Collected Papers of
S. R. S. Varadhan

Volume 1
Limit Theorems,
Review Articles

Collected Papers of S. R. S. Varadhan

Volume 1: Limit Theorems,
Review Articles

Volume 2: PDE, SDE, Diffusions,
Random Media

Volume 3: Large Deviations

Volume 4: Particle Systems and their Large Deviations

Collected Papers of S. R. S. Varadhan

Volume 1
Limit Theorems, Review Articles

Editors

Rajendra Bhatia
Abhay G. Bhatt
K. R. Parthasarathy

HINDUSTAN BOOK AGENCY

Published by:

Hindustan Book Agency (India)
P 19 Green Park Extension
New Delhi 110 016
India

email: info@hindbook.com
www.hindbook.com

ISBN 978-93-80250-39-7 (4 volume set)

Printed and bound by Gopsons Papers Ltd., Noida

Preface

Srinivasa Varadhan began his research career at the Indian Statistical Institute (ISI), Calcutta, where he came in 1959 as a graduate student. His first paper appeared in *Sankhyā*, the Indian Journal of Statistics in 1962. The erudite founder of the Institute, and of the Journal, traced the origins of probabilistic thinking to the Indian-Jaina philosophy of *syādvāda*, or the "assertion of possibilities". The school of logic that propagated this was active around the fourth century BCE, and a full exposition of this theory is given in a sixth century text called *Saptabhaṅgīnaya* (the dialectic of seven-fold prediction).

In the middle of the twentieth century, research work on all aspects of statistics, practical and theoretical, was started in all earnestness at ISI. Varadhan was there in the golden period of the Institute. Among his teachers were C. R. Rao and R. R. Bahadur, and just a little older than him were fellow students V. S. Varadarajan, R. Ranga Rao and K. R. Parthasarathy. Together with this latter group Varadhan began the study of probability on topological groups and on Hilbert spaces, and quickly made an international reputation.

At this time Varadhan realised that there are strong connections between Markov processes and differential equations, and in 1963 he came to the Courant Institute in New York where he has stayed ever since. (According to him "here you do not have to make an effort to learn PDE, you just have to breathe it".) Here he came under the influence of probabilists Monroe Donsker and Marc Kac, and of a young graduate student Daniel Stroock. With Stroock he wrote a series of papers on the Martingale Problem and Diffusions, and with Donsker another series of papers on Large Deviations. With this work Varadhan's reputation as one of the leading mathematicians of the time was firmly established. Since then he has contributed to several other topics in probability, analysis and physics, and collaborated with several distinguished mathematicians.

These Collected Works contain all his research papers over the half-century from 1962 to the time of going to print in early 2012. It is befitting that they are being published in India. We are happy and proud to be associated with this project.

Editors

Contents

Volume 1

Limit Theorems

Volume 3

Large Deviations

Volume 4

Particle Systems and their Large Deviations

Autobiography: S. R. S. Varadhan

According to my school records I was born on Jan 2, 1940, in the city of Madras, in the state of Madras in India, which was then a British colony. The city is now called Chennai and the state has become Tamil Nadu in the Republic of India.

My father was born in the last year of the nineteenth century, in 1899 and he married my mother in 1917, when he was eighteen and she was ten. I am an only child and my parents had been married for nearly twenty five years when I was born. Both my parents were the eldest siblings in rather large families and I have always received special attention from all my uncles, aunts, cousins, grandmothers and other assorted relatives. I was born so late that I did not really get to know either of my grandfathers.

As a child I grew up in several small towns not far from Madras. My father was a high school teacher and later principal in the county school system and was periodically transferred from one town to another within the county or district as it is called in India. That was both good and bad for me. I was treated with consideration by all my teachers. But I could not do any mischief at school without my father finding out right away.

Growing up in these small towns was fun. There was plenty of time after school to play with friends on the riverbed that was mostly dry, or play indoors on rainy days. There was very little homework, and in fact there was only minimal learning at school. I did well relative to my class, but was not challenged in any sense. It was only in the last year of the high school, that my mathematics teacher took a special interest in a small group of us and would ask us to come to his house on weekends to do some mathematics problems for fun. More than anything else he taught me that solving mathematical problems or puzzles can be fun.

I had some vague ideas of becoming a doctor as a child. But once, with a group of fellow students from the high school, I went to a medical exhibition at the local

S.R.S. Varadhan (✉)
Courant Institute of Mathematical Sciences, New York University, New York, NY 10012, USA
e-mail: varadhan@cims.nyu.edu

H. Holden, R. Piene (eds.), *The Abel Prize*,
DOI 10.1007/978-3-642-01373-7_12,

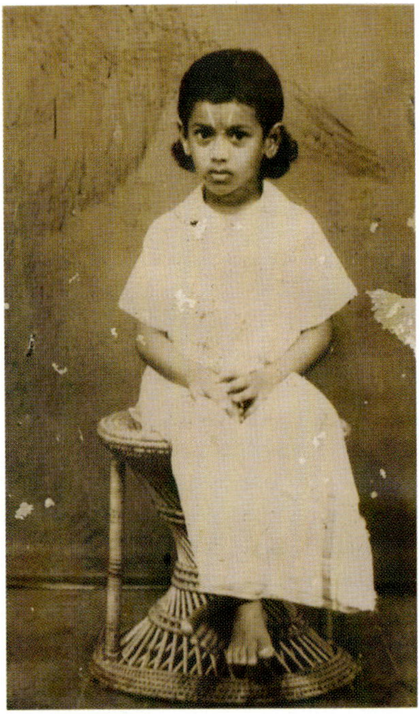

(a) As a seven year old in traditional Brahmin attire

(b) As a nine year old in a group with teachers in school. I am there because my father was the principal. I am sitting at his feet. He is the one with the head dress (turban)

medical college where the medical students demonstrated their surgical skills on cadavers. That cured me of the desire to enter the medical profession.

In 1954, I moved out to live with my uncle in a suburb of Madras to attend a local college where we were required to study for two years before entering the University. This was a difficult transition. From a rather parochial school system where everything was taught in Tamil, the vernacular, I had to grasp new ideas that were being taught in English. Some of the professors were from UK and I could not quite understand their accent. I studied English and Tamil as my two languages. These were the harder subjects at college since we were expected to write critical essays, where as in school we were only required to remember facts and memorize poetry.

In addition to languages I studied Mathematics, Physics and Chemistry. I did well in all. I liked Physics the most, and I began slowly to see the connections between them.

(c) High school graduation picture. I am standing on the top row. My father as principal is there too, sitting in the front row wearing a turban

It was time to enter the University and the seats were limited and the competition was tough. I came from a Brahmin community, viewed by the government as privileged, and there was reverse discrimination. The goal was to get into a Honors program that led directly to a Master's degree in three years and saved you a year.

It was at this time that I heard of Statistics for the first time. I was told that there was only one college in the state that offered an Honours program in Statistics and they admitted only fourteen students each year. My father knew somebody who knew somebody who could help and luckily I was admitted. I had a choice between Physics and Statistics and I opted for Statistics.

Basically it was a three year program devoted to pure mathematics, probability and statistics. The Presidency College where the program was offered, was situated on the beach. You looked out the window and it was a sandy beach that was as nice as Ipanema. Meanwhile, my father had retired and moved out to the same town as my uncle and I stayed with my parents and commuted for a year. I wanted my parents to let me stay in the dorm for the last two years and they did, after much persuasion on my part.

I really enjoyed those two years. Long walks on the beach with a dozen or so close friends from my class. Lots of movies, arguments, discussions and a generally carefree student life. We are still a close group and see each other from time to time. Although I am the only academic, the others have gone on to do well in government, business and other areas. The studies themselves were relatively easy. I was learning a lot of new concepts and they all felt natural and I could do exceptionally well without much effort. The trouble with the system was that the expectations were not high and it was too rigid. We had only set courses and no electives.

(d) Picture with Kolmogorov during his visit to India (1962). I am third from the left behind Kolmogorov. With me are K.R. Parthasarathy (graduate student), B.P. Adhikari (Professor), me, J. Sethuraman (graduate student), C.R. Rao (Head of the Institute and my advisor) and P.K. Pathak (graduate student)

I graduated after three years having established a new record for grades obtained in the final examination where you are tested in one week on everything that you have learned in three years.

The usual career path at this point is to write a competitive civil service examination that is used to recruit high level government officials. My parents expected me to sit for this. If you succeed, you are set for life. I wanted to go for research. When my father saw that I was firm, he supported me and I went off to Calcutta to study at the Indian Statistical Institute. I had no idea what I was supposed to do. When I arrived in August, 1959, they gave me a desk and expected me to write a thesis in three years. No graduate courses were offered. There were seminars that were optional. I took one on point set topology from Varadarajan and one on measure theory from Bahadur. I did not know what to do and so learned to play bridge and I played a lot. But slowly over the year I met up with a few other graduate students who were there already for a year or two, and we organized our own seminars and programs of study. We lectured to each other, formulated our own problems and tried to solve them. It was a wonderful learning experience and I started to do "Research".

The atmosphere at ISI was very stimulating. We had a steady stream of distinguished visitors, during the pleasant winter months. Sir Ronald Fisher came every year. I had just missed Norbert Wiener, who had come the year before I went there. We had tea, twice a day, when we all met at the tea room and talked informally. J.B.S. Haldane was a regular at tea, constantly puffing on his cigar. There was always some excitement in the air and some thing or other was always happening. During my second and third year I learned a lot of functional analysis and more or less finished my dissertation. I had begun to learn Markov Processes and wanted to start working on it. During my third year Kolmogorov visited us for two months and I gave a seminar on my thesis and he was in the audience asking questions. He spoke no English, none of us knew any Russian, and so we talked through an interpreter who knew French. A group of graduate students accompanied him on a two week tour of parts of India. He was a member of my thesis committee and brought a copy of my thesis with him, to Moscow, promising to send a report from there. The report came after six months, but only after Parthasarathy, a colleague who went to Moscow on an exchange program, provided a steady daily reminder.

At this point a year had gone by after my dissertation was finished, I formally got my Ph.D., and it was time to go abroad, for a postdoctoral study. In those post Sputnik days that meant USA. Varadarajan, who had just returned to India, after three years in the US, suggested that I go to NYU, to Courant Institute. He wrote a letter on my behalf that went unacknowledged for nearly three months. In the end I was offered a postdoctoral position and I arrived in New York in the Fall of 1963.

There was an international organization that greeted foreign students upon arrival, and they offered to find a place for me to stay for my first night. They met me at the old Idlewild airport and arranged for me to stay in a hotel on Times Square. When I showed up at Courant the next day and told them where I was staying, they were horrified and found me a place to stay on Tenth and University Place. Later I found myself a studio in the West Village.

Before leaving India I became engaged to be married to Vasundara and the wedding was to take place the following June, in Madras. The year went by very fast. I met a small group of Indian academics and over the years we have remained close.

Work went well at Courant. I saw a lot of Mony Donsker, who was the principal probabilist at Courant. There were lots of visitors, seminars and other activities that kept me busy during the week and I socialized with my Indian group during weekends.

Donsker and I, discussed a lot. He had many students working on an assortment of problems. This provided me with a broad perspective and I was beginning to view things from a slightly different angle that made a lot of sense to me. Donsker was interested in specific problems that he wanted to be solved and I was always interested in developing techniques that would enable one to solve a class of problems. We complemented each other and worked closely for over twenty-five years. I went to India to get married in the summer and we found a nice apartment in Washington Square Village. My wife, who was only sixteen, started undergraduate studies at Washington Square College. When my post doctoral fellowship was renewed for two more years it proved ideal for both of us. Then, I joined the faculty in the Fall of 1966, as an assistant professor.

Donsker had a student named Schilder who did a very fine thesis on Laplace asymptotics on Wiener space. I thought a lot about it, and felt that one could develop it

(e) This picture is at the airport on my way to Courant as a postdoc (Fall 64)

(f) This picture is at the airport, returning to USA with Vasu immediately after our wedding in the summer of 1964

considerably. This has been a major thrust of my research over the years and goes by the name of Large Deviation Theory. It is the technique of estimating precisely, in a logarithmic scale, how small the probabilities of certain rare events are. I have found this to be a problem that crops up in many different contexts, and have returned to it again and again during my career.

We used to have a joint seminar with Rockefeller Institute, where Mark Kac was a professor. We would go up to 68th and York often. Once on the way back, in a taxi, someone commented on a result by Cieselski, a Polish mathematician and I immediately saw connections with large deviations and diffusions. The diffusion processes are Markov processes with continuous paths, closely related to second order elliptic partial differential operators. I was able to work out this connection and it proved to be a nice result.

There was a graduate student at Rockefeller University, Dan Stroock, who came by to talk to me about this and we became close friends. He came to Courant in 1966 and stayed for six years. We worked very closely during this period and I believe we changed the way Markov processes were viewed. We introduced systematically methods based on martingales that have become more or less standard today. We jointly wrote a book that appeared in 1979 and has been received well.

In the fall of 1973, we (the family at this time included two sons Gopal born in 69 and we lost him on 9/11/01, and Ashok born 1972) had just returned after a sabbatical from Sweden and India. Donsker as usual was interested in a specific problem about a probabilistic explanation of the variational formula for the ground state energy of the Schrödinger operator. I remember going to Durham to give a talk at Duke, and sitting in their library before my talk, I saw the connection. This led to a whole series of results on large deviation theory that Donsker and I worked on, till his untimely death in 1991.

I had become an Associate professor in 1968, won an Alfred P. Sloan Fellowship in 1969, and became a Professor in 1972. This was also a tough time at NYU, involving a financial crisis and the sale of the Bronx campus. We moved to Stanford for a year long visit in 1976–77, enjoyed the outdoor life and the open skies. But professionally I could not wait to return to New York. I also started working occasionally with George Papanicolau, who had lots of applied problems that often required new techniques. I was invited to give a talk at the International Congress of Mathematicians, in Helsinki in 1978 which was a kind of recognition.

I served for four years from 1980–84 as Director of Courant and survived the experience. I was surprised to find that the central administration consisted for the most part of talented individuals doing their best at a difficult task. I began to appreciate for the first time that we at Courant were part of a larger University. My research slowed down some due to the demands of administrative work, but I was able to continue with it. I had started to attract a steady stream of graduate students by this time and enjoyed working with them. I became recognized as an expert on Large Deviations, lectured often on it and wrote a set of lecture notes on the topic.

This was also the time when Wall Street discovered probability. The Black and Scholes model was suddenly very popular. I started some consulting for a small company that managed pension funds and worked closely with Harry Markowitz, who was also a consultant. Harry went on to win the Nobel Prize in Economics. Richard Brignoli, who was the CEO of the company was a maverick and that made the experience all the more enjoyable. It lasted several years, until the company went under. Not because it performed poorly, but the partners had a fallout, fought in court and the lawyers ended up getting everything. Our son Gopal got his first experience working for the company in the summer and later part time during his senior year at college. He chose to stay with the financial industry and his younger brother Ashok has followed him.

I took a sabbatical in 84–85, after serving for four years as Director of the Courant Institute supported partly on a Guggenheim fellowship. Around this time I went to a conference in Marseilles and we usually walked down to the sea and back up the hill, to the campus at Luminy, after lunch. During one of the walks, George explained to me a problem of establishing what he called bulk diffusion under rescaling for interacting diffusions. The problem intrigued me and I thought it would be a simple problem. I spent my sabbatical year thinking about it and found it rather difficult. I found out later that there were a whole class of such problems generally referred to as problems of hydrodynamic scaling. I worked on the problem a lot, but made no progress. Although I had some ideas, they were not sufficient to solve the problem. We had a seminar at which the speaker Josef Fritz from Budapest, proved a result of similar type for a different model. I thought about the new model and found to my surprise, that my ideas worked well for this model and could provide a better result than Fritz. George, Guo who was a student of George, and I worked feverishly on this model and completed our work. This led me to a whole set of problems requiring new methods and I worked closely with a very talented younger colleague, H.T. Yau. He has since developed the subject and has taken it in a variety of different directions.

The period from 1984–94 was personally very hard for me. My parents grew old and infirm. Since I had no siblings, I had the responsibility for their welfare. This meant frequent and prolonged trips back and forth between New York and Madras. This was hard on my wife, who had stayed in school part time, while raising the two boys, and earned a Ph.D. degree (her fourth degree at NYU) in 1985. My father died in 1990 and my mother in 1994. My wife lost her mother in 1991. The generation above us started thinning out slowly.

There was a bit of a problem at the office in 1992. George, who was to have been the director, left to go to Stanford and there was no time to find a replacement. I was drafted to serve for two years during which period a search committee was appointed and found an excellent choice of director in Dave McLaughlin.

This was also a time when I was receiving honors and recognition and that was gratifying. I was elected a Fellow of American Academy of Arts and Sciences and an Associate Fellow of the Third World Academy of Sciences in 1988, a Fellow of the Institute of Mathematical Statistics in 1991, a member of National Academy of Sciences in 1995, and a Fellow of The Royal Society in 1998. I received the George D. Birkhoff Prize from AMS and SIAM in 1994 and the Leroy P. Steele prize of the AMS that I shared with Dan Stroock in 1996. NYU gave me the Margaret and Herman Sokol award in 1995. I was invited to give a plenary address at the International Congress of Mathematicians in Zurich, in 1994. I was appointed as Frank J. Gould Professor of Science at NYU. In 2001, I was elected to serve as the President of Institute of Mathematical Sciences for 2002–03. I received honorary degrees from University of Paris in 2003 and Indian Statistical Institute in 2004.

Tragedy hit on 9/11/01 when we lost our older son Gopal who was working at the World Trade Center that day. I was in Paris, visiting IHP for a month and was told of the crash by my wife who called me from New York. It took four days to get back to New York. We miss him. Our grandson Gavin, born to our younger son Ashok and his wife Maggie helps take some of the edge off our grief.

Professionally I consider myself to have been fortunate, in that my career spanned a period, when science and mathematics were generously funded by the public. I had a stimulating ambience at Courant that has traditionally provided a very supportive environment for their younger faculty. I have enjoyed working closely with my colleagues as well as everyone of my nearly thirty doctoral students. Finally, I hope to have several more years of productive academic life.

Varadhan with M. D. Donsker

Varadhan with G. C. Papanicolaou

Varadhan with D. W. Stroock

Varadhan receiving the Abel Prize from HM King Harald V, Oslo, 2007

Varadhan receiving the National Medal for Science from Mr. Barack Obama, President of the United States, 2011

Introduction

S. R. S. Varadhan

These volumes of collected works begin with the work done at the Indian Statistical Institute, with R. Ranga Rao and K. R. Parthasarathy. We were learning functional analysis, particularly topics around weak convergence of probability distributions, an area that had been investigated by V.S. Varadarajan earlier at ISI. We wanted to apply it to study convolution properties of probability distributions on topological groups. While some initial work was carried out directly, other parts dealing with locally compact abelian groups needed the use of characteristic functions. These methods were later modified to cover the infinite dimensional case of Hilbert Space, which was a significant part of my Ph.D thesis. Later developments include the work with D. Stroock [12] on the analog of the Central Limit Theorem on Lie Groups and extension to limit theorems of products of independent random elements on a Lie Group carried out by P. Feinsilver [4] in his thesis. These works use martingale techniques developed with Stroock in the sixties at the Courant Institute.

The work done at the Courant Institute falls mainly into four interrelated areas; large deviations, diffusion processes, interacting particle systems and processes in random environments. The work on large deviations began in 1964 when, Michael Schilder, a student of Monroe Donsker proved in his thesis [11] a result that can be interpreted as a large deviation result for Brownian motion. He showed that if P is the Wiener measure on $C[0,1]$, then for any bounded continuous function $V(\cdot)$

$$\lim_{\epsilon\to 0} \epsilon \log E^P\left[\exp[\frac{1}{\epsilon}\int_0^1 V(\sqrt{\epsilon}x(s))ds\right] = \sup_{\substack{f(\cdot)\\ f(0)=0}} \left[\int_0^1 V(f(t))dt - \frac{1}{2}\int_0^1 [f'(t)]^2 dt\right].$$

My interest in large deviation theory began with trying to interpret this result as the combination of a Laplace formula and a large deviation result directly in $C[0,1]$. This point of view was developed in [14]. Some time in 1965 while returning after a seminar from Rockefeller University in a taxi, some one remarked about a theorem of Czieselski [1] that compared the ratio of the two fundamental solutions of the heat equation

$$r(t,x,y) = \frac{p_D(t,x,y)}{p(t,x,y)}$$

where p_D corresponds to the process killed at the boundary of a domain D and p to the whole space. The ratio $r \to 1$ as $t \to 0$ for all $x, y \in D$, if and only if

D is essentially convex. The explanation is that if a Brownian path goes from x to y, in a short time, it would most likely go along the straight line from x to y. The analogue for diffusions would be that the straight line has to be replaced by the geodesic in the Riemannian metric associated to the diffusion. This, along with my interest in large deviations, led to the study of small time behavior of diffusions. Michael Glass in his Ph.D. thesis [7], extended this to small second order perturbations of vector fields. This topic was developed profoundly and independently by Freidlin and Wentzell in their work [5] on small random perturbations of dynamical systems.

One of the things that Dan Stroock, who was then finishing up his thesis at Rockefeller University, and I noticed was that in proving the large deviation results for diffusions the main tool was some exponential martingales that were associated with the diffusion process. We then started wondering if the martingales characterized the process. We quickly worked out that if they did, then one can use the machinery of weak convergence to take a fresh look at many of the convergence theorems in Markov Processes. For the next few years Dan and I explored many aspects of Markov Process theory through their associated martingales. This continued till Dan left for Colorado in 1972, and we later wrote our book on multidimensional diffusion processes [13]. Martingale methods have since become an important tool in proving limit theorems for stochastic processes.

I was on a sabbatical leave during 1972-73, and when I returned, Donsker asked me a question. Can we explain the variational formula

$$\lambda(V) = \inf_{\|f\|_2=1} \Big[\int V(x)[f(x)]^2 dx + \frac{1}{2} \int |\nabla f|^2 dx \Big],$$

for the principal eigenvalue $\lambda(V)$ of $-\frac{1}{2}\Delta + V$ as the Laplace formula that corresponds to some large deviation result? After all there is the Feynman-Kac formula and its consequence

$$\lambda(V) = -\lim_{t\to\infty} \frac{1}{t} \log E[\exp[-\int_0^t V(x(s))ds]$$

where E is the expectation with respect to Brownian motion. I was thinking of this sitting in the library of the Mathematics Department at Duke University, where I was visiting to give a talk, and I saw how there was a simple answer. We can rewrite the variational formula in terms of $g = f^2$ as

$$\lambda(V) = \inf_{\substack{g\geq 0 \\ \int g dx=1}} \Big[\int V(x)g(x)dx + \frac{1}{8} \int \frac{|\nabla g|^2}{g} dx \Big],$$

and the Feynman-Kac formula as

$$\lambda(V) = -\lim_{t\to\infty} \frac{1}{t} \log E[\exp[-t \int_R V(x)\, L_t(dx)],$$

where $L_t(A) = \frac{1}{t}\int_0^t \mathbf{1}_A(x(s))ds$ is the empirical distribution, or occupation time, i.e., the proportion of time spent by the Brownian path in the set A. This suggests there is a large deviation result for the behavior of the distribution of L_t on the space $\mathcal{M}$ of probability distributions in R, with the rate function for the distribution with density g being

$$I(g) = \frac{1}{8}\int \frac{|\nabla g|^2}{g} dx.$$

We, in fact, established such a result for Brownian motion [2] and for other Markov processes. At this time Mark Kac raised a problem about what he named the Wiener Sausage: if $|D_\epsilon(t)|$ is the volume of the range of a Brownian path in R^d enlarged by a neighborhood of size ϵ, show that

$$\lim_{t\to\infty} \frac{1}{t^{\frac{d}{d+2}}} \log E[\exp[-\nu D_\epsilon(t)]] = -k(d,\nu)$$

exists and calculate it explicitly. This prompted us to investigate the role of topology in large deviations. While the usual theory of large deviations involved weak topology for measures, to answer the question raised by Kac, we need the strong or L_1 topology. The ϵ neighborhood provides a certain level of smoothing, and the issue was to show that was enough. Mark had suggested that a similar result for random walk on the number of distinct sites visited could be more tractable. But it turns out that the random walk case is proved by a combination of the central limit theorem and the result for Brownian motion [3].

Similar large deviation results were obtained independently by Jurgen Gärtner [6]. It turns out that one can prove the existence of the limit

$$\lambda(V) = \lim_{t\to\infty} \frac{1}{t} \log E\big[\exp[\int_0^t V(x(s))ds]\big]$$

for a wide class of Markov processes. While according to the ergodic theorem L_t converges to the invariant distribution μ_0, the dual

$$I(\mu) = \sup_V \left[\int V(x)\mu(dx) - \lambda(V)\right]$$

of $\lambda(V)$ is the rate function for large deviations of L_t from μ_0.

We adopted a different method where the upper bound was established by the use of exponential martingales, and the lower bound was obtained through "tilting" which provides a direct connection to entropy that controls, in some sense, all large deviations probabilities. The idea is that if we have a model and a rare phenomenon, whose probability we want to estimate, we can change the model so that the the rare phenomenon is normal in the new model. The amount of change is measured by relative entropy. Since there may be many possible "tilts" that can produce the phenomenon in question, it is natural to choose the optimal tilt, i.e., one with the least relative entropy. This then ends up as the large deviation rate.

Over the years I have worked with George Papanicolaou on problems involving random media of different kinds. These problems have their origins in problems of homogenization of periodic and random media that arise in the study of composite materials. The theory of interacting particle systems is closely related, in the sense that the random medium felt by any one particle keeps changing randomly due to the movement of other particles. Some of these problems are related to the study of non-equilibrium statistical mechanics, where a system with multiple equilibria evolves from non-equilibrium. Local equilibria are quickly reached, parametrized by variables that correspond to conserved quantities, that evolve in space and time when suitably rescaled. The limiting equations that describe their evolution are often a non-linear system of conservation laws. While attending a conference in Luminy, near Marseille, during a walk to the calanque after lunch, George posed the question of studying a large collection of $N\rho$ interacting Brownian motions, rescaled as

$$dx_i(t) = -N \sum V'(N(x_i(t) - x_j(t)) + d\beta_i(t)$$

in a periodic one dimensional space, i.e., the unit circle. In the limit the local density of particles $\rho(t, x)$ will evolve according to a nonlinear heat equation

$$\rho_t = (D(\rho)\rho_x)_x.$$

$D(\rho)$ is determined through the statistical mechanical equilibrium distribution, or the Gibbs state, μ_ρ at density ρ. I thought about this problem for a while. Initially I was confident it could be solved easily, but that did not turn out to be so. I did make some progress but was stuck at a crucial point. Some time later Josef Fritz from Hungary was giving a seminar at Courant in which he was proving a similar result for a different model. To my surprise I realized that the difficulty I had for the model involving interacting Brownian motions, could be resolved easily in this model dealing with interacting random fields at lattice locations. This resulted in a joint paper with Guo and Papanicolaou [8]. This involved field variables at lattice locations that represented charges that were randomly exchanged between neighboring sites. The random exchange was driven by a nonlinear force, modified by a random noise, which was Brownian motion

$$d\xi_i(t) = [\phi'(x_{i-1}(t)) - 2\phi'(x_i(t)) + \phi(x_{i+1}(t))]\,dt + [d\beta_{i-1,i}(t) - d\beta_{i,i+1}(t)].$$

The rescaled limiting equation for the averages of $\xi_i(t)$ is again a nonlinear diffusion equation of the form

$$m_t = (D(m)m_x)_x$$

with an explicit formula for D in terms of ϕ. I was then curious to see how some inhomogeneity will change the answer. Replace $d\beta_{i-1,i}(t)$ by $\sigma(x_{i-1}(t), x_i(t))d\beta_{i-1,i}(t)$. To my surprise I found the answer to be much more complicated, and it involved some very heavy analysis in infinite dimensional spaces [15]. In the end I could get a variational formula for $a(\rho)$ and the limiting equation

$$m_t = (a(m)D(m)m_x)_x.$$

I remember being in L'Aquila, Italy, and explaining to Herbert Spohn what I could do. He was quick to recognize the significance of what I had done, and explained to me the difference between gradient and non-gradient models. The diffusion coefficient $a(m)$ is, in general, given by a Green-Kubo formula, in terms of the entire evolution in time of the equilibrium, and it simplifies only in gradient models. What I had done was perhaps the first example of a non-gradient model whose diffusive scaling limit had been rigorously established. I quickly understood the significance and with a group of students, started working on problems of this elk, referred to as hydrodynamic scaling limits. Around this time I was lucky to have H. T. Yau, join our group. He had an excellent background in mathematical physics (and everything else!) and together with many students and postdocs we worked on several problems in this area. It turned out that there is a subtle connection between these problems and large deviations that proved to be useful in sorting out many things.

During the last decade I have been studying large deviations for random walks in a random environment. I was always of the opinion that one should consider the environment as seen by the moving particle, and with many students we explored this. It turned out that certain nonlinear homogenization problems are related to large deviations of random walks or diffusions in a random environment. With Rezakhanlou and Kosygina we worked on these issues [9],[10].

I have been lucky to have found excellent collaborators, throughout my career. The Mathscinet lists nearly thirty coauthors. I have enjoyed working with all of them: from fellow graduate students Ranga Rao and Parthasarathy to colleagues Donsker, Stroock, Papanicolaou, Nirenberg, Yau and Chatterjee at Courant; visitors and post-docs Sznitman, Olla, Kipnis and Kifer. There is even an unpublished but widely circulated manuscript with P. L. Lions. I have continued to work with many of my past students, Rezakhanlou, Quastel, Kosygina and Sethuraman. All of them have enriched my professional life and I am greatly indebted to them.

References

[1] Ciesielski, Z. Heat conduction and the principle of not feeling the boundary. Bull. Acad. Polon. Sci. Sr. Sci. Math. Astronom. Phys. 14 1966 435–440.

[2] Donsker, M. D.; Varadhan, S. R. S. Asymptotic evaluation of certain Wiener integrals for large time. Functional Integration and its Applications (Proc. Internat. Conf., London, 1974), pp. 15–33. Clarendon Press, Oxford, 1975.

[3] Donsker, M. D.; Varadhan, S. R. S. Asymptotics for the Wiener sausage. Comm. Pure Appl. Math. 28 (1975), no. 4, 525–565.

[4] Feinsilver, Philip. Processes with independent increments on a Lie group. Trans. Amer. Math. Soc. 242 (1978), 73–121.

[5] Freidlin, Mark I.; Wentzell A.D. On Small random perturbations of dynamical systems. Russian Mathematical Surveys Volume 25(1970) , number 1 pages 1-56

[6] Gärtner, Jurgen. Large deviation theorems for a class of random processes. (Russian) Teor. Verojatnost. i Primenen. 21 (1976), no. 1, 95–106.

[7] Glass, Michael Steven. Perturbation of a first order equation by a small diffusion. Thesis (Ph.D.)New York University. ProQuest LLC, Ann Arbor, MI, 1970. 53 pp..

[8] Guo, M. Z.; Papanicolaou, G. C.; Varadhan, S. R. S. Nonlinear diffusion limit for a system with nearest neighbor interactions. Comm. Math. Phys. 118 (1988), no. 1, 31–59.

[9] Kosygina, Elena; Rezakhanlou, Fraydoun; Varadhan, S. R. S. Stochastic homogenization of Hamilton-Jacobi-Bellman equations. Comm. Pure Appl. Math. 59 (2006), no. 10, 1489–1521.

[10] Kosygina, Elena; Varadhan, S. R. S. Homogenization of Hamilton-Jacobi-Bellman equations with respect to time-space shifts in a stationary ergodic medium. Comm. Pure Appl. Math. 61 (2008), no. 6, 816–847.

[11] Schilder, M. Some asymptotic formulas for Wiener integrals. Trans. Amer. Math. Soc. 125 (1966) 63–85.

[12] Stroock, Daniel W.; Varadhan, S. R. S. Limit theorems for random walks on Lie groups. Sankhya Ser. A 35 (1973), no. 3, 277–294.

[13] Stroock, Daniel W.; Varadhan, S. R. S. Multidimensional diffusion processes. Grundlehren der Mathematischen Wissenschaften [Fundamental Principles of Mathematical Sciences], 233. Springer-Verlag, Berlin-New York, 1979.

[14] Varadhan, S. R. S. Asymptotic probabilities and differential equations. Comm. Pure Appl. Math. 19 (1966) 261–286.

[15] Varadhan, S. R. S. Nonlinear diffusion limit for a system with nearest neighbor interactions. II. Asymptotic problems in probability theory: stochastic models and diffusions on fractals (Sanda/Kyoto, 1990), 75–128, Pitman Res. Notes Math. Ser., 283, Longman Sci. Tech., Harlow, 1993.

Prize Citations

National Medal of Science, 2011.

To Srinivasa S.R. Varadhan, for his work in probability theory, especially his work on large deviations from expected random behavior, which has revolutionized this field of study during the second half of the twentieth century and become a cornerstone of both pure and applied probability. The mathematical insights he developed have been applied in diverse fields including quantum field theory, population dynamics, finance, econometrics, and traffic engineering.

Abel Prize, 2007.

To S. R. S. Varadhan, for his fundamental contributions to probability theory and in particular for creating a unified theory of large deviations.

Probability theory is the mathematical tool for analyzing situations governed by chance. The law of large numbers, discovered by Jacob Bernoulli in the eighteenth century, shows that the average outcome of a long sequence of coin tosses is usually close to the expected value. Yet the unexpected happens, and the question is: how? The theory of large deviations studies the occurrence of rare events. This subject has concrete applications to fields as diverse as physics, biology, economics, statistics, computer science, and engineering.

The law of large numbers states that the probability of a deviation beyond a given level goes to zero. However, for practical applications, it is crucial to know how fast it vanishes. For example, what capital reserves are needed to keep the probability of default of an insurance company below acceptable levels? In analyzing such actuarial "ruin problems", Harald Cramér discovered in 1937 that standard approximations based on the Central Limit Theorem (as visualized by the bell curve) are actually misleading. He then found the first precise estimates of large deviations for a sequence of independent random variables. It took 30 years before Varadhan discovered the underlying general principles and began to demonstrate their tremendous scope, far beyond the classical setting of independent trials.

In his landmark paper "Asymptotic probabilities and differential equations" in 1966 and his surprising solution of the polaron problem of Euclidean quantum field theory in 1969, Varadhan began to shape a general theory of large deviations that was much more than a quantitative improvement of convergence rates. It addresses a fundamental question: what is the qualitative behaviour of a stochastic system if it deviates from the ergodic behaviour predicted by some law of large numbers or if it arises as a small perturbation of a deterministic system? The key to the answer is a powerful variational principle that describes the unexpected behaviour in terms of a new probabilistic mode minimizing a suitable entropy distance to the initial probability measure. In a series of joint papers with Monroe D. Donsker exploring the hierarchy of large deviations in the context of Markov processes, Varadhan demonstrated the relevance and the power of this new approach. A striking application is their solution of a conjecture of Mark Kac concerning large time asymptotics of a tubular neighbourhood of the Brownian motion path, the so-called "Wiener sausage".

Varadhan's theory of large deviations provides a unifying and efficient method for clarifying a rich variety of phenomena arising in complex stochastic systems, in fields as diverse as quantum field theory, statistical physics, population dynamics, econometrics and finance, and traffic engineering. It has also greatly expanded our ability to use computers to simulate and analyze the occurrence of rare events. Over the last four decades, the theory of large deviations has become a cornerstone of modern probability, both pure and applied.

Varadhan has made key contributions in several other areas of probability. In joint work with Daniel W. Stroock, he developed a martingale method for characterizing diffusion processes, such as solutions of stochastic differential equations. This new approach turned out to be an extremely powerful way of constructing new Markov processes, for example infinite-dimensional diffusions arising in population genetics.

Another major theme is the analysis of hydrodynamical limits describing the macroscopic behavior of very large systems of interacting particles. A first breakthrough came in joint work with Maozheng Guo and George C. Papanicolaou on gradient models. Varadhan went even further by showing how to handle non-gradient models, greatly extending the scope of the theory. His ideas also had a strong influence on the analysis of random walks in a random environment. His name is now attached to the method of "viewing the environment from the travelling particle", one of the few general tools in the field.

Varadhan's work has great conceptual strength and ageless beauty. His ideas have been hugely influential and will continue to stimulate further research for a long time.

Leroy P. Steele Prize for Seminal Contribution to Research, 1996.

To Daniel Stroock and S.R.S. Varadhan for their four papers: *Diffusion processes with continuous coefficients I and II*, Comm. Pure Appl. Math. 22 (1969), 345-400, 479-530; *On the support of diffusion processes with applications to the strong maximum principle*, Sixth Berkeley Sympos. Math. Statist. Probab., vol. III, 1970, pp. 333-360; *Diffusion processes with boundary conditions*, Comm. Pure Appl. Math. 34 (1971), 147-225; *Multidimensional diffusion processes*, Springer-Verlag, 1979, in which they introduced the new concept of a martingale solution to a stochastic differential equation, enabling them to prove existence, uniqueness, and other important properties of solutions to equations which could not be treated before by purely analytic methods; their formulation has been widely used to prove convergence of various processes to diffusions.

Birkhoff Prize in Applied Mathematics (AMS-SIAM George David Birkhoff Prize), 1994.

To S. R. S. Varadhan for important contributions to the martingale characterization of diffusion processes, to the theory of large deviations for functionals of occupation times of Markov processes, and to the study of random media.

Varadhan's Work on Diffusion Theory

Daniel W. Stroock*

Given a second order, (possibly degenerate) elliptic differential operator

$$L = \frac{1}{2}\sum_{i,j=1}^{N} a_{ij}(x)\partial x_i \partial_{x_j} + \sum_{i=1}^{N} b_i(x)\partial_{x_i} \tag{1}$$

acting on $C^2(\mathbb{R};\mathbb{C})$ and a Borel probability measure μ on $\mathbb{R}^N$, what does it mean for a Borel probability measure $\mathbb{P}_\mu$ on $C([0,\infty);\mathbb{R}^N)$ to be a diffusion process determined by L with initial μ? When Varadhan and I asked ourselves this question in the mid 1960's, we found no answer which satisfied us.

It was not that there were no answers. Indeed, already in the 1930's Kolmogorov (cf. [6]) had mapped out a path which started from L and ended at a measure $\mathbb{P}_\mu$. Namely, he began by solving what is now called Kolmogorov's forward equation

$$\partial_t p(t,x,\cdot) = L^* p(t,x,\cdot) \quad \text{with } p(0,x,\cdot) = \delta_x, \tag{2}$$

where L^* is the formal adjoint of L given by

$$L^*\psi(y) = \frac{1}{2}\sum_{i,j=1}^{N} \partial_{y_i}\partial_{y_j}\big(a_{ij}(y)\psi(y)\big) - \sum_{i=1}^{N} \partial_{y_i}\big(b_i(y)\psi(y)\big).$$

He then showed that, under suitable regularity and non-degeneracy conditions on the coefficients a and b, there is one and only solution to (2) which is the density of a probability measure on $\mathbb{R}^N$. Further he showed that $p(t,x,y)$ is a continuous function of $(t,x,y) \in (0,\infty)\times\mathbb{R}^N\times\mathbb{R}^N$ and that it satisfies the Chapman–Kolmogorov equation

$$p(s+t,x,y) = \int p(t,\xi,y)p(s,x,\xi)\,d\xi. \tag{3}$$

*Department of Mathematics, Massachusetts Institute of Technology, Cambridge, MA 02139

Knowing (3), he could check that, for any $\mu \in M_1(\mathbb{R}^N)$, the family of probability measures given by

$$P_\mu\big((t_1,\dots,t_n);\Gamma\big) = \int\limits_\Gamma \cdots \int \prod_{m=1}^{n} p(t_m - t_{m-1}, y_{m-1}, y_m)\, \mu(dy_0)dy_1 \cdots dy_n$$
$$\text{for } n \geq 0,\ 0 = t_0 < t_1 < \cdots < t_n, \text{ and } \Gamma \in \mathcal{B}_{(\mathbb{R}^N)^{n+1}}$$

is consistent in the sense that if $\{t'_1,\dots,t'_{n'}\} \subset \{t_1,\dots,t_n\}$ then $P_\mu\big((t'_1,\dots,t'_{n'});\cdot\big)$ is the marginal distribution of $P_\mu\big((t_1,\dots,t_n);\cdot\big)$ on the coordinates corresponding to $\{t'_1,\dots,t'_{n'}\}$. Having established consistency, Kolmogorov applied his Consistency Theorem to produce a measure $\tilde{\mathbb{P}}_\mu$ on $(\mathbb{R}^N)^{[0,\infty)}$ whose finite dimensional marginals are given by

$$\tilde{\mathbb{P}}_\mu\big(\{\omega : \big(\omega(t_0),\dots,\omega(t_n)\big) \in \Gamma\}\big) = P_\mu\big((t_1,\dots,t_n);\Gamma\big).$$

Finally, using his continuity criterion, he showed that $C\big([0,\infty);\mathbb{R}^N\big)$ has outer measure 1 under $\tilde{\mathbb{P}}_\mu$ and therefore that the restriction of $\tilde{\mathbb{P}}_\mu$ to $C\big([0,\infty);\mathbb{R}^N\big)$ determines a unique Borel probability measure $\mathbb{P}_\mu$ there.

The trail which Kolmogorov blazed was widened and smoothed by many people, including Wm. Feller [4], J. Doob[2], G. Hunt [5], R. Blumenthal and R. Getoor [11], E.B. Dynkin [3], and a host of others. What emerged, especially after the work of Hunt, from this line of research was an isomorphism between Markov processes and potential theory. However, none of these really addressed the question which was bothering Varadhan and me. Namely, from our point of view, the route mapped out by Kolmogorov was too circuitous: one started with L, constructed from it either a transition probability function or resolvent operator, and only then characterized the resulting Markov processes in terms of these ancillary objects. What we were looking for is a characterization of $\mathbb{P}_\mu$ directly in terms of L, one from which it would be clear that if, for each $n \geq 1$, $\mathbb{P}_n$ is related to L_n with initial distribution μ_n, and if $\mu_n \longrightarrow \mu$ weakly and $L_n\varphi \longrightarrow L\varphi$ for smooth φ's with compact support, then $\mathbb{P}_n$ tends weakly to $\mathbb{P}_\mu$. Obviously, the best way to achieve this goal is to phrase the characterization in terms $\mathbb{P}_\mu$-integrals involving $L\varphi$.

Formulating such a characterization was relatively easy. Namely, no matter how one goes about its construction, $\mathbb{P}_\mu$ should have the property that[1]

$$\mathbb{E}^{\mathbb{P}_\mu}\big[\varphi\big(\omega(t_2)\big)\,\big|\,\mathcal{B}_{t_1}\big] - \varphi\big(\omega(t_1)\big) = \mathbb{E}^{\mathbb{P}_\mu}\left[\int_{t_1}^{t_2} L\varphi\big(\omega(\tau)\big)\,d\tau\,\middle|\,\mathcal{B}_{t_1}\right]$$
$$\text{for all } \varphi \in C_c^\infty(\mathbb{R}^N;\mathbb{C}), \tag{4}$$

[1]Throughout, I will be somewhat casual about integrability questions. In the following, no integrability problems arise unless the coefficients are unbounded, in which case special considerations have to be taken into account.

where $\mathcal{B}_t$ is the σ-algebra generated by $\{\omega(\tau) : \tau \in [0,t]\}$. More succinctly,

$$\Big(\varphi\big(\omega(t)\big) - \int_0^t L\varphi\big(\omega(\tau)\big)\,d\tau, \mathcal{B}_t, \mathbb{P}_\mu\Big)$$
$$\text{is a martingale for all } \varphi \in C_c^\infty(\mathbb{R}^N;\mathbb{C}). \tag{5}$$

Thus, we said that $\mathbb{P}_\mu$ *solves the martingale problem for* L *with initial distribution* μ if μ is the $\mathbb{P}_\mu$-distribution of $\omega \rightsquigarrow \omega(0)$ and (5) holds.

Having formulated the martingale problem, our first task was to prove that, in great generality, solutions exist. Of course, seeing as all roads lead to solutions to the martingale problem, in situations to which they applied, we could have borrowed results from earlier work (e.g., of either the Kolmogorov or Itô schools). However, not only would that have made no contribution to known existence results, it would have been very inefficient. At least when a and b are bounded and continuous, existence of solutions to the martingale problem is an easy application of compactness, completely analogous to the proof of existence of solutions to ordinary differential equations with bounded continuous coefficients. Our second task was to show that, under appropriate conditions, solutions are unique, and we knew that uniqueness posed a challenge of an entirely different order from the one posed by existence. Indeed, if all roads lead to solutions to the martingale problem, then uniqueness would prove that all roads must have the same destination.

Our approach to uniqueness depended on the assumptions being made about the coefficients a and b. One approach turned on the observation that any solution to the martingale is the distribution of a solution to an Itô stochastic integral equation. When the coefficients are reasonably smooth, this observation allowed us to prove uniqueness for the martingale problem as a consequence of uniqueness for the associated stochastic integral equation. However, when the coefficients are not smooth, Itô's theory cannot cope. Specifically, because his theory ignores ellipticity properties, it is difficult to see how one goes about bolstering it with powerful results from the vast and beautiful P.D.E. literature.

The key to overcoming the problem just raised is to realize that uniqueness for the martingale problem is a purely distributional question, whereas the uniqueness found in Itô's theory is a path-by-path statement about a certain transformation (the "Itô map") of Brownian paths. Thus, to take advantage of the P.D.E. existence theory, we abandoned Itô and adopted an easy duality argument which shows that uniqueness for the martingale problem requires only a sufficiently good existence statement about the initial value problem for the backward equation $\partial_t u = Lu$. Namely, if u is a classical solution[2] to the

[2]By "classical solution," I mean that the solution is twice continuously differentiable in space and once in time.

backward equation with initial data f, then one can show that, for any solution $\mathbb{P}$ to the martingale problem for L,

$$\left(u\big((t_2-t)^+,\omega(t)\big),\mathcal{B}_t,\mathbb{P}\right)$$

is a martingale and therefore that, for all $0 \le t_1 < t_2$,

$$\mathbb{E}^{\mathbb{P}}\left[f\big(\omega(t_2)\big)\,\middle|\,\mathcal{B}_{t_1}\right] = u\big(t_2-t_1,\omega(t_1)\big).$$

Hence, if such classical solutions exist for enough f's, then the conditional distribution of $\omega \rightsquigarrow \omega(t_2)$ under $\mathbb{P}$ given $\mathcal{B}_{t_1}$ is uniquely determined for all $0 \le t_1 < t_2$, and from this, together with the condition that the $\mathbb{P}$-distribution of $\omega \rightsquigarrow \omega(0)$ is μ, it follows that there is only one solution to the martingale problem. This argument works without a hitch when, for example, a and b are bounded and uniformly Hölder continuous and a is uniformly elliptic (i.e., $a \ge \epsilon I$ for some $\epsilon > 0$), since in that case P.D.E. theorists had long ago provided the necessary existence statement about the backward equation.

When uniform Hölder continuity is replaced by simple uniform continuity, the existence theory for the backward equation is less satisfactory. To be precise, although solutions continue to exist, they are no longer classical. Instead, their second derivatives exist only in $L^p(\mathbb{R}^N;\mathbb{C})$ for $p \in (1,\infty)$. As a consequence, the preceding duality argument breaks down unless one can show *a priori* that every solution to the martingale problem is sufficiently regular to tolerate limit procedures involving $L^p(\mathbb{R}^N)$-convergence of the integrands. Without question, the proof of such an *a priori* regularity result was the single most intricate part of our program. In broad outline, our proof went as follows. First, we developed machinery which allowed us to reduce everything to the case when $b = 0$ and a is an arbitrarily small perturbation of the identity. Second, we again used the connection with Itô stochastic integral equations to show that any solution $\mathbb{P}$ can be approximated by $\mathbb{P}_n$'s which, conditioned on $\mathcal{B}_{\frac{m}{n}}$, are Gaussian during the time interval $\left[\frac{m}{n},\frac{m+1}{n}\right]$ and therefore have the required regularity properties. Finally, as an application of the Calderon–Zygmund theory of singular integral operators, for each $p \in (1,\infty)$ we were able to show that, when the perturbation is sufficiently small, the $L^p(\mathbb{R}^N)$-regularity properties of the approximating $\mathbb{P}_n$'s can be controlled independent of n and are therefore enjoyed by $\mathbb{P}$ itself.

The steps outlined above led us to a proof of the following theorem, which should be considered the centerpiece of [12].

Theorem. *Let L be given by* (1)*, where a and b are bounded, a is continuous, b is Borel measurable, and $a(x)$ is symmetric and strictly positive definite for each $x \in \mathbb{R}^N$. Then, for each Borel probability measure μ on $\mathbb{R}^N$, there is precisely* *one solution to the martingale for L with initial distribution μ. Moreover, if δ_x is the unit mass at x and $\mathbb{P}_x = \mathbb{P}_{\delta_x}$, then $x \rightsquigarrow \mathbb{P}_x$ is weakly continuous and $\{\mathbb{P}_x : x \in \mathbb{R}^N\}$ is a strong Markov family.*

Concluding Remarks: There are several additional comments which may be helpful.

(1) Although, with 20-20 hindsight, the characterization given by (4) is the most obvious one, it is not the one which we chose initially. Instead, because we were influenced at the time both by Henry McKean's treatment in [9] of Itô's theory of stochastic integral equations and by a possible analogy with characteristic functions, we chose to characterize $\mathbb{P}_\mu$ by saying that

$$\left(\exp\left[\left(\xi, \omega(t) - \int_0^t b(\omega(\tau))\, d\tau\right)_{\mathbb{R}^N} - \tfrac{1}{2}\int_0^t \left(\xi, a(\tau)\xi\right)_{\mathbb{R}^N} d\tau\right], \mathcal{B}_t, \mathbb{P}_\mu\right)$$
$$\text{is a martingale for all } \xi \in \mathbb{C}^N. \tag{6}$$

The equivalence of (4) and (5) is a quite easy application of a elementary Fourier analysis and the observation that, aside from integrability issues,

$$\left(M(t)V(t) - \int_0^t M(\tau)\, dV(\tau), \mathcal{B}_t, \mathbb{P}\right)$$

is a martingale when $\left(M(t), \mathcal{B}_t, \mathbb{P}\right)$ is a continuous martingale and $V(t)$ is a continuous, $\{\mathcal{B}_t : t \geq 0\}$-adapted process of locally bounded variation.

(2) The preceding theorem is not the one for which we had originally hoped that we would have to settle. When we started, we had hoped for a statement in which a need only be Borel measurable and locally uniformly elliptic. Using special, heavily dimension dependent arguments, we did prove such a result when $N \in \{1, 2\}$, and we somewhat naïvely assumed that it was only the limits on our own technical powers which prevented us from doing so in higher dimensions. However, a few years ago Nikolai Nadirashvili gave in [?] a beautiful example which showed that the limits on our powers were not responsible for our inability to go to higher dimensions. Namely, he produced a uniformly elliptic, bounded, Borel measurable a on $\mathbb{R}^3$ for which the associated martingale problem admits more than one solution.

(3) In retrospect, there are many aspects of this work which look rather crude and unnecessarily cumbersome. For instance, around the same time, first P.A. Meyer [?] and then N. Kunita and S. Watenabe [8] gave a much better and cleaner treatment of the stochastic integral calculus which we developed. More significant, had we been aware of the spectacular work being done by N. Krylov, in particular the estimate in [7], we could have vastly simplified the regularity proof just described.

(4) In a later work, we extended the martingale problem approach to cover diffusions with boundary conditions. The resulting paper [13] has, for good reasons, been much less influential: it is nearly unreadable. On the other hand, it contains information which, so far as I know, is available nowhere else. In particular, I do not think the theorems there dealing with approximations by discrete Markov chains have ever been improved.

(5) Varadhan and my joint work on diffusion theory culminated in our papers [14] and [15] on degenerate diffusions. The first of these contains our initial version of "the support theorem," which has been given several other proofs over the years. Among other things, the second paper contains an extension of "the support theorem" to cover situations in which the diffusion matrix a is smooth but does not admit a smooth square root. It seems that most probabilists are not particularly impressed by this extension, but we were very proud of it at the time.

References

[1] R.M. Blumenthal and R.K. Getoor, *Markov Processes and Potential Theory,* Academic Press, 1961.

[2] J.L. Doob, *Stochastic Processes,* J. Wiley, N.Y., 1950.

[3] E.B. Dynkin, *Markov Processes, volumes I & II,* Springer–Verlag translation of 1963 Russian edition, 1965.

[4] Wm. Feller, *Zur Theorie der Stochastichen Prozesse,* Math. Ann. **133** (1936) 113–160.

[5] G. Hunt, *Markov processes and potentials I, II, III,* Illinois J. Math. **1** (1957) 44-93; **1** (1957) 316-369; **2** (1958) 151-213.

[6] A.N. Kolmogorov, *Uber die Analytischen Methoden in der Wahrscheinlichkeitsrechnung,* Math. Ann. **104** (1931) 415–458.

[7] Krylov, N., *On quasi diffusion processes,* Theory Prob. and Appl. **11** (1966) 373-389.

[8] Kunita, H. and Watanabe, S. , *On square integrable martingales,* Nagoya Math. J. **30** (1967) 209–245.

[9] H.P. McKean, *Stochastic Integrals,* Academic Press and now available in the AMS Chelsea Series, 1973.

[10] 0P.A. Meyer, *Probability and Potentials,* Blaisdell, Boston, 1966.

[11] 1N.S. Nadirashvili, *Nonuniqueness in the martingale problem and the Dirichlet problem,* Ann. Scuola Norm. Sup. Pisa Cl. Sci. (4) **24** (1997) 537–550.

[12] Stroock, D. and Varadhan, S. , *Diffusion processes with continuous coefficients, I and II,* CPAM **XXII** (1969) 345–400 and 479–30.

[13] ———, *Diffusion processes with boundary conditions,* CPAM **XXIV** (1971) 147–225.

[14] ———, *On the support of diffusion processes, with applications to the strong maximum principle,* Proceedings of the Sixth Berkeley Symposium on Probability and Statistics, 1970333–359 .

[15] ———, *On degenerate elliptic-parabolic operators of the second order and their associated diffusions,* CPAM **XXV** (1972) 651–713.

Varadhan's Work on Large Deviations

Daniel W. Stroock*

Suppose that $\{\mu_n : n \in \mathbb{N}\}$ is a sequence of Borel probability measures on a Polish space Ω, and assume that, as $n \to \infty$, μ_n degenerates to the point mass δ_{ω_0} at $\omega_0 \in \Omega$. Then it is reasonable to say that, at least for large n's, neighborhoods of ω_0 represent "typical" behavior and that their complements represent "deviant" behavior, and it is often important to know how fast their complements are becoming deviant. Finding a detailed solution to such a problem usually entails rather intricate analysis. However, if one's interest is in behavior which is "highly deviant" in the sense that it is dying out at an exponential rate, and if one is satisfied with finding the exponential rate at which it is disappearing, then one is studying *large deviations* and life is much easier. Indeed, instead of trying to calculate the asymptotic limit of quantities like[1] $\mu_n\big(B(\omega_0, r)^c\big)$, one is trying to calculate $\lim_{n\to\infty} \frac{1}{n} \log \mu_n\big(B(\omega_0, r)^c\big)$, which is an inherently simpler task.

To develop intuition for this type of analysis, remember that we are dealing here with probability measures. Thus, the only way that the μ_n's can be degenerating to δ_{ω_0} is that more and more of their mass is getting concentrated in a neighborhood of ω_0. In the nicest situation, this concentration is taking place because $\mu_n(d\omega) = \frac{1}{Z_n} e^{-nI(\omega)} \lambda(d\omega)$, where $I(\omega) > I(\omega_0) \geq 0$ for $\omega \neq \omega_0$ and λ is some reference measure. Indeed, assuming that $\lim_{n\to\infty} \frac{1}{n} \log Z_n = 0$, then

$$\begin{aligned} \lim_{n\to\infty} \frac{1}{n} \log \mu_n(\Gamma) &= \lim_{n\to\infty} \log \|1_\Gamma e^{-I}\|_{L^n(\lambda)} \\ &= \log \|1_\Gamma e^{-I}\|_{L^\infty(\lambda)} = -\operatorname*{essinf}_{\omega\in\Gamma} I(\omega). \end{aligned}$$

That is,

$$\lim_{n\to\infty} \frac{1}{n} \log \mu_n(\Gamma) = -\operatorname*{essinf}_{\omega\in\Gamma} I(\omega). \tag{1}$$

Of course, for many applications (e.g., to number theory, geometry, or statistical mechanics), the non-appearance of Z_n in the answer would mean that one has thrown out the baby with the wash. On the other hand, because it is so crude,

*Department of Mathematics, Massachusetts Institute of Technology, Cambridge, MA 02139

[1]Here, $B(\omega, r)$ denotes the ball of radius r centered at ω.

the type of thinking used in the preceding predicts correct results even when it has no right to. To wit, suppose that $\Omega = \{\omega \in C([0,1];\mathbb{R}) : \omega(0) = 0\}$ and μ_n is the distribution of $\omega \in \Omega \longmapsto n^{-\frac{1}{2}}\omega \in \Omega$ under standard Wiener measure. Clearly the μ_n's are degenerating to the point mass at the path $\mathbf{0}$ which is identically 0. Moreover, Feynman's representation of μ_n is

$$\mu_n(d\omega) = \frac{1}{Z_n} e^{-nI(\omega)}\, \lambda(d\omega),$$

where

$$I(\omega) = \frac{1}{2}\int_0^1 |\dot{\omega}(t)|^2\, dt$$

and λ is Lebesgue measure on Ω. Ignoring the fact that none of this is mathematically kosher and proceeding formally, one is led to the guess that (1) may nonetheless be true, at least after one has taken into account some of its obvious flaws. In particular, there are two sources of concern. The first of these is the almost sure non-differentiability of Wiener paths. However this objection is easily overcome by simply defining $I(\omega) = \infty$ unless ω has a square integrable derivative. The second objection is that λ does not exist and therefore the "ess" before the "inf" has no meaning. This objection is more serious, and its solution requires greater subtlety. In fact, it was his solution to this problem which was one of Varadhan's seminal contributions to the whole field of large deviations. Namely, our derivation of (1) was purely measure theoretic: we took no account of topology. On the other hand, not even the sense in which the μ_n's are degenerating can be rigorously described in purely measure theoretic terms. The best that one can say is that they are tending weakly to $\delta_{\mathbf{0}}$. Thus, one should suspect that (1) must be amended to reflect the topology of Ω and that topology should appear in exactly the same way that it does in weak convergence. With this in mind, one can understand Varadhan's answer that (1) should be replaced by

$$\begin{aligned} -\inf_{\omega\in\Gamma^\circ} I(\omega) &\le \varliminf_{n\to\infty} \frac{1}{n}\log\mu_n(\Gamma) \\ &\le \varlimsup_{n\to\infty} \frac{1}{n}\log\mu_n(\Gamma) \le -\inf_{\omega\in\overline{\Gamma}} I(\omega). \end{aligned} \tag{2}$$

Monroe Donsker provided the original inspiration for this type of analysis of rescaled Wiener measure, and his student Schilder was the first to obtain rigorous results. However, it was Varadhan in [1] who first realized that Schilder's work could be viewed in the context of large deviations, and it was he who gave and proved the validity of the formulation in (2). Indeed, a strong case can be made for saying that the modern theory of large deviations was born in [1]. In particular, (2) quickly became the archetype for future results, and families

$\{\mu_n : n \in \mathbb{N}\}$ for which (2) are now said to satisfy *the large deviation principle with rate function I*. In addition, it was in the same article that Varadhan proved how to pass from (2) to the sort of results which Schilder had proved. Namely, he proved that if (2) holds with a rate function I which has compact level sets[2] and $F : \Omega \longrightarrow \mathbb{R}$ is a bounded, continuous function, then

$$\lim_{n\to\infty} \mathbb{E}^{\mu_n}\left[e^{nF}\right] = \sup_{\omega\in\Omega}\big(F(\omega) - I(\omega)\big). \tag{3}$$

This result, which is commonly called *Varadhan's Lemma*, is exactly what one would expect from the model case when $\mu_n(d\omega) = \frac{1}{Z_n} e^{-nI}\,\lambda(d\omega)$, and its proof in general is quite easy, but one would be hard put to overstate its importance. Not only is it a practical computational tool, it provides a link between the theory of large deviations and convex analysis. Specifically, when Ω is a closed, convex subset of a topological vector space E, then, under suitable integrability assumptions, Varadhan's Lemma combined with the inversion formula for Legendre transforms often can be used to identify the rate function I as the Legendre transform

$$\Lambda^*(\omega) = \sup_{\lambda\in E^*}\big\{\langle\omega,\lambda\rangle - \Lambda(\lambda) : \lambda \in E^*\big\},$$

where

$$\lambda \in E^* \longrightarrow \Lambda(\lambda) \equiv \lim_{n\to\infty} \frac{1}{n}\log \mathbb{E}^{\mu_n}\left[e^{n\lambda(\omega)}\right].$$

Had he only laid the foundation for the field, Varadhan's impact on the study of large deviations would have been already profound. However, he did much more. Perhaps his deepest contributions come from his recognition that large deviations underlie and explain phenomena in which nobody else even suspected their presence. The depth of his understanding is exemplified by his explanation of Marc Kac's famous formula for the principal eigenvalue of a Schrödinger operator. Donsker had been seeking such an explanation for years, but it was not until he joined forces with Varadhan that real progress was made on the problem. Prior to their article [2], all applications (Schilder's theorem, including its extensions and improvements by Varadhan, as well as the many beautiful articles by Freidlin and Wentcel) of large deviations to diffusion theory had been based on the observation that, during a short time interval, "typical" behavior of a diffusion is given by the solution to an ordinary differential equation. Thus, the large deviations in these applications come from the perturbation of an ordinary differential equation by a Gaussian-like noise term. The large deviations in [2] have an entirely different origin. Instead of short time behavior of the diffusion paths themselves, the quantity under consideration is

[2] That is, $\{\omega : I(\omega) \le R\}$ is compact for each $R \in [0,\infty)$.

the long time behavior of their empirical distribution. In this case, "typical" behavior is predicted by ergodic theory, and the large deviations are those of the empirical distribution from ergodic behavior. The situation in [2] is made more challenging by the fact that there really is no proper ergodic behavior of Brownian motion on the whole of space since, in so far as possible, the empirical distribution of a Brownian path is trying to become normalized Lebesgue measure. What saves the day is the potential term in the Schrödinger operator, whose presence penalizes paths which attempt to spread out too much.

The upshot of Donsker and Varadhan's analysis in [2] is a new variational formula for the principal eigenvalue. Although their formula reduces to classical one in the case of self-adjoint operators, it has the advantage that it relies entirely on probabilistic reasoning (i.e., the minimum principle) and, as they showed in [3], is therefore equally valid for operators which are not self-adjoint. More important, it launched a program which produced a spectacular sequence of articles. The general theory was developed in [3] and [5], each one raising the level[3] of abstraction and, at the same time, revealing more fundamental principles. However, they did not content themselves with general theory. On the contrary, they applied their theory to the solve a remarkably varied set of problems, ranging from questions about the range of a random walk in [6] to questions coming from mathematical physics about function space integrals in [5] and [7], and each abstraction was designed to tackle a specific problem.

As is nearly always the case when breaking new ground, the applications required ingenious modifications of the general theory. To give an indication of just how ingenious these modifications had to be, consider the "Wiener sausage" calculation in [4]. The problem there, which grew out of a question posed by the physicist J. Luttinger, was to find the asymptotic volume of the tubular neighborhood of a Brownian path as the time goes to infinity and the diameter of the neighborhood goes to 0. If one thinks about it, one realizes that this volume can be computed by looking at a neighborhood in the space of measures of the empirical distribution. However, the neighborhood that one needs is the one determined by the variation norm, whereas their general theory deals with the weak topology. Thus, except in one dimension where local time comes to the rescue, they had to combine their general theory with an intricate approximation procedure in order to arrive at their goal. Their calculation in [5] is a true *tour de force*, only exceeded by their solution to the polaron problem in [7].

[3]Among their disciples, these levels are designated as levels I, II, and III. Because it reminds me, depending on my mood, of the voice one hears on the elevator as one ascends from "boys clothing" to "women's lingerie" at Macy's or of the levels of hell through which Dante descended, I have never subscribed to this terminology.

In conclusion, it should be emphasized that Varadhan's contributions to the theory of large deviations were to both its foundations and its applications. Because of his work, the subject is now seen as basic tool of analysis, not simply an arcane branch of probability and statistics. With twenty-twenty hindsight, it has become clear that large deviations did not always provide the most efficient or best approach to some of the problems which he solved, but there can be no doubt that his insights have transformed the field forever.

References

[1] Varadhan, S.R.S., *Diffusion processes in a small time interval* , CPAM **20** (1967) 659–685.

[2] Donsker, M. and Varadhan, S.R.S., *On a variational formula for the principal eigenvalue for operators with maximum principle,* PNAS **72** (1975) 780–783.

[3] ———, *Asymptotic evaluation of certain Markov process expectations for large time, I & II,* CPAM **28** (1975) 1–47 & 279–301.

[4] ———, *Asymptotics for the Wiener sausage* CPAM **28#4** (1975) 525–565.

[5] ———, *Asymptotic evaluation of certain Markov process expectations for large time, III* , CPAM **29 #4** (1976) 389–461.

[6] ———, *On the number of distinct sites visited by a random walk,* CPAM **32 #6** (1979) 721–747.

[7] ———, *The polaron problem and large deviations* , Physics Reports. A Review Section of Physics Letters **77** (No. 3) (1981) 235–237.

Varadhan's Work on Large Deviations and Homogenization

Fraydoun Rezakhanlou*

The main theme of this expository article is to review some of Raghu Varadhan and his collaborators' contributions to the question of homogenization for the following stochastic models:

- Stationary Hamilton-Jacobi (HJ) and Hamilon-Jacobi-Bellman (HJB) equation.
- Random walk in random environment (RWRE).
- Simple exclusion process (SEP).

All the above models share similar scaling behaviors and in some sense represent evolving height functions which are governed by local and random growth rules. In fact the law of a RWRE satisfies an equation which resembles a discrete HJB equation, and growth rates of the particle currents in SEP are described by a nonlinear function of the height differences. Reviewing Raghu Varadhan's fundamental contributions sheds light on some universal behavior of stochastic growth models.

1. Hamilton-Jacobi and Hamilon-Jacobi-Bellman Equation

To introduce the basic idea behind homogenization, we first consider (inhomogeneous) Hamilton–Jacobi (HJ) equation

$$u_t = H(x, u_x), \tag{1.1}$$

where H is stationary and ergodic in the first variable x. More precisely, we have a probability space $(\Omega, \mathcal{F}, \mathbb{P})$ with $\mathcal{F}$ a Borel σ-field on Ω and $\mathbb{P}$ a probability measure on $(\Omega, \mathcal{F})$, which is invariant with respect to a family of translation operators, i.e., for every $x \in \mathbb{R}^d$, there exists a measurable function $\tau_x : \Omega \to \Omega$

*This work is supported in part by NSF grant DMS07-07890. Department of Mathematics, UC Berkeley

so that $\tau_x \circ \tau_y = \tau_{x+y}$ and $\mathbb{P}(\tau_x A) = \mathbb{P}(A)$ for every $A \in \mathcal{F}$ and $x, y \in \mathbb{R}^d$. We also assume that τ_x is ergodic; i.e., $\tau_x A = A$ for all $x \in \mathbb{R}^d$ implies that either $\mathbb{P}(A) = 1$ or 0.

Now $H(x,p,\omega) = H_0(\tau_x\omega, p)$ where $H_0 : \Omega \times \mathbb{R}^d \to \mathbb{R}$ is a measurable function. We think of (x,t,u) as the *microscopic coordinates* with the graph of $u(\cdot,t)$ representing a random interface. To switch to *macroscopic coordinates*, we set

$$u^\varepsilon(x,t;\omega) = \varepsilon u\left(\frac{x}{\varepsilon}, \frac{t}{\varepsilon}; \omega\right). \tag{1.2}$$

We now have

$$u_t^\varepsilon = H\left(\frac{x}{\varepsilon}, u_x^\varepsilon\right). \tag{1.3}$$

We note that the right-hand side of (1.3) fluctuates greatly over macroscopic shifts in position x. The huge fluctuation in H, though, does not necessarily imply correspondingly huge fluctuations in u^ε. This is the homogenization phenomenon, i.e., we expect $u^\varepsilon \to \bar{u}$ as $\varepsilon \to 0$ with $\bar{u}$ solving a *homogenized HJ* equation

$$\bar{u}_t = \bar{H}(\bar{u}_x), \tag{1.4}$$

where $\bar{H} : \mathbb{R}^d \to \mathbb{R}$ is the *homogenized Hamiltonian* and does not depend on ω.

As our second example, we consider Hamilton–Jacobi–Bellmann equation

$$u_t = H(x, u_x) + \frac{1}{2}\Delta u \tag{1.5}$$

with $H(x,p) = H(x,p,\omega)$ as before. We define u^ε as in (1.2) and (1.3) becomes

$$u_t^\varepsilon = H\left(\frac{x}{\varepsilon}, u_x^\varepsilon\right) + \frac{\varepsilon}{2}\Delta u^\varepsilon. \tag{1.6}$$

Again we expect to have $u^\varepsilon \to \bar{u}$ with $\bar{u}$ satisfying an equation of the form (1.4) for a different homogenized Hamiltonian $\bar{H}$. Indeed the homogenization for both (1.3) and (1.6) have been achieved by Souganidis [S], Rezakhanlou–Tarver [RT], Lions-Souganidis and Kosygina–Rezakhanlou–Varadhan [KRV] provided that $H(x,p)$ is convex in p, and satisfies suitable technical assumptions that we do not elaborate in this article. (See also Kosygina-Varadhan [KR] when H is allowed to depend on the time variable.) Most notably [KRV] obtains a variational formula for $\bar{H}$. In the case of (1.6), $\bar{H}$ is given by

$$\bar{H}(p) = \inf_g \operatorname*{esssup}_\omega \left[H_0(p + g(\omega), \omega) + \frac{1}{2}\nabla \cdot g(\omega)\right] \tag{1.7}$$

where essential supremum is taken with respect to the probability measure $\mathbb{P}$ and the infimum is taken over functions $g : \Omega \to \mathbb{R}^d$ such that $\mathbb{E}g = 0$ and

$\nabla \cdot g = 0$ weakly. Here ∇ is the generator of the group $\{\tau_x\}$. That is,

$$\nabla f(\omega) \cdot v = \lim_{t \to 0} \frac{1}{t}(f(\tau_{tv}\omega) - f(\omega)), \tag{1.8}$$

whenever the limit exists. We expect a similar formula to hold in the case of (1.3), namely

$$\bar{H}(p) = \inf_g \operatorname*{esssup}_w H_0(p + g(\omega), \omega). \tag{1.9}$$

Before we turn to our next model, let us make an observation regarding the homogenization of (1.6). Note that if

$$H(x, p, \omega) = \frac{1}{2}|p|^2 + b(x, \omega) \cdot p + V(x, \omega), \tag{1.10}$$

and u is a solution of (1.5), then by Hopf–Cole transform, the function $w = e^u$ solves

$$w_t = \frac{1}{2}\Delta w + b(x, \omega) \cdot \nabla w + V(x, \omega) w(x, \omega). \tag{1.11}$$

By Feynmann–Kac Formula, there is a probabilistic representation for w using a diffusion with a drift b. More precisely, if $X(t, x; \omega)$ denotes the solution to

$$\begin{aligned} dX(t) &= b(X(t), \omega)dt + d\beta(t), \\ X(0) &= x, \end{aligned} \tag{1.12}$$

then

$$w(x, t; \omega) = E^\omega w(X(t, x; \omega), 0) e^{\int_0^t V(X(s, x; \omega), \omega) ds}. \tag{1.13}$$

Here β is a standard Brownian motion and E^ω denotes the expected value for the process $X(t)$. The function V is the potential and if $V \leq 0$, then $-V$ may be interpreted as a killing rate for the diffusion X. With this interpretation, $w(x, t; \omega)$ is the expected value of $w(\hat{X}(t), 0)$ with $\hat{X}$ denoting the diffusion with the killing. We now would like to use our probabilistic representation to rewrite u^ε. If

$$u^\varepsilon(x, 0; \omega) = f(x) \tag{1.14}$$

for a deterministic initial condition f, then

$$u^\varepsilon(x, t; \omega) = \varepsilon \log E^\omega \exp\left[\varepsilon^{-1} f(\varepsilon X(t/\varepsilon, x/\varepsilon; \omega)) + \int_0^{t/\varepsilon} V(X(s, x/\varepsilon; \omega), \omega) ds\right]. \tag{1.15}$$

In particular

$$u^\varepsilon(0, 1; \omega) = \varepsilon \log E^\omega e^{\varepsilon^{-1} f(X(\varepsilon^{-1}; \omega)) + \int_0^{\varepsilon^{-1}} V(X(s; \omega), \omega) ds} \tag{1.16}$$

where $X(s;\omega) := X(s,0;\omega)$ is the diffusion starting from the origin. On the other hand, since $\bar{H}$ is convex (which is evident from (1.7)) we may use Hope–Lax–Oleinik formula to write

$$\bar{u}(x,t) = \sup_y \left(f(y) - t\bar{L}\left(\frac{y-x}{t}\right)\right) \tag{1.17}$$

where $\bar{L}$ is the convex conjugate of $\bar{H}$. In particular,

$$\lim_{\varepsilon\to 0} u^{\varepsilon}(0,1;\omega) = \bar{u}(0,1) = \sup_y (f(y) - \bar{L}(y)). \tag{1.18}$$

By a celebrated lemma of Varadhan, (1.18) is equivalent to saying that for almost all ω, the diffusion $\hat{X}$ satisfies a *large deviation principle* with the rate function $\bar{L}$. When $b \equiv 0$ and

$$-V(x,\omega) = \sum_{j\in I} V_0(x - x_j) \tag{1.19}$$

with $\omega = \{x_j : j \in I\}$ a Poisson point process and V_0 a continuous function of compact support, the large deviation principle for $\hat{X}$ was earlier established by Sznitman [Sz].

In words, large deviation principle for the diffusion $\hat{X}(\cdot;\omega)$ is equivalent to homogenization for the equation (1.6). Let us write P^{ω} for the law of the process $\hat{X}(\cdot;\omega)$. What we have in (1.18) is an example of a *quenched* large deviation principle. We may also consider the annealed law

$$\bar{P} = \int P^{\omega}\, \mathbb{P}(d\omega) \tag{1.20}$$

and wonder whether an annealed large deviation principle is true for the process $\hat{X}$. More precisely, whether or not

$$\begin{aligned}\lim_{\varepsilon\to 0} \varepsilon \log \int E^{\omega} \exp\left[\varepsilon^{-1} f(X(\varepsilon^{-1},\omega)) + \int_0^{\varepsilon^{-1}} V(X(s,\omega),\omega)ds\right] \mathbb{P}(d\omega)\\ = \sup_y (f(y) - J(y))\end{aligned} \tag{1.21}$$

for a suitable rate function J. In terms of u^{ε}, this is equivalent to saying

$$\lim_{\varepsilon\to 0} \varepsilon \log \int e^{\varepsilon^{-1} u^{\varepsilon}(0,1;\omega)} \mathbb{P}(d\omega) = \sup_y (f(y) - J(y)). \tag{1.22}$$

This would follow if we can establish a large deviation principle for the convergence of u^{ε} to $\bar{u}$. That is, if we can find a function $K_f(y;x,t)$ such that

$$\lim_{\varepsilon\to 0} \varepsilon \log \int e^{\varepsilon^{-1}\lambda u^{\varepsilon}(x,t;\omega)} \mathbb{P}(d\omega) = \sup_y (\lambda y - K_f(y;x,t)). \tag{1.23}$$

The annealed large deviation (1.22) in the case of $b \equiv 0$ and V of the (1.19) can be found in the manuscript [S] but (1.23) remains open even when $b = 0$.

It is worth mentioning that there is also a variational description for the large deviation rate function $\bar{L}$, namely

$$\bar{L}(v) = \inf_a \inf_{\mu \in \Gamma_{a,v}} \int L_0(\omega, a(\omega))\mu(d\omega) \tag{1.24}$$

where $L_0(\omega, v)$ is the convex conjugate of $H_0(\omega, p)$ and $\Gamma_{a,v}$ is the set of invariant measures for the diffusions $\mathcal{A}_a = a(\omega) \cdot \nabla + \frac{1}{2}\Delta$ with $\int a(\omega)\mu(d\omega) = v$. In the case of (1.3) the generator $\mathcal{A}_a$ takes the form $a \cdot \nabla$ and when H is periodic in x (i.e., Ω is the d-dimensional torus with $\mathbb{P}$ the uniform measure), the formula (1.24) is equivalent to a formula of Mather for the averaged Lagrangian and our homogenization is closely related to the weak KAM theory. See Fathi [F] and Evans–Gomez [EG] for more details.

2. Random Walk in Random Environment

As our second class of examples, we consider a discrete version of the diffusion (1.12). This is simply a random walk in a random environment (RWRE). To this end, let us write $\mathcal{P}$ for the space of probability densities on the d-dimensional lattice $\mathbb{Z}^d$, i.e., $p \in \mathcal{P}$ if $p : \mathbb{Z}^d \to [0, 1]$ with $\sum_z p(z) = 1$. We set $\Omega = \mathcal{P}^{\mathbb{Z}^d}$ and $\omega \in \Omega$ is written as

$$\omega = (p_a : a \in \mathbb{Z}^d).$$

Given $\omega \in \Omega$, we write $X(n, a; \omega)$ to denote a random walk at time n with starting point $a \in \mathbb{Z}^d$ and transition probabilities $p_a, a \in \mathbb{Z}^d$. More precisely

$$P^\omega(X(n+1) = y \mid X(n) = x) = p_x(y - x).$$

Given a function $g : \mathbb{Z}^d \to \mathbb{R}$, we write

$$T_n g(x) = E^\omega g(X(n, x; \omega)),$$

so that

$$T_1 g(x) = \sum_{y \in \mathbb{Z}^d} g(y) p_x(y - x).$$

To compare with (1.11) in the case of $V \equiv 0$, we also write

$$w(x, n) = T_n g(x)$$

for a given initial g. This trivially solves

$$w(x, n+1) = (T_1 w(\cdot, n))(x).$$

To compare with (1.5), we set $u = \log w$ so that

$$u(x, n+1) - u(x, n) = (Au(\cdot, n))(x)$$

where

$$Ag(x) = \log T_1 e^g(x) - g(x) = \log \sum_z e^{g(x+z)-g(x)}\ p_x(z).$$

Now homogenization means that we are interested in

$$\bar{u}(x, t) = \lim_{\varepsilon \to 0} \varepsilon u \left(\left[\frac{x}{\varepsilon} \right], \left[\frac{t}{\varepsilon} \right]; \omega \right)$$

provided that ω is distributed according to an ergodic stationary probability measure $\mathbb{P}$ where

$$\tau_x \omega = (p_{y+x} : y \in \mathbb{Z}^d).$$

(Here by $[a]$ we mean the integer part of a.) Again $\bar{u}$ solves (1.4) provided that $\lim_{\varepsilon \to 0} u^\varepsilon(x, 0) = \bar{u}(x, 0) = f(x)$ exists initially. The function $\bar{L}$, the convex conjugate of $\bar{H}$, is the quenched large deviation rate function for $X(n; \omega)$. More precisely, for any bounded continuous $f : \mathbb{R}^d \to \mathbb{R}$,

$$\lim_{n \to \infty} n^{-1} \log E e^{f(n^{-1} X(n,0;\omega))} = \sup_y (f(y) - \bar{L}(y)).$$

This has been established under an ellipticity condition on p_x by Varadhan [V1]. See Bolthausen and Sznitman [BS] for a survey on earlier results. The analog of (1.7) is the following formula of Rosenbluth [Ro] :

$$\bar{H}(p) = \inf_g \operatorname*{esssup}_\omega \sum_y p_0(y) e^{p \cdot z + g(\omega, z)} \tag{2.1}$$

with infimum over functions $(g(\cdot, z) : \Omega \to \mathbb{R} : z \in \mathbb{Z}^d)$ such that $\mathbb{E} g(\cdot, z) = 0$ and g is a "closed 1-form". By the latter we mean that for every loop $x_0, x_1, \ldots, x_{k-1}, x_k = x_0$, we have that $\sum_{r=0}^{k-1} g(\tau_{x_r} \omega, x_{k+1} - x_k) = 0$.

We now turn to the annealed large deviations for a RWRE. For this, we need to select a tractable law for the environment. Pick a probability measure β on $\mathcal{P}$ and set $\mathbb{P}$ to be the product of β to obtain a law on $\mathcal{P}^{\mathbb{Z}^d}$. The annealed measure $\bar{P} = \int P^\omega\ \mathbb{P}(d\omega)$ has a simple description. For this, we write $Z(n) = X(n+1) - X(n)$ for the jump the walk performs at time n. We also define

$$N_{x,z}(n) = \#\{i \in \{0, 1, 2, \ldots, n\} : X(i) = x, Z(i) = z\}.$$

We certainly have

$$P(X(1; \omega) = x_1, \ldots, X(n, \omega) = x_n) = \prod_{z, x \in \mathbb{Z}^d} (p_x(z))^{N_{x,z}(n)},$$

$$\bar{P}(X(1) = x_1, \ldots, X(n) = x_n) = \prod_{x \in \mathbb{Z}^d} \int \prod_z (p(z))^{N_{x,z}(n)} \beta(dp),$$

where now

$$N_{x,z}(n) = \#\{i \in \{0, 1, \ldots, n-1\} : x_i = x, x_{i+1} - x_i = z\}.$$

Evidently $\bar{P}$ is the law of a non-Markovian walk in $\mathbb{Z}^d$. Varadhan in [V1] establishes the annealed large deviations principle under a suitable ellipticity condition on β. The method relies on the fact that the environment seen from the walker is a Markov process for which Donsker–Varadhan Theory may apply if we have enough control on the transition probabilities.

If we set

$$\begin{aligned} W_n &= (0 - X(n), X(1) - X(n), \ldots, X(n-1) - X(n), X(n) - X(n)) \\ &= (s_{-n}, \ldots, s_{-1}, s_0 = 0) \end{aligned}$$

for the chain seen from the location $X(n)$, then we obtain a walk of length n that ends at 0. The space of such walks is denoted by $\mathbf{W}_n$. Under the law $\bar{P}$, the sequence $W_1, W_2, \ldots$ is a Markov chain with the following rule:

$$\bar{P}(W_{n+1} = T_z W_n \mid W_n) = \frac{\bar{P}(T_z W_n)}{\bar{P}(W_n)} = \frac{\int_{\mathcal{P}} p(z) \prod_a p(a)^{N_{0,a}} \beta(dp)}{\int_{\mathcal{P}} \prod_a p(a)^{N_{0,a}} \beta(dp)}, \tag{2.2}$$

where $N_{0,a} = N_{0,a}(W_n)$ is the number of jumps of size a from 0 for the walk W_n. Here $T_z W_n$ denotes a walk of size $n+1$, which is formed by translating the walk W_n by $-z$ so that it ends at $-z$ instead of 0, and then making the new jump of size z so that it ends at 0.

We wish to establish a large deviation principle for the Markov chain with transition probability $q(W, z)$ given by (2.2) where $W = W_n \in \bigcup_{m=0}^{\infty} \mathbf{W}_m$ and z is the jump size. We assume that with probability one, the support of $p_0(\cdot)$ is contained in the set $D = \{z : |z| \leq C_0\}$. Naturally q extends to those infinite walks $W \in \mathbf{W}_\infty$ with $N_{0,a} < \infty$ for every $a \in D$. If we let $\mathbf{W}_\infty^{tr}$ denote the set of transient walks, then the expression $q(W, z) = \mathbf{q}(W, T_z W)$ given by (2.2) defines the transition probability for a Markov chain in $\mathbf{W}_\infty^{tr}$. Donsker–Varadhan Theory suggests that the empirical measure

$$\frac{1}{n} \sum_{m=0}^{n-1} \delta_{W_m}$$

satisfies a large deviation principle with a rate function

$$\mathcal{I}(\mu) = \int_{\mathbf{W}_\infty^{tr}} \mathbf{q}_\mu(W, z) \log \frac{\mathbf{q}_\mu(W, z)}{\mathbf{q}(W, z)} \mu(dW)$$

where μ is any T-invariant measure on $\mathbf{W}_\infty^{tr}$ and $\mathbf{q}_\mu(W,z)$ is the conditional probability of a jump of size z, given the past history. We then use the contraction principle to come up with a candidate for the large deviation rate function

$$H(v) = \inf\left\{I(\mu) : \int z_0\mu(dW) = v\right\}$$

where $(z_j : j \in \mathbb{Z})$ denotes the jumps of a walk W. Several technical difficulties arise as one tries to apply Donsker–Varadhan Theory because of non-compactness of the state space and the fact that the transition probabilities are not continuous. These issues are handled masterfully in [V1].

3. Simple Exclusion Process

We now turn to our final model. This time our environment $\omega = (p_i(t) : i \in \mathbb{Z})$ is a collection of independent Poisson clocks. More precisely p_i, $i \in \mathbb{Z}$ are independent and each p_i is a Poisson process of rate 1; $p_i(t) = k$ for $t \in [\tau_1^i + \cdots + \tau_k^i, \tau_1^i + \cdots + \tau_{k+1}^i)$ with τ_j^i independent mean 1 exponential random variables. Given a realization of ω and an initial height function $h^0 \in \Gamma = \{h : \mathbb{Z} \to \mathbb{Z}$ such that $0 \le h(i+1) - h(i) \le 1\}$, we construct $h(i,t) = h(i,t;\omega)$ such that $h(\cdot,t;\omega) \in \Gamma$ for all t. More precisely, at each Poisson time $t = \tau_1^i + \cdots + \tau_k^i$, the height $h(i,t)$ increases by one unit provided that the resulting height function h^i belongs to Γ; otherwise the increase is suppressed.

The process $h(\cdot,t)$ is a Markov process with the following rules: $h \to h^i$ with rate $\eta(i+1)(1-\eta(i))$ where $\eta(i) = h(i) - h(i-1)$. The process $(\eta(i,t;\omega) : i \in \mathbb{Z})$ is also Markovian with the interpretation that $\eta(i,t) = 1$ if the site i is occupied by a particle and $\eta(i,t) = 0$ if the site i is vacant. Now the growth $h \to h^i$ is equivalent to jumping a particle from site $i+1$ to i provided that the site i is vacant. Since $h \in \Gamma$ is nondecreasing, we may define its inverse $x \in \Gamma'$, where

$$\Gamma' = \{x : \mathbb{Z} \to \mathbb{Z} \text{ such that } x(h+1) > x(h)\}.$$

Since h increases at a site $i+1$ if the site i is occupied by a particle, we may regard $x(h)$ as the position of a particle of label h. Equivalently we may interpret $h(i)$ as the label of a particle at an occupied site i.

The process $x(h,t;\omega)$ is also a Markov process with the following rules: $x(h,t) \to x(h,t)-1$ with rate $1\!\!1(x(h,t)-x(h-1,t) > 1)$. In words $x(h)$ decreases by one unit with rate 1 provided that the resulting configuration x_h is still in Γ'. For the construction of $x(h,t;\omega)$ we may use the clocks ω or equivalently we may use clocks that are assigned to sites $h \in \mathbb{Z}$. More precisely, if $\omega' = (p'_h(t) : h \in \mathbb{Z})$ is a collection of independent Poisson processes of rate 1, then we decrease $x(h)$ by one unit when the clock p'_h rings. The processes $x(h,t;\omega)$ and $x(h,t;\omega')$ have the same distribution. If we define $\zeta(h,t) = x(h,t) - x(h-1,t) - 1$, then $\zeta(h,t)$ represents the gap between the h-th and $(h-1)$-th particles in the exclusion process. The process $(\zeta(h,t) : h \in \mathbb{Z})$ is the celebrated *zero range process* and can be regarded as the occupation number at site h. The ζ-process is also Markovian where a ζ-particle at site h jumps to site $h+1$ with rate $1\!\!1(\zeta(h) > 0)$.

As in the previous sections, we set

$$u^\varepsilon(x,t;\omega) = \varepsilon h\left(\left[\frac{x}{\varepsilon}\right], \frac{t}{\varepsilon};\omega\right), \quad x^\varepsilon(u,t;\omega) = \varepsilon x\left(\left[\frac{u}{\varepsilon}\right], \frac{t}{\varepsilon};\omega\right)$$

and as a homogenization we expect to have $u^\varepsilon \to \bar{u}$, $x^\varepsilon \to \bar{x}$ with $\bar{u}$ and $\bar{x}$ deterministic solutions to Hamilton-Jacobi equations

$$\bar{u}_t = \bar{H}_1(\bar{u}_x) = \bar{u}_x(1 - \bar{u}_x) \tag{3.1}$$

$$\bar{x}_t = \bar{H}_2(\bar{x}_u) = \frac{1}{\bar{x}_u} - 1. \tag{3.2}$$

(See [R].) As for the large deviations, we will be interested in

$$\lim_{\varepsilon\to 0} \varepsilon \log \mathbb{P}(u^\varepsilon(x,t) \geq u) = \lim_{\varepsilon\to 0} \varepsilon \log \mathbb{P}(x^\varepsilon(u,t) \geq x) =: -W(x,u,t). \tag{3.3}$$

Evidently $W(x,u,t) = 0$ if $u \leq \bar{u}(x,t)$ or $x \leq \bar{x}(u,t)$. However we have that $W(x,u,t) > 0$, whenever $u > \bar{u}(x,t)$ or $x > \bar{x}(u,t)$. As it turns out,

$$\lim_{\varepsilon\to 0} \varepsilon \log \mathbb{P}(u^\varepsilon(x,t) \leq u) = -\infty \tag{3.4}$$

for $u < \bar{u}(x,t)$ because for such a number u,

$$\liminf_{\varepsilon\to 0} \varepsilon^2 \log \mathbb{P}(u^\varepsilon(x,t) \leq u) > 0$$

as was demonstrated by Jensen and Varadhan [JV]. Quoting from [JV], the statement (3.3) has to do with the fact that one may slow down $x(h,t)$ for $h \leq h_0$ in a time interval of order $O(\varepsilon^{-1})$ by simply slowing down $x(h_0,t)$. This can be achieved for an entropy price of order $O(\varepsilon^{-1})$. However for $x^\varepsilon(u,t) \leq \bar{x}(u,t) - \delta$, $\delta > 0$, we need to speed up $O(\varepsilon^{-1})$ many particles for a time interval of order $O(\varepsilon^{-1})$. This requires an entropy price of order $O(\varepsilon^{-2})$.

As was observed by Seppäläinen [S1], both h and x processes enjoy a strong monotonicity property. More precisely, if we write $x(h,t;\omega) = T_t^\omega x^0(h)$ for the x-process starting from the initial configuration $x^0 \in \Gamma'$, then $T_t^\omega(\sup_\alpha x_\alpha^0) = \sup_\alpha T_t^\omega(x_\alpha^0)$. In words, if the initial height $x^0 = \sup_\alpha x_\alpha^0$ is the supremum of a family of height functions x_α^0, then it suffices to evolve each x_α^0 separately for a given realization of ω and take the supremum afterwards. From this, it is not hard to show that such a strong monotonicity must be valid for W and this in turn implies that W solves a H-J equation of the form

$$W_t = K(W_x, W_u). \tag{3.5}$$

Here the initial data $W(x, u, 0)$ is the large deviation rate function at initial time. Of course we assume that there is a large deviation rate function initially and would like to derive a large deviation principle at later times. In the case of exclusion or zero-range process, it is not hard to guess what K is because when the process is at equilibrium, then the height function at a given site has a simple description. To construct the equilibrium measures for the x-process, let us pick a number $b \in (0, 1)$ and define a random initial height function $x(\cdot, 0)$ by the requirement that $x(\cdot, 0) = 0$ and that $(x(h + 1, 0) - x(h, 0) - 1 :$ $h \in \mathbb{Z})$ are independent geometric random variables of parameter b. That is $x(h + 1, 0) - x(h, 0) = k + 1$ with probability $(1 - b)b^k$. Let us write $\mathbb{P}^b$ for the law of the corresponding process $x(h, t; \omega)$. Using Cramer's large-deviation theorem, we can readily calculate that for u positive,

$$W(x, u, 0) = -\lim_{\varepsilon \to 0} \varepsilon \log \mathbb{P}^b(x^\varepsilon(u, 0) \geq x) = u\left(I_1\left(\frac{x}{u} + 1, b\right)\right)^+ \tag{3.6}$$

where $I_1(r, b) = r \log \frac{r}{b(1+r)} - \log[(1 - b)(1 + r)]$. As is well-known (see for example Chapter VIII, Corollary 4.9 of Liggett [L]), $-x(0, t)$ is a Poisson process which decreases one unit with rate b. Again Cramer's theorem yields

$$W(x, 0, t) = bt\left(I_2\left(\frac{-x}{bt}\right)\right)^+ \tag{3.7}$$

where $I_2(r) = r \log r - r + 1$. The expressions (3.5)–(3.7) provide us with enough information to figure out what K is. We refer to [S2] for a large deviation principle of the form (3.3) for a related particle system known as Hammersley's model.

Alternatively, we may study the large derivation of the particle densities. For this purpose, we define the empirical measure by

$$\pi^\varepsilon(t, dx) = \pi^\varepsilon(t, dx; \omega) = \varepsilon \sum_i \delta_{\varepsilon i}(dx)\eta(i, t/\varepsilon; \omega).$$

We regard π^ε as an element of the Skorohod space $X = \mathcal{D}([0, T]; M)$ where M is the space of locally bounded measures. The law of $\omega \mapsto \pi^\varepsilon(t, dx; \omega)$ induces a probability measure $\mathcal{P}^\varepsilon$ on X.

Hydrodynamic limit for the exclusion process means that $\mathcal{P}^\varepsilon \Rightarrow \mathcal{P}$ where $\mathcal{P}$ is concentrated on the single *entropy* solution of

$$\bar{\rho}_t = (\bar{\rho}(1 - \bar{\rho}))_x, \tag{3.8}$$

for a given initial data $\bar{\rho}(x,0) = \bar{\rho}^0(x)$. The function $\hat{\rho}$ is related to the macroscopic height function $\bar{u}$ by $\bar{\rho} = \bar{u}_x$. In Jensen–Varadhan [JV], a large deviation principle has been established for the convergence of $\mathcal{P}^\varepsilon$. Roughly

$$\mathcal{P}^\varepsilon\left(\pi^\varepsilon(t,dx) \text{ is near } \mu(t,dx)\right) \approx e^{-\varepsilon^{-1}\mathcal{I}(\mu)} \tag{3.9}$$

with the following rate function $\mathcal{I}$. First $\mathcal{I}(\mu) = +\infty$ unless $\mu(t,dx) = m(x,t)dx$ and m is a weak solution of (3.8). However when $0 < \mathcal{I}(m) < \infty$, then m is a non-entropic solution of (3.8). In fact $\mathcal{I}(\mu) = \mathcal{I}_0(\mu) + \mathcal{I}_{dym}(\mu)$ where $\mathcal{I}_0(\mu)$ is the large-deviation rate function coming from the initial deviation and depends only on our choice of initial configurations, and $\mathcal{I}_{dym}(\mu)$ is the contribution coming from dynamics and quantitatively measures how the entropy condition is violated. By the entropy condition we mean that for a pair (ϕ, q) with ϕ convex and $\phi'\bar{H}_1' = q'$ for $\bar{H}_1(p) = p(1-p)$, we have

$$\phi(\bar{\rho})_t + q(\bar{\rho})_x \leq 0 \tag{3.10}$$

in the weak sense. The left-hand side is a negative distribution which can only be a negative measure. As our discussions around (3.5) and (3.6) indicate, the invariant measures play an essential role in determining the large-deviations rate function. As it turns out the relevant ϕ to choose is simply the large deviation rate function for the invariant measure which is given by

$$\phi(m) = m\log m + (1-m)\log(1-m) + \log 2.$$

Here for the invariant measure we choose a Bernoulli measure ν under which $(\eta(i) : i \in \mathbb{Z})$ are independent and $\nu(\eta(i) = 1) = 1/2$. To measure the failure of the entropy solution, we take a weak solution m for which the corresponding

$$\phi(m)_t + q(p)_x = \gamma = \gamma^+ - \gamma^-$$

is a measure with γ^+ and γ^- representing the positive and negative parts of γ. We now have

$$\mathcal{I}_{dyn}(\mu) = \gamma^+(\mathbb{R} \times [0,T]).$$

It is customary in equilibrium statistical mechanics to represent a state as a probability measure with density $\frac{1}{Z}e^{-\beta H}$ with H some type of energy and Z the normalizing constant. In non-equilibrium statistical mechanics a large deviation principle of the form (3.9) offers an analogous expression with $\mathcal{I}(\mu)$ playing the role of "effective" energy (or rather potential). What we learn from [JV] is that after the entropy solution, the most frequently visited configurations are those associated with non-entropic solutions and the entropic price for such visits are measured by the amount the inequality (3.10) fails. Even though the entropy solutions for scalar conservation laws are rather well-understood, our understanding of non-entropic solutions is rather poor perhaps because we had no reason to pay attention to them before. The remarkable work [JV] urges us to look more deeply into non-entropic solutions for gaining insight into the way the microscopic densities deviate from the solution to the macroscopic equations.

References

[BS] Bolthausen, E., Sznitman, A-S., Ten lectures on random media. DMV Seminar, 32. Birkhuser Verlag, Basel, 2002.

[EG] Evans, L. C., Gomes, D., Effective Hamiltonians and averaging for Hamiltonian dynamics. I. Arch. Ration. Mech. Anal. 157 (2001), 1–33.

[F] Fathi,A., The Weak Kam Theorem in Lagrangian Dynamics (Cambridge Studies in Advanced Mathematics), to appear.

[JV] Jensen, L. H., and Varadhan, S. R. S., Large deviations of the asymmetric exclusion process in one dimension, preprint.

[KR] Kosygina, E., Varadhan S.R.S. Homogenization of Hamilton-Jacobi-Bellman equations with respect to time-space shifts in a stationary ergodic medium, Comm. Pure Appl. Math. **61** (2008), 816847.

[KRV] Kosygina, E., Rezakhanlou, F., and Varadhan, S. R. S., Homogenization of stochastic Hamilton–Jacobi–Bellmann equations, Comm. Pure Appl. Math. **59**(2006), 1489–1521.

[LS] Lions, P-L,, Souganidis, P. E., Homogenization of "viscous" Hamilton-Jacobi equations in stationary ergodic media, Comm. Partial Differential Equations **30** (2005), no. 1-3, 335–375.

[L] Liggett,T. "Interacting particle systems." Springer-Verlag, Berlin, 2005.

[R] Rezakhanlou, F., Hydrodynamic limit for attractive particle systems on $\mathbb{Z}^d$, Comm. Math. Phys. **140** (1991), 417–448.

[RT] Rezakhanlou, F., and Tarver, J. E., Homogenizations for stochastic Hamilton–Jacobi equations, Arch. Ration. Mech. Anal. **151**(4) (2000), 227–309.

[Ro] Rosenbluth, J., Quenched large deviations for multidimensional random walk in random environment: A variational formula. Ph.D thesis, New York University, 2006.

[S1] Seppäläinen, T., Existence of hydrodynamics for the totally asymmetric simple K-exclusion process, Ann. Probab. **27** (1999), 361–415.

[S2] Seppäläinen, T., Large deviations for increasing sequences on the plane, Probab. Theory Related Fields, **112** (1998), 221–244.

[S] Souganidis, P. E., Stochastic homogenization of Hamilton–Jacobi equations and some applications, Asymptot. Anal. **20**(1) (1999), 1–11.

[Sz] Sznitman, A. S., Brownian Motion, Obstacles and Random Media, Springer, 1998.

[V1] Varadhan, S. R. S., Large deviations for the asymmetric exclusion process, 1–27, Adv. Stud. Pure Math., **39**, Math. Soc. Japan, Tokyo, 2004.

[V2] Varadhan, S. R. S., Large deviations for random walks in a random environment, Comm. Pure Appl. Math. **56** (2003), 1222-1245.

Varadhan's Work on Hydrodynamical Limits

Jeremy Quastel* and Horng-Tzer Yau†

In the centuries since the discovery of Newton equation, the quest to solve the many-body problem has been one of the most persistent endeavours of mathematics and physics. Although progress was made in approximating it when the number of particles is small, a solution for large numbers of particles in any useful form seems simply impossible. The fundamental observation of Boltzmann was that the typical behavior for classical Hamiltonian systems in equilibrium is governed by ensemble averages – Gibbs states in today's language. This avoided the difficulty of directly solving the Newton equations by postulating statistical ensembles, and lead to modern statistical physics and ergodic theory. Boltzmann's formulation concerned systems in equilibrium; in other words, behavior of systems as the time approaches infinity.

At the other end is the kinetic theory for short time behavior. Classical dynamics are exactly solvable when there is no interactions among particles, i.e., free dynamics. In short time classical dynamics can be understood by supplementing the free dynamics with collisions. The fundamental observation of kinetic theory, the idealization of the collision processes, is again due to Boltzmann in his celebrated work on the Boltzmann equation.

For systems neither in equilibrium nor near free dynamics, i.e., for time scales too short for equilibrium theory, but longer than the scale for kinetic theory, the most useful descriptions are still the classical macroscopic equations, e.g., the Euler and Navier-Stokes equations. These are continuum formulations of conservation of mass and momentum, and also contain some phenomenological concepts such as viscosity. They are equations for macroscopic quantities such as density, velocity (momentum) and energy, while the Boltzmann equation is an equation for the probability density of finding a particle at a fixed position and velocity. The classical Hamiltonian plays no active role in either formulation, and all the microscopic effects are summarized by the viscosity in the Navier-Stokes equations, or the collision operator in the Boltzmann equation. But the central theoretical question, i.e., understanding the connection

*Departments of Mathematics and Statistics, University of Toronto, Toronto, Canada M5S 2E4

†Department of Mathematics, Harvard University, Cambridge, MA 02138

between large particle systems and their continuum approximations, remained unsolved and is still one of the fundamental questions in non-equilibrium statistical physics.

Since classical dynamics of large systems are all but impossible to solve, a more feasible goal is to replace the classical dynamics with stochastic dynamics. From the sixties to early eighties, tremendous effort was made by Dobrushin, Lebowitz, Spohn, Presutti, Spitzer, Liggett, and others to understand large stochastic particle systems. A key focus was to derive rigorously the classical phenomenological equations from the interacting particle systems in suitable scaling limits. The methods at the time were based on coupling and perturbative arguments; the systems which can be treated rigorously were restricted to special one dimensional systems, perturbations of the symmetric simple exclusion or mean-field type interactions. Together with Guo and Papanicoloau, Varadhan [3] introduced the first general approach, the entropy method. The key ideas are the dissipative nature of the entropy, and large deviations. As long as the equilibrium measures of the dynamics are known, and the scaling is diffusive, this approach is very effective and now has been applied to many systems. We now sketch the approach in [3]. The method is most transparent in a model (the hydrodynamical limit of this model was first proved under some more restrictive assumptions in [1]) where the particle number is replaced by a real valued scalar field $\phi_x \in \mathbb{R}$, $x \in \{1, \cdots, N\}$ with periodic boundary condition so that $N+1=1$. These evolve as interacting diffusions. Denote by μ_0 the product measure such that the law of ϕ_x is given by $e^{-V(\phi_x)}$. Let $f_t\mu_0$ be distribution of the field ϕ at the time t. The dynamics of ϕ will be given by the evolution equation for f_t:

$$\begin{aligned} \partial_t f_t = L f_t, \quad L \;&=\; \sum_{j=1}^{N} \left(\frac{\partial}{\partial \phi_j} - \frac{\partial}{\partial \phi_{j+1}} \right)^2 \\ &\quad - \sum_{j=1}^{N} \left(V'(\phi_j) - V'(\phi_{j+1})\right) \left(\frac{\partial}{\partial \phi_j} - \frac{\partial}{\partial \phi_{j+1}} \right) \end{aligned} \tag{1}$$

This dynamics is reversible with respect to the invariant measure μ_0 and the Dirichlet form is given by

$$D(f) := -\int f L f \, d\mu_0 = \sum_{j=1}^{N} \int \left(\frac{\partial f}{\partial \phi_j} - \frac{\partial f}{\partial \phi_{j+1}} \right)^2 d\mu_0 \tag{2}$$

We rescale the time diffusively so that the evolution equation becomes

$$\partial_t f_t = \varepsilon^{-2} L f_t, \tag{3}$$

where $\varepsilon = N^{-1}$ is the scaling parameter.

The dynamics (3) can be written as a conservation law,

$$d\phi_i = (w_{i+1} - w_i)dt + dM_i, \qquad w_i = N^2 V'(\phi_i) - N^2 V'(\phi_{i-1}) \tag{4}$$

where M_i are martingales and w_i is the microscopic current. The current is itself a gradient and our dynamics is formally diffusive, i.e., for any smooth test function J

$$dN^{-1}\sum_i J(\varepsilon i)\phi_i \sim N^{-1}\sum_i J''(\varepsilon i)V'(\phi_i)dt + dM. \tag{5}$$

Denote by $\rho(x,t)$ the local average of ϕ around $x = \varepsilon i$,

$$\rho(x,t) = \lim_{\delta\to 0}\lim_{\varepsilon\to 0}\frac{1}{2\varepsilon\delta+1}\sum_{|j-i|\le\delta\varepsilon^{-1}}\phi_j \tag{6}$$

Let $p(\lambda)$ denote the pressure

$$p(\lambda) = \log\int_{\mathbb{R}} d\phi e^{\lambda\phi - V(\phi)} \tag{7}$$

and $h(m)$ denote the free energy, the Legendre transform of p. It is not hard to check that the martingale term in (5) vanishes in the limit and hence the main task to establish the hydrodynamic limit of the form

$$\partial_t\rho = \partial_{xx}h'(\rho) \tag{8}$$

is to prove that we can replace $V'(\phi_i)$ in (5) by $h'(\rho(x,t))$ in the sense of a law of large numbers with respect to the distribution f_t satisfying (3).

Consider the local Gibbs measure with a chemical potential λ,

$$d\mu_\lambda(\phi_1,\cdots,\phi_N) = \exp\left[\sum_j \lambda(\varepsilon j)\phi_j\right]d\mu_0 \tag{9}$$

where the chemical potential λ is allowed to depend on the site slowly. If f_t is a local Gibbs state, then certainly we can replace $V'(\phi_i)$ in (5) by $h'(\rho(x,t))$ in the sense of law of large numbers. The key observation of GPV is to consider the evolution of the entropy:

$$\partial_t S(f_t) = -D(\sqrt{f_t}), \qquad S(f) = \int f\log f d\mu_0 \tag{10}$$

For a typical system of N variables given by a density f with respect to μ_0, the entropy $S(f)$ is of order N. This implies that

$$\int_0^t D(\sqrt{f_s})ds \le CN \tag{11}$$

This information alone is sufficient to establish that the solution of (3) is close enough to a local Gibbs state that the law of large numbers continues to hold.

Systems where the current is itself a gradient of some other function are known as *gradient systems*. For such systems, [3] provides a general framework to establish the hydrodynamical limit. However many systems are of nongradient type. A simple illustrative example is to modify the dynamics (1) so that the generator L becomes the symmetric operator with Dirichlet form,

$$D(f) := -\int fLf d\mu_0 = \sum_{j=1}^{N} \int a(\phi_j, \phi_{j+1}) \left(\frac{\partial f}{\partial \phi_j} - \frac{\partial f}{\partial \phi_{j+1}} \right)^2 d\mu_0 \tag{12}$$

The current can be computed easily and is given (up to scale factors) by

$$w_j = -\frac{\partial a(\phi_j, \phi_{j+1})}{\partial \phi_j} + \frac{\partial a(\phi_j, \phi_{j+1})}{\partial \phi_{j+1}} + a(\phi_j, \phi_{j+1})[\phi_1 - \phi_{j+1}]. \tag{13}$$

To establish the hydrodynamical limit, it is now required to prove that in some sense w_j can be replaced by ($x = \varepsilon j$)

$$w_j = D(\rho(x,t))[\phi_{j+1} - \phi_j] \tag{14}$$

for some function D, which will be the diffusion coefficient of the hydrodynamical equation. The key observation of Varadhan's work on nongradient systems is that

$$w_j = D(\rho(x,t))[\phi_{j+1} - \phi_j] + LF \tag{15}$$

where F is some local function of ϕ_j and L is the generator given by (12). The idea is that functions of the type LF represent incoherent rapid fluctuations which vanish over the long time scale of the hydrodynamic limit. This fluctuation is indeed in the system and the hydrodynamical limit can be established only if we properly account for its effect.

The sense in which (15) holds is in the H_{-1} sense, corresponding to the vanishing of the variance in the central limit theorem for the corresponding additive functional. This goes back to Varadhan's earlier work on *tagged particles* [4]. The problem of proving the convergence of tagged particles to appropriate diffusions is somewhat complementary to the hydrodynamic limit. Varadhan introduced the martingale method in this context so that the idea of viewing the system from the point of view of the particle can be implemented. These ideas have had broad influence not only in hydrodynamic limits, but also in homogenization theory and in random walk in random environment. In fact, there is even a more explicit connection between the tagged particle problems and the non-gradient systems. Suppose that one gives each particle one of m different labels, and watches the evolution of the m different densities in the hydrodynamic limit. The corresponding particle systems are usually of non-gradient form [8] as long as $m \geq 2$. This is a weak form of tagging, and the large m limit of this system is a (weak) way to keep track of individual particles. It can be proved, via nongradient system method, that each species of particles evolves according to a diffusion equation and thus the hydrodynamical limit of tagged particles in nonequilibrium is established [9]. The advantage of this approach is that it can be done in nonequilibrium, identifying the collective drift imposed by the flow of the bulk towards equilibrium. However, it is strictly speaking not the behavior of a single tagged particle, but the average behavior of tagged particles with vanishing density.

The equation (15) is quite difficult to solve as it involves the full generator L. In order to solve it, Varadhan developed a method which can be viewed as an infinite dimensional version of Hodge theory. This is a deep theory and we shall only attempt to give some flavor. First note that because of the entropy bound (11), one only has to solve (15) in equilibrium. So the diffusion coefficient can be treated as a constant. Now for simplicity take $j = 0$. The current w_0 has the property that

$$\int w_0 d\mu_\lambda = 0 \tag{16}$$

for any Gibbs state with constant chemical potential λ. The space of functions with this property corresponds to a space of *closed forms*. A subspace of *exact forms* corresponds to the fluctuation terms LF. The deep result is that the exact forms are of codimension one in the space of closed forms with orthogonal complement corresponding to $\phi_{j+1} - \phi_j$. This solves (15). This approach, as it stands, is based on the integration by parts nature of L and applies only to reversible dynamics. It is possible to formulate it also for nonreversible dynamics and the formal analogy between this equation and the Hodge theory can be strengthened [15].

The two fundamental papers [3, 16] of Varadhan ushered in an era of hydrodynamical limits based on the idea of entropy. The developments following these two papers are astonishing and we shall only mention a few. The approach of [3] was successfully applied to many systems, including interacting Brownian motions [14], interacting Ornstein-Uhlenbeck processes [6] and Ginzburg-Landau models [11]. The interacting Brownian motions and interacting Ornstein-Uhlenbeck processes are continuum systems with no lattice structure. The hydrodynamical limit for the Ginzburg-Landau models was proved for all temperatures, including the phase transition region—a remarkable result. Furthermore, the approach of GPV was successfully applied to kinetic scaling and led to the derivation of the Boltzmann equation from stochastic particle systems [13]. The idea that the solution of (3) is heuristically a local Gibbs states goes back many decades. But the estimates obtained in GPV are in fact strong enough to prove it. It is observed in [17] that one can bypass many technical difficulties in GPV and prove directly that the local Gibbs states are in fact an approximate solution to (3) in the sense of relative entropy. The assumptions needed in this approach are: 1. Some ergodic properties of the dynamics. 2. Smoothness of solutions to the hydrodynamical equations. This method is more restrictive than GPV for diffusive systems, but it essentially relies only on the identification of the invariant measures of the dynamics and it applies also to hyperbolic systems before the formation of shocks. It was adapted in [7] to derive the classical Euler equation from Hamiltonian systems with vanishing noise. This is the most significant advance since Morrey stated this problem in the 60's. Once the hyperbolic equations develop shocks, a very different method is needed, see [12, 2] for references and related results.

Varadhan's work on nongradient systems requires a spectral gap of order ℓ^{-2} for the system in a box of side length ℓ. This inspired work on the estimates of spectral gaps of conservative dynamics and it led to the development of martingale methods for estimating spectral gap for conservative dynamics. Using this spectral estimate, Varadhan and his coworker [15] established the hydrodynamical limit of lattice gas in the high temperature phase.

The idea that the current can be decomposed into a dissipative term $\phi_j - \phi_{j+1.}$ and a fluctuation term LF is a deep idea and is really a rigorous statement of the so-called fluctuation-dissipation theorem in physics. In a sense, the insight that this equation is fundamental to the hydrodynamical limit is at least as significant as the solution of this equation for the specific model considered in [16]. Although the fluctuation-dissipation equation was solved in [16] only for reversible dynamics, it was realized that one can develop a method to solve this equation for nonreversible dynamics provided that the spatial dimension is larger than two [5]. This led to the derivation of the incompressible Navier-Stokes (INS) equations from stochastic lattice gases for dimension $d = 3$ [10]. The result obtained in [10] is very strong; it identifies the large deviation rate that the hydrodynamical equation is not a Leray solution and does not assume that the INS equations have classical solutions. The physical significance is the following: The first principles equation governing a classical fluid is the Newton equation, which is time reversible and has no dissipation. The INS equations possess viscosity and are time irreversible. Therefore, a derivation of the INS equations from classical mechanics would have to answer the fundamental question relating to the origin of dissipation and the breaking of the time reversibility in classical dynamics. Although the underlying dynamics in [10] is stochastic, it was proved that the viscosity in the INS equations was *strictly* larger than the original viscosity of the underlying stochastic dynamics. In other words, the deterministic part of the dynamics makes a nontrivial contribution to the viscosity. We remark that the condition $d = 3$ is critical. For dimension $d = 2$, it was proved that the hydrodynamic limit equations for such lattice gas models are not the INS equations. Indeed, even the diffusive scaling is incorrect and there are logarithmic corrections. Although these works do not answer directly the fundamental question regarding the derivation of the incompressible Navier-Stokes equations from the classical dynamics, it is the first time we understand the generation of the viscosity from many particle dynamics. These developments are largely attributed to Varadhan's insight of the importance of the fluctuation-dissipation equations (15).

References

[1] Fritz, J.: On the hydrodynamic limit of a scalar Ginzburg-Landau lattice model: the resolvent approach. Hydrodynamic behavior and interacting particle systems (Minneapolis, Minn., 1986), 75–97, IMA Vol. Math. Appl., 9, Springer, New York, 1987.

[2] Fritz, J.; Tooth, B.: Derivation of the Leroux system as the hydrodynamic limit of a two-component lattice gas. Comm. Math. Phys. 249 (2004), no. 1, 1–27.

[3] Guo, M. Z.; Papanicolaou, G. C.; Varadhan, S. R. S. Nonlinear diffusion limit for a system with nearest neighbor interactions. Comm. Math. Phys. 118 (1988), no. 1, 31–59.

[4] Kipnis, C.; Varadhan, S. R. S.: Central limit theorem for additive functionals of reversible Markov processes and applications to simple exclusions. Comm. Math. Phys. 104 (1986), no. 1, 1–19.

[5] Landim, C.; Yau, H. T. Fluctuation-dissipation equation of asymmetric simple exclusion processes. Probab. Theory Related Fields 108 (1997), no. 3, 321–356.

[6] Olla, Stefano; Varadhan, S. R. S.: Scaling limit for interacting Ornstein-Uhlenbeck processes. Comm. Math. Phys. 135 (1991), no. 2, 355–378.

[7] Olla, S.; Varadhan, S. R. S.; Yau, H.-T. Hydrodynamical limit for a Hamiltonian system with weak noise. Comm. Math. Phys. 155 (1993), no. 3, 523–560.

[8] Quastel, J.: Diffusion of color in the simple exclusion process. Comm. Pure Appl. Math. 45 (1992), no. 6, 623–679

[9] Quastel, J.; Rezakhanlou, F.; Varadhan, S. R. S. Large deviations for the symmetric simple exclusion process in dimensions $d \geq 3$. Probab. Theory Related Fields 113 (1999), no. 1, 1–84.

[10] Quastel, J.; Yau, H.-T. : Lattice gases, large deviations, and the incompressible Navier-Stokes equations. Ann. of Math. (2) 148 (1998), no. 1, 51–108.

[11] F. Rezakhanlou: Hydrodynamic limit for a system with finite range interactions, Comm. Math. Phys. 129 (1990) 445–480.

[12] Rezakhanlou, F. : Hydrodynamic limit for attractive particle systems on Z^d. Comm. Math. Phys. 140 (1991), no. 3, 417–448.

[13] Rezakhanlou, F.: Boltzmann-Grad limits for stochastic hard sphere models. Comm. Math. Phys. 248 (2004), no. 3, 553–637.

[14] Varadhan, S. R. S. Scaling limits for interacting diffusions. Comm. Math. Phys. 135 (1991), no. 2, 313–353.

[15] Varadhan, S. R. S.; Yau, H.-T.: Diffusive limit of lattice gas with mixing conditions. Asian J. Math. 1 (1997), no. 4, 623–678.

[16] Varadhan, S. R. S. Nonlinear diffusion limit for a system with nearest neighbor interactions. II. Asymptotic problems in probability theory: stochastic models and diffusions on fractals (Sanda/Kyoto, 1990), 75–128, Pitman Res. Notes Math. Ser., 283, Longman Sci. Tech., Harlow, 1993.

[17] Yau, H.T.: Relative entropy and hydrodynamics of Ginsburg-Landau models, Letters Math. Phys., **22**, 63–80,(1991)

Book Review

BULLETIN (New Series) OF THE
AMERICAN MATHEMATICAL SOCIETY
Volume 2, Number 3, May 1980

0002-9904/80/0000-0219/$02.75

Multidimensional diffusion processes, by D. W. Stroock and S. R. S. Varadhan, Die Grundlehren der mathematischen Wissenschaften, vol. 233, Springer-Verlag, Berlin and New York, 1979, xii + 338 pp., $34.80.

1. *Is it best to think of a 'diffusion' as meaning* (i) *a continuous strong Markov process*, (ii) *a strong solution of an Itô stochastic differential equation, or* (iii) *a solution of a martingale problem*? Both the Markov-process approach and the Itô approach (which holds so special a place in the hearts of probabilists after the appearance of McKean's wonderful book [7]) have been immensely successful in diffusion theory. The Stroock-Varadhan book, developed from the historic 1969 papers by its authors, presents the martingale-problem approach as a more powerful–and, in certain regards, more intrinsic–means of studying the foundations of the subject.

The martingale-problem method has been applied with great success to other problems in Markov-process theory, both 'pure' (Stroock [**10**], . . .) and applied (Holley and Stroock [**3**], [**4**], . . .). It has conditioned our whole way of thinking about still-more-general processes (Jacod [**5**], . . .). Moreover, the method's ideas and results now feature largely in work on filtering and control (Davis [**1**], . . .).

I 'batter' you with the preceding paragraph because the authors make the uncompromising decision not 'to proselytize by intimidating the reader with myriad examples demonstrating the full scope of the techniques', but rather to persuade the reader 'with a careful treatment of just one problem to which they apply'. Halmos's doctrine 'More is less, and less is more' is thereby thoroughly tested; but if one had to choose a single totally-integrated piece of work which in depth and importance shows that probability theory has 'come of age', it would surely be the theorem towards which so much of this book is directed–or perhaps Stroock's extension of it [**10**]. Most of the main tools of stochastic-process theory are used, after first having been honed to a sharper edge than usual. But it is the formidable combination of probability theory with analysis (in the form of deep estimates from the theory of singular integrals) which is the core of the work.

The book's importance has persuaded me to write a review accessible to the general reader–I even define *continuous martingale*! I therefore concentrate on background material, with just a few clues (some for the sharp-eyed!) on how the book's results are proved.

2. The following purely-analytic (and very special) corollary of the main Stroock-Varadhan theorem can help set the scene; and the fact that, in spite of so much research, no analytic proof of its uniqueness assertion has been discovered, can serve as a first indication of the depth of the theorem.

Let L be a second-order elliptic operator on $C_K^\infty(\mathbf{R}^d)$ of the form

$$L = \frac{1}{2}\sum_{i,j \leqslant d}\sum a_{ij}(x)\partial_i\partial_j + \sum_{i \leqslant d} b_i(x)\partial_i, \tag{2.1}$$

where ∂_i denotes $\partial/\partial x_i$, $\{a_{ij}(x)\}$ is a strictly positive-definite real symmetric matrix for each x, and where each $a_{ij}(\cdot)$ and each $b_i(\cdot)$ is a bounded continuous real-valued function on $\mathbf{R}^d$. No Lipschitz conditions are assumed. Then there is one and only one Feller-Dynkin semigroup $\{P_t\colon t \geqslant 0\}$[1] with infinitesimal generator extending L. (Chapter 10 of the book examines closely the extent to which the 'boundedness' condition on a and b may be relaxed. It cannot be relaxed completely because one must preclude 'explosion'.)

3. With a general Feller-Dynkin semigroup $\{P_t\}$ is associated an $\mathbf{R}^d$ valued Markov process $X = \{X_t\colon t \geqslant 0\}$ with *right-continuous* paths, such that the law P^x of X started from x is given by the usual recipe: for $0 \leqslant t_1 \leqslant t_2 \leqslant \cdots \leqslant t_n$, we have[2]

$$\begin{aligned} &P^x\left[X_{t_1} \in dx_1; X_{t_2} \in dx_2; \ldots ;X_{t_n} \in dx_n\right] \\ &= P_{t_1}(x, dx_1)P_{t_2-t_1}(x_1, dx_2) \ldots P_{t_n-t_{n-1}}(x_{n-1}, dx_n). \end{aligned} \tag{3.1}$$

But if $\{P_t\}$ has generator extending our operator L, then, because L is *local*, we can take X to be (path-)*continuous*.[3]

---

[1]A 'Feller-Dynkin' semigroup $\{P_t\colon t > 0\}$ is a family of bounded linear maps $P_t\colon C_0(\mathbf{R}^d) \to C_0(\mathbf{R}^d)$, where $C_0(\mathbf{R}^d)$ is the Banach space of continuous functions on $\mathbf{R}^d$ which 'vanish at ∞', such that

$$f > 0 \Rightarrow P_t f > 0;\quad P_t 1 = 1;\quad P_s P_t = P_{s+t};\quad P_\varepsilon f \to f \quad (\varepsilon \downarrow 0).$$

('Pointwise' and 'strong-topology' convergence in the last statement are equivalent under the remaining hypotheses.) The statement that $\{P_t\}$ has *infinitesimal generator extending* L amounts to the statement that

$$P_t f - f = \int_0^t P_s Lf\, ds, \quad \forall f \in C_K^\infty(\mathbf{R}^d). \tag{2.2}$$

Note that the differentiated form $(P_t f)' = P_t Lf$ of (2.2) corresponds to the Fokker-Planck equation.

[2]Here, the transition probability measure $P_t(x, \cdot)$ is of course that associated with the map $f \mapsto P_t f(x)$ on $C_0(\mathbf{R}^d)$ by the Riesz representation theorem.

[3]Conversely, if each P^x measure associated with a Feller-Dynkin semigroup $\{P_t\}$ assigns mass 1 to the set of continuous paths, then the generator G of $\{P_t\}$ must be *local* and must obey the *maximum principle*: if $f \in D(G)$, then $Gf \leqslant 0$ at a maximum of f. Hence if $C_K(\mathbf{R}^d) \subset D(G)$, then G acts on $C_K^\infty(\mathbf{R}^d)$ as a (possibly singular) *second-order* elliptic operator with continuous coefficients. The fact that it can happen that $D(G) \cap C_K^\infty(\mathbf{R}^d) = \{0\}$ is one reason why the definition of a diffusion as a continuous strong Markov process, though intrinsic, is too wide.

A Markov process is a much richer structure than its transition semigroup, and it may be subjected to all sorts of transformations which have no analytic counterpart. This explains why probability theory can obtain analytic results which escape analytic proof (and the literature contains many very striking instances of this). But in the richer setting, the purely-analytic questions are of secondary interest.

4. Continuous P-martingales. We discuss martingale theory only in the special context appropriate to diffusion theory.

Let $\Omega = C([0, \infty), \mathbf{R}^d)$ be the space of all continuous functions ω from $[0, \infty)$ to $\mathbf{R}^d$. For $t \in [0, \infty)$ and $\omega \in \Omega$, write X for the coordinate process with $X_t(\omega) = \omega(t)$. Put $\mathfrak{F}_\infty^\circ = \sigma\{X_s\colon s \geqslant 0\}$, the smallest σ-algebra of subsets of Ω such that every map X_s from Ω to $\mathbf{R}^d$ is $\mathfrak{F}_\infty^\circ$ measurable; and, for $t \geqslant 0$, put $\mathfrak{F}_t^\circ = \sigma\{X_s\colon s \leqslant t\}$.

Let M be a continuous adapted real-valued process. Thus, the map M: $[0, \infty) \times \Omega \to \mathbf{R}$ is such that $t \mapsto M_t(\omega)$ is continuous for every ω, and M is $\{\mathfrak{F}_t^\circ\}$ *adapted* in that for each t, $M_t(\cdot)$ is $\mathfrak{F}_t^\circ$ measurable.[4]

Let P be a probability measure on $(\Omega, \mathfrak{F}_\infty^\circ)$. Then M is called a (continuous) *P-martingale* if $M_t \in \mathfrak{L}^1(\Omega, \mathfrak{F}_t^\circ, P)$, $\forall t$, and, whenever $s \leqslant t$ and $\Lambda \in \mathfrak{F}_s^\circ$,

$$\int_\Lambda M_t(\omega)P(d\omega) = \int_\Lambda M_s(\omega)P(d\omega).$$[5]

5. Martingale problems for L. Let L be as at (2.1). Let P be a probability measure on $(\Omega, \mathfrak{F}_\infty^\circ)$. Let $x \in \mathbf{R}^d$.

Definition. We say that *P solves the martingale problem for L starting from x* if $P[X_0 = x] = 1$ and, $\forall f \in C_K^\infty(\mathbf{R}^d)$, the process C^f, where

$$C_t^f = f(X_t) - \int_0^t Lf(X_s)\, ds$$

is a P-martingale.

Assume for the moment that a Feller-Dynkin semigroup $\{P_t\}$ exists with infinitesimal generator extending L. As explained in §3, it is possible to define a probability measure P^x on $(\Omega, \mathfrak{F}_\infty^\circ)$ via (3.1). It is almost trivial to use (2.2) to show that P^x solves the martingale problem for L starting from x.

In the special context corresponding to §2, the main Stroock-Varadhan theorem takes the following form.

Theorem. *Let L be a second-order elliptic operator on $C_K^\infty(\mathbf{R}^d)$ satisfying the hypotheses described at* (2.1). *Then, for each x in $\mathbf{R}^d$, there is one and only one solution of the martingale problem for L starting from x. Moreover, this solution is of the Markovian form P^x derived as at* (3.1) *from the unique Feller-Dynkin semigroup $\{P_t\}$ with generator extending L.*

In fact, for the existence-and-uniqueness part of the result, Stroock and Varadhan allow the matrices $a(\cdot)$ and $b(\cdot)$ to be *time-dependent*, and impose

---

[4]The intuitive thrust of this requirement is that $M_t(\omega)$ is determined when the values $X_s(\omega)$ for $s \in [0, t]$ are known.

[5]This is the full statement of the requirement that the conditional P-expectation of M_t, given the information $\mathfrak{F}_s^\circ$ about what has happened up to time s, is M_s: under P, the 'game' M is fair.

only a *measurability* requirement on b. This generalisation is not merely an academic matter: it is just what engineers need (see Davis [1], . . .).

Before thinking briefly about how the theorem is proved, let us spend some time in getting an intuitive feeling for continuous martingales and semimartingales.[6]

6. Continuous P-semimartingales. Again let P be a probability measure on $(\Omega, \mathcal{F}_\infty^\circ)$. The integrability condition ($M_t \in \mathcal{L}^1$) in the definition of 'martingale' is a nuisance. Call a continuous adapted process M a (continuous) *local P-martingale* if, for every n, M^n is a P-martingale, where

$$M_t^n = M_{t\wedge\tau(n)} - M_0, \quad \text{and} \quad \tau(n) = \inf\{s: |M_s - M_0| > n\}.$$

(This is one of many instances where I use simple definitions and/or results which work only for continuous processes. Meyer [8] has the 'correct'–that is, *generalisable*–versions).

We now arrive at one of the central concepts of Strasbourg theory. Call a continuous adapted process Z a (continuous) *P-semimartingale* if Z may be written as follows:

$$Z_t = Z_0 + M_t + V_t, \tag{6.1}$$

where M is a continuous local P-martingale with $M_0 = 0$, and V is a (continuous adapted) process with paths of finite variation and with $V_0 = 0$. It is surprising that *the decomposition* (6.1) *is P-unique*: if $Z = Z_0 + M_t^* + V_t^*$, then $P[M_t = M_t^*, \forall t] = 1$.[7]

7. Generalised Itô formula: continuous case. Let Z be a continuous P-semimartingale with decomposition as at (6.1). You can see that V represents the 'drift' of Z away from the fair (local martingale) situation.

By the celebrated *Meyer decomposition theorem*, there is a P-unique continuous adapted process $\langle M, M\rangle$ with nondecreasing paths such that $\langle M, M\rangle_0 = 0$ and $M^2 - \langle M, M\rangle$ is a local P-martingale. Moreover, by a theorem of Kunita, Watanabe, and Doléans, there exists a sequence (n_k) along which, with P-probability 1,

$$\langle M, M\rangle_t = \lim \sum_i \left[M_{t\wedge i2^{-n}} - M_{t\wedge(i-1)2^{-n}}\right]^2 \quad \forall t,$$

[6]The theory of martingales, and of the still-more-important semimartingales, has been developed to a very advanced level by the French school of probabilists led by Meyer. (But we must not forget massive Japanese contributions.) I myself have on occasion been mischievous in comments on 'Strasbourg theory', and I think that Stroock and Varadhan are perhaps just a little mischievous in this regard. Strasbourg publications are generally inclined to be somewhat abstract in spirit. However, they do contain a profound analysis of intuitive thinking; and, with sufficient poetic licence, they may be regarded as in the direct tradition of Lévy's work. Meyer [8] is the definitive work on the Strasbourg theory of stochastic integrals. Dellacherie, Doléans, Letta, and Meyer [2] is a most helpful 'Strasbourg' look at part of the Stroock-Varadhan theory. These two papers are my guides for the next few sections. Incidentally, I am using the long-thought-about Strasbourg notation throughout the review.

[7]An immediate consequence of the P-uniqueness just asserted is that a *continuous* local P-martingale M cannot have paths of finite variation unless it is constant. (*You* supply the 'almost surely' qualification.) This explains why we need a *stochastic integral*. It is important that the mildest regularity requirements on a 'stochastic integral' force us to consider only stochastic integrals relative to (possibly discontinuous) semimartingales.

the limit existing uniformly on compact t-intervals. Infinitesimally,

$$(dZ_t)^2 = (dM_t)^2 = d\langle M, M\rangle_t$$

(in some sense!), so Taylor's theorem leads us to anticipate the generalised Itô formula for the continuous case:

$$df(Z) = f'(Z)\,dZ + \tfrac{1}{2}f''(Z)d\langle M, M\rangle. \qquad (7.1)$$

The moral is that a C^2 function f of a continuous P-semimartingale Z is again a continuous P-semimartingale with decomposition

$$f(Z_t) = f(Z_0) + \int_0^t f'(Z)\,dM + \left\{\int_0^t f'(Z_s)\,dV_s + \frac{1}{2}\int_0^t f''(Z_s)\,d\langle M, M\rangle_s\right\}, \qquad (7.2)$$

the first of the integrals, a stochastic integral, yielding a local P-martingale.[8] The final two integrals in (7.2) are Stieltjes integrals over s for each ω.

8. Diffusions as semimartingales. If Z is our continuous P-semimartingale as at (6.1), and f is a C_K^∞ function on $\mathbf{R}$, then, by (7.2),

$$f(Z_t) - \left\{\int_0^t f'(Z_s)\,dV_s + \frac{1}{2}\int_0^t f''(Z_s)\,d\langle M, M\rangle_s\right\}$$

is a local P-martingale. (More general martingale problems loom into view; but back to diffusions . . . .)

Take $d = 1$, and $L = \frac{1}{2}a(x)d^2/dx^2 + b(x)d/dx$, where $a(\cdot)$ and $b(\cdot)$ are continuous (but 'measurable' will do!) and $a(\cdot) > 0$. We see that *a measure P on $(\Omega, \mathcal{F}_\infty^\circ)$ solves the martingale problem for L starting from x if and only if $P[X_0 = x] = 1$ and the coordinate process X is a P-semimartingale with decomposition $X = X_0 + M + V$ satisfying*

$$V_t = \int_0^t b(X_s)\,ds, \qquad \langle M, M\rangle_t = \int_0^t a(X_s)\,ds.$$

'Vector' generalisations to $d > 1$ are obvious. See footnote 8!

9. The martingale problem for Wiener measure. Continue with $d = 1$. Take $b \equiv 0$, $a \equiv 1$, $x = 0$. If P satisfies the corresponding martingale problem, then $P[X_0 = 0] = 1$, and X is a local P-martingale with $\langle X, X\rangle_t = t$. A trivial modification of (7.1) shows that $dY_t^\theta = i\theta Y_t^\theta\,dX$, where $Y_t^\theta = \exp(i\theta X_t + \frac{1}{2}\theta^2 t)$. Hence Y is a local P-martingale, and indeed a *P-martingale* because Y_t^θ is bounded by $\exp(\frac{1}{2}\theta^2 t)$. It is immediate to anyone who recognizes the characteristic function of the normal distribution that *X is an $(\{\mathcal{F}_t^\circ\}, P)$ Brownian motion: X_t is $\mathcal{F}_t^\circ$ measurable $(\forall t)$, and, under P, $X_{t+h} - X_t$ is independent of $\mathcal{F}_t^\circ$ and has the normal distribution of mean 0 and variance h.*

[8] With U standing for $f'(Z)$, the stochastic integral $\int U\,dM$ is a limit of Riemann sums $\sum U_s(M_{s+h} - M_s)$, *with each M-increment pointing into the future of the U-value*. This is why $\int U\,dM$ is a local martingale. The super-elegant modern (Kunita, Motoo, Watanabe, . . . , Meyer) theory *defines* $\int U\,dM$ as that P-unique continuous local P-martingale starting at 0 such that for every continuous local P-martingale N with $N_0 = 0$, $\langle \int U\,dM, N\rangle_t = \int_0^t U_s\,d\langle M, N\rangle$ up to P-uniqueness, where $\langle M, N\rangle = \frac{1}{2}(\langle M+N, M+N\rangle - \langle M, N\rangle - \langle N, N\rangle)$. See Meyer [8].

Thus P is forced to be Wiener measure. This result, discovered by Lévy (who else?!), is what gave hope for the martingale problem method. Splendid as the Kunita-Watanabe proof just given is, it is tied to this 'freak' case, and so is superseded by the proof in §12 below.

10. Martingale problems and stochastic differential equations. Again take $d = 1$, and assume that P solves the martingale problem starting from x for $L = \frac{1}{2}a(x)d^2/dx^2 + b(x)d/dx$, where a and b are continuous (or just measurable) and $a(\cdot) > 0$.

Put $\sigma(x) = a(x)^{1/2}$, and write

$$dB = \sigma(X)^{-1}\,dM = \sigma(X)^{-1}\left[dX - b(X)\,dt\right], \qquad B_0 = 0. \tag{10.1}$$

Then B is a local P-martingale and $(dB)^2 = a(X)^{-1}d\langle M, M\rangle = dt$. By Lévy's theorem, *$B$ is an $(\{\mathcal{F}_t^\circ\}, P)$ Brownian motion. Moreover, the formally-obvious consequence*:

$$dX = \sigma(X)\,dB + b(X)\,dt \tag{10.2}$$

of (10.1) *is easily proved.* (It is obvious from the modern theory in footnote 8.) Note that *we constructed B from X.*

This is the reverse of the situation in Itô theory, where the aim is to study (10.2) *as an equation for an 'unknown' X for a given Brownian motion B.* Under Lipschitz conditions on a and b, Itô solved (10.2) via successive approximation, Picard-style, obtaining X as a B-adapted process (X_t is $\sigma\{B_s$: $s \leqslant t\}$ measurable). It is not known whether a B-adapted solution X of (10.2) exists when a mere continuity hypothesis is imposed on a and b: *X may require more randomness than B can provide.* To understand what might be going on here, study the still-rather-mystifying *Tsirel'son counterexample* in Liptser and Shiryayev [**6**].

Many aspects of the relationship between Itô theory and S-V theory are subtle and difficult. Chapter 8 of the S-V book is a careful study, following on from important work of Watanabe and Yamada. One of the main results is that 'Itô uniqueness' (watch the formulation!) implies uniqueness in the martingale-problem sense.

11. Why has the martingale-problem method succeeded where others have failed? In regard to *existence theorems*, the reason lies in the 'solidarity' of probability measures on Polish spaces under 'weak' convergence, and in the fact that *the martingale-problem method is ideally suited for establishing 'weak' compactness of families of measures.* This is because martingale inequalities automatically provide modulus-of-continuity properties for application of Arzéla-Ascoli-Prohorov criteria. (That 'practical' weak-convergence results are handled effectively is evidenced by the book's Chapter 11.) Itô's results are 'dense', and we can take limits!

We now turn to the deeper *uniqueness results*. These are deeper than analysis can *formulate*! But the cunning martingale-problem method *stays sufficiently close to analysis to be able to utilise its estimates.* By contrast, the more probabilistic *Itô theory is too subtle for analysis to be of any help*!

12. Thumbnail sketch of proof of the uniqueness result. Let L be as at (2.1) but with $b \equiv 0$. The *Cameron-Martin-Girsanov change-of-measure theorem* allows us to 'add in (measurable) b' later. Since $a(\cdot)$ is continuous, it is locally almost constant. *Localisation techniques* (*probabilistic*–with no analytic counterpart) make it enough to deal with the case: *$a(\cdot)$ is always close to the identity matrix, so L is 'appoximately'* $\frac{1}{2}\Delta$. You can very easily derive directly from the definition that if P^x (*not* known to be Markovian) solves the martingale problem for L starting from x, then, for $\lambda > 0$, and $f \in C_K^\infty(\mathbf{R}^d)$,

$$R_\lambda^x(\lambda f - Lf) = f(x) \quad \text{where } R_\lambda^x(h) = \int_0^\infty dt\, e^{-\lambda t} \int_\Omega h \circ X_t(\omega) P^x(d\omega). \tag{12.1}$$

The idea of perturbation theory is to try to show that R_λ^x is therefore determined because $R_\lambda^x g = (\lambda - L)^{-1} g(x)$, where we try to define

$$(\lambda - L)^{-1} = V_\lambda (I - K_\lambda)^{-1}, \tag{12.2}$$

where $K_\lambda = \frac{1}{2}\Sigma\Sigma[a_{ij}(\cdot) - \delta_{ij}]\partial_i\partial_j V_\lambda$, and $V_\lambda = (\lambda - \frac{1}{2}\Delta)^{-1}$ is the usual Brownian resolvent. The key fact with which we can work is provided by Littlewood-Paley theory (see Meyer [**9**] for a fine probabilistic proof), namely: *for $p > 1$, the map $\partial_i\partial_j V_\lambda$ from $C_K^\infty(\mathbf{R}^d)$ to $L^p(\mathbf{R}^d)$ extends (uniquely) to a bounded linear operator on $L^p(\mathbf{R}^d)$*. By holding $a(\cdot)$ everywhere close to the identity matrix, we can arrange that, for all λ in an interval, $\|K_\lambda\|_p < 1$. While (12.2) is now meaningful in $L^p(\mathbf{R}^d)$ terms, we need to make a much more careful analysis (not done in this review!) to show that *for $p > \frac{1}{2}d$, R_λ^x is a bounded linear functional on $L^p(\mathbf{R}^d)$*. Only then can we say that R_λ^x is determined.[9] By 'inversion of Laplace transforms', *$P^x \circ X_t^{-1}$ is uniquely determined for all t (and this for every x). On conditioning the martingale-problem formulation relative to $\mathcal{F}_t^\circ$ in the sense of regular conditional probabilities*, we find that

$$\left[P^x \circ X_{t+h}^{-1} | \mathcal{F}_t^\circ\right] = P^{X(t)} \circ X_h^{-1} \qquad (P^x \text{ almost surely}).$$

The 'full' uniqueness of P^x and its Markov property (the two are inextricably linked) follow immediately. (For *Markov selection principles* in the presence of nonuniqueness, see Chapter 12 of the book.)

For their more general and much fuller results, Stroock and Varadhan use much deeper inequalities due to Jones, Fabes, and Rivière. The book's Appendix proves these estimates, and links such results to the theory of *BMO*, etc.

13. I have emphasised the great importance of the Stroock-Varadhan book. It contains a lot more than I have indicated; in particular, its many exercises contain much interesting material.

For immediate confirmation of the subject's sparkle, virtuosity, and depth, see Mozart–sorry, I mean *McKean*–[7]. The Stroock-Varadhan book proceeds on its inexorable way like a massive Bach fugue. 'Too much counterpoint; and, what is worse, *Protestant* counterpoint', said Beecham of Bach. But old

[9] Argue (via localisation) that we can assume that $a(\cdot)$ is constant far out. Hence we can approximate $a(\cdot)$ uniformly by smooth $a_k(\cdot)$. Now it is fairly easy to show that if $g \in C_K^\infty$, then we can find f_n in C_K^∞ with $\lambda f_n - Lf_n \to g$ in L^p. Thus, since R_λ^x is continuous on L^p, $R_\lambda^x g$ may be determined from (12.1).

J. S. can be something of a knockout if his themes get hold of you. And his influence on what followed was (you may say) substantial.

REFERENCES

1. M. H. A. Davis, *A pathwise solution of the equations of non-linear filtering*, Teor. Veroyatnost i Primenen (to appear).

2. C. Dellacherie, C. Doléans-Dade, G. Letta and P. A. Meyer, *Diffusions à coefficients continus d'après D. W. Stroock et S. R. S. Varadhan*, Séminaire de Probabilités IV, Lecture Notes in Math., vol. 124, Springer-Verlag, Berlin and New York, 1970, pp. 241–282.

3. R. A. Holley and D. W. Stroock, *A martingale approach to infinite systems of interacting processes*, Ann. Probability **4** (1976), 195–228.

4. ______, *Nearest neighbor birth and death processes on the real live*, Acta Math. **140** (1978), 103–154.

5. J. Jacod, *Calcul stochastique et problèmes de martingales*, Lecture Notes in Math., vol. 714, Springer-Verlag, Berlin and New York, 1979.

6. R. S. Liptser and A. N. Shiryayev, *Statistics of random processes*. I: *general theory* (English translation from the Russian), Applications of Mathematics, vol. 5, Springer-Verlag, Berlin and New York, 1977.

7. H. P. McKean, *Stochastic integrals*, Academic Press, New York, 1969.

8. P. A. Meyer, *Un cours sur les intégrales stochastiques*, Séminaire de Probabilités X, Lecture Notes in Math., vol. 511, Springer-Verlag, Berlin and New York, 1976, pp. 245–400.

9. ______, *Démonstration probabiliste de certaines inégalités de Littlewood-Paley*, Séminaire de Probabilités X, Lecture Notes in Math., vol. 511, Springer-Verlag, Berlin and New York, 1976, pp. 125–183.

10. D. W. Stroock, *Diffusion processes associated with Lévy generators*, Z. Wahrscheinlichkeitstheorie ver. Gebiete **32** (1975), 209–244.

DAVID WILLIAMS

Limit Theorems

Sankhya Ser. A 24 (1962) 213-238

LIMIT THEOREMS FOR SUMS OF INDEPENDENT RANDOM VARIABLES WITH VALUES IN A HILBERT SPACE

By S. R. S. VARADHAN

Indian Statistical Institute

SUMMARY. In this article distributions on a Real Separable Hilbert Space are considered. Limit distributions are derived for sums of infinitesimal random variables.

A representation similar to the Levy-Khintchine representation is derived for infinitely divisible distributions. Necessary and sufficient conditions for compactness are obtained in terms of the quantities occurring in the representation.

1. Introduction

The celebrated Levy-Khintchine representation for the characteristic function $\phi(t)$ of an infinitely divisible distribution on the real line is given by

$$\phi(t) = \exp\left[i\nu t - \frac{\sigma^2 t^2}{2} + \int_{-\infty}^{\infty} \left(e^{itx} - 1 - \frac{itx}{1+x^2} \right) \frac{1+x^2}{x^2}\, dG(x) \right] \quad \ldots \quad (1.1)$$

where ν and σ are real constants, $\sigma \geqslant 0$ and $G(x)$ is a bounded non-decreasing function of x which is continuous at the origin. ν, σ and G are uniquely determined by $\phi(t)$. Conversely, any function of the type (1.1) is the characteristic function of an infinitely divisible distribution. Khintchine and Bawly went further and proved that the limit distributions of sums of uniformly infinitesimal random variables are infinitely divisible and that they can be obtained as limits of certain accompanying infinitely divisible distributions. For all historical and other details concerning these we refer to Gnedenko and Kolmogorov (1954). Attempts have been made by several authors to extend these results to more general situations. We mention, in particular, the works of Bochner (1955) and (1958), Hunt (1956), Kloss (1961), Levy (1939), Takano (1955) and Vorobev (1954). Takano has generalised both the representation and the limit theorems to the case of a finite dimensional vector space. Levy has considered the circle group (the multiplicative group of complex numbers of modulus unity). The axiomatic development of the concept of characteristic functions by Bochner (1958) throws some light on the nature of the problem. Hunt has obtained the representation for one parametric semigroups of distributions on a Lie group. Vorobev has considered finite groups and Kloss a wide class of compact groups.

The case of a general locally compact abelian group was considered very recently by K. R. Parthasarathy, R. Rangarao and the present author (1962a). The results obtained are mentioned briefly in the next section and we will have occasion to use them in the later sections.

The case of a non-locally compact group offers considerable difficulty. However, the particular case of a separable Hilbert space can be tackled using some compactness criteria of Prohorov (1956). In this paper we will consider the case of a Hilbert space, and the analysis will throw some light on the precise nature of the difficulties in extending these results to non-locally compact groups.

2. The locally compact case

In the present section we will briefly mention the results obtained for the locally compact case. We will be using some of these results in the latter sections.

Let X be a locally compact abelian group which is separable metric and let Y be its character group. For $x \epsilon X$ and $y \epsilon Y$ we write (x, y) to denote the value of the character y at the point x.

Let $\mathcal{M}$ denote the convolution semi-group of probability measures on X, endowed with the weak topology. A sequence μ_n in $\mathcal{M}$ is shift compact whenever the sequence formed by some translates of these distributions is compact. Throughout the article $*$ denotes the convolution operation. When x is an element of the group and λ a distribution $\lambda * x$ denotes the convolution of λ with the distribution degenerate at x. λ^n denotes the n-th convolution of λ. $|\lambda|^2$ denotes $\lambda * \bar{\lambda}$ where $\bar{\lambda}$ is the distribution defined by $\bar{\lambda}(A) = \lambda(-A)$. $-A$ consists of the set of inverses of A. The following theorem was proved by Parthasarathy *et al* (1962).

Theorem 2.1 : *Let $\mu_n = \alpha_n * \beta_n$ for each n. If μ_n is compact then α_n and β_n are shift compact.*

Definition 2.1 : A distribution μ is said to be infinitely divisible if for each n there are elements x_n in X and $\lambda_n \epsilon \mathcal{M}$ such that

$$\mu = \lambda_n^n * x_n.$$

Remark 2.1 : This modification of the classical definition is intended to avoid the consequence of the presence of elements that are not divisible. In a divisible group this definition is equivalent to the classical definition as can be easily seen.

Theorem 2.2 : *The totality of infinitely divisible distributions is a closed subsemigroup of $\mathcal{M}$.*

Definition 2.2 : If F is any totally finite measure on X the distribution $e(F)$ associated with it is defined as follows.

$$e(F) = e^{-F(X)} \left[1 + F + \frac{F^2}{2!} + \ldots \right].$$

For $\mu \epsilon \mathcal{M}$ we denote by $\hat{\mu}(y)$ its characteristic function. We then have

$$\hat{e}(F)(y) = [\int [(x, y) - 1] dF].$$

Theorem 2.3 : *If μ is infinitely divisible and if its characteristic function vanishes at some point then μ has an idempotent factor.*

Theorem 2.4 : *Let $\mu_n = e(F_n)$. Then the necessary and sufficient conditions that*

(a) *μ_n is shift compact,*

(b) *if μ is any limit of shifts of μ_n then μ has no idempotent factors, are*

(i) *for each neighbourhood N of identity e the family F_n restricted to $X - N$ is weakly conditionally compact;*

(ii) *for each $y \epsilon Y$*

$$\sup_n \int [1 - \text{Real } (x, y)] dF_n < \infty.$$

Definition 2.3 : A double sequence $\alpha_{nj} : j = 1, 2, \ldots, k_n;\ n = 1, 2, \ldots$ of distributions is said to be uniformly infinitesimal if

$$\underset{n\to\infty}{\text{lt}} \sup_{1\leqslant j\leqslant k_n} \sup_{y\epsilon K} |\alpha_{nj}(y)-1| = 0$$

for every compact subset $K \subseteq Y$.

Lemma 2.1 : *There exists a real valued function $g(x, y)$ on $X\times Y$ which is continuous in both the variables x and y and such that*

(i) $g(x, y_1+y_2) = g(x, y_1)+g(x, y_2)$ *for each* $x\epsilon X$ *and* $y_1, y_2 \epsilon Y$,

(ii) *for any compact subset* $K \subseteq Y$

$$\sup_{x\epsilon X} \sup_{y\epsilon K} |g(x, y)| < +\infty,$$

(iii) *for each compact subset* $K \subseteq Y$ *there exists a neighbourhood* N_K *of the identity in* X *such that*

$$(x, y) = e^{ig(x,y)}$$

holds for $x \epsilon N_K$ *and* $y \epsilon K$,

(iv) $g(x, y) = 0$ *whenever* $x = e$ *for any* y.

Theorem 2.5 : *Let α_{nj} be a uniformly infinitesimal sequence and let*

$$\mu_n = \prod_{j=1}^{k_n} \alpha_{nj}.$$

Suppose μ_n is shift compact and that no limit of shifts of μ_n has an idempotent factor. Let $\beta_{nj} = e[\alpha_{nj}*g_{nj}]$ where g_{nj} is that element of the group X defined by the equality

$$(g_{nj}, y) = \exp\{-i \textstyle\int g(x, y)d\alpha_{nj}\}.$$

If
$$\lambda_n = \prod_{j=1}^{kn} \beta_{nj}*g_n \text{ where } g_n = -\sum_j g_{nj},$$

then
$$\overline{\lim_{n\to\infty}} \sup_{y\epsilon K} |\lambda_n(y)-\mu_n(y)| = 0$$

for every compact set K of Y.

Corollary 2.1 : *Limit distribution of sums infinitesimal independent random variables is infinitely divisible.*

Definition 2.4 : A distribution μ is said to be Gaussian if it has the following properties,

(i) μ is infinitely divisible,

(ii) $\mu = e(F) *\alpha$, where α is infinitely divisible implies that F vanishes outside the identity.

Therem 2.6 : *A distribution μ is Gaussian if and only if $\mu(y)$ is of the form*

$$\mu(y) = (x, y) \exp[-\phi(y)]$$

where x is a fixed element of X and $\phi(y)$ a continuous non-negative function of y satisfying the equation

$$\phi(y_1+y_2)+\phi(y_1-y_2) = 2[\phi(y_1) + \phi(y_2)]$$

for every pair y_1, y_2 in Y.

Theorem 2.7 : *A distribution μ on X is infinitely divisible without an idempotent factor if and only if $\hat{\mu}(y)$ is of the form*

$$\hat{\mu}(y) = (x_0, y) \exp [\int [(x, y) - 1 - ig(x, y)] dM - \phi(y)]$$

where

(i) *x_0 is a fixed element of X,*

(ii) *$\phi(y)$ is a function as in Theorem 2.6,*

(iii) *M is a σ-finite measure giving finite mass outside every neighbourhood of the identity,*

(iv) *$\int [1 - \text{Real}\,(x, y)] dM < +\infty$ for each $y \in Y$.*

Remark 2.2 : This representation is not in general unique. Examples of non-uniqueness and conditions for uniqueness are discussed by Parthasarathy *et al* (1962a) . It turns out that if the character group is connected then the representation is unique.

Corollary 2.2 : *Every one parametric weakly continuous semigroup μ_t has a unique representation*

$$\hat{\mu}_t(y) = (x_t, y) \;\; \exp [t \int [(x, y) - 1 - ig(x, y)] dM - t\phi(y)]$$

where x_t is a continuous semigroup in X and M and ϕ are as in Theorem 2.7.

Remark 2.3 : Of the results mentioned in this section Theorems 2.1, 2.2 and the necessity part of Theorem 2.3 are valid in any complete separable metric group. We will also need the following corollary and theorem deducible from Theorem 2.2.

Corollary 2.3 : *λ_n is shift compact if and only if $|\lambda_n|^2$ is compact.*

Theorem 2.8 : *Let λ_n be a sequence such that λ_n is a factor of λ_{n+1} for all n. Then if λ_n is shift compact there are translates λ_n' of λ_n which converge.*

Remark 2.4 : We can assume instead that λ_{n+1} is a factor of λ_n for each n and then also the theorem will be valid.

3. Preliminaries

X is a real separable Hilbert space, (x, y) denotes the inner product and $||x||$ the norm. With vector addition as group operation X becomes a complete separable metric group. We will denote by $\mathcal{M}$ the semigroup of all distributions.

For every $\mu \in \mathcal{M}$ its characteristic function is defined on X by the formula

$$\hat{\mu}(y) = \int e^{i(x,\, y)} \; d\mu(x).$$

We will, in this section, mention some results obtained by Prohorov (1956) concerning compactness criteria for distributions on X.

Definition 3.1 : A positive semi-definite Hermitian operator A is called an S-operator if it has finite trace. The class of sets of the type $[x : (Sx, x) < t]$ where S runs over S-operators and t over positive numbers forms a neighbourhood system at the origin for a certain topology which is called the S-topology. A net x_α converges to zero in S-topology if and only if (Sx_α, x_α) converges to zero for every S-operator S.

LIMIT DISTRIBUTION FOR RANDOM VARIABLES

We have the following theorem concerning characteristic functions and S-topology which was obtained by Sazanov (1958).

Theorem 3.1 : *In order that a function $\mu(y)$ may be the characteristic function of a distribution on X it is necessary and sufficient that $\mu(0) = 1$, $\mu(y)$ be positive definite and continuous at $y = 0$ in the S-topology. (Here and elsewhere in the article 0 will denote the null element of X or the identity of the group and will be called the origin).*

We also have the following theorem of Prohorov (1956).

Theorem 3.2 : *In order that a positive definite function $\mu(y)$ with $\mu(0) = 1$ be the characteristic function of a distribution on X it is necessary and sufficient that for every $\epsilon > 0$ there exists an S-operator S_ϵ such that*

$$1-R\mu(y) \leqslant (S_\epsilon y, y)+\epsilon$$

where R denotes the real parts.

Definition 3.2 : Let μ be a distribution on X for which $\int (x, y)^2 d\mu < \infty$ for each y. Then the covariance operator S of μ is that Hermitian operator for which

$$(Sy, y) = \int (x, y)^2 d\mu(x).$$

This operator S will be positive semi-definite and will be an S-operator if and only if

$$\int \|x\|^2 \, d\mu(x) < \infty.$$

Definition 3.3 : A sequence S_n of S-operators will be called compact if and only if the following two conditions are satisfied

$$\text{(i)} \qquad \sup_n \text{ Trace } (S_n) < \infty$$

$$\text{(ii)} \qquad \operatorname*{lt}_{N \to \infty} \sup_n \sum_{j=N}^{\infty} (S_n e_j, e_j) \to 0$$

for some orthonormal sequence $e_1, e_2, \ldots e_j, \ldots$

When S is the covariance operator of a distribution μ on X for which

$$\int \|x\|^2 d\mu(x) < \infty$$

we have the relation

$$\sum_{j=N}^{\infty} (Se_j, e_j) = \int r_N^2(x) d\mu(x)$$

where

$$r_N^2(x) = \sum_{j=N}^{\infty} (x, e_j)^2$$

and $e_1, e_2, \ldots$ any orthonormal basis.

The following theorems concerning conditions for compactness of a sequence μ_n of distributions on X were obtained by Prohorov (1956).

Theorem 3.3 : *In order that a sequence μ_n of distributions on X be weakly conditionally compact it is necessary and sufficient that for every $\epsilon > 0$, there exists a compact sequence S_n^ϵ of S-operators such that*

$$1-R\,\mu_n(y) \leqslant (S_n^\epsilon y, y)+\epsilon$$

for all n and y. Here $\mu_n(y)$ is the characteristic function of μ_n.

Theorem 3.4 : *Let μ_n be a sequence of distributions for which the covariance operators S_n exist and are S-operators. Let further S_n be compact. Then μ_n is weakly conditionally compact.*

Theorem 3.5 : *In order that a sequence μ_n of distributions on X may be weakly conditionally compact it is necessary that for any $\epsilon > 0$*

$$\underset{N \to \infty}{\text{lt}} \sup_n \mu_n[r_N^2(X) > \epsilon] = 0.$$

In defining the covariance operator in Definition 3.2. we assumed that μ is a distribution. Actually, if M is any σ-finite measure for which

$$\int ||x||^2 \, dM < \infty$$

then S is well defined as an S-operator by the relation

$$(Sy, y) = \int (x, y)^2 \, dM(x).$$

We will need also the following.

Theorem 3.6 : (Rangarao, 1960). *Let $\mu_n \to \mu$. Let $\{f_a\}$ be a family of continuous functions equicontinuous and uniformly bounded. Then*

$$\underset{n \to \infty}{\text{lt}} \sup_a \left| \int f_a d\mu_n - \int f_a d\mu \right| = 0.$$

4. An estimate of the variance

Let $X_1, X_2, \ldots, X_n$, be n symmetric independent random variables in the Hilbert space X. We will give in this section an estimate for the variance $E||X_1 + \ldots X_n||^2$ when each X_j is bounded uniformly in norm by a constant C independent of j.

To this end we introduce the concentration function $Q_\mu(t)$ following Levy [(1954), page 138].

Definition 4.1 : The concentration function $Q_\mu(t)$ of a distribution μ in the Hilbert space X is defined for $0 < t < \infty$ as

$$Q_\mu(t) = \sup_{x \epsilon X} \mu(S_t + x)$$

where S_t denotes the sphere $[x : ||x|| \leqslant t]$ and $S_t + x$ its translate by the element x of X.

We now list a few elementary properties of these functions.

Theorem 4.1 : (1) *$Q_\mu(t)$ is a nondecreasing function of t and* $\lim_{t \to \infty} Q_\mu(t) = 1$. (2) *If $\mu_1 * \mu_2 = \mu$ then, for every t, $Q_\mu(t) \leqslant$* min $[Q_{\mu_1}(t), Q_{\mu_2}(t)]$. (3) *If μ_n is shift compact then*

$$\lim_{t \to \infty} \inf_n Q_{\mu_n}(t) = 1.$$

Proof : (1) and (2) are obvious and (3) is a consequence of tightness.

Theorem 4.2 : *Let $X_1, X_2, \ldots, X_n$ be n mutually independent symmetric random variables. Let $S_j = X_1 + \ldots + X_j$. Further, let $Q(t)$ denote the concentration function of the sum $S_n = X_1 + \ldots X_n$. If T is defined as*

$$T = \sup_{1 \leqslant j \leqslant n} || S_j ||$$

then one has for any $t > 0$ $\quad P\{T > 4t\} \leqslant 2[1 - Q(t)].$

Proof : This remarkable result of Levy, which is a refinement of Kolmogorov's inequality in terms of the variance, although proved by him for the real line, offers no difficulty for generalisation. Let the events E_k be defined as follows :

$$E_k = [\|S_1\| \leqslant 4t, \ldots, \|S_{k-1}\| \leqslant 4t, \ \|S_k\| > 4t]$$

$$\{T > 4t\} = E_1 \cup E_2 \ldots \cup E_n, \quad E_i \cap E_j = \phi.$$

By $Pr\{ \ \}$ we will denote the probability of the event within the brackets given that the event E_r has occurred. We then have

$$Pr\{\|S_n\| \leqslant 2t\} \leqslant Pr\{\|S_n - S_r\| > 2t\} = P\{\|S_n - S_r\| > 2t\}. \quad \ldots (4.1)$$

This is because E_r and $\|S_n\| \leqslant 2t$ imply that $\|S_n - S_r\| > 2t$ and $S_n - S_r$ is distributed independently of E_r. Let us further suppose that $Q(t) > \frac{1}{2}$. We now consider the distribution μ_{rn} of $S_n - S_r$ whose concentration function is denoted by $Q_{rn}(t)$. Since μ_{rn} is a factor of μ we have $Q_{rn}(t) > \frac{1}{2}$. This implies the existence of a point x in the space X such that

$$\mu_{rn}(S_t + x) > Q_{rn}(t) - \epsilon > \tfrac{1}{2}. \quad \ldots (4.2)$$

Since μ_{rn} is a symmetric distribution

$$\mu_{rn}(S_t - x) = \mu_{rn}(-S_t + x) = \mu_{rn}(S_t + x) > \tfrac{1}{2}.$$

Therefore, $\qquad S_t - x \cap S_t + x \neq \phi.$

In other words there exists a point y such that

$$\|x + y\| \leqslant t, \ \|x - y\| \leqslant t.$$

These two imply that $\|x\| \leqslant t$ and hence

$$x + S_t \subseteq S_{2t}. \quad \ldots (4.3)$$

From (4.2) and (4.3) follows $\qquad \mu_{rn}(S_{2t}) > Q_{rn}(t) - \epsilon.$

Since ϵ is arbitrary we have $\qquad \mu_{rn}(S_{2t}) \geqslant Q_{rn}(t) \quad \ldots (4.4)$

which is the same as $\quad P\{\|S_n - S_r\| > 2t\} \leqslant 1 - Q_{rn}(t) \leqslant 1 - Q(t). \quad \ldots (4.5)$

We have further, using (4.1) and (4.5)

$$P\{T > 4t, \|S_n\| \leqslant 2t\} = \sum_{r=1}^{n} Pr\{\|S_n\| \leqslant 2t\} P[E_r] \leqslant \Big(\sum_{r=1}^{n} P[E_r](1 - Q(t)) = P\{T > 4t\}(1 - Q(t)), \quad \ldots (4.6)$$

we also have $\quad P\{T \leqslant 4t, \|S_n\| \leqslant 2t\} \leqslant P\{T \leqslant 4t\} = 1 - P\{T > 4t\}. \quad \ldots (4.7)$

Adding (4.6) and (4.7) we get $\quad P\{\|S_n\| \leqslant 2t\} \leqslant 1 - P\{T > 4t\} Q(t). \quad \ldots (4.8)$

From (4.4) putting $r = 0$, we obtain

$$P\{\|S_n\| \leqslant 2t\} \geqslant Q(t), \quad \ldots (4.9)$$

(4.8) and (4.9) imply, since $Q(t) > \frac{1}{2}$,

$$P\{T > 4t\} \leqslant \frac{1 - Q(t)}{Q(t)} \leqslant 2[1 - Q(t)]. \quad \ldots (4.10)$$

However, if $Q(t) \leqslant \frac{1}{2}$ the theorem is trivially true. Thus the proof of the theorem is complete.

We have the following theorem which gives an estimate of the variance. This is well known for the real line. We reproduce the proof for the real line, which can be found in Halmos [(1950) page 198], replacing however $|x|$ by $\|x\|$.

Theorem 4.3 : *Let $X_1, X_2, \ldots, X_n$ be n independent random variables in a Hilbert space, uniformly bounded by a constant c in norm. Let each X_i have zero expectation. In addition let*

$$P\{\sup_{1\leqslant j\leqslant n} \|S_j\| \leqslant d\} \geqslant \epsilon > 0$$

where
$$S_j \doteq X_1 + \ldots X_j$$

Then
$$E\,\|S_n\|^2 \leqslant \frac{d^2+(c+d)^2}{\epsilon}\ .$$

Proof: Let the sets E_k be defined as follows

$$E_k = \sup_{1\leqslant j\leqslant k} \|S_j\| \leqslant d.$$

Then $E_1 \supset E_2 \ldots \supset E_n$ and $P(E_n) \geqslant \epsilon > 0$. By P we mean the product measure of $X_1, X_2, \ldots, X_n$. Let us define

$$F_k = E_{k-1} - E_k$$

and
$$\alpha_k = \int_{E_k} \|S_k\|^2\, dP.$$

We will take E_0 as the whole space and α_0 as zero. We then have

$$\begin{aligned}
\alpha_k - \alpha_{k-1} &= \int_{E_k} \|S_k\|^2\, dP - \int_{E_{k-1}} \|S_{k-1}\|^2\, dP \\
&= \int_{E_{k-1}} \|S_k\|^2\, dP - \int_{F_k} \|S_k\|^2\, dP - \int_{E_{k-1}} \|S_{k-1}\|^2\, dP \\
&= \int_{E_{k-1}} \|S_{k-1}\|^2\, dP - 2\int_{E_{k-1}} (S_{k-1}, X_k) dP + \int_{E_{k-1}} \|X_k\|^2\, dP \\
&\quad - \int_{F_k} \|S_k\|^2\, dP - \int_{E_{k-1}} \|S_{k-1}\|^2\, dP \\
&= \int_{E_{k-1}} \|X_k\|^2\, dP - \int_{F_k} \|S_k\|^2\, dP \\
&\geqslant P(E_{k-1}) E\|X_k\|^2 - (c+d)^2 P(F_k). \qquad \ldots (4.11)
\end{aligned}$$

We notice here that $\int_{E_{k-1}} (S_{k-1}, X_k)\, dP = 0$ since X_k is independent of S_{k-1}, has zero expectation and is independent of E_{k-1} as well. Moreover, $F_k \subset E_{k-1}$ hence $\|S_k\| \leqslant \|S_{k-1}\| + \|X_k\| \leqslant (c+d)$ over F_k. Since $P(E_n) \leqslant P(E_k)$ for any k we have

$$\alpha_k - \alpha_{k-1} \geqslant P(E)\, E\|X_k\|^2 - (c+d)^2 P(F_k) \quad \text{for } k = 1, 2, \ldots \qquad \ldots (4.12)$$

Since $F_1, F_2, \ldots, F_n$ are all disjoint adding (4.12) for $k = 1, 2, \ldots, n$, we have

$$\alpha_n \geqslant P(E) E\,\|S_n\|^2 - (c+d)^2. \qquad \ldots (4.13)$$

However, since $\|S_n\| \leqslant d$ on E_n $\alpha_n = \int_{E_n} \|S_n\|^2\, dP \leqslant d^2,$ ... (4.14)

(4.13) and (4.14) give
$$E\|S_n\|^2 \leqslant \frac{(d+d)^2+d^2}{P(E)} \leqslant \frac{(c+d)^2+d^2}{\epsilon}\ .$$

This completes the proof.

Theorem 4.4 : *Let* $X_1, X_2, \ldots, X_n$ *be symmetric independent random variables in the Hilbert space such that* $\| X_i \| \leqslant c$ *for* $i = 1, 2, \ldots, n$. *Let* $Q(t)$ *denote the concentration function of* $S_n = X_1 + \ldots X_n$. *Then*

$$E\|S_n\|^2 \leqslant \frac{16t^2+(c+4t)^2}{2Q(t)-1}$$

for any t *such that* $Q(t) > \frac{1}{2}$.

Proof : This follows at once from Theorems 4.2 and 4.3 and the fact that a bounded symmetric random variable has zero expectation.

5. Infinitely divisible distributions

In this section we will obtain a representation for infinitely divisible distributions. The definition of an infinitely divisible distribution is the same as Definition 2.1, but since the Hilbert space as a group is divisible it is equivalent to the classical definition which requires that μ be written as λ_n^n for each n.

As we have already remarked at the end of Section 2 that some of the results mentioned in that Section are valid for the Hilbert space we will now state them. We will keep in mind also that the Hilbert space has no nontrivial compact subgroups and hence there are no idempotent distributions.

Theorem 5.1 : *The infinitely divisible distributions form a closed subsemigroup among all distributions.*

Theorem 5.2 : *If* μ *is an infinitely divisible distribution and* $\mu(y)$ *its characteristic function then* $\mu(y)$ *is nonvanishing.*

For every finite measure F the infinitely divisible distribution $e(F)$ is associated in the same way as in Definition 2.2. We then have the following theorem.

Theorem 5.3 : *Let* $\mu_n = e(F_n)$. *In order that* μ_n *may be shift compact it is necessary that*

(i) *for each neighbourhood* N *of the identity* F_n *restricted to* N' *is weakly conditionally compact,*

(ii) $\sup\limits_n \int [1-\cos(x, y)] dF_n < \infty$ *for each* y.

Theorem 5.4 : *Let* $\mu_n \Longrightarrow \mu$. *Then* $\mu_n(y) \to \mu(y)$ *uniformly over every bounded sphere.*

Proof : Since the set of functions (x, y) as y varies over a bounded sphere forms a equicontinuous family of functions in x the theorem follows.

Theorem 5.5 : *Let* μ_n *be shift compact and* $\mu_n(y) \to \mu(y)$ *uniformly over bounded spheres. Then* $\mu_n \Longrightarrow \mu$.

Proof : Since μ_n is shift compact let x_n in X be chosen such that $\mu_n * x_n$ is compact. We will now show that x_n in X is compact. If x_n is not compact then we can produce a subsequence from x_n which has no further convergent subsequence. We will denote the subsequence by x_n itself. Since $\mu_n * x_n$ is weakly compact it has a subsequence $\mu_{n_j} * x_{n_j}$ converging weakly. Thus $\mu_{n_j}(y)e^{i(x_{n_j}, y)}$ as well as $\mu_{n_j}(y)$ converge uniformly over bounded spheres. Since from the compactness of $|\mu_n|^2$ one can conclude the existence of a sphere S such that $\inf_n |\mu_n(y)| \geqslant \epsilon > 0$ for all y in S, it follows that $e^{i(x_{n_j}, y)}$ converges uniformly over S and hence x_{n_j} converges in norm. This proves the theorem.

Before obtaining the representation we will show that if $e(F_n)$ is shift compact then

$$\sup_n \int_{\|x\| \leqslant 1} \|x^2\| dF_n < \infty.$$

To this end we consider the following lemma.

Lemma 5.1 : *Let $f(y)$ be a non-negative function on X such that $f(2y) \leqslant 4f(y)$ for all values of y. If $f(y) \leqslant \epsilon$ whenever $(Sy, y) \leqslant \delta$ where S is some S-operator then*

$$f(y) \leqslant (S_1 y, y) + \epsilon \quad \textit{for all } y$$

where

$$S_1 = 4\epsilon\delta^{-1}S.$$

Proof : Defining $S_0 = \epsilon\delta^{-1}S$,

we see that
$$f(y) \leqslant \epsilon \text{ whenever } (S_0 y, y) \leqslant \epsilon. \quad \ldots (5.1)$$

Further, if $(S_0 y, y) \leqslant 4^n\epsilon$ where n is a positive integer, then denoting by y_n the element $2^{-n}y$, we have

$$(S_0 y_n, y_n) = 4^{-n}(S_0 y, y) \leqslant \epsilon$$

consequently $f(y_n) \leqslant \epsilon$. But since $f(2y) \leqslant 4f(y)$,

we have
$$f(y) \leqslant 4^n f(y_n) \leqslant 4^n\epsilon.$$

So from (5.1) we have

$$f(y) \leqslant 4^n\epsilon \text{ whenever } (S_0 y, y) \leqslant 4^n\epsilon. \quad \ldots (5.2)$$

Let y be any element of X and let $(S_0 y, y) = t$.

Case (i) : Let $t > \epsilon$. If n is a non-negative integer such that

$$4^n \epsilon < t \leqslant 4^{n+1}\epsilon. \quad \ldots (5.3)$$

We have, since $(S_0 y, y) = t \leqslant 4^{n+1}\epsilon$, using (5.2) and (5.3)

$$f(y) \leqslant 4^{n+1}\epsilon \leqslant 4t = 4(S_0 y, y). \quad \ldots (5.4)$$

Case (ii) : Let $t \leqslant \epsilon$. Then from (4.1) we have

$$f(y) \leqslant \epsilon. \quad \ldots (5.5)$$

(5.4) and (5.5) give at once $\quad f(y) \leqslant \max[\epsilon, (4S_0 y, y)] \leqslant \epsilon + (S_1 y, y)$.

We shall need while proving the next theorem the following inequality. If $a_1, \ldots a_m$ are any m real numbers such that $|a_j| \leqslant 1$ for $1 \leqslant j \leqslant m$, then

$$1 - a_1 a_2 \ldots a_m \leqslant \sum_{j=1}^{m} (1 - a_j). \quad \ldots (5.6)$$

This inequality is proved by induction if all the a's are positive. If at least one of them say a_r is negative

$$1 - a_1 \ldots a_m \leqslant 1 + |a_1 \ldots a_m| \leqslant 1 + |a_r| = 1 - a_r \leqslant \sum_{j=1}^{m} (1 - a_j).$$

We will now prove the following theorem.

Theorem 5.6 : *Let F_n be a sequence of finite measures such that $e(F_n)$ is shift compact. Then*

$$\sup_n \int_{\|x\| \leqslant 1} \|x\|^2 dF_n < \infty.$$

Proof : We assume without any loss of generality that each F_n vanishes outside the unit sphere. Otherwise we can consider the restriction of F_n to the unit sphere instead of F_n. Let $M_n = F_n + \bar{F}_n$. Then $e(M_n) = |e(F_n)|^2 = \lambda_n$ is compact. We will show that $\int \|x\|^2 dM_n$ is uniformly bounded. To this end we assume that the total mass of M_n is an integer for every n. If this were not so we can write $M_n = M_n^{(1)} + M_n^{(2)}$ where $M_n^{(1)}$ is symmetric with an integral total mass and $M_n^{(2)}$ has total mass less than unity. Consequently

$$\int_{\|x\| \leqslant 1} \|x\|^2 dM_n^{(2)} \leqslant 1.$$

Since our aim is to prove that $\sup_n \int_{\|x\| \leqslant 1} \|x\|^2 dM_n < \infty$

it suffices to show that $\sup_n \int_{\|x\| \leqslant 1} \|x\|^2 dM_n^{(1)} < \infty.$

Now since the total mass of M_n is an integer say k_n we will write

$$M_n = F_{n1} + \dots F_{nk_n}$$

where F_{nj} for $j = 1, 2, \dots, k_n$, $n = 1, 2, \dots$ is a symmetric probability measure in the unit sphere. Let us now denote by μ_n the convolution

$$\mu_n = F_{n1} * F_{n2} \dots * F_{nk_n}.$$

Since each F_{nj} is symmetric and has zero expectation

$$\int \|x\|^2 d\mu_n = \sum_{j=1}^{k_n} \int \|x\|^2 dF_{nj} = \int \|x\|^2 dM_n.$$

Hence it suffices to show that $\sup_n \int \|x\|^2 d\mu_n < \infty.$

If $Q_n(t)$ denotes the concentration function of μ_n from Theorem 4.4 we have

$$\int \|x\|^2 \, d\mu_n \leqslant \frac{16t^2 + (4t+1)^2}{2Q_n(t) - 1}$$

whenever $Q_n(t) > \frac{1}{2}$. Therefore, it is enough to prove that

$$\inf_n Q_n(t) \geqslant \tfrac{3}{4} \text{ for some } t$$

which will follow from Theorem 4.1 if we prove that μ_n is weakly conditionally compact.

Since each $F_{nj}(y)$ is a real characteristic function and

$$\mu_n(y) = \prod_{j=1}^{k_n} F_{nj}(y) \qquad \dots \quad (5.7)$$

it follows from (5.6) that $1 - \mu_n(y) \leqslant \sum_{j=1}^{k_n} [1 - F_{nj}(y)] = \sum_{j=1}^{k_n} \int [1 - \cos(x, y)] dF_{nj}(x)$

$$= \int [1 - \cos(x, y)] dM_n(x) = f_n(y) \text{ say.} \qquad \dots \quad (5.8)$$

We also have $\lambda_n(y) = \exp[-f_n(y)]$ and hence for any given ϵ there exists a δ depending only on ϵ such that

$$f_n(y) \leqslant \epsilon \quad \text{if} \quad 1-\lambda_n(y) \leqslant \delta. \qquad \ldots \quad (5.9)$$

Since λ_n is compact we have from Theorem 3.5 for any given $\delta > 0$ a compact sequence S_n of S-operators depending on δ only such that

$$1-\lambda_n(y) \leqslant (S_n y, y)+\frac{\delta}{2}. \qquad \ldots \quad (5.10)$$

From (5.9) and (5.10) it follows that whenever $(S_n y, y) \leqslant \delta/2, f_n(y) \leqslant \epsilon$ and hence from Lemma 5.1 we have

$$f_n(y) \leqslant (S'_n y, y)+\epsilon \qquad \ldots \quad (5.11)$$

where $$S'_n = 8\epsilon\delta^{-1}S_n.$$

Since S_n is compact so is S'_n and Theorem 3.3 and (5.11) imply that μ_n is weakly conditionally compact. The proof of the theorem is now complete.

We will denote by $K(x, y)$ the following function

$$K(x, y) = e^{i(x,y)}-1-\frac{i(x, y)}{1+\|x\|^2}.$$

Theorem 5.7 : *Let μ_n for each n be of the form $e(F_n)$ where F_n is a finite measure. Let $\mu_n * x_n \Longrightarrow \mu$ for some suitably chosen elements x_n in X. We further assume that F_n is increasing. Then F_n increases to a measure F which may be σ-finite but gives finite mass outside every neighbourhood of the origin and for which*

$$\int_{\|x\|\leqslant 1} \|x\|^2 dF < \infty.$$

In addition $$\mu(y) = \exp[i(x_0, y)+\int K(x, y)dM(x)]$$

where x_0 is a fixed element of X.

Proof: Let $\lambda_n(y)$ be defined as

$$\lambda_n(y) = \exp[\int K(x, y)dF_n].$$

Then $\lambda_n(y)$ is the characteristic function of λ_n which is the shift of μ_n by the element

$$Z_n = -\int \frac{x}{1+\|x\|^2} dF_n.$$

We will now show that $\lambda_n(y)$ converges uniformly in y over bounded spheres. For this purpose we write

$$\int K(x, y)dF_n = \int_{\|x\|\leqslant 1} K(x, y)dF_n+\int_{\|x\|>1} K(x, y)dF_n. \qquad \ldots \quad (5.12)$$

Let F be the limit of F_n. From Theorems 5.3 and 5.6 it follows that F is finite outside every neighbourhood of the origin and

$$\int_{\|x\|\leqslant 1} \|x\|^2 dF < \infty.$$

Since $$|K(x, y)| \leqslant 2+\frac{\|x\|\,\|y\|}{1+\|x\|^2} \leqslant 2+\|y\|$$

it follows that

$$\sup_{y\varepsilon S} \Big| \int_{||x||>1} K(x, y)dF_n - \int_{||x||>1} K(x, y)dF \Big| \rightarrow 0 \text{ as } n\rightarrow\infty \qquad \ldots \quad (5.13)$$

for every bounded sphere S. On the other hand if $||x|| \leqslant 1$

$$|K(x, y)| \leqslant |e^{i(x,y)}-1-i(x, y)| + \frac{|(x, y)|\, ||x||^2}{1+||x||^2}$$

$$\leqslant \tfrac{1}{2}(x, y)^2 + |(x, y)|\, ||x||^2 \leqslant \tfrac{1}{2}||x||^2||y||^2 + ||y||\; ||x||^2.$$

Since $\int\limits_{||x||\leqslant 1} ||x||^2\, dF$ is finite it follows that

$$\sup_{y\varepsilon S} \Big| \int_{||x||\leqslant 1} K(x, y)dF_n - \int_{||x||\leqslant 1} K(x, y)dF \Big| \rightarrow 0 \text{ as } n\rightarrow\infty \qquad \ldots \quad (5.14)$$

for every bounded sphere S. (5.12), (5.13) and (5.14) imply that

$$\sup_{y\varepsilon S} |\lambda_n(y) - \lambda(y)| \rightarrow 0 \quad \text{as } n\rightarrow\infty$$

for every bounded sphere S where $\lambda(y) = \exp [\int K(x, y)dF]$.

Since λ_n is shift compact from Theorem 5.5 it follows that $\lambda_n \Longrightarrow \lambda$ and λ has to be a shift of μ. Hence the theorem follows.

Theorem 5.8 : *Let $\mu(y)$ be a function of the form*

$$\mu(y) = \exp [\int K(x, y)dF(x)]$$

where F is a σ-finite measure giving finite mass outside every neighbourhood of the identity and for which

$$\int_{||x||\leqslant 1} ||x||^2\, dF < \infty.$$

Then $\mu(y)$ is the characteristic function of an infinitely divisible distribution.

Proof : Let N_n denote the sphere of radius $\frac{1}{n}$ around the origin and F_n the restriction of F to N_n'. Then F_n increases to F. Let

$$\mu_n(y) = \exp [\int K(x, y)dF_n].$$

From the proof of Theorem 5.1 it follows that

$$\sup_{y\,\varepsilon\, V} |\mu_n(y) - \mu(y)| \rightarrow 0 \text{ as } n\rightarrow\infty$$

for every bounded sphere V. In view of Theorems 5.1 and 5.5 it is enough to show that μ_n is shift compact. We will now show that $\lambda_n = |\mu_n|^2$ is compact.

$$\lambda_n = |\mu_n|^2 = |e(F_n)|^2 = e(M_n) \text{ where } M_n = F_n + \bar{F}_n.$$

Since F_n increases to F it follows that M_n increases to M where $M = F + \bar{F}$. Without any loss of generality we can assume that F and hence M vanishes outside the sphere $||x|| \leqslant 1$. We further have

$$1-\lambda_n(y) = 1-\exp [\int [\cos (x, y)-1]dM_n] = 1-\exp [-\int [1-\cos (x, y)]dM_n]$$

$$\leqslant \int [1-\cos (x, y)]\, dM_n \leqslant \int [1-\cos (x, y)]\, dM$$

$$\leqslant \tfrac{1}{2} \int_{||x||\leqslant 1} (x, y)^2\, dM = \tfrac{1}{2}(Sy, y).$$

Since
$$\int_{\|x\| \leqslant 1} \|x\|^2 \, dM = 2 \int_{\|x\| \leqslant 1} \|x\|^2 \, dF < \infty$$

it follows that S is an S-operator. Since S is a fixed S-operator independent of n it follows from Theorem 3.5 that λ_n is compact. Consequently μ_n is shift compact and the theorem is proved.

Gaussian distributions are defined in the Hilbert Space in exactly the same manner as in Definition 2.4. We shall now prove

Theorem 5.9 : *A distribution μ on X is Gaussian if and only if $\mu(y)$ is of the form*
$$\mu(y) = \exp \{i(x_0, y) - (Sy, y)\}$$
where X_0 is a fixed element and S an S-operator.

Proof: Let us take a countable dense subset $y_1, y_2, \ldots, y_n \ldots$ in X and consider the map τ from X to Z^{∞}, the countable product of the circle groups, defined as
$$\tau(x) = [e^{i(x, y_1)}, \ldots e^{i(x, y_n)} \ , \ldots].$$
Let H be the image of X under τ in Z^{∞}. Then τ is a both ways measurable isomorphism of the two groups X and H. If μ is Gaussian on X then $\mu\tau^{-1}$ is Gaussian in H and hence in Z^{∞}. From Theorem 2.5 we have
$$\mu\tau^{-1}(\theta) = \theta(Z) \exp [-\phi(\theta)] \qquad \ldots \quad (5.15)$$
where θ is a character on Z^{∞}, Z is a fixed element of Z^{∞} and ϕ a function with properties specified in Theorem 2.5. $\theta(Z)$ denotes the value of the character θ at the point Z. Further any character θ on Z^{∞} is of the form
$$\theta(Z) = Z_1^{n_1} \ldots Z_k^{n_k} \qquad \ldots \quad (5.16)$$
where n_1, n_k are integer and $Z_1, \ldots Z_k$ the first k coordinates of Z in Z^{∞}.

Therefore,
$$\mu\tau^{-1}(\theta) = \mu(n_1 y_1 + \ldots n_k y_k)$$
where $\mu(y)$ is the characteristic function on X and $n_1, \ldots n_k$ are related to θ by the relation (5.16). Hence
$$|\mu\tau^{-1}(\theta)| = |\mu| (n_1 y_1 + \ldots n_k y_k) = e^{-\phi(\theta)}.$$
Since for any θ and θ'
$$\phi(\theta+\theta') + \phi(\theta-\theta') = 2[\phi(\theta) + \phi(\theta')]$$
it follows that
$$h(y+y') + h(y-y') = 2[h(y') + h(y)] \qquad \ldots \quad (5.17)$$
whenever y, y' are of the form $n_1 y_1 + \ldots n_k y_k$, where
$$h(y) = -\log |\mu(y)|.$$
Since $y_1, y_2, \ldots$ are dense and h is continuous in the norm topology of X it follows that the relation (5.17) is valid for any pair y, y'. This implies that $h(y)$ can be put as
$$h(y) = (Sy, y)$$
where S is a positive semi-definite Hermitian operator. From the continuity of $\mu(y)$ in the S-topology it follows that S is an S-operator. If we now consider the distribution λ on X defined by the equation
$$\lambda(y) = \exp [-(Sy, y)].$$
We have
$$\lambda\tau^{-1}(\theta) = e^{-\phi(\theta)}.$$

Hence $\lambda\tau^{-1}$ is a shift of $\mu\tau^{-1}$. Since both $\lambda\tau^{-1}$ and $\mu\tau^{-1}$ give unit mass for the subgroup H of Z^{∞} the element Z by which $\lambda\tau^{-1}$ is shifted to obtain $\mu\tau^{-1}$ belongs to H and hence $Z = \tau(x_0)$ for some $x_0 \epsilon X$. Consequently,

$$\mu = \lambda * x_0$$

or

$$\mu(y) = \exp\,[i(x_0, y) - (Sy, y)].$$

Conversely, if $\mu(y)$ is of the form $\mu(y) = \exp\,[i(x_0, y) - (Sy, y)]$ the distribution $\mu\tau^{-1}$ in Z^{∞} is Gaussian and hence so is μ.

We now prove the representation theorem for infinitely divisible distributions.

Theorem 5.10 : *A function $\mu(y)$ is the characteristic function of an infinitely divisible distribution μ on X if and only if it is of the form*

$$\mu(y) = \exp\,[i\,(x_0, y) - (Sy, y) + \int K(x, y)\, dM(x)] \qquad \ldots \quad (5.18)$$

where x_0 is a fixed element of X, S an S-operator and M a σ-finite measure giving finite-mass outside every neighbourhood of the origin and for which

$$\int_{\|x\| \leqslant 1} \|x\|^2\, dM < \infty.$$

Here $K\,(x, y)$ is the function $K(x, y) = \left[e^{i(x,y)} - 1 - \dfrac{i(x, y)}{1 + \|x\|^2} \right].$

The representation (5.18) of $\mu(y)$ is unique.

Proof : Let $\mu(y)$ be the characteristic function of an infinitely divisible distribution μ. Then in the same manner as in the proof of Theorem 7.1 of Parthasarathy *et al* (1962a) we can construct a sequence of distributions λ_n such that (i) $\lambda_n = e(M_n)$, (ii) M_n increases to M, (iii) $\lambda_n * x_n \Longrightarrow \lambda$, (iv) $\mu = \lambda * \mu_0$ where μ_0 is Gaussian.

Now using Theorems 5.7 and 5.9 we can complete the proof. Sufficiency is immediate from Theorems 5.8 and 5.9. Uniqueness is proved in an exactly same manner as Theorem 8.1 of Parthasarathy *et al* (1962a) but keeping in mind that the space X playing the role of the character group is connected.

6. Compactness criteria

In the present section we will find out the necessary and sufficient conditions in order that a sequence μ_n of infinitely divisible distributions may be weakly conditionally compact.

If μ is any infinitely divisible distribution by $\mu = [x, S, M]$ we will mean that the three quantities occurring in the representation of μ according to Theorem 5.10 are respectively x, S and M. We will associate with any such $\mu = [x, S, M]$ another S-operator which we will denote by T.

$$(Ty, y) = 2(Sy, y) + \int_{\|x\| \leqslant 1} (x, y)^2\, dM(x) \qquad \ldots \quad (6.1)$$

T is an S-operator since $\displaystyle\int_{\|x\| \leqslant 1} \|\, x \,\|^2\, dM(x) < \infty.$

Lemma 6.1 : *In order that a sequence μ_n of Gaussian distributions with covariance operators S_n be shift compact it is necessary and sufficient that S_n should be compact. [If μ is Gaussian with covariance operator S its characteristic function is* $\exp[i(x_0, y)-\frac{1}{2}(Sy, y)]$.

Proof : Sufficiency is immediate from Theorem 3.4. We will prove necessity. If μ_n is shift compact then $|\mu_n|^2$ is compact. But $|\mu_n|^2$ is Gaussian with mean zero and covariance operator $2S_n$.

$$|\mu_n|^2(y) = \exp[-(S_n y, y)].$$

From Theorem 3.3 it follows that there exists a compact sequence U_n of S-operators such that

$$1-|\mu_n|^2(y) \leqslant (U_n y, y)+\epsilon.$$

Hence there is a δ such that for any n, $(U_n y, y) \leqslant \delta$ implies $(S_n y, y) \leqslant 1$. From this we can deduce that

$$(S_n y, y) \leqslant \delta^{-1}(U_n y, y).$$

Since δ is independent of n and U_n is compact, S_n is also compact.

Lemma 6.2 : *Let μ be a symmetric infinitely divisible distribution with* $\mu = [0, 0, M]$. *Further, let M be concentrated in the unit sphere. Then*

$$\int \|x\|^4 \, d\mu \leqslant \int \|x\|^4 \, dM+3[\int \|x\|^2 \, dM]^2 < \infty.$$

Proof : It is enough to prove the theorem when M is finite since the other case can be obtained by limit transition. Let $M(x) = t$ and F be the distribution such that $F(X) = 1$ and $M = tF$. Then

$$\mu = e^{-t} \Sigma \frac{t^r F^r}{r!}$$

$$\begin{aligned}\int \|x\|^4 \, dF^r &= E \, \|X_1+\dots X_r\|^4 \\ &= r\, E\, \|X_1\|^4+r(r-1)[E\, \|X_1\|^2]^2+2r(r-1)\, E(X_1, X_2)^2 \\ &= rE\, \|X_1\|^4+3r(r-1)(E\, \|X_1\|^2)^2.\end{aligned}$$

Expectation is with respect to F and $X_1, \dots X_r$ are independent random variables in X with the same distribution F. Terms with zero expectation have been omitted.

$$\begin{aligned}\int \|x\|^4 \, d\mu &\leqslant e^{-t} \Sigma \frac{rt^r \, E\, \|X_1\|^4}{r!}+3\, e^{-t} \Sigma \frac{r(r-1)t^r(E\, \|X_1\|^2)^2}{r!} \\ &= tE\, \|X_1\|^4+3t^2(E\, \|X_1\|^2)^2 \\ &= t\int \|X\|^4 \, dF+3t^2[\int \|X\|^2 \, dF]^2 \\ &= \int \|X\|^4 \, dM+3[\int \|X\|^2 \, dM]^2.\end{aligned}$$

Lemma 6.3 : *Let μ_n be symmetric infinitely divisible distributions such that*

$$\mu_n = [0, 0, M_n]$$

with M_n vanishing outside the sphere $\|X\| \leqslant 1$ for all n. If μ_n is compact then

$$\sup_n \int \|X\|^4 \, d\mu_n < \infty.$$

Proof: Since M_n is concentrated in the unit sphere

$$\int \|X\|^4 \, dM_n \leqslant \int \|X\|^2 \, dM_n.$$

Theorem 5.6 implies that $\sup_n \int \|X\|^2 \, dM_n < \infty$

and hence an application of Lemma 6.2 will complete the proof.

Remark 6.1 : In the same manner as Lemma 6.2 it can be shown that if M is symmetric and vanishes outside the sphere $\|X\| \leqslant 1$ for

$$\mu = [0, 0, M].$$

We have $\int (x, y)^2 d\mu = \int (x, y)^2 dM$ for all y.

Lemma 6.4 : *Let μ_n be a weakly conditionally compact sequence of symmetric distribution such that*

$$\sup_n \int \|X\|^4 \, d\mu_n < \infty.$$

Then if S_n is the covariance operator of μ_n, $\{S_n\}$ is compact.

Proof: If θ_n is a sequence of distributions on the real line such that $\theta_n \Longrightarrow \theta$ and $\int x^2 \, d\theta_n$ is uniformly bounded then

$$\int x d\theta_n \rightarrow \int x d\theta \text{ as } n \rightarrow \infty.$$

Theorem 3.5 can be applied now and the lemma follows at once.

Theorem 6.1 : *Let μ_n be symmetric distributions that are infinitely divisible with representations*

$$\mu_n = [0, S_n, M_n].$$

Then in order that μ_n be compact it is necessary and sufficient that

(i) *M_n restricted outside any neighbourhood of the identity is weakly conditionally compact,*

(ii) *T_n as defined in* (6.1) *is compact.*

Proof: We will first prove sufficiency. Let unit sphere be chosen as the neighbourhood and let us write $M_n = M_n^{(1)} + M_n^{(2)}$ where $M_n^{(1)}$ and $M_n^{(2)}$ are respectively the restrictions of M_n inside the unit sphere and outside the unit sphere. Since $M_n^{(2)}$ is weakly conditionally compact and $F_n \Longrightarrow F$ implies $e(F_n) \Longrightarrow e(F)$ it is enough to show that the distributions

$$\lambda_n = [0, S_n, M_n^{(1)}]$$

form a compact sequence

$$\begin{aligned} \int (x, y)^2 d\lambda_n(x) &= 2(S_n y, y) + \int (x, y)^2 \, dM_n^{(1)}(x) \\ &= 2(S_n y, y) + \int_{\|x\| \leqslant 1} (x, y)^2 \, dM_n(x) \\ &= (T_n y, y). \end{aligned}$$

Since T_n is compact sufficiency follows from Theorem 3.4. Necessity of (i) is a consequence of Theorem 5.3 and (ii) follows from Lemmas 6.1, 6.2 and 6.4.

Theorem 6.2 : *In order that a sequence μ_n of infinitely divisible distributions with representations*

$$\mu_n = [x_n, S_n, M_n]$$

be shift compact it is necessary and sufficient that

(i) *M_n restricted to the complement of any neighbourhood N of the origin is weakly conditionally compact,*

(ii) *T_n as defined in (6.1) is compact.*

Proof : Since μ_n is shift compact if and only $|\mu_n|^2$ is compact what we need are conditions for the compactness of

$$|\mu_n|^2 = [0, 2S_n, M_n+\bar{M}_n].$$

In addition we have $\int\limits_{\|x\| \leqslant 1} (x, y)^2 \, d(M_n+\bar{M}_n) = 2 \int\limits_{\|x\| < 1} (x, y)^2 dM_n.$

Hence the theorem follows from 6.1.

Theorem 6.3 : *In order that μ_n with the representations*

$$\mu_n = [x_n, S_n, M_n]$$

be weakly conditionally compact it is necessary and sufficient that in addition to the conditions of Theorem 6.2 x_n should be compact in X.

Proof : In order to prove the theorem it suffices to show that whenever

$$\mu_n = [x_n, S_n, M_n]$$

is shift compact, $\qquad \lambda_n = [0, S_n, M_n]$

is compact. Let $F_n \Longrightarrow F$. Then the following convergence takes place in norm.

$$\int \frac{x}{1+\|x\|^2} \, dF_n \rightarrow \int \frac{x}{1+\|x\|^2} \, dF.$$

Hence we can assume that M_n vanishes for all n outside the sphere $\|x\| \leqslant 1$. Let us now consider

$$f_n(y) = \int K(x, y) \, dM_n(x)$$

$$|f_n(y)| \leqslant \tfrac{1}{2} \int\limits_{\|x\| \leqslant 1} (x, y)^2 \, dM_n(x) + \int\limits_{\|x\| > 1} \frac{|(x, y)| \, \|x\|^2}{1+\|x\|^2} \, dM_n(x)$$

$$\leqslant \tfrac{1}{2}(T_n y, y) + \int |(x, y)| \; \|x\|^2 \, dM_n(x)$$

$$\leqslant \tfrac{1}{2}(T_n y, y) + [\int (x, y)^2 \|x\|^2 \, dM_n \int \|x\|^2 \, dM_n]^{\frac{1}{2}}$$

$$\leqslant \tfrac{1}{2}(T_n y, y) + C(T_n y, y)^{\frac{1}{2}}.$$

Hence given any $\rho > 0$ there is a $\delta > 0$ such that $|f_n(y)| \leqslant \rho$ whenever $(T_n y, y) < \delta$. But since $\lambda_n(y) = \exp[f_n(y)]$, for any $\epsilon > 0$ there exists a $\rho > 0$ such that $1 - R\lambda_n(y) \leqslant \epsilon$ whenever $|f_n(y)| \leqslant \rho$. Consequently, for any $\epsilon > 0$ there exists a $\delta > 0$ such that whenever $(T_n y, y) \leqslant \delta$ it follows that $1 - R\lambda_n(y) \leqslant \epsilon$. Here R denotes the real part. Now Lemma 5.1 and Theorem 3.5 imply the validity of the theorem since $\{T_n\}$ is compact.

Remark 6.2 : In defining the operator T we could have taken any bounded sphere around the origin instead of the unit sphere. When M_n restricted outside any neighbourhood of the origin is known to be weakly conditionally compact, the compactness of T_n when it is based on some sphere implies the compactness of T_n when it is based on any finite sphere.

Since the representation is unique one can give conditions for the convergence of

$$\mu_n = [x_n, S_n, M_n]$$

to the distribution $\mu = [x, S, M]$

in terms of $[x_n, S_n, M_n]$ and $[x, S, M]$. However, we will mention only the following.

Theorem 6.4 : *Let μ_n be a sequence such that μ_n has the representation*

$$\mu = [x_n, S_n, M_n].$$

If $\mu_n \Longrightarrow \mu$, μ is Gaussian if and only if

$$M_n(N') \rightarrow 0 \text{ as } n \rightarrow \infty$$

for every neighbourhood N of the origin.

Proof: Let μ be Gaussian. Since μ cannot be written as $e(F)*\lambda$ with λ infinitely divisible it follows that $M_n(N') \rightarrow 0$ as $n \rightarrow \infty$ for every neighbourhood N of the origin, conversely, if $M_n(N') \rightarrow 0$ for every neighbourhood N in exactly the same manner as in the proof of Theorem 6.1 of Parthasarathy *et al* (1962a) it can be shown that $\mu(y)$ is of the form

$$\mu(y) = \exp [i (x_0, y) - (Sy, y)]$$

which shows that μ is Gaussian.

7. Accompanying laws

Definition 7.1 : A sequence α_{nj} of distributions $j = 1, 2, \ldots, k_n$, $n = 1, 2, \ldots$ is said to be uniformly infinitesimal if for any neighbourhood N of the origin

$$\operatorname*{lt}_{n \to \infty} \inf_{1 \leqslant j \leqslant k_n} \alpha_{nj}(N) = 1.$$

Theorem 7.1 : *In order that α_{nj} be uniformly infinitesimal it is necessary that*

$$\operatorname*{lt}_{\to \infty} \sup_{1 \leqslant j \leqslant k_n} \sup_{\|y\| \leqslant K} |\alpha_{nj}(y) - 1| = 0$$

for every constant K.

Proof: This is immediate from Theorem 5.4.

Theorem 7.2 : *Let α_{nj} be uniformly infinitesimal symmetric distributions with non-negative characteristic function. Let*

$$\mu_n = \prod_{j=1}^{k_n} \alpha_{nj}$$

and λ_n be defined as

$$\lambda_n = \prod_{j=1}^{k_n} e(\alpha_{nj}).$$

In order that $\mu_n \Longrightarrow \mu$ it is necessary and sufficient that $\lambda_n \Longrightarrow \mu$.

Proof : Let μ_n be compact. Since the inequality $e^{x-1} \geqslant x$ is valid for all x

we have $$e(\alpha_{nj})(y) \geqslant \alpha_{nj}(y) \quad \text{for } j = 1, 2, \ldots, k_n, \quad n = 1, 2 \ldots$$

since $\alpha_{nj}(y)$ is non-negative $$\lambda_n(y) \geqslant \mu_n(y).$$

Or $$1-\lambda_n(y) \leqslant 1-\mu_n(y).$$

From the compactness of μ_n and Theorem 3.3 it follows that λ_n is compact. Now let λ_n be compact. It follows from Theorem 6.2 and the remark made after Theorem 6.3 that

(i) F_n restricted to N' is weakly conditionally compact for every neighbourhood N;

(ii) the sequence S_n of operators defined by

$$(S_n\, y, y) = \int_{\|x\| \leqslant t} (x, y)^2 dF_n(x)$$

is compact for every t.

Here F_n denotes the sum $\alpha_{n1}+\alpha_{n2}+\ldots\alpha_{nk_n}$. We will now show that μ_n is compact. We have

$$1-\mu_n(y) \leqslant \sum_{j=1}^{k_n} [1-\alpha_{nj}(y)] = \int [1-\cos(x, y)]dF_n(x)$$

$$= \int_{\|x\|<t} [1-\cos(x, y)]dF_n(x) + \int_{\|x\|>t} [1-\cos(x,y)\, dF_n(x)$$

$$= \tfrac{1}{2} \int_{\|x\|\leqslant t} (x, y)^2 dF_n(x) + 2F_n[\|x\| > t] = \tfrac{1}{2}(S_n y, y) + 2F_n[\|x\| > t].$$

Since F_n is weakly compact outside any neighbourhood we can choose t such that for all n

$$F_n[\|x\| > t] < \epsilon/2.$$

Since for that fixed t, S_n is compact Theorem 3.3 shows that μ_n is compact.

We will now complete the proof by showing that whenever λ_n is compact, for every costant k

$$\sup_{\|y\|\leqslant K} |\lambda_n(y) - \mu_n(y)| \to 0.$$

To this end it is enough to show that

$$\sup_{\|y\|\leqslant K} \sup_n \sum_{j=1}^{k_n} [1-\alpha_{nj}(y)] < \infty.$$

But the expression is equal to

$$\sup_{\|y\|\leqslant K} \sup_n \int [1-\cos(x, y)]dF_n \leqslant \sup_{\|y\|\leqslant K} \sup_n \int_{\|x\|\leqslant 1} [1-\cos(x, y)]dF_n + \sup_n 2F_n(\|x\|>1)$$

$$\leqslant \tfrac{1}{2}K^2 \sup_n \int_{\|x\|\leqslant 1} \|x\|^2 dF_n + 2 \sup_n F_n(\|x\| > 1)$$

$$< \infty.$$

The last step follows since $\lambda_n = e(F_n)$ is compact.

Lemma 7.1 : *Let α_{nj} be uniformly infinitesimal. Then if x_{nj} is defined by the relation*

$$x_{nj} = \int_{\|x\| \leqslant 1} x \, d\, \alpha_{nj}$$

then
$$\sup_{1 \leqslant j \leqslant k_n} \|x_{nj}\| \to 0 \text{ as } n \to \infty.$$

Proof : Let ϵ be arbitrary and V the sphere $\|x\| \leqslant \epsilon$. Then

$$\|x_{nj}\| \leqslant \|\int_V d\alpha_{nj}\| + \| \int_{\epsilon < \|x\| \leqslant 1} x d\alpha_{nj}\| \leqslant \epsilon + \alpha_{nj}(V').$$

Hence
$$\limsup_{n \to \infty} \sup_{1 \leqslant j \leqslant k_n} \|x_{nj}\| \leqslant \epsilon.$$

Since ϵ is arbitrary the lemma follows.

Lemma 7.2 : *Let α_{nj} be uniformly infinitesimal. Let x_{nj} be defined as in Lemma 4.6. Then if $\theta_{nj} = \alpha_{nj} * (-x_{nj})$ there exists a n_0 such that for all $1 \leqslant j \leqslant k_n$ and $n \geqslant n_0$ we have*

$$\| \int_{\|x\| \leqslant 1} x d\theta_{nj}\| \leqslant 2\theta_{nj}[\|x_{nj}\| > 1].$$

Proof : Let n_0 be so chosen that

$$\sup_{1 \leqslant j \leqslant k_n} \|x_{nj}\| \leqslant 1/4 \quad \text{for all } n > n_0.$$

Then
$$\int_{\|x\| \leqslant 1} x d\theta_{nj} = \int_{\|x - x_{nj}\| \leqslant 1} (x - x_{nj}) d\alpha_{nj}$$

$$= \int_{\|x - x_{nj}\| \leqslant 1} x d\alpha_{nj} - \int_{\|x\| \leqslant 1} x d\alpha_{nj} + x_{nj}\alpha_{nj}[\|x - x_{nj}\| > 1].$$

Therefore, for $n > n_0$ and $1 \leqslant j \leqslant k_n$,

$$\| \int_{\|x\| \leqslant 1} x d\theta_{nj}\| \leqslant \int_{[\|x - x_{nj}\| \leqslant 1 \Delta \|x\| \leqslant 1]} \|x\| d\alpha_{nj} + \|x_{nj}\| \theta_{nj} [\|x\| > 1]$$

$$\leqslant \int_{\frac{3}{4} \leqslant \|x\| \leqslant \frac{5}{4}} \|x\| d\alpha_{nj} + \tfrac{1}{4} \, \theta_{nj} [\|x\| > 1]$$

$$\leqslant \tfrac{5}{4} \, \alpha_{nj} [\|x\| > 3/4] + \tfrac{1}{4} \, \theta_{nj} [\|x\| > 1]$$

$$\leqslant \tfrac{5}{4} \, \theta_{nj} [\|x\| > 1] + \tfrac{1}{4} \, \theta_{nj} [\|x\| > 1]$$

$$\leqslant 2\theta_{nj} [\|x\| > 1].$$

Lemma 7.3 : *Let F_n be a sequence of σ-finite measures such that F_n restricted to N' is finite and weakly conditionally compact for every neighbourhood N of the origin. Then for any $\epsilon > 0$ there exists a compact set K such that*

$$F_n(K') \leqslant \epsilon \text{ for } n = 1, 2, \ldots$$

Proof : Let ϵ be any positive number. Choose a sequence N_r of neighbourhoods of the origin decreasing to the origin. Let A_r be defined as $N_{r-1} - N_r$, N_0 being taken as the whole space. From the conditions of the lemma it is possible to find a compact set K_r in A_r such that for all n,

$$F_n(A_r - K_r) \leqslant \frac{\epsilon}{2^r}.$$

Let K be defined as $$K = \bigcup_{r=1}^{\infty} K_r \cup \{0\}.$$

Since $$K \cap N'_r = K_1 \cup K_2 \ldots \cup K_r$$
and N_r decreases to the origin it follows that K is compact and

$$F_n(K') \leqslant \sum_{r=1}^{\infty} F_n(A_r - K_r) \leqslant \epsilon.$$

We now proceed to prove the main theorem of the section.

Let α_{nj} be uniformly infinitesimal sequence of distributions on X. Let μ_n, x_n, θ_{nj}, λ_n be defined as follows.

$$\mu_n = \prod_{j=1}^{k_n} \alpha_{nj}$$

$$x_{nj} = \int_{||x|| \leqslant 1} x d\alpha_{nj}$$

$$\theta_{nj} = \alpha_{nj} * (-x_{nj})$$

$$\lambda_n = \prod_{j=1}^{k_n} e(\theta_{nj}) * \left(\sum_{j=1}^{k_n} x_{nj} \right)$$

In what follows we will adopt the above notation.

Theorem 7.3 : *If μ_n is shift compact so is λ_n.*

Proof : Since μ_n is shift compact $|\mu_n|^2$ is compact. But

$$|\mu_n|^2 = \prod_{j=1}^{k_n} |\alpha_{nj}|^2.$$

It follows now from Theorem 6.2 that if one defines

$$F_n = \sum_{j=1}^{k_n} |\alpha_{nj}|^2 = \sum_{j=1}^{k_n} |\theta_{nj}|^2. \qquad \ldots \quad (7.1)$$

Then $e(F_n)$ is compact. We can now apply Theorem 6.2 and Lemma 7.3 and deduce that for any $\epsilon > 0$ there exists a compact set K such that

$$F_n(K') \leqslant \epsilon \quad \text{for } n = 1, 2, \ldots \qquad \ldots \quad (7.2)$$

Let us now define the S-operators T_n by the formula

$$(T_n y, y) = \int_{||x|| \leqslant t} (x, y)^2 \, dF_n(x).$$

Then for any finite t the sequence T_n is compact.
Let us now define G_n as

$$G_n = \sum_{j=1}^{k_n} \theta_{nj}. \qquad \ldots \quad (7.3)$$

In order to complete the theorem we have to show that $e(G_n)$ is shift compact or

(a) G_n is weakly conditionally compact when restricted outside any neighbourhood of the identity;

(b) if the S-operators S_n are defined as

$$(S_n y, y) = \int_{\|x\| \leqslant 1} (x, y)^2 dG_n$$

then S_n is compact.

Since α_{nj} is uniformly infinitesimal by Lemma 7.1 θ_{nj} are also uniformly infinitesimal. Hence for any $\epsilon > 0$ there exists a compact set C such that

$$\theta_{nj}(C) \geqslant 1-\epsilon \text{ for all } n \text{ and } 1 \leqslant j \leqslant k_n.$$

In the same manner as in the proof of Theorem 5.1 of Parthasarathy *et al* (1962a) we have for all n and $1 \leqslant j \leqslant k_n$

$$|\theta_{nj}|^2(K') = \int \theta_{nj}(K'+x)d\theta_{nj}(x) \geqslant \int_C \theta_{nj}(K'+x)d_{nj}(x)$$

$$\geqslant (1-\epsilon) \inf_{x\epsilon C}\theta_{nj} (K'+x) = (1-\epsilon)[1-\sup_{x\epsilon C} \theta_{nj} (K+x)]$$

$$= (1-\epsilon)\theta_{nj}(\overline{K+C}') = (1-\epsilon)\, \theta_{nj}(K_1'). \qquad \dots (7.4)$$

where K_1 is another compact set.

In a similar manner it can be shown that if V and N are two neighbourhoods of the origin such that $V+V \subseteq N$ and

$$\theta_{nj}(V) \geqslant 1-\epsilon \text{ for all } n \geqslant n_0,\ 1 \leqslant j \leqslant k_n \qquad \dots (7.5)$$

then for $\quad n \geqslant n_0$ and $1 \leqslant j \leqslant k_n$ $|\theta_{nj}|^2(V') \geqslant (1-\epsilon)\theta_{nj} (N') \qquad \dots (7.6)$

(7.2), (7.3) and (7.4) imply that for any $\epsilon > 0$ there exists a compact set K_1 such that

$$G_n(K_1') \leqslant \epsilon \text{ for all } n. \qquad \dots (7.7)$$

On the other hand (7.1), (7.3), (7.6) and the weak compactness of F_n when restricted to the complement of any neighbourhood of the origin imply that for any neighbourhood N of the origin,

$$\sup_n G_n(N') < \infty. \qquad \dots (7.8)$$

(7.7) and (7.8) prove that G_n is weakly conditionally compact when restricted outside any neighbourhood of the origin. To prove (b) let us consider

$$\int_{\|x\| \leqslant 2} (x, y)^2\, d|\theta_{nj}|^2(x) = \iint_{\|x_1-x_2\| \leqslant 2} (x_1-x_2, y)^2 d\theta_{nj}(x_1)d\theta_{nj}(x_2)$$

$$\geqslant \iint_{\|x_1\| \leqslant 1,\ \|x_2\| \leqslant 1} (x_1-x_2, y)^2 d\theta_{nj}(x_1)d\theta_{nj}(x_2)$$

$$= 2\theta_{nj}(\|x\| \leqslant 1) \int_{\|x\| \leqslant 1} (x, y)^2 d\theta_{nj}(x) - 2[\int_{\|x\| \leqslant 1} (x, y)d\theta_{nj}]^2.$$

Since θ_{nj} is uniformly infinitesimal we can assume that $\theta_{nj}(\|x\| \leqslant 1) \geqslant \frac{1}{2}$ for all suitably large n and for all $1 \leqslant j \leqslant k_n$. Hence

$$(S_n y, y) \leqslant (T_n y, y) + 2(U_n y, y)$$

where $$(U_n y, y) = \sum_{j=1}^{k_n} [\int (x, y) d\theta_{nj}(x)]^2.$$

Since we know that T_n is compact in order to show that S_n is compact it is enough to prove that U_n is compact. We will show now that trace of U_n tends to zero as $n \to \infty$, and this would complete the proof.

Let us put $$y_{nj} = \int_{\|x\| \leqslant 1} x d\theta_{nj}$$

Then for the trace of U_n we have

trace of $$U_n = \sum_{j=1}^{k_n} \|y_{nj}\|^2 \leqslant (\sup_{1 \leqslant j \leqslant k_n} \|y_{nj}\|) \left(\sum_{j=1}^{k_n} |y_{nj}| \right).$$

From Lemmas 7.1, 7.2 and 7.8 it follows that trace $U_n \to 0$ as $n \to \infty$.

Theorem 7.4 : *If λ_n is shift compact so is μ_n.*

Proof : $$1 - |\mu_n|^2(y) = 1 - \prod_{j=1}^{k_n} |\theta_{nj}|^2(y) \leqslant \sum_{j=1}^{k_n} [1 - |\theta_{nj}|^2(y)]$$

$$\leqslant 2 \sum_{j=1}^{k_n} [1 - \text{Real } \theta_{nj}(y)] = 2 \sum_{j=1}^{k_n} \int [1 - \cos (x, y)] d\theta_{nj}$$

$$= 2 \int [1 - \cos (x, y)] dG_n(x)$$

$$= 2 \int_{\|x\| \leqslant t} [1 - \cos (x, y)] dG_n(x) + 2G_n[\|x\| > t]$$

$$\leqslant \int_{\|x\| \leqslant t} (x, y)^2 dG_n(x) + 2G_n[\|x\| > t]$$

$$= (S_n y, y) + 2G_n[\|x\| > t]. \quad \ldots \quad (7.9)$$

From the shift compactness of λ_n for any $\epsilon > 0$ we can choose t such that

$$G_n[\|x\| > t] \leqslant \frac{\epsilon}{2} \text{ for all } n \quad \ldots \quad (7.10)$$

and the sequence of operators S_n defined by

$$(S_n y, y) = \int_{\|x\| \leqslant t} (x, y)^2 dG_n(x)$$

is compact. From (7.9), and (7.10) the compactness of S_n and Theorem 3.3 the present theorem follows.

Theorem 7.5 : *Let λ_n be shift compact. Then for any finite number t we have*

$$\operatorname{lt}_{n \to \infty} \sup_{\|y\| \leqslant t} |\lambda_n(y) - \mu_n(y)| = 0.$$

Proof : Since θ_{nj} is uniformly infinitesimal it follows from Theorem 7.1 that

$$\operatorname{lt}_{n \to \infty} \sup_{\|y\| \leqslant t} \sup_{1 \leqslant j \leqslant k_n} |\theta_{nj}(y) - 1| = 0.$$

Hence it is enough to show that $\sup_n \sup_{\|y\| \leqslant t} \sum_{j=1}^{k_n} |\theta_{nj}(y) - 1| < \infty.$ ... (7.11)

To this end we have

$$\theta_{nj}(y) - 1 = \int [e^{i(x,y)} - 1] d\theta_{nj}$$

$$= \int_{\|x\| \leqslant 1} [e^{(ix,y)} - 1 - i(x, y)] d\theta_{nj} + i \int_{\|x\| \leqslant 1} (x, y) d\theta_{nj} + \int_{\|x\| > 1} [e^{i(x,y)} - 1] d\theta_{nj}$$

$$|\theta_{nj}(y) - 1| \leqslant \tfrac{1}{2} \int_{\|x\| \leqslant 1} (x, y)^2 d\theta_{nj} + \|y\| \, \Big\| \int_{\|x\| \leqslant 1} x d_{nj} \Big\| + 2\theta_{nj}[\|x\| > 1].$$

(7.11) follows at once from the compactness of S_n, Lemma 7.2 and (7.8).

Theorem 7.6 : *In order that $\mu_n * x_n \Longrightarrow \mu$, where x_n are arbitrary points in X, it is necessary and sufficient that $\lambda_n * x_n \Longrightarrow \mu$.*

Proof : This is an immediate consequence of Theorems 7.3, 7.4, 7.5 and 5.5.

Theorem 7.7 : *Limit distribution of sums of independent uniformly infinitesimal random variables in X is infinitely divisible.*

Proof : Theorems 7.6 and 5.1 imply the present theorem since λ_n is infinitely divisible for each n.

Theorem 7.8 : *Let $\mu_n * x_n \Longrightarrow \mu$. In order that μ may be Gaussian it is necessary and sufficient that for each neighbourhood N*

$$\operatorname{lt}_{n \to \infty} G_n(N') = 0,$$

where

$$G_n = \sum_{j=1}^{k_n} \theta_{nj}.$$

Proof : This follows immediately from Theorems 7.6 and 6.4.

The author wishes to express his thanks to K. R. Parthasarathy and R. Rangarao for several valuable comments and discussions.

REFERENCES

BOCHNER, S. (1955) : *Harmonic Analysis in Probability*, University of California Press.

——— (1958) : General analytical setting for the central limit theory of probability. Golden Jubilee Commemoration Volume of the Calcutta Mathematical Society, Part I, pp. 111-128.

GNEDENKO, B. V. and KOLMOGOROV, A. N. (1954) : *Limit Distributions for Sums of Independent Random Variables*, Addison-Wesley, Cambridge.

HALMOS, P. R. (1950) : *Measure Theory*, Van Nostrand, Princeton.

HUNT, G. A. (1956) : Semigroups of measures on Lie groups. *Trans. Amer. Math. Soc.*, **81**, 264-293.

KLOSS, B. M. (1961) : Limiting distributions on bicompact abelian groups. *Teorya Veroyot i Premen.*, **6**, 392-421.

LEVY, P. (1939) : L'addition des variables aleatoires des variables definies sur une cerconference. *Bull. Soc. Math.*, **67**, 1-41.

——— (1954) : *Theorie de L'addition des Variables Aleatoires*, Gauthier—Villars.

PARTHASARATHY, K. R., RANGARAO, R. and VARADHAN, S. R. S. (1962) : On the category of indecomposable distributions on topological groups. *Trans. Amer. Math. Soc.*, **102**, No. 2, 200-217.

——— (1962a) : On the semigroup of distributions on topological groups. To appear in *Illinois J. Math.*

PROHOROV, YU. V. (1956) : Convergence of random processes and limit theorems in the theory of probability. *Teorya Veroyot. i. Premen.*, **1**, 177-238.

RANGARAO, R. (1960) : Some problems in probability theory. Thesis submitted to the Calcutta University (unpublished).

SAZANOV, V. (1958) : On characteristic functionals. *Teorya Veroyot i. Premen.*, **3**, 201-205.

TAKANO, K. (1955) : Central convergence criterion in the multidimensional case. *Ann. Inst. Stat. Math.*, **7** 95-102.

VARADARAJAN, V. S. (1958) : Weak convergence of measures on separable metric spaces. *Sankhyā*, **19**, 15-22.

VOROBEV, N. N. (1954) : Addition of independent random variables on finite abelian groups. *Mat. Sbornik*, **34**, 89-216.

Paper received : June, 1962.

Trans. Amer. Math. Soc. 102 (1962) 200-217

ON THE CATEGORY OF INDECOMPOSABLE DISTRIBUTIONS ON TOPOLOGICAL GROUPS

BY

K. R. PARTHASARATHY, R. RANGA RAO AND S. R. S. VARADHAN

1. Introduction. According to a theorem of A. I. Khinchine [1] any distribution on the real line can be written as the convolution of two distributions one of which is the convolution of a finite or a countable number of indecomposable distributions and the other is an infinitely divisible distribution without indecomposable factors. Further, any distribution which is not infinitely divisible has an indecomposable component. This result gives an indication of the existence of a large collection of indecomposable distributions. It is, however, not clear from this result alone that there exists a nonatomic or absolutely continuous indecomposable distribution. This question was raised by H. Cramér [2] and an answer in the affirmative was given by P. Lévy [3]. However, what is available is only a meagre supply of examples [3; 4; 5] even in the case of the real line. In this connection there arises naturally the question of the "size" of the class $\mathfrak{M}_1$, of indecomposable distributions. Or more precisely, what is the category of $\mathfrak{M}_1$? The object of this paper is to answer questions of this type.

These questions are studied here under the general framework of a complete separable metric group G. We consider three classes of distributions: (1) all indecomposable distributions, (2) all nonatomic indecomposable distributions and (3) all indecomposable distributions which are absolutely continuous with respect to the Haar measure in the case of a locally compact abelian group. It is shown that, under appropriate conditions on G, the indecomposable distributions form a dense G_δ in the class of all distributions in cases (1) and (2) and in the class of all absolutely continuous distributions in case (3). This lends substance to the statement that, in general, a distribution is indecomposable. It may be added that most of these results seem to be new even in the case when G is the real line.

In this connection it is interesting to compare these results with that of W. Rudin [6]. Rudin considers the convolution algebra $L_1(G)$ of all functions integrable with respect to the Haar measure in a locally Euclidean group G and shows that every element in $L_1(G)$ is decomposable. The decomposability of all elements is made possible by the fact that the factors are not confined to non-negative elements of $L_1(G)$ as we have done in our work.

Our analysis incidentally throws some light on the problem of existence of nonatomic measures in a separable metric space. It follows as a consequence of our work that the set of all nonatomic measures in a complete

Received by the editors May 15, 1961.

separable metric space without isolated point is a dense G_δ in the set of all measures under the weak topology.

2. **Preliminaries.** Throughout the paper we suppose that G denotes a complete separable metric group. Additional assumptions on G will be specifically mentioned as and when necessary. We employ the customary notation of denoting the group operation as xy, $x, y \in G$ in the case of general groups and as $x+y$ if G is abelian. e always denotes the unit in G. For any two subsets A, B of G we write $AB = [z: z = xy, x \in A, y \in B]$ and $A^{-1} = [z: z^{-1} \in A]$ (in case the group is abelian we use instead the symbols $A+B$ and $-A$ respectively).

The convolution operation. By a measure (or distribution) we mean a probability measure defined on the σ-field $\mathcal{B}$ of Borel subsets of G. Let $\mathfrak{M}$ denote the collection of all probability measures on $\mathcal{B}$. For $\mu, \nu \in \mathfrak{M}$ the convolution $\mu * \nu$ is defined as follows

$$(\mu*\nu)(A) = \int \mu(Ax^{-1})d\nu(x). \tag{2.1}$$

With this operation $\mathfrak{M}$ becomes a semigroup which is abelian if and only if G is so. It should be noted that $\mu * \nu$ in (2.1) can be written in the equivalent form

$$(\mu*\nu)(A) = \int \nu(x^{-1}A)d\mu(x).$$

For each $g \in G$ and $\mu \in \mathfrak{M}$, $\mu * g$ denotes the right translate of μ by g i.e., the measure $\mu(Eg^{-1})$. $g * \mu$ is defined similarly. By a translate of μ we mean a measure of the form $\mu * g$ or $g * \mu$. Regarding the measure-theoretic terminology used but unaccompanied by explanation, we follow Halmos [7].

DEFINITION 2.1. A measure λ is *decomposable* if and only if there exist two nondegenerate measures μ and ν such that $\lambda = \mu * \nu$. In the contrary case λ is said to be *indecomposable*.

DEFINITION 2.2. A nondegenerate measure α is said to be a factor of a measure β if and only if there exists a measure γ such that either $\beta = \alpha * \gamma$ or $\beta = \gamma * \alpha$.

We shall denote by $\mathfrak{M}_0$ the set of all decomposable measures and $\mathfrak{M}_1$ the set of all indecomposable measures.

DEFINITION 2.3. The spectrum of a measure μ is the smallest closed set $A \subset G$ such that $\mu(A) = 1$.

The existence of the spectrum is well known and it is also easy to see that if A, B, and C are the spectra of the measures μ, ν and $\mu * \nu$ respectively then C = closure of (AB).

Topologies in $\mathfrak{M}$. In the sequel we shall be mainly concerned with the weak topology in $\mathfrak{M}$. It is defined through convergence as follows:

DEFINITION 2.4. A sequence of measures $\{\mu_n\}$ converges weakly to a measure μ if and only if, for every real valued bounded continuous function f defined on G, $\int f d\mu_n \to \int f d\mu$.

It is clear that the class of subsets of $\mathfrak{M}$, of the form

$$\left[\mu: \left|\int f_i d\mu - \int f_i d\mu_0\right| < \epsilon_i, i = 1, 2, \cdots, k\right]$$

where $(f_1, \cdots, f_k)$ is any finite set of bounded continuous functions and $(\epsilon_1, \cdots, \epsilon_k)$ is any finite set of positive numbers forms a neighbourhood system for the weak topology in $\mathfrak{M}$. It is useful to note that the sets of the type

$$[\mu: \mu(V_i) > \mu_0(V_i) - \epsilon_i, i = 1, 2, \cdots, k]$$

where $\epsilon_i > 0$ for all i and V_i are open subsets of G, are open in the weak topology.

Now we shall gather a few results about the weak topology in $\mathfrak{M}$ which we need in the sequel.

THEOREM 2.1 (PROHOROV [8], VARADARAJAN [9]). *If G is a complete separable metric space, the space $\mathfrak{M}$ of measures on G becomes a complete separable metric space under the weak topology.*

THEOREM 2.2 (PROHOROV [8]). *If G is a complete separable metric space, a subset $M \subset \mathfrak{M}$ is conditionally compact in the weak topology if and only if, for every $\epsilon > 0$, there exists a compact set $K_\epsilon \subset G$ such that $\mu(K_\epsilon) > 1 - \epsilon$ for every $\mu \in M$.*

THEOREM 2.3 (RANGA RAO [10]). *In a complete separable metric space G a sequence $\mu_n \in \mathfrak{M}$ converges weakly to $\mu \in \mathfrak{M}$ if and only if the following holds: For every class $\mathfrak{A}$ of continuous functions on G such that*

(i) *$\mathfrak{A}$ is uniformly bounded,*

(ii) *$\mathfrak{A}$ is compact in the topology of uniform convergence on compacta,*

$$\lim_{n \to \infty} \sup_{f \in \mathfrak{A}} \left| \int f d\mu_n - \int f d\mu \right| = 0.$$

All topological notions in $\mathfrak{M}$ used in §§1–6 refer to the weak topology. Only in the last section, we find it necessary to consider the strong topology induced by the norm $\|\mu\| = \sup_{A \in \mathcal{B}} |\mu(A)|$, $\mu \in \mathfrak{M}$.

Indecomposable distributions on the real line. We now state two results due to P. Lévy [3] concerning absolutely continuous indecomposable distributions on the real line. We shall have occasion to use the latter one in the last section.

Let Z be any real-valued random variable which takes values in a bounded interval. We denote by $[Z]$ the integral part of Z and the conditional distribution of Z given that $[Z] = n$, by μ_n.

THEOREM 2.4. *Let Z be any real-valued random variable taking values in a bounded interval and satisfying the following properties:*

(a) *$[Z]$ is even with probability one,*

(b) *the distribution of $[Z]$ is indecomposable,*

(c) *the family of distributions μ_n, n running over possible values of $[Z]$, has no common factor.*

Then the distribution of Z is indecomposable.

THEOREM 2.5. *Let μ_1, μ_2 denote the uniform distributions on the intervals $[a, b]$ and $[c, d]$ respectively. If $(b-a)/(d-c)$ is irrational then μ_1 and μ_2 have no common factor.*

3. **Shift compactness in $\mathfrak{M}$.** In the theory of sums of independent random variables we often come across the situation where a sequence of distributions fails to converge to any limit but actually does converge when suitably centered (see Gnedenko and Kolmogorov [11]). In this section we shall make a systematic analysis of this phenomenon in relation to the convolution operation between distributions on groups. To this end, it is convenient to introduce the following

DEFINTION 3.1. A family $\mathfrak{N}$ is said to be shift compact if, for every sequence $\mu_n \in \mathfrak{N}$ $(n=1, 2, \cdots)$, there is a sequence of measures ν_n such that (1) ν_n is a translate of μ_n and (2) ν_n has a convergent subsequence.

The main result of this section is the following theorem which reveals an important structural property of the topological semigroup in relation to the notion of shift compactness. Its applications in the theory of factorisation of distributions in groups will be discussed in a separate paper.

THEOREM 3.1. *Let $\{\lambda_n\}$, $\{\mu_n\}$, $\{\nu_n\}$ be three sequences of measures on G such that*

$$\lambda_n = \mu_n * \nu_n \qquad (n = 1, 2, \cdots). \tag{3.1}$$

If the sequence $\{\lambda_n\}$ is conditionally compact, then each one of the sequences $\{\mu_n\}$ and $\{\nu_n\}$ is shift compact.

As an immediate consequence we have the following

COROLLARY 3.1. *For any $\lambda \in \mathfrak{M}$, the family $\mathfrak{F}(\lambda)$ of all factors of λ is shift compact.*

Before proceeding to the proof of Theorem 3.1 we shall establish the following

LEMMA 3.1. *Let $\{\lambda_n\}$, $\{\mu_n\}$, $\{\nu_n\}$ be three sequences of measures on G such that $\lambda_n = \mu_n * \nu_n$ for each n. If the sequences $\{\lambda_n\}$ and $\{\mu_n\}$ are conditionally compact then so is the sequence $\{\nu_n\}$.*

Proof. Since the sequences λ_n and μ_n are conditionally compact it follows from Theorem 2.2 that, given $\epsilon>0$, there exists a compact set K_ϵ such that

$$\lambda_n(K_\epsilon) > 1 - \epsilon, \qquad \mu_n(K_\epsilon) > 1 - \epsilon$$

for all n. Then we have

$$\begin{aligned} 1 - \epsilon < \lambda_n(K_\epsilon) &= \int \nu_n(x^{-1}K_\epsilon)d\mu_n(x) \\ &\leqq \int_{K_\epsilon} \nu_n(x^{-1}K_\epsilon)d\mu_n(x) + \epsilon \end{aligned} \tag{3.2}$$

or

$$\int_{K_\epsilon} \nu_n(x^{-1}K_\epsilon)d\mu_n(x) > 1 - 2\epsilon. \tag{3.3}$$

(3.3) implies the existence of a point $x_n \in K_\epsilon$ with the property

$$\nu_n(x_n^{-1}K_\epsilon) > 1 - 3\epsilon$$

and consequently we have

$$\nu_n(K_\epsilon^{-1}K_\epsilon) > 1 - 3\epsilon$$

for all n. Since $K_\epsilon^{-1}K_\epsilon$ is compact and independent of n, another application of Theorem 2.2 leads to the fact that the sequence $\{\nu_n\}$ is conditionally compact. This completes the proof of the lemma.

Proof of Theorem 3.1. We choose a sequence ϵ_r of positive numbers such that $\sum \epsilon_r < \infty$. Then the conditional compactness of the sequence $\{\lambda_n\}$ implies, by Theorem 2.2, that there exists a sequence of compact sets K_r such that

$$\lambda_n(K_r) > 1 - \epsilon_r, \qquad r = 1, 2, \cdots$$

for all n. Now choose a positive sequence η_r descending to zero and satisfying

$$\sum_{r=1}^{\infty} \epsilon_r \eta_r^{-1} \leqq \frac{1}{2}\cdot$$

Let

$$E_{nr} = [x: \mu_n(K_r x^{-1}) > 1 - \eta_r], \qquad F_n = \bigcap_{r=1}^{\infty} E_{nr}. \tag{3.4}$$

Then, from (3.1) and (3.4), we have

$$\begin{aligned} 1 - \epsilon_r \leqq \lambda_n(K_r) &= \int_{E_{nr}} \mu_n(K_r x^{-1})d\nu_n(x) + \int_{E'_{nr}} \mu_n(K_r x^{-1})d\nu_n(x) \\ &\leqq \nu_n(E_{nr}) + (1 - \eta_r)\nu_n(E'_{nr}) \end{aligned}$$

where E'_{nr} denotes the complement of the set E_{nr}. Thus we obtain

$$\nu_n(E'_{nr}) \leqq \frac{\epsilon_r}{\eta_r}$$

and consequently

$$\nu_n(F'_n) \leqq \sum_n \epsilon_r \eta_r^{-1} \leqq \frac{1}{2}\cdot$$

Hence $F_n \neq \phi$. Let x_n be any element in F_n. Then, from the definition of F_n, we have

$$\mu_n(K_r x_n^{-1}) > 1 - \eta_r \tag{3.5}$$

for all n and all r.

We now write $\alpha_n = \mu_n * x_n$ (the right translate of μ_n by x_n) and $\beta_n = x_n^{-1} * \nu_n$. Then, obviously, $\lambda_n = \alpha_n * \beta_n$ and from (3.5) and Theorem 2.2 it follows that both the sequences λ_n and α_n are conditionally compact. Lemma 3.1 now implies that β_n is conditionally compact. The fact that α_n and β_n are translates of μ_n and ν_n respectively completes the proof of the theorem.

4. The class $\mathfrak{M}_1$ is a G_δ. The purpose of this section is simply to prove the following.

THEOREM 4.1. *Let G be a complete separable metric group. Then the class $\mathfrak{M}_1$ of all indecomposable distributions forms a G_δ.*

Before proceeding to the proof of this theorem we make a digression in order to pick up a few auxiliary facts.

Let $f_1, f_2, \cdots$, be a sequence of bounded functions on G with the following properties

(a) for each j, $f_j(x)$ is uniformly continuous in both the right and left uniformities of G.

(b) the sequence $\{f_j\}$ separates points of G.

The existence of such a sequence of functions may be seen as follows. Since G is a separable metric group there is a sequence of neighbourhoods $\{N_i\}$ of e such that $\bigcap_{i=1}^{\infty} N_i = \{e\}$. Then by a well known result (cf. A. Weil [16, pp. 13–14]) there exists a sequence of functions $f_i(x)$ $(i=1, 2, \cdots)$ which are uniformly continuous in the two-sided uniformity (i.e., in both the right and left uniformities) and such that $f_i(e)=0$ and $f_i(x)=1$ for $x \notin N_i$. Let $\{x_n\}$ be a sequence dense in G. Then the countable family $S=\{\phi_{ij}\}$ where $\phi_{ij}(x)=f_i(xx_j)$ possesses both the properties. It is only necessary to prove that the family S separates points of G. If not let a and b be two distinct elements of G such that $\phi_{ij}(a)=\phi_{ij}(b)$ for all i and j or equivalently $f_i(ax_j) = f_i(bx_j)$ for all i and j. Since $\{x_j\}$ is dense it follows that $f_i(ab^{-1})=f_i(e)=0$ for all i. But $ab^{-1} \notin N_i$ for some i and hence $f_i(ab^{-1})=1$ for some i. This contradiction shows that $S=\{\phi_{ij}\}$ separates points of G.

In all that follows, $S=\{f_j\}$ is a fixed sequence with the above properties.

It is then clear that a measure μ on G is degenerate if and only if the induced measure μf_j^{-1} on the real line is degenerate for each j. For any real valued bounded continuous function f and any measure μ, we write

$$V(f,\mu) = \sup_{a\in G} \max\left\{\left[\int f^2(ax)d\mu - \left(\int f(ax)d\mu\right)^2\right], \left[\int f^2(xa)d\mu - \left(\int f(xa)d\mu\right)^2\right]\right\}. \tag{4.1}$$

It is obvious that a measure μ is degenerate if and only if $V(f_j,\mu)=0$ for all j.

LEMMA 4.1. *If f is bounded and uniformly continuous in both the right and left uniformities of the group G and μ_n is a sequence of measures converging weakly to μ, then*

$$\lim_{n\to\infty} V(f,\mu_n) = V(f,\mu).$$

Proof. For each f which is bounded and uniformly continuous in the right as well as the left uniformity, it is clear that each one of the families of functions $\{f(xa),\ a\in G\}$, $\{f^2(xa),\ a\in G\}$, $\{f(ax),\ a\in G\}$, $\{f^2(ax),\ a\in G\}$ is uniformly bounded and equi-continuous at each point of G. Consequently, they are conditionally compact in the topology of uniform convergence on compacta. The lemma is then an immediate consequence of Theorem 2.3.

Let now $E_{ij}(\epsilon)$ be defined as follows

$$E_{ij}(\epsilon) = [\mu : \mu = \alpha * \beta,\ V(f_i,\alpha) \geqq \epsilon,\ V(f_j,\beta) \geqq \epsilon] \tag{4.2}$$

where f_i and f_j are any two functions from S. Then we have

LEMMA 4.2. *For any $\epsilon>0$ and each i, j the set $E_{ij}(\epsilon)$ is closed.*

Proof. Let μ_n be a sequence of measures in $E_{ij}(\epsilon)$ converging to some measure μ. Then by (4.2) there exist measures α_n and β_n such that

$$\begin{aligned} \mu_n &= \alpha_n * \beta_n \\ V(f_i,\alpha_n) &\geqq \epsilon, \qquad V(f_j,\beta_n) \geqq \epsilon. \end{aligned} \tag{4.3}$$

From Theorem 3.1, it follows that there exists a sequence $a_n\in G$ such that the sequences of measures $\{\alpha_n * a_n\}$ and $\{a_n^{-1} * \beta_n\}$ are conditionally compact. Thus we can choose subsequences $\alpha_{n_k} * a_{n_k}$ and $a_{n_k}^{-1} * \beta_{n_k}$ converging to some measures α_0 and β_0 respectively. Since μ_n converges to μ and $\mu_{n_k} = \alpha_{n_k} * a_{n_k} * a_{n_k}^{-1} * \beta_{n_k}$ we have $\mu = \alpha_0 * \beta_0$. It is clear from the definition of $V(f,\mu)$ (see (4.1)) that

$$V(f,\alpha_{n_k} * a_{n_k}) = V(f,\alpha_{n_k}). \tag{4.4}$$

From Lemma 4.1 it follows immediately that

$$\lim_{k\to\infty} V(f_i, \alpha_{n_k}) = V(f_i, \alpha_0).$$

Similarly

$$\lim_{k\to\infty} V(f_j, \beta_{n_k}) = V(f_j, \beta_0).$$

Thus from (4.3) we have

$$V(f_i, \alpha_0) \geqq \epsilon, \qquad V(f_j, \beta_0) \geqq \epsilon$$

or $\mu \in E_{ij}(\epsilon)$. This completes the proof.

We shall now prove Theorem 4.1 by showing that the set of all decomposable measures is an F_σ. In fact

$$\mathfrak{M}_0 = \bigcup_{j=1}^{\infty} \bigcup_{i=1}^{\infty} \bigcup_{r=1}^{\infty} E_{ij}(r^{-1}). \tag{4.5}$$

It is clear that any measure belonging to the right side of (4.5) is decomposable and hence belongs to $\mathfrak{M}_0$. Let now μ be any measure in $\mathfrak{M}_0$. Then there exist two nondegenerate measures α and β such that $\mu = \alpha * \beta$. Since α and β are nondegenerate it follows from the remarks made earlier that there exist two functions f_i and f_j belonging to S, with the property

$$V(f_i, \alpha) > 0, \qquad V(f_j, \beta) > 0.$$

Let $\epsilon = \min[V(f_i, \alpha), V(f_j, \beta)]$. Then for $r > 1/\epsilon$, $\mu \in E_{ij}(r^{-1})$. Thus $\mathfrak{M}_0$ is contained in the right side of (4.5). An application of Lemma 4.2 completes the proof of the theorem.

REMARKS. 1. Let $\mathfrak{M}_c$ denote the class of nonatomic measures. It may be noted that the class of all indecomposable nonatomic measures is a G_δ in $\mathfrak{M}_c$ under the relative topology. At the moment, it is not clear whether even a single nonatomic measure exists in G. These points will be clarified in §6.

2. Let G be a locally compact abelian separable metric group and $\mathfrak{A}(G)$ the class of all absolutely continuous distributions in G. Since the norm topology in $\mathfrak{A}(G)$ is stronger than the weak topology it is clear that the set of all indecomposable absolutely continuous distributions is a G_δ in $\mathfrak{A}(G)$ (the relevant topology being the norm topology).

To determine the category of the various classes of indecomposable distributions, it is thus sufficient to find their closures. We note one case where the class $\mathfrak{M}_1$ is of first category. This is the situation when the group G is finite, as is implied in the work of Vorobev [12].

In the rest of the paper we will study the closures of three classes of indecomposable distributions

(1) the general case—the class $\mathfrak{M}_1$ itself;

(2) the nonatomic case—the class of all nonatomic measures in $\mathfrak{M}_1$;

(3) the absolutely continuous case—the class of indecomposable distributions absolutely continuous with respect to the Haar measure in a locally compact abelian group G.

5. **The general case.** Before we state the main theorem of this section we begin with some lemmas, the purpose of which is to construct indecomposable distributions in G. To this end we introduce the following definition.

DEFINITION 5.1. A subset $A \subset G$ is said to be *decomposable* if there exist two sets $A_1, A_2 \subset G$ such that (a) each of A_1, A_2 contains at least two elements and (b) $A_1A_2 = A$; a set $A \subset G$ is said to be *indecomposable* if it is not decomposable.

LEMMA 5.1. *Let A be any countable indecomposable set and μ a measure such that $\mu(A)=1$ and $\mu(\{g\})>0$ for every $g \in A$. Then μ is indecomposable.*

Proof. Let us suppose that μ is decomposable. Then $\mu=\mu_1 * \mu_2$ where μ_1 and μ_2 are nondegenerate measures with mass concentrated at a countable or finite number of points. Let

$$A_i = [g: g \in G, \mu_i(g) > 0], \qquad i = 1, 2.$$

From the conditions of the lemma it follows that $A = A_1A_2$, which contradicts the fact that A is indecomposable.

LEMMA 5.2. *Let B be an infinite countable set $\{g_1, g_2, \cdots\}$ with the following property: $g_rg_s^{-1} \neq g_tg_u^{-1}$ for every set of distinct integers r, s, t, u no two of which are equal. If F is any finite subset of G then the set $B \cup F$ is indecomposable.*

Proof. Suppose the lemma is not true. Then there exist two sets A_1, A_2, at least one of which contains an infinite number of elements and such that $B \cup F = A_1A_2$. Let $A_1 = (x_1, x_2 \cdots x_n \cdots)$ and $y_1, y_2 \in A_2$. Since the elements x_ry_1, $r=1, 2, \cdots$ are all distinct all but a finite number of them belong to B. Thus there exists a finite set N of integers such that

$$x_ry_1 \in B \qquad \text{for } r \notin N.$$

Take any integer $m \notin N$. For at most one integer s, say $s=k_1$, x_sy_1 can be equal to x_my_2. Similarly, for at most one integer, say $s=k_2$, x_sy_2 can be equal to x_my_1. Choose any integer $n \notin N$ and different from k_1 and k_2. Then x_my_1, x_my_2, x_ny_1 and x_ny_2 are all distinct and belong to B. But

$$x_my_1(x_ny_1)^{-1} = x_my_2(x_ny_2)^{-1}$$

which contradicts the defining property of B.

LEMMA 5.3. *If G is an infinite group, then there exists a set B with the property described in Lemma* 5.1.

Proof. Let g_1, g_2, g_3 be any three distinct elements of the group G. Suppose $g_1, g_2, \cdots, g_n$ have been chosen. Consider the set A_n of all elements of the form $g_{i_1}^{\pm 1} g_{i_2}^{\pm 1} g_{i_3}^{\pm 1}$ where i_1, i_2, i_3 are any three positive integers less than or equal to n. Since A_n is finite and the group G is infinite A_n' is nonempty. Choose any element g_{n+1} from A_n'. The sequence $g_1, g_2, \cdots$ chosen in this way has the required property.

THEOREM 5.1. *If the group G is infinite, then* $\mathfrak{M}_1$ *is a dense* G_δ.

Proof. Any measure in G is a weak limit of measures concentrated at a finite number of points. From Lemmas 5.1–5.3 it is clear that any measure with a finite spectrum is a weak limit of indecomposable distributions. Thus indecomposable distributions are dense in $\mathfrak{M}$. In view of Theorem 4.1, this completes the proof.

6. **The nonatomic case.** To start with we shall investigate the existence of a nonatomic measure in an arbitrary complete separable metric space. This, in itself, seems to be an interesting problem (cf. [13]).

THEOREM 6.1. *Let X be any complete separable metric space without any isolated points. Then there exists a nonatomic measure on X.*

Proof. Let $\mathfrak{M}$ be the class of all probability measures on X. Then by Theorem 2.1 $\mathfrak{M}$ is a complete separable metric space under the weak topology. For any given $\epsilon > 0$, we denote by $C(\epsilon)$ the class of all measures which have at least one atom of mass greater than or equal to ϵ. Then the class of all measures with atomic components can be represented as $\bigcup_{r=1}^{\infty} C(1/r)$. If there does not exist any nonatomic measure, we have

$$\mathfrak{M} = \bigcup_{r=1}^{\infty} C\left(\frac{1}{r}\right).$$

It is not difficult to verify by making use of Theorem 2.2 that $C(\epsilon)$ is closed in the weak topology. Thus by Baire's category theorem, at least one $C(1/r)$ has interior. Hence there exists a measure μ_0 with an atom of positive mass $> \delta > 0$ such that, whenever a sequence of measures μ_n converges weakly to μ_0, μ_n has an atom of mass at least δ for sufficiently large n. Since measures with finite spectrum (i.e., for which spectrum is a finite set) are everywhere dense we can, without loss of generality, assume that μ_0 is a measure with masses $p_1, p_2, \cdots, p_k$ at the points $x_1, x_2, \cdots, x_k$ respectively.

Let $N_n(x_1), N_n(x_2), \cdots, N_n(x_k)$ be sequences of neighbourhoods shrinking to $x_1, x_2, \cdots, x_k$ respectively. We can and do assume that these neighbourhoods are disjoint. Since by assumption X has no isolated points each of these neighbourhoods contains an infinite number of points. We distribute the mass p_i among the points of $N_n(x_i)$ such that the mass at each point is less than $\delta/2$. By doing this for every i and every n we obtain a sequence of measures μ_n converging weakly to μ_0 and such that the mass of μ_n at any point is $\leqq \delta/2$. This contradicts the defining property of μ_0 and shows that $C(1/r)$ has no interior for any r. The proof is complete.

COROLLARY 6.1. *Let X be any complete separable metric space with an uncountable number of points. Then there exists a nonatomic measure on X.*

Proof. Let Y denote the set of all accumulation points of X. Then it is well known (cf. Hausdorff [14, p. 146]) that X can be written in the form $X=Y\cup N$ where

(a) Y is closed and dense in itself,

(b) N is countable.

By Theorem 6.1 there exists a nonatomic measure on Y and hence on X.

The proof of Theorem 6.1 actually yields something more. In fact we have

COROLLARY 6.2. *Let X be a complete separable metric space without isolated points. Then the set of all nonatomic measures which give positive mass to each open subset of X is a dense G_δ in $\mathfrak{M}$.*

If G is any nondiscrete complete metric group, then it is clear that it is necessarily uncountable and cannot have any isolated point. Consequently we have the following:

THEOREM 6.2. *Let G be any nondiscrete complete separable metric group (not necessarily abelian). Then the set of all nonatomic indecomposable distributions which give positive mass to each open set is a dense G_δ in $\mathfrak{M}$.*

7. **The absolutely continuous case.** In this section we suppose that G is a locally compact abelian group and consider measures absolutely continuous with respect to the Haar measure on G. Let $\mathfrak{A}=\mathfrak{A}(G)$ denote the collection of these measures. The convergence notion that is appropriate for $\mathfrak{A}$ is the norm convergence of measures or the L_1 convergence of their densities. The main object here is to show that in the sense of this convergence the indecomposable measures in $\mathfrak{A}$ are dense.

In the first instance we develop in the following lemmas a general method of constructing absolutely continuous indecomposable distributions in G.

LEMMA 7.1. *Let A_1, A_2, A_3 be three closed subsets of G satisfying the following conditions:*

$$(1)\qquad (A_i - A_i)\cap(A_j - A_k) = \emptyset \qquad \text{for } i=1,2,3 \text{ and } j\neq k;$$

$$(2)\qquad (A_1 - A_2)\cap(A_2 - A_3) = (A_2 - A_3)\cap(A_3 - A_1) = (A_3 - A_1)\cap(A_1 - A_2) = \emptyset.$$

Let μ_1, μ_2, μ_3 be measures with $\mu_i(A_i)=1$ and $\lambda=p_1\mu_1+p_2\mu_2+p_3\mu_3$ where $p_i>0$ ($i=1, 2, 3$) and $p_1+p_2+p_3=1$. Then λ is decomposable if and only if μ_1, μ_2, μ_3 have a nondegenerate common factor.

Proof. If μ_1, μ_2, μ_3 have a nondegenerate common factor, it is obvious that λ is decomposable. Conversely let us suppose that λ is decomposable. Then there exist two nondegenerate measures α and β such that $\lambda=\alpha*\beta$. Let C

and D denote the spectra of α and β respectively. It is obvious that

$$C + D \subset A_1 \cup A_2 \cup A_3 = A. \tag{7.1}$$

For each $c \in G$, we write

$$D_i(c) = [d: d \in D \text{ and } c + d \in A_i] = D \cap (A_i - c) \tag{7.2}$$

for $i=1, 2, 3$. The rest of the proof depends on an analysis of the nature of decomposition $\{D_i(c)\}$ of D. It is convenient to divide it into three steps.

I. The sets $D_i(c)$ possess the following properties.

(i) $\bigcup_{i=1}^{3} D_i(c) = D$ for each c.

(ii) $D_i(c) \cap D_j(c) = \emptyset$ for $i \neq j$.

(iii) For any two distinct c_1 and c_2, $D_i(c_1) = D_i(c_2)$ for some i, implies that $D_j(c_1) = D_j(c_2)$ for $j=1, 2, 3$.

(iv) if $c_1 \neq c_2$, $D_i(c_1) \cap D_j(c_2) \neq \emptyset$ implies that $D_i(c_1) = D_j(c_2)$.

The first three properties are very simple. We shall prove (iv). Let us suppose that $D_i(c)$ and $D_j(c)$ have a common point d and $D_i(c_1) \neq D_j(c_2)$. Then there exists a point $d' \in D_i(c_1)$ which is not in $D_j(c_2)$. From (i) above it follows that there is a $k(\neq j)$ such that $d' \in D_k(c_2)$. From these facts we have

$$\begin{aligned} c_1 + d &\in A_i, & c_1 + d' &\in A_i, \\ c_2 + d &\in A_j, & c_2 + d' &\in A_k. \end{aligned} \tag{7.3}$$

Consequently

$$d - d' \in (A_i - A_i) \cap (A_j - A_k).$$

This contradicts the assumption (1) of the lemma and proves (iv). It should be noted that the property (iv) implies that the decompositions $\{D_i(c)\}$ for $c \in C$ are only permutations of each other.

II. One of the following relations is always satisfied. Either

(a) for each $c \in C$, all but one of $D_i(c)$ are empty, i.e., $D_i(c) = D$ for some i, or

(b) for any two $c_1, c_2 \in C$, $D_i(c_1) = D_i(c_2)$ for $i=1, 2, 3$.

The proof of this is quite straightforward and is similar to that of (iv) above. We shall not go into details.

III. Now we suppose that case (a) obtains. Let $C_i = [c: D_i(c) \neq \emptyset]$. It is then easily verified that (1) C_i's are mutually disjoint and their union is C, (2) $C_i + D \subset A_i$ for each i. Let the measures α_i ($i=1, 2, 3$) be defined as follows:

$$\alpha_i(E) = \alpha(E \cap C_i)/\alpha(C_i).$$

(Note that $\alpha(C_i) > 0$.) It is then not difficult to verify that

$$\alpha_i * \beta = \mu_i \qquad \text{for } i = 1, 2, 3.$$

Thus for each i, β is a factor of μ_i.

In case (b), let $D_i = D_i(c)$. Obviously the D_i's are mutually disjoint and $C+D_i \subset A_i$ for each i. Writing $\beta_i(E) = \beta(E \cap D_i)/\beta(D_i)$ we get as before

$$\alpha * \beta_i = \mu_i \qquad \text{for } i = 1, 2, 3.$$

In this case α is the required common factor. This completes the proof of the lemma.

LEMMA 7.2. *Let G be a noncompact group. Then for any given compact set K, there exist elements g, $h \in G$ such that the sets K, $K+g$, $K+h$ satisfy the conditions* (1) *and* (2) *of Lemma* 7.1.

Proof. It may be verified that conditions (1) and (2) of Lemma 7.1 in this case reduce to choosing g and h such that none of the elements g, h, $g-h$, $g+h$, $2g+h$, $2h-g$ belong to the compact set $C = (K-K)-(K-K)$. Let

$$F = [x\colon x = 2y,\ y \in G].$$

Then there are two possibilities.

Case 1. F has compact closure. In this case we can choose an element g such that $g \notin C$ and $\bar{F} \cap (C+g) = \emptyset$. Since G is noncompact such elements exist. Let h be any element such that $h \notin C \cup (C+g) \cup (C-g) \cup (C+2g)$. The pair g, h satisfies our requirements.

Case 2. The closure of F is not compact. Let $g \notin C$ be arbitrary. Since $\bar{F}$ is not compact we can find an $h \in G$ such that $2h \notin C+g$ and $h \notin C \cup (C+g) \cup (C-g) \cup (C-2g) \cup (C+2g)$. As is easily verified the pair g, h serves our purpose. This completes the proof.

LEMMA 7.3. *Let G be an infinite compact metric abelian group. Let A be a subset such that*

$$\text{(1)} \qquad 0 < \lambda(A) < 1,$$

$$\text{(2)} \qquad \int_A \chi_j(x)\,d\lambda(x) \neq 0 \qquad \text{for } j = 0, 1, 2, \cdots,$$

where λ is the normalized Haar measure on G and $\chi_0, \chi_1, \cdots$ are the characters of G. If λ_1, λ_2 are defined by

$$\lambda_1(E) = \lambda(E \cap A)/\lambda(A),$$
$$\lambda_2(E) = \lambda(E \cap A')/\lambda(A'),$$

then λ_1 and λ_2 do not have a common factor.

Proof. Let χ_0 be the identity character. Since for the Haar measure $\int \chi_j d\lambda = 0$ for $j \neq 0$ and $0 < \lambda(A) < 1$, we have

$$\text{(7.4)} \qquad \int_{A'} \chi_j d\lambda \neq 0 \qquad \text{for every } j.$$

We shall now prove that the measures λ_1 and λ_2 cannot have a common factor. If this is not true, then let μ be a common factor. Then there exist measures α_1, α_2 such that

$$\lambda_1 = \alpha_1 * \mu, \qquad \lambda_2 = \alpha_2 * \mu. \tag{7.5}$$

From the definitions of λ_1 and λ_2 and (7.5) we have

$$\lambda(A)\lambda_1 + \lambda(A')\lambda_2 = \lambda = (\lambda(A)\alpha_1 + \lambda(A')\alpha_2) * \mu. \tag{7.6}$$

Taking the characteristic functionals on both sides of (7.5) and (7.6) we get

$$\begin{aligned} &\int \chi_j d\lambda_1 = \left(\int \chi_j d\alpha_1\right)\left(\int \chi_j d\mu\right), \\ &\int \chi_j d\lambda_2 = \left(\int \chi_j d\alpha_2\right)\left(\int \chi_j d\mu\right), \\ &\left(\lambda(A)\int \chi_j d\alpha_1 + \lambda(A')\int \chi_j d\alpha_2\right)\left(\int \chi_j d\mu\right) = 0 \qquad \text{for } j \neq 0. \end{aligned} \tag{7.7}$$

From condition (2) of the lemma and (7.7) we deduce that

$$\int \chi_j d\mu \neq 0 \qquad \text{for all } j.$$

Thus from (7.7) we have

$$\lambda(A)\int \chi_j d\alpha_1 + \lambda(A')\int \chi_j d\alpha_2 = 0 \qquad \text{for all } j \neq 0,$$

which is the same as saying

$$\lambda(A)\alpha_1 + \lambda(A')\alpha_2 = \lambda. \tag{7.8}$$

From (7.5) and the definition of λ_1 and λ_2 we get

$$\int \alpha_1(A' - x)d\mu(x) = \lambda_1(A') = 0,$$

$$\int \alpha_2(A - x)d\mu(x) = \lambda_2(A) = 0.$$

Consequently

$$\alpha_1(A' - x) = 0 \;\; a\cdot e(\mu),$$
$$\alpha_2(A - x) = 0 \;\; a\cdot e(\mu).$$

Thus there exists a point x_0 such that

$$\alpha_1(A' - x_0) = \alpha_2(A - x_0) = 0. \tag{7.9}$$

(7.8) and (7.9) imply that

$$\alpha_1(E) = \frac{\lambda(E \cap [A - x_0])}{\lambda(A)} = [\lambda_1 * (-x_0)](E),$$

$$\alpha_2(E) = \frac{\lambda(E \cap [A' - x_0])}{\lambda(A')} = [\lambda_2 * (-x_0)](E).$$

Thus from (7.5) and (7.10) we obtain

$$\lambda_1 = \lambda_1 * (-x_0) * \mu, \qquad \lambda_2 = \lambda_2 * (-x_0) * \mu. \tag{7.11}$$

Taking characteristic functionals on both sides of (7.11) we have

$$\int \chi_j d((-x_0) * \mu) = 1 \qquad \text{for all } j.$$

Thus μ is degenerate at the point x_0. The proof of the lemma is complete.

LEMMA 7.4. *In any infinite compact group G there exists a set A possessing the properties* (1) *and* (2) *of Lemma* 7.3.

Proof. Let $S(\lambda)$ be the measure ring obtained by considering the space of Borel subsets of G modulo λ-null sets. This is a complete metric space with the distance $d(E, F) = \lambda(E \triangle F)$ where E and F belong to $S(\lambda)$ (cf. [7, pp. 165–169]). Let $\chi_0, \chi_1, \cdots$ be the characters of G, χ_0 being the identity. We consider the following mapping from $S(\lambda)$ to the complex plane. For any $E \in S(\lambda)$, we write

$$f_j(E) = \int_E \chi_j d\lambda.$$

The mapping f_j is obviously continuous. Hence the sets

$$V_j = \left[E\colon \int_E \chi_j d\lambda \neq 0\right]$$

are open in $S(\lambda)$. We shall now prove that each V_j is dense in $S(\lambda)$. Let $A \in S(\lambda)$ and

$$\int_A \chi_j d\lambda = 0.$$

Let $\lambda(A) = c > 0$. Since λ is nonatomic, for any $0 < \epsilon < c$ there exists a set $B \subset A$ such that $\epsilon/2 < \lambda(B) < \epsilon$. Let C be any subset of B for which

$$\int_C \chi_j d\lambda \neq 0.$$

Such a C exists, for otherwise χ_j will vanish in B almost everywhere but at the same time $|\chi_j| = 1$. The set $A \cap C'$ has the property

$$d(A \cap C', A) = \lambda((A \cap C') \triangle A) = \lambda(C) < \epsilon.$$

Since this is true for any sufficiently small ϵ it is possible to get A as a limit of elements belonging to V_j. Since the class of sets A with $\lambda(A) > 0$ is dense in the ring $S(\lambda)$ it follows that the sets V_j are actually dense in $S(\lambda)$. By the Baire category theorem it follows that $\bigcap_{j=1}^{\infty} V_j$ is dense in $S(\lambda)$. Thus there exist Borel sets with the required properties.

LEMMA 7.5. *In any locally compact separable metric abelian group G there exist two absolutely continuous measures with compact supports which do not have a common factor.*

We shall prove this lemma in two steps. First of all let us assume that G is a finite dimensional vector space. Let A_1 and A_2 be two cubes in G such that the ratio of the lengths of their sides is irrational.

Then the uniform distributions μ_1 and μ_2, concentrated in A_1 and A_2 respectively, cannot have a common factor. For, if they have, then at least one of the one-dimensional marginal distributions of μ_1 must have a common factor with the corresponding marginal distribution of μ_2. Since the corresponding marginal distributions of μ_1 and μ_2 are rectangular distributions in the real line with the ratio of the lengths of their supports irrational, it follows from Theorem 2.5 that they cannot have a common factor. This proves the lemma in the case when G is a vector space.

If G is an infinite compact group, we have by Lemma 7.3 two absolutely continuous measures which do not have a common factor.

Now a result of Pontrjagin [15] states that for any general locally compact group G there exists an open subgroup H such that

$$H = V \oplus Z$$

where V is a vector group, Z a compact group and $\oplus$ denotes the direct sum. In V we take any two absolutely continuous measures μ_1 and μ_2 without any common factor. If Z is infinite we take two absolutely continuous measures ν_1 and ν_2 in Z without common factor. If Z is finite we take ν_1 and ν_2 to be any two degenerate measures. We form the product measures

$$\lambda_1 = \mu_1 \times \nu_1, \qquad \lambda_2 = \mu_2 \times \nu_2$$

in H. Since H is open λ_1 and λ_2 are absolutely continuous with respect to the Haar measure in G. Since none of the marginals of λ_1 and λ_2 have a common factor we conclude that λ_1 and λ_2 themselves cannot have a common factor. This completes the proof of the lemma.

THEOREM 7.1. *In any locally compact noncompact complete separable metric* *abelian group G the set of all absolutely continuous indecomposable distributions is a dense G_δ in $\mathfrak{A}(G)$.*

That the set under consideration is a G_δ follows from the remarks made in §4. It remains to be proved that it is dense. It is clear that the set of all absolutely continuous measures with compact supports is everywhere dense. Thus it remains only to prove that any absolutely continuous measure μ with compact support is a limit of a sequence of absolutely continuous indecomposable measures. Let the support of μ be K_0. Let μ_1 and μ_2 be two absolutely continuous measures with compact supports K_1 and K_2 and having no common factor. Such measures exist because of Lemma 7.5. Let

$$K = K_0 \cup K_1 \cup K_2.$$

By using Lemma 7.3 we choose two points $g, h \in G$ such that $K, K+g, K+h$ satisfy conditions (1) and (2) of Lemma 7.1. We write

$$\alpha_1 = \mu, \qquad \alpha_2 = \mu_1 * g, \qquad \alpha_3 = \mu_2 * h$$

and

$$\mu_n = \left(1 - \frac{2}{n}\right)\alpha_1 + \frac{1}{n}\,\alpha_2 + \frac{1}{n}\,\alpha_3, \qquad n \geqq 2.$$

From Lemma 7.1 it follows that μ_n is indecomposable. It is obvious that μ_n converges in norm to α_1 which is the same as μ. This completes the proof of the theorem.

REMARK. In the above theorem the assumption of noncompactness of G has played a crucial role. The question arises—is this assumption necessary? Or, more precisely, if G is an infinite compact group, is the collection of indecomposable distributions in $\mathfrak{A}(G)$ dense in $\mathfrak{A}(G)$? The answer is not known.

REFERENCES

1. A. I. Khinchine, *Contribution à l'arithmétique des lois de distribution*, Bull. Math. Univ. Moscou 1 (1937), 6–17.

2. H. Cramér, *Problems in probability theory*, Ann. Math. Statist. **18** (1948), 165–193.

3. P. Lévy, *Sur une classe de lois de probabilité indécomposables*, C. R. Acad. Sci. Paris **235** (1952), 489–492.

4. D. Dugué, *Sur certains exemples de décomposition en arithmétique des lois des probabilités*, Ann. Inst. H. Poincaré **12** (1951), 159–181.

5. D. Dugué et R. A. Fisher, *Un résultat assez inattendu d'arithmétique des lois des probabilités*, C. R. Acad. Sci. Paris **227** (1948), 1205–1207.

6. W. Rudin. *Representation of functions by convolutions*, J. Math. Mech. **7** (1958), 103–116.

7. P. R. Halmos, *Measure theory*, Van Nostrand, New York, 1950.

8. Yu. V. Prohorov, *Convergence of random processes and limit theorems in the theory of probability*, Teor. Veroyatnost. i Primenen. **1** (1956), 177–238. (Russian)

9. V. S. Varadarajan, *Weak convergence of measures on separable metric spaces*, Sankhyā **19** (1958), 15–22.

10. R. Ranga Rao, *Some problems in probability theory*, thesis submitted to Calcutta University, 1960.

11. B. V. Gnedenko and A. N. Kolmogorov, *Limit theorems for sums of independent random variables*, Addison-Wesley, Cambridge, Mass., 1954.

12. N. N. Vorobev, *Addition of independent random variables on finite abelian groups*, Mat. Sbornik **34** (1954), 89–126. (Russian)

13. R. Sikorski, *Boolean algebras*, Springer, Berlin, 1960.

14. F. Hausdorff, *Set theory*, Chelsea, New York, 1957.

15. L. S. Pontrjagin, *Topological groups*, Princeton Univ. Press, Princeton, N. J., 1939.

16. André Weil, *Sur les espaces à structure uniforme et sur la topologie générale*, Hermann, Paris, 1937.

INDIAN STATISTICAL INSTITUTE,
CALCUTTA, INDIA

Illinois J. Math. 7 (1963) 337-369

PROBABILITY DISTRIBUTIONS ON LOCALLY COMPACT ABELIAN GROUPS

BY

K. R. PARTHASARATHY, R. RANGA RAO, AND S. R. S. VARADHAN

1. Introduction

For probability distributions on the real line there are three main theorems on which the entire study of limit theorems for sums of independent random variables is based. These are (1) the Lévy-Khinchin representation of an infinitely divisible distribution, (2) the criteria for weak convergence of such distributions, and (3) Khinchin's theorem on sums of infinitesimal summands stating that these converge weakly if and only if certain associated infinitely divisible laws converge. For a precise statement of these results we refer to Kolmogorov and Gnedenko [3].

During the last two decades or so these results have been extended by many authors to varying degrees of generality. We mention in particular the works of Lévy [12], Kawada and Itô [17], Takano [9], Bochner [1], [2], Hunt [4], Urbanik [13], [14], Kloss [16]. In this paper we study probability distributions on a locally compact abelian (separable) group and obtain definitive extensions of all the three main results mentioned above.

The preliminaries are developed in Section 2. We mention in particular the concept of shift-compactness introduced therein and the important role that Theorem 2.1 plays in our study. A slightly modified notion of an infinitely divisible law is given in this paper to take into account the fact that the group may not be divisible.

The main results of the paper are the following. Weak limits of sums of uniformly infinitesimal random variables (with values in a group) are infinitely divisible. These limits can be obtained from certain accompanying infinitely divisible distributions if they have no idempotent factors. If μ is any infinitely divisible distribution without an idempotent factor, then its characteristic functional $\hat{\mu}(y)$, defined on the character group, has the form

$$(x_0, y) \exp\left\{\int [(x, y) - 1 - ig(x, y)]\, dF(x) - \phi(y)\right\}$$

where (x, y) is the value of the character y at x, x_0 is a fixed element of the group, $g(x, y)$ is a fixed function independent of μ, F is a σ-finite measure which integrates the function $(x, y) - 1 - ig(x, y)$ and has finite mass outside each neighborhood of the identity, and $\phi(y)$ is a nonnegative continuous function satisfying the equality

$$\phi(y_1 + y_2) + \phi(y_1 - y_2) = 2[\phi(y_1) + \phi(y_2)].$$

Received December 18, 1961.

Distributions μ with characteristic function of the form

$$(x_0, y) \exp(-\phi(y)),$$

where $\phi(y)$ has the above mentioned property, are precisely those characterized by the following algebraic property: $\mu = \alpha * \beta$, where α is a (generalized) Poisson distribution and β is infinitely divisible, implies that α is degenerate at the identity. We call them Gaussian since they are really so in the case of finite-dimensional Euclidean spaces.

One interesting feature of the representation of an infinitely divisible distribution without idempotent factors is its nonuniqueness. However, the Gaussian component is always unique. If F_1 and F_2 are two measures which occur in two different representations, their difference is always concentrated in a subgroup H of X characterized by the following property: H is the smallest closed subgroup containing all compact subgroups of X. Thus nonuniqueness is due to the prevalence of compact subgroups. If, however, we take a weakly continuous one-parameter convolution semigroup of measures $\{\mu_t\}$, then there exists a unique representation

$$\hat{\mu}_t(y) = (x_t, y) \exp\left[t\left(\int [(x, y) - 1 - ig(x, y)]\, dF - \phi(y)\right)\right].$$

In the last section we prove the following generalization of a result due to Khinchin [5] on the factorization of arbitrary probability distributions. Any distribution is the convolution of the normalized Haar measure of a compact subgroup, a finite or a countable number of indecomposable distributions, and an infinitely divisible distribution without indecomposable factors.

2. Preliminaries

2.1. All groups considered in this paper are locally compact abelian separable metric groups. Let X denote such a group, and let Y be its character group. For $x \in X$ and $y \in Y$, let (x, y) denote the value of the character y at x. By duality theory the relation between X and Y is perfectly symmetric, i.e., X is the character group of Y. Further, if G is a closed subgroup of X and H is the annihilator of G in Y, i.e., the set

$$H = [y: (x, y) = 1 \text{ for all } x \in G],$$

then G and Y/H are character groups of each other. These facts and some well-known results on the structure of locally compact abelian groups will be freely used in the sequel. For these details we refer to A. Weil [10], and Pontrjagin [18].

2.2. By a measure on X we shall mean a nonnegative completely additive set function defined on the Borel σ-field of subsets of X. We shall refer to probability measures as distributions. Let $\mathfrak{M}$ denote the class of all distributions. For $\lambda, \mu \in \mathfrak{M}$ we write

$$(\lambda * \mu)(E) = \int \mu(E - x)\, d\lambda$$

for any Borel set E. $\lambda * \mu$ is a distribution obtained by the convolution operation. With this operation $\mathfrak{M}$ becomes a commutative semigroup. If μ is the distribution degenerate at a point $x \in X$, then we write $\lambda * x$ for $\lambda * \mu$. We call $\lambda * x$ the shift or translate of λ by the element x. If $\lambda_1, \lambda_2, \cdots, \lambda_n$ are some n distributions, then the distribution $\mu = \lambda_1 * \cdots * \lambda_n$ has an obvious meaning. We denote μ by $\prod_{i=1}^{n} \lambda_i$. If all the distributions λ_i are identically equal to a single distribution λ, we write $\mu = \lambda^n$. These definitions can obviously be extended to all measures, and we shall have occasion to use them.

For any measure μ we write $\bar{\mu}(A) = \mu(-A)$ where $-A$ is the set of all inverses of elements in A. Then $\bar{\mu}$ is also a measure. We denote by $| \mu |^2$ the measure $\mu * \bar{\mu}$.

We introduce in $\mathfrak{M}$ the weak topology, defined as follows: For $\mu_n, \mu \in \mathfrak{M}$, μ_n is said to converge to μ (or $\mu_n \Rightarrow \mu$) if $\int f\, d\mu_n \to \int f\, d\mu$ for each bounded continuous function f on X. We assume some familiarity with this convergence and in particular the description of compact subsets in this topology.

2.3. For each $\mu \in \mathfrak{M}$ its characteristic function $\hat{\mu}(y)$ is a function on the character group Y, defined as follows:

$$\hat{\mu}(y) = \int_X (x, y)\, d\mu(x).$$

Some of the basic properties of the characteristic function $\hat{\mu}(y)$ are given below.

(1) $\hat{\mu}(y)$ is a uniformly continuous function of y,

(2) $\hat{\mu}(y)$ determines μ uniquely,

(3) $(\mu * \lambda)^{\wedge}(y) = \hat{\mu}(y)\hat{\lambda}(y)$ for all $y \in Y$, and $\mu, \lambda \in \mathfrak{M}$,

(4) $\hat{\bar{\mu}}(y) = \overline{\hat{\mu}(y)}$,

(5) $\mu_n \Rightarrow \mu$ if and only if $\hat{\mu}_n(y) \to \hat{\mu}(y)$ uniformly over compact subsets of Y,

(6) if $\hat{\mu}_n(y)$ converges to a limit uniformly on each compact subset of Y, then there is a $\mu \in \mathfrak{M}$ such that

$$\text{(i)}\quad \hat{\mu}(y) = \lim_{n\to\infty} \hat{\mu}_n(y) \qquad \text{and} \qquad \text{(ii)}\quad \mu_n \Rightarrow \mu.$$

2.4. A distribution μ is said to be idempotent if $\mu^2 = \mu * x$ for some $x \in X$. If we write $\lambda = \mu * (-x)$, then it is clear that $\lambda^2 = \lambda$, so that $\hat{\lambda}(y) = 0$ or 1. From the inequality

$$(2.1)\qquad 1 - R(x, y_1 + y_2) \leq 2[(1 - R(x, y_1)) + (1 - R(x, y_2))]$$

(R denoting the real part) it is clear that the set of all y for which $\hat{\lambda}(y) = 1$ is a both open and closed subgroup of Y. It is not difficult to see that the annihilator G of this subgroup is compact and λ is the normalized Haar measure of G. Thus μ is the translate of the Haar distribution of a compact subgroup.

2.5. For $\alpha, \mu \in \mathfrak{M}$, α is said to be a factor of μ ($\alpha \prec \mu$ in symbols) if there is a $\beta \in \mathfrak{M}$ such that $\mu = \alpha * \beta$. A distribution μ whose only factors are either degenerate distributions or translates of μ is said to be indecomposable. We denote by $F(\mu)$ the collection of all factors of μ.

Two distributions α, β are said to be equivalent ($\alpha \sim \beta$ in symbols) if each is a shift of the other. "$\sim$" is an equivalence relation. For each α, $\tilde{\alpha}$ denotes the equivalence class containing α. The collection $\tilde{\mathfrak{M}}$ of equivalence classes forms a semigroup. $\tilde{\mathfrak{M}}$ will be endowed with the quotient topology.

Definition 2.1. A subset $\mathfrak{N} \subset \mathfrak{M}$ is said to be *shift-compact* if its image $\tilde{\mathfrak{N}}$ in $\tilde{\mathfrak{M}}$ is conditionally compact.

The following theorem proved in [8] (see Section 3) plays a fundamental role in our study.

Theorem 2.1. *Let $\mathfrak{N}$ be a conditionally compact subset of $\mathfrak{M}$. Then $F(\mathfrak{N})$ consisting of the totality of all factors of elements belonging to $\mathfrak{N}$ is shift-compact.*

Corollary 1. *A subset $\mathfrak{N} \subset \mathfrak{M}$ is shift-compact if and only if the set $|\mathfrak{N}|^2$ consisting of all elements $\alpha \in \mathfrak{M}$ of the form $|\mu|^2$ with $\mu \in \mathfrak{N}$, is conditionally compact.*

Corollary 2. *For any distribution μ, $F(\mu)$ is shift-compact.*

Corollary 3. *Suppose $\alpha_1 \prec \alpha_2 \prec \cdots \prec \alpha_n \prec \cdots$ and $\alpha_n \prec \mu$ for all n. Then there is a translate α_n' of α_n for every n such that α_n' is weakly convergent.*

Proof of Corollary 3. Since $\alpha_n \in F(\mu)$ for each n, $\{\alpha_n\}$ is shift-compact. Suppose that β and β' are any two limits of shifts of α_n. Then the fact that $\alpha_n \prec \alpha_{n+1}$ implies that β and β' are translates of each other. The corollary is an easy consequence of this fact.

The following corollary is proved in a similar manner.

Corollary 4. *Suppose $\alpha_1 \succ \alpha_2 \cdots$. Then there exists a translate α_n' of α_n for each n such that α_n' weakly converges to a limit.*

Remark. Corollaries 2–4 have been obtained earlier by Itô [11] for distributions in the real line.

3. Two auxiliary lemmas

Lemma 3.1. *For each compact set $C \subset Y$ there are a neighborhood N_C of the identity in X and a finite set $E \subset C$ such that*

$$\sup_{y \in C} [1 - R(x, y)] \leqq M \cdot \sup_{y \in E} [1 - R(x, y)]$$

for all $x \in N_C$, *where* M *is a finite constant depending on* C. (*Here* $R(x, y)$ *denotes the real part of* (x, y).)

Proof. From the inequality

$$1 - R(x_1 + x_2, y) \leqq 2[(1 - R(x_1, y)) + (1 - R(x_2, y))]$$

it is clear that if the lemma is valid in two groups X_1 and X_2, it is valid for their direct sum $X_1 \oplus X_2$. Let now Y' denote the closed subgroup generated by C, and Φ its annihilator in X. If τ denotes the canonical homomorphism from X onto $X' = X/\Phi$, it is obvious that $R(x, y) = R(\tau(x), y)$ for all $x \in X$ and $y \in Y'$. It is thus sufficient to prove the lemma when the groups concerned are X' and Y' instead of X and Y. Since Y' is compactly generated, it is of the form $V \oplus C \oplus I^r$ where V is a finite-dimensional vector group, C is a compact group, and I^r is the product of r copies of the integer group. Hence X' is of the form $V \oplus D \oplus K^r$ where D is a discrete group and K^r is the product of r copies of the circle group. Since the lemma is trivially valid in the case of the real line, discrete group, and compact group, the proof of the lemma is complete.

LEMMA 3.2. *For any* $y \in Y$ *there is a continuous function* $h_y(x)$ *on* X *with the following properties*:

(1) $|h_y(x)| \leqq \pi$ *for all* $x \in X$, *and* $h_y(-x) = -h_y(x)$,

(2) $(x, y) = \exp(ih_y(x))$ *for all* $x \in N_y$ *where*

$$N_y = [x: |(x, y) - 1| \leqq \tfrac{1}{2}].$$

Proof. Let $(x, y) = \exp(i\phi(x))$ where $-\pi \leqq \phi(x) < \pi$. Then it is not difficult to verify that $\phi(x)$ is a continuous function of x in the closed set N_y. Now choose any continuous extension of $\phi(x)$ to X such that the first condition is fulfilled. This will serve our purpose.

LEMMA 3.3. *There is a function* $g(x, y)$ *defined on the product space* $X \times Y$ *possessing the following properties*:

(1) $g(x, y)$ *is a continuous function of both the variables* x *and* y.

(2) $\sup_{x \in X} \sup_{y \in C} |g(x, y)| < \infty$ *for each compact set* $C \subset Y$.

(3) $g(x, y_1 + y_2) = g(x, y_1) + g(x, y_2)$ *for each* $x \in X$ *and* $y_1, y_2 \in Y$, *and* $g(-x, y) = -g(x, y)$.

(4) *If* C *is any compact subset of* Y, *then there is a neighborhood* N_C *of the identity in* X *such that* $(x, y) = \exp[ig(x, y)]$ *for all* $x \in N_C$ *and* $y \in C$.

(5) *If* C *is any compact subset of* Y, *then* $g(x, y)$ *tends to zero uniformly in* $y \in C$ *as* x *tends to the identity of the group* X.

Proof. We shall reduce the proof of the proposition to the case of certain simple groups by making use of the structure theory. Suppose that the proposition is true for an open subgroup G of X. Let H and Y be the character groups of G and X respectively. Since H can be obtained as a quotient group of Y by taking the quotient with respect to the annihilator of G in Y, there is a canonical homomorphism τ from Y to H. Suppose $g(x, h)$ has been defined for $x \in G$ and $h \in H$ with the required properties. We extend the definition of g as follows. For $x \in G$ and $y \in Y$ we define

$$g(x, y) = g(x, \tau(y)).$$

For $x \notin G$, we define

$$g(x, y) = 0 \quad \text{for all } y \in Y.$$

Since an open subgroup is closed, the continuity of $g(x, y)$ follows immediately. The rest of the properties of $g(x, y)$ are immediate consequences of their validity in $G \times H$.

In the case of a general group X we take G to be the group generated by a compact neighborhood of the identity. This is both open and closed in X. This group G has the simple structure, $V \oplus C \oplus I^r$ where V is a finite-dimensional vector group, C is a compact group, and I^r is the product of the integer group taken r times. We now observe that if functions $g_1(x, y)$ and $g_2(u, v)$ with the properties mentioned in the lemma exist in groups X and U with character groups Y and V respectively, then a function $g(\xi, \eta)$ with the same properties exists for $\xi \in X \oplus U$ and $\eta \in Y \oplus V$. We have only to define

$$g(\xi, \eta) = g_1(x, y) + g_2(u, v)$$

where x and u are projections of ξ into X and U respectively, and y and v are projections of η into Y and V respectively. Thus it is enough to construct $g(x, y)$ in the case of a real line, a compact group, and the integer group. In the case of the integer group we can take $g(x, y)$ to be identically zero. In the case of the real line we can take $g(x, y) = \theta(x)y$ where

$$\begin{aligned} \theta(x) &= x, & x &\in [-1, 1], \\ &= 1, & x &> 1, \\ &= -1, & x &< -1. \end{aligned}$$

(Note that the character group of the real line is itself.) Thus, in order to complete the proof of the lemma it is enough to consider the case of a compact group X.

Let X be a compact group with Y as character group. Let X_0 be the component of identity in X, Y_1 the annihilator of X_0 in Y, $X_1 = X/X_0$, and $Y_0 = Y/Y_1$. Then Y_0 is the character group of X_0, and Y_1 is the character group of X_1. Since X_0 is connected and compact, Y_0 is a discrete torsion-free group. Let $\{d_\alpha\}$ be a maximal family of mutually independent elements in Y_0. Then for $d \in Y_0$ there exist elements $d_{\alpha_1}, \cdots, d_{\alpha_k}$ from the maximal family and integers $n, n_1, \cdots, n_k$ $(n > 0)$ such that

$$nd = n_1 d_{\alpha_1} + \cdots + n_k d_{\alpha_k}. \tag{3.1}$$

This representation is unique except for multiplication by an integer on both sides.

Each element of Y_0 is a coset of Y_1 in Y. We take the coset d_α and pick out an element y_α of Y from this coset. We fix the elements y_α. We define

$$g(x, y_\alpha) = h_{y_\alpha}(x)$$

for every α where $h_{y_\alpha}(x)$ is as in Lemma 3.2. Let now $y \in Y$ be arbitrary. Then y belongs to some coset of Y_1 which is an element of Y_0. If this element is denoted by d, then there exist integers $n, n_1, \cdots, n_k$ $(n > 0)$ and elements $d_{\alpha_1}, \cdots, d_{\alpha_k}$ from the collection $\{d_\alpha\}$ such that equation (3.1) is satisfied. We define

$$g(x, y) = (n_1/n)g(x, y_{\alpha_1}) + \cdots + (n_k/n)g(x, y_{\alpha_k}).$$

We shall now prove that the function $g(x, y)$ constructed in this way has all the required properties.

Since $g(x, y)$ is continuous in x for each fixed y and Y is discrete, the continuity of $g(x, y)$ in both the variables follows immediately. Properties (2) and (3) are obvious from the nature of the construction.

Since compact sets in Y are finite sets, it is enough to prove property (4) for each $y \in Y$. For any $y \in Y$, let $[y]$ denote the coset of Y_0 to which y belongs. Then $[y]$ is an element of Y_0. If we write $[y]$ for d and $[y_\alpha]$ for d_α, then equation (3.1) can be written as

$$n[y] = n_1[y_{\alpha_1}] + \cdots + n_k[y_{\alpha_k}]. \tag{3.2}$$

For any two elements $y_1, y_2 \in [y]$ it is clear that $y_1 - y_2 \in Y_1$. Since Y_1 is the character group of X/X_0 which is totally disconnected, every element of Y_1 is of finite order. Hence, for any $y \in Y_1$ there exists a neighborhood of the identity in X where $(x, y) = 1$. Thus for any two elements $y_1, y_2 \in [y]$ there exists a neighborhood of the identity in X where $(x, y_1) = (x, y_2)$.

Making use of the remarks made in the previous paragraph we shall complete the proof of the lemma. From the construction of $g(x, y)$ and Lemma 3.2 it is clear that, for each y_α, there exists a neighborhood of the identity in X where $e^{ig(x,y_\alpha)} = (x, y_\alpha)$. Let now $y \in Y$ be arbitrary. From (3.2) it is clear that there exist elements $y_{\alpha_j 1}, \cdots y_{\alpha_j n_j}$ in $[y_{\alpha_j}]$, for $j = 1, 2, \cdots k$, such that

$$ny = \textstyle\sum_{j=1}^{k} (y_{\alpha_j 1} + \cdots + y_{\alpha_j n_j}). \tag{3.3}$$

From the remarks made in the previous paragraph it follows that there exists a neighborhood of the identity in X (depending on $y_{\alpha_j r}$ and y_{α_j}) where

$$(x, y_{\alpha_j r}) = (x, y_{\alpha_j}).$$

Denoting by N the intersection of all the neighborhoods corresponding to $y_{\alpha_j r}$ $(r = 1, 2, \cdots, n_j$ and $j = 1, 2, \cdots, k)$ we have

$$(x, y_{\alpha_j r}) = (x, y_{\alpha_j}) \qquad \text{for } x \in N,\ r = 1, 2, \cdots, n_j,\ j = 1, 2, \cdots, k. \tag{3.4}$$

From (3.3) and (3.4) we have

$$(x, y)^n = (x, ny) = \textstyle\prod_{j=1}^{k} (x, y_{\alpha_j})^{n_j} \qquad \text{for } x \in N.$$

Since there are neighborhoods of the identity where $(x, y_{\alpha_j}) = \exp\{ig(x, y_{\alpha_j})\}$, it follows that there exists a neighborhood of the identity where

$$(x, y)^n = e^{ing(x,y)}.$$

Since (x, y) and $e^{ig(x,y)}$ are continuous functions nonvanishing at the identity of X, there exists a neighborhood of the identity where

$$(x, y) = e^{ig(x,y)}.$$

Property (5) is obvious from property (1) and the fact that $g(x, y)$ vanishes when either x or y is the identity of the corresponding group. This completes the proof of the lemma.

In the following paragraphs we give examples of the function $g(x, y)$ for some particular groups.

Example 1. Let $X = Y = R^n$. If $x = (x_1, \cdots, x_n) \in X$ and $y = (y_1, y_2, \cdots, y_n) \in Y$, then

$$g(x, y) = \sum_{i=1}^n \phi_i(x_i) y_i$$

where $\phi_i(t)$ $(i = 1, 2, \cdots, n)$ are bounded continuous functions on the real line such that $\phi_i(t) = t$ in a neighborhood of $t = 0$ and $\phi_i(-t) = -\phi_i(t)$.

Example 2. Let K denote the circle group, and let $X = K^n$ and $Y = I^n$, I being the integer group. Let

$$X = [(x_1, \cdots, x_n): -1 < x_i \leqq 1 \text{ for } i = 1, 2, \cdots, n]$$

addition being taken modulo 2. Then if $y = (y_1, \cdots, y_n) \in I^n$,

$$g(x, y) = \sum_{i=1}^n \phi_i(x_i) y_i$$

where the functions $\phi_i(t)$ are as in Example 1.

Example 3. Y = additive group of rationals, and X is its character group. Let $\phi(x)$ be a bounded continuous function on X such that $\exp(i\phi(x)) = (x, y_0)$ for x in a neighborhood of the identity in X, where y_0 is a fixed element of Y, other than the identity. Then

$$g(x, y) = \phi(x) y / y_0 .$$

Example 4. If X is totally disconnected, then every homomorphism of Y into the real line is trivial, so that in this case $g(x, y) \equiv 0$ for all $x \in X$ and $y \in Y$.

4. Infinitely divisible distributions

We shall now introduce the definition of infinitely divisible distributions and study some of their elementary properties.

DEFINITION 4.1. A distribution μ is said to be *infinitely divisible* if, for each n, there are elements $x_n \in X$ and $\lambda_n \in \mathfrak{M}$ such that $\mu = \lambda_n^n * x_n$.

We remark that the definition given here is slightly different from the classical definition in the case of the real line. Such a modification is necessary if we want to avoid the role of divisibility of elements from the group. As yet it is not clear whether there exists a single element x of the group X with the property $\mu = \lambda_n^n * x$ for every n. That such is the case for every infinitely divisible distribution will be obvious from the representation which we shall give later in Section 7.

THEOREM 4.1. *The infinitely divisible distributions form a closed subsemigroup of* $\mathfrak{M}$.

Proof. If λ and μ are infinitely divisible, it is obvious from the definition that $\lambda * \mu$ is also infinitely divisible. Let now μ_k, $k = 1, 2, \cdots$, be a sequence of infinitely divisible distributions weakly converging to μ. For any fixed integer n, let

$$\mu_k = \lambda_{kn}^n * x_{kn}. \tag{4.1}$$

From Theorem 2.1 it is clear that there exists a subsequence of λ_{kn} which after a suitable shift converges to a distribution λ_n. Since $\mu_k \Rightarrow \mu$, it is obvious from (4.1) that there exists an element x_n such that $\mu = \lambda_n^n * x_n$. This completes the proof.

The normalized Haar measure of a compact subgroup is an example of an infinitely divisible distribution. We shall now prove a result concerning the absence of zeros for the characteristic function of an infinitely divisible distribution without idempotent factors.

THEOREM 4.2. *Let $\hat{\mu}(y)$ be the characteristic function of an infinitely divisible distribution μ. If $\hat{\mu}(y_0) = 0$ for some character y_0, then μ has an idempotent factor.*

Proof. From the definition of infinite divisibility it follows that, for each n, there exist an element $x_n \in X$ and a distribution λ_n such that $\mu = \lambda_n^n * x_n$. Since $\hat{\mu}(y_0) = 0$, $\hat{\lambda}_n(y_0)$ also vanishes for every n. By Theorem 2.1, λ_n is shift-compact. Let λ be a limit of shifts of λ_n. Then $\hat{\lambda}(y_0) = 0$ and is hence a nondegenerate distribution. It is also clear that every power of λ is a factor of μ. Thus the sequence λ^n is shift-compact, and any limit of shifts is a nondegenerate idempotent factor of μ. This completes the proof.

DEFINITION 4.2. If F is any totally finite measure on X the *distribution* $e(F)$ *associated with* F is defined as follows:

$$e(F) = e^{-F(X)}[1 + F + F^2/2! + \cdots + F^n/n! + \cdots]$$

where 1 is used to denote the measure with unit mass and degenerate at the identity.

$e(F)$ is obviously an infinitely divisible distribution since $e(F) = [e(F/n)]^n$.

Its characteristic function is given by

$$(e(F))\hat{}(y) = \exp\left[\int ((x, y) - 1)\, dF(x)\right].$$

Suppose F_n is a sequence of totally finite measures and we form the sequence $e(F_n)$. We shall now obtain a necessary condition (which will be shown to be sufficient in Section 9) for the shift-compactness of $e(F_n)$.

THEOREM 4.3. *Let $\mu_n = e(F_n)$ where F_n is a sequence of totally finite measures. Then, in order that*

(a) *μ_n be shift-compact,*

(b) *if μ is any limit of shifts of $\{\mu_n\}$, then μ have no idempotent factor,*

the following conditions are necessary:

(i) *For each neighborhood N of the identity the family $\{F_n\}$ restricted to $X - N$ is weakly conditionally compact.*

(ii) *For each $y \in Y$*

$$\sup_n \int [1 - R(x, y)]\, dF_n < \infty,$$

$R(x, y)$ denoting the real part of (x, y).

Before proceeding to the proof of this theorem we shall prove the following.

LEMMA 4.1. *Let F_n be as in Theorem 4.3, and suppose $\sup_n F_n(N') \leqq k$ where N' denotes the complement of a symmetric neighborhood N of the identity. If the sequence $e(F_n)$ is shift-compact, then the sequence of measures F_n is tight when restricted to N'.*

Proof. Let G_n denote the restriction of F_n to N'. Then $e(G_n)$ is a factor of $e(F_n)$. Since $\{e(F_n)\}$ is shift-compact, so is the sequence $\{e(G_n)\}$ by Theorem 2.1. Let $H_n = G_n + \bar{G}_n$. Then, by Corollary 1 of Theorem 2.1, the sequence $\{e(H_n)\}$ is compact. Hence, for any $\varepsilon > 0$ there exists a compact set C such that $e(H_n)(C') < \varepsilon$. Since $e(H_n) = e^{-H_n(X)}[\sum_{r=0}^{\infty} H_n^r/r!]$, we have

$$\begin{aligned}\varepsilon > e(H_n)(C') &\geqq e^{-H_n(X)} H_n(C') \geqq e^{-2k} H_n(C') \\ &\geqq e^{-2k} G_n(C')\end{aligned}$$

for all n. Since k is a constant not depending on n and ε is arbitrary, it follows that the family $\{G_n\}$ is tight.

Proof of Theorem 4.3. Since any neighborhood of the identity contains a symmetric neighborhood, we can assume that N is symmetric. Suppose (a) and (b) are valid. Let, if possible, $\sup_n F_n(N') = \infty$. We can then choose a subsequence for which

$$F_{n_k}(N') \geqq 2k \qquad \text{for } k = 1, 2, \cdots. \tag{4.2}$$

Let L_k, $k = 1, 2, \cdots$, be measures such that

$$(4.3) \qquad \begin{aligned} L_k(A) &\leqq (1/k)F_{n_k}(A) \qquad \text{for every Borel set } A, \\ L_k(N) &= 0, \\ L_k(N') &= 1. \end{aligned}$$

The distribution $\lambda_k = e(L_k)$ is a factor of $e(F_{n_k})$, and the shift-compactness of $e(F_n)$ implies the shift-compactness of $\{\lambda_k\}$ by Theorem 2.1. Let λ be any limit of shifts of λ_k. From (4.2) and (4.3) it follows that any power of λ is a factor of μ. Thus the sequence λ^n is shift-compact, and any limit of shifts of λ^n is a factor of μ. Since these limits will be idempotent and μ has no idempotent factors, it follows that any such limit must be degenerate. Since λ^n is an increasing sequence (in the order $\prec$), it follows that λ itself must be degenerate. Thus the sequence $|\lambda_k|^2$ converges to the distribution degenerate at the identity. Hence

$$e(L_k + \bar{L}_k)(N') \to 0 \quad \text{as} \quad k \to \infty.$$

But

$$e(L_k + \bar{L}_k)(N') = e^{-(L_k+\bar{L}_k)(X)} \sum \frac{(L_k + \bar{L}_k)^r}{r!} (N') \geqq e^{-2} L_k(N') = e^{-2},$$

which is a contradiction. Thus we have $\sup_n F_n(N') < \infty$. Now an application of Lemma 4.1 shows that condition (i) of the theorem is necessary.

In order to prove the necessity of (ii) we observe that $e(F_n + \bar{F}_n) = |e(F_n)|^2$ is a compact sequence, and an application of Theorem 4.2 shows that any limit of $|e(F_n)|^2$ has a nonvanishing characteristic function. Thus, for any y,

$$\lim_n \exp\left\{\int [R(x, y) - 1]\, d(F_n + \bar{F}_n)\right\} \neq 0,$$

which implies condition (ii). The proof is thus complete.

5. General limit theorems for sums of infinitesimal summands

In the case of the real line a well-known result due to Bawly and Khinchin (see [3, Chapter 4]) asserts that the limit of sums of infinitesimal random variables is infinitely divisible, and it can be obtained as the limit of a certain accompanying sequence of infinitely divisible distributions. The purpose of this section is to introduce the notion of infinitesimal distributions in a group and prove a generalized version of the above-mentioned result in the case when the limiting distribution has no idempotent factor.

DEFINITION 5.1. A triangular sequence $\{\alpha_{nj}\}$, $j = 1, 2, \cdots, k_n$ of distributions is said to be *uniformly infinitesimal* if

$$\lim_{n\to\infty} \sup_{1\leqq j\leqq k} \sup_{y\in K} |\hat{\alpha}_{nj}(y) - 1| = 0$$

for each compact set $K \subset Y$.

Before going to the statement of the main result of this section we shall prove a lemma which will be often used in the sequel.

LEMMA 5.1. *Let* $\mu_n = \prod_{j=1}^{k_n} \alpha_{nj}$ *where the sequence* $\{\alpha_{nj}\}$ *is uniformly infinitesimal. If the distribution* μ *is a limit of shifts of* μ_n, *then the set of characters* $[y:\hat{\mu}(y) \neq 0]$ *is an open subgroup of* Y, *and consequently the normalized Haar measure of the annihilator of this subgroup in* X *is a factor of* μ.

Proof. Since μ is a limit of shifts of μ_n, it is clear that for a subsequence (which we shall denote by μ_n itself) $|\mu_n|^2 \Rightarrow |\mu|^2$, and hence

$$\lim_{n\to\infty} \prod_{j=1}^{k_n} |\hat{\alpha}_{nj}(y)|^2 = |\hat{\mu}(y)|^2. \tag{5.1}$$

If $\hat{\mu}(y) \neq 0$, it is obvious that $\hat{\mu}(-y) \neq 0$. (5.1) implies that a necessary and sufficient condition that $\hat{\mu}(y) \neq 0$ is that

$$\sup_n \sum_{j=1}^{k_n} (1 - |\hat{\alpha}_{nj}(y)|^2) < \infty. \tag{5.2}$$

Thus if we make use of the inequality

$$1 - \phi(y_1 + y_2) \leqq 2[(1 - \phi(y_1)) + (1 - \phi(y_2))]$$

for any real-valued characteristic function ϕ, it is clear that the validity of (5.2) for y_1 and y_2 implies its validity for $y_1 + y_2$. The continuity of $\hat{\mu}(y)$ now implies that the set $[y:\hat{\mu}(y) \neq 0]$ is an open subgroup of Y.

We choose and fix a function $g(x, y)$ defined on $X \times Y$ and satisfying all the properties mentioned in Lemma 3.3. The main theorem of this section can now be stated as follows:

THEOREM 5.1. *Let* $\{\alpha_{nj}\}$ *be a uniformly infinitesimal sequence of distributions, and let*

$$\mu_n = \prod_{j=1}^{k_n} \alpha_{nj}.$$

Suppose that $\{\mu_n\}$ *is shift-compact such that no limit of shifts of* μ_n *has an idempotent factor. Let*

$$\beta_{nj} = e(\alpha_{nj} * x_{nj})$$

where x_{nj} *is that element of the group* X *defined by the equality*

$$(x_{nj}, y) = \exp\left[-i \int g(x, y)\, d\alpha_{nj}\right].$$

If $\lambda_n = (\prod_{j=1}^{k_n} \beta_{nj}) * x_n$, *where* $x_n = -\sum_j x_{nj}$, *then*

$$\lim_{n\to\infty} \sup_{y \in K} |\hat{\lambda}_n(y) - \hat{\mu}_n(y)| = 0$$

for each compact set K *of* Y.

Proof. During the course of the proof of the theorem we shall adopt the following conventions: We denote by $C_1, C_2, \cdots$ constants depending only on the compact set K (and not on n). All the statements that we make are for sufficiently large n. By N we denote any arbitrarily small neighborhood of the identity in X.

Turning to the proof of the theorem, we observe that the elements x_{nj} are well defined, since, from the properties of $g(x, y)$, it follows that $\exp [-i \int g(x, y)\, d\alpha]$ is a character on Y for any distribution α. Further, for any neighborhood N of the identity all the points x_{nj} are in N for sufficiently large n. Therefore the uniform infinitesimality of $\{\alpha_{nj}\}$ implies the uniform infinitesimality of the sequence $\{\beta_{nj}\}$. Thus $\hat{\lambda}_n(y)$ and $\hat{\mu}_n(y)$ are nonvanishing in K, and hence we use the logarithmic notation freely. Since no limit of shifts of μ_n has an idempotent factor, it follows from Lemma 5.1 that the sequence $\hat{\mu}_n(y)$ is uniformly bounded away from zero for all $y \in K$. Thus it is enough to prove that

$$\lim_{n\to\infty} \sup_{y\in K} | \log \hat{\lambda}_n(y) - \log \hat{\mu}_n(y) | = 0.$$

We have

$$\begin{aligned}\log \hat{\lambda}_n(y) &= \textstyle\sum_j \log \hat{\beta}_{nj}(y) - \sum_j \log (x_{nj}, y)\\ &= \textstyle\sum_j \log \hat{\beta}_{nj}(y) + i \sum_j \displaystyle\int g(x, y)\, d\alpha_{nj}(x)\\ &= \textstyle\sum_j [(\alpha_{nj} * x_{nj})^\wedge(y) - 1] + i \sum_j \displaystyle\int g(x, y)\, d\alpha_{nj}(x),\end{aligned}$$

and

$$\log \hat{\mu}_n(y) = \textstyle\sum_j \log \hat{\alpha}_{nj}(y).$$

Writing $\theta_{nj} = \alpha_{nj} * x_{nj}$, we obtain

$$\begin{aligned}| \log \hat{\lambda}_n(y) - \log \hat{\mu}_n(y) | &= \Big| \textstyle\sum_j (\hat{\theta}_{nj}(y) - 1) + i \sum_j \displaystyle\int g(x, y)\, d\alpha_{nj}\\ &\qquad - \textstyle\sum_j \log \hat{\theta}_{nj}(y) + \sum_j \log (x_{nj}, y) \Big|\\ &= | \textstyle\sum_j (\hat{\theta}_{nj}(y) - 1) - \sum_j \log \hat{\theta}_{nj}(y) |\\ &\leqq C_1(\textstyle\sum_j | 1 - \hat{\theta}_{nj}(y)|) \sup_j | 1 - \hat{\theta}_{nj}(y)| .\end{aligned}$$

Since $\{\theta_{nj}\}$ is uniformly infinitesimal, it is clear from the above inequality that it is enough to prove that

$$\sup_n \sup_{y\in K} [\textstyle\sum | 1 - \hat{\theta}_{nj}(y)|] < \infty. \tag{5.3}$$

We have, for any neighborhood N of the identity in X

$$\begin{aligned}| 1 - \hat{\theta}_{nj}(y)| &\leqq \left| \int_N (1 - (x, y))\, d\theta_{nj} \right| + \left| \int_{N'} (1 - (x, y))\, d\theta_{nj} \right|\\ &\leqq \left| \int_N (1 - (x, y))\, d\theta_{nj} \right| + 2\theta_{nj}(N').\end{aligned} \tag{5.4}$$

From property (4) of $g(x, y)$ in Lemma 3.3 it follows that there exists a neighborhood of the identity in X where

$$(x, y) = e^{ig(x,y)} \qquad \text{for } y \in K.$$

In such a neighborhood we have, for $y \in K$,

$$|1 - (x, y) + ig(x, y)| \leqq C_2 g^2(x, y). \tag{5.5}$$

(5.4) and (5.5) imply

$$|1 - \hat{\theta}_{nj}(y)| \leqq \left| \int_N g(x, y)\, d\theta_{nj} \right| + C_2 \int_N g^2(x, y)\, d\theta_{nj} + 2\theta_{nj}(N') \tag{5.6}$$

for all $y \in K$. By property (2) of $g(x, y)$ in Lemma 3.3

$$\begin{aligned} \left| \int_X g(x, y)\, d\theta_{nj} \right| &= \left| \int_X g(x + x_{nj}, y)\, d\alpha_{nj}(x) \right| \\ &\leqq \left| \int_N g(x + x_{nj}, y)\, d\alpha_{nj}(x) \right| + C_3\, \alpha_{nj}(N'). \end{aligned} \tag{5.7}$$

Since all the x_{nj} will be in any small neighborhood of the identity after a certain stage, and since $e^{ig(x,y)} = (x, y)$ for $x \in N$ and $y \in K$, we conclude, by making use of property (5) of $g(x, y)$ in Lemma 3.3, that

$$g(x + x_{nj}, y) = g(x, y) + g(x_{nj}, y) \tag{5.8}$$

for $x \in N$ and $y \in K$. Further

$$e^{ig(x_{nj}, y)} = (x_{nj}, y) = \exp\left\{ - i \int g(x, y)\, d\alpha_{nj} \right\}$$

for all $y \in K$ and sufficiently large n. By property (5) of $g(x, y)$ in Lemma 3.3, we get

$$g(x_{nj}, y) = - \int g(x, y)\, d\alpha_{nj}. \tag{5.9}$$

(5.8) and (5.9) imply

$$\begin{aligned} \left| \int_N g(x + x_{nj}, y)\, d\alpha_{nj} \right| &= \left| \int_N (g(x, y) + g(x_{nj}, y))\, d\alpha_{nj} \right| \\ &= \left| \int_N g(x, y)\, d\alpha_{nj} - \alpha_{nj}(N) \int g(x, y)\, d\alpha_{ni} \right| \\ &= \left| \alpha_{nj}(N') \int_N g(x, y)\, d\alpha_{nj} - \alpha_{nj}(N) \int_{N'} g(x, y)\, d\alpha_{nj} \right| \\ &\leqq C_4\, \alpha_{nj}(N'). \end{aligned}$$

The above inequality together with (5.7) implies that

$$\left| \int g(x, y)\, d\theta_{nj} \right| \leqq C_5\, \alpha_{nj}(N') \qquad \text{for } y \in K. \tag{5.10}$$

(5.6), (5.10), and property (2) of Lemma 3.3 give

$$| 1 - \hat{\theta}_{nj}(y) | \leqq C_2 \int g^2(x, y)\, d\theta_{nj} + C_6\, \theta_{nj}(N') + C_7\, \alpha_{nj}(N')$$

for $y \in K$. Thus, in order to complete the proof of the theorem we have only to show that

$$\limsup_n \textstyle\sum_j \theta_{nj}(N') < \infty, \tag{5.11}$$

$$\limsup_n \textstyle\sum_j \alpha_{nj}(N') < \infty, \tag{5.12}$$

$$\limsup_n \sup_{y \in K} \textstyle\sum_j \int_N g^2(x, y)\, d\theta_{nj} < \infty. \tag{5.13}$$

To this end we consider the distribution

$$| \mu_n |^2 = \textstyle\prod_{j=1}^{k_n} | \alpha_{nj} |^2.$$

Since $| \mu_n |^2$ is compact and no limit of $| \mu_n |^2$ has an idempotent factor, according to Lemma 5.1, $| \hat{\mu}_n(y) |^2$ is bounded away from zero uniformly for $y \in K$ and in n. Thus

$$\limsup_n \sup_{y \in K} \textstyle\sum_j (1 - | \hat{\alpha}_{nj}(y) |^2) < \infty.$$

This is the same as (5.3) with $| \alpha_{nj} |^2$ replacing θ_{nj}, and hence

$$\lim_{n \to \infty} \sup_{y \in K} [\exp (\textstyle\sum_j (| \hat{\alpha}_{nj}(y) |^2 - 1)) - | \hat{\mu}_n(y) |^2] = 0.$$

Thus the sequence $e(\sum_j | \alpha_{nj} |^2)$ is compact. We now appeal to Theorem 4.3. Then

$$\limsup_n \textstyle\sum_j | \alpha_{nj} |^2(N') < \infty, \tag{5.14}$$

$$\limsup_n \textstyle\sum_j \int (1 - R(x, y))\, d\, | \alpha_{nj} |^2 < \infty. \tag{5.15}$$

We now choose a neighborhood V of the identity such that $V + V \subset N$. Then

$$\begin{aligned}
\textstyle\sum_j \alpha_{nj}(N') &\leqq \textstyle\sum_j \alpha_{nj}((V + V)') \\
&\leqq \textstyle\sum_j \inf_{x \in V} \alpha_{nj}((V + x)') = \textstyle\sum_j \inf_{x \in V} \alpha_{nj}(V' + x) \\
&\leqq \textstyle\sum_j [\alpha_{nj}(V)]^{-1} \int_V \alpha_{nj}(V' + x)\, d\alpha_{nj} \\
&\leqq \textstyle\sum_j [\alpha_{nj}(V)]^{-1} \int \alpha_{nj}(V' + x)\, d\alpha_{nj} \\
&\leqq (\sup_j [\alpha_{nj}(V)]^{-1}) \textstyle\sum_j | \alpha_{nj} |^2(V').
\end{aligned}$$

Since $\{\alpha_{nj}\}$ is uniformly infinitesimal, $\sup_j [\alpha_{nj}(V)]^{-1} < 1 + \varepsilon$, for any given $\varepsilon > 0$ and all sufficiently large n depending on ε. The above inequality and the validity of (5.14) for any neighborhood N of the identity imply (5.12). Since $| \alpha_{nj} |^2 = | \theta_{nj} |^2$ and $\{\theta_{nj}\}$ is uniformly infinitesimal, the same argument leads to (5.11).

From (5.15) we have, for any neighborhood V of the identity in X,

$$(5.16) \qquad \limsup_n \sum_j \iint_{V \times V} [1 - R(x_1 - x_2, y)]\, d\theta_{nj}(x_1)\, d\theta_{nj}(x_2) < \infty.$$

We now choose V such that $V - V \subset N$. Then

$$R(x_1 - x_2, y) = \cos g(x_1 - x_2, y).$$

Since $1 - \cos \theta > \theta^2/4$ for sufficiently small θ, we have from property (5) of $g(x, y)$ in Lemma 3.3,

$$1 - R(x_1 - x_2, y) \geqq \tfrac{1}{4} g^2(x_1 - x_2, y)$$

for $y \in K$. Since $e^{ig(x,y)} = (x, y)$ for $x \in N$ and $y \in K$, the same property of $g(x, y)$ gives

$$g(x_1 - x_2, y) = g(x_1, y) - g(x_2, y), \qquad x_1, x_2 \in V, \quad y \in K.$$

Thus, for $x_1, x_2 \in V$, $y \in K$,

$$(5.17) \quad 1 - R(x_1 - x_2, y) \geqq \tfrac{1}{4}[g^2(x_1, y) + g^2(x_2, y) - 2g(x_1, y)g(x_2, y)].$$

(5.16) and (5.17) imply

$$(5.18) \qquad \limsup_n \sup_{y \in K} \left\{ \sum_j \int_V g^2(x, y)\, d\theta_{nj}(x) - \left(\int_V g(x, y)\, d\theta_{nj} \right)^2 \right\} < \infty.$$

(5.10), (5.12), and (5.18) imply (5.13). This completes the proof of the theorem.

THEOREM 5.2. *If $\{\alpha_{nj}\}$ is uniformly infinitesimal, $\mu_n = \prod_j \alpha_{nj}$, and $\mu_n \Rightarrow \mu$, then μ is infinitely divisible.*

Proof. If μ has no idempotent factor, then it is also a limit of the sequence λ_n where λ_n is constructed as in Theorem 5.1. Since λ_n is infinitely divisible for each n, μ is also infinitely divisible.

Now let us consider the general case. By Lemma 5.1 the set $[y : \hat{\mu}(y) \neq 0]$ is an open subgroup H. If G is the annihilator of this subgroup in X, then the normalized Haar measure of the compact group G is a factor of μ. Let τ be the canonical homomorphism from X to X/G. Then the sequence $\{\alpha_{nj}\, \tau^{-1}\}$ is uniformly infinitesimal in X/G, and $\mu_n \tau^{-1} \Rightarrow \mu\tau^{-1}$. $\mu\tau^{-1}$ has no idempotent factor and hence is infinitely divisible. For $y \in H$, $\mu(y) = \hat{\mu\tau^{-1}}(y)$ and $\hat{\mu}(y) = 0$ for $y \notin H$. Thus μ itself is infinitely divisible.

Remark. In the statement of Theorem 5.2 we have assumed that $\{\alpha_{nj}\}$ is uniformly infinitesimal. However, it is enough to assume the existence of a sequence $\{x_{nj}\}$ of elements from the group with the property that $\{\alpha_{nj} * x_{nj}\}$ is uniformly infinitesimal. This is equivalent to the statement that any limit of shifts of α_{nj} is degenerate.

6. Gaussian distributions

In this section we give an algebraic definition of a Gaussian distribution and obtain its representation. This definition is also consistent with the classical definition of Gaussian laws in the case of the finite-dimensional vector spaces. Another definition of a Gaussian law is due to Urbanik [14] who also discussed some of its properties.

From the point of view of limit theorems, the Gaussian laws arise very naturally as follows. Suppose F_n is a sequence of finite measures on the group X such that (1) outside each neighborhood of the identity $F_n \to 0$ as $n \to \infty$, and (2) $e(F_n)$ converges to a limit after a suitable shift. If the total mass of F_n is not uniformly bounded, $e(F_n)$ may actually converge to a nondegenerate distribution. These are precisely the Gaussian laws.

DEFINITION 6.1. A distribution μ is said to be *Gaussian* if it has the following properties: (i) μ is infinitely divisible, and (ii) if $\mu = e(F) * \alpha$ where α is infinitely divisible, then F is degenerate at the identity.

THEOREM 6.1. *A function on Y is the characteristic function of a Gaussian distribution on X if and only if it has the form*

$$(x, y) \exp [-\phi(y)]$$

where x is a fixed point of X and $\phi(y)$ is a continuous, nonnegative function on Y satisfying the equality

$$\phi(y_1 + y_2) + \phi(y_1 - y_2) = 2[\phi(y_1) + \phi(y_2)] \tag{6.1}$$

for all y_1, y_2 in Y.

Proof. Let μ be Gaussian. Then μ cannot have a nondegenerate idempotent factor. For, otherwise, the Haar measure of some compact subgroup will be a factor of μ, and hence if F is any measure concentrated in that subgroup, then $\mu = e(F) * \mu$. This contradicts property (ii) of Definition 6.1. From the definition of infinite divisibility it follows that, for each n, there exist a distribution α_n and an element g_n of X such that $\mu = \alpha_n^n * g_n$.

Since μ has no idempotent factors, any limit of shifts of α_n is degenerate, and hence α_n's can be shifted so as to converge to the distribution degenerate at the identity. All these shifts may be absorbed into g_n, so that α_n itself can be assumed to converge to the distribution degenerate at the identity. We now write (as in the proof of Theorem 5.1)

$$\theta_n = \alpha_n * x_n, \qquad \beta_n = e(\theta_n), \qquad \lambda_n = \beta_n^n * (-nx_n) * g_n,$$

where the element x_n is determined by the identity

$$(x_n, y) = \exp\left[-i\int g(x,y)\,d\alpha_n\right].$$

The absence of idempotent factors for μ implies

$$\lim\sup_n \sup_{y\epsilon K} |\hat{\lambda}_n(y) - \hat{\mu}(y)| = 0$$

by Theorem 5.1.

Thus

$$|\hat{\mu}(y)| = \lim_{n\to\infty} \exp\left(n\int [R(x,y) - 1]\,d\theta_n\right).$$

We shall first show that the function

$$\phi(y) = \lim_{n\to\infty}\left(n\int [1 - R(x, y)]\,d\theta_n\right)$$

satisfies (6.1). We write $P_n = n\theta_n$. Then $e(P_n)$ is a shift of λ_n, and hence $e(P_n)$ is shift-compact. Now Theorem 4.3 implies that P_n restricted to N' is tight for every neighborhood N of the identity. But any limit P of P_n restricted to N' will be such that $\mu = e(P) * \alpha$ where α is also infinitely divisible. From condition (ii) of Definition 6.1 it follows that the mass of P_n outside every neighborhood of the identity tends to zero. Thus

$$\phi(y) = \lim_{n\to\infty}\int_N [1 - R(x,y)]\,dP_n \tag{6.2}$$

for every neighborhood N of the identity. (6.2) and the following identity[1]

$$\lim_{x\to e}\frac{[1 - R(x, y_1 + y_2)] + [1 - R(x, y_1 - y_2)]}{2[1 - R(x, y_1)] + 2[1 - R(x, y_2)]} = 1$$

(e denoting the identity element of the group X) imply

$$\phi(y_1 + y_2) + \phi(y_1 - y_2) = 2[\phi(y_1) + \phi(y_2)].$$

Thus, in order to complete the proof of the theorem it suffices to show that $\hat{\mu}(y)/|\hat{\mu}(y)|$ is a character on Y. Let us denote this by $\chi(y)$. It is not difficult to verify that, for every neighborhood N of the identity,

$$\begin{aligned}\chi(y_1 + y_2)[\chi(y_1)\chi(y_2)]^{-1} \\ = \exp\left(\lim_{n\to\infty}\int [I(x, y_1 + y_2) - I(x, y_1) - I(x, y_2)]\,dP_n\right)\end{aligned} \tag{6.3}$$

where $I(x, y)$ denotes the imaginary part of (x, y). For any given $\varepsilon > 0$, we

[1] Since both the numerator and denominator can vanish, this limiting relation is to be interpreted as follows: the numerator lies between $(1 - \varepsilon)$ and $(1 + \varepsilon)$ times the denominator, if x is near enough to e.

choose a neighborhood N of the identity such that

$$| I(x, y_1)| < \varepsilon, \qquad | I(x, y_2)| < \varepsilon$$

for $x \in N$. Since

$$| I(x, y_1 + y_2) - I(x, y_1) - I(x, y_2)| \leqq | I(x, y_1)| \, | 1 - R(x, y_2)| + | I(x, y_2)| \, | 1 - R(x, y_1)|$$

and

$$\limsup \int [1 - R(x, y)] \, dP_n < \infty$$

(by Theorem 4.3), we have

$$\left| \int_N [I(x, y_1 + y_2) - I(x, y_1) - I(x, y_2)] \, dP_n \right| < C \cdot \varepsilon$$

where C is a constant depending only on y_1 and y_2. Since ε is arbitrary, the right side of (6.3) is equal to unity. The continuity of $\chi(y)$ is obvious. This shows that χ is a character on Y. Thus there exists an element $x \in X$ such that

$$\hat{\mu}(y) = (x, y) \exp(-\phi(y)).$$

This proves the necessity part.

Conversely, let $\hat{\mu}(y) = (x, y) \exp(-\phi(y))$ where $\phi(y)$ is a nonnegative continuous function of y satisfying (6.1). Let $y_1, \cdots, y_k$ be some k characters. Then it is easily verified that $\exp[-\phi(n_1 y_1 + \cdots + n_k y_k)]$, considered as a function of integers $n_1, \cdots, n_k$, is positive-definite in the product of integer group taken k times. This implies the positive-definiteness of $\exp(-\phi(y))$, and hence this function is the characteristic function of a measure. Since $\hat{\mu}(y) = 1$ at the identity of Y, the measure is a distribution. The infinite divisibility of μ is obvious. We shall now prove property (ii) of Definition 6.1. Let, if possible, $\hat{\mu}(y) = \hat{\mu}_1(y)\hat{\mu}_2(y)$, where $\mu_1 = e(F)$ and μ_2 is infinitely divisible. Since $\hat{\mu}(y)$ does not vanish for any y, $\hat{\mu}_2(y)$ also does not vanish, and hence by Theorem 5.1, μ_2 is a limit of distributions of the type $e(H)$. From (2.1) it is clear that for any finite measure H

$$\begin{aligned} -\log | (e(H))^\wedge(y_1 + y_2)| - \log | (e(H))^\wedge(y_1 - y_2)| \\ \leqq 2[-\log |(e(H))^\wedge(y_1)| - \log |(e(H))^\wedge(y_2)|]. \end{aligned} \tag{6.4}$$

Thus (6.4) is also valid when $e(H)$ is replaced by either μ_1 or μ_2. Substituting μ_1 and μ_2 for $e(H)$ in (6.4) and adding, we get $\phi(y_1 + y_2) + \phi(y_1 - y_2) \leqq 2[\phi(y_1) + \phi(y_2)]$. Since equality holds good in this case, we must have

$$\int ([1 - R(x, y_1 + y_2)] + [1 - R(x, y_1 - y_2)]) \, dF = 2 \int ([1 - R(x, y_1)] + [1 - R(x, y_2)]) \, dF$$

for each $y_1, y_2 \in Y$, i.e.,

$$\int [1 - R(x, y_1)][1 - R(x, y_2)]\, dF = 0.$$

Since F is a measure, this implies that F must be degenerate at the origin. This completes the proof.

Remark. 1. Consider real-valued continuous functions $\psi(y_1, y_2)$ defined for $y_1, y_2 \in Y$ possessing the following properties:

(i) $\psi(y_1, y_2) = \psi(y_2, y_1)$,
(ii) $\psi(y_1 + y_2, y_3) = \psi(y_1, y_3) + \psi(y_2, y_3)$,
(iii) $\psi(y, y) \geqq 0$.

Clearly for any such function ψ, $\phi(y) = \psi(y, y)$ satisfies the identity (6.1). Conversely, any nonnegative continuous function satisfying (6.1) can be obtained in this way. In fact, $\psi(y_1, y_2)$ can be recovered from $\phi(y)$ by the relation

$$\psi(y_1, y_2) = \tfrac{1}{2}[\phi(y_1 + y_2) - \phi(y_1) - \phi(y_2)].$$

Remark 2. If X_0 is the component of the identity in X, then its annihilator Y_0 is the smallest closed subgroup containing all compact subgroups of Y. Consequently any $\psi(y_1, y_2)$ with the properties stated above vanishes identically in Y_0, i.e., $\psi(y_1, y_2) = 0$ if y_1 or $y_2 \in Y_0$. Thus if μ is a symmetric Gaussian distribution on X, then $\hat{\mu}(y) = 1$ for $y \in Y_0$. In other words μ is necessarily concentrated in X_0. Theorem 6.1 and Remark 1 above also show that every connected locally compact group has nontrivial Gaussian measures defined on it.

7. Representation of infinitely divisible distributions

As we have mentioned in the introduction, the characteristic function of any infinitely divisible distribution on the real line possesses the famous Lévy-Khinchin representation [3]. The purpose of this section is to obtain such a canonical representation in the case of a general locally compact abelian group.

DEFINITION 7.1. An infinitely divisible distribution λ is said to be a *proper factor* of another infinitely divisible distribution μ, if $\mu = \lambda * \alpha$ and α is infinitely divisible.

LEMMA 7.1. *The set of proper factors of an infinitely divisible distribution is closed.*

LEMMA 7.2. *If $e(F_n)$ converges to the distribution degenerate at the identity, then $F_n(N') \to 0$ as $n \to \infty$ for every neighborhood N of the identity.*

The proof of both the lemmas is quite elementary and is left to the reader.

THEOREM 7.1. *If μ is an infinitely divisible distribution without idempotent factors, then $\hat{\mu}(y)$ has a representation*

$$\hat{\mu}(y) = (x_0, y) \exp\left[\int [(x, y) - 1 - ig(x, y)]\, dF(x) - \phi(y)\right] \tag{7.0}$$

where x_0 is a fixed element of X, $g(x, y)$ is a function on $X \times Y$ which is independent of μ and has the properties mentioned in Lemma 3.3, F is a σ-finite measure with finite mass outside every neighborhood of the identity in X which satisfies

$$\int [1 - R\,(x, y)]\, dF < \infty \qquad \textit{for every } y,$$

and $\phi(y)$ is a nonnegative continuous function satisfying

$$\phi(y_1 + y_2) + \phi(y_1 - y_2) = 2[\phi(y_1) + \phi(y_2)]$$

for each y_1, $y_2 \in Y$. Conversely, any function of the type (7.0) is the characteristic function of an infinitely divisible distribution.

Proof. Let μ be any infinitely divisible distribution without an idempotent factor. Choose and fix a sequence $\{N_k\}$ of neighborhoods of the identity in X descending to the identity. Let μ_1 be that proper factor of μ which is of the type $e(F)$ and for which $F(N_1) = 0$ and $F(N_1') = 0$ and $F(N_1')$ is maximum. Such an $e(F)$ exists because of Theorems 2.1 and 4.3 and Lemma 7.1. Let the F at which the maximum is attained be F_1, and let $\mu = \mu_1 * \lambda_1$ and $\mu_1 = e(F_1)$. Since λ_1 is infinitely divisible and without idempotent factors, the same argument can be applied to λ_1 and the neighborhood N_2. Thus there exists a measure F_2 for which $F_2(N_2) = 0$, $F_2(N_2')$ is a maximum, $\mu_2 = e(F_2)$, $\lambda_1 = \mu_2 * \lambda_2$, and λ_2 is infinitely divisible. Repeating this procedure we can write

$$\mu = \mu_1 * \mu_2 * \cdots * \mu_n * \lambda_n, \tag{7.1}$$

$$\lambda_{n-1} = \mu_n * \lambda_n, \tag{7.2}$$

$$\mu_n = e(F_n), \qquad F_n(N_n) = 0. \tag{7.3}$$

λ_n is infinitely divisible, and $F_n(N_n')$ is a maximum in the sense explained earlier. Thus by Theorem 2.1 there exist shifts of $\mu_1 * \cdots * \mu_n$ and λ_n converging to ν and λ respectively, and $\mu = \nu * \lambda$. We now assert that λ cannot have a proper factor of the type $e(F)$. Suppose, on the contrary, $e(F)$ is a proper factor of λ. Then it will have a positive mass outside some N_k. Further, since the sequence λ_n is descending (in the order $\prec$), $e(F)$ is a proper factor of λ_k. $\lambda_k = e(F) * \theta$ where θ is infinitely divisible. If F' is the restriction of F to N_k', then (7.2) and (7.3) imply that

$$\lambda_{k-1} = e(F_k + F') * e(F - F') * \theta.$$

This is a contradiction since the total mass of $F_k + F'$ exceeds that of F_k. Thus λ has no proper factor of the type $e(F)$ and is therefore a Gaussian distribution. An application of Theorem 6.1 leads to the existence of a function $\phi(y)$ and an element $x \in X$ for which

$$\hat{\lambda}(y) = (x, y) \exp[-\phi(y)] \tag{7.4}$$

and

$$\phi(y_1 + y_2) + \phi(y_1 - y_2) = 2[\phi(y_1) + \phi(y_2)], \qquad y_1, y_2 \in Y.$$

Now we write $H_n = F_1 + F_2 + \cdots + F_n$. From the construction of the distributions $\nu_n = \mu_1 * \cdots * \mu_n$, it is clear that

$$\hat{\nu}_n(y) = (x_n, y) \exp\left[\int [(x, y) - 1]\, dH_n(x)\right]$$

for some element $x_n \in X$. Since $\exp[i\int g(x, y)\, dH_n(x)]$ is a character on Y, it can be considered as an element of X. Thus there exists an element $z_n \in X$ such that

$$\hat{\nu}_n(y) = (z_n, y) \exp\left[\int [(x, y) - 1 - ig(x, y)]\, dH_n(x)\right]. \tag{7.5}$$

Since $e(H_n)$ is a factor of μ and H_n increases as $n \to \infty$, it follows from the shift-compactness of $e(H_n)$, Theorem 4.3, and Lemma 2.1 that H_n increases to a σ-finite measure H for which $H(N') < \infty$ for every neighborhood N of the identity and

$$\int \sup_{y \in K} [1 - R(x, y)]\, dH < \infty,$$

for every compact $K \subset Y$. Since $[(x, y) - 1 - ig(x, y)]$ is bounded uniformly in $y \in K$, by property (2) of $g(x, y)$ in Lemma 3.3, we have, for every neighborhood N of the identity,

$$\lim_{n\to\infty} \int_{N'} [(x, y) - 1 - ig(x,y)]\, dH_n = \int_{N'} [(x, y) - 1 - ig(x, y)]\, dH,$$

uniformly in $y \in K$. When N is sufficiently small, we have, by the properties (4) and (5) of $g(x, y)$ in Lemma 3.3,

$$(x, y) = e^{ig(x,y)} \qquad \text{for} \quad x \in N, \quad y \in K,$$

$$g^2(x, y) \leqq C_1[1 - R(x, y)], \qquad x \in N, \quad y \in K,$$

where C_1 is a constant depending on K only. Thus

$$|(x, y) - 1 - ig(x, y)| \leqq C_2[1 - R(x, y)], \quad x \in N, \quad y \in K, \tag{7.6}$$

where C_2 is a constant depending on K only. Thus

$$\int_N \sup_{y \in K} |(x, y) - 1 - ig(x, y)|\, dH_n \leqq C_2 \int_N \sup_{y\in K} [1 - R(x, y)]\, dH.$$

The above inequality implies the convergence of

$$\int [(x, y) - 1 - ig(x, y)] \, dH_n \quad \text{to} \quad \int [(x, y) - 1 - ig(x, y)] \, dH$$

uniformly in $y \in K$. Now (7.5) implies that $\hat{\nu}_n(y)$, after a suitable shift, converges uniformly over compact sets to $\exp \int [(x, y) - 1 - ig(x, y)] \, dH$. This completes the proof of the first part since ν_n converges to ν after a suitable shift, $\mu = \nu * \lambda$, and λ satisfies (7.4).

To prove the converse, we first observe that if F is a totally finite measure, then $\exp[i \int g(x, y) \, dF]$ is a character, and hence $\hat{\mu}(y)$ given by (7.0) is the characteristic function of an infinitely divisible distribution. In the general case we consider a sequence F_n of totally finite measures increasing to F. If K is any compact subset of Y, then, according to Lemma 2.1, $1 - R(x, y)$ is uniformly integrable with respect to F for $y \in K$. (7.6) implies the uniform integrability of $[(x, y) - 1 - ig(x, y)]$ for $y \in K$. This shows that the function

$$(x_0, y) \exp\left(\int [(x, y) - 1 - ig(x, y)] \, dF_n - \phi(y)\right)$$

converges uniformly over compact sets to $\hat{\mu}(y)$. Thus, by Theorem 5.2' $\hat{\mu}(y)$ is infinitely divisible.

Remark. If the group X is totally disconnected, then the representation takes a simpler form. For in such groups $\phi(y) = 0$ (see Remark 2 following Theorem 6.1), and $g(x, y) = 0$ (see Example 4 at the end of Section 3). Thus every infinitely divisible distribution without idempotent factors has the representation

$$\hat{\mu}(y) = (x_0, y) \exp\left[\int [(x, y) - 1] \, dF\right]$$

where F is a σ-finite measure which has finite mass outside each neighborhood of the identity.

8. Uniqueness of the representation

In the case of the real line it is well known that the canonical representation of an infinitely divisible distribution is unique. However, this is not true in the case of a general group. We shall show that the nonuniqueness is essentially due to the presence of compact subgroups in the original group or equivalently due to the disconnectedness of the character group.

Before proceeding to the statement of the main result of this section we shall explain a few conventions and prove an elementary lemma. If μ is any infinitely divisible distribution without idempotent factors, we say that μ has the representation (x_0, F, ϕ) where x_0, F, and ϕ are as in Theorem 7.1. If F is any signed measure, we denote by F_y the measure given by

$$F_y(A) = \int_A [1 - R(x, y)] \, dF.$$

LEMMA 8.1. *Let μ be a totally finite signed measure. If $\hat{\mu}(y)$ is constant on the cosets of a closed subgroup Y_0 of Y, then μ vanishes identically on the complement of the annihilator of Y_0 in X.*

Proof. Let $Y_1 = Y/Y_0$, and let X_1 be the annihilator of Y_0 in X. $\hat{\mu}(y)$, being constant on cosets of Y_0, can be considered as a function on Y_1. Then $\hat{\mu}(y_1)$, $y_1 \in Y_1$ is the characteristic function of a signed measure on X_1. Since, for $x \in X$, (x, y) remains constant on cosets of Y_1 and $\hat{\mu}(y)$ has the same property, we can write

$$\hat{\mu}(y) = \int_{X_1} (x, y_1)\, d\lambda = \int_{X_1} (x, y)\, d\lambda$$

where y_1 denotes that coset of Y_1 to which y belongs. This shows that the signed measures μ and λ are identical, and hence μ vanishes identically on the complement of X_1.

THEOREM 8.1. *If (x_1, F_1, ϕ_1) and (x_2, F_2, ϕ_2) are two representations of the same infinitely divisible distribution without idempotent factors, then* (i) $\phi_1 = \phi_2$, *and* (ii) *the signed measure $F_1 - F_2$ vanishes identically on the complement of the annihilator of the component of identity of the character group Y.*

Proof. Writing $F = F_1 - F_2$, $\phi = \phi_1 - \phi_2$, and $x_0 = x_2 - x_1$, we have

$$\exp \int [(x, y) - 1 - ig(x, y)]\, dF = (x_0, y) \exp \phi(y), \tag{8.1}$$

$$\phi(y_1 + y_2) + \phi(y_1 - y_2) - 2[\phi(y_1) + \phi(y_2)] = 0, \quad y_1, y_2 \in Y. \tag{8.2}$$

Equating the logarithm of the absolute value on both sides of (8.1) we obtain

$$\phi(y) = \int [R(x, y) - 1]\, dF. \tag{8.3}$$

Substituting the values of the above expression at $y_1 + y_2$, $y_1 - y_2$, y_1, and y_2 in (8.2), we get

$$\int (1 - R(x, y_1))(1 - R(x, y_2))\, dF = 0. \tag{8.4}$$

(8.4) can be rewritten as

$$\int [1 - R(x, y_1)]\, dF_{y_2} = 0, \qquad \text{for} \quad y_1, y_2 \in Y.$$

Since F_{y_2} is totally finite, we have

$$\int (x, y)\, d(F_{y_2} + \bar{F}_{y_2}) = 2F_{y_2}(X). \tag{8.5}$$

Since the right side of (8.5) is constant when y_2 is fixed, we conclude that the signed measure $F_{y_2} + \bar{F}_{y_2}$ is degenerate at the identity. But the mass of F_{y_2} at the identity is zero. Thus $F_{y_2} + \bar{F}_{y_2} = 0$. In particular $F_{y_2}(X) = 0$, i.e.,

$$\int [1 - R(x, y_2)] \, dF = 0 \qquad \text{for every } y_2 . \tag{8.6}$$

(8.3) and (8.6) imply the equality of ϕ_1 and ϕ_2.

In order to prove the second part of the theorem we make use of the equality of ϕ_1 and ϕ_2 and rewrite (8.1) as

$$\exp \int [(x, y) - 1 - ig(x, y)] \, dF = (x_0, y), \qquad y \in Y. \tag{8.7}$$

Substituting $y = y_1 + y_2$, $y_1 - y_2$, and y_1 successively in (8.7) and dividing the product of the first two by the square of the third, we obtain

$$\exp \int (x, y_1)[1 - R(x, y_2)] \, dF = 1, \qquad y_1, y_2 \in Y, \tag{8.8}$$

or equivalently

$$\int (x, y_1)[1 - R(x, y_2)] \, dF = 2\pi n(y_1, y_2)$$

where $n(y_1, y_2)$ is an integer-valued continuous function of y_1 and y_2. We fix y_2 for the present. Then $n(y_1, y_2)$ remains constant on every connected subset of Y and, in particular, on the cosets of the component of identity in Y. This implies, by Lemma 8.1, that the signed measure F_{y_2} vanishes identically on the complement of the annihilator in X of the component of the identity of Y. Since this is true for each y_2, it follows that F itself vanishes identically outside this annihilator. This completes the proof of the theorem.

Remark 1. It is not difficult to show that the annihilator of the component of the identity of Y is the smallest closed subgroup containing all compact subgroups of X. This reflects the role of compact subgroups in making the representation nonunique. In particular, if the group X has no compact subgroups, then the representation is unique.

Remark 2. It was shown in the course of the proof of Theorem 8.1 that the measure F_y is antisymmetric for each character y, i.e., $F_y(A) = -F_y(-A)$ for every Borel set. But if every element in the group were of order two, then such a measure would be identically zero. Coupling this with Remark 1 we can say that if the group X is such that every compact subgroup of X consists only of elements of order two, then the representation is unique.

Remark 3. Conversely, if X is a compact group such that not all elements of X are of order two, then the representation is not unique as can be seen from the following example. We take an element y_0 in the character group which is not of order two and consider the function

$$f(x) = 2\pi i[\overline{(x, y_0)} - (x, y_0)].$$

$f(x)$ is real and not identically zero. If h denotes the normalized Haar measure of X, we have

$$\begin{aligned}\int (x, y)f(x)\, dh(x) &= 2\pi i \qquad \text{if} \quad y = y_0\,,\\ &= -2\pi i \qquad \text{if} \quad y = -y_0\,,\\ &= 0 \qquad \text{otherwise.}\end{aligned}$$

Writing f^+ and f^- to denote the positive and negative parts of $f(x)$, we define the two measures

$$F_1(A) = \int_A f^+(x)\, dh, \qquad F_2(A) = \int_A f^-(x)\, dh.$$

Then $F_1 \neq F_2$, but

$$\exp \int [(x, y) - 1 - ig(x, y)]\, dF_1 = (x_0\,, y) \exp \int [(x,y) - 1 - ig(x, y)]\, dF_2$$

where $(x_0\,, y) = \exp [i \int g(x, y)\, d(F_2 - F_1)]$ is a character on Y and hence an element of X. Thus $(e, F_1\,, 0)$ and $(x_0\,, F_2\,, 0)$ are two representations of the same infinitely divisible distribution.

9. Compactness criteria

When the group is the real line, necessary and sufficient conditions that a sequence of infinitely divisible distributions may converge to a given infinitely divisible distribution have been obtained in terms of their representations (cf. Gnedenko and Kolmogorov [3]). Such a result fails to be valid in the general case because of the nonuniqueness of the representations. However, it is possible to obtain conditions for the compactness of a family of infinitely divisible distributions in terms of their representations.

Before proceeding to state the main result of this section we shall investigate what happens to the representation when we pass over from a group to its quotient group. Let $G \subset X$ be some closed subgroup of X, and $X' = X/G$ the quotient group. Let τ denote the canonical homomorphism from X to X'. If Y' is the character group of X', we choose and fix a function $g'(x', y')$ defined on $X' \times Y'$ and satisfying all the properties of Lemma 3.3. We observe that Y' is the annihilator of G in Y and hence a subgroup of Y. Any infinitely divisible distribution μ' on X' without idempotent factors has a representation (x', F', ϕ') (with g replaced by g') according to Theorem 7.1.

LEMMA 9.1. *Let μ be an infinitely divisible distribution on X with a representation (x, F, ϕ). If $\mu' = \mu\tau^{-1}$, $x' = \tau x$, $F' = F\tau^{-1}$, and ϕ' is the restriction of ϕ to Y', then μ' is an infinitely divisible distribution on X' and is a shift of the distribution represented by (x', F', ϕ').*

Proof. Let $\mu = \mu_1 * \mu_2$ where μ_2 is the unique Gaussian component of μ. Clearly $\mu' = \mu\tau^{-1} = \mu_1\tau^{-1} * \mu_2\tau^{-1}$. Since $e(\alpha)\tau^{-1} = e(\alpha\tau^{-1})$ for every finite measure α, it follows easily that $\mu_1\tau^{-1}$ and the distribution represented by $(x', F', 0)$ are shifts of each other. Thus $\mu_2\tau^{-1}$ is the Gaussian component of μ'. This completes the proof.

We shall now prove the following.

THEOREM 9.1. *Let $\{\mu_\alpha\}$ be a family of infinitely divisible distributions without idempotent factors and with representations $\{(x_\alpha, F_\alpha, \phi_\alpha)\}$. Necessary and sufficient conditions that $\{\mu_\alpha\}$ be shift-compact and any limit of shifts of $\{\mu_\alpha\}$ be devoid of idempotent factors, are*

(1) *The family $\{F_\alpha\}$ of measures is compact when restricted to N' for every neighborhood N of the identity.*

$$(2) \quad \sup_\alpha\left[\int [1 - R(x, y)]\, dF_\alpha + \phi_\alpha(y)\right] < \infty \quad \textit{for all } y.$$

Proof. The necessity of the above two conditions is obvious in view of Theorem 4.3. Regarding sufficiency we first observe that if $\{\mu_\alpha\}$ is a shift-compact family, then condition (2) is sufficient to ensure the absence of idempotent factors in any limit of shifts of μ_α. We shall now prove shift-compactness.

We have only to prove the compactness of the family $\{|\mu_\alpha|^2\}$. We now observe that if (x, F, ϕ) is a representation of an infinitely divisible distribution μ in a group X and τ is a continuous homomorphism of X onto another group X', then $\mu\tau^{-1}$ is a shift of the distribution represented by $(\tau x, F\tau^{-1}, \phi')$ where ϕ' is the restriction of ϕ to the character group of X' (which is a subgroup of Y). Further if $\{(x_\alpha, F_\alpha, \phi_\alpha)\}$ satisfies conditions (1) and (2), so does the family $\{(\tau x_\alpha, F_\alpha\tau^{-1}, \phi'_\alpha)\}$. Making use of these remarks we shall reduce the proof of the general case to that of certain simple groups.

In order that a family of measures be compact, it is necessrry and sufficient that the family be tight. If C is a compact subgroup of X, and if the family of measures induced by the canonical homomorphism on X/C is tight, then the original family itself is tight. We now choose the group C in such a manner that X/C has the structure $V \oplus D \oplus K^r$ where V is a vector group, D a discrete group, and K^r the r-dimensional torus. The existence of such a compact subgroup is well known. But a family of measures in the product of two topological spaces is tight as soon as the two marginal families are so. Thus it is enough to prove the sufficiency of (1) and (2) in the case of the real line, discrete groups, and compact groups. In the case of the real line the boundedness of $\int [1 - R(x, y)]\, dF_\alpha$ implies the boundedness of $\int_{|x|<\varepsilon} x^2\, dF_\alpha$ for a suitable ε, which together with condition (1) implies the equicontinuity of the family of functions $\exp\{-\int (1 - R(x, y))\, dF_\alpha\}$. Since $\phi_\alpha(y)$ assumes the form $\sigma_\alpha^2 y^2$ and hence σ_α^2 is bounded, it is clear that $\{|\mu_\alpha(y)|^2\}$ is equicontinuous. But equicontinuity implies compactness. In the case of a discrete group, identity itself is an open set, and hence the family $\{F_\alpha\}$ is compact outside the identity. This, together with the fact that every infinitely divisible distribution without idempotent factors is a shift of $e(F)$ for some F with zero mass at the identity, implies the required result. In the case of compact groups any family of measures is compact. This completes the proof of the theorem.

COROLLARY 9.1. *In addition to the conditions* (1) *and* (2) *of Theorem* 9.1 *the condition that* $\{x_\alpha\}$ *be a conditionally compact set is necessary and sufficient to ensure the conditional compactness of* $\{\mu_\alpha\}$ *with representations*

$$(x_\alpha, F_\alpha, \phi_\alpha).$$

Proof. From (7.6) we have

$$|(x, y) - 1 - ig(x, y)| \leqq C[1 - R(x, y)]$$

for all $x \in N$, $y \in K$ where C is a constant depending on K only, N is a sufficiently small neighborhood of the identity in X, and K is a compact subset of Y. This implies the equicontinuity of the family of functions

$$\exp\left[\int [(x, y) - 1 - ig(x, y)]\, dF_\alpha - \phi_\alpha(y)\right].$$

Hence $\{x_\alpha\}$ is conditionally compact if and only if $\{\mu_\alpha\}$ is so.

For any σ-finite measure F which has finite mass outside every neighborhood of the identity and integrates the function $1 - R(x, y)$ for every y, we write

$$(E(F))^\wedge(y) = \exp \int [(x, y) - 1 - ig(x, y)]\, dF. \tag{9.1}$$

Theorem 7.1 implies that $(E(F))^\wedge(y)$ is the characteristic function of an infinitely divisible distribution. By proceeding along the same lines as in the proof of Theorem 9.1 it is possible to prove the following.

THEOREM 9.2. *Let* $\{\mu_n\}$ *be a sequence of infinitely divisible distributions with representations* (x_n, F_n, ϕ_n). *Let* μ_n *converge to* μ *after a suitable shift, and* F_n *to* F *outside each neighborhood of the identity. Then* μ *has a representation* (x, F, ϕ) *for a suitable choice of* x *and* ϕ.

10. Representation of convolution semigroups

We have observed earlier that the representation of an infinitely divisible distribution is not unique. We shall now consider the representation problem for a one-parameter convolution semigroup of distributions. By such a semigroup we mean a family $\{\mu_t\}$ of distributions indexed by $t \geqq 0$ such that $\mu_t * \mu_s = \mu_{t+s}$. We shall further assume that μ_t converges weakly to the distribution degenerate at the identity when $t \to 0$. Obviously, for such semigroups $\hat{\mu}_t(y) \neq 0$ for any $t > 0$ and $y \in Y$. That such a semigroup has a unique canonical representation is the content of the following.

THEOREM 10.1. *Let $\{\mu_t\}$ be a one-parameter convolution semigroup of distributions such that μ_t converges weakly to the distribution degenerate at the identity as $t \to 0$. Then $\hat{\mu}_t(y)$ has the canonical representation*

$$\hat{\mu}_t(y) = (x_t, y) \exp\left(t \int [(x, y) - 1 - ig(x, y)]\, dF - t\phi(y)\right)$$

where F and ϕ are as in Theorem 7.1 and $\{x_t\}$ is a continuous one-parameter semigroup in X. Moreover, $\{x_t\}$, F, and ϕ are uniquely determined by $\{\mu_t\}$.

Proof. Since $\mu_t = (\mu_{t/n})^n$, μ_t is infinitely divisible. As remarked at the beginning, $\hat{\mu}_t(y)$ is nonvanishing at any point and hence has no idempotent factor. Thus by Theorem 7.1, μ_t has a representation (z_t, F_t, ϕ_t). The uniqueness of ϕ_t implies that $\phi_t = t\phi_1$. We write $\phi = \phi_1$ and $\hat{\lambda}_t(y) = \hat{\mu}_t(y)e^{t\phi}$. Then $\{\lambda_t\}$ is a weakly continuous convolution semigroup, and λ_t has neither idempotent nor Gaussian factors. For any σ-finite measure F which has finite mass outside every neighborhood of the identity and integrates the function $1 - R(x, y)$, we define the distribution $E(F)$ as in (9.1). We now observe that the distribution $E(n!F_{1/n!})$ is a shift of λ_1 for every n. By Theorem 4.3 the sequence of measures $n!F_{1/n!}$ is compact outside every neighborhood of the identity. Thus we can choose a subsequence $n_k!F_{1/n_k!}$ which converges to a measure F outside every neighborhood of the identity. Thus by Theorem 9.2, λ_1 has a representation $(z, F, 0)$ for some $z \in X$. We now define

$$\hat{P}_t(y) = \exp\left(t \int [(x, y) - 1 - ig(x, y)]\, dF\right).$$

If $t = p/q$ is rational, then $E((p/q)n!\, F_{1/n!})$ is a shift of $\lambda_{p/q}$ for all sufficiently large n. Since $(p/q)n_k!F_{1/n_k!}$ converges to $(p/q)F$ outside every neighborhood of the identity, another application of Theorem 9.2 shows that $\lambda_{p/q}$ is a shift of $P_{p/q}$. By the continuity of the semigroups it is clear that, for every t, λ_t is a shift of P_t. Thus $\hat{\lambda}_t(y)$ can be written as $(x_t, y)\hat{P}_t(y)$. Then x_t automatically becomes a continuous one-parameter semigroup in X.

If now $(x_t, tF, t\phi)$ and $(x_t', tF', t\phi)$ are two representations of the semigroup $\{\mu_t\}$, then by proceeding in the same way as in the proof Theorem 8.1 we obtain

$$\exp\left[t \int (x, y_1)[1 - R(x, y)]\, d(F - F')\right] = 1$$

for every t. But this can happen only if $F - F' = 0$. Thus the representation of the semigroup is always unique.

COROLLARY. *If μ_t is symmetric for each t, i.e., $\mu_t(A) = \mu_t(-A)$ for each Borel set A, then the representation stated in Theorem* 10.1 *takes the simpler form*

$$\hat{\mu}_t(y) = \exp\left[t \int [(x, y) - 1]\, dF - t\phi(y)\right]$$

where F is a symmetric measure which integrates $[1 - (x, y)]$ for each y, and $\phi(y)$ is continuous, nonnegative, and satisfies (6.1).

Remark. The above corollary could be used to give a representation for the "negative definite" functions of Herz [15, p. 198]. (The problem of obtaining such a representation for "negative definite" functions is raised by Herz on page 207 of [15].) A function $\nu(y)$ on Y is said to be "negative definite" if

(i) $\nu(y)$ is continuous, $\nu(y) = \nu(-y)$, and $\nu(y) \geqq \nu(0) = 0$, and

(ii) if $y_1, \cdots, y_n$ are arbitrary elements of Y and $c_1, c_2, \cdots, c_n$ are real numbers such that $\sum c_j = 0$, then

$$\sum \sum c_i c_j \nu(y_i - y_j) \leqq 0.$$

From the corollary above and Theorem 3.1 of [15, p. 198], it follows that $\nu(y)$ is negative definite if and only if it is of the form

$$\nu(y) = \int_X [1 - (x, y)]\, dF + \phi(y)$$

where F and ϕ have properties stated in the above corollary.[2]

11. A decomposition theorem

According to a theorem of Khinchin [5] any distribution on the real line can be written as the convolution of two distributions one of which is the convolution of a finite or countable number of indecomposable distributions, and the other of which is a distribution without indecomposable factors. Further, any distribution which is not infinitely divisible has an indecomposable factor. The object of this section is to extend this result to the general case with a slight modification to counteract the existence of idempotent factors.

THEOREM 11.1. *Let μ be any distribution on X. Then it can be written as $\lambda_G * \lambda$ where λ is a distribution without any idempotent factor and λ_G is the maximal idempotent factor of μ.*

Proof. Let H be the group generated by the set of all characters y at which $\hat{\mu}(y) \neq 0$. H is open, and hence its annihilator G in X is compact. The normalized Haar measure λ_G of G is the required maximal idempotent factor. If we denote by τ the canonical homomorphism from X to X/G, then the distribution $\mu\tau^{-1}$ in X/G has no idempotent factors. We now choose a Borel

[2] The authors wish to thank the referee for pointing out that this representation could be derived from the results of the paper.

set $A \subset X$ such that τ restricted to A maps A onto X/G in a one-to-one manner. The existence of such a Borel set follows from a result due to Mackey [7, Lemma 1.1, p. 102]. By a result of Kuratowski [6, p. 251] it follows that the inverse of τ from X/G to A is a measurable map. Hence $\mu\tau^{-1}$ induces a measure λ on A. This λ satisfies the requirements of the theorem.

We shall now introduce a function $\theta(\alpha)$ defined for all factors of a distribution μ and similar to the function introduced by Khinchin. Let μ be a distribution without any idempotent factor. Then there is a sequence $y_1, y_2, \cdots$ of characters in Y such that $\hat{\mu}(y_i) \neq 0$ for $i = 1, 2, \cdots$ and the smallest closed subgroup generated by this sequence is Y. Since $-\log | \hat{\mu}(y_i) |$ is well defined, we can choose a sequence $\varepsilon_n > 0$ such that

$$-\sum_{n=1}^{\infty} \varepsilon_n \log | \hat{\mu}(y_n) | < \infty .$$

This implies at once that, for every factor α of μ, the function

$$\theta(\alpha) = -\sum_{n=1}^{\infty} \varepsilon_n \log | \hat{\alpha}(y_n) |$$

is well defined and has the following obvious properties:

(11.1)

(i) $\theta(\alpha) \geqq 0$.

(ii) $\theta(\alpha) = 0$ if and only if α is degenerate.

(iii) $\theta(\alpha_1 * \alpha_2) = \theta(\alpha_1) + \theta(\alpha_2)$.

(iv) If $\alpha_n \Rightarrow \alpha$, then $\theta(\alpha_n) \to \theta(\alpha)$.

(v) $\theta(\alpha) = \theta(\beta)$ if α is a shift of β.

THEOREM 11.2. *Let μ be a distribution without any indecomposable or idempotent factors. Then μ is infinitely divisible.*

Proof. In view of the remark made under Theorem 5.2 it is enough to factorize μ in the form

$$\mu = \alpha_{n1} * \cdots * \alpha_{n2^n} \tag{11.2}$$

where $\{\alpha_{nj}\}$ is such that any limit of their shifts is degenerate. From the properties (i)–(v) of the θ function it is clear that it is sufficient to factorize μ in the form (11.2) with $\theta(\alpha_{nj}) = 2^{-n}\theta(\mu)$. If μ satisfies the conditions stated in the theorem, then any factor of μ also satisfies them. Thus it suffices to prove that μ can be written as $\mu_1 * \mu_2$ with $\theta(\mu_1) = \theta(\mu_2) = \frac{1}{2}\theta(\mu)$. A repetition of this argument will then complete the proof. In order to do this we first observe that

$$\inf_{\alpha \epsilon F(\mu), \theta(\alpha) \neq 0} \theta(\alpha) = 0$$

where $F(\mu)$ is the class of all factors of μ. For, otherwise, the class $F(\mu)$ being shift-compact, the infimum would be attained at an indecomposable distribution. But this is a contradiction. Thus there are factors of μ with arbitrarily small θ-values. We now take two distributions μ_1, μ_2 for which $\mu =$

$\mu_1 * \mu_2$ and $|\theta(\mu_1) - \theta(\mu_2)|$ is minimum. From the shift-compactness of $F(\mu)$ it follows that the minimum is attained. This minimum has to be zero. For, otherwise, by transferring a factor of μ_1 or μ_2 with an arbitrarily small θ-value to μ_2 or μ_1 we can make $|\theta(\mu_1) - \theta(\mu_2)|$ smaller. This completes the proof of the theorem.

THEOREM 11.3. *Any distribution μ on X can be written as $\lambda_H * \lambda_1 * \lambda_2$ where λ_H is the maximal idempotent factor of μ, λ_1 is the convolution of a finite or a countable number of indecomposable distributions, and μ_2 is an infinitely divisible distribution without indecomposable or idempotent factors.*

Proof. An application of Theorem 11.1 shows that μ can be written as $\lambda_H * \lambda$ where λ_H is the maximal idempotent factor and λ has no idempotent factors. Thus we can define a θ function on $F(\lambda)$ satisfying the properties (i)–(v) of (11.1). Let now δ_1 be the maximum of $\theta(\alpha)$ as α varies over the indecomposable factors of λ. If δ_1 is greater than zero, we write $\lambda = \alpha_1 * P_1$ where α_1 is indecomposable and $\theta(\alpha_1) \geqq \delta_1/2$. We now denote by δ_2 the maximum of $\theta(\alpha)$ as α varies over the indecomposable factors of P_1. If $\delta_2 > 0$, then we write $P_1 = \alpha_2 * P_2$ where α_2 is indecomposable and $\theta(\alpha_2) \geqq \delta_2/2$. We repeat this argument. If the process terminates at the nth stage, then $\lambda = \alpha_1 * \cdots * P_n$ and $\delta_{n+1} = 0$, which means that P_n has no indecomposable factors. Otherwise the process continues ad infinitum. Since $\sum_1^\infty \theta(\alpha_i)$ is convergent, $\theta(\alpha_n) \to 0$ as $n \to \infty$. The sequence $\alpha_1 * \cdots * \alpha_n$ being monotonic (in the order $\prec$) converges after a suitable shift. Absorbing this shift in α_n we can assume that $\alpha_1 * \cdots * \alpha_n$ converges to a distribution λ_1. Automatically P_n will converge to a distribution λ_2. If λ_2 has an indecomposable factor α, then it is a factor of P_n. Thus $\theta(\alpha) \leqq \delta_n$ for each n. But $\delta_n \leqq 2\theta(\alpha_n)$ and hence tends to zero. Therefore $\theta(\alpha) = 0$, or equivalently λ_2 has no indecomposable factors.

12. Concluding remarks

In obtaining the representation for an infinitely divisible distribution as well as in the proof of the theorem on accompanying laws (Theorems 5.1 and 7.1), we have assumed that the distributions under consideration do not have any idempotent factors. If a distribution μ has an idempotent factor, then, as in the proof of Theorem 11.1, we can construct a compact subgroup G such that the measure induced by the canonical homomorphism onto X/G has no idempotent factors. Since uniform infinitesimality and infinite divisibility are preserved by this map, these results can now be discussed in the quotient group X/G.

Another assumption we have made in the paper is that the group is separable (i.e., satisfies the second axiom of countability). All the results of Sections 3–10 are easily extended, with little essential modification in the proofs, if in place of this restriction on the group, we suppose that the measures under consideration have supports contained in a σ-compact subset.

References

1. S. Bochner, *Harmonic analysis and the theory of probability*, Berkeley, University of California Press, 1955.
2. ———, *General analytical setting for the central limit theory of probability*, Calcutta Mathematical Society Golden Jubilee Commemoration Volume (1958), Part I, pp. 111–128.
3. B. V. Gnedenko and A. N. Kolmogorov, *Limit distributions for sums of independent random variables*, trans. from the Russian, Cambridge, Mass., Addison-Wesley Pub. Co., 1954.
4. G. A. Hunt, *Semi-groups of measures on Lie groups*, Trans. Amer. Math. Soc., vol. 81 (1956), pp. 264–293.
5. A. Khinchin, *Contribution à l'arithmétique des lois de distribution*, Bull. Univ. État Moscou, Sér. Int., Sect. A: Math. et Mécan., vol. 1 (1937), pp. 6–17.
6. C. Kuratowski, *Topologie I*, 2nd ed., Warsaw, 1948.
7. G. W. Mackey, *Induced representations of locally compact groups. I*, Ann. of Math. (2), vol. 55 (1952), pp. 101–139.
8. K. R. Parthasarathy, R. Ranga Rao, and S. R. S. Varadhan, *On the category of indecomposable distributions on topological groups*, Trans. Amer. Math. Soc., vol. 102 (1962), pp. 200–217.
9. K. Takano, *Central convergence criterion in the multidimensional case*, Ann. Inst. Statist. Math., Tokyo, vol. 7 (1956), pp. 95–102.
10. A. Weil, *L'intégration dans les groupes topologiques et ses applications*, 2 éd., Paris, Hermann, 1951.
11. K. Itô, *On stochastic processes (I)*, Japanese J. Math., vol. 18 (1942), pp. 261–301.
12. P. Lévy, *L'addition des variables aléatoires définies sur une circonférence*, Bull. Soc. Math. France, vol. 67 (1939), pp. 1–41.
13. K. Urbanik, *Poisson distributions on compact Abelian topological groups*, Colloq. Math., vol. 6 (1958), pp. 13–24.
14. ———, *Gaussian measures on locally compact Abelian topological groups*, Studia Math., vol. 19 (1960), pp. 77–88.
15. C. Herz, *The spectral theory of bounded functions*, Trans. Amer. Math. Soc., vol. 94 (1960), pp. 181–232.
16. B. M. Kloss, *Limiting distributions on bicompact Abelian groups*, Teor. Veroyatnost. i Primenen., vol. 6 (1961), pp. 392–421 (in Russian with English summary).
17. Y. Kawada and K. Itô, *On the probability distribution on a compact group. I*, Proc. Phys.-Math. Soc. Japan (3), vol. 22 (1940), pp. 977–998.
18. L. Pontrjagin, *Topological groups*, trans. from the Russian, Princeton University Press, 1946.

Indian Statistical Institute
Calcutta, India

ТЕОРИЯ ВЕРОЯТНОСТЕЙ

Том IX И ЕЕ ПРИМЕНЕНИЯ Выпуск 1

1964

EXTENSION OF STATIONARY STOCHASTIC PROCESSES

K. R. PARTHASARATHY, S. R. S. VARADHAN

1. Introduction. In this paper the following problem is solved. Suppose x_t is a stochastically continuous and relatively stationary process defined in the unit interval [0, 1]. Does there exist a stationary process X_t defined on the entire real line with the property that the distributions of X_t and x_t coincide when $t \in [0, 1]$? The answer to this question is always in the affirmative. However we do not know when the extension is unique and when does there exist an ergodic extension.

As a particular case we deduce the well known result of Krein [1] that a continuous positive definite function defined on the interval [—1,1] has a positive definite extension to the entire real line.

2. A preliminary lemma on the extension of invariant linear functionals. Let X be any topological space and $C(X)$ the space of all real valued bounded continuons functions on X. $C(X)$ becomes a Banach lattice under the usual supremum norm and ordering. Let $B_0 \subset C(X)$ be a subspace of $C(X)$ containing constants. Let further T_g, $g \in G$ be a group of homeomorphisms of X into itself. Each T_g, $g \in G$ induces an automorphism U_g of the Banach lattice $C(X)$ in a natural manner. We are interested in the problem of extending a nonnegative invariant linear functional on B_0 to the entire space X. For our purpose it is enough to consider the special case when G is the group of rationals under addition. In what follows we tacitly assume that G is the group of rationals under addition. However, we may remark that the problem mentioned above is meaningful when $C(X)$ is replaced by an arbitrary Banach lattice and U_g, $g \in G$ is an arbitrary group of order-preserving automorphisms of the Banach lattice into itself.

Lemma 2.1. *Let Λ_0 be a non-negative invariant linear functional defined on B_0 and invariant under U_g for all $g \in G$. Then there exists an extension Λ of Λ_0 to the whole of $C(X)$ which is non-negative and invariant.*

Proof. Let f be any fixed element outside B_0. We shall then show that there is an extension Λ_1 of Λ_0 to the smallest closed invariant subspace B_1 containing B_0 and f. A routine zornification will then complete the proof.

To this end we define

$$\theta = \sup_{l} \sup_{\substack{a_1+a_2+\dots+a_l=1 \\ g_1, g_2, \dots, g_l \in G}} \sup_{\substack{\varphi \in B_0 \\ \varphi \leqslant a_1 U_{g_1} f + \dots + a_l U_{g_l} f}} \Lambda_0(\varphi), \tag{2.1}$$

$$\theta' = \inf_{m} \inf_{\substack{b_1+b_2+\ldots+b_m=1 \\ h_1, h_2, \ldots, h_m \in G}} \inf_{\substack{\psi \in B_0 \\ \psi \geqslant b_1 U_{h_1} f + \ldots + b_m U_{h_m} f}} \Lambda_0(\psi). \tag{2.2}$$

Suppose now that φ and ψ are any two functions in B_0, $a_1, a_2, \ldots, a_l$; $b_1, b_2, \ldots, b_m$ are real numbers and $g_1, g_2, \ldots, g_l$; $h_1, h_2, \ldots, h_m$ are elements of G (i. e. rational numbers) with the following properties:

$$\begin{aligned} &\text{(I)} && a_1 + a_2 + \ldots + a_l = 1, \\ &\text{(II)} && b_1 + b_2 + \ldots + b_m = 1, \\ &\text{(III)} && \varphi \leqslant a_1 U_{g_1} f + \ldots + a_l U_{g_l} f, \\ &\text{(IV)} && \psi \geqslant b_1 U_{h_1} f + \ldots + b_m U_{h_m} f. \end{aligned} \tag{2.3}$$

We shall now show that $\Lambda_0(\varphi) \leqslant \Lambda_0(\psi)$. First of all, we observe tha there exists a rational α such that $g_1, g_2, \ldots, g_l, h_1, \ldots, h_m$ are all integral multiples of α.

We define the operator A_n in $C(X)$ as follows:

$$A_n = \frac{U_\alpha + U_{2\alpha} + \ldots + U_{n\alpha}}{n}. \tag{2.4}$$

Obviously, for any integer k, $U_{k\alpha} = U_\alpha^k$. It is not difficult to verify that

$$\lim_{n \to \infty} \| A_n f - A_n U_g f \| = 0$$

whenever g is any one of the elements $g_1, g_2, \ldots, g_l, h_1, h_2, \ldots, h_m$. Further A_n is an order-preserving operator for every n. Thus properties (III) and (IV) of (2.3) imply

$$A_n \varphi \leqslant A_n \left[\sum_{r=1}^{l} a_r U_{g_r} f \right], \tag{2.5}$$

$$A_n \psi \geqslant A_n \left[\sum_{r=1}^{m} b_r U_{h_r} f \right]. \tag{2.6}$$

From (2.5), (2.6) and properties (I), (II) of (2.3) we get

$$A_n \varphi - A_n f \leqslant \left(\sum_{1}^{l} | a_r | \right) \sup_{1 \leqslant r \leqslant l} \| A_n f - A_n U_{g_r} f \|,$$

$$A_n \psi - A_n f \geqslant - \left(\sum_{1}^{m} | b_r | \right) \sup_{1 \leqslant r \leqslant m} \| A_n f - A_n U_{h_r} f \|.$$

Thus, subtracting both the sides of the second inequality from the corresponding sides of the first above we obtain

$$A_n \varphi - A_n \psi \leqslant \varepsilon_n$$

where ε_n is a sequence of positive numbers tending to zero as $n \to \infty$. Since $\varphi, \psi \in B_0$ we have $\Lambda_0(A_n\varphi) = \Lambda_0(\varphi)$, $\Lambda_0(A_n\psi) = \Lambda_0(\psi)$. The fact that constants belong to B_0 and Λ_0 is non-negative on B_0 implies that

$$\Lambda_0(\varphi) \leqslant \Lambda_0(\psi). \tag{2.7}$$

The definition of θ and θ' together with (2.7) implies

$$\theta \leqslant \theta'. \tag{2.8}$$

Now choose and fix a t such that

$$\theta \leqslant t \leqslant \theta'.$$

Let now $\varphi \in B_0$ and $a_1, a_2, \ldots, a_l$ be constants such that

$$\varphi + a_1 U_{g_1} f + \ldots + a_l U_{g_l} f \geqslant 0, \tag{2.9}$$

$$a_1 + a_2 + \ldots + a_l \neq 0. \tag{2.10}$$

We shall now show that

$$\Lambda_0(\varphi) + (a_1 + a_2 + \ldots + a_l)\, t \geqslant 0. \tag{2.11}$$

Without loss of generality we can assume that $a_1 + a_2 + \ldots + a_l > 0$. As before we choose an $\alpha \in G$ such that $g_1, g_2, \ldots, g_l$ are integral multiples of α and define A_n as in (2.4). From (2.9) we have

$$A_n\varphi + a_1 A_n U_{g_1} f + \ldots + a_l A_n U_{g_l} f \geqslant 0.$$

Thus

$$A_n\varphi + (a_1 + \ldots + a_l)\, A_n f \geqslant \varepsilon_n \tag{2.12}$$

where

$$\varepsilon_n = -\left(\sum_i |a_i|\right) \sup_{1 \leqslant i \leqslant l} \| A_n f - A_n U_{g_i} f \|. \tag{2.13}$$

Inequality (2.12) can be rewritten as

$$A_n f \geqslant \frac{\varepsilon_n}{a_1 + \ldots + a_l} - A_n \left(\frac{\varphi}{a_1 + \ldots + a_l} \right). \tag{2.14}$$

The function on the right side of (2.14) is an element of B_0. Thus by the definition of θ we have

$$\Lambda_0 \left(\frac{\varepsilon_n}{a_1 + \ldots + a_l} \right) - \frac{\Lambda_0(\varphi)}{a_1 + \ldots + a_l} \leqslant \theta \leqslant t.$$

Since $\varepsilon_n \to 0$ as $n \to \infty$ we have, in the limit,

$$\Lambda_0(\varphi) + t\,(a_1 + \ldots + a_l) \geqslant 0.$$

A similar argument can be given when $a_1 + \ldots + a_l < 0$. In that case we have to use the inequality $t \leqslant \theta'$. If $a_1 + a_2 + \ldots + a_l = 0$, we have by the same analysis as in the derivation of (2.12),

$$A_n \varphi \geqslant \varepsilon_n$$

where ε_n is given by (2.13). Applying Λ_0 we have

$$\Lambda_0(\varphi) = \Lambda_0(A_n \varphi) \geqslant \Lambda_0(\varepsilon_n).$$

Since $\varepsilon_n \to 0$ as $n \to \infty$ we have in the limit

$$\Lambda_0(\varphi) \geqslant 0$$

which can be rewritten as

$$\Lambda_0(\varphi) + (a_1 + \ldots + a_l)\, t \geqslant 0. \tag{2.15}$$

Thus (2.15) holds good whenever

$$\varphi + a_1 U_{g_1} f + \ldots + a_l U_{g_l} f \geqslant 0$$

for some $g_1, g_2, \ldots, g_l \in G$.

Now we define, for any $\varphi \in B_0$, $g_1, \ldots, g_l \in G$ and constants $a_1, \ldots, a_l$

$$\Lambda_1(\varphi + a_1 U_{g_1} f + \ldots + a_l U_{g_l} f) = \Lambda_0(\varphi) + (a_1 + \ldots + a_l)\, t.$$

If

$$\varphi + a_1 U_{g_1} f + \ldots + a_l U_{g_l} f = \varphi' + b_1 U_{h_1} f + \ldots + b_m U_{h_m} f$$

then it follows from the non-negativity of Λ_1 that

$$\Lambda_1(\varphi + a_1 U_{g_1} f + \ldots + a_l U_{g_l} f) = \Lambda_1(\varphi' + b_1 U_{h_1} f + \ldots + b_m U_{h_m} f).$$

Thus, Λ_1 is unambiguously defined on the smallest invariant linear manifold generated by B_0 and f. Non-negativity implies boundedness. Thus Λ_1 can be extended to the closure of this linear manifold without the destruction of the properties of invariance and non-negativity. This completes the proof of the lemma.

3. Extension of stationary stochastic processes. Suppose x_t is a stochastic process with the following properties: a) x_t is defined for $t \in [0, 1]$; b) x_t takes values in the real line; c) x_t is stochastically continuous; d) x_t is relatively stationary in $[0, 1]$, i. e., whenever $t_1, t_2, \ldots, t_n$; $t_1 + h, t_2 + h, \ldots, t_n + h$ belong to $[0, 1]$ the joint distributions of the random vectors $(x_{t_1}, x_{t_2}, \ldots, x_{t_n})$ and $(x_{t_1+h}, \ldots, x_{t_n+h})$ are same. If the above conditions are satisfied then the question arises as to whether there exists a stationary process x_t, $t \in R$ (real line) whose restriction to the unit interval is the given process. That such an extension always exists is the content of the following

Theorem 3.1. *Let x_t be a real valued, stochastically continuous and relatively stationary process defined in the interval $t \in [0, 1]$. Then there exists a stochastically continuous stationary process X_t, $t \in R$ (real line) whose restriction to the unit interval is x_t.*

Remark. x_t is assumed to be real valued only for convenience. The theorem is valid even for complex valued processes.

Proof of theorem 3.1. Let $G \subset R$ be the set of all rationals and $G_1 \subset [0, 1]$ be the set of rationals in the unit interval. We consider the process x_t, $t \in G_1$. Suppose there exists an extension X_t, $t \in G$. Then we can get the extension to the entire real line by using stationarity and stochastic continuity. Thus it is enough to get the extension X_t for $t \in G$ with the following properties: if $t_1, \ldots, t_n \in G$, $t_1 + r, \ldots, t_n + r \in G$ then the joint distributions of $(X_{t_1}, \ldots, X_{t_n})$ and $(X_{t_1+r}, \ldots, X_{t_n+r})$ are same; X_t coincides with x_t when $t \in G_1$.

Further it is enough to consider the case when x_t takes values in a bounded interval, say $[0, 1]$. For, if not, we can compactify the real line and consider the process x_t in this space. We can make use of the following fact to ultimately get the process in the real line: if (X, Y) is a random vector such that the marginal distributions of X aud Y have no mass at infinity then the joint distribution of (X, Y) also has no mass at infinity.

Thus we shall now consider a process x_t, $t \in G_1$, which is relatively stationary in G_1. We shall get a stationary extension for $t \in G$. Let $\mathfrak{X} = I^G$ be the space of all real valued functions on G taking values in the unit interval $I = [0, 1]$. $\mathfrak{X}$ is a compact topological space under the product topology. A point $z \in \mathfrak{X}$ can be written as $\{z_t, t \in G\}$ where z_t is a real number in the interval $[0, 1]$ corresponding to each t. Suppose $E \subset G$ is some subset and the value of an element $f \in C(\mathfrak{X})$ (the space of continuous functions on $\mathfrak{X}$) is completely determined by the coordinates z_t, $t \in E$. Then we shall write $f(E)$ for f.

Let C_0 be the subspace of all those bounded continuous functions which are completely determined by the coordinates z_t, $t \in G_1$. The stochastic process x_t, $t \in G_1$ can be viewed upon as a nonnegative bounded linear functional Λ_0 on the subspace C_0.

For any $t \in G$ let T_t denote the shift by t in the space $\mathfrak{X}$. If $z = \{z_t, t \in G\}$ and $y = T_{t_0} z = \{y_t, t \in G\}$ then $y_t = z_{t+t_0}$. The relative-stationarity of the given process x_t can then be described as follows: if the functions $f(z)$ and $g(z) = f(T_t z)$ belong to C_0 then $\Lambda_0(f) = \Lambda_0(g)$. It is clear that T_t induces an automorphism U_t in $C(\mathfrak{X})$ in a natural way. C_0 is not invariant under T_t. If Λ_0 is extended to the smallest closed invariant subspace generated by C_0 then an application of lemma 2.1 and the famous Riesz representation theorem will complete the proof. Thus we shall now concentrate our attention on the extension of Λ_0 to the smallest closed invariant subspace generated by C_0.

For any $f_1, f_2, \ldots, f_n \in C_0$ and $t_1, t_2, \ldots, t_n \in G$ define

$$\Lambda(U_{t_1} f_1 + U_{t_2} f_2 + \ldots + U_{t_n} f_n) = \Lambda_0(f_1) + \Lambda_0(f_2) + \ldots + \Lambda_0(f_n).$$

If we show the non-negativity of Λ then this will establish the unambiguity of the definition of Λ on the smallest invariant linear manifold generated by C_0. In order to establish non-negativity of Λ we have to only show that

$$\Lambda_0(f_1) + \Lambda_0(f_2) + \ldots + \Lambda_0(f_n) \geqslant 0 \tag{3.1}$$

whenever

$$U_{t_1}f_1 + U_{t_2}f_2 + \ldots + U_{t_n}f_n \geqslant 0 \tag{3.2}$$

for $f_1, f_2, \ldots, f_n \in C_0$, $t_1 < t_2 < \ldots < t_n \in G$. We shall do this by induction. Suppose (3.1) is true whenever (3.2) is true for n. We shall show that the same is true for $n+1$. Let

$$f = U_{t_1}g_1 + U_{t_2}g_2 + \ldots + U_{t_{n+1}}g_{n+1} \geqslant 0, \tag{3.3}$$

$$t_1 < t_2 < \ldots < t_{n+1} \in G, \tag{3.4}$$

$$g_1, g_2, \ldots, g_{n+1} \in C_0. \tag{3.5}$$

Let $E_r = G_1 + t_r$ and $h_r = U_{t_r}g_r$. Then the value of h_r is completely determined by the coordinates corresponding to E_r. Thus we can write $h_r = h_r(E_r)$. Let Δ_{n+1} be the set of all points in E_{n+1} but not in E_n. Then Δ_{n+1} is disjoint from $E_1, E_2, \ldots, E_n$. The inequality (3.3) can be written as

$$h_1(E_1) + h_2(E_2) + \ldots + h_{n+1}(E_{n+1}) \geqslant 0.$$

Thus

$$\min_{\Delta_{n+1}} \{h_1(E_1) + h_2(E_2) + \ldots + h_{n+1}(E_{n+1})\} \geqslant 0 \tag{3.6}$$

where the minimum is taken over all coordinates corresponding to Δ_{n+1}. Since Δ_{n+1} is disjoint from $E_1 \cup E_2 \cup \ldots \cup E_n$, (3.6) can be written as

$$h_1(E_1) + h_2(E_2) + \ldots + h_n(E_n) + \min_{\Delta_{n+1}} h_{n+1}(E_{n+1}) \geqslant 0. \tag{3.7}$$

Let

$$u_n(E_n) = \min_{\Delta_{n+1}} h_{n+1}(E_{n+1}), \quad v_n(E_n) = h_n(E_n) + u_n(E_n).$$

Obviously u_n and v_n are determined completely by the coordinates corresponding to E_n. From the compactness of the product space it also follows that u_n and v_n are continuous functions. Thus (3.7) can be written as

$$h_1(E_1) + h_2(E_2) + \ldots + h_{n-1}(E_{n-1}) + v_n(E_n) \geqslant 0.$$

From the validity of non-negativity for n, we obtain

$$\Lambda_0(g_1) + \Lambda_0(g_2) + \ldots + \Lambda_0(g_{n-1}) + \Lambda_0(U_{-t_n}v_n) \geqslant 0. \tag{3.8}$$

But

$$\Lambda_0(U_{-t_n}v_n) = \Lambda_0(g_n) + \Lambda_0(U_{-t_n}u_n). \tag{3.9}$$

Since $U_{-t_{n+1}}u_n$ and $U_{-t_n}u_n$ are in C_0 it follows from relative stationarity that

$$\Lambda_0(U_{-t_{n+1}}u_n) = \Lambda_0(U_{-t_n}u_n).$$

Since

$$h_{n+1}(E_{n+1}) \geqslant u_n(E_n)$$

it follows that

$$g_{n+1} \geqslant U_{-t_{n+1}} u_n. \tag{3.10}$$

Both the sides of (3.10) are in C_0. From the non-negativity of Λ_0 in C_0 it follows that

$$\Lambda_0 (g_{n+1}) \geqslant \Lambda_0 (U_{-t_{n+1}} u_n) = \Lambda_0 (U_{-t_n} u_n). \tag{3.11}$$

Thus (3.8), (3.9) and (3.11) imply that

$$\Lambda_0 (g_1) + \Lambda_0 (g_2) + \ldots + \Lambda_0 (g_{n+1}) \geqslant 0.$$

Since Λ_0 is non-negative the statement is valid for $n = 0$. This completes the proof of non-negativity and unambiguity of the functional Λ. Now Λ can be extended to the smallest closed invariant subspace containing C_0 by limiting process. By the very definition Λ is invariant under U_t, $t \in G$. This completes the proof of theorem 3.1.

Remark 1. As a particular case of theorem 3.1 it follows that if $\varphi(t)$ nis a continuous positive definite function in the interval [—1,1] then it can be extended as continuous positive definite function on the real line. To do this we have to consider the Gaussian process in $[-{}^1/_2, {}^1/_2]$ with $\varphi(t)$ as the covariance function, extend the process to $(-\infty, +\infty)$ and take the covariance function of the extended process as the required extension.

Remark 2. The following questions arise in this connection. When is the extension of the process unique? When is an ergodic extension possible?

Remark 3. It is clear from the proofs that theorem 3.1 is valid in the case when the process x_t takes values in any locally compact separable metric space. In particular, if $x_t \, (0 \leqslant t \leqslant 1)$ is a relatively stationary finite-dimensional Gaussian process then there exists a stationary Gaussian extension in $-\infty < t < \infty$.

Indian Statistical Institute, Calcutta.

Поступила в редакцию 10.12.62

REFERENCES

[1] М. Г. Крейн, О проблеме продолжения эрмитово положительных непрерывных функций, ДАН СССР, **XXVI**, 1 (1940), 17—21.

Math. Student 32 (1964) 17-21

LIMIT THEOREMS IN PROBABILITY

by

S. R. S. VARADHAN

The collection of all probability distributions on the real line admits of two important structures one algebraic and the other topological. The interplay between these two structures has been studied in some detail.

Let $\mathcal{M}$ denote the collection of all probability distributions on the real line. For any two elements α, β of $\mathcal{M}$ their convolution $\alpha*\beta$ is defined by the relation

$$\alpha*\beta(A)=\int_R \alpha(A-x)d\beta(x)$$

Under this operation $\mathcal{M}$ becomes a commutative semi-group. One can also define a notion of convergence in $\mathcal{M}$: A sequence α_n in $\mathcal{M}$ converges to α in $\mathcal{M}$ if and only if

$$\int f(x)d\alpha_n(x) \to \int f(x)d\alpha(x)$$

for every bounded continuous function $f(x)$ on R. One can verify that this convergence is actually induced by a metric on $\mathcal{M}$. The topology on $\mathcal{M}$ is compatible with the semi-group operation and $\mathcal{M}$ is actually a topological semi-group.

The main tool in the study of this semi-group is the notion of Fourier transform. For any $\alpha\varepsilon\mathcal{M}$ its 'characteristic function' is defined by the relation

$$\alpha(t)=\int e^{itx}d\alpha(x).$$

It is well known that $\alpha(t)$ determines α uniquely, $\alpha*\beta(t)=\alpha(t).\beta(t)$ and $\alpha_n\to\alpha$ if and only if $\alpha_n(t)\to\alpha(t)$ uniformly over all t in every bounded interval.

The theory revolves around the following main problem : Let, for each n, there be given k_n elements of $\mathcal{M}$, $\{\alpha_{nj}\}$ $j=1, 2,\ldots k_n$. Form the convolution

$$\mu_n=\alpha_{n_1}*\alpha_{n_2}\ldots*\alpha_{nk_n}.$$

Under the assumption that for sufficiently large n the various α_{nj} are nearly the unit element of $\mathcal{M}$ (the distribution giving mass 1 to the point 0) what can one say about the asymptotic behaviour of μ_n? The assumption known as uniform infinitesimality is formulated in precise terms as follows :

$$\lim_{n\to\infty} \sup_{|t|\leqslant T} \sup_{1\leqslant j\leqslant k_n} |\alpha_{nj}(t)-1| = 0$$

for every finite T.

It turns out that every limit (along any subsequence) of μ_n is infinitely divisible in the sense that if μ is the limit it can be put in the form λ_n^n for every n. One then obtains a complete characterization of all infinitely divisible distributions and the conditions for the convergence of μ_n to a specified infinitely divisible distribution μ.

We shall examine now what happens when the real line is replaced by a more general object. Let X be a locally compact abelian group which is separable, Y its character group. Then the class $\mathcal{M}$ of distribution on X becomes in a natural manner a topological semigroup. The convolution operation is

$$\alpha*\beta(A)=\int_X \alpha(A-x)d\beta(x)$$

and the convergence notion is

$$\int_X f(x)d\alpha_n(x) \to \int_X f(x)d\alpha(x)$$

for every bounded continuous function $f(x)$ on X. For every $\alpha\varepsilon\mathcal{M}$ the characteristic function is defined on Y by the formula

$$\alpha(y)=\int_X (x,y)d\alpha(x)$$

where (x, y) denotes the value of the character y at the point x. The relations between α and $\alpha(y)$ valid for the real line continue to be valid even in this set up.

We shall now state the various results known concerning the semigroup. A sequence μ_n in $\mathcal{M}$ is called shift compact if there are elements x_n in such that μ_n translated by x_n is compact. We shall denote this translation by $\mu_n * x_n$. If $\mu=\alpha*\beta$ for some β then α is called a factor of μ. α is idempotent if $\alpha^2=\alpha$. Idempotent distributions are precisely normalised Haar measures of compact sub-groups.

Theorem 1. Let $\mu_n=\alpha_n*\beta_n$ for each n. If μ_n is compact then α_n and β_n are shift compact.

Definition. A distribution μ is said to be infinitely divisible if for each n there are elements x_n in X and $\lambda_n\varepsilon\mathcal{M}$ such that

$$\mu=\lambda_n^n * x_n.$$

Remark. This modification of the classical definition is intended to avoid the consequence of the presence of elements that are not divisible. In a divisible group this definition is equivalent to the classical definition as can be easily seen.

Theorem 2. The totality of infinitely divisible distributions is a closed sub-semigroup of $\mathcal{M}$.

Definition. If F is any totally finite measure on X the distribution $e(F)$ associated with it is defined as follows.

$$e(F)=e^{-F(X)}[1+F+\frac{F^2}{2!}+\ldots].$$

For $\mu\varepsilon\mathcal{M}$ we denote by $\mu(y)$ its characteristic function. We then have

$$e(F)(y)=[\int[(x, y)-1]dF].$$

Theorem 3. If μ is infinitely divisible and if its characteristic function vanishes at some point then μ has an idempotent factor.

Theorem 4. Let $\mu_n=e(F_n)$. Then the necessary and sufficient conditions that

(a) μ_n is shift compact,

(b) if μ is any limit of shifts of μ_n then μ has no idempotent factors,

are

(i) for each neighbourhood N of identity ε the family F_n restricted to $X-N$ is weakly conditionally compact,

(ii) for each $y\varepsilon Y$

$$\sup_n \int [1-\text{Real } (x, y)]dF_n<\infty.$$

Definition. A double sequence α_{nj} : $j=1, 2,\ldots, k_n$; $n=1, 2,\ldots$ of distributions is said to be uniformly infinitesimal if

$$\lim_{n\to\infty} \sup_{1\leqslant j\leqslant k_n} \sup_{y\varepsilon K} |\alpha_{nj}(y)-1| =0$$

for every compact subset $K\subset Y$.

Lemma. There exists a real valued function $g(x, y)$ on $X\times Y$ which is continuous in both the variables x and y and such that

(i) $g(x, y_1+y_2)=g(x, y_2)$ for each $x\varepsilon X$ and $y_1, y_2\varepsilon Y$.

(ii) for any compact subset $K\subset Y$.

$$\sup_{x\varepsilon X} \sup_{y\varepsilon K} | g(x, y) | <+\infty.$$

(iii) for each compact subset $K\subset Y$ there exists a neighbourhood N_K of the identity in X such that

$$(x, y)=e^{ig(x,y)}$$

holds for $x\varepsilon N_K$ and $y\varepsilon K$,

(*iv*) $g(x, y)=0$ whenever $x=e$ for any y.

Theorem 5. Let α_{nj} be a uniformly infinitesimal sequence and let

$$\mu_n=\prod_{j=1}^{k_n}\alpha_{nj}.$$

Suppose $\beta_{nj}=e[\alpha_{nj}*g_{nj}]$ where g_{nj} is that element of the group X defined by the equality

$$(g_{nj}, y)=\exp-i\int g(x, y)d\alpha_{nj}.$$

If

$$\lambda_n=\prod_{j=1}^{k_n}\beta_{nj}*g_n \text{ where } g_n=-\sum_j g_{nj},$$

then

$$\lim_{n\to\infty}\ \sup_{y\varepsilon K}\ |\lambda_n(y)-\mu_n(y)|=0$$

for every compact set K of Y.

Corollary. Limit distribution of sums of uniformly infinitesimal independent random variables is infinitely divisible.

Definition. A distribution μ is said to be Gaussian if it has the following properties :

(*i*) μ is infinitely divisible,

(*ii*) $\mu=e(F)*\alpha$, where α is infinitely divisible implies that F vanishes outside the identity.

Theorem 6. A distribution μ is Gaussian if and only if $\mu(y)$ is of the form

$$\mu(y)=(x. y)\exp[-\phi(y)]$$

where x is a fixed element of X and $\phi(y)$ a continuous non-negative function of y satisfying the equation

$$\phi(y_1+y_2)+\phi(y_1-y_2)=2[\phi(y_1)+\phi(y_2)]$$

for every pair y_1, y_2 in Y.

Theorem 7. A distribution μ on X is infinitely divisible without an idempotent factor if and only if $\mu(y)$ is of the form

$$\mu(y)=(x_0, y)\exp\left[\int[(x, y)-1-ig(x, y)]dM-\phi(y)\right]$$

where

(*i*) x_0 is a fixed element of X,

(*ii*) $\phi(y)$ is a function as in Theorem 6,

(*iii*) μ is a σ-finite measure giving finite mass outside every neighbourhood of the identity,

(*iv*) $\int[1-\text{Real }(x, y)]d\mu<+\infty$ for each $y\varepsilon Y$.

Remark. This representation is not in general unique. If the group X has no compact sub-groups or more generally if every compact sub-group consists only of elements of order two then the representation is unique. In the contrary case examples can be constructed to exhibit non-uniqueness.

Corollary. Every one parametric weakly continuous semi-group μ_t has a unique representation

$$\mu_t(y)=(x_t, y) \exp [t \int [(x, y)-1-ig(x, y)]dM-t\phi(y)]$$

where x_t is a continuous semi-group in X and M and ϕ are as in Theorem 7.

REFERENCES

1. B, V, GNEDENKO and A, N. KOLMOGOROV, Limit distributions for sums of independent random variables, trans., from the Russian, Cambridge, Mass., Addison Wesley Pub. Co., 1954.

2. K. R. PARTHASARATHY, R. RANGA RAO, and S.R.S. VASADHAN, Probability distributions on locally compact abelian groups, Illinois, J. Math. Vol. 7, No. 2, 1963, p.p. 337–369.

Indian Statistical Institute,
Calcutta.

COMMUNICATIONS ON PURE AND APPLIED MATHEMATICS, VOL. XXVI, 497–524 (1973)

A Limit Theorem with Strong Mixing in Banach Space and Two Applications to Stochastic Differential Equations*

G. C. PAPANICOLAOU AND S. R. S. VARADHAN

1. Introduction

In [1] the asymptotic behavior of the expected value of the solution of an abstract stochastic equation was investigated. This work was motivated by the results of Khasminskii [2] and Stratonovich [3] for stochastic ordinary differential equations and other works (cf. [4], [5], [6]) concerned with operator equations. The results obtained in [1] were limited by severe restrictions on the allowed from of the stochastic perturbation.

Recently, Cogburn and Hersh [7] have generalized the results of [1] considerably by allowing a much broader class of stochastic perturbations and requiring only a strong mixing condition. Our aim here is to improve the results of [7] by giving an estimate for the error committed in the asymptotic approximation. If ε denotes the small parameter of the problem we show that the error is $O(\varepsilon)$ as $\varepsilon \to 0$. The results in [1], [2], [3] and [7] show only that the error is $o(1)$. Our estimate is best possible since it is achieved for the classical central limit theorem which is a special case of our Theorem 2.

In Section 2 we state and prove an abstract limit theorem, Theorem 1. In Section 3 we apply the theorem of Section 2 to obtain a result similar to that of Khasminskii in [2] with error estimate $O(\varepsilon)$, Theorem 2. In Section 4 we apply Theorem 1 to obtain a limit theorem for random motion in a Lie group. This is the content of Theorem 3. Theorem 4 contains a result which is needed in Theorem 3 but it is stated as a theorem since it concerns the behavior of the solution of a diffusion equation on the group and is of independent interest. Theorem 3 has a number of interesting applications (cf. [8], [9]).

The nature of the hypotheses of the abstract theorem, Theorem 1, can best be understood by the form they take in the concrete applications of Theorems 2 and 3. In fact, Theorem 1 is a simple abstraction of the situation encountered in the other theorems. It has, however, other interesting applications in connection with random evolution equations (cf. [7], [10], [11], [12], [13]).

---

*Research performed at the Courant Institute of Mathematical Sciences and supported by NSF Grant #GP-32996X2 and Air Force Grant #AF-AFOSR-72-2307.

The first named author wishes to thank R. Hersh, D. Stroock and E. B. Davies for several enlightening conversations concerning the present problem.

2. A Limit Theorem in Banach Space

We begin with some definitions. Then we state the problem under consideration and the main result.

Let $(\Omega, \mathscr{F}, P)$ be a probability space and $\mathscr{F}_s^t \subset \mathscr{F}$, $0 \leqq s \leqq t \leqq \infty$, a family of σ-algebras such that $\mathscr{F}_{s_1}^{t_1} \subset \mathscr{F}_s^t$, $0 \leqq s \leqq s_1 \leqq t_1 \leqq t \leqq \infty$. Let the conditional probabilities $P(A \mid \mathscr{F}_0^s)$ have a regular version (cf. [15], p. 354) i.e., there exists a function $P_s(A \mid \omega)$ on $\mathscr{F} \times \Omega$ such that

(i) $P_s(A \mid \cdot)$ is $\mathscr{F}_0^s$ measurable for every $A \in \mathscr{F}$,

and

(ii) $P_s(\cdot \mid \omega)$ is a probability measure on $\mathscr{F}$ for every $\omega \in \Omega$.

We assume that the strong mixing property holds as follows:

$$\sup_{s \geqq 0} \sup_{\substack{A \in \mathscr{F}_{t+s}^{\infty} \\ B \in \mathscr{F}_0^s}} |P(A \mid B) - P(A)| = \sup_{s \geqq 0} \operatorname*{ess.\,sup}_{\substack{\omega \in \Omega \\ A \in \mathscr{F}_{t+s}^{\infty}}} |P_s(A \mid \omega) - P(A)| \tag{2.1}$$

$$= \rho(t) \downarrow 0\,, \qquad t \to \infty\,.$$

Let L_0 be a separable Banach space, its norm being denoted by $\| \ \|_0$. Let L_k, $1 \leqq k \leqq 4$, be dense linear subspaces of L_0 such that $L_{k-1} \supset L_k$. Let $\| \ \|_k$ be a norm on L_k, $1 \leqq k \leqq 4$, so that the L_k spaces normed this way are also separable Banach spaces. We assume that $\|f\|_{k-1} \leqq \|f\|_k$, $f \in L_k$, $1 \leqq k \leqq 4$.

For each $t \in [0, \infty)$ and $\omega \in \Omega$, let $V(t, \omega)$ be a bounded linear operator from L_k into L_{k-1}, $1 \leqq k \leqq 4$. We shall assume that the operators $V(t, \omega)$ are strongly measurable (cf. [14], p. 72) with respect to the σ-algebra generated by $B \times \mathscr{F}$, where B denotes the Borel sets on $[0, \infty)$. We assume further that, for $f \in L_k$, $1 \leqq k \leqq 4$, $V(t)f = V(t, \omega)f \in L_{k-1}$ is strongly $\mathscr{F}_t^t$ measurable as a function of $\omega \in \Omega$, t fixed.

Let $\|V(t, \omega)\|_k$ denote the norm of V as an operator on $L_k \to L_{k-1}$, $1 \leqq k \leqq 4$, i.e.,

$$\|V(t, \omega)\|_k = \sup_{f \in L_k, f \neq 0} \frac{\|Vf\|_{k-1}}{\|f\|_k}\,. \tag{2.2}$$

Since the spaces L_k are separable, $\|V(t, \omega)\|_k$ is an $\mathscr{F}_t^t$ measurable function of ω and jointly measurable in t and ω. We shall assume that $\|V(t, \omega)\|_k$ satisfies one of the following two hypotheses:

(V.i) There exist constants c_k such that, for almost all $\omega \in \Omega$,

$$\| V(t, \omega) \|_k \leqq c_k , \qquad 1 \leqq k \leqq 4, t \geqq 0 . \tag{2.3}$$

(V.ii) There exist constants $\tilde{c}_k$ such that

$$E^{1/5}\{\| V(t, \omega) \|_k^5\} \leqq \tilde{c}_k , \qquad 1 \leqq k \leqq 4 , t \geqq 0 . \tag{2.4}$$

Here and below $E\{\cdot\}$ stands for integration over Ω relative to P. Note also that

$$\| Vf \|_{k-1} \leqq \| V \|_k \, \| f \|_k , \tag{2.5}$$

where we omit the $\omega \in \Omega$ for convenience.

The object of our study are the asymptotic properties of the expected value of the solution of the Cauchy problem

$$\frac{dy(t)}{dt} = \varepsilon V(t)\, y(t) , \qquad t \geqq 0 , y(0) = f . \tag{2.6}$$

Here ε is a small real parameter. For the asymptotic analysis it is best to rescale the problem (2.6) by defining

$$\tau = \varepsilon^2 t , \qquad y^{(\varepsilon)}(\tau) = y(\tau/\varepsilon^2) , \tag{2.7}$$

so that

$$\frac{dy^{(\varepsilon)}(\tau)}{d\tau} = \frac{1}{\varepsilon} V(\tau/\varepsilon^2)\, y^{(\varepsilon)}(\tau) , \qquad y^{(\varepsilon)}(0) = f . \tag{2.8}$$

In order to make the sense in which (2.8) is to be taken precise, we introduce the solution operators $U^{(\varepsilon)}(\tau, \sigma)$ by

$$y^{(\varepsilon)}(\tau) = U^{(\varepsilon)}(\tau, \sigma)\, y^{(\varepsilon)}(\sigma) , \qquad \tau \geqq \sigma \geqq 0 . \tag{2.9}$$

We call the operators $U^{(\varepsilon)}(\tau, \sigma)$ propagators and assume that the following hypotheses hold.

(P.i) $U^{(\varepsilon)}(\tau, \sigma)$ are contraction operators on $L_0 \to L_0$ and $U^{(\varepsilon)}(\tau, \sigma) f$, $f \in L_0$, is strongly $\mathscr{F}_{\sigma/\varepsilon^2}^{\tau/\varepsilon^2}$ measurable. The $U^{(\varepsilon)}(\tau, \sigma)$ are bounded linear operators on $L_k \to L_k$, $k = 1, \cdots, 4$.

(P.ii) The finite propagator property holds:

$$U^{(\varepsilon)}(\tau,\sigma)U^{(\varepsilon)}(\sigma,s)=U^{(\varepsilon)}(\tau,s)\,,\qquad 0\leqq s\leqq\sigma\leqq\tau \tag{2.10}$$
$$U^{(\varepsilon)}(\tau,\tau)=I\,.$$

(P.iii) The infinitesimal forward and backward propagator properties hold:

$$U^{(\varepsilon)}(\tau,\sigma)f=f+\frac{1}{\varepsilon}\int_\sigma^\tau V(s/\varepsilon^2)U^{(\varepsilon)}(s,\sigma)f\,ds\,, \tag{2.11}$$
$$U^{(\varepsilon)}(\tau,\sigma)f=f+\frac{1}{\varepsilon}\int_\sigma^\tau U^{(\varepsilon)}(\tau,s)V(s/\varepsilon^2)f\,ds\,.\qquad f\in L_1\,,$$

Integrals of L_k-valued functions are in Bochner's sense and for their existence it is sufficient that the integrand be strongly measurable and its norm integrable (cf. [14], p. 80). Other properties required in the calculations below, such as Fubini's theorem for vector-valued functions, are examined in detail in [14]. The use of such properties is not stated explicitly in what follows.

In the applications of Sections 3 and 4 the role of τ and σ in (P.i)–(P.iii) is reversed. However, we adhere to the present format because this is the way results are frequently presented in the physical literature (cf. [4]).

We adopt the convention that C denotes a constant independent of ε, t, τ, σ, etc. In particular situations we distinguish different constants by using small c's with primes etc. in order to make an argument more explicit.

We now state the main result of this section.

THEOREM 1. *Let* $U^{(\varepsilon)}(\tau,\sigma)$, $0\leqq\sigma\leqq\tau$, *be defined as above and have properties* (P.i)–(P.iii). *Let the following conditions hold:*

(1.i) *For* $f\in L_1$,

$$E\{V(t)f\}=0\,. \tag{2.12}$$

(1.ii) *For* $f\in L_2$, *the strong limit in* L_0,

$$\lim_{t\to\infty}\frac{1}{t}\int_{t_0}^{t_0+t}\int_{t_0}^{s}E\{V(s)V(\sigma)f\}\,d\sigma\,ds\,, \tag{2.13}$$

exists uniformly in t_0 *and defines a bounded operator* $\bar V$ *on* $L_2\to L_0$. *We assume specifically that*

$$\sup_{t_0\geqq 0}\left\|\bar Vf-\frac{1}{t}\int_{t_0}^{t_0+t}\int_{t_0}^{s}E\{V(s)V(\sigma)f\}\,d\sigma\,ds\right\|_0\leqq C\frac{\|f\|_2}{t}\,,\qquad t\to\infty\,, \tag{2.14}$$

and that $\bar{V}$ *has a closed extension in* L_0 *with domain* $\mathscr{D}_{\bar{V}} \supset L_2$ *which generates a strongly continuous contraction semigroup on* L_0 *denoted by* $e^{\tau \bar{V}}$, $\tau \geqq 0$. *The set* $L_4 \cap \mathscr{D}^2_{\bar{V}}$ *is assumed dense in* L_0. ($\mathscr{D}^2_{\bar{V}}$ *is the domain of* $\bar{V}^2$).

(1.iii) $V(t)$, $t \geqq 0$, *satisfies* (V.1).

(1.iv) *The following inequalities are satisfied*:

$$\| U^{(\varepsilon)}(\tau, \sigma) f \|_1 \leqq c_1' \, e^{\alpha(\tau-\sigma)/\varepsilon} \, \| f \|_1 \qquad a.s., \tag{2.15}$$

$$\| U^{(\varepsilon)}(\tau, \sigma) f \|_2 \leqq c_2' \left(1 + \frac{\tau - \sigma}{\varepsilon}\right) e^{\alpha(\tau-\sigma)/\varepsilon} \, \| f \|_2 \qquad a.s., \tag{2.16}$$

where α, c_1' *and* c_2' *are positive constants.*

(1.v) *The inequality*

$$\sup_{0 \leqq \tau \leqq \tau_0} \| e^{\tau \bar{V}} f \|_k \leqq c_k'' \, \| f \|_k \, , \qquad 1 \leqq k \leqq 4 \, , \tag{2.17}$$

is satisfied, where τ_0 *is a fixed point in* $[0, \infty)$ *and* c_k'', $k = 1, 2, \cdots$, *are positive constants depending on* τ_0.

(1.vi) *In addition*,

$$\int_0^\infty \rho^{1/2}(s) \, ds < \infty \, . \tag{2.18}$$

Under these hypotheses,

$$\sup_{0 \leqq \tau \leqq \tau_0} \| E\{U^{(\varepsilon)}(\tau, 0) f\} - e^{\tau \bar{V}} f \|_0 \leqq C(f; \tau_0)\varepsilon \, , \qquad f \in L_4 \cap \mathscr{D}^2_{\bar{V}} \, . \tag{2.19}$$

$C(f; \tau_0)$ *denotes a positive constant that depends on* f *and* τ_0. *Thus* $E\{U^{(\varepsilon)}(\tau, 0) f\}$ *converges strongly in* L_0 *to* $e^{\tau \bar{V}} f$ *as* $\varepsilon \to 0$, $0 \leqq \tau \leqq \tau_0$, *and the error in the approximation is* $O(\varepsilon)$.

Before proceeding with the proof we insert a lemma which we shall use often. This lemma is a generalization of a result in [16], p. 222, to L_k-valued random variables; it is also used in [7].

LEMMA 1. *Let* $\mathscr{L}$, $\mathscr{M}$, $\mathscr{N}$ *be Banach spaces and let* $W(\omega)$ *and* $U(\omega)$, $\omega \in \Omega$, *be bounded operators from* $\mathscr{N} \to \mathscr{M}$ *and* $\mathscr{M} \to \mathscr{L}$, *respectively. Let* U *be strongly* $\mathscr{F}^\infty_{t+s}$ *measurable and* W *strongly* $\mathscr{F}^s_0$ *measurable. Then, for any* $f \in \mathscr{N}$,

$$\| E\{UWf\} - E\{UE\{Wf\}\} \|_{\mathscr{L}} \leqq 2\rho(t) \sup_{\omega, \omega' \in \Omega} \| U(\omega) W(\omega') f \|_{\mathscr{L}} \, . \tag{2.20}$$

Proof of Lemma 1: Set $\mu_s(A \mid \omega) = P_s(A \mid \omega) - P(A)$. For fixed ω and s, $\mu_s(A \mid \omega)$ is a signed measure. Let $|\mu_s|\,(A \mid \omega)$ denote the variation of μ_s on A. From the Hahn decomposition theorem we know that

$$\sup_{A \in \mathscr{F}_{t+s}^{\infty}} |\mu_s|\,(A \mid \omega) \leqq 2\rho(t)\,, \tag{2.21}$$

where we have used (2.1). On using Fubini's theorem for vector-valued functions, we obtain

$$\begin{aligned}
\|E\{UWf\} - E\{UE\{Wf\}\}\|_{\mathscr{L}} &= \|E\{UE\{Wf \mid \mathscr{F}_0^s\}\} - E\{UE\{Wf\}\}\|_{\mathscr{L}} \\
&= \left\| \iint U(\omega')W(\omega) f \mu_s(d\omega \mid \omega') P(d\omega') \right\|_{\mathscr{L}} \\
&\leqq \iint \|U(\omega')W(\omega)f\|_{\mathscr{L}} |\mu_s|\,(d\omega \mid \omega') P(d\omega') \\
&\leqq \int \sup_{\omega} \|U(\omega')W(\omega)f\|_{\mathscr{L}} \sup_{A \in \mathscr{F}_{t+s}^{\infty}} |\mu_s|\,(A \mid \omega') P(d\omega') \\
&\leqq 2\rho(t) \sup_{\omega,\omega' \in \Omega} \|U(\omega')W(\omega)f\|_{\mathscr{L}}\,.
\end{aligned}$$

Since we omit the ω's throughout, we shall also omit the sup in (2.20) when we apply it below. It is, of course, understood that the supremum is taken.

Proof of Theorem 1: Fix $\tau \in [0, \tau_0]$ and let ε go to zero through values that render $m = \tau/\varepsilon$ an integer going to infinity.

We use the finite propagator property (P.ii) to obtain the decomposition

$$\begin{aligned}
&E\{(U^{(\varepsilon)}(\tau, 0) - e^{\tau\bar{V}})f\} \\
&\quad = \sum_{k=0}^{m-1} E\{U^{(\varepsilon)}(\tau, (k+1)\varepsilon)[U^{(\varepsilon)}((k+1)\varepsilon, k\varepsilon) - e^{\varepsilon\bar{V}}]e^{k\varepsilon\bar{V}} f\}\,.
\end{aligned} \tag{2.22}$$

We decompose the right side of (2.22) into two parts and use property (P.i) to obtain the estimate

$$\begin{aligned}
&\|E\{U^{(\varepsilon)}(\tau, 0)f\} - e^{\tau\bar{V}} f\|_0 \\
&\quad \leqq \sum_{k=0}^{m-1} \|[E\{U^{(\varepsilon)}((k+1)\varepsilon, k\varepsilon)\} - e^{\varepsilon\bar{V}}]e^{k\varepsilon\bar{V}} f\|_0 \\
&\qquad + \sum_{k=0}^{m-1} \|E\{U^{(\varepsilon)}(\tau, (k+1)\varepsilon) U^{(\varepsilon)}((k+1)\varepsilon, k\varepsilon) e^{k\varepsilon\bar{V}} f\} \\
&\qquad - E\{U^{(\varepsilon)}(\tau, (k+1)\varepsilon) E\{U^{(\varepsilon)}((k+1)\varepsilon, k\varepsilon) e^{k\varepsilon\bar{V}} f\}\}\|_0 \\
&\quad = I_1 + I_2\,.
\end{aligned} \tag{2.23}$$

Let us estimate I_2 first. Set $g_k = e^{k\varepsilon\bar{V}} f$. Using (2.11), we have

$$
\begin{aligned}
I_2^{(k)} &\equiv \| E\{U^{(\varepsilon)}(\tau, (k+1)\varepsilon) U^{(\varepsilon)}((k+1)\varepsilon, k\varepsilon) g_k\} \\
&\qquad - E\{U^{(\varepsilon)}(\tau, (k+1)\varepsilon) E\{U^{(\varepsilon)}((k+1)\varepsilon, k\varepsilon) g_k\}\}\|_0 \\
&= \frac{1}{\varepsilon^2} \left\| E\left\{\int_{(k+1)\varepsilon}^{\tau} U^{(\varepsilon)}(\tau, s) V(s/\varepsilon^2)\, ds \int_{k\varepsilon}^{(k+1)\varepsilon} V(\sigma/\varepsilon^2) U^{(\varepsilon)}(\sigma, k\varepsilon) g_k\, d\sigma\right\}\right. \\
&\qquad \left. - E\left\{\int_{(k+1)\varepsilon}^{\tau} U^{(\varepsilon)}(\tau, s) V(s/\varepsilon^2)\, ds E\left\{\int_{k\varepsilon}^{(k+1)\varepsilon} V(\sigma/\varepsilon^2) U^{(\varepsilon)}(\sigma, k\varepsilon) g_k\, d\sigma\right\}\right\}\right\|_0 \qquad (2.24) \\
&= \varepsilon^2 \left\| \int_{(k+1)/\varepsilon}^{\tau/\varepsilon^2} \int_{k/\varepsilon}^{(k+1)/\varepsilon} [E\{U^{(\varepsilon)}(\tau, \varepsilon^2 s) V(s) V(\sigma) U^{(\varepsilon)}(\varepsilon^2 \sigma, k\varepsilon) g_k\}\right. \\
&\qquad \left. - E\{U^{(\varepsilon)}(\tau, \varepsilon^2 s) V(s) E\{V(\sigma) U^{(\varepsilon)}(\varepsilon^2 \sigma, k\varepsilon) g_k\}\}]\, d\sigma\, ds \right\|_0 .
\end{aligned}
$$

Note that in the first equality above the identity terms of (2.11) cancel and that in the second equality a simple change of variables and a rearrangement have been made. We may now use Lemma 1 in the last expression in (2.24), since of the four factors $U^{(\varepsilon)}, V(s), V(\sigma), U^{(\varepsilon)}$ the first two are $\mathscr{F}_s^\infty$ measurable and the last two $\mathscr{F}_0^\sigma$ measurable. Thus,

$$
\begin{aligned}
I_2^{(k)} &\leqq \varepsilon^2 \int_{(k+1)/\varepsilon}^{\tau/\varepsilon^2} \int_{k/\varepsilon}^{(k+1)/\varepsilon} \rho(s-\sigma)\, d\sigma\, ds \\
&\qquad \times \sup_{\substack{s \geqq 0 \\ k/\varepsilon \leqq \sigma \leqq (k+1)/\varepsilon}} \| U^{(\varepsilon)}(\tau, \varepsilon^2 s) V(s) V(\sigma) U^{(\varepsilon)}(\varepsilon^2 \sigma, k\varepsilon) g_k \|_0 . \qquad (2.25)
\end{aligned}
$$

We use hypotheses (P.i), (1.iii) twice, (1.iv) once, and (1.v) once to obtain

$$
\begin{aligned}
&\sup_{\substack{s \geqq 0 \\ k/\varepsilon \leqq \sigma \leqq (k+1)/\varepsilon}} \| U^{(\varepsilon)}(\tau, \varepsilon^2 s) V(s) V(\sigma) U^{(\varepsilon)}(\varepsilon^2 \sigma, k\varepsilon) g_k \|_0 \\
&\qquad \leqq \sup_{\substack{s \geqq 0 \\ k/\varepsilon \leqq \sigma \leqq (k+1)/\varepsilon}} \| V(s) V(\sigma) U^{(\varepsilon)}(\varepsilon^2 \sigma, k\varepsilon) g_k \|_0 \\
&\qquad \leqq c_1 \sup_{k/\varepsilon \leqq \sigma \leqq (k+1)/\varepsilon} \| V(\sigma) U^{(\varepsilon)}(\varepsilon^2 \sigma, k\varepsilon) g_k \|_1 \qquad (2.26) \\
&\qquad \leqq c_1 c_2 \sup_{k/\varepsilon \leqq \sigma \leqq (k+1)/\varepsilon} \| U^{(\varepsilon)}(\varepsilon^2 \sigma, k\varepsilon) g_k \|_2 \\
&\qquad \leqq 2 c_1 c_2 c_2' e^{\alpha} \|g_k\|_2 \leqq c_1 c_2 c_2\, 2 e^{\alpha} C_2'' \|f\|_2 .
\end{aligned}
$$

Thus

$$I_2^{(k)} \leqq C\varepsilon^2 \|f\|_2 \int_{(k+1)/\varepsilon}^{\tau/\varepsilon^2} \int_{k/\varepsilon}^{(k+1)/\varepsilon} \rho(s-\sigma)\, d\sigma\, ds\,. \tag{2.27}$$

We show next that the double integral in (2.27) is bounded independently of k and $\varepsilon \in [0, 1]$, say. We have

$$\begin{aligned} \int_{(k+1)/\varepsilon}^{\tau/\varepsilon^2} \int_{k/\varepsilon}^{(k+1)/\varepsilon} \rho(s-\sigma)\, d\sigma\, ds &\leqq \int_{(k+1)/\varepsilon}^{\infty} \int_{k/\varepsilon}^{(k+1)/\varepsilon} \rho(s-\sigma)\, d\sigma\, ds \\ &= \int_0^{1/\varepsilon} \sigma\rho(\sigma)\, d\sigma + \frac{1}{\varepsilon}\int_{1/\varepsilon}^{\infty} \rho(\sigma)\, d\sigma\,. \end{aligned} \tag{2.28}$$

By the monotonicity of $\rho(s)$ and (1.vi) we have further

$$\begin{aligned} \int_0^{1/\varepsilon} \sigma\rho(\sigma)\, d\sigma = \int_0^{1/\varepsilon} \rho(\sigma)\int_0^{\sigma} ds\, d\sigma &\leqq \int_0^{1/\varepsilon} \rho^{1/2}(\sigma)\, d\sigma \int_0^{\sigma} \rho^{1/2}(s)\, ds \\ &\leqq \left[\int_0^{\infty} \rho^{1/2}(\sigma)\, d\sigma\right]^2. \end{aligned} \tag{2.29}$$

Similarly,

$$\frac{1}{\varepsilon}\int_{1/\varepsilon}^{\infty} \rho(\sigma)\, d\sigma \leqq \frac{1}{\varepsilon}\, \rho^{1/2}(1/\varepsilon) \int_0^{\infty} \rho^{1/2}(\sigma)\, d\sigma\,. \tag{2.30}$$

Since ρ decreases to zero and (1.vi) holds, it follows that $(1/\varepsilon)\rho^{1/2}(1/\varepsilon)$ is uniformly bounded for $\varepsilon \in [0, 1]$. Consequently, (2.30), (2.29), and (2.27) yield

$$I_2^{(k)} \leqq C\varepsilon^2 \|f\|_2\,. \tag{2.31}$$

Using this last estimate in (2.23), we get finally the desired estimate

$$I_2 \leqq mC\varepsilon^2 \|f\|_2 \leqq \tau_0 C \|f\|_2 \cdot \varepsilon\,. \tag{2.32}$$

The next step in the proof is the estimation of I_1. For this purpose we proceed as follows. We use the backward and forward infinitesimal propagator property (2.11) iterated four times to obtain the expansion

$$
\begin{aligned}
&U^{(\varepsilon)}((k+1)\varepsilon, k\varepsilon)f \\
&\quad = f + \frac{1}{\varepsilon}\int_{k\varepsilon}^{(k+1)\varepsilon} V(s/\varepsilon^2) f\, ds \\
&\qquad + \frac{1}{\varepsilon^2}\int_{k\varepsilon}^{(k+1)\varepsilon}\int_{s}^{(k+1)\varepsilon} V(\sigma/\varepsilon^2)V(s/\varepsilon^2) f\, d\sigma\, ds \\
&\qquad + \frac{1}{\varepsilon^3}\int_{k\varepsilon}^{(k+1)\varepsilon}\int_{s}^{(k+1)\varepsilon}\int_{\sigma}^{(k+1)\varepsilon} V(p/\varepsilon^2)V(\sigma/\varepsilon^2)V(s/\varepsilon^2) f\, dp\, d\sigma\, ds \qquad (2.33)\\
&\qquad + \frac{1}{\varepsilon^4}\int_{k\varepsilon}^{(k+1)\varepsilon}\int_{s}^{(k+1)\varepsilon}\int_{\sigma}^{(k+1)\varepsilon} V(p/\varepsilon^2)\int_{\sigma}^{p} V(q/\varepsilon^2)U^{(\varepsilon)}(q,\sigma)\, dq \\
&\qquad \times V(\sigma/\varepsilon^2)V(s/\varepsilon^2) f\, dp\, d\sigma\, ds\,, \qquad f\in L_4\,.
\end{aligned}
$$

We take expected values in (2.33), rearrange and take norms. This yields

$$
\begin{aligned}
&\| E\{U^{(\varepsilon)}((k+1)\varepsilon, k\varepsilon)f\} - f - \varepsilon\bar{V}f\}\|_0 \\
&\quad \leqq \varepsilon\left\|\bar{V}f - \varepsilon\int_{k/\varepsilon}^{(k+1)/\varepsilon}\int_{k/\varepsilon}^{\sigma} E\{V(\sigma)V(s)f\}\, d\sigma\, ds\right\|_0 \\
&\qquad + \varepsilon^3\left\|\int_{k/\varepsilon}^{(k+1)/\varepsilon}\int_{s}^{(k+1)/\varepsilon}\int_{\sigma}^{(k+1)/\varepsilon} E\{V(p)V(\sigma)V(s)f\}\, dp\, d\sigma\, ds\right\|_0 \qquad (2.34)\\
&\qquad + \varepsilon^4\left\|\int_{k/\varepsilon}^{(k+1)/\varepsilon}\int_{s}^{(k+1)/\varepsilon}\int_{\sigma}^{(k+1)/\varepsilon}\int_{\sigma}^{p} E\{V(p)V(q)U^{(\varepsilon)}(\varepsilon^2 q, \varepsilon^2\sigma) \right. \\
&\qquad\qquad \left. \times V(\sigma)V(s)f\}\, dq\, dp\, d\sigma\, ds\right\|_0 .
\end{aligned}
$$

The first term on the right side of the inequality (2.34) is less than or equal to a constant times $\varepsilon^2 \|f\|_2$ independently of k. This follows from (2.14). The other two terms are estimated in Lemmas 2 and 3 which are stated and proved at the end of this proof. On using (2.38) and (2.42) and the above observation, we have the estimate

$$
\| E\{U^{(\varepsilon)}((k+1)\varepsilon, k\varepsilon)f\} - f - \varepsilon\bar{V}f\|_0 \leqq C\varepsilon^2\|f\|_4 . \qquad (2.35)
$$

We recall here the convention about the constants C. By elementary considerations and the contractiveness of $e^{\tau\bar{V}}$, we find that

$$
\| e^{\varepsilon\bar{V}}f - f - \varepsilon\bar{V}f\|_0 \leqq \tfrac{1}{2}\varepsilon^2\|\bar{V}^2 f\|_0\,, \qquad f\in\mathscr{D}^2_{\bar{V}} \qquad (2.36)
$$

When we combine (2.35) and (2.36) we obtain the desired estimate of I_1 :

$$
\begin{aligned}
I_1 &= \sum_{k=0}^{m-1} \|E\{U^{(\varepsilon)}((k+1)\varepsilon, k\varepsilon)e^{k\varepsilon\bar{V}}f - e^{\varepsilon\bar{V}} \cdot e^{k\varepsilon\bar{V}}f\|_0 \\
&\leqq \sum_{k=0}^{m-1} \varepsilon^2[C\,\|e^{k\varepsilon\bar{V}}f\|_4 + \tfrac{1}{2}\,\|\bar{V}^2 e^{k\varepsilon\bar{V}}f\|_0] \qquad (2.37)\\
&\leqq \varepsilon\tau_0(Cc_4''\,\|f\|_4 + \tfrac{1}{2}\,\|\bar{V}^2 f\|_0]\,, \qquad f \in L_4 \cap \mathscr{D}_{\bar{V}}^2 .
\end{aligned}
$$

In the last inequality in (2.37) we have used the hypothesis (1.v) and the fact that $\bar{V}^2$ and $e^{s\bar{V}}$ commute on $\mathscr{D}_{\bar{V}}^2$. The estimates (2.37) and (2.32) imply that (2.19) holds. Since $U^{(\varepsilon)}$ and $e^{\tau\bar{V}}$ are contractions, and $L_4 \cap \mathscr{D}_{\bar{V}}^2$ is dense in L_0, we also have the strong convergence in L_0. This completes the proof of the theorem.

LEMMA 2. *For* $f \in L_3$, $k = 0, 1, 2, \cdots, 0 \leqq \varepsilon \leqq 1$,

$$
J = \left\| \int_{k/\varepsilon}^{(k+1)/\varepsilon} \int_s^{(k+1)/\varepsilon} \int_\sigma^{(k+1)/\varepsilon} E\{V(p)V(\sigma)V(s)f\}\, dp\, d\sigma\, ds \right\|_0 \leqq \frac{C}{\varepsilon}\, \|f\|_3 . \qquad (2.38)
$$

Proof: By a simple change of variables and the triangle inequality, we have

$$
\begin{aligned}
J \leqq \tilde{J} \equiv \int_0^{1/\varepsilon} \int_0^{(1/\varepsilon)-s} \int_0^{(1/\varepsilon)-s-\sigma} & \left\| E\left\{ V\left(s + \frac{k}{\varepsilon} + \sigma + p\right) V\left(s + \frac{k}{\varepsilon} + \sigma\right) \right.\right. \qquad (2.39)\\
& \times \left.\left. V\left(s + \frac{k}{\varepsilon}\right) f\right\}\right\|_0 dp\, d\sigma\, ds .
\end{aligned}
$$

In order to estimate the integral in (2.39) by using Lemma 1 and hypothesis (1.i), we first rewrite it as follows:

$$
\begin{aligned}
\tilde{J} \leqq \int_0^{1/\varepsilon} \Bigg[\left[\int_0^{1/2\varepsilon} \int_0^p + \int_{1/2\varepsilon}^{1/\varepsilon} \int_0^{(1/\varepsilon)-p}\right] & \left\| E\left\{ V\left(s + \frac{k}{\varepsilon} + \sigma + p\right) V\left(s + \frac{k}{\varepsilon} + \sigma\right) \right.\right. \\
& \times \left.\left. V\left(s + \frac{k}{\varepsilon}\right) f\right\}\right\|_0 d\sigma\, dp \\
+ \int_0^{1/2\varepsilon} \int_0^\sigma + \int_{1/2\varepsilon}^{1/\varepsilon} \int_0^{(1/\varepsilon)-\sigma} & \left\| E\left\{ V\left(s + \frac{k}{\varepsilon} + \sigma + p\right) V\left(s + \frac{k}{\varepsilon} + \sigma\right) \right.\right. \qquad (2.40)\\
& \times \left.\left. V\left(s + \frac{k}{\varepsilon}\right) f\right\}\right\|_0 dp\, d\sigma \Bigg]\, ds .
\end{aligned}
$$

Now we use Lemma 1, (1.i), (1.iii) (three times) and obtain

$$
\begin{aligned}
J &\leqq C \|f\|_3 \int_0^{1/\varepsilon}\Bigg[\Bigg[\int_0^{1/2\varepsilon}\int_0^{p} + \int_{1/2\varepsilon}^{1/\varepsilon}\int_0^{(1/\varepsilon)-p}\Bigg]\rho(p)\, d\sigma\, dp \\
&\qquad + \Bigg[\int_0^{1/2\varepsilon}\int_0^{\sigma} + \int_{1/2\varepsilon}^{1/\varepsilon}\int_0^{(1/\varepsilon)-\sigma}\Bigg]\rho(\sigma)\, dp\, d\sigma\Bigg] ds \\
&\leqq 2C \|f\|_3 \frac{1}{\varepsilon}\Bigg[\int_0^{1/2\varepsilon}\sigma\rho(\sigma)\, d\sigma + \int_{1/2\varepsilon}^{1/\varepsilon}\rho(\sigma)\Big(\frac{1}{\varepsilon} - \sigma\Big)\, d\sigma\Bigg] \qquad (2.41)\\
&\leqq 2C \|f\|_3 \frac{1}{\varepsilon}\Bigg[\Big(\int_0^{\infty}\rho^{1/2}(\sigma)\, d\sigma\Big)^2 + \frac{1}{\varepsilon}\,\rho^{1/2}\Big(\frac{1}{2\varepsilon}\Big)\int_0^{\infty}\rho^{1/2}(\sigma)\, d\sigma\Bigg].
\end{aligned}
$$

Remarks identical to those following (2.30) show that the quantity in the square brackets in the last inequality in (2.41) is bounded independently of $\varepsilon \in [0, 1]$. This then completes the proof of the lemma.

LEMMA 3. *For $f \in L_4$, $k = 0, 1, 2, \cdots, 0 \leqq \varepsilon \leqq 1$,*

$$
\begin{aligned}
\Bigg\| \int_{k/\varepsilon}^{(k+1)/\varepsilon}\int_s^{(k+1)/\varepsilon}\int_\sigma^{(k+1)/\varepsilon}\int_\sigma^{p} & E\{V(p)V(q)U^{(\varepsilon)}(\varepsilon^2 q, \varepsilon^2 \sigma) \\
& \times V(\sigma)V(s)f\}\, dq\, dp\, d\sigma\, ds \Bigg\|_0 \qquad (2.42)\\
& \leqq \frac{C\|f\|_4}{\varepsilon^2}.
\end{aligned}
$$

Proof: The proof of this lemma is very similar to the previous one. However, now in addition to the hypotheses used above, we need (P.i) concerning the measurability of $U^{(\varepsilon)}$ and also the inequality (2.16) concerning the growth of the $\| \ \|_2$ norm of $U^{(\varepsilon)}$. In the decomposition analogous to (2.40), the additional integral is simply carried along. We shall omit the details of the derivation.

THEOREM 1′. *Let all hypotheses of Theorem* 1 *hold with the following changes*:
(1.iii)′ $V(t)$ *satisfies* (V.ii).
(1.iv)′ $U^{(\varepsilon)}(\tau, \sigma)$, $\varepsilon \geqq 0$, $0 \leqq \sigma \leqq \tau$, *satisfies*

$$E^{1/5}\{\|U^{(\varepsilon)}(\tau, \sigma)f\|_1^5\} \leqq c_1' e^{\alpha(\tau-\sigma)/\varepsilon}\|f\|_1, \tag{2.43}$$

$$E^{1/5}\{\|U^{(\varepsilon)}(\tau, \sigma)f\|_2^5\} \leqq c_2'\Big(1 + \frac{\tau - \sigma}{\varepsilon}\Big)e^{\alpha(\tau-\sigma)/\varepsilon}\|f\|_2. \tag{2.44}$$

(1.vi)′ *Moreover, assume*

$$\int_0^\infty \rho^{1/10}(s)\,ds < \infty . \tag{2.45}$$

Then the conclusion of Theorem 1 *is again valid.*

The proof of Theorem 1′ is very similar to that of Theorem 1. We require now the following analogue of Lemma 1 whose proof we omit since it is entirely analogous to the one for random variables given in [16], p. 222 (cf. also [7]).

LEMMA 1′. *Let* $\mathscr{L}$, $\mathscr{M}$, $\mathscr{N}$ *be separable Banach spaces and let* $W(\omega)$ *and* $U(\omega)$, $\omega \in \Omega$, *be bounded operators from* $\mathscr{N} \to \mathscr{M}$ *and* $\mathscr{M} \to \mathscr{L}$, *respectively. Let* U *be strongly* $\mathscr{F}_{t+s}^{\infty}$ *measurable and* W *strongly* $\mathscr{F}_0^{s}$ *measurable. Set*

$$\begin{aligned} \|U(\omega)\|_{\mathscr{M},\mathscr{L}} &= \sup_{\substack{g\neq 0\\ g\in\mathscr{M}}} \frac{\|U(\omega)g\|_{\mathscr{L}}}{\|g\|_{\mathscr{M}}}, \\ \|W(\omega)\|_{\mathscr{N},\mathscr{M}} &= \sup_{\substack{g\neq 0\\ g\in\mathscr{N}}} \frac{\|W(\omega)g\|_{\mathscr{M}}}{\|g\|_{\mathscr{N}}}, \end{aligned} \tag{2.46}$$

and suppose that

$$E\{\|U(\omega)\|^p_{\mathscr{M},\mathscr{L}}\} < \infty , \qquad E\{\|W(\omega)\|^q_{\mathscr{N},\mathscr{M}}\} < \infty , \qquad \frac{1}{p}+\frac{1}{q} = 1 . \tag{2.47}$$

Then

$$\begin{aligned} &\|E\{UWf\} - E\{UE\{Wf\}\}\|_{\mathscr{L}} \\ &\qquad \leqq 2\rho^{1/p}(t)E^{1/p}\{\|U\|^p_{\mathscr{M},\mathscr{L}}\}E^{1/q}\{\|W\|^q_{\mathscr{N},\mathscr{M}}\}\,\|f\|_{\mathscr{N}} . \end{aligned} \tag{2.48}$$

Proof of Theorem 1′: The proof is quite similar to the one for Theorem 1. We now use Lemma 1′ in place of Lemma 1. Specifically, we use Lemma 1′ on (2.24) with $p = q = 2$ so that in (2.25) ρ is replaced by $\rho^{1/2}$. Thus (2.45) is not used fully at this point. Similarly, hypotheses (1.iii)′ and (1.iv)′ are used to obtain the analogue of (2.26) (but only fourth moments are needed at this point). Lemma 1′ is used again to obtain the analogues of Lemmas 2 and 3. In Lemma 2 the new hypotheses are not used fully but in Lemma 3 they are. For Lemma 3 we first use Lemma 1′ with $p = 5$, $q = \frac{5}{4}$, so that ρ is replaced by $\rho^{1/5}$ and then use Hölder's inequality repeatedly. This brings about the appearance of fifth moments.

3. Application to Systems of Nonlinear Stochastic Ordinary Differential Equations

In this section we shall apply Theorem 1 to obtain a limit theorem for the solution of a system of stochastic equations.

Let $z(t) = z(t, \omega) \in R^n$, $t \geqq 0$, $\omega \in \Omega$, be defined as the solution of the equations

$$\frac{dz(t)}{d} = \varepsilon F(z(t), t, \omega)\,, \qquad z(0) = z \in R^n\,. \tag{3.1}$$

Here ε is a real parameter and $F(z, t, \omega)$ is a mapping from $R^n \times [0, \infty) \times \Omega$ into R^n which has bounded and continuous derivatives in z up to order four for all $t \geqq 0$, $\omega \in \Omega$. For fixed $z \in R^n$, F is measurable in t and ω. Thus the solution of (3.1) exists and is unique and the process $z(t, \omega)$ is well defined. To study the process we let

$$\tau = \varepsilon^2 t\,, \qquad z^{(\varepsilon)}(\tau) = z(\tau/\varepsilon^2)\,, \tag{3.2}$$

so that

$$\frac{dz^{(\varepsilon)}(\tau)}{d\tau} = \frac{1}{\varepsilon} F(z^{(\varepsilon)}(\tau), \tau/\varepsilon^2, \omega)\,, \qquad z^{(\varepsilon)}(0) = z\,. \tag{3.3}$$

In order to apply Theorem 1 we must introduce appropriate function spaces and operators on them. The space $C(R^n)$ of bounded continuous functions on R^n is a natural candidate for L_0 of Section 2. However, this space is not separable and the measurability hypotheses we need are conveniently verified in a separable space. We introduce therefore the one-point compactification of R^n, denoted by R^n_c, and $C(R^n_c) = L_0$, the space of bounded continuous functions on R^n_c with $\| \ \|_0$ the maximum norm. This space is a separable Banach space. The space $C(R^n_c)$ is sufficiently rich from the point of view of weak convergence of measures on R^n. On the other hand, because of this choice of L_0 (and $L_1, \cdots, L_4$) certain other hypotheses are unnecessarily restrictive (compare with the results in [2]). The example of this section gives a good indication of some of the disadvantages of formulating the limit theorem associated with (3.3) in terms of spaces and operators rather than following a more probabilistic approach.

Assume now that the hypotheses concerning F above hold with $z \in R^n_c$ instead of R^n. We have the following result.

THEOREM 2. *Let* $z^{(\varepsilon)}(\tau)$, $0 \leqq \tau \leqq \tau_0$, *be the process defined by* (3.3) *and assume the following conditions hold:*

(2.i) *For all* $z \in R_c^n$, $t \geqq 0$, $\omega \in \Omega$,

$$|F_i(z, t, \omega)| \leqq C\,, \qquad \left|\frac{\partial F_i}{\partial z_j}(z, t, \omega)\right| \leqq C\,,$$

$$\left|\frac{\partial^2 F_i}{\partial z_j\,\partial z_k}(z, t, \omega)\right| \leqq C\,, \qquad \left|\frac{\partial^3 F_i}{\partial z_j\,\partial z_k\,\partial z_l}(z, t, \omega)\right| \leqq C\,. \tag{3.4}$$

$$i, j, k, l = 1, \cdots, n\,,$$

(2.ii) *For each fixed* $z \in R_c^n$ *and* $t \geqq 0$,

$$E\{F_i(z, t, \omega)\} = 0\,, \qquad i = 1, \cdots, n\,. \tag{3.5}$$

(2.iii) $F(z, t, \omega)$ *as a function of* ω *is* $\mathscr{F}_t^t$ *measurable for fixed* t *and all* $z \in R_c^n$ *and the* σ*-algebras* $\mathscr{F}_s^t$ *satisfy the conditions stated above* (2.1). *In addition,*

$$\int_0^\infty \rho^{1/2}(s)\,ds < \infty\,.$$

(2.iv) *The limits*

$$a_{ij}(z) = \lim_{t\to\infty}\frac{1}{t}\int_{t_0}^{t_0+t}\int_{t_0}^{s} E\{F_i(z, \sigma, \omega)\,F_j(z, s, \omega)\}\,d\sigma\,ds\,, \tag{3.6}$$

$$i, j = 1, 2, \cdots, n\,,$$

$$b_j(z) = \lim_{t\to\infty}\frac{1}{t}\int_{t_0}^{t_0+t}\int_{t_0}^{s}\sum_{i=1}^{n} E\left\{F_i(z, \sigma, \omega)\,\frac{\partial F_j}{\partial z_i}(z, s, \omega)\right\}d\sigma\,ds\,, \tag{3.7}$$

exist uniformly in $z \in R_c^n$ *and* $t_0 \geqq 0$ *and the rate of approach of the double integrals to* a_{ij}, b_j *is not slower than* $1/t$, $t \to \infty$. *Assume further that* $a_{ij}(z)$ *and* $b_j(z)$, $i, j = 1, \cdots, n$, *have bounded continuous derivatives in* R_c^n *up to order* 4.

Under these hypotheses and with $f \in C^4(R_c^n)$ *and* $w^{(\varepsilon)}(\tau, z) = E\{f(z^{(\varepsilon)}(\tau))\}$, *we have*

$$\sup_{0\leqq\tau\leqq\tau_0}\ \sup_{z\in R_c^n}|w^{(\varepsilon)}(\tau, z) - w^{(0)}(\tau, z)| \leqq C(f; \tau_0)\cdot\varepsilon\,. \tag{3.8}$$

Here $C(f; \tau_0)$ *denotes a constant depending on* f *and* τ_0 *and* $w^{(0)}$ *is the solution of the Cauchy problem*

$$\frac{\partial w^{(0)}}{\partial \tau} = \sum_{i,j=1}^{n} a_{ij}\frac{\partial^2 w^{(0)}}{\partial z_i\,\partial z_j} + \sum_{j=1}^{n} b_j\frac{\partial w^{(0)}}{\partial z_j} = \bar{V}w^{(0)}\,, \qquad \tau > 0\,, \tag{3.9}$$

$$w^{(0)}(0, z) = f(z)\,.$$

Proof: We prove this theorem by reducing it to a special case of Theorem 1.

The Banach spaces L_k are $C^k(R_c^n)$, $k = 0, 1, \cdots$, and $\| \ \|_k$ is the sum of the maximum norm of the function and its derivatives up to order k in the usual manner. The spaces L_k are separable and the L_k, $k \geqq 1$, are dense in L_0. The operator $V(t)$ is defined on L_1 by

$$V(t, \omega) = \sum_{j=1}^{n} F_j(z, t, \omega) \frac{\partial}{\partial z_j} . \tag{3.10}$$

Since the dual space L_0^* of L_0 is the space of finite signed Borel measures on R_c^n, the strong measurability of $V(t)$ is implied by the $\mathscr{F}_t^t$ measurability of

$$\int (V(t, \omega)f)(z)\mu(dz) , \qquad f \in L_1 , \quad \mu \in L_0^* .$$

The latter follows from our hypotheses on F.

Let $T_\sigma^\tau(\omega)z = T_\sigma^\tau z$ denote the solution of (3.3) for $\tau \geqq \sigma \geqq 0$ with $T_\sigma^\sigma z = z$. From our hypotheses it follows that $T_\sigma^\tau(\omega)z$ is a differentiable transformation of R^n into itself for fixed σ, τ, and ω, and $\mathscr{F}_\sigma^\tau$ measurable as a function of ω for all z (if z is the point at infinity, then $T_\sigma^\tau(\omega)z$ is the point at infinity for all $\tau \geqq \sigma$, $\omega \in \Omega$). Using the same argument as above to check measurability we define the operator $U^{(\varepsilon)}$ by

$$(U^{(\varepsilon)}(\sigma, \tau)f)(z) = f(T_\sigma^\tau z) , \qquad f \in C(R_c^n) . \tag{3.11}$$

These propagators do not satisfy (P.i)–(P.iii) because the roles of σ and τ are interchanged. This is however an entirely superficial difference; it only changes the form of $\bar{V}$ since now σ and s are interchanged in (2.13). With this explanation in mind we record the new form of (P.i)–(P.iii):

(P.i)′ The $U^{(\varepsilon)}$ of (3.11) are contraction operators on $L_0 \to L_0$ and they are strongly $\mathscr{F}_{\sigma/\varepsilon^2}^{\tau/\varepsilon^2}$ measurable. $U^{(\varepsilon)}(\tau, \sigma)$: $L_k \to L_k$, $k \geqq 1$.

(P.ii)′ The following property holds:

$$U^{(\varepsilon)}(\sigma, s)U^{(\varepsilon)}(s, \tau) = U^{(\varepsilon)}(\sigma, \tau) . \tag{3.12}$$

(P.iii)′ Moreover,

$$U^{(\varepsilon)}(\sigma, \tau)f = f + \frac{1}{\varepsilon}\int_\sigma^\tau V(s/\varepsilon^2)U^{(\varepsilon)}(s, \tau)f \, ds , \tag{3.13}$$

$$U^{(\varepsilon)}(\sigma, \tau)f = f + \frac{1}{\varepsilon}\int_\sigma^\tau U^{(\varepsilon)}(\sigma, s)V(s/\varepsilon^2)f \, ds , \qquad f \in L_1 . \tag{3.14}$$

We proceed next to the verification of hypotheses (1.i)–(1.vi) of Theorem 1. Hypothesis (1.i) follows from (2.ii). Hypothesis (1.ii) follows from (2.iii), the fact that the Cauchy problem (3.9) is well posed (cf. [17]), and the maximum principle. Note that (a_{ij}) is positive semidefinite as can be verified from its definition (3.6). The operator $\bar{V}$ is the second order (possibly degenerate) elliptic differential operator on the right side of (3.9), $\mathscr{D}_{\bar{V}}$ contains $C^2(R_c^n)$ and, since a_{ij}, b_j are smooth, $\mathscr{D}_{\bar{V}}^2$ contains $C^4(R_c^n)$. Thus $L_4 \cap \mathscr{D}_{\bar{V}}^2 \supset C^4(R_c^n)$ which is dense in $L_0 = C(R_c^n)$.

Hypothesis (1.iii) follows from (2.i). Hypothesis (1.vi) is included in (2.iii). Hypothesis (1.v) follows from the smoothness of the coefficients a_{ij}, b_j, and the results of Oleinik (cf. [17], Theorem 6, and [23], Appendix). It remains to show that (1.iv) holds. This means that we must show that

$$\|f(T_\sigma^\tau z)\|_1 \leqq C_1' e^{\alpha(\tau-\sigma)/\varepsilon} \|f\|_1 \quad \text{a.s.}, \tag{3.15}$$

and

$$\|f(T_\sigma^\tau z)\|_2 \leqq C_2'\left(1+\frac{\tau-\sigma}{\varepsilon}\right) e^{\alpha(\tau-\sigma)/\varepsilon} \|f\|_2 \quad \text{a.s.}. \tag{3.16}$$

We demonstrate (3.15) only, since (3.16) can be derived in the same manner.

From the definition of the $\| \; \|_1$ norm it follows that it suffices to estimate

$$\left\|\frac{\partial}{\partial z_j} f(T_\sigma^\tau z)\right\|_0, \qquad j = 1, 2, \cdots, n. \tag{3.17}$$

Now for $f \in C^1[R_c^n]$ we have

$$\begin{aligned} \left\|\frac{\partial}{\partial z_j} f(T_\sigma^\tau z)\right\| &= \left\|\sum_{i=1}^n f_i(T_\sigma^\tau z) \frac{\partial (T_\sigma^\tau z)_i}{\partial z_j}\right\| \\ &\leqq \|f\|_1 \sup_{z\in R^n} \left\|\frac{\partial (T_\sigma^\tau z)}{\partial z}\right\|. \end{aligned} \tag{3.18}$$

In (3.18), f_j, $j = 1, 2, \cdots, n$, denotes the partial derivative of f with respect to its j-th argument and $\| \; \|$ denotes some matrix norm. Also, $\partial(T_\sigma^\tau z)/\partial z$ denotes the matrix with entries $\partial(T_\sigma^\tau z)_i/\partial z_j$, $i, j = 1, 2, \cdots, n$. From (3.3) we have

$$T_\sigma^\tau z = z + \frac{1}{\varepsilon}\int_\sigma^\tau F(T_\sigma^s z, s/\varepsilon^2, \omega)\, ds, \tag{3.19}$$

and hence

$$\frac{\partial(T_\sigma^\tau z)}{\partial z} = I + \frac{1}{\varepsilon}\int_\sigma^\tau \frac{\partial(T_\sigma^s z)}{\partial z} F^{(1)}(T_\sigma^s z, s/\varepsilon^2, \omega)\, ds\,. \tag{3.20}$$

Here I denotes the identity matrix and $F^{(1)}$ the matrix whose entries are the partial derivatives of F_i, $i = 1, \cdots, n$, with respect to its various arguments in the first (the z) position. From (3.20) and (2.i) it follows that

$$\left\|\frac{\partial(T_\sigma^\tau z)}{\partial z}\right\| \leqq 1 + \frac{c}{\varepsilon}\int_\sigma^\tau \left\|\frac{\partial(T_\sigma^s z)}{\partial z}\right\| ds\,, \tag{3.21}$$

so that

$$\left\|\frac{\partial(T_\sigma^\tau z)}{\partial z}\right\| \leqq e^{c(\tau-\sigma)/\varepsilon}\,. \tag{3.22}$$

The estimate (3.22) and (3.18) together imply (3.15). This completes the verification of the hypotheses and hence the proof of Theorem 2.

Let us observe the following.

1. If the operator $\bar{V}$ in (2.13) is unaltered by the interchange of s and σ in $E\{V(s)V(\sigma)f\}$, then $\bar{V}$ is symmetric in the sense that, for example, the coefficients $a_{ij}(z)$ in the concrete application of Theorem 2 are symmetric: $a_{ij} = a_{ji}$. In general, the fact that s and σ in (2.13) are interchangeable is called the microscopic reversibility (detailed balance) and the resulting symmetry of coefficients an Onsager relation. Some further comments on this can be found in [11].

2. If $F(z, t, \omega) = x(t, \omega) = x(t)$ independently of z, where $x(t)$ is an $\mathscr{F}_t^t$ measurable R^n-valued process, then

$$z^{(\varepsilon)}(\tau) = z + \frac{1}{\varepsilon}\int_0^\tau x(s/\varepsilon^2, \omega)\, ds\,, \tag{3.23}$$

and

$$V(t) = \sum_{j=1}^n x_j(t)\frac{\partial}{\partial z_j}\,. \tag{3.24}$$

Clearly then

$$\bar{V} = \sum_{i,j=1}^n \left\{\lim_{t\to\infty}\frac{1}{t}\int_{t_0}^{t_0+t}\int_{t_0}^{s} E\{x_i(\sigma)x_j(s)\}\, d\sigma\, ds\right\}\frac{\partial^2}{\partial z_i\,\partial z_j}, \tag{3.25}$$

and Theorem 2 implies that $z^{(\varepsilon)}(\tau)$ converges weakly to a Gaussian random vector with covariance matrix

$$\lim_{t\to\infty} \frac{1}{t} \int_{t_0}^{t_0+t} \int_{t_0}^{t_0+t} E\{x_i(\sigma)x_j(s)\}\, d\sigma\, ds .$$

This is the classical central limit theorem for random vectors satisfying a strong mixing condition (cf. [16]). It is clearly a special case of Theorem 2. Since ε corresponds to $1/\sqrt{n}$ in the classical scaling (in discrete time), the error $O(\varepsilon)$ corresponds to $O(1/\sqrt{n})$ which is best possible since $O(1/\sqrt{n})$ is achieved for sums of independent Bernoulli variables.

4. Application to Random Motion on a Lie Group

Let G denote a connected n-dimensional Lie group, I the identity element of G, and Λ the Lie algebra of G; that is, the tangent space of G at the identity. For concreteness (specifically so that we may write (4.6) directly) we assume that G is a matrix Lie group with elements g and the bracket operation in Λ is the matrix commutator, i.e., if $m_1, m_2 \in \Lambda$, then

$$[m_1, m_2] = m_1 m_2 - m_2 m_1 . \tag{4.1}$$

Let $Y_1, Y_2, \cdots, Y_n$ denote a fixed basis in Λ and c_{ij}^k the structure constants

$$[Y_i, Y_j] = \sum_{k=1}^{n} c_{ij}^k Y_k . \tag{4.2}$$

Let G_c denote the one-point compactification of G and $C(G_c)$ the separable Banach space of bounded continuous functions on G_c with the maximum norm $\|\ \|_0$. Let $C^1(G_c)$ denote the linear subspace of $C(G_c)$ for which the strong limits

$$D_j f(g) \equiv \lim_{h\downarrow 0} \frac{f(e^{Y_j h} g) - f(g)}{h}, \qquad j = 1, 2, \cdots, n, \tag{4.3}$$

exist. Here $e^{Y_j h}$ denotes the exponential mapping of Λ into G which is the ordinary exponential matrix since we are dealing with matrix groups. The operators D_j, $j = 1, 2, \cdots, n$, form a basis for the right invariant vector fields on G. The space $C^1(G_c)$ equipped with the norm

$$\|f\|_1 = \|f\|_0 + \sum_{j=1}^{n} \|D_j f\|_0 \tag{4.4}$$

becomes a Banach space which is dense in $C(G_c)$ (cf. [18]). Spaces $C^k(G_c)$, $k = 2, 3, \cdots$, can be defined along with norms $\| \|_k$ in a similar fashion. All spaces $C^k(G_c)$, $k = 1, 2, \cdots$, are dense in $C(G_c)$.

We wish to study a random motion on G. For this purpose let

$$\tilde{m}(t, \omega) = \sum_{j=1}^{n} \tilde{m}_j(t, \omega) Y_j , \qquad \omega \in \Omega , \tag{4.5}$$

be a process with values in Λ. The R^n-valued process $(\tilde{m}_1(t), \tilde{m}_2(t), \cdots, \tilde{m}_n(t))$ has several properties which we state below. Let $A \in \Lambda$ be a fixed matrix and consider the matrix equation

$$\frac{dg(t)}{dt} = Ag(t) + \varepsilon\tilde{m}(t)g(t) , \qquad g(0) = g . \tag{4.6}$$

In order to study the asymptotic properties of the process $g(t)$ when $\varepsilon \to \infty$, $t \to \infty$, we let

$$\tau = \varepsilon^2 t , \qquad g^{(\varepsilon)}(\tau) = e^{-A\tau/\varepsilon^2} g(\tau/\varepsilon^2) . \tag{4.7}$$

Then $g^{(\varepsilon)}(\tau)$ satisfies the matrix equation

$$\frac{dg^{(\varepsilon)}(\tau)}{d\tau} = \frac{1}{\varepsilon} m(\tau/\varepsilon^2)g^{(\varepsilon)}(\tau) , \qquad g^{(\varepsilon)}(0) = g . \tag{4.8}$$

$$m(t) = e^{-At}\tilde{m}(t)e^{At} = \sum_{j=1}^{n} m_j(t, \omega) Y_j . \tag{4.9}$$

We have the following result.

THEOREM 3. *Let $g^{(\varepsilon)}(\tau)$ be the process defined by* (4.8) *and let $0 \leqq \tau \leqq \tau_0$. Assume that the following conditions hold:*

(3.i) *$|\tilde{m}_j(t)| \leqq C$ a.s., $j = 1, \cdots, n$, and A belongs to the compact part of Λ so that e^{At} remains in a compact subset of G containing the identity for all $t \in (-\infty, \infty)$.*

(3.ii) *For $j = 1, 2, \cdots, n$,*

$$E\{\tilde{m}_j(t)\} = 0 . \tag{4.10}$$

(3.iii) *$\tilde{m}_j(t)$, $j = 1, \cdots, n$, are $\mathscr{F}_t^t$ measurable and the σ-algebras $\mathscr{F}_s^t$ satisfy all hypotheses above* (2.1). *In addition,*

$$\int_0^\infty \rho^{1/2}(s)\, ds < \infty .$$

(3.iv) *There exist constants* a_{ij}, $i, j = 1, \cdots, n$, *such that*

$$\sup_{t_0 \geqq 0} \left| a_{ij} - \frac{1}{t} \int_{t_0}^{t_0+t} \int_{t_0}^{s} E\{m_i(\sigma) m_j(s)\}\, d\sigma\, ds \right| \leqq \frac{C}{t}, \tag{4.11}$$

$$t \to \infty, \; i, j = 1, \cdots, n.$$

Let $w^{(\varepsilon)}(\tau, g) = E\{f(g^{(\varepsilon)}(\tau))\}, f \in C^4(G_c)$. *Then*

$$\sup_{0 \leqq \tau \leqq \tau_0} \sup_{g \in G_c} |w^{(\varepsilon)}(\tau, g) - w^{(0)}(\tau, g)| \leqq C \|f\|_4 \cdot \varepsilon, \tag{4.12}$$

where C *is a constant depending on* τ_0, *and* $w^{(0)}$ *satisfies the Cauchy problem*

$$\frac{\partial w^{(0)}}{\partial \tau} = \sum_{i,j=1}^{n} a_{ij} D_i D_j w^{(0)} \equiv \bar{V} w^{(0)}, \qquad w^{(0)}(0, g) = f(g). \tag{4.13}$$

Proof: As for Theorem 2 we transform the present problem so that Theorem 1 applies.

Let T_σ^τ denote the solution matrix of (4.8), $\tau \geqq \sigma$, such that $T_\sigma^\sigma = I$. Define $U^{(\varepsilon)}$ and $V(t)$ by

$$(U^{(\varepsilon)}(\sigma, \tau) f)(g) = f(T_\sigma^\tau g), \qquad f \in C(G_c), \tag{4.14}$$

$$(V(t) f)(g) = \sum_{j=1}^{n} m_j(t) D_j f(g), \qquad f \in C^1(G_c). \tag{4.15}$$

It is easily verified that (P.i)′–(P.iii)′ of Section 3 hold again here. We may proceed therefore with the verification of the hypotheses and the identification of the objects with the ones of Theorem 1.

Clearly, $L_k = C^k(G_c)$, $k = 0, 1, 2, \cdots$. Hypothesis (1.i) follows from (3.ii). Hypothesis (1.ii) follows from (3.iv) and the results of K. Itô (cf. [19], [20]); a detailed treatment is given in [21]. In terms of semigroups, Itô's results are also given in [18]. $\mathscr{D}_{\bar{V}} \supset L_2$ and $L_4 \cap \mathscr{D}_{\bar{V}}^2 \supset C^4(G_c)$ which is dense in $C(G_c)$. Hypotheses (1.iii) and (1.vi) follow from (3.i) and (3.iii), respectively.

Next we verify hypothesis (1.iv). As in Theorem 2, it suffices to obtain (2.16) since (2.17) follows by similar considerations. We shall show that, for each $j = 1, 2, \cdots, n$,

$$\|D_j U^{(\varepsilon)}(\sigma, \tau) f\|_0 \leqq C e^{\alpha(\tau - \sigma)/\varepsilon} \|f\|_1, \tag{4.16}$$

which suffices for (2.16) to hold. We have

$$
\begin{aligned}
D_j U^{(\varepsilon)}(\sigma,\tau)f(g) &= \lim_{h\downarrow 0} \frac{U^{(\varepsilon)}(\sigma,\tau)f(e^{Y_jh}g) - U^{(\varepsilon)}(\sigma,\tau)f(g)}{h} \\
&= \lim_{h\downarrow 0} \frac{f(T_\sigma^\tau e^{Y_jh}g) - f(T_\sigma^\tau g)}{h} \qquad (4.17)\\
&= \lim_{h\downarrow 0} \frac{f(T_\sigma^\tau e^{Y_jh}(T_\sigma^\tau)^{-1} T_\sigma^\tau g) - f(T_\sigma^\tau g)}{h} .
\end{aligned}
$$

For $g \in C$, let $Ad(g)$ be the linear transformation on Λ defined by

$$
Ad(g)\,Y = \frac{d}{dh}(ge^{hY}g^{-1})\Big|_{h=0} = gYg^{-1}, \tag{4.18}
$$

and let $(Ad(g))_{ij}$ denote the matrix elements of $Ad(g)$ relative to the basis $Y_1, Y_2, \cdots, Y_n$. Then we have

$$
D_j U^{(\varepsilon)}(\sigma,\tau)f(g) = \sum_{l=1}^{n} (Ad(T_\sigma^\tau))_{jl} D_l f(T_\sigma^\tau g), \tag{4.19}
$$

and hence

$$
\|D_j U^{(\varepsilon)}(\sigma,\tau)f\|_0 \leqq \|f\|_1 \, \|Ad(T_\sigma^\tau)\| . \tag{4.20}
$$

Here $\|\ \|$ denotes some matrix norm.

From (4.20) it follows that the desired estimate has been reduced to estimating $\|Ad(T_\sigma^\tau)\|$. For this purpose we proceed as follows. From (4.18) and (4.8) we obtain

$$
\frac{\partial}{\partial\tau} Ad(T_\sigma^\tau) = \frac{1}{\varepsilon}[m(\tau/\varepsilon^2), Ad(T_\sigma^\tau)], \tag{4.21}
$$

which in terms of the matrix elements of $Ad(T_\sigma^\tau)$ takes the form

$$
\begin{aligned}
&\frac{\partial}{\partial\tau}(Ad(T_\sigma^\tau))_{ij} = \frac{1}{\varepsilon}\sum_{k=1}^{n}(Ad(T_\sigma^\tau))_{ik} Q_{kj}(\tau/\varepsilon^2), \\
&Q_{kj}(t) = \sum_{l=1}^{n} m_l(t) c_{lk}^j, \qquad Ad(T_\sigma^\sigma) = I .
\end{aligned} \tag{4.22}
$$

In integral form, (4.22) becomes

$$Ad(T_\sigma^\tau) = I + \frac{1}{\varepsilon}\int_\sigma^\tau Ad(T_\sigma^s)\cdot Q(s/\varepsilon^2)\, ds\,, \tag{4.23}$$

so that

$$\|Ad(T_\sigma^\tau)\| \leqq 1 + \frac{C}{\varepsilon}\int_\sigma^\tau \|Ad(T_\sigma^s)\|\, ds\,. \tag{4.24}$$

Here we have used hypothesis (3.i) concerning $\tilde{m}$ and $A \in \Lambda$ and the boundedness of the structure constants. Clearly, (4.24) yields the desired estimate for $\|Ad(T_\sigma^\tau)\|$ and then (4.20) in turn yields hypothesis (1.iv).

We come now to the verification of hypothesis (1.v). From the results of Itô [20] and Hunt [18] it follows that there is a uniquely determined family of probability measures p_τ, $\tau \geqq 0$, on the Borel sets of G_c such that

$$e^{\tau\bar{V}} f(g) = \int_{G_c} f(g'g)\, p_\tau(dg')\,, \qquad f \in C(G_c)\,. \tag{4.25}$$

Assuming that integrals exist we have now

$$D_i\, e^{\tau\bar{V}} f(g) = \int_{G_c} \sum_{j=1}^{n} (Ad(g'))_{ji}\, D_j\, f(g'g)\, p_\tau(dg')\,, \qquad g \in C^1(G_c)\,. \tag{4.26}$$

Thus, for any $j = 1, 2, \cdots, n$ and $f \in C^1(G_c)$,

$$\|D_j\, e^{\tau\bar{V}} f\|_0 \leqq \|f\|_1 \int_{G_c} \|Ad(g')\|\, p_\tau(dg')\,. \tag{4.27}$$

Here $\|\cdot\|$ denotes some matrix norm. Since the function $\|Ad(g)\|$ does not belong to the class $C(G_c)$, we must show that the integral on the right side of (4.27) exists. Specifically we shall show that

$$\sup_{0 \leqq \tau \leqq \tau_0} \int_{G_c} \|Ad(g)\|\, p_\tau(dg) \leqq c(\tau_0) < \infty\,, \tag{4.28}$$

where $c(\tau_0)$ is a constant depending on τ_0. Estimates for $\|D_i\, D_j\, e^{\tau\bar{V}} f\|_0$ are obtained in the same manner. We proceed now with the proof of (4.28).

Let us observe that for any matrix norm there is a constant c such that if $A = (A_{ij})$, A_{ij} real, then

$$\|A\| \leqq c\left(\sum_{i,j=1}^{n} A_{ij}^2\right)^{1/2} \equiv c\,\|A\|_E\,. \tag{4.29}$$

It suffices therefore to show (4.28) with $\|\ \|$ replaced by $\|\ \|_E$. Now, if $A \otimes A$ denotes the tensor (Kronecker) product of A with itself, then

$$\|A\|_E = \{H(A \otimes A)\}^{1/2}, \tag{4.30}$$

where $H(A \otimes A)$ is defined as $\sum_{i,j} A_{ij}^2$. Let us also note that the linear transformations $Ad(g)$, $g \in G$, constitute a finite-dimensional representation of G, the adjoint representation. The desired estimate (4.28), and hence the completion of the proof of Theorem 3, follow directly from the following result which is also of independent interest.

THEOREM 4. *Let $\pi(g)$, $g \in G$, be a finite-dimensional representation of G. Then*

$$\int_{G_c} \pi(g)\, p_\tau(dg) = \exp\{\tau \nabla \pi(g)|_{g=I}\}, \tag{4.31}$$

in the sense that the integral on the left exists and equals the expression on the right.

Proof: First note that $\pi(g)$ is analytic in g, being a finite-dimensional representation (cf. [22], p. 107). Thus the expression on the right side of (4.31) is well defined. We shall give the remainder of the proof after proving two lemmas.

LEMMA 4.1. *Let $g(\tau)$ be the diffusion (Brownian motion) corresponding to ∇ starting at time 0 from the identity. Let K_1 be a compact neighborhood of the identity and let σ_1 be the first exit time from K_1 of the process $g(\tau)$. Then, with probability one,*

$$\lim_{K_1 \uparrow G} \sigma_1 = \infty \tag{4.32}$$

Proof of Lemma 4.1: Define sets $K_2, K_3, \cdots$ as follows:

$$\begin{aligned} K_2 &= \{g \in G \mid g = g_1 g_2, g_1, g_2 \in K_1\}, \\ K_3 &= \{g \in G \mid g = g_1 g_2, g_1, g_2 \in K_2\}, \cdots. \end{aligned} \tag{4.33}$$

Clearly, $K_1 \subset K_2 \subset K_3 \subset \cdots$ and $K_n \uparrow G$. Let $\sigma_1, \sigma_2, \cdots$ denote the exit times of $g(\tau)$ from $K_1, K_2, \cdots$ and let $g^{(j)} = g(\sigma_j)$, $j = 1, 2, \cdots$. Define $K_1^{(j)}$ by

$$K_1^{(1)} = K_1, \quad K_1^{(2)} = K_1 g^{(1)}, \quad K_1^{(3)} = K_1 g^{(2)}, \quad \cdots. \tag{4.34}$$

The exit time of $g(\tau)$ from $K_1^{(j)}$, conditional on $g^{(j-1)}$ $(g^{(0)} = I), j = 1, 2, \cdots$, at $\tau = 0$, is denoted by $\sigma^{(1)}, \sigma^{(2)}, \cdots (\sigma^{(1)} = \sigma_1)$. In view of the right invariance of $g(\tau)$ and the strong Markov property, the positive random variables $\sigma^{(1)}, \sigma^{(2)}, \cdots$ are independent and identically distributed. Furthermore, since $K_1^{(2)} \subset K_2$ we have

$$\sigma_2 \geqq \sigma^{(1)} + \sigma^{(2)}, \tag{4.35}$$

and similarly

$$\sigma_n \geqq \sigma^{(1)} + \sigma^{(2)} + \cdots + \sigma^{(n)}, \qquad n = 3, 4, \cdots. \tag{4.36}$$

From the strong law of large numbers it follows now that $\sigma^{(1)} + \sigma^{(2)} + \cdots + \sigma^{(n)} \uparrow \infty$ with probability one and hence, from (4.36), $\sigma_n \uparrow \infty$ with probability one. This then completes the proof of the lemma.

LEMMA 4.2. *Let $\pi(g)$ be any finite-dimensional representation of G and let $\pi(g) \otimes \pi(g)$ be the Kronecker product representation. Then*

$$\begin{aligned} &H\{\exp\{\tau \tilde{V}\pi(g) \otimes \pi(g)|_{g=I}\}\pi(g) \otimes \pi(g)\} \\ &\qquad \geqq H\{\exp\{\tau\tilde{V}\pi(g)|_{g=I}\}\pi(g) \otimes \exp\{\tau\tilde{V}\pi(g)|_{g=I}\}\pi(g)\} \geqq 0. \end{aligned} \tag{4.37}$$

Proof of Lemma 4.2: First we note that there exists a family of probability measures $R_\varepsilon(dg)$ all having support in a compact neighborhood of I and such that

$$(\tilde{V}f)(I) = \lim_{\varepsilon \downarrow 0} \frac{1}{\varepsilon} \int [f(g) - f(I)] R_\varepsilon(dg). \tag{4.38}$$

With $\varepsilon > 0$ fixed, define the operator $\tilde{V}_\varepsilon$ by

$$(\tilde{V}_\varepsilon f)(g') = \frac{1}{\varepsilon} \int [f(gg') - f(g')] R_\varepsilon(dg). \tag{4.39}$$

$\tilde{V}_\varepsilon$ generates a Markov semigroup of Poisson type; for any function $f(g)$,

$$\begin{aligned} (e^{\tilde{V}_\varepsilon \tau} f)(g') &= \int f(gg') P_\varepsilon(\tau, dg) = \sum_{n=0}^{\infty} \frac{\tau^n}{n!} (\tilde{V}_\varepsilon^n f)(g') \\ &= \sum_{n=0}^{\infty} e^{-\tau/\varepsilon} \frac{(\tau/\varepsilon)^n}{n!} E\{f(g_n g_{n-1} \cdots g_1 g')\}. \end{aligned} \tag{4.40}$$

Here $g_1, g_2, \cdots, g_n$ are independent random elements in G with common distribution $R_\varepsilon(dg)$.

If $\pi(g)$ is any finite-dimensional representation, then

$$\sup_{g\in K} \|\pi(g)\|_E \leqq M < \infty , \tag{4.41}$$

where K is the support of the measures $R_\varepsilon(dg)$. Clearly then, if f is replaced by π,

$$\|(e^{\mathcal{V}_\varepsilon \tau} \pi)(g')\|_E \leqq \|\pi(g')\|_E \sum_{n=0}^{\infty} e^{-\tau/\varepsilon}\left(\frac{\tau}{\varepsilon}\right)^n \frac{M^n}{n!} < \infty , \qquad g' \in K . \tag{4.42}$$

In view of (4.42) we find, by direct computation, that (4.31) holds with $\mathcal{V}_\varepsilon$ in place of $\mathcal{V}$. But then (4.37), with $\mathcal{V}_\varepsilon$ in place of $\mathcal{V}$, is just Schwartz's inequality:

$$0 \leqq \|E\{\pi(g_\varepsilon(\tau))\}\|_E^2 \leqq E\{\|\pi(g_\varepsilon(\tau))\|_E^2\} . \tag{4.43}$$

Here $g_\varepsilon(\tau)$ denotes the Markov process corresponding to $\mathcal{V}_\varepsilon$ and such that $g_\varepsilon(0) = I$. On letting $\varepsilon \to 0$ in (4.43) (which is (4.37) with $\mathcal{V}$ replaced by $\mathcal{V}_\varepsilon$) we obtain the desired result.

Proof of Theorem 4: Define

$$\begin{aligned} U(\tau, g) &= \exp\{\tau \mathcal{V}\pi(g)|_{g=I}\}\pi(g) , \\ W(\tau, g) &= \exp\{\tau \mathcal{V}\pi(g) \otimes \pi(g)|_{g=I}\}\pi(g) \otimes \pi(g) , \\ w(\tau, g) &= H(W(\tau, g)) . \end{aligned} \tag{4.44}$$

From Lemma 4.2 we have

$$w(\tau, g) \geqq \|U(\tau, g)\|_E^2 \geqq 0 . \tag{4.45}$$

Recalling that $\mathcal{V}$ is a right invariant differential operator and that $\pi(g)$ is a representation, we find by direct computation that

$$\frac{\partial \theta(\tau, g)}{\partial \tau} = \mathcal{V}\theta(\tau, g) , \qquad \tau > 0 , \tag{1.46}$$

where θ is either w or any entry of U or W. Let K be any compact neighborhood of I and let σ_K be the first exit time for the process $g(\tau)$ corresponding to $\mathcal{V}$ with $g(0) = I$. Then for any $\tau \geqq 0$ we have, by Itô's formula,

$$\theta(\tau, I) = E\{\theta(0, g(\tau)), \tau \leqq \sigma_K\} + E\{\theta(\tau - \sigma_K, g(\sigma_K)), \tau > \sigma_K\} . \tag{4.47}$$

Taking $\theta = w$ in (4.47) and using (4.45) yields the inequality

$$w(\tau, I) \geqq E\{w(0, g(\tau)), \tau \leqq \sigma_K\}. \tag{4.48}$$

We now let $K \uparrow G$, use Lemma 4.1, and Fatou's lemma, and obtain

$$\infty > w(\tau, I) \geqq E\{W(0, g(\tau))\} = E\{\|\pi(g(\tau))\|_E^2\}. \tag{4.49}$$

Therefore, $E\{\|\pi(g(\tau))\|_E\} < \infty$.

For any entry $\pi_{ij}(g)$ of $\pi(g)$ we have

$$\begin{aligned} \left|\int \pi_{ij}(g'g)\, P_\tau(dg')\right| &\leqq \int |\pi_{ij}(g'g)|\, P_\tau(dg') \\ &\leqq \left(\int \pi_{ij}^2(g'g)\, p_\tau(dg')\right)^{1/2} \leqq w^{1/2}(\tau, g) < \infty. \end{aligned} \tag{4.50}$$

Thus,

$$E\{\pi_{ij}(g(\tau))\} = \int \pi_{ij}(g)\, p_\tau(dg) \tag{4.51}$$

is finite.

We shall now show that (4.31) holds, i.e.,

$$U(\tau, I) = E\{\pi(g(\tau))\}. \tag{4.52}$$

For this purpose we let $K \uparrow G$ in (4.47) with θ any of the entries of U. Since $P\{\tau \leqq \sigma_K\} \to 1$, $K \uparrow G$, and $\pi(g)$ is integrable with respect to $p_\tau(dg)$, the first term on the right side of (4.47) converges to

$$\int \pi(g)\, p_\tau(dg).$$

It is therefore sufficient to show that if θ is any entry of U, then

$$E\{\theta(\tau - \sigma_K, g(\sigma_K)), \tau > \sigma_K\} \to 0 \qquad \text{as} \qquad K \uparrow G. \tag{4.53}$$

But

$$\begin{aligned} |E\{\theta(\tau - \sigma_K, g(\sigma_K)), \tau > \sigma_K\}|^2 & \\ &\leqq E\{|\theta(\tau - \sigma_K, g(\sigma_K))|^2, \tau > \sigma_K\} P\{\tau > \sigma_K\} \\ &\leqq E\{w(\tau - \sigma_K, g(\sigma_K)), \tau > \sigma_K\} P\{\tau > \sigma_K\} \\ &\leqq w(\tau, I) P\{\tau > \sigma_K\} \to 0, \qquad K \uparrow G. \end{aligned} \tag{4.54}$$

The last inequality in (4.54) follows from (4.47), with $\theta = w$, in the same manner that (4.48) was obtained. This completes the proof of Theorem 4.

The integrability of $\pi(g)$ and hence (4.31) can also be shown by the following simple argument due to D. Stroock. Recall the definition of the sets

$$K_1, K_2, \cdots$$

in (4.33) and the first exit times $\sigma_1, \sigma_2, \sigma_3, \cdots$ from these sets. From (4.36) it follows that $P\{\sigma_k \leqq T\} \leqq (P\{\sigma_1 \leqq T\})^k$, $k = 1, 2, \cdots$. Let $M = \sup_{g \in K_1} \|\pi(g)\|$ and choose $T > 0$ such that $P\{\sigma_1 \leqq T\} \leqq 1/2M$. Then, with $\sigma_0 \equiv 0$, $g(0) = I$, and the representation property of $\pi(g)$,

$$E\{\|\pi(g(T))\|\} = \sum_{k=1}^{\infty} E\{\|\pi(g(T))\|, \sigma_{k-1} \leqq T < \sigma_k\}$$

$$\leqq \sum_{k=1}^{\infty} \frac{M^k}{(2M)^{k-1}} = 2M .$$

From the representation property again it follows that, for any $0 \leqq \tau < \infty$,

$$E\{\|\pi(g(\tau))\|\} \leqq (2M)^n , \qquad n = [\tau/T] + 1 .$$

This proves the integrability of $\pi(g)$ from which (4.31) follows easily.

Let us comment briefly on the role of the term Ag in (4.6). Because of this term the process $m(t)$ in (4.9) is not stationary and one must allow for the generality we have here. Many applications, for example in [8], [9], present problems in the form (4.6). Frequently the presence of A, which is called the fast varying part of the process, simplifies the analysis of (4.13) considerably (see [8], [9]).

A central limit theorem on Lie groups due to Wehn is presented in [24]. A recent treatment of limit theorems on Lie groups is [25].

Bibliography

[1] Papanicolaou, G. C., and Hersh, R., *Some limit theorems for stochastic equations and applications*, Indiana Univ. Math. J., Vol. 21, 1972, pp. 815–840.

[2] Khasminskii, R. Z., *A limit theorem for the solutions of differential equations with random right-hand sides*, Theory Prob. Applications, Vol. 11, 1966, pp. 390–406.

[3] Stratonovich, R. L., *Conditional Markov Processes and their Application to the Theory of Optimal Control*, Elsevier, New York, 1968.

[4] Kubo, R., *Stochastic Liouville equation*, J. Math. Phys., Vol. 4, 1963, pp. 174–183.

[5] Lax, M., *Classical noise IV; Langevin methods*, Rev. Modern Phys., Vol. 38, 1966, 561–566.

[6] Papanicolaou, G., and Keller, J. B., *Stochastic differential equations with applications to random harmonic oscillators and wave propagation in random media*, SIAM J. Appl. Math., Vol. 20, 1971, pp. 287–305.

[7] Cogburn, R., and Hersh, R., *Two limit theorems for random equations*, Indiana Univ. Math. Jour., Vol. 22, 1973, pp. 1067–1089.
[8] Burridge, R., and Papanicolaou, G., *The geometry of coupled mode propagation in one dimensional random media*, Comm. Pure Appl. Math., Vol. 25, 1972, pp. 715–757.
[9] Papanicolaou, G., *A kinetic theory for power transfer in stochastic systems*, J. Math. Phys., Vol. 13, 1972, pp. 1912–1918.
[10] Griego, R., and Hersh, R., *Theory and application of random evolutions*, Trans. Amer. Math. Soc., Vol. 156, 1971, pp. 405–418.
[11] Hersh, R., and Papanicolaou, G., *Non-commuting random evolution and an operator valued Feynman-Kac formula*, Comm. Pure Appl. Math., Vol. 25, 1972, pp. 337–367.
[12] Pinsky, M., *Multiplicative operator functionals and their asymptotic properties*, in Advances in Probability, to appear.
[13] Kertz, R., Northwestern University Thesis, 1972.
[14] Hille, E., and Phillips, R., *Functional Analysis and Semigroups*, Am. Math. Soc. Coll. Publ. Vol. XXXI, Providence, R.I., 1957.
[15] Loève, M., *Probability Theory*, Van Nostrand, Princeton, 1963.
[16] Doob, J. L., *Stochastic Processes*, J. Wiley, New York, 1953.
[17] Oleinik, O., *Alcuni risultati sulle equazioni lineari e quasi lineari ellitticoparaboliche a derivate parziali del secondo ordine*, Rendiconti Accad. Naz., Vol. 40, 1966, pp. 775–784.
[18] Hunt, G. A., *Semigroups of measures on Lie groups*, Trans. Am. Math. Soc., Vol. 31, 1956, pp. 264–293.
[19] Itô, K., *Stochastic differential equations in a differentiable manifold*, Nagoya Math. J., Vol. 1, 1950, pp. 35–47.
[20] Itô, K., *Brownian Motion in a Lie Group*, Proc. Japan Acad., Vol. 26 (8), 1950, pp. 4–10.
[21] McKean, H. P. Jr., *Stochastic Integrals*, Academic Press, New York, 1969.
[22] Helgason, S., *Differential Geometry and Symmetric Spaces*, Academic Press, New York, 1962.
[23] Stroock, D. S., and Varadhan, S. R. S., *On degenerate elliptic-parabolic operators of second order and their associated diffusions*, Comm. Pure Appl. Math., Vol. 25, 1972, pp. 651–714.
[24] Grenander, U., *Probabilities on Algebraic Structures*, J. Wiley and Sons, New York, 1963.
[25] Stroock, D. S., and Varadhan, S. R. S., *Limit theorems on Lie groups*, to appear.

Received February, 1973.

Sankhya Ser. A 35 (1973) 277-294

LIMIT THEOREMS FOR RANDOM WALKS ON LIE GROUPS*

By DANIEL W. STROOCK

University of Colorado

and

S. R. S. VARADHAN

New York University, Courant Institute of Mathematical Sciences

SUMMARY. Limit Theorems for Random walks on Lie groups are prooved. Necessary and sufficient conditions analogous to Lindeberg's conditions are obtained.

1. Introduction

The aim of this paper is to prove the analogue of the classical central limit theorem under Lindeberg's conditions, when the random variables take values in a Lie group. Actually we consider the convergence of the associated random walks to the corresponding "Gaussian" process with independent increments. There have been various generalizations arrived at earlier. For instance the locally compact abelian group has been treated in Parthasarathy, Rao and Varadhan (1963). One can find some partial results in the noncommutative case in Parthasarathy (1964) and Wehn (1962). A survey of the results known until then appears in Sazanov and Tutubalin (1966). But there seems to be no complete theory.

In Stroock and Varadhan (1969b) we developed methods for examining the convergence of Markov chains to diffusion processes. These methods are applicable in the current context. One can exploit the special situation at hand and obtain complete results. We determine all "Gaussian" processes with independent increments on a Lie group G and obtain necessary and sufficient conditions for the convergence of random walks to any given "Gaussian" process with independent increments. Unlike the earlier works cited, the techniques do not require an intimate knowledge of the Lie group. In particular the representation theory is not used at all.

2. Notation and summary

Let G be a connected Lie group of dimension d. A continuous process with independent left increments is a stochastic process $X(t) : 0 \leqslant t \leqslant T$, with values in G, satisfying the following.

A1. $X(0) = e$ with probability 1.

A2. For $0 \leqslant t_1 \leqslant t_2 \leqslant \ldots \leqslant t_n \leqslant T$ and every n, the increments $X(t_1)$, $[X(t_1)]^{-1}X(t_2), \ldots, [X(t_j)]^{-1}X(t_{j+1}), \ldots$ are mutually independent.

* The research in this paper was supported by U.S. Air Force, Grant AF-AFOSR-72-2307.

A3. $X(t)$ is a continuous function of t with probability 1. Such a process will be said to be homogeneous in time if in addition to A1, A2 and A3 the process satisfies :

A4. The distribution of the increment $[X(t_1)]^{-1}X(t_2)$ for $0 \leqslant t_1 \leqslant t_2 \leqslant T$, as a function of t_1 and t_2 depends only on the difference $t_2 - t_1$.

Let Ω_c denote the space of continuous functions on $[0\ T]$ with values in G endowed with uniform convergence and the standard Borel structure. Let $x(s)$ denote the evaluation map at time s. Let M_t be the σ-field generated by $x(s)$ for $0 \leqslant s \leqslant t$. The total σ-field M is of course M_T. Let $\mathcal{M}$ denote the totality of all probability measures on $[\Omega_c, M]$. The convergence notion on $\mathcal{M}$ will be weak convergence. Processes satisfying assumptions A1, A2 and A3 will form a closed subset of $\mathcal{P}$ of $\mathcal{M}$. Those that satisfy in addition A4, will be a closed subset $\mathcal{P}_h$ of $\mathcal{P}$. One of the aims of this paper is to describe all the elements of $\mathcal{P}$ and parametrize them in a canonical fashion.

The theorem of Hunt (1956) describes $\mathcal{P}_h$ in a natural fashion. Let $\mathcal{G}$ be the Lie algebra of G. We shall view the elements ξ of $\mathcal{G}$ as left invariant vector fields on G. They act on the space $\mathcal{S}$ of C^∞ functions on $\mathcal{G}$ with compact support.

$$(\xi f)(x) = \frac{d}{dt} f(xe^{t\xi})\Big|_{t=0} \qquad \ldots \quad (2.1)$$

Let $\xi_1, \xi_2, \ldots, \xi_d$ be a basis for $\mathcal{G}$. Let $a = \{a_{ij}\}$ be any $d \times d$ symmetric nonnegative definite matrix and b_j any d vector. Let us denote by $L_{a,b}$ the operator

$$L_{a,b} = \frac{1}{2} \Sigma\, a_{ij}\xi_i\xi_j + \Sigma b_j\xi_j.$$

Then $L_{a,b}$ is the infinitesimal generator of a homogeneous Markov process, which is also a process with independent left increments. Fixing the starting point always at the identity e of the group G these processes exhaust all of $\mathcal{P}_h$ in a one to one manner as a and b are allowed to vary over their respective classes.

In some sense if one chose a and b to be time dependent, the processes corresponding to these should exhaust the bulk of $\mathcal{P}$. In the case of the additive group of real numbers the most general element of $\mathcal{P}$ is characterized by the variance function $\sigma^2(t)$ which is an arbitrary continuous nondecreasing function and the mean function $m(t)$ which is an arbitrary continuous function. The relation to a and b is

$$\sigma^2(t) = \int_0^t a(s)\, ds$$

$$m(t) = \int_0^t b(s)\, ds.$$

This shows that if $a(\cdot)$ and $b(\cdot)$ are chosen to be arbitrary integrable choices, then one gets most of $\mathcal{P}$ but not all of it. Hence by analogy, we define the following :

(i) Let V be the totality of all continuous mappings $A(t)$ from $[0, T]$ into nonnegative definite symmetric matrices such that $A(t) \geqslant A(s)$ for $t \geqslant s$ and $A(0) = 0$.

(ii) V_0 consists of absolutely continuous elements of V.

(iii) $\mathscr{F}$ consists of all continuous maps of $[0, T]$ into G.

(iv) $\mathscr{F}_0$ consists of absolutely continuous elements of $\mathscr{F}$. ... (2.2)

The relevant topologies on V and $\mathscr{F}$ are those induced by uniform convergence of the elements. Let us take a function $A(t)$ in V_0 and a function $g(t) \in \mathscr{F}_0$. We shall write

$$A(t) = \int_0^t a(s)\, ds$$

$$\dot{g}(t) = \Sigma b_j(t)\, \xi_j(g(t)). \qquad \text{... (2.3)}$$

Corresponding to this we shall write down the family of operators L_s, $0 \leqslant s \leqslant T$ given by

$$L_s = \frac{1}{2}\, \Sigma a_{ij}(s) \xi_i \xi_j + \Sigma b_j(s) \xi_j.$$

We define a measure P on Ω_c by the relation

(i) $P\{x(0) = e\} = 1$

(ii) For each $f \in \mathcal{S}$

$$f(x(t)) - \int_0^t (L_s f)(x(s))\, ds \qquad \text{... (2.4)}$$

is a martingale relative to $[\Omega_c, M_t, P]$ for $0 \leqslant t \leqslant T$.

This determines a unique measure $P = P_{A(\cdot), g(\cdot)}$ for $A(\cdot) \in V_0$ and $g(\cdot) \in \mathscr{F}_0$. We have the following.

Theorem 2.1 : *The map* $(A(\cdot), g(\cdot)) \rightarrow P_{A(\cdot), g(\cdot)}$ *from* $V_0 \times \mathscr{F}_0 \rightarrow \mathscr{P}$ *extends by continuity into a map from all of* $V \times \mathscr{F} \rightarrow \mathscr{P}$, *which is one to one, onto and bicontinuous.*

This is the analog of the description in the simplest case of Gaussian processes with independent increments in terms of their mean functions and covariance functions. In such a case, a process $X(t)$ with variance $\sigma^2(t)$ and mean $m(t)$ can be written as

$$X(t) = Y(t) + m(t)$$

where $Y(t)$ is a process with mean 0 and variance $\sigma^2(t)$. The analog of this decomposition will yield a slightly better description of $\mathscr{P}$ than that provided by Theorem 2.1.

For each $A(\cdot) \in V$, let us associate directly a process Q as follows :

(i) $Q[x(0) = e] = 1$

(ii) For each $f \in \mathcal{S}$

$$f(x(t)) = \frac{1}{2}\, \Sigma \int_0^t (\xi_i \xi_j f)(x(s))\, dA_{ij}(s) \qquad \text{... (2.5)}$$

is a martingale relative to $[\Omega_c, M_t, Q]$ for $0 \leqslant t \leqslant T$.

We can show that $Q = Q_{A(\cdot)}$ exists and is unique. If $e(\cdot)$ is the element of $\mathcal{F}_0$ identically equal to e then one can easily identify $Q_{A(\cdot)}$, with $P_{A(\cdot),\, e(\cdot)}$ [if $A(\cdot)$ happens to be an element of $\mathcal{F}_0$]. If $X(t)$ is a process in $\mathcal{P}$ then so is $X(t)\, g(t)$ where $g(t)$ is an arbitrary element of $\mathcal{F}$. If $X(t)$ has distribution $Q_{A(\cdot)}$ we shall denote by $Q_{A(\cdot), g(\cdot)}$ the distribution of $X(t) \cdot g(t)$. We then have

Theorem 2.2 : *The map* $(A(\cdot), g(\cdot)) \to Q_{A(\cdot), g(\cdot)}$ *from* $V \times \mathcal{F} \to \mathcal{P}$ *is bicontinuous, one to one and onto.*

This of course leaves the question of reconciling the two parametrizations. Let $g \epsilon G$ be arbitrary. Consider the induced inner automorphism

$$x \to gxg^{-1}$$

of G onto G. This in turn induces an automorphism Ad_g of the Lie algebra $\mathcal{G}$ on $\mathcal{G}$ by

$$e^{t(A_d g)\xi} = ge^{t\xi}g^{-1}.$$

We can express this linear transformation by a matrix $\tau_{ij}(g)$ where

$$Ad_g \xi_j = \Sigma \tau_{ij}(g) \xi_j.$$

The following theorem links $P_{A(\cdot),\, g(\cdot)}$ to $Q_{B(\cdot), h(\cdot)}$.

Theorem 2.3 : $P_{A(\cdot),\, g(\cdot)} = Q_{B(\cdot), h(\cdot)}$ *if and only if*

(i) $g(t) \equiv h(t)$ *for all* $t \epsilon [0, T]$.

and

(ii) $B_{rl}(t) = \Sigma \int\limits_0^t \tau_{ir}(g(s)) \tau_{jl}(g(s))\, dA_{ij}(s)$.

We next turn to the convergence of random walks to processes with independent increments. For each n let k_n independent random variables $X_{n1}, X_{n2}, \ldots X_{nk_n}$ with values in G be given. Let $\alpha_{n1}, \ldots, \alpha_{nk_n}$ be their respective distributions. For each n, divide the interval $[0, T]$ into $k_n + 1$ parts with $0 = t_{n0} < t_{n1} < t_{n2} < \ldots < t_{nk_n} < t_{n,k_n+1} = T$ being the points of division. We assume that $\sup\limits_{0 \leqslant j \leqslant k_n} | t_{nj} - t_{nj+1} |$ tends to zero with n.

Let us construct a process $X_n(t)$ by the following recipe :

$$\begin{aligned} X_n(t) &= e && \text{for } 0 \leqslant t < t_{n1} \\ &= X_{n1} X_{n2} \ldots X_{nj} && \text{for } t_{nj} \leqslant t < t_{n,j+1}, \quad j = 1, 2, \ldots, k_{n-1} \\ &= X_{n1}\, X_{n2} \ldots X_{nk_n} && \text{for } t_{n,k_n} \leqslant t \leqslant T. \end{aligned}$$

Let Ω_d be the space of functions on $[0, T]$ with values in G which have discontinuities only of the first kind. The space is provided with the Skorohod topology. The process $X_n(t)$ induces a measure P_n on the space Ω_d. Since $\Omega_c \subset \Omega_d$ we can view any measure

$P \epsilon \mathcal{P}$ as a measure on Ω_d concentrated on Ω_c. Given such a P we want to find the necessary and sufficient conditions for P_n to converge weakly to P. One necessary condition is obviously the Lindeberg condition

$$\lim_{n\to\infty} \sum_{j=1}^{k_n} \alpha_{nj}(U^c) = 0 \qquad \dots (2.6)$$

for any neighbourhood U of e in G.

Let us take a coordinate system $\phi = (\phi_1, \dots, \phi_d)$ given in a neighbourhood N of e. We shall assume that the coordinate system satisfies

(i) $\phi(e) = 0$

and

(ii) $(\xi_i \phi_j)(e) = 1$ for $i = j$ and 0 for $i \neq j$. ... (2.7)

Let us take a neighbourhood V with a compact closure contained in N and define the mean and covariance of α_{nj}. The mean g_{nj} is defined by

$$\phi(g_{nj}) = \int_V \phi(x)\, \alpha_{nj}(dx). \qquad \dots (2.8)$$

It is well defined because (2.7) implies

$$\lim_{n\to\infty} \sup_{1 \leqslant j \leqslant k_n} \alpha_{nj}(U^c) = 0 \qquad \dots (2.9)$$

for any neighbourhood U of e.

The covariance matrix a_{pq}^{nj} is defined by

$$a_{pq}^{nj} = \int_V [\phi_p(x) - \phi_p(g_{nj})][\phi_q(x) - \phi_q(g_{nj})]\alpha_{nj}(dx). \qquad \dots (2.10)$$

We next define the cumulated means and variances by

$$h_{nj} = g_{n1} g_{n2} \cdots g_{nj} \qquad \dots (2.11)$$

and

$$A^{nj} = a^{n1} + a^{n2} + \dots + a^{nj}. \qquad \dots (2.12)$$

We superimpose these on the time scale by defining

$$\begin{aligned} \tau_n(t) &= 0 && \text{for } 0 \leqslant t \leqslant t_{n1} \\ &= j && \text{for } t_{nj} \leqslant t < t_{n,j+1},\ j = 1, \dots, k_n - 1 \\ &= k_n && \text{for } t_{nk_n} \leqslant t \leqslant T, \end{aligned} \qquad \dots (2.13)$$

$$h_n(t) = h_{n,\tau_n}(t), \qquad \dots (2.14)$$

and

$$A^n(t) = A^{n,\tau_n(t)}. \qquad \dots (2.15)$$

Given an element $P \epsilon \mathcal{P}$, according to Theorem 2.1, we can write it in the form $P_{A(\cdot),\, g(\cdot)}$. In order to state the condition on $A^n(t)$ and $h_n(t)$ so that P_n may converge to P we need the following considerations. The means g_{nj} are sensitive to the coordinate system used. Equations (2.7) alone are not enough to determine g_{nj} uniquely. The representation $P_{A(.),g(.)}$ of P on the other hand is canonical. So if the means $h_n(t)$ in the ϕ coordinates converge to $h(t)$, there is a rule which connects $h(t)$ with $g(t)$. Let us define

$$\rho_i^{pq} = (\xi_p \xi_q \phi_i)(e) \qquad \dots \quad (2.16)$$

$$b_i(t) = -\frac{1}{2} \Sigma \ \rho_i^{pq} A_{pq}(t). \qquad \dots \quad (2.17)$$

Given $h(t)$, consider the solution $k(t)$ of the equation

$$dk(t) = \Sigma \ db_i(t) \tau_{il}(h(s)) \xi_l(k(t))$$

$$k(0) = e. \qquad \dots \quad (2.18)$$

Then

$$g(t) = k(t)\, h(t). \qquad \dots \quad (2.19)$$

To invert this, given $g(t)$, consider the solution $y(t)$ of the equation

$$dy(t) = -\Sigma \ db_i(t) \tau_{il}(g(t))\, \xi_l(y(t))$$

$$y(0) = e. \qquad \dots \quad (2.20)$$

Then

$$h(t) = y(t)\, g(t). \qquad \dots \quad (2.21)$$

One can verify that $y(t) = k^{-1}(t)$. Indeed, one has only to remember the formula for left invariant differentials

$$dk^{-1}(t) = -\tau(k(t))\, dk(t).$$

Therefore (2.20)-(2.21) and (2.18)-(2.19) express the same relation. It should be pointed out that if one were to use the canonical exponential coordinates, then $\rho_i^{pq} + \rho_i^{qp} = 0$ for all i, p and q. Therefore $h(t) \equiv g(t)$ and things are much simpler. We are now ready to state the theorem on the convergence of P_n.

Theorem 2.4 : *In order that P_n may converge to $P_{A(.),g(.)}$ it is necessary and sufficient that besides* (2.6) *the following hold.*

$$\lim_{\to\infty} A^n(t) = A(t) \text{ uniformly on } [0, T] \qquad \dots \quad (2.22)$$

$$\lim_{n\to\infty} h_n(t) = h(t) \text{ uniformly on } [0, T]. \qquad \dots \quad (2.23)$$

$h(t)$ and $g(t)$ are related by (2.18)-(2.19) *or equivalently by* (2.20)-(2.21).

3. Proofs

The first order of business is to describe how for any given set of integrable functions $\{a_{ij}(s)\}$ and $\{b_j(s)\}$ on $[0, T]$, one associates a process with independent increments. Recall that we define the operator

$$L_s = \frac{1}{2} \Sigma a_{ij}(s) \xi_i \xi_j + \Sigma b_j(s) \xi_j$$

and then look for a measure P on Ω_c such that

$$P[x(0) = e] = 1$$

and for each $f \in \mathcal{S}$

$$f(x(t)) = \int_0^t (L_s f)(x(s))\, ds \quad \dots \quad (3.1)$$

is a martingale relative to $[\Omega_c, M_t, P]$ for $0 \leqslant t \leqslant T$.

To construct such a process globally on the manifold we construct the process locally and piece together. Let U be a neighbourhood of e in G, which admits a coordinate system and which has in addition a compact closure. Let V, W be neighbourhoods such that

$$V \subset \overline{V} \subset W \subset \overline{W} \subset U.$$

Let $y_1, y_2, \dots, y_d$ be the coordinate system and let $\widetilde{V}$, $\widetilde{W}$ and $\widetilde{U}$ be the respective images of V, W and U in R^d. Choose ψ to be a smooth nonnegative cut-off function which is 1 on $\widetilde{V}$ and zero off $\widetilde{W}$. We can write L_s in local coordinates as

$$\tilde{L}_s = \frac{1}{2} \Sigma \tilde{a}_{ij}(s, y) \frac{\partial^2}{\partial y_i \partial y_j} + \Sigma \tilde{b}_j(s, y) \frac{\partial}{\partial y_j}$$

valid in the neighbourhood $\widetilde{U}$ of 0, which is assumed to be the image of e under the coordinate map. Since ξ_i are invariant operators, the $\tilde{a}_{ij}$ and $\tilde{b}_j$ are smooth functions of y for each fixed s. Thus $\psi \tilde{a}_{ij}$ and $\psi \tilde{b}_j$ are global C^∞ functions on R^d. Set

$$\hat{L} = \psi L_s = \frac{1}{2} \Sigma\, \hat{a}_{ij}(s, y) \frac{\partial^2}{\partial y_i \partial y_j} + \Sigma\, \hat{b}_j(s, y) \frac{\partial}{\partial y_j}.$$

We denote by $\varepsilon(s)$ the global bound in y for the $\hat{a}_{ij}$, their first and second order derivatives with respect to $y_1, \dots, y_d$, $\hat{b}_j$ and their first order derivatives with respect to $y_1, \dots, y_d$. We can estimate

$$\varepsilon(s) \leqslant C\, \Sigma[a_{ii}(s) + |b_i(s)|].$$

The constant C depends only on the neighbourhoods and coordinate system chosen. In the smaller neighbourhood $\widetilde{V}$, $\hat{L}_s$ is a representation of L_s in the given coordinates.

If we now take the positive semidefinite square root $\hat{\sigma}$ of $\hat{A}$, by a theorem of Phillips and Sarason (1967)

$$|\hat{\sigma}_{ij}(s, y) - \hat{\sigma}(s, y')| \leqslant C\sqrt{\varepsilon(s)}\, |y - y'|. \quad \dots \quad (3.2)$$

In addition

$$\begin{aligned} &|\hat{\sigma}_{ij}(s, y)| \leqslant C\sqrt{\varepsilon(s)} \\ &|\hat{b}_j(s, y) - \hat{b}_j(s, y')| \leqslant C\varepsilon(s)\, |y - y'| \\ &|\hat{b}_j(s, y)| \leqslant C\varepsilon(s). \end{aligned} \quad \dots \quad (3.3)$$

In view of (3.2) and (3.3) the Itô equations

$$\begin{aligned} dy(t) &= \hat{\sigma}(t, y(t))\, d\beta(t) + \hat{b}(t, y(t))dt, \ t \geqslant t_0 \\ y(t_0) &= y_0 \end{aligned} \qquad \dots \quad (3.4)$$

have a unique solution for any starting point y_0 and starting time t_0. By a theorem of Watanabe and Yamada [Proposition 1 on page 158, 1971] this implies the uniqueness of the solution to the martingale problem for $\hat{L}_s$. By the results in Stroock and Varadhan (1969a) (specifically Lemma 5.6 on page 386), this in turn implies that the solution to the martingale problem

$$\begin{cases} P\{x(t_0) = e\} = 1 \\ \text{for } f \in \mathcal{S} \\ f(x(t)) - \int_{t_0}^{t} (L_s f)(x(s))\, ds, \ t_0 \leqslant t \leqslant T. \\ \text{is a martingale relative to } [\Omega_c, M_t^{t_0}, P] \end{cases} \qquad \dots \quad (3.5)$$

exists and is unique until the first exit time $\tau_U^{t_0}$ from the set U after time t_0. If the starting point is any other point g of G, then the same observations hold by just translating by g on the left. The neighbourhood U_g of g will be gU. The same coordinates and the same cut-off function can be used. The resulting process is stochastically equivalent to $gx(t)$ where $x(t)$ is the process starting at the identity. We can now piece together the local solutions in the manner described in Stroock and Varadhan (1969a, Lemma 3.6 and Theorem 3.4 on pages 367-368), and we get a unique solution until the explosion time. That is, if K is a compact set and τ_K is the first exit time from K then the explosion time is

$$\zeta = \sup_K \tau_K.$$

We have to show that the explosion cannot occur prior to time T. In the compact case this is obvious. In general one has to make an argument. Let us define U_n to be the set of all elements which can be expressed as the product of n elements each of which is from U. Of course U_1 is the same as U. Let τ_n^s be the exit time from U_n for the process starting at the identity at time s. If the process does not exit before time T, τ_n^s is taken to be $+\infty$. Showing that the explosion time is infinite amounts to

$$\lim_{n \to \infty} E\left[e^{-\tau_n^s}\right] = 0 \quad \text{for } 0 \leqslant s \leqslant T. \qquad \dots \quad (3.6)$$

We notice the obvious Markov inequality

$$E[e^{-\tau_n^s}] \leqslant \rho\, E[e^{-\tau_{n-1}^s}] \qquad \dots \quad (3.7)$$

where

$$\rho = \sup_{s \leqslant t \leqslant T} E[e^{-\left(\tau_1^t - 1\right)}].$$

This is obvious because in order to get out of U_n the process has to get out of U_{n-1} at some point g and then get out of gU. Therefore, proving (3.6), amounts to proving that $\rho < 1$.

Let ϕ be a function in $\mathcal{S}$ which is 1 at the identity and zero off U. Then

$$\phi(x(t))-\int_{t_0}^{t} (L_s\phi)(x(s))\,ds \qquad \dots \quad (3.8)$$

is a martingale. Given the function ϕ let

$$\theta(s) = \sup_{X\in G} |(L_s\phi)(x)|. \qquad \dots \quad (3.9)$$

One can verify that $\theta(s)$ is integrable on $[0, T]$.

It follows from (3.8) and (3.9) that

$$\phi(x(t))+\int_{t_0}^{t} \theta(s)\,ds$$

is a submartingale. It is easy to deduce from this, using Doob's stopping theorem for submartingales, that

$$\text{Prob } [\tau_1^t \leqslant t+h] \leqslant \int_{t}^{t+h} \theta(s)\,ds. \qquad \dots \quad (3.10)$$

From (3.10) it follows that $\rho < 1$. We have therefore proved

Lemma 3.1 : *Given $\{a_{ij}(s)\}$ and $\{b_j(s)\}$ which are integrable on $[0, T]$ the solution to the martingale problem* (3.1) *exists and is unique.*

Lemma 3.2 : *Given a sequence $\{a_{ij}^{(n)}(s)\}$ and $\{b_j^{(s)}(s)\}$ of integrable functions such that the family*

$$E^n(t) = \int_0^t \Sigma[a_{ii}^{(n)}(s)+|b_i^{(s)}(s)|]\,ds$$

is equicontinuous, the corresponding family of solutions $P^{(n)}$ to the martingale problem (3.1) *is weakly compact on Ω_c.*

Proof : In order to prove the compactness of $P^{(n)}$ we have to check that for every neighbourhood V of e

$$\lim_{\delta\downarrow 0}\ \sup_{0\leqslant t\leqslant T}\ \sup_{n} P^{(n)}[\tau_V^t - t < \delta] = 0 \qquad \dots \quad (3.11)$$

and

$$\sup_{n}\ \sup_{0\leqslant t\leqslant T}\ E^{P^{(n)}}\left[e^{-\left(\tau_1^t - t\right)}\right] < 1. \qquad \dots \quad (3.12)$$

It is clear that we can estimate

$$\theta^{(n)}(s) \leqslant C\Sigma[a_{ii}^{(n)}(s)+|b_i^{(n)}(s)|]$$

and therefore (3.10) enables one to prove (3.11) and (3.12).

Lemma 3.3 : *Let P correspond to the operator*

$$L_s = \frac{1}{2} \Sigma a_{ij}(s)\xi_i\xi_j + \Sigma b_j(s)\xi_j$$

and Q correspond to

$$L_s^* = \frac{1}{2} \Sigma c_{ij}(s)\xi_i\xi_j$$

where

$$c_{ij}(t) = \Sigma\, \tau_{pi}(g(t))\tau_{qj}(g(t))a_{pq}(t) \quad \ldots \quad (3.13)$$

and $g(t)$ *is the solution of*

$$g(t) = \Sigma\, b_j(t)\xi_j(g(t)) : g(0) = e. \quad \ldots \quad (3.14)$$

If $X(t)$ *is a process having P as its distribution then* $x(t)$ *can be represented as*

$$X(t) = Y(t)\; g(t)$$

where $Y(t)$ *is distributed according to Q.*

Proof : $Y(t)$ has the property that for $f \epsilon s$

$$f(Y(t)) - \int_0^t (L_s^* f)(Y(s))\, ds$$

is a martingale. It is easily seen that this implies

$$f(t,\, Y(t)) - \int_0^t \left(\frac{\partial}{\partial s} + L_s^*\right)(f)(s,\, Y(s))\, ds \quad \ldots \quad (3.15)$$

is a martingale for functions $f(s, x)$ which are smooth. Let $h(x)$ be a function in the class $\mathcal{S}$ and let us take

$$f(t, y) = h(y\; g(t))$$

substituting in (3.15) we get

$$R(Y(t)\; g(t)) - \int_0^t \Phi(s,\, Y(s))\, ds \quad \ldots \quad (3.16)$$

is a martingale. Here Φ is computed in terms of h as

$$\Phi(s, y) = \left(\frac{\partial}{\partial s} + L_s^*\right) h(y\; g(s)).$$

We can rewrite the martingales (3.16) as

$$h(x(t)) - \int_0^t \psi(s,\, x(s))\, ds \quad \ldots \quad (3.17)$$

where

$$X(t) = Y(t)\; g(t).$$

A straightforward computation of ψ in terms of h is seen to lead to

$$\psi(s, x) = (L_s h)(x). \qquad \ldots \quad (3.28)$$

One has to keep in mind the definition of $\tau_{ij}(g)$ in terms of the inner automorphisms. The formula (3.18) identifies $X(t)$ as having distribution P.

Lemma 3.4 : *Let $A(t)$ be an element of V. Then there is a unique measure Q on Ω_c such that*

(i) $Q[x(0) = e] = 1$

(ii) *for* $f \in \mathcal{S}$,

$$f(x(t)) - \frac{1}{2} \Sigma \int_0^t (\xi_i \xi_j f)(x(s))\, dA_{ij}(s), \quad 0 \leqslant t \leqslant T$$

is a martingale relative to $[\Omega_c, M_t, Q]$.

Proof: By a transformation of the time scale one can always reduce the case of arbitrary $\{A_{ij}(t)\}$ to one where all the functios $\{A_{ij}(t)\}$ are absolutely continuous functions of t. Then Lemma 3.4 is exactly the same as Lemma 3.1.

Let us denote by $Q_{A(\cdot)}$ the measure corresponding to $A(\cdot)$ provided by Lemma 3.4. Recall that if $x(t)$ is distributed according to $Q_{A(\cdot)}$ we denote by $Q_{A(\cdot),\, g(\cdot)}$ the distribution of $X(t)\, g(t)$.

Lemma 3.5 : *The mapping* $(A(\cdot), g(\cdot)) \to Q_{A(\cdot),\, g(\cdot)}$ *is a continuous map of* $V \times \mathcal{F} \to \mathcal{P}$.

Proof: It is sufficient to prove that $Q_{A(\cdot)}$ is a continuous map of $V \to \mathcal{P}$. It follows easily from Lemma 3.2 that if $A_{n(\cdot)} \to A(\cdot)$ uniformly then $Q_{A^{n}(\cdot)}$ is compact. It is then obvious that any limit point Q is characterized by Lemma 3.4 to be $Q_{A(\cdot)}$.

Lemma 3.6 : *The map* $(A(\cdot),\ g(\cdot)) \to P_{A(\cdot),\, g(\cdot)}$ *from* $V_0 \times \mathcal{F}_0 \to \mathcal{P}$ *defined by* (2.3) *and Lemma* 3.1 *extends by continuity to a map of* $V \times \mathcal{F} \to \mathcal{P}$. *Moreover*

$$P_{A(\cdot),\, g(\cdot)} = Q_{B(\cdot),\, g(\cdot)}$$

provided

$$B_{k,\, l}(t) = \Sigma \int_0^t \tau_{ir}(g(s))\, \tau_{jl}(g(s))\, dA_{ij}(s).$$

Proof: This lemma follows immediately from Lemmas 3.3 and 3.5.

We now turn to the problem of proving the convergence of random walks to processes with independent increments. To simplify the notation we take $T = 1$. We shall also assume that for each n we are given $n-1$ random variables $X_{n1}, X_{n2}, \ldots, X_{n,\, n-1}$ with values in G which are mutually independent. We construct a process $X_n(t)$ by defining

$$X_n(t) = X_{n1} X_{n2} \ldots X_{n, j_n}(t)$$

where

$$j_n(t) = \begin{cases} [nt] & \text{for } 0 \leqslant t < 1 \\ n-1 & \text{for } t = 1. \end{cases}$$

Although this is a special situation, the proof for the general case described in Section 2 is only different in terms of notation. The reason for not taking $j_n(1) = n$, is that convention dictates the functions in Ω_d be left continuous at the right end. We have the means g_{nj} defined by (2.8) and the covariances a^{nj}_{pq} defined by (2.9) . Let us define

$$X'_{nj} = X_{nj}g^{-1}_{nj} \quad \dots \quad (3.19)$$

and

$$Y_{nj} = h_{n,j-1}X'_{nj}h^{-1}_{n,j-1}. \quad \dots \quad (3.20)$$

Let α_{nj} be the distribution of X_{nj} and β_{nj} the distribution of Y_{nj}. Define the process $Y_n(t)$ by

$$Y_n(t) = Y_{n1}Y_{n2} \dots Y_{nj_n}(t). \quad \dots \quad (3.21)$$

Let us recall that

$$h_{nj} = g_{n1}\, g_{n2} \dots g_{nj} \quad \dots \quad (3.22)$$

and

$$h_n(t) = h_{n\, j_n}(t). \quad \dots \quad (3.23)$$

One can easily verify that

$$X_n(t) = Y_n(t)h_n(t).. \quad \dots \quad (3.24)$$

Let us also remember that

$$A^{nj} = a^{n1}+\dots+a^{nj}, \quad \dots \quad (3.25)$$

$$A^n(t) = A^{nj_n(t)} \quad \dots \quad (3.26)$$

and Q_n, P_n are the distributions of $Y_n(\cdot)$, $X_n(\cdot)$ respectively We will now assume that

$$\lim_{n\to\infty} A^n(t) = A(t) \qquad \text{uniformly on } [0, T]$$

and

$$\lim_{n\to\infty} h_n(t) = h(t) \qquad \text{uniformly on } [0, T]. \quad \dots \quad (3.27)$$

Lemma 3.7 : *Let* (3.27) *and* (2.7) *hold. Define the operators*

$$(\delta_{nj}f)(x) = \int [f(xy)-f(x)]\beta_{nj}(dy)$$

and

$$(L_{nj}f)(x) = \left(\frac{1}{2}\,\Sigma\, a^{nj}_{pq}\tau_{pr}(h_{n\,j-1})\tau_{ql}(h_{n\,j-1})\xi_\gamma\xi_l\right)(f)(x) - \left(\frac{1}{2}\,\Sigma\, a^{nj}_{pq}\rho^{pq}_i\tau_{il}(h_{n\,j-1})\xi_l\right)(f)(x).$$

Then for any $f\epsilon\mathcal{S}$ and for any $\varepsilon > 0$ there is a constant C, a neighbourhood N of e and an integer n_0 such that for $n \geqslant n_0$ and for all j

$$\sup_{x\varepsilon G} |(\delta_{nj}f)(x)-(L_{nj}f)(x)| \leqslant \varepsilon \operatorname{tr} a^{nj}+c\,\alpha_{nj}(N^c).$$

In particular since $\varepsilon > 0$ *is arbitrary and* $\sum_{j=1}^{n-1} \operatorname{Tr} a^{nj}$ *is bounded,*

$$\lim_{n\to\infty} \sum_{j=1}^{n-1} \sup_{x\in G} |(\delta_{nj}f)(x)-(L_{nj}f)(x)| = 0.$$

Proof: To avoid cumbersome notation in the proof of the lemma we shall denote h_{nj-1} by h and g_{nj} by g. We shall denote $(\delta_{nj}f)(e)$ by $\delta_{nj}(f)$ and $(L_{nj}f)(e)$ by $L_{nj}(f)$.

$$\delta_{nj}(f) = \int [f(hyg^{-1}h^{-1})-f(e)]\alpha_{nj}(dy).$$

We shall expand $f(hyg^{-1}h^{-1})$ around $y = g$ using $\phi_1, \ldots, \phi_d$ as coordinates. We shall use the Taylor expansion up to the second order terms.

$$f(hyg^{-1}h^{-1})-f(e)=\Sigma\theta_i[\phi_i(y)-\phi_i(g)]+\frac{1}{2}\Sigma\,\theta_{pq}[\phi_p(y)-\phi_p(g)][\phi_q(y)-\phi_q(y)]+R. \quad \ldots \quad (3.28)$$

This is possible because for large n, g is uniformly close to the identity. In the above equation

$$\theta_i = \frac{\partial}{\partial\phi_i} f(hyg^{-1}h^{-1})\Big|_{y=g}$$

$$\theta = \frac{\partial^2}{\partial\phi_p\partial\phi_q} f(hyg^{-1}h^{-1})\Big|_{y=g}$$

Let us also define

$$\lambda_{pq} = \frac{\partial^2}{\partial\phi_p\partial\phi_q} f(hyh^{-1})\Big|_{y=e}$$

We first note that because h ranges over a compact set for any $\varepsilon > 0$ there is a neighbourhood N of e such that for all n and j

$$|R| \leqslant \varepsilon\Sigma|\phi_p(y)-\phi_p(g)|^2.$$

Integrating (3.28) by splitting G into N and N^c yields

$$\left|\delta_{nj}(f)-\frac{1}{2}\,\Sigma\theta_{pq}a^{nj}_{pq}\right| \leqslant \varepsilon\operatorname{tr} a^{nj}+C\alpha_{nj}(N^c).$$

Since h ranges over a compact set and g is uniformly close to the identity,

$$|\theta_{pq}-\lambda_{pq}| < \varepsilon \text{ for } n \geqslant n_0 \text{ and for all } j.$$

We notice also that a computation of λ_{pq} in terms of $\xi_1, \ldots, \xi_d$, using (2.7) and (2.16), yields

$$\lambda_{pq} = (\Sigma\,\tau_{pr}(h)\,\tau_{ql}(h)\xi_r\xi_l-\Sigma\,\rho_i^{pq}\tau_{il}(h)\xi_l)(f)(e).$$

Finally if $f \in \mathcal{S}$, the family $f_x(y) = f(xy)$ for $x \in G$ is uniformly bounded and uniformly smooth around e. Hence all the above considerations hold uniformly for them. We identify

$$(\delta_{nj} f)(x) = d_{nj}(f_x)$$

and the

$$(L_{nj} f)(x) = L_{nj}(f_x)$$

and this completes the proof of the lemma.

Lemma 3.8 : *Let* (3.27) *and* (2.7) *hold. Then Q_n converges weakly to a limit Q. Q is characterized by the fact that*

(i) $Q[x(0) = e] = 1$

and

(ii) *for* $f \in \mathcal{S}$

$$f(x(t)) - \Sigma \frac{1}{2} [\tau_{pr}(h(s))\tau_{ql}(h(s))\xi_\gamma \xi_l - \rho_i^{pq} \tau_{il}(h)\xi_l](f)(x(s))\, dA_{pq}(s)$$

is a martingale.

Proof: First we verify the compactness of Q_n. Let k be an integer $0 \leqslant k \leqslant n-1$. Consider the process $Y_n(t)$ and let Π_U^k be the number of steps needed for the Markov chain $\left[Y_n\left(\frac{k}{n}\right)\right]^{-1} Y_n\left(\frac{j}{n}\right) : j \geqslant k$, to exit from a neighbourhood U of e. We have to show that for any U

$$\lim_{\delta \downarrow 0} \limsup_{n \to \infty} \sup_{0 \leqslant t \leqslant 1} Q_n[\Pi_U^{[nt]} \leqslant n\delta] = 0. \qquad \ldots \quad (3.29)$$

This is proved in the same way as Lemma 3.2. For any $f \in s$ and k_0, denoting by $y(t)$ the function $[x(k_0/n)]^{-1} X(t)$

$$f\left(y\left(\frac{k}{n}\right)\right) - \sum_{k_0+1}^{k} \int \left[f\left(y\left(\frac{j-1}{n}\right) y\right) - f\left(y\left(\frac{j-1}{n}\right)\right)\right] d\beta_{nj}(dy) \qquad \ldots \quad (3.30)$$

is a Q_n martingale. We can estimate

$$\left| \int [f(zy) - f(z)] \beta_{nj}(dy) \right| \leqslant \sup_{X \varepsilon G} |(L_{nj} f)(x)| + \varepsilon_{nj}$$

where $\lim_{n=1} \sum_{j=1}^{n-1} \varepsilon_{nj} = 0$. Moreover

$$\sup_{,} |(L_{nj} f)(x)| \leqslant C \operatorname{Tr} a^{nj}.$$

Therefore there are numbers Δ_{nj} such that

$$f\left(y\left(\frac{k}{n}\right)\right) + \sum_{k_0+1}^{k} \Delta_{nj} \text{ is a } Q_n \text{ submartingale}$$

and

$$\lim_{\delta \downarrow 0} \limsup_{n \to \infty} \sup_{|t_1 - t_2| \leqslant \delta} \sum_{[nt_1] \leqslant j \leqslant [nt_2]} \Delta_{nj} = 0.$$

Equation (3.29) is now proved in the same manner as Lemma 3.2 by taking f to be a function which is 1 at e and 0 off U. We then apply Doob's stopping theorem. Let Q be any limit point of the sequence $\{Q_n\}$. By taking $k_0 = 0$ in (3.30),

$$f\left(x\left(\frac{k}{n}\right)\right) - \sum_{1}^{k} \int \left[f\left(x\left(\frac{j-1}{n}\right)y\right) - f\left(x\left(\frac{j-1}{n}\right)\right)\right] \beta_{nj}(dy) \qquad \dots \quad (3.31)$$

is a martingale relative to $[\Omega_d, M_{k/q}, Q_n]$. From (2.7) it is clear that Q is concentrated on Ω_c. We look at the expression (3.31) as a functional $F_n(t, w)$ defined for t of the form k/n on Ω_d. From Lemma 3.7 it is clear that $F_n(t, w)$ is converging uniformly on a compact subset of Ω_d to the limit

$$F(t, w) = f(x(t)) - \Sigma \frac{1}{2} \int_0^t [\tau_{pr}(h(s))\tau_{ql}(h(s))\xi_\gamma \xi_l - \rho_i^{pq} \tau_{il}(h(s))\xi_l] \,.\, (f)(x(s))\, dA_{pq}(s).$$

Therefore for any $f \epsilon \mathfrak{S}$, $F(t, w)$ is a martingale relative to $[\Omega_c, M_t, Q]$. One can then verify that such a Q is unique. Hence the entire sequence Q_n converges to the limit Q.

Lemma 3.9 : *P_n converges weakly to $P_{A(\cdot),g(\cdot)}$ where $g(\cdot)$ is given by* (2.18) *and* (2.19).

Proof : From the convergence of Q_n to Q it follows that P_n converges to P where P is the distribution of $Y(t)h(t)$ and $Y(t)$ is distributed according to Q. We can use Lemmas 3.3 and 3.6 to compute what P is and verify that it is $P_{A(\cdot),\, g(\cdot)}$.

We now prove that the representations $P_{A(\cdot),\, g(\cdot)}$ and $Q_{A(\cdot),\, g(\cdot)}$ are unique.

Lemma 3.10 : *If $P_{A(\cdot),\, g(\cdot)} = P_{B(\cdot),\, h(\cdot)}$ then $A(\cdot) = B(\cdot)$ and $g(\cdot) = h(\cdot)$. If $Q_{A(\cdot),\, g(\cdot)} = Q_{B(\cdot),\, h(\cdot)}$ then $A(\cdot) = B(\cdot)$ and $g(\cdot) = h(\cdot)$.*

Proof : Since we can use Lemma 3.6 to go back and forth between "P" and the "Q" representations, it suffices to prove that if $Q_{A(\cdot),\, g(\cdot)} = Q_{B(\cdot),\, h(\cdot)}$ then $A(\cdot) = B(\cdot)$ and $g(\cdot) = h(\cdot)$. From the definition of $Q_{A(\cdot),\, g(\cdot)}$ find $Q_{B(\cdot),\, h(\cdot)}$ it follows that it is sufficient to prove that if $Q_{A(\cdot),\, g(\cdot)} = Q_{B(\cdot),\, e(\cdot)}$ then $g(\cdot) = e(\cdot)$ and $A(\cdot) = B(\cdot)$. Here $e(\cdot)$ is the function identically equal to e. For the moment let us suppose that $g(\cdot)$ is of bounded variation. Then we can write $Q_{A(\cdot),\, g(\cdot)}$ as $P_{C(\cdot),\, g(\cdot)}$ and $Q_{B(\cdot),\, e(\cdot)} = P_{B(\cdot),\, e(\cdot)} \cdot P_{C(\cdot),\, g(\cdot)}$ has the property that for $f \epsilon \mathfrak{S}$,

$$f(x(t)) - \int_0^t \frac{1}{2} \Sigma(\xi_p \xi_q f)(x(s))\, dC_{pq}(s) - \int_0^t \Sigma\, (\xi_i f)(x(s))\, db_1(s)$$

is a martingale. Here

$$dg(t) = \Sigma db_i(t)\, \xi_i(g(t)).$$

On the other hand with respect to $P_{B(\cdot),\, e(\cdot)}$

$$f(x(t)) - \int_0^t \frac{1}{2}\, \Sigma\, (\xi_p \xi_q f)(x(s))\, dB_{p_l}(s)$$

is a martingale. If the two measures are the same, by the uniqueness of the Meyer decomposition for quasi-martingales $C(\cdot) = B(\cdot)$ and $db_i(\cdot) \equiv 0$. Therefore $A(\cdot) = B(\cdot)$. So all we have to do is prove that if $Q_{A(\cdot),\, g(\cdot)} = Q_{B(\cdot),\, e(\cdot)}$ then $g(\cdot)$ is of bounded variation. Let $X(t)$ be any stochastic process with P as its distribution. We can call P "regular" if for any $f \in \mathcal{S}$, $f(X(t))$ is a quasi-martingale, that is there is a process $F(t)$ such that $F(t)$ is of bounded variation and

$$f(X(t)) - F(t)$$

is a square integrable martingale. Suppose $X(t)$, $Y(t)$ are two "regular" processes which are possibly dependent, then Kunita-Watanabe's theory of square integrable martingales (Kunita and Watanabe, 1967) gurantees that the two processes are jointly regular, i.e., for any smooth function ϕ of two variables, $\phi(X(t), Y(t))$ is a quasi-martingale. Going back to our problem the process corresponding to $Q_{A(\cdot),\, g(\cdot)}$ is of the form $X(t)g(t)$ where $X(t)$ is regular. The process corresponding to $Q_{B(\cdot),\, e(\cdot)}$ is of course always regular. Since both processes are the same we have that $X(t)$ as well as $X(t)g(t)$ are regular processes. Therefore for smooth ϕ, $\phi(X(t), Y(t))$ is a quasi-martingale. In particular taking $\phi(x, y) = f(x^{-1}y)$ we get $f(g(t))$ to be a quasi-martingale. This forces $g(\cdot)$ to be of bounded variation.

Lemma 3.10 : *Let $Q_{A^n(\cdot),\, g^n(\cdot)}$ be compact on Ω_c, where $A^n(\cdot)$, $g^n(\cdot)$ are arbitrary sequences in V and $\mathcal{F}$. Then $A^n(\cdot)$, $g^n(\cdot)$ are equicontinuous at every point of* $[0, T]$.

Proof : Let us prove equicontinuity at 0 for $A^n(\cdot)$. Suppose the sequence is not equicontinuous. By taking a subsequence if need be we can assume that there is a sequence δ_n of positive numbers converging to 0 such that

$$\text{Tr } A^n(\delta_n) = \varepsilon > 0.$$

Let us make a monotone continuous transformation ϕ_n of $[0, \varepsilon]$ onto $[0, \delta_n]$ such that $A^n(\phi_n(t))$ is equicontinuous on $[0, \varepsilon]$. Denote $A^n(\phi_n(t))$ by $B^n(t)$, $g^n(\phi_n(t))$ by $h^n(t)$ and the measure corresponding to $x(\phi_n(t))$ under $Q_{A^n(\cdot),\, g^n(\cdot)}$ on $[0, \varepsilon]$ by Q^n. Then Q^n on $[0, \varepsilon]$ is $Q_{B^n(\cdot),\, h^n(\cdot)}$. Since $B^n(\cdot)$ is equicontinuous we can assume that it converges to $B(\cdot)$ uniformly. Then $Q_{B^n(\cdot),\, e^n(\cdot)}$ will converge to $Q_{B(\cdot),\, e(\cdot)}$. Moreover since $\delta_n \to 0$ Q^n will converge to the process constantly equal to e. This forces $h^n(\cdot)$ to be equicontinuous and if $h^n(\cdot)$ were to converge to $h(\cdot)$ then $Q_{B(\cdot),\, h(\cdot)}$ is the trivial process. By the uniqueness of Lemma 3.9 this forces $B(\cdot)$ to be zero. But $\text{Tr } B(\varepsilon) = \lim \text{Tr } A^n(\phi_n(\varepsilon))$ $= \lim \text{Tr } A^n(\delta_n) = \varepsilon > 0$. Thus $A^n(\cdot)$ is equicontinuous at 0. By a similar proof it is equicontinuous at every point. Thus $Q_{A^n(\cdot),\, e^n(\cdot)}$ is compact. Since $Q_{A^n(\cdot),\, h^n(\cdot)}$ is also compact, $h^n(\cdot)$ is equicontinuous too.

Lemma 3.11 : *Let P_n be as in Lemma 3.9. If P_n is compact then $A^n(\cdot)$ and $h_n(\cdot)$ are equicontinuous on* $[0, T]$.

Proof : Again we shall prove equicontinuity only at $t = 0$. The same proof will work at any other point. Let U be a neighbourhood of e with a compact closure and δ_n be the first time $h_n(t)$ exits from U. Let $\varepsilon > 0$ be an arbitrary positive number and let θ_n be the first time Tr $A^n(t)$ exceeds ε. We have to show that δ_n and θ_n are bounded below. Let us suppose that they are not. Then by a suitable choice of a subsequence we are reduced to one of the following cases.

Case 1 : $\theta_n \leqslant \delta_n$ and $\theta_n \to 0$.

Case 2 : $\delta_n \leqslant \theta_n$ and $\delta_n \to 0$.

Let us consider case 1. We can find monotone continuous maps $\phi_n(t)$ of $[0, \varepsilon]$ onto $[0, \theta_n]$ such that (a) $A_n(\phi_n(t))$ is equicontinuous on $[0, \varepsilon]$ and (b) the points t_{nj} where the jumps of the process $X(\phi_n(t))$ occur have the property $\lim_{n\to\infty} \sup_j |t_{nj}-t_{nj-1}| = 0$. We can write $X(\phi_n(t))$ as $Y(\phi_n(t))\cdot h_n(\phi(t))$ and we shall denote by $\tilde{P}_n$ and $\tilde{Q}_n$ the measures corresponding to $X(\phi_n(t))$ and $Y(\phi_n(t))$ on $[0, \varepsilon]$. Since $\delta_n \geqslant \theta_n$, we have that $h_n(\phi_n(t))$ is in U and therefore in a compact set during $[0, \varepsilon]$. Since $A^{(n)}(\phi_n(t))$ is equicontinuous and $h_n(\phi_n(t))$ ranges over a compact set $B^{(n)}(\phi_n(t))$ is also equicontinuous where

$$B_{rl}^{(n)}(t) = \sum_{j\leqslant[nt]} a_{pq}^{nj}\tau_{pr}(h_{n\,j-1})\tau_{ql}(h_{n\,j-1}).$$

Moreover Tr $B^{(n)}(\phi_n(\varepsilon))$ is bounded from below because Tr $A^{(n)}(\phi_n(\varepsilon)) = \varepsilon$ and $h_n(\phi_n(t))$ ranges over a compact set. If we now choose a further subsequence and assume that $B^{(n)}(t) \to B(t)$ uniformly then $\tilde{Q}_n$ converges to a nontrivial limit $\tilde{Q}$. By that we man that the variance term is nonzero. Since $\theta_n \to 0$ we have $\tilde{P}_n$ converges to the trivial process. Since $\tilde{Q}_n$ and $\tilde{P}_n$ are compact so is $h_n(\phi_n(t))$. If we assume that $h_n(\phi_n(t))$ converges to a limit $h(t)$ then the trivial process $X(t) \equiv e$ is expressed as $Y(t)h(t)$ where $Y(t)$ has a nontrivial variance. This is a contradiction.

Case 2 : We proceed exactly as in case 1.

We conclude that the trivial process $X(t) \equiv e$ is expressed as $Y(t)h(t)$ where $h(\varepsilon)\epsilon U^c$ and therefore $h(\cdot)$ is a nontrivial continuous function. We are however unable to assert that the $Y(\cdot)$ process has nontrivial variance. However if the $Y(t)$ process has trivial variance then $Y(\cdot) \equiv e$ with probability 1. (See Lemma 3.8 for the types of processes that come up.] So there is never any possibility of the $Y(\cdot)$ process exactly cancelling the function $h(t)$. This is again a contradiction.

We have now proved all the results claimed in the summary. Characterization of the compactness and uniqueness of the representation prove the continuity of the inverse maps in Theorems 2.1 and 2.2 and the necessity part of Theorem 2.4. Theorem 2.3 follows from the uniqueness of the representations and Lemma 3.6. The only remaining part is to prove that the maps in Theorems 2.1 and 2.2 are onto. This is proved below.

Lemma 3.12 : *If* $P \in \mathcal{P}$ *then* $P = P_{A(\cdot),\, g(\cdot)}$ *for some* $A(\cdot)$ *and* $g(\cdot)$.

Proof : Let $X(t)$ be the process corresponding to P_0.

If we define random variables

$$X_{nj} = X\left(\frac{j-1}{n}\right)^{-1} X\left(\frac{j}{n}\right), \qquad j = 1, 2, ..., n-1$$

then we can use Lemma 3.11 to conclude that $A^n(\cdot)$ and $h_n(\cdot)$ constructed as in Theorem 2.4 are equicontinuous because P_n in this context converges to P. By an appropriate choice of a subsequence we can assume that $A^n(\cdot) \to A(\cdot)$ and $h_n(\cdot) \to h(\cdot)$. Therefore $P = P_{A(\cdot), g(\cdot)}$ for some $A(\cdot)$ and $g(\cdot)$.

REFERENCES

HUNT, G. A. (1956) : Semi-groups of measures on Lie groups. *Trans. Amer. Math. Soc.*, **81**(2), 264-293.

KUNITA, S. and WATANABE, S (1967) : On square integrable martingales. *Nagoya Math. J.*, **30**, 209-245.

PATHASARATHY, K. R. (1964) : The central limit theorem for the rotation group. *Theory Prob. Applications*, **9**, 248-257 (English translation).

PARTHASARATHY, K. R., RANGA RAO, R. and VARADHAN, S. R. (1963) : Probability distributions on locally compact abelian groups. *Illinois J. Math.*, **7**, 337369.

PHILLIPS, R. S. and SARASON, L. (1967) : Elliptic-parabolic equations of the second order. *J. Math. Mech.* **17**, 891-917.

SAZANOV, V. V. and TUTUBALIN, V. N. (1966) : Probability distributions on Topological groups. *Theory Prob. Applications*, **11**, 1-45 (English translation).

STROOCK, D. W. and VARADHAN, S. R. S. (1969a) : Diffusion processes with continuous coefficients. I. *Comm. Pure Appl. Math.*, **XII**, 345-400.

——— (1969b) : Diffusion processes with continuous coefficients, II. *Comm. Pure Appl. Math.*, **XII**, 479-530.

YAMADA, T. and WATANABE, S. (1971) : On the uniqueness of solutions of stochastic differential equations. *J. Math. Kyoto Univ.*, **11**, No. 1, 155-167.

WEHN, D. (1968) : Probabilities on Lie groups. *Proc. Nat. Zcad. Sci.*, U.S.A. **48**, No. 5, 791-795.

Paper received : December, 1972.

Martingale Approach to Some Limit Theorems

G. C. Papanicolaou†

D. Stroock††

S. R. S. Varadhan†

---

†Courant Institute of Mathematical Sciences, New York University. Research supported by the National Science Foundation under Grant No. MCS75-09837.

††Department of Mathematics, University of Colorado.

This is an expanded version of a lecture given by one of us (G. P.) at the Conference on Turbulence at Duke University in April 1976.

Contents

1. Introduction.

The asymptotic behavior of stochastic processes under various circumstances is a central problem in probability theory and its applications. The theory of limiting distributions of sums of independent (or weakly dependent) random variables is among the best known examples.

Diffusion approximations for suitably normalized Markov chains have also been studied extensively starting with Khinchine [1] who employed a differential equations approach and continuing with Prohorov [2] and Skorohod [3] who employed compactness criteria that they developed. By formulating diffusion theory in terms of martingale problems, very general diffusion approximations were obtained in [4] under minimal regularity hypotheses. In [5] corresponding problems with boundary conditions were treated.

In a variety of applications one encounters asymptotic problems for stochastic processes that are somewhat different from the above. These problems are to some extent analogous to problems in differential equations that are suitable for application of the method of averaging [6] and can be described roughly as follows.

There are two processes given that vary on two different time scales. One, the rapidly varying one, is driving the other by entering as a coefficient in the equation defining the slowly varying process. The problem is to characterize the asymptotic behavior of the driven process as the fluctuations of the driving process become more and more rapid. In this work we analyze such problems, under a variety of different circumstances, by adapting the martingale approach of [4,5] (cf. also [7]).

Among the many places where asymptotic problems of the above form appear we mention the work of Stratonovich [8] (on noise phenomena) and subsequent developments [9-12,14], the diffusion limit of transport theory [13, 15 and references cited there], random evolutions [16,17,18], waves in random media [19], etc.

The following is organized in two parts. One deals with the case of a driving process that has strong ergodic properties. Sections 2.1 and 2.2 contain the formulation of a theorem that is typical of what follows. Sections 2.4-2.8 contain results that may be considered as variations of the main theme but are also of independent interest.

The second part deals with the case of a not so strongly ergodic driving process and as a consequence the nature of the limit process changes. Typical examples here are occupation time problems for Brownian motion [20,21,22]. Sections 3.1 and 3.4 contain the main theorems.

2. Ergodic driving processes.

2.1 Formulation of the problem.

We shall first formulate a general problem and then give some examples which explain why we are interested in this problem.

Let $\varepsilon > 0$ be a parameter and let $(x^\varepsilon(t), y^\varepsilon(t))$ be the diffusion process in $R^n \times R^m$ defined by the following system of stochastic differential equations

$$
\begin{aligned}
(2.1.1)\qquad dx_i^\varepsilon(t) &= \left[\frac{1}{\varepsilon} F_i^{(1)}(x^\varepsilon(t),y^\varepsilon(t)) + G_i^{(1)}(x^\varepsilon(t),y^\varepsilon(t))\right]dt \\
&\quad + \sum_{j=1}^{p} \sigma_{ij}^{(1)}(x^\varepsilon(t),y^\varepsilon(t))\,d\beta_j^{(1)}(t) , \\
&\quad x_i^\varepsilon(0) = x_i , \qquad i = 1,2,\dots,n , \\
dy_k^\varepsilon(t) &= \left[\frac{1}{\varepsilon^2} F_k^{(2)}(x^\varepsilon(t),y^\varepsilon(t)) + \frac{1}{\varepsilon} G_k^{(2)}(x^\varepsilon(t),y^\varepsilon(t))\right. \\
&\quad \left. + H_k^{(2)}(x^\varepsilon(t),y^\varepsilon(t))\right]dt \\
&\quad + \sum_{\ell=1}^{q} \left[\frac{1}{\varepsilon} \sigma_{k\ell}^{(2)}(x^\varepsilon(t),y^\varepsilon(t)) \right. \\
&\quad \left. + \sigma_{k\ell}^{(3)}(x^\varepsilon(t),y^\varepsilon(t))\right] d\beta_\ell^{(2)}(t) , \\
&\quad y_k^\varepsilon(0) = y_k , \qquad k = 1,2,\dots,m .
\end{aligned}
$$

Here $(\beta_1^{(1)}(t),\beta_2^{(1)}(t),\dots,\beta_p^{(1)}(t))$ and $(\beta_1^{(2)}(t),\beta_2^{(2)}(t),\dots,\beta_q^{(2)}(t))$ are p-dimensional and q-dimensional standard Brownian motions, respectively, and

(2.1.2) $$E\{\beta_j^{(1)}(t)\beta_\ell^{(2)}(s)\} = \min(t,s)b_{j\ell} ,$$

$$j = 1,\dots,p , \quad \ell = 1,\dots,q .$$

The functions $F_i^{(1)}$, $G_i^{(1)}$, $\sigma_{ij}^{(1)}$, $i = 1,\dots,n$, $j = 1,\dots,p$ and $F_k^{(2)}$, $G_k^{(2)}$, $H_k^{(2)}$, $\sigma_{k\ell}^{(2)}$, $\sigma_{k\ell}^{(3)}$, $k = 1,\dots,m$, $\ell = 1,\dots,q$ are assumed to be bounded for $(x,y) \in R^n \times R^m$ and smooth i.e., C^∞ in both x and y. These hypotheses can be weakened considerably as we indicate briefly later. However, we assume them here in order to simplify the statement of results.

For $\varepsilon > 0$ and fixed the existence theory of Itô equations [23] provides a unique solution to (2.1.1) which is a continuous strong Markov process $(x^\varepsilon(t), y^\varepsilon(t))$. The infinitesimal generator $\mathscr{L}^\varepsilon$ of this process has the following form

(2.1.2) $$\mathscr{L}^\varepsilon = \frac{1}{\varepsilon^2}\mathscr{L}_1 + \frac{1}{\varepsilon}\mathscr{L}_2 + \mathscr{L}_3$$

where

(2.1.3) $$\mathscr{L}_1 = \sum_{k=1}^{m} F_k^{(2)}(x,y)\frac{\partial}{\partial y_k} + \frac{1}{2}\sum_{k,\ell=1}^{m} a_{k\ell}^{(2)}(x,y)\frac{\partial^2}{\partial y_k \partial y_\ell} ,$$

(2.1.4) $$\mathscr{L}_2 = \sum_{i=1}^{n} F_i^{(1)}(x,y)\frac{\partial}{\partial x_i} + \sum_{k=1}^{m} G_k^{(2)}(x,y)\frac{\partial}{\partial y_k}$$

$$+ \sum_{i=1}^{n}\sum_{k=1}^{m} a_{ik}^{(12)}(x,y)\frac{\partial^2}{\partial x_i \partial y_k}$$

$$+ \sum_{k,\ell=1}^{m} a_{k\ell}^{(23)}(x,y)\frac{\partial^2}{\partial y_k \partial y_\ell} ,$$

$$(2.1.5)\qquad \mathscr{L}_3 = \sum_{i=1}^{n} G_i^{(1)}(x,y)\frac{\partial}{\partial x_i} + \sum_{k=1}^{m} H_k^{(2)}(x,y)\frac{\partial}{\partial y_k}$$

$$+ \frac{1}{2}\sum_{i,j=1}^{n} a_{ij}^{(1)}(x,y)\frac{\partial^2}{\partial x_i \partial x_j}$$

$$+ \sum_{i=1}^{n}\sum_{k=1}^{m} a_{ik}^{(13)}(x,y)\frac{\partial^2}{\partial x_i \partial y_k}$$

$$+ \frac{1}{2}\sum_{k,\ell=1}^{m} a_{k\ell}^{(3)}(x,y)\frac{\partial^2}{\partial y_k \partial y_\ell},$$

and

$$a_{ij}^{(1)}(x,y) = \sum_{r=1}^{p} \sigma_{ir}^{(1)}(x,y)\sigma_{jr}^{(1)}(x,y), \quad i,j = 1,\dots,n,$$

$$a_{k\ell}^{(2)}(x,y) = \sum_{r=1}^{q} \sigma_{kr}^{(2)}(x,y)\sigma_{\ell r}^{(2)}(x,y),$$

$$a_{k\ell}^{(3)}(x,y) = \sum_{r=1}^{q} \sigma_{kr}^{(3)}(x,y)\sigma_{\ell r}^{(3)}(x,y), \quad k,\ell = 1,\dots,m,$$

$$(2.1.6)\qquad a_{ik}^{(12)}(x,y) = \sum_{r=1}^{p}\sum_{s=1}^{q} \sigma_{ir}^{(1)}(x,y)\sigma_{ks}^{(2)}(x,y)b_{rs},$$

$$a_{ik}^{(13)}(x,y) = \sum_{r=1}^{p}\sum_{s=1}^{q} \sigma_{ir}^{(1)}(x,y)\sigma_{ks}^{(3)}(x,y)b_{rs},$$

$$i = 1,\dots,n, \quad k = 1,\dots,m,$$

$$a_{k\ell}^{(23)}(x,y) = \sum_{r=1}^{q} \sigma_{kr}^{(2)}(x,y)\sigma_{\ell r}^{(3)}(x,y),$$

$$k,\ell = 1,2,\dots,m.$$

We now give some examples that motivate the above formulation.

Example 1. $F^{(1)}$, $F^{(2)}$, $H^{(2)}$, $\sigma^{(2)}$ and $\sigma^{(3)}$ are identically zero (we omit subscripts for simplicity), m = 1 and $G^{(1)}$ and $\sigma^{(1)}$ are almost periodic functions of $y \in R^1$. Thus, (2.1.1) reduces to

$$
\begin{aligned}
(2.1.7)\qquad dx_i^{\varepsilon}(t) &= G_i^{(1)}(x^{\varepsilon}(t), y + t/\varepsilon)\,dt \\
&\quad + \sum_{j=1}^{p} \sigma_{ij}^{(1)}(x^{\varepsilon}(t), y + t/\varepsilon)\,d\beta_j^{(1)}(t) , \\
x_i^{\varepsilon}(0) &= x , \qquad i = 1,2,\ldots,n .
\end{aligned}
$$

The asymptotic analysis of the process $x^{\varepsilon}(t)$ defined by (2.1.7) constitutes the method of averaging extended to Itô stochastic differential equations [9 ,24]. This example is considered further in Section 2.8.

Example 2. $F^{(1)} \equiv 0$ in (2.1.1). This example corresponds to what may be called generalized averaging for Itô stochastic differential equations. The deterministic equivalent was analyzed by Volosov, Tichonov [25,26] and others, and the stochastic problem by Khasminskii [12]. We return to this example in Section 2.8.

Example 3. n = m, $F^{(1)} = F^{(2)}$, $G^{(1)} = G^{(2)}$, $\sigma^{(1)} = \sigma^{(2)}$, $\beta^{(1)}(t) \equiv \beta^{(2)}(t)$ so that $b_{j\ell} = \delta_{j\ell}$ in (2.1.2), and all quantities as functions of y_j are periodic functions of period one, j = 1,2,...,n. This example corresponds to homogenization [27] and is considered again in Section (2.7). If the initial data are related by $y = x/\varepsilon$ then, in this example,

$$y^{\varepsilon}(t) \equiv x^{\varepsilon}(t)/\varepsilon \tag{2.1.8}$$

and (2.1.1) becomes (cf. [37])

$$dx_i^{\varepsilon}(t) = \left[\frac{1}{\varepsilon} F_i^{(1)}\left(x^{\varepsilon}(t),\frac{x^{\varepsilon}(t)}{\varepsilon}\right) + G_i^{(1)}\left(x^{\varepsilon}(t),\frac{x^{\varepsilon}(t)}{\varepsilon}\right)\right]dt + \sum_{j=1}^{p} \sigma_{ij}^{(1)}\left(x^{\varepsilon}(t),\frac{x^{\varepsilon}(t)}{\varepsilon}\right)d\beta_j^{(1)}(t) , \tag{2.1.9}$$

$$x_i^{\varepsilon}(0) = x , \quad i = 1,\dots,n .$$

<u>Example 4</u>. $n = m$, $F^{(1)}(x,y) \equiv y$, $\sigma^{(1)} \equiv 0$ and $y^{\varepsilon}(t) = y(t/\varepsilon^2)$ where

$$dy_k(t) = -y_k(t)dt + d\beta_k(t) , \tag{2.1.10}$$

$$y_k(0) = y_k , \qquad k = 1,2,\dots,n .$$

The asymptotic limit of (2.1.1) in this case is the Smoluchovski-Kramers limit of Brownian motion [28] and was analyzed by Ilin and Khasminskii [13] and Bromberg [29]. In the latter paper the process $x^{\varepsilon}(t)$ is confined to a bounded region in space with appropriate boundary conditions which give rise to boundary layers as $\varepsilon \to 0$.

From these examples it is clear that scaling by ε in (2.1.1) is intended to articulate the fact that in the pair $(x^{\varepsilon}(t),y^{\varepsilon}(t))$, $x^{\varepsilon}(t)$ is the <u>slowly varying</u> component of the process and $y^{\varepsilon}(t)$ the <u>rapidly varying</u> component. We also refer to $y^{\varepsilon}(t)$ as the <u>driving process</u> and to $x^{\varepsilon}(t)$ as the driven process. Note that, in general, the driven process, in turn, influences the driving process since the functions with superscript (2) and (3) depend on y. Thus, when $\varepsilon > 0$ is not small, the component processes $x^{\varepsilon}(t)$ and $y^{\varepsilon}(t)$ are on an equal footing; it is only when ε is small that the above terminology is meaningful. Problem (2.1.1) can also be considered as a random evolution problem [16,17]. In that context the driven component can be a very general object, not necessarily an R^n valued process. On the other hand the driving component is frequently taken to be a Markov chain. We return to this in Section (2.6) in connection with transport problems.

The process $(x^{\varepsilon}(t), y^{\varepsilon}(t))$, $\varepsilon > 0$ fixed, is also characterized by its infinitesimal generator $\mathscr{L}^{\varepsilon}$ acting on smooth functions. Under our hypotheses, almost any method can be used to construct the process from $\mathscr{L}^{\varepsilon}$ of (2.1.2)-(2.1.6). In particular the martingale method of [4]. In the martingale approach the measures $P^{\varepsilon}_{x,y}$ in the space of continuous trajectories in $R^n \times R^m$, with (x,y) initial point, corresponding to $(x^{\varepsilon}(t), y^{\varepsilon}(t))$, are characterized as follows:

(i) $P^{\varepsilon}_{x,y}\{x(0) = x, y(0) = y\} = 1$,

(ii) for any function $f(x,y)$ which is smooth and with compact support, the expression

$$M_f(t) = f(x(t),y(t)) - \int_0^t \mathscr{L}^{\varepsilon} f(x(s),y(s))ds \tag{2.1.11}$$

is a $P^{\varepsilon}_{x,y}$ martingale.

Here we do not use superscripts on $(x(t),y(t))$ since they appear as arguments of functionals while the measures are labeled by ε. If $\mathcal{F}_t$, $t \geq 0$, denotes the σ-algebras of events generated by the paths $(x(s),y(s))$ for $s \leq t$, then (ii) above says that (P^ε a.s.)

$$E^{P^\varepsilon}\{M_f(t) \mid \mathcal{F}_s\} = M_f(s) \ , \quad 0 \leq s \leq t < \infty \ . \tag{2.1.12}$$

In the following we employ systematically the martingale approach. The reasons for this are that (i) it is well suited to asymptotics centering on weak convergence and provides a reasonably general and convenient scheme for a large class of problems, (ii) it is well suited to problems involving boundary conditions [5,30] (the characterization being formulated as a submartingale problem [5]) and (iii) it provides results with minimal regularity hypotheses on the coefficients of the processes under consideration [4].

2.2 The limit theorem.

In addition to the hypotheses introduced above regarding (2.1.1), we shall require the following hypotheses for the asymptotic analysis.

Let $Y(t;x)$ be the diffusion process in R^m generated by $\mathscr{L}_1$ of (2.1.3) with $x \in R^n$ regarded as a parameter. From (2.1.1) it is clear that, roughly, one has $y^\varepsilon(t) \sim Y(t/\varepsilon^2;x)$ at least for times t that are not too large. The asymptotic analysis of (2.1.1) depends in an essential way on the ergodic properties of $Y(t;x)$. Let $P(t,y,A;x)$ be the transition function of $Y(t;x)$

$$P(t,y,A;x) = P\{Y(t;x) \in A | Y(0;x) = y\} , \tag{2.2.1}$$

where $y \in R^m$ and A is a Borel subset of R^m. From (2.1.3) it follows that $P(t,y,A;x)$ is a Feller transition function, depending smoothly upon $x \in R^n$, and such that the semigroup associated with P maps functions that vanish at infinity to functions that vanish at infinity.

We assume that $Y(t;x)$ is ergodic i.e., there exists a unique invariant probability measure $\bar{P}(A;x)$, $A \subset R^m$, $x \in R^n$, depending smoothly upon x, such that

$$\lim_{t\uparrow\infty} P(t,y,A;x) = \bar{P}(A;x)$$

uniformly in (x,y) on compact sets and $A \subset R^m$. Define the recurrent potential kernel

$$(2.2.2) \qquad \psi(y,A;x) = \int_0^\infty [P(t,y,A;x) - \bar{P}(A;x)]dt .$$

We assume that the kernel ψ exists, depends smoothly on x and transforms bounded smooth functions of y into themselves. These are the hypotheses regarding $Y(t;x)$.

We assume further that

$$(2.2.3) \qquad \int_{R^m} F^{(1)}(x,y)\bar{P}(dy;x) \equiv 0 ,$$

and this is called the centering condition.

Let $f(x) \in C_o^\infty(R^n)$ and define $\psi_f^{(1)}$ by

$$(2.2.4) \qquad \psi_f^{(1)}(x,y) = \int_{R^m} \psi(y,dz;x) \sum_{j=1}^{n} F_j^{(1)}(x,z) \frac{\partial f(x)}{\partial x_j} .$$

We define a diffusion operator $\bar{\mathscr{L}}$ as follows

$$(2.2.5) \qquad \bar{\mathscr{L}} f(x) = \int_{R^m} \bar{P}(dy;x)A_f(x,y) ,$$

$$(2.2.6) \qquad A_f(x,y) = \sum_{k=1}^{m} G_k^{(2)}(x,y) \frac{\partial \psi_f^{(1)}(x,y)}{\partial y_k} + \sum_{i=1}^{n}\sum_{k=1}^{m} a_{ik}^{(12)}(x,y) \frac{\partial^2 \psi_f^{(1)}(x,y)}{\partial x_i \partial y_k} + \sum_{k,\ell=1}^{m} a_{k\ell}^{(23)}(x,y) \frac{\partial^2 \psi_f^{(1)}(x,y)}{\partial y_k \partial y_\ell} + \sum_{i=1}^{n} F_i^{(1)}(x,y) \frac{\partial \psi_f^{(1)}(x,y)}{\partial x_i}$$

$$+ \sum_{i=1}^{n} G_i^{(1)}(x,y) \frac{\partial f(x)}{\partial x_i}$$

$$+ \frac{1}{2} \sum_{i,j=1}^{n} a_{ij}^{(1)}(x,y) \frac{\partial f(x)}{\partial x_i \partial x_j} .$$

As a consequence of our smoothness requirements, the coefficients of $\bar{\mathcal{L}}$ are smooth and bounded functions of x. Thus, $\bar{\mathcal{L}}$ generates a diffusion Markov process on R^n which is denoted by $x(t)$, $t \geq 0$. In particular, the martingale problem associated with $\bar{\mathcal{L}}$ has a unique solution P_x which for each x are probability measures in the space of continuous trajectories in R^n starting at x.

Theorem 2.1.

Under the above hypotheses the process $x^{\varepsilon}(t)$, $t \geq 0$ defined by (2.1.1) converges weakly in $C([0,T];R^n)$, $T < \infty$ but arbitrary, as $\varepsilon \to 0$ to the diffusion Markov process $x(t)$, $t \geq 0$ generated by $\bar{\mathcal{L}}$ of (2.2.5).

Remark 1.

We clarify now the role of smoothness in the above. Let $\mathcal{L}^{\varepsilon}$, $\varepsilon > 0$, be given by (2.1.2)-(2.1.6) and assume that the martingale problem for it has a unique solution. If $\mathcal{L}^{\varepsilon}$ is uniformly elliptic (in $R^n \times R^m$), then [4] it is enough for this that the diffusion coefficients be bounded and continuous and the drift coefficients bounded and measurable. If $\mathcal{L}^{\varepsilon}$ is not uniformly elliptic some smoothness is sufficient [31].

Suppose that $Y(t;x)$ is ergodic so that (2.2.2) is well defined and that (2.2.3) holds. Assume that for f smooth bounded, $\psi_f^{(1)}$ is bounded and has bounded derivatives of first order in x and second order in y. Assume further that $\bar{\mathscr{L}}$ of (2.2.5) is uniformly elliptic with bounded continuous coefficients. If not uniformly elliptic smoothness is sufficient.

Then Theorem 2.1 is valid again.

Remark 2.

Concerning boundedness of the coefficients of $\mathscr{L}^{\varepsilon}$ and $\bar{\mathscr{L}}$ we remark that one can have unbounded coefficients provided that $x(t)$, associated with $\bar{\mathscr{L}}$ of (2.2.5), has infinite explosion time with probability one. In this generality, however, $x^{\varepsilon}(t)$ may exist only up to a finite explosion time $\tau^{\varepsilon} \to \infty$ as $\varepsilon \to 0$.

If convergence of moments of $x^{\varepsilon}(t)$ to moments of $x(t)$ is required, a convenient sufficient hypothesis is that the x-derivatives of all terms in (2.1.1) be uniformly bounded as functions of x and y.

Remark 3.

Concerning the strong ergodicity hypothesis about $Y(t;x)$ embodied in the existence of ψ in (2.2.2), we note that it can be weakened somewhat, as for example is done in Section (2.8) where the situation encountered is typical. The results of Part 3 show that distinctly different results arise without (2.2.2).

Specifically, Theorem 2.1 states that in the limit $\varepsilon \to 0$ the driven process $x^{\varepsilon}(t)$ decouples from the driving process $y^{\varepsilon}(t)$ and becomes a Markov process by itself. This is not true in the examples of Part 3.

2.3 Proof of Theorem 2.1.

The proof is divided into two steps. In Step 1 we prove that the processes $x^{\varepsilon}(\cdot)$ of (2.1.1) are relatively weakly compact in $D([0,T];R^n)$, $T < \infty$ but arbitrary. Since both $x^{\varepsilon}(\cdot)$ and the limit $x(\cdot)$ are supported in $C([0,T];R^n)$ this suffices for weak convergence in C. In Step 2 we identify the limit measure by showing that it solves the martingale problem for $\bar{\mathcal{L}}$ of (2.2.5).

Both Step 1 and Step 2 are proved in a somewhat more general manner than required by the form (2.1.1). The advantage is that this method, which uses only the martingale formulation and not the Itô equations, applies equally well to processes with jumps (cf. Section 2.6) and/or boundary conditions (cf. Section 2.5).

Let f(x) be a smooth, bounded function of x and define $\psi_f^{(1)}(x,y)$ by (2.2.4). Let

$$f^{\varepsilon}(x,y) = f(x) + \varepsilon\psi_f^{(1)}(x,y) . \tag{2.3.1}$$

By construction, $\psi_f^{(1)}$ satisfies

$$\mathcal{L}_1\psi_f^{(1)} + \mathcal{L}_2 f = 0 \tag{2.3.2}$$

since ψ of (2.2.2) is the kernel of the operator $-\mathcal{L}_1^{-1}$ acting on functions whose integrals relative to $\bar{P}$ are zero (Fredholm alternative). Thus, from (2.1.2)-(2.1.5) and the above we find that

(2.3.3) $\mathscr{L}^{\varepsilon} f^{\varepsilon}(x,y) = A_f(x,y) + \varepsilon \mathscr{L}_3 \psi_f^{(1)}(x,y)$

where $A_f(x,y)$ is given by (2.2.6).

From the martingale characterization of $(x^{\varepsilon}(t), y^{\varepsilon}(t))$ (cf. (2.1.11)) it follows that

(2.3.4)
$$M_{f^{\varepsilon}}(t) = f^{\varepsilon}(x^{\varepsilon}(t), y^{\varepsilon}(t)) - f^{\varepsilon}(x,y) - \int_0^t \mathscr{L}^{\varepsilon} f^{\varepsilon}(x^{\varepsilon}(s), y^{\varepsilon}(s))ds$$
$$= f(x^{\varepsilon}(t)) + \varepsilon\psi_f^{(1)}(x^{\varepsilon}(t), y^{\varepsilon}(t)) - f(x) - \varepsilon\psi_f^{(1)}(x,y)$$
$$- \int_0^t \left[A_f(x^{\varepsilon}(s), y^{\varepsilon}(s)) + \varepsilon \mathscr{L}_3 \psi_f^{(1)}(x^{\varepsilon}(s), y^{\varepsilon}(s))\right]ds$$

is a bounded zero-mean martingale. Of course, we can express $M_{f^{\varepsilon}}(t)$ explicitly as a Brownian stochastic integral but, as remarked above, we prefer not to make use of this.

The increasing process associated with $M_{f^{\varepsilon}}$ and denoted by $\langle M_{f^{\varepsilon}}(t), M_{f^{\varepsilon}}(t)\rangle$ is given by

(2.3.5)
$$\langle M_{f^{\varepsilon}}(t), M_{f^{\varepsilon}}(t)\rangle = \int_0^t H_{f^{\varepsilon}}(x^{\varepsilon}(s), y^{\varepsilon}(s))ds$$

where

(2.3.6)
$$H_{f^{\varepsilon}}(x,y) = \left\{\mathscr{L}^{\varepsilon} (f^{\varepsilon})^2 - 2f^{\varepsilon}\mathscr{L}^{\varepsilon} f^{\varepsilon}\right\}(x,y)$$

and, as can be verified by direct computation, $H_{f^{\varepsilon}}(x,y)$ is a bounded function of x and y independently of ε (it is the quadratic form associated with second derivatives in (2.1.3)-(2.1.5)).

To prove compactness in $D([0,T];R^n)$ for $x^\varepsilon(\cdot)$, it suffices to show that [5]

$$\text{(2.3.7)}\qquad \overline{\lim_{\delta\downarrow 0}}\ \overline{\lim_{\varepsilon\downarrow 0}}\ \sup_{|t-s|\le\delta}\ \sup E\{|x^\varepsilon(t)-x^\varepsilon(s)|^2 \big| \mathcal{F}_s\} = 0 ,$$

where the sup next to the expectation is over the past up to time s i.e., over $x^\varepsilon(s) = x$, $y^\varepsilon(s) = y$. This follows immediately from (2.3.4)-(2.3.6) upon taking $f(x) = x_i$, $i = 1,2,\ldots,n$, successively and squaring, etc. in (2.3.4). Note that this choice of f is an unbounded function but $\psi_f^{(1)}$, A_f, $\mathcal{L}_3\psi_f^{(1)}$ and H_{f^ε} involve only drivatives of f and hence are bounded. Specifically, using vector notation, we have,

$$\begin{aligned}|x^\varepsilon(t)-x^\varepsilon(s)|^2 &\le 4\varepsilon\left[\psi_x^{(1)}(x^\varepsilon(t),y^\varepsilon(t)) - \psi_x^{(1)}(x^\varepsilon(s),y^\varepsilon(s))\right]^2 \\ &+ 4\left|\int_s^t \left[A_x(x^\varepsilon(\sigma),y^\varepsilon(\sigma)) + \varepsilon\mathcal{L}_3\psi_x^{(1)}(x^\varepsilon(\sigma),y^\varepsilon(\sigma))\right]d\sigma\right|^2 \\ &+ 4|M^\varepsilon(t)-M^\varepsilon(s)|^2 ,\end{aligned}$$

where $M^\varepsilon(t)$ stands for $M_{f^\varepsilon}(t)$ with f^ε the vector function $x + \varepsilon\psi_x^{(1)}(x,y)$. Upon using

$$E\left\{|M^\varepsilon(t)-M^\varepsilon(s)|^2 \big| \mathcal{F}_s\right\} = \int_s^t E\left\{|H^\varepsilon(x^\varepsilon(\sigma),y^\varepsilon(\sigma))|^2 \big| \mathcal{F}_s\right\}d\sigma ,$$

with H^ε the vector function H_{f^ε} and f^ε as above, the result (2.3.7) follows and Step 1 is complete.

For Step 2, the identification of the limit, we rewrite (2.3.4), with f(x) a smooth bounded function, as follows

$$(2.3.8)\qquad f(x^\varepsilon(t)) - f(x^\varepsilon(s)) - \int_s^t \bar{\mathcal{L}} f(x^\varepsilon(\sigma))d\sigma$$

$$= -\varepsilon\left[\psi_f^{(1)}(x^\varepsilon(t),y^\varepsilon(t)) - \psi_f^{(1)}(x^\varepsilon(s),y^\varepsilon(s))\right]$$

$$+ \int_s^t \left[A_f(x^\varepsilon(\sigma),y^\varepsilon(\sigma)) - \bar{\mathcal{L}} f(x^\varepsilon(\sigma))\right]d\sigma$$

$$+ \varepsilon \int_s^t \mathcal{L}_3\psi_f^{(1)}(x^\varepsilon(\sigma),y^\varepsilon(\sigma))d\sigma$$

$$+ M_{f^\varepsilon}(t) - M_{f^\varepsilon}(s) \ .$$

Multiplying (2.3.8) by any bounded continuous functional Φ_s of $x(\cdot)$ which is $\mathcal{F}_s$ measurable, taking expectations and passing to the limit $\varepsilon \to 0$ we obtain

$$(2.3.9)\qquad \lim_{\varepsilon\downarrow 0} E\left\{\Phi_s\left[f(x^\varepsilon(t)) - f(x^\varepsilon(s)) - \int_s^t \bar{\mathcal{L}} f(x^\varepsilon(\sigma))d\sigma\right]\right\}$$

$$= \lim_{\varepsilon\downarrow 0} E\left\{\Phi_s \int_s^t \left[A_f(x^\varepsilon(\sigma),y^\varepsilon(\sigma)) - \mathcal{L} f(x^\varepsilon(\sigma))\right]d\sigma\right\} \ .$$

The right hand side of (2.3.9) is zero by the definition (2.2.5) of $\bar{\mathcal{L}} f(x)$ and the ergodicity of $Y(t;x)$ i.e., (2.2.2). Since the coefficients are smooth here, this fact can be seen immediately by defining first

$$(2.3.10) \quad \psi_f^{(2)}(x,y) = \int \psi(y,dz)\, A_f(x,z)$$

so that

$$(2.3.11) \quad \mathcal{L}_1 \psi_f^{(2)} + A_f - \bar{\mathcal{L}} f = 0 \ .$$

Then we apply (2.3.4) with $f^\varepsilon = \varepsilon^2 \psi_f^{(2)}$ (using the smoothness of the coefficients), multiply by Φ_s, take expectations and pass to the limit $\varepsilon \to 0$ to obtain the result that the right hand side of (2.3.9) is zero. If the coefficients are only continuous, then the above is preceded by a preliminary smoothing of the coefficients which is removed again at the end.

In view of continuity of the functional on the left side of (2.3.9) and the compactness it follows that for any limit measure of the processes $x^\varepsilon(\cdot)$, the expression

$$f(x(t)) - f(x) - \int_0^t \bar{\mathcal{L}} f(x(s))ds \ ,$$

is a martingale for all $f(x) \in C^\infty$ and bounded. But the solution to this martingale problem is unique by hypothesis hence Step 2 and proof of the theorem is complete.

Remark.

Assuming smoothness as we do, the essential ingredient of the above proof is the following formal computation.

Given f(x) smooth and bounded let $\psi_f^{(1)}(x,y)$ and $\psi_f^{(2)}(x,y)$ be two smooth functions to be chosen appropriately. Let

(2.3.12) $$f^{\varepsilon} = f + \varepsilon\psi_f^{(1)} + \varepsilon^2\psi_f^{(2)} .$$

Then

(2.3.13)
$$\begin{aligned}\mathcal{L}^{\varepsilon}f^{\varepsilon} &= \left(\frac{1}{\varepsilon^2}\mathcal{L}_1 + \frac{1}{\varepsilon}\mathcal{L}_2 + \mathcal{L}_3\right)\left(f + \varepsilon\psi_f^{(1)} + \varepsilon^2\psi_f^{(2)}\right) \\ &= \frac{1}{\varepsilon}\left(\mathcal{L}_1\psi_f^{(1)} + \mathcal{L}_2 f\right) + \left(\mathcal{L}_1\psi_f^{(2)} + \mathcal{L}_2\psi_f^{(1)} + \mathcal{L}_3 f - \bar{\mathcal{L}}f\right) \\ &\quad + \bar{\mathcal{L}}f + \varepsilon\left(\mathcal{L}_2\psi_f^{(2)} + \mathcal{L}_3\psi_f^{(1)}\right) + \varepsilon^2\mathcal{L}_3\psi_f^{(2)} \\ &= \bar{\mathcal{L}}f + 0(\varepsilon) ,\end{aligned}$$

provided that

(2.3.14) $$\mathcal{L}_1\psi_f^{(1)} + \mathcal{L}_2 f = 0$$

and

(2.3.15) $$\mathcal{L}_1\psi_f^{(2)} + \mathcal{L}_2\psi_f^{(1)} + \mathcal{L}_3 f - \bar{\mathcal{L}}f = 0 .$$

This is exactly how $\psi_f^{(1)}$ and $\psi_f^{(2)}$ were defined by (2.2.4) (cf. (2.3.2)) and (2.3.11) (since $A_f \equiv \mathcal{L}_2\psi_f^{(1)} + \mathcal{L}_3 f$).

2.4 Stability and related questions.

In this section we discuss briefly how one can adapt, very simply in the present context, the usual Lyapounov methods for Markov processes (cf. for example [24]) to answer questions regarding the behavior of $x^{\varepsilon}(t)$ as $t \to \infty$, with ε fixed, $0 < \varepsilon \leq \varepsilon_o$ and ε_o sufficiently small. The objective here is to answer such questions by using only corresponding properties of the limit process $x(t)$ as $t \to \infty$. We assume, of course, that Theorem 2.1 holds.

Let $(x^{\varepsilon}(t), y^{\varepsilon}(t))$ be defined by (2.1.1) and assume that the hypotheses of Theorem 2.1 hold. Assume further that

$$F_i^{(1)}(0,y) = G_i^{(1)}(0,y) = 0 , \tag{2.4.1}$$

$$\sigma_{ij}^{(1)}(0,y) = 0 , \quad i = 1,2,\dots,n , \quad j = 1,2,\dots,p , \quad y \in R^m ,$$

which means that the origin is an equilibrium point of $x^{\varepsilon}(t)$. From (2.4.1) and (2.2.4)-(2.2.6) it follows that the origin is also an equilibrium point of $x(t)$ generated by $\bar{\mathcal{L}}$. A typical result is now the following.

Theorem 2.2.

Let $V(x)$ be a smooth positive definite function on R^n i.e.,

$$V(x) \geq 0 , \quad V(x) = 0 \Rightarrow , \quad x = 0 . \tag{2.4.2}$$

Suppose that there is a constant $\gamma > 0$ and an open neighborhood K of the origin in R^n having compact closure, such that

$$\bar{\mathcal{L}}V(x) \leq -\gamma V(x) \ , \qquad x \in K \ . \tag{2.4.3}$$

Then for $0 < \varepsilon \leq \varepsilon_o$, ε fixed and ε_o sufficiently small, $x^\varepsilon(t)$ is uniformly stochastically asymptotically stable as $t \to \infty$ i.e., for any $\eta_1 > 0$ and $\eta_2 > 0$, there is a $\delta > 0$ such that if $|x(0)| = |x| < \delta$ then

$$P_{x,y}\{|x^\varepsilon(t)| \leq \eta_2 e^{-\tilde{\gamma}t} \ , \quad t \geq 0\} \geq 1 - \eta_1 \tag{2.4.4}$$

for all y and with $\tilde{\gamma} > 0$ a constant.

Proof.

Let V(x) be as in the statement above and define $\psi_V^{(1)}(x,y)$ and $\psi_V^{(2)}(x,y)$ by (2.2.4) and (2.3.11), respectively, with $f = V$. Let

$$V^\varepsilon(x,y) = V(x) + \varepsilon\psi_V^{(1)}(x,y) + \varepsilon^2\psi_V^{(2)}(x,y) \ . \tag{2.4.5}$$

From the smoothness and boundedness of the coefficients in (2.1.1) and (2.4.1), it follows that there is an ε_o sufficiently small and constants $\tilde{c}_1 > 0$ and $\tilde{c}_2 > 0$ such that for all $y \in R^m$ and $0 < \varepsilon \leq \varepsilon_o$

$$\tilde{c}_1 V(x) \leq V^\varepsilon(x,y) \leq \tilde{c}_2 V(x) \ , \qquad x \in K \ . \tag{2.4.6}$$

Let $\hat{\gamma} > 0$ be a constant. From (2.3.13) with $f^\varepsilon = V^\varepsilon$ and (2.4.6) we have

$$(2.4.7) \qquad (\mathcal{L}^{\varepsilon} + \hat{\gamma})V^{\varepsilon} = \hat{\gamma}V^{\varepsilon} + \bar{\mathcal{L}}V(x)$$

$$+ \varepsilon\left[\mathcal{L}_2\psi_V^{(2)} + \mathcal{L}_3\psi_V^{(1)}\right] + \varepsilon^2\mathcal{L}_3\psi_V^{(2)}$$

$$\leq \left[\bar{\mathcal{L}} + \tilde{c}_2\hat{\gamma} + \varepsilon_o\tilde{c}_3\right]V \ , \quad x \in K \ ,$$

where $\tilde{c}_3$ is some constant. If we choose $\hat{\gamma}$ so that $\tilde{c}_2\hat{\gamma} + \varepsilon_o\tilde{c}_3 \leq \gamma$ (ε_o being sufficiently small), we find that (2.4.3) yields

$$(2.4.8) \qquad (\mathcal{L}^{\varepsilon} + \hat{\gamma})V^{\varepsilon}(x,y) \leq 0 \ , \qquad x \in K \ , \quad y \in R^m \ .$$

Suppose that $x^{\varepsilon}(0) = x \in K$ and let τ be the first exit time of $x^{\varepsilon}(t)$ from K. We use the identity (2.3.4) with $f^{\varepsilon} = e^{t\hat{\gamma}}V^{\varepsilon}$ and the upper time limit equal to $\tau \wedge t = \min(t,\tau)$ to obtain

$$(2.4.9) \qquad e^{\hat{\gamma}\tau\wedge t}V^{\varepsilon}(x^{\varepsilon}(t\wedge\tau), y^{\varepsilon}(t\wedge\tau)) = V^{\varepsilon}(x,y)$$

$$+ \int_0^{t\wedge\tau} e^{s\hat{\gamma}}(\mathcal{L}^{\varepsilon} + \hat{\gamma})V^{\varepsilon}(x^{\varepsilon}(s), y^{\varepsilon}(s))ds$$

$$+ \tilde{M}_{V^{\varepsilon}}(t\wedge\tau) \ .$$

Here $\tilde{M}_{V^{\varepsilon}}(t\wedge\tau)$ is a zero mean bounded martingale with increasing process as in (2.3.5) with $f^{\varepsilon} = V^{\varepsilon}$, an extra factor $e^{2\hat{\gamma}t}$ multiplying $H_{V^{\varepsilon}}$ and $t \wedge \tau$ for upper limit of intergration in (2.3.5).

Using (2.4.6) and (2.4.8) in (2.4.9) we obtain

$$(2.4.10)\quad 0 \le e^{\gamma\tau\wedge t}\, \tilde{c}_1 V(x^{\varepsilon}(t\wedge\tau) \le \tilde{c}_2 V(x) + \tilde{M}_{V^{\varepsilon}}(t\wedge\tau) .$$

Thus, $\tilde{c}_2 V + \tilde{M}_{V^{\varepsilon}}(t\wedge\tau)$ is a nonnegative bounded martingale and by Kolmogorov's inequality we have that for each $\tilde{\eta}_2 > 0$

$$(2.4.11)\quad P_{x,y}\left\{\sup_{0\le t\le T} e^{\hat{\gamma}t\wedge\tau}\, \tilde{c}_1 V(x^{\varepsilon}(t\wedge\tau)) > \tilde{\eta}_2\right\}$$

$$\le P_{x,y}\left\{\sup_{0\le t\le T} (\tilde{c}_2 V(x) + \tilde{M}_{V^{\varepsilon}}(t\wedge\tau)) > \tilde{\eta}_2\right\}$$

$$\le \frac{\tilde{c}_2 V(x)}{\tilde{\eta}_2} .$$

Since the right hand side in (2.4.11) is independent of the set K and T, it follows that

$$(2.4.12)\quad P_{x,y}\left\{\sup_{t\ge 0} e^{\hat{\gamma}t}\tilde{c}_1 V(x^{\varepsilon}(t)) > \tilde{\eta}_2\right\} \le \frac{\tilde{c}_2 V(x)}{\tilde{\eta}_2} .$$

The positive definiteness (2.4.2) of V(x) and estimate (2.4.12) yield, in the usual way, the result (2.4.4). The proof is complete.

Some more details regarding stability and related questions can be found in [32].

2.5 Boundary conditions.

Let $\phi(x)$, $x \in R^n$, be a smooth real-valued function such that

$$\mathcal{D} = \{x \in R^n \mid \phi(x) > 0\} \tag{2.5.1}$$

is an open connected set in R^n with compact closure. We denote the boundary of $\mathcal{D}$ by $\partial\mathcal{D}$ and the unit inner normal at $x \in \partial\mathcal{D}$ by $\hat{n}(x)$. We assume that ϕ is such that $\nabla\phi(x) \equiv \hat{n}(x)$ for $x \in \partial\mathcal{D}$.

We shall consider the process $(x^\varepsilon(t), y^\varepsilon(t))$ with generator $\mathcal{L}^\varepsilon$ given by (2.1.2)-(2.1.5) in $\mathcal{D} \times R^m$ and we shall prescribe reflecting boundary conditions for $x^\varepsilon(t)$ on $\partial\mathcal{D}$. Throughout this section $\mathcal{L}^\varepsilon$ is assumed to be uniformly elliptic in $\bar{\mathcal{D}} \times R^m$.

Boundary conditions are given as follows. Let $B(x,y)$, $x \in \partial\mathcal{D}$, $y \in R^m$, be a smooth n-vector function pointing inward at the boundary. Specifically we assume that there is a $\delta > 0$ such that

$$\inf_{\substack{x \in \partial\mathcal{D} \\ y \in R^m}} \sum_{i=1}^{n} B_i(x,y) \frac{\partial\phi(x)}{\partial x_i} \geq \delta . \tag{2.5.2}$$

Boundary conditions are prescribed so that $x^\varepsilon(t)$ is reflected obliquely at $\partial\mathcal{D}$ in the direction B. According to the results of [5] the reflected process in $\bar{\mathcal{D}} \times R^m$, again denoted by $(x^\varepsilon(t), y^\varepsilon(t))$, is under our hypotheses the unique solution of the following submartingale problem $(\mathcal{L}^\varepsilon, B)$. The measures $P^{\varepsilon}_{x,y}$ in the space of continuous functions in $\bar{\mathscr{D}} \times R^m$ corresponding to $(x^{\varepsilon}(t), y^{\varepsilon}(t))$ satisfy

(i) $P^{\varepsilon}_{x,y}\{x(0) = x, y(0) = y\} = 1$,

$$x \in \bar{\mathscr{D}}, \quad y \in R^m,$$

(2.5.3) (ii) for all bounded smooth functions on $\bar{\mathscr{D}} \times R^m$ such that

$$\sum_{i=1}^{n} B_i(x,y) \frac{\partial f(x,y)}{\partial x_i} \geq 0,$$

$$x \in \partial\mathscr{D}, \quad y \in R^m,$$

the expression

$$f(x(t),y(t)) - \int_0^t \mathscr{L}^{\varepsilon} f(x(s),y(s)ds$$

is a submartingale for $P^{\varepsilon}_{x,y}$.

We shall assume, with no loss in generality in view of (2.5.2), that the vector field $B(x,y)$ is normalized so that

$$(2.5.4) \qquad B_{\hat{n}}(x,y) \equiv \sum_{i=1}^{m} B_i(x,y) \frac{\partial \phi(x)}{\partial x_i} \equiv 1 .$$

The corresponding increasing process of $x^{\varepsilon}(t)$ at the boundary will be denoted by $N^{\varepsilon}(t)$. It is defined by the property that for all $f(x,y)$ smooth and bounded on $\bar{\mathscr{D}} \times R^m$,

$$(2.5.5) \qquad f(x^{\varepsilon}(t),y^{\varepsilon}(t)) - \int_0^t \mathscr{L}^{\varepsilon} f(x^{\varepsilon}(s),y^{\varepsilon}(s))$$

$$- \int_0^t \sum_{i=1}^{n} B_i(x^{\varepsilon}(s),y^{\varepsilon}(s)) \frac{\partial f(x^{\varepsilon}(s),y^{\varepsilon}(s))}{\partial x_i} dN^{\varepsilon}(s)$$

$$\equiv M_f(t)$$

is a martingale (when superscript ε appears on the paths we do not refer explicitly to the measure P^{ε} as in (2.5.3)). We refer to [5] for the details of the existence and uniqueness theory.

According to Theorem 2.1 the process $x^{\varepsilon}(t)$ converges, away from boundaries, to the diffusion process $x(t)$ generated by $\bar{\mathcal{L}}$ of (2.2.5), (2.2.6). The question is: What happens to the boundary conditions as $\varepsilon \to 0$? Stated another way, we wish to characterize the diffusion process in $\bar{\mathcal{D}}$ to which the reflected process $x^{\varepsilon}(t)$ converges weakly (under our hypotheses) and, of course, to prove weak convergence.

It is clear that the behavior of the reflected process $x^{\varepsilon}(t)$ must be analyzed in detail near the boundary $\partial\mathcal{D}$. For this purpose we introduce stretched boundary layer coordinates as follows.

Since $\partial\mathcal{D}$ is a compact (n-1)-dimensional submanifold of R^n, it can be covered by a finite system of neighborhoods in each of which we have a given coordinate patch. We fix attention to a single patch and let $x = \xi(\gamma)$, $\gamma \in R^{n-1}$ be the parametric equation of the surface $\partial\mathcal{D}$.

We introduce stretched coordinates $x \to (\gamma, \varepsilon\eta)$ near $\partial\mathcal{D}$ by setting

$$x = \xi(\gamma) + \varepsilon\eta\hat{n}(\gamma) \quad . \tag{2.5.6}$$

Here $\hat{n}(x)$, $x \in \partial\mathcal{D}$ is the unit inner normal and we think of it, alternatively, as a function $x \in \partial\mathcal{D}$ or of γ abbreviating $\hat{n}(\xi(\gamma)) = \hat{n}(\gamma)$. The scalar parameter $\eta \geq 0$ is confined to an interval depending on the size of the patch and goes to $+\infty$ as $\varepsilon \to 0$. In fact we have

$$\eta = \frac{(x-\xi(x))\cdot\hat{n}(x)}{\varepsilon}, \tag{2.5.7}$$

where, in returning from $(\gamma,\varepsilon\eta)$ to x, we write $\xi(x)$ instead of $\xi(\gamma(x))$, with $\gamma(x)$ being uniquely defined if the patch is small enough.

With $\xi \in \partial\mathcal{D}$ a fixed parameter we define the operator $\mathcal{L}_{BL}$, acting on functions of $y \in R^m$ and $\eta \geq 0$ as follows (recall that $\hat{n}(x) \equiv \nabla\phi(x)$ by hypothesis):

$$\begin{aligned}\mathcal{L}_{BL} = {} & \sum_{k=1}^{m} F_k^{(1)}(\xi,y)\frac{\partial}{\partial y_k} + \frac{1}{2}\sum_{k,\ell=1}^{m} a_{k\ell}^{(2)}(\xi,y)\frac{\partial^2}{\partial y_k \partial y_\ell} \\ & + \sum_{i=1}^{n} F_i^{(1)}(\xi,y)\frac{\partial\phi(\xi)}{\partial x_i}\frac{\partial}{\partial\eta} \\ & + \sum_{k=1}^{m}\left[\sum_{i=1}^{n} a_{ik}^{(12)}(\xi,y)\frac{\partial\phi(\xi)}{\partial x_i}\right]\frac{\partial^2}{\partial\eta\partial y_k} \\ & + \frac{1}{2}\left[\sum_{i,j=1}^{n} a_{ij}^{(1)}(\xi,y)\frac{\partial\phi(\xi)}{\partial x_i}\frac{\partial\phi(\xi)}{\partial x_j}\right]\frac{\partial^2}{\partial\eta^2}.\end{aligned} \tag{2.5.8}$$

This operator, the boundary layer operator, is characterized uniquely by the following property. With $\mathcal{L}^\varepsilon$, $\mathcal{L}_1$, $\mathcal{L}_2$ and $\mathcal{L}_3$ defined by (2.1.2)-(2.1.5) and with the change of variables $x \to (\gamma,\varepsilon\eta)$ of (2.5.6),

$$\varepsilon\left[\mathcal{L}^\varepsilon - \frac{1}{\varepsilon^2}\mathcal{L}_{BL}\right]f(\gamma,\eta) \tag{2.5.9}$$

is well defined near the boundary on smooth functions and is bounded independently of ε.

We shall assume, as in Section 2.2, that $\mathcal{L}_{BL}$ has the following ergodic properties. Their validity (and of the ones in Section 2.2) is discussed in the Appendix.

There exists a unique probability measure $\bar{U}(A;\xi)$, A a Borel subset of R^m, depending on $\xi \in \partial\mathcal{D}$ such that for any bounded smooth function $\psi(y)$ on R^m with

$$\int_{R^m} \psi(y)\bar{U}(dy;\xi) = 0 , \tag{2.5.10}$$

the boundary value problem

$$\begin{aligned} &\mathcal{L}_{BL} v(\eta,y) = 0 , \quad \eta > 0 , \quad y \in R^m , \\ &\left.\frac{\partial v(\eta,y)}{\partial \eta}\right|_{\eta=0} = \psi(y) , \quad y \in R^m , \end{aligned} \tag{2.5.11}$$

has a smooth solution which can be made unique by imposing the condition

$$\lim_{\eta \to +\infty} v(\eta,y) = 0 . \tag{2.5.12}$$

The limit zero in (2.5.12) is approached exponentially fast, uniformly† in y, and so do derivatives of v.

Some additional information regarding $\mathcal{L}_{BL}$ is given in the appendix. We comment briefly on the probabilistic significance of the above.

The operator $\mathcal{L}_{BL}$ with Neumann boundary condition at $\eta = 0$ is the generator on $[0,\infty) \times R^m$ of a reflected process

---

†Uniformly on compact sets suffices in all considerations.

$(\eta(t), y(t))$. Let $N(t)$ be the increasing process of $y(t)$ at $\eta = 0$. Then the solution $v(\eta,y)$ of (2.5.11) has the representation

$$v(\eta,y) = \lim_{t\uparrow\infty} E\left\{\int_0^t \psi(y(s))\,dN(s)\right\}, \tag{2.5.13}$$

where $\eta(0) = \eta$ and $y(0) = y$ and provided the limit exists.

Let

$$\tau_\theta = \sup\{s: s \geq 0,\ N(s) \leq \theta\}.$$

We show in the appendix that $\tau_\theta \to \infty$ as $\theta \to \infty$ a.s. and that we can write

$$v(\eta,y) = \lim_{\theta\uparrow\infty} E\left\{\int_0^\theta \psi(\tilde{y}(\zeta))\,d\zeta\right\} \tag{2.5.14}$$

where $\tilde{y}(\zeta) = y(\tau_\zeta)$ is the boundary process [5]. This process is ergodic and the $\bar{U}$ introduced above is its stationary measure. This explains (2.5.10).

We return now to our main objective in this section which is the asymptotic analysis.

For each $x \in \partial\mathcal{D}$ we define the vector field

$$\bar{B}(x) = \int_{R^m} B(x,y)\bar{U}(dy;x), \qquad x \in \partial\mathcal{D}. \tag{2.5.15}$$

In view of (2.5.2), there is a $\delta > 0$ such that

$$\inf_{x\in\partial\mathcal{D}} \sum_{i=1}^n \bar{B}_i(x)\,\frac{\partial\phi(x)}{\partial x_i} \geq \delta, \tag{2.5.16}$$

and $\bar{B}(x)$ is smooth. Let $\bar{\mathcal{L}}$ be defined by (2.2.5) and (2.2.6).

Theorem 2.3.

The reflected process $x^{\varepsilon}(t)$ under the above hypotheses converges weakly in $C([0,T];\bar{\mathcal{D}})$, $T < \infty$ but arbitrary, as $\varepsilon \to 0$ to the reflected diffusion process $x(t)$ in $\bar{\mathcal{D}}$ which is the unique solution of the submartingale problem $(\bar{\mathcal{L}}, \bar{B})$.

Proof.

As with Theorem 2.1, the proof is divided into two steps: (1) compactness in $D([0,T];\bar{\mathcal{D}})$ and (2) identification of the limit with $(\bar{\mathcal{L}}, \bar{B})$.

Let $f(x)$ be a smooth function on $\bar{\mathcal{D}}$ and define $\psi_f^{(1)}$ and ψ_f^{BL} as follows. The function $\psi_f^{(1)}(x,y)$ is defined by (2.2.4), as before, so that (2.3.2) holds. The function ψ_f^{BL} is a function of $y \in R^m$ and $\eta \geq 0$ with $\xi \in \partial\mathcal{D}$ a parameter (actually, a function of $\gamma \in R^{n-1}$ i.e. we think of ξ as $\xi(\gamma)$). It satisfies the problem

$$\mathcal{L}_{BL}\psi_f^{BL}(\eta,y;\xi) = 0\ , \qquad \eta > 0\ , \quad y \in R^m\ , \quad \xi \in \partial\mathcal{D},$$

$$\frac{\partial}{\partial\eta}\psi_f^{BL}(\eta,y;\xi)\Big|_{\eta=0} = -\sum_{i=1}^{n} B_i(\xi,y)\frac{\partial f(\xi)}{\partial x_i} + \sum_{i=1}^{n} \bar{B}_i(\xi)\frac{\partial f(\xi)}{\partial x_i}\ , \tag{2.5.17}$$

$$\lim_{\eta\to+\infty} \psi_f^{BL}(\eta,y;\xi) = 0\ .$$

In view of definition (2.5.15) and (2.5.10)-(2.5.12), ψ_f^{BL} is well defined and, along with derivatives, decays to zero exponentially as $\eta \to \infty$.

Define

$$(2.5.18)\quad f^{\varepsilon}(x,y) = f(x) + \varepsilon\psi_f^{(1)}(x,y) + \varepsilon\zeta(x)\psi_f^{BL}\left(\frac{(x-\zeta(x))\cdot\hat{n}(x)}{\varepsilon},y;\xi(x)\right),$$

where $\zeta(x)$ is a C^{∞} function identically equal to one in a fixed neighborhood θ_1 of $\partial\mathcal{D}$, covered by the coordinate patches, and identically equal to zero outside some fixed neighborhood $\theta_2 \supset \theta_1$. From the construction of $\psi_f^{(1)}$ and ψ_f^{BL} we have that

$$(2.5.19)\quad \mathcal{L}^{\varepsilon}f^{\varepsilon}(x,y) = A_f(x,y) + \varepsilon\mathcal{L}_3\psi_f^{(1)}(x,y) + g_f^{\varepsilon}(x,y), \qquad x \in \bar{\mathcal{D}},\quad y \in R^m,$$

where A_f is defined by (2.2.6) and, because of (2.5.17),

$$(2.5.20)\quad g_f^{\varepsilon} = \varepsilon\left(\mathcal{L}^{\varepsilon}\zeta\psi_f^{BL} - \zeta\mathcal{L}^{\varepsilon}\psi_f^{BL}\right) + \varepsilon\zeta\left(\mathcal{L}^{\varepsilon}\psi_f^{BL} - \frac{1}{\varepsilon^2}\mathcal{L}_{BL}\psi_f^{BL}\right).$$

The first term on the right side of (2.5.20) is identically equal to zero for $x \in \theta_1 \cap \bar{\mathcal{D}}$ and $x \notin \theta_2 \cap \bar{\mathcal{D}}$. In the rest of $\bar{\mathcal{D}}$ it decays to zero exponentially since ψ_f^{BL} and its derivatives decay to zero as $\eta \to \infty$. The second term on the right side of (2.5.20) is identically zero for $x \notin \theta_2 \cap \bar{\mathcal{D}}$ and uniformly bounded as $\varepsilon \downarrow 0$ in the rest of $\bar{\mathcal{D}}$; this follows from the characterization (2.5.9).

For $x \in \partial\mathcal{D}$ we have, by construction again and by (2.5.4),

$$(2.5.21)\qquad \sum_{i=1}^{n} B_i(x,y)\left.\frac{\partial f^{\varepsilon}(x,y)}{\partial x_i}\right|_{x\in\partial\mathcal{D}} = \sum_{i=1}^{n} \bar{B}_i(x)\left.\frac{\partial f(x)}{\partial x_i}\right|_{x\in\partial\mathcal{D}} + h_f^{\varepsilon}(x,y) \ .$$

Here $h_f^{\varepsilon} = 0(\varepsilon)$ because $\psi_f^{(1)}(x,y)$, $\psi_f^{BL}(\eta,y;\xi)$ and derivatives are bounded.

The above constructions and (2.5.5) with $f = f^{\varepsilon}$ lead to the following:

$$\begin{aligned}(2.5.22)\qquad & f(x^{\varepsilon}(t)) + \varepsilon\psi_f^{(1)}(x^{\varepsilon}(t),y^{\varepsilon}(t)) \\ & + \varepsilon\zeta(x^{\varepsilon}(t))\psi_f^{BL}\left(\frac{(x^{\varepsilon}(t)-\xi(x^{\varepsilon}(t)))\ \cdot\hat{n}(x^{\varepsilon}(t))}{\varepsilon},y^{\varepsilon}(t);\xi(x^{\varepsilon}(t))\right) \\ & - f(x) - \varepsilon\psi_f^{(1)}(x,y) - \varepsilon\zeta(x)\psi_f^{BL}\left(\frac{(x-\xi(x))\cdot\hat{n}(x)}{\varepsilon},y;\xi(x)\right) \\ & - \int_0^t \bar{\mathcal{L}}f(x^{\varepsilon}(s))ds - \int_0^t (A_f - \bar{\mathcal{L}}f)(x^{\varepsilon}(s),y^{\varepsilon}(s))ds \\ & - \int_0^t \varepsilon\mathcal{L}_3\psi_f^{(1)}(x^{\varepsilon}(s),y^{\varepsilon}(s))ds - \int_0^t g_f(x^{\varepsilon}(s),y^{\varepsilon}(s))ds \\ & - \int_0^t \bar{B}(x^{\varepsilon}(s))\cdot\frac{\partial f(x^{\varepsilon}(s))}{\partial x}\, dN^{\varepsilon}(s) \\ & - \int_0^t h_f(x^{\varepsilon}(s),y^{\varepsilon}(s))dN^{\varepsilon}(s) \\ & \qquad \equiv M_{f^{\varepsilon}}(t) \ ,\end{aligned}$$

where $M_{f^\varepsilon}(t)$ is a zero-mean martingale and $\bar{B}\cdot\frac{\partial f}{\partial x}$ stands for the inner product of $\bar{B}$ and the gradient of f (we shall see shortly that $N^\varepsilon(t)$ has finite second moments).

For Step 1, compactness, it suffices to show that (2.3.7) holds again. Suppose that f(x) = x, the vector of coordinate functions (letting f be a vector is all right, of course) in (2.5.22) and that the lower limit is not zero but s with $s \le t$. It is clear that (2.3.7) will hold provided we show that

$$\overline{\lim_{\delta\downarrow 0}}\ \overline{\lim_{\varepsilon\downarrow 0}}\ \sup_{|t-s|\le\delta}\ \sup E\{(N^\varepsilon(t)-N^\varepsilon(s))^2|\mathcal{F}_s\} = 0 , \tag{2.5.23}$$

where the sup next to the expectation is over all $x^\varepsilon(s) = x \in \bar{\mathcal{D}}$ and $y^\varepsilon(s) = y \in R^m$.

To prove (2.5.23) we note that the increasing process corresponding to $M_f(t)$ in (2.5.5) is given by

$$<M_f(t),M_f(t)> = \int_0^t (\mathcal{L}^\varepsilon f^2 - 2f\mathcal{L}^\varepsilon f)(x^\varepsilon(s),y^\varepsilon(s))ds . \tag{2.5.24}$$

At this point, we do not know that $N^\varepsilon(t)$ has second moments; this will follow later. If we let τ_ν be the first time N(t) equals $\nu < \infty$, then τ_ν is a stopping time and the above is true with t replaced by $t \wedge \tau_\nu$. If the same is done with the calculations that follow, the desired estimate will be independent of ν so letting $\nu \to \infty$ in the end will produce the results. After this remark we continue with (2.5.23).

We apply (2.5.22) with $f(x) = \phi(x)$, the support function of the region $\mathcal{D}$. In view of (2.5.16) there is a $\delta_1 > 0$ so that for ε small enough

$$N^{\varepsilon}(t) - N^{\varepsilon}(s) \leq \frac{1}{\delta_1}\int_s^t \left(\overline{B}\cdot\frac{\partial\phi}{\partial x} + h_{\phi}^{\varepsilon}\right)(x^{\varepsilon}(\sigma),y^{\varepsilon}(\sigma))\,dN^{\varepsilon}(\sigma) \tag{2.5.25}$$

$$= \frac{1}{\delta_1}\left[\phi(x^{\varepsilon}(t)) - \phi(x^{\varepsilon}(s)) - \left(M_{\phi^{\varepsilon}}(t) - M_{\phi^{\varepsilon}}(s)\right)\right]$$

$$+ 0(t-s) + 0(\varepsilon) .$$

Hence,

$$E\{(N^{\varepsilon}(t)-N^{\varepsilon}(s))^2|\mathcal{F}_s\} \tag{2.5.26}$$

$$\leq \frac{4}{\delta_1} E\{(\phi(x^{\varepsilon}(t))-\phi(x^{\varepsilon}(s)))^2|\mathcal{F}_s\}$$

$$+ \frac{4}{\delta_1} E\{(M_{\phi^{\varepsilon}}(t)-M_{\phi^{\varepsilon}}(s))^2|\mathcal{F}_s\}$$

$$+ 0(t-s) + 0(\varepsilon) .$$

Using (2.5.24) with $f = \phi^{\varepsilon}$ (i.e., the f^{ε} that results in (2.5.18) when $f = \phi$) it follows that the second term on the right of (2.5.26) is $0(t-s)$. Hence it remains to consider the first term on the right.

Using (2.5.22) with $f = \phi^2$ it follows that

$$E\{(\phi(x^{\varepsilon}(t))-\phi(x^{\varepsilon}(s))^2|\mathcal{F}_s\} \tag{2.5.27}$$

$$= E\{\phi^2(x^{\varepsilon}(t))-\phi^2(x^{\varepsilon}(s))\,|\,\mathcal{F}_s\}$$

$$- 2\phi(x^{\varepsilon}(s))E\{\phi(x^{\varepsilon}(t))-\phi(x^{\varepsilon}(s))\,|\,\mathcal{F}_s\}$$

$$= 0(t-s) + 0(\varepsilon) - 2\phi(x^\varepsilon(s))E\left\{\int_s^t \left(\bar{B}\cdot\frac{\partial\phi}{\partial x} + h_\phi^\varepsilon\right)(x^\varepsilon(s),y^\varepsilon(s))dN^\varepsilon(s)\right\}.$$

But then the last term on the right is nonpositive for ε sufficiently small. Combining (2.5.27) and (2.5.26) yields (2.5.23).

Now we pass to Step 2 the idenfication. With f(x) a smooth function on $\bar{\mathcal{D}}$ we rewrite (2.5.22) as follows

$$f(x^\varepsilon(t)) - f(x) - \int_0^t \bar{\mathcal{L}}f(x^\varepsilon(s)ds - \int_0^t \bar{B}(x^\varepsilon(s))\cdot\frac{\partial f(x^\varepsilon(s))}{\partial x}\,dN^\varepsilon(s) \tag{2.5.28}$$

$$= \int_0^t (A_f - \bar{\mathcal{L}}f)(x^\varepsilon(s),y^\varepsilon(s))ds + \int_0^t g_f^\varepsilon(x^\varepsilon(s),y^\varepsilon(s)ds + M_{f^\varepsilon}(t) + 0(\varepsilon)\ .$$

As with Step 2 of Theorem 2.1, it suffices to show that the expectation of the right hand side of (2.5.28) tends to zero as $\varepsilon \to 0$. This implies that for any continuous $\mathcal{F}_s$ measurable functional Φ_s

$$\lim_{\varepsilon\downarrow 0} E\left\{\Phi_s\left[f(x^\varepsilon(t)) - f(x^\varepsilon(s)) - \int_s^t \mathcal{L}f(x^\varepsilon(\sigma))d\sigma\right]\right\} \geq 0\ , \tag{2.5.29}$$

for all f smooth with $\bar{B}\cdot\frac{\partial f}{\partial x} \geq 0$. Hence the limit coincides with the solution $(\bar{\mathcal{L}},\bar{B})$ of the submartingale problem.

The first term on the right side of (2.5.28) has zero expectation as $\varepsilon \to 0$ by exactly the same argument following (2.3.9). Hence it remains to examine the second term.

Recall that g_f^ε is defined by (3.5.20) and is uniformly bounded in $\bar{\mathcal{D}}$ for all ε and goes to zero exponentially fast as $\varepsilon \to 0$ uniformly in any compact subset of $\mathcal{D}$. Let $\delta > 0$ be fixed and define

$$\mathcal{D}_\delta = \{x \in \bar{\mathcal{D}} \,|\, |x - \partial\mathcal{D}| < \delta\} .$$

Now from the remarks just made we have

$$\text{(2.5.30)} \qquad \overline{\lim_{\delta\downarrow 0}}\ \overline{\lim_{\varepsilon\downarrow 0}}\ E\left\{\int_0^t g_f^\varepsilon(x^\varepsilon(s),y^\varepsilon(s))\,ds\right\}$$
$$\leq c\ \overline{\lim_{\delta\downarrow 0}}\ \overline{\lim_{\varepsilon\downarrow 0}}\ E\left\{\int_0^t \chi_{\mathcal{D}_\delta}(x^\varepsilon(s))\,ds\right\} ,$$

where c is a constant and $\chi_{\mathcal{D}_\delta}(x)$ is the characteristic function of the set $\mathcal{D}_\delta$.

To show that the right hand side of (2.5.30) is zero we proceed in a manner that is a bit more general than needed. Namely, we obtain the desired result using the uniform ellipticity of $\bar{\mathcal{L}}$ (the limit operator) only and not the uniform ellipticity of $\mathcal{L}^\varepsilon$ (which was used only to get existence and uniqueness of $(x^\varepsilon(t),y^\varepsilon(t))$).

Define $\eta_\delta(r)$ by

$$\text{(2.5.31)} \qquad \eta_\delta(r) = \begin{cases} \frac{1}{\delta}(\delta-r)^3 , & 0 \leq r \leq \delta \\ 0 , & r \geq \delta \end{cases}$$

and

$$\hat{\phi}_\delta(x) = \eta_\delta(\phi(x)) = \eta_\delta \circ \phi(x) , \tag{2.5.32}$$

where $\phi(x)$ is the support function of $\mathcal{D}$. Let $\psi^{(1)}_{\hat{\phi}_\delta}(x,y)$ be the function defined by (2.2.4) with $f = \hat{\phi}$ and put

$$\hat{\phi}^\varepsilon_\delta(x,y) = \hat{\phi}_\delta(x) + \varepsilon\psi^{(1)}_{\hat{\phi}_\delta}(x,y) . \tag{2.5.33}$$

Then, as in (2.5.19) with $\psi^{BL} \equiv 0$,

$$\mathcal{L}^\varepsilon\hat{\phi}^\varepsilon_\delta = \overline{\mathcal{L}}\hat{\phi}_\delta + A_{\hat{\phi}_\delta} - \overline{\mathcal{L}}\hat{\phi} + O(\varepsilon) . \tag{2.5.34}$$

Also, let us write $\overline{\mathcal{L}}$ of (2.2.5) in the usual form

$$\overline{\mathcal{L}} = \frac{1}{2}\sum_{i,j=1}^{n} \bar{a}_{ij}(x)\frac{\partial^2}{\partial x_i \partial x_j} + \sum_{j=1}^{n} \bar{b}_j(x)\frac{\partial}{\partial x_j}$$

so that

$$\overline{\mathcal{L}}\hat{\phi}_\delta(x) = \eta''_\delta(\phi(x))\,\frac{1}{2}\sum_{i,j=1}^{n} \bar{a}_{ij}(x)\frac{\partial\phi}{\partial x_i}\frac{\partial\phi}{\partial x_j} + \eta'_\delta(\phi(x))\overline{\mathcal{L}}\phi(x) . \tag{2.5.35}$$

We now use the above in the martingale (2.5.5) with $f = \hat{\phi}^\varepsilon_\delta$ to obtain the estimate

$$E\left\{\int_0^t \left(\eta''_\delta \circ \phi \sum_{i,j=1}^{n} \bar{a}_{ij}\frac{\partial\phi}{\partial x_i}\frac{\partial\phi}{\partial x_j}\right)(x^\varepsilon(s))\,ds\right\} = O(\delta) + O(\varepsilon) + c_1\left|E\left\{\int_0^t (A_{\hat{\phi}_\delta} - \overline{\mathcal{L}}_{\hat{\phi}_\delta})(x^\varepsilon(s),y^\varepsilon(s))\,ds\right\}\right| + c_2\delta\, E\{N^\varepsilon(t)\} . \tag{2.5.36}$$

From the remark following (2.5.28) and the estimate on moments of $N^{\varepsilon}(t)$, it follows that

$$\overline{\lim_{\delta\downarrow 0}}\ \overline{\lim_{\varepsilon\downarrow 0}}\ E\left\{\int_0^t \eta''_{\delta}\circ\phi \sum_{i,j=1}^{n} \bar{a}_{ij} \frac{\partial\phi}{\partial x_i}\frac{\partial\phi}{\partial x_j}(x^{\varepsilon}(s))ds\right\} = 0 . \tag{2.5.37}$$

Thus, uniform ellipticity of $\bar{\mathscr{L}}$, $|\nabla\phi| = 1$ and the construction of η_{δ} in (2.5.31) imply that for $t < \infty$

$$\overline{\lim_{\delta\downarrow 0}}\ \overline{\lim_{\varepsilon\downarrow 0}}\ E\left\{\int_0^t \chi_{\mathscr{D}_{\delta}}(x^{\varepsilon}(s))ds\right\} = 0 .$$

The proof of Theorem 2.3 is complete.

<u>Remark 1</u>.

The smoothness of the coefficients is a convenient assumption to avoid counting derivatives. The method used above and the results of [5] yield results under minimal regularity hypotheses.

<u>Remark 2</u>.

When smoothness is available, then one can construct full asymptotic expansions to solutions of the backward equation associated with the reflected $(x^{\varepsilon}(t),y^{\varepsilon}(t))$ process. In the context of transport problems (see next section) this is carried out in [30]. The methods used there extend directly to the problems of this section.

Remark 3.

More general boundary conditions, not just oblique reflection, can be treated by using the results of Anderson [33].

2.6 Transport problems.

This is a class of processes $(x^{\varepsilon}(t), y^{\varepsilon}(t))$ scaled in the same manner as (2.1.1) but with $y^{\varepsilon}(t)$ being a jump process and not a diffusion. We shall consider a typical situation, not the most general, and we shall concentrate on the asymptotics.

Let $q(x,y) \geq 0$ be a bounded function smooth in x and let $\pi(x,y,A)$, A a Borel subset of R^m, be a probability measure for each $(x,y) \in R^n \times R^m$, smoothly depending on x. Define on bounded continuous functions on R^m the bounded operator Q by

$$Qf(y) = q(x,y) \int_{R^m} \pi(x,y,dz) f(x) - q(x,y) f(y) \,. \tag{2.6.1}$$

Here $x \in R^n$ is merely a parameter. Let $F(x,y)$ and $H(x,y)$ be smooth in x and bounded n-vector functions on $R^n \times R^m$ and define a differential operator $\mathscr{L}^{\varepsilon}$ as follows

$$\mathscr{L}^{\varepsilon} = \frac{1}{\varepsilon^2} Q + \frac{1}{\varepsilon} F \cdot \frac{\partial}{\partial x} + H \cdot \frac{\partial}{\partial x} \,, \tag{2.6.2}$$

where

$$F \cdot \frac{\partial}{\partial x} = \sum_{i=1}^{n} F_i(x,y) \frac{\partial}{\partial x_i} \,.$$

Note that $\mathscr{L}^{\varepsilon}$ has the form (2.1.2) but $\mathscr{L}_1$, $\mathscr{L}_2$ and $\mathscr{L}_3$ are a bit different here.

It is elementary to show that $\mathscr{L}^{\varepsilon}$ of (2.6.2) is the infinitesimal generator of a Markov process $(x^{\varepsilon}(t), y^{\varepsilon}(t))$ with $\varepsilon > 0$ fixed. This process has very simple structure which

which motivates the name transport process: $x^{\varepsilon}(t)$ moves along the orbits of the vector field $\varepsilon^{-1}F + H$ in between jumps of $y^{\varepsilon}(t)$ which occur with frequency proportional to ε^{-2}. The scaling corresponds to the situation described after example 4 in Section 2.1 and is the usual diffusion limit of transport theory. The sample space of the process can be taken to be $C([0,\infty),R^n) \times D([0,\infty) \times R^m)$. It is a Feller process and hence a strong Markov process.

Let $Y(t;x)$ be the jump process in R^m generated by Q of (2.6.1) with $x \in R^n$ a parameter. We introduce here the same hypotheses regarding ergodicity of $Y(t;x)$ as in Section 2.2.

Assume that there are constants q_{ℓ} and q_u such that

$$(2.6.3) \qquad 0 < q_{\ell} \leq q(x,y) \leq q_u < \infty ,$$

and that $\pi(x,y,A)$ has a density $\tilde{\pi}(x,y,z)$ relative to a non-trivial probability measure on R^m such that

$$(2.6.4) \qquad 0 < \pi_{\ell} \leq \tilde{\pi}(x,y,z) \leq \pi_u < \infty .$$

Let $P(t,y,A;x)$ be the transition probability function of $Y(t;x)$. It is easily seen that (2.6.3) and (2.6.4) imply that there exists a unique invariant probability measure $\bar{P}(A;x)$ and that

$$|P(t,y,A;x) - \bar{P}(A;x)| \leq e^{-\alpha t}$$

for some $\alpha > 0$, for t large enough and for all $y \in R^m$, $A \in R^n$ as well as $x \in R^n$. Consequently, the recurrent potential kernel

$$(2.6.5)\qquad \psi(y,A;x) = \int_0^\infty [P(t,y,A;x) - \bar{P}(A;x)]dt$$

is well defined and takes bounded continuous functions of y to themselves. Smoothness in x persists in the parametric dependence.

As in (2.2.3) we assume that $F(x,y)$ satisfies the centering condition

$$(2.6.6)\qquad \int F(x,y)\bar{P}(dy;x) \equiv 0 ,$$

and for $f \in C_o^\infty(R^n)$ we define

$$(2.6.7)\qquad \psi_f^{(1)}(x,y) = \int_{R^m} \psi(y,dz;x) \sum_{j=1}^n F_j(x,y) \frac{\partial f(x)}{\partial x_j} .$$

We define a diffusion operator $\bar{\mathscr{L}}$ (cf. (2.2.5) and (2.2.6)) as follows

$$(2.6.8)\qquad \bar{\mathscr{L}} f(x) = \int_{R^m} \bar{P}(dy;x) A_f(x,y)$$

$$(2.6.9)\qquad A_f(x,y) = \sum_{i=1}^n F_i(x,y) \frac{\partial \psi_f^{(1)}(x,y)}{\partial x_i} + \sum_{i=1}^n H_i(x,y) \frac{\partial f(x)}{\partial x_i} .$$

From our smoothness requirements above, the coefficients of $\bar{\mathscr{L}}$ are smooth bounded functions of x and hence $\bar{\mathscr{L}}$ generates a diffusion process $x(t)$ in R^n.

Theorem 2.4.

Under the above hypotheses the process $x^{\varepsilon}(t)$, $t \geq 0$ of the pair $(x^{\varepsilon}(t), y^{\varepsilon}(t))$ generated by (2.6.2) converges weakly in $C([0,T];R^n)$, $T < \infty$ but arbitrary, as $\varepsilon \to 0$ to the diffusion Markov process $x(t)$, $t \geq 0$, generated by $\bar{\mathscr{L}}$ of (2.6.8).

The proof of this theorem is identical to the one for Theorem 2.1. Regarding regularity assumptions, we refer to Remarks 1 and 2 following the statement of that theorem. Regarding the ergodicity of $Y(t;x)$, see Remark 3.

It is easily seen that one can consider more general pairs $(x^{\varepsilon}(t), y^{\varepsilon}(t))$ with generators $\mathscr{L}^{\varepsilon}$ having the same general structure as (2.6.2) and (2.1.2)-(2.1.5). The results do not change except for notation.

Let us now consider transport processes with boundary conditions. We limit discussion to one typical problem.

Let $\mathscr{D} \subset R^n$ be a bounded open set with smooth boundary $\partial\mathscr{D}$ and let $\phi(x)$ be its support function (cf. (2.5.1)) as described in the first two paragraphs of Section 2.5.

For each $x \in \partial\mathscr{D}$ let

$$S^{\pm} = \left\{ y \in R^m \,|\, F(x,y)\cdot\hat{n}(x) \lessgtr 0 \right\}, \tag{2.6.10}$$

where $\hat{n}(x) \equiv \nabla\phi(x)$ is the unit inward normal to $\partial\mathscr{D}$. In the usual moving particle model of a transport process where $x^{\varepsilon}(t)$ is position and $y^{\varepsilon}(t)$ is velocity, the spaces $S^{\pm}$ correspond to outward directed (+) and inward directed (-) velocities at the boundary. For each $x \in \partial\mathscr{D}$ and $y \in S^{+}$ let

$B(x,y,A)$, A a Borel subset of S^-, be a probability measure S^- depending smoothly on x. B is the reflection probability at the boundary. Let

$$G = (\mathcal{D} \times R^m) \cup (\partial\mathcal{D} \times S^-) \tag{2.6.11}$$

$$\partial G = \partial\mathcal{D} \times S^+ \tag{2.6.12}$$

and consider the following operator on $\bar{G} = G \cup \partial G$ corresponding to instantaneously reflected† transport processes.

$$\mathcal{L}^\varepsilon = \frac{1}{\varepsilon^2} Q + \frac{1}{\varepsilon} F\cdot\frac{\partial}{\partial x}, \qquad (x,y) \in G, \tag{2.6.13}$$

$$\text{boundary operator } (B - I), \qquad (x,y) \in \partial G,$$

where B is the operator with kernel $B(x,y,A)$

$$Bf(x,y) = \int_{S^-} B(x,y,dz)f(x,z), \qquad (x,y) \in \partial G, \tag{2.6.14}$$

and I is the identity operator. It is easily seen that there exist processes on $\bar{G}$ with instantaneous reflection generated by $\mathcal{L}^\varepsilon$ of (2.1.13) with $\varepsilon > 0$ fixed. Moreover, in the interior G we have an ordinary transport process uniquely defined as above by (2.6.2) (with $H \equiv 0$ for simplicity here). To assure uniqueness for (2.6.13) we assume that given $\delta_1 > 0$ there is a $\delta_2 > 0$ such that

$$\inf_{x\in\partial\mathcal{D}} \inf_{y\in S^+} \int_{F(x,y)\cdot\hat{n}(x)<-\delta_1} B(x,y,dz) \geq \delta_2 . \tag{2.6.15}$$

---

†More precisely, the process characterized by (2.6.17) below.

The meaning of this condition is that upon reflection at the boundary the process has nontangential (grazing) velocity with high probability in a uniform sense.

Since the process is uniquely defined in the interior let for each $(x,y) \in G$, τ_1 be the first time ∂G is reached. It is instantaneously reflected there according to the law $B(x,y,A)$. Let τ_2 be the next time ∂G is reached, etc. Since τ_n is increasing we define

$$\tau_\infty = \lim_{n \uparrow \infty} \tau_n .$$

For uniqueness we must show $\tau_\infty = \infty$ with probability 1. It is easily shown using (2.1.15) that indeed, for $(x,y) \in G$,

$$\lim_{n \uparrow \infty} P_{x,y}\{\tau_n < t\} = 0 , \qquad 0 \leq t < \infty .$$

Now we define the increasing process $N^\varepsilon(t)$ as follows (cf. (2.5.5)). With $\tau_0 = 0$, $\tau_n \uparrow \infty$,

$$(2.6.16) \quad N^\varepsilon(t) = n , \qquad \tau_n \leq t < \tau_{n+1} , \qquad n = 0,1,2,\ldots .$$

With this definition and with $f(x,y)$ a smooth bounded function on $\bar{G}$,

$$(2.6.17) \quad M_f(t) = f(x^\varepsilon(t),y^\varepsilon(t)) - f(x,y) - \int_0^t \chi_G \mathscr{L}^\varepsilon f(x^\varepsilon(s),y^\varepsilon(s))ds - \int_0^t \chi_{\partial G} \mathscr{L}^\varepsilon f(x^\varepsilon(s),y^\varepsilon(s))dN^\varepsilon(s)$$

is a right-continuous, zero mean martingale. Here χ_G is the characteristic function of the set G. Note that the existence and uniqueness theory for the reflected transport process $(x^{\varepsilon}(t),y^{\varepsilon}(t))$ on $\bar{G}$ is elementary by comparison to the theory needed in Section 2.5.

For the asymptotics we shall introduce one more hypothesis about the reflection operator B as follows. There is a nontrivial probability measure on S^- such that B(x,y,A) has a density $\tilde{B}(x,y,z)$ relative to this measure and there are constants b_ℓ and b_u such that

$$(2.6.18)\quad 0 < b_\ell \le \tilde{B}(x,y,z) \le b_u < \infty ,$$

$$x \in \partial\mathscr{D}, \qquad y \in S^+, \qquad z \in S^- .$$

To characterize the limit of $x^{\varepsilon}(t)$ in $\bar{\mathscr{D}}$ we note that, clearly, in the interior the limit process should be governed by $\bar{\mathscr{L}}$ of (2.6.8), (2.6.9) (with $H \equiv 0$). Thus, we must find the limiting form of the reflecting boundary conditions. As in Section 2.5, we must look closely at the appropriate boundary layer problem.

We introduce stretched coordinates $x \to (\gamma,\varepsilon\eta)$ as in (2.5.6) and carry over the various definitions and remarks made there. The operator $\mathscr{L}_{BL}$ and the boundary layer problem (2.5.11) take the following form here

$$(2.6.19)\quad \mathscr{L}_{BL} = \sum_{i=1}^{n} F_i(\xi,y)\, \frac{\partial\phi(\xi)}{\partial x_i}\, \frac{\partial}{\partial\eta} + Q \Big|_{x=\zeta}$$

where Q is defined by (2.6.1) and $x = \zeta \in \partial\mathcal{D}$ is a fixed parameter, and (with $\zeta \in \partial\mathcal{D}$ fixed)

$$\mathcal{L}_{BL} v(\eta,y) = 0 , \quad \eta > 0 , \quad y \in R^m ,$$

$$\eta = 0 , \quad y \in S^- ,$$

(2.6.20)

$$\int_{S^-} B(\zeta,y,dz)v(0,z) - v(0,y) = \psi(y;x) , \quad y \in S^+ ,$$

$$\lim_{\eta\to+\infty} v(\eta,y) = 0 .$$

By results similar to the ones in the appendix (cf. [30] for details), it follows that under the above hypotheses, there is a unique probability measure $\bar{U}(A;x)$, $A \in S^+$, $x \in \partial\mathcal{D}$, such that if

$$\int_{S^+} \bar{U}(dy;x)\psi(y;x) \equiv 0 , \qquad x \in \partial\mathcal{D} , \tag{2.6.21}$$

then (2.6.20) has a unique solution that decays to zero as $\eta \to \infty$ exponentially fast, uniformly in y.

Let $\psi(y,A;x)$ be defined by (2.6.5) and define the vector field $\bar{B}(x)$, $x \in \partial\mathcal{D}$ by

$$\bar{B}(x) = \int_{S^+} \bar{U}(dy;x)\left[\int_{S^-} B(x,y,dz) \int_S \psi(z,dz';x)F(x,z') - \int_S \psi(y,dz;x)F(x,z)\right] . \tag{2.6.22}$$

From our smoothness requirements in x it follows that $\bar{B}(x)$ is indeed a smooth vector field on $\partial\mathscr{D}$ and, also, it points inwards. We need to assume, however, the following stronger hypothesis. There is a $\delta > 0$ such that

$$(2.6.23)\qquad \sum_{i=1}^{n} \bar{B}_i(x)\,\frac{\partial\phi(x)}{\partial x_i} \geq \delta\,, \qquad x \in \partial\mathscr{D}.$$

This means that $\bar{B}$ is uniformly nontangential.

Theorem 2.5.

Under the above hypotheses the reflected process $x^\varepsilon(t)$ in $\bar{\mathscr{D}}$, of the pair $(x^\varepsilon(t), y^\varepsilon(t))$ in $\bar{G}$, converges weakly in $C([0,T];\bar{\mathscr{D}})$, $T < \infty$ but arbitrary, as $\varepsilon \to 0$ to the reflected diffusion process $x(t)$ which is the unique solution of the submartingale problem $(\bar{\mathscr{L}},\bar{B})$.

The proof of this theorem is nearly identical, step by step, to the one in Section 2.5. Only minor differences arise in showing that (2.5.23) holds here. We emply the theory of [5] to obtain the characterization of the limit problem $(\bar{\mathscr{L}},\bar{B})$. The details are actually given in [30] where, in addition, absorbing boundary conditions and full expansions (not just weak convergence) are considered.

2.7 Homogenization.

An example of this was given in Section 2.1 (Example 3). Here we examine briefly the relation of homogenization problems to the ones of previous sections. We refer to Babuska's papers [27] for examples from continuum mechanics and many other nonprobabilistic stituations, to the work of De Giorgi-Spagnolo and Spagnolo [34, 35] for connections with the theory of G-convergence (other references can be found there) and to [36] for problems in variational inequalities and other areas, including probabilistic ones (see also [37]). A systematic treatment of these problems, boundary layers, etc., is given in [38].

Consider the following operator on $C_o^\infty(R^n)$

$$(2.7.1) \quad \mathcal{L}^\varepsilon = \frac{1}{2}\sum_{i,j=1}^{n} a_{ij}(x,\tfrac{x}{\varepsilon})\,\frac{\partial^2}{\partial x_i \partial x_j} + \sum_{j=1}^{n}\left(\frac{1}{\varepsilon}\, b_j(x,\tfrac{x}{\varepsilon}) + c_j(x,\tfrac{x}{\varepsilon})\right)\frac{\partial}{\partial x_j},$$

where we assume the following

(2.7.2) $(a_{ij}(x,y))$ is symmetric smooth and uniformly elliptic for $(x,y) \in R^n \times R^n$.

(2.7.3) $b_j(x,y)$ and $c_j(x,y)$ are smooth bounded functions on $R^n \times R^n$, $j = 1,2,\ldots,n$.

Clearly, these conditions imply immediately that $\mathcal{L}^\varepsilon$, $\varepsilon > 0$, generates a diffusion Markov process $x^\varepsilon(t)$, $t \geq 0$.

Let $\mathscr{L}_1$ be defined on $C_o^\infty(R^n)$ by

$$(2.7.4) \qquad \mathscr{L}_1 = \frac{1}{2} \sum_{i,j=1}^{n} a_{ij}(x,y) \frac{\partial^2}{\partial y_i \partial y_j} + \sum_{j=1}^{n} b_j(x,y) \frac{\partial}{\partial y_j} .$$

This operator is the generator of a process $Y(t;x) \in R^n$ where $x \in R^n$ is a parameter. As in Section 2.2 we assume that this process is ergodic with $\bar{P}(A;x)$ its invariant measure and $\psi(y,A;x)$, $y \in R^n, A \in R^n$, $x \in R^n$, as in (2.2.2) and well defined as described there.

Assume that

$$(2.7.5) \qquad \int_{R^n} b_j(x,y)\bar{P}(dy;x) \equiv 0 , \qquad x \in R^n , \quad j = 1,2,\ldots,n$$

and for $f(x) \in C_o^\infty(R_n)$ define $\psi_f^{(1)}(x,y)$ by

$$(2.7.6) \qquad \psi_f^{(1)}(x,y) = \int_{R^n} \sum_{j=1}^{n} \psi(y,dz;x) b_j(x,y) .$$

Define $\bar{\mathscr{L}}$ on $C_o^\infty(R^n)$ by

$$(2.7.7) \qquad \bar{\mathscr{L}} f(x) = \int_{R^n} \bar{P}(dy;x) A_f(x,y) ,$$

where

$$(2.7.8) \qquad A_f(x,y) = \sum_{i,j=1}^{n} a_{ij}(x,y) \frac{\partial^2 \psi_f^{(1)}(x,y)}{\partial x_i \partial y_i} + \sum_{j=1}^{n} b_j(x,y) \frac{\partial \psi_f^{(1)}(x,y)}{\partial x_j} + \sum_{j=1}^{n} c_j(x,y) \frac{\partial \psi_f^{(1)}(x,y)}{\partial y_j}$$

$$+ \frac{1}{2} \sum_{i,j=1}^{n} a_{ij}(x,y) \frac{\partial^2 f(x)}{\partial x_i \partial x_j}$$

$$+ \sum_{j=1}^{n} c_j(x,y) \frac{\partial f(x)}{\partial x_j} .$$

Clearly the coefficients of $\bar{\mathscr{L}}$ are smooth bounded functions of x.

Theorem 2.6.

Under the above hypotheses the process $x^\varepsilon(t)$ generated by $\mathscr{L}^\varepsilon$ in (2.7.1) converges weakly on $C([0,T];R^n)$, $T < \infty$ but arbitrary, as $\varepsilon \to 0$ to the diffusion process $x(t)$ generated by $\bar{\mathscr{L}}$ of (2.7.7), (2.7.8).

Remark 1.

Concerning smoothness of the coefficients see Remark 1 and 2 following Theorem 2.1.

Remark 2.

If

$$b_j(x,y) = \frac{1}{2} \sum_{i=1}^{n} \frac{\partial a_{ij}(x,y)}{\partial y_i} , \tag{2.7.9}$$

then $\mathscr{L}^\varepsilon$ in (2.7.1) can be put in divergence form

$$\mathscr{L}^\varepsilon = \frac{1}{2} \sum_{i,j=1}^{n} \frac{\partial}{\partial x_i}\left(a_{ij}(x,\tfrac{x}{\varepsilon}) \frac{\partial}{\partial x_j}\right) + \sum_{j=1}^{n} \left(c_j(x,y) - \frac{1}{2} \sum_{i=1}^{n} \frac{\partial a_{ij}(x,\frac{x}{\varepsilon})}{\partial x_i}\right)\frac{\partial}{\partial x_j} . \tag{2.7.10}$$

In this case one can analyze the limit (in the appropriate sense) of the operator $\mathcal{L}^{\varepsilon}$ in a Hilbert space framework with $(a_{ij}(x,y))$ being only bounded measurable functions (and uniformly elliptic); see [34-36]. In particular, the case where $(a_{ij}(x,y))$ and $(c_j(x,y))$ are periodic functions of y has received most attention. For the case (2.7.10) with periodic y dependence, $\bar{P}(dy;x)$ is Lebesque measure, as can be seen easily, (2.7.5) is automatically satisfied and (2.7.7) simplifies a lot.

Proof of Theorem 2.6.

The proof is identical to the one of Theorem 2.1 after various objects are identified with the corresponding ones there.

The idea is to define

$$y^{\varepsilon}(t) = \frac{x^{\varepsilon}(t)}{\varepsilon} \tag{2.7.11}$$

and then treat $(x^{\varepsilon}(t), y^{\varepsilon}(t))$ as a pair in the framework of Section 2.1. After all constructions are carried out we re-establish (2.7.11) which is only an analytical device. Implementing this we find that the generator of $(x^{\varepsilon}(t), y^{\varepsilon}(t))$ is of the form (2.1.2) with $\mathcal{L}_1$ as in (2.7.4) and

$$\mathcal{L}_2 = \sum_{i,j=1}^{n} a_{ij}(x,y) \frac{\partial^2}{\partial x_i \partial y_i} + \sum_{j=1}^{n} b_j(x,y) \frac{\partial}{\partial x_j} + \sum_{j=1}^{n} c_j(x,y) \frac{\partial}{\partial y_j} , \tag{2.7.12}$$

$$(2.7.13)\qquad \mathscr{L}_3 = \frac{1}{2}\sum_{i,j=1}^{n} a_{ij}(x,y)\frac{\partial^2}{\partial x_i \partial x_j} + \sum_{j=1}^{n} c_j(x,y)\frac{\partial}{\partial x_j}\,.$$

From here on we proceed as in Theorem 2.1.

Remark 3.

Using the same device (2.7.11) one can treat processes $x^{\varepsilon}(t)$ with generator $\mathscr{L}^{\varepsilon}$ of (2.7.1) in $\mathscr{D}$, a bounded open set in R^n, and boundary conditions on $\partial\mathscr{D}$. The results are formally the same as the ones in Section 2.5. In the case of reflection, the only difference is that in the boundary layer problem (2.5.11) (or (2.5.17)) instead of having the normal derivative given, an appropriate oblique derivative is given at $\eta = 0$. The proof of the results analagous to the ones in the appendix is, however, different.

2.8 Averaging.

In the first two theorems that follow we deal with a variant of the process $(x^\varepsilon(t), y^\varepsilon(t))$ defined at the beginning of Section 2.6. Then we deal with a variant (2.1.1).

Let Q be defined by (2.6.1) and assume that (2.6.3) and (2.6.4) hold so that ψ in (2.6.5) is well defined as discussed there. Let $F(x,y,\tau)$ and $H(x,y,\tau)$ be n-vector functions defined on $R^n \times R^m \times R^1$, bounded, smooth in x and almost periodic in τ. Define for $t \geq 0$

$$(2.8.1) \quad \mathscr{L}_t^\varepsilon = \frac{1}{\varepsilon^2} Q + \frac{1}{\varepsilon} F(x,y,t/\varepsilon^2)\cdot\frac{\partial}{\partial x} + H(x,y,t/\varepsilon^2)\cdot\frac{\partial}{\partial x} .$$

This is a time inhomogeneous operator which, it is easily seen, generates a Markov process $(x^\varepsilon(t), y^\varepsilon(t))$, $t \geq 0$ in $R^n \times R^m$.

For the asymptotic analysis of $x^\varepsilon(t)$ we introduce the following hypotheses concerning $F(x,y,\tau)$ and $H(x,y,\tau)$.

$$(2.8.2) \qquad \lim_{T\uparrow\infty} \frac{1}{T}\int_0^T d\tau \int \bar{P}(dy;x)F(x,y,\tau) \equiv 0 ,$$

and, more specifically, if

$$(2.8.3) \qquad \bar{F}(x,\tau) \equiv \int \bar{P}(dy;x)F(x,y,\tau) ,$$

then

$$(2.8.4) \qquad \int_0^\tau ds\, \bar{F}(x,s) \leq C , \qquad x \in R^n , \qquad \tau \geq 0 ,$$

where C is a constant. For $f \in C_o^\infty(R^n)$ let

$$(2.8.5)\qquad \psi_f^{(1)}(x,y,\tau) = \int_0^\infty ds \int_{R^m} [P(s,y,dz;x)-\bar{P}(dz;x)]$$

$$\cdot \sum_{j=1}^n \left(F_j(x,z,s+\tau)-\bar{F}_j(x,s+\tau)\right)\frac{\partial f(x)}{\partial x_j}$$

$$- \int_0^\tau ds \sum_{j=1}^n \bar{F}_j(x,s)\ \frac{\partial f(x)}{\partial x_j}\ .$$

Let $A_f(x,y,\tau)$ be defined by

$$(2.8.6)\qquad A_f(x,y,\tau) = \sum_{j=1}^n \left(F_j(x,y,\tau)\ \frac{\partial \psi_f^{(1)}(x,y,\tau)}{\partial x_j} + H_j(x,y,\tau)\ \frac{\partial f(x)}{\partial x_j} \right) .$$

We assume that for $f \in C_o^\infty(R^n)$ fixed,

$$(2.8.7)\qquad \bar{\mathscr{L}} f(x) \equiv \lim_{T\uparrow\infty} \frac{1}{T} \int_0^T ds \int \bar{P}(dy;x) A_f(x,y,\tau+s)$$

exists uniformly in x and τ, it is independent of τ and the coefficients of the second order operator $\bar{\mathscr{L}}$ (time homogeneous) are smooth and bounded functions of $x \in R^n$.

Theorem 2.7.

Under the above hypotheses $x^\varepsilon(t)$, $t \geq 0$ converges weakly on $C([0,T];R^n)$, $T < \infty$, as $\varepsilon \to 0$ to the time homogeneous diffusion process generated by $\bar{\mathscr{L}}$ of (2.8.7).

Remark 1.

In the case F and H are independent of y we are dealing with the case of averaging for ordinary differential equations [6], including second order effects (the effect of $\varepsilon^{-1}F$). The orbits of the vector field $[\varepsilon^{-1}F(x,t/\varepsilon^2) + H(x,t/\varepsilon^2)]\cdot\frac{\partial}{\partial x}$ converge to the orbits of the vector field ($F = \bar{F}$ here)

$$\bar{\mathscr{L}} = -\lim_{T\uparrow\infty}\frac{1}{T}\int_0^T ds\int_0^s d\sigma\sum_{i,j=1}^n F_j(x,s)\,\frac{\partial F_i(x,\sigma)}{\partial x_j}\,\frac{\partial}{\partial x_i}$$

$$+\lim_{T\uparrow\infty}\frac{1}{T}\int_0^T ds\sum_{i=1}^n H_i(x,s)\,\frac{\partial}{\partial x_i}\,.$$

Note that the coefficients of second order derivatives in (2.8.7) are identically zero in this case. Clearly one cannot get a diffusion process from deterministic vector fields even when they oscillate rapidly (in an almost periodic fashion).

Remark 2.

In some applications [19] the interaction of averaging with the diffusion approximation plays an important role. This is one reason for presenting Theorem 2.7.

Proof of Theorem 2.7.

The proof is nearly the same as the one of Theorem 2.1. It suffices to show how the scheme (2.3.12)-(2.3.15) must be modified here.

In the spirit of the device (2.7.11) we introduce the "process"

(2.8.8) $\tau^{\varepsilon}(t) = \tau + t/\varepsilon^2 , \quad \tau \in R^1 , \quad t \geq 0 .$

The augmented process $(x^{\varepsilon}(t), y^{\varepsilon}(t), \tau^{\varepsilon}(t))$ on $R^n \times R^m \times R^1$ is time homogeneous and its generator is given by

(2.8.9) $$\frac{1}{\varepsilon^2}(Q + \frac{\partial}{\partial\tau}) + \frac{1}{\varepsilon} F(x,y,\tau)\cdot\frac{\partial}{\partial x} + H(x,y,\tau)\cdot\frac{\partial}{\partial x}$$

which is of the form (2.1.2).

Let $f(x) \in C_o(R^n)$, let $\psi_f^{(1)}(x,y,\tau)$ and $\psi_f^{(2,\lambda)}(x,y,\tau)$, $\lambda > 0$, be functions to be defined shortly and put

(2.8.10) $$f^{\varepsilon}(x,y,\tau) = f(x) + \varepsilon\psi_f^{(1)}(x,y,\tau) + \varepsilon^2\psi_f^{(2,\lambda)}(x,y,\tau) .$$

Then we have

(2.8.11) $$\left[\frac{1}{\varepsilon^2}(Q + \frac{\partial}{\partial\tau}) + \frac{1}{\varepsilon} F\cdot\frac{\partial}{\partial x} + H\cdot\frac{\partial}{\partial x}\right]\left(f + \varepsilon\psi_f^{(1)} + \varepsilon^2\psi_f^{(2,\lambda)}\right)$$

$$= \frac{1}{\varepsilon}\left[(Q + \frac{\partial}{\partial\tau})\psi_f^{(1)} + F\cdot\frac{\partial f}{\partial x}\right]$$

$$+ \left[(Q + \frac{\partial}{\partial\tau})\psi_f^{(2,\lambda)} + F\cdot\frac{\partial\psi_f^{(1)}}{\partial x} + H\cdot\frac{\partial f}{\partial x} - \bar{\mathscr{L}}f - \lambda\tilde{\psi}_f^{(2,\lambda)}\right]$$

$$+ \lambda\tilde{\psi}_f^{(2,\lambda)} + \bar{\mathscr{L}}f + \varepsilon\left[F\cdot\frac{\partial\psi_f^{(2,\lambda)}}{\partial x} + H\cdot\frac{\partial\psi_f^{(1)}}{\partial x}\right]$$

$$+ \varepsilon^2 H\cdot\frac{\partial\psi_f^{(2,\lambda)}}{\partial x} .$$

We choose $\psi_f^{(1)}(x,y,\tau)$ so that

(2.8.12) $$(Q + \frac{\partial}{\partial\tau})\psi_f^{(1)} + F\cdot\frac{\partial f}{\partial x} = 0 .$$

With $\bar{F}$ defined by (2.8.3) we rewrite (2.8.12) as follows

$$(2.8.13) \quad (Q + \frac{\partial}{\partial \tau})\psi_f^{(1)}(x,y,\tau) + \left[F(x,y,\tau)-\bar{F}(x,\tau)\right]\cdot\frac{\partial f(x)}{\partial x} + \bar{F}(x,\tau)\cdot\frac{\partial f(x)}{\partial x} = 0 .$$

Because of the ergodic properties of Q (cf. (2.6.3)-(2.6.5)) it follows from (2.8.13) that $\psi_f^{(1)}(x,y,\tau)$ given by (2.8.5) satisfies (2.8.13) and it is bounded.

With $\psi_f^{(1)}$ so determined consider the second term on the right side of (2.8.11). We rewrite it as

$$(2.8.14) \quad (Q + \frac{\partial}{\partial \tau})\psi_f^{(2,\lambda)}(x,y,\tau) + A_f(x,y,\tau) - \bar{\mathscr{L}}_\tau f(x) + \bar{\mathscr{L}}_\tau f(x) - \bar{\mathscr{L}} f(x) - \lambda\tilde{\psi}_f^{(2,\lambda)} = 0 .$$

where

$$(2.8.15) \quad \bar{\mathscr{L}}_\tau f(x) = \int \bar{P}(dy;x)A_f(x,y,\tau) ,$$

with A_f given by (2.8.6). From (2.8.14) it is clear that

$$(2.8.16) \quad \psi_f^{(2,\lambda)}(x,y,\tau) = \tilde{\psi}_f^{(2,\lambda)}(x,y,\tau) + \int_0^\infty ds \cdot \int [P(s,y,dz;x)-\bar{P}(dz;x)] \cdot \left[A_f(x,z,s+\tau) - \bar{\mathscr{L}}_{s+\tau} f(x)\right]$$

$$(2.8.17) \quad \tilde{\psi}_f^{(2,\lambda)}(x,\tau) = \int_0^\infty e^{-\lambda s}\left[\bar{\mathscr{L}}_{\tau+s} f(x) - \bar{\mathscr{L}} f(x)\right] d\sigma , \quad \lambda > 0 .$$

Collecting the above formulas we find that with $\mathcal{L}_t^\varepsilon$ given by (2.8.1) we have $(f \in C_o^\infty(R^n))$,

$$\left[\frac{\partial}{\partial t} + \mathcal{L}_t^\varepsilon\right]\left[f(x) + \varepsilon\psi_f^{(1)}(x,y,\frac{t}{\varepsilon^2}) + \varepsilon^2\psi_f^{(2,\lambda)}(x,y,\frac{t}{\varepsilon^2})\right] \tag{2.8.18}$$

$$= \overline{\mathcal{L}}f(x) + \lambda\tilde{\psi}_f^{(2,\lambda)}(x,\frac{t}{\varepsilon^2})$$

$$+ \varepsilon\left[F\cdot\frac{\partial\psi_f^{(2,\lambda)}}{\partial x} + H\cdot\frac{\partial\psi_f^{(1)}}{\partial x}\right](x,y,\frac{t}{\varepsilon^2})$$

$$+ \varepsilon^2 H\cdot\frac{\partial\psi_f^{(2,\lambda)}}{\partial x}(x,y,\frac{t}{\varepsilon^2}) \ .$$

Now because of (2.8.7) it follows that $\lambda\tilde{\psi}_f^{(2,\lambda)}(x,\tau)$ goes to zero as $\lambda \to 0$, uniformly in x and τ, while for $\lambda > 0$, $\psi_f^{(2,\lambda)}(x,y,t/\varepsilon^2)$ and x derivatives are bounded.

These are the analogs of the necessary facts (2.3.13)-(2.3.15) for the proof of Theorem 2.7. The rest follows as in Theorem 2.1 and the proof is complete.

A very similar situation to the above, which arises in a problem similar to the one in [19], is the following.

The time inhomogeneous process $(x^\varepsilon(t), y^\varepsilon(t))$ has generator

$$\mathcal{L}_t^\varepsilon = \frac{1}{\varepsilon^2}Q + \frac{1}{\varepsilon}F(x,y,t/\varepsilon)\cdot\frac{\partial}{\partial x} + H(x,y,t/\varepsilon)\cdot\frac{\partial}{\partial x} \ , \tag{2.8.19}$$

the only difference being that F and H are functions of t/ε and not t/ε^2 as in (2.8.1). We now assume that

$$\int \overline{P}(dy;x)F(x,y,\tau) \equiv 0 \ , \qquad x \in R^n \ , \qquad \tau \in R^1 \ , \tag{2.8.20}$$

and that $F(x,y,\tau)$ and $H(x,y,\tau)$ are also smooth functions of $\tau \in R^1$.

For $f \in C_o^\infty(R^n)$ let

$$\psi_f^{(1)}(x,y,\tau) = \int_{R^m} \psi(y,dz;x) \sum_{j=1}^{n} F_j(x,y,\tau) \cdot \frac{\partial f(x)}{\partial x} \tag{2.8.21}$$

where ψ is defined by (2.6.5). Let

$$A_f(x,y,\tau) = \sum_{j=1}^{n} \left[F_j(x,y,\tau) \frac{\partial \psi_f^{(1)}(x,y,\tau)}{\partial x} + H_j(x,y,\tau) \frac{\partial f(x)}{\partial x_j} \right] \tag{2.8.22}$$

and

$$\bar{\mathscr{L}}_\tau f(x) = \int_{R^m} \bar{P}(dy;x) A_f(x,y,\tau) . \tag{2.8.23}$$

The limit operator $\bar{\mathscr{L}}$ is defined for $f \in C_o^\infty(R^n)$ by

$$\bar{\mathscr{L}} f(x) = \lim_{T \uparrow \infty} \frac{1}{T} \int_0^T ds \bar{\mathscr{L}}_{s+\tau} f(x) , \tag{2.8.24}$$

where the limit is assumed to exist uniformly in (x,τ) and to be independent of τ as shown.

Theorem 2.8.

Under the above hypotheses the process $x^\varepsilon(t)$ associated with (2.8.19) converges weakly in $C([0,T];R^n)$, $T < \infty$ but arbitrary, as $\varepsilon \to 0$ to the diffusion process $x(t)$ generated by $\bar{\mathscr{L}}$ of (2.8.24).

<u>Proof</u>.

As in Theorem 2.7 we indicate the essential differences in the argument from Theorem 2.1.

With $f \in C_o^\infty(R^n)$ fixed define $\psi_f^{(1)}$ by (2.8.21) and

$$\tilde{\psi}_f^{(1,\lambda)}(x,\tau) = \int_0^\infty e^{-\lambda s}\left(\bar{\mathscr{L}}_{\tau+s}f(x) - \bar{\mathscr{L}}f(x)\right)ds \ , \qquad \lambda > 0 \ , \tag{2.8.25}$$

$$\psi_f^{(2,\lambda)}(x,y,\tau) = \int_{R^m} \psi(y,dz;x)\left[A_f(x,z,t) - \bar{\mathscr{L}}_\tau f(x) + \frac{\partial\psi_f^{(1)}(x,z,\tau)}{\partial\tau} + F(x,z,\tau)\cdot\frac{\partial\tilde{\psi}^{(1,\lambda)}(x,\tau)}{\partial x}\right] . \tag{2.8.26}$$

From these definitions it follows that

$$\begin{aligned}(\partial_t + \mathscr{L}_t^\varepsilon)&\left(f(x) + \varepsilon\psi_f^{(1)}(x,y,\tfrac{t}{\varepsilon}) + \varepsilon\tilde{\psi}_f^{(1,\lambda)}(x,\tfrac{t}{\varepsilon}) + \varepsilon^2\psi_f^{(2,\lambda)}(x,y,\tfrac{t}{\varepsilon})\right) \\ &= \bar{\mathscr{L}}f(x) + \lambda\tilde{\psi}_f^{(1,\lambda)}(x,\tfrac{t}{\varepsilon}) + \varepsilon\left[\left(F\cdot\frac{\partial}{\partial x} + \frac{\partial}{\partial\tau}\right)\psi_f^{(2,\lambda)}(x,y,\tfrac{t}{\varepsilon}) \right.\\ &\quad \left. + H\cdot\frac{\partial\psi_f^{(1)}}{\partial x}(x,y,\tfrac{t}{\varepsilon}) + H(x,y,\tfrac{t}{\varepsilon})\cdot\frac{\partial\tilde{\psi}_f^{(1,\lambda)}}{\partial x}(x,\tfrac{t}{\varepsilon})\right] \\ &\quad + \varepsilon^2 H\cdot\frac{\partial\tilde{\psi}_f^{(2,\lambda)}}{\partial x}(x,y,\tfrac{t}{\varepsilon}) \ .\end{aligned} \tag{2.8.27}$$

We note that coefficients of ε and ε^2 on the right side of (2.8.27) are bounded for $\lambda > 0$ and that, because of the uniform existence of (2.8.24), $\lambda\tilde{\psi}_f^{(1,\lambda)}(x,\tau) \to 0$ as $\lambda \to 0$, uniformly in x and τ. With these facts the proof goes exactly as the one of Theorem 2.1 and we omit further details.

As a final problem we consider the following system of stochastic equations, a special case of (2.1.1) (cf. example 2 in Section 2.1), also considered in [12].

$$(2.8.28)\quad dx_i^\varepsilon(t) = G_i^{(1)}(x^\varepsilon(t),y^\varepsilon(t))dt + \sum_{j=1}^{p} \sigma_{ij}^{(1)}(x^\varepsilon(t),y^\varepsilon(t))d\beta_j^{(1)}(t) ,$$

$$x_i^\varepsilon(0) = x_i , \qquad i = 1,2,\dots,n ,$$

$$dy_k^\varepsilon(t) = \frac{1}{\varepsilon^2} F_k^{(2)}(x^\varepsilon(t),y^\varepsilon(t)) + \frac{1}{\varepsilon}\sum_{\ell=1}^{q} \sigma_{k\ell}^{(2)}(x^\varepsilon(t),y^\varepsilon(t))d\beta_\ell^{(2)}(t)$$

$$y_k^\varepsilon(0) = y_k , \qquad k = 1,2,\dots,m .$$

Here $(\beta_1^{(1)}(t),\beta_2^{(1)}(t),\dots,\beta_p^{(1)}(t))$ and $(\beta_1^{(2)}(t),\beta_2^{(2)}(t),\dots,\beta_q^{(2)}(t))$ are p-dimensional and q-dimensional standard Brownian motions, respectively, and (2.1.2) holds. The generator of $(x^\varepsilon(t),y^\varepsilon(t))$ has the form

$$(2.8.29)\qquad \mathscr{L}^\varepsilon = \frac{1}{\varepsilon^2}\mathscr{L}_1 + \frac{1}{\varepsilon}\mathscr{L}_2 + \mathscr{L}_3$$

where $\mathscr{L}_1$ is given by (2.1.3) and

$$(2.8.30)\qquad \mathscr{L}_2 = \sum_{i=1}^{n}\sum_{k=1}^{m} a_{ik}^{(12)}(x,y)\frac{\partial^2}{\partial x_i \partial y_k} ,$$

$$(2.8.31)\qquad \mathscr{L}_3 = \sum_{i=1}^{n} G_i^{(1)}(x,y)\frac{\partial}{\partial x_i} + \frac{1}{2}\sum_{i,j=1}^{n} a_{ij}^{(1)}(x,y)\frac{\partial^2}{\partial x_i \partial x_j} .$$

The a's are defined by (2.1.6) and the smoothness and boundedness conditions of Section 2.1 hold here also.

Let $Y(t;x)$ be the process as R^m generated by $\mathscr{L}_1$ with $x \in R^n$ a fixed parameter as in Section 2.2. Let $P(t,y,A;x)$ be defined by (2.2.1). We shall not assume the strong ergodicity assumptions that lead to the existence of ψ in (2.2.2). Instead, we assume that for each $f(x) \in C_o^\infty(R^n)$,

$$(2.8.32) \qquad \bar{\mathscr{L}}f(x) = \lim_{T\uparrow\infty} \frac{1}{T}\int_0^T dt \int_{R^m} P(t,y,dz;x)\mathscr{L}_3 f(x,z) \ ,$$

exists uniformly in $x \in R^n$, $y \in R^m$, is independent of y as shown and the coefficients of $\bar{\mathscr{L}}$ are smooth and bounded functions of x.

Theorem 2.9.

Under the above hypotheses the process $x^\varepsilon(t)$ defined by (2.8.28) converges weakly on $C([0,T];R^n)$, $T < \infty$, but arbitrary, as $\varepsilon \to 0$ to the diffusion process $x(t)$ with generator $\bar{\mathscr{L}}$ given by (2.8.32).

Proof.

The proof is simpler than the one of Theorem 2.1 since compactness of $x^\varepsilon(t)$ is now immediate. For the identification, we define for $f(x) \in C_o^\infty(R^n)$, $\lambda > 0$,

$$(2.8.33) \qquad \psi_f^{(1,\lambda)}(x,y) = \int_0^\infty e^{-\lambda s} \int_{R^m} P(s,y,dz;x)(\mathscr{L}_3 f(x,z) - \bar{\mathscr{L}} f(x)) .$$

From this definition and ($\mathscr{L}^\varepsilon$ given by (2.8.29))

$$\mathscr{L}^\varepsilon\left[f(x) + \varepsilon^2 \psi_f^{(1,\lambda)}(x,y)\right]$$

$$= \left[\mathscr{L}_1 \psi_f^{(1,\lambda)}(x,y) + \mathscr{L}_3 f(x,y) - \bar{\mathscr{L}} f(x) - \lambda\psi_f^{(1,\lambda)}(x,y)\right]$$

$$+ \bar{\mathscr{L}} f(x) + \lambda\psi_f^{(1,\lambda)}(x,y) ,$$

it follows that

$$\mathscr{L}^\varepsilon(f + \varepsilon^2\psi_f^{(1,\lambda)}) = \bar{\mathscr{L}} f + \lambda\psi_f^{(1,\lambda)} .$$

In view of (2.8.32), $\lambda\psi_f^{(1,\lambda)}(x,y) \to 0$ as $\lambda \to 0$, uniformly in $(x,y) \in R^n \times R^m$ and this is all that is needed for the proof as usual.

Remark.

The special case $m = 1$, $F^{(2)} \equiv 1$, $\sigma^{(2)} \equiv 0$, is example 1 of Section 2.1 i.e., the method of averaging [6,25].

3. Non-ergodic driving process.

3.1 First order limit theorem.

We shall formulate first a special problem that displays the features that are characteristic in general and can be contrasted easily with those of previous problems, i.e., the case of ergodic driving processes.

Let $F_i(x,y)$, $x \in R^n$, $y \in R^1$, $i = 1,2,\ldots,n$ be smooth functions with compact support in y and let a(y) be a smooth function on R^1 such that

$$0 < \alpha_\ell \leq a(y) \leq \alpha_u < \infty , \quad \alpha_\ell, \alpha_u \text{ constants,} \tag{3.1.1}$$

$$\lim_{y \to \pm\infty} a(y) = a_\pm > 0 . \tag{3.1.2}$$

Consider the system of stochastic differential equations

$$\frac{dx_i^\varepsilon(t)}{dt} = \frac{1}{\varepsilon} F_i\left(x^\varepsilon(t), \frac{y^\varepsilon(t)}{\varepsilon}\right) , \tag{3.1.3}$$

$$x_i^\varepsilon(0) = x_i , \qquad i = 1,2,\ldots,n ,$$

$$dy^\varepsilon(t) = a^{1/2}\left(\frac{y^\varepsilon(t)}{\varepsilon}\right) d\beta(t) ,$$

$$y^\varepsilon(0) = y$$

Here $\beta(t)$ is the standard Brownian motion process on R^1.

The essential difference between system (3.1.3) and (2.1.1) is that the driving process $y^\varepsilon(t)$ is Brownian motion on R^1 with a time substitution and hence it is not ergodic.

It is null recurrent, however, and its local time is well defined. This is an important feature of (3.1.3). Note also that the scaling in (3.1.3) is different from (2.1.1).

Define the vector field $\bar{F}(x)$ by

$$\bar{F}_i(x) = \int_{-\infty}^{\infty} \frac{F_i(x,y)}{a(y)}\, dy \,, \qquad i = 1,2,\ldots,n \,, \tag{3.1.4}$$

and assume that it is not identically equal to zero (the case $\bar{F}(x) \equiv 0$ is considered in Section 3.4). Let $\bar{x}(t)$ be the solution of

$$\frac{d\bar{x}(t)}{dt} = \bar{F}(\bar{x}(t)) \,, \qquad \bar{x}(0) = x \,. \tag{3.1.5}$$

Let

$$\tilde{a}(y) = a_+\chi_{y>0} + a_-\chi_{y\leq 0} \tag{3.1.6}$$

and let $y(t)$ be the process satisfying

$$dy(t) = \tilde{a}^{1/2}(y(t))\,d\tilde{\beta}(t) \,, \qquad y(0) = y \tag{3.1.7}$$

where $\tilde{\beta}(t)$ is another standard Brownian motion. Let $\ell_o(t)$ denote the local time at $y = 0$ of this process. The local time at zero is defined by

$$\ell_o(t) = 2y_+(t) - 2y_+ - 2\int_0^t H(y(s))\,dy(s) \,, \tag{3.1.8}$$

where

$$y_+ = \max(0,y) \; ,$$

(3.1.9)
$$H(y) = \begin{cases} 0 \; , & y < 0 \; , \\ 1 \; , & y \geq 0 \; . \end{cases}$$

It is clear that the left hand side of (3.1.8) is formally

$$\int_0^t \tilde{a}(y(s))\,\delta(y(s))ds$$

which explains the terminology. The right hand side of (3.1.8) is well defined as it stands and is referred to as Tanaka's representation of $\ell_o(t)$ [ ].

Define the process $x(t)$ in R^n by

(3.1.10) $x(t) = \bar{x}(\ell_o(t))$.

Theorem 3.1.

Under the above hypotheses $(x^\varepsilon(t), y^\varepsilon(t))$ converges weakly in $C([0,T];\, R^n \times R^1)$, $T < \infty$ but arbitrary, as $\varepsilon \to 0$ to the process $(x(t), y(t))$ defined by (3.1.7) and (3.1.5), (3.1.10).

Remark 1.

When $n = 1$ and $F(x,y) = \chi_A(y)$, A a bounded subset of R^1, then

$$\bar{F} = \int \frac{1}{a(y)}\, dy \; ,$$

is independent of x, and

$$x(t) = x + \bar{F}\ell_o(t) .$$

Thus,

$$\frac{1}{\varepsilon}\int_0^t \chi_A\left(\frac{y^\varepsilon(s)}{\varepsilon}\right)ds$$

converges weakly as $\varepsilon \to 0$ to $\bar{F}\ell_o(t)$ where $\ell_o(t)$ is the local time of $y(t)$ in (3.1.7). This is a well known result [20].

Remark 2.

Instead of (3.1.3), consider

$$\begin{aligned} dx_i^\varepsilon(t) &= \frac{1}{\varepsilon} F_i\left(x^\varepsilon(t), \frac{y^\varepsilon(t)}{\varepsilon}\right)dt \\ &\quad + \frac{1}{\sqrt{\varepsilon}}\sum_{j=1}^q \sigma_{ij}\left(x^\varepsilon(t), \frac{y^\varepsilon(t)}{\varepsilon}\right)d\beta_j^{(1)}(t) \\ x_i^\varepsilon(0) &= x_i , \qquad i = 1,2,\ldots,n , \\ dy^\varepsilon(t) &= a^{1/2}\left(\frac{y^\varepsilon(t)}{\varepsilon}\right)d\beta(t) , \\ y^\varepsilon(0) &= y . \end{aligned} \tag{3.1.11}$$

Here $(\sigma_{ij}(x,y))$ are bounded smooth functions on $R^n \times R^1$ and $(\beta_1^{(1)}(t), \beta_2^{(1)}(t), \ldots, \beta_q^{(1)}(t))$ is a q-dimensional Brownian motion, not necessarily independent of $\beta(t)$.

Define

$$a_{ij}^{(1)}(x,y) = \sum_{k=1}^{q} \sigma_{ik}(x,y)\sigma_{jk}(x,y) ,$$

and, assuming $a_{ij}^{(1)}(x,y)$ vanish for y outside a bounded set,

$$\bar{a}_{ij}(x) = \int \frac{a_{ij}^{(1)}(x,y)}{a(y)} dy , \qquad i,j = 1,2,\ldots,n . \tag{3.1.12}$$

Let $\bar{\mathscr{L}}$ be defined on $C_o^\infty(R^n)$ by

$$\bar{\mathscr{L}} f(x) = \frac{1}{2} \sum_{i,j=1}^{n} \bar{a}_{ij}(x) \frac{\partial^2 f(x)}{\partial x_i \partial x_j} + \sum_{j=1}^{n} \bar{F}_j(x) \frac{\partial f(x)}{\partial x_j} , \tag{3.1.13}$$

where $\bar{F}(x)$ is defined by (3.1.4). From the smoothness of the coefficients of $\bar{\mathscr{L}}$ it follows that it generates a diffusion Markov process $\bar{x}(t)$ in R^n. Let

$$x(t) = \bar{x}(\ell_o(t)) \tag{3.1.14}$$

where $\ell_o(t)$ is the local time at $y = 0$ of $y(t)$ in (3.1.7).

The proof of Theorem 3.1 extends immediately to the above situation so that $(x^\varepsilon(t), y^\varepsilon(t))$ of (3.1.11) converges weakly as $\varepsilon \to 0$ to $(x(t),y(t))$ of (3.1.14) and (3.1.7). In fact, the proof of Section 3.2 deals with this example.

Remark 3.

The conclusion of Theorem 3.1 and Remark 2 should be contrasted with (2.1.1) and the theorems of Part 2. The driving process $y^\varepsilon(t)$ does not decouple in the limit and continues to drive the process $x(t)$ through its local time at the origin.

Other examples and remarks are given in Section 3.3 after the proof.

3.2 Proof of Theorem 3.1.

As pointed out in Remark 2 we shall give the proof for (3.1.11) with $\bar{\mathscr{L}}$ defined by (3.1.13).

Before going into the actual proof, which is similar to the one of Theorem 2.1, we prove the following five lemmas that we shall need.

Lemma 3.1.

Let

(3.2.1) $$\ell_o^\alpha(t) = \frac{1}{\alpha}\int_0^t \tilde{a}(y(s))\chi_{[0,\alpha]}(y(s))ds ,$$

where $\alpha > 0$ is a parameter and $\chi_A(y)$ is the characteristic function of the set $A \subset R^1$. Let $\ell_o(t)$ be defined by (3.1.8). Then

(3.2.2) $$E\{(\ell_o^\alpha(t) - \ell_o(t))^2\} \leq c_5\alpha^2 + c_6\alpha\sqrt{t}$$

and

(3.2.3) $$E\{\ell_o^\alpha(t)\} \leq c_1\sqrt{t} , \quad E\{(\ell_o^\alpha(t))^2\} \leq c_2 t ,$$
$$E\{e^{\lambda\ell_o^\alpha(t)}\} \leq c_3 e^{c_4\lambda^2 t} , \quad \lambda > 0 ,$$

where the c_i are constants independent of t, λ and α. Note that in view of (3.2.2), (3.2.3) holds for $\ell_o(t)$ also.

Proof of Lemma.

Define $g^{\alpha}(y)$ by

$$(3.2.4)\qquad g^{\alpha}(y) = \frac{2}{\alpha}\int_{-\infty}^{y} d\eta \int_{-\infty}^{\eta} d\zeta \chi_{[0,\alpha]}(\zeta) .$$

By Itô's formula and (3.2.1) we have

$$(3.2.5)\qquad \ell_o^{\alpha}(t) = g^{\alpha}(y(t)) - g^{\alpha}(y) - \int_0^t \frac{dg^{\alpha}(y(s))}{dy}\, dy(s) .$$

Thus

$$(3.2.6)\qquad E\{\ell_o^{\alpha}(t)\} \le 2E\{|y(t)|\} + E^{1/2}\left\{\left(\int_0^t \frac{dg^{\alpha}}{dy}\, dy(s)\right)^2\right\} .$$

$$\le C_1\sqrt{t} ,$$

where C_1 depends only on $\tilde{a}(y)$ of (3.1.6). Similary we obtain the bound for $E\{(\ell_o^{\alpha}(t))^2\}$.

From (3.2.5) it follows that

$$(3.2.7)\qquad E\left\{e^{\lambda\ell_o^{\alpha}(t)}\right\} = e^{-\lambda g^{\alpha}(y)}E\left\{e^{\lambda g^{\alpha}(y(t))}e^{-\lambda\int_0^t \frac{dg^{\alpha}}{dy}(y(s))dy(s)}\right\}$$

$$= e^{-\lambda g^{\alpha}(y)}E\left\{e^{\lambda g^{\alpha}(y(t)) + \lambda^2\int_0^t \tilde{a}(y(s))\left(\frac{dg^{\alpha}}{dy}(y(s))\right)^2 ds}\right.$$

$$\left.\cdot\, e^{-\lambda\int_0^t \frac{dg^{\alpha}}{dy}(y(s))dy(s) - \lambda^2\int_0^t \tilde{a}(y(s))\left(\frac{dg^{\alpha}}{dy}(y(s))\right)^2 ds}\right\}$$

$$\le E^{1/2}\left\{e^{4\lambda y_+(t) + 4\tilde{c}\lambda^2 t}\right\} .$$

Here $\tilde{c}$ is the largest of a_+ and a_-. Clearly the desired exponential estimate in (3.2.3) follows from (3.2.7).

Subtracting (3.1.8) and (3.2.5) and squaring we obtain

$$\left(\ell_o^\alpha(t) - \ell_o(t)\right)^2 = 4\left(2y_+(t) - g^\alpha(y(t))\right) + 4\left(2y_+ - g^2(y)\right)^2$$

$$+ 4\left(\int_0^t \left(2H(y(s)) - \frac{dg^\alpha}{dy}(y(s))\right)dy(s)\right)^2 .$$

Hence

$$(3.2.8) \qquad E\left\{\left(\ell_o^\alpha(t) - \ell_o(t)\right)^2\right\} \le 8\alpha^2 + 4E\left\{\int_0^t \tilde{c}\chi_{[0,\alpha]}(y(s))ds\right\}$$

$$= 8\alpha^2 + 4\tilde{c}\alpha E\left\{\ell_o^\alpha(t)\right\}$$

$$\le 8\alpha^2 + 4\tilde{c}C_1\alpha\sqrt{t} ,$$

which is (3.2.2). The proof is complete.

Lemma 3.2.

Let $y(t)$ be defined by (3.1.7) and $\ell_o(t)$ by (3.1.8) and let $\bar{x}(t)$ be the diffusion in R^n generated by $\bar{\mathcal{L}}$ of (3.1.13) (it is uniquely defined by smoothness of its coefficients). Let $x(t)$ be defined by (3.1.14). The measures $P_{x,y}$ on $C([0,\infty), R^n \times R^1)$ corresponding to $(x(t),y(t))$ are characterized by the following martingale problem.

(i) $P_{x,y}\{x(0) = x, y(0) = y\} = 1$.

(ii) <u>For each</u> $f(x,y) \in C_o^\infty(R^n \times R^1)$,

$$(3.2.9) \qquad f(x(t),y(t)) - \frac{1}{2}\int_0^t \tilde{a}(y(s)) \frac{\partial^2 f(x(s),y(s))}{\partial y^2} ds - \int_0^t \bar{\mathscr{L}} f(x(s),y(s)) d\ell_o(s)$$

<u>is a</u> $P_{x,y}$ <u>martingale</u>.

In (ii) $\ell_o(t)$ is the local time at zero of the y component of the process as defined by (3.1.8).

<u>Proof</u>.

If $(x(t),y(t))$ is defined by (3.1.14) and (3.1.7) and if $f(x,y) \in C_o^\infty(R^n \times R^1)$, then we have

$$(3.2.10) \qquad E_{x,y}\{f(x(t),y(t))\} = f(x,y) + E_{x,y}\left\{\int_0^t \frac{1}{2} a(y(s) \frac{\partial^2 f(x(s),y(s))}{\partial y^2} ds\right\} + E_{x,y}\left\{\int_0^t \bar{\mathscr{L}} f(x(s),y(s)) d\ell_o(s)\right\} .$$

Here we use the fact that $\ell_o(t)$ is almost surely bounded, increasing and continuous $(t < \infty)$. The equation (3.2.10) implies (3.2.9).

Now suppose (3.2.9) holds. According to Theorem 2.1 in [5] and Lemma 3.1 above, if (3.2.9) is a P martingale then for any f(x,y) smooth, bounded and strictly positive,

$$f(x(t),y(t))e^{-\int_0^t \frac{Lf(x(s),y(s))}{f(x(s),y(s))}ds - \int_0^t \frac{\overline{\mathscr{L}}f(x(s),y(s))}{f(x(s),y(s))}d\ell_o(s)} \tag{3.2.11}$$

is also a P martingale and conversely. In (3.2.11) we have defined

$$L = \frac{1}{2}\tilde{a}(y)\frac{\partial^2}{\partial y^2}. \tag{3.2.12}$$

Thus, it suffices to show that (3.2.11) being a martingale for each f positive, bounded and smooth, implies that the process $(x(\cdot),y(\cdot))$ is as defined by (3.1.14) and (3.1.7).

First, by taking f = f(y) in (3.2.11) and using the uniqueness of the process associated with L, we conclude that the restriction of the measure P to M_t, $t \geq 0$, the σ-algebra generated by events associated with y(s), $s \leq t$, is the one defined by (3.1.7).

Let $\mathcal{F}_t$ be the σ-algebra of events generated by (x(s),y(s)), $s \leq t$ and let $M = \sigma(\bigcup_{t>0} M_t)$ be the σ-algebra of events generated by y(s), $s \geq 0$. Let $Q_{y(\cdot)}$ be a regular conditional probability of P given M. We shall show that if for every $f(x) \in C_o^\infty(R^n)$ and strictly positive,

$$(3.2.13) \quad X_f(t) = f(x(t))e^{-\int_0^t \frac{\overline{\mathscr{L}}f(x(s))}{f(x(s))} d\ell_o(s)},$$

then $(X_f(t), \mathscr{F}_t, Q_{y(\cdot)})$ is a martingale for almost all (relative to P) paths $y(\cdot)$. Here, as above, $\ell_o(t)$ is the functional of $y(\cdot)$ defined by (3.1.8). The uniqueness of the process $\bar{x}(t)$ associated with $\mathscr{L}$ and (3.2.13), identify $x(t)$ as in (3.1.14) (or (3.1.10)).

To show that $(X_f(t), \mathscr{F}_t, Q_{y(\cdot)})$ is P a.s. a martingale we proceed as follows. Let $f(x)$ and $g(y)$ be strictly positive and smooth functions on R^n and R^1 respectively. Define $X_f(t)$ by (3.2.13) and

$$(3.2.14) \quad Y_g(t) = g(y(t))e^{-\int_0^t \frac{Lg(y(s))}{g(y(s))} ds}.$$

By (3.2.11) we have that $(X_f(t)Y_g(t), \mathscr{F}_t, P)$ is a martingale. For $s \geq 0$ fixed define a measure $P_{f,s}$ on $\mathscr{F}_t$, $t \geq s$ so that

$$(3.2.15) \quad \left.\frac{dP_{f,s}}{dP}\right|_{\mathscr{F}_t} = \frac{X_f(t)}{X_f(s)}, \qquad t \geq s .$$

With this definition, $(Y_g(t), \mathscr{F}_t, P_{f,s}, t \geq s)$ is a martingale. Since however $Y_g(t)$ is M_t measurable it follows that $(Y_g(t), M_t, P_{f,s}, t \geq s)$ is also a martingale. Hence

$$E^{P_{f,s}}\{Y_g(t_2)|M_{t_1}\} = Y_g(t_1) \quad ,$$

P a.s. for each g, f and $s \geq 0$. Taking $g \equiv 1$ it follows that

$$(3.2.16) \qquad E^{P_{f,s}}\{1|M\} = 1 \ .$$

If $Q_{y(\cdot)}$ is a regular conditional probability of P given M, then (3.2.16) and (3.2.15) imply that

$$(3.2.17) \qquad E^{Q_{y(\cdot)}}\{X_f(t)|\mathcal{F}_s\} = X_f(s) \quad , \qquad t \geq s \ ,$$

P a.s.

We note that the null set involved in (3.2.17) depends on f, s and t. Since a countable dense set of f's suffices and $X_f(t)$ is P a.s. continuous it follows that (3.2.17) holds for a determining class of f's and hence (3.2.13) is a martingale as was to be shown.

Lemma 3.3.

The following are equivalent.

(i) For every $f(x,y) \in C_o^\infty(R^n \times R^1)$,

$$(3.2.18) \qquad f(x(t),y(t)) - \frac{1}{2}\int_0^t \tilde{a}(y(s)) \frac{\partial^2 f(x(s),y(s))}{\partial y^2}\, ds - \int_0^t \bar{\mathcal{L}} f(x(s),y(s))\, d\ell_o(s)$$

is an $\mathcal{F}_t$, P martingale with $\ell_o(t)$ defined by (3.1.8).

(ii) For every $f(x,y) \in C_o^\infty(R^n \times R^1)$,

$$f(x(t),y(t)) - 2y_+(t)\bar{\mathscr{L}}f(x(t),y(t)) - \int_0^t \tilde{a}(y(s)) \cdot \left[\frac{1}{2}\frac{\partial^2 f(x(s),y(s))}{\partial y^2} - 2H(y(s))\bar{\mathscr{L}}\frac{\partial f(x(s),y(s))}{\partial y} - y_+(s)\bar{\mathscr{L}}\frac{\partial^2 f(x(s),y(s))}{\partial y^2}\right]ds \tag{3.2.19}$$

is an $\mathfrak{F}_t$, P martingale where y_+ and $H(y)$ are as in (3.1.9).

Proof.

First we show that (i) implies (ii). Let $g^\alpha(y)$ be defined by (3.2.4). The operator $\bar{\mathscr{L}}$ of (3.1.13) is assumed to have smooth coefficients. In what follows we make use of this smoothness in a superficial way which can be avoided by a preliminary smoothing.

For f(x,y) smooth and bounded let

$$f^\alpha(x,y) = f(x,y) - g^\alpha(y)\bar{\mathscr{L}}f(x,y) . \tag{3.3.20}$$

With f replaced by f^α in (3.2.18) we find that

$$f^\alpha(x(t),y(t)) - \int_0^t \frac{1}{2}\tilde{a}(y(s))\frac{\partial^2 f(x(s),y(s))}{\partial y^2}ds + \int_0^t \frac{1}{2}\tilde{a}(y(s))\left[\frac{d^2 g^\alpha(y(s))}{dy^2}\bar{\mathscr{L}}f(x(s),y(s)) + 2\frac{dg^\alpha(y(s))}{dy}\bar{\mathscr{L}}\frac{\partial f(x(s),y(s))}{dy} + g^\alpha(y(s))\bar{\mathscr{L}}\frac{\partial^2 f(x(s),y(s))}{\partial y^2}\right] \tag{3.2.21}$$

$$- \int_0^t \bar{\mathscr{L}} f(x(s),y(s))d\ell_o(s)$$

$$+ \int_0^t g^\alpha(y(s))\bar{\mathscr{L}}^2 \; f(x(s),y(s))d\ell_o(s)$$

is an $\mathcal{F}_t$, P martingale.

Using (3.2.4) and (3.2.1) we rewrite (3.2.21) as follows

$$(3.2.22) \quad f(x(t),y(t)) - 2y_+(t)\bar{\mathscr{L}} f(x(t),y(t))$$

$$- \int_0^t \tilde{a}(y(s)) \Bigg[\frac{1}{2} \frac{\partial^2 f(x(s),y(s))}{\partial y} - 2H(y(s))\bar{\mathscr{L}} \frac{\partial f(x(s),y(s))}{\partial y} - y_+(s)\bar{\mathscr{L}} \frac{\partial^2 f(x(s),y(s))}{\partial y^2}\Bigg] ds$$

$$- (g^\alpha(y(t)) - 2y_+(t))\bar{\mathscr{L}} f(x(t),y(t))$$

$$+ \int_0^t \bar{\mathscr{L}} f(x(s),y(s))d(\ell_o^\alpha(s) - \ell_o(s))$$

$$+ \int_0^t \tilde{a}(y(s)) \Bigg[\Big(\frac{dg^\alpha(y(s))}{dy} - 2H(y(s))\Big)\bar{\mathscr{L}} \frac{f(x(s),y(s))}{\partial y} + \Big(\frac{1}{2} g^\alpha(y(s)) - y_+(s)\Big)\bar{\mathscr{L}} \frac{\partial^2 f(x(s),y(s))}{\partial y^2}\Bigg] ds$$

$$+ \int_0^t (g^\alpha(y(s)) - 2y_+(s))\bar{\mathscr{L}}^2 f(x(s),y(s))d\ell_o(s)$$

$$+ \int_0^t 2y_+(s)\bar{\mathscr{L}}^2 f(x(s),y(s))d\ell_o(s) \;,$$

is an $\mathcal{F}_t$, P martingale. Now we note first that the last term in (3.2.22) is identically zero since $\ell_o(t)$ increases only when $y(t) = 0$. The expectation of the absolute value of the other terms that involve α goes to zero as $\alpha \downarrow 0$. Thus from (3.2.22) we immediately get that (3.2.19) is a martingale by passing to the limit $\alpha \downarrow 0$. Since the $y(\cdot)$ process has finite moments, this limit is elementary. Only the term

$$\int_0^t \overline{\mathcal{L}} f(x(s),y(s))\, d(\ell_o^\alpha(s) - \ell_o(s)) \tag{3.2.23}$$

requires comment since $\ell_o^\alpha(t) - \ell_o(t)$ does not converge in variation (over t) in probability to zero as $\alpha \downarrow 0$. However, we have (3.2.2) and (3.2.3) and this is easily seen to imply that the expectation of (3.2.23) tends to zero as $\alpha \downarrow 0$.

To prove that (ii) implies (i) we proceed as follows. Let $g^\alpha(y)$ be defined by (3.2.4). For $f(x,y)$ smooth and bounded let $f^\alpha(x,y)$ be defined by

$$f^\alpha(x,y) - g^\alpha(y)\overline{\mathcal{L}} f^\alpha(x,y) = f(x,y) \; . \tag{3.2.24}$$

The function $f^\alpha(x,y)$ is well defined by (3.2.24) since $g^\alpha(y) \geq 0$ and $\overline{\mathcal{L}}$ is a generator; f^α is a smooth and bounded function of (x,y) for $\alpha > 0$ (y is a parameter in (3.2.24)).

We now use (3.2.19) with f replaced by f^α of (3.2.24) to find that

$$(3.2.25)\quad f^\alpha(x(t),y(t)) - 2y_+(t)\bar{\mathscr{L}}f^\alpha(x(t),y(t))$$
$$- \int_0^t a(y(s))\Big[\frac{1}{2}\frac{\partial^2 f(x(s),y(s))}{\partial y^2}$$
$$- (2H(y(s)) - \frac{dg^\alpha}{dy}(y(s)))\bar{\mathscr{L}}\frac{\partial f^\alpha(x(s),y(s))}{\partial y}$$
$$+ \frac{1}{2}(2y_+(s) - g^\alpha(y(s)))\bar{\mathscr{L}}\frac{\partial^2 f^\alpha(x(s),y(s))}{\partial y^2}$$
$$+ \frac{1}{2}\frac{d^2 g^\alpha}{dy^2}(y(s))\bar{\mathscr{L}}f^\alpha(x(s),y(s))\Big]ds$$

is an $\mathscr{F}_t$, P martingale. Note that $\bar{\mathscr{L}}$ acts on x dependence alone. Let R_{g^α} be defined by

$$(3.2.26)\qquad R_{g^\alpha} = (I - g^\alpha\bar{\mathscr{L}})^{-1} ,$$

so that $||R_{g^\alpha}|| \le 1$ and

$$(3.2.27)\qquad f^\alpha = R_{g^\alpha}f \quad .$$

Using (3.2.24) and (3.2.5) we may rewrite (3.2.25) as follows

$$(3.2.28)\quad f(x(t),y(t)) - \int_0^t \frac{1}{2}\tilde{a}(y(s))\frac{\partial^2 f(x(s),y(s))}{\partial y^2}ds$$
$$- \int_0^t \bar{\mathscr{L}}f(x(s),y(s))d\ell_o(s)$$
$$- (2y_+(t) - g^\alpha(y(t)))\bar{\mathscr{L}}R^\alpha f(x(t),y(t))$$

$$+ \int_0^t \tilde{a}(y(s)) \left[\left[2H(y(s)) - \frac{dg^\alpha}{dy}(y(s))\right] \bar{\mathscr{L}} \frac{\partial f^\alpha(x(s),y(s))}{\partial y} \right.$$

$$\left. + \frac{1}{2}(2y_+(s) - g^\alpha(y(s))) \bar{\mathscr{L}} \frac{\partial^2 f^\alpha(x(s),y(s))}{\partial y^2}\right] ds$$

$$- \int_0^t \bar{\mathscr{L}} f(x(s),y(s)) d(\ell_o^\alpha(s) - \ell_o(s))$$

$$- \int_0^t g^\alpha(y(s)) \bar{\mathscr{L}}^2 f^\alpha(x(s),y(s)) d\ell_o^\alpha(s)$$

is an $\mathscr{F}_t$, P martingale. We must show that the expectation of the absolute value of the last four terms goes to zero as $\alpha \to 0$. This goes exactly as in the first part of the proof. Only the term that involves $\partial^2 f^\alpha/\partial y^2$ requires special consideration.

Differentiating (3.2.24) with respect to y we obtain

$$f_y^\alpha - g^\alpha \bar{\mathscr{L}} f_y^\alpha = f_y + \frac{dg^\alpha}{dy} \bar{\mathscr{L}} f^\alpha ,$$

hence

$$f_y^\alpha = R_{g^\alpha} f_y + \frac{dg^\alpha}{dy} (R_{g^\alpha})^2 \bar{\mathscr{L}} f . \tag{3.2.29}$$

Differentiating once again we obtain

$$f_{yy}^\alpha = R_{g^\alpha} f_{yy} + 2 \frac{dg^\alpha}{dy^2} R_{g^\alpha} \bar{\mathscr{L}} \left(R_{g^\alpha} f_y + \frac{dg^\alpha}{dy} (R_{g^\alpha})^2 \bar{\mathscr{L}} f\right) + \frac{d^2 g^\alpha}{dy^2} (R_{g^\alpha})^2 \bar{\mathscr{L}} f . \tag{3.2.30}$$

We note next that

$$(3.2.31) \qquad (2y_+ - g^\alpha(y))\ \frac{1}{2}\ \frac{d^2 g^\alpha(y)}{dy^2}$$

is bounded by a linear function of y independently of $\alpha > 0$ and $\alpha \to 0$ it tends to zero pointwise.

Substituting (3.2.30) into the fifth term in (3.2.28) and using what was just pointed out about (3.2.31) we obtain the desired result (3.2.18) by taking the limit $\alpha \to 0$ in (3.2.28). The proof of the lemma is complete.

Lemma 3.4.

Let X be a complete separable metric space and let P_n be sequence of probability measures on the Borel sets of X that converge weakly to a probability measure P. Let $f_n(x)$, $x \in X$ be a sequence of measurable real-valued functions on X which are uniformly bounded. Let $\mathcal{D} \subset X$ be such that

$$(3.2.32) \qquad f_n(x) \to f(x)\ , \quad \text{as} \quad n \to \infty\ , \quad x \in \mathcal{D},$$

and

$$(3.2.33) \qquad P(\mathcal{D}) = 1\ ,$$

where f(x) is a bounded measurable function on X. Assume in addition the following

$$(3.2.34) \qquad \text{If} \quad x_n \in X\ , \quad x_n \to x \quad \text{and} \quad x \in \mathcal{D}\ , \quad \text{then} \quad f_n(x_n) \to f(x)\ .$$

Then

$$(3.2.35) \qquad \int_X f_n(x)dP_n(x) \to \int_X f(x)dP(x) \quad \text{as} \quad n \to \infty .$$

Proof.

According to a Lemma of Skorohod [3], there exists a probability space (Ω,Σ,μ) and random variables $\xi_n(\omega)$, $\xi(\omega)$ with values in X such that

(i) $\mu\{\xi_n \in A\} = P_n(A)$, $\mu\{\xi \in A\} = P(A)$,

for each Borel set $A \subset X$,

and

(ii) $\xi_n(\omega) \to \xi(\omega)$ for μ almost all ω.

It follows that if $E\{\cdot\}$ denotes integration relative to μ, (3.2.35) takes the form

$$(3.2.36) \qquad E\{f_n(\xi_n)\} \to E\{f(\xi)\} .$$

By (3.2.34), (3.2.33) and (ii), the set of points $\omega \in \Omega$ for which $f_n(\xi_n(\omega)) \to f(\xi(\omega))$ has μ measure equal to one. Since f_n and f are uniformly bounded, (3.2.36) follows by the bounded convergence theorem.

Lemma 3.5.

Let Y be a locally compact metric space and let $g_n(y)$ and $g(y)$ be uniformly bounded and measurable real-valued functions _on_ Y. _Let_ G _be an open subset of_ Y _and suppose that the_ g_n _are continuous on_ G _and_

$$g_n(y) \to g(y) \text{ uniformly for } y \in K \subset G \,, \tag{3.2.37}$$

for each compact subset K _of_ G. _Let_ $X = C([0,T];Y)$ _and let_

$$\mathcal{D} = \left\{ y(\cdot) \in X \mid m\{t \in [0,T] \mid y(t) \in G^c\} = 0 \right\} , \tag{3.2.38}$$

where m is Lebesque measure on the line.

_If_ $y_n(\cdot) \in X$ _and_ $y_n(\cdot) \to y(\cdot) \in \mathcal{D}$ _then_

$$\int_0^T g_n(y_n(t))dt \to \int_0^T g(y(t))dt \,, \qquad \text{as } n \to \infty \,. \tag{3.2.39}$$

Proof.

Fix a path $y(\cdot) \in \mathcal{D}$ and let $t \in [0,T]$ be such that $y(t) \in G$. The set of such points $t \in [0,T]$ has Lebesque measure T, by definition of $\mathcal{D}$. Since $y_n(t) \to y(t)$ and $y(t) \in G$, it follows that for n sufficiently large all $y_n(t)$ belong to a compact subset K of G. From the uniform convergence (3.2.37) it follows that $g_n(y_n(t)) \to g(y(t))$. Thus, the bounded convergence theorem yields (3.2.39).

Proof of Theorem 3.1.

Let $\zeta^\varepsilon(t)$ be defined by

$$\zeta^\varepsilon(t) = \frac{y^\varepsilon(t)}{\varepsilon} \,, \tag{3.2.40}$$

and consider jointly the process $(x^\varepsilon(t), y^\varepsilon(t), \zeta^\varepsilon(t))$ whose generator $\mathcal{L}^\varepsilon$ has the form

$$(3.2.41)\quad \mathcal{L}^\varepsilon = \frac{1}{\varepsilon^2}\left[\frac{1}{2}\, a(\zeta)\, \frac{\partial^2}{\partial y^2}\right] + \frac{1}{\varepsilon}\left[a(\zeta)\, \frac{\partial^2}{\partial \zeta \partial y} + \sum_{k=1}^{n} F_k(x,\zeta)\, \frac{\partial}{\partial x_k} + \frac{1}{2} \sum_{k,j=1}^{n} a_{kj}^{(1)}(x,\zeta)\, \frac{\partial^2}{\partial x_k \partial x_j}\right] + \frac{1}{2}\, a(\zeta)\, \frac{\partial^2}{\partial y^2} .$$

Here we have used the notation following (3.1.11) and we have assumed that $(\beta_1^{(1)}(t), \ldots, \beta_q^{(1)}(t))$ and $\beta(t)$ in (3.1.11) are independent. This entails no loss in generality and does not affect the limit. Note that $\mathcal{L}^\varepsilon$ has the form

$$(3.2.42)\quad \mathcal{L}^\varepsilon = \frac{1}{\varepsilon^2}\, \mathcal{L}_1 + \frac{1}{\varepsilon}\, \mathcal{L}_2 + \mathcal{L}_3$$

as in (2.1.2), with $\mathcal{L}_1, \mathcal{L}_2$ and $\mathcal{L}_3$ identified from (3.2.41).

Let $f(x,y)$ be a bounded smooth function and define $\psi_f^{(1)}(x,y,\zeta)$ and $\psi_f^{(2)}(x,y,\zeta)$ as follows

$$(3.2.43)\quad \psi_f^{(1)}(x,y,\zeta) = -2 \int_0^\zeta d\eta \int_{-\infty}^{\eta} d\xi\, \frac{1}{a(\xi)}\, A_f(x,y,\xi) ,$$

$$A_f(x,y,\zeta) = \sum_{k=1}^{n} F_k(x,\zeta)\, \frac{\partial f(x,y)}{\partial x_k} + \frac{1}{2} \sum_{j,k=1}^{n} a_{kj}^{(1)}(x,\zeta)\, \frac{\partial^2 f(x,y)}{\partial x_k \partial x_j} ,$$

$$(3.2.44)\quad \psi_f^{(2)}(x,y,\zeta) = -2 \int_0^\zeta d\eta \int_{-\infty}^{\eta} d\xi\, \frac{1}{a(\xi)}\, A_{\psi_f^{(1)}}(x,y,\xi) .$$

By construction we have

$$(3.2.45)\quad \mathscr{L}^{\varepsilon}(f + \varepsilon\psi_f^{(1)} + \varepsilon^2\psi_f^{(2)})(x,y,\zeta)$$

$$= a(\zeta)\left[\frac{1}{2}\frac{\partial^2 f(x,y)}{\partial y^2} + \frac{\partial^2\psi_f^{(1)}(x,y,\zeta)}{\partial\zeta\partial y} + \frac{\varepsilon}{2}\frac{\partial^2\psi_f^{(1)}(x,y,\zeta)}{\partial y^2}\right]$$

$$+ \varepsilon\left[\sum_{k=1}^{n} F_k(x,\zeta)\frac{\partial\psi_f^{(2)}(x,y,\zeta)}{\partial x_k}\right.$$

$$+ \frac{1}{2}\sum_{k,j=1}^{n} a_{kj}^{(1)}(x,\zeta)\frac{\partial^2\psi_f^{(2)}(x,y,\zeta)}{\partial x_k\partial x_j}$$

$$\left.\cdot\ a(\zeta)\frac{\partial^2\psi_f^{(2)}(x,y,\zeta)}{\partial\zeta\partial y}\right]$$

$$+ \varepsilon^2\frac{1}{2}a(\zeta)\frac{\partial^2\psi_f^{(2)}(x,y,\zeta)}{\partial y^2}.$$

With $\bar{\mathscr{L}}$ defined by (3.1.13) we note that

$$(3.2.46)\quad \lim_{\varepsilon\downarrow 0}\varepsilon\psi_f^{(1)}(x,y,\tfrac{y}{\varepsilon}) = -2y_+\bar{\mathscr{L}}f(x,y),$$

uniformly on compact sets in $R^n \times R^1$. Also,

$$(3.2.47)\quad \lim_{\varepsilon\downarrow 0}\frac{\partial\psi_f^{(1)}(x,y,\frac{y}{\varepsilon})}{\partial\zeta} = -2H(y)\bar{\mathscr{L}}f(x,y),$$

uniformly on compact subsets of $R^n \times (R^1 - \{0\})$. Similarly,

$$(3.2.48)\quad \lim_{\varepsilon\downarrow 0}\varepsilon^2\psi_f^{(2)}(x,y,\tfrac{y}{\varepsilon}) = 0, \qquad \lim_{\varepsilon\downarrow 0}\varepsilon^2\frac{\partial^2\psi_f^{(2)}}{\partial y^2}(x,y,\tfrac{y}{\varepsilon}) = 0,$$

$$\lim_{\varepsilon\downarrow 0}\varepsilon\frac{\partial^2\psi_f^{(2)}(x,y,\frac{y}{\varepsilon})}{\partial\zeta\partial y} = 0,$$

$$\lim_{\varepsilon\downarrow 0} \varepsilon \sum_{k=1}^{n} F_k(x,\tfrac{y}{\varepsilon}) \frac{\partial\psi_f^{(2)}(x,y,\frac{y}{\varepsilon})}{\partial x_k} + \frac{\varepsilon}{2} \sum_{k,j=1}^{n} a_{kj}^{(1)}(x,\tfrac{y}{\varepsilon}) \frac{\partial^2\psi_f^{(2)}(x,y,\frac{y}{\varepsilon})}{\partial x_k \partial x_j} = 0 ,$$

uniformly on compact subsets of $R^n \times R^1$. Here we use the fact that $F_k(x,y)$ and $a_{kj}^{(1)}(x,y)$ have compact support in y and they are smooth.

On the basis of the above constructions and the fact that $y^{\varepsilon}(t)$ has bounded moments of all orders, for t in a finite interval, with bounds independent of ε, it follows as in Section 2.3 that

$$\begin{aligned}(3.2.49)\quad M_f^{\varepsilon}(t) &= f(x^{\varepsilon}(t),y^{\varepsilon}(t)) + \varepsilon\psi_f^{(1)}(x^{\varepsilon}(t),y^{\varepsilon}(t),\tfrac{y^{\varepsilon}(t)}{\varepsilon}) \\ &+ \varepsilon^2\psi_f^{(2)}(x^{\varepsilon}(t),y^{\varepsilon}(t),\tfrac{y^{\varepsilon}(t)}{\varepsilon}) - f(x,y) \\ &- \varepsilon\psi_f^{(1)}(x,y,\tfrac{y}{\varepsilon}) - \varepsilon^2\psi_f^{(2)}(x,y,\tfrac{y}{\varepsilon}) \\ &- \int_0^t a\left(\frac{y^{\varepsilon}(s)}{\varepsilon}\right)\left[\frac{1}{2}\frac{\partial^2 f}{\partial y^2} + \frac{\partial^2\psi_f^{(1)}}{\partial\zeta\partial y} + \frac{\varepsilon}{2}\frac{\partial^2\psi_f^{(1)}}{\partial y^2}\right] \\ &\cdot (x^{\varepsilon}(s),y^{\varepsilon}(s),\tfrac{y^{\varepsilon}(s)}{\varepsilon})\,ds \\ &- \varepsilon\int_0^t a(\tfrac{y^{\varepsilon}(s)}{\varepsilon})\left[\sum_{k=1}^{n} F_k \frac{\partial\psi_f^{(2)}}{x_k} + \frac{1}{2}\sum_{k,j=1}^{n} a_{kj}^{(1)} \frac{\partial^2\psi_f^{(2)}}{\partial x_k\partial x_j} \right. \\ &\left. + \frac{\varepsilon}{2}\frac{\partial^2\psi_f^{(2)}}{\partial y^2}\right](x^{\varepsilon}(s),y^{\varepsilon}(s),\tfrac{y^{\varepsilon}(s)}{\varepsilon})\,ds\end{aligned}$$

is an integrable zero-mean martingale. The increasing process corresponding to $M_f^\varepsilon(t)$ has the form

$$(3.2.50)\quad \langle M_f^\varepsilon(t), M_f^\varepsilon(t)\rangle = \int_0^t H_f^\varepsilon(x^\varepsilon(s), y^\varepsilon(s), \frac{y^\varepsilon(s)}{\varepsilon})\, ds\ ,$$

where, one can verify it easily by direct computation, $H_f^\varepsilon(x,y,\frac{y}{\varepsilon})$ is bounded by a constant times one plus a power of $|y|$ independently of ε.

As in Section 2.3 we conclude from the above that the process $(x^\varepsilon(\cdot), y^\varepsilon(\cdot))$ is relatively weakly compact in $D([0,T];\ R^n \times R^1)$. It remains now to identify the limit process.

We rewrite (3.2.49) as follows.

$$(3.2.51)\quad f(x^\varepsilon(t),y^\varepsilon(t)) - 2y_+^\varepsilon(t)\bar{\mathscr{L}}f(x^\varepsilon(t),y^\varepsilon(t))$$
$$- f(x^\varepsilon(\sigma),y^\varepsilon(\sigma)) + 2y_+^\varepsilon(\sigma)\bar{\mathscr{L}}f(x^\varepsilon(\sigma),y^\varepsilon(\sigma))$$
$$- \int_\sigma^t \tilde{a}(y^\varepsilon(s))\left[\frac{1}{2}\frac{\partial^2 f}{\partial y^2} - 2H(y)\bar{\mathscr{L}}\frac{\partial f}{\partial y} - y_+\bar{\mathscr{L}}\frac{\partial^2 f}{\partial y^2}\right](x^\varepsilon(s),y^\varepsilon(s))\,ds$$

$$= M_f^\varepsilon(t) - M_f^\varepsilon(\sigma) - \left[2y_+^\varepsilon(t)\bar{\mathscr{L}}f(x^\varepsilon(t),y^\varepsilon(t)) - \varepsilon\psi_f^{(1)}(x^\varepsilon(t),y^\varepsilon(t),\frac{y^\varepsilon(t)}{\varepsilon})\right]$$
$$+ \left[2y_+^\varepsilon(\sigma)\bar{\mathscr{L}}f(x^\varepsilon(\sigma),y^\varepsilon(\sigma)) - \varepsilon\psi_f^{(1)}(x^\varepsilon(\sigma),y^\varepsilon(\sigma),\frac{y^\varepsilon(\sigma)}{\varepsilon})\right]$$
$$- \int_\sigma^t \left[a(\frac{y^\varepsilon(s)}{\varepsilon}) - \tilde{a}(y^\varepsilon(s))\right]\left[\frac{1}{2}\frac{\partial^2 f}{\partial y^2} - 2H(y)\bar{\mathscr{L}}\frac{\partial f}{\partial y} - y_+\bar{\mathscr{L}}\frac{\partial^2 f}{\partial y^2}\right]$$
$$\cdot\ (x^\varepsilon(s),y^\varepsilon(s))\,ds$$

$$- \int_\sigma^t a\left(\frac{y^\varepsilon(s)}{\varepsilon}\right) \left[\frac{\partial^2 \psi_f^{(1)}}{\partial \zeta \partial y} + 2H(y)\overline{\mathscr{L}}\frac{\partial f}{\partial y} + \frac{\varepsilon}{2}\frac{\partial^2 \psi_f^{(1)}}{\partial y^2} + y_+ \overline{\mathscr{L}}\frac{\partial^2 f}{\partial y^2}\right] \cdot \left(x^\varepsilon(s), y^\varepsilon(s), \frac{y^\varepsilon(s)}{\varepsilon}\right) ds$$

$$- \varepsilon \int_\sigma^t a\left(\frac{y^\varepsilon(s)}{\varepsilon}\right) \left[\sum_{k=1}^n F_k \frac{\partial \psi_f^{(2)}}{\partial x_k} + \frac{1}{2}\sum_{k,j=1}^n a_{kj}^{(1)} \frac{\partial^2 \psi_f^{(2)}}{\partial x_k \partial x_j} + \frac{\varepsilon}{2}\frac{\partial^2 \psi_f^{(2)}}{\partial y^2}\right] \cdot \left(x^\varepsilon(s), y^\varepsilon(s), \frac{y^\varepsilon(s)}{\varepsilon}\right) ds .$$

We intend to show that the absolute value of the terms after M_f^ε on the right hand side have expectation that tends to zero as $\varepsilon \to 0$ along a subsequence for which P^ε, the measure corresponding to $(x^\varepsilon(\cdot), y^\varepsilon(\cdot))$, converges weakly to a measure P (by compactness).

Let Φ_σ be a bounded continuous and $\mathscr{F}_\sigma$ measurable functional of $(x(\cdot), y(\cdot)) \in C([0,T]; R^n \times R)$. We multiply both sides of (3.2.51) by Φ_σ, take expectations relative to P^ε and let $\varepsilon \to 0$ along the chosen subsequence. We have

$$(3.2.52) \quad \lim_{\varepsilon \downarrow 0} E^{P^\varepsilon}\Big\{\Phi_\sigma\Big[f(x(t),y(t)) - 2y_+(t)\overline{\mathscr{L}} f(x(t),y(t)) - f(x(\sigma),y(\sigma)) + 2y_+(\sigma)\overline{\mathscr{L}} f(x(\sigma),y(\sigma)) - \int_\sigma^t \tilde{a}(y(s))\left(\frac{1}{2}\frac{\partial^2 f}{\partial y^2} - 2H(y)\overline{\mathscr{L}}\frac{\partial f}{\partial y} - y_+\overline{\mathscr{L}}\frac{\partial^2 f}{\partial y^2}\right)(x(s),y(s))ds\Big]\Big\}$$

$$= \lim_{\varepsilon \downarrow 0} E^{P^\varepsilon}\Big\{-\Phi_\sigma\Big[\int_\sigma^t \left(a\left(\frac{y(s)}{\varepsilon}\right) - \tilde{a}(y(s))\right)\left(\frac{1}{2}\frac{\partial^2 f}{\partial y^2} - 2H(y)\overline{\mathscr{L}}\frac{\partial f}{\partial y} - y_+\overline{\mathscr{L}}\frac{\partial^2 f}{\partial y^2}\right) \cdot (x(s),y(s))ds\Big]\Big\}$$

$$+ \lim_{\varepsilon \downarrow 0} E^{P^\varepsilon}\left\{ -\Phi_\sigma \left[\int_\sigma^t a\left(\frac{y(s)}{\varepsilon}\right) \left(\frac{\partial^2 \psi_f^{(1)}}{\partial \zeta \partial y} + 2H(y) \mathscr{Z} \frac{\partial f}{\partial y} \right. \right. \right.$$

$$\left. \left. \left. + \frac{\varepsilon}{2} \frac{\partial^2 \psi_f^{(1)}}{\partial y} + y_+ \mathscr{Z} \frac{\partial^2 f}{\partial y^2} \right) (x(s), y(s), \frac{y(s)}{\varepsilon}) ds \right] \right\} .$$

Here we have used the compactness of the measures P^ε and (3.2.46), (3.2.48) to drop some terms on the right side of (3.2.52) by the dominated convergence theorem.

In the notation of Lemma 3.5 we let $X = C([0,T],Y)$, $Y = R^n \times R$, $G = R - \{0\}$ and

$$\mathscr{D} = \{(x(\cdot), y(\cdot)) \in X \mid m\{t \in [0,T] \mid y(t) = 0\} = 0\} .$$

From the existence on the local time for $y^\varepsilon(t)$ it follows that for any $\varepsilon > 0$

(3.2.53) $\quad P^\varepsilon(\mathscr{D}) = 0$,

and hence $P(\mathscr{D})$ for any weak limit of the measures P^ε.

The functional inside the expectation on the left hand side of (3.2.52) is P a.s. continuous for any limit measure by Lemma 3.5 (with $g_n \equiv g$). By Lemma 3.4 we can equate the left side to the expectation of the functional with respect to the weak limit P. On the right side we use the compactness of the measures P^ε, (3.2.53), (3.2.46), (3.2.47) and Lemmas 3.5 and 3.4 to conclude it is zero.

From the fact that the right side of (3.2.52) is zero, Lemmas 3.3 and 3.2 and the uniqueness of the limit process described in Theorem 3.1 (and Remark 2) the identification step is complete. The proof of Theorem 3.1 is complete.

Remark.

As in the case of ergodic driving, the above proof extends immediately to a variety of similar situations of merely changing (3.2.43), (3.2.44) and (3.2.45) appropriately. Examples are given in the following sections.

3.3 Some simple consequences.

Instead of (3.1.3) (or (3.1.11)) we may consider the asymptotic behavior as $\varepsilon \to 0$ of the following system

$$(3.3.1) \qquad dx_i^\varepsilon(t) = \frac{1}{\varepsilon}\sum_{p=1}^{N} F_i^{(p)}\left(x^\varepsilon(t), \frac{y^\varepsilon(t)-y_p}{\varepsilon}\right),$$

$$x_i^\varepsilon(0) = x_i, \qquad i = 1,2,\ldots,n$$

$$dy^\varepsilon(t) = a^{1/2}\left(\frac{y^\varepsilon(t)}{\varepsilon}\right)d\beta(t), \qquad y^\varepsilon(0) = y.$$

Here $F^{(1)}, F^{(2)},\ldots,F^{(N)}$ are smooth bounded and with compact support in y n-vector functions of (x,y) and $y_1,y_2,\ldots,y_N$ are fixed points on the line. The function a(y) satisfies (3.1.1) and (3.1.2) and is smooth.

It is easily seen that the methods of Section 3.2 apply immediately to (3.3.1). Let

$$(3.3.2) \qquad \bar{F}^{(p)}(x) = \int_{-\infty}^{\infty} \frac{F^{(p)}(x,y)}{a(y)}\,dy,$$

and

$$\bar{\mathscr{L}}_p = \sum_{i=1}^{n} \bar{F}_i^{(p)}(x)\frac{\partial}{\partial x_i}, \qquad p = 1,2,\ldots,N.$$

Under our hypotheses, there is a unique family of probability measures $P_{(x,y)}$ on $C([0,\infty), R^n \times R^1)$ for $(x,y) \in R^n \times R^1$ such that

(i) $P_{(x,y)}\{x(0) = x, y(0) = y\}$

(ii) For each $f(x,y) \in C_o^\infty(R^n \times R^1)$

$$f(x(t),y(t)) - \int_0^t \frac{1}{2}\,\tilde{a}(y(s))\,\frac{\partial^2 f(x(s),y(s))}{\partial y^2}\,ds \tag{3.3.3}$$

$$- \sum_{p=1}^{N} \int_0^t \bar{\mathscr{L}}_p f(x(s),y(s))\,d\ell(s,y_p)$$

is a P martingale. Here $\tilde{a}(y)$ is defined by (3.1.6) and $\ell(t,\bar{y})$ is the local time of $y(t)$ of (3.1.7) at $\bar{y} \in R^1$. The local time is defined by

$$\ell(t,\bar{y}) = 2(y(t)-\bar{y})_+ - 2(y-\bar{y})_+ \tag{3.3.4}$$

$$- 2\int_0^t H(y(s)-\bar{y})\,dy(s)\ .$$

This is the natural generalization of (3.1.8).

Theorem 3.2.

Under the above hypotheses $(x^\varepsilon(t), y^\varepsilon(t))$ *of* (3.3.1) *converges weakly in* $C([0,T]; R^n \times R^1)$, $T < \infty$, *to the process* $(x(t),y(t))$ *characterized by the martingale problem* (3.3.3).

As an application consider the following problem which is similar to one considered by Isaacson [39].

Let $y(t)$ be Brownian motion on R^1 and $F(y)$ be a smooth nonnegative function of compact support such that

$$\int_{-\infty}^{\infty} F(y)\,dy = 1 \,. \tag{3.3.5}$$

Consider the Cauchy problem (f(y) smooth bounded)

$$\begin{aligned} \frac{\partial u^{\varepsilon}(t,y)}{\partial t} &= \frac{1}{2}\frac{\partial^2 u^{\varepsilon}(t,y)}{\partial y^2} - \frac{1}{2\varepsilon}\left[F\left(\frac{y+1}{\varepsilon}\right) + F\left(\frac{y-1}{\varepsilon}\right)\right]u^{\varepsilon}(t,y) \,, \\ u^{\varepsilon}(0,y) &= f(y) \,. \end{aligned} \tag{3.3.6}$$

It follows from Theorem 3.2 that

$$\lim_{\varepsilon\downarrow 0} u^{\varepsilon}(t,y) = E_y\left\{e^{-\frac{1}{2}[\ell(t,+1)+\ell(t,-1)]} f(y(t))\right\} . \tag{3.3.7}$$

If u(t,y) stands for the right hand side of (3.3.7), formally, it satisfies the Cauchy problem

$$\begin{aligned} \frac{\partial u(t,y)}{\partial t} &= \frac{1}{2}\frac{\partial^2 u(t,y)}{\partial y^2} - \frac{1}{2}\Big(\delta(y+1)+\delta(y-1)\Big)u(t,y) \,, \\ u(0,y) &= f(y) \,. \end{aligned} \tag{3.3.8}$$

To see in more detail how (3.3.7) follows introduce $x^{\varepsilon}(t)$ by

$$x^{\varepsilon}(t) = x + \frac{1}{2\varepsilon}\int_0^t \left(F\left(\frac{y(s)+1}{\varepsilon}\right) + F\left(\frac{y(s)-1}{\varepsilon}\right)\right) ds$$

so that

$$\begin{aligned} \frac{dx^{\varepsilon}(t)}{dt} &= \frac{1}{2\varepsilon}\left(F\left(\frac{y(s)+1}{\varepsilon}\right) + F\left(\frac{y(s)-1}{\varepsilon}\right)\right) \\ x^{\varepsilon}(0) &= x \,. \end{aligned} \tag{3.3.9}$$

By the Feynman-Kac formula,

$$u^{\varepsilon}(t,y) = E_y\left\{e^{-(x^{\varepsilon}(t)-x)} f(y(t))\right\}$$
$$= e^x E_{x,y}\left\{e^{-x^{\varepsilon}(t)} f(y(t))\right\} .$$

Thus, the quantity of interest $u^{\varepsilon}(t,y)$ is a bounded continuous functional of $(x^{\varepsilon}(t),y(t))$ to which Theorem 3.2 applies directly.

Theorem 3.1 (or Theorem 3.2) can be extended to the case where $y^{\varepsilon}(t)$, the driving process, is not, essentially, Brownian motion as we have assumed (cf. (3.1.3)).

One possibility is to allow a(y) to violate (3.1.1) so that $\int_0^y \frac{1}{a(\zeta)} d\zeta$ tends to ∞ as $y \to \pm\infty$ at a rate other than $|y|$ (the Brownian case). Another is to allow $y^{\varepsilon}(t)$ to be a symmetric stable process and to adjust the scaling appropriately. When F(x,y) in (3.1.3) is a function of y only, such problems are considered in [20,21]. In connection with results analogous to the one of the next section see [22].

We note finally that if F(x,y) is a periodic function of y (or an appropriate almost periodic function) in (3.1.3), then, even though the driving process is not ergodic, we are effectively in the ergodic case; in fact in a situation similar to the one of Section 2.7.

3.4 Second order limit theorem.

We shall now consider a rescaled form of (3.1.3) (or (3.1.11)) where $\bar{F}(x) \equiv 0$ in (3.1.4). We call the asymptotic limit a second order limit because the first order result, Theorem 3.1, is trivial in case $\bar{F}(x) \equiv 0$ and we have to go to higher order (rescale) to get the first nontrivial result.

Let $F_i^{(1)}(x,y)$, $F_i^{(2)}(x,y)$ and $\sigma_{ij}(x,y)$, $i = 1,2,\ldots,n$, $j = 1,2,\ldots,q$, $x \in R^n$, $y \in R^1$ be smooth bounded functions with compact support in y and let $a(y)$ satisfy (3.1.1) and (3.1.2). Consider the system of stochastic differential equations

$$(3.4.1) \qquad dx_i^\varepsilon(t) = \left[\frac{1}{\varepsilon^3} F_i^{(1)}\left(x^\varepsilon(t), \frac{y^\varepsilon(t)}{\varepsilon^2}\right) + \frac{1}{\varepsilon^2} F^{(2)}\left(x^\varepsilon(t), \frac{y^\varepsilon(t)}{\varepsilon^2}\right)\right] dt$$

$$+ \frac{1}{\varepsilon} \sum_{j=1}^{q} \sigma_{ij}\left(x^\varepsilon(t), \frac{y^\varepsilon(t)}{\varepsilon^2}\right) d\beta_j^{(1)}(t)$$

$$x_i^\varepsilon(0) = x_i \ , \qquad i = 1,2,\ldots,n \ ,$$

$$dy^\varepsilon(t) = a^{1/2}\left(\frac{y^\varepsilon(t)}{\varepsilon^2}\right) d\beta(t) \ , \qquad y^\varepsilon(0) = y \ .$$

Here $(\beta_1^{(1)}(t), \beta_2^{(1)}(t), \ldots, \beta_q^{(1)}(t))$ is a q-dimensional Brownian motion, not necessarily independent of the Brownian motion $\beta(t)$.

The distinguishing hypothesis about (3.4.1) is that

$$(3.4.2) \qquad \int_{-\infty}^{\infty} \frac{F^{(1)}(x,y)}{a(y)}\, dy \equiv 0 \ .$$

The scaling of (3.4.1) should be compared with (3.1.3) (or (3.1.11)); replacing ε by $\sqrt{\varepsilon}$ in (3.4.1) identifies the $F^{(1)}$ drift as a more singular term than usual which explains the necessity of (3.4.2).

Define

$$a_{ij}^{(1)}(x,y) = \sum_{k=1}^{q} \sigma_{ik}(x,y)\sigma_{jk}(x,y)$$

and

$$\bar{a}_{ij}(x) = \int_{-\infty}^{\infty} \frac{a_{ij}^{(1)}(x,y)}{a(y)}\, dy \,, \tag{3.4.3}$$

$$\bar{F}_i^{(2)}(x) = \int_{-\infty}^{\infty} \frac{\bar{F}_i^{(2)}(x,y)}{a(y)}\, dy \,, \qquad i,j = 1,2,\dots,n \,.$$

For $f(x) \in C_o^\infty(R^n)$, define $\tilde{\psi}_f^{(1)}(x,\zeta)$, $\zeta \in R^1$, by

$$\tilde{\psi}_f^{(1)}(x,\zeta) = -2\int_0^{\zeta} d\eta \int_{-\infty}^{\eta} d\zeta \sum_{j=1}^{n} \frac{F_j^{(1)}(x,\xi)}{a(\xi)} \frac{\partial f(x)}{\partial x_j} \,. \tag{3.4.4}$$

Similary define

$$\begin{aligned} \tilde{\psi}_f(x) &= \int_{-\infty}^{\infty} d\zeta \sum_{i=1}^{n} \frac{F_i^{(1)}(x,\zeta)}{a(\zeta)} \frac{\partial \psi_f^{(1)}(x,\zeta)}{\partial x_i} \\ &= 2\int_{-\infty}^{\infty} d\zeta \int_{-\infty}^{\zeta} d\eta \sum_{i=1}^{n} \frac{F_i^{(1)}(x,\eta)}{a(\eta)} \frac{\partial}{\partial x_i} \\ &\quad \cdot \left[\int_{-\infty}^{\zeta} d\xi \sum_{j=1}^{n} \frac{F_j^{(1)}(x,\xi)}{a(\xi)} \frac{\partial f(x)}{\partial x_j} \right] . \end{aligned} \tag{3.4.5}$$

Note that because of (3.4.2), (3.4.5) is well defined and smooth for $f(x) \in C_o^\infty(R^n)$.

Let $\bar{\mathscr{L}}$ be the second order elliptic operator

$$\bar{\mathscr{L}}f(x) = \tilde{\psi}_f(x) + \frac{1}{2}\sum_{i,j=1}^{n} \bar{a}_{ij}(x)\,\frac{\partial^2 f(x)}{\partial x_i \partial x_j} + \sum_{j=1}^{n} \bar{F}_j^{(1)}(x)\,\frac{\partial f(x)}{\partial x_j}\,. \tag{3.4.6}$$

By hypothesis it has smooth coefficients and so it generates a diffusion process $\bar{x}(t)$ on R^n.

Let $y(t)$ be the process defined by (3.1.7) and let $\ell_o(t)$ be the local time of $y(t)$ at zero, defined by (3.1.8). Let $x(t)$ be the diffusion process $\bar{x}(\cdot)$ run at the local time of $y(\cdot)$ i.e., let (3.1.10) hold.

Theorem 3.3.

Under the above hypotheses, the process $(x^\varepsilon(t), y^\varepsilon(t))$ *defined by* (3.4.1) *converges weakly on* $C([0,T]; R^n \times R^1)$, $T < \infty$, *to the process* $(x(t), y(t))$ *described above*.

Proof.

The proof of this theorem is nearly identical to the one of Theorem 3.1. It suffices to show how the constructions centering around (3.2.45) change here.

To simplify writing we shall assume that $F^{(2)} \equiv 0$, $\sigma \equiv 0$ and $F^{(1)} = F$ with $n = 1$.

As in (3.2.40) put

$$(3.4.7) \qquad \zeta^{\varepsilon}(t) = \frac{y^{\varepsilon}(t)}{\varepsilon^2} ,$$

and consider $(x^{\varepsilon}(t), y^{\varepsilon}(t), \zeta^{\varepsilon}(t))$ jointly so that its generator $\mathscr{L}^{\varepsilon}$ has the form

$$(3.4.8) \qquad \mathscr{L}^{\varepsilon} = \frac{1}{\varepsilon^4}\left[\frac{1}{2}\, a(\zeta)\, \frac{\partial^2}{\partial \zeta^2}\right] + \frac{1}{\varepsilon^3}\, F(x,\zeta)\, \frac{\partial}{\partial x} + \frac{1}{\varepsilon^2}\, a(\zeta)\, \frac{\partial^2}{\partial \zeta \partial y} + \frac{1}{2}\, a(\zeta)\, \frac{\partial^2}{\partial y^2} .$$

Define $\psi_f^{(1)}(x,y,\zeta)$, $\psi_f^{(2)}(x,y,\zeta)$ and $\psi_f^{(3)}(x,y,\zeta)$ for each $f \in C_o^{\infty}(R^n \times R^1)$ as follows

$$(3.4.9) \qquad \psi_f^{(1)}(x,y,\zeta) = -2 \int_0^{\zeta} d\eta \int_{-\infty}^{\eta} d\xi\, \frac{F(x,\xi)}{a(\xi)}\, \frac{\partial f(x,y)}{\partial x} ,$$

$$(3.4.10) \qquad \psi_f^{(2)}(x,y,\zeta) = -2 \int_0^{\zeta} d\eta \int_{-\infty}^{\eta} d\xi\, \frac{F(x,\xi)}{a(\xi)}\, \frac{\partial \psi_f^{(1)}(x,y,\xi)}{\partial x} ,$$

$$(3.4.11) \qquad \psi_f^{(3)}(x,y,\zeta) = -2 \int_0^{\zeta} d\eta \int_{-\infty}^{\eta} d\xi \left[\frac{F(x,\xi)}{a(\xi)}\, \frac{\partial \psi_f^{(2)}(x,y,\xi)}{\partial x} + \frac{\partial^2 \psi_f^{(1)}(x,y,\xi)}{\partial \zeta \partial y}\right] .$$

We have the following

$$(3.4.12)\quad \mathscr{L}^{\varepsilon}\left[f + \varepsilon\psi_f^{(1)} + \varepsilon^2\psi_f^{(2)} + \varepsilon^3\psi_f^{(3)}\right](x,y,\zeta)$$

$$= a(\zeta)\left[\frac{1}{2}\frac{\partial^2 f(x,y)}{\partial y^2} + \frac{\partial^2\psi_f^{(2)}(x,y,\zeta)}{\partial\zeta\partial y} + \frac{\varepsilon^2}{2}\frac{\partial^2\psi_f^{(2)}(x,y,\zeta)}{\partial y^2}\right]$$

$$+ F(x,\zeta)\frac{\partial\psi_f^{(3)}(x,y,\zeta)}{\partial x} + \varepsilon a(\zeta)\frac{\partial^2\psi_f^{(3)}(x,y,\zeta)}{\partial\zeta\partial y}$$

$$+ \frac{\varepsilon}{2}a(\zeta)\frac{\partial^2\psi_f^{(1)}(x,y,\zeta)}{\partial y^2} + \frac{\varepsilon^3}{2}a(\zeta)\frac{\partial^2\psi_f^{(3)}(x,y,\zeta)}{\partial y^2}\,.$$

From (3.4.2) it follows that $\psi_f^{(1)}(x,y,\zeta)$ is bounded and it is easily seen that the following limits hold.

$$(3.4.13)\quad \lim_{\varepsilon\downarrow 0}\varepsilon^2\psi_f^{(2)}(x,y,\tfrac{y}{\varepsilon^2}) = -2y_+\overline{\mathscr{L}}f(x,y)$$

uniformly on compact subsets of R × R,

$$(3.4.14)\quad \lim_{\varepsilon\downarrow 0}\frac{\partial\psi_f^{(2)}(x,y,\frac{y}{\varepsilon^2})}{\partial\zeta} = -H(y)\overline{\mathscr{L}}f(x,y)$$

uniformly on compact subsets of R × (R-[0]),

$$(3.4.15)\quad \lim_{\varepsilon\downarrow 0}F(x,\tfrac{y}{\varepsilon^2})\frac{\partial\psi_f^{(3)}(x,y,\frac{y}{\varepsilon^2})}{\partial x} = 0\,,$$

$$\lim_{\varepsilon\downarrow 0}\varepsilon\frac{\partial\psi_f^{(3)}(x,y,\frac{y}{\varepsilon^2})}{\partial\zeta} = 0\,,$$

$$\lim_{\varepsilon\downarrow 0}\varepsilon^3\psi_f^{(3)}(x,y,\tfrac{y}{\varepsilon^2}) = 0\,,$$

uniformly on compact subsets of R × R.

The above parts correspond to (3.2.46)-(3.2.48). The proof now proceeds exactly as in Theorem 2.1.

3.5 A jump driving process.

We shall restrict attention to the situation that is comparable to Theorem 3.1. It is easily extended to the one of Theorem 3.3.

Let $Y_0 = 0$ and $Y_1, Y_2, \ldots,$ be independent identically distributed random variables with distribution $\phi(dz)$ such that

$$(3.5.1) \qquad \int_{-\infty}^{\infty} z\phi(dz) = 0 \ , \qquad \int_{-\infty}^{\infty} z^2\phi(dz) = 1$$

and for some $\alpha > 0$

$$(3.5.2) \qquad \int_{-\infty}^{\infty} e^{\alpha z}\phi(dz) < \infty \ .$$

Let $N(t)$ be an independent Poisson process with parameter 1 and let $N^{\varepsilon}(t)$ denote the same process rescaled so that its parameter is $1/\varepsilon^2$. Define

$$(3.5.3) \qquad y^{\varepsilon}(t) = y + \varepsilon \int_0^t Y_{N^{\varepsilon}(s)} \, dN^{\varepsilon}(s) \ ,$$

which is simply the sum of a Poisson number of random variables. The scaling is such that $y^{\varepsilon}(t)$ converges weakly to Brownian motion as $\varepsilon \to 0$. This can be verified very easily.

Let $F(x,y)$ be a smooth and bounded n-vector function on $R^n \times R$ with compact support in y. Consider the process $x^{\varepsilon}(t)$ on R^n defined by

$$(3.5.4) \qquad \frac{dx^{\varepsilon}(t)}{dt} = \frac{1}{\varepsilon} F(x^{\varepsilon}(t), \frac{y^{\varepsilon}(t)}{\varepsilon}) , \qquad x^{\varepsilon}(0) = x .$$

We shall study the asymptotic behavior of $(x^{\varepsilon}(t), y^{\varepsilon}(t))$ as $\varepsilon \to 0$.

Let

$$(3.5.5) \qquad \bar{F}(x) = \int_{-\infty}^{\infty} F(x,y)$$

and define $\bar{x}(t)$ by

$$(3.5.6) \qquad \frac{d\bar{x}(t)}{dt} = \bar{F}(\bar{x}(t)) , \qquad \bar{x}(0) = x .$$

Let $y(t)$ be Brownian motion with $y(0) = y$ and let $\ell_o(t)$ be its local time at $y = 0$. Define $x(t)$ by

$$(3.5.7) \qquad x(t) = \bar{x}(\ell_o(t)) .$$

Theorem 3.4.

Under the above hypotheses, $(x^{\varepsilon}(t), y^{\varepsilon}(t))$ *converges weakly in* $C([0,T];R^n \times R)$, $T < \infty$, *to the process* $(x(t), y(t))$ *defined above.*

Proof.

Again the proof is very similar to the one of Theorem 3.1. We outline the constructions centering around (3.2.45). We take $n = 1$ for simplicity.

Define

$$(3.5.8) \qquad \zeta^{\varepsilon}(t) = \frac{y^{\varepsilon}(t)}{\varepsilon}$$

so that $(x^\varepsilon(t), y^\varepsilon(t), \zeta^\varepsilon(t))$ are jointly Markov with generator

$$(3.5.9)\quad \mathcal{L}^\varepsilon f(x,y,\zeta) = \frac{1}{\varepsilon^2}\left[\int_{-\infty}^{\infty} f(x,y+\zeta z,\varepsilon+z)\phi(dz) - f(x,y,\zeta)\right] + \frac{1}{\varepsilon} F(x,\zeta)\frac{\partial}{\partial x} .$$

We rewrite this as follows:

$$(3.5.10)\quad \mathcal{L}^\varepsilon f(x,y,\zeta) = \frac{1}{\varepsilon^2}\left[\int_{-\infty}^{\infty} f(x,y,\zeta+z)\phi(dz) - f(x,y,\zeta)\right] + \frac{1}{\varepsilon} F(x,\zeta)\frac{\partial f(x,y,\zeta)}{\partial x} + \frac{1}{\varepsilon}\int \frac{\partial f(x,y,\zeta+z)}{\partial y} z\phi(dz) + \frac{1}{2}\int \frac{\partial^2 f(x,y,\zeta+z)}{\partial y^2} z^2\phi(dz) + S^\varepsilon f(x,y,z) .$$

Here

$$(3.5.11)\quad S^\varepsilon f(x,y,\zeta) = \frac{1}{\varepsilon^2}\int_{-\infty}^{\infty}\left[f(x,y+\varepsilon z,\zeta+z) - f(x,y,\zeta+z) - \varepsilon\frac{\partial f(x,y,\zeta+z)}{\partial y} z - \frac{\varepsilon^2}{2}\frac{\partial^2 f(x,y,\zeta+z)}{\partial y^2} z^2\right]\phi(dz)$$

and it easily seen that for f smooth $S^\varepsilon f \to 0$ as $\varepsilon \to 0$ uniformly on compact sets.

Now we construct $\psi_f^{(1)}(x,y,\zeta)$ for each $f(x,y) \in C_o^\infty$ so that

$$(3.5.12)\quad \int \psi_f^{(1)}(x,y,\zeta+z)\phi(dz) - \psi_1(x,y,\zeta) + F(x,\zeta)\frac{\partial f(x,y)}{\partial x} = 0 .$$

If $\hat{\phi}(k)$ denotes the characteristic function of $\phi(dz)$ and $\hat{}$, in general, Fourier transform, we have

(3.5.13) $$\hat{\psi}_f^{(1)}(x,y,k) = \frac{\hat{F}(x,k)\frac{\partial f(x,y)}{\partial x}}{1 - \hat{\phi}(k)}$$

which is well defined by (3.5.1), (3.5.2) and the smoothness and compact support of F(x,y). Furthermore, we have

(3.5.14) $$\lim_{\varepsilon \downarrow 0} \varepsilon\psi_f^{(1)}(x,y,\frac{y}{\varepsilon}) = -2y_+ \,\bar{F}(x)\, \frac{\partial f(x,y)}{\partial x}$$

uniformly on compact sets,

and

(3.5.15) $$\lim_{\varepsilon \downarrow 0} \frac{\partial \psi_f^{(1)}(x,y,\frac{y}{\varepsilon})}{\partial y} = -\, 2H(y)\bar{F}(x)\, \frac{\partial f(x,y)}{\partial x}$$

uniformly on compact subsets of $R \times (R-[0])$.

We define, similarly, $\psi_f^{(2)}(x,y,\zeta)$ as solutions of

(3.5.16) $$\int_{-\infty}^{\infty} \psi_f^{(2)}(x,y,\zeta+z)\phi(dz) - \psi_f^{(2)}(x,y,\zeta) + F(x,\zeta)\, \frac{\partial \psi_f^{(1)}(x,y,\zeta)}{\partial x} = 0$$

and let $\mathscr{L}^\varepsilon$ of (3.5.10) act on $f(x,y) + \varepsilon\psi_f^{(1)}(x,y,\zeta) + \varepsilon^2\psi_f^{(2)}(x,y,\zeta)$ as usual. Upon collecting terms, using (3.5.12), (3.5.16), the smallness of S^ε and (3.5.14), (3.5.15) we arrive, after some additional simple manipulations, to the usual set-up of the proof of Theorem 3.1.

3.6 Interaction with ergodic driving.

Clearly one can consider combined forms of Theorem 2.1 and Theorem 3.1. The situation is somewhat analogous to the one of Section 2.8. We shall confine attention to one example.

Let $y(t)$ be Brownian motion on R^1 with $y(0) = y$ and let $z(t)$ be an ergodic Markov process on R^m with exponentially convergent recurrent potential kernel (cf. Section 2.1 where $z(t)$ corresponds to $Y(t)$). Let $F(x,y,z)$ be a smooth and bounded n-vector function on $R^n \times R \times R^m$ with compact support in y. Let

$$z^{\varepsilon}(t) = z(t/\varepsilon^2) \tag{3.6.1}$$

and define $x^{\varepsilon}(t)$ by

$$\frac{dx^{\varepsilon}(t)}{dt} = \frac{1}{\varepsilon} F(x^{\varepsilon}(t), \frac{y(t)}{\varepsilon}, z^{\varepsilon}(t)) , \qquad x^{\varepsilon}(0) = x . \tag{3.6.2}$$

In view of the Brownian scaling $y(t/\varepsilon^2) \sim y(t)/\varepsilon$. If $y^{\varepsilon}(t) = y(t/\varepsilon^2)$ we may write (3.6.2) as

$$\frac{dx^{\varepsilon}(t)}{dt} = \frac{1}{\varepsilon} F(x^{\varepsilon}(t), y^{\varepsilon}(t), z^{\varepsilon}(t)) , \qquad x^{\varepsilon}(0) = x . \tag{3.6.3}$$

Let

$$\bar{F}(x) = \int_{-\infty}^{\infty} dy \int_{R^m} \bar{P}(z) F(x,y,z) , \tag{3.6.4}$$

where $\bar{P}(A)$, $A \subset R^m$ is the invariant probability measure of $z(t)$. Let $\bar{x}(t)$ be defined by

$$(3.6.5) \qquad \frac{d\bar{x}(t)}{dt} = \bar{F}(\bar{x}(t)) , \qquad \bar{x}(0) = x ,$$

and let

$$(3.6.6) \qquad x(t) = \bar{x}(\ell_o(t))$$

where $\ell_o(t)$ is the local time of $y(t)$ at $y = 0$.

Theorem 3.5.

The process $x^\varepsilon(t)$ converges weakly on $C([0,T];R^n)$, $T < \infty$, to the process defined by (3.6.5) and (3.6.6) (relative to a Brownian motion $y(t)$).

The proof of this result is again entirely analogous to the one of Theorem 3.1. We omit further details.

Appendix. Ergodic Properties of $\mathscr{L}_{BL}$.

We shall discuss briefly here the hypotheses concerning ergodicity of the process generated by $\mathscr{L}_1$ in Section (2.2) ($\mathscr{L}_1$ is defined by (2.1.3)) and the process generated by $\mathscr{L}_{BL}$ in Section (2.5) (defined by (2.5.8)). These hypotheses were sufficient for the asymptotic analysis.

It is convenient to restrict attention to the case where the diffusion process Y(t) generated[†] by $\mathscr{L}_1$ is confined to a compact subset S of R^m. This can be done by imposing reflecting boundary conditions on ∂S, assumed smooth. In the sequel Y(t) will be this process with $\mathscr{L}_1$ its generate in the interior (the boundary conditions will not be introduced explicitly) and with transition function $P(t,y,A)$, $y \in S$, $A \subset S$.

A simple sufficient condition for ergodicity and existence of the recurrent potential (2.2.2) is the following well-known Doeblin's condition: there is a reference probability measure μ on S a $t_o > 0$ and a $c > 0$ such that for all Borel subsets A of S,

$$\text{(A.1)} \qquad P(t_o,y,A) \geq c\mu(A) , \qquad y \in S .$$

One easily deduces from this that there exists a unique invariant measure $\bar{P}(A)$ on S and that it is approached exponentially fast by $P(t,y,A)$, as $t \to \infty$, uniformly in y and A.

[†]Parametric dependence on other, frozen variables will not be shown in the Appendix.

A slightly different formulation of this result, useful in the context of differential equations and boundary value problems for $P(t,y,A)$ as here, is the following: $P(t,y,A)$ is uniformly Feller continuous i.e., for each $t > 0$

$$\text{(A.2)} \qquad \sup_{A \subset S} |P(t,y,A)-P(t,y',A)| \to 0 \quad \text{as } |y-y'| \to 0 ,$$

and for any x and y in S and some $t_o > 0$ the measures $P(t_o,x,\cdot)$ and $P(t_o,y,\cdot)$ are not orthogonal.

To see how this implies ergodicity with exponential approach let x and y be points in S and let $H_{x,y}$ be a positive Hahn decomposition of S for $P(t_o,x,\cdot) - P(t_o,y,\cdot)$ then,

$$\sup_{A \subset S} |P(t_o,x,A)-P(t_o,y,A)|$$

$$= P(t_o,x,H_{x,y}) - P(t_o,y,H_{x,y}) \equiv \rho(x,y)$$

and by the orthogonality hypothesis $\rho(x,y) < 1$ for each $x,y \in S$. By the uniform Feller property $\rho(x,y)$ is continuous on $S \times S$ which is compact. Hence

$$\sup_{(x,y) \in S\times S} \; \sup_{A \subset S} |P(t_o,x,A)-P(t_o,y,A) = \bar{\rho} < 1 ,$$

from which exponential ergodicity follows.

In the context of diffusions on S that are uniformly elliptic, the uniform Feller property follows from interior in t estimates of the y derivatives of $P(t,y,A)$ and the orthogonality property from the strong maximum principle.

Let us next consider the operator $\mathcal{L}_{BL}$ of (2.5.8) which we rewrite as follows.

$$(A.3) \qquad \mathcal{L}_{BL} = \mathcal{L}_1 + F(y)\frac{\partial}{\partial\eta} + \sum_{k=1}^{n} \bar{b}_k(y)\frac{\partial^2}{\partial\eta\partial y_k} + \frac{1}{2}\bar{a}(y)\frac{\partial^2}{\partial\eta^2} .$$

Here $\mathcal{L}_1$ is defined by (2.1.3) (with x-dependence omitted) and acts on functions on S; the reflecting boundary conditions are again not written explicitly. The coefficients F(y), $\bar{b}_k(y)$ and $\bar{a}(y)$ are identified from (2.5.8) and their form is not important here except that

$$(A.4) \qquad \int_S F(y)\bar{P}(dy) = 0 \qquad \text{(inherited from (2.2.3))}$$

and that they are smooth functions of y, independent of η, with $\bar{a}(y)$ strictly positive on S.

The generator $\mathcal{L}_{BL}$ above defines uniquely a process $(Y(t),H(t))$ on $S \times [-\infty,\infty]$, with Y(t) in S as before. Suppose that $H(0) = \eta > 0$ and let τ be the first time $H(0) = 0$. We want to show that τ is a finite random variable with probability one so that for $A \subset S$

$$(A.5) \qquad Q(\eta,y,A) = P\{Y(\tau) \in A | Y(0) = y,\ H(0) = \eta\} , \qquad \eta \geq 0 , \quad y \in S ,$$

is a well defined transition probability function. Moreover, Q satisfies[†]

[†]The reflecting boundary conditions on ∂S are again not shown.

(A.6) $$\mathscr{L}_{BL}Q = 0\ , \qquad \eta > 0\ , \qquad y \in S\ ,$$
$$Q(0,y,A) = \chi_A(y)\ .$$

Let $N > 0$ be an integer and let τ_N be the first time $H(t)$ equals N, starting from the left of N. The function[†]

(A.7) $$Q_N(\eta,y,A) = P_{\eta,y}\{Y(\tau) \in A,\ \tau < \tau_N\}$$

satisfies the boundary value problem

(A.8) $$\mathscr{L}_{BL}Q_N = 0\ , \qquad 0 < \eta < N\ , \qquad y \in S$$
$$Q_N(0,y,A) = \chi_A(y)\ , \qquad Q_N(N,y,A) = 0$$

as can be seen by Itô's formula applied to the smooth solution of (A.8) (the latter follows from results in differential equations). Now

$$Q(\eta,y,A) = Q_N(\eta,y,A)$$
$$+ E_{\eta,y}\{\chi_A(Y(\tau))\,|\,\tau \ge \tau_N\}P_{\eta,y}\{\tau \ge \tau_N\}$$

so that if we show that $P_{\eta,y}(\tau \ge \tau_N) \to 0$ as $N \to \infty$ then, we wil have (A.5) and (A.6) and $Q(\eta,y,A)$ will be, in fact a transition probability function on S for $\eta \ge 0$.

Since (A.4) holds (cf. (2.2.2))

(A.9) $$\psi_F(y) = \int \psi(y,dz)F(z)$$

[†] Subscripts denote the starting points as usual.

is well defined and

$$\text{(A.10)} \qquad \mathcal{L}_1 \psi_F + F = 0 .$$

It follows that $\eta + \psi_F(y)$ is a harmonic function for $\mathcal{L}_{BL}$

$$\text{(A.11)} \qquad \mathcal{L}_{BL}(\eta + \psi_F(y)) = 0 .$$

Thus,

$$E_{\eta,y}\{H(\tau \wedge \tau_N) + \psi_F(Y(\tau \wedge \tau_N))\} = \eta + \psi_F(y)$$

and also

$$NP_{\eta,y}\{\tau_N \geq \tau\} + E_{\eta,y}\{\psi_F(Y(\tau \wedge \tau_N))\} = \eta + \psi_F(y)$$

from which we conclude that $P_{\eta,y}\{\tau_N \geq \tau\} \to 0$ as $N \to \infty$.

Since the operator $\mathcal{L}_{BL}$ is uniformly elliptic, its coefficients are smooth and S is compact the previous considerations imply that $Q(\eta,y,A)$ is an ergodic transition function with a unique invariant measure $\bar{Q}(A)$ approached exponentially fast as $\eta \to \infty$ uniformly in $y \in S$ and $A \in S$. $\bar{Q}(A)$ is the limiting distribution of $Y(\tau)$ as $\eta \to \infty$.

The exponential ergodicity implies the validity of the Fredholm alternative for the Dirichlet problem

$$\text{(A.11)} \qquad \begin{cases} \mathcal{L}_{BL} u(\eta,y) = 0 , & \eta > 0 , \quad y \in S , \\ u(0,y) = g(y) , & y \in S \\ \lim\limits_{\eta\to\infty} u(\eta,y) = 0 . \end{cases}$$

Namely, (A.11) has a unique solution u provided

$$(A.12) \qquad \int_S g(y)\bar{Q}(dy) = 0$$

and u decays to zero exponentially fast as $\eta \to \infty$, uniformly in $y \in S$.

The Fredholm alternative for the Neumann problem

$$(A.13) \qquad \begin{cases} \mathcal{L}_{BL} v(\eta,y) = 0 , & \eta > 0 , \quad y \in S \\ \dfrac{\partial v}{\partial \eta}(0,y) = g(y) , & \\ \lim\limits_{\eta\to\infty} v(\eta,y) \to 0 , & \end{cases}$$

can be obtained simply[†] from (A.11). Indeed, if $u(\eta,y)$ is the solution of (A.11) then

$$(A.14) \qquad v(\eta,y) = -\int_\eta^\infty u(\zeta,y)\,d\zeta = \int_0^\infty u(\eta+\zeta,y)\,d\zeta ,$$

satisfies (A.13) so that (A.12) is its solvability condition.

The reason why (A.12) serves as solvability condition for both (A.11) and (A.13) can be understood better perhaps by noting that if $T_Q(\eta)$ denotes the semigroup generated by $Q(\eta,y,A)$ on continuous functions on S then, the infinitesimal generator of $T_Q(\eta)$ is the generator of the boundary process associated with $(Y(t),H^R(t))$, the reflected process at $\eta = 0$

[†] In the present special case of nonoblique derivative condition at $\eta = 0$.

corresponding to $(Y(t), H(t))$, on $S \times [0,\infty)$ (cf. [5]). This observation and the remarks concerning (2.5.13) and (2.5.14) explain the relation between (A.11) and (A.13).

References

[1] A. Khinchine, Asymptotische Gesetze der Wahrscheinlichkeitsrechnung, Springer, Berlin, 1933.

[2] Yu. V. Prohorov, Convergence of random processes and limit theorems, Theor. Probability Appl., 9 (1956), pp. 157-214.

[3] A. V. Skorohod, Studies in the theory of random processes, Addison Wesley, Reading, Mass., 1964.

[4] D. W. Stroock and S. R. S. Varadhan, Diffusion processes with continuous coefficients I, II, Comm. Pure Appl. Math., 22 (1969), pp. 345-400 and 479-530.

[5] D. W. Stroock and S. R. S. Varadhan, Diffusion processes with boundary conditions, Comm. Pure Appl. Math., 24 (1971), pp. 147-225.

[6] N. N. Bogoliubov and Ju. A. Mitropolski, Asymptotic methods in the theory of nonlinear oscillations, Gordon and Breach, New York, 1961.

[7] T. G. Kurtz, Semigroups of conditioned shifts and approximation of Markov processes, Annals of Probab., 3 (1975), pp. 618-642.

[8] R. L. Stratonovich, Topics in the theory of random noise, Vols. I, II, Gordon and Breach, New York, 1963.

[9] R. Z. Khasminskii, The averaging principle for parabolic and elliptic differential equations and Markov processes with small diffusion. Theor. Prob. Appl., 8 (1963), pp. 1-21.

[10] R. Z. Khasminskii, On stochastic processes defined by differential equations with a small parameter, Theor. Prob. Appl., 11 (1966), pp. 211-228.

[11] R. Z. Khasminskii, A limit theorem for solutions of differential equations with a random right hand side, Theor. Prob. Appl., 11 (1969), pp. 211-228.

[12] R. Z. Khasminskii, On the principle of averaging for Itô's stochastic differential equations, Kybernetica (Prague), 4 (1968), pp. 260-279.

[13] A. M. Ilin and R. Z. Khasminskii, On the equations of Brownian motion, Theor. Prob. Appl., 9 (1964), pp. 421-444.

[14] G. C. Papanicolaou and W. Kohler, Asymptotic theory of mixing stochastic ordinary differential equations, Comm. Pure and Appl. Math., 27 (1974), pp. 641-668.

[15] G. C. Papanicolaou, Asymptotic analysis of transport processes, Bull. A.M.S., 81 (1975), pp. 330-392.

[16] R. Hersh, Random evolutions: a survey of results and problems, Rocky Mountain Math. Journal, 14 (1974), pp. 443-496.

[17] M. Pinsky, Multiplicative operator functionals and their asymptotic properties, in Advances in Probability III, Marcel Dekker, New York, 1974.

[18] T. G. Kurtz, A limit theorem for perturbed operator semigroups with applications to random evolution, J. Funct. Analysis, 12 (1973), pp. 55-67.

[19] W. Kohler and G. C. Papanicolaou, Power statistics for waves in one dimension I, II, J. Math. Phys., 14 (1973), pp. 1733-1745 and 15 (1974), pp. 2186-2197.

[20] D. A. Darling and M. Kac, On occupation times for Markov processes, Trans. Amer. Math. Soc., 84 (1957), pp. 444-458.

[21] C. Stone, Limit theorems for random walks, birth and death processes and diffusion processes, Ill. J. Math., 7 (1963), pp. 638-660.

[22] Y. Kasahara, A limit theorem for occupation times for Markov processes, to appear.

[23] H. P. McKean, Stochastic Integrals, Academic Press, New York, 1968.

[24] I. I. Gihman and A. V. Skorohod, Stochastic Differential Equations, Springer, Berlin-Heidelberg-New York, 1972.

[25] V. M. Volosov, Averaging in systems or ordinary differential equations, Uspehi Mat. Nauk, 17 (1962), #6 (108), pp. 3-126 (Russian Math. Surveys, 17 (1962), pp. 1-126).

[26] A. N. Tichonov, Systems of differential equations containing a small parameter multiplying the highest derivatives, Mat. Sb., 73 (1952), pp. 575-585.

[27] I. Babuška, Several reports from the Department of Math., University of Maryland, College Park, Maryland.

[28] H. Kramers, Brownian motion in a field of force and the diffusion model of chemical reactors, Physica, 7 (1940), pp. 284-304.

[29] N. Bromberg, N.Y.U. dissertation, 1974.

[30] A. Bensoussan, J. L. Lions and G. C. Papanicolaou, Boundary layers and homogenization of transport processes, to appear.

[31] D. W. Stroock and S. R. S. Varadhan, On degenerate elliptic-parabolic operators of second order and their associated diffusions, Comm. Pure Appl. Math., 25 (1972), pp. 651-713.

[32] G. Blankenship and G. C. Papanicolaou, Stability and control of stochastic systems with wide-band noise disturbances I, II, SIAM J. Appl. Math., to appear.

[33] R. F. Anderson, Diffusions with second order boundary I, II, Indiana Univ. Math. J., 25, pp. 367-395 and 403-441.

[34] E. De Giorgi and S. Spagnolo, Sulla convergenza degli integrali dell' energia per operatsii ellittici del secondo ordine, Boll. U.M.I., 8 (1973), pp. 391-411.

[35] S. Spagnolo, Convergence in energy for elliptic operators, in Numerical solution of partial differential equations III, Academic Press, New York, 1976.

[36] A. Bensoussan, J. L. Lions, G. C. Papanicolaou, C. R. Acad. Sc. Paris, 281 (A) (1975), pp. 89-94 and pp. 317-322; also, 282 (A) (1976), pp. 143-147 and pp. 1277-1282.

[37] A. Bensoussan, J. L. Lions, G. C. Papanicolaou, Homogenization and ergodic theory, to appear.

[38] A. Bensoussan, J. L. Lions, G. C. Papanicolaou, book in preparation.

[39] D. Isaacson, N.Y.U. dissertation, 1976.

Commun. Math. Phys. 104, 1–19 (1986)

Communications in Mathematical Physics

Central Limit Theorem for Additive Functionals of Reversible Markov Processes and Applications to Simple Exclusions

C. Kipnis[1] and S. R. S. Varadhan[2]*

1 Centre de Mathematiques Appliqueés, Ecole Polytechnique, F-91128 Palaiseau, France
2 Courant Institute, New York University, New York 10012, USA

Abstract. We prove a functional central limit theorem for additive functionals of stationary reversible ergodic Markov chains under virtually no assumptions other than the necessary ones. We use these results to study the asymptotic behavior of a tagged particle in an infinite particle system performing simple excluded random walk.

Introduction

In a recent work Lebowitz and Spohn [4] proved that diffusion of color for "mechanically" identical particles with two colors as well as convergence to local equilibrium in the hydrodynamical limit were related to the asymptotics, after proper rescaling, of the movements of one or more tagged particles of the system.

Since for the moment purely mechanical systems seem to be out of reach, several models which are to lesser or greater extent stochastic have been proposed. In [3] Kipnis et al. considered the case of a one dimensional system of hard rods with stochastic collisions. In this paper we consider the case of the so-called simple exclusion process.

The intuitive description of the process is the following: Infinitely many particles move on $\mathbb{Z}^d$ according to a simple random walk with exponential (mean one) holding time at each site and jump law $p(x)$ which is symmetric, i.e. satisfies $p(x) = p(-x)$ for all x's. However if a particle attempts a transition to a site already occupied, the jump is suppressed.

The key remark is that, due to invariance of the mechanics under translations, the evolution of the rest of the medium *seen from an observer sitting on a tagged particle* follows a Markovian evolution (which possesses many reversible probabilities) and that the movement of the tagged particle (the observer) in the absolute frame is in a certain sense "driven" by this Markov process.

* Supported by NSF Grant MCS-8301364, ONR Contract N00014-81-K-0012 and a Fellowship from John S. Guggenheim Memorial Foundation

In Sect. 1 we prove an abstract theorem on weak convergence to Brownian motion of a process driven by a reversible Markov process.

In Sect. 2 we prove that our tagged particle satisfies the conditions described in Sect. 1 to conclude that if x_t denotes the position of the tagged particle then $\varepsilon x_{t\varepsilon^{-2}}$ converges to a Brownian motion with finite diffusion constant D. (Notice that before taking the limit the random process x_t is not Markovian due to the interactions!)

However it has been shown by Arratia [1] that when $d=1$ and $p(1)=p(-1)=1/2$, the diffusion constant D is equal to zero and the correct scaling needed to obtain a nontrivial limit is $\varepsilon x_{t\varepsilon^{-4}}$. We therefore prove that if $d \geqq 2$ or $d=1$ and $p(x)$ is concentrated on more than two points then D is strictly positive.

In Sect. 3. we prove that two different tagged particles become independent in the limit contrary to what happens in [3] where they remain correlated.

In Sect. 4. we apply our theorem to a jump process in random environment obtained by first choosing (randomly) rates for each bond of nearest-neighbor points of the lattice $\mathbb{Z}^d$ and letting the particle jump across a bond with the aforementioned rate.

1. Convergence to Brownian Motion of an Arbitrary Reversible Velocity Process

The question of proving a central limit theorem for the partial sums

$$X_n = \sum_{j=1}^{n} Z_j$$

of a sequence $Z_1, Z_2, \ldots, Z_n, \ldots$ of random variables has been studied exhaustively in probability theory for at least fifty years. We will limit our discussion here to the situation where $\{Z_j\}$ forms a stationary sequence in the strict sense. If the Z_j's are independent with mean zero and variance σ^2 then the classical central limit theorem asserts that the distribution of $X_n/\sqrt{n}$ converges to a normal distribution with mean 0 and variance σ^2. The invariance principle asserts that the distribution of the stochastic process $X_n(t) = 1/\sqrt{n}\, X_{[nt]}$ converges weakly in the Skorohod space to the distribution of a Brownian motion with variance σ^2.

These results have been generalized to the situation where the independence of the random variables $\{Z_j\}$ is replaced by asymptotic independence of one type or other. The assumptions are often difficult to check in specific circumstances.

One special situation where analogs of the classical results are valid without additional assumptions is the following:

Theorem 1.1. *Let* $\{Z_j\colon -\infty < j < \infty\}$ *be a stationary ergodic process such that* $E[Z_{n+1} | F_n] = 0$ a.e., *where* F_n *is the* σ*-field generated by* Z_j *for* $j \leqq n$. *For such a martingale difference sequence the distribution of* $X_n(t) = 1/\sqrt{n}\,[Z_1 + \cdots + Z_{[nt]}]$ *converges weakly to the distribution of Brownian motion with variance* σ^2 *provided* $EZ_n^2 = \sigma^2 < \infty$.

Proof. The above theorem is deduced easily from Theorem 3.2 of [2] where condition (b) can be checked easily in our situation.

Remark 1.2. If $\{Z_j\}$ is a vector valued sequence one has an identical theorem where one obtains in the limit a multidimensional Brownian motion with a covariance matrix identical to that of any Z_j.

Suppose $\{y_j\}$, $-\infty < j < \infty$ is a stationary Markov chain on a state space X and $Z_j = V(y_j)$. Let us suppose that $\pi(dx)$ is the common invariant distribution and $q(x, dy)$ is the transition probability of the chain. Let us assume that the chain is ergodic. We denote by the symbol q the operator $(qf)(x) = \int f(y)\, q(x, dy)$ as well. Suppose $V(x) = f(x) - (qf)(x)$ for some bounded function $f(x)$. Then $V(x)$ has necessarily mean 0 with respect to π and it is natural to ask if the central limit theorem is valid for

$$X_n = \sum_{j=1}^{n} V(y_j).$$

If we define

$$Z'_j = f(y_j) - (qf)(y_{j-1}),$$

then an elementary computation shows that

$$E[Z'_{n+1} | F_n] = E[f(y_{n+1}) | F_n] - (qf)(y_n) = (qf)(y_n) - (qf)(y_n) = 0.$$

Moreover if

$$X'_n = \sum_{j=1}^{n} Z'_j,$$

then

$$X_n = \sum_{j=1}^{n} f(y_j) - \sum_{j=1}^{n} (qf)(y_j) = \sum_{j=1}^{n} Z'_j + (qf)(y_0) - (qf)(y_n).$$

We write $X_n = X'_n + X''_n$, where $X''_n = (qf)(y_0) - (qf)(y_n)$. Since $X''_n/\sqrt{n}$ is negligible as $n \to \infty$, the central limit theorem for X_n follows from the central limit theorem for X'_n which is an application of Theorem 1.1.

Although the approach is quite elementary, when the underlying Markov chain $\{y_j\}$ is reversible the method is powerful enough to establish the central limit theorem and the invariance principle for the most general V that one can expect the theory to hold.

Let $q(x, dy)$ be a translation probability on a state space X with $\pi(dx)$ as a reversible stationary probability. We will assume that the stationary Markov process P with π as marginal distribution and $q(x, dy)$ with transition probability is ergodic. Let $V(y)$ be a function on X with $\int V^2(x)\, \pi(dx) < \infty$, and we will assume that $\int V(x)\, \pi(dx) = 0$ and

$$\lim_{n \to \infty} \frac{1}{n} E^P[(V(y_1) + \cdots + V(y_n))^2] = \sigma^2 < \infty.$$

If we denote by $e_V(d\lambda)$, the spectral measure of V corresponding to the selfadjoint operator q on $L_2(X, \pi)$, we obtain

$$\sigma^2 = \int \frac{1+\lambda}{1-\lambda} e_V(d\lambda) < \infty. \tag{1.1}$$

The condition that $\sigma^2 < \infty$ is therefore equivalent to the condition $\int 1/1-\lambda\, e_V(d\lambda) < \infty$ or $V \in \text{Range}(I-q)^{1/2}$. This can be verified by checking the inequality

$$|\int V(y)\phi(y)\pi(dy)| \leqq C(\int((I-q)\phi)(y)\phi(y)\pi(dy))^{1/2} \tag{1.2}$$

for all test functions ϕ in $L_2(X,\pi)$. If we introduce the inner products

$$\langle f_1, f_2 \rangle = \int f_1(y) f_2(y)\pi(dy),$$
$$\langle f_1, f_2 \rangle_1 = \int((I-q)f_1)(y) f_2(y)\pi(dy),$$

then our assumption can be restated as $\langle V, V\rangle_{-1} < \infty$, where $\langle\ \ \rangle_{-1}$ is dual to $\langle\ \ \rangle_1$ under the pairing $\langle\ \ \rangle$.

We denote by Ω the space of all X-valued sequences $\{y_j\}$, $-\infty < j < \infty$ and by F_n the σ-field generated by y_j for $j \leqq n$. We have the measure P on (Ω, F), where $F = \bigvee_n F_n$.

Theorem 1.3. *Let V in $L_2(x,\pi)$ satisfy condition (1.1). Then the sequence*

$$X_n = \sum_{j=1}^{n} V(y_j)$$

can be written as $X_n = M_n + \xi_n$, where M_n is a martingale relative to (Ω, F_n, P) and

$$\lim_{n\to\infty} \frac{1}{\sqrt{n}} \sup_{1\leqq j\leqq n} |\xi_j| = 0 \text{ in probability } (P). \tag{1.3}$$

Moreover

$$\lim_{n\to\infty} \frac{1}{n} E^P|\xi_n|^2 = 0. \tag{1.4}$$

Proof. We denote by u_ε the solution of the equation

$$(1+\varepsilon)u_\varepsilon - qu_\varepsilon = V \tag{1.5}$$

for $\varepsilon > 0$. We will investigate the behavior of u_ε as $\varepsilon \to 0$. We can rewrite (1.5) as

$$u_\varepsilon - qu_\varepsilon = V - \varepsilon u_\varepsilon. \tag{1.6}$$

Therefore for every $n \geqq 0$,

$$E^P\{u_\varepsilon(y_{n+1}) - u_\varepsilon(y_n) - \varepsilon u_\varepsilon(y_n) + V(y_n)|F_n\} = 0.$$

If we define

$$M_n^\varepsilon = \sum_{j=0}^{n-1} [u_\varepsilon(y_{j+1}) - u_\varepsilon(y_j) + V_\varepsilon(y_j)],$$

where $V_\varepsilon = V - \varepsilon u_\varepsilon$, then for each $\varepsilon > 0$, M_n^ε if a martingale relative to (Ω, F_n, P). We can now express for each $\varepsilon > 0$, $X_n = M_n^\varepsilon + \xi_n^\varepsilon + \eta_n^\varepsilon$, where

$$\xi_n^\varepsilon = -\sum_{j=0}^{n-1} [u_\varepsilon(y_{j+1}) - u_\varepsilon(y_j)],$$

and

$$\eta_n^\varepsilon = \sum_{j=0}^{n-1} (\varepsilon u_\varepsilon)(y_j).$$

We shall now establish that for each $n \geqq 1$,

$$\lim_{\varepsilon \to 0} M_n^\varepsilon = M_n \text{ exists in } L_2(\Omega, P), \tag{1.7}$$

$$\lim_{\varepsilon \to 0} \xi_n^\varepsilon = \xi_n \text{ exists in } L_2(\Omega, P), \tag{1.8}$$

$$\lim_{\varepsilon \to 0} \eta_n^\varepsilon = 0 \text{ in } L_2(\Omega, P). \tag{1.9}$$

It follows then that $X_n = M_n + \xi_n$. We will then establish (1.3) and (1.4) for the sequence $\{\xi_n\}$. Since M_n^ε is a martingale with stationary increments in order to show that M_n^ε has a limit in $L_2(\Omega, P)$, it is sufficient to check that

$$\lim_{\varepsilon \to 0} \{u_\varepsilon(y_2) - u_\varepsilon(y_1) + V_\varepsilon(y_1\} \text{ exists in } L_2(\Omega, P). \tag{1.10}$$

From relation (1.6) the above limit is of the form

$$\lim_{\varepsilon \to 0} \{u_\varepsilon(y_2) - (qu_\varepsilon)(y_1)\}.$$

An easy computation yields

$$E^P[f(y_2) - (qf)(y_1)]^2 = \langle f, (I - q^2)f \rangle.$$

Therefore we need to check only that

$$\lim_{\substack{\varepsilon_1 \to 0 \\ \varepsilon_2 \to 0}} \langle u_{\varepsilon_1} - u_{\varepsilon_2}, (I - q^2)(u_{\varepsilon_1} - u_{\varepsilon_2}) \rangle = 0. \tag{1.11}$$

Equation (1.11) can be calculated in terms of spectral measure $e_V(d\lambda)$ and the formula reads

$$\begin{aligned} \langle u_{\varepsilon_1} - u_{\varepsilon_2}, (I - q^2)(u_{\varepsilon_1} - u_{\varepsilon_2}) \rangle &= \int_{-1}^{1} (1 - \lambda^2)\left(\frac{1}{1 + \varepsilon_1 - \lambda} - \frac{1}{1 + \varepsilon_2 - \lambda}\right)^2 e_V(d\lambda) \\ &= \int_{-1}^{1} \frac{(\varepsilon_2 - \varepsilon_1)^2(1 - \lambda^2)}{(1 + \varepsilon_1 - \lambda)^2(1 + \varepsilon_2 - \lambda)^2} e_V(d\lambda). \end{aligned}$$

If we assume that $\varepsilon_2 \geqq \varepsilon_1 > 0$ the integrand is dominated by $(\varepsilon_2^2(1-\lambda^2)/(1-\lambda)^2\varepsilon_2^2) = (1 + \lambda)/(1 - \lambda)$. Since our basic hypothesis is that $\int (1 + \lambda)/(1 - \lambda)\, e_V(d\lambda) < \infty$, we have (1.7) by the dominated convergence theorem. By a similar argument we also obtain

$$\lim_{\substack{\varepsilon_1 \to 0 \\ \varepsilon_2 \to 0}} \langle u_{\varepsilon_1} - u_{\varepsilon_2}, (I - q)(u_{\varepsilon_1} - u_{\varepsilon_2}) \rangle = 0. \tag{1.11a}$$

A similar calculation also yields

$$\lim_{\varepsilon\to 0} \langle u_\varepsilon, (I-q^2)u_\varepsilon\rangle = \int \frac{1+\lambda}{1-\lambda} e_V(d\lambda) = \sigma^2.$$

Equation (1.9) involves proving

$$\lim_{\varepsilon\to 0} \varepsilon \| u_\varepsilon \| = 0.$$

We will actually prove the stronger version,

$$\lim_{\varepsilon\to 0} \varepsilon \langle u_\varepsilon, u_\varepsilon\rangle = 0. \tag{1.12}$$

Again by spectral calculation

$$\varepsilon\langle u_\varepsilon, u_\varepsilon\rangle = \int_{-1}^{1} \frac{\varepsilon}{(1+\varepsilon-\lambda)^2} e_V(d\lambda).$$

The integrand goes to zero and is dominated by $1/(1-\lambda)$, which is integrable with respect to $e_V(d\lambda)$. Since $X_n = M_n^\varepsilon + \xi_n^\varepsilon + \eta_n^\varepsilon$ and is independent of $\varepsilon > 0$, (1.7) and (1.9) clearly imply (1.8).

We now concentrate on proving (1.3) and (1.4) for the sequence $\{\xi_n\}$. We first prove (1.4). Clearly for every $\varepsilon > 0$,

$$\xi_n = M_n^\varepsilon - M_n + \xi_n^\varepsilon + \eta_n^\varepsilon.$$

Since $M_n^\varepsilon - M_n$ is a martingale with stationary increments,

$$\begin{aligned}\frac{1}{n}E^P|\xi_n|^2 &\leqq \frac{3}{n}E^P|M_n^\varepsilon - M_n|^2 + \frac{3}{n}E^P|\xi_n^\varepsilon|^2 + \frac{3}{n}E^P|\eta_n^\varepsilon|^2 \\ &= 3E^P|M_1^\varepsilon - M_1|^2 + \frac{3}{n}E^P|\xi_n^\varepsilon|^2 + \frac{3}{n}E^P|\eta_n^\varepsilon|^2.\end{aligned}$$

We have

$$\lim_{\varepsilon\to 0} E^P|M_1^\varepsilon - M_1|^2 = 0.$$

Therefore it is enough to choose $\varepsilon = 1/n$ and then show that

$$\lim_{n\to\infty} \frac{1}{n}E^P|\xi_n^{1/n}|^2 = 0 \quad \text{and} \quad \lim_{n\to\infty}\frac{1}{n}E^P|\eta_n^{1/n}|^2 = 0.$$

Clearly

$$\begin{aligned}E^P|\xi_n^{1/n}|^2 &= E^P|u_{1/n}(y_n) - u_{1/n}(y_0)|^2 \leqq 4E^P|u_{1/n}(y_0)|^2 \\ &= 4\langle u_{1/n}, u_{1/n}\rangle = o(n) \quad \text{by (1.12).}\end{aligned}$$

Moreover

$$\begin{aligned}E^P|\eta_n^{1/n}|^2 &= E^P\left[\frac{1}{n}\sum_{j=0}^{n-1} u_{1/n}(y_j)\right]^2 \leqq E^P|u_{1/n}(y_0)|^2 \\ &= \langle u_{1/n}, u_{1/n}\rangle = o(n) \quad \text{again by (1.12).}\end{aligned}$$

Now we turn to (1.3). We need Lemma 1.4 which we will use but prove after completing the argument;

$$\sup_{1\leqq j\leqq n} |\xi_j| \leqq \sup_{1\leqq j\leqq n} |M_j^{1/n} - M_j| + \sup_{1\leqq j\leqq n} |\xi_j^{1/n}| + \sup_{1\leqq j\leqq n} |\eta_j^{1/n}|.$$

Now by Doob's inequality

$$P[\sup_{1\leqq j\leqq n} |M_j^{1/n} - M_j| \geqq \delta\sqrt{n}] \leqq \frac{1}{n\delta^2} E^P[M_n^{1/n} - M_n]^2$$
$$= \frac{1}{\delta^2} E^P[M_1^{1/n} - M_1]^2 = o(1) \quad \text{as} \quad n\to\infty.$$

$$P[\sup_{1\leqq j\leqq n} |\eta_j^{1/n}| \geqq \delta\sqrt{n}] \leqq P\left[\frac{1}{n}\sum_0^{n-1} |u_{1/n}(y_j)| \geqq \delta\sqrt{n}\right]$$
$$\leqq \frac{1}{n\delta^2} E^P\left[\frac{1}{n}\sum_0^{n-1} |u_{1/n}(y_j)|\right]^2 \leqq \frac{1}{n\delta^2}\langle u_{1/n}, u_{1/n}\rangle$$
$$= o(1) \quad \text{as} \quad n\to\infty.$$

Finally

$$\sup_{1\leqq j\leqq n} |\xi_j^{1/n}| = \sup_{1\leqq j\leqq n} |u_{1/n}(y_j) - u_{1/n}(y_0)| \leqq \sup_{1\leqq j\leqq n} |u_{1/n}(y_j)| + |u_{1/n}(y_0)|,$$

$$P[|u_{1/n}(y_0)| \geqq \delta\sqrt{n}] \leqq \frac{1}{n\delta^2} E^P[u_{1/n}(y_0)]^2 = \frac{1}{n\delta^2}\langle u_{1/n}, u_{1/n}\rangle$$
$$= o(1) \quad \text{as} \quad n\to\infty.$$

As for $\sup_{1\leqq j\leqq n} |u_{1/n}(y_j)|$, we write

$$\sup_{1\leqq j\leqq n} |u_{1/n}(y_j)| \leqq \sup_{1\leqq j\leqq n} |u_\varepsilon(y_j) - u_{1/n}(y_j)| + \sup_{1\leqq j\leqq n} |u_\varepsilon(y_j)|$$

for some $\varepsilon > 0$.

$$P[\sup_{1\leqq j\leqq n} |u_\varepsilon(y_j)| \geqq \delta\sqrt{n}] \leqq nP[|u_\varepsilon(y_0)| \geqq \delta\sqrt{n}] = o(1),$$

because $E^P|u_\varepsilon(y_0)|^2 < \infty$.

We use Lemma 1.4 to estimate

$$P[\sup|u_\varepsilon(y_j) - u_{1/n}(y_j)| \geqq \delta\sqrt{n}]$$

$$\leqq \frac{3}{\delta\sqrt{n}}(n\langle u_\varepsilon - u_{1/n}, (I-q)(u_\varepsilon - u_{1/n})\rangle + \langle u_\varepsilon - u_{1/n}, u_\varepsilon - u_{1/n}\rangle)^{1/2},$$

letting $n\to\infty$,

$$\limsup_{n\to\infty} P[\sup_{1\leqq j\leqq n} |u_\varepsilon(y_j) - u_{1/n}(y_j)| \geqq \delta\sqrt{n}]$$
$$\leqq \frac{3}{\delta}\lim_{n\to\infty} \langle u_\varepsilon - u_{1/n}, (I-q)(u_\varepsilon - u_{1/n})\rangle^{1/2}.$$

If we let $\varepsilon \to 0$, we obtain

$$\lim_{\varepsilon\to 0} \limsup_{n\to\infty} P[\sup_{1\leqq j\leqq n} |u_\varepsilon(y_j) - u_{1/n}(y_j)| \geqq \delta\sqrt{n}]$$
$$\leqq \frac{3}{\delta} \lim_{\varepsilon\to 0} \lim_{n\to\infty} \langle u_\varepsilon - u_{1/n}, (I-q)(u_\varepsilon - u_{1/n})\rangle$$
$$= 0 \text{ by (1.1a)},$$

and we are done.

Lemma 1.4. *Let $f(x)$ be an arbitrary function in $L_2(X,\pi)$. Let*

$$\langle f,f\rangle = a, \quad \langle f,(I-q)f\rangle = b.$$

Then

$$P\{\sup_{0\leqq j\leqq n} |f(y_j)| \geqq l\} \leqq \frac{3}{l}\sqrt{a+nb}. \tag{1.13}$$

Proof. Let $G = \{x: |f(x)| \geqq l\}$ and let τ be the hitting time of G, i.e. $\tau = \{\inf j: y_j \in G\}$. Then

$$P\{\sup_{0\leqq j\leqq n} |f(y_j)| \geqq l\} = P\{\tau \leqq n\}.$$

Let us consider the function $\phi_\rho(x) = E^{P_x}\{\rho^\tau\}$, where P_x is the Markov measure on the space of sequences $\{y_j\}$ for $j \geqq 0$ initialized to start from x at time 0. Then

$$P\{\tau \leqq y\} \leqq \rho^{-n} \int \phi_\rho(x)\pi(dx) \leqq \rho^{-n}[\int \phi_\rho^2(x)\pi(dx)]^{1/2}.$$

To estimate the last term we note that

$$\phi_\rho = \rho q \phi_\rho \text{ on } G^c, \quad \phi_\rho = 1 \text{ on } G.$$

By the Dirichlet principle

$$\int \phi_\rho^2(x) d\pi \leqq \langle \phi_\rho, \phi_\rho\rangle + \frac{\rho}{1-\rho}\langle \phi_\rho, (I-q)\phi_\phi\rangle \leqq \langle \omega,\omega\rangle + \frac{\rho}{1-\rho}\langle \omega,(I-q)\omega\rangle$$

for any ω which is 1 on G. The function $(|f(x)| \wedge l/l) = \bar\omega$ is an admissible choice;

$$\langle \bar\omega,\bar\omega\rangle \leqq \frac{1}{l^2}\langle f,f\rangle = \frac{a}{l^2}, \quad \langle \bar\omega,(I-q)\bar\omega\rangle \leqq \frac{1}{l^2}\langle f,(I-q)f\rangle = \frac{b}{l^2}.$$

Therefore

$$P[\tau \leqq n] \leqq \rho^{-n}\left(\frac{a}{l^2} + \frac{\rho}{1-\rho}\frac{b}{l^2}\right)^{1/2}.$$

Taking $\rho = e^{-1/n}$ we obtain (1.13).

Corollary 1.5. *If V satisfies (1.1), then the distribution of $X_n(t) = 1/\sqrt{n}\ X_{[nt]} = (1/\sqrt{n}) \sum_{j=1}^{[nt]} V(y_j)$ under P converges to Brownian motion with variance σ^2 weakly in the Skorohod space on any finite interval. Moreover σ^2 is given by (1.1).*

Remark 1.6. If V is vector valued then essentially the same result is valid. The vector valued process $X_n(t)$ converges weakly to a multidimensional Brownian motion with the corresponding covariance matrix.

Remark 1.7. If we replace P by P_x the Markov measure starting from x at time 0, then the distribution of $X_n(t)$ under P_x will again converge to the correct Brownian motion but the convergence is in measure as functions of x relative to the measure π. The question of almost sure convergence is open.

In the case of continuous time Markov processes we have the transition probabilities $p(t, x, dy)$ and an invariant probability measure $\pi(dx)$ with respect to which the process is assumed to be reversible and ergodic. We also have the infinitesimal generator L of the process. L is self adjoint on $L_2(\pi)$ with a nonpositive spectrum and a simple eigenvalue of 0, constants being the eigenspace corresponding to that eigenvalue.

If one is interested in proving a central limit theorem for an additive functional of the form

$$X(t) = \int_0^t V(y(s))ds,$$

where $\int V\, d\pi = 0$, then

$$\begin{aligned}\lim_{t\to\infty} \frac{1}{t} E^P[X^2(t)] &= 2\int_0^\infty E^P[V(y(s))V(y(0))]ds = 2\int_0^\infty \langle T_s V, V\rangle ds \\ &= 2\langle -L^{-1}V, V\rangle,\end{aligned}$$

where T_s is the semigroup induced by the transition probabilities.

The natural assumption on V then is that $V \in D((-L)^{-1/2})$ or equivalently an estimate of the form

$$|\langle V, \phi\rangle| \leqq c\langle -L\phi, \phi\rangle^{1/2} \tag{1.14}$$

for all ϕ in the domain of L. We will assume that V in $L_2(\pi)$ satisfies (1.14). When we say domain of L we mean the domain as the generator of a semigroup in $L_2(\pi)$.

We now state the analog of Theorem 1.3.

Theorem 1.8. *Let $y(t)$ be a Markov process, reversible with respect to a probability measure π, and let us suppose that the reversible stationary process P with π as invariant measure is ergodic. Let V be a function on the state space in $L_2(\pi)$ satisfying $\int V\, d\pi = 0$ and condition (1.14). Let F_t be the σ-field generated by the process up to time t. There exists a square integrable Martingale M_t relative to (Ω, F_t, P) for $f \geqq 0$ such that M_t has stationary increments and*

$$\lim_{t\to\infty} \frac{1}{\sqrt{t}} \sup_{0\leqq s\leqq t} |x(s) - M_s| = 0$$

in probability with respect to P, where $X(0) = M_0 = 0$ and

$$X(t) = \int_0^t V(y(s))ds.$$

Moreover

$$\lim_{t\to\infty} \frac{1}{t} E^P |X(t) - M_t|^2 = 0.$$

Corollary 1.9. *It follows now from results is Helland [2] that $1/\sqrt{\lambda}\, X(\lambda t)$ satisfies a functional central limit theorem relative to P and the limiting variance σ^2 is given by*

$$\sigma^2 = \lim_{t\to\infty} \frac{1}{t} EX^2(t) = 2 \int_0^\infty \langle T_t V, V\rangle dt = 2\langle -L^{-1}V, V\rangle.$$

Remark 1.10. If we denote by P_x the Markov process starting from a point x in the state space and by $Q_{\lambda,x}$ the measure on the space of continuous functions induced by $1/\sqrt{\lambda}\, X(\lambda t)$ from P_x and by Q the Brownian motion measure with variance σ^2 on the same function space, then we have besides $\int Q_{\lambda,x}\pi(dx) \Rightarrow Q$ actually $Q_{\lambda,x} \Rightarrow Q$ in measure with respect to π. The question of almost sure convergence is open.

Remark 1.11. If V is a vector of functions one has analogous results regarding convergence to the corresponding vector Brownian motion with appropriate covariance matrix. The basic approximation in Theorem 1.8 can be done for each component and in [2] the multidimensional version of the central limit theorem for vector martingales can be found.

Comments on Proofs. The proof follows the discrete time situation very closely. The only tricky point is the analog of Lemma 1.4. Since the $X(t)$ process is almost surely continuous and M_t can be taken to be almost surely right continuous, one replaces the supremum in

$$\sup_{0\le s\le t} |X(s) - M_s|$$

by a supremum over diadic points in $[0, t]$. Denoting by D_N the set of diadic points of the form $j/2^N$, it is clearly sufficient to prove

Lemma 1.12. *Let f satisfy $\langle f, f\rangle = a$, $\langle f, -Lf\rangle = b$. Then for every t, N and l*

$$P[\sup_{0\le s\le t, s\in D_N} |f(y(s)| \ge l] \le \frac{3}{l}\sqrt{a+tb}.$$

Proof. If we denote by $q_N(x, dy)$ the transition probability $p(1/2^N, x, dy)$ and by q_N the operator $T_{1/2^N}$, then by the spectral theorem,

$$\langle f, (I - q_N) f\rangle = \int (1 - e^{-\lambda/2^N}) E_f(dy) \le \frac{1}{2^N} \langle f, -Lf\rangle = \frac{b}{2^N}.$$

We now apply Lemma 1.4 for the discretely sampled process with $n = [t2^N]$, and we obtain the validity of the lemma.

2. Asymptotics for a Single Test Particle in Simple Exclusion

In this section we study the behavior of one test particle in the symmetric simple exclusion process. We will denote by $\xi \in \{0, 1\}^{\mathbb{Z}^d}$ the state of such a process

$$\xi(u)=\begin{cases}1 & \text{if the site } u \text{ has a particle}\\ 0 & \text{if the site } u \text{ is free.}\end{cases}$$

For any state ξ and any two sites u, v we denote by $\xi^{u,v}$ the new state obtained by

$$\xi^{u,v}(z)=\begin{matrix}\xi(z) & \text{if} & z\neq u \text{ or } v\\ \xi(u) & \text{if} & z=v\\ \xi(v) & \text{if} & z=u.\end{matrix}$$

The evolution of the process $\xi(t,\cdot)$ is governed by its infinitesimal generator L

$$(LF)=\sum_{u\neq v}[F(\xi^{u,v})-F(\xi)]p(u-v). \tag{2.1}$$

The summation here is over unordered pairs (u,v) with $u\neq v$. For the process $\xi(t,\cdot)$ the Bernoulli measures P_θ, where θ is the probability that any site is occupied (independently of all the other sites) are reversible ergodic measures provided the random walk $p(x-y)$ is irreducible.

If we have a single test particle in the system, it is more convenient to describe the system in terms of the location x of the test particle and the configuration η of the entire system seen from the test particle:

$$\eta(u)=\xi(x+u)\quad\text{for}\quad u\neq 0, \tag{2.2}$$

and the state space for the system is $x\in\mathbb{Z}^d$ and $\eta\in\{0,1\}^{\mathbb{Z}^d-\{0\}}$. The infinitesimal generator for the $(x,\eta)_t$ process is given by

$$\begin{aligned}(\tilde{L}F)(x,\eta)=&\sum_{z\neq 0}p(z)(1-\eta(z))[F(x+z,\tau_{-z}\eta)-F(x,\eta)]\\ &+\sum_{u\neq v\neq 0}[F(x,\eta^{u,v})-F(x,\eta)]p(u-v),\end{aligned} \tag{2.3}$$

where

$$\begin{aligned}(\tau_{-z}\eta)(u)&=\eta(u+z)\quad\text{for}\quad u\neq -z\\ &=0\quad\text{for}\quad u=-z.\end{aligned} \tag{2.4}$$

From the form of (2.3) we see that η by itself is a Markov process with generator

$$\begin{aligned}(L_0F)(\eta)&=\sum_z p(z)(1-\eta(z))[F(\tau_{-z}\eta)-F(\eta)]+\sum_{u\neq v\neq 0}[F(\eta^{u,v})-F(\eta)]p(u-v)\\ &=(L_1F)(\eta)+(L_2F)(\eta).\end{aligned} \tag{2.5}$$

The invariant measures for the η process will turn out to be Bernoulli product measures P_θ with common probability θ, and these will be ergodic and reversible for L_0. [Although for the η process there is no site at 0 we will continue to denote by P_θ the Bernoulli measure over all the other sites.]

Lemma 2.1. The η_t process is reversible and ergodic with respect to P_θ.

Proof. We need only check that

$$\int fL_0g\,dP_\theta=\int gL_0f\,dP_\theta,$$

where f and g are of the form

$$f(\eta)=\prod_{u\in A}\eta(u),\quad g(\eta)=\prod_{u\in B}\eta(u),$$

where $A,\ B\subset \mathbb{Z}^d-\{0\}$.

From the form of the generator (2.5) we need only check that

$$\int f(\eta^{u,v})g(\eta)dP_\theta=\int g(\eta^{u,v})f(\eta)dP_\theta, \tag{2.6}$$

$$\int f(\eta)g(\tau_{-z}\eta)(1-\eta(z))dP_\theta=\int g(\eta)f(\tau_z\eta)(1-\eta(-z))dP_\theta. \tag{2.7}$$

Equation (2.6) follows from the invariance of P_θ under the map $\eta\to\eta^{u,v}$ and (2.7) follows from the fact that $\eta\to\tau_z\eta$ maps $(1-\eta(z))dP_\theta$ to $(1-\eta(-z))dP_\theta$. It is easy to see that P_θ is reversible and ergodic for L_2. Since any harmonic function ϕ such that $(L_1+L_2)\phi=0$ has the property $\int\phi L_1\phi dP_\theta=-\int\phi L_2\phi\, dP_\theta$ and $\int\phi L_i\phi\, dP_\theta\leqq 0$ for $i=1,\ 2$ we have $\int\phi L_1\phi dP_\theta=\int\phi L_2\phi dP_\theta=0$. But L_2 is reversible and ergodic. Therefore ϕ is a constant almost surely with respect to P_θ. This completes the proof of the lemma.

The x_t process is driven by the η_t process. If we pick a vector l and consider $(x\cdot l)$, then

$$L_0(x\cdot l)=\sum(z\cdot l)p(z)(1-\eta(z))=\phi_l(\eta),$$

so that

$$(x_t\cdot l)-\int_0^t\phi_l(\eta_s)ds=N_t^l \tag{2.7}$$

is a Martingale relative to the natural σ-fields. We now apply Theorem 1.3 to $\phi_l(\eta)$ and write

$$\int_0^t\phi_l(\eta_s)ds=M_t^l+E_t^l,$$

where M_t^l is another martingale and E_t^l is negligible. Taking l to be the various basis vectors, we can approximate each component of x_t by a sum of two martingales. The central limit theorem now follows from what we said earlier. We need to show that the conditions of Theorem 1.3 apply and that the limiting covariance of the sum M_t+N_t is nondegenerate. We assume that

$$\sum|z|^2p(z)<\infty. \tag{2.8}$$

Lemma 2.2. *For any l the function*

$$\phi_l(\eta)=\sum(z\cdot l)(1-\eta(z))p(z)$$

satisfies the condition

$$\left|\int\phi_l(\eta)F(\eta)dP_\theta\right|\leqq C\left(-\int(L_0F)(\eta)F(\eta)dP_\theta\right)^{1/2}.$$

Proof. We can take F to be a function depending on a finite number of coordinates and then

$$\begin{aligned}\int \phi_l(\eta)F(\eta)dP_\theta &= \int(\sum(z,l)(1-\eta(z))p(z))F(\eta)dP_\theta \\ &= \tfrac{1}{2}\int(\sum_z(z\cdot l)[(1-\eta(z))-(1-\eta(-z))]F(\eta)p(z))dP_\theta \\ &= \tfrac{1}{2}\int\sum_z(z\cdot l)(1-\eta(z))[F(\tau_{-z}\eta)-F(\eta)]p(z)dP_\theta \\ &\leqq \tfrac{1}{2}(\sum p(z)(z\cdot l)^2)^{1/2}\,(\int\sum(1-\eta(z))[F(\tau_{-z}\eta)-F(\eta)]^2p(z)dP_\theta)^{1/2} \\ &\leqq C(\int(-L_1F)(\eta)F(\eta)dP_\theta)^{1/2} \\ &\leqq C(-\int(L_0F)(\eta)F(\eta)dP_\theta)^{1/2}. \end{aligned} \tag{2.9}$$

For the next lemma which is required to prove the nondegeneracy of the limiting covariance we will assume that under the random walk $p(u-v)$ one can get from any nonzero site to any other nonzero site without going through the origin. Notice that this rules out only the nearest neighbor random walk in one dimension.

Lemma 2.3. *Under the above assumption we have an estimate of the form*

$$|\int\phi_l(\eta)F(\eta)dP_\theta| \leqq C(-\int(L_2F)(\eta)F(\eta)dP_\theta)^{1/2}.$$

Proof. It is clearly sufficient to prove (see (2.9))

$$|\int[(1-\eta(u))-(1-\eta(v))]F(\eta)dP_\theta| \leqq C|u-v|(-\int(L_2F)(\eta)F(\eta)dP_\theta)^{1/2}.$$

In view of the triangle inequality is to enough to establish

$$\sup_{|u-v|\leqq k}|\int[(1-\eta(u))-(1-\eta(v))]F(\eta)dP_\theta| \leqq C(-\int(L_2F)(\eta)F(\eta)dP_\theta)^{1/2} \tag{2.10}$$

The estimate is going to be derived through a random walk path from u to v that avoids zero. Since for a fixed path of finite length, translations will produce only at most a finite number that will go through zero the supremum in (2.10) comes for free provided we establish (2.10) for each fixed u and v (with "translation invariant" C, see below). Let

$$u=z_0,\quad z_1,\dots z_n=v$$

be a path that avoids 0, and let $p(z_i-z_{i-1})>0$ for $i=1,2,\dots,n$. Again because of the triangle inequality we can concentrate on

$$\begin{aligned}&|\int[(1-\eta(z_i))-(1-\eta(z_{i-1}))]F(\eta)dP_\theta| = |\int(1-\eta(z_{i-1}))(F(\eta^{z_{i-1},z_i})-F(\eta))dP_\theta| \\ &\leqq(\int[F(\eta^{z_{i-1},z_i})-F(\eta)]^2dP_\theta)^{1/2} \\ &\leqq\left(\frac{1}{p(z_i-z_{i-1})}\right)^{1/2}(-\int(L_2F)(\eta)F(\eta)dP_\theta)^{1/2}. \end{aligned}$$

□

We are finally ready to state and prove our main theorem: Let us consider a test particle in simple exclusion where the underlying random walk probabilities $p(u-v)$ are assumed to be symmetric and satisfying $\sum' |z|^2\, p(z)<\infty$. Moreover the random walk is assumed to be irreducible and *not* the one dimensional nearest neighbor random walk. The test particle is assumed to start at 0. The rest of the configuration is distributed according to the invariant Bernoulli product measure P_θ. Then

Theorem 2.4. *The distribution of* $(1/\sqrt{\lambda})x_{\lambda t}$ *converges to a nondegenerate diffusion with covariance* σ *in the Skorohod space. Moreover the covariance* σ *is also the limit of the covariances of* $(1/\sqrt{\lambda})x_\lambda$ as $\lambda \to \infty$.

Proof. In view of the earlier comments we need only prove nondegeneracy. There are many martingales associated with the $(x, \eta)_t$ process;

$$v_t^{u,v} = \sum_{s \leq t} \chi_{(\eta_{s-} = \eta_s^{u,v})} |\eta_s(u) - \eta_s(v)| - \int_0^t |\eta_s(u) - \eta_s(v)| - \int_0^t |\eta_s(u) - \eta_s(v)| p(u-v) ds,$$

$$v_t^z = \sum_{s \leq t} \chi_{(\eta_s = \tau_{-z}\eta_{s-})} - \int_0^t p(z)(1 - \eta_s(z)) ds$$

are basic martingales, and any other martingale is expressible in terms of these. For instance N_t given by (2.7) can be expressed as

$$N_t = \sum_{z \neq 0} z v_t^z,$$

For any function F the martingale

$$M_t^F = F(\eta(t) - F(\eta(0)) - \int_0^t (L_0 F)(\eta(s)) ds$$

can be expressed as

$$M_t^F = \int_{u,v \neq 0} \int_0^t [F(\eta^{u,v}(s)) - F(\eta(s))] dv_s^{u,v} + \sum_z \int_0^t [F(\tau_{-z}\eta(s)) - F(\eta(s))] dv_s^z.$$

Moreover the martingales $\{v_t^z v_t^{u,v}\}$ are mutually orthogonal in the sense that the product of any two distinct ones is again a martingale. For any l the martingale $\langle N_t, l \rangle$ is of the form

$$\langle N_t, l \rangle = \sum_{z \neq 0} (z, l) v_t^z,$$

and $\langle M_t, l \rangle$ of the decomposition

$$\int_0^t \phi_l(\eta_s) ds = M_t^l + E_t^l$$

comes from $M_t^{F_\lambda}$ as $\lambda \to 0$, where $F_\lambda = (\lambda I - L_0)^{-1} \phi_l$. From the orthogonality of these martingales it is clear that the nondegeneracy of $M_t + N_t$ is assured if we establish that M_t should involve at least some $v_t^{u,v}$ component so that it cannot cancel N_t. In other words we need to prove

$$\lim_{\lambda \to 0} E^{P_\theta} \{ \sum_{u \neq v \neq 0} p(u-v) [F_\lambda(\eta^{u,v}(s)) - F_\lambda(\eta(s)]^2 \} \neq 0,$$

i.e.

$$\lim_{\lambda \to 0} \langle -L_2(\lambda I - L_0)^{-1} \phi_l, \ (\lambda I - L_0)^{-1} \phi_l \rangle \neq 0.$$

Lemma 2.3 assures us that

$$|\langle\phi_l,\psi\rangle|\leqq C\langle -L_2\psi,\psi\rangle^{1/2}$$

or

$$\langle -L_2^{-1}\phi_l,\phi_l\rangle\leqq C.$$

We can estimate

$$\begin{aligned}\langle\phi_l,(\lambda I-L_0)^{-1}\phi_l\rangle &= \langle(-L_2^{-1/2}\phi_l,(-L_2)^{1/2}(\lambda I-L_0)^{-1}\phi_l\rangle\\ &\leqq\langle -L_2^{-1}\phi_l,\phi_l\rangle^{1/2}\langle -L_2(\lambda I-L_0)^{-1}\phi_l,(\lambda I-L_0)^{-1}\phi_l\rangle\\ &\leqq C\langle -L_2(\lambda I-L_0)^{-1}\phi_l,(\lambda I-L_0)^{-1}\phi_l\rangle.\end{aligned}$$

Since $\langle\phi_l,(\lambda I-L_0)^{-1}\phi_l\rangle$ can go to zero only if $\phi_l=0$, we are done.

Corollary 2.5. *Suppose the initial configuration in the sites $u\neq 0$ is different from P_θ but is some measure $Q\ll P_\theta$. Then the basic central limit theorem is still valid and the limiting distribution is the same as before. If dQ/dP_θ is bounded, then the covariance of $(1/\sqrt{\lambda})x_\lambda$ converges to the covariance of the limiting Brownian motion.* The proof follows easily from Remark 1.10. The boundedness of dQ/dP_θ guarantees that the error is still negligible in the sense of mean square.

3. Asymptotic Independence of Two Test Particles

In this section we study the behavior of two test particles located initially at points x and y.

Theorem 3.1. *The scaled positions of two test particles converge jointly to two independent Brownian motions.*

Proof. First of all we note that we are working with an initial distribution conditioned to having sites x and y occupied. Since the conditioning set has nonzero measure by Corollary 2.5, the two individual positions converge to Brownian motions separately and only asymptotic independence has to be proved. In view of the results in Helland [2] in order to prove that two Martingales satisfying the central limit theorem separately converge jointly, after rescaling, to independent Brownian motions we need only show that

$$\lim_{t\to\infty}\frac{1}{t}EM_{i,t}^{(1)}M_{j,t}^{(2)}=0,$$

where $M_{i,t}^{(j)}$ is the ith component of the jth martingale. By standard formula one can compute

$$EM_{1t}^{(1)}M_{jt}^{(2)}=E\int_0^t\psi_{ij}(s)ds$$

for some suitable function $\psi_{ij}(s)$.

We will carry out these computations in the next several lemmas and establish that $E\psi_{ij}(s)\to 0$ as $s\to\infty$. This will prove the result.

First some notation. The probability that a typical site is occupied is of course θ and is fixed. We will not specify it hereafter. We have the initial measures P, P_x, P_y,

P_{xy} corresponding to no conditions, conditioned for site x, site y and both sites x and y to be occupied respectively at time 0. We also have the measures for the entire process ξ_t of the infinite particle system corresponding to these initial conditions which we denote by Q, Q_x, Q_y and Q_{xy} respectively.

Lemma 3.2. *The positions x_t and y_t of two tagged particles starting from x and y at time $t=0$ can be represented as*

$$x_t = x + M_t^{(1)} + E_t^{(1)}, \quad y_t = y + M_t^{(2)} + E_t^{(2)},$$

where $M_t^{(1)}$ and $M_t^{(2)}$ are martingales relative to Q_x and Q_y respectively. They are both martingales relative to $Q_{x,y}$. Further $E_t^{(1)}$, $E_t^{(2)}$ are negligible relative to $Q_{x,y}$ in the mean square sense.

Proof. The representation of x_t relative to Q_x and y_t relative to Q_y follow from Sect. 2. Since all martingale properties are valid relative to almost all initial starting points and $P_{x,y} \ll P_x$ as well as $P_{x,y} \ll P_y$, the martingale property holds true for $Q_{x,y}$ as well. Moreover $dQ_{x,y}/dQ_x$ and $dQ_{x,y}/dQ_y$ are bounded. So all the negligibility results are valid in the mean square sense as well.

Let us look at the complete class of martingales for the ξ_t process for any $a, b \in \mathbb{Z}^d$ and $a \neq b$:

$$n(t,a,b) = N_t(a,b) - \int_0^t p(b-a)\xi_s(a)(1-\xi_s(b))\,ds$$

is a martingale where $N_t(a,b)$ is the number of transitions in time $[0,t]$ from site a to site b. Since any martingale is expressible in terms of these basic ones

$$M_{i,t}^{(1)} = \sum_{a \neq b} \int_0^t \phi_{i,a,b}^{(1)}(s)\,dn(s,a,b) \tag{3.1}$$

and

$$M_{i,t}^{(2)} = \sum_{a \neq b} \int_0^t \phi_{i,a,b}^{(2)}(s)\,dn(s,a,b). \tag{3.2}$$

Lemma 3.3. *For each $t>0$*

$$E^{Q_{x,y}} M_{i,t}^{(1)} M_{j,t}^{(2)} = E^{Q_{x,y}} \int_0^t \psi_{ij}(s)\,ds,$$

where

$$\psi_{ij}(s) = \sum_{a \neq b} \phi_{i,a,b}^{(1)}(s)\phi_{j,a,b}^{(2)}(s)\eta_s(a)(1-\eta_s(b))p(b-a).$$

Proof. Follows from the fact that unless $a=a'$ and $b=b'$ the martingales $n(t,a,b)$ and $n(t,a',b')$ are orthogonal and

$$En^2(t,a,b) - \int_0^t \eta_s(a)(1-\eta_s(b))p(b-a)\,ds$$

is a martingale.

Lemma 3.4. *For any* $L < \infty$,

$$\lim_{t\to\infty} Q_{x,y}[|x(t)-y(t)| \leqq L] = 0.$$

Proof. Let us consider the system consisting of the first tagged particle being the origin of reference and the rest of the world including the second tagged particle being viewed from the first tagged particle. For this system there is a σ-finite invariant measure with mass one corresponding to each location of the second tagged particle. Given the location of the second tagged particle the rest of the universe (excluding also the origin) is provided a Bernoulli product measure. One checks that except for the one dimensional nearest neighbor model the system is ergodic. Moreover the system is reversible so that if T_t is the evolution semigroup for the system $\langle f, T_t g\rangle \to 0$ by the spectral theorem provided f and g are square integrable with respect to the invariant measure. Picking $g = \chi_{\{|u|\leqq L\}}(\cdot)$ and $f = \chi_y$, i.e. the indicators of the sets where the second tagged particle is within a distance L of the origin and at y respectively we obtain our result.

Lemma 3.5.

$$\lim_{t\to\infty} E^{Q_{x,y}}\psi_{ij}(t) = 0.$$

Proof. We use the simple inequality

$$|\psi_{ij}(t)| \leqq \chi E_1(t)\cap E_2(t) \neq \phi \cdot [Q_i^{(1)}(t)]^{1/2}[Q_j^{(2)}(t)]^{1/2} + [Q_i^{(1)}(t)]^{1/2}[Q_{j,E_2^c(t)}^{(2)}(t)]^{1/2} + [Q_{i,E_1^c(t)}^{(1)}(t)]^{1/2}[Q_j^{(2)}(t)]^{1/2}, \tag{3.3}$$

where $L < \infty$ is fixed, and

$$E_1(t) = \{(u,v)\in \mathbb{Z}^d \times \mathbb{Z}^d,\ u \neq v \text{ and } \max[|u-x_t|, |v-x_t|] \leqq L\},$$

$$E_2(t) = \{(u,v)\in \mathbb{Z}^d \times \mathbb{Z}^d,\ u \neq v \text{ and } \max[|u-y_t|, |v-y_t|] \leqq L\},$$

$$Q_i^{(r)}(t) = \sum_{u,v} |\phi_{i,u,v}^{(r)}(t)|^2 \eta_t(u)(1-\eta_t(v))p(v-u),$$

$$Q_{i,A}^{(r)}(t) \sum_{(u,v)\in A} |\phi_{i,u,v}^{(r)}(t)|^2 \eta_t(u)(1-\eta_t(v))p(v-u).$$

We write for each $L < \infty$

$$E^{Q_{x,y}}|\psi_{ij}(t)| \leqq a_L(t) + b_L(t) + c_L(t),$$

corresponding to the estimate (3.3), which we think of as

$$\psi_{ij}(t) \leqq T_{1,L}(t) + T_{2,L}(t) + T_{3,L}(t).$$

Step 1.

$$a_L(t) = E^{Q_{x,y}} T_{1,L}(t) \leqq \int_{|x_t-y_t|\leqq 2L} [Q_i^{(1)}(t)Q_j^{(2)}(t)]^{1/2} dQ_{x,y}$$

$$\leqq \Big(\int_{|x_t-y_t|\leqq 2L} Q_i^{(1)}(t)dQ_{x,y}\Big)^{1/2}\Big(\int_{|x_t-y_t|\leqq 2L} Q_j^{(2)}(t)dQ_{x,y}\Big)^{1/2}.$$

To show that $a_L(t)\to 0$ as $t\to 0$, it suffices to know that $Q_i^{(r)}(t)$ are uniformly integrable with respect to $Q_{x,y}$. But $dQ_{x,y}/dQ_x$ and $dQ_{x,y}/dQ_y$ are bounded and $Q_i^{(1)}(t)$ and $Q_i^{(2)}(t)$ are stationary processes relative to Q_x and Q_y respectively.

Step 2.

$$\begin{aligned} b_L(t) = E^{Q_{x,y}}T_{2,L}(t) &\leq (E^{Q_{x,y}}[Q_i^{(1)}(t)])^{1/2}(E^{Q_{x,y}}[Q^{(2)}_{j,E_2^\varepsilon(t)}(t)])^{1/2} \\ &\leq C(E^{Q_x}[Q_i^{(1)}(t)])^{1/2}(E^{Q_y}[Q^{(2)}_{j,E_2^\varepsilon(t)}(t)]). \end{aligned}$$

$E^{Q_x}[Q_i^{(1)}(t)]$ is independent of t and so is $E^{Q_y}[Q^{(2)}_{j,E_2^\varepsilon(t)}(t)]$. Moreover the last quantity tends to zero as $L\to\infty$, and we are done.

Step 3. The estimate for $c_L(t)$ is identical to step 2 with the roles of the two tagged particles reversed.

4. Application to a Bond Diffusion Process

In this section we want to study the movement of a particle in a random environment. Consider for each bond b of Z^d a positive random variable a_b and suppose that these random variables are i.i.d.. Once this random environment is chosen, let a particle evolve according to the jump process with generator $Lf(x)=\sum_{y\sim x} a_{yx}[f(y)-f(x)]$, where $y\sim x$ means that y and x are nearest-neighbors and a_{yx} is the rate of the corresponding bond. We also assume that the a_b's have a uniform upper and lower bound. Call x_t the position of the particle at time t, and $\varepsilon x_{t/\varepsilon^2}$ the rescaled position. Then

Theorem 4.1. *After rescaling the position of the particle converges to Brownian motion.*

Proof. The only quantities that matter in this problem are the values of the rates given to each bond (which we call an environment).

It is also clear that the process ξ_t of the environment seen from the position of the particle at time t is a Markov process whose generator is

$$\tilde{L}f(\xi)=\sum_e a_e(\xi)[f(\tau_e\xi)-f(\xi)].$$

Therefore to prove our central limit theorem we only have to check that, for any $\vec{\alpha}\in R^d$, the function

$$\phi(\xi)=\sum a_e(\xi)(\vec{e}\cdot\vec{\alpha})$$

satisfies the condition of Theorem 1.3, because we easily check that the original product measure (say $m(d\xi)$) is reversible for this process;

$$\begin{aligned} \int a_e(\xi)(e\cdot\alpha)g(\xi)m(d\xi) &= \int a_{-e}(\tau_e\xi)(\vec{e}\cdot\vec{\alpha})g(\xi)m(d\xi) \\ &= -\int a_{-e}(\eta)(-\vec{e}\cdot\vec{\alpha})g(\tau_{-e}\eta)m(d\eta). \end{aligned}$$

Therefore

$$\sum_e \int g(\xi)a_e(\xi)(\vec{e}\cdot\vec{\alpha})m(d\xi)=\tfrac{1}{2}\int\sum_e a_e(\xi)(\vec{e}\cdot\vec{\alpha})(f(\tau_e\xi)-f(\xi))m(d\xi)$$

$$\leq \tfrac{1}{2}\Big[\sum_e \int a_e(\xi)(\vec{e}\cdot\vec{\alpha})^2 m(d\xi)\Big]\Big[\sum_e \int a_e(\xi)(f(\tau_e\xi)-f(\xi))^2 m(d\xi)\Big],$$

which is the bound that we were looking for.

Remark 4.2. With some additional work one can prove the nondegeneracy of the limiting Brownian motion.

Acknowledgements. C. K. acknowledges the warm hospitality of the CIMS where this work was done. The authors wish to thank S. Goldstein and H. Spohn for valuable discussions as well as detailed comments on an earlier version of the paper which are responsible for this improved version.

References

1. Arratia, R.: The motion of a tagged particle in the simple symmetric exclusion system in Z. Ann. Probab. **11** 362–373 (1983)
2. Helland, I.: Central limit theorems for martingales with discrete or continuous time. Scand. J. Stat., **9**, 79–94 (1982)
3. Kipnis, C., Lebowitz, J. L., Presutti, E., Spohn, H.: J. Stat. Phys. **30**, 107–121 (1983)
4. Lebowitz, J. L., Spohn, H.: J. Stat. Phys. **28**, 539–556 (1982)

Communicated by J. L. Lebowitz

Received August 10, 1984; in revised form July 2, 1985

Adv. Appl. Prob. **21**, 842–860 (1989)
Printed in N. Ireland

BOUNDING FUNCTIONS OF MARKOV PROCESSES AND THE SHORTEST QUEUE PROBLEM

JOHN A. GUBNER,* *University of Maryland*
B. GOPINATH,** *Bell Communications Research*
S. R. S. VARADHAN,*** *Courant Institute of Mathematical Sciences*

Abstract

We prove a theorem which can be used to show that the expectation of a non-negative function of the state of a time-homogeneous Markov process is uniformly bounded in time. This is reminiscent of the classical theory of non-negative supermartingales, except that our analog of the supermartingale inequality need not hold almost surely. Consequently, the theorem is suitable for establishing the stability of systems that evolve in a stabilizing mode in most states, though from certain states they may jump to a less stable state. We use this theorem to show that 'joining the shortest queue' can bound the expected sum of the squares of the differences between all pairs among N queues, *even under arbitrarily heavy traffic.*

STABILITY; LYAPUNOV FUNCTION; SUPERMARTINGALE; LOAD BALANCING; PARALLEL QUEUES

1. Introduction

Controlled Markov processes often arise when modeling the behavior of asynchronous algorithms, queuing systems, and computer and communication networks. In these models it is often the case that the control policy considers only the immediate future of the process in an attempt to stabilize it. While in general a myopic policy will not guarantee the long-term stability of the process, in this paper we shall give sufficient conditions on the one-step behavior of a process to guarantee its long-term stability.

Suppose we are given a time-homogeneous Markov process, $\{X_n\}$. Consider an analysis in which we choose a non-negative function u such that if the expected value of $u(X_n)$ is suitably bounded, then the process is stable in some sense. For example, in Section 3 our process will be the state of a system of N parallel queues. Our control policy will direct the traffic arriving at time $n+1$ to the queue whose

Received 10 December 1987; revision received 21 October 1988.

* Present address: Department of Electrical and Computer Engineering, University of Wisconsin, Madison, WI 53706, USA.

** Present address: Department of Electrical Engineering, Rutgers University, New Brunswick, NJ 08903, USA.

*** Postal address: Courant Institute of Mathematical Sciences, New York University, 251 Mercer Street, New York, NY 10012, USA.

This work was supported by Bell Communications Research. The research of the first author was also supported in part by the University of Maryland Systems Research Center under NSF Grant OIR-85-00108.

length was shortest at time n. We shall show that this policy can stabilize the system in the sense that the expected sum of the squares of the differences between all pairs among the N queues is uniformly bounded for all time.

To motivate our results, consider the following example of a single-server queue. Let $\{Y_n, n=0, 1, 2, \cdots\}$ be a discrete-time Markov chain whose transition probabilities are as follows. For $i=0$,

$$\mathbb{P}(Y_{n+1}=j \mid Y_n=i)=\begin{cases}\lambda, & \text{if } j=1,\\ 1-\lambda, & \text{if } j=0,\end{cases}$$

and for $i\geqq 1$,

$$\mathbb{P}(Y_{n+1}=j \mid Y_n=i)=\begin{cases}\lambda, & \text{if } j=i+1,\\ 1-(\lambda+\mu), & \text{if } j=i,\\ \mu, & \text{if } j=i-1,\end{cases}$$

where $0\leqq\lambda<\mu\leqq 1$ and $\lambda+\mu\leqq 1$. Our concern here is the long-term behavior of $\mathbb{E}[Y_n \mid Y_0=i]$. In particular, we should like to know if

$$\limsup_{n\to\infty} \mathbb{E}[Y_n \mid Y_0=i]<\infty.$$

In this simple example, since $\lambda<\mu$, it is very easy to compute the stationary distribution (see Billingsley (1979), Theorem 8.6, p. 106),

$$\lim_{n\to\infty} \mathbb{P}(Y_n=j \mid Y_0=i)=\left(\frac{\lambda}{\mu}\right)^j\left(1-\frac{\lambda}{\mu}\right).$$

Recalling that for $|z|<1$, $\sum_{j=1}^{\infty} jz^{j-1}=(1-z)^{-2}$, it is not hard to compute that indeed,

$$\lim_{n\to\infty} \mathbb{E}[Y_n \mid Y_0=i]=\frac{\lambda}{\mu-\lambda}<\infty,$$

assuming that we can take the limit inside the expectation.

How could we study the behavior of $\mathbb{E}[Y_n \mid Y_0=i]$ if the stationary distribution were not available to us? Recall that in the field of differential equations, a 'Lyapunov function' can sometimes be used to study the behavior of solutions to a differential equation without actually solving it. In this spirit, our paper generalizes the preceding example and gives conditions on the one-step behavior of a process that will yield useful information about its long-run (and short-run) behavior.

We now make a precise statement of our model. Let $\{X_n, n=0, 1, 2, \cdots\}$ be a time-homogeneous Markov process defined on a probability space $(\Omega, \mathcal{F}, \mathbb{P})$ and taking values in a measurable space $(\mathbb{X}, \mathcal{X})$. Let $u:\mathbb{X}\to[0,\infty)$ be a given measurable function. (Here $[0,\infty)$ is equipped with the usual Borel σ-field.) For $x\in\mathbb{X}$, let

(1.1) $$\hat{u}_n(x)\triangleq\mathbb{E}[u(X_n) \mid X_0=x], \qquad n=0, 1, 2, \cdots.$$

In this paper we give sufficient conditions to answer the following question: 'Does there exist a positive, finite constant D such that

$$\limsup_{n\to\infty} \hat{u}_n(x) \leqq D$$

for all $x \in \mathbb{X}$?' To answer this question, we construct a sequence of functions $\{u_n\}$ such that for all $n \geqq 0$,

$$\hat{u}_n(x) \leqq u_n(x), \qquad \forall x \in \mathbb{X}, \tag{1.2}$$

and such that $\{u_n\}$ is easy to analyze. To suggest how one might establish (1.2), consider the following result.

Proposition 1. *Set* $u_0(x) \triangleq u(x)$, *and let* $\{u_n\}_{n\geqq 1}$ *be given. If one can show that for all* $n \geqq 0$, *and all* $x \in \mathbb{X}$,

$$\mathbb{E}[u_n(X_1) \mid X_0 = x] \leqq u_{n+1}(x), \tag{1.3}$$

then $\hat{u}_n(x) \leqq u_n(x)$ *for all* $n \geqq 0$, *and all* $x \in \mathbb{X}$.

Proof. We proceed by induction. Clearly,

$$\hat{u}_0(x) \triangleq \mathbb{E}[u(X_0) \mid X_0 = x] = u(x) = u_0(x),$$

and so (1.2) holds for $n = 0$. Suppose (1.2) holds for some $n \geqq 0$. Then

$$\begin{aligned}
\hat{u}_{n+1}(x) &\triangleq \mathbb{E}[u(X_{n+1}) \mid X_0 = x] \\
&= \mathbb{E}[\mathbb{E}[u(X_{n+1}) \mid X_1, X_0] \mid X_0 = x] \\
&= \mathbb{E}[\mathbb{E}[u(X_{n+1}) \mid X_1] \mid X_0 = x], \quad \text{by the Markov property,} \\
&= \mathbb{E}[\hat{u}_n(X_1) \mid X_0 = x], \quad \text{by time homogeneity,} \\
&\leqq \mathbb{E}[u_n(X_1) \mid X_0 = x], \quad \text{by the induction hypothesis,} \\
&\leqq u_{n+1}(x), \quad \text{by (1.3).}
\end{aligned}$$

The difficulty is to establish conditions under which (1.3) holds. If we let $p(x, B)$ denote a regular version of $\mathbb{P}(X_1 \in B \mid X_0 = x)$, then we can rewrite (1.3) as

$$\int_{\mathbb{X}} u_n(y) p(x, dy) \leqq u_{n+1}(x). \tag{1.4}$$

Remark. The careful reader will see later that since $u_n(x)$ will have the form $u_n(x) = f_n(u(x))$, in order to derive the results in this paper, we do not need a regular version of $\mathbb{P}(X_1 \in B \mid X_0 = x)$, but only a regular version of $\mathbb{P}(u(X_1) \in C \mid X_0 = x)$. Since u is real valued, such a regular version always exists; see Billingsley (1979), Theorem 33.3, p. 390.

In Section 2 we give sufficient conditions under which (1.4) will hold. These conditions are summarized in Theorem 2. In Section 3 we describe the problem of 'joining the shortest queue'. We then apply Theorem 2 to show that the servers are equally loaded, even in heavy traffic. In Section 4 we prove Theorem 2, in Section 5 we present an extension of Theorem 2, and in Section 6 we compare Theorem 2 to a result of Hajek (1982).

2. Statement of main result

As pointed out in Section 1, the problem of finding a sequence $\{u_n\}$ satisfying (1.2) can be reduced to the problem of finding a sequence satisfying (1.4). We now present our main result, Theorem 2, which gives conditions under which this can be done. The proof is given in Section 4.

Theorem 2. *Fix* $b \geqq 0$ *and* $0 < \varepsilon < 1$. *Suppose that there exist positive constants* C *and* v *satisfying*

$$\frac{(b+1)(b+2)}{2}\frac{(1+\varepsilon)^b}{(1-\varepsilon)^{b+3}}C \leqq 2v, \tag{2.1}$$

and such that for some L,

$$|u(y)-u(x)| \leqq \varepsilon u(x),\ p(x,\cdot)\text{-a.s.},\quad \textit{if}\quad u(x) > L, \tag{2.2}$$

$$\mathbb{E}[u(X_{n+1})-u(x)\mid X_n = x] \leqq -v,\quad \textit{if}\quad u(x) > L, \tag{2.3}$$

and

$$\mathbb{E}[|u(X_{n+1})-u(x)|^2 \mid X_n = x] \leqq Cu(x),\quad \textit{if}\quad u(x) > L. \tag{2.4}$$

If one can find a finite constant A, *depending on* b *and* L, *such that*

$$\mathbb{E}[u(X_{n+1})^{b+2}\mid X_n = x]^{1/(b+2)} \leqq A,\quad \textit{if}\quad 0 \leqq u(x) \leqq L, \tag{2.5}$$

then the sequence $\{u_n\}$ *given by*

$$u_n(x) \triangleq \frac{u(x)^{b+2}}{\left[u(x)+\dfrac{v}{b+1}n\right]^{b+1}} + A + \frac{A^{b+2}}{\left(\dfrac{v}{b+1}\right)^{b+1}}\sum_{k=1}^{n-1}\frac{1}{k^{b+1}}, \tag{2.6}$$

satisfies (1.4), *and so by Proposition* 1,

$$\mathbb{E}[u(X_n)\mid X_0 = x] \leqq u_n(x),\qquad n \geqq 0,\ x \in \mathbb{X}. \tag{2.7}$$

Of course, when $n=0$, *Equation* (2.6) *is to be interpreted as* $u_0(x) \triangleq u(x)$, *and when* $n=1$, *it is to be interpreted as*

$$u_1(x) \triangleq \frac{u(x)^{b+2}}{\left[u(x)+\dfrac{v}{b+1}\right]^{b+1}} + A.$$

Remark. When $b = 1$, (2.6) becomes

$$u_n(x) = \frac{u(x)^3}{\left[u(x) + \frac{v}{2}n\right]^2} + A + 4\frac{A^3}{v^2}\sum_{k=1}^{n-1}\frac{1}{k^2}. \tag{2.8}$$

In fact, since $\sum_{k=1}^{\infty} k^{-2} = \pi^2/6$ (see Papadimitriou (1973) for an elementary proof of this fact; see Stark (1974) for more general series), it is easy to see that

$$\lim_{n\to\infty} u_n(x) = A\left(1 + \frac{2}{3}\left(\frac{\pi A}{v}\right)^2\right), \tag{2.9}$$

and that

$$u_n(x) \leqq u(x) + A\left(1 + \frac{2}{3}\left(\frac{\pi A}{v}\right)^2\right). \tag{2.10}$$

Note that the limit in (2.9) is independent of x, whereas the bound in (2.10) is independent of n.

Remark. The inequality (2.3) is very similar to the definition of a Markovian supermartingale. To see this, recall that if $\mathcal{F}_n \triangleq \sigma(X_0, \cdots, X_n)$, then $\{u(X_n)\}$ is an $\{\mathcal{F}_n\}$-supermartingale if and only if for all $n \geqq 0$,

$$\mathbb{E}[u(X_{n+1}) \mid \mathcal{F}_n] \leqq u(X_n). \tag{2.11}$$

Since $\{X_n\}$ is Markovian, this can be rewritten as

$$\mathbb{E}[u(X_{n+1}) \mid X_n] \leqq u(X_n),$$

or

$$\mathbb{E}[u(X_{n+1}) - u(x) \mid X_n = x] \leqq 0, \qquad \forall x \in \mathbb{X}. \tag{2.12}$$

To compare (2.3) with (2.12), observe that (2.3) is a stronger inequality, but holds only when $u(x) > L$. On the other hand, (2.12) is a weaker inequality, but holds for *every* $x \in \mathbb{X}$. Now, observe that if (2.11) were to hold, then we could immediately write

$$\mathbb{E}[u(X_{n+1}) \mid X_0] \leqq \mathbb{E}[u(X_n) \mid X_0],$$

from which it would follow by induction that for all $n \geqq 0$,

$$\mathbb{E}[u(X_n) \mid X_0] \leqq u(X_0),$$

which we can rewrite as

$$\mathbb{E}[u(X_n) \mid X_0 = x] \leqq u(x).$$

Compare this with the conclusion of Theorem 2 when $b = 1$; substituting (2.10) into (2.7) yields

$$\mathbb{E}[u(X_n) \mid X_0 = x] \leqq u(x) + A\left(1 + \frac{2}{3}\left(\frac{\pi A}{v}\right)^2\right).$$

Example. Let us apply Theorem 2 to the single-server queue described in Section 1. Since Y_n takes values in $[0, \infty)$, we take $u(Y_n) = Y_n$. Now, set $b = 1$ and fix $0 < \varepsilon < 1$. Make the following observations. First, it is easy to calculate

$$\mathbb{E}[Y_{n+1} \mid Y_n = i] = \begin{cases} \lambda, & \text{if } i = 0, \\ i - (\mu - \lambda), & \text{if } i > 0; \end{cases}$$

rewrite this as

$$\mathbb{E}[Y_{n+1} - i \mid Y_n = i] = \begin{cases} \lambda, & \text{if } i = 0, \\ -(\mu - \lambda), & \text{if } i > 0. \end{cases} \tag{2.13}$$

Clearly, we should take $\nu = \mu - \lambda > 0$. Set

$$C = \tfrac{2}{3}\nu \frac{(1-\varepsilon)^4}{(1+\varepsilon)}.$$

Now note that whenever $Y_n = i$,

$$|Y_{n+1} - i| \leqq 1, \tag{2.14}$$

and so

$$\mathbb{E}[|Y_{n+1} - i|^2 \mid Y_n = i] \leqq 1. \tag{2.15}$$

To put (2.14) and (2.15) into the form of (2.2) and (2.4), we can require that

$$1 \leqq \varepsilon i \quad \text{and} \quad 1 \leqq Ci.$$

This will happen whenever

$$i \geqq \max\left\{\frac{1}{\varepsilon}, \frac{3(1+\varepsilon)}{2\nu(1-\varepsilon)^4}\right\}. \tag{2.16}$$

Let L denote the right-hand side of (2.16). Finally, note that since (2.14) implies

$$Y_{n+1} \leqq i + 1,$$

we have

$$\begin{aligned} \mathbb{E}[Y_{n+1}^3 \mid Y_n = i]^{\frac{1}{3}} &\leqq i + 1 \\ &\leqq L + 1, \quad \text{if } i \leqq L. \end{aligned}$$

Hence, we may take $A = L + 1$. It should now be clear that Theorem 2 applies to the single-server queue if $\mu > \lambda$. In this case,

$$\limsup_{n\to\infty} \mathbb{E}[Y_n \mid Y_0 = i] \leqq (L+1) + \frac{2}{3}\left(\frac{\pi}{\mu - \lambda}\right)^2 (L+1)^3 < \infty.$$

We remark here that the preceding bound depends on L, which by (2.16), depends on ε. To reduce L, we should try to choose $0 < \varepsilon < 1$ so that

$$\frac{1}{\varepsilon} = \frac{3(1+\varepsilon)}{2\nu(1-\varepsilon)^4},$$

or equivalently,

$$2\nu(1-\varepsilon)^4 - 3\varepsilon(1+\varepsilon) = 0. \tag{2.17}$$

In other words, to reduce L, we must find the zeros of a fourth-degree polynomial in ε. Observe that there is at least one solution of (2.17) in $(0, 1)$.

3. Joining the shortest queue

In this section we investigate the balancing of N queues to N identical servers under a range of traffic conditions. Consider a system of N identical servers, each with its own infinite waiting room, indexed by $i = 1, \cdots, N$. Suppose also that we are given $N+1$ arrival processes indexed by $i = 1, \cdots, N, N+1$. Customers from arrival processes $i = 1, \cdots, N$ are assigned to the corresponding waiting room for server i. Customers from the $(N+1)$th arrival process may in general be assigned to any waiting room; however, we consider only what happens when we employ the policy of assigning these customers to the waiting room with the fewest waiting customers. In case of a tie, the waiting room of smallest index is selected.

Let $Q_n^{(i)}$ denote the size of the queue for server i at time $n = 0, 1, 2, \cdots$. This includes customers in the waiting room as well as the customer being served. Let

$$Q_n \triangleq (Q_n^{(1)}, \cdots, Q_n^{(N)}).$$

We want to show that the policy of joining the shortest queue will ensure, even under arbitrarily heavy traffic conditions, that for all $i \neq j$,

$$\mathbb{E}[|Q_n^{(j)} - Q_n^{(i)}| \mid Q_0 = q] \tag{3.1}$$

is uniformly bounded for all time. To apply the theory of Section 2, we need to find a suitable function u. Below we show that the function defined in (3.2) will suffice.

Before proceeding, we point out that the continuous-time analog of this model has been studied by a number of authors, though, to our knowledge, none has investigated the behavior of the continuous-time analog of (3.1). For example, Haight (1958), Kingman (1961), Flatto and McKean (1977), and Conolly (1984) characterized the stationary distribution of the state of the system. Foschini and Salz (1978) studied a diffusion approximation. Blanc (1987) outlined a numerical method for calculating the state probabilities and moments of the queue-length distribution. Brumelle (1971) bounded the expected waiting time. Wolff (1977) bounded moments of the delay distribution (see also Wolff (1987) for corrections and comments). Halfin (1985) bounded the probability distribution of the number of customers in the system, and its expected value in equilibrium. Winston (1977) and Weber (1978) proved that the policy of 'joining the shortest queue' is optimal in the sense of maximizing the customer throughput. Ephremides et al. (1980) proved that this policy is also optimal in the sense of minimizing the expected total time for completing service on all customers which arrive before a fixed time.

We now make a precise statement of our model. Using the N-fold cartesian product of the non-negative integers as the state space for $\{Q_n\}$, we take $\{Q_n\}$ to be a Markov process with transition probabilities defined as follows. Suppose that

$$Q_n = q = (q^{(1)}, \cdots, q^{(N)}),$$

where the $q^{(\cdot)}$ are non-negative integers. Let $\mu, \lambda, \lambda_1, \cdots, \lambda_N$ be a sequence of elements from the interval $[0, 1]$ such that $\lambda + \sum_{j=1}^{N} \lambda_j + N\mu \leqq 1$. Suppose that the N-vector Δq_j is defined by $(0, \cdots, 0, 1, 0, \cdots, 0)$, where the 1 occupies the jth coordinate. Let i denote the smallest index such that $q^{(i)} \leqq q^{(j)}$ holds for $j = 1, \cdots, N$. Let $\delta_i(j) = 1$ if $j = i$, and 0 otherwise. Let $\mu(t) = \mu$ if $t \geqq 1$ and $\mu(t) = 0$ if $t = 0$. We assume that

$$\mathbb{P}(Q_{n+1} = q' \mid Q_n = q) = \begin{cases} \lambda_j + \lambda \delta_i(j), & \text{if } q' = q + \Delta q_j, \\ \mu(q^{(j)}), & \text{if } q' = q - \Delta q_j, \\ 1 - \lambda - \sum_{j=1}^{N} [\lambda_j + \mu(q^{(j)})], & \text{if } q' = q. \end{cases}$$

Let $u(q)$ denote the sum of the squares of the differences between all pairs of components of $q = (q^{(1)}, \cdots, q^{(N)})$. More precisely,

$$u(q) \triangleq \sum_{\alpha=1}^{N-1} \sum_{\beta=\alpha+1}^{N} |q^{(\beta)} - q^{(\alpha)}|^2. \tag{3.2}$$

Theorem 3. *Without loss of generality, assume that* $\lambda_1 \leqq \cdots \leqq \lambda_N$. *If*

$$\lambda > \sum_{k=1}^{N-1} (\lambda_N - \lambda_k), \tag{3.3}$$

then there exist finite positive constants v *and* A *such that*

$$\mathbb{E}[u(Q_n) \mid Q_0 = q] \leqq \frac{u(q)^3}{\left[u(q) + \frac{v}{2} n\right]^2} + A\left(1 + 4\left(\frac{A}{v}\right)^2 \sum_{k=1}^{n-1} \frac{1}{k^2}\right),$$

from which it follows that

$$\mathbb{E}[u(Q_n) \mid Q_0 = q] \leqq u(q) + A\left(1 + \frac{2}{3}\left(\frac{\pi A}{v}\right)^2\right), \tag{3.4}$$

and

$$\limsup_{n \to \infty} \mathbb{E}[u(Q_n) \mid Q_0 = q] \leqq A\left(1 + \frac{2}{3}\left(\frac{\pi A}{v}\right)^2\right) < \infty.$$

Remarks. (i) If $\lambda > (N-1)(\lambda_N - \lambda_1)$ then (3.3) holds. (ii) Since the inequality in (3.3) does not depend on μ, the interpretation of (3.4) is that the policy of 'joining the shortest queue' will ensure that even under arbitrarily heavy traffic, $\mathbb{E}[u(Q_n) \mid Q_0 = q]$ does not 'blow up'. By Jensen's inequality, (3.4) also implies that for every $i \neq j$,

$$\begin{aligned} \mathbb{E}[|Q_n^{(j)} - Q_n^{(i)}| \mid Q_0 = q] &\leqq \mathbb{E}[|Q_n^{(j)} - Q_n^{(i)}|^2 \mid Q_0 = q]^{\frac{1}{2}} \\ &\leqq \mathbb{E}[u(Q_n) \mid Q_0 = q]^{\frac{1}{2}} \end{aligned}$$

is bounded for all time. In this sense, each queue has 'approximately' the same number of customers; that is, the queues are 'equally loaded' or 'balanced'. The reader may find it interesting to contemplate various special cases such as $\mu = 0$, $\lambda_N = \lambda_1 > 0$, or $\lambda_N = \lambda_1 = 0$.

Proof. We show below that (3.3) implies

$$|u(q \pm \Delta q_j) - u(q)| \leqq (N-1)(1 + 2\sqrt{u(q)}), \tag{3.5}$$

$$\mathbb{E}[u(Q_{n+1}) - u(q) \mid Q_n = q] \leqq (N-1) - 2\Big(\lambda - \sum_{k=1}^{N-1} (\lambda_N - \lambda_k)\Big)\sqrt{\frac{u(q)}{2N-1}}, \tag{3.6}$$

and

$$\mathbb{E}[|u(Q_{n+1}) - u(q)|^2 \mid Q_n = q] \leqq (N-1)^2(1 + 4\sqrt{u(q)} + 4u(q)). \tag{3.7}$$

First, however, we use (3.3) and (3.5)–(3.7) to show that the hypotheses of Theorem 2 are satisfied. Set $b = 1$, and fix any $0 < \varepsilon < 1$. Now make the following observations. Since

$$\lim_{v \to \infty} \frac{(N-1)(1 + 2\sqrt{v})}{v} = 0,$$

it is clear that for all sufficiently large v,

$$(N-1)(1 + 2\sqrt{v}) \leqq \varepsilon v.$$

Now, fix any $C > 4(N-1)^2$. Since

$$\lim_{v \to \infty} \frac{(N-1)^2(1 + 4\sqrt{v} + 4v)}{v} = 4(N-1)^2,$$

for all sufficiently large v,

$$(N-1)^2(1 + 4\sqrt{v} + 4v) \leqq Cv.$$

Next, choose (recall (2.1))

$$\nu \geqq \tfrac{3}{2} C \frac{(1+\varepsilon)}{(1-\varepsilon)^4}; \tag{3.8}$$

since (3.3) holds, we have

$$\lim_{v \to \infty} \Big\{(N-1) - 2\Big(\lambda - \sum_{k=1}^{N-1} (\lambda_N - \lambda_k)\Big)\sqrt{\frac{v}{2N-1}}\Big\} = -\infty.$$

It follows that for all sufficiently large v, we also have

$$(N-1)-2\left(\lambda-\sum_{k=1}^{N-1}(\lambda_N-\lambda_k)\right)\sqrt{\frac{v}{2N-1}}\leqq -\nu.$$

It should now be clear that if $b=1$, $0<\varepsilon<1$, $C>4(N-1)^2$, and ν satisfies (3.8), then there is some L such that (2.2)–(2.4) hold. To find a finite constant A satisfying (2.5), observe that if $u(q)\leqq L$, then (3.5) implies

$$|u(q\pm\Delta q_j)-u(q)|\leqq(N-1)(1+2\sqrt{L}),$$

and so

$$u(q\pm\Delta q_j)\leqq(N-1)(1+2\sqrt{L})+L.$$

Hence, we may take

$$A=(N-1)(1+2\sqrt{L})+L.$$

We conclude that if (3.3) holds, (3.5)–(3.7) are sufficient to apply Theorem 2.

We now establish (3.5)–(3.7). Clearly, (3.5) implies (3.7). To establish (3.5) and (3.6), we proceed as follows. We denote by i the smallest integer in $\{1,\cdots,N\}$ such that $q^{(i)}\leqq q^{(j)}$ holds for $j=1,\cdots,N$. Setting $l_j\triangleq q^{(j)}-q^{(i)}$, we can write

$$q=(q^{(i)}+l_1,\cdots,q^{(i)}+l_N).$$

Note that each $l_j\geqq 0$, and $l_i=0$. Fixing any $j\in\{1,\cdots,N\}$, we have

$$\begin{aligned}u(q)&=\sum_{\beta\neq j}|q^{(\beta)}-q^{(j)}|^2+\sum_{\substack{\alpha=1\\ \alpha\neq j}}^{N-1}\sum_{\substack{\beta=\alpha+1\\ \beta\neq j}}^{N}|q^{(\beta)}-q^{(\alpha)}|^2\\ &=\sum_{\beta\neq j}|l_\beta-l_j|^2+\sum_{\substack{\alpha=1\\ \alpha\neq j}}^{N-1}\sum_{\substack{\beta=\alpha+1\\ \beta\neq j}}^{N}|l_\beta-l_\alpha|^2,\end{aligned}\tag{3.9}$$

and

$$\begin{aligned}u(q\pm\Delta q_j)&=\sum_{\beta\neq j}|q^{(\beta)}-(q^{(j)}\pm 1)|^2+\sum_{\substack{\alpha=1\\ \alpha\neq j}}^{N-1}\sum_{\substack{\beta=\alpha+1\\ \beta\neq j}}^{N}|q^{(\beta)}-q^{(\alpha)}|^2\\ &=\sum_{\beta\neq j}|(l_\beta-l_j)\mp 1|^2+\sum_{\substack{\alpha=1\\ \alpha\neq j}}^{N-1}\sum_{\substack{\beta=\alpha+1\\ \beta\neq j}}^{N}|l_\beta-l_\alpha|^2,\end{aligned}$$

so that

$$u(q\pm\Delta q_j)-u(q)=\sum_{\beta\neq j}(1\mp 2(l_\beta-l_j)).\tag{3.10}$$

It follows from (3.9) that for any $\beta\neq j$,

$$|l_\beta-l_j|\leqq\sqrt{u(q)}.$$

This with (3.10) implies (3.5).

To establish (3.6), use (3.10) to show that

$$\begin{aligned}\mathbb{E}[u(Q_{n+1})-u(q)\mid Q_n=q] &= \sum_{j=1}^{N}\Big\{\Big(\sum_{\beta\neq j}(1-2(l_\beta-l_j))\Big)(\lambda_j+\lambda\delta_i(j)) \\ &\qquad +\Big(\sum_{\beta\neq j}(1+2(l_\beta-l_j))\Big)\mu(q^{(i)}+l_j)\Big\} \\ &= \sum_{j=1}^{N}\Big\{(N-1)[\mu(q^{(i)}+l_j)+\lambda_j+\lambda\delta_i(j)] \\ &\qquad +2\sum_{\beta\neq j}(l_\beta-l_j)[\mu(q^{(i)}+l_j)-\lambda_j-\lambda\delta_i(j)]\Big\} \\ &\leqq \nu_0+2\sum_{j=1}^{N}\sum_{\beta\neq j}(l_\beta-l_j)[\mu(q^{(i)}+l_j)-\lambda_j-\lambda\delta_i(j)],\end{aligned}$$

where $\nu_0 \triangleq (N-1)[N\mu+\sum_{j=1}^{N}\lambda_j+\lambda]\leqq(N-1)$. For the second term, write

$$\nu_1 \triangleq \sum_{j=1}^{N}\sum_{\beta\neq j}(l_\beta-l_j)[\mu(q^{(i)}+l_j)-\lambda_j-\lambda\delta_i(j)].$$

Break this into two terms. In the first, change the order of summation so that

$$\nu_1=\sum_{\beta=1}^{N}l_\beta\sum_{j\neq\beta}[\mu(q^{(i)}+l_j)-\lambda_j-\lambda\delta_i(j)]-\sum_{j=1}^{N}l_j\sum_{\beta\neq j}[\mu(q^{(i)}+l_j)-\lambda_j-\lambda\delta_i(j)].$$

Using the fact that $l_i=0$, that $\mu(\cdot)\leqq\mu$, and that the minimum value of $\sum_{j\neq\beta}\lambda_j$ occurs when $\beta=N$,

$$\begin{aligned}\nu_1 &\leqq \sum_{\beta\neq i}l_\beta\sum_{j\neq\beta}[\mu-\lambda_j-\lambda\delta_i(j)]-\sum_{j\neq i}l_j[\mu(q^{(i)}+l_j)-\lambda_j-\lambda\delta_i(j)](N-1) \\ &= \sum_{\beta\neq i}l_\beta\Big[(N-1)\mu-\lambda-\sum_{j\neq\beta}\lambda_j\Big]-\sum_{j\neq i}l_j[\mu(q^{(i)}+l_j)-\lambda_j](N-1) \\ &\leqq \sum_{\beta\neq i}l_\beta\Big[(N-1)\mu-\lambda-\sum_{k=1}^{N-1}\lambda_k\Big]-\sum_{j\neq i}l_j[\mu(q^{(i)}+l_j)-\lambda_j](N-1) \\ &= \sum_{j\neq i}l_j\Big[(N-1)\mu-\lambda-\sum_{k=1}^{N-1}\lambda_k\Big]-\sum_{j\neq i}l_j[\mu-\lambda_j](N-1),\end{aligned}$$

where the last step follows by observing that $l_j\mu(q^{(i)}+l_j)=l_j\mu$ for all j. Continuing, we see that

$$\begin{aligned}\nu_1 &\leqq \Big[(N-1)\lambda_N-\sum_{k=1}^{N-1}\lambda_k-\lambda\Big]\sum_{j\neq i}l_j \\ &= -\Big[\lambda-\sum_{k=1}^{N-1}(\lambda_N-\lambda_k)\Big]\sum_{j\neq i}l_j.\end{aligned}\tag{3.11}$$

Now, observe that according to (3.9), if we set $j = i$ and use the fact that $l_i = 0$, then

$$\begin{aligned} u(q) &= \sum_{\beta \neq i} l_\beta^2 + \sum_{\substack{\alpha=1 \\ \alpha \neq i}}^{N-1} \sum_{\substack{\beta=\alpha+1 \\ \beta \neq i}}^{N} |l_\beta - l_\alpha|^2 \\ &\leqq \sum_{\beta \neq i} l_\beta^2 + \sum_{\substack{\alpha=1 \\ \alpha \neq i}}^{N-1} \sum_{\substack{\beta=\alpha+1 \\ \beta \neq i}}^{N} (l_\beta^2 + l_\alpha^2) \\ &\leqq (2N-1) \sum_{\beta \neq i} l_\beta^2, \end{aligned}$$

and so

$$\sqrt{\frac{u(q)}{2N-1}} \leqq \sum_{\beta \neq i} l_\beta. \tag{3.12}$$

If (3.3) holds, substituting (3.12) into (3.11) now yields

$$v_1 \leqq -\left(\lambda - \sum_{k=1}^{N-1} (\lambda_N - \lambda_k)\right) \sqrt{\frac{u(q)}{2N-1}}. \tag{3.13}$$

Finally, we have

$$\begin{aligned} \mathbb{E}[u(Q_{n+1}) - u(q) \mid Q_n = q] &\leqq v_0 + 2v_1 \\ &\leqq (N-1) - 2\left(\lambda - \sum_{k=1}^{N-1} (\lambda_N - \lambda_k)\right) \sqrt{\frac{u(q)}{2N-1}}, \end{aligned}$$

which is exactly (3.6). This completes the proof of Theorem 3.

We now briefly discuss the continuous-time analog of this model. Suppose that $\{Q_n\}$ is the jump chain corresponding to a continuous-time queuing process $\{\hat{Q}_t\}$ with instantaneous arrival rates $\hat{\lambda}, \hat{\lambda}_1, \cdots, \hat{\lambda}_N$, and instantaneous departure rate $\hat{\mu}$ for each of the N servers. If we let $d(q) \triangleq \hat{\lambda} + \sum_{j=1}^N [\hat{\lambda}_j + \hat{\mu}(q^{(j)})]$, then the jump chain parameters are

$$\begin{aligned} \lambda &= \hat{\lambda}/d(q), \\ \lambda_j &= \hat{\lambda}_j/d(q), \\ \mu(q^{(j)}) &= \hat{\mu}(q^{(j)})/d(q). \end{aligned} \tag{3.14}$$

Clearly, the transition probabilities given that $Q_n = q$ now depend on q in a slightly different way. However, assuming that $\hat{\lambda}_1 \leqq \cdots \leqq \hat{\lambda}_N$, a review of the proof of Theorem 3 through (3.11) when (3.14) is in force shows that the crucial question is whether or not

$$\frac{\hat{\lambda}}{d(q)} > \frac{1}{d(q)} \sum_{k=1}^{N-1} (\hat{\lambda}_N - \hat{\lambda}_k).$$

This condition obviously holds if and only if

$$\hat{\lambda} > \sum_{k=1}^{N-1} (\hat{\lambda}_N - \hat{\lambda}_k). \tag{3.15}$$

If (3.15) holds, we can express (3.13) in terms of the infinitesimal rates:

$$v_1 \leqq -\frac{\hat{\lambda} - \sum_{k=1}^{N-1} (\hat{\lambda}_N - \hat{\lambda}_k)}{d(q)} \sqrt{\frac{u(q)}{2N-1}}.$$

Since $\max_q d(q) = \hat{\lambda} + \sum_{j=1}^{N} \hat{\lambda}_j + N\hat{\mu}$,

$$v_1 \leqq -\frac{\hat{\lambda} - \sum_{k=1}^{N-1} (\hat{\lambda}_N - \hat{\lambda}_k)}{\hat{\lambda} + \sum_{j=1}^{N} \hat{\lambda}_j + N\hat{\mu}} \sqrt{\frac{u(q)}{2N-1}}.$$

Thus, for jump chains, (3.6) should be replaced by

$$\mathbb{E}[u(Q_{n+1}) - u(q) \mid Q_n = q] \leqq (N-1) - 2\frac{\hat{\lambda} - \sum_{k=1}^{N-1} (\hat{\lambda}_N - \hat{\lambda}_k)}{\hat{\lambda} + \sum_{j=1}^{N} \hat{\lambda}_j + N\hat{\mu}} \sqrt{\frac{u(q)}{2N-1}},$$

and (3.3) should be replaced by (3.15).

4. Proof of Theorem 2

Before proving Theorem 2, we introduce the following notation. Let

$$s \triangleq \frac{v}{b+1}, \tag{4.1}$$

and set

$$B \triangleq \frac{A^{b+2}}{s^{b+1}}$$

so that we can use (2.5) to write

$$\frac{1}{s^{b+1}} \int_{\mathbb{X}} u(y)^{b+2} p(x, dy) \leqq B, \quad \text{if} \quad 0 \leqq u(x) \leqq L. \tag{4.2}$$

Now, for $v \geqq 0$, let $\varphi_0(v) \equiv v$, and for $n \geqq 1$ set

$$\varphi_n(v) \triangleq \frac{v^{b+2}}{[v+sn]^{b+1}}. \tag{4.3}$$

With this notation we can write (recall (2.6))

$$(4.4)\qquad \begin{aligned} u_0(x) &= \varphi_0(u(x)), \\ u_1(x) &= \varphi_1(u(x)) + A, \\ u_n(x) &= \varphi_n(u(x)) + A + B\sum_{k=1}^{n-1}\frac{1}{k^{b+1}}, \qquad n \geqq 2. \end{aligned}$$

For later reference, note that differentiating φ_n with respect to v yields

$$(4.5)\qquad \varphi_n'(v) = \frac{v^{b+2} + sn(b+2)v^{b+1}}{[v+sn]^{b+2}},$$

and

$$(4.6)\qquad \varphi_n''(v) = \frac{(b+1)(b+2)s^2n^2v^b}{[v+sn]^{b+3}}.$$

Proof of Theorem 2. Fix $x \in \mathbb{X}$. We first show that (1.4) holds if $0 \leqq u(x) \leqq L$. Then we show that (1.4) holds if $u(x) > L$. Suppose $0 \leqq u(x) \leqq L$ and $n = 0$. Use Jensen's inequality and (2.5) to get

$$\begin{aligned} \int_{\mathbb{X}} u_0(y)p(x, dy) - u_1(x) &= \int_{\mathbb{X}} u(y)p(x, dy) - \varphi_1(x) - A \\ &\leqq \left(\int_{\mathbb{X}} u(y)^{b+2}p(x, dy)\right)^{1/(b+2)} - \varphi_1(x) - A \\ &\leqq A - \varphi_1(x) - A \leqq 0. \end{aligned}$$

Now for $n \geqq 1$, keep in mind (4.2) while using (4.4) to write

$$\begin{aligned} \int_{\mathbb{X}} u_n(y)p(x, dy) - u_{n+1}(x) &= \int_{\mathbb{X}} \varphi_n(u(y))p(x, dy) - \varphi_{n+1}(u(x)) - \frac{B}{n^{b+1}} \\ &= \int_{\mathbb{X}} \frac{u(y)^{b+2}}{[u(y)+sn]^{b+1}}p(x, dy) - \varphi_{n+1}(u(x)) - \frac{B}{n^{b+1}} \\ &\leqq \frac{1}{(sn)^{b+1}}\int_{\mathbb{X}} u(y)^{b+2}p(x, dy) - \varphi_{n+1}(u(x)) - \frac{B}{n^{b+1}} \\ &\leqq \frac{B}{n^{b+1}} - \varphi_{n+1}(u(x)) - \frac{B}{n^{b+1}} \\ &\leqq 0. \end{aligned}$$

We next establish (1.4) when $u(x) > L$ and $n \geqq 1$. The case $n = 0$ follows the same pattern (without Taylor's theorem, since $\varphi_0(v) \equiv v$) and is left to the reader. Using Taylor's theorem, we have

$$(4.7)\qquad \begin{aligned} \int_{\mathbb{X}} u_n(y)p(x, dy) &= u_n(x) + \int_{\mathbb{X}} [\psi_n(u(y)) \mid \varphi_n(u(x))]p(x, dy) \\ &\leqq u_n(x) + \int_{\mathbb{X}} [\varphi_n'(u(x))(u(y) - u(x)) + K(u(y) - u(x))^2]p(x, dy), \end{aligned}$$

where K is the upper bound on $\frac{1}{2}\varphi_n''$ given by

$$K \triangleq \frac{(b+1)(b+2)}{2} \frac{(1+\varepsilon)^b}{(1-\varepsilon)^{b+3}} \frac{u(x)^b s^2 n^2}{[u(x)+sn]^{b+3}}.$$

To verify that K is an upper bound, observe that (2.2) implies that $p(x, \cdot)$-a.s.,

$$u(x)(1-\varepsilon) \leqq u(y) \leqq u(x)(1+\varepsilon).$$

Now, using (4.6),

$$\begin{aligned} \sup_{u(x)(1-\varepsilon) \leqq v \leqq u(x)(1+\varepsilon)} \tfrac{1}{2}\varphi_n''(v) &\leqq \frac{(b+1)(b+2)s^2n^2u(x)^b(1+\varepsilon)^b}{2[u(x)(1-\varepsilon)+sn]^{b+3}} \\ &\leqq K, \end{aligned}$$

since $[u(x)(1-\varepsilon)+sn] \geqq (1-\varepsilon)[u(x)+sn]$. Returning to (4.7), we use (2.3) and (2.4) to see that

$$\int_{\mathbb{X}} u_n(y)p(x, dy) \leqq u_n(x) + \varphi_n'(u(x))(-v) + KCu(x).$$

Using the definition of K and (2.1) together with (4.5),

$$\begin{aligned} \int_{\mathbb{X}} u_n(y)p(x, dy) &\leqq u_n(x) - v\frac{u(x)^{b+2} + sn(b+2)u(x)^{b+1}}{[u(x)+sn]^{b+2}} + \frac{2vu(x)^{b+1}s^2n^2}{[u(x)+sn]^{b+3}} \\ &= u_n(x) - \frac{vu(x)^{b+2}}{[u(x)+sn]^{b+2}} - \frac{vsnu(x)^{b+1}}{[u(x)+sn]^{b+3}}[(b+2)u(x)+bsn] \\ &\leqq u_n(x) - \frac{vu(x)^{b+2}}{[u(x)+sn]^{b+2}} \\ &= u_{n+1}(x) + u_n(x) - u_{n+1}(x) - \frac{vu(x)^{b+2}}{[u(x)+sn]^{b+2}}. \end{aligned} \tag{4.8}$$

Observe that

$$\begin{aligned} u_n(x) - u_{n+1}(x) &= \varphi_n(u(x)) - \varphi_{n+1}(u(x)) - \frac{B}{n^{b+1}} \\ &\leqq \varphi_n(u(x)) - \varphi_{n+1}(u(x)) \\ &= u(x)^{b+2}\left(-\int_n^{n+1} \frac{\partial}{\partial\theta}\left(\frac{1}{[u(x)+s\theta]^{b+1}}\right) d\theta\right) \\ &= u(x)^{b+2}(b+1)s\int_n^{n+1} \frac{d\theta}{[u(x)+s\theta]^{b+2}} \\ &\leqq \frac{vu(x)^{b+2}}{[u(x)+sn]^{b+2}}, \end{aligned} \tag{4.9}$$

where the last step follows by setting $\theta = n$ in the integrand and then using (4.1). Combining (4.8) and (4.9) yields (1.4) when $u(x) > L$.

5. Extension of Theorem 2

It is possible to weaken the hypotheses of Theorem 2 by replacing (2.2)–(2.4) with (5.1)–(5.4) below.

Suppose that for each $x \in \mathbb{X}$, there is a subset, B_x, with

$$B_x \subset \{y \in \mathbb{X} : u(y) \leqq u(x)\}, \tag{5.1}$$

and such that

$$I_{B_x^c}(y)\,|u(y) - u(x)| \leqq \varepsilon u(x), \qquad p(x, \cdot)\text{-a.s., if } u(x) > L, \tag{5.2}$$

$$\mathbb{E}[I_{B_x^c}(X_{n+1})(u(X_{n+1}) - u(x)) \mid X_n = x] \leqq -\nu, \quad \text{if } u(x) > L, \tag{5.3}$$

and

$$\mathbb{E}[I_{B_x^c}(X_{n+1})\,|u(X_{n+1}) - u(x)|^2 \mid X_n = x] \leqq Cu(x), \quad \text{if } u(x) > L. \tag{5.4}$$

Here, B_x^c denotes the complement of B_x and $I_{B_x^c}$ denotes the indicator function of the set B_x^c.

To prove that we can weaken the hypotheses as claimed, we proceed as follows. Observe that (4.5) implies $\varphi_n' \geqq 0$, and so φ_n is a non-decreasing function on $[0, \infty)$. Hence, (5.1) implies

$$\int_{B_x} [\varphi_n(u(y)) - \varphi_n(u(x))]p(x, dy) \leqq 0.$$

From this and Taylor's theorem,

$$\begin{aligned}\int_{\mathbb{X}} u_n(y)p(x, dy) &= u_n(x) + \int_{\mathbb{X}} [\varphi_n(u(y)) - \varphi_n(u(x))]p(x, dy)\\ &\leqq u_n(x) + \int_{B_x^c} [\varphi_n(u(y)) - \varphi_n(u(x))]p(x, dy) \qquad (5.5)\\ &\leqq u_n(x) + \int_{B_x^c} [\varphi_n'(u(x))(u(y) - u(x)) + K(u(y) - u(x))^2]p(x, dy).\end{aligned}$$

If we replace (4.7) with (5.5), then the derivation given in Section 4 will prove the modification of Theorem 2 stated above.

6. Relation to other work

In this section, we consider systems to which we can apply both Theorem 2 and the results of Hajek (1982). We point out that Theorem 2 applies only to non-negative functions of time-homogeneous Markov processes while Hajek's results apply to real-valued functions of non-Markov processes. The purpose of this section is to show that when considering a non-negative function of a time-homogeneous Markov process such that (2.2) holds, whenever Hajek's results apply, so does Theorem 2. Further, if the parameter ρ below is greater than or equal to 1, then Hajek's results do not yield useful asymptotic information, while Theorem 2 still does.

To apply Hajek's results to $\{u(X_n)\}$ requires that the following hold.

There exist positive constants ν, η, ρ, and δ such that for some $L<\infty$,

$$\mathbb{E}[u(X_{n+1})-u(x)\mid X_n=x]\leqq -\nu, \quad \text{if} \quad u(x)>L, \tag{6.1}$$

$$\mathbb{E}[\exp(\eta(u(X_{n+1})-u(x)))\mid X_n=x]\leqq \rho, \quad \text{if} \quad u(x)>L, \tag{6.2}$$

and

$$\mathbb{E}[\exp(\eta(u(X_{n+1})-L))\mid x_n=x]\leqq \delta, \quad \text{if} \quad u(x)\leqq L. \tag{6.3}$$

Remarks. (i) If $\rho<1$, applying Jensen's inequality with the natural logarithm function to (6.2) yields (6.1) with $-\nu=(1/\eta)\ln\rho$. (ii) Observe that if (6.1) and (6.2) hold for some L, they also hold for any $L'\geqq L$. As for (6.3), rewrite (6.2) and (6.3) as

$$\mathbb{E}[\exp(\eta u(X_{n+1}))\mid X_n=x]\leqq \rho e^{\eta u(x)}, \quad \text{if} \quad u(x)>L,$$

and

$$\mathbb{E}[\exp(\eta u(X_{n+1}))\mid X_n=x]\leqq \delta e^{\eta L}, \quad \text{if} \quad u(x)\leqq L.$$

For any $L'\geqq L$,

$$\begin{aligned}\mathbb{E}[\exp(\eta u(X_{n+1}))\mid X_n=x]&\leqq \max\{\delta e^{\eta L}, \rho e^{L'}\}, && \text{if} \quad u(x)\leqq L',\\ &\leqq \max\{\delta,\rho\}e^{\eta L'}, && \text{if} \quad u(x)\leqq L'.\end{aligned}$$

It now follows that if (6.1)–(6.3) hold for some L, they hold with L replaced by any $L'\geqq L$ and δ replaced by $\max\{\delta,\rho\}$.

Hajek proves by a simple induction argument that (6.2) and (6.3) imply

$$\mathbb{E}[\exp(\eta u(X_n))\mid X_0=x]\leqq \rho^n e^{\eta u(x)}+\frac{1-\rho^n}{1-\rho}\delta e^{\eta L}. \tag{6.4}$$

To relate this to $\mathbb{E}[u(X_n)\mid X_0=x]$, apply Jensen's inequality to (6.4) to get

$$\mathbb{E}[u(X_n)\mid X_0=x]\leqq \frac{1}{\eta}\ln\left[\rho^n e^{\eta u(x)}+\frac{1-\rho^n}{1-\rho}\delta e^{\eta L}\right], \tag{6.5}$$

which yields both

$$\mathbb{E}[u(X_n)\mid X_0=x]\leqq \frac{1}{\eta}\ln\left[e^{\eta u(x)}+\frac{1}{1-\rho}\delta e^{\eta L}\right], \quad \text{if} \quad \rho<1, \tag{6.6}$$

and

$$\limsup_{n\to\infty}\mathbb{E}[u(X_n)\mid X_0=x]\leqq \frac{1}{\eta}\ln\left(\frac{\delta}{1-\rho}\right)+L, \quad \text{if} \quad \rho<1. \tag{6.7}$$

Under the hypotheses of Theorem 2 (with $b=1$ and $0<\varepsilon<1$),

$$\mathbb{E}[u(X_n)\mid X_0=x]\leqq \frac{u(x)^3}{[u(x)+(v/2)n]^2}+A+4\frac{A^3}{v^2}\sum_{k=1}^{n-1}\frac{1}{k^2}, \tag{6.8}$$

$$\mathbb{E}[u(X_n)\mid X_0=x]\leqq u(x)+A\left(1+\frac{2}{3}\left(\frac{\pi A}{v}\right)^2\right), \tag{6.9}$$

and

$$\limsup_{n\to\infty}\mathbb{E}[u(X_n)\mid X_0=x]\leqq A\left(1+\frac{2}{3}\left(\frac{\pi A}{v}\right)^2\right)<\infty. \tag{6.10}$$

Clearly, both approaches yield very similar results. We now compare the respective assumptions which each approach requires. First, (6.1) is identical to (2.3). Second, we show below that (6.3) implies the existence of a *finite* constant A satisfying (2.5). Third, we show below that if (2.2) holds, then (6.2) implies (2.4) with a constant C satisfying (2.1). It follows that if there exist positive constants v, η, ρ, δ, and $0<\varepsilon<1$ such that for some $L<\infty$, (6.1), (6.2), and (6.3) hold, and if we have $p(x,\cdot)$-a.s.,

$$|u(y)-u(x)|\leqq \varepsilon u(x), \quad \text{if} \quad u(x)>L,$$

then both (6.5) and (6.8) hold. However, if $\rho\geqq 1$, the uniform bounds (6.6) and (6.7) are not available, while the uniform bounds (6.9) and (6.10) still hold.

The remainder of this section is devoted to establishing the preceding claims. Suppose that (6.3) holds. Rewrite it as

$$\int_{\mathbb{X}} e^{\eta u(y)}p(x,dy)\leqq \delta e^{\eta L}, \quad \text{if} \quad u(x)\leqq L.$$

Now, for any *integer* $b\geqq 0$, since $\eta u(y)\geqq 0$ implies

$$\frac{[\eta u(y)]^{b+2}}{(b+2)!}\leqq e^{\eta u(y)},$$

we have

$$\int_{\mathbb{X}} u(y)^{b+2}p(x,dy)\leqq \frac{(b+2)!}{\eta^{b+2}}\delta e^{\eta L}, \quad \text{if} \quad u(x)\leqq L,$$

providing a finite constant A satisfying (2.5).

Suppose that (6.2) and (2.2) hold. Rewrite (2.2) as

$$(1-\varepsilon)u(x)\leqq u(X_{n+1})\leqq (1+\varepsilon)u(x).$$

Taylor's theorem implies

$$\exp(\eta(u(X_{n+1})-u(x)))=1+\eta[u(X_{n+1})-u(x)]+\tfrac{1}{2}\eta^2 e^{\eta v^*}(u(X_{n+1})-u(x))^2,$$

where $(1-\varepsilon)u(x) \leqq v^* \leqq (1+\varepsilon)u(x)$. Thus,

$$\exp(\eta(u(X_{n+1}) - u(x))) \geqq 1 - \eta\varepsilon u(x) + \tfrac{1}{2}\eta^2 \exp(\eta(1-\varepsilon)u(x))(u(X_{n+1}) - u(x))^2,$$

and then (6.2) implies

$$\mathbb{E}[|u(X_{n+1}) - u(x)|^2 \mid X_n = x] \leqq [\eta\varepsilon u(x) - 1 + \rho]\frac{2}{\eta^2}\exp(-\eta(1-\varepsilon)u(x)).$$

Now, setting

$$a(v) \triangleq [\eta\varepsilon v - (1-\rho)]\frac{2}{\eta^2}\exp(-\eta(1-\varepsilon)v),$$

we have $\lim_{v\to\infty} a(v)/v = 0$. If we choose any positive C satisfying (2.1) we can then take L large enough so that whenever $v > L$, $a(v) \leqq Cv$.

Acknowledgement

We thank Professor A. M. Makowski for his careful reading of an earlier version of this paper and his suggestions for improving its presentation. We thank R. B. Sowers for his suggestions for improving the revised version of this paper.

References

BILLINGSLEY, P. (1979) *Probability and Measure.* Wiley, New York.

BLANC, J. P. C. (1987) A note on waiting times in systems with queues in parallel. *J. Appl. Prob.* **24,** 540–546.

BRUMELLE, S. L. (1971) Some inequalities for parallel-server queues. *Operat. Res.* **19,** 402–413.

CONOLLY, B. W. (1984) The autostrada queueing problem. *J. Appl. Prob.* **21,** 394–403.

EPHREMIDES, A., VARAIYA, P., AND WALRAND, J. (1980) A simple dynamic routing problem. *IEEE Trans. Autom. Contr.* **25,** 690–693.

FLATTO, L. AND MCKEAN, H. P. (1977) Two queues in parallel. *Comm. Pure Appl. Math.* **30,** 255–263.

FOSCHINI, G. J. AND SALZ, J. (1978) A basic dynamic routing problem and diffusion. *IEEE Trans. Comm.* **26,** 320–327.

HAIGHT, F. A. (1958) Two queues in parallel. *Biometrika* **45,** 401–410.

HAJEK, B. (1982) Hitting-time and occupation-time bounds implied by drift analysis with applications. *Adv. Appl. Prob.* **14,** 502–525.

HALFIN, S. (1985) The shortest queue problem. *J. Appl. Prob.* **22,** 865–878.

KINGMAN, J. F. C. (1961) Two similar queues in parallel. *Ann. Math. Statist.* **32,** 1314–1323.

PAPADIMITRIOU, I. (1973) A simple proof of the formula $\sum_{k=1}^{\infty} k^{-2} = \pi^2/6$. *Amer. Math. Monthly* **80,** 424–425.

STARK, E. L. (1974) The series $\sum_{k=1}^{\infty} k^{-s}$, $s = 2, 3, 4, \cdots$, once more. *Mathematics Mag.* **47,** 197–202.

WEBER, R. R. (1978) On the optimal assignment of customers to parallel servers. *J. Appl. Prob.* **15,** 406–413.

WINSTON, W. (1977) Optimality of the shortest line discipline. *J. Appl. Prob.* **14,** 181–189.

WOLFF, R. W. (1977) An upper bound for multi-channel queues. *J. Appl. Prob.* **14,** 884–888.

WOLFF, R. W. (1987) Upper bounds on work in system for multichannel queues. *J. Appl. Prob.* **24,** 547–551.

Rev. Math. Phys. 6 (1994) 621-648

FINITE APPROXIMATIONS TO QUANTUM SYSTEMS

TROND DIGERNES*
Department of Mathematics
University of Trondheim
7034 Trondheim-NTH
Norway

V. S. VARADARAJAN
Department of Mathematics
University of California at Los Angeles
Los Angeles, CA 90024
USA

S. R. S. VARADHAN**
Courant Institute of Mathematical Sciences
New York University
New York, N. Y. 10012
USA

Received 8 November 1993

1. Introduction

1.1. In his studies on the foundations of quantum mechanics in the 1960's Julian Schwinger had considered quantum systems in finite dimensional spaces as approximations to quantum systems in infinite dimensional spaces. For example, to study the motion of a single particle in the one dimensional space $\mathbf{R}$, Schwinger's idea [1], which is already present in Weyl [2] (pp. 277–280), consists in replacing the configuration space $\mathbf{R}$ by a large finite grid $X(N) \subset \mathbf{R}$:

$$X(N) = \{r\varepsilon_N \mid r = 0, \pm 1, \pm 2, \ldots, \pm N^0\} \qquad \varepsilon_N = \left(\frac{2\pi}{N}\right)^{1/2}, \quad N = 2N^o + 1\ .$$

The Hilbert space of the system is $L^2(X(N))$; the position operator q_N is multiplication by the coordinate; the momentum operator p_N is the Fourier transform of q_N, the Fourier transform being the one defined for the finite group $\mathbf{Z}(N) = \mathbf{Z}\varepsilon_N/N\cdot\mathbf{Z}\varepsilon_N$ with which $X(N)$ can be identified. Schwinger's idea is to regard the system with the Hamiltonian $H = \frac{1}{2}(p^2 + V(q))$ on $L^2(\mathbf{R})$ as approximated for large N by the system on $L^2(X(N))$ with Hamiltonian $H_N^{(s)} = \frac{1}{2}(p_N^2 + V_N)$ ($V_N = V(q_N)$).

*Supported by the Research Council of Norway. The first two authors would like to express their gratitude to the departments of Mathematics of the University of Trondheim, Norway, and the University of California at Los Angeles, Ca, USA, for their hospitality.
**Supported by NSF grant DMS–920122 and ARO grant DAAL 03–92–G–0317.

Schwinger did not analyse this approximation in detail. For the harmonic oscillator where $H = \frac{1}{2}(p^2 + q^2)$ and the spectrum of H simple, consisting of

$$\frac{1}{2}, \frac{3}{2}, \frac{5}{2}, \ldots$$

a computation with *Mathematica* showed that even for relatively small values of N, the eigenvalues and eigenfunctions of $H_N^{(s)}$ were remarkably close to those of H: for $N = 41$, the first 10 eigenvalues of $H_N^{(s)}$ agree with those of H up to 12 decimal places, and the graphs of the eigenfunctions could hardly be differentiated. This suggested that one should try to prove that for a large class of Hamiltonians, the spectral data of $H_N^{(s)}$ converge in a very strong fashion to the spectral data of H.

In this paper we shall treat this problem in arbitrary dimension d for Hamiltonians $H = -\Delta + V$ where Δ is the Laplacian in $\mathbf{R}^d$ and V is a potential which is continuous, ≥ 0, and $V(x) \longrightarrow \infty$ as $|x| \longrightarrow \infty (x \in \mathbf{R}^d)$. The resolvent operators are then compact so that H has a pure discrete spectrum $0 \leq h_1 < h_2 < \ldots, h_n, \ldots, h_n \longrightarrow \infty$, with finite multiplicities. We then prove, using analytic arguments in Sec. 2, that the eigenvalues and eigenfunctions of H_N converge to those of H; here $H_N = -\Delta_N + V_N$ is defined as above using finite grids in the lattice $(\mathbf{Z}\varepsilon_N)^d$, and Δ_N is either a discrete Laplacian (i.e., a second difference operator) with suitable modifications at the boundary of the grid, or the Schwinger Laplacian associated to the grid by finite Fourier transform. In Secs. 3 through 6 we use path integral methods based on the Feynman–Kac formula to prove that if the potential V blows up at infinity fast enough so that the operators $e^{-tH}(t > 0)$ are of trace class, then, for each $t > 0$ the operator e^{-tH_N} converges to the operator e^{-tH} in the very strong trace norm.

Both the analytic and probabilistic techniques used in this paper are well-known to specialists so that the paper as a semi-expository character. The analytic problem is one of studying the stability and convergence of difference schemes for solving eigenvalue problems. Kreiss [3] has treated this question for ODE's ($d = 1$) on a finite interval. The present case, when the potential becomes infinite at infinite distances, does not go very much beyond what is done in [3] since the large size of the potential at infinity introduces a natural cutoff. The Schwinger approximation is however not a true difference scheme; but it has good comparison properties with periodic difference schemes, and so can be handled. Nevertheless it is in some sense the best approximation to the infinite system. It is desirable to explore this aspect more.

In the probabilistic approach $\frac{1}{2}\Delta$ is the infinitesimal generator of Brownian motion in $\mathbf{R}^d$ while $\frac{1}{2}\Delta_N$ and its variants are infinitesimal generators of random walks on the lattice $(\mathbf{Z}\varepsilon_N)^d$ and its finite grids. In the continuum limit when the interlattice distance ε_N goes to 0, the random walks will converge to Brownian motion in the sense of measures in function space. The trace convergence of the dynamical operators will then follow from this on applying the Feynman–Kac formula. As before the Schwinger Hamiltonians are handled by comparison with the periodic ones.

1.2. Notation. For any real nonnegative continuous function V on $\mathbf{R}^d$ the operator $H = -\Delta + V$ is essentially self-adjoint on $C_c^\infty(\mathbf{R}^d)$, hence on $C_c^\infty(\mathbf{R})\otimes\cdots\otimes C_c^\infty(\mathbf{R})$ ([4], Theorem X. 28, pp. 183–184). Suppose now that $V(x) \longrightarrow \infty$ as $|x| \longrightarrow \infty$. The formula

$$(Hf,f) = \sum_{1\leq i\leq d} ||\partial f/\partial x_i||^2 + ||V^{1/2}f||^2 \qquad (f \in C_c^\infty(\mathbf{R}^d))$$

shows that $H \geq 0$ and moreover yields the compactness of the resolvent $(I+H)^{-1}$, as is classically known. Indeed, if $(I+H)f = g$ where $f \in C_c^\infty(\mathbf{R}^d)$ and $||g|| \leq 1$, then $||f|| \leq 1$ and so $((I+H)f,f) \leq 1$, giving $||\partial f/\partial x_i|| \leq 1 (1 \leq i \leq d), ||V^{1/2}f|| \leq 1$. The first estimate implies, in view of the Rellich imbedding [5], that f varies in a compact set in each $L^2(K)$, K compact in $\mathbf{R}^d$, while the second estimate shows that

$$\int_{|x|>a} |f|^2 dx \leq \left(\inf_{|x|>a} V(x)\right)^{-1} \longrightarrow 0 \qquad (a \longrightarrow \infty)$$

thus proving that f varies in a compact set in $L^2(\mathbf{R}^d)$. The analytical proof of the main approximation theorem given in Sec. 2 is based on the fact that the above argument remains uniform when ε goes to 0.

For any $\varepsilon > 0$, let $G(\varepsilon)$ denote the lattice $(\mathbf{Z}\varepsilon)^d$. For $h \in G(\varepsilon)$ the translation operator T_h acts on functions on $G(\varepsilon)$ by $(T_h f)(x) = f(x+h)$. If $e_i (1 \leq i \leq d)$ are the standard basis vectors in $\mathbf{R}^d$, the partial difference operators $D_i(\varepsilon)^\pm$ are defined by

$$D_i(\varepsilon)^+ = \varepsilon^{-1}(T_{\varepsilon e_i} - I), \qquad D_i(\varepsilon)^- = \varepsilon^{-1}(I - T_{-\varepsilon e_i}) \quad (1 \leq i \leq d)$$

and the discrete Laplacian $\Delta(\varepsilon)$ by

$$\Delta(\varepsilon) = \sum_{1\leq i\leq d} D_i(\varepsilon)^- D_i(\varepsilon)^+ .$$

On $L^2(G(\varepsilon))$ the T_h are unitary, $D_i(\varepsilon)^- = -(D_i(\varepsilon)^+)^\dagger$ and $\Delta(\varepsilon)$ are bounded, and

$$(-\Delta(\varepsilon)f,f) = \sum_{1\leq i\leq d} ||D_i(\varepsilon)^+ f||^2 \qquad (f \in L^2(G(\varepsilon))) .$$

The Hamiltonian $H(\varepsilon)$ on $L^2(G(\varepsilon))$ is given by $H(\varepsilon) = -\Delta(\varepsilon) + V_\varepsilon$, the operator V_ε acting as multiplication by the restriction of V to $G(\varepsilon)$. It is clear that $H(\varepsilon) \geq 0$ and $(I + H(\varepsilon))^{-1}$ is compact.

Let $N^0 = N^0(\varepsilon)$ be an integer ≥ 1 depending on ε; $\varepsilon N^0 \longrightarrow \infty$ as $\varepsilon \longrightarrow 0$; $N = N(\varepsilon) = 2N^0 + 1$. $G(\varepsilon)^0$ is the grid $X(\varepsilon)^d$ in $G(\varepsilon)$ where $X(\varepsilon)$ is the set $\{r\varepsilon \mid r = 0, \pm1, \ldots, \pm N^0\}$. There are several variants of $\Delta(\varepsilon)$ depending on the treatment of boundary terms.

The periodic case: $H(\varepsilon)^{(p)} = -\Delta(\varepsilon)^{(p)} + V_\varepsilon$. We identify $G(\varepsilon)^0$ with the finite group $G(\varepsilon)/N(\varepsilon)\cdot G(\varepsilon)$ and hence $L^2(G(\varepsilon)^0)$ with the space of functions on $G(\varepsilon)$

invariant under $T_h, h \in N(\varepsilon)\cdot G(\varepsilon)$; $D_i(\varepsilon)^{\pm(p)}$ are the restrictions of $D_i(\varepsilon)^{\pm}$ to this space; and

$$\Delta(\varepsilon)^{(p)} = \sum_{1\leq i\leq d} D_i(\varepsilon)^{-(p)} D_i(\varepsilon)^{+(p)}, \quad (-\Delta(\varepsilon)^{(p)} f, f) = \sum_{1\leq i\leq d} \|D_i(\varepsilon)^{+(p)} f\|^2.$$

The Hamiltonians $H(\varepsilon)^{(0)} = -\Delta(\varepsilon)^{(0)} + V_\varepsilon$. We denote by $\Delta(\varepsilon)^{(0)}$ any member of a class of operators obtained from $\Delta(\varepsilon)$ by boundary modifications. The simplest of these is defined as $\Delta(\varepsilon)^0 = P_\varepsilon \Delta(\varepsilon) P_\varepsilon$ where P_ε is multiplication by the characteristic function of $G(\varepsilon)^0$ and we identify $L^2(G(\varepsilon)^0)$ as a subspace of $L^2(G(\varepsilon))$ in the obvious way. More generally, $\Delta(\varepsilon)^{(0)}$ is defined so that it satisfies the following conditions:

(a) $\Delta(\varepsilon)^{(0)}$ is symmetric, real, and ≥ 0.

(b) For any $b > 0$ let $C(b)$ be the cube of all $x = (x_1, \ldots, x_d) \in \mathbf{R}^d$ with $|x_i| < b$ for all i; if ε is small enough, $C(b) \cap G(\varepsilon) \subset G(\varepsilon)^0$. We then have, for all $f \in L^2(G(\varepsilon)^0)$ with support contained in $C(b) \cap G(\varepsilon)$,

$$\Delta(\varepsilon)^{(0)} f = \Delta(\varepsilon) f \ .$$

This will be true if for a suitable $N' = N'(\varepsilon) < N^0, N'/N^0 \longrightarrow 1$ as $\varepsilon \longrightarrow 0$, the matrix elements of $\Delta(\varepsilon)^{(0)}$ and $\Delta(\varepsilon)$ (in the delta function basis) satisfy

$$\Delta(\varepsilon)^{(0)}(r,s) = \Delta(\varepsilon)(r,s) \qquad (|r|, |s| \leq N^0, \min(|r|,|s|) \leq N') \ .$$

(c) For any $b > 0$ and all sufficiently small ε, with the suffix indicating that the norm is computed over $C(b) \cap G(\varepsilon)$,

$$(-\Delta(\varepsilon)^{(0)} f, f) \geq \sum_{1\leq i\leq d} \|D_i(\varepsilon)^{+} f\|_b^2 \ .$$

It is clear that $\Delta(\varepsilon)^0 = P_\varepsilon \Delta(\varepsilon) P_\varepsilon$ satisfies (a)–(c). To get another example let $d = 1$ and define D_ε^+ by

$$(D_\varepsilon^+ f)(k\varepsilon) = \varepsilon^{-1}[f((k+1)\varepsilon) - f(k\varepsilon)] \ (-N^0 \leq k < N^0), \ (D_\varepsilon^+ f)(N^0\varepsilon) = 0 \ .$$

We then take $-\Delta(1:\varepsilon) = (D_\varepsilon^+)^\dagger D_\varepsilon^+$. For $d \geq 1, \Delta(d:\varepsilon) := \sum_{1\leq i\leq d} \Delta_i(1:\varepsilon)$ where $\Delta_i(1:\varepsilon)$ is just $\Delta(1:\varepsilon)$ acting only on the ith variable. It is clear that $\Delta(d:\varepsilon)$ verifies (a)–(c).

The Schwinger Hamiltonians $H(\varepsilon)^{(s)} = -\Delta(\varepsilon)^{(s)} + V_\varepsilon$. In this case we relate ε to N by the equation $\varepsilon = (2\pi/N)^{1/2}$ and write $\mathbf{F}_\varepsilon$ for the finite Fourier transform on $L^2(G(\varepsilon)/N\cdot G(\varepsilon))$ transferred to $L^2(G(\varepsilon)^0)$. It is a unitary operator with matrix elements $\mathbf{F}_\varepsilon(rs)(r,s \in G(\varepsilon)^0)$ in the basis $(u_\varepsilon^r)_{r\in G(\varepsilon)^0}$, u_ε^r being the delta function at r, given by

$$\mathbf{F}_\varepsilon(rs) = N^{-d/2}\, e^{ir\cdot s}.$$

The operator $q_{i,\varepsilon}$ is the multiplication by the ith coordinate and $p_{i,\varepsilon}$ is its Fourier transform, while $-\Delta(\varepsilon)^{(s)}$ is the sum of squares of the $p_{i,\varepsilon}$:

$$(q_{i,\varepsilon}f)(x) = x_i f(x) \quad (x \in G(\varepsilon)), \quad p_{i,\varepsilon} = \mathbf{F}_\varepsilon q_{i,\varepsilon} \mathbf{F}_\varepsilon^{-1}, \quad -\Delta(\varepsilon)^{(s)} = \sum_{1 \le k \le d} p_{i,\varepsilon}^2 .$$

From our remarks it is clear that for $* = p, 0, s$,

$$(-\Delta(\varepsilon)^{(*)}f, f) \ge \sum_{1 \le i \le d} \|D_i(\varepsilon)^+ f\|_b^2 \qquad (f \in L^2(G(\varepsilon)^0)) .$$

The validity of this for $* = s$ is due to the fact that *the Schwinger Laplacian (with the negative sign) is bigger than the periodic version*:

$$(-\Delta(\varepsilon)^{(s)}f, f) \ge (-\Delta(\varepsilon)^{(p)}f, f) \qquad (f \in L^2(G(\varepsilon)^0)) .$$

Indeed, it is enough to verify this when $d = 1$ and in that case it follows from the inequality $1 - \cos x \le \frac{1}{2}x^2 (x \in \mathbf{R})$ and the formula

$$\mathbf{F}_\varepsilon^{-1}(-(1/2)\Delta(\varepsilon)^{(p)})\mathbf{F}_\varepsilon = \varepsilon^{-2}(1 - \cos q_\varepsilon \varepsilon) .$$

Since it is necessary to work in a single Hilbert space to formulate the approximation theorems neatly we shall introduce certain natural imbeddings of the $L^2(G(\varepsilon))$ (hence also of the $L^2(G(\varepsilon)^0)$)) into $L^2(\mathbf{R}^d)$. For any $x \in \mathbf{R}^d$ let

$$R(x) = \{y = (y_1, \ldots, y_d) \mid x_i - \varepsilon/2 \le y_i < x_i + \varepsilon/2, 1 \le i \le d\} .$$

Then the map $f \mapsto f^\#$ is an isometric imbedding of $L^2(G(\varepsilon))$ into $L^2(\mathbf{R}^d)$ where

$$f^\# = \varepsilon^{-d/2} \sum_{x \in G(\varepsilon)} f(x)\chi_{R(x)} \qquad (\chi_E \text{ is the characteristic function of } E) .$$

Thus we may view operators on $L^2(G(\varepsilon))$ as operators on $L^2(\mathbf{R}^d)$ by the identification $A \to Q_\varepsilon A Q_\varepsilon$ where Q_ε is the orthogonal projection on the image of $L^2(G(\varepsilon))$. If $\theta_\varepsilon = \varphi_\varepsilon \otimes \cdots \otimes \varphi_\varepsilon, \varphi_\varepsilon = \varepsilon^{-1}\chi_{[-\varepsilon/2, \varepsilon/2]}$, then we have,

$$Q_\varepsilon f =: f_\varepsilon = \sum_{x \in G(\varepsilon)} (\theta_\varepsilon * f)(x)\chi_{R(x)} \qquad (f \in L^2(\mathbf{R}^d)) .$$

If Q_ε^0 is the orthogonal projection on the image of $L^2(G(\varepsilon)^0)$, then $Q_\varepsilon^0 f \longrightarrow f$ for all $f \in L^2(\mathbf{R}^d)$; this is trivial for $f \in C_c^1(\mathbf{R}^d)$ and follows for all f by approximation.

2. Analytic Method: Convergence of Eigenvalues and Eigenfunctions When $V(x) \longrightarrow \infty$ For $|x| \longrightarrow \infty$

2.1. Let $\Delta(\varepsilon)^{(*)}$ be any of $\Delta(\varepsilon), \Delta(\varepsilon)^{(p)}, \Delta(\varepsilon)^{(0)}, \Delta(\varepsilon)^{(s)}$, and $H(\varepsilon)^{(*)} = -\Delta(\varepsilon)^{(*)} + V_\varepsilon$. Assuming that $V(x)$ is continuous, ≥ 0, and $\longrightarrow \infty$ as $|x| \longrightarrow \infty$ we shall give in this section an analytic proof that the eigenvalues and eigenfunctions of $H(\varepsilon)^{(*)}$ converge to those of $H = -\Delta + V$, as $\varepsilon \longrightarrow 0$. There are two steps to the proof. The first proves that the resolvents $(I + H(\varepsilon)^{(*)})^{-1}$ converge in the strong operator topology to the resolvent $(I + H)^{-1}$. The second step is to show that the operators $(I + H(\varepsilon)^{(*)})^{-1}$ are uniformly compact in a suitable sense.

2.2. The strong convergence of the resolvents and exponentials will follow if we show that $H(\varepsilon)^{(*)}f \longrightarrow Hf$ for all f in a common core for all the operators $H, H(\varepsilon)^{(*)}$ ([6], Theorem VIII. 20 and VIII. 25). It follows from [4] (Theorem X. 28) that $C_c^\infty(\mathbf{R}^d)$ is a common core, hence also $C_c^\infty(\mathbf{R}) \otimes \cdots \otimes C_c^\infty(\mathbf{R})$. This is true even if V has singularities of a certain type and is in $L^2(\mathbf{R}^d)^{\text{loc}}$, for instance if V is an atomic potential ([4], Example 2, p. 167 and Theorem X. 26). For the strong convergence of the resolvents and exponentials it is enough to suppose only the following.

(a) $-\Delta + V$ is essentially self-adjoint on $C_c^\infty(\mathbf{R}^d)$.

(b) $V = \sum_j V_j$, where each V_j is continuous away from a linear subspace L_j of codimension ≥ 3 and V_j is bounded by constant times (distance from L_j)$^{-1}$ on compact sets of $\mathbf{R}^d$.

Then $C_c^\infty(\mathbf{R}) \otimes \cdots \otimes C_c^\infty(\mathbf{R})$ is a common core for all $H, H(\varepsilon)^{(*)}$.

Proposition 1. *Let V be as above. Then the resolvents of $H(\varepsilon)^{(*)}$ converge strongly to those of H. In particular, if V is ≥ 0 and continuous, $(I + H(\varepsilon)^{(*)})^{-1}$ converges strongly to $(I+H)^{-1}$, and we also have the strong convergence of $e^{-tH(\varepsilon)^{(*)}}$ to e^{-tH} for all $t > 0$.*

Proof. Since we deal only with f with compact support, boundary modifications do not matter. Hence there are only two cases to be treated: $H(\varepsilon)$ and $H(\varepsilon)^{(s)}$. For the convergence of the potential part we may suppose that $V = V_j$. If V_ε is the function which is $V(x)$ if the cube $R(x)(x \in G(\varepsilon))$ does not meet L_j and is 0 otherwise, we have the estimate $|V_\varepsilon(x)| \leq$ const. (distance of x to L_j)$^{-1}$ for all x, and so, as $V_\varepsilon f_\varepsilon \longrightarrow Vf$ pointwise away from L_j, we have, by the dominated convergence theorem the same limit relation in $L^2(\mathbf{R}^d)$. □

To prove that $\Delta(\varepsilon)f_\varepsilon \longrightarrow \Delta f$ for f in $L^2(\mathbf{R}^d)$ it is clear that we need only work in dimension 1. Then $\Delta(\varepsilon)f_\varepsilon$ is the function $\sum_{x \in G(\varepsilon)} \theta_\varepsilon^{[3]} * f''(x)\chi_{R(x)}$, where $\theta_\varepsilon^{[3]} = \theta_\varepsilon * \theta_\varepsilon * \theta_\varepsilon$. It is then easy to check that $\Delta(\varepsilon)f_\varepsilon \longrightarrow f''$.

To show that $\Delta(\varepsilon)^{(s)}f_\varepsilon \longrightarrow \Delta f$ requires a little more effort. As before it is enough to treat the case $d = 1$. Let $\mathbf{F}_\varepsilon$ denote the operator which is identified with the finite Fourier transform on the image of $L^2(G(\varepsilon)^0)$ and the identity on the orthogonal complement. Let $\mathbf{F}$ be the *unitary* Fourier transform on $\mathbf{R}$ so that

$$\mathbf{F}f = \frac{1}{\sqrt{2\pi}}\hat{f}, \qquad \hat{f}(x) = \int e^{ixy} f(y)dy \ .$$

It is easily seen to be a question of showing that, as $\varepsilon \longrightarrow 0$, for all $f \in C_c^\infty(\mathbf{R})$,

$$q_\varepsilon^r(\mathbf{F}_\varepsilon^{-1}f_\varepsilon - (\mathbf{F}^{-1}f)_\varepsilon) \longrightarrow 0 \qquad (r \text{ an integer} \geq 0)$$

(we need this only for $r = 0, 2$). Recall that $\varepsilon = (2\pi/N)^{1/2}, N \longrightarrow \infty, N^0 = (N-1)/2$. An elementary calculation gives

$$\|q_\varepsilon^r \mathbf{F}_\varepsilon^{-1} f_\varepsilon - q_\varepsilon^r (\mathbf{F}_\varepsilon^{-1} f)_\varepsilon\|^2 = (2\pi)^{-1} \sum_{|k| \le (N-1)/2} (k\varepsilon)^{2r} |\alpha_{k\varepsilon} - A_{k\varepsilon}|^2$$

where, writing $h = \theta_\varepsilon * f$,

$$\alpha_{k\varepsilon} = \varepsilon^{3/2} \sum_{|m| \le (N-1)/2} e^{-ik\varepsilon m\varepsilon} h(m\varepsilon) = \varepsilon^{3/2} \sum_{m \in \mathbf{Z}} e^{-ik\varepsilon m\varepsilon} h(m\varepsilon)$$

$$A_{k\varepsilon} = \varepsilon^{-1/2} \int_{(k-\frac{1}{2})\varepsilon}^{(k+\frac{1}{2})\varepsilon} \hat{f}(-x)dx = \varepsilon^{1/2} \int e^{-ik\varepsilon y} f(y) a(\varepsilon y) dy$$

with $a(u) = \sin(u/2)/(u/2)$. Applying the *Poisson Summation Formula* to the function $y \mapsto h(y)e^{-ik\varepsilon y}$ we get

$$\alpha_{k\varepsilon} = \varepsilon^{1/2} a(-k\varepsilon^2) \hat{f}(-k\varepsilon) + \varepsilon^{1/2} \sum_{|m| \ge 1} \hat{h}(Nm\varepsilon - k\varepsilon) .$$

We next rewrite $A_{k\varepsilon}$ as

$$A_{k\varepsilon} = \varepsilon^{1/2} \hat{f}(-k\varepsilon) + \varepsilon^{1/2} \int e^{-ik\varepsilon y} (a(\varepsilon y) - 1) f(y) dy .$$

Then $A_{k\varepsilon} - \alpha_{k\varepsilon} = \sum_{p=0}^{p=2} E_{k\varepsilon}^{(p)}$ where

$$E_{k\varepsilon}^{(0)} = \varepsilon^{1/2} \hat{f}(-k\varepsilon)(1 - a(-k\varepsilon^2)), \qquad E_{k\varepsilon}^{(1)} = \varepsilon^{1/2} \int e^{-ik\varepsilon y} (a(\varepsilon y) - 1) f(y) dy$$

$$E_{k\varepsilon}^{(2)} = -\varepsilon^{1/2} \sum_{|m| \ge 1} \hat{h}(Nm\varepsilon - k\varepsilon) .$$

It is to be shown that as $\varepsilon \longrightarrow 0$,

$$J^{(p)} := \sum_{|k| \le (N-1)/2} (k\varepsilon)^{2r} |E_{k\varepsilon}^{(p)}|^2 \longrightarrow 0 \qquad (p = 0, 1, 2) .$$

$p = 0$. $J^{(0)}$ is seen to go to 0 in view of the estimate (uniform in g)

$$\varepsilon \sum_{|k| \le (N-1)/2} g(k\varepsilon) - \int_{|x| \le N\varepsilon/2} g dx = O(N^{-1/2} \|g''\|_\infty)$$

for $g(x) = |\hat{f}(-x)|^2 x^{2r} |(1 - a(-\varepsilon x))|^2$, and the remark that the error term is $O(N^{-1/2})$ while the main term goes to 0 by the dominated convergence theorem. Here we must note that a and all its derivatives are bounded on $\mathbf{R}$.

$p = 1$. We argue as before with $g = u_\varepsilon$ where

$$u_\varepsilon(x) = x^{2r}\int e^{-ixy}(a(\varepsilon y)-1)f(y)dy .$$

It is clear that $u_\varepsilon \longrightarrow 0$ pointwise. Since

$$(u_\varepsilon(x)x^m)^{(p)} = (i)^{2r+m}\int e^{-ixy}(a(\varepsilon y)-1)f(y))^{(2r+m)}(-iy)^p dy$$

it is clear that $u_\varepsilon^{(p)}(x)x^m = O(1)$ as $|x| \longrightarrow \infty$, uniformly in ε, for each p, m.

$p = 2$. We have $|\hat{h}| = |a\hat{f}| \leq$ const. $|\hat{f}|$, so that

$$(k\varepsilon)^{2r}|E_{k\varepsilon}^{(2)}|^2 \leq \text{const.}\varepsilon(k\varepsilon)^{2r}\left(\sum_{|m|\geq 1}|\hat{f}(Nm\varepsilon - k\varepsilon)|\right)^2 .$$

Since $|Nm\varepsilon - k\varepsilon| \geq N|m|\varepsilon/2(|k| \leq (N-1)/2)$, the expression within the parentheses is $O(N^{-t})$ for every $t > 0$ uniformly in k and so $J^{(2)} \longrightarrow 0$.

Corollary 2. *For any $t > 0$ and any operator K in $L^2(\mathbf{R}^d)$ of finite rank, as $\varepsilon \longrightarrow 0$*

$$\mathrm{Tr}(e^{-tH(\varepsilon)^{(*)}}K) \longrightarrow \mathrm{Tr}(e^{-tH}K) .$$

2.3. We now suppose that V is continuous, ≥ 0 and goes to infinity at infinite distances and prove the uniform compactness of $(I + H(\varepsilon)^{(*)})^{-1}$ as $\varepsilon \longrightarrow 0$. This is based on the uniformity of the lattice versions of the Rellich imbeddings as $\varepsilon \longrightarrow 0$, which is formulated as follows.

Lemma 1. *Suppose $f_\varepsilon \in L^2(G(\varepsilon))$ satisfy the following conditions: for some $a > 0$ there is a constant $A > 0$ such that for all sufficiently small ε,*

$$||f_\varepsilon||_a \leq A, \qquad ||D_i(\varepsilon)^+ f_\varepsilon||_a \leq A \quad (1 \leq i \leq d)$$

where the suffix means that the norm is computed over $C(a) \cap G(\varepsilon)$, $C(a)$ being the open cube where all coordinates are $< a$ in absolute value. If (ε_n) is a sequence tending to 0, then for any $b > 0$ with $0 < b < a$, the sequence $(f_{\varepsilon_n}^{\#})$ is relatively compact in $L^2(C(b))$.

Proof. We first reduce the proof to the case where all the f_ε vanish outside a cube $C(c)$ with $b < c < a$. Select $u \in C^1(\mathbf{R}^d)$ with $u = 1$ on $C(b), u = 0$ outside $C(c)$, and $0 \leq u \leq 1$. It is then immediate that $||(uf_\varepsilon)|| \leq A', ||D_i(\varepsilon)^+(uf_\varepsilon)|| \leq A'$ for all sufficiently small ε, A' being a suitable constant. So we assume that $f_\varepsilon = 0$ outside $C(c)$. For small ε we see that the $f_\varepsilon^{\#}$ vanish outside $C(a)$ and hence their Fourier transforms $F_\varepsilon(\xi)$ are entire functions of ξ; moreover for any $t > 0$ and all ξ with $|\mathrm{Im}\ \xi| \leq t$ and all sufficiently small ε we have the estimate $|F_\varepsilon(\xi)| \leq B_t$ where $B_t > 0$ is a constant depending only on A, a, t. So passing to a subsequence we may suppose that $F_\varepsilon \longrightarrow g$ uniformly on compact subsets of $\mathbf{C}^d$ for some entire function g. The Fourier transform of $D_k(\varepsilon)^+ f_\varepsilon^\#$ is

$$-i\xi_k e^{-i\varepsilon\xi_k/2}\Phi(\varepsilon\xi_k)F_\varepsilon(\xi)\ , \qquad \Phi(z) = \frac{\sin(z/2)}{(z/2)}$$

Hence, for all sufficiently small $\varepsilon > 0$,

$$\int_{\mathbb{R}^d} |\xi|^2 |\Phi(\varepsilon\xi_1)\ldots\Phi(\varepsilon\xi_d)F_\varepsilon(\xi)|^2 d\xi \leq A^2 d\ .$$

If $g_\varepsilon(\xi) = \Phi(\varepsilon\xi_1)\ldots\Phi(\varepsilon\xi_d)F_\varepsilon(\xi)$ it now follows that (g_{ε_n}) is relatively compact in $L^2(\mathbf{R}^d)$. As g_ε is the Fourier transform of $\theta_\varepsilon * f_\varepsilon^\#$ we see that $\theta_{\varepsilon_n} * f_{\varepsilon_n}^\#$ is relatively compact in $L^2(\mathbf{R}^d)$. But, it is easy to see that there is a fixed finite set $F \subset \mathbb{Z}^d$ such that for all $\varepsilon > 0$ and $x \in \mathbb{R}^d, \theta_\varepsilon * f_\varepsilon^\#(x)$ differs from $f_\varepsilon^\#(x)$ by at most the sum of terms $|f_\varepsilon^\#(\varepsilon m + x) - f_\varepsilon^\#(x)|, m \in F$. As $||T_{\varepsilon e_k} f_\varepsilon^\# - f_\varepsilon^\#|| = ||T_{\varepsilon e_k} f_\varepsilon - f_\varepsilon|| = O(\varepsilon)$ for each k, it follows that $||\theta_\varepsilon * f_\varepsilon^\# - f_\varepsilon^\#|| = O(\varepsilon)$. □

We shall now consider the resolvents $(I + H(\varepsilon_n)^{(*)})^{-1}, (\varepsilon_n)$ a sequence tending to 0. Let $f_{\varepsilon_n}, g_{\varepsilon_n} \in L^2(G(\varepsilon_n)^{(*)})$ with $||g_{\varepsilon_n}|| \leq 1, (I + H(\varepsilon_n)^{(*)})f_{\varepsilon_n} = g_{\varepsilon_n}$. Then $||f_{\varepsilon_n}|| \leq 1$ also.

Lemma 2. *With the notations as above, $(f_{\varepsilon_n}^\#)$ is relatively compact in $L^2(\mathbf{R}^d)$.*

Proof. The equation $((I + H(\varepsilon_n)^{(*)} f_{\varepsilon_n}, f_{\varepsilon_n}) = (g_{\varepsilon_n}, f_{\varepsilon_n})$ leads to the estimates

$$||f_{\varepsilon_n}|| \leq 1, \quad ||V_{\varepsilon_n}^{1/2} f_{\varepsilon_n}|| \leq 1, \quad ||D_k(\varepsilon_n)^+ f_{\varepsilon_n}||_b \leq 1 \quad (1 \leq k \leq d)$$

for all $b > 0$. By Lemma 1, $(f_{\varepsilon_n}^\#)$ is relatively compact in $L^2(C(b))$ for any $b > 0$. Since

$$\int_{\mathbf{R}^d \setminus C(b)} |f_{\varepsilon_n}^\#|^2 dx \leq \left(\inf_{x \notin C(b)} V(x)\right)^{-1} \longrightarrow 0$$

as $b \longrightarrow \infty$ uniformly in n we see that $(f_{\varepsilon_n}^\#)$ is relatively compact in $L^2(\mathbf{R}^d)$. This proves the Lemma. □

A formal functional analytic argument is all that is now needed to prove the approximation of H by $H(\varepsilon_n)^{(*)}$. Let $\mathcal{H}$ be a Hilbert space; $K, K_n (n = 1, 2, \ldots)$ self-adjoint operators on $\mathcal{H}$ with $0 \leq K, K_n \leq I$, K compact; we suppose that $K_n \longrightarrow K$ strongly as $n \longrightarrow \infty$. Further there are closed subspaces $L_n \subset \mathcal{H}$ stable under K_n such that (i) K_n is compact on L_n and $K_n = I$ on $L_n^\perp$. (ii) $P_{L_n} \longrightarrow I$ strongly where P_{L_n} is the orthogonal projection on L_n. (iii) If $K_n f_n = g_n$ where $f_n \in L_n$ and $||f_n|| \leq 1$, the sequence (g_n) is relatively compact. We then have the following.

Lemma 3. *Let $1 \geq k_1 > k_2 > \cdots > 0$ be the positive eigenvalues of K and S_j the eigensubspace corresponding to k_j. Then* (i) *if J is a compact subset of* $(0, 1]$ *not* *containing any eigenvalues of K, then no eigenvalue of $K_n|_{L_n}$ belongs to J if n is large enough* (ii) *if J is a compact neighborhood of k_j not containing any $k_i, i \neq j$, all the eigenvalues of $K_n|_{L_n}$ that belong to J converge to k_j; if S_{nj} is the span of the corresponding eigenspaces,* $\dim(S_{nj}) = \dim(S_j)$ *for n large enough, and there is an orthonormal basis of S_{nj} that converges to an orthonormal basis of S_j.*

Proof. If $f_r (r = 1, 2, \ldots)$ are unit vectors in L_{n_r} such that $K_{n_r} f_r = k'_r f_r$ where $k'_r \longrightarrow k' > 0$ as $r \longrightarrow \infty$, (f_r) is relatively compact by our assumptions, and for any limit point f^* of (f_r) we have $Kf^* = k'f^*$. So k' must be one of the k_j. this proves the first statement and the first part of the second; and further shows that $\dim(S_{nj}) \leq \dim(S_j)$ if n is large enough. The remaining assertions are proved by induction on j. Assume that the result is true for indices $< j$. It is enough to prove that if f is a unit vector in S_j and f_{nj} is its projection on S_{nj}, $f_{nj} \longrightarrow f$. Fix $k, k_j > k > k_{j+1}$ and let f_n be the projection of f on L_n. Then $f_n = f_{n1} + \cdots + f_{nj} + g_n$ where g_n is the projection of f_n on the span of the eigenspaces of $K_n|_{L_n}$ of eigenvalues $\leq k$. Now $f_n \longrightarrow f$ while the induction hypothesis shows that $f_{nr} \longrightarrow 0, r < j$. Passing to a subsequence we may assume that $f_{nj} \longrightarrow f^*$ so that $g_n \longrightarrow g^*$ and $f = f^* + g^*$. But $g^* = f - f^* \in S_j$ and $(Kg^*, g^*) \leq k(g^*, g^*)$, showing that $g^* = 0$. Hence $f^* = f$. This proves the Lemma. □

It is now enough to take $K_n = (I + H(\varepsilon_n)^{(*)})^{-1}, K = (I + H)^{-1}$ and L_n as the image of $L^2(G(\varepsilon_n)^{(*)})$ to obtain the main theorem of this section.

Theorem 4. *Let (ε_n) be a sequence tending to 0 and $* = p, s, 0$. Let $0 < h_1 < h_2 < \ldots$ be the eigenvalues of H and T_j the eigensubspace corresponding to h_j. Then* (i) *if J is a compact subset of $[0, \infty)$ not containing any eigenvalues of H, then no eigenvalue of $H(\varepsilon_n)^{(*)}$ belongs to J if n is large enough;* (ii) *if J is a compact neighborhood of h_j not containing any $h_i, i \neq j$, all the eigenvalues of $H(\varepsilon_n)^{(*)}$ that belong to J converge to h_j; if T_{nj} is the span of the corresponding eigenspaces,* $\dim(T_{nj}) = \dim(T_j)$ *for n large enough, and there is an orthonormal basis of T_{nj} that converges to an orthonormal basis of T_j.*

3. Convergence of Random Walks on Lattices to Brownian Motion

3.1. In this section we shall write $\varepsilon = c^{-1/2}$ where c is a real parameter ≥ 1 which will go to ∞. We write

$$\mathbf{Z}_c = \mathbf{Z}c^{-1/2}, \quad \mathbf{Z}_c^d = (\mathbf{Z}_c)^d, \quad j_c = jc^{-1/2} \quad (j \in \mathbf{Z}^d)$$

and consider the Markov process $\{X_c(t)\}$ with state space $\mathbf{Z}_c^d$ and infinitesimal generator $Q_c = \frac{1}{2}\Delta(\varepsilon)$. Thus, for $j = (j_1, \ldots, j_d), k = (k_1, \ldots, k_d) \in \mathbf{Z}^d$,

$$Q_c(j_c, k_c) = \begin{cases} c/2, & \text{if for some } s, k_r = j_r \pm \delta_{rs} \\ -cd, & \text{if } k = j \\ 0, & \text{in all other cases .} \end{cases}$$

The Q_c are elements of the Banach algebra $\mathbf{M}$ of matrices $M = (M(j_c, k_c))_{j,k \in \mathbf{Z}}$ with

$$||M|| = \sup_j \sum_k |M(j_c, k_c)| < \infty .$$

It therefore makes sense to consider the one parameter group in $\mathbf{M}$ of matrices

$$p_c(t) = \exp tQ_c .$$

It is a standard result in the theory of Markov chains that $(p_c(t))_{t \geq 0}$ is a stochastic semigroup and therefore defines a Markov chain [7], and that the associated probability measures are defined on the space of right continuous step functions on $[0, \infty)$ with values in $\mathbf{Z}_c^d$. Since we intend to vary c it is necessary to view these measures as defined on the single larger space of all right continuous maps of $[0, \infty)$ into $\mathbf{R}^d$ whose discontinuities are only of the first kind, namely the *Skorohod space* $D[0, \infty)^d$ [8, 9], and, when we want to consider a fixed time t, in the Skorohod space $D[0, t]^d$. We wish to prove the following basic limit theorem.

Theorem 1. *Let $x, y \in \mathbf{R}^d$ and $t > 0$ be fixed. Let $a_c, b_c \in \mathbf{Z}_c^d$ vary in such a way that $a_c \longrightarrow x, b_c \longrightarrow y$ as $c \longrightarrow \infty$. Let $\mathbf{P}_{c,a_c,b_c}$ be the probability measure on $D[0, t]^d$ corresponding to the process X_c that starts from a_c at time 0 and is conditioned to reach b_c at time t. Then, in the sense of convergence of probability measures in $D[0, t]^d$,*

$$\mathbf{P}_{c,a_c,b_c} \Longrightarrow W_{x,y}^d$$

where $W_{x,y}^d$ is the probability measure defined by the Brownian motion process in $D[0, t]^d$ that starts from x at time 0 and is conditioned to reach y at time t.

Here convergence of measures is the usual weak convergence (cf. [8, 9]). The conditional Brownian measures (*Brownian bridges*) are defined in the usual way to depend continuously on x and y (cf. [10]). When $d = 1$ we omit the superscript d. For arbitrary d we have

$$W_{x,y}^d = W_{x_1,y_1} \times \cdots \times W_{x_d,y_d} \qquad (x = (x_1, \ldots, x_d), y = (y_1, \ldots, y_d) \in \mathbf{R}^d) .$$

They are of course defined on the space $C[0, t]^d$ of continuous paths but we shall view them as defined on the larger space $D[0, t]^d$.

Theorem 1 already contains in itself a *uniformity* of the convergence. Let us suppose that x and y vary in compact sets of $\mathbf{R}^d$ and that $a_c = a_c(x)$ and $b_c = b_c(y)$ depend on x and y. The following corollary is immediate.

Corollary 2. *Suppose that the convergences $a_c(x) \longrightarrow x, b_c(y) \longrightarrow y$ are uniform as $c \longrightarrow \infty$ when x, y vary over compact sets. Then the convergence*

$$\mathbf{P}_{c,a_c(x),b_c(y)} \Longrightarrow W_{x,y}^d$$

is uniform in the following sense. If g is any bounded continuous functional on $D[0,t]^d$, the convergence

$$\int g d\mathbf{P}_{c,a_c(x),b_c(y)} \longrightarrow \int g dW^d_{x,y}$$

is uniform when x, y vary over compact sets of $\mathbf{R}^d$.

3.2. The first step is to calculate the transition probabilities for the X_c-process. Under the Fourier transform that takes $f \in L^2(\mathbf{Z}_c^d)$ to the function $\hat{f}$ in $L^2((S^1)^d)$ given by

$$\hat{f} = \sum_j f(j_c)e^{ij.\theta} \qquad (j.\theta = j_1\theta_1 + \cdots + j_d\theta_d)$$

Q_c goes over to the operator of multiplication by $-c\sum_{1\le k\le d}(1 - \cos\theta_k)$, so that $\exp tQ_c$ goes over to the operator of multiplication by $\exp\{-ct\sum_{1\le k\le d}(1-\cos\theta_k)\}$. This implies that the transition probability $p_c(t : j_c, k_c)$ of going from state j_c to state k_c in time t is given by

$$p_c(t : j_c, k_c) = p_c(t : k - j), \quad p_c(t : r) = \prod_{1\le k\le d} p_c(t : r_k) \qquad (r = (r_1, \ldots, r_d) \in \mathbf{Z}^d)$$

where

$$p_c(t : r) = p_c(t : -r) = \frac{1}{2\pi}\int_{-\pi}^{\pi} e^{-ct(1-\cos\theta)}e^{ir\theta}d\theta \quad (r \in \mathbf{Z}) .$$

It follows from this that X_c is a process of independent increments. It is also clear from the product structure of the transition probabilities that the components of X_c are mutually independent. The basic result on p_c is the following.

Lemma 1. *We have the following.*

(i) $p_c(s : r) > 0$ *for all* $r \in \mathbf{Z}^d$ and $s > 0$.

(ii) *For any fixed $s > 0$ there is a constant $B > 0$ such that*

$$p_c(s : r) \le Bc^{-d/2} \text{ for all } r \in \mathbf{Z}^d, c \ge 1 .$$

(iii) *For fixed $s > 0$,*

$$p_c(s : r) = \frac{1}{(2\pi cs)^{d/2}}\left(e^{-|r_c|^2/2s} + \eta(r : c)\right)$$

where $|\cdot|$ is the usual Euclidean norm in $\mathbf{R}^d$, and $\eta(r : c) \longrightarrow 0$ as $c \longrightarrow \infty$, uniformly for $r \in \mathbf{Z}^d$.

Proof. It is clearly enough to prove these when $d = 1$.

(i) We have $p_c'(0 : \pm 1) = c/2$ while $p_c(0 : \pm 1) = 0$. So $p_c(s/n : \pm 1) > 0$ if $n >> 1$, s being fixed. Also $p_c(s : 0) > 0$ for all $s \ge 0$. Hence, if $r > 0$, choosing $n > r$,

$$p_c(s : r) > p_c(s/n : 1)^r p_c(s/n : 0)^{n-r} > 0 .$$

(ii) For a suitably small $a > 0$ we can find $b > 0$ such that for all $r \in \mathbf{Z}, c \geq 1$,

$$p_c(s:r) \leq \frac{1}{2\pi}\int_{|\theta|\leq a} e^{-cs\theta^2/4}d\theta + \frac{1}{2\pi}\int_{a\leq|\theta|\leq\pi} e^{-bcs}d\theta .$$

This gives (ii).

(iii) We refine the argument in (ii) but work with the new variable $z = \theta\sqrt{cs}$. Then

$$\begin{aligned}\sqrt{2\pi cs}p_c(s:r) &= \frac{1}{\sqrt{2\pi}}\int_{-\pi\sqrt{cs}}^{\pi\sqrt{cs}} e^{-cs(1-\cos(z/\sqrt{cs}))}e^{irz/\sqrt{cs}}dz \\ &= \frac{1}{\sqrt{2\pi}}\int_{-\infty}^{\infty} e^{-z^2/2+irz/\sqrt{cs}}dz + \eta \\ &= e^{\frac{-r_c^2}{2s}} + \eta .\end{aligned}$$

To prove that $\eta \to 0$ as $c \to \infty$ uniformly in $r \in \mathbf{Z}$ we split the integrals involved in the definition of η into the regions where $|z| < a\sqrt{cs}$ and $a\sqrt{cs} \leq |z| \leq \pi\sqrt{cs}$. We omit the details. This proves (iii).

For later use we collect the following information on the moments of the X_c-process. Let E_0 denote the expectation value of the 1-dimensional X_c-process starting from 0 at time 0. It is obvious from the space symmetry that $E_0(X_c(t)^k) = 0$ for odd k. Since

$$E_0(X_c(t)^k) = i^k c^{-k/2} f^{(k)}(0), \qquad f(\theta) = e^{-ct(1-\cos\theta)}$$

we find, remembering that the process has independent increments,

$$E_0\left((X_c(s_1) - X_c(s_2))^2\right) = s_1 - s_2, \quad (0 < s_2 < s_1)$$
$$E_0\left((X_c(s_1) - X_c(s_2))^2(X_c(s_2) - X_c(s_3))^2\right) = (s_1 - s_2)(s_2 - s_3) \quad (0 < s_3 < s_2 < s_1)$$

By expanding the exponential and noting that $(1 - \cos\theta)^r$ vanishes to the order $2r$ at $\theta = 0$ we find that for each $t > 0$ and $k \geq 1$,

$$\sup_{c\geq 1} E_0\left(X_c(t)^{2k}\right) < \infty .$$

3.3. *Convergence of the unconditioned process.* In view of the independence of the components of the processes in question it is clearly enough to prove Theorem 3.1.1 when $d = 1$. So we shall assume that $d = 1$ from now on till the end of this section. Let $\mathbf{P}_{c,a_c}$ be the probability measure on $D[0,t]$ defined by the X_c-process starting from $a_c \in \mathbf{Z}_c$ at time 0.

Proposition 1. *Let W_x be the probability measure of the Wiener process starting from x at time 0. If $a_c \longrightarrow x$ as $c \longrightarrow \infty$, then*

$$\mathbf{P}_{c,a_c} \Longrightarrow W_x \qquad (c \longrightarrow \infty) .$$

Proof. We first verify the convergence of the finite dimensional distributions at time points $t_i, 0 < t_1 < t_2 < \cdots < t_N < t$:

$$\mathbf{P}_{c,a_c}^{t_1,t_2,\ldots,t_N} \Longrightarrow W_x^{t_1,t_2,\ldots,t_N} .$$

If $J_i, 1 \leq i \leq N$ are compact intervals of $\mathbf{R}, \mathbf{P}_{c,a_c}(X_c(t_i) \in J_i, 1 \leq i \leq N)$ is equal to

$$\sum_{b_{ic} \in J_i, 1 \leq i \leq N} p_c(t_1 : b_1 - a) p_c(t_2 - t_1 : b_2 - b_1) \ldots p_c(t_N - t_{N-1} : b_N - b_{N-1}) .$$

The uniform estimate for $p_c(t : r)$ given in Lemma 3.1.1 easily implies that this sum converges to

$$\int_{z_i \in J_i, 1 \leq i \leq N} p(t_1 : z_1 - x) p(t_2 - t_1 : z_2 - z_1) \ldots p(t_N - t_{N-1} : z_N - z_{N-1}) dz_1 dz_2 \ldots dz_N$$

where

$$p(s : u) = (2\pi s)^{-1/2} \exp\{-u^2/2s\} .$$

The limit is clearly

$$W_x(X_c(t_i) \in J_i, 1 \leq i \leq N) .$$

To complete the proof we must show that the measures $\mathbf{P}_{c,a_c}$ are relatively compact. This is immediate from the criterion of Chentsov [11] because, for $0 < s_3 < s_2 < s_1 \leq t$, we have the uniform bound

$$E_{\mathbf{P}_{c,a_c}}\{(X_c(s_1) - X_c(s_2))^2 (X_c(s_2) - X_c(s_3))^2\} \leq (s_1 - s_3)^2$$

for all $c \geq 1$. This proves the proposition. □

3.4. *Convergence of the conditioned processes.* To get the convergence of the conditioned processes from the convergence of the unconditioned processes established in the preceding paragraph we should divide by the probability of the conditioning event, namely, Prob $\{X_c(t) = b_c\}$, and we can take care of this using the precise asymptotics of this probability proved in (iii) of Lemma 3.2.1.

We shall fix $t > 0$ and prove that if $a_c \longrightarrow x, b_c \longrightarrow y$, then $\mathbf{P}_{c,a_c,b_c} \Longrightarrow W_{x,y}$. The method of proof is the same as before; first we establish as in the previous case that the finite dimensional distributions converge, and then verify the relative compactness. The latter is somewhat harder to do in this case. The idea for doing it is as follows. It is a question of showing that if A is the set where the modulus of continuity of the path is large, then $\mathbf{P}_{c,a_c,b_c}(A)$ is small uniformly in c. If the modulus of continuity is computed over the interval $[0, 3t/4]$, this is easy to deduce from the convergence of the unconditioned process as we have to only divide by the probability of the conditioning event, and this is easy because the conditioning takes place at t, which is "far away" from $3t/4$. We also have to show the uniform smallness when the modulus of continuity is calculated on $[t/4, t]$; but this can be deduced from the previous case by using the time reversal $s \longrightarrow t - s$ that takes $\mathbf{P}_{c,a_c,b_c}$ to $\mathbf{P}_{c,b_c,a_c}$. We now proceed to give a few more details of this argument.

Let us write, for any $X \in D[0,t]$ and any $\delta > 0$,

$$\omega(X : \delta) = \sup_{0 \leq s' < s < s'' \leq t, |s' - s''| \leq \delta} \min(|X(s) - X(s')|, |X(s) - X(s'')|)$$

and for any $\eta > 0$ let

$$A(\delta, \eta) = \{X : \omega(X : \delta) > \eta\} .$$

The compactness criterion [8, 9] comes down to showing that for any $\eta > 0$,

$$\lim_{\delta \to 0} \mathbf{P}_{c,a_c,b_c}(A(\delta, \eta)) = 0 \quad \text{uniformly in } c$$

as $\delta \longrightarrow 0$. Let ω_1 and ω_2 be the analogues of ω for the intervals $[0, 3t/4]$ and $[t/4, t]$ respectively. If $\delta < t/4$ and s', s'' are two points in $[0, t]$ with $|s' - s''| \leq \delta$, then they are both in one of the two intervals. Hence

$$\omega(X : \delta) = \max\{\omega_1(X : \delta), \omega_2(X : \delta)\}$$

so that

$$A(\delta, \eta) = A_1(\delta, \eta) \cup A_2(\delta, \eta), \quad A_j(\delta, \eta) = \{\omega_j(X : \delta) > \eta\}$$

Hence it is enough to prove that

$$\lim_{\delta \to 0} \mathbf{P}_{c,a_c,b_c}(A_j(\delta, \eta)) = 0 \quad \text{uniformly in } c \quad (j = 1, 2)$$

Case of $A_1(\delta, \eta)$. Write A_1 for $A_1(\delta, \eta)$. Then

$$\mathbf{P}_{c,a_c,b_c}(A_1) = \frac{\mathbf{P}_{c,a_c}(A_1 \cap (X(t) = b_c))}{\mathbf{P}_{c,a_c}(X(t) = b_c)} .$$

Considering the values of X at $\frac{3t}{4}$ and we see that the numerator is

$$\sum_j \mathbf{P}_{c,a_c}(A_1 \cap (X(3t/4) = j_c)) \cdot \mathbf{P}_{c,0}(X(t/4) = b_c - j_c)$$

which, in view of the Markov property, is majorized by

$$\mathbf{P}_{c,a_c}(A_1) \cdot \sup_k p_c(\frac{t}{4} : k) .$$

Hence

$$\mathbf{P}_{c,a_c,b_c}(A_1) \leq \mathbf{P}_{c,a_c}(A_1) \cdot \frac{\sup_k p_c(\frac{t}{4} : k)}{p_c(t : b - a)} .$$

By Lemma 3.2.1, for fixed $s > 0$, if $k_c \longrightarrow u$, then $\sqrt{c}p_c(s : k) \geq \gamma > 0$ for some constant $\gamma > 0$, while, $\sup_m \sqrt{c}p_c(s : m) \leq \gamma' < \infty$ for some other constant γ'. Hence, for some constant $B > 0$ independent of c, δ, η,

$$\mathbf{P}_{c,a_c,b_c}(A_1) \leq B\mathbf{P}_{c,a_c}(A_1) .$$

But the right side goes to 0 as $\delta \longrightarrow 0$ uniformly in c because of the convergence $\mathbf{P}_{c,a_c} \Longrightarrow W_x$ that we have already established. Hence the left side goes to 0 as $\delta \longrightarrow 0$ uniformly in c.

Case of $A_2(\delta,\eta)$. As we remarked earlier we use time reversal. Since the elements of the Skorohod space are in general only right continuous we define the time reversal map as follows:

$$x^*(s) = x(t-s-0), \qquad 0 \le s \le t .$$

It is easy to see that the map $x \longrightarrow x^*$ is Borel and involutive. For any probability measure P on $D[0,t]$ let P^* be the probability measure defined by $P(E) = P^*(E^*)$. It is then easily seen from the stochastic continuity of the processes and the symmetry $p_c(t:r) = p_c(t:-r)$ that

$$\mathbf{P}_{c,a_c,b_c}{}^* = \mathbf{P}_{c,b_c,a_c} .$$

If we now define $\omega_j{}^*$ as the analogue of ω_j where $X(s')$ etc are replaced by $X(s'-0)$ etc, it is then obvious that

$$A_2(\delta,\eta)^* = \{\omega_1{}^*(X:\delta) > \eta\}$$

and further that

$$\{\omega_1{}^*(X:\delta) > \eta\} \subset \{\omega_1(X:2\delta) > \eta\} .$$

From these remarks we find that

$$\mathbf{P}_{c,a_c,b_c}(A_2(\delta,\eta)) = \mathbf{P}_{c,b_c,a_c}(A_2(\delta,\eta)^*) \le \mathbf{P}_{c,b_c,a_c}(A_1(2\delta,\eta)) .$$

and the proof for the case of A_2 is now reduced to the case of A_1. This completes the proof of the theorem.

4. Convergence of Traces: Continuum Limit of Infinite Lattices

4.1. We shall now prove that in the infinite lattice approximation the traces of the dynamical operators converge. Let $\mathcal{V}$ be the class of potentials on $\mathbf{R}^d$ with the following properties.

(i) V is continuous and ≥ 0;

(ii) $V(x)/\log|x| \longrightarrow \infty$ as $|x| \longrightarrow \infty$.
It is not difficult to show (see also the remarks at the end of the section) that in this case we have

(iii) $e^{t(\frac{1}{2}\Delta - V)}$ is of trace class for every $t > 0$.

When $d = 1$, the harmonic oscillator potential $V(x) = x^2$ is in $\mathcal{V}$ and so is any potential V on $\mathbf{R}^d$ for which $V(x) \ge |x|^2$ for $|x|$ sufficiently large. We shall prove

Proposition 1. *Let* $V \in \mathcal{V}$. *Then*

(i) $e^{t(Q_c - V)}$ *is of trace class on* $L^2(\mathbf{Z}_c^d)$ *for each* $t > 0$.

(ii) *For each $t > 0$, as $c \longrightarrow \infty$, we have*

$$\mathrm{Tr}\left(e^{t(Q_c - V)}\right) \longrightarrow \mathrm{Tr}\left(e^{t(\frac{1}{2}\Delta - V)}\right) .$$

4.2. Recall that under the Fourier transform $L^2(\mathbf{Z}_c^d) \longrightarrow L^2((S^1)^d)$ the operator Q_c goes over to the operator of multiplication by $-c\sum_{1 \leq k \leq d}(1 - \cos\theta_k)$, thus $Q_c \leq 0$. Hence, $Q_c - V$, self adjoint on the domain of multiplication by V, is ≤ 0. As $e^{t(Q_c - V)} \geq 0$ for each $t > 0$ it will be of trace class if and only if

$$\sum_{j \in \mathbf{Z}^d} e^{t(Q_c - V)}(j_c, j_c) < \infty .$$

This fact will emerge during the course of the proof itself.

By the Feynman–Kac formula for the X_c-process

$$e^{t(Q_c - V)}(j_c, k_c) = \int e^{-\int_0^t V(X(s))ds} d\mathbf{P}_{c, j_c, k_c} \cdot p_c(t : k - j) \geq 0 .$$

Lemma 1. *There is a constant $B_t > 0$ such that for all $c \geq 1$,*

$$\sup_j e^{t(Q_c - V)}(j_c, j_c) \leq B_t c^{-d/2} .$$

Furthermore, let

$$g(x) = \frac{1}{(2\pi t)^{d/2}} \int e^{-\int_0^t V(X(s))ds} dW_{x,x}^d$$

Then

$$\mathrm{Tr}\left(e^{t(\frac{1}{2}\Delta - V)}\right) = \int_{\mathbf{R}^d} g(x)dx .$$

Proof. The Feynman–Kac formula shows that

$$\sup_j e^{t(Q_c - V)}(j_c, j_c) \leq p_c(t : 0)$$

and the first assertion follows from Lemma 3.2.1. We observe next by the Feynman–Kac formula again that $e^{t(\frac{1}{2}\Delta - V)}$ is an integral operator on $L^2(\mathbf{R}^d)$ with kernel

$$K_V(x, y) = K_{V,t}(x, y) = \int e^{-\int_0^t V(X(s))ds} dW_{x,y} \cdot p(t : x - y) .$$

Since the operator is ≥ 0, the trace evaluation follows from Mercer's theorem [12].

Lemma 2. *Suppose $c \longrightarrow \infty$. If $j_c \longrightarrow x, k_c \longrightarrow y$, then,*

$$e^{t(Q_c - V)}(j_c, k_c)c^{d/2} \longrightarrow K_V(x, y) .$$

If the convergence of j_c to x and k_c to y are uniform when x and y vary over compact sets of $\mathbf{R}^d$, the above convergence is also uniform over compact sets for x, y.

Proof. Since, by Lemma 3.2.1,

$$c^{d/2} p_c(t : k - j) = \frac{1}{(2\pi t)^{d/2}} \left(e^{-|k_c - j_c|^2/2t} + \mathrm{o}(1) \right)$$

with uniform o(1), it is a question of showing that

$$\int e^{-\int_0^t V(X(s))ds} d\mathbf{P}_{c,j_c,k_c} \longrightarrow \int e^{-\int_0^t V(X(s))ds} dW^d_{x,y} \ .$$

This follows from Theorem 3.1.1 since the integrand is a continuous functional on $D[0,t]^d$.

To prove Proposition 1 we must prove that as $c \longrightarrow \infty$,

$$\sum_{j \in \mathbf{Z}} e^{t(Q_c - V)}(j_c, j_c) \longrightarrow \int_{\mathbf{R}^d} g(x) dx$$

where the statement includes the assertion that the left side is finite for all $c > 0$.

Lemma 3. *For any $m > 0$,*

$$\sum_{|j_c| \leq m} e^{t(Q_c - V)}(j_c, j_c) \longrightarrow \int_{|x| \leq m} g(x) dx \ .$$

Proof. Define the function g_c on $\mathbf{R}^d$ by

$$g_c(u) = e^{t(Q_c - V)}(j_c, j_c) c^{d/2} \qquad \text{if } j_r c^{-1/2} \leq u_r < (j_r + 1) c^{-1/2} \text{ for } 1 \leq r \leq d \ .$$

It then follows from Lemma 2 that g_c converges uniformly to g so that

$$\int_{|x| \leq m} g_c dx \longrightarrow \int_{|x| \leq m} g dx$$

for any $m > 0$. Since

$$\int_{|x| \leq m} g_c dx = \sum_{|j_c| \leq m} e^{t(Q_c - V)}(j_c, j_c) + \mathrm{O}(c^{-1/2})$$

the lemma is proved.

To finish the proof of Proposition 1 it is now sufficient to prove that

$$\lim_{m \to \infty} \sum_{|j_c| \geq m} e^{t(Q_c - V)}(j_c, j_c) = 0 \quad \text{uniformly in } c \quad (*) \ .$$

Lemma 4. *There exists a constant $B > 0$ such that for all integers $p \geq 1$, and $|j_c| \geq p$ and all $c \geq 1$, we have the estimate*

$$e^{t(Q_c - V)}(j_c, j_c) \leq Bc^{-d/2} \int_{p-1}^{p} \left(e^{-\frac{t}{2}V^*(\frac{x}{2})} + \frac{1}{x^4} \right) dx$$

where $V^*(x) = \inf_{|y| \geq x} V(y) (x \geq 0)$.

Proof. Before starting the proof we recall the classical fact ([13], pp. 21–24) which itself is a consequence of P. Levy's inequality for sums of independent random variables: if $X(t)$ is a separable stochastic process with right continuous paths and independent increments, then

$$P(|X(s) - X(0)| > 2a \text{ for some } s \in [0, u]) \leq \frac{2}{a^{2k}} \sup_{s,t \in [0,u]} E(X(s) - X(t))^{2k} .$$

Applying this to each component of X_c it follows that for each $u > 0$ and integer $k \geq 1$ there is a constant $B_k = B_k(u) > 0$ such that for all $a > 0, c \geq 1, j_c \in \mathbf{Z}_c^d$,

$$\mathbf{P}_{c,j_c}(|X(s) - X(0)| > 2a \text{ for some } s \in [0, u]) \leq B_k a^{-2k} .$$

With this said, we shall prove the lemma. We write

$$e^{t(Q_c - V)}(j_c, j_c) = I_1 + I_2$$

where

$$I_i = \int_{R_i} e^{-\int_0^t V(X(s))ds} \chi_{j_c}(X(t)) d\mathbf{P}_{c,j_c} \quad (i = 1, 2) .$$

Here χ_{j_c} is the characteristic function of the singleton $\{j_c\}$ and $R_i (i = 1, 2)$ are the regions defined by

$$R_1 = \{X | X(0) = j_c, |X(s)| \geq p/2 \text{ for all } s \in [0, t/2]\}$$
$$R_2 = \{X | X(0) = j_c, |X(s)| < p/2 \text{ for some } s \in [0, t/2]\} .$$

To estimate I_1 we note that for $X \in R_1$,

$$\int_0^t V(X(s))ds \geq \int_0^{\frac{t}{2}} V(X(s))ds \geq tV^*(p/2)/2 .$$

Hence

$$\begin{aligned} I_1 &\leq e^{-tV^*(p/2)/2} \mathbf{P}_{c,j_c}(R_1 \cap (X(t) = j_c)) \\ &\leq e^{-tV^*(p/2)/2} \mathbf{P}_{c,j_c}(R_1) \sup_k p_c(t/2 : k) \\ &\leq Bc^{-d/2} e^{-tV^*(p/2)/2} \end{aligned}$$

where $B > 0$ is a constant > 0 independent of c, j_c, p. For I_2 we replace the integrand by 1 and observe that for $X \in R_2$, we have $|X(s) - X(0)| > p/2$ for some $s \in [0, t/2]$. Hence we get

$$\begin{aligned} I_2 &\leq \mathbf{P}_{c,j_c}(R_2) \cdot \sup_k p_c(t/2 : k) \\ &\leq Bc^{-d/2} \mathbf{P}_{c,j_c}(|X(s) - X(0)| > p/2 \text{ for some } s \in [0, t/2]) \\ &\leq Bc^{-d/2} \frac{1}{p^4}. \end{aligned}$$

This proves the lemma.

To get $(*)$ we sum over j_c in the range $p \leq |j_c| \leq p + 1$, the number of terms in the sum being $O(c^{d/2})$, and then over $p = m, m+1, m+2, \ldots$. Thus

$$\sum_{|j_c| \geq m+1} e^{t(Q_c - V)}(j_c, j_c) \leq B \int_m^\infty \left(e^{-\frac{t}{2} V^*(\frac{x}{2})} + \frac{1}{x^4} \right) dx \ .$$

Since $V^*(x)/\log x \longrightarrow \infty$ the integrand is $O(x^{-4})$ and so the right side goes to zero as $m \longrightarrow \infty$. As it is independent of c we are finished with the proof of the Proposition.

Remark. It is a consequence of the above proof that the integral of $K_{V,t}(x, x)$ over $\mathbf{R}^d$ is finite. This integral is the Hilbert–Schmidt norm of $K_{V,t/2}$ since the kernels are continuous, symmetric and we have a semi-group. So $K_{V,t}$ is of trace class for every $t > 0$. Also the finiteness of the integral on the diagonal can be established by splitting the integration as in the proof of Lemma 4 and estimating the parts directly.

5. Convergence of Traces: Continuum Limit of Finite Lattices

5.1. We shall now truncate our infinite lattice $\mathbf{Z}_c^d$ to finite grids $\mathbf{Z}_c^{*d} = (\mathbf{Z}_c^*)^d$ where

$$\mathbf{Z}_c^* = \{k_c \| k| \leq N(c)\}$$

and $N(c)$ are integers that tend to ∞ as $c \longrightarrow \infty$. We assume that for some constant $a > 0$,

$$N(c) \geq a \cdot c \ .$$

We write $||x|| = \max_{1 \leq r \leq d} |x_r|, x \in \mathbf{R}^d$ and suppose that for suitable integers $N^*(c)$,

$$N^*(c) < N(c), \quad N^*(c)/N(c) \longrightarrow 1 \quad (c \longrightarrow \infty)$$

we have

$$Q_c^*(j_c, k_c) = Q_c(j_c, k_c), \qquad ||j|| \leq N^*(c), ||k|| \leq N(c)$$

and further that the Q_c^* generate Markov processes. We may suppose

$$N^*(c) \geq (3/4)N(c) \ .$$

We intend to prove

Proposition 1. *Suppose $V \in \mathcal{V}$. Then, for each $t > 0$, as $c \longrightarrow \infty$,*

$$\operatorname{Tr}\left(e^{t(Q_c^* - V)}\right) = \sum_{\|j\| \le N(c)} e^{t(Q_c^* - V)}(j_c j_c) \longrightarrow \operatorname{Tr}\left(e^{t(\frac{1}{2}\Delta - V)}\right) .$$

5.2. We write $\mathbf{P}^*_{c,j_c}$ for the probability measure of the Markov process corresponding to the generator Q_c^* starting from j_c at time 0. We have the Feynman–Kac formula

$$e^{t(Q_c^* - V)}(j_c, k_c) = \int e^{-\int_0^t V(X(s))ds} \chi_{k_c}(X(t)) d\mathbf{P}^*_{c,j_c} .$$

To prove the proposition it is enough, in view of Proposition 4.1.1, to show that

$$\sum_{\|j\| \le N(c)} e^{t(Q_c^* - v)}(j_c, j_c) - \sum_{j \in \mathbf{Z}^d} e^{t(Q_c - V)}(j_c, j_c) \longrightarrow 0 \qquad (*)$$

as $c \longrightarrow \infty$. Since $N(c)c^{-1/2} \longrightarrow \infty$ and

$$\lim_{m \to \infty} \sum_{\|j_c\| > m} e^{t(Q_c - V)}(j_c, j_c) = 0 \quad \text{uniformly in } c$$

we have

$$\lim_{c \to \infty} \sum_{\|j\| > N(c)/2} e^{t(Q_c - V)}(j_c, j_c) = 0 .$$

So, to prove $(*)$, it is enough to prove the following two limit relations:

$$\lim_{c \to \infty} \sum_{\|j\| \le N(c)/2} \left| e^{t(Q_c^* - V)}(j_c, j_c) - e^{t(Q_c - V)}(j_c, j_c) \right| = 0 \quad \text{(I)}$$

$$\lim_{c \to \infty} \sum_{\|j\| > N(c)/2} e^{t(Q_c^* - V)}(j_c, j_c) = 0 . \quad \text{(II)}$$

We have already noted that $e^{t(Q_c^* - V)}(j_c, k_c)$ is always ≥ 0 by the Feynman–Kac formula. Since each sum has $O(N(c)^d)$ terms, it is enough to verify the following replacements for (I) and (II):

$$\sup_{\|j\| \le N(c)/2} \left| e^{t(Q_c^* - V)}(j_c, j_c) - e^{t(Q_c - V)}(j_c, j_c) \right| = o(N(c)^{-d}) \quad \text{(I')}$$

$$\sup_{\|j\| \ge N(c)/2} e^{t(Q_c^* - V)}(j_c, j_c) = o(N(c)^{-d}) . \qquad \text{(II')}$$

5.3. *Proof of* (I'). The key to proving both (I') and (II') is the following lemma. Let

$$W_{c,j_c} = \left\{ X \in D[0,t]^d \mid X(0) = j_c, \|X(u)\| \le N^*(c)c^{-1/2} \text{ for all } u \in [0,t] \right\} .$$

Lemma 1. *For $||j|| \leq N^*(c)$ the measures $\mathbf{P}_{c,j_c}$ and $\mathbf{P}^*_{c,j_c}$ coincide on the σ-algebra of subsets of W_{c,j_c}. If W'_{c,j_c} denotes the complement of W_{c,j_c}, then there are constants $B > 0, c_1 \geq 1$ such that for all $c \geq c_1$ and $||j|| \leq N(c)/2$,*

$$\mathbf{P}_{c,j_c}(W'_{c,j_c}) = \mathbf{P}^*_{c,j_c}(W'_{c,j_c}) \leq BN(c)^{-(d+1)}.$$

Proof. The first assertion is more or less obvious if we note that the Markov measures are determined by the waiting times and the jump probabilities. The relation between Q_c and Q^*_c implies that the jump probabilities from j_c to k_c are the same as long as $||j_c|| \leq N^*(c)$. This shows that the measures coincide on W_{c,j_c} (see [7]).

For the second assertion note that both measures take the same value for W_{c,j_c} and so for its complement also; this common value is the probability under $\mathbf{P}_{c,j_c}$ that X, starting from j_c at time 0, visits some k_c, $||k|| > N^*(c)$, during $[0,t]$. So, if $||j|| \leq N(c)/2$, then,

$$\begin{aligned}\mathbf{P}_{c,j_c}(W'_{c,j_c}) &\leq \mathbf{P}_{c,j_c}\left\{||X(s) - X(0)|| > N(c)c^{-1/2}/4 \text{ for some } s \in [0,t]\right\}\\ &\leq BN(c)^{-(d+1)}\end{aligned}$$

where B is a constant independent of c, j_c. This proves the lemma.

Lemma 2. *We have, uniformly for $||j|| \leq N(c)/2$,*

$$\sum_{||k|| \leq N(c)} \left|\left(e^{t(Q^*_c - V)} - e^{t(Q_c - V)}\right)(j_c, k_c)\right| \leq BN(c)^{-(d+1)}$$

where $B > 0$ is a constant independent of c, j_c.

Proof. It follows from Lemma 1 that

$$\left|\left(e^{t(Q^*_c - V)} - e^{t(Q_c - V)}\right)(j_c, k_c)\right|$$

is majorized by

$$\left|\int_{W'_{c,j_c} \cap \{X(t) = k_c\}} e^{-\int_0^t V(X(s))ds} \chi_{j_c}(X(t)) d(\mathbf{P}_{c,j_c} - \mathbf{P}^*_{c,j_c})\right|$$

which is

$$\leq (\mathbf{P}_{c,j_c} + \mathbf{P}^*_{c,j_c})(W'_{c,j_c} \cap \{X(t) = k_c\}) .$$

Hence

$$\begin{aligned}\sum_{||k|| \leq N(c)} \left|\left(e^{t(Q^*_c - V)} - e^{t(Q_c - V)}\right)(j_c, k_c)\right| &\leq (\mathbf{P}_{c,j_c} + \mathbf{P}^*_{c,j_c})(W'_{c,j_c})\\ &\leq BN(c)^{-(d+1)}.\end{aligned}$$

The proof of (I') is now quite simple. We have, for $||j|| \leq N(c)/2$,

$$\left|\left(e^{t(Q_c^*-V)} - e^{t(Q_c-V)}\right)(j_c, j_c)\right| \leq 2BN(c)^{-(d+1)}$$

where the O is uniform in $||j|| \leq N(c)/2$. This proves (I′).

5.4. *Proof of* (II′). We consider $j, ||j|| \geq N(c)/2$. We split the integral expressing $e^{t(Q_c^*-V)}(j_c, j_c)$ into two parts, over the region R_1 and over its complement R_1', where

$$R_1 = \left\{X \mid X(0) = j_c, |X(u)| \geq N(c)c^{-1/2}/4 \text{ for all } u \in [0, t/2]\right\} .$$

Lemma 1. *We have, as $c \longrightarrow \infty$, uniformly for $N(c)/2 \leq ||j|| \leq N(c)$,*

$$\int_{R_1} e^{-\int_0^t V(X(s))ds} \chi_{j_c}(X(t))dP^*_{c,j_c} \leq o(N(c)^{-d}) .$$

Proof. For $X \in R_1$,

$$\int_0^t V(X(s))ds \geq \int_0^{t/2} V(X(s))ds \geq t(V^*(N(c)c^{-1/2}/4)/2$$

so that

$$\int_{R_1} e^{-\int_0^t V(X(s))ds} \chi_{j_c}(X(t))dP^*_{c,j_c} \leq e^{-t(V^*(N(c)c^{-1/2}/4)/2}.$$

For any $b > 0$, $V^*(x) \geq b \log x$ for all sufficiently large $x > 0$, and so the right side of the above estimate is $O(N(c)^{-tb/4})$ which is $o(N(c)^{-d})$ if $tb > 4d$.

Lemma 2. *We have, as $c \longrightarrow \infty$, uniformly for $N(c)/2 \leq ||j|| \leq N(c)$,*

$$\int_{R_1'} e^{-\int_0^t V(X(s))ds} \chi_{j_c}(X(t))dP^*_{c,j_c} \leq o(N(c)^{-d}) .$$

Proof. The integral is obviously majorized by $\mathbf{P}^*_{c,j_c}(R_1')$ and so we must estimate this probability. We have

$$R_1' = \left\{X | X(0) = j_c, ||X(u)|| < N(c)c^{-1/2}/4 \text{ for some } u \in [0, t/2]\right\} .$$

Let

$$M_c = \left\{k_c \mid ||k_c|| = [N(c)/2]c^{-1/2}\right\}$$

and let us introduce the stopping time τ of the first visit of the X_c^*-process to the set M_c. Then, writing A for the event

$$A = \left\{X \mid \tau < t/2, |X(\tau + u)| < N(c)c^{-1/2}/4 \text{ for some } u \in [0, t/2]\right\}$$

we have

$$\sup_{j_c} \mathbf{P}^*_{c,j_c}(R_1') \le \sup_{j_c} \mathbf{P}^*_{c,j_c}(A) = \sup_{j_c} \sum_{k_c} \mathbf{P}^*_{c,j_c}(A \cap (X(\tau) = k_c)) \ .$$

In view of the strong Markov property ([7], p. 232), this is majorized by

$$\sup_{j_c} \sum_{k_c} \mathbf{P}^*_{c,j_c}(\tau < t/2, X(\tau) = k_c)\mathbf{P}^*_{c,k_c}(\|X(u)\| < N(c)c^{-1/2}/4 \quad \text{for some } u \in [0,t/2]) \ .$$

Hence

$$\sup_{j_c} \mathbf{P}^*_{c,j_c}(R_1') \le \sup_{k_c} \mathbf{P}^*_{c,k_c}(\|X(u)\| < N(c)c^{-1/2}/4 \text{ for some } u \in [0,t/2]) \ .$$

Let

$$C = \left\{\|X(u)\| < N(c)c^{-1/2}/4 \text{ for some } u \in [0,t/2]\right\}$$
$$A = \left\{\|X(u)\| \le N^*(c)c^{-1/2} \text{ for all } u \in [0,t/2]\right\}$$
$$B = \left\{\|X(u)\| > N^*(c)c^{-1/2} \text{ for some } u \in [0,t/2]\right\}$$

then

$$\mathbf{P}^*_{c,k_c}(C) = \mathbf{P}^*_{c,k_c}(C \cap A) + \mathbf{P}^*_{c,k_c}(C \cap B) \ .$$

By Lemma 3.1

$$\mathbf{P}^*_{c,k_c}(C \cap A) = \mathbf{P}_{c,k_c}(C \cap A) \le \mathbf{P}_{c,k_c}(C), \quad \mathbf{P}^*_{c,k_c}(C \cap B) \le \mathbf{P}^*_{c,k_c}(B) = \mathbf{P}_{c,k_c}(B)$$

so that

$$\mathbf{P}^*_{c,k_c}(C) \le \mathbf{P}_{c,k_c}(C) + \mathbf{P}_{c,k_c}(B) \ .$$

Furthermore,

$$(X(0) = k_c) \cap (B \cup C) \subset \left\{X \mid \|X(u) - X(0)\| > [N(c)/4]c^{-1/2} \quad \text{for some } u \in [0,t/2]\right\} \ .$$

Hence

$$\begin{aligned} \mathbf{P}^*_{c,k_c}(C) &\le 2\mathbf{P}_{c,k_c}(\|X(u) - X(0)\| > [N(c)/4]c^{-1/2} \text{ for some } u \in [0,t/2]) \\ &\le BN(c)^{-(d+1)} \end{aligned}$$

for a constant $B > 0$ independent of c, k_c, M_c. Lemmas 1 and 2 imply (II′) and thus the proof of Proposition 1 is finished.

6. Convergence of Dynamical Groups in Trace Norm

6.1. Let us denote by $\mathcal{H}_2$ the Hilbert space of operators of Hilbert-Schmidt class on $L^2(\mathbf{R}^d)$ with scalar product and norm given by

$$(S,T)_2 = \mathrm{Tr}(ST^\dagger), \quad |||T|||_2^2 = \mathrm{Tr}\,(TT^\dagger) \ .$$

We also consider the trace norm $|||\cdot|||_1$; we have

$$|||AB|||_1 \leq |||A|||_2 |||B|||_2 \,.$$

Recall the class $\mathcal{V}$ of potentials defined in Sec. 4. We consider the Hamiltonians $H(\varepsilon)^{(*)} = -\Delta(\varepsilon)^{(*)} + V_\varepsilon$ where the notation is as before except that for the truncated Laplacian we use $2Q_c^*$ as in Sec. 5, with Q_c^* symmetric so that the Hamiltonians are self-adjoint.

Theorem 1. *Let $V \in \mathcal{V}$. Then for any $t > 0$,*

$$|||e^{-tH(\varepsilon)^{(*)}} - e^{-tH}|||_1 \longrightarrow 0 \quad (\varepsilon \longrightarrow 0) \,.$$

Proof. We do this first with respect to $|||\cdot|||_2$. We note that for this it is sufficient to prove that

$$\mathrm{Tr}\left(e^{-tH(\varepsilon)^{(*)}}\right) \longrightarrow \mathrm{Tr}\left(e^{-tH}\right) \quad (*)$$

for each $t > 0$. For this means that

$$|||e^{-tH(\varepsilon)^{(*)}}|||_2^2 = \mathrm{Tr}(e^{-2tH(\varepsilon)^{(*)}}) \longrightarrow \mathrm{Tr}(e^{-2tH}) = |||e^{-tH}|||_2^2 \,.$$

By Corollary to Proposition 2.2 and the density of the operators of finite rank in $\mathcal{H}_2$ we have

$$e^{-tH(\varepsilon)^{(*)}} \longrightarrow e^{-tH}$$

weakly in $\mathcal{H}_2$ while the norms also converge. Hence we have convergence in $\mathcal{H}_2$. The convergence $(*)$ is now immediate from the results of Secs. 4 and 5 except for the Schwinger Hamiltonians. However for $H(\varepsilon)^{(s)}$ we use as before the comparison with the periodic Hamiltonians. Indeed it is clear from the inequality

$$\mathrm{Tr}(e^{-tH(\varepsilon)^{(s)}}) \leq \mathrm{Tr}(e^{-tH(\varepsilon)^{(p)}})$$

that

$$\mathrm{Tr}\left(e^{-tH}\right) = \lim_\varepsilon \mathrm{Tr}\left(e^{-tH(\varepsilon)^{(p)}}\right) \geq \limsup_\varepsilon \mathrm{Tr}\left(e^{-tH(\varepsilon)^{(s)}}\right) \geq \liminf_\varepsilon \mathrm{Tr}\left(e^{-tH(\varepsilon)^{(s)}}\right).$$

On the other hand, $e^{-tH(\varepsilon)^{(s)}} \longrightarrow e^{-tH}$ strongly. So if $f_1, f_2, \ldots,$ is an orthonormal basis for $L^2(\mathbf{R}^d)$,

$$\mathrm{Tr}\left(e^{-tH(\varepsilon)^{(s)}}\right) \geq \sum_{1\leq j\leq k} \left(e^{-tH(\varepsilon)^{(s)}} f_j, f_j\right) \longrightarrow \sum_{1\leq j\leq k} (e^{-tH} f_j, f_j)$$

so that, letting $k \longrightarrow \infty$,

$$\liminf_\varepsilon \mathrm{Tr}\left(e^{-tH_N^{(s)}}\right) \geq \mathrm{Tr}\left(e^{-tH}\right)$$

which proves (*). For furnishing the proof with $|||\cdot|||_1$ we compare trace norms at t with Hilbert–Schmidt norms at $t/2$ and use the relation

$$|||A^2 - B^2|||_1 \leq |||A + B|||_2 \cdot |||A - B|||_2 + 2|||B|||_2 \cdot |||A - B|||_2$$

which follows from the identity

$$A^2 - B^2 = (A + B)(A - B) + (A - B)B - B(A - B)$$

6.2. The trace convergence implies of course convergence in the uniform operator topology and hence the convergence of eigenvalues and eigenfunctions. The results in Secs. 4 and 5 allow us, as we shall show now, to refine the L^2-convergence of the eigenfunctions to uniform convergence on compacta at least *for the stochastic Hamiltonians* (this means that the Laplacian is approximated by $\Delta(\varepsilon)$ for the infinite lattice and by the stochastic Q_c^* in the finite grid). We treat first the case of the infnite lattice approximation $H(\varepsilon)$. The proof for it is based on the following two simple lemmas.

Lemma 1. *Fix $t > 0$. Then there is a constant $C = C(t) > 0$ such that for all $f \in L^2(\mathbf{R}^d)$ and all ε,*

$$||e^{-tH(\varepsilon)}f||_\infty \leq C||f||_2, \qquad ||e^{-tH}f||_\infty \leq C||f||_2 \,.$$

Proof. The kernel K_V that defines e^{-tH} as an integral operator satisfies

$$0 \leq K_V(x, y) \leq (2\pi t)^{-d/2} e^{-|x-y|^2/2t}$$

so that

$$|(e^{-tH}f)(x)| \leq (4\pi t)^{-d/4}||f||_2 \,.$$

Further

$$||e^{-tH(\varepsilon)}f||_\infty^2 \leq \sup_j \sum_k |e^{-tH(\varepsilon)}(j_c, k_c)|^2 \cdot ||f||_2^2 \cdot \varepsilon^{-d} \,.$$

But, as $e^{-tH(\varepsilon)}$ is self-adjoint, Lemma 4.2.1 gives

$$\sum_k |e^{-tH(\varepsilon)}(j_c, k_c)|^2 = e^{-2tH(\varepsilon)}(j_c, j_c) \leq B_{2t}\varepsilon^d \,.$$

Lemma 2. *Fix $t > 0$. Then, for any $f \in L^2(\mathbf{R}^d)$,*

$$e^{-tH(\varepsilon)}f \longrightarrow e^{-tH}f$$

uniformly on compacta.

Proof. In view of Lemma 1 it is enough to do this for a dense set of f, say $f \in C_c(\mathbf{R}^d)$. Now it is obvious that $e^{-tH(\varepsilon)}$ is the integral operator whose defining kernel is $K(\varepsilon)$ where

$$K(\varepsilon)(x,y) = e^{-tH(\varepsilon)}(j_c, k_c)\varepsilon^{-d} \quad (x \in R(j_c), y \in R(k_c))$$

where for $r \in \mathbf{Z}^d$, $R(r_c)$ is the cube centered at r_c of sides ε. Lemma 4.2.2 implies that $K(\varepsilon) \longrightarrow K_V$ uniformly on compact sets of $\mathbf{R}^d \times \mathbf{R}^d$. The present lemma is now immediate. □

Let f_j be an orthonormal sequence of eigenfunctions of H and $f(\varepsilon)_j$ be a corresponding family for $H(\varepsilon)$ converging in $L^2(\mathbf{R}^d)$.

Proposition 3. *We have*

$$f(\varepsilon)_j \longrightarrow f_j$$

uniformly on compact sets of $\mathbf{R}^d$.

Proof. It is enough to show that

$$e^{-tH(\varepsilon)} f(\varepsilon)_j \longrightarrow e^{-tH} f_j$$

uniformly on compacta. But

$$e^{-tH(\varepsilon)} f(\varepsilon)_j - e^{-tH} f_j = e^{-tH(\varepsilon)}(f(\varepsilon)_j - f_j) + (e^{-tH(\varepsilon)} f_j - e^{-tH} f_j)$$

and the result follows from Lemmas 1 and 2.

We shall now treat the case of the finite grid stochastic approximation. It is clearly enough to establish the analogues of the two lemmas above for $H(\varepsilon)^{(*)}$ in place of $H(\varepsilon)$.

Lemma 1. *Fix $t > 0$. Then for any compact set $E \subset \mathbf{R}^d$ there is a constant $C = C(E,t) > 0$ such that for all $f \in L^2(\mathbf{R}^d)$,*

$$\|e^{-tH(\varepsilon)^{(*)}} f\|_{E,\infty} \le C\|f\|_2$$

where $\|\ \|_{E,\infty}$ is the L^∞-norm on E.

Proof. As before it is enough to show that for any $a > 0, u > 0$ there is a constant $B_{a,u} > 0$ such that for all ε

$$\sup_{|j_c| \le a} e^{-uH(\varepsilon)^{(*)}}(j_c, j_c) \le B_{a,u}\varepsilon^d .$$

This is immediate from Lemma 5.3.2 and Lemma 4.2.1.

Lemma 2. *We have, for all $f \in L^2(\mathbf{R}^d)$ and stochastic $H(\varepsilon)^{(*)}$,*

$$e^{-tH(\varepsilon)^{(*)}} f \longrightarrow e^{-tH} f$$

uniformly on compacta.

Proof. It is a question of showing that

$$e^{-tH(\varepsilon)^{(*)}} f - e^{-tH(\varepsilon)} f \longrightarrow 0$$

uniformly on compact sets for any $f \in C_c(\mathbf{R}^d)$. This comes to showing that for each $a > 0$

$$\sup_{|j_c| \leq a} \sum_{|k_c| \leq a} \left| e^{-tH(\varepsilon)^{(*)}}(j_c, k_c) - e^{-tH(\varepsilon)}(j_c, k_c) \right| = o(1)$$

This is immediate from Lemma 5.3.2.

The following is now clear.

Proposition 3. *Denoting by $f(\varepsilon)^*_j$ an orthonormal sequence of eigenfunctions for stochastic $H(\varepsilon)^{(*)}$ converging in L^2-norm to f_j, we have*

$$f(\varepsilon)^*_j \longrightarrow f_j$$

uniformly on compacta.

References

[1] J. Schwinger, *Proc. Nat. Acad. Sci. USA* **46** (1960) 570–579, 1401–1405.
[2] H. Weyl, *Group Theory and Quantum Mechanics*, Dover, 1931.
[3] H. Kreiss, "Difference approximations for boundary and eigenvalue problems for ordinary differential equations", *Math. Comp.* **26** (1972) 605–624.
[4] M. Reed, and B. Simon, *Methods of Modern Mathematical Physics*, II, Academic Press, 1975.
[5] R. Narasimhan, *Analysis on Real and Complex Manifolds*, North Holland, 1968.
[6] M. Reed, and B. Simon, *Methods of Modern Mathematical Physics, I*, Academic Press, 1980.
[7] D. Freedman, *Markov Chains*, Springer Verlag, 1971.
[8] P. Billingsley, *Weak Convergence of Probability Measures*, Wiley, 1968.
[9] K. R. Parthasarathy, *Probability Measures on Metric Spaces*, Academic Press, 1967.
[10] B. Simon, *Functional Integration and Quantum Physics*, Academic Press, 1979.
[11] N. N. Chentsov, *Theory of Probab. App.* **1** (1956) 140–144.
[12] F. Riesz and B. Sz. Nagy, *Lecons d'Analyse Fonctionnelle*, Budapest, 1952.
[13] S. R. S. Varadhan, *Diffusion Problems and Partial Differential Equations*, Tata Institute Lectures, Bombay, 1980.

Ann. Inst. Henri Poincaré, (In Memoriam C. KIPNIS)

Vol. 31, n° 1, 1995, p. 273-285. *Probabilités et Statistiques*

Self Diffusion of a tagged particle in equilibrium for asymmetric mean zero random walk with simple exclusion

by

S. R. S. VARADHAN (*)

Courant Institute of Mathematical Sciences,
New York University, New York, U.S.A.

Dedicated to the memory of Claude Kipnis.

ABSTRACT. – We consider a tagged particle in a simple exclusion model where the probability distribution of the jump sizes has zero mean, but is not necessarily symmetric. We establish for the tagged particle, a central limit theorem under the usual scaling.

RÉSUMÉ. – Nous considérons une particule marquée dans un modèle d'exclusion simple où la loi de probabilité des sauts est de moyenne nulle mais pas nécessairement symétrique. Nous démontrons pour le déplacement de la particule marquée un théorème central limite avec le changement d'échelle usuel.

1. INTRODUCTION

We consider a random walk with simple exclusion. This means that we have a collection of particles in $Z^d, d \geq 1$ with at most one particle per site. Each particle waits for an exponential time with mean 1 and at the end of this time picks a random site to jump to. The probability that a

(*) Supported by NSF grant DMS-9201222 and ARO grant DAAL03-92-G-0317.
Classification A.M.S.: 60 K 35, 60 F 17.

particle, located at x, picks the site y to jump to is given by $p(y-x)$. However the jump can be executed only if the site y is free. Otherwise the particle remains in the original site and waits for a new exponential time. All the particles are doing this simultaneously and independently of each other. Since we are dealing with processes in continuous time ties will never occur and this describes the evolution completely.

For a more mathematical definition of the model we start with the state space Ω consisting of functions η on Z^d taking values either 0 or 1. If $\eta(x)=1$, then there is a particle at x and if $\eta(x)=0$, the site x is free.

We next define certain transformations on the space Ω.

$$(\sigma_{x,y}\eta)(a) = \eta(a) \quad if \quad a \neq x \; or \; y$$
$$(\sigma_{x,y}\eta)(x) = \eta(y)$$
$$(\sigma_{x,y}\eta)(y) = \eta(x)$$

The transformation $\sigma_{x,y}$ for two sites x, y in Z^d with $x \neq y$ could be either a particle jumping from x to y or one jumping from y to x. Of course if $\eta(x)=\eta(y)$ then $\sigma_{x,y}$ does nothing and is identity. Given $p(x) \geq 0$ with $p(0)=0$ and $\Sigma p(x)=1$, *i.e.* the distribution of jumps, we can define a formal infinitesimal generator acting on suitable functions F on Ω

$$(LF)(\eta) = \Sigma_{x,y}\eta(x)(1-\eta(y))p(y-x)[F(\sigma_{x,y}\eta) - F(\eta)]$$

It is known [3] that this generator defines a good Markov Process on Ω . For each ρ in $0 \leq \rho \leq 1$, we have on Ω the Bernoulli product measure P_ρ with $P_\rho[\eta(x)=1]=\rho$ for all x in Z^d. Each P_ρ is an invariant measure for our evolution. Corresponding to each ρ we have a system of interacting particles with initial distribution P_ρ which is in equilibrium. Of $p(x)$ we assume that it is zero outside a finite set. The case when $p(x)=p(-x)$ is the symmetric case and then the process is reversible with respect to any one of the invariant measures. But in this article we will assume only that

$$\Sigma < x, \theta > p(x) = 0$$

or that the jumps have mean zero. Now the process is not necessarily reversible. We will also assume an irreducibility condition namely that $\{x : p(x) > 0\}$ generates the whole group Z^d .

If we start our evolution under the assumption that there is a particle at 0 and that at other sites we have a random Bernoulli configuration with density ρ , we can tag the particle at 0 as it moves around and denote its location at time t by z_t . We are interested in proving a Brownian scaling limit for z_t , *i.e.* to establish that $\lambda^{-\frac{1}{2}} z_{\lambda t}$ converges as $\lambda \to \infty$ to a

Brownian motion in distribution. In [2], with Claude Kipnis we proved that this was indeed so in the symmetric case. The case $p(1) = p(-1) = \frac{1}{2}$ in one dimension is special and leads to Brownian motion with zero variance. In all other cases it is nondegenerate. In this article we will extend these results to the asymmetric mean zero case.

2. EVOLUTION OF THE TAGGED PARTICLE

In order to follow the motion of a tagged particle the state space has to be changed. The state space Ω cannot distinguish between particles. We shall find it convenient to describe the current state of the system by giving the location of the tagged particle and the environment around the tagged particle. In other words the state space is $\hat{\Omega} = Z^d \times \Omega_0$. Here Ω_0 is the space of configurations on $Z^d - \{0\}$. In order to describe the evolution in the space $\hat{\Omega}$ we need to describe a transformation τ_x that acts on Ω_0 for each $x \neq 0$.

$$(\tau_x\eta)(a) = \eta(x+a) \quad if \quad a \neq 0 \;\; or \quad -x$$
$$(\tau_x\eta)(-x) = 0$$

$\tau_x\eta$ is meaningful only if $\eta(x) = 0$ and describes the effect of the tagged particle, which by definition is always at zero, jumping to x. We relocate the origin at x the new location of the tagged particle. The transformations $\sigma_{x,y}$ for $x, y \neq 0$ are well defined on Ω_0. The evolution of the tagged particle on $\hat{\Omega}$ is governed by the following generator

$$(\hat{L}F)(z,\eta) = \Sigma_{x,y\neq 0}\eta(x)(1-\eta(y))p(y-x)[F(z,\sigma_{x,y}\eta) - F(z,\eta)]$$
$$+ \Sigma_x(1-\eta(x))p(x)[F(z+x,\tau_x\eta) - F(z,\eta)].$$

If F were a function of η only, then

$$(L_0F)(\eta) = \Sigma_{x,y\neq 0}\eta(x)(1-\eta(y))p(y-x)[F(\sigma_{x,y}\eta) - F(\eta)]$$
$$+ \Sigma_x(1-\eta(x))p(x)[F(\tau_x\eta) - F(\eta)].$$

In other words if (z_t, η_t) is the tagged system , η_t by itself is a Markov Process with generator L_0 . The Bernoulli measures P^0_ρ on Ω_0 defined by

$$P^0_\rho[\eta(x) = 1] = \rho \quad for\ all\ x \in Z^d - \{0\}$$

are invariant measures for L_0. $\hat{L}$ on the other hand does not have an invariant measure because the z_t part will wander away in Z^d.

Our initial distribution on $\hat{\Omega}$ is $\delta_0 \times P^0_\rho$, *i.e.* we start from $z = 0$ and η is in equilibrium . Then under the $\hat{L}$ evolution η is always in equilibrium. z_t is the location of the tagged particle at time t and the scaling limit is to be established for it.

3. OUTLINE OF PROOF

A direct calculation yields

$$\hat{L}z = \Sigma_x(1 - \eta(x))xp(x)$$

or equivalently for each θ in R^d

$$\begin{aligned}\hat{L} < z, \theta > &= \Sigma_x(1 - \eta(x)) < x, \theta > p(x) \\ &= < \psi(\eta), \theta > .\end{aligned}$$

Therefore,

$$< z_t, \theta > = \int_0^t < \psi(\eta_s), \theta > ds + < M(t), \theta > .$$

Here $M(t)$ is a Martingale with stationary increments. The basic idea in [2] was to replace the first term on the right in the above equation by a Martingale term and an error term.

$$\int_0^t < \psi(\eta_s), \theta > ds = < N(t), \theta > + < E(t), \theta > .$$

Our ability to do so depended on the following procedure.

Step 1.

Solve the equation

$$\lambda u_\lambda - L_0 u_\lambda = \psi$$

For simplicity let us take just one component of ψ . Based on an estimate

$$|E^{P^0_\rho}[\psi(\eta)G(\eta)]| \leq C_\rho [D_\rho(G)]^{\frac{1}{2}}$$

where $D_\rho(u) = - < L_0 u, u >_\rho$ is the Dirichlet form of u, we get for the solution u_λ

$$E[\lambda u_\lambda^2 + D_\rho(u_\lambda)] = < \psi, u_\lambda >_\rho$$
$$\leq C_\rho [D_\rho(u_\lambda)]^{\frac{1}{2}}.$$

This gives us bounds

$$\lambda E u_\lambda^2 \leq C_\rho^2$$
$$D_\rho(u_\lambda) \leq C_\rho^2$$

holding uniformly as $\lambda \to 0$.

Step 2.

Using either spectral theory or a more direct argument one establishes in fact that

$$\lambda E[u_\lambda^2] \to 0 \;\; as \;\; \lambda \to 0$$

and

$$D_\rho(u_{\lambda_1} - u_{\lambda_2}) \to 0 \;\; as \;\; \lambda_1, \lambda_2 \to 0.$$

This was enough to reduce the Brownian scaling property for z_t to the same thing for a Martingale with stationary increments which is elementary and standard.

Step 3.

We prove compactness by establishing the tightness of the distributions of

$$\int_0^t \psi(\eta_s) ds$$

under Brownian scaling.

Step 4.

Establish the nondegeneracy of the Brownian motion obtained as the scaling limit.

4. DETAILS OF PROOF

Due to the asymmetry one has to make changes in the proof along the way.

Step 1.

$$\begin{aligned}\psi(\eta) &= \Sigma(1-\eta(x)) < x, \theta > p(x) \\ &= \Sigma_{x\neq 0}\eta(x)\omega_\theta(x)\end{aligned}$$

where $\omega_\theta(x)$ are weights with $\Sigma\omega_\theta(x) = 0$. We can therefore rewrite

$$\psi(\eta) = \Sigma_{x,y\neq 0}[\eta(x) - \eta(y)]\tilde{\omega}(x, y)$$

for some weights $\tilde{\omega}(x, y)$. Because of irreducibility we can assume that $\tilde{\omega}(x, y) \neq 0$ only if $p(y - x) > 0$. Since $\eta(x) - \eta(y) = -((\sigma_{x,y}\eta)(y) - (\sigma_{x,y}\eta)(x))$ we can write,

$$\begin{aligned}E[G(\eta)\psi(\eta)] &= \frac{1}{2}E[\Sigma\tilde{\omega}(x, y)[\eta(x) - \eta(y)][G(\sigma_{x,y}\eta) - G(\eta)] \\ &\leq C[D_1(G)]^{1/2}.\end{aligned}$$

where

$$D_1(G) = E[\Sigma_{x,y\neq 0}p(y - x)[G(\sigma_{x,y}\eta) - G(\eta)]^2.$$

Here and in what follows E is expectation relative to P_ρ^0 for some fixed $0 < \rho < 1$. If we denote by

$$D_2(G) = E[\Sigma_x[G(\tau_x\eta) - G(\eta)]^2]$$

then the full Dirichlet form for L_0 is

$$D_0(G) = D_1(G) + D_2(G)$$

Step 2.

If now one solves the equation

$$\lambda u_\lambda - L_0 u_\lambda = \psi$$

one can get the estimates

$$\begin{aligned}\lambda E u_\lambda^2 + D_0(u_\lambda) &= E\psi u_\lambda \leq C[D_1(u_\lambda)]^{\frac{1}{2}}, \\ \lambda E u_\lambda^2 &\leq C, \\ D_0(u_\lambda) &\leq C.\end{aligned} \tag{4.1}$$

We can as before take a subsequence of u_λ converging to a limit u in H_1. Here H_1 is the completion of nice functions with respect to the

Dirichlet norm $D_0(u)$. It contains a lot of generalized functions and consists technically of limits of equivalence classes of functions modulo constants. Let us suppose that the following estimate holds. For all functions F and G,

$$E[FL_0G] \leq C[D_0(F)]^{\frac{1}{2}}[D_0(G)]^{\frac{1}{2}} \tag{4.2}$$

then L_0 will be a bounded operator from H_1 into H_{-1} and $\lambda u_\lambda \to 0$ weakly in H_{-1}. Therefore

$$L_0 u = -\psi \quad in \quad H_{-1}$$

From this we obtain,

$$D_0(u) = < u, \psi >$$

where $<,>$ is the pairing between H_1 and H_{-1}, the latter being the formal dual of the former. Taking limits as $\lambda \to 0$ in (4.1) if $d_1 = lim\ \lambda E u_\lambda^2$ and $d_2 = lim\ D_0(u_\lambda)$ along suitable subsequences then

$$d_1 + d_2 = < u, \psi > = D_0(u)$$

By lower semicontinuity $D_0(u) \leq d_2$. Therefore $d_1 = 0$ and $d_2 = D_0(u)$. This proves the strong convergence of u_λ in H_1 and also establishes

$$lim\ \lambda E u_\lambda^2 = 0.$$

It is now routine to prove the uniqueness of the limit point. So the inequality (4.2) which is obvious with $C = 1$ in the symmetric case is the crucial step in the nonsymmetric case. We shall prove it in the last section. The rest of the details are identical to the symmetric case.

Step 3.

Whereas in [2], for the symmetric case, compactness was a consequence of the estimate in step 1, in our more general context we need to use some special properties of ψ. We start with the estimate

$$\frac{1}{t} log E\left\{exp\left[\lambda \int_0^t \psi(\eta_s)ds\right]\right\} \leq \sup_{||G||_2=1} [\lambda E[\psi G^2] - D_0(G)]$$

As in step 1

$$E[G^2\psi] = \frac{1}{2} E\Sigma[\eta(x) - \eta(y)]\tilde{\omega}(x,y)[G^2(\sigma_{x,y}\eta) - G^2(\eta)]$$

and

$$\begin{aligned}
&E[(\eta(x)-\eta(y))(G^2(\sigma_{x,y}\eta)-G^2(\eta))]\\
&=E[[\eta(x)-\eta(y)][G(\sigma_{x,y}\eta)-G(\eta)][G(\sigma_{x,y}(\eta)+G(\eta)]]\\
&\le E[|G(\sigma_{x,y}\eta)-G(\eta)||G(\sigma_{x,y}\eta)+G(\eta)|]\\
&\le [E[G(\sigma_{x,y}\eta)-G(\eta)]^2]^{\frac{1}{2}}[E[G(\sigma_{x,y}\eta)+G(\eta)]^2]^{\frac{1}{2}}\\
&\le C[E[G(\sigma_{x,y}\eta)-G(\eta)]^2]^{\frac{1}{2}}.
\end{aligned}$$

We have used the facts that $|\eta(x)-\eta(y)|\le 1$ and $EG^2(\sigma_{x,y}\eta)=EG^2(\eta)=1$. Therefore

$$\begin{aligned}
\sup_{||G||_2=1}[\lambda E\psi G^2 \quad - D_0(G)] &\le \sup_{||G||_2=1}[C\lambda(D_0(G))^{\frac{1}{2}}-D_0(G)]\\
&\le \frac{A\lambda^2}{2}
\end{aligned}$$

In other words

$$E\left[exp\left[\lambda\int_s^t \psi(\eta_s)ds\right]\right]\le exp\left[A(t-s)\frac{\lambda^2}{2}\right].$$

Compactness is now a consequence of Garsia-Rodemick-Rumsey estimate which can be found in [4].

Step 4

We now establish the nondegeneracy of the limiting Brownian motion. From the outline of the proof it is clear that what we need to establish is that the martingale produced by $\int_0^t <\psi(\eta_s),\theta> ds$ namely $<N(t),\theta>$ cannot cancel the martingale $<M(t),\theta>$. There are two classes of martingales generated by the underlying Poisson events: jumps that involve the tagged particle, identified through τ_x and those involving other particles identified through $\sigma_{x,y}$ with $x,y\neq 0$. These are orthogonal martingales because they relate to non simultaneous Poisson jumps. If $N(t)$ were to totally cancel $M(t)$, which involves only the first type of jumps, its orthogonal projection on to the second type of martingales should go to zero. This can happen only if for the corresponding u_λ, $D_1(u_\lambda)\to 0 \ as \ \lambda\to 0$. But if $D_1(u_\lambda)\to 0$ then it follows from our estimates that $D_0(u_\lambda)\to 0$ as well and this forces $N(t)$ to be identically zero. With a small bit of extra calculation one can turn this into a lower bound for the variance of the limiting Brownian motion.

With only the basic estimate (4.2) remaining to be proved in the next section we have now established our main result.

THEOREM 4.1. – *If z_t is the location of the tagged particle in Z^d then as $k \to \infty$ the distribution of the stochastic process $\frac{z_{kt}}{\sqrt{k}}$ converges in Skorohod space to a Brownian motion on R^d with a nondegenerate covariance matrix.*

5. THE BASIC ESTIMATE

For the operator L_0 acting on functions defined on Ω_0

$$\begin{aligned}(L_0F)(\eta) = \Sigma_x(1-\eta(x))[F(\tau_x\eta) - F(\eta)]p(x) \\ + \Sigma_{x,y\neq 0}\eta(x)(1-\eta(y))p(y-x)[F(\sigma_{x,y}\eta) - F(\eta)]\end{aligned}$$

we wish to establish the following estimate.

THEOREM 5.1. - *For functions $F(\eta)$ and $G(\eta)$*

$$| < F, L_0G >_\rho | \leq C_\rho[D_0(F)]^{\frac{1}{2}}[D_0(G)]^{\frac{1}{2}}$$

for some constant C_ρ depending only on ρ.

Here $<,>_\rho$ represents the inner product in $L_2(P^0_\rho)$ and $D_0(F)$ is the Dirichlet form

$$\begin{aligned}D_0(F) = \frac{1}{2}E^{P^0_\rho}\{\Sigma_x(1-\eta(x))p(x)[F(\tau_x\eta) - F(\eta)]^2 \\ + \Sigma_{x,y\neq 0}\eta(x)(1-\eta(y))p(y-x)[F(\sigma_{x,y}\eta) - F(\eta)]^2\}\end{aligned}$$

Before we prove Theorem 5.1 let us make some observations.

1. If L_0 were symmetric in $L_2(P^0_\rho)$ then the inequality is valid with $C_\rho = 1$.

2. Suppose A is a Markov generator on a finite state space with an invariant measure μ and A is not symmetric. Let $\bar{A}$ be the symmetrization $\frac{A+A^*}{2}$ in $L_2(\mu)$. Since A and $\bar{A}$ have the same range, if we have $Ag = h$ we also have for some $\bar{g}$, $\bar{A}\bar{g} = h$ and the map $g \to \bar{g}$ is well defined and the two Dirichlet forms are related by

$$D(\bar{g}) \leq CD(g)$$

Let us first prove a similar estimate for the untagged system L on Ω.

$$(LF)(\eta) = \Sigma_{x,y}\eta(x)(1-\eta(y))p(y-x)[F(\sigma_{x,y}\eta) - F(\eta)]$$

LEMMA 5.2. – *For any two functions F,G on* Ω *and for any* ρ

$$| < F, LG >_\rho | \leq C_\rho [D(F)]^{1/2} [D(G)]^{1/2}.$$

Proof. – The proof depends on the following considerations. Suppose $a_1, a_2, \cdots a_k$ are k points in Z^d such that $a_1 + a_2 + \cdots + a_k = 0$, then the probability distribution $\pi(x)$ on Z^d defined by $\pi(a_j) = \frac{1}{k}$ for $j = 1, 2 \cdots k$ and $\pi(x) = 0$ for all other x has clearly mean 0. The sequence $y_0 = 0, y_1 = a_1, y_2 = a_1 + a_2, \cdots, y_k = a_1 + a_2 + \cdots + a_k = 0$ defines a cycle $C = \{y_0, y_1, \cdots, y_k\}$ and a π can be associated to each such cycle by taking $a_j = y_j - y_{j-1}$. Moreover one can assume that the cycle has no double points. Otherwise the cycle decomposes into two or more cycles and π_C is a convex combination of π_{C_i} corresponding to the component cycles. We can therefore limit ourselves to irreducible cycles. If $C_1, C_2, \cdots, C_l$ are l irreducible cycles and $w_1, w_2, \cdots, w_l$ are nonnegative weights adding up to 1, the convex combination $p(x) = \Sigma w_i \pi_{C_i}(x)$ is a mean zero probability distribution on Z^d. We have the converse

LEMMA 5.3. – *Any* $p(x)$ *with finite support and mean 0 has a representation*

$$p(x) = \Sigma w_i \pi_{C_i}(x)$$

for some weights $w_1, w_2, \cdots, w_l$ *and irreducible cycles* $C_1, C_2, \cdots, C_l$.

Proof. – A proof of this lemma can be found in [5].

We continue with the proof of lemma 5.2. Each cycle C introduces an operator A_C on Ω by

$$(A_C F)(\eta) = (1/k) \Sigma_{i=0}^{k-1} \eta(y_i)(1 - \eta(y_{i+1}))[F(\sigma_{y_i, y_{i+1}} \eta) - F(\eta)]$$

One can verify that A_C has P_ρ for an invariant measure and is in general nonreversible unless $k = 2$. For any cycle C we can consider the cycle $C + x$ described by $x, x + y_1, \cdots, x + y_k = x$ starting and ending at x rather than at 0 and define

$$(A_{C+x} F)(\eta) = (1/k) \Sigma_{i=0}^{k-1} \eta(x + y_i)(1 - \eta(x + y_{i+1}))$$
$$[F(\sigma_{x+y_i, x+y_{i+1}} \eta) - F(\eta)]$$

We can make a translation invariant generator on Ω by defining

$$L_C = \Sigma_{x \in Z^d} A_{C+x}$$

L_C is a typical generator for asymmetric mean zero random walk with simple exclusion. A consequence of lemma 5.3. is the representation

LEMMA 5.4. – *Our generator L for asymmetric mean zero simple exclusion has a representation*

$$L = \Sigma_{i=0}^{l} w_i L_{C_i}$$

for some $C_1, C_2, \cdots, C_l$ which are irreducible cycles.

Now we can complete the proof of lemma 5.2. The Dirichlet form for L is easily seen to be weighted sum of the forms for each C_i. Therefore there is no loss of generality in assuming $l = 1$ or $L = A_C$ for some irreducible C.

$$L = \Sigma_x A_{C+x}$$

The Dirichlet form for L_C is the sum of Dirichlet forms for each A_{C+x}. By Schwartz's inequality it is enough to prove the estimate for each A_{C+x} and this is essentially observation 2. The idea is that instead of detailed balance that produces reversibility we now have local balance that yields bounds.

Now we have to extend these considerations to the tagged system. We need another basic idea. Suppose $(\Omega, \mathcal{F}, P)$ is a probability space and $T_1, T_2, \cdots, T_k$ are k measure preserving transformations. Let us assume $T_k \cdots T_1 = I$, and define

$$(LF)(\omega) = \Sigma_{i=0}^{k}[F(T_i\omega) - F(\omega)].$$

Our claim is that our basic estimate is valid in this context.

LEMMA 5.5. – *We have for all F and G,*

$$|E[FLG]| \leq C[D(F)]^{1/2}[D(F)]^{1/2}.$$

Proof.

$$\begin{aligned}\int FLGdP &= \Sigma \int F(\omega)[G(T_i\omega) - G(\omega)]dP \\ &= \Sigma \int F(T_{i-1} \cdots T_1\omega)[G(T_i \cdots T_1\omega) - G(T_{i-1} \cdots T_1\omega)]dP \\ &= \Sigma \int [F(T_{i-1} \cdots T_1\omega) - F(\omega)] \\ &\qquad [G(T_i \cdots T_1\omega) - G(T_{i-1} \cdots T_1\omega)]dP\end{aligned}$$

because

$$\begin{aligned}&\Sigma_i F(\omega)[G(T_i \cdots T_1\omega) - G(T_{i-1} \cdots T_1\omega)] \\ &\quad = F(\omega)[G(T_k \cdots T_1\omega) - G(\omega)] = 0.\end{aligned}$$

Rest is Schwartz.

Finally we return to the proof of Theorem 5.1. The operator

$$\begin{aligned}(L_0F)(\eta) &= \Sigma p(x)(1-\eta(x))[F(\tau_x\eta) - F(\eta)] \\ &\quad + \Sigma p(y-x)\eta(x)(1-\eta(y))[F(\sigma_{x,y}\eta) - F(\eta)]\end{aligned}$$

decomposes into cycles and we can assume without loss of generality that $p(x) = \pi_C(x)$ for some irreducible C. The problem is that 0 plays a special role and translates $C + x$ of the cycle C that touch 0 create trouble. We can write

$$L_0F = L_1F + L_2F$$

where L_2F involves only jumps of the untagged particles and in addition to full cycles that are estimated without any trouble there are incomplete cycles because they go through the origin. L_1 on the other hand involves only jumps of the tagged particle.

Let us remark that

$$\eta(x)(1-\eta(y))[F(\sigma_{x,y}\eta) - F(\eta)] = (1-\eta(y))[F(\sigma_{x,y}\eta) - F(\eta)]$$

so that the terms of L_1 and L_2 look similar. Let our cycle be $0 = y_0, y_1, \cdots, y_k = 0$ with $a_i = y_i - y_{i-1}$ for $i = 1, \cdots, k$. Then ignoring constants 1/k and disregarding the full cycles,

$$\begin{aligned}(L_1F)(\eta) &= \Sigma_i(1-\eta(a_i))[F(\tau_{a_i}\eta) - F(\eta)] \\ (L_2F)(\eta) &= \Sigma_{(x,y)\in\Delta}(1-\eta(y))[F(\sigma_{x,y}\eta) - F(\eta)]\end{aligned}$$

The rest of the proof can be best described in words. Suppose there is an empty site in the loop in front of the tagged particle, *i.e.* $\eta(a_1) = 0$. Then the tagged particle can move, that is we can apply τ_{a_1}. Now the empty site created at the origin is really at $-a_1 = (a_2 + \cdots + a_k)$ because the origin has shifted with the tagged particle. $-a_1$ is the last site of the loop that starts from the new origin *i.e.* the old a_1. One can effect $\sigma_{a_2+\cdots+a_{k-1}, a_2+\cdots+a_k}$ and get a free site at $a_2 + \cdots + a_{k-1}$. We can proceed in this fashion till we get a free site at a_2. Now the tagged particle can jump to a_2 and the whole process starts again. After several steps the tagged particle will return to its original starting point with an empty site in front and all other particles in exactly the same position that we started from. This takes exactly $n = k(k-1)$ steps and we use up every term of L_1 and L_2 exactly once. In other words

$$(L_1 + L_2)(F)(\eta) = \Sigma\chi_{E_j}[F(S_j\eta) - F(\eta)]$$

where $E_{j+1} = S_j S_{j-1} \cdots S_1 E_1$ and $S_n S_{n-1} \cdots S_1 = I \ on \ E_1$. We can use lemma 5.5 at this point and we are done.

6. REMARKS

The basic estimate (4.2) for the untagged system was derived and used by Lin Xu in his New York University PhD dissertation to establish a hydrodynamic scaling limit for mean zero random walks with simple exclusion. A central limit theorem for the position of a tagged particle in the case of asymmetric one dimensional simple exclusion (non zero mean) was considered in [1] by C. Kipnis.

ACKNOWLEDGEMENTS

The author wishes to thank F. Rezakhanlou for valuable discussions.

REFERENCES

[1] C. Kipnis, Central limit theorems for infinite series of queues and applications to simple exclusion, *Ann. Prob.* Vol. **14**, 1986, pp. 397-408.

[2] C. Kipnis and S. R. S. Varadhan, Central limit theorem for additive functionals of reversible Markov processes and application to simple exclusions, *Comm. Math. Phys.* Vol. **104**, 1986, pp. 1-19.

[3] T. Liggett, *Interacting Particle Systems*, Springer-Verlag, 1985.

[4] D. W. Stroock and S. R. S. Varadhan, *Multidimensional Diffusion Processes*, Springer-Verlag, Berlin-Heidelberg-New York, 1979.

[5] L. Xu, *Ph. D Dissertation,* New York University, 1993.

(Manuscript received April 9, 1994; revised August 14, 1994.)

Communications on Pure and Applied Mathematics, Vol. LIII, 0972–1006 (2000)

Diffusive Limit of a Tagged Particle in Asymmetric Simple Exclusion Processes

SUNDER SETHURAMAN
Iowa State University
S. R. S. VARADHAN
Courant Institute
AND
HORNG-TZER YAU
Courant Institute

Dedicated to Dr C. R. Rao on his eightieth birthday.

Abstract

Invariance principles are proved under diffusive scaling for the centered position of a tagged particle in the simple exclusion process with asymmetric nonzero drift jump probabilities in dimensions $d \geq 3$. The method of proof is by martingale techniques which rely on the fact that symmetric random walks are transient in high dimensions.

1 Introduction

In the study of interacting particle systems, an important component is the motion of a tagged particle as it moves through other particles that are viewed as indistinguishable among themselves. An initial step is to understand the behavior of a tagged particle in equilibrium, in particular, the fluctuation behavior. The simple exclusion processes are a natural class of models where one can study this problem. In the case of simple exclusion processes on $\mathbb{Z}^d$ when the underlying random walk has a symmetric distribution, this question is answered in [1] for $d = 1$ nearest-neighbor walks, and in [6] for all other distributions and dimensions. It was generalized to the case of random walks with mean 0 in [11]. When the mean is nonzero, the only case studied previously is the asymmetric nearest-neighbor $d = 1$ case in [5]. Here we consider the general case where the mean is not zero. We need to assume, however, that *the dimension d of the lattice $\mathbb{Z}^d$ is at least* 3.

If X is a countable set and $p(x, y)$ is the transition probability matrix of a Markov chain, then one can define the simple exclusion process on X with transition probabilities $p(\cdot,\cdot)$ on X. This is actually a Markov process whose state space is the set $\Omega(X)$ of all possible subsets of X. The set $A \subset X$ signifies that there are particles present at every point of A and that the sites in $X \setminus A$ are empty. One can also view $\Omega(X)$ as the set of maps $\eta : X \to \{0,1\}$ where $\eta(x) = 1$ means that $x \in A$ or that there is a particle at the site x. $\eta(x) = 0$ signifies that the site is empty. Each particle waits for an exponential time and then jumps to a new site or at least tries to. The new site y is selected with probability $p(x,y)$ where the current site is x. If the site y is empty, then the jump is completed and things start afresh. If the site is occupied with a particle already there, the jump is forbidden and the original particle remains at x. Again things start afresh. All the particles are doing this simultaneously, and because we are dealing with continuous time, there will be no ties to resolve. However, it requires some work to make sure that the process is well-defined, especially if there are an infinite number of particles to begin with.

Suppose $u : \Omega \to \mathbb{R}$ is a function that depends only on a finite number of coordinates, i.e., on $\{\eta(x) : x \in F\}$ for some finite set F; then the infinitesimal generator of the process is defined by

$$(\mathcal{L}u)(\eta) = \sum_{x,y} \eta(x)(1-\eta(y))p(x,y)[u(\eta^{x,y}) - u(\eta)]$$

where

$$\eta^{x,y}(z) = \begin{cases} \eta(z) & \text{if } z \neq x \text{ or } y \\ \eta(y) & \text{if } z = x \\ \eta(x) & \text{if } z = y. \end{cases} \tag{1.1}$$

Under some mild conditions on $p(\cdot,\cdot)$ (for instance, it is sufficient to assume that $\sup_y \sum_x p(x,y) < \infty$), there is a well-defined stochastic process starting from any initial configuration in Ω. The details can be found in [8]. This is referred to as the simple exclusion process on X with transition probability $p(\cdot,\cdot)$. A special case of particular interest is when $p(\cdot,\cdot)$ is doubly stochastic, i.e.,

$$\sum_x p(x,y) = 1 \;\text{ for all } y \quad \text{in addition to} \quad \sum_y p(x,y) = 1 \;\text{ for all } x.$$

In this case the process is always well-defined. A more special case is when $p(\cdot,\cdot)$ is symmetric. This case is referred to as symmetric simple exclusion.

In the doubly stochastic case, the uniform measure on X is a σ-finite invariant measure for the Markov chain. One can verify that, for any $0 \le \rho \le 1$, the Bernoulli product measure P_ρ on Ω defined by $P_\rho[\eta(x) = 1] = \rho$ for every x, with $\{\eta(x)\}$ being mutually independent for different x, is an ergodic invariant measure for the evolution on Ω with generator L. If $p(\cdot,\cdot)$ is symmetric, the evolution is reversible with respect to each P_ρ and has for its Dirichlet form the quantity

$$D^\rho(u) = \frac{1}{2} \int_\Omega \sum_{x,y} p(x,y)[u(\eta^{x,y}) - u(\eta)]^2 dP_\rho .$$

Our interest is mainly in the case where $X = \mathbb{Z}^d$ for some $d \ge 3$ and $p(x,y) = p(y-x)$ for some probability distribution $p(\cdot)$ on $\mathbb{Z}^d$. We can assume without loss of generality that $p(0)=0$. We are interested in the case when

$$m=\sum_x xp(x)\neq 0. \tag{1.2}$$

We shall assume for simplicity that $p(x)=0$ outside a finite set F, although it will not matter that much. We shall start the process in equilibrium, that is to say, with the initial distribution being some P_ρ but conditioned to have a particle at the origin 0, which will be tagged. As the system evolves, we wish to follow the trajectory of the tagged particle. It is convenient to change our coordinates in $\mathbb{Z}^d$ so that the tagged particle is always seen at the origin. In other words, our description of the current state consists of the position of the tagged particle, which we denote by z, and the environment seen from the tagged particle, which can be viewed as a point in $\Omega_0=\{\eta:\mathbb{Z}^d\setminus\{0\}\to\{0,1\}\}$.

There are two types of motion. When an untagged particle jumps, it is from some x to a y, neither of which can be 0. When the tagged particle jumps from 0 to an empty site x, the origin shifts with it, so what we see is a shift of the environment by $-x$. The tagged particle at 0 is not part of the shift, and so we always end up with $-x$ being empty. In other words, if we define on the set $\eta(x)=0$, the map τ_x,

$$(\tau_x\eta)(y)=\begin{cases}\eta(x+y) & \text{for } y\neq 0,\ y\neq -x\\ 0 & \text{for } y=-x,\end{cases}$$

then the generator of our process is given by $\mathcal{L}=\mathcal{L}^{\mathrm{sh}}+\mathcal{L}^{\mathrm{ex}}$ where

$$(\mathcal{L}^{\mathrm{sh}}u)(z,\eta)=\sum_x(1-\eta(x))p(x)[u(z+x,\tau_x\eta)-u(z,\eta)]$$

and

$$(\mathcal{L}^{\mathrm{ex}}u)(z,\eta)=\sum_{x,y\neq 0}\eta(x)(1-\eta(y))p(y-x)\left[u(z,\eta^{x,y})-u(z,\eta)\right].$$

The environment by itself is a Markov process, and the generator is given by $L=L^{\mathrm{sh}}+L^{\mathrm{ex}}$ where

$$(L^{\mathrm{sh}}u)(\eta)=\sum_x(1-\eta(x))p(x)[u(\tau_x\eta)-u(\eta)] \tag{1.3}$$

$$(L^{\mathrm{ex}}u)(\eta)=\sum_{x,y\neq 0}\eta(x)(1-\eta(y))p(y-x)\left[u(\eta^{x,y})-u(\eta)\right]. \tag{1.4}$$

We have adopted the convention that the generators associated with the original process are denoted by script $\mathcal{L}$, while those associated with the environment process are denoted by L. The Bernoulli product measure P_ρ restricted to points in $\mathbb{Z}^d\setminus\{0\}$ is an ergodic invariant measure for the environment process (see proposition 3 of [9]). An elementary computation shows that

$$\mathcal{L}z=g(\eta)=\sum_x xp(x)(1-\eta(x)) \tag{1.5}$$

so that

$$z(t) - z(0) - \int_0^t g(\eta(s))ds = M(t)$$

is a martingale with stationary increments. One can see almost surely [9],

$$\lim_{t\to\infty} \frac{z(t)}{t} = \lim_{t\to\infty} \frac{1}{t} \int_0^t g(\eta(s))ds = \int_{\Omega_0} g(\eta)dP_\rho = m(1-\rho).$$

Let

$$\xi(t) = \frac{1}{\sqrt{t}}(z(t) - m(1-\rho)t); \tag{1.6}$$

in this article we prove a functional central limit theorem for ξ. As remarked earlier, the symmetric case for nearest-neighbor $d = 1$ walks was considered in [1], and all the other symmetric cases in [6]. The more general case $m = 0$ was covered in [11]. Also, the $d = 1$ asymmetric nearest-neighbor $m \neq 0$ case was done in [5]. We are now interested in the case when $m \neq 0$ in high dimensions, which requires different methods. Our main result is the following theorem:

THEOREM 1.1 *As $\alpha \to \infty$, the distribution of $(1/\sqrt{\alpha})(z(\alpha t) - m(1-\rho)\alpha t)$ converges weakly in Skorohod space to a nondegenerate Brownian motion with covariance $\mathbf{C}(\rho)$ given in formula* (2.23).

In the next section, we outline the proof of this theorem. Our proof follows the approach in [6] of considering the environment process as seen by the tagged particle and to explore the associated martingales. The key step in this approach is an estimate of the resolvent equation. The methods used here for solving the resolvent equation are similar to the methods used for solving the fluctuation-dissipation equation of the hydrodynamical limit of the simple exclusion processes in [7] and of lattice gases in [2]. Notice that both the fluctuation-dissipation equation and the resolvent equation require estimates on the Green's function of the generator, though in somewhat different contexts. More technical comments can be found at the end of the next section.

2 Outline of the Proof of Theorem 1.1

The tagged particle process has a lot of martingales associated with it. They are of the form

$$M_x(t) = N_x(t) - \int_0^t p(x)(1-\eta(s,x))ds \tag{2.1}$$

and

$$M_{x,y}(t) = N_{x,y}(t) - \int_0^t p(y-x)\eta(s,x)(1-\eta(s,y))ds, \tag{2.2}$$

corresponding to the number of jumps $N_x(t)$ of size x for the tagged particle or the number of jumps $N_{x,y}(t)$ of untagged particles from x to y. The compensators of their squares are given by

$$d\langle [M_{x,y}(t)]^2\rangle = p(y-x)\eta(t,x)(1-\eta(t,y))dt\,,$$
$$d\langle [M_x(t)]^2\rangle = p(x)(1-\eta(t,x))dt\,.$$

These are the basic martingales, and every other martingale is a combination of these. For example, the position $z(t)$ of the tagged particle satisfies

$$z(t) = \sum_x xN_x(t) = \sum_x xM_x(t) + \int_0^t \sum_x xp(x)(1-\eta(s,x))ds\,.$$

Subtracting the term $m(1-\rho)t$, we have

$$\xi(t) = z(t) - m(1-\rho)t = \xi^{(1)}(t) + \int_0^t g(\eta(s))ds \tag{2.3}$$

where

$$\xi^{(1)}(t) = \sum_x xM_x(t) \tag{2.4}$$

and g is the vector-valued function given by (1.5), i.e.,

$$g(\eta) = \sum_x xp(x)(\rho-\eta(x))\,.$$

The problem now reduces to proving the central limit theorem for the centered additive functional

$$A(t) = \int_0^t g(\eta(s))ds\,.$$

We would like to do it by the martingale method, which means finding a square integrable martingale $\xi^{(2)}(t)$ such that

$$A(t) = \xi^{(2)}(t) + B(t)$$

with $B(\cdot)$ becoming negligible under rescaling. Then the central limit theorem for square integrable martingales with stationary increments applies to the sum $\xi^{(1)}(t)+\xi^{(2)}(t)$, and this will establish the result. We may write the generator of the process that describes the environment of the tagged particle as the sum of its symmetric and antisymmetric pieces,

$$L = L_{\text{sym}} + L_{\text{skew}}\,.$$

Associated with the symmetric part, we have the Dirichlet form

$$D^\rho(u) = D^{ex,\rho}(u) + D^{sh,\rho}(u) \tag{2.5}$$

where

$$D^{ex,\rho}(u) = \frac{1}{2}\int_{\Omega_0} \sum_{x,y\neq 0} a(y-x)[u(\eta^{x,y}) - u(\eta)]^2 dP_\rho \tag{2.6}$$

and

$$D^{sh,\rho}(u) = \frac{1}{2}\int_{\Omega_0} \sum_x a(x)(1-\eta(x))[u(\tau_x\eta) - u(\eta)]^2 dP_\rho \tag{2.7}$$

with $a(x) = \frac{1}{2}[p(x) + p(-x)]$. The associated Dirichlet norm is $\|\cdot\|_1 = \sqrt{D_\rho(u)}$. The dual norm $\|\cdot\|_{-1}$ is defined by

$$\|g\|_{-1} = \inf\left[C : \left|\int g u\, dP_\rho\right| \le C\|u\|_1 \text{ for all local functions } u\right].$$

We define $\mathbb{H}_1$ and $\mathbb{H}_{-1}$ as the Hilbert spaces generated by local functions with respect to the norms $\|\cdot\|_1$ and $\|\cdot\|_{-1}$. We shall drop the index ρ, which is fixed throughout the whole paper.

2.1 Step 1: $\mathbb{H}_{-1}$ Estimate and Tightness

Our first step is to show that for the function g defined in (1.1), $\|g\|_{-1} < \infty$. This follows from the next lemma.

LEMMA 2.1 *Let f be a mean zero local function, i.e., a function that depends only on a finite number of coordinates. Then for $d \ge 3$, there is a bound*

$$|\langle f, v\rangle| \le C_f \|v\|_1 .$$

Lemma 2.1 will be proved in Section 3. Recall that $\sqrt{t}\,\xi$ is a sum of a martingale $\xi^{(1)}$ and $A(t)$. The martingale $\xi^{(1)}$ satisfies Doob's inequality. Hence the tightness of ξ in Skorohod space is deduced from the following general theorem:

THEOREM 2.2 *Let P be a stationary and ergodic Markov process with marginal μ on some state space. Let A and A^* be the infinitesimal generators in $L_2(\mu)$ of the original process and its time reversal, respectively. Let $S = \frac{1}{2}[A + A^*]$ be the symmetrized operator defined on* $\mathrm{dom}(A) \cap \mathrm{dom}(A^*)$ *with the associated Dirichlet form*

$$D(u) = \langle -Su, u\rangle . \tag{2.8}$$

Let f be a function on the state space belonging to $L_2(\mu)$ such that for some sequence $u_n \in$ $\mathrm{dom}(A) \cap \mathrm{dom}(A^*)$ *we have*

$$\|Su_n - f\|_{L_2(\mu)} \to 0$$

while

$$\sup_n D(u_n) \le C . \tag{2.9}$$

Then for all $T \ge 0$

$$E^P\left\{\sup_{0\le t\le T}\left|\int_0^t f(x(s))ds\right|^2\right\} \le 8CT$$

with the same constant C as in equation (2.9).

PROOF: We deal with both the forward and backward filtrations.

$$u_n(x(t)) - u_n(x(s)) - \int_s^t (Au_n)(x(\sigma))d\sigma = M_n^+(t) - M_n^+(s)$$

are martingales adapted to the forward filtration for $t \geq s$.

$$u_n(x(s)) - u_n(x(t)) + \int_s^t (A^*u_n)(x(\sigma))d\sigma = M_n^-(s) - M_n^-(t)$$

are martingales adapted to the backward filtration for $s \leq t$. In any case

$$\int_s^t (Su_n)(x(\sigma))d\sigma = \frac{1}{2}[M_n^+(t) - M_n^+(s) + M_n^-(s) - M_n^-(t)].$$

Since

$$E^P\{[M_n^+(t) - M_n^+(s)]^2\} = E^P\{[M_n^-(t) - M_n^-(s)]^2\} \leq 2C|t-s|,$$

Doob's inequality applied separately to the two martingales provides an estimate that is uniform in n. Since $Su_n \to f$ in $L_2(\mu)$, we are done. □

Remark. If $\mathrm{dom}(A) \cap \mathrm{dom}(A^*)$ is large enough to be a core for the Dirichlet space, from an estimate of the form

$$|\langle f, u\rangle| \leq \sqrt{C}\sqrt{D}(u)$$

valid for all $u \in \mathrm{dom}(A) \cap \mathrm{dom}(A^*)$ we can construct by standard resolvent techniques a sequence u_n with the desired properties, with the same exact constant C appearing in equation (2.9). In our case of tagged versions of simple exclusion processes, the space of local functions, included in all domains, is always large enough to form a core.

2.2 Step 2: Resolvent Estimate

Consider the resolvent equation

$$\lambda u_\lambda - Lu_\lambda = f. \tag{2.10}$$

The key input for the proof of Theorem 1.1 is the following estimate for the solution of the resolvent equation (2.10) to be proved in Section 6.

THEOREM 2.3 *For any local f with mean 0 and $d \geq 3$,*

$$\sup_\lambda \|Lu_\lambda\|_{-1} < \infty, \qquad \sup_\lambda \|\lambda u_\lambda\|_{-1} < \infty. \tag{2.11}$$

2.3 Step 3: Approximation via the Resolvent Equation

THEOREM 2.4 *If f is a local function with mean 0, then for the solution u_λ of the resolvent equation we have*

$$\lim_{\lambda\to 0} \|u_\lambda - w\|_1 = 0 \tag{2.12}$$

for some $w \in \mathbb{H}_1$ and

$$\lim_{\lambda\to 0} \lambda\|u_\lambda\|_0^2 = 0. \tag{2.13}$$

PROOF: Multiplying the resolvent equation by u_λ and integrating, we get

$$\lambda\|u_\lambda\|_0^2 + \|u_\lambda\|_1^2 = \langle u_\lambda, f\rangle . \tag{2.14}$$

From the estimate of Lemma 2.1, we see that $\sup_{0\le\lambda\le 1}\|u_\lambda\|_1 < \infty$. Hence we can choose a subsequence $\lambda_n \to 0$ such that along this subsequence $u_{\lambda_n} = u_n$ has a weak limit w in the Dirichlet space $\mathbb{H}_1$. Because $\lambda\|u_\lambda\|_0^2$ is bounded, $\lambda u_\lambda \to 0$ as $\lambda \to 0$ in $L_2 = \mathbb{H}$ and consequently, from equation (2.11), converges to 0 in $\mathbb{H}_{-1}$. Together with the resolvent equation, this implies that Lu_n converges weakly to $-f$ in $\mathbb{H}_{-1}$. By standard functional analysis there are convex combinations v_n of $u_1,\dots,u_n$ such that v_n and Lv_n converge strongly to w and $-f$ in $\mathbb{H}_1$ and $\mathbb{H}_{-1}$, respectively. It is easy to see, from equation (2.11), that $\|v_n\|_1 \to \|w\|_1$ and $\langle v_n, Lv_n\rangle$ converges to $-\langle w, f\rangle$, thus proving

$$\langle w, f\rangle = \|w\|_1^2 .$$

From equation (2.14), we have, in particular,

$$\limsup_{n\to\infty}\|u_{\lambda_n}\|_1^2 \le \limsup_{n\to\infty}[\lambda_n\|u_{\lambda_n}\|_0^2 + \|u_{\lambda_n}\|_1^2] = \langle w, f\rangle = \|w\|_1^2 . \tag{2.15}$$

Since $\|\cdot\|_1$ is lower semicontinuous and $u_{\lambda_n} \to w$ weakly,

$$\|w\|_1^2 \le \liminf_{n\to\infty}\|u_{\lambda_n}\|_1^2 .$$

We have thus proved that the equality holds in equation (2.15). This in turns implies that u_{λ_n} converges to w strongly. The proof of Theorem 2.4 is completed by showing that the limits w_1 and w_2 along any two subsequences λ_n and λ_n' are necessarily equal. To see this, we observe that $L(u_{\lambda_n} - u_{\lambda_n'}) \to 0$ weakly in $\mathbb{H}_{-1}$ while $u_{\lambda_n} - u_{\lambda_n'} \to w_1 - w_2$ strongly in $\mathbb{H}_1$, proving that

$$\|u_{\lambda_n} - u_{\lambda_n'}\|_1^2 = -\langle L(u_{\lambda_n} - u_{\lambda_n'}), (u_{\lambda_n} - u_{\lambda_n'})\rangle \to 0 .$$

□

We now have the following martingale decomposition theorem:

THEOREM 2.5 *There is a square integrable martingale $M(t)$ with stationary increments and an additive functional $\Omega(t)$ such that*

$$\int_0^t f(\eta(s))ds = M(t) + \Omega(t)$$

with

$$T^{-1}EM(T)^2 \le \|f\|_{-1}^2 \tag{2.16}$$

and

$$\lim_{T\to\infty}\frac{1}{T}E[[\Omega(T)]^2] = 0 . \tag{2.17}$$

PROOF: The proof is based on Theorem 2.4. From Ito's formula

$$u_\lambda(\eta(t)) - u_\lambda(\eta(0)) = \int_0^t Lu_\lambda(\eta(s))ds + M_\lambda(t),$$

where M_λ denotes the martingale part with quadratic variation given by $EM_\lambda(t)^2 = t\|u_\lambda\|_1^2$. From the resolvent equation (2.10),

$$\int_0^t f(\eta(s))ds = M_\lambda(t) + \Omega_\lambda(t)$$

where

$$\Omega_\lambda(t) = u_\lambda(\eta(t)) - u_\lambda(\eta(0)) - \int_0^t \lambda u_\lambda(\eta(s))ds.$$

By Theorem 2.4, the martingale part converges as $\lambda \to 0$, to, say, $M(t)$. Clearly, $M(t)$ satisfies the estimate (2.16). Hence we obtain

$$\int_0^t f(\eta(s))ds = M(t) + \Omega(t)$$

where

$$\Omega(t) = \Omega_\lambda(t) + [M_\lambda(t) - M(t)].$$

If we pick $\lambda = 1/T$, then together with (2.13)

$$\begin{aligned}\frac{1}{T}E^P[(\Omega(T))^2] &\le \frac{12}{T}\|u_{1/T}\|_0^2 + \frac{4}{T}E^P\Big[[M_{1/T}(T) - M(T)]^2\Big] \\ &= \frac{12}{T}\|u_{1/T}\|_0^2 + 4E^P\Big[[M_{1/T}(1) - M(1)]^2\Big] \to 0\end{aligned}$$

as $T \to \infty$. This concludes the proof. □

We can now apply Theorem 2.5 to our setting with f taken to be the function g defined in equation (1.5). Hence

$$\xi(t) = \xi^{(1)}(t) + \xi^{(2)}(t) + \Omega(t) \tag{2.18}$$

where $\xi^{(1)}$ is defined in equation (2.4), $\xi^{(2)}(t)$ is the martingale obtained from applying Theorem 2.5, and $\Omega(t)$ is the error term satisfying equation (2.17). Hence, up to a negligible error, we have two square integrable martingales $\xi_1(t)$ and $\xi_2(t)$ with stationary increments adapted to the environment process. From the martingale convergence theorem (see theorem 3.2 of [3] where condition (b) can be checked in this situation), $\xi(\alpha t)/\sqrt{\alpha}$, as $\alpha \uparrow \infty$, converges to a Brownian motion with the covariance matrix $\mathbf{C}(\rho)$ characterized by

$$\langle \mathbf{C}(\rho)a, a\rangle = E^{P_\rho}\{\langle \xi_1(1) + \xi_2(1), a\rangle^2\}.$$

Hence our final task is to estimate the variance.

2.4 Step 4: Bounds on the Variance

For upper bounds on the variance, we can estimate each term separately, where, as for lower bounds, we have to worry about possible cancellations. We start with the upper bound. From the definition of $\xi^{(1)}$ in (2.4), we have immediately

$$E^{P_\rho}\{\langle\xi_1(1),a\rangle^2\} = (1-\rho)\sum_x p(x)\langle x,a\rangle^2 . \tag{2.19}$$

An upper bound of $\xi^{(2)}$ can also be easily obtained from equation (2.16)

$$E\{\langle\xi_2(1),a\rangle^2\} \le C\rho(1-\rho)\sum_x p(x)\langle x,a\rangle^2 . \tag{2.20}$$

Therefore, we have the upper bound that follows:

THEOREM 2.6 $\langle \mathbf{C}(\rho)a,a\rangle \le C(1-\rho)\sum_x p(x)\langle x,a\rangle^2 .$

The lower bound can be reduced to the symmetric case. The limiting variance of the tagged particle in the asymmetric case is bounded below by the limiting variance of the tagged particle in the corresponding symmetric case. A lower bound for the symmetric case was derived in [10].

THEOREM 2.7 $\langle \mathbf{C}a,a\rangle \ge C_3(1-\rho)\langle a,a\rangle .$

PROOF: The martingale $\xi^{(2)}(t)$ can be expressed in terms of the solution w obtained in Therorem 2.4

$$\xi^{(2)}(t) = \sum_x w_x M_x(t) + \sum_{x,y} w_{x,y} M_{x,y}(t) \tag{2.21}$$

where $\{M_x(\cdot)\}$ and $\{M_{x,y}(\cdot)\}$ are the basic martingales introduced in equations (2.1) and (2.2). The functions $\{w_x(\cdot)\}$ and $\{w_{x,y}(\cdot)\}$ are the limits of $u_{\lambda,x} = (u_\lambda(\tau_x\eta) - u_\lambda(\eta))$ and $u_{\lambda,x,y} = (u_\lambda(\eta^{x,y}) - u_\lambda(\eta))$, respectively, in $L_2(P_\rho)$. The convergence of u_λ to w in $\mathbb{H}_1$ guarantees the convergence to 0 of

$$E^{P_\rho}\left[\sum_x p(x)(1-\eta(x))(u_{\lambda,x}-w_x)^2 + \sum_{x,y\neq 0} p(y-x)(u_{\lambda,x,y}-w_{x,y})^2\right]$$

as $\lambda \to 0$. From (2.4), $\xi^{(1)}$ has the representation

$$\xi^{(1)}(t) = \sum_x x M_x(t) . \tag{2.22}$$

Combining equations (2.21) and (2.22), we get the representation

$$\xi^{(1)}(t) + \xi^{(2)}(t) = \sum_x (w_x + x) M_x(t) + \sum_x w_{x,y} M_{x,y}(t) ,$$

from which we can easily deduce that

$$\langle C(\rho)a,a\rangle = E^{P_\rho}\left[\sum_x p(x)(1-\eta(x))(w_x^a + \langle x,a\rangle)^2 + \sum_{x,y} p(y-x) w_{x,y}^2\right] . \tag{2.23}$$

Here we have denoted by w^a the solution that corresponds to $f = \langle a, g\rangle$. The right-hand side of formula (2.23), being essentially a Dirichlet form, is clearly insensitive to replacing $p(\cdot)$ by its symmetrized version. The corresponding formula for the symmetric case, proved in [6], involves solving a different equation and therefore a different w, but for the symmetric case the correct w is given by a variational formula that minimizes the expression on the right-hand side of formula (2.23). Therefore the quantity $\langle C(\rho)a, a\rangle$ is bounded below by the corresponding value for the symmetric case for which a lower bound of the form claimed in Theorem 2.7 was proved in [10]. □

We have based our proof on estimating the resolvent equation (2.10) with the key estimate equation (2.11). In fact, we can also base it on the following estimate: *For any $\varepsilon > 0$ there is a local function u_ε such that*

$$\|Lu_\varepsilon - f\|_{-1} \le \varepsilon .$$

From this estimate we can obtain a result analogous to Theorem 2.4 ([7] or section 6 of [4]) and thus the martingale decomposition Theorem 2.5. This was the estimate established in [2, 7] for the fluctuation-dissipation equation. In a sense these two estimates on the Green's functions can be substituted for each other in many contexts; see section 6 of the lecture [4].

3 Estimates Related to Simple Exclusion Processes

Simple exclusion processes describe dynamics of infinitely many simple random walks on $\mathbb{Z}^d$ with the exclusion that no two particles are allowed to occupy the same site. The simple random walk on $\mathbb{Z}^d$ has the generator given by

$$(Af)(x) = \sum_y p(y-x)[f(y) - f(x)] .$$

The corresponding symmetric generator is

$$(\mathbf{S}f)(x) = \sum_y a(y-x)[f(y) - f(x)] \tag{3.1}$$

where

$$p(x) = a(x) + b(x) ,$$

$a(\cdot)$ and $b(\cdot)$ being the symmetric and asymmetric components of $p(\cdot)$, respectively. Note that

$$|b(x)| = \frac{1}{2}|p(x) - p(-x)| \le \frac{1}{2}[p(x) + p(-x)] = a(x) .$$

The Dirichlet form is

$$\mathcal{D}(f) = \frac{1}{2}\sum_{x,y} a(y-x)[f(y) - f(x)]^2 . \tag{3.2}$$

Since we are mainly interested in the tagged particle process, the state space is $\mathbb{Z}^d \setminus \{0\}$ rather than $\mathbb{Z}^d$. The random walk on $\mathbb{Z}^d \setminus \{0\}$ can be viewed as the random walk on $\mathbb{Z}^d$ with jumps into the origin disallowed. The main result in this section was obtained in collaboration with C. Landim and was reported in [7] by a purely analytic proof. Here we give a more probabilistic proof and for more general processes.

We can consider the general setting of an irreducible transient Markov process on a countable state space X that is symmetric with respect to the counting or uniform measure on X. The generator and the Dirichlet form are given by (3.1) and (3.2) with obvious interpretation. There is a subset $E \subset X$ that is excluded, and transitions into E are disallowed. On the state space $Y = X \setminus E$ we have the induced generator

$$(\tilde{\mathbf{S}}u)(x) = \sum_{y \in Y} a(x,y)[u(y) - u(x)]$$

for $x \in Y$. The corresponding Dirichlet form is

$$\tilde{\mathcal{D}}(u) = \frac{1}{2} \sum_{x,y \in Y} a(x,y)[u(y) - u(x)]^2.$$

For the special case of a random walk, E consists of just the origin. We shall keep the set E general because we want to apply it to other settings as well. We shall assume that the basic process on X is transient, which means that the Green's function exists, i.e.,

$$g(x,y) = \int_0^\infty p(t,x,y)dt < \infty \quad \text{for all } x,y \in X$$

where $p(t,x,y)$ is the transition probability function. The Green's function $g(x,y)$ satisfies

$$g(x,y) = g(y,x) \leq g(x,x) \quad \text{for all } x,y \in X$$

and

$$(\mathbf{S}g(x,\cdot))(y) = \sum_z [g(x,z) - g(x,y)]a(y,z) = -\delta_x(y) \quad \text{for all } x,y \in X$$

where $\delta_x(y) = 1$ if $x = y$ and 0 otherwise.

We want to show that the induced process on Y with generator $\tilde{\mathbf{S}}$ is again transient and compare its Green's function $\tilde{g}(x,y)$ to the original Green's function $g(x,y)$. We will assume that the probability

$$\theta(x) = P_x\{x(t) \text{ visits } E \text{ for some } t \geq 0\} < 1$$

for each $x \in Y$. This is clearly satisfied for any random walk in $d \geq 3$ with $E = \{0\}$. It is well-known that in the transient case there is an estimate of the form

$$|u(x)| \leq C(x)\sqrt{\mathcal{D}(u)}$$

valid uniformly for all functions u that vanish outside a finite set and by completion for all functions that belong to the Dirichlet space $\mathbb{H}_1$, which we recall was defined as the completion of the space of finitely supported functions under the norm $\|u\|_1 = \sqrt{\mathcal{D}(u)}$. A precise estimate on the constant $C(x)$ is given in the following lemma:

LEMMA 3.1 *Suppose $V(x)$ is a nonnegative, compactly supported function. Then*

$$\sum_x u^2(x)V(x) \le \sup_x \Big[\sum_y g(x,y)V(y)\Big]\mathcal{D}(u)$$

for all u. In particular, taking V to be 1 at x and 0 elsewhere,

$$|u(x)| \le \sqrt{g(x,x)}\sqrt{\mathcal{D}(u)}\,.$$

PROOF: Let us define $W(x) = \sum_y g(x,y)V(y)$ and $C = \sup_x W(x)$. Since $0 \le W \le C$ and $V \ge 0$,

$$\begin{aligned}
\sum_x u^2(x)V(x) &\le C\sum_x u^2(x)\frac{V(x)}{W(x)} \\
&= -C\sum_x \frac{u^2(x)}{W(x)}(SW)(x) \\
&= \frac{C}{2}\sum_{x,y}\left[\frac{u^2(y)}{W(y)} - \frac{u^2(x)}{W(x)}\right][W(y)-W(x)]a(x,y) \\
&= \frac{C}{2}\sum_{x,y}\left[u^2(y)+u^2(x)-u^2(x)\frac{W(y)}{W(x)} - u^2(y)\frac{W(x)}{W(y)}\right]a(x,y) \\
&\le \frac{C}{2}\sum_{x,y}[u^2(y)+u^2(x)-2u(x)u(y)]a(x,y) = C\mathcal{D}(u)\,.
\end{aligned}$$

□

Notice that $\theta(x)$ solves the equation

$$\begin{cases} (\mathbf{S}\theta)(x) = \sum_{y\in X} a(x,y)[\theta(y)-\theta(x)] = 0 & \text{for } x \in Y \\ \theta(x) = 1 & \text{for } x \in E\,. \end{cases}$$

Therefore

$$(\tilde{\mathbf{S}}\theta)(x) = -\sum_{y\in E} a(x,y)[\theta(y)-\theta(x)] = -\sum_{y\in E} a(x,y)[1-\theta(x)] = -A(x)(1-\theta(x))$$

where

$$A(x) = \sum_{y\in E} a(x,y)\,. \tag{3.3}$$

Because of irreducibility, $A(x)$ cannot vanish identically on Y. Thus $\theta(x)$ is a bounded, nonconstant superharmonic function, and so the process is transient. We will actually assume that

$$\sup_{x\in Y}\theta(x)=\beta<1\,. \tag{3.4}$$

The following lemma is a quantified version of the transience:

LEMMA 3.2 *For the Green's function* $\tilde{g}(x,y)$, *we have*

$$\sum_{y\in Y}\tilde{g}(x,y)A(y)\leq\frac{\theta(x)}{(1-\beta)}\,.$$

PROOF: Since

$$-(\tilde{\mathbf{S}}\theta)(x)\geq(1-\beta)A(x)\,,$$

the lemma follows from the maximum principle. □

LEMMA 3.3 *Let* $U\geq 0$ *be supported on* Y*, that is, it vanishes on* E*. Then*

$$\begin{aligned}\sup_{x\in Y}\sum_{y\in Y}\tilde{g}(x,y)U(y)&\leq\frac{1}{(1-\beta)}\sup_{x\in X}\sum_{y\in X}g(x,y)U(y)\\&=\frac{1}{(1-\beta)}\sup_{x\in Y}\sum_{y\in Y}g(x,y)U(y)\end{aligned} \tag{3.5}$$

Furthermore, for all $x,y\in Y$,

$$\tilde{g}(x,y)\leq g(x,y)+\frac{1}{(1-\beta)}g(y,y)\theta(x) \tag{3.6}$$

and

$$\tilde{g}(x,x)\leq\frac{1}{(1-\beta)}g(x,x)\,. \tag{3.7}$$

PROOF: Let

$$W(x)=\sum_{y\in X}g(x,y)U(y).$$

The function W is nonnegative and solves $\mathbf{S}W=-U$. A computation shows that for $x\in Y$,

$$\begin{aligned}(\tilde{\mathbf{S}}W)(x)&=(\mathbf{S}W)(x)-\sum_{y\in E}a(x,y)[W(y)-W(x)]\\&\leq-U(x)+\sum_{y\in E}a(x,y)W(x)=-U(x)+A(x)W(x)\leq-U(x)+CA(x)\end{aligned}$$

where $C=\sup_{x\in Y}W(x)$. Then

$$\sum_{y\in Y}\tilde{g}(x,y)U(y)\leq W(x)+C\sum_{y\in Y}\tilde{g}(x,y)A(y)\leq W(x)+\frac{C\theta(x)}{(1-\beta)}\,. \tag{3.8}$$

Taking the supremum over x, we get

$$\sup_{x\in Y}\sum_{y\in Y}\tilde{g}(x,y)U(y)\le C+\frac{C\beta}{(1-\beta)}=\frac{C}{(1-\beta)},$$

which proves equation (3.5). Taking $U(x)$ to be $\delta_y(x)$, we have $W(x)=g(x,y)$ and $\sup_x W(x)=g(y,y)$. Hence equation (3.6) follows from equation (3.8). Taking $x=y$, we get equation (3.7). □

Remark. If $a(x,y)$ is local, then we can do better if we define $\bar{E}=\bigcup_{x\in E}\{y:a(x,y)>0\}$ and take $C=\sup_{x\in\bar{E}}W(x)$.

We can combine Lemmas 3.1 and 3.2 to obtain the next lemma.

LEMMA 3.4 *For any function u*

$$\sum_{x\in Y}A(x)u^2(x)\le\frac{\beta}{(1-\beta)}\tilde{\mathcal{D}}(u).$$

PROOF: According to Lemma 3.1

$$\sum_{x\in Y}A(x)u^2(x)\le C\tilde{\mathcal{D}}(u)$$

where the constant can be taken to be

$$C=\sup_{x\in Y}\sum_{y\in Y}\tilde{g}(x,y)A(y).$$

In Lemma 3.2 we established that for $x\in Y$

$$\sum_{y\in Y}\tilde{g}(x,y)A(y)\le\frac{\theta(x)}{(1-\beta)}.$$

Taking the supremum over x, clearly we can take C to be $\beta/(1-\beta)$. □

The following results concerning two random walks will be useful:

LEMMA 3.5 *For the symmetric random walk on* $\{\mathbb{Z}^d\setminus\{0\}\}\times\{\mathbb{Z}^d\setminus\{0\}\}$ *with generator*

$$\sum_z a(z)[f(x_1+z,x_2)-f(x_1,x_2)]+\sum_z a(z)[f(x_1,x_2+z)-f(x_1,x_2)],$$

the probability $\theta(x_1,x_2)$ *of hitting the excluded set* $E_2=\{x_1=0\}\cup\{x_2=0\}\cup\{x_1=x_2\}$ *has the property*

$$\sup_{(x_1,x_2)\in(\mathbb{Z}^d\times\mathbb{Z}^d)\setminus E_2}\theta(x_1,x_2):=\alpha<1.$$

PROOF: If we denote by $\delta(x)$ the probability that the continuous time random walk in $\mathbb{Z}^d$ hits 0 for some $t\ge 0$, then $\theta(x_1,x_2)$ can be bounded by

$$\theta(x_1,x_2)\le\delta(x_1)+\delta(x_2)+\delta(x_1-x_2).$$

We know that $\sup_{x\neq 0}\delta(x) < 1$ and $\delta(x) \to 0$ as $x \to \infty$. This is enough to conclude that $\theta(x_1,x_2)$ stays away from 1 near ∞, and since it is strictly less than 1 for each (x_1,x_2), we are done. □

The number α is a constant of the underlying random walk and $0 < \alpha < 1$. The same α is also a bound for the hitting probability of 0 for a single random walk on $\mathbb{Z}^d$. From Lemma 3.3 we get for the random walk on $\mathbb{Z}^d$ that excludes the origin the following bound for the Green's function:

$$g_0(x,x) \le \frac{1}{(1-\alpha)}g(x,x) = \frac{g(0,0)}{(1-\alpha)}.$$

We now consider the symmetric simple exclusion processes on $\mathbb{Z}^d \setminus \{0\}$. The state space Ω consists of maps η from the countable set $\mathbb{Z}^d \setminus \{0\}$ into $\{0,1\}$. Fix $0 < \rho < 1$ and denote the Bernoulli product measure with density ρ by P_ρ. For $x \in \mathbb{Z}^d$, we define

$$\xi_x(\eta) = \frac{\eta(x)-\rho}{\sqrt{\rho(1-\rho)}}. \tag{3.9}$$

For $A \subset \mathbb{Z}^d \setminus \{0\}$ we define

$$\begin{cases} \xi_A(\eta) = \prod_{x\in A}\xi_x(\eta) & \text{if } A \text{ is nonempty} \\ \xi_A(\eta) = 1 & \text{if } A \text{ is empty.} \end{cases} \tag{3.10}$$

Then $\{\xi_A(\cdot)\}$ is an orthonormal basis for $\mathbf{H} = L_2(\Omega)$. It comes naturally graded as $\mathbf{H} = \bigoplus_{n\ge 0} H_n$ where H_n is the span of ξ_A over sets of cardinality n.

If we write

$$u = \sum_A \tilde{u}(A)\xi_A,$$

then the Dirichlet form can be calculated explicitly, and we obtain

$$D^{\text{ex}}(u) = \frac{1}{2}\sum_{n\ge 1}\sum_{|A|=n}\sum_{x,y} a(x,y)[\tilde{u}(A^{x,y}) - \tilde{u}(A)]^2 := D^{\text{ex}}(\tilde{u})$$

where

$$A^{x,y} = \begin{cases} A & \text{if either } x,y \in A \text{ or } x,y \notin A \\ (A\setminus x)\cup y & \text{if } x \in A \text{ and } y \notin A \\ (A\setminus y)\cup x & \text{if } x \notin A \text{ and } y \in A. \end{cases} \tag{3.11}$$

The computation depends on the simple observation that, for all x,y in X, $\xi_A(\eta^{x,y}) = \xi_{A^{x,y}}(\eta)$.

Let us denote by $\mathcal{X}$ the space of all finite subsets of $\mathbb{Z}^d \setminus \{0\}$, and we write $\mathcal{X}$ as the natural union $\bigcup \mathcal{X}_n$ of spaces of subsets of cardinality n. We will consider functions on $\mathcal{X}$ or $\mathcal{X}_n$, and we assume initially that they are zero outside a finite set of points in $\mathcal{X}$ or $\mathcal{X}_n$. Then $\tilde{u}$ is a function on $\mathcal{X}$, and we can write it as $\tilde{u} = \sum_n \tilde{u}_n$ with $\tilde{u}_n$ being the restriction of $\tilde{u}$ to $\mathcal{X}_n$. Clearly,

$$D^{\mathrm{ex}}(\tilde{u}) = \sum_{n\geq 1} D_n^{\mathrm{ex}}(\tilde{u}_n),$$

where for $f : \mathcal{X}_n \to R$

$$\begin{aligned} D_n^{\mathrm{ex}}(f) &= \frac{1}{2} \sum_{A\in\mathcal{X}_n} \sum_{x,y} a(x,y)[f(A^{x,y}) - f(A)]^2 \\ &= \sum_{A\in\mathcal{X}_n} \sum_{\substack{x\in A\\ y\notin A}} a(x,y)[f((A\setminus x)\cup y) - f(A)]^2 . \end{aligned}$$

The Dirichlet form D^{sh} associated with the shift operator L^{sh} is not graded naturally according to our decomposition. A related Dirichlet form allowing the tagged particle to jump to an occupied site,

$$\hat{D}^{\mathrm{sh}}(u) = \frac{1}{2} \int \sum_x a(x)[u(\tau_x\eta) - u(\eta)]^2 \, dP_\rho , \tag{3.12}$$

is naturally graded with

$$\hat{D}_n^{\mathrm{sh}}(u) = \hat{D}_n^{\mathrm{sh}}(\tilde{u}_n) = \frac{1}{2} \sum_{x,|A|=n} a(x)[\tilde{u}_n(\tau_x A) - \tilde{u}_n(A)]^2 .$$

We shall prove later on an estimate of $\hat{D}^{\mathrm{sh}}$ in terms of D^{sh} and D^{ex}.

We need the following concept later on: Suppose $n > m$ and we are given a function g on $\mathcal{X}_n$. The function f on $\mathcal{X}_m$ is defined by $f \geq 0$ and

$$f^2(B) = \sum_{\substack{A:A\supset B\\ |A|=n}} g^2(A) .$$

Notice that if we consider g as a wave function, then f^2 is simply the m point function associated with the wave function. A very important relation between f and g is the connection between their kinetic energies, especially the special case $m = 1$, which leads to the semiclassical limit for the kinetic energy of Bose gases. Notice that in our setting there is an interaction between particles, namely, the exclusion rule. Nevertheless, it is still valid.

LEMMA 3.6 *For any $n > m$*

$$D_m^{\mathrm{ex}}(f) \leq \binom{n-1}{m-1} D_n^{\mathrm{ex}}(g) .$$

In particular, for $m = 1$ we have for the function

$$f^2(x) = \sum_{A:A\ni x} g^2(A) ,$$

that

$$D_1^{\text{ex}}(f) \le D_n^{\text{ex}}(g).$$

PROOF: If B_1 and B_2 are two subsets of size m that differ by one point, namely, $B_2 = B_1 \setminus x \cup y$, then one has the following obvious estimate

$$|f(B_1) - f(B_2)|^2 \le \sum_{\substack{A:|A|=n \\ A \supset B_1 \\ x \in A,\, y \notin A}} |g(A \setminus x \cup y) - g(A)|^2.$$

One multiplies this inequality by $a(x,y)$ and sums over everything in sight. One has to be careful and count the number of times a term like $|g(A \setminus x \cup y) - g(A)|^2$ occurs in the summation on the right-hand side. That number is clearly the number of different subsets of A of size m that include a given $x \in A$, and this yields the combinatorial prefactor $\binom{n-1}{m-1}$. □

The following lemma is the main result of this section:

LEMMA 3.7 *Let $n \ge 2$. For any set A in $\mathcal{X}_n$, i.e., a subset of $\mathbb{Z}^d \setminus \{0\}$ of cardinality n, that consists of n distinct points $x_1, x_2, \dots, x_n$, we define*

$$W_1(A) = \sum_i a(x_i) \quad \text{and} \quad W_2(A) = \sum_{i \ne j} a(x_i - x_j).$$

Then for any u defined on $\mathcal{X}_n$ that is in the Dirichlet space,

$$\sum_A W_1(A) u^2(A) \le \frac{\alpha}{(1-\alpha)} D_n^{\text{ex}}(u)$$

and

$$\sum_A [W_1(A) + W_2(A)] u^2(A) \le \frac{\alpha}{(1-\alpha)} n D_n^{\text{ex}}(u).$$

PROOF: Define v on $\mathcal{X}_1 = \mathbb{Z}^d \setminus 0$ by

$$v^2(x) = \sum_{A: A \ni x} u^2(A).$$

By Lemma 3.6

$$D_1^{\text{ex}}(v) \le D_n^{\text{ex}}(u).$$

By definition of v,

$$\sum_{|A|=n} W_1(A) u^2(A) = \sum_{x \ne 0} a(x) v^2(x).$$

By Lemma 3.4 and the remark at the end of Lemma 3.6,

$$\sum_{x \ne 0} a(x) v^2(x) \le \frac{\alpha}{(1-\alpha)} D_n^{\text{ex}}(v) = \frac{\alpha}{(1-\alpha)} D_n^{\text{ex}}(u).$$

This proves the estimate involving W_1.

To prove the second inequality, define the function w on $\mathcal{X}_2$ by

$$w^2(B) = \sum_{A:A\supset B} u^2(A).$$

We can view w as a symmetric function on $(\mathbb{Z}^d \times \mathbb{Z}^d)\setminus E_2$ and compute its Dirichlet form,

$$\begin{aligned} D_2^{\text{ex}}(w) = &\frac{1}{2} \sum_{\substack{x_1,x_2,y_1 \in \mathbb{Z}^d\setminus\{0\} \\ x_1\neq y_1}} a(y_1-x_1)[w(y_1,x_2)-w(x_1,x_2)]^2 \\ &+ \frac{1}{2} \sum_{\substack{x_1,x_2,y_2 \in \mathbb{Z}^d\setminus\{0\} \\ x_2\neq y_2}} a(y_2-x_2)[w(x_1,y_2)-w(x_1,x_2)]^2 ; \end{aligned}$$

here the factor $\frac{1}{2}$ appears because each set B is counted twice. Again by Lemma 3.6

$$D_2^{\text{ex}}(w) \le (n-1)D_n^{\text{ex}}(u). \tag{3.13}$$

Clearly,

$$\sum_A [W_1(A)+W_2(A)]u^2(A) = \frac{1}{2} \sum_{\substack{x_1,x_2 \in \mathbb{Z}^d\setminus\{0\} \\ x_1\neq x_2}} H(x_1,x_2)w^2(x_1,x_2)$$

where

$$H(x_1,x_2) = a(x_1)+a(x_2)+2a(x_1-x_2)$$

is the sum of all the transition rates into the excluded set for the two random walks.

An easy calculation using Lemma 3.5 establishes

$$\sum_B H(B)w^2(B) = \frac{1}{2}\sum_{x_1\neq x_2} H(x_1,x_2)w^2(x_1,x_2) \le \frac{1}{2}\frac{\alpha}{(1-\alpha)} 2D_2^{\text{ex}}(w).$$

This fact together with equation (3.13) proves the lemma. □

Another important Markovian evolution, referred to as Glauber dynamics, will be useful later on. We now collect some of its properties here. The Glauber dynamics has each site flipping back and forth between 0 and 1 as a two-state Markov process. The flip rates from $1 \to 0$ and $0 \to 1$ are taken to be $1/\rho$ and $1/(1-\rho)$, respectively, so that the generator for a single site is

$$(Gv)(0) = \frac{1}{(1-\rho)}(v(1)-v(0)), \quad (Gv)(1) = \frac{1}{\rho}(v(0)-v(1)),$$

with the corresponding Dirichlet form, $[v(0)-v(1)]^2$. For the full process on Ω, we need to define the flip operator σ_x at site x,

$$\begin{cases} (\sigma_x\eta)(y) = \eta(y) & \text{if } x \neq y \\ (\sigma_x\eta)(x) = 1-\eta(x) & \text{for } x \in X. \end{cases}$$

The Glauber generator then has the Dirichlet form

$$D_g(u) = \sum_x D_{g,x}(u)$$

where

$$D_{g,x}(u) = \int_\Omega [u(\sigma_x\eta) - u(\eta)]^2 dP_\rho . \tag{3.14}$$

Our next lemma is an estimate of $D_{g,x}(u)$ in terms of $D^{\text{ex}}(u)$ in the transient case. The transience is important because in the Glauber dynamics "particles" are not conserved, whereas in exclusion processes they are conserved. The only way to kill a particle in the exclusion model is to send it to infinity, and transience plays an important role in that.

LEMMA 3.8 *Assume transience. Let $u : \Omega \to \mathbb{R}$ depend on a finite set of coordinates. Then for any $x \in X$,*

$$\int_{\eta(x)=1} [u(\sigma_x\eta) - u(\eta)]^2 dP_\rho \le \frac{1}{(1-\rho)} g(x,x)\, D^{\text{ex}}(u)$$

and

$$\int_{\eta(x)=0} [u(\sigma_x\eta) - u(\eta)]^2 dP_\rho \le \frac{1}{\rho} g(x,x)\, D^{\text{ex}}(u) .$$

By adding the two inequalities,

$$D_{g,x}(u) \le \frac{1}{\rho(1-\rho)} g(x,x) D^{\text{ex}}(u) .$$

PROOF: An elementary calculation shows that for a function

$$u(\eta) = \sum_A \tilde{u}(A) \xi_A(\eta)$$

we have

$$\int_{\eta(x)=1} [u(\sigma_x\eta) - u(\eta)]^2 dP_\rho = \frac{1}{(1-\rho)} \sum_{A:A\ni x} \tilde{u}^2(A),$$

$$\int_{\eta(x)=0} [u(\sigma_x\eta) - u(\eta)]^2 dP_\rho = \frac{1}{\rho} \sum_{A:A\ni x} \tilde{u}^2(A) .$$

Adding the two equalities, we get

$$D_{g,x}(u) = \frac{1}{\rho(1-\rho)} \sum_{A:A\ni x} \tilde{u}^2(A) .$$

If we define $\tilde{u}_n$ to be the restriction of $\tilde{u}$ to $\mathcal{X}_n$,

$$\sum_{A:A\ni x} \tilde{u}^2(A) = \sum_n v_n^2(x) \quad \text{where} \quad v_n^2(x) = \sum_{\substack{A:|A|=n \\ A\ni x}} \tilde{u}_n^2(A).$$

From Lemma 3.1 $v_n^2(x) \le g(x,x)D_1(v_n)$, and by Lemma 3.6 $D_1^{\text{ex}}(v_n) \le D_n^{\text{ex}}(\tilde{u}_n)$. The lemma follows. □

As corollaries of previous lemmas, we prove Lemma 2.1.

PROOF OF LEMMA 2.1: Clearly, to get a bound of the form

$$|\langle f, v\rangle| \le C_f \sqrt{\sum_{x\in F} D_{g,x}(v)},$$

where f depends only on the coordinates x from F, is just finite-dimensional matrix algebra. We can now use Lemma 3.8 to complete the proof. □

Next we obtain an estimate of $\hat{D}^{\text{sh}}$. We already know that $D^{\text{sh}}(u) \le \hat{D}^{\text{sh}}(u)$, since the $\hat{D}^{\text{sh}}$ form is stronger than the D^{sh} form because it corresponds to the tagged particle exchanging places with an untagged particle, which is not allowed. However, we could make the particle at x disappear, let the tagged particle jump to x, which is now empty, and have our old disappeared particle reappear at $-x$ and accomplish our goal. This means that the $\hat{D}^{\text{sh}}$ form can be estimated in terms of the D^{sh} form and the Glauber forms at x and $-x$. More precisely,

$$\begin{aligned}
&\int_{\eta(x)=1} [u(\tau_x\eta) - u(\eta)]^2 dP_\rho \\
&\quad \le 3 \int_{\eta(x)=1} \Big[[u(\sigma_{-x}\tau_x\sigma_x\eta) - u(\tau_x\sigma_x\eta)]^2 + [u(\tau_x\sigma_x\eta) - u(\sigma_x\eta)]^2 \\
&\qquad\qquad + [u(\sigma_x\eta) - u(\eta)]^2 \Big] dP_\rho \\
&\quad = 3\frac{\rho}{(1-\rho)} \int_{\eta(-x)=0} [u(\sigma_{-x}\eta) - u(\eta)]^2 + 3\frac{\rho}{(1-\rho)} \int_{\eta(x)=0} [u(\tau_x\eta) - u(\eta)]^2 dP_\rho \\
&\qquad + 3 \int_{\eta(x)=1} [u(\sigma_x\eta) - u(\eta)]^2 dP_\rho .
\end{aligned}$$

Note that the maps σ_x are not measure preserving when $\rho \ne \frac{1}{2}$, and we therefore pick up the factor $\rho/(1-\rho)$. We can now use Lemma 3.8 to conclude that for some constant C not depending on ρ

$$\hat{D}^{\text{sh}}(u) \le \frac{C}{1-\rho}[D^{\text{ex}}(u) + D^{\text{sh}}(u)]. \tag{3.15}$$

4 Some Estimates on the Generator

Recall the generator $L = L^{\text{ex}} + L^{\text{sh}}$ of the environment process defined in equations (1.3) and (1.4). Note also that $L_2 = \mathbf{H} = \bigoplus_{n\geq 0} H_n$ where H_n is spanned by the orthonormal basis $\{\xi_A : |A| = n\}$ with ξ_A defined in equation (3.10). Notice that L is a bounded operator on each H_n and maps it into $H_{n-1} \oplus H_n \oplus H_{n+1}$. There are thus bounded operators $B_{n,n-1} : H_n \to H_{n-1}$, $B_{n,n} : H_n \to H_n$, and $B_{n,n+1} : H_n \to H_{n+1}$ such that for $u \in H_n$

$$Lu = B_{n,n-1}u + B_{n,n}u + B_{n,n+1}u\,.$$

To compute these operators explicitly, we start with the action of L^{ex} on ξ_A,

$$L^{\text{ex}}\xi_A = \sum_{x,y\neq 0} p(y-x)\eta(x)(1-\eta(y))[\xi_{A^{x,y}} - \xi_A]\,.$$

Recall $p(x) = a(x) + b(x)$, the sum of its symmetric and asymmetric parts. Hence we obtain

$$L^{\text{ex}}\xi_A = \frac{1}{2}\sum_{x,y\neq 0} a(y-x)[\xi_{A^{x,y}} - \xi_A] + \frac{1}{2}\sum_{x,y\neq 0} b(y-x)(\eta(x)-\eta(y))[\xi_{A^{x,y}} - \xi_A]\,.$$

To proceed further, we write

$$\eta(x) - \eta(y) = \sqrt{\rho(1-\rho)}[\xi_x - \xi_y]$$

and use the rule

$$\xi_x{}^2 = 1 + \frac{1-2\rho}{\sqrt{\rho(1-\rho)}}\xi_x$$

to obtain

$$[\eta(x)-\eta(y)][\xi_{A^{x,y}} - \xi_A] = \begin{cases} 2\sqrt{\rho(1-\rho)}[\xi_{A\cup y} - \xi_{A\setminus x}] + \frac{2\rho-1}{\sqrt{\rho(1-\rho)}}[\xi_A + \xi_{A\setminus x\cup y}] & \text{if } x\in A, y\notin A,\\ 2\sqrt{\rho(1-\rho)}[\xi_{A\setminus y} - \xi_{A\cup x}] - \frac{2\rho-1}{\sqrt{\rho(1-\rho)}}[\xi_A + \xi_{A\setminus y\cup x}] & \text{if } x\notin A, y\in A,\\ 0 & \text{in all other cases.}\end{cases}$$

This yields

(4.1)
$$
\begin{aligned}
L^{\mathrm{ex}}\xi_A &= \frac{1}{2}\sum_{\substack{x,y\neq 0\\ x\in A, y\notin A}} a(y-x)[\xi_{A\backslash x\cup y}-\xi_A] + \frac{1}{2}\sum_{\substack{x,y\neq 0\\ x\notin A, y\in A}} a(y-x)[\xi_{A\backslash y\cup x}-\xi_A]\\
&\quad + \frac{2\rho-1}{2}\sum_{\substack{x,y\neq 0\\ x\in A, y\notin A}} b(y-x)[\xi_{A\backslash x\cup y}+\xi_A]\\
&\quad - \frac{2\rho-1}{2}\sum_{\substack{x,y\neq 0\\ x\notin A, y\in A}} b(y-x)[\xi_{A\backslash y\cup x}+\xi_A]\\
&\quad + \sqrt{\rho(1-\rho)}\sum_{\substack{x,y\neq 0\\ x\in A, y\notin A}} b(y-x)\xi_{A\cup y} - \sqrt{\rho(1-\rho)}\sum_{\substack{x,y\neq 0\\ x\notin A, y\in A}} b(y-x)\xi_{A\cup x}\\
&\quad - \sqrt{\rho(1-\rho)}\sum_{\substack{x,y\neq 0\\ x\in A, y\notin A}} b(y-x)\xi_{A\backslash x} + \sqrt{\rho(1-\rho)}\sum_{\substack{x,y\neq 0\\ x\notin A, y\in A}} b(y-x)\xi_{A\backslash y}.
\end{aligned}
$$

We have arranged the terms in equation (4.1) so that the first line is a symmetric piece of $B_{n,n}$, the second two are an asymmetric piece again of $B_{n,n}$, and the last two lines are parts of $B_{n,n+1}$ and $B_{n,n-1}$. In fact, there is a symmetry relative to the interchange of x and y and the first and second terms on each line are equal. We use this to rewrite

$$
\begin{aligned}
&L^{\mathrm{ex}}\xi_A =\\
&\sum_{\substack{x,y\neq 0\\ x\in A, y\notin A}} a(y-x)[\xi_{A\backslash x\cup y}-\xi_A] + (2\rho-1)\sum_{\substack{x,y\neq 0\\ x\in A, y\notin A}} b(y-x)[\xi_{A\backslash x\cup y}+\xi_A]\\
&+2\sqrt{\rho(1-\rho)}\sum_{\substack{x,y\neq 0\\ x\in A, y\notin A}} b(y-x)\xi_{A\cup y} - 2\sqrt{\rho(1-\rho)}\sum_{\substack{x,y\neq 0\\ x\in A, y\notin A}} b(y-x)\xi_{A\backslash x}.
\end{aligned}
\tag{4.2}
$$

Now we compute in a similar fashion the term $L^{\mathrm{sh}}\xi_A$. Although the tagged particle cannot jump to a site where there is already a particle present, it is convenient to extend the definition of τ_x to the set $\{x : \eta(x) = 1\}$ by allowing the tagged particle to jump to the site $-x$, which now becomes the origin, and move the particle that was originally at $-x$ back to where the tagged particle was, which is now x. This defines a version of τ_x with $\tau_x\xi_A = \xi_{\tau_x A}$ where

$$
\tau_x A = \begin{cases} A+x & \text{if } -x\notin A\\ (A+x)\setminus 0\cup x & \text{if } -x\in A. \end{cases}
$$

Note that $x \in \tau_x A$ if and only if $-x \in A$. After careful calculation we obtain

$$
\begin{aligned}
L^{\mathrm{sh}}\xi_A = {} & (1-\rho)\sum_{-x\notin A} a(x)[\xi_{\tau_x A}-\xi_A]+\rho\sum_{-x\in A} a(x)[\xi_{\tau_x A}-\xi_A] \\
& +(1-\rho)\sum_{-x\notin A} b(x)[\xi_{\tau_x A}+\xi_A]+\rho\sum_{-x\in A} b(x)[\xi_{\tau_x A}+\xi_A] \\
& +\sqrt{\rho(1-\rho)}\sum_{x\notin A} p(x)\xi_{A\cup x}-\sqrt{\rho(1-\rho)}\sum_{-x\notin A} p(x)\xi_{(A+x)\cup x} \\
& +\sqrt{\rho(1-\rho)}\sum_{x\in A} p(x)\xi_{A\backslash x}-\sqrt{\rho(1-\rho)}\sum_{-x\in A} p(x)\xi_{(A+x)\backslash 0}.
\end{aligned}
\tag{4.3}
$$

Again the first line is the symmetric part of $B_{n,n}$, followed by its asymmetric part and then the pieces of $B_{n,n+1}$ and $B_{n,n-1}$, respectively.

Putting together L^{ex} and L^{sh}, we have

$$
\begin{aligned}
& B_{n,n}\xi_A \\
& \quad = \sum_{\substack{x,y\neq 0\\ x\in A,y\notin A}} a(y-x)[\xi_{A\backslash x\cup y}-\xi_A]+(1-\rho)\sum_{-x\notin A} a(x)[\xi_{\tau_x A}-\xi_A] \\
& \qquad +\rho\sum_{-x\in A} a(x)[\xi_{\tau_x A}-\xi_A]+(2\rho-1)\sum_{\substack{x,y\neq 0\\ x\in A,y\notin A}} b(y-x)[\xi_{A\backslash x\cup y}+\xi_A] \\
& \qquad +(1-\rho)\sum_{-x\notin A} b(x)[\xi_{\tau_x A}+\xi_A]+\rho\sum_{-x\in A} b(x)[\xi_{\tau_x A}+\xi_A]
\end{aligned}
\tag{4.4}
$$

$$
\begin{aligned}
B_{n,n+1}\xi_A = \sqrt{\rho(1-\rho)}\Bigg[& 2\sum_{\substack{x,y\neq 0\\ x\in A,y\notin A}} b(y-x)\xi_{A\cup y} \\
& +\sum_{x\notin A} p(x)\xi_{A\cup x}-\sum_{-x\notin A} p(x)\xi_{(A+x)\cup x}\Bigg],
\end{aligned}
\tag{4.5}
$$

$$
\begin{aligned}
B_{n+1,n}\xi_A = \sqrt{\rho(1-\rho)}\Bigg[& -2\sum_{\substack{x,y\neq 0\\ x\in A,y\notin A}} b(y-x)\xi_{A\backslash x} \\
& +\sum_{x\in A} p(x)\xi_{A\backslash x}-\sum_{-x\in A} p(x)\xi_{(A+x)\backslash 0}\Bigg].
\end{aligned}
\tag{4.6}
$$

First, we will provide some estimates on $B_{n,n+1}$ and $B_{n+1,n}$. If we separate out the odd and even parts,

$$B_{n,n+1} = \sqrt{\rho(1-\rho)}[B^{\mathrm{odd}}_{n,n+1}+B^{\mathrm{even}}_{n,n+1}]$$

where the odd and even parts are given by

$$
\begin{aligned}
B^{\text{odd}}_{n,n+1}(\xi_A) &= \Big[2 \sum_{\substack{x,y\neq 0\\ x\in A, y\notin A}} b(y-x)\xi_{A\cup y} + \sum_{x\notin A} b(x)\xi_{A\cup x} - \sum_{-x\notin A} b(x)\xi_{(A+x)\cup x}\Big], \\
B^{\text{even}}_{n,n+1}(\xi_A) &= \Big[\sum_{x\notin A} a(x)\xi_{A\cup x} - \sum_{-x\notin A} a(x)\xi_{(A+x)\cup x}\Big].
\end{aligned} \tag{4.7}
$$

The even part of $B_{n+1,n}$ is the adjoint of the even part of $B_{n,n+1}$, and for the odd part an extra change of sign is involved. So if we bound the odd and even pieces of $B_{n,n+1}$ separately, the dual bounds are valid for $B_{n+1,n}$. Our basic estimate is the following:

LEMMA 4.1 *For all $n \geq 1$ and $u \in H_n$ and $v \in H_{n+1}$,*

$$
|\langle B_{n,n+1}u, v\rangle| \leq C\sqrt{\rho(1-\rho)}\sqrt{D^{\text{ex}}_{n+1}(v)}\Big[\sqrt{n}\sqrt{D^{\text{ex}}_n(u)} + \sqrt{\hat{D}^{\text{sh}}_n(u)}\Big]
$$

and

$$
|\langle B_{n+1,n}v, u\rangle| \leq C\sqrt{\rho(1-\rho)}\sqrt{D^{\text{ex}}_n(u)}\Big[\sqrt{n}\sqrt{D^{\text{ex}}_{n+1}(v)} + \sqrt{\hat{D}^{\text{sh}}_{n+1}(v)}\Big]
$$

where C is a constant that is independent of n and ρ.

PROOF: We shall only prove the first estimate. The second one follows in a similar way or by invoking duality. Write functions u and v in H_n and H_{n+1}, respectively, with expansions $u = \sum_{|A|=n} \tilde{u}(A)\xi_A$ and $v = \sum_{|D|=n+1} \tilde{v}(D)\xi_D$. From equation (4.7), the estimate for $B^{\text{odd}}_{n,n+1}$ reduces to the estimation of

$$
\begin{aligned}
&\Big|2\sum_{|A|=n}\sum_{x\in A, y\notin A} b(y-x)\tilde{u}(A)\tilde{v}(A\cup y)\Big| \\
&\quad + \Big|\sum_{|A|=n}\sum_{x\notin A} b(x)\tilde{u}(A)\tilde{v}(A\cup x) - \sum_{|A|=n} l \sum_{-x\notin A} b(x)\tilde{u}(A)\tilde{v}((A+x)\cup x))\Big|.
\end{aligned}
$$

The first term is estimated by

$$\begin{aligned}
&\left|2\sum_{|D|=n+1}\sum_{x,y\in D} b(y-x)\tilde{u}(D\setminus y)\tilde{v}(D)\right| \\
&\quad=\left|\sum_{|D|=n+1}\sum_{x,y\in D} b(y-x)(\tilde{u}(D\setminus y)-\tilde{u}(D\setminus x))\tilde{v}(D)\right| \quad (b \text{ is odd}) \\
&\quad\le \sum_{|D|=n+1}\sum_{x,y\in D} a(y-x)|\tilde{u}(D\setminus y)-\tilde{u}(D\setminus x)||\tilde{v}(D)| \quad (\text{because } |b|\le a) \\
&\quad\le \left[\sum_{|D|=n+1}\sum_{x,y\in D} a(y-x)|\tilde{u}(D\setminus y)-\tilde{u}(D\setminus x)|^2\right]^{\frac{1}{2}} \\
&\qquad\times\left[\sum_{|D|=n+1}\sum_{x,y\in D} a(y-x)|\tilde{v}(D)|^2\right]^{\frac{1}{2}}.
\end{aligned} \tag{4.8}$$

The first expression

$$\left[\sum_{|D|=n+1}\sum_{x,y\in D} a(y-x)|\tilde{u}(D\setminus y)-\tilde{u}(D\setminus x)|^2\right]$$

is seen to equal $2D_{ex,n}(u)$. In view of Lemma 3.7, the second expression is bounded by $CnD_{ex,n+1}(v)$.

We turn to the second term. This leads to

$$\begin{aligned}
&\left|\sum_{\substack{x,D\\ x\in D,|D|=n+1}} b(x)[\tilde{u}(D\setminus x)-\tilde{u}((D\setminus x)-x)]\tilde{v}(D)\right| \\
&\quad\le\left[\sum_{\substack{x,D\\ x\in D,|D|=n+1}} a(x)[\tilde{u}(D\setminus x)-\tilde{u}((D\setminus x)-x)]^2\right]^{\frac{1}{2}}\left[\sum_{\substack{x,D\\ x\in D,|D|=n+1}} a(x)[\tilde{v}(D)]^2\right]^{\frac{1}{2}} \\
&\quad=\left[\sum_{\substack{x,A\\ x\notin A,|A|=n}} a(x)[\tilde{u}(A)-\tilde{u}(A-x)]^2\right]^{\frac{1}{2}}\left[\sum_{\substack{x,D\\ x\in D,|D|=n+1}} a(x)[\tilde{v}(D)]^2\right]^{\frac{1}{2}}.
\end{aligned} \tag{4.9}$$

We have used the inequality $|b(x)| \le a(x)$ in the last step and used $A = D \setminus x$ as the summation variable. The first expression is estimated by

$$\sum_{\substack{x,A \\ x\notin A, |A|=n}} a(x)[\tilde{u}(A) - \tilde{u}(A-x)]^2 = \sum_{\substack{x,A \\ x\notin A, |A|=n}} a(x)[\tilde{u}(\tau_x A) - \tilde{u}(A)]^2$$
$$\le \sum_{\substack{x,A \\ x\notin A, |A|=n}} a(x)[\tilde{u}(\tau_x A) - \tilde{u}(A)]^2 = 2\hat{D}_n^{\text{sh}}(\tilde{u})$$

and the second by

$$\sum_{\substack{x,D \\ x\in D, |D|=n+1}} a(x)[\tilde{v}(D)]^2 \le \frac{\alpha}{(1-\alpha)} D_{n+1}^{\text{ex}}(\tilde{v}) .$$

The estimate for $B_{n,n+1}^{\text{even}}$ is similar. So we have proved the first estimate of Lemma 4.1. The second one follows from duality. This concludes the proof of the lemma. □

5 Some Estimates on the Resolvent

The main theorem of this section is the following:

THEOREM 5.1 *Let $0 < \rho < 1$ and a local mean zero function f be given. Then the solution u_λ of the resolvent equation* (2.10) *satisfies the following estimates*: *For every $k \ge 0$ there is a constant C_k independent of n and ρ such that*

$$\sup_{\lambda>0} \sum_n n^{2k} [D_n^{\text{ex}}(u_\lambda) + (1-\rho)\hat{D}_n^{\text{sh}}(u_\lambda)] \le C_k C_f , \quad \sup_{\lambda>0} \lambda \sum_n n^{2k} \|u_{\lambda,n}\|_0^2 \le C_k C_f .$$

PROOF: We first recall standard estimates. Multiplying equation (2.10) by u_λ and integrating,

$$\lambda \|u_\lambda\|_0^2 + D^{\text{ex}}(u_\lambda) + D^{\text{sh}}(u_\lambda) = \langle f, u_\lambda \rangle \le C_f \sqrt{D^{\text{ex}}(u_\lambda)} .$$

This leads immediately to the estimates

$$\sup_{\lambda>0} \lambda \|u_\lambda\|_0^2 \le C_f , \tag{5.1}$$

$$\sup_{\lambda>0} D^{\text{ex}}(u_\lambda) \le C_f , \tag{5.2}$$

$$\sup_{\lambda>0} D^{\text{sh}}(u_\lambda) \le C_f . \tag{5.3}$$

Because of estimate (3.15), we also have

$$\sup_{\lambda>0} \hat{D}^{\text{sh}}(u_\lambda) \le \frac{1}{(1-\rho)} C_f . \tag{5.4}$$

Let us define the operator T as multiplication by a scalar $t(n)$ on each H_n. The sequence $t(n)$ is assumed to be positive, increasing, and eventually constant. Since L is bounded on each H_n and T is a multiple of I except on a finite number of H_n, it is easily verified that T leaves the domain of L invariant and the commutator $[T,L] = TL - LT$ is a bounded operator from $\mathbf{H} \to \mathbf{H}$. For $u = \sum_n u_n$ with $u_n \in H_n$, an explicit calculation yields

$$[TL - LT]u = \sum_n [(t(n+1) - t(n))B_{n,n+1}u_n + (t(n-1) - t(n))B_{n,n-1}u_n]$$

and

$$\begin{aligned}\langle [TL - LT]u, Tu\rangle &= \sum_n t(n+1)(t(n+1) - t(n))\langle B_{n,n+1}u_n, u_{n+1}\rangle \\ &\quad + \sum_n t(n-1)(t(n-1) - t(n))\langle B_{n,n-1}u_n, u_{n-1}\rangle .\end{aligned} \tag{5.5}$$

From Lemma 4.1, we have

$$\begin{aligned}|\langle [LT - TL]u, Tu\rangle| \le C\sum_n & t(n+1)|t(n+1) \\ & - t(n)|\sqrt{\rho(1-\rho)}\sqrt{D^{\mathrm{ex}}_{n+1}(u)}\left[\sqrt{n}\sqrt{D^{\mathrm{ex}}_n(u)} + \sqrt{\hat{D}^s_n(u)}\right] \\ & + C\sum_n t(n-1)|t(n-1) \\ & - t(n)|\sqrt{\rho(1-\rho)}\sqrt{D^{\mathrm{ex}}_{n-1}(u)}\left[\sqrt{n}\sqrt{D^{\mathrm{ex}}_n(u)} + \sqrt{\hat{D}^{\mathrm{sh}}_n(u)}\right].\end{aligned}$$

Since $D^{\mathrm{ex}}_n(u) = t(n)^{-2}D^{\mathrm{ex}}_n(Tu)$, the last term is equal to

$$\begin{aligned}& C\sum_n \left|\frac{t(n+1)}{t(n)} - 1\right|\sqrt{\rho(1-\rho)}\sqrt{D^{\mathrm{ex}}_{n+1}(Tu)}\left[\sqrt{n}\sqrt{D^{\mathrm{ex}}_n(Tu)} + \sqrt{\hat{D}^{\mathrm{sh}}_n(Tu)}\right] \\ & \quad + \sum_n \left|\frac{t(n-1)}{t(n)} - 1\right|\sqrt{\rho(1-\rho)}\sqrt{D^{\mathrm{ex}}_{n-1}(Tu)}\left[\sqrt{n}\sqrt{D^{\mathrm{ex}}_n(Tu)} + \sqrt{\hat{D}^{\mathrm{sh}}_n(Tu)}\right].\end{aligned}$$

We now make the choice of $t(n)$ such that for every n, $t(n)$ satisfies

$$\frac{C(1+\sqrt{n})}{2}\left\{\left|\frac{t(n+1)}{t(n)} - 1\right| + \left|\frac{t(n-1)}{t(n)} - 1\right|\right\} \le \delta \tag{5.6}$$

where δ will be chosen soon. This can be achieved by making $t(n) = n_0^k$ for $1 \le n \le n_0$ and $t(n) = n^k$ for $n_0 \le n \le n_1$ and $t(n) = n_1^k$ for $n \ge n_1$. Whereas the choice of n_0 will be governed by δ and k, n_1 can be totally arbitrary and the estimates will be uniform in n_1. In the end n_1 can be allowed to go to infinity, providing us with estimates for the choice of $t(n) = (n_0 \vee n)^k$.

From the factor $\sqrt{\rho(1-\rho)}$, we keep only $\sqrt{(1-\rho)}$ and use it only with $\hat{D}_n^{\text{sh}}(u)$ terms. Then

$$
\begin{aligned}
&|\langle [LT-TL]u, Tu\rangle| \\
&\leq \delta \sum_n [D_{n-1}^{\text{ex}}(Tu) + D_{ex,n}(Tu) + D_{n+1}^{\text{ex}}(Tu) + (1-\rho)\hat{D}_n^{\text{sh}}(Tu)] \qquad (5.7)\\
&= \delta[3D^{\text{ex}}(Tu) + (1-\rho)\hat{D}^{\text{sh}}(Tu)] \leq \delta C D(Tu)
\end{aligned}
$$

where the constant C comes from equation (3.15). We pick δ so that $C\delta < \frac{1}{4}$. Let us remark that the estimates depend on T only through δ.

From the resolvent equation we have

$$\lambda T u_\lambda - LTu_\lambda = [T, L]u_\lambda + Tf\,.$$

Multiply both sides by Tu_λ and integrate. From equation (5.7) and Lemma 5.1,

$$\lambda \|Tu_\lambda\|_0^2 + D^{env}(Tu_\lambda) \leq \frac{1}{4} D^{env}(Tu_\lambda) + C_{Tf}\sqrt{D^{\text{ex}}(u_\lambda)}\,.$$

Clearly this is sufficient to give the estimates

$$\sup_\lambda \lambda \|Tu_\lambda\|_0^2 \leq C_{Tf}\,, \quad \sup_\lambda \sum_n [t(n)]^2 [D_n^{\text{ex}}(u_\lambda) + (1-\rho)\hat{D}_n^{\text{sh}}(u)] \leq C_0 C_{Tf}^2\,.$$

Since f is local, C_{Tf} is easily controlled. □

6 Proof of Theorem 1.1

We first rewrite $B_{n,n}$ as follows:

LEMMA 6.1 *The operator $B_{n,n}$ from $H_n \to H_n$ can be rewritten as*

$$
\begin{aligned}
B_{n,n}\xi_A = &\sum_{\substack{x,y\neq 0\\ x\in A, y\notin A}} q(y-x)[\xi_{A\setminus x\cup y} - \xi_A] + (1-\rho)\sum_{-x\notin A} p(x)[\xi_{\tau_x A} - \xi_A] \\
&+ \rho \sum_{-x\in A} p(x)[\xi_{\tau_x A} - \xi_A]
\end{aligned}
$$

where $q(x) = q_\rho(x) = a(x) + (2\rho - 1)b(x) = \rho p(x) + (1-\rho)p(-x)$.

PROOF: We have to compare with the earlier expression and make sure that the difference is zero. For this we need to show that

$$(2\rho - 1)\sum_{\substack{x,y\neq 0\\ x\in A, y\notin A}} b(y-x) + (1-\rho)\sum_{-x\notin A} b(x) + \rho \sum_{-x\in A} b(x) = 0\,.$$

If we use the antisymmetry of $b(\cdot)$ as well as its consequence $\sum_x b(x) = 0$, the above relation is easily seen to be true. □

Our next step is to make $B_{n,n}$ look more like a convolution operator. The full translation symmetry is not available because the set $\mathcal{X}_n$ corresponds only to distinct nonzero n-tuples. A subset $A \subset \mathbb{Z}^d \setminus 0$ of cardinality n is really an equivalence class of $n!$ points in $(\mathbb{Z}^d)^n$. All the functions that we consider on $(\mathbb{Z}^d)^n$ will be symmetric under permutation. Let us denote by $\mathcal{G}_n \subset (\mathbb{Z}^d)^n$ the collection of distinct ordered nonzero n-tuples. A function on $\mathcal{X}_n$ can be considered as a symmetric function on $\mathcal{G}_n$ and then extended to all of $(\mathbb{Z}^d)^n$ by defining it to be 0 on the complement $(\mathbb{Z}^d)^n \setminus \mathcal{G}_n$. We will decompose this complement into three parts:

$$\mathcal{B}_n^1 = \{(x_1, \ldots, x_n) : x_i \neq 0 \text{ for } 1 \leq i \leq n \text{ and } x_i = x_j \text{ for just one pair}\},$$
$$\mathcal{B}_n^2 = \{(x_1, \cdots, x_n) : x_i \neq x_j \text{ for } 1 \leq i \neq j \leq n \text{ and for just one } i,\ x_i = 0\},$$

and

$$\mathcal{B}_n^3 = \mathcal{G}_n^c \setminus (\mathcal{B}_n^1 \cup \mathcal{B}_n^2).$$

We want to replace the operator $B_{n,n}$ acting on functions defined on $\mathcal{G}_n$ by the following operator $C_{n,n}$ of convolution type acting on the space of functions defined on all of $(\mathbb{Z}^d)^n$. Note that $\mathcal{B}_n^1$ and $\mathcal{B}_n^2$ are the boundary of $\mathcal{G}_n$, and transitions are possible from $\mathcal{G}_n$ only into $(\mathcal{G}_n \cup \mathcal{B}_n^1 \cup \mathcal{B}_n^2)$.

$$\begin{aligned}
&(C_{n,n}v)(x_1, \ldots, x_n) \\
&\quad = \sum_{x,j} q(x)[v(x_1, \ldots, x_j + x, \ldots, x_n) - v(x_1, \ldots, x_n)] \\
&\qquad + (1-\rho) \sum_{x} p(x)[v(x_1 - x, \ldots, x_j - x, \ldots, x_n - x) - v(x_1, \ldots, x_n)].
\end{aligned}$$

A comparison has to be made with $B_{n,n}$. Given a function u on $\mathcal{X}_n$, we can view it as a symmetric function defined on $\mathcal{G}_n$ and extend it to all of $(\mathbb{Z}^d)^n$ as a symmetric function by making it zero outside $\mathcal{G}_n$. We will abuse the notation somewhat and not distinguish between the three versions of the same function on $\mathcal{X}_n$, on $\mathcal{G}_n$, and on $(\mathbb{Z}^d)^n$. Since we will deal only with symmetric functions, it will not matter. We can also extend in the same fashion the function $f = B_{n,n}u$.

$$\begin{aligned}
&(B_{n,n}u)(x_1, \ldots, x_n) \\
&\quad = \sum_{j} \sum_{(x_1, \ldots, y, \ldots, x_n) \in \mathcal{G}_n} q(y - x_j)[u(x_1, \ldots, y, \ldots, x_n) - u(x_1, \ldots, x_n)] \\
&\qquad + (1-\rho) \sum_{x \notin \{x_1, \ldots, x_n\}} p(x)[u(x_1 - x, \ldots, x_n - x) - u(x_1, \ldots, x_n)] \\
&\qquad + \rho \sum_{j} p(x_j)[u(x_1 - x_j, \ldots, -x_j, \ldots, x_n - x_j) - u(x_1, \ldots, x_n)].
\end{aligned}$$

We define h by

$$C_{n,n}u = B_{n,n}u + h$$

and try to estimate h in terms of u. First, let us compute h explicitly. On $\mathcal{B}_n^3$ the function h is identically zero. For a point in $\mathcal{B}_n^2$, consisting of n distinct points $x_1,\dots,x_n$, exactly one of which is zero,

$$h(x_1,\dots,x_{n-1},0) = \sum_{x_n:(x_1,\dots,x_n)\in\mathcal{G}_n} q(x_n)u(x_1,\dots,x_{n-1},x_n) + (1-\rho)\sum_{x\neq x_1,\dots,x_n} p(x)u(x_1-x,\dots,x_n-x).$$

On $\mathcal{B}_n^1$ where a typical point is $(x_1,\dots,x_{n-1},x_{n-1})$ with distinct nonzero $x_1,\dots,x_{n-1}$,

$$h(x_1,\dots,x_{n-1},x_{n-1}) = 2\sum_{x_n:(x_1,\dots,x_n)\in\mathcal{G}_n} q(x_n-x_{n-1})u(x_1,\dots,x_{n-1},x_n).$$

Finally on $\mathcal{G}_n$,

$$\begin{aligned} &h(x_1,\dots,x_{n-1},x_n) \\ &= -\Bigg[\sum_{i\neq j} q(x_i-x_j)+\sum_i q(-x_i)]u(x_1,\dots,x_n)-(1-\rho)[\sum_j p(x_j)]u(x_1,\dots,x_n) \\ &\quad -\rho\sum_j p(x_j)[u(x_1-x_j,\dots,-x_j,\dots,x_n-x_j)-u(x_1,\dots,x_n)\Bigg].\end{aligned}$$

There are two new Dirichlet forms. The forms $D_n^{\text{ex}}(\cdot)$ that we already saw on functions defined on $\mathcal{X}_n$ as well as the new Dirichlet forms

$$\bar{D}_n(u) = \frac{1}{2}\sum_{\substack{i,x_i' \\ x_1,\dots,x_n}} a(x_i-x_i')\,[u(x_1,\dots,x_i,\dots x_n)-u(x_1,\dots,x_i',\dots x_n)]^2,$$

which corresponds to n free random walks, and

$$\bar{D}_n^{\text{sh}}(u) = \frac{1}{2}\sum_{x,x_1,\dots,x_n} a(x)[u(x_1-x,\dots,x_n-x)-u(x_1,\dots,x_n)]^2,$$

which corresponds to shifts. We have the following estimates:

LEMMA 6.2 *If u is any symmetric function of $x_1,\dots,x_n$ on $(\mathbb{Z}^d)^n$*

$$D_n^{\text{ex}}(u) \le \frac{1}{n!}\bar{D}_n(u).$$

If, in addition, $u\equiv 0$ outside $\mathcal{G}_n$, then for some constant C independent of n and u,

$$\frac{1}{n!}\bar{D}_n(u) \le CD_{\text{ex},n}(u) \quad \text{and} \quad \left|\frac{1}{n!}\bar{D}_n^{\text{sh}}(u)-\hat{D}_n^{\text{sh}}(u)\right| \le CD_n^{\text{ex}}(u).$$

PROOF: The first part is obvious. The factorial is just the number of times each term is counted. Clearly

$$\bar{D}_n(u) \le n!D_n^{\text{ex}}(u)+2(n-1)!\sum_{x_1,\dots,x_n} a(x_n-x_{n-1})u^2(x_1,\dots,x_n),$$

and by Lemma 3.7

$$\sum_{x_1,\dots,x_n}\sum_{i\neq j} a(x_i-x_j)u^2(x_1,\dots,x_n)\leq C n D_n^{\mathrm{ex}}(u),$$

and now the second part follows. The third part also follows from Lemma 3.7 and is nearly identical:

$$\left|\frac{1}{n!}\bar{D}_n^{\mathrm{sh}}(u)-\hat{D}_n^{\mathrm{sh}}(u)\right|\leq C\sum_{A\in\mathcal{X}_n}\left(\sum_{x\in A}a(x)\right)[u(A)]^2,$$

where we have identified A with $(x_1,\dots,x_n)$. □

We now return to the estimation of h. If we define

$$\langle h,w\rangle=\frac{1}{n!}\sum h(x_1,\dots,x_n)\,w(x_1,\dots,x_n),$$

then $|\langle h,w\rangle|$ is less than

$$\begin{aligned}|\langle h,w\rangle| &\leq C\left[\frac{1}{n!}\sum_{x_1,\dots,x_n}A(x_1,\dots,x_n)u^2(x_1,\dots,x_n)\right]^{\frac{1}{2}}\\ &\quad\times\left[\frac{1}{n!}\sum_{x_1,\dots,x_n}A(x_1,\dots,x_n)w^2(x_1,\dots,x_n)\right]^{\frac{1}{2}} \qquad (6.1)\\ &\leq Cn[D_n^{\mathrm{ex}}(u)]^{\frac{1}{2}}\left[\frac{1}{n!}\bar{D}_n(w)\right]^{\frac{1}{2}}.\end{aligned}$$

We have defined

$$A(x_1,\dots,x_n)=\sum_{i\neq j}a(x_i-x_j)+\sum_j a(x_j).$$

We have used the following facts: The symmetric part q is also a, and p and q are dominated by $2a$.

The next lemma is a simple consequence of Fourier analysis.

LEMMA 6.3 *Let u be a symmetric function of n variables satisfying*

$$\lambda u - C_{n,n}u = v$$

where v satisfies the bound

$$\left|\sum_{x_1,\dots,x_n}v(x_1,\dots,x_n)w(x_1,\dots,x_n)\right|\leq C[\bar{D}_n(w)]^{\frac{1}{2}}.$$

Then λu satisfies the same bound with the same constant,

$$\left|\sum_{x_1,\dots,x_n}\lambda u(x_1,\dots,x_n)w(x_1,\dots,x_n)\right|\leq C[\bar{D}_n(w)]^{\frac{1}{2}}.$$

PROOF: Denoting by $\mathbb{T}_d^n$ the n-fold product of the d-torus and by $\hat{u}(\theta)$ and $\hat{v}(\theta)$ the Fourier transforms of u and v, respectively, we have

$$[\lambda + \Phi(\theta) + i\Psi(\theta)]\hat{u}(\theta) = \hat{v}(\theta)$$

where

$$\begin{aligned}\Phi(\theta) + i\Psi(\theta) &= \sum_j \sum_x q(x)[1 - \cos(\theta_j x) - i \sin(\theta_j x)] \\ &\quad + (1-\rho)\sum_x p(x)\left[1 - \cos\left(x \sum_j \theta_j\right) + i \sin\left(x \sum_j \theta_j\right)\right].\end{aligned}$$

Φ and Ψ are real, and $\Phi(\theta) \geq 0$. Moreover,

$$\|v\|_{-1}^2 = \int_{\mathbb{T}_d^n} \frac{|\hat{v}(\theta)|^2}{H(\theta)} d\theta \quad \text{where } H(\theta) = \sum_j \sum_x a(x)[1 - \cos(\theta_j x)]$$

and

$$\begin{aligned}\|\lambda u\|_{-1}^2 &= \int_{\mathbb{T}_d^n} \frac{|\lambda \hat{u}(\theta)|^2}{H(\theta)} d\theta = \int_{\mathbb{T}_d^n} \frac{|\hat{v}(\theta)|^2}{H(\theta)} \frac{|\lambda|^2}{|\lambda + \Phi(\theta) + i\Psi(\theta)|^2} d\theta \\ &\leq \int_{\mathbb{T}_d^n} \frac{|\hat{v}(\theta)|^2}{H(\theta)} d\theta,\end{aligned}$$

which proves the lemma. □

PROOF OF THEOREM 1.1: We are now ready to prove our main theorem. We start with a solution u_λ of (2.10) with a local f. If we decompose and write $u = \sum_n u_n$ and $f = \sum_n f_n$ with u_n and f_n from H_n, following the notation of (4.5) and (4.6),

$$\lambda u_{\lambda,n} - B_{n,n} u_{\lambda,n} = f_n + B_{n-1,n} u_{\lambda,n-1} + B_{n+1,n} u_{\lambda,n+1} = g_n.$$

As before, we rewrite

$$\lambda u_{\lambda,n} - C_{n,n} u_{\lambda,n} = g_n + h_n.$$

From (6.1) and Theorem 5.1, we have the estimate

$$|\langle h_n, w\rangle| \leq C_f C_k n^{-k} \left[\frac{1}{n!} \bar{D}(w)\right]^{\frac{1}{2}}.$$

From Lemmas 4.1 and 6.3 and Theorem 5.1, we have

$$|\langle w, B_{n-1,n} u_{\lambda,n-1} + B_{n+1,n} u_{\lambda,n+1}\rangle| \leq C_f C_k n^{-k} \left[\frac{1}{n!} \bar{D}(w)\right]^{\frac{1}{2}}.$$

If f is local, $f_n = 0$ for large enough n. By Lemma 6.4, we now conclude that for all large n,

$$|\langle \lambda u_{\lambda,n}, w\rangle| \leq C_f C_k \, n^{-k} \left[\frac{1}{n!} \bar{D}(w)\right]^{\frac{1}{2}}.$$

For small n where $f_n \neq 0$, we have from the proof of Lemma 2.1 and Lemma 6.3 that

$$|\langle w, f_n\rangle| \leq C_f\sqrt{D^e_n} \leq C_f\left[\frac{1}{n!}\bar{D}(w)\right]^{\frac{1}{2}}$$

and so for small n by Lemma 6.4,

$$|\langle \lambda u_{\lambda,n}, w\rangle| \leq C_f\left[\frac{1}{n!}\bar{D}(w)\right]^{\frac{1}{2}}.$$

Finally, by the use of Lemma 6.3, adding up contributions with a larger constant,

$$\|\lambda u_\lambda\|_{-1} \leq C_f C_k \sum_n n^{-k},$$

and we are done. □

Acknowledgments. The authors wish to express their thanks to the referee. The original proof of Theorem 2.7 was by direct estimation and the referee's suggestion to bound it below by the corresponding variance for the symmetric situation simplified the proof considerably.

The authors also wish to acknowledge the support of the National Science Foundation through grants DMS-9703811 for Sethuraman, DMS-9503419 for Varadhan, and DMS-9403462 as well as DMS-9703752 for Yau. In addition, Varadhan wants to acknowledge the support of ARO through Grant ARO-DAAH04-95-1-0666, while Yau wishes to acknowledge the support of the David and Lucile Packard Foundation.

Bibliography

[1] Arratia, R. The motion of a tagged particle in the simple symmetric exclusion system on $\mathbb{Z}$. *Ann. Probab.* **11** (1983), no. 2, 362–373.

[2] Esposito, R.; Marra, R.; Yau, H. T. Navier-Stokes equations for stochastic particle systems on the lattice. *Comm. Math. Phys.* **182** (1996), no. 2, 395–456.

[3] Helland, I. S. Central limit theorems for martingales with discrete or continuous time. *Scand. J. Statist.* **9** (1982), no. 2, 79–94.

[4] Jensen, L.; Yau, H.-T. Hydrodynamical scaling limits of simple exclusion models. *Probability theory and applications (Princeton, N.J., 1996)*, 167–225. IAS/Park City Math. Ser., 6. Amer. Math. Soc., Providence, R.I., 1999.

[5] Kipnis, C. Central limit theorems for infinite series of queues and applications to simple exclusion. *Ann. Probab.* **14** (1986), no. 2, 397–408.

[6] Kipnis, C.; Varadhan, S. R. S. Central limit theorem for additive functionals of reversible Markov processes and applications to simple exclusions. *Comm. Math. Phys.* **104** (1986), no. 1, 1–19.

[7] Landim, C.; Yau, H. T. Fluctuation-dissipation equation of asymmetric simple exclusion processes. *Probab. Theory Related Fields* **108** (1997), no. 3, 321–356.

[8] Liggett, T. M. *Interacting particle systems.* Grundlehren der Mathematischen Wissenschaften, 276. Springer, New York-Berlin, 1985.

[9] Saada, E. A limit theorem for the position of a tagged particle in a simple exclusion process. *Ann. Probab.* **15** (1987), no. 1, 375–381.

[10] Varadhan, S. R. S. Regularity of self-diffusion coefficient. *The Dynkin Festschrift*, 387–397. Progr. Probab., 34. Birkhäuser Boston, Boston, 1994.

[11] Varadhan, S. R. S. Self-diffusion of a tagged particle in equilibrium for asymmetric mean zero random walk with simple exclusion. *Ann. Inst. H. Poincaré Probab. Statist.* **31** (1995), no. 1, 273–285.

SUNDER SETHURAMAN
Iowa State University
Department of Mathematics
400 Carver Hall
Ames, IA 50011
E-mail: sethuram@math.iastate.edu

S. R. S. VARADHAN
Courant Institute
251 Mercer Street
New York, NY 10012
E-mail: varadhan@cims.nyu.edu

HORNG-TZER YAU
Courant Institute
251 Mercer Street
New York, NY 10012
E-mail: yau@cims.nyu.edu

Received March 1999.

Electronic Journal of Probability

Vol. 10 (2005), Paper no. 36, pages 1221-1235.

Journal URL
http://www.math.washington.edu/~ejpecp/

A Martingale Proof of Dobrushin's Theorem for Non-Homogeneous Markov Chains [1]

S. Sethuraman and **S.R.S. Varadhan**

Department of Mathematics, Iowa State University
Ames, IA 50011, USA
Email: sethuram@iastate.edu

and

Courant Institute, New York University
New York, NY 10012, USA
Email: varadhan@cims.nyu.edu

Abstract. In 1956, Dobrushin proved an important central limit theorem for non-homogeneous Markov chains. In this note, a shorter and different proof elucidating more the assumptions is given through martingale approximation.

Keywords and phrases: non-homogeneous Markov, contraction coefficient, central limit theorem, martingale approximation

AMS Subject Classification (2000): Primary 60J10; secondary 60F05.

Submitted to EJP on March 21, 2005. Final version accepted on September 6, 2005.

[1]Research supported in part by NSF/DMS-0071504, NSA-H982300510041 and NSF/DMS-0104343.

1 Introduction and Results

Nearly fifty years ago, R. Dobrushin proved in his thesis [2] an important central limit theorem (CLT) for Markov chains in discrete time that are not necessarily homogeneous in time. Previously, Markov, Bernstein, Sapagov, and Linnik, among others, had considered the central limit question under various sufficient conditions. Roughly, the progression of results relaxed the state space structure from 2 states to an arbitrary set of states, and also the level of asymptotic degeneracy allowed for the transition probabilities of the chain.

After Dobrushin's work, some refinements and extensions of his CLT, some of which under more stringent assumptions, were proved by Statuljavicius [16] and Sarymsakov [13]. See also Hanen [6] in this regard. A corresponding invariance principle was also proved by Gudinas [4]. More general references on non-homogeneous Markov processes can be found in Isaacson and Madsen [7], Iosifescu [8], Iosifescu and Theodorescu [9], and Winkler [18].

We now define what is meant by "degeneracy." Although there are many measures of "degeneracy," the measure which turns out to be most useful to work with is that in terms of the contraction coefficient. This coefficient has appeared in early results concerning Markov chains, however, in his thesis, Dobrushin popularized its use, and developed many of its important properties. [See Seneta [14] for some history.]

Let $(\mathbf{X}, \mathcal{B}(\mathbf{X}))$ be a Borel space, and let $\pi = \pi(x, dy)$ be a Markov transition probability on $(\mathbf{X}, \mathcal{B}(\mathbf{X}))$. Define the contraction coefficient $\delta(\pi)$ of π as

$$
\begin{aligned}
\delta(\pi) &= \sup_{x_1, x_2 \in \mathbf{X}} \|\pi(x_1, \cdot) - \pi(x_2, \cdot)\|_{\mathrm{Var}} \\
&= \sup_{\substack{x_1, x_2 \in \mathbf{X} \\ A \in \mathcal{B}(\mathbf{X})}} |\pi(x_1, A) - \pi(x_2, A)| \\
&= \frac{1}{2} \sup_{\substack{x_1, x_2 \in \mathbf{X} \\ \|f\|_{L^\infty} \leq 1}} \left| \int f(y)[\pi(x_1, dy) - \pi(x_2, dy)] \right|.
\end{aligned}
$$

Also, define the related coefficient $\alpha(\pi) = 1 - \delta(\pi)$.

Clearly, $0 \leq \delta(\pi) \leq 1$, and $\delta(\pi) = 0$ if and only if $\pi(x, dy)$ does not depend on x. It makes sense to call π "non-degenerate" if $0 \leq \delta(\pi) < 1$. We use the standard convention and denote by $\mu\pi$ and πu the transformations induced by π on countably additive measures and bounded measurable functions respectively,

$$
(\mu\pi)(A) = \int \mu(dx)\, \pi(x, A) \qquad \text{and} \qquad (\pi u)(x) = \int \pi(x, dy)\, u(y).
$$

One can see that $\delta(\pi)$ has the following properties.

$$
\delta(\pi) = \sup_{\substack{x_1, x_2 \in \mathbf{X} \\ u \in \mathcal{U}}} |(\pi u)(x_1) - (\pi u)(x_2)| \tag{1.1}
$$

with $\mathcal{U} = \{u : \sup_{y_1,y_2} |u(y_1) - u(y_2)| \leq 1\}$. It is the operator norm of π with respect to the Banach (semi-) norm $\mathrm{Osc}(u) = \sup_{x_1,x_2} |u(x_1) - u(x_2)|$, namely the oscillation of u. In particular, for any transition probabilities π_1, π_2 we have

$$\delta(\pi_1\,\pi_2) \leq \delta(\pi_1)\,\delta(\pi_2) \tag{1.2}$$

where $\pi_1\pi_2$ is the two-step transition probability $\pi_1\pi_2(x,\cdot) = \int \pi_1(x,dy)\pi_2(y,\cdot)$.

By a non-homogeneous Markov chain of length n on state space $(\mathbf{X}, \mathcal{B}(\mathbf{X}))$ corresponding to transition operators $\{\pi_{i,i+1} = \pi_{i,i+1}(x,dy) : 1 \leq i \leq n-1\}$ we mean the Markov process P on the product space $(\mathbf{X}^n,\ \mathcal{B}(\mathbf{X}^n))$,

$$P[X_{i+1} \in A | X_i = x] = \pi_{i,i+1}(x, A),$$

where $\{X_i : 1 \leq i \leq n\}$ are the canonical projections. In particular, under the initial distribution $X_1 \sim \mu$, the distribution at time $k \geq 1$ is $\mu\pi_{1,2}\pi_{2,3}\cdots\pi_{k-1,k}$. For $i < j$ we will define

$$\pi_{i,j} = \pi_{i,i+1}\pi_{i+1,i+2}\cdots\pi_{j-1,j}.$$

We denote by $E[Z]$ and $V(Z)$ the expectation and variance of the random variable Z with respect to P.

Dobrushin's theorem concerns the fluctuations of an array of non-homogeneous Markov chains. For each $n \geq 1$, let $\{X_i^{(n)} : 1 \leq i \leq n\}$ be n observations of a non-homogeneous Markov chain on $\mathbf{X}$ with transition matrices $\{\pi_{i,i+1}^{(n)} = \pi_{i,i+1}^{(n)}(x,dy) : 1 \leq i \leq n-1\}$ and initial distribution $\mu^{(n)}$. Let also

$$\alpha_n = \min_{1\leq i\leq n-1} \alpha(\pi_{i,i+1}^{(n)}).$$

In addition, let $\{f_i^{(n)} : 1 \leq i \leq n\}$ be real valued functions on $\mathbf{X}$. Define, for $n \geq 1$, the sum

$$S_n = \sum_{i=1}^{n} f_i^{(n)}(X_i^{(n)}).$$

Theorem 1.1 *Suppose that for some finite constants C_n,*

$$\sup_{1\leq i\leq n}\sup_{x\in\mathbf{X}} |f_i^{(n)}(x)| \leq C_n.$$

Then, if

$$\lim_{n\to\infty} C_n^2\alpha_n^{-3}\left[\sum_{i=1}^{n} V(f_i^{(n)}(X_i^{(n)}))\right]^{-1} = 0, \tag{1.3}$$

we have the standard Normal convergence

$$\frac{S_n - E[S_n]}{\sqrt{V(S_n)}} \Rightarrow \mathrm{N}(0,1). \tag{1.4}$$

Also, there is an example where the result is not true if condition (1.3) is not met.

In [2], Dobrushin also states the direct corollary which simplifies some of the assumptions.

Corollary 1.1 *When the functions are uniformly bounded, i.e.* $\sup_n C_n = C < \infty$ *and the variances are bounded below, i.e.* $V(f_i^{(n)}(X_i^{(n)})) \geq c > 0$, *for all* $1 \leq i \leq n$ *and* $n \geq 1$, *then we have the convergence (1.4) provided*

$$\lim_{n\to\infty} n^{1/3}\alpha_n = \infty.$$

We remark that in [2] (e.g. Theorems 3, 8) there are also results where the boundedness condition on $f_i^{(n)}$ is replaced by integrability conditions. As these results follow from truncation methods and Theorem 1.1 for bounded variables, we only consider Dobrushin's theorem in the bounded case.

Also, for the ease of the reader, and to be complete, we will discuss in the next section an example, given in [2] and due to Bernstein and Dobrushin, of how the convergence (1.4) may fail when the condition (1.3) is not satisfied.

We now consider Dobrushin's methods. The techniques used in [2] to prove the above results fall under the general heading of the "blocking method." The condition (1.3) ensures that well-separated blocks of observations may be approximated by independent versions with small error. Indeed, in many remarkable steps, Dobrushin exploits the Markov property and several contraction coefficient properties, which he himself derives, to deduce error bounds sufficient to apply CLT's for independent variables. However, in [2], it is difficult to see, even at the technical level, why condition (1.3) is natural.

The aim of this note is to provide a different, shorter proof of Theorem 1.1 which explains more why condition (1.3) appears in the result. The methods are through martingale approximations and martingale CLT's. These methods go back at least to Gordin [3] in the context of homogeneous processes, and have been used by others in mostly "stationary" situations (e.g. Kifer [10], Kipnis and Varadhan [11], Pinsky [12], and Wu and Woodroofe [19]). The approximation with respect to the non-homogeneous setting of Theorem 1.1 makes use of three ingredients: (1) negligibility estimates for individual components, (2) a law of large numbers (LLN) for conditional variances, and (3) lower bounds for the variance $V(S_n)$. Negligibility bounds and a LLN are well known requirements for martingale CLT's (cf. Hall-Heyde [5, ch. 3]), and in fact, as will be seen, the sufficiency of condition (1.3) is transparent in the proofs of these two components (Lemma 3.2, and Lemmas 3.3 and 3.4). The variance lower bounds which we will use (Proposition 3.2) were as well derived by Dobrushin in his proof. However, using some martingale and spectral tools, we give a more direct argument for a better estimate.

We note also, with this martingale approximation, that an invariance principle for the partial sums holds through standard martingale propositions, Hall-Heyde [5], among other results. In fact, from the martingale invariance principle, it should be possible to derive Gudynas's theorems [4] although this is not done here.

We now explain the structure of the article. In section 2, we give the Bernstein-Dobrushin example of a Markov chain with anomalous behavior. In section 3, we state a martingale

CLT and prove Theorem 1.1 assuming a lower bound on the variance $V(S_n)$. Last, in section 4, we prove this variance estimate.

2 Bernstein-Dobrushin Example

Here, we summarize the example in Dobrushin's thesis, attributed to Bernstein, which shows that condition (1.3) is sharp.

Example 2.1 Let $\mathbf{X} = \{1, 2\}$, and consider the 2×2 transition matrices on $\mathbf{X}$,

$$Q(p) = \begin{pmatrix} 1-p & p \\ p & 1-p \end{pmatrix}$$

for $0 \leq p \leq 1$. The contraction coefficient $\delta(Q(p))$ of $Q(p)$ is $|1-2p|$. Note that $\delta(Q(p)) = \delta(Q(1-p))$. The invariant measures for all the $Q(p)$ are the same $\mu(1) = \mu(2) = \frac{1}{2}$. We will be looking at $Q(p)$ for p close to 0 or 1 and the special case of $p = \frac{1}{2}$. However, when p is small, the homogeneous chains behave very differently under $Q(p)$ and $Q(1-p)$. More specifically, when p is small there are very few switches between the two states whereas when $1-p$ is small it switches most of the time. In fact, this behavior can be made more precise (see Dobrushin [1], or from direct computation). Let $T_n = \sum_{i=1}^n \mathbf{1}_{\{1\}}(X_i)$ count the number of visits to state 1 in the first n steps.

Case A. Consider the homogeneous chain under $Q(p)$ with $p = \frac{1}{n}$ and initial distribution $\mu(1) = \mu(2) = \frac{1}{2}$. Then,

$$\frac{T_n}{n} \Rightarrow G \quad \text{and} \quad \lim_{n\to\infty} n^{-2}\, V(T_n) \;=\; V_A \tag{2.1}$$

where $0 < V_A < \infty$ and G is a non-degenerate distribution supported on $[0, 1]$.

Case B. Consider the homogeneous chain run under $Q(p)$ with $p = 1 - \frac{1}{n}$ and initial distribution $\mu(1) = \mu(2) = \frac{1}{2}$. Then,

$$T_n - \frac{n}{2} \Rightarrow F \quad \text{and} \quad \lim_{n\to\infty} V(T_n) \;=\; V_B \tag{2.2}$$

where $0 < V_B < \infty$ and F is a non-degenerate distribution.

Let a sequence $\alpha_n \to 0$ with $\alpha_n \geq n^{-\frac{1}{3}}$ be given . To construct the anomalous Markov chain, it will be helpful to split the time horizon $[1, 2, \ldots, n]$ into roughly $n\alpha_n$ blocks of size α_n^{-1}. We interpose a $Q(\frac{1}{2})$ between any two blocks that has the effect of making the blocks independent of each other. More precisely let $k_i^{(n)} = i[\alpha_n^{-1}]$ for $1 \leq i \leq m_n$ where $m_n = [n/[\alpha_n^{-1}]]$. Also, define $k_0^{(n)} = 0$, and $k_{m_n+1}^{(n)} = n$.

Define now, for $1 \le i \le n$,

$$\pi^{(n)}_{i,i+1} = \begin{cases} Q(\alpha_n) & \text{for } i = 1, 2, \ldots, k^{(n)}_1 - 1 \\ Q(\frac{1}{2}) & \text{for } i = k^{(n)}_1, k^{(n)}_2, \ldots, k^{(n)}_{m_n} \\ Q(1-\alpha_n) & \text{for all other } i. \end{cases}$$

Consider the non-homogeneous chain with respect to $\{\pi^{(n)}_{i,i+1} : 1 \le i \le n-1\}$ starting from equilibrium $\mu^{(n)}(0) = \mu^{(n)}(1) = \frac{1}{2}$. ¿From the definition of the chain, one observes, as $Q(\frac{1}{2})$ does not distinguish between states, that the process in time horizons $\{(k^{(n)}_i + 1, k^{(n)}_{i+1}) : 0 \le i \le m_n\}$ are mutually independent. For the first time segment 1 to $k^{(n)}_1$, the chain is in regime A, while for the other segments, the chain is in case B.

Once again, let us concentrate on the number of visits to state 1. Denote by $T^{(n)} = \sum_{i=1}^n \mathbf{1}_{\{1\}}(X^{(n)}_i)$ and $T^{(n)}(k,l) = \sum_{i=k}^l \mathbf{1}_{\{1\}}(X^{(n)}_i)$ the counts in the first n steps and in steps k to l respectively. It follows from the discussion of independence above that

$$T^{(n)} = \sum_{i=0}^{m_n} T^{(n)}(k^{(n)}_i + 1, k^{(n)}_{i+1})$$

is the sum of independent sub-counts where, additionally, the sub-counts for $1 \le i \le m_n - 1$ are identically distributed, the last sub-count perhaps being shorter. Also, as the initial distribution is invariant, we have $V(\mathbf{1}_{\{1\}}(X^{(n)}_i)) = 1/4$ for all i and n. Then, in the notation of Corollary 1.1, $C = 1$ and $c = 1/4$.

¿From (2.1), we have that

$$V(T^{(n)}(1, k^{(n)}_1)) \sim \alpha_n^{-2} V_A \quad \text{as } n \uparrow \infty.$$

Also, from (2.2) and independence of m_n sub-counts, we have that

$$V(T^{(n)}(k^{(n)}_1 + 1, n)) \sim n\alpha_n V_B \quad \text{as } n \uparrow \infty.$$

¿From these calculations, we see if $n^{1/3}\alpha_n \to \infty$, then $\alpha_n^{-2} << n\alpha_n$, and so the major contribution to $T^{(n)}$ is from $T^{(n)}(k^{(n)}_1 + 1, n)$. However, since this last count is (virtually) the sum of m_n i.i.d. sub-counts, we have that $T^{(n)}$, properly normalized, converges to N$(0,1)$, as predicted by Dobrushin's Theorem 1.1.

On the other hand, if $\alpha_n = n^{-1/3}$, we have $\alpha_n^{-2} = n\alpha_n$, and count $T^{(n)}(1, k^{(n)}_1)$, independent of $T^{(n)}(k^{(n)}_1, n)$, also contributes to the sum $T^{(n)}$. After centering and scaling, then, $T^{(n)}$ approaches the convolution of a non-trivial non-normal distribution and a normal distribution, and therefore is not Gaussian.

3 Proof of Theorem 1.1

The CLT for martingale differences is a standard tool. We quote the following form of the result implied by Corollary 3.1 in Hall and Heyde [5].

Proposition 3.1 *For each $n \geq 1$, let $\{(W_i^{(n)}, \mathcal{G}_i^{(n)}) : 0 \leq i \leq n\}$ be a martingale relative to the nested family $\mathcal{G}_i^{(n)} \subset \mathcal{G}_{i+1}^{(n)}$ with $W_0^{(n)} = 0$. Let $\xi_i^{(n)} = W_i^{(n)} - W_{i-1}^{(n)}$ be their differences. Suppose that*

$$\begin{array}{rcll} \max_{1\leq i\leq n} \|\xi_i^{(n)}\|_{L^\infty} & \to & 0 & \text{and} \\ \sum_{i=1}^n E[(\xi_i^{(n)})^2 | \mathcal{G}_{i-1}^{(n)}] & \to & 1 & \text{in } L^2. \end{array}$$

Then,

$$W_n^{(n)} \Rightarrow \mathrm{N}(0,1).$$

The first and second limit conditions are the so called "negligibility" assumption on the sequence, and LLN for conditional variances mentioned in the introduction.

Consider now the non-homogeneous setting of Theorem 1.1. To simplify notation, we will assume throughout that the functions $\{f_i^{(n)}\}$ are mean-zero, $E[f_i^{(n)}(X_i^{(n)})] = 0$ for $1 \leq i \leq n$ and $n \geq 1$. Define

$$Z_k^{(n)} = \sum_{i=k}^{n} E[f_i^{(n)}(X_i^{(n)}) | X_k^{(n)}]$$

so that

$$Z_k^{(n)} = \begin{cases} f_k^{(n)}(X_k^{(n)}) + \sum_{i=k+1}^n E[f_i^{(n)}(X_i^{(n)}) | X_k^{(n)}] & \text{for } \ 1 \leq k \leq n-1 \\ f_n^{(n)}(X_n^{(n)}) & \text{for } \ k = n. \end{cases} \tag{3.1}$$

Remark 3.1 Before going further, we remark that the sequence $\{Z_k^{(n)}\}$ can be thought of as a type of "Poisson-resolvent" sequence often seen in martingale approximations Namely, when the array $\{X_i^{(n)}\}$ is formed from the sequence $\{X_i\}$, $f_i^{(n)} = f$ for all i and n, and the chain is homogeneous, $P_n = P$ for all n, then indeed $Z_k^{(n)}$ reduces to $Z_k^{(n)} = f(X_k) + \sum_{i=1}^{n-k}(P^i f)(X_k)$ which approximates the Poisson-resolvent solution $\sum_{i=0}^{\infty}(P^i f)(X_k) = [(I - P)^{-1} f](X_k)$ usually used to prove the CLT in this case (cf. p. 145-6 Varadhan [17]).

Returning to the full non-homogeneous setting of Theorem 1.1, by rearranging terms in (3.1), we obtain for $1 \leq k \leq n-1$ that

$$f_k^{(n)}(X_k^{(n)}) \;\; = \;\; Z_k^{(n)} - E[Z_{k+1}^{(n)} | X_k^{(n)}] \tag{3.2}$$

which for $2 \leq k \leq n-1$ further equals $[Z_k^{(n)} - E[Z_k^{(n)} | X_{k-1}^{(n)}]] + [E[Z_k^{(n)} | X_{k-1}^{(n)}] - E[Z_{k+1}^{(n)} | X_k^{(n)}]]$. Then, we have the decomposition,

$$\begin{aligned} S_n &= \sum_{k=1}^{n} f_k^{(n)}(X_k^{(n)}) \\ &= \sum_{k=2}^{n} [Z_k^{(n)} - E[Z_k^{(n)} | X_{k-1}^{(n)}]] + Z_1^{(n)} \end{aligned} \tag{3.3}$$

and so in particular $V(S_n) = \sum_{k=2}^{n} V(Z_k^{(n)} - E[Z_k^{(n)}|X_{k-1}^{(n)}]) + V(Z_1^{(n)})$. Let us now define the scaled differences

$$\xi_k^{(n)} = \frac{1}{\sqrt{V(S_n)}}[Z_k^{(n)} - E[Z_k^{(n)}|X_{k-1}^{(n)}]] \tag{3.4}$$

and the martingale $M_k^{(n)} = \sum_{l=2}^{k} \xi_l^{(n)}$ with respect to $\mathcal{F}_k^{(n)} = \sigma\{X_l^{(n)} : 1 \le l \le k\}$ for $2 \le k \le n$. The plan to obtain Theorem 1.1 will now be to approximate $S_n/\sqrt{V(S_n)}$ by $M_n^{(n)}$ and use Proposition 3.1. Condition (1.3) will be a sufficent condition for "negligibility" (Lemma 3.2) and "LLN" (Lemmas 3.3 and 3.4) with regard to Proposition 3.1.

Lemma 3.1 *We have, for $1 \le i < j \le n$,*

$$\|\pi_{i,j} f_j^{(n)}\|_{L^\infty} \le 2C_n(1-\alpha_n)^{j-i} \quad \text{and} \quad \mathrm{Osc}\big(\pi_{i,j}(f_j^{(n)})^2\big) \le 2C_n^2(1-\alpha_n)^{j-i}$$

and, for $1 \le l < i < j \le n$,

$$\mathrm{Osc}\big(\pi_{l,i}(f_i^{(n)}\pi_{i,j}f_j^{(n)})\big) \le 6C_n^2\,(1-\alpha_n)^{i-l}(1-\alpha_n)^{j-i}.$$

Proof. As $\|f_j^{(n)}\|_{L_\infty} \le C_n$ its oscillation $\mathrm{Osc}(f_j^{(n)}) \le 2C_n$. From definition of $\delta(\cdot)$ (cf. (1.1)) and (1.2),

$$\mathrm{Osc}(\pi_{i,j}f_j^{(n)}) \le \mathrm{Osc}(f_j^{(n)})\,\delta(\pi_{i,j}) \le 2C_n(1-\alpha_n)^{j-i}.$$

Because $E[(\pi_{i,j}f_j^{(n)})(X_i^{(n)})] = E[f_j^{(n)}(X_j^{(n)})] = 0$, the first bound follows as

$$\|\pi_{i,j}f_j^{(n)}\|_{L_\infty} \le \mathrm{Osc}(\pi_{i,j}f_j^{(n)}) \le 2C_n(1-\alpha_n)^{j-i}.$$

The second bound is analogous. For the third bound, write

$$\begin{aligned}
\mathrm{Osc}\big(\pi_{l,i}(f_i^{(n)}\pi_{i,j}f_j^{(n)})\big) &\le (1-\alpha_n)^{i-l}\,\mathrm{Osc}\big(f_i^{(n)}\pi_{i,j}f_j^{(n)}\big)\\
&\le (1-\alpha_n)^{i-l}\Big[\mathrm{Osc}(f_i^{(n)})\,\|\pi_{i,j}f_j^{(n)}\|_{L^\infty}\\
&\qquad + \|f_i^{(n)}\|_{L^\infty}\,\mathrm{Osc}(\pi_{i,j}f_j^{(n)})\Big]\\
&\le 6C_n^2\,(1-\alpha_n)^{i-l}(1-\alpha_n)^{j-i}.
\end{aligned}$$

□

We now state a lower bound for the variance proved in the next section. For comparison, we remark that in [2] the bound $V(S_n) \ge (\alpha_n/8)\sum_{i=1}^{n} V(f_i^{(n)}(X_i^{(n)}))$ is given (see also section 1.2.2 [9]).

Proposition 3.2 *For $n \geq 1$,*

$$V(S_n) \geq \frac{\alpha_n}{4} \sum_{i=1}^{n} V\big(f_i^{(n)}(X_i^{(n)})\big). \tag{3.5}$$

The next estimate shows that the asymptotics of $S_n/\sqrt{V(S_n)}$ depend only on the martingale approximant $M_n^{(n)}$, and that the differences $\xi_k^{(n)}$ are negligible.

Lemma 3.2 *Under condition (1.3), we have that*

$$\lim_{n\to\infty} \sup_{1\leq k\leq n} \frac{\|Z_k^{(n)}\|_{L^\infty}}{\sqrt{V(S_n)}} = 0.$$

Proof. By Lemma 3.1,

$$\|Z_k^{(n)}\|_{L^\infty} \leq \sum_{i=k}^{n} \|E[f_i^{(n)}(X_i^{(n)})|X_k^{(n)}]\|_{L^\infty} \leq 2C_n \sum_{i=k}^{n} (1-\alpha_n)^{i-k} \leq 2C_n\alpha_n^{-1}.$$

Then, by Proposition 3.2,

$$\sup_{1\leq k\leq n} \frac{\|Z_k^{(n)}\|_{L^\infty}}{\sqrt{V(S_n)}} \leq 4C_n\left(\alpha_n^3 \sum_{i=1}^{n} V(f_i^{(n)}(X_i^{(n)}))\right)^{-1/2}$$

which in turn by (1.3) is $o(1)$. □

The next two lemmas help prove the LLN part of Proposition 3.1 for array $\{M_k^{(n)}\}$. By the oscillation of a random variable η we mean $\mathrm{Osc}(\eta) = \sup_{\omega,\omega'} |\eta(\omega) - \eta(\omega')|$.

Lemma 3.3 *Let $\{Y_l^{(n)} : 1 \leq l \leq n\}$ and $\{\mathcal{G}_l^{(n)} : 1 \leq l \leq n\}$, for $n \geq 1$, be respectively an array of non-negative variables and σ-fields such that $\sigma\{Y_1^{(n)}, \ldots, Y_l^{(n)}\} \subset \mathcal{G}_l^{(n)}$. Suppose that*

$$\lim_{n\to\infty} E\Big[\sum_{l=1}^{n} Y_l^{(n)}\Big] = 1 \quad \text{and} \quad \sup_{1\leq i\leq n} \|Y_i^{(n)}\|_{L^\infty} \leq \epsilon_n$$

where $\lim_{n\to\infty} \epsilon_n = 0$. In addition, assume

$$\lim_{n\to\infty} \sup_{1\leq l\leq n-1} \mathrm{Osc}\left(E\Big[\sum_{j=l+1}^{n} Y_j^{(n)} | \mathcal{G}_l^{(n)}\Big]\right) = 0.$$

Then,

$$\lim_{n\to\infty} \sum_{l=1}^{n} Y_l^{(n)} = 1 \quad \text{in } L^2.$$

Proof. Write

$$E[(\sum_{l=1}^{n} Y_l^{(n)})^2] = \sum_{l=1}^{n} E[(Y_l^{(n)})^2] + 2\sum_{l=1}^{n-1} E[Y_l^{(n)}(\sum_{j=l+1}^{n} Y_j^{(n)})].$$

The first sum on the right-hand side is bounded as follows. ¿From non-negativity,

$$\sum_{l=1}^{n} E[(Y_l^{(n)})^2] \;\le\; \epsilon_n \sum_{l=1}^{n} E[Y_l^{(n)}] \;=\; \epsilon_n \cdot (1+o(1)) \;\to 0 \;\text{ as }\quad n \uparrow \infty.$$

For the second sum, write

$$\sum_{l=1}^{n-1} E[Y_l^{(n)}(\sum_{j=l+1}^{n} Y_j^{(n)})] = \sum_{l=1}^{n-1} E[Y_l^{(n)} E[\sum_{j=l+1}^{n} Y_j^{(n)} | \mathcal{G}_l^{(n)}]].$$

¿From the oscillation assumption, we have that

$$\sup_{1\le l\le n-1} \sup_{\omega} \left| E[\sum_{j=l+1}^{n} Y_j^{(n)} | \mathcal{G}_l^{(n)}](\omega) - E[\sum_{j=l+1}^{n} Y_j^{(n)}] \right| = o(1).$$

Therefore,

$$\begin{aligned} 2\sum_{l=1}^{n-1} E[Y_l^{(n)}(\sum_{j=l+1}^{n} Y_j^{(n)})] &= 2\sum_{l=1}^{n-1} E[Y_l^{(n)}]E[\sum_{j=l+1}^{n} Y_j^{(n)}] + o(1)\cdot \sum_{l=1}^{n-1} E[Y_l^{(n)}] \\ &= \Big(\sum_{l=1}^{n} E[Y_l^{(n)}]\Big)^2 - \sum_{l=1}^{n} E[(Y_l^{(n)})^2] + o(1) \\ &= 1 + o(1) \end{aligned}$$

finishing the proof. □

To apply later this result to $v_j^{(n)} = E[(\xi_j^{(n)})^2 | \mathcal{F}_{j-1}^{(n)}]$ measureable with respect to $\mathcal{G}_j^{(n)} = \mathcal{F}_{j-1}^{(n)}$ for $2 \le j \le n$ we will need the following oscillation estimate.

Lemma 3.4 *Under condition (1.3), we have*

$$\sup_{2\le l\le n-1} \text{Osc}\left(E[\sum_{j=l+1}^{n} v_j^{(n)} | \mathcal{F}_{l-1}^{(n)}](\omega) \right) \;=\; o(1).$$

Proof. From the martingale property, $E[\xi_r^{(n)}\xi_s^{(n)}|\mathcal{F}_u^{(n)}] = 0$ for $r > s > u$, (3.4) and (3.2), we have

$$\begin{aligned}
E\Big[\sum_{j=l+1}^{n} v_j^{(n)}|\mathcal{F}_{l-1}^{(n)}\Big] &= E\Big[\sum_{j=l+1}^{n} (\xi_j^{(n)})^2\,|\mathcal{F}_{l-1}^{(n)}\Big] \\
&= E\Big[\Big(\sum_{j=l+1}^{n} \xi_j^{(n)}\Big)^2\,|X_{l-1}^{(n)}\Big] \\
&= V(S_n)^{-1}E\Big[\,\Big(\sum_{j=l+1}^{n} f_j^{(n)}(X_j^{(n)}) - E\big[Z_{l+1}^{(n)}|X_l^{(n)}\big]\Big)^2\,|X_{l-1}^{(n)}\Big] \\
&= V(S_n)^{-1}E\Big[\,\Big(\sum_{j=l+1}^{n} f_j^{(n)}(X_j^{(n)})\Big)^2\,|X_{l-1}^{(n)}\Big] \\
&\quad -V(S_n)^{-1}E\big[E\big[Z_{l+1}^{(n)}|X_l^{(n)}\big]^2|X_{l-1}^{(n)}\big]. \qquad (3.6)
\end{aligned}$$

By Lemma 3.2, the last term in (3.6) is bounded $\sup_{2\leq l\leq n-1} V(S_n)^{-1}\|Z_{l+1}^{(n)}\|_{L^\infty}^2 = o(1)$, and so its oscillation is also uniformly $o(1)$.

To estimate oscillation of the first term on right-side of (3.6), we write

$$\begin{aligned}
&\mathrm{Osc}\Big(V(S_n)^{-1}E\Big[\Big(\sum_{j=l+1}^{n} f_j^{(n)}(X_j^{(n)})\Big)^2|X_{l-1}^{(n)}\Big]\Big) \qquad (3.7) \\
&\leq V(S_n)^{-1}\sum_{l+1\leq j,m\leq n} \mathrm{Osc}\Big(E\big[f_j^{(n)}(X_j^{(n)})f_m^{(n)}(X_m^{(n)})|X_{l-1}^{(n)}\big]\Big).
\end{aligned}$$

But, for $l+1 \leq j \leq m \leq n$, we have from Lemma 3.1 that

$$\mathrm{Osc}\Big(E\big[f_j^{(n)}(X_j^{(n)})f_m^{(n)}(X_m^{(n)})|X_{l-1}^{(n)}\big]\Big) \leq 6C_n^2\,(1-\alpha_n)^{j-l+1}(1-\alpha_n)^{m-j}.$$

Then, (3.7) is bounded, uniformly in l, on order $V(S_n)^{-1}C_n^2\alpha_n^{-2}$ which from Proposition 3.2 and (1.3) is $o(1)$. □

Proof of Theorem 1.1. From Lemma 3.2, we need only show that $M_n^{(n)}/\sqrt{V(S_n)} \Rightarrow$ N$(0,1)$. This will follow from martingale convergence (Proposition 3.1) as soon as we show (1) $\sup_{2\leq k\leq n}\|\xi_k^{(n)}\|_{L^\infty} \to 0$ and (2) $\sum_{k=2}^{n} E[(\xi_k^{(n)})^2|\mathcal{F}_{k-1}^{(n)}] \to 1$. However, (1) follows from the negligibility estimate Lemma 3.2, and (2) from LLN Lemmas 3.3 and 3.4 since "negligibility" (1) holds and $\sum_{k=2}^{n} E[(\xi_k^{(n)})^2] = 1+o(1)$ (from variance decomposition after (3.3) and Lemma 3.2). □

4 Proof of Variance Lower Bound

Let λ be a probability measure on $\mathbf{X}\times\mathbf{X}$ with marginals α and β respectively. Let $\pi(x_1, dx_2)$ and $\widehat{\pi}(x_2, dx_1)$ be the corresponding transition probabilities in the two directions so that $\alpha\pi = \beta$ and $\beta\widehat{\pi} = \alpha$.

Lemma 4.1 *Let $f(x_1)$ and $g(x_2)$ be square integrable with respect to α and β respectively. If*

$$\int f(x_1)\alpha(dx_1) = \int g(x_2)\beta(dx_2) = 0$$

then,

$$\left|\int f(x_1)g(x_2)\lambda(dx_1, dx_2)\right| \leq \sqrt{\delta(\pi)}\, \|f\|_{L_2(\alpha)}\|g\|_{L_2(\beta)}.$$

Proof. Let us construct a measure on $\mathbf{X}\times\mathbf{X}\times\mathbf{X}$ by starting with λ on $\mathbf{X}\times\mathbf{X}$ and using reversed $\widehat{\pi}(x_2, dx_3)$ to go from x_2 to x_3. The transition probability from x_1 to x_3 defined by

$$Q(x_1, A) = \int \pi(x_1, dx_2)\widehat{\pi}(x_2, A)$$

satisfies $\delta(Q) \leq \delta(\pi)\delta(\widehat{\pi}) \leq \delta(\pi)$. Moreover $\alpha Q = \alpha$ and the operator Q is self-adjoint and bounded with norm 1 on $L_2(\alpha)$. Then, if f is a bounded function with $\int f(x)\alpha(dx) = 0$ (and so $E_\alpha[Q^n f] = 0$), we have for $n \geq 1$,

$$\|Q^n f\|_{L_2(\alpha)} \leq \|Q^n f\|_{L_\infty} \leq (\delta(Q))^n \mathrm{Osc}(f). \tag{4.1}$$

Hence, as bounded functions are dense, on the subspace of functions, $M = \{f \in L_2(\alpha) : \int f(x)\alpha(dx) = 0\}$, the top of the spectrum of Q is less than $\delta(Q)$ and so $\|Q\|_{L_2(\alpha,M)} \leq \delta(Q)$. Indeed, suppose the spectral radius of Q on M is larger than $\delta(Q)+\epsilon$ for $\epsilon > 0$, and $f \in M$ is a non-trivial bounded function whose spectral decomposition is with respect to spectral values larger than $\delta(Q)+\epsilon$. Then, $\|Q^n f\|_{L^2(\alpha)} \geq \|f\|_{L^2(\alpha)}(\delta(Q)+\epsilon)^n$ which contradicts the bound (4.1) when $n \uparrow \infty$. [cf. Thm. 2.10 [15] for a proof in discrete space settings.]

Then,

$$\|\widehat{\pi} f\|^2_{L_2(\beta)} = \langle \pi\widehat{\pi} f, f\rangle_{L_2(\alpha)} = \langle Qf, f\rangle_{L_2(\alpha)} \leq \|Q\|_{L_2(\alpha,M)}\|f\|^2_{L_2(\alpha)} \leq \delta(Q)\|f\|^2_{L_2(\alpha)}.$$

Finally,

$$|\int f(x_1)g(x_2)\lambda(dx_1, dx_2)| = |\langle \widehat{\pi} f, g\rangle_{L_2(\beta)}| \leq \sqrt{\delta(\pi)}\, \|f\|_{L_2(\alpha)}\|g\|_{L_2(\beta)}.$$

□

Lemma 4.2 *Let $f(x_1)$ and $g(x_2)$ be square integrable with respect to α and β respectively. Then,*

$$E\big[\big(f(x_1) - g(x_2)\big)^2\big] \ \geq \ (1-\delta(\pi))\ V\big(f(x_1)\big)$$

as well as

$$E\big[\big(f(x_1) - g(x_2)\big)^2\big] \ \geq \ (1-\delta(\pi))\ V\big(g(x_2)\big).$$

Proof. To get lower bounds, we can assume without loss of generality that f and g have mean 0 with respect to α and β respectively. Then by Lemma 4.1

$$\begin{aligned}E\big[\big(f(x_1) - g(x_2)\big)^2\big] &= E\big[[f(x_1)]^2\big] + E\big[[g(x_2)]^2\big] - 2E\big[f(x_1)g(x_2)\big] \\ &\geq E\big[[f(x_1)]^2\big] + E\big[[g(x_2)]^2\big] - 2\sqrt{\delta(\pi)}\,\|f\|_{L_2(\alpha)}\|g\|_{L_2(\beta)} \\ &\geq (1-\delta(\pi))\|f\|^2_{L_2(\alpha)}.\end{aligned}$$

The proof of the second half is identical. □

Proof of Proposition 3.2. Applying Lemma 4.2 to the Markov pairs $\{(X_k^{(n)}, X_{k+1}^{(n)}) : 1 \leq k \leq n-1\}$ with $f(X_k^{(n)}) = E[Z_{k+1}^{(n)}|X_k^{(n)}]$ and $g(X_{k+1}^{(n)}) = Z_{k+1}^{(n)}$, we get

$$E\big[\big(Z_{k+1}^{(n)} - E[Z_{k+1}^{(n)}|X_k^{(n)}]\big)^2\big] \geq \alpha_n E\big[\big(Z_{k+1}^{(n)}\big)^2\big].$$

On the other hand from (3.2), for $1 \leq k \leq n-1$, we have

$$\begin{aligned}V(f_k^{(n)}(X_k^{(n)})) &\leq E\big[\big(f_k^{(n)}(X_k^{(n)})\big)^2\big] \\ &\leq 2E\big[\big(Z_k^{(n)}\big)^2\big] + 2E\big[\big(E\big[Z_{k+1}^{(n)}|X_k^{(n)}\big]\big)^2\big] \\ &\leq 2E\big[\big(Z_k^{(n)}\big)^2\big] + 2E\big[\big(Z_{k+1}^{(n)}\big)^2\big].\end{aligned}$$

Summing over k, and noting $f_n^{(n)}(X_n^{(n)}) = Z_n^{(n)}$ and variance decomposition after (3.3),

$$\begin{aligned}\sum_{k=1}^{n} V(f_k^{(n)}(X_k^{(n)})) &\leq 4\sum_{k=1}^{n} E\big[\big(Z_k^{(n)}\big)^2\big] \\ &\leq \frac{4}{\alpha_n}\left[\sum_{k=1}^{n-1} E\big[\big(Z_{k+1}^{(n)} - E[Z_{k+1}^{(n)}|X_k^{(n)}]\big)^2\big] + E[(Z_1^{(n)})^2]\right] = \frac{4}{\alpha_n}V(S_n).\end{aligned}$$

□

Acknowledgement. We would like to thank the referees for their comments.

References

[1] Dobrushin, R. (1953) Limit theorems for Markov chains with two states. (Russian) *Izv. Adad. Nauk SSSR* **17:4** 291-330.

[2] Dobrushin, R. (1956) Central limit theorems for non-stationary Markov chains I,II. *Theory of Probab. and its Appl.* **1** 65-80, 329-383.

[3] Gordin, M.I. (1969) The central limit theorem for stationary processes. *Soviet Math. Dokl.* **10** 1174-1176.

[4] Gudynas, P. (1977) An invariance principle for inhomogeneous Markov chains. *Lithuanian Math. J.* **17** 184-192.

[5] Hall, P. and Heyde, C.C. (1980) *Martingale Limit Theory and Its Application.* Academic Press, New York.

[6] Hanen, A. (1963) Théorèmes limites pour une suite de châines de Markov. *Ann. Inst. H. Poincaré* **18** 197-301.

[7] Isaacson, D.L., and Madsen, R.W. (1976) *Markov Chains. Theory and Applications.* John Wiley and Sons, New York.

[8] Iosifescu, M. (1980) *Finite Markov Processes and Their Applications.* John Wiley and Sons, New York.

[9] Iosifescu, M., and Theodorescu, R. (1969) *Random Processes and Learning. Springer,* Berlin.

[10] Kifer, Y. (1998) Limit theorems for random transformations and processes in random environments. *Trans. Amer. Math. Soc.* **350** 1481-1518.

[11] Kipnis, C., Varadhan, S. R. S. (1986) Central limit theorem for additive functionals of reversible markov processes. *Commun. Math. Phys.* **104** 1-19.

[12] Pinsky, M. (1991) *Lectures on Random Evolution.* World Scientific, Singapore.

[13] Sarymsakov, T.A. (1961) Inhomogeneous Markov chains. *Theor. Probability Appl.* **6** 178-185.

[14] Seneta, E. (1973) On the historical development of the theory of finite inhomogeneous Markov chains. *Proc. Cambridge Philos. Soc.* **74** 507-513.

[15] Seneta, E. (1981) *Non-negative Matrices and Markov Chains.* Second Edition, Springer-Verlag, New York.

[16] Statuljavicius, V. (1969) Limit theorems for sums of random variables connected in Markov chains.(Russian) *Litovsk. Mat. Sb.* **9** 345-362; ibid. **9**, 635-672; ibid. **10** 161–169.

[17] Varadhan, S.R.S. (2001) *Probability Theory.* Courant Lecture Notes **7**, American Mathematical Society, Providence, R.I.

[18] Winkler, G. (1995) *Image Analysis, Random Fields and Dynamic Monte Carlo Methods. A Mathematical Introduction.* Applications of Mathematics **27**, Springer-Verlag, Berlin.

[19] Wu, Wei Biao, Woodroofe, M. (2004) Martingale approximations for sums of stationary processes. *Ann. Probab.* **32** 1674–1690.

Review Articles

Handbook of Statistics 19, North-Holland, 2001, 853-872

Diffusion Processes

S. R. S. Varadhan

1. Brownian motion

Brownian motion is the prototypical example of a diffusion process. The standard Brownian motion on R is a stochastic process $x(t)$ defined for $0 \leq t < \infty$. It has independent increments, i.e. the increments over disjoint intervals are stochastically independent. The distribution of the increment $x(t) - x(s)$, over the interval $[s, t]$ with $s \leq t$, is normally distributed with mean 0 and variance $t - s$. The Brownian motion is usually normalized so that $x(0) = 0$ with probability 1. It can be equivalently defined as a Gaussian Process with mean 0 and covariance $E[x(s)x(t)] = s \wedge t$. The joint distribution of $x(t_1), \ldots, x(t_n)$ at n time points $0 < t_1 < \cdots < t_n < \infty$ is multivariate normal with mean 0 and covariance $C_{i,j} = t_i \wedge t_j$. Its distribution on R^n has the joint density

$$p_n(t_1, \ldots, t_n; x_1, \ldots, x_n) = \left[\prod_{j=1}^{n} (2\pi(t_j - t_{j-1}))^{-\frac{1}{2}}\right] \times \exp\left[-\sum_{j=1}^{n} \frac{(x_j - x_{j-1})^2}{2(t_j - t_{j-1})}\right] \quad (1.1)$$

with respect to the Lebesgue measure. Here we adopt the convention that $t_0 = x_0 = 0$. This defines a consistent family of finite dimensional distributions, and by Kolmogorov's theorem defines a stochastic process. It is a theorem Norbert Wiener that this stochastic process can be supported on the space $\Omega = C[0, \infty)$ of continuous functions on $[0, \infty)$. The space Ω comes with a natural σ-field of Borel sets $\mathscr{F}$ and canonical sub-σ-fields $\mathscr{F}_t$ of events generated by $[x(s) : 0 \leq s \leq t]$. Clearly $\mathscr{F}_s \subset \mathscr{F}_t$ for $s \leq t$ and $\mathscr{F} = \sigma(\cup_t \mathscr{F}_t)$, i.e. the smallest σ-field containing $\mathscr{F}_t$ for all t. The Wiener measure P_0 on Ω is the unique measure such that for every positive integer n, time points $0 < t_1 < \cdots < t_n$

$$P_0[x(t_1), \ldots, x(t_n) \in A] = \int_A p_n(t_1, \ldots, t_n; x_1, \ldots, x_n) \mathrm{d}x_1 \ldots \mathrm{d}x_n$$

for all Borel sets $A \subset R^n$. It has the following additional properties.

1. If we denote by H_{α,C,T_1,T_2} the set of functions $x(\cdot)$ that satisfy a local Hölder condition of the form

$$|x(s) - x(t)| \leq C|t - s|^\alpha$$

for $T_1 \leq s \leq t \leq T_2$ then for every $\alpha \geq 1/2$, $T_1 < T_2$ and $C < \infty$

$$P_0[H_{\alpha,C,T_1,T_2}] = 0 \ .$$

2. On the other hand for every $\alpha < 1/2$, and $T_1 < T_2 < \infty$,

$$\lim_{C\to\infty} P_0[H_{\alpha,C,T_1,T_2}] = 1 \ .$$

3. If D_t is the set of paths that are differentiable at the point t, then

$$P_0[\cup_t D_t] = 0$$

4. For any positive integer k, if $\{j/2^k; j \geq 0\}$ is the discrete set of points with spacing $1/2^k$, then

$$P_0\left[\lim_{k\to\infty} \sum_{1\leq j\leq[2^k t]} \left(x\left(\frac{j}{2^k}\right) - x\left(\frac{j-1}{2^k}\right)\right)^2 = t \quad \text{for all } t \geq 0\right] = 1 \qquad (1.2)$$

It is clear from these properties that a typical Brownian path, being nowhere differentiable, is not of bounded variation in any finite interval. However it has a definite 'quadratic variation' in any interval that is almost surely equal to the length of the interval.

One of the ways of understanding the behavior of a Brownian path is to start with a random walk $\{S_n : n \geq 1\}$ which, for each n, is the sum

$$S_n = X_1 + X_2 + \cdots + X_n$$

of the first n terms of a sequence $\{X_i\}$ of independent identically distributed random variables that take the value ± 1 with probability $1/2$ each. While the central limit theorem asserts that for large n, the distribution $S_n/\sqrt{n}$ is asymptotically the standard normal, Brownian motion is to be thought of as the limit in distribution of $S_{[nt]}/\sqrt{n}$ as a function of t.

The Wiener measure P_0 or the Brownian motion $x(\cdot)$ is not only the prime example of a diffusion process but it is also the building block out of which other diffusion processes are constructed. A discussion of Brownian motion and its properties can be found, among other places, in the following books (Stroock, 1993; Revuz and Yor, 1999). Usually one refers to the measure P_0 as the Wiener measure and a random path $x(\cdot)$ that is 'distributed' according to the Wiener measure as Brownian motion.

Brownian motion derives its name from the botanist Robert Brown who observed in 1828, that, in water, pollen from plants dispersed in an irregular swarming motion. From that time many scientists have examined the Brownian phenomenon including Einstein. Norbert Wiener was the first one who essentially constructed the measure corresponding to Brownian motion, and proved the

Hölder continuity of it paths in Wiener (1923) and Wiener (1924). It was the first rigorous example of integration in an infinite dimensional function space.

Just as any multivariate normal random vector can be generated by a linear transformation $y = Tx$ from one with covariance equal to the Identity matrix, one can start with any orthonormal basis $\{f_j(\cdot)\}$ in $L_2[0,\infty]$ and represent the Brownian motion as

$$x(t) = \sum_j Z_j F_j(t)$$

where $\{Z_j\}$ are independent standard normal random variables and

$$F_j(t) = \int_0^t f_j(s)\mathrm{d}s \ .$$

While formally the expansion is always valid its convergence is far more delicate. Such expansions have been considered for Brownian motion and more generally for Gaussian processes in Itô and Nisio (1968).

2. Brownian motion as a Markov process

The probability density p_n in (1.1) can be rewritten as

$$p_n(t_1,\ldots,t_n;x_1,\ldots,x_n) = \prod_{j=1}^{n} p(t_j - t_{j-1}, x_j - x_{j-1}) \tag{2.1}$$

where

$$p(t,x) = \frac{1}{\sqrt{2\pi t}}\exp\left[-\frac{x^2}{2t}\right] \tag{2.2}$$

In other words the process P_0 is a Markov process with transition probability

$$\begin{aligned} P_0[x(t) \in A | x(\sigma) : 0 \le \sigma \le s] &= \int_A p(t-s, y - x(s))\mathrm{d}y \\ &= \int_A q(s, x(s), t, y)\mathrm{d}y \end{aligned}$$

where $q(s,x;t,y) = p(t-s, y-x)$.

We can in fact initialize the Brownian motion to start from any point x_0 at time 0 to get P_{x_0} which is just the distribution of $x_0 + x(\cdot)$ under P_0. The transition probability density as we saw depends only on the differences $y - x$ and $t - s$. It is because Brownian motion as a Markov process is homogeneous in time and has independent increments, i.e. is it is invariant under translations in space as well as time.

In the theory of Markov processes the concept of stopping times is extremely important. These are random variables $\tau(\omega) \ge 0$ that are $\mathscr{F}$ measurable that have the additional property that for every $t \ge 0$, the set $[\omega : \tau(\omega) \le t]$ is $\mathscr{F}_t$ measurable. Examples of stopping times are the first time some thing happens, for instance

$$\tau(\omega) = \inf[t : x(t) \geq a]$$

In order to determine if $\tau(\omega) \leq t$, we only need to observe the path up to time t. On the other hand, the last time some thing happens like

$$\sup\left[t : \int_0^t V(x(s))\mathrm{d}s \leq 0\right]$$

are not stopping times. An important fact is that for Brownian motion as well as for most other diffusion processes, the Markov property extends to stopping times and this is called the strong Markov property and was first considered by Hunt (1956).

Associated with any stopping time τ there is a natural σ-field $\mathscr{F}_\tau$ defined by

$$\mathscr{F}_\tau = \{A : A \cap [\tau \leq t] \in \mathscr{F}_t \quad \text{for all } t \geq 0\}$$

Roughly speaking if we observe the process only up to time τ and then stop, then $\mathscr{F}_\tau$ represents the information that is available. The strong Markov property for Brownian motion, first established by G. A. Hunt [H], states that with respect to any P_{x_0}, the future increments $y_\tau(t) = x(\tau + t) - x(\tau)$ is again a Brownian motion, i.e. distributed according to P_0 and more over is stochastically independent of the events in $\mathscr{F}_\tau$. This can be restated in the form

$$P_{x0}[x(\tau + t) \in A|\mathscr{F}_\tau] = P_{x(\tau)}[x(t) \in A]$$

An easy consequence of this is the reflection principle of Bachelier that computes for $\ell \geq 0$,

$$P_0\left[\sup_{0 \leq s \leq t} x(s) \geq \ell\right] = 2P[x(t) \geq \ell] = \sqrt{\frac{2}{\pi t}} \int_\ell^\infty \exp\left[-\frac{x^2}{2t}\right]\mathrm{d}x$$

3. Semigroups and generators

An important functional analytical tool in the study of Markov processes is the notion of semigroups associated with them. For any bounded measurable function f on R, we define $T_t f$ by

$$(T_t f)(x) = E[f(x(t))|x(0) = x] = \int_R f(y)p(t, y - x)\mathrm{d}y$$

The Chapman–Kolmogorov equations which in our case reduce to

$$\int_R p(t, y - x)p(s, z - y)\mathrm{d}y = p(t + s, z - x)$$

yield

$$T_t T_s f = T_s T_t f = T_{t+s} f$$

for all f and $s, t \geq 0$. Such semigroups are characterized by their infinitesimal generators

$$A = \left.\frac{\mathrm{d}T_t}{\mathrm{d}t}\right|_{t=0}$$

and can be recovered in some sense as

$$T_t = \exp[tA]$$

For our Brownian motion semigroup it is an easy calculation to derive that for any smooth function f

$$(Af)(x) = \lim_{t\to 0}\frac{1}{t}\int [f(x+y) - f(x)]p(t,y)\mathrm{d}y = \frac{1}{2}\frac{\mathrm{d}^2 f(x)}{\mathrm{d}x^2}$$

It is therefore natural to expect connections between the operator $1/2D_x^2$ and Brownian motion. The first such relation is that the transition probability $p(t,x)$ satisfies the heat equation

$$\frac{\partial p}{\partial t} = \frac{1}{2}\frac{\partial^2 p}{\partial x^2}$$

and consequently

$$(T_t f)(x) = u(t,x) = \int f(y)p(t, y-x)\mathrm{d}y$$

solves the Cauchy problem

$$\frac{\partial u}{\partial t} = \frac{1}{2}\frac{\partial^2 u}{\partial x^2}$$

with the initial condition $\lim_{t\to 0} u(t,x) = f(x)$.

Taking Laplace transforms in t, for $\lambda > 0$, the resolvent equations

$$\lambda u - \tfrac{1}{2}u_{xx} = f$$

are solved by

$$u(x) = \int_0^\infty \mathrm{e}^{-\lambda t}(T_t f)(x)\mathrm{d}t = E^{P_x}\left[\int_0^\infty \mathrm{e}^{-\lambda t} f(x(t))\mathrm{d}t\right]$$

The Feynman–Kac formula provides a representation for solutions of

$$\frac{\partial u}{\partial t} = \frac{1}{2}\frac{\partial^2 u}{\partial x^2} + V(x)u$$

with the initial condition $\lim_{t\to 0} u(t,x) = f(x)$ in the form

$$u(t,x) = E^{P_x}\left[f(x(t))\exp\left[\int_0^t V(x(s))\mathrm{d}s\right]\right]$$

4. Stochastic integrals

Since $x(t)$ is almost surely a continuous function of t we can define $Y = \int_0^\infty f(t)x(t)\mathrm{d}t$ for smooth functions f with compact support and will give us a Gaussian random variable with mean 0 and variance $\iint f(s)f(t)s\wedge t\,\mathrm{d}s\,\mathrm{d}t$. In fact, if f is a smooth function we can even define $\int f(t)\mathrm{d}x(t)$ as $-\int f'(t)x(t)\mathrm{d}t$ after an integration by part and calculate

$$E\left[\left|\int f(t)\mathrm{d}x(t)\right|^2\right] = \iint f'(s)f'(t)s\wedge t\,\mathrm{d}s\,\mathrm{d}t = \int [f(t)]^2\,\mathrm{d}t$$

By completion in $L_2[P]$, we can now define $\int f(t)\mathrm{d}x(t)$ for functions f in $L_2[0,\infty]$ and will give us a mean zero normal random variable with variance $\|f\|_2^2$. This was carried out already by Wiener and yields the following result known as the Cameron–Martin formula [see Cameron and Martin (1953)].

Let $F(t)$ be of the form $F(t) = \int_0^t f(s)\mathrm{d}s$ for some f in $L_2[0,\infty]$. Let $x(\cdot)$ be Brownian Motion distributed according to the Wiener measure P_0. Then the distribution P_F of $y(\cdot) = x(\cdot) + F(\cdot)$ is absolutely continuous with respect to P_0 and has the Radon–Nikodym derivative

$$\frac{\mathrm{d}P_F}{\mathrm{d}P_0}(\omega) = \exp\left[\int_0^\infty f(t)\mathrm{d}x(t) - \frac{1}{2}\int_0^\infty [f(t)]^2\,\mathrm{d}t\right].$$

Formally,the relative density of a Gaussian vector ξ with mean μ and covariance C with respect the Gaussian with mean 0 and the same covariance is given by

$$\exp\left[\langle\xi\,,C^{-1}\mu\rangle - \tfrac{1}{2}\langle\mu\,,C^{-1}\mu\rangle\right]$$

and our formula is just a special case.

In the definition of Wiener's stochastic integral the integrand f is nonrandom, i.e. does depend on the path ω. Itô extended Wiener's definition in a far reaching way. Since the calculation is essentially one involving two moments, in an approximation of the form

$$S = \sum_j f(t_j)(x(t_{j+1}) - x(t_j))$$

if we let $f(t) = f(t,\omega)$ depend on ω, but insist that for every t, $f(t,\omega)$ be measurable with respect to $\mathscr{F}_t$ then the increment $x(t_{j+1}) - x(t_j)$ is independent of $f(t_j,\omega)$. One still gets

$$E[S] = 0, \quad E[S^2] = E\left[\sum_j [f(t_j, \omega)]^2\right]$$

leading to Itô's definition of the stochastic integral for the class of 'progressively measurable' functions $f(t, \omega)$ satisfying

$$E\left[\int_0^\infty [f(t, \omega)]^2 \mathrm{d}t\right] < \infty$$

Since we do not necessarily have to integrate all the way to ∞, it suffices that f is progressively measurable and satisfies

$$E\left[\int_0^t [f(s, \omega)]^2 \mathrm{d}s\right] < \infty \quad \text{for all} \quad t < \infty$$

Itô's stochastic integral

$$y(t) = \int_0^t f(s, \omega)\mathrm{d}x(s)$$

is well defined under these conditions and has the following properties.

(1) The process $y(t)$ is almost surely continuous in t and is a progressively measurable martingale relative to $\mathscr{F}_t$.
(2) If τ is any stopping time, $y(\tau \wedge t)$ is also a martingale and

$$y(\tau \wedge t) = \int_0^t [\mathbf{1}_{\tau > s}(\omega) f(s, \omega)]\mathrm{d}x(s) = \int_0^t f_\tau(s, \omega)\mathrm{d}x(s) \ .$$

(3) The process $z(t) = y^2(t) - \int_0^t [f(s, \omega)]^2 \mathrm{d}s$ is a progressively measurable martingale.
(4) If f is in addition uniformly bounded, then

$$\exp\left[y(t) - \frac{1}{2}\int_0^t [f(s, \omega)]^2 \mathrm{d}s\right]$$

is again a martingale.

5. Itô calculus

The stochastic integral $y(t) = \int_0^t f(s, \omega)\mathrm{d}x(s)$ can be thought of as $\mathrm{d}y = f\mathrm{d}x$ and the first step in developing any calculus is the chain rule. What is $\mathrm{d}\phi(y(t))$? The problem exists already if $f = 1$ i.e. what is $\mathrm{d}\phi(x(t))$? If we take $\phi(x) = x^2$ one might think that $\mathrm{d}x^2(t) = 2x(t)\mathrm{d}x(t)$ or

$$x^2(t) - x^2(0) = x^2(t) = \int_0^t 2x(t)\mathrm{d}x(t)$$

This is clearly wrong because the left hand side has expectation t while the right hand side has expectation 0. A better guess is

$$x^2(t) - x^2(0) = x^2(t) = \int_0^t 2x(t)\mathrm{d}x(t) + t$$

This can indeed be seen to be correct by using (1.2). More generally Itô's formula asserts that

$$f(x(t)) - f(x(0)) = \int_0^t f'(x(s))\mathrm{d}x(s) + \frac{1}{2}\int_0^t f''(x(s))\mathrm{d}s$$

Since $x(\cdot)$ is not of bounded variation but has finite quadratic variation, formally $(\mathrm{d}x)^2 = \mathrm{d}t$ and $(\mathrm{d}x)^i(\mathrm{d}t)^j = 0$ if $i + 2j \geq 3$. We have to take an additional term in the Taylor expansion. We can now compute

$$\mathrm{d}\phi(t, y(t)) = \phi_t \mathrm{d}t + \phi_y \mathrm{d}y + \tfrac{1}{2}\phi_{yy}(\mathrm{d}y)^2 = [\phi_t + \tfrac{1}{2}\phi_{yy}f^2]\mathrm{d}t + \phi_y f\,\mathrm{d}x$$

or

$$\begin{aligned}
&\phi(t, y(t)) - \phi(0, y(0)) \\
&= \int_0^t \left[\phi_s(s, y(s)) + \frac{1}{2}\phi_{yy}(s, y(s))[f(s,\omega)]^2\right]\mathrm{d}s \\
&\quad + \int_0^t \phi_y(s, y(s))f(s,\omega)\mathrm{d}x(s)
\end{aligned}$$

In particular if $f = 1$

$$\begin{aligned}
\phi(t, x(t)) - \phi(0, x(0)) &= \int_0^t \left[\phi_s(s, x(s)) + \frac{1}{2}\phi_{xx}(s, x(s))\right]\mathrm{d}s \\
&\quad + \int_0^t \phi_x(s, x(s))\mathrm{d}x(s)
\end{aligned}$$

One can use Itô's formula to establish a connection with the infinitesimal generator $1/2D_x^2$. For example if $u(t,x)$ solves

$$u_t = \tfrac{1}{2}D_x^2 u$$

with $u(0,x) = f(x)$, Itô's formula for $v(t,x) = u(T - t, x)$ yields

$$\mathrm{d}v(t, x(t)) = \int_0^t v_x(s, x(s))\mathrm{d}x(s)$$

with no $\mathrm{d}t$ term and therefore $v(t, x(t))$ is a martingale. Equating expectations at times 0 and T we get

$$v(0,x) = u(T,x) = E^{P_x}[v(T, x(T))] = E^{P_x}[f(x(T))] = \int f(y)p(T, y - x)\mathrm{d}y$$

Itô's formula in particular provides a proof of the uniqueness of the Cauchy problem for the heat equation!

A more interesting example is to consider a smooth function $u = u(t,x)$ that solves

$$u_t + \tfrac{1}{2}D_x^2 u + V(x,t)u = 0 \quad \text{in } [0,T] \times [a,b]$$

with $u(T,x) = f(x)$. Then by a similar calculation

$$u(t,x(t)) \exp\left[\int_0^t V(s,x(s))\mathrm{d}s\right]$$

is a martingale until the exit time τ from the interval $[a,b]$, and we get

$$u(0,x) = E^{P_x}[u(\tau \wedge T, x(\tau \wedge T))] \exp\left[\int_0^{\tau\wedge T} V(s,x(s))\mathrm{d}s\right]$$

expressing $u(0,x)$ in terms of the boundary values of u along $t = T$ and $0 < t < T, x = a$ or b. Itô's theory of stochastic integrals and stochastic differential equations appeared in [Itô (1942)]. We can now find a treatment in various texts in probability, engineering and in recent times finance.

6. Brownian motion with a drift

Brownian motion is characterized as a process with independent increments with the distribution of the increment $x(t) - x(s)$ being normal with mean 0 and variance $t - s$. We can build in a drift by requiring the increment to have a normal distribution with mean $\mu(t-s)$ and variance $t - s$. Such a process is very easily defined as $y(t) = x(t) + \mu t$. It is again a Markov process with a transition probability density

$$q_\mu(s,x,t,y) = \frac{1}{\sqrt{2\pi(t-s)}} \exp\left[-\frac{(y - x - \mu(t-s))^2}{2(t-s)}\right]$$

Symbolically $\mathrm{d}y = \mathrm{d}x + \mu \mathrm{d}t$ and there are analogous results. But the basic difference is that the infinitesimal generator $1/2D_x^2$ has now to be replaced by $1/2D_x^2 + \mu D_x$.

Often, in considering stochastic models, we come across situations where the drift is a restoring force that is proportional to the displacement. Formally

$$\mathrm{d}y = -\mu y \, \mathrm{d}t + \mathrm{d}x \ . \tag{6.1}$$

Such an example will not be a process with independent increments, but will still be Markov. We can solve (6.1) explicitly with initial condition $y(0) = y$, to obtain

$$y(t) = \mathrm{e}^{-\mu t} y(0) + \int_0^t \mathrm{e}^{-\mu(t-s)} \, \mathrm{d}x(s) \tag{6.2}$$

From the representation (6.2), we see that given $y(0) = y$, $y(t)$ is again normally distributed, but now with mean $e^{-\mu t}y$ and variance $1/2\mu(1 - e^{-2\mu t})$. We therefore have a time homogeneous Markov process with transition probability density

$$q(s,x,t,y) = \frac{1}{\sqrt{\frac{\pi}{\mu}(1 - e^{-2\mu(t-s)})}} \exp\left[-\frac{\mu(y - e^{-\mu(t-s)}x)^2}{(1 - e^{-2\mu(t-s)})}\right]$$

As $t \to \infty$ this has a limit which is a stationary density for the process and is in fact the normal distribution with mean 0 and variance $1/2\mu$. For this Markov process, the generator is $1/2D_y^2 - \mu y D_y$. The way to compute the generator quickly is by applying Itô's formula

$$\begin{aligned} du(y(t)) &= u'dy(t) + \frac{1}{2}u''(dy(t))^2 = u'[-\mu y dt] + u'dx + \frac{1}{2}u'' \, dt \\ &= \left[\frac{1}{2}D_y^2 u - \mu y D_y u\right] dt + D_y u \, dx \end{aligned}$$

The generator is given by the coefficient of the dt term. More generally we can have models of the form $dy = b(y)dt + dx$, written as

$$y(t) = y(0) + x(t) + \int_0^t b(y(s))ds$$

If $b(y)$ satisfies a Lipschitz condition in y, i.e., a condition of the form $|b(y) - b(z)| \le C|y - z|$ which is valid for some constant C for all y and z, the equation can be solved uniquely by Picard iteration. The solution $y(t)$ will be a function $y(t) = \phi(t, y(0), \omega)$ that will be a measurable function of ω relative to the σ-field $\mathscr{F}_t$. The autonomous nature of the equation means that

$$\phi(t, y(0), \omega) = \phi(t - s, y(s), \theta_s\omega)$$

where $\theta_s\omega$ is the new Brownian path $x(\sigma) - x(s)$ for $\sigma \ge s$. This guarantees that $y(t)$ is still a Markov process and Itô's formula computes the generator as $1/2D_y^2 + b(y)D_y$. The transition probability is no longer explicit. It is of course given by

$$q(s,x,t,A) = P_0[\phi(t - s, x, \omega) \in A]$$

Hopefully this has a density. The conditional distribution of the increment $z = y(t+h) - y(t)$ of a Brownian motion with drift $b(y)$ in an interval $[t, t+h]$ is roughly normal with mean $hb(y(t))$ and variance h. The relative density with respect to the distribution of the increment of Brownian motion over the same interval is given by

$$\exp\left[b(y(t))z - \frac{h}{2}[b(y(t))]^2\right]$$

This suggests the formula

$$\exp\left[\int_0^t b(x(s))\mathrm{d}x(s) - \frac{1}{2}\int_0^t [b(x(s))]^2\,\mathrm{d}s\right]$$

for the Radon–Nikodym derivative $(\mathrm{d}P_{b(\cdot),x})/\mathrm{d}P_x$ of the Brownian motion with drift $b(\cdot)$ relative to Brownian motion starting at the same point x, but without any drift. This formula, known some times as Girsanov's formula, can be proved again using Itô's formula, at least when b is bounded. It can become quite technical if b is unbounded for reasons that have to do with the possible explosion of the process, i.e. becoming infinite at a finite time. The formula in a sense is still true, but needs to be carefully interpreted. See Stroock and Varadhan (1997) for a general discussion on explosion.

7. One-dimensional diffusions

The idea of a one dimensional diffusion process is to have a Markov process $x(t)$ with continuous paths such that given the past history $[x(s) : 0 \le s \le t\,]$, the future increment $x(t+h) - x(t)$ has a conditional distribution that for small h is approximately normal with mean hb and variance ha. The mean b and the variance a can be functions $b(t, x(t))$ and $a(t, x(t))$ respectively. The case $b \equiv 0$, $a \equiv 1$ is clearly Brownian motion. If $b(t, x) \equiv \mu$ and $a \equiv 1$ we have the Brownian motion with a constant drift. If a and b are purely functions of t we have a Gaussian process with independent increments, with the increment over $[s, t]$ being normally distributed with mean $\int_s^t b(\sigma)\mathrm{d}\sigma$ and variance $\int_s^t a(\sigma)\mathrm{d}\sigma$. The case when a and b are functions of x will give us the time homogeneous case. Another interesting class of examples are given by $a = a(t)$ and $b = c(t)x + d(t)$. These are the Gauss Markov processes.

The aim is to explore the relationship between the processes and the coefficients that define them. There are several possible avenues to make the connection.

a. Stochastic differential equations

If we want to generate a Gaussian random variable Y with mean b and variance a we can do that by means of a linear transformation $Y = \sigma X + b$ from a standard normal X, with the choice of $\sigma = \sqrt{a}$. With this analogy in mind, in Itô (1942), Itô's treatment of the problem was to construct the increment of a 'Diffusion Process' $y(t)$ by

$$y(t+h) - y(t) \simeq b(t, y(t))h + \sigma(t, y(t))(x(t+h) - x(t))$$

where $x(t)$ is a standard Brownian motion. Here $\sigma(t, y) = \sqrt{a(t, y)}$. This is formally written as

$$\mathrm{d}y = \sigma\,\mathrm{d}x + b\,\mathrm{d}t$$

Mathematically one tries to solve the above in integrated form

$$y(t) = y(0) + \int_0^t b(s, y(s))\mathrm{d}s + \int_0^t \sigma(s, y(s))\mathrm{d}x(s) \tag{7.1}$$

Let us, following Itô, make the following assumptions.

(A) The functions σ and b satisfy a Lipschitz condition.

$$|\sigma(t,x) - \sigma(t,y)| + |b(t,x) - b(t,y)| \leq C|x - y|$$

for some constant C.

(B) They satisfy linear growth conditions

$$|\sigma(t,x)| + |b(t,x)| \leq C(1 + |x|)$$

for some constant C.

Then within the class of almost surely continuous (in s) progressively measurable functions $\xi(s,\omega)$ on the Wiener space $(\Omega, \mathcal{F}_s, P_0)$ that satisfy

$$\sup_{0 \leq s \leq t} E^{P_0}[(\xi(s,\omega))^2] \leq C(t) < \infty$$

for every t, there exists a unique solution $y(\cdot,\cdot)$ to (7.1). Moreover the solution is a strong Markov process with transition probability function $p(s,y,t,A)$ described below. The equation can be solved for $t \geq s$ in the form

$$y(t) = x + \int_s^t b(v, y(v))\mathrm{d}v + \int_s^t \sigma(v, y(v))\mathrm{d}x(v)$$

and if we denote the solution by $y(s,x,t,\omega)$, then

$$p(s,y,t,A) = P_0[y(s,x,t,\omega) \in A]$$

This gives us a family $Q_{s,x}$ of probability measures on the space of paths $\Omega_{s,x}$ that start from x at time s.

It can be shown [see McKean (1969)] that under additional conditions of smoothness on the coefficients σ and b the solutions $y(s,x,t,\omega)$ depend smoothly on x and in fact for any smooth function $f(x)$

$$u(s,x) = \int f(y)p(s,x,t,\mathrm{d}y) = E^{Q_{s,x}}[f(x(t))]$$

is a smooth function of s and x and satisfies the equation

$$\frac{\partial u}{\partial s} + \frac{1}{2}a(s,x)\frac{\partial^2 u}{\partial x^2} + b(s,x)\frac{\partial u}{\partial x} = 0 \tag{7.2}$$

with the boundary condition $u(t,x) = f(x)$. The relation between a, b and $p(\cdot,\cdot,\cdot,\cdot)$ can also expressed by

$$\lim_{t\to s}\frac{1}{t-s}\int(y-x)p(s,x,t,\mathrm{d}y)=b(s,x)$$

$$\lim_{t\to s}\frac{1}{t-s}\int(y-x)^2p(s,x,t,\mathrm{d}y)=a(s,x)$$

$$\lim_{t\to s}\frac{1}{t-s}\int(y-x)^4p(s,x,t,\mathrm{d}y)=0$$

In fact this was the way Kolmogorov [see Kolmogorov (1931)], formulated the problem of constructing the transition probabilities from given coefficients.

b. Partial differential equations

One can try to construct $p(s,x,t,y)$ as a probability density directly and then construct a Markov process with these transition probability densities. To do this we need to find a nonnegative p that solves the equation

$$\frac{\partial p}{\partial s}+\frac{1}{2}a(s,x)\frac{\partial^2 p}{\partial x^2}+b(s,x)\frac{\partial p}{\partial x}=0$$

with the boundary condition

$$p(s,x,t,\cdot)\to\delta_x(\cdot)$$

as $s\uparrow t$. Under the assumptions

(**C**) For some constant C and $0<\alpha\leq 1$

$$|a(s,x)-a(t,y)|+|b(s,x)-b(t,y)|\leq C[|s-t|^\alpha+|x-y|^\alpha]$$

(**D**) For some constants $C<\infty$ and $c>0$

$$c\leq a(s,x)\leq C,\quad\text{and}\quad |b(s,x)|\leq C$$

the existence uniqueness of p can be established. See for instance Friedman (1964) for the details. From p one constructs the Markov process in a canonical manner via finite dimensional distributions. The almost sure continuity of the paths follows from estimates on p that are obtained along the way.

If we drop the lower bound $a\geq c>0$ but strengthen the smoothness assumptions by demanding additional regularity on a and b as functions its variables of s and x, one can prove the existence of smooth solutions u for the Equation (7.2), provided the boundary data f is smooth. The value of the solution $u(s,x)$ can be shown for each fixed t to be a nonnegative bounded linear functional of f and a suitable version of Riesz representation theorem will give

$$u(s,x)=\int f(y)p(s,x,t,\mathrm{d}y)$$

and we get p in this manner. No matter what assumptions we use the final goal is always to get the family of measures $Q_{s,x}$ on the space of paths. One can show, as one should, that if more than one method works in a given situation they lead to the same measures $Q_{s,x}$.

c. Martingale formulation

No matter how the measures are constructed it is always true that for any smooth f

$$f(x(t)) - f(x(s)) - \int_s^t (L_{v,x}f)(x(v))dv \tag{7.3}$$

is a martingale for times $t \geq s$ relative to $(\Omega_s, \mathscr{F}_t^s, Q_{s,x})$ and

$$Q_{s,x}[\omega : x(s,\omega) = x] = 1$$

Here Ω_s is the space of continuous maps $\omega = x(\cdot)$ from $[s,\infty]$ into R. $\mathscr{F}_t^s$ is the σ-field generated by $\{x(v) : s \leq v \leq t\}$. And

$$(L_{v,x}f)(x) = \tfrac{1}{2}a(v,x)f''(x) + b(v,x)f'(x)$$

In the martingale formulation, that was introduced in Stroock and Varadhan (1969), for given a and b and starting point (s,x) one asks if $Q_{s,x}$ exists such that (7.3) is a martingale for every smooth f and if such a $Q_{s,x}$ is unique. If we assume that the coefficients are bounded and continuous, existence is easy. Uniqueness is more difficult and under assumptions **(A)**, **(B)** or **(C)**, **(D)** uniqueness can be shown. Uniqueness implies that the processes $Q_{s,x}$ are strong Markov processes. In the martingale formulation, by establishing the corresponding Girsanov formula the consideration for any bounded measurable b can be reduced to the case of $b \equiv 0$ provided a has a uniform positive lower bound.

Using the martingale formulation one can establish the following random time change relation in the time homogeneous case. If $V(x)$ is positive function with $0 < c \leq V(x) \leq C < \infty$ we can define stopping times τ_t by the relation

$$\int_0^{\tau_t} \frac{1}{V(x(s))} ds = t$$

and use the stopping times to change the time scale of the process through the following map $\Phi : \Omega \to \Omega$

$$\Phi(\omega)(t) = x(\tau_t(\omega), \omega)$$

If Q is a solution to the martingale problem for $[a(x), b(x)]$ then $Q\Phi_V^{-1}$ solves the martingale problem for $[V(x)a(x), V(x)b(x)]$. One can therefore transfer existence and uniqueness results from one to the other. For the matingale formulation Stroock and Varadhan (1997) is a good source.

In the one dimensional case there are some special results. For instance in the Stochastic Differential Equations formulation the Lipschitz condition can be weakened to a Hölder condition in x of exponent $1/2$. Using a combination of random time change and Girsanov's formula one can reduce the case of any bounded $[a, b]$ with a positive lower bound for a to the case of $a \equiv 1, b \equiv 0$, which is the case of Brownian motion and enjoys both existence and uniqueness. This gives a way of directly constructing such processes from Brownian motion. These results can be found in Dynkin (1959) or Stroock and Varadhan (1997) in full detail.

d. Limit theorems

Suppose that we have a Markov chain with transition probability $\pi_h(x, \mathrm{d}y)$ and this represents a transition in a time step of duration h. We can construct a piecewise constant stochastic process $X_h(t)$ by defining first, a Markov chain ξ_n^h with transition probability π_h and then making a stochastic process out of it by defining,

$$X_h(t) = \xi_n^h \quad \text{for } nh \le t < (n+1)h$$

We can, for any initial condition $X_h(0) = x$ define the distribution of this process as a measure $Q_{x,h}$ on the Skorohod space of paths with discontinuities only of the first kind. If one assumes that the limits

$$\lim_{h \to 0} \frac{1}{h} \int (y - x)\pi_h(x, \mathrm{d}y) = b(x)$$

$$\lim_{h \to 0} \frac{1}{h} \int (y - x)^2 \pi_h(x, \mathrm{d}y) = a(x)$$

and

$$\lim_{h \to 0} \frac{1}{h} \int |y - x|^{2+\delta} \pi_h(x, \mathrm{d}y) = 0$$

hold locally uniformly in x for some bounded continuous functions $[a(x), b(x)]$ and that there is unique family Q_x that is a solution to the martingale problem for these coefficients that start at time 0 from the point x, then it follows that

$$\lim_{h \to 0} Q_{h,x} = Q_x$$

in the sense of weak convergence. The basic idea in this approach is that with $\pi_h f$ defined as $\int f(y)\pi_h(x, \mathrm{d}y)$,

$$f(X_h(nh)) - f(X_h(0)) - \sum_{j=0}^{n-1} (\pi_h f - f)(X_h((n-1)h))$$

is a martingale with respect to $(\Omega, \mathscr{F}_{nh}, Q_{h,x})$ and under our assumptions

$$\lim_{h \to 0} \frac{1}{h}(\pi_h f - f) = Lf$$

exists locally uniformly, with

$$(Lf)(x) = \tfrac{1}{2}a(x)f''(x) + b(x)f'(x)$$

This link is enough to establish that any limit point of $Q_{h,x}$ as $h \to 0$ is a solution of the martingale problem for $[a(x), b(x)]$. If we have uniqueness the limit is identified. See for instance Stroock and Varadhan (1997) and Ethier and Kurtz (1986) for results of this type.

8. The multidimensional case

To start with, we note that one can define Brownian motion in d dimensions by just taking d independent Brownian motions and making them the d components of a vector valued process $\{x_j(\cdot)\}$. We get a Markov process (actually independent increments) with transition probability density given by

$$p(s,x,t,y) = \frac{1}{[2\pi(t-s)]^{d/2}} \exp\left[-\frac{\|y-x\|^2}{2(t-s)}\right]$$

This diffusion process has the generator

$$L = \frac{1}{2}\sum_{j=1}^{d} \frac{\partial^2}{\partial x_j^2}$$

and the transition probability density itself is the fundamental solution of the heat equation

$$\frac{\partial p}{\partial t} = \frac{1}{2}\sum_{j=1}^{d} \frac{\partial^2 p}{\partial x_j^2}$$

For our Brownian motion we have taken the identity $\mathscr{I}$ as the covariance matrix. We can instead take any positive definite matrix A for the covariance and if $A = \sigma\sigma^*$ with $*$ representing the adjoint operation, we can represent the new process as the linear transform $y(t) = \sigma x(t)$. The new transition probabilities will be

$$p(s,x,t,y) = \frac{1}{[2\pi(t-s)]^{d/2}} \frac{1}{[\det A]^{1/2}} \exp\left[-\frac{\langle (y-x), A^{-1}(y-x)\rangle}{2(t-s)}\right]$$

and the new generator

$$L = \frac{1}{2}\sum_{i,j=1}^{d} a_{i,j} \frac{\partial^2}{\partial x_i \partial x_j}$$

We can add a constant drift vector vector b and then the corresponding transition probability density and generator will be given respectively by

$$p(s,x,t,y) = \frac{1}{[2\pi(t-s)]^{d/2}}\frac{1}{[\det A]^{1/2}} \times \exp\left[-\frac{\langle (y-x-b(t-s)), A^{-1}(y-x-b(t-s))\rangle}{2(t-s)}\right]$$

$$L = \frac{1}{2}\sum_{i,j=1}^{d} a_{i,j}\frac{\partial^2}{\partial x_i \partial x_j} + \sum_j b_j \frac{\partial}{\partial x_j}$$

The general time dependent diffusion process in d dimensions corresponds to the generator

$$L = \frac{1}{2}\sum_{i,j=1}^{d} a_{i,j}(t,x)\frac{\partial^2}{\partial x_i \partial x_j} + \sum_j b_j(t,x) \frac{\partial}{\partial x_j}$$

and can be constructed by any of the methods outlined in the one dimensional case. The essential intuition is that we have a vector valued process $x(t)$ such that the conditional distribution of $x(t+h) - x(t)$, given the past history $\{x(s) : 0 \leq s \leq t\}$, is approximately normally distributed in R^d with mean $hb(t,x(t))$ and covariance $ha(t,x(t))$. Here $b(t,x) = \{b_j(t,x)\}$ is an R^d valued function on $[0,\infty] \times R^d$ while $a(t,x) = a_{i,j}(t,x)$ is a symmetric positive semidefinite $d \times d$ matrix for each $(t,x) \in [0,\infty) \times R^d$.

Just as in the one dimensional case if we assume that $a(t,x)$ can be written as $\sigma(t,x)\sigma(t,x)^*$ and that both $\sigma(t,x)$ and $b(t,x)$ satisfy a Lipschitz condition of the form

$$\sum_j |b_j(t,x) - b_j(t,y)| + \sum_{i,j} |\sigma_{i,j}(t,x) - \sigma_{i,j}(t,y)| \leq C|x-y|$$

as well as a bound

$$\sum_j |b_j(t,x)| + \sum_{i,j} |\sigma_{i,j}(t,x)| \leq C(1+|x|)$$

with a constant C that does not depend on t, then we can solve the stochastic differential equation

$$\mathrm{d}y(t) = \sigma(t,y(t))\mathrm{d}x(t) + b(t,y(t))\mathrm{d}t$$

and construct a map from the d dimensional Brownian motion to the d dimensional diffusion $y(\cdot)$ corresponding to $[a(\cdot,\cdot), b(\cdot,\cdot)]$.

If we assume that $a(t,x)$ is uniformly elliptic, i.e. for some constants $0 < c < C < \infty$ we have

$$c\sum_{j=1}^{d} \xi_j^2 \leq \sum_{i,j=1}^{d} a_{i,j}(t,x)\xi_i\xi_j \leq C\sum_{j+1}^{d} \xi_j^2 \ ,$$

$b_j(\cdot,\cdot)$ are uniformly bounded by some constant C, and that $a(\cdot,\cdot)$ and $b(\cdot,\cdot)$ satisfy a Hölder condition

$$\sum_j |b_j(s,x) - b_j(t,y)| + \sum_{i,j} |a_{i,j}(s,x) - a_{i,j}(t,y)| \leq C[|t-s|^\alpha + |x-y|^\alpha]$$

for some exponent α and constant C, then just as in the one dimensional case we can get a fundamental solution $p(s,x,t,y)$ that can serve as the transition probability density for the diffusion. The heat operator is of course replaced by $\partial/\partial s + L_s$ where

$$L_s = \frac{1}{2}\sum_{i,j=1}^{d} a_{i,j}(s,x)\frac{\partial^2}{\partial x_i \partial x_j} + \sum_{j+1}^{d} b_j(s,x)\frac{\partial}{\partial x_j}$$

The corresponding diffusion processes $Q_{s,x}$ can again be characterized through the martingale formulation as one for which $Q_{s,x}[x(s) = x] = 1$, and

$$f(x(t)) - f(x(s)) - \int_s^t (L_v f)(x(v))dv$$

is a martingale for every smooth function f.

An important observation is the connection with the Dirichlet problem. We will state it for the Brownian motion and of course there are analogs for the general time homogeneous case.

If G is a connected open set in R^d with a smooth boundary, for example the open ball $B = \{x : |x| < 1\}$ and if P_x is the d dimensional Brownian motion starting from the point $x \in B$ we can define the stopping time

$$\tau(\omega) = \inf[t : x(t) \notin B]$$

By the continuity of paths one can see that $|x(\tau)| = 1$ and it is not difficult to see that $P_x[\omega : \tau(\omega) < \infty] = 1$ for $x \in B$. Then for any continuous data f on $|x| = 1$

$$u(x) = E^{P_x}[f(x(\tau))]$$

solves the Dirichlet problem

$$\sum_j \frac{\partial^2 u}{\partial x_j^2} = 0$$

in B and for $b \in \partial B$,

$$\lim_{\substack{x\to b\\ x\in B}} u(x) = f(b)$$

Although, Kolmogorov [see Kolmogorov (1931)], introduced the connection between Markov processes and parabolic partial differential equations it was not till much later that advances in PDE Friedman (1964) made it possible to use the results from PDE for constructing diffusions whereas Itô had made considerable progress using stochastic differential equations. For additional results on one dimensional diffusions Dynkin (1959) is an excellent source.

9. Diffusions with reflection

If we try to construct the one dimensional Brownian motion on the half line rather than the full line we run into trouble. The Brownian path eventually reaches 0 and wants to get out to the negative side. We must do some thing to keep it nonnegative. We may decide to give it enough of a kick to keep it nonnegative. This is easier to see in the discrete setting of a symmetric random walk or a gambler's ruin situation. Every time the gambler loses his entire fortune (i.e. reaches 0) we provide him with enough to make one more bet (i.e. move him from 0 to 1). He may lose it again in which case we make another contribution. Every time the random walk comes to zero, at the next step it is moved to 1.

The continuous analog is the set of relations $y(t) = x(t) + F(t)$, $F(t)$ is nondecreasing and continuous in t, $y(t) \geq 0$ and $x(t)$ is a given Brownian path. We are interested in the minimal F that achieves this. Alternately F is allowed to increase only when $y(t) = 0$. It turns out that such a pair of function $y(t), F(t)$ exist and is unique for every continuous path $x(t)$ and in fact are given explicitly by

$$F(t) = -\left[\inf\left[0, \inf_{0 \leq s \leq t} x(s)\right]\right]$$

and

$$y(t) = x(t) + F(t)$$

Moreover $y(t)$ is a Markov process whose distribution as a process is the same as that of $|x(t)|$. The corresponding expectations

$$u(t, y) = E[f(y(t))|y(0) = y]$$

now solve the equation

$$\frac{\partial u}{\partial t} = \frac{1}{2}\frac{\partial^2 u}{\partial y^2} \quad \text{on } [0, \infty) \times [0, \infty)$$

with the Neumann boundary condition $u_y(t, 0) = 0$ for all $t > 0$.

There are analogs for general diffusions in R^d. It becomes more complicated because the push into the region from the boundary which can be administered in only one direction in the one dimensional case can now be administered in any nontangential direction pointing towards the interior. Wentzell in (1959) wrote down the most general boundary condition that can arise. The one dimensional case for general diffusions was considered by Feller in (1957). For a martingale formulation of the problem of constructing diffusions with boundary conditions see Stroock and Varadhan (1971).

References

Cameron, R. H. and W. T. Martin (1953). The transformation of Wiener integrals by non linear transformations. *Trans. Amer. Math. Soc.* **75**, 552–575.

Dynkin, E. B. (1959). One dimensional continuous strong Markov processes. *Theor. Prob. Appl.* **4**, 3–54.

Ethier, S. N. and T. G. Kurtz (1986). *Markov processes. Characterization and convergence.* Wiley Series in Probability and Mathematical Statistics: John Wiley & Sons, Inc.

Feller, W. (1957). Generalized second order differential operators and their lateral conditions. *Illinois J. Math.* **1**, 495–504.

Friedman, A. (1964). *A Partial Differential Equations of Parabolic Type.* Prentice Hall, New Jersey.

Hunt, G. A. (1956). Some theorems concerning Brownian motion. *Trans. Amer. Math. Soc.* **81**, 294–319.

Itô, K. (1942). Differential equations determining a Markoff process (original Japanese title: Zenkoku Sizyo Sugaku Danwakai-si). *J. Pan-Japan Math. Coll.* 1077.

Itô, K. and M. Nisio (1968). On the Convergence of Sums of Independent Banach Space Valued Random Variables, Osaka. *J. Math.* **5**, 35–48.

Kolmogorov, A. N. (1931). Uber die Analytischen Methoden in der Wahrschienlichtkeitsrechnung. *Math. Ann.* **104**, 415–458.

McKean, H. P. (1969). *Stochastic Integrals.* Academic Press.

Revuz, D. and M. Yor (1999). *Continuous Martingale and Brownian Motion.* 3rd edn., Springer-Verlag.

Stroock, D. W. (1993). Probability Theory. An Analytic View. Cambridge University Press.

Stroock, D. W. and S. R. S. Varadhan (1997). *Multidimensional Diffusion Processes.* 2nd edn., Springer-Verlag.

Stroock, D. W. and S. R. S. Varadhan (1969). Diffusion Processes with continuous coefficients, I, II. *Comm. Pure Appl. Math.* **XXII**, 345–400, 479–530.

Stroock, D. W. and S. R. S. Varadhan (1971). Diffusion Processes with boundary conditions. *Comm. Pure Appl. Math.* **XXIV**, 147–225.

Wentzel, A. D. (1959). On the boundary conditions for multidimensional diffusion processes. *Theor. Prob. Appl.* **4**, 164–177.

Wiener, N. (1923). N. Differential-Space. *J. Math. Phys.* **2**, 131–174.

Wiener, N. (1924). The average value of a functional. *Proc. Lon. Math. Soc.* **22**, 454–467.

Bull. Amer. Math. Soc. (N.S.) 40 (2003) 89-97

STOCHASTIC ANALYSIS AND APPLICATIONS

S. R. S. VARADHAN

1. Introduction

The world we live in has never been very predictable, and randomness has always been part of our lives. There is ample evidence that our ancestors did enjoy playing games of chance, and the early development of probability theory had to do with these games. The study of probability has always been motivated by potential applications. Besides gambling, the stimulus has come from life insurance, pricing of annuities, statistical modeling of errors in observations, genetics, etc.

By 1900, there had developed a fairly well understood body of work, although it was not within an axiomatic framework. This prompted Hilbert to pose the axiomatization of probability theory as one of his famous problems. Although it was not until 1933 in [13] that Kolmogorov axiomatized probability theory by making it part of measure theory, the subject continued to blossom in the early part of the twentieth century at the hands of people like Lévy, Khintchin, and Wiener.

2. Infinite dimensional analysis

Wiener, in [27], was the first person to construct a measure, corresponding to what we now call Brownian motion, on the space of R^d valued continuous functions on $[0, 1]$, thereby making it possible to integrate legitimately in function spaces.

The basic ingredient in the construction of the Wiener measure is the kernel

$$p(t, x, y) = \frac{1}{(2\pi t)^{\frac{d}{2}}} \exp\left[-\frac{\|y - x\|^2}{2t}\right]$$

and its relationship to the Wiener measure P through the formula

$$P[x(\cdot) : x(t_1) \in A_1, \ldots, x(t_n) \in A_n] \\ = \int_{A_1} \cdots \int_{A_n} p(t_1, 0, x_1) p(t_2 - t_1, x_1, x_2) \cdots p(t_n - t_{n-1}, x_{n-1}, x_n) dx_1 \cdots dx_n.$$

The function p is of course the fundamental solution of the heat equation

$$\frac{\partial p}{\partial t} = \frac{1}{2} \Delta p,$$

and this creates a basic connection between the Wiener measure and the Laplace operator. Kolmogorov observed, in [12], that this relationship can be extended to a large class of probability measures (Markov or diffusion processes) that generalize

Received by the editors September 7, 2001, and, in revised form, February 21, 2002.
2000 *Mathematics Subject Classification.* Primary 60-02.

the Wiener measure on the one hand, and a large class of differential operators (second order parabolic) that generalize the heat equation on the other:

$$\frac{\partial p}{\partial t} = \frac{1}{2}\sum_{i,j} a_{i,j}(x)\frac{\partial^2 p}{\partial x_i \partial x_j} + \sum_j b_j(x)\frac{\partial p}{\partial x_j}$$

or

$$\frac{\partial p}{\partial t} = \frac{1}{2}\sum_{i,j} \frac{\partial^2}{\partial y_i \partial y_j}[a_{i,j}(y)p] - \sum_j \frac{\partial}{\partial y_j}[b_j(y)p].$$

These are referred to respectively as Kolmogorov's backward and forward equations. Itô, in 1944, described a mapping of the space of trajectories that will directly transform the Wiener measure into the more general measure that corresponds to a given operator

$$\mathcal{L} = \frac{1}{2}\sum_{i,j} a_{i,j}(x)\frac{\partial^2}{\partial x_i \partial x_j} + \sum_j b_j(x)\frac{\partial}{\partial x_j}.$$

This transformation is defined in [9] by means of a system of stochastic differential equations

$$dy_i(t) = \sum_j \sigma_{i,j}(y(t))dx_j(t) + b_i(y(t))dt$$

with $y(0) = z$, where

$$\sigma(x)\sigma^*(x) \equiv a(x).$$

One big advantage of Itô's construction is that it bypasses the PDE, thereby opening up the possibility of saying something about the solution of the PDE by studying the properties of the SDE.

It is not as simple as it sounds. The functions $x_j(t)$ are not of bounded variation (almost surely with respect to the Wiener measure), and therefore even in integrated form it is a task to make sense of the dx integrals

$$y_i(t) = z_i + \int_0^t \sum_j \sigma_{i,j}(y(s))dx_j(s) + \int_0^t b_i(y(s))ds.$$

Itô developed his theory of stochastic integration to take care of this sticky point, including a stochastic calculus with special rules

$$dx_i(t)dx_k(t) = \delta_{i,k}dt,$$
$$dx_i(t)dt = 0,$$

and

$$dy_i(t)dy_j(t) = a_{i,j}(y(t))dt.$$

In integrated form the identity

$$f(y(t)) - f(y(0)) = \int_0^t \langle \sigma^*(y(s))\nabla f(y(s)), dx(s)\rangle + \int_0^t (\mathcal{L}f)(y(s))ds$$

is known as Itô's formula. An advantage of Itô's approach is that the matrix of coefficients $\{a_{i,j}(x)\}$ can be semi-definite. Often in PDE, nondegeneracy is needed.

Another big advantage of Itô's approach is the possibility of dealing with infinite systems just as easily as finite ones. With suitable regularity and proper definition of norms, this can be realized. The PDE, on the other hand, is hard to study directly in infinite dimensions.

By a combination of methods from analysis (PDE) and probability (SDE), diffusion processes in finite dimensional spaces were thoroughly studied during 1950–1980.

The Wiener measure provides us with a reasonable integration theory in the infinite dimensional space $\Omega = C_0[[0,1];R^d]$. However it is not invariant with respect to translations on Ω, and in fact is not even quasi-invariant with respect to all translations. It is however quasi-invariant with respect to translations from the following dense subspace $\mathcal{H} \subset \Omega$:

$$\mathcal{H} = \left\{ h : h(t) = \int_0^t f(s)ds, \int_0^1 \|f(s)\|^2 ds < \infty \right\}.$$

One can therefore talk about the smoothness of a function, defined only almost everywhere, in the direction $h \in \mathcal{H}$. The gradient operator ∇_h exists, and it makes sense to talk of $\|\nabla_h u\|_p$ existing in the L_p sense. Since $\mathcal{H}$ has a natural Hilbert structure, it also makes sense to talk about the squared gradient

$$\sum_i \|\nabla_{e_i} u\|_2^2,$$

where $\{e_i\}$ is an orthonormal basis of $\mathcal{H}$. There are natural Sobolev spaces and corresponding spaces of distributions. Since we are in infinite dimensions, no amount of L_p regularity will yield continuity, but a very nice theory of differential analysis can still be built up. This was initiated by Malliavin in [16] and developed further by others, including Kusuoka and Stroock in [14], [15], and [20].

A particularly interesting byproduct is Hörmander's theory of hypoellipticity, which can be explained in terms of differential calculus on Wiener space and an integration by parts formula. Using Itô's theory, one can write the fundamental solution $p(t,z,y)$ of

$$\frac{\partial}{\partial t} - \mathcal{L}$$

as

$$E^P[\delta(y_z(t) - y)].$$

The idea is to differentiate as many times as we need to with respect to z and y, integrate by parts to get rid of unwanted derivatives, and express the answers as Wiener integrals and estimate them.

The general method bears the name of Malliavin calculus and is an important tool in the study of regularity of maps defined as stochastic integrals from one infinite dimensional space to another. Clearly the study of infinite dimensional spaces locally modeled after Wiener spaces will rely heavily on such a calculus.

One area from where problems are likely to come is nonlinear analysis in an infinite dimensional setting.

Probability theory has always generated its problems by its contact with other areas. There are very few problems that are generated by its own internal structure. This is partly because, once stripped of everything else, a probability space is essentially the unit interval with Lebesgue measure. If you have seen one you have seen them all.

3. Large discrete random structures

Perhaps the most studied example of a discrete random structure is the Ising model [8]. In its simplest version we have a finite lattice $\Lambda_N \subset Z^d$, a state space

$\Omega_N = \{-1,1\}^{\Lambda_N}$ of maps $\xi(\cdot) : \Lambda_N \to \{-1,1\}$, and a probability measure depending on a small number of parameters. Given the two parameters J and H, the Ising measure on Ω_N is the probability measure defined by

$$p^g_{N,J,H}(\xi) = \frac{1}{Z^g_N(J,H)} \exp[-JE^g_N(H,\xi)],$$

where

$$E^g_N(H,\xi) = \sum_{\substack{x,y\in\Lambda_N \\ |x-y|=1}} |\xi(x)-\xi(y)|^2 + 2 \sum_{\substack{x\in\Lambda_N \\ y\notin\Lambda_N \\ |x-y|=1}} |\xi(x)-g(y)|^2 + H\sum_{x\in Z_N} \xi(x).$$

The interesting questions concern the behavior of the measures p_N on Ω_N as $N \to \infty$ and the absence or persistence of the influence of the boundary conditions g, after the infinite volume limit has been taken.

This is an area that has been very well studied as a statistical model for phase transitions. While a lot is known, there are still some basic issues that are unresolved, even in the simplest model. There are issues of universality, etc., that are still open. The infinite volume limits of these types of models provide natural examples of random fields in Z^d that are correlated. The recent results concerning surface tension and the Wulff construction are excellent examples of interesting detailed analysis of such systems in the strongly correlated region.

The subject of random graphs had its originals in a paper of Erdös and Rényi [3]. The basic model is of a graph with n vertices in which some randomly chosen edges are turned on while the others are shut down. The subtle interplay between the probability with which the individual edges are turned on and the qualitative behavior of the resulting graph for large values of n has been a fascinating subject for study ever since the paper of Erdös and Rényi.

Percolation. If we take the standard square lattice in Z^d with $d \geq 2$ and select the nearest neighbor edges independently with probability p, there is a qualitative change in the nature of the resultant infinite graph as p changes. For small p the graph is just a collection of finite clusters, whereas for p large enough, with probability 1, there is an infinite cluster. There are detailed studies regarding the nature of this transition at a critical value $0 < p_c < 1$ of p. See [5].

The random cluster models of Fortuin and Kasteleyn [4] provide a unified point of view, connecting percolation and Ising type models.

Recently there have been exciting developments [24], [1], and [11] in situations that appear at first glance to be different, but lead to the same phenomenon. These have to do with (non-Gaussian) asymmetric fluctuations of certain extremal problems.

1. Random matrix theory. We look at the largest eigenvalue λ_N of a random symmetric matrix with an orthogonally invariant Gaussian distribution.

2. We look at the length l_N of the longest increasing subsequence of a random permutation of $1, 2, \ldots, N$.

3. Random growth models. We have a collection of independent random variables $X_{i,j}$ at each site in Z^2 and a path π connecting $(0,0)$ and (M,N) that moves only either up or to the right, and we define

$$G(M,N) = \max_{\pi} \sum_{(i,j)\in\pi} X_{i,j}.$$

All these random variables exhibit non-Gaussian asymmetric fluctuations and after rescaling have the same limiting distribution, known as the Tracy-Widom distribution. Its distribution function is given by

$$F(s) = \exp\left[-\int_s^\infty (x-s)[u(x)]\right]^2 dx,$$

where u is the solution of the Painlevé II equation

$$u'' = 2u^3 + xu$$

with $u \simeq \mathrm{Ai}(x)$ as $x \to \infty$.

The theory, as developed by P. Deift and others, depends on exact formulas and a precise analysis of the asymptotic behavior of solutions of associated Riemann-Hilbert problems.

To a probabilist there is a mystery here, and it is very likely that these results will one day be part of a larger class of limit theorems.

Disordered systems. Random environments, etc. One of the earliest examples of a problem involving disorder is a random walk in a random environment. Let us take a family of independent random variables π_j, $-\infty < j < \infty$, with $0 < \pi_j < 1$, and consider a random walk on Z with transition probabilities $p_{x,x+1} = \pi_x$ and $p_{x,x-1} = 1 - \pi_x$. The random choices are made once and fixed for all time. The asymptotic behavior of such a walk can be studied in detail for the one dimensional nearest neighbor case. Other problems of a similar nature range from the difficult to the impossibly difficult.

Roughly speaking, these problems deal with probability measures $\mu_{\mathbf{a}}$, where $\mathbf{a}$ is a large or an infinite set of parameters $\{a_x\}$ indexed by $x \in \Lambda$. If $a_x \equiv a$ (a constant), the situation is well understood. In problems of disorder, $\{a_x\}$ are assumed to be random, chosen independently for each x and with a common distribution. One is interested in properties that are shared by $\mu_{\mathbf{a}}$ for almost all $\mathbf{a}$. A random walk in a random environment is one example. See [21].

The Sherrington-Kirkpatrick or S-K model consists of a family of probability measures $\mu_N = \mu_N^{\beta,h,\{g_{i,j}\}}$ on $\{0,1\}^N$ depending on parameters $\beta, h, \{g_{i,j}\}$:

$$\mu_N\{\epsilon\} = \frac{1}{Z_N}\exp[-\beta H_N(\epsilon)],$$

where

$$H_N(\epsilon) = -\frac{1}{\sqrt{N}}\sum_{1\le i<j\le N} g_{i,j}\epsilon_i\epsilon_j + h\sum_{1\le i\le N}\epsilon_i;$$

$\{g_{i,j}\}$ is a realization of i.i.d. standard Gaussian random variables.

If $h = 0$ and $\beta < 1$, the direction of the vector ϵ is rather random, so that

$$E^{\mu_N\times\mu_N}\left[\frac{(\epsilon.\epsilon')^2}{N^2}\right]$$

is small for most $\{g_{i,j}\}$. This is not expected to be true for $\beta > 1$, although what is known is a slightly weaker form of the result for $\beta \gg 1$.

This is just one of several similar problems. See the recent work of Talagrand [22] and [23] for detailed comments.

The similar problem in the case of short range interactions, where

$$H_N(\epsilon) = -\sum_{(x,y):|x-y|=1} g_{x,y}\epsilon_x\epsilon_y + h\sum_x \epsilon_x$$

and the summation extends over nearest neighbors in Z^d, is even less understood. See for instance the article [17] by Newman and Stein.

4. Scaling limits

Another area that has been active is the dynamical behavior of large or infinite interacting systems. The evolution may have conserved quantities, in which case the evolution of slow modes can be studied under hydrodynamical scaling, averaging out the fast modes with respect to appropriate invariant distributions.

Fairly general methods involving entropy, rate of entropy production and relative entropy have been developed.

In the case of systems without conserved quantities, one of the issues is whether the rate of approach to equilibrium slows down as the system size increases. This can be measured in terms of either the spectral gap or the log-Sobolev constant. Conditions for these estimates to hold uniformly are known. They can be used to estimate the accuracy of simulations. See Stroock and Zegarliński [19].

In the presence of conserved quantities, the spectral gaps and the log-Sobolev constants become degenerate with increasing volume, and the rate can be guessed by looking at the slow modes. That this is in fact the case is a useful tool in the analysis of these systems and is known for several models. See Yau [29], or Cancrini and Martinelli [2].

A typical class of models that illustrates the scaling behavior is known as simple exclusion processes. See [25], [26] and [28].

One has, as physical space, either Z^d or Z^d_N, the periodic lattice in Z^d of N^d elements. Each site can have either one particle or no particle. The state is described by $\{\eta(x) : x \in Z^d_N\}$.

The dynamics is specified by the generator

$$(\mathcal{A}f)(\eta) = \sum_{x,y} \eta(x)(1-\eta(y))p(y-x)[f(\eta^{x,y}) - f(\eta)].$$

The scaling $x \to \frac{x}{N}$ of space imbeds Z^d_N into the d-torus $\mathcal{T}^d$. The micro-state η can be replaced by a macro-state

$$r_N(t,\cdot) = \frac{1}{N^d}\sum_x \delta_{\frac{x}{N}},$$

which as $n \to \infty$ turns into the particle density $\rho(\theta)$.

Assuming that the initial configuration leads to the macro-state $\rho_0(\theta)$, the dynamical behavior of the large system should be describable in terms of $\rho_0(\cdot)$.

There are two cases. If $m = \sum zp(z)$, either $m = 0$ or $m \neq 0$. Let us first look at $m \neq 0$. The special case $d = 1$, $p(1) = 1$ is typical. In this case one needs to speed up the time scale by a factor of N as well, and the micro-state at time t is, with probability nearly 1, close to the macro-state $\rho(t,\theta)$, which is the unique entropic solution of the Burgers equation

$$\rho_t + [\rho(1-\rho)]_\theta = 0.$$

The secondary issues of probabilities of large deviations can also be studied for this model. See Jensen [10]. Given $\rho(\cdot,\cdot)$, the question of estimating the probability

$$P[r_N(\cdot,\cdot) \simeq \rho(\cdot,\cdot)] = \exp[-N\ I(\rho(\cdot,\cdot)) + o(N)]$$

with an explicit I has a simple answer: $I(\rho(\cdot,\cdot)) = \infty$ unless $\rho(\cdot,\cdot)$ is also a weak solution, and then $I(\rho(\cdot,\cdot))$ is the precise amount by which the entropy condition is violated for the convex functional

$$h(\rho) = \rho \log \rho + (1-\rho)\log(1-\rho);$$

that is,

$$I(\rho(\cdot,\cdot)) = \int_0^T \int [[h(\rho)]_t + [g(\rho)]_x]^+ dx dt.$$

In particular, one can show that

$$\lim_{N\to\infty} \frac{1}{N} \log E[\exp\{hN(t)\}] = \begin{cases} \frac{h}{4} & \text{if } h \geq 0, \\ \tanh \frac{h}{4} & \text{if } h \leq 0, \end{cases}$$

and one can compare the fluctuation theory with the large deviation estimate.

The case $m = 0$ exhibits more complex behavior. There exists $A(\rho)$ such that the equality

$$\rho_t = \frac{1}{2} \nabla A(\rho) \nabla \rho$$

describes the evolution of the macroscopic state in the new time scale, which now needs to be speeded up by a factor of N^2. The calculation of $A(\rho)$ is in general rather complicated. However, if p is symmetric, $A(\rho)$ simplifies to $A = \sum z \otimes z p(z)$.

Again the problems of large deviation as well as the motion of a tracer particle have satisfactory solutions. $S(\rho)$ is the self-diffusion coefficient in equilibrium, i.e., $\frac{x(t)}{\sqrt{t}} \simeq \beta$, and β has dispersion $S(\rho)$. Instead of just looking at the evolution of the macro-state, we may wish to consider the random process

$$R_N = \frac{1}{N^d} \sum_j \delta_{x_j(\cdot)},$$

which describes the collective history of all the particles. The limit

$$\lim_{N\to\infty} R_N = Q$$

exists. Both Q and the rate function $I(R)$ in the large deviation estimate

$$P[R_N \simeq R] = \exp[-N^d I(R) + o(N)]$$

have explicit exact descriptions. See Quastel, Rezakhanlou, and Varadhan [18].

The models that we have described have a fairly regular structure. The physical space is the lattice Z^d with connections between nearby sites. The rates are translation invariant.

One can imagine disordered versions of these problems. There have been some studies where the disorder is only in the rates. See [6] and [7]. But a more challenging problem is when the underlying structure itself is some sort of a random graph.

Whereas problems coming from physics present a more regular structure, problems from the social sciences present far more disorder.

Today more sophisticated models are being used in finance, insurance and other areas involving an analysis of uncertainty and risk. If we wish to fully understand how large systems behave and why, one possible approach is to specify the model at the microscopic level and develop methods that allow us to make predictions of macroscopic behavior.

References

1. Jinho Baik, Percy Deift, and Kurt Johansson, *On the distribution of the length of the longest increasing subsequence of random permutations*, J. Amer. Math. Soc. **12** (1999), 1119–1178. MR **2000e**:05006
2. N. Cancrini and F. Martinelli, *On the spectral gap of Kawasaki dynamics under a mixing condition revisited. Probabilistic techniques in equilibrium and nonequilibrium statistical physics*, J. Math. Phys. **41** (2000), 1391–1423.
3. P. Erdös and A. Renyi, *On the evolution of random graphs*, Magyar Tud. Akad. Mat. Kutató Int. Közl. **5** (1960), 17–61. MR **23**:A2338
4. C. M. Fortuin and P. W. Kasteleyn, *On the random cluster model. I: Introduction and relation to other models*, Physica **59** (1972), 535–564. MR **50**:12107
5. Geoffrey Grimmett, *Percolation*, 2nd ed., Springer-Verlag, Berlin, 1999. MR **2001a**:60114
6. Alice Guionnet and Boguslaw Zegarliński, *Decay to equilibrium in random spin systems on a lattice*, Comm. Math. Phys. **181** (1996), 703–732. MR **97j**:82105
7. ———, *Decay to equilibrium in random spin systems on a lattice. II*, J. Statist. Phys. **86** (1997), 899–904. MR **98h**:82046
8. E. Ising, *Beitrag zur Theorie des Ferromagnitismus*, Z. Phys. **31** (1926), 253–258.
9. Kiyosi Itô, *Stochastic integral*, Proc. Imp. Acad. Tokyo **20** (1944), 519–524. MR **7**:313c
10. Leif Jensen, *Large deviations of the asymmetric simple exclusion process*, Ph.D. thesis, Courant Inst., New York Univ., New York, 2000.
11. Kurt Johansson, *Shape fluctuations and random matrices*, Comm. Math. Phys. **209** (2000), 437–476. MR **2001h**:60177
12. A. N. Kolmogorov, *Über die analyttischen Methoden in der Wahrscheinlichkeitsrechnung*, Math. Ann. **104** (1931), 415–458.
13. ———, *Grundbegriffe der Wahrscheinlichkeitsrechnung*, Springer-Verlag, Berlin, 1933.
14. S. Kusuoka and D. Stroock, *Applications of the Malliavin calculus.* II, J. Fac. Sci. Univ. Tokyo Sect. IA Math. **32** (1985), 1–76. MR **86k**:60100b
15. ———, *Applications of the Malliavin calculus.* III, J. Fac. Sci. Univ. Tokyo Sect. IA Math. **34** (1987), 391–442. MR **89c**:60093
16. Paul Malliavin, *Stochastic calculus of variation and hypoelliptic operators*, Proc. Internat. Sympos. Stochastic Differential Equations (Res. Inst. Math. Sci., Kyoto Univ., Kyoto, 1976), Wiley, New York, 1978, pp. 195–263. MR **81f**:60083
17. C. M. Newman and D. L. Stein, *Equilibrium pure states and nonequilibrium chaos*, J. Statist. Phys. **94** (1999), 709–722. MR **2000d**:82040
18. Jeremy Quastel, Fraydoun Rezakhnalou, and S. R. S. Varadhan, *Large deviations for the symmetric simple exclusion process in dimensions* $d \geq 3$, Probab. Theory Related Fields **113** (1999), 1–84. MR **2000c**:60162
19. D. Stroock and B. Zegarliński, *On the ergodic properties of Glauber dynamics*, J. Statist. Phys. **81** (1995), 1007–1019. MR **96i**:60109
20. Daniel Stroock, *Applications of the Malliavin calculus.* I, Stochastic Analysis (Katata/Kyoto, 1982), North-Holland Math. Library, vol. 32, North-Holland, Amsterdam, 1984, pp. 271–306. MR **86k**:60100a
21. Alain-Sol Sznitman, *Brownian motion, obstacles and random media*, Springer-Verlag, Berlin, 1998. MR **2001h**:60147
22. Michel Talagrand, *Huge random structures and mean field models for spin glasses*, Proc. Internat. Congr. Math. (Berlin, 1998), Vol. I, Doc. Math. Extra Vol. I (1998), 507–536 (electronic). MR **2000c**:60164
23. ———, *The Sherrington-Kirkpatrick model: a challenge for mathematicians*, Probab. Theory Related Fields **110** (1998), 109–176. MR **99a**:60117
24. Craig A. Tracy and Harold Widom, *Level spacing distributions and the Bessel kernel*, Comm. Math. Phys. **161** (1994), 289–309. MR **95e**:82004
25. S. R. S. Varadhan, *The complex story of simple exclusion*, Itô's Stochastic Calculus and Probability Theory, Springer, Tokyo, 1996, pp. 385–400. MR **98g**:60058

26. ______, *Large deviations for interacting particle systems*, Perplexing Problems in Probability, 373–383, Progr. Probab., vol. 44, Birkhauser Boston, Boston, MA, 1999, pp. 373–383. MR **2001h:**60044

27. Norbert Wiener, *Differential space*, J. Math. and Phys. **2** (1923), 131–174.

28. Horng-Tzer Yau, *Scaling limit of particle systems, incompressible Navier-Stokes equation and Boltzmann equation*, Proc. Internat. Congr. Math. (Berlin, 1998), Vol. III Doc. Math. Extra Vol. III, 193–202 (electronic). MR **2000h:**82061

29. ______, *Logarithmic Sobolev inequality for lattice gases with mixing conditions*, Comm. Math. Phys. **181** (1996), 367–408. MR **98e:**82021

COURANT INSTITUTE, NEW YORK UNIVERSITY, NEW YORK, NEW YORK 10012-1110
E-mail address: Varadhan@cims.nyu.edu

Entropy, Princeton Univ. Press, 2003, 199-214

Large Deviations and Entropy

S. R. S. Varadhan

Courant Institute of Mathematical Sciences

We describe how entropy plays a central role in the calculation of probabilities of large deviations. This is illustrated by several examples.

9.1 Where Does Entropy Come From?

The entropy functional $h(p) = h(p_1, \dots, p_m) = -\sum_j p_j \log p_j$, defined on the simplex of probability distributions on m points, i.e. on the simplex

$$\left\{p = (p_1, \dots, p_m) : p_i \geqslant 0, \ \sum_i p_i = 1\right\},$$

is usually characterized abstractly as the only function that satisfies certain natural properties (Khinchin 1957). However, for a probabilist the natural way to meet entropy is in a combinatorial computation, while approximating factorials with the use of Stirling's formula (Feller 1957).

Let us a toss the mythical fair coin n times and observe the number of heads. The number of ways of getting k heads is $\binom{n}{k}$. If we use Stirling's approximation for factorials, then we obtain

$$\begin{aligned}\binom{n}{k} &\simeq \frac{\sqrt{2\pi}\mathrm{e}^{-n}n^{n+1/2}}{\sqrt{2\pi}\mathrm{e}^{-k}k^{k+1/2}\sqrt{2\pi}\mathrm{e}^{-(n-k)}(n-k)^{n-k+1/2}} \\ &= \frac{\sqrt{n}}{\sqrt{2\pi k(n-k)}}\exp\left[n\left[-\frac{k}{n}\log\frac{k}{n} - \frac{n-k}{n}\log\frac{n-k}{n}\right]\right].\end{aligned}$$

If we fix the proportion of heads and tails at approximately p and $q = 1 - p$, then

$$\binom{n}{k} \simeq \frac{1}{\sqrt{2\pi npq}}\mathrm{e}^{nh(p,q)}.$$

The entropy is the factor in the exponential growth rate of the 'volume,' or in this case the number of distinct outcomes in the space of coin tosses that corresponds to a given number k of heads. The actual probability in the context of tossing a fair coin is

$$P_n(k) \simeq \frac{1}{\sqrt{2\pi npq}}\mathrm{e}^{n[h(p,q)-\log 2]}.$$

Let us look at the situation when an 'unfair' coin with probabilities α for heads and $1-\alpha$ for tails is tossed. Then

$$P_n(\alpha, k) \simeq \frac{1}{\sqrt{2\pi npq}} e^{n[h(p,q)+p\log\alpha+(1-p)\log(1-\alpha)]}.$$

The exponential constant $h(p,q) + p\log\alpha + (1-p)\log(1-\alpha)$ is a sum of two terms, both depending on $p = 1-q = k/n$. The first term $h(p,q)$ is the 'volume' term that we have already seen and does not depend on α. The second term is the 'energy' term and is a function of p and α. The most likely state for k corresponds to the value of p that maximizes the combination $h(p,q) - E(p,q)$, which is $p = \alpha$. Deviations of p away from α are called 'large deviations.' Their probabilities decay exponentially as

$$\exp[-nI(\alpha; p)],$$

where

$$I(\alpha; p) = p\log\frac{p}{\alpha} + (1-p)\log\frac{1-p}{1-\alpha} = E(p,q) - h(p,q). \tag{9.1}$$

The large-deviation rate function is the rate of decay given by (9.1). $I(\alpha; p) \geqslant 0$ and in fact $I(\alpha; p) > 0$ unless $p = \alpha$.

While it is difficult to extend the definition of the entropy functional $h(p)$ to arbitrary probability distributions, the 'relative entropy' $I_\alpha(p)$ has a definition that makes sense for any pair of probability distributions α and β on an arbitrary measurable space. If $\beta \ll \alpha$, with a Radon–Nikodym derivative

$$f(x) = \frac{\mathrm{d}\beta}{\mathrm{d}\alpha} \quad \text{and} \quad \int f(x)\log f(x)\,\mathrm{d}\alpha(x) < \infty,$$

then

$$I(\alpha; \beta) = \int f(x)\log f(x)\,\mathrm{d}\alpha(x) = \int \log f(x)\,\mathrm{d}\beta(x). \tag{9.2}$$

Otherwise, $I_\alpha(\beta) = +\infty$.

There are large-deviation rate functions that at first glance do not seem to come directly from entropy. Cramér's Theorem (Cramér 1938) is an example. Consider a sequence $X_1, X_2, \ldots, X_n, \ldots$ of i.i.d. random variables. Assume that the moment-generating function

$$M(\theta) = E[\exp[\theta X]]$$

is finite for all $\theta \in \mathbb{R}$. Let $\phi(a)$ be the Legendre transform of $\psi(\theta) = \log M(\theta)$:

$$\phi(a) = \sup_\theta[a\theta - \psi(\theta)] = \sup_\theta[a\theta - \log M(\theta)]. \tag{9.3}$$

Then $\phi(a) \geqslant 0$ and $\phi(a) = 0$ if and only if $a = E[X]$. The 'law of large numbers' tells us that the probabilities

$$P\left[\frac{X_1 + \cdots + X_n}{n} \sim a\right]$$

of large-deviations decay, away from $a = E[X]$. In fact, they do so exponentially, with an explicit rate constant

$$P\left[\frac{X_1 + \cdots + X_n}{n} \sim a\right] \sim \exp[-\phi(a)]$$

with $\phi(a)$ given by (9.3).

The precise formulation of the above relation is as follows. A sequence of probability distributions $\{P_n\}$ on a Polish space (complete separable metric space) $\mathcal{X}$ is said to satisfy a 'large-deviation principle' with rate n and rate function $I(x)$ if the following are valid.

1. The rate function $I(x)$ is lower semicontinuous, nonnegative and has level sets $K_\ell = \{x : I(x) \leqslant \ell\}$ that are compact in $\mathcal{X}$.

2. Moreover,

$$\limsup_{n\to\infty} \frac{1}{n} \log P_n(C) \leqslant - \inf_{x\in C} I(x) \quad \text{for closed } C \subset \mathcal{X}, \tag{9.4}$$

$$\liminf_{n\to\infty} \frac{1}{n} \log P_n(G) \geqslant - \inf_{x\in G} I(x) \quad \text{for open } G \subset \mathcal{X}. \tag{9.5}$$

It looks like that, in the context of Cramér's Theorem, Legendre transforms control large-deviation rate functions rather than entropy. But, as we shall see in the next section, this is deceptive.

One of the consequences of the validity of a large-deviation principle is a Laplace asymptotic formula for certain types of integrals. If P_n satisfies a large-deviation principle with a rate function $I(x)$ on a Polish space $\mathcal{X}$ and if $F : \mathcal{X} \to \mathbb{R}$ is a bounded continuous function, then

$$\lim_{n\to\infty} \frac{1}{n} \log E^{P_n}[\exp[nF(x)]] = \sup_{x\in\mathcal{X}}[F(x) - I(x)]. \tag{9.6}$$

9.2 Sanov's Theorem

In the multinomial case, $X_1, X_2, \ldots, X_n$ are n independent random variables that take values $1, 2, \ldots, m$ with probabilities $\pi_1, \pi_2, \ldots, \pi_m$ and the frequencies $f_1, f_2, \ldots, f_m$ are observed. The probabilities of large deviations for the empirical probabilities $\{f_j/n\}$ are

$$\begin{aligned} P_n\left[\left\{\frac{f_j}{n}\right\} \sim \{p_j\}\right] &= \exp\left[-n\sum_j p_j \log\frac{p_j}{\pi_j} + o(n)\right] \\ &= \exp[-nI(\{\pi_j\}; \{p_j\}) + o(n)]. \end{aligned}$$

This extends to the empirical distribution

$$\nu_n = \frac{1}{n}\sum_{r=1}^{n} \delta_{X_r} \tag{9.7}$$

of n identically distributed random variables on $\mathbb{R}$ with a common distribution μ viewed as a random measure on $\mathbb{R}$. The large-deviation rate function takes the form

$$P[\nu_n \simeq \nu] = \exp[-n\, I(\mu; \nu) + o(n)],$$

where $I(\mu; \nu)$ was defined in (9.2). The connection to Cramér's Theorem is provided by the contraction principle

$$\inf_{\nu:\int x\, \mathrm{d}\nu = a} I(\mu; \nu) = \phi(a)$$

and the infimum is attained when

$$\mathrm{d}\nu = \frac{\mathrm{e}^{\theta x}}{M(\theta)}\, \mathrm{d}\mu$$

and θ is chosen so that

$$\psi'(\theta) = \frac{M'(\theta)}{M(\theta)} = a.$$

9.3 What about Markov Chains?

Let $\pi_{i,j}$ be the transition probabilities of a Markov chain on a finite state space with k points. We want to compute the probabilities of large deviations for the empirical distribution ν_n of the proportion of times the chain spends at different sites. The rate function is easily computed as

$$\mathcal{I}_\pi(\nu) = \sup_V \left[\sum_j V(j)\nu(j) - \lambda(V) \right],$$

where the supremum is taken over all vectors $V = \{V(1), V(2), \ldots, V(k)\}$. Here

$$\lambda(V) = \log \rho(V)$$

and $\rho(V)$ is the eigenvalue with the largest modulus of the matrix

$$\pi^V_{i,j} = \pi_{i,j}\mathrm{e}^{V(j)}$$

(which is positive because the matrix has nonnegative entries). It is not clear what the spectral radius of nonnegative matrices has to do with entropy. But again there is a connection. Instead of considering just the one-dimensional marginals $\nu_n(j)$, we could consider the frequencies of sites visited at consecutive times,

$$\nu_n^{(2)}(i, j) = \frac{1}{n} \sum_{r=0}^{n-1} \delta_{X_r}(i)\delta_{X_{r+1}}(j), \tag{9.8}$$

where we have adopted the convention that $X_n = X_0$. If F is the state space, then $\nu_n^{(2)}$ is a probability measure on $F \times F$ such that both marginals are equal to the empirical distribution ν_n defined in (9.7). If $\mu(i)$ is the invariant measure for the chain, then $\nu_n^{(2)} \to \mu^{(2)}$ given by $\mu^{(2)}(i, j) = \mu(i)\pi_{i,j}$. We want to compute the rate function $\mathcal{I}_2(q)$ defined for measures q on $F \times F$ with equal marginals that governs the probabilities of large deviations of the empirical frequencies $\nu_n^{(2)}(\cdot, \cdot)$ away from $\mu^{(2)}(\cdot, \cdot)$:

$$P[\nu_n^{(2)} \sim q] = \exp[-n\mathcal{I}_2(q) + o(n)].$$

For each q define q_1 to be its marginal and $\hat{q}$ on $F \times F$ by $\hat{q}(i, j) = q_1(i)\pi_{i,j}$. Then

$$\mathcal{I}_2(q) = I(\hat{q}; q) = \sum_{i,j} q(i, j) \log \frac{q(i, j)}{\hat{q}(i, j)} \tag{9.9}$$

and we are back with our relative entropy. In particular,

$$\inf_{q:q_1=\nu} \mathcal{I}_2(q) = \mathcal{I}_\pi(\nu).$$

9.4 Gibbs Measures and Large Deviations

All of this can be understood in terms of large deviations of Gibbs measures in one dimension. Let F be a finite set and $\Omega = F^{\mathbb{Z}}$ be the set of maps $\omega : \mathbb{Z} \to F$ of the set of integers $\mathbb{Z}$ into F. $T : \Omega \to \Omega$ is the shift operator defined by $(T\omega)(n) = \omega(n+1)$. If $\varphi(\omega)$ is a local function of ω, i.e. a function depending only on a finite number of coordinates $\{\omega(j) : |j| \leqslant k\}$, the Gibbs measure P_φ is the (unique) limit as $N \to \infty$ of normalized probability measures $P_{N,\varphi}$ on $F^{[-N,N]}$, defined by

$$P_{N,\varphi}[\omega(-N) \dots, \omega(N)] \simeq \exp\left[-\sum_{j=-N}^{N} \varphi(T^j\omega)\right].$$

In principle, for finite N, these measures depend on some boundary effects, arising from the edges in the summation $\sum_{j=-N}^{N} \varphi(T^j\omega)$. But, in one dimension as $N \to \infty$, these effects disappear and we get a unique, well-defined, stationary probability measure P_φ on $\Omega = F^{\mathbb{Z}}$ corresponding to any local function φ. A special P, which we denote by P_0, corresponds to $\varphi \equiv 0$ and is the product measure on Ω of the uniform distribution on the finite set F. If Q is any stationary measure on Ω, we have its marginal distribution Q_N on $\Omega_N = F^{[-N,N]}$ with entropy $h_N(Q)$ given by

$$-\sum_{\omega_N \in \Omega_N} Q_N(\omega_N) \log Q(\omega_N)$$

and the Kolmogorov–Sinai entropy $H(Q)$ of Q, defined by

$$H(Q) = \lim_{N\to\infty} \frac{1}{2N+1} h_N(Q),$$

can be easily shown to exist (by considerations of subadditivity). The largest possible value for $H(Q)$ is $\log k$ and is achieved only when $Q = P_0$, the product of uniform distributions.

The 'free energy' $\psi_0(g)$ for a local function g (relative to P_0) is defined by

$$\psi_0(g) = \lim_{N\to\infty} \frac{1}{2N+1} \log E^{P_0}\left[\exp\left[\sum_{j=-N}^{N} g(T^j\omega)\right]\right]. \tag{9.10}$$

The empirical process $R_n(\omega)$ is defined by

$$R_n(\omega) = \frac{1}{n}[\delta_\omega + \delta_{T\omega} + \cdots + \delta_{T^{n-1}\omega}]$$

and is nearly stationary for large n. Moreover, the marginals of R_n are precisely the empirical distributions $\nu_n^{(1)}$, $\nu_n^{(2)}$, ... with a negligible correction coming from the edges for large n. By the law of large numbers $R_n(\omega) \to P_0$ and the probabilities of large deviations from the limit P_0 are governed by the rate function

$$I_0(Q) = [\log k - H(Q)].$$

There is also a variational formula for the free energy, arising from the Laplace asymptotic formula (9.6)

$$\psi_0(\varphi) = \sup_{Q\in\mathcal{S}}[E^Q[\varphi(\omega)] - I_0(Q)],$$

where the supremum is taken over all translation-invariant probability measures Q on Ω. It is attained uniquely at some $Q = P_\varphi$ which is called the Gibbs measure with energy φ. The free energy relative to P_φ is defined as

$$\psi_\varphi(g) = \lim_{N\to\infty} \frac{1}{2N+1} \log E^{P_\varphi}\left[\exp\left[\sum_{j=-N}^{N} g(T^j\omega)\right]\right]$$

and is seen to equal

$$\psi_0(\varphi + g) - \psi_0(\varphi) = \sup_{Q\in\mathcal{S}}[E^{P_\varphi}[\varphi(\omega)] - I_\varphi(Q)]$$

with

$$I_\varphi(Q) = \psi_0(\varphi) - [E^Q[\varphi(\omega)] - I_0(Q)].$$

It is now clear that the large deviations of $R_n(\omega)$ under P_φ will be controlled by the rate function $I_\varphi(Q)$. If we are only interested in the marginals $\nu_n^{(1)}$ or $\nu_n^{(2)}$, the rate function will have to be contracted to yield

$$I_\varphi^{(1)}(\{p_j\}) = \inf_{Q\in\mathcal{S}_1(\{p_j\})} I_\varphi(Q), \tag{9.11}$$

where the infimum is taken over stationary (i.e. translation-invariant) probability measures Q on Ω with one-dimensional marginal $\{p_j\}$ or

$$I_\varphi^{(2)}(\{q_{i,j}\}) = \inf_{Q \in \mathcal{S}_2(\{q_{ij}\})} I_\varphi(Q), \tag{9.12}$$

where the infimum is taken over stationary processes Q with two-dimensional marginal $\{q_{i,j}\}$.

If $\phi(\omega) = V(\omega(0))$, then P_φ is a product measure with $\pi_j \simeq \mathrm{e}^{V(j)}$. The infimum in attained at the product measure $Q = \otimes^{\mathbb{Z}}\{p_j\}$ and $I_\varphi^{(1)}(\{p_j\})$ becomes the relative entropy $I(\{\pi_j\}; \{p_j\})$. We are back at Sanov's Theorem.

If $\phi(\omega) = V(\omega(0), \omega(1))$, then P_φ is Markov with transition a transition probability π that can be determined. Moreover, $\{q_{i,j}\}$ determines a unique Markov process Q_0 and $I_\varphi^{(2)}(\{q_{i,j}\})$, with the infimum attained at Q_0, is seen to equal $\mathcal{I}_2(q)$, which was defined in (9.9). See Lanford (1973) for more details.

9.5 Ventcel–Freidlin Theory

There are other large-deviation investigations, e.g. the Ventcel–Freidlin theory (Ventcel and Freidlin 1970). Here the context is very different. Consider a Markov process with generator

$$L_\varepsilon = \tfrac{1}{2}\varepsilon\Delta + b(x)\cdot\nabla$$

acting on smooth functions on $\mathbb{R}^d$. If the paths start from $x \in \mathbb{R}^d$, then they are given by the solution to the SDE

$$x_\varepsilon(t) = x + \int_0^t b(x_\varepsilon(s))\,\mathrm{d}s + \sqrt{\varepsilon}\,\beta(t),$$

where $\beta(t)$ is the d-dimensional Brownian motion. As $\varepsilon \to 0$, $x_\varepsilon(t)$ converges to the solution of the ODE

$$x(t) = x + \int_0^t b(x(s))\,\mathrm{d}s.$$

We are interested in the probabilities of large deviations, i.e.

$$P[x_\varepsilon(\cdot) \sim f(\cdot)],$$

where f is an arbitrary function $f(\cdot) : [0, T] \to \mathbb{R}^d$ with $f(0) = x$. These probabilities decay like $\exp[-(1/\varepsilon)\mathcal{I}(f)]$ and $\mathcal{I}(f)$ is calculated as

$$\mathcal{I}(f) = \frac{1}{2}\int_0^T \|f'(t) - b(f(t))\|^2\,\mathrm{d}t.$$

This is also essentially an entropy calculation. One way to make sure that the solution converges to $f(t)$ is to replace the original generator by the new time-dependent generator

$$L_{\varepsilon,f} = \tfrac{1}{2}\varepsilon\Delta + f'(t)\cdot\nabla$$

corresponding to the SDE

$$y_\varepsilon(t) = x + \int_0^t f'(s)\,\mathrm{d}s + \sqrt{\varepsilon}\,\beta(t).$$

The solution $y_\varepsilon(t)$ converges to $f(t)$. A formula, known to probabilists as Girsanov's formula, gives us the Radon–Nikodym derivative (on the σ-field up to time T) of the distribution μ_ε (on path space) of the solution of

$$\mathrm{d}z(t) = b(t, z(t))\,\mathrm{d}t + \sqrt{\varepsilon}\,\mathrm{d}\beta(t)$$

relative to the distribution of λ_ε of $z(\cdot)$ given by $z(t) = \sqrt{\varepsilon}\beta(t)$, as

$$\frac{\mathrm{d}\mu_\varepsilon}{\mathrm{d}\lambda_\varepsilon}(z(\cdot)) = \exp\left[\frac{1}{\varepsilon}\int_0^T \langle b(t, z(t)), \mathrm{d}z(t)\rangle - \frac{1}{2\varepsilon}\int_0^T \|b(t, z(t))\|^2\,\mathrm{d}t\right],$$

where $\int_0^T \langle b(t, z(t)), \mathrm{d}z(t)\rangle$ is the Itô stochastic integral. The relative entropy of the two measures P_ε of $x_\varepsilon(\cdot)$ and Q_ε of $y_\varepsilon(\cdot)$, which correspond to the choice of $b(t, x) = b(x)$ and $b(t, x) = f'(t)$ respectively, can now be easily computed as

$$\begin{aligned} I(P_\varepsilon; Q_\varepsilon) &= E^{Q_\varepsilon}\left[\frac{1}{\varepsilon}\int_0^T \langle b(x(s)), \mathrm{d}x(s)\rangle - \frac{1}{2\varepsilon}\int_0^T \|b(x(s))\|^2\,\mathrm{d}s \right. \\ &\qquad \left. - \frac{1}{\varepsilon}\int_0^T \langle f'(s), \mathrm{d}x(s)\rangle + \frac{1}{2\varepsilon}\int_0^T \|f'(s)\|^2\,\mathrm{d}s\right] \\ &\simeq \frac{1}{2\varepsilon}\int_0^T |f'(s) - b(f(s))|^2\,\mathrm{d}s. \end{aligned}$$

9.6 Entropy and Large Deviations

In fact, it is not hard to see that any 'large deviation' has to be related to 'entropy.' If P is a measure and we want to estimate $P(A)$, then we can do it if we can estimate

$$\inf_{Q: Q(A)=1} I(P; Q) = \ell.$$

We have the inequality (Jensen) that says

$$E^Q[g] \leqslant I(P; Q) + \log E^P[\exp[g]]. \tag{9.13}$$

Taking $g = c\chi_A$ in (9.13), with $c > 0$, we get

$$Q(A) \leqslant \frac{1}{c}[I(P; Q) + \log[\mathrm{e}^c P(A) + (1 - P(A))]$$

or

$$\ell \geqslant c - \log[1 - P(A) + \mathrm{e}^c P(A)].$$

Letting $c \to \infty$ we get

$$\ell \geqslant -\log P(A)$$

or

$$P(A) \leqslant e^{-\ell}.$$

There is a partial converse. If $Q \ll P$ and $E^Q|\log dQ/dP - \ell|$ is small and $Q(A) = q$, then

$$\begin{aligned} P(A) &\geqslant \int_A \frac{dP}{dQ}\, dQ = Q(A)\frac{1}{Q(A)}\int_A \exp\left[-\log\frac{dQ}{dP}\right] dQ \\ &\geqslant q\exp\left[\frac{1}{q}\int_A \left[-\log\frac{dQ}{dP}\right] dQ\right] \\ &\geqslant q\exp\left[-\ell - \frac{1}{q}E^Q\left|\log\frac{dQ}{dP} - \ell\right|\right]. \end{aligned}$$

Large deviations is then essentially estimating entropies, but the new measure Q has to be guessed. In a dynamical situation the dynamics has to be perturbed in order to get a Q with the right property. There are lots of choices and the entropy can be calculated by a Girsanov formula. Then the optimal choice becomes an optimal control problem with entropy as the cost functional (Dupuis and Ellis 1997; Fleming 1985).

Let us illustrate this by means of the examples in Sections 9.2–9.4.

If $X_1, \dots, X_n, \dots$ is a sequence of i.i.d. random variables with a common distribution α and we want the empirical distribution $(1/n)\sum_{r=1}^{n}\delta_{X_r}$ to be close to β, we can achieve this by changing the joint distribution of the $\{X_r\}$ from the product measure $P = \otimes^{\mathbb{Z}}\alpha$ to any stationary process Q with one-dimensional marginals equal to β. The relative entropy $I_n(P; Q)$ of Q with respect to P on the σ-field generated by the first n coordinates grows linearly in n and

$$\lim_{n\to\infty}\frac{1}{n}I_n(P; Q) = \mathcal{I}(P; Q)$$

exists and can be evaluated as

$$\mathcal{I}(P; Q) = E^Q[h(q_\omega, \alpha)],$$

where q_ω is the conditional distribution of X_1 given the past history $\{X_j : j \leqslant 0\}$ calculated under Q. The control problem reduces to the variational problem

$$\mathcal{I}(\beta) = \inf_{Q\in\mathcal{M}_\beta}[\mathcal{I}(P; Q)],$$

where $\mathcal{M}_\beta$ is the set of stationary processes Q, with one-dimensional marginals β. Self-consistency requires that for $Q \in \mathcal{M}_\beta$,

$$\int q_\omega\, dQ = \beta.$$

It is not hard to see at this point that, from the convexity of $\mathcal{I}(\cdot, \alpha)$, the infimum is attained when $Q = \otimes^{\mathbb{Z}}\beta$, and we arrive at Sanov's Theorem.

The advantage of this somewhat complex view point is that the principle is universal. If we replace the product measure P by a nicely mixing Markov chain with transition probabilities $\pi(x, \mathrm{d}y)$, nothing changes except the formula for $\mathcal{I}(P; Q)$, which now reads

$$\mathcal{I}(P; Q) = E^Q[h(q_\omega, p_\omega)],$$

where, by our Markovian assumptions, $p_\omega = \pi(X_{-1}, \mathrm{d}y)$. Let us carry out the optimization

$$\inf_{Q \in \mathcal{M}_\beta} \mathcal{I}(P; Q)$$

in two steps. We let Γ_β be the set of bivariate distributions γ with common marginals β and denote by $\mathcal{M}_\gamma^{(2)}$ the set of stationary processes Q such that its bivariate marginal at two consecutive times is γ. Then the infimum above can be carried out as

$$\inf_{\gamma \in \Gamma_\beta} \inf_{Q \in \mathcal{M}_\gamma} \mathcal{I}(P; Q).$$

Any $\gamma \in \Gamma_\beta$ can be disintegrated as $\beta(\mathrm{d}x) \otimes \hat{\pi}(x, \mathrm{d}y)$ through its marginal and conditional distribution. $\hat{\pi}(\cdot, \cdot)$ is a transition probability with β as an invariant measure, and therefore defines a canonical stationary Markov process Q_γ with (consecutive) bivariate marginals γ and one-dimensional marginals β. One can show that the inner infimum is attained at $Q = Q_\gamma$,

$$\mathcal{I}(Q_\gamma; P) = \mathcal{I}^{(2)}(\gamma) = \int I(\pi(x, \cdot), \hat{\pi}(x, \cdot))\, \beta(\mathrm{d}x),$$

and the rate function becomes

$$\mathcal{I}(\beta) = \inf_{\gamma \in \Gamma_\beta} \mathcal{I}^{(2)}(\gamma).$$

The duality relation

$$\sup_\beta \left[\int V(x)\, \beta(\mathrm{d}x) - \mathcal{I}(\beta) \right] = \log \lambda(V),$$

where $\lambda(V)$ is the spectral radius of $\pi(x, \mathrm{d}y)\mathrm{e}^{V(x)}$, is just a piece of convex analysis and establishes the connection with Section 9.3. It goes more or less like this. We denote by

$$\Gamma = \bigcup_\beta \Gamma_\beta$$

the set of all bivariate distributions with equal marginals:

$$
\begin{aligned}
\sup_V &\left[\int V(x)\,\mathrm{d}\beta - \mathcal{I}(\beta)\right] \\
&= \sup_\beta \sup_{\gamma\in\Gamma_\beta}\left[\int V(x)\,\gamma(\mathrm{d}x,\mathrm{d}y) - \mathcal{I}^{(2)}(\gamma)\right] \\
&= \sup_{\gamma\in\Gamma}\left[\int V(x)\,\gamma(\mathrm{d}x,\mathrm{d}y) - \int \log\frac{\hat\pi(x,\mathrm{d}y)}{\pi(x,\mathrm{d}y)}\,\gamma(\mathrm{d}x,\mathrm{d}y)\right] \\
&= \sup_\gamma \inf_U\left[\int [V(x)+U(x)-U(y)]\,\gamma(\mathrm{d}x,\mathrm{d}y)\right. \\
&\qquad\qquad \left. - \int \log\frac{\hat\pi(x,\mathrm{d}y)}{\pi(x,\mathrm{d}y)}\,\gamma(\mathrm{d}x,\mathrm{d}y)\right] \\
&= \inf_U \sup_\gamma\left[\int [V(x)+U(x)-U(y)]\,\gamma(\mathrm{d}x,\mathrm{d}y)\right. \\
&\qquad\qquad \left. - \int \log\frac{\hat\pi(x,\mathrm{d}y)}{\pi(x,\mathrm{d}y)}\,\gamma(\mathrm{d}x,\mathrm{d}y)\right] \\
&= \inf_U\left[\sup_x [V(x)+U(x)+\log\int \mathrm{e}^{-U(y)}\pi(x,\mathrm{d}y)]\right] \\
&= \lambda(V).
\end{aligned}
$$

Finally, the connection with the example in Section 9.5 is straightforward. We replace $b(x)$ with a new $c(t,x)$ and the SDE is

$$\mathrm{d}x(t) = c(t,x(t))\,\mathrm{d}t + \sqrt{\varepsilon}\,\beta(t).$$

We need c to be such that f is a solution of $\dot{x} = c(t,x(t))$ or $c(t,f(t)) = f'(t)$, which is really what we did.

9.7 Entropy and Analysis

Entropy plays a role in analysis as well. Let $p(t,x,y)$ be the transition probability density (relative to an invariant distribution q) of a Markov process. We have the Markov semigroup

$$(T_t f)(x) = \int f(y)p(t,x,y)q(\mathrm{d}y)$$

acting on bounded measurable functions and the adjoint semigroup

$$(T_t^* g)(y) = \int g(x)p(t,x,y)q(\mathrm{d}x)$$

acting on $L_1(q)$. Let L and L^* be the infinitesimal generators of the semigroups T_t and T_t^*, respectively. The function

$$u(t, x) = (T_t f)(x)$$

satisfies the evolution equation

$$\frac{\partial u}{\partial t} = Lu$$

and $g_t = T_t^* g_0$ satisfies the Fokker–Planck (Kolmogorov's forward) equation

$$\frac{\partial g}{\partial t} = L^* g$$

with L^* being the adjoint of L with respect to q, i.e.

$$\int (Lf)g \, \mathrm{d}q = \int f(L^* g) \, \mathrm{d}q.$$

The H-theorem asserts that the relative entropy $H(t) = \int g_t \log g_t \, \mathrm{d}q$ is non-increasing in t. If we know that $H_0 = \int g_0 \log g_0 \, \mathrm{d}q < \infty$, then

$$\int_0^\infty \left[-\frac{\mathrm{d}H_t}{\mathrm{d}t} \right] \mathrm{d}t \leqslant H_0,$$

giving us a decay rate on $I(t) = -\mathrm{d}H_t/\mathrm{d}t$. An elementary calculation shows that

$$-\frac{\mathrm{d}}{\mathrm{d}t} H_t = -\frac{\mathrm{d}}{\mathrm{d}t} \int g_t \log g_t \, \mathrm{d}q = -\int (L^* g_t) \log g_t \, \mathrm{d}q = -\int g_t (L \log g_t) \, \mathrm{d}q$$
$$= -2 \int g_t (L \log \sqrt{g_t}) \, \mathrm{d}q \geqslant -2 \int g_t \frac{L\sqrt{g_t}}{\sqrt{g_t}} \, \mathrm{d}q = 2\mathcal{D}(\sqrt{g_t}).$$

Here $\mathcal{D}$ is the Dirichlet form on $L_2(q)$, defined as

$$\mathcal{D}(g) = \langle g, -Lg \rangle_q = -\int g(x)(Lg)(x)q(\mathrm{d}x) = -\int g(x)(L^* g)(x)q(\mathrm{d}x),$$

and this provides us with a control

$$\int_0^\infty \mathcal{D}(\sqrt{g_t}) \, \mathrm{d}t \leqslant \tfrac{1}{2} H_0.$$

Since usually $D(g) = 0$ only if g is a constant, a small value of the Dirichlet form implies that the state is close to equilibrium. This can be used to show that large systems with multiple equilibria approach local equilibria over large volumes with a length scale that is related to the time scale. If a log-Sobolev estimate is available, then this can be used to establish hydrodynamic scaling results. These typically derive a partial differential equation for the parameters that define the local equilibria as functions of rescaled space and time. We shall illustrate this in the next section. Even if there are no such inequalities available, one can still use the estimate on the Dirichlet form in very many ways (see, for instance, Guo et al. 1988).

9.8 Hydrodynamic Scaling: an Example

Hydrodynamic scaling refers to a wide class of phenomena where a large interacting system with conserved quantities and local interactions evolves over time. Because of the conserved quantities there will be multiple equilibria, which can usually be parametrized by the average values of the conserved quantities. The system will reach local equilibria first before evolving towards the global equilibrium rather slowly. If we speed up time, the local equilibria will be established over larger domains. At the appropriate rescaling of space and time, the evolution of the complex system can be fully described by describing the parametric label of the equilibrium as a function of space and time, which is determined for later times by an evolution equation from its initial values. This evolution equation, usually a parabolic or hyperbolic transport equation, is called the hydrodynamic limit of the complex interacting system.

Let us illustrate this by a familiar class of examples called simple exclusion processes. We will limit ourselves to the one-dimensional case. We start with the lattice $\mathbb{Z}$. At any given time we have particles at some of the sites and the associated variables $\eta(x)$, which are 1 if there is a particle at x and 0 otherwise. The term 'exclusion' signifies that we are not allowed to have more than one particle at the same time at any site. The particles wait for a random exponential waiting time (with mean 1) and then pick a random site to jump to. If the particle is at site x, then the probability of picking site y to jump to is $p(y-x)$. Of course, $\sum_{z\neq 0} p(z) = 1$. If the site y is selected, the particle jumps to the new site if it is empty. If the chosen site has a particle and is therefore unavailable, the jump is disallowed and the particle waits at the original site for a new exponential time. All of this can be specified by writing down the generator L acting on functions $f(\eta)$ that depend on the configuration η:

$$(Lf)(\eta) = \sum_{x,y} p(y-x)\eta(x)(1-\eta(y))[f(\eta^{x,y}) - f(\eta)].$$

Here $\eta^{x,y}$ is the new configuration obtained by exchanging the situations at x and y, i.e.

$$\eta^{x,y}(z) = \begin{cases} \eta(x) & \text{if } z = y, \\ \eta(y) & \text{if } z = x, \\ \eta(z) & \text{otherwise.} \end{cases} \tag{9.14}$$

For simplicity, let us look at the periodic case where the lattice $\mathbb{Z}$ is replaced by $\mathbb{Z}_N$ of N sites with periodic boundary conditions. Let us also suppose that $p(z)$ is symmetric, i.e. $p(z) = p(-z)$. Let the initial configuration, which depends on N and could be random, be such that

$$\lim_{N\to\infty} \frac{1}{N} \sum_x J\left(\frac{x}{N}\right)\eta_0(x) = \int_T J(z)\rho_0(z)\,\mathrm{d}z \tag{9.15}$$

for some deterministic $0 \leqslant \rho_0(z) \leqslant 1$. In the random case the above limit is interpreted as limit in probability. Here the continuum limit of the rescaled lattice is the circle $\boldsymbol{T}$, which is viewed as the unit interval with end points identified. Then the evolution has the property that, at time $N^2 t$,

$$\lim_{N\to\infty} \frac{1}{N} \sum_x J\left(\frac{x}{N}\right)\eta_{N^2 t}(x) = \int_{\boldsymbol{T}} J(z)\rho(t,z)\,\mathrm{d}z$$

in probability, where $\rho(\cdot,\cdot)$ is the solution of the heat equation

$$\rho_t = \tfrac{1}{2}\sigma^2 \rho_{xx} \tag{9.16}$$

with initial condition $\rho(0,z) = \rho_0(z)$. Here $\sigma^2 = \sum_z z^2 p(z)$. In this context the heat equation (9.16) is called the hydrodynamic limit of the symmetric simple exclusion process.

It is fairly easy to establish this fact by a direct computation of two moments remembering that the speeded-up process has generator $N^2 L$. One expects that in the speeded-up time scale, for any local function $f(\eta)$, denoting by $f_x(\eta)$ its spatial translate by the lattice variable x,

$$\lim_{N\to\infty} \frac{1}{N} \int_0^T \left[\sum_x f_x(\eta_{N^2 s})\right] \mathrm{d}s = \int_0^T \int_{\boldsymbol{T}} \hat{f}(\rho(t,z))\,\mathrm{d}z\,\mathrm{d}t \tag{9.17}$$

with $\hat{f}(\rho) = E^{\rho}[f(\eta)]$. Here E^{ρ} refers to the expectation with respect to the Bernoulli product measure on $\{0,1\}^{\mathbb{Z}}$ with $E^{\rho}[\eta(x)] = \rho$. This result will attest to the fact that in the time scale N^2 the system is locally in a Bernoulli state over spatial scales that involve N lattice sites. This can be deduced from the estimate on the Dirichlet form in the following way: if we use the uniform distribution on 2^N sites as our q, then the relative entropy of any initial state with respect to q is at most $N \log 2$. Therefore, in the speeded-up time scale, since the Dirichlet form inherits a factor of N^2,

$$\int_0^{\infty} \mathcal{D}(\sqrt{g_t})\,\mathrm{d}t \leqslant \frac{\log 2}{2N}.$$

By convexity, if

$$\bar{g} = \frac{1}{T}\int_0^T g_t\,\mathrm{d}t,$$

then

$$\mathcal{D}(\sqrt{\bar{g}}) \leqslant \frac{\log 2}{2NT}.$$

We want to use this estimate to show that on a large block of εN sites the conditional distribution under $\bar{q}$ of the configurations given the mean number m of particles in that block is close to the Bernoulli measure μ_m on that block. This can be controlled if we can control the relative entropy of the restriction of $\bar{q}$ to a block of size εN to a mixture of μ_m on that block. We have to use a mixture, because the average density m can be random under $\bar{q}$. Since the Dirichlet form is additive in volume, the restriction of $\bar{q}$ to a typical block of length εN will have a Dirichlet form of about $(\varepsilon \log 2)/N$. The log-Sobolev inequality is known to hold with a constant proportional to the square of the block size. Therefore, the relative entropy of the restriction of $\bar{q}$ to a typical block of size εN will have a relative entropy of order $\varepsilon^3 N$ relative to a suitable mixture of Bernoulli distribution in that block. The specific entropy is roughly ε^3 per site. This can be controlled uniformly in N provided ε is small. This is enough to prove (9.17).

We now turn to the totally asymmetric case where $p(1) = 1$ and $p(z) = 0$ for $z \neq 1$. In this case time is rescaled by a factor of N and we assume that the initial condition (9.15) holds as before. Then

$$\lim_{N\to\infty} \frac{1}{N} \sum_x J\left(\frac{x}{N}\right)\eta_{Nt}(x) = \int_T J(z)\rho(t,z)\,\mathrm{d}z, \tag{9.18}$$

where $\rho(t, z)$ is now a weak solution of Burgers equation

$$\frac{\partial \rho}{\partial t} + \frac{\partial[\rho(1-\rho)]}{\partial z} = 0 \tag{9.19}$$

on $[0, T] \times \boldsymbol{T}$ with initial condition

$$\rho(0, z) = \rho_0(z).$$

There is no uniqueness for solutions of Burgers equation (9.19) in the class of weak solutions. See Chapter 6 by Dafermos (2003) in this book. While there is uniqueness within the class of regular solutions, in general, regular solutions do not exist or exist only up to a certain time when shocks develop. The hydrodynamic limit (9.18) is singled out as the unique weak solution satisfying the entropy condition (Rezakhanlou 1991), i.e. for every convex 'entropy' h and the corresponding 'flux' g, defined by

$$h'(\rho)(1 - 2\rho) = g'(\rho),$$

one has

$$\frac{\partial h(\rho)}{\partial t} + \frac{\partial g(\rho)}{\partial z} \leqslant 0 \tag{9.20}$$

as a distribution. A special case is the entropy

$$h(\rho) = \rho \log \rho + (1-\rho)\log(1-\rho)$$

with corresponding flux

$$g(\rho) = \rho - \rho(1-\rho)\log\frac{\rho}{1-\rho}.$$

It turns out (Jensen 2000) that this particular entropy controls the probabilities of large deviations. If we want the empirical density to be close to an arbitrary

$\rho(\cdot,\cdot)$ in $[0,T]$, which may not be the entropy solution, this will be a large deviation. The rate function $\mathcal{I}(\rho)$ for such a large deviation is finite only if $\rho=\rho(t,z)$ is a weak solution of equation (9.19), and the distribution

$$\mu=\frac{\partial h(\rho)}{\partial t}+\frac{\partial g(\rho)}{\partial z}$$

is a signed measure of bounded variation. While the condition for the weak solution $\rho(\cdot,\cdot)$ to be an entropy solution is $\mu\leqslant 0$, the large-deviation rate function is given by

$$\mathcal{I}(\rho(\cdot,\cdot))=\mu^{+}[[0,T]\times\boldsymbol{T}].$$

References

Cramér, H. 1938 Sur un noveau théorème—limites de la théorie des probabilités. *Actualités Scientifiques et Industrielles* **736**, 5–23. *Colloque consacré à la théorie des probabilités*, Vol. 3. Hermann, Paris

Dafermos, C. 2003 Entropy for hyperbolic conservation laws. In *Entropy* (ed. A. Greven, G. Keller and G. Warnecke), pp. 107–120. Princeton University Press.

Dupuis, P. and Ellis, R. S. 1997 *A Weak Convergence Approach to the Theory of Large Deviations*. Wiley Series in Probability and Statistics (Probability and Statistics Section). Wiley-Interscience.

Feller, W. 1957 *An Introduction to Probability Theory and Its Applications*, Vol. I, 2nd edn. Wiley, Chapman and Hall.

Fleming, W. H. 1985 A stochastic control approach to some large deviations problems. *Recent Mathematical Methods in Dynamic Programming* (Rome, 1984), pp. 52–66. Lecture Notes in Mathematics, vol. 1119. Springer.

Guo, M. Z., Papanicolaou, G. C. and Varadhan, S. R. S. 1988 Nonlinear diffusion limit for a system with nearest neighbor interactions. *Commun. Math. Phys.* **118**(1), 31–59.

Jensen, L. 2000 Large deviations of the asymmetric simple exclusion process. PhD thesis, New York University.

Khinchin, A. I. 1957 *Mathematical Foundations of Information Theory* (translated by R. A. Silverman and M. D. Friedman). Dover, New York.

Lanford III, O. E. 1973 Entropy and equilibrium states in classical statistical mechanics. In *Statistical Mechanics and Mathematical Problems* (ed. A. Lenard), pp. 1–113. Lecture Notes in Physics, 20. Springer.

Rezakhanlou, F. 1991 Hydrodynamic limit for attractive particle systems on Z^d. *Commun. Math. Phys.* **140**, 417–448.

Ventcel, A. D. and Freidlin, M. I. 1970 Small random perturbations of dynamical systems. *Russ. Math. Surv.* (translation of *Usp. Mat. Nauk*) **25**, 3–55.

The Annals of Probability
2008, Vol. 36, No. 2, 397–419
DOI: 10.1214/07-AOP348

SPECIAL INVITED PAPER

LARGE DEVIATIONS[1]

BY S. R. S. VARADHAN

New York University

This paper is based on Wald Lectures given at the annual meeting of the IMS in Minneapolis during August 2005. It is a survey of the theory of large deviations.

1. Large deviations for sums.

The role of "large deviations" is best understood through an example. Suppose that $X_1, X_2, \dots, X_n, \dots$ is a sequence of i.i.d. random variables, for instance, normally distributed with mean zero and variance 1. Then,

$$E\big[e^{\theta(X_1+\cdots+X_n)}\big] = E[e^{\theta X_1}]^n = e^{n(\theta^2/2)}.$$

On the other hand,

$$E\big[e^{\theta(X_1+\cdots+X_n)}\big] = E\big[e^{n\theta(S_n/n)}\big].$$

Since, by the law of large numbers, $\frac{S_n}{n}$ is nearly zero, we have

$$E\big[e^{\theta(X_1+\cdots+X_n)}\big] = E\big[e^{o(n)}\big] \neq e^{o(n)}.$$

There is, of course, a very simple explanation for this. In computing expectations of random variables that can assume large values with small probabilities, contributions from such values cannot be ignored. After all, a product of something big and something small can still be big! In our case, assuming $\theta > 0$, for any $a > 0$,

$$E[e^{\theta S_n}] \geq e^{n\theta a} P\left[\frac{S_n}{n} \geq a\right] = e^{n\theta a} e^{-(na^2/2)+o(n)} = e^{n(a\theta - a^2/2)+o(n)}.$$

Since $a > 0$ is arbitrary,

$$E[e^{\theta S_n}] \geq e^{n \sup_{a>0}(\theta a - a^2/2)+o(n)} = e^{n\theta^2/2+o(n)},$$

which is the correct answer.

The simplest example for which one can calculate probabilities of large devia-

Received April 2007; revised April 2007.

[1]Supported in part by the NSF Grants DMS-01-04343 and DMS-06-04380.

AMS 2000 subject classifications. 60-02, 60F10.

Key words and phrases. Large deviations, rare events.

tions is coin tossing. The probability of k heads in n tosses of a fair coin is

$$P(n,k)=\binom{n}{k}2^{-n}=\frac{n!2^{-n}}{k!(n-k)!},$$

which using Stirling's approximation, is

$$\simeq \frac{\sqrt{2\pi}\,e^{-n}n^{n+1/2}2^{-n}}{\sqrt{2\pi}\,e^{-(n-k)}(n-k)^{n-k+1/2}\sqrt{2\pi}\,e^{-k}k^{k+1/2}},$$

$$\begin{aligned}\log P(n,k)&\simeq -\frac{1}{2}\log(2\pi)\left(n+\frac{1}{2}\right)\log n-\left(n-k+\frac{1}{2}\right)\log(n-k)\\&\quad-\left(k+\frac{1}{2}\right)\log k-n\log 2\\&=-\frac{1}{2}\log(2\pi)-\frac{1}{2}\log n-\left(n-k+\frac{1}{2}\right)\log\left(1-\frac{k}{n}\right)\\&\quad-\left(k+\frac{1}{2}\right)\log\frac{k}{n}-n\log 2.\end{aligned}$$

If $k\simeq nx$, then

$$\begin{aligned}\log P(n,k)&\simeq -n[\log 2+x\log x+(1-x)\log(1-x)]+o(n)\\&=-nH(x)+o(n),\end{aligned}$$

where $H(x)$ is the Kullback–Leibler information or relative entropy of Binomial$(x,1-x)$ with respect to Binomial$(\frac{1}{2},\frac{1}{2})$.

This is not a coincidence. In fact, if f_i are the observed frequencies in n trials of a multinomial with probabilities $\{p_i\}$ for the individual cells, then

$$P(n,p_1,\ldots,p_k;f_1,\ldots,f_k)=\frac{n!}{f_1!\cdots f_k!}p_1^{f_1}\cdots p_k^{f_k}.$$

A similar calculation using Stirling's approximation yields, assuming $f_i\simeq nx_i$,

$$\log P(n,p_1,\ldots,p_k;f_1,\ldots,f_k)=-nH(x_1,\ldots,x_k;p_1,\ldots,p_k)+o(n),$$

where $H(x,p)$ is again the Kullback–Leibler information number

$$H(x)=\sum_{i=1}^{k}x_i\log\frac{x_i}{p_i}.$$

Any probability distribution can be approximated by one that is concentrated on a finite set and the empirical distribution from a sample of size n will then have a multinomial distribution. One therefore expects that the probability $P(n,\alpha,\beta)$ that the empirical distribution

$$\frac{1}{n}\sum_{i=1}^{n}\delta_{X_i}$$

of n independent observations from a distribution α is close to β should satisfy

$$\log P(n,\alpha,\beta) = -nH(\beta,\alpha) + o(n),$$

where $H(\beta,\alpha)$ is again the Kullback–Leibler information number

$$\int \log \frac{d\beta}{d\alpha}\, d\beta = \int \frac{d\beta}{d\alpha} \log \frac{d\beta}{d\alpha}\, d\alpha.$$

This theorem, proven by Sanov, must be made precise. This requires a formal definition of what is meant by large deviations. We have a family $\{P_n\}$ of probability distributions on some space X which we assume to be a complete separable metric space. There is a sequence of numbers $a_n \to \infty$ which we might as well assume to be n. Typically, P_n concentrates around a point $x_0 \in X$ and, for sets A away from x_0, $P_n(A)$ tends to zero exponentially rapidly in n, that is,

$$\log P_n(A) \simeq -nc(A),$$

where $c(A) > 0$ if $x_0 \notin A$.

We say that a *large deviation principle* holds for a sequence of probability measures P_n defined on the Borel subsets of a Polish (complete separable metric) space X, with a *rate function* $H(x)$, if:

(1) $H(x) \geq 0$ is a lower semicontinuous function on X with the property that $K_\ell = \{x : H(x) \leq \ell\}$ is a compact set for every $\ell < \infty$;

(2) for any closed set $C \subset X$,

$$\limsup_{n\to\infty} \frac{1}{n} \log P_n(C) \leq -\inf_{x\in C} H(x);$$

(3) for any open set $U \subset X$,

$$\liminf_{n\to\infty} \frac{1}{n} \log P_n(U) \geq -\inf_{x\in U} H(x).$$

While condition (1) is not really necessary for the validity of the large deviation principle, it is a useful condition on the rate function that will allow us to reduce the analysis to what happens on compact sets. Rate functions with this property are referred to as "good" rate functions.

For Sanov's theorem, the i.i.d. random variables will be taking values in a complete separable metric space X. Their common distribution will be a probability measure $\alpha \in \mathcal{M} = \mathcal{M}(X)$, the set of probability measures on X. The space $\mathcal{M}$ under weak convergence is a complete separable metric space. The empirical distributions will be random variables with values in $\mathcal{M}$, and P_n will be their distribution, which will therefore be a probability measure on $\mathcal{M}$. The rate function

$$H(\beta,\alpha) = \int \frac{d\beta}{d\alpha} \log \frac{d\beta}{d\alpha}\, d\alpha$$

will be defined to be $+\infty$ unless $\beta \ll \alpha$ and the Radon–Nikodym derivative $f = \frac{d\beta}{d\alpha}$ is such that $f \log f$ is integrable with respect to α.

2. Rate functions, duality and generating functions. We start with i.i.d. random variables and look at the sample mean

$$Y_n = \frac{S_n}{n} = \frac{X_1 + \cdots + X_n}{n}.$$

According to a theorem of Cramér, its distribution P_n satisfies a large deviation principle with a rate function $h(a)$ given by

$$h(a) = \sup_\theta [\theta a - \log E[e^{\theta X}]].$$

In general, if a large deviation principle is valid for probability measures P_n on a space X with some rate function $H(x)$, it is not hard to see that, under suitable conditions on the function F (boundedness and continuity will suffice),

$$\frac{1}{n} \log \int e^{nF(x)}\, dP_n \to \sup_x [F(x) - H(x)].$$

The basic idea is just

$$\frac{1}{n} \log \sum e^{na_i} \to \sup_i a_i.$$

In other words, the logarithms of generating functions are dual to, or Legendre transforms of, large deviation rate functions. For instance,

$$\log\left[\int e^{V(x)}\, d\alpha\right] = \sup_\beta \left[\int V(x)\, d\beta - H(\beta, \alpha)\right].$$

If the rate function is convex, as it is in the case of sums of independent identically distributed random variables, the duality relationship is invertible and

$$H(\beta, \alpha) = \sup_{V(\cdot)} \left[\int V(x)\, d\beta - \log E\left[e^{V(x)}\, d\alpha\right]\right],$$

where the supremum on V in either case is taken over all bounded measurable functions or bounded continuous functions.

3. Markov processes. This relationship for i.i.d. sequences can be extended to the Markovian context. For simplicity, let us assume that we have a finite state space X and transition probabilities $\pi(x, y)$ of a Markov chain on X. Let us suppose that $\pi(x, y) > 0$ for all x, y. If $V(\cdot) : X \to R$ is a function on X, then

$$E_x\left[\exp[V(X_1) + V(X_2) + \cdots + V(X_n)]\right]$$

can be explicitly evaluated as

$$\sum_y \pi_V^n(x, y),$$

where $\pi_V(x, y) = \pi(x, y)e^{V(y)}$ and π_V^n is the nth power of π_V. Since π_V is a matrix with positive entries,

$$\frac{1}{n}\log\sum_y \pi_V^n(x, y) \to \log\lambda_\pi(V),$$

where $\lambda_\pi(V)$ is the principal eigenvalue of π_V. The analog of Sanov's theorem in this context establishes a large deviation result for the empirical distributions of a long sequence $\{x_1, x_2, \ldots, x_n\}$. They belong to the space $\mathcal{M}$ of probability measures on X. Let us denote by P the Markov process with transition probability $\pi(x, y)$ and by Q_n the measure on $\mathcal{M}$ which is the distribution of the empirical distribution $\{x_1, x_2, \ldots, x_n\}$. The large deviation upper bound for Q_n, with a rate function given by

$$H_\pi(q) = \sup_V\left[\sum_x V(x)q(x) - \log\lambda_\pi(V)\right],$$

is an easy consequence of estimates on the generating function.

There is a more direct way of approaching H_π through lower bounds. Let us pretend that our Markov chain exhibits "atypical" behavior and behaves like a different Markov chain, one with transition probability $\tilde{\pi}(x, y)$. In other words, the empirical distribution of visits to different sites of X is close to q, which turns out to be the invariant distribution for a different chain, one with transition probabilities $\tilde{\pi}(x, y)$. We can estimate the probability of this event. Let U be an open set around q in the space $\mathcal{M}$ of probability measures on X. Let A_n be the set of all realizations of $\{x_1, \ldots, x_n\}$ with empirical distributions belonging to U. We can estimate $P(A_n)$, the probability of A_n under the original $\pi(x, y)$ chain, by

$$\begin{aligned}
P(A_n) &= \sum_{x_1,x_2,\ldots,x_n\in A_n} \pi(x, x_1)\cdots\pi(x_{n-1}, x_n) \\
&= \sum_{x_1,x_2,\ldots,x_n\in A_n} \tilde{\pi}(x, x_1)\cdots\tilde{\pi}(x_{n-1}, x_n)\exp\left[-\sum_{i=1}^n \log\frac{\tilde{\pi}(x_{i-1}, x_i)}{\pi(x_{i-1}, x_i)}\right] \\
&= \int_{A_n}\exp\left[-\sum_{i=1}^n \log\frac{\tilde{\pi}(x_{i-1}, x_i)}{\pi(x_{i-1}, x_i)}\right] d\tilde{P},
\end{aligned}$$

where $\tilde{P}$ is the Markov chain with transition probability $\tilde{\pi}(x, y)$. By the ergodic theorem,

$$\lim_{n\to\infty}\tilde{P}(A_n) = 1$$

as $n \to \infty$. An application of Jensen's inequality yields

$$\liminf_{n\to\infty}\frac{1}{n}\log Q_n(A) \geq -\sum_{x,y}\tilde{\pi}(x, y)\log\frac{\tilde{\pi}(x, y)}{\pi(x, y)}q(x).$$

We can pick any $\tilde{\pi}$, provided q is invariant for $\tilde{\pi}$, that is, $q\tilde{\pi} = q$ and we will get an upper bound for $H_\pi(q)$. Therefore,

$$H_\pi(q) \le \inf_{\tilde{\pi}:\tilde{q}=q} \sum_{x,y} \tilde{\pi}(x, y) \log \frac{\tilde{\pi}(x, y)}{\pi(x, y)} q(x).$$

In fact, there is equality here and $H_\pi(q)$ is dual to $\log \lambda_\pi(V)$.

$$H_\pi(q) = \sup_{V(\cdot)} \left[\sum_x V(x)q(x) - \log \lambda_\pi(V) \right],$$

$$\log \lambda_\pi(V) = \sup_{q(\cdot)} \left[\sum_x V(x)q(x) - H_\pi(q) \right].$$

If we are dealing with a process in continuous time, we will have a matrix A of transition rates $\{a(x, y)\}$ with $a(x, y) \ge 0$ for $x \ne y$ and $\sum_y a(x, y) = 0$. The transition probabilities $\{\pi(t, x, y)\}$ will be given by $\pi(t, x, y) = (\exp tA)(x, y)$. If a is symmetric, that is, $a(x, y) = a(y, x)$, then so is $\pi(t, \cdot, \cdot)$. The uniform distribution on X will, in this case, be the invariant measure. The role of $\log \lambda(V)$ in the previous discrete situation will now be played by the principal eigenvalue $\lambda_a(V)$ of $A + V$, where $(A + V)(x, y) = a(x, y) + \delta(x, y)V(y)$. Here, $I = \{\delta(x, y)\}$ is the identity matrix. This is the conjugate of the rate function $H(q)$.

$$H_a(q) = \sup_{V(\cdot)} \left[\sum_x V(x)q(x) - \lambda_a(V) \right],$$

$$\lambda_a(V) = \sup_{q(\cdot)} \left[\sum_x V(x)q(x) - H_a(q) \right].$$

Note that, in the symmetric case, we have the usual variational formula

$$\begin{aligned}
\lambda_a(V) &= \sup_{\substack{u(\cdot) \\ \sum_x [u(x)]^2 = 1}} \left[\sum_x V(x)u^2(x) + \sum_{x,y} a(x, y)u(x)u(y) \right] \\
&= \sup_{\substack{u(\cdot) \\ \sum_x [u(x)]^2 = 1}} \left[\sum_x V(x)u^2(x) - \tfrac{1}{2} \sum_{x,y} a(x, y)\big(u(x) - u(y)\big)^2 \right] \\
&= \sup_{\substack{q(\cdot) \ge 0 \\ \sum_x q(x) = 1}} \left[\sum_x V(x)u^2(x) - \tfrac{1}{2} \sum_{x,y} a(x, y)\big(\sqrt{q(x)} - \sqrt{q(y)}\big)^2 \right].
\end{aligned}$$

It is therefore not very surprising that

$$H_a(q) = \tfrac{1}{2} \sum_{x,y} a(x, y)\big(\sqrt{q(x)} - \sqrt{q(y)}\big)^2.$$

4. Small random perturbations. The exit problem. The context for large deviations is a probability distribution that is nearly degenerate. So far, this has come from the law of large numbers or the ergodic theorem. But it can also come from small random perturbations of a deterministic system.

Consider, for instance, the Brownian motion or Wiener measure on the space of continuous functions $C_0[[0, T]; R]$ that are zero at 0, that is, $x(0) = 0$. We can make the variance of the Brownian motion at time t equal to ϵt instead of t. We can consider $x_\epsilon(t) = \sqrt{\epsilon}\, x(t)$. Perhaps consider $\epsilon = \frac{1}{n}$ and $x_n(t) = \frac{1}{n}\sum_i y_i(t)$, the average of n independent Brownian motions. In any case, the measure P_ϵ of x_ϵ is nearly degenerate at the path $f(\cdot) \equiv 0$.

We have a large deviation principle for P_ϵ with a rate function

$$H(f) = \frac{1}{2}\int_0^T [f'(t)]^2\, dt.$$

There are various ways of seeing this. We will exhibit two. As a Gaussian process, the generating function for Brownian motion is

$$\log E\left[\exp\left[\int_0^T x(t)g(t)\, dt\right]\right] = \frac{1}{2}\int_0^T\int_0^T \min(s,t)g(s)g(t)\, ds\, dt.$$

Its dual is given by

$$\sup_{g(\cdot)}\left[\int_0^T f(t)g(t)\, dt - \frac{1}{2}\int_0^T\int_0^T \min(s,t)g(s)g(t)\, ds\, dt\right]$$
$$= \frac{1}{2}\int_0^T [f'(t)]^2\, dt.$$

Or, we can perturb $\sqrt{\epsilon}\, x(t)$ by $f(t)$ and change the measure from P_ϵ to Q_ϵ that concentrates near f rather that at 0. Of course, there are many measures that do this. Q_ϵ is just one such. The relative entropy is easily calculated for our choice. It is

$$H(Q_\epsilon, P_\epsilon) = \frac{1}{2\epsilon}\int_0^T [f'(t)]^2\, dt.$$

This is also a lower bound for the rate function $H(f)$. In general, the rate function has a lower bound which is the "entropy cost" of changing the measure to do what we want it to do, namely, to concentrate on the "wrong spot." The cheapest way of achieving this is invariably the actual rate function. Let us now start with an ODE in R^n,

$$dx(t) = b(x(t))\, dt, \qquad x(0) = x_0,$$

and perturb it with a small noise,

$$dx_\epsilon(t) = b(x_\epsilon(t))\, dt + \sqrt{\epsilon}\beta(t), \qquad x(0) = x_0,$$

where $\beta(t)$ is the Brownian motion in R^n. As $\epsilon \to 0$, the distribution P_ϵ of $x_\epsilon(\cdot)$ will concentrate on the unique solution $x_0(\cdot)$ of the ODE. But there will be a large deviation principle with a rate function

$$H(f) = \frac{1}{2}\int_0^T \|f'(t) - b(f(t))\|^2\,dt$$

which will happen if $\sqrt{\epsilon}\beta(\cdot)$ concentrates around $g(\cdot)$ with $g'(\cdot) = f'(\cdot) - b(f(\cdot))$.

One application of this is to the "exit problem" and the resulting interpretation of "punctuated equilibria." Let us take the ODE to be a gradient flow:

$$dx(t) = -(\nabla V)(x(t))\,dt.$$

The system will move toward a minimum of V. If V has multiple local minimum or valleys surrounded by mountains, the solutions of the ODE could be trapped near a local minimum, depending on the starting point. They will not move from one local minimum to a deeper minimum. On the other hand, with even a small noise, they develop "wanderlust." While, most of the time, they follow the path dictated by the ODE, they will, from time to time, deviate sufficiently to be able to find and reach a lower minimum. If one were to sit at a deeper minimum and wait for the path to arrive, then trace its history back, one would be apt to find that it did not wander at all, but made the most efficient beeline from the previous minimum to this one, as if it were guided by a higher power. The large deviation interpretation of this curious phenomenon is simply that the system will experiment with every conceivable path, with probabilities that are extremely small. The higher the rate function, that is, the less efficient the path, the smaller will be the probability, hence, less frequent the attempts involving that path. Therefore, the first attempt to take the path to its new location will be the most efficient path, one that climbs the lowest mountain pass to get to the new fertile land!

5. Gibbs measures and statistical mechanics. Let us look at point processes on R. The simplest is the Poisson point process. One can imagine it as starting from a finite interval $[-\frac{n}{2}, \frac{n}{2}]$ and placing $k_n = \rho n$ particles there randomly, independently and uniformly. Their joint density is

$$\frac{1}{n^{k_n}} dx_1 \cdots dx_{k_n}.$$

We can ignore the labeling and think of it as a point process P_n. The density is then

$$\frac{k_n!}{n^{k_n}} dx_1 \cdots dx_{k_n}.$$

As $n \to \infty$, we obtain a point process P which is the Poisson point process with intensity ρ. We can try to modify P_n into Q_n given by

$$dQ_n = \frac{1}{Z_n} \exp\left[-\sum_{i,j} V(x_i - x_j)\right] dP_n,$$

where $V \geq 0$ has compact support. Here, Z_n is the normalizing constant

$$Z_n = \int \exp\left[-\sum_{i,j} V(x_i - x_j)\right] dP_n.$$

The limit

$$\lim_{n\to\infty} \frac{1}{n} \log Z_n = \Psi(V)$$

is called the *free energy*. If Q is any stationary, that is, translation-invariant point process on R, its restriction to $[a, b]$ is a symmetric (i.e., permutation-invariant) probability distribution $q_{[a,b]}$ on

$$\bigcup_{k\geq 0} [a, b]^k.$$

Similarly, the Poisson point process with intensity ρ generates $p_{[a.b]}$, a convex combination of uniform distributions on $[a, b]^k$ with Poisson weights

$$e^{-\rho(b-a)} \frac{\rho^k (b-a)^k}{k!}.$$

The relative entropy $H(q_{[a,b]}, p_{[a,b]}) = H_\ell(Q, P)$ depends only on the length $\ell = (b - a)$ and is superadditive, that is,

$$H_{\ell_1+\ell_2}(Q, P) \geq H_{\ell_1}(Q, P) + H_{\ell_2}(Q, P).$$

Therefore, the limit

$$\lim_{\ell\to\infty} \frac{1}{\ell} H_\ell(Q, P) = \sup_\ell H_\ell(Q, P) = \widehat{H}(Q, P)$$

exists and is called the *specific relative entropy*. There is the expectation

$$E^Q[V] = \lim_{n\to\infty} \frac{1}{n} E^Q\left[\sum_{x_\alpha, x_\beta \in [-n/2, n/2]} V(x_\alpha - x_\beta)\right].$$

Then,

$$\Psi(V) = -\inf_Q [E^Q[V] + \widehat{H}(Q, P)]$$

over all stationary point processes and the minimizer (unique) is the limit of Q_n. Such considerations play a crucial role in the theory of equilibrium statistical mechanics and thermodynamics [16].

6. Interacting particle systems. Interacting particle systems offer an interesting area where methods of large deviation can be applied. Let us look at some examples. Consider the lattice $\mathbf{Z}_N$ of integers modulo N. We think of them as points arranged uniformly on the circle of unit circumference or the unit interval with endpoints identified. For each $x \in \mathbf{Z}_N$, there corresponds a point $\xi = \frac{x}{N}$ on the interval (really, the circle). There are a certain number of particles in the system that occupy some of the sites. The particles wait for an exponential time and then decide to jump. They pick the adjacent site either to its left or right with equal probability and then jump. In this model, the particles do not interact and, after a diffusive rescaling of space and time, the particles behave like independent Brownian motions. Since there is a large number of particles, by an application of the law of large numbers, the empirical density will be close to a solution $\rho(t, \xi)$ of the heat equation

$$\rho_t = \tfrac{1}{2}\rho_{\xi\xi},$$

with a suitable initial condition, depending on how the particles were initially placed.

We can introduce an interaction in this model, by imposing a limit of at most one particle per site. When the particle decides to jump to a randomly chosen site after a random exponential waiting time, it can jump only when the site is unoccupied. Otherwise, it must wait for a new exponential time before trying to jump again. All of the particles wait and attempt to jump independently of each other. Since we are dealing with continuous-time and exponential distributions, two particle will not try to jump at the same time and we do not have to resolve ties.

One can express all of this by simply writing down the generator. η is a configuration and $X_N = \{0, 1\}^N$is the set of all configurations. $\eta(x) = 1$ if the site x is occupied and $\eta(x) = 0$ if the site x is free. A point in X_N is just a map $\eta : \mathbf{Z}_N \to \{0, 1\}$. The generator A_N acting on functions $f : X_N \to R$ is given by

$$(A_N f)(\eta) = \tfrac{1}{2}\sum_x \eta(x)\big[(1 - \eta(x+1))\big(f(\eta^{x,x+1}) - f(\eta)\big) \\ + (1 - \eta(x-1))\big(f(\eta^{x,x-1}) - f(\eta)\big)\big],$$

where $\eta^{x,y}$ is defined by

$$\eta^{x,y}(z) = \begin{cases} \eta(x), & \text{if } z = y, \\ \eta(y), & \text{if } z = x, \\ \eta(z), & \text{otherwise.} \end{cases}$$

This system has some interesting features. The total number $\sum_x \eta(x)$ of particles does not change over time. Particles just diffuse over time. It will take time of order N^2 for the effect to be felt at a distance of order N. The equilibrium distributions $\mu_{N,k}$ are uniform over all possible $\binom{N}{k}$ configurations, where k is the number of particles. In particular, there are multiple equilibria. Such systems can be locally in equilibrium while approaching the global equilibrium rather slowly. For example, we initially place our particles in such a way that one half of the circle C_1 has particle density $\frac{1}{4}$, while the other half C_2 has density $\frac{3}{4}$. The system will locally stay near the two different equilibria at the two intervals for quite some time, with just some smoothing near the edges. If we wait for times of order N^2, that is, time $N^2 t$ with $t > 0$, the density will be close to $\rho(t, \frac{x}{N})$. A calculation that involves applying the speeded up generator $N^2 \mathcal{A}_N$ to expressions of the form

$$\frac{1}{N} \sum J\left(\frac{x}{N}\right) \eta(x)$$

shows that $\rho(t, \xi)$ will be the solution of the heat equation

$$\rho_t(t, \xi) = \tfrac{1}{2} \rho_{\xi\xi}(t, \xi); \qquad \rho(0, \xi) = \tfrac{1}{4} \mathbf{1}_{C_1}(\xi) + \tfrac{3}{4} \mathbf{1}_{C_2}(\xi).$$

In this case, this behavior is the same as if the particles moved independently without any interaction. While this is true at the level of the law of large numbers, we shall see that for large deviation rates, the interaction does matter.

7. Superexponential estimates. Let us consider the process in equilibrium, with $k = \rho N$ particles. We let N, k go to ∞ while ρ is kept fixed. It is easy to see, for any fixed site x, that

$$\mu_{N,k}[\eta(x) = 1] \to \rho$$

and $\{\eta(x_j)\}$ become independent for any finite set of distinct sites. If $f(\eta)$ is a local function and $f_x(\eta) = f(\tau_x \eta)$ its translation, then

$$\frac{1}{N} \sum_x f_x(\eta) \to E^{P_\rho}[f(\eta)]$$

as $N \to \infty$. The probabilities of deviations decay exponentially fast. The exponential rate is given by

$$H(a) = \inf_{Q \in \mathcal{M}(a,\rho)} H(Q, P_\rho),$$

where $\mathcal{M}(a, \rho)$ consists of stationary measures Q with density ρ and $E^Q[f(\eta)] = a$, and P_ρ is Bernoulli with density ρ. $H(Q, P)$ is the specific entropy, calculated as the limit

$$\lim_{n \to \infty} \frac{1}{n} H(Q_n, P_n),$$

where Q_n and P_n are restrictions of Q and P to a block of n sites. This is, of course, an equilibrium calculation done at a fixed time. What about space–time averages

$$\frac{1}{NT} \int_0^T \sum_x f_x(\eta(t))\, dt?$$

The rate function is now some $H_T(a)$ that depends on T. While this is hard to calculate, one can show that

$$\lim_{T\to\infty} H_T(a) = \inf_{\rho(\cdot)\in\mathcal{P}(a,\rho)} H(P_{\rho(\xi)}, P_\rho)\, d\xi,$$

where $\mathcal{P}(a,\rho)$ consists of density profiles $\rho(\xi)$ that satisfy

$$\int \rho(\xi)\, d\xi = \rho \quad \text{and} \quad \int E^{P_{\rho(\xi)}}[f(\eta)]\, d\xi = a.$$

Note that the rate increases to a finite limit as $T \to \infty$. The reason is that large-scale fluctuations in density can occur with exponentially small probability and these fluctuations do not necessarily decay when T is large. Particles diffuse slowly and it takes times of order N^2 to diffuse through order N sites.

To resolve this difficulty, we define the approximate large-scale empirical density by

$$r_{N,\epsilon}(\xi,\eta) = \frac{1}{N\epsilon} \sum_{x:|x/N-\xi|\le\epsilon} \eta(x)$$

and compare

$$\frac{1}{N^3}\int_0^{N^2T} \sum_x f_x(\eta(t))\, dt = Y_{N,f}(T)$$

with

$$\frac{1}{N^3}\int_0^{N^2T} \sum_x \widehat{f}\Big(r_{N,\epsilon}\Big(\frac{x}{N}, \eta(t)\Big)\Big)\, dt = \widehat{Y}_{N,f,\epsilon}(T),$$

where

$$\widehat{f}(\rho) = E^{P_\rho}[f(\eta)].$$

One can show that

$$\limsup_{\epsilon\to 0}\limsup_{N\to\infty} \frac{1}{N}\log P[|Y_{N,f}(T) - \widehat{Y}_{N,f,\epsilon}(T)| \ge \delta] = -\infty$$

for any $\delta > 0$. In other words, in time scale N^2, the large scale density fluctuations are responsible for all large deviations with a logarithmic decay rate of N. Such estimates, called *superexponential estimates*, play a crucial role in the study of interacting particle systems where approach to equilibrium is slow in large scales.

8. Hydrodynamical limits. If we consider a large system with multiple equilibria, the system can remain in a local equilibrium for a long time. Let the different equilibria be labeled by parameter ρ. If the system lives on a large spatial domain and if ξ denotes a point in the macroscopic scale of space, a function $\rho(\xi)$ could describe a system that is locally in equilibrium, albeit at different ones at macroscopically different points. If t is time measured in a suitably chosen faster scale, then $\rho(\xi) = \rho(t, \xi)$ may evolve gently in time and converge as $t \to \infty$ to a constant ρ that identifies the global equilibrium.

For instance, in our example, let us initially start with $k = N\rho$ particles and distribute them in such a way that we achieve a density profile $\rho_0(\xi)$ as $N \to \infty$. Technically, this means

$$\lim_{N\to\infty} \frac{1}{N} \sum_x J\left(\frac{x}{N}\right)\eta(x) = \int J(\xi)\rho_0(\xi)\, d\xi$$

for bounded continuous test functions J. What will we see after time tN^2, especially when N is large?

The answer is very easy. Consider the sum

$$f(\eta) = \frac{1}{N} \sum_x J\left(\frac{x}{N}\right)\eta(x)$$

and apply the speeded up operator $N^2 A_N$ to it.

$$\begin{aligned}
f(\eta^{x,y}) - f(\eta) &= \frac{1}{N}[\eta(x) - \eta(y)]\left[J\left(\frac{y}{N}\right) - J\left(\frac{x}{N}\right)\right], \\
(N^2 A_N f)(\eta) &= \frac{N}{2} \sum_x \eta(x)(1 - \eta(x+1))\left[J\left(\frac{x+1}{N}\right) - J\left(\frac{x}{N}\right)\right] \\
&\quad + \eta(x)(1 - \eta(x-1))\left[J\left(\frac{x-1}{N}\right) - J\left(\frac{x}{N}\right)\right] \\
&= \frac{N}{2} \sum_x [\eta(x)(1 - \eta(x+1)) - \eta(x+1)(1 - \eta(x))] \\
&\quad \times \left(J\left(\frac{x+1}{N}\right) - J\left(\frac{x}{N}\right)\right) \\
&= \frac{N}{2} \sum_x [\eta(x) - \eta(x+1)]\left[J\left(\frac{x+1}{N}\right) - J\left(\frac{x}{N}\right)\right] \\
&= \frac{N}{2} \sum_x \left[J\left(\frac{x+1}{N}\right) + J\left(\frac{x-1}{N}\right) - 2J\left(\frac{x}{N}\right)\right]\eta(x) \\
&\simeq \frac{1}{2N} \sum_x J''\left(\frac{x}{N}\right)\eta(x),
\end{aligned}$$

leading to the limiting heat equation

$$\rho_t = \tfrac{1}{2}\rho_{\xi\xi}, \qquad \rho(0, \xi) = \rho_0(\xi).$$

Let us change the problem slightly. Introduce a slight bias. The probabilities to right and left are $\frac{1}{2} \pm \frac{1}{2N} b$, where $b > 0$ is the bias to the right. This introduces an extra term, so that now

$$N^2 A_N \frac{1}{N}\sum_x J\Big(\frac{x}{N}\Big)\eta(x) = \frac{1}{2N}\sum_x J''\Big(\frac{x}{N}\Big)\eta(x) + F_N(\eta),$$

where

$$F_N(\eta) = \frac{b}{2N}\sum J'\Big(\frac{x}{N}\Big)[\eta(x)(1-\eta(x+1)) + \eta(x)(1-\eta(x-1))].$$

This is a problem because F_N is nonlinear in η and cannot be replaced by a simple expression involving ρ. If we were locally in equilibrium, F_N would be replaced by

$$G_{N,\epsilon}(\eta) = \frac{b}{N}\sum J'\Big(\frac{x}{N}\Big) r_{N,\epsilon}\Big(\frac{x}{N},\eta\Big)\Big(1 - r_{N,\epsilon}\Big(\frac{x}{N},\eta\Big)\Big),$$

where $r_{N,\epsilon}(\frac{x}{N},\eta)$ is the empirical density. If we were in a global equilibrium, in the faster time scale,

$$\Big|\int_0^T F_N(\eta(t)) - G_{N,\epsilon}(\eta(t))\,dt\Big| \geq \delta,$$

with superexponentially small probability. The importance of superexponential estimates lies in the following elementary, but universal, inequality:

$$Q(A) \leq \frac{2 + H(Q,P)}{\log(1/P(A))}.$$

If $P(A)$ is superexponentially small and $H(Q,P)$ is linear in N, then $Q(A)$ is small. Perturbation of the process by a slight bias of order $\frac{1}{N}$ produces a relative entropy of order $\frac{1}{N^2}$ per site per unit time. With N sites and time of order N^2, this is still only linear in N. Changes in initial conditions can also only contribute linear relative entropy. So, we can carry out the approximation and end up with

$$\rho_t = \tfrac{1}{2}\rho_{\xi\xi} - [b\rho(1-\rho)]_\xi.$$

Note that without exclusion, the limit would have been

$$\rho_t = \tfrac{1}{2}\rho_{\xi\xi} - [b\rho]_\xi,$$

which is just the CLT when there is a small mean.

9. Large deviations in hydrodynamical limits. We saw earlier that in our model, if we started with an initial profile with density ρ_0, in a speeded up time scale, the system will evolve with a time-dependent profile which is the solution of the heat equation

$$\rho_t = \tfrac{1}{2}\rho_{\xi\xi}.$$

This is, of course, true with probability nearly 1. We can make the initial condition deterministic in order that no initial deviation is possible. Still, our random evolution could produce, with small probability, strange behavior. If we limit ourselves to deviations with only exponentially small probabilities, what deviations are possible and what are their exponential rates?

We can doctor the system with a bias b which is not constant, but is a function $b(t, \frac{x}{N})$ of x and fast time t with $b(t, \xi)$ being a nice function of t and ξ. This will produce a solution of

$$\rho_t(t, \xi) = \tfrac{1}{2} D_\xi^2 \rho(t, \xi) - D_\xi\big[b(t, \xi)\rho(t, \xi)(1 - \rho(t, \xi))\big], \qquad \rho(0, \xi) = \rho_0(\xi).$$

This can be done with an entropy cost that can be calculated as

$$\Psi(b) = \tfrac{1}{2} \int_0^T \int |b(t, \xi)|^2 \rho(t, \xi)(1 - \rho(t, \xi))\, dt\, d\xi,$$

for the duration $[0, T]$. If $\rho(t, \xi)$ is given, then we can minimize $\Psi(b)$ over all compatible b. This is a lower bound for the rate. One can match this bound in the other direction.

10. Large deviations for random walks in random environments. We will start with a probability space (Ω, Σ, P) on which $\mathbf{Z}^d$ acts ergodically as a family τ_z of measure-preserving transformations. We are given $\pi(\omega, z)$, which is a probability distribution on $\mathbf{Z}^d$ for each ω, and is a measurable functions of ω for each z. One can then generate random transition probabilities $\pi(\omega, z', z)$ by defining

$$\pi(\omega, z', z' + z) = \pi(\tau_{z'}\omega, z).$$

For each ω, $\pi(\omega, z', z)$ can serve as the transition probability of a Markov process on $\mathbf{Z}^d$, and the measure corresponding to this process, starting from 0, is denoted by Q^ω. This is, of course, random (it depends on ω and is called the *random walk in the random environment* ω). One can ask the usual questions about this random walk and, in some form, they may be true for almost all ω with respect to P. The law of large numbers, if valid, will take the form

$$P\left[\omega \colon \lim_{n\to\infty} \frac{S_n}{n} = m(P) \text{ a.e. } Q^\omega\right] = 1.$$

Such statement concerning the almost sure behavior under Q^ω for almost all ω with respect to P are said to deal with the “quenched” version. Sometimes one wishes to study the behavior of the “averaged” or “annealed” measure,

$$\overline{Q} = \int Q^\omega P(d\omega).$$

The law of large numbers is the same because it is equivalent to

$$\overline{Q}\left[\omega \colon \lim_{n\to\infty} \frac{S_n}{n} = m(P)\right] = 1.$$

On the other hand, questions on the asymptotic behavior of probabilities, like the central limit theorem or large deviations, could be different for the quenched and the averaged cases.

A special environment, which is called the *product environment*, is one in which $\pi(\omega, z', z'+z)$ are independent for different z' and have a common distribution β which is a probability measure on the space $\mathcal{M}$ of all probability measures on $\mathbf{Z}^d$. In this case, the canonical choice for (Ω, Σ, P) is the countable product of $\mathcal{M}$ and the product measure P with marginals β.

There are large deviation results regarding the limits

$$\lim_{n\to\infty} \frac{1}{n} \log Q^{\omega}\left[\frac{S_n}{n} \simeq a\right] = I(a)$$

and

$$\lim_{n\to\infty} \frac{1}{n} \log \overline{Q}\left[\frac{S_n}{n} \simeq a\right] = \overline{I}(a).$$

The difference between I and $\overline{I}$ has a natural explanation. In the special case of Z, in terms of the large deviation behavior in the one-dimensional random environment, it is related to the following question. If, at time n, we see a particle reach an improbable value na, what does it say about the environment in $[0, na]$? Did the particle behave strangely in the environment or did it encounter a strange environment? It is probably a combination of both.

Large deviation results exists, however, in much wider generality, both in the quenched and the averaged cases. The large deviation principle is essentially the existence of the limits

$$\lim_{n\to\infty} \frac{1}{n} \log E[\exp[\langle\theta, S_n\rangle]] = \Psi(\theta).$$

The expectation is with respect to Q^{ω} or $\overline{Q}$, which could produce different limits for Ψ. The law of large numbers and central limit theorem involve the differentiability of Ψ at $\theta = 0$, which is harder. In fact, large deviation results have been proven by general subadditivity arguments for the quenched case. Roughly speaking, fixing ω, we have

$$Q^{\omega}[S_{k+\ell} \simeq (k+\ell)a] \geq Q^{\omega}[S_k \simeq ka] \times Q^{\tau_{ka}\omega}[S_\ell \simeq \ell a].$$

It is harder for the averaged case. We want to prove that the limits

$$\lim_{n\to\infty} \frac{1}{n} \log \overline{Q}\left[\frac{S_n}{n} \simeq a\right] = -\overline{I}(a)$$

or, equivalently,

$$\lim_{n\to\infty} \frac{1}{n} \log E^{\overline{Q}}[\exp[\langle\theta, S_n\rangle]] = \overline{\Psi}(\theta),$$

exist. The problem is that the measure $\overline{Q}$ is not very nice. As the random walk explores $\mathbf{Z}^d$, it learns about the environment and in the case of the product environment, when it returns to a site that it has visited before, the experience has not been forgotten and leads to long term correlations. However, if we are interested in the behavior $S_n \simeq na$ with $a \neq 0$, the same site is not visited too often and the correlations should rapidly decay.

One can use Bayes' rule to calculate the conditional distribution

$$\overline{Q}[S_{n+1} = S_n + z | S_1, S_2, \ldots, S_n] = q(z|w),$$

where w is the past history of the walk. Before we do that, it is more convenient to shift the origin as we go along so that the current position of the random walk is always the origin and the current time is always 0. An n step walk then looks like $w = \{S_0 = 0, S_{-1}, \ldots, S_{-n}\}$. We pick a z with probability $q(z|w)$. We obtain a new walk of $n+1$ steps $w' = \{S'_0 = 0, S'_{-1}, \ldots, S'_{-(n+1)}\}$ given by $S'_{-(k+1)} = S_{-k} - z$ for $k \geq 0$. We can now calculate $q(z|w)$. We need to know all the quantities $\{k(w, x, z)\}$, the numbers of times the walk has visited x in the past and jumped from x to $x + z$. It is not hard to see that the a posteriori probability can be calculated as

$$q(z|w) = \frac{\int \pi(z) \Pi_{z'} \pi(z')^{k(w,0,z')} \beta(d\pi)}{\int \Pi_{z'} \pi(z')^{k(w,0,z')} \beta(d\pi)}.$$

While this initially makes sense only for walks of finite length, it can clearly be extended to all transient paths. Note that although we only use $k(w, 0, z')$, in order to obtain the new $k(w', 0, z')$, we would need to know the collection $\{k(w, z, z')\}$.

Now, suppose that R is a process with stationary increments $\{z_j\}$. Then, we can again make the current position the origin and if the process is transient, as it would be if the increments were ergodic and had a nonzero mean a, the conditional probabilities $q(z|w)$ would exist a.e. R and can be compared to the corresponding conditional probabilities $r(z|w)$ under R. The relative entropy

$$H(R) = E^R\left[\sum_z r(z|w) \log \frac{r(z|w)}{q(z|w)}\right]$$

is then well defined.

The function

$$\overline{I}(a) = \inf_{\substack{R: \int z_1 dR = a \\ R \text{ ergodic}}} H(R),$$

defined for $a \neq 0$, extends as a convex function to all of R^d and, with this $\overline{I}$ as rate function, a large deviation result holds.

We now turn to the quenched case. Although a proof using the subadditive ergodic theorem exists, we will provide an alternate approach that is more appealing.

We will illustrate this in the context of Brownian motion with a random drift. Instead of the action of $\mathbf{Z}^d$, we can have $\mathbf{R}^d$ acting on (Ω, Σ, P) ergodically and consider a diffusion on $\mathbf{R}^d$ with a random infinitesimal generator

$$(\mathcal{L}^\omega u)(x) = \tfrac{1}{2}(\Delta u)(x) + \langle b(\omega, x), (\nabla u)(x)\rangle$$

acting on smooth functions on $\mathbf{R}^d$. Here, $b(\omega, x)$ is generated from a map $b(\omega) : \Omega \to \mathbf{R}^d$ by the action of $\{\tau_x : x \in \mathbf{R}^d\}$ on ω:

$$b(\omega, x) = b(\tau_x \omega).$$

Again, there is the quenched measure Q^ω that corresponds to the diffusion with generator $\mathcal{L}^\omega$ that starts from 0 at time 0, and the averaged measure that is given by a similar formula. This model is referred to as *diffusion with a random drift*. Exactly the same questions can be asked in this context. We can define a diffusion on Ω with generator

$$\mathcal{L} = \tfrac{1}{2}\Delta + \langle b(\omega), \nabla\rangle,$$

where $\nabla = \{D_i\}$ are the generators of the translation group $\{\tau_x : x \in \mathbf{R}^d\}$. This is essentially the image of lifting the paths $x(t)$ of the diffusion on $\mathbf{R}^d$ corresponding to $\mathcal{L}^\omega$ to Ω by

$$\omega(t) = \tau_{x(t)}\omega.$$

While there is no possibility of having an invariant probability measure on $\mathbf{R}^d$, on Ω, one can hope to find an invariant probability density $\phi(\omega)$, that is, to find $\phi(\omega) \geq 0$ in $L_1(P)$ with $\int \phi \, dP = 1$ which solves

$$\tfrac{1}{2}\Delta\phi = \nabla \cdot (b\phi).$$

If such a ϕ exists, then we have an ergodic theorem for the diffusion process Q^ω corresponding to $\mathcal{L}$ on Ω,

$$\lim_{t\to\infty} \frac{1}{t}\int_0^t f(\omega(s))\, ds = \int f(\omega)\phi(\omega)\, dP \qquad \text{a.e. } P.$$

This also translates to an ergodic theorem on R^d. If we define the stationary process g by $g(\omega, x) = f(\tau_x\omega)$, then

$$\lim_{t\to\infty} \frac{1}{t}\int_0^t f(\omega, x(s))\, ds = \int f(\omega)\phi(\omega)\, dP \qquad \text{a.e. } Q^\omega, \text{ a.e. } P,$$

where Q^ω is now the quenched process in the random environment. Since

$$x(t) = \int_0^t b(\omega, x(s))\, ds + \beta(t),$$

where $\beta(\cdot)$ is the Brownian motion, it is clear that

$$\lim_{t\to\infty} \frac{x(t)}{t} = \int b(\omega)\phi(\omega)\, dP \qquad \text{a.e. } Q^\omega, \text{ a.e. } P,$$

providing a law of large numbers for $x(t)$. While we cannot be sure of finding ϕ for a given b, it is easy to find a b for a given ϕ. For instance, if $\phi > 0$, we could take $b = \frac{\nabla\phi}{2\phi}$. Or, more generally, $b = \frac{\nabla\phi}{2\phi} + \frac{c}{\phi}$ with $\nabla \cdot c = 0$. If we change b to b' which satisfies $\frac{1}{2}\Delta\phi = \nabla \cdot (b'\phi)$, the new process $Q^{b',\omega}$ with drift b' will, in the time interval $[0, t]$, have relative entropy

$$E^{Q^{b',\omega}}\left[\frac{1}{2}\int_0^t \|b(\omega(s)) - b'(\omega(s))\|^2\,ds\right]$$

and, by the ergodic theorem, one can see that, a.e. with respect to P,

$$\lim_{t\to\infty} \frac{1}{t} E^{Q^{b',\omega}}\left[\frac{1}{2}\int_0^t \|b(\omega(s)) - b'(\omega(s))\|^2\,ds\right] = \frac{1}{2}\int \|b(\omega) - b'(\omega)\|^2 \phi(\omega)\,dP.$$

Moreover, for almost all ω with respect to P, almost surely with respect to $Q^{b',\omega}$,

$$\lim_{t\to\infty} \frac{x(t)}{t} = \int b'(\omega)\phi(\omega)\,dP.$$

If we fix $\int b'(\omega)\phi(\omega)\,dP = a$, the bound

$$\liminf_{t\to\infty} \frac{1}{t} \log Q^{\omega}\left[\frac{x(t)}{t} \simeq a\right] \geq -\frac{1}{2}\int \|b - b'\|^2 \phi\,dP$$

is easily obtained. If we define

$$I(a) = \inf_{\substack{b',\phi \\ (1/2)\Delta\phi = \nabla\cdot(b'\phi) \\ \int b'\phi\,dP = a}} \frac{1}{2}\int \|b - b'\|^2 \phi\,dP,$$

then

$$\liminf_{t\to\infty} \frac{1}{t} \log Q^{\omega}\left[\frac{x(t)}{t} \simeq a\right] \geq -I(a).$$

Of course, these statements are valid a.e. P. One can check that I is convex and that the upper bound amounts to proving the dual estimate

$$\lim_{t\to\infty} \frac{1}{t} \log E^{Q^{\omega}}[e^{\langle\theta, x(t)\rangle}] \leq \Psi(\theta),$$

where

$$\Psi(\theta) = \sup_a [\langle a, \theta\rangle - I(a)].$$

We need a bound on the solution of

$$u_t = \tfrac{1}{2}\Delta u + \langle b, \nabla u\rangle$$

with $u(0) = \exp[\langle\theta, x\rangle]$. By the Hopf–Cole transformation $v = \log u$, this reduces to estimating

$$v_t = \tfrac{1}{2}\Delta v + \tfrac{1}{2}\|\nabla v\|^2 + \langle b, \nabla v\rangle$$

with $v(0) = \langle \theta, x \rangle$. This can be done if we can construct a subsolution

$$\tfrac{1}{2}\nabla \cdot w + \tfrac{1}{2}\|\nabla w\|^2 + \langle b, w \rangle \le \psi(\theta)$$

on Ω, where $w : \Omega \to \mathbf{R}^d$ satisfies $\int w\, dP = \theta$ and $\nabla \times w = 0$ in the sense that $D_i w_j = D_j w_i$. The existence of the subsolution comes from convex analysis.

$$\begin{aligned}
\psi(\theta) = & \sup_{\substack{b',\phi \\ (1/2)\Delta\phi = \nabla\cdot(b'\phi)}} \Big[\int \langle b', \theta\rangle \phi\, dP - \tfrac{1}{2}\int \|b - b'\|^2 \phi\, dP\Big] \\
&= \sup_{\phi}\sup_{b'}\inf_{u}\Big[\int \langle b', \theta\rangle \phi\, dP - \tfrac{1}{2}\int \|b - b'\|^2 \phi\, dP \\
&\qquad + \tfrac{1}{2}[\Delta u + \langle b', \nabla u\rangle]\phi\, dP\Big] \\
&= \sup_{\phi}\inf_{u}\sup_{b'}\Big[\int \langle b', \theta\rangle \phi\, dP - \tfrac{1}{2}\int \|b - b'\|^2 \phi\, dP \\
&\qquad + \int \tfrac{1}{2}[\Delta u + \langle b', \nabla u\rangle]\phi\, dP\Big] \\
&= \sup_{\phi}\inf_{u}\int \sup_{b'}\Big[\langle b', \theta + \nabla \cdot u\rangle \phi\, dP - \tfrac{1}{2}\int \|b - b'\|^2 \phi\, dP + \int \tfrac{1}{2}\Delta u \phi\, dP\Big] \\
&= \sup_{\phi}\inf_{u}\int [\tfrac{1}{2}\Delta u + \langle b, \theta + \nabla u\rangle + \tfrac{1}{2}\|\theta + \nabla u\|^2]\phi\, dP \\
&= \sup_{\phi} \inf_{\substack{\nabla\times w = 0 \\ \int w\, dP = \theta}} \int [\tfrac{1}{2}\nabla \cdot w + \langle b, w\rangle + \tfrac{1}{2}\|w\|^2]\phi\, dP \\
&= \inf_{\substack{\nabla\times w = 0 \\ \int w\, dP = \theta}} \sup_{\phi} \int [\tfrac{1}{2}\nabla \cdot w + \langle b, w\rangle + \tfrac{1}{2}\|w\|^2]\phi\, dP \\
&= \inf_{\substack{\nabla\times w = 0 \\ \int w\, dP = \theta}} \sup_{\omega} [\tfrac{1}{2}\nabla \cdot w + \langle b, w\rangle + \tfrac{1}{2}\|w\|^2],
\end{aligned}$$

which proves the existence of a subsolution. One needs to justify the free interchange of sup and inf. In passing from the first line to the second, the restriction on b' and ϕ is replaced by the Lagrange multiplier u.

This can be viewed as showing the existence of a limit as $\epsilon \to 0$ (homogenization) of the solution of

$$u_t^\epsilon = \frac{\epsilon}{2}\Delta u^\epsilon + \frac{1}{2}\|\nabla u^\epsilon\|^2 + \Big\langle b\Big(\frac{x}{\epsilon}, \omega\Big), \nabla u^\epsilon \Big\rangle$$

with $u^\epsilon(0,x) = f(x)$. The limit satisfies

$$u_t = \Psi(\nabla u)$$

with $u(0,x) = f(x)$.

11. Homogenization of Hamilton–Jacobi–Bellman equations. This can be generalized to equations of the form

$$u_t^\epsilon = \frac{\epsilon}{2}\Delta u^\epsilon + H\left(\frac{x}{\epsilon}, \nabla u^\epsilon, \omega\right), \qquad u(0,x) = f(x),$$

where $H(x,p,\omega) = H(\tau_x\omega, p)$ is a stationary process of convex functions in p. By the changes of variables $x = \epsilon y$, $u = \epsilon v$, $t = \epsilon\tau$, this reduces to the behavior of $\epsilon v^\epsilon(\frac{T}{\epsilon}, \frac{x}{\epsilon})$, where v^ϵ solves

$$v_\tau^\epsilon = \tfrac{1}{2}\Delta v^\epsilon + H(y, \nabla v^\epsilon, \omega), \qquad v^\epsilon(0,y) = \epsilon^{-1} f(\epsilon y).$$

From the principle of dynamic programming, we can represent the solution as the supremum of a family of solutions of linear equations,

$$v^\epsilon(\tau, y) = \sup_{b(\cdot,\cdot)\in\mathcal{B}} w^\epsilon\big(b(\cdot,\cdot), T-\tau, y\big),$$

where w^ϵ solves

$$w_\tau^\epsilon + \tfrac{1}{2}\Delta w^\epsilon + \langle b(\tau,y), \nabla w^\epsilon\rangle - L(\tau_y\omega, b(\tau,y)) = 0,$$
$$w^\epsilon(\epsilon^{-1}T, y) = \epsilon^{-1} f(\epsilon y),$$

L being the convex dual of H with respect to the variable p. If b is chosen as $b(\tau_y\omega)$, for suitable choices of $b(\omega) \in \mathfrak{C}$ that admit positive integrable solutions ϕ to

$$\tfrac{1}{2}\Delta\phi(\tau_x\omega) = \nabla\cdot b(\tau_x\omega)\phi(\tau_x,\omega),$$

then it is not hard to see that

$$\epsilon w^\epsilon(0,0) \to f\left(T\int b(\omega)\phi(\omega)\,dP\right) - T\int L(b(\omega),\omega)\phi(\omega)\,dP.$$

This provides a lower bound

$$\liminf_{\epsilon\to 0} u^\epsilon(T,0) \geq \sup_{b\in\mathfrak{C}}\left[f\left(T\int b(\omega)\phi(\omega)\,dP\right) - T\int L(b(\omega),\omega)\phi(\omega)\,dP\right]$$

which can be shown to be an upper bound as well.

12. History and references. The origin of large deviation theory goes back to Scandinavian actuaries [10] who were interested in the analysis of risk in the insurance industry. For sums of independent random variables, the general large deviations result was established by Cramér in [1]. The result for empirical distributions of independent identically distributed random variables is due to Sanov [18]. The generalization to Markov chains and processes can be found in several papers of Donsker and Varadhan [3–6] and Gärtner [13]. The results concerning small random perturbations of deterministic systems goes back to the work of Varadhan [20], as well as Vencel and Freidlin [12]. Several monographs have appeared on the subject. Lecture notes by Varadhan [21], texts by Deuschel and Stroock [7], Dembo and Zeitouni [2], Schwartz and Weiss [19], Ellis [9], Dupuis and Ellis [8] and, most recently, by Feng and Kurtz [11]. They cover a wide spectrum of topics in large deviation theory. For large deviations in the context of hydrodynamic scaling, there is the text by Kipnis and Landim [14], as well as an exposition by Varadhan [23]. As for large deviations for random walks in a random environment, see [24], as well as references in Zeitouni's article [25]. For a general survey on large deviations and entropy, see [22]. The results on homogenization of random Hamilton–Jacobi–Bellman equations and its application to large deviations has appeared in [15] and [17]. Undoubtedly, there are many more references. A recent Google search on "Large Deviations" produced 3.4 million hits.

REFERENCES

[1] CRAMÈR, H. (1938). Sur un nouveau théorème-limite de la théorie des probabilités. *Actualités Scientifiques et Industrialles* **736** 5–23. *Colloque Consecré à la Théorie des Probabilités* **3**. Hermann, Paris.

[2] DEMBO, A. and ZEITOUNI, O. (1998). *Large Deviations Techniques and Applications*, 2nd ed. Springer, New York. MR1619036

[3] DONSKER, M. D. and VARADHAN, S. R. S. (1975). Asymptotic evaluation of certain Markov process expectations for large time. I. *Comm. Pure Appl. Math.* **28** 1–47. MR0386024

[4] DONSKER, M. D. and VARADHAN, S. R. S. (1975). Asymptotic evaluation of certain Markov process expectations for large time. II. *Comm. Pure Appl. Math.* **28** 279–301. MR0386024

[5] DONSKER, M. D. and VARADHAN, S. R. S. (1976). Asymptotic evaluation of certain Markov process expectations for large time. III. *Comm. Pure Appl. Math.* **29** 389–461. MR0428471

[6] DONSKER, M. D. and VARADHAN, S. R. S. (1983). Asymptotic evaluation of certain Markov process expectations for large time. IV. *Comm. Pure Appl. Math.* **36** 183–212. MR0690656

[7] DEUSCHEL, J. D. and STROOCK, D. W. (1989). *Large Deviations*. Academic Press, Boston. MR0997938

[8] DUPUIS, P. and ELLIS, R. S. (1997). *A Weak Convergence Approach to the Theory of Large Deviations*. Wiley, New York. MR1431744

[9] ELLIS, R. S. (2006). *Entropy, Large Deviations, and Statistical Mechanics*. Springer, Berlin. MR2189669

[10] ESSCHER, F. (1932). On the probability function in the collective theory of risk. *Skandinavisk Aktuarietidskrift* **15** 175–195.

[11] FENG, J. and KURTZ, T. G. (2006). *Large Deviations for Stochastic Processes*. Amer. Math. Soc., Providence, RI. MR2260560

[12] FREIDLIN, M. I. and WENTZELL, A. D. (1998). *Random Perturbations of Dynamical Systems*, 2nd ed. Springer, New York. MR1652127

[13] GÄRTNER, J. (1977). On large deviations from an invariant measure (in Russian). *Teor. Verojatnost. i Primenen.* **22** 27–42. MR0471040

[14] KIPNIS, C. and LANDIM, C. (1999). *Scaling Limits of Interacting Particle Systems*. Springer, Berlin. MR1707314

[15] KOSYGINA, E., REZAKHANLOU, F. and VARADHAN, S. R. S. (2006). Stochastic homogenization of Hamilton–Jacobi–Bellman equations. *Comm. Pure Appl. Math.* **59** 1489–1521. MR2248897

[16] LANFORD, O. E. (1973). Entropy and equilibrium states in classical statistical mechanics. In *Statistical Mechanics and Mathematical Problems* (A. Lenard, ed.) 1–113. *Lecture Notes in Phys.* **20**. Springer, Berlin.

[17] LIONS, P. L. and SOUGANIDIS, P. E. (2005). Homogenization of "viscous" Hamilton–Jacobi equations in stationary ergodic media. *Comm. Partial Differential Equations* **30** 335–375. MR2131058

[18] SANOV, I. N. (1957). On the probability of large deviations of random magnitudes (in Russian). *Mat. Sb. N. S.* **42 (84)** 11–44. MR0088087

[19] SHWARTZ, A. and WEISS, A. (1995). *Large Deviations for Performance Analysis*. Chapman and Hall, London. MR1335456

[20] VARADHAN, S. R. S. (1967). Diffusion processes in a small time interval. *Comm. Pure Appl. Math.* **20** 659–685. MR0217881

[21] VARADHAN, S. R. S. (1984). *Large Deviations and Applications*. SIAM, Philadelphia. MR0758258

[22] VARADHAN, S. R. S. (2003). Large deviations and entropy. In *Entropy. Princeton Ser. Appl. Math.* 199–214. Princeton Univ. Press. MR2035822

[23] VARADHAN, S. R. S. (1996). The complex story of simple exclusion. In *Itô's Stochastic Calculus and Probability Theory* 385–400. Springer, Tokyo. MR1439538

[24] VARADHAN, S. R. S. (2003). Large deviations for random walks in a random environment. Dedicated to the memory of Jürgen K. Moser. *Comm. Pure Appl. Math.* **56** 1222–1245. MR1989232

[25] ZEITOUNI, O. (2002). Random walks in random environments. In *Proceedings of the International Congress of Mathematicians* **III** (*Beijing, 2002*) 117–127. Higher Ed. Press, Beijing. MR1957524

COURANT INSTITUTE OF MATHEMATICAL SCIENCES
NEW YORK UNIVERSITY
NEW YORK, NEW YORK 10012
USA
E-MAIL: varadhan@cims.nyu.edu

Contemporary Mathematics
Volume **490**, 2009, **3-10**

The Role of Weak Convergence in Probability Theory

S.R.S.Varadhan

1. Introduction

The concept of limit theorems in probability goes way back. The first limit theorem, the weak law of large numbers was proved by Jacob Bernoulli [**1**] in the early eighteenth century. This was quickly followed by De Moivre [**3**] who proved the central limit theorem establishing the approximation of the binomial distribution by a normal distribution. Further work by, Laplace, Gauss, Levy, Khintichine, Kolmogorov, Gnedenko and others put limit theorems at the center of probability theory [**11**].

The connection between random walks and Brownian motion was understood by Levy and others along with the idea that distributions of quantities like the maximum etc, based on random walks, converges to the corresponding distributions derived from Brownian motion. Doob [**6**] formulated this more precisely in his paper on 'Heuristic approach to the Kolmogorov-Smirnov Theorems'. Donsker [**5**], in his thesis, established the first general theorem to the effect that Doob's heuristic proof can in fact be justified. However his approach was too dependent on finite dimensional approximations.

At this point the study of stochastic processes as probability distributions on function spaces began. Contributions were made by LeCam [**13**] in the United States, Kolmogorov [**12**], Prohorov [**15**], Skorohod [**17**] and others in USSR, as well as Varadarajan [**20**] in India. Alexandrov in the 1940's had studied set functions on topological spaces and now powerful techniques from functional analysis could be used to study stochastic processes as measures on function spaces. A random walk or any stochastic process induces a probability distribution on the space of paths. By interpolation or some such simple device both the approximating and the limiting distributions can be put on the same space of paths. The question then reduces to the investigation of the convergence of a sequence μ_n of probability measures on a space X of paths to a limit μ. It is clear that the measures μ_n, in the case of random walks, look qualitatively different from Brownian paths and hence $\mu_n \perp \mu$. It is not going to be true that $\mu_n(A) \to \mu(A)$ for all measurable sets $A \in X$.

Functional analysis now provides a useful window. The space X of paths comes with a topology. A probability measure μ defines a normalized non-negative linear functional

$$\Lambda_\mu(f) = \int f(x)\mu(dx)$$

on $\mathbf{B} = C(X)$ and they form a convex subset $\mathcal{M}$ in the dual $\mathbf{B}^*$ of $\mathbf{B}$. The natural weak$*$ topolgy on this subset is called 'weak convergence' by probabilists, and has proved to be a very useful tool. The following questions arise naturally and were investigated.

1. Which linear functionals $\Lambda(f)$ have the above representation as Λ_μ for some probability measure μ, i.e. when does $\lambda \in \mathcal{M}$ and when is μ uniquely determined by Λ?

2. What can one say about the space $\mathcal{M}$ as a topological space?

3. Is it metrizable? What are its compact subsets?

Viewed in this manner the classical limit theorems tell us that if we place random walks as well as Brownian Motion on the same space $X = C[0,T]$ of continuous paths, then the only possible limit for the measures μ_n coming from random walks is the Wiener measure μ. Weak convergence as elements in $C(X)^*$ would provide justification for the convergence of distributions of continuous functionals on X under μ_n to the corresponding distribution under μ. The issue then is that of compactness.

Prohorov, Skorohod and others in the Russian school worked mostly under the assumption that X is a complete separable metric space. They obtained characterizations of compact subsets of $\mathcal{M}$, and provided useful general sufficient conditions to verify compactness in several useful function spaces. Varadarajan had independently worked out similar results, in a more general context, in his thesis during 1956-57 at the Indisn Statistical Institute. These tools were used by Parthasarathy, Ranga Rao and Varadhan [**14**] , [**21**] to study limit theorems in different contexts.

2. The Martingale Problem.

The work of Stroock and Varadhan [**18**], [**19**] on the Martingale approach to the study of Markov process is also motivated by these considerations. In approximating diffusion processes by Markov chains we start with an approximation that is valid at the infinitesimal level. If $h > 0$ is the discretized time unit and $\pi_h(x, dy)$ is the single step transition probability, it is natural to assume that

$$\int [f(y) - f(x)]\pi_h(x, dy) = h(Lf)(x) + o(h)$$

where L is the generator of the semigroup associated with the limiting process. Our aim is to show $v_n(x) = \int f(y)\pi_h^n(x, dy) \simeq (e^{n\,h\,L} f)(x)$. The standard analytical method is to to solve the evolution equation

$$u_t = Lu, u(0, x) = f(x)$$

and estimate the difference $\Delta_n(x) = v_n(x) - u(n\,h, x)$. Let $nh = t$.

$$\begin{aligned}\int u(k\,h, y)\pi_h(x, dy) &\simeq u(k\,h, x) + h(Lu(k\,h, \cdot))(x) + o(h)\\ &= u(k\,h, x) + hu_t(k\,h, x)(x) + o(h)\\ &= u((k+1)\,h, x) + o(h)\end{aligned}$$

Hence

$$\left|u(n\,h,x) - \int u(0,y)\pi_h^n(x,dy)\right| \le n\,o(h) = o(1)$$

and the proof depends on the regularity of the solution $u(t,x)$. The actual random paths of the Markov chain did not play any role. Only the iterates of π_h were needed.

The martingale method, on the other hand, starts with the measures $P_{h,x}$ on a function space of paths, that corresponds to the given Markov chain starting from x and shows that sequence has a limit P_x. Then if $nh = t$,

$$v_n(x) = E^{P_{h,x}}[f(x(t)] \to E^{P_x}[f(x(t)] = u(t,x)$$

To achieve this one needs to verify compactness and then characterize the possible limit uniquely. It is done as the unique measure on $C[0,T]$ (or $D[0,T]$ which is a space of paths that admits simple jumps), with certain properties:

$$P[x(0) = x] = 1$$

and

$$f(x(t)) - f(x(0) - \int_0^t (Lf)(x(s))ds = Z_f(t)$$

is a martingale with respect to $(\Omega, \mathcal{F}_t, P)$ for a wide class of functions f. From the definition of a Markov chain, if P_h is the probability distribution of the chain and

$$P_h[x(0) = x] = 1$$

then

$$f(x(n\,h)) - f(x(0)) - \sum_{j=0}^{n-1} \int [f(y) - f(x(j\,h))]\pi_h(x(j\,h), dy) = Z_f^h$$

is a martingale. One now uses ideas from weak convergence to show that $\{P_h\}$ is compact and that any limit satisfies the properties.

3. Large Deviations

In limit theorems one is primarily interested in the behavior of $P_n(A)$ as $n \to \infty$. We look at the larger problem of weak convergence of P_n and establish such convergence to the limit P, in a some topological space X that contains A as a subset. If A is a continuity set for P, i.e $P[\delta A] = 0$, then $P_n(A) \to P(A)$.

In large deviations, we are dealing with a situation where P_n tends to a distribution that is degenerate at some point, i.e. $P_n \to \delta_{x_0}$ for some $x_0 \in X$. Then if $x \notin \bar{A}$, $P_n(A) \to 0$ and we wish to know how fast. In particular we expect the rate to be exponential and we wish to examine

$$\lim_{n\to\infty} \frac{1}{n} \log P_n(A) = -c(A)$$

It will turn out that

$$c(A) = \inf_{x\in A} I(x)$$

for some non-negative function $I(x)$ with $I(x_0) = 0$.

In the theory of weak convergence it is established that in metric spaces weak convergence is equivalent to

$$\limsup_{n\to\infty} P_n(C) \le P(C) \qquad \text{for closed sets} \quad C$$
$$\liminf_{n\to\infty} P_n(G) \ge P(G) \qquad \text{for open sets} \quad G$$

and for continuity sets i.e sets with $P(A^o) = P(\bar{A})$, one has

$$\lim_{n\to\infty} P_n(A) = P(A)$$

An important criterion for compactness of $\{P_n\}$, is the uniform tightness condition: given $\epsilon > 0$, there is a compact set K_ϵ such that

$$\inf_n P_n(K_\epsilon) \ge 1 - \epsilon$$

In large deviation theory a sequence P_n is said to satisfy the large deviation principle with the rate function $I(x)$ if

$$\limsup_{n\to\infty} \frac{1}{n} \log P_n(C) \le -\inf_{x\in C} I(x) \qquad \text{for closed sets} \quad C \tag{3.1}$$

$$\liminf_{n\to\infty} \frac{1}{n} \log P_n(G) \ge -\sup_{x\in G} I(x) \qquad \text{for open sets} \quad G \tag{3.2}$$

$I(x) : X \to [0,\infty]$ is a lower semi continuous function. There is a similar tightness condition.

1. $I(x)$ has compact level sets, i.e $C_\ell = \{x : I(x) \le \ell\}$ is compact for each $\ell < \infty$.

2. For any ℓ there is a compact set K_ℓ such that

$$\sup_n P_n(K_\ell^c) \le e^{-n\ell}$$

Just as weak convergence implies

$$\lim_{n\to\infty} P_n(A) = P(A)$$

for continuity sets, in large deviation theory it follows easily from (3.1) and (3.2) that for sets A such that

$$\inf_{x\in A^o} I(x) = \inf_{x\in A} I(x) = \inf_{x\in \bar{A}} I(x)$$

we will have

$$\lim_{n\to\infty} \frac{1}{n} \log P_n(A) = -c(A) = -\inf_{x\in A} I(x)$$

In weak convergence we will have

$$\lim_{n\to\infty} \int f(x) dP_n = \int f(x) dP$$

for bounded continuous functons f. In large deviation theory for such functions the analogous result is

$$\lim_{n\to\infty} \frac{1}{n} \log \int e^{n\,f(x)} dP_n = \sup_x [f(x) - I(x)]$$

In weak convergence when $f_n \hookrightarrow f$, in order to make sure that $\int f_n(x)dP_n \mapsto \int f(x)dP$ when P_n converges weakly to P, one assumes the uniform convergence of f_k to f on compact sets. This is used together with tightness to control

$$\limsup_{k\to\infty} \limsup_{n\to\infty} P_n[|f_k - f| \geq \epsilon] = 0$$

for every $\epsilon > 0$. In large deviation theory super exponential estimates play a similar role. For every $\epsilon > 0$,

$$\limsup_{k\to\infty} \limsup_{n\to\infty} \frac{1}{n} \log P_n[|f_k - f| \geq \epsilon] = -\infty \tag{3.3}$$

The methods used in large deviation theory are often similar to the ones used in studying weak convergence of probability measures. There are several sources [**22**], [**4**], [**2**] that describe these. Let us look at an example.

Schilder's theorem [**16**]. P_n is the distribution of $x(t) = \frac{1}{\sqrt{n}}\beta(t)$, where β is the standard Brownian motion. There is a large deviation principle on $C[0,1]$ with rate function

$$I(f) = \frac{1}{2}\int_0^1 [f'(t)]^2 dt$$

on functions f with $f(0) = 0$ that have a square integrable derivative. Otherwise $I(f) = \infty$.

The proof proceeds by approximation. If $x_k(\cdot)$ is the piecewise linear approximation with intervals $\frac{1}{k}$.

$$P_n[\|x_k(\cdot) - x(\cdot)\| \geq \epsilon] \leq C\, n\, exp[-\frac{n\, k\, \epsilon^2}{2}].$$

This provides the estimate (3.3) and enables us to interchange the limits on k and n. In a recent book [**7**] Ellis and Dupuis emphasize the weak convergence aspect of large deviations. If one wants to prove a large deviation upper bound for P_n with rate function $I(x)$, it suffices to show that when ever $Q_n << P_n$ and $Q_n \to \delta_x$ weakly, then

$$\liminf_{n\to\infty} \frac{1}{n} H(Q_n|P_n) \geq I(x).$$

Here $H(Q|P)$ is the relative entropy $\int \frac{dQ}{dP} \log \frac{dQ}{dP}\, dP$.

4. Scaling Limits

In studying scaling limits of large systems of interacting processes ideas from weak convergence play a crucial role. For simplicity consider as in [**8**], a family $\{x_i(t)\}$ of processes indexed by points $i = 1, 2, \ldots, N$ arranged (periodically) on the unit circle that satisfy the stochastic differential equation

$$dx_i(t) = [\phi(x_{i-1}(t)) - 2\phi(x_i(t)) + \phi(x_{i+1}(t))]dt + [d\beta_{i,i+1}(t) - d\beta_{i-1,i}(t)].$$

Assume initially that

$$\frac{1}{N}\sum_i \delta_{\frac{i}{N}} x_i(0) = \mu_N(0) \to u_0(x)dx$$

in probability. Then show that

$$\frac{1}{N}\sum_i \delta_{\frac{i}{N}} x_i(N^2 t) = \mu_N(N^2 t) \to u(t,x)dx$$

where $u(t,x)$ satisfies a certain partial differential equation with $u(0,x) = u_0(x)$. Let P_N be the distribution of the process $\mu_N(N^2\cdot)$ on the space $C[[0,T];\mathcal{M}(S^1)]$. Then,

1. P_N is tight.

2. If P is any limit then P is supported on paths $\mu(t)$, that satisfy

(i) they have a density for every t, i.e $\mu(t,dx) = u(t,x)dx$

(ii) if $\psi' = \phi$ then with

$$h(x) = \sup[\theta\, x - \log \int e^{\theta y - \psi(y)} dy]$$

$u(t,x)$ is a weak solution of

$$u_t = [h(u(t,x))]_{xx}$$

that satisfies

$$\int_{S^1}\int_0^T [[h'(u(t,x))]_x]^2 dx dt < \infty$$

(iii) $u(0,x) = u_0(x)$

3. The solution satisfying (i), (ii) and (iii) is unique.

The measure P therefore has to be the δ-measure at this unique solution.

5. Large Deviations for the Simple Exclusion Process.

The totally asymmetric simple exclusion model (TASEP) is a Markov process of interacting particles [**10**]. These particles live on Z and are restricted to at most one particle per site. The current state then is a map $\eta : Z \to \{0,1\}$. If there is a particle at site x then $\eta(x) = 1$, otherwise it is $\eta(x) = 0$. The generator of the process is given by

$$(\mathcal{L}f)(\eta) = \sum_z \eta(z)(1 - \eta(z+1))[f(\eta^{z,z+1}) - f(\eta)]$$

which means that particles wait for a random exponential time at the end of which they try to jump to the next site on its right. They jump if the site is free. If not they wait again for the next opportunity. If the density of particles is not too high, then they march to the right, in a haphazard fashion. We rescale space by $\frac{1}{N}$ and speed up time by a factor of N. The density profile $\rho(t,x)$ that provides a macroscopic view, at time Nt is defined by

$$\frac{1}{N}\sum_z J(\frac{z}{N})\eta(Nt,z) \to \int J(x)\rho(t,x)dx.$$

Given $\rho(0,x) = \rho_0(x)$ at time $t = 0$, for $t > 0$, $\rho(t,x)$ is determined as the weak solution of

$$\rho_t + [\rho(1-\rho)]_x = 0$$

with initial condition $\rho(0,x) = \rho_0(x)$. However the weak solution is not unique and an 'entropy condition' has to be imposed in order to make the solution unique. If ϕ is a convex function and h is determined by $h'(r) = \phi'(r)(1-2r)$, then for smooth solutions we will have

$$[\phi(\rho)]_t + [h(\rho)]_x = 0.$$

The entropy entropy condition for weak solutions is that

$$[\phi(\rho)]_t + [h(\rho)]_x \leq 0$$

At a simple shock discontinuity this requires the density on the left to be lower than the density on the right. Non entropic solutions can occur with small probability. For example one particle may not jump for a long time and will hold up every particle behind it, producing a shock, i.e. discontinuity. We want to see how small this probability is. It turns out that the rate function $I(\rho(\cdot,\cdot))$ is infinite unless $\rho(\cdot,\cdot)$ is a weak solution and in such a case the rate function is the total mass of the positive part (can be infinite) of $[\phi(\rho)]_t + [h(\rho)]_x$ viewed as a distribution on $R \times [0,T]$ for the special convex function

$$\phi(r) = r \log r + (1-r)\log(1-r).$$

The proof by Jensen [**9**], [**23**] depends on the method of Ellis-Dupuis that uses weak convergence. If we perturb the dynamics from $\mathcal{L}$ to

$$\mathcal{L}_c = \sum_z c_z(t,\eta)\eta(z)(1-\eta(z+1))[f(\eta^{z,z+1}) - f(\eta)]$$

so that the macroscopic profile converges to $\rho(t,x)$ that may not be entropic then, denoting by Q_n the perturbed process and by P_n the unperturbed one, it suffices to prove that we must necessarily have

$$\liminf_{N\to\infty} \frac{1}{N} H(Q_N|P_N) \geq I(\rho(\cdot,\cdot).)$$

This can be done.

6. Conclusion

I have tried to demonstrate how ideas from 'weak convergence' of probability measures on abstract spaces, that at the first glance appear to be too abstract, do in fact provide an essential tool to study some hard and concrete problems. Of course my view is biased, in that I have found the tool very useful in my own work. I learnt these tools as a graduate student in Calcutta from Varadarajan through my assocaition with Ranga Rao and Parthasarathy. You can not teach an old dog new tricks!

Acknowledgments. This work was supported by a grant from the National Science Foundation DMS-0604380.

References

[1] Bernoulli, J. Ars Conjectandi, Basel, 1713

[2] Dembo, A.; Zeitouni, O. Large deviations techniques and applications. Jones and Bartlett Publishers, Boston, MA, 1993.

[3] DeMoivre, A. The Doctrine of Chances, London, 1718

[4] Deuschel, J.D.; Stroock, D. W. Large deviations. Pure and Applied Mathematics, 137. Academic Press, Inc., Boston, MA, 1989.

[5] Donsker, M. D. Justification and extension of Doob's heuristic approach to the Komogorov-Smirnov theorems. Ann. Math. Statistics 23, (1952). 277–281.

[6] Doob, J. L. Heuristic approach to the Kolmogorov-Smirnov theorems. Ann. Math. Statistics 20, (1949). 393–403.

[7] Dupuis, P.; Ellis, R. S. A weak convergence approach to the theory of large deviations. Wiley Series in Probability and Statistics: Probability and Statistics. A Wiley-Interscience Publication. John Wiley & Sons, Inc., New York, 1997.

[8] Guo, M. Z.; Papanicolaou, G. C.; Varadhan, S. R. S. Nonlinear diffusion limit for a system with nearest neighbor interactions. Comm. Math. Phys. 118 (1988), no. 1, 31–59.

[9] Jensen, L. Large deviations of TASEP. PhD thesis, NYU, 2000.

[10] Kipnis, C.; Landim, C. Scaling limits of interacting particle systems. Grundlehren der Mathematischen Wissenschaften [Fundamental Principles of Mathematical Sciences], 320. Springer-Verlag, Berlin, 1999.

[11] Gnedenko, B.V.; Kolmogorov, A.N. Limit Distributions for Sums of Independent Random Variables. Moscow, 1949.

[12] Kolmogoroff, A.N. ; Prochorow, Yu.V. Zufllige Funktionen und Grenzverteilungsstze. (German) Bericht ber die Tagung Wahrscheinlichkeitsrechnung und mathematische Statistik in Berlin, Oktober, 1954, pp. 113–126. Deutscher Verlag der Wissenschaften, Berlin, 1956.

[13] LeCam, L. Convergence in distribution of stochastic processes. Univ. Calif. Publ. Statist. 2 (1957), 207–236.

[14] Parthasarathy, K. R.; Ranga Rao, R.; Varadhan, S. R. S. Probability distributions on locally compact abelian groups. Illinois J. Math. 7 1963 337–369.

[15] Prohorov, Yu. V. Convergence of random processes and limit theorems in probability theory. (Russian) Teor. Veroyatnost. i Primenen. 1 (1956), 177–238.

[16] Schilder, M. Some asymptotic formulas for Wiener integrals. Trans. Amer. Math. Soc. 125 1966 63–85.

[17] Skorohod, A. V. Limit theorems for stochastic processes. (Russian) Teor. Veroyatnost. i Primenen. 1 (1956), 289–319.

[18] Stroock, D.W.; Varadhan, S. R. S. Diffusion processes with continuous coefficients. I. Comm. Pure Appl. Math. 22 1969 345–400.

[19] Stroock, Daniel W.; Varadhan, S. R. S. Diffusion processes with continuous coefficients. II. Comm. Pure Appl. Math. 22 1969 479–530.

[20] Varadarajan, V. S. Convergence of stochastic processes. Bull. Amer. Math. Soc. 67 1961 276–280.

[21] Varadhan, S. R. S. Limit theorems for sums of independent random variables with values in a Hilbert space. Sankhyā Ser. A 24 1962 213–238.

[22] Varadhan, S. R. S. Large deviations and applications. CBMS-NSF Regional Conference Series in Applied Mathematics, 46. Society for Industrial and Applied Mathematics (SIAM), Philadelphia, PA, 1984.

[23] Varadhan, S. R. S. Large deviations for the asymmetric simple exclusion process. Stochastic analysis on large scale interacting systems, 1–27, Adv. Stud. Pure Math., 39, Math. Soc. Japan, Tokyo, 2004.

Courant Institute of Mathematical Sciences, New York University, New York, NY, 10012, USA. E-mail: varadhan@cims.nyu.edu

PDE, SDE, Diffusions

COMMUNICATIONS ON PURE AND APPLIED MATHEMATICS, VOL. XXII, 345–400 (1969)

Diffusion Processes with Continuous Coefficients, I*

DANIEL W. STROOCK AND S. R. S. VARADHAN

1. Introduction

The aim of this article is to study diffusion processes on R^d corresponding to coefficients $a = a_{ij}(t, x)$, $1 \leqq i, j \leqq d$, and $b = b_i(t, x)$, $1 \leqq i \leqq d$. The first question which naturally arises is: what do we mean by a diffusion process corresponding to these coefficients? This is intimately connected with analytic questions about the existence and uniqueness of solutions to certain parabolic equations. Indeed, it is traditional to initiate the study of a diffusion process by producing transition probability distributions $P(s, x; t, \Gamma)$, $t \geqq s \geqq 0$, $x \in R^d$, and $\Gamma \in \mathscr{B}(R^d)$. The function $P(s, x; t, \Gamma)$ plays a central role for both the probabilist and analyst. For the probabilist, $P(s, x; t, \Gamma)$ is the probability of being in Γ at time t, given that the process started from x at time s. In the analytic context, $P(s, x; t, \Gamma)$ is most easily described as the measure whose density is the solution to the *forward equation*

$$(1.1) \qquad \frac{\partial u}{\partial t}(t, y) = \frac{1}{2} \sum_{i,j=1}^{d} \frac{\partial^2}{\partial y_i \, \partial y_j} (a_{ij}(t, y)u(t, y)) - \sum_{i=1}^{d} \frac{\partial}{\partial y_i} (b_i(t, y)u(t, y))$$

for $t > s$ with the boundary condition $u(s, \cdot) = \delta_x(\cdot)$. Of course, equation (1.1) as it stands does not make very much sense unless the coefficients are smooth. For this reason, one is forced in general to adopt a modified analytic description of $P(s, x; t, \Gamma)$. One such modification is to look at $P(s, x; t, \Gamma)$ as the solution to the *backward equation*

$$(1.2) \qquad -\frac{\partial u}{\partial s}(s, x) = \frac{1}{2} \sum_{i,j=1}^{d} a_{ij}(s, x) \frac{\partial^2 u}{\partial x_i \, \partial x_j}(s, x) + \sum_{i=1}^{d} b_i(s, x) \frac{\partial u}{\partial x_i}(s, x)$$

for $0 \leqq s < t$ with the initial condition $u(t, \cdot) = \chi_\Gamma(\cdot)$. This description has the obvious advantage that it makes perfectly good sense even for badly behaved coefficients. Furthermore, when the coefficients are bounded and uniformly

* This paper represents results obtained at the Courant Institute of Mathematical Sciences, New York University. This work was supported in part by a Ford Foundation Grant for Statistics and in part by a Sloan Foundation Grant for Statistics. Reproduction in whole or in part is permitted for any purpose of the United States Government.

Hölder continuous and the matrix a is uniformly positive definite, then a great deal is known about the solutions to (1.2). In particular, it is known that (1.2) possesses a nice fundamental solution $p(s, x; t, y)$; and starting with this fact one can generate a diffusion process having $P(s, x; t, \Gamma) = \int_\Gamma p(s, x; t, y)\, dy$ as its transition probability function (cf. Dynkin [5]).

However, there is not very much known about equation (1.2) when the coefficients are not Hölder continuous, and therefore one must adopt slightly less classical tools. In particular, if the coefficients do not depend on time, it can be useful to phrase the analytic description of $P(s, x; t, \Gamma)$ in the language of semi-group theory. Loosely speaking, the idea is to start out with the operator

$$L = \frac{1}{2} \sum_{i,j=1}^{d} a_{ij}(x) \frac{\partial^2}{\partial x_i\, \partial x_j} + \sum_{i=1}^{d} b_i(x) \frac{\partial}{\partial x_i}$$

and to end up with a semi-group $\{T_t\}_{t>0}$ which has L as its infinitesimal generator. The reason why we say "loosely speaking" is that one knows *a priori* that L acting on smooth functions cannot be the generator of a semi-group, because it is not closed. This seemingly pedantic point turns out to be a most difficult one to overcome. In fact, what is involved here is the extension of L to an $\hat{L}$ for which the equations

$$(\lambda - \hat{L})f_\lambda = g\,, \qquad \lambda > 0\,,$$

have a solution for sufficiently many g. Although this problem is trivial when a nice fundamental solution to (1.2) is known to exist, it is far from trivial when such a fundamental solution is not available. Nonetheless, with the aid of some estimates from the theory of elliptic differential equations, Tanaka [17] and Krylov [12], independently, found a good extension of L and were thereby enabled to construct an appropriate semi-group in the case when the $a_{ij}(x)$ are bounded and uniformly continuous, the $b_i(x)$ are bounded and measurable, and the matrix a is uniformly positive definite. However, although their extension of L is clearly the natural one, neither Tanaka nor Krylov was able to show, in general, that it is the only one. Thus, it is impossible, on the basis of their results, to identify the process that they construct as *the* diffusion corresponding to the given coefficients.

The principal goal of the present article is to clear up the sort of ambiguity just mentioned. In order to do this we must provide a more probabilistic setting for the problem. What a probabilist means by a continuous Markov process on R^d is a family of probability measures $P_{s,x}$, $s \geqq 0$ and $x \in R^d$, on $\Omega = C([0, \infty), R^d)$ [1] such that $P_{s,x}(A \mid M_t^s) = P_{t,x(t)}(A)$ a.s. $P_{s,x}$, $A \in M_\infty^t$, where

[1] $C([0, \infty), R^d)$ stands for the space of continuous functions on $[0, \infty)$ into R^d.

M_t^s is the smallest σ-algebra on Ω with respect to which all the functions $x(u) = x(u, \cdot)$: $\Omega \to R^d$, $s \leqq u \leqq t$, are measurable. In order for the process to correspond to the coefficients a and b, one wants, at the very least, that the measure $P_{s,x}$ satisfies

$$(1.3) \qquad E_{s,x}[f(x(t))] - f(x) = E_{s,x}\left[\int_s^t L_u f(x(u))\, du\right], \qquad t \geqq s \text{ and } f \in C_0^\infty(R^d),^2$$

where

$$L_u = \frac{1}{2} \sum_{i,j=1}^{d} a_{ij}(u, y) \frac{\partial^2}{\partial y_i\, \partial y_j} + \sum_{i=1}^{d} b_i(u, y) \frac{\partial}{\partial y_i}.$$

Such a process is called a *quasi-diffusion* by Krylov. The requirement imposed by (1.3) can be interpreted as saying that the transition probability distribution $P(s, x; t, \Gamma) = P_{s,x}(x(t) \in \Gamma)$ satisfies the forward equation (1.1) in the sense that

$$(1.1') \qquad \frac{\partial}{\partial t} \int f(y) P(s, x; t, dy) = \int L_t f(y) P(s, x; t, dy).$$

Thus, one possible approach to the problem of constructing a quasi-diffusion corresponding to a and b is to start by studying the analytic problem of finding $P(s, x; t, \Gamma)$ satisfying (1.1′). However, such an approach will at best lead to the same lack of uniqueness encountered by Tanaka and Krylov. An alternative method is to resort to K. Itô's theory of stochastic differential equations. In Itô's theory, the idea is to define the measure $P_{s,x}$ directly on Ω by making a certain nonlinear transformation of Brownian motion. Itô's transformation is defined in the following way. Let $\beta(t, w)$, $t \geqq 0$, be d-dimensional Brownian motion, and denote by $\sigma = \sigma(t, x)$ the non-negative definite square root of a. Then the Brownian path $\beta(t, w)$, $t \geqq 0$, is transformed into the path $\xi(t, w)$, $t \geqq s$, such that $\xi(s, w) = x$ and

$$(1.4) \qquad d\xi(t, w) = \sigma(t, \xi(t, w))\, d\beta(t, w) + b(t, \xi(t, w))\, dt, \qquad t \geqq s,$$

where $d\xi(t, w)$ denotes the infinitesimal increment in ξ during the time interval $[t, t + dt]$, and $d\beta(t, w)$ denotes the corresponding increment in β. When σ and b are Lipschitz continuous, Itô [9] shows that for a set of w having full Wiener measure, there exists a unique continuous solution $\xi(t, w)$ satisfying these requirements. Hence we can define the measure $P_{s,x}$ on Ω by setting

$$P_{s,x}(x(t_1) \in \Gamma_1, \cdots, x(t_n) \in \Gamma_n) = W(\xi(t_1) \in \Gamma_1, \cdots, \xi(t_n) \in \Gamma_n)$$

[2] $C_0^\infty(R^d)$ is the set of C^∞ functions having compact support.

for $s \leqq t_1 < \cdots < t_n$, $n \geqq 1$, and $\Gamma_1, \cdots, \Gamma_n \in \mathscr{B}(R^d)$, where W denotes Wiener measure. Further, Itô proves that this recipe gives a quasi-diffusion corresponding to a and b. Obviously, Itô's approach, when it works, gives us everything we need to know. However, as it stands, this method seems to be restricted to Lipschitz continuous coefficients.

In order to go further, we reformulate what is meant by a solution to the stochastic differential equation (1.4). According to this new formulation, a solution to (1.4) is a probability measure P on Ω such that

$$dx(t, w) = \sigma(t, x(t, w))\, d\beta(t, w) + b(t, x(t, w))\, dt\,, \quad t \geqq s\,, \text{ a.s. } P\,, \tag{1.4'}$$

where $\beta(t, w)$, $t \geqq s$, is a Brownian motion with respect to P. In [16], Skorokhod shows that if a and b are continuous, then for each $s \geqq 0$ and $x \in R^d$ there exists a $P_{s,x}$ on Ω satisfying (1.4′) such that $P_{s,x}(x(s) = x) = 1$. However, Skorokhod did not prove the uniqueness of $P_{s,x}$, and without uniqueness we cannot even prove the Markov property or any properties of $P_{s,x}$ as s and x vary, much less that there is only one such quasi-diffusion. Later, Girsanov pointed out in [8] that if there is a nice fundamental solution to the backward equation (1.2), then $P_{s,x}$ is unique. In particular, this is the case when the coefficients are bounded and uniformly Hölder continuous and the matrix a is uniformly positive definite. Moreover, Girsanov shows in [7] that positive definiteness is necessary in the sense that there is an example of Hölder continuous coefficients for which there is a continuum of the $P_{s,x}$.

In the present article it is proved that there is at most one $P_{s,x}$ whenever a is bounded and continuous, b is bounded and measurable, and the matrix a is positive definite. However, we prefer to phrase the problem a little differently. Instead of considering probability measures P on Ω satisfying (1.4′), we look at P's on Ω such that

$$\begin{aligned} X_\theta^s(t) = \exp\Big\{\langle \theta, x(t) - x(s)\rangle &- \frac{1}{2}\int_s^t \langle \theta, a(u, x(u))\theta\rangle\, du \\ &- \int_s^t \langle \theta, b(u, x(u))\rangle\, du\Big\} \end{aligned} \tag{1.5}$$

is a martingale for all $\theta \in R^d$. As we shall see, this description of P is equivalent to the one in equation (1.4′). In fact, the martingale formulation bears the same sort of relation to the stochastic differential equation as an integral equation does to a differential equation. This analogy is deeper than may be apparent at first sight. Indeed, as with an integral equation, the relations implied by the martingale formulation are much more stable than those implied by (1.4′). In particular, they are stable under the convergence of measures in function space. We shall exploit this stability systematically, not only in proving existence but also in studying $P_{s,x}$ as a function of s and x. Also, it is this stability

which enables us to prove an invariance principle for Markov chains like the one established by Donsker [2] for sums of independent random variables. A principle of this sort has been studied by Skorokhod in [16] under different conditions.

It is important to see where the sort of uniqueness which we prove stands in relation to the other kinds that we have mentioned. We shall show that the uniqueness of $P_{s,x}$ is stronger than any statement of uniqueness that can be formulated in terms of semi-groups or diffusion processes. On the other hand, it is not so strong as the path-wise uniqueness which is natural in Itô's original formulation of stochastic differential equations, but we have not been able to settle whether our uniqueness holds in some case when Itô's does not. Finally, our uniqueness implies that of weak solutions to the backward equation (1.2) in a certain Sobolev space.

A synopsis of the results proved in this article appears in Section 2. In that section we have also summarized some of the more important, less familiar facts which we shall use. The actual work begins with Section 3.

In conclusion, we wish to express our gratitude to Professors H. P. McKean, L. Nirenberg, and H. Weinberger, with whom we had many helpful discussions in the course of our research.

2. Preliminaries and Summary

In this section we review some of the results that we use later on and give a section by section summary of the results proved.

The processes that we construct will have continuous trajectories. Therefore, we consider the space Ω of trajectories w consisting of continuous functions defined on $[0, \infty)$ with values in R^d. The value of the function w at time t will be denoted by $x(t, w)$ or $x_t(w)$. The argument w may be suppressed sometimes. The space Ω will be viewed as a complete separable metric space. The relevant topology is that of uniform convergence on compact subsets of $[0, \infty)$. M^0 is the Borel σ-field of Ω and can be identified as the smallest σ-field with respect to which $x(t, w)$ are measurable for $0 \leqq t < \infty$. M_β^α will denote the σ-field generated by the functions $x(t, w)$ for $\alpha \leqq t \leqq \beta$. When $\beta = +\infty$, this σ-field will be denoted by M^α. C_β^α will be the collection of sets of the form $\{x(t, w) \in \Gamma_1, \cdots, x(t_n, w) \in \Gamma_n\}$, where $\Gamma_1, \cdots, \Gamma_n$ are Borel subsets of R^d and $t_1, \cdots, t_n$ are points in the interval $\alpha \leqq t \leqq \beta$. Clearly, C_β^α generates M_β^α and, if two measures agree on C_β^α, they agree on M_β^α. As before, C^α will denote C_β^α with $\beta = +\infty$.

A function $\tau(w)$ defined on Ω is called an s-Markov time or an s-stopping time if $\tau(w) \geqq s$ and the set $\{w : \tau(w) \leqq t\}$ is M_t^s-measurable for every $t \geqq s$. Associated with these stopping times are the σ-fields M_τ^s defined as the collection of sets

$$\{A \in M^s : A \cap \{\tau \leqq t\} \in M_t^s \text{ for all } t \geqq s\}.$$

Any constant $t \geqq s$ is an s-stopping time and the corresponding σ-field coincides with M_t^s. If we define C_τ^s as the collection of sets of the form

$$\{x(\tau \wedge t_1, w) \in \Gamma_1, \cdots, x(\tau \wedge t_n, w) \in \Gamma_n\}, \qquad s \leqq t_1 < \cdots < t_n,$$

then C_τ^s generates M_τ^s.

These σ-fields are nice in the sense that they have regular conditional probabilities. The following theorem, which states this fact, is proved in the appendix (Theorem A.2).

THEOREM 2.1. *Let P be a probability measure on (Ω, M^s). Let τ be an s-stopping time with the associated σ-field M_τ^s. Then there exist probability distributions $Q(w, A)$ such that*

(i) *$Q(w, A)$ is a probability measure in $A \in M^s$ for each $w \in \Omega$,*
(ii) *$Q(w, A)$ is M_τ^s-measurable for each $A \in M^s$,*
(iii) *$Q(w, A_w) = 1$, where*

$$A_w = [w' : x(t, w') = x(t, w) \text{ for } s \leqq t \leqq \tau(w)],$$

(iv) *$Q(w, A) = P(A \mid M_\tau^s)$ for almost all w with respect to P.*

A stochastic process defined for $t \geqq s$ is a probability measure P on (Ω, M^s). Such a process is said to start from a point x of R^d at time s provided

$$P\{w : x(s, w) = x\} = 1.$$

The notion of weak convergence of a sequence P_n of measures on (Ω, M^0) is familiar and is defined as follows: $P_n \to P$ weakly if for every bounded and continuous function $f(w)$ on Ω

$$\lim_{n\to\infty} \int_\Omega f(w)\, dP_n = \int_\Omega f(w)\, dP.$$

A detailed study of weak convergence can be found in [14] or [15]. For instance, it is proved there that weak convergence arises from a metric, and a family $\{P_\alpha\}$ is conditionally compact in that metric if and only if for every $\varepsilon > 0$ there exists a compact subset K_ε of Ω such that

$$P_\alpha(K_\varepsilon) \geqq 1 - \varepsilon \quad \text{for all} \quad \alpha.$$

We are interested in the following consequences of this result for later use.

THEOREM 2.2. *In order that a family $\{P_\alpha\}$ of measures on (Ω, M^0) be weakly conditionally compact, the following two conditions are necessary and sufficient:*

(i) *For every* $T < \infty$ *and* $\varepsilon > 0$,

$$\limsup_{\delta \downarrow 0} \sup_{\alpha} P_\alpha[\sup_{\substack{0 \leqq t_1 < t_2 \leqq T \\ |t_2 - t_1| \leqq \delta}} |x(t_2) - x(t_1)| \geqq \varepsilon] = 0 ;$$

(ii)
$$\lim_{l \uparrow \infty} \sup_{\alpha} P_\alpha[|x(0)| \geqq l] = 0 .$$

THEOREM 2.3. *Let* $\{P_\alpha\}$ *be a family of measures on* (Ω, M^0) *such that, for every* $T < \infty$ *and* $0 \leqq t_1 < t_2 \leqq T$, *there are numbers* $n > 0$, $\delta > 0$ *and constants* C *and* $C(T)$, *all independent of* α, *satisfying*

$$E_\alpha[|x(0)|^n] \leqq C ,$$
$$E_\alpha[|x(t_2) - x(t_1)|^n] \leqq C(T)\, |t_1 - t_2|^{1+\delta} .$$

Then $\{P_\alpha\}$ *is conditionally compact.*

Remark 1. Theorem 2.2 can be found in [14] or [15] and Theorem 2.3 in [15].

Remark 2. We can also apply these theorems for measures defined on (Ω, M^s). M^s can be identified as the Borel σ-field of the space of R^d-valued continuous functions on $[s, \infty)$.

The following inequalities for martingales are standard and can be found in [3]. They are stated here for convenient future reference.

THEOREM 2.4. *Let* $X_n, n = 1, 2, \cdots, N$, *be a martingale with respect to a measure* P *on some measure space. Then, for every* $r \geqq 1$ *and* $l > 0$,

$$P\{\sup_{1 \leqq j \leqq N} |X_j| \geqq l\} \leqq \frac{1}{l^r} E[|X_N|^r] .$$

COROLLARY 2.1. *Let* $X(t)$, $t \in [0, T]$, *be a separable martingale with respect to some measure* P. *Then, for every* $r \geqq 1$ *and* $l > 0$,

$$P\{\sup_{0 \leqq t \leqq T} |X(t)| \geqq l\} \leqq \frac{1}{l^r} E[|X(T)|^r] .$$

We need the following theorem on stochastic integrals which can be found in [10]. Let $\beta(t)$ be a d-dimensional Brownian motion and $x(t)$ a stochastic integral of the form

$$x(t) = x(s) + \int_s^t b(u, w)\, du + \int_s^t \sigma(u, w)\, d\beta(u)$$

with the usual assumption of independence of the increments of $\beta(s)$ for $s \geqq t$, on the one hand, and $b(s, w)$ and $\sigma(s, w)$ for $s \leqq t$, on the other. Of course, b is a d-vector of functions and σ is a $d \times d$ matrix. Let $F(u, x)$ be a smooth function on $[s, t] \times R^d$ and let us define

$$G(u, w) = \frac{\partial F}{\partial u}(u, x(u)) + \sum \frac{\partial F}{\partial x_j}(u, x(u)) b_j(u, w)$$

$$+ \frac{1}{2} \sum \frac{\partial^2 F}{\partial x_i \, \partial x_j}(u, x(u)) \sigma_{ik}(u, w) \sigma_{jk}(u, w) ,$$

$$H_k(u, w) = \sum \frac{\partial F}{\partial x_j}(u, x(u)) \, \sigma_{jk}(u, w) .$$

Then Itô's formula states:

THEOREM 2.5.

$$F(t, x(t)) = F(s, x(s)) + \int_s^t G(u, w) \, du + \int_s^t H(u, w) \, d\beta(u) ,$$

where H is the vector of functions H_k defined above.

We need also the following theorem from differential equations. The form used is different from the one in the literature (cf. [11]), and the transition is carried out in the appendix (Theorem A.1).

THEOREM 2.6. *Let Δ be the Laplacian in d-dimensions. $U(s, x)$ and $g(s, x)$ are C^∞ functions defined for $s \geqq 0$ and $x \in R^d$ satisfying the following conditions:*

(i) $\dfrac{\partial U}{\partial s} + \dfrac{1}{2} \Delta U = g,$

(ii) *U and g vanish exponentially as $|x| \to \infty$ and are identically zero for large s.*

Then, for every p in the range $1 < p < \infty$, there exists a constant $C_{d,p}$ such that

$$\left\| \frac{\partial^2 U}{\partial x_i \, \partial x_j} \right\|_p \leqq C_{d,p} \, \|g\|_p .^3$$

As was already mentioned in the introduction, the basic problem is to examine

---

[3] $\| \quad \|_p$ denotes the L_p norm in s and x.

the existence and uniqueness of a measure P on (Ω, M^s) such that $P\{x(s) = x\} = 1$ and

$$\exp\left\{\langle\theta, x(t) - x\rangle - \left\langle\theta, \int_s^t b(u, x(u))\, du\right\rangle - \frac{1}{2}\int_s^t \langle\theta, a(u, x(u))\theta\rangle\, du\right\}$$

is a martingale for $\theta \in R^d$ and $t \geqq s$.

In Section 3, properties of such measures P are studied. Under mild conditions it is proved that such a relation is equivalent to $x(t)$ being a solution of the stochastic integral equation

$$x(t) = x + \int_s^t b(u, x(u))\, du + \int_s^t \sigma(u, x(u))\, d\beta(u)\,,$$

where β is a Brownian motion with respect to P and is constructable from $x(t)$. The increments of $\beta(u)$ for $u \geqq t$ are independent of M_t^s, and σ is a square root of a. Some properties of the conditional distributions of P are established. These properties will later ensure that P is a strong Markov process.

In Section 4 the existence of a solution P to the basic problem is proved under the hypotheses that $b \equiv 0$ and a is continuous and bounded.

In Section 5, the uniqueness of the solution constructed in Section 4 is established under the additional hypothesis that a is strictly elliptic.

In Section 6 the existence and uniqueness established in Sections 4 and 5 for the case $b \equiv 0$ is used to prove existence and uniqueness for any bounded measurable b. Once uniqueness is established, one can easily check that the processes are all Markov processes and in fact strong Markov processes. The Cameron-Martin formula (cf. [1]), which gives the Radon-Nikodym derivatives with respect to one another of two processes with the same a, is derived.

The next section is an appendix which starts from some known results and modifies them to suit our current needs.

In Section 7,[4] the continuity of the transition probabilities $P(s, x; t, dy)$ in s and x is considered. It is proved, for example, that the processes are all strong Feller. That is, for each function $\phi(y)$ which is bounded and measurable and each fixed t,

$$U(s, x) = \int \phi(y) P(s, x; t, dy)$$

is continuous in s and x for $s < t$. In fact, these functions are equicontinuous provided $\phi(y)$ stays uniformly bounded.

Section 8 is concerned with estimating $P(s, x; t, dy)$. It is proved that for almost all t there is a density $p(s, x; t, y)$. Moreover, as a function of t and y,

[4] Sections 7–11 of this paper will appear as Part II in the next issue of this journal.

the density is locally in L_α for every $1 < \alpha < \infty$. The only possible exception is the diagonal $s = t$, $x = y$. These estimates hold uniformly over suitable classes of coefficients. In the homogeneous case, the transition probability is of the form $P(t - s, x, dy)$ and it is proved that $P(u, x, dy)$ has a density for every $u > 0$. In fact, the density $p(u, x, y)$ is locally in L_α as a function of y for every $1 < \alpha < \infty$.

Section 9 deals with the continuity of the solution with respect to the coefficients. It is proved that if $a^{(n)}$ converges to a uniformly on compact sets and $b^{(n)}$ converges to b in measure on compact sets, then the solutions converge (with some additional assumptions regarding boundedness and ellipticity). In the homogeneous case, this convergence is strengthened by the convergence of the densities in L_α norm for $1 \leqq \alpha < \infty$.

Section 10 deals with the invariance principle (i.e., the convergence of Markov chains to diffusion processes). The analogues of the results in [2] and generalizations of the results in [16] are established.

Section 11 connects the semi-group solution with our solution. It also identifies weak solutions of parabolic solutions (cf. [13]) as being the same as the solutions discussed by us.

We wish to point out, in conclusion, that we are concerned here with a new type of uniqueness theorem for stochastic integral equations. This is not the classical path-wise uniqueness valid for the Lipschitz continuous case. This uniqueness in the sense of the probability distributions holds very generally and is adequate for a profitable and meaningful discussion of the diffusion processes. Such a possibility was considered in [8] by Girsanov, but did not lead to new results because the uniqueness still depended too heavily on differential equations.

3. Some Properties of Martingales

Let $I^s = [s, \infty)$ and $J \subset I^s$ be a subset. A function $\eta(t, w)$ on $J \times \Omega$ is said to be s-non-anticipating if η is $\mathscr{B}(J) \times M^s$-measurable and $\eta(t, w)$ is M_t^s-measurable for each fixed $t \in J$. Given a probability measure P on (Ω, M^s) and a function $X(t, w)$ on $J \times \Omega$ with values in R (or C), we say that X is a P-martingale if X is s-non-anticipating and

$$E^P[X(t_2) \mid M_{t_1}^s] = X(t_1) \text{ a.s. } P,$$

for all $t_1, t_2 \in J$, $t_1 \leqq t_2$. If $X(t)$, $t \in J$, is a separable P-martingale, then from Theorem 2.4 it follows that

$$P[\sup_{u \in J \cap I_t^s} |X(u)| \geqq c] \leqq \frac{1}{c} E[|X(t)|] \tag{3.1}$$

provided $t \in J$ and $c > 0$. I_t^s is the interval $[s, t]$; (3.1) will be referred to as the *martingale inequality*.

LEMMA 3.1. *Let P be a probability measure on (Ω, M^s) and let $\xi : I^s \times \Omega \to R^d$ and $a : I^s \times \Omega \to R^d \otimes R^d$ be s-non-anticipating functions such that*

(i) *$\xi(t, w)$ is continuous in t for almost all $w \in \Omega$,*

(ii) *$0 \leqq \sum_{i,j=1}^{d} a_{ij}(t, w)\theta_i\theta_j \leqq A\,|\theta|^2$ for all $t \in I^s$, $w \in \Omega$ and $\theta \in R^d$.*

If, for some $J \subset I^\alpha$, $\alpha \geqq s$, and some $\theta \in R^d$,

$$X_\theta^\alpha(t) = \exp\left\{\langle\theta, \xi(t) - \xi(\alpha)\rangle - \frac{1}{2}\int_\alpha^t \langle\theta, a(u)\theta\rangle\, du\right\}$$

is a P-martingale for $t \in J$, then for all $\gamma \geqq \alpha$

$$X_\theta^\gamma(t) = \exp\left\{\langle\theta, \xi(t) - \xi(\gamma)\rangle - \frac{1}{2}\int_\gamma^t \langle\theta, a(u)\theta\rangle\, du\right\}, \qquad t \in J \cap I^\gamma,$$

is a P-martingale. Moreover, if J is dense in I^α and $[X_\theta^\alpha(t) : t \in J]$ is a P-martingale for $\theta \in \Theta$, a dense subset of R^d, then for all $\gamma \geqq \alpha$ and $l > 0$

$$P(\sup_{\gamma \leqq u \leqq t} |\xi(u) - \xi(\gamma)|\,\phi \geqq l) \leqq 2d \exp\left\{-\frac{l^2}{2A(t-\gamma)}\right\}, \tag{3.2}$$

and $[X_\theta^\gamma(t) : t \in I^\gamma]$ is a P-martingale for all $\theta \in R^d$.

Proof: To prove the first assertion, let $\gamma \in I^\alpha$ be given and suppose that $t_1, t_2 \in J \cap I^\gamma$ with $t_1 \leqq t_2$. Then,

$$E^P[X_\theta^\gamma(t_2) \mid M_{t_1}^s] = \exp\left\{-\langle\theta, \xi(\gamma) - \xi(\alpha)\rangle + \frac{1}{2}\int_\gamma^\alpha \langle\theta, a(u)\theta\rangle\, du\right\} E^P[X_\theta^\alpha(t_2) \mid M_{t_1}^s]$$

$$= X_\theta^\gamma(t_1) \quad \text{a.s. } P.$$

The inequality (3.2) is now a consequence of the martingale inequality. In fact, if $[X_\theta^\alpha(t) : t \in J]$ is a P-martingale for all $\theta \in \Theta$, then $[X_\theta^\gamma(t) : t \in J \cap I^\gamma]$ is a P-martingale for all $\theta \in \Theta$ and $\gamma \in I^\alpha$. Hence, for $\theta \in \Theta$,

$$P\left[\sup_{\gamma \leqq u \leqq t}\left\langle\frac{\theta}{|\theta|}, \xi(u) - \xi(\gamma)\right\rangle \geqq l\right] \leqq P[\sup_{\gamma \leqq u \leqq t} X_\theta^\gamma(u) \geqq \exp\{|\theta|\, l - \tfrac{1}{2}|\theta|^2 A(t-\gamma)\}]$$

$$\leqq \exp\{-|\theta|\, l + \tfrac{1}{2}|\theta|^2 A(t-\gamma)\}.$$

We have used here the fact that $X_\theta^\gamma(t)$ is almost surely continuous in t and J is dense in I^α. Since Θ is dense in R^d, the above inequality continues to be valid for all $\theta \in R^d$. In particular, if $\hat{\theta} = \theta/|\theta|$ and $|\theta| = l/A(t-\gamma)$, we find that

$$P[\sup_{\gamma \leqq u \leqq t} \langle\hat{\theta}, \xi(u) - \xi(\gamma)\rangle \geqq l] \leqq \exp\left\{-\frac{l^2}{2A(t-\gamma)}\right\}.$$

The inequality (3.2) is immediate from this. Finally, in order to show that $[X_\theta^\gamma(t) : t \in I^\gamma]$ is a P-martingale for all $\theta \in R^d$ and $\gamma \in I^\alpha$, it suffices to show that $E^P[|X_{\theta_n}^\gamma(t_n) - X_\theta^\gamma(t)|] \to 0$ as $n \to \infty$ for sequences $\{t_n\}$ from $J \cap I^\gamma$ and $\{\theta_n\}$ from Θ, converging, respectively, to t and θ. We have continuity of $X_\theta^\gamma(t)$ in t and θ, and (3.2) establishes uniform integrability.

THEOREM 3.1. *Let P be a probability measure on (Ω, M^s) and suppose that the functions $\xi : I^s \times \Omega \to R^d$ and $a : I^s \times \Omega \to R^d \otimes R^d$ are s-non-anticipating and satisfy*

(i) *$\xi(t, w)$ is continuous in t for almost all $w \in \Omega$,*

(ii) $0 \leqq \sum_{i,j=1}^d a_{ij}(t, w)\theta_i\theta_j \leqq A\,|\theta|^2$ *for all $t \in I^s$, $w \in \Omega$ and $\theta \in R^d$,*

(iii) $X_\theta^s(t) = \exp\left\{\langle\theta, \xi(t) - \xi(s)\rangle - \frac{1}{2}\int_s^t \langle\theta\, a(u)\theta\rangle\, du\right\}$ *is a P-martingale for $t \in I^s$ and for all $\theta \in R^d$.*

If τ_1 and τ_2 are two s-stopping times such that $0 \leqq \tau_2 - \tau_1 \leqq C$, then for all $\theta \in R^d$

$$E[X_\theta^s(\tau_2) \mid M_{\tau_1}^s] = X_\theta^s(\tau_1) \text{ a.s. } P. \tag{3.3}$$

Moreover, if τ is an s-stopping time and Q is a regular conditional probability of P given M_τ^s (Theorem 2.1), then there exists a set $N \in M_\tau^s$ with $P(N) = 0$ such that, for $w \notin N$ and $\theta \in R^d$,

$$X_\theta^{\tau(w)}(t) = \exp\left\{\langle\theta, \xi(t) - \xi(\tau(w))\rangle - \frac{1}{2}\int_{\tau(w)}^t \langle\theta, a(u)\theta\rangle\, du\right\}, \quad t \in I^{\tau(w)},$$

is a Q_w-martingale.

Proof: We begin by proving the second statement under the assumption that the first assertion is true. It should be observed that the derivation will hold for bounded τ even if it is assumed that (3.3) holds only for bounded τ_1 and τ_2.

Given $0 \leqq t_1 \leqq t_2$, $B \in M_\tau^s$ and $A \in M_{\tau+t_1}^s$, we have

$$\begin{aligned} &E^P[\chi_B(w)E^{Q_w}[\chi_A X_\theta^{\tau(w)}(\tau(w) + t_2)]] \\ &\quad = E^P\left[\chi_{A\cap B}\exp\left\{-\langle\theta, \xi(\tau) - \xi(s)\rangle + \frac{1}{2}\int_s^\tau \langle\theta, a(u)\theta\rangle\, du\right\}X_\theta^s(\tau + t_2)\right] \\ &\quad = E^P\left[\chi_{A\cap B}\exp\left\{-\langle\theta, \xi(\tau) - \xi(s)\rangle + \frac{1}{2}\int_s^\tau \langle\theta, a(u)\theta\rangle\, du\right\}X_\theta^s(\tau + t_1)\right] \\ &\quad = E^P[\chi_B(w)E^{Q_w}[\chi_A X_\theta^{\tau(w)}(\tau(w) + t_1)]]\,.^5 \end{aligned}$$

[5] When there is a possibility of confusion, we emphasize to which measure an expectation refers by inserting the appropriate measure as a superscript.

Thus for each pair $0 \leqq t_1 \leqq t_2$, each $\theta \in R^d$ and $A \in M^s_{\tau+t_1}$, there is a set $N^{A,\theta}_{t_1,t_2} \in M^s_\tau$ such that $P(N^{A,\theta}_{t_1,t_2}) = 0$ and

$$E^{Q_w}[\chi_A X^{\tau(w)}_\theta(\tau(w) + t_2)] = E^{Q_w}[\chi_A X^{\tau(w)}_\theta(\tau(w) + t_1)]$$

for all $w \notin N^{A,\theta}_{t_1,t_2}$. Since $M^s_{\tau+t_1}$ is countably generated, we can find $N^\theta_{t_1,t_2} \in M^s_\tau$ such that $P(N^\theta_{t_1,t_2}) = 0$ and

$$E^{Q_w}[X^{\tau(w)}_\theta(\tau(w) + t_2) \mid M^s_{\tau+t_1}] = X^{\tau(w)}_\theta(\tau(w) + t_1) \text{ a.s. } Q_w ,$$

for all $w \notin N^\theta_{t_1,t_2}$.

Let J be a countable dense set in I^s and Θ a countable dense set in R^d. Define

$$N = \bigcup_{\substack{t_1, t_2 \in J \\ \theta \in \Theta}} N^\theta_{t_1,t_2} .$$

Then $P(N) = 0$, and for $w \notin N$, $[X^{\tau(w)}_\theta(t) : t \in J \cap I^{\tau(w)}]$ is a Q_w-martingale for $\theta \in \Theta$. We can apply Lemma 3.1 and conclude that $[X^{\tau(w)}_\theta(t) : t \in I^{\tau(w)}]$ is a Q_w-martingale for all $\theta \in R^d$ and $w \notin N$.

We now turn to the proof of equation (3.3) in the case when there exists an integer $L > 1$, such that $\tau_1 \leqq \tau_2 \leqq L - 1$. For each $n \geqq 1$, define $\tau_i^{(n)} = ([n\tau_i] + 1)/n$ for $i = 1, 2$. Then $\tau_1^{(n)} \leqq \tau_2^{(n)} \leqq L$ for $n = 1, 2, \cdots$, and $\tau_1^{(n)}$ and $\tau_2^{(n)}$ are s-Markov times.

Since $\tau_i^{(n)} \geqq \tau_i$ for $i = 1, 2$, $M^s_{\tau_i} \subseteq M^s_{t_i^{(n)}}$. Thus, if $A \in M^s_{\tau_1}$, then

$$A_k = A \cap \left\{\tau_1^{(n)} = \frac{k}{n}\right\} \in M^s_{k/n} ,$$

$$A_{k,l} = A_k \cap \left\{\tau_2^{(n)} = \frac{l}{n}\right\} \in M^s_{l/n} ,$$

$$B_{k,l} = A_k \cap \left\{\tau_2^{(n)} > \frac{l}{n}\right\} \in M^s_{l/n} ,$$

for $1 \leqq k \leqq l \leqq nL$. Hence,

$$\int_{B_{K,l}} X^s_\theta\left(\frac{l}{n}\right) dP = \int_{B_{K,l}} X^s_\theta\left(\frac{l+1}{n}\right) dP = \int_{A_{K,l+1}} X^s_\theta\left(\frac{l+1}{n}\right) dP + \int_{B_{l,l+1}} X^s_\theta\left(\frac{l+1}{n}\right) dP ,$$

and so

$$\int_{A_K} X^s_\theta(\tau_1^{(n)})\, dP = \sum_{l=k}^{nL} \int_{A_{K,l}} X^s_\theta\left(\frac{l}{n}\right) dP = \int_{A_K} X^s_\theta(\tau_2^{(n)})\, dP .$$

Summing over k, we obtain

$$\int_A X^s_\theta(\tau_1^{(n)})\, dP = \int_A X^s_\theta(\tau_2^{(n)})\, dP , \qquad n \geqq 1 .$$

Because $\tau_1^{(n)} \leqq \tau_2^{(n)} \leqq L$ for all n, using inequality (3.2), one finds that $X_\theta^s(\tau_i^{(n)})$, $i = 1, 2$ and $n \geqq 1$, are uniformly integrable. Now the almost sure continuity of $X_\theta^s(t)$ as a function of t allows us to let $n \to \infty$ and conclude that equation (3.3) is valid for bounded τ_1 and τ_2. In particular, if τ_1 is bounded and $0 \leqq \tau_2 - \tau_1 \leqq C$, then by the preceding paragraph and Lemma 3.1

$$Q_w[|\xi(\tau_2) - \xi(\tau_1)| \geqq l] \leqq Q_w[\sup_{\tau_1(w) \leqq u \leqq \tau_1(w)+C} |\xi(u) - \xi(\tau_1(w))| \geqq l]$$

$$\leqq 2d \exp\left\{-\frac{l^2}{2AC}\right\} \text{ a.s. } P,$$

where Q is a regular conditional probability distribution of P given $M_{\tau_1}^s$. Thus,

$$P[|\xi(\tau_2) - \xi(\tau_1)| \geqq l] \leqq 2d \exp\left\{-\frac{l^2}{2AC}\right\}.$$

Let τ_1 be any s-Markov time and suppose τ_2 is a second s-Markov time such that $0 \leqq \tau_2 - \tau_1 \leqq C$. Then, for all $t > s$,

$$P[|\xi(\tau_2 \wedge t) - \xi(\tau_1 \wedge t)| \geqq l] \leqq 2d \exp\left\{-\frac{l^2}{2AC}\right\}.$$

Therefore, the random variables $[X_\theta^{\tau_1 \wedge t}(\tau_2 \wedge t) : t > s]$ are uniformly integrable. Also, for $t \geqq s$,

$$E[X_\theta^{\tau_1 \wedge t}(\tau_2 \wedge t) \mid M_{\tau_1 \wedge t}^s] = 1 \text{ a.s. } P.$$

Thus, if $A \in M_{\tau_1 \wedge t_0}^s$ for some $t_0 \geqq s$, then

$$\int_A X_\theta^{\tau_1}(\tau_2)\, dP = \lim_{t \to \infty} \int_A X_\theta^{\tau_1 \wedge t}(\tau_2 \wedge t)\, dP = P(A).$$

Since $M_{\tau_1}^s$ is the smallest σ-algebra generated by $\bigcup_{t \geqq s} M_{\tau_1 \wedge t}^s$, we see that

$$E^P[X_\theta^{\tau_1}(\tau_2) \mid M_{\tau_1}^s] = 1 \text{ a.s. } P.$$

Hence, (3.3) is proved.

LEMMA 3.2. *Let P be a probability measure on (Ω, M^s) and suppose that the functions $\eta : I^s \times \Omega \to R$ and $e : I^s \times \Omega \to R$ are s-non-anticipating. If η is almost surely continuous, $0 \leqq e \leqq A$ and*

$$X_\lambda^s(t) = \exp\left\{\lambda(\eta(t) - \eta(s)) - \tfrac{1}{2}\lambda^2 \int_s^t e(u)\, du\right\}, \qquad t \in I^s,$$

is a P-martingale for all $\lambda \in R$, then,

(i) $\eta(t) - \eta(t_0)$, $t \in I^{t_0}$, *is a P-martingale for each* $t_0 \geqq s$,

(ii) $E[(\eta(t_2) - \eta(t_1))^2] = E\left[\int_{t_1}^{t_2} e(u)\,du\right]$ *for all* $t_2 \geqq t_1 \geqq s$,

(iii) $E[(\eta(t_2) - \eta(t_1))^4] \leqq C\,|t_2 - t_1|^2$ *for all* $t_2 \geqq t_1 \geqq s$, *where C depends only on A.*

Proof: From the argument given to prove (3.2) in Lemma 3.1, we see that $[X_\lambda^{t_0}(t): t \geqq t_0]$ is a P-martingale. We see also that

$$P[\sup_{t_1 \leqq t \leqq t_2} |\eta(t) - \eta(t_1)| \geqq l] \leqq 2 \exp\left\{-\frac{l^2}{2A(t_2 - t_1)}\right\}.$$

Therefore, we can differentiate the equality

$$\int_A X_\lambda^{t_1}(t_2)\,dP = \int_A X_\lambda^{t_1}(t_1)\,dP\,, \qquad A \in M_{t_1}^s\,,\ s \leqq t_1 \leqq t_2\,,$$

under the integral sign as many times as we want with respect to λ. The statements (i), (ii), (iii) are obtained as immediate consequences of differentiating these expressions once, twice and four times, respectively, and then setting $\lambda = 0$; q.e.d.

We now want to develop a theory of stochastic integration which is closely related to our martingales. Suppose P, ξ and a are as in Theorem 3.1 and let $\theta: I^s \times \Omega_t \to R^d$ be an s-non-anticipating function such that

$$E^P\left[\int_s^t |\theta(u)|^2\,du\right] < \infty$$

for $t \geqq s$. We define an integral $\int_{t_1}^{t_2} \langle \theta(u), d\xi(u)\rangle$ such that if

$$\theta(u) = \theta(s + [n(u - s)]/n)$$

for $u \geqq s$, then

$$\int_{t_1}^{t_2} \langle \theta(u), d\xi(u)\rangle = \sum_{k=[n(t_1-s)]}^{[n(t_2-s)]} \left\langle \theta\left(s + \frac{k}{n}\right), \xi\left(\left[s + \frac{k+1}{n}\right] \wedge t_2\right) - \xi\left(\left[s + \frac{k}{n}\right] \vee t_1\right)\right\rangle. \tag{3.4}$$

From now on we call an s-non-anticipating function θ such that $\theta(t) = \theta(s + [n(t - s)]/n)$, $t \geqq s$, an n-simple function.

LEMMA 3.3. *Let P, ξ and a be as in Theorem* 3.1. *Suppose $\theta: I^s \times \Omega \to R^d$ is an n-simple function such that*

$$E\left[\int_s^t |\theta(u)|^2 \, du\right] < \infty$$

for $t \geqq s$. Then, if $\int_{t_1}^{t_2} \langle \theta(u), d\xi(u)\rangle$ is defined as in (3.4) *for $t_2 \geqq t_1 \geqq s$, it has the following properties*:

$$\text{(i)} \qquad \int_{t_1}^{t_2} \langle \theta(u), d\xi(u)\rangle + \int_{t_2}^{t_3} \langle \theta(u), d\xi(u)\rangle = \int_{t_1}^{t_3} \langle \theta(u), d\xi(u)\rangle,$$

(ii) $\int_{t_0}^{t} \langle \theta(u), d\xi(u)\rangle$ *is an almost surely continuous P-martingale for $t \geqq t_0$ and $t_0 \geqq s$*,

$$\text{(iii)} \qquad X_\theta^s(t) = \exp\left\{\int_s^t \langle \theta(u), d\xi(u)\rangle - \frac{1}{2}\int_s^t \langle \theta(u), a(u)\theta(u)\rangle \, du\right\}$$

is a P-martingale for $t \geqq s$,

$$\text{(iv)} \quad E\left[\left(\int_{t_1}^{t_2} \langle \theta(u), d\xi(u)\rangle\right)^2\right] = E\left[\int_{t_1}^{t_2} \langle \theta(u), a(u)\theta(u)\rangle \, du\right] \textit{ for } t_2 \geqq t_1 \geqq s\,,$$

(v) *for $t_2 \geqq t_1 \geqq s$ and $\varepsilon > 0$*,

$$P\left[\sup_{t_1 \leq t \leq t_2} \left|\int_{t_1}^{t_2} \langle \theta(u), d\xi(u)\rangle\right| \geqq \varepsilon\right] \leqq \frac{1}{\varepsilon^2} E^P\left[\int_{t_1}^{t_2} \langle \theta(u), a(u)\theta(u)\rangle \, du\right].$$

Proof: Clearly, (i) is immediate from (3.4). It is also obvious that, for $t \geqq t_0 \geqq s$, $\int_{s_0}^{t} \langle \theta(u), d\xi(u)\rangle$ is s-non-anticipating and almost surely continuous in t. To prove (iii), it suffices to show that, if $t_2 \geqq t_1 \geqq s + [n(t_2 - s)]/n$, then $E^P[X_\theta^s(t_2) \mid M_{t_1}^s] = X_\theta^s(t_1)$ a.s. P. To this end, let Q be a regular conditional probability of P given $M_{t_1}^s$. Then,

$$\begin{aligned} E^P[X_\theta^s(t_2) \mid M_{t_1}^s] &= X_\theta^s(t_1, w) E^{Q_w}[X_\theta^{t_1}(t_2)] \\ &= X_\theta^s(t_1, w) E^{Q_w}\Big[\exp\Big\{\langle \theta(t_1, w), \xi(t_2) - \xi(t_1)\rangle \\ &\qquad - \frac{1}{2}\int_{t_1}^{t_2} \theta(t_1, w), a(u)\theta(t_1, w) \, du\Big\}\Big] \\ &= X_\theta^s(t_1, w) \text{ a.s. } P\,. \end{aligned}$$

This follows because

$$\exp\left\{\langle \theta(t_1, w), \xi(t) - \xi(t_1)\rangle - \frac{1}{2}\int_{t_1}^{t} \langle \theta(t_1, w), a(u)\theta(t_1, w)\rangle \, du\right\}$$

is a Q_w-martingale for $t \geqq t_1$ and almost all w. Thus (iii) is proved. The assertions (ii) and (iv) are immediate from (iii) by Lemma 3.2 when θ is bounded. In that case, (iv) and the martingale inequality give (v). A straightforward argument allows one to pass to the unbounded case from the bounded case.

LEMMA 3.4. *Let P be a probability measure on (Ω, M^s). Let $\theta: I^s \times \Omega \to R^d$ be an s-non-anticipating function such that $E\left[\int_s^t |\theta(u)|^2\, du\right] < \infty$ for $t \geqq s$. Then there exists a sequence $\{\theta_m\}_1^\infty$ of s-non-anticipating $\theta_m: I^s \times \Omega \to R^d$ such that*

(i) $E\left[\int_s^t |\theta_m(u)|^2\, du\right] < \infty$,

(ii) *θ_m is n_m-simple for some $n_m \geqq 1$,*

(iii) $E\left[\int_s^t |\theta_m(u) - \theta(u)|^2\, du\right] \to 0$ *as $m \to \infty$ for each $t \geqq s$.*

Furthermore, if θ is bounded, then θ_m can be chosen to be uniformly bounded by the same bound.

Proof: This lemma is proved in Itô's memoir [9] (cf. Lemma 7.1 on page 15).

We can now define stochastic integrals in general. Let P, ξ and a be as in Theorem 3.1 and let $\theta: I^s \times \Omega \to R^d$ be an s-non-anticipating function such that $E\left[\int_s^t |\theta(u)|^2\, du\right] < \infty$ for each $t > s$. Choose a sequence $\{\theta_m\}$ as in Lemma 3.4, and define $\int_{t_1}^{t_2} \langle \theta_m(u), d\xi(u)\rangle$ according to (3.4). Then, by Lemma 3.3,

$$P\left[\sup_{t_1 \leqq t \leqq t_2} \left|\int_{t_1}^{t_0} \langle \theta_m(u), d\xi(u)\rangle - \int_{t_1}^{t_2} \langle \theta_{m'}(u), d\xi(u)\right| \geqq \varepsilon\right]$$

$$\leqq \frac{A}{\varepsilon^2} E^P\left[\int_{t_1}^{t_2} |\theta_m(u) - \theta_{m'}(u)|^2\, du\right]$$

for all $m, m' \geqq 1$ and $t_2 \geqq t_1 \geqq s$. Hence, $\int_{t_1}^{t} \langle \theta_m(u), d\xi(u)\rangle$ converges uniformly in probability to a limit which is almost surely continuous. This limit, which we shall denote by $\int_{t_1}^{t} \langle \theta(u), d\xi(u)\rangle$, is independent of the choice of the sequence $\{\theta_m\}$ and depends only on θ. We list some elementary properties of this stochastic integral.

THEOREM 3.2. *Let P, ξ and a be as in Theorem 3.1 and suppose $\theta: I^s \times \Omega \to R^d$ is an s-non-anticipating function such that $E^P\left[\int_s^t |\theta(u)|^2 \, du\right] < \infty$ for $t \geqq s$. Then, the following assertions hold:*

(i) $\int_{t_1}^{t_2} \langle \theta(u), d\xi(u)\rangle + \int_{t_2}^{t_3} \langle \theta(u), d\xi(u)\rangle = \int_{t_1}^{t_3} \langle \theta(u), d\xi(u)\rangle$ a.s. P.

(ii) $\int_{t_0}^{t} \langle \theta(u), d\xi(u)\rangle$ *is an almost surely continuous P-martingale for $t \geqq t_0$ and for all $t_0 \geqq s$,*

(iii) $E\left[\left(\int_{t_1}^{t_2} \langle \theta(u), d\xi(u)\rangle\right)^2\right] = E\left[\int_{t_1}^{t_2} \langle \theta(u), a(u)\theta(u)\rangle \, du\right]$ *for $t_2 \geqq t_1 \geqq s$.*

(iv) $$P\left(\sup_{t_1 \leqq t \leqq t_2} \left|\int_{t_1}^{t} \langle \theta(u), d\xi(u)\rangle\right| \geqq \varepsilon\right) \leqq \frac{1}{\varepsilon^2} E\left[\int_{t_1}^{t_2} \langle \theta(u), a(u)\theta(u)\rangle \, du\right], \qquad t_2 \geqq t_1 \geqq s.$$

(v) *If θ is bounded, then for each $t_0 \geqq s$*

$$X_\theta^{t_0}(t) = \exp\left\{\int_{t_0}^{t} \langle \theta(u), d\xi(u)\rangle - \frac{1}{2}\int_{t_0}^{t} \langle \theta(u), a(u)\theta(u)\rangle \, du\right\}, \qquad t \geqq t_0, \tag{3.5}$$

is a P-martingale.

(vi) *Finally, if θ_1 and θ_0 are two s-non-anticipating functions on $I^s \times \Omega \to R^d$ such that $E^P\left[\int_s^t |\theta_i(u)|^2 \, du\right] < \infty$ for $i = 1, 2$ and $t > s$, then for $t_2 \geqq t_1 \geqq s$ and almost all w*

$$\int_{t_1}^{t_2} \langle \theta_1(u), d\xi(u)\rangle + \int_{t_1}^{t_2} \langle \theta_2(u), d\xi(u)\rangle = \int_{t_1}^{t_2} \langle \theta_1(u) + \theta_2(u), d\xi(u)\rangle .$$

Proof: The only assertion requiring comment is (v). Assuming θ is bounded, we can find a sequence $\{\theta_m\}$ as in Lemma 3.4 bounded by the same bound. For each $m \geqq 1$ and $\lambda \in R$, $[X^s_{\lambda\theta_m}(t) : t \geqq s]$ is a P-martingale. Therefore, by the argument at the beginning of the proof of Lemma 3.2, for $t_2 \geqq t_1 \geqq s$,

$$P\left[\sup_{t_1 \leqq t \leqq t_2} \left|\int_{t_1}^{t_2} \langle \theta_m(u), d\xi(u)\rangle\right| \geqq l\right] \leqq 2 \exp\left\{-\frac{l^2}{2A\,\|\theta\|^2(t_2 - t_1)}\right\},$$

where $\|\theta\|$ is the bound of θ. In particular, the $X^s_{\theta_m}(t)$ are all uniformly integrable for $t \geqq s$. Therefore, by letting $m \to \infty$, $X^s_\theta(t)$ inherits the martingale property; q.e.d.

Let P, ξ and a be as in Theorem 3.1 and suppose that $\sigma: I^s \times \Omega \to R^d$ is s-non-anticipating and $E\left[\int_s^t \langle \theta, \sigma^2(u)\theta \rangle\, du\right] < \infty$ for $t \geqq s$ and $\theta \in R^d$, where $\sigma^2(u) = \sigma^*(u)\sigma(u)$.[6] Then we define $\int_s^t \sigma(u)\, d\xi(u)$ to be the s-non-anticipating function on $I^s \times \Omega \to R^d$ such that, for each $\theta \in R^d$,

$$\left\langle \theta, \int_s^t \sigma(u)\, d\xi(u) \right\rangle = \int_s^t \langle \sigma^*(u)\theta, d\xi(u) \rangle .$$

Clearly for $t \geqq s$, $\int_s^t \sigma(u)\, d\xi(u)$ is almost surely continuous and is a vector-valued P-martingale.

An s-non-anticipating function $\beta: I^s \times \Omega \to R^d$ is said to be an s-Brownian motion with respect to a measure P on (Ω, M^s) if, for $t \geqq s$, $\beta(t)$ is almost surely continuous and, for all $t_2 \geqq t_1 \geqq s$ and almost all w,

$$P[\beta(t_2) \in \Gamma \mid M^s_{t_1}] = \frac{1}{[2\pi(t_2 - t_1)]^{d/2}} \int_\Gamma \exp\left\{ -\frac{|y - \beta(t_1)|^2}{2(t_2 - t_1)} \right\} dy .$$

It is easy to see that if $[\beta(t): t \geqq s]$ is an s-Brownian motion with respect to P, then the increments $\beta(t_2) - \beta(t_1)$ for $t_2 \geqq t_1 \geqq s$ are independent of $M^s_{t_1}$. Furthermore, it is simple to check that an s-non-anticipating function β: $I^s \times \Omega \to R^d$ is an s-Brownian motion with respect to P on (Ω, M^s) if and only if $[\beta(t): t \geqq s]$ is almost surely continuous, $P[\beta(s) = 0] = 1$, and

$$\exp\{\langle \theta, \beta(t) \rangle - \tfrac{1}{2} |\theta|^2 (t - s)\}, \qquad t \geqq s, \tag{3.6}$$

is a P-martingale for each $\theta \in R^d$. In particular, if $[\beta(t): t \geqq s]$ is an s-Brownian motion with respect to P, then we can define $\int_{t_1}^{t_2} \langle \theta(u), d\beta(u) \rangle$ for all s-non-anticipating functionals $\theta: I^s \times \Omega \to R^d$ and all $t_2 \geqq t_1 \geqq s$, provided

$$E\left[\int_{t_1}^{t_2} |\theta(u)|^2\, du\right] < \infty .$$

The properties of these integrals can be read off from Theorem 3.2 by replacing ξ by β and a by the identity throughout.

LEMMA 3.5. *Let P, ξ and a be as in Theorem 3.1 and let $\sigma: I^s \times \Omega \to R^d \otimes R^d$ be an s-non-anticipating function such that*

$$\left| \sum_{i,j=1}^d \sigma_{ij}(u, w)\theta_i\theta_j \right| \leqq B\, |\theta|^2 \quad \textit{for } \theta \in R^d ,$$

[6] If A is a matrix, then A^* denotes the transpose of A.

where B is a constant. Let

$$\eta(t) = \int_s^t \sigma(u)\, d\xi(u)\,.$$

Then,

$$Y_\theta^s(t) = \exp\left\{\langle \theta, \eta(t)\rangle - \frac{1}{2}\int_s^t \langle \theta, \sigma^*(u)a(u)\sigma(u)\rangle\, du\right\}$$

is a P-martingale for $t \geqq s$ and for each $\theta \in R^d$. Moreover, if $\theta: I^s \times \Omega \to R^d$ is an s-non-anticipating function with $E\left[\int_s^t |\theta(u)|^2\, du\right] < \infty$, then $\int_{t_1}^{t_2} \langle \theta(u), d\eta(u)\rangle$ is well defined, and for $t \geqq s$ and almost all w

$$\int_s^t \langle \theta(u), d\eta(u)\rangle = \int_s^t \langle \sigma^*(u)\theta(u), d\xi(u)\rangle\,.$$

Proof: Observe that the first assertion is an immediate consequence of Theorem 3.2, and therefore that P, η and $\sigma^* a\sigma$ satisfy the hypothesis of Theorem 3.1. Thus we know how to define the integrals $\int_{t_1}^{t_2} \langle \theta(u), d\eta(u)\rangle$ for s-non-anticipating $\theta: I^s \times \Omega \to R^d$ with $E\left[\int_s^t |\theta(u)|^2\, du\right] < \infty$. These integrals will have the properties listed in Theorem 3.2. In particular, $\int_s^t \langle \theta(u), d\eta(u)\rangle$ is almost surely continuous. Therefore, it suffices to prove that

$$\int_s^{t_1} \langle \theta(u), d\eta(u)\rangle = \int_s^{t_1} \langle \sigma^*(u)\theta(u), d\xi(u)\rangle \quad \text{a.s.} \quad P\,,$$

for each $t_1 \geqq s$. But this will follow if we can prove it in the case when $\theta(u)$ is n-simple. For such a θ we have

$$\begin{aligned}
\int_s^{t_1} \langle \theta(u), d\eta(u)\rangle &= \sum_{k=0}^{[n(t_1-s)]} \left\langle \theta\left(s + \frac{k}{n}\right), \eta\left(\left(s + \frac{k+1}{n}\right) \wedge t_1\right) - \eta\left(s + \frac{k}{n}\right)\right\rangle \\
&= \sum_{k=0}^{[n(t_1-s)]} \left\langle \theta\left(s + \frac{k}{n}\right), \int_{k/n}^{[(k+1)/n]\wedge t_1} \sigma(u)\, d\xi(u)\right\rangle \\
&= \sum_{k=0}^{[n(t_1-s)]} \int_{k/n}^{[(k+1)/n]\wedge t_1} \langle \sigma^*(u)\theta(u), d\xi(u)\rangle \\
&= \int_s^{t_1} \langle \sigma^*(u)\theta(u), d\xi(u)\rangle\,.
\end{aligned}$$

THEOREM 3.3. *Let P, ξ, a be as in Theorem 3.1. We assume in addition that, for each $u \geqq s$ and $w \in \Omega$, the matrix $a(u, w)$ is symmetric and satisfies*

$$\sum_{i,j=1}^d a_{ij}(u, w)\theta_i\theta_j \geqq A' |\theta|^2$$

for all $\theta \in R^d$, where $A' > 0$ is a constant. Let σ be the positive definite symmetric square root of a and σ^{-1} the inverse of σ. Then for all $\theta \in R^d$,

$$\exp\left\{\int_s^t \langle \sigma^{-1}(u)\theta, d\xi(u)\rangle - \tfrac{1}{2}|\theta|^2(t-s)\right\}, \qquad t \geqq s,$$

is a P-martingale, and for $t \geqq s$

$$\beta(t) = \int_s^t \sigma^{-1}(u)\, d\xi(u) \tag{3.7}$$

is an s-Brownian motion with respect to P. In particular, for $t \geqq s$ and almost all w,

$$\xi(t) - \xi(s) = \int_s^t \sigma(u)\, d\beta(u). \tag{3.8}$$

Proof: The first two assertions follow from Theorem 3.2 and the remarks preceding Lemma 3.5. To prove (3.8), observe that by (3.7) and Lemma 3.5, for $t \geqq s$ and $\theta \in R^d$,

$$\left\langle \theta, \int_s^t \sigma(u)\, d\beta(u)\right\rangle = \int_s^t \langle \sigma(u)\theta, d\beta(u)\rangle = \int_s^t \langle \sigma^{-1}(u)\sigma(u)\theta, d\xi(u)\rangle$$

$$= \langle \theta, \xi(t) - \xi(s)\rangle \text{ a.s. } P.$$

COROLLARY 3.1. *Let P be a probability measure on (Ω, M^s) and let $\xi : I^s \times \Omega \to R^d$ and $a : I^s \times R^d \to R^d \otimes R^d$ be s-non-anticipating. Assume that ξ is almost surely continuous and that a is symmetric and satisfies*

$$A'|\theta|^2 \leqq \langle \theta, a\theta\rangle \leqq A|\theta|^2, \qquad \theta \in R^d,$$

where $0 < A' < A < \infty$ are constants. Then

$$X_\theta^s(t) = \exp\left\{\langle \theta, \xi(t) - \xi(s)\rangle - \frac{1}{2}\int_s^t \langle \theta, a(u)\theta\rangle\, du\right\}, \qquad t \geqq s,$$

is a P-martingale for all $\theta \in R^d$ if and only if there is an s-Brownian motion β with respect to P such that

$$\xi(t) - \xi(s) = \int_s^t \sigma(u)\, d\beta(u), \qquad t \geqq s, \text{ a.s. } P,$$

where σ is the positive definite symmetric square root of a.

Of particular interest for us is the case when $\xi(t) = x(t) - \int_s^t b(u)\,du$, where $b: I^s \times \Omega \to R^d$ is a bounded s-non-anticipating function. Suppose that P, ξ and a are as in Theorem 3.1, where ξ has the above form. Then for $t_2 \geqq t_1 \geqq s$ we know how to define $\int_{t_1}^{t_2} \langle \theta(u), d\xi(u)\rangle$ whenever θ is an s-non-anticipating function such that $E\left[\int_s^t |\theta(u)|^2\,du\right] < \infty$, $t \geqq s$. We also know how to define $\int_{t_1}^{t_2} \langle \theta(u), b(u)\rangle\,du$ for such θ. Therefore, we can introduce the integral

$$\int_{t_1}^{t_2} \langle \theta(u), dx(u)\rangle = \int_{t_1}^{t_2} \langle \theta(u), d\xi(u)\rangle - \int_{t_1}^{t_2} \langle \theta(u), b(u)\rangle\,du\,.$$

It is easy to see that if $\theta(u)$ is n-simple for some $n \geqq 1$, then

$$\int_{t_1}^{t_2} \langle \theta(u), dx(u)\rangle = \sum_{k=[n(t_1-s)]}^{[n(t_2-s)]} \left\langle \theta\left(s + \frac{k}{n}\right), x\left(\left(s + \frac{k+1}{n}\right) \wedge t_2\right) - x\left(\left(s + \frac{k}{n}\right) \vee t_1\right)\right\rangle. \tag{3.9}$$

Moreover, it is clear from the properties of $\int_{t_1}^{t_2} \langle \theta(u), d\xi(u)\rangle$ and $\int_{t_1}^{t_2} \langle \theta(u), b(u)\rangle\,du$ that

$$P\left[\sup_{t_1 \leqq t \leqq t_2} \left|\int_{t_1}^{t_2} \langle \theta(u), dx(u)\rangle - \int_{t_1}^{t_2} \langle \theta_m(u), dx(u)\rangle\right| \geqq \varepsilon\right] \to 0$$

as $m \to \infty$ if $E\left[\int_{t_1}^{t_2} |\theta(u) - \theta_m(u)|^2\,du\right] \to 0$. Thus we could have defined $\int_{t_1}^{t_2} \langle \theta(u), dx(u)\rangle$ by starting with equation (3.9) and then passing to limits as we did before.

COROLLARY 3.2. *Let P be a probability measure on (Ω, M^s) and let $b: I^s \times \Omega \to R^d$ and $a: I^s \times \Omega \to R^d \otimes R^d$ be s-non-anticipating functions such that b is bounded and a is a symmetric matrix satisfying*

$$A'\,|\theta|^2 \leqq \langle \theta, a\theta\rangle \leqq A\,|\theta|^2\,, \qquad \theta \in R^d\,,$$

where $0 < A' \leqq A < \infty$ are constants. Then,

$$X_\theta^s(t) = \exp\left\{\langle \theta, x(t) - x(s)\rangle - \int_s^t \langle \theta, b(u)\rangle\,du - \frac{1}{2}\int_s^t \langle \theta, a(u)\theta\rangle\,du\right\}$$

is a P-martingale for each $\theta \in R^d$ if and only if there is an s-Brownian motion β with respect to P such that

$$x(t) - x(s) = \int_s^t \sigma(u)\, d\beta(u) + \int_s^t b(u)\, du\,, \qquad t \geqq s\,, \text{ a.s. } P,$$

where $\sigma(u, w)$ is the symmetric positive definite square root of $a(u, w)$.

We now prove a theorem concerning the patching together of measures.

LEMMA 3.6. *Let P be a probability measure on (Ω, M^s) and let τ be an s-Markov time. For each $w \in \Omega$ let P_w be a probability measure on $(\Omega, M^{\tau(w)})$ such that*

(i) $P_w[w' : x(\tau(w), w) = x(\tau(w), w')] = 1$ *for each* $w \in \Omega$,

(ii) $w \to P_w(A)$ *is* M_t^s*-measurable on* $\{\tau \leqq t\}$ *for all* $t \geqq s$ *and* $A \in M^t$.

Then there is a unique probability measure P' on (Ω, M^s) such that P' and P agree on M_τ^s and Q'_w (the regular conditional probability distribution of P' given M_τ^s) agrees with P_w on $M^{\tau(w)}$ a.s. *P*.

Proof: For each $w \in \Omega$ define P'_w on (Ω, M^s) such that

$$P'_w[x(u_0) \in \Gamma_0, \cdots, x(u_n) \in \Gamma_n]$$
$$= \begin{cases} \chi_{\Gamma_0}(x(u_0, w)) \cdots \chi_{\Gamma_k}(x(u_k, w)) P_w[x(u_{k+1}) \in \Gamma_{k+1}, \cdots, x(u_n) \in \Gamma_k] \\ \qquad\qquad \text{for } u_k \leqq \tau(w) < u_{k+1}, \\ \chi_{\Gamma_0}(x(u_0, w)) \cdots \chi_{\Gamma_n}(x(u_n, w)) \qquad \text{for } \tau(w) \geqq u_n, \end{cases}$$

where $s = u_0 < u_1 < u_2 \cdots < u_n$, $n \geqq 1$, and $\Gamma_0, \cdots, \Gamma_n \in \mathscr{B}(R^d)$. Clearly, $w \to P'_w(A)$ is M_τ-measurable for all $A \in C^s$, and therefore $w \to P'_w(A)$ is M_τ-measurable for all $A \in M^s$. Also, $P'_w(A) = \chi_A(w)$ for $A \in C_\tau^s$ and, therefore, for all $A \in M_\tau^s$. Define P' on (Ω, M^s) by

$$P'(A) = E^P[P'_w(A)]\,, \qquad A \in M^s\,.$$

Then $P'(A) = P(A)$ for $A \in M_\tau^s$. Moreover, if $A \in M_\tau^s$ and $B \in M^s$, then

$$P'(A \cap B) = E^P[P'_w(A \cap B)] = E^P[\chi_A P'_w(B)] = \int_A P'_w(B) P(dw)\,.$$

Therefore, if $Q''(w, A) = P'_w(A)$, then $Q'' : \Omega \times M^s \to R$ is a regular conditional probability distribution of P' given M_τ^s. Clearly, if $t \geqq s$, $A \in M^t$ and $\tau(w) \leqq t$, then $Q''(w, A) = P'_w(A) = P_w(A)$. Hence if Q' is any regular conditional probability distribution of P' given M_τ^s, then there is a set $N \in M^s$ with $P(N) = P'(N) = 0$ and $Q''_w = Q'_w$ for $w \notin N$. Therefore, P' has the desired properties.

To prove the uniqueness of P', it suffices to note that the conditions of the lemma uniquely determine P' on M^s and also uniquely determine any regular conditional probability distribution of P given M_τ^s up to a M_τ^s-null set.

THEOREM 3.4. *Let P be a probability measure on (Ω, M^s) and let τ be a bounded s-Markov time. Suppose $a^{(1)}$ and $a^{(2)}$ are measurable functions on $I^s \times R^d$ with values in $R^d \otimes R^d$ and $b^{(1)}$ and $b^{(2)}$ are measurable functions on $I^s \times R^d$ with values in R^d such that*

(i) $0 \leqq \langle\theta, a^k\theta\rangle \leqq A\,|\theta|^2, \qquad \theta \in R^d,\ k = 1, 2,$

(ii) $a^{(1)}(u, x(u)) = a^{(2)}(u, x(u))$ *for* $s \leqq u \leqq \tau(w)$ a.s. P,

(iii) $b^{(1)}(u, x(u)) = b^{(2)}(u, x(u))$ *for* $s \leqq u \leqq \tau(w)$ a.s. P,

(iv) $$X_\theta^s(t) = \exp\left\{\langle\theta, x(t) - x(s)\rangle - \int_s^t \langle\theta, b^{(1)}(u, x(u))\theta\rangle\, du - \frac{1}{2}\int_s^t \langle\theta, a^{(1)}(u, x(u))\theta\rangle\, du\right\}$$

is a P-martingale for each $\theta \in R^d$.

For each $w \in \Omega$, let P_w be a probability measure on $(\Omega, M^{\tau(w)})$ such that

(v) $P_w[w' : x(\tau(w), w') = x(\tau(w), w)] = 1$ *for all* $w \in \Omega$,

(vi) $w \to P_w(A)$ *is M_t^s-measurable on* $\{\tau \leqq t\}$ *for all* $A \in M^t$ *and* $t \geqq s$,

(vii) $$Y^{\tau(w)} = \exp\left\{\langle\theta, x(t) - x(\tau(w))\rangle - \int_{\tau(w)}^t \langle\theta, b^{(2)}(u, x(u))\rangle\, du - \frac{1}{2}\int_{\tau(w)}^t \langle\theta, a^{(2)}(u, x(u))\theta\rangle\, du\right\}, \qquad t \geqq \tau(w),$$

is a P_w-martingale for all $\theta \in R^d$ and $w \in \Omega$.

Then there is a probability measure P' on (Ω, M^s) such that $P' = P$ on M_τ^s and

$$Y_\theta^s(t) = \exp\left\{\langle\theta, x(t) - x(s)\rangle - \int_s^t \langle\theta, b^{(2)}(u, x(u))\rangle\, du - \frac{1}{2}\int_s^t \langle\theta, a^{(2)}(u, x(u))\theta\rangle\, du\right\}, \qquad t \geqq s,$$

is a P'-martingale for all $\theta \in R^d$.

Proof: Let P' be the probability measure on (Ω, M^s) such that $P' = P$ identically on M_τ^s, with the regular conditional probability distribution Q'_w of P' given M_τ^s taken as P_w on $M^{\tau(w)}$ for each $w \in \Omega$. We shall show that

$$\int_A Y_\theta^s(t_2)\, dP' = \int_A Y_\theta^s(t_1)\, dP'$$

for $t_2 \geqq t_1 \geqq s$ and $A \in M^s_{t_1}$. In fact, let

$$A = \{x(u_0) \in \Gamma_0, \cdots, x(u_n) \in \Gamma_n\},$$

$$B_k = \{x(u_0) \in \Gamma_0, \cdots, x(u_k) \in \Gamma_n\},$$

$$C_k = \{x(u_{k+1}) \in \Gamma_{k+1}, \cdots, x(u_n) \in \Gamma_n\},$$

where $s = u_0 < u_1 \cdots < u_n = t_1$, $n > k \geqq 0$ and $\Gamma_0, \cdots, \Gamma_n \in \mathscr{B}(R^d)$. Then,

$$\begin{aligned}\int_{\{u_k \leqq \tau < u_{k+1}\}} E^{Q_w'}[\chi_A Y^s_\theta(t_2)]P(dw) &= \int_{\{u_k \leqq \tau < u_{k+1}\} \cap B_k} X^s_\theta(\tau(w), w) E^{P_w}[\chi_{C_k} Y^{\tau(w)}_\theta(t_2)]P(dw) \\ &= \int_{\{u_k \leqq \tau < u_{k+1}\} \cap B_k} X^s_\theta(\tau(w), w) E^{P_w}[\chi_{C_k} Y^{\tau(w)}_\theta(t_1)]P(dw) \\ &= \int_{\{u_k \leqq \tau < u_{k+1}\}} E^{Q_w'}[\chi_A Y^s_\theta(t_1)]P(dw)\end{aligned}$$

for $0 \leqq k \leqq n-1$. Thus,

$$\int_{\{\tau < t_1\}} E^{Q_w'}[\chi_A Y^s_\theta(t_2)]P(dw) = \int_{\{\tau < t_1\}} E^{Q_w'}[\chi_A Y^s_\theta(t_1)]P(dw).$$

Next we observe that

$$\begin{aligned}\int_{\{t_1 \leqq \tau < t_2\}} E^{Q_w'}[\chi_A Y^s_\theta(t_2)]P(dw) &= \int_{\{t_1 \leqq \tau < t_2\} \cap A} X^s_\theta(\tau(w), w) E^{P_w}[Y^{\tau(w)}_\theta(t_2)]P(dw) \\ &= \int_{\{t_1 \leqq \tau < t_2\} \cap A} X^s_\theta(\tau(w), w) P(dw),\end{aligned}$$

and

$$\begin{aligned}\int_{\{\tau \geqq t_2\}} E^{Q_w'}[\chi_A Y^s_\theta(t_2)]P(dw) &= \int_{\{\tau \geqq t_2\} \cap A} X^s_\theta(t_2) P(dw) \\ &= \int_{\{\tau \geqq t_2\} \cap A} X^s_\theta(\tau(w), w) P(dw),\end{aligned}$$

since $\{\tau \geqq t_2\} \cap A \in M_{\tau \wedge t_2}$. Combining these equations, we get

$$\int_{\{\tau \geqq t_1\}} E^{Q_w'}[\chi_A Y_\theta^s(t_2)]P(dw) = \int_{\{\tau \geqq t_1\} \cap A} X_\theta^s(\tau(w), w)P(dw)$$

$$= \int_{\{\tau \geqq t_1\} \cap A} X_\theta^s(\tau \wedge t_1)P(dw)$$

$$= \int_{\{\tau \geqq t_1\} \cap A} Y_\theta^s(t_1)P(dw)$$

$$= \int_{\{\tau \geqq t_1\}} E^{Q_w'}[\chi_A Y_\theta^s(t_1)]P(dw) .$$

Hence,

$$\int_A Y_\theta^s(t_2)\, dP' = \int E^{Q_w'}[\chi_A Y_\theta^s(t_2)]P(dw)$$

$$= \int E^{Q_w'}[\chi_A Y_\theta^s(t_1)]P(dw)$$

$$= \int_A Y_\theta^s(t_1)\, dP' .$$

4. The Existence of Solutions to the Martingale Problem

Let $a:[s, \infty) \times \Omega \to R^d \otimes R^d$ be an s-non-anticipating function which is symmetric and satisfies

$$0 \leqq \langle \theta, a\theta \rangle \leqq A\, |\theta|^2 , \qquad \theta \in R^d . \tag{4.1}$$

We say that a probability measure P on (Ω, M^s) is a solution to the *martingale problem for a, starting from x at time s,* if

(I) $P(x(s) = x) = 1$,

(II) $X_\theta^s(t) = \exp\left\{\langle \theta, x(t) - x(s) \rangle - \frac{1}{2}\int_s^t \langle \theta, a(u)\theta \rangle\, du\right\}$, $t \geqq s$, is a P-martingale for all $\theta \in R^d$.

As we saw in the preceding section, if a is positive definite, then P is a solution to this problem if and only if there is an s-Brownian motion β with respect to P such that

$$x(t) - x = \int_s^t \sigma(u)\, d\beta(u) , \qquad t \geqq s , \text{ a.s. } P . \tag{4.2}$$

In the case when $a(u) = a(x(u))$, where $a: R^d \to R^d \otimes R^d$, it is clear that the problem of finding a P satisfying (I) and (II) is related to the problem of finding a Markov semigroup $\{T_t\}_{t>0}$ such that, for all $f \in C_0^2(R^d)$,

$$T_t f(x) - f(x) = \int_0^t T_u Lf(x)\, du\,, \qquad t \geqq 0,\ x \in R^d\,, \tag{4.3}$$

where

$$L = \frac{1}{2} \sum_{i,j=1}^{d} a_{ij}(x) \frac{\partial^2}{\partial x_i\, \partial x_j}\,.$$

However, the two problems are not identical. Indeed, we shall see that the existence of such a semigroup implies the simultaneous existence of a whole Markov family of solutions to the martingale; whereas, even if we construct a martingale solution for each $x \in R^d$, we cannot assert the existence of such a semigroup unless we have uniqueness as well.

THEOREM 4.1. *Let $a: R^d \to R^d \otimes R^d$ be a continuous function such that*

$$0 \leqq \langle \theta, a\theta \rangle \leqq A\, |\theta|^2\,, \qquad \theta \in R^d\,,$$

and let $L = \frac{1}{2} \sum_{i,j=1}^d a_{ij}(x)\, \partial^2/\partial x_i\, \partial x_j$. Suppose $\{T_t\}_{t>0}$ is a Markov semigroup on $\mathscr{B}(R^d)$ such that

$$T_t f(x) - f(x) = \int_0^t T_u L f(x)\, du\,, \qquad t \geqq 0,\ x \in R^d\,, \tag{4.4}$$

for all $f \in C_0^\infty(R^d)$. Then there is a unique probability measure P_x on (Ω, M^0) for each $x \in R^d$ such that P_x is a solution to the martingale problem for $a = a(x(u))$ starting from x, and

$$\begin{aligned} &P_x(x_{u_1} \in \Gamma_1, \cdots, x_{u_n} \in \Gamma_n) \\ &\quad = \int_{\Gamma_1} \cdots \int_{\Gamma_n} P(u_1, x, dy_1) P(u_2 - u_1, dy_2) \cdots P(u_n - u_{n-1}, y_{n-1}, dy_n)\,, \end{aligned} \tag{4.5}$$

for $0 \leqq u_1 < \cdots < u_n$, $n \geqq 1$, and $\Gamma_1, \cdots, \Gamma_n \in \mathscr{B}(R^d)$, where $P(t, x, dy)$ is the transition probability function associated with $\{T_t\}_{t>0}$.

Proof: It is easy to check that (4.4) must remain true for $f \in C_b^2(R^d)$.[7] From this fact one can see that, for any $\varepsilon > 0$, $\sup_{x \in R^d} P(t, x, R^d - B(x, \varepsilon)) \to 0$ as $t \downarrow 0$.[8]

[7] $C_b^k(R^d)$, $k \geqq 0$, stands for the set of functions on R^d which have k bounded continuous derivatives.

[8] $B(x, r) = \{y \in R^d : |y - x| < r\}$.

Hence, by Theorem 6.3 in Dynkin [4], there is a right continuous Markov process $(\bar{x}(t), \bar{M}_t, \bar{P}_x)$ such that $\bar{P}_x(\bar{x}(t) \in \Gamma) = P(t, x, \Gamma)$, $t \geqq 0$, $x \in R^d$, and $\Gamma \in \mathscr{B}(R^d)$. Given $f \in C_b^2(R^d)$ such that f is uniformly bounded below by some $\alpha > 0$, consider

$$u(t, x) = E^{\bar{P}_x}\left[f(\bar{x}(t)) \exp\left\{-\int_0^t \frac{Lf(\bar{x}_u)}{f(\bar{x}_u)}\, du\right\}\right].$$

It is easy to see that $\partial u(t, x)/\partial t = 0$, $t \geqq 0$, and therefore $u(t, x) = u(0, x) = f(x)$ for all $t \geqq 0$ and $x \in R^d$. Thus, if we choose a sequence of functions $f_n \in C_b^2(R^d)$ such that each f_n is bounded below by some positive α_n and $f_n(x)$, together with its first and second derivatives, tend to $e^{\langle\theta, x\rangle}$ and its derivatives, then we see from Fatou's lemma that

$$e^{\langle\theta, x\rangle} = \lim_{n\to\infty} f_n(x) \geqq E^{\bar{P}_x}\left[\exp\left\{\langle\theta, \bar{x}(t)\rangle - \frac{1}{2}\int_0^t \langle\theta, a(\bar{x}(u))\theta\rangle\, du\right\}\right], \quad t \geqq 0.$$

Since this is true for all $\theta \in R^d$, an application of Chebychev's inequality yields

$$\bar{P}_x(|\bar{x}(t) - x| \geqq l) \leqq C \exp\left\{\frac{l^2}{2At}\right\}, \qquad l \geqq 0.$$

In particular, this shows that we could have chosen the preceding sequence of the f_n so that

$$E^{\bar{P}_x}\left[f_n(\bar{x}(t)) \exp\left\{-\int_0^t \frac{Lf_n(\bar{x}(u))}{f_n(\bar{x}(u))}\, du\right\}\right]$$
$$\to E^{\bar{P}_x}\left[\exp\left\{\langle\theta, \bar{x}(t)\rangle - \frac{1}{2}\int_0^t \langle\theta, a(\bar{x}(u))\theta\rangle\, du\right\}\right],$$

and therefore

$$E^{\bar{P}_x}\left[\exp\left\{\langle\theta, \bar{x}(t) - x\rangle - \frac{1}{2}\int_0^t \langle\theta, a(\bar{x}(u))\theta\rangle\, du\right\}\right] = 1,$$

$$t \geqq 0,\ x \in R^d \text{ and } \theta \in R^d.$$

Hence, if $t_2 \geqq t_1 \geqq 0$, then for $A \in \bar{M}_{t_1}^0$

$$\int_A \exp\left\{\langle\theta, \bar{x}(t_2) - x\rangle - \frac{1}{2}\int_0^{t_2} \langle\theta, a(\bar{x}(u))\theta\rangle\, du\right\}\, d\bar{P}_x$$
$$= E^{\bar{P}_x}\left[\chi_A \exp\left\{\langle\theta, \bar{x}(t_1) - x\rangle - \frac{1}{2}\int_0^{t_1} \langle\theta, a(\bar{x}(u)\theta\rangle\, du\right\} F(\bar{x}(t_1))\right],$$

where

$$F(\bar{x}(t_1)) = E^{\bar{P}_{\bar{x}(t_1)}}\left[\exp\left\{\langle\theta, \bar{x}(t_2 - t_1)\rangle - \frac{1}{2}\int_0^{t_2 - t_1} \langle\theta, a(\bar{x}(u)\theta\rangle\, du\right\}\right] = 1.$$

Consequently,

$$\exp\left\{\langle\theta, \bar{x}(t) - x\rangle - \frac{1}{2}\int_0^t \langle\theta, a(\bar{x}(u)\theta\rangle\, du\right\}, \qquad t \geqq 0,$$

is a $\bar{P}_x$-martingale for all $\theta \in R^d$. Using the same reasoning as in Lemma 3.2, it follows that

$$E^{\bar{P}_x}[|\bar{x}(t_2) - \bar{x}(t_1)|^4] \leqq C\,|t_2 - t_1|^2 .$$

Hence, by Kolmogorov's theorem, there is a unique probability measure P_x on (Ω, M^0) satisfying (4.5). Clearly this P_x fulfills (I) and (II); q.e.d.

Theorem 4.1 shows that one method of constructing solutions to the martingale problem in the case $a(u) = a(x(u))$ is by a semigroup approach. Thus, for instance, if $a: R^d \to R^d \otimes R^d$ is uniformly continuous and satisfies

$$A'\,|\theta|^2 \leqq \langle\theta, a\theta\rangle \leqq A\,|\theta|^2, \qquad \theta \in R^d,$$

then the results of Tanaka [17] and Krylov [12] assure us the existence of solutions to the corresponding martingale problem. This method could also be extended to the case when $a(u) = a(u, x(u))$ by considering operators T_{t_1,t_2}, $0 \leqq t_1 \leqq t_2$, such that

$$T_{t_1,t_3} = T_{t_1,t_2}T_{t_2,t_3}, \qquad 0 \leqq t_1 \leqq t_2 \leqq t_3,$$

$$T_{t_1,t_3}f(x) - T_{t_2,t_3}f(x) = \int_{t_1}^{t_2} T_{u,t_3}L_u f(x)\, du, \quad 0 \leqq t_1 \leqq t_2 \leqq t_3,$$

for $f \in C_0^\infty(R^d)$. However, the existence of such operators is not easily proved, and in any case the operator theory approach to our problem cannot work in the general non-anticipating case. Hence we shall abandon this tack and turn instead to one which exploits the connection with stochastic integral equations.

Let $a: [s, \infty) \times \Omega \to R^d \otimes R^d$ be a continuous s-non-anticipating function satisfying (4.1). Our aim is to construct a sequence of probability measures $P^{(m)}$ on (Ω, M^s) such that, for each $m \geqq 1$, $P^{(m)}$ is a solution to the martingale problem for $a(s + [m(u - s)]/m)$, $u \geqq s$, starting from x. Once this is done, it will be easy to find a subsequence of $\{P^{(m)}\}$ which converges to a solution for a.

Let W^s be the probability measure on (Ω, M^s) such that

$$W^s(\{x(u_1) \in \Gamma_1, \cdots, x(u_n) \in \Gamma_n\})$$
$$= \int_{\Gamma_1} \cdots \int_{\Gamma_n} g(s, 0, u_1, y_1) \cdots g(u_{n-1}, y_{n-1}, u_n, y_n)\, dy_1 \cdots dy_n,$$

where $s \leqq u_1 < u_2 < \cdots < u_n$, $n \geqq 1$, $\Gamma_1, \cdots, \Gamma_n \in \mathscr{B}(R^d)$ and

$$g(s, x, u, y) = [2\pi(u - s)]^{-d/2} \exp\left\{-\frac{|y - x|^2}{2(u - s)}\right\}.$$

The measure W^s is called *s-Wiener measure* and its existence is well known. Notice that W^s could equally well be described as the measure on (Ω, M^s) with respect to which the non-anticipating function $x(t)$, $t \geqq s$, is an s-Brownian motion. Given a symmetric positive semi-definite matrix a and an $x \in R^d$, $W^s_{a,x}$ is defined to be the probability measure on (Ω, M^s) such that

$$W^s_{a,x}(x(u_1) \in \Gamma_1, \cdots, x(u_n) \in \Gamma_n)$$
$$= W^s(\sigma(x(u_1) - x) \in \Gamma_1, \cdots, \sigma(x(u_n) - x) \in \Gamma_n)$$

for $s < u_1 < \cdots < u_n$, $n \geqq 1$, and $\Gamma_1, \cdots, \Gamma_n \in \mathscr{B}(R^d)$, where σ is the positive semi-definite symmetric square root of a. It is easily checked that

$$\exp\{\langle\theta, x(t) - x\rangle - \tfrac{1}{2}(t - s)\langle\theta, a\theta\rangle\}, \qquad t \geqq s,$$

is a $W^s_{a,x}$-martingale for all $\theta \in R^d$.

LEMMA 4.1. *Let $a: [s, \infty) \times \Omega \to R^d \otimes R^d$ be an s-non-anticipating function such that*

(i) *a is symmetric and satisfies* (4.1),

(ii) *a is m-simple for some $m \geqq 1$.*

Then there is a probability measure P on (Ω, M^s) for each $x \in R^d$ which solves the martingale problem for a, starting from x.

Proof: We shall define a sequence of measures P'_k, $k \geqq 1$, on (Ω, M^s) by induction. P'_1 is equal to $W^s_{a_0,x}$, where a_0 is the matrix $a(s, w_0)$ and w_0 is any element of Ω such that $x(s, w_0) = x$. If P'_k has been defined, then P'_{k+1} is the uniquely determined measure on (Ω, M^s) such that $P'_{k+1} = P'_k$ identically on $M^s_{s+k/m}$ and the regular conditional probability distribution Q_w of P'_{k+1} on $M^{s+k/m}$, given $M^s_{s+k/m}$, is $W^{s+k/m}_{a(s+k/m,w),x(k/m,w)}$ a.s. P'_k. The existence and uniqueness of P'_{k+1} is proved in Lemma 3.6.

Now take $P_k = P'_k$ on $M^s_{s+k/m}$ and observe that $P_{k+1} = P_k$ on $M^s_{s+k/m}$. Thus, by Theorem V.4.1 in Parthasarathy [14], the P_k determine a probability measure P on (Ω, M^s) such that $P = P_k$ on $M^s_{s+k/m}$ for all k. We shall show that P is the desired measure. Clearly, $P(x(s) = x) = 1$. In order to see that

$$X_\theta(t) = \exp\left\{\langle\theta, x(t) - x\rangle - \frac{1}{2}\int_s^t \langle\theta, a(u)\theta\rangle\, du\right\}, \qquad t \geqq s,$$

is a P-martingale, it suffices to prove that if

$$s + \frac{k}{m} \leqq t_1 \leqq t_2 \leqq s + \frac{k+1}{m},$$

then

$$E^{P_{k+1}}[X_\theta(t_2) \mid M^s_{t_1}] = X_\theta(t_1) \quad \text{a.s.} \quad P'_{k+1}.$$

To this end, let $A = B \cap C$, where $B \in M^s_{s+k/m}$ and $C \in M^{s+k/m}_{t_1}$. Then,

$$\int_A X_\theta(t_2)\, dP'_{k+1} = \int_B X_\theta\left(s + \frac{k}{m}, w\right)\left(\int_C \exp\left\{\left\langle \theta, x(t_2) - x\left(s + \frac{k}{m}, w\right)\right\rangle\right.\right.$$

$$\left.\left. - \frac{1}{2}\left(t_2 - \left(s + \frac{k}{m}\right)\right)\left\langle \theta, a\left(s + \frac{k}{m}, w\right)\theta\right\rangle\right\} dW^{s+k/m}_{a(s+k/m,w),x(s+k/m,w)}\right) P(dw)$$

$$= \int_B X_\theta\left(s + \frac{k}{m}, w\right)\left(\int_C \exp\left\{\left\langle \theta, x(t_1) - x\left(s + \frac{k}{m}, w\right)\right\rangle\right.\right.$$

$$\left.\left. - \frac{1}{2}\left(t_1 - \left(s + \frac{k}{m}\right)\right)\left\langle \theta, a\left(s + \frac{k}{m}, w\right)\theta\right\rangle\right\} dW^{s+k/m}_{a(s+k/m,w),x(s+k/m,w)}\right) P(dw)$$

$$= \int_A X_\theta(t_1)\, dP\,.$$

Since $M^s_{t_1}$ is generated by sets A of this form, it follows that

$$E^{P'_{k+1}}[X_\theta(t_2) \mid M^s_{t_1}] = X_\theta(t_1) \text{ a.s. } P'_{k+1}\,.$$

THEOREM 4.2. *Let $a:[s, \infty) \times \Omega \to R^d \otimes R^d$ be a symmetric continuous s-non-anticipating function satisfying (4.1). Then for each $x \in R^d$ there is a probability measure P on (Ω, M^s) such that*

$$P(x(s) = x) = 1$$

and

$$X_\theta(t) = \exp\left\{\langle \theta, x(t) - x\rangle - \frac{1}{2}\int_s^t \langle \theta, a(u)\theta\rangle\, du\right\}, \qquad t \geqq s\,,$$

is a P-martingale for all $\theta \in R^d$.

Proof: For each $m \geqq 1$, define $a^{(m)}(u) = a(s + [m(u - s)]/m)$, $u \geqq s$, and let $P^{(m)}$ be a probability measure on (Ω, M^s) such that

$$P^{(m)}(x(s) = x) = 1$$

and

$$X^{(m)}_\theta(t) = \exp\left\{\langle \theta, x(t) - x\rangle - \frac{1}{2}\int_s^t \langle \theta, a^{(m)}(u)\theta\rangle\, du\right\}, \qquad t \geqq s\,,$$

is a $P^{(m)}$-martingale for all $\theta \in R^d$. Then by Lemma 3.2,

$$E^{P^{(m)}}[|x(t_2) - x(t_1)|^4] \leqq C(t_2 - t_1)^2\,, \qquad t_2 \geqq t_1 \geqq s\,,$$

where C depends only on the constant A in equation (4.1). Thus, by Theorem 2.3, there is a subsequence $P^{(m_k)}$, $k \geqq 1$, of $P^{(m)}$, $m \geqq 1$, converging to a probability measure P on (Ω, M^s) in the sense that, if $t \geqq s$, then

$$E^P[\eta] = \lim_{k \to \infty} E^{P^{(m_k)}}[\eta]$$

for any bounded continuous M_t^s-measurable function η. In particular, since $\{x(s) = x\}$ is closed and M_s^s-measurable,

$$P(x(s) = x) \geqq \overline{\lim_{k \to \infty}} P^{(m_k)}\{(x(s) = x)\} = 1 .$$

We must still show that $X_\theta(t)$, $t \geqq s$, is a P-martingale. In order to do this it suffices to show that, if $t \geqq s$ and if η is a bounded continuous M_t^s-measurable function, then

$$E^P[\eta X_\theta(t)] = \lim_{k \to \infty} E^{P^{(m_k)}}[\eta X_\theta^{(m_k)}(t)] .$$

Note that for each $N > 0$, the functions $\eta \cdot (X^{(m_k)}(t) \wedge N)$, $k \geqq 1$, are uniformly bounded, M_t^s-measurable, and equicontinuous at each $w \in \Omega$. Thus, since $X_\theta^{(m_k)}(t) \to X_\theta(t)$ as $k \to \infty$, it follows from Theorem 6.8 in Parthasarathy [14] that

$$E^P[\eta \cdot (X_\theta(t) \wedge N)] = \lim_{k \to \infty} E^{P^{(m_k)}}[\eta \cdot (X_\theta^{(m_k)}(t) \wedge N)] ,$$

for each $N > 0$. Hence, we need only show that

$$E^P[|X_\theta(t) - (X_\theta(t) \wedge N)|]$$

and

$$\sup_{k \geqq 1} E^{P^{(m_k)}}[|X_\theta^{(m_k)}(t) - (X_\theta^{(m_k)}(t) \wedge N)|]$$

tend to 0 as $N \to \infty$. But by Lemma 3.1,

$$P^{(m_k)}(\sup_{s \leqq u \leqq t} |x(u) - x(s)| > l) \leqq 2d \exp\left\{-\frac{l^2}{2A(t-s)}\right\},$$

and so

$$\sup_{k \geqq 1} E^{P^{(m_k)}}[|X_\theta^{(m_k)}(t) - X_\theta^{(m_k)}(t) \wedge N|] \to 0$$

as $N \to \infty$. Moreover, $\{|x(t) - x(s)| > l\}$ is open and M_t^s-measurable, and therefore

$$\begin{aligned} P(|x(t) - x(s)| > l) &\leqq \underline{\lim_{k \to \infty}} P^{(m_k)}(|x(t) - x(s)| > l) \\ &\leqq 2d \exp\left\{-\frac{l^2}{2A(t-s)}\right\}. \end{aligned}$$

Thus $E^P[|X_\theta(t) - (X_\theta(t) \wedge N)|] \to 0$ as $N \to \infty$.

5. Uniqueness

Let us consider the fundamental solution

$$g(s, x, t, y) = [2\pi(t-s)]^{-d/2} \exp\left\{-\frac{|y-x|^2}{2(t-s)}\right\}, \qquad t \geqq s \geqq 0,$$

of the heat equation

$$\frac{\partial u}{\partial s} + \frac{1}{2}\sum \frac{\partial^2 u}{\partial x_1^2} = f$$

in d dimensions. We define the operators G_λ acting on suitable functions $f(t, y)$ on $[0, \infty) \times R^d$ by

$$(G_\lambda f)(s, x) = \int_s^\infty \int_{R^d} g(s, x, t, y) e^{-\lambda(t-s)} f(t, y)\, dt\, dy, \qquad \lambda > 0.$$

If $h = G_\lambda f$, then h formally satisfies the equation

$$\lambda h - \frac{\partial h}{\partial s} - \frac{1}{2}\sum \frac{\partial^2 h}{\partial x_i^2} = f. \tag{5.1}$$

The following facts regarding G_λ are well known and are easily verified.

1°. G_λ maps $L_p([0, \infty) \times R^d)$ into itself and has norm at most λ^{-1}. This is true for $1 \leqq p \leqq \infty$.

2°. The integral

$$\int_s^\infty \int_{R^d} |g(s, x, t, y)|^\alpha e^{-\lambda\alpha(t-s)}\, dt\, dy = \|G_\lambda\|_\alpha^\alpha$$

can be computed and shown to be finite provided $\alpha < (d+2)/d$. Together with Young's inequality, this shows that G_λ maps L_{p_1} to L_{p_2} boundedly, provided $p_2 > p_1$ is such that

$$\frac{1}{p_1} - \frac{1}{p_2} = 1 - \frac{1}{\alpha}$$

for some $\alpha < (d+2)/d$.

3°. The integral

$$\int_s^\infty \int_{R^d} \left\| \frac{\partial g}{\partial x}(s, x, t, y) \right\|^\alpha e^{-\lambda\alpha(t-s)}\, dt\, dy = \|\nabla G_\lambda\|^\alpha$$

can be computed and shown to be finite provided $\alpha < (d+2)/(d+1)$.

4°. A consequence of 2° is that there exists a p_B such that $1 < p_B < \infty$ and G_λ maps L_{p_B} into bounded continuous functions and

$$\sup_{s,x} |(G_\lambda f)(s, x)| \leqq C_{\lambda, p_B} \|f\|_{p_B};$$

p_B of course must be chosen so that

$$\frac{1}{p_B} + \frac{1}{\alpha} = 1 \quad \text{for some} \quad \alpha < \frac{d+2}{d}.$$

5°. A consequence of 3° is that there exists a p_L such that $1 < p_L < \infty$ and G_λ maps L_{p_L} into Lipschitz continuous functions and

$$|(G_\lambda f)(s, x_1) - (G_\lambda f)(s, x_2)| \leqq C_{\lambda, p_L} \|f\|_{p_L} |x_1 - x_2| ;$$

p_L must be chosen so that

$$\frac{1}{p_L} + \frac{1}{\alpha} = 1 \quad \text{for some} \quad \alpha < \frac{d+2}{d+1}.$$

Moreover, $G_\lambda f$ has uniformly continuous first derivatives in x and

$$\begin{aligned} |(G_\lambda f)(s, x_1 + h) - (G_\lambda f)(s, x_1) - (G_\lambda f)(s, x_2 + h) + (G_\lambda f)(s, x_2)| \\ \leqq w_0(|h|) C_{\lambda, p_L} \|f\|_{p_L} |x_1 - x_2| , \end{aligned}$$

where $w_0(|h|) \downarrow 0$ as $|h| \downarrow 0$.

6°. If f is a C^∞ function having compact support, then the function $h_\lambda = G_\lambda f$ is again C^∞, vanishes exponentially fast as $|x| \to \infty$, is identically zero for large s, and satisfies equation (5.1).

From equation (5.1) it follows that $h = h_\lambda$ satisfies the equation

$$\frac{\partial h}{\partial s} + \frac{1}{2} \sum \frac{\partial^2 h}{\partial x_i^2} = \lambda h - f,$$

and moreover

$$h(s, x) = \int_s^\infty \int_{R^d} (\lambda h - f)(t, y) g(s, x, t, y)\, dt\, dy .$$

We know further that for any p in $1 \leqq p \leqq \infty$, $\|h\|_p \leqq \lambda^{-1} \|f\|_p$. Therefore, using a result of B. F. Jones (cf. Theorem 2.6), we conclude that

$$\left\| \frac{\partial^2 h}{\partial x_i\, \partial x_j} \right\|_p \leqq C_{d,p} \|\lambda h - f\|_p \leqq 2C_{d,p} \|f\|_p , \tag{5.2}$$

where $C_{d,p}$ is an absolute constant depending only on d and p and well defined for $1 < p < \infty$.[9]

[9] This inequality of B. F. Jones is an extension to parabolic equations of the renowned *a priori* estimate of Calderon and Zygmund, who considered elliptic equations.

Let $a_{ij}(s, x)$ be symmetric and positive definite. Let us further assume that $|a_{ij} - \delta_{ij}| \leqq \varepsilon$, everywhere. Moreover, we shall assume that ε is so small that any such coefficients are always uniformly elliptic. A more stringent choice of ε will be made later on. We shall denote by $\varepsilon_{ij}(s, x)$ the difference $a_{ij}(s, x) - \delta_{ij}$, thus giving us $|\varepsilon_{ij}| \leqq \varepsilon$. We define the operator T_λ on smooth functions $f(t, y)$ with compact support by

$$T_\lambda f = \frac{1}{2} \sum \varepsilon_{ij} \frac{\partial^2 (G_\lambda f)}{\partial x_i \, \partial x_j} .$$

From (5.2) it follows that T_λ is a bounded operator from L_p to L_p for all $1 < p < \infty$ and in fact can be extended to all of L_p. An elementary computation yields

$$\|T_\lambda\|_p \leqq d^2 \varepsilon C_{d,p} .$$

Let $1 < p_1 < p_2 \cdots < p_N < \infty$ be any finite set of p-values. Then one can clearly choose an $\varepsilon > 0$ such that

$$d^2 \varepsilon C_{d,p} \leqq \tfrac{1}{2} \quad \text{for} \quad p = p_1, \cdots, p_N, \tag{5.3}$$

implying that

$$\|T_\lambda\|_p \leqq \tfrac{1}{2} \quad \text{for} \quad p = p_1, \cdots, p_N .$$

Let us now define the operators K_λ by

$$K_\lambda = G_\lambda (I - T_\lambda)^{-1} .$$

$(I - T_\lambda)^{-1}$ is well defined as bounded operator with norm at most 2 from L_{p_i} into L_{p_i} for $i = 1, 2, \cdots, N$. Let $p_1 > 1$ be arbitrary but fixed. Then, in view of property 2° of G_λ, we can construct a finite sequence $p_2, \cdots, p_N$ such that G_λ maps L_{p_i} boundedly into $L_{p_{i+1}}$ for $i = 1, 2, \cdots, N - 1$ and p_N itself satisfies

$$\frac{1}{p_N} + \frac{1}{\alpha_0} = 1 \quad \text{for some} \quad \alpha_0 < \frac{d+2}{d+1} .$$

N depends only on the dimension. If we now choose $\varepsilon > 0$ small enough according to (5.3), then we can conclude that K_λ has the same properties as G_λ: that K_λ maps L_{p_i} into $L_{p_{i+1}}$ for $i = 1, 2, \cdots, N - 1$, and that

$$\sup_{(s,x)} |(K_\lambda f)(s, x)| \leqq C_\lambda \|f\|_{p_N} ,$$

$$|(K_\lambda f)(s, x_1) - (K_\lambda f)(s, x_2)| \leqq C_\lambda \|f\|_{p_N} |x_1 - x_2| ,$$

where C_λ is some constant. Although it is obvious on heuristic grounds, we shall prove the following consistency lemma.

LEMMA 5.1. *K_λ is defined uniquely, independent of the choice of the L_p space used.*

Proof: We need prove this only for $(I - T_\lambda)^{-1}$. Since T_λ was obtained by extention from smooth functions, the various extensions of T_λ are obviously consistent. $(I - T_\lambda)^{-1} = \sum_{r=0}^{\infty} T_\lambda^r$, therefore, has the same property; q.e.d.

If instead of $((\delta_{ij}))$ we choose any positive definite symmetric matrix $((C_{ij}))$, then we can define the analogous Green's function $g_C(s, x, t, y)$ and the corresponding operator G_λ. For coefficients $a_{ij}(s, x) = C_{ij} + \varepsilon_{ij}(s, x)$ we can define again the operator

$$T_\lambda f = \frac{1}{2} \sum \varepsilon_{ij} \frac{\partial^2(G_\lambda f)}{\partial x_i \, \partial x_j}$$

and the corresponding $K_\lambda = G_\lambda(I - T_\lambda)^{-1}$. These of course will behave exactly in the same manner except that the choice of ε, and the various bounds, will be governed by the size of $((C_{ij}))$, or more precisely by the lowest and highest eigenvalues of $((C_{ij}))$.

We summarize these remarks in the following theorem.

THEOREM 5.1. *There exists an $\varepsilon > 0$ depending on the positive definite matrix $((C_{ij}))$ such that if a_{ij} are symmetric coefficients satisfying $|a_{ij} - C_{ij}| \leqq \varepsilon$, then a is uniformly elliptic and the corresponding operator K_λ maps L_{p_i} boundedly into $L_{p_{i+1}}$ for $i = 1, 2, \cdots, N-1$. Moreover,*

$$\sup_{(s,x)} |K_\lambda f(s, x)| \leqq C_\lambda \, \|f\|_{p_N}$$

and

$$|K_\lambda f(s, x_1) - K_\lambda f(s, x_2)| \leqq C_\lambda \, \|f\|_{p_N} \, |x_1 - x_2| \, .$$

The choice of ε, the bounds for the various norms of K_λ from $L_{p_i} \to L_{p_{i+1}}$ and the constant C_λ can be made independent of $((C_{ij}))$ provided the eigenvalues of $((C_{ij}))$ stay in an interval $[\rho_-, \rho_+]$ with $0 < \rho_- < \rho_+ < \infty$.

Let a_{ij} and b_{ij} be two symmetric coefficients with $|a_{ij} - C_{ij}| \leqq \varepsilon$ and $|b_{ij} - C_{ij}| \leqq \varepsilon$, where $\varepsilon > 0$ is chosen according to Theorem 5.1. Moreover, let $|a_{ij} - b_{ij}| \leqq \eta$. Then let K_λ^a and K_λ^b be the operators for a and b constructed as in Theorem 5.1. Then we have

THEOREM 5.2. *$\|K_\lambda^a - K_\lambda^b\| \leqq A_\lambda \eta$ for all norms from $L_{p_i} \to L_{p_{i+1}}$, $i = 1, 2, \cdots, N-1$. For $i = N-1$ the uniform and the Lipshitz norms are considered on the range space. A_λ again depends only on the extreme eigenvalues of $((C_{ij}))$.*

Proof: It is obvious that $\|T_\lambda^a - T_\lambda^b\|$ can be estimated, and therefore $\|(I - T_\lambda^a)^{-1} - (I - T_\lambda^b)^{-1}\|$ can be; q.e.d.

We shall use certain easy corollaries of Theorems 5.1 and 5.2 and we now state them for future reference. We shall replace all constants by C_λ, which depends on $((C_{ij}))$ as before.

COROLLARY 5.1. *In the earlier notation,*

(i) $|K_\lambda f(s, x)| \leqq C_\lambda \|f\|_{p_N}$ *for all* s, x,

(ii) $|K_\lambda^N f(s, x)| \leqq C_\lambda \|f\|_{p_1}$ *for all* s, x,

(iii) $|K_\lambda^N f(s, x_1) - K_\lambda^N f(s, x_2)| \leqq C_\lambda \|f\|_{p_1} |x_1 - x_2|$ *for all* s, x_1, x_2,

(iv) $|(K_\lambda^a)^N f(s, x) - (K_\lambda^b)^N f(s, x)| \leqq C_\lambda \|f\|_{p_1} \eta$ *for all* s, x,

(v) $|(K_\lambda^a)^N f(s, x_1) - (K_\lambda^b)^N f(s, x_1) - (K_\lambda^a)^N f(s, x_2) + (K_\lambda^b)^N f(s, x_2)|$
$\leqq C_\lambda \|f\|_{p_1} \eta |x_1 - x_2|$ *for all* s, x_1, x_2, *where* $\eta = \sup |a_{ij} - b_{ij}|$.

Now let $((a_{ij}))$ be continuous coefficients which are uniformly elliptic and satisfy

$$\|a_{ij} - \delta_{ij}\| \leqq \varepsilon \quad \text{for all} \quad i, j, \tag{5.4}$$

where ε is chosen according to Theorem 5.1; in particular,

$$2C_{d, p_N} \varepsilon d^2 \leqq 1. \tag{5.5}$$

From Theorem 4.2 we know that for any s, x_0 there exists a solution to the martingale problem. Under hypotheses (5.4) and (5.5) we shall prove the uniqueness of this solution. Up to Theorem 5.3 we shall assume that the coefficients satisfy (5.4) and (5.5).

LEMMA 5.2. *Let s and x_0 be arbitrary and fixed. For any solution $P = P_{s,x_0}$ of the martingale problem starting from x_0 at time s, let us define the measures μ_λ on $I^s \times R^d$ by*

$$\mu_\lambda(f) = E\left[\int_s^\infty e^{-\lambda(t-s)} f(t, x(t))\, dt\right]. \tag{5.6}$$

Then,

$$|\mu_\lambda(f)| \leqq 2C_\lambda \|f\|_{P_N}.$$

Proof: According to Corollary 3.2, $x(t)$ satisfies the equation

$$x(t) = x_0 + \int_s^t \sigma(\sigma, x(\sigma))\, d\beta(\sigma),$$

where σ is the positive definite symmetric square root of a. Let $\pi_n(\sigma)$ be defined as follows:

$$\pi_n(\sigma) = \begin{cases} \sigma + \dfrac{j-1}{n} & \text{for } s + \dfrac{j-1}{n} \leqq \sigma < s + \dfrac{j}{n}, \qquad j = 1, 2, \cdots, n^2, \\ \sigma + n & \text{for } \sigma \geqq s + n . \end{cases}$$

Then $\lim_{n\to\infty} \pi_n(\sigma) = \sigma$ for all $\sigma \geqq s$. Define

$$x_n(t) = x_0 + \int_s^t \sigma(\pi_n(\sigma), x(\pi_n(\sigma))\, d\beta(\sigma) . \tag{5.7}$$

Since $x(\sigma)$ and $\sigma(\sigma, x)$ are continuous, we conclude that $x_n(t)$ converges to $x(t)$ in the mean for each $t > 0$. Moreover, it is easily seen that

$$\lim_{n\to\infty} E^P\left[\int_s^\infty e^{-\lambda(t-s)} f(t, x_n(t))\, dt\right] = E^P\left[\int_s^\infty e^{-\lambda(t-s)} f(t, x(t))\, dt\right]$$

for bounded and continuous functions $f(t, x)$. Thus if we denote

$$\mu_\lambda^{(n)}(f) = E\left[\int_s^\infty e^{-\lambda(t-s)} f(t, x_n(t))\, dt\right],$$

then $\mu_\lambda^{(n)}$ converges weakly to μ_λ. We shall conclude the proof of the lemma by getting a uniform bound for $\mu_\lambda^{(n)}$. From equation (5.7) it follows by using Itô's formula (Theorem 2.5) that, for any function $f(t, x)$ which is sufficiently smooth,

$$f(t, x_n(t)) = f(s, x_0) + \int_s^t \langle \nabla f(\sigma, x_n(\sigma))\sigma(\pi_n(\sigma), x(\pi_n(\sigma))\, d\beta(\sigma)\rangle$$
$$+ \int_s^t \left[f_\sigma + \frac{1}{2}\sum a_{ij}(\pi_n(\sigma), x(\pi_n(\sigma)))f_{ij}\right](\sigma, x_n(\sigma))\, d\sigma .^{10}$$

Multiply both sides by $\lambda e^{-\lambda(t-s)}$ and integrate from s to ∞. Integrating by parts and using the fact that

$$E^P\int_s^t \langle(\nabla f)(\sigma, x_n(\sigma))\sigma(\pi_n(\sigma), x(\pi_n(\sigma)))\, d\beta(\sigma)\rangle = 0 ,$$

we get

$$\lambda\mu_\lambda^{(n)}(f) = f(s, x_0)$$
$$+ E^P\int_s^\infty e^{-\lambda(t-s)}[f_t + \tfrac{1}{2}\sum a_{ij}(\pi_n(t), x(\pi_n(t))f_{ij}](t, x_n(t))\, dt .$$

[10] Here and elsewhere ∇ stands for gradient with respect to the space variables. We use f_s to denote $\partial f/\partial s$, and f_{ij} to stand for $\partial^2 f/\partial x_i\, \partial x_j$.

This can be rewritten as

$$\begin{aligned}(5.8)\qquad &\mu_\lambda^{(n)}[\lambda f - f_s - \tfrac{1}{2}\textstyle\sum f_{ii}] \\ &= f(s, x_0) + \tfrac{1}{2}\, E^P\left[\int_s^\infty e^{-\lambda(t-s)}[\textstyle\sum \varepsilon_{ij}(\pi_n(t), x(\pi_n(t))f_{ij}](t, x_n(t))\, dt\right],\end{aligned}$$

where $\varepsilon_{ij} = (a_{ij} - \delta_{ij})$. We now choose $f = G_\lambda h$, where h is C^∞ with compact support. Then f is smooth enough and vanishes fast enough at ∞ so that (5.8) is valid, and this yields

$$\begin{aligned}(5.9)\qquad |\mu_\lambda^n(h)| &\leqq C_\lambda \|h\|_{p_N} + \tfrac{1}{2}\varepsilon_0\, E^P \int_s^\infty e^{-\lambda(t-s)} \textstyle\sum |f_{ij}|\,(t, x_n(t))\, dt \\ &= C_\lambda \|h\|_{p_N} + \tfrac{1}{2}\varepsilon_0\, \mu^n(H)\,,\end{aligned}$$

where $H(t, x) = \Sigma\, |(G_\lambda h)_{ij}|$.

From (5.3) and (5.5) it follows that

$$(5.10)\qquad \|H\|_{p_N} \leqq 2d^2 C_{d, p_N} \|h\|_{p_N}\,.$$

Looking at the process $x_n(t)$, we see that it is given by a stochastic integral of a simple function and is therefore piecewise Browanian motion. An elementary computation shows that

$$|\mu_\lambda^{(n)}(h)| \leqq A_\lambda^{(n)} \|h\|_{p_N}\,.$$

Let $A_\lambda^{(n)}$ be the best possible constant. Of course, $A_\lambda^{(n)} = \|d\mu_\lambda^{(n)}/dm\|_{q_N}$, where m is the Lebesgue measure on $I^s \times R^d$ and q_N is the conjugate of p_N (i.e., $p_N^{-1} + q_N^{-1} = 1$). From (5.5), (5.8) and (5.9) it follows that

$$A_\lambda^{(n)} \leqq C_\lambda + \tfrac{1}{2} A_\lambda^{(n)}\,.$$

Since $A_\lambda^{(n)}$ is finite, we must have

$$A_\lambda^{(n)} \leqq 2C_\lambda\,,$$

or $\sup_n A_\lambda^{(n)} \leqq 2C_\lambda < \infty$. Since $\mu_\lambda^{(n)}$ converges weakly to μ_λ, it follows that

$$|\mu_\lambda(h)| \leqq 2C_\lambda \|h\|_{p_N}\,.$$

Lemma 5.3. *Let P be any solution of the martingale problem starting at time s from x_0. Define*

$$\mu_\lambda(f) = E^P \int_s^\infty e^{-\lambda(t-s)} f(t, x(t))\, dt\,.$$

Then $\mu_\lambda(f) = (K_\lambda f)(s, x_0)$, where K_λ is the operator constructed in Theorem 5.1.

Proof: Using Corollary 3.2 and Itô's formula (Theorem 2.5) we derive for μ_λ the equation analogous to (5.8):

$$\mu_\lambda[\lambda f - f_s - \tfrac{1}{2}\sum f_{ii}] = f(s, x_0) + \mu_\lambda(\tfrac{1}{2}\sum \varepsilon_{ij} f_{ij}) .$$

Choosing $f = G_\lambda h$ for smooth enough h with compact support, we have

$$\mu_\lambda(h) = (G_\lambda h)(s, x_0) + \mu_\lambda(T_\lambda h) .$$

In other words,

$$\mu_\lambda(h - T_\lambda h) = (G_\lambda h)(s, x_0) .$$

But K_λ also satisfies $K_\lambda(I - T_\lambda) = G_\lambda$ or

$$[K_\lambda(h - T_\lambda h)](s, x_0) = (G_\lambda h)(s, x_0) .$$

Therefore, $\mu_\lambda(f) = (K_\lambda f)(s, x_0)$ for $f = (I - T_\lambda)h$ with h smooth. Since such functions are dense in L_{p_N} and both functionals are bounded on L_{p_N}, the lemma is true.

LEMMA 5.4. *Let P be any solution of the martingale problem starting at time s from x_0. Then,*

$$P[x(t) \in A] = P(s, x_0 \; ; \; t, A)$$

is uniquely determined for $t > s$.

Proof: From Lemma 5.3, the Laplace transform

$$\int_{s_0}^{\infty} e^{-\lambda(t-s)} P(s, x_0 \; ; \; t, A) \, dt$$

is uniquely determined. Moreover, $P(s, x_0 \; ; \; t, \cdot)$ is weakly continuous in t because of the continuity of paths. Consequently, $P(s, x_0 \; ; \; t, A)$ is uniquely determined for all $t > s$.

LEMMA 5.5. *Let P be any solution of the martingale problem starting at time s from x_0. Then P is a Markov process. Moreover, the solution to the problem is unique (on M^s).*

Proof: Let $r > s$ be any time point. Then by Theorem 3.1, we know that the regular conditional probability of P given the σ-field M_r^s exists and is again a solution to the martingale problem starting at time r from $x(r, w)$ for almost all w with respect to P. We denote by Q_w this conditional probability distribution.

Let $t > r$ be another time point. Then by the uniqueness established in Lemma 5.4,

$$Q_w[x(t) \in A] = P(r, x(r);\ t, A) \text{ a.s. } P\,,$$

which implies

$$P[x(t) \in A \mid M^s] = P(r, x(r);\ t, A) \text{ a.s. } P\,.$$

This proves that P is a Markov process and has $P(r, x; t, A)$ as its transition probabilities. Consequently, P is unique given that it starts at s, x_0.

THEOREM 5.3. *Under hypotheses* (5.4) *and* (5.5), *the martingale problem has a unique solution for each starting time and point s, x. Denoting this unique solution by $P_{s,x}$, we get a family of Markov processes depending continuously on s and x in the obvious sense.*

Proof: We know from uniqueness, that we get a family $\{P_{s,x}\}$, $P_{s,x}$ being a measure on the space $C([0, \infty), R^d)$ on the σ-field M^s. If necessary, we can treat these as defined on all of M^0 by just imposing the condition

$$P_{s,x}[x(u) \equiv x \text{ for } x \leqq s] = 1\,.$$

Then, for $s_n \to s$ and $x_n \to x$, the family P_{s_n,x_n} is compact, and if Q is any weak limit we see, arguing along the lines of the proof of Theorem 4.2, that

$$Q[x(u) \equiv x \text{ for } u \leqq s] = 1\,.$$

Q is again a solution to the martingale problem starting from x at time s. Therefore, any limit Q is equal to $P_{s,x}$ by uniqueness. Consequently,

$$\lim_{n\to\infty} P_{s_n,x_n} = P_{s,x}\,,$$

that is $P_{s,x}$ depends continuously on s and x.

THEOREM 5.4. *Let $0 < \rho_- \leqq \rho_+ < \infty$. There exists an $\varepsilon > 0$ such that for any matrix C_{ij} with $\rho_- I \leqq ((C_{ij})) \leqq \rho_+ I$ and for any continuous coefficients $((a_{ij}))$ with*

$$\|a_{ij} - C_{ij}\| \leqq \varepsilon \quad \textit{for} \quad \textit{all} \quad i, j\,,$$

the matrix $((a_{ij}))$ is uniformly elliptic, and the solution to the martingale problem with these coefficients yields a unique solution $P_{s,x}$ depending continuously on s and x.

Proof: Because of Theorem 5.1, the case of any C_{ij} is no different from $C_{ij} = \delta_{ij}$.

COROLLARY 5.2. *A consequence of continuity is that the uniquely defined $P(s, x;\ t, A)$ is a measurable function of s and x for $s < t$ and $x \in R^d$, whenever A is a Borel subset of R^d.*

From now on we shall assume that the coefficients $((a_{ij}))$ are bounded, continuous and uniformly elliptic. Let s, x_0 be arbitrary. Define the stopping

time τ_s by the formula

$$\tau_s = \inf_{s+1 \geqq t \geqq s} \{t : \sup_{i,j} |a_{ij}(s, x_0) - a_{ij}(t, x(t))| \geqq \varepsilon\}.$$

If the t-set is empty, τ_s is defined as $s + 1$. This ε is as in Theorem 5.2.

LEMMA 5.6. *Let P_1, P_2 be two solutions of the martingale problem starting from s, x_0. Then P_1 and P_2 agree on the σ-algebra $M^s_{\tau_s}$.*

Proof: Let $G \subset I^s \times R^d$ be defined as

$$G = \{t, x : |a_{ij}(t, x) - a_{ij}(s, x_0)| < \varepsilon \text{ for all } i, j\}.$$

Let D be the closure of G. Let $C_{ij} = a_{ij}(s, x_0)$. Then we have $|a_{ij} - C_{ij}| \leqq \varepsilon$ on D, and therefore the a_{ij} can be modified to a continuous function b_{ij} defined on all of $[I^s \times R^d]$ satisfying $\|b_{ij} - C_{ij}\| \leqq \varepsilon$. Moreover, $a_{ij} = b_{ij}$ on D, and $((b_{ij}))$ can be chosen to be symmetric. From the definition of τ_s we have $\tau_s \leqq s + 1 < \infty$, and $a_{ij} = b_{ij}$ for $s \leqq t \leqq \tau_s$. For coefficients b_{ij} we know by Theorem 5.2 that there exists a solution $Q_{t,x}$ which is measurable in t and x (in fact continuous). Let us define two solutions Q_1 and Q_2 as follows: Q_i is P_i on $M^s_{\tau_s}$ and the conditional distribution of P_i given M_{τ_s} is defined for each w by $Q_{\tau_s, x(\tau_s)}$ on $M^{\tau_s(w)}$. This is explained in Theorem 3.4 and Lemma 3.6, and it is shown there that these are also solutions. It follows that Q_1 and Q_2 are solutions for b_{ij} starting from s, x_0. Hence by uniqueness they must agree, i.e., $Q_1 \equiv Q_2$. But $Q_i \equiv P_i$ on $M^s_{\tau_s}$ by the construction, and therefore $P_1 \equiv P_2$ on $M^s_{\tau_s}$.

LEMMA 5.7. *Let P_1, P_2 be two measures on M^0 such that $P_1[w : x(t, w) \equiv f(t)$ for $t \leqq s] = P_2[w : x(t, w) \equiv f(t)$ for $t \leqq s] = 1$. Assume they are solutions of our problem for $t > s$ starting from s, x_0, where $x_0 = f(s)$ ($f(t)$ is some continuous function from $[0, s]$ to R^d). Then $P_1 \equiv P_2$ on $M^0_{\tau_s}$, where τ_s is as in Lemma 5.6.*

Proof: From Lemma 5.6 we have $P_1 \equiv P_2$ on $M^s_{\tau_s}$. However, in the interval $0 \leqq t \leqq s$, $x(t, w) \equiv f(t)$ with probability 1 with respect to P_1 as well as P_2; q.e.d.

Now let P_1, P_2 be two solutions to our problem starting at time 0 from the point x_0. Let us define successive stopping times and places:

$$\tau_1 = \inf_{1 \geqq t > 0} [t : \sup_{i,j} |a_{ij}(t, x(t) - a_{ij}(0, x_0)| \geqq \varepsilon_1],$$

$$x_1 = x(\tau_1),$$

. . . .

$$\tau_n = \inf_{\tau_{n-1} \leqq t \leqq \tau_{n-1}+1} [t : \sup_{i,j} |a_{ij}(t, x(t)) - a_{ij}(\tau_{n-1}, x_{n-1})| \geqq \varepsilon],$$

$$x_n = x(\tau_n).$$

Obviously $\tau_n \leqq n$.

LEMMA 5.8. *$P_1 = P_2$ on $M^0_{\tau_n}$ for each n.*

Proof: We prove this by induction. Lemma 5.6 assures the validity for $n = 1$. Let us suppose $P_1 \equiv P_2$ on $M^0_{\tau_{n-1}}$. We denote by $Q^{(i)}_w(\cdot)$ the conditional distribution of P_i given $M^0_{\tau_{n-1}}$. Then for $i = 1, 2$,

$$P_i(A) = \int_\Omega Q^{(i)}_w(A)\, dP_i .$$

Since $Q^{(i)}_w(A)$ is $M^0_{\tau_{n-1}}$-measurable and the P_i agree on $M^0_{\tau_{n-1}}$, it suffices to show that, for $A \in M^0_{\tau_n}$, $Q^{(1)}_w(A) = Q^{(2)}_w(A)$ a.s. P. Let $w = w_0$ be fixed. Assuming that w_0 is not in an exceptional set of measure zero, we know by Theorem 3.1 that $Q^{(i)}_{w_0}$ has the two properties

(a) $Q^{(i)}_{w_0}[w : x(t, w) = x(t, w_0) \text{ for } 0 \leqq t \leqq \tau_{n-1}(w_0)] = 1$,

(b) $Q^{(i)}_{w_0}$ is a solution to our problem starting from t_0, x_{n-1}, where $t_0 = \tau_{n-1}(w_0)$.

By Lemma 5.7 this implies that $Q^{(1)}_{w_0} \equiv Q^{(2)}_{w_0}$ on the σ-field M^0_τ, where

$$\tau = \inf_{t_0+1 \geqq t \geqq t_0} \{t : \sup_{i,j} |a_{ij}(t_0, x_{n-1}) - a_{ij}(t, x(t)| \geqq \varepsilon\} .$$

Now if $A \in M^0_{\tau_n}$, let us denote by A_{w_0} the set

$$A_{w_0} = A \cap \{w : x(t, w) = x(t, w_0) \text{ for } 0 \leqq t \leqq \tau_{n-1}(w_0)\} .$$

Clearly, $A_{w_0} \in M^0_{\tau_n}$ and $\tau_n = \tau$ on A_{w_0}. Therefore, $A_{w_0} \in M^0_\tau$. Since

$$Q^{(i)}_{w_0}(A) = Q^{(i)}_{w_0}(A_{w_0}) ,$$

it follows that $Q^{(1)}_{w_0}(A)$ and $Q^{(2)}_{w_0}(A)$ agree for $A \in M_{\tau_n}$ for almost all w_0.

THEOREM 5.5. *Let P_1, P_2 be two solutions starting at time 0 from x_0 to the martingale problem. Then $P_1 \equiv P_2$.*

Proof: We know by Lemma 5.8 that $P_1 \equiv P_2$ on M_{τ_n} for every n. Also, since $x(t, w)$ is continuous in t, we must have $\lim_{n\to\infty} \tau_n = \infty$ a.e. with respect to both P_1 and P_2. Let $A \in M_t$ for some $t < \infty$. Then,

$$A = [A \cap \{\tau_n < t\}] \cup [A \cap \{\tau_n \geqq t\}] ,$$

and $\{A \cap \{\tau_n \geqq t\}\} \in M^0_{\tau_n}$. Since P_1, P_2 agree on $M^0_{\tau_n}$,

$$|P_1(A) - P_2(A)| \leq P_1[A \cap \{\tau_n < t\}] + P_2[A \cap \{\tau_n < t\}] .$$

Letting $n \to \infty$, we have $P_1 \equiv P_2$ on M^0_t. Since t is arbitrary, $P_1 \equiv P_2$ on M^0.

COROLLARY 5.3. *The solution $P_{s,x}$ to the martingale problem starting from x at time s is unique.*

Proof: This was established in Theorem 5.5 for $s = 0$. The general case is just the change of origin for time; q.e.d.

We now drop the hypothesis of uniform ellipticity and replace it by strict ellipticity at each point. This of course implies uniform ellipticity of the coefficients on compact subsets.

THEOREM 5.6. *Let the coefficients $((a_{ij}))$ be continuous and let us suppose that $\|a_{ij}\| \leqq M$ and that $((a_{ij}))$ is strictly elliptic (i.e., positive definite at each point). Then the solution $P_{s,x}$ to the martingale problem starting from x at time s is unique.*

Proof: We know already that $P_{s,x}$ exists (Theorem 4.2). Let P_1, P_2 be two possible solutions. Then one can take a large compact set $s_0 \leqq t \leqq R$, $\|y\| \leqq l$, and change the coefficients $((a_{ij}))$ outside to get coefficients $((b_{ij}))$ which are symmetric, uniformly elliptic and continuous. There is a unique solution for the coefficients $((b_{ij}))$, say $Q_{t,y}$, starting from y at time t. Combining P_1 and P_2 with $Q_{t,y}$ we can show, as in Lemma 5.6, that P_1 and P_2 agree on M_{τ_l}, where

$$\tau_l = \inf_{s \leqq t \leqq R} \{t: |x(t, w)| \geqq l\} .$$

Again, since $\tau_l \to \infty$ as $l \to \infty$, we have that $P_1 \equiv P_2$.

THEOREM 5.7. *$P_{s,x}$ is a Markov process. More precisely,*

$$P_{s,x}[x(t_2) \in A \mid M_{t_1}^s] = P_{t_1,x(t_1)}[x(t_2) \in A] \text{ a.s. } P_{s,x}.$$

It is, moreover, a strong Markov process.

Proof: By Theorem 3.1, the conditional distribution of $P_{s,x}$ given $M_{t_1}^s$ is again a solution of the martingale problem starting from $x(t_1)$ at time t_1 almost surely. By uniqueness this must be $P_{t_1,x(t_1)}$. For strong Markov property one uses a Markov time instead of t_1; q.e.d.

Let $((a_{ij}))$ be as in Theorem 5.6. Then the unique solution $P_{s,x}$ defines the transition probabilities $P(s, x; t, A)$ by the relation

$$P(s, x; t, A) = P_{s,x}[x(t) \in A] \quad \text{for} \quad t > s .$$

The Markov property of $P_{s,x}$ ensures that $P(s, x, t, A)$ satisfies the Chapman-Kolmogorov equations

$$P(s, x; u, A) = \int P(s, x; t, dy) P(t, y; u, A) .$$

By an argument similar to Theorem 5.1, one can show from uniqueness and compactness that $P_{s,x}$ depends continuously on s and x (i.e.,

$$\int \phi(y)P(s, x; t, dy) = u(s, x)$$

is continuous in s and x for $s \leqq t$), provided $\phi(y)$ is a bounded continuous function on R^d. This of course is the Feller property. The strong Feller property will assert that $u(s, x)$ is continuous for $s < t$ if $\phi(y)$ is bounded and measurable. We shall prove later that actually u is equicontinuous at each s, x, $s < t$, provided ϕ stays uniformly bounded.

6. The Case with an Additional First Order Term

In this section we consider diffusion processes with a drift term. We shall suppose that the drift terms $b_j(s, x)$ are uniformly bounded and measurable.

PROBLEM 6.1. For given s and x we must prove the existence and uniqueness of a measure $Q_{s,x}$ on M^s such that

$$\exp\left\{\langle\theta, x(t) - x(s)\rangle - \left\langle\theta, \int_s^t b(\sigma, x(\sigma))\, du\right\rangle - \frac{1}{2}\int_s^t \langle\theta, a(\sigma, x(\sigma))\theta\rangle\, du\right\}$$

is a martingale and $Q_{s,x}(x(s) = x) = 1$.

We shall suppose for a while that $b(s, x) = 0$ outside a region $0 \leqq s \leqq l$, $|x| \leqq l$.

LEMMA 6.1. *Let P be a solution to Problem* 6.1 *with $b \equiv 0$ starting at s, x. Then the measure Q defined by*

$$\left.\frac{dQ}{dP}\right|_{M_t^s} = \exp\left\{\int_s^t \langle b(\sigma, x(\sigma)), a^{-1}(\sigma, x(\sigma))\, dx(\sigma)\rangle - \frac{1}{2}\int_s^t \langle b(\sigma, x(\sigma)), a^{-1}(\sigma, x(\sigma))b(\sigma, x(\sigma))\rangle\, d\sigma\right\} \tag{6.1}$$

is a solution to Problem 6.1.

Proof: From Theorem 3.2 we know that

$$\exp\left\{\int_s^t \langle\theta(\sigma, w), dx(\sigma)\rangle - \frac{1}{2}\int_s^t \langle\theta(\sigma, w), a(\sigma, x(\sigma))\theta(\sigma, w)\rangle\, d\sigma\right\}$$

is a P-martingale for all bounded s-non-anticipating functionals $\theta(\sigma, w)$. Let us choose

$$\theta(\sigma, w) = \theta + a^{-1}(\sigma, x(\sigma))b(\sigma, x(\sigma)) ,$$

where θ is a constant vector. This is allowed because a is uniformly elliptic on compact sets and $b \equiv 0$ outside a compact set. With this choice we see that

$$\exp\left\{\langle\theta, x(t) - x(s)\rangle - \left\langle\theta, \int_s^t b(\sigma, x(\sigma))\, d\sigma\right\rangle - \frac{1}{2}\int_s^t \langle\theta, a(\sigma, x(\sigma))\theta\rangle\, d\sigma\right\} R_t$$

is a P-martingale for all θ, where

$$R_t = \exp\left\{\int_s^t \langle b(\sigma, x(\sigma))a^{-1}(\sigma, x(\sigma)), dx(\sigma)\rangle - \frac{1}{2}\int_s^t \langle b(\sigma, x(\sigma)), a^{-1}(\sigma, x(\sigma))b(\sigma, x(\sigma))\rangle\, d\sigma\right\}. \tag{6.2}$$

Since $R_t(w)$ is itself a P-martingale, the Q given by (6.1) is well defined, and

$$\exp\left\{\langle\theta, x(t) - x(s)\rangle - \left\langle\theta, \int_s^t b(\sigma, x(\sigma))\, d\sigma\right\rangle - \frac{1}{2}\int_s^t \langle\theta, a(\sigma, x(\sigma))\theta\rangle\, d\sigma\right\}$$

is a Q-martingale. Hence Q solves Problem 6.1.

LEMMA 6.2. *If Q solves Problem* 6.1, *then Q is given by* (6.1), *where P solves Problem* 6.1 *with $b \equiv$* 0.

Proof: If Q solves Problem 6.1, we know that

$$\exp\left\{\int_s^t \langle\theta(\sigma, w), dx(\sigma)\rangle - \int_s^t \langle\theta(\sigma, w), b(\sigma, x(\sigma))\rangle\, d\sigma - \frac{1}{2}\int_s^t \langle\theta(\sigma, w), a^{-1}(\sigma, x(\sigma))\theta(\sigma, w)\rangle\, d\sigma\right\}$$

is a Q-martingale for bounded s-non-anticipating θ. Let us choose

$$\theta(\sigma, w) = \theta - a^{-1}(\sigma, x(\sigma))b(\sigma, x(\sigma)) .$$

Then,

$$\exp\left\{\langle\theta, x(t) - x(s)\rangle - \frac{1}{2}\int_s^t \langle\theta, a(\sigma, x(\sigma))\theta\rangle\, d\sigma\right\} S_t(w)$$

is a Q-martingale for all θ, where $S_t(w) = R_t^{-1}(w)$ (cf. (6.2)). In particular, this implies that $dQ = R_t(w)\, dP$, where P is a solution of Problem 6.1 with $b \equiv 0$.

We have therefore proved

THEOREM 6.1. *If b has compact support, then Problem 6.1 has a unique solution starting at s, x, and the solution $Q_{s,x}$ is related to the solution $P_{s,x}$ starting at s, x for $b \equiv 0$ by the formula*

$$\left.\frac{dQ_{s,x}}{dP_{s,x}}\right|_{M_t^s} = \exp\left\{\int_s^t \langle b(\sigma, x(\sigma)), a^{-1}(\sigma, x(\sigma))\, dx(\sigma)\rangle - \frac{1}{2}\int_s^t \langle b(\sigma, x(\sigma)), a^{-1}(\sigma, x(\sigma)) b(\sigma, x(\sigma))\rangle\, d\sigma\right\}.$$

Remark. This formula gives the density of M_t^s for every $s \leqq t < \infty$. Of course Q and P may be inequivalent on M^s.

We now turn to the case when b need not vanish outside a compact set.

LEMMA 6.3. *Let b be bounded by M. Then the solution to Problem 6.1 exists.*

Proof: Let b_l be the function agreeing with b in $0 \leqq t \leqq l$ and $|x| \leqq l$ and zero outside. Then the solution $Q_{s,x}^l$ for b_l exists. We shall construct the solution $Q_{s,x}$ by taking the limit as $l \to \infty$. If $\hat{\tau}_l^s$ is the exit time after s from the sphere $\{|x| < l\}$ and $\tau_l^s = \hat{\tau}_l^s \wedge l$, then it is clear that $Q_{s,x}^{l'}$ agrees with $Q_{s,x}^l$ on $M_{\tau_l^s}^s$ for $l' > l$. Therefore, the limit $Q_{s,x}$ of $Q_{s,x}^l$, as $l \to \infty$, exists on each σ-field $M_{\tau_l^s}^s$. Now if $A \in M_t^s$, then $A \cap \{\tau_l^s > t\} \in M_{\tau_l^s}^s$, and we can estimate

$$Q_{s,x}^{l'}[\tau_l^s \leqq t] \leqq \exp\left\{-\frac{\alpha l^2}{2t}\right\}.$$

Therefore, $Q_{s,x}$ is a countably additive measure on each M_t^s and hence on M^s. That it is again a solution of the martingale problem follows easily.

LEMMA 6.4. *The solution $Q_{s,x}$ to Problem 6.1 is unique and bears the same relation to $P_{0,x}$ as described in Lemma 6.1.*

Proof: We have a unique solution $Q_{s,x}^l$ for b_l, and we can see that $Q_{s,x}^l = Q_{s,x}$ on $M_{\tau_l^s}^s$. Therefore, $Q_{s,x}$ is unique on $M_{\tau_l^s}^s$ for every l. Letting l tend to infinity

we have uniqueness. Since $Q_{s,x} = \lim_{l\to\infty} Q^l_{s,x}$, we can take

$$\frac{dQ_{s,x}}{dP_{s,x}} = \lim_{l\to\infty} \frac{dQ^l_{s,x}}{dP_{s,x}},$$

provided the latter exists. The existence of the preceding limit follows from a standard uniform integrability argument.

We shall summarize the result as:

THEOREM 6.2. *The solution to the martingale problem with bounded drift coefficients b exists and is unique. The solution has the form*

$$\left.\frac{dQ}{dP}\right|_{M_t^s} = \exp\left\{\int_s^t \langle b(\sigma, x(\sigma)), a^{-1}(\sigma, x(\sigma))\, dx(\sigma)\rangle - \frac{1}{2}\int_s^t \langle b(\sigma, x(\sigma)), a^{-1}(\sigma, x(\sigma)) b(\sigma, x(\sigma))\rangle\, d\sigma\right\},$$

where P is the corresponding solution for $b \equiv 0$. Moreover, $\{Q_{s,x}\}$ is a strong Markov process.

Proof: The only thing that remains to be shown is the Markov property. This follows from uniqueness and Theorem 3.1.

A similar approach for computing the Radon-Nikodym derivative was carried out in [6] by Girsanov. However, without uniqueness one can get only a partial statement.

Appendix

The estimate of B. F. Jones. Let $p(t, x)$, $t > 0$ and $x \in R^d$, be defined by

$$p(t, x) = \frac{1}{(2\pi t)^{d/2}} e^{-|x|^2/2t}.$$

Given $g \in C_0^\infty([0, \infty) \times R^d)$ define

$$G_0^* g(t, x) = \int_0^t \int_{R^d} p(t - s, x - y) g(s, y)\, dy\, ds.$$

Then the estimate of B. F. Jones [11] says that there is a constant A_p, $1 < p < \infty$, such that

$$\text{(A.1)} \qquad \|D^2 G_0^* g\|_p \leqq A_p \|g\|_p, \qquad g \in C_0^\infty([0, \infty) \times R^d),$$

where D^2 stands for the generic second spacial derivative and

$$\|\cdot\|_p = \|\cdot\|_{L_p([0,\infty)\times R^d)} .$$

We use the result to prove the following theorem.

THEOREM A.1. *For each* $1 < p < \infty$, *there exists a* C_p *such that if*

$$u \in C_b^\infty([0, \infty) \times R^d)$$

satisfies

(i) $\lim_{t\to\infty} \sup_{|x|\leq l} |u(t, x)| = 0$ *for each* $l > 0$,

(ii) $\dfrac{\partial u}{\partial t} + \frac{1}{2}\Delta u = -f \in L_p([0, \infty) \times R^d)$ *for some* $1 < p < \infty$, *and* $e^t\,|f(t, x)|$ *is bounded, then*

$$\|D^2 u\|_p \leqq C_p \|f\|_p . \tag{A.2}$$

Proof: We first show that if u satisfies (i) and (ii), then

$$u(t, x) = \int_t^\infty \int_{R^d} p(s - t, x - y) f(s, y)\, dy\, ds . \tag{A.3}$$

To do this, let $[x_t, M_t^0, P_x : t \geqq 0,\ x \in R^d]$ be d-dimensional Brownian motion. Then, by Itô's formula,

$$\begin{aligned} E^{P_x}[u(t + s, x_s)] - u(t, x) &= -E^{P_x}\left[\int_0^s f(t + \alpha, x_\alpha)\, d\alpha\right] \\ &= -\int_t^{t+s} \int_{R^d} p(\alpha - t, x - y) f(\alpha, y)\, dy\, d\alpha . \end{aligned}$$

Letting $s \to \infty$, we obtain (A.3).

Now let $g \in C_0^\infty([0, \infty) \times R^d)$. Then,

$$\begin{aligned} &\int_0^\infty \int_{R^d} g(t, x) D^2 u(t, x)\, dx\, dt \\ &= \int_0^\infty \int_{R^d} D^2 g(t, x)\, dx\, dt \int_t^\infty \int_{R^d} p(s - t, x - y) f(s, y)\, dy\, ds \\ &= \int_0^\infty \int_{R^d} f(s, y)\, dy\, ds \int_0^s \int_{R^d} p(s - t, y - x) D^2 g(t, x)\, dx\, dt \\ &= \int_0^\infty \int_{R^d} f(s, y) D^2[G_0^* g](s, y)\, dy\, ds . \end{aligned}$$

Hence,

$$\left| \int_0^\infty \int_{R^d} g(t, x) D^2 u(t, x)\, dx\, dt \right| \leqq A_q \, \|f\|_p \, \|g\|_q ,$$

where $1/p + 1/q = 1$, and therefore

$$\|D^2 u\|_p \leqq C_p \, \|f\|_p ,$$

where $C_p = A_q$.

Some facts from measure theory. Throughout this section we shall use $\mathscr{B}_X$ to denote the Borel field of a topological space X. We remind the reader that a measurable space $\langle X, \mathscr{B} \rangle$ is said to be *standard* if there exists a complete separable metric space Y such that $\mathscr{B}$ is σ-isomorphic to $\mathscr{B}_Y$. In particular, if $X \in \mathscr{B}_Y$ where Y is a complete separable metric space, then $\langle X, \mathscr{B}_X \rangle$ is a standard measurable space.

Let $\Omega = C([0, \infty) \times R^d)$ and endow Ω with the topology of uniform convergence on compact sets. Then Ω is a complete separable metric space and it is simple to see that $M^0 = \mathscr{B}_\Omega$.

LEMMA A.1. *Let $s \geqq 0$ and suppose τ is an s-Markov time. Define C_τ^s to be the system of sets*

$$\{x_{t_1 \wedge \tau} \in \Gamma_1, \cdots, x_{t_n \wedge T} \in \Gamma_n\},$$

where $s \leqq t_1 < \cdots < t_n$, $n \geqq 1$, and $\Gamma_1, \cdots, \Gamma_n \in \mathscr{B}_{R^d}$. Then M_τ^s is the σ-field generated by C_τ^s.

Proof: First observe that $\{x_{t \wedge \tau} \in \Gamma\} \in M_\tau^s$ for all $t \geqq s$ and $\Gamma \in \mathscr{B}_{R^d}$. Hence, $\sigma(C_\tau^s) \subseteq M_\tau^s$. On the other hand, if ξ is M_τ^s-measurable, then by Lemma 1.5 in Dynkin [4] there is a $(\mathscr{B}_{R^d})^\infty$-measurable function f and a sequence

$$s \leqq t_1 < \cdots < t_n < \cdots$$

such that

$$\xi(w) = f(x_{t_1}(w), \cdots, x_{t_n}(w), \cdots), \qquad w \in \Omega .$$

Hence, since $\xi \cdot \chi_{\tau \leqq T}$ is M_T^s-measurable for each $T \geqq s$,

$$\xi \cdot \chi_{\tau \leqq T}(w) = f(x_{t_1 \wedge T}(w), \cdots, x_{t_n \wedge T}(w), \cdots), \qquad w \in \Omega ,$$

for all $T \geqq s$. From this we see that

$$\xi(w) = f(x_{t_1 \wedge \tau(w)}(w), \cdots, x_{t_n \wedge \tau(w)}(w), \cdots), \qquad w \in \Omega ,$$

and so ξ is $\sigma(C_\tau^s)$-measurable.

LEMMA A.2. *Given $s \geqq 0$, let $\Omega^s = C([s, \infty), R^d)$ and give Ω^s the topology of uniform convergence on compact sets. Define $\phi^s:\Omega \to \Omega^s$ so that if $w^s = \phi(w)$, then $x(t, w^s) = x(t, w)$, $t \geqq s$. If τ is an s-Markov time, then $\tau(w) = \tau(w')$ whenever $\phi^s(w) = \phi^s(w')$ and so we can define τ^s on Ω^s so that $\tau^s(\phi^s(w)) = \tau(w)$, $w \in \Omega$. Given an s-Markov time τ, define*

$$\Omega^s(\tau^s) = \{w^s \in \Omega^s : x(t, w^s) = x(t \wedge \tau^s(w^s), w^s) \text{ for all } t \geqq s\} .$$

Then, $\Omega^s(\tau^s) \in \mathscr{B}_{\Omega^s}$. Moreover, if $\phi_\tau^s:\Omega \to \Omega^s(\tau)$ in such a way that

$$x(t, \phi_\tau^s(w)) = x(t \wedge \tau(w), w) , \qquad t \geqq s,$$

then $M_\tau^s = \{(\phi_\tau^s)^{-1}(A) : A \in \mathscr{B}_{\Omega^s(\tau^s)}\}$.

Proof: Since an s-Markov time is M^s-measurable, it is obvious that $\tau(w) = \tau(w')$ if $\phi^s(w) = \phi^s(w')$, and therefore τ^s is well-defined on Ω^s.

To prove that $\Omega^s(\tau^s) \in \mathscr{B}_{\Omega^s}$, we first show that $\Omega^s(t) \in \mathscr{B}_{\Omega^s}$ for each $t \geqq s$. Note that

$$\Omega^s(t) = \bigcap_{n=1}^{\infty} \bigcap_{k=0}^{\infty} \left\{w^s \in \Omega^s : x\left(t + \frac{k}{n}, w^s\right) = x(t, w^s)\right\} ,$$

and

$$\left\{w^s \in \Omega^s : x\left(t + \frac{k}{n}, w^s\right) = x(t, w^s)\right\} \in \mathscr{B}_{\Omega^s}$$

for all $n \geqq 1$ and $k \geqq 0$. Next observe that

$$\Omega^s(\tau^s) = \{\tau^s = \infty\} \cup \bigcap_{n=1}^{\infty} \bigcap_{k=0}^{\infty} \left\{s + \frac{k}{n} \leqq \tau^s < s + \frac{k+1}{n}\right\} \cap \Omega^s\left(s + \frac{k+1}{n}\right),$$

and so $\Omega^s(\tau^s) \in \mathscr{B}_{\Omega^s}$, since τ^s is clearly $\mathscr{B}_{\Omega^s}$-measurable.

Finally, observe that $\mathscr{B}_{\Omega^s(\tau^s)} = \sigma(C_{\tau^s})$, where C_{τ^s} is the system of sets

$$\{w^s \in \Omega^s : x(t_1 \wedge \tau^s(w^s)) \in \Gamma_1 , \cdots , x(t_n \wedge \tau^s(w^s)) \in \Gamma_n\} ,$$

where $s \leqq t_1 < \cdots < t_n$, $n \geqq 1$, and $\Gamma_1 , \cdots , \Gamma_n \in \mathscr{B}_{R^d}$. Furthermore, it is clear that

$$\{w \in \Omega : x_{t\wedge\tau(w)} \in \Gamma\} = (\phi_\tau^s)^{-1}(\{w^s \in \Omega^s : x(t \wedge \tau^s(w^s) \in \Gamma\}) ,$$

for all $t \geqq s$ and $\Gamma \in \mathscr{B}_{R^d}$. Hence, by Lemma A.1, we have

$$M_\tau^s = \{(\phi_\tau^s)^{-1}(A) : A \in \mathscr{B}_{\Omega^s(\tau^s)}\} ;$$

q.e.d.

We are now ready to put the Kuratowski fiber theorem into the form in which we want to apply it. The usual statement of this theorem says that if ϕ is a measurable map from one standard measurable space $\langle X, \mathscr{B}\rangle$ onto a second standard measurable space $\langle Y, \mathscr{C}\rangle$ and if P is a probability measure on $\langle X, \mathscr{B}\rangle$ and P' is the corresponding measure induced on $\langle Y, \mathscr{C}\rangle$ by ϕ, then there exists, for each $y \in Y$, a probability measure P_y on $\langle X, \mathscr{B}\rangle$ such that

(i) $P_y(A)$ is $\mathscr{C}$-measurable for each $A \in \mathscr{B}$,

(ii) there is a set $N \in \mathscr{C}$ for which $P'(N) = 0$ and $P_y(\{x : \phi(x) = y\}) = 1$ for $y \notin N$,

(iii) for all $A \in \mathscr{B}$ and $C \in \mathscr{C}$,

$$P(A \cap \phi^{-1}(C)) = \int_C P_y(A)P'(dy) .$$

The measures P_y constitute a *regular conditional probability distribution of P given* ϕ. It is easy to see that if P_y and Q_y are both regular conditional probability distributions of P given Q, then there is an $N \in \mathscr{C}$ such that $P'(N) = 0$ and $P_y = Q_y$ for $y \notin N$.

THEOREM A.2. *Let P be a probability measure on (Ω, M^s) and let τ be an s-Markov time. Then there exists a map $Q:\Omega \times M^s \to R$ such that*

(i) *for each $A \in M^s$, $Q_w(A) = Q(w, A)$ is M^s_τ-measurable,*

(ii) *for each $w \in \Omega$, $Q_w(\cdot)$ is a probability measure on (Ω, M^s),*

(iii) *there is an $N \in M^s_\tau$ such that $P(N) = 0$ and for all $w \notin N$*

$$Q_w(\{x_t = x_t(w), s \leqq t \leqq \tau(w)\}) = 1 ,$$

(iv) *for all $A \in M^s$ and $B \in M^s_\tau$,*

$$P(A \cap B) = \int_B Q_w(A)P(dw) .$$

Moreover, if $Q':\Omega \times M^s \to R$ is any other map satisfying (i)–(iv), *then there is an $N \in M^s_\tau$ such that $P(N) = 0$ and $Q_w = Q'_w$ for $w \notin N$.*

Proof: Define $\phi^s_\tau:\Omega \to \Omega^s(\tau^s)$ as in Lemma A.2. Then ϕ^s_τ is measurable from (Ω, M^s) onto $\langle \Omega^s(\tau^s), \mathscr{B}_{\Omega^s(\tau^s)}\rangle$. Note that M^s is σ-isomorphic to $\mathscr{B}_{\Omega^s}$, and therefore (Ω, M^s) is standard. Also, $\Omega^s(\tau^s) \in \mathscr{B}_{\Omega^s}$, and therefore $\langle \Omega^s(\tau^s), \mathscr{B}_{\Omega^s(\tau^s)}\rangle$ is standard. Thus we can find probability measures

$$[P_{w^s} : w^s \in \Omega^s(\tau^s)]$$

on (Ω, M^s) such that

(i′) $P_{w^s}(A)$ is $\mathscr{B}_{\Omega^s(\tau^s)}$-measurable for each $A \in M^s$,

(ii′) there is an $N \in \mathscr{B}_{\Omega^s(\tau^s)}$ such that $P'(N) = 0$ and

$$P_{w^s}(\{w' \in \Omega : \phi^s_\tau(w') = w^s\}) = 1$$

for $w^s \notin N$,

(iii') for $A \in M^s$ and $B \in \mathscr{B}_{\Omega^s(\tau^s)}$,

$$P(A \cap (\phi_\tau^s)^{-1}(B)) = \int_B P_{w^s}(A) P'(dw^s) ,$$

where P' is the probability measure on $\langle \Omega^s(\tau^s), \mathscr{B}_{\Omega^s(\tau^s)} \rangle$ corresponding to P induced by ϕ_τ^s. If we now define $Q(w, A) = P_{\phi_\tau^s(w)}(A)$, $w \in \Omega$ and $A \in M^s$, our assertion follows.

We refer to the mapping $Q: \Omega \times M^s \to R$ described in Theorem A.2 as a *regular conditional probability distribution of* P *given* M_τ^s.

An estimate of Kolmogorov type.[11]

THEOREM A.3. *Let* $\langle \Omega, \mathscr{B}, P \rangle$ *be an arbitrary probability space and let* Y *be a Banach space. Suppose* $F: R^d \times \Omega \to Y$ *satisfies:*

(i) $F(x, \cdot)$ *is* $\mathscr{B}$*-measurable for each* $x \in R^d$,

(ii) $l(F(\cdot, w))$ *is continuous for almost all* $w \in \Omega$ *for each* l *in a dense subset of* Y^*,

(iii) *there exists an* $\alpha > d$ *and a* $B < \infty$ *such that*

$$E^P[\|F(x) - F(y)\|^\alpha] \leqq B\,|x - y|^\alpha , \qquad x, y \in R^d ,$$

$$E^P[\|F(x)\|^\alpha] \leqq B , \qquad x \in R^d .$$

Then $F(\cdot, w)$ *is strongly continuous for almost all* $w \in \Omega$. *Moreover, for each* $\beta > 1 + 1/\alpha$, *there exists a constant* C *depending only on* d, α, *and* β *such that*

$$P(\|F(x)\| \geqq L(1 + |x|)^\beta \textit{ for some } x \in R^d) \leqq \frac{CB}{L^\alpha} , \qquad L > 0 .$$

Proof: Given $t > 0$ and $n \geqq 0$ let us define functions $F_n^{(t)}(x)$ on

$$\square(t) = \{y \in R^d : |y_i| \leqq t \text{ for } i = 1, 2, \cdots, d\}$$

as follows. At the n-th stage the domain $\square(t)$ is split up into 2nd cubes by a regular partition of each coordinate segment into 2^n equal parts. Inside any single cube, $F_n^{(t)}(x)$ is multilinear in $x_1, \cdots, x_n$ and is chosen to coincide with $F(x)$ at the corners of that cube. There is a unique $F_n^{(t)}(x)$ which achieves this. Let us define

$$\Delta_n^{(t)} = \sup_{x \in \square(t)} \|F_{n+1}^{(t)}(x) - F_n^{(t)}(x)\|$$

[11] The results of this sub-section will be used only in Part II of this article.

and take

$$\delta_n^{(t)} = \sup_{1\leq j\leq d} \sup_{-2^n\leq l_1\leq 2^n} \cdots \sup_{-2^n\leq l_j\leq 2^n-1} \cdots \sup_{-2^n\leq l_d\leq 2^n-1} \left\|F\left(\frac{l_1 t}{2^n}, \ldots, \frac{l_d t}{2^n}\right) - F\left(\frac{l_1 t}{2^n}, \ldots, \frac{l_{j-1} t}{2^n}, \frac{(l_j+1)t}{2^n}, \frac{l_{j+1} t}{2^n}, \ldots, \frac{l_d t}{2^n}\right)\right\| .$$

Then,

$$\Delta_n^{(t)} \leqq d\, \delta_{n+1}^{(t)}$$

and so

$$\text{(A.4)} \qquad P(\Delta_n^{(t)} \geqq 2^{-(n+1)\gamma} A) \leqq \frac{Bd^{\alpha+1}t^\alpha 2^d}{A^\alpha} 2^{(n+1)(d+\alpha\gamma-\alpha)}, \quad A > 0 \text{ and } \gamma \in R .$$

Since $\alpha > d$, we can choose $\gamma > 0$ so that $-\varepsilon = d + \alpha\gamma - \alpha < 0$. Thus,

$$\sum_{n=1}^{\infty} P(\Delta_n^{(t)} \geqq 2^{-(n+1)\gamma}) < \infty ,$$

and therefore, by the Borel-Cantelli lemma,

$$P(\lim_{n\to\infty} \Delta_n^{(t)} \geqq 2^{-(n+1)\gamma}) = 0 .$$

In particular, we see that there is a $G: \square(t) \times \Omega \to Y$ such that

$$\sup_{x\in\square(t)} \|F_n^{(t)}(x, w) - G(x, w)\| \to 0 \quad \text{as} \quad n \to \infty$$

for almost all $w \in \Omega$. Using (ii) and the definition of the $F_n^{(t)}$, one easily sees that $F(x, w) = G(x, w)$, $x \in \square(t)$, for almost all $w \in \Omega$. Hence, $F(x, w)$, is continuous in $x \in \square(t)$ for almost all $w \in \Omega$. Since $t > 0$ was arbitrary, the first assertion now follows.

A consequence of the preceding is that

$$\sup_{x\in\square(t)} \|F(x) - F_0^{(t)}(x)\| \leqq \sum_0^\infty \Delta_n^{(t)} \text{ a.s. } P .$$

Hence, if we define $C_\eta = \sum_0^\infty 2^{-(n+1)\eta}$ for $\eta > 0$, then

$$P(\sup_{x\in\square(t)} \|F(x) - F_0^{(t)}(x)\| \geqq A) \leqq P\left(\Delta_n^{(t)} \geqq \frac{A}{C_\gamma} 2^{-(n+1)\gamma} \text{ for some } n \geqq 0\right)$$

$$\leqq \frac{Bd^{\alpha+1}t^\alpha 2^d C_\gamma^\alpha}{A^\alpha} \sum_0^\infty 2^{-(n+1)\varepsilon} = \frac{Bd^{\alpha+1}t^\alpha 2^d C_\gamma^\alpha C_\varepsilon}{A^\alpha},$$

where γ and ε are chosen as in the preceding paragraph. Also,

$$P(\sup_{x\in\square(t)} \|F_0^{(t)}(x)\| \geqq A) \leqq \frac{2^d B}{A^\alpha}.$$

Thus if $\beta > 1 + 1/\alpha$, then

$$\begin{aligned}
&P(\sup_{x\in\square(t)} \|F(x)\| \geqq L(1+t)^\beta \text{ for some } t > 0) \\
&\quad \leqq P(\sup_{x\in\square(1)} \|F(x)\| \geqq L) + P(\sup_{x\in\square(t)} \|F(x)\| \geqq L(t+1)^\beta \text{ for some } t \geqq 1) \\
&\quad \leqq P(\sup_{x\in\square(1)} \|F_0^{(1)}(x)\| \geqq \tfrac{1}{2}L) + P(\sup_{x\in\square(1)} \|F(x) - F_0^{(1)}(x)\| \geqq \tfrac{1}{2}L) \\
&\qquad + P(\sup_{x\in\square(t)} \|F_0^{(t)}(x)\| \geqq \tfrac{1}{2}L(1+t)^\beta \text{ for some } t \geqq 1) \\
&\qquad + P(\sup_{x\in\square(t)} \|F(x) - F_0^{(t)}(x)\| \geqq \tfrac{1}{2}L(1+t)^\beta \text{ for some } t \geqq 1) \\
&\quad \leqq \frac{2^{d+\alpha}B}{L^\alpha} + \frac{2^\alpha C'B}{L^\alpha} + \frac{2^\alpha C'B}{L^\alpha}\sum_1^\infty \frac{1}{N^{\alpha(1-\beta)}} + \frac{2^{d+\alpha}B}{L^\alpha}\sum_1^\infty \frac{1}{N^\alpha} \\
&\quad \leqq \frac{2^\alpha B}{L^\alpha}\left(2^d + C' + C'\sum_1^\infty \frac{1}{N^{\alpha(1-\beta)}} + 2^d \sum_1^\infty \frac{1}{N^\alpha}\right) = \frac{CB}{L^\alpha},
\end{aligned}$$

where $C' = 2^d d^{\alpha+1} C_\gamma^\alpha C_\varepsilon$ and

$$C = 2^\alpha\left(2^d + C' + C'\sum_1^\infty \frac{1}{N^{\alpha(1-\beta)}} + 2^d \sum_1^\infty \frac{1}{N^\alpha}\right).$$

Bibliography

[1] Cameron, R. H., and Martin, W. T., *The transformation of Wiener integrals by non-linear transformations*, Trans. Amer. Math. Soc., Vol. 75, 1953, pp. 552–575.

[2] Donsker, M. D., *Four Papers in Probability*, Mem. Amer. Math. Soc., No. 6, 1951.

[3] Doob, J. L., *Stochastic Processes*, John Wiley and Sons, New York, 1953.

[4] Dynkin, E. B., *Foundations of the Theory of Markov Processes*, Pergamon Press, New York, 1960.

[5] Dynkin, E. B., *Markov Processes*, Vols. 1 and 2, Academic Press, New York, 1965.

[6] Girsanov, I. V., *On transforming a class of stochastic processes by absolutely continuous substitution of measures*, Theor. Probability Appl., Vol. 5, 1960, pp. 285–301.

[7] Girsanov, I. V., *An example of non-uniqueness of the solution of the stochastic equation of K. Itô*, Theor. Probability Appl., Vol. 7, 1962, pp. 325–331.

[8] Girsanov, I. V., *On Itô's stochastic integral equations*, Doklady Akad. Nauk USSR, Vol. 138, 1961, pp. 18–21. (In Russian.)

[9] Itô, K., *On Stochastic Differential Equations*, Mem. Amer. Math. Soc., No. 4, 1951.

[10] Itô, K., *On a formula concerning stochastic differentials*, Nagoya Math. J., Vol. 3, 1951, pp. 55–65.

[11] Jones, B. F., *A class of singular integrals*, Amer. J. Math., Vol. 86, 1964, pp. 441–462.

[12] Krylov, N. V., *On quasi-diffusion processes*, Theor. Probability Appl., Vol. 11, 1966, pp. 373–389.
[13] Ladyzenskaia, O. A., Solonnikov, V. A., and Ural'tseva, N. N., *Linear and Quasilinear Equations of Parabolic Type*, Izd. Nauka, Moscow, 1967. (In Russian.)
[14] Parthasarathy, K. R., *Probability Measures on Metric Spaces*, Academic Press, New York, 1967.
[15] Prokhorov, Yu. V., *Convergence of random processes and limit theorems*, Theor. Probability Appl., Vol. 9, 1956, pp. 157–214.
[16] Skorokhod, A. V., *Investigations in the Theory of Random Processes*, Izd. Kievskovo Universiteta, Kiev, 1961. (In Russian.)
[17] Tanaka, H., *Existence of diffusions with continuous coefficients*, Mem. Fac. Sci. Kyushu Univ. Ser. A., Vol. 18, 1964, pp. 89–103.

Received October, 1968.

COMMUNICATIONS ON PURE AND APPLIED MATHEMATICS, VOL. XXII, 479–530 (1969)

Diffusion Processes with Continuous Coefficients, II*

DANIEL W. STROOCK AND S. R. S. VARADHAN

7. Strong Feller Property[1]

Let $P(s, x, t, dy)$ be the transition probabilities of the Markov process corresponding to the coefficients $a_{ij}(s, x)$ and $b_j(s, x)$. We wish to show that, for $s < t$ and arbitrary x, $P(s, x, t, dy)$ is continuous in s, x in the uniform sense. This amounts to showing that for bounded functions $\Phi(y)$ the functions

$$U(s, x) = \int \Phi(y) P(s, x, t_0, dy)$$

are continuous in s and x for $s < t_0$, and are equicontinuous provided Φ stays uniformly bounded. Because of the continuity of paths, it suffices to investigate the continuity of U in x for each fixed $s < t_0$. All continuity properties will depend only on the bound of Φ, provided s stays away from t_0.

LEMMA 7.1. *Let $b \equiv 0$, $a_{ij}(s, x) = C_{ij} + \varepsilon_{ij}(s, x)$, where C_{ij} is constant, symmetric and positive definite, and $|\varepsilon_{ij}(s, x)| \leqq \varepsilon$, where ε is chosen as in Theorem* 5.1. *Then,*

$$U(s, x) = \int \Phi(y) P(s, x, t_0, dy)$$

is continuous in x.

Proof:

$$U(s, x) = \int U(t, y) P(s, x, t, dy) \quad \text{for} \quad s < t < t_0 .$$

Therefore,

$$U(s, x) = \frac{1}{h} \iint_{t_0 - h}^{t_0} U(t, y) P(s, y, t, dy)\, dt$$

$$= \frac{1}{h} \int_{t_0 - h}^{t_0} U(t, y) e^{-(t-s)} e^{(t-s)} P(s, x, t, dy)\, dt ,$$

$$s \leqq t_0 - h .$$

---

* This paper represents results obtained at the Courant Institute of Mathematical Sciences. New York University. This work was supported in part by a Ford Foundation Grant for Statistics and in part by a Sloan Foundation Grant for Statistics. Reproduction in whole or in part is permitted for any purpose of the United States Government.

[1] This is the continuation of the paper by the same title which appeared in the previous issue of this journal. The notation and the numbering of sections, theorems, lemmas, equations, etc., are continued.

In view of Theorem 5.1,

$$|U(s, x_1) - U(s, x_2)| \leqq C\,|x_1 - x_2|\;|||U|||_{p_N}\,,$$

where

$$|||U|||_{p_N}^{p_N} = \frac{1}{h^{p_N}}\int\int_{t_0-h}^{t_0} |U(t, y)e^{(t-s)}|^{p_N}\,dt\,dy\,.$$

If $|\Phi(y)| \leqq M$ and vanishes outside a sphere of radius l, then $U(t, y)$ is bounded by M and vanishes exponentially fast at ∞. Therefore,

$$|||U|||_{p_N} \leqq A(l, M)\,.$$

On the other hand, if $|\Phi(y)| \leqq M$, the contribution to

$$U(s, x) = \int P(s, x, t_0\,, dy)\Phi(y)$$

from outside $|y| \geqq l$ is uniformly small for large l, provided x varies over a compact set and $t_0 - s$ is bounded.

LEMMA 7.2. *Let a_{ij} be as in Lemma 7.1 and let b_j be bounded. Then,*

$$U(s, x) = \int \Phi(y)P(s, x, t_0\,, dy)$$

is continuous in x for $s < t_0$.

Proof: Let $P_{s,x}$ be the process corresponding to these coefficients and $Q_{s,x}$ the process with the same a_{ij} but with $b_j \equiv 0$. Then,

$$U(s, x) = \int \Phi(x(t_0))\,dP_{s,x}$$

$$= \int \Phi(x(t_0))\,\exp\{S_s^{t_0}\}\,dQ_{s,x}\,,$$

where

$$S_s^t = \int_s^t \langle b(u, x(u)), a^{-1}(u, x(u))\,dx(u)\rangle$$

$$- \tfrac{1}{2}\int \langle b(u, x(u), a^{-1}(u, x(u))b(u, x(u))\rangle\,du\,.$$

Let

$$U_\varepsilon(s, x) = \int \Phi(x(t_0)) \exp \{S_{s+\varepsilon}^{t_0}\} dQ_{s,x} .$$

Then

$$U_\varepsilon(s, x) = \int \psi_\varepsilon(y) Q(s, x, s + \varepsilon, dy) ,$$

where

$$\psi_\varepsilon(y) = \int \Phi(x(t_0)) \exp \{S_{s+\varepsilon}^{t_0}\} dQ_{s+\varepsilon,y} ,$$

and $Q(s, x, s + \varepsilon, dy)$ is the transition probability of the process with $b \equiv 0$. Moreover, $|\Phi(y)| \leqq M$ implies $|\psi_\varepsilon(y)| \leqq M$. Therefore, $U_\varepsilon(s, x)$ is continuous for every $\varepsilon > 0$. It is therefore sufficient to estimate

$$\Delta_\varepsilon = \int \exp \{S_{s+\varepsilon}^{t_0}\} |\exp \{S_s^{s+\varepsilon}\} - 1| \, dQ_{s,x} .$$

Now,

$$\Delta_\varepsilon^2 \leqq \left(\int \exp \{2S_{s+\varepsilon}^{t_0}\} dQ_{s,x}\right)\left(\int [\exp \{S_s^{s+\varepsilon}\} - 1]^2 \, dQ_{s,x}\right) .$$

The first factor is at most $\exp \{c(t_0 - s + \varepsilon)\}$, where c is the bound for $ba^{-1}b$, which exists in view of the boundedness of b and the uniform ellipticity of a. The second factor equals

$$\begin{aligned} \int (\exp \{2S_s^{s+\varepsilon}\} - 2 \exp \{S_s^{s+\varepsilon}\} + 1) \, dQ_{s,x} &\leqq \exp \{c\varepsilon\} - 2 + 1 \\ &= \exp \{c\varepsilon\} - 1 . \end{aligned}$$

Therefore, $\Delta_\varepsilon \to 0$ as $\varepsilon \to 0$, uniformly with respect to x and s.

LEMMA 7.3. *Take a and b to be general (i.e., a continuous, bounded and strictly elliptic and b bounded). Consider a point (s_0, x_0) such that $s_0 < t_0$. Then, $U(s, x)$ is continuous in x at x_0.*

Proof: Let us take a neighborhood N_δ of the form $[s_0, s_0 + \delta]$, $|x - x_0| \leqq \delta$. We can choose this neighborhood in such a way that the coefficients a can be altered outside to yield new coefficients $\bar{a}$ satisfying the conditions in Lemma 7.2. Then for the coefficients $\bar{a}$ and b we have transition probabilities $\bar{P}(s, x, t, dy)$ for which Lemma 7.2 is applicable. Also,

$$\begin{aligned} U(s, x) &= \int \Phi(y) P(s, x, t_0 , dy) \\ &= \int U(t, y) P(s, x, t, dy) \quad \text{for} \quad s < t < t_0 . \end{aligned}$$

Let

$$\bar{U}_\varepsilon(s, x) = \int U(s + \varepsilon, y)\bar{P}(s, x, s + \varepsilon, dy) .$$

We know from Lemma 7.2 that $\bar{U}_\varepsilon(s, x)$ is continuous. Moreover, $U(s, x)$ is bounded by M if Φ is bounded by M. It is therefore enough to examine the difference

$$\bar{U}_\varepsilon(s, x) - U(s, x) = \int U(s + \varepsilon, y)[\bar{P}(s, x, s + \varepsilon, dy) - P(s, x, s + \varepsilon, dy)] .$$

Since U is bounded by M for $\varepsilon \leqq \delta$, we get

$$|\bar{U}_\varepsilon(s, x) - U(s, x)| \leqq 2MP_{s,x}[\tau^s_{N_\delta} \leqq \varepsilon] \leqq 2M \exp\left\{-\frac{\alpha\delta_x^2}{2\varepsilon}\right\},$$

where $\delta_x = \delta - |x - x_0|$. Thus, the difference goes to zero uniformly over $|x - x_0| \leqq \frac{1}{2}\delta$, and this proves the lemma.

THEOREM 7.1. *Let $U(s, x) = \int \Phi(y)P(s, x, t_0, dy)$, where $P(s, x, t_0, dy)$ is the transition probability corresponding to general a and b. Then $U(s, x)$ is continuous in s and x for $s < t_0$ and for every bounded Φ. Moreover, the modulus of continuity of $U(s, x)$ at a point (s, x) with $s < t_0$ depends only on the bound of Φ.*

Proof: Lemmas 7.1 to 7.3 establish the continuity in x and the uniformity of the continuity provided s stays away from t_0. To estimate $U(s_1, x) - U(s_2, x)$ we proceed as follows. Let $s_1 < s_2$. Then,

$$|U(s_1, x) - U(s_2, x)| \leqq \int |U(s_2, y) - U(s_2, x)| \, P(s_1, x, s_2, dy) .$$

We have the estimate $P(s_1, x, s_2, |y - x| \geqq \varepsilon) \leqq \exp\{-\alpha\varepsilon^2/2(s_2 - s_1)\}$, and the continuity of $U(s_2, x)$. The continuity in s follows from this.

8. Estimates on the Transition Probabilities

The coefficients a are continuous and strictly elliptic. We shall denote by $\{\psi_K(\delta)\}$ and $\{\lambda_K\}$ the modulus of continuity and the lowest eigenvalue of a on compact subsets K of $I^0 \times R^d$. By m we denote the common upper bound for all a_{ij} and b_j. The aim of this section is to obtain some estimates on $P(s, x, t, dy)$ which will depend only on $\{\psi_K(\delta)\}$, $\{\lambda_K\}$ and m. This will be clear from the method of estimation and we shall not keep track of the constants.

We shall first work out a method of localization. Consider the following sets

$$J_0 = [s:t_1 \leqq s \leqq t_2], \qquad J_1 = J_2 = [s:t_1 - \delta \leqq s \leqq t_2],$$

$$B_j = [x:|x - x_0| \leqq R + j\delta], \qquad j = 0, 1, 2,$$

$$S_j = J_j \times B_j, \qquad j = 0, 1, 2,$$

$$D_j = \text{boundary of } S_j \text{ for } j = 1, 2.$$

Let $\bar{a}$, $\bar{b}$ be new coefficients agreeing with a and b in S_2. Let $\bar{P}(s, x, t, dy)$ and $P(s, x, t, dy)$ be the transition probabilities corresponding to the coefficients $\bar{a}$, $\bar{b}$ and a, b, respectively, $\bar{P}_{s,x}$, $P_{s,x}$ the processes, and $\bar{E}_{s,x}$, $E_{s,x}$ the corresponding expectations with respect to these processes. The function $f(t, y)$ is bounded and measurable in t and y, and is zero outside S_0. It can also be a function of the form $\delta(t - t_2)f(y)$, where $f(y)$ is a bounded measurable function of y which is zero outside B_0 and $\delta(t)$ is the Dirac delta function. We shall treat the first case. The second case will be a limiting form.

Let us define the quantities

$$T(s, x) = E_{s,x}\int_s^{t_2} e^{-(t-s)}f(t, x(t))\, dt,$$

$$\bar{T}(s, x) = \bar{E}_{s,x}\int_s^{t_2} e^{-(t-s)}f(t, x(t))\, dt,$$

$$T_j = \sup_{(s,x)\in D_j} |T(s, x)|, \qquad j = 1, 2,$$

$$\bar{T}_j = \sup_{(s,x)\in D_j} |\bar{T}(s, x)|, \qquad j = 1, 2,$$

$$\bar{T} = \max(\bar{T}_1, \bar{T}_2),$$

$$\tau_1 = \begin{cases} \text{first entrance time into } B_1 \text{ after time } t_1 - \delta \\ \qquad \text{if such a time is less than } t_2, \\ t_2 \text{ otherwise,} \end{cases}$$

$$\tau_2 = \begin{cases} \text{first exit time from the interior of } B_2 \\ \qquad \text{if such a time is less than } t_2, \\ t_2 \text{ otherwise,} \end{cases}$$

$$H_1(s, x) = E_{s,x}[e^{(-\tau_1-s)}; \tau_1 < t_2], \qquad (s, x) \notin S_1,$$

$$H_2(s, x) = E_{s,x}[e^{-(\tau_2-s)}; \tau_2 < t_2], \qquad (s, x) \in S_2,$$

$$\rho = \sup_{(s,x)\in D_1} H_2(s, x).$$

LEMMA 8.1. *$\rho \leqq \rho(m, \delta) < 1$ for every $\delta > 0$.*

Proof:

$$\begin{aligned} P_{s,x}[(\tau_2 - s) \leqq \Delta, \tau_2 < t_2] &\leqq P_{s,x}[|x(t) - x| \geqq \delta \text{ for some } t \text{ in } (s, s + \Delta)] \\ &\leqq e^{-c\delta^2/\Delta}, \end{aligned}$$

where c depends on m only. This estimate, valid for $(s, x) \in D_1$, gives a bound for $H_2(s, x)$.

LEMMA 8.2. (a) *For $(s, x) \notin S_1$,*

$$|T(s, x)| \leqq H_1(s, x) T_1 .$$

(b) *For $(s, x) \in S_2$,*

$$|T(s, x)| \leqq |\bar{T}(s, x)| + \bar{T}_2 + H_2(s, x) T_2 .$$

Proof: (a) Let $(s, x) \notin S_1$. Since f has support in S_0, we have

$$\begin{aligned} T(s, x) &= E_{s,x} \int_s^{t_2} e^{-(t-s)} f(t, x(t))\, dt \\ &= E_{s,x}\left[\int_{\tau_1}^{t_2} e^{-(t-s)} f(t, x(t))\, dt, \tau_1 < t_2\right] \\ &= E_{s,x} e^{-(\tau_1 - s)} \{\tau_1 < t_2\} E_{\tau_1, x(\tau_1)}\left[\int_{\tau_1}^{t_2} e^{-(t-\tau_1)} f(t, x(t))\, dt\right] \\ &= E_{s,x}[e^{-(\tau_1 - s)}\{\tau_1 < t_2\} T(\tau_1 , x(\tau_1))] . \end{aligned}$$

Hence,

$$|T(s, x)| \leqq H_1(s, x) T_1 .$$

(b) Let $(s, x) \in S_2$. Then,

$$\begin{aligned} T(s, x) &= E_{s,x} \int_s^{t_2} e^{-(t-s)} f(t, x(t))\, dt \\ &= E_{s,x} \int_s^{\tau_2} e^{-(t-s)} f(t, x(t)) dt \\ &\quad + E_{s,x} \int_{\tau_2}^{t_2} e^{-(t-s)} f(t, x(t))\, dt . \end{aligned}$$

Since $P_{s,x}$ and $\bar{P}_{s,x}$ agree on $M^s_{\tau_2}$, we have

$$
\begin{aligned}
T(s, x) &= \bar{E}_{s,x}\int_s^{\tau_2} e^{-(t-s)}f(t, x(t))\, dt + E_{s,x}\int_{\tau_2}^{t_2} e^{-(t-s)}f(t, x(t))\, dt \\
&= \bar{E}_{s,x}\int_s^{t_2} e^{-(t-s)}f(t, x(t))\, dt + E_{s,x}\int_{\tau_2}^{t_2} e^{-(t-s)}f(t, x(t))\, dt \\
&\qquad - \bar{E}_{s,x}\int_{\tau_2}^{t_2} e^{-(t-s)}f(t, x(t))\, dt \\
&= \bar{T}(s, x) + E_{s,x}[e^{-(\tau_2-s)}\{\tau_2 < t_2\}T(\tau_2, x(\tau_2))] \\
&\qquad - \bar{E}_{s,x}[e^{-(\tau_2-s)}\{\tau_2 < t_2\}\bar{T}(\tau_2, x(\tau_2))] .
\end{aligned}
$$

Hence,

$$|T(s, x)| \leqq |\bar{T}(s, x)| + H_2(s, x)T_2 + \bar{T}_2 .$$

LEMMA 8.3. *We have the following inequality:*

$$T_1 \leqq 2(1 - \rho)^{-1}\bar{T} .$$

Proof: From Lemma 8.2(b) we know that, for $(s, x) \in D_1$,

$$|T(s, x)| \leqq \bar{T}_1 + \bar{T}_2 + \rho T_2$$

or, what is the same,

$$T_1 \leqq 2\bar{T} + \rho T_2 .$$

From Lemma 8.2(a), we have $T_2 \leqq T_1$. Therefore,

$$T_1 \leqq 2\bar{T} + \rho T_1$$

that is,

$$T_1 \leqq 2(1 - \rho)^{-1}\bar{T} .$$

LEMMA 8.4. *There is a constant C_δ depending on m and δ such that, for $(s, x) \notin S_1$,*

$$|T(s, x)| \leqq C_\delta H_1(s, x)\bar{T} ,$$

and, for $(s, x) \in S_2$,

$$|T(s, x)| \leqq \bar{T}(s, x) + C_\delta \bar{T} .$$

Proof: The statement of this lemma is an immediate consequence of the earlier lemmas.

As a consequence of Lemma 8.4 one can use an estimate of the transition probability $\bar{P}(s, x, t, dy)$ to obtain an estimate for $P(s, x, t, dy)$. The term $H_1(s, x)$

yields a good estimate for large values of x and large values of $(t_1 - s)$. Let a and b be as before and (s_0, x_0) a fixed point. We choose a neighborhood $J_0 = [s_0 - \delta, s_0 + \delta]$ around s_0 and call it $[t_1, t_2]$, and set $J_1 = J_2 = [t_1 - \delta, t_2]$. We let B_0 be a neighborhood around x_0 of the form $B_0 = \{x: |x - x_0| \leqq \delta\}$, and B_1 and B_2 the slightly enlarged neighborhoods $\{x: |x - x_0| \leqq 2\delta\}$ and $\{x: |x - x_0| \leqq 3\delta\}$, respectively. S_j and D_j are defined as before in terms of T_j and B_j. The δ is chosen small enough so that on S_2 we have $|a_{ij} - C_{ij}| \leqq \varepsilon$, where $C_{ij} = a_{ij}(s_0, x_0)$; ε is to be chosen later. The $\bar{a}$ are uniformly continuous coefficients satisfying $|\bar{a}_{ij} - C_{ij}| \leqq \varepsilon$ everywhere and agreeing with a on S_2. Moreover, the modulus of continuity of $\bar{a}$ can be assumed to be no worse than that of a on S_2. We now consider three sets of coefficients $[a, b]$, $[\bar{a}, b]$, and $[\bar{a}, 0]$. The transition probabilities will be denoted by $P(s, x, t, dy)$, $\bar{P}(s, x, t, dy)$ and $\bar{Q}(s, x, t, dy)$, respectively. If they have densities, we shall denote them by $p(s, x, t, y)$, $\bar{p}(s, x, t, y,)$ and $\bar{q}(s, x, t, y)$. The processes will be denoted by $P_{s,x}$, $\bar{P}_{s,x}$ and $\bar{Q}_{s,x}$. We want to estimate $\iint_{S_0} |p(s, x, t, y)|^\alpha \, dt\, dy$. We do this by first estimating $\iint_{S_0} |\bar{q}(s, x, t, y)|^\alpha \, dt\, dy$ and then $\iint_{S_0} |\bar{p}(s, x, t, y)|^\alpha \, dt\, dy$. The final α will be a little less than the initial α. But since the initial α is arbitrary, this does not affect the final result. Fix $1 < \alpha < \infty$. Choose p_1 so that $p_1^{-1} + \alpha^{-1} = 1$. Corresponding to this p_1 there exists an integer N and a number ε stated in Theorem 5.1. This will be the choice for our ε.

LEMMA 8.5. *$\bar{Q}(s, x, t, dy)$ has a density for almost all t. Denoting it by $\bar{q}(s, x, t, y)$, we have*

$$\sup_{(s,x)\in L} \iint |e^{-(t-s)}\bar{q}(s, x, t, y)|^{\alpha'} \, dt\, dy \leqq C_\delta < \infty$$

for every $\alpha' < \alpha$, where $L = D_1 \cup D_2$.

Proof: $p_1^{-1} + \alpha^{-1} = 1$. Using Theorem 5.1 with $\lambda = 1$, we get an estimate:

$$|K_1^N f(s, x)| \leqq C_1 \, \| f(s, x) \|_{p_1}. \tag{8.1}$$

It is easily verified that

$$K_1^N f(s, x) = \iint_s^\infty e^{-(t-s)} \frac{(t - s)^{N-1}}{(N - 1)!} f(t, y) \bar{Q}(s, x, t, dy)\, dt,$$

and therefore (8.1) tells us that $\bar{Q}$ has a density for almost all t and that

$$\iint_s^\infty |e^{-(t-s)}\bar{q}(s, x, t, y)|^\alpha \, (t - s)^{(N-1)\alpha} \, dy\, dt \leqq C_1^\alpha.$$

Since $J_0 = [t_1, t_2]$, we have, for $s < t_1$,

$$\iint_{S_0} |e^{-(t-s)}\bar{q}(s, x, t, y)|^\alpha \, dt \, dy \leqq C_1^\alpha (t_1 - s)^{-(N-1)\alpha} . \tag{8.2}$$

We need worry only about the case where s is arbitrary but x is away from B_0 by at least a distance δ. Let s be arbitrary. Let $J_h = J \cap \{t \geqq s + h\}$. Then,

$$\iint_{J_h \times B_0} |e^{-(t-s)}\bar{q}(s, x, t, y)|^\alpha \, dt \, dy \leqq C_1^\alpha h^{-(N-1)\alpha} .$$

On the other hand,

$$\begin{aligned} \iint_{(J-J_h)\times B_0} e^{-(t-s)}\bar{q}(s, x, t, y) \, dt \, dy &\leqq \int_s^{s+h} \bar{Q}(s, x, t, B_0) \, dt \\ &\leqq h \exp\left\{-\frac{c\delta^2}{2h}\right\}, \end{aligned}$$

where c is a positive constant depending on m, and δ is the distance of x from B_0. Let

$$\Phi_h(t, y) = \begin{cases} e^{-(t-s)}\bar{q}(s, x, t, y) & \text{for } t \in J_h, y \in B_0 , \\ 0 & \text{otherwise} , \end{cases}$$

$$\Phi(t, y) = \begin{cases} e^{-(t-s)}\bar{q}(s, x, t, y) & \text{for } t \in J, y \in B_0 , \\ 0 & \text{otherwise} . \end{cases}$$

Our estimates are

$$\|\Phi_h(t, y)\|_\alpha \leqq C_1 h^{-(N-1)} ,$$

$$\|\Phi - \Phi_h\|_1 \leqq h e^{-c\delta^2/2h} .$$

We claim that this gives us an estimate of the form

$$\|\Phi\|_{\alpha'} \leqq C_2 ,$$

for every $\alpha' < \alpha$ with a constant, C_2, depending on C_1, N, c, δ and α'. To see this let us take $h = 1/n$. Then, for $\Delta_n = \Phi_n - \Phi_{n+1}$, we have

$$\|\Delta_n\|_\alpha \leqq 2C_1(n + 1)^{N-1} ,$$

and

$$\|\Delta_n\|_1 \leqq C_2 e^{-\nu n} ,$$

where $\nu > 0$ is a constant $(\nu = \frac{1}{2}c\delta)^2$. If we now use the fact that $\log \|\Delta_n\|_\beta$ is a convex function of $1/\beta$, we find that for every $1 < \alpha' < \alpha$ there exists a θ such that $0 < \theta < 1$ and

$$\|\Delta_n\|_{\alpha'} \leqq \|\Delta_n\|_\alpha^{(1-\theta)} \|\Delta_n\|_1^\theta$$
$$\leqq [2C_1(n+1)]^{(N-1)\theta}[C_2 e^{-\nu n}]^\theta$$

for $n \geqq 1$. This gives us an estimate for $\sum_n \|\Delta_n\|_{\alpha'}$ and therefore for $\|\Phi\|_{\alpha'}$.

LEMMA 8.6. *$\bar{P}(s, x, t, dy)$ has a density $\bar{p}(s, x, t, y)$ for almost all t which satisfies*

$$\sup_{(s,x)\in L} \iint_{S_0} |e^{-(t-s)}\bar{p}(s, x, t, y)|^{\alpha''} \, dy \, dt \leqq C_\delta < \infty$$

for every $\alpha'' < \alpha'$, where $L = D_1 \cup D_2$ is as in Lemma 8.5.

Proof: According to Theorem 6.2, $\bar{P}_{s,x}$ and $\bar{Q}_{s,x}$ are absolutely continuous with respect to one another and

$$\left.\frac{d\bar{P}_{s,x}}{d\bar{Q}_{s,x}}\right|_{M_t^s} = \exp\left\{\int_s^t b(\sigma, x(\sigma))\bar{a}^{-1}(\sigma, x(\sigma)) \, dx(\sigma) - \frac{1}{2}\int_s^t b(\sigma, x(\sigma))\bar{a}^{-1}(\sigma, x(\sigma))b(\sigma, x(\sigma)) \, d\sigma\right\}.$$

We denote the expression on the right by $\exp\{R(t, w)\}$. Then,

$$[b(\sigma, x(\sigma))\bar{a}^{-1}(\sigma, x(\sigma))b(\sigma, x(\sigma))]$$

has an upper bound c depending on the ellipticity of $\bar{a}$ and the bound for b. We now estimate

$$\int_\Omega \exp\{\rho R(t, w)\} \, d\bar{Q}_{s,x} = N_\rho .$$

Clearly,

$$N_\rho = \int \exp\left\{\rho \int_s^t b\bar{a}^{-1} \, dx - \tfrac{1}{2}\rho \int_s^t b\bar{a}^{-1}b \, d\sigma\right\} d\bar{Q}_{s,x}$$
$$\leqq \exp\{\tfrac{1}{2}c(t-s)\rho(\rho-1)\}$$
$$\times\left[\int_\Omega \exp\left\{\rho \int_s^t b\bar{a}^{-1} \, dx - \tfrac{1}{2}\rho^2 \int_s^t b\bar{a}^{-1}b \, d\sigma\right\} d\bar{Q}_{s,x}\right]$$
$$\leqq \exp\{\tfrac{1}{2}c(t-s)\rho(\rho-1)\}.$$

It follows that $\bar{P}(s, x, t, dy) = \bar{Q}(s, x, t, dy) r(s, x, t, y)$; in fact,

$$r(s, x, t, y) = E^{\bar{Q}_{s,x}}[\exp\{R(t, w)\} \mid x(t) = y] .$$

Therefore, $\bar{P}(s, x, t, dy)$ has a density whenever $\bar{Q}(s, x, t, dy)$ has one and

$$\bar{p}(s, x, t, y) = r(s, x, t, y)\bar{q}(s, x, t, y) .$$

Since conditional expectation does not increase L_p norm, we have

$$\int |r(s, x, t, y)|^p \bar{q}(s, x, t, y)\, dy \leqq N_p .$$

Consequently,

$$\iint_{S_0} |r(s, x, t, y)|^p e^{-(t-s)}\bar{q}(s, x, t, y)\, dt\, dy \tag{8.3}$$
$$\leqq (t_2 - t_1) \exp\{\tfrac{1}{2}c(t_2 - s)\rho(\rho - 1)\} .$$

Lemma 8.5 implies that

$$\sup_{(s,x)\in L} \iint_{S_0} |\bar{q}(s, x, t, y)e^{-(t-s)}|^{\alpha'}\, dt\, dy \leqq C_\delta ,$$

and from equation (8.3) we see that

$$\iint_{S_0} |e^{-(t-s)}\bar{p}(s, x, t, y)|^p\, |e^{-(t-s)}\bar{q}(s, x, t, y)|^{1-p}\, dy\, dt$$
$$\leqq (t_2 - t_1) \exp\{\tfrac{1}{2}c(t_2 - s)\rho(\rho - 1)\} .$$

Combining these two, we have by Hölder's inequality

$$\iint_{S_0} |e^{-(t-s)}\bar{p}(s, x, t, y)|^{\alpha''}\, dt\, dy \leqq C_\delta ,$$

provided $(s, x) \in L$ and $\alpha'' < \alpha'$.

THEOREM 8.1. *Let $1 < \alpha < \infty$ be arbitrary. Assume K is a compact subset of $I \times R^d$. Then, for every $\delta_0 > 0$,*

$$\iint_K |p(s, x, t, y)|^\alpha\, dy\, dt \leqq C_{\delta_0} < \infty ,$$

provided (s, x) stays away from K by a distance δ_0. Moreover, if a should be uniformly continuous and uniformly elliptic, then

$$\int_{t_0}^{\infty}\int_{R^d} |p(s, x, t, y)|^\alpha e^{-(t-t_0)\alpha}\, dy\, dt < \infty ,$$

provided $s < t_0$.

Proof: Let (s_0, x_0) be a point of K and let J_0, J_1, J_2, B_0, B_1, B_2, S_0, S_1, S_2, D_1, D_2, be as in the paragraph preceding Lemma 8.5. We are now ready to apply Lemma 8.4. We let $f(t, y)$ be a function with support S_0, and choose δ so that (s, x) is outside S_1. Lemma 8.6 provides us with an estimate of the form

$$\bar{T} \leqq C_\delta \|f\|_{p_1},$$

where $p_1^{-1} + \alpha^{-1} = 1$. Since $(s, x) \notin S_1$, we have from Lemma 8.4,

$$|T(s, x)| \leqq A_\delta \|f\|_{p_1} H_1(s, x) .$$

This in turn yields

$$\iint_{B_0} |p(s, x, t, y)|^\alpha e^{-\alpha(t-s)}\, dt\, dy \leqq A_\delta[H_1(s, x)]^\alpha .$$

It is trivially seen that $H_1(s, x) \leqq \exp\{-(s_0 - s - \delta)\}$. Therefore,

$$\iint_{B_0} |p(s, x, t, y)|^\alpha\, dt\, dy \leqq A'_\delta .$$

Since we have such an estimate around each point (s_0, x_0) we get an estimate for every compact set K. If we want to replace K by an infinite region, under the hypothesis of uniform continuity and uniform ellipticity on a, A_δ stays bounded provided $s < t_0$, and H_1 decays exponentially in space and time directions. This enables one to sum the integrals over a lattice of neighborhoods (which are of uniform size).

We now turn to the homogeneous case (i.e., we assume that a and b depend only on x). Here we shall use the notation $a_{ij}(x)$ and $b_j(x)$ to denote the coefficients. The transition probability will now be time homogeneous and will be denoted by $P(s, x, dy)$. The aim of the rest of this section is to get L_p estimates on $P(s, x, dy)$ for each s. As before, let x_0 be an arbitrary point. Then the $\bar{a}_{ij}(x)$ are chosen so as to be uniformly continuous, uniformly elliptic and to satisfy $|\bar{a}_{ij}(x) - C_{ij}| \leqq \varepsilon$ everywhere; ε will be chosen later. Moreover, $\bar{a}_{ij}(x)$ will

agree with $a_{ij}(x)$ in a neighborhood of the form $|x - x_0| \leqq 3\delta$. Such a neighborhood exists for every $\varepsilon > 0$ because of continuity of the coefficients. Let $1 < \alpha < \infty$ be arbitrary. Then the integers N and ε are chosen according to Theorem 5.1 to serve for p_1 with $p_1^{-1} + \alpha^{-1} = 1$. B_2 will denote $\{x: |x - x_0| \leqq 3\delta\}$ and B_1, B_0 will denote $\{x: |x - x_0| \leqq 2\delta\}$ and $\{x: |x - x_0| \leqq \delta\}$, respectively. The transition probabilities corresponding to $\bar{a}_{ij}(x)$, $b_j(x)$ will be denoted by $\bar{P}(t, x, dy)$ and those corresponding to $\bar{a}_{ij}(x)$ with zero drift terms by $\bar{Q}(t, x, dy)$.

LEMMA 8.7. *There exists a nonincreasing function $\rho(t)$, which, for t near zero, is no worse than $At^{-\nu}$ for some ν, such that $\bar{Q}(t, x, dy)$ has a density $\bar{q}(t, x, y)$, for each $t > 0$, satisfying*

$$\sup_{x \in R^d} \|\bar{q}(t, x, \cdot)\|_\alpha \leqq \rho(t) .$$

Proof: From Corollary 5.1 we know that

$$\int_0^\infty e^{-t\alpha} t^{(N-1)\alpha} \int_{R^d} |\bar{q}(t, x_1, y) - \bar{q}(t, x_2, y)|^\alpha \, dy \, dt \leqq A |x_1 - x_2|^\alpha .$$

We can assume without loss of generality that $\alpha > 1$ since we have a bound on the L_1 norm. If dG denotes the mass distribution on $(0, \infty)$ with $dG = e^{-t\alpha} t^{(N-1)\alpha} dt$ and $F(t, x)$ denotes the function $\bar{q}(t, x, \cdot)$, as a map into L_α, then $F(t, x)$ is in L_α for almost all t and

$$\int_0^\infty \|F(t, x_1) - F(t, x_2)\|^\alpha \, dG \leqq A |x_1 - x_2|^\alpha ,$$

$$\int_0^\infty \|F(t, x)\|^\alpha \, dG \leqq A .$$

We now refer to Theorem A.3 of the Appendix of Part I. Since $F(t, x)$ is weakly continuous, the set

$$\{t : \|F(t, x)\| \geqq M(1 + |x|)^\beta\}$$

has G measure at most CA/M^α. Since the G measure of the set $(0, \varepsilon)$ decreases like $A\varepsilon^k$ as $\varepsilon \downarrow 0$ in the interval $[0, t]$, there must be a point t_0 such that

$$\|F(t_0, x)\| \leqq M(t)(1 + |x|^\beta) ,$$

where $M(t)$ can be chosen to be $At^{-\nu}$ for small enough t with some ν depending on k and α. For $0 < t_0 < t$, we have

$$\bar{q}(t, 0, y) = \int_{R^d} \bar{q}(t_0, x, y) \bar{Q}(t - t_0, 0, dx) .$$

Therefore,

$$\|\bar{q}(t, 0, \cdot)\| \leqq \int M(t)[1 + |x|^{\beta}]\bar{Q}(t - t_0, 0, dx)$$
$$\leqq A'M(t) ;$$

since the hypothesis is homogeneous we can replace 0 by any other point. Therefore, we get

$$\sup_{x \in R^d} \|\bar{q}(t, x, \cdot)\|_{\alpha} \leqq \rho(t) .$$

The left-hand side is obviously nonincreasing.

LEMMA 8.8. *There exists a nonincreasing function $\rho(t)$ which for small t is no worse than $At^{-\nu}$ for some ν, such that $\bar{P}(t, x, dy)$ has a density $\bar{p}(t, x, y)$, for each $t > 0$, satisfying*

$$\sup_{x \in R^d} \|\bar{p}(t, x, y)\|_{\alpha'} \leqq \rho(t)$$

for every $\alpha' < \alpha$.

Proof: The idea is the same as that of Lemma 8.6. We can go from $\bar{q}(t, x, y)$ to $\bar{p}(t, x, y)$ using the Radon-Nikodym derivative $r(t, x, y)$,

$$r(t, x, y) = \frac{\bar{p}(t, x, y)}{\bar{q}(t, x, y)} ,$$

where

$$r(t, x, y) = E^{Q_x}\left[\exp\left\{\int_0^t b\bar{a}^{-1}\, dx - \frac{1}{2}\int_0^t b\bar{a}^{-1}b\, ds\right\} \,\Big|\, x(t) = y\right] .$$

We then get an estimate of the form

$$\int_{R^d} [r(t, x, y)]^{\rho}\bar{q}(t, x, y)\, dy \leqq \exp\{\tfrac{1}{2}ct\rho(\rho - 1)\} .$$

This together with Lemma 8.7 yields Lemma 8.8.

THEOREM 8.2. *Let K be a compact subset of R^d. Then, for every $1 < \beta < \infty$ and $\delta_0 > 0$,*

$$\int_K |p(s, x, y)|^{\beta}\, dy \leqq A_{\delta_0} < \infty ,$$

provided either $s \geqq \delta_0 > 0$ or x stays away from K by a distance $\delta_0 > 0$. Moreover, if the coefficients $a_{ij}(x)$ are uniformly continuous and uniformly elliptic, then K can be replaced

by R^d and we have

$$\sup_{x\in R^d}\int_{R^d}|p(s,x,y)|^\beta\,dy \leqq A_{\delta_0} < \infty$$

provided $s \geqq \delta_0 > 0$.

Proof: Let us take $x_0 \in K$ and neighborhoods B_0, B_1, B_2 as in the paragraph preceding Lemma 8.7. We shall suppose that the size δ of the neighborhoods is such that x is outside B_1. We shall also assume that $\delta < \delta_0$. Let $f(x)$ be a function with support B_0. We want to estimate

$$T(s,x) = e^{-s}\int_{B_0} P(s,x,dy)\,f(y)$$

in terms of the corresponding L_θ norm of $f(y)$. We want to use Lemma 8.4 for the special case $f(t,y) = \delta(t-s)f(y)$. For that, we need an estimate of

$$\bar{T}(s,x) = e^{-s}\int_{B_0}\bar{P}(s,x,dy)\,f(y)$$

when x is in $B_1 \cup B_2$ or $s \geqq \delta > 0$. Therefore, we have to estimate

$$\int_{B_0}|\bar{q}(s,x,y)|^\beta\,dy$$

for $s \geqq \delta > 0$ or $x \in B_1 \cup B_2$. We have two types of estimates. One is given by Lemma 8.8,

$$\|\bar{p}(s,x,\cdot\|_{\alpha'} \leqq \rho(s)\,, \tag{8.4}$$

and the other one is obtained from the exponential bound

$$\int_{B_0}\bar{p}(s,x,y)\,dy \leqq \exp\left\{-\frac{c\delta^2}{2s}\right\} \tag{8.5}$$

which holds for x which are a distance greater than δ away from B_0. Inequalities (8.4) and (8.5) show that for every $\beta < \alpha'$ we can estimate $\bar{T}(s,x)$ for $s \geqq \delta > 0$ and $x \in B_1 \cup B_2$. Therefore, we can estimate

$$\int_{B_0}|p(s,x,y)|^\beta\,dy$$

and consequently we can bound

$$\int_K |p(x,s,y)|^\beta\,dy$$

for every compact K. If the coefficients were uniformly continuous and uniformly elliptic, then the term $H_1(s, x)$ in Lemma 8.4 again would enable one to replace K by R^d. Since $\beta < \alpha' < \alpha$ and α was arbitrary, β can be arbitrary in the range $1 < \beta < \infty$; q.e.d.

We now treat the special case $b \equiv 0$ and wish to estimate the modulus of continuity

$$\int_K |p(s, x, y) - p(s, x, y + h)|^\beta \, dy$$

as $|h| \to 0$. We denote by $p(s, x, y)$ the transition probability density and by $\bar{p}(s, x, y)$ the transition probability density for the coefficients $\bar{a}_{ij}(x)$ localized around a point x_0. Since the method of proof is very similar to the one used to prove Lemma 8.7 and Theorem 8.2, we shall only give a sketch.

THEOREM 8.3. *Let K be a compact subset of R^d and take $1 < \beta < \infty$. Then,*

$$\int_K |p(s, x, y) - p(s, x, y + h)|^\beta \, dy \leqq \Phi_{\delta_0}(|h|) \,,$$

provided either x is away from K by a distance δ_0 or $s \geqq \delta_0 > 0$. $\Phi_{\delta_0}(|h|)$ is a function decreasing to zero as $|h| \to 0$ and depends only on $\{\psi_K(\delta)\}$, $\{\lambda_K\}$ and m.

Proof: Let $x_0 \in K$ and let B_0, B_1, B_2 be as in the paragraph preceding Lemma 8.7. We shall take an even smaller neighborhood $\hat{B}$ of radius $\frac{1}{2}\delta$ around x_0. Let g be a function with support in $\hat{B}$. Then the function $g(y + h) - g(y) = f(y)$ has support in B_0 provided $|h| \leqq \frac{1}{2}\delta$. To estimate

$$\int_K |p(s, x, y) - p(s, x, y + h)|^\beta \, dy \,,$$

it suffices to estimate

$$\int_{\hat{B}} |p(s, x, y) - p(s, x, y + h)|^\beta \, dy$$

which means we need an estimate of the form

$$\left| \int_{B_0} p(s, x, y) f(y) \, dy \right| \leqq \Phi(|h|) \, \|g\|_\theta \,,$$

where $\Phi(|h|) \to 0$ as $|h| \to 0$ and $\theta^{-1} + \beta^{-1} = 1$. In view of Lemma 8.4, it is sufficient to get an estimate of the form

$$\left| \int_{B_0} \bar{p}(s, x, y) f(y) \, dy \right| \leqq \Phi(|h|) \, \|g\|_\theta \,,$$

or equivalently to estimate

$$\int_K |\bar{p}(s, x, y) - \bar{p}(s, x, y + h)|^\beta \, dy \tag{8.6}$$

as $|h| \to 0$ with $s \geqq \delta > 0$ or x away from K by a distance $\delta > 0$. Suppose that for some $\alpha > \beta$ we have

$$\left(\int |\bar{p}(s, x, y) - \bar{p}(s, x, y + h)^\alpha \, dy\right)^{1/\alpha} \leqq \Phi(|h|)\rho(s) , \tag{8.7}$$

where $\rho(s)$ is nonincreasing and no worse than $As^{-\nu}$ for some ν. Then

$$\int_K |\bar{p}(s, x, y) - \bar{p}(s, x, y + h)|^\alpha \, dy \leqq [\Phi(|h|)\rho(s)]^\alpha ,$$

and

$$\int_K |\bar{p}(s, x, y) - \bar{p}(s, x, y + h)| \, dy \leqq 2 \exp\left\{-\frac{c\delta^2}{2s}\right\} .$$

These inequalities imply that, for $1 < \beta < \alpha$,

$$\int_K |\bar{p}(s, x, y) - \bar{p}(s, x, y + h)|^\beta \, dy \leqq [\Phi(|h|)\rho(s)]^{\theta_1} 2^{\theta_2} \exp\left\{-\frac{c\delta^2\theta_2}{2s}\right\} ,$$

where θ_1, θ_2 are greater than zero. This yields an estimate for (8.6). We shall choose α first, and ε and N will be as in Theorem 5.1. We shall complete the proof by establishing an estimate of the form (8.7).

Let us denote by $f(s, x)$ the function $\bar{p}(s, x, \cdot)$ as an element of L_α-space. Let $f_h(s, x)$ be $\bar{p}(s, x, y + h)$, Δ_h the difference $f_h - f$, $\bar{a}_h(x)$ the coefficients $\bar{a}_{ij}(x + h)$, and $\bar{p}_h(s, x, y)$ the transition probabilities of $\bar{a}_h(x)$. It is obvious that

$$\bar{p}_h(s, x, y) = \bar{p}(s, x + h, y + h) .$$

Therefore,

$$\begin{aligned} &\|\Delta_h(s, x_1) - \Delta_h(s, x_2)\| \\ &\qquad \leqq \|\bar{p}(s, x_1, \cdot) - \bar{p}_h(s, x_1, \cdot) - \bar{p}(s, x_2, \cdot) + p_h(s, x_2, \cdot)\| \\ &\qquad\quad + \|\bar{p}(s, x_1 + h, \cdot) - \bar{p}(s, x_1, \cdot) - \bar{p}(s, x_2 + h, \cdot) + \bar{p}_h(s, x_2, \cdot)\| \\ &\qquad = F_1 + F_2 . \end{aligned}$$

If $dG = e^{-\alpha s} s^{(N-1)\alpha}\, ds$, then as before

$$\int_0^\infty \|\Delta_h(s, x_1) - \Delta_h(s, x_2)\|^\alpha \, dG(s) \leqq 2^\alpha \int_0^\infty F_1^\alpha \, dG + 2^\alpha \int_0^\infty F_2^\alpha \, dG .$$

Corollary 5.1 gives the estimates with $\lambda = 1$:

$$\int_0^\infty F_1^\alpha \, dG \leqq C_1 [w(|h|)]^\alpha \, |x_1 - x_2|^\alpha , \tag{8.8}$$

where $w(|h|)$ is the modulus of continuity of $\bar{a}(x)$, as well as

$$\int_0^\infty F_2^\alpha \, dG \leqq C_2 [w_0(|h|)]^\alpha \, |x_1 - x_2|^\alpha , \tag{8.9}$$

where $w_0(|h|) \to 0$ as $|h| \to 0$. Combining these estimates, we have

$$\int_0^\infty \|\Delta_h(s, x_1) - \Delta_h(s, x_2)\|^\alpha \, dG \leqq w_1(|h|) \, |x_1 - x_2|^\alpha , \tag{8.10}$$

where again $w_1(|h|) \to 0$ as $|h| \to 0$. Moreover, by the same corollary with $\lambda = 1$,

$$\int_0^\infty \|\Delta_h(s, x)\|^\alpha \, dG \leqq C_3 w(|h|) . \tag{8.11}$$

Now we can again use Theorem A.3 of the Appendix of Part I, and just as in the proof of Lemma 8.7, we obtain an estimate of the form (8.7), where $\Phi(|h|)$ is some simple combination of $w(|h|)$ and $w_1(|h|)$. This proves the theorem.

9. Continuous Dependence on the Coefficients

We have seen in Sections 5 and 6 that, under certain conditions, for time dependent coefficients a and b, there are Markov processes $P_{s,x}$. In this section we shall investigate the continuity of the dependence of $P_{s,x}$ on the coefficients.

THEOREM 9.1. *Let a^n and b^n be continuous and uniformly bounded, and assume they converge uniformly on compact subsets of $I \times R^d$ to limits a and b, respectively. Let all the coefficients a^n and the limit a be strictly elliptic. Then the processes $P^n_{s,x}$, corresponding to the n-th set of coefficients, converge in the sense of weak convergence to $P_{s,x}$, corresponding to the limiting coefficients. Moreover, the convergence is uniform on compact subsets of s and x.*

Proof: From the assumptions there exists a constant m such that

$$|a^n_{ij}| \leqq m < \infty ,$$

$$|b^n_j| \leqq m < \infty .$$

Under these hypotheses, for each fixed s and x, the family $P^n_{s,x}$ of measures on M^s is weakly compact. This follows from the fact that

$$\exp\left\{\langle\theta, x(t) - x(s)\rangle - \left\langle\theta, \int_s^t b^n(\sigma, x(\sigma))\right\rangle d\sigma - \frac{1}{2}\int_s^t \langle\theta, a^n(\sigma, x(\sigma))\theta\rangle\, d\sigma\right\}$$

is a martingale with respect to $P^n_{s,x}$. Moreover,

$$P^n_{s,x}[x(s) = x] = 1 .$$

In addition, there are uniform estimates of the form

$$P^n_{s,x}[|x(t) - x| \geqq l] \leqq \exp\{-cl^2\} ,$$

c depending only on m and t. Therefore, it follows that any limit Q of any subsequence of $P^n_{s,x}$ is a solution to the martingale problem with coefficients a and b, starting from x at time s. Hence, by the uniqueness theorem, $Q = P_{s,x}$. This proves that $P^n_{s,x}$ converges to $P_{s,x}$. If we had chosen s and x depending on n with $s^n \to s$ and $x^n \to x$, we would have obtained uniformity as well. There are some details missing, but these are elementary and identical to those in the proof of Theorem 4.2.

Since we have processes corresponding to bounded measurable b, the uniformity and the continuity demanded in Theorem 9.1 are not realistic. We shall weaken this by demanding only that b^n converge to b in measure (Lebesgue measure on $I \times R^d$) on compact subsets of $I \times R^d$. For this purpose we need a lemma.

LEMMA 9.1. *Let a be continuous coefficients with modulus of continuity $\{\psi_k(|h|)\}$ on compact sets K and $\{\lambda_K\}$ the lower bound for ellipticity on K. Let m be finite, $|a_{ij}| \leqq m$, (s, x) a given point, and $t > s$ fixed. Take b^1 and b^2 to be two sets of drift coefficients such that $|b^1_j| \leqq m$ and $|b^2_j| \leqq m$. The processes $Q^1_{s,x}$ and $Q^2_{s,x}$ correspond to $[a, b^1]$ and $[a, b^2]$, respectively. For any $\varepsilon > 0$ there exist l and δ, depending only on $\{\psi_K(|h|)\}$, $\{\lambda_K\}$ and m, such that the variation $\|Q^1_{s,x} - Q^2_{s,x}\|_t$ of $Q^1_{s,x} - Q^2_{s,x}$ on the σ-field M^s_t is less than ε, provided that*

$$\textit{Lebesgue}\ \text{mes}\ \{\sigma, z : s \leqq \sigma \leqq t;\ |z - x| \leqq l,\ |b^1_j(\sigma, z) - b^2_j(\sigma, z)| \geqq \delta \ \textit{for some}\ j\} \leqq \delta .$$

Proof: Let $P_{s,x}$ be the process corresponding to $[a, 0]$. Then from Section 6 we know that on M_t^s, $Q_{s,x}^1$ and $Q_{s,x}^2$ are absolutely continuous with respect to $P_{s,x}$ and the densities Φ_1, Φ_2 are

$$\Phi_1 = \exp\{T_1\},$$

where

$$T_1 = \int_{s_0}^t b^1 a^{-1}\, dx - \frac{1}{2}\int_{s_0}^t b^1 a^{-1} b^1\, d\sigma,$$

and

$$\Phi_2 = \exp\{T_2\},$$

where

$$T_2 = \int_{s_0}^t b^2 a^{-1}\, dx - \frac{1}{2}\int_{s_0}^t b^2 a^{-1} b^2\, d\sigma.$$

The problem is to estimate

$$\Delta = \int |\exp\{T_1\} - \exp\{T_2\}|\, dP_{s,x}.$$

Let b^{1l}, b^{2l} be coefficients obtained by replacing b^1 and b^2 by zero outside a sphere of radius l around x. The processes corresponding to these coefficients will be denoted by Q^{1l} and Q^{2l}. (We omit.the suffix s, x.) Then,

$$\|Q^1 - Q^{1l}\|_t \leqq Q^1\Big[\sup_{s\leqq\sigma\leqq t} |x(s) - x(0)| \geqq l\Big]$$
$$\leqq \exp\{-\alpha(t)l^2\},$$

where $\alpha(t)$ is a constant depending only on t and m. Similarly,

$$\|Q^2 - Q^{2l}\|_t \leqq \exp\{-\alpha(t)l^2\}.$$

Let us choose l so that $\exp\{-\alpha(t)l^2\} \leqq \frac{1}{4}\varepsilon$. Then we need estimate only

$$\Delta_l = \int |\exp\{T_1^l\} - \exp\{T_2^l\}|\, dP,$$

where T_1^l, T_2^l are obtained by replacing b^1, b^2 by b^{1l}, b^{2l}, respectively, in the expressions for T_1 and T_2. Let us estimate

$$D_l = \int (T_1^l - T_2^l)^2\, dP$$

(9.1)
$$\leqq 2E^P\left[\left(\int_s^t (b^{1l} - b^{2l})a^{-1}\, dx\right)^2 + \frac{1}{4}\left(\int_s^t b^{1l}a^{-1}b^{1l}\, du - \int_s^t b^{2l}a^{-1}b^{2l}\, du\right)^2\right]$$
$$\leqq C_l E^P\left[\int_s^t (b^{1l} - b^{2l})a^{-1}(b^{1l} - b^{2l})\, du\right],$$

where C_l is a constant depending on $\{\lambda_K\}$ and m. To see this, we note that $(b^{1l}a^{-1}b^{1l})$ can be bounded by λ_K^{-1} and m. Having chosen l, we need only estimate

$$(9.2) \qquad E^P\left[\int_s^t (b^{1l} - b^{2l})a^{-1}(b^{1l} - b^{2l})\, du\right].$$

Since a^{-1} is again bounded above, where b^{1l} and b^{2l} have support, (9.2) can be replaced by

$$(9.3) \qquad E^P\int_s^t \|b^{1l} - b^{2l}\|^2\, d\sigma = \int_s^t \|b^{1l}(u, y) - b^{2l}(u, y)\|^2\, P(s, x, u, dy)\, du\,,$$

where $P(s, x, u, dy)$ is the transition probability corresponding to P. Since we have a bound m on both b^{1l} and b^{2l}, we only need the uniform absolute continuity of $P(s, x, u, dy)du$ on $[s, t] \times \{|z - x| \leqq l\}$ with respect to the Lebesgue measure. This is provided by Theorem 8.1. If we can prove that for any $\varepsilon > 0$ and $l < \infty$ there exists an $\eta > 0$ such that $D_l \leqq \eta$ implies $\Delta_l \leqq \frac{1}{2}\varepsilon$, then the lemma will be proved. This depends only on estimating

$$\int \exp\{2T_1^l\}\, dP \leqq C$$

and

$$\int \exp\{2T_2^l\}\, dP \leqq C\,.$$

These estimates are given in Lemma 8.6. C depends only on the lower bound λ_l of ellipticity, on $\{|z - x| \leqq l\}$ and on m. What remains is taken care of in the following lemma.

LEMMA 9.2. *Let x_1, x_2 be random variables with respect to some probability measure P such that $E[e^{2x_1}] \leqq C$ and $E[e^{2x_2}] \leqq C$. Further, let $E(x_1 - x_2)^2 \leqq \alpha$. Then,*

$$E|\, e^{x_1} - e^{x_2}| \leqq \alpha^{1/4}(1 + 2C)\,.$$

Proof: Truncating x_1 and x_2 at a point ρ yields

$$\int |e^{x_1} - e^{x_2}|\, dP \leqq \int_{\substack{x_1 \leqq \rho \\ x_2 \leqq \rho}} |e^{x_1} - e^{x_2}|\, dP + \int_{x_1 > \rho} e^{x_1}\, dP + \int_{x_2 > \rho} e^{x_2}\, dP$$

$$\leqq e^{\rho}\int |x_1 - x_2|\, dP + 2C\, e^{-\rho}$$

$$\leqq e^{\rho}\sqrt{\alpha} + 2Ce^{-\rho}\,.$$

Choosing $\rho = -\frac{1}{4}\log\alpha$, we have

$$\int |e^{x_1} - e^{x_2}|\, dP \leqq (1 + 2C)\alpha^{1/4}.$$

THEOREM 9.2.* *Let a^n converge uniformly on compact sets to a and let b^n converge in measure to b on each compact set. Moreover, assume that all the coefficients a^n and a are continuous and strictly elliptic. Furthermore let $|a^n_{ij}|$ and $|b^n_j|$ be all uniformly bounded by a constant m. Then $P^n_{s,x}$ converges weakly to the process $P_{s,x}$ corresponding to the limiting coefficients. Moreover, this convergence is uniform over compact sets.*

Proof: Let $\varepsilon > 0$ and $t > s$ be given. According to Lemma 9.1, there exist l and δ independent of n such that if c^n are coefficients with $|c^n_j| \leqq m$ and

$$\text{mes}\,\{s \leqq \sigma \leqq t,\ |z - x| \leqq l\,; \quad |b^n_j(\sigma, z) - c^n_j(\sigma, z)| \geqq \delta \text{ for some } j\} \leqq \delta\,, \tag{9.4}$$

then

$$\|Q^n_{s,x} - P^n_{s,x}\|_t \leqq \varepsilon \quad \text{for all} \quad n\,,$$

$Q^n_{s,x}$ being the process corresponding to $[a^n, c^n]$. We shall choose c^n to be continuous, bounded by m and converging uniformly on compact sets to a limit c satisfying

$$\text{mes}\,\{s \leqq \sigma \leqq t,\ |z - x| \leqq l\,; \quad |b_j(\sigma, z) - c_j(\sigma, z)| \geqq \delta \text{ for some } j\} \leqq \delta\,. \tag{9.5}$$

This in turn yields

$$\|Q_{s,x} - P_{s,x}\|_t \leqq \varepsilon\,,$$

where $Q_{s,x}$ is the process corresponding to $[a, c]$. From Theorem 9.1, we know that $Q^n_{s,x}$ converges weakly to $Q_{s,x}$; since $\varepsilon > 0$ and t are arbitrary, it follows that $P^n_{s,x}$ converges to $P_{s,x}$. Moreover, the uniformity in Theorem 9.1 easily carries over to this case.

COROLLARY 9.3. *If under the same conditions on the coefficients, a is fixed and b^n converges in measure to b, then the convergence $P^n_{s,x}$ to $P_{s,x}$ is in the sense of variation on each σ-field M^s_t.*

* *Note added in proof:* The convergence in measure of the b^n to b can be replaced by weak convergence. The same remark holds for Corollary 9.3. This observation was motivated by the article of M. Freidlin, *On small perturbations in the coefficients of a diffusion process*, Theory Prob. Appl., Vol. 12, 1967, pp. 487–490.

Proof: This is contained in Lemma 9.1.

We now consider the homogeneous case. Let the $a^n_{ij}(x)$ be continuous strictly elliptic coefficients converging uniformly on compact subsets to $a_{ij}(x)$ which is again continuous and strictly elliptic. We shall assume that $a^n_{ij}(x)$ and $a_{ij}(x)$ all have a common bound m. Let Q^n_x be the process corresponding to the n-th set of coefficients and Q_x the one corresponding to the limiting coefficients, when we take the drift terms identically zero for all processes; $q_n(s, x, y)$ and $q(s, x, y)$ are the transition probability densities. Let $0 \leqq s \leqq t$ be a fixed time interval and $\pi = \{0 = s_0 < s_1 < \cdots < s_k = t\}$ a finite partition of $0 \leqq s \leqq t$. Let $b^n(x)$ be a sequence of drift coefficients which are all zero outside a compact set, continuous and converging uniformly to $b(x)$. We consider the following stochastic integrals and approximating sums:

$$T^n = \int_0^t b^n(a^n)^{-1}\,dx - \frac{1}{2}\int_0^t b^n(a^n)^{-1}b^n\,d\sigma\,,$$

$$T = \int_0^t ba^{-1}\,dx - \frac{1}{2}\int_0^t ba^{-1}b\,d\sigma\,,$$

$$T^n_\pi = \sum_{j=1}^k b^n(x(s_{j-1}))(a^n)^{-1}(x(s_{j-1}))[x(s_j) - x(s_{j-1})]$$

$$- \frac{1}{2}\sum_{j=1}^k b^n(x(s_{j-1}))(a^n)^{-1}(x(s_{j-1}))b^n(x(s_{j-1}))[s_j - s_{j-1}]\,,$$

$$T_\pi = \sum_{j=1}^k b(x(s_{j-1}))a^{-1}(x(s_{j-1}))[x(s_j) - x(s_{j-1})]$$

$$- \frac{1}{2}\sum_{j=1}^k b(x(s_{j-1}))a^{-1}(x(s_{j-1}))b(x(s_{j-1}))[s_j - s_{j-1}]\,.$$

By $\pi \downarrow 0$ we shall mean that the partition becomes dense in $[0, t]$.

LEMMA 9.3.

$$\lim_{\pi\downarrow 0}\ \sup_n \int (T^n - T^n_\pi)^2\,dQ^n_x = 0\,,$$

$$\lim_{\pi\downarrow 0} \int (T - T_\pi)^2\,dQ_x = 0\,,$$

uniformly for x in a compact set.

Proof: Let us compare the stochastic integral parts of the two expressions first:

$$e_\pi^n = \int\left\{\int_0^t b^n(a^n)^{-1}\,dx - \sum_{j=1}^k b^n(x(s_{j-1}))(a^n)^{-1}(x(s_{j-1}))\,[x(s_j) - x(s_{j-1})]\right\}^2 dQ_x^n\,.$$

If we put $\Phi_n(x) = b^n(x)(a^n)^{-1}(x)$, then e_π^n equals

$$\sum_{i=1}^k \int\left\{\int_{s_{j-1}}^{s_j} [\Phi_n(x(s_{j-1})) - \Phi_n(x(\sigma))]a^n(x(\sigma))\,[\Phi_n(x(s_{j-1})) - \Phi_n(x(\sigma))]\,d\sigma\right\} dQ_x^n.$$

From the facts that $\|\Phi_n(x)\|$ is uniformly bounded, that Φ_n converges uniformly to a Φ, that everything is continuous, and that Q_x^n is weakly compact, it follows that

$$\sup_n \varepsilon_\pi^n \to 0 \qquad \text{as} \qquad \pi \downarrow 0\,.$$

As for the regular integral terms, the uniform continuity of $b^n(a^n)^{-1}b^n$ implies that

$$\int_0^t b^n(a^n)^{-1}b^n\,d\sigma - \sum_{j=1}^k b^n(x(s_{j-1}))(a^n)^{-1}(x(s_{j-1}))b^n(x(s_{j-1}))[s_j - s_{j-1}]$$

tends to zero as $\pi \to 0$ uniformly in n and uniformly over compact subsets of paths in $[0, t]$. Since Q_x^n is weakly compact, the contribution to the integrals from these terms tends to zero uniformly in n. The limiting case is identical.

LEMMA 9.4. *Let us define measures on* M_t *by*

$$P_x^{n,\pi} = \int_A \exp\{T_\pi^n\}\,dQ_x^n \qquad \textit{for} \qquad A \in M_t\,,$$

$$P_x^\pi = \int_A \exp\{T_\pi\}\,dQ_x \qquad \textit{for} \qquad A \in M_t\,,$$

$$P_x^n = \int_A \exp\{T^n\}\,dQ_x^n\,,$$

$$P_x = \int_A \exp\{T\}\,dQ_x\,.$$

Then,

$$\lim_{\pi\downarrow 0}\sup_n \|P_x^{n,\pi} - P_x^n\| = 0\,,$$

and

$$\lim_{\pi\downarrow 0}\|P_x^\pi - P_x\| = 0\,,$$

uniformly for x in a compact set.

Proof: Since

$$\|P_x^{n,\pi} - P_x^n\| = \int |\exp\{T_\pi^n\} - \exp\{T^n\}| \, dQ_x^n ,$$

we can use Lemmas 9.2 and 9.3. We therefore need only show that

$$\int \exp\{2T_\pi^n\} \, dQ_x^n$$

and similar integrals are uniformly bounded in n for x in a compact set. To see this we note that $b^n(a^n)^{-1}b^n$ has a uniform bound which is guaranteed by the uniform ellipticity of a^n on any compact set, the vanishing of b^n outside a compact set, and their uniform bound on compact sets. The rest of the estimation is as in Lemma 8.6.

LEMMA 9.5. *Let $p_n(s, x, y)$ and $p(s, x, y)$ be the transition probability densities for $[a^n, b^n]$ and $[a, b]$, respectively. Then,*

$$\lim_{n\to\infty} \int |p_n(s, x, y) - p(s, x, y)| \, dy = 0 ,$$

uniformly for x in a compact set and $0 < t_1 \leqq s \leqq t_0 < \infty$.

Proof: We know that $q_n(s, x, y)$ converges in L_1 to $q(s, x, y)$. This is a consequence of weak convergence (cf. Theorem 9.1), uniform modulus of continuity estimates (cf. Theorem 8.3), and uniform exponential tail estimates for the probability when y is in the complement of a compact set K. If one now considers the measures Q_x^n on the σ-field M_π generated by $x(s_0), \cdots, x(s_k)$, then Q_x^n converges in variation. Since T_π^n is M_π measurable and T_π^n converges to T_π as n tends to infinity, we have

$$\|P_x^{n,\pi} - P_x^\pi\|_{M_\pi} \to 0$$

for each π as $n \to \infty$. Let s belong to $[t_1, t_0]$ and π contain the point s. Then,

$$\begin{aligned} \Delta_n &= \int |p_n(s, x, y) - p(s, x, y)| \, dy \\ &\leqq \|P_x^n - P_x\|_{M_\pi} \\ &\leqq \|P_x^n - P_x^{n,\pi}\|_{M_\pi} + \|P_x^{n,x} - P_x^\pi\|_{M_\pi} + \|P_x^\pi - P_x\|_{M_\pi} . \end{aligned}$$

Taking lim sup as $n \to \infty$, we get

$$\limsup_{n\to\infty} \Delta_n \leqq \sup_n [\|P_x^n - P_x^{n,\pi}\| + \|P_x^\pi - P_x\|] .$$

Since this is true for every π containing s and Δ_n is independent of π, we have, letting $\pi \downarrow 0$,

$$\limsup_{n\to\infty} \Delta_n = 0 ,$$

in view of Lemma 9.4. The uniformity in the convergence of $q_n(s, x, y)$ naturally carries over to this case.

LEMMA 9.6. *Let b^n be uniformly bounded and converge in measure to b on compact subsets. Under the usual assumptions on a^n and a this gives*

$$\lim_{n\to\infty} \sup_{0<t_1\leqq s\leqq t_0<\infty} \sup_{x\in K} \int |p_n(s, x, y) - p(s, x, y)| \, dy = 0 .$$

Proof: We can find functions $\hat{b}^n(x)$, which are continuous and vanish outside a compact set $\{|x| \leqq l\}$, and converge uniformly to a limit $\hat{b}(x)$, such that

$$\text{mes}\,[x\colon |x| \leqq l,\ |\hat{b}^n(x) - b^n(x)| \geqq \delta] \leqq \delta \quad \text{for all} \quad n ,$$

and

$$\text{mes}\,[x\colon |x| \leqq l,\ |\hat{b}(x) - b(x)| \geqq \delta] \leqq \delta ,$$

mes being the Lebesgue measure and l, δ being given in advance. If we denote by $\hat{p}_n(s, x, y)$ and $\hat{p}(s, x, y)$ the densities corresponding to $[a^n, \hat{b}^n]$ and $[a, \hat{b}]$, respectively, then we get, from Lemma 9.5,

$$\int |\hat{p}_n(s, x, y) - \hat{p}(s, x, y)| \, dy \to 0$$

uniformly for s, x in appropriate sets. If we now use Lemma 9.1, and a proper choice of l and δ, we can be sure that

$$\int |\hat{p}_n(s, x, y) - p_n(s, x, y)| \, dy$$

and

$$\int |\hat{p}(s, x, y) - p(s, x, y)| \, dy$$

are uniformly small. Hence the lemma follows.

THEOREM 9.3. *Let a^n be continuous, strictly elliptic and converge uniformly on compact sets to a continuous strictly elliptic a. Let b^n converge in measure on compact sets to b. Assume a_{ij}^n and b_j^n have a uniform bound m. If $p_n(s, x, y)$, $p(s, x, y)$ are the transition probability densities of the homogeneous processes corresponding to $[a^n, b^n]$, $[a, b]$, respectively, then*

(i)
$$\lim_{n\to\infty} \sup_{0<t_1\leq s\leq t_0<\infty} \sup_{x\in K} \int |p_n(s, x, y) - p(s, x, y)|\, dy = 0$$

for every compact K.

(ii) *Moreover, for every α in the range $1 < \alpha < \infty$ and any two compact sets K_1 and K_2,*

$$\lim_{n\to\infty} \sup_{0<t_1\leq s\leq t_0<\infty} \sup_{x\in K_1} \int_{K_2} |p_n(s, x, y) - p(s, x, y)|^\alpha\, dy = 0 .$$

Proof: (i) is just a restatement of Lemma 9.6; (ii) follows from (i) and Theorem 8.2 because boundedness of L_α norm and convergence in L_1 imply convergence in L_β norm for every $\beta < \alpha$.

Remark 1. K_1 in Theorem 9.3 (ii) can be replaced by R^d. This is due to the fact that $H_1(s, x)$, in Lemma 8.4, gives a uniform exponentially dying estimate for $\int |p_n(s, x, y)|^\alpha\, dy$ for large $|x|$.

Remark 2. If, in addition, for a^n we had a uniform lower bound of ellipticity and a uniform global modulus of continuity (uniform in n and x), then K_2 could be replaced by R^d. See the end of the proof of Theorem 8.2.

Remark 3. Replacing both K_1 and K_2 by R^d is more involved. We require, in addition to the assumptions in Remark 2, a^n to converge to a uniformly in R^d. On b^n we need two types of assumptions. The convergence in measure on compact subsets should be uniform with respect to translations of these compact sets. Some kind of L_α modulus of continuity of b^n on compact sets should be uniform in n and the location of the compact sets. These conditions ensure that the argument of replacing b^n by $\hat{b}^n$ in Lemma 9.6 works uniformly in the entire space.

10. An Invariance Principle for Markov Processes

The purpose of this section is to determine how our uniqueness result gives rise to an invariance principle for Markov chains analogous to that for sums of independent random variables (Donsker [2]). The idea is the following. Suppose that for each $n \geqq 1$ we are given a Markov chain $X^{(n)}(k)$, $k \geqq 0$, on R^d with transition probabilities $\Pi_k^{(n)}(x, dy)$. That is, if $\Gamma_1, \cdots, \Gamma_m \in \mathscr{B}(R^d)$, then the conditional probability that $X^{(n)}(k+1) \in \Gamma_1, \cdots,$ and $X^{(n)}(k+m) \in \Gamma_m$, given that $X^{(n)}(k) = x$, is

$$\int_{\Gamma_m} \cdots \int_{\Gamma_1} \Pi_k^{(n)}(x, dy_1)\Pi_{k+1}^{(n)}(y_1, dy_2) \cdots \Pi_{k+m-1}^{(n)}(y_{m-1}, dy_m) .$$

Further, assume that, as $n \to \infty$,

$$n \int_{|y-x| \leq 1} (y - x) \otimes (y - x) \Pi^{(n)}_{[nt]} (x, dy) \to a(t, x) \,,$$

$$n \int_{|y-x| \leq 1} (y - x) \Pi^{(n)}_{[nt]} (x, dy) \to b(t, x) \,,$$

where $a\colon [0, \infty) \times R^d \to R^d \otimes R^d$ and $b\colon [0, \infty) \times R^d \to R^d$ are bounded continuous functions such that $a(t, x)$ is strictly positive definite at each point and symmetric. Finally, assume that $\Pi^{(n)}_k(x, \{|y - x| \geqq \varepsilon\}) \to 0$ sufficiently fast for each $\varepsilon > 0$. Given such a family of Markov chains, define for each $n \geqq 1$ the probability measure $P^{(n)}_x$ on (Ω, M^0) by

$$P^{(n)}_x(x(t_1) \in \Gamma_1 , \cdots , x(t_m) \in \Gamma_m) = \Pi^{(n)}_x(Y^{(n)}(t_1) \in \Gamma_1 , \cdots , Y^{(n)}(t_m) \in \Gamma_m),$$

where $\Pi^{(n)}_x$ is the probability measure on $R^d \times \cdots \times R^d \times \cdots$ induced by the process $X^{(n)}(k)$, $k \geqq 0$, conditioned to start at x and

$$Y^{(n)}(t) = ([nt] + 1 - t)X^{(n)}([nt]) + (t - [nt])X^{(n)}([nt] + 1) \,.$$

What we shall show is that $P^{(n)}_x \to P_x$, where P_x is the unique probability measure on (Ω, M^0) such that

(i) $P_x(x(0) = x) = 1$,

and

$$\text{(ii) } X_\theta(t) = \exp\left\{\langle\theta, x(t) - x\rangle - \frac{1}{2}\int_0^t \langle\theta, a(u, x(u))\theta\rangle\, du - \int_0^t \langle\theta, b(u, x(u))\rangle\, du\right\}, \quad t \geqq 0 \,,$$

is a P_x-martingale for all $\theta \in R^d$.

As in the proof of the classical invariance principle, our demonstration is accomplished in two steps. The first of these is quite standard and consists of showing that the family $\{P^{(n)}_x\}_{n=1}^\infty$ is weakly compact. The second step involves identifying the limit points of $\{P^{(n)}_x\}_{n=1}^\infty$ as solutions to (i) and (ii) above. For technical reasons, it is convenient in carrying out the second step to replace $X_\theta(t)$, $t \geqq s$, by $X_{i\theta}(t)$, $t \geqq s$. In the present context this replacement is analogous to considering the characteristic function instead of the moment generating function when studying random variables. We are justified in making this change by the following theorems.

THEOREM 10.1. *Let* $a:[s, \infty) \times \Omega \to R^d \otimes R^d$ *be an s-nonanticipating function such that*

$$\alpha |\theta|^2 \leqq \sum_{i,j=1}^{d} a_{ij}(u, w)\theta_i\theta_j \leqq A |\theta|^2$$

$$\text{for all} \quad u \geqq s, \quad w \in \Omega \quad \text{and} \quad \theta \in R^d,$$

where $0 < \alpha < A$, *and suppose* $\xi:[s, \infty) \times \Omega \to R^d$ *is an s-nonanticipating function. If P is a probability measure on* (Ω, M^s), *then a necessary and sufficient condition for*

$$X_\theta(t) = \exp\left\{\langle \theta, \xi(t) - \xi(s)\rangle - \frac{1}{2}\int_s^t \langle \theta, a(u)\theta\rangle\, du\right\}, \qquad t \geqq s,$$

to be a P-martingale for all $\theta \in R^d$ *is that*

$$X_{i\theta}(t) = \exp\left\{i\langle \theta, \xi(t) - \xi(s)\rangle + \frac{1}{2}\int^t \langle \theta, a(u)\theta\rangle\, du\right\}, \qquad t \geqq s,$$

is a P-martingale for all $\theta \in R^d$.

Proof: First suppose that $[X_\theta(t), t \geqq s]$ is a P-martingale for all $\theta \in R^d$. Then, by the estimate derived in Lemma 3.1,

$$\int_B X_\theta(t)\, dP$$

is an entire function of $\theta \in C^d$ for each $B \in M^s$ and $t \geqq s$. Hence, if $t_2 \geqq t_1 \geqq s$ and $B \in M^s_{t_1}$, then the fact that

$$\int_B X_\theta(t_2)\, dP = \int_B X_\theta(t_1)\, dP$$

for all $\theta \in R^d$ implies that this equality holds for all $\theta \in C^d$. In particular, $[X_{i\theta}(t), t \geqq s]$ is a P-martingale for all $\theta \in R^d$.

Now suppose that $[X_{i\theta}(t), t \geqq s]$ is a P-martingale for all $\theta \in R^d$. We shall show that the estimate in Lemma 3.1 is true in this case also. Then it will follow by the argument given in the preceding paragraph that $[X_\theta(t), t \geqq s]$ is a P-martingale for all $\theta \in C^d$.

Given $\theta_0 \in R^d$ such that $|\theta_0| = 1$, define $\tau_t(w)$, $t \geqq s$ and $w \in \Omega$, by the equation

$$\int_s^{\tau_t(w)} \langle \theta_0, a(u, w)\theta_0\rangle\, du = t - s.$$

Then the τ_t are s-Markov times and

$$\tau_t = s + \int_s^t \frac{1}{\langle\theta_0, a(\tau_u)\theta_0\rangle} du .$$

In particular, $\tau_{t_1} \leqq \tau_{t_2}$ if $s \leqq t_1 \leqq t_2$, and

$$s + \frac{t-s}{A} \leqq \tau_t \leqq s + \frac{t-s}{\alpha} .$$

Hence, by an argument exactly like that used to prove equation (3.3) in Theorem 3.1,

$$E^P[X_{i\theta}(\tau_{t_2}) \mid M^s_{\tau_1}] = X_{i\theta}(\tau_{t_1}) \qquad \text{a.s.} \quad P ,$$

for all $t_2 \geqq t_1 \geqq s$ and $\theta \in R^d$. But this means that

$$E^P[\exp\{i\lambda\langle\theta_0, \xi(\tau_{t_2}) - \xi(\tau_{t_1})\rangle \mid M^s_{\tau_1}\}] = e^{-\lambda^2/2(t_2-t_1)}$$

for all $t_2 \geqq t_1 \geqq s$ and $\lambda \in R$. In other words, $\langle\theta_0, \xi(\tau_t)\rangle$ is a one-dimensional Brownian motion. It follows then from well known estimates that

$$P(\sup_{s\leqq u\leqq t} |\langle\theta_0, \xi(\tau_u) - \xi(s)\rangle| \geqq l) \leqq 2 \exp\left\{-\frac{l^2}{2(t-s)}\right\},$$

$$l \geqq 0 \quad \text{and} \quad t \geqq s .$$

Since $\tau_t - s \geqq (t-s)/A$, we have therefore

$$P(\sup_{s\leqq u\leqq s+\delta} |\langle\theta_0, \xi(u) - \xi(s)\rangle| \geqq l) \leqq P(\sup_{s\leqq u\leqq s+A\delta} |\langle\theta_0, \xi(\tau_u) - \xi(s)\rangle| \geqq l)$$

$$\leqq 2 \exp\left\{-\frac{l^2}{2A(t-s)}\right\} .$$

Hence, because θ_0 was an arbitrary element of S^{d-1},

$$P(\sup_{s\leqq u\leqq t} |\xi(u) - \xi(s)| \geqq l) \leqq 2d \exp\left\{-\frac{l^2}{2A(t-s)}\right\} .$$

COROLLARY 10.1. *Let $\{a_{ij}(s, x)\}$, $\{b_j(s, x)\}$ be bounded functions such that $\{a_{ij}(s, x)\}$ is continuous symmetric and strictly elliptic at each point. The $\{b_j(s, x)\}$ are measurable functions. Then P is a solution of the martingale problem starting from x_0 at time s_0 if and only if*

$$P[x(s_0) = x_0] = 1$$

and

$$\exp\left\{i\langle\theta, x(t) - x(s_0)\rangle - i\left\langle\theta, \int_{s_0}^{t} b(u, x(u))\,du\right\rangle + \frac{1}{2}\int_{s_0}^{t}\langle\theta, a(u, x(u))\theta\rangle\,du\right\}$$

is a P-martingale for all $\theta \in R^d$.

Proof: The necessary part follows from Theorem 10.1 because only the upper bound for $\{a_{ij}(s, x)\}$ was used in the proof. To prove sufficiency, we shall choose a large sphere S in R^d and alter the coefficients outside $S \times [0, l]$ so that the altered coefficients $\{\overline{a_{ij}}(s, x)\}$, in addition to being bounded, continuous and symmetric, are also uniformly elliptic. To these coefficients corresponds a process $\bar{Q}_{s,x}$. Let τ' be the exit time form S and $\tau = \tau' \wedge l$. Then as in Theorem 3.4 we can construct a measure $\bar{P}$ which is a solution to the martingale problem starting from x_0 at time s_0, and agrees with P on M_τ. Therefore, if $\tau_t = \tau \wedge t$, then

$$\exp\left\{\langle\theta, x(\tau_t) - x(s_0)\rangle - \int_{s_0}^{\tau_t}\langle\theta, b(u, x(u))\rangle\,du - \frac{1}{2}\int_{s_0}^{\tau_t}\langle\theta, a(u, x(u))\theta\rangle\,du\right\}$$

is a P-martingale. Letting S and l increase to whole space and ∞, respectively, we find that P is a solution of the martingale problem.

A function Π: $R^d \times \mathscr{B}(R^d) \to R$ is called a *transition probability on* R^d if $\Pi(x, \cdot)$ is a probability measure for each $x \in R^d$ and $\Pi(\cdot, \Gamma)$ is $\mathscr{B}(R^d)$-measurable for each $\Gamma \in \mathscr{B}(R^d)$. Let $\bar{\Omega} = R^d \times \cdots \times R^d \times \cdots$, and for each $k \geqq 0$ let $X(k, \bar{w})$ be the k-th coordinate of $\bar{w}$. Given $l \geqq k \geqq 0$, define $\bar{M}_l^k$ to be the σ-algebra on $\bar{\Omega}$ generated by the sets

$$\{X(j_1) \in \Gamma_1, \cdots, X(j_m) \in \Gamma_m\},$$

where $k \leqq j_1 < \cdots < j_m \leqq l$, $1 \leqq m \leqq k - l$, and $\Gamma_1, \cdots, \Gamma_m \in \mathscr{B}(R^d)$, and define $\bar{M}^k$ as the smallest σ-field containing $\bar{M}_l^k$ for $l \geqq k$. If $\Pi_k(x, \cdot)$, $k \geqq 0$, are transition probabilities, define $\nu_{k,x}$, $k \geqq 0$ and $x \in R^d$, to be the probability measure on $\langle\bar{\Omega}, \bar{M}^k\rangle$ given by

$$\nu_{k,x}(X(k) \in \Gamma_0, \cdots, X(k+m) \in \Gamma_m)$$

$$= \int_{\Gamma_m} \cdots \int_{\Gamma_0} \delta_x(dy_0)\Pi_k(y_0, dy_1) \cdots \Pi_{k+m-1}(y_{m-1}, dy_m)$$

for $m \geqq 1$ and $\Gamma_0, \cdots, \Gamma_m \in \mathscr{B}(R^d)$. We call $\nu_{k,x}$ *the measure induced by* Π_l, $l \geqq k$, *on* $\langle\bar{\Omega}, \bar{M}^k\rangle$ *conditioned to start at* x.

THEOREM 10.2. *Let Π_k, $k \geqq 0$, be a sequence of transition probabilities on R^d and let $\theta \in R^d$ be such that*

$$\sup_{\substack{x \in R^d \\ k \geqq 0}} \left| 1 - \int e^{i\langle \theta, y-x\rangle} \Pi_k(x, dy) \right| < 1 .$$

Define

$$\phi_{k,\theta}(x) = \log \left(\int e^{i\langle \theta, y-x\rangle} \Pi_k(x, dy) \right) ,$$

and let

$$\bar{X}_\theta^k(l) = \exp \left\{ i\langle \theta, X(l) - X(k)\rangle - \sum_{j=k}^{l-1} \phi_{j,\theta}(X(j)) \right\} , \qquad l \geqq k .$$

Then $[\bar{X}_\theta^k(l), \bar{M}_l^k, \nu_{k,x}\colon\ l \geqq k]$ is a martingale, where $\nu_{k,x}$ is the measure induced by Π_l, $l \geqq k$, on $(\bar{\Omega}, \bar{M}_l^k)$ conditioned to start at x.

Proof: Observe that

$$\int e^{i\langle \theta, y_{m+1} - y_0\rangle} \Pi_{k+m}(y_m, dy_{m+1}) = \exp \{ i\langle \theta, y_m - y_0\rangle + \phi_{m,\theta}(y_m) \} ,$$

and that therefore

$$E^{\nu_{k,x}} \left[\exp \left\{ i\langle \theta, X(k+m+1) - X(k)\rangle - \sum_{j=k}^{k+m} \phi_{j,\theta}(X(j)) \right\} \middle| \bar{M}_{k+m}^k \right]$$

$$= \exp \left\{ - \sum_{j=k}^{k+m} \phi_{j,\theta}(X(j)) \right\} \int_{R^d} e^{i\langle \theta, y_{m+1} - X(k)\rangle} \Pi_{k+m}(X(k+m), dy_{m+1})$$

$$= \exp \left\{ i\langle \theta, X(k+m) - X(k)\rangle - \sum_{j=k}^{k+m-1} \phi_{j,\theta}(X(j)) \right\} .$$

Thus,

$$E^{\nu_{k,x}}[\bar{X}_\theta^k(k+m+1) \mid \bar{M}_{m+k}^k] = \bar{X}_\theta^k(k+m) \qquad \text{a.s.} \quad \nu_{k,x} .$$

LEMMA 10.1. *For each $n \geqq 1$, let $\Pi_k^{(n)}$, $k \geqq 0$, be a sequence of transition probabilities on R^d. Assume that*

$$\Pi_k^{(n)}(x, \{|y - x| > \varepsilon\}) = \eta_\varepsilon(n; k, x) ,$$

$$\int_{|y-x| \leqq 1} (y - x)\, \Pi_k^{(n)}(x, dy) = \frac{1}{n} b\left(\frac{k}{n}, x\right) + \beta(n; k, x) ,$$

$$\int_{|y-x| \leqq 1} (y - x) \otimes (y - x)\, \Pi_k^{(n)}(x, dy) = \frac{1}{n} a\left(\frac{k}{n}, x\right) + \alpha(n; k, x) ,$$

where $b\colon [0, \infty) \times R^d \to R^d$ *and* $a\colon [0, \infty) \times R^d \to R^d \otimes R^d$ *are bounded, and*

$$\sup_{\substack{k\ge 0\\ |x|\le l}} \eta_\varepsilon(n; k, x), \qquad \sup_{\substack{k\ge 0\\ |x|\le l}} |\beta(n; k, x)|, \qquad \sup_{\substack{k\ge 0\\ |x|\le l}} |\alpha(n; k, x)|$$

are $o(1/n)$ *as* $n \to \infty$ *for each* $\varepsilon > 0$ *and* $l \geqq 0$. *Then for each* $l > 0$ *there exists a function* $\gamma(n; l)$ *such that* $\gamma(n; l) = o(1/n)$ *as* $n \to \infty$ *and, for large* n,

$$\sup_{\substack{k\ge 0\\ |x|\le l\\ |\theta|\le l}} \left| \phi_{k,\theta}^{(n)}(x) - i\left\langle \theta, \frac{1}{n} b\left(\frac{k}{n}, x\right)\right\rangle + \frac{1}{2}\left\langle \theta, \frac{1}{n} a\left(\frac{k}{n}, x\right) \theta\right\rangle \right| \leqq \gamma(n; l),$$

where

$$\phi_{k,\theta}^{(n)}(x) = \log\left(\int e^{i\langle\theta, y-x\rangle}\, \Pi_k^{(n)}(x, dy)\right).$$

Proof: Observe that for each $l > 0$, there is an N such that, for $n \geqq N$,

$$\sup_{\substack{k\ge 0\\ |x|\le l\\ |\theta|\le l}} \left| \int e^{i\langle\theta, y-x\rangle}\, \Pi_k^{(n)}(x, dy) - 1 \right| < 1.$$

If $n \geqq N$, $|x| \leqq l$, and $|\theta| \leqq l$,

$$\phi_{k,\theta}^{(n)}(x) = \int e^{i\langle\theta, y-x\rangle}\, \Pi_k^{(n)}(x, dy) - 1 + o\left(\int e^{i\langle\theta, y-x\rangle}\, \Pi_k^{(n)}(x, dy) - 1\right).$$

Thus, it suffices to show that

$$\sup\left| \left(\int e^{i\langle\theta, y-x\rangle}\, \Pi_k^{(n)}(x, dy) - 1\right) - \frac{i}{n}\left\langle \theta, b\left(\frac{k}{n}, x\right)\right\rangle + \frac{1}{2n}\left\langle \theta, a\left(\frac{k}{n}, x\right)\theta\right\rangle \right| = o\left(\frac{1}{n}\right).$$

Taylor's theorem shows that, for each $0 < \varepsilon \leqq 1$,

$$\begin{aligned}\int e^{i\langle\theta, y-x\rangle}\, \Pi_k^{(n)}(x, dy) - 1 &= i\left\langle \theta, \int_{|y-x|\le\varepsilon} (y - x)\, \Pi_k^{(n)}(x, dy)\right\rangle \\ &\quad - \frac{1}{2}\left\langle \theta, \int_{|y-x|\le\varepsilon} (y - x) \otimes (y - x)\Pi_k^{(n)}(x, dy)\theta\right\rangle \\ &\quad + \Delta_\varepsilon(n; k, x, \theta) + \int_{|y-x|>\varepsilon} e^{i\langle\theta, y-x\rangle}\Pi_k^{(n)}(x, dy),\end{aligned}$$

where

$$\Delta_\varepsilon(n; k, x, \theta) = \int_{|y-x|\leqq\varepsilon} \langle\theta, (y-x)\otimes(y-x)\theta\rangle\delta_\theta(x-y)\Pi_k^{(n)}(x, dy) ,$$

and $|\delta_\theta(x-y)| \to 0$ uniformly for $|\theta| \leqq l$ as $|x-y| \to 0$. Also,

$$\left|\left\langle\theta, \int_{|y-x|\leqq\varepsilon}(y-x)\Pi_k^{(n)}(x, dy)\right\rangle - \left\langle\theta, \frac{1}{n}b\left(\frac{k}{n}, x\right)\right\rangle\right|$$
$$\leqq |\theta|\,\beta(n; k, x) + |\theta|\,\eta_\varepsilon(n; k, x) ,$$

$$\left|\left\langle\theta, \int_{|y-x|\leqq\varepsilon}(y-x)\otimes(y-x)\Pi_k^{(n)}(x, dy)\theta\right\rangle - \left\langle\theta, \frac{1}{n}a\left(\frac{k}{n}, x\right)\theta\right\rangle\right|$$
$$\leqq |\theta|^2\,\alpha(n; k, x) + |\theta|^2\,\eta_\varepsilon(n; k, x) ,$$

and

$$\left|\int_{|y-x|>\varepsilon} e^{i\langle\theta, y-x\rangle}\Pi_k^{(n)}(x, dy)\right| \leqq \eta_\varepsilon(n; k, x) .$$

Hence,

$$\sup_{\substack{k\geqq 0\\ |x|\leqq l\\ |\theta|\leqq l}}\left|\left(\int e^{i\langle\theta, y-x\rangle}\Pi_k^{(n)}(x, dy) - 1\right) - i\left\langle\theta, \frac{1}{n}b\left(\frac{k}{n}, x\right)\right\rangle + \frac{1}{2}\left\langle\theta, \frac{1}{n}a\left(\frac{k}{n}, x\right)\theta\right\rangle\right|$$
$$\leqq f_\varepsilon(n; l) + \sup_{\substack{|x-y|\leqq\varepsilon\\ |\theta|\leqq l}}|\delta_\theta(x-y)|\,g(n; l) ,$$

where $f_\varepsilon(n; l) = o(1/n)$ for each $\varepsilon > 0$ and $g(n; l) = O(1/n)$. Since we can find $\varepsilon_n \downarrow 0$ such that $f_{\varepsilon_n}(n; l) = o(1/n)$, this completes the proof.

For each $n \geqq 1$, let $\Pi_k^{(n)}$, $k \geqq 0$, be a sequence of transition probabilities on R^d and let $[\nu_{k,x}^{(n)} : k \geqq 0, x \in R^d]$ be the associated measures on $(\bar\Omega, \bar M^k)$. Define Φ_n: $\bar\Omega \to \Omega$ so that

$$x(t, \Phi_n(\bar w)) = ([nt] + 1 - t) \times ([nt], \bar w) + (t - [nt]) \times ([nt] + 1, \bar w).$$

Now define $P_x^{(n)}$ on (Ω, M^0) by

$$P_x^{(n)}(B) = \nu_{0,x}^{(n)}(\Phi_n^{-1}(B)), \qquad B \in M_0 . \tag{10.1}$$

LEMMA 10.2. *For each $n \geqq 1$, let $\Pi_k^{(n)}$, $k \geqq 0$, be a sequence of transition probabilities on R^d and define $\nu_{k,x}^{(n)}$, $k \geqq 0$ and $x \in R^d$, and $P_x^{(n)}$, $x \in R^d$, as above. Assume*

that

(i) $\Pi_k^{(n)}(x, \{|y - x| > R\}) = 0$, $n \geqq 1$, $k \geqq 0$, *and* $x \in R^d$,

(ii) $\lim_{n\to\infty} \sup_{\substack{k\geqq 0\\ x\in R^d}} n\Pi_k^{(n)}(x, \{|y - x| > \varepsilon\}) = 0$ *for each* $\varepsilon > 0$,

(iii) $\sup_{\substack{k\geqq 0\\ x\in R^d}} \left| \int (y - x)\Pi_k^{(n)}(x, dy) \right| \leqq \frac{C_1}{n}$,

(iv) $\sup_{\substack{k\geqq 0\\ x\in R^d}} \left| \int (y - x) \otimes (y - x)\Pi_k^{(n)}(x, dy) \right| \leqq \frac{C_2}{n}$.

Then for each $L > 0$ *the set* $\{P_x^{(n)}: n \geqq 1, |x| \leqq L\}$ *is compact in the sense that it is weakly compact on* M_T^0 *as a collection of measures on* $C([0, T], R^d)$ *for any* $T > 0$.

Proof: Let $T > 0$ be given and define

$$Q_\delta^{(n)}(\varepsilon, x) = P_x^{(n)}\left(\sup_{\substack{s,t\in[0,T]\\ |t-s|<\delta}} |x(t) - x(s)| \geqq \varepsilon \right).$$

What we must show is that, for each $\varepsilon > 0$,

$$\lim_{\delta\downarrow 0} \lim_{n\to\infty} \sup_{x\in R^d} Q_\delta^{(n)}(\varepsilon, x) = 0 .$$

Let $I_\delta^k = [k\delta, (k + 2)\delta]$, $k \geqq 0$ and $\delta > 0$. Then,

$$\begin{aligned} Q_\delta^{(n)}(\varepsilon, x) &\leqq P_x^{(n)}\left(\sup_{0\leqq k\leqq[T/\delta]} \sup_{s,t\in I_\delta^k} |x(t) - x(s)| \geqq \varepsilon \right) \\ &\leqq \left(\left[\frac{T}{\delta}\right] + 1 \right) \sup_{0\leqq k\leqq[T/\delta]} P_x^{(n)}\left(\sup_{x,t\in I_\delta^k} |x(t) - x(s)| \geqq \varepsilon \right). \end{aligned}$$

Observe that if we split I_δ^k into three intervals

$$I_1 = \left[k\delta, \frac{[nk\delta] + 1}{n} \right],$$

$$I_2 = \left[\frac{[nk\delta] + 1}{n}, \frac{[nk\delta] + 2[n\delta] + 1}{n} \right],$$

and

$$I_3 = \left[\frac{[nk\delta] + 2[n\delta] + 1}{n}, (k + 2)\delta \right],$$

then

$$P_x^{(n)}\left(\sup_{s,t\in I_\delta^k}|x(t)-x(s)|\geqq\varepsilon\right)\leqq P_x^{(n)}\left(\sup_{t\in I_1}\left|x(t)-x\left(\frac{[nk\delta]+1}{n}\right)\right|\geqq\tfrac{1}{3}\varepsilon\right)$$
$$+P_x^{(n)}\left(\sup_{t\in I_2}\left|x\left(\frac{[nk\delta]+1}{n}\right)-x(t)\right|\geqq\tfrac{1}{3}\varepsilon\right)$$
$$+P_x^{(n)}\left(\sup_{t\in I_3}\left|x\left(\frac{[nk\delta]+2[n\delta]+1}{n}\right)-x(t)\right|\geqq\tfrac{1}{3}\varepsilon\right),$$

Since

$$P_x^{(n)}\left(\sup_{t\in I_1}\left|x(t)-x\left(\frac{[nk\delta]+1}{n}\right)\right|\geqq\tfrac{1}{3}\varepsilon\right)\leqq\Pi_{[k\delta]}^{(n)}(x,\{|y-x|\geqq\tfrac{1}{3}\varepsilon\})$$

and

$$P_x^{(n)}\left(\sup_{k\in I_3}\left|x\left(\frac{[nk\delta]+2[n\delta]+1}{n}\right)-x(t)\right|\geqq\tfrac{1}{3}\varepsilon\right)$$
$$\leqq\Pi_{[nk\delta]+2[n\delta]+1}^{(n)}(x,\{|y-x|\geqq\tfrac{1}{3}\varepsilon\}),$$

it follows that

$$\overline{\lim_{n\to\infty}}\sup_{x\in R^d}Q_\delta^{(n)}(\varepsilon,x)\leqq\left(\left[\frac{T}{\delta}\right]+1\right)\overline{\lim_{n\to\infty}}\sup_{x\in R^d}\rho_{[n\delta]}^{(n)}(\tfrac{1}{3}\varepsilon,x),$$

where

$$\rho_l^{(n)}(\varepsilon,x)=\sup_{0\leqq k\leqq[nT]+1-l}\Pi_{0,x}^{(n)}(\sup_{1\leqq j\leqq l}|X(k+j)-X(k)|\geqq\varepsilon).$$

Thus it suffices to show that

$$\lim_{\delta\downarrow 0}\left(\left[\frac{T}{\delta}\right]+1\right)\lim_{n\to\infty}\sup_{x\in R^d}\rho_{[n\delta]}^{(n)}(\tfrac{1}{3}\varepsilon,x)=0.$$

In order to estimate $\rho_l^{(n)}(\varepsilon,x)$, we consider the process $Z^{(n)}(k)$, $k\geqq 0$, which is given by

$$Z^{(n)}(0)=\langle\theta,x\rangle$$

and

$$Z^{(n)}(k+1)=\langle\theta,X(k+1)\rangle-\sum_{l=0}^{k}\int\langle\theta,(y-X(l)\rangle\Pi_l^{(n)}(X(l),dy),$$

where θ is some fixed element of S^{d-1}. It is easy to check that $[Z^{(n)}(k),\bar{M}_k^0,\nu_{0.x}^{(n)}:k\geqq 0]$ is a martingale for each $n\geqq 1$. Hence, by the martingale inequality,

$$\nu_{0,x}^{(n)}(\sup_{1\leqq j\leqq l}|Z^{(n)}(k+j)-Z^{(n)}(k)|\geqq\varepsilon)\leqq\frac{1}{\varepsilon^4}E^{\nu_{x,0}^{(n)}}[(Z^{(n)}(k+l)-Z^{(n)}(k))^4].$$

Define

$$T_k^{(n)}(l) = Z^{(n)}(k+l) - Z^{(n)}(k)$$

and

$$\Delta_k^{(n)}(l+1) = Z^{(n)}(k+l+1) - Z^{(n)}(k+l) .$$

Then

$$E^{\nu_{0,x}^{(n)}}[(T_k^{(n)}(l+1))^4] = E^{\nu_{0,x}^{(n)}}[(T_k^{(n)}(l))^4] + 6E^{\nu_{0,x}^{(n)}}[(T_k^{(n)}(l))^2(\Delta_k^{(n)}(l+1))^2]$$
$$+ 4E^{\nu_{x,0}^{(n)}}[T_k^{(n)}(l)(\Delta_k^{(n)}(l+1)^3] + E^{\nu_{0,x}^{(n)}}[(\Delta_k^{(n)}(l+1))^4] .$$

Observe that

$$E^{\nu_{k+l,X(k+l)}^{(n)}}[\Delta_k^{(n)}(l+1)^2] \leqq 2(E^{\nu_{k+l,X(k+l)}^{(n)}}[\langle\theta, X(k+l+1) - X(k+l)\rangle^2]$$
$$+ E^{\nu_{k+l,X(k+l)}^{(n)}}\left[\left(\int \langle\theta, y - X(k+l)\rangle \Pi_{k+l}^{(n)}(X(k+l), dy)\right)^2\right]$$
$$\leqq 2\left(\frac{C_2}{n} + \frac{C_1^2}{n^2}\right) \leqq \frac{C}{n}$$

for all $k, l \geqq 0$, and therefore,

$$E^{\nu_{0,x}^{(n)}}[(T_k^{(n)}(l))^2] = \sum_{j=0}^{l-1} E^{\nu_{0,x}^{(n)}}[(\Delta_k^{(n)}(j))^2] = \sum_{j=0}^{l-1} E^{\nu_{0,x}^{(n)}}[E^{\nu_{k+j,X(k+j)}^{(n)}}[(\Delta_k^{(n)}(j))^2]] \leqq \frac{lC}{n}$$

for all $k, l \geqq 0$ and $x \in R^d$. Thus,

$$E^{\nu_{0,x}^{(n)}}[(T_k^{(n)}(l+1))^4] \leqq E^{\nu_{0,x}^{(n)}}[(T_k^{(n)}(l))^4] + 6C^2 \frac{l}{n^2}$$
$$+ 8C^{3/2}R \frac{l^{1/2}}{n^{3/2}} + E^{\nu_{0,x}}[(\Delta_k^{(n)}(l+1))^4]$$

and so

$$E^{\nu_{0,x}^{(n)}}[(T_k^{(n)}(l+1))^4] \leqq A\left(\frac{l}{n}\right)^2 + B\left(\frac{l}{n}\right)^{3/2} + \sum_{j=0}^{l} E^{\nu_{0,x}^{(n)}}[(\Delta_k^{(n)}(j+1))^4] ,$$

where A and B do not depend on $k, l \geqq 0$, $x \in R^d$, or $\theta \in S^{d-1}$. Also,

$$E^{\nu_{0,x}^{(n)}}[(\Delta^{(n)}(j+1))^4] \leqq 8\Big(E^{\nu_{0,x}^{(n)}}[\langle\theta, X(k+j+1) - X(k+j)\rangle^4]$$
$$+ E^{\nu_{0,x}^{(n)}}\left[\left(\int \langle 0, y - X(k+j)\rangle \Pi_{k+j}^{(n)}(X(k+j), dy)\right)^4\right]\Big)$$
$$\leqq 8\left(\alpha^2 \frac{C}{n} + R^4 \sup_{\substack{i \geqq 0 \\ y \in R^d}} \Pi_i^{(n)}\left(y, \{|z-y| \geqq \alpha\}\right) + \frac{C_1^4}{n^4}\right)$$

for each $\alpha > 0$. Hence,

$$E^{\nu_{0,x}^{(n)}}([\Delta_k^{(n)}(j+1))^4] \leqq \frac{f(n)}{n},$$

where f is independent of $x \in R^d$, $\theta \in S^{d-1}$ and $j, k \geqq 0$, and $\lim_{n\to\infty} f(n) = 0$. Combining these inequalities, we have

$$\sup_{x\in R^d} \rho_l^{(n)}\left(\varepsilon + C_1 \frac{l}{n}, x\right) \leqq \frac{d}{\varepsilon^4}\left(A\left(\frac{l}{n}\right)^2 + B\left(\frac{l}{n}\right)^{3/2} + f(n)\frac{l}{n}\right)$$

and hence

$$\lim_{n\to\infty} \sup_{x\in R^d} \frac{1}{\delta}\, \rho_{[n\delta]}^{(n)}(\varepsilon + C_1\delta, x) \leqq \frac{d}{\varepsilon^4}(A\delta + B\delta^{1/2}) .$$

LEMMA 10.3. *Let $\Pi_k^{(n)}$, $k \geqq 0$, be a sequence of transition probabilities for each $n \geqq 1$, and define $\nu_{k,x}^{(n)}$, $k \geqq 0$ and $x \in R^d$, and $P_x^{(n)}$, $x \in R^d$, as in Lemma 10.2. Assume that*

(i) $\displaystyle\sup_{\substack{k\geqq 0\\ x\in R^d}} \Pi_k^{(n)}(x, \{|y-x| \geqq l\}) \leqq \frac{f(l)}{n}$, *where* $\displaystyle\lim_{l\to\infty} f(l) = 0$,

(ii) $\displaystyle\lim_{n\to\infty} \sup_{\substack{k\geqq 0\\ |x|\leqq L}} n\, \Pi_k^{(n)}(x, \{|y-x| \geqq \varepsilon\}) = 0$ *for all* $L < \infty$ *and* $\varepsilon > 0$,

(iii) $\displaystyle\sup_{\substack{k\geqq 0\\ x\in R^d}} \left| \int_{|y-x|\leqq 1} (y-x)\Pi_k^{(n)}(x, dy) \right| \leqq \frac{C_1}{n}$,

(iv) $\displaystyle\sup_{\substack{k\geqq 0\\ x\in R^d}} \left| \int_{|y-x|\leqq 1} (y-x)\otimes(y-x)\Pi_k^{(n)}(x, dy) \right| \leqq \frac{C_2}{n}$.

Then $[P_x^{(n)}\colon n \geqq 1, |x| \leqq L]$ is compact in the sense described in Lemma 10.2 for each $L < \infty$.

Proof: The proof of this lemma turns on two applications of the following remark. In order that $[P_x^{(n)}\colon n \geqq 1, |x| \leqq L]$ be compact in the required sense it is sufficient to find, for each $T > 0$, $L < \infty$ and $\varepsilon > 0$, a family $[\bar{P}_x^{(n)}\colon n \geqq 1, |x| \leqq L]$ such that $[\bar{P}_x^{(n)}\colon n \geqq 1, |x| \leqq L]$ is weakly compact on $C([0, T], R^d)$ and $|P_x^{(n)}(B) - \bar{P}_x^{(n)}(B)| < \varepsilon$ for all $B \in M_T^0$ and $|x| \leqq L$.

Given $T > 0$ and $\varepsilon > 0$, choose $l \geqq 1$ so that $f(l) < \frac{1}{2}\varepsilon(T+1)$ and define

$$\bar{\Pi}_k^{(n)}(x, \Gamma) = \Pi_k^{(n)}(x, \Gamma \cap \{|y-x| \leqq l\}) + \Pi_k^{(n)}(x, \{|y-x| \geqq l\})\delta_x(\Gamma) .$$

Then $[\bar{\Pi}_k^{(n)}\colon k \geqq 0]$ is a sequence of transition probabilities for each $n \geqq 1$. Define $\bar{\nu}_{k,x}^{(n)}$ and $\bar{P}_x^{(n)}$ from the $\bar{\Pi}_k^{(n)}$ in the same way as the $\nu_{k,x}^{(n)}$ and $P_k^{(n)}$ were defined from the $\Pi_k^{(n)}$. For each $n \geqq 1$ let

$$A^{(n)} = \{|X(k+1) - X(k)| \leqq l, 0 \leqq k \leqq [nT]\}.$$

Then, for $\bar{B} \in \bar{M}^0_{[nT]+1}$, we have

$$0 \leqq \bar{\nu}_{0,x}^{(n)}(\bar{B} \cap A^{(n)}) - \nu_{0,x}^{(n)}(\bar{B} \cap A^{(n)}) \leqq \bar{\nu}_{0,x}^{(n)}(A^{(n)}) - \nu_{0,x}^{(n)}(A^{(n)})$$

$$\leqq \nu_{0,x}^{(n)}(\bar{\Omega} - A^{(n)}) \leqq ([nT]+1)\frac{f(l)}{n} < \tfrac{1}{2}\varepsilon .$$

Thus, if $B \in M_T^0$, then

$$|P_x^{(n)}(B) - \bar{P}_x^{(n)}(B)| = |\nu_{0,x}^{(n)}(\Phi_n^{-1}(B)) - \bar{\nu}_{0,x}^{(n)}(\Phi_n^{-1}(B))|$$

$$\leqq \tfrac{1}{2}\varepsilon + \nu_{0,x}^{(n)}(\Phi_n^{-1}(B) \cap (\bar{\Omega} - A^{(n)})) < \varepsilon ,$$

since $\Phi_n^{-1}(B) \in \bar{M}^0_{[nT]+1}$.

In view of the preceding paragraph, we see that it suffices to prove compactness in the case when the $\Pi_k^{(n)}$, $k \geqq 0$, satisfy

(i′) $\Pi_k^{(n)}(x, \{|y - x| \geqq l\}) = 0, \quad n \geqq 1, \quad k \geqq 0, \quad \text{and} \quad x \in R^d,$

(ii′) $\lim_{n\to\infty} \sup_{\substack{k \geqq 0 \\ |x| \leqq L}} n\Pi_k^{(n)}(x, \{|y - x| \geqq \varepsilon\}) = 0 \quad \text{for each} \quad L < \infty,$

(iii′) $\sup_{\substack{k \geqq 0 \\ x \in R^d}} \left| \int (y - x)\Pi_k^{(n)}(x, dy) \right| \leqq \frac{C_1}{n},$

(iv′) $\sup_{\substack{k \geqq 0 \\ x \in R^d}} \left| \int (y - x) \otimes (y - x)\Pi_k^{(n)}(x, dy) \right| \leqq \frac{C_2}{n}.$

What we shall do is to show that if the $\Pi_k^{(n)}$, $k \geqq 0$, satisfy (i′)–(iv′), then for each $T > 0$, $L < \infty$ and $\varepsilon > 0$ there exist transition probabilities $\bar{\Pi}_k^{(n)}$, $k \geqq 0$, satisfying the conditions of Lemma 10.2 such that $|P_x^{(n)}(B) - \bar{P}_x^{(n)}(B)| < \varepsilon$ for all $B \in M_T^0$ and $|x| \leqq L$, where the $\bar{P}_x^{(n)}$ are defined from the $\bar{\Pi}_k^{(n)}$ as above.

Given $T > 0$, $L < \infty$, and $\varepsilon > 0$, choose $L' > L$ such that

$$\frac{2(dC_2 + C_1^2)(T+1)}{[(L' - L) - (T+1)C_1]^2} < \tfrac{1}{2}\varepsilon$$

and define

$$\bar{\Pi}_k^{(n)}(x, \Gamma) = \begin{cases} \Pi_k^{(n)}(x, \Gamma) & \text{if } |x| \leqq L', \\ \delta_x(\Gamma) & \text{if } |x| > L'. \end{cases}$$

It is easy to see that the $\bar{\Pi}_k^{(n)}$, $k \geqq 0$, satisfy the conditions of Lemma 10.2. Moreover, if $\bar{B} \in \bar{M}_k^0$ and $|x| \leqq L'$, then

$$\bar{\nu}_{0,x}^{(n)}(\bar{B} \cup \{\sup_{0 \leqq l \leqq k} |X(l)| \leqq L'\}) = \nu_{0,x}^{(n)}(\bar{B} \cap \{\sup_{0 \leqq l \leqq k} |X(l)| \leqq L'\}) .$$

Hence,

$$|P_x^{(n)}(B) - \bar{P}_x^{(n)}(B)| \leqq \nu_{0,x}^{(n)}(\sup_{0 \leqq l \leqq [nT]+1} |X(l)| > L') + \bar{\nu}_{0,x}^{(n)}(\sup_{0 \leqq l \leqq [nT]+1} |X(l)| > L')$$

for all $B \in M_T^0$. Since

$$\bar{\nu}_{0,x}^{(n)}(\sup_{0 \leqq l \leqq [nT]+1} |X(l)| > L') = \nu_{0,x}^{(n)}(\sup_{0 \leqq l \leqq [nT]+1} |X(l)| \geqq L')$$

for $x \leqq L'$, we shall be done once we show that

$$\sup_{|x| \leqq L} \nu_{0,x}^{(n)}(\sup_{0 \leqq l \leqq [nT]+1} |X(l)| > L') < \tfrac{1}{2}\varepsilon .$$

But proceeding as we did in the proof of Lemma 10.2, we find that

$$\begin{aligned}\nu_{0,x}^{(n)}(\sup_{0 \leqq l \leqq [nT]+1} |X(l)| \geqq L') &\leqq \frac{2n(T+1)}{[(L'-L)-(T+1)C_1]^2} \\ &\quad \times \sup_{l \geqq 0} \left(E^{\nu_{0,x}^{(n)}}[|X(l+1) - X(l)|^2] + E^{\nu_{0,x}^{(n)}}\left[\left|\int (y - X(l))\Pi_l^{(n)}(X(l), dy)\right|^2\right]\right) \\ &\leqq \frac{2(T+1)(dC_2 + C_1^2)}{[(L'-L)-(T+1)C_1]^2} < \tfrac{1}{2}\varepsilon .\end{aligned}$$

THEOREM 10.3. *Let a: $[0, \infty) \times R^d \to R^d \otimes R^d$ and b: $[0, \infty) \times R^d \to R^d$ be bounded continuous functions, and assume that a is symmetric and is strictly positive definite at each point. Denote by P_x, $x \in R^d$, the unique probability measure on (Ω, M^0) for which*

(a) $P_x(x(0) = x) = 1$,

(b) $X_\theta(t) = \exp \{\langle \theta, x(t) - x\rangle - \frac{1}{2} \int \langle \theta, a(u, x(u))\theta\rangle\, du - \int_0^t \langle \theta, b(u, x(u))\rangle\, du\}$,

$t \geqq 0$,

is a P_x-martingale for all $\theta \in R^d$.

Suppose $[\Pi_k^{(n)}: k \geqq 0]$ *is a sequence of transition probabilities on* R^d *and define* $\nu_{k,x}^{(n)}$, *and* $P_x^{(n)}$ *from the* $\Pi_k^{(n)}$ *as in Lemma* 10.2. *Assume*

(i) $\sup_{\substack{n \geqq 1 \\ k \geqq 0 \\ x \in R^d}} n\Pi_k^{(n)}(x, \{|y - x| \geqq l\}) \to 0 \quad as \quad l \to \infty$,

(ii) $\lim_{n \to \infty} \sup_{\substack{k \geqq 0 \\ |x| \leqq L}} n\Pi_k^{(n)}(x, \{|y - x| \geqq \varepsilon\}) = 0$ *for each* $\varepsilon > 0$ *and* $L < \infty$,

(iii) $$\int_{|y-x| \leqq 1} (y - x)\Pi_k^{(n)}(s, dy) = \frac{1}{n} b\left(\frac{k}{n}, x\right) + \beta(n; k, x),$$

where

$$\sup_{\substack{k \geqq 0 \\ x \in R^d}} |\beta(n; k, x)| \leqq \frac{C_1}{n} \quad and \quad \lim_{n \to \infty} \sup_{\substack{k \geqq 0 \\ |x| \leqq L}} n\,|\beta(n; k, x)| = 0 \quad for\ each \quad L < \infty,$$

(iv) $$\int_{|y-x| \leqq 1} (y - x) \otimes (y - x)\Pi_k^{(n)}(x, dy) = \frac{1}{n} a\left(\frac{k}{n}, x\right) + \alpha(n; k, x),$$

where

$$\sup_{\substack{k \geqq 0 \\ x \in R^d}} |\alpha(n; k, x)| \leqq \frac{C_2}{n} \quad and \quad \lim_{n \to \infty} \sup_{\substack{k \geqq 0 \\ |x| \leqq L}} n\,|\alpha(n; k, x)| = 0 \quad for\ each \quad L < \infty.$$

If $x_n \to x$ *in* R^d, *then* $P_{x_n}^{(n)} \to P_x$ *in the sense of weak convergence on* (Ω, M^0).

Proof: By Lemma 10.3, we can find a subsequence of $[P_{x_n}^{(n)}: n \geqq 1]$ which converges to a probability measure on (Ω, M^0) in the required sense. Hence, all that we need to show is that any probability measure which is a limit of $[P_{x_n}^{(n)}: n \geqq 1]$, in this sense, satisfies (a) and (b).

Let P_m be a subsequence of $[P_{x_n}^{(n)}: n \geqq 1]$ which converges in the required sense to a limit point P on $\langle \Omega, M^0 \rangle$. We first observe that $P(x(0) = x) = 1$. In fact,

$$P(|x(0) - x| \leqq \delta) \geqq \lim_{m \to \infty} P_m(|x(0) - x| \leqq \delta) = 1.$$

We next want to prove that, for each $\theta \in R^d$,

$$X_{i\theta}(t) = \exp\left\{i\langle\theta, x(t) - x\rangle + \frac{1}{2}\int \langle\theta, a(u, x(u))\theta\rangle\, du - i\int_0^t \langle\theta, b(u, x(u))\rangle\, du\right\}, \quad t \geqq 0,$$

is a P-martingale. By Theorem 10.1, it will then follow that P satisfies (b). Without loss of generality, we may assume that

$$\sup_{\substack{m\geqq 1\\ k\geqq 0\\ x\in R^d}}\left|1-\int e^{i\langle\theta,y-x\rangle}\,\Pi_k^{(n_m)}(x,dy)\right|<1\,.$$

Define

$$\phi_{k,\theta}^{(m)}(x)=\log\left(\int e^{i\langle\theta,y-x\rangle}\,\Pi_k^{(n_m)}(x,dy)\right)$$

and

$$\bar{X}_\theta^{(m)}(k)=\exp\left\{i\langle\theta,X(k)-X(0)\rangle-\sum_{j=0}^{k-1}\phi_{j,\theta}^{(m)}(X(j))\right\},\qquad k\geqq 0\,.$$

Then, by Theorem 10.2, $[\bar{X}_\theta^{(m)}(k),\ \bar{M}_k^0,\ \nu_{0,x_{n_m}}^{(n_m)}:\ k\geqq 0]$ is a martingale. Thus, if

$$X_\theta^{(m)}(t)=\exp\left\{i\langle\theta,x(t)-x(0)\rangle-n_m\int_0^{[n_m t]}\phi_{[n_m u],\theta}^{(m)}\left(x\left(\frac{[n_m u]}{n_m}\right)\right)du\right\},$$

then $[X_\theta^{(m)}(k/n_m),\ M_{k/n_m}^0,\ P_m:\ k\geqq 0]$ is a martingale for all $m\geqq 1$. Hence, in order to prove that $[X_{i\theta}(t):\ t\geqq 0]$ is a P-martingale, it suffices to show that if ξ is a bounded continuous M_T^0-measurable function on Ω, then

$$\lim_{m\to\infty}E^{P_m}\left[\xi\cdot X_\theta^{(m)}\left(\frac{[n_m t]}{n_m}\right)\right]=E^P[\xi\cdot X_{i\theta}(t)]$$

for each $t\geqq 0$. To do this, we first observe that, by Lemma 10.1,

$$\lim_{m\to\infty}\sup_{w\in K}\left|X_\theta^{(m)}\left(\frac{[n_m t]}{n_m},w\right)-X_{i\theta}(t,w)\right|=0$$

for each $K\subseteq\Omega$ which is compact as a subset of $C([0,t],R^d)$. Now suppose that $\varepsilon>0$ is given and choose $K\subseteq\Omega$ so that K is compact as a subset of $C([0,t],R^d)$ and

$$\inf_{m\geqq 1}P_m(K)\geqq 1-\frac{\varepsilon}{\Delta},$$

where $\Delta=\sup_{m\geqq 1}\|\xi\cdot(X_\theta^{(m)}([n_m t]/n_m)-X_{i\theta}(t))\|$. Next, choose N such that

$$\sup_{\substack{m\geqq N\\ w\in K}}\left|X_\theta^{(m)}\left(\frac{[n_m t]}{n_m},w\right)-X_{i\theta}(t,w)\right|<\varepsilon$$

and

$$\sup_{m\geqq N}|E^{P_m}[\xi\cdot X_{i\theta}(t)]-E^P[\xi\cdot X_{i\theta}(t)]|<\varepsilon\,.$$

Then for $m \geqq N$,

$$\left| E^{P_m}\left[\xi \cdot X_\theta^{(m)}\left(\frac{[n_m t]}{n_m}\right)\right] - E^P[\xi \cdot X_{i\theta}(t)]\right| \leqq \left| E^{P_m}\left[\xi \cdot \left(X_\theta^{(m)}\left(\frac{[n_m t]}{n_m}\right) - X_{i\theta}(t)\right)\right]\right|$$
$$+ |E^{P_m}[\xi \cdot X_{i\theta}(t)] - E^P[\xi \cdot X_{i\theta}(t)]| < 3\varepsilon .$$

COROLLARY 10.2. *Let everything be defined as in Theorem* 10.3. *If f is a bounded continuous function on R^d, then, for each $t > 0$, $E^{P_x^{(n)}}[f(x(t))] \to E^{P_x}[f(x(t))]$ uniformly for x in compact sets.*

In conclusion, we point out that the same development could be carried out for Markov chains with values on a lattice. In this way we could get the appropriate generalization of the random walk approximation to the Brownian motion.

11. Connections with Semi-Groups and Parabolic Equations

Let $a\colon R^d \to R^d \otimes R^d$ be a continuous and $b\colon R^d \to R^d$ a bounded measurable function. Further, assume that a is symmetric and strictly positive definite for each $x \in R^d$ and that there are constants $A < \infty$ and $B < \infty$ such that

$$(11.1) \qquad \begin{aligned} |a_{ij}| &\leqq A , \\ |b_j| &\leqq B . \end{aligned}$$

What we want to show is that there is at most one Markov semi-group $\{T_t\}_{t>0}$ on $\mathbf{B}(R^d)$ such that $\{T_t\}_{t>0}$ is strongly continuous on $C_0(R^d)$ and

$$(11.2) \qquad T_t f = f + \int_0^t T_s Lf\, ds , \qquad t \geqq 0 ,$$

for all $f \in C_0^\infty(R^d)$, where

$$(11.3) \qquad L = \frac{1}{2}\sum_{i,j=1}^d a_{ij}(x)\frac{\partial^2}{\partial x_i\,\partial x_j} + \sum_{i=1}^d b_i(x)\frac{\partial}{\partial x_i} .$$

The idea of our proof is to show that if $\{T_t\}_{t>0}$ is such a semi-group, then for each $x \in R^d$ there is a probability measure P_x on $\langle \Omega, M^0\rangle$ such that

$$\text{(i)} \quad X_\theta(t) = \exp\left\{\langle \theta, x(t) - x\rangle = \frac{1}{2}\int_0^t \langle \theta, a(x(u))\theta\rangle\, du - \int_0^t \langle \theta, b(x(u))\rangle\, du\right\}$$
$$t \geqq 0 ,$$

is a P-martingale for each $\theta \in R^d$,

(ii) $P_x(x(0) = x) = 1$,

(iii) $T_t f(x) = E^{P_x}[f(x(t))]$ for all $t \geqq 0$ and $f \in B(R^d)$.

Since the first two of these conditions uniquely determine P_x , the uniqueness of $\{T_t\}_{t>0}$ will follow immediately from (iii).

LEMMA 11.1. *Let* $u \in C_b^{1,2}([0, \infty) \times R^d)$.[2] *Then*

$$[T_t u(t, \cdot)](x) - u(0, x) = \int_0^t [T_s(u_1(s, \cdot) + Lu(s, \cdot))](x)\, ds , \tag{11.4}$$

where $u_1(s,y) = \partial u(s,y)/\partial s$.

Proof: It suffices to prove this assertion for $u \in C_b^\infty([0, \infty) \times R^d)$ with compact support for each $t \geqq 0$.[3] Given such a u, we have

$$\begin{aligned} T_t u(t, \cdot) - u(0, \cdot) &= T_t (u(t, \cdot) - u(0, \cdot)) + T_t u(0, \cdot) - u(0, \cdot) \\ &= T_t \int_0^t u_1(s, \cdot)\, ds + \int_0^t T_\alpha Lu(0, \cdot)\, d\alpha . \end{aligned}$$

Observe that

$$\begin{aligned} T_t \int_0^t u_1(s, \cdot)\, ds - \int_0^t T_s u_1(s, \cdot)\, ds &= \int_0^t T_s(T_{t-s} u_1(s, \cdot) - u_1(s, \cdot))\, ds \\ &= \int_0^t T_s \int_0^{t-s} T_\alpha Lu_1(s, \cdot)\, d\alpha\, ds , \\ &= \int_0^t \int_s^t T_\alpha Lu_1(s, \cdot)\, d\alpha\, ds , \end{aligned}$$

and that

$$\begin{aligned} \int_0^t T_\alpha Lu(0, \cdot)\, d\alpha - \int_0^t T_\alpha Lu(\alpha, \cdot)\, d\alpha &= - \int_0^t T_\alpha \int_0^\alpha Lu_1(s, \cdot)\, ds\, d\alpha \\ &= - \int_0^t \int_0^\alpha T_\alpha Lu_1(s, \cdot)\, ds\, d\alpha \\ &= - \int_0^t \int_s^t T_\alpha Lu_1(s, \cdot)\, d\alpha\, ds . \end{aligned}$$

Combining these equations we obtain (11.4).

[2] $C_b^{1,2}([0, \infty) \times R^d)$ stands for the space of continuous functions on $[0, \infty) \times R^d$, which are bounded and have one bounded continuous derivative in t and two bounded continuous partial derivatives in x.

[3] $C_b^\infty([0, \infty) \times R^d)$ denotes the space of bounded continuous functions on $[0, \infty) \times R^d$ having infinitely many bounded continuous derivatives in t and x.

LEMMA 11.2. *There exist constants C_1 and C_2 such that*

$$\sup_{x\in R^d} P(t, x, \{|y - x| \leqq l\}) \leqq d \cdot \exp\left\{-\frac{(l - c_1 t)^2}{2c_2 t}\right\}, \qquad t \geqq \frac{l}{c_1}, \tag{11.5}$$

where $P(t, x, \Gamma)$ is the transition probability function associated with $\{T_t\}_{t>0}$.

Proof: Let $\phi_n^{(\theta)} \in C_b^\infty(R^d)$, $n \geqq 1$ and $\theta \in S^{d-1}$, be a family of functions such that

(a) $\phi_n^{(\theta)}(y) \to \langle\theta, y\rangle$ uniformly on compact sets for each $\theta \in S^{d-1}$,

(b) $\displaystyle\sup_{\substack{n\geqq 1\\ y\in R^d\\ |\theta|=1}} \sum_{i,j=1}^{d} \left(\frac{\partial^2\phi^{(\theta)}}{\partial y_i\, \partial y_j}(y)\right)^2 \leqq L_2^2,$

(c) $\displaystyle\sup_{\substack{n\geqq 1\\ y\in R^d\\ |\theta|=1}} \sum_{i=1}^{d} \left(\frac{\partial\phi_n^{(\theta)}}{\partial y_i}(y)\right)^2 \leqq L_1^2.$

Extend the definition of $\phi_n^{(\theta)}$ to all $\theta \in R^d$ by setting

$$\phi_n^{(\theta)}(y) = \phi_n^{(\theta/|\theta|)}(|\theta| y).$$

Then (a) continues to hold. Moreover,

$$\sup_{\substack{n\geqq 1\\ y\in R^d}} \sum_{i,j=1}^{d} \left(\frac{\partial^2\phi_n^{(\theta)}}{\partial x_i\, \partial x_j}(y)\right)^2 \leqq |\theta|^4 L_2^2, \qquad \theta \in R^d,$$

$$\sup_{\substack{n\geqq 1\\ y\in R^d}} \sum_{i=1}^{d} \left(\frac{\partial\phi_n^{(\theta)}}{\partial x_i}(y)\right)^2 \leqq |\theta|^2 L_1^2, \qquad \theta \in R^d.$$

Consider the function

$$u_n^{(\theta)}(t, y) = \exp\{\phi_n^{(\theta)}(y) - \tfrac{1}{2}C_2 |\theta|^2 t - C_1 |\theta|\, t\},$$

where

$$C_2 = \sup_{y\in R^d}\left(\sum_{i,j=1}^{d} (a_{ij}(y))^2\right) L_2 + AL_1^2,$$

$$C_1 = BM_1.$$

Then $u_n^{(\theta)} \in C_b^{1,2}([0, \infty) \times R^d)$ and $\partial u_n^{(\theta)}/\partial t + Lu_n^{(\theta)} \leqq 0$. Hence, by Lemma 11.1,

$$[T_t u_n^{(\theta)}(t, \cdot)](x) \leqq u_n^{(\theta)}(0, x).$$

Letting $n \to \infty$, we can use Fatou's lemma to conclude that

$$\int e^{\langle\theta, y-x\rangle} P(t, x, dy) \leqq \exp\{\tfrac{1}{2}C_2 |\theta|^2 t + C_1 |\theta| t\}.$$

Therefore, if $\theta \in S^{d-1}$, then

$$P(t, x, \langle\theta, y - x\rangle \geqq l) \leqq \exp\{-\lambda l + \tfrac{1}{2}\lambda^2 C_2 t + \lambda C_1 t\}, \qquad \lambda \geqq 0.$$

In particular, if $t \leqq l/C_1$, then by taking $\lambda = (l - C_1 t)/C_2 t$, we obtain

$$\sup_{\substack{x \in R^d \\ \theta \in S^{d-1}}} P(t, x, \{\theta, y - x \geqq l\}) \leqq \exp - \frac{(l - C_1 t)^2}{2C_2 t}, \qquad t \leqq \frac{l}{C_1}.$$

Equation (11.5) is now immediate.

LEMMA 11.3. *There exists a continuous Markov process* $[x(t), M_t^0, P_x \colon t \geqq 0, x \in R^d]$ *such that*

$$T_t f(x) = E^{P_x}[f(x(t))], \qquad t \geqq 0, \; x \in R^d, \; f \in B(R^d), \tag{11.6}$$

and, for all $\theta \in R^d$ *and* $x \in R^d$,

$$\xi_\theta(t) = \langle\theta, x(t) - x(0)\rangle - \int_0^t \langle\theta, b(x(u))\rangle\, du \tag{11.7}$$

and

$$\eta_\theta(t) = \xi_\theta^2(t) - \int_0^t \langle\theta, a(x(u))\theta\rangle\, du \tag{11.8}$$

are P_x*-martingales.*

Proof: That there exists a continuous Markov process $[x(t), M_t^0, P_x \colon t \geqq 0, x \in R^d]$ satisfying (11.6) is immediate from Lemma 11.2 above and Theorem 6.5 in Dynkin [5].

Observe that, by Lemmas 11.1 and 11.2, we have

$$E^{P_x}[\langle\theta, x(t) - x(0)\rangle] = E^{P_x}\left[\int_0^t \langle\theta, b(x(u))\rangle\, du\right],$$

$$\begin{aligned} E^{P_x}[(\langle\theta, x(t) - x(0)\rangle)^2] &= E^{P_x}\left[\int_0^t \langle\theta, a(x(u))\theta\rangle\, du\right] \\ &\quad + E^{P_x}\left[\int_0^t \langle\theta, b(x(u))\rangle\langle\theta, x(u)\rangle\, du\right]. \end{aligned}$$

Hence,

$$E^{P_x}[\xi_\theta(t+s) \mid M_t^0] = E^{P_x}\left[\langle\theta, x(t+s) - x(t)\rangle - \int_t^{t+s} \langle\theta, b(x(u))\rangle\, du \,\Big|\, M_t^0\right] + \xi_\theta(t)$$

$$= E^{P_{x(t)}}[\langle\theta, x(s) - x(0)\rangle] - E^{P_{x(t)}}\left[\int_0^s \langle\theta, b(x(u))\rangle\, du\right] + \xi_\theta(t)$$

$$= \xi_\theta(t) \qquad \text{a.s. } P_x,$$

and so $[\xi_\theta(t), M_t^0, P_x\colon t \geqq 0]$ is a martingale. Also,

$$\begin{aligned}
&E^{P_x}[\xi_\theta^2(t+s) - \xi_\theta^2(t) \mid M_t^0] \\
&= E^{P_x}[(\xi_\theta(t+s) - \xi(t))^2 \mid M_t^0] + 2E^{P_x}[\xi(t+s)\xi(t) \mid M_t^0] - 2\xi^2(t) \\
&= E^{P_{x(t)}}[\xi_\theta^2(s)] \\
&= E^{P_{x(t)}}[(\langle\theta, x(s) - x(0)\rangle)^2] \\
&\quad - 2E^{P_{x(t)}}\left[x(s)\int_0^s \langle\theta, b(x(u))\rangle\, du\right] + E^{P_{x(t)}}\left[\left(\int_0^s \langle\theta, b(x(u))\rangle\, du\right)^2\right] \\
&= E^{P_{x(t)}}\left[\int_0^s \langle\theta, a(x(u))\theta\rangle\, du\right] \\
&\quad - 2E^{P_{x(t)}}\left[\int_0^s (\langle\theta, b(x(u))\rangle)(\langle\theta, x(s) - x(u)\rangle)\, du\right] \\
&\quad + E^{P_{x(t)}}\left[\left(\int_0^s \langle\theta, b(x(u))\rangle\, du\right)^2\right] \\
&= E^{P_{x(t)}}\left[\int_0^s \langle\theta, a(x(u))\theta\rangle\, du\right] \\
&\quad - 2E^{P_{x(t)}}\left[\int_0^s \langle\theta, b(x(u))\rangle\left(\int_u^s \langle\theta, b(x(\alpha))\rangle\, d\alpha\right)\right] du \\
&\quad + E^{P_{x(t)}}\left[\left(\int_0^s \langle\theta, b(x(u))\rangle\, du\right)^2\right] \\
&= E^{P_x}\left[\int_t^{t+s} \langle\theta, a(x(u))\theta\rangle\, du \,\Big|\, M_t^0\right] \qquad \text{a.s. } P_x.
\end{aligned}$$

In other words, $[\eta_\theta(t), M_t^0, P_x\colon t \geqq 0]$ is a martingale.

This fact gives us our result. Indeed, Doob proves a theorem in [3] (cf. Theorem 5.3 on page 449) which shows that if ξ: $[0, \infty) \times \Omega \to R^d$ is a non-anticipating function which is continuous with respect to $t \geqq 0$ and if P is a probability measure on $\langle \Omega, M^0 \rangle$ such that

(i) $\langle \theta, \xi(t) \rangle$, $t \geqq 0$, is a P-martingale for all $\theta \in R^d$,

(ii) $$E^{P_x}[(\langle \theta, \xi(t) - \xi(s) \rangle)^2 \mid M_s^0] = E^{P_x}\left[\int_s^t \langle \theta, a(x(u))\theta \rangle \, du \mid M_s^0\right], \quad t \geqq s \geqq 0, \quad \theta \in R^d,$$

then there exists a 0-Brownian motion β with respect to P such that

$$\xi(t) - \xi(0) = \int_0^t \sigma(x(u)) \, d\beta(u) \quad \text{a.s. } P, \qquad t \geqq 0,$$

where $\sigma(y)$ is the non-negative definite square root of $a(y)$. Combining Doob's result with Lemma 11.3 above, we find that, for each $x \in R^d$,

(i) $$X_\theta(t) = \exp\left\{\langle \theta, x(t) - x(0) \rangle - \frac{1}{2}\int_0^t \langle \theta, a(x(u))\theta \rangle \, du - \int_0^t \langle \theta, b(x(u)) \rangle \, du\right\}, \ t \geqq 0,$$

is a P_x-martingale for all $\theta \in R^d$,

(ii) $P_x(x(0) = x) = 1$.

We have therefore proved the following theorem.

THEOREM 11.1. *Let $\{T_t\}_{t>0}$ be a Markov semi-group of the sort described in the first paragraph of this section. Then $\{T_t\}_{t>0}$ is unique.*

Before moving on, it is interesting to look at Doob's result in terms of our uniqueness theorem. Indeed, the argument preceding the above theorem shows that there is only one probability measure P on $\langle \Omega, M^0 \rangle$ such that

(i) $\langle \theta, \xi(t) \rangle$, $t \geqq 0$, is a P-martingale for all $\theta \in R^d$,

(ii) $(\langle \theta, \xi(t) \rangle)^2 - \int_0^t \langle \theta, a(x(u))\theta \rangle \, du$, $t \geqq 0$, is a P-martingale for all $\theta \in R^d$,

(iii) $P(x(0) = x) = 1$,

where

$$\xi(t) = x(t) - x(0) - \int_0^t b(x(u)) \, du .$$

Of course, everything that we have said so far in this section extends to the inhomogeneous case. That is, if the coefficients a and b depend on t as well as on x, then the appropriate analogue of Theorem 11.1 is the following: there is only one inhomogeneous Markov semi-group $\{T_{s,t}\}_{t \geqq s > 0}$ such that

$$T_{s,t}f - f = \frac{1}{2}\int_s^t T_{s,u} \sum_{i,j=1}^{a} a_{ij}(u, \cdot) \frac{\partial^2 f}{\partial y_i \, \partial y_j} \, du + \int_s^t T_{s,u} \sum_{i=1}^{d} b_i(u, \cdot) \frac{\partial f}{\partial y_i} \, du$$

for all $t \geqq s > 0$ and $f \in C_0^\infty(R^d)$. The proof of this statement is the same as that of Theorem 11.1. In particular, we use Doob's result and our uniqueness theorem to prove that there is only one probability measure P on $\langle \Omega, M^s \rangle$ such that

(i) $\langle \theta, \xi(t) \rangle$, $t \geqq s$, is a P-martingale for all $\theta \in R^d$,

(ii) $(\langle \theta, \xi(t) \rangle)^2 - \int_s^t \langle \theta, a(u, x(u))\theta \rangle \, du$, $t \geqq s$, is a P-martingale for all $\theta \in R^d$,

(iii) $P(x(s) = x) = 1$,

where

$$\xi(t) = (x(t) - x(s)) - \int_s^t b(u, x(u)) \, du .$$

We shall now connect the stochastic process $P_{s,x}$ with solutions of the Cauchy problem for parabolic differential equations.

Let us consider a strip $\{s_0 \leqq s \leqq T\} \times R^d$, which we denote by $\bar{D}$. The strip $\{s_0 < s < T\} \times R^d$ will be denoted by D. For any function $v = v(s, x)$ which is infinitely differentiable in s and x and vanishes outside a compact subset of $\bar{D}$ we denote by $F = F(s, x)$ and $G = G(s, x)$ the following scalar and vector functions

$$\begin{aligned} F &= \frac{\partial v}{\partial s} + \frac{1}{2}\sum_{i,j} a_{ij} \frac{\partial^2 v}{\partial x_i \, \partial x_j} + \sum_j b_j \frac{\partial v}{\partial x_j}, \\ G &= \sum_i \frac{\partial v}{\partial x_i} \sigma_{ij}, \end{aligned} \tag{11.9}$$

where σ is the positive definite square root of a, and $a = \{a_{ij}(s, x)\}$, and $b = \{b_j(s, x)\}$ are given coefficients satisfying our general hypothesis (both are bounded, and a is strictly elliptic and continuous).

Let v_n be a sequence of functions on $\bar{D}$ each of which vanishes outside a compact subset and suppose that they satisfy the following hypotheses:

(I) $|v_n| \leqq L < \infty$ for all n,

(II) $v_n \to U$ as $n \to \infty$ uniformly on compact subsets of $\bar{D}$,

(III) $F_n \to 0$ in $L_p(D_0)$ over every compact subset D_0 of the open strip D; p is some index greater than 1. (F_n is constructed from v_n as in (11.2)). Then we have the following theorem.

THEOREM 11.2. *If P_{s_0,x_0} is the process starting from x_0 at time s_0 corresponding to these coefficients, then $u(t, x(t))$ is a martingale with respect to P_{s_0,x_0} in the time range $s_0 \leqq t \leqq T$.*

Proof: Combining Itô's formula with the stochastic integral representation of $x(t)$, we have, for each n and $s \geqq s_0$,

$$v_n(s, x(s)) = v_n(s_0, x_0) + \int_{s_0}^{s} \langle G_n(u, x(u)), d\beta(u)\rangle + \int_{s_0}^{s} F_n(u, x(u))\, du . \tag{11.10}$$

We might as well assume that $T > s_0$. Let us choose h so small that $s_0 < s_0 + h < T - h < T$. We take K to be a region $\{|x - x_0| < l\}$ in R^d and define the following stopping time τ. If $x(s_0 + h)$ is not in K, then $\tau = s_0 + h$. If $x(s_0 + h)$ is in K and if the process exits from K between times $s_0 + h$ and $T - h$, then τ is the first exit time. If the process stays in K from time $s_0 + h$ to $T - h$, then τ equals $T - h$. Put $\tau_t = \tau \wedge t$ for $s_0 + h \leqq t \leqq T - h$. Let us define

$$\Phi_n^h(s, w) = v_n(s, x(s)) - v_n(s_0 + h, x(s_0 + h)) - \int_{s_0+h}^{s} F_n(u, x(u))\, du ,$$

and

$$\psi_n^h(s, w) = v_n(\tau_s, x(\tau_s)) - v_n(s_0 + h, x(s_0 + h)) - \int_{s_0+h}^{\tau_s} F_n(u, x(u))\, du .$$

From (11.10) it follows that $\Phi_n^h(s, w)$, and therefore $\psi_n^h(s, w)$, is a martingale for $s_0 + h \leqq s \leqq T - h$. Our method of proof is to let $n \to \infty$, $l \to \infty$, $h \to 0$, in that order.

Let $E_l = \{w: |x(s_0 + h) - x_0| < l\}$. Then $E_l \in M_{s_1}^{s_0}$ for every $s_1 \geqq s_0 + h$. Let us now take $s_0 + h \leqq s_1 < s_2 \leqq T - h$. Then by the martingale equations, for $A \in M_{\tau_{s_1}}^{s_0}$,

$$\int_{A \cap E_l} \psi_n^h(s_2, w)\, dP_{s_0,x_0} = \int_{A \cap E_l} \psi_n^h(s_1, w)\, dP_{s_0,x_0} . \tag{11.11}$$

Clearly, $\psi_n^h(s, w) \to U(\tau_s, x(\tau_s)) - U(s_0 + h, x(s_0 + h))$, as $n \to \infty$. Because of condition (I), we need only check that

$$\lim_{n\to\infty} \int_{E_l} \left[\int_{s_0+h}^{\tau_s} F_n(\sigma, x(\sigma))\, d\sigma\right] dP_{s_0,x_0} = 0 . \tag{11.12}$$

This follows from the fact that, on E_l, the paths never get out of K and therefore

$$\int_{E_l}\left[\int_{s_0+h}^{\tau_s}|F_n(\sigma, x(\sigma))|\,d\sigma\right]dP_{s_0,x_0} \leqq \|F_n(s,x)\|_p\,\|p(s_0, x_0, s, y)\|_q,$$

where

$$\|F_n(s,x)\|_p = \left(\int_K\int_{s_0+h}^{T-h}|F_n(s,x)|^p\,ds\,dx\right)^{1/p}$$

and

$$\|p(s_0, x_0, s, y)\|_q = \left(\int_K\int_{s_0+h}^{T-h}|p(s_0, x_0, s, y)|^q\,dy\,ds\right)^{1/q}$$

with $p^{-1}+q^{-1}=1$. From condition (III) and the estimates in Theorem 8.1, it follows that (11.12) is true. Therefore, we can let $n \to \infty$ in (11.11) and conclude that, for $A \in M_{\tau_{s_1}}^{s_0}$,

$$\int_{A\cap E_l} U(\tau_{s_2}, x(\tau_{s_2}))\,dP_{s_0,x_0} = \int_{A\cap E_l} U(\tau_{s_1}, x(\tau_{s_1}))\,dP_{s_0,x_0}. \tag{11.13}$$

As $l \to \infty$, E_l increases to the entire space, τ_s tends to s and $M_{\tau_s}^{s_0}$ increases to $M_s^{s_0}$ Moreover, U is uniformly bounded. Therefore, letting $l \to \infty$ in (11.13), we arrive at

$$\int_A U(s_2, x(s_2))\,dP_{s_0,x_0} = \int_A U(s_1, x(s_1))\,dP_{s_0,x_0}. \tag{11.14}$$

This shows that $U(s, x(s))$ is a martingale for $s_0+h \leqq s \leqq T-h$. Since $h>0$ is arbitrary and $U(s, x(s))$ tends to a limit as $s \to s_0$ or $s \to T$ in a bounded manner, we conclude that $U(s, x(s))$ is a martingale for $s_0 \leqq s \leqq T$.

Remark 1. This gives the representation

$$U(s_0, x_0) = \int P(s_0, x_0, T, dy)\,U(T, y).$$

Remark 2. Instead of $U_n(s, x)$ being uniformly bounded, we might have allowed it a certain exponential rate of growth and with a little more work could have proved the uniqueness theorem under conditions similar to the classical ones for the heat equation.

COROLLARY 11.1. *Let $U(s, x)$ be a solution of the equation*

$$\frac{\partial U}{\partial s} + \frac{1}{2}\sum a_{ij}\frac{\partial^2 U}{\partial x_i\,\partial x_j} + \sum b_j\frac{\partial U}{\partial x_j} = 0, \qquad s_0 \leqq s \leqq T,$$

with the boundary condition $U(T, x) = f(x)$, *in the class* $W_p^{1,2}$ *with* $p > p_0 = \frac{1}{2}(d+2)$.[4] *Then*

$$U(s_0, x_0) = \int f(y)p(s_0, x_0, T, dy) .$$

Proof: By the definition of $W_p^{1,2}$, we can find a sequence $U_n(s, x)$ satisfying the conditions of Theorem 11.2. The fact that $p > \frac{1}{2}(d+2)$ implies that the convergence of $U_n(s, x)$ to $U(s, x)$ is uniform. Therefore Remark 1 yields the representation

$$U(s_0, x_0) = \int f(y)p(s_0, x_0, T, dy) .$$

(For definition and properties of solutions in the class $W_p^{1,2}$ we refer to [13].)

Bibliography

[1] Cameron, R. H., and Martin, W. T., *The transformation of Wiener integrals by non-linear transformations*, Trans. Amer. Math. Soc., Vol. 75, 1953, pp. 552–575.

[2] Donsker, M. D., *Four Papers in Probability*, Mem. Amer. Math. Soc., No. 6, 1951.

[3] Doob, J. L., *Stochastic Processes*, John Wiley and Sons, New York, 1953.

[4] Dynkin, E. B., *Foundations of the Theory of Makov Processes*, Pergamon Press, New York, 1960.

[5] Dynkin, E. B., *Markov Processes*, Vols. 1 and 2, Academic Press, New York, 1965.

[6] Girsanov, I. V., *On transforming a certain class of stochastic processes by absolutely continuous substitution of measures*, Theory Prob. Applications, Vol. 5, 1960, pp. 285–301.

[7] Girsanov, I. V., *An example of non-uniqueness of the solution of the stochastic equation of K. Itô*, Theor. Probability Appl., Vol. 7, 1962, pp. 325–331.

[8] Girsanov, I. V., *On Itô's stochastic integral equations*, Doklady Akad. Nauk USSR, Vol. 138, 1961, pp. 18–21. (In Russian).

[9] Itô, K., *On Stochastic Differential Equations*, Mem. Amer. Math. Soc., No. 4, 1951.

[10] Itô, K., *On a formula concerning stochastic differentials*, Nagoya Math. J., Vol. 3, 1951, pp. 55–65.

[11] Jones, B. F., *A class of singular integrals*, Amer. J. Math., Vol. 86, 1964, pp. 441–462.

[12] Krylov, N. V., *On quasi-diffusion processes*, Theor. Probability Appl., Vol. 11, 1966, pp. 373–389.

[13] Ladyzenskaia, O. A., Solonnikov, V. A., and Ural'tseva, N. N., *Linear and Quasilinear Equations of Parabolic Type*, Izd. Nauka, Moscow, 1967. (In Russian).

[14] Parthasarathy, K. R., *Probability Measures on Metric Spaces*, Academic Press, New York, 1967.

[15] Prokhorov, Yu. V., *Convergence of random processes and limit theorems*, Theor. Probability Appl., Vol. 9, 1956, pp.118–122.

[16] Skorohod, A. V., *Investigations in the Theory of Random Processes*, Izdatelstvo Kievskovo Universiteta, Kiev, 1961. (In Russian).

[17] Tanaka, H., *Existence of diffusions with continuous coefficients*, Mem. Fac. Sci. Kyushu Univ., Ser. A., Vol. 18, 1964, pp. 89–103.

Received October, 1968.

---

[4] $W_p^{1,2}$ stands for L_p-functions which have one L_p-derivative in t and two L_p-derivatives in x. (See [13] for more details).

COMMUNICATIONS ON PURE AND APPLIED MATHEMATICS, VOL. XXIV, 147–225 (1971)

Diffusion Processes with Boundary Conditions*

DANIEL W. STROOCK AND S. R. S. VARADHAN

1. Introduction

In this paper we adapt the approach that we took in [8] to the study of diffusions with boundary conditions. In [8] the idea was to describe the diffusion P corresponding to

$$L = \frac{1}{2}\sum_{i,j} a_{ij}(t, x)\frac{\partial^2}{\partial x_i\,\partial x_j} + \sum b_j(t, x)\frac{\partial}{\partial x_j}$$

by saying that P is determined by an initial condition plus the condition that

$$f(t, x(t)) - \int_0^t \left(\frac{\partial f}{\partial u} + Lf\right)(u, x(u))\,du$$

is a P-martingale for all smooth f with compact support. To explain the analogous description for a diffusion with boundary conditions, suppose we consider the reflecting Brownian motion in a smooth region G. Then we want to describe P by an initial condition plus the condition that

$$f(t, x(t)) - \int_0^t \left(\frac{\partial f}{\partial u} + \frac{1}{2}\Delta f\right)(u, x(u))\,du$$

is a P-submartingale whenever f is a smooth function whose inner normal derivative on ∂G is non-negative.

The class of problems that we have treated by this method is as follows. The region G is an open set in R^d such that $G = \{x \in R^d : \phi(x) > 0\}$ and $\partial G = \{x \in R^d : \phi(x) = 0\}$, where ϕ is bounded, has two bounded continuous derivatives on R^d and $|\nabla\phi| \geqq 1$ on ∂G. In G we are given bounded continuous diffusion coefficients $a_{ij}(t, x)$, which are strictly elliptic, and bounded measurable drift coefficients $b_i(t, x)$. On ∂G there is a non-negative, bounded, Lipschitz continuous function $\rho(t, x)$ and a bounded Lipschitz continuous vector field $\gamma(t, x)$ satisfying $\langle\gamma, \nabla\phi\rangle \geqq \beta > 0$. If ρ is either identically zero

---

* Results obtained at the Courant Institute of Mathematical Sciences, New York University; this research was sponsored by the U.S. Air Force Office of Scientific Research, Contract AF-49(638)-1719.

or uniformly positive, then we can show that for each $x \in \bar{G}$ there is one and only one P on $C([0, \infty), \bar{G})$ such that $P(x(0) = x) = 1$ and

$$f(t, x(t)) - \int_0^t \left(\frac{\partial f}{\partial u} + Lf\right)(u, x(u))\chi_G(x(u))\, du$$

is a P-submartingale whenever f is a smooth function with compact support and satisfies

$$\rho \frac{\partial f}{\partial t} + \langle \gamma, \nabla f \rangle \geqq 0 \qquad \text{on} \qquad \partial G .$$

Moreover, if P is such a measure, then there is a continuous non-decreasing process $\xi(t)$ which increases only when $x(t) \in \partial G$ and has the property that

$$f(t, x(t)) - \int_0^t \chi_G\left(\frac{\partial f}{\partial u} + Lf\right)(u, x(u))\, du - \int_0^t \left(\rho \frac{\partial f}{\partial u} + \langle \gamma, \nabla f \rangle\right)(u, x(u))\, d\xi(u)$$

is a P-martingale for all smooth f. In the case when the coefficients are independent of time, we can handle general non-negative ρ.

As a consequence of our uniqueness result, we are able to prove an invariance principle for Markov chains converging to diffusions with boundary conditions. The conditions for such convergence coincide with the intuitively natural ones.

2. Some Martingale Manipulations

This section contains an assortment of results which are in one way or another related to martingales.

LEMMA 2.1. *Let $(\Omega, \mathcal{M}, P)$ be a probability space and let $\{\mathcal{M}_t : t \geqq 0\}$ be a non-decreasing family of sub σ-algebras of $\mathcal{M}$. Suppose ϕ and ψ are complex-valued functions on $[0, \infty) \times \Omega$ which are non-anticipating with respect to the $\mathcal{M}_t$. Further, assume that*

(i) *$\phi(t)$ is a continuous martingale which is uniformly bounded on finite time intervals,*

(ii) *$\psi(t)$ is a continuous function of bounded variation such that for each t*

$$E[|\psi|(t)] < \infty ,$$

where $|\psi|(t)$ stands for the total variation of $\psi(s)$ on $[0, t]$.

Under these conditions,

$$\phi(t)\psi(t) - \int_0^t \phi(u)\, d\psi(u)$$

is a martingale.

Proof: Consider

$$E\left[\phi(t)\psi(t) - \phi(s)\psi(s) - \int_s^t \phi(u)\psi(u) \mid \mathcal{M}_s\right]$$

$$= E\left[\int_s^t (\phi(t) - \phi(u))\, d\psi(u) \mid \mathcal{M}_s\right] + E[(\phi(t) - \phi(s))\psi(s) \mid \mathcal{M}_s]$$

$$= I_1 + I_2 .$$

Clearly, $I_2 = 0$. To prove that $I_1 = 0$, we approximate

$$\int_s^t (\phi(t) - \phi(u))\, d\psi(u)$$

by Riemann sums of the form

$$\sum_1^n (\phi(t) - \phi(u_j))(\psi(u_j) - \psi(u_{j-1})) ,$$

where $s = u_0 < \cdots < u_n$; q.e.d.

From now on we assume that Ω is the space of continuous functions on $[0, \infty)$ with values in R^d. We denote by $x(t, \omega)$ the value of the function ω at time t. Ω will be viewed as a complete separable metric space with uniform convergence on compact time intervals defining the basic topology. $\mathcal{M}_t^s$ will be the σ-field generated by $x(u, \omega)$ for $s \leqq u \leqq t$. If $s = 0$, $\mathcal{M}_t^s$ will be denoted by $\mathcal{M}_t$. If $t = \infty$, $\mathcal{M}_t^s$ will be denoted by $\mathcal{M}^s$.

THEOREM 2.1. *Let* $a : [t_0 , \infty) \times \Omega \to S_N$[1] *and* $b : [t_0 , \infty) \times \Omega \to R^N$ *be bounded non-anticipating functions. Define*

$$K_u = \tfrac{1}{2} \sum_{i,j=1}^N a_{ij}(u) \frac{\partial^2}{\partial x_i\, \partial x_j} + \sum_{i=1}^N b_i(u) \frac{\partial}{\partial x_i} .$$

Let P *be a probability measure on* $\langle \Omega, \mathcal{M}^{t_0} \rangle$, *and suppose that* $\xi : [t_0 , \infty) \times \Omega \to R^1$ *is a continuous, non-decreasing non-anticipating function satisfying*:

(i) $\int_{t_0}^t \langle \theta, a(u)\theta \rangle\, d\xi(u)$ *is bounded for each* $t \geqq t_0$ *and* $\theta \in R^N$,

(ii) $E^P[e^{\lambda(\xi(t)-\xi(t_0))}] < \infty$ *for all* $\lambda \geqq 0$ *and* $t \geqq t_0$.

---

[1] S_N denotes the class of symmetric, non-negative definite $N \times N$-matrices. S_N^+ is the subclass of S_N consisting of positive definite matrices.

Finally, let $\alpha : [t_0, \infty) \times \Omega \to R^N$ *be a continuous, non-anticipating function. Then the following are equivalent:*

(a) *for all* $f \in C_0^\infty(R^N)$,[2]

$$f(\alpha(t)) - \int_{t_0}^{t} K_u f(\alpha(u))\, d\xi(u)$$

is a P-martingale,

(b) *for all* $f \in C_b^{1,2}([t_0, \infty] \times R^N)$,[3]

$$f(t, \alpha(t)) - \int_{t_0}^{t} f_u(u, \alpha(u))\, du - \int_{t_0}^{t} K_u f(u, \alpha(u))\, d\xi(u) \quad {}^4$$

is a P-martingale,

(c) *for all* $f \in C_b^{1,2}([t_0, \infty) \times R^N)$ *which are uniformly positive,*

$$f(t, \alpha(t)) \exp\left\{-\int_{t_0}^{t} \frac{f_u(u, \alpha(u))}{f(u, \alpha(u))}\, du - \int_{t_0}^{t} \frac{K_u f(u, \alpha(u))}{f(u, \alpha(u))}\, d\xi(u)\right\}$$

is a P-martingale,

(d) *for all* $g \in C_b^{1,2}([t_0, \infty) \times R^N$ *and* $\theta \in R^N$,

$$\begin{aligned}\exp\Big\{g(t, \alpha(t)) - g(t_0, \alpha(t_0)) + \langle \theta, \alpha(t) - \alpha(t_0)\rangle - \int_{t_0}^{t} g_u(u, \alpha(u))\, du \\ - \tfrac{1}{2}\int_{t_0}^{t} \langle \nabla g(u, \alpha(u)), a(u)\nabla g(u, \alpha(u))\rangle\, d\xi(u) \\ - \int_{t_0}^{t} \langle \theta, a(u)\nabla g(u, \alpha(u))\rangle\, d\xi(u) \\ - \tfrac{1}{2}\int_{t_0}^{t} \langle \theta, a(u)\theta\rangle\, d\xi(u) - \int_{t_0}^{t} \langle \theta, b(u)\rangle\, d\xi(u)\Big\}\end{aligned}$$

is a P-martingale,

(e) *for all* $\theta \in R^N$, *the function* $X_\theta^{t_0}(t)$ *defined by*

$$X_\theta^{t_0}(t) = \exp\left\{\langle \theta, \alpha(t) - \alpha(t_0)\rangle - \tfrac{1}{2}\int_{t_0}^{t} \langle \theta, a(u)\theta\rangle\, d\xi(u) - \int_{t_0}^{t} \langle \theta, b(u)\rangle\, d\xi(u)\right\}$$

is a P-martingale,

---

[2] $C_0^\infty(R^N)$ is the class of infinitely differentiable functions on R^N having compact support.

[3] $C_b^{1,2}([t_0, \infty) \times R^N)$ denotes the class of functions on $[t_0, \infty) \times R^N$ which together with their first t-derivative and first two x-derivatives are bounded and continuous.

[4] f_u denotes $\partial f/\partial u$.

(f) *for all* $\theta \in R^N$, *the function* $X_{i\theta}^{t_0}$ *defined by*

$$X_{i\theta}^{t_0}(t) = \exp\left\{i\langle\theta, \alpha(t) - \alpha(t_0)\rangle + \tfrac{1}{2}\int_{t_0}^{t}\langle\theta, a(u)\theta\rangle\, d\xi(u) - i\int_{t_0}^{t}\langle\theta, b(u)\rangle\, d\xi(u)\right\}$$

is a P-martingale.

Moreover, (e) *implies that*

$$P\left(\sup_{t_0 \leq t \leq T}\left|\alpha(t) - \alpha(t_0) - \int_{t_0}^{t} b(u)\, d\xi(u)\right| \geqq R\right) \leqq 2Ne^{-R^2/2A_T N^{1/2}}, \tag{2.1}$$

where

$$A_T = \sup_{\substack{\theta \in R^N \\ \omega \in \Omega}} \frac{1}{|\theta|^2}\int_{t_0}^{T}\langle\theta, a(u, \omega)\theta\rangle\, d\xi(u)\,.$$

Proof: We may take $t_0 = 0$. Assume (a) and observe that it suffices to prove (b) for $f \in C_0^\infty([0, \infty) \times R^N)$. Given such an f, we have

$$\begin{aligned} E[f(t, \alpha(t)) - f(s, \alpha(s)) \mid \mathcal{M}_s] &= E\left[\int_s^t f_u(u, \alpha(u))\, du + \int_s^t K_u f(u, \alpha(u))\, d\xi(u) \mid \mathcal{M}_s\right] \\ &\quad + E\left[\int_s^t (f_u(u, \alpha(t)) - f_u(u, \alpha(u)))\, du \mid \mathcal{M}_s\right] \\ &\quad + E\left[\int_s^t (K_u f(s, \alpha(u)) - K_u f(u, \alpha(u)))\, d\xi(u) \mid \mathcal{M}_s\right] \\ &= I_1 + I_2 + I_3\,. \end{aligned}$$

But

$$\begin{aligned} I_2 &= E\left[\int_s^t du \int_u^t d\xi(v) K_v f_u(u, \alpha(v)) \mid \mathcal{M}_s\right] \\ &= E\left[\int_s^t d\xi(u) \int_s^u dv K_u f_v(v, \alpha(u)) \mid \mathcal{M}_s\right] \\ &= E\left[\int_s^t K_u f(u, \alpha(u)) - K_u f(s, \alpha(u)) \mid \mathcal{M}_s\right] \\ &= -I_3\,. \end{aligned}$$

This proves that $I_1 = E[f(t, \alpha(t)) - f(s, \alpha(s)) \mid \mathcal{M}_s^0]$, and therefore

$$f(t, \alpha(t)) - \int_0^t f_u(u, \alpha(u))\, du - \int_0^t K_u f(u, \alpha(u))\, d\xi(u)$$

is a P-martingale. Hence (a) implies (b).

That (b) implies (c) follows from Lemma 2.1. Indeed, simply take

$$\phi(t) = f(t, \alpha(t)) - \int_0^t f_u(u, \alpha(u))\, du - \int_0^t K_u f(u, \alpha(u))\, d\xi(u)$$

and

$$\psi(t) = \exp\left\{-\int_0^t \frac{f_u(u, \alpha(u))}{f(u, \alpha(u))}\, du - \int_0^t \frac{K_u f(u, \alpha(u))}{f(u, \alpha(u))}\, d\xi(u)\right\},$$

where $f \in C_b^{1,2}([0, \infty) \times R^N)$ is uniformly positive.

Assume (c) and let $\theta \in R^N$ and $g \in C_b^{1,2}([0, \infty) \times R^N)$ be given. Choose $\phi'_n \in C_0^\infty(R')$ so that $0 \leqq \phi'_n \leqq 1$, $\phi'_n(s) = 1$ for $|s| \leqq n$, and $\phi'_n(s) = 0$ for $|s| \geqq n + 1$. Take $\phi_n(t) = \int_0^t \phi'_n(s)\, ds$, define Φ_n to be the R^N-valued function whose i-th component is $\phi_n(x_i)$, and set

$$f_n(t, x) = \exp\{g(t, x) + \langle \theta, \Phi_n(x)\rangle\}.$$

Then $f_n \in C_b^{1,2}([0, \infty) \times R^N)$, f_n is uniformly positive, and

$$\frac{K_u f_n(u, x)}{f_n(u, x)} = K_u g + \tfrac{1}{2}\langle \nabla g, a\nabla g\rangle + \langle \theta\Phi'_n, a\nabla g\rangle + \tfrac{1}{2}\langle \theta\Phi'_n, a\Phi'_n\rangle + \langle \theta\Phi'_n, b\rangle,$$

$$\frac{K_u f_n^2(u, x)}{f_n^2(u, x)} = 2K_u g + 2\langle \nabla g, a\nabla g\rangle + 4\langle \theta\Phi'_n, a\nabla g\rangle + 2\langle \theta\Phi'_n, a\theta\Phi'_n\rangle + 2\langle \theta\Phi'_n, b\rangle,$$

where $\theta\Phi'_n(x)$ is the N-vector whose i-th component is $\theta_i\Phi'_n(x_i)$. Thus,

$$\begin{aligned}\int_0^t \frac{(f_n^2)_u(u, \alpha(u))}{f_n^2(u, \alpha(u))}\, du &+ \int_0^t \frac{K_u f_n^2(u, \alpha(u))}{f_n^2(u, \alpha(u))}\, d\xi(u) \\ &- 2\int_0^t \frac{(f_n)_u(u, \alpha(u))}{f_n(u, \alpha(u))}\, du - \int_0^t \frac{K_u f_n(u, \alpha(u))}{f_n(u, \alpha(u))}\, d\xi(u) \\ &= \int_0^t \langle \nabla g, a\nabla g\rangle\, d\xi(u) + 2\int_0^t \langle \theta\Phi'_n, a\nabla g\rangle\, d\xi(u) + \int_0^t \langle \theta\Phi'_n, a\theta\Phi'_n\rangle\, d\xi(u) \\ &\leqq C,\end{aligned}$$

where C is independent of n. Using this fact, we now see that

$$E\left[\left(\frac{f_n(t,\alpha(t))}{f_n(0,\alpha(0))}\exp\left\{-\int_0^t\frac{(f_n)_u(u,\alpha(u))}{f_n(u,\alpha(u))}\,du-\int_0^t\frac{K_uf_n(u,\alpha(u))}{f_n(u,\alpha(u))}\,d\xi(u)\right\}\right)^2\right]$$
$$\leqq e^C E\left[\frac{f_n^2(t,\alpha(t))}{f_n^2(0,\alpha(0))}\exp\left\{-\int_0^t\frac{(f_n^2)_u(u,\alpha(u))}{f_n^2(u,\alpha(u))}\,du-\int_0^t\frac{K_uf_n^2(u,\alpha(u))}{f_n^2(u,\alpha(u))}\,d\xi(u)\right\}\right]$$
$$= e^C,$$

because the integrand in the last expression is a P-martingale. It follows that the sequence

$$X_n(t)=\frac{f_n(t,\alpha(u))}{f_n(0,\alpha(0))}\exp\left\{-\int_0^t\frac{(f_n)_u(u,\alpha(u))}{f_n(u,\alpha(u))}\,du-\int_0^t\frac{K_uf_n(u,\alpha(u))}{f_n(u,\alpha(u))}\,d\xi(u)\right\}$$

is uniformly P-integrable. Moreover, $X_n(t)$ tends to the expression in (d) pointwise, and therefore (d) follows from (c).

Obviously (e) is just the special case of (d) when $g\equiv 0$. If (e) obtains, then, by the argument used in Lemma 3.1 of [8], we get the estimate (2.1). From this it is clear that $E[X_\theta^0(t)Y]$ is an analytic function of $\theta\in C^N$ for each bounded measurable function Y. Thus the equality

$$E[X_\theta^0(t)\mid\mathscr{M}_s^0]=X_\theta^0(s)\qquad\text{a.s. } P$$

for $\theta\in R^N$ extends to all $\theta\in C^N$.

It remains to show that (f) implies (a). To that end, take $\phi(t)=X_{i\theta}^0(t)$ and

$$\psi(t)=\exp\left\{-\tfrac{1}{2}\int_0^t\langle\theta,a(u)\theta\rangle\,d\xi(u)+i\int_0^t\langle\theta,b(u)\rangle\,d\xi(u)\right\}.$$

Applying Lemma 2.1 to ϕ and ψ, we see that

$$\exp\{i\langle\theta,\alpha(t)-\alpha(0)\rangle\}+\int_0^t(\tfrac{1}{2}\langle\theta,a(u)\theta\rangle-i\langle\theta,b(u)\rangle)\exp\{i\langle\theta,\alpha(t)-\alpha(0)\rangle\}\,d\xi(u)$$

is a P-martingale for all $\theta\in R^N$. Hence,

$$f(\alpha(t))-\int_0^t K_uf(\alpha(u))\,d\xi(u)$$

is a P-martingale for all f which are finite linear combinations of functions of the form $\exp\{i\langle\theta,x\rangle\}$, $\theta\in R^N$. It is easy to pass from this to (a); q.e.d.

We continue using the notation of Theorem 2.1 with $t_0=0$. Assume that

$\xi(t)$ is such that

$$\int_0^t \langle \theta, a(u)\theta \rangle \, d\xi(u) = \int_0^t \langle \theta, a(u)\theta \rangle \, du \,, \qquad t \geqq 0 \,, \quad \theta \in R^1 \,, \tag{2.2}$$

and let P satisfy (a). From (e) we see that

$$X_\theta(t) = \exp\left\{\langle \theta, \bar{\alpha}(t) - \bar{\alpha}(0) \rangle - \tfrac{1}{2}\int_0^t \langle \theta, a(u)\theta \rangle \, du\right\}$$

is a P-martingale for all $\theta \in R^N$, where

$$\bar{\alpha}(t) = \alpha(t) - \int_0^t b(u) \, d\xi(u) \,.$$

In view of Theorem 3.2 in [8], we can now define a stochastic integral with respect to $\bar{\alpha}(t)$. Indeed, given a non-anticipating function $\theta : [0, \infty) \times \Omega \to R^N$ satisfying $E\left[\int_0^t |\theta(u)|^2 \, du\right] < \infty$, the expression $\int_0^t \langle \theta(u), d\bar{\alpha}(u) \rangle$ is well-defined and enjoys all the properties listed in that theorem (for a summary of these properties see Theorem 2.2 below). We now define stochastic integrals with respect to $\alpha(t)$ by setting

$$\int_0^t \langle \theta(u), d\alpha(u) \rangle = \int_0^t \langle \theta(u), d\bar{\alpha}(u) \rangle + \int_0^t \langle \theta(u), b(u) \rangle \, d\xi(u)$$

when the terms on the right are defined. In particular, $\int_0^t \langle \theta(u), d\alpha(u) \rangle$ is well-defined for bounded non-anticipating $\theta : [0, \infty) \times \Omega \to R^N$.

We want next to develop an Itô formula for $d\alpha(u)$-stochastic integrals. The method which we employed in [8] was to relate $d\alpha(u)$ integrals to Brownian integrals (cf. Theorem 3.3 of [8]). However, that approach breaks down here because we do not want to assume that $a(u)$ is strictly positive definite. Thus, in order to construct a Brownian motion $\beta(t)$ such that $d\bar{\alpha}(t) = a^{1/2}(t) \, d\beta(t)$, we would be forced to go outside our original sample space (cf. Theorem 5.3. in Doob [2]). We could avoid such problems by invoking the extremely general results along these lines of H. Kunita and S. Watanabe [4]. However, the reasoning which seems most consistent with our overall philosophy is the following.

Let $g \in C_b^{1,2}([0, \infty) \times R^N)$, and define $\tilde{\alpha} : [0, \infty) \times \Omega \to R^{N+1}$ by the recipe

$$\tilde{\alpha}_i(t) = \bar{\alpha}_i(t) \,, \qquad 1 \leqq i \leqq N \,,$$

$$\tilde{\alpha}_{N+1}(t) = g(t, \alpha(t)) - \int_0^t g_u(u, \alpha(u)) \, du - \int_0^t K_u \, g(u, \alpha(u)) \, d\xi(u) \,.$$

Define $\tilde{\alpha} : [0, \infty) \times \Omega \to S_{N+1}$ by

$$\begin{aligned} \tilde{\alpha}_{ij}(t) &= a_{ij}(t)\,, && 1 \leqq i,j \leqq N\,, \\ \tilde{\alpha}_{N+1,i}(t) &= \tilde{\alpha}_{i,N+1}(t) = (a(t)\nabla g(t, \alpha(t)))_i\,, && 1 \leqq i \leqq N\,, \\ \tilde{\alpha}_{N+1,N+1}(t) &= \langle \nabla f(t, \alpha(t)), a(t)\nabla f(t, \alpha(t))\rangle\,. \end{aligned}$$

Then, by (e),

$$X_{\tilde{\theta}}(t) = \exp\left\{\langle \tilde{\theta}\, \tilde{\alpha}(t) - \tilde{\alpha}(0)\rangle - \tfrac{1}{2}\int_0^t \langle \tilde{\theta}, \tilde{a}(u)\tilde{\theta}\rangle\, du\right\}$$

is a P-martingale for all $\tilde{\theta} \in R^{N+1}$. Hence we can define $d\tilde{\alpha}(t)$-stochastic integrals, and, by Theorem 3.2 of [8], we know that

$$\exp\left\{\int_0^t \langle \tilde{\theta}(u), d\tilde{\alpha}(u)\rangle - \frac{1}{2}\int_0^t \langle \tilde{\theta}(u), \tilde{a}(u)\tilde{\theta}(u)\rangle\, du\right\}$$

is a P-martingale for bounded non-anticipating $\tilde{\theta} : [0, \infty) \times \Omega \to R^{N+1}$. In particular, if

$$\tilde{\theta}_i(u) = -\lambda \frac{\partial g}{\partial x_i}(u, \alpha(u))$$

for $1 \leqq i \leqq N$ and $\tilde{\theta}_{N+1}(u) = \lambda$, then

$$\exp\left\{\lambda(g(t, \alpha(t)) - g(0, \alpha(0)) - \int_0^t g_u(u, \alpha(u))\, du \right.$$
$$\left. - \int_0^t K_u g(u, \alpha(u))\, d\xi(u) - \int_0^t \langle \nabla g(u, \alpha(u)), d\bar{\alpha}(u)\rangle\right\}$$

is a P-martingale. Since this is true for all $\lambda \in R^1$, we conclude that

$$\begin{aligned} g(t, \alpha(t)) - g(0, \alpha(0)) &= \int_0^t \langle \nabla g(u, \alpha(u)), d\bar{\alpha}(u)\rangle \\ &\quad + \int_0^t g_u(u, \alpha(u))\, du + \int_0^t K_u g(u, \alpha(u))\, d\xi(u)\,. \end{aligned} \tag{2.3}$$

If K_u^0 denotes the second order part of K_u, then (2.3) can be rewritten as

$$\begin{aligned} g(t, \alpha(t)) - g(0, \alpha(0)) &= \int_0^t \langle \nabla g(u, \alpha(u)), d\alpha(u)\rangle \\ &\quad + \int_0^t (g_n + K_u^0 g)(u, \alpha(u))\, du\,. \end{aligned} \tag{2.4}$$

For future reference, we summarize these findings and some of their consequences in the following theorem, using the same notation as in Theorem 2.1.

THEOREM 2.2. *In addition to the assumptions in Theorem 2.1, let us assume that*

$$(2.5)\qquad \int_{t_0}^{t} \langle\theta, a(u)\theta\rangle\, d\xi(u) = \int_{t_0}^{t} \langle\theta, a(u)\theta\rangle\, du \qquad \text{for} \qquad t \geqq t_0\,, \quad \theta \in R^N\,,$$

and that P satisfies any of the equivalent martingale formulations (a)–(f) of that theorem. Then stochastic integrals with respect to $\alpha(t)$ are well-defined for non-anticipating $\theta : [t_0, \infty) \times \Omega \to R^N$ satisfying

$$(2.6)\qquad E\left[\int_{t_0}^{t} |\theta(u)|^2\, du + \int_{t_0}^{t} |\langle\theta(u), b(u)\rangle|\, d\xi(u)\right] < \infty\,.$$

The $d\alpha(t)$-stochastic integrals have the following properties:

(i) $\displaystyle\int_{t_1}^{t_2} \theta(u)\, d\alpha(u) + \int_{t_2}^{t_3} \theta(u)\, d\alpha(u) = \int_{t_1}^{t_3} \theta(u)\, d\alpha(u), \qquad t_0 \leqq t_1 \leqq t_2 \leqq t_3\,,$

(ii) $\displaystyle\int_{t_0}^{t} \theta(u)\, d\alpha(u)$ *is continuous in $t \geqq t_0$ and*

$$\int_{t_0}^{t} \theta(u)\, d\alpha(u) - \int_{t_0}^{t} \langle\theta(u), b(u)\rangle\, d\xi(u)$$

is a P-martingale,

(iii) $$E\left[\left(\int_{t_1}^{t_2} \theta(u)\, d\alpha(u) - \int_{t_1}^{t_2} \langle\theta(u), b(u)\rangle\, d\xi(u)\right)^2\right] = E\left[\int_{t_1}^{t_2} \langle\theta(u), a(u), \theta(u)\rangle\, du\right],$$

(iv) *if θ is bounded, then*

$$X_\theta^{t_0}(t) = \exp\left\{\int_{t_0}^{t} \theta(u)\, d\alpha(u) - \int_{t_0}^{t} \langle\theta(u), b(u)\rangle\, d\xi(u) - \tfrac{1}{2}\int_{t_0}^{t} \langle\theta(u), a(u)\theta(u)\rangle\, du\right\}$$

is a P-martingale,

(v) $\displaystyle\int_{t_1}^{t_2} \theta_1(u)\, d\alpha(u) + \int_{t_1}^{t_2} \theta_2(u)\, d\alpha(u) = \int_{t_1}^{t_2} (\theta_1(u) + \theta_2(u))\, d\alpha(u)\,,$

(vi) *if $g \in C_b^{1,2}([0, \infty) \times R^N)$, then the Itô formula given in (2.4) holds.*

Finally, if τ *is a* t_0*-stopping time*[5] *and* Q_ω *is the regular conditional probability distribution*[6] *of* P *given* $\mathcal{M}_\tau^{t_0}$ *(cf. Theorem* 2.1 *of* [8]), *then there is an* $\mathcal{N} \in \mathcal{M}_\tau^{t_0}$ *such that* $P(\mathcal{N}) = 0$ *and, for* $\omega \notin \mathcal{N}$, Q_ω *satisfies* (a) *of Theorem* 2.1 *above with* $\tau(\omega)$ *replacing* t_0.

Proof: The only assertion that has not been explained is the last. However, since (a) is equivalent to (e) in Theorem 2.1, this assertion follows from Theorem 3.1 of [8].

THEOREM 2.3. *Let* $a : [0, \infty) \times R^d \to S_d^+$ *be bounded and continuous, and let* $b : [0, \infty) \times R^d \to R^d$ *and* $\gamma : [0, \infty) \times R^d \to R^d$ *be bounded and measurable. Suppose* P *is a probability measure on* $(\Omega, \mathcal{M}^{t_0})$ *and that* $\xi : [0, \infty) \times \Omega \to R^1$ *is a continuous, non-decreasing, non-anticipating function which satisfies*

$$\text{(i)}\ \ \xi(t) - \xi(t_0) = \int_{t_0}^t \chi_{\partial G}(x(s))\, d\xi(s) \qquad \textit{for} \qquad t \geqq t_0\,,$$

$$\text{(ii)}\ \ E[\exp\{\lambda(\xi(t) - \xi(t_0))\}] < \infty \qquad \textit{for} \qquad t \geqq t_0 \qquad \textit{and} \qquad \lambda \geqq 0\,,$$

where ∂G *is the boundary of some open region* $G \subseteq R^d$. *Furthermore, assume that, for* $f \in C_0^\infty(R^d)$,

$$f(x(t)) - \int_{t_0}^t \chi_G L_u f(x(u))\, du - \int_{t_0}^t \langle \gamma(u, x(u)), \nabla f(x(u))\rangle\, d\xi(u)$$

is a P-martingale, where

$$L_u = \tfrac{1}{2} \sum_{i,j=1}^d a_{ij}(u, x) \frac{\partial^2}{\partial x_i\, \partial x_j} + \sum_{i=1}^d b_i(u, x) \frac{\partial}{\partial x_i}\,.$$

Finally, let τ *be a bounded* t_0*-stopping time, define*

$$\tau' = \inf\{t \geqq \tau : x(t) \notin G\}\,,$$

and let P_ω *be the r.c.p.d. of* P *given* $\mathcal{M}_\tau^{t_0}$. *Then there is an* $\mathcal{N} \in \mathcal{M}_\tau^{t_0}$ *such that* $P(\mathcal{N}) = 0$ *and, for* $\omega \notin \mathcal{N}$, P_ω *equals* $Q_{\tau(\omega), x(\tau(\omega), \omega)}$ *on* $\mathcal{M}_{\tau'_\omega}^{\tau(\omega)}$, *where* $Q_{s,x}$ *is the unique solution to the martingale problem for* a *and* b *starting from* x *at time* s,[7] *and* $\tau'_\omega = \inf\{t \geqq \tau(\omega) : x(t) \notin G\}$.

Proof: Choose $T \geqq \tau$ and let P^T be the probability measure on $\langle \Omega, \mathcal{M}^{t_0} \rangle$ which equals P on $\mathcal{M}_{\tau'}^{t_0}$ and whose r.c.p.d. given $\mathcal{M}_{\tau' \wedge T}^{t_0}$ is $Q_{\tau' \wedge T, x(\tau' \wedge T)}$ (cf. Lemma 3.6 of [8]). Let P_ω^T be the r.c.p.d. of P^T given $\mathcal{M}_\tau^{t_0}$. Then, by

---

[5] A t_0-stopping time is a function $\tau : \Omega \to [t_0, \infty)$ such that $\{\tau \leqq t\} \in \mathcal{M}_t^{t_0}$ for $t \geqq t_0$.

[6] In the future we shall use r.c.p.d.

[7] This terminology is explained in [8].

Theorem 2.1 above and Theorem 3.6 of [8], for almost all ω, P^T_ω solves the martingale problem for a and b starting from $x(\tau(\omega), \omega)$ at time $\tau(\omega)$. Hence, $P^T_\omega = Q_{\tau(\omega), x(\tau(\omega), \omega)}$ on $\mathscr{M}^{\tau(\omega)}$ a.s. P^T. Since $P_\omega = P^T_\omega$ a.s. P, we have $P_\omega = Q_{\tau(\omega), x(\tau(\omega), \omega)}$ on $\mathscr{M}^{\tau(\omega)}_{\tau'(\omega)\wedge T}$. By letting $T \uparrow \infty$, we get our result.

COROLLARY 2.1. *(The notation is the same as that in Theorem 2.3.) If*

$$f \in W^{1,2}_p([0, \infty) \times R^d)\,,^8$$

where $p > d + 2$, *and if* $f_u + L^0_u f = g \in B([0, \infty) \times G)$,[9] *where* L^0_u *is the principal part of* L_u, *then for* $T \geqq \tau$

$$f(\tau' \wedge T, x(\tau' \wedge T)) - f(\tau, x(\tau)) = \int_\tau^{\tau'\wedge T} g(u, x(u))\, du + \int_\tau^{\tau'\wedge T} \nabla_x f(u, x(u))\, dx(u)$$

a.s. P.

Proof: In view of Theorem 2.3, it suffices to prove that

$$f(t, x(t)) - f(s, x(s)) = \int_s^t g(u, x(u))\, du + \int_s^t \nabla_x f(u, x(u))\, dx(u) \qquad a.s.\ Q_{s,x}\,.$$

This is done in the same way as Theorem 11.2 of [9] was proved.

Given an open set G in R^d and a probability measure P on $\langle \Omega, \mathscr{M}^{t_0} \rangle$, we say that a non-anticipating function $X: [t_0, \infty) \times \Omega \to R^1$ is a *local P-martingale in* G if, for every t_0-stopping time τ,

$$X(t \wedge \tau') - X(t \wedge \tau)$$

is a P-martingale, where $\tau' = \inf\{t \geqq \tau : x(t) \notin G\}$.

We now introduce the general set-up with which we shall be concerned in the rest of this paper. Let G be a non-empty open set in R^d. We assume that there is associated with G a function ϕ, called the defining function of G, such that

(i) $\phi \in C^2_b(R^d)$,

(ii) $G = \{x \in R^d : \phi(x) > 0\}$ and $\partial G = \{x \in R^d : \phi(x) = 0\}$,

(iii) $|\nabla \phi(x)| \geqq 1$ for all $x \in \partial G$.

---

[8] The space $W^{1,2}_p([0, \infty) \times R^d)$ is the completion of $C^\infty_0([0, \infty) \times R^d)$ with respect to the norm

$$|||f||| = \|f\|_p + \|f_t\|_p + \sum_{i=1}^d \|f_{x_i}\|_p + \sum_{i,j=1}^d \|f_{x_i x_j}\|_p,$$

where $\|\cdot\|_p$ is the ordinary L_p norm.

[9] $B([0, \infty) \times G)$ is the class of bounded measurable functions on $[0, \infty) \times G$.

The following quantities will also be given

(i′) $a : [0, \infty) \times G \to S_d^+$ which is bounded and continuous,

(ii′) $b : [0, \infty) \times G \to R^d$ which is bounded and measurable,

(iii′) $\gamma : [0, \infty) \times \partial G \to R^d$ which is bounded, continuous, and satisfies $\langle \gamma(t, x), \nabla\phi(x)\rangle \geqq \beta > 0$ for $t \geqq 0$ and $x \in \partial G$,

(iv′) $\rho : [0, \infty) \times \partial G \to [0, \infty)$ which is bounded and continuous.

Define

$$L_u = \frac{1}{2} \sum_{i,j=1}^{d} a_{ij}(u, x) \frac{\partial^2}{\partial x_i \partial x_j} + \sum_{i=1}^{d} b_i(u) \frac{\partial}{\partial x_i}$$

and

$$J_u = \sum_{i=1}^{d} \gamma_i(u, x) \frac{\partial}{\partial x_i}.$$

We say that a probability measure P on $\langle \Omega, \mathscr{M}^{t_0}\rangle$ solves the *sub-martingale problem on* G *for coefficients* a, b, γ *and* ρ *if*

(I) $\qquad P(x(t) \in \bar{G}) = 1, \qquad t \geqq t_0,$

(II) $f(t, x(t)) - \int_{t_0}^{t} [\chi(f_u {}_G + L_u f)](u, x(u))\, du$ is a P-submartingale for any $f \in C_0^{1,2}([0, \infty) \times R^d)$ satisfying

$$\rho f_t + J_t f \geqq 0 \quad \text{on} \quad [t_0, \infty) \times \partial G.$$

Unless it is stated to the contrary, the P in the following is assumed to be such a solution.

LEMMA 2.2. *If* $f \in C_b^{1,2}([0, \infty) \times R^d)$, *then*

$$X_f(t) = f(t, x(t)) - \int_{t_0}^{t} \chi_G(f_u + L_u f)(u, x(u))\, du$$

is a local P-martingale on G. If $\rho f_t + J_t f \geqq 0$ *on* $[t_0, \infty) \times \partial G$, $X_f(t)$ *is a P-submartingale.*

PROOF. Suppose f has compact support in $[t_0, \infty) \times G$. Then

$$\rho f_t + J_t f = 0$$

on $[t_0, \infty) \times \partial G$, and so $X_f(t)$ is a P-martingale. In general, choose a non-decreasing sequence of open sets G_n such that

$$\bar{G}_n \subset G_{n+1} \subset \bar{G}_{n+1} \subset\subset G$$

and $G_n \uparrow G$.[10] Choose $\eta_n \in C_0^\infty([0, \infty) \times R^d)$ so that $\eta_n = 1$ on $[0, t_0 + n] \times G_n$ and $\eta_n \equiv 0$ off $[0, t_0 + n + 1] \times G_{n+1}$. Take $f_n = \eta_n f$. Then $X_{f_n}(t)$ is a P-martingale. Given a t_0-stopping time τ, let

$$\tau_n = \tau \wedge (t_0 + n) \qquad \text{and} \qquad \tau'_n = \inf\{t \geqq \tau_n : x(t) \in G_n\} \wedge (t_0 + n) .$$

Then

$$X_{f_n}(t \wedge \tau'_n) - X_{f_n}(t \wedge \tau_n)$$

is a P-martingale. But

$$X_f(t \wedge \tau'_n) - X_f(t \wedge \tau_n) = X_{f_n}(t \wedge \tau'_n) - X_{f_n}(t \wedge \tau_n) ,$$

$\tau_n \wedge t \uparrow \tau \wedge t$, and τ'_n increases to

$$\tau' \wedge t = \inf\{s \geqq \tau : x(s) \in G\} \wedge t .$$

Hence, $X_f(t \wedge \tau') - X_f(t \wedge \tau)$ is a P-martingale.

Now suppose that $f \in C_b^{1,2}([0, \infty) \times R^d)$ and that $f_t + J_t f \geqq 0$ on $[t_0, \infty) \times \partial G$. Assume that $f(t, \cdot) \equiv 0$ for $t \geqq T$, where $T > t_0$. For each $N \geqq 1$, choose $\eta_N \in C_0^\infty(R^d)$ such that $0 \leqq \eta_N \leqq 1$, $\eta_N = 1$ on $\{x : |x| \leqq N\}$, $|J_t \eta_N| \leqq 1/N$ on $[0, \infty) \times \partial G$, and all derivatives of η_N up to the second order are uniformly bounded. Let $\bar{\eta}_N$ be an element of $C_0^\infty(R^d)$ having the properties

$$0 \leqq \bar{\eta}_N \leqq 1 ,$$

$$\bar{\eta}_N \equiv 1 \qquad \text{on} \qquad \text{supp.}\,(\eta_N) ,$$

and all derivatives up to the second order are uniformly bounded. Set

$$f_N = \eta_N \cdot f + \frac{\|f\|}{N} \bar{\eta}_N \cdot \phi .$$

Then f_N is in $C_0^{1,2}([0, \infty) \times R^d)$ and $(\rho(\partial/\partial t) + J_t) f_N \geqq 0$ on $[0, \infty) \times \partial G$. Hence, X_{f_N} is a P-submartingale. Clearly, $X_{f_N}(t) \to X_f(t)$ boundedly, and so $X_f(t)$ is a P-submartingale. Finally, we can drop the assumption that $f(t, \cdot) = 0$ for $t > T$, because we can do the same truncation argument in time; q.e.d.

---

10 $K \subset\subset U$ means that K is a compact subset of U.

Suppose that $f \in C_b^{1,2}([0, \infty) \times R^d)$ and that $\rho f_t + J_t f \geqq 0$ on $[t_0, \infty) \times \partial G$. Then $X_f(t)$ is a locally bounded, continuous P-submartingale. Hence, by the Doob-Meyer decomposition theorem (cf. [10]), there exists an integrable, non-decreasing, non-anticipating continuous function $\xi_f : [t_0, \infty) \times \Omega \to [0, \infty)$ such that $\xi_f(t_0) = 0$ and $X_f(t) - \xi_f(t)$ is a P-martingale. In general, if $f \in C_b^{1,2}([0, \infty) \times R^d)$, we can find an α such that

$$\rho \tilde{f}_t + J_t \tilde{f} \geqq 0 \qquad \text{on} \qquad [t_0, \infty) \times \partial G,$$

where $\tilde{f} = f + \alpha\phi$. Hence we can choose a $\xi_{\tilde{f}}$ for $\tilde{f}$. If we let $\xi_f = \xi_{\tilde{f}} - \alpha\xi_\phi$, then we see that $\xi_f(t)$ is an non-anticipating continuous function of bounded variation such that $\xi_f(t_0) = 0$, $E[|\xi_f|(t)] < \infty$ for $t \geqq t_0$, and $X_f(t) - \xi_f(t)$ is a P-martingale. These remarks show that $C_b^{1,2}([0, \infty) \times R^d)$ is contained in the class F described below.

The class F consists of those functions f which satisfy:

(i) $f \in C_b([t_0, \infty) \times R^d) \cap C_b^{1,1}([t_0, \infty) \times \partial G)$, [11]

(ii) there is a function $Kf \in B([0, \infty) \times G)$ such that

$$X_f(t) = f(t, x(t)) - \int_{t_0}^t \chi_G \, Kf(u, x(u)) \, du$$

is a local P-martingale in G,

(iii) there is a continuous non-anticipating function $\xi_f : [t_0, \infty) \times \Omega \to R$ such that

(a) $\xi_f(t_0) = 0$, $\xi_f(t)$ is of locally bounded variation, and $E[|\xi_f|(t)] < \infty$ for $t \geqq t_0$,

(b) $X_f(t) - \xi_f(t)$ is a P-martingale,

(iv) if $g \in C_b^{1,2}([0, \infty) \times R^d)$, then $\tilde{f} = f + g$, satisfies (i), (ii) and (iii), and if $\rho \tilde{f}_t + J_t \tilde{f} \geqq 0$ on $[t_0, \infty) \times \partial G$, then $\xi_{\tilde{f}}(t)$ can be chosen to be non-decreasing.

LEMMA 2.3. *For each $f \in F$, there is at most one ξ_f satisfying* (iii). *Moreover,*

$$\int_{t_0}^{t_1} \chi_G(x(u)) \, d\,|\xi_f|\,(u) = 0 .$$

Proof: Suppose ξ_1 and ξ_2 satisfy (iii) for some $f \in F$. Then $\eta = \xi_1 - \xi_2$

[11] If $f \in C_b^{1,2}[(t_0, \infty) \times \partial G)$, we mean that f_t and f_{x_i}, $1 \leqq i \leqq d$, exist and are continuous on $[t_0, \infty) \times \partial G$.

is an integrable, continuous martingale with paths of bounded variation. Hence, by an elementary argument, $\eta(t)$ is almost surely constant, and therefore almost surely zero.

Finally, let $f \in F$. We must show that $\int_{t_0}^{t_1} \chi_G(x(u))\, d\,|\xi_f|\,(u) = 0$. Since $\xi_f(t) = \xi_{f+\alpha\phi}(t) - \xi_{\alpha\phi}(t)$, by choosing large enough α, it suffices to treat the case when $\rho f_t + J_t f \geqq 0$ on $[t_0, \infty) \times \partial G$. In this case $\xi_f(t)$ is non-decreasing and so we need only show that $\int_{t_0}^{t_1} \chi_K(x(u))\, d\xi_f(u) = 0$ for all compact $K \subset G$. For such K, define

$$\tau_0 = \inf\{t \geqq t_0 : x(t) \in K\} \wedge t_1\,,$$

$$\tau_{2n+1} = \inf\{t \geqq \tau_{2n} : x(t) \in \partial U\} \wedge t_1\,,$$

$$\tau_{2n} = \inf\{t \geqq \tau_{2n-1} : x(t) \in K\} \wedge t_1\,.$$

Then $X_f(\tau_{2n+1} \wedge t) - X_f(\tau_{2n} \wedge t)$ is a P-martingale, and therefore so is $\xi_f(\tau_{2n+1} \wedge t) - \xi_f(\tau_{2n} \wedge t)$. Thus, $\xi_f(\tau_{2n+1} \wedge t) = \xi_f(\tau_{2n} \wedge t)$ a.s. P, and from this we get

$$\int_{t_0}^{t_1} \chi_K(x(u))\, d\xi_f(u) = 0 \qquad \text{a.s. } P\,.$$

LEMMA 2.4. *Take $t_0 \leqq a < b$. If $f \in F$ and U is an open neighborhood of a point $x \in \partial G$ such that $f \equiv 0$ on $(a, b) \times U$, then*

$$\int_a^b \chi_U(x(u))\, d\,|\xi_f|\,(u) = 0\,.$$

Proof: Let $K \subset\subset U$ and take $a < c < d < b$. Define

$$\tau_0 = \inf\{t \geqq c : x(t) \in K\} \wedge d\,,$$

$$\tau_{2n+1} = \inf\{t \geqq \tau_{2n} : x(t) \in \partial U\} \wedge d\,,$$

$$\tau_{2n} = \inf\{t \geqq \tau_{2n-1} : x(t) \in K\} \wedge d\,.$$

Then,

$$X_f(t \wedge \tau_{2n+1}) - X_f(t \wedge \tau_{2n})$$
$$= \int_{t\wedge\tau_{2n}}^{t\wedge\tau_{2n+1}} \chi_G(x(u))\, Kf(u, x(u))\, du - (\xi_f(t \wedge \tau_{2n+1}) - \xi_f(t \wedge \tau_{2n}))$$

is a continuous integrable P-martingale of bounded variation. Hence,

$$\int_{t\wedge\tau_{2n}}^{t\wedge\tau_{2n+1}} \chi_G(x(u))\, Kf(u, x(u))\, du = -(\xi_f(t \wedge \tau_{2n}) - \xi_f(t \wedge \tau_{2n+1}))\,.$$

Since $d\xi_f(t)$ is supported on $\{t : x(t) \in \partial G\}$, it follows that $\xi_f(t \wedge \tau_{2n+1}) = \xi_f(\tau_{2n})$ for $t \geqq \tau_{2n}$. Therefore,

$$\int_c^d \chi_K(x(u))\, d\xi_f(u) = \sum_0^\infty \int_{\tau_{2n}}^{\tau_{2n+1}} \chi_K(x(u))\, d\xi_f(u) = 0\,.$$

LEMMA 2.5. *Take $t_0 \leqq a < b$. Let $f \in F$ and let U be a neighborhood of a point $x \in \partial G$ such that $\rho f_t + J_t f \geqq 0$ on $(a, b) \times (U \cap \partial G)$. Then*

$$\int_a^b \chi_U(x(u))\, d\xi_f(u) \geqq 0\,.$$

Proof: Let $a < c < d < b$ and let N be an open set such that $\bar{N} \subset\subset U$. Choose $\eta \in C_b^\infty([0, \infty) \times R^d)$ such that $0 \leqq \eta \leqq 1$, $\eta = 0$ on $(c, d) \times N$, and $\eta = 1$ off $(a, b) \times U$. Define $\tilde{f} = f + \alpha\eta \cdot \phi$. Then, for larger α, $\rho\tilde{f} + J_t\tilde{f} \geqq 0$ on $[t_0, \infty) \times \partial G$ and $\tilde{f} = f$ on $(c, d) \times N$. Thus

$$\int_c^d \chi_N(x(u))\, d\xi_f(u) = \int_c^d \chi_N(x(u))\, d\xi_{\tilde{f}}(u) \geqq 0\,.$$

THEOREM 2.4. *There is a unique, continuous, non-decreasing, non-anticipating function $\xi_0 : [t_0, \infty) \times \Omega \to [0, \infty)$ such that $\xi_0(t_0) = 0$, $E[\xi_0(t)] < \infty$,*

$$\xi_0(t) = \int_{t_0}^t \chi_{\partial G}(x(u))\, d\xi_0(u)$$

and

$$f(t, x(t)) - \int_{t_0}^t \chi_G\, Kf(u, x(u))\, du - \int_{t_0}^t (\rho f_u + J_u f)(u, x(u))\, d\xi_0(u)$$

is a P-martingale for all $f \in F$.

Proof: We first show that if $f \in F$, then $d\xi_f(t)$ is absolutely continuous with respect to $d\xi_\phi(t)$. Indeed, let $\tilde{f} = \alpha\phi - f$. Then, for large α, $\rho\tilde{f}_t + J_t\tilde{f} \geqq 0$ on $[t_0, \infty) \times \partial G$, and so $\alpha d\xi_\phi(t) \geqq d\xi_f(t) \geqq -\alpha d\xi_\phi(t)$. Let $d\xi_f(t)/d\xi_\phi(t) = \alpha(t)$. Given $s \in (t_0, \infty)$ and $x \in \partial G$, let

$$\beta = \frac{(\rho f_s + J_s f)(s, x)}{\langle \gamma(s, x), \nabla\phi(x)\rangle}\,.$$

If $\varepsilon > 0$, choose $s \in (a, b) \subset [t_0, \infty)$ and an open set U containing x such that

$$(\beta - \varepsilon)\langle \gamma, \nabla\phi\rangle \leqq (\rho f_u + J_u f) \leqq (\beta + \varepsilon)\langle \gamma, \nabla\phi\rangle$$

in $(a, b) \times U$. Then

$$(\beta - \varepsilon)\int_{a'}^{b'} \chi_U(x(u))\, d\xi_\phi(u) \leqq \int_{a'}^{b'} \chi_U(x(u))\alpha(u)\, d\xi_\phi(u)$$
$$\leqq (\beta + \varepsilon)\int_{a'}^{b'} \chi_U(x(u))\, d\xi_\phi(u)$$

for $(a', b') \subset (a, b)$ and $U' \subset U$. Hence,

$$(\beta - \varepsilon)\chi_U(x(u)) \leqq \chi_U(x(u))\alpha(u) \leqq (\beta + \varepsilon)\chi_U(x(u)) .$$

It follows that

$$\alpha(u) = \frac{\rho f_u + J_u f}{\langle \gamma, \nabla\phi\rangle} .$$

We now define

$$\xi_0(t) = \int_{t_0}^{t} \frac{1}{\langle \gamma, \nabla\phi\rangle}\, d\xi_\phi(u) .$$

Clearly $\xi_0(t)$ has the desired properties. Finally, if ξ is any other function with these properties, then

$$\int_{t_0}^{t} \langle \gamma, \nabla\phi\rangle\, d\xi_0(u) = \xi_\phi(t) = \int_{t_0}^{t} \langle \gamma, \nabla\phi\rangle\, d\xi(u) ,$$

and so $\xi_0(t) = \xi(t)$.

LEMMA 2.6 *For all* $t \geqq t_0$ *and* $\lambda > 0$,

$$E[e^{\lambda\xi_0(t)}] < \infty .$$

Proof: Clearly, it suffices to prove that $E[e^{\lambda\xi_\phi(t)}] < \infty$. We know that

$$Z_\lambda(t) = e^{i\lambda\phi(x(t))} - i\lambda\int_{t_0}^{t} e^{i\lambda\phi(x(u))}\, d\bar{\xi}(u) + \tfrac{1}{2}\lambda^2\int_{t_0}^{t} \beta(u)e^{i\lambda\phi(x(u))}\, du$$

is a P-martingale, where

$$\bar{\xi}(t) = \int_{t_0}^{t} \chi_G\, L_u\, \phi(x(u))\, du + \xi_\phi(t)$$

and

$$\beta(u) = \chi_G(x(u))\langle \nabla\phi(x(u)), a(u, x(u))\, \nabla\phi(x(u))\rangle .$$

Take $\eta(t) = \phi(x(t)) - \bar{\xi}(t)$. We shall show that

$$\exp\left\{i\lambda\eta(t) + \tfrac{1}{2}\lambda^2\int_{t_0}^t \beta(u)\,du\right\}$$

is a P-martingale. Indeed, simply apply Lemma 2.1 to the martingale $Z_\lambda(t)$ and the function of bounded variation

$$\exp\left\{-i\lambda\bar{\xi}(t) + \tfrac{1}{2}\lambda^2\int_{t_0}^t \beta(u)\,du\right\}.$$

Hence, by Theorem 2.1,

$$P\left(\sup_{t_0\leqq s\leqq t}|\eta(s)| \geqq R\right) \leqq 2e^{-R^2/2At}.$$

In particular, $E[e^{\lambda|\eta(t)|}] < \infty$ for all $t > t_0$ and $\lambda > 0$. Since $\phi \in C_b^2(R^d)$, it is immediate from this that $E[e^{\lambda\xi\phi(t)}] < \infty$.

THEOREM 2.5. *The measure P on $\langle\Omega, \mathscr{M}^{t_0}\rangle$ solves the submartingale problem for a, b, ρ and γ if and only if there exists a continuous, non-decreasing, non-anticipating function $\xi_0 : [t_0, \infty) \times \Omega \to [0, \infty)$ such that*

(i) $\xi_0(t_0) = 0$, $E[e^{\lambda\xi_0(t)}] < \infty$ *for all* $\lambda \geqq 0$ *and* $t \geqq t_0$,

(ii) $\xi_0(t) = \int_{t_0}^t \chi_{\partial G}(x(u))\,d\xi_0(u)$ *for* $t \geqq t_0$,

(iii) *for all* $\lambda \in R$ *and* $\theta \in R^d$,

$$X_{\lambda,\theta}^{t_0}(t) = \exp\left\{\langle\theta, x(t) - x(t_0)\rangle - \tfrac{1}{2}\int_{t_0}^t \chi_G\langle\theta, a\theta\rangle\,du - \int_{t_0}^t \chi_G\langle\theta, b\rangle\,du - \int_{t_0}^t \langle\theta, \gamma\rangle\,d\xi_0(u)\right\}\exp\left\{\lambda\left(\int_{t_0}^t \chi_{\partial G}(x(u))\,du - \int_{t_0}^t \rho\,d\xi_0(u)\right)\right\}$$

is a P-martingale.

Furthermore, if P is such a solution, then $\xi_0(t)$ is uniquely determined, up to P-equivalence, by the condition that

$$\phi(x(t)) - \int_{t_0}^t \chi_G\,L_u\,\phi(u, x(u))\,du - \int_{t_0}^t \langle\nabla\phi, \gamma\rangle\,d\xi_0(u)$$

is a P-martingale.

Finally, if P is a solution, then we can define stochastic integrals with respect to $x(t)$.

The $dx(t)$ integrals have the following properties:

1. $$X_\theta(t) = \exp\left\{\int_{t_0}^t \theta(u)\, dx(u) - \tfrac{1}{2}\int_{t_0}^t \chi_G\langle\theta(u), a\theta(u)\rangle\, du - \int_{t_0}^t \chi_G\langle\theta(u), b\rangle\, du - \int_{t_0}^t \langle\theta(u), \gamma\rangle\, d\xi_0(u)\right\}$$

is a P-martingale for bounded non-anticipating $\theta : [t_0, \infty) \times \Omega \to R^d$.

2. *If* $f \in C_b^{1,2}([0, \infty) \times R^d)$, *then, for* $t_0 \leqq s \leqq t$,

$$f(t, x(t)) - f(s, x(s)) = \int_s^t \nabla_x f(u, x(u))\, dx(u) + \int_s^t (f_u + \chi_G L_u^0 f)(u, x(u))\, du\,,$$

where L_u^0 *is the principal part of* L_u.

3. $$\int_{t_0}^t \chi_{\partial G}(x(u))\, dx(u) = \int_{t_0}^t \gamma(u, x(u))\, d\xi_0(u)\,, \qquad t \geqq t_0\,, \quad a.s.\ P\,.$$

4. *Moreover,*

$$\int_{t_0}^t \chi_{\partial G}(x(u))\, du = \int_{t_0}^t \rho(u, x(u))\, d\xi_0(u)\,, \qquad t \geqq t_0\,, \quad a.s.\ P\,.$$

5. *Finally, if* $\rho = 0$, *then*

$$P\left[\int_{t_0}^t \chi_{\partial G}(x(u))\, du > 0\right] = 0\,.$$

Proof: The first few assertions are immediate from Lemma 2.6 and Theorems 2.1, 2.2 and 2.3. To prove properly 3, take $\theta(u) = \chi_{\partial G}(x(u))\theta$, where $\theta \in R^d$. Then

$$\exp\left\{\left\langle \theta, \int_{t_0}^t \chi_{\partial G}(x(u))\, dx(u) - \int_{t_0}^t \gamma(u, x(u))\, d\xi_0(u)\right\rangle\right\}$$

is a P-martingale. Since this is true for all $\theta \in R^d$, property 3 follows immediately.

COROLLARY 2.2 *If* P *solves the submartingale problem and if* τ *is a* t_0*-stopping time, then there is a set* $N \in \mathcal{M}_\tau^{t_0}$ *such that* $P(N) = 0$, *and for all* $\omega \notin N$ *the r.c.p.d.* P_ω *of* P *given* $\mathcal{M}_\tau^{t_0}$ *is again a solution and* $P_\omega\{x(\tau(\omega)) = x(\tau(\omega), \omega)\} = 1$. *Moreover, if* $\tau'_\omega = \inf\{t > \tau(\omega) : x(t) \in \partial G\}$, *then for* $\omega \notin N$, $P_\omega = Q_{\tau(\omega), x(\tau(\omega)), \omega)}$ *on*

$\mathscr{M}_{\tau\omega'}^{\tau(\omega)}$, *where* $Q_{t,x}$ *solves the martingale problem for* a *and* b *starting from* x *at time* t.

Proof: These assertions are easy consequences of the preceding.

COROLLARY 2.3. *Let* P *be a solution to the submartingale problem starting from* $x \in \partial G$ *at time* t_0. *Then* $P(\xi_0(t) > 0, t > t_0) = 1$.

Proof: By Theorem 2.1, we know that

$$X_\lambda(t) = \exp\left\{\lambda\phi(x(t)) - \lambda\int_{t_0}^t \chi_G\, L_u\, \phi(x(u))\, du \right.$$
$$\left. - \lambda\int_{t_0}^t \langle\gamma, \nabla\phi\rangle\, d\xi_0(u) - \tfrac{1}{2}\lambda^2\int_{t_0}^t \chi_G \langle\nabla\phi, a\nabla\phi\rangle\, du\right\}$$

is a P-martingale relative to $\mathscr{M}_t^{t_0}$. Since $X_\lambda(t)$ is continuous and locally bounded, it is easy to see that $X_\lambda(t)$ is a P-martingale relative to $\mathscr{M}_{t+0}^{t_0} = \bigcap_{\varepsilon>0} \mathscr{M}_{t+\varepsilon}^{t_0}$. Define $\tau_0 = \sup\{s \geqq t_0 : \xi_0(s) = 0\}$. Then τ_0 is a stopping time relative to the $\mathscr{M}_{t+0}^{t_0}$. Moreover, since

$$\int_{t_0}^{\tau_0} \chi_{\partial G}(x(u))\, du = \int_{t_0}^{\tau_0} \rho(u, x(u))\, d\xi_0(u) = 0\,,$$

we see that

$$X_\lambda(\tau_0 \wedge t) = \exp\left\{\lambda\phi(x(t\wedge\tau_0)) - \lambda\int_{t_0}^{t\wedge\tau_0} L_u\,\phi\, du - \tfrac{1}{2}\lambda^2\int_{t_0}^{t\wedge\tau_0} \langle\nabla\phi, a\,\nabla\phi\rangle\, du\right\}$$

is a P-martingale relative to $\mathscr{M}_{t\wedge\tau_0+0}^{t_0}$. Choose $\delta > 0$ so that

$$\langle\nabla\phi, a\nabla\phi\rangle \geqq \alpha > 0$$

in $B(x, \delta) \cap \bar{G}$ and let $\tau_\delta = \inf\{t \geqq t_0 : |x(t) - x| \geqq \delta\}$.[12] Then

$$E[X_\lambda(\tau_0 \wedge \tau_\delta \wedge t)] = 1$$

for $t \geqq t_0$. Let us fix $t > t_0$ and choose $\lambda < 0$. Then, since $\phi \geqq 0$ on $\bar{G}$, we have

$$E\left[\exp\left\{-\lambda^2\left(\tfrac{1}{2}\alpha - \frac{1}{\lambda}\beta\right)(\tau - t_0)\right\}\right] \geqq 1\,,$$

[12] $B(x, \delta) = \{y : |y - x| < \delta\}$.

where

$$\beta = \sup_{\substack{u \geq t_0 \\ x \in G}} |L_u \phi(x)| \qquad \text{and} \qquad \tau = \tau_0 \wedge \tau_\delta \wedge t .$$

As $\lambda \to -\infty$, we obtain $P(\tau - t_0 = 0) = 1$. Since $\tau_\delta > t_0$ a.s. P, it follows that $P(\tau_0 = t_0) = 1$.

COROLLARY 2.4. *Suppose a and b are such that $a^{-1} b$ is uniformly bounded on $[t_0, \infty) \times G$. Define*

$$R_t^{t_0} = \exp\left\{\int_{t_0}^t \chi_G \langle b(u, x(u)), a^{-1}(u, x(u))\, dx(u)\rangle \right.$$
$$\left. - \tfrac{1}{2}\int_{t_0}^t \chi_G \langle b(u, x(u)), a^{-1}(u, x(u)) b(u, x(u))\rangle\, du\right\}.$$

If P^0 is a solution to the submartingale problem for a, ρ and γ (i.e., $b \equiv 0$), then $R_t^{t_0}$ is a P^0-martingale and the measure P defined by

$$\frac{dP}{dP^0} = R_t^{t_0} \quad \text{on} \quad \mathscr{M}_t^{t_0}, \qquad t \geq t_0 ,$$

is a solution to the submartingale problem for a, b, ρ, and γ. Conversely, if P is a solution for a, b, ρ and γ, then $(R_t^{t_0})^{-1}$ is a P-martingale and the measure P^0 defined by

$$\frac{dP^0}{dP} = (R_t^{t_0})^{-1} \quad \text{on} \quad \mathscr{M}_t^{t_0}, \qquad t \geq t_0 ,$$

is a solution for a, ρ and γ.

Proof: By Theorem 2.5, $R_t^{t_0}$ is a P^0-martingale. Hence the measure P is consistently defined by

$$\frac{dP}{dP^0} = R_t^{t_0} \quad \text{on} \quad \mathscr{M}_t^{t_0} .$$

Moreover, by setting $\theta(u) = \theta + [\chi_G a^{-1} b](u, x(u))$, we see from Theorem 2.5 that

$$\exp\left\{\langle\theta, x(t) - x(t_0)\rangle - \tfrac{1}{2}\int_{t_0}^t \chi_G\langle\theta, a\theta\rangle\, du - \int_{t_0}^t \chi_G\langle\theta, b\rangle\, du - \int_{t_0}^t \langle\theta, \gamma\rangle\, d\xi_0(u)\right\}$$
$$\cdot R_t^{t_0} \exp\left\{\lambda\left(\int_{t_0}^t \chi_{\partial G}\, du - \int_{t_0}^t \rho\, d\xi_0(u)\right)\right\}$$

is a P^0-martingale. This is equivalent to saying that

$$\exp\left\{\langle\theta, x(t) - x(t_0)\rangle - \tfrac{1}{2}\int_{t_0}^{t} \chi_G\langle\theta, a\theta\rangle\, du - \int_{t_0}^{t} \chi_G\langle\theta, b\rangle\, du - \int_{t_0}^{t}\langle\theta, \gamma\rangle\, d\xi_0(u)\right\}$$
$$\cdot \exp\left\{\lambda\left(\int_{t_0}^{t} \chi_{\partial G}\, du - \int_{t_0}^{t} \rho\, d\xi_0(u)\right)\right\}$$

is a P-martingale. The converse is proved in a similar fashion.

So far the only functions which we know to be in F are functions in $C_b^{1,2}([0, \infty) \times R^d)$. Later on we shall need to know that F also includes functions f of the form

$$f(t, x) = E^{Q_{t,x}}\left[\int_t^{\tau_G} g(s, x(s))\, ds\right] + E^{Q_{t,x}}[h(\tau_G, x(\tau_G))]\ ^{13}$$

such that $f \in C_b^{1,1}([0, \infty) \times \partial G)$, where $Q_{t,x}$ is as in Corollary 2.2 and

$$g, h \in B_0([0, \infty) \times R^d).\ ^{14}$$

The rest of the present section is devoted to proving this inclusion.

LEMMA 2.7. *Let P be any solution to the submartingale problem for a, b, ρ and γ starting from $x \in \partial G$ at time t_0. If a is uniformly elliptic and if we define*

$$\tau_\delta = \inf\{t \geqq t_0 : |x(t) - x(t_0)| \geqq \delta\},$$
$$T_\delta = E\left[\int_{t_0}^{\tau_\delta} \chi_G(x(u))\, du\right], \qquad T'_\delta = E\left[\left(\int_{t_0}^{\tau_\delta} \chi_G(x(u))\, du\right)^2\right],$$
$$\Xi_\delta = E[\xi_0(\tau_\delta)], \qquad \Xi'_\delta = E[\xi_0^2(\tau_\delta)],$$

then there exist positive numbers C, α, and δ_0 such that

$$E[\tau_\delta - t_0] \leqq C\delta, \qquad E[(\tau_\delta - t_0)^2] \leqq C\delta^2, \qquad T_\delta \leqq C\delta^2, \qquad T'_\delta \leqq C\delta^4,$$
$$\alpha\delta \leqq \Xi_\delta \leqq C\delta, \qquad \Xi'_\delta \leqq C\delta^2,$$

for $0 < \delta < \delta_0$. The quantities C, α and δ_0 can be chosen independent of $x \in \partial G$, $t_0 \geqq 0$, and the particular choice of P.

Proof: We use C to denote "the existence of a constant such that". It is

---

[13] $\tau_G = \inf\{t > 0 : x(t) \notin G\}$.

[14] $B_0(\delta)$ is the set of bounded measurable functions having compact support in δ.

easy to check that all the estimates obtained are uniform in the relevant parameters.

Denote by $T_{\delta,t}$ and $T'_{\delta,t}$ the quantities

$$E\left[\int_{t_0}^{\tau_\delta\wedge t} \chi_G(x(u))\,du\right] \quad \text{and} \quad E\left[\left(\int_{t_0}^{\tau_\delta\wedge t} \chi_G(x(u))\,du\right)^2\right],$$

respectively. We know that

$$\phi^2(x(t)) = 2\int_{t_0}^{t} \chi_G\,\phi\nabla\phi\,dx + 2\int_{t_0}^{t} \chi_G\,\phi L_u^0\,\phi\,du + \int_{t_0}^{t} \chi_G\langle\nabla\phi, a\nabla\phi\rangle\,du\,.$$

Since $|\nabla\phi| \geqq 1$ on ∂G, we have for small δ

$$T_{\delta,t} \leqq C(E[\phi^2(x(t\wedge\tau_\delta))] + \delta T_{\delta,t})\,,$$

where C is independent of $t \geqq t_0$. Hence

$$T_{\delta,t} \leqq CE[\phi^2(x(\tau_\delta\wedge t))] \leqq C\delta^2\,, \qquad t \geqq t_0\,. \tag{2.7}$$

Similarly

$$T'_{\delta,t} \leqq CE[\phi^4(x(t\wedge\tau_\delta))] + \delta^2 T'_{\delta,t} + \delta^2 T_{\delta,t}\,,$$

and so

$$T'_{\delta,t} \leqq C\delta^2 E[\phi^2(x(t\wedge\tau_\delta))] \leqq C\delta^4\,, \qquad t \geqq t_0\,. \tag{2.8}$$

Next we use (2.8) and the equation

$$\phi(x(t)) = \int_{t_0}^{t} \chi_G\,\nabla\phi\,dx + \int_{t_0}^{t} \chi_G\,L_u^0\,\phi\,du + \int_{t_0}^{t} \chi_G\langle\nabla\phi, \gamma\rangle\,d\xi_0 \tag{2.9}$$

to obtain

$$E[\xi_0(\tau_\delta)] \leqq C\delta\,.$$

Since

$$\tau_\delta = \int_0^{\tau_\delta} \chi_G(x(u))\,du + \int_0^{\tau_\delta} \rho(u, x(u))\,d\xi_0(u)\,,$$

we get $E[\tau_\delta] \leqq C\delta$. Also, we have from (2.9)

$$E[\xi_0^2(\tau_\delta)] \leqq C(E[\phi^2(x(\tau_\delta))] + T_\delta + T'_\delta)\,,$$

and so

$$E[\xi_0^2(\tau_\delta)] \leqq CE[\phi^2(x(\tau_\delta))] \leqq C\delta^2\,. \tag{2.10}$$

From (2.8) and (2.10) we obtain $E[\tau_\delta^2] \leqq C\delta^2$, and from (2.7) and (2.10),

$$\delta^2 = E[|x(\tau_\delta) - x|^2] \leqq 2E\left[\left|\int_{t_0}^{\tau_\delta} \chi_G(x(u))\, dx(u)\right|^2 + \left|\int_{t_0}^{\tau_\delta} \gamma\, d\xi_0\right|^2\right]$$
$$\leqq CE[\phi^2(x(\tau_\delta))]\,.$$

Since $\phi(x(\tau_\delta)) \leqq \delta$, this implies that $E[\phi(x(\tau_\delta))] \geqq \alpha'\delta$ for some positive α'. Combining the preceding with (2.7) and (2.9), we conclude that $E[\phi(\tau_\delta)] \geqq \alpha\delta$ for some $\alpha > 0$.

LEMMA 2.8. *The notation and hypotheses are the same as in Lemma 2.7. If* $f \in C_b([0, \infty) \times \bar{G}) \cap C_b^{1,1}([0, \infty) \times \partial G)$ *and* $f_t + J_t f \geqq \beta$ *for some* $\beta > 0$, *then there exists a* $\delta(t_0, x)$ *such that* $0 < \delta(t_0, x) \leqq \delta_0$ *and*

$$E[f(\tau_\delta, x(\tau_\delta)) - f(t_0, x_0)] \geqq 0$$

for $0 < \delta < \delta(t_0, x)$. *The choice of* $\delta(t_0, x)$ *depends only on* δ_0 *and the modulus of continuity of* ρ, γ, f_u *and* $\nabla_x f$ *at* (t_0, x). *In particular,* $\delta(t, x)$ *can be chosen uniformly positive on compact sets.*

Proof: We have

$$E[f(\tau_\delta, x(\tau_\delta)) - f(t_0, x)] = E[\langle\nabla_x f(\tau_\delta, x(t_0)), x(\tau_\delta) - x(t_0)\rangle]$$
$$+ E\left[\int_{t_0}^{\tau_\delta} f_u(u, x(t_0))\, du\right] + o(\delta)$$
$$= I_1 + I_2 + o(\delta)\,.$$

Now

$$I_1 = E[\langle\nabla_x f(\tau_\delta, x(t_0)), \gamma(\tau_\delta, x(t_0))\rangle\xi_0(\tau_\delta)]$$
$$+ E\left[\left\langle\nabla_x f(\tau_\delta, x(t_0)), \int_{t_0}^{\tau_\delta}(\gamma(u, x(u)) - \gamma(\tau_\delta, x(t_0)))\, d\xi_0(u)\right\rangle\right]$$
$$+ E\left[\left\langle\nabla_x f(t_0, x(t_0)), \int_{t_0}^{\tau_\delta}\chi_G(x(u))b(u, x(u))\, du\right\rangle\right]$$
$$+ E\left[\left\langle\nabla_x f(\tau_\delta, x(t_0)) - \nabla_x f(t_0, x(t_0)), \int_{t_0}^{\tau_\delta}\chi_G(x(u))\, dx(u)\right\rangle\right]$$
$$= I_{11} + I_{12} + I_{13} + I_{14}\,.$$

Using the estimates from Lemma 2.7, we get

$$I_{12} \leqq \|\nabla_x f\| \, E\left[\int_{t_0}^{\tau_\delta} |\gamma(u, x(u)) - \gamma(\tau_\delta, x(t_0))| \, d\xi_0(u)\right] = o(\delta) ,$$

$$I_{13} \leqq \|\nabla_x f\| \, E\left[\int_{t_0}^{\tau_\delta} \chi_G(x(u)) \, |b(u, x(u))| \, du\right] = o(\delta) ,$$

and

$$I_{14} \leqq CE[|\nabla_x f(\tau_\delta, x(t_0)) - \nabla_x f(t_0, x(t_0))|^2]^{1/2} T_\delta^{1/2} = o(\delta) .$$

Also,

$$\begin{aligned} I_2 &= E[\rho(\tau_\delta, x(t_0)) f_t(\tau_\delta, x(t_0)) \xi_0(\tau_\delta)] \\ &\quad + E\left[\int_{t_0}^{\tau_\delta} [\rho(u, x(u)) f_u(u, x(t_0)) - \rho(\tau_\delta, x(t_0)) f_t(\tau_\delta, x(t_0))] \, d\xi_0(u)\right] \\ &\quad + E\left[\int_{t_0}^{\tau_\delta} \chi_G f_u(u, x(t_0)) \, du\right] \\ &= I_{21} + I_{22} + I_{23} . \end{aligned}$$

Arguing as above, we find that I_{22} and I_{23} are $o(\delta)$. Combining these estimates, we have

$$E[f(\tau_\delta, x(t_\varepsilon)) - f(t_0, x(t_0))] = I_{11} + I_{12} + o(\delta) .$$

Since $I_{11} + I_{12} \geqq \alpha\beta\delta$ for $0 < \delta < \delta_0$, where α and δ_0 are as in Lemma 2.7, our assertion follows.

LEMMA 2.9. *We still use the notation and hypotheses of Lemma* 2.7, *except that here we do not assume that P starts at a boundary point. The contention is*:

$$E^P[(\tau_\delta - t_0)^2] \leqq C\delta^2 \qquad \textit{for} \qquad 0 < \delta \leqq \delta_0 ,$$

where C is independent of the starting point and the particular solution P.

Proof: If P starts on ∂G, then this estimate follows from Lemma 2.7. In general, let

$$\tau = \inf\{t \geqq t_0 : x(t) \notin G\} \qquad \text{and} \qquad \tau'_{2\delta} = \inf\{t \geqq \tau : |x(t) - x(\tau)| \geqq 2\delta\} .$$

Then

$$\begin{aligned} E^P[(\tau_\delta - t_0)^2] &\leqq E^P[(\tau_\delta - t_0)^2\chi_{\tau_\delta<\tau}] + E^P[(\tau_\delta - t_0)^2\chi_{\tau_\delta\geqq\tau}] \\ &\leqq E^{Q_{t_0,x}}[(\tau_\delta - t_0)^2] + 2E^P[(\tau - t_0)^2\chi_{\tau_\delta\geqq\tau}] \\ &\quad + 2E^P[(\tau'_{2\delta} - \tau)^2\chi_{\tau_\delta\geqq\tau}] \\ &\leqq E^{Q_{t_0,x}}[(\tau_\delta - t_0)^2] + 2E^{Q_{t_0,x}}[(\tau_\delta \wedge \tau - t_0)^2] \\ &\quad + 2E^{Q_{t_0,x}}[\chi_{\tau_\delta\geqq\tau}E^{P^\tau_\omega}[(\tau'_{2\delta} - t_0)^2]] \\ &\leqq C\delta^2 , \end{aligned}$$

where P^τ_ω is the r.c.p.d. of P given $\mathscr{M}^{t_0}_\tau$ and $Q_{t_0,x}$ is as in Corollary 2.2.

LEMMA 2.10. *(Notation and hypotheses are as in Lemma 2.9.) Assume that* $g, h \in B_0([0, \infty) \times R^d)$ *and* $y \in \bar{G}$. *Define*

$$f(s, y) = -E^{Q_{s,y}}\left[\int_s^{\tau_s} g(u, x(u))\, du\right] + E^{Q_{s,y}}[h(\tau_s , x(\tau_s))] ,$$

where $\tau_s = \inf\{t \geqq s : x(t) \notin G\}$ *and* $Q_{s,y}$ *is as in Corollary 2.2. Let*

$$X_f(t) = f(t, x) - \int_{t_0}^t \chi_G\, g(u, x(u))\, du .$$

Then $X_f(t)$ *is a local P-martingale in G. If, in addition,*

$$f \in C_b([0, \infty) \times \bar{G}) \cap C_b^{1,1}([0, \infty) \times \partial G),$$

then, for any $\psi \in C_b^{1,2}([0, \infty) \times R^d)$ *satisfying* $(\rho\, \partial/\partial t + J_t)(f + \psi) \geqq \beta > 0$ *on* $[t_0 , \infty) \times \partial G$,

$$X_{\tilde{f}}(t) = \tilde{f}(t, x(t)) - \int_{t_0}^t \chi_G (g + \psi_u + L_u\, \psi)(u, x(u))\, du$$

is a P-submartingale, where $\tilde{f} = f + \psi$.

Proof: That $X_f(t)$ is a local P-martingale in G is immediate from Corollary 2.2 and the definition of f.

To prove the second assertion, let $0 < r < 1$ and define

$$\begin{aligned} \tau^r_0 &= t_0 , \\ \tau^r_{2n+1} &= \inf\{t \geqq \tau^r_{2n} : x(t) \notin G\} , \\ \tau^r_{2n} &= \inf\{t \geqq \tau^r_{2n-1} : |x(t) - x(\tau^r_{2n-1})| \geqq r\delta(\tau_{2n-1} , x(\tau_{2n-1}))\} , \end{aligned}$$

where $\delta(t, x)$ is chosen for f as in Lemma 2.9. Define

$$\sigma_r = \begin{cases} T & \text{if there is no } n \text{ such that } \tau_{2n}^r \leqq T < \tau_{2n-1}^r , \\ \tau_{2n}^r & \text{if} \quad \tau_{2n-1}^r \leqq T < \tau_{2n}^r , \end{cases}$$

and let

$$Y(t) = X_f(t) - f(t_0, x(t_0)) .$$

We shall show that the family $\{Y(\sigma_r)\}_{0<r<1}$ is uniformly integrable and that $E[Y(\sigma_r)] \geqq 0$ for $0 < r < 1$. Since $\sigma_r \to T$ as $r \downarrow 0$, it will follow that $E[Y(T)] \geqq 0$.

To show that $\{Y(\sigma_r)\}_{0<r<1}$ is uniformly integrable, it suffices to check that $E[(\sigma_r - t_0)^2] \leqq C < \infty, 0 < r < 1$. But

$$\begin{aligned} E[(\sigma_r - t_0)^2] &\leqq 2(T - t_0)^2 + 2E[\sum \chi_{\tau_{2n-1}^r < T < \tau_{2n}^r}(\tau_{2n}^r - T)^2] \\ &= 2(T - t_0)^2 + 2E[\sum \chi_{\tau_{2n-1}^r \leqq T < \tau_{2n}^r}(\omega) E^{P_\omega^T}[\tau_{2n}^r - T)^2]] \\ &\leqq 2(T - t_0)^2 + 2E[\sum \chi_{\tau_{2n-1}^r \leqq T < \tau_{2n}^r}(\omega) E^{P_\omega^T}[(\tau_{2r\delta_0} - T)^2]] \\ &\leqq 2(T - t_0)^2 + 8(r\delta_0)^2 , \end{aligned}$$

where $\tau_{2r\delta_0} = \inf\{u \geqq T : |x(u) - x(T)| \geqq \{2r\delta_0\}$ and P_ω^T is the r.c.p.d. of P given $\mathscr{M}_T^{t_0}$. To prove that $E[Y(\sigma_r)] \geqq 0$, note that $Y(\sigma_r) = \lim_{N\to\infty} Y(\sigma_r \wedge \tau_{2N}^r)$ and that $E[|Y(\sigma_r \wedge \tau_{2N}^r)|^2] \leqq C_1 + C_2 E[\sigma_r^2]$. Hence it suffices to show that $E[Y(\sigma_r \wedge \tau_{2N}^r)] \geqq 0$. But

$$\begin{aligned} E[Y(\sigma_r \wedge \tau_{2N}^r)] &= E\left[\sum_{n=0}^{N-1} \chi_{\tau_{2n}^r \leqq T}(X_f(\tau_{2n+1}^r \wedge T) - X_f(\tau_{2n}^r))\right] \\ &\quad + E\left[\sum_{n=1}^{N} \chi_{\tau_{2n-1}^r \leqq T}(X_{f'}(\tau_{2n}^r) - X_f(\tau_{2n-1}^r))\right], \end{aligned}$$

$$E[\chi_{\tau_{2n}^r \leqq T}(X_f(\tau_{2n+1}^r \wedge T) - X_f(\tau_{2n}^r))] = 0 ,$$

and

$$\begin{aligned} &E[\chi_{\tau_{2n-1}^r \leqq T}(X_f(\tau_{2n}^r) - X_f(\tau_{2n-1}^r))] \\ &\qquad = E[\chi_{\tau_{2n-1}^r \leqq T}(\omega) E^{P_\omega^{\tau_{2n-1}^r}}[X_f(\tau_{2n}^r) - X_f(\tau_{2n}^r(\omega), \omega)]] \geqq 0 , \end{aligned}$$

where $P_\omega^{\tau_{2n-1}^r}$ is the r.c.p.d. of P given $\mathscr{M}_{\tau_{2n-1}^r}^{t_0}$.

We know now that, for any solution of the submartingale problem starting at time t_0,

$$E[X_f(T)] \geqq E[X_f(t_0)] \qquad \text{for} \qquad T > t_0 .$$

Given $t_0 \leqq s < T$, apply this result to the r.c.p.d. P^s_ω of P given $\mathcal{M}^{t_0}_s$. Then

$$E^{P^s_\omega}[X_f(T)] \geqq E^{P^s_\omega}[X_f(x)] \qquad \text{a.s. } P,$$

since P^s_ω is almost surely a solution to the submartingale problem starting at time s. In other words, $X_f(t)$ is a P-submartingale.

THEOREM 2.6. *Let P be a solution to the submartingale problem for a, b, ρ and γ starting at time t_0. Assume that a is uniformly elliptic. Define f as in Lemma 2.10 and suppose $f \in C_b([0, \infty) \times \bar{G}) \cap C_b^{1,1}([0, \infty) \times \partial G)$. Then $f \in F$.*

Proof: We must check that f fulfills conditions (ii) and (iii) defining the class F, since condition (i) is part of the hypothesis.

From the preceding we know that

$$X_f(t) = f(t, x(t)) - \int_{t_0}^t \chi_G g(u, x(u))\, du$$

is a local P-martingale in G. Hence f satisfies (ii) with $Kf = g$. To show that f satisfies (iii), choose α such that $\rho \tilde{f}_t + J_t \tilde{f} \geqq \beta > 0$ on $[t_0, \infty) \times \partial G$, where $\tilde{f} = f + \alpha\phi$. Then, by the preceding,

$$\tilde{f}(t, x) - \int_{t_0}^t \chi_G (g + \alpha L_u \phi)(u, x(u))\, du$$

is a P-submartingale. Hence there is a $\xi_{\tilde{f}}$ satisfying (iii) for $\tilde{f}$. Clearly $\xi_f = \xi_{\tilde{f}} - \alpha\xi_\phi$ satisfies (iii) for f. Finally, if $\psi \in C_b^{1,2}([0, \infty) \times R^d)$ satisfying

$$(\rho \frac{\partial}{\partial t} + J_t)(f + \psi) \geqq 0 \quad \text{on} \quad [t_0, \infty) \times \partial G,$$

then, for every $\beta > 0$,

$$(f + \psi + \beta\phi)(t, x(t)) - \int_{t_0}^t \chi_G (g + L_u \psi + \beta L_u \phi)(u, x(u))\, du$$

is a P-submartingale. Let $\beta \downarrow 0$, we see that

$$(f + \psi)(t, x(t)) - \int_{t_0}^t \chi_G (g + L_u \psi)(u, x(u))\, du$$

is also a P-submartingale. Therefore, $\xi_{f+\psi}(t)$ must be non-decreasing.

3. Existence

Let G be an open subset of R^d and assume there exists a $\phi \in C_b^2(R^d)$ such that $G = \{x : \phi(x) > 0\}$, $\bar{G} = \{x : \phi(x) \geqq 0\}$, $|\nabla\phi| \geqq 1$ on ∂G. Let

$$a : [0, \infty) \times \bar{G} \to S_d^+$$

be bounded and continuous, let $\gamma : [0, \infty) \times \partial G \to R^d$ be a bounded continuous function such that $\langle\gamma, \nabla\phi\rangle \geqq \beta > 0$ on $[0, \infty) \times \partial G$, and let $\rho : [0, \infty) \times \partial G \to R^1$ be a uniformly positive, bounded, continuous function. Given $t_0 \geqq 0$ and $x_0 \in \bar{G}$, we are going to construct a measure P on $\langle\Omega, \mathscr{M}^{t_0}\rangle$ such that $P(x(t_0) = x_0) = 1$ and P solves the submartingale problem for a, ρ and γ. Since ρ is bounded and uniformly positive, we may replace γ by γ/ρ and thereby reduce the problem to the case when $\rho = 1$. Our construction will make use of the following simple lemma.

LEMMA 3.1. *Let P be a measure on $\langle\Omega, \mathscr{M}^{t_0}\rangle$. Let*

$$t_0 \equiv \tau_{-1} \leqq \tau_0 \leqq \tau_1 \leqq \cdots \leqq \tau_n \leqq \cdots$$

be a non-decreasing sequence of t_0-stopping times such that $\tau_n \to \infty$ a.s. P. If $X : [t_0, \infty) \times \Omega \to R$ is a locally bounded non-anticipating function such that $X(t \wedge \tau_n) \vee \tau_{n-1})$ is almost surely a $P_\omega^{\tau_{n-1}}$-submartingale for each $n \geqq 0$, where $P_\omega^{\tau_{n-1}}$ is the r.c.p.d. of P given $\mathscr{M}_{\tau_{n-1}}^{t_0}$, then the process $X(t)$ is a P-submartingale.

Proof: Given $t_0 \leqq s < t$ and $A \in \mathscr{M}_s^{t_0}$, we have

$$\begin{aligned} E[(X(t) - X(s))\chi_A] &= E\left[\sum_0^\infty (X(\tau_n \vee s) \wedge t) - X((\tau_{n-1} \vee s) \wedge t))\chi_A\right] \\ &= \lim_{N\to\infty} \sum_0^N E[(X((\tau_n \vee s) \wedge t) - X((\tau_{n-1} \vee s) \wedge t))\chi_A]\,. \end{aligned}$$

But

$$\begin{aligned} E[(X((\tau_n &\vee s) \wedge t) - X((\tau_{n-1} \vee s) \wedge t))\chi_A] \\ &= E[(X(\tau_n \wedge t) - X((\tau_{n-1} \vee s) \wedge t))\chi_{A\cap\{\tau_n > s\}}] \\ &= E[(X(\tau_n \wedge t) - X(\tau_{n-1} \wedge t))\chi_{A\cap\{\tau_{n-1} > s\}}] \\ &\quad + E[(X(\tau_n \wedge t) - X(s))\chi_{A\cap\{\tau_n > s \geqq \tau_{n-1}\}}] \\ &= E[\chi_{A\cap\{s<\tau_{n-1}<t\}}(\omega) E^{P_\omega^{\tau_{n-1}}}[X(\tau_n \wedge t) - X(\tau_{n-1}(\omega))]] \\ &\quad + E[\chi_{\{\tau_{n-1}\leqq s\}}(\omega) E^{P_\omega^{\tau_{n-1}}}[(X(\tau_n \wedge t) - X(s))\chi_{A\cap\{\tau_n > s\}}]] \\ &\geqq 0\,; \end{aligned}$$

q.e.d.

Now choose $\delta_0 > 0$ such that $x + \delta\gamma(u, x) \in G$ for all $x \in \partial G$, $u \geqq 0$, and $0 \leqq \delta \leqq \delta_0$. Given $0 < \delta < \delta_0$, define

$$Q^\delta(s, x; t, \Gamma) = \begin{cases} \chi_\Gamma(x)e^{-(t-s)/\delta} + \dfrac{1}{\delta}\displaystyle\int_s^t e^{-(u-s)/\delta}\chi_\Gamma(x + \delta\gamma(u, x))\, du & \text{if} \quad x \in \partial G\,, \\ \chi_\Gamma(x) \quad \text{if} \quad x \in G\,. \end{cases}$$

Clearly there is a Markov process $Q_{s,x}$, $s \geqq 0$ and $x \in \bar{G}$, on $D([0, \infty), \bar{G})$ [15] for which $Q^\delta(s, x;\ t, \Gamma)$ is the transition probability. Moreover, if f belongs to $C_b^{1,2}([0, \infty) \times R^d)$, then

$$f(t, x(t)) - \int_s^t \chi_{\partial G}(x(u)) K_\delta f(u, x(u))\, du - \int_s^t f_u(u; x(u))\, du$$

is a $Q^\delta_{s,x}$-martingale, where

$$K_\delta f(t, x) = \frac{1}{\delta}\left[f(t, x + \delta\gamma(t, x)) - f(t, x)\right].$$

Let $P^{(a)}_{s,x}$, $s \geqq 0$ and $x \in R^d$, be the Markov process on $D([s, \infty), R^d)$ associated with the diffusion coefficients a. [16] Define

$$\tau_{-1} = s\,,$$
$$\tau_{2n} = \inf\{t \geqq \tau_{2n-1} : x(t) \notin G\}\,,$$
$$\tau_{2n+1} = \inf\{t \geqq \tau_{2n} : |x(t) - x(\tau_{2n})| \geqq \tfrac{1}{2}\beta\delta\}\,.$$

For each fixed $\delta > 0$ we shall define a family of measures $P^\delta_{s,x}$, $s \geqq 0$ and $x \in \bar{G}$. First we define a sequence $P^{(n)}_{s,x}$ as follows:

$$P^{(-1)}_{s,x} = P^{(a)}_{s,x}\,.$$

Having defined $P^{(2n-1)}_{s,x}$, $P^{(2n)}_{s,x}$ is defined to be equal to $P^{(2n-1)}_{s,x}$ on $\mathscr{M}^s_{\tau_{2n-1}}$, and the r.c.p.d. of $P^{(2n)}_{s,x}$ given $\mathscr{M}^s_{\tau_{2n-1}}$ to be equal to $P^{(a)}_{\tau_{2n-1}, x(\tau_{2n-1})}$. Having defined $P^{(2n)}_{s,x}$, $P^{(2n+1)}_{s,x}$ is defined to be equal to $P^{(2n)}_{s,x}$ on $\mathscr{M}^s_{\tau_{2n}}$, and the r.c.p.d. of $P^{(2n+1)}_{s,x}$ given $\mathscr{M}^s_{\tau_{2n}}$ to be equal to $Q^\delta_{\tau_{2n}, x(\tau_{2n})}$. Because of the consistency

[15] $D(0, T)$ is the space of functions with no discontinuities of the second kind. We formalize the elements of $D(0, T)$ to be right-continuous and endow the space with the Skorokhod metric.

[16] We may assume that the coefficients a are globally defined.

of the sequence $P^{(n)}_{s,x}$, we can define $P^{\delta}_{s,x}$ to be equal to $P^{(n)}_{s,x}$ on $\mathcal{M}^{s}_{\tau_n}$. Moreover, by Lemma 3.1, if $f \in C_b^{1,2}([0, \infty) \times R^d)$, then

$$
\begin{aligned}
X^{\delta}_f(t) &= f(t, x(t)) - \int_s^t \chi_G \cdot (f_u + L_u f)(u, x(u))\, du \\
&\quad - \int_s^t \chi_{\partial G} \cdot (f_u + K_\delta f)(u, x(u))\, du
\end{aligned} \tag{3.1}
$$

is a $P^{\delta}_{s,x}$-martingale, where

$$
L_u f(u, x) = \sum a_{ij}(u, x) \frac{\partial^2 f}{\partial x_i\, \partial x_j} .
$$

We want to show that, for given t_0, x_0, there is a sequence $\{\delta_m\}$ decreasing to zero such that $P^{\delta_m}_{t_0,x_0}$ converges in $D[[t_0, \infty), \bar{G}]$ to a measure P fulfilling our requirements.

LEMMA 3.2. *There exists a constant C_T, independent of s, x and δ, such that*

$$
E^{P^{\delta}_{s,x}}[|x(t) - x(s)|^2] \leqq C_T(t - s), \qquad T \geqq t \geqq s,
$$

$$
E^{P^{\delta}_{s,x}}\left[\int_s^t \chi_{G_\varepsilon}(x(u))\, du\right] \leqq C_T\, \varepsilon, \qquad t_0 \leqq t \leqq T,
$$

where $G_\varepsilon = \{y \in G : |y - \partial G| < \varepsilon\}$.

Proof: The first assertion is immediate from (3.1). To prove the second estimate, define

$$
\eta_\varepsilon(r) = \begin{cases} \dfrac{1}{\varepsilon}(\varepsilon - r)^3 & \text{if} \quad |r| < \varepsilon, \\ 0 & \text{if} \quad |r| \geqq \varepsilon, \end{cases}
$$

and take $\psi_\varepsilon = \eta_\varepsilon \circ \phi$. Then

$$
L_u \psi_\varepsilon = \eta''_\varepsilon \circ \phi \cdot \langle \nabla\phi, a \nabla\phi \rangle + \eta''_\varepsilon \circ \phi L_u \phi, \qquad |K_\delta \psi_\varepsilon| \leqq C\varepsilon .
$$

Hence, from (3.1),

$$
E^{P_{s,x}}\left[\int_s^t [\chi_G \eta''_\varepsilon \circ \phi \langle \nabla\phi, a \nabla\phi\rangle](u, x(u))\, du\right] \leqq \varepsilon^2 + C(t - s)\varepsilon .
$$

Since a is uniformly elliptic, $|\nabla\phi| \geqq 1$ in ∂G, and $\eta''_\varepsilon \geqq \frac{1}{2}$ in $G_{\varepsilon/2}$, our estimate follows.

For each $\delta > 0$, the family $P^{\delta}_{s,x}$, $s \geqq 0$ and $x \in \bar{G}$, forms a Markov process. Consequently, the first estimate in Lemma 3.2 is enough to guarantee

weak compactness on $D([t_0, \infty), \bar{G})$ for the family $P^{\delta}_{t_0,x_0}$, $0 < \delta < \delta_0$. Choose $\delta_m \downarrow 0$ such that $P^{\delta_m}_{t_0,x_0}$ converges weakly on $D([t_0, \infty), \bar{G})$ to a measure P. Clearly, $P(x(t_0) = x_0) = 1$. Furthermore,

$$P((\exists t \geqq t_0)\ |x(t) - x(t-)| \geqq \delta_m \|\gamma\|) = 0$$

for all m, and therefore P is concentrated on $C([t_0, \infty), \bar{G})$. Finally, we must show that P solves the submartingale problem. Given $f \in C_0^{1,2}([0, \infty) \times R^d)$ such that $f_u + \langle \gamma, \nabla_x f \rangle \geqq \alpha > 0$ on $[t_0, \infty) \times \partial G$, choose $0 < \delta_1 \leqq \delta_0$ so that $f_u + K_\delta f \geqq 0$ on $[t_0, \infty) \times \partial G$ for $0 < \delta \leqq \delta_1$. Given $\varepsilon > 0$, choose $\eta_\varepsilon \in C_b^\infty(R^d)$ such that $0 \leqq \eta_\varepsilon \leqq 1$, $\eta_\varepsilon \equiv 1$ on $G - \bar{G}_\varepsilon$, and $\eta_\varepsilon \equiv 0$ off G. Let $t > s \geqq t_0$ and suppose Φ is a bounded, non-negative, continuous $\mathscr{M}_s^{t_0}$-measurable function. Then there is a constant C such that

$$\begin{aligned}
E^P&\left[\Phi\{(f(t, x(t)) - f(s, x(s)) - \int_s^t [\chi_G \cdot (f_u + L_u f)](u, x(u))\, du)\}\right] \\
&\geqq -C\varepsilon + E^P\left[\Phi\{(f(t, x(t)) - f(s, x(s))) \right. \\
&\qquad\qquad \left. - \int_s^t [\eta_\varepsilon \cdot (f_u + L_u f)](u, x(u))\, du\}\right] \\
&= -C\varepsilon + \lim_{m\to\infty} E^{P^{\delta_m}_{t_0,x_0}}\left[\Phi\{(f(t, x(t)) - f(s, x(s))) \right. \\
&\qquad\qquad \left. - \int_s^t [\eta_\varepsilon \cdot (f_u + L_u f)](u, x(u))\, du\}\right] \\
&\geqq -2C\varepsilon + \lim_{m\to\infty} E^{P^{\delta_m}_{t_0,x_0}}\left[\Phi\{(f(t, x(t)) - f(s, x(s))) \right. \\
&\qquad\qquad \left. - \int_s^t [\chi_G \cdot (f_u + L_u f)](u, x(u))\, du\}\right] \\
&\geqq -2C\varepsilon\,.
\end{aligned}$$

Letting $\varepsilon \downarrow 0$, we see that

$$f(t, x(t)) - \int_{t_0}^t [\chi_G \cdot (f_u + L_u f)](u, x(u))\, du$$

is a P-submartingale. It is easy to pass from this to the fact that P solves the submartingale problem.

We have now constructed a solution to the submartingale problem starting from x_0 at time t_0 in the case when a is uniformly elliptic, $b \equiv 0$, and ρ is uniformly positive. We next extend the construction to the case when ρ

is merely non-negative. To do this, choose uniformly positive ρ_n such that $\rho_n \to \rho$ uniformly. For each n, let P^n denote a solution starting from x_0 at time t_0 for a, ρ_n and γ. We shall show that $\{P^n\}_{n\geq 1}$ is weakly compact on $C([t_0, \infty), R^d)$ and that any limit point P of $\{P^n\}_{n\geq 1}$ solves the submartingale problem for a, ρ, and γ.

LEMMA 3.3. *There exists a constant C_T, independent of n, such that*

$$E^{P^n}[|x(t_3) - x(t_2)|^2 \, |x(t_2) - x(t_1)|^2] \leqq C_T(t_3 - t_1)^2 ,$$

$$t_0 \leqq t_1 \leqq t_2 \leqq t_3 \quad \textit{and} \quad t_3 - t_1 \geqq T ,$$

$$E^{P^n}\left[\int_{t_0}^t \chi_{G_\varepsilon}(x(u))\, du\right] \leqq C_T\, \varepsilon , \qquad t_0 \leqq t \leqq T ,$$

where $G_\varepsilon = \{x \in G : |x - \partial G| < \varepsilon\}$.

Proof: Let P on $\langle \Omega, \mathscr{M}^s\rangle$ be a solution to the submartingale problem for a, ρ and γ. Denote by ξ the associated increasing function described in Theorem 2.4. Then,

$$\phi(x(t)) - \phi(x(s)) = \int_s^t \chi_G \nabla\phi\, dx(u) + \int_s^t \chi_G\, L_u\, \phi\, du + \int_s^t \langle \gamma, \nabla\phi\rangle\, d\xi(u)$$

and

$$\phi^2(x(t)) - \int_s^t \chi_G \langle \nabla\phi, a\,\nabla\phi\rangle - 2\int_s^t \chi_G\, \phi L_u\, \phi\, du$$

are P-martingales. From these two facts, we see that

$$\begin{aligned} &E^P[(\phi(x(t)) - \phi(x(s)))^2] \\ &\quad = E^P[\phi^2(x(t)) - \phi^2(x(s))] - 2E^P[\phi(x(s))(\phi(x(t)) - \phi(x(s)))] \\ &\quad = E^P\left[\int_s^t \chi_G \langle \nabla\phi, a\nabla\phi\rangle\, du\right] + 2E^P\left[\int_s^t \chi_G\, \phi L_u\, \phi\, du\right] \\ &\qquad - 2E^P\left[\phi(x(s))\int_s^t \chi_G\, L_u\, \phi\, du\right] - 2E^P\left[\int_s^t \langle \gamma, \nabla\phi\rangle\, d\xi(u)\right] \\ &\quad \leqq C(t-s) , \end{aligned}$$

since $\phi(x(s)) \displaystyle\int_s^t \langle \gamma, \nabla\phi\rangle\, d\xi(u) \geqq 0$. Substituting this back into the first relation, we get

$$E^P\left[\left(\int_s^t \langle \gamma, \nabla\phi\rangle\, d\xi(u)\right)^2\right] \leqq C_T(t-s) , \qquad 0 \leqq t - s \leqq T,$$

where C_T does not depend on ρ or the particular choice of P. Since $\langle\gamma, \nabla\phi\rangle \geqq \beta > 0$, this proves that $E[\xi^2(t)] \leqq C_T\, t$, $s \leqq t \leqq T$. Hence,

$$E^P[|x(t) - x(s)|^2] \leqq 2E^P\left[\left|\int_s^t \chi_G(x(u))\, dx(u)\right|^2\right]$$

$$+ 2E^P\left[\left|\int_s^t \gamma(u, x(u))\, d\xi(u)\right|^2\right] \leqq C_T(t-s)\,,$$

$$0 \leqq t - s \leqq T\,.$$

Now take $t_0 \leqq t_1 \leqq t_2 \leqq t_3$, where $t_3 - t_1 \leqq T$. Applying the preceding result to the r.c.p.d. of P^n given $\mathscr{M}_{t_2}^{t_0}$, we get

$$E^{P^n}[|x(t_3) - x(t_2)|^2 \mid \mathscr{M}_{t_2}^{t_0}] \leqq C_T(t_3 - t_2) \qquad \text{a.s. } P^n\,.$$

From this our first assertion is immediate.

The second assertion is proved in very much the same way as we proved the second estimate in Lemma 3.2. The only difference is that here one must use the estimate $E^{P^n}[\xi^{(n)}(t)] \leqq C_T(t - t_0)^{1/2}$, $0 \leqq t - t_0 \leqq T$, which has just been found in the preceding paragraph.

The first estimate in Lemma 2.3 shows that $\{P^n\}_{n\geqq 1}$ is compact on $D([t_0, \infty), \bar{G})$. Using the second estimate in Lemma 2.3 and arguing in just the same way as we did before, we see that every limit point of $\{P^n\}_{n\geqq 1}$ solves our problem. Combining this with Corollary 2.3, we have proved:

THEOREM 3.1. *Let G, ϕ, a and γ be as described at the beginning of this section. Let $b : [0, \infty) \times \partial G \to R^d$ be bounded and measurable and $\rho : [0, \infty) \times \partial G \to [0, \infty)$ bounded and continuous. Then, for each $t_0 \geqq 0$ and $x_0 \in \bar{G}$, there is a solution P, starting from x_0 at time t_0, to the submartingale problem for a, b, ρ and γ.*

Remark. If one applies the uniform estimate

$$P\left(\sup_{t_0 \leqq u \leqq T} |x(u) - x(t_0)| \geqq R\right) \leqq \frac{C(T - t_0)}{R^2}\,,$$

one can replace the assumption of uniform ellipticity by strict ellipticity using an obvious piecing-together argument (cf. [8]).

4. Boundary Processes

Suppose that we are given a region G in R^d and coefficients a, b, ρ and γ of the type discussed previously. Consider only the coefficients a and b. Let $Q_{t,x}$ be the unique solution to the martingale problem for a and b, starting

from (t, x) up to the first exit time τ after t from the region G. We call $Q_{t,x}$, defined on $\mathscr{M}_\tau^t$, the interior process.

We now proceed to define a process called the boundary process. The importance of this process is that the entire process is built out of the interior and the boundary processes in a concrete way.

Let P be a solution to the submartingale problem corresponding to a, b, ρ and γ, starting from a boundary point x at time t_0. Then by Theorem 2.4, there exists a nondecreasing continuous function $\xi(t)$ with $\xi(t_0) = 0$, whose increases are only at those times when the trajectory $x(t)$ is at the boundary, such that, for smooth functions $u(t, x)$,

$$u(t, x(t)) - \int_{t_0}^t u_s(s, x(s))\, ds - \int_{t_0}^t (L_s u)(s, x(s))\chi_G(x(s))\, ds$$
$$- \int_{t_0}^t (J_s u)(s, x(s))\, d\xi(s)$$

is a martingale. Let us define stopping times τ_θ as follows:

$$\tau_\theta = [\sup s : s \geqq t_0 \text{ and } \xi(s) \leqq \theta] .$$

Let $T(\omega) = \lim_{t\to\infty} \xi(t)$. Then $\tau_\theta < \infty$ if and only if $T(\omega) > \theta$. Although τ_θ is not strictly a stopping time, for any $\varepsilon > 0$, $\tau_\theta + \varepsilon$ is a stopping time. Or equivalently, τ_θ is a stopping time relative to $\mathscr{M}_{t+0}^{t_0}$.[17] Moreover, since the starting point is on the boundary, we have $\xi(t) > 0$ for $t > 0$ almost surely (i.e., $\tau_0 = t_0$ with probability one (see Corollary 2.3)).

Also, if $\theta_n \downarrow \theta$ we have $\tau_{\theta_n} \downarrow \tau_\theta$. Since $\xi(\tau_\theta + \varepsilon) > \xi(\tau_\theta) = \theta$ for every $\varepsilon > 0$, $x(\tau_\theta)$ must be on the boundary. Assume, for simplicity, that $T(\omega) = \infty$ a.s. Then the process

$$y(\theta) = x(\tau_\theta)$$

is defined for all $0 \leqq \theta < \infty$, and it is a right continuous process living on the boundary. Moreover, the left limit exists for almost all trajectories. Also, $y(0) = x(t_0) = x$. We find it more convenient to work with the $(d + 1)$-dimensional process $(\tau_\theta, y(\theta))$. This is what we call the boundary process starting from x at time t_0. It is defined relative to the σ-fields $\mathscr{M}_{\tau_\theta}^{t_0}$. (If $T(\omega) < \infty$ with positive probability, it may be convenient to adjoin a point at ∞ and define $(\tau(\theta), y(\theta)) = \infty$ for $\theta \geqq T(\omega)$.)

In order to compute explicitly the "generator" of the boundary process, we make certain assumptions on the coefficients a and b and the region G. For any two functions f in $C_0^{1,2}([0, \infty) \times \partial G)$ and h in $C_0([0, \infty) \times G)$, we

[17] $\mathscr{M}_{t+0}^{t_0} = \bigcap_{s>t} \mathscr{M}_s^t$.

assume that the function

$$u(t, x) = E^{Q_{t,x}}\left[\int_t^\tau h(s, x(s))\, ds + f(\tau, x(\tau))\right]$$

has bounded continuous spatial derivatives of the first order throughout $[0, \infty) \times \bar{G}$.

From certain results in partial differential equations (cf. [5]) it is possible to show that this assumption is justified. However, we shall use the results of this section only in special cases where the assumption will be verified without appealing to the general theory.

Let us take h to be zero and set

$$u(t, x) = E^{Q_{t,x}}[f(\tau, x(\tau)] \,.$$

Since the spatial derivatives of the first order exist everywhere and since u equals f on the boundary,

$$(Kf)(s, x) = (J_s u)(s, x) + \rho(s, x) u_s(s, x) \tag{4.1}$$

is well defined as a bounded continuous function $[0, \infty) \times \partial G$.

THEOREM 4.1. *For any function $f(t, x)$ on the boundary in $C_0^{1,2}([t_0, \infty) \times \partial G)$,*

$$f(\tau(\theta), y(\theta)) - \int_0^\theta (Kf)(\tau(s), y(s))\, ds$$

is a martingale relative to $\mathcal{M}_{\tau\theta}^{t_0}$ and P. Moreover,

$$P[(\tau(0), y(0)) = (t_0, x)] = 1 \,.$$

Proof: We know that $u \in F$. Therefore, since u is harmonic in the interior,

$$u(t, x(t)) - \int_{t_0}^t [u_s(s, x(s))] \chi_{\partial G}(x(s))\, ds - \int_{t_0}^t (J_s u)(s, x(s))\, d\xi(s)$$

is a martingale. We can write

$$\chi_{\partial G}(x(s))\, ds = \rho(s, x(s))\, d\xi(s) \,,$$

and therefore

$$u(t, x(t)) - \int_{t_0}^t (Kf)(s, x(s))\, d\xi(s)$$

is a martingale. Let us assume for the moment that we can justify substituting τ_θ for t. Then,

$$u(\tau_\theta, x(\tau_\theta)) - \int_{t_0}^{\tau_\theta} (Kf)(s, x(s))\, d\xi(s)$$

is a martingale relative to $\mathcal{M}_{\tau_\theta}^{t_0}$. Clearly, $x(\tau_\theta) = y(\theta)$ is a boundary point and $u = f$ on the boundary. If we change variables in the integration from s to τ_s, then

$$f(\tau(\theta), y(\theta)) - \int_0^\theta (Kf)(\tau(s), y(s))\, ds$$

is a martingale relative to $\mathcal{M}_{\tau_\theta}^{t_0}$, and that will prove the theorem.

To justify the substitution of random stopping times, we need to take care of two points. First, τ_θ is a stopping time only relative to $\mathcal{M}_{t+0}^{t_0}$, and second, the τ_θ are unbounded. The first point is overcome by considering $\tau_\theta + \varepsilon$ and the second point by considering $(\tau_\theta + \varepsilon) \wedge N$. We first let $N \to \infty$ and then let $\varepsilon \to 0$. Up to τ_θ, $\xi(s)$ can be at most θ. Therefore, for $t_0 \leqq t \leqq \tau_\theta + \varepsilon$, there exist constants A and B such that

$$\begin{aligned}\left| u(t, x(t)) - \int_0^t (Kf)(s, x(s))\, d\xi(s) \right| &\leqq A + B\xi(t) \\ &\leqq A + B[\theta + (\xi(\tau_\theta + \varepsilon) - \xi(\tau_\theta))] .\end{aligned}$$

$E^P[|\xi(\tau_\theta + \varepsilon) - \xi(\tau_\theta)|^2]$ can be easily seen to be bounded. In fact, by Lemma 2.6, $E\xi^2(t)$ is finite for all t and therefore, for any stopping time τ,

$$E(\xi(\tau + t) - \xi(\tau))^2$$

can be estimated by taking r.c.p.d. We have uniform integrability by the same estimate, and the martingale is right continuous. This completes the justification.

We say that the uniqueness theorem is valid for the boundary process if, for any given point x on the boundary and for any starting time t_0, there is only one solution Q to the problem of finding a measure on the space

$$D([0, \infty), [t_0, \infty) \times \partial G)\ ^{18}$$

such that

1. $Q[\tau(0) = t_0, y(0) = x] = 1$,
2. ∞ is an absorbing point,

[18] See footnote 15.

3. for $f(t, x)$ in $C_0^{1,2}([t_0, \infty) \times \partial G)$,

$$f(\tau(\theta), y(\theta)) - \int_0^\theta (Kf)(s, y(s))\, ds$$

is a martingale relative to the natural σ-algebras $\mathscr{M}_\theta$ in $D([0, \infty), [t_0, \infty) \times \partial G)$ and the measure Q.

Remark 1. Of course it is clear that conditions 1, 2 and 3 are borrowed from Theorem 4.1. In particular, the validity of the uniqueness theorem for the boundary process implies that if P_1 and P_2 are two solutions to the submartingale problem for the same set of coefficients a, b, ρ and γ, and if $\xi_1(t)$ and $\xi_2(t)$ are defined accordingly, then

$$E^{P_1}\int_{t_0}^\infty f(s, x(s))\, d\xi_1(s) = E^{P_2}\int_{t_0}^\infty f(s, x(s))\, d\xi_2(s)$$

for any bounded function on $[t_0, \infty) \times \partial G$ with compact support.

Remark 2. The point ∞ is only a device to overcome the fact that $T(\omega) < \infty$ may occur with positive probability. This has to do with the fact that the harmonic measure may be substochastic (i.e., with positive probability the process may stay away from the boundary, with the result that the boundary process will be nonconservative). In Section 5, where we use this result to prove uniqueness for the submartingale problem, $T(\omega)$ will be ∞ with probability 1.

Remark 3. Let y be a point on the boundary and R a solution to the submartingale problem starting from y at time t. By Remark 1, if the uniqueness theorem holds for the boundary process, then

$$E^R\int_t^\infty f(s, x(s))\, d\xi(s) = q(t, y)$$

is uniquely determined. Further, if x in $\bar{G}$ is any point, on or off the boundary, and P is any solution to the submartingale problem starting from x at time t_0, then

$$\begin{aligned} E^P\left[\int_{t_0}^\infty f(s, x(s))\, d\xi(s)\right] &= E^P\left[\int_\tau^\infty f(s, x(s))\, d\xi(s)\right] \\ &= E^P\left[E^P\left[\int_\tau^\infty f(s, x(s))\, d\xi(s) \mid \mathscr{M}_\tau^{t_0}\right]\right] \\ &= E^P[q(\tau, x(\tau))] \\ &= E^{Q_{t_0,x}}[q(\tau, x(\tau))] . \end{aligned}$$

Here τ is the first exit time from G and $Q_{t_0,x}$ is the interior process corresponding to the given coefficients starting from x at time t_0. Hence the uniqueness of the boundary process implies that

$$E^P\left[\int_{t_0}^{\infty} f(s, x(s))\, d\xi(s)\right]$$

is uniquely determined for all starting points on or off the boundary.

THEOREM 4.2. *Let a, b, ρ and γ be coefficients such that the uniqueness theorem holds for the boundary process. Then the solution to the submartingale problem for any starting time and place is unique.*

Proof: Let x_0 be the starting point and let t_0 be the starting time. Let P_1, P_2 be two solutions. Let $Q_{t,x}$ be the process corresponding to a and b in the interior and let $h(t, x)$ be a bounded continuous function with compact support in $[t_0, \infty) \times G$. Set

$$u(t, x) = E^{Q_{t,x}}\left[\int_t^{\tau} h(s, x(s))\, ds\right],$$

where τ is the first exit time from G after time t. We define

$$g(s, x) = (J_s\, u)(s, x) + \rho(s, x)\, u_s(s, x)\,.$$

Then $u \in F$ and

$$u(t, x(t)) - \int_{t_0}^{t} h(s, x(s))\, ds - \int_{t_0}^{t} g(s, x(s))\, d\xi_i(s)$$

is a martingale relative to P_i.

Since $u(t, x)$ has compact support in t, we have

$$u(t_0, x_0) = E^{P_i}\left[\int_{t_0}^{\infty} h(s, x(s))\, ds + \int_{t_0}^{\infty} g(s, x(s))\, d\xi_i(s)\right].$$

In view of Remark 3,

$$E^{P_1}\left[\int_{t_0}^{\infty} h(s, x(s))\, ds\right] = E^{P_2}\left[\int_{t_0}^{\infty} h(s, x(s))\, ds\right].$$

Since this is true for all bounded continuous functions $h(s, x)$ with compact support in $[t_0, \infty) \times G$, it is also valid for all functions vanishing identically for large t which are bounded and supported in G. For all bounded functions g on the boundary with compact support,

$$E^{P_1}\left[\int_{t_0}^{\infty} g(s, x(s))\, d\xi_1(s)\right] = E^{P_2}\left[\int_{t_0}^{\infty} g(s, x(s))\, d\xi_2(s)\right].$$

Taking g to be a function with support in $\{(s, x) \in [t_0, \infty) \times \partial G \cap \{\rho > 0\}\}$, we have

$$E^{P_1}\left[\int_{t_0}^{\infty} g(s, x(s))\rho(s, x(s))\, ds\right] = E^{P_2}\left[\int_{t_0}^{\infty} g(s, x(s))\rho(s, x(s))\, ds\right].$$

Therefore,

$$E^{P_1}\left[\int_{t_0}^{\infty} g(s, x(s))\, ds\right] = E^{P_2}\left[\int_{t_0}^{\infty} g(s, x(s))\, ds\right]$$

for functions $g(s, x)$ with support on the boundary. Hence,

$$E^{P_1}\Big[\int_{t_0}^{\infty} F(s, x(s))\, ds = E^{P_2}\Big[\int_{t_0}^{\infty} F(s, x(s))\, ds\Big]$$

for all $F(s, x)$ bounded and vanishing for all large s. This in turn implies that

$$P_1[x(t) \in A] = P_2[x(t) \in A] \quad \text{for} \quad \text{all} \quad t > t_0 .$$

Our result now follows by taking r.c.p.d.

5. Uniqueness

Let $\bar{D}$ be a closed half-space, D its interior and ∂D its boundary. That is,

$$\bar{D} = \{x_1, \cdots, x_d : x_1 \geqq 0\},$$
$$D = \{x_1, \cdots, x_d : x_1 > 0\},$$
$$\partial D = \{x_1, \cdots, x_d : x_1 = 0\}.$$

Let us consider $[0, \infty) \times \bar{D}$ and suppose $\{a_{ij}(t, x)\}$ is a symmetric, positive definite continuous coefficient matrix such that $a_{ij}(t, x) = \delta_{ij}$ for (t, x) lying outside a bounded neighborhood of (t_0, x_0) and

$$|a_{ij}(t, x) - \delta_{ij}| \leqq \varepsilon \quad \text{for} \quad \text{all} \quad i, j, t, x,$$

where $\varepsilon > 0$ will be chosen later on. $Q_{t,x}$ will denote the interior process corresponding to $\{a_{ij}(t,x)\}$ starting from an interior point x at time t. $Q^0_{t,x}$ is the similar process for $\{\delta_{ij}\}$ (i.e., Brownian motion in D). The harmonic extension operators corresponding to these processes are defined by

$$(Hf)(t, x) = E^{Q_{t,x}}\{f(\tau, x(\tau))\}$$

and

$$(H_0 f)(t, x) = E^{Q^0_{t,x}}\{f(\tau, x(\tau))\},$$

$f(\cdot, \cdot)$ is a function on $[0, \infty) \times \partial D$. Clearly, Hf and $H_0 f$ are functions defined in $[0, \infty) \times \bar{D}$ and agree with f on the boundary.

By ∇ we denote the gradient in D and by $\tilde{\nabla}$ the gradient with respect to the tangential components on the boundary. From Theorem 7.1 of the appendix we know that, for any $\delta > 0$ and $1 < p < \infty$, there is an ε such that

$$\|(H - H_0)f\|_p^{(2)} \leqq \delta \,|||f|||_{\sim p}^{(0,1)} .$$

The definition of these norms are given in the appendix. We now introduce two boundary operators acting on functions f on the boundary such that $|||f|||_{\sim p}^{(0,1)}$ and $|||\tilde{\nabla} f|||_{\sim p}^{(0,1)}$ are finite. We take $p > d + 2$ and fix it. Then, by Remark 3, following Theorem 7.3 of the appendix,

$$(Kf)(t, x) = \langle \gamma(t, x), \nabla Hf(t, x)\rangle \quad \text{for} \quad x \in \partial D$$

is well defined and so is

$$(K_0 f)(t, x) = \langle \gamma_0 , \nabla H_0 f(t, x)\rangle \quad \text{for} \quad x \in \partial D .$$

Here γ_0 is a constant vector with $\langle \gamma_0 , e_1\rangle = 1$, where e_1 is the vector $(1, 0, \cdots, 0)$ or the unit normal into D; γ is also normalized so that

$$\langle \gamma(t, x), e_1\rangle = 1 .$$

THEOREM 5.1. *Let $\delta > 0$ and $p > d + 2$ be given. Suppose that*

$$|\gamma(t, x) - \gamma(t', x')| \leqq A(|t - t'| + |x - x'|) \tag{5.1}$$

for some finite A. Then there is an $\varepsilon > 0$, depending on δ, p, and A, such that the inequalities

$$|a_{ij}(t, x) - \delta_{ij}| \leqq \varepsilon \tag{5.2}$$

and

$$|\gamma(t, x) - \gamma_0| \leqq \varepsilon \tag{5.3}$$

imply that

$$|||Kf - K_0 f|||_{\sim p} \leqq \delta \,|||f|||_{\sim p}^{(0,1)} .$$

Proof:

$$\begin{aligned} Kf - K_0 f &= \langle \gamma, \nabla Hf\rangle - \langle \gamma_0 , \nabla h_0 f\rangle \\ &= \langle \gamma_0 , \nabla(H - H_0)f\rangle - \langle \gamma - \gamma_0 , \nabla Hf\rangle . \end{aligned}$$

Therefore,

$$|||Kf - K_0 f|||_{\sim p} \leqq \|\gamma_0\| \,|||\nabla(H - H_0)f|||_{\sim p} + |||\langle \gamma - \gamma_0 , \nabla Hf\rangle|||_{\sim p} .$$

The first term is estimated by

$$|||\nabla(H - H_0)f|||_{\sim p} \leqq C_1 \,\|(H - H_0)f\|_p^{(2)} \leqq C_2\, \varepsilon\, |||f|||_{\sim p}^{(0,1)} ,$$

because $(H - H_0)f$ is identically zero for large enough t. Moreover, since the first component of $\gamma - \gamma_0$ is identically zero and $\gamma - \gamma_0$ is Lipschitz continuous, by Theorem 7.3, there is an $\varepsilon > 0$ such that if $\|\gamma - \gamma_0\| \leqq \varepsilon$, then

$$|||\langle \gamma - \gamma_0 , \nabla Hf\rangle|||_{\sim p} \leqq \tfrac{1}{2}\delta\, |||\tilde{\nabla} f|||_{\sim p} ;$$

q.e.d.

For each function g which is C^∞ with compact support in $[0, \infty) \times \partial D$ and for each $\lambda > 0$, we know that the equation

$$\lambda f - K_0 f = g$$

has a solution f which is bounded and is again C^∞. Moreover, from Theorem 7.2,

$$|||f|||_{\sim p}^{(0,1)} \leqq C_\lambda\, |||g|||_{\sim p}$$

and

$$|||f|||_{\sim p} \leqq C_\lambda\, |||g|||_{\sim p} ,$$

where C_λ is bounded for $\lambda \geqq 1$. Therefore, by a perturbation argument,

$$(\lambda - K)^{-1} = (\lambda - K_0)^{-1}[I + (K_0 - K)(\lambda - K_0)^{-1}]^{-1} ,$$

and for $\lambda \geqq 1$

$$|||(K_0 - K)(\lambda - K_0)^{-1}\, g|||_{\sim p} \leqq C_4\, \delta\, |||g|||_{\sim p} .$$

Hence, for suitably small δ, the equation

$$\lambda f - Kf = g$$

has a solution for $\lambda \geqq 1$ such that

$$|||f|||_{\sim p}^{(0,1)} \leqq C\, |||g|||_{\sim p}$$

and

$$|||f|||_{\sim p} \leqq C\, |||g|||_{\sim p} .$$

THEOREM 5.2. *Let, for a suitably small* ε, a_{ij} *satisfy* (5.2) *and be equal to* δ_{ij} *outside a bounded set. Let* $b = 0$ *and* $\rho = 0$. *If* γ *satisfies* (5.1) *and* (5.3), *then the uniqueness theorem holds for the corresponding boundary process.*

Proof: Let t_0, x_0 be a starting point with $x_0 \in \partial D$. We have to show that there is only one measure R_{t_0,x_0} on $D([0, \infty), [t_0, \infty) \times \partial D)$ such that

$$f(\tau(t), y(t)) - \int_0^t (Kf)(\tau(s), y(s))\, ds$$

is a martingale with respect to R_{t_0,x_0}, with

$$R_{t_0,x_0}[\tau(0) = t_0, y(0) = x_0] = 1\,.$$

The martingale property, which is valid to start with for smooth functions, is easily extended to functions for which $|||f|||_{\sim p}$ and $|||f|||_{\sim p}^{(0,1)}$ are finite, provided $p > d + 2$. This is because such functions can be approximated by smooth functions f_n with compact support in such a manner that

$$f_n \to f \quad \text{and} \quad Kf_n \to Kf \quad \text{uniformly}\,.$$

(See Remark 4 after Theorem 7.3 of the appendix.) Hence, for any solution R,

$$E^R[f(\tau(t), y(t))] = f(t, x) + E^R\left[\int_0^t (Kf)(\tau(s), y(s))\, ds\right].$$

From this one passes easily to

$$\int_0^\infty e^{-\lambda s} E^R[g(\tau(s), y(s))]\, ds = f(t, x)\,,$$

provided

$$\lambda f - Kf = g\,.$$

Therefore,

$$\int_0^\infty e^{-\lambda s} E^R[g(\tau(s), y(s))]\, ds$$

is uniquely determined for g in $W_p([t_0, \infty) \times \partial D)$.[19] If one uses stochastic continuity, the uniqueness theorem for Laplace transforms guarantees the uniqueness of the marginals of R. By taking r.c.p.d. one gets the uniqueness of R; q.e.d.

We now turn to the elastic case (i.e., when $\rho > 0$). H_0, H are as before. The new operators K_0 and K are defined by

$$(Kf)(t, x) = \langle \gamma(t, x), \nabla Hf(t, x)\rangle + \rho(t, x) f_t(t, x)\,,$$
$$(K_0 f)(t, x) = \langle \gamma_0, \nabla H_0 f(t, x)\rangle + \rho_0 f_t(t, x)\,.$$

[19] $W_p([t_0, \infty) \times \partial D)$ is defined before Lemma 7.2 in the appendix.

In addition to the earlier assumptions on γ and γ_0, we now suppose that ρ satisfies a Lipschitz condition and that

$$|\rho(t, x) - \rho_0| \leqq \varepsilon , \tag{5.4}$$

where $\rho_0 > 0$. Of course, ε is so small that $\rho(t, x) > 0$. As before, we can estimate

$$|||(K - K_0)f|||_{\sim p} \leqq \delta \, |||f|||^{(1,1)}_{\sim p}$$

provided ε is chosen small depending on δ, the Lipschitz constant A and $\rho_0 > 0$. We can also solve the equation

$$\lambda f - K_0 f = g$$

for g such that $g \in W_p([t_0 , \infty) \times \partial D)$, and $\lambda \geqq 1$. Moreover,

$$|||f|||^{(1,1)}_{\sim p} \leqq C \, |||g|||_{\sim p}$$

with C independent of λ.

We can again use a perturbation argument and solve

$$\lambda f - Kf = g$$

for $g \in W_p$ with an f such that

$$|||f|||^{(1,1)}_{\sim p} \leqq C \, |||g|||_{\sim p},$$

provided ε is chosen small enough. Armed with this, we can prove the following theorem in exactly the same manner as Theorem 5.2.

THEOREM 5.3. *If a, b, and γ are as in Theorem 5.2 and ρ satisfies (5.4), then for small ε the uniqueness theorem holds for the corresponding boundary process.*

THEOREM 5.4. *Let a, b, ρ, and γ satisfy the assumptions of Theorem 5.2 or of Theorem 5.3. Then the solution to the submartingale problem for a, b, ρ, and γ is unique and depends continuously on t and x.*

Proof: The uniqueness follows from the preceding theorems and from Section 4. Now let $t_n \to t$ and $x_n \to x$. We consider $P_n = P_{t_n, x_n}$. The proof consists of establishing the compactness of P_n and checking that any limit P of a subsequence of the P_n is a solution for the submartingale problem starting at t and x. We need the uniqueness theorem to prove convergence. Compactness

is elementary, because we have the inequalities

$$E^{P_n}[|x(t_2) - x(t_1)|^2 \, |x(t_3) - x(t_2)|^2] \leqq C \, |t_3 - t_1|^2 \quad \text{for} \quad 0 \leqq t_1 \leqq t_2 \leqq t_3 < \infty .$$

The only other estimate needed is that, for the strip $G_\delta = \{x : 0 < x_1 < \delta\}$,

$$\sup_n E^{P_n}\left[\int_0^T \chi_{G_\delta}(x(s)) \, ds\right] \to 0 \quad \text{as} \quad \delta \to 0$$

for each $T < \infty$. This is proved in the same way as in Theorem 3.1. The rest of the proof also follows that of Theorem 3.1.

THEOREM 5.5 *Let a, ρ, and γ satisfy the hypotheses of Theorem* 5.2 *or Theorem* 5.3 *and let b be a bounded measurable function. Then the uniqueness theorem holds for the submartingale problem with these coefficients and, moreover, the solution $P_{t,x}$ depends measurably on t and x.*

Proof: By Corollary 2.4, if $Q_{t,x}$ is the unique solution to the problem corresponding to a, $(b = 0)$, ρ and γ, then $Q_{t,x}$ and $P_{t,x}$ are equivalent to one another and

$$\left.\frac{dP_{t,x}}{dQ_{t,x}}\right|_{\mathscr{M}_s^t} = \exp\left\{\int_t^s \langle b(u, x(u)) a^{-1}(u, x(u)) \chi_D(x(u)), dx(u)\rangle \right.$$

$$\left. - \tfrac{1}{2} \int_t^s \langle b(u, x(u)) a^{-1}(u, x(u)), b(u, x(u))\rangle \chi_D(x(u)) \, du \right\} .$$

This proves uniqueness. The continuity of $Q_{t,x}$ in (t, x) implies the measurability of $P_{t,x}$ in t and x; q.e.d.

Let $t_0 > 0$ and $x_0 \in \partial D$. Let S be a spherical neighborhood of x_0 in $\bar{D}$. Let $\varepsilon > 0$ be arbitrary and take τ to be the minimum of $t_0 + \varepsilon$ and the first exit time from S after time t_0. A measure P on $(\Omega, \mathscr{M}_\tau^{t_0})$ is said to solve the *submartingale problem for a, b, ρ, and γ up to time τ starting from (t_0, x_0) if*

(i) $P(x(t_0) = x_0) = 1$,

(ii) $f(t \wedge \tau, x(t \wedge \tau)) - \int_{t_0}^{t \wedge \tau} (f_u + Lf)(u, x(u)) \chi_D(x(u)) \, du$

is a P-submartingale for $f \in C_0^\infty([t_0, \infty) \times D)$ satisfying $\rho f_u + \langle \gamma, \nabla f \rangle \geqq 0$ on $[t_0, \infty) \times \partial D$.

THEOREM 5.6. *Let a, b, ρ, and γ be defined on $[t_0, t_0+\varepsilon] \times S$ and have extensions a', b', ρ', and γ' to $[0, \infty) \times \bar{D}$ satisfying the hypotheses of Theorem 5.5. Then the solution to the corresponding submartingale problem starting from (t_0, x_0) up to time τ is unique.*

Proof: If P is a solution up to time τ for a, b, ρ, and γ, then by adjoining to P a solution for a', b', ρ', and γ', we can obtain a solution P' for a', b', ρ' and γ' which agrees with P on $\mathscr{M}_\tau^{t_0}$. Since P' is unique, P must be unique on $\mathscr{M}_\tau^{t_0}$; q.e.d.

We now turn to the region G and coefficients a, b, ρ and γ defined on $[0, \infty) \times \bar{G}$. We suppose

(i) $\{a_{ij}(t, x)\}$ is bounded, continuous, symmetric and strictly positive definite for each (t, x),

(ii) b is bounded and measurable on $[0, \infty) \times \bar{G}$,

(iii) γ is a locally Lipschitz, bounded vector function on $[0, \infty) \times \partial G$ and $\langle \gamma(t, x), \nabla\phi(x)\rangle \geqq \beta > 0$,

(iv) either $\rho \equiv 0$, or ρ is a bounded locally Lipschitz function which is strictly positive at each point of $[0, \infty) \times \partial G$.

THEOREM 5.7. *Under the above assumptions, the solution to the submartingale problem for a, b, ρ and γ is unique for each starting point (t, x).*

Proof: Let us suppose that the following conditions hold:

1. For each $t_0 \geqq 0$ and $x_0 \in \partial G$, there is an $\varepsilon > 0$ and a neighborhood N_{x_0} of x_0 in $\bar{G}$ such that, if τ is the minimum of the exit time from N_{x_0} and $t_0 + \varepsilon$, the solution to the submartingale problem for a, b, ρ and γ up to time τ starting from (t_0, x_0) is unique.

2. ε and the size of N_{x_0} stay uniformly positive as (t_0, x_0) varies over a compact subset of $[0, \infty) \times \partial G$.

Then obviously the uniqueness theorem is valid for all time. Indeed, since uniqueness is valid up to the first exit time from G, by taking r.c.p.d., we need only prove uniqueness for starting points on the boundary. Then we use the stopping time described in condition 1 and we get uniqueness up to τ. Repeating this process indefinitely, we get a sequence of stopping times $\tau_1, \tau_2, \cdots, \tau_n, \cdots$ such that the r.c.p.d. of P given $\mathscr{M}_{\tau_j}^t$ is unique on $\mathscr{M}_{\tau_{j+1}}^{\tau_j}$ for each j. We also have uniqueness on $\mathscr{M}_{\tau_1}^{t_0}$. Since condition 2 guarantees that $\lim_{j\to\infty}\tau_j = \infty$ a.s., we have uniqueness on $\mathscr{M}^{t_0}$.

We need only prove that (i) and (ii) are valid.

Let $x_0 \in \partial G$ be given. Then there exists a neighborhood N'_{x_0} of x_0 which can be mapped by a twice continuously differentiable map Φ onto a spherical

neighborhood S_0 in $\bar{D}$ of a point y_0 on ∂D. The transformed boundary is part of a hyperplane. By choosing a linear map, if necessary, we can always assume that the transformed coefficients a', b', ρ' and γ' are such that

$$a'_{ij}(t_0, y_0) = \delta_{ij}.$$

Moreover, since the ratio of ρ' and γ' is the only relevant quantity, we may always assume that

$$\langle \gamma', e_1 \rangle = 1.$$

By computing a', b', ρ' and γ' in terms of the old coefficients, we check that they satisfy the hypotheses of Theorem 5.6. We may have to choose a smaller neighborhood S and consider everything there. Since the basic transformation Φ takes solutions to the submartingale problem into solutions for the transformed coefficients in a one-to-one manner, we get uniqueness for a, b, ρ and γ up to the time τ, which is the smaller of $t_0 + \varepsilon$ and the escape time from the inverse image N_{x_0} of S.

That condition 2 is valid is obvious because these various regions are uniformly large, since, on compact sets, everything is uniform; q.e.d.

We now turn to the homogeneous case (i.e., we assume that a, b, ρ and γ are functions of x alone). In this case, the uniqueness theorem can be strengthened by weakening the assumptions on ρ. We assume that a, b and γ satisfy the same assumptions as before, but $\rho(x)$ is only assumed to be non-negative, continuous and bounded on ∂G. Since the transformation taking a small neighborhood of a point on the boundary of the half-space is time independent, the time homogeneity is not lost by such a transformation. Therefore, we must prove uniqueness only in the case which is analogous to Theorems 5.2 and 5.3.

The clue is the fact that, in this case, K commutes with time translations. Let us use the earlier notation and define

$$(Kf)(t, x) = \langle \gamma(x), (\nabla Hf)(t, x) \rangle + \rho(x) f_t(t, x),$$
$$(K'f)(t, x) = \langle \gamma(x), \nabla Hf(t, x) \rangle.$$

We have the uniqueness theorem for K' and we want to prove the uniqueness theorem for K. Let $R_{t,x}$ be a solution to the martingale problem for K. Then, if $(\tau(s), y(s))$ is a trajectory, we define

$$\tau'(s) = \tau(s) - \int_0^s \rho(y(u))\, du$$

and consider the process $R'_{t,x}$ corresponding to $(\tau'(s), y(s))$. We show now

that $R'_{t,x}$ is a solution for K' and, therefore, is uniquely determined. Since

$$\tau(s) = \tau'(s) + \int_0^s \rho(y'(u))\, du$$

and

$$y(s) = y'(s)\,,$$

this means that $R_{t,x}$ is also unique.

We start with a measure R such that for suitable functions f,

$$f(\tau(t), y(t)) - \int_0^t (Kf)(\tau(s), y(s))\, ds$$

is a martingale. It is easy to generalize this formula to functions g depending on time t as well. In fact,

$$g(t, \tau(t), y(t)) - \int_0^t (g_s + Kg)(s, \tau(s), y(s))\, ds$$

is a martingale. We choose g to be of the form

$$g(t, \tau, y) = f(\tau - \rho t, y)\,.$$

Since K commutes with translations in the τ direction,

$$f(\tau(t) - \rho t, y(t)) - \int_0^t (Kf)(\tau(s) - \rho s, y(s))\, ds + \int_0^t f_s(\tau(s) - \rho s, y(s))\, ds$$

is a martingale. Since this is true for any constant ρ, it must also be true for simple non-anticipating functions $\rho(t)$. In particular,

$$f(\tau(t) - \eta(t), y(t)) - \int_0^t (Kf)(\tau(s) - \eta(s), y(s))\, ds + \int_0^t f_s(\tau(s) - \eta(s), y(s))\, ds$$

is a martingale, where

$$\eta(t) = \int_0^t \rho(s, \omega)\, ds\,,$$

$$\rho(s, \omega) = \rho(y(v_j)) \qquad \text{for} \qquad \sigma_j \leqq s < \sigma_{j+1}\,,$$

and the σ_j are the successive exit times from small neighborhoods. By letting

these neighborhoods shrink, we obtain

$$\eta(t) = \int_0^t \rho(y(s))\, ds$$

and see that

$$f(\tau'(t), y(t)) - \int_0^t (K'f)(\tau'(s), y(s))\, ds$$

is a martingale. We have, therefore, proved

THEOREM 5.8. *If a, b, ρ and γ are time independent, then the following conditions are sufficient for the uniqueness of the solution to the submartingale problem:*

(i) *$a(x)$ is continuous, symmetric and positive definite at each point of $\bar{G}$,*

(ii) *$b(x)$ is bounded and measurable,*

(iii) *γ is bounded, locally Lipschitz and $\langle\gamma(x), \nabla\phi(x)\rangle \geqq \beta > 0$ on ∂G,*

(iv) *$\rho(x)$ is bounded, continuous and non-negative.*

To complete this section, we wish to state a few results which can be proved in a routine manner (cf. [8]).

1. By taking r.c.p.d., one sees that any time uniqueness holds, the solutions are strong Markov processes.

2. The process $P_{t,x}$ corresponding to a given a, b, ρ and γ for which the uniqueness holds is a Feller process. This is proved by establishing compactness, taking a subsequence that converges, and identifying any limiting process as the unique solution corresponding to the limiting starting point.

3. A similar theorem could be proved on the convergence of $P^n_{t,x}$ to $P_{t,x}$ if a_n, b_n, ρ_n and γ_n converge suitably to a, b, ρ and γ. Essentially, we need compactness and uniqueness for the coefficients.

6. An Invariance Principle

G is a region in R^d with boundary ∂G. $\bar{G}$ stands for $G \cup \partial G$. G is such that there exists a function $\phi(x)$ in $C_b^2(R^d)$ with

$$G = [x : \phi(x) > 0]\,,$$

$$\partial G = [x : \phi(x) = 0]\,,$$

$$\|\nabla\phi(x)\| \geqq 1 \qquad \text{on} \qquad \partial G\,.$$

For each $h > 0$ we are given a Markov chain with $\bar{G}$ as its state space. The transitions of this Markov chain occur at times that are multiples of h.

Let $\Pi_j^h(x, dy)$ denote the transition probabilities of this chain at time jh. We assume that the chain is defined for $0 \leqq jh \leqq T$.

Let P_z^h denote the measure corresponding to this chain starting from the point $z \in \bar{G}$ at time 0 on the space $\Omega = D([0, T], \bar{G})$. If a transition occurs at time T it will be ignored. P_z^h is therefore defined by the following relations:

$$P_z^h[x(0) = z] = 1 ,$$

$$P_z^h[x(t) = x(jh) \quad \text{for} \quad jh \leqq t < (j + h)h] = 1 ,$$

$$P_z^h[x((j + 1)h) \in A \mid \mathcal{M}_{jh}] = \Pi_j^h(x(jh), A) .$$

Here $x(t)$ is the coordinate function at time t and $\mathcal{M}_t$ stands for the σ-field generated by $x(s)$ for $0 \leqq s \leqq t$. The customary demand that $x(T - 0)$ be equal to $x(T)$ for elements of $D([0, T], \bar{G})$ forces us to disregard any transition occurring at time T.

We consider the following functions defined in terms of the $\Pi_j^h(x, dy)$:

$$\Delta^h(jh, x) = \frac{1}{h}\int \|y - x\|^{2+\alpha}\, \Pi_j^h(x, dy) ,$$

$$a^h(jh, x) = \frac{1}{h}\int (y - x) \otimes (y - x)\Pi_j^h(x, dy) ,$$

$$b^h(jh, x) = \frac{1}{h}\int (y - x)\Pi_j^h(x, dy) .$$

Here $\alpha > 0$ is arbitrary but fixed. The b^h are of course vectors and the a^h are positive semi-definite matrices. We extend the definition of these functions from points of the form jh to all times t by defining

$$\Delta^h(t, x) = \Delta^h(jh, x) \quad \text{for} \quad jh \leqq t < (j + 1)h ,$$

$$a^h(t, x) = a^h(jh, x) \quad \text{for} \quad jh \leqq t < (j + 1)h ,$$

$$b^h(t, x) = b^h(jh, x) \quad \text{for} \quad jh \leqq t < (j + 1)h .$$

We shall make the following assumptions regarding these functions:

A.
$$\sup_{0 \leqq t \leqq T} \sup_{x \in \bar{G}} \Delta^h(t, x) = \Delta(h) ,$$

$$\Delta(h) \to 0 \quad \text{as} \quad h \to 0 .$$

B. There exists a constant $M < \infty$ and a constant $c > 0$ such that

$$\|b^h(t, x)\| > M$$

implies

$$\langle \nabla\phi(x), b^h(t, x)\rangle \geqq c\, \|b^h(t, x)\| \,.$$

C. For every $\delta > 0$, there exists a constant $M_\delta < \infty$ such that

$$\|b^h(t, x)\| > M_\delta$$

implies

$$\phi(x) < \delta \,.$$

D. There exists a constant $M < \infty$ such that

$$\sup_{0 \leqq t \leqq T}\ \sup_{x \in \bar{G}} \|a^h(t, x)\| \leqq M \,.$$

Remark. The norm for matrices shall be the square root of d times the sum of squares of all the entries. In particular, $\operatorname{Tr} a \leqq \|a\|$. Conditions B and C guarantee that if there are large drifts at all, they must occur close to the boundary and be directed uniformly inward and never outward or tangentially.

THEOREM 6.1. *Under hypotheses* A–D, *for any compact set* $K \subset \bar{G}$, *the family*

$$\{P_z^h : h > 0, z \in K\}$$

is conditionally compact in $D([0, T], \bar{G})$. *Moreover, any weak limit of these is concentrated on the subset* $C([0, T], \bar{G}) \subseteq D([0, T], \bar{G})$.

The proof of Theorem 6.1 will follow after a few lemmas.

LEMMA 6.1. *There exists a constant* $A < \infty$ *such that, with respect to any* P_z^h,

$$h\sum_{j=0}^{k-1} \|b^h(jh, x(jh))\| - A\phi(x(kh)) - Akh$$

is a supermartingale relative to the σ-algebras $\mathcal{M}_{kh}$.

Proof: From the relation of the process P_z^h to the transition probabilities $\{\Pi_j^h(x, dy)\}$, it follows that

$$\phi(x(kh)) - \sum_{j=0}^{k-1} \int [\phi(y) - \phi(x(jh))]\Pi_j^h(x(jh), dy)$$

is a martingale relative to $\mathcal{M}_{kh}$ and P_z^h. Let

$$\theta(jh, x) = \int [\phi(y) - \phi(x)]\Pi_j^h(x, dy) \,.$$

Since $\phi \in C_b^2(\bar{G})$, there exists a $C < \infty$ such that

$$|\phi(y) - \phi(x) - \langle y - x, \nabla\phi\rangle| \leqq C_1 |y - x|^2 . \tag{6.1}$$

By integrating (6.1) with respect to $\Pi_j^h(x, dy)$, we get

$$\begin{aligned} |\theta(jh, x) - h\langle b^h(jh, x), \nabla\phi(x)\rangle| &\leqq C_1 h \operatorname{Tr} a^h(jh, x) \\ &\leqq C_2 h . \end{aligned}$$

Therefore,

$$\theta(jh, x) \geqq h\langle b^h(jh, x), \nabla\phi(x)\rangle - C_2 h . \tag{6.2}$$

If $\|b^h(jh, x)\| \leqq M$, we have

$$\langle b^h(jh, x), \nabla\phi(x)\rangle \geqq -MC , \tag{6.3}$$

since $|\nabla\phi(x)| \leqq C$. If $\|b^h(jh, x)\| > M$, then, by hypothesis, we have

$$\langle b^h(jh, x), \nabla\phi(x)\rangle \geqq c\, \|b^h(jh, x)\| . \tag{6.4}$$

Therefore, combining (6.3) and (6.4), there exists a constant C_3 such that

$$\langle b^h(jh, x), \nabla\phi(x)\rangle \geqq c\, \|b^h(jh, x)\| - C_3 .$$

Using this and (6.2), we can find a constant A such that

$$A\theta(jh, x) \geqq h\, \|b^n(jh, x)\| - Ah .$$

Since

$$\sum_{j=0}^{k-1} \theta(jh, x(jh)) - \phi(x(kh))$$

is a martingale,

$$h \sum_{j=0}^{k-1} \|b^h(jh, x)\| - Ah - A\phi(x(kh))$$

is a supermartingale.

LEMMA 6.2. *For any $\delta > 0$, there exists a constant $M_\delta < \infty$ such that, relative to $\mathcal{M}_{kh}$ and any P_z^h,*

$$\phi^2(x(kh)) - \delta h \sum_{j=0}^{k-1} \|b^h(jh, x(jh)\| - M_\delta\, kh$$

is a supermartingale.

Proof: Let

$$\theta(jh, x) = \int [\phi^2(y) - \phi^2(x)] \Pi_j^h(x, dy) .$$

Since ϕ^2 is in $C_b^2(\bar{G})$, there exists a constant C_1 such that

$$|\phi^2(y) - \phi^2(x) - 2\phi(x)\langle y - x, \nabla\phi(x)\rangle| \leqq C_1 |y - x|^2 .$$

Therefore, as in Lemma 6.1, we have

$$\theta(jh, x) \leqq C_2 h + 2\phi(x)\langle b^h(jh, x), \nabla\phi(x)\rangle h . \tag{6.5}$$

Since $\|\nabla\phi(x)\|$ is bounded, given $\delta > 0$, there exists $M'_\delta < \infty$ such that

$$\|b^h(jh, x)\| > M'_\delta \Rightarrow 2\phi(x) \|\nabla\phi(x)\| < \delta .$$

Now if $\|b^h(jh, x)\| \leqq M'_\delta$, we have

$$2\phi(x)\langle b^h(jh, x), \nabla\phi(x)\rangle \leqq 2C_3 M'_\delta , \tag{6.6}$$

since $2\phi(x) \|\nabla\phi(x)\| \leqq C_3$. On the other hand, if $\|b^h(jh, x)\| > M'_\delta$,

$$2\phi(x)\langle b^h(jh, x), \nabla\phi(x)\rangle \leqq \delta \|b^h(jh, x)\| . \tag{6.7}$$

In any case, combining (6.6) and (6.7), we have

$$2\phi(x)\langle b^h(jh, x), \nabla\phi(x)\rangle \leqq (1 + 2C_3)M'_\delta + \delta \|b^h(jh, x)\| . \tag{6.8}$$

Therefore, there is a constant $M_\delta < \infty$ such that

$$\theta(jh, x) \leqq M_\delta h + \delta h \|b^h(jh, x)\| .$$

Since

$$\phi^2(x(kh)) - \sum_{j=0}^{k-1} \theta(jh, x(jh))$$

is a martingale, the proof of the lemma is complete.

Proof of Theorem 6.1: We want to estimate the quantity

$$E^{P_x^h}[\|x(kh) - x(rh)\| \mid \mathcal{M}_{rh}] .$$

Let us define

$$y(kh) = x(kh) - \sum_{l=0}^{k-1} hb^h(lh, x(lh)) .$$

Then $y(kh)$ is a martingale. Hence,

$$\begin{aligned} E^{P_z^h}[\|y(kh) - y(rh)\|^2 \mid \mathcal{M}_{rh}] &= \sum_{l=r}^{k-1} E^{P_z^h}[\|y(l+1)h) - y(lh)\|^2 \mid \mathcal{M}_{rh}] \\ &\leqq \sum_{l=r}^{k-1} E^{P_z^h}[\|x(l+1)h) - x(lh)\|^2 \mid \mathcal{M}_{rh}] \\ &\leqq (k-r)Mh\,. \end{aligned}$$

Therefore,

$$E^{P_z^h}[\|y(kh) - y(rh)\| \mid \mathcal{M}_{rh}] \leqq [M(k-r)h]^{1/2}\,.$$

It follows from this that

$$\begin{aligned} &E^{P_z^h}[\|x(kh) - x(rh)\| \mid \mathcal{M}_{rh}] \\ &\quad \leqq [M(k-r)h]^{1/2} + E^{P_z^h}\left[h \sum_{l=r}^{k-1} \|b^h(lh, x(lh)\|\right] \\ &\quad \leqq [M(k-r)h]^{1/2} + [A(k-r)h] + AE^{P_z^h}[\phi(x(kh)) - \phi(x(rh)|\mathcal{M}_{rh}]\,. \end{aligned}$$

To complete the proof we estimate

$$\begin{aligned} E^{P_z^h}[\phi(x(kh)) \mid \mathcal{M}_{rh}] &\leqq \{E^{P_z^h}[\phi^2(x(kh) \mid \mathcal{M}_{rh}]\}^{1/2} \\ &\leqq \phi^2(x(rh)) + M_\delta(k-r)h + \delta C^{1/2} \\ &\leqq \phi(x(rh)) + M_\delta^{1/2}((k-r)h)^{1/2} + \delta^{1/2} C^{1/2}\,. \end{aligned}$$

Consequently, for any $\delta > 0$, there exists a constant C_δ such that

$$E^{P_z^h}[\|x(kh) - x(rh)\| \mid \mathcal{M}_{rh}] \leqq \delta + C_\delta[(k-r)h]^{1/2}\,.$$

According to Theorem 7.5 of the appendix, this implies compactness in $D([0, T], \bar{G})$. To check that the limits are in $C([0, T], \bar{G})$, we note that

$$P_z^h\left[\sup_{0 \leqq jh \leqq T} \|x((j+1)h) - x(jh)\| \geqq \varepsilon\right] \leqq \left(\left[\frac{T}{h}\right] + 1\right)\frac{\Delta(h)}{\varepsilon^{2+\rho}} \to 0 \quad \text{as} \quad h \to 0\,,$$

by hypothesis.

THEOREM 6.2. *Let* $f \in C_b^{1,2}([0, T] \times \bar{G})$. *Define*

$$(6.9)\qquad \begin{aligned} F_j^h(x) = \int &[f((j+1)h, y) - f(jh, x)]\Pi_j^h(x, dy) - hf_t(jh, x) \\ &- h\langle b^h(jh, x) \cdot \nabla f(jh, x)\rangle - \tfrac{1}{2}h \operatorname{Tr}[a^h(jh, x) \cdot D^2 f(jh, x)]\,.^{20} \end{aligned}$$

[20] $D^2 f$ stands for the Hessian matrix $((\partial^2 f / \partial x_i\, \partial x_j))$.

Then, for $K \subset\subset G$,

$$\lim_{h\to 0}\sup_{z\in K} E^{P_z^h}\left[\sum_{0\leq jh\leq T} |F_j^h(x(jh))|\right] = 0 .$$

Proof: Since the family P_z^h, $h > 0$ and $z \in K$, is compact, we can find for any $\varepsilon > 0$ an M such that

$$\sup_{z\in K} P_z^h\left[\sup_{0\leq t\leq T} |x(t)| \geq M\right] \leq \varepsilon .$$

Clearly,

$$F_j^h(x) = \int H_j^h(x, y)\Pi_j^h(x, dy) ,$$

where

$$H_j^h(x, y) = f((j+1)h, y) - f(jh, x) - hf_t(jh, x) - (y - x)\nabla f(jh, x) - \tfrac{1}{2}\operatorname{Tr}[(y-x)^2\cdot D^2 f(jh, x)] .$$

From our hypotheses we know that

$$|H_j^h(x, y)| \leq C_1 |x - y|^2 + C_2 h ,$$

and so

$$|F_j^h(x)| \leq Ch .$$

To complete the proof of the theorem, it suffices to prove that

$$\sup_{\substack{x\in G\\ |x|\leq M}} |F_j^h(x)| = o(h) \quad \text{as} \quad h \to 0 \quad \text{for every} \quad M < \infty .$$

By Taylor's theorem, $|H_j^h(x, y)| = o(h) + o(|x - y|^2)$ uniformly for $x \in \bar{G}$ and $|x| \leq M$. Hence, for any $\varepsilon > 0$, there exists a $\delta > 0$ such that

$$|H_j^h(x, y)| \leq \varepsilon h + \varepsilon |x - y|^2 ,$$

provided $|x - y| \leq \delta$ and $h \leq \delta$. Therefore, if $h \leq \delta$,

$$\begin{aligned} |F_j^h(x)| &\leq \left[\int_{|x-y|\leq\delta} H_j^h(x, y)\Pi_j^h(x, dy)\right] + \left[\int_{|x-y|>\delta} H_j^h(x, y)\Pi_j^h(x, dy)\right] \\ &\leq \varepsilon\int |y - x|^2\, \Pi_j^h(x, dy) + \left[\int_{|x-y|>\delta} |x - y|^2\, \Pi_j^h(x, dy)\right] + o(h) \\ &\leq \varepsilon Ch + o(h) . \end{aligned}$$

Thus,

$$\sup_{\substack{x \in G \\ |x| \leqq M}} |F_j^h(x)| = o(h) \qquad \text{as} \qquad h \to 0\,;$$

q.e.d.

We assume the following concerning the behavior of the coefficients $a^h(t, x)$ and $b^h(t, x)$ defined for $(t, x) \in [0, T] \times \bar{G}$ as $h \to 0$.

1. $a^h \to a$ and $b^h \to b$ uniformly on compact subsets $K \subset [0, T] \times G$, where a and b are continuous diffusion and drift coefficients defined on $[0, T] \times G$.

2. There is a bounded continuous function $\rho(t, y) \geqq 0$ and a bounded continuous vector function $\gamma(t,y)$ on $[0, T] \times \partial G$ such that $\langle \gamma(t,y), \nabla\phi(y) \rangle = 1$ on $[0, T] \times \partial G$. (This is a convenient normalization. It is only the ratio of γ to ρ which is relevant.) The a^h and b^h are related to ρ and γ as $h \to 0$ in the following manner. Let

$$J_0 = [(t,y) : \rho(t,y) = 0] \quad \text{and} \quad J_1 = [(t,y) : \rho(t,y) > 0]\,.$$

(i) Given $(t, y) \in J_1$ and $\varepsilon > 0$, there exist $h_0 > 0$, $\delta_0 > 0$ such that if $|t - s| < \delta_0$, $|x - y| < \delta_0$, $h < h_0$ and $\langle \nabla\phi(x), a^h(s, x)\nabla\phi(x) \rangle < \delta_0$, we have

$$\|a^h(s, x)\| < \varepsilon \qquad \text{and} \qquad \|b^h(s, x) - \rho^{-1}(t, y)\gamma(t, y)\| < \varepsilon\,.$$

(ii) Given $(t, y) \in J_1$, there exist $\delta_0 > 0$ and $M_0 < \infty$ such that, if $|s - t| < \delta_0$ and $|y - x| < \delta_0$, we have

$$\|b^h(s, x)\| \leqq M_0 \quad \text{for all} \quad h\,.$$

(iii) Given $(t, y) \in J_0$ and $M < \infty$, there exist $\delta_0 > 0$ and $h_0 > 0$ such that, if

$$|t - s| < \delta_0\,, \quad |x - y| < \delta_0\,, \quad h < h_0 \quad \text{and} \quad \langle \nabla\phi(x), a^h(s, x)\nabla\phi(x) \rangle < \delta_0\,,$$

we have

$$\|b^h(s, x)\| \geqq M\,.$$

(iv) Given $(t, y) \in J_0$ and $\varepsilon > 0$, there exist $\delta_0 > 0$, $h_0 > 0$ and $N_0 < \infty$ such that, if

$$|s - t| < \delta_0\,, \quad |x - y| < \delta_0\,, \quad h < h_0 \quad \text{and} \quad \|b^h(s, x)\| > N_0\,,$$

we have

$$\left\| \frac{b^h(s, x)}{\langle b^h(s, x), \nabla\phi(x) \rangle} - \gamma(t, y) \right\| < \varepsilon\,.$$

THEOREM 6.3. *Let $h_n \to 0$, $z_n \to z$ and let $P^{h_n}_{z_n}$ converge weakly to a limit P. Then P has the following properties:*

(i) $P\{x(0) = z\} = 1$.

(ii) *For any function f in $C_0^{1,2}([0, T] \times \bar{G})$ with $\rho(\partial f/\partial t) + \langle \gamma, \nabla f\rangle \geqq 0$ on $[0, T] \times \partial G$,*

$$f(t, x(t)) - \int_0^t (f_s + L_s f)(s, x(s))\chi_G(x(s))\, ds$$

is a submartingale, where

$$(L_s f) = \tfrac{1}{2}\sum_{i,j=1}^{d} a_{ij}\frac{\partial^2 f}{\partial x_i\, \partial x_j} + \sum_{j=1}^{d} b_j \frac{\partial f}{\partial x_j}.$$

The proof of the theorem will be deferred until a few lemmas have been proved. We denote $P^{h_n}_{z_n}$ simply by P_n and a^{h_n} and b^{h_n} by a_n and b_n, respectively.

LEMMA 6.3. *There is a constant C such that*

$$E^{P_n}\left[\int_0^T \|b_n(s, x(s))\|\, ds\right] \leqq C.$$

Proof: By Lemma 6.1, if k_n is the largest integer such that $k_n h_n \leqq T$, then

$$h_n \sum_{j=0}^{k_n - 1} \|b_n(jh, x(jh))\| = \int_0^{k_n h_n} \|b_n(s, x(s))\|\, ds \qquad \text{a.s. } P_n,$$

and the expectation of this expression is bounded by that of

$$A\phi(x(k_n h_n)) + Ak_n h_n.$$

Since ϕ is bounded by C_1, we have

$$E^{P_n}\left[\int_0^{h_n k_n} \|b_n(s, x))\|\, ds\right] \leqq A(C_1 + T).$$

Moreover, it is obvious that

$$\int_{k_n h_n}^T \|b_n(s, x(s))\|\, ds \leqq h_n \sup_{s, x} \|b_n(s, x)\|.$$

Since $\int \|y - x\|^2 \, \Pi_j^h(x, dy) \leqq Mh$, we also have

$$\left\| \int (y - x)\Pi_j^h(x, dy) \right\| \leqq \sqrt{Mh}\,.$$

Therefore,

$$\|b_n(s, x)\| \leqq \frac{C_2}{\sqrt{h_n}}\,.$$

This proves Lemma 6.3.

LEMMA 6.4. *Let B be a compact subset of the boundary $[0, T] \times \partial G$, let U be a neighborhood of B, and take $\delta > 0$. Define*

$$f_n(s, x) = \begin{cases} 1 \text{ if } \langle \nabla\phi(x), a_n(s, x)\nabla\phi(x)\rangle \geqq \delta\,, \\ 0 \text{ otherwise}\,. \end{cases}$$

Then

$$\lim_{U \downarrow B} \limsup_{n \to \infty} E^{P_n}\left[\int_0^T f_n(s, x(s))\chi_U(x(s))\, ds\right] = 0\,.$$

Proof: Let $\rho > 0$ be arbitrary. Consider the function $\psi_\rho(x)$ defined by

$$\psi_\rho(x) = \begin{cases} 0 & \text{for} \quad \phi(x) \geqq \rho\,, \\ \dfrac{1}{\rho}\,(\rho - \phi(x))^3 & \text{for} \quad \phi(x) \leqq \rho\,; \end{cases}$$

$\psi_\rho(x)$ is clearly in $C^2(\bar{G})$ and by differentiating we find that there is a constant C such that

$$\|\nabla\psi_\rho(x)\| \leqq C\rho \tag{6.10}$$

and

$$\left\| D_i\, D_j\, \psi_\rho(x) - \left(1 - \frac{\phi(x)}{\rho}\right)(D_i\, \phi(x))(D_j\, \phi(x)) \right\| \leqq C\rho\,.$$

Also we have $0 \leqq \psi_\rho(x) \leqq \rho^2$, which implies that

$$E^{P_n}[\psi_\rho(x(T)) - \psi_\rho(x(0))] \leqq \rho^2\,.$$

From Theorem 6.2 and the definition of P_n, this implies that

$$\overline{\lim_{n \to \infty}}\, E^{P_n}\left[\int_0^T [\langle b_n(s, x(s)), \nabla\psi_\rho(x(s))\rangle + \operatorname{Tr} a_n(s, x(s)) \cdot D^2\, \psi_\rho(x(s))]\, ds\right] \leqq \rho^2\,,$$

which in turn implies that there exists a constant C_1 with

$$\overline{\lim_{n\to\infty}} E^{P_n}\left[\int_0^T 6\left(1-\frac{\phi(x)}{\rho}\right)\langle\nabla\phi(x(s)),\, a_n(s, x(s))\,\nabla\phi(x(s))\rangle\, ds\right] \leqq \rho^2 + \rho C_1 .$$

We have used here the bound (6.10) for $\|\nabla\psi_\rho(x)\|$, Lemma 6.3, and the bounds for $\|a_n(s, x)\|$. Since $\phi(x) = 0$ on ∂G, $U \subset \{x : \phi(x) < \frac{1}{2}\rho\}$ implies that, on U,

$$6\left(1-\frac{\phi(x)}{\rho}\right) \geqq 3$$

and $\langle\nabla\phi(x),\, a_n(s, x(s))\,\nabla\phi(x)\rangle \geqq \delta f_n(x)$. Therefore we have

$$\overline{\lim_{n\to\infty}} E^{P_n}\left[\int_0^T f_n(s, x)\chi_U(x(s))\, ds\right] \leqq \frac{\rho^2 + 3C_1}{3\delta}$$

which proves the lemma.

Let Φ be a bounded non-negative continuous function on $D([0, T], \bar{G})$. Let us define the measures Q_n and Q by

$$dQ_n = \Phi\, dP_n$$

and

$$dQ = \Phi\, dP .$$

Then Q_n converges to Q weakly. Let us define the following measures μ_n and μ on $[t_1, t_2] \times \bar{G}$, where $0 \leqq t_1 < t_2 \leqq T$, by setting

$$\int u(t, x)\, d\mu_n = E^{Q_n}\left[h_n \sum_{t_1 \leqq jh < t_2} u(jh_n, x(jh_n))\right],$$

$$\int u(t, x)\, d\mu = E^{Q}\left[\int_{t_1}^{t_2} u(s, x(s))\, ds\right],$$

for $u \in C([t_1, t_2] \times \bar{G})$. Then clearly μ_n converges weakly to μ. Given $f(t, x)$ in $C_0^{1,2}([0, T] \times \bar{G})$, we define the functions

$$k_n(s, x) = f_t(s, x) + \tfrac{1}{2}\operatorname{Tr}(a_n(s, x)\cdot D^2 f(s, x)) + \langle b_n(s, x), \nabla f(s, x)\rangle ,$$

$$k(s, x) = f_t(s, x) + \tfrac{1}{2}\operatorname{Tr}(a(s, x)\cdot D^2 f(s, x)) + \langle b(s, x), \nabla f(s, x)\rangle .$$

Then k_n and k vanish outside a fixed compact set $K \subset [0, T] \times \bar{G}$, and $k_n \to k$ uniformly on compact subsets of $[0, T] \times G$, and therefore on closed

subsets of $[0, T] \times G$. Let $f(t, x)$ have the further property that

$$\rho(s, x) f_t(s, x) + \langle \gamma(s, x), \nabla f(s, x) \rangle \geqq 0$$

on $[0, T] \times \partial G$.

LEMMA 6.5. *Let B be a compact subset of $[0, T] \times \partial G$ and $B_0 = B \cap J_0$. Given any $\varepsilon > 0$ and $L < \infty$, there exist $n_0 < \infty$, $\delta_0 > 0$, $N_0 > L$ and a neighborhood U_0 of B_0 such that*

(i) *if $(s, x) \in U_0$, $n \geqq n_0$ and $\langle \nabla \phi(x), a_n(s, x) \nabla \phi(x) \rangle < \delta_0$, then*

$$\|b_n(s, x)\| \geqq N_0,$$

(ii) *if $(s, x) \in U_0$, $n \geqq n_0$ and $\|b_n(s, x)\| \geqq N_0$, then*

$$\langle b_n(s, x), \nabla f(s, x) \rangle \geqq -\varepsilon \|b_n(s, x)\|.$$

Proof: Let $(t, y) \in B_0$ and $\varepsilon' > 0$ be given. Then, by hypotheses, we can find N_0', n_0' and a neighborhood U_0' of (t, y) such that, if

$$(s, x) \in U_0', \qquad n \geqq n_0', \qquad \|b_n(s, x)\| \geqq N_0',$$

then

$$\left\| \frac{b_n(s, x)}{\langle b_n(s, x), \nabla \phi(x) \rangle} - \gamma(t, y) \right\| < \varepsilon'.$$

We assume that the neighborhood U_0' is so small that for $(s, x) \in U_0'$ we have

$$|f(t, y) - f(s, x)| \leqq \varepsilon',$$
$$|\nabla f(t, y) - \nabla f(s, x)| \leqq \varepsilon'.$$

From this we conclude that

$$\frac{\langle b_n(s, x), \nabla f(s, x) \rangle}{\langle b_n(s, x), \nabla \phi(x) \rangle} \geqq -C\varepsilon',$$

and therefore that

$$\langle b_n(s, x), \nabla f(s, x) \rangle \geqq -C \|\nabla \phi(x)\| \, \varepsilon' \, \|b_n(s, x)\|.$$

Since ε' was arbitrary, we can achieve that

$$C \|\nabla \phi(x)\| \, \varepsilon' < \varepsilon.$$

If we consider the covering $\{U_0'\}$ as (t, y) varies over B_0, we can extract a finite subcovering for B_0. The union of these open sets will be denoted by V_1. Then $V_1 \supset B_0$ and there exist N_0 and n_0 such that, if

$$(s, x) \in V_1, \qquad n \geqq n_0 \qquad \text{and} \qquad \|b_n(s, x)\| \geqq N_0,$$

then

$$\langle b_n(s, x), \nabla f(s, x)\rangle \geqq -\varepsilon \|b_n(s, x)\|.$$

We can assume that $N_0 > L$ without any loss of generality. Also for each $(t, y) \in B_0$, there is a neighborhood U_0'' of (t, y), a $\delta_0'' > 0$ and an n_0'' such that if

$$n \geqq n_0'', \qquad (s, x) \in U_0'', \qquad \text{and} \qquad \langle \nabla\phi(x), a_n(s, x)\nabla\phi(x)\rangle < \delta_0'',$$

then

$$\|b_n(s, x)\| \geqq N_0.$$

We can again extract a finite subcovering. The union of these open sets will be denoted by V_2. Clearly, $U_0 = V_1 \cap V_2$ and the natural choices of n_0 and δ_0 work.

LEMMA 6.6. *Let U_0 be as in Lemma 6.5 and $B_1 = B \cap U_0^c$. Then $B_1 \subset J_1$ is compact. Given any $\varepsilon > 0$, there exist a neighborhood U_1 of B_1, a $\delta_0 > 0$ and an $n_0 < \infty$ such that if*

$$(s, x) \in U_1, \quad n \geqq n_0 \quad \textit{and} \quad \langle \nabla\phi(x), a_n(s, x)\nabla\phi(x)\rangle < \delta_0,$$

then

$$f_t(s, x) + \langle b_n(s, x), \nabla f(s, x)\rangle \geqq -\varepsilon$$

and

$$\|a_n(s, x)\| \leqq \varepsilon.$$

Moreover, there exists a constant M, independent of ε, such that the choice of U_1 can always be made so that, for $(s, x) \in U_1$ and for all n,

$$\|b_n(s, x)\| \leqq M.$$

Proof: The proof is exactly like that of Lemma 6.5 in that it uses our hypotheses locally and the compactness of B_1 to get a global bound.

LEMMA 6.7.

$$\liminf_{U \downarrow B} \liminf_{n\to\infty} \int_{U \cap I} k_n(s, x)\, d\mu_n \geqq 0,$$

where

$$I = \{(s, x) : t_1 \leqq s \leqq t_2\}.$$

Proof: Let $\varepsilon > 0$ be arbitrary. We choose U_0, U_1, δ_0 and n_0 according to Lemmas 6.5 and 6.6. Set $U = U_0 \cup U_1$. Let $n \geqq n_0$ and

$$U_{1,1}^{(n)} = U_1 \cap \{(s, x) : \langle \nabla\phi(x), a_n(s, x)\, \nabla\phi(x)\rangle < \delta_0\},$$
$$U_{1,2}^{(n)} = U_1 \cap \{(s, x) : \langle \nabla\phi(x), a_n(s, x)\, \nabla\phi(x)\rangle \geqq \delta_0\},$$
$$U_{0,1}^{(n)} = U_0 \cap U_1^c \cap \{(s, x) : \|b_n(s, x)\| \geqq N_0\},$$
$$U_{0,2}^{(n)} = U_0 \cap U_1^c \cap \{(s, x) : \|b_n(s, x)\| < N_0\}.$$

Then, for each n,

$$U = U_{1,1}^{(n)} \cup U_{1,2}^{(n)} \cup U_{0,1}^{(n)} \cup U_{0,2}^{(n)}.$$

We consider the behavior of k_n on each of these sets separately.

Case 1. By Lemma 6.6,

$$\int_{U_{1,1}^{(n)} \cap I} k_n(s, x)\, d\mu_n(s, x) \geqq -C\varepsilon, \tag{6.11}$$

where C depends on the bounds on $D^2 f$.

Case 2. Again by Lemma 6.6,

$$\int_{U_{1,2}^{(n)} \cap I} k_n(s, x)\, d\mu_n(s, x) \geqq -M'\mu_n(U_{1,2}^{(n)}), \tag{6.12}$$

where M' depends on the bound on the k_n.

Case 3. By Lemma 6.5,

$$\int_{U_{0,1}^{(n)} \cap I} k_n(s, x)\, d\mu_n(s, x) \geqq -\varepsilon \int_{U_{0,1}^{(n)}} \|b_n(s, x)\|\, d\mu_n - C\mu_n(U_{0,1}^{(n)}), \tag{6.13}$$

where C depends on the bound on $k_n - \langle b_n, \nabla f\rangle$.

Case 4. Finally, using the bound on k_n restricted to $U_{0,2}^{(n)}$, we have

$$\int_{U_{0,2}^{(n)} \cap I} k_n(s, x)\, d\mu_n(s, x) \geqq -(A_1 + A_2 N_0)\mu_n(U_{0,2}^{(n)}). \tag{6.14}$$

Combining (6.11) through (6.14), one obtains

$$\begin{aligned} \int_{U \cap I} k_n(s, x)\, d\mu_n \geqq -C\varepsilon - M'\mu_n\{U_{1,2}^{(n)}\} - \varepsilon \int_{[0,T]\times G} \|b_n(s, x)\|\, d\mu_n \\ - C\mu_n[U_{0,1}^{(n)}] - (A_1 + A_2 N_0]\mu_n\{U_{0,2}^{(n)}\}. \end{aligned} \tag{6.15}$$

Now let $U \downarrow B$. Since $U_{0,2}^{(n)} \subset \{s, x : \langle \nabla\phi(x) a_n(s, x) \nabla\phi \rangle \geqq \delta\}$, we have, by Lemma 6.4,

$$\limsup_{U \downarrow B} \limsup_{n\to\infty} \mu_n[U_{0,2}^{(n)} \cup U_{1,2}^{(n)}] = 0 . \tag{6.16}$$

By Lemma 6.3,

$$\int_{[0,T]\times\bar{G}} \|b_n(s, x)\| \, d\mu_n \leqq C ,$$

and since, in $U_{0,1}^{(n)}$, $\|b_n(s, x)\| \geqq N_0 > L$ we have

$$\mu_n[U_{0,1}^{(n)}] \leqq \frac{C}{L} . \tag{6.17}$$

Therefore, using (6.16) and (6.17) in (6.15), we see that

$$\liminf_{\downarrow B} \liminf_{n\to\infty} \int_{U\cap J} k_n(s, x) \, d\mu_n \geqq -C'\varepsilon - C'\varepsilon - \frac{C'}{L} .$$

Since ε and L are arbitrary, we have

$$\liminf_{U \downarrow B} \liminf_{n\to\infty} \int_{U\cap J} k_n(s, x) \, d\mu_n \geqq 0 .$$

LEMMA 6.8.

$$\liminf_{n\to\infty} \int_{[t_1 \leqq s \leqq t_2]\times\bar{G}} k_n(s, x) \, d\mu_n \geqq \int_{[t_1 \leqq s \leqq t_2]\times\bar{G}} k(s, x) \, d\mu .$$

Proof: Since in the t-component, μ is Lebesgue measure, $[t_1, t_2] \times \bar{G}$ is a μ-continuity set. We can, therefore, restrict our attention to $[t_1, t_2] \times \bar{G}$. Let B be the support of f and its derivatives on $[t_1, t_2] \times \partial G$. Choose U to be an open neighborhood of B so that $\mu(\partial U) = 0$. Since $k_n \to k$ uniformly on U^c,

$$\lim_{n\to\infty} \int_{U^c} k_n(s, x) \, d\mu_n = \int_{U^c} k(s, x) \, d\mu .$$

By Lemma 6.7, for any given $\varepsilon > 0$, we can choose U so that

$$\liminf_{n\to\infty} \int_{U} k_n(s, x) \, d\mu_n \geqq -\varepsilon .$$

Therefore,

$$\liminf_{n\to\infty} \int_{[t_1,t_2]\times \bar{G}} k_n(s, x)\, d\mu_n \geqq \int_{U^c} k(s, x)\, d\mu - \varepsilon .$$

Letting $U \downarrow B$ and letting $\varepsilon \to 0$, we have the lemma.

Proof of Theorem 6.3: Set

$$Y(t) = \int_0^t (f_s + L_s f)(s, x(s))\chi_G(x(s))\, ds$$
$$= \int_0^t k(s, x(s))\chi_G(x(s))\, ds ,$$

$$Y_n(t) = h_n \sum_{0\leqq jh_n<t} k_n(jh_n , x(jh_n)) ,$$

and

$$Z_n(t) = \sum_{0\leqq jh_n<t} \int [f((j+1)h_n , y) - f(jh_n , x(jh_n))]\Pi_j^{h_n}(x(jh_n), dy) .$$

Relative to $\mathscr{M}_{kh_n}$ and P_n,

$$f(kh_n , x(kh_n)) - Z_n(kh_n)$$

is a martingale. Let us choose t_0 such that $0 \leqq t_0 < T$. Let Φ be a non-negative bounded continuous function on $D([0, T], \bar{G})$ which is $\mathscr{M}_{t_0}$ measurable. Let t be an arbitrary element of (t_0 , T). Find k_n and k'_n such that $k_n h_n \to t_0$ from the right and $k'_n h_n \to t$. Then we have, for each n,

$$\int [f(k_n h_n , x(k_n h_n)) - Z_n(k_n h_n) - f(k'_n h_n , x(k'_n h_n)) + Z_n(k'_n h_n)]\Phi\, dP_n = 0 .$$

From Theorem 6.2 we know that

$$E^{P_n} \sup_{0\leqq t\leqq T} |Z_n(t) - Y_n(t)| \to 0 \qquad \text{as} \qquad n \to \infty .$$

Also,

$$f(k_n h_n , x(k_n h_n)) \to f(t, x(t))$$

and

$$f(k'_n h_n , x(k'_n h_n)) \to f(t_0 , x(t_0)) .$$

Hence,

$$E^P[\{f(t, x(t)) - f(t_0, x(t_0))\}\Phi]$$

$$\begin{aligned} &= \lim_{n\to\infty} \int [Z_n(t) - Z_n(t_0)]\Phi\, dP_n \\ &= \lim_{n\to\infty} \int [Z_n(t) - Z_n(t_0)]\, dQ_n \\ &= \lim_{n\to\infty} \int_{t_0}^{t} \int_{\bar G} k_n(s, x)\, d\mu_n(s, x) \\ &\geqq \int_{t_0}^{t} \int_{G} k(s, x)\, d\mu(s, x) \\ &= E^Q\left[\int_{t_0}^{t} (f_s + L_s f)(s, x(s))\chi_G(x(s))\, ds\right] \\ &= E^P\left[\Phi \int_{t_0}^{t} (f_s + L_s f)(s, x(s))\chi_G(x(s))\, ds\right]. \end{aligned}$$

Therefore,

$$E^P[f(t, x(t) - f(t_0, x(t_0)) - \int_{t_0}^{t} \chi_G(f_s + L_s f)(s, x(s))\, ds \mid \mathscr{M}_{t_0}] \geqq 0.$$

This proves property (ii) of the theorem. Property (i) is trivial.

Remark. If the uniqueness theorem is valid for the limiting coefficients a, b, ρ, and γ, then

$$\lim_{\substack{h\to 0\\ x\to z}} P_x^h = P_z ,$$

where P_z is the unique solution to the submartingale problem for the limiting coefficients.

7. Appendix

Let $D = \{x \in R^d : x_1 > 0\}$. Given $u \in C_0^\infty([0, T] \times \bar D)$, define

$$\|u\|_{p,T}^{(2)} = \|u\|_{p,T} + \|D_t u\|_{p,T} + \sum_{|\alpha|\leqq 2} \|D_x^\alpha u\|_{p,T} ,$$

where $\|\cdot\|_{p,T}$ is the ordinary L_p-norm on $[0, T] \times \bar D$. Denote by $W_p^{(2)}([0, T) \times \bar G)$ the completion of $C_0^\infty([0, T) \times \bar G)$ with respect to $\|\cdot\|_{p,T}^{(2)}$. If $T = \infty$, we drop the subscript T. Given a function f on $[0, T) \times \bar D$, let f^* denote the anti-asymmetric extension of f across $x_1 = 0$. We define,

for $f \in C_0^\infty([0, T) \times \bar{D})$,

$$G_0 f(s, x) = \int_{-\infty}^{\infty} dt \int_{R^d} dy\, p(t - s, x - y) f^*(t, y) , \tag{7.1}$$

where

$$p(t, y) = \chi_{t \geq 0} \frac{1}{(2\pi t)^{d/2}} e^{-|y|^2/2t} . \tag{7.2}$$

LEMMA 7.1. *If $f \in L_p([0, T) \times \bar{D})$, then for each $T < \infty$ and $1 \leqq p \leqq \infty$ there is a constant $C_{p,T}$ independent of f such that*

$$\|G_0 f\|_{p,T} \leqq C_{p,T} \|f\|_{p,T} , \tag{7.3}$$

$$\sum_{|\alpha|=1} \|D_x^\alpha G_0 f\|_{p,T} \leqq C_{p,T} \|f\|_{p,T} . \tag{7.4}$$

Moreover, for each $1 < p < \infty$ there is a constant C_p independent of T and f such that

$$\sum_{|\alpha|=2} \|D_x^\alpha G_0 f\|_p \leqq C_p \|f\|_p . \tag{7.5}$$

In particular, for each finite T the operator G_0 maps $L_p([0, T) \times \bar{D})$ boundedly into $W_p^{(2)}([0, \infty) \times D)$.

Proof: The only nontrivial assertion is (7.5). However, by the estimate of B. Frank Jones (cf. [3]), for $|\alpha| = 2$ the $L_p([0, \infty) \times R^d)$-norm of $D_x^\alpha G_0 f$ is bounded by the $L_p([0, \infty) \times R^d)$-norm of f^*, and (7.5) follows immediately from this.

The estimates (7.3), (7.4) can be proved by checking that the $L_1([0, T) \times R^d)$-norms of $p(t, x)$ and $D_{x_i} p(t, x)$ are finite for each finite T.

Next let

$$L = \tfrac{1}{2} \sum_{i,j=1}^{d} a_{ij}(t, x) \frac{\partial^2}{\partial x_i \partial x_j}$$

be a second order elliptic operator. Define $\varepsilon_{ij}(t, x) = a_{ij}(t, x) - \delta_{ij}$ and let

$$T_\varepsilon f(t, x) = \tfrac{1}{2} \sum_{i,j=1}^{d} \varepsilon_{ij}(t, x) \frac{\partial^2 G_0 f}{\partial x_i \partial x_j}(t, x)$$

for $f \in C_0^\infty([0, \infty) \times \bar{D})$. By the estimate (7.4), we see that for each $1 < p < \infty$

and $\delta > 0$ there is an $\varepsilon_{p,\delta} < 0$ such that

$$\|T_\varepsilon f\|_p \leqq \delta \|f\|_p \tag{7.6}$$

if

$$\sup_{\substack{t\geqq 0\\ x\in \bar{D}}} \sum_{i,j} |\varepsilon_{ij}(t, x)| \leqq \varepsilon_{p,\delta} .$$

In particular, if

$$\sup_{\substack{t\geqq 0\\ x\in \bar{D}}} \sum_{i,j} |\varepsilon_{ij}(t, x)| \leqq \varepsilon_{p,\delta} ,$$

where $\delta < 1$, then $(I - T_\varepsilon)^{-1}$ is bounded from $L_p([0, \infty) \times \bar{D})$ into $L_p([0, \infty) \times \bar{D})$, and $\|(I - T_\varepsilon)^{-1}\|_p^p \leqq 1/(1 - \delta)$. Hence we can define

$$G_L = G_0(I - T_\varepsilon)^{-1} , \tag{7.7}$$

and we can assert the existence of constants $C_{p,T,L}$ such that

$$\|G_L f\|_p \leqq C_{p,T,L} \|f\|_{p,T} , \qquad 1 \leqq p \leqq \infty , \quad 0 \leqq T < \infty , \tag{7.8}$$

$$\|D_x^\alpha f\|_p \leqq C_{p,T,L} \|f\|_{p,T} , \qquad 1 \leqq p \leqq \infty , \quad 0 \leqq T < \infty , \quad |\alpha| = 1 , \tag{7.9}$$

for $f \in L_p([0, T) \times \bar{D})$, and constants $C_{p,L}$ such that

$$\|D_x^\alpha f\|_p \leqq C_{p,L} \|f\|_p , \qquad 1 < p < \infty , \quad |\alpha| = 2 , \tag{7.10}$$

for $f \in L_p([0, \infty) \times \bar{D})$. Of course, the dependence of $C_{p,T,L}$ and $C_{p,L}$ on L is through the size of the ε_{ij}.

We next introduce certain L_p-type norms on $[0, \infty) \times \partial D$. In order to avoid confusion, we use $\| \cdot \|_{\sim p}$ to denote the ordinary L_p-norm on $[0, \infty) \times \partial D$. Given $f \in C_0^\infty([0, \infty) \times \partial D)$, define

$$\langle\langle f \rangle\rangle_{\sim p,x} = \left[\int_0^\infty dt \int_{\partial D} \int_{\partial D} \frac{|f(t, x) - f(t, y)|^p}{|x - y|^{d+p-2}} \, dx \, dy \right]^{1/p} ,$$

$$\langle\langle f \rangle\rangle_{\sim p,t} = \left[\int_{\partial D} dy \int_0^\infty \int_0^\infty \frac{|f(t, y) - f(s, y)|^p}{|t - s|^{(p+1)/2}} \, ds \, dt \right]^{1/p} ,$$

$$|||f|||_{\sim p} = \|f\|_{\sim p} + \langle\langle f \rangle\rangle_{\sim p,x} + \langle\langle f \rangle\rangle_{\sim p,t},$$

$$|||f|||_{\sim p}^{(0,1)} = |||f|||_{\sim p} + \sum_2^d |||D_{x_j} f|||_{\sim p} ,$$

$$|||f|||_{\sim p}^{(1,1)} = |||f|||_{\sim p}^{(0,1)} + |||D_t f|||_{\sim p} .$$

Associated with these norms we introduce the spaces $W_p([0, \infty) \times \partial D)$, $W_p^{(0,1)}([0, \infty) \times \partial D)$, and $W_p^{(1,1)}([0, \infty) \times \partial D)$. Before putting these spaces to work, we state the following simple facts.

LEMMA 7.2. *Let T be a bounded, linear, translation invariant map of $L_p([0,\infty) \times \partial D)$ into itself. Then*

$$\langle\langle Tf\rangle\rangle_{\sim p,x} \leqq \|T\|_{\sim p}^{p}\langle\langle f\rangle\rangle_{\sim p,x} \quad \text{and} \quad \langle\langle Tf\rangle\rangle_{\sim p,t} \leqq \|T\|_{\sim p}^{p}\langle\langle f\rangle\rangle_{\sim p,t}.$$

In particular, T maps each of the spaces $W_p([0, \infty) \times \partial D)$, $W_p^{(0,1)}([0, \infty) \times \partial D)$, and $W_p^{(1,1)}([0, \infty) \times \partial D)$ into itself with norm $\|T\|_{\sim p}^{p}$.

The basic reason for dealing with these spaces is the following. Define

$$h_{x_1}^{(0)}(t, \tilde{y}) = \chi_{t \geqq 0} \frac{x_1}{t(2\pi t)^{d/2}} \exp\{-(x_1^2 + |\tilde{y}|^2)/2t\}, \qquad x_1 > 0, \quad y \in D,$$

where $\tilde{y}$ is the projection of y on ∂D. For $f \in C_0^\infty([0, \infty) \times \partial D)$, let

$$H_0 f(s, x) = H_{x_1}^{(0)} f(s, \tilde{x}) = \int_s^\infty dt \int_D d\tilde{y}\, h_{x_1}^{(0)}(t - s, x - \tilde{y}) f(t, \tilde{y}). \tag{7.11}$$

It is easily checked that $H_0 f$ is the time-space harmonic extension for $(\partial/\partial s + \frac{1}{2}\Delta)$ of f in $[0, \infty) \times \bar{D}$. The crucial fact which we need is that, for each $1 < p < \infty$, there is a constant C_p such that

$$\sum_{|\alpha|=2} \|D_x^\alpha f\|_p \leqq C_p \, |||f|||_{\sim p}^{(0,1)}. \tag{7.12}$$

This estimate is proved on p. 294 of [5].

THEOREM 7.1. *Let $L = \frac{1}{2}\sum_{i,j} a_{ij}(t, x)\, \partial^2/\partial x_i\, \partial x_j$ be an elliptic operator and let $\varepsilon_{ij}(t, x) = a_{ij}(t, x) - \delta_{ij}$. Define*

$$D_\varepsilon = \tfrac{1}{2}\sum \varepsilon_{ij}(t, x)\, \partial^2/\partial x_i\, \partial x_j.$$

Given $f \in C_0^\infty([0, \infty) \times \partial D)$, let $H_L f$ denote the time-space harmonic extension of f to $[0, \infty) \times \bar{D}$. If for some $T < \infty$ the ε_{ij} vanish when $t \geqq T$, then for each $1 < p < \infty$ and $\delta > 0$ there is an $\varepsilon_{p,\delta,T} > 0$ such that

$$\sup_{\substack{x \geqq 0 \\ x \in \bar{D}}} \sum_{i,j} \varepsilon_{ij}(t, x) \leqq \varepsilon_{p,\delta,T}$$

implies

(7.13) $$\|(H_L - H_0)f\|_p^{(2)} \leqq \delta \,|||f|||_p^{(0,1)} .$$

Proof: Assume that the ε_{ij} are small enough for (7.8), (7.9) and (7.10) to hold. A simple computation yields

$$H_L - H_0 = G_L D_\varepsilon H_0 .$$

From (7.12) we see that $D_\varepsilon H_0 f \in L_p([0, T) \times \bar{D})$ and that

$$\|D_\varepsilon H_0 f\|_{p,T} \leqq C_p \varepsilon \,|||f|||_{\sim p}^{(0,1)},$$

where

$$\varepsilon = \sup_{\substack{t \geq 0 \\ x \in D}} \sum_{i,j} |\varepsilon_{ij}(t, x)| .$$

Thus, by (7.8), (7.9) and (7.10),

$$\|(H_L - H_0)f\|_p^{(2)} \leqq C\varepsilon \,|||f|||_{\sim p}^{(0,1)} .$$

We want to study next the operators $\{H_{x_1}^{(0)}\}_{x_1>0}$ as transformations from $C_0^\infty([0, \infty) \times \partial D)$ into $C_b([0, \infty) \times \partial D)$. First note that

(7.14) $$\widehat{h_{x_1}^{(0)}}(\mu, \tilde{\theta}) = \exp\{-x_1(2i\mu + |\tilde{\theta}|^2)^{1/2}\} .$$

This shows that $\{H_{x_1}^{(0)}\}_{x_1>0}$ is a semigroup. Let K_0 denote its generator. Then

(7.15) $$\widehat{K_0 f}(\mu, \tilde{\theta}) = -(2i\mu + |\tilde{\theta}|^2)^{1/2} \hat{f}(\mu, \tilde{\theta}) .$$

Denoting by $R_\lambda^{(0)}$ the resolvent operators $(\lambda - K_0)^{-1}$ associated with $\{H_{x_1}^{(0)}\}_{x_1>0}$, we have

(7.16) $$\widehat{R_\lambda^{(0)} f}(\mu, \tilde{\theta}) = \frac{1}{\lambda + (2i\mu + |\tilde{\theta}|^2)^{1/2}} \hat{f}(\mu, \tilde{\theta}) .$$

Using the representation

$$R_\lambda^{(0)} = \int_0^\infty e^{-\lambda x_1} H_{x_1}^{(0)}\, dx_1 ,$$

we see that

(7.17) $$\|R_\lambda^{(0)} f\|_{\sim p} \leqq \frac{1}{\lambda} \|f\|_{\sim p} , \qquad 1 \leqq p \leqq \infty$$

Since $\{H_{x_1}^{(0)}\}_{x_1>0}$ is a convolution semigroup, in order to find the semigroup $\{H_{x_1}^{(\rho,\tilde{\gamma})}\}_{x_1>0}$ generated by $K_{\rho,\tilde{\gamma}} = K_0 + \langle\tilde{\gamma}, \tilde{\nabla}\rangle + \rho\,\partial/\partial t$, where $\rho \geqq 0$ and $\tilde{\gamma} \in R^{d-1}$ are constants, we simply have to take

$$h_{x_1}^{(\rho,\tilde{\gamma})}(t, \tilde{x}) = h_{x_1}^{(0)}(t - \rho x, \tilde{x} - x_1\tilde{\gamma}) \tag{7.18}$$

and set

$$H_{x_1}^{(\rho,\tilde{\gamma})}f(s, \tilde{x}) = \int_s^\infty dt \int_{\partial G} d\tilde{y}\, h_{x_1}^{(\rho,\tilde{\gamma})}(t - s, \tilde{x} - \tilde{y}) f(t, \tilde{y})\,. \tag{7.19}$$

We can now read off the following facts about $\{H_{x_1}^{(\rho,\tilde{\gamma})}\}_{x_1>0}$, $K_{\rho,\tilde{\gamma}}$, and $R_\lambda^{(\rho,\tilde{\gamma})}$ from the corresponding statements about $\{H_{x_1}^{(0)}\}_{x_1>0}$, K_0, and $R_\lambda^{(0)}$:

$$K_{\rho,\tilde{\gamma}}f(\mu, \tilde{\theta}) = -((2i\mu + |\tilde{\theta}|^2)^{1/2} + i\langle\tilde{\gamma}, \tilde{\theta}\rangle + i\mu)\hat{f}(\mu, \tilde{\theta})\,, \tag{7.20}$$

$$R_\lambda^{(\rho,\tilde{\gamma})}f(\mu, \tilde{\theta}) = \frac{1}{\lambda + (2i\mu + |\tilde{\theta}|^2)^{1/2} + i\langle\tilde{\gamma}, \tilde{\theta}\rangle + i\rho\mu}\hat{f}(\mu, \tilde{\theta})\,, \tag{7.21}$$

$$\|R_\lambda^{(\rho,\tilde{\gamma})}f\|_p \leqq \frac{1}{\lambda}\|f\|_p\,, \qquad 1 \leqq p \leqq \infty\,. \tag{7.22}$$

THEOREM 7.2. *For each* $1 < p < \infty$, $\rho \geqq 0$, $\tilde{\gamma} \in R^{d-1}$, *there is a* $C_{p,\rho,\tilde{\gamma}}$ *such that*

$$\sum_{j=2}^{d} |||D_{x_j} R_\lambda^{(\rho,\tilde{\gamma})}f|||_{\sim p} \leqq C_{p,\rho,\tilde{\gamma}} |||f|||_{\sim p}\,, \qquad \lambda \geqq 0\,. \tag{7.23}$$

In particular, if $\lambda > 0$, *then*

$$|||R_\lambda^{(\rho,\tilde{\gamma})}f|||_{\sim p}^{(0,1)} \leqq C_{p,\rho,\tilde{\gamma},\lambda} |||f|||_{\sim p}\,. \tag{7.24}$$

Moreover, if $\rho < 0$, *then*

$$|||D_t R_\lambda^{(\rho,\tilde{\gamma})}f|||_{\sim p} \leqq C_{p,\rho,\tilde{\gamma}} |||f|||_{\sim p}\,, \tag{7.25}$$

and therefore, if $\lambda > 0$, *then*

$$|||R_\lambda^{(\rho,\tilde{\gamma})}f|||_{\sim p}^{(1,1)} \leqq C_{p,\rho,\tilde{\gamma},\lambda} |||f|||_{\sim p}\,. \tag{7.26}$$

Proof: Let $\rho = 0$. In order to prove (7.23) for $\lambda = 0$, it suffices to know that $\theta_j/[2i\mu + |\tilde{\theta}|^2)^{1/2} + i\langle\tilde{\gamma}, \tilde{\theta}\rangle]$ is an L_p-multiplier (cf. [3] or [6]), $1 < p < \infty$. That this is the case follows from the work of Fabes and Riviere [3] on multipliers of mixed homogeneity. To prove (7.23) when $\lambda > 0$, we use the resolvent

equation

$$R_\lambda^{(0,\tilde{\gamma})} = R_0^{(0,\tilde{\gamma})} - \lambda R_0^{(0,\tilde{\gamma})} R_\lambda^{(0,\tilde{\gamma})}.$$

From this we see that

$$\| D_{x_j} R_\lambda^{(0,\tilde{\gamma})} \|_p \leqq 2 \, \| D_{x_j} R_0^{(0,\tilde{\gamma})} f \|_p ,$$

and so (7.23) follows immediately.

We next turn to the case when $\rho > 0$. Now we must know that

$$\frac{\mu}{[2i\mu + \tilde{\theta}^2)^{1/2} + i\langle \tilde{\gamma}, \tilde{\theta} \rangle + i\rho\mu]}$$

is an L_p-multiplier. This multiplier no longer fits into the category treated in [3]. However, it does satisfy the conditions of Theorem 4.5 in [6], and is therefore an L_p-multiplier for $1 < p < \infty$. Arguing as in the preceding paragraph, we now have (7.25) for all $\rho > 0$.

In order to prove (7.23) for $\rho > 0$, we must show that

$$\frac{\theta_j}{(2i\mu + |\tilde{\theta}|^2)^{1/2} + i\langle \tilde{\gamma}, \tilde{\theta} \rangle + i\rho\mu}$$

is an L_p-multiplier. But, according to the preceding paragraph,

$$\frac{\mu}{(2i\mu + |\tilde{\theta}|^2)^{1/2} + i\langle \tilde{\gamma}, \tilde{\theta} \rangle + i\rho\mu}$$

and therefore also

$$\frac{(2i\mu + |\tilde{\theta}|^2)^{1/2} + i\langle \tilde{\gamma}, \tilde{\theta} \rangle}{(2i\mu + |\tilde{\theta}|^2)^{1/2} + i\langle \tilde{\gamma}, \tilde{\theta} \rangle + i\rho\mu} = 1 - \frac{i\rho\mu}{(2i\mu + |\tilde{\theta}|^2)^{1/2} + i\langle \tilde{\gamma}, \tilde{\theta} \rangle + i\rho\mu}$$

are L_p-multipliers. By the Fabes-Riviere result, we know that

$$\frac{\theta_j}{(2i\mu + |\tilde{\theta}|^2)^{1/2} + i\langle \tilde{\gamma}, \tilde{\theta} \rangle}$$

is an L_p-multiplier. Combining this with the above and remembering that the product of two L_p-multipliers is again an L_p-multiplier, we see that

$$\frac{\theta_j}{(2i\mu + |\tilde{\theta}|^2)^{1/2} + i\langle \tilde{\gamma}, \tilde{\theta} \rangle + i\rho\mu}$$

is an L_p-multiplier.

The last result that we prove in this section concerns the possibility of perturbing in $W_p([0, \infty) \times \partial D)$.

THEOREM 7.3. *Let* α *be a function on* $[0, \infty) \times \partial D$ *such that*

$$\sup_{\substack{t \geqq 0 \\ x \in \partial G}} |\alpha(t, x)| \leqq \varepsilon \leqq 1$$

and

$$|\alpha(s, x) - \alpha(t, y)| \leqq A(|t - s| + |x - y|) .$$

Then there is a constant C_q *for each* $1 < p < \infty$ *such that*

$$|||\alpha f|||_{\sim p} \leqq C_p \, \varepsilon^{1/(p+1)} \, |||f|||_{\sim p} .$$

Proof: Clearly, $\|\alpha f\|_{\sim p} \leqq \varepsilon \, \|f\|_{\sim p}$. We shall show that

$$\langle\langle \alpha f \rangle\rangle_{\sim p, x} \leqq C_p \, \varepsilon^{1/(p+1)} \, |||f|||_{\sim p} .$$

The proof that

$$\langle\langle \alpha f \rangle\rangle_{\sim p, t} \leqq C_p \, \varepsilon^{1/(p+1)} \, |||f|||_{\sim p}$$

is similar. We have

$$\begin{aligned}
(\langle\langle \alpha f \rangle\rangle_{\sim p, x})^p &= \int_0^\infty dt \int_{\partial D} \int_{\partial D} \frac{|\alpha f(t, x) - \alpha f(t, y)|^p}{|x - y|^{d+p-2}} \, dx \, dy \\
&= \int_0^\infty dt \iint_{|x-y| \geqq 1} \frac{|\alpha f(t, x) - \alpha f(t, y)|^p}{|x - y|^{d+p-2}} \, dx \, dy \\
&\quad + \int_0^\infty dt \iint_{|x-y| < 1} \frac{|\alpha f(t, x) - \alpha f(t, y)|^p}{|x - y|^{d+p-2}} \, dx \, dy \\
&= I_1 + I_2 .
\end{aligned}$$

Note that

$$I_1 \leqq 2\varepsilon^p \int_0^\infty dt \int_{\partial D} |f(t, x)|^p \, dx \int_{|y-x| \geqq 1} \frac{1}{|y - x|^{d+p-2}} \, dy \leqq C_p \, \varepsilon^p \, \|f\|_{\sim p}^p .$$

We treat I_2 next:

$$\begin{aligned}
I_2 &\leqq C_p \int_0^\infty dt \int_{|x-y| < 1} \frac{|\alpha(t, x)|^p \, |f(t, x) - f(t, y)|^p}{|x - y|^{d+p-2}} \, dx \, dy \\
&\quad + C_p \int_0^\infty dt \int_{|x-y| < 1} \frac{|f(t, y)|^p \, |\alpha(t, x) - \alpha(t, y)|^p}{|x - y|^{d+p-2}} \, dx \, dy \\
&= C_p (I_3 + I_4) .
\end{aligned}$$

Clearly,

$$I_3 \leqq \varepsilon^p(\langle\langle f\rangle\rangle_{p,x})^p .$$

Observe now that

$$I_4 = \int_0^\infty dt \iint_{|x-y|<\varepsilon^\mu} \frac{|f(t, y)|^p\,|\alpha(t, x) - \alpha(t, y)|^p}{|x - y|^{d+p-2}}\, dx\, dy$$

$$+ \int_0^\infty dt \iint_{\varepsilon^\mu \leqq |x-y| \leqq 1} \frac{|\ (t, y)|^p\,|\alpha(t, x) - \alpha(t, y)|^p}{|x - y|^{d+p-2}}\, dx\, dy$$

$$= I_5 + I_6 ,$$

$$I_5 \leqq A \int_0^\infty dt \int_{\partial D} f(t, y)^p\, dy \int_{|x-y|\leqq\varepsilon^\mu} \frac{1}{|x - y|^{d-2}}\, dx \leqq C\varepsilon^\mu\, \|f\|^p_{\sim p} ,$$

$$I_6 = \int_0^\infty dt \int_{\partial D} dy\, f(t, y)^p \int_{\varepsilon^\mu \leqq |x-y| \leqq 1} \frac{|\alpha(t, x) - \alpha(t, y)|}{|x - y|^{d+p-2}}\, dx$$

$$\leqq C\frac{\varepsilon^p}{\varepsilon^{p\mu}} \int_0^\infty dt \int_{\partial D} |f(t, y)|^p \int_{|x-y|\leqq 1} \frac{1}{|x - y|^{d-2}}\, dx \leqq C\varepsilon^{p(1-\mu)}\, \|f\|^p_{\sim p} .$$

Combining these and taking $\mu = p/(p + 1)$, we get our result.

Remark 1. Suppose L is as in Theorem 7.1. Given $h \in C_0^\infty([0, \infty) \times \partial D)$, we write

$$H_L f = H_0 h + G_L D_\varepsilon H_0 h .$$

Then $H_0 h \in C_b^{1,2}([0, \infty) \times D)$ and $G_L D_\varepsilon H_0 h \in W_p^{(2)}([0, \infty) \times \bar{D})$. Thus if $Q_{s,x}$ is a solution to the martingale problem for a starting from $x \in G$ at times s, then for $p > d + 2$

$$H_L h(t, x(t)) - \int_s^t \left(\frac{\partial}{\partial u} + L_u\right) H_L h(u, x(u))\, du$$

is a local $Q_{s,x}$-martingale in D. But $((\partial/\partial u) + L_u)H_L f = 0$ on D, and so $H_L f(t, x(t))$ is a local $Q_{s,x}$-martingale in D. In particular,

$$H_L h(s, x) = E^{Q_{s,x}}[h(\tau_s , x(\tau_s))] , \tag{7.27}$$

where $\tau_s = \inf\{t \geqq s : x(t) \notin D\}$.

Remark 2. Let L be as in Remark 1, and $p > d + 2$. If $g \in C_0^\infty([0, \infty) \times \bar{D})$,

$$G_L g \in W_p^{(2)}([0, \infty) \times \bar{D})$$

and

$$\left(\frac{\partial}{\partial t} + L\right) G_L g = -g .$$

Hence,

$$G_L g(t, x(t)) + \int_s^t g(u, x(u))\, du$$

is a local $Q_{s,x}$-martingale in D. Combining this with the fact that $G_L g \to 0$ at $[0, \infty) \times \partial D$, we have

$$G_L g(s, x) = E^{Q_{s,x}}\left[\int_s^{\tau_s} g(u, x(u))\, du\right]. \tag{7.28}$$

Remark 3. Let L be as before, $p > d + 2$, and let h and g be chosen as in Remarks 1 and 2, respectively. Define $u(s, x) = -G_L g(s, x) + H_L(s, x)$. Then,

$$u(s, x) = -E^{Q_{s,x}}\left[\int_s^{\tau_s} g(u, x(u))\, du\right] + E^{Q_{s,x}}[h(\tau_s, x(\tau_s))] .$$

Moreover, $u \in W_p^{(2)}([0, \infty) \times \bar{D})$. In particular, u has a bounded continuous gradient in all directions on the boundary, and so Theorem 2.5 applies to u.

Remark 4. For $p > d + 2$, one can prove a Sobolev type theorem bounding the maximum of $|f|$ by $|||f|||_p$.

CONDITIONS FOR COMPACTNESS IN $D([0, T], R^d)$.

THEOREM 7.4. *Let P_n be a sequence of measures on $D([0, T], R^d)$ such that*

(i) *$\{P_n\}$ is uniformly stochastically continuous in the interval $[0, T]$,*

(ii) *there exists a constant A independent of n such that*

$$E^{P_n}[|x(t_1) - x(t_2)|^2\, |x(t_3) - x(t_2)|^2] \leqq A\, |t_1 - t_3|^2$$

for $0 \leqq t_1 < t_2 < t_3 \leqq T$. Then the sequence $\{P_n\}$ is relatively compact.

This theorem is found in Chentsov [1].

Let $h > 0$ be arbitrary. For each h let a stochastic process P_h be defined on $D([0, T], R^d)$ in the following manner. P_h is concentrated on step functions

which are constant on intervals of the form $jh \leqq t < (j+1)h$. There is never a jump at T. The chains $x(jh)$ constitute a Markov chain in j for each $h > 0$. For each $\delta > 0$, there is a constant C_δ such that

$$E^{P_h}[\|x(kh) - x(rh)\| \mid \mathcal{M}_{rh}] \leqq \delta + C_\delta |(k-r)h|^{1/2}$$

for $k \geqq r$. Moreover,

$$P_h[x(0) = z_h] = 1$$

and z_h varies over a compact set as $h \to 0$. Then we can state

THEOREM 7.5. *The family* $\{P_h\}$ *is relatively compact in* $D([0, T], R^d)$ *as* $h \to 0$.

Proof: From the hypothesis it follows that

$$E^{P_h}[\|x(t)\|] \leqq A \quad \text{for} \quad 0 \leqq t \leqq T \quad \text{and} \quad h \to 0\,.$$

Therefore, if $\{t_j\}$ is a dense set in $[0, T]$, for given $\eta > 0$, we can find constants C_j such that

$$(7.29) \qquad P_h[\|x(t_j)\| \leqq C_j \text{ for all } j] \geqq 1 - \tfrac{1}{3}\eta \qquad \text{for} \qquad 0 < h \leqq 1\,.$$

Moreover, we can find $\varepsilon_j \downarrow 0$, $\alpha_j \downarrow 0$ and $\beta_j \downarrow T$ such that

$$(7.30) \qquad P_h[\|x(\alpha_j) - x(0)\| \leqq \varepsilon_j \text{ for all } j] \geqq 1 - \tfrac{1}{6}\eta \qquad \text{for} \qquad 0 < h \leqq 1\,,$$

$$(7.31) \qquad P_h[\|x(\beta_j) - x(T)\| \leqq \varepsilon_j \text{ for all } j] \geqq 1 - \tfrac{1}{6}\eta \qquad \text{for} \qquad 0 < h \leqq 1\,.$$

We shall now prove that

$$\lim_{\rho \to 0} \limsup_{h \to 0} P_h[\omega_x^*(\rho) \geqq \varepsilon] = 0\,,$$

where

$$\omega_x^*(\rho) = \sup_{\substack{0 \leq t < u < s \leq T \\ |t-s| < \rho}} [\inf\{|x(t) - x(u)|, |x(u) - x(s)|\}]\,.$$

Then the theorem will be proved in view of (7.29), (7.30), (7.31) and Theorem 7.2 of Parthasarathy [7].

Let us take $h > 0$ and set $x_j = x(jh)$, for convenience. Then x_0, x_1, x_2, $\cdots$ form a Markov chain. Let us define

$$\tau_1 = \inf[j : j > 0 \, |x_j - x_0| > \varepsilon]\,,$$

$$\tau_2 = \inf[j : j > 0 \, |x_{\tau_1} - x_{\tau_1 + j}| > \varepsilon]\,,$$

$$\vdots$$

$$\tau_k = \inf[j : j > 0 \, |x_{\tau_1 + \cdots + \tau_{k-1}} - x_{\tau_1 + \cdots + \tau_{k-1} + j}| > \varepsilon]\,.$$

Of course, after a certain stage, τ_k does not exist. In fact, since $\tau_{k+1} \geqq \tau_k + 1$, the process ends after T/h steps. We wish actually to estimate

$$P_h[\min(\tau_1, \cdots, \tau_{k_0}) \leqq \rho h^{-1}] = \phi_h(\rho),$$

where k_0 is the last j for which τ_j exists:

$$\begin{aligned}
&P_h[\min(\tau_1, \cdots, \tau_k) \leqq \rho h^{-1}] \\
&\qquad \leqq P_h[\tau_1 \leqq \rho h^{-1}, \tau_1 h \leqq T] + \cdots \\
&\qquad\quad + P_h[\tau_k \leqq \rho h^{-1}, (\tau_1 + \cdots + \tau_k)h \leqq T] \\
&\qquad\quad + P_h[(\tau_1 + \cdots + \tau_{k+1})h \leqq T] \\
&\qquad \leqq P_h[\tau_1 \leqq \rho h^{-1}] + \cdots + P_h[\tau_k \leqq \rho h^{-1}] + P_h[(\tau_1 + \cdots + \tau_{k+1})h \leqq T] \\
&\qquad = P_h[h\tau_1 \leqq \rho] + \cdots + P_h[h\tau_k \leqq \rho] + P_h[h(\tau_1 + \cdots + \tau_{k+1}) \leqq T].
\end{aligned}$$

Here k is an arbitrary integer. We note that

$$P_h[h\tau_1 \leqq \rho] = P_h\Big[\sup_{0 \leq t \leq 0} |x(t) - x(0)| > \varepsilon\Big].$$

If $x_0, x_1, \cdots, x_n$ is a Markov chain and if

$$f_h(n, \varepsilon) = \sup_{\substack{x \\ 0 \leq r < s < n}} P_h[|x_r - x_s| \geqq \tfrac{1}{2}\varepsilon \,|x_r = x],$$

we see, by a standard estimate, that

$$P_h\Big[\sup_{0 \leqq j \leqq n} |x_0 - x_j| \geqq \varepsilon\Big] \leqq \frac{f_h(n, \varepsilon)}{1 - f_n(n, \varepsilon)}.$$

Therefore,

$$P_h[h\tau_1 \leqq \rho] \leqq \frac{f_h(n, \varepsilon)}{1 - f_h(n, \varepsilon)},$$

where $n = [\rho h^{-1}]$. To estimate $f_h(n, \varepsilon)$ we note that, by hypothesis, for any $\delta > 0$

$$E^{P_h}[\|x_r - x_s\| \mid x_s = x] \leqq \delta + C_\delta |(r - s)h|^{1/2}.$$

Therefore,

$$\begin{aligned}
f_h(n, \varepsilon) &\leqq \frac{2}{\varepsilon}[\delta + C_\delta(nh)^{1/2}] \\
&\leqq \frac{2}{\varepsilon}[\delta + C_\delta \rho^{1/2}].
\end{aligned}$$

Hence, for each $\varepsilon > 0$,

$$\sup_h P_h[h\tau_1 \leqq \rho] = f(\rho, \varepsilon) ,$$

where $f(\rho, \varepsilon) \to 0$ as $\rho \to 0$, for each $\varepsilon > 0$.

In view of the fact that we have the same estimate in terms of $f(\rho, \varepsilon)$ for all conditional distributions involved and because the strong Markov property is valid for any chain, we have

$$P_h[h\tau_j \leqq \rho] \leqq f(\rho, \varepsilon) ,$$

since all the τ_j are stopping times for the chain. Moreover,

$$P_h[h\tau_j \leqq \rho \mid \mathscr{M}_{h\tau_{j-1}}] \leqq f(\rho, \varepsilon) .$$

This means that

$$P_h[h(\tau_1 + \cdots + \tau_k) \leqq T] \leqq \psi_k(\varepsilon) ,$$

where $\psi_k(\varepsilon)$ is estimated in terms of $f(\rho, \varepsilon)$ and

$$\psi_k(\varepsilon) \to 0 \quad \text{as} \quad k \to \infty \quad \text{for} \quad \text{each} \quad \varepsilon > 0 .$$

Therefore,

$$\phi_k(\rho) \leqq kf(\rho, \varepsilon) + \psi_{k+1}(\varepsilon)$$

or

$$\sup_h \phi_h(\rho) = \phi(\rho) \to 0 \quad \text{as} \quad \rho \to 0 \quad \text{for} \quad \text{each} \quad \varepsilon > 0 .$$

It is easily verified that if $\min(\tau_1, \cdots, \tau_{k_0}) \leqq \rho h^{-1}$, then

$$\omega_x^*(\rho) \leqq 2\varepsilon .$$

Consequently,

$$\sup_h P_h[\omega_x^*(\rho) \geqq 2\varepsilon] \leqq \phi(\rho) ,$$

and this completes the proof of the theorem.

Bibliography

[1] Chentsov, N. N, *Weak convergence of stochastic processes whose trajectories have no discontinuities of the second kind and the "heuristic" approach to the Kolmogorov-Smirnov tests*, Theor. Prob. Appl., Vol. 1, 1956, pp. 140–144.

[2] Doob, J. L., *Stochastic Processes*, John Wiley and Sons, New York, 1953.

[3] Fabes, E., and Rivière, N. M., *Singular integrals with mixed homogeneity*, Studia Math., Vol. 27, 1966, pp. 19–38.

[4] Kunita, H., and Watanabe, S., *On square integrable martingales*, Nagoya Math., J., Vol. 30, 1967, pp. 209–245.

[5] Ladyzhenskaya, O. A., Solonnikov, V. A., and Ural'tseva, N. N., *Linear and Quasi-linear Equations of Parabolic Type*, Amer. Math. Soc. Translations of Mathematical Monographs, No. 23, Amer. Math. Soc., Providence, R.I., 1968.

[6] Littman, W., McCarthy, C., and Rivière, N., *L^p multiplier theorems*, Studia Math., Vol. 30, 1968, pp. 193–217.

[7] Parthasanathy, K. R., *Probability Measures on Metric Spaces*, Academic Press, New York, 1967.

[8] Stroock, D. W., and Varadhan, S. R. S., *Diffusion processes with continuous coefficients, I*, Comm. Pure Appl. Math., Vol. 22, 1969, pp. 345–400.

[9] Stroock, D. W., and Varadhan, S. R. S., *Diffusion processes with continuous coefficients, II*, Comm. Pure Appl. Math., Vol. 22, 1969, pp. 479–530.

[10] Meyer, P. A., *Probability and Potentials*, Blaisdell, Waltham, Toronto-London, 1966.

Received July, 1970.

COMMUNICATIONS ON PURE AND APPLIED MATHEMATICS, VOL. XXV, 651–713 (1972)

On Degenerate Elliptic-Parabolic Operators of Second Order and Their Associated Diffusions*

D. STROOCK AND S. R. S. VARADHAN

1. Introduction

This paper consists of two parts. In the first part we extend our earlier results [10] on the strong maximum principle to a broader class of operators, namely degenerate parabolic operators

$$\frac{\partial}{\partial t} + L_t ,$$

where $L_t = \frac{1}{2}\nabla \cdot (a(t, x)\nabla) + b(t, x) \cdot \nabla$ with a and b suitably smooth. This leads to a generalization of the results of M. Bony [1] that was sought by C. D. Hill [3]. It is also related to a recent result of M. Redheffer [8].

The second part of the paper is devoted to the study of the first boundary value problem for degenerate elliptic operators

$$L = \frac{1}{2}\nabla \cdot (a(x)\nabla) + b(x) \cdot \nabla - k(x)$$

in smooth regions. For example we show that if G is a smooth region in R^d and if g is a bounded continuous function on $\Sigma_2 \cup \Sigma_3$ (see [2], [4] or [5] for the definition of Σ_2 and Σ_3), then there is a unique $u \in L^\infty(G)$ such that

$$\text{(i)} \qquad \int uL^*v \, dx = 0 \quad \text{for all} \quad v \in C_0^\infty(G) ,$$

$$\text{(ii)} \qquad \operatorname*{ess\,lim}_{\substack{x \to a \\ x \in G}} u(x) = g(a) , \qquad a \in \Sigma_2 \cup \Sigma_3 ,$$

provided $\inf_{x \in G} k(x) > 0$. Under an additional assumption on G, a and b, we need assume only that $k(x) \geqq 0$. Moreover, there is a set F of measure zero such that for $x \notin F$ the solution $u(x)$ is the limit of solutions of perturbed problems. One can use a wide class of perturbations and the set F of measure zero depends

* The work for this paper was carried out at the Courant Institute of Mathematical Sciences and supported by Air Force Contract No. AFOSR-71-2055.

only on the coefficients a, b and the region G. We also give sufficient conditions which ensure that F is empty.

The second half of Section 2 contains a description of the main results from the analytical as well as the probabilistic points of view.

The first part, consisting of Sections 3 and 4, is almost independent of Sections 5 through 8, which constitute the second part. Either part can be read independently of the other. Results from the first part are used only twice in the second part, and there in a rather weak form. Remarks following Theorem 5.1 and Theorem 6.3 explain how the results from the first part are used.

2. Background and Summary

Let $\Omega = C[[0, \infty), R^d]$ be the space of R^d-valued continuous functions on $[0, \infty)$. The value of the function $\omega \in \Omega$ at t is $x(t, \omega) = x_t(\omega) = \omega(t)$. M_t^s is the σ-field generated by the functions $x_u(\cdot)$ for $s \leqq u \leqq t$. M^s is the smallest σ-field containing M_t^s for all $t \geqq s$. M_t denotes M_t^0. Given

$$a:[0, \infty) \times R^d \to S_d \quad \text{and} \quad b:[0, \infty) \times R^d \to R^d,$$

we define

$$L_t = \frac{1}{2} \sum_{i,j=1}^{d} a^{ij}(t, x) \frac{\partial^2}{\partial x_i \, \partial x_j} + \sum_{i=1}^{d} b^i(t, x) \frac{\partial}{\partial x_i} . \tag{2.1}$$

A solution to the martingale problem for L_t starting at $(s, x) \in [0, \infty) \times R^d$ is a probability measure P on $\langle \Omega, M^s \rangle$ satisfying

$$P[x(s) = x] = 1 , \tag{2.2}$$

and

$$f(x(t)) - \int_s^t (L_u f)(x(u)) \, du \tag{2.3}$$

is a P-martingale for all $f \in C_0^\infty(R^d)$. We shall also refer to P as a solution to the martingale problem for a and b starting at (s, x).

THEOREM 2.1. *Let* $a:[0, \infty) \times R^d \to S_d$ *and* $b:[0, \infty) \times R^d \to R^d$ *be bounded measurable functions. Assume that* $a \in C^{1,2}[[0, \infty) \times R^d]$ *and that* b *is uniformly Lipschitz continuous with respect to the space variables on* $[0, \infty) \times R^d$. *Let* L_t *be as in* (2.1). *Then, for each* $(s, x) \in [0, \infty) \times R^d$, *there is exactly one probability*

measure $P_{s,x}$ *on* $\langle \Omega, M^s \rangle$ *satisfying* (2.2) *and* (2.3). *Moreover, the family*

$$\{P_{s,x}: (s, x) \in [0, \infty) \times R^d\}$$

is a Feller continuous strong Markov process. Finally there is an enlargement $\hat{\Omega}$ *of* Ω *and an extension* $\hat{P}_{s,x}$ *of* $P_{s,x}$ *to* $\hat{\Omega}$ *such that*

$$x(t) = x + \int_s^t a^{1/2}(u, x(u))\, d\hat{\beta}(u) + \int_s^t b(u, x(u))\, du \qquad a.s.\ P_{s,x},$$

where $\hat{\beta}(t)$ *is a* $\hat{P}_{s,x}$*-Brownian motion.*

Theorem 2.1 is proved in [10]. Another fact proved in [10] which we want to use is

THEOREM 2.2. *If* a *and* b *are as in Theorem* 2.1 *and* $c:[0, \infty) \times R^d \to R^d$ *is bounded and measurable, then there is exactly one solution* $Q_{s,x}$ *to the martingale problem for* a *and* $b + a^{1/2} c$ *starting at* (s, x). *Moreover, the support of* $Q_{s,x}$ *is independent of* c.

COROLLARY 2.1. *Suppose* G *is an open subset of* $[0, \infty) \times R^d$. *Let* $a: G \to S_d$ *and* $b: G \to R^d$ *be bounded measurable functions such that* $a \in C^{1,2}(G)$ *and* b *satisfies a Lipschitz condition with respect to* x *uniformly on compact subsets of* G. *Given a bounded measurable function* $c: G \to R^d$, *set*

$$L_u = \tfrac{1}{2}\nabla \cdot a\nabla + (b + a^{1/2} c) \cdot \nabla\ ;$$

then for each $(s, x) \in G$ *there is a unique probability measure* $Q_{s,x}$ *on* $\langle \Omega, M^s_{\tau_s} \rangle$, *where* $\tau_s = \inf\{t \geqq s\colon\ (t, x(t)) \notin G\}$, *such that*

$$Q_{s,x}[x(s) = x] = 1\,, \tag{2.4}$$

and

$$f(t \wedge \tau_s, x(t \wedge \tau_s)) - \int_s^{t \wedge \tau_s} L_u f(u, x(u))\, du \tag{2.5}$$

is a $Q_{s,x}$ *martingale for all* $f \in C_b^{1,2}(G)$.

In [10] we proved (cf. Section 5) that if $\beta(t)$ is a W-Brownian motion on R^d and if

$$\eta(t) = x + \int_s^t \sigma(u, \eta(u))\, d\beta(u) + \int_s^t b(u, \eta(u))\, du\,,$$

then under certain assumptions on σ,

$$W\left(\sup_{s\leq t\leq T}\left|\int_s^t \sigma(u,\eta(u))\,d\beta(u) + \tfrac{1}{2}\int_s^t (\sigma'\sigma)(u,\eta(u))\,du\right| \geqq \varepsilon \,\middle|\, |||\beta(\cdot)|||_s^T < \delta\right)$$

tends to 0 as $\delta \to 0$ for each $\varepsilon > 0$, where

$$(\sigma'\sigma)^i = \sigma^{il}_{\ \ ,j}\,\sigma^{lj}\,,$$

and $|||\beta(\cdot)|||_s^T = \max_{1\leq i\leq d}\sup_{s\leq t\leq T}|\beta_i(t)|$. By an easy extension of the same argument we can prove the following result.

THEOREM 2.3. *Let $\sigma, \tilde{\sigma}:[0,\infty)\times R^d \to R^d \otimes R^d$ and $b:[0,\infty)\times R^d \to R^d$ be bounded measurable functions. Assume that $\sigma \in C_b^{1,2}([0,\infty)\times R)$ and suppose that*

$$\eta(t) = x + \int_s^t \sigma(u,\eta(u))\,d\beta(u) + \int_s^t \tilde{\sigma}(u,\eta(u))\,d\tilde{\beta}(u) + \int_s^t b(u,\eta(u))\,du\,,$$

where $\beta(\cdot)$ and $\tilde{\beta}(\cdot)$ are independent W-Brownian motions on R^d. Then, for all $\varepsilon > 0$ and $T > s$,

$$\lim_{\delta\to 0} W\left(\sup_{s\leq t\leq T}\left|\int_s^t \sigma(u,\eta(u))\,d\beta(u) + \tfrac{1}{2}\int_s^t (\sigma'\sigma)(u,\eta(u))\,du\right| \geqq \varepsilon \,\middle|\, |||\beta(\cdot)|||_s^T < \delta\right) = 0.$$

THEOREM 2.4. *Let $a_i:[0,\infty)\times R^d \to S_d$ and $b_i:[0,\infty)\times R^d \to R^d$ satisfy the conditions of Theorem 2.1 for $i = 1, 2$. Suppose G is an open set in $[0,\infty)\times R^d$ such that $a_1 \equiv a_2$ and $b_1 \equiv b_2$ in G. Denote by $P^{(i)}_{s,x}$, $i = 1, 2$, the solution to the martingale problem for a_i and b_i starting from $(s,x) \in \bar{G}$. Define*

$$\tau_s' = \inf\{t \geqq s : (t, x(t)) \notin \bar{G}\}\,.$$

Then

$$P^{(1)}_{s,x}[A] = P^{(2)}_{s,x}[A]$$

for all $A \in M^s_{\tau_s'}$.

Proof: The proof of the theorem is an easy application of Theorem 3.4 in [9]. Indeed let Q be the measure which results from patching $P^{(1)}_{s,x}$ and $P^{(2)}_{\tau_s',x(\tau_s')}$ together at time τ_s' as in Lemma 3.6 of [9]. One must show that $Q = P^{(2)}_{s,x}$. To do this let $\{G_n\}_1^\infty$ be a decreasing sequence of open sets containing $\bar{G}$ such

that $\bar{G} = \bigcap_1^\infty G_n$. For each $n \geqq 1$, define

$$\tau^{(n)} = \inf\{t \geqq s : (t, x(t)) \notin G_n\}$$

and let $Q^{(n)}$ be the result of patching $P_{s,x}^{(1)}$ and $P_{\tau^{(n)},x(\tau^{(n)})}^{(2)}$ at time $\tau^{(n)}$. Then $\tau^{(n)} \downarrow \tau'$ and it is easy to check that $Q^{(n)}$ tends to both Q and $P_{s,x}^{(2)}$ as $n \to \infty$.

We shall now give a summary of the results contained in this paper.

In [10] we proved that the support in function space of the solution to the martingale problem $P_{s,x}$ corresponding to

$$L_t = \tfrac{1}{2}\sigma^*(t, x)\nabla \cdot \sigma^*(t, x)\nabla + b(t, x) \cdot \nabla$$

starting from (s, x) consists of the closure of the set of trajectories of the form

$$\phi(t) = x + \int_s^t \sigma(u, \phi(u))\psi(u)\, du + \int_s^t b(u, \phi(u))\, du\,, \qquad t \geqq s\,,$$

where ψ runs over smooth functions. In Section 3 this result is extended to operators of the form

$$L_t = \tfrac{1}{2}\nabla \cdot (a(t, x)\nabla) + b(t, x) \cdot \nabla \tag{2.6}$$

and it is shown (Theorem 3.2) that the support of the corresponding solution $P_{s,x}$ consists of the closure of the set of trajectories of the form

$$\phi(t) = x + \int_s^t a(u, \phi(u))\psi(u)\, du + \int_s^t b(u, \phi(u))\, du\,, \qquad t \geqq s\,, \tag{2.7}$$

where ψ runs over smooth functions. This enables us to prove a general version of the strong maximum principle (Theorem 4.1). Namely, if L_t is as in (2.6) and v is a function on $G \subset [0, \infty) \times R^d$ such that

$$\frac{\partial v}{\partial t} + L_t v \geqq 0 \tag{2.8}$$

and

$$v(s, x) = \sup_{(t,y) \in G} v(t, y)\,,$$

then $v(s, x) = v(s', x')$ for all $(s', x') \in A_{s,x}$. $A_{s,x}$ is the set in G containing (s, x) which is described as follows:

$$A_{s,x} = \text{closure}\,\{(t, \phi(t)) : t \geqq s,\ \phi(u) \in G \quad \text{for all} \quad s \leqq u \leqq t\}\,,$$

ϕ being as in (2.7) with ψ running over smooth functions. We also prove that $A_{s,x}$ is maximal. For this purpose we have to construct counterexamples. These are in general only generalized solutions of (2.8). From the probabilistic point of view the natural notion of a generalized solution is in terms of martingales. Solving the equation

$$\frac{\partial u}{\partial t} + L_t u = f$$

should be the same as finding a function u such that

$$u(t, x(t)) - \int_s^t f(\sigma, x(\sigma))\, d\sigma$$

is a $P_{s,x}$ martingale for all s, x. In Sections 4 and 5 we adopt this as our notion of a solution. Later in Section 8, we prove that it is equivalent to extending L_t by certain closure operations starting from L_t acting on smooth functions. See for instance Theorem 8.1.

The rest of the paper is devoted to the study of the elliptic first boundary value problem:

$$Lu - ku = -f \quad \text{in} \quad G,$$
$$u = g \quad \text{in} \quad \delta G,$$

where

$$L = \tfrac{1}{2} \sum a^{ij}(x) \frac{\partial^2}{\partial x_i\, \partial x_j} + \sum b^j(x) \frac{\partial}{\partial x_j},$$

$G \subset R^d$, and δG is the boundary of G. From the probabilistic point of view, once the problem is properly formulated one would expect the solution to be given by the formula

(2.9)
$$u(x) = E^{P_x}\left[\int_0^\tau \exp\left\{-\int_0^t k(x(s))\, ds\right\} f(x(t))\, dt + \exp\left\{-\int_0^\tau k(x(s))\, ds\right\} g(x(\tau))\right],$$

where $\tau(\omega)$ is the first exit time of the path $x(\cdot, \omega)$ from G. This is because (2.9) holds in the nondegenerate case. It is easy to verify that, under suitable assumptions, $u(x)$ is well defined by (2.9) and satisfies

$$Lu - ku = -f$$

in the martingale sense. The manner in which the boundary value is taken on is more difficult to describe. Obviously the boundary function has to be specified only at those points at which $x(\cdot, \omega)$ can exit. An additional difficulty is that

although g, k and f are smooth functions, u need not even be continuous, due to the discontinuous nature of the functional $\tau(\omega)$.

Actually, instead of considering τ one can consider τ', the first exit time of the path $x(\cdot, \omega)$ from $\bar{G}$. There is no *a priori* reason why one should prefer one to the other. We have a similar formula for the solution

(2.10)

$$v(x) = E^{P_x}\left[\int_0^{\tau'} \exp\left\{-\int_0^t k(x(s))\, ds\right\} f(x(t))\, dt + \exp\left\{-\int_0^{\tau'} k(x(s))\, ds\right\} g(x(\tau'))\right]$$

when one uses τ'. The function $v(x)$ has as much right to be a solution as $u(x)$. In the nondegenerate case, $P_x[\tau = \tau'] = 1$ for all $x \in G$ so that the two solutions coincide. In the general case, this need not be true. If one wants the boundary value to be taken on continuously at a point $x_0 \in \delta G$, one needs

$$\lim_{\substack{x \to x_0 \\ x \in G}} P_x[\tau' \geqq \varepsilon] = 0 \quad \text{for} \quad \text{any} \quad \varepsilon > 0\,. \tag{2.11}$$

If one defines Γ as the set of points x_0 on δG such that (2.11) is true, one can show that, for g which are continuous and bounded,

$$\lim_{x \to x_0} v(x) = g(x_0) \qquad \text{for} \qquad x_0 \in \Gamma\,.$$

In order to show that it is enough to prescribe g on Γ, one has to show that

$$P_x[x(\tau') \notin \Gamma] = 0 \quad \text{for} \quad \text{all} \quad x \in G\,.$$

This is also relatively easy. The analogous properties for τ are not true in general. Thus the τ' problem is better behaved than the τ problem.

It is nonetheless possible to formulate both problems in analytical terms so that the solutions are $u(x)$ and $v(x)$, respectively. This is done in Section 5.

If $\Sigma_2 \cup \Sigma_3 \subset \delta G$ is defined as in [2], [4] or [5], then we have

$$\Sigma_2 \cup \Sigma_3 \subset \Gamma \subset \overline{\Sigma_2 \cup \Sigma_3}\,.$$

Sections 6 and 7 deal with the above inclusions and give some criteria which help determine the set Γ precisely. We also prove in Section 7 that $P_x[\tau = \tau'] = 1$ for almost all x in G. This means that although $u(x)$ and $v(x)$ can be different they must agree almost everywhere. Moreover, at any point x where

$$P_x[\tau = \tau'] = 1$$

both u and v must be continuous and the solution is stable with respect to various perturbations. Even if one cannot determine Γ exactly, it is still enough to prescribe the boundary function on $\Sigma_2 \cup \Sigma_3$. This is made possible by the fact that

$$P_x[\tau = \tau' \quad \text{and} \quad x(\tau) \in \Sigma_2 \cup \Sigma_3] = 1$$

for almost all x in G. Therefore, $\Sigma_2 \cup \Sigma_3$ is a big enough set on the boundary to determine u almost everywhere. This is the basis of the uniqueness proved in Theorem 8.2 and Corollary 8.2.

3. The Support

It was proved in [10] that the support of the diffusion process starting from (t_0, x_0) corresponding to the operator

$$L_t = \tfrac{1}{2}\sigma^*(t, x)\nabla \cdot \sigma^*(t, x)\nabla + b(t, x) \cdot \nabla \tag{3.1}$$

consists of the closure of the set of trajectories having the form

$$\phi(t) = x_0 + \int_{t_0}^{t} \sigma(s, \phi(s))\psi(s)\, ds + \int_{t_0}^{t} b(s, \phi(s))\, ds\,, \tag{3.2}$$

where ψ runs over nice functions. Of course, if L_t is given in the form

$$L_t = \tfrac{1}{2}\sum a^{ij}(t, x)\frac{\partial^2}{\partial x_i\, \partial x_j} + \sum b^j(t, x)\frac{\partial}{\partial x_j}\,, \tag{3.3}$$

then one cannot always write it in the form (3.1) with a smooth choice of $\sigma(t, x)$. The purpose of this section is to extend the result proved in [10] for operators in the form (3.1) to those in the form (3.3).

The main idea behind the proof is that we notice first that σ need not be a $d \times d$ matrix in (3.1). It could be any $m \times d$ matrix so long as $\sigma^*\sigma = a$. While it is not true that an arbitrary operator with smooth coefficients in the form (3.3) can be always written in the form (3.1) with m equal d or with any other finite m, we prove that it can be written in the form (3.1) with smooth coefficients provided one takes $m = \infty$. This of course introduces technical difficulties regarding convergence. The approach we take overcomes these problems.

We start with a few lemmas.

Lemma 3.1. *Assume that* $\Omega^{(i)} = C[[0, \infty), R^{d_i}]$, $i = 1, 2$, *and let us identify*

$$\Omega = C[[0, \infty), R^{d_1+d_2}]$$

with $\Omega^{(1)} \times \Omega^{(2)}$ *(i.e.,* $z(t) = (x(t), y(t))$*); let*

$$a:[0, \infty) \times \Omega \to S_{d_2} \quad \text{and} \quad b:[0, \infty) \times \Omega \to R^{d_2}$$

be bounded nonanticipating functions. Define

$$\bar{a}:[0, \infty) \times \Omega \to S_{d_1+d_2} \quad \text{and} \quad \bar{b}:[0, \infty) \times \Omega \to R^{d_1+d_2}$$

by

$$\langle \theta, \bar{a}\theta \rangle = |\lambda|^2 + \langle \mu, a\mu \rangle ,$$

$$\langle \theta, \bar{b} \rangle = \langle \mu, b \rangle ,$$

where $\theta = (\lambda, \mu)$ *is any element of* $R^{d_1+d_2}$*. Suppose* P *is a probability measure on* Ω *such that* $P[z(0) = 0] = 1$ *and*

$$Z_\theta(t) = \exp\left\{\left\langle \theta, z(t) - \int_0^t \bar{b}(u)\, du \right\rangle - \tfrac{1}{2}\int_0^t \langle \theta, \bar{a}(u)\theta \rangle\, du\right\}$$

is a P-martingale for all $\theta \in R^{d_1+d_2}$*. Denote by* P_w *the r.c.p.d. of* P *given the* σ*-field* $\mathcal{M}^{(1)}$ *generated by* $x(s)$ *for* $0 \leqq s < \infty$*. Then there is a set* $N \in \mathcal{M}^{(1)}$ *with* $P(N) = 0$ *such that, for* $w \notin N$*,*

(i) $$P_w[y(0) = 0] = 1 ,$$

(ii) *for all* $\mu \in R^{d_2}$,

$$Y(t) = \exp\left\{\left\langle \mu, y(t) - \int_0^t b(u)\, du \right\rangle - \tfrac{1}{2}\int_0^t \langle \mu, a(u)\mu \rangle\, du\right\}$$

is a P_w*-martingale.*

Proof: First we shall prove that for each $t_0 \geqq 0$

$$E^{P_w}[Y_\mu(t_0)] = 1 \qquad \text{a.s.}$$

Let us denote the function above by $H(w)$. We define a measure Q on $\mathcal{M}^{(1)}$ by

$$\frac{dQ}{dP} = H(w) .$$

Since $H(w)$ is obviously measurable with respect to $\mathcal{M}^{(1)}$ it suffices to show that Q on $\mathcal{M}^{(1)}$ is the same as P on $\mathcal{M}^{(1)}$, i.e., the Wiener measure. Since

$Q[x(0) = 0] = 1$, we have to establish that

$$X_\lambda(t) = \exp\{\langle \lambda, x(t)\rangle - \tfrac{1}{2}t|\lambda|^2\}$$

is a Q-martingale. Assume $0 \leqq t_1 \leqq t_2$ and let $A \in \mathscr{M}_{t_1}^{(1)}$. We have to prove that

$$E^Q[\chi_A X_\lambda(t_2)] = E^Q[\chi_A X_\lambda(t_1)]. \tag{3.4}$$

Case i. $0 \leqq t_1 \leqq t_2 \leqq t_0$.

$$\begin{aligned} E^Q[\chi_A X_\lambda(t_2)] &= E^P[Y_\mu(t_0) X_\lambda(t_2)\chi_A] \\ &= E^P[Y_\mu(t_0) X_\lambda(t_0)\chi_A] \\ &= E^P[Z_\theta(t_0)\chi_A] \\ &= E^P[Z_\theta(t_1)\chi_A] \\ &= E^P[X_\lambda(t_1) Y_\mu(t_1)\chi_A] \\ &= E^P[X_\lambda(t_1) Y_\mu(t_0)\chi_A] \\ &= E^Q[X_\lambda(t_1)\chi_A]. \end{aligned}$$

Here $\theta = (\lambda, \mu)$ and $X_\lambda(t)$, $Y_\mu(t)$ and $Z_\theta(t)$ are all P-martingales.

Case ii. $0 \leqq t_1 \leqq t_0 \leqq t_2$.

$$\begin{aligned} E^Q[\chi_A X_\lambda(t_2)] &= E^P[\chi_A X_\lambda(t_2) Y_\mu(t_0)] \\ &= E^P[\chi_A X_\lambda(t_0) Y_\mu(t_0)] \\ &= E^P[\chi_A Z_\theta(t_0)] \\ &= E^P[\chi_A Z_\theta(t_1)] \\ &= E^P[\chi_A X_\lambda(t_1) Y_\mu(t_1)] \\ &= E^P[\chi_A X_\lambda(t_1) Y_\mu(t_0)] \\ &= E^Q[\chi_A X_\lambda(t_1)]. \end{aligned}$$

Case iii. $0 \leqq t_1 \leqq t_0 \leqq t_2$.

$$\begin{aligned} E^Q[\chi_A X_\lambda(t_2)] &= E^P[\chi_A X_\lambda(t_2) Y_\mu(t_0)] \\ &= E^P[\chi_A X_\lambda(t_1) Y_\mu(t_0)] \\ &= E^Q[\chi_A X_\lambda(t_1)]. \end{aligned}$$

This proves (3.4) in all cases. Let P_w^s be the r.c.p.d. of P given $\mathscr{M}_s$. Then $X_\lambda(t)$, $Y_\mu(t)$ and $Z_\theta(t)$ are all martingales with respect to P_w^s for almost all w. Hence the same argument as before yields

$$E^{P_w^s}[Y_\mu(t) \mid \mathscr{M}^{(1)}] = Y_\beta(s) \qquad \text{for} \qquad t \geqq s\,.$$

By the usual argument one can end up with one exceptional set of measure zero which works for all times and μ simultaneously.

LEMMA 3.2. *Suppose $\beta(\cdot)$ and $\tilde{\beta}(\cdot)$ are two independent d-dimensional Q-Brownian motions and let $\sigma, \tilde{\sigma}:[0, \infty) \times R^d \to R^d \otimes R^d$ and $b:[0, \infty) \times R^d \to R^d$ be bounded measurable functions which are uniformly Lipschitz continuous with respect to space variables on $[0, \infty) \times R^d$. Further, assume that $\sigma \in C_b^{1,2}([0, \infty) \times R^d)$. Let $\eta(\cdot)$ be the solution of*

$$\eta(t) = x + \int_s^t \sigma(u, \eta(u))\, d\beta(u) + \int_s^t \tilde{\sigma}(u, \eta(u))\, d\tilde{\beta}(u) + \int_s^t b(u, \eta(u))\, du$$

for $s \leqq t \leqq T$, where $T > s$ is fixed. For any $\delta > 0$, let Q_δ be defined so that

$$Q_\delta(A) = \frac{Q(A \cap |||\beta|||_T^s < \delta)}{Q(|||\beta|||_T^s < \delta)} = Q(A \mid |||\beta|||_T^s < \delta)\,,$$

where

$$|||\beta|||_T^s = \sup_{\substack{s \leqq t \leqq T \\ 1 \leqq i \leqq d}} |\beta_i(t)|\,.$$

Denote by P_δ the distribution of $\eta(t)$ under Q_δ for $s \leqq t \leqq T$. Then

$$\lim_{\delta \to 0} P_\delta = P_0\,,$$

where P_0 is the unique solution of the martingale problem in $s \leqq t \leqq T$ corresponding to the coefficients $\tilde{a} = \tilde{\sigma}\tilde{\sigma}^*$ and $\tilde{b} = b - \frac{1}{2}\sigma'\sigma$, starting at time s from the point x.

Proof: There is no loss of generality in assuming that $x = 0$ and $s = 0$. We make the following definitions:

$$\xi(t) = \int_0^t \sigma(u, \eta(u))\, d\beta(u)\,,$$

$$\tilde{\xi}(t) = \int_0^t \tilde{\sigma}(u, \eta(u))\, d\tilde{\beta}(u)\,,$$

$y(t)$ is the solution of

$$y(t) = \int_0^t \tilde{\sigma}(u, y(u))\, d\tilde{\beta}(u) + \int_0^t \tilde{b}(u, y(u))\, du\,,$$

$$x(t) = -\tfrac{1}{2}\int_0^t (\sigma'\sigma)(u, y(u))\, du\,,$$

$$\tilde{x}(t) = \int_0^t \tilde{\sigma}(u, y(u))\, d\tilde{\beta}(u)\,,$$

and

$$\bar{x}(t) = -\tfrac{1}{2}\int_0^t (\sigma'\sigma)(u, \eta(u))\, du\,.$$

By Lemma 3.1, conditioning with respect to β does not affect the distributions of $y(\cdot)$ or $\tilde{\beta}(\cdot)$. Hence they have the same distribution under Q_δ as under Q. In particular, $y(\cdot)$ has P_0 for its distribution. It is therefore sufficient to prove that, for any $\varepsilon > 0$,

$$\lim_{\delta\to 0} Q_\delta(\|\eta - y\|_T^0 \geqq \varepsilon) = 0\,.$$

First let us observe that, by Theorem 2.3, for any $\varepsilon > 0$,

$$\lim_{\delta\to 0} Q_\delta(\|\xi - \bar{x}\|_T^0 \geqq \varepsilon) = 0\,.$$

Let us choose a large number M so that it dominates σ, $\tilde{\sigma}$, b, $\sigma'\sigma$ and their Lipschitz constants. Define

$$A_\delta(t) = E^{Q_\delta}[\{\min(\|\eta - y\|_T^0, M)\}^2]\,,$$

$$B_\delta(t) = E^{Q_\delta}[\{\min(\|\xi - x\|_T^0, M)\}^2]\,,$$

$$\tilde{B}_\delta(t) = E^{Q_\delta}(\{\min(\|\tilde{\xi} - \tilde{x}\|_T^0, M)\}^2]\,,$$

$$C_\delta(t) = E^{Q_\delta}[\{\min(\|\xi - \bar{x}\|_T^0, M)\}^2]\,.$$

Then

$$A_\delta(t) \leqq 8B_\delta(t) + 8\tilde{B}_\delta(t) + 8M^2 t\int_0^t A_\delta(u)\,du\,,$$

$$B_\delta(t) \leqq 2C_\delta(t) + 2M^2 t\int_0^t A_\delta(u)\,du\,,$$

$$\tilde{B}_\delta(t) \leqq E^{Q_\delta}[\{\|\xi - \tilde{x}\|_T^0\}^2]$$

$$\leqq 2E^{Q_\delta}[|\xi(\tau) - \tilde{x}(\tau)|^2]$$

$$\leqq 2M^2\int_0^t A_\delta(u)\,du\,.$$

Combining the above inequalities, we obtain

$$A_\delta(t) \leqq 16\,C_\delta(t) + k\int_0^t A_\delta(u)\,du\,,$$

where k is some constant. Since $C_\delta(t) \to 0$ as $\delta \to 0$, it follows that $A_\delta(t) \to 0$ as $\delta \to 0$. This concludes the proof.

LEMMA 3.3. *Let a, σ, $\tilde{\sigma}:[0, \infty) \times \Omega \to S_d$ and $b:[0, \infty) \times \Omega \to R^d$ be bounded nonanticipating functions. Assume that, for each t and w, $a(t, w)$, $\sigma(t, w)$ and $\tilde{\sigma}(t, w)$ commute. Further, let $a = \sigma^2 + \tilde{\sigma}^2$. Let P be a probability measure on Ω with the property that*

$$f(x(t)) - \int_0^t (L_s f)(x(s))\,ds$$

is a P-martingale for all $f \in C_0^\infty(R^d)$, where

$$L_s = \frac{1}{2}\sum_{i,j=1}^d a^{ij}(s)\frac{\partial^2}{\partial x_i\,\partial x_j} + \sum_{i=1}^d b^i(s)\frac{\partial}{\partial x_i}\,.$$

Then there is an enlargement $\hat{\Omega}$ of Ω and an extension $\hat{P}$ of P to $\hat{\Omega}$ such that

$$x(t) = x(0) + \int_0^t \sigma(s)\,d\beta(s) + \int_0^t \tilde{\sigma}(s)\,d\tilde{\beta}(s) + \int_0^t b(s)\,ds \qquad a.s.\ \hat{P}\,,$$

where β and $\tilde{\beta}$ are independent d-dimensional $\hat{P}$-Brownian motions.

Proof: There is an enlargement Ω' of Ω and an extension P' of P to Ω' such that

$$x(t) - x(0) = \int_0^t a^{1/2}(s)\, d\alpha(s) + \int_0^t b(s)\, ds\,,$$

where $\alpha(s)$ is a d-dimensional P'-Brownian motion. Define

$$\rho = \lim_{\varepsilon \downarrow 0} a^{1/2}(a + \varepsilon I)^{-1}\,.$$

Then $\rho a^{1/2} = a^{1/2}\rho = \pi_R$, where π_R is the orthogonal projection onto the range of a. Let $\pi_N = I - \pi_R$ be the orthogonal projection onto the null space of a. Define

$$U^{11} = U^{22} = \rho\sigma + \frac{1}{\sqrt{2}}\pi_N\,,$$

$$U^{12} = -U^{21} = \rho\tilde{\sigma} + \frac{1}{\sqrt{2}}\pi_N\,;$$

then the $2d \times 2d$ matrix

$$U = \begin{bmatrix} U^{11} & U^{12} \\ U^{21} & U^{22} \end{bmatrix}$$

is an orthogonal matrix. Let $\hat{\Omega} = \Omega' \times \tilde{\Omega}$ and $\hat{P} = P' \times \tilde{P}$, where $(\tilde{\Omega}, \tilde{P})$ admits the Brownian motion $\tilde{\alpha}(t)$. Let us define

$$\begin{bmatrix} \beta(t) \\ \tilde{\beta}(t) \end{bmatrix} = \int_0^t U(s) \begin{bmatrix} d\alpha(s) \\ d\tilde{\alpha}(s) \end{bmatrix}.$$

Then β and $\tilde{\beta}$ are independent $\hat{P}$-Brownian motions. Moreover,

$$\int_0^t \sigma(s)\, d\beta(s) = \int_0^t \sigma(s) U^{11}(s)\, d\alpha(s) + \int_0^t \sigma(s) U^{12}(s)\, d\tilde{\alpha}(s)\,,$$

$$\int_0^t \tilde{\sigma}(s)\, d\tilde{\beta}(s) = \int_0^t \tilde{\sigma}(s) U^{21}(s)\, d\alpha(s) + \int_0^t \tilde{\sigma}(s) U^{22}(s)\, d\tilde{\alpha}(s)\,.$$

Therefore,

$$\int_0^t \sigma(s)\, d\beta(s) + \int_0^t \tilde{\sigma}(s)\, d\tilde{\beta}(s)$$

$$= \int_0^t (\sigma U^{11} + \tilde{\sigma} U^{21})(s)\, d\alpha(s) + \int_0^t (\sigma U^{12} + \tilde{\sigma} U^{22})(s)\, d\tilde{\alpha}(s)$$

$$= \int_0^t a^{1/2}(s)\, d\alpha(s)\,.$$

Indeed, one need only note that

$$\sigma U^{11} + \tilde{\sigma} U^{21} = a^{1/2}$$

and

$$\sigma U^{12} + \tilde{\sigma} U^{22} = 0\,.$$

THEOREM 3.1. *Let a, σ, $\tilde{\sigma}: [0, \infty) \times R^d \to S_d$ and b, $c: [0, \infty) \times R^d \to R^d$ be bounded measurable functions which are uniformly Lipschitz continuous in the space variables on $[0, \infty) \times R^d$. Let us assume further that $a, \sigma \in C_b^{1,2}([0, \infty) \times R^d)$, that a, σ and $\tilde{\sigma}$ commute and that $a = \sigma^2 + \tilde{\sigma}^2$. Let $\tilde{a} = \tilde{\sigma}^2$ and $\tilde{b} = b + a^{1/2} c - \frac{1}{2}\sigma' \sigma$. If $P_{s,x}$ and $\tilde{P}_{s,x}$ are the solutions to the martingale problems for $[a, b]$ and $[\tilde{a}, \tilde{b}]$, respectively, starting from x at time s, then*

$$\operatorname{supp}(\tilde{P}_{s,x}) \subset \operatorname{supp}(P_{s,x})\,.$$

Proof: Since the support of $P_{s,x}$ is the same as the support of the solution $Q_{s,x}$ to the martingale problem for $[a, b + a^{1/2} c]$, we can assume that $c = 0$. In that case the theorem is an immediate consequence of Lemmas 3.2 and 3.3.

COROLLARY 3.1. *Let $a \in C_b^{1,2}([0, \infty) \times R^d, S_d)$ and let $b: [0, \infty) \times R^d \to R^d$ be bounded and uniformly Lipschitz continuous in space. Given a bounded measurable*

$$\psi: [0, \infty) \to R^d\,,$$

define $\phi(\cdot)$ for $t \geqq s$ by

$$\phi(t) = x + \int_s^t a(u, \phi(u))\psi(u)\, du + \int_s^t \tilde{b}(u, \phi(u))\, du\,,$$

where $\tilde{b} = b - \frac{1}{2}a'$. Then $\phi \in \operatorname{supp}(P_{s,x})$, where $P_{s,x}$ solves the martingale problem for $[a, b]$ starting from x at time s.

Proof: Clearly we can assume without loss of generality that the trace of a is at most $\frac{1}{2}$. Define $\sigma_0 = a$ and, for $n \geqq 1$,

$$\sigma_n = \sigma_{n-1} - \sigma_{n-1}^2 \quad \text{and} \quad b_n = b + a\psi - \tfrac{1}{2}\sum_0^{n-1} (\sigma_j^2)' .$$

Let $P_{s,x}^{(n)}$ be the solution to the martingale problem for $[\sigma_n, b_n]$ starting from x at time s. Then, by induction, we see from the preceding theorem that for all n

$$\operatorname{supp}(P_{s,x}^{(n)}) \subset \operatorname{supp}(P_{s,x}) .$$

Moreover, by Theorem 10.2 of the appendix, $\sigma_n \to 0$ and $b_n \to b + a\psi - \frac{1}{2}a'$ uniformly on compacts as $n \to \infty$. Therefore, $P_{s,x}^{(n)}$ converges weakly as $n \to \infty$ to the distribution degenerate at the path ϕ. In particular,

$$\phi \in \operatorname{supp}(P_{s,x}) .$$

Given coefficients $[a, b]$ as in the preceding corollary, let us define $S_{a,b}(s, x)$ to be the set of $\phi:[s, \infty) \to R^d$ such that

$$\phi(t) = x + \int_s^t a(u, \phi(u))\psi(u)\, du + \int_s^t b(u, \phi(u))\, du , \qquad t \geqq s,$$

for some bounded measurable function $\psi:[s, \infty) \to R^d$.

THEOREM 3.2. *Let $[a, b]$ satisfy the conditions of Corollary 3.1. Given a bounded measurable $c:[0, \infty) \times R^d \to R^d$, let*

$$L_t = \tfrac{1}{2}\nabla \cdot a\nabla + (b + a^{1/2} c) \cdot \nabla ,$$

and denote by $P_{s,x}$ the solution to the martingale problem for L_t starting from x at time s. Then

$$\operatorname{supp}(P_{s,x}) = \overline{S_{a,b}(s, x)} .$$

Proof: We can assume that $c = 0$. The inclusion $\overline{S_{a,b}(s, x)} \subseteq \operatorname{supp}(P_{s,x})$ follows at once from Corollary 3.1. To prove the opposite inclusion, let us assume that the trace of a is at most $\frac{1}{2}$ and define σ_n, for $n \geqq 0$, as in Corollary 3.1. Let $a_n = \sum_0^n \sigma_j^2$ and

$$L_t^{(n)} = \tfrac{1}{2}\nabla \cdot a_n \nabla + b \cdot \nabla .$$

If $P_{s,x}^{(n)}$ is the solution to the martingale problem corresponding to $L_t^{(n)}$, starting

from x at time s, then we know that

$$\operatorname{supp}(P_{s,x}^{(n)}) \subset \overline{S_{a_n,b}(s, x)} .$$

Moreover,

$$\overline{S_{a_n,b}(s, x)} \subset \overline{S_{a,b}(s, x)} \quad \text{for all} \quad n .$$

Since $P_{s,x}^{(n)}$ converges weakly to $P_{s,x}$ as $n \to \infty$, we have

$$\operatorname{supp}(P_{s,x}) \subset \overline{S_{a,b}(s, x)} .$$

Remark 3.1. Suppose that G is an open set in $[0, \infty) \times R^d$ and that a, b and c are coefficients satisfying the conditions of Corollary 2.1. Let τ_s and $Q_{s,x}$, $(s, x) \in G$, be defined as in that corollary. Given $T > s$ such that

$$Q_{s,x}(\tau_s \geqq T) > 0 ,$$

let $Q_{s,x}^T$ denote the measure on $C([s, T], R^d)$ obtained by conditioning $Q_{s,x}$ with respect to $\{\tau_s \geqq T\}$. Then an easy consequence of the results of this section is that $\operatorname{supp}(Q_{s,x}^T)$ coincides with the closure in $C([s, T], R^d)$ of the paths $\phi:[s, T] \to R^d$ whose graph lies in G and for which there is a bounded measurable $\psi:[s, T] \to R^d$ such that

$$\phi(t) = x + \int_s^t a(u, \phi(u))\psi(u)\, du + \int_s^t b(u, \phi(u))\, du , \qquad t \in [s, T] .$$

4. The Strong Maximum Principle

Let G be an open set in $[0, \infty) \times R^d$ and suppose a, b and c are coefficients satisfying the conditions of Corollary 2.1. Define

$$L_u = \tfrac{1}{2}\nabla \cdot a\nabla + (b + a^{1/2} c) \cdot \nabla$$

and define τ_s and $Q_{s,x}$, $(s, x) \in G$, as in Corollary 2.1. Given $(s, x) \in G$, consider the measures

$$q(s, x;\ t, \Gamma) = Q_{s,x}(x(t) \in \Gamma ,\quad \tau_s > t). \tag{4.1}$$

Define $A(s, x)$ to be the closure in G of the points $(t, \phi(t))$, where $t \geqq s$ and $\phi:[s, t] \to R^d$ is a path whose graph lies in G and for which there is a bounded measurable $\psi:[s, t] \to R^d$ such that

$$\phi(u) = x + \int_s^u a(\alpha, \phi(\alpha))\psi(\alpha)\, d\alpha + \int_s^u b(\alpha, \phi(\alpha))\, d\alpha , \qquad s \leqq u \leqq t .$$

As a consequence of Remark 3.1, we see that $y \in \operatorname{supp}(q(s, x; t, \cdot))$ if and only if $(t, y) \in A(s, x)$.

The purpose of the present section is to show that if u is a "subsolution" for $\partial/\partial t + L_t$ in G and if $u(s, x) = \sup_G u(t, x)$, then $u \equiv u(s, x)$ on $A(s, x)$. Moreover, we want to show that $A(s, x)$ is the largest subset of G having this property. Before proving this result, it is necessary to explore what we mean by a "subsolution". Certainly any function $u \in C_b^{1,2}(G)$ satisfying

$$\frac{\partial u}{\partial t} + L_t u \geqq 0 \qquad \text{in} \qquad G$$

must be covered by our definition. However, in order to show that $A(s, x)$ is maximal, we need to extend our notion of subsolution to include functions which, in the case $L_t = \frac{1}{2}\Delta$, are called subparabolic functions on G.

Let $k: G \to [0, \infty)$ be a bounded measurable function. Given a function $u: G \to R$, we shall say that u is a *k-subsolution* if u is bounded above, u is upper semicontinuous, and, for any $(s, x) \in G$,

$$u(t \wedge \tau_s, x(t \wedge \tau_s)) \exp\left\{-\int_0^{t\wedge\tau_s} k(u, x(u))\, du\right\}$$

is a $Q_{s,x}$-submartingale. We shall first show that if $u \in C_b^{1,2}(G)$ satisfies

$$\frac{\partial u}{\partial t} + L_t u - ku \geqq 0 \qquad \text{in} \qquad G, \tag{4.2}$$

then u is a k-subsolution.

LEMMA 4.1. *Let $\xi: [0, \infty) \times \Omega \to R$ and $\eta: [0, \infty) \times \Omega \to R$ be bounded nonanticipating functions. Assume that $\xi(t)$ is a P-martingale for some probability measure P on Ω, and define $\psi(t) = \int_0^t \eta(s)\, ds$. Then*

$$\xi(t)\psi(t) - \int_0^t \xi(s)\eta(s)\, ds$$

is again a P-martingale.

Proof: Note that

$$E\left[\xi(t)\phi(t) - \xi(s)\phi(s) - \int_s^t \xi(u)\eta(u)\, du \,\middle|\, M_s\right]$$

$$= E[(\xi(t) - \xi(s))\phi(s) \mid M_s] + E\left[\int_s^t (\xi(t) - \xi(u))\eta(u)\, du \,\middle|\, M_s\right].$$

The first term obviously vanishes. Moreover, if $A \in M_s$, then

$$E\left[\chi_A \int_s^t (\xi(t) - \xi(u))\eta(u)\, du\right] = \int_s^t E[\chi_A(\xi(t) - \xi(u))\eta(u)]\, du = 0\,,$$

and this completes the proof.

LEMMA 4.2. *If* $u \in C_b^{1,2}(G)$ *satisfies* (4.2), *then* u *is a k-subsolution.*

Proof: Let

$$\xi(t) = u(t \wedge \tau_s, x(t \wedge \tau_s)) - \int_0^{t\wedge\tau_s} \left(\frac{\partial u}{\partial \alpha} + L_\alpha u\right)(\alpha, x(\alpha))\, d\alpha\,,$$

and

$$\eta(t) = -\chi_{\tau_s > t} k(t, x(t)) \exp\left\{-\int_s^{t\wedge\tau_s} k(\alpha, x(\alpha))\, d\alpha\right\}.$$

Then an application of Lemma 4.1 shows that

$$u(t \wedge \tau_s, x(t \wedge \tau_s)) \exp\left\{-\int_s^{t\wedge\tau_s} k(\alpha, x(\alpha))\, d\alpha\right\} - \int_t^{t\wedge\tau_s} \left(\frac{\partial u}{\partial \alpha} + L_\alpha u - ku\right) \exp\left\{-\int_s^\alpha k(\gamma, x(\gamma))\, d\gamma\right\} d\alpha$$

is a $Q_{s,x}$-martingale. Since the integrand in the second term is non-negative, this proves that the first term is a $Q_{s,x}$-submartingale.

THEOREM 4.1. *If* u *is a k-subsolution for some non-negative bounded measurable* k *and if* $0 = u(s, x) = \sup_G u(t, y)$ *for some* $(s, x) \in G$, *then* $u \equiv u(s, x)$ *on* $A(s, x)$. *Conversely, if* $(t_0, y_0) \in G - A(s, x)$, *then there is a* 0*-subsolution* u *such that* $u \leqq 0$, $u(s, x) = 0$, *and* $u(t_0, y_0) < 0$.

Proof: Choose $\lambda \geqq \sup_G k(t, y)$. Then u is a λ-subsolution, and so

$$\begin{aligned} e^{-\lambda(t-s)} \int q(s, x; t, dy) u(t, y) &= E^{Q_{s,x}}(e^{-\lambda(t-s)} u(t, x(t))\chi_{\tau_s > t}] \\ &\geqq E^{Q_{s,x}}(e^{-\lambda(t\wedge\tau_s - s)} u(t \wedge \tau_s, x(t \wedge \tau_s))] \geqq u(s, x) = 0\,. \end{aligned}$$

Thus, since u is upper semicontinuous and at most 0, $u(t, \cdot) \equiv 0$ on

$$\operatorname{supp}(q(s, x;\ t, \cdot)) = A(s, x)\,.$$

To prove the converse, suppose that $(t_0, y_0) \in G - A(s, x)$ and choose $\psi \in C_0^\infty(G - A(s, x))$ such that $\psi \leqq 0$ and $\psi(t_0, y_0) = -1$. Define

$$u(t, y) = E^{Q_{t,y}}\left[\int_t^{\tau_t} \psi(\alpha, x(\alpha))\, d\alpha\right].$$

Then $u \leqq 0$ and $u(t_0, y_0) < 0$. Moreover, using Remark 3.1 and the fact that $\psi = 0$ on $A(s, x)$, one sees that $u(s, x) = 0$. Finally, the weak continuity of the $Q_{t,y}$ and standard Markov arguments imply that u is a 0-subsolution.

Remark 4.1. It is obvious from the proof of Theorem 4.1 that u need be a subsolution only on $A(s, x)$ and $0 = u(s, x) = \sup_{A(s,x)} u(t, y)$ in order for the conclusion to obtain.

Remark 4.2. The analytic meaning of the extension of $\partial/\partial t + L_t$ which is implicit in our definition of subsolutions will be discussed in Section 8. In particular, it will be described there in what sense the counterexample constructed in Theorem 4.1 satisfies $\partial u/\partial t + L_t u \geqq 0$ on G. However, it should be mentioned here that if $G = (t_1, t_2) \times R^d$, then our counterexample is essentially as smooth as the coefficients in L_t. This fact follows from the theorem of Oleinik proved in the appendix.

Remark 4.3. Because our notion of subsolution is basically a mean-value property, our class of subsolutions is closed under the same operations as the class of subparabolic functions. Hence it is considerably larger than the set of smooth functions satisfying (4.2).

5. The First Boundary Value Problem

Let $a: R^d \to S_d$ and $b: R^d \to R^d$ be bounded measurable functions such that $a \in C^2(R^d)$ and $b \in C^1(R^d)$. Let

$$L = \frac{1}{2}\sum_{i,j=1}^d a^{ij}(x)\frac{\partial^2}{\partial x_i\, \partial y_j} + \sum_{i=1}^d b^i(x)\frac{\partial}{\partial x_i},$$

and denote by P_x the solution to the martingale problem for L starting at x. Given an open set G in R^d, define

$$\tau = \inf\{t \geqq 0\colon\ x(t) \notin G\},$$

$$\tau' = \inf\{t \geqq 0\colon\ x(t) \notin \bar{G}\}.$$

If L is strictly elliptic and G is reasonably nice (e.g., satisfies an exterior cone condition), then $P_x(\tau \neq \tau') = 0$ for all $x \in G$. However, if L is degenerate,

then, as we shall see, $P_x(\tau \neq \tau')$ need not vanish for all $x \in G$ no matter how smooth G is. This fact leads us to two formulations of the first boundary value problem, one corresponding to τ and the other to τ'. As will be shown later, these two formulations correspond analytically to approximating the solution from inside and from the outside, respectively.

We shall use Γ to denote the set of $x \in \delta G$ such that $P_x(\tau' > 0) = 0$. By the $0-1$ law, if $x \in \delta G$, then $x \notin \Gamma$ if and only if $P_x(\tau' > 0) = 1$.

LEMMA 5.1. *The set* Γ *is a* G_δ *subset of* δG. *Moreover, for all* $x \in G$,

$$P_x(x(\tau') \notin \Gamma\,, \quad \tau' < \infty) = 0\,.$$

Finally, if $x \in \Gamma$, *then* $\lim_{\substack{y \to x \\ y \in G}} P_y(\tau' \geqq \varepsilon) = 0$ *for all* $\varepsilon > 0$.

Proof: Note that $\{\tau' < \varepsilon\}$ is open. Since P_x is weakly continuous in x, $P_x(\tau' < \varepsilon) \leqq \underline{\lim}_{y \to x} P_y(\tau' < \varepsilon)$. Hence $\{x \in \delta G\colon\ P_x(\tau' < \varepsilon) > 0\}$ is open in δG, and so

$$\Gamma = \{x \in \delta G : P_x(\tau' = 0) > 0\} = \bigcap_1^\infty \{x \in \delta G : P_x(\tau' < 1/n) > \tfrac{1}{2}\}$$

is a G_δ.

To prove the second assertion, observe that $x(\tau') \in \delta G$ a.s. P_x for $x \in G$ since the paths are continuous. Hence,

$$P_x(x(\tau') \notin \Gamma, \tau' < \infty) = E_x(\chi_{\tau' < \infty} P_{x(\tau')}(\tau' > 0)] = P_x(\tau' > \tau') = 0\,,$$

by the strong Markov property.

Finally, suppose $x \in \Gamma$. Then, for all $\varepsilon > 0$, $P_x(\tau' \geqq \varepsilon) = 0$. Since $\{\tau' \geqq \varepsilon\}$ is closed, $\overline{\lim}_{y \to x} P_y(\tau' \geqq \varepsilon) \leqq P_x(\tau' \geqq \varepsilon) = 0$; q.e.d.

Given bounded measurable functions u and f on $\bar{G}$, we shall say that

$$\tilde{L}u = f \quad \text{in} \quad \bar{G}$$

if

$$u(x(t \wedge \tau')) - \int_0^{t \wedge \tau'} f(x(s))\, ds$$

is a P_x-martingale for all $x \in \bar{G}$.

Remark 5.1. If $u \in C^2(\bar{G})$ and $Lu = f$ on $\bar{G}$, then clearly $\tilde{L}u = f$. Hence $\tilde{L}$ is an extension of L. Moreover, it follows from Theorem 2.4 that the definition of $\tilde{L}$ really depends only on the restriction of the coefficients of L to G. Finally,

although the operator $\tilde{L}$ is not strictly single-valued, nonetheless if $\tilde{L}u = f_i$ in $\bar{G}$, $i = 1, 2$, then $f_1 = f_2$ at common points of continuity.

We are now ready to state the first boundary value problem associated with τ'. Let k be a bounded, continuous non-negative function on $\bar{G}$. Given bounded measurable functions f and g on $\bar{G}$ and Γ, respectively, we shall say that u solves the *τ'-first boundary value problem* for f and g if

$$
\begin{aligned}
\tilde{L}u - ku &= f \quad \text{on} \quad \bar{G}, \\
u\big|_{\Gamma} &= g .
\end{aligned}
\tag{5.1}
$$

THEOREM 5.1. *If either*

$$\sup_{x\in\bar{G}} E_x[\tau'] < \infty \tag{5.2}$$

or

$$\inf_{x\in\bar{G}} k(x) > 0 , \tag{5.3}$$

the τ'-first boundary value problem has exactly one solution u for each f and g. Moreover, $\lim_{\substack{y\to x\\ y\in G}} u(y) = g(x)$ *for* $x \in \Gamma$ *which are points of continuity of* g.

Proof: Standard Markov arguments show that

$$
\begin{aligned}
u(x) = E_x\Big[-\int_0^{\tau'} \exp\Big\{ -\int_0^t k(x(s))\, ds\Big\} f(x(t))\, dt \\
+ \exp\Big\{ -\int_0^{\tau'} k(x(s))\, ds\Big\} g(x(\tau'))\Big]
\end{aligned}
\tag{5.4}
$$

is a solution (note that all the indicated integrals exist under either (5.2) or (5.3)).

Conversely, if u is a solution, then

$$u(x(t\wedge\tau')) - \int_0^{t\wedge\tau'} (ku + f)(x(s))\, ds$$

is a P_x-martingale for all $x \in \bar{G}$. Hence, by an application of Lemma 4.1,

$$u(x(t\wedge\tau')) \exp\Big\{ -\int_0^{t\wedge\tau'} k(x(s))\, ds\Big\} - \int_0^{t\wedge\tau'} \exp\Big\{ -\int_0^{\theta} k(x(s))\, ds\Big\} f(x(\theta))\, d\theta$$

is also a P_x-martingale for all $x \in \bar{G}$, and so

$$u(x) = E_x\left[-\int_0^{t\wedge\tau'} \exp\left\{-\int_0^\theta k(x(s))\,ds\right\} f(x(\theta))\,d\theta + \exp\left\{-\int_0^{t\wedge\tau'} k(x(s))\,ds\right\} u(x(t\wedge\tau'))\right],$$

for all $t \geqq 0$ and $x \in G$. Using (5.2) or (5.3) and the fact that

$$u(x(t\wedge\tau')) = g(x(\tau')) \qquad \text{on} \qquad \{\tau' \leqq t\},$$

one sees upon letting $t \uparrow \infty$ that $u(x)$ is given by (5.4).

To prove that $\lim_{\substack{y\to x\\ y\in G}} u(y) = g(x)$ if $x \in \Gamma$ is a point of continuity of g, observe that by Lemma 5.1 and either (5.2) or (5.3)

$$\lim_{\substack{y\to x\\ y\in \bar{G}}} E_y\left[\int_0^{t\wedge\tau'} \exp\left\{-\int_0^t k(x(s))\,ds\right\} dt\right] = 0 .$$

This is obvious if (5.3) obtains. To prove it under (5.2) note that

$$\begin{aligned}\overline{\lim_{\substack{y\to x\\ y\in G}}}\, E_y[\tau'] &\leqq \varepsilon + \overline{\lim_{\substack{y\to x\\ y\in G}}}\, E_y[\tau', \tau' > \varepsilon] = \varepsilon + \overline{\lim_{\substack{y\to x\\ y\in G}}}\, E_y(\chi_{\tau'>\varepsilon} E_{x(\varepsilon)}[\tau']] \\ &\leqq \varepsilon + c\,\overline{\lim_{\substack{y\to x\\ y\in G}}}\, P_y(\tau' > \varepsilon) = \varepsilon ,\end{aligned}$$

where $c = \sup_{x\in \bar{G}} E_x[\tau']$. This implies that

$$\lim_{\substack{y\to x\\ y\in G}} u(y) = \lim_{\substack{y\to x\\ y\in G}} E_y[g(x(\tau')), \tau' < \infty] .$$

Our assertion now follows from another application of Lemma 5.1 and use of the fact that

$$\lim_{\delta\searrow 0}\, \sup_{y\in \bar{G}} P_y\left(\sup_{0\leqq t\leqq \delta} |x(t) - y| \geqq \varepsilon\right) = 0$$

for all $\varepsilon > 0$.

Remark 5.2. The condition (5.2) is difficult to verify in general. However, one sufficient condition for (5.2) to hold can be obtained from our characterization of supp (P_x). Indeed, suppose $\bar{G}$ is compact and that there is a T such that

for each $x \in \bar{G}$ there is a $\phi \in S_{a,b}(x)$ such that $\phi(T) \notin \bar{G}$. Then $P_x(\tau' < T) > 0$ for all $x \in \bar{G}$. Hence, since $P_x(\tau' < T)$ is lower semi-continuous,

$$\inf_{x \in \bar{G}} P_x(\tau' < T) \geqq 1 - \alpha > 0 .$$

Thus by induction and the strong Markov property, $P_x(\tau' \geqq nT) \leqq \alpha^n$ for all $x \in \bar{G}$. Hence, in this case, we even have $\sup_{x \in \bar{G}} E_x[e^{\lambda \tau'}] < \infty$ for λ such that $\alpha e^{\lambda T} < 1$.

We now want to describe the first boundary problem associated with τ. To do this we must first say what extension of L we are going to use. Given bounded measurable functions u and f on G, we write

$$\hat{L}u = f \quad \text{in} \quad G ,$$

if

$$u(x(t \wedge \tau)) - \int_0^{t \wedge \tau} f(x(s))\, ds$$

is a P_x-martingale for all $x \in G$. Observe that $\hat{L}u = f$ on G if and only if for every compact $K \subset G$, $\tilde{L}u = f$ on K. In order to describe how we want the boundary data to be taken on, we introduce the "barrier" function,

$$\psi(x) = 1 - E_x[e^{-\tau}]$$

and define

$$\Gamma_0 = \left\{ x \in \delta G : \varliminf_{\substack{y \to x \\ y \in G}} \psi(y) = 0 \right\} .$$

Let k be a non-negative bounded continuous function on G. Given a bounded measurable function f on G and a bounded continuous function g on Γ_0, we say that a bounded measurable function u solves the *τ-first boundary value problem* for f and g if

$$(5.5) \qquad \begin{aligned} \hat{L}u - ku &= f \quad &&\text{in} \quad G , \\ u(y) &\xrightarrow{\psi} g(x) \quad &&\text{as} \quad y \to x \in \Gamma_0 , \end{aligned}$$

where $u(y) \xrightarrow{\psi} g(x)$ means that $u(y_n) \to u(x)$ if $\langle y_n \rangle_1^\infty \subseteq G$ tends to x and $\lim_{n \to \infty} \psi(y_n) = 0$.

Remark 5.3. It is easy to check that $\hat{L}\psi = \psi - 1$ in G. However, ψ is lower semicontinuous and not in general upper semicontinuous. It is this fact that prevents ψ from being a good barrier and forces us to give such a clumsy

description of the way in which u takes on its boundary data. If one were to take the lower semicontinuous regularization of ψ, then for nice regions G one would end up with $1 - E_x[e^{-\tau'}]$, which is the appropriate barrier for the τ'-first boundary value problem but not for the τ-problem. In this connection, it should be noted that if $G_n \nearrow G$ and $\bar{G}_n \subset G$, then $1 - E_x[e^{-\tau'_n}] \searrow \psi(x)$ on G.

LEMMA 5.2. *For all* $x \in G$, $\chi_{\tau<\infty}\psi(x(t \wedge \tau)) \to 0$ *a.s.* P_x *as* $t \nearrow \infty$.

Proof: Clearly it is enough to show that $\xi(t) = e^{-t\wedge\tau}\psi(x(t \wedge \tau)) \to 0$ a.s. P_x as $t \to \infty$. Using Lemma 4.1, one can easily check that $\xi(t)$ is a P_x-supermartingale, and therefore, since it is bounded, $\xi(t)$ converges a.s. P_x. Thus, to prove that $\xi(t) \to 0$ a.s. P_x, it is enough to show that $E_x[\xi(t)] \to 0$ as $t \to \infty$. But

$$\begin{aligned} E_x(e^{-t\wedge\tau}\psi(x(t \wedge \tau))] &= E_x(e^{-t\wedge\tau}E_{x(t\wedge\tau)}[1 - e^{-\tau}]] \\ &= E_x(e^{-t\wedge\tau}(1 - e^{-(\tau-t\wedge\tau)})] \\ &= E_x(e^{-t}(1 - e^{-(\tau-t)})\chi_{\tau>t}] \to 0 \quad \text{as} \quad t \nearrow \infty . \end{aligned}$$

Remark 5.4. As a consequence of Lemma 5.2, we see that

$$P_x(x(\tau) \notin \Gamma_0 , \tau < \infty) = 0 , \qquad x \in G .$$

In the next section we shall show that $P_x(x(\tau) \notin \bar{\Gamma}, \tau < \infty) = 0, x \in G$. From this it follows that $\Gamma_0 \subseteq \bar{\Gamma}$. Indeed, if $x \in \Gamma_0 - \bar{\Gamma}$, then we can find an $\varepsilon > 0$ so that $|y - x| \leqq \varepsilon$ implies $y \notin \bar{\Gamma}$. But $\underset{\substack{y \to x \\ y \in G}}{\underline{\lim}} P_x(\tau \geqq t) = 0$ for all $t > 0$, and so there is a $y \in G$ such that $P_y(|x(\tau) - x| < \varepsilon, \tau < \infty) > 0$, which contradicts $P_y(x(\tau) \in \bar{\Gamma}, \tau < \infty) = 0$. Also notice that, for $x \in \Gamma$,

$$\lim_{\substack{y \to x \\ y \in G}} P_y(\tau \geqq t) \leqq \lim_{\substack{y \to x \\ y \in G}} P_y(\tau' \geqq t) = 0$$

for $t > 0$ and so $\Gamma \subseteq \Gamma_0$. Hence the only points where the bad convergence may occur lie in $\bar{\Gamma} - \Gamma$.

THEOREM 5.2. *Let k be a non-negative continuous bounded function on G. Assume that either*

$$\sup_{x \in G} E_x[\tau] < \infty \tag{5.6}$$

or

$$\inf_{x\in G} k(x) > 0 . \tag{5.7}$$

Then for each bounded measurable function f on G and bounded continuous function g on Γ_0 there is exactly one bounded measurable u which solves the τ-first boundary value problem for f and g.

Proof: To prove existence, set

$$\begin{aligned} u(x) = E_x\Big[-\int_0^\tau \exp\Big\{-\int_0^t k(x(\alpha))\, d\alpha\Big\} f(x(t))\, dt \\ + \exp\Big\{-\int_0^\tau k(x(\alpha))\, d\alpha\Big\} g(x(\tau))\Big] . \end{aligned} \tag{5.8}$$

Then, by the same type of argument as that given in Theorem 5.1, one can show that $\hat{L}u - ku = f$ in G and that $u(y_n) \to g(x)$ if $\{y_n\}_1^\infty \subseteq G$ tends to x in such a way that $\psi(y_n) \to 0$.

To prove uniqueness, note that if u is a solution, then

$$\begin{aligned} u(x) = E_x\Big[-\int_0^{t\wedge\tau} \exp\Big\{-\int_0^s k(x(\alpha))\, d\alpha\Big\} f(x(s))\, ds \\ + \exp\Big\{-\int_0^{t\wedge\tau} k(x(\alpha))\, d\alpha\Big\}\, u(x(t\wedge\tau))\Big] . \end{aligned}$$

By Lemma 5.2, $u(x(t\wedge\tau)) \to g(x(\tau))$ a.s. P on $\{\tau < \infty\}$, and so we see, upon letting $t \nearrow \infty$, that u must be given by (5.8).

Remark 5.5. Although the τ-first boundary value problem has many drawbacks, it has one major advantage over the τ'-problem. Consider coefficients $a \in C^2(G)$ and $b \in C^1(G)$ which cannot be extended smoothly outside of G. (For example, let $G = (0, \infty)$ and suppose $a(x) = x$ near 0.) Then one can no longer discuss the τ'-problem, but the τ-problem still makes sense and Theorem 5.2 remains true.

6. Infinitesimal Processes and Conditions for Regularity

Let $G \subset R^d$ be a region with $\phi(x)$ as its defining function. That is,

$$G = \{x:\ \phi(x) > 0\} ,$$

$$\bar{G} = \{x:\ \phi(x) = 0\} ,$$

$$|\nabla\phi| \neq 0 \quad \text{on} \quad \delta G .$$

Let $x^0 \in \delta G$. We would like to determine, in terms of the behavior of the coefficients of L and the function ϕ, when $x^0 \in \Gamma$, the set of τ' regular points described in Section 5. For this purpose we look at the process $x(t)$ corresponding to L, starting at time 0 from the point x^0. Let $\xi(t) = \phi(x(t))$. We have to decide when $\xi(t)$ becomes immediately negative. Let us define

$$\alpha(t) = E\xi(t)\,,$$
$$\beta(t) = E\xi^2(t)\,,$$
$$s(t) = [\beta(t) - \alpha^2(t)]^{1/2}\,;$$

$s(t)$ is of course non-negative and is the standard deviation of $\xi(t)$.

We shall assume that

$$\alpha(t) = \alpha t^r + o(t^r) \qquad \text{as} \qquad t \downarrow 0\,,$$

and

$$\beta(t) = \beta t^s + o(t^s) \qquad \text{as} \qquad t \downarrow 0\,,$$

where α, β are different from zero and r, s are some integers. Clearly, $s \leqq 2r$ and if $s = 2r$, then $\beta \geqq \alpha^2$. Three cases arise naturally in trying to compare $\alpha(t)$ with $s(t)$.

Case 1: $s < 2r$.

In this case, $s(t)$ is much larger than $\alpha(t)$. So the diffusion must dominate and this indicates that $\xi(t)$ takes on immediately positive and negative values. Therefore, the point in question is probably regular.

Case 2: $s = 2r$, $\beta = \alpha^2$.

In this case, $s(t)$ is negligible compared to $\alpha(t)$. So the behavior of $\alpha(t)$ must depend on the sign of α; $\alpha > 0$ should imply irregularity and $\alpha < 0$ should imply regularity.

Case 3: $s = 2r$, $\beta > \alpha^2$.

In this case, $s(t)$ and $\alpha(t)$ are of the same order of magnitude. The situation is unclear and one must look more carefully.

To illustrate these cases let us consider some examples.

EXAMPLE 1. If, at a boundary point, $\langle a\nabla\phi,\ \nabla\phi\rangle > 0$, then that point must be regular. In this case, $s = 1$ and no matter what r is we always have $s < 2r$.

EXAMPLE 2. Consider the region $G \subset R^2$ consisting of

$$0 < x < 1 , \qquad -\infty < y < \infty .$$

Let the operator be

$$\frac{\partial}{\partial y^2} + y\frac{\partial}{\partial x} .$$

Take the boundary point (0, 0). For the function $\phi(x) = x$, $s = 3$ and $r = \infty$. So the origin is regular.

EXAMPLE 3. If the operator is purely of first order or at least if the second order terms have a strong enough zero at the point under consideration, then we are in Case 2, so that the regularity is determined purely by the sign of $E\phi(x(t))$ for small t.

EXAMPLE 4. Consider the operator

$$\frac{1}{2}\frac{\partial^2}{\partial x^2} + x\frac{\partial^2}{\partial x\,\partial y} + \tfrac{1}{2}x^2\frac{\partial}{\partial y^2} + (\alpha + \tfrac{1}{2})\frac{\partial}{\partial y}$$

in either $y > 0$ or $y < 0$. A solution starting from (0, 0) is easily seen to be

$$y(t) = \tfrac{1}{2}\beta^2(t) + \alpha t ,$$

where $\beta(t)$ is Brownian motion. If the region is $y > 0$, we have regularity of (0, 0) provided $\alpha < 0$. If the region is $y < 0$ we have regularity for all α. All these cases come under Case 3 with $s = 2$ and $r = 1$.

Theorem 6.1 deals with Case 1 and under some additional assumptions proves what is indicated. Theorem 6.2 treats Case 2 and establishes what was suggested before. Theorem 6.3 looks in detail at Case 3 when $r = 1$ and $s = 2$. Each theorem is preceded by a lemma or two which cover some preliminary material.

Remark. If the coefficients and the function ϕ are sufficiently smooth, we can determine α, β, r and s by the relations

$$(L^k\phi)(x^0) = 0 \quad \text{for} \quad k = 1, 2, \cdots, r-1 ,$$

$$(L^r\phi)(x^0) = (r!)\alpha ,$$

$$(L^k\phi^2)(x^0) = 0 \quad \text{for} \quad k = 1, 2, \cdots, s-1 ,$$

$$(L^r\phi^2)(x^0) = (s!)\beta .$$

We first consider Case 1. Here we have to assume in addition that $a = \sigma\sigma^*$, where $\sigma \in C^\infty[R^d, R^d \otimes R^N]$ for some N.

LEMMA 6.1. *If* $a = \sigma\sigma^*$, *where* $\sigma \in C^\infty[R^d, R^d \otimes R^N]$, *then for* $f \in C_d^\infty[R^d]$

$$E^{P_{x^0}}[f^2(x(t))] \leqq Ct^n \quad for \quad small \quad t$$

implies that

$$E^{P_{x^0}}[f^4(x(t))] \leqq C't^{2n} \quad for \quad small \quad t.$$

Proof: We shall drop P_{x^0} and denote the expectations by just E. We shall prove by induction on k that if $g = \sum_1^m f_i^2$, where $f_i \in C_b^\infty[R^d]$, for any arbitrary m, then

$$Eg(x(t)) \leqq Ct^k \quad \text{implies that} \quad Eg^2(x(t)) \leqq C't^{2k}.$$

The assertion is trivial for $k = 0$. Suppose it is true for $k \leqq n$ and let $g = \sum_1^m f_i^2$ satisfy $Eg(x(t)) \leqq Ct^{n+1}$. Then $Ef_i^2(x(t)) \leqq Ct^{n+1}$ for $1 \leqq i \leqq m$. This means that $L^r f_i^2(x^0) = 0$ for $0 \leqq r \leqq n$. In particular, $L^r < \nabla f_i, a\nabla f_i > (x^0) = 0$ for $0 \leqq r \leqq n - 1$ (by Theorem 10.3 of the appendix) and $L^r(Lf)^2(x^0) = 0$ for $0 \leqq r \leqq n - 2$. Hence,

$$\begin{aligned} E[f_i^4(x(t))] &= E\int_0^t (Lf_i^4)(x(s))\, ds \\ &= 6E\int_0^t f_i^2\langle\nabla f_i, a\nabla f_i\rangle(x(s))\, ds + 4E\int_0^t f_i^3(Lf_i)(x(s))\, ds \\ &\leqq 6\int_0^t [Ef_i^4(x(s))]^{1/2}(E\,|\sigma^*\nabla f_i|^4\,(x(s)))]^{1/2}\, ds \\ &\quad + 4\int_0^t (Ef_i^4(x(s))]^{3/4}(E(Lf)^4(x(s))]^{1/4}\, ds. \end{aligned}$$

Note that

$$E[|\sigma^*\nabla f_i|^2(x(t))] \leqq Ct^n,$$

and

$$E[(Lf_i)^2(x(t))] \leqq Ct^{n-1}.$$

Since $|\sigma^*\nabla f_i|^2$ is a sum of squares, the induction hypothesis yields

$$E[|\sigma^*\nabla f_i|^4(x(t))] \leqq C't^{2n},$$

and

$$E[(Lf_i)^4(x(t))] \leqq C' t^{2n-2}.$$

Thus

$$E[f_i^4(x(t))] \leqq A \int_0^t [Ef_i^4(x(s))]^{1/2} s^n \, ds + B \int_0^t [Ef_i^4(x(s))]^{3/4} s^{(2n-2)/4} \, ds.$$

We can now show that $E[f_i^4(x(t))] \leqq Ct^{2n+2}$. Indeed, if $L^r f_i^4(x^0) = 0$ for all r, there is nothing to prove. Let r_0 be the first r for which $L^r f_i^4(x^0)$ is nonzero. Then there are positive numbers λ_1 and λ_2 such that

$$\lambda_1 t^{r_0} \leqq E[f^4(x(t))] \leqq \lambda_2 t^{r_0}.$$

But this means that

$$\begin{aligned}\lambda_1 t^{r_0} &\leqq \lambda_2^{1/2} A \int_0^t s^{r_0/2+n} \, ds + \chi_2^{3/4} B \int_0^t s^{(3r_0/4)+(2n-2)/4} \, ds \\ &\leqq A' t^{(r_0+2n+2)/2} + B' t^{(3r_0+2n+2)/4},\end{aligned}$$

and this is possible only if $r_0 \geqq 2n+2$. It follows that

$$E[g^2(x(t))] \leqq C \sum_1^m E[f^4(x(t))] \leqq C' {}^{2n+2}.$$

LEMMA 6.2. *Let* X *be a random variable with mean zero. Let* $\alpha = EX^2$ *and* $\beta^4 = EX^4$. *Then*

$$P\left[X^\pm \geqq \frac{\alpha^{3/2}}{4\beta^2}\right] \geqq \frac{\alpha^2}{64\beta^4}.$$

Proof: By Hölder's inequality,

$$\alpha = E[|X|^2] \leqq E[|X|]^{2/3} E[|X|^4]^{1/3},$$

and therefore $E[|X|] \geqq \alpha^{3/2}/\beta^2$

Since $E[X^+] = E[X^-]$, we now have that $E[X^\pm] \geqq \alpha^{3/2}/2\beta^2$. Next observe that

$$\frac{\alpha^{3/2}}{2\beta^2} \leqq E[X^\pm] \leqq \frac{\alpha^{3/2}}{4\beta^2} + RP\left(X^\pm \geqq \frac{\alpha^{3/2}}{4\beta^2}\right) + \frac{\alpha}{R}$$

for all $R > 0$. Hence,

$$P\left(X^{\pm} \geqq \frac{\alpha^{3/2}}{4\beta^2}\right) \geqq \frac{\alpha^{3/2}}{4\beta^2}\frac{1}{R} - \frac{\alpha}{R^2}.$$

Taking $R = 8\beta^2/\alpha^{1/2}$, we get our result.

THEOREM 6.1. *Assume that* $a = \sigma\sigma^*$, *where* $\sigma \in C^\infty(R^d, R^d \otimes R^N)$. *Suppose that* $L^r \phi^2(x^0) \neq 0$ *for some* r *and let* n *denote the first such* r. *If* $L^r \phi(x^0) = 0$ *for* $r \leqq \frac{1}{2}n$, *then* $P_{x^0}(\tau' > 0) = 0$.

Proof: Let

$$X(t) = \frac{\phi(x(t)) - E[\phi(x(t))]}{t^{n/2}}.$$

Then there is a $\lambda > 0$ such that $1/\lambda \leqq E[|X(t)|^2] \leqq \lambda$ and there is a $C < \infty$ such that $E[|X(t)|^4] \leqq C$. Hence, we can find positive numbers ε, γ such that $P(X^-(t) \geqq \varepsilon) \geqq \gamma$. Since $E[\phi(x(t))]/t^{n/2} \to 0$ as $t \to 0$, this proves that, for small t,

$$P_{x^0}(\phi(x(t)) \leqq -\tfrac{1}{2}\varepsilon t^{n/2}) \geqq \gamma.$$

Hence $P_{x^0}(\tau' \leqq t) \geqq \gamma$ for all $t > 0$, and so, by the $0-1$ law, $P_{x^0}(\tau' > 0) = 0$. We next consider Case 2.

THEOREM 6.2. *If* $L^r \phi(x^0) = 0$ *for* $r \leqq n-1$, $L^r \phi^2(x^0) = 0$ *for* $r \leqq 2n-1$ *and*

$$\frac{1}{(2n)!} L^{2n} \phi^2(x^0) = \left(\frac{1}{n!} L^n \phi(x^0)\right)^2 \neq 0.$$

Then $P_{x^0}(\tau' > 0) = 0$ *if and only if* $L^n \phi(x^0) < 0$.

Proof: Let $\psi(t, x) = \phi(x) - (t^n/n!)L^n \phi(x^0)$ and $\bar{L} = \partial/\partial t + L$. Then clearly $\bar{L}^r \psi(0, x^0) = 0$ for $r \leqq n$, and $\bar{L}^r\psi^2(0, x^0) = 0$ for $r \leqq 2n$. Consequently,

$$E^{P_{x^0}}\left[\left|\frac{\phi(x(t))}{t^n} - \frac{(L^n \phi)(x^0)}{n!}\right|^2\right] = o(t) \qquad \text{as} \qquad t \to 0.$$

It follows that $\phi(x(t))/t^n$ converges to $L^n \phi(x^0)/n!$ in mean-square as $t \searrow 0$. We want to show now that this convergence is almost sure. By enlarging the sample space, if necessary, we can produce a one-dimensional Brownian motion

$\beta(\cdot)$ such that

$$\psi(t, x(t)) = \int_0^t \langle \nabla\phi, a\nabla\phi\rangle^{1/2}(x(s))\, d\beta(s) + \int_0^t L\psi(s, x(s))\, ds \qquad \text{a.s. } P_{x^0}.$$

Thus,

$$\frac{\phi(x(t))}{t^n} - \frac{\phi(x(s))}{s^n} = \int_s^t \frac{\langle \nabla\phi, a\nabla\phi\rangle^{1/2}(x(u))}{u^n}\, d\beta(s) + \int_s^t \frac{L\psi(u, x(u))}{u^n}\, du - n\int_s^t \frac{\psi(u, x(u))}{u^{n+1}}\, du \qquad \text{a.s. } P.$$

Observe that

$$E^{P_{x^0}}\left[\left(\int_s^t \frac{\langle \nabla\phi, a\nabla\phi\rangle^{1/2}(x(u))}{u^n}\, d\beta(u)\right)^2\right] = \int_s^t \frac{E^{P_{x^0}}[\langle \nabla\phi, a\nabla\phi\rangle(x(u))]}{u^{2n}}\, du \leqq C(t-s),$$

$$E^{P_{x^0}}\left[\int_s^t \left|\frac{L\psi(u, x(u))}{u^n}\right| du\right] \leqq \int_s^t \frac{E^{P_{x^0}}[|\psi(u, x(u))|^2]^{1/2}}{u^n}\, du \leqq C\int_s^t u^{-1/2}\, du,$$

$$E^{P_{x^0}}\left[\int_s^t \left|\frac{\psi(u, x(u))}{u^{n+1}}\right| du\right] \leqq \int_s^t \frac{E^{P_x}[|\psi(u, x(u))|^2]^{1/2}}{u^{n+1}}\, du \leqq C\int_s^t u^{-1/2}\, du,$$

where we have used the facts that

$$L^r\langle \nabla\phi, a\nabla\phi\rangle(x^0) = 0, \qquad r \leqq 2n-1,$$
$$L^r\psi^2(0, x^0) = 0, \qquad r \leqq 2n,$$
$$L^r(L\psi)^2(0, x^0) = 0, \qquad r \leqq 2n-1.$$

Using these facts together with Doob's martingale inequality, we see that

$$P_{x^0}\left(\sup_{0<s<t<\delta}\left|\frac{\phi(x(t))}{t^n} - \frac{\phi(x(s))}{s^n}\right| \geqq \varepsilon\right) = o(1)$$

as $\delta \searrow 0$ for all $\varepsilon > 0$. Hence, $\phi(x(t))/t^n \to L^n\phi(x^0)/n!$ a.s. P_{x^0}. From here it is easy to see that $P_{x^0}(\tau' > 0) = 0$ if $L^n\phi(x^0) < 0$ and $P_{x^0}(\tau' > 0) = 1$ if $L^n\phi(x^0) > 0$.

We turn finally to Case 3.

LEMMA 6.3. *Let* $a \in C_b^2[R^d, S_d]$ *and* $b \in C_b^1[R^d, R^d]$. *Assume that* $a^{ii}(0) = 0$ *for* $1 \leqq i \leqq r$. *Suppose that* P *solves the martingale problem for* a *and* b *starting at* 0. *For each* $\varepsilon > 0$ *let* $P^{(\varepsilon)}$ *be the distribution of the process*

$$x^{\varepsilon}(t) = \left(\frac{x_1(\varepsilon t)}{\varepsilon}, \cdots, \frac{x_r(\varepsilon t)}{\varepsilon}, \frac{x_{r+1}(\varepsilon t)}{\varepsilon^{1/2}}, \cdots, \frac{x_d(\varepsilon t)}{\varepsilon^{1/2}}\right),$$

where $x(t) = (x_1(t), \cdots, x_n(t))$ *has* P *for its distribution. Then* P^{ε} *is the solution to the martingale problem for* a_ε *and* b_ε *starting at* 0, *where*

$$a_\varepsilon^{ij}(x) = \frac{a^{ij}(\varepsilon x_1, \cdots, \varepsilon x_r, \varepsilon^{1/2} x_{r+1}, \cdots, \varepsilon^{1/2} x_d)}{\varepsilon^{\alpha_i+\alpha_j-1}},$$

$$b_\varepsilon^{i}(x) = \frac{b^{i}(\varepsilon x_1, \cdots, \varepsilon x_r, \varepsilon^{1/2} x_{r+1}, \cdots, \varepsilon^{1/2} x_d)}{\varepsilon^{\alpha_i-1}},$$

and $\alpha_i = 1$ *if* $1 \leqq i \leqq r$ *and* $\alpha_i = \frac{1}{2}$ *if* $r < i \leqq d$. *Moreover,* $P^{(\varepsilon)}$ *converges weakly as* $\varepsilon \to 0$ *to* $P^{(0)}$ *which is the solution to the martingale problem for* a_0 *and* b_0, *starting from* 0, *where*

$$\begin{aligned}
a_0^{ij}(x) &= \tfrac{1}{2} \sum_{k,l>r} a^{ij}_{,kl}(0) x_k x_l && \textit{for} \quad i, j \leqq r, \\
a_0^{ij}(x) &= \sum_{k>r} a^{ij}_{,k}(0) x_k && \textit{for} \quad i \leqq r < j, \\
a_0^{ij}(x) &= a^{ij}(0) && \textit{for} \quad i, j > r, \\
b_0^{i}(x) &= b^{i}(0) && \textit{for} \quad i \leqq r, \\
b_0^{i}(x) &= 0 && \textit{for} \quad i > r.
\end{aligned}$$

Proof: Note that

$$x_\theta(t) = \exp\left[\left\langle \theta, x(t) - \int_0^t b(x(s))\, ds \right\rangle - \tfrac{1}{2}\int_0^t \langle \theta, a(x(s))\theta \rangle\, ds\right]$$

is a P-martingale for all $\theta \in R$. Hence if

$$\theta^{(\varepsilon)} = \left(\frac{\theta_1}{\varepsilon}, \cdots, \frac{\theta_r}{\varepsilon}, \frac{\theta_{r+1}}{\varepsilon^{1/2}}, \cdots, \frac{\theta_d}{\varepsilon^{1/2}}\right),$$

then

$$\begin{aligned} X_\theta^{(\varepsilon)}(t) &= X_{\theta^{(\varepsilon)}}(\varepsilon t) \\ &= \exp\left\{\left\langle \theta, x^{(\varepsilon)}(t) - \frac{1}{\varepsilon}\int_0^{\varepsilon t} b_\varepsilon\left(x^{(\varepsilon)}\left(\frac{s}{\varepsilon}\right)\right) ds\right\rangle - \frac{1}{2\varepsilon}\int_0^{\varepsilon t}\left\langle \theta, a_\varepsilon\left(x^{(\varepsilon)}\left(\frac{s}{\varepsilon}\right)\right)\theta\right\rangle ds\right\} \\ &= \exp\left\{\left\langle \theta, x^{(\varepsilon)}(t) - \int_0^t b_\varepsilon(x^{(\varepsilon)}(s))\right\rangle ds - \tfrac{1}{2}\int_0^t \langle \theta, a_\varepsilon(x^{(\varepsilon)}(s))\theta\rangle\, ds\right\} \end{aligned}$$

is a P-martingale. This proves that the distribution of $x^{(\varepsilon)}(\cdot)$ under P is $P^{(\varepsilon)}$.

Finally, since $a_\varepsilon \to a_0$ and $b_\varepsilon \to b_0$ uniformly on compacts, we shall see that $P^{(\varepsilon)} \Rightarrow P^{(0)}$ once we have shown that $\{P^{(\varepsilon)}\}_{\varepsilon>0}$ is relatively compact. Note that $\operatorname{tr} a_\varepsilon(x) \leqq C|x|^2$. Hence, if $a_{\varepsilon,R}(x) = a_\varepsilon(x \wedge R)$ and $P^{(\varepsilon,R)}$ solves the martingale problem for $a_{\varepsilon,R}$ and b_ε starting at 0, then $\{P^{(\varepsilon,R)}\}_{\varepsilon>0}$ is relatively compact for each $R > 0$. Moreover, $P^{(\varepsilon)}|_{M_{\tau_R}} = P^{(\varepsilon,R)}|_{M_{\tau_R}}$, where τ_R is the first exit time from $B(0, R)$. Thus it suffices to show that

$$\lim_{R\to\infty} \sup_{\varepsilon>0} P^{(\varepsilon)}(\tau_R \leqq t) = 0$$

for all $t \geqq 0$. Observe that

$$E^{P^{(\varepsilon)}}[|x(t)|^2] \leqq 2CE^{P^{(\varepsilon)}}\left[\int_0^t |x(s)|^2\, ds\right] + 2\,\|b\|_\infty^2\, t^2,$$

and therefore

$$E^{P^{(\varepsilon)}}[|x(t)|^2] \leqq 2\,\|b\|_\infty^2\left(t^2 + 2C\int_0^t e^{2C(t-s)} s^2\, ds\right).$$

Since

$$x(t) - \int_0^t b_\varepsilon(x(s))\, ds$$

is a $P^{(\varepsilon)}$-martingale, it follows from this that

$$\lim_{R\to\infty} \sup_{\varepsilon>0} P^{(\varepsilon)}(\tau_R \leqq t) = 0.$$

LEMMA 6.4. *Let*

$$L = \tfrac{1}{2}\sum_{i,j=1}^d a^{ij}(x)\frac{\partial^2}{\partial x_i\, \partial x_j} + \sum_1^d b^i(x)\frac{\partial}{\partial x_i}$$

and assume $\phi \in C_b^2(R^d)$. *If* $\langle \nabla\phi, a\nabla\phi\rangle(x^0) = 0$, *then* $\langle b - \frac{1}{2}a', \nabla\phi\rangle$ *is an invariant scalar and*

$$aH_{\langle\nabla\phi, a\nabla\phi\rangle}a(x^0) \equiv \left(\left(a^{ik}\frac{\partial^2}{\partial x_k\,\partial x_l}\,a^{lj}\right)\right)_{1\leq i,j\leq d},$$

$$a(\nabla(a\nabla\phi))^*(x^0) \equiv ((a^{ik}(a^{lj}\,\phi_{,l})_{,k}))_{1\leq i,j\leq d},$$

and

$$(\nabla(a\nabla\phi))a(\nabla(a\nabla\phi))^*(x^0) \equiv (((a^{li}\,\phi_{,l})_{,k}\,a^{kk'}(a^{l'j}\phi_{,l'})_{,k'}))_{1\leq i,j\leq d}$$

are invariant as elements of $T_{x^0}\otimes T_{x^0}$.

Proof: Let $F\colon R^d \to R^d$ be a C^2 diffeomorphism and set $y = F(x)$. Then $L(f\circ F) = (\bar{L}f)\circ F$, where

$$\bar{L} = \tfrac{1}{2}\sum_{i,j=1}^{d}\bar{a}^{ij}\frac{\partial}{\partial y_i\,\partial y_j} + \sum_{i=1}^{d}\bar{b}^i\frac{\partial}{\partial y_i},$$

$$\bar{a}^{ij}\circ F = F^i_{,k}\,a^{kl}\,F^j_{,l},$$

$$\bar{b}^i\circ F = F^i_{,k}\,b^k + \tfrac{1}{2}F^i_{,kl}\,a^{kl}.$$

Hence,

$$\begin{aligned}\bar{a}^{ij}_{,j}\circ F &= (F^{-1})^r_{,j}\circ F(F^i_{,k}\,a^{kl}\,F^j_{,l})_{,r}\\ &= (F^{-1})^r_{,j}\circ FF^i_{,kr}\,a^{kl}\,F^j_{,l} + (F^{-1})^r_{,j}\circ FF^i_{,k}\,a^{kl}_{,r}\,F^j_{,l}\\ &\quad + (F^{-1})^r_{,j}\circ FF^i_{,k}\,a^{kl}\,F^j_{,lr}\\ &= F^i_{,kr}\,a^{kr} + F^i_{,k}\,a^{kr}_{,r} + (F^{-1})^r_{,j}\circ FF^i_{,k}\,a^{kl}F^j_{,lr}.\end{aligned}$$

Thus

$$(\bar{b}^i - \tfrac{1}{2}\bar{a}^{i,j}_{,j})\circ F = F^i_{,k}\,b^k - \tfrac{1}{2}F^i_{,k}\,a^{kr}_{,r} - \tfrac{1}{2}(F^{-1})^r_{,j}\circ FF^i_{,k}\,a^{kl}F^j_{,lr}.$$

Since $(\nabla\bar{\phi})^i\circ F = (F^{-1})^r_{,i}\circ F\phi_{,r}$,

$$\langle\bar{b} - \tfrac{1}{2}\bar{a}', \nabla\bar{\phi}\rangle\circ F = \langle b - \tfrac{1}{2}a', \nabla\phi\rangle - \tfrac{1}{2}(F^{-1})^r_{,j}\,F^j_{,lr}\,a^{rl}\phi_{,r}.$$

In particular, if $\langle\nabla\phi, a\nabla\phi\rangle(x^0) = 0$, then

$$\langle\bar{b} - \tfrac{1}{2}\bar{a}', \nabla\bar{\phi}\rangle(y^0) = \langle b - \tfrac{1}{2}a', \nabla\phi\rangle(x^0).$$

Since a transforms like an element of $T_{x^0}\otimes T_{x^0}$, it suffices to prove that $H_{\langle\nabla\phi, a\nabla\phi\rangle}(x^0) \in T^*_{x^0}\otimes T^*_{x^0}$ in order to show that $aH_{\langle\nabla\phi, a\nabla\phi\rangle}a(x^0) \in T_{x^0}\otimes T_{x^0}$.

But

$$\frac{\partial^2\langle\nabla\bar\phi, \bar a\nabla\bar\phi\rangle}{\partial y_i\,\partial y_j}\circ F = (F^{-1})^k_{,ij}\circ F(\langle\nabla\phi, a\nabla\phi\rangle)_{,k} + (F^{-1})^k_{,j}\circ F(\langle\nabla\phi, a\nabla\phi\rangle)_{,kl}(F^{-1})^k_{,i}\circ F,$$

and $(\langle\nabla\phi, a\nabla\phi\rangle)_{,k}(x^0) = 0$, since x^0 is a minimum of $\langle\nabla\phi, a\nabla\phi\rangle$.

Next note that

$$(\nabla(\bar a\nabla\bar\phi))^i_j\circ F = (F^{-1})^r_{,j}\circ F(F^i_{,k}\,a^{lk}\phi_{,l})_{,r} = F^i_{,k}(a^{lk}\phi_{,l})_{,r}(F^{-1})^r_{,j}$$

when $\langle\nabla\phi, a\nabla\phi\rangle = 0$. Hence,

$$(\bar a(\nabla(\bar a\nabla\phi))^*)^{ij}(y^0) = F^i_{,k}\,a^{kr}(a^{lk'}\phi_{,l})_{,r}\,F^j_{,k'}(x^0) = F^i_{,k}(a(\nabla(a\nabla\phi))^*)^{kk'}F^j_{,k'}(x^0)\,.$$

Finally,

$$((\nabla(\bar a\nabla\bar\phi))\bar a(\nabla\bar a\nabla\bar\phi)^*)^{ij}(y^0) = F^i_{,k}(a^{lk}\phi_{,l})_{,r}\,a^{rs}(a^{l'k'}\phi_{,l'})_{,s}\,F^j_{,k'}(x^0) = F^i_{,k}[(\nabla a(a\nabla\phi))a(\nabla a(a\nabla\phi))^*]^{kk'}F^j_{,k'}(x^0)\,.$$

THEOREM 6.3. *Assume $x^0\in\delta G$ is given; then for $P_{x^0}(\tau'>0)=1$ to hold it is necessary that each of the following conditions be satisfied:*

(i) $\langle\nabla\phi, a\nabla\phi\rangle(x^0) = 0\,,$

(i′) $L\phi^2(x^0) = 0\,,$

(ii) $\langle b-\tfrac12 a', \nabla\phi\rangle(x^0)\geqq 0\,,$

(iii) $L\phi(x^0)\geqq 0\,,$

(iv) $a(\nabla(a\nabla\phi))^*(x^0)\in S_d\,,$

(v) $(\nabla(a\nabla\phi))a(\nabla(a\nabla\phi))^*(x^0) = \tfrac12 aH_{\langle\nabla\phi.a\nabla\phi\rangle}a(x^0).$

Proof: First observe that $L\phi^2(x^0) = \langle\nabla\phi, a\nabla\phi\rangle(x^0)$. Hence (i) and (i′) are equivalent. Next note that conditions (i)–(v) are invariant under C^2 changes of coordinates. Therefore we shall assume that $x^0=0$ and that, near 0, $\phi(x) = x_1$. Further, we can choose the coordinates so that $((a^{ij}(0)))_{2\le i,j\le d}$

is diagonal. In such a coordinate system our conditions become:

(i) $a^{11}(0) = 0$,

(ii) $b^1(0) - a^{1j}_{,j}(0) \geqq 0$,

(iii) $b^1(0) \geqq 0$,

(iv) $((a^{ii}a^{1j}_{,i}(0)))_{2\leq i,j\leq d} \in S_{d-1}$,

(v) $a^{1i}_{,k}\, a^{kk}\, a^{1j}_{,k}(0) = \frac{1}{2}a^{ii}a^{11}_{,ij}\, a^{ij}(0)$, $\qquad 2 \leqq i, j \leqq d$.

To prove (i), let $Q^{(\varepsilon)}$ be the distribution of $x(\varepsilon t)/\varepsilon^{1/2}$ under P_0. Then $Q^{(\varepsilon)}$ tends weakly to $Q^{(0)}$, the solution of the martingale problem for $a(0)$ and 0 starting at 0. In particular, $x_1(\cdot)$ is distributed under $Q^{(0)}$ like $(a^{11}(0))^{1/2}\beta(\cdot)$, where $\beta(\cdot)$ is a one-dimensional Brownian motion. Since $P_0(\tau' > 0) = 1$ implies $Q^{(0)}(\tau' = \infty) = 1$, it follows that $a^{11}(0) = 0$. Condition (iii) is proved by observing that $x_1(\cdot)$ is distributed under P_0 like a process $\xi(\cdot)$ having the form

$$\xi(t) = \int_0^t \alpha(s)\, d\beta(s) + \int_0^t b^1(\xi(s))\, ds,$$

where $\alpha : [0, \infty) \times \Omega \to R$ is a bounded nonanticipating function relative to the one-dimensional Brownian motion $\beta(\cdot)$. Thus,

$$P_0\left(x(t) = \int_0^t b^1(x(s))\, ds \text{ for some } t \in (0, \delta) \text{ and every } \delta > 0\right) = 1 .$$

We turn now to the proof of (ii), (iv) and (v). Let $P^{(\varepsilon)}$ denote the distribution of

$$x^{(\varepsilon)}(t) = \frac{x_1(\varepsilon t)}{\varepsilon}, \frac{x_2(\varepsilon t)}{\varepsilon^{1/2}}, \ldots, \frac{x_d(\varepsilon t)}{\varepsilon^{1/2}}$$

under P_0. Since $a^{11}(0) = 0$, $P^{(\varepsilon)}$ tends weakly to the solution of the martingale problem for a_0 and b_0 starting at 0, where

$$a^{11}(x) = \tfrac{1}{2}\sum_{k,l\geqq 2} a^{11}_{,kl}(0)x_k\, x_l ,$$

$$a_0^{1i}(x) = \sum_{k\geqq 2} a^{1i}_{,k}\, x_k , \qquad i \geqq 2 ,$$

$$a_0^{ij}(x) = a^{ij}(0) , \qquad i, j \geqq 2 ,$$

$$b_0^1(x) = b^1(0) \quad \text{and} \quad b_0^i(x) = 0 , \qquad i \geqq 2 .$$

In particular, supp $(P^0) = \overline{S_{a_0, b_0 - \frac{1}{2}a'_0}(0)}$. Let

$$A^{ij} = a^{ij}(0)\,,$$
$$B^i_j = a^{1i}_{,j}(0)\,,$$
$$H_{ij} = a^{11}_{,ij}(0)\,,$$

for $2 \leqq i, j \leqq d$. Then $P_0(\tau' > 0) = 1$ implies that, for $\psi \in C([0, \infty), R^d)$ and all $t \geqq 0$,

$$\tfrac{1}{2}\int_0^t \langle \tilde{\eta}(s), H\tilde{\eta}(s)\rangle \psi_1(s)\, ds + \int_0^t \langle \tilde{\psi}(s), B\tilde{\eta}(s)\rangle\, ds + \int_0^t (b - \tfrac{1}{2}a')'(0)\, ds \geqq 0\,,$$

where

$$\tilde{\eta}(t) = \int_0^t B\tilde{\eta}(s)\psi_1(s)\, ds + A\int_0^t \tilde{\psi}(s)\, ds$$

and $\tilde{x} = (x_2, \cdots, x_d)$. Taking $\psi \equiv 0$, we see that $(b - \frac{1}{2}a')'(0) \geqq 0$. Taking $\psi_1 \equiv 0$, we have

$$\int_0^t \langle \tilde{\psi}(s), BA\dot{\tilde{\psi}}(s)\rangle\, ds + \int_0^t (b - \tfrac{1}{2}a')'(0)\, ds \geqq 0$$

for all $\tilde{\psi} \in \Psi = \{f \in C^1([0, \infty), R^{d-1})\colon f(0) = 0\}$. Thus,

$$\int_0^t \langle \tilde{\psi}(s), BA\dot{\tilde{\psi}}(s)\rangle\, ds \geqq 0$$

for all $\tilde{\psi} \in \tilde{\Psi}$. Now suppose that $\tilde{\psi} \in \tilde{\Psi}$ such that $\tilde{\psi}(1) = 0$ and

$$\int_0^1 \langle \tilde{\psi}(s), BA\dot{\tilde{\psi}}(s)\rangle\, ds > 0\,.$$

Then,

$$\int_0^1 \langle \tilde{\psi}(1 - s), BA\dot{\tilde{\psi}}(1 - s)\rangle\, ds < 0\,,$$

which is a contradiction. It follows that the vector field $BA\tilde{x}$ is an exact differential, and therefore $BA = (BA)^* = AB^*$. In particular,

$$0 \leqq \int_0^t \langle \tilde{\psi}(s), BA\dot{\tilde{\psi}}(s)\rangle\, ds = \langle \tilde{\psi}(t), BA\tilde{\psi}(t)\rangle$$

for all $\tilde{\psi} \in \tilde{\Psi}$, and so $BA \in S_{d-1}$. This proves (iv).

To prove (v), suppose $w \in R^d$ is given and define $\psi_1(t) = w_1$ and

$$\tilde{\psi}(t) = \tilde{w} - tw_1 B^* \tilde{w} .$$

Then, using the fact that $BA = AB^*$, one can easily show that the corresponding $\tilde{\eta}(\cdot)$ is given by $\tilde{\eta}(t) = tA\tilde{w}$. Thus,

$$\tfrac{1}{2}w_1 \int_0^t s^2 \langle \tilde{w}, AHA\tilde{w} \rangle \, ds + \int_0^t s \langle \tilde{w}, BA\tilde{w} \rangle \, ds - w_1 \int_0^t s^2 \langle B^* \tilde{w}, BA\tilde{w} \rangle \, ds$$
$$+ \int_0^t (b - \tfrac{1}{2}a')'(0) \, ds \geqq 0$$

for all $w \in R^d$ and $t \geqq 0$. Clearly this is possibly only if

$$\tfrac{1}{2}\langle \tilde{w}, AHA\tilde{w} \rangle = \langle \tilde{w}, B^2 A\tilde{w} \rangle$$

for all $\tilde{w} \in R^{d-1}$. Since $B^2 A = BAB^*$, we have now proved (v).

Remark. It is interesting to note that although in the proof of Theorem 6.3 we used the characterization of supp $(P^{(0)})$, we could have proved the theorem without our previous result. Indeed, the stochastic integral equations associated with a_0 and b_0 are sufficiently simple to prove (i) and (ii) directly. The importance of this observation is that Theorem 6.3 can be used to give an alternate derivation of our characterization of the support of a degenerate diffusion. This alternate route would resemble the method used by Bony [1], Hill [3] and Redheffer [8] in their work on the strong maximum principle. In their scheme, Theorem 6.3 would play the role of their lemma which states that if x^0 is a characteristic boundary point of a ball and if the Fichera drift at x^0 points into the ball, then a "barrier" exists for x^0. With minor modifications, our method can be made to cover the operators studied by Redheffer. Namely, if

$$L = \tfrac{1}{2} \sum_{i,j=1}^{n} a^{ij}(t, x) X_i X_j + Y ,$$

where $X_1, \cdots, X_n$, Y are d-dimensional C^1-vector fields and $((a^{ij}(t, x)))_{1 \leqq i,j \leqq n}$ is bounded, continuous, and uniformly positive definite, then the support of any solution to the martingale problem for L starting at (s, x) is independent of $((a^{ij}(t, x)))_{1 \leqq i,j \leqq n}$. Unfortunately, unless $((a^{ij}(t, x)))$ is smooth enough, we have no uniqueness result for the martingale problem.

7. On the Set of Regular Points

Let us define the following sets on the boundary δG of G:

$$\Sigma_3 = \{x \in \delta G : \langle \nabla\phi, a\nabla\phi\rangle(x) > 0\},$$
$$\Sigma_2 = \{x \in \delta G - \Sigma_3 : \langle b - \tfrac{1}{2}a', \nabla\phi\rangle(x) < 0\},$$
$$\Sigma_2^* = \{x \in \delta G - \Sigma_3 : (L\phi)(x) < 0\},$$
$$\Sigma_0 = \{x \in \delta G - \Sigma_3 : \langle b - \tfrac{1}{2}a', \nabla\phi\rangle(x) = 0\},$$
$$\Sigma_1 = \{x \in \delta G - \Sigma_3 : \langle b - \tfrac{1}{2}a', \nabla\phi\rangle(x) > 0\}.$$

Clearly, the sets $\Sigma = \Sigma_2 \cup \Sigma_3$ and $\Sigma^* = \Sigma_2^* \cup \Sigma_3$ are both open sets of δG. Moreover, in view of Theorem 6.3, the set Γ of τ'-regular points contains Σ and Σ^*.

From the earlier works of Fichera, Kohn-Nirenberg and Oleinik it is clear that the set of τ'-regular points Γ and the set Σ cannot be essentially different.

We prove in this section that $\Gamma \subset \bar{\Sigma}$ so that one always has

$$\Sigma \cup \Sigma^* \subset \Gamma \subset \bar{\Sigma} = \overline{\Sigma^*}.$$

Moreover, one should also try to answer the following questions: Are the τ and τ' solutions for the boundary value problems really different? If the boundary value is specified only on Σ (or Σ^*), which can be described more easily than Γ, is the problem meaningful? It is proved in this section that

1. the two solutions corresponding to τ and τ' agree almost everywhere, that is,
$$P_x[\tau = \tau'] = 1 \quad \text{a.e. } x \text{ in } G;$$

2. it is sufficient to provide the boundary values only on Σ (or Σ^*), if one is satisfied with having the solution only almost everywhere, that is,

$$P_x[x(\tau) \in \Sigma \cap \Sigma^*, \tau < \infty] = 1 \quad \text{a.e. } x \text{ in } G,$$
$$P_x[x(\tau') \in \Sigma \cap \Sigma^*, \tau' < \infty] = 1 \quad \text{a.e. } x \text{ in } G.$$

LEMMA 7.1. *The sets Σ_0, Σ_1, Σ_2, Σ_2^* and Σ_3 are all invariantly defined and $\bar{\Sigma} = \overline{\Sigma^*}$.*

Proof: The invariance is obvious from the earlier results. To prove that $\bar{\Sigma} = \overline{\Sigma^*}$, it suffices to show that $L\phi(x^0) = \langle b - \tfrac{1}{2}a', \nabla\phi\rangle(x^0)$ for $x^0 \in \delta G - \bar{\Sigma}_3$. Given $x^0 \in \delta G - \overline{\Sigma}_3$, choose coordinates in such a way that $x^0 = 0$ and

$x_1 = \phi(x)$ near the point x^0. In terms of this coordinate system, we have to prove that $a^{1j}_{,j}(0) = 0$. But $a^{11}(0, x_2, \cdots, x_d) = 0$ for x near 0. Thus, $a^{ij}(0, x_2, \cdots, x_d) = 0$ for x near 0 and so $a^{1j}_{,l}(0) = 0$ for $j \geqq 2$ and $l \geqq 2$. Moreover, a^{11} is a non-negative function vanishing at 0. Thus $a^{11}_{,1} = 0$.

LEMMA 7.2. *For any* $x \in G$, $P_x(x(\tau) \notin \overline{\Sigma}, \tau < \infty) = 0$.

Proof: Assume that $x^0 = 0$ and that $\phi(x) = x_1$ in some neighborhood $|x| \leqq R$. Choose $0 < 2\delta < R$ so that $\{x : |x| \leqq 2\delta\} \cap \overline{\Sigma} = \phi$. Then

$$a^{11}(0, x_2, \cdots, x_d) = 0 \qquad \text{and} \qquad b^1(0, x_2, \cdots, x_d) \geqq 0$$

for $|x| \leqq 2\delta$. Hence, there is a $\lambda > 0$ such that for $|x| \leqq 2\delta$

$$a^{11}(x) \leqq \lambda x_1^2 ,$$

and

$$b^1(x) \geqq -\lambda |x_1| .$$

Given $0 < \varepsilon < \delta$, define $V_\varepsilon(x) = \varepsilon/x_1$. Then

$$LV_\varepsilon(x) = \varepsilon a^{11}(x) \frac{1}{x_1^3} - \varepsilon b^1(x) \frac{1}{x_1^2} \leqq 2\lambda V_\varepsilon(x) \qquad \text{if} \qquad |x| \leqq 2\delta .$$

Thus if

$$\sigma_T = (T \wedge \inf\{t \geqq 0 : |x(t)| \geqq 2\delta\})$$

and

$$\tau_\varepsilon = \inf\{t \geqq 0 : x_1(t) = \varepsilon\} ,$$

then

$$\begin{aligned} 1 &= E^{P_x}\left[\frac{V_\varepsilon(x(\tau_\varepsilon \wedge \sigma_T))}{V_\varepsilon(x)} \exp\left\{-\int_0^{\tau_\varepsilon \wedge \sigma_T} \frac{LV_\varepsilon}{V_\varepsilon}(x(s))\,ds\right\}\right] \\ &\geqq E^{P_x}\left[\frac{V_\varepsilon(x(\tau_\varepsilon \wedge \sigma_T))}{V_\varepsilon(x)} \exp\{-2\lambda\tau_\varepsilon \wedge \sigma_T\}\right] \\ &\geqq \frac{x_1}{\varepsilon} E^{P_x}[e^{-2\lambda\tau_\varepsilon}, \tau_\varepsilon \leqq \sigma_T] \end{aligned}$$

for $|x| \leqq \delta$ such that $x_1 \geqq \varepsilon$. Letting $\varepsilon \to 0$ and then letting $T \to \infty$, we conclude that

$$P_x[\tau < \sigma_\infty] = 0$$

for x such that $x_1 > 0$ and $|x| < \delta$. Now define

$$\tau_0 = \inf\{t \geqq 0 : |x(t)| \leqq \delta\}, \cdots, \tau_{2n+1} = \{\inf t \geqq \tau_{2n} : |x(t)| \geqq 2\delta\}$$

and

$$\tau_{2n} = \inf\{t \geqq \tau_{2n-1} : |x(t)| \leqq \delta\}.$$

Then,

$$\begin{aligned} P_x(|x(\tau)| < \delta, \tau < \tau_{2n+1}) &= \sum_{k=0}^{n} P_x(|x(\tau)| < \delta, \tau_{2k} \leqq \tau < \tau_{2k+1}) \\ &= \sum_{k=0}^{n} E^{P_x}[\chi_{\tau \geqq \tau_{2k}} P_{x(\tau_{2k})}(\tau < \tau_1)] = 0 \end{aligned}$$

for all n. Since $\tau_{2n+1} \to \infty$ as $n \to \infty$, we see that

$$P_x(|x(\tau)| < \delta, \ \tau < \infty) = 0.$$

LEMMA 7.3. *If $x^0 \in \delta G - \bar{\Sigma}$, then there is an $\alpha > 0$ such that*

$$\varliminf_{\substack{x \to x_0 \\ x \in G}} P_x[\tau \geqq \alpha] \geqq \alpha.$$

In particular, $\Gamma \subset \bar{\Sigma}$.

Proof: If $x^0 \in \delta G - \bar{\Sigma}$, then there is a $\delta > 0$ such that, for $|x - x^0| \leqq \delta$, $\{y : |y - x| \leqq \delta\} \cap \bar{\Sigma} = \varnothing$. Thus, by standard estimates and Lemma 7.2, there is an $\alpha > 0$ such that

$$P_x[\tau \geqq \alpha] \geqq P_x\left[\sup_{0 \leqq t \leqq \alpha} |x(t) - x| < \delta\right] \geqq \alpha \qquad \text{if} \qquad |x - x^0| < \delta.$$

Finally, if $x^0 \in \Gamma$, then $P_{x^0}(\tau' > 0) = 0$. Hence $P_{x^0}(\tau' \geqq t) = 0$ for any $t > 0$. But $[\tau' \geqq t]$ is a closed set and therefore

$$\varlimsup_{\substack{x \to x^0 \\ x \in G}} P_x[\tau' \geqq t] \leqq P_{x^0}[\tau' \geqq t] = 0 \quad \text{for all} \quad t > 0.$$

Since $\tau \leqq \tau'$, this proves that

$$\lim_{\substack{x \to x^0 \\ x \in G}} P_x[\tau' \geqq t] = 0 \quad \text{for all} \quad t > 0.$$

In particular, $x^0 \in \bar{\Sigma}$.

LEMMA 7.4. *Let* $\Delta = \delta G - \Sigma_3$ *and* $\Delta_1 = \{x \in \Delta : L\langle \nabla\phi, a\nabla\phi\rangle(x) \neq 0\}$. *Then* $L\phi(x) = \langle b - \frac{1}{2}a', \nabla\phi\rangle(x)$ *for* $x \in \Delta - \Delta_1$ *and* Δ_1 *has zero boundary measure.*

Proof: First suppose that $x^0 \in \Delta - \Delta_1$ and choose a coordinate system so that $x^0 = 0$, $\phi(x) = x_1$ near x^0, and $((a^{ij}(0)))_{i,j\leqq 2}$ is diagonal. Then

$$0 = La^{11}(0) = \tfrac{1}{2}\,\mathrm{tr}\,(a(0)H_{a^{11}}(0))\,.$$

Since $H_{a^{11}}(0)$ is non-negative definite, $a^{il}(0)a^{11}_{,lj}(0) = 0$ for $1 \leqq i, j \leqq d$. In particular, $a^{ii}(0) \neq 0$ implies $a^{11}_{,ii}(0) = 0$ for $2 \leqq i \leqq d$. From this and the inequality $|a^{1i}(x)|^2 \leqq a^{11}(x)a^{ii}(x)$, it follows that $a^{1i}_{,i}(0) = 0$, $2 \leqq i \leqq d$. Since $a^{11}_{,i}(0) = 0$, we see that $a_1'(0) = 0$.

Next suppose that $x^0 \in \Delta_1$. Again choose coordinates such that $x^0 = 0$ and $\phi(x) = x_1$ near x^0. Then, locally on Δ_1, $La^{11} \neq 0$. Since a^{11} is zero, we must have $a^{11}_{,ij} \neq 0$ for some $i, j \geqq 2$.

If A is any open set in a Euclidean space and f is C^2 and non-negative, then on the set A_0 where $f = 0$ we must also have $D^2 f = 0$ a.e.

This means that Δ_1 must be of boundary measure zero.

LEMMA 7.5. *Let* $\Delta_2 = \{x \in \Delta - \Delta_1 : L\phi(x) = 0\}$. *Then there is a set* $E \subseteq G$ *of zero Lebesgue measure such that* $P_x(x(\tau) \in \Delta_1 \cup \Delta_2,\ \tau < \infty) = 0$ *for all* $x \notin E$.

Proof: Given a non-negative $g \in C_0^\infty(G)$, choose $\lambda > 0$ large enough so that $(\lambda - L^*)u = g$ has a C_b^2-solution. Let $f \in C_0^\infty(R^d)$ be non-negative. Then,

$$\begin{aligned}\int u(x)f(x)\,dx &= \int g(x)E^{P_x}\left[\int_0^\infty e^{-\lambda t}f(x(t))\,dt\right]dx\\ &\geqq \int g(x)E^{P_x}\left[e^{-\lambda\tau}E^{P_{x(\tau)}}\left[\int_0^\infty e^{-\lambda t}f(x(t))\,dt\right]\right]dx\\ &= \int_{\partial G}\Pi_g^\lambda(dz)E^{P_z}\left[\int_0^\infty e^{-\lambda t}f(x(t))\,dt\right],\end{aligned}$$

where

$$\Pi_g^\lambda(F) = \int g(x)E^{P_x}(e^{-\lambda\tau}\chi_F(x(\tau))]\,dx\,, \qquad F \in B[\delta G].$$

Suppose $x^0 \in \Delta$ and choose coordinates so that $x^0 = 0$ and $\phi(x) = x_1$ for x in a neighborhood N of x^0. Then

$$\int u(x)f(x)\,dx \geqq \int_{\Delta\cap T}\Pi_g^\lambda(dz)E^{P_z}\left[\int_0^\infty e^{-\lambda t}f(x(t))\,dt\right].$$

Making the obvious substitution, one now gets

$$\int u(\varepsilon x_1, x_2, \cdots, x_d) f(x)\, dx \geqq \int_{\Delta \cap T} \Pi_g^\lambda(dz) E^{P_z^{(0)}}\left[\int_0^\infty e^{-\lambda \varepsilon t} f(x(t))\, dt\right],$$

where $P_z^{(\varepsilon)}$ denotes the distribution of $[x_1(\varepsilon t)/\varepsilon, x_2(\varepsilon t), \cdots, x_d(\varepsilon t)]$ under P_z. Using our earlier result, Lemma 6.3, one can easily see that, for $z \in \Delta \cap N$, $P_z^{(\varepsilon)}$ tends weakly to a measure $P_z^{(0)}$ under which $x_i(0) \equiv z_i$, $2 \leqq i \leqq d$. Moreover, $x_i(\cdot) \equiv z_i$ a.s. $P_z^{(0)}$ if $z \in \Delta_2 \cap N$. Hence,

$$\int u(0, x_2, \cdots, x_d) f(x)\, dx \geqq \int_{\Delta \cap N} \Pi_g^\lambda(dz) E^{P_z^{(0)}}\left[\int_0^\infty f(x_1(t), z_2, \cdots, z_d)\, dt\right],$$

where $x_1(\cdot) \equiv z_1$ a.s. $P_z^{(0)}$ for $z \in \Delta_2 \cap N$. Since this is true for all non-negative $f \in C_0^\infty(R^d)$, it is also true for $f(x) = h(x_1)\bar{h}(x_2, \cdots, x_d)$, where h and $\bar{h}$ are bounded, non-negative measurable functions having compact support. Thus

$$\left(\int h(x_1)\, dx_1\right)\int u(0, \bar{x})\bar{h}(z)\, dz \geqq \int_{\Delta \cap T} \Pi_g^\lambda(dz)\bar{h}(z) E^{P_z^{(0)}}\left[\int_0^\infty h(x_1(t))\, dt\right].$$

But this is possible only if $\Pi_g^\lambda|_{\Delta \cap N}$ is absolutely continuous with respect to Lebesgue measure on $\Delta \cap N$ and if $\Pi_g^\lambda(\Delta_2 \cap N) = 0$. Hence,

$$\Pi_g^\lambda((\Delta_1 \cup \Delta_2) \cap N) = 0.$$

We see now that

$$\int g(x) P_x(x(\tau) \in \Delta_1 \cup \Delta_2, \tau < \infty)\, dx = 0$$

for all non-negative $g \in C_0^\infty(G)$, and our assertion follows from this.

THEOREM 7.1. *There is a set $E \subseteq G$ of zero Lebesgue measure such that*

$$P_x(x(\tau) \notin \Sigma, \tau < \infty) = P(x(\tau) \notin \Sigma^*, \ \tau < \infty) = 0$$

for $x \notin E$. In particular, $P_x(\tau < \tau') = 0$ for $x \notin E$.

Proof: Since $\delta G - \Sigma^* \subseteq \Delta_1 \cup \Delta_2 \cup \Delta_3$, where

$$\Delta_3 = \{x \in \Delta - \Delta_1 : L\phi(x) > 0\},$$

and $(\Sigma - \Sigma^*) \cup (\Sigma^* - \Sigma) \subseteq \Delta_1$, we need only show that

$$P_x(x(\tau) \in \Delta_3, \ \tau < \infty) = 0 \qquad \text{a.e.} \qquad \text{in} \qquad G.$$

To do this, let V be a non-negative, bounded, smooth function which is positive off $\bar{G}$. Given a non-negative $g \in C_0^\infty(G)$, choose $\mu > 0$ in such a way that $(\mu + \lambda V - L^*)u_\lambda = g$ has a C_b^2-solution u_λ for all $\lambda > 0$. Then for any non-negative $f \in C_0^\infty(R^d)$ we have

$$\int u_\lambda(x)f(x)\,dx = \int g(x)E^{P_x}\left[\int_0^\infty e^{-\mu t}\exp\left\{-\lambda\int_0^t V(x(s))\,ds\right\}f(x(t))\,dt\right]dx$$
$$\geqq \int_{\partial G}\Pi_g^\mu(dz)E^{P_z}\left[\int_0^\infty e^{-\mu t}\exp\left\{-\lambda\int_0^t V(x(s))\,ds\right\}f(x(t))\,dt\right],$$

where Π_g^μ is defined as in the proof of Lemma 7.5. Letting $\lambda \to \infty$, we now get

$$\int u_\infty(x)f(x)\,dx \geqq \int_{\partial G}\Pi_g^\mu(dz)E^{P_z}\left[\int_0^{\tau'} e^{-\mu t}f(x(t))\,dt\right],$$

where u_∞ is the non-increasing limit of the u_λ. The rest of the proof parallels that of Lemma 7.5. Given $x^0 \in \Delta_3$, choose coordinates so that $x^0 = 0$ and $\phi(x) = x$ in a neighborhood N of x^0. Substituting as in Lemma 7.5, we find that

$$\int u_\infty(\varepsilon x_1, x_2, \cdots, x_d)f(x)\,dx \geqq \int_{\Delta_3\cap N}\Pi_g^\mu(dz)E^{P_z^{(\varepsilon)}}\left[\int_0^{\tau'} e^{-\mu\varepsilon t}f(x(t))\,dt\right],$$

where $P_z^{(\varepsilon)}$ is as in Lemma 7.5. Next observe that, for $z \in \Delta_3$, $L\phi(z) > 0$, $L\phi^2(z) = 0$ and $L^2\phi^2(z) = 2(L\phi)^2(z)$. Hence $P_z(\tau' > 0) = 1$ for $z \in \Delta_3$; and so for every $z \in \Delta_3$ and $0 < \alpha < 1$ there is a $t > 0$ such that

$$P_z^{(\varepsilon)}(\tau' > t/\varepsilon) \geqq 1 - \alpha$$

for all $\varepsilon > 0$. Therefore,

$$\lim_{\varepsilon\searrow 0}\int u(\varepsilon x_1, x_2, \cdots, x_d)f(x)\,dx$$
$$\geqq \int_{\Delta_3\cap N}\Pi_g^\mu(dz)E^{P_z^{(0)}}\left[\int_0^\infty f(x_1(t), z_2, \cdots, z_d)\,dt\right].$$

This inequality holds for all bounded, non-negative, measurable f with compact support. In particular, if $f(x) = h(x_1)\bar{h}(x_2, \cdots, x_d)$, where $\operatorname{supp}\bar{h} \subseteq \Delta_3 \cap N$, then, since points in Δ_3 are regular for the L^*-process, we find that

$$\int_{\Delta_3\cap N}\Pi_g^\mu(dz)E^{P_z^{(0)}}\left[\int_0^\infty h(x_1(t))\,dt\right]\bar{h}(z) = 0,$$

and therefore $\Pi_G^\mu(\Delta_3 \cap N) = 0$. We have used here the facts that u_∞ vanishes at regular points of the L^*-process and that u_∞ is upper semi-continuous. This proves the first part of the theorem.

To prove the second part, note that if $P_x(x(\tau) \notin \Sigma, \tau < \infty) = 0$, then

$$P_x(\tau < \tau') = E^{P_x}[\chi_{\tau<\infty} P_{x(\tau)}(\tau' > 0)] = 0,$$

since $P_x(\tau' > 0) = 0$ for $x \in \Sigma$.

8. On Generalized Solutions

Let G be an open set in R^d and $\bar{G}$ its closure. In connection with the solution of the boundary value problem we introduced, in Section 5, the extension $\tilde{L}$ of L. Although from the probabilistic point of view this extension is very natural, from the analytical point of view it is rather obscure. The first part of this section provides a characterization of $\tilde{L}$ in analytic terms.

There is also the standard notion of a weak solution to the equation $Lu = f$. This uses twice differentiable test functions. The second part of the section is devoted to a study of the relation of weak solutions to our solutions of the boundary value problem.

According to the definition in Section 5, we say that $\tilde{L}u = f$ in $\bar{G}$ if

$$u(x(\tau' \wedge t)) - \int_0^{\tau' \wedge t} f(x(s))\, ds$$

is a P_x-martingale for all $x \in \bar{G}$.

Let us denote by X the Banach space of pairs of bounded measurable functions (u, f) on $\bar{G}$. We say that a sequence (u_n, f_n) in X is $*$-convergent to a point (u, f) in X if

$$\lim_{n\to\infty} u_n(x) = u(x) \quad \text{for each} \quad x \in \bar{G},$$

$$\lim_{n\to\infty} f_n(x) = f(x) \quad \text{for each} \quad x \in \bar{G},$$

$$\sup_n |u_n(x)| < \infty,$$

and

$$\sup_n |f_n(x)| < \infty.$$

We denote by $Y \subset X$ pairs (u, f) such that u is smooth and $Lu = f$. We denote by $\bar{Y}$ the smallest $*$-closed subset of X containing Y. We then have the following characterization of $\tilde{L}$.

THEOREM 8.1. *$\tilde{L}u = f$ in $\bar{G}$ if and only if $(u, f) \in \bar{Y}$.*

The above theorem will be proved in several steps, each of which is stated as a lemma.

LEMMA 8.1. *If $(u, f) \in \bar{Y}$, then $\tilde{L}u = f$ in $\bar{G}$.*

Proof: Clearly if $(u, f) \in Y$, then by Itô's formula $\tilde{L}u = f$. The set of pairs (u, f) such that $\tilde{L}u = f$ is $*$-closed because limits of martingales are again martingales. Therefore, $\tilde{L}u = f$ if $(u, f) \in \bar{Y}$.

For technical reasons it is easier to work with the operator $L_1 \equiv L - I$ rather than L itself. By the relation

$$\tilde{L}_1 u = f$$

we mean that

$$e^{-t\wedge\tau'} u(x(t \wedge \tau')) - \int_0^{t\wedge\tau'} e^{-s} f(x(s))\, ds$$

is a P_x-martingale for all $x \in G$. Letting Y_1 be the pairs $(u, L_1 u)$ with smooth u, we take $\bar{Y}_1$ to be the $*$-closure of Y_1.

LEMMA 8.2. *$\tilde{L}u = f$ if and only if $\tilde{L}_1 u = f - u$, and $(u, f) \in \bar{Y}$ if and only if $(u, f - u) \in \bar{Y}_1$.*

Proof: Let $\tilde{L}u = f$. Set

$$\xi(t) = u(x(t \wedge \tau')) - \int_0^{t\wedge\tau'} f(x(s))\, ds\,.$$

We apply Lemma 4.1 with $\eta(t) = -\chi_{\tau'>t}\, e^{-t}$ and conclude that

$$\xi(t) e^{-t\wedge\tau'} - \int_0^t \xi(s)\eta(s)\, ds$$

is a P_x-martingale for all $x \in G$. After simplification this yields

$$\tilde{L}_1 u = f - u\,.$$

The converse is proved by retracing the steps. To prove the second part, we note that $(u, f) \in Y$ if and only if $(u, f - u) \in Y_1$. Hence $(u, f) \in \bar{Y}$ if and only if $(u, f - u) \in \bar{Y}_1$.

It is now clear that in order to prove Theorem 8.1 we need show only that $\tilde{L}_1 u = f$ if and only if $(u, f) \in \bar{Y}_1$.

LEMMA 8.3. *If* $\tilde{L}_1 u = f$, *then*

$$u(x) = E^{P_x}\left[e^{-\tau'} u(x(\tau')) - \int_0^{\tau'} e^{-s} f(x(s))\, ds\right].$$

Proof: From the fact that

$$e^{-t\wedge\tau'} u(x(t \wedge \tau')) - \int_0^{t\wedge\tau'} e^{-s} f(x(s))\, ds$$

is a P_x-martingale we know that

$$u(x) = E^{P_x}\left[e^{-t\wedge\tau'} u(x(t \wedge \tau')) - \int_0^{t\wedge\tau'} e^{-s} f(x(s))\, ds\right]$$

for $t \geqq 0$. By letting $t \to \infty$ we get the desired result. There is no difficulty regarding convergence because

$$\left| e^{-t\wedge\tau'} u(x(t \wedge \tau')) - \int_0^{t\wedge\tau'} e^{-s} f(x(s))\, ds \right| \leqq \|u\|_\infty + \|f\|_\infty .$$

We do not need any continuity regarding u because if $\tau' < \infty$, then

$$u(x(t \wedge \tau')) = u(x(\tau'))$$

for $t \geqq \tau'$ and if $\tau' = \infty$, then $e^{-t\wedge\tau'} \to 0$ as $t \to \infty$.

LEMMA 8.4. *If f and g are bounded measurable functions on R^d and u is given by*

$$u(x) = E^{P_x}\left[\bar{e}^{\tau'} g(x(\tau')) - \int_0^{\tau'} \bar{e}^{s} f(x(s))\, ds\right],$$

then $(u, f) \in \bar{Y}_1$.

Proof: Observe that the class of functions f, g for which $(u, f) \in \bar{Y}_1$ is $*$-closed. Hence we can and do assume that f and g are in $C_b^\infty[R^d]$. Let $V \in C_b^2[R^d]$ be a non-negative function which is zero on $\bar{G}$ and is strictly positive

off $\bar{G}$. Define

$$w_n(t, x) = E^{P_x}\left[\int_0^t (ngV - f)(x(s)) \exp\left\{-s - n\int_0^s V(x(\sigma))\, d\sigma\right\} ds\right].$$

Using Oleinik's result (cf. Theorem 10.1 of the appendix), one can easily check that $w_n \in C^{1,2}([0, \infty) \times R^d)$ and that

$$\frac{\partial w_n}{\partial t} = (L_1 - nV)w_n + ngV - f.$$

Clearly,

$$\lim_{t\to\infty} w_n(t, x) = w_n(x)$$

$$= E^{P_x}\left[\int_0^\infty (ngV - f)(x(s)) \exp\left\{-s - n\int_0^s V(x(\sigma))\, d\sigma\right\} ds\right]$$

uniformly in x. Moreover, $\partial w_n(t, x)/\partial t \to 0$ uniformly in x as $t \to \infty$. Hence, $(w_n, f) \in \bar{Y}_1$ for each n. But it is clear that $w_n(x) \to u(x)$ boundedly for $x \in \bar{G}$ and therefore $(u, f) \in \bar{Y}_1$.

This completes the proof of Theorem 8.1.

Corollary 8.1. *If $\bar{L}u = f$, then, for each $x \in \bar{G}$, the P_x-martingale*

$$u(x(t \wedge \tau')) - \int_0^{t\wedge\tau'} f(x(s))\, ds$$

is a.s. continuous. In particular, $u(x(t \wedge \tau'))$, $0 \leqq t < \infty$, is a.s. continuous.

Proof: The assertion is obvious if $u \in C(\bar{G})$. Moreover, by Doob's inequality, if (u, f) is the $*$-limit of $\{(u_n, f_n)\}_1^\infty \subseteq \bar{Y}$ and

$$u_n(x(t \wedge \tau')) - \int_0^{t\wedge\tau'} f_n(x(s))\, ds$$

is a.s. continuous, then

$$u(x(t \wedge \tau')) - \int_0^{t\wedge\tau'} f(x(s))\, ds$$

is a.s. continuous. Hence

$$u(x(t \wedge \tau')) - \int_0^{t \wedge \tau'} f(x(s))\, ds$$

is a.s. continuous for all $(u, f) \in \bar{Y}$; q.e.d.

We now consider the standard notion of a weak solution. Given $u, f \in L_\infty(G)$, we say that

$$Lu \overset{W}{=} f \quad \text{in} \quad G$$

if

$$\int uL^*\phi\, dx = \int f\phi\, dx$$

for all $\phi \in C_0^\infty(G)$. One way of formulating the boundary value problem is to demand that

$$Lu \overset{W}{=} f \quad \text{in} \quad G,$$

$$u = g \quad \text{on} \quad \Sigma_2 \cup \Sigma_3 .$$

We proceed to prove that the above formulation can be reduced to the earlier one.

LEMMA 8.5. *Suppose $\psi \in C_b^2[R^d]$ is a non-negative function which is strictly positive on G and vanishes off G. Denote by L^ψ the operator ψL and let P_x^ψ denote the solution to the martingale problem for L^ψ starting at x. Then $P_x^\psi[\tau < \infty] = 0$ for $x \in G$. Define*

$$\sigma_t = \inf\left\{ s \geqq 0 : \int_0^s \frac{du}{\psi(x(u))} \geqq t \right\},$$

and let Q_x be the distribution of $x(\sigma_t)$, $t \geqq 0$, under P_x. Then $P_x^\psi = Q_x$. In particular, $P_x(\sigma_t < \tau) = 1$ for all $t \geqq 0$ and $x \in G$. Finally, $P_x\left(\sigma_\infty = \lim_{t \nearrow \infty} \sigma_t = \tau\right) = 1$ for all x.

Proof: That $P_x^\psi(\tau < \infty) = 0$, $x \in G$, is an immediate consequence of the fact that $\Sigma_2 \cup \Sigma_3 = \varnothing$ for L^ψ. The equality $P_x^\psi = Q_x$ follows from the uniqueness of P_x^ψ. To see that $P_x(\sigma_t < \tau) = 1$ for $x \in G$ and $t \geqq 0$, note that

$$P_x(\sigma_t < \tau) = P_x(x(\sigma_s) \in G, 0 \leqq s \leqq t) = Q_x(\tau > t) = 1 .$$

Finally,

$$P_x(\sigma_\infty < \tau) = P_x\left(\int_0^{\tau - \varepsilon} \frac{1}{\psi(x(s))}\, ds = \infty,\ \varepsilon > 0 \right) = 0 .$$

Lemma 8.6. *If V is a non-negative function in $C_b^2(R^d)$ and $u \in L^\infty(R^d)$ satisfies*

$$\int u(L - V)^* \phi \, dx = 0 \tag{8.1}$$

for all $\phi \in C_0^\infty(R^d)$, then

$$E^{P_g}\left[u(x(t)) \exp\left\{-\int_0^t V(x(s)) \, ds\right\} dt\right] = \int g \, u \, dx \, , \qquad t \geqq 0 \, , \tag{8.2}$$

for all non-negative $g \in L^1(R^d)$.

Proof: It suffices to prove (8.2) when $g \in C_0^\infty(R)$. First note that (8.1) must remain true when $\phi \in C_b^2(R^d)$ and $(L - V)^*\phi \in L^1(R^d)$. In particular, one can use Oleinik's result to show that for large λ there is a ϕ_λ such that

$$\lambda \int u\phi_\lambda \, dx = \int g \, u \, dx \, .$$

Moreover,

$$\int u\phi_\lambda \, dx = \int_0^\infty e^{-t} E^{P_g}\left[\exp\left\{-\int_0^t V(x(s)) \, ds\right\} u(x(t))\right] dt \, ,$$

and $E^{P_g}\left[\exp\left\{-\int_0^t V(x(s)) \right\} ds \, u(x(t))\right]$ is continuous in t. Hence, by the uniqueness of the Laplace transform, (8.2) follows.

We now have to describe the manner in which u takes on the boundary value g on $\Sigma_2 \cup \Sigma_x$. Let us suppose that

$$\begin{gathered} Lu \overset{W}{=} f \, , \\ \operatorname*{ess\,lim}_{x \to a} u(x) = g(a) \qquad \text{for} \qquad a \in \Sigma_2 \cup \Sigma_3 \, . \end{gathered} \tag{8.3}$$

It is easy to check that g is continuous on $\Sigma_2 \cup \Sigma_3$ and u can be changed on a set of measure zero so that

$$\begin{gathered} Lu \overset{W}{=} f \, , \\ \lim_{x \to a} u(x) = g(a) \qquad \text{for} \qquad a \in \Sigma_2 \cup \Sigma_3 \, . \end{gathered} \tag{8.4}$$

THEOREM 8.2. *Let u, f and g satisfy* (8.4). *Then there are bounded measurable versions u' and f' of u and f so that*

$$\tilde{L}u' = f'$$

and

$$u'(a) = g(a) = \lim_{\substack{x \to a \\ x \in G}} u'(x) \qquad \text{for} \qquad a \in \Sigma_2 \cup \Sigma_3 .$$

Proof: Let $\tilde{f}$ be a bounded measurable version of f. Let $\bar{g}$ be a bounded measurable extension of g from $\Sigma_2 \cup \Sigma_3$ to Γ. Define

$$v(x) = E^{P_x}\left[e^{-\tau'} \bar{g}(x(\tau')) - \int_0^{\tau'} e^{-s} (\tilde{f} - u)(x(s))\, ds\right].$$

Then, by Lemma 8.4, $(v, \tilde{f} - u) \in \bar{Y}_1$. Hence, for $\phi \in C_0^\infty(G)$,

$$\int v(L - I)^* \phi\, dx = \int (\tilde{f} - u)\phi\, dx .$$

Also for such ϕ

$$\int u(L - I)^* \phi\, dx = \int (\tilde{f} - u)\phi\, dx .$$

Therefore, if $w = v - u$,

(8.5)
$$\int w(L - I)^* \phi\, dx = 0$$

for $\phi \in C_0^2(G)$. It is easily seen that (8.5) holds for $\phi \in C_0^2(R^d)$ which are zero off G. In particular, if $\psi \in C_b^2(R^d)$ is non-negative and zero off G, then

$$\int w(L^\psi - \psi)^* \phi\, dx = 0$$

for all $\phi \in C_0^\infty(R^d)$. Thus by Lemmas 8.5 and 8.6, if $g \in L^1(R^d)$ is non-negative and is supported in G, then

$$\int gw\, dx = E^{P_g}[e^{-\sigma_t} w(x(\sigma_t))] \qquad \text{for} \qquad t \geqq 0 ,$$

where σ_t is as in Lemma 8.6. We let $t \to \infty$; then $\sigma_t \to \tau$. We know that for almost all paths with respect to P_g, $\tau = \tau'$ and $x(\tau) \in \Sigma_2 \cup \Sigma_3$. But in such

a case, $w(x(\sigma_t)) \to 0$ because both v and u approach $g(a)$ as $x \to a$. All this is when $\tau < \infty$. Otherwise, $e^{-\sigma_t} \to 0$. So in any case $\int gw\,dx = 0$ for all $g \in L^1(R^d)$ which are supported in G. Therefore,

$$v(x) = u(x) \qquad \text{a.e.} \qquad \text{in} \qquad G\,.$$

If we now take $u' = v$ and $f' = f + w$, then it is easily checked that

$$\tilde{L}u' = f'$$

and

$$u'(a) = \lim_{\substack{x \to a \\ x \in G}} u'(x) = g(a) \qquad \text{for} \qquad a \in \Sigma_2 \cup \Sigma_3\,.$$

Corollary 8.2. *Let k be a bounded non-negative function on G and assume that either $\sup_{x\in G} E_x[\tau'] < \infty$ or that k is uniformly positive. Given*

$$f \in L^\infty(G) \qquad \textit{and} \qquad g \in L^\infty(\Gamma) \cap C(\Sigma_2 \cup \Sigma_3)\,,$$

the function

$$u(x) = E^{P_x}\Big[g(x(\tau'))\exp\Big\{-\int_0^{\tau'} k(x(s))\,ds\Big\} - \int_0^{\tau'} \exp\Big\{-\int_0^t k(x(s))\,ds\Big\} f(x(t))\,dt\Big]$$

satisfies

(i) $$Lu \overset{W}{=} ku + f,$$

(ii) $$\lim_{\substack{x \to a \\ x \in G}} u(x) = g(a) \qquad \textit{for} \qquad a \in \Sigma_2 \cup \Sigma_3\,.$$

Moreover, if $v \in L^\infty(G)$ satisfies (i) *and* (ii), *then $v = u$ a.e. in G.*

Proof: That u satisfies (i) and (ii) is obvious. Suppose v is any solution of (i) and (ii). Then by Theorem 8.2 we can find v', f' such that they agree almost everywhere with v, f,

$$\tilde{L}v' = kv' + f'\,,$$

$$v'(a) = \lim_{\substack{x \to a \\ x \in G}} v'(x) = g(a) \qquad \text{for} \qquad a \in \Sigma_2 \cup \Sigma_3\,.$$

But from Section 5, we know that

$$v'(x) = E^{P_x}\Big[v'(x(\tau'))\exp\Big\{-\int_0^{\tau'} k(x(s))\,ds\Big\} - \int_0^{\tau'} \exp\Big\{-\int_0^t k(x(s))\,ds\Big\} f'(x(t))\,dt\Big]\,.$$

Since $f = f'$ almost everywhere and $v' = g$ on $\Sigma_2 \cup \Sigma_3$, it follows that $v' = u$ a.e.

9. Some Additional Remarks

Consider once again the first boundary value problem

$$\tilde{L}u - ku = f \quad \text{in} \quad \bar{G},$$
$$u = g \quad \text{on} \quad \Gamma,$$

under the assumption that k is either uniformly positive or that $\sup_{x\in\bar{G}} E_x[\tau'] < \infty$. We saw in Section 5 that

$$u(x) = E_x[\xi(\omega)], \tag{9.1}$$

where

$$\xi(\omega) = g(x(\tau')) \exp\left\{-\int_0^{\tau'} k(x(s))\, ds\right\} - \int_0^{\tau'} \exp\left\{-\int_0^t k(x(s))\, ds\right\} f(x(t))\, dt. \tag{9.2}$$

Now we want to use this representation of u to study certain stability questions associated with the first boundary value problem. In the ensuing discussion we shall assume that k and f are continuous on $\bar{G}$ and that g is bounded and continuous on Γ. There are three closely related questions to which we seek the answer:

A. Is $u(x)$ continuous as a function of x?

B. Is u continuous as a function of the coefficients of L?

C. Can u be approximated by the solution of a suitable scheme of difference equations?

It is easy to check that the measures P_x are weakly continuous under the various kinds of convergence associated with A, B and C. Hence what we need to study are conditions under which $\xi(\omega)$ is a.s. continuous.

LEMMA 9.1. *If $P_x[\tau = \tau'] = 1$, then $\xi(\omega)$ is continuous almost surely with respect to P_x.*

Proof: τ is upper semicontinuous and τ' is lower semicontinuous; so if $P_x[\tau = \tau'] = 1$, then τ' is almost surely continuous. Therefore, $\xi(\omega)$ is continuous almost surely.

LEMMA 9.2. *Let Γ be closed. Then, for any $x \in G$, $P_x[\tau = \tau'] = 1$.*

Proof: We know that $\Gamma = \overline{\Sigma_2 \cup \Sigma_3}$. Therefore, for any $x \in G$,

$$P_x[x(\tau) \in \Gamma] = 1 .$$

This in turn implies that $P_x[\tau = \tau'] = 1$. Combining Lemmas 9.1 and 9.2 we have

THEOREM 9.1. *Let* Γ *be closed. Then the answers to questions* A, B, *and* C *raised earlier are in the affirmative.*

Remark 9.1. It is now obviously important to know when Γ is closed. This means that we want to know when the boundary points of $\Sigma_2 \cup \Sigma_3$ are regular. The results of Section 6 are relevant in this context.

If Γ is not closed or if we do not know that it is closed, we consider the set

$$E = [x \in G : P_x[x(\tau) \in \Sigma_2 \cup \Sigma_3] = 1] .$$

$G - E$ is a set of measure zero. For $x \in E$, $P_x[\tau = \tau'] = 1$. Consequently, the answers to questions A, B, C are in the affirmative for points $x \in E$. We combine all of this in the following theorem.

THEOREM 9.2. *There is a set* F *of measure zero in* G *such that*

(i) $u(x)$ *is continuous at* $x \in G - F$;

(ii) *if the operator* L *is approximated, then the solutions converge continuously on* $G - F$;

(iii) *if the operator* L *is approximated by suitable difference schemes, then the solutions converge continuously at points of* $G - F$;

(iv) *if* Γ *is closed, then* F *is empty.*

Remark 9.2. One consequence of Theorem 9.1 is that our solution to the first boundary value problem is almost everywhere the limit of the solution to the corresponding problem for the operators $\varepsilon\Delta + L$ as $\varepsilon \searrow 0$.

10. Appendix

We begin this section with a lemma of Phillips and Sarason (cf. Lemma 1.1 in [7]) and some variations on their result.

LEMMA 10.1. *Let* $a : R \to S_d$ *and assume* $a^{ij} \in C_b^2(R)$, $1 \leqq i, j \leqq d$. *Let* $a^{1/2}$ *denote the non-negative definite square root of* a. *Then* $\alpha = a^{1/2}$ *is uniformly Lipschitz continuous with a Lipschitz constant depending only on the bounds on the second derivatives of the* a^{ij}. *Furthermore, there is a constant* C *depending only on the bounds on*

the second derivatives of the a^{ij} *such that, for all* $\lambda, \mu \in R^d$,

$$|\langle \lambda, \dot{a}(x)\mu\rangle| \leqq C(|\lambda|\, \langle \mu, a(x)\mu\rangle^{1/2} + |\mu|\, \langle \lambda, a(x)\lambda\rangle^{1/2}) , \tag{10.1}$$

and for all symmetric $((u_{ij})) \in R^d \otimes R^d$:

$$(\mathrm{tr}\,(\dot{a}u))^2 \leqq C\,\mathrm{tr}\,(u\,a\,u) . \tag{10.2}$$

Proof: Our proof mimics that of Phillips and Sarason. Clearly we can assume that $a(\cdot)$ is uniformly positive definite. Further, we may assume that $a(x^0)$ is diagonal. Then Loewner's formula tells us that

$$\dot{\alpha}^{ij}(x^0) = \frac{\dot{a}^{ij}(x^0)}{(a^{ii}(x^0))^{1/2} + (a^{jj}(x^0))^{1/2}} .$$

Given $\lambda \in R^d$, note that $\langle \lambda, a(\cdot)\lambda\rangle$ is a non-negative C^2-function. Hence,

$$\langle \lambda, \dot{a}(x)\lambda\rangle^2 \leqq 2\, \|\langle \lambda, \ddot{a}(\cdot)\lambda\rangle\|_\infty \langle \lambda, a(x)\lambda\rangle .$$

Applying this fact to the vectors e_i, and $(e_i + e_j)$, where e_i is the i-th coordinate vector, we obtain

$$|\dot{a}^{ij}(x^0)| \leqq A(a^{ii}(x^0) + a^{jj}(x^0))^{1/2} ,$$

where A depends only on the bounds on the $\ddot{a}^{ij}$. This together with Loewner's formula proves the first assertion. To prove (10.1), observe that

$$\begin{aligned}
|\langle \lambda, \dot{a}(x^0)\mu\rangle|^2 &= (\lambda_i \dot{a}^{ij}(x^0)\mu_j)^2 \leqq d^4 A^2 \lambda_i^2 (a^{ii}(x^0) + a^{jj}(x^0))\mu_j^2 \\
&= d^4 A^2 (|\mu|^2 \langle \lambda, a(x^0)\lambda\rangle + |\lambda|^2 \langle \mu, a(x^0)\mu\rangle) \\
&\leqq d^4 A^2 (|\mu|\, \langle \lambda, a(x^0)\lambda\rangle^{1/2} + |\lambda|\, \langle \mu, a(x^0)\mu\rangle^{1/2})^2 .
\end{aligned}$$

Finally, (10.2) follows from

$$\begin{aligned}
(\mathrm{tr}\,(\dot{a}\,u)(x^0))^2 &= (\dot{a}^{ij}(x^0)u_{ij})^2 \leqq d^4 A^2 (a^{ii}(x^0) + a^{jj}(x^0))\,u_{ij}^2 \\
&= 2d^4 A^2\, \mathrm{tr}\,(u\,a(x^0)u) .
\end{aligned}$$

We turn now to the proof of a theorem due to Oleinik [6].

Let $a : [0, \infty) \times R^d \to S_d$, $b : [0, \infty) \times R^d \to R^d$, and $k : [0, \infty) \times R^d \to R$ be bounded smooth functions. Assume for the present that a is uniformly positive

definite and let u be the unique smooth solution of

$$(10.3)\qquad \begin{aligned} &u_t + a^{ij}u_{,ij} + b^i u_{,i} + ku = f, \qquad 0 \leqq t < T, \\ &\lim_{t \nearrow T} u(t,\cdot) = g, \end{aligned}$$

where f and g are bounded smooth functions. We want to derive estimates on the derivatives of u which are independent of the ellipticity of a.

Let $D_x^\alpha = \partial^{|\alpha|}/\partial x_1 \cdots \partial x_d$ and set $u^\alpha = D^\alpha u$, $f^\alpha = D^\alpha f$, and $g^\alpha = D^\alpha g$. Then

$$u_t^\alpha + a^{ij}u^\alpha_{,ij} + b^i u^\alpha_{,i} + \sum_{l=1}^{d} \alpha_l a^{ij}_{,l} u^{\hat\alpha l}_{,ij} + \sum_{|\beta| \leqq |\alpha|} c_\beta u^\beta = f,$$

$$\lim_{t \nearrow T} v^\alpha(t,\cdot) = g^\alpha,$$

where $\hat\alpha^l$ is defined so that

$$\frac{\partial}{\partial x_l} D^{\hat\alpha l} = D^\alpha$$

and the c_β are linear combinations of the coefficients a, b, and k and their spacial derivatives up to order $|\alpha|$. Define $w = \sum_{|\alpha|=n} (u^\alpha)^2$. Then

$$w_t + a^{ij}w_{,ij} + b^i w_{,i} - 2\sum_{|\alpha|=n} \langle \nabla_x u^\alpha, a\nabla_x u^\alpha \rangle$$

$$+ 2\sum_{l=1}^{d} \sum_{|\alpha|=n} \alpha_l u^\alpha a^{ij}_{,l} u^{\hat\alpha l}_{,ij} + 2\sum_{|\beta| \leqq n} c_\beta u^\alpha u^\beta = 2\sum_{|\alpha|=n} u^\alpha f^\alpha,$$

$$\lim_{t \nearrow T} w(t,\cdot) = \sum_{|\alpha|=n} (g^\alpha)^2.$$

Put $A = \sum_{|\alpha|=n} \langle \nabla_x u^\alpha, a\nabla_x u^\alpha \rangle$. Then, applying first Schwartz's inequality and then (10.2), we have

$$\left| \sum_{|\alpha|=n} \alpha_l u^\alpha a^{ij}_{,l} u^{\hat\alpha l}_{,ij} \right| \leqq n w^{1/2} \left(\sum_{|\alpha|=n} (a^{ij}_{,l} u^{\hat\alpha l}_{,ij})^2 \right)^{1/2}$$

$$\leqq C_1 w^{1/2} A^{1/2},$$

where C_1 depends only on the second spacial derivatives of the a^{ij}. Hence,

$$w_t + a^{ij}w_{,ij} + (C_1 w^{1/2} A^{1/2} - 2A) + 2\sum_{|\beta| \leqq n} \sum_{|\alpha|=n} C_\beta u^\alpha u^\beta - 2\sum_{|\alpha|=n} u^\alpha f^\alpha \geqq 0.$$

First note that $C_1 w^{1/2} A^{1/2} - 2A \leqq C_1^2 w/8$. Also

$$\left| 2 \sum_{|\beta|=n} \sum_{|\alpha|=n} c_\beta u^\alpha u^\beta \right| \leqq C_2 w ,$$

$$\left| 2 \sum_{|\beta|<n} \sum_{|\alpha|=n} c_\beta u^\alpha u^\beta \right| \leqq C_3(1 + w) \|u\|^{(n-1)} ,$$

$$\left| 2 \sum_{|\alpha|=n} u^\alpha f^\alpha \right| \leqq C_4(1 + w) \|f\|^{(n)} ,$$

where the constants depend only on the bounds on the derivatives, up to order n, of the coefficients in equation (10.3), and $\|h\|^{(m)} = \sum_{|\alpha|\leqq m} \|D_x^\alpha h\|_\infty$. Hence there exist constants C_5 and C_6, depending only on $C_1 - C_4$, such that

$$w_t + a^{ij} w_{,ij} + b^i w_{,i} + C_5(1 + \|u\|^{(n-1)} + \|f\|^{(n)})w \\ + C_6(1 + \|u\|^{(n-1)} + \|f\|^{(n)}) \geqq 0 .$$

Using induction on n and the weak maximum principle, one arrives at the following theorem.

THEOREM 10.1 (Oleinik). *Let*

$$a : [0, \infty) \times R^d \to S_d , \qquad b : [0, \infty) \times R^d \to R^d , \qquad k : [0, \infty) \times R^d \to R$$

be bounded continuous functions. Assume that

$$a^{ij} \in C_b^{0,m}([0, \infty) \times R^d) , \qquad b^i \in C^{0,n}([0, \infty) \times R^d) , \qquad k \in C^{0,n}([0, \infty) \times R^d) .$$

Let $f \in C^{0,m\wedge n}([0, T) \times R^d)$ *and* $g \in C^{m\wedge n}(R^d)$, *and suppose* u *solves* (10.3). *If* $m \geqq 2$, *then for* $l \leqq m \wedge n$

$$\|u(t, \cdot)\|^{(l)} \leqq A(1 + \|f\|^{(l)} + \|g\|^{(l)}) e^{B(T-t)} , \tag{10.4}$$

where A *and* B *depend only on the bounds on the second spacial derivatives of the* a^{ij} *and the spacial derivatives up to order* l *of all the coefficients in* (10.3).

COROLLARY 10.1. *If* m *and* n *in Theorem* 10.1 *are at least* 2, *then, for each* $T > 0$, $f \in C_b^{0,2}([0, T) \times R^d)$ *and* $g \in C_b^2(R^d)$, *there is a unique* u *satisfying* (10.3). *Moreover, if the coefficients in* (10.3) *are independent of* t *and if* $l = m$, $n \geqq 2$, *then,*

for each $f \in C_b(R^d)$, *there is a* $\lambda_0 > 0$ *such that, for all* $\lambda \geqq \lambda_0$,

$$\lambda u - a^{ij} u_{,ij} - b^i u_{,i} - ku = f \tag{10.5}$$

has a unique solution in $C_b^l(R^d)$.

THEOREM 10.2. *Let* $a : [0, \infty) \times R^d \to S_d$ *and assume that* $\operatorname{tr} a(t, x) \leqq \frac{1}{2}$ *for all* (t, x). *If* $a^{ij} \in C_b^{1,2}([0, \infty) \times R^d)$, $1 \leqq i, j \leqq d$, *then* $\sum_0^\infty \sigma_n^2$ *converges to* a *uniformly on compacts, where* $\sigma_0 = a$ *and* $\sigma_n = \sigma_{n-1} - \sigma_{n-1}^2$. *Moreover,* $\sum_0^\infty (\sigma_n^2)'$ *converges to* a' *uniformly on compacts.*

Proof: Define σ_n, $n \geqq 0$, as indicated and note that $\sigma_n \leqq \sigma_{n-1}$, in the sense of non-negative definite matrices, for all $n \geqq 1$. Moreover, we know that $\sum_0^N \sigma_n^2 = a - \sigma_{N+1} \leqq a$. Hence, for any $\theta \in R^d$, $\sum_0^N \langle \theta, \sigma_n^2 \theta \rangle \leqq \langle \theta, a\theta \rangle$, and so $\sum_0^\infty \langle \theta, \sigma_n^2 \theta \rangle$ is absolutely convergent. But this means $\langle \theta, \sigma_{N+1}^2 \theta \rangle \searrow 0$ as $N \to \infty$. Hence, by Dini's theorem, $\langle \theta, \sigma_{N+1}^2 \theta \rangle \searrow 0$ uniformly on compacts. Using the fact that the σ_n are non-negative definite, one sees that $\sigma_n \searrow 0$ uniformly on compacts. The first assertion is now proved.

To prove the second assertion, note that

$$\dot{\sigma}_{n+1} = \dot{\sigma}_n - \dot{\sigma}_n \sigma_n - \sigma_n \dot{\sigma}_n = \tfrac{1}{2}[\dot{\sigma}_n(I - 2\sigma_n) + (I - 2\sigma_n)\dot{\sigma}_n],$$

where the dot indicates that one takes a generic spacial derivative. Continuing by induction, one arrives at

$$\dot{\sigma}_{n+1} = \frac{1}{2^{n+1}} \sum_{A \in \mathscr{P}_n} (I - 2\sigma)_A \, \dot{a} (I - 2\sigma)_{A^c},$$

where $\mathscr{P}_n$ denotes the set of subsets of $\{0, \cdots, n\}$ and $(I - 2\sigma)_A = \prod_{j \in A} (I - 2\sigma_j)$. Hence, if $\lambda, \mu \in R^d$, then

$$|\langle \lambda, \dot{\sigma}_{n+1} \mu \rangle| \leqq \frac{1}{2^{n+1}} \sum_{A \in \mathscr{P}_n} |\langle (I - 2\sigma)_A \lambda, \dot{a}(I - 2\sigma)_{A^c} \mu \rangle|.$$

Using (10.1), we have

$$\begin{aligned} &|\langle (I - 2\sigma)_A \lambda, \dot{a}(I - 2\sigma)_{A^c} \mu \rangle| \\ &\quad \leqq C(|(I - 2\sigma)_A \lambda| \, \langle (I - 2\sigma)_{A^c} \mu, a(I - 2\sigma)_{A^c} \mu \rangle^{1/2} \\ &\qquad + |(I - 2\sigma)_{A^c} \mu| \, \langle (I - 2\sigma)_A \lambda, a(I - 2\sigma)_A \lambda \rangle^{1/2}), \end{aligned}$$

where C depends on the bounds on a and its first two spacial derivatives.

Note that $|(I-2\sigma)_A \lambda| \leqq |\lambda|$. Thus it remains only to estimate quantities of the form

$$I_n = \frac{1}{2^{n+1}} \sum_{A \in \mathscr{S}_n} \langle (I-2\sigma)_A \lambda, a(I-2\sigma)_A \lambda \rangle^{1/2} .$$

By Schwarz's inequality,

$$I_n^2 \leqq \frac{1}{2^{n+1}} \sum_{A \in \mathscr{S}_n} \langle (I-2\sigma)_A \lambda, a(I-2\sigma)_A \lambda \rangle = \frac{1}{2^{n+1}} \sum_{A \in \mathscr{S}_n} \langle \lambda, a(I-2\sigma)_A^2 \lambda \rangle .$$

But

$$\frac{1}{2^{n+1}} \sum_{A \in \mathscr{S}_n} (I-2\sigma)_A^2 = \prod_{j=0}^{n} \frac{I+(I-2\sigma_j)}{2},$$

and so

$$I_n^2 \leqq \left\langle \lambda, a \prod_{j=0}^{n} \frac{I+(I-2\sigma_j)}{2} \lambda \right\rangle = \left\langle \lambda, a \prod_0^n ((I-\sigma_j)^2 + \sigma_j^2) \lambda \right\rangle$$

$$\leqq \left\langle \lambda, a \exp\left\{2 \sum_0^n \sigma_j^2\right\} \prod_0^n (I-\sigma_j)^2 \lambda \right\rangle .$$

Observe that $\sigma_{n+1} = a \prod_0^n (I-\sigma_j)$, and therefore

$$a \prod_0^n (I-\sigma_j)^2 = \sigma_{n+1} \prod_0^n (I-\sigma_j) \leqq a \prod_0^n (I-\sigma_j) = \sigma_{n+1} .$$

Hence,

$$I_{n-1}^2 \leqq \langle \lambda, e^{2a} \sigma_n \lambda \rangle \to 0$$

uniformly on compacts; q.e.d.

Let L be a second order elliptic-parabolic operator of the form

$$Lu = \tfrac{1}{2} \sum a^{ij}(x) \frac{\partial^2 u}{\partial x_i \partial x_j} + \sum b^j(x) \frac{\partial u}{\partial x_j}$$

with smooth coefficients. Let ϕ be a smooth function.

LEMMA 10.2. *For any integer* $m > 0$,

$$L^m \phi^2 = \sum_{\substack{p+q \leqq m-1 \\ p \geqq 0 \\ q \geqq 0}} \binom{p+q}{p} L^{m-p-q-1} \langle a \nabla L^p \phi, \nabla L^q \phi \rangle + \sum_{0 \leqq \alpha \leqq m} \binom{m}{\alpha} (L^\alpha \phi)(L^{m-\alpha} \phi) .$$

Proof: This Lemma is proved easily by induction on m.

THEOREM 10.3. *Let ϕ be such that, at a certain point x^0,*

$$L^r\phi^2 = 0 \qquad \textit{for} \qquad r = 0, 1, 2, \cdots, n.$$

Then at the same point x^0

$$\text{(i)} \qquad L^s\phi = 0 \qquad \textit{for} \qquad 0 \leqq s \leqq \left[\frac{n}{2}\right],$$

$$\text{(ii)} \qquad L^r\langle a\nabla L^p\phi, \nabla L^q\phi\rangle = 0 \qquad \textit{if} \qquad p + q + r \leqq n - 1 .$$

Proof: First we prove (i). Since L generates a non-negative semigroup, our assumption implies that

$$(T_t\phi^2)(x^0) = o(t^n) \qquad \text{as} \qquad t \to 0 .$$

Moreover, $T_t\phi^2 \geqq (T_t\phi)^2$ by the Cauchy Schwartz inequality. Therefore,

$$(T_t\phi)(x^0) = o(t^{n/2}) \qquad \text{as} \qquad t \to 0 .$$

This implies that $\phi, L\phi, \cdots, L^{[n/2]}\phi$ are zero at x^0. We now turn to the proof of (ii). First, in view of Lemma 10.1 and the first part of the theorem, we have

$$\sum_{\substack{p+q\leqq m-1 \\ p\geqq 0 \\ q\geqq 0}} \binom{p+q}{q} L^{m-p-q-1}\langle a\nabla L^p\phi, \nabla L^q\phi\rangle = 0 \tag{10.6}$$

for $m = 1, 2, \cdots, n$. Let

$$\psi_{pq}(x) = \langle a\nabla L^p\phi, \nabla L^q\phi\rangle , \quad 0 \leqq p \leqq n, \quad 0 \leqq q \leqq n;$$

$\psi_{pq}(x)$ is a positive semidefinite matrix for each x. Therefore,

$$A_{pq}(t) = (T_t\psi_{pq})(x^0)$$

is a positive semidefinite matrix for each $t \geqq 0$. We shall prove our result by induction on $k = p + q + r$. We note that, for $k = 0$, (10.6) yields

$$\langle a\nabla\phi, \nabla\phi\rangle = 0 \qquad \text{at} \qquad x^0 ,$$

with $m = 1$. Let us suppose that (ii) is valid for $p + q + r$ equal to $0, 1, 2, \cdots, l - 1$. This implies that at x^0

$$L^r\langle a\nabla L^p\phi, \nabla L^q\phi\rangle = 0 \qquad \text{for} \qquad 0 \leqq r \leqq l - p - q - 1$$

or

$$A_{pq}(t) = \frac{a_{pq}}{(l-p-q)\,!}\, t^{l-p-q} + o(t^{l-p-q}),$$

where

$$a_{p,q} = (L^{l-p-q}\langle a\nabla L^p\phi, \nabla L^q\phi\rangle)(x^0).$$

Since $A_{pq}(t)$ is a positive semidefinite matrix for every $t \geqq 0$, so is the matrix

$$B_{pq}(t) = A_{p,q}(t)t^{p+q-l}.$$

In particular, by taking $t = 0$, we see that

$$b_{p,q} = \begin{cases} \dfrac{a_{pq}}{(l-p-q)\,!} & \text{for} \quad p+q \leqq l, \\ 0 & \text{for} \quad p+q > l, \end{cases}$$

is a positive semidefinite matrix. Our assumptions imply that

$$\sum \binom{p+q}{p}(l-p-q)\,!\, b_{pq} = 0,$$

and what we want to conclude is that $b_{pq} = 0$ for $p + q \leqq l$. For this it is sufficient to prove that

$$\rho_{pq} = \binom{p+q}{p}(l-p-q)\,!$$

is positive definite. We know that $\rho_{pq} = (p + q)\,!\,(l - p - q)\,!\,/p\,!\,q\,!$ and so it is positive definite if $(r!)(l - r)!$ is a moment sequence. We have by computation

$$(r\,!)(l-r)\,! = c_l\int_0^1 \left(\frac{u}{1-u}\right)^r (1-u)^l\, du.$$

This completes the proof.

Bibliography

[1] Bony, J. M., *Principe due maximum, inégalité de Harnack et unicité du probleme de Cauchy pour les operateurs elliptiques dégénérés*, Ann. Inst. Fourier, XIX, No. 1, 1969, pp. 277–304.

[2] Fichera, G., *Sulla equazioni differenziali lineari ellittico-paraboliche del secondo ordine*, Atti. Naz. Lincei Mem., Ser. 8, Vol. 5, 1956, pp. 1–30.

[3] Hill, C. D., *A sharp maximum principle for degenerate elliptic parabolic equations*, Indiana Univ. Math. Jour., Vol. 20, No. 3, 1970, pp. 213–230.

[4] Kohn, J. J., and Nirenberg, L., *Degenerate elliptic-parabolic equations of second order*, Comm. Pure Appl. Math., Vol. 20, 1967, pp. 797–872.

[5] Oleinik, O. A., *A problem of Fichera*, Doklady Acad. Nauk SSSR, Vol. 157, 1964, pp. 1297–1301. (In Russian.) English translation in Soviet Math., Vol. 5, 1964, pp. 1129–1133.

[6] Oleinik, O. A., *Alcuni risultati sulle equazioni lineari e quasi lineari ellitico-paraboliche a derivate parziali del secondo ordine*, Rend. Classe, Sci. Fis. Mat. Nat. Acad. Naz. Lincei, Ser. 8, Vol. 40, 1966, pp. 775–784.

[7] Phillips, R. S., and Sarason, L., *Elliptic-parabolic equations of the second order*, J. Math. Mech. 17, 1967, pp. 891–917.

[8] Redheffer, R., *The sharp maximum principle for nonlinear inequalities*. (Preprint.)

[9] Stroock, D. W., and Varadhan, S. R. S., *Diffusion processes with continuous coefficients*, I, Comm. Pure Appl. Math., Vol. 22, 1969, pp. 345–400.

[10] Stroock, D. W., and Varadhan, S. R. S., *On the support of diffusion processes, with applications to the strong maximum principle*, to appear in the Proceedings of the Sixth Berkeley Symposium in Probability and Statistics.

Received March, 1972.

Proceedings of the Sixth Berkeley Symposium on Mathematical Statistics and Probability,
Vol. III: Probability theory, Univ. California Press, 1972, 333-359

ON THE SUPPORT OF DIFFUSION PROCESSES WITH APPLICATIONS TO THE STRONG MAXIMUM PRINCIPLE

DANIEL W. STROOCK and S. R. S. VARADHAN
COURANT INSTITUTE, NEW YORK UNIVERSITY

1. Introduction

Let $a: [0, \infty) \times R^d \to S_d$ and $b: [0, \infty) \times R^d \to R^d$ be bounded continuous functions, where S_d denotes the class of symmetric, nonnegative definite $d \times d$ matrices. From a and b form the operator

$$L_t = \frac{1}{2} \sum_{i,j=1}^{d} a^{ij}(t, x) \frac{\partial^2}{\partial x_i \partial x_j} + \sum_{i=1}^{d} b_i(t, x) \frac{\partial}{\partial x_i}. \tag{1.1}$$

A strong maximal principle for the operator $(\partial/\partial t) + L_t$ is a statement of the form: "for each open $\mathscr{G} \subseteq [0, \infty) \times R^d$ and each $(t_0, x_0) \in \mathscr{G}$ there is a set $\mathscr{G}(t_0, x_0) \subseteq \mathscr{G}$ with the property that $(\partial f/\partial t) + L_t f \geqq 0$ on $\mathscr{G}(t_0, x_0)$ and $f(t_0, x_0) = \sup_{\mathscr{G}(t_0, x_0)} f(t, x)$ imply $f \equiv f(t_0, x_0)$ on $\mathscr{G}(t_0, x_0)$." Of course, in order for a strong maximum principle to be very interesting it must describe the set $\mathscr{G}(t_0, x_0)$. Further, it should be possible to show that $\mathscr{G}(t_0, x_0)$ is maximal. That is, one wants to know that if $(t_1, x_1) \in \mathscr{G} - \mathscr{G}(t_0, x_0)$, then there is an f satisfying $(\partial f/\partial t) + L_t f \geqq 0$ on $\mathscr{G}$ (perhaps in a generalized sense) such that $f(t_0, x_0) = \sup f(t, x)$, and $f(t_1, x_1) < f(t_0, x_0)$.

In the case when $a(t, x)$ is positive definite for all (t, x), L. Nirenberg [6] has shown that $\mathscr{G}(t_0, x_0)$ can be taken as the closure in $\mathscr{G}$ of the set of $(t_1, x_1) \in \mathscr{G} \cap ([t_0, \infty) \times R^d)$ such that there exists a continuous map $\phi: [t_0, t_1] \to R^d$ with the properties that $\phi(t_0) = x_0$, $\phi(t_1) = x_1$, and $(t, \phi(t)) \in \mathscr{G}$ for all $t \in (t_0, t_1)$. We will give a probabilistic proof of the Nirenberg maximum principle in Section 3. Moreover, we will also prove there that Nirenberg's $\mathscr{G}(t_0, x_0)$ is maximal in the desired sense.

If a is only nonnegative definite, the problem of finding a suitable maximum principle is more difficult. Results in this direction have been proved by J.-M. Bony [1] and C. D. Hill [3]. Both of these authors employ a modification of the technique originally introduced by E. Hopf for elliptic operators and later adapted by Nirenberg for parabolic ones. The major drawback to Bony's

Results obtained at the Courant Institute of Mathematical Sciences, New York University, this research was sponsored by the U.S. Air Force Office of Scientific Research. Contract AF-49(638)-1719.

work is his restrictive assumptions on the coefficients of L_t. As we will point out later, our own approach has not removed all his restrictions.

Before going into the details, we will conclude this section with an outline of our method. Suppose P_{t_0,x_0} is a probability measure on $C([t_0, \infty), R^d)$ with the properties that $P_{t_0,x_0}(x(t_0) = x_0) = 1$ and

$$\int_{t_0}^{t} \left(\frac{\partial}{\partial u} + L_u\right) f(u, x(u))\, du \tag{1.2}$$

is a martingale for all $f \in C_0^\infty([t_0, \infty) \times R^d)$ (here and in what follows $C_0^\infty(S)$ denotes the space of infinitely differentiable functions having compact support in S). Given an open $\mathscr{G} \subseteq [0, \infty) \times R^d$, let $\tau = \inf\{t \geqq t_0: (t_0, x(t)) \notin \mathscr{G}\}$. Then it is easy to see that $f(t \wedge \tau, x(t \wedge \tau))$ is a submartingale if $f \in C_b^{1,2}(\mathscr{G})$ and $(\partial f/\partial t) + L_t \geqq 0$. (We use $C_b^{1,2}(S)$ to denote the class of bounded f having one bounded continuous t derivative and two bounded continuous x derivatives on S.) Hence,

$$f(t_0, x_0) \leqq E^{P_{t_0,x_0}}[f(t \wedge \tau, x(t \wedge \tau))]. \tag{1.3}$$

In particular, if $f(t_0, x_0) = \sup_{\mathscr{G}} f(t, x)$, then $f(t_1, x_1) = f(t_0, x_0)$ at all (t_1, x_1) for which there exists a path $\phi \in \operatorname{supp}(P_{t_0,x_0})$ such that $\phi(t_1) = x_1$ and $(t, \phi(t)) \in \mathscr{G}$ for $t \in [t_0, t_1]$. Thus, for example, if

$$\operatorname{supp}(P_{t_0,x_0}) = \{\phi \in C([0, \infty), R^d): \phi(t_0) = x_0\}, \tag{1.4}$$

then $(\partial/\partial t) + L_t$ satisfies the Nirenberg maximum principle. What we are going to do is study the measure P_{t_0,x_0} and try to describe its support.

2. Background material

In this section, we discuss diffusions from the point of view introduced in [9]. Our notation throughout is the same as it was in that paper. Namely,

$$\begin{gathered} \Omega = C([0, \infty), R^d), \\ x(t, \omega) = x_t(\omega) \text{ is the position of } \omega \text{ at time } t, \\ \mathscr{M}_t^s = \mathscr{B}[x_u: s \leqq u \leqq t], \qquad \mathscr{M}^s = \mathscr{B}[x_u: u \geqq s]. \end{gathered} \tag{2.1}$$

In order to discuss the weak convergence of measures on $\langle \Omega, \mathscr{M}^s \rangle$, we will sometimes think of a measure on $\langle \Omega, \mathscr{M}^s \rangle$ as defined on $C([s, \infty), R^d)$. A useful criterion for the relative compactness of a set Γ of probability measures P on $\langle \Omega, \mathscr{M}^s \rangle$ is the following:

$$\lim_{R \to \infty} \sup_{P \in \Gamma} P(|x(s)| \geqq R) = 0, \tag{2.2}$$

$$\sup_{P \in \Gamma} E^P[|x(t_2) - x(t_1)|^4] \leqq C_T(t_2 - t_1)^2, \quad s \leqq t_1 \leqq t_2 \leqq T, \quad T > 0. \tag{2.3}$$

A proof of this fact may be found in [7].

A function η on $[s, \infty) \times \Omega$ into a measurable space is said to be s *nonanticipating* if η is $\mathscr{B}_{[s,\infty)} \times \mathscr{M}^s$ measurable and $\eta(t)$ is $\mathscr{M}_t^s$ measurable for each $t \geqq s$. If P is a probability measure on $\langle \Omega, \mathscr{M}^s \rangle$, then η is a *P martingale* if η is complex valued, s nonanticipating, and

$$E^P[\eta(t_2) \mid \mathscr{M}_{t_1}^s] = \eta(t_1) \qquad \text{a.s. } P \tag{2.4}$$

for $s \leqq t_1 < t_2$.

Let $a\colon [0, \infty) \times R^d \to S_d$ and $b\colon [0, \infty) \times R^d \to R^d$ be bounded and measurable. Define

$$L_t = \frac{1}{2} \sum_{i,j=1}^{d} a^{ij}(t, x) + \sum_{i=1}^{d} b_i(t, x). \tag{2.5}$$

A probability measure P on $\langle \Omega, \mathscr{M}^{t_0} \rangle$ is said to solve the *martingale problem for L_t starting at (t_0, x_0) if $P(x(t_0) = x_0) = 1$ and $f(x(t)) - \int_{t_0}^t L_s f(x(u))\, du$ is a P* martingale for all $f \in C_0^\infty(R^d)$. In [9] it was shown that if a is continuous and $a(t, x)$ is positive definite for all (t, x), then there is exactly one solution $P_{t,x}$ to the martingale problem for L_t starting at (t, x). Moreover, we proved there that the family $\{P_{t,x}\colon (t, x) \in [0, \infty) \times R^d\}$ forms a strong Feller, strong Markov process. The purpose of the present section is to extend this result to the case when a and b are smooth in x and $a(t, x)$ is only nonnegative definite. The idea is to reduce the martingale problem to a stochastic differential equation which can be solved by the techniques introduced by K. Itô in [4]. For purposes of easy reference, we state here the following theorem, whose proof may be found in [10].

THEOREM 2.1. *Let $a\colon [t_0, \infty) \times \Omega \to S_d$ and $b\colon [t_0, \infty) \times \Omega \to R^d$ be bounded t_0 nonanticipating functions and define*

$$L_t = \frac{1}{2} \sum_{i,j=1}^{d} a^{ij}(t) \frac{\partial^2}{\partial x_i \partial x_j} + \sum_{i=1}^{d} b_i(t) \frac{\partial}{\partial x_i}. \tag{2.6}$$

Suppose $\alpha\colon [t_0, \infty) \times \Omega \to R^d$ is a continuous (in time) t_0 nonanticipating function and that P is a probability measure on $\langle \Omega, \mathscr{M}^{t_0} \rangle$. Then the following are equivalent:

(i) *$f(\alpha(t)) - \int_{t_0}^t L_u f(\alpha(u))\, du$ is a P martingale for all $f \in C_0^\infty(R^d)$;*

(ii) *$f(t, \alpha(t)) - \int_{t_0}^t [(\partial/\partial u) + L_u] f(u, \alpha(u))\, du$ is a P martingale for all $f \in C_b^{1,2}([t_0, \infty) \times R^d)$;*

(iii) *$\exp\{\langle \theta, \alpha(t) - \alpha(t_0) - \int_{t_0}^t b(u)\, du \rangle - \frac{1}{2}\int_{t_0}^t \langle \theta, a(u)\, \theta \rangle\, du\}$ is a P martingale for all $\theta \in R^d$.*

Moreover, if P satisfies one of these and if $\bar{\alpha}(t) - \int_0^t b(u)\, du$, then $d\bar{\alpha}(t)$ stochastic integrals $\int_{t_0}^t \langle \theta(u), d\bar{\alpha}(u) \rangle$ can be defined when the integrand $\theta\colon [t_0, \infty) \times \Omega \to R^d$ is t_0 nonanticipating and satisfies

$$E^P\left[\int_{t_0}^t \langle \theta(u), a(u)\theta(u) \rangle\, du\right] < \infty, \qquad t \geqq t_0. \tag{2.7}$$

The process $\int_{t_0}^t \langle\theta(u), d\bar{\alpha}(u)\rangle$ *is a continuous P martingale; and if* $\langle\theta(u), a(u)\theta(u)\rangle$ *is bounded, then*

$$\exp\left\{\int_{t_0}^t \langle\theta(u), d\bar{\alpha}(u)\rangle - \frac{1}{2}\int_{t_0}^t \langle\theta(u), a(u)\theta(u)\rangle\, du\right\} \tag{2.8}$$

is a P martingale. Finally, for $f \in C_b^{1,2}([t_0, \infty) \times R^d)$, *one has*

$$f(t, \alpha(t)) - f(t_0, \alpha(t_0)) = \int_{t_0}^t \langle\nabla_x f(u, \alpha(u)), d\bar{\alpha}(u)\rangle + \int_{t_0}^t \left(\frac{\partial}{\partial u} + L_u\right) f(u, \alpha(u))\, du, \tag{2.9}$$

where $\nabla_x f(t, x) = [(\partial f/\partial x_1)(t, x), \cdots, (\partial f/\partial x_d)(t, x)]$.

In [9], we showed that if the a in Theorem 2.1 is uniformly positive definite, then $\beta(t) = \int_{t_0}^t a^{-1/2}(u)\, d\bar{\alpha}(u)$ is a P Brownian motion (that is, $P(\beta(t_0) = 0) = 1$ and $\exp\{\langle\theta, \beta(t)\rangle - \frac{1}{2}|\theta|^2(t - t_0)\}$ is a P martingale for all $\theta \in R^d$), where $a^{1/2}$ is the positive definite, symmetric square root of a. Hence,

$$\alpha(t) - \alpha(t_0) = \int_{t_0}^t a^{1/2}(u)\, d\beta(u) + \int_{t_0}^t b(u)\, du \quad \text{a.s. } P \tag{2.10}$$

for some P Brownian motion β. We will now extend this result to nonnegative definite a. As we will see, this entails enlarging the sample space.

THEOREM 2.2. *Let a, b, and α be as in Theorem 2.1 and suppose P is a probability measure of* $\langle\Omega, \mathscr{M}^{t_0}\rangle$ *satisfying one of the conditions* (i), (ii), *or* (iii) *of that theorem. Assume* $\sigma\colon [t_0, \infty) \times \Omega \to R^d \times R^d$ *is t_0 nonanticipating and satisfies* $a(u) = \sigma(u)\sigma^*(u)$. *Then there is an extension* $\hat{P}$ *of P to* $\langle\Omega \times \Omega, \mathscr{M}^{t_0} \times \mathscr{M}^{t_0}\rangle$ *and a* $\hat{P}$ *Brownian motion* $\hat{\beta}$ *such that*

$$\alpha(t) - \alpha(t_0) = \int_{t_0}^t \sigma(u)\, d\hat{\beta}(u) + \int_{t_0}^t b(u)\, du \quad \text{a.s. } \hat{P}. \tag{2.11}$$

PROOF. It suffices to treat the case when $\sigma = a^{1/2}$. Indeed, if $a = \sigma\sigma^*$ and $U = (a + \varepsilon I)^{-1/2}\sigma$, then $U_\varepsilon \to U_0$ as $\varepsilon \downarrow 0$, where U_0 is an orthogonal transformation such that $a^{1/2} = U_0\sigma$. Hence, if

$$\alpha(t) - \alpha(t_0) = \int_{t_0}^t a^{1/2}(u)\, d\hat{\beta}(u) + \int_{t_0}^t b(u)\, du \quad \text{a.s. } \hat{P}, \tag{2.12}$$

then

$$\alpha(t) - \alpha(t_0) = \int_{t_0}^t \sigma(u)\, d\hat{\hat{\beta}}(u) + \int_{t_0}^t b(u)\, du \quad \text{a.s. } \hat{P}, \tag{2.13}$$

where $\hat{\hat{\beta}}(t) = \int_{t_0}^t U_0^*(u)\, d\hat{\beta}(u)$ is again a $\hat{P}$ Brownian motion.

To prove the assertion when $\sigma = a^{1/2}$, define $\tilde{a}(u) = \lim_{\varepsilon\downarrow 0} a^{1/2}(u)(a(u) + \varepsilon I)^{-1}$. Then $\tilde{a}(u)a^{1/2}(u) = a^{1/2}(u)\tilde{a}(u) = E_R(u)$, where $E_R(u)$ is the orthogonal projec-

tion onto the range of $a(u)$. Let $\beta(t)$ be a W Brownian motion on $\langle \Omega, \mathscr{M}^{t_0}\rangle$ and define $\hat{P} = P \times W$. Then, by Theorem 2.1,

$$\hat{\beta}(t) = \int_{t_0}^{t} \tilde{a}(u)\, d\bar{\alpha}(u) + \int_{t_0}^{t} E_N(u)\, d\beta(u) \tag{2.14}$$

is a $\hat{P}$ Brownian motion, where $E_N(u) = I - E_R(u)$. Moreover,

$$\begin{aligned}\int_{t_0}^{t} a^{1/2}(u)\, d\hat{\beta}(u) &= \int_{t_0}^{t} E_R(u)\, d\bar{\alpha}(u) + \int_{t_0}^{t} a^{1/2}(u) E_N(u)\, d\beta(u) \\ &= \int_{t_0}^{t} E_R(u)\, d\bar{\alpha}(u),\end{aligned} \tag{2.15}$$

and

$$\begin{aligned} &E^{\hat{P}}\left[\left|\bar{\alpha}(t) - \bar{\alpha}(t_0) - \int_{t_0}^{t} E_R(u)\, d\bar{\alpha}(u)\right|^2\right] \\ &= E^{\hat{P}}\left[\left|\int_{t_0}^{t} E_N(u)\, d\bar{\alpha}(u)\right|^2\right] \\ &= E^{\hat{P}}\left[\left|\int_{t_0}^{t} \operatorname{tr}\left(E_N(u) a(u) E_N(u)\right) du\right|\right] = 0,\end{aligned} \tag{2.16}$$

where tr means trace. *Q.E.D.*

Theorem 2.2 is the multidimensional analogue of Theorem 5.3 in J. L. Doob's book [2].

THEOREM 2.3. *Let $a\colon [0, \infty) \times R^d \to S_d$ and $b\colon [0, \infty) \times R^d \to R^d$ be bounded measurable functions. Assume that there is a bounded measurable $\sigma\colon [0, \infty) \times R^d \to S_d$ such that $a = \sigma\sigma^*$ and*

$$\sup_{0 \leq t \leq T, |x| + |y| \leq R} \left(\|\sigma(t, x) - \sigma(t, y)\| + |b(t, x) - b(t, y)|\right) \leq C(T, R)|x - y| \tag{2.17}$$

for all $T, R > 0$. Then for each (t_0, x_0) there is exactly one solution P_{t_0, x_0} to the martingale problem for

$$L_t = \frac{1}{2}\sum_{i,j=1}^{d} a^{ij}(t, x)\frac{\partial^2}{\partial x_i \partial x_j} + \sum_{i=1}^{d} b_i(t, x)\frac{\partial}{\partial x_i} \tag{2.18}$$

starting at (t_0, x_0). Moreover, the family $\{P_{t_0, x_0}\colon (t_0, x_0) \in [0, \infty) \times R^d\}$ forms a Feller continuous, strong Markov process.

PROOF. We will only prove the first assertion, since the second one follows by standard methods used in [9]. Moreover, we will restrict ourselves to the case when $C(T, R)$ is independent of T and R, because the general case can then be handled by the techniques employed in Theorem 5.6 of [9].

To prove existence, let β be a W Brownian motion on $\langle \Omega, \mathcal{M}^{t_0} \rangle$. Define $\xi_0(t) \equiv x_0$ and

$$\xi_{n+1}(t) = x_0 + \int_{t_0}^{t} \sigma(u, \xi_n(u))\, d\beta(u) + \int_{t_0}^{t} b(u, \xi_n(u))\, du. \tag{2.19}$$

Following H. P. McKean [5], it is easy to show that

$$W(\sup_{t_0 \leqq t \leqq T} |\xi_n(t) - \xi(t)| > \varepsilon) \to 0 \tag{2.20}$$

for all $T > t_0$, where

$$\xi(t) = x_0 + \int_{t_0}^{t} \sigma(u, \xi(u))\, d\beta(u) + \int_{t_0}^{t} b(u, \xi(u))\, du. \tag{2.21}$$

Letting P be the distribution of $\xi(t)$ (that is, P is defined on $\langle \Omega, \mathcal{M}^{t_0} \rangle$ by

$$P(x(t_1) \in \Gamma_1, \cdots, x(t_n) \in \Gamma_n) = W(\xi(t_1) \in \Gamma_1, \cdots, \xi(t_n) \in \Gamma_n), \tag{2.22}$$

and using Theorem 2.1, one sees that P solves the martingale problem for L_t starting at (t_0, x_0).

Turning to uniqueness, suppose that P is a solution. Choose $\hat{P}$ and $\hat{\beta}(t)$ as in Theorem 2.2. Using the technique of the preceding paragraph, we can find $\xi_n(t)$ such that

$$\xi_{n+1}(t) = x_0 + \int_{t_0}^{t} a^{1/2}(u, \xi_n(u))\, d\hat{\beta}(u) + \int_{t_0}^{t} b(u, \xi_n(u))\, du, \tag{2.23}$$

and $\xi_n(t) \to x(t)$ in probability uniformly on finite intervals. Because the distribution of each $\xi_n(t)$ is uniquely determined, P is unique. *Q.E.D.*

We next state another theorem for reference purposes. Its proof can be found in [9].

THEOREM 2.4 (Cameron-Martin). *Let $\beta(t)$ be a W Brownian motion on $\langle \Omega, \mathcal{M}^{t_0} \rangle$ and suppose $c: [t_0, \infty) \times \Omega \twoheadrightarrow R^d$ is bounded and t_0 nonanticipating. Then*

$$R(t) = \exp\left\{ \int_{t_0}^{t} \langle c(u), d\beta(u) \rangle - \frac{1}{2} \int_{t_0}^{t} |c(u)|^2\, du \right\} \tag{2.24}$$

is a W martingale. In particular, there is a unique probability measure Q on $\langle \Omega, \mathcal{M}^{t_0} \rangle$ such that

$$\frac{dQ}{dW} = R(t) \text{ on } \mathcal{M}_t^{t_0}, \qquad t \geqq t_0. \tag{2.25}$$

Moreover, $\bar{\beta}(t) = \beta(t) - \int_{t_0}^{t} c(u)\, du$ is a Brownian motion.

COROLLARY 2.1. *Let a, b, L_t, and σ be as in Theorem 2.3, and let $c: [0, \infty) \times R^d \twoheadrightarrow R^d$ be bounded and measurable. Suppose P_{t_0, x_0} is the solution to the martingale problem for L_t starting at (t_0, x_0), and choose $\hat{P}_{t_0, x_0}$ and $\hat{\beta}(t)$ as in Theorem 2.2 so that*

$$(2.26)\qquad x(t) = x_0 + \int_{t_0}^{t} \sigma(u, x(u))\, d\hat{\beta}(u) + \int_{t_0}^{t} b(u, x(u))\, du \qquad a.s.\ \hat{\Gamma}_{t_0,x_0}.$$

Define

$$(2.27)\qquad R(t) = E^{P_{t_0,x_0}}\left[\exp\left\{\int_{t_0}^{t} \langle c(u, x(u)), d\hat{\beta}(u)\rangle - \frac{1}{2}\int_{t_0}^{t} |c(u)|^2\, du\right\}\middle|\, \mathcal{M}_t^{t_0}\right],$$

and determine Q_{t_0,x_0} *by the relations*

$$(2.28)\qquad \frac{dQ_{t_0,x_0}}{dP_{t_0,x_0}} = R(t) \text{ on } \mathcal{M}_t^{t_0}, \qquad t \geqq t_0.$$

Then Q_{t_0,x_0} *is the only solution to the martingale problem for*

$$(2.29)\qquad L_t^c = \frac{1}{2}\sum_{i,j=1}^{d} a^{ij}(t, x)\frac{\partial^2}{\partial x_i \partial x_j} + \sum_{i=1}^{d} (b + \sigma c)_i(t, x)\frac{\partial}{\partial x_i}$$

starting at (t_0, x_0). *Moreover, the family* $\{Q_{t,x}\colon (t, x) \in [0, \infty) \times R^d\}$ *forms a strong Markov process which is Feller continuous when* c *is continuous in* x.

Remark 2.1. If $a\colon [0, \infty) \times R^d \to S_d$ is bounded and measurable, then

$$(2.30)\qquad \sup_{t \leqq T, |x| + |y| \leqq R} \|a^{1/2}(t, x) - a^{1/2}(t, y)\| \leqq C(T, R)|x - y|,$$

if $a(t, x)$ is twice continuous differentiable in x and

$$(2.31)\qquad \max_{1 \leqq i,j \leqq d}\ \sup_{t \leqq T, |x| + |y| \leqq R} \left\|\frac{\partial^2 a}{\partial x_i \partial x_j}(t, x)\right\| \leqq C(T, R).$$

This fact is proved by R. S. Philips and L. Sarason [8].

Remark 2.2. Unfortunately there is no nice criterion on $a\colon [0, \infty) \times R^d \to S_d$ which guarantees the existence of a smooth $\sigma\colon [0, \infty) \times R^d \to R^d \otimes R^d$ such that $a = \sigma\sigma^*$. Nonetheless, we will often assume in what follows that L_t can be written in the form

$$(2.32)\qquad L_t = \tfrac{1}{2}\sigma^*\nabla_x \cdot \sigma^*\nabla_x + b \cdot \nabla_x,$$

where

$$(2.33)\qquad \sigma^*\nabla_x \cdot \sigma^*\nabla_x = \sum_{i,j=1}^{d} \sigma^{i\ell}\frac{\partial}{\partial x_i}\left(\sigma^{j\ell}\frac{\partial}{\partial x_j}\right).$$

(Here, and in what follows, repeated indices are summed.) Notice that (2.32) can be written as

$$(2.34)\qquad L_t = \frac{1}{2}\sum_{i,j=1}^{d} a^{ij}\frac{\partial^2}{\partial x_i \partial x_j} + \sum_{i=1}^{d}\left(b + \frac{1}{2}\sigma'\sigma\right)_i \frac{\partial}{\partial x_i},$$

where $a = \sigma\sigma^*$ and the vector $\sigma'\sigma$ is defined by

$$(2.35)\qquad (\sigma'\sigma)_i = \sigma^{ij}_{,\ell}\sigma^{\ell j}.$$

(We have used here, and will continue to use, the notation $f_{,\ell}$ to stand for $\partial f/\partial x_\ell$.) What specific assumptions are made about the smoothness σ will depend on our immediate needs. But in any case, it will be necessary to assume that σ is once differentiable in x in order to even define $\sigma^*\nabla_x \cdot \sigma^*\nabla_x$.

3. The nondegenerate case

Throughout this section $a: [0, \infty) \times R^d \to S_d$ will be bounded, continuous, positive definite valued function and $b: [0, \infty) \times R^d \to R^d$ will be bounded and measurable. We will use P_{t_0, x_0} to denote the unique solution to the martingale problem for

$$L_t = \frac{1}{2} \sum_{i,j=1}^{d} a^{ij}(t, x) \frac{\partial^2}{\partial x_i \partial x_j} + \sum_{i=1}^{d} b_i(t, x) \frac{\partial}{\partial x_i} \tag{3.1}$$

starting at (t_0, x_0), and $P^0_{t_0, x_0}$ will denote the unique solution for

$$L^0_t = \frac{1}{2} \sum_{i,j=1}^{d} a^{ij}(t, x) \frac{\partial^2}{\partial x_i \partial x_j} \tag{3.2}$$

starting at (t_0, x_0). Our aim is to prove that

$$\operatorname{supp}(P_{t_0, x_0}) = \Omega(t_0, x_0), \tag{3.3}$$

where $\Omega(t_0, x_0)$ is the set of $\omega \in \Omega$ such that $x(t_0, \omega) = x_0$.

Before proceeding, we make two simplifying observations. First, since P_{t_0, x_0} and $P^0_{t_0, x_0}$ are equivalent (that is, mutually absolutely continuous) on $\mathscr{M}^{t_0}_t$ for all $t \geqq t_0$, it suffices to work with $P^0_{t_0, x_0}$. Second, by an obvious transformation, we can always assume that $t_0 = 0$ and $x_0 = 0$. Thus, what we need to show is that

$$\operatorname{supp}(P_0) = \Omega_0, \tag{3.4}$$

where $P_0 = P^0_{0,0}$ and $\Omega_0 = \Omega(0, 0)$.

LEMMA 3.1. *Let $\phi: [0, \infty) \to R^d$ be once continuously differentiable such that $\phi(0) = 0$. Then for all $T > 0$ and $\varepsilon > 0$, $P_0(\|x(t) - \phi(t)\|^0_T < \varepsilon) > 0$, where $\|\cdot\|^s_T$ denotes the sup norm on the interval $[s, T]$.*

PROOF. Let $\psi(t) = \chi_{[0,T]}(t)\dot{\phi}(t)$ and define Q_0 by

$$\frac{dQ_0}{dP_0} = R(t) = \exp\left\{\int_0^t \langle a^{-1}(u, x(u)), \psi(u)\, dx(u)\rangle - \frac{1}{2}\int_0^t \langle \psi(u), a^{-1}(u, x(u)), \psi(u)\rangle\, du\right\} \tag{3.5}$$

on $\mathscr{M}^0_t$, $t \geqq 0$. Then Q_0 is the unique solution to the martingale problem for

$$L^\psi_t = \frac{1}{2} \sum_{i,j=1}^{d} a^{ij}(t, x) \frac{\partial^2}{\partial x_i \partial x_j} + \sum_{i=1}^{d} \psi_i(t) \frac{\partial}{\partial x_i} \tag{3.6}$$

starting at (0, 0). In particular, there is a Q_0 Brownian motion $\beta(t)$ such that

$$\bar{x}(t) = \int_0^t a^{1/2}(u, x(u))\, d\beta(u), \tag{3.7}$$

where $\bar{x}(t) = x(t) - \int_0^t \psi(u)\, du = x(t) - \phi(t \wedge T)$. Hence, by Theorem 1 in [11], $Q_0(\|\bar{x}(t)\|_T^0 < \varepsilon) > 0$ for all $\varepsilon > 0$. Since Q_0 and P_0 are equivalent on $\mathscr{M}_T^0$, this implies $P_0(\|x(t) - \phi(t)\|_T^0 < \varepsilon) > 0$ for all $\varepsilon > 0$. *Q.E.D.*

Using Lemma 3.1, we see that supp(P_0) contains all differentiable paths which start at 0. Because these are dense in Ω_0, equation (3.4) is now proved. We have therefore proved the following theorem.

THEOREM 3.1. *If P_{t_0,x_0} is the solution to the martingale problem for L_t starting at (t_0, x_0), then equation* (3.3) *holds.*

As we saw in Section 1, equation (3.3) implies Nirenberg's strong maximum principle. In fact, it implies more. Let $\mathscr{G}$ be an open set in $[0, \infty) \times R^d$ and let $(t_0, x_0) \in \mathscr{G}$. Define $\mathscr{H}_{L_t}^-(t_0, x_0, \mathscr{G})$ to be the set of $f: \mathscr{G} \to R \cup \{-\infty\}$ which are upper semicontinuous, bounded above, and have the property that $f(t \wedge \tau, x(t \wedge \tau))$ is a P_{t_0,x_0} submartingale, that is,

$$f(t_1 \wedge \tau, x(t_1 \wedge \tau)) \leqq E^{P_{t_0,x_0}}[f(t_2 \wedge \tau, x(t_2 \wedge \tau)) | \mathscr{M}_{t_1 \wedge \tau}^{t_0}], \quad t_0 \leqq t_1 \leqq t_2, \tag{3.8}$$

where $\tau = \inf\{t \geqq t_0: (t, x(t)) \notin \mathscr{G}\}$. Note that $\mathscr{H}_{L_t}^-(t_0, x_0, \mathscr{G})$ is closed under nonincreasing limits, multiplication by nonnegative constants, and maximums (that is, if $f_1, f_2 \in \mathscr{H}_{L_t}^-(t_0, x_0, \mathscr{G})$, then so is $f_1 \vee f_2$). Next define (t_0, x_0) to be the closure in $\mathscr{G}$ of the set of $(t_1, x_1) \in \mathscr{G} \cap ([t_0, \infty) \times R^d)$ such that $x_0 = \phi(t_0)$ and $x_1 = \phi(t_1)$ for some $\phi \in C([t_0, t_1], R^d)$ satisfying $(t, \phi(t)) \in \mathscr{G}$, $t \in [t_0, t_1]$. Observe that if $f \in C_b^{1,2}(\mathscr{G})$ and $((\partial/\partial t) + L_t)f \geqq 0$ on $\mathscr{G}(t_0, x_0)$, then $f \in \mathscr{H}_{L_t}^-(t_0, x_0)$. Finally, define

$$\mathscr{H}_{L_t}^-(\mathscr{G}) = \bigcap_{\mathscr{G}} \mathscr{H}_{L_t}^-(t_0, x_0, \mathscr{G}). \tag{3.9}$$

THEOREM 3.2. *If $f \in \mathscr{H}_{L_t}^-(t_0, x_0, \mathscr{G})$ and $f(t_0, x_0) = \sup f(t, x)$, then $f \equiv f(t_0, x_0)$ on $\mathscr{G}(t_0, x_0)$. Moreover, if $(t_1, x_1) \in \mathscr{G} - \mathscr{G}(t_0, x_0)$, then there is an $f \in \mathscr{H}_{L_t}^-(\mathscr{G})$ such that $f(t_0, x_0) = \sup f(t, x)$ and $f(t_1, x_1) < f(t_0, x_0)$.*

PROOF. The first assertion follows from the argument given in Section 1. To prove the second assertion, let $(t_1, x_1) \in \mathscr{G} - \mathscr{G}(t_0, x_0)$. If $t_1 < t_0$, take $f(t, x) = t \wedge t_0$. If $t_1 \geqq t_0$, choose an open neighborhood N of (t_1, x_1) such that $N \subseteqq \mathscr{G}$ and $N \cap \mathscr{G}(t_0, x_0) = \phi$. Let $h \in C_0^\infty(N)$ such that $-1 \leqq h \leqq 0$ and $h(t_1, x_1) = -1$. Define

$$f(t, x) = E^{P_{t,x}}\left[\int_t^{\tau_t} e^{-u} h(u, x(u))\, du\right], \tag{3.10}$$

where $\tau_t = \inf\{u \geqq t: (u, x(u)) \notin \mathscr{G}\}$. Since $\{P_{t,x}: (t, x) \in [0, \infty) \times R^d\}$ is strongly Feller continuous, f is continuous. Moreover,

$$(3.11) \qquad E^{P_{s,x}}[f(t \wedge \tau_s, x(t \wedge \tau_s))] = E^{P_{s,x}}\left[\chi_{\tau_s > t} \int_t^{\tau_t} e^{-u} h(u, x(u))\, du\right] \geqq f(s, x),$$

for $(s, x) \in \mathscr{G}$ and $t \geqq s$; and therefore

$$(3.12) \qquad \begin{aligned} E^{P_{s,x}}[f(t_2 \wedge \tau_s, x(t_2 \wedge \tau_s)) | \mathscr{M}^s_{t_1 \wedge \tau_s}] &= E^{P_{t_1 \wedge \tau_s, x(t_1 \wedge \tau_s)}}[f(t_2 \wedge \tau_{t_1}, x(t_2 \wedge \tau_{t_1}))] \\ &\geqq f(t_1 \wedge \tau_3, x(t_1 \wedge \tau_s)) \end{aligned}$$

for $s \leqq t_1 \leqq t_2$. This proves that $f \in \mathscr{H}^-_{L_t}(\mathscr{G})$. Clearly, $f \leqq 0$, $f(t_0, x_0) = 0$, and $f(t_1, x_1) < 0$. *Q.E.D.*

REMARK 3.1. It is important to know in what sense $\mathscr{H}^-_{L_t}(\mathscr{G})$ is an extension of the class of $f \in C^{1,2}_b(\mathscr{G})$ satisfying $(\partial f/\partial t) + L_t f \geqq 0$ on $\mathscr{G}$. Using the estimates obtained in [9], one can show that $\mathscr{H}^-_{L_t}(\mathscr{G})$ contains the class of $f \in W^{1,2}_p(\mathscr{G})$ (see [9] for the definition of $W^{1,2}$) satisfying $(\partial f/\partial t) + L_t f \geqq 0$ when $p > (d + 2)/2$. To give a complete analytic description of $\mathscr{H}^-_{L_t}(\mathscr{G})$, consider the transition function $\hat{P}(t, (s, x), \cdot)$ defined by

$$(3.13) \qquad \hat{P}(t, (s, x), \Delta \times \Gamma) = \chi_\Delta(s + t) P_{s,t}(x(s + t) \wedge \tau_s) \in \Gamma)$$

for $\Delta \in \mathscr{B}_{[0,\infty)}$ and $\Gamma \in \mathscr{B}[\bar{\mathscr{G}}]$, where $\tau_s = \inf\{t \geqq s: (t, x(t)) \notin \mathscr{G}\}$. It is easy to see that

$$(3.14) \qquad \hat{P}(t_1 + t_2, (s, x), \cdot) = \int_0^\infty \int_{\mathscr{G}} \hat{P}(t_1, (s, x), du \times dy) \hat{P}(t_2, (u, y), \cdot).$$

Thus, we can define a Markov semigroup $\{\hat{T}_t\}_{t \geqq 0}$ on $\mathscr{B}(\mathscr{G})$ by setting

$$(3.15) \qquad \hat{T}_t f(s, x) = \int_0^\infty \int_{\mathscr{G}} \hat{P}(t, (s, x), du \times dy) f(u, y).$$

Furthermore, $\{\hat{T}_t\}_{t \geqq 0}$ is the only Markov semigroup having the property that

$$(3.16) \qquad \hat{T}_t f(s, x) - f(s, x) = \int_0^t \left(\hat{T}_u\left[\chi_{\mathscr{G}} \cdot \left(\frac{\partial f}{\partial u} + L_u f\right)\right]\right)(u, x)\, du$$

for all $f \in C^{1,2}_b(\bar{\mathscr{G}})$. Finally, $\mathscr{H}^-_{L_t}(\mathscr{G})$ coincides with the class of f on $\bar{\mathscr{G}}$ such that f is bounded above and $\hat{T}_t f(s, x) \downarrow f(s, x)$ for $(s, x) \in \mathscr{G}$. When $L_t = \frac{1}{2}\Delta$, it is easy to see that $\mathscr{H}^-_{L_t}(\mathscr{G})$ is just the class of subparabolic functions on $\mathscr{G}$.

4. The degenerate case, part I

Let $\sigma: [0, \infty) \times R^d \twoheadrightarrow R^d \otimes R^d$ and $b: [0, \infty) \times R^d \twoheadrightarrow R^d$ be bounded measurable functions. In this section, we will assume that b and the first spatial

derivatives of σ are uniformly Lipschitz continuous in x. From σ and b, we form the operator

$$L_t = \tfrac{1}{2}\sigma^*\nabla_x \cdot \sigma^*\nabla_x + b \cdot \nabla_x. \tag{4.1}$$

by the prescription given in Remark 2.2. Under the above assumptions, we know from Theorem 2.3 that there is exactly one solution P_{t_0,x_0} to the martingale problem for L_t starting at (t_0, x_0). The purpose of this section is to prove that

$$\operatorname{supp}(P_{t_0,x_0}) \subseteqq \overline{\mathscr{S}_{\sigma,b}(t_0, x_0)}, \tag{4.2}$$

where $\overline{\mathscr{S}_{\sigma,b}(t_0, x_0)}$ is the class of $\phi \in C([0, \infty), R^d)$ for which there exists a piecewise constant $\psi: [t_0, \infty) \to R^d$ such that

$$\phi(t) = x_0 + \int_{t_0}^t \sigma(u, \phi(u))\psi(u)\,du + \int_{t_0}^t b(u, \phi(u))\,du, \qquad t \geqq t_0. \tag{4.3}$$

Clearly, it suffices to treat the case when $t_0 = 0$ and $x_0 = 0$.

Let $\beta(t)$ be a W Brownian motion on $\langle \Omega, \mathscr{M}^0 \rangle$. Given $n \geqq 0$, define $t_n = [2^n t]/(2^n)$, $t_n^+ = ([2^n t] + 1)/2^n$, and

$$\beta^{(n)}(t) = 2^n(\beta(t_n^+) - \beta(t_n)). \tag{4.4}$$

Next, let $\xi^{(n)}(t)$ be the stochastic process determined by the ordinary integral equation

$$\xi^{(n)}(t) = \int_0^t \sigma(u, \xi^{(n)}(u))\beta^{(n)}(u)\,du + \int_0^t b(u, \xi^{(n)}(u))\,du; \tag{4.5}$$

and denote by P_n the distribution of $\xi^{(n)}(t)$. Clearly, $\operatorname{supp}(P_n) \subseteqq \overline{\mathscr{S}_{\sigma,b}(0, 0)}$ for all $n \geqq 0$. Hence, if we show that P_n tends weakly to $P_{0,0}$ as $n \to \infty$, then it will follow that $\operatorname{supp}(P_{0,0}) \subseteqq \overline{\mathscr{S}_{\sigma,b}(0, 0)}$. Thus, we must prove that $P_n \Rightarrow P_{0,0}$. Results of this sort are familiar in various branches of applied mathematics when $d = 1$ (see E. Wong and M. Zakai [12]). However, to the best of our knowledge, the proof which follows is the first complete one for $d > 1$.

The procedure which we will use consists of two steps. The first of these is to prove that $\{P_n\}_{n\geqq 1}$ is relatively weakly compact. Once this is done, we will then show that every convergent subsequence of $\{P_n\}_{n\geqq 1}$ converges to a solution of the martingale problem for L_t starting at $(0, 0)$. For convenience in writing, we will use the following notation:

$$\eta^{(n)}(t) = \int_0^t \sigma(u, \xi^{(n)}(u))\beta^{(n)}(u)\,du, \tag{4.6}$$

$$\alpha^{(n)}(t) = \sigma\left(t, \xi^{(n)}\left(\frac{[2^n t]}{2^n}\right)\right)\beta^{(n)}(t), \tag{4.7}$$

$$(\sigma'\sigma)_i^{\ell,\ell'}(t, x) = \left(\frac{\partial}{\partial x_j}\sigma^{i\ell}(t, x)\right)\sigma^{j\ell'}(t, x), \tag{4.8}$$

$$\Delta_k^{(n)} = \beta\left(\frac{k+1}{2^n}\right) - \beta\left(\frac{k}{2^n}\right). \tag{4.9}$$

LEMMA 4.1. *The set $\{P_n\}_{n\geqq 1}$ is relatively weakly compact.*

PROOF. It suffices to prove that

$$\sup_n E^{P_n}[|x(t) - x(s)|^4] \leqq C_T|t_2 - s|^2, \qquad 0 \leqq s \leqq t \leqq T, \qquad T > 0. \tag{4.10}$$

To do this, first observe that

$$\begin{aligned} &E^{P_n}|x(t) - x(s)|^4 \\ &\quad \leqq 8\left(E^W[|\eta^{(n)}(t) - \eta^{(n)}(s)|^4] + E^W\left[\left|\int_s^t b(u, \xi^{(n)}(u))\,du\right|^4\right]\right). \end{aligned} \tag{4.11}$$

Hence, it suffices to examine

$$E^W[|\eta^{(n)}(t) - \eta^{(n)}(s)|^4]. \tag{4.12}$$

But

$$\begin{aligned} &\eta^{(n)}(t) - \eta^{(n)}(s) \\ &\quad = \int_s^t \alpha^{(n)}(u)\,du + \int_s^t du \int_{u_n}^u dv(\sigma'\sigma)^{\ell,\ell'}(u, \xi^{(n)}(v))\dot\beta_\ell^{(n)}(v)\dot\beta_{\ell'}^{(n)}(v), \end{aligned} \tag{4.13}$$

and

$$\begin{aligned} E^W\left[\left|\int_s^t \alpha^{(n)}(u)\,du\right|^4\right] &= E^W\left[\left|\int_s^t \sigma(u, \xi^{(n)}(u_n))\dot\beta^{(n)}(u)\,du\right|^4\right] \\ &= E^W\left[\left|\int_{s_n}^{t_n} \sigma^{(n)}(u)\,d\beta(u)\right|^4\right] \leqq C_1(t - s)^2, \end{aligned} \tag{4.14}$$

where

$$\sigma^{(n)}(u) = 2^n \int_{u_n \vee s}^{u_n^+ \wedge t} \sigma(v, \xi^{(n)}(u_n))\,dv. \tag{4.15}$$

Finally,

$$\begin{aligned} &E^W\left[\left|\int_s^t du \int_{u_n}^u dv(\sigma'\sigma)^{\ell,\ell'}(u, \xi^{(n)}(v))\dot\beta_\ell^{(n)}(v)\dot\beta_{\ell'}^{(n)}(v)\right|^4\right] \\ &\quad \leqq (t - s)^3 E^W\left[\int_s^t du \left|\int_{u_n}^u dv(\sigma'\sigma)^{\ell,\ell'}(u, \xi^{(n)}(v))\dot\beta_\ell^{(n)}(v)\dot\beta_{\ell'}^{(n)}(v)\right|^4\right] \\ &\quad \leqq C_2(t - s)^3 \sum_{k=[2^n s]}^{[2^n t]} 2^{8n} \int_{k/2^n}^{k+1/2^n} du\left(u - \frac{k}{2^n}\right)^4 E^W[|\Delta_k^{(n)}|^8] \\ &\quad \leqq C_3(t - s)^3. \end{aligned} \tag{4.16}$$

Q.E.D.

We now have to show that if P is the limit of a convergent subsequence of $\{P_n\}$, then

$$E^P[F\cdot(f(x(t))-f(x(s)))]=E^P\left[F\cdot\int_s^t L_u f(x(u))\,du\right] \tag{4.17}$$

for all $f\in C_0^\infty(R^d)$, $0\leqq s<t$, and bounded $\mathscr{M}_s^0$ measurable $F:\Omega\twoheadrightarrow R$. Clearly, it will suffice to do this when s and t have the form $k/2^N$ and F is continuous as well as bounded and $\mathscr{M}_s^0$ measurable. For the sake of convenience, we will use $\{P_n\}_{n\geqq 1}$ to denote the subsequence which converges to P. Observe that

$$\begin{aligned} E^{P_n}[F\cdot(f(x(t))-f(x(s)))] &= E^{P_n}\left[F\cdot\int_s^t\langle\nabla_x f(x(u)),b(u,x(u))\rangle\,du\right]\\ &\quad+E^W\left[F\cdot\int_s^t\langle\nabla_x f(\xi^{(n)}(u)),\alpha^{(n)}(u)\rangle\,du\right]\\ &\quad+E^W\left[F\cdot\int_s^t\langle\nabla_x f(\xi^{(n)}(u)),\dot\eta^{(n)}(u)-\alpha^{(n)}(u)\rangle\,du\right]\\ &=I_1^{(n)}+I_2^{(n)}+I_3^{(n)}. \end{aligned} \tag{4.18}$$

Obviously, $I_1^{(n)}\to E^P[F\cdot\int_s^t\langle\nabla_x f(x(u)),b(u,x(u))\rangle\,du]$. Before examining the limiting behavior of the other two terms, we need the following simple lemma.

LEMMA 4.2. *If ϕ and ψ are bounded measurable functions on $[s,t]$, then*

$$2^n\int_s^t du\,\phi(u)\int_{u_n}^u dv\,\psi(v)\to\frac{1}{2}\int_s^t\phi(u)\psi(u)\,du. \tag{4.19}$$

PROOF. Let $\psi_n(u)=2^n\int_{u_n}^u\psi(v)\,dv$. Then $\{\psi_n\}_{n\geqq 1}$ is obviously relatively compact in the weak topology on $\mathsf{L}^1([s,t])$ induced by $\mathsf{L}^\infty([s,t])$. Hence, it suffices to prove the result when ϕ is continuous. Next observe that

$$2^n\int_s^t du\,\phi(u)\int_{u_n}^u dv\,\psi(v)=2^n\int_s^t du\,\psi(u)\int_u^{u_n^+}dv\,\phi(v). \tag{4.20}$$

Thus, by the argument just used, we may assume that both ϕ and ψ are continuous. But the result is obvious for continuous ϕ and ψ, and so we are done. *Q.E.D.*

LEMMA 4.3. *Let $L_u^0=\frac{1}{2}(\sigma\sigma^*)^{ij}(t,x)(\partial^2/\partial x_i\partial x_j)$. Then*

$$I_2^{(n)}\to E^P\left[F\cdot\int_s^t L_u^0 f(x(u))\,du\right]. \tag{4.21}$$

PROOF. We will use $H(x)$ to denote the Hessian matrix of f. Note that

$$E^W[\alpha^{(n)}(u)\,|\,\mathscr{M}_{u_n}]=0, \tag{4.22}$$

and therefore

$$(4.23)\qquad I_2^{(n)} = E^W\left[F\cdot\int_s^t \langle\nabla_x f(\xi^{(n)}(u)) - \nabla_x f(\xi^{(n)}(u_n)), \alpha^{(n)}(u)\rangle\,du\right]$$
$$= E^W\left[F\cdot\int_u^t du\int_{u_n}^u dv\langle\dot\eta^{(n)}(v), H(\xi^{(n)}(v))\alpha^{(n)}(u)\rangle\right]$$
$$+ E^W\left[F\cdot\int_s^t du\int_{u_n}^u dv\langle b(v,\xi^{(n)}(v)), H(\xi^{(n)}(v))\alpha^{(n)}(u)\rangle\right]$$
$$= J_1^{(n)} + J_2^{(n)}.$$

Clearly, $|J_2^{(n)}| \to 0$ and

$$(4.24)\qquad J_1^{(n)} = E^W\left[F\cdot\int_s^t du\int_{u_n}^u dv\langle\alpha^{(n)}(v), H(\xi^{(n)}(v)\alpha^{(n)}(u)\rangle\right]$$
$$+ E^W\left[F\cdot\int_s^t du\int_{u_n}^u dv\int_{u_n}^v dw\langle(\sigma'\sigma)^{\ell,\ell'}(v,\xi^{(n)}(w))\dot\beta_\ell^{(n)}\beta_{\ell'}^{(n)},\ H[\xi^{(n)}(v)]\alpha^{(n)}(u)\rangle\right]$$
$$= J_3^{(n)} + J_4^{(n)}.$$

Again, it is obvious that $|J_4^{(n)} \to 0$ and that

$$(4.25)\qquad J_3^{(n)} = E^W\left[F\cdot\int_s^t du\int_{u_n}^u dv\langle\alpha^{(n)}(v), H(\xi^{(n)}(u_n))\alpha^{(n)}(u)\rangle\right]$$
$$+ E^W\left[F\cdot\int_s^t du\int_{u_n}^u dv\langle\alpha^{(n)}(v), (H(\xi^{(n)}(v)) - H(\xi^{(n)}(u_n)))\alpha^{(n)}(u)\rangle\right]$$
$$= J_5^{(n)} + J_6^{(n)}.$$

Since $|J_6^{(n)}| \to 0$, it remains to examine $J_5^{(n)}$. Clearly,

$$(4.26)\qquad J_5^{(n)} = E^W\left[2^nF\cdot\int_s^t du\int_{u_n}^u dv\ \mathrm{tr}\,[\sigma^*(v,\xi^{(n)}(v_n))H(\xi^{(n)}(u_n))\sigma(u,\xi^{(n)}(u_n))]\right]$$
$$= E^{P_n}\left[2^nF\cdot\int_s^t du\int_{u_n}^u dv\ \mathrm{tr}\,[\sigma^*(v,x(v_n))H(x(u_n))]\sigma(u,x(u_n))\right].$$

Hence, by Lemma 4.2 and the fact that $P_n \Rightarrow P$, we have that

$$(4.27)\qquad J_5^{(n)} \to \frac{1}{2}E^P\left[F\cdot\int_s^t \mathrm{tr}\,[\sigma^*(u,v(u))H(x(u))\sigma(u,x(u))]\,du\right].$$

Q.E.D.

LEMMA 4.4. *Let* $(\sigma'\sigma)_i(t,x) = (\sigma'\sigma)_i^{\ell\ell}(t,x)$. *Then*

$$(4.28)\qquad I_3^{(n)} \to \frac{1}{2}E^P\left[F\cdot\int_s^t \nabla_x f(x(u)), (\sigma'\sigma)(u,x(u))\,du\right].$$

PROOF. Note that

$$(4.29)\quad I_3^{(n)} = E^W\left[F\cdot\int_s^t du\int_{u_n}^u dv\,\nabla_x f(\xi^{(u)}(u)),\,(\sigma\sigma')^{\ell\ell'}(u,\xi^{(n)}(v))\dot\beta_\ell(v)\dot\beta_{\ell'}(v)\right]$$

$$= E^W\left[F\cdot\int_s^t \nabla_x f(\xi^{(n)}(u_n)),\,(\sigma\sigma')(u,\xi^{(n)}(u_n))(u-u_n)\,du\right]$$

$$+ E^W\left[F\cdot\int_s^t du\int_{u_n}^u dv\,\nabla_x f(\xi^{(n)}(u_n)),\right.$$

$$\left.[(\sigma\sigma')^{\ell\ell'}(u,\xi^{(n)}(v)) - (\sigma\sigma')^{\ell\ell'}(u,\xi^{(n)}(u_n))]\dot\beta_\ell(v)\dot\beta_{\ell'}(v)\right]$$

$$+ E^W\left[F\cdot\int_s^t du\int_{u_n}^u dv\langle\nabla_x f(\xi^{(n)}(u)) - \nabla_x f(\xi^{(n)}(u_n)),\right.$$

$$\left.(\sigma'\sigma)^{\ell\ell'}(u,\xi^{(n)}(v))\dot\beta_\ell(v)\dot\beta_{\ell'}(v)\rangle\right]$$

$$= J_1^{(n)} + J_2^{(n)} + J_3^{(n)}.$$

Clearly, $|J_2^{(n)}|$ and $|J_3^{(n)}|$ tend to zero. Moreover,

$$(4.30)\quad J_1^{(n)} = E^{P_n}\left[F\cdot\int_s^t \langle\nabla_x f(x(u_n)),\,(\sigma\sigma')(u,x(u_n))\rangle\,(u-u_n)\,du\right]$$

$$\to \frac{1}{2}E^P\left[F\cdot\int_s^t \nabla_x f(x(u)),\,(\sigma\sigma')(u,x(u))\,du\right].$$

Q.E.D.

THEOREM 4.1. *Let P_n be the distribution of the process $\xi^{(n)}(t)$ defined in* (4.5). *Then P_n converges weakly to $P_{0,0}$ as $n\to\infty$. In particular, equation* (4.2) *is valid for all (t_0, x_0).*

COROLLARY 4.1. *If $c\colon [0,\infty)\times R^d \to R^d$ is bounded and measurable, then for each (t_0, x_0) the unique solution $P^c_{t_0,x_0}$ to the martingale problem for*

$$(4.31)\quad L_t^c = \tfrac{1}{2}\sigma^*\nabla_x\cdot\sigma^*\nabla_x + (b+\sigma c)\cdot\nabla_x$$

starting at (t_0, x_0) satisfies

$$(4.32)\quad \operatorname{supp}(P^c_{t_0,x_0}) \subseteq \overline{\mathscr{S}_{\sigma,b}(t_0,x_0)}.$$

PROOF. According to Corollary 2.1, $P^c_{t_0,x_0}$ and P_{t_0,x_0} are equivalent on $\mathscr{M}_t^{t_0}$ for all $t\geq t_0$. Hence, $\operatorname{supp}(P^c_{t_0,x_0}) = \operatorname{supp}(P_{t_0,x_0})$. *Q.E.D.*

REMARK 4.1. Let $a_i\colon [0,\infty)\times R^d \to S_d$, $i=1,2$, and $b\colon [0,\infty)\times R^d\to R^d$ be bounded measurable functions which are uniformly Lipschitz continuous in x. Define $\mathscr{S}_{a_i,b}(t_0,x_0)$ to be the class of $\phi\in C([0,\infty), R^d)$ for which there exists a piecewise constant $\psi\colon [0,\infty)\times R^d\to R^d$ such that

$$(4.33)\quad \phi(t) = x_0 + \int_{t_0}^t a_i(u,\phi(u))\psi(u)\,du + \int_{t_0}^t b(u,\phi(u))\,du,\qquad t\geq t_0.$$

Then Range $(a_1(t, x)) \subseteqq$ Range $(a_2(t, x))$, $(t, x) \in [t_0, \infty) \times R^d$, implies

$$\overline{\mathscr{S}_{a_1,b}(t_0, x_0)} \subseteqq \overline{\mathscr{S}_{a_2,b}(t_0, x_0)}. \tag{4.34}$$

REMARK 4.2. Suppose $a: [0, \infty) \times R^d \twoheadrightarrow S_d$ satisfies the conditions stated in Remark 2.1. Then for each (t_0, x_0) there is exactly one solution P_{t_0,x_0} to the martingale problem for $L_t = \frac{1}{2}\nabla_x \cdot (a\nabla_x) + b \cdot \nabla_x$, where

$$\nabla_x \cdot (a\nabla_x) = \sum_{i,j=1}^{d} \frac{\partial}{\partial x_i}\left(a^{ij}(t, x)\frac{\partial}{\partial x_j}\right) \tag{4.35}$$

and $b: [0, \infty) \times R^d \twoheadrightarrow R^d$ is bounded, measurable, and uniformly Lipschitz continuous in s. Moreover, if $a^{1/2}(t, x)$ possesses first spatial derivatives which are uniformly Lipschitz continuous in x, then L_t can be written in the form

$$L_t = \tfrac{1}{2} a^{1/2}\nabla_x \cdot a^{1/2}\nabla_x + (b + a^{1/2}c) \cdot \nabla_x; \tag{4.36}$$

and therefore

$$\text{supp}\,(P_{t_0,x_0}) \subseteqq \overline{\mathscr{S}_{a^{1/2},b}(t_0, x_0)}. \tag{4.37}$$

By Remark 4.1, $\overline{\mathscr{S}_{a^{1/2},b}(t_0, x_0)} = \overline{\mathscr{S}_{a,b}(t_0, x_0)}$, and so we have

$$\text{supp}\,(P_{t_0,x_0}) \subseteqq \overline{\mathscr{S}_{a,b}(t_0, x_0)}. \tag{4.38}$$

Using a localization procedure, it is possible to prove (4.38) under the assumptions on a stated in Remark 2.1, without any further conditions on $a^{1/2}$.

REMARK 4.3. Suppose $\sigma: [0, \infty) \times R^d \twoheadrightarrow R^d \times R^d$ and $b: [0, \infty) \times R^d \twoheadrightarrow R^d$ are bounded infinitely differentiable functions. Define vector fields

$$X_\ell = \sum_{i=1}^{d} \sigma^{i\ell}(t, x)\frac{\partial}{\partial x_i}, \qquad 1 \leqq \ell \leqq d, \tag{4.39}$$

and

$$Y = \sum_{i=1}^{d} b_i(t, x)\frac{\partial}{\partial x_i}. \tag{4.40}$$

Then

$$\tfrac{1}{2}\sigma^*\nabla_x \cdot \sigma^*\nabla_x + b \cdot \nabla_x = \tfrac{1}{2}\sum_{\ell=1}^{d} X_\ell^2 + Y. \tag{4.41}$$

Using the techniques of Bony [1], one can show that $\overline{\mathscr{S}_{\sigma,b}(t_0, x_0)}$ contains all $\phi \in C([0, \infty), R^d)$ such that

$$\phi(t) = x_0 + \int_{t_0}^{t} Z(u, \phi(u))\,du + \int_{t_0}^{t} Y(u, \phi(u))\,du, \qquad t \geqq t_0, \tag{4.42}$$

where Z is an element of the Lie algebra $\mathscr{L}(X_1, \cdots, X_d)$ generated by $X_1, \cdots, X_d$. In particular, if $\mathscr{L}(X_1, \cdots, X_d)$ has rank d at every point, then $\overline{\mathscr{S}_{\sigma,b}(t_0, x_0)}$ coincides with the set of $\phi \in C([0, \infty), R^d)$ such that $\phi(t_0) = x_0$.

5. The degenerate case, part II

Throughout this section $\sigma\colon [0, \infty) \times R^d \to R^d \otimes R^d$ will satisfy $\sigma^{ij} \in C_b^{1,2}([0, \infty) \times R^d)$, $1 \leq i, j \leq d$, and $b\colon [0, \infty) \times R^d \to R^d$ will be a bounded measurable function which is uniformly Lipschitz continuous in x. For each (t_0, x_0), P_{t_0,x_0} will denote the unique solution to the martingale problem for $L_t = \frac{1}{2}\sigma^*\nabla_x \cdot \sigma^*\nabla_x + b \cdot \nabla_x$, starting at (t_0, x_0). Our aim is to prove that

$$\text{supp}\,(P_{t_0,x_0}) = \overline{\mathscr{S}_{\sigma,b}(t_0, x_0)}, \tag{5.1}$$

where $\mathscr{S}_{\sigma,b}(t_0, x_0)$ is defined as in Section 4. In view of Theorem 4.1, equation (5.1) will be proved once we have shown that for all $T > t_0$, $\varepsilon > 0$, and ϕ in a dense subset of $\mathscr{S}_{\sigma,b}(t_0, x_0)$,

$$P_{t_0,x_0}\left(\|x(t) - \phi(t)\|_T^{t_0} < \varepsilon\right) > 0 \tag{5.2}$$

and $x_0 = 0$.

Using Theorem 2.2, one sees that the desired result is equivalent to proving that, for a dense set of $\phi \in \mathscr{S}_{\sigma,b}(0, 0)$,

$$W\left(\|\eta(t) - \phi(t)\|_T^0 < \varepsilon\right) > 0, \qquad T > 0, \varepsilon > 0, \tag{5.3}$$

where

$$\eta(t) = \int_0^t \sigma(u, \eta(u))\, d\beta(u) + \int_0^t \tilde{b}(u, \eta(u))\, du \qquad \text{a.s. } W, \tag{5.4}$$

$\beta(t)$ being a W Brownian motion and $\tilde{b}$ standing for $b + \frac{1}{2}\sigma'\sigma$ (see equation (2.35) for the definition of $\sigma'\sigma$). Actually, what we will show is more; namely, we will prove that if $\psi \in C^2([0, \infty) \times R^d)$, $\psi(0) = 0$, and

$$\phi(t) = x_0 + \int_0^t \sigma(u, \phi(u))\dot{\psi}(u)\, du + \int_0^t b(u, \phi(u))\, du, \tag{5.5}$$

then, for all $\varepsilon > 0$ and $T > 0$,

$$W\left(\|\eta(t) - \phi(t)\|_T^0 < \varepsilon \,\big|\, \||\beta(t) - \psi(t)\||_T^0 < \delta\right) \to 1 \tag{5.6}$$

as $\delta \downarrow 0$, where $\||\alpha(t)\||_T^0 = \max_{1 \leq i \leq d} \|\alpha_i(t)\|_T^0$. We will first prove (5.6) when $\psi \equiv 0$.

After some easy manipulations involving Itô's formula (see McKean [5]), one can show that (5.4) is equivalent to

$$\eta(t) = \int_0^t b(u, \eta(u))\, du + \sigma(t, \eta(t))\beta(t) - \int_0^t \left[\left(\frac{\partial}{\partial u} + L_u\right)\sigma\right](u, \eta(u))\beta(u)\, du - \Delta(t), \tag{5.7}$$

where

$$\Delta_i(t) = \sum_{j\neq k} \int_t^t (\sigma_\ell^{ij}\sigma^{\ell k})(u, \eta(u))\beta_j(u)\, d\beta_k(u) \tag{5.8}$$

$$+ \sum_j \int_0^t (\sigma_\ell^{ij}\sigma^{\ell j})(u, \eta(u))\, d\beta_j^2(u).$$

Thus, in order to prove that

$$W(\|\eta(t) - \phi(t)\|_T^0 < \varepsilon \,|\, \|\|\beta(t)\|\|_T^0 < \delta) \to 1 \tag{5.9}$$

as $\delta \downarrow 0$, where $\phi(t) = \int_0^t b(u, \phi(u))\, du$, it suffices to show that

$$W(\|\Delta(t)\|_T^0 < \varepsilon \,|\, \|\|\beta(t)\|\|_T^0 < \delta) \to 1 \tag{5.10}$$

as $\delta \downarrow 0$.

LEMMA 5.1. *Let $\eta(t)$ be given by (5.4) and suppose $f\colon [0, \infty) \times R^d \twoheadrightarrow R$ is bounded and uniformly continuous. Then for all $\varepsilon > 0$*

$$W\left(\left\| \int_0^t f(u, \eta(u))\, d\beta_i^2(u)\right\|_T^0 < \varepsilon \,\Big|\, \|\|\beta(t)\|\|_T^0 < \delta\right) \to 1 \tag{5.11}$$

as $\delta \downarrow 0$, and

$$W\left(\left\| \int_0^t f(u, \eta(u))\beta_i(u)\, d\beta_j(u)\right\|_T^0 < \varepsilon \,\Big|\, \|\|\beta(t)\|\|_T^0 < \delta\right) \to 1 \tag{5.12}$$

as $\delta \downarrow 0$, where $1 \leqq i \leqq d$ and $j \neq i$.

PROOF. We will first prove (5.11) under the assumption that $f \in C_b^\infty([0, \infty) \times R^d)$. Applying Itô's formula, we have

$$\int_0^t f(u, \eta(u))\, d\beta_i^2(u) \tag{5.13}$$

$$= f(t, \eta(t))\beta_i^2(t) - \int_0^t \beta_i^2(u)(f_u + L_u f)(u, \eta(u))\, du$$

$$- 2\int_0^t \beta_i(u)(\sigma^* \nabla_x f)_i(u, \eta(u))\, du$$

$$- \int_0^t \beta_i^2(u)\langle(\sigma^* \nabla_x f)(u, \eta(u)), d\beta(u)\rangle.$$

Clearly, the first three terms on the right tend to 0 as $\|\|\beta(t)\|\|_T^0 \to 0$. Moreover, by standard estimates,

$$W\left(\left\| \int_0^t \beta_i^2(u)\langle(\sigma^* \nabla_x f)(u, \eta(u)), d\beta(u)\rangle\right\|_T^0 > \varepsilon, \|\|\beta(t)\|\|_T^0 < \delta\right) \tag{5.14}$$

$$\leqq A_1 \exp\left\{\frac{-B_1\varepsilon^2}{\delta^4 T}\right\}$$

and

$$W(\|\beta(t)\|_T^0 < \delta) \geqq A_2 \exp\left\{\frac{-B_2T}{\delta^2}\right\}. \tag{5.15}$$

Hence, (5.11) is proved in the case when $f \in C_b^\infty([0, \infty) \times R^d)$. Now suppose f is uniformly continuous and choose $\{f_n\}_0^\infty \subseteq C_b^\infty([0, \infty) \times R^d)$ so that $f_n \to f$ uniformly. Then

$$\begin{aligned}\int_0^t f(u, \eta(u))\, d\beta_i^2(u) &- \int_0^t f_n(u, \eta(u))\, d\beta_i^2(u) \\ &= 2\int_0^t (f - f_n)(y, \eta(u))\beta_i(u)\, d\beta_i(u) + \int_0^t (f - f_n)(u, \eta(u))\, du.\end{aligned} \tag{5.16}$$

Given $\varepsilon > 0$, and $T > 0$, it is clear that for large n,

$$\begin{aligned}&W\left(\left\|\int_0^t f(u, \eta(u))\, d\beta_i^2(u) - \int_0^t f_0(u, \eta(u))\, d\beta_i^2(u)\right\|_T^0 > \varepsilon, \|\|\beta(t)\|\|_T^0 < \delta\right) \\ &\leqq W\left(\left\|\int_0^t (f - f_n)(u, \eta(u))\beta_i(u)\, d\beta_i(u)\right\|_T^0 > \tfrac{1}{2}\varepsilon, \|\|\beta(t)\|\|_T^0 < \delta\right) \\ &\leqq A \exp\left\{\frac{-B_1\varepsilon^2}{\|f - f_n\|\delta^2 T}\right\}.\end{aligned} \tag{5.17}$$

Hence, for all $\varepsilon > 0$ and $\alpha > 0$, there is an $n(\varepsilon, \alpha)$ such that

$$\begin{aligned}W\Bigg(\Bigg\|\int_0^t f(u, \eta(u))\, d\beta_i^2(u) \\ - \int_0^t f_n(u, \eta(u))\, d\beta_i^2(u)\Bigg\|_T^0 > \varepsilon \,\Big|\, \|\|\beta(t)\|\|_T^0 < \delta\Bigg) < \alpha\end{aligned} \tag{5.18}$$

independent of $0 < \delta \leqq 1$. Combining this with our original result, (5.11) follows.

Next we will prove (5.12) under the assumption that $f \in C_b^\infty([0, \infty) \times R^d)$. Again the general result follows from this by the technique used above. For convenience, we will let $\xi(t) = \int_0^t \beta_i(u)\, d\beta_j(u)$. Using Itô's formula, we have

$$\begin{aligned}\int_0^t f(u, \eta(u))\beta_i(u)\, d\beta_j(u) &\\ = f(t, \eta(t))\zeta(t) &- \int_0^t \xi(u)(f_n + L_n f)(u, \eta(u))\, du \\ &- \int_0^t (\sigma^* \nabla_x f)_j(u, \eta(u))\beta_i(u)\, du \\ &- \int_0^t \xi(u)\langle(\sigma^* \nabla_x f)(u, \eta(u)), d\beta(u)\rangle.\end{aligned} \tag{5.19}$$

Applying Theorem A.1 of the Appendix, we see that only the last term on the right need be examined. Let $\phi = (\sigma^* \nabla_x f)_\ell$. Then

$$\begin{aligned}(5.20)\qquad &\int_0^t \xi(u)(\sigma^*\nabla_x f)_\ell(u, \eta(u))\, d\beta_\ell(u)\\ &\quad = \int_0^t \phi(u, \eta(u))\, d(\xi(u)\beta_\ell(u))\\ &\quad\quad - \int_0^t \phi(u, \eta(u))\beta_i(u)\beta_\ell(u)\, d\beta_j(u) - \delta_{j,\ell}\int_0^t \phi(u, \eta(u))\beta_i(u)\, du.\end{aligned}$$

The third term on the right gives no trouble. Moreover, the second term can be handled by the estimates used to prove (5.11). Thus, we need only worry about the first term. But

$$\begin{aligned}(5.21)\qquad &\int_0^t \phi(u, \eta(u))\, d(\xi(u)\beta_\ell(u))\\ &\quad = \phi(t, \eta(t))\xi(t)\beta_\ell(t) - \int_0^t \xi(u)\beta_\ell(u)(\phi_u + L_u\phi)(u, \eta(u))\, du\\ &\quad\quad - \int_0^t \xi(u)(\sigma^*\nabla_x)_\ell(u, \eta(u))\, du\\ &\quad\quad - \int_0^t \beta_\ell(u)\beta_i(u)(\sigma^*\nabla_x\phi)_j(u, \eta(u))\, du\\ &\quad\quad - \int_0^t \langle \xi(u)\beta_\ell(u)(\sigma^*\nabla_x\phi)(u, \eta(u)), d\beta(u)\rangle.\end{aligned}$$

Again, only the final term need be examined. Using $\gamma(t)$ to denote this term, we have

$$\begin{aligned}(5.22)\qquad &W(\|\gamma(t)\|_T^0 \geqq \varepsilon, \||\beta(t)\||_T^0 < \delta)\\ &\quad = W(\|\gamma(t)\|_T^0 \geqq \varepsilon, \||\beta(t)\||_T^0 < \delta, \|\xi(t)\|_T^0 < M\delta)\\ &\quad\quad + W(\|\gamma(t)\|_T^0 \geqq \varepsilon, \||\beta(t)\||_T^0 < \delta, \|\xi(t)\|_T^0 \geqq M\delta).\end{aligned}$$

By the standard estimates,

$$(5.23)\qquad W(\|\gamma(t)\|_T^0 \geqq \varepsilon, \|\beta(t)\|_T^0 \leqq \delta, \|\xi(t)\|_T^0 < M\delta) \leqq A \exp\left\{-\frac{B\varepsilon^2}{M^2\delta^4}\right\}$$

By Theorem A.1,

$$(5.24)\qquad \lim_{M\to\infty} \sup_{0<\delta\leqq 1} \frac{W(\|\gamma(t)\|_T^0 \geqq \varepsilon, \||\beta(t)\||_T^0 < \delta, \|\xi(t)\|_T^0 \geqq M\delta)}{W(\||\beta(t)\||_T^0 < \delta)} = 0.$$

Combining these, we see that

$$(5.25)\qquad W(\|\gamma(t)\|_T^0 \geqq \varepsilon \mid \||\beta(t)\||_T^0 < \delta) \to 0$$

as $\delta \downarrow 0$. *Q.E.D.*

Clearly, (5.10) is an immediate consequence of Lemma 5.1. Hence, we have proved (5.6) when $\psi \equiv 0$.

THEOREM 5.1. *Let $\eta(t)$ be given by*

$$\eta(t) = x_0 + \int_{t_0}^{t} \sigma(u, \eta(u))\, d\beta(u) + \int_{t_0}^{t} \tilde{b}(u, \eta(u))\, du \quad a.s.\ W, \qquad t \geqq t_0, \tag{5.26}$$

where $\beta(t)$ is a W Brownian motion. Given $\psi \in C^2([0, \infty), R^d)$ satisfying $\psi(t_0) = 0$, define $\phi(t)$ by

$$\phi(t) = x_0 + \int_{t_0}^{t} \sigma(u, \phi(u))\dot{\psi}(u)\, du + \int_{t_0}^{t} b(u, \phi(u))\, du, \qquad t \geqq t_0. \tag{5.27}$$

Then for all $\varepsilon > 0$

$$W(\|\eta(t) - \phi(t)\|_T^{t_0} < \varepsilon \mid \|\|\beta(t) - \psi(t)\|\|_T^{t_0} < \delta) \to 1 \tag{5.28}$$

as $\delta \downarrow 0$.

PROOF. We may assume that $t_0 = 0$ and $x_0 = 0$. The case when $\psi \equiv 0$ has just been proved. To handle the general case, define $\bar{W}$ so that

$$\frac{d\bar{W}}{dW} = R(t) = \exp\left\{\int_0^t \langle \dot{\psi}(u), d\beta(u)\rangle - \frac{1}{2}\int_0^t |\dot{\psi}(u)|^2\, du\right\}, \tag{5.29}$$

on $\mathscr{M}_t^0$, $t \geqq 0$. Then, by Theorem 2.4, $\bar{\beta}(t) = \beta(t) - \psi(t)$ is a $\bar{W}$ Brownian motion and

$$\eta(t) = \int_0^t \sigma(u, \eta(u))\, d\bar{\beta}(u) + \int_0^t \tilde{c}(u, \eta(u))\, du \qquad \text{a.s. } \bar{W}, \qquad t \geqq 0, \tag{5.30}$$

where $c = b + \sigma\dot{\psi}$ and $\tilde{c} = c + \frac{1}{2}\sigma'\sigma$. Hence,

$$\bar{W}(\|\eta(t) - \phi(t)\|_T^0 < \varepsilon \mid \|\|\bar{\beta}(t)\|\|_T^0 < \delta) \to 1 \tag{5.31}$$

as $\delta \downarrow 0$, where $\phi(t) = \int_0^t c(u, \phi(u))\, du$. But this means that

$$\begin{aligned} &\lim_{\delta \downarrow 0} W(\|\eta(t) - \phi(t)\|_T^0 < \varepsilon \mid \|\|\beta(t) - \psi(t)\|\|_T^0 < \delta) \\ &= \lim_{\delta \downarrow 0} \frac{W(\|\eta(t) - \phi(t)\|_T^0 < \varepsilon, \|\|\beta(t) - \psi(t)\|\|_T^0 < \delta)}{E^W[R(T)\chi_{[0,\varepsilon)}(\|\eta(t) - \phi(t)\|_T^0)\chi_{[0,\delta)}(\|\|\beta(t) - \psi(t)\|\|_T^0)]} \\ &\qquad \cdot \frac{E^W[R(T)\chi_{[0,\delta)}(\|\|\beta(t) - \psi(t)\|\|_T^0)]}{W(\|\|\beta(t) - \psi(t)\|\|_T^0 < \delta)} \\ &= 1, \end{aligned} \tag{5.32}$$

since

$$R(t) = \exp\left\{\psi(t)\beta(t) - \int_0^t \langle \beta(u), \ddot{\psi}(u)\rangle\, du - \frac{1}{2}\int_0^t |\dot{\psi}(u)|^2\, du\right\}, \tag{5.33}$$

is a continuous functional of β. *Q.E.D.*

THEOREM 5.2. *Let* $c: [0, \infty) \times R^d \to R^d$ *be bounded and measurable, and set*

$$L_t = \tfrac{1}{2}\sigma^* \nabla_x \cdot \sigma^* \nabla_x + (b + \sigma c) \cdot \nabla_x \tag{5.34}$$

(as before). If P_{t_0,x_0} *is the solution to the martingale problem for* (5.34) *starting at* (t_0, x_0), *then equation* (5.1) *obtains.*

PROOF. By Corollary 2.1, we may take $c \equiv 0$. Choose $\hat{P}_{t_0,x_0}$ and $\hat{\beta}(t)$ as in Theorem 2.2. Then

$$x(t) + \int_{t_0}^{t} \sigma(u, x(u))\, d\hat{\beta}(u) + \int_{t_0}^{t} \tilde{b}(u, x(u))\, du \quad \text{a.s. } \hat{P}_{t_0,x_0}, \qquad t \geqq t_0. \tag{5.35}$$

Hence, by Theorem 5.1, for all $T \geqq t_0$ and $\varepsilon > 0$,

$$\hat{P}_{t_0,x_0}\left(\|x(t) - (t)\|_T^{t_0} < \,|\, \||\hat{\beta}(t) - \psi(t)\||_T^{t_0} < \delta\right) \to 1 \tag{5.36}$$

as $\delta \downarrow 0$, where $\psi \in C^2([t_0, \infty), R^d)$ satisfies $\psi(t_0) = 0$ and (5.27). By Theorem 3.1, $\hat{P}_{t_0,x_0}(\||\hat{\beta}(t) - \psi(t)\||_T^{t_0} < \delta) > 0$ for all $T > t_0$ and $\delta > 0$. In particular, supp (P_{t_0,x_0}) contains a dense subset of $\mathscr{S}_{\sigma,b}(t_0, x_0)$.

REMARK 5.1. Let $\langle X, \rho \rangle$ be a metric space and suppose that μ is a probability measure on X. Given a μ measurable transformation $T: X \to X$, we say that $x \in X$ is a *continuity point* of the transformation T if for all $\varepsilon > 0$,

$$\lim_{\delta \downarrow 0} \frac{\mu\left(T^{-1}(B(T(x), \varepsilon)) \cap B(x, \delta)\right)}{\mu(B(x, \delta))} = 1, \tag{5.37}$$

where $B(y, \alpha) = \{z \in X: \rho(y, z) < \alpha\}$. In the terminology of continuity points, Theorem 5.1 says the following. Define $T: C^2([0, \infty), R^d) \to \Omega$ so that $T(\psi) = \phi$, where

$$\phi(t) = \int_0^t \sigma(u, \phi(u))\dot{\psi}(u)\, du + \int_0^t b(u, \phi(u))\, du, \qquad t \geqq 0, \tag{5.38}$$

and let W be Wiener measure on Ω (that is, $x(t)$ is a W Brownian motion). Then the transformation $\hat{T}$ such that $\hat{T}(\omega)(t) = \eta(t, \omega)$, where

$$\eta(t) = \int_0^t \sigma(u, \eta(u))\, dx(u) + \int_0^t \tilde{b}(u, \eta(u))\, du \quad \text{a.s. } W, \tag{5.39}$$

is a W measurable extension of T of Ω with the property that all elements in $C^2([0, \infty), R^d)$ are continuity points of $\hat{T}$.

REMARK 5.2. It seems likely that the hypotheses under which Theorem 5.1 was proved are close to the best possible. However, the authors believe that Theorem 5.2 is valid under weaker assumptions. In particular, it seems likely that if $a: [0, \infty) \times R^d \to S_d$ satisfies $a^{ij} \in C_b^{1,2}([0, \infty) \times R^d)$, $1 \leqq i, j \leqq d$ and $b: [0, \infty) \times R^d \to R^d$ is a bounded, measurable function which is uniformly Lipschitz continuous in x, then

$$\text{supp}\,(P_{t_0,x_0}) = \overline{\mathscr{S}_{a,b}(t_0, x_0)}, \tag{5.40}$$

where P_{t_0,x_0} is the solution to the martingale problem for $L_t = \frac{1}{2}\nabla_x \cdot (a\nabla_x) + b \cdot \nabla_x$ starting at (t_0, x_0).

6. Applications

Let σ and b be as in Section 5 and define $\mathscr{S}_{\sigma,b}(t_0, x_0)$, $(t_0, x_0) \in [0, \infty) \times R^d$, accordingly. Given an open $\mathscr{G} \subseteq [0, \infty) \times R^d$, take $\mathscr{G}_{\sigma,b}(t_0, x_0)$ to be the closure in $\mathscr{G}$ of $\{(t, \phi(t)): t \geqq t_0, \phi \in \mathscr{S}_{\sigma,b}(t_0, x_0), \text{ and } (s, \phi(s)) \in \mathscr{G} \text{ for } t_0 \leqq s \leqq t\}$. Let $c: [0, \infty) \times R^d \rightarrow R^d$ be bounded, measurable, and continuous in x. Denote by P_{t_0,x_0} the solution to the martingale problem for $L_t = \frac{1}{2}\sigma^* \nabla_x \cdot \sigma^* \nabla_x + (b + \sigma c) \cdot \nabla_x$, starting at (t_0, x_0). We will use $\mathscr{H}_{L_t}^-(t_0, x_0, \mathscr{G})$ to denote the class of upper semicontinuous functions f on $\mathscr{G}$ into $R \cup \{-\infty\}$ which are bounded above and have the property that $f(t \wedge \tau, x(t \wedge \tau))$, $t \geqq t_0$, is a P_{t_0,x_0} submartingale, where $\tau = \inf\{t \geqq t_0: (t, x(t)) \notin \mathscr{G}\}$. We use $\mathscr{H}_{L_t}^-(\mathscr{G})$ to denote the class $\bigcap_{\mathscr{G}} \mathscr{H}_{L_t}^-(t_0, x_0, \mathscr{G})$. Observe that $\mathscr{H}_{L_t}^-(t_0, x_0, \mathscr{G})$ and $\mathscr{H}_{L_t}^-(\mathscr{G})$ have the same closure properties as the analogously defined classes in Section 3. Further, note that if $f \in C_b^{1,2}(\mathscr{G})$, then $f \in \mathscr{H}_{L_t}^-(t_0, x_0, \mathscr{G})$ if and only if $(\partial f/\partial t) + L_t f \geqq 0$ on $\mathscr{G}(t_0, x_0)$.

THEOREM 6.1. *If $f \in \mathscr{H}_{L_t}^-(t_0, x_0, \mathscr{G})$ and $f(t_0, x_0) = \sup_{\mathscr{G}} f(t, x)$, then $f \equiv f(t_0, x_0)$ on $\mathscr{G}(t_0, x_0)$. Moreover, if $(t_1, x_1) \in \mathscr{G} - \mathscr{G}(t_0, x_0)$, then there is an $f \in \mathscr{H}_{L_t}^-(\mathscr{G})$ such that $f(t_0, x_0) = \sup_{\mathscr{G}} f(t, x)$ and $f(t_1, x_1) < f(t_0, x_0)$.*

PROOF. The proof is exactly the same as that of Theorem 3.2. The only difference lies in the fact that the $P_{t,x}$ are no longer strongly Feller continuous and therefore $E^{P_{t,x}}[\int_t^{\tau_t} e^{-u} h(u, x(u))\, du]$ will not, in general, be continuous. Nonetheless, if $h \leqq 0$, then it will still be upper semicontinuous, by virtue of the Feller continuity of the $P_{t,x}$. *Q.E.D.*

REMARK 6.1. The class $\mathscr{H}_{L_t}^-(\mathscr{G})$ admits the same semigroup interpretation as we gave in Remark 3.1, only it is no longer true that every measurable function f which is bounded above and satisfies $\hat{T}_t f \downarrow f$ is upper semicontinuous in the ordinary sense. However, it will be upper semicontinuous in the "intrinsic topology" of the time-space process; and one can still show that such an f will be constant on the intrinsic closure of $\{(t, \phi(t)): t \geqq t_0, \phi \in \mathscr{S}_{\sigma,b}(t_0, x_0), \text{ and } (s, \phi(s)) \in \mathscr{G} \text{ for } t_0 \leqq s \leqq t\}$ if $f(t_0, x_0) = \sup_{\mathscr{G}} f(t, x)$.

◇ ◇ ◇ ◇ ◇

APPENDIX

Let $\beta(t)$ be a W Brownian motion and set $\xi(t) = \int_0^t \beta_i(u)\, d\beta_j(u)$, where $i \neq j$. The purpose of this section is to prove that for all $T > 0$,

$$\lim_{M \to \infty} \inf_{0 < \delta \leqq 1} W(\|\xi(t)\|_T^0 < M\delta \mid \|\beta(t)\|_T^0 < \delta) = 1. \tag{A.1}$$

It is clear that we may assume that $d = 2$, $i = 1$, and $j = 2$.

For each $(s, x) \in [0, \infty) \times R^2$, let $W_{s,x}$ be Wiener measure starting at (s, x) (that is, $W_{s,x}$ is the measure on $\langle \Omega, \mathscr{M}^s \rangle$ such that $x(t) - x$ is a $W_{s,x}$ Brownian motion). Define

$$\tau_s = \inf \{t \geqq s: |x_1(t)| \vee |x_2(t)| \geqq \delta\}, \tag{A.2}$$

and let

$$Q_{s,x}(A) = W_{s,x}(A \mid \{\tau_s > T\}), \qquad A \in \mathscr{M}^s. \tag{A.3}$$

LEMMA A.1. *Let* $g(s, x) = W_{s,x}(\tau_s > T)$ *and* $h(s, a) = W_{s,(a,0)}(\|x_1(t)\|_T^s < \delta)$. *Then* $g(s, x) = h(s, x_1) \cdot h(s, x_2)$, *and* $h \in C^\infty[(0, T) \times (-\delta, \delta)]$. *Next, let* $b(s, a) = h_a(s, a)/h(s, a)$ *in* $(0, T) \times (-\delta, \delta)$. *Then* $b(s, \cdot)$ *is nonincreasing and* $b(s, -a) = -b(s, a)$. *Finally, define* $B(s, x) = (b(s, x_1), b(s, x_2))$. *Then for all* $s \in [0, \tau)$ *and* x *such that* $|x_1| \vee |x_2| < \delta$,

$$\beta(t) = x(t) - x - \int_s^{t \wedge \tau} B(u, x(u))\, du \tag{A.4}$$

is a $Q_{s,x}$ *Brownian motion.*

PROOF. The first assertion is trivial. To prove the second assertion, note that $h_s + \frac{1}{2} h_{aa} = 0$ and that $h_s \geqq 0$. Hence,

$$b_a(s, a) = \frac{h h_{aa} - h_a^2}{h^2} \leqq 0, \tag{A.5}$$

and so $b(s, \cdot)$ is nonincreasing. Also, note that $h(s, -a) = h(s, a)$, and therefore $h_a(s, -a) = -h_a(s, a)$. Hence, $b(s, -a) = -b(s, a)$.

Finally, to prove the last assertion, let $\{\delta_n\}_1^\infty \subseteqq (0, \delta)$ such that $\delta_n \uparrow \delta$ and define

$$\tau_s^{(n)} = (\inf \{t \geqq s: |x_1(t)| \vee |x_2(t)| \geqq \delta_n\}) \wedge T. \tag{A.6}$$

Then, by Itô's formula,

$$\begin{aligned} &g(t \wedge \tau_s^{(n)}, x(t \wedge \tau_s^{(n)})) - g(s, x) \\ &\qquad = \int_s^{t \wedge \tau_s^{(n)}} g(u, x(u)) \langle B(u, x(u)), dx(u) \rangle \quad \text{a.s. } W_{s,x}. \end{aligned} \tag{A.7}$$

Hence,

$$\begin{aligned} &\frac{g(t \wedge \tau_s^{(n)}, x(t \wedge \tau_s^{(n)}))}{g(s, x)} \\ &\quad = R^{(n)}(t) \\ &\quad = \exp\left\{\int_s^{t \wedge \tau_s^{(n)}} \langle B(u, x(u)), dx(u) \rangle - \frac{1}{2} \int_s^{t \wedge \tau_s^{(n)}} |B(u, x(u))|^2\, du\right\} \\ &\qquad\qquad \text{a.s. } W_{s,x}. \end{aligned} \tag{A.8}$$

Given $A \in \mathscr{M}^s_{t \wedge \tau_s^{(n)}}$, we now have

$$(A.9)\qquad Q_{s,x}(A) = \frac{W_{s,x}(A \cap \{\tau_s > T\})}{g(s, x)} = \frac{E^{W_{s,x}}[\chi_A g(t \wedge \tau_s^{(n)}, x(t \wedge \tau_s^{(n)})]}{g(s, x)}$$
$$= E^{W_{s,x}}[R^{(n)}(t)\chi_A].$$

Thus, by Theorem 2.1,

$$(A.10)\qquad X_\theta^{(n)}(t) = \exp\left\{\langle \theta, x(t \wedge \tau_s^{(n)}) - x - \int_s^{t \wedge \tau_s^{(n)}} B(u, x(u))\, du\rangle - \tfrac{1}{2}|\theta|^2(t \wedge \tau_s^{(n)} - s)\right\},$$

is a $Q_{s,x}$ martingale for all $\theta \in R^2$. Since $\tau_s^{(n)} \uparrow T$ as $n \to \infty$ and

$$(A.11)\qquad E^{Q_{s,x}}[|X_\theta^{(n)}(t)|^2] = E^{Q_{s,x}}[X_{2\theta}^{(n)}(t) \exp\{|\theta|^2(t \wedge \tau_s^{(n)} - s)\}]$$
$$\leqq \exp\{|\theta^{2(T-s)}\},$$

it follows that

$$(A.12)\qquad \exp\left\{\langle \theta, x(t \wedge T) - x - \int_s^{t \wedge T} B(u, x(u))\, du\rangle - \tfrac{1}{2}|\theta|^2(t \wedge T - s)\right\},$$

is a $Q_{s,x}$ martingale for all $\theta \in R^2$. Combining these, we have that

$$(A.13)\qquad x(t) - x - \int_s^{t \wedge T} B(u, x(u))\, du$$

is a $Q_{s,x}$ Brownian motion. *Q.E.D.*

LEMMA A.2. *Let* $\xi(t) = \int_0^t x_1(u)\, dx_2(u)$. *Then*

$$(A.14)\qquad \sup_{0 \leqq s \leqq T, |x_1| \vee |x_2| < \delta} E^{Q_{s,x}}[|\xi(T) - \xi(s)|^2] \leqq C\delta^2.$$

PROOF. Note that by Lemma A.1

$$(A.15)\qquad \xi(t) = \int_s^t x_1(u)\, d\beta_1(u) + \int_s^{t \wedge T} x_1(u) b(u, x_2(u))\, du,$$

where $\beta_1(t) = x_1(t) - x_1 - \int_s^{t \wedge T} b(u, x_1(u))\, du$ is a one dimensional $Q_{s,x}$ Brownian motion. Hence,

$$(A.16)\qquad E^{Q_{s,x}}[|\xi(T) - \xi(s)|^2]$$
$$\leqq 2E^{Q_{s,x}}\left[\left|\int_s^T x_1(u)\, d\beta_1(u)\right|^2\right] + 2E^{Q_{s,x}}\left[\left|\int_s^T x_1(u) b(u, x_2(u))\, du\right|^2\right].$$

Since $x_1(u) < \delta$ a.s. $Q_{s,x}$ for $s \leqq u \leqq T$,

$$(A.17)\qquad E^{Q_{s,x}}\left[\left|\int_s^T x_1(u)\, d\beta_1(u)\right|^2\right] \leqq \delta^2(T - s).$$

Using the independence of $x_1(\cdot)$ and $x_2(\cdot)$ under $Q_{s,x}$, we have

$$(\text{A.18})\qquad E^{Q_{s,x}}\left[\left|\int_s^T x_1(u)b(u, x_2(u))\,du\right|^2\right] = 2\iint\limits_{s\leqq u_1\leqq u_2\leqq T} E^{Q_{s,x}}[x_1(u_1)x_1(u_2)] \cdot E^{Q_{s,x}}\cdot[b(u_1, x_2(u_1))b(u_2, x_2(u_2))]\,du_1du_2.$$

Observe that

$$(\text{A.19})\qquad E^{Q_{s,x}}[b(u_1, x_2(u_1))b(u_1, x_2(u_2))] = E^{Q_{s,x}}[b(u_1, x_2(u_1))E^{Q_{u_1,(0,x_2(u_1))}}[b(u_2, x_2(u_2))]].$$

Since $(b(u, \cdot)$ is nonincreasing and antisymmetric, it is easy to see that $v(u_1, \cdot) = E^{Q_{u_1,(0,\cdot)}}[b(u_2, x_2(u_2))]$ has the same properties. In particular,

$$(\text{A.20})\qquad E^{Q_{s,x}}[b(u_1, x_2(u_1))b(u, x_2(u_2))] \geqq 0.$$

Hence,

$$(\text{A.21})\qquad E^{Q_{s,x}}\left[\left|\int_s^T x_1(u)b(u, x_2(u))\,du\right|^2\right] \leqq \delta^2 E^{Q_{s,x}}\left[\left|\int_s^T b(u, x_2(u))\,du\right|^2\right] = \delta^2 E^{Q_{s,x}}[|\beta_2(T) - x_2(T) - x_2|^2] \leqq C\delta^2.$$

Q.E.D.

THEOREM A.1. *Define* $\xi(t)$ *as in Lemma* A.2 *and let* $W = W_{0,0}$. *Then*

$$(\text{A.22})\qquad \lim_{M\to\infty}\inf_{0<\delta\leqq 1} W(\|\xi(t)\|_T^0 < M\delta \,|\, \||x(t)\||_T^0 < \delta) = 1.$$

PROOF. Let σ be a stopping time which is dominated by T. Then for $A \in \mathcal{M}_\sigma^0$,

$$(\text{A.23})\qquad \begin{aligned} E^W[\chi_{A\cap\{\tau_s>T\}}|\xi(T) - \xi(\sigma)|^2] &= E^W[\chi_{A\cap\{\tau_s>\sigma\}}E^{W_{\sigma,x(\sigma)}}(\chi_{\{\tau_0>T\}}|\xi(T) - \xi(\sigma)|^2)] \\ &= E^W[\chi_{A\cap\{\tau_s>\sigma\}}W_{\sigma,x(\sigma)}(\tau_\sigma > T)E^{Q_{\sigma,x(\sigma)}}[|\xi(T) - \xi(\sigma)|^2]] \\ &\leqq C\delta^2 W(A\cap\{\tau_s > T\}). \end{aligned}$$

Thus if $Q = Q_{0,0}$, then

$$(\text{A.24})\qquad E^Q[|\xi(T) - \xi(\sigma)|^2 \,|\, \mathcal{M}_\sigma^0] \leqq C\delta^2.$$

Now let $\sigma_\ell = (\inf\{t \geqq 0: |\xi(t)| \geqq \ell\delta) \wedge T$. Then

$$(\text{A.25})\qquad \begin{aligned} Q(\{\sigma_{2\ell} < T\}\cap\{|\xi(T)| \leqq \ell\delta\}) &\leqq Q(\{\sigma_{2\ell} < T\}\cap\{|\xi(T) - \xi(\sigma_{2\ell})| \geqq \ell\delta\}) \\ &\leqq \frac{C}{\ell^2}Q(\sigma_{2\ell} < T). \end{aligned}$$

Thus,

$$Q(\sigma_{2\ell} < T) \leqq \frac{Q(|\xi(T)| > \ell\delta)}{1 - (C/\ell^2)} \leqq \frac{C}{\ell^2}\left(1 - \frac{C}{\ell^2}\right)^{-1}. \tag{A.26}$$

Q.E.D.

REFERENCES

[1] J. M. Bony, "Principe du maximum, unicité du problème de Cauchy et inégalité de Harnack pour les operateurs elliptiques degenérés," *Ann. Inst. Fourier*, Vol. 19 (1969), pp. 277–304.

[2] J. L. Doob, *Stochastic Processes*, New York, Wiley, 1953.

[3] C. D. Hill, "A sharp maximum principle for degenerate elliptic-parabolic equations," *J. Math. and Mech.*, Vol. 20 (1970), pp. 213–228.

[4] K. Itô, "On stochastic differential equations," *Mem. Amer. Math. Soc.*, No. 4 (1951).

[5] H. P. McKean, *Stochastic Integrals*, New York, Academic Press, 1969.

[6] L. Nirenberg, "A strong maximum principle for parabolic equations," *Comm. Pure Appl. Math.*, Vol. 6 (1953), pp. 167–177.

[7] Yu. V. Prokhorov, "Convergence of random processes and limit theorems," *Theory Probability Appl.*, Vol. 1 (1956), pp. 118–122.

[8] R. S. Philips and L. Sarason, "Elliptic-parabolic equations of the second order," *J. Math. Mech.*, Vol. 17 (1968), pp. 891–917.

[9] D. W. Stroock and S. R. S. Varadhan, "Diffusion processes with continuous coefficients," *Comm. Pure Appl. Math.* II, Vol. 22 (1969), pp. 345–400; 479–530.

[10] ———, "Diffusion processes with boundary conditions," *Comm. Pure Appl. Math.*, Vol. 24 (1971), pp. 147–225.

[11] D. W. Stroock, "On the growth of stochastic integrals," *Z. Wahrscheinlichkeitstheorie und Verww. Gebiete*, Vol. 18 (1971), pp. 340–344.

[12] E. Wong and M. Zakai, "Riemann-Stieltjes approximation of stochastic integrals," *Z. Wahrscheinlichkeitstheorie und Verww. Gebiete*, Vol. 12 (1969), pp. 87–97.

Proceedings of the Sixth Berkeley Symposium on Mathematical Statistics and Probability,
Vol. III: Probability theory, Univ. California Press, 1972, 361-368

DIFFUSION PROCESSES

DANIEL W. STROOCK and S. R. S. VARADHAN
COURANT INSTITUTE, NEW YORK UNIVERSITY

1. Introduction

One of the major problems in the theory of diffusion processes is to construct the process for a given set of diffusion coefficients. A diffusion process in R^d is hopefully determined by the two sets of coefficients

$$\begin{aligned} a &= a(t, x) = \{a_{ij}(t, x)\}, \qquad 1 \leqq i, j \leqq d,\ t \in [0, \infty),\ x \in R^d, \\ b &= b(t, x) = \{b_j(t, x)\}, \qquad 1 \leqq j \leqq d,\ t \in [0, \infty),\ x \in R^d. \end{aligned} \tag{1.1}$$

Here a is a positive semidefinite symmetric matrix for each t and x, and b is a d vector for each t and x. There are various ways of describing exactly what we mean by a diffusion process corresponding to the specified set of coefficients. We shall adopt the following approach.

Let Ω be the space of R^d valued continuous functions on $[0, \infty)$. The value of a function $\omega = x(\cdot)$ in Ω at time t will be denoted by $x(t)$. The σ-field generated by $x(s)$ for $t_1 \leqq s \leqq t_2$ will be denoted by $M_{t_2}^{t_1}$. If $t_1 = 0$, we will denote this by M_{t_2} and by M^{t_1} in case $t_2 = \infty$, where M is the σ-field generated by $x(s)$ for $0 \leqq s < \infty$. The space Ω can be viewed as a complete separable metric space, with uniform convergence on bounded intervals defining the topology. Then M is the Borel σ-field in Ω. A stochastic process with values in R^d, defined for $t \geqq t_0$, is a probability measure on (Ω, M^{t_0}).

Given the coefficients $\{a_{ij}(t, x)\}$ and $\{b_j(t, x)\}$, we define an operator L_t acting on functions $f(x) \in C_0^\infty(R^d)$ by

$$(L_t f)(x) = \frac{1}{2} \sum a_{ij}(t, x) \frac{\partial^2 f}{\partial x_i \partial x_j} + \sum b_j(t, x) \frac{\partial f}{\partial x_j}. \tag{1.2}$$

We say that a measure P is a solution to the Martingale problem corresponding to the given coefficients, starting at time t_0 from the point x_0 if

(a) P is a probability measure on (Ω, M^{t_0}) such that $P[x(t_0) = x_0] = 1$, and

(b) for each $f \in C_0^\infty(R^d)$, $f(x(t)) - \int_{t_0}^t (L_s f)(x(s))\, ds$ is a martingale relative to $(\Omega, M_t^{t_0}, P)$.

Under suitable conditions on the coefficients a and b, one should attempt to answer the following questions:

(1) For each t_0 and x_0, does a solution P_{t_0, x_0} exist?

Results obtained at the Courant Institute of Mathematical Sciences, New York University; this research was sponsored by the US Air Force Office of Scientific Research, Contract AF-49(638)-1719.

(2) Is it unique?

(3) Is the solution a strong Markov process?

(4) Does the solution depend continuously, in some suitable sense, on t_0, x_0, a and b?

(5) If a diffusion process can be constructed in some other natural manner, does it coincide with the above construction?

Let us take up question (5) first. There are at least two other possible ways of constructing diffusion processes for given coefficients. If we use the stochastic differential equations of Itô, we take a matrix $\sigma(t, x)$ such that

$$\sigma(t, x)\sigma^*(t, x) = a(t, x). \tag{1.3}$$

If $\beta(t)$ is Brownian motion in d dimensions, one can set up the stochastic differential equation

$$dx(t) = \sigma(t, x(t))\, d\beta(t) + b(t, x(t))\, dt. \tag{1.4}$$

If σ and b are assumed to be bounded and uniformly (with respect to t) Lipschitz continuous in x, one can show that the above equation has a unique solution $x(t)$ for $t \geqq t_0$, for each initial condition $x(t_0) = x_0$. The process so obtained is the diffusion process corresponding to a and b starting from the point x_0 at time t_0.

One can prove that if the solution to the stochastic differential equation exists, then the solution to the martingale problem also exists. Moreover, if the former is unique, then so is the latter. Furthermore, one shows that the two solutions are the same in the sense that the solution to the martingale problem is the distribution of the solution to the stochastic differential equation.

Another possibility is to use the theory of partial differential equations. We consider the fundamental solution $p(s, x, t, y)$ of the equation

$$\frac{\partial p}{\partial s} + \frac{1}{2} \sum a_{ij}(s, x) \frac{\partial^2 p}{\partial x_i \partial x_j} + \sum b_j(s, x) \frac{\partial p}{\partial x_j} = 0, \tag{1.5}$$

where $p(s, x, t, y)$ serves as the transition probability density of a Markov process. This is taken as the diffusion process corresponding to a and b. One can verify that, in this case also, the solution to the martingale problem exists, is unique, and coincides with the process constructed through the fundamental solution. The existence of a fundamental solution is proved under the assumption that a and b are bounded, satisfy a Hölder condition and a is uniformly positive definite.

2. Existence and uniqueness: the general case

Let us now assume that $a(t, x)$ is bounded continuous and positive definite for each t and x. Assume $b(t, x)$ bounded and measurable. Under these assumptions we can answer the questions (1) through (4) raised in the introduction, affirmatively.

THEOREM 2.1. *The solution $P_{s,x}$ to the martingale problem, starting from the point x at some s, exists and is unique for every s and x. Moreover, $P_{s,x}$ is a strong Markov process with transition probabilities $P(s, x, t, A) = P_{s,x}[x(t) \in A]$. Further, the solution depends continuously on $s, x, a(\cdot, \cdot)$, and $b(\cdot, \cdot)$ in the following sense: if $a_n(\cdot, \cdot) \to a(\cdot, \cdot)$ uniformly on compact sets, $b_n(\cdot, \cdot) \to b(\cdot, \cdot)$ in measure, $s_n \to s$, $x_n \to x$ and if $a_n(\cdot, \cdot)$, $b_n(\cdot, \cdot)$ are uniformly bounded, then the solution P_n corresponding to a_n, b_n starting from x_n at time s_n converges weakly to the solution P for the limiting coefficients starting from x at time s.*

This theorem can be found in [2] and [3].

3. Boundary conditions

Let us suppose that $G \subset R^d$ is a smooth region. More precisely, there is a function $\phi(x) \in C_b^2(R^d)$ such that

$$(3.1)\qquad \begin{aligned} G &= [x: \phi(x) > 0], \\ \delta G &= [x: \phi(x) = 0], \\ \|\nabla G\| &\geqq 1 \text{ on } \delta G. \end{aligned}$$

As before we shall assume that a is bounded, positive definite for each t and x and is continuous on $[0, \infty) \times \bar{G}$. Assume b is bounded and measurable. One can check that some sort of boundary condition on δG is needed to describe what happens to the process when it reaches the boundary δG. A class of these boundary conditions are of the form

$$(3.2)\qquad \rho(s, x)\frac{\partial}{\partial s} + \sum_{j=1}^{d} \gamma_j(s, x)\frac{\partial}{\partial x_j} = 0,$$

where ρ and γ are suitable function on the boundary $[0, \infty) \times \delta G$. We shall assume the following regarding ρ and γ:

(i) γ is Lipschitz continuous in t and x, is bounded, and there is a constant $\beta > 0$, such that $\langle\gamma, \nabla\phi\rangle \geqq \beta > 0$; for all $t, x \in [0, \infty) \times \delta G$;

(ii) either ρ is identically zero or it is everywhere positive, is bounded, and satisfies a Lipschitz condition in t and x.

We formulate the problem in the following manner. A solution corresponding to a, b, ρ, and γ, starting a time t_0 from the point x_0, is a measure P on (Ω, M^{t_0}), where now $\Omega = C[[0, \infty), \bar{G}]$ and M^{t_0} is the natural σ-field as before. The measure P is such that

(a) $P[x(t_0) = x_0] = 1$, and

(b) for every $u \in C_0^\infty[[0, \infty) \times \bar{G}]$ with $\rho(\partial u/\partial s) + \gamma. \nabla u \geqq 0$ on $[0, \infty) \times \delta G$,

$$(3.3)\qquad u(t, x(t)) - \int_{t_0}^{t} (U_s + L_s u)(s, x(s))\chi_G(x(s))\, ds,$$

is a submartingale relative to $(\Omega, M_t^{t_0}, P)$. We will call P, if it exists, a solution to the submartingale problem.

THEOREM 3.1. *Under the assumptions* (a) *and* (b) *on* ρ *and* γ, *along with the assumptions on* a, b *mentioned in the earlier section, for each* $x \in \bar{G}$ *and* $s \geqq 0$, *there is a unique solution* $P_{s,x}$ *to the submartingale problem. The solution is a strong Markov process. Moreover, the solution depends continuously on* s, x, a, b, ρ, *and* γ.

REMARK 3.1. In the homogeneous case, that is, when a, b, ρ, and γ are independent of t, assumption (b) on ρ can be replaced by the assumption that ρ is bounded, continuous and nonnegative.

These results can be found in [4].

4. Invariance principle

Using the results of Sections 2 and 3, one can prove general theorems concerning the convergence of Markov chains to diffusion processes. The conditions for convergence are very natural and involve essentially the first two moments. For simplicity, we shall treat the homogeneous case and assume further that $d = 1$. The general case is quite similar. Let us suppose that there is no boundary.

For each $\delta > 0$, $\pi_\delta(x, dy)$ is the transition probability of a Markov chain with R as its state space. The transitions of the chain occur in multiples of time δ. Let us define

$$
\begin{aligned}
b_\delta(x) &= \frac{1}{\delta} \int (y - x)\pi_\delta(x, dy), \\
a_\delta(x) &= \frac{1}{\delta} \int (y - x)^2 \pi_\delta(x, dy), \\
\theta_\delta(x) &= \frac{1}{\delta} \int |y - x|^\mu \pi_\delta(x, dy), \qquad \mu > 2.
\end{aligned}
\tag{4.1}
$$

We assume:

(a) $|b_\delta(x)| \leqq C$ and $b_\delta(x)$ converges as δ tends to zero to a continuous limit $b(x)$ uniformly on bounded intervals;

(b) $|a_\delta(x)| \leqq C$ and $a_\delta(x)$ converges as δ tends to zero to a continuous limit $a(x)$ uniformly on bounded intervals;

(c) $\theta_\delta(x) \to 0$, as δ tends to zero, uniformly on bounded intervals;

(d) the solution to the martingale problem for any starting point, corresponding to the coefficients a and b is unique.

THEOREM 4.1. *Under the above assumptions the Markov chain converges as* δ *tends to zero, to the diffusion corresponding to* a *and* b.

The general case for the d dimensional case, where a and b could depend on t is treated in [3]. There are similar results when a boundary is involved and these can be found in [4].

5. Special case: $d = 2$

The method of the main theorem mentioned in Section 2, involves several steps. The hypothesis of continuity of $a(t, x)$ is superfluous. What one needs is that the discontinuities of $a(t, x)$ should be small compared to the eigenvalues of $a(t, x)$. The actual constants are hard to follow through in general. But when $d = 2$, in the homogeneous case, this leads to the following theorem.

THEOREM 5.1. *If $a(x)$ is uniformly positive definite on compact sets and if trace $a(x) \equiv 1$, then for any bounded measurable $b(x)$, existence and uniqueness hold for the martingale problem.*

By a random time change, the condition that trace $a(x) \equiv 1$ can be removed and replaced by the boundedness of a. Once uniqueness is established, one shows by a standard reasoning that the process is strongly Markovian. With a little more work, one can establish the continuous dependence of the solution on the coefficients and the starting point.

These results can be found essentially in [1].

6. Special case: $d = 1$

When $d = 1$, the results are even stronger. If $a(t, x)$ is bounded measurable and uniformly positive on compact sets and if $b(t, x)$ is bounded and measurable, then the solution to the martingale problem exists and is unique for every starting point. Since a proof has not appeared anywhere, we will give a quick sketch. From the way uniqueness is proved in [2] for the general case the basic step is a perturbation in L_p. When $d = 1$, p can be taken to be 2. We write the operator

$$\frac{1}{2}a(t, x)\frac{\partial}{\partial x^2} + \frac{\partial}{\partial t}, \tag{6.1}$$

as

$$\frac{1}{2}\ell\frac{\partial}{\partial x^2} + \frac{1}{2}\big(a(t, x) - \ell\big)\frac{\partial}{\partial x^2} + \frac{\partial}{\partial t}, \tag{6.2}$$

where ℓ is chosen to be a suitable large number. For this to work in L_2, where the norms of the various operators are explicitly computable, it suffices that

$$\big|a(t, x) - \ell\big| \leqq \ell' < \ell. \tag{6.3}$$

If a is such that $0 < \alpha_1 \leqq a \leqq \alpha_2 < \infty$, then ℓ can be taken as α_2 and ℓ' can be $\alpha_2 - \alpha_1$. Since uniqueness is purely a local property, our assertion follows.

We fix $0 < \alpha_1 < \alpha_2 < \infty$ and consider the totality of all bounded measurable $a(t, x)$ satisfying $\alpha_1 \leqq a(t, x) \leqq \alpha_2$. For simplicity, we shall assume that b is identically zero. By existence and uniqueness mentioned above, we know that there are transition probabilities $\{p_a(s, x, t, dy)\}$ depending on a. As a varies over the above class, one can check that $\{p_a\}$ varies over a compact family. The

convergence notion is weak convergence in dy which is uniform over compact sets of s, x and t. The limits are all again transition probabilities corresponding to some a in the same class.

This means that there is a notion of "weak" convergence such that $a_n \to a$ "weakly" if and only if $p_{a_n} \to p_a$ as described above. It will of course be very interesting to know precisely what this convergence is. There are two special cases worth noticing. If $a_n(t, x)$ are purely functions of x, then $p_{a_n} \to p_a$ if and only if

$$\int_{\ell_1}^{\ell_2} \frac{dx}{a_n(x)} \to \int_{\ell_1}^{\ell_2} \frac{dx}{a(x)} \tag{6.4}$$

for $-\infty < \ell_1 < \ell_2 < \infty$. If $a_n(t, x)$ are purely functions of t, then $p_{a_n} \to p_a$ if and only if

$$\int_{\ell_1}^{\ell_2} a_n(s)\, ds \to \int_{\ell_1}^{\ell_2} a(s)\, ds \tag{6.5}$$

for $0 \leqq \ell_1 \leqq \ell_2 < \infty$.

For the general case when a_n depends on both t and x, these special cases provide conflicting clues. The problem is more involved than one expects to begin with.

7. An example

Let us define the coefficients

$$a_+^h(x) = \begin{cases} \alpha & \text{if } \left[\frac{x}{h}\right] \text{ is even,} \\ \beta & \text{if } \left[\frac{x}{h}\right] \text{ is odd,} \end{cases} \tag{7.1}$$

and

$$a_-^h(x) = \begin{cases} \alpha & \text{if } \left[\frac{x}{h}\right] \text{ is odd,} \\ \beta & \text{if } \left[\frac{x}{h}\right] \text{ is even} \end{cases} \tag{7.2}$$

$\beta > \alpha > 0$. (For $x < 0$, $[x] = -[-x] - 1$.)
Let $\pi_\pm^h(t, x, dy)$ be the transition probabilities corresponding to the homogeneous coefficients $a_\pm^h(x)$. Let us define

$$a^{k,h}(t, x) = \begin{cases} a_+^h(x) & \text{if } \left[\frac{t}{k}\right] \text{ is even,} \\ a_-^h(x) & \text{if } \left[\frac{t}{k}\right] \text{ is odd.} \end{cases} \tag{7.3}$$

Let $\pi^{k,h}(s, x, t, dy)$ be the transition probability corresponding to $a^{k,h}$. We denote by $P^{k,h}_{s,x}$, the process corresponding to $a^{k,h}$ starting off at time s from x.

THEOREM 7.1. *Let h_n and k_n tend to zero such that $\rho_n = k_n h_n^{-2}$ converges to a limit $\rho, 0 \leqq \rho \leqq \infty$. Then π^{k_n,h_n} converges to a limit π which is Brownian motion*

$$\pi(s, x, t, dy) = [2\pi(t - s)\sigma^2]^{-1/2} \exp\left\{-\frac{|y - x|^2}{2\sigma^2(t - s)}\right\}, \tag{7.4}$$

where $\sigma^2 = \sigma^2(\rho)$ is a continuous function of ρ with $\sigma^2(0) = \frac{1}{2}(\alpha + \beta)$ and $\sigma^2(\infty) = 2\alpha\beta/(\alpha + \beta)$.

PROOF. There are always convergent subsequences because of compactness. The limits are uniform in s, x, and t, so that the periodicity of $\pi^{k,h}$ leads to the invariance of the limit π, with respect to space and time translations. Therefore, any possible limit π is Brownian motion. We only have to compute σ^2.

Let us denote by π_n the transition probability π^{k_n,h_n}, by a_n the corresponding coefficient and by P_n the process starting from 0 at time 0. Let us define

$$\sigma_n^2 = E^{P_n}|x(1)|^2 = E^{P_n}\int_0^1 a_n(s, x(s))\,ds. \tag{7.5}$$

The computation of σ_n^2 really involves an idea of how much time, on the average, the process spends in the regions where $a = \alpha$ and $a = \beta$. We shall suppose that N is such that $2Nk_n = 1$. This involves at most an error of magnitude k_n in the computation of σ_n^2. Instead of considering the process $x(s)$, let us consider the process

$$y(s) = h_n^{-1}x(k_n s). \tag{7.6}$$

We denote by θ_n the measure corresponding to it. The generator of the $y(s)$ process can be written as

$$\frac{1}{2}\rho_n a(t, x)\frac{\partial^2}{\partial x^2}, \tag{7.7}$$

where $p_n = k_n h_n^{-1}$ and $a(t, x) = a^{1,1}(t, x)$. Now for the y process it is a question of how much time is spent in regions $\{a = \alpha\}$ and $\{a = \beta\}$ relatively up to time $2N$. Since the problem is periodic, we consider the reduced problem on the circle. A^+ and A^- are the upper and lower semicircles. π^+ and π^- are the transition probabilities for the homogeneous processes corresponding to $a^+(x)$ and $a^-(x)$, where

$$a^+(x) = \begin{cases}\alpha & \text{on } A^+,\\ \beta & \text{on } A^-.\end{cases} \tag{7.8}$$

and

$$a^-(x) = \begin{cases}\beta & \text{on } A^+,\\ \alpha & \text{on } A^-.\end{cases} \tag{7.9}$$

If $\pi^{\pm}(t)$ denotes the semigroup corresponding to $a^{\pm}$, then the ρ_n in front of the generator changes $\pi^{\pm}(t)$ to $\pi^{\pm}(t\rho_n)$. We can now write

$$(7.10)\qquad \sigma_n^2 = E^{P_n}\left[\int_0^1 a_n(s, x(s))\,ds\right]$$
$$= k_n E^{Q_n}\left[\int_0^{2N} a(s, y(s))\,ds\right]$$
$$= k_n \sum_{r=1}^{N} E^{Q_n}\int_{(2r-1)}^{2r} a(s, y(s))\,ds + k_n \sum_{r=0}^{N-1} E^{Q_n}\int_{2r}^{(2r+1)} a(s, y(s))\,ds$$
$$= k_n \sum_{r=1}^{N} E^{Q_n}\int_{(2r-1)}^{2r} a^-(y(s))\,ds + k_n \sum_{r=0}^{N-1} E^{Q_n}\int_{2r}^{(2r+1)} a^+(y(s))\,ds.$$

Since k_n is nearly $(2N)^{-1}$, it suffices to look at $E^{Q_n}\int_{2r}^{(2r+1)} a^+(y(s))\,ds$ and $E^{Q_n}\int_{(2r-1)}^{2r} a^-(y(s))\,ds$ for large r and n.

We have

$$(7.11)\qquad E^{Q_n}\int_{2r}^{(2r+1)} a^+(y(s))\,ds = [\pi^+(\rho_n)\pi^-(\rho_n)]^r\left(\int_0^1 \pi^+(s\rho_n)a^+\,ds\right)$$

and

$$(7.12)\qquad E^{Q_n}\int_{(2r-1)}^{2r} a^-(y(s))\,ds = [\pi^-(\rho_n)\pi^+(\rho_n)]^{r-1}\left(\int_0^1 \pi^-(s\rho_n)a^-\,ds\right).$$

As ρ_n tends to ρ and $n \to \infty$, it is clear that we have to look at the invariant measures μ_ρ^+ and μ_ρ^- solving $\mu_\rho^+\pi^+(\rho)\pi^-(\rho) = \mu_\rho^+$ and $\mu_\rho^-\pi^-(\rho)\pi^+(\rho) = \mu_\rho^-$. We can then write

$$(7.13)\qquad \lim_{n\to\infty} \sigma_n^2 = \sigma^2 = \frac{1}{2}\left[\int_0^1 \mu_\rho^+\pi^+(s\rho)a^+\,ds + \int_0^1 \mu_\rho^-\pi^-(s\rho)a^-\,ds\right],$$

where σ^2 is of course only a function of ρ. By standard techniques, one can justify all the steps and even compute $\sigma^2(\rho)$ when $\rho = 0$ or ∞.

REFERENCES

[1] N. V. KRYLOV, "On the first boundary value problem for elliptic equations of second order," *Differentsial'nye Uravneniya*, Vol. 3 (1967), pp. 315–326. (In Russian.)

[2] D. W. STROOCK and S. R. S. VARADHAN, "Diffusion processes with continuous coefficients I," *Comm. Pure Appl. Math.*, Vol. 22 (1969), pp. 345–400.

[3] ———, "Diffusion processes with continuous coefficients II," *Comm. Pure Appl. Math.*, Vol. 22 (1969), pp. 479–530.

[4] ———, "Diffusion processes with boundary conditions," *Comm. Pure Appl. Math.*, Vol. 24 (1971), pp. 147–225.

TRANSACTIONS OF THE
AMERICAN MATHEMATICAL SOCIETY
Volume 192, 1974, 245-260

A PROBABILISTIC APPROACH TO $H^p(R^d)$ (1)

BY
D. STROOCK AND S.R.S. VARADHAN

ABSTRACT. The relationship between $H^p(R^d)$, $1 \le p < \infty$, and the integrability of certain functionals of Brownian motion is established using the connection between probabilistic and analytic notions of functions with bounded mean oscillation. An application of this relationship is given in the derivation of an interpolation theorem for operators taking $H^1(R^d)$ to $L^1(R^d)$.

Introduction. The purpose of this paper is to give a "soft" analytic treatment of the connection between Brownian motion and the theory of $H^p(R^d)$, $1 \le p \le \infty$. The original relationship between these two subjects was discovered by Burkholder, Gundy, and Silverstein [2] for S^1. Moreover, we have learned that Burkholder and Gundy have recently extended their earlier result to R^d. The route they take is more refined than ours and establishes the connection for all $0 < p < \infty$, whereas our technique restricts us to $p \ge 1$. Nonetheless, our method avoids the difficult estimates on which theirs turns and has some nice dividends of its own.

There are three contexts in which we discuss $H^p(R^d)$. The first of these (cf. §1) is described in terms of Wiener martingales having certain integrability properties. This space is denoted by $\mathcal{H}^p(\Omega)$. Second, we talk about those harmonic functions in R_+^{d+1} which become members of $\mathcal{H}^p(\Omega)$ when evaluated along Wiener paths. This space is denoted by $\mathcal{H}^p(R_+^{d+1})$. Finally, in §2 we discuss $H^p(R^d)$ itself, described in terms of Riesz transforms. Theorem (3.3) and its corollaries establish $H^p(R^d)$ as the boundary values of $\mathcal{H}^p(R_+^{d+1})$, and Theorem (3.1) shows that $H^p(R^d)$ arises as the result of "conditioning" members of $\mathcal{H}^p(\Omega)$ with respect to the first place that a Wiener path exits from R_+^{d+1}.

§4 is concerned with a Marcinkiewicz-type interpolation theorem (cf. Theorem (4.1)) whose proof takes advantage of the above relationships between the various ways of defining $H^p(R^d)$. As a consequence of this theorem, we are able to show that $H^p(R^d) = L^p(R^d)$ for $1 < p < \infty$.

It has been pointed out to us by E. M. Stein that essentially the same interpolation theorem was discovered much earlier by S. Igari [7] using entirely different methods.

Received by the editors September 25, 1972 and, in revised form, February 2, 1973.

AMS (MOS) subject classifications (1970). Primary 60G45; Secondary 42A32.

Key words and phrases. Martingale, Wiener measure, harmonic functions, interpolation.

(1) The research in this paper was supported by the U.S. Air Force, Contract AFOSR-72-2307, while the authors were at Courant Institute of Mathematical Sciences, New York University.

1. **Probabilistic background.** Let $\Omega = C([0,\infty), R^{d+1})$ (2) and let $z(t,\omega)$ denote the position of the path ω at time t. We will use $x(t,\omega)$ for the first d-coordinates of $z(t,\omega)$ and $y(t,\omega)$ for the $(d+1)$st coordinate. Let $\mathfrak{M}_t = \mathcal{B}[z(s): 0 \le s \le t]$, for $t \ge 0$, and put $\mathfrak{M} = \sigma(\bigcup_{t\ge 0}\mathfrak{M}_t)$. Given $z = (x,y) \in R^{d+1}$, we will denote by P_z the Wiener measure on $\langle \Omega, \mathfrak{M}\rangle$ starting at z. That is, $P_z(z(0) = z) = 1$ and

$$P_z(z(t+s) \in \Gamma \mid \mathfrak{M}_s) = \frac{1}{(2\pi t)^{(d+1)/2}} \int_\Gamma \exp(-|\xi - z(s)|^2/2t)\, d\xi \quad (\text{a.s.}, P_z)$$

for all $s, t \ge 0$ and $\Gamma \in \mathcal{B}[R^{d+1}]$.

Let $\zeta = \inf\{t \ge 0: y(t) \le 0\}$. Given $z \in R_+^{d+1} = \{(x,y) \in R^{d+1}: y > 0\}$, we will call the triple $\langle \eta(t), \mathfrak{M}_t, P_z\rangle$ a *continuous local* martingale on $[0,\zeta)$ if η is a real valued, measurable function on $\{(t,\omega) \in [0,\infty) \times \Omega: \zeta(\omega) > t\}$, $\eta(\cdot,\omega)$ is continuous on $[0,\zeta(\omega))$ for P_z-almost all ω, and there exists a sequence of stopping times $\tau_n < \zeta$ such that $\tau_n \nearrow \zeta$ (a.s., P_z) and $\langle \eta(t \wedge \tau_n), \mathfrak{M}_t, P_z\rangle$ is a bounded martingale for each n. If $\sup_n E_z[|\eta(\tau_n)|] < \infty$, then

$$\eta(\zeta) \equiv \lim_{t \nearrow \zeta} \eta(t)$$

exists for P_z-almost all ω. Moreover, if $\{\eta(\tau_n)\}_{n\ge 1}$ is uniformly P_z-integrable, then $\langle \eta(t \wedge \zeta), \mathfrak{M}_t, P_z\rangle$ is a uniformly P_z-integrable, continuous martingale.

Examples of continuous local martingales on $[0,\zeta)$ are plentiful. For instance, if $h \in C^2(R_+^{d+1})$, then $\langle h(z(t)) - \frac{1}{2}\int_0^t \Delta h(z(s))\, ds, \mathfrak{M}_t, P_z\rangle$ is a continuous local martingale on $[0,\zeta)$ for all $z \in R_+^{d+1}$. This can be easily seen by combining the basic property defining P_z with the Doob stopping time theorem applied to $\tau_n = \inf\{t \ge 0: y(t) \notin (1/n, n) \text{ or } |x(t)| \ge n\} \wedge n$. See [12] for details.

The continuous version of the Doob decomposition theorem, proved by Meyer (cf. M. Rao [9]), says that if $\langle \eta(t), \mathfrak{M}_t, P_z\rangle$ is a continuous local martingale on $[0,\zeta)$, then there is a measurable function $\Psi_\eta: [0,\infty) \times \Omega \to R$ such that $\Psi_\eta(0,\cdot) = 0$, $\Psi_\eta(\cdot,\omega)$ is continuous and nondecreasing on $[0,\zeta(\omega))$ for P_z-almost all ω, $\Psi_\eta(t,\cdot)$ is $\mathfrak{M}_t$-measurable for all $t \ge 0$, and $\langle \eta^2(t) - \Psi_\eta(t), \mathfrak{M}_t, P_z\rangle$ is a continuous local martingale on $[0,\zeta)$. Moreover, $\Psi_\eta(t \wedge \zeta)$ is unique up to a set of P_z-measure zero. In the example cited above, $\Psi_\eta(t \wedge \zeta) = \int_0^{t\wedge\zeta} |\nabla h(z(s))|^2\, ds$.

The Burkholder-Gundy inequality says that if $0 < p < \infty$, then there exist universal constants $0 < a_p \le A_p < \infty$ such that

$$a_p E_z[|\Psi_\eta(\zeta)|^{p/2}] \le E_z\left[\sup_{0\le t<\zeta} |\eta(t) - \eta(0)|^p\right] \le A_p E_z[|\Psi_\eta(\zeta)|^{p/2}] \tag{1.1}$$

for all continuous local martingales $\langle \eta(t), \mathfrak{M}_t, P_z\rangle$ on $[0,\zeta)$. A nice proof of (1.1) can be found in the paper of Getoor and Sharpe [5].

Let $1 \le p < \infty$. We will say that $\eta \in \mathcal{K}^p(\Omega)$ if $\langle \eta(t) - \eta(0), \mathfrak{M}_t, P_z\rangle$ is a continuous local martingale on $[0,\zeta)$ for all $z \in R_+^{d+1}$ and

(2) $C([0,\infty), R^{d+1})$ is the space of continuous functions on $[0,\infty)$ with values in R^{d+1}.

(1.2) $$\|\eta\|_{\mathcal{H}^p(\Omega)} \equiv \left(\sup_{y>0} \int_{R^d} E_{x,y}\left[\sup_{0\le t<\zeta} |\eta(t) - \eta(0)|^p\right] dx\right)^{1/p} < \infty.$$

Notice that if $\eta \in \mathcal{H}^p(\Omega)$ then, for each $y > 0$, $\langle \eta(t \wedge \zeta) - \eta(0), \mathfrak{M}_t, P_{x,y}\rangle$ is a continuous uniformly integrable martingale for almost every $x \in R^d$. There is a subspace of $\mathcal{H}^p(\Omega)$ which will be of particular interest to us. Namely, let $\mathcal{H}^p(R^{d+1}_+)$ be the space of harmonic functions u on R^{d+1}_+ such that

(1.3) $$\|u\|_{\mathcal{H}^p(R^{d+1}_+)} \equiv \left(\sup_{y>0} \int_{R^d} E_{x,y}\left[\sup_{0\le t<\zeta} |u(z(t))|^p\right] dx\right)^{1/p} < \infty.$$

Obviously $u \in \mathcal{H}^p(R^{d+1}_+)$ implies $\eta_u \in \mathcal{H}^p(\Omega)$ and $\|\eta_u\|_{\mathcal{H}^p(\Omega)} \le 2\|u\|_{\mathcal{H}^p(R^{d+1}_+)}$, where $\eta_u(t) = u(z(t \wedge \zeta))$.

There is one more class of martingales with which we will have to deal. We will say that $\theta \in \mathfrak{B}.\mathfrak{M}.\mathfrak{O}.(\Omega)$ if $\langle \theta(t \wedge \zeta) - \theta(0), \mathfrak{M}_t, P_z\rangle$ is a continuous square integrable martingale for every $z \in R^{d+1}_+$ and

(1.4) $$\|\theta\|_{\mathfrak{B}.\mathfrak{M}.\mathfrak{O}.(\Omega)} \equiv \sup_{z;t\ge 0} \operatorname{ess\,sup} E_z[(\theta(\zeta) - \theta(t \wedge \zeta))^2 | \mathfrak{M}_t]^{1/2} < \infty.$$

As we will see later, the notion of $\mathfrak{B}.\mathfrak{M}.\mathfrak{O}.$ martingales is a natural probabilistic outgrowth of the analytic notion of a function of bounded mean oscillation. The importance of this class of martingales to us is contained in the following theorem.

Theorem (1.1). *Let $\eta \in \mathcal{H}^1(\Omega) \cap \mathcal{H}^2(\Omega)$ and $\theta \in \mathfrak{B}.\mathfrak{M}.\mathfrak{O}.(\Omega)$. Then for $z \in R^{d+1}_+$:*

$$|E_z[(\eta(\zeta) - \eta(0))\theta(\zeta)]| \le 2^{1/2}\|\theta\|_{\mathfrak{B}.\mathfrak{M}.\mathfrak{O}.(\Omega)} E_z[(\Psi_\eta(\zeta))^{1/2}].$$

In particular, there is a universal constant F such that

$$\int_{R^d} |E_{x,y}[(\eta(\zeta) - \eta(0))\theta(\zeta)]|\, dx \le F\|\theta\|_{\mathfrak{B}.\mathfrak{M}.\mathfrak{O}.(\Omega)} \int_{R^d} E_{x,y}\left[\sup_{0\le t<\zeta} |\eta(t) - \eta(0)|\right] dx$$
$$\le F\|\eta\|_{\mathcal{H}^1(\Omega)}\|\theta\|_{\mathfrak{B}.\mathfrak{M}.\mathfrak{O}.(\Omega)}.$$

A proof of this theorem may be found in [5]. It is the inequality of Fefferman in the present context.

2. **Analytic background.** For each $1 \le j \le d$, define R_j on $L^2(R^d)$ by the relation $(R_j f)^\wedge(\xi) = i(\xi_j/|\xi|)\hat{f}(\xi)$. Clearly R_j is a bounded, translation invariant, skew-adjoint operator on $L^2(R^d)$ into itself. One can also express the action of R_j by the use of principal value integrals. Namely,

(2.1) $$R_j f(x) = \lim_{\varepsilon \searrow 0} R_j^{(\varepsilon)} * f(x),$$

where $R_j^{(\varepsilon)}(x) = a_d(x_j/|x|^{d+1})\mathfrak{X}_{[\varepsilon,\infty)}(|x|)$. The convergence takes place in $L^2(R^d)$.

These operators have an extensive theory and are known as the *Riesz transformations*. The book of Stein and Weiss [11] is an excellent reference for this material.

For each $1 \le p < \infty$, let $\|\cdot\|_{H^p(R^d)}$ be defined on $L^p(R^d) \cap L^2(R^d)$ by

$$\|f\|_{H^p(R^d)} = \sum_0^d \|R_j f\|_{L^p(R^d)},$$

where R_0 is the identity. Denote by $H^p(R^d)$ the completion with respect to $\|\cdot\|_{H^p(R^d)}$ of the class of $f \in L^p(R^d) \cap L^2(R^d)$ such that $\|f\|_{H^p(R^d)} < \infty$. Observe that $\|\cdot\|_{H^2(R^d)}$ is commensurate with $\|\cdot\|_{L^2(R^d)}$ and that $H^2(R^d) = L^2(R^d)$. We will see later that this is also the case for all $1 < p < \infty$, but it is false for $p = 1$. Obviously, the R_j determine unique bounded operators from $H^p(R^d)$ to $L^p(R^d)$. We will use R_j to denote these operators as well. Note that R_j is determined on $H^1(R^d)$ by the relation $(R_j f)^\wedge(\xi) = i(\xi_j/|\xi|)\hat{f}(\xi)$.

Lemma (2.1). *The space $H^1(R^d)$ coincides with the class of $f \in L^1(R^d)$ such that, for each $1 \le j \le d$, $i(\xi_j/|\xi|)\hat{f}(\xi)$ is the Fourier transform of a function in $L^1(R^d)$.*

It is immediately apparent from this lemma why $H^1(R^d)$ is smaller than $L^1(R^d)$. Indeed, if $f \in H^1(R^d)$, then $(\xi_j/|\xi|)\hat{f}(\xi)$ is continuous, since it is the Fourier transform of an $L^1(R^d)$-function. Hence a necessary condition for $f \in H^1(R^d)$ is that $\int_{R^d} f(x)\,dx = \hat{f}(0) = 0$.

There is another class of functions with which we will be concerned; namely, the John-Nirenberg class of functions having bounded mean oscillation. A locally integrable function ϕ on R^d will be said to have bounded mean oscillation if

$$\|\phi\|_{\text{B.M.O.}(R^d)} \equiv \sup_Q \frac{1}{|Q|}\int_Q |\phi(x) - \phi_Q|\,dx < \infty, \tag{2.2}$$ (3)

where Q runs over the cubes contained in R^d. The space of all such functions is denoted by B.M.O.(R^d). Actually, it is convenient to think of functions in B.M.O. (R^d) as being defined only up to additive constants. Then one can easily check that $\|\cdot\|_{\text{B.M.O.}(R^d)}$ is a complete norm on B.M.O.(R^d). One of the nice facts about B.M.O.(R^d) is that many maps which do not map $L^\infty(R^d)$ into itself continuously map $L^\infty(R^d)$ continuously into B.M.O.(R^d). The basic result in this direction is the following (cf. Fefferman and Stein [3]).

Lemma (2.2). *Let $K \in L^1(R^d)$ be given and suppose B is a constant satisfying $\|\hat{K}\|_{L^\infty(R^d)} \le B$ and*

$$\sup_{x \in R^d} \int_{|\xi| \ge 2|x|} |K(x - \xi) - K(-\xi)|\,d\xi \le B.$$

Then there is a constant C depending only on d and B such that $\|K * \phi\|_{\text{B.M.O.}(R^d)} \le C\|\phi\|_{L^\infty(R^d)}$, $\phi \in L^\infty(R^d)$.

(3) Here and in what follows, if f is a function on R^d and $S \subseteq R^d$, then f_S denotes that mean value of f on S.

Following Fefferman and Stein [3], one can show from Lemma (2.2) that the Riesz transforms determine a continuous map of $L^\infty(R^d)$ into B.M.O.(R^d). Indeed, the mapping with which they work is given by:

$$\lim_{\varepsilon \searrow 0} \int_{R^d} (R_j^{(\varepsilon)}(x-\xi) - R_j^{(1)}(-\xi))\phi(\xi)\,d\xi. \tag{2.3}$$

Notice that the definition here differs from that in (2.1), but the difference is contained in an additive constant and so does not matter in B.M.O.(R^d).

An immediate consequence of the preceding is the next lemma.

Lemma (2.3). *Let $f \in L^2(R^d)$ and assume that*

$$\left|\int_{R^d} f(x)\phi(x)\,dx\right| \le A\|\phi\|_{\mathrm{B.M.O.}(R^d)}$$

for all $\phi \in L^2(R^d) \cap$ B.M.O.(R^d). Then $f \in H^1(R^d)$ and $\|f\|_{H^1(R^d)} \le CA$, where C depends only on d.

We next want to give a characterization of B.M.O.(R^d) which will enable us to relate this class of functions to $\mathfrak{B}.\mathfrak{M}.\mathfrak{O}.(\Omega)$. Essential for this is the following inequality of John and Nirenberg [8].

Theorem (2.1). *Suppose $\phi \in$ B.M.O.(R^d) and that $\|\phi\|_{\mathrm{B.M.O.}(R^d)} \le 2^{-(d+1)}$. Then for all cubes Q, $(1/|Q|)\int_Q \exp[\alpha|\phi(x) - \phi_Q|]\,dx \le 2e^{4\alpha}/(2 - e^{4\alpha})$ for $\alpha < \frac{1}{4}\log 2$. In particular, there exist universal constants B_p, $1 \le p < \infty$, such that*

$$\left(\frac{1}{|Q|}\int_Q |\phi(x) - \phi_Q|^p\,dx\right)^{1/p} \le B_p\|\phi\|_{\mathrm{B.M.O.}(R^d)}.$$

A nice probabilistic derivation of the above appears in Garsia [4]. Using this result we are now able to prove the characterization of B.M.O.(R^d) which we will need. This characterization was first discovered by Gundy for S^1. The proof we give below is based on an idea due to Garsia.

Theorem (2.2). *If $\phi \in$ B.M.O.(R^d), then $\phi(x)/(1 + |x|^{d+1}) \in L^1(R^d)$ and*

$$\sup_{y>0;\,x\in R^d} [p_y * (\phi - u_\phi(x,y))^2](x) \le C\|\phi\|^2_{\mathrm{B.M.O.}(R^d)},\ (^4)$$

where C depends only on d. Conversely, if $\phi(x)/(1 + |x|^{d+1}) \in L^1(R^d)$, then

$$\|\phi\|^2_{\mathrm{B.M.O.}(R^d)} \le C' \sup_{y>0;\,x\in R^d} [p_y * (\phi - u_\phi(x,y))^2](x),$$

where C' depends only on d.

(4) Here and in what follows, if f is a function on R^d, then $u_f(x,y) = p_y * f(x)$ is its harmonic extension to R^{d+1}_+, where $p_y(x) = b_d(y/(y^2 + |x|^2))^{(d+1)/2}$.

Proof. We first observe that there are positive constants a and A depending only on d such that for all $p \geq 1$:

$$\frac{1}{2}\left(\sup_B \frac{a}{|B|}\int_B |\phi(x) - \phi_B|^p\,dx\right)^{1/p} \leq \left(\sup_Q \frac{1}{|Q|}\int_Q |\phi(x) - \phi_Q|^p\,dx\right)^{1/p}$$
$$\leq 2\left(\sup_B \frac{A}{|B|}\int_B |\phi(x) - \phi_B|^p\,dx\right)^{1/p}$$

where B runs over the balls in R^d. Combining this with Theorem (2.1), we see in particular that

$$\sup_B \frac{1}{|B|}\int_B |\phi(x) - \phi_B|^p\,dx \leq C_p \|\phi\|^p_{B.M.O.(R^d)}$$

for $1 \leq p < \infty$.

We next note that

$$p_y(x) = \int_0^\infty \frac{1}{|B(0,r)|}\chi_{B(0,r)}(x)\rho_y(r)\,dr$$

where

$$\rho_y(r) = c_d\frac{yr^{d+1}}{(r^2+y^2)^{(d+3)/2}}$$

and c_d is a positive constant. Integrating with respect to x, one sees that $\int_0^\infty \rho_y(r)\,dr = 1$.

Now suppose $\phi \in$ B.M.O.(R^d). In [3] it is shown that $\phi(x)/(1+|x|^{d+1}) \in L^1(R^d)$. In particular, $u_\phi(x,y)$ is well defined. Using the preceding paragraph, we have:

$$\begin{aligned}
[p_y * (\phi - u_\phi(x,y))^2](x) &= p_y * \phi^2(x) - (p_y * \phi(x))^2 \\
&= \int_0^\infty \rho_y(r)(\phi^2)_{B(x,r)}\,dr - \left(\int_0^\infty \rho_y(r)\phi_{B(x,r)}\,dr\right)^2 \\
&= \int_0^\infty \rho_y(r)((\phi - \phi_{B(x,r)})^2)_{B(x,r)}\,dr \\
&\quad + \int_0^\infty \rho_y(r)\left(\phi_{B(x,r)} - \int_0^\infty \rho_y(s)\phi_{B(x,s)}\,ds\right)^2 dr \\
&\leq \int_0^\infty \rho_y(r)((\phi - \phi_{B(x,r)})^2)_{B(x,r)}\,dr \\
&\quad + \int_0^\infty \rho_y(s)\,ds \int_0^\infty \rho_y(r)(\phi_{B(x,s)} - \phi_{B(x,r)})^2\,dr.
\end{aligned}$$

We know that $((\phi - \phi_{B(x,r)})^2)_{B(x,r)} \le C_1 \|\phi\|^2_{\text{B.M.O.}(R^d)}$. Moreover, $0 < r < s$,

$$\begin{aligned}|\phi_{B(x,s)} - \phi_{B(x,r)}|^{2d+2} &= |(\phi - \phi_{B(x,s)})_{B(x,r)}|^{2d+2} \\ &\le \frac{1}{|B(x,r)|}\int_{B(x,s)} |\phi(\xi) - \phi_{B(x,s)}|^{2d+2}\, d\xi \\ &= \left(\frac{s}{r}\right)^d \frac{1}{|B(x,s)|}\int_{B(x,s)} |\phi(\xi) - \phi_{B(x,s)}|^{2d+2}\, d\xi \\ &\le C_2\left(\frac{s}{r}\right)^d \|\phi\|^{2d+2}_{\text{B.M.O.}(R^d)}.\end{aligned}$$

Hence

$$\begin{aligned}&\int_0^\infty \rho_y(s)\, ds \int_0^s \rho_y(r)(\phi_{B(x,s)} - \phi_{B(x,r)})^2\, dr \\ &\qquad\le C_2^{1/(d+1)} \|\phi\|^2_{\text{B.M.O.}(R^d)} \int_0^\infty \rho_y(s)\, ds \int_0^s \rho_y(r)\left(\frac{s}{r}\right)^{d/(d+1)} dr.\end{aligned}$$

Note that

$$\int_0^\infty \rho_y(s)\, ds \int_0^s \rho_y(r)\left(\frac{s}{r}\right)^{d/(d+1)} dr = \int_0^\infty \rho_1(s)\, ds \int_0^s \rho_1(r)\left(\frac{s}{r}\right)^{d/(d+1)} dr < \infty,$$

and therefore the first assertion is proved.

To prove the converse statement, let $B = B(x,r)$ be given. Note that

$$\frac{r}{(r^2 + |x|^2)^{(d+1)/2}} = \frac{1}{r^d}\frac{1}{(|x|^2/r^2 + 1)^{(d+1)/2}} \ge \frac{1}{2^{(d+1)/2} r^d}$$

for $|x| \le r$. Hence

$$\begin{aligned}\frac{1}{|B(x,r)|}\int_{B(x,r)} (\phi(\xi) - \phi_{B(x,r)})^2\, d\xi &\le \frac{1}{|B(x,r)|}\int_{B(x,r)} (\phi(\xi) - u_\phi(x,r))^2\, d\xi \\ &\le C\int_{R^d} |\phi(\xi) - u_\phi(x,r)|^2 p_r(x - \xi)\, d\xi. \quad \text{Q.E.D.}\end{aligned}$$

We conclude this section with an important theorem due to Stein and Weiss. Let $f \in L^2(R^d)$ and define $u_j(\cdot,y) = p_y * R_j f$ for $0 \le j \le d$ and $y > 0$. Then it is easy to check that the u_j satisfy the generalized Cauchy-Riemann equations of M. Riesz (cf. p. 91 of Stein and Weiss [11]). Hence by Theorem (4.14) of [8], $(\sum_0^d u_j^2)^{\alpha/2}$ is subharmonic in R_+^{d+1} for $\alpha \ge (d-1)/d$. We will not go through the proof of this fact but will only mention that it can be checked by hand when $d = 1$ and $\alpha = \frac{1}{2}$. The general proof involves a clever application of elementary matrix algebra.

3. A projection theorem.

Lemma (3.1). *Suppose* $\phi \in$ B.M.O. (R^d) *and let* $\theta_\phi(t) = u_\phi(z(t))$, $0 < t < \zeta$. *Then* $\theta_\phi \in \mathfrak{B}.\mathfrak{M}.\mathfrak{O}.(\Omega)$ (5) *and there is a universal constant C depending only on d such that*

$$\|\theta_\phi\|_{\mathfrak{B}.\mathfrak{M}.\mathfrak{O}.(\Omega)} \leq C\|\phi\|_{\mathrm{B.M.O.}(R^d)},$$

Proof. Observe that for $z \in R_+^{d+1}$:

$$\begin{aligned} E_z[(\theta_\phi(\zeta) - \theta_\phi(t \wedge \zeta))^2 \mid \mathfrak{M}_t] &= \mathfrak{X}_{\{\zeta > t\}} E_{z(t)}[(\phi(x(\zeta)) - u_\phi(z(0)))^2] \\ &= \mathfrak{X}_{\{\zeta > t\}} p_{y(t)} * [(\phi - u_\phi(z(t)))^2](x(t)). \end{aligned}$$

Hence the lemma follows from Theorem (2.2). Q.E.D.

Lemma (3.2). *If* $\eta \in \mathcal{H}^1(\Omega) \cap \mathcal{H}^2(\Omega)$ *and* $\phi \in$ B.M.O.(R^d), *then*

$$\begin{aligned} &\int_{R^d} |E_{x,y}[(\eta(\zeta) - \eta(0))\phi(x(\zeta))]|\, dx \\ &\qquad \leq CF\|\phi\|_{\mathrm{B.M.O.}(R^d)} \int_{R^d} E_{x,y}\left[\sup_{0 \leq t < \zeta} |\eta(t) - \eta(0)|\right] dx, \end{aligned}$$

where C is the constant in Lemma 3.1 *and F is the constant in Theorem* 1.1.

Theorem (3.1). *For each* $1 \leq p < \infty$ *and* $y > 0$ *there is a continuous linear map* $\Pi_y: \mathcal{H}^p(\Omega) \to L^p(R^d)$, *such that* $\|\Pi_y \eta\|_{L^p(R^d)} \leq \|\eta\|_{\mathcal{H}^p(R^d)}$, *given by*

$$\int_{R^d} (\Pi_y \eta)(x)\phi(x)\, dx = \int_{R^d} E_{x,y}[(\eta(\zeta) - \eta(0))\phi(x(\zeta))]\, dx \tag{3.1}$$

for all $\phi \in L^q(R^d)$, $1/p + 1/q = 1$. *Moreover, if* $\eta \in \mathcal{H}^1(\Omega)$, *then* $\Pi_y \eta \in H^1(R^d)$ *and there is a universal constant* C_d, *depending only on d, such that*

$$\|\Pi_y \eta\|_{H^1(R^d)} \leq C_d \int_{R^d} E_{x,y}\left[\sup_{0 \leq t < \zeta} |\eta(t) - \eta(0)|\right] dx. \tag{3.2}$$

Finally, if η *is bounded, then* $\Pi_y \eta$ *is dominated by twice the bound on* η.

Proof. When $1 < p < \infty$ the assertion is trivial. Indeed:

$$\begin{aligned} \left|\int_{R^d} E_{x,y}[(\eta(\zeta) - \eta(0))\phi(x(\zeta))]\, dx\right| &\leq \|\eta\|_{\mathcal{H}^p(\Omega)} \left(\int_{R^d} E_{x,y}[|\phi(x(\zeta))|^q]\, dx\right)^{1/q} \\ &= \|\eta\|_{\mathcal{H}^p(\Omega)} \left(\int_{R^d} dx \int_{R^d} p_y(x - \xi)|\phi(\xi)|^q\, d\xi\right)^{1/q} \\ &= \|\eta\|_{\mathcal{H}^p(\Omega)} \|\phi\|_{L^q(R^d)}, \end{aligned}$$

(5) Note that $\langle u_\phi(z(t)), \mathfrak{M}_t, P_z\rangle$ is a continuous local martingale on $[0,\zeta)$, since $\Delta u_\phi = 0$. In particular, $u_\phi(z) = E_z[\phi(x(\zeta))]$.

and therefore the right-hand side of (3.1) determines a continuous linear functional on $L^q(R^d)$.

To treat the case when $p = 1$, we first suppose that $\eta \in \mathcal{H}^1(\Omega) \cap \mathcal{H}^2(\Omega)$. Then $\Pi_y\eta \in L^2(R^d)$. Moreover, if $\phi \in L^2(R^d) \cap$ B.M.O.(R^d), then by Lemma (3.2):

$$\left|\int_{R^d} (\Pi_y\eta)(x)\phi(x)\,dx\right| \leq CF\left(\int_{R^d} E_{x,y}\left[\sup_{0\leq t<\zeta} |\eta(t) - \eta(0)|\right] dx\right) \|\phi\|_{\text{B.M.O.}(R^d)}.$$

Hence, by Lemma (2.3), $\Pi_y\eta \in H^1(R^d)$ and (3.2) obtains. Given a general $\eta \in \mathcal{H}^1(\Omega)$, define $\eta^{(\alpha)}(t) = \eta(t \wedge \tau_\alpha)$ for $0 \leq t < \zeta$, where $\tau_\alpha = \inf\{t \geq 0: |\eta(t \wedge \zeta) - \eta(0)| \geq \alpha\}$. Clearly $\eta^{(\alpha)} \in \mathcal{H}^1(\Omega) \cap \mathcal{H}^2(\Omega)$ for all $\alpha > 0$. Also,

$$\int_{R^d} E_{x,y}\left[\sup_{0\leq t<\zeta} |\eta(t) - \eta^{(\alpha)}(t)|\right] dx$$
$$\leq 2\int_{R^d} E_{x,y}\left[\sup_{0\leq t<\zeta} |\eta(t) - \eta(0)|; \sup_{0\leq t<\zeta} |\eta(t) - \eta(0)| \geq \alpha\right] dx \to 0$$

as $\alpha \to \infty$. From this and the preceding it follows that $\Pi_y\eta^{(\alpha)}$ tends in $H^1(R^d)$ to a $\Pi_y\eta$ satisfying (3.1) and (3.2). Q.E.D.

Lemma (3.3). *Let $1 \leq p \leq \infty$ and suppose $u \in \mathcal{H}^p(R^{d+1}_+)$. Given $h > 0$, let $w(x,y) = u(x,y + h)$ in R^{d+1}_+. Then $w \in \mathcal{H}^p(R^{d+1}_+)$ and $\|w\|_{\mathcal{H}^p(R^{d+1}_+)} \leq \|u\|_{\mathcal{H}^p(R^{d+1}_+)}$.*

Proof. Observe that

$$\begin{aligned}\int_{R^d} E_{x,y}\left[\sup_{0\leq t<\zeta} |w(z(t))|^p\right] dx &= \int_{R^d} E_{x,y}\left[\sup_{0\leq t<\zeta} |u(x(t),y(t) + h)|^p\right] dx\\ &= \int_{R^d} E_{x,y+h}\left[\sup_{0\leq t<\zeta_h} |u(z(t))|^p\right] dx\\ &\leq \int_{R^d} E_{x,y+h}\left[\sup_{0\leq t<\zeta} |u(z(t))|^p\right] dx\\ &\leq \|u\|^p_{\mathcal{H}^p(R^{d+1}_+)},\end{aligned}$$

where $\zeta_h = \inf\{t \geq 0: y(t) \leq h\}$. Q.E.D.

Theorem (3.2). *Let $1 \leq p < \infty$ and $u \in \mathcal{H}^p(R^{d+1}_+)$. Set $\eta_u(t) = u(z(t))$, $0 \leq t < \zeta$. Then $\eta_u \in \mathcal{H}^p(\Omega)$ and the function $f = u(\cdot,2h) + \Pi_h\eta_u$ is independent of $h > 0$. Moreover, $u = u_f$ and $\|f\|_{L^p(R^d)} \leq \|u\|_{\mathcal{H}^p(R^{d+1}_+)}$. Finally, if $p = 1$, then $f \in H^1(R^d)$ and $\|f\|_{H^1(R^d)} \leq C\|u\|_{\mathcal{H}^1(R^{d+1}_+)}$, where C depends only on d.*

Proof. Let $h > 0$ be given and set $f = u(\cdot,2h) + \Pi_h\eta_u$. Clearly $f \in L^p(R^d)$. Moreover, if $\phi \in L^q(R^d)$, then

$$\begin{aligned}\int_{R^d} f(x)\phi(x)\,dx &= \int_{R^d} u(x,2h)\phi(x)\,dx + \int_{R^d} E_{x,h}[(\eta_u(\zeta) - \eta_u(0))\phi(x(\zeta))]\,dx\\ &= \int_{R^d} E_{x,h}[\eta_u(\zeta)\phi(x(\zeta))]\,dx,\end{aligned}$$

since

$$\int_{R^d} E_{x,h}[\eta_u(0)\phi(x(\zeta))]\,dx = \int_{R^d} u(x,h)u_\phi(x,h)\,dx = \int_{R^d} u(x,2h)\phi(x)\,dx.^{(6)}$$

Define $\zeta_n = \inf\{t \geq 0: y(t) \leq h/n\}$. Then

$$\int_{R^d} E_{x,h}[\eta_u(\zeta)\phi(x(\zeta))]\,dx$$
$$= \lim_{n\to\infty} \int_{R^d} E_{x,h}[u(x(\zeta_n), h/n)u_\phi(x(\zeta_n), h/n)]\,dx$$
$$= \lim_{n\to\infty} \int_{R^d} u(x, 2h/n)\phi(x)\,dx.$$

Taking $\phi(x) = p_y(x^0 - x)$, we obtain

$$u_f(x^0,y) = \int_{R^d} f(x)p_y(x^0 - x)\,dx = \lim_{n\to\infty} \int_{R^d} u(x,2h/n)p_y(x^0 - x)\,dx$$
$$= \lim_{n\to\infty} u(x^0, y + 2h/n) = u(x^0,y).$$

This proves that $u = u_f$, and, in particular, that $u(\cdot,2h) + \Pi_h\eta_u$ is independent of $h > 0$. Obviously, from $u = u_f$ it follows that $\|f\|_{L^p(R^d)} \leq \|u\|_{\mathcal{H}^p(R^{d+1}_+)}$, $f = \lim_{y\searrow 0} p_y * f = \lim_{y\searrow 0} u_f$.

Now suppose that $p = 1$. Since $f = u(\cdot,2) + \Pi_1\eta_u$, and $\Pi_1\eta_u \in H^1(R^{d+1}_+)$ satisfies (3.2), it suffices to prove that $u(\cdot,2) \in H^1(R^{d+1}_+)$ and $\|u(\cdot,2)\|_{H^1(R^d)} \leq C\|u\|_{\mathcal{H}^1(R^{d+1}_+)}$. Let $g = u(\cdot,2)$ and $w = u_g$. Then $w(x,y) = u(x,y+2)$ and therefore $\|w\|_{\mathcal{H}^1(R^{d+1}_+)} \leq \|u\|_{\mathcal{H}^1(R^{d+1}_+)}$. Also we have just seen that $g = w(\cdot,2h) + \Pi_h\eta_w$ for all $h > 0$. Clearly $g \in L^2(R^d)$. Given $\phi \in L^2(R^d) \cap$ B.M.O.(R^d), we have

$$\left|\int_{R^d} g(x)\phi(x)\,dx\right| = \left|\int_{R^d} w(x,2h)\phi(x)\,dx + \int_{R^d} E_{x,h}[(\eta_w(\zeta) - \eta_w(0))\phi(x(\zeta))]\,dx\right|$$
$$\leq \|w(\cdot,2h)\|_{L^2(R^d)}\|\phi\|_{L^2(R^d)} + C\|u\|_{\mathcal{H}^1(R^{d+1}_+)}\|\phi\|_{\text{B.M.O.}(R^d)}.$$

Since $\|w(\cdot,2h)\|_{L^2(R^d)} \to 0$ as $h \to \infty$, we see from Lemma (2.3) that $g \in H^1(R^d)$ and $\|g\|_{H^1(R^d)} \leq C'\|u\|_{\mathcal{H}^1(R^{d+1}_+)}$. Q.E.D.

Theorem (3.3). *If $1 < p < \infty$ and $f \in L^p(R^d)$, then $u_f \in \mathcal{H}^p(R^{d+1}_+)$ and $\|u_f\|_{\mathcal{H}^p(R^{d+1}_+)} \leq C_p\|f\|_{L^p(R^d)}$, where C_p depends only on p. If $f \in H^1(R^d)$ and $u_j = u_{R_jf}$, $0 \leq j \leq d$, then $u_j \in \mathcal{H}^1(R^{d+1}_+)$ and $\|u_j\|_{\mathcal{H}^1(R^{d+1}_+)} \leq C\|f\|_{H^1(R^d)}$, where C depends only on d.*

Proof. First suppose $f \in L^p(R^d)$ and $1 < p < \infty$. Then, by Doob's inequality,

$$E_z\left[\sup_{0\leq t<\zeta} |u_f(z(t))|^p\right] \leq C_p^p E_z[|f(x(\zeta))|^p]$$

(6) We have used the fact that convolution with p_y is selfadjoint.

for all $z \in R_+^{d+1}$. Hence

$$\int_{R^d} E_{x,y}\left[\sup_{0\le t<\zeta} |u_f(z(t))|^p\right] dx < C_p^p \int_{R^d} p_y * |f|^p(x)\,dx = C_p^p \|f\|_{L^p(R^d)}^p.$$

Next suppose $f \in H^1(R^d) \cap L^2(R^d)$ and set $u_j = u_{R_jf}$, $0 \le j \le d$. By the theorem of Stein and Weiss mentioned at the end of §2, $F = (\sum_0^d u_j^2)^{\alpha/2}$ is subharmonic in R_+^{d+1} for some $0 < \alpha < 1$. Hence, by Doob's inequality,

$$E_z\left[\sup_{0\le t<\zeta} |u_j(z(t))|\right] \le E_z\left[\sup_{0\le t<\zeta} |F(z(t))|^{1/\alpha}\right] \le C_{1/\alpha}^{1/\alpha} E_z[|F(z(\zeta))|^{1/\alpha}]$$

and therefore

$$\int_{R^d} E_{x,y}\left[\sup_{0\le t<\zeta} |u_j(z(t))|\right] dx \le C \int_{R^d} E_{x,y}\left[\sum_0^d |R_jf(x(\zeta))|\right] dx$$

$$= C\|f\|_{H^1(R^d)}.$$

Since $H^1(R^d) \cap L^2(R^d)$ is dense in $H^1(R^d)$, this completes the proof. Q.E.D.

Corollary (3.3.1). *For each* $1 < p < \infty$, *the mapping* $f \to u_f$ *is an isomorphism from* $L^p(R^d)$ *onto* $\mathcal{H}^p(R_+^{d+1})$. *Also,* $f \to u_f$ *is an isomorphism from* $H^1(R^d)$ *onto* $\mathcal{H}^1(R_+^{d+1})$. *Finally,* R_j, $1 \le j \le d$, *maps* $H^1(R^d)$ *continuously into itself.*

Corollary (3.3.2). *For each* $1 < p < \infty$, $H^1(R^d) \cap L^\infty(R^d)$ *is dense in* $L^p(R^d)$. *Moreover, if* $f \in L^p(R^d) \cap L^\infty(R^d)$, *then there exists a sequence* $\{f_n\} \subseteq H^1(R^d) \cap L^\infty(R^d)$ *such that* $f_n \to f$ *in* $L^p(R^d)$ *and* $\sup_n \|f_n\|_{L^\infty(R^d)} \le 2\|f\|_{L^\infty(R^d)}$.

Proof. Let $f \in L^p(R^d)$ for some $1 < p < \infty$. Define $\eta(t) = u_f(z(t))$, $0 \le t < \zeta$, and $\eta_\alpha(t) = \eta(t \wedge \tau_\alpha)$, $0 \le t < \zeta$, where $\tau_\alpha = (\inf\{t \ge 0\colon |\eta(t) - \eta(0)| \vee |x(t)| \ge \alpha\}) \wedge \zeta$. Clearly $\Pi_y\eta_\alpha \in H^1(R^d) \cap L^\infty(R^d)$ for all $\alpha > 0$ and $y > 0$. Moreover, $f - \Pi_y\eta_\alpha = u_f(\cdot, 2y) + \Pi_y(\eta - \eta_\alpha)$. Since $u_f(\cdot, 2y) \to 0$ in $L^p(R^d)$ (7) as $y \to \infty$, it remains only to show that $\|\Pi_y(\eta - \eta_\alpha)\|_p \to 0$ as $\alpha \to \infty$ for each $y > 0$. But

$$\|\Pi_y(\eta - \eta_\alpha)\|_p^p \le 2^p \int_{R^d} E_{x,y}\left[\sup_{0\le t<\zeta} |\eta(t) - \eta(0)|^p, \tau_\alpha < \zeta\right] dx \to 0 \quad \text{as } \alpha \to \infty$$

since $\tau_\alpha \nearrow \zeta$ (a.s., P_z) for all $z \in R_+^{d+1}$. Q.E.D.

Remark (3.1). A slight variation of the reasoning just given yields the following multi-dimensional analogue of the F. and M. Riesz theorem. Let u_0 be a harmonic function in R_+^{d+1}. Then $u_0 = u_f$ for some $f \in H^1(R^d)$ if and only if there exist functions $u_1, \ldots, u_d$ which are harmonic in R_+^{d+1} such that the $(d + 1)$-triple $(u_0, \ldots, u_d)$ satisfies the generalized Cauchy-Riemann equation of M. Riesz in R_+^{d+1} and

(7) This is obvious, since $p_y \to 0$ in L^q and $\|p_y\|_{L^1(R^d)} = 1$.

$$\sup_{y>0} \int_{R^d} \left(\sum_0^d u_j^2(x,y) \right)^{1/2} dx < \infty.$$

The "only if" statement is obvious; simply take $u_j = u_{R_j f}$. To prove the "if" assertion it is enough to show that $u_0 \in \mathcal{H}^1(R_+^{d+1})$. But by the theorem of Stein and Weiss at the end of §2, we know that $(\sum_0^d u_j^2(x,y))^{\alpha/2}$ is subharmonic in R_+^{d+1} for some $0 < \alpha < 1$. Hence, reasoning just as we did in the proof of Theorem (3.3),

$$\int_{R^d} E_{x,y}\left[\sup_{0 \le t < \zeta_h} |u_0(z(t))| \right] dx < C_{1/\alpha}^{1/\alpha} \int_{R^d} E_{x,y}\left[\left(\sum_0^d u_j^2(x(\zeta_h), h) \right)^{1/2} \right] dx$$

$$= C_{1/\alpha}^{1/\alpha} \int_{R^d} \left(\sum_0^d u_j^2(x,h) \right)^{1/2} dx$$

for all $0 < h < y$, where $\zeta_h = \inf\{t \ge 0 : y(t) \le h\}$.

Remark (3.2). We have shown that if $f \in L^1(R^d)$ satisfies

$$(*) \qquad \sup_{y>0} \int_{R^d} E_{x,y}\left[\sup_{0 \le t < \zeta} |u_f(z(t))| \right] dx < \infty,$$

then $f \in H^1(R^d)$. In particular, $\int_{R^d} f(x)\,dx = 0$. In fact, using the lemma on p. 225 of Stein [10], it is easy to show that $u_f(\cdot,y) \to 0$ in $L^1(R^d)$ as $y \to \infty$, and therefore, since the R_j are continuous on $H^1(R^d)$ into itself and commute with convolution by p_y, $u_f(\cdot,y) \to 0$ in $H^1(R^d)$. However, we have found no direct probabilistic proof of any of these properties of f directly from $(*)$. It would be interesting to find such a proof.

4. **An interpolation theorem.** Let $\mathcal{H}^\infty(\Omega)$ stand for the class of $\eta: [0,\infty) \times \Omega \to R$ such that $\langle \eta(t) - \eta(0), \mathfrak{M}_t, P_z \rangle$ is a continuous local martingale on $[0,\zeta)$ for each $z \in R_+^{d+1}$ and for which there exists a number $B < \infty$ with the property that

$$\sup_{z \in R_+^{d+1}} P_z\left(\sup_{0 \le t < \zeta} |\eta(t) - \eta(0)| > B \right) = 0.$$

Lemma (4.1). *Let $\langle X, \mathcal{B}, \lambda \rangle$ be a measure space and suppose $\mathcal{T}$ is a subadditive mapping of $\mathcal{H}^1(\Omega) \cap \mathcal{H}^\infty(\Omega)$ into a measurable function on $\langle X, \mathcal{B} \rangle$. Further, assume that there exist numbers A,B, and $1 < p < \infty$ such that*

$$\lambda(|\mathcal{T}\eta| \ge \gamma) \le \frac{A}{\gamma} \int_{R^d} E_{x,y}\left[\sup_{0 \le t < \zeta} |\eta(t) - \eta(0)| \right] dx$$

and

$$\lambda(|\mathcal{T}\eta| \ge \gamma) \le \frac{B}{\gamma^p} \int_{R^d} E_{x,y}\left[\sup_{0 \le t < \zeta} |\eta(t) - \eta(0)|^p \right] dx,$$

for all $\gamma > 0$ *and some* $y > 0$. *Then for each* $1 < r < p$, $\mathfrak{T}$ *has a unique extension to* $\mathcal{K}^r(\Omega)$ *such that*

$$\|\mathfrak{T}\eta\|_{L^r(\lambda)} \le C\left(\int_{R^d} E_{x,y}\left[\sup_{0\le t<\zeta}|\eta(t)-\eta(0)|^r\right]dx\right)^{1/r}. \tag{4.1}$$

The constant C in (4.1) *depends only on p, r, A, and B.*

Proof. Since $\mathcal{K}^1(\Omega) \cap \mathcal{K}^\infty(\Omega)$ is dense in $\mathcal{K}^r(\Omega)$ with respect to the norm on the right-hand side of (4.1) (cf. the proof of Corollary (3.3.2)), it suffices to prove (4.1) for $\eta \in \mathcal{K}^1(\Omega) \cap \mathcal{K}^\infty(\Omega)$. Let such an η be given. For convenience we will use $\|\eta\|$ to denote $\sup|\eta(t) - \eta(0)|$. By subadditivity

$$\lambda(|\mathfrak{T}\eta| \ge \gamma) \le \lambda(|\mathfrak{T}\eta^{(\gamma)}| \ge \gamma/2) + \lambda(|\mathfrak{T}\eta_{(\gamma)}| \ge \gamma/2),$$

where $\eta^{(\gamma)}(t) = \eta(t \wedge \tau_\gamma)$, $0 \le t < \zeta$, $\eta_{(\gamma)} = \eta - \eta^{(\gamma)}$, and $\tau_\gamma = \inf\{t \ge 0$: $|\eta(t$

$$\begin{aligned}\lambda(|\mathfrak{T}\eta^{(\gamma)}| \ge \gamma/2) &\le \frac{2^pB}{\gamma^p}\int_{R^d} E_{x,y}[\|\eta\|^p \wedge \gamma^p]\,dx\\ &\le \frac{2^pB}{\gamma^p}\int_{R^d} E_{x,y}[\|\eta\|^p, \|\eta\| \le \gamma]\,dx + 2^pB\int_{R^d} P_{x,y}(\|\eta\| \ge \gamma)\,dx.\end{aligned}$$

Hence

$$\begin{aligned} &r\int_0^\infty \gamma^{r-1}\lambda(|\mathfrak{T}\eta^{(\gamma)}| \ge \gamma/2)\,d\gamma\\ &\quad\le 2^prB\int_{R^d}dx\int_0^\infty \gamma^{r-p-1}E_{x,y}[\|\eta\|^p, \|\eta\| \le \gamma]\,d\gamma\\ &\qquad + 2^prB\int_{R^d}dx\int_0^\infty \gamma^{r-1}P_{x,y}(\|\eta\| \ge \gamma)\,d\gamma\\ &\quad\le 2^prpB\int_{R^d}dx\int_0^\infty \gamma^{r-p-1}\,d\gamma\int_0^\gamma \alpha^{p-1}P_{x,y}(\|\eta\| \ge \alpha)\,d\alpha\\ &\qquad + 2^pB\int_{R^d} E_{x,y}[\|\eta\|^r]\,dx\\ &\quad= \frac{2^prpB}{p-r}\int_{R^d}dx\int_0^\infty \alpha^{r-1}P_{x,y}(\|\eta\| \ge \alpha)\,d\alpha\\ &\qquad + 2^pB\int_{R^d} E_{x,y}[\|\eta\|^r]\,dx\\ &\quad= 2^pB\frac{2p-r}{p-r}\int_{R^d} E_{x,y}[\|\eta\|^r]\,dx.\end{aligned}$$

Also,

$$\lambda(|\mathfrak{T}\eta_{(\gamma)}| \ge \gamma/2) \le \frac{4A}{\gamma}\int_{R^d} E_{x,y}[\|\eta\|, \|\eta\| \ge \gamma]\,dx,$$

and so

$$r\int_0^\infty \gamma^{r-1}\lambda(|\mathfrak{T}\eta_{(\gamma)}| \geq \gamma/2)\,d\gamma$$
$$\leq 4Ar\int_{R^d} dx \int_0^\infty \gamma^{r-2}E_{x,y}[\|\eta\|, \|\eta\| \geq \gamma]\,d\gamma$$
$$= 4Ar\int_{R^d} dx \int_0^\infty d\gamma\gamma^{r-2}\left[\int_\gamma^\infty P_{x,y}(\|\eta\| \geq \alpha)\,d\alpha + \gamma P_{x,y}(\|\eta\| \geq \gamma)\right]$$
$$\leq \frac{8Ar}{r-1}\int_{R^d} dx \int_0^\infty \alpha^{r-1}P_{x,y}(\|\eta\| \geq \alpha)\,d\alpha$$
$$= \frac{8A}{r-1}\int_{R^d} E_{x,y}[\|\eta\|^r]\,dx.$$

Combining these, we arrive at (4.1). Q.E.D.

Theorem (4.1). *Let $\langle X, \mathfrak{B}, \lambda\rangle$ be a σ-finite measure space and suppose T is a subadditive mapping of $H^1(R^d) \cap L^\infty(R^d)$ into measurable functions on $\langle X, \mathfrak{B}\rangle$. If there exist numbers A, B, and $1 < p < \infty$ such that*

$$\lambda(|Tf| \geq \gamma) \leq (A/\gamma)\|f\|_{H^1(R^d)}$$

and

$$\lambda(|Tf| \geq \gamma) \leq (B/\gamma^p)\|f\|^p_{L^p(R^d)}$$

for all $\gamma > 0$, then for $1 < r < p$, T has a unique extension to $L^r(R^d)$ such that

$$\|Tf\|_{L^r(\lambda)} \leq C\|f\|_{L^r(R^d)}. \tag{4.2}$$

The constant C depends only on A, B, p, r, and d.

Proof. By Corollary (3.3.2), $H^1(R^d) \cap L^\infty(R^d)$ is dense in $L^r(R^d)$. Hence it suffices to prove (4.2) for $f \in H^1(R^d) \cap L^\infty(R^d)$. Also, we may assume λ is finite.

Given $y > 0$, define $\mathfrak{T}_y = T \circ \Pi_y$ on $\mathcal{H}^1(\Omega) \cap \mathcal{H}^\infty(\Omega)$. Then it is easily checked that $\mathfrak{T}_y$ satisfies the hypotheses of Lemma (4.1). Hence $\mathfrak{T}_y$ has a unique extension to $\mathcal{H}^r(\Omega)$ such that

$$\|\mathfrak{T}_y\eta\|_{L^r(\lambda)} \leq C\left(\int_{R^d} E_{x,y}\left[\sup_{0\leq t<\zeta}|\eta(t) - \eta(0)|^r\right]dx\right)^{1/r}. \tag{4.3}$$

The constant C in (4.3) is independent of $y > 0$.

Now let $f \in H^1(R^d) \cap L^\infty(R^d)$ be given. Then $f = u_f(\cdot, 2y) + \Pi_y\eta_f$, and so $|Tf| \leq |Tu_f(\cdot, 2y)| + |\mathfrak{T}_y\eta_f|$. Since $u_f(\cdot, 2y) \to 0$ in $L^p(R^d)$, the hypotheses of the theorem guarantee that $Tu_f(\cdot, 2y) \to 0$ in $L^r(\lambda)$ as $y \to \infty$. Hence, by Fatou's lemma,

$$\|Tf\|_{L^r(\lambda)} \leq \liminf_{y\to\infty}\|\mathfrak{T}_y\eta_f\|_{L^r(R^d)} \leq C\|\eta_f\|_{\mathcal{H}^r(\Omega)}$$
$$\leq 2C\|u_f\|_{\mathcal{H}^r(R^{d+1}_+)} \leq C'\|f\|_{L^r(R^d)}. \quad \text{Q.E.D.}$$

Corollary (4.1.1). *For each* $1 \leq j \leq d$ *and* $1 < p < \infty$, R_j *has a unique continuous extension to* $L^p(R^d)$ *into itself. In particular,* $H^p(R^d)$ *is isomorphic to* $L^p(R^d)$ *for* $1 < p < \infty$.(8)

Corollary (4.1.2). *Suppose* T *is a linear map on* $H^1(R^d) \cap L^2(R^d)$ *into measurable functions on* R^d. *Assume that*

$$\|Tf\|_{L^2(R^d)} \leq A\|f\|_{L^2(R^d)}$$

and

$$\|T^*f\|_{\text{B.M.O.}(R^d)} \leq B\|f\|_{L^\infty(R^d)}$$

where T^* *is the adjoint of* T *on* $L^2(R^d)$. *Then* T *has a unique bounded extension as a map from* $H^r(R^d)$ *into* $L^r(R^d)$ *for* $1 \leq r \leq 2$. *Moreover, if* T *commutes with all the* R_j, $1 \leq j \leq d$, *then* T *maps* $H^1(R^d)$ *continuously into itself.*

Proof. Suppose $f \in H^1(R^d) \cap L^2(R^d)$ is given. Given $\phi \in L^2(R^d) \cap L^\infty(R^d)$, we have

$$\left|\int_{R^d} Tf(x)\phi(x)\,dx\right| \leq C\|f\|_{H^1(R^d)}\|\phi\|_{L^\infty(R^d)},$$

and therefore T extends uniquely as a bounded map of $H^1(R^d)$ into $L^1(R^d)$. We now apply Theorem (4.1) to conclude that T has a unique bounded extension as a map from $H^r(R^d)$ into $L^r(R^d)$, $1 < r \leq 2$. Finally, if T commutes with all the R_j, then

$$\|R_j Tf\|_{L^1(R^d)} = \|TR_j f\|_{L^1(R^d)} \leq C\|R_j f\|_{H^1(R^d)} \leq C'\|f\|_{H^1(R^d)}$$

for all $f \in H^1(R^d) \cap L^2(R^d)$ and $1 \leq j \leq d$. Q.E.D.

Corollary (4.1.3). *If* $K \in L^1(R^d)$ *satisfies the conditions of Lemma* (2.2), *then* $\|K * f\|_{H^r(R^d)} \leq C_r(B)\|f\|_{H^r(R^d)}$, $1 \leq r < \infty$, *where* $C_r(B)$ *depends only on* r *and* B.

Remark (4.1). Corollary (4.1.3) contains the hard part of some of Hörmander's results in [5]. In particular, his result about "almost L^1" operators follows from Corollary (4.1.3) and the type of reasoning used in showing that R_j is bounded on $L^\infty(R^d)$ to B.M.O.(R^d).

References

1. D. Burkholder and R. Gundy, *Extrapolation and interpolation of quasi-linear operators on martingales*, Acta. Math. **124** (1970) 249–304.
2. D. Burkholder, R. Gundy and M. Silverstein, *A maximal function characterization of* H^p, Trans. Amer. Math. Soc. **157** (1971), 137–153. MR **43** #527.
3. C. Fefferman and E. Stein, H^p *spaces of several variables* (to appear).

(8) To see the result for $2 < p < \infty$, one uses a trivial duality argument.

4. A. Garsia, *Notes on B.M.O. martingales and related topics* (to appear).

5. R. Getoor and M. Sharpe, *Conformal martingales* (to appear).

6. L. Hörmander, *Estimates for translation invariant operators in L^p spaces*, Acta. Math. **104** (1960) 93–140. MR **22** #12389.

7. S. Igari, *An extension of the interpolation theorem of Marcinkiewicz*, Proc. Japan Acad. **38** (1962), 731–734. MR **26** #5328.

8. F. John and L. Nirenberg, *On functions of bounded mean oscillation*, Comm. Pure Appl. Math. **14** (1961), 415–426. MR **24** #A1348.

9. M. Rao, *On decomposition theorems of Meyer*, Math. Scand. **24** (1969), 66–78. MR **43** #1264.

10. E. Stein, *Singular integrals and differentiability properties of functions*, Princeton Math. Series, no. 30, Princeton Univ. Press, Princeton, N.J., 1970. MR **44** #7280.

11. E. Stein and G. Weiss, *Introduction to Fourier analysis on Euclidean spaces*, Princeton Univ. Press, Princeton, N.J., 1971.

12. D. Stroock and S. R. S. Varadhan, *Diffusion processes with continuous coefficients*. I, Comm. Pure Appl. Math. **22** (1969), 345–400. MR **40** #6641.

COURANT INSTITUTE OF MATHEMATICAL SCIENCES, NEW YORK UNIVERSITY, NEW YORK, NEW YORK 10012 (Current address of S. R. S. Varadhan)

Current address (D. Stroock): Department of Mathematics, University of Colorado, Boulder, Colorado 80302

STUDIA MATHEMATICA, T. LXVIII. (1980)

Kac functional and Schrödinger equation

by

KAI LAI CHUNG* and S. R. S. VARADHAN (Stanford, Calif.)

1. Introduction. Let $X = \{x(t),\ t \geqslant 0\}$ be a strong Markov process with continuous paths on $\boldsymbol{R} = (-\infty, +\infty)$. Such a process is often called a *diffusion*. For each real b, we define the hitting time τ_b as follows:

$$\tau_b = \inf\{t > 0 \mid x(t) = b\}. \tag{1}$$

Let P_a and E_a denote as usual the basic probability and expectation associated with paths starting from a. It is assumed that for every a and b, we have

$$P_a\{\tau_b < \infty\} = 1. \tag{2}$$

Now let q be a bounded Borel measurable function on $\boldsymbol{R}$, and write for brevity

$$e(t) = \exp \int_0^t q\big(x(s)\big)\,ds. \tag{3}$$

This is a multiplicative functional introduced by M. Kac in [3]. In this paper we study the quantity

$$u(a, b) = E_a\{e(\tau_b)\}. \tag{4}$$

Since q is bounded below, (2) implies that $u(a, b) > 0$ for every a and b, but it may be equal to $+\infty$. A fundamental property of u is given by

$$u(a, b)u(b, c) = u(a, c), \tag{5}$$

valid for $a < b < c$, or $a > b > c$. This is a consequence of the strong Markov property (SMP).

2. Probabilistic investigation. We begin by defining two abscissas of finiteness, one for each direction.

$$\begin{aligned} \beta &= \inf\{b \in \boldsymbol{R} \mid \exists a < b:\ u(a, b) = \infty\} \\ &= \sup\{b \in \boldsymbol{R} \mid \forall a < b:\ u(a, b) < \infty\}; \\ \alpha &= \sup\{a \in \boldsymbol{R} \mid \exists b > a:\ u(b, a) = \infty\} \\ &= \inf\{a \in \boldsymbol{R} \mid \forall b > a:\ u(b, a) < \infty\}. \end{aligned} \tag{6}$$

* Research supported in part by NSF Grant MCS77-01319

It is possible, e.g., that $\beta = -\infty$ or $+\infty$. The first case occurs when X is the standard Brownian motion, and $q(x) \equiv 1$; for then, $u(a, b) \geqslant E_a(\tau_b) = \infty$, for any $a \neq b$.

LEMMA 1. *We have*

$$\beta = \inf\{b \in \boldsymbol{R} |\ \forall a < b:\ u(a, b) = \infty\}$$
$$= \sup\{b \in \boldsymbol{R} |\ \exists a < b:\ u(a, b) < \infty\};$$
$$\alpha = \sup\{a \in \boldsymbol{R} |\ \forall b > a:\ u(b, a) = \infty\}$$
$$= \inf\{a \in \boldsymbol{R} |\ \exists b > a:\ u(b, a) < \infty\}.$$

Proof. It is sufficient to prove the first equation above for β, because the second is trivially equivalent to it, and the equations for α follow by similar arguments. Suppose $u(a, b) = \infty$; then for $x < a < b$ we have $u(x, b) = \infty$ by (5). For $a < x < b$ we have by SMP,

$$u(x, b) \geqslant E_x\{e(\tau_a);\ \tau_a < \tau_b\} u(a, b) = \infty$$

since $P_x\{\tau_a < \tau_b\} > 0$ in consequence of (2).

The next lemma is a martingale argument. Let $\mathfrak{J}_t$ be the σ-field generated by $\{x_s,\ 0 \leqslant s \leqslant t\}$ and all null sets, so that $\mathfrak{J}_{t+} = \mathfrak{J}_t$ for $t \geqslant 0$; and for any optional τ let $\mathfrak{J}_{\tau+}$, $\mathfrak{J}_\tau$, $\mathfrak{J}_{\tau-}$ have the usual meanings.

LEMMA 2. *If* $a < b < \beta$, *then*

$$\lim_{a \uparrow b} u(a, b) = 1; \tag{7}$$

$$\lim_{b \downarrow a} u(a, b) = 1. \tag{8}$$

Proof. Let $a < b_n \uparrow b$ and consider

$$E_a\{e(\tau_b) |\ \mathfrak{J}(\tau_{b_n})\}, \qquad n \geqslant 1. \tag{9}$$

Since $b < \beta$, $u(a, b) < \infty$ and the sequence in (9) forms a martingale. As $n \uparrow \infty$, $\tau_{b_n} \uparrow \tau_b$ a.s. and $\mathfrak{J}(\tau_{b_n}) \uparrow \mathfrak{J}(\tau_b^-)$. Since $e(\tau_b) \in \mathfrak{J}(\tau_b^-)$, the limit of the martingale is a.s. equal to $e(\tau_b)$. On the other hand, the conditional probability in (9) is also equal to

$$E_a\left\{e(\tau_b) \exp\left(\int_{\tau_{b_n}}^{\tau_b} q\big(x(s)\big)\, ds\right) \Big|\ \mathfrak{J}(\tau_{b_n})\right\} = e(\tau_{b_n}) u(b_n, b).$$

As $n \uparrow \infty$, this must then converge to $e(\tau_b)$ a.s.; since $e(\tau_{b_n})$ converges to $e(\tau_b)$ a.s., we conclude that $u(b_n, b) \to 1$. This establishes (7).

Now let $\beta > b > a_n \downarrow a$, and consider

$$E_a\{e(\tau_b) |\ \mathfrak{J}(\tau_{a_n})\}, \qquad n \geqslant 1. \tag{10}$$

This is again a martingale. Although $a \to \tau_a$ is a.s. left continuous, not right continuous, for each fixed a we do have $\tau_{a_n} \downarrow \tau_a$ and $\mathfrak{J}(\tau_{a_n}) \downarrow \mathfrak{J}(\tau_a)$.

Hence we obtain as before $u(a_n, b) \to u(a, b)$ and consequently

$$u(a, a_n) = \frac{u(a, b)}{u(a_n, b)} \to 1.$$

This establishes (8).

The next result illustrates the basic probabilistic method.

THEOREM 1. *The following three propositions are equivalent:*

(i) $\beta = +\infty$;

(ii) $\alpha = -\infty$;

(iii) *For every* a *and* b, *we have*

$$u(a, b)\,u(b, a) \leqslant 1. \tag{11}$$

Proof. Suppose $x(0) = b$ and let $a < b < c$. If (i) is true, then $u(b, c) < \infty$ for every $c > b$. Define a sequence of successive hitting times T_n as follows (where θ denotes the usual shift operator):

$$S = \begin{cases} \tau_a & \text{if} \quad \tau_a < \tau_c, \\ \infty & \text{if} \quad \tau_c < \tau_a; \end{cases}$$

$$T_0 = 0, \quad T_1 = S, \tag{12}$$

$$T_{2n} = T_{2n-1} + \tau_b \circ \theta_{T_{2n-1}}, \qquad T_{2n+1} = T_{2n} + S \circ \theta_{T_{2n}},$$

for $n \geqslant 1$. Define also

$$N = \min\{n \geqslant 0 \mid T_{2n+1} = \infty\}. \tag{13}$$

It follows from $P_b\{\tau_c < \infty\} = 1$ that $0 \leqslant N < \infty$ a.s. For $n \geqslant 0$, we have

$$E_b\{e(\tau_c)\colon N = n\} = E_b\Big\{\exp\Big(\sum_{k=0}^{2n} \int_{T_k}^{T_{k+1}} q\big(x(s)\big)\,ds\Big)\Big\} \tag{14}$$

$$= E_b\{e(\tau_a);\ \tau_a < \tau_c\}^n\, E_a\{e(\tau_b)\}^n E_b\{e(\tau_c);\ \tau_c < \tau_a\}.$$

Since the sum of the first term in (14) over $n \geqslant 0$ is equal to $u(b, c) < \infty$, the sum of the last term in (14) must converge. Thus we have

$$E_b\{e(\tau_a);\ \tau_a < \tau_c\}\, u(a, b) < 1. \tag{15}$$

Letting $c \to \infty$ we obtain (11). Hence $u(b, a) < \infty$ for every $a < b$ and so (ii) is true. Exactly the same argument shows that (ii) implies (iii) and so also (i).

We are indebted to R. Durrett for ridding the next lemma of a superfluous condition.

LEMMA 3. *Given any* $a \in \boldsymbol{R}$ *and* $Q > 0$, *there exists an* $\varepsilon = \varepsilon(a, Q)$ *such that*

$$E_a\{e^{Q\sigma_\varepsilon}\} < \infty \tag{16}$$

where $\sigma_\varepsilon = \inf\{t > 0 | \; x(t) \notin (a-\varepsilon, a+\varepsilon)\}$.

Proof. Since X is strong Markov and has continuous paths, there is no "stable" point. This implies $P_a\{\sigma_\varepsilon \geqslant 1\} \to 0$ as $\varepsilon \to 0$ and so there exists ε such that

$$P_a\{\sigma_\varepsilon \geqslant 1\} < e^{-(Q+1)}. \tag{17}$$

Now σ_ε is a terminal time, so $x \to P_x\{\sigma_\varepsilon \geqslant 1\}$ is an excessive function for the process X killed at σ_ε. Hence by standard theory it is finely continuous. By for a diffusion under hypothesis (2) it is clear that fine topology coincides with the Euclidean. Thus $x \to P_x\{\sigma_\varepsilon \geqslant 1\}$ is in fact continuous. (1) It now follows that we have, further decreasing ε if necessary:

$$\sup_{|x-a|<\varepsilon} P_x\{\sigma_\varepsilon \geqslant 1\} < e^{-(Q+1)}. \tag{18}$$

A familiar inductive argument then yields for all $n \geqslant 1$

$$P_a\{\sigma_\varepsilon \geqslant n\} < e^{-n(Q+1)} \tag{19}$$

and (16) follows.

Lemma 4. *For any* $a < \beta$ *we have*

$$u(a, \beta) = \infty; \tag{20}$$

for any $b > a$ *we have* $u(b, a) = \infty$.

Proof. We will prove that if $u(a, b) < \infty$, then there exists $c > b$ such that $u(b, c) < \infty$. This implies (20) by Lemma 1, and the second assertion is proved similarly.

Let $Q = \|q\|$. Given b we choose a and b so that $a < b < d$ and

$$E_b\{e^{Q(\tau_a \wedge \tau_d)}\} < \infty. \tag{21}$$

This is possible by Lemma 3. Now let $b < c < d$; then as $c \downarrow b$ we have

$$E_b\{e(\tau_a); \; \tau_a < \tau_c\} \leqslant E_b\{e^{Q(\tau_a \wedge \tau_d)}; \; \tau_a < \tau_c\} \to 0 \tag{22}$$

because $P_b\{\tau_a < \tau_c\} \to 0$. Hence there exists c such that

$$E_b\{e(\tau_a); \; \tau_a < \tau_c\} < \frac{1}{u(a, b)}. \tag{23}$$

This is just (15) above, and so reversing the argument there, we conclude that the sum of the first term in (14) over $n \geqslant 0$ must converge. Thus $u(b, c) < \infty$, as was to be shown.

To sum up:

Theorem 2. *The function* $(a, b) \to u(a, b)$ *is continuous in the region* $a \leqslant b < \beta$ *and in the region* $a < b \leqslant a$. *Furthermore, extended continuity*

(1) This fact can also be proved in an elementary way.

holds in $a \leqslant b \leqslant \beta$ *and* $a \leqslant b \leqslant a$, *except at* (β, β) *when* $\beta < \infty$, *and at* (a, a) *when* $a > -\infty$.

Proof. To see that there is continuity in the extended sense at (a, β), where $a < \beta$, let $a < b_n \uparrow \beta$. Then we have by Fatou's lemma

$$\varliminf_{n\to\infty} u(a, b_n) \geqslant E_a\{\lim_{n\to\infty} e(\tau_{b_n})\} = E_a\{e(\tau_\beta)\} = u(a, \beta) = \infty.$$

If $\beta < \infty$, then $u(\beta, \beta) = 1$ by definition, but $u(a, \beta) = \infty$ for all $a < \beta$; hence u is not-continuous at (β, β). The case for a is similar.

3. The Schrödinger equation. From now on the process X will be the standard Brownian motion on $\boldsymbol{R}$ and q will be bounded and continuous in $\boldsymbol{R}$. The Schrödinger equation

$$\tfrac{1}{2}\varphi''(x) + q(x)\varphi(x) = 0, \qquad x \in \boldsymbol{R} \tag{24}$$

will be referred to as "the equation", and any of its solutions "a solution". The fundamental existence and uniqueness theorem for linear differential equations (with a Lipshitz condition) is applicable and guarantees a unique solution for given initial values $\varphi(a)$ and $\varphi'(a)$ for any $a \in \boldsymbol{R}$. A fortiori, it guarantees the unique extension of any solution given in a non-empty interval to a solution in $\boldsymbol{R}$.

LEMMA 5. *Let* φ *be a solution in* $[a, b]$ *with* $\varphi(a) = \varphi(b) = 0$; *then* $b - a \geqslant (2Q)^{-1/2}$, *where* $Q = \|q\|$. *Let* $0 < b - a < (2Q)^{-1/2}$, $A > 0$, $B > 0$; *then there is a unique solution* φ *satisfying*

$$\varphi(a) = A, \qquad \varphi(b) = B, \tag{25}$$

and $\varphi > 0$ *in* $[a, b]$.

Proof. The first assertion is known as de la Vallée Poussin's theorem. The second assertion then follows from a well-known criterion for the solvability of the equation with given boundary conditions. For both results see [5], pp. 91–92. The solution cannot vanish more than once by the first assertion. Nor can it vanish just once for then it must assume its minimum there and so Φ' must also vanish which is impossible by the uniqueness theorem.

LEMMA 6. *Let* Φ *be a solution in* $\boldsymbol{R}$. *For* $-\infty < a < b < \infty$ *define*

$$\tau = \inf\{t \geqslant 0 \mid x(t) \notin [a, b]\}\tag{26}$$

and

$$M(t) = \Phi(x(t)) \exp \int_0^t q(x(s))\, ds. \tag{27}$$

Then $\{M(t \wedge \tau), \mathfrak{F}(t), t \geqslant 0\}$ *is a martingale.*

Proof. Using Ito's formula (see [2]), we have

$$(28)\qquad dM(t) = \exp\int_0^t q(x(s))\,ds \times$$
$$\times\Big[\Phi'(x(t))\,dx(t) + \big(\tfrac{1}{2}\Phi''(x(t)) + q(x(t))\,\Phi(x(t))\big)dt\Big],$$

namely, for $t \geqslant 0$:

$$(29)\qquad M(t) - M(0) = \int_0^t \Big[\exp\int_0^s q(x(r))\,dr\Big]\Phi'(x(s))\,dx(s) +$$
$$+\int_0^t \Big[\exp\int_0^s q(x(r))\,dr\Big][\tfrac{1}{2}\Phi'' + q\Phi](x(s))\,ds.$$

If we substitute $t\wedge\tau$ for t in the above, the second term on the right vanishes because Φ is a solution. The first term is then of the form $\int_0^t f(s)\,dx(s)$ where

$$f(s) = \chi(s)\Big[\exp\int_0^s q(x(r))\,dr\Big]\,\Phi'(x(s)),$$

$$(30)\qquad \chi(s, w) = \begin{cases} 1 & \text{if} \quad s < \tau(w), \\ 0 & \text{if} \quad s \geqslant \tau(w). \end{cases}$$

Clearly, $f(s, w)$ is progressively measurable, being right continuous, and

$$(31)\qquad E_x\Big\{\int_0^t f(x)^2\,ds\Big\} < \infty, \qquad x \in \mathbf{R},$$

because φ' as well as φ is bounded in $[a, b]$. Thus the first term on the right side of (29) is an Ito integral, hence a martingale.

THEOREM 3. *Suppose $u(x, b) < \infty$ for some, hence all, $x < b$. Then $u(\cdot, b)$ is a solution in $(-\infty, b)$.*

Proof. Let $x_1 < x < x_2 < b$ where $x_2 - x_1 < (2Q)^{-1/2}$. Then by Lemma 5, there is a solution Φ satisfying

$$(32)\qquad \Phi(x_i) = u(x_i, b), \qquad i = 1, 2.$$

Let

$$\sigma = \inf\{t \geqslant 0 \mid x(t) \notin [x_1, x_2]\}$$

and M be as in (27). Then $M(t\wedge\sigma)$ is a martingale by Lemma 6. Hence for $t \geqslant 0$ we have

$$(33)\qquad \Phi(x) = E_x\{M(0)\} = E_x\{M(t\wedge\sigma)\}$$
$$= E_x\{M(\sigma);\ \sigma < t\} + E_x\{M(t);\ \sigma \geqslant t\}.$$

By Lemma 5, we can choose $x_2 - x_1$ so small that

$$E\,\{e^{Q\sigma}\} < \infty. \tag{34}$$

Then

$$E_x\{M(t);\ \sigma \geqslant t\} \sup_{x_1 \leqslant x \leqslant x_2} |\Phi| \int_{\{\sigma \geqslant t\}} e^{Q\sigma} dP_x$$

converges to zero as $t \to \infty$ by (34), since $\sigma < \infty$ almost surely. Therefore, if we let $t \to \infty$ in (33) and use (32) we obtain

$$\Phi(x) = E_x\{M(\sigma)\} = E_x\{u(x(\sigma), b)\, e(\sigma)\}. \tag{35}$$

On the other hand, by the strong Markov property applied at σ $(< \tau_b)$, we see that

$$\begin{aligned} u(x, b) &= E_x\{e(\sigma) E_{x(\sigma)}[e(\tau_b)]\} \\ &= E_x\{e(\sigma)\, u(x(\sigma), b)\}. \end{aligned} \tag{36}$$

Comparison of (35) with (36) yields $u(x, b) = \Phi(x)$. This being true for each $x < b$, Theorem 3 is established. See the Remarks at the end of the paper.

THEOREM 4. *Let Φ be any solution such that $\Phi(x) > 0$ for $x \in (-\infty, b]$. Then $b < \beta$, and we have*

$$u(x, b) \leqslant \frac{\Phi(x)}{\Phi(b)}, \qquad -\infty < x \leqslant b. \tag{37}$$

In other words, $u(\cdot, b)$ is the minimal positive solution in $(-\infty, b)$ with $\lim_{x \uparrow b} u(x, b) = 1$.

Proof. Consider the M in (27) but write $\tau_{[a,b]}$ for the τ in (26). Then for each $x \in \boldsymbol{R}$ and $t \geqslant 0$, we have the martingale relation

$$\Phi(x) = E_x\{M(0)\} = E_x\{M(t \wedge \tau_{[a,b]})\}. \tag{38}$$

If we keep b fixed but let $a \to -\infty$, then under P_x for $x \in (-\infty, b)$ we have $\tau_{[a,b]} \uparrow \tau_b$ and $M(t \wedge \tau_{[a,b]}) \to M(t \wedge \tau_b)$ by continuity. Since $M(t \wedge \tau_{[a,b]}) > 0$ because $\Phi > 0$ in $(-\infty, b]$, it follows from (38) by Fatou's lemma that

$$\Phi(x) \geqslant E_x\{M(t \wedge \tau_b)\}, \qquad -\infty < x < b. \tag{39}$$

Letting $t \to \infty$, we obtain by another application of Fatou's lemma

$$\Phi(x) \geqslant E_x\{M(\tau_b)\} = \Phi(b)\, u(x, b). \tag{40}$$

Thus (37) is true for $x < b$; for $x = b$ it reduces to a trivial equation.

As a corollary to Theorem 4, we can relate the two abscissas β and α to the solutions $u(\cdot, b)$. Suppose $\beta > -\infty$. For each $b < \beta$, $u(\cdot, b)$ is a solution in $(-\infty, b)$ by Theorem 3. By the fundamental existence and

uniqueness theorem for linear differential equations, there is a solution Φ_b in $\boldsymbol{R}$ satisfying

$$\Phi_b(x) = u(x, b), \qquad -\infty < x < b. \tag{41}$$

It follows from (7) that $\Phi_b(b) = 1$. Furthermore, the uniqueness theorem implies that for $b < c < \beta$ we have

$$\Phi_c(x) = \Phi_b(x)u(b, c), \qquad x \in \boldsymbol{R}. \tag{42}$$

Thus the family of solutions $\{\Phi_b,\ b < \beta\}$ are linearly dependent.

Similarly if $\alpha < +\infty$, then there is a family of linearly dependent solutions $\{{}_a\Phi,\ a > \alpha\}$ satisfying

$$ {}_a\Phi(x) = u(x, a), \qquad a < x < +\infty. \tag{43}$$

Corollary to Theorem 4. *For each $b < \beta$, β is the smallest root of Φ_b. For each $a > \alpha$, α is the largest root of ${}_a\Phi$.*

Proof. Fix $b < \beta$ and denote the smallest root of Φ_b by z. Since $\Phi_b > 0$ in $(-\infty, z)$, we must have $z \leqslant \beta$ by Theorem 4. On the other hand, for $b < c < \beta$ we have $\Phi_b(x) = \varphi_c(x)/u(b, c) > 0$ for $x \in (-\infty, c)$ by (42).

Hence $\beta \leqslant z$ and so $\beta = z$ as asserted. The assertion about α is proved similarly.

Theorem 5. *The following propositions are all equivalent:*

(i) *There is a solution which is positive in $\boldsymbol{R}$;*

(ii) $\beta = +\infty$;

(iii) $\alpha = -\infty$;

(iv) *for every pair of real numbers a and b we have*

$$u(a, b)u(b, a) \leqslant 1; \tag{44}$$

(v) *for a pair of real numbers a and b we have* (44).

Proof. The equivalence of (ii), (iii) and (iv) has already been proved for any diffusion in Theorem 1.

Let Φ be a positive solution in $\boldsymbol{R}$. Then Theorem 4 applies for every b in $\boldsymbol{R}$ and yields $\beta = +\infty$. By symmetry $\alpha = -\infty$.

If $\beta = +\infty$, then for any $b \in \boldsymbol{R}$, Φ_b has no root by Corollary to Theorem 4; hence it is a positive solution in $\boldsymbol{R}$. Similarly if $\alpha = -\infty$, then for any $a \in \boldsymbol{R}$, ${}_a\Phi$ is a positive solution in $\boldsymbol{R}$. We have thus proved the equivalence of (i) with (ii), (iii), and (iv).

It remains to prove that (v) implies (i). Let a and b satisfy (44) and $a < b$. Then

$$\alpha < a < b < \beta \tag{45}$$

by the definition of α and β and Lemma 4. Bisecting the interval $[a, b]$

we deduce from (44) and (5) that $u(a_1, b_1)u(b_1, a_1) \leqslant 1$ for some a_1 and b_1 where $a \leqslant a_1 \leqslant b_1 \leqslant b$ and $b_1 - a_1 = (b-a)/2$. Continuing this process in the grand tradition of Bolzano–Weierstrass we see that there exist a_n and b_n such that $a_n \uparrow c$, $b_n \downarrow c$ where $c \in [a, b]$, and $u(a_n, b_n)u(b_n, a_n) \leqslant 1$. Since $a \leqslant a_n \leqslant b_n \leqslant b$ we may use (41), and (43) and (5) to obtain

$$\Phi_b(a_n)\,{}_a\Phi(b_n) \leqslant \Phi_b(b_n)\,{}_a\Phi(a_n). \tag{46}$$

Writing, e.g., $\Phi_b(b_n) = \Phi_b(a_n) + \Phi_b'(c_n)(b_n - a_n)$ where $a_n \leqslant c_n \leqslant b_n$ by the mean value theorem, substituting into (46) and letting $n \to \infty$, we obtain

$$W(c) = {}_a\Phi(c)\Phi_b'(c) - \Phi_b(c)\,{}_a\Phi'(c) \geqslant 0, \tag{47}$$

where W is the Wronskian of ${}_a\Phi$ and Φ_b. Since ${}_a\Phi$ and Φ_b are both solutions of the equation, it is an elementary fact that W is a constant. Now suppose for the sake of contradiction that Φ_b has a root; then by Corollary to Theorem 4 its smallest root is β. Since $\Phi_b > 0$ in $(-\infty, \beta)$, it is obvious that $\Phi_b'(\beta) < 0$; since $\beta > a$, we have ${}_a\Phi(\beta) > 0$ by the definition of ${}_a\Phi$. Thus

$$W(\beta) = {}_a\Phi(\beta)\Phi_b'(\beta) < 0, \tag{48}$$

which contradicts (47). Therefore, φ_b is a positive solution in $\boldsymbol{R}$ and (i) is proved.

Acknowledgement. Under the assumption (49) below, van Moerbeke proved that condition (v) implies that the solution w_1 in (50) has no zero. The proof above is modelled after his.

4. A particular case. In the analytical study of the Schrödinger equation (24), the following condition on q is often assumed:

$$\int_{-\infty}^{\infty} |xq(x)|\,dx < \infty. \tag{49}$$

It is known (see e.g. [1], p. 284) that there are then two solutions w_1 and w_2 such that

$$\lim_{x\to-\infty} w_1(x) = 1, \quad \lim_{x\to+\infty} w_2(x) = 1. \tag{50}$$

Any solution v which tends to a finite limit as $x \to -\infty$ $[+\infty]$ must be a constant multiple of w_1 $[w_2]$. For v' must tend to zero at $-\infty$ $[+\infty]$, so that Wronskian of v and w_1 $[w_2]$ must vanish.

The probabilistic counterpart of the result above is given below.

THEOREM 6. *Under the assumption* (49) *we have for any* $b < \beta$,

$$\lim_{x\to-\infty} u(x, b) < \infty \tag{51}$$

and for any $\alpha > a$,

(52) $$\lim_{x\to+\infty} u(x, \alpha) < \infty.$$

Proof. We use the trivial inequality, for $x \leqslant b$,

(53) $$|u(x,b)-1| \leqslant \sum_{n=1}^{\infty} \frac{1}{n!} E_x\left\{\left[\int_0^{\tau_b} |q|(x(s))\,ds\right]^n\right\} = \sum_{n=1}^{\infty} M^{(n)}(x, b),$$

where ($t_0 = 0$)

(54) $$M^{(n)}(x, b) = E_x\left\{\prod_{j=1}^{n} \int_{t_{j-1}}^{\tau_b} |q|(x(t_j))\,dt_j\right\}.$$

Put also

$$M^{(n)}(b) = \sup_{x\leqslant b} M^{(n)}(x, b),$$

then $M^{(n)}(b)$ is nondecreasing in b. Using the Markov property of x at t_{n-1} in (54), we obtain

(55) $$M^{(n)}(x, b) \leqslant M^{(n-1)}(x, b)\, M^{(1)}(b) \leqslant [M^{(1)}(b)]^n.$$

Let τ be as in (26). Then it is part of the standard theory of Brownian motion that for $x \in [a, b]$ we have

(56) $$E_x\left\{\int_0^{\tau} |q|(x(t))\,dt\right\} = \int_a^b G(x, y)|q|(y)\,dy,$$

where G is the Green's function for $[a, b]$, specifically (see, e.g. [2]),

$$G(x, y) = \begin{cases} \dfrac{(x-a)(b-y)}{b-a} & \text{if } a \leqslant x \leqslant y \leqslant b; \\[2ex] \dfrac{(y-a)(b-x)}{b-a} & \text{if } a \leqslant y \leqslant x \leqslant b. \end{cases}$$

Letting $a \to -\infty$ and using (49), we have by dominated convergence

(57) $$M^{(1)}(x, b) = E_x\left\{\int_0^{\tau_b} |q|(x(t))\,dt\right\} = \int_{-\infty}^{b} [(b-x)\wedge(b-y)]\,|q|(y)\,dy.$$

Hence

(58) $$M^{(1)}(b) \leqslant \int_{-\infty}^{b} (b-y)\,|q|(y)\,dy$$

which tends to zero as $b \to -\infty$ by (49). Choose b_0 so that

(59) $$M^{(1)}(b_0) = \eta < 1.$$

Then we have by (53), (55) and (59) for $b < b_0$:

$$\sup_{x\leqslant b} |u(x,\ b)-1| \leqslant \frac{M^{(1)}(b)}{1-\eta},$$

and consequently the left member above tends to zero as $b\to-\infty$. This and the fundamental relation (5) for u implies (51) for $b=b_0$, hence also for all $b<\beta$ by the meaning of β (Lemma 1). This proves (51), and (52) is entirely similar.

COROLLARY. *We have the following bounds for β and a:*

$$\beta \geqslant \sup\Big\{ b\in \boldsymbol{R} \Big|\ \int_{-\infty}^{b} (b-y)|q|(y)\,dy \leqslant 1\Big\},$$

(60)

$$a \leqslant \inf\Big\{ a\in \boldsymbol{R} \Big|\ \int_{a}^{\infty} (y-a)|q|(y)\,dy \leqslant 1\Big\}.$$

5. Remarks. In [3], Kac proved that if $p(t;\ a,b)$ denotes the fundamental solution of the partial differential equation:

$$\frac{\partial\varphi}{\partial t} = \frac{1}{2}\frac{\partial^2\varphi}{\partial x^2} + q(x)\varphi,$$

then

$$p(t;\ a,\ b)\,db = E_a\{e(t);\ x(t)\in db\},$$

where x is the standard Brownian motion on $\boldsymbol{R}$, and $e(t)$ is defined in (3). The associated semigroup has been called the *Kac semigroup*. It appears that Theorem 3 can be derived by Kac's original method using Laplace transforms (resolvents). This idea is due to Moerbeke, but a difficulty arises because $x\to u(x,b)$ in Theorem 3 is not necessarily bounded, so that a crucial dominated convergence required by the method appears missing. This was pointed out by Murali Rao. Although we have managed to overcome this difficulty by a moderate detour, the present approach to Theorems 3 and 4 throws more light on the "connections between probability theory and differential and integral equations" — Kac's original theme. We are indebted to Moerbeke for some stimulating discussions.

References

[1] G. Birkhoff and G. C. Rota, *Ordinary differential equations*, Ginn & Co., 1962.

[2] K. Ito and H. P. McKean, *Diffusion processes and their sample path*, Springer-Verlag, 1965.

[3] M. Kac, *On some connections between probability theory and differential and integral equations*, Proc. Second Berkeley Symposium (1951), pp. 189–215.
[4] H. P. McKean, Jr., *Stochastic integrals*, Academic Press, 1969.
[5] F. G. Tricomi, *Differential equations*, Blackie & Sons, 1961.

Added in proof: For results in $\boldsymbol{R}^d$, $d \geqslant 1$ see K. L. Chung and K. M. Rao, *Sur la théorie du potentiel avec la fonctionnelle de Feynman–Kac,* Comp. Rend. Acad. Sci., Paris 290 (31 mars 1980).

DEPARTMENT OF MATHEMATICS
STANFORD UNIVERSITY
STANFORD, CALIFORNIA

Received December 15, 1977 (1385)

Comm. Pure Appl. Math. 38 (1985) 405-443

Brownian Motion in a Wedge with Oblique Reflection

S.R.S. VARADHAN
Courant Institute

AND

R.J. WILLIAMS
University of California, San Diego

Abstract

This work is concerned with the existence and uniqueness of a strong Markov process that has continuous sample paths and the following additional properties:

(i) The state space is an infinite two-dimensional wedge, and the process behaves in the interior of the wedge like an ordinary Brownian motion.

(ii) The process reflects instantaneously at the boundary of the wedge, the angle of reflection being constant along each side.

(iii) The amount of time that the process spends at the corner of the wedge is zero (i.e., the set of times for which the process is at the corner has Lebesgue measure zero).

Hereafter, let ξ be the angle of the wedge ($0 < \xi < 2\pi$), let θ_1 and θ_2 be the angles of reflection on the two sides of the wedge, measured from the inward normals, the positive angles being toward the corner ($-\frac{1}{2}\pi < \theta_1, \theta_2 < \frac{1}{2}\pi$), and set $\alpha = (\theta_1 + \theta_2)/\xi$.

The question of existence and uniqueness is recast as a submartingale problem in the style used by Stroock and Varadhan (*Diffusion processes with boundary conditions*, Comm. Pure Appl. Math. 24, 1971, pp. 147–225), for diffusions on smooth domains with smooth boundary conditions. It is shown that no solution exists if $\alpha \geqq 2$. In this case, there is a unique continuous strong Markov process satisfying (i) and (ii) above; it reaches the corner of the wedge almost surely and it remains there. If $\alpha < 2$, however, then there is a unique continuous strong Markov process satisfying (i)–(iii). It is shown that starting away from the corner this process does not reach the corner of the wedge if $\alpha \leqq 0$, and does reach the corner if $0 < \alpha < 2$.

The general theory of multi-dimensional diffusions does not apply to the above problem because in general the boundary of the state space is not smooth and there is a discontinuity in the direction of reflection at the corner. For some values of α, the process arises from diffusion approximations to storage systems and queueing networks.

1. Introduction

1.1 Overview. The basic problem considered here is the existence and uniqueness of a strong Markov process that has continuous sample paths and the following three properties:

(1.1) The state space is an infinite two-dimensional wedge, and the process behaves in the interior of the wedge like ordinary Brownian motion.

(1.2) The process reflects instantaneously at the boundary of the wedge, the direction of reflection being constant along each side.

(1.3) The amount of time that the process spends at the corner of the wedge is zero (i.e., the set of times for which the process is at the corner has Lebesgue measure zero).

Motivation for the study of this basic problem comes from several sources. First, for some wedges and directions of reflection, processes with the above properties arise as diffusion approximations in the theory of tandem queues. To facilitate study of the properties of these diffusions and thereby queueing models, it is useful to have an alternative characterization for them. This characterization should encompass existence and uniqueness and not depend on being able to verify conditions of limit theorems, such as those for heavy traffic given in [13] or those in [20] which require that the diffusion does not reach the corner almost surely. The second reason for studying the problem is that it does not fit within the realm of the general theory of multi-dimensional diffusions because in general the boundary of the state space is not smooth and the direction of reflection has a discontinuity at the corner. Indeed, no systematic treatment of such processes has been given previously, although a few special cases have been examined in the literature. These range from the much studied processes with normal reflection at the boundary (see Bellot [3] for a short survey and a path-by-path construction), to processes for which the wedge is a half-plane (cf. [2], [7], [11], [16], [17]), to those arising as diffusion approximations in the theory of tandem queues (cf. [6]) for which the state space is a quadrant and the directions of reflection are restricted to point inward, but not too sharply. Finally, study of this concrete example increases the detailed knowledge about the possible boundary behavior of the multi-dimensional diffusions. In particular, the intriguing behavior at the corner and the techniques used to investigate this problem should provide impetus for the study of other multi-dimensional diffusions with nonsmooth boundaries and singular boundary conditions.

The above formulation of the basic problem is a little heuristic in that one can ask what it means for the process to behave like Brownian motion in the interior of the wedge, or to reflect instantaneously at the boundary in a given direction. The basic problem is formulated in precise mathematical terms as a question of the existence and uniqueness of a family $\{P_x\}$ of solutions of a submartingale problem (see (1.4)–(1.6) below). Here x runs through all points in the wedge. This formulation is in the style used by Stroock and Varadhan [14] for smooth domains and boundary conditions. By

means of a systematic approach, this question is settled for all nondegenerate wedges and directions of reflection.

Let S denote the wedge with corner at the origin, ξ denote the angle of the wedge $(0 < \xi < 2\pi)$, and θ_1, θ_2 denote the angles that the directions of the reflection on the two sides of the wedge make with the inward normals to those sides, positive angles being toward the corner $(-\frac{1}{2}\pi < \theta_1, \theta_2 < \frac{1}{2}\pi)$. See Figure 1.1 for an example of an acute wedge with $\theta_1 > 0$ and $\theta_2 > 0$. The key characteristic quantity turns out to be $\alpha \equiv (\theta_1 + \theta_2)/\xi$. Loosely speaking, α is a normalized measure of the total "push" towards the corner caused by reflection at the boundary.

Given $x \in S$, a solution of the submartingale problem starting from x is a probability measure P_x on the space of continuous paths w in S such that

$$P_x(w(0) = x) = 1, \tag{1.4}$$

$$f(w(t)) - \frac{1}{2}\int_0^t \Delta f(w(s))\,ds \tag{1.5}$$

is a P_x submartingale whenever f is a function which together with its first and second partial derivatives is continuous and bounded on S, and such that f is constant in a neighborhood of the corner and has non-negative directional derivative in the direction of reflection on each side of the wedge, and

$$E^{P_x}\left(\int_0^\infty 1_{\{0\}}(w(s))\,ds\right) = 0. \tag{1.6}$$

If there is a unique solution of the submartingale problem starting from each $x \in S$, then the family $\{P_x\}$ has the strong Markov property.

If $\alpha < 2$, then for each $x \in S$ there is a unique solution P_x of the submartingale problem starting from x, and the family $\{P_x, x \in S\}$ has the strong Markov property. For $x \neq 0$, $w(\cdot)$ reaches the corner of the wedge with P_x probability zero or one, respectively, as $\alpha \leqq 0$ or $0 < \alpha < 2$.

If $\alpha \geqq 2$, then for each $x \in S$ there is no P_x satisfying (1.4)–(1.6). However, there is a unique P_x satisfying (1.4) and (1.5). Moreover, in this case, $\{P_x, x \in S\}$ has the strong Markov property and, for each x, P_x is concentrated on those paths which reach the corner and terminate there, i.e., it corresponds to the process starting at x which is almost surely absorbed at the corner.

This paper is divided into three sections. In Section 2, a process with properties (1.1) and (1.2) is defined, which instead of satisfying (1.3) is absorbed at the corner, if it ever gets there. It is shown that the corner is reached almost surely if $\alpha > 0$, and is almost surely not reached when starting away from it if $\alpha \leqq 0$. Estimates of some expected occupation times and hitting times for this process with absorption at the corner are also obtained. In particular, it is shown that the expected time of hitting the corner, when starting away from it, is finite if and only if $\alpha > 2$. Furthermore, the estimates obtained for the expected times of occupation of neighborhoods of the corner are vital to the proof given in subsection 2.4 of the existence for $\alpha < 2$ of a continuous process which heuristically has properties (1.1)–(1.3), and more precisely

gives a solution of (1.4)–(1.6). This process is obtained as a weak limit point of an approximating family of processes indexed by points δ in the wedge. The approximating process associated with δ behaves like the absorbed process prior to hitting the corner, but instead of being absorbed there, it jumps instantaneously to the point δ and continues on from there as if it had started there. In Section 3, the question of uniqueness of a solution of (1.4)–(1.6) is addressed. Uniqueness is the key to the proof of the strong Markov property and moreover implies that the process obtained in subsection 2.4 does not depend on the particular choice of a sequence of δ's tending to 0. For $\alpha < 2$, the solution is unique. For $\alpha \geqq 2$, it is shown that there is a unique solution of (1.4) and (1.5), which instead of satisfying (1.6) is concentrated on those paths which absorb at the corner in finite time.

For those wedges and directions of reflection for which the process exists and is unique, i.e., for $\alpha < 2$, the following auxiliary questions are treated in separate papers by Williams. Is the process a semimartingale? Is the process recurrent? If so, what is its invariant measure?

1.2 Notations and terminology. The following are used extensively in the sequel. The σ-algebras on the Euclidean spaces R^k, $k \geqq 1$, are taken to be the Borel σ-algebras. The Lebesgue measure on R is denoted by λ, and the Euclidean norm in R^k, $k \geqq 1$, is denoted by $|\cdot|$.

Polar coordinates in R^2 are denoted by (r,θ), $r \geqq 0$, $\theta \in [0,2\pi)$. Let $\xi \in (0,2\pi)$ and

$$S = \{(r,\theta) : 0 \leqq \theta \leqq \xi, r \geqq 0\}.$$

The boundary ∂S of the wedge S consists of the two sides $\partial S_1 = \{(r,\theta) : \theta = 0, r \geqq 0\}$ and $\partial S_2 = \{(r,\theta) : \theta = \xi, r \geqq 0\}$. Let $S^0 = S \backslash \partial S$ be the interior of S. The origin 0 is often referred to as the corner of the wedge. For $j = 1, 2$, let n_j denote the unit vector normal to ∂S_j that points into S^0 and let v_j be a vector in R^2 such that $v_j \cdot n_j = 1$. The angle between n_j and v_j, measured such that it is positive if and only if v_j points towards the origin, is denoted by θ_j. Note that $-\frac{1}{2}\pi < \theta_j < \frac{1}{2}\pi$. An acute wedge with $\theta_1 > 0$ and $\theta_2 > 0$ is shown in Figure 1.1. The set of continuous functions $w : [0,\infty) \to S$ is denoted by C_S. Then $\mathcal{M}_t \equiv \sigma\{w(s) : 0 \leqq s \leqq t\}$ denotes the smallest σ-algebra of subsets of C_S which makes each of the coordinate maps $w \to w(s)$ from C_S to R^2 $\mathcal{M}_t$-measurable for $0 \leqq s \leqq t$. Equivalently, $\mathcal{M}_t$ is characterized as the Borel σ-algebra associated with the topology on C_S of uniform convergence on $[0,t]$. This follows from Stroock and Varadhan [15], page 30. Let $\mathcal{M} \equiv \sigma\{w(s) : 0 \leqq s < \infty\}$. A function $X : [0,\infty) \times C_S \to R^k$ for some $k \geqq 1$ is called adapted if $X(t,\cdot)$ is $\mathcal{M}_t$-measurable for each $t \geqq 0$, and it is called (right) continuous if $X(\cdot,w)$ is (right) continuous for each $w \in C_S$. If P is a probability measure on $(C_S,\mathcal{M})$, then X is a P (sub/semi)-martingale if and only if $\{X_j(t,\cdot), t \geqq 0\}$ is a real-valued (sub/semi)-martingale on the filtered probability space $(C_S,\mathcal{M},\{\mathcal{M}_t\},P)$ for $1 \leqq j \leqq k$. A function $T : C_S \to [0,\infty]$ is called a stopping time if $\{T \leqq t\} \equiv \{w : T(w) \leqq t\} \in \mathcal{M}_t$ for all $t \geqq 0$. The σ-algebra $\mathcal{M}_T$ associated with a stopping time T is defined by:

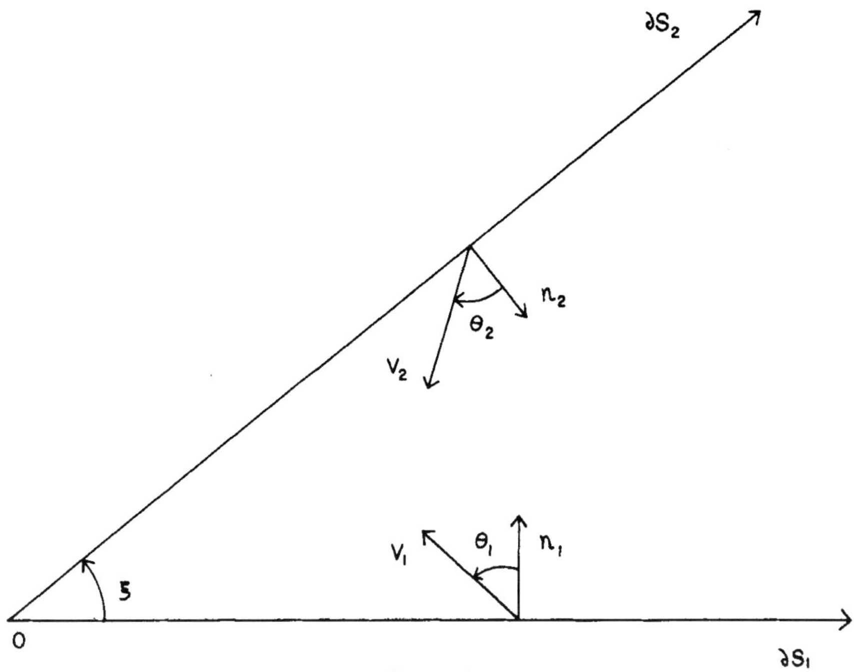

Figure 1.1

$$\mathcal{M}_T = \{A \in \Omega_F : A \cap \{T \leqq t\} \in \mathcal{M}_t \quad \text{for all} \quad t \leqq 0\}.$$

In fact,

$$\mathcal{M}_T = \sigma\{w(t \wedge T) : t \geqq 0\}.$$

The Laplace differential operator will be denoted by

$$\Delta \equiv \sum_{k=1}^{2} \frac{\partial^2}{\partial x_k^2},$$

and for $j = 1, 2$ we define

$$D_j \equiv v_j \cdot \nabla,$$

the directional derivative operator associated with the direction of reflection v_j on the side ∂S_j.

The set of continuous real-valued functions on S is denoted by $C(S)$, and $C_b(S)$ denotes those functions in $C(S)$ that are bounded on S. For each $n \geqq 1$, $C^n(S)$ denotes the set of real-valued functions that are n-times continuously differentiable in some domain containing S, and $C_b^n(S)$ denotes those functions in $C^n(S)$ that together with their partial derivatives, up to and including those of order n, are bounded on S. The

set of real-valued functions $f = f(t,x)$ that together with their first t-derivative and first two x-derivatives are continuous and bounded on $[0,\infty) \times R^2$ is denoted by $C_b^{1,2}([0,\infty) \times R^2)$.

2. Existence

2.1 Absorbed process. For $w \in C_S$ we define

$$\tau_0(w) = \inf\{t \geqq 0 : w(t) = 0\}. \tag{2.1}$$

For stopping times like τ_0, the explicit dependence on w will often be suppressed in the sequel.

THEOREM 2.1. *For each $x \in S$ there exists a unique measure P_x^0 on C_S such that*

(i) $P_x^0(w(0) = x) = 1$;

(ii) $f(w(t \wedge \tau_0)) - \frac{1}{2}\int_0^{t\wedge\tau_0}(\Delta f)(w(s))\, ds$

is a P_x^0 submartingale for all $f \in C_b^2(S)$ satisfying the boundary conditions

$$D_j f \geqq 0 \quad on \quad \partial S_j \quad for \quad j = 1,2;$$

(iii) $P_x^0(w(t) = 0 \text{ for } t \geqq \tau_0) = 1$.

We shall call the process associated with P_x^0 the "absorbed process" (starting from x).

Proof: It is clear that a Brownian motion starting at x can be reflected obliquely from a single side. By successive conditioning a reflected Brownian motion can be built up until the time when the path undergoes an infinite number of successive reflections from the two sides. If this time is finite, then it is easy to check that the path must hit the corner at this time. By forcing the process to remain at the corner after that time, we obtain a solution P_x^0 of (i)–(iii).

To prove uniqueness it suffices because of (iii) to show, for each $x \neq 0$, $n \geqq 1/|x|$ and $\sigma_n = \inf\{t \geqq 0 : |w(t)| \leqq 1/n\}$, that there is a unique probability measure P_x^n on $\mathcal{M}_{\sigma_n}$ such that (i)–(iii) hold with σ_n in place of τ_0. Indeed (cf. [19], subsection 3.6), by virtue of the time homogeneity of the data of our problem, it is equivalent to show this with (ii) replaced by:

$$\text{(ii)}' \quad f(t \wedge \sigma_n, w(t \wedge \sigma_n)) - \int_0^{t\wedge\sigma_n} 1_{S^0}(w(s))\left(\frac{\partial f}{\partial s} + \frac{1}{2}\Delta f\right)(s,w(s))\, ds$$

is a P_x^n submartingale for all $f \in C_b^{1,2}([0,\infty) \times R^2)$ satisfying $D_j f \geqq 0$ on ∂S_j for $j = 1, 2$. But the uniqueness of such a P_x^n follows from the theory of Stroock and Varadhan [14].

Remark. By construction, the family $\{P_x^0, x \in S\}$ has the strong Markov property and it is Feller continuous on $S\backslash\{0\}$ (cf. Stroock and Varadhan [14], page 196).

2.2 Is the corner reached? We look next at the question of whether the corner is reached. It is proved in this subsection that for $x \neq 0$, $P_x^0(\tau_0 < \infty) = 0$ or 1 according to whether $\alpha \equiv (\theta_1 + \theta_2)/\xi \leqq 0$ or > 0. First we note the following elementary lemma.

LEMMA 2.1. *Let* $x \in S\backslash\{0\}$ *and*

$$\sigma_r = \inf\{t \geqq 0 : |w(t)| = r\} \quad \textit{for each} \quad r \geqq 0. \tag{2.2}$$

Suppose $0 \leqq \varepsilon \leqq |x| \leqq K \leqq \infty$. *Then*

$$P_x^0(\sigma_\varepsilon \wedge \sigma_K < \infty) = 1. \tag{2.3}$$

Proof: The process $w(\cdot)$ under P_x^0 is equivalent in law to a continuous process Z obtained by a path-by-path construction from a two-dimensional Brownian motion B with starting point x, such that $Z(t) \in S$ for all $t \geqq 0$ and

$$Z(t) = \begin{cases} B(t) + v_1 Y_1(t) + v_2 Y_2(t) \\ \qquad \text{for } 0 \leqq t < \tau_0^Z \equiv \inf\{s \geqq 0 : Z(s) = 0\}, \\ 0 \quad \text{for } t \geqq \tau_0^Z, \end{cases} \tag{2.4}$$

where Y_1 and Y_2 are continuous increasing processes adapted to B, with $Y_1(0) = Y_2(0) = 0$ and $Y_1(t)$, $Y_2(t)$ finite for all $t < \tau_0^Z$; moreover, for each $j = 1, 2$, Y_j increases only when Z is on $\partial S_j\backslash\{0\}$ (for further details see [19]). Thus, it suffices to show that $\sigma_\varepsilon^Z \wedge \sigma_K^Z$ is finite a.s. (for σ_r^Z defined by (2.2) with Z in place of w). For this, let v be a unit vector in R^2 such that $v_1 \cdot v \geqq 0$ and $v_2 \cdot v \geqq 0$. Then on $\{\sigma_\varepsilon^Z = \infty, Z(0) = x\}$, we have $\tau_0^Z = \infty$ and, by (2.4),

$$\sup_t |Z(t)| \geqq \sup_t (Z(t) \cdot v) \geqq \sup_t (B(t) \cdot v).$$

The right member above is a.s. equal to $+\infty$ by the (null) recurrence of the one-dimensional Brownian motion "$B \cdot v$". Hence $\sigma_K^Z < \infty$ a.s. on $\{\sigma_\varepsilon^Z = \infty\}$.

THEOREM 2.2. *Suppose* $x \in S\backslash\{0\}$. *Then*

$$P_x^0(\tau_0 < \infty) = \begin{cases} 1 & \textit{if} \quad \alpha > 0, \\ 0 & \textit{if} \quad \alpha \leqq 0. \end{cases} \tag{2.5}$$

Proof: To prove this theorem, we shall apply the submartingale property (ii) of P_x^0 to a function (actually to $C_b^2(S)$ approximations of it) that is harmonic in $S\backslash\{0\}$ and has zero directional derivative in the direction of reflection on each side of $\partial S\backslash\{0\}$. Such a function Φ can be found by seeking a solution of Laplace's equation using

separation of variables in polar coordinates, and choosing the separation constant and one arbitrary constant to satisfy the directional derivative boundary conditions:

$$D_j\Phi = 0 \quad \text{on} \quad \partial S_j\backslash\{0\} \quad \text{for} \quad j = 1,2.$$

For $\alpha \neq 0$, the appropriate function Φ that can be found in this way is defined on $S\backslash\{0\}$ by

$$\Phi(r,\theta) = r^\alpha \cos(\alpha\theta - \theta_1). \tag{2.6}$$

It can be verified to have the desired properties as follows. Since Φ is a linear combination of the real and imaginary parts of z^α, for $z = (r,\theta)$, it follows that (2.6) actually defines a twice continuously differentiable harmonic extension of Φ to some domain containing $S\backslash\{0\}$. We then have

$$\Delta\Phi = 0 \quad \text{in} \quad S\backslash\{0\}. \tag{2.7}$$

In polar coordinates we have:

$$\nabla\Phi = \left(\frac{\partial\Phi}{\partial r}, \frac{1}{r}\frac{\partial\Phi}{\partial\theta}\right) = \alpha r^{\alpha-1}(\cos(\alpha\theta - \theta_1), -\sin(\alpha\theta - \theta_1)), \tag{2.8}$$

$$v_1 = (-\tan\theta_1, 1), \tag{2.9}$$

$$v_2 = (-\tan\theta_2, -1). \tag{2.10}$$

Then it is easily verified that

$$\begin{aligned} D_1\Phi &= v_1\cdot\nabla\Phi = 0 \quad \text{on} \quad \partial S_1\backslash\{0\} = \{(r,\theta) : \theta = 0, r>0\}, \\ D_2\Phi &= v_2\cdot\nabla\Phi = 0 \quad \text{on} \quad \partial S_2\backslash\{0\} = \{(r,\theta) : \theta = \xi, r>0\}. \end{aligned} \tag{2.11}$$

Let σ_r be defined as in Lemma 2.1 and let $0 < \varepsilon \leqq |x| \leqq K < \infty$. By applying the submartingale property of P_x^0 to $C_b^2(S)$ extensions of Φ outside $F_{\varepsilon K} \equiv \{z \in S : \varepsilon \leqq |z| \leqq K\}$ and then using Doob's stopping theorem, it follows from (2.7) and (2.11) that $\Phi(w(\cdot \wedge \sigma_\varepsilon \wedge \sigma_K))$ is a P_x^0 martingale. Hence, for all $t \geqq 0$,

$$E^{P_x^0}[\Phi(w(t\wedge\sigma_\varepsilon\wedge\sigma_K))] = \Phi(x). \tag{2.12}$$

Since Φ is bounded on $F_{\varepsilon K}$ and $\sigma_\varepsilon \wedge \sigma_K < \infty$ P_x^0 a.s. by Lemma 2.1, it follows on letting $t \to \infty$ in (2.12) that

$$E^{P_x^0}[\Phi(w(\sigma_\varepsilon\wedge\sigma_K))] = \Phi(x). \tag{2.13}$$

Let $p_{\varepsilon K} = P_x^0(\sigma_\varepsilon < \sigma_K)$. Then by (2.13) and (2.6) we have

$$(\varepsilon^\alpha p_{\varepsilon K} + K^\alpha(1 - p_{\varepsilon K}))c \leqq |x|^\alpha, \tag{2.14}$$

where

$$c \equiv \min_{\theta \varepsilon [0,\xi]} \cos(\alpha\theta - \theta_1) \geqq \cos(|\theta_1| \vee |\theta_2|) > 0. \tag{2.15}$$

Hence,

$$p_{\varepsilon K} \geqq (K^\alpha - c^{-1}|x|^\alpha)/(K^\alpha - \varepsilon^\alpha) \quad \text{if} \quad \alpha > 0, \tag{2.16}$$

$$p_{\varepsilon K} \leqq (c^{-1}|x|^\alpha - K^\alpha)/(\varepsilon^\alpha - K^\alpha) \quad \text{if} \quad \alpha < 0. \tag{2.17}$$

By Lemma 2.1, $P_x^0(\sigma_0 \wedge \sigma_K < \infty) = 1$, and so combining this with the continuity of the paths $w(\cdot)$ we obtain $\{\sigma_0 < \sigma_K\} = \cap_{\varepsilon>0}\{\sigma_\varepsilon < \sigma_K\}$ P_x^0 a.s. By letting $\varepsilon \downarrow 0$ in (2.16), it follows that, if $\alpha > 0$,

$$P_x^0(\sigma_0 < \sigma_K) \geqq 1 - c^{-1}|x/K|^\alpha.$$

By the continuity of $w(\cdot)$, $\sigma_K \to \infty$ as $K \to \infty$. Hence, on letting $K \to \infty$ in the above we obtain, $P_x^0(\sigma_0 < \infty) = 1$. On the other hand, for $\alpha < 0$, we obtain, by letting $\varepsilon \downarrow 0$ in (2.17), $P_x^0(\sigma_0 < \sigma_K) = 0$ and then, on letting $K \to \infty$, $P_x^0(\sigma_0 < \infty) = 0$. This proves the theorem for $\alpha \neq 0$, since $\sigma_0 \equiv \tau_0$.

For the case $\alpha = 0$, define Φ on $S\backslash\{0\}$ by

$$\Phi(r,\theta) = \log r + \theta \tan \theta_1. \tag{2.18}$$

Now (2.18) actually defines a twice continuously differentiable harmonic extension of Φ to a domain containing $S\backslash\{0\}$, and then $\nabla\Phi = (r^{-1}, r^{-1} \tan \theta_1)$ in polar coordinates. Using $\theta_1 = -\theta_2$, it can be verified that (2.7) and (2.11) still hold when $\alpha = 0$, and then so does (2.13). Consequently, for $p_{\varepsilon K}$ as defined before we have

$$p_{\varepsilon K} \log \varepsilon + (1 - p_{\varepsilon K}) \log K + \xi|\tan \theta_1| \geqq \log|x| - \xi|\tan \theta_1|,$$

and therefore

$$p_{\varepsilon K} \leqq \frac{\log K - \log|x| + 2\xi|\tan \theta_1|}{\log K - \log \varepsilon}.$$

Letting $\varepsilon \downarrow 0$ and then $K \uparrow \infty$, we obtain

$$P_x^0(\sigma_0 < \infty) = 0 \quad \text{if} \quad \alpha = 0.$$

If, in the above proof, σ_ε and σ_K are replaced by the times of hitting the arcs where Φ equals ε or K, respectively, then a precise formula for the corresponding $p_{\varepsilon K}$ is obtained. For example, by doing this for $\alpha > 0$ and letting $\varepsilon \to 0$ we obtain the following

COROLLARY 2.1. *Suppose $x \in S$ and $\alpha > 0$. Let Φ be given by* (2.6) *and set $\Phi(0) = 0$. For each $r \geqq 0$, let*

$$\tau_r = \inf\{t \geqq 0 : \Phi(w(t)) = r\}.$$

Then for all $K \geqq \Phi(x)$ we have

$$P_x^0(\tau_K < \tau_0) = \Phi(x)/K.$$

2.3 Some estimates of expected occupation and hitting times. Let Φ be defined by (2.6) if $\alpha \neq 0$ or by (2.18) if $\alpha = 0$. Define Ψ on $S\backslash\{0\}$ by

$$\Psi = \begin{cases} \Phi & \text{if } \alpha > 0, \\ e^{\Phi} & \text{if } \alpha = 0, \\ 1/\Phi & \text{if } \alpha < 0, \end{cases} \tag{2.19}$$

and let $\Psi(0) = 0$. Then Ψ is continuous on S and $\Psi(r,\theta)$ is increasing with r for each fixed θ. An estimate of the expected amount of time that $w(\cdot)$ spends in the neighborhood $\{z \in S : 0 < \Psi(z) < \varepsilon\}$ of the corner prior to hitting the corner or the arc$\{z : \Psi(z) = K\}$ is given by Theorem 2.3 and Corollary 2.2 below. As a further corollary, it is proved that the time at which $w(\cdot)$ first hits the corner, starting from $x \neq 0$, has finite expectation under P_x^0 if and only if $\alpha > 2$.

The idea behind the proof of the aforementioned estimate is as follows. If we were to try to exactly calculate the expected amount of time $u(x)$ that $w(\cdot)$ starting from x spends in the neighborhood $U = \{z \in S : 0 < \Psi(z) < \varepsilon\}$ of the corner prior to hitting the corner or the arc $\Gamma = \{z \in S : \Psi(z) = K\}$, we would need to solve the following difficult two-dimensional boundary value problem:

$$\begin{aligned} \tfrac{1}{2}\Delta u &= -1_U \quad \text{in} \quad S\backslash\{0\}, \\ D_j u &= 0 \quad \text{on} \quad \partial S_j\backslash\{0\} \quad \text{for} \quad j = 1,2, \\ u &= 0 \quad \text{on} \quad \Gamma \cup \{0\}, \end{aligned}$$

However, for our purposes it is enough to have a sufficiently precise estimate of this expectation, which can be obtained by solving a boundary value problem "close" to the above one in which 1_U is replaced by $1_U h$, where h is a function bounded away from zero and infinity on $S\backslash\{0\}$. A solution of such a modified problem is obtained by considering functions of the form $g(\Phi)$. Such functions automatically satisfy the same oblique derivative boundary conditions as Φ, and $\Delta g(\Phi) = g''(\Phi)|\nabla\Phi|^2$, where $|\nabla\Phi|^2$ can be written as a function of Φ times a function bounded away from zero and infinity on $S\backslash\{0\}$. The function g is chosen to solve a one-dimensional boundary value problem such that $g(\Phi)$ solves a problem "close" to that for u (in the above sense).

For each $r \geqq 0$, let

$$\tau_r = \inf\{t \geqq 0 : \Psi(w(t)) = r\}. \tag{2.20}$$

For each $K > 0$, let $\tau_{0K} = \tau_0 \wedge \tau_K$.

THEOREM 2.3. *Let $x \in S\backslash\{0\}$. Then for any $K \geqq \Psi(x)$ and $0 \leqq \varepsilon \leqq K$, we have*

$$E^{P_x^0}\int_0^{\tau_{0K}} 1_{(0,\varepsilon)}(\Psi(w(s))h(w(s))\,ds) = g(\Phi(x)), \tag{2.21}$$

where, for $z = (r,\theta) \in S\backslash\{0\}$,

(2.22)
$$h(z) = \begin{cases} (\cos(\alpha\theta - \theta_1))^{(2/\alpha)-2} & \textit{if} \quad \alpha \neq 0, \\ (1 + \tan^2 \theta_1)\exp\{2\theta \tan \theta_1\} & \textit{if} \quad \alpha = 0, \end{cases}$$

and if $\alpha \in (0,\infty)\backslash\{2\}$,

(2.23)
$$g(y) = \begin{cases} 0 & \textit{for} \quad y = 0, \\ y(\varepsilon^{(2/\alpha)-1}((2/\alpha)(1 - \varepsilon/K) + \varepsilon/K) - y^{(2/\alpha)-1})/(2 - \alpha) & \textit{for} \quad 0 < y \leqq \varepsilon, \\ \alpha^{-1}\varepsilon^{2/\alpha}(1 - y/K) & \textit{for} \quad y \geqq \varepsilon, \end{cases}$$

if $\alpha = 2$,

(2.24)
$$g(y) = \begin{cases} 0 & \textit{for} \quad y = 0, \\ \tfrac{1}{2}y((\log \varepsilon + 1 - \varepsilon/K) - \log y) & \textit{for} \quad 0 < y \leqq \varepsilon, \\ \tfrac{1}{2}\varepsilon(1 - y/K) & \textit{for} \quad y \geqq \varepsilon, \end{cases}$$

if $\alpha = 0$,

(2.25)
$$g(y) = \begin{cases} \tfrac{1}{2}(\varepsilon^2 - e^{2y}) + \varepsilon^2(\log K - \log \varepsilon) & \textit{for} \quad -\infty < y \leqq \log \varepsilon, \\ \varepsilon^2(\log K - y) & \textit{for} \quad y \geqq \log \varepsilon, \end{cases}$$

if $\alpha < 0$,

(2.26)
$$g(y) = \begin{cases} -2\varepsilon^{1-2/\alpha}(y - K^{-1})/\alpha(2 - \alpha) & \textit{for} \quad 0 \leqq y \leqq \varepsilon^{-1} \\ (2 - \alpha)^{-1}(\varepsilon^{-2/\alpha} - y^{2/\alpha}) - 2\varepsilon^{1-2/\alpha}(\varepsilon^{-1} - K^{-1})/\alpha(2 - \alpha) & \\ & \textit{for} \quad y \geqq \varepsilon^{-1}. \end{cases}$$

Proof: It is easily verified that, for $\alpha \neq 0$, g is continuous on $[0,\infty)$, continuously differentiable once in $(0,\infty)$, and twice in $(0,\infty)\backslash\{\varepsilon\}$ if $\alpha > 0$ or in $(0,\infty)\backslash\{\varepsilon^{-1}\}$ if $\alpha < 0$. For $\alpha = 0$, g is continuously differentiable once in $(-\infty,\infty)$ and twice in $(-\infty,\infty)\backslash\{\log \varepsilon\}$.

Moreover, if $\alpha > 0$,

(2.27)
$$g''(y) = \begin{cases} -2\alpha^{-2}y^{(2/\alpha)-2} & \text{for} \quad 0 < y < \varepsilon, \\ 0 & \text{for} \quad y > \varepsilon, \end{cases}$$

if $\alpha = 0$,

$$g''(y) = \begin{cases} -2e^{2y} & \text{for} \quad -\infty < y < \log \varepsilon, \\ 0 & \text{for} \quad y > \log \varepsilon, \end{cases} \tag{2.28}$$

and, if $\alpha < 0$,

$$g''(y) = \begin{cases} 0 & \text{for} \quad 0 < y < \varepsilon^{-1}, \\ -2\alpha^{-2} y^{(2/\alpha)-2} & \text{for} \quad y > \varepsilon^{-1}. \end{cases} \tag{2.29}$$

Extend Φ to be harmonic in some domain containing $S\backslash\{0\}$ and define $f = g(\Phi)$ there. Then $f \in C^1(S\backslash\{0\})$ and $f \in C^2(S\backslash\{0\} \cup \Psi^{-1}(\varepsilon)))$. In particular, for $z \in S\backslash(\{0\} \cup \Psi^{-1}(\varepsilon))$ we have

$$\begin{aligned} \Delta f(z) &= g''(\Phi(z))|\nabla\Phi(z)|^2 + g'(\Phi(z))\Delta\Phi(z) \\ &= g''(\Phi(z))|\nabla\Phi(z)|^2. \end{aligned} \tag{2.30}$$

Now $|\nabla\Phi|^2$ can be written as a product of a function of Φ and a function that is bounded away from zero and infinity on $S\backslash\{0\}$, i.e.,

(2.31)

$$|\nabla\Phi(z)|^2 = \begin{cases} \alpha^2 r^{2\alpha-2} = \alpha^2(\Phi(z)/\cos(\alpha\theta - \theta_1))^{2-2/\alpha} & \text{if} \quad \alpha \neq 0, \\ r^{-2}(1 + \tan^2\theta_1) = e^{-2\Phi(x)}\exp\{2\theta\tan\theta_1\}(1 + \tan^2\theta_1) & \text{if} \quad \alpha = 0. \end{cases}$$

Then, since g was chosen to satisfy one of the one-dimensional differential equations (2.27)–(2.29), depending on the value of α, it follows that f satisfies a two-dimensional problem "close" to $\frac{1}{2}\Delta f = -1_{\{0<\Psi<\varepsilon\}}$, namely,

$$\tfrac{1}{2}\Delta f(z) = \begin{cases} -h(z) & \text{for} \quad 0 < \Psi(z) < \varepsilon, \\ 0 & \text{for} \quad \Psi(z) > \varepsilon, \end{cases} \tag{2.32}$$

where h defined by (2.22) is bounded away from zero and infinity on $S\backslash\{0\}$. Since Φ satisfies the first-order homogeneous oblique derivative boundary conditions (2.11), so does f, i.e., for $j = 1, 2$,

$$D_j f(z) = (v_j \cdot \nabla f)(z) = (v_j \cdot \nabla\Phi(z))g'(\Phi(z)) = 0 \quad \text{for} \quad z \in \partial S_j\backslash\{0\}. \tag{2.33}$$

In addition, it can be verified that f is non-negative and bounded on $\{z \in S : 0 < \Psi(z) < K\}$, $f(z) = 0$ on $\{z \in S : \Psi(z) = K\}$, and $\lim_{z\to 0} f(z) = 0$ if $\alpha > 0$. Consistent with the last property, we define $f(0) \equiv 0$ if $\alpha > 0$.

For $0 < r < \Psi(x)$, let $\tau_{rK} = \tau_r \wedge \tau_K$ and $F_{rK} = \{z \in S : r \leqq \Psi(z) \leqq K\}$. Although f is not twice continuously differentiable across $\{z \in S : \Psi(z) = \varepsilon\}$, there is an approximating family of functions $\{f_n\}$ in $C_h^2(S)$ such that f_n satisfies (2.33), f_n and ∇f_n converge uniformly on F_{rK} to f and ∇f, respectively, and $\{\Delta f_n\}$ is bounded on F_{rK} and converges pointwise to Δf on $F_{rK}\backslash\{z : \Psi(z) = \varepsilon\}$. By applying the submartingale property of P_x^0 to each f_n, using Doob's stopping theorem, and then letting $n \to \infty$ we may conclude that

$$f(w(t\wedge\tau_{rK})) - \frac{1}{2}\int_0^{t\wedge\tau_{rK}} \Delta f(w(s))\,ds \tag{2.34}$$

is a P_x^0 martingale. The exceptional set $\{z \in S : \Psi(z) = \varepsilon\}$, where Δf is not defined, is of no consequence because P_x^0 a.s. $w(\cdot)$ spends zero time there in the sense of Lebesgue measure (cf. [14]). By taking expectations in (2.34) and using (2.32) we obtain

$$E^{P_x^0}[f(w(t\wedge\tau_{rK}))] + E^{P_x^0}\left[\int_0^{t\wedge\tau_{rK}} 1_{(0,\varepsilon)}(\Psi(w(s)))h(w(s))\,ds\right] = f(x). \tag{2.35}$$

Now, P_x^0 almost surely, $\tau_{rK} \uparrow \tau_{0K}$ as $r \downarrow 0$. Here we have P_x^0 a.s. $\tau_{0K} < \infty$ by Lemma 2.1, and if $\alpha \leqq 0$, $\tau_{0K} = \tau_K$ by Theorem 2.2. Letting $t \uparrow \infty$ and then $r \downarrow 0$ in (2.35), we obtain (2.21) by using bounded and monotone convergence, and the fact that P_x^0 almost surely $f(w(\tau_{0K})) = 0$.

COROLLARY 2.2. *Under the hypothesis of Theorem* 2.3 *there are constants* a_1 *and* a_2 *depending on* θ_1, θ_2, *and* α, *such that* $0 < a_1 \leqq a_2 < \infty$ *and*

$$a_1 g(\Phi(x)) \leqq E^{P_x^0}\left[\int_0^{\tau_{0K}} 1_{(0,\varepsilon)}(\Psi(w(s)))\,ds\right] \leqq a_2 g(\Phi(x)). \tag{2.36}$$

Proof: This follows from (2.21), (2.22). We have $0 < a_1 \leqq a_2 < \infty$ because, for $\theta \in [0,\xi]$,

$$0 < \cos(|\theta_1| \vee |\theta_2|) \leqq \cos(\alpha\theta - \theta_1) \leqq 1 \quad \text{if} \quad \alpha \neq 0,$$

or

$$0 < \exp\{-2\xi|\tan\theta_1|\} \leqq \exp\{2\theta\tan\theta_1\} \leqq \exp\{2\xi|\tan\theta_1|\} < \infty \quad \text{if} \quad \alpha = 0.$$

COROLLARY 2.3. *For* $\alpha \leqq 2$ *and* $x \in S\backslash\{0\}$,

$$E^{P_x^0}(\tau_0) = \infty. \tag{2.37}$$

For $\alpha > 2$, *there are constants* b_1 *and* b_2 *depending on* θ_1, θ_2, *and* α, *such that* $0 < b_1 \leqq b_2 < \infty$ *and, for each* $x \in S$,

$$b_1|x|^2 \leqq E^{P_x^0}(\tau_0) \leqq b_2|x|^2. \tag{2.38}$$

Proof: If $\alpha \leqq 0$, $P_x^0(\tau_0 < \infty) = 0$ for $x \in S\backslash\{0\}$, by Theorem 2.2. Thus we may suppose $\alpha > 0$. By setting $\varepsilon = K$ in the definition of g in (2.23), (2.24), we obtain, for $0 \leqq y \leqq K$,

$$g(y) = \begin{cases} y(K^{(2/\alpha)-1} - y^{(2/\alpha)-1})/(2-\alpha) & \text{if} \quad \alpha \neq 2, \\ \frac{1}{2}y(\log K - \log y) & \text{if} \quad \alpha = 2. \end{cases}$$

Thus, for each $y \in (0,\infty)$, as $K \to \infty$:

$$g(y) \to \begin{cases} \infty & \text{if } \alpha \leqq 2, \\ y^{(2/\alpha)}/(\alpha - 2) & \text{if } \alpha > 2. \end{cases}$$

Since $\tau_{0K} \uparrow \tau_0$ as $K \to \infty$, it follows from (2.36) that, for each $x \in S\backslash\{0\}$,

$$E^{P_x^0}(\tau_0) = \infty \quad \text{if} \quad \alpha \leqq 2, \tag{2.39}$$

$$\frac{a_1}{\alpha - 2}(\Phi(x))^{2/\alpha} \leqq E^{P_x^0}(\tau_0) \leqq \frac{a_2}{\alpha - 2}(\Phi(x))^{2/\alpha} \quad \text{if} \quad \alpha > 2. \tag{2.40}$$

Here $\Phi(x) = |x|^\alpha \cos(\alpha\theta_x - \theta_1)$, where $\theta = \arg(x) \in [0,\xi]$. It follows that (2.37), (2.38) hold for $x \in S\backslash\{0\}$. Since (2.38) holds trivially for $x = 0$, the proof is complete.

2.4. Existence. We now prove that, if $\alpha < 2$, a solution to the submartingale problem with "reflection" (rather than absorption) at the corner exists for any starting point $x \in S$. Let D_S denote the space of all functions $w : [0,\infty) \to S$ which are right continuous on $[0,\infty)$ and have finite left limits on $(0,\infty)$. We endow D_S with the Skorokhod topology (cf. Kurtz [8], page 7). The Borel σ-algebra $\mathcal{M}^D$ associated with this metric topology on D_S is the same as that generated by the coordinate maps, so $\mathcal{M}^D = \sigma\{w(s) : 0 \leqq s < \infty\}$ (cf. [10], page 117). The restriction of $\mathcal{M}^D$ to C_S is $\mathcal{M}$ and so we may think of P_x^0 as a probability measure on $(D_S, \mathcal{M}^D)$, concentrated on C_S. Let $\mathcal{M}_t^D = \sigma\{w(s) : 0 \leqq s \leqq t,\ w \in D_S\}$ for each $t \geqq 0$.

We define an approximating family of processes with a jump at the corner as follows. For each $x \in S$ and $\delta \in S\backslash\{0\}$ we define a process which induces a probability measure P_x^δ on $(D_S,\mathcal{M}^D)$ as follows (see Stroock and Varadhan [14], page 177, for details on how to define P_x^δ using successive conditioning). The process (P_x^δ) behaves like the absorbed process (P_x^0) starting from x, prior to the time that the corner is reached. If

$$P_x^0(\tau_0 < \infty) = 1,$$

we move the process instantaneously to the point $\delta \in S\backslash\{0\}$ as soon as the absorbed process hits the corner. The process is then allowed to start afresh (i.e., behaves according to P_δ^0). This procedure is repeated forever. Of course, if $P_x^0(\tau_0 < \infty) = 0$, then $P_x^\delta = P_x^0$ for $x \neq 0$ and P_x^δ is defined to be P_δ^0. For each $\delta \in S\backslash\{0\}$, the family $\{P_x^\delta,\ x \in S\}$ has the strong Markov property.

THEOREM 2.4. *For each $x \in S$, the family $\{P_x^\delta\}$ is weakly relatively compact as $|\delta| \to 0$.*

Proof: Since S is locally compact and the trajectories have jumps of size at most $|\delta|$, it is sufficient (cf. Prokhorov [12], page 182) to prove that

$$\lim_{t\to 0} \limsup_{|\delta|\to 0} \sup_{x\in S} P_x^\delta(\tau_x(\eta) \leqq t) = 0$$

for each $\eta > 0$, where $\tau_x(\eta)$ is the exit time of $w(\cdot)$ from a ball of radius η centered at x.

Let $\eta > 0$ be fixed. We assume that $|\delta|$ is so small that $|\delta| \leqq 3\eta/4$. Pick $\beta > 0$ so small that $\beta \leqq \eta/4$ and the ball of radius β centered at any $x \in S \cap \{z : |z| \geqq 2\eta/4\}$ intersects at most one side of S. Then clearly,

$$\begin{aligned}\sup_{x\in S} P_x^\delta(\tau_x(\eta) \leqq t) &\leqq \sup_{|x|\geqq 3\eta/4} P_x^\delta(\tau_x(\beta) \leqq t)\\ &= \sup_{|x|\geqq 3\eta/4} P_x^0(\tau_x(\beta) \leqq t)\\ &= o(1) \quad \text{as} \quad t \to 0.\end{aligned}$$

Next we show that any weak limit point P_x of P_x^δ as $|\delta| \to 0$ is a solution to the submartingale problem (1.4)–(1.6) provided $\alpha < 2$.

THEOREM 2.5. *Let $x \in S$ and suppose P_x is a weak limit point of $\{P_x^\delta\}$ as $|\delta| \to 0$. Then P_x is a probability measure on $(C_S, \mathcal{M})$ with the following two properties:*

(i) $P_x(w(0) = x) = 1$.

(ii) *For each $f \in C_b^2(S)$ which is constant in a neighborhood of the origin and satisfies*

$$D_j f \geqq 0 \quad on \quad \partial S_j \quad for \quad j = 1, 2, \tag{2.41}$$

we know that

$$f(w(t)) - \frac{1}{2}\int_0^t \Delta f(w(s))\, ds \tag{2.42}$$

is a P_x submartingale.

Moreover, if $\alpha < 2$, we also have

(iii) $$E^{P_x}\left(\int_0^\infty 1_{\{0\}}(w(s))\, ds\right) = 0.$$

We call a probability measure on $(C_S, \mathcal{M})$ satisfying (i)–(iii) a *solution of the submartingale problem starting from x*.

Proof: The fact that P_x is concentrated on the space of continuous paths follows from the property that P_x^δ only allows jumps of size at most $|\delta|$ from the origin. Property (i) is clearly inherited from that for P_x^δ, even when $x = 0$.

For any f as described in (ii), $f(0) = f(\delta)$ for all sufficiently small $|\delta|$ and it follows that, for such δ, (2.42) is a P_x^δ submartingale on $(D_S, \mathcal{M}^D, \{\mathcal{M}_t^D\})$. This submartingale property is preserved in the limit as $|\delta| \to 0$.

For τ_K defined by (2.20), to prove (iii) it suffices to show that, if $\alpha < 2$,

$$\lim_{\varepsilon \downarrow 0} E^{P_x}\left(\int_0^{\tau_K} 1_{[0,\varepsilon)}(\Psi(w)(s)))\, ds\right) = 0 \tag{2.43}$$

for each $K > \Psi(x)$. The proof of this is computational and relies on the following two lemmas. Here the definition of $\tau_r(w)$, $r \geqq 0$, is extended to $w \in D_S$:

$$\tau_r(w) = \inf\{t \geqq 0 : \Psi(w(t)) = r\}.$$

LEMMA 2.2. *Suppose $K > \Psi(x)$ and $0 < \varepsilon \leqq K$. Let $\{\delta_n\}$ be a sequence of points in $S\backslash\{0\}$ such that $|\delta_n| \to 0$ and $\{P_x^{\delta_n}\}$ converges weakly to P_x. Then*

$$E^{P_x}\left[\int_0^{\tau_K} 1_{[0,\varepsilon)}(\Psi(w(s)))\,ds\right] \leqq \liminf_{n\to\infty} E^{P_x^{\delta_n}}\left[\int_0^{\tau_K} 1_{[0,\varepsilon)}(\Psi(w)(s)))\,ds\right]. \tag{2.44}$$

Proof: For each $s \in (0,\infty)$, the function $\chi : D_S \to [0,\infty)^2$ defined by

$$\chi(w) = (\max_{0\leqq u\leqq s} \Psi(w(u)), \Psi(w(s)))$$

is continuous in the Skorokhod topology of D_S at each $w \in C_S$. Since $P_x(C_S) = 1$, it follows by the continuous mapping theorem (see Kurtz [8], Proposition 2.5) that $\chi(w)$ under $P_x^{\delta_n}$ converges weakly to $\chi(w)$ under P_x, and hence since $[0,K) \times [0,\varepsilon)$ is open in $[0,\infty)^2$:

$$P_x(\chi(w) \in [0,K) \times [0,\varepsilon)) \leqq \liminf_{n\to\infty} P_x^{\delta_n}(\chi(w) \in [0,K) \times [0,\varepsilon)). \tag{2.45}$$

Since $K > \Psi(x)$, we have P_x a.s.:

$$\{s < \tau_k\} = \{\max_{0\leqq u\leqq s} (\Psi(w(u))) \in [0,K)\}. \tag{2.46}$$

Furthermore, since the paths of $P_x^{\delta_n}$ are continuous outside $\{z : |z| \leqq \delta_n\}$ and $\Psi(\delta_n) \to 0$ as $n \to \infty$, (2.46) holds $P_x^{\delta_n}$ a.s. for all sufficiently large n. Thus, (2.45) is equivalent to

$$P_x(s < \tau_K;\ \Psi(w(s)) \in [0,\varepsilon)) \leqq \liminf_{n\to\infty} P_x^{\delta_n}(s < \tau_K;\ \Psi(ws)) \in [0,\varepsilon)).$$

To obtain (2.44) from this, first integrate over $s \in (0,\infty)$, then apply Fatou's lemma to the resulting right member, and finally interchange the expectation and time integration.

LEMMA 2.3. *Suppose $\delta \in S\backslash\{0\}$, $x \in S$, $K > \Psi(x) \vee \Psi(\delta)$, and $0 < \varepsilon \leqq K$. Let g and h be defined as in Theorem 2.3. Then, if $x \neq 0$,*

$$E^{P_x^\delta}\left[\int_0^{\tau_K} 1_{(0,\varepsilon)}(\Psi(w(s)))h(w(s))\,ds\right] = g(\Phi(x)) + \frac{P_x^0(\tau_0 < \tau_K)}{P_\delta^0(\tau_K < \tau_0)} g(\Phi(\delta)), \tag{2.47}$$

or if $x = 0$,

$$E^{P_0^\delta}\left[\int_0^{\tau_K} 1_{(0,\varepsilon)}(\Psi(w(s)))h(w(s))\,ds\right] = \frac{g(\Phi(\delta))}{P_\delta^0(\tau_K < \tau_0)}. \tag{2.48}$$

Proof: For each $z \in S$, let

$$\rho(z) = 1_{(0,\varepsilon)}(\Psi(z))h(z). \tag{2.49}$$

For each $w \in D_S$ define

$$\tau_{0K}(w) = \tau_0(w) \wedge \tau_K(w),$$

$$T_1(w) = \sup_m \inf\{t \geqq 0 : |w(t)| \leqq 1/m\},$$

where the sup is over all positive integers m; and, by induction on n, define

$$T_{n+1}(w) = \sup_m \inf\{t > T_n(w) : |w(t)| \leqq 1/m\} \quad \text{for all} \quad n \geqq 1.$$

The supinf definition of the T_n is needed because P_x^δ a.s. the paths $w(\cdot)$ have an instantaneous jump to δ at the times T_n and by right continuity are never at the origin. Now by the definition of P_x^δ we have, for $x \neq 0$,

$$\begin{aligned} E^{P_x^\delta}\left[\int_0^{\tau_K} \rho(w(s))\, ds\right] &= E^{P_x^0}\left[\int_0^{\tau_{0K}} \rho(w(s))\, ds\right] \\ &+ \sum_{n=1}^{\infty} E^{P_x^\delta}\left[1_{\{T_n<\tau_K\}} E^{P_\delta^0}\left(\int_0^{\tau_{0K}} \rho(w(s))\, ds\right)\right]. \end{aligned} \tag{2.50}$$

By Theorem 2.3, the right side of equation (2.50) is equal to

$$g(\Phi(x)) + \sum_{n=1}^{\infty} P_x^\delta(T_n < \tau_K) g(\Phi(\delta)), \tag{2.51}$$

where, by induction on n, we have

$$P_x^\delta(T_n < \tau_K) = P_x^0(\tau_0 < \tau_K)(P_\delta^0(\tau_0 < \tau_K))^{n-1}.$$

Then (2.47) follows by summing the geometric series in (2.51).

For $x = 0$ we have $P_0^\delta = P_\delta^\delta$ and so (2.48) is obtained by setting $x = \delta$ in (2.47).

To verify (2.43), we suppose $\alpha < 2$ and $\Psi(x) < K$. Let $\{\delta_n\}$ be as in Lemma 2.2. Since $P_x^{\delta_n}(w(t) = 0 \text{ for some } t \geqq 0) = 0$ (the $P_n^{\delta_n}$ process jumps instantaneously to δ_n), we can replace $1_{[0,\varepsilon)}$ by $1_{(0,\varepsilon)}$ in the right member of (2.44). By combining this with Lemma 2.3, we obtain

$$\begin{aligned} &cE^{P_x}\left[\int_0^{\tau_K} 1_{[0,\varepsilon)}(\Psi(w(s)))\, ds\right] \\ &\leqq \begin{cases} \liminf\limits_{n\to\infty} \left[g(\Phi(x)) + \dfrac{P_x^0(\tau_0 < \tau_K)}{P_{\delta_n}^0(\tau_K < \tau_0)} g(\Phi(\delta_n))\right] & \text{for} \quad x \neq 0, \\ \liminf\limits_{n\to\infty} \dfrac{g(\Phi(\delta_n))}{P_{\delta_n}^0(\tau_K < \tau_0)} & \text{for} \quad x = 0, \end{cases} \end{aligned} \tag{2.52}$$

where

$$c \equiv \inf_{z \in S\setminus\{0\}} h(z) > 0.$$

If $\alpha \leqq 0$, $P_x^0(\tau_0 < \tau_K) = 0$ for $x \neq 0$, and $P_{\delta_n}^0(\tau_K < \tau_0) = 1$, by Theorem 2.2. Then, using the expressions for g given in Theorem 2.3, we see that the right member of (2.52) is equal to:

$$(2.53)\quad \begin{cases} g(\Phi(x)) & \text{if } x \neq 0 \text{ and } \alpha \leqq 0, \\ \varepsilon^2(\tfrac{1}{2} + \log K - \log \varepsilon) & \text{if } x = 0 \text{ and } \alpha = 0, \\ (2-\alpha)^{-1}(\varepsilon^{-2/\alpha} - 2\alpha^{-1}\varepsilon^{1-2/\alpha}(\varepsilon^{-1} - K^{-1})) & \text{if } x = 0 \text{ and } \alpha < 0. \end{cases}$$

Since each of the expressions in (2.53) tends to zero as $\varepsilon \downarrow 0$ (recall that g depends on ε), it follows that (2.43) holds if $\alpha \leqq 0$.

Suppose $\alpha > 0$. Then the top expression in the right member of (2.52) can be used for all x, because when x is set to zero in it, it equals the bottom expression. By Corollary 2.1,

$$P_y^0(\tau_K < \tau_0) = \Phi(y)/K \quad \text{for all} \quad y : \Psi(y) \leqq K,$$

and therefore the right member of (2.52) equals

$$(2.54)\quad \begin{cases} g(\Phi(x)) + \left(1 - \dfrac{\Phi(x)}{K}\right)\dfrac{K}{2-\alpha} \\ \quad \times \left(\varepsilon^{(2/\alpha)-1}\left(\dfrac{2}{\alpha}\left(1 - \dfrac{\varepsilon}{K}\right) + \dfrac{\varepsilon}{K}\right) - \liminf\limits_{n\to\infty} \Phi(\delta_n)^{(2/\alpha)-1}\right) & \text{if } \alpha \neq 2, \\ g(\Phi(x)) + \left(1 - \dfrac{\Phi(x)}{K}\right)\dfrac{K}{2} \\ \quad \times \left(\left(\log \varepsilon + 1 - \dfrac{\varepsilon}{K}\right) - \liminf\limits_{n\to\infty} \log \Phi(\delta_n)\right) & \text{if } \alpha = 2. \end{cases}$$

In the above, the term involving the "$-\liminf$" is zero, $+\infty$, or $-\infty$, respectively, as $\alpha < 2$, $\alpha = 2$, or $\alpha > 2$. Thus, for $0 < \alpha < 2$, it follows by the dependence of g on ε that (2.54) tends to zero as $\varepsilon \downarrow 0$. Hence (2.43) holds in this case also.

3. Uniqueness

In subsection 3.1 it is proved that when $\alpha < 2$, the submartingale problem starting from x has at most one solution for each $x \in S$. By combining this with the existence Theorem 2.5, it follows that there is a unique solution of the submartingale problem starting from each $x \in S$ when $\alpha < 2$. On the other hand, it is shown in subsection 3.2 that if $\alpha \geqq 2$, there is no solution of the submartingale problem starting from any $x \in S$. Indeed, in this case, any probability measure on $(C_S, \mathcal{M})$ satisfying properties (i) and (ii) of the submartingale problem coincides with P_x^0, the probability measure associated with the absorbed process. Finally, in subsection 3.3 it is shown that the

family $\{P_x, x \in S\}$ of solutions of the submartingale problem (one for each $x \in S$) is Feller continuous (i.e., $P_{x_n} \to P_x$ weakly if $x_n \to x$), and it has the strong Markov property.

In this section, Ψ and τ_r are defined by (2.19) and (2.20).

3.1 Uniqueness. Let $C_c^\infty(S)$ denote the set of functions that are infinitely differentiable in a domain containing S and have compact support. The following theorem shows that, in order to prove the uniqueness of a solution of the submartingale problem starting from x, it suffices to show uniqueness of the resolvent operator acting on the class of functions in $C_c^\infty(S)$ which vanish in a neighborhood of the origin. This requires proof because we do not know *a priori* that the process associated with a family of solutions of the submartingale problem (one solution for each $x \in S$) is a Markov process.

THEOREM 3.1. *Suppose that, for every $x \in S$, $\gamma > 0$, and $h \in C_c^\infty(S)$ which vanishes in a neighborhood of the origin, we have*

$$E^{P^1}\left(\int_0^\infty e^{-\gamma t} h(w(t))\,dt\right) = E^{P^2}\left(\int_0^\infty e^{-\gamma t} h(w(t))\,dt\right) \tag{3.1}$$

whenever P^1 and P^2 both solve the submartingale problem starting from x. Then starting from each x there is at most one solution to the submartingale problem.

Proof: This follows by a simple modification of the proofs of Theorem 6.2.3 and Corollary 6.2.4 in Stroock and Varadhan [15], pages 147–8. In particular, the expressions: "martingale", "R^d", and "determining class Φ", used there, should be replaced, respectively, by "submartingale", "S", and "functions in $C_c^\infty(S)$ which vanish in a neighborhood of the origin".

In the sequel, the hypotheses of the above theorem are verified by first showing that for each fixed $x \in S$, and for γ and h as in Theorem 3.1, $E^P(\int_0^{\tau_0} e^{-\gamma t} h(w(t))\,dt)$ is the same for all solutions P of the submartingale problem starting from x. Then, for $\alpha < 2$, it is shown that $E^P(\int_0^\infty e^{-\gamma t} h(w(t))\,dt)$ is the same for all solutions P of the submartingale problem starting from the origin. A conditioning argument is then used to show that (3.1) holds whenever P^1 and P^2 both solve the submartingale problem starting from some $x \in S$, and $\alpha < 2$. The following lemma is used several times in proving these results.

LEMMA 3.1. *Suppose $x \in S$ and P is a solution of the submartingale problem starting from x. Let τ be a stopping time and let $\{P_w^\tau\}$ be a regular conditional probability distribution (r.c.p.d.) of $P|\mathcal{M}_\tau$. For each $w \in \{\tau < \infty\}$, define P_w^τ on $(C_S, \mathcal{M})$ by*

$$\hat{P}_w^\tau\{\omega \in C_S : \omega \in \mathcal{A}\} = P_w^\tau(\omega(\cdot + \tau) \in \mathcal{A}) \quad \textit{for all} \quad \mathcal{A} \in \mathcal{M}. \tag{3.2}$$

Then there is a P-null set $N \in \mathcal{M}_\tau$ such that, for $w \notin N \cup \{\tau = \infty\}$, $\hat{P}_w^\tau$ is a solution of the submartingale problem starting from $w(\tau)$.

Proof: Let Γ denote the set of $f \in C_b^2(S)$ that are constant in a neighborhood of the origin and satisfy (2.41). There is a countable set Γ^* dense in Γ with respect to bounded pointwise convergence of functions together with their first and second partial derivatives (cf. [19], Lemma 4.6). For each $f \in \Gamma^*$, (2.42) is a P submartingale, and since Theorem 1.2.10 of [15], page 28, holds with the hypothesis of non-negativity of the submartingale replaced by that of boundedness on each finite time interval, it follows that there is a P-null set $N_f \in \mathcal{M}_\tau$ such that, for all $w \notin N_f$,

$$\chi_f^w(t,\omega) = f(\omega(t)) - f(\omega(t\wedge\tau)) - \frac{1}{2}\int_{t\wedge\tau}^{t} \Delta f(\omega(s))\,ds \tag{3.3}$$

defines a P_w^τ submartingale. The abbreviation $\chi_f^w(t)$ will often be used for $\chi_f^w(t,\cdot)$. Let $N = \cup_{f\epsilon\Gamma^*} N_f$. Since the submartingale property is preserved under bounded convergence for each time t, it follows by the density of Γ^* in Γ that χ_f^w is a P_w^τ submartingale for each $w \notin N$ and $f \in \Gamma$. Then by Doob's submartingale stopping theorem, for each positive integer n, $\{\chi_f^w(t + (\tau \wedge n)), \mathcal{M}_{t+(\tau\wedge n)}; t \geqq 0\}$ is a P_w^τ submartingale. On letting $n \to \infty$, it follows in view of continuity and bounded convergence that $\{1_{(\tau<\infty)}\chi_f^w(\cdot + \tau), \mathcal{M}_{\cdot+\tau}\}$ is a P_w^τ submartingale. Then, by using (3.2), it is easy to verify that P_w^τ satisfies properties (i)–(iii) of the submartingale problem starting from $w(\tau)$ whenever $w \notin N \cup \{\tau = \infty\}$.

THEOREM 3.2. *Suppose P solves the submartingale problem starting from some $x \in S$. Then $P = P_x^0$ on $(C_S, \mathcal{M}_{\tau_0})$, where P_x^0 was defined in Theorem* 2.1.

Proof: Since

$$\mathcal{M}_{\tau_0} = \bigvee_{0<r<\Psi(z)} \mathcal{M}_{\tau_r} \vee \sigma\{w(\tau_0)\cdot 1_{\{\tau_0<\infty\}}\}, \tag{3.4}$$

where $w(\tau_0)1_{\{\tau_0<\infty\}} = 0$, it suffices to prove $P = P_x^0$ on $\mathcal{M}_{\tau_r}$ for $0 < r < \Psi(x)$. But this follows by the uniqueness proof in Theorem 2.1.

For each $\gamma > 0$ and $h \in C_b(S)$, consider the stopped resolvent:

$$R_\gamma^0 h(x) = E^{P_x^0}\left(\int_0^{\tau_0} E^{-\gamma t} h(w(t))\,dt\right). \tag{3.5}$$

A closed form expression can be given for this; it is in terms of a Green's function G given by Van Dantzig [18], page 393, for the operator $\Delta - 1$ in S^0 with boundary conditions obtained by taking equality in (2.41). Furthermore, G vanishes at the origin if $\alpha > 0$. The qualitative properties that follow from this representation of (3.5) are vital to the uniqueness proof. For $(r,\theta) \neq (r_0,\theta_0)$ let, if $\alpha \leqq 0$,

$$G(r,\theta;r_0,\theta_0) = \frac{1}{2\tau\xi}\int_{-\infty}^{\infty} du \int_{-\infty}^{\infty} du_0 \cos(r\sinh u)\sin(r_0 \sinh u_0)$$

$$\cdot\frac{\rho(u+i\theta)}{\rho(u_0+i\theta_0)}\frac{\sinh \pi\xi^{-1}(u_0+i\theta_0)}{(\cosh(\pi\xi^{-1}(u_0+i\theta_0)) - \cosh(\pi\xi^{-1}(u+i\theta)))}, \tag{3.6}$$

or if $\alpha > 0$ let

$$G(r,\theta;r_0,\theta_0) = \frac{1}{2\pi\xi}\int_{-\infty}^{\infty} du \int_{-\infty}^{\infty} du_0 \sin(r\sinh u)\cos(r_0 \sinh u_0)$$

$$\cdot \frac{\rho(u+i\theta)}{\rho(u_0+i\theta_0)} \frac{\sinh \pi\xi^{-1}(u+i\theta)}{(\cosh(\pi\xi^{-1}(u+i\theta)) - \cosh(\pi\xi^{-1}(u_0+i\theta_0)))}, \tag{3.7}$$

where

$$\rho(z) \equiv e(z - i\xi, -\theta_1)e(z,\theta_2), \tag{3.8}$$

$$e(z,\theta) \equiv \prod_{m,n}\left\{1 + z^2\left(m\xi + \frac{1}{2}\pi n + \theta\right)^{-2}\right\} \cdot \left\{1 + z^2\left(m\xi + \frac{1}{2}\pi n - \theta\right)^{-2}\right\}^{-1}, \tag{3.9}$$

the product being over all m and n that are odd positive integers. This Green's function has the following properties (see Van Dantzig [18] and Lauwerier [9], II).

The function G is continuous off the diagonal in $S \times S$, and the first and second partial derivatives with respect to (r,θ) are continuous off the diagonal in $(S\times S)\backslash\{(r,\theta;r_0,\theta_0) : r = 0 \text{ or } r_0 = 0\}$. For $r \to \infty$ and $\theta \in [0,\xi]$, $G(r,\theta;r_0,\theta_0) = O(e^{-\beta r})$ for some $\beta > 0$ which may be chosen independent of (r_0,θ_0) lying in a compact subset of S. Similarly, for $r_0 \to \infty$, $\theta_0 \in [0,\xi]$, $G(r,\theta;r_0,\theta_0) = O(\exp\{-\beta r_0\})$ for some $\beta > 0$ which may be chosen independent of (r,θ) lying in a compact subset of S. For fixed $(r_0,\theta_0) \in S\backslash\{0\}$, as a function on $S\backslash(r_0,\theta_0)$, $G(\cdot,\cdot;r_0,\theta_0)$ satisfies:

$$(\Delta - 1)G(r,\theta;r_0,\theta_0) = -r_0^{-1}\delta(r-r_0)\delta(\theta-\theta_0) \quad \text{for} \quad (r,\theta) \in S^0 = S\backslash\partial S, \tag{3.10}$$

$$D_j G(r,\theta;r_0,\theta_0) = 0 \quad \text{for} \quad (r,\theta) \in \partial S_j\backslash\{0\}, \qquad j = 1, 2, \tag{3.11}$$

$$G(0,\theta;r_0,\theta_0) = 0 \quad \text{if} \quad \alpha > 0. \tag{3.12}$$

In polar coordinates the Laplacian is given here by

$$\Delta = \frac{\partial^2}{\partial r^2} + \frac{1}{r}\frac{\partial}{\partial r} + \frac{1}{r^2}\frac{\partial^2}{\partial \theta^2}. \tag{3.13}$$

The right member of (3.10) represents a two-dimensional point source of unit strength written as a Schwartz distribution. In particular, this means that the behavior of $G(\cdot,\cdot;r_0,\theta_0)$ near (r_0,θ_0) is:

$$G(r,\theta;r_0,\theta_0) = -\frac{1}{2\pi}\log\{r^2 + r_0^2 - 2rr_0\cos(\theta-\theta_0)\}^{1/2} + O(1). \tag{3.14}$$

By a minor rescaling of G and integration against h, the desired resolvent (3.5) is obtained for $x \neq 0$, it equals zero since $\tau_0 = 0$ P_0^0 a.s.

THEOREM 3.3. *Suppose $\gamma > 0$ and $h \in C_b(S)$. For each $(r,\theta) \in S$, let*

$$f(r,\theta) = 2\int_0^\xi\int_0^\infty G(\sqrt{2\gamma}r,\theta;\sqrt{2\gamma}r_0,\theta_0)h(r_0,\theta_0)r_0\,dr_0\,d\theta_0. \tag{3.15}$$

Then, for each $x \in S\backslash\{0\}$,

$$R_\gamma^0 h(x) = f(x). \tag{3.16}$$

Proof: By approximation, it suffices to consider $h \in C_c^\infty(S)$. Then by the properties of G, especially its behavior near the singularity $(r,\theta) = (r_0,\theta_0)$, it follows that $f \in C_b(S) \cap C_b^2(S_\varepsilon)$ for each $\varepsilon > 0$, where $S_\varepsilon = S\backslash\{z : \Psi(z) < \varepsilon\}$, and, furthermore,

$$(\tfrac{1}{2}\Delta - \gamma)f = -h \quad \text{in} \quad S^0, \tag{3.17}$$

$$D_j f = 0 \quad \text{on} \quad \partial S_j\backslash\{0\} \quad \text{for} \quad j = 1,2, \tag{3.18}$$

$$f(0,\theta) = 0 \quad \text{if} \quad \alpha > 0. \tag{3.19}$$

Fix $x \in S\backslash\{0\}$. By applying the submartingale property of P_x^0 to suitable approximations to f and using Doob's stopping theorem we see that

$$f(w(t \wedge \tau_\varepsilon)) - \frac{1}{2}\int_0^{t\wedge\tau_\varepsilon} (\Delta f)(w(s))\,ds \tag{3.20}$$

is a P_x^0 martingale. Then, since $\exp\{-\gamma(t \wedge \tau_\varepsilon)\}$ is a process of bounded variation, stochastic calculus (see for example [4]) yields that

$$\exp\{-\gamma(t\wedge\tau_\varepsilon)\}f(w(t\wedge\tau_\varepsilon)) + \int_0^{t\wedge\tau_\varepsilon} e^{-\gamma s}(\gamma f - \tfrac{1}{2}\Delta f)(w(s))\,ds \tag{3.21}$$

is a P_x^0 martingale. The integrand $\gamma f - \frac{1}{2}\Delta f$ may be replaced by h, by (3.17) and the continuity of f, Δf and h on S_ε. Then taking expectations in (3.21) and letting $\varepsilon \downarrow 0$, we obtain

$$E^{P_x^0}[\exp\{-\gamma(t\wedge\tau_0)\}f(w(t\wedge\tau_0))] + E^{P_x^0}\left[\int_0^{t\wedge\tau_0} e^{-\gamma s}h(w(s))\,ds\right] = f(x). \tag{3.22}$$

If $\alpha \leqq 0$, then $P_x^0(\tau_0 = \infty) = 1$, by Theorem 2.2. On the other hand, if $\alpha > 0$, then $f(0) = 0$ by (3.19). In either case, as $t \to \infty$ the first expectations in (3.22) tends to zero and (3.16) follows.

COROLLARY 3.1. *Suppose $\alpha \leqq 0$ and P is a solution of the submartingale problem starting from some $x \in S\backslash\{0\}$. Then, for γ, h and f as defined in Theorem 3.3, we have*

$$E^P\left[\int_0^\infty e^{-\gamma t}h(w(t))\,dt\right] = f(x). \tag{3.23}$$

Proof: By Theorem 3.2, $P = P_x^0$ on $\mathcal{M}_{\tau_0}$. Then, since $\tau_0 = \infty$ P_x^0 a.s., it follows that this also holds P a.s. Hence (3.23) follows from (3.16).

THEOREM 3.4. *Suppose $\alpha \leqq 0$. Then starting from each $x \in S$ there is at most one solution to the submartingale problem.*

Proof: By Theorem 3.1 and Corollary 3.1, it suffices to show for γ and h as in Theorem 3.1 that

$$E^P\left[\int_0^\infty e^{-\gamma t} h(w(t))\,dt\right] \tag{3.24}$$

has the same value for all solutions P of the submartingale problem starting from the origin. For this, define

$$\sigma_\varepsilon = \inf\{t \geqq 0 : \Psi(w(t)) \geqq \varepsilon\} \tag{3.25}$$

for each $\varepsilon > 0$. By condition (iii) of the submartingale problem, $\sigma_\varepsilon \downarrow 0$ P a.s. as $\varepsilon \downarrow 0$. Furthermore, since h is zero in a neighborhood of the origin, there is $\varepsilon_0 > 0$ such that

$$h = 0 \quad \text{on} \quad \{z \in S : \Psi(z) \leqq \varepsilon_0\}.$$

It follows from Lemma 3.1 and Corollary 3.1 that, for each $0 < \varepsilon < \varepsilon_0$, there ia a P-null set $N_\varepsilon \in M_{\sigma_\varepsilon}$ such that, for $w \notin N_\varepsilon \cap \{\sigma_\sigma = \infty\}$,

$$E^{\hat{P}^\varepsilon_w}\left[\int_0^\infty e^{-\gamma t} h(w(t))\,dt\right] = f(w(\sigma_\varepsilon)).$$

Here $\hat{P}^\varepsilon_w \equiv \hat{P}^{\sigma_\varepsilon}_w$ is defined by (3.2) and f is given by (3.15). Hence, by conditioning, the expression in (3.24) equals

$$E^P[\exp\{-\gamma\sigma_\varepsilon\} f(w(\sigma_\varepsilon))].$$

Since $\sigma_\varepsilon \downarrow 0$ P a.s., and f is continuous at the origin and bounded, the above converges to $f(0)$ as $\varepsilon \downarrow 0$. Hence this is the value of (3.24) for all P.

It remains to prove uniqueness for $0 < \alpha < 2$. The main argument for this appears in Theorem 3.10, but first some preliminaries must be established. These are also important in their own right. In the next theorem, for $\alpha > 0$, the submartingale property is extended to those functions f which instead of being constant in a neighborhood of the origin are such that $f(z) - f(0)$ tends to zero sufficiently rapidly as $z \to 0$. Define $S_0 = S\backslash\{0\}$.

THEOREM 3.5. *Suppose $\alpha > 0$ and $h \in C_b(S\backslash\{0\})$ and $f \in C_b(S)$, $C^2(S\backslash\{0\})$ such that*

$$\tfrac{1}{2}\Delta f = h \quad \textit{in} \quad S\backslash\{0\}, \tag{3.26}$$

$$D_j f \geqq 0 \quad \textit{on} \quad \partial S_j\backslash\{0\} \quad \textit{for} \quad j = 1, 2, \tag{3.27}$$

$$|f(z) - f(0)| = o(\Psi(z)) \quad \textit{as} \quad z \to 0. \tag{3.28}$$

Let $x \in S$ and P be a solution of the submartingale problem starting from x. Then,

$$f(w(t)) - \int_0^t (1_{S_0}h)(w(s))\,ds \tag{3.29}$$

is a P submartingale.

Proof: Let $\eta > 0$. Define T_0 and a sequence $\{T_n\}_{n=1}^{\infty}$ of times at which w alternately hits the origin or the arc$\{z : \Psi(z) = \eta\}$ in the following way:

$$T_0 \equiv 0,$$

$$T_{2n+1} = \inf\{t \geqq T_{2n} : w(t) = 0\},$$

$$T_{2n+2} = \inf\{t \geqq T_{2n+1} : \Psi(w(t)) = \eta\},$$

for $n = 0, 1, 2, \cdots$. Then,

$$\begin{aligned} f(w(t)) &- f(w(0)) - \int_0^t (1_{S_0}h)(w(s))\,ds \\ &= \sum_{n=0}^{\infty}\left\{ f(w(t\wedge T_{2n+1})) - f(w(t\wedge T_{2n})) - \int_{t\wedge T_{2n}}^{t\wedge T_{2n+1}} (1_{S_0}h)(w(s))\,ds \right\} \\ &+ \sum_{n=0}^{\infty}\left\{ f(w(t\wedge T_{2n+2})) - f(w(t\wedge T_{2n+1})) - \int_{t\wedge T_{2n+1}}^{t\wedge T_{2n+2}} (1_{S_0}h)(w(s))\,ds \right\}. \end{aligned} \tag{3.30}$$

By stopping w at the first time it hits

$$F_m \equiv \{z \in S : \Psi(z) \leqq m^{-1} \quad \text{or} \quad \Psi(z) \geqq m\}$$

after T_{2n}, m a positive integer, and then applying the submartingale property to functions $f_m \in C_b^2(S)$ which satisfy (3.27), are constant near the origin, and agree with f on F_m, and finally letting $m \to \infty$, it follows that, for each n,

$$f(w(t\wedge T_{2n+1})) - f(w(t\wedge T_{2n})) - \frac{1}{2}\int_{t\wedge T_{2n}}^{t\wedge T_{2n+1}} (1_{S_0}\Delta f)(w(s))\,ds \tag{3.31}$$

is a P submartingale.

Since an L^1 limit of submartingales is again a submartingale, to prove the theorem it suffices to show that the sum over n of the expressions in (3.31) is a P submartingale and that, as $\eta \downarrow 0$, the second sum in (3.30) converges to zero in $L^1 \equiv L^1(C_S, \mathcal{M}, P)$. For this we need an estimate of

$$E^P\left[\sum_{n=1}^{\infty} 1_{\{T_{2n} \leqq t\}}\right].$$

Now, for each $\varepsilon \geqq \eta$,

$$
\begin{aligned}
t &= E^P\left[\sum_{n=0}^{\infty} 1_{\{T_n \leqq t\}} \int_{T_n}^{T_{n+1}\wedge t} ds\right] \\
&\geqq E^P\left[\sum_{n=1}^{\infty} 1_{\{T_{2n} \leqq t\}} \int_{T_{2n}}^{T_{2n+1}\wedge t} 1_{(0,\varepsilon)}(\Psi(w(s)))\, ds\right] \\
&= E^P\left[\sum_{n=1}^{\infty}\left\{ 1_{\{T_{2n} \leqq t\}} \int_{T_{2n}}^{T_{2n+1}} 1_{(0,\varepsilon)}(\Psi(w(s)))\, ds \right.\right. \\
&\qquad \left.\left. - 1_{\{T_{2n} \leqq t < T_{2n+1}\}} \int_{t}^{T_{2n+1}} 1_{(0,\varepsilon)}(\Psi(w(s)))\, ds \right\}\right] \\
&= E^P\left[\sum_{n=1}^{\infty}\left\{ 1_{\{T_{2n} \leqq t\}} E^P\left(\int_{T_{2n}}^{T_{2n+1}} 1_{(0,\varepsilon)}(\Psi(w(s)))\, ds\, \mathcal{M}_{T_{2n}}\right)\right.\right. \\
&\qquad \left.\left. - 1_{\{T_{2n} \leqq t < T_{2n+1}\}} E^P\left(\int_{t}^{T_{2n+1}} 1_{(0,\varepsilon)}(\Psi(w(s)))\, ds\, \mathcal{M}_t\right\}\right].
\end{aligned} \tag{3.32}
$$

It is shown below that the conditional expectations in the last line above are P a.s. finite on $\{T_{2n} \leqq t\}$. This justifies the last two equalities in (3.32), i.e., P a.s. the illegal operation $\infty - \infty$ is not being performed. By Lemma 3.1 and Theorem 3.2, on $[T_{2n} \leqq t\}$ we have P a.s.:

$$
E^P\left(\int_{T_{2n}}^{T_{2n+1}} 1_{(0,\varepsilon)}(\Psi(w(s)))\, ds \mathcal{M}_{T_{2n}}\right) = E^{P^0_{w(T_{2n})}}\left(\int_0^{\tau_0} 1_{(0,\varepsilon)}(\Psi(\omega(s)))\, ds\right). \tag{3.33}
$$

Since $\Psi(w(T_{2n})) = \eta$ P a.s. on $\{T_{2n} < \infty\}$ for $n \geqq 1$, it follows, letting $K \to \infty$ in (2.36), that P a.s. the right member of (3.33) is finite and is greater than or equal to

$$
C_{\varepsilon\eta} \equiv \begin{cases} a_1\eta(2\alpha^{-1}\varepsilon^{(2/\alpha)-1} - \eta^{(2/\alpha)-1})/(2-\alpha) & \text{if} \quad \alpha \neq 2, \\ a_1\eta(\log\varepsilon + 1 - \log\eta)/2 & \text{if} \quad \alpha = 2. \end{cases} \tag{3.34}
$$

In a similar manner, but using the maximum value of g to obtain the upper bound on $\{T_{2n} \leqq t < T_{2n+1}\}$ we have P a.s.:

$$
\begin{aligned}
E^P\left(\int_t^{T_{2n+1}} 1_{(0,\varepsilon)}(\Psi(w(s)))\, ds \mathcal{M}_t\right) &= E^{P^0_{w(t)}}\left(\int_0^{\tau_0} 1_{(0,\varepsilon)}(\Psi(\omega(s)))\, ds\right) \\
&\leqq \alpha^{-1}a_2\varepsilon^{2/\alpha}.
\end{aligned} \tag{3.35}
$$

By combining (3.32)–(3.35) we obtain

$$
E^P\left(\sum_{n=1}^{\infty} 1_{\{T_{2n} \leqq t\}}\right) \leqq (t + \alpha^{-1}a_2\varepsilon^{2/\alpha})/C_{\varepsilon\eta}. \tag{3.36}
$$

It follows that the sum over n of (3.31) is convergent in L^1 and consequently this sum

is a P submartingale. It remains to prove that the second sum in (3.30) converges to zero in L^1 as $\eta \downarrow 0$. For this let $\|h\| = \sup_{z\in S\setminus\{0\}}|h(z)|$. By the definition of the T_n and (3.28), the L^1 norm of the second sum in (3.30) is dominated by:

$$\beta(\eta) + \sum_{n=1}^{\infty} E^P[1_{\{T_{2n+1}\leqq t\}}]\beta(\eta) + \|h\|E^P\left[\int_0^t 1_{(0,\eta)}(\Psi(w(s)))\,ds\right], \tag{3.37}$$

where $\beta(\eta) = o(\eta)$ as $\eta \downarrow 0$. Then by (3.36) and dominated convergence, (3.37) tends to zero as $\eta \downarrow 0$.

Next, for $0 < \alpha < 2$, an estimate is given for the expected amount of time that the process associated with a solution of the submartingale problem starting from the origin takes to hit the arc $\{z \in S : \Psi(z) = \varepsilon\}$.

THEOREM 3.6. *Suppose $0 < \alpha < 2$ and P is a solution of the submartingale problem starting from the origin. For each $\varepsilon > 0$, let*

$$\sigma_\varepsilon \equiv \inf\{t \geqq 0 : \Psi(w(t)) \geqq \varepsilon\}. \tag{3.38}$$

Then there are positive constants a_1 and a_2, independent of ε, such that

$$a_1\varepsilon^{2/\alpha} \leqq E^P(\sigma_\varepsilon) \leqq a_2\varepsilon^{2/\alpha}. \tag{3.39}$$

Proof: For $y \in [0,\infty)$, define

$$g(y) = \alpha^2(\varepsilon^{2/\alpha} - y^{2/\alpha})/(2 - \alpha).$$

Let $f(z) = g(\Psi(z))$ for each $z \in S$. By using (2.30)–(2.33) (recall $\Psi = \Phi$ when $\alpha > 0$), it can be readily verified, since $2/\alpha > 0$, that $f \in C(S) \cap C^2(S\setminus\{0\})$ and satisfies (3.26)–(3.28) with

$$h(z) = -\alpha^2(\cos(\alpha\theta - \theta_1))^{(2/\alpha)-2} \quad \text{for} \quad z = (r,\theta), \tag{3.40}$$

and with equality in (3.27). Furthermore, f can be modified off $B_\varepsilon \equiv \{z \in S : \Psi(z) \leqq \varepsilon\}$, to be bounded on S, by multiplying it by a function in $C_b^2(S)$ which equals one on B_ε and satisfies (3.27) with equality there. Then it follows from Theorem 3.5 and Doob's stopping theorem that (3.29) with $t \wedge \sigma_\varepsilon$ in place of t is a P martingale. Hence,

$$E^P[f(w(t\wedge\sigma_\varepsilon))] - f(0) = E^P\left[\int_0^{t\wedge\sigma_\varepsilon} (1_{S_0}h)(w(s))\,ds\right]. \tag{3.41}$$

Since h is bounded away from zero on $S_0 \equiv S\setminus\{0\}$ and f is bounded on B_ε, letting $t \to \infty$ in (3.41) we may first conclude using property (iii) of the submartingale problem that

$$\sigma_\varepsilon < \infty \quad P\text{ a.s.}, \tag{3.42}$$

and then by bounded and monotone convergence that

$$E^P[f(w(\sigma_\varepsilon))] - f(0) = E^P\left[\int_0^{\sigma_\varepsilon} (1_{S_0}h)(w(s))\,ds\right]. \tag{3.43}$$

Now, $f(w(\sigma_\varepsilon)) = 0$ P a.s. and $f(0) = \alpha^2\varepsilon^{2/\alpha}/(2-\alpha)$. Thus, setting

$$\begin{aligned} a_1 &= \frac{\alpha^2}{2-\alpha}\left\{\sup_{z\in S\setminus\{0\}}(-h(z))\right\}^{-1}, \\ a_2 &= \frac{\alpha^2}{2-\alpha}\left\{\inf_{z\in S\setminus\{0\}}(-h(z))\right\}^{-1}, \end{aligned} \tag{3.44}$$

and using property (iii) of the submartingale problem, (3.39) can be obtained from (3.43).

COROLLARY 3.2. *Let α, P, and σ_ε be as in Theorem 3.6. Then, for each $\gamma > 0$,*

$$1 - E^P[\exp\{-\gamma\sigma_\varepsilon\}] = o(\varepsilon) \quad as \quad \varepsilon \downarrow 0. \tag{3.45}$$

Proof: Since $2/\alpha > 1$ and

$$1 - \exp\{-\gamma\sigma_\varepsilon\} = \gamma\int_0^{\sigma_\varepsilon} e^{-\gamma t}\,dt \leqq \gamma\sigma_\varepsilon,$$

the result follows immediately from (3.39).

In the following theorem, for the definition of weak convergence of probability measures defined on C_S, the space C_S is endowed with the Skorokhod topology inherited from D_S.

THEOREM 3.7. *Let $0 < \alpha < 2$ and $x \in S$. The family of all probability measures on C_S which solve the submartingale problem starting from x is weakly compact.*

Proof: Weak relative compactness is proved by similar arguments to those in Theorem 2.4, using Theorem 3.2. Suppose P^* is a weak limit point of the family. By similar reasoning to that in the proof of Theorem 2.5, P^* is a probability measure on $(C_S, \mathcal{M})$ satisfying properties (i) and (ii) of the submartingale problem starting from x. To show that property (iii) holds for P^*, fix ε and K such that $0 < \varepsilon < K$ and $0 \leqq \Psi(x) < K$. Define

$$g(y) = \begin{cases} (2\alpha\varepsilon^{(2/\alpha)-1}(K-\varepsilon) + \alpha^2(\varepsilon^{2/\alpha} - y^{2/\alpha}))/(2-\alpha) & \text{for } 0 \leqq y \leqq \varepsilon, \\ 2\alpha\varepsilon^{(2/\alpha)-1}(K-y)/(2-\alpha) & \text{for } y \geqq \varepsilon. \end{cases} \tag{3.46}$$

Then $g \in C^1([0,\infty)) \cap C^2((0,\infty)\setminus\{\varepsilon\})$ and

$$\frac{1}{2}g''(y) = \begin{cases} -y^{(2/\alpha)-2} & \text{for } 0<y<\varepsilon, \\ 0 & \text{for } y>\varepsilon, \end{cases} \tag{3.47}$$

$$|g(y) - g(0)| = o(y) \quad \text{as} \quad y \downarrow 0, \tag{3.48}$$

$$g(K) = 0. \tag{3.49}$$

Let $\{g_n\}$ be a sequence in $C_b^1([0,\infty)) \cap C_b^2((0,\infty))$ which converges uniformly to g on $[0,K]$ and is such that $g_n = g$ on $[0,\varepsilon)$ and $g_n'' \leqq 0$ on $[0,K]$. Define $f_n = g_n(\Psi)$ on S. Then by similar reasoning to that in Theorem 3.6, it follows that, for σ_K defined by setting $\varepsilon = K$ in (3.38), and any solution P of the submartingale problem starting from x, and $t \geqq 0$, we have

$$E^P[f_n(w(t\wedge\sigma_K))] - f_n(x) = \frac{1}{2}E^P\left[\int_0^{t\wedge\sigma_K} (1_{S_0}\Delta f_n)(w(s))\,ds\right]. \tag{3.50}$$

From (2.30) and the choice of g_n, it follows that

$$-\Delta f_n \geqq 0 \quad \text{on} \quad B_K \equiv \{z\in S : 0 \leqq \Psi(z) \leqq K\}.$$

Thus, by letting $t \to \infty$ in (3.50) and using monotone convergence, we conclude that

$$-E^P\left[\frac{1}{2}\int_0^{\sigma_K} (1_{S_0}\Delta f_n)(w(s))\,ds\right] \leqq f_n(z) + \|f_n\|_K, \tag{3.51}$$

where $\|f_n\|_K = \sup\{|f_n(z)| : 0 \leqq \Psi(z) \leqq K\}$. Furthermore, by (3.40) and (3.44), we have

$$-\tfrac{1}{2}\Delta f_n(z) = -h(z) \geqq \alpha^2[(2-\alpha)a_2\}^{-1} \quad \text{when} \quad 0<\Psi(z)<\varepsilon.$$

Thus, (3.51) implies

$$E^P\left[\int_0^{\sigma_K} 1_{(0,\varepsilon)}(\Psi(w(s)))\,ds\right] \leqq a_2\left(\frac{2-\alpha}{\alpha^2}\right)[f_n(x) + \|f_n\|_K]. \tag{3.52}$$

By property (iii) of the submartingale problem, we can replace $1_{(0,\varepsilon)}$ by $1_{[0,\varepsilon)}$ in (3.52). Then, by the uniform convergence of f_n to $g(\Psi) \leqq g(0)$ on B_K, it follows, letting $n \to \infty$ in (3.52), that

$$E^P\left[\int_0^{\sigma_K} 1_{[0,\varepsilon)}(\Psi(w(s)))\,ds\right] \leqq 2a_2\{2\alpha^{-1}\varepsilon^{(2/\alpha)-1}(K-\varepsilon) + \varepsilon^{2/\alpha}\}. \tag{3.53}$$

Similar reasoning to that in Lemma 2.2 implies that (3.53) holds with P^* in place of P. Then property (iii) of the submartingale problem follows for P^* by letting $\varepsilon \downarrow 0$ and then $K \to \infty$, since $2/\alpha > 1$. Hence P^* is a solution of the submartingale problem starting from x. Thus, the family of probability measures on C_S which solve the

submartingale problem starting from x is weakly relatively compact and any weak limit point is a member of the family. Hence the family is weakly compact.

The following proof of uniqueness for $0 < \alpha < 2$ will rely heavily on knowing the asymptotic behavior as $x \to 0$ of $R_\gamma^0 h(x)$, which is defined by (3.5). This behavior is not easily obtained from the representation (3.16) due to the complicated nature of G given by (3.7). However, it can be derived using the Green's function G_0 for the operator $\frac{1}{2}\Delta$ (in place of $\Delta - 1$ for G), with the same boundary conditions as for G and such that G_0 vanishes at the origin.

For $\alpha > 0$ this Green's function was obtained by Lauwerier [9], I, pages 477–478, and for $(r,\theta) \neq (r_0,\theta_0)$ in S it is given by

$$\begin{aligned} G_0(r,\theta;r_0,\theta_0) = 2\pi\Bigg[&\mathcal{B}\left(\left(\frac{r}{r_0}\right)^{\nu} \exp\{i\nu(\theta-\theta_0)\},\mu\right) \\ &+ \mathcal{B}\left(\left(\frac{r}{r_0}\right)^{\nu} \exp\{-i\nu(\theta-\theta_0)\},\mu\right) \\ &+ \exp\{-2i\theta_1\}\mathcal{B}\left(\left(\frac{r}{r_0}\right)^{\nu} \exp\{i\nu(\theta+\theta_0)\},\mu\right) \\ &+ \exp\{2i\theta_1\}\mathcal{B}\left(\left(\frac{r}{r_0}\right)^{\nu} \exp\{-i\nu(\theta+\theta_0)\},\mu\right)\Bigg], \end{aligned} \tag{3.54}$$

where $\nu = \pi/\xi$, $\mu = \alpha/\nu$, and $\mathcal{B}$ is the incomplete Beta function defined by

$$\mathcal{B}(z,\mu) = \mu^{-1} z^{\mu} F(\mu,1;\mu+1;z). \tag{3.55}$$

Here $z^\mu \equiv r^\mu e^{i\mu\theta}$ for $z = (r,\theta)$, $r \geqq 0$, $\theta \in (-\infty,\infty)$, and F is the hypergeometric function such that, for z lying in the complex plane cut along the real axis from 1 to $+\infty$, we have

$$F(\mu,1;\mu+1;z) = \mu\int_0^1 t^{\mu-1}(1-tz)^{-1}\,dt.$$

Consequently,

$$F(\mu,1;\mu+1;z) = \begin{cases} \mu\sum_{k=0}^{\infty}\dfrac{z^k}{\mu+k} & \text{for } |z|<1, \\ \mu\sum_{k=1}^{\infty}\dfrac{z^{-k}}{k-\mu} + \mu z^{-\mu}\pi \text{ for} \operatorname{cosec}(\mu\pi)\exp\{i\mu\pi\operatorname{sgn}(\arg z)\} & \\ & \text{for } |z|>1, z\notin(1,\infty). \end{cases} \tag{3.56}$$

At $z = 1$, F has a logarithmic singularity and the change in F upon crossing the branch cut is a multiple of the complex number i. It follows that $\mathcal{B}(z,\mu) + \mathcal{B}(\bar{z},\mu)$ (where $\bar{z} = (r,-\theta)$ if $z = (r,\theta)$) is real and can be defined continuously across the

branch cut from 1 to ∞. Thus G_0 is real and well defined off the diagonal in $S \times S$. For fixed $(r_0,\theta_0) \in S\backslash\{0\}$, as a function on $S\backslash(r_0,\theta_0)$, $G(\cdot,\cdot;r_0,\theta_0)$ satisfies

$$\tfrac{1}{2}\Delta G_0(r,\theta;r_0,\theta_0) = -r_0^{-1}\delta(r-r_0)\delta(\theta-\theta_0) \quad \text{for} \quad (r,\theta)\in S^0, \tag{3.57}$$

$$D_jG_0(r,\theta;r_0,\theta_0) = 0 \quad \text{for} \quad (r,\theta) \in \partial S_j\backslash\{0\}, \qquad j = 1, 2, \tag{3.58}$$

$$G_0(0,\theta;r_0,\theta_0) = 0. \tag{3.59}$$

The following theorem can be proved in a similar manner to Theorem 3.3. Here h is assumed to vanish sufficiently rapidly at infinity so that the right side of (3.60) is well defined.

THEOREM 3.8. *Suppose $\alpha > 0$ and $h \in C_b(S)$ such that $h(z) = O(\exp\{-\beta|z|\})$ as $|z| \to \infty$ for some $\beta > 0$. For each $x = (r,\theta) \in S$,*

$$E^{P_x^0}\left[\int_0^{\tau_0} h(w(t))\,dt\right] = \int_0^{\xi}\int_0^{\infty} G_0(r,\theta;r_0,\theta_0)h(r_0,\theta_0)r_0\,dr_0\,d\theta_0. \tag{3.60}$$

The asymptotic result that will be needed is the following.

THEOREM 3.9. *Suppose $0<\alpha<2$ and $h \in C_b(S)$ is such that $h(z) = O(\exp\{-\beta|z|\})$ as $|z| \to \infty$ for some $\beta > 0$. Then, uniformly, for $\theta \in [0,\xi]$,*

$$\lim_{r\downarrow 0}(\Psi(r,\theta))^{-1}E^{P_{(r,\theta)}^0}\left[\int_0^{\tau_0} h(w(t))\,dt\right] = c(h). \tag{3.61}$$

Here $c(h)$ is a constant which is strictly positive whenever $h \geqq 0$ and $h \not\equiv 0$.

Proof: Let $x = (r,\theta) \in S\backslash\{0\}$. For $0 < r_0 < r$, we have, by (3.54)–(3.56) and since $\mu\pi = \theta_1 + \theta_2$,

$$\begin{aligned} G_0&(r,\theta;r_0,\theta_0) \\ &= 4\pi\left\{\sum_{k=1}^{\infty}\frac{1}{k-\mu}\left(\frac{r}{r_0}\right)^{\alpha-k\nu}[\cos((\alpha-k\nu)(\theta-\theta_0)) \right.\\ &\qquad + \cos((\alpha-k\nu)(\theta+\theta_0)-2\theta_1)] \\ &\qquad \left. + \tau\operatorname{cosec}(\mu\pi)[\cos(\mu\pi)+\cos(\mu\pi-2\theta_1)]\right\}, \\ &= 8\pi\left\{\sum_{k=1}^{\infty}\frac{1}{k-\mu}\left(\frac{r}{r_0}\right)^{\alpha-k\nu}[\cos((\alpha-k\nu)\theta-\theta_1)\cos((\alpha-k\nu)\theta_0-\theta_1) \right.\\ &\qquad \left. + \pi\operatorname{cosec}(\theta_1+\theta_2)\cos\theta_2\cos\theta_1]\right\}. \end{aligned} \tag{3.62}$$

Consequently, by setting $C = \pi\operatorname{cosec}(\theta_1+\theta_2)\cos\theta_2\cos\theta_1$ we obtain

$$\int_0^r |G_0(r,\theta;r_0,\theta_0)| r_0\, dr_0 \leqq 8\pi r^2 \left[\sum_{k=1}^{\infty} \frac{1}{(k-\mu)(2-\alpha+k\nu)} + \frac{1}{2} C \right] = O(r^2), \tag{3.63}$$

by comparison with $\Sigma_{k=1}^{\infty} k^{-2}$. Similar to (3.62), for $r_0 > r$,

$$G_0(r,\theta;r_0,\theta_0) = 4\pi \left\{ \sum_{k=0}^{\infty} \frac{1}{\mu + k} \left(\frac{r}{r_0} \right)^{\alpha + k\nu} [\cos((\alpha + k\nu)\theta - \theta_1)\cos((\alpha + k\nu)\theta_0 - \theta_1)] \right\}. \tag{3.64}$$

By combining this with the boundedness of $h(z)$ and its exponential decay as $|z| \to \infty$, we may conclude that as $r \to 0$, the principal contribution to the following integral involving G_0 comes from integrating against the $k = 0$ term in the above series. More precisely, we have

$$\begin{aligned} &\int_0^{\xi} \int_r^{\infty} G_0(r,\theta;r_0,\theta_0) h(r_0,\theta_0) r_0\, dr_0\, d\theta_0 \\ &= 4\pi \left\{ \mu^{-1} r^{\alpha} \cos(\alpha\theta - \theta_1) \times \int_0^{\xi} \int_r^{\infty} r_0^{1-\alpha} \cos(\alpha\theta_0 - \theta_1) h(r_0,\theta_0) dr_0\, d\theta_0 + \chi(r,\theta) \right\}, \end{aligned} \tag{3.65}$$

where

$$\chi(r,\theta) = \begin{cases} O(r^{(\alpha+\nu)\wedge 2}) & \text{if} \quad \alpha + \nu \neq 2 \\ O(r^2 \log r) & \text{if} \quad \alpha + \nu = 2, \end{cases}$$

as $r \to 0$. From (3.60), (3.63), and (3.65) it follows, since $\Psi(r,\theta) = r^{\alpha} \cos(\alpha\theta - \theta_1)$ and $\alpha < 2$, that (3.61) holds with

$$c(h) = \frac{4\pi^2}{\theta_1 + \theta_2} \int_0^{\xi} \int_0^{\infty} r_0^{1-\alpha} \cos(\alpha\theta_0 - \theta_1) h(r_0,\theta_0)\, dr_0\, d\theta_0. \tag{3.66}$$

If h is non-negative and not identically zero, then the integrand in (3.66) also has these properties and consequently $c(h) > 0$.

LEMMA 3.2. *Suppose $\alpha > 0$. Then for each $\gamma > 0$ there is a constant $\beta(\gamma) > 0$ such that*

$$E^{P_x^0}[e^{-\gamma\tau_0}] = O(\exp\{-\beta(\gamma)|x|\}) \quad as \quad |x| \to \infty. \tag{3.67}$$

Proof: For $\upsilon > 0$, let K_{υ} denote the modified Bessel function of order υ which tends to zero at infinity. For the properties of K_{υ}, see Abramowitz and Stegun [1], pages 374–378. Define

$$g_\gamma(y) = \begin{cases} by^{1/2}K_{\alpha/2}(\sqrt{2\gamma}\alpha y^{1/\alpha}) & \text{for } y > 0, \\ 1 & \text{for } y = 0, \end{cases} \tag{3.68}$$

where $b = 2^{1-\alpha/4}\gamma^{\alpha/4}\alpha^{\alpha/2}(\Gamma(\alpha/2))^{-1}$ and $\Gamma(\cdot)$ is the usual Gamma function. Then $g_\gamma \in C_b([0,\infty)) \cap C^2((0,\infty))$ and

$$g''_\gamma(y) = 2\gamma y^{(2/\alpha)-2}g_\gamma(y) \quad \text{for} \quad y>0. \tag{3.69}$$

Define $f_\gamma(z) = g_\gamma(\Psi(z))$ for $z \in S$. Then $f_\gamma \in C_b(S) \cap C^2(S\backslash\{0\})$, and since $\Psi = \Phi$ for $\alpha > 0$, we have, by (2.30), (2.31) and (3.69),

$$\tfrac{1}{2}\Delta f_\gamma(z) = \gamma(\theta)f_\gamma(z) \quad \text{for} \quad z = (r,\theta) \in S\backslash\{0\}, \tag{3.70}$$

where $\gamma(\theta) = \gamma\alpha^2(\cos(\alpha\theta - \theta_1))^{(2/\alpha)-2}$. Furthermore,

$$D_j f_\gamma = 0 \quad \text{on} \quad \partial S_j\backslash\{0\} \quad \text{for} \quad j = 1,2. \tag{3.71}$$

The asymptotic behavior of K_v at infinity is given by (see [1], page 378)

$$K_v(y) = O(y^{-1/2}e^{-y}) \quad \text{as} \quad y \to \infty. \tag{3.72}$$

Replacing y by $\sqrt{2\gamma}\alpha y^{1/\alpha}$ and setting $v = \alpha/2$ in the above, we see that there is a constant $d(\gamma) > 0$:

$$f_\gamma(z) = O(\exp\{-d(\gamma)|z|\}) \quad \text{as} \quad |z| \to \infty. \tag{3.73}$$

Then by similar arguments to those in the proof of Theorem 3.3, in particular using (3.71), (3.70), and (3.73), we conclude that

$$f_\gamma(w(t\wedge\tau_\varepsilon))\exp\left\{-\int_0^{t\wedge\tau_\varepsilon}\gamma(\theta_s)\,ds\right\} + \int_0^{t\wedge\tau_\varepsilon}\left\{\gamma(\theta_s)f_\gamma(w(s)) - \frac{1}{2}\Delta f(w(s))\right\}\exp\left\{-\int_0^s\gamma(\theta_u)\,du\right\}ds$$

is a P_x^0 martingale, where $\theta_s \equiv \arg(w(s))$. Here the internal term is zero P_x^0 a.s. by (3.70). Then by taking expectations and letting $\varepsilon \downarrow 0$ and then $t \to \infty$, it follows by bounded convergence that

$$E^{P_x^0}\left[f_\gamma(w(\tau_0))\exp\left\{-\int_0^{\tau_0}\gamma(\theta_s)\,ds\right\}\right] = f_\gamma(x).$$

Now $f_\gamma(w(\tau_0)) = f_\gamma(0) = 1$ P_x^0 a.s. Furthermore,

$$\sup_{\theta\in[0,\xi]}\gamma(\theta) \leqq \gamma a, \quad \text{where} \quad a \equiv \max_{\theta\in[0,\xi]}\left[\alpha^2(\cos(\alpha\theta-\theta_1))^{(2/\alpha)-2}\right] < \infty.$$

Hence

$$E^{P_x^0}[\exp\{-\gamma a\tau_0\}] \leqq f_\gamma(x).$$

Therefore, (3.67) holds with $\beta(\gamma) = d(\gamma a^{-1})$, by (3.73).

THEOREM 3.10. *Suppose $0 < \alpha < 2$. Then starting from each $x \in S$ there is at most one solution of the submartingale problem.*

Proof: By Theorem 3.2 and Lemma 3.1, for each $\gamma > 0$, $h \in C_c^\infty(S)$ which vanishes in a neighborhood of the origin, and solution P of the submartingale problem starting from $x \in S$, we have

$$E^P\left[\int_0^\infty e^{-\gamma t}h(w(t))\,dt\right] = E^{\hat{P}_x^0}\left[\int_0^{\tau_0} e^{-\gamma t}h(w(t))\,dt\right] + E^P\left[\exp\{-\gamma\tau_0\}E^{\hat{P}_w^0}\left(\int_0^\infty e^{-\gamma t}h(\omega(t))\,dt\right)\right]. \tag{3.74}$$

Here $\hat{P}_w^0 \equiv \hat{P}_w^{\tau_0}$ is a solution of the submartingale problem starting from the origin, for P_x^0 almost every $w \in \{\tau_0 < \infty\}$, by Lemma 3.1. From (3.74) and Theorem 3.1, it follows that it is enough to show that the left member of (3.74) has the same value for all solutions P of the submartingale problem starting from the origin. Let P be such a solution.

Let σ_ε, ε_0 and $\hat{P}_w^\varepsilon$ be as in the proof of Theorem 3.4. Suppose $0 < \varepsilon < \varepsilon_0$. Then P a.s., $w(t) \in h^{-1}(0)$ for $0 \leqq t \leqq \sigma_\varepsilon$, and consequently we have

$$E^P\left[\int_0^\infty e^{-\gamma t}h(w(t))\,dt\right] = E^P\left[\exp\{-\gamma\sigma_\varepsilon\}E^{\hat{P}_w^\varepsilon}\left\{\int_0^\infty e^{-\gamma t}h(\omega(t))\,dt\right\}\right]. \tag{3.75}$$

Similarly to (3.74) we have P a.s. on $\{\sigma_\varepsilon < \infty\}$,

$$E^{\hat{P}_\cdot^\varepsilon}\left\{\int_0^\infty e^{-\gamma t}h(w)(t))\,dt\right\} = E^{\hat{P}_\cdot^\varepsilon}\left\{\int_0^{\tau_0} e^{-\gamma t}h(w(t))\,dt\right\} + E^{\hat{P}_\cdot^\varepsilon}\left\{\exp\{-\gamma\tau_0\}E^{\hat{Q}_w^0}\left(\int_0^\infty e^{-\gamma t}h(\omega(t))\,dt\right)\right\}, \tag{3.76}$$

where $\cdot$ denotes a generic element of $\{\sigma_\varepsilon < \infty\}$ and, for $Q \equiv \hat{P}_\cdot^\varepsilon$, $\hat{Q}_w^0 \equiv \hat{Q}_w^{\tau_0}$ is a solution of the submartingale problem starting from the origin for $\hat{P}_\cdot^\varepsilon$ almost every $w \in \{\tau_0 < \infty\}$.

Define

$$M = \sup_P E^P\left(\int_0^\infty e^{-\gamma t}h(w(t))\,dt\right)$$

where the supremum is over all solutions P of the submartingale problem starting

from the origin. By Theorem 3.7, the collection of such P's is weakly compact. Furthermore, the expectation $E^P(\int_0^\infty e^{-\gamma t}h(w(t))\,dt)$ is weakly continuous in P. Hence the above supremum is attained by some P. For this P we have, by (3.75) and (3.76),

$$\begin{aligned} M &= E^P\left[\int_0^\infty e^{-\gamma t}h(w(t))\,dt\right] \\ &\leqq E^P\left[\exp\{-\gamma\sigma_\varepsilon\}E^{\hat{P}_w^\varepsilon}\left\{\int_0^{\tau_0} e^{-\gamma t}h(\omega(t))\,dt\right\}\right] \\ &\quad + E^P\left[\exp\{-\gamma\sigma_\varepsilon\}E^{\hat{P}_w^\varepsilon}\{\exp-\gamma\tau_0\}M\right] \\ &= E^P\left[\exp\{-\gamma\sigma_\varepsilon\}E^{P^0_{w(\sigma_\varepsilon)}}\left\{\int_0^{\tau_0} e^{-\gamma t}h(\omega(t))\,dt\right\}\right] \\ &\quad + ME^P\left[\exp\{-\gamma\sigma_\varepsilon\}E^{P^0_{w(\sigma_\varepsilon)}}\{\exp\{-\gamma\tau_0\}\right]. \end{aligned} \tag{3.77}$$

The last equality above follows because, for P almost everywhere $w \in \{\sigma_\varepsilon < \infty\}$, $\hat{P}_w^\varepsilon$ is a solution of the submartingale problem starting from $w(\sigma_\varepsilon)$ and hence agrees with $P^0_{w(\sigma_\varepsilon)}$ on $\mathcal{M}_{\tau_0}$. By rearranging (3.77), since $\sigma_\varepsilon > 0$ P a.s. (by (3.39)), we have

$$M \leqq \frac{E^P\left[\exp\{-\gamma\sigma_\varepsilon\}E^{P^0_{w(\sigma_\varepsilon)}}\left\{\int_0^{\tau_0} e^{-\gamma t}h(\omega(t))\,dt\right\}\right]}{1 - E^P\left[\exp\{-\gamma\sigma_\varepsilon\}E^{P^0_{w(\sigma_\varepsilon)}}\{\exp\{-\gamma\tau_0\}\}\right]}. \tag{3.78}$$

For each $x \in S$, let

$$u(x) = E^{P^0_x}\left\{\int_0^{\tau_0} e^{-\gamma t}h(w(t))\,dt\right\} \tag{3.79}$$

and

$$v(x) = E^{P^0_x}\left\{\int_0^{\tau_0} e^{-\gamma t}\,dt\right\}. \tag{3.80}$$

Then (3.78) may be rewritten as:

$$M \leqq \frac{E^P[\exp\{-\gamma\sigma_\varepsilon\}u(w(\sigma_\varepsilon))]}{1 - E^P[\exp\{-\gamma\sigma_\varepsilon\}] + \gamma E^P[\exp\{-\gamma\sigma_\varepsilon\}v(w(\sigma_\varepsilon))]}. \tag{3.81}$$

By Theorem 3.3, for each $x \in S\backslash\{0\}$, $u(x) = R^0_\gamma h(x)$ satisfies (3.17)–(3.19) with u in place of f. Moreover, $u(x) = O(\exp\{-\beta_0(\gamma)|x|\})$ as $|x| \to \infty$ for some $\beta_0(\gamma) > 0$, since $G(\cdot,\cdot;r_0,\theta_0)$ has the property uniformly for (r_0,θ_0) lying in the (compact) support of h. By a similar proof to that in Theorem 3.3, using $u(w(t \wedge \tau_\varepsilon))$ in place of $\exp\{-\gamma(t \wedge \tau_\varepsilon)\}f(w(t \wedge \tau_\varepsilon))$ we have, since $u(0) = 0$,

$$u(x) = E^{P_x^0}\left[\int_0^{-\tau_0} (h - \gamma u)(w(s))\,ds\right].$$

The right side above is well defined because the hypotheses of Theorem 3.8 are satisfied with $h - \gamma u$ in place of h. Then, by Theorem 3.9, on $\{\gamma_\varepsilon < \infty\}$ we have

$$\text{(3.82)}\qquad \begin{aligned}\varepsilon^{-1}u(w(\sigma_\varepsilon)) &= (\Psi(w(\gamma_\varepsilon)))^{-1}E^{P^0_{w(\sigma_\varepsilon)}}\left[\int_0^{\tau_0} (h - \gamma u)(\omega(s))\,ds\right]\\ &\to c(h - \gamma u) \quad \text{as} \quad \varepsilon \downarrow 0.\end{aligned}$$

By similar reasoning to that in Theorem 3.3, $v = R_\gamma^0 1$ satisfies

$$\tfrac{1}{2}\Delta v = \gamma v - 1 \quad \text{in} \quad S^0,$$

and (3.18), (3.19) hold with $f = v$. By Lemma 3.2, we have, for $x \in S\setminus\{0\}$,

$$0 < 1 - \gamma v(x) = E^{P_x^0}[\exp\{-\gamma\tau_0\}] = O(\exp\{-\beta(\gamma)|x|\})$$

as $|x| \to \infty$. Then, as for u above, it follows that

$$\text{(3.83)}\qquad \varepsilon^{-1}v(w(\sigma_\varepsilon)) \to c(1 - \gamma v) > 0 \quad \text{as} \quad \varepsilon \downarrow 0.$$

Thus, by dividing the numerator and denominator of (3.81) by ε, and using (3.45), (3.82), (3.83), and bounded convergence, we may conclude on letting $\varepsilon \downarrow 0$ that

$$M \leqq C(\gamma,h) \equiv \frac{c(h - \gamma u)}{\gamma c(1 - \gamma v)} < \infty.$$

By similar reasoning we can obtain

$$m = \inf_P E^P\left(\int_0^\infty e^{-\gamma t}h(w(t))\,dt\right) \geqq C(\gamma,h).$$

Hence,

$$E^P\left(\int_0^\infty e^{-\gamma t}h(w(t))\,dt\right) = M = m = C(\gamma,h)$$

for all solutions P of the submartingale problem starting from the origin, as desired.

3.2. Non-existence for $\alpha \geqq 2$.

THEOREM 3.11. *Suppose $\alpha \geqq 2$ and $x \in S$. Then there is no solution of the submartingale problem starting from x. Indeed P_x^0 is the unique probability measure on $(C_S,\mathcal{M})$ satisfying properties* (i) *and* (ii) *of the submartingale problem.*

Proof: Suppose P is a probability measure on $(C_S,\mathcal{M})$ satisfying properties (i) and (ii) of the submartingale problem. Since the proof of Theorem 3.2 does not require property (iii) of the submartingale problem, it follows that $P = P_x^0$ on $\mathcal{M}_{\tau_0}$. Consequently, $\tau_0 < \infty$ P a.s. By omitting the verification of property (iii) from the proof of Lemma 3.1, it can be shown that there is a P-null set $N \in \mathcal{M}_{\tau_0}$ such that, for $w \notin N \cup \{\tau_0 = \infty\}$, $\hat{P}_w^{\tau_0}$ satisfies properties (i) and (ii) of the submartingale problem starting from the origin. We prove below that, for such w,

$$\hat{P}_w^{\tau_0}[\omega(t) = 0 \text{ for all } t \geqq 0] = 1. \tag{3.84}$$

Here ω denotes a generic element of C_S. It follows from this that $P = P_x^0$ on $\mathcal{M}$ and consequently it cannot satisfy property (iii) of the submartingale problem. Thus it suffices to verify (3.84).

Let $g : [0,\infty) \to [0,1]$ be a twice continuously differentiable function such that

$$g(y) = \begin{cases} 0 & \text{for} \quad 0 \leqq y \leqq \frac{1}{2}, \\ 1 & \text{for} \quad y \geqq 1. \end{cases}$$

Define

$$M = \max_{y \in [0,\infty)} \{|g'(y)| + |g''(y)|\}.$$

For each $r > 0$ and $z \in S$, define

$$h_r(z) = g(r^{-1}\Psi(z)).$$

For $0 < \varepsilon < K < \infty$, let

$$f_{\varepsilon K} = h_\varepsilon(1 - h_{2K})\Psi.$$

Then $f_{\varepsilon K} \in C_b^2(S)$ and $f = f_{\varepsilon K}$ satisfies (2.41) with equality there. Moreover, since $\Delta\Psi = 0$ in $S\backslash\{0\}$, we have

$$\Delta f_{\varepsilon K} = \begin{cases} 0 & \text{for} \quad 0 \leqq \Psi(z) \leqq \frac{1}{2}\varepsilon, \\ (\varepsilon^{-2}g''\Psi + 2\varepsilon^{-1}g')|\nabla\Psi|^2(z) & \text{for} \quad \frac{1}{2}\varepsilon \leqq \Psi(z) \leqq \varepsilon, \\ 0 & \text{for} \quad \varepsilon \leqq \Psi(z) \leqq K. \end{cases} \tag{3.85}$$

Let $\sigma_K = \inf\{t \geqq 0 : \Psi(w(t)) \geqq K\}$. Then by Doob's stopping theorem and property (ii) of the submartingale problem satisfied by $\hat{P}_w^{\tau_0}$, we see that, for each $w \notin N \cup \{\tau_0 = \infty\}$,

$$f_{\varepsilon K}(\omega(t \wedge \sigma_K)) - \frac{1}{2}\int_0^{t \wedge \sigma_K} \Delta f_{\varepsilon K}(\omega(s))\, ds \tag{3.86}$$

is a $\hat{P}_w^{\tau_0}$ martingale. Now on $\{z \in S : 0 \leqq \Psi(z) \leqq K\}$:

(a) $$f_{\varepsilon K} \to \Psi \quad \text{uniformly as} \quad \varepsilon \downarrow 0,$$

(b) $$\Delta f_{\varepsilon K} \to 0 \qquad \text{as} \quad \varepsilon \downarrow 0,$$

and

(c) $$|\Delta f_{\varepsilon K}| \leqq b\{(\varepsilon^{-2}M)\varepsilon + 2\varepsilon^{-1}M\}\varepsilon^{2-2/\alpha}.$$

Here $b < \infty$ is a constant such that $|\nabla \Psi|^2 \leqq b\Psi^{2-2/\alpha}$ on $S\backslash\{0\}$ (see equation (2.31)). Since $\alpha \geqq 2$, the right side of the inequality in (c) is a bounded function of $\varepsilon \in (0,K)$. Since martingales are preserved under bounded convergence at each time t, it follows, by letting $\varepsilon \downarrow 0$ in (3.86), that $\Psi(\omega(\cdot \wedge \sigma_K))$ is a $\hat{P}_w^{\tau_0}$ martingale. Consequently,

$$E^{\hat{P}_w^{\tau_0}}[\Psi(\omega(t \wedge \sigma_K))] = \Psi(0) = 0 \quad \text{for all} \quad t \geqq 0.$$

Since $\Psi > 0$ on $S\backslash\{0\}$ it follows that $\omega(\cdot \wedge \sigma_K) \equiv 0$ $\hat{P}_w^{\tau_0}$ a.s. Letting $K \to \infty$, we see that (3.84) holds.

By Theorem 2.5, any weak limit point P_x of the $\{P_x^\delta\}$ satisfies as $|\delta| \to 0$ properties (i) and (ii) of the submartingale problem. By the above we have $P_x = P_x^0$ if $\alpha \geqq 2$.

3.3 Feller continuity and the strong Markov property. The existence Theorem 2.5 may be combined with the uniqueness Theorems 3.4, 3.10, and 3.11, and Corollary 3.1 to yield the following.

THEOREM 3.12. *For each $x \in S$, there is a unique weak limit P_x of $\{P_x^\delta\}$ as $|\delta| \to 0$. If $\alpha < 2$, it is the unique solution of the submartingale problem starting from x. Moreover, if $\alpha \leqq 0$ and $x \in S\backslash\{0\}$, then $P_X = P_x^0$. On the other hand, if $\alpha \geqq 2$, then $P_x = P_x^0$ and it is the unique probability measure on $(C_S,\mathcal{M})$ satisfying properties* (i) *and* (ii) *of the submartingale problem starting from x.*

This uniqueness theorem is the key to the proofs given below that $\{P_x, x \in S\}$ is Feller continuous and has the strong Markov property.

THEOREM 3.13. *Suppose $\{x_n\}$ is a sequence in S which converges to $x \in S$. Then $\{P_{x_n}\}$ converges weakly to P_x.*

Proof: By similar arguments to those in the first part of the proof of Theorem 3.7, it follows that $\{P_{x_n}\}$ is weakly relatively compact and any weak limit point P^* is concentrated on C_S and satisfies properties (i) and (ii) of the submartingale problem statting from x. If $\alpha \geqq 2$, it then follows from Theorem 3.12 that $P^* = P_x$, and since P^* was an arbitrary limit point of $\{P_{x_n}\}$, we can conclude that $\{P_{x_n}\}$ converges weakly to P_x.

If $0 < \alpha < 2$, it follows by the same reasoning as in Theorem 3.7 that if $K > \Psi(x_n)$ for all n, then (3.53) holds with P^* in place of P. Then, by letting $\varepsilon \downarrow 0$ and then $K \to \infty$, we see that P^* also satisfies property (iii) of the submartingale problem starting from x. Thus, as for $\alpha \geqq 2$, we may conclude that $P^* = P_x$ and $\{P_{x_n}\}$ converges weakly to P_x, if $0 < \alpha < 2$.

Finally, suppose $\alpha \leqq 0$. By the same reasoning as for $0 < \alpha < 2$, it suffices to show that P^* satisfies property (iii) of the submartingale problem starting from x. For this, by selecting a subsequence if necessary, we suppose that $x_n \neq 0$ for all n, even if $x = 0$. Since $\alpha \leqq 0$, we have $P_{x_n} = P_{x_n}^0$ for all n. Then it follows by similar reasoning to that in the proof of Theorem 2.5 (iii) that P^* satisfies property (iii) of the submartingale problem starting from x.

COROLLARY 3.3. *Suppose $h : C_S \to R$ is bounded and $\mathcal{M}$-measurable. Then $Ph : S \to R$ defined by*

$$(Ph)(x) \equiv E^{P_x}[h(w(\cdot))] \quad \textit{for all} \quad x \in S$$

is a Borel measurable function on S.

Proof: By the weak convergence proved in Theorem 3.13 and the weak metrizability of $\{P_x, x \in S\}$, it follows that if h is continuous and bounded on C_S (endowed with the Skorokhod topology), then Ph is continuous and bounded on S. The desired result follows from this by a monotone class argument (see Dellacherie and Meyer [5], Section I-21, pages 14-I, for the appropriate monotone class theorem).

THEOREM 3.14. *The family $\{P_x, x \in S\}$ has the strong Markov property, i.e., for each stopping time τ, each $x \in S$, and each bounded $\mathcal{M}$-measurable function $h : C_S \to R$, we have*

$$E^{P_x}[1_{\{\tau<\infty\}}h(w(\cdot + \tau))\mathcal{M}_\tau] = 1_{\{\tau<\infty\}}E^{P_{w(\tau)}}[h(\omega(\cdot))].$$

Proof: If $\alpha < 2$, this follows immediately from Lemma 3.1, and the uniqueness Theorem 3.12 and Corollary 3.3, which enable us to use $P_{w(\tau)}$ for $\hat{P}_w^\tau$. If $\alpha \geqq 2$, the result follows in a similar manner, except that a modified form of Lemma 3.1 must be used in which the words "a solution of the submartingale problem" are replaced by "a solution of properties (i) and (ii) of the submartingale problem". This modified form can be proved in the same manner as Lemma 3.1.

Acknowledgment. This research was supported by the National Science Foundation under Grants MCS-80-02568 and MCS-81-17526.

Bibliography

[1] Abramowitz, M., and Stegun, I. A., *Handbook of Mathematical Functions*, National Bureau of Standards, Applied Mathematics Series 55, 1972.

[2] Anderson, R. F., and Orey, S., *Small random perturbations of dynamical systems with reflecting boundary*, Nagoya Math. J. 60, 1976, pp. 189–216.

[3] Bellot, B. A., *On Reflected Brownian Motion in Two Dimensions*, Ph.D. Dissertation, Department of Statistics, University of California at Berkeley, California, 1975.

[4] Chung, K. L., and Williams, R. J., *An Introduction to Stochastic Integration*, Birkhäuser, Boston, 1983.

[5] Dellacherie, C., and Meyer, P. A., *Probabilities and Potential,* Vol. I, North-Holland, Amsterdam, 1978.

[6] Harrison, J. M., and Reiman, M. I., *Reflected brownian Motion on an orthant,* Annals Prob., 9, 1981, pp. 302–308.

[7] Keller, J. B., *Oblique derivative boundary conditions and the image method,* SIAM J. Appl. Math., 41, 1981, pp. 294–300.

[8] Kurtz, T. G., *Approximation of Population Processes,* CBMS-NSF Regional Conference Series in Applied Mathematics, SIAM, 1981.

[9] Lauwerier, H. A., *Solutions of the equation of Helmholtz in an angle,* I and II. I. Kon. Ned. Akad. v. Wet. Proc. A62, No. 5, 1959, pp. 475–488; II. *Ibid.*, A63, No. 4, 1960, pp. 355–372.

[10] Lindvall, T., *Weak convergence of probability measures and random functions in the function space* $D[0,\infty)$, J. Appl. Prob. 10, 1973, pp. 109–121.

[11] Motoo, M., *Brownian motions in the half-plane with singular inclined periodic boundary conditions,* Topics in Probability Theory, ed. D. W. Stroock and S. R. S. Varadhan, New York University, 1973, pp. 163–179.

[12] Prokhorov, Yu, V., *Convergence of random processes and limit theorems in probability theory,* Theor. Prob. and Applic. 1, 1956, pp. 157–214.

[13] Reiman, M. I., *Open queueing networks in heavy traffic,* to appear in Mathematics of Operations Research.

[14] Stroock, D. W., and Varadhan, S. R. S., *Diffusion processes with boundary conditions,* Comm. Pure Appl. Math. 24, 1971, pp. 147–225.

[15] Stroock, D. W., and Varadhan, S. R. S., *Multi-Dimensional Diffusion Processes,* Springer-Verlag, New York, 1979.

[16] Tsuchiya, M., *On the stochastic differential equation for a Brownian motion with oblique reflection on the half-plane,* Proc. of International Symposium on Stochastic Differential Equations, Kyoto, 1976, K. Itô, Ed., Wiley-Interscience, New York, 1978, pp. 427–436.

[17] Tsuchiya, M., *On the stochastic differential equation for a two-dimensional Brownian motion with boundary condition,* J. Math. Soc. Japan 32, 1980, pp. 233–249.

[18] Van Dantzig, D., *Solutions of the equation of Helmholz in an angle with vanishing directional derivatives along each side,* Kon. Ned. Akad. v. Wet. Proc. A61, No. 4, 1958, pp. 384–398.

[19] Williams, R. J., *Brownian motion in a wedge with oblique reflection at the boundary,* Ph.D. Dissertation, Stanford University, 1983.

[20] Zelicourt, Catherine de, *Une méthode de martingales pour la convergence d'une suite de processus de sauts markoviens vers une diffusion associée à une condition frontière. Application aux systèmes de files d'attente,* Ann. Inst. Henri Poincaré 17, 1981, pp. 351–375.

Received July, 1984.

Probab. Th. Rel. Fields 71, 553–579 (1986)

Probability Theory and Related Fields

A Multidimensional Process Involving Local Time

A.S. Sznitman[1] and S.R.S. Varadhan[2]

[1] Laboratoire de Probabilités, Université Paris VI Associé C.N.R.S. n° 224, 4 place Jussieu, F-75005 Paris, France

[2] Courant Institute of Mathematical Sciences, New York University, 251 Mercer Street New York 10012, USA

I. Introduction

In his paper [20], McKean posed the problem of the construction of a system of N particles on $\mathbb{R}$ with formal generator:

$$Lf(x^1, \dots, x^N) = \tfrac{1}{2}\Delta f(x^1, \dots, x^N) + \frac{1}{2(N-1)} \sum_{i<j} \delta(x^i - x^j) \left(\frac{\partial}{\partial x^i} + \frac{\partial}{\partial x^j}\right) f(x^1, \dots, x^N), \tag{1.1}$$

where $\delta(x^i - x^j)$ is the Dirac mass on the hyperplane $x^i = x^j$. Gutkin-Kac [11], Kotani-Osada [15] have given answers to this question but the methods they use are quite analytical, and do not give an insight on the trajectorial behavior of the process constructed.

In the paper presented here, we adopt a rather probabilistic approach. We investigate the solutions of a certain multidimensional stochastic differential equation with local time. Multidimensional S.D.E. with a local time supported on a smooth hypersurface have already been considered (see for instance Anulova [2], Portenko [24]). The difficulty of the case we study here comes from the fact that the hypersurface we have to consider is not smooth (it is a reunion of hyperplanes). More precisely, we study the solutions of

$$\begin{cases} dX_t = dB_t + \sum_k V_k \, dL^0(n_k \cdot X)_t \\ X_0 = x(o) \end{cases} \tag{1.2}$$

where B_t is an $\mathbb{R}^d$ valued Brownian motion independant of $x(0)$, n_k are unit vectors generating pairwise distinct hyperplanes H_k, V_k are constant vectors satisfying

$$V_k \cdot n_k = 0 \tag{1.3}$$

for every k (that is V_k belongs to H_k), and $L^0(n_k \cdot X)$ is the symmetric local time in zero of $n_k \cdot X$, the Euclidean scalar product of n_k with X.

Formally, using the identification $\delta(X_t^i - X_t^j) \times 2dt = dL^0(X^i - X^j)_t$ one recovers "solutions" to (1.1) by setting

$$d=N, \quad n_{ij}=\frac{1}{\sqrt{2}}(e_i-e_j), \quad V_{ij}=\frac{\sqrt{2}}{4(N-1)}(e_i+e_j), \quad i<j, \tag{1.4}$$

if (e_i) is the canonical basis of $\mathbb{R}^d$.

One should point out that this formal type of identification can hide a certain number of difficulties.

For instance, in dimension one, one might guess that the solutions of

$$\begin{cases} dX_t = dB_t + a\,\phi_\alpha(X_t)\,dt, & \text{where } \phi_\alpha = \frac{1}{\alpha}\phi\left(\frac{\cdot}{\alpha}\right) \text{ is a mollifier,} \\ X_0 = x, \end{cases} \tag{1.5}$$

converge as α goes to zero towards the solution of

$$\begin{cases} dX_t = dB_t + a\,dL^0(X)_t \\ X_0 = x. \end{cases} \tag{1.6}$$

But the solution of (1.6) does not exist if $|a|>1$ (see Harrison-Shepp [12]). In fact it can be shown that the solutions of (1.5) converge when α goes to zero to the solution of (1.6) with a replaced by $\tanh a$ (see Le Gall [18]). So one can very well question the validity of the choice (1.4) to recover (1.1), or even the existence of solutions of (1.2) when V_k are large.

In fact, as we are going to see, condition (1.3) prevents this type of phenomena. As this appears in Lemma 2.12, the "formal generator", that one should associate with (1.2) is of divergence type

$$L = \operatorname{div}[(\tfrac{1}{2}I + A)\,\nabla], \tag{1.7}$$

where I is the identity matrix and A is bounded skew symmetric. (For instance if $V=e_2$, $n=e_1$, $d=2$, one has $\delta(x)\frac{\partial}{\partial y} = \operatorname{div}(A\cdot\nabla)$, if $A(x)=\begin{pmatrix} 0 & H(x) \\ -H(x) & 0 \end{pmatrix}$, with $H(x)=1$ for $x>0$, 0 for $x\leqq 0$.)

As a consequence of this divergence structure, one can either apply P.D.E. results (see Aronson-Serrin [4], Ladyzhenskaya-Solonnikov-Ural'ceva [17], Moser [21], Nash [22]) in the same fashion as Kotani-Osada [15] did in the case of (1.1), or as well apply the theory of non-symmetric Dirichlet forms (see Carillo Menendez [6], Le Jan [19]). But we adopt here a rather trajectorial method. Let us describe the results presented here. It is not difficult to see that a solution of (1.2) starting at a point $x\in H_{i_0}\setminus \bigcup_{j\neq i_0} H_j$, because of (1.3) is equal to

$$x + B_t + V_{i_0}\cdot L^0(n_{i_0}\cdot B)_t, \tag{1.8}$$

as long as this process does not reach an hyperplane different of H_{i_0}.

So starting outside of the singular set $\bigcup_{i\neq j} H_i \cap H_j$, a.s. the process (1.8) will remain outside of $\bigcup_{i\neq j} H_i \cap H_j$ (Proposition 2.4), and our solution will have the structure of (1.8) between the successive times (T_n) of visit to different hyperplanes.

The question is now is $\lim T_n = T_\infty$ infinite or not. The same type of question arises in the study of the oblique reflection of a Brownian motion in a two dimensional wedge (see Varadhan-Williams [29]), in which case it may very well happen that $T_\infty < \infty$ a.s. In our case T_∞ turns out to be a.s. infinite, as a consequence one obtains that when

$$P(x(0) \in H_i \cap H_j) = 0, \quad i \neq j, \tag{1.9}$$

there is trajectorial existence and uniqueness for (1.2) and also uniqueness in law, and the process constructed is strong Markov in

$$\mathbb{R}^d \setminus \bigcup_{i\neq j} H_i \cap H_j.$$

The solution has other nice properties such as a Brownian scaling:

$$\lambda X^x_{\cdot/\lambda^2} \text{ has same law as } X^{\lambda x}_{\cdot}, \quad \lambda > 0, \tag{1.10}$$

and a uniform approximation property on compact sets of $\mathbb{R}_+$, by the solution of

$$dX_t = dB_t + \sum_i V_i \phi_\alpha(n_i \cdot X)\, dt,$$

when α goes to zero, (in contrast with (1.5), (1.6)).

Let us now try to explain the reason why T_∞ is infinite a.s. This has to do with some "symmetries" of the process. In the case of dimension 2, one has to worry about the possibility that X^x_t converges in a finite time towards zero. Roughly speaking, what prevents this phenomenon is the fact that one has a zero-one law for this event and that if p is the inversion of pole zero and radius 1, $-p(X^x_{\tau_t})$ has the same law as $X_t^{-p(x)}$, if τ_t is the inverse of $A_t = \int_0^t \frac{ds}{r_s^4}$ (see Proposition 3.1). Along these lines one sees that the probability of converging towards zero is the same as the probability of converging towards infinity, these events being clearly disjoint, by the zero-one law, have to be of probability zero.

In the case of a dimension $d \geqq 3$, by induction one can see that the only possibility for T_∞ to be finite is that the process converges in a finite time towards zero. One uses the fact that the process X_t constructed with the (n, V_i) is in duality with the process $\bar{X}_t$ constructed with $(n_i, -V_i)$, with respect to Lebesgue measure.

This property is related to the fact that because of (1.3), the "vector field" $\sum \delta(n_i \cdot X) \cdot V_i$ is divergence free, so the adjoint vector field is $-\sum \delta(n_i \cdot X) \cdot V_i$.

If one time changes X_t and $\bar{X}_t$ respectively by the inverse of $A_t = \int_0^t \frac{ds}{|X_s|^2}$,

$t<T_\infty$, $\bar{A}_t=\int_0^t \frac{ds}{|\bar{X}_s|^2}$, $t<\bar{T}_\infty$, the result Y_t, $\bar{Y}_t$ are now in duality with respect to $\mu(dx)=\frac{1}{r^2}dx$.

Using a Brownian scaling property, one has the fact that A_{T_∞} and $\bar{A}_{T_\infty}$ are a.s. infinite, so that the semi groups associated with the Y_t, $\bar{Y}_t$ are truly Markovian, moreover $\mu(dx)=\frac{dx}{r^2}$ is Radon ($d\geqq 3$) and invariant, and by a simple argument on the Hopf decomposition, into the dissipative and conservative part, this prevents the possibility that $\lim_{t\to\infty} Y_t=0$, (and that $T_\infty<\infty$).

At some point we use a moment estimate of Nash [22] to obtain a tightness property (Proposition 2.13, Corollary 2.14); it should be pointed out that in the case of the system:

$$dX_t^i=dB_t^i+\frac{c}{N}\sum_{j\neq i} dL^0(X^i-X^j)_t, \qquad (\text{and } 2c<N),$$

which was the original motivation for this work, in view of a propagation of chaos result, one can instead use the probabilistic estimates in Sznitman [27]. As an other possible approach, one can rely more heavily on analytical methods to show that the singular set is polar. Although our approach is less general, it gives some trajectorial understanding of the process, and may also be helpful in the construction of certain obliquely reflected Brownian motions in multi-dimensional wedges, where analytical results are not necessarily as well developped. To conclude we give at the end of this paper a few remarks about what happens to the laws of the solution when one goes near the set of singularities.

II. Characterization of the Process up to the Time T_∞

Our goal in this section, is to give a characterization of a process $(X_t)_{t\geqq 0}$ satisfying (1.2), (1.9), up to the first "accumulation" of the passage times through the various hyperplanes where the process feels the effect of a local time, and then to derive some consequences of this characterization.

More precisely, we consider a filtered probability space $(\Omega, F, (F_t)_{t\geqq 0}, P)$ (F is complete, $(F_t)_{t\geqq 0}$ is right continuous, and each F_t contains the P-negligible subsets of Ω), endowed with a d-dimensional Brownian motion $(B_t)_{t\geqq 0}$, N unit vectors n_j, $1\leqq j\leqq N$, of R^d that determine pairwise distinct hyperplanes H_j, $1\leqq j\leqq N$:

$$H_j=\{x\in R^d/n_j\cdot x=0\}, \tag{2.1}$$

and N vectors of R^d, V_j, $1\leqq j\leqq N$, satisfying:

$$V_j\cdot n_j=0, \quad 1\leqq j\leqq N. \tag{2.2}$$

We suppose that we are given a continuous semimartingale $(X_t)_{t\geqq 0}$, satisfying:

$$X_t=X_0+B_t+\sum_{j=1}^N V_j L^0(n_j\cdot X)_t, \tag{2.3}$$

(where $L^0(n_j \cdot X)_t$ denotes the symmetric local time in zero of the continuous semimartingale $(n_j \cdot X_t)_{t \geqq 0}$, and

$$\text{for } i \neq j, \quad P(X_0 \in H_i \cap H_j) = 0. \tag{2.4}$$

We are now going to study the process $(X_t)_{t \geqq 0}$, in a number of steps.

Let us first start with the

Remark 2.1. Let S be a finite (F_t)-stopping time, such that X_S does not belong to $H_i \cap H_j$, P a.s., then it is immediate to check that with respect to $(\Omega, F, (F_{S+t})_{t \geqq 0}, (B_{S+t} - B_S)_{t \geqq 0}, P)$, $(X_{S+t})_{t \geqq 0}$ satisfies (2.3), (2.4). □

Define

$$\begin{aligned} T_0 &= \inf\{t \geqq 0,\ \exists j \in [1, N],\ n_j \cdot (X_0 + B_t) = 0\}, \\ J_0 &= \{j \in [1, N],\ n_j \cdot (X_0 + B_{T_0}) = 0\}, \end{aligned} \tag{2.5}$$

we have *Lemma* 2.2:

$$P\text{-a.s., for } t \geqq 0, \quad X_{t \wedge T_0} = X_0 + B_{t \wedge T_0}, \tag{2.6}$$

$T_0 < \infty$, and J_0 is reduced to a single point $\{j_0(\omega)\}$.

Proof. (2.6) is immediate from the equation satisfied by $(X_t)_{t \geqq 0}$, and from the fact that the intersection of two distinct hyperplanes is polar for Brownian motion. □

Now define the "translated" filtration and Brownian motion:

$$(F_t^0)_{t \geqq 0} = (F_{T_0 + t})_{t \geqq 0}, \quad (B_t^0)_{t \geqq 0} = (B_{T_0 + t} - B_{T_0})_{t \geqq 0}, \tag{2.7}$$

similarly the "translated" process:

$$(X_t^0)_{t \geqq 0} = (X_{T_0 + t})_{t \geqq 0}, \tag{2.8}$$

and

$$\tau_1(\omega) = \inf\{t \geqq 0,\ \exists j \neq j_0(\omega),\ n_j \cdot (X_0^0 + B_t^0 + V_{j_0(\omega)} \cdot L^0(n_{j_0(\omega)} \cdot B^0)_t) = 0\}, \tag{2.9}$$

$$T_1(\omega) = T_0(\omega) + \tau_1(\omega), \tag{2.10}$$

$$J_1(\omega) = \{j \in [1, N],\ n_j \cdot (X_0^0 + B_{\tau_1}^0 + V_{j_0(\omega)} \cdot L^0(n_{j_0(\omega)} \cdot B^0)_{\tau_1}) = 0\}. \tag{2.11}$$

Remark 2.3. $j_0(\omega)$ is F_0^0 measurable, consequently $(n_{j_0} \cdot B_t^0)_{t \geqq 0}$ is an (F_t^0) continuous semimartingale, in fact it is an F_t^0-Brownian motion, and so the expression $L^0(n_{j_0} \cdot B^0)_t$ makes sense, it is in fact equal to $L(n_k \cdot B_{T_0}, n_k \cdot B)_{T_0+t} - L(n_k \cdot B_{T_0}, n_k \cdot B)_{T_0}$ on the set $\{j_0 = k\}$, if $L(x, n_k \cdot B)_t$ is a continuous version in (x, t) of the local time of $n_k \cdot B$.

We have the following characterization of $(X_t^0)_{t \geqq 0}$ up to time τ_1 given by:

Proposition 2.4.

$$P \text{ a.s., } X_{t \wedge \tau_1}^0 - X_0^0 + B_{t \wedge \tau_1}^0 + V_{j_0} L^0(n_{j_0} \cdot B^0)_{t \wedge \tau_1} \quad \textit{for } t \geqq 0, \tag{2.12}$$

moreover when $N \geqq 2$, *P-a.s.* $\tau_1 < \infty$, *and* J_1 *is reduced to a single point* $\{j_1(\omega)\}$. (2.13)

Proof. Let us first prove (2.12). Introduce temporarily

$$\bar{\tau}_1 = \inf\{t \geqq 0,\ \exists j \neq j_0(\omega),\ n_j \cdot X_t^0 = 0\},$$

we have:

$$X_{T_0+(t\wedge\bar{\tau}_1)} = X_{T_0} + B^0_{t\wedge\bar{\tau}_1} + \sum_j V_j(L^0(n_j\cdot X)_{T_0+(t\wedge\bar{\tau}_1)} - L^0(n_j\cdot X)_{T_0}),$$

but the set of $j\in[1,N]$ such that there exists $0\leqq s<\bar{\tau}_1$ where $n_j\cdot X_{T_0+s}=0$ is reduced to $\{j_0(\omega)\}$ by definition of $\bar{\tau}_1$, as a consequence

$$X_{T_0} + (t\wedge\bar{\tau}_1) = X_{T_0} + B^0_{t\wedge\bar{\tau}_1} + \sum_k 1\{j_0 = k\} \times V_k \cdot L^0(n_k\cdot X^0)_{t\wedge\bar{\tau}_1}, \tag{2.14}$$

projecting (2.14) on the vector n_{j_0} we obtain, using (2.2) and Lemma 2.2: $n_{j_0}\cdot X^0_{t\wedge\bar{\tau}_1} = n_{j_0}\cdot B^0_{t\wedge\bar{\tau}_1}$, as a consequence $X^0_{t\wedge\bar{\tau}_1} = X^0_0 + B^0_{t\wedge\bar{\tau}_1} + V_{j_0}\cdot L^0(n_{j_0}\cdot B^0)_{t\wedge\bar{\tau}_1}$, and from which we deduce $\tau_1 = \bar{\tau}_1$ and (2.12).

Remark 2.5. In the case when $N=1$, we have obtained the characterization

$$X_t = X_0 + B_t + V_1 L^0(n_1\cdot(B+X_0))_t. \tag{2.15}$$

It is immediate to check that the process defined by (2.15) is indeed a solution of (2.3). (Condition (2.4) is empty when $N=1$.) We will suppose now that $N\geqq 2$.

Suppose first we know that

$$\tau_1 < \infty \text{ a.s.} \quad \text{and} \quad P(X^0_{\tau_1}\in H_{j_0}) = 0. \tag{2.16}$$

and let us see how we get (2.13) from it. Consider $i \neq j$ and set

$$T = \tau_1 \quad \text{on} \quad X^0_{\tau_1}\in H_i\cap H_j, +\infty \text{ else}.$$

We have $\{T<\infty,\ n_{j_0}\cdot X_T^0 \neq 0\} \subset \bigcup_{r\in Q_+} \{r<T<\infty,\ \inf_{s\geqq 0}|n_{j_0}\cdot X_{(r+s)\wedge T}|>0\}$ $\subset \bigcup_{r\in Q_+}\{r<T<\infty,\ X_r^0 + B_T^0 - B_r^0 \in H_i\cap H_j\}$, this last event is negligible because, conditionally on $\{r<T\}$ the Brownian motion $X_r^0 + B^0_{r+s} - B_r^0$ starts outside $H^i \cap H^j$ and will never hit the polar set $H_i\cap H_j$. As a consequence $\{T<\infty\}$ is included P a.s. in $\{T<\infty,\ n_{j_0}\cdot X_T^0 = 0\}$ which has zero probability by (2.16). Let us then prove (2.16). In view of (2.12) and of the definition of τ_1, it is sufficient to show that

$\tau_1<\infty$, and for $K(\omega)$ a $\sigma(j_0)$-adapted r.v. with values in $[1,N]\backslash\{j_0(\omega)\}$ the probability that $X_0^0 + B_t^0 + V_{j_0} L^0(n_{j_0}\cdot B^0)_t$ lies in $H_{j_0}\cap H_{K(\omega)}$ for some t is zero. (2.17)

In order to do this, consider the plane generated by n_K, n_{j_0}, let e_1, e_2, be an orthonormal basis for this plane, such that $e_2 = n_{j_0(\omega)}$. Define:

$$\begin{aligned} Z_t^1 &= e_1(\omega)\cdot(X_0^0 + B_t^0 + V_{j_0} L^0(n_{j_0}\cdot B^0)_t), \\ Z_t^2 &= |e_2(\omega)\cdot(X_0^0 + B_t^0 + V_{j_0} L^0(n_{j_0}\cdot B^0)_t)|, \end{aligned} \tag{2.18}$$

let $W_t = (W_t^1, W_t^2)$ be the F_t^0-two dimensional Brownian motion $(e_1(\omega) \cdot B_t^0, e_2(\omega) \cdot B_t^0)$. Using the definition of $j_0(\omega)$, we find

$$Z_t^1 = Z_0^1 + W_t^1 + (V_{j_0} \cdot e_1)\, L^0(W^2)_t \tag{2.19}$$

$$Z_t^2 = |W_t^2| = \int_0^t \operatorname{sign}(W_s^2)\, dW_s^2 + L^0(W^2)_t. \tag{2.20}$$

Let D be the half plane $\{(x^1, x^2) \in R^2,\ x^2 \geqq 0\}$. Denote by $F(x, a, w^1, w^2)$ the continuous mapping from the space $D \times \mathbb{R} \times C_0(\mathbb{R}_+, \mathbb{R}) \times C_0(\mathbb{R}_+, \mathbb{R})$ into $C(\mathbb{R}_+, D)$, ($C_0(\mathbb{R}_+, \mathbb{R})$ is the set $\{w \in C(\mathbb{R}_+, \mathbb{R}),\ w(0) = 0\}$), sending (x, a, w^1, w^2) on the unique solution of

$$\begin{aligned} z_t^1 &= x^1 + w_t^1 + a\, k_t \\ z_t^2 &= x^2 + w_t^2 + k_t \end{aligned} \tag{2.21}$$

with $z_t^2 \geqq 0$, k_t continuous increasing and $\int_0^t 1\{z_s^2 \neq 0\}\, dk_s = 0$. (For this see El-Karoui-Chaleyat-Maurel [8] or Ikeda-Watanabe [13], p. 120.)

From (2.19) (2.20), we see that

$$(Z_\cdot^1, Z_\cdot^2) = F((Z_0^1, 0), V_{j_0} \cdot e_1, W_\cdot^1, \bar{W}_\cdot^2)\, P \text{ a.s.},$$

where $\bar{W}_\cdot^2$ is the Brownian motion $\int_0^\cdot \operatorname{sign}(W_s^2)\, dW_s^2$, in other words conditionally on F_0^0, $(Z_\cdot^1, Z_\cdot^2)$ is distributed as an oblique reflected Brownian motion in the half plane D, along the vector $v = (V_{j_0} \cdot e_1, 1)$ and starting at $(Z_0^1, 0)$.

We are now going to show that any such oblique reflected Brownian motion in D, when starting away from the origin never hits the origin, but eventually hits any straight line going through the origin. This will prove our claim (2.17) and Proposition 2.4.

This result is contained in the next lemma, that we will use later,

Lemma 2.6. *Let $(B_t)_{t \geqq 0}$ be a two dimensional Brownian motion, and $(Z_t)_{t \geqq 0}$, be defined by* (2.21), *(B replacing W), suppose $x = (x^1, x^2) \neq 0$, ($x \in D$), then*

$$\textit{the probability that } (Z_\cdot) \textit{ hits the origin is zero}, \tag{2.22}$$

$$\int_0^\infty \frac{ds}{|Z_s|^2} = \infty \quad \textit{and} \quad \int_0^\infty \frac{ds}{|Z_s|^4} = \infty \ \textit{a.s.}; \tag{2.23}$$

if τ_t is the continuous increasing inverse of $\int_0^t \frac{ds}{|Z_s|^2}$, and (r_t, θ_t) are the polar coordinates of Z_t, ($\theta_t \in [0, \pi]$),

$$\theta_{\tau_t} \textit{ is a reflected Brownian motion on } [0, \pi]. \tag{2.24}$$

Proof. For any $c \geqq 0$, define $T_c = \inf\{t \geqq 0,\ |Z_t| = c\}$, using the fact that the analytic function $\operatorname{Log} r + i\theta$ is analytic in a conic neighborhood of $D - \{0\}$, (θ determination of the angle in $[0, \pi]$), and Ito's formula, we obtain for $0 < \varepsilon \leqq |x|$

$$\operatorname{Log} r_{t\wedge T_\varepsilon} = \operatorname{Log} r_0 + \int_0^{t\wedge T_\varepsilon} \frac{Z_s^1\,dB_s^1 + Z_s^2\,dB_s^2}{r_s^2} + a\int_0^{t\wedge T_\varepsilon} \frac{\operatorname{sign} Z_s^1}{|Z_s^1|}\,dk_s, \tag{2.25}$$

$$\theta_{t\wedge T_\varepsilon} = \theta_0 + \int_0^{t\wedge T_\varepsilon} \frac{Z_s^1\,dB_s^2 - Z_s^2\,dB_s^1}{r_s^2} + \int_0^{t\wedge T_\varepsilon} \frac{\operatorname{sign}(Z_s^1)}{|Z_s^1|}\,dk_s, \tag{2.26}$$

as a consequence,

$$M_t = \operatorname{Log} r_{t\wedge T_\varepsilon} - a\,\theta_{t\wedge T_\varepsilon} \tag{2.27}$$

is a local martingale with increasing process

$$\langle M\rangle_t = \int_0^{t\wedge T_\varepsilon} \frac{ds}{r_s^2} \times (1+a^2). \tag{2.28}$$

Pick $R>|x|$, and set $T_{\varepsilon,R} = T_\varepsilon \wedge T_R$, from the fact that $M_{t\wedge T_{\varepsilon,R}}$ is bounded, and $M^2_{t\wedge T_{\varepsilon,R}} - \langle M\rangle_{t\wedge T_{\varepsilon,R}}$ is a martingale we get easily that $T_{\varepsilon,R}$ is finite almost surely (in fact integrable), and $E[M_{T_{\varepsilon,R}}] = E[M_0]$, that is

$$\operatorname{Log}\varepsilon \times P(T_\varepsilon < T_R) + \operatorname{Log} R \times P(T_R < T_\varepsilon) = E[M_0] + a\,E[\theta_{T_{\varepsilon,R}}] \tag{2.29}$$

keeping R fixed and letting ε go to zero in (2.30), we find $0 = \lim_{\varepsilon\to 0} P(T_\varepsilon < T_R) = P(T_0 < T_R)$, now letting R go to infinity, we obtain $P(T_0 < \infty) = 0$, which proves (2.22).

Let us notice also that

$$\operatorname{Log}\frac{r_{t\wedge T_\varepsilon}}{\varepsilon} - a\,\theta_{t\wedge T_\varepsilon} + |a|\,\pi \tag{2.30}$$

is a positive supermartingale and consequently converges a.s. to a finite limit, so on the set $\{T_\varepsilon = \infty\}$ we find that $\operatorname{Log}\frac{r_t}{\varepsilon} - a\,\theta_t$ converges to a finite limit, but this contradicts the fact that $T_{\varepsilon,R} = T_R$ (on $\{T_\varepsilon = \infty\}$) is finite for every $R>|x|$, a.s., so $T_\varepsilon < \infty$ a.s., $\varepsilon \leqq |x|$, in the same way, one proves that $T_R < \infty$ a.s., $R \geqq |x|$.

Let us show (2.23). Define P_x to be the law on $C(\mathbb{R}_+, D)$ of the solution of (2.21), with initial condition $x \in D$. It is known (see for instance Ikeda-Watanabe [13]) that P_x determines a strong Markov process, and it is easy to see from (2.21) that the image of P_x under the map $z_. \to \lambda\, z_{./\lambda^2}$ is $P_{\lambda x}$ (Brownian scaling).

Define for $x \neq 0 \in D$,

$$\phi(x) = E_{P_x}\left[\exp - \int_0^\infty \frac{1}{|Z_s|^4}\,ds\right], \qquad \text{pick } \lambda > 0,$$

$$\phi(\lambda x) = E_{P_x}\left[\exp - \int_0^\infty \frac{1}{\lambda^4}\,\frac{1}{|Z_{s/\lambda^2}|^4}\,ds\right\} = E_{P_x}\left[\exp - \frac{1}{\lambda^2}\int_0^\infty \frac{ds}{|Z_s|^4}\right].$$

As a consequence, $\lambda \to \phi(\lambda x)$ is an increasing function.

Suppose there exists x such that $\phi(x) \neq 0$, then set $R = |x|$, for $y \in D$, by strong Markov property, we have

$$\phi(y) = E_{P_y}\left[\phi(Z_{\tau_{R/2}})\exp - \int_0^{\tau_{R/2}} \frac{ds}{|Z_s|^4}\right] \tag{2.31}$$

(with the notation $\tau_a = \inf\{s \geqq 0,\ |Z_s| = a\}$) and we know that $\tau_{R/2}$ is P_y-a.s. finite.

Define

$$a = \sup_{|y|=R} \phi(y), \quad \text{then} \quad \sup_{|y|=\frac{R}{2}} \phi(y) \leqq a. \tag{2.32}$$

Now as a consequence of (2.31), (2.32), for $|y|=R$,

$$\phi(y) \leqq a \times E_y\left[\exp - \int_0^{\tau_{R/2}} \frac{ds}{|Z_s|^4}\right]. \tag{2.33}$$

But it is easy to see from the "explicit" formula for the solution of (2.21) (see previously cited references) that P_y, $|y|=R$, is a tight family of probabilities, as a consequence, there will be with a uniformity strictly positive probability a modulus of continuity of (Z) when $|y|=R$. So one can find $\alpha>0$, $\beta>0$, such that for $|y|=R$

$$P_y(\tau_{2R} \wedge \tau_{R/2} \geqq \alpha) \geqq \beta > 0.$$

As a consequence,

$$\sup_{|y|=R} E_y\left[\exp - \int_0^{\tau_{R/2}} \frac{ds}{|Z_s|^4}\right] < 1$$

and we have a contradiction with (2.33), so $\phi(x)$ is identically zero.

In the same way setting $\psi(x) = E_{P_x}\left[\exp - \int_0^\infty \frac{ds}{|Z_s|^2}\right]$, $x \in D\setminus\{0\}$ one finds that ψ is homogeneous of degree zero, and in fact ψ is identically zero. So we have obtained (2.23). Let us prove (2.24),

$$N_t = \int_0^t \frac{-Z_s^2\, dB_s^1 + Z_s^1\, dB_s^2}{r_s^2}$$

is a local martingale with increasing process $\int_0^t \frac{ds}{r_s^2}$, so N_{τ_t} is a Brownian motion, and

$$\theta_{\tau_t} = \theta_0 + N_{\tau_t} + \int_0^t \frac{\operatorname{sign}(Z_{\tau_s}^1)}{|Z_{\tau_s}^1|}\, dk_{\tau_s} \tag{2.34}$$

but $s \to \int_0^{\cdot} \frac{dk_{\tau_s}}{|Z_{\tau_s}^1|}$ is a continuous increasing process with support contained in the set where $\theta_{\tau_s}=0$ or π, this and (2.34) implies (see Tanaka [28]) that θ_{τ_s} is a reflected Brownian motion on $[0, \pi]$. □

We are now going to make a repeated use of Proposition 2.4 and Remark 2.1, to characterize $(X_t)_{t \geqq 0}$ up to the first time of accumulation of successive visits of (X_t) to different hyperplanes H_i.

Define by induction, using Proposition 2.4

$$T_0, \tau_1, T_1, \ldots, \tau_n, T_n \ldots \quad \text{and}$$
$$X_0^0, X_0^1, \ldots, X_0^n, \ldots, j_0(\omega), \ldots, j_n(\omega), \ldots \quad \text{satisfying:} \tag{2.35}$$

$$\infty > T_{n+1} = T_n + \tau_{n+1} \quad (n \geqq 0),$$
$$\infty > \tau_{n+1} = \inf\{s \geqq 0, \exists j \neq j_n(\omega), X_0^n + B_s^n + V_{j_n(\omega)} L^0(n_{j_n} \cdot B^n)_s \in H_j\},$$

(with the notation $B^n_s = B_{T_n+s} - B_{T_n}$, which is a (F^n_t) $(=F_{T_n+t})$ Brownian motion),

$$X^0_{n+1} = X^n_0 + B^n_{\tau_{n+1}} + V_{j_n(\omega)} L^0(n_{j_n} \cdot B^n)_{\tau_{n+1}},$$

$$\{j_{n+1}(\omega)\} = \{j \in [1,N],\ X^0_{n+1} \in H_j\} \quad \text{(which is a singleton)},$$

we obtain immediately using Remark 2.1 and Proposition 2.4:

Proposition 2.7.

$$X_{T_n+s\wedge\tau_{n+1}} = X^n_0 + B^n_{s\wedge\tau_{n+1}} + V_{j_n(\omega)} \cdot L^0(B^n \cdot n_{j_n})_{s\wedge\tau_{n+1}}, \quad \textit{for } s \geqq 0,\ P\text{-a.s.} \quad \square \tag{2.36}$$

If we set $T_\infty = \lim\uparrow T_n$, we have characterized $(X_t)_{t\geqq 0}$ for $t < T_\infty$.

It will be convenient now to introduce a "canonical" way of constructing the laws of the process X_t, $t < T_\infty$; in order to do that, consider the space $C_0(\mathbb{R}_+, \mathbb{R}^d)$ of continuous functions on $\mathbb{R}_+$ with values in $\mathbb{R}^d$, such that their value at time 0 is 0. Denote by $(W_s)_{s\geqq 0}$, the canonical process, Q the Wiener measure on $C_0(\mathbb{R}_+, \mathbb{R}^d)$, $(G_s)_{s\geqq 0}$ the canonical right continuous completed filtration, for the Wiener measure on $C_0(\mathbb{R}_+, \mathbb{R}^d)$, and $L(y,j,t,w)$ a version of the local time of $W \cdot n_j$, at y at time t, Q a.s. continuous in (y,t).

We define for $x \in \mathbb{R}^d \setminus \bigcup_{i\neq j} H_i \cap H_j$, the random variables $(T^x_n,\ n\geqq 0)$, $(\tau^x_n,\ n\geqq 1)$, $(Z^x_n,\ n\geqq 0)$, $(j_n,\ n\geqq 0)$, satisfying:

$$T^x_0 = \inf\{t \geqq 0,\ \exists j \in [1,N],\ n_j \cdot (x + W_t) = 0\}, \tag{2.37}$$

$$T^x_{n+1} = T^x_n + \tau^x_{n+1}, \quad n \geqq 0,$$

$$\tau^x_{n+1} = \inf\{s \geqq 0,\ \exists j \neq j_n,\ Z^x_n + W_{T^x_n+s} - W_{T^x_n} + V_{j_n}(L(W_{T^x_n} \cdot n_{j_n}, j_n, T^x_n + s) - L(W_{T^x_n} \cdot n_{j_n}, j_n, T^x_n)) \in H_j\}, \quad n \geqq 0,$$

$$Z^x_0 = x + W_{T^x_0},$$

$$Z^x_{n+1} = Z^x_n + W_{T^x_{n+1}} - W_{T^x_n} + V_{j_n}(L(W_{T^x_n} \cdot n_{j_n}, j_n, T^x_{n+1}) - L(W_{T^x_n} \cdot n_{j_n}, j_n, T^x_n)), \quad n \geqq 0;$$

$$\{j_n\} = \{j \in [1,N],\ Z^x_n \in H_j\}, \quad n \geqq 0.$$

We set $T^x_\infty = \lim_n \uparrow T^x_n$, we consider an isolated point δ, and define our "canonical" version

$$Z^x_t = x + W_t, \quad \text{on } t \leqq T^x_0, \tag{2.38}$$

$$= Z^x_n + W_t - W_{T^x_n} + V_{j_n}(L(W_{T_n} \cdot n_{j_n}, j_n, t) - L(W_{T_n} \cdot n_{j_n}, j_n, T^x_n)),$$

on $T^x_n \leqq t \leqq T^x_{n+1}$,

$$= \delta, \text{ on } t \geqq T^x_\infty, \text{ and we set } Z^\delta_t = \delta \text{ for all } t.$$

So $(X^x_t)_{t\geqq 0}$ is a process valued in the space of continuous trajectories in $\mathbb{R}^d$ with finite lifetime (see Dellacherie-Meyer [7], p. 148).

Remark. If we choose a version of $L(y,j,t,w)$ which is continuous in (y,t), for every $w \in C_0$, and we define $Z(x,t,w) \in \mathbb{R}^d \cup \{\delta\}$, using (2.38), and the sup-

plementary requirement that Z is stopped once it reaches the set of singularities (this is a negligible set for every fixed x), it is easy to check that the version $Z(x, t, w)$ obtained is $B(\mathbb{R}^d \cup \{\delta\} \times \mathbb{R}_+ \times C_0)$ measurable, and that $x \to Z(x, \cdot, \cdot)$ (valued in the space of continuous trajectories with finite lifetime), is also measurable.

We are now ready to state the strong Markov property satisfied by X_t, $t \geqq 0$.

Let S be a finite (F_t)-stopping time, such that X_S does not belong to $H_i \cap H_j$, $i \neq j$, $i, j \in [1, N]$. We define the translated process $X_t^S = X_{t+S}$, $t \geqq 0$, we can define (see Remark 2.1) $T_\infty^S = \lim_n \uparrow T_n^S$, for the translated process.

Set $\hat{X}_t = X_t^S$, $t < T_\infty^S$,

δ, $t \geqq T_\infty^S$,

we have:

Proposition 2.8. a) *For any bounded measurable function f*

$$E[f(\hat{X}_\cdot(\omega))/F_S] = \int_{C_0} f(Z^{X_S(\omega)}(w))\, dQ(w), \qquad P\text{-a.s.} \tag{2.39}$$

b) *Let T be a finite (G_s)-stopping time, for f bounded measurable function*

$$E_Q[f(Z^x_{T+\cdot}(w))/G_T] = \int_{C_0} f(Z^{Z_T(w)}(w'))\, dQ(w'), \quad Q.\ a.s. \tag{2.40}$$

Proof. a) is a simple consequence of the fact that for $\Lambda \in F_S$, on the restricted space $\Omega \cap \Lambda$, endowed with $\dfrac{P(\cdot \cap \Lambda)}{P(\Lambda)}$ the translated process $X_{S+\cdot}$ is a solution of (2.3), (2.4) (with the translated Brownian motion $B_{S+t} - B_S$, so, by Proposition 2.7,

$$E[1_\Lambda f(\hat{X}_\cdot)[= E[1_\Lambda(\omega) \int_{C_0} f(Z_\cdot^{X_S(\omega)}(w'))\, dQ(w')],$$

which proves (2.39).

b) is a consequence of the fact that Q. a.s.

$$Z(x, t+T, w) = Z(Z_T^x(w), t, W_{T+\cdot} - W_T)$$

as one checks on the formulas (2.38). □

The process Z_t^x, also satisfies a "Brownian scaling" property:

Proposition 2.9. *Let* $\lambda > 0$, *for* $x \notin \bigcup_{i \neq j} H_i \cap H_j$, $\lambda Z^x_{\cdot/\lambda^2}$ *has the same law as* $Z_\cdot^{\lambda x}$.

Proof. This is a consequence of the fact that

$$\lambda Z_{\cdot/\lambda^2}(x, \omega) = Z_\cdot(\lambda x, \lambda \omega_{\cdot/\lambda^2}), \quad Q \text{ a.s.}, \tag{2.41}$$

as can be seen on (2.37), (2.38). □

Our next step is now to prove a type of stability result for the processes $(Z_\cdot^x)$.

Consider ϕ a smooth function on $\mathbb{R}$, positive, symmetric, with compact support in $[-1, +1]$, such that $\int \phi(x)\,dx=1$, for $\alpha>0$, we define $\phi_\alpha(x) = \frac{1}{\alpha}\phi\left(\frac{x}{\alpha}\right)$.

We can now construct the processes $(Z_{\cdot}^{\alpha,x})$ satisfying

$$Z_t^{\alpha,x} = x + W_t + \int_0^t \sum_{i=1}^N V_i \cdot \phi_\alpha(Z_s^{\alpha,x}\cdot n_i)\,ds. \tag{2.42}$$

Remark 2.10. (2.42) is the equation obtained by smoothing the local times that appear in Eq. (2.3) (with initial condition x). We are going to show that the process $(Z_{\cdot}^{\alpha,x})$ approximate $Z_{\cdot}^x$ when α goes to zero. This result is not only of technical interest, it shows a special feature of the Eq. (2.3) due to condition (2.2).

If we were trying to the same thing, in one dimension, with the equation

$$dX_t = dB_t + a\,L^0(X)_t \tag{2.43}$$

$(L^0(X)_\cdot$ symmetric local time of X in 0, and $|a|<1)$,

(see Harrison-Shepp [12]) and the approximating processes

$$dX_t^\alpha = dB_t + b\,\phi_\alpha(X_t^\alpha)\,dt. \tag{2.44}$$

We would find that when α goes to zero X^α goes to the solution of (2.43) with $a=\tanh b$ (see for instance Le Gall [18]).

Proposition 2.11. *For* $x\notin\bigcup_{i\neq j} H_i\cap H_j$, *on* $\{t<T_\infty^x\}$, Q *a.s.* $\lim_{\alpha\to 0}\sup_{s\leq t}|Z_s^{\alpha,x}-Z_s^x|=0.$

(2.45)

Proof. We will show by induction that

$$\lim_{\alpha\to 0}\sup_{s\leq T_n^x}|Z_s^{\alpha,x}-Z_s^x|=0. \tag{2.46}$$

In the course of the proof we will use several times the density occupation formula which says that for w outside of N negligible in C_0, for $t\geq 0$, the image by $W_s\cdot n_j$ of Lebesgue measure on $[0,t]$ is $L(y,j,t,w)\,dy$. As a consequence, for $f(\cdot,\cdot)$ bounded measurable on $\mathbb{R}\times C_0$, and $w\notin N$, for every $t\geq 0$, $j\in[1,N]$, $\int_0^t f(W_s\cdot n_j, w)\,ds = \int_{\mathbb{R}} f(y,w)\,L(y,j,t,w)\,dy$ (notice that $f(W_s\cdot n_j,w)$ need not be adapted).

Let us prove (2.46) by induction on n. We will suppress the superscript x in what follows to simplify the notations. Start with the case $n=0$. If $x\in\bigcup_j H_j$, (2.46) is immediate, $(T_0=0)$, so suppose $x\notin\bigcup_j H_j$. We have:

$$Z_{s\wedge T_0} = x + W_{s\wedge T_0}, \qquad Z_{s\wedge T_0}^\alpha = x + W_{s\wedge T_0} + \int_0^{s\wedge T_0}\sum_i V_i\,\phi_\alpha(Z_s^\alpha\cdot n_i)\,ds.$$

Set $H_j^\alpha=\{x\in\mathbb{R}^d,\ |n_j\cdot x|\leq\alpha\}$; until Z_s^α reaches one of the sets H_j^α, $j\neq j_0(w)$, (notice that this way we do not define a stopping time since j_0 is anticipating),

we have:

$$Z_s^\alpha = x + W_s + \int_0^s V_{j_0} \cdot \phi_\alpha(Z_u^\alpha \cdot n_{j_0})\, du \tag{2.47}$$

as a consequence, until Z_s^α reaches $\bigcup_{j \neq j_0} H_j^\alpha$,

$$n_{j_0} \cdot Z_s^\alpha = n_{j_0} \cdot (x + W_s) \quad \text{(using (2.2)), so}$$

$$Z_{s\wedge T_0}^\alpha = x + W_{s\wedge T_0} + \int_0^{s\wedge T_0} V_{j_0}\, \phi_\alpha(n_{j_0}(x + W_u))\, du \tag{2.48}$$

$$= x + W_{s\wedge T_0} + V_{j_0} \int_{\mathbb{R}} \phi_\alpha(y + n_{j_0} \cdot x)\, L(y, j_0, s\wedge T_0)\, dy \tag{2.48'}$$

(occupation density formula) until $Z_{s\wedge T_0}^\alpha$ reaches $\bigcup_{j \neq j_0} H_j^\alpha$.

If T^α denotes the first time $x + W_s$ reaches one of the H_j^α, $j \in [1, n]$, we have $\lim_{\alpha\to 0} T^\alpha = T_0$, now if we consider the expression defined by (2.48)', for any s we can replace the quantity $L(y, j_0, s\wedge T_0)$ by $L(y, j_0, s\wedge T_0) - L(y, j_0, s\wedge T^\alpha)$, leaving (2.48)' unchanged. By a.s. uniform continuity of the function L, we find for α small enough that,

$$x + W_{s\wedge T_0} + V_{j_0} \times \int_R \phi_\alpha(y + n_{j_0} \cdot x)\,(L(y, j_0, s\wedge T_0) - L(y, j_0, s\wedge T^\alpha)),$$

does not reach $\bigcup_{j \neq j_0} H_j^\alpha$, so (2.48) is matter of factly valid for all positive s, and also $\lim_{\alpha\to 0} \sup_{s \leq T_0} |Z_s^\alpha - Z_s| = 0$, Q a.s.

Suppose (2.46) proved for $n \geq 0$, let us prove it for $n+1$. The idea of the proof is basically the same, but now we have to take into account the terms corresponding to the hyperplanes H_{j_n} and $H_{j_{n+1}}$ when α is small.

With the notations

$$W_s^n = W_{T_n + s} - W_{T_n}, \quad Z_s^n = Z_{T_n + s}, \quad Z_n = Z_0^n, \quad Z_s^{\alpha, n} = Z_{T_n + s}^\alpha, \quad Z_n^\alpha = Z_0^{\alpha, n},$$

we have:

$$Z_{s\wedge \tau_{n+1}}^n = Z_n + W_{s\wedge\tau_{n+1}}^n + V_{j_n} \times (L(W_{T_n} \cdot n_{j_n}, j_n, (\tau_{n+1} \wedge s) + T_n) - L(W_{T_n} \cdot n_{j_n}, j_n, T_n)). \tag{2.49}$$

Set $\tau^\alpha = \inf\{s \geq 0,\ Z_s^{\alpha, n} \in \bigcup_{j \neq j_n} H_j^\alpha\}$, for $s \in [0, \tau^\alpha]$ we find $Z_s^{\alpha, n} \cdot n_{j_n} = Z_n^\alpha \cdot n_{j_n} + W_s^n \cdot n_{j_n}$ (because of (2.42) and (2.2)), using this fact and the occupation density formula, from (2.42) we get Q a.s.:

$$Z_s^{\alpha, n} = Z_n^\alpha + W_s^n + V_{j_n} \int_{\mathbb{R}} \phi_\alpha(Z_n^\alpha \cdot n_{j_n} + y)\,(L(W_{T_n} \cdot n_{j_n} + y, j_n, T_n + s) - L(W_{T_n} \cdot n_{j_n} + y, j_n, T_n))\, dy, \quad 0 \leq s \leq \iota^\alpha. \tag{2.50}$$

From the Q.a.s. uniform continuity of L on sets bounded in the time variable, and (2.50), we find that Q a.s. $Z_{\cdot\wedge\tau^\alpha\wedge\tau_{n+1}}^{\alpha, n}$ have a uniform modulus of

continuity. Set $\tau = \liminf_{\alpha \to 0} \tau^\alpha \wedge \tau_{n+1}$, by the previous remark, we have Q a.s. $\tau > 0$. Using again the a.s. uniform continuity of L, we find that

$$Q \text{ a.s., } Z_\cdot^{\alpha, n} \text{ converges towards } Z_\cdot^n \text{ uniformly on compact sets of } [0, \tau[. \tag{2.51}$$

Let us now observe that Q-a.s., $\tau = \tau_{n+1}$. Indeed if $\{\tau < \tau_{n+1}\}$ had positive measure, on this set we could write, with η small enough $Z_\tau^n = Z_\tau^n - Z_{\tau-\eta}^n + Z_{\tau-\eta}^n - Z_{\tau-\eta}^{n,\alpha} + Z_{\tau-\eta}^{n,\alpha} - Z_{\tau^\alpha}^{n,\alpha} + Z_{\tau^\alpha}^{n,\alpha}$. Now using the a.s. uniform continuity of the $Z_{\cdot \wedge \tau^\alpha}^{n,\alpha}$, (2.51)′

we would obtain $Z_\tau^n \in \bigcup_{j \neq j_n} H_j$, contradicting the definition of τ_{n+1}.

So we have obtained that Q.a.s. $Z_\cdot^{\alpha,n}$ converges uniformly on compact intervals of $[0, \tau_{n+1}[$. (2.52)

Let I_η be the interval $[(\tau_{n+1} - \eta)_+, \tau_{n+1}]$ for $\eta > 0$, there just remains to show that Q a.s. there exists $\eta > 0$ such that $Z_\cdot^{\alpha,n}$ converges uniformly to $Z_\cdot^n$ on I_η.

Q a.s. there exists η_0, a, small enough such that Z_s^n does not meet $\bigcup_{j \neq j_{n+1}} H_j$ for $s \in I_{\eta_0}$, and $\inf_{s \in I_{\eta_0}} |Z_s^n \cdot n_j| \geqq a > 0$, $j \neq j_{n+1}$, (as a consequence $\eta_0 < \tau_{n+1}$).

We are basically in the same situation as for the case $n = 0$. Fix $\eta \leqq \eta_0$, for α small enough $Z_{\tau_{n+1}-\eta}^{\alpha,n}$ is not in $\bigcup_{j \neq j_{n+1}} H_j^\alpha$, and before $Z_s^{\alpha,n}$, $s \in I_\eta$ reaches one of the sets H_j^α, $j \neq j_{n+1}$ we have

$$Z_s^{\alpha,n} = Z_{\tau_{n+1}-\eta}^{\alpha,n} + W_s^n - W_{\tau_{n+1}-\eta}^n + V_{j_{n+1}} \int_{\tau_{n+1}-\eta}^{s} \phi_\alpha(Z_u^{\alpha,n} \cdot n_{j_{n+1}})\, du \tag{2.53}$$

$$\begin{aligned} &= Z_{\tau_{n+1}-\eta}^{\alpha,n} + W_s^n - W_{\tau_{n+1}-\eta}^n + V_{j_{n+1}} \int_{\mathbb{R}} dy\, \phi_\alpha(Z_{\tau_{n+1}-\eta}^{\alpha,n} \cdot n_{j_{n+1}} + y) \\ &\quad \cdot (L(y + W_{T_{n+1}-\eta} \cdot n_{j_{n+1}}, j_{n+1}, s + T_n) - L(y + W_{T_{n+1}-\eta} \cdot n_{j_{n+1}}, j_{n+1}, \tau_{n+1} - \eta + T_n)), \end{aligned} \tag{2.53'}$$

(we have used (2.53) to compute $Z_u^{\alpha,n} \cdot n_{j_{n+1}}$, and the occupation density formula).

Now pick $\eta_1 \leqq \eta_0$ such that the last term of (2.53)′, with $\eta = \eta_1$, is less in module than $a/3$, $s \in I_{\eta_1}$, and α small enough so that

$$|Z_{\tau_{n+1}-\eta_1}^{\alpha,n} - Z_{\tau_{n+1}-\eta_1}^n| \leqq \frac{a}{3} \quad \text{and} \quad \alpha < \frac{a}{3}; \tag{2.54}$$

as a consequence for α satisfying (2.54), the right member of (2.53)′ will have projection on n_j, $j \neq j_{n+1}$, of modulus bigger than $\frac{a}{3}$, for $s \in I_{\eta_1}$. This shows that the right member of (2.53)′ does not meet $\bigcup_{j \neq j_{n+1}} H_j^\alpha$ for $s \in I_{\eta_1}$ and so (2.53), (2.53)′ is valid for $s \in I_{\eta_1}$.

But in (2.53)′ we can replace in the last term of local time $\tau_{n+1} - \eta_1$ by $\bar{\tau}^\alpha = (\tau_{n+1} - \eta_1) \vee (\tau^\alpha \wedge \tau_{n+1})$ so that we have proved that Q a.s. there exists

$0<\eta_1 \leqq \eta_0$ such that for $s \in I_{\eta_1}$, and α such that (2.54) is satisfied:

$$Z_s^{\alpha,n} = Z_{\tau_{n+1}-\eta_1}^{\alpha,n} + W_s^n - W_{\tau_{n+1}-\eta_1}^n + V_{j_{n+1}} \times \int dy\, \phi_\alpha(y) \times [L(y + W_{T_{n+1}-\eta_1} \cdot n_{j_{n+1}} - Z_{\tau_{n+1}-\eta_1}^{\alpha,n} \cdot n_{j_{n+1}}, j_{n+1}, T_n+.)]_{\tilde{\tau}^\alpha}^{s \vee \tilde{\tau}^\alpha}.$$

Using the a.s. uniform continuity of L for bounded times, (2.52), we obtain since $\tilde{\tau}^\alpha$ converges towards τ_{n+1} when α goes to zero, that $\lim_{\alpha \to 0} \sup_{s \in I_{\eta_1}} |Z_s^{\alpha,n} - Z_s^n| = 0$, and this terminates the proof of Proposition 2.11. □

So far we have considered Z_t^x as a process with possibly finite lifetime, we are going to show that on the set $\{T_\infty^n < \infty\}$, when t goes to T_∞^x (from the left), Z_t^x converges to a finite limit in $\bigcup_{i \neq j} H_i \cap H_j$, (this will imply that if we consider δ as the point at infinity of $R^d \setminus \bigcup_{i \neq j} H_i \cap H_j$, Z_t^x valued in $(R^d \setminus \bigcup_{i \neq j} H_i \cap H_j) \cup \{\delta\}$ is a.s. continuous).

In order to obtain this result, we first prove the following

Lemma 2.12. *The laws of the processes $Z_\cdot^{\alpha,x}$ where $\alpha > 0$ and x remains in a ball of $\mathbb{R}^d$, are tight.*

Proof. Fix x and α, the law of $Z_\cdot^{\alpha,n}$ is solution of the martingale problem (x, L), where L is the differential operator:

$$L = \tfrac{1}{2}\Delta + \sum_i \phi_\alpha(n_i \cdot x)\, V_i \cdot \nabla$$

introduce the function $H_\alpha(x) = \int_{-\infty}^{x/\alpha} \phi(u)\, du$, so that $H'_\alpha(x) = \phi_\alpha(x)$, and for every $i \in [1, N]$ an orthonormal matrix P_i such that $P_i e_1 = n_i$, $P_i e_2 = \frac{V_i}{|V_i|}$, $(e_k, k \in [1, d]$, is the canonical basis of $\mathbb{R}^d$). Then set $a_i = |V_i| \times P_i \cdot \bar{a}({}^tP_i x) \cdot {}^tP_i$, where $\bar{a}(x)$ is the matrix with coefficients $\bar{a}_{12}(x) = H_\alpha(x \cdot e_1)$, $\bar{a}_{21}(x) = -H_\alpha(x \cdot e_1)$, and zero coefficients everywhere else.

Our first claim is that

$$L = \tfrac{1}{2}\Delta + \sum_i \operatorname{div}(a_i \nabla). \tag{2.56}$$

This follows from the following computation, introduce f, g smooth function with compact support on $\mathbb{R}^d$:

$$\begin{aligned} \int g \operatorname{div}(a_i \nabla f)\, dx &= -\int {}^t\nabla g\, a_i \nabla f\, dx \\ &= -|V_i| \times \int ({}^tP_i \nabla g)^t\, \bar{a}({}^tP_i x)({}^tP_i \nabla f)\, dx| \\ &= -|V_i| \int \nabla \bar{g}_i\, \bar{a} \nabla \bar{f}_i\, dx \quad \text{(change of variable)}, \end{aligned}$$

with the notation $\bar{g}_i = g(P_i x)$, $\bar{f}_i = f(P_i x)$, (so that $\nabla \bar{g}_i(x) = {}^tP_i(\nabla g)(P_i x)$ and the same for f).

But the last quantity is:

$$-|V_i|\cdot\int H_\alpha(x_1)\frac{\partial\bar g_i}{\partial x_1}\cdot\frac{\partial\bar f_i}{\partial x_2}-H_\alpha(x_1)\cdot\frac{\partial\bar g_i}{\partial x_2}\cdot\frac{\partial\bar f_i}{\partial x_1}dx=|V_i|\int\phi_\alpha(x_1)\cdot\bar g_i\cdot\frac{\partial\bar f_i}{\partial x_2}dx,$$

after integration by parts. But,

$$\begin{aligned}\int\phi_\alpha(n_i\cdot x)\,g\,V_i\,\nabla f\,dx&=|V_i|\times\int\phi_\alpha(e_1\cdot{}^tP_ix)\times\langle\nabla f(x),P_ie_2\rangle g(x)\,dx\\&=|V_i|\int\phi_\alpha(x_1)\cdot\bar g_i\cdot\frac{\partial\bar f_i}{\partial x_2}dx,\end{aligned}$$

this proves that: $\operatorname{div}(a_i\,\nabla\cdot)=\phi_\alpha(n_i\cdot x)\,V_i\cdot\nabla$ and (2.56) follows.

So we have just showed that $L=\operatorname{div}(A\nabla)$ where $A=\frac{1}{2}\mathrm{Id}+\sum_i a_i$ but the matrix A is bounded independently of α, and satisfies the uniform coercivity assumption $\langle y,Ay\rangle=\frac{1}{2}|y|^2$, it turns out that we can apply (using the connection between transition densities and fundamental solutions of parabolic equations), the moment bounds in Nash [22], Part I, p. 938, which show that $E[|Z_t^{\alpha,x}-x|]\leqq K\cdot\sqrt{t}$, where K is independant of α, and x. One can then use the strong Markov property of the processes $Z_\cdot^{\alpha,x}$ and obtain that for $T_{\alpha_n}(G_t)$-stopping times uniformly bounded by C, one has for η, $\delta>0$,

$$\sup_{0\leqq\theta\leqq\delta}P[|Z^{\alpha,x}_{T_{\alpha_n}+\theta}-Z^{\alpha,x}_{T_{\alpha_n}}|>\eta]\leqq\frac{K}{\eta}\sqrt{\delta},$$

applying now Aldous criterion [1], (see also Joffe-Metivier [14]) one obtains Lemma 2.12. One simply has to make some easy modification in Nash's proof because it is written in the case of a symmetric matrix A (basically one has to majorize $|(x,Ay)|$ by $\frac{1}{2}|x|^2+\frac{1}{2}(y,{}^tAAy)$). □

Remark. For further properties of the fundamental solution of divergence form equations (comparison with the Brownian densities), we refer the reader to Aronson [3]. From these estimates, which require more work, one can obtain the assertion of Lemma 2.12, by the application of a classical tightness criterion, see Billingsley [5].

Proposition 2.13. *For* $x\in\mathbb{R}^d\setminus\bigcup_{i\neq j}H_i\cap H_j$, *Q.a.s. on the set* $\{T_\infty^x<\infty\}$, $\lim_{t\uparrow T_\infty^x}Z_t^x$ *exists and belongs to* $\bigcup_{i\neq j}H_i\cap H_j$.

Proof. Fix $x\in\mathbb{R}^d\setminus\bigcup_{i\neq j}H_i\cap H_j$, by Lemma 2.11 we have for every $\varepsilon>0$ and $T>0$, an increasing function g_T on $\mathbb{R}_+$ with limit 0 in 0 (modulus of continuity) such that for every α

$$Q\Big(\bigcap_{\delta>0}\{\sup_{\substack{s,t\leqq T\\|s-t|\leqq\delta}}|Z_s^\alpha-Z_t^\alpha|\leqq g_T(\delta)\}\Big)\geqq1-\varepsilon\tag{2.57}$$

(we suppress here and in the following superscript x).

So for every $n\geqq0$,

$$Q\Big(\bigcap_{\delta>0}\{\sup_{\substack{s,t\leqq T\\|s-t|\leqq\delta}}|Z^\alpha_{s\wedge T_n}-Z^\alpha_{t\wedge T_n}|\leqq g_T(\delta)\}\Big)\geqq1-\varepsilon,$$

by Proposition 2.11, we get, letting α go to zero, (Fatou's lemma),

$$Q\Big(\bigcap_{\delta>0}\{\sup_{\substack{s,t\leqq T\\|s-t|\leqq\delta}}|Z_{s\wedge T_n}-Z_{t\wedge T_n}|\leqq g_T(\delta)\}\Big)\geqq 1-\varepsilon$$

for every n. But the sets

$$\bigcap_{\delta>0}\{\sup_{\substack{s,t\leqq T\\|s-t|\leqq\delta}}|Z_{s\wedge T_n}-Z_{t\wedge T_n}|\leqq g_T(\delta)\}$$

decrease with n and we find

$$Q\Big(\bigcap_{\delta>0}\{\sup_{\substack{s,t<T\wedge T_\infty\\|s-t|\leqq\delta}}|Z_s-Z_t|\leqq g_T(\delta)\}\Big)\geqq 1-\varepsilon$$

since T and ε are arbitrary, we have obtained that Q a.s. Z_t, $t<T_\infty$ are Cauchy in the neighborhood of T_∞, on the set $\{T_\infty<\infty\}$, this proves that Q a.s. $\lim_{t\uparrow T_\infty} Z_t$ exists on $\{T_\infty<\infty\}$. Since this limit is also the limit of the sequence Z_{T_n} which belong to H_{j_n} and j_n and j_{n+1} are distinct, we find that $\lim_{t\uparrow T_\infty} Z_t\in\bigcup_{i\neq j} H_i\cap H_j$ on $\{T_\infty<\infty\}$ this proves Proposition 2.13. □

We can now replace the value δ in (2.38) by the limit asserted in Proposition 2.13, by defining for $x\in\mathbb{R}^d\setminus\bigcup_{i\neq j} H_i\cap H_j$

$$\bar{Z}_t^x = \begin{cases} Z_t^x, & t<T_\infty^x,\\ \lim_{s\uparrow T_\infty^x} Z_s^x, & t\geqq T_\infty^x.\end{cases}$$

In the course of the Proof of Proposition 2.13 we have obtained that:

Corollary 2.14. *The laws of the processes*

$$\text{$\bar{Z}^x_\cdot$ on $C(\mathbb{R}_+,\mathbb{R}^d)$, for $|x|\leqq c$, and $x\notin\bigcup_{i\neq j} H_i\cap H_j$ are tight.} \tag{2.58}$$

Remark 2.15. In order to prove Proposition 2.13 and Corollary 2.14, we have used Nash's moment bounds. In the case of the system:

$$dX_t^i = dB_t^i + \frac{c}{N}\sum_{j\neq i} dL^0(X^i-X^j)_t, \qquad (2c<N),$$

which was the original motivation of this work, one obtains by probabilistic means a modulus of continuity estimate, and consequently Proposition 2.13 and Corollary 2.14, by the application to the processes $X_{t\wedge T_n}$, of the method of Lemma 3.3, in Sznitman [27].

III. The Existence and Uniqueness Result

In the second part we have characterized the process X_t solution of (2.3), (2.4), up to the time T_∞. The main goal of this section is to show that as a matter of fact $T_\infty=\infty$ a.s. This will be performed by proving this result on the canonical

T_∞^x defined on C_0. As a consequence, we will obtain an existence and uniqueness result for the solutions of (2.3), (2.4).

We will first start by investigating the case of dimension 2, which needs a special treatment. One has:

Proposition 3.1. *In the case of dimension* 2, *for* $x \neq 0$, $T_\infty^x = \infty$, Q *a.s.*

Proof. We will only consider the case where there are $N \geqq 2$ lines $H_i = \mathbb{R}\, v_i$ with $v_i = V_i/|V_i|$, $i \in [1, N]$ (see Remark 2.5).

Consider the set $D = \{\pm v_i,\ i \in [1, N]\} \subset S_1$, we can identify the set $\bigcup H_i \backslash \{0\}$ with $]0, \infty[\times D$, by the map

$$\psi\colon\ x \in \bigcup H_i \backslash \{0\} \to \left(|x|, \frac{x}{|x|}\right). \tag{3.1}$$

Now by Proposition 2.8, the sequence $\left(|Z_{T_n}|, \dfrac{Z_{T_n}}{|Z_{T_n}|},\ n \geqq 0\right)$ (we omit the superscript x) is a Markov chain on $]0, \infty[\times D$, with transition probability R defined by

$$\int R(r, y, dr'\, dy')\, f(r', y') = \int_{C_0} f(\psi(r \cdot y + W_{T_1} + V_j L(0, j, T_1)))\, dQ(w) \tag{3.2}$$

(where ψ is defined by (3.1)), $T_1 = T_1^{r \cdot y}$ with the notation of (2.37), and j is the unique integer in $[1, N]$ such that $n_j \cdot y = 0$). Observe on formula (3.2) by scaling that

$$\int R(r, y, dr'\, dy')\, f(r', y') = \int R(1, y, dr'\, dy')\, f(r \cdot r', y') \tag{3.3}$$

Consider now for $y \in D$ the process defined on C_0:

$$Y(y, t) = y + W_t + V_j L(0, j, t) \quad (\text{with } n_j \cdot y = 0). \tag{3.4}$$

Set $$Y^1(y, t) = Y(y, t) \cdot y, \qquad Y^2(y, r) = Y(y, t) \cdot n_y.$$

The computations in the proof of Proposition 2.4 show immediately that the law of the process $Y^1(y, t)$, $|Y^2(y, t)|$ is the image under the application $F((1, 0), V_j \cdot y, \cdot, \cdot)$ (see (2.21)) of a two dimensional Brownian motion. From Lemma 2.6 we obtain that for $r_t = |Y(y, t)|$ we have:

$$\forall t \geqq 0,\ r_t \neq 0, \qquad \int_0^\infty \frac{ds}{r_s^2} = +\infty \quad \text{and} \quad \int_0^\infty \frac{ds}{r_s^4} = +\infty, \quad Q\text{-a.s.} \tag{3.5}$$

By the same computation as in (2.27), on the angular motion of $Y(y, t)$ we see that $\dfrac{Y(y, T_1)}{|Y(y, T_1)|}$ has the same distribution on D, has a Brownian motion on S_1 starting at y and stopped at the first time it reaches $D - \{y\}$, (that is to say concentrated on the two nearest neighbors $y-$, $y+$ of y, with weights corresponding to the angular distance of (y, y^+) and (y^-, y) suitably normalized). Let us already observe that if $P_{(r, y)}$ is the canonical law of the Markov chain starting at (r, y) with transition probability R, $\phi(r, y) = P_{(r, y)}[\lim_n R_n = 0]$ is either identically zero or 1.

This comes from the fact that the set $\{\lim_n R_n = a\}$ is invariant for the chain, so ϕ is a bounded harmonic function (see Neveu [23], Revuz [25], p. 82) and $\phi(r, y) = \int R(1, y, dr', dy')\, \phi(r'r, y)$.

But observe because of (3.3) that $P_{(r,y)}$ is the image under $(r \cdot R_n, Y_n)_{n \geq 0}$ of the probability $P_{(1,y)}$. So $\phi(r, y) = \phi(1, y)$, and $\phi(1, y)$ is harmonic for $R(y, dy')$ (with obvious notations) and is constant, but this constant has to be 0 or 1 because $\phi(R_n, Y_n)$ is a bounded martingale which converges $P_{(r,y)}$ a.s. (for every (r, y)) towards $1\{\lim R_n = 0\}$. The same method shows that $P_{(r,y)}(\lim R_n = \infty)$ is identically 0 or 1.

Our next step is to show that

$$P_{(1,y)}(\lim R_n = 0) = P_{(1,y)}(\lim R_n = \infty), \tag{3.6}$$

since clearly $\{\lim R_n = 0\}$ and $\{\lim R_n = \infty\}$ are disjoint events this will show that $P_{(r,y)}(\lim R_n = 0) = 0$ identically, since we know by (2.13) that $\{T_\infty < \infty\}$ is Q a.s. included in the set $\{\lim_n Z_{T_n} = 0\}$ (because $\{0\} = H_i \cap H_j$, for $i \neq j$) this will prove that Q a.s. $\{T_\infty = \infty\}$. In order to prove (3.6), let us first show

Lemma 3.2. *Define* $A_t = \int_0^t \frac{ds}{r_s^4}$, τ_t *the continuous increasing inverse of* A_t. *If* $p(x)$: $\mathbb{R}^2 \setminus \{0\} \to \mathbb{R}^2 \setminus \{0\}$ *is the inversion of pole* 0 *and radius* 1,

$$-p(Y(y, \tau_t)) \text{ has the same law as } Y(-y, t). \tag{3.7}$$

Proof. Without any loss of generality, we can assume that $y = 1$, then we have

$$\begin{cases} Y_t^1 = 1 + W_t^1 + a\, L^0(W^2)_t, & (a = V_j \cdot y), \\ Y_t^2 = W_t^2 \end{cases}$$

then $-p(Y_t) = H_t$, with $H_t^1 = -\frac{Y_t^1}{r_t^2}$, $H_t^2 = -\frac{Y_t^2}{r_t^2}$, $r_t = |Y_t|$. Using Ito's formula, we find:

$$H_t^1 = -1 + \int_0^t \frac{(Y_s^1)^2 - (Y_s^2)^2}{r_2^4}\, dW_s^1 + \int_0^t 2\frac{Y_s^1\, Y_s^2}{r_s^4}\, dW_s^2 + a \int_0^t \frac{(Y_s^1)^2 - (Y_s^2)^2}{r_s^4}\, dL^0(W^2)_s, \tag{3.8}$$

$$H_t^2 = \int_0^t 2\frac{Y_s^1\, Y_s^2}{r_s^4}\, dW_s^1 + \int_0^t \frac{(Y_s^2)^2 - (Y_s^1)^2}{r_s^4}\, dW_s^2. \tag{3.9}$$

From (3.8), (3.9), we deduce that $\langle H^1, H^2 \rangle_t = 0$, $\langle H^1 \rangle = \langle H^2 \rangle_t = \int_0^t \frac{ds}{r_s^4}$. Moreover H_t^2 is a local martingale and we know that its local time in 0, L_t satisfies

$$|H_t^2| = \int_0^t \operatorname{sign} H_s^2\, dH_s^2 + L_t. \tag{3.10}$$

But we can also compute the semimartingale decomposition of $|H_t^2|$ by applying Ito's formula to $|H_t^2| = \frac{|Y_t^2|}{r_t^2}$, the finite variation terms that will appear

will be, (using (3.10)), $dL_s = \frac{1}{r_s^2} dL^0(W^2)_s + |W_s^2|\, dC_s$ where C_t is some bounded variation process which is absolutely continuous with respect to $dL^0(W^2)_s + ds$. But $1\{H_s^2 = 0\}\, dL_s = dL_s$ and we have $|W_s^2| \times 1\ \{H_s^2 = 0\} \times (ds + dL^0(W_s^2)) = 0$, as a consequence we find:

$$L_t = \int_0^t \frac{dL^0(W^2)_s}{r_s^2}. \tag{3.11}$$

Now it is easy to see that the last term in (3.8) is $a \times L_t$, (because $dL^0(W^2)_t$ is concentrated on the set where $W_t^2 = 0$). As a consequence in the filtration G_{τ_t}, we have

$$H_{\tau_t}^1 = -1 + \bar{W}_t^1 + aL_{\tau_t}, \qquad H_{\tau_t}^2 = \bar{W}_t^2$$

and $\bar{W}_t^1, \bar{W}_t^2$ are independent G_{τ_t}-Brownian motions, moreover $|H_{\tau_t}^2| - L_{\tau_t}$ is a G_{τ_t} martingale, so L_{τ_t} is the local time of $H_{\tau_t}^2$ in 0, and this proves that $-p(Y(y, \tau_t))$ has the same law as $Y(-y, t)$. □

As a consequence of Lemma 3.2, we find that $Y(-y, T_1^{-y})$ has the same law as $-p(Y(y, \tau_S))$ if S is the first time $Y(y, \tau_s)$ reaches $\bigcup H_i \backslash \mathbb{R} \times y$, but clearly $Y(y, \tau_S) = Y(y, T_1^y)$. So we have obtained that

$$Y(-y, T_1^{-y}) \quad \text{has same law as} \quad -p(Y(y, T_1^y)). \tag{3.12}$$

Now consider the sequence $U_n = (\bar{R}_n, \bar{Y}_n)$, $n \geqq 0$, with $\bar{R}_n = \frac{1}{R_n}$, $\bar{Y}_n = -Y_n$, under the law $P_{(1,y)}$ we find that

$$E_{P_{(1,0)}}[f(U_{n+1})/(R_i, Y_i), i \leqq n] = \int f\left(\frac{1}{R_n \cdot r'}, -y'\right) R(1, Y_n, dr', dy')$$

by (3.12) this is equal to

$$\int f\left(\frac{r'}{R_n}, y'\right) R(1, -Y_n, dr', dy') = \int f(\bar{R}_n r', y')\, R(1, \bar{Y}_n, dr', dy'),$$

so as a consequence

$$E_{P_{(1,y)}}[f(U_{n+1})/U_1, \ldots, U_n] = \int f(\bar{R}_n r', y')\, R(1, \bar{Y}_n, dr', dy'), \tag{3.13}$$

moreover $U_1 = (1, -y)$, $P_{(1,y)}$ a.s.

So we have obtained that the law of $\left(\frac{1}{R_n}, -Y_n\right)$ under $P_{(1,y)}$ is $P_{(1,-y)}$ from which we deduce (3.6), which proves Proposition 3.1. □

Let us now turn to the case of dimension $d \geqq 3$. It is not difficult to see in an informal way that the processes Z_t^x (constructed with (n_j, V_j), $1 \leqq j \leqq N$, satisfying (2.2)), are in duality with the processes $\bar{Z}_t^x$ (constructed with $(n_j, -V_j)$, $1 \leqq j \leqq N$) with respect to Lebesgue measure, this can be guessed on the generators $L^\alpha = \frac{1}{2}\Delta + \sum_{i=1}^N \operatorname{div}(a_i^\alpha V)$ (see (2.56)), $\bar{L}^\alpha = \frac{1}{2}\Delta + \sum_{i=1}^N \operatorname{div}({}^t a_i^\alpha V)$, for the approximating processes $Z_t^{\alpha,x}$ and $\bar{Z}_t^{\alpha,x}$ respectively.

As a result one should get that the processes $Z^x_{\tau^x_t}$, where τ^x_t is the continuous inverse of $A^x_t = \int_0^t \frac{ds}{|Z^x_s|^2}$, are in duality with the processes $\bar{Z}^x_{\bar{\tau}^x_t}$, where the duality measure is now $\mu(dx) = \frac{1}{r^2} dx$, (observe that since $d \geqq 3$, μ is a Radon measure on $\mathbb{R}^d$).

Our goal is now to establish this result more precisely. Let c belong to $]0, \infty[$, and h_c be a smooth bounded function from $\mathbb{R}_+$ into $\mathbb{R}_+$ such that,

$$h(x) \leqq x, h(x) = x \quad \text{for } x \leqq c, \quad \text{and} \quad h_c(x) = \text{constant} \quad \text{for } x \geqq c+1. \tag{3.14}$$

We define
$$A^{x,c}_t = \int_0^t \frac{ds}{h_c(|Z^x_s|^2)}, \quad x \notin \bigcup_{i \neq j} H_i \cap H_j, \ t \in [0, T^x_\infty[\tag{3.15}$$

$$A^{\alpha,x,c}_t = \int_0^t \frac{ds}{h_c(|Z^{\alpha,x}_s|^2)}, \quad x \neq 0, \ t \geqq 0. \tag{3.16}$$

((3.16) is well defined, since starting away from 0, almost surely $Z^{\alpha,x}_s$ does not reach 0, moreover $A^{\alpha,x,c}_\infty = \infty$).

We will now omit the superscript x and c except when there might be any ambiguity. We can define the continuous increasing inverse: τ_t, $t \in [0, A_{T_\infty}[$, (resp. $\tau^\alpha_t, t \in [0, \infty[$) of $A_.$ (resp. $A^\alpha_.$).

We can consider the time changed processes

$$Y_t = Z_{\tau_t}, \quad t \in [0, A_{T_\infty}[, \tag{3.17}$$

$$Y^\alpha_t = Z^\alpha_{\tau^\alpha_t}, \quad t \in [0, \infty[, \tag{3.18}$$

and denote by $\bar{Y}_t$ and $\bar{Y}^\alpha_t$ the corresponding processes obtained when replacing (n_j, V_j), by $(n_j, -V_j)$, $1 \leqq j \leqq N$. We have

Lemma 3.3.

a) *Let* $x \notin \bigcup_{i \neq j} H_i \cap H_j$ *be fixed, Q-a.s. on* $\{t < A_{T_\infty}\}$

$$\lim_{\alpha \to 0} \sup_{s \leqq t} |Y^\alpha_s - Y_s| = 0, \tag{3.19}$$

b) *for* f, g *continuous functions with compact support in* $\mathbb{R}^d$, *and* $t \geqq 0$,

$$\int_{\mathbb{R}^d} d\mu_c(x) E_Q[f(Y^{\alpha,x,c}_t)] g(x) = \int_{\mathbb{R}^d} d\mu_c(x) f(x) E_Q[g(\bar{Y}^{\alpha,x,c}_t)], \tag{3.20}$$

$$\left(\text{where } \mu_c(dx) = \frac{1}{h_c(r^2)} dx, d \geqq 3\right).$$

Proof. a) Q-a.s. $A^\alpha_.$ converges uniformly on compact sets of $[0, T_\infty[$, when α goes to zero, towards $A_.$ (x and c are fixed), using Proposition 2.11. One checks easily that as a consequence, τ^α_t converge uniformly to τ_t on compact sets of $[0, A_{T_\infty}[$, (by Dini's theorem it is enough to check simple convergence).

But $Y_t = Z_{\tau_t}$, $Y^\alpha_t = Z^\alpha_{\tau^\alpha_t}$ and Q a.s. on $\{t < A_{T_\infty}\}$, for α small enough, $\tau^\alpha_s \leqq \tau^\alpha_t < T_\infty$ for $0 \leqq s \leqq t$. Then we have $|Z_{\tau_s} - Z^\alpha_{\tau^\alpha_s}| \leqq |Z_{\tau_s} - Z_{\tau^\alpha_s}| + \sup_{u \leqq \tau^\alpha_t} |Z_u - Z^\alpha_u|$, and (3.19) follows easily.

b) Notice that for $f \in C_K^\infty(\mathbb{R}^d)$, $M_t^f = f(Z_t^{\alpha,x}) - f(x) - \int_0^t L^\alpha f(Z_s^{\alpha,x})\,ds$ is a G_t-martingale and $\langle M^f\rangle_t = \int_0^t |\nabla f(Z_s^{\alpha,x})|^2\,ds \leq K_f\, t$.

So for $x \neq 0$, $\langle M^f\rangle_{\tau_t^{\alpha,x}} \leq K_f\, \tau_t^{\alpha,x} \leq K_f(c+1) \int_0^{\tau_t^{\alpha,x}} \frac{ds}{h_c(|Z_s^{\alpha,x}|^2)} = K_f(c+1)\,t$.

As a consequence we find that

$$f(Y_t^{\alpha,x}) - f(x) - \int_0^t \underline{L}^\alpha(Y_s^{\alpha,x})\,ds \quad \text{is a } G_{\tau_t^{\alpha,x}}\text{-martingale}$$

$$\text{if } \underline{L}^\alpha = h_c(r^2) \times L^\alpha. \tag{3.21}$$

Define P_x to be the unique solution of the martingale problem associate to x and $\underline{L}^\alpha$, $(x \in R^d)$, on $C(\mathbb{R}_+, \mathbb{R}^d)$, and $T_t f(x) = E_{P_x}[f(X_t)]$.

We have the fact that the law of $(Y_\cdot^{\alpha,x})$, $x \neq 0$, on $C(\mathbb{R}_+, \mathbb{R}^d)$ is P_x, and to prove (3.20) it is clearly enough to show that

$$\int d\mu_c\, T_t f\, g = \int d\mu_c\, f\, \bar{T}_t g \tag{3.22}$$

with the obvious notation $\bar{T}_t$, and f, g continuous with compact support in $\mathbb{R}^d$. $\sqrt{h_c}$ is simply a Lipschitz function, we are going to approximate the generators $\underline{L}^\alpha$, in order to prove (3.22). Observe first that for $x \neq 0$, P a.s. X_t does not reach the origin, this can be seen from the fact that P_x is the law of $(Z_{\tau_t^\alpha}^{\alpha,x})$ which never reaches the origin, or from the fact that P_x is the law of the solution of an S.D.E. with Lipschitz coefficients, for which the constant trajectory 0 is the solution starting from zero (see Kunita [16], Chapter II.4).

If $h_{c',c}(c' < c)$ is a positive smooth function on $\mathbb{R}_+$ equal to h_c on $[c', +\infty[$, which remains bounded away from 0, the solution to the martingale problem $\underline{L}_{c',c}^\alpha = h_{c',c}(r^2)\, L^\alpha$ starting at x, coincides with P_x up to the first time the process reaches the ball centered at 0 of radius c'. From these remarks it follows that if one shows for f, g continuous with compact support in $\mathbb{R}^d \setminus \{0\}$,

$$\int d\mu_{c',c}\, T_t f\, g = \int d\mu_{c',c}\, f\, \bar{T}_t g, \tag{3.23}$$

where $\mu_{c',c}(dx) = dx/h_{c',c}(r^2)$, and T_t, $\bar{T}_t$ are the semigroups associated to the solutions of the martingale problems $\underline{L}_{c',c}^\alpha$, $\bar{\underline{L}}_{c',c}^\alpha$, letting c' go to zero and using a monotone class argument one will obtain (3.22).

We will use the fact that for $f \in C_K^\infty(R^d)$, $T_t f$ is a $C^2(\mathbb{R}^d)$ function which decays rapidly with its first two derivatives, uniformly when t remains bounded. This is a consequence of the estimates on the L^p - derivatives of the solution of an S.D.E. having smooth coefficients with respect to its initial condition (see Gikhman-Skorohod [10], Kunita [16]) and of the exponential estimates for the diffusion (Stroock-Varadhan [26]).

So for $f \in C_K^\infty(\mathbb{R}^d)$, we have $\frac{d}{dt} T_t f = \underline{L}_{c',c}^\alpha T_t f = T_t \underline{L}_{c',c}^\alpha f$, and we find for $f, g \in C_K^\infty(\mathbb{R}^d)$

$$\frac{d}{ds} \int T_{t-s} f\, \bar{T}_s g\, d\mu_{c',c} = \int (-\underline{L}_{c',c}^\alpha T_{t-s} f\, \bar{T}_s g + T_{t-s} f\, \bar{\underline{L}}_{c',c}^\alpha \bar{T}_s g)\, d\mu_{c',c} = 0,$$

(integrating by parts).

As a consequence $\int T_t f\, g\, d\mu_{c',c} = \int f\, \bar{T}_t g\, d\mu_{c',c}$ for $f, g \in C_K^\infty$, and one gets (3.23) by approximation.

We are now ready to show

Proposition 3.4.

$$\text{for } x \notin \bigcup_{i \neq j} H_i \cap H_j, \quad Q \text{ a.s. } T_\infty^x = \infty. \tag{3.24}$$

Proof. We will prove (3.24) by induction on the dimension d. The result is already proved for $d=2$, and is trivial for $d=1$. Suppose that (3.24) has been proved for all dimensions $\leqq d$, consider the case of dimension $d+1$. If $N=1$ (3.24) is trivial, suppose $N \geqq 2$. Observe that we can suppose that $\bigcap_i H_i = \{0\}$, else denote by p the orthogonal projection on $H^\perp$, if $H = \bigcap_i H_i$, for $y \notin \bigcup_{i \neq j} H_i \cap H_j$, $p(y) \notin \bigcup_{i \neq j} p(H_i) \cap p(H_j)$, denote by Y_t the projection $p(Z_t^y)$. Using the fact that for $x \in R^{d+1}$, $p(x) \cdot n_i = x \cdot n_i$, for $1 \leqq i \leqq N$, it is easy to check on the formula (2.37) that Y_t is the process in $H^\perp$ starting from $p(y)$ constructed with (2.37) with the Brownian motion $p(W_t)$ and the data $n_j, p(V_j), j \in [1, n]$ (note that $p(n_j) = n_j$ for every j, since $n_j \in H^\perp$ by definition), and that $T_n^y = T_n^{p(y)}$, $n \geqq 0$. But now by the induction hypothesis we get $T_\infty^y = \infty Q$ a.s.

So we can suppose $\bigcap_{i=1}^N H_i = \{0\}$; our next step is to prove that necessarily

$$\text{on } \{T_\infty^x < \infty\}, \quad \lim_{t \uparrow T_\infty^x} Z_t^x = 0 \tag{3.25}$$

(the existence of this limit comes from Proposition (2.13)). Let I be a subset of $[1, N]$, $x \notin \bigcup_{i \neq j} H_i \cap H_j$, it is immediate from (2.37) that Z_t^x coincides with the process $Z_t^{x,I}$ constructed with (n_j, V_j), $j \in I$, until the first time T_n^x such that $j_n \notin I$. As a consequence on the set

$$\bigcap_{n \geqq 0} \{j_n \in I\}, \quad Q \text{ a.s. } Z_\cdot^{x,I} = Z_\cdot^x. \tag{3.26}$$

Let us now denote by $\underline{I}$ the family of subsets I of $[1, N]$ such that $\bigcap_{j \in I} H_j \neq \{0\}$, then Q a.s. on $\{T_\infty^x < \infty\} \cap \{\lim_{t \uparrow T_\infty^x} Z_t^x \neq 0\}$, there exists $n \geqq 0$ and $I \in \underline{I}$ such that for $p \geqq n$, $j_p \in I$. But

$$Z_{T_n^x + t}(x, w) = Z_t(Z_{T_n^x}, W_{T_n^x + \cdot} - W_{T_n^x}) = Z_t^I(Z_{T_n^x}, W_{T_n^x + \cdot} - W_{T_n^x}) \tag{3.27}$$

a.s. on the set $A_{n,I} = \bigcap_{p \geqq n} \{j_p \in I\}$.

Now notice that for $Z_\cdot^{y,I}$ we know that $T_\infty^{y,I} = \infty$ a.s. because $\bigcap_{j \in I} H_j \neq \{0\}$, using (3.27) we find that $\{T_\infty^x < \infty\} \cap \{\lim_{t \uparrow T_\infty^x} Z_t^x \neq 0\}$ is negligible, which proves (3.25).

We are now going to show that for $x \notin \bigcup_{i \neq j} (H_i \cap H_j)$,

$$Q \text{ a.s.,} \quad \int_0^{T_\infty^x} \frac{ds}{|Z_s^x|^2} = \infty. \tag{3.28}$$

The method is very similar to the proof in Lemma 2.6. Set

$$\phi(x)=E_Q\left[\exp-\int_0^{T_\infty^x}\frac{ds}{|Z_s^x|^2}\right],$$

by Proposition 2.9, we find for $\lambda>0$,

$$\phi(\lambda x)=E_Q\left[\exp-\int_0^{\lambda^2 T_\infty^x}\frac{ds}{\lambda^2|Z_{s/\lambda^2}^x|^2}\right]=\phi(x).$$

Now for $|x|=2$ we have, using strong Markov property (2.40)

$$\phi(x)=E_Q\left[\exp-\int_0^{\tau_1\wedge t}\frac{ds}{|Z_s^x|^2}\phi(X_{\tau_1\wedge t})\right]\qquad(t>0).$$

if $\tau_1=\inf\{s\geqq 0; |Z_s^x|=1\}$ $(\leqq T_\infty^x)$. If a denotes $\sup\{\phi(x), x\notin\bigcup_{i\neq j}H_i\cap H_j\}$, we have

$$a\leqq\sup_{|x|=2}E_Q\left[\exp-\int_0^{\tau_1\wedge t}\frac{ds}{|Z_s^x|^2}\right]\times a,$$

but

$$\sup_{|x|=2}E_Q\left[\exp-\int_0^{\tau_1\wedge t}\frac{ds}{|Z_s^x|^2}\right]\leqq\sup_{|x|=2}E_Q\left[\exp-\int_0^{\tau_1\wedge\tau_3\wedge t}\frac{ds}{|Z_s^x|^2}\right]<1$$

because of the tightness asserted in (2.58). This proves (3.28).

Define now for $x\notin\bigcup_{i\neq j}H_i\cap H_j$, $(\tau_t^x)_{t\geqq 0}$ the continuous increasing inverse of A_t^x, $Y_t^x=Z_{\tau_t^x}^x$, $t\geqq 0$, the processes Y_t^x as well as $Y_t^{x,c}$ are defined on $\mathbb{R}_+$, because of (3.28) $(A_{T_\infty^x}^{x,c}=\infty)$.

Letting α go to zero in (3.20), by (3.19) we obtain for $t\geqq 0$,

$$\int_{\mathbb{R}^{d+1}}d\mu_c(x)E_Q[f(Y_t^{x,c})]g(x)=\int_{\mathbb{R}^{d+1}}d\mu_c(x)f(x)E_Q[g(\bar{Y}_t^{x,c})]\tag{3.29}$$

for f,g continuous function with compact support in $\mathbb{R}^{d+1}$. Notice that $A_t^{x,c}=A_t^x$ for $t\leqq T_c^x\wedge T_\infty^x$, if T_c^x is the first time Z_t^x reaches the complement of the ball of radius c, centered at the origin; as a consequence we have $Y_t^{x,c}=Y_t^x$ for $t\leqq A_{T_c^x\wedge T_\infty^x}^x$. Since μ and μ_c coincide when restricted to $B(0,c)$, we find letting c go to infinity in (3.29), that for $t\geqq 0$, f,g, continuous with compact support in $\mathbb{R}^{d+1}$

$$\int_{R^{d+1}}d\mu(x)E_Q[f(Y_t^x)]g(x)=\int_{R^{d+1}}d\mu(x)E_Q[g(\bar{Y}_t^x)].\tag{3.30}$$

Now consider the canonical Markov chain on $(\mathbb{R}^{d+1})^N=\Omega$ with transition probability $P(x,dy)=E_Q[Y_1^x\in dy]$, for $x\notin\bigcup_{i\neq j}H_i\cap H_j$ and $=\delta_x(dy)$ for $x\in\bigcup_{i\neq j}H_i\cap H_j$. By (3.30), we see that μ is an invariant measure for $P(x,dy)$. Denote by P_μ the law of the chain on Ω, with initial measure μ.

Now Ω can be decomposed into a conservative part C and dissipative part D (see Neveu [23], Revuz [25]).

Since $d+1\geqq 3$, every ball $B(o,r)$ has finite μ measure.

$$\text{On } D,\ P_\mu \text{ a.s.,} \quad \sum_{m\geqq 0} 1_{B(0,1)}(X_m)<\infty, \tag{3.31}$$

$$\text{on } C,\ P_\mu \text{ a.s.,} \quad \sum_{m\geqq 0} 1_{B\left(0,\frac{1}{n}\right)^c \cap B(0,1)}(X_m)=0 \quad \text{or } +\infty$$

for every $n\geqq 2$ (see Revuz [25] Proposition 2.3, p. 112).

As a consequence C is included P_μ a.s. in the set

$$\{\forall n\geqq 2, \sum_m 1_{B\left(0,\frac{1}{n}\right)^c \cap B(0,1)}(X_m)=0\} \cup \{\exists n\geqq 2, \sum_m 1_{B\left(0,\frac{1}{n}\right)^c \cap B(0,1)}(X_m)=+\infty\}. \tag{3.32}$$

From (3.31) and (3.32) and $P_\mu(\exists n\geqq 0,\ X_n=0)=0$, we conclude that

$$P_\mu[\lim_n X_n=0]=0. \tag{3.33}$$

So by (3.25), we see that for $\mu\cdot$a.e. $x\in\mathbb{R}^{d+1}\setminus\bigcup_{i\neq j} H_i\cap H_j$, $T_\infty^x=\infty$.

Let us show that the result is indeed true for every

$$x\in\mathbb{R}^{d+1}\setminus\bigcup_{i\neq j} H_i\cap H_j.$$

For such an x we can find a ball small enough centered at x, such that until the exit time τ of this ball Z_t^x coincides with $x+W_t$ if $x\notin\bigcup_i H_i$, or with $x+W_t+V_{i_0}L^0(W_{\cdot}n_{i_0})_t$ else.

Denote by A the set where $\psi(x)=Q(T_\infty^x=\infty)$ has value 1. Since the laws of $x+W_t$ or $x+W_t+V_{i_0}L^0(W_{\cdot}n_{i_0})_t$ for $t>0$ admit densities with respect to Lebesgue measure, and since A^c is negligible with respect to Lebesgue measure, the optional set $D=\{(w,t)\in C_0\times\mathbb{R}_+,\ t<\tau(w) \text{ and } Z_t^x\in A\}$ has a projection on C_0 equal to C_0 up to a negligible set.

Applying the section theorem (see Dellacherie-Meyer [7], p. 220), for every $\varepsilon>0$, we can find a G_t - stopping time T such that for w where $T(w)<\infty$, $(w,T(w))\in D$ and $Q(T<\infty)\geqq 1-\varepsilon$.

Now considering the stopping time $\tau\wedge T$, by (2.40), we obtain $\phi(x)=E_Q[\phi(X_{\tau\wedge T})]\geqq 1-\varepsilon$, letting ε go to zero we find that $T_\infty^x=\infty$, Q-a.s. and this proves Proposition 3.4. □

We can now state as a consequence of our results:

Theorem 3.5. *There exists a unique solution on* $(\Omega,F,(F_t)_{t\geqq 0},P)$ *of* (2.3) *with an initial condition* X_0 *satisfying* (2.4), *this solution is given by the formulas* (2.5), (2.6), (2.35), (2.36), *moreover the laws of the solutions do not depend on the particular choice of the space* $(\Omega,F,(F_t)_{t\geqq 0},P)$.

Remark 3.6. We have also obtained that Q a.s. $Z_\cdot^{\alpha,x}$ converges to $Z_\cdot^x$ on the compact sets of $\mathbb{R}_+$. One also obtains easily the duality relation

$$\text{for } t\geqq 0, \quad \int E_Q[f(Z_t^x)]\,g(x)\,dx=\int f(x)\,E_Q[g(\bar{Z}_t^x)]\,dx$$

for fg bounded measurable with compact support, and the fact that the laws Q_x, $x \in \mathbb{R}^d \setminus \bigcup_{i \neq j} H_i \cap H_j$, on $C(\mathbb{R}_+, \mathbb{R}^d)$ of the $Z^x_.$, form a strong Markov process, with state space $\mathbb{R}^d \setminus \bigcup_{i \neq j} H_i \cap H_j$. □

Remark 3.7. One can wonder if one can define Q_x for x in the set of singularities $\bigcup_{i \neq j} H_i \cap H_j$. Observe by Remark 2.14, that the law Q_x are tight when x remains bounded ($x \in \mathbb{R}^d \setminus \bigcup_{i \neq j} H_i \cap H_j$). As a consequence of Aronson-Serrin [4], Moser [20], Nash [21] results there exists a continuous transition density $p(t,x,y), (t,x,y) \in (0, +\infty) \times \mathbb{R}^d \times \mathbb{R}^d$, such that for

$$x \notin \bigcup_{i \neq j} H_i \cap H_j, \qquad Q_x(f(X_t)) = \int_{\mathbb{R}^d} p(t,x,y) f(y) dy.$$

This determines the finite dimensional laws of any limit point of the Q_x when x converges to $y \in \bigcup_{i \neq j} H_i \cap H_j$, so one can extend the family Q_x, $x \in \mathbb{R}^d \setminus \bigcup_{i \neq j} H_i \cap H_j$, into a family Q_x, $x \in \mathbb{R}^d$ which is strong Markov (because of the continuity of $p(t,x,y)$). □

The first author wishes to thank the Courant Institute of Mathematical Sciences for its hospitality. Research partially supported by grant n° AFOSR-85-0017 from the Air Force Office of Scientific Research.

Bibliography

1. Aldous, D.: Stopping times and tightness. Ann. Probab. **6**, 335–340 (1978)
2. Anulova, S.V.: Diffusion processes with singular characteristics. Stochastic Differential Systems, Proc. I.F.I.P.-W.G. Conf. Vilnius, 264–269, Lecture notes in Control **25**. Berlin Heidelberg New York: Springer 1978
3. Aronson, D.G.: Bounds for the fundamental solution of a parabolic equation. Bull. Am. Math. Soc. **73**, 890–896 (1967)
4. Aronson, D.G., Serrin, J.: Local behaviour of solutions of quasilinear parabolic equations. Arch. Rational Mech. Anal. **25**, 81–122 (1967)
5. Billingsley, P.: Convergence of probability measures. New York: Wiley 1968
6. Carillo-Menendez, S.: Processus de Markov Associé à une forme de Dirichlet non Symétrique. Z. Wahrscheinlichkeitstheorie Verw. Geb. **33**, 139–154 (1975)
7. Dellacherie, C., Meyer, P.A.: Probabilités et Potentiels, cha. I a IV. Paris: Hermann 1975
8. El-Karoui, N., Chaleyat-Maurel, M.: Un Problème de réflexion et ses applications au temps local et aux équations différentielles stochastiques, sur R, cas continu. Temps locaux. Astérisque 52–54, (1978)
9. Friedman, A.: Partial differential equations of parabolic type. London: Prentice Hall 1964
10. Gimman, I.I., Skorohod, A.V.: Stochastic differential equations. Berlin: Heidelberg New York: Springer 1972
11. Gutkin, E., Kac, M.: Propagation of chaos and the Burgers' equations. SIAM J. Appl. Math., **43**, 972–980 (1983)
12. Harrison, J.M., Shepp, L.A.: On skew Brownian motion. Ann. Probab. **9**, 309–313 (1981)
13. Ikeda, N., Watanabe, S.: Stochastic differential equations and diffusion processes. Amsterdam Kodansha Tokyo: North Holland 1981
14. Joffe, A., Métivier, M.: On tightness in $D(\mathbb{R}_+, H)$, and a few applications to limits of branching processes. Sep. 1981, CRMA-1054, Université de Montreal

15. Kotani, S., Osada, H.: Propagation of chaos for Burgers'equation. J. Math. Soc. Japan **37**, 275–294 (1985)
16. Kunita, H.: Stochastic flows. École d'été de probabilités de S^T flour (1982). Lecture notes in Math. **1097**. Berlin Heidelberg New York Tokyo: Springer 1985
17. Ladyzhenskaya, O.A., Solonnikov, V.A., Uralceva, N.N.: Linear and quasilinear equations of parabolic type. Providence, R.I.: Amer. Math. Soc. 1968
18. Le Gall, J.F.: Doctorat de 3eme cycle, Paris VI (1982)
19. Le Jan, Y.: Mesures associées à une forme de Dirichlet et applications. Bull. Soc. Math. France, **106**, 61–112 (1978)
20. McKean, H.P.: Propagation of chaos for a class of nonlinear parabolic equations. Lecture series in Differential Equations, Vol. 7, 41–57, Catholic Univ., Washington, D.C., (1967)
21. Moser, J.: A Harnack inequality for parabolic differential equations. Comm. Pure Appl. Math. **17**, 101–134 (1964), and **20**, 231–236 (1967)
22. Nash, J.: Continuity of solutions of parabolic and elliptic equations. Am. J. Math. **80**, 931–954 (1958)
23. Neveu, J.: Bases mathématiques des probabilités. Paris: Masson 1970
24. Portenko, N.I.: On stochastic differential equations with generalized trend vectors, II Vilnius Conf., abstracts of communications, Vol. 1, Vilnius (1977)
25. Revuz, D.: Markov chains. Amsterdam: North Holland 1975
26. Stroock, D.W., Varadhan, S.R.S.: Multidimensional diffusion processes. Berlin Heidelberg New York: Springer 1979
27. Sznitman, A.S.: A propagation of chaos result for Burgers' equation. Probab. Th. Rel. Fields **71**, 581–613 (1986)
28. Tanaka, H.: Stochastic differential equations with reflecting boundary conditions in convex regions. Hiroshima Math. J. **9**, 163–177 (1979)
29. Varadhan, S.R.S., Williams, R.J.: Brownian motion in a wedge with oblique reflection. Comm. Pure Appl. Math. **38**, 405–443 (1985)

Received November 8, 1984; in revised form September 2, 1985

Équations aux dérivées partielles/*Partial Differential Equations*

État fondamental et Principe du Maximum pour les opérateurs elliptiques du second ordre dans des domaines généraux

Henri BERESTYCKI, Louis NIRENBERG et Srinivasa VARADHAN

Résumé – Pour un opérateur elliptique L dans un domaine borné général $\Omega\subset\mathbb{R}^N$ (sans hypothèse de régularité) on définit la valeur propre principale λ_1 en posant

(1) $$\lambda_1=-\inf_{\{\phi>0\}}\sup_{x\in\Omega}\{L\phi(x)/\phi(x)\}=\sup\{\lambda;\ \exists\varphi>0 \text{ tel que } L\varphi+\lambda\varphi\leqq 0 \text{ dans } \Omega\}.$$

On montre que la théorie de Krein-Rutman s'étend à ce cadre général. Il existe en effet une fonction propre $\phi\in L^\infty(\Omega)$, unique, à une constante multiplicative près, telle que $(L+\lambda_1)\phi_1=0$ dans Ω et telle que ϕ_1 s'annule sur $\partial\Omega$ en un sens convenablement défini. D'autre part, L satisfait le Principe du Maximum (en un sens convenablement généralisé) si et seulement si $\lambda_1>0$. Nous obtenons diverses propriétés (notamment relatives à la dépendance par rapport aux coefficients ou au domaine) et estimations de λ_1 en en dégageant l'aspect « structurel » (indépendamment de l'opérateur L). Nous montrons en particulier que le Principe du Maximum est vérifié dans des domaines suffisamment « étroits » ou de mesure suffisamment petite.

The Ground State and Maximum Principle for 2nd order elliptic operators in general domains

Abstract – *For an elliptic operator* L *in a general bounded domain* $\Omega\subset\mathbb{R}^N$ (*no assumption of smoothness is made here*), *we define the principal eigenvalue by* (1) *above. We show that the Krein-Rutman theory extends to this general setting. Indeed, we show that there exists a function* $\phi_1\in L^\infty(\Omega)$ *such that* $(L+\lambda_1)\phi_1=0$ *in* Ω, ϕ_1 *vanishes on* $\partial\Omega$ *in a sense which is made precise. This function is unique up to a multiplicative constant. Furthermore, the Maximum Principle* (*in a conveniently refined formulation*) *holds for* L *in* Ω *if and only if* $\lambda_1>0$. *We establish several properties of* λ_1 *about the dependence on the coefficients, the domain, etc. and several estimates which are new – even in the case of a regular domain* Ω. *In deriving these estimates we emphasize the* structural *aspect of the various constants – independently of the particular operator under consideration. In particular we show that the maximum principle holds for domain which are sufficiently "narrow" or have small measure.*

Abridged English Version – We consider a general bounded domain $\Omega\subset\mathbb{R}^N$. No smoothness condition whatsoever is being assumed here. Let L be an elliptic operator in Ω of the form

(2) $$L=M+c(x)=a_{ij}(x)\,\partial_{ij}+b_i(x)\,\partial_i+c(x),$$

with

(3) $$\begin{cases} c_0|\xi|^2\leqq a_{ij}(x)\xi_i\xi_j\leqq C_0|\xi|^2, & \forall\xi\in\mathbb{R}^N,\quad \forall x\in\Omega\\ a_{ij}\in C(\Omega);\qquad b_i, c\in L^\infty(\Omega); & \|b_i\|_{L^\infty},\ \|c\|_{L^\infty}\leqq b,\end{cases}$$

where c_0, $C_0>0$ and $b\geqq 0$ are constants. The functions to which we apply L are always supposed to be in $W^{2,N}_{loc}$.

In this general framework we define the principal eigenvalue (P.E.) of L by (1).

In the regular case, it is known ([4], [5]) that the lowest eigenvalue of $-L$ with Dirichlet condition is given by the minimax formula (1).

This work has a triple scope: (i) To extend the Krein-Rutman theory for the P.E. in the general case. (ii) To investigate the relationships of λ_1 with the maximum principle. (iii) To derive qualitative properties and *a priori* estimates emphasizing their dependence on Ω

Note présentée par Louis NIRENBERG.

and the constants c_0, C_0, b. Proofs of the results stated here as well as of others are given in [3].

In the following, we will use a function u_0 which was constructed by Stroock and Varadhan [12]. Let $\Omega_j \subset \Omega$ be an increasing sequence of smooth domains with $\bigcup_j \Omega_j = \Omega$.

Let u_j denote the solution of $M u_j + 1 = 0$ in Ω_j, $u_{j\,|\,\partial\Omega_j} = 0$. Then u_j converges to a function u_0 which satisfies $M u_0 + 1 = 0$ in Ω and such that $u_0 \to 0$ at "good" points of the boundary (u_0 is independent of the chosen sequences of set Ω_j's). We say that

$$u \overset{u_0}{=} 0 \qquad (\text{respectively } u \overset{u_0}{\leqq} 0) \quad \text{on } \partial\Omega$$

if for any sequence $x_j \to \partial\Omega$ with $u_0(x_j) \to 0$, we have $u(x_j) \to 0$ (resp. $\limsup_{j \to \infty} u(x_j) \leqq 0$).

We say that the refined maximum principle "MP" holds for L in Ω if any function w which is bounded from above and such that $L w \geqq 0$ in Ω, $w \overset{u_0}{\leqq} 0$ on $\partial\Omega$ satisfies: $w \leqq 0$ in Ω.

THEOREM 1. — L *satisfies MP in* Ω *if and only if* $\lambda_1(L, \Omega) > 0$.

THEOREM 2. — *Suppose* $\lambda_1(L, \Omega) > 0$. *Then, for all* $f \in L^N(\Omega)$, *there exists a unique solution* $u \in L^\infty(\Omega)$ *of* $L u = f$ *in* Ω, $u \overset{u_0}{=} 0$ *on* $\partial\Omega$. *Furthermore, there is a constant* A *such that* $\|u\|_{L^\infty} \leqq A \|f\|_{L^N}$.

This result allows one to extend the classical Alexandrov-Bakelman-Pucci estimate (*see* [6] for the statement when $c \leqq 0$).

We say that a function ψ is an eigenfunction of $-L$ with eigenvalue λ if $\psi \not\equiv 0$, $\psi \in L^\infty(\Omega)$, and $(L + \lambda)\psi = 0$ in Ω, with $\psi \overset{u_0}{=} 0$ in $\partial\Omega$. In our next theorem, we show that the Krein-Rutman theory extends to this general framework.

THEOREM 3. — *There exists an eigenfunction* $\phi_1 > 0$ *associated to* λ_1. *The associated eigenspace has dimension* 1.

(ii) λ_1 *is the unique eigenvalue which has a positive eigenfunction.*

(iii) *For any eigenvalue* $\lambda \neq \lambda_1$, *one has* $\operatorname{Re} \lambda > \lambda_1$.

(iv) *There exists a constant* C *such that* $\phi_1 \leqq C u_0$. *In addition,* $u_0 \overset{\phi_1}{=} 0$ *on* $\partial\Omega$, *though, in general it is not true that there exists a constant* C *such that* $u_0 \leqq C \phi_1$.

Results related to Theorem 3 have been obtained by Agmon [1], Pinchover [8], Nussbaum and Pinchover [7].

In [3] we derive several further qualitative properties related to λ_1 and to the maximum principle MP. Let us just mention the following one concerning "narrow" domains.

THEOREM 4. — *Let* $d_0 > 0$ *be fixed and* $0 < \rho \leqq d_0$. *Assume that* $\exists\, \delta > 0$ *such that* $\forall\, x \in \Omega$, *there exists a ball* $B(x, \rho_x)$ *of radius* $\rho_x \leqq \rho$ *with*

$$|B(x, \rho_x) \cap \Omega^c| \geqq \delta\, |B(x, \rho_x)|.$$

Then, there exists a constant $\tau = \tau(N, c_0, C_0, b, d_0, \delta) > 0$ *such that*

$$\lambda_1(M, \Omega) \geqq \frac{\tau}{\rho^2}.$$

COROLLARY 5. — *There exists* $\eta > 0$ *depending only on* N, c_0, C_0 *and* b *such that if* meas $(\Omega) < \eta$, *then MP holds for* L *in* Ω.

This extends a similar result of Bakelman [2] where, however, η also depends on the diameter of Ω.

---

1. CADRE GÉNÉRAL ET VALEUR PROPRE PRINCIPALE D'UN OPÉRATEUR ELLIPTIQUE. — On considère un domaine borné $\Omega \subset \mathbb{R}^N$ sur lequel *aucune restriction* de régularité n'est imposée. Soit L un opérateur elliptique dans Ω de la forme (2) et qui vérifie (3).

Un de nos objectifs dans les énoncés qui suivent est de préciser dans les diverses estimations *a priori* la dépendance des constantes par rapport à l'opérateur L. En particulier, nous disons que C est une *constante structurelle* si C est une constante que l'on peut choisir uniquement à partir de Ω, N, c_0, C_0, b et non de l'opérateur L — ce que l'on écrira sous la forme $C = C(\Omega, N, c_0, C_0, b)$, ou en restreignant ou augmentant le nombre d'arguments dans les cas où c'est possible ou bien nécessaire. De ce point de vue, plusieurs de nos résultats sont nouveaux même dans le cas où $\partial\Omega$ est régulier.

Dans ce cadre général, on définit la *valeur propre principale* (on écrira dans la suite V.P.P.) de l'opérateur L en posant (1), c'est-à-dire

$$\lambda_1 = \sup\{\lambda;\ \exists \phi > 0, \text{ avec } (L+\lambda)\phi \leqq 0 \text{ dans } \Omega\} = -\inf_{\{\varphi > 0\}} \sup_{x \in \Omega} \{L\varphi(x)/\varphi(x)\}. \tag{1.1}$$

On convient dans toute la Note que l'opérateur L est appliqué à des fonctions de classe $W^{2,N}_{loc}$. On notera $\lambda_1 = \lambda_1(L, \Omega)$.

On sait que dans le cas classique, lorsque Ω est *régulier* et $a_{ij} \in C(\bar{\Omega})$, λ_1 coïncide avec la définition classique, à savoir la plus petite valeur propre de $-L$ avec condition de Dirichlet, qui est encore caractérisée par (Théorie de Krein-Rutman) l'existence de ϕ telle que :

$$\begin{cases} (L+\lambda_1)\phi = 0, & \phi > 0 \text{ dans } \Omega \\ \phi \in C(\bar{\Omega}), & \phi_{|\partial\Omega} = 0. \end{cases} \tag{1.2}$$

Le fait que, dans le cas régulier, λ_1 *vérifie la propriété* (1.1) a été prouvé par Donsker et Varadhan ([4], [5]). D'autres auteurs avaient obtenu des inégalités liées à (1.1) (*voir* [9], [10], [11] et références citées). Notons, qu'à l'inverse de ces travaux, (1.1) constitue ici *la définition* de λ_1.

L'objet de ce travail est de comprendre le rôle de λ_1 dans le cas général, d'éclairer le lien de λ_1 avec le Principe du Maximum, d'étendre à ce cadre la Théorie de Krein-Rutman, d'obtenir diverses propriétés qualitatives de λ_1 et estimations *a priori* en s'attachant à en dégager le caractère structurel.

Nous n'énonçons ici que quelques-uns des principaux résultats. On en trouvera les démonstrations détaillées dans [3] ainsi que divers autres résultats et généralisations.

2. PRINCIPE DU MAXIMUM ET V.P.P. — Il convient dans un premier temps de raffiner dans le cadre d'un domaine général la notion de Principe du Maximum. Pour cela, on utilise la construction d'une fonction u_0 due à Stroock et Varadhan [12]. Soit Ω_j une suite croissante de domaines réguliers $\Omega_j \subset \Omega$, $\bigcup \Omega_j = \Omega$. Soit u_j la solution de $M u_j + 1 = 0$ dans Ω_j, $u_{j\,|\,\partial\Omega} = 0$. Alors u_j converge (en croissant) vers une fonction u_0 vérifiant $M u_0 + 1 = 0$ dans Ω et qui s'annule sur une partie du bord (par exemple, en tout point $x_0 \in \partial\Omega$ tel qu'il existe un cône extérieur, on peut montrer que $\lim_{x \to x_0} u_0(x) = 0$); u_0 est indépendant du choix des Ω_j. Il est facile de voir que si Ω est régulier, alors $u_0 \in C(\bar{\Omega})$ et $u_{0\,|\,\partial\Omega} = 0$. Dans le cas général cependant u_0 ne s'annule pas sur tout $\partial\Omega$.

On a également une estimation utile

$$u_0 \leqq K(N, c_0, C_0, b|\Omega|^{1/N})|\Omega|^{2/N}$$

où $|\Omega|$ désigne la mesure de Ω.

Dans ce contexte, pour une suite de points telle que $(x_j) \subset \Omega$, on dit que $x_j \overset{u_0}{\to} \partial\Omega$ si $u_0(x_j) \to 0$.

On écrira $u \overset{u_0}{=} 0$ sur $\partial\Omega$ (*resp.* $u \overset{u_0}{\leqq} 0$) si pour toute suite $x_j \overset{u_0}{\to} \partial\Omega$, on a $u(x_j) \to 0$ [*resp.* $\limsup u(x_j) \leqq 0$].

Nous pouvons indiquer le sens généralisé du Principe du Maximum, noté PM. On dit qu'un opérateur L vérifie PM dans Ω, si pour toute fonction $w \in W^{2, N}_{loc}(\Omega)$ telle que

(2.1) $\qquad Lw \geqq 0$ dans Ω, $\quad w$ bornée supérieurement

(2.2) $\qquad \limsup w(x_j) \leqq 0, \quad \forall (x_j), \quad x_j \overset{u_0}{\to} \partial\Omega$

on a nécessairement $w \leqq 0$ dans Ω.

Théorème 1. — *Le Principe du Maximum PM est vérifié par* L *dans* Ω *si et seulement si* $\lambda_1 > 0$.

Lorsque $\lambda_1 > 0$ on peut résoudre le problème de Dirichlet :

Théorème 2. — *On suppose* $\lambda_1(L, \Omega) > 0$. *Pour tout* $f \in L^N(\Omega)$, *il existe une solution unique* $u \in L^\infty(\Omega)$ *de*

(2.3)
$$L u = f \quad \text{dans } \Omega$$
$$u \overset{u_0}{=} 0 \quad \text{sur } \partial\Omega.$$

De plus, il existe une constante $A = A(\Omega, c_0, C_0, b, \lambda_1)$ *telle que*

(2.4) $$\|u\|_{L^\infty} \leqq A \|f\|_{L^N}$$

On obtient également une généralisation de l'estimation L^∞ d'Alexandrov, Bakelman, Pucci (*voir* [6]) pour des opérateurs L avec un terme c de signe quelconque mais tels que $\lambda_1 > 0$. C'est l'objet de l'énoncé suivant.

Théorème 3. — *Soit* $\lambda_1 > 0$. *On suppose que* v *est bornée supérieurement et vérifie* $Lv \geqq f$ *dans* Ω, $f \leqq 0$, $f \in L^n(\Omega)$,

$$\limsup v(x_j) \leqq \beta, \quad \forall x_j \overset{u_0}{\to} \partial\Omega \quad (\textit{avec } \beta \geqq 0)$$

Alors

$$\sup_\Omega v \leqq A(\|f\|_{L^N} + \beta(\sup c^+)|\Omega|^{1/n})$$

où A *est la constante dans* (2.4).

3. Extension de la Théorie de Krein-Rutman à des domaines généraux. — On dit qu'une fonction $\psi \in W^{2, N}_{loc}(\Omega)$ est une fonction propre de $-L$ associée à la valeur propre λ si $\psi \not\equiv 0$, $|\psi|$ est bornée et ψ vérifie

$$(L + \lambda)\psi = 0 \quad \text{dans } \Omega, \qquad \psi \overset{u_0}{=} 0 \quad \text{sur } \partial\Omega.$$

Théorème 4. — *Il existe une fonction propre* (*état fondamental*) $\phi_1 > 0$ *associée à* λ_1 *satisfaisant*

$$(L + \lambda_1)\phi_1 = 0.$$

Il existe une constante structurelle $C>0$ *telle que* $\phi_1 \leqq C u_0$. *De plus,* $u_0 \overset{\phi_1}{=} 0$ *sur* $\partial\Omega$.

Dans [3], on construit un contre-exemple montrant que l'on n'a pas nécessairement $u_0 \leqq \text{Cte}\, \phi_1$.

Théorème 5. — (i) *Si* ψ *est une fonction propre de* $-L$ *associée à* λ_1, *alors* $\psi = \sigma \phi_1$ *pour une constante* σ.

(ii) *Si* ψ *est une fonction propre associée à* λ, *telle que* $\psi > 0$ *dans* Ω, *alors* $\lambda = \lambda_1$ *et* $\psi = \sigma \phi_1$ *(*σ *constante).*

(iii) *Si* $\lambda \neq \lambda_1$ *est valeur propre, alors* $\operatorname{Re} \lambda > \lambda_1$.

(iv) *Il n'existe pas de fonction* ψ *telle que*

$$(L+\lambda_1)\psi = \phi_1 \quad \textit{dans } \Omega, \qquad \psi \overset{u_0}{\leqq} 0 \quad \textit{sur } \partial\Omega.$$

On trouvera dans [3] des résultats plus précis et des généralisations. Les deux théorèmes montrent donc que la théorie de Krein-Rutman se généralise intégralement au cadre d'un domaine Ω quelconque. On remarquera en particulier que nous n'avons pas recours au lemme de Hopf.

Remarque. — L'existence de ϕ_1 dans le théorème 4 s'obtient à l'aide de l'inégalité de Harnack dans la version générale de Krylov et Safonov (*voir* [6]) et de l'estimation d'Alexandrov, Bakelman et Pucci (*ibid.*). Un résultat dans le même esprit que le théorème 4 a été montré avec la définition

$$\lambda_1(\Omega) \equiv \inf_{\Omega_0 \subset \Omega} \lambda_1(\Omega_0)$$

par Agmon [1]. Dans [1], Agmon considère également des variétés et des domaines non bornés. Agmon [1] obtient d'autres résultats liés à certains résultats de [3]. On trouvera des travaux reliés au théorème 4 dans Pinchover [8]. Diverses autres propriétés de l'état fondamental ont également été étudiées par Nussbaum et Pinchover [7].

On sait dans le cas régulier que le Principe du Maximum est vérifié s'il existe une fonction $g>0$ dans $\bar{\Omega}$ telle que $Lg \leqq 0$. Cette propriété s'étend au cas général.

Théorème 6. — *Soit* u *une fonction vérifiant* $u>0$ *et* $Lu \leqq 0$ *dans* ω. *Alors, ou bien* $\lambda_1 > 0$, *ou bien* $\lambda_1 = 0$ *et* u *est un multiple constant de* ϕ_1.

4. Propriétés qualitatives. — Le résultat suivant étend au cadre général des propriétés classiques de la dépendance de λ_1 par rapport à c.

Proposition 7. — *Dans sa dépendance par rapport au coefficient* c, λ_1 *est une fonction concave et lipschitzienne de rapport* 1 *(pour la norme* L^∞ *sur* c*). De plus, si* $\tilde{c} \geqq c$, $\tilde{c} \neq c$, *alors* $\lambda_1(M+c) > \lambda_1(M+\tilde{c})$.

Le théorème suivant est un résultat de monotonie en *un sens uniforme* pour la dépendance de λ_1 par rapport au domaine Ω.

Théorème 8. — *Soit* Γ *un sous-ensemble fermé de* Ω *avec* $|\Gamma| \geqq \delta_0 > 0$. *Alors il existe* $\bar{\eta} = \bar{\eta}(c_0, C_0, b, \Omega, \delta_0)$ *tel que*

$$\lambda_1(\Omega \setminus \Gamma) - \lambda_1(\Omega) \geqq \bar{\eta} > 0.$$

$|\Gamma|$ désigne la mesure de Γ.

Des variantes et des versions plus générales de ce résultat sont indiquées dans [3]. Le résultat suivant fournit une borne inférieure de λ_1, lorsque le domaine est *étroit*.

THÉORÈME 9. — *Soient* $d_0>0$ *fixé et* $0<\rho\leqq d_0$. *On suppose qu'il existe* $\delta>0$ *tel que pour tout* $x\in\Omega$, *il existe une boule* $B(x, \rho_x)$ *de rayon* $\rho_x\leqq\rho$ *telle que*

$$|B(x, \rho_x)\cap\Omega^c|\geqq\delta|B(x, \rho_x)|.$$

Il existe une constante positive τ *dépendant de* N, c_0, C_0, b, d_0 *et telle que*

$$\lambda_1(M, \Omega)\geqq\frac{\tau}{\rho^2}.$$

On peut dire que ρ est un paramètre indiquant l'étroitesse (en mesure) de Ω. En effet, plus ρ est petit, plus le domaine Ω est étroit. Une conséquence du théorème 9 est que pour un opérateur L donné, le Principe du Maximum PM est vérifié dès que le domaine est suffisamment « étroit ».

COROLLAIRE 10. — *Le Principe du Maximum est vérifié par* L *dans* Ω *dès que* $|\Omega|<\eta$ *où* η *ne dépend que de* N, c_0, C_0 *et* b.

Remarque. — On ne sait pratiquement rien sur les autres valeurs propres. Nous conjecturons l'existence d'une infinité de valeurs propres, dans le cadre général.

Note remise le 12 octobre 1992, acceptée le 12 novembre 1992.

RÉFÉRENCES BIBLIOGRAPHIQUES

[1] S. AGMON, On positivity and decay of solutions of second order elliptic equations on Riemannian manifolds. *Methods of functional analysis and theory of elliptic equations*, D. GRÉCO éd., Liguori Napoli, 1983, p. 19-52.

[2] I. YA BAKELMAN, *Convex functions and elliptic equations*, manuscript of book.

[3] H. BERESTYCKI, I. NIRENBERG et S. R. S. VARADHAN, The principal eigenvalue and maximum principle for second order elliptic operators in general domains, *Comm. Pure Applied Math.* (à paraître).

[4] M. DONSKER et S. R. S. VARADHAN, On a variational formula for the principal eigenvalue for operators with maximum principle, *Proc. Nat. Acad. Sci. U.S.A.*, 72, 1975, p. 780-783.

[5] M. DONSKER et S. R. S. VARADHAN, On the principal eigenvalue of second-order elliptic differential operators, *Comm. Pure Appl. Math.*, 29, 1976, p. 595-621.

[6] D. GILBARG et N. S. TRUDINGER, *Elliptic partial differential equations of second order*, Grundlehren der math. Wiss., Springer Verlag, 2e éd., 1983.

[7] R. D. NUSSBAUM et Y. PINCHOVER, *On variational principles for the generalized principal eigenvalue of second order elliptic operators and some applications*, Preprint.

[8] Y. PINCHOVER, Large scale properties of multiparameter oscillation problems, *Comm. Part. Diff. Equ.*, 15, 1990, p. 647-673.

[9] M. H. PROTTER et H. W. WEINBERGER, On the spectrum of general second order operators, *Bull. A.M.S.*, 72, 1966, p. 251-255.

[10] M. H. PROTTER et H. W. WEINBERGER, *Maximum principle for differential equations*, Prentice-Hall, N.J., 1967.

[11] C. PUCCI, Maximum and minimum first eigenvalues for a class of elliptic operators, *Proc. A.M.S.*, 17, 1966, p. 788-795.

[12] D. STROOCK et S. R. S. VARADHAN, On degenerate elliptic-parabolic operators of second order and their associated diffusions, *Comm. Pure Appl. Math.*, 25, 1972, p. 651-713.

H. B. : *Laboratoire d'Analyse numérique, Université Pierre-et-Marie-Curie, 4, place Jussieu, 75252 Paris Cedex 05;*

L. N. et S. R. S. V. : *Courant Institute of Mathematical Sciences, N.Y.U., 251, Mercer Street, New York, NY 10012.*

Comm. Pure Appl. Math. 47 (1994) 47-92

The Principal Eigenvalue and Maximum Principle for Second-Order Elliptic Operators in General Domains

H. BERESTYCKI
Université de Paris VI

L. NIRENBERG AND S. R. S. VARADHAN
Courant Institute

Dedicated to Henry McKean.

Chapter 1. Introduction and Main Results

1. Introduction

(a) The maximum principle. Let L be an elliptic operator in a bounded domain $\Omega \subset \mathbb{R}^n$, of the form

$$L = M + c(x) = a_{ij}(x)\partial_{ij} + b_i(x)\partial_i + c(x) \; ; \tag{1.1}$$

we will only consider uniformly elliptic ones, i.e., satisfying, for some positive constants c_0, C_0,

$$c_0|\xi|^2 \leqq a_{ij}(x)\xi_i\xi_j \leqq C_0|\xi|^2 \quad \forall \; \xi \in \mathbb{R}^n \; . \tag{1.2}$$

In addition we will always assume that

$$\begin{gathered} a_{ij} \in C(\Omega) \, , \quad b_i \, , \quad c \in L^\infty \, , \\ (\textstyle\sum b_i^2)^{\frac{1}{2}} \, , \quad |c| \leqq b \end{gathered} \tag{1.3}$$

for some constant $b \geqq 0$.

This paper is concerned with the maximum principle, and the existence of a principal eigenvalue and eigenfunction, for such operators. We first recall a classical form of the maximum principle:

DEFINITION. We say that the maximum principle holds for the operator L in Ω if

$$Lw \geqq 0 \quad \text{in } \Omega \tag{1.4}$$

and

$$\limsup_{x \to \partial\Omega} w(x) \leqq 0 \tag{1.5}$$

imply $w \leqq 0$ in Ω.

To be precise, we will assume throughout the paper that all functions w on which L acts belong to $W^{2,n}_{\text{loc}}(\Omega)$. In fact, in almost all cases, they will belong to $W^{2,p}_{\text{loc}}(\Omega)$ for all finite $p > n$.

We recall also three well-known conditions, each of which is sufficient for the maximum principle to hold. Indeed, these are classical in case the functions w satisfying (1.4), (1.5) are in $C^2(\Omega)$; for w in $W^{2,n}_{\text{loc}}$, see [7], [13], [4], and Chapter 9 in [10]. The sufficient conditions are:

(i) $c(x) \leqq 0$ in Ω,

(ii) There is a function $g \in C(\overline{\Omega})$ which is positive in $\overline{\Omega}$ and satisfies $Lg \leqq 0$.

(iii) Ω is narrow in, say, the x_1-direction. More precisely, there is an $\varepsilon > 0$ depending only on c_0, C_0 and b, such that Ω lies in a narrow band $a < x_1 < a + \varepsilon$.

The proof that (ii) is sufficient follows from (i) by considering the function $z = w/g$. A simple calculation shows that

$$\left(M' + \frac{Lg}{g}\right) z \geqq 0 \tag{1.6}$$

where $M' = M + 2a_{ij}g_i\partial_j$. Since $Lg/g \leqq 0$, we may apply the sufficient condition (i) to conclude that z, and hence w, are nonpositive. That condition (iii) is sufficient, is easily proved by constructing a function $g = g(x_1)$ in $C(\overline{\Omega})$ satisfying $Lg \leqq 0$. Then condition (ii) yields sufficiency.

Good general references for various forms of the maximum principle are [17] and [10].

In case the coefficients of L, and also $\partial\Omega$, have some degree of regularity, the maximum principle as formulated above, holds if and only if the principal eigenvalue λ_1 of $-L$, for the Dirichlet problem, is positive. Recall that a number λ, and function ψ (both possibly complex), are, respectively, an eigenvalue and eigenfunction for the Dirichlet problem for $-L$, if

$$\begin{aligned} (L+\lambda)\psi &= 0 \text{ in } \Omega \\ \psi &= 0 \text{ in } \partial\Omega\,. \end{aligned} \tag{1.7}$$

In case the operator L is self adjoint (with respect to $L^2(\Omega)$), i.e., L has the form

$$L = \partial_i a_{ij}(x)\partial_j + c(x)\,, \tag{1.8}$$

the eigenvalues and eigenfunctions are always real, but this is not necessarily true for L of the general form (1.1). In case the coefficients, and $\partial\Omega$, are smooth, however, there is a principal eigenvalue λ_1 and associated (principal) eigenfunction

ϕ_1 which are real and satisfy:

$$\text{(1.9)}\quad \begin{aligned}&\text{(a)}\ \ \phi_1 \text{ is a simple eigenfunction, i.e., } \phi_1 \text{ spans } \ker(L+\lambda_1) \text{ under}\\ &\quad\ \ \text{Dirichlet boundary conditions; furthermore, } \phi_1 > 0 \text{ in } \Omega,\\ &\text{(b)}\ \ \text{if } \psi \text{ is a positive eigenfunction with eigenvalue } \lambda,\\ &\quad\ \ \text{then } \lambda = \lambda_1;\\ &\text{(c)}\ \ \text{for any eigenvalue } \lambda,\ \operatorname{Re}\lambda \geqq \lambda_1 .\end{aligned}$$

The existence of such a pair (λ_1, ϕ_1) is usually proved with the aid of the Krein-Rutman Theorem (see, for example, [19], Sections VIII-3 and VIII-7), and the proof requires $\partial\Omega$ to be smooth. Very recently, using some results of this paper, I. Birindelli, [6], was able to use the Krein-Rutman Theorem (in a modified form) to prove the existence of (λ_1, ϕ_1) in the general case, thereby giving an alternate proof of Theorem 2.1 below.

In a recent paper, [5], the authors relied on improved sufficient conditions (see Proposition 1.1 below) for the maximum principle to hold — in particular when Ω has irregular boundary. It seemed natural therefore to find corresponding principal eigenvalue and eigenfunction (λ_1, ϕ_1) for *any* bounded domain (i.e., open connected set) in $\mathbb{R}^n$, and to study the relationship with the maximum principle.

(b) The principal eigenvalue. In this paper we define such a pair (λ_1, ϕ_1) for any bounded domain Ω and we show that they have properties (1.9). The eigenvalue λ_1 is defined by a min-max expression. (For L of the form (1.8) one has of course the well-known expression

$$\lambda_1 = \inf \int_\Omega (a_{ij}(x) u_i u_j + c(x) u^2)\, dx$$

where the inf is taken over all functions in $C_0^\infty(\Omega)$ with L^2 norm equal to 1.)

In addition, for any bounded domain Ω, we formulate a refined version of the maximum principle above and prove (Theorem 1.1) that the refined maximum principle holds for $L \Leftrightarrow \lambda_1 > 0$.

In Section 5 we study the dependence of λ_1 on Ω and on the coefficients b_i and c: λ_1 strictly decreases if we enlarge Ω; λ_1 is decreasing and concave in its dependence on the coefficient $c(x)$, and is Lipschitz continuous with respect to c, in the L^∞ norm, with Lipschitz constant 1; λ_1 is also locally Lipschitz continuous with respect to the b_i — in the L^∞ norm. In later sections we obtain lower bounds for λ_1 in various situations.

An essential point in this paper is that in *all our estimates, the constants depend only on the constants c_0, C_0 and b, as well as on the domain Ω*. Consequently, in this aspect, many of our results are new even when $\partial\Omega$ and the coefficients of L are smooth.

Here is our definition of λ_1:

$$\text{(1.10)}\qquad \lambda_1 = \sup\{\lambda \mid \exists \phi > 0 \text{ in } \Omega \text{ satisfying } (L+\lambda)\phi \leqq 0\} .$$

Clearly

$$\lambda_1(\Omega) \leqq \lambda_1(\Omega') \text{ if } \Omega \supset \Omega' . \tag{1.11}$$

As we will see, strict inequality holds if $\Omega \supsetneqq \Omega'$; in fact Theorem 2.4 is a much stronger result. We point out below that if L has smooth coefficients, and if $\partial\Omega$ is smooth, then λ_1 defined by (1.10) is indeed the principal eigenvalue of L. It is easy to derive from (1.10) a min-max characterization of λ_1. Note first that if $\phi > 0$ and $(L + \lambda)\phi \leqq 0$, then

$$\lambda \leqq \inf_{\Omega} \left(-\frac{L\phi}{\phi} \right) = -\sup_{\Omega} \frac{L\phi}{\phi} .$$

Consequently

$$\lambda_1 = \sup_{\phi} \inf_{\Omega} \left(-\frac{L\phi}{\phi} \right) = \sup_{\phi} \left(-\sup_{\Omega} \frac{L\phi}{\phi} \right), \tag{1.12}$$

where in $\sup_\phi$, the supremum is taken over all positive functions ϕ in $W^{2,n}_{\text{loc}}(\Omega)$. (1.12) may be written as

$$\lambda_1 = -\inf_{\phi} \sup_{\Omega} \frac{L\phi}{\phi} . \tag{1.13}$$

It is useful to rewrite this in another form, as follows: Any positive $\phi \in W^{2,n}_{\text{loc}}(\Omega)$ can be written as $\phi = e^z$ with $z \in W^{2,n}_{\text{loc}}$. Then

$$\frac{L\phi}{\phi} = Mz + a_{ij}z_i z_j .$$

Let A denote the set of functions $z \in W^{2,p}_{\text{loc}}(\Omega)$ such that

$$F(z)(x) := Mz + a_{ij}z_i z_j$$

is bounded from above in Ω. Since F is convex in z, the set A is convex. Then

$$\begin{aligned} \lambda_1 &= \sup_{z\in A}\{-\sup_\Omega[F(z) + c]\} \\ &= -\inf_{z\in A} \sup_{x\in\Omega}[F(z) + c(x)] . \end{aligned} \tag{1.14}$$

We will sometimes denote λ_1 by $\lambda_1(\Omega)$, $\lambda_1(L, \Omega)$, or $\lambda_1(c)$ if M and Ω are fixed and the coefficient c is allowed to vary.

In [8]–[9] Donsker and Varadhan gave a variational formula for the principal eigenvalue of a general operator with maximum principle. In [9] (pp. 598–599) they showed, furthermore, that for L of the form (1.1) with continuous coefficients in a bounded domain Ω with smooth boundary, the principal eigenvalue λ_1 is given by (1.12) and (1.13). If $\partial\Omega$ is smooth, a different proof of (1.13), where, however,

the functions ϕ are required to be positive in $\bar{\Omega}$, was given in [22]; it makes use of earlier results of M. Chicco.

We are taking (1.10), i.e., (1.12)–(1.14), as *our definition* of λ_1 and we will construct a corresponding principal eigenfunction ϕ_1.

In the course of the paper, various bounds on λ_1 will be established. In earlier work, a number of authors have obtained bounds on λ_1 — in case $\partial\Omega$ is smooth. Already in 1937, Barta, [3], showed that in case $L = \Delta$, the Laplace operator, for any positive C^2 function ϕ in Ω,

$$\lambda_1 \geqq \inf_{\Omega} \left(-\frac{L\phi}{\phi} \right).$$

This result was then extended to other self adjoint operators (see [17], p. 92, for other references). Protter and Weinberger, [16], proved Barta's result for general operators (1.1) — even under more general boundary conditions, and also in unbounded domains. They showed that for any eigenvalue λ, not necessarily real, and any positive $\phi \in C^2(\Omega)$

$$\text{Re}\,\lambda \geqq \inf_{\Omega} \left(-\frac{L\phi}{\phi} \right). \tag{1.15}$$

Through a not very direct proof, they also established an upper bound for λ_1: If $q \in C^2(\Omega) \cap C(\overline{\Omega})$, $q > 0$ in Ω, $q = 0$ in $\partial\Omega$, then

$$\lambda_1 \leqq \sup_{\Omega} \left(-\frac{Lq}{q} \right). \tag{1.15$'$}$$

From (1.12) it is, in fact, very easy to obtain an upper bound on $\lambda_1(\Omega)$ in case Ω contains a ball B_R of radius R.

LEMMA 1.1. *Suppose $B_R = \{|x| < R\}$ lies in Ω, with $R \leqq 1$. Then*

$$\lambda_1(\Omega) \leqq \frac{C}{R^2}$$

where C depends only on n, c_0, C_0 and b.

Proof: With $r = R/2$, $B_r = \{|x| < r\}$, consider the function

$$\sigma = \frac{1}{4}(r^2 - |x|^2)^2 . \tag{1.16}$$

Direct calculation shows that

$$\sup_{\Omega} \left(-\frac{L\sigma}{\sigma} \right) \leqq \frac{C}{R^2} . \tag{1.17}$$

Namely

$$\sigma_i = -x_i(r^2 - |x|^2) , \qquad \sigma_{ij} = -\delta_{ij}(r^2 - |x|^2) + 2x_i x_j ,$$

and hence

$$
\begin{aligned}
-\frac{L\sigma}{4\sigma} &= \frac{1}{r^2-|x|^2}\sum a_{ii} - \frac{2a_{ij}x_ix_j}{(r^2-|x|^2)^2} + \frac{b_ix_i}{r^2-|x|^2} - c \\
&\leqq \frac{nC_0+br}{r^2-|x|^2} - \frac{2c_0|x|^2}{(r^2-|x|^2)^2} + b\,. \qquad (1.18)
\end{aligned}
$$

In the region where

$$|x|^2(2c_0+nC_0+br) > r^2(nC_0+br)\,,$$

the second term on the right of (1.18) dominates the first, and hence $-L\sigma/\sigma \leqq b$. In the remainder of B_r,

$$
\begin{aligned}
-\frac{L\sigma}{4\sigma} &\leqq \frac{nC_0+br}{r^2-|x|^2} + b \\
&\leqq b + \frac{1}{2c_0r^2}(nC_0+br)\,(nC_0+br+2c_0)\,.
\end{aligned}
$$

Thus (1.17) holds.

To prove the lemma it suffices to show that if for some positive function ϕ and some λ, $(L+\lambda)\phi \leqq 0$, then $\lambda < \sup(-L\sigma/\sigma)$. Consider such a pair ϕ, λ, and suppose $\lambda \geqq \sup(-L\sigma/\sigma) = \tau$. Then $(L+\tau)\sigma \geqq 0$ in B_r while $(L+\tau)\phi \leqq (L+\lambda)\phi \leqq 0$. By the sufficient condition (ii), the maximum principle holds for $L+\tau$ in B_r. But then, since $\sigma = 0$ on ∂B_r, this would imply that $\sigma \leqq 0$ in B_r. This is impossible; the lemma is proved.

Corollary 1.1 below is a stronger form of (1.15)$'$.

For operators of the form (1.1), with $b_i = c = 0$, and for Ω a ball, Pucci, [18], gave sharp upper and lower bounds for λ_1.

Recently, we learned that R. D. Nussbaum and Y. Pinchover introduced in [15] a definition of principal eigenvalue for an operator L in terms of a positive weight function W, i.e., they consider $L+\lambda W$, in any domain Ω (not necessarily bounded) in $\mathbb{R}^n$. Their definition is

$$\lambda_0 = \sup\{\lambda \in \mathbb{R} \,|\, \exists\, u \in C^2(\Omega),\ u > 0 \text{ and } (L+\lambda W)u = 0\}\,.$$

They show that this is the usual principal eigenvalue in case Ω is a bounded domain with smooth boundary, and they compare this with other variational formulas for the principal eigenvalue. [15] also contains other results and references to earlier work on related topics. If Ω is a bounded domain, and $W \equiv 1$, their λ_0 coincides with our λ_1. They do not prove the existence of a principal eigenfunction.

(c) More on the maximum principle; the refined form. This paper was initiated by the following observation which was first noted by Bakelman in [2] and used extensively in [5].

PROPOSITION 1.1. *Let L satisfy conditions* (1.2) *and* (1.3) *in* Ω. *Assume* diam $\Omega \leqq d$. *There exists $\delta > 0$ depending only on n, c_0, b, and d such that the maximum*

principle holds for L in Ω if the measure of Ω, $|\Omega|$, satisfies

$$|\Omega| < \delta . \tag{1.19}$$

In the proof, which we include here, concerning c, we will only require the one-sided inequality

$$c \leqq b . \tag{1.3$'$}$$

The proof is a direct consequence of a basic inequality of Alexandroff, Bakelman, Pucci. We shall use it in the following form:

THEOREM ABP. *If $c \leqq 0$ and w satisfies $Lw \geqq f$ in Ω, and* (1.5):

$$\limsup_{x \to \partial\Omega} w(x) \leqq 0 ,$$

then

$$\sup_{\Omega} w \leqq B\|f\|_{L^n} . \tag{1.20}$$

Here the constant B depends only on n, c_0, b, and d.

In Theorem 9.1 in [10" this result is proved in a more general form. It is assumed there that $w \in C(\overline{\Omega})$, but it is also pointed out that the proof works just under condition (1.5). Theorem 2.5 below is an extension of Theorem ABP to operators with $\lambda_1 > 0$ and without any condition on the sign of c.

Proof of Proposition 1.1: Consider a function w satisfying (1.4) and (1.5). Writing (1.4) in the form, here $c = c^+ - c^-$,

$$(M - c^-)w \geqq -c^+ w^+$$

we may apply Theorem ABP and infer from (1.20) that

$$\sup_{\Omega} w \leqq Bb \sup_{\Omega} w^+ |\Omega|^{\frac{1}{n}}$$

and it follows that $w \leqq 0$ in Ω in case, say,

$$Bb\delta^{\frac{1}{n}} = \frac{1}{2} . \tag{1.21}$$

In Theorem 2.6 we give a much stronger form of Proposition 1.1.

Now we come to our refinement of the maximum principle. In an arbitrary bounded domain Ω, the condition (1.5) in the maximum principle is too strong

a hypothesis; in general one cannot prescribe boundary values, or know precise behaviour, of solutions at *every* point of $\partial\Omega$. In place of (1.5) we will work with a refined version which goes back to [21]. It makes use of a solution $u_0 > 0$ in Ω of $Mu_0 = -1$, which vanishes, in a suitable sense, on $\partial\Omega$. In place of (1.5) we will require that for every sequence $x_j \to \partial\Omega$ such that $u_0(x_j) \to 0$,

$$\limsup w(x_j) \leqq 0 .$$

In general we will use the following

Notation. Given $u \in C(\Omega)$, for a sequence $x_j \to \partial\Omega$ we use the notation

$$x_j \overset{u}{\to} \partial\Omega \tag{1.22}$$

to mean $u(x_j) \to 0$. We also say that

$$u \overset{u_0}{=} 0 \quad \text{on} \quad \partial\Omega , \tag{1.22$'$}$$

if $x_j \overset{u_0}{\to} \partial\Omega$ implies $u(x_j) \to 0$. The function u_0 plays an essential role in our considerations. In Section 3 we will construct it and study its behaviour near the boundary.

REFINED MAXIMUM PRINCIPLE. Using u_0 we now formulate the refined maximum principle — for want of a better name. It concerns functions which are bounded from above.

DEFINITION. We say that L satisfies the *refined maximum principle* in Ω if

$$Lw \geqq 0 \quad \text{in} \quad \Omega , \; w \text{ bounded above} , \tag{1.23}$$

and

$$\limsup w(x_j) \leqq 0 \quad \text{if} \quad x_j \overset{u_0}{\to} \partial\Omega , \tag{1.24}$$

imply $w \leqq 0$ in Ω.

Remark 1.1. The requirement in (1.23) that w be bounded from above cannot be omitted. For example, suppose $L = \Delta$ and

$$\Omega = B_1 \setminus \{0\} \quad \text{in } \mathbb{R}^3 ,$$

where B_1 is the unit ball. The function u_0 is easily seen to be

$$u_0 = \frac{1}{6}(1 - |x|^2) ;$$

thus u_0 is not sensitive to the origin. On the other hand, the function $w = |x|^{-1} - 1$ satisfies the conditions (1.23), (1.24), except for the boundedness from above, and it is positive.

We now state a few of the main results of the paper — showing that the use of the notion $x_j \overset{u_0}{\to} \partial\Omega$ is not unnatural. In these results no assumption is made about the sign of c.

THEOREM 1.1. *The refined maximum principle holds for $L \Leftrightarrow \lambda_1(L,\Omega) > 0$.*

This, and Corollary 2.2 below, yield a much more general form of (1.15)′:

COROLLARY 1.1. *Let q be bounded above, $q \not\equiv 0$, in Ω, and satisfy for some constant τ,*

$$\begin{aligned}(L+\tau)q &\geqq 0 \quad in\ \Omega\\ \limsup q(x_j) &\leqq 0 \quad if\ x_j \overset{u_0}{\longrightarrow} \partial\Omega\,.\end{aligned}$$

If, in addition, q is non-negative somewhere in Ω then $\lambda_1(L) \leqq \tau$. Furthermore, if q vanishes somewhere in Ω, then $\lambda_1 < \tau$.

Proof: If $\lambda_1 > \tau$ then, by Theorem 1.1, the refined maximum principle holds for $L+\tau$ — which implies $q \leqq 0$ in Ω. Since $q = 0$ somewhere in Ω, by a well-known form of the maximum principle, $q \equiv 0$; impossible. If $\lambda_1 = \tau$ then we obtain a contradiction with the aid of Corollary 2.2.

THEOREM 1.2. *Assume $\lambda_1(L,\Omega) > 0$. Then, given $f \in L^n(\Omega)$, there is a unique solution u, with $|u|$ bounded, of*

$$\begin{aligned}Lu &= f \quad in\ \Omega\\ u &\overset{u_0}{=} 0 \quad on\ \partial\Omega\,.\end{aligned} \tag{1.25}$$

Furthermore, for a constant A depending only on Ω, c_0, C_0, b, and λ_1

$$\|u\|_{L^\infty} \leqq A\|f\|_{L^n}\,. \tag{1.26}$$

Theorems 1.1 and 1.2 yield an extension of Theorem ABP.

THEOREM 1.3 (REFINED ABP). *Assume $\lambda_1 > 0$. Suppose v is a function, bounded from above, satisfying*

$$Lv \geqq f \quad in\ \Omega\,, \quad f \leqq 0\,, \quad f \in L^n(\Omega)\,,$$

$$\limsup v(x_j) \leqq \beta \quad if \quad x_j \overset{u_a}{\to} \partial\Omega\,, \quad \beta \geqq 0\,.$$

Then

$$\sup v \leqq A(\|f\|_{L^n} + \beta(\sup c^+)|\Omega|^{1/n}) + \beta\,. \tag{1.27}$$

where A is the constant in (1.26).

Plan of the paper. In Section 2 we state some of the other principal results of the paper and prove several corollaries. In Section 3, the function u_0 is constructed and its behaviour near the boundary is studied. It is shown that u_0 can be extended as a continuous function to all boundary points "admitting a strong barrier" — with $u_0 = 0$ at those points. In addition, it is shown (Lemma 3.1) that the notion $x_j \overset{u_0}{\to} \partial\Omega$ depends only on $\Omega \cap$ any neighbourhood of $\partial\Omega$.

In Section 4 we prove the existence of a principal eigenfunction ϕ_1, and establish some of its properties, in particular $\phi_1 \leqq Cu_0$. Section 5, as we said earlier, is devoted to proving some simple facts concerning the dependence of λ_1 on Ω, c, and the b_i. In addition, in Remark 5.1 we show that in general u_0 is not bounded by a constant ϕ_1. But we establish in Theorem 5.1 that

$$u_0 \overset{\phi_1}{=} 0 \quad \text{on} \quad \partial\Omega .$$

Section 6 is a key section; we prove several results in case $\lambda_1 > 0$, including Theorems 1.1–1.3.

In Section 7 we prove Theorem 2.2, using results of Section 6. Corollary 2.5 and Theorem 2.3 are proved in Section 8. In Section 9, Theorem 2.4 is obtained as an immediate corollary of a more general result, Proposition 9.3. Some variants of that proposition are also given. Finally in Section 10 we present the proof of Theorem 2.5.

The following remark, which is easily proved by an indirect argument, will be used repeatedly.

Remark 1.2. If w satisfies (1.24) then $\forall\, \varepsilon > 0$, $\exists\, \delta > 0$ such that

$$w(x) \leqq \varepsilon \quad \text{if } x \in \Omega , \quad \text{and} \quad u_0(x) < \delta .$$

If also $w \leqq N$, then

$$w(x) \leqq \varepsilon + \frac{N}{\delta} u_0(x) \quad \text{in } \Omega . \tag{1.28}$$

It has been pointed out to us that the paper by Agmon, [1", already contains some of the ideas developed here. It deals, however, with a slightly different class of operators L on a manifold M: L is written in semi-divergence form:

$$L = \partial_i(a_{ij}(x))\partial_j + \tilde{b}_i(x)\partial_i + c(x) ,$$

and it acts on functions $u \in H^1_{\text{loc}}(\Omega)$, where Ω is any open connected set in M — in some results, connectivity is not required. In case $\overline{\Omega}$ is compact, λ_1 is defined as the infimum of Re λ, for all λ in the spectrum of $-L$. (The proof of this is different from the proof of the corresponding result, Theorem 2.3 here.) Furthermore, it is shown that the maximum principle for functions in H^1_0 holds $\Leftrightarrow \lambda_1 > 0$ — as in our Theorem 1.1.

Results are also derived for noncompact domains Ω, e.g., a domain in $\mathbb{R}^n$. In particular, the analogue of λ_1 is defined as

$$\inf_{\Omega_0 \subset\subset \Omega} \lambda_1(\Omega_0) ,$$

and connections concerning the spectrum of L on Ω, and the existence of positive solutions of $Lu = 0$ are presented. A notion of positive solutions of $Lu = 0$ on $\Omega \backslash K$, for some compact K in Ω, "having minimal growth at infinity" is introduced and a related notion of ground state, namely a positive solution in Ω of $(L+\lambda)u = 0$ having minimal growth at infinity. Our Theorem 2.2 and its corollaries (and their proofs) have some relationship with Theorem 5.2 and Lemma 5.3 in [1] (and their proofs). An interesting result is Theorem 5.4 which asserts that there is a positive ground state (with eigenvalue λ) $\Leftrightarrow$ $L + \lambda$ does not admit a positive fundamental solution in Ω with pole at some given point in Ω. [1] contains still further results.

Added March 28, 1993. We wish to thank E. Calabi and M. Gromov for a suggestion which improved the formulation of Theorem 2.5.

2. More Principal Results

We say that ψ is an eigenfunction of $-L$, with eigenvalue λ (both possibly complex) if $\psi \not\equiv 0$, $|\psi|$ is bounded, and

$$\begin{aligned} (L + \lambda)\psi &= 0 \quad \text{in } \Omega , \\ \psi &\overset{u_0}{=} 0 \quad \text{on } \partial\Omega . \end{aligned}$$

We begin with the existence of a principal eigenfunction.

THEOREM 2.1.

(i) *There exists a positive function* ϕ_1 *in* $W^{2,p}_{\text{loc}}(\Omega)$, $\forall\, p < \infty$, *called a principal eigenfunction, satisfying*

$$(L + \lambda_1)\phi_1 = 0 .$$

(ii) *If we normalize* ϕ_1 *so that* $\phi_1(x_0) = 1$ *at a fixed point* $x_0 \in \Omega$, *then* $\phi_1 \leqq k$ *in* Ω *where* k *is a constant depending only on* x_0, Ω, c_0, C_0, *and* b.

(iii) ϕ_1 *is bounded by a constant times the function* u_0.

Clearly the results of Theorem 2.1 hold for the problem with a weight $W \in C(\Omega)$ which is bounded above and below by positive constants: $(L + \lambda W)\phi = 0$ in Ω.

As we will see, (iii) will imply that ϕ_1 can be extended as a continuous function to every point $y \in \partial\Omega$ where $\partial\Omega$ satisfies an exterior cone condition (see Section 3) and $\phi_1(y) = 0$. In the proof we will also see that λ_1 is the limit of the principal eigenvalues μ_j for a sequence $\Omega_j \subset \Omega_{j+1}$, with smooth boundaries, and $\cup\Omega_j = \Omega$.

That (ii) holds seems surprising.

THEOREM 2.2.[1] *Assume there is a function* $u > 0$ *in* Ω *satisfying* $Lu \leqq 0$. *Assume also that the refined maximum principle does not hold, i.e., that there is a function* v, *bounded above, and satisfying* (1.23) *and* (1.24), *which is positive somewhere. Then*

$$Lu \equiv 0\,, \quad \textit{and} \quad u \overset{u_0}{=} 0 \quad \textit{on } \partial\Omega\,.$$

Furthermore, v *is a constant multiple of* u.

This has immediate corollaries; since their proofs are short we include them here. The first is another, perhaps more explicit, way of stating Theorem 2.2.

COROLLARY 2.1. *Assume that there is a function* $u > 0$ *with* $Lu \leqq 0$ *in* Ω. *Then either* $\lambda_1 > 0$ *or* $\lambda_1 = 0$ *and* u *is a constant multiple of our principal eigenvector* ϕ_1.

Note that in Theorem 2.2 and Corollary 2.2, u is not assumed to be bounded from above.

Proof: Clearly $\lambda_1 \geqq 0$. If $\lambda_1 = 0$, then apply Theorem 2.2 with $v = \phi_1$.

The next two corollaries show that the eigenfunction ϕ_1 is algebraically simple in a strong sense.

COROLLARY 2.2. *If* v *is bounded above, satisfies*

$$(L + \lambda_1)v \geqq 0 \quad \textit{in } \Omega\,,$$

and

$$\limsup v(x_j) \leqq 0 \quad \textit{if } x_j \overset{u_0}{\to} \partial\Omega\,,$$

then v *is a constant multiple of the principal eigenfunction* ϕ_1.

The same example as in Remark 1.1, namely $w = |x|^{-1} - 1$ in $B_1 \setminus \{0\}$, $L = \Delta$, shows that the assumption that v is bounded above may not be omitted.

Proof: We may suppose $\lambda_1 = 0$. Suppose v is not a constant multiple of ϕ_1. Then, applying Theorem 2.2, with $u = \phi_1$, we infer that $v \leqq 0$. Since $v \not\equiv 0$, it follows from the usual maximum principle that $v < 0$. But then we may apply Theorem 2.2 again, in which the function $-v$ plays the role of u, and ϕ_1 plays the role of v. This implies that ϕ_1 is a constant multiple of $-v$. Contradiction.

COROLLARY 2.3. *There is no function* ψ, *bounded above, and satisfying*

$$\begin{aligned} (L + \lambda_1)\psi &= \phi_1 && \text{in } \Omega \\ \limsup \psi(x_j) &\leqq 0 && \text{if } x_j \overset{u_0}{\to} \partial\Omega\,. \end{aligned}$$

[1] A weaker form is given in Theorem 2 of [23].

Proof: If such a function ψ were to exist then, by Corollary 2.2, ψ would be a constant multiple of ϕ_1, which implies that $(L + \lambda_1)\psi = 0$. Impossible.

Remark 2.1. Theorem 2.2 is a substantial improvement of the standard sufficient condition (ii) of Section 1 for the maximum principle to hold for L in Ω. Serrin weakened (ii) somewhat in [20], in case $\partial\Omega$ is smooth.

We have namely the following corollary of Theorem 2.2:

COROLLARY 2.4. *The refined maximum principle holds if there exists a positive function ϕ (which need not even be bounded) satisfying*

$$\begin{aligned} L\phi &\lneqq 0\,, && \textit{or} \\ L\phi &\equiv 0 && \textit{and } \phi(x_j) \geqq \delta > 0 \textit{ for some sequence } x_j \overset{u_0}{\to} \partial\Omega\,. \end{aligned} \tag{2.1}$$

For related results when M is the Laplacian see [24].

THEOREM 2.3. *Suppose ψ is an eigenfunction of L with eigenvalue $\lambda \neq \lambda_1$. Then* (i) Re $\lambda > \lambda_1$. (ii) *If ψ is real then it changes sign in Ω.*

Theorems 2.1, Corollaries 2.2 and 2.3, and Theorem 2.3, show that (1.9) holds for any bounded domain Ω.

We continue with further properties.

PROPOSITION 2.1. *In its dependence on the coefficient function c, λ_1 is concave. Furthermore, it is Lipschitz continuous in c (using the L^∞ norm) with Lipschitz constant* 1. *In addition, if $\tilde{c} \gneqq c$ then $\lambda_1(M + c) > \lambda_1(M + \tilde{c})$.*

We shall make use of a simple corollary of Theorem 1.1, in which we consider elliptic operators L and $\tilde{L}$ acting, respectively, on bounded domains $\Omega \subset \mathbb{R}^n$, $\tilde{\Omega} \in \mathbb{R}^m$. Points in the product domain $\hat{\Omega} = \Omega \times \tilde{\Omega} \in \mathbb{R}^{n+m}$ are represented by (x, y), $x \in \Omega$, $y \in \tilde{\Omega}$. In $\hat{\Omega}$, consider the operator $\hat{L} = L + \tilde{L}$, where L acts only in the x variables and $\tilde{L}$ only in the y variables.

COROLLARY 2.5. *Consider a pair (λ, ϕ) — eigenvalue and eigenfunction for L in Ω — and a similar pair $(\tilde{\lambda}, \tilde{\phi})$ for $\tilde{L}$ in $\tilde{\Omega}$.* (i) *Then* $(\lambda + \tilde{\lambda}, \phi(x)\tilde{\phi}(y)))$ *is a corresponding pair for $\hat{L}$ in $\hat{\Omega}$.* (ii) *Furthermore if λ and $\tilde{\lambda}$ are the principal eigenvalues for L and $\tilde{L}$, then so is $\lambda + \tilde{\lambda}$ for $\hat{L}$.*

The next two results are concerned with bounds for $\lambda_1(\Omega)$.

THEOREM 2.4. *Let Γ be a closed set in Ω with $|\Gamma| \geqq \delta_0 > 0$. Then*

$$\lambda_1(\Omega \setminus \Gamma) - \lambda_1(\Omega) \geqq \bar{\delta} > 0\,,$$

where $\bar{\delta}$ depends only on c_0, C_0, b, Ω, δ_0.

Variants, and a stronger form, Proposition 9.3, of this theorem will be given in Section 9. Our next result, Theorem 2.5, yields a lower bound for λ_1 in case Ω is narrow. The sufficient condition (iii) (of Section 1) for the maximum principle to hold is a very particular case of this.

THEOREM 2.5. *Let δ, ρ, and d_0 be fixed positive numbers, $0 < \rho \leqq d_0$. Assume that for any $x \in \Omega$ there is a ball $B(x, \rho_x)$, centred at x with radius $\rho_x \leqq \rho$, such that if Ω_x is the component of $\Omega \cap B(x, \rho_x)$ containing x,*

$$|B(x, \rho_x) \backslash \Omega_x| \geqq \delta |B(x, \rho_x)| \,. \tag{2.2}$$

Then there is a positive constant τ depending only on n, c_0, C_0, b, d_0, and δ, such that

$$\lambda_1(M, \Omega) \geqq \frac{\tau}{\rho^2} \,. \tag{2.3}$$

Note that nothing is assumed about diam Ω. Theorem 2.5 implies that $\lambda_1(L, \Omega) > 0$ in case ρ is so small that $\tau/\rho^2 > \sup c$. Theorem 2.5 yields immediately the following stronger form of Proposition 1.1.

THEOREM 2.6. *Let L satisfy conditions* (1.2) *and* (1.3) *in Ω. There exists $\eta > 0$ depending only on n, c_0, C_0, and b such that the refined maximum principle holds for L in Ω if*

$$|\Omega| < \eta \,.$$

Proof: Suppose $|\Omega| < \eta$, with η so small that the ball $B(0, \rho)$ with volume 2η has radius $\rho \leqq 1$. Then for every $x \in \Omega$ the ball $B(x, \rho)$ satisfies (2.2) with $\delta = \frac{1}{2}$. Applying Theorem 2.5, with $d_0 = 1$, we find

$$\lambda_1(M, \Omega) \geqq \tau \rho^{-2} \,, \tag{2.4}$$

where τ depends only on n, c_0, C_0, and b. If η is so small that $\tau\rho^{-2} > \sup c$, the refined maximum principle then holds for L.

Remark 2.2. Inequality (2.4) shows that if $|\Omega|$ is small, namely, if the ball $B(0, \rho)$, with $|B(0, \rho)| = 2|\Omega|$, has radius $\rho \leqq 1$, then

$$\lambda_1(M) \geqq \tau \rho^{-2} = \tau_1 |\Omega|^{-2/n} \,. \tag{2.5}$$

This is a much sharper result than that given (later) by Lemma 4.1:

$$\lambda_1(M) \geqq B^{-1} |\Omega|^{-1/n} \,.$$

For $L = \Delta$, Lieb (see Corollary 2 in [12]) has given a short, elegant proof of Theorem 2.5 (in case $\rho_x \equiv \rho$), with an explicit expression for τ. His argument makes use of energy (the Dirichlet integral) associated with Δ — as well as the fact that the operator Δ has constant coefficient. Since our operator L has nondivergence form, arguments involving energy cannot be used.

3. Construction of u_0, and Boundary Behaviour

We now construct the function u_0 satisfying $Mu_0 = -1$ in Ω and $u_0 = 0$, in some sense, on $\partial\Omega$.

For convenience we will always suppose

$$\inf_{\Omega} x_1 = 0 . \tag{3.1}$$

Let H_j be a sequence of open subsets of Ω having smooth boundaries and such that $H_j \subset \overline{H}_j \subset H_{j+1}, \cup H_j = \Omega$. Now $a_{ik} \in C(\overline{H}_j)$ and hence in H_j, by the standard existence theory, there is a solution $u_j \in W^{2,p}(H_j)$ of

$$\begin{aligned} Mu_j &= -1 \quad \text{in } H_j \\ u_j &= 0 \quad \text{in } \partial H_j \end{aligned}$$

for any $p > n$. By the usual maximum principle, $u_j > 0$ in H_j, and the functions u_j are increasing in j. Furthermore, they are all bounded from above. Indeed, if $\sigma > 0$ is such that

$$c_0\sigma^2 - b\sigma - b = 1 , \tag{3.2}$$

then

$$Me^{\sigma x_1} = e^{\sigma x_1}(a_{11}\sigma^2 + b_1\sigma) \geqq e^{\sigma x_1} \geqq 1 \tag{3.3}$$

and (for later use),

$$Le^{\sigma x_1} \geqq 1 . \tag{3.3'}$$

Using the maximum principle again we find, if $d = \operatorname{diam} \Omega$,

$$u_j \leqq e^{\sigma d} - e^{\sigma x_1} \leqq e^{\sigma d} \quad \text{in } H_j .$$

Consequently $u_j \nearrow u_0$ weakly in $W^{2,p}(J)$, strongly in $C^1(J)$, for any compact set $J \subset \Omega$ and

$$Mu_0 = -1 \quad \text{in } \Omega , \quad 0 < u_0 \leqq e^{\sigma d} .$$

Furthermore we claim that u_0 "vanishes on the boundary" in the following sense: u_0 can be extended as a continuous function to every point y of $\partial\Omega$ "admitting a strong barrier" by setting $u_0(y) = 0$.

DEFINITION. A point $y \in \partial\Omega$ is said to admit a strong barrier if for some ball $B_r(y) = \{|x-y| < r\}$ there is in $B_r(y) \cap \Omega = U$ a positive function $h \in W^{2,n}_{\text{loc}}(U)$ satisfying

$$Mh \leqq -1 ,$$

which can be extended continuously to the point y with $h(y) = 0$.

The claim is easily proved: Set $V = H_j \cap B_{r/2}$. We may fix $\varepsilon > 0$ depending only on n, C_0, b, and r, so that $\varepsilon M|x-y|^2 \leqq \frac{1}{2}$ in U; then $\tilde{h} = h + \varepsilon|x-y|^2$ satisfies $M\tilde{h} \leqq -\frac{1}{2}$ in U. Furthermore if $|x-y| = r/2$ and $x \in \overline{H}_j$,

$$\tilde{h}(x) \geqq \delta = \varepsilon r^2/4 ;$$

we may suppose $\delta \leqq 1$. Hence the function

$$\zeta = e^{\sigma d}\frac{\tilde{h}}{\delta} - u_j \geqq 0 \quad \text{on } \partial V$$

and

$$M\zeta \leqq 0 \quad \text{in } V .$$

The usual maximum principle implies that $\zeta \geqq 0$ in V, i.e.,

$$u_j(x) \leqq e^{\sigma d}\frac{1}{\delta}\tilde{h}(x) \quad \text{in } V .$$

Keeping x fixed and letting $j \to \infty$ we find

$$u_0(x) \leqq e^{\sigma d}\frac{1}{\delta}\tilde{h}(x) \quad \text{for} \quad x \in H_j \cap B_{r/2} ,$$

and the claim follows.

Remark 3.1. It is easily seen that the function u_0 defined above is independent of the choice of subsets H_j.

The following is a simple but useful corollary of Th orem ABP:

COROLLARY 3.1. *Given $f \in L^\infty(\Omega)$, there is a solution z of*

$$Mz = f$$

satisfying

$$|z(x)| \leqq \|f\|_{L^\infty} u_0(x) \quad \text{in } \Omega ,$$

and

$$\|z\|_{L^\infty} \leqq B\|f\|_{L^n} .$$

Furthermore, if $f \leqq 0$ then $z \geqq 0$ in Ω.

Proof: As in the construction of u_0, let z_j be the solution in H_j of

$$\begin{aligned} Mz_j &= f \quad \text{in } H_j \\ z_j &= 0 \quad \text{on } \partial H_j \,. \end{aligned}$$

By the maximum principle, $|z_j| \leqq u_0\|f\|_{L^\infty}$, while by ABP, $|z_j| \leqq B\|f\|_{L^n}$, in Ω. The result follows easily on letting $j \to \infty$. Note that if $f \leqq 0$ then $z_j \geqq 0$ in H_j; hence $z \geqq 0$ in Ω.

Remark 3.2. We recall that $\partial\Omega$ satisfies an exterior cone condition at $y \in \partial\Omega$ if there is a truncated right spherical cone with vertex at y:

$$\{x \mid |x-y|\cos\theta \leqq \langle x-y\,,\,\nu\rangle \leqq \gamma\,,\ \text{for some unit vector } \nu\}\,,$$

lying outside $\overline{\Omega}$. Here $0 < \theta < \frac{\pi}{2}$, $\gamma > 0$. K. Miller, [14", has shown that every point $y \in \partial\Omega$ where $\partial\Omega$ satisfies an exterior cone condition, admits a strong barrier. In a neighbourhood of any such point y he constructed a function h which is positive in $B_r(y) \setminus \{\text{the cone}\}$, $h(y) = 0$, h is homogeneous in $|x-y|$ of some positive degree $\tau < 2$, $h \geqq c_1|x-y|^\tau$, $Mh \leqq -c_1|x-y|^{\tau-2}$ for some $c_1 > 0$ in $B_r \cap \Omega$.

Next we point out that the condition

$$x_j \overset{u_0}{\to} \partial\Omega$$

depends only on a neighbourhood of $\partial\Omega$:

LEMMA 3.1. *Let H be an open set with smooth boundary, $H \subset \overline{H}_j$ where H_j is one of the sets used in the construction of u_0; j will be fixed. Let $\tilde{u}_0$ be the function u_0 constructed earlier — but now for the set $\Omega \setminus \overline{H}$. Then*

$$\tilde{u}_0 \leqq u_0 \leqq C\tilde{u}_0 \quad \text{in } \Omega \setminus H_j \tag{3.4}$$

where C depends only on H, H_j, c_0, C_0, and b.

Proof: The first inequality in (3.4) is obvious by the maximum principle; we have only to prove the second. First we use the simple fact that if $w \geqq 0$ and $Mw \leqq -1$ in a ball $B_r(y)$, then

$$w(y) \geqq \alpha > 0 \tag{3.5}$$

where α depends only on r, C_0, and b. Indeed, for $\kappa > 0$ the function $\zeta = \kappa(r^2 - |x-y|)^2)$ satisfies in $B_r(y)$,

$$\begin{aligned} M\zeta &= -2\kappa \textstyle\sum(a_{ii} + b_i(x_i - y_i)) \\ &\geqq -2\kappa(nC_0 + br) \\ &\geqq -1 \qquad \text{if } \kappa \text{ is small}\,. \end{aligned}$$

Then $w \geqq \zeta$ in $B_r(y)$ and $w(y) \geqq \kappa r^2$.

It is easily seen that our function $\tilde{u}_0$ is obtained in this case as the (increasing) limit of $\tilde{u}_i$ satisfying

$$\begin{aligned} M\tilde{u}_i &= -1 \quad \text{in } H_i \setminus \overline{H} \\ \tilde{u}_i &= 0 \quad \text{on } \partial(H_i \setminus \overline{H}) . \end{aligned}$$

For $i \geqq$ some i_0, and for some fixed r satisfying $r < d(x, \overline{H})$ and $r < d(x, \partial\Omega)$ $\forall\, x \in \partial H_j$, H_i contains an r-neighbourhood of H_j. By (3.5),

$$\tilde{u}_i \geqq \alpha \quad \text{on } \partial H_j .$$

Recalling that $u_0 \leqq e^{\sigma d} = C'$ we see that for any such Ω_i,

$$\frac{C'}{\alpha}\tilde{u}_i \geqq u_0 \quad \text{on } \partial H_j .$$

We may suppose $C' \geqq \alpha$.

Since $u_i \nearrow u_0$ where u_i is the solution of

$$\begin{aligned} Mu_i &= -1 \quad \text{in } H_i \\ u_i &= 0 \quad \text{on } \partial H_i , \end{aligned}$$

we see that

$$\frac{C'}{\alpha}\tilde{u}_i \geqq u_i \quad \text{on } \partial H_j$$

and $\tilde{u}_i = u_i = 0$ on ∂H_i. Consequently

$$\frac{C'}{\alpha}\tilde{u}_i \geqq u_i \quad \text{in } H_i \setminus H_j , \quad \text{for } i \geqq i_0 .$$

Letting $i \to \infty$ we obtain the desired inequality

$$\frac{C'}{\alpha}\tilde{u}_0 \geqq u_0 .$$

We conclude this section with two results which make essential use of Remark 1.2. First, a particularly simple result which will be used later.

LEMMA 3.2. *M satisfies the refined maximum principle in Ω.*

Proof: Suppose w satisfies (1.23), (1.24), with the operator M rather than L, and $w \leqq N$. Then given $\varepsilon > 0$, w satisfies (1.28). In particular — recall the construction of u_0 — (1.28) holds for $x \in \partial H_j$. Since $Mu_0 = Mu_j = -1$ in H_j, and $u_j = 0$ on ∂H_j, it follows from the maximum principle that

$$w \leqq \varepsilon + \frac{N}{\delta}(u_0 - u_j) \quad \text{in } H_j .$$

Fixing $x \in H_j$ and letting $j \to \infty$, so that $u_j(x) \to u_0(x)$, we find

$$w(x) \leqq \varepsilon \quad \text{in } \Omega .$$

Now let $\varepsilon \to 0$.

Finally, we prove a special case of Theorem 1.2 (which is related to Corollary 3.1); it will be used in proving the general result.

LEMMA 3.3. *Given $f \in L^n(\Omega)$ there is a unique L^∞ solution z of*

$$Mz = f \quad \text{in } \Omega \tag{3.6}$$

$$z \overset{u_0}{=} 0 \quad \text{on } \partial\Omega . \tag{3.7}$$

If $f \leqq 0$ then $z \geqq 0$ in Ω. Furthermore

$$\|z\|_{L^\infty} \leqq B\|f\|_{L^n} .$$

Proof: The uniqueness follows from Lemma 3.2. To prove existence, let f_i be a sequence of bounded functions converging to f in L^n. Let z_i be the solutions of $Mz_i = f_i$ given in Corollary 3.1. In particular,

$$z_i \overset{u_0}{=} 0 \quad \text{on } \partial\Omega ,$$

and $z_i \geqq 0$ if $f \leqq 0$. By the corollary,

$$\|z_i\|_{L^\infty} \leqq B\|f_i\|_{L^n} .$$

Using standard interior $W^{2,n}$ estimates, we see that a subsequence of the z_i converges (uniformly on compact subsets) to a solution z of (3.6). We have only to verify (3.7). Applying Corollary 3.1 again, we find

$$\|z_i - z_j\|_{L^\infty} \leqq B\|f_i - f_j\|_{L^n} ,$$

and it follows that the z_i converge uniformly to z in Ω. Consider a sequence $x_j \overset{u_0}{\to} \partial\Omega$. One now sees easily that (3.7) holds.

Chapter 2. The Principal Eigenfunction

4. Proof of Theorem 2.1

Proof of Theorem 2.1: Suppose Ω contains the closure of a ball B_R of radius R. Let C be the constant in Lemma 1.1 and fix $\delta > 0$ so that

$$\left(b + \frac{C}{R^2}\right) B\delta^{1/n} \leqq \frac{1}{2} , \tag{4.1}$$

where B is the constant in (1.20). Let K be a closed subset of Ω such that $U = \Omega \backslash K$ satisfies $|U| < \delta$, and suppose K contains x_0 and B_R. Let Ω_j be a sequence of expanding subdomains of Ω with $\partial\Omega_j$ smooth and

$$K \subset \Omega_j \subset \overline{\Omega}_j \subset \Omega_{j+1} \ldots \subset \Omega\,, \quad \text{and} \quad \bigcup_j \Omega_j = \Omega\,.$$

Now $a_{ik} \in C(\overline{\Omega}_j)$, and hence in Ω_j by the standard existence theory, there is a principle eigenvalue μ_j and eigenfunction $\phi_j \in W^{2,p}(\Omega_j)$ $\forall\, p < \infty$, i.e., satisfying $\phi_j > 0$ in Ω_j, $\phi_j(x_0) = 1$,

$$\begin{aligned} (L + \mu_j)\phi_j &= 0 \quad \text{in } \Omega_j \\ \phi_j &= 0 \quad \text{on } \partial\Omega_j\,. \end{aligned}$$

Since $\phi_{j+1} > 0$ in $\overline{\Omega}_j$ we see by the usual maximum principle, that $\mu_j > \mu_{j+1} > \lambda_1$. Set $\mu = \lim \mu_j \geqq \lambda_1$.

Now we make use of our next main tool: the Krylov-Safonov Harnack inequality. It asserts that if $u > 0$ satisfies $Lu = 0$ in a bounded domain Ω, then for any closed subset $K \subset \Omega$, there is a constant C_1 depending only on K, Ω, c_0, C_0, and b, such that

$$\max_K u \leqq C_1 \min_K u$$

(see Corollary 9.25 in [10]). Applying the inequality in Ω_1 (note that the μ_j are bounded) we may infer that

$$\phi_j \leqq C_1 \quad \text{on } K\,, \quad j = 1, 2, \ldots$$

In $U_j = \Omega_j \backslash K$, $v = \phi_j - C_1$ satisfies

$$\begin{aligned} (M - c^-)v &= -c^+\phi_j + c^- C_1 - \mu_j \phi_j \\ &\geqq -b\phi_j - \mu_j \phi_j, \end{aligned}$$

and

$$\limsup_{x \to \partial U_j} v \leqq 0\,.$$

By Lemma 1.1, $\mu_j \leqq CR^{-2}$. With the aid of (1.20) we may infer as before that

$$\begin{aligned} \max\nolimits_{\bar{U}_j} \phi_j - C_1 &\leqq (b + CR^{-2})B(\max\nolimits_{\bar{U}_j} \phi_j)\delta^{\frac{1}{n}} \\ &\leqq \tfrac{1}{2} \max\nolimits_{\bar{U}_j} \phi_j \qquad \text{by (4.1)}\,. \end{aligned}$$

Hence

$$\max_{\bar{\Omega}_j} \phi_j \leqq 2C_1 =: k\,.$$

Standard interior $W^{2,p}$ estimates yield

$$\|\phi_k\|_{W^{2,p}(\Omega_j)} \leqq C(j) \qquad \forall\, k \geqq j+1 .$$

Consequently a subsequence of the ϕ_j converges to a positive function ϕ_1 in Ω, $\phi_j \to \phi_1$ weakly in $W^{2,p}_{\mathrm{loc}}$, and strongly in $W^{1,\infty}_{\mathrm{loc}}$, i.e., in every compact subset of Ω. Clearly ϕ_1 is a solution of

$$L\phi_1 + \mu\phi_1 = 0 \quad \text{in } \Omega\,, \quad \phi_1(x_0) = 1\,, \tag{4.2}$$

and $\phi_1 \leqq k$ everywhere. By our definition of λ_1 we see that $\mu = \lambda_1$. Statements (i) and (ii) of Theorem 2.1 are proved.

To prove (iii) we show that

$$\phi_1 \leqq k(\lambda_1^+ + b)u_0\,. \tag{4.3}$$

Here $\lambda_1^+ = \max(0, \lambda_1)$. Namely, $\phi_1 = \lim \phi_j$, $\phi_j \leqq k$. $M\phi_j = -\mu_j\phi_j - c\phi_j$, $\phi_j = 0$ on $\partial\Omega_j$, and so by the maximum principle,

$$\phi_j \leqq k(\mu_j^+ + b)u_0 \quad \text{in } \Omega_j\,.$$

Letting $j \to \infty$ we obtain (4.3).

The proof of Theorem 2.1 yields also the following simple result (see, however, Remark 2.2).

LEMMA 4.1. *Under the conditions of Theorem* 2.1, *if* B *is the constant in* (1.20), *then*

$$\lambda_1 \geqq \frac{1}{B|\Omega|^{1/n}} - \sup c^+\,. \tag{4.4}$$

This shows in particular that $\lambda_1(L) > 0$ for any operator L with $c \leqq 0$.

Proof: In our proof of Theorem 2.1 we had $\phi_1 = \lim \phi_j$, with

$$\begin{aligned} (L+\mu_j)\phi_j &= 0 \quad \text{in } \Omega_j \\ \phi_j &= 0 \quad \text{on } \partial\Omega_j\,. \end{aligned}$$

and $\mu_j \searrow \lambda_1$. Thus

$$(M - c^-)\phi_j = -(c^+ + \mu_j)\phi_j\,.$$

Setting $t^+ = \max(0, t)$ and $\gamma = \sup c^+$, we see that

$$(M - c^-)\phi_j \geqq -(\gamma + \mu_j)^+\phi_j\,.$$

Applying (1.20) once more we find

$$\max_{\Omega_j} \phi_j \leqq B(\gamma + \mu_j)^+ \max_{\Omega_j} \phi_j |\Omega|^{1/n} \tag{4.5}$$

or,

$$1 \leqq B|\Omega|^{1/n}(\gamma + \mu_j)^+ .$$

Letting $j \to \infty$ we find

$$1 \leqq B|\Omega|^{1/n}(\gamma + \lambda_1)^+$$

and (4.4) follows.

5. Dependence of λ_1 on c and the b_i

In this section we prove Proposition 2.1 and some related results.

Proof of Proposition 2.1: Consider the expression $F(z)$ defined in the formula before (1.14). According to (1.14),

$$-\lambda_1(c) = \inf_{z\in A} \sup_{x\in\Omega} \left[F(z) + c(x)\right] . \tag{5.1}$$

Now suppose $c_1, c_2 \in L^\infty(\Omega)$. For functions z_1, z_2 in A we find from the convexity of $F(z)$ in z that for $0 \leqq t \leqq 1$,

$$\begin{aligned} F(tz_1 + (1-t)z_2) &+ tc_1 + (1-t)c_2 \\ &\leqq t(F(z_1) + c_1(x)) + (1-t)(F(z_2) + c_2(x)) . \end{aligned}$$

Taking sup with respect to x we find from (5.1):

$$\begin{aligned} -\lambda_1(tc_1 + 1 - tc_2) \quad \leqq \quad & t \sup_{x\in\Omega}[F(z_1)(x) + c_1(x)\text{``} \\ & + (1-t)\sup_{x\in\Omega}[F(z_2)(x) + c_2(x)\text{``} . \end{aligned}$$

Taking, now, inf with respect to both z_1 and z_2 we obtain

$$-\lambda_1(tc_1 + (1-t)c_2) \leqq -t\lambda_1(c_1) - (1-t)\lambda_1(c_2) .$$

This shows that λ_1 is concave in its dependence on c.

That λ_1 is Lipschitz continuous in c with Lipschitz constant 1, follows immediately from the definition of λ_1. To complete the proof of the proposition, suppose $c \lneqq \tilde{c}$. we may suppose $\lambda_1(M + c) = 0$, and have to show that $\tilde{\lambda} := \lambda_1(M + \tilde{c}) < 0$. Clearly $\tilde{\lambda} \leqq 0$.

Let ϕ_1 and ψ be the corresponding positive principal eigenfunctions of Theorem 2.1:

$$(M + c)\phi_1 = (M + \tilde{c} + \tilde{\lambda})\psi = 0 .$$

Suppose $\tilde{\lambda} = 0$. Then $(M + c)\psi \lneqq 0$ in Ω. It follows from Remark 2.1 and Theorem 1.1 (both of which will be proved later) that $\lambda_1(M + c) > 0$. Contradiction.

Next we describe the dependence of λ_1 on the first-order coefficients b_i.

PROPOSITION 5.1. *In its dependence on the coefficients b_i, λ_1 is locally Lipschitz continuous. If the coefficients b_i lie in the region $\sum b_i^2 \leqq b^2$, then the Lipschitz constant can be chosen to depend only on Ω, c_0, C_0, and b.*

Proof: Suppose b_i', $i = 1, \dots, n$ are L^∞ functions satisfying

$$\sum (b_i' - b_i)^2 \leqq \delta^2 \leqq bc_0 .$$

Set

$$M' = a_{ij}\partial_{ij} + b_i'\partial_i , \quad L' = M' + c .$$

We will show that λ_1 for L', call it $\lambda_1(L')$, satisfies

$$\lambda_1(L') \geqq \lambda_1(L) - \sqrt{\frac{b}{c_0}}\,\delta . \tag{5.2}$$

This implies the desired Lipschitz continuity.

We may suppose $\lambda_1(L) = 0$. Let u be our principal eigenfunction, of Theorem 2.1, satisfying $Lu = 0$. For $0 < \alpha < 1$ set $v = u^\alpha$. Then

$$\begin{aligned}
L'v &= \alpha u^{\alpha-1}M'u + \alpha(\alpha-1)u^{\alpha-2}a_{ij}u_iu_j + cu^\alpha \\
&= \alpha u^{\alpha-1}Mu + \alpha u^{\alpha-1}(b_i' - b_i)u_i + \alpha(\alpha-1)u^{\alpha-2}a_{ij}u_iu_j + cu^\alpha \\
&\leqq -\alpha cu^\alpha + \alpha u^{\alpha-1}\delta|\nabla u| - \alpha(1-\alpha)u^{\alpha-2}c_0|\nabla u|^2 + cu^\alpha \\
&\leqq c(1-\alpha)u^\alpha + \frac{\alpha}{4(1-\alpha)c_0}u^\alpha\delta^2 \\
&\leqq \left[b(1-\alpha) + \frac{\alpha}{4(1-\alpha)c_0}\delta^2\right]v .
\end{aligned}$$

The expression in the brackets is minimized if

$$1 - \alpha = \frac{\delta}{2\sqrt{bc_0}} .$$

With this choice of α, we obtain

$$\begin{aligned}
L'v &\leqq \left[\frac{b\delta}{2\sqrt{bc_0}} + \frac{\delta}{4c_0}\left(2\sqrt{bc_0} - \delta\right)\right]v \\
&\leqq \sqrt{\frac{b}{c_0}}\,\delta v .
\end{aligned}$$

This implies (5.2).

We turn now to a comparison of ϕ_1 and u_0.

Remark 5.1. Inequality (4.3) asserts that $\phi_1 \leqq Cu_0$. It is not true, in general, that $u_0 \leqq C\phi_1$ — which does hold if $\partial\Omega$ is smooth. This may be seen by a simple example. Let $\Omega = (0,\pi) \times (0,\pi)$ in $\mathbb{R}^2$ and let $L = \Delta$ in Ω. Then

$$\phi_1 = \sin x \sin y\ .$$

On the other hand, the function

$$\sigma = \sin x \sin y \log r\ ,$$

where $r^2 = x^2 + y^2$, vanishes on $\partial\Omega$ and $|\Delta\sigma|$ is bounded, as one easily verifies. Hence $|\sigma| \leqq$ const. u_0. But near the origin $|\sigma|$ is not bounded by a constant times ϕ_1.

We now prove, however:

THEOREM 5.1.

$$u_0 \overset{\phi_1}{=} 0 \quad on \quad \partial\Omega\ .$$

First two simple lemmas.

LEMMA 5.1. *If ϕ, ψ are positive functions in an open set $U \subset \Omega$, and $0 < t <$ 1, then*

$$w = \phi^t \psi^{1-t} \tag{5.3}$$

satisfies

$$Mw \leqq t\left(\frac{\psi}{\phi}\right)^{1-t} M\phi + (1-t)\left(\frac{\phi}{\psi}\right)^{t} M\psi\ . \tag{5.4}$$

Proof: Set $\phi = e^{z_1}$, $\psi = e^{z_2}$, then $w = e^z$, $z = tz_1 + (1-t)z_2$. Hence

$$e^{-z}Mw = Mz + a_{ij}z_i z_j =: F(z) \tag{5.5}$$

is convex in z, and so it is

$$\begin{aligned} &\leqq tF(z_1) + (1-t)F(z_2) \\ &= te^{-z_1}M\phi + (1-t)e^{-z_2}M\psi \end{aligned}$$

and (5.4) follows.

Remark 5.2. Observe that if $c = tc_1 + (1-t)c_2$ then (5.4) yields

$$(M + c)w \leqq t\left(\frac{\psi}{\phi}\right)^{1-t}(M + c_1)\phi + (1-t)\left(\frac{\phi}{\psi}\right)^{t}(M + c_2)\psi\,. \tag{5.6}$$

This yields the following result; we use the notation above.

LEMMA 5.2. *If*

$$(M + c_1)\phi \leqq f\,, \quad (M + c_2)\psi \leqq f\,, \quad \textit{with} \quad f \leqq 0\,,$$

then

$$(M + c)w \leqq f\,. \tag{5.6$'$}$$

Proof: By (5.6),

$$(M + c)w \leqq \left[t\left(\frac{\psi}{\phi}\right)^{1-t} + (1-t)\left(\frac{\phi}{\psi}\right)^{t}\right]f\,,$$

(5.6)$'$ then follows from the fact that

$$t\,a^{1-t} + (1-t)a^{-t} \geqq 1 \quad \text{for} \quad 0 \leqq t \leqq 1\,, \quad a > 0\,.$$

Now we turn to the

Proof of Theorem 5.1: We may suppose $\lambda_1(L, \Omega) = 0$. We will denote ϕ_1 by ϕ — which is normalized by $\phi(x_0) = 1$, for some fixed $x_0 \in \Omega$. Assume that the claim in the theorem is false, i.e., that there is a sequence x_i such that

$$x_i \xrightarrow{\phi} \partial\Omega \quad \text{and} \quad u_0(x_i) \geqq \mu > 0\,,$$

we will obtain a contradiction. Let K be a closed set containing x_0 such that

$$B|\Omega \setminus K|^{1/n} \leqq \mu/2\,.$$

The function ϕ satisfies $(M + c)\phi = 0$. By the Krylov-Safonov Harnack theorem,

$$\phi \geqq \rho \quad \text{on} \quad K\,, \quad \text{for some constant } \rho > 0\,.$$

With $q = 1 + \sup|c|$, ϕ satisfies $M\phi - q\phi \leqq -\phi$. Hence

$$(M - q)\phi \leqq -\rho\chi_K$$

where χ_K is the characteristic function of K.

It is convenient at this point to work with the expanding sequence of subdomains $\Omega_j \supset K$, with $\partial\Omega_j$ smooth. For $\delta = (2B|\Omega|^{1/n})^{-1}$, we see, as in the proof of Proposition 1.1, that the maximum principle holds for $M + \delta$ in each Ω_j. Recall that $u_0 = \lim u_j$, where the u_j are the solutions of

$$\begin{aligned} Mu_j &= -1 \quad \text{in } \Omega_j \\ u_j &= 0 \quad \text{on } \partial\Omega_j \,. \end{aligned}$$

Let v_j and ψ_j be the solutions of

$$\begin{aligned} Mv_j = -\chi_K \,, \quad (M+\delta)\psi_j &= -\rho\chi_K \quad \text{in } \Omega_j \,, \\ v_j = \psi_j &= 0 \quad \text{on } \partial\Omega_j \,. \end{aligned}$$

With

$$t = \frac{\delta}{q+\delta}\,, \quad \text{so that} \quad -tq + (1-t)\delta = 0\,,$$

we now apply Lemma 5.2, in Ω_j, to the functions ϕ and ψ_j — with $c_1 = -q$, $c_2 = \delta$, and $c = 0$ — and infer that

$$M(\phi^t\psi_j^{1-t}) \leqq -\rho\chi_K \quad \text{in } \Omega_j \,.$$

The maximum principle then yields

$$\phi^t\psi_j^{1-t} \geqq \rho v_j \,. \tag{5.7}$$

Next we observe that $\psi_j \leqq C$ independent of j. This follows immediately from Theorem ABP applied to the equation for ψ_j. Namely, we find

$$\sup\psi_j \leqq (\delta\sup\psi_j + \rho)B|\Omega|^{1/n}\,,$$

or

$$\sup\psi_j \leqq 2\rho B|\Omega|^{1/n} =: C\,.$$

Inserting this in (5.7) we find

$$v_j \leqq \rho^{-1}C^{1-t}\phi^t\,. \tag{5.8}$$

Clearly

$$M(u_j - v_j) = -\chi_{\Omega_j - K} \quad \text{in } \Omega_j\,.$$

Applying Theorem ABP once more we find

$$u_j - v_j \leqq B|\Omega\setminus K|^{1/n} \leqq \mu/2$$

and, using (5.8),

$$u_j \leqq \rho^{-1}C^{1-t}\phi^t + \mu/2\,.$$

Now let $j \to \infty$. It follows in particular that

$$u_0(x_i) \leqq \rho^{-1} C^{1-t} \phi^t(x_i) + \mu/2 .$$

Since $\phi(x_i) \to 0$ we find that for i large,

$$u_0(x) \leqq \frac{3}{4}\mu$$

— which is impossible.

Remark (added January 25, 1993). Let U be an open set, with $\overline{U} \subset \Omega$. Let v be the minimal positive solution of $Mv = 0$ in $\Omega \backslash \overline{U}$, $v = 1$ on ∂U. Namely, for j large, our domains Ω_j contain $\overline{U}$; then, $v = \lim v_j$ where v_j is the solution of $Mv_j = 0$ in $\Omega_j \backslash \overline{U}, v_j = 1$ on $\partial U, v_j = 0$ on $\partial \Omega_j$. Clearly $v \leqq$ constant $\cdot\, u_0$ in $\Omega \backslash \overline{U}$. A proof similar to that of Theorem 5.1 shows that $u_0 \overset{v}{=} 0$ on $\partial \Omega$.

6. Some Results for L in Case $\lambda_1 > 0$; Proofs of Theorems 1.1–1.3

In case $\lambda_1 > 0$, then for any subdomain $U \subset \overline{U} \subset \Omega$, the maximum principle holds for L in U. This follows from the fact that if $\phi_1 > 0$ is the principal eigenfunction of L in Ω, then $L\phi = -\lambda_1 \phi < 0$ and $\phi > 0$ in $\overline{U}$ — the sufficient condition (ii) of page 48 then applies. This remark will be used repeatedly in this section.

We begin with two preliminary results.

Proposition 6.1. *Assume $\lambda_1(\Omega) > 0$. Then there is a function v in Ω satisfying $Lv \leqq 0$ and*

$$1 \leqq v \leqq C$$

where the constant C depends only on c_0, C_0, b, Ω, and λ_1.

Proof: Choose a closed set K in Ω such that $|\Omega \setminus K| \leqq \delta = (2bB)^{-n}$. Here B is the constant in Theorem ABP. By Lemma 3.3, there exists a unique L^∞ solution w of

$$Mw = \begin{cases} -2b & \text{in } \Omega \setminus K \\ 0 & \text{in } K \end{cases} .$$

$$w \overset{u_0}{=} 0 \quad \text{on } \partial\Omega .$$

Furthermore, we know that

$$0 < w < B 2 b \delta^{1/n} = 1 \quad \text{in } \Omega .$$

Then we have in $\Omega \setminus K$:

$$\begin{aligned} L(1 + w) &= c(x)(1 + w(x)) - 2b \\ &\leqq 0 . \end{aligned}$$

Next, by the Krylov-Safonov Harnack inequality we know that $\exists\,\delta > 0$ such that $\phi_1 \geqq \delta$ on K. Set

$$A = \frac{2b}{\lambda_1 \delta} \quad \text{and} \quad v = 1 + w + A\phi_1 \, .$$

Then

$$Lv = L(1+w) - \lambda_1 A\phi_1 \leqq \begin{cases} L(1+w) & \leqq \; 0 \quad \text{in } \Omega \setminus K \\ 2b - \lambda_1 A\delta & = \; 0 \quad \text{in } K \end{cases} .$$

Thus $Lv \leqq 0$ in Ω and

$$1 \leqq v \leqq 2 + \frac{2b}{\lambda_1 \delta} \sup \phi_1 =: C$$

(C depends on the appropriate quantities). Proposition 6.1 is proved.

With the aid of Proposition 6.1 we now present a weak form of Theorem 1.2:

PROPOSITION 6.2. *Assume* $\lambda_1(\Omega) > 0$. *There is a constant* C_1 *depending only on* c_0, C_0, b, Ω, *and* λ_1, *such that for any* $f \in L^\infty(\Omega)$, *there is a solution* w *of*

$$\begin{aligned} Lw &= f \quad \text{in } \Omega \\ w &\overset{u_0}{=} 0 \quad \text{on } \partial\Omega \end{aligned}$$

satisfying, in addition,

$$\|w\|_{L^\infty} \leqq C_1 \|f\|_{L^\infty} \, .$$

Furthermore, if $f \leqq 0$ *then* $w \geqq 0$ *in* Ω.

Proof: We may suppose $\|f\|_{L^\infty} = 1$. Using the sequence of subdomains H_j which was used in constructing u_0, let w_j be the solution in Ω_j of

$$\begin{aligned} Lw_j &= f \quad \text{in } H_j \\ w_j &= 0 \quad \text{on } \partial H_j \, . \end{aligned} \tag{6.1}$$

That the solution exists is a consequence of the fact that the homogeneous problem has only zero as a solution — since the maximum principle holds for L in H_j. The maximum principle also shows that $w_j \geqq 0$ if $f \leqq 0$.

We now apply Proposition 6.1 to the operator $L + \frac{1}{2}\lambda_1$. Then there is a function v in Ω satisfying

$$\left(L + \frac{\lambda_1}{2}\right) v \leqq 0 \quad \text{in } \Omega$$

and $1 \leqq v \leqq C$. Consequently in H_j with j fixed, the functions

$$w_{\pm} := \frac{2}{\lambda_1} v \pm w_j \tag{6.2}$$

satisfy

$$Lw_{\pm} \leqq 0 .$$

They are clearly positive on ∂H_j. The maximum principle holds in H_j; indeed, the function v itself satisfies $v > 0$, $Lv < 0$ in H_j. Hence $w_{\pm} \geqq 0$ in H_j. Thus

$$|w_j| \leqq \frac{2}{\lambda_1} v \leqq \frac{2}{\lambda_1} C .$$

Consequently, in H_j,

$$\begin{aligned} |Mw_j| &= |f - cw_j| \\ &\leqq \|f\|_{L^\infty} + \frac{2}{\lambda_1} Cb . \end{aligned}$$

Using the maximum principle for M, it follows that

$$|w_j| \leqq (1 + \frac{2}{\lambda_1} Cb) u_0 \quad \text{in} \quad H_j . \tag{6.3}$$

Letting $j \to \infty$ we obtain, in the usual way, the conclusion of the proposition.

Next, a useful result.

THEOREM 6.1. *Assume* $\lambda_1(L, \Omega) > 0$. *There is a solution* w_0 *in* Ω *of*

$$Lw_0 = -1 \quad \text{in} \ \Omega$$

satisfying

$$C^{-1} u_0 \leqq w_0 \leqq C u_0 \tag{6.4}$$

where C *depends only on* Ω, c_0, C_0, b, *and* $\lambda_1(\Omega)$.

Proof: Let w_0 be the solution constructed in Proposition 6.2 — for $f = -1$. It was obtained as a limit of the solutions w_j of (6.1), and it was proved that $|w_0| \leqq C_1$. Hence $|Mw_0| \leqq 1 + bC_1$. The right inequality of (6.4) then follows from Corollary 3.1; indeed, the construction of the solution in Proposition 6.2 is the same as the one in Corollary 3.1.

To prove the left inequality, recall that u_0 was obtained as the limit of solutions u_j of

$$\begin{aligned} Mu_j &= -1 \quad \text{in } H_j , \\ u_j &= 0 \quad \text{on } \partial H_j . \end{aligned}$$

Then

$$|(M+c)u_j| \leqq 1 + bu_j \leqq 1 + b\sup u_0 .$$

Thus

$$(M+c)\left[b(\sup u_0)w_j + (w_j - u_j)\right] \leqq 0 .$$

Applying once more the maximum principle for L in H_j, as in the proof of Proposition 6.2, we find

$$u_j \leqq (1 + b\sup u_0)w_j .$$

Letting $j \to \infty$, we obtain

$$u_0 \leqq (1 + b\sup u_0)w_0 .$$

The following important observation is an immediate corollary of Lemma 3.1 and Theorem 6.1.

COROLLARY 6.1. *Let H be an open set in Ω with smooth boundary, and $\overline{H} \subset \Omega$. Assume $\lambda_1(L, \Omega\backslash\overline{H}) > 0$. Let $\tilde{w}_0$ be the function of Theorem 6.1 for the region $\Omega\backslash\overline{H}$. Let u_0 be our solution of $Mu_0 = -1$ constructed in Ω, and $\tilde{u}_0$ the corresponding solution of $M\tilde{u}_0 = -1$ in $\Omega \setminus \overline{H}$. Then for any sequence $x_j \to \partial\Omega$ the following three statements are equivalent:*

$$u_0(x_j) \to 0 , \quad \tilde{u}_0(x_j) \to 0 , \quad \tilde{w}_0(x_j) \to 0 .$$

We turn now to the proofs of Theorems 1.1–1.3.

Proof of Theorem 1.1: If the refined maximum principle holds for L in Ω then necessarily $\lambda_1 > 0$. For if $\lambda_1 \leqq 0$, the principal eigenfunction $\phi_1 > 0$ would be a counterexample to the refined maximum principle. Thus we have only to prove that $\lambda_1 > 0 \Rightarrow$ the refined maximum principle holds.

Suppose then that $\lambda_1 > 0$ and that v is a function, bounded above, which satisfies $Lv \geqq 0$ in Ω and

$$\limsup v(x_j) \leqq 0 \quad \text{if} \quad x_j \overset{u_0}{\to} \partial\Omega .$$

We have to prove $v \leqq 0$ in Ω. We may suppose $N = \sup v > 0$, otherwise we are through. Once more we make use of Remark 1.2, rather, (1.25), according to which, for any $\varepsilon > 0$, $\exists\, \delta > 0$ such that

$$v \leqq \varepsilon + \frac{N}{\delta}u_0 \leqq \varepsilon + C_4\frac{N}{\delta}w_0 \quad \text{in } \Omega .$$

The last inequality relies on (6.4) in Theorem 6.1. Recall that w_0 was obtained as the limit of functions w_j in H_j satisfying (6.7) in H_j, for $f = -1$.

In H_j consider the function

$$u := v - \varepsilon - C_4 \frac{N}{\delta}(w_0 - w_j) .$$

On ∂H_j, $u \leqq 0$, while in H_j

$$Lu \geqq -c\varepsilon \geqq -b\varepsilon .$$

The maximum principle for L in H_j then yields

$$\begin{aligned} u &\leqq b\varepsilon w_j \\ &\leqq \varepsilon b\left(1 + \frac{2}{\lambda_1}Cb\right)\sup u_0 \end{aligned}$$

by (6.3). Thus $u \leqq \varepsilon C_2$. For any fixed point x in H_j, therefore,

$$v(x) \leqq \varepsilon + C_4 \frac{N}{\delta}(w_0(x) - w_j(x)) + \varepsilon C_2 .$$

Letting $j \to \infty$, so that $w_0(x) - w_j(x) \to 0$, we find

$$v(x) \leqq \varepsilon(1 + C_2) .$$

Letting $\varepsilon \to 0$ we conclude the proof.

Proof of Theorem 1.2: Here $\lambda_1 > 0$, and for any $f \in L^n$ we wish to prove existence and uniqueness of a bounded solution u of

$$\begin{aligned} Lu &= f \quad \text{in } \Omega \\ u &\overset{u_0}{=} 0 \quad \text{on } \partial\Omega . \end{aligned}$$

The uniqueness follows from Theorem 1.1. To prove existence we first let z be the solution — given by Lemma 3.3 — of

$$\begin{aligned} Mz &= f \quad \text{in } \Omega \\ z &\overset{u_0}{=} 0 \quad \text{on } \partial\Omega . \end{aligned}$$

By that lemma,

$$\|z\|_{L^\infty} \leqq B\|f\|_{L^n} .$$

Writing our desired solution u as $u = z + v$ we see that we have only to find v satisfying

$$\begin{aligned} Lv &= -cz \quad \text{in } \Omega \\ v &\overset{u_0}{=} 0 \quad \text{on } \partial\Omega . \end{aligned}$$

Since $cz \in L^\infty$, by Proposition 6.2, this problem is solvable for v, with

$$\begin{aligned}\|v\|_{L^\infty} &\leqq C_1 b \|z\|_{L^\infty} \\ &\leqq C_1 bB \|f\|_{L^n} .\end{aligned}$$

Theorem 1.2 is proved.

Proof of Theorem 1.3: By considering $\tilde{v} = v - \beta$ in place of v we may suppose $\beta = 0$. (Note that $L\tilde{v} \geqq f - \beta \sup c^+$.) From now on we suppose $\beta = 0$ and that $Lv \geqq f$. By Theorem 1.2 there is a solution u of

$$\begin{aligned}Lu &= f \quad \text{in } \Omega \\ u &\overset{u_0}{=} 0 \quad \text{on } \partial\Omega ,\end{aligned}$$

with

$$\|u\|_{L^\infty} \leqq A \|f\|_{L^n} .$$

Then $w = v - u$ is bounded above and satisfies

$$\begin{aligned}Lw &\geqq 0 \quad \text{in } \Omega \\ \limsup w(x_j) &\leqq 0 \quad \text{if } x_j \overset{u_0}{\to} \partial\Omega .\end{aligned}$$

But then $w \leqq 0$ — by Theorem 1.1. Consequently

$$v \leqq u \leqq A \|f\|_{L^n} .$$

7. Proof of Theorem 2.2

Proof of Theorem 2.2: Let K be the closure of an open set H in Ω with smooth boundary, $K \subset \Omega$, such that

$$Bb|\Omega \setminus K|^{1/n} \leqq \frac{1}{2} . \tag{7.1}$$

For $0 < t$ large, $z = v - tu < 0$ on K. Then $Lz \geqq 0$ in $\Omega \setminus K$. We will prove that $z \leqq 0$ in all of Ω. By Lemma 4.1, $\lambda_1(L, \Omega \setminus K) > 0$. Let $\tilde{w}_0$ be the function of Corollary 6.1. By that corollary,

$$\limsup v(x_j) \leqq 0 \quad \text{if} \quad x_j \overset{\tilde{w}_0}{\longrightarrow} \partial\Omega .$$

Since $z < 0$ on K, we conclude that

$$\limsup z(x_j) \leqq 0 \quad \text{if} \quad x_j \overset{\tilde{w}_0}{\longrightarrow} \partial(\Omega \setminus K) .$$

But then, since $\lambda_1(L, \Omega \setminus K) > 0$, the maximum principle holds for L in $\Omega \setminus K$ — by Theorem 1.1. Consequently $z \leqq 0$ in $\Omega \setminus K$, and hence in all of Ω.

Now set

$$\tau = \inf\{t \mid v - tu \leqq 0 \text{ in } \Omega\}.$$

Clearly $v - \tau u \leqq 0$ in Ω, and since $v > 0$ somewhere, τ must be positive. If $v - \tau u \equiv 0$ then u and v satisfy the assertion in Theorem 2.2 and the theorem is proved. If not, then by the usual maximum principle, $v - \tau u < 0$ in Ω. But then for some positive $s < \tau$, $\xi = v - (\tau - s)u < 0$ on K. Since $L\xi \geqq 0$, we find as before that $\xi \leqq 0$ in $\Omega \setminus K$ and so in Ω. This contradicts the definition of τ.

8. Proofs of Corollary 2.5 and Theorem 2.3

Proof of Corollary 2.5: We prove part (i) first. Clearly $\hat{\phi}(x,y) = \phi(x)\tilde{\phi}(y)$, and $\hat{\lambda} = \lambda + \tilde{\lambda}$, satisfy

$$\left(\hat{L} + \hat{\lambda}\right)\hat{\phi} = 0. \tag{8.1}$$

We have only to verify that

$$\hat{\phi} \overset{\hat{u}_0}{=} 0 \quad \text{on } \partial\hat{\Omega} \tag{8.2}$$

where $\hat{u}_0$ is our solution (see Section 3) of

$$\begin{aligned} \hat{M}\hat{u}_0 &= -1 && \text{in } \hat{\Omega} \\ \text{“}\hat{u}_0 &= 0 && \text{on } \partial\hat{\Omega}\,\text{.”} \end{aligned}$$

Here $L = M + C$, $\tilde{L} = \tilde{M} + \tilde{C}$ and $\hat{M} = M + \tilde{M}$. If u_0 and $\tilde{u}_0$ are the corresponding functions for M and $\tilde{M}$ (as in Section 3) then

$$\begin{aligned} \hat{M}(u_0 \cdot \tilde{u}_0)(x,y) &= -u_0(x) - \tilde{u}_0(y) \\ &\geqq -\overline{C}, \end{aligned}$$

for some constant $\overline{C}$, since u_0 and $\tilde{u}_0$ are bounded. From the way we constructed u_0 and $\tilde{u}_0$, via interior domains with smooth boundaries, we find, with the aid of the usual maximum principle, that

$$u_0(x)\,\tilde{u}_0(y) \leqq \overline{C}\hat{u}_0(x,y).$$

Since ϕ and $\tilde{\phi}$ are bounded in absolute value we see, in addition, that for some constants, $C, \tilde{C}$,

$$|\phi| \leqq Cu_0 \quad \text{in}\,, \Omega\,, \qquad |\tilde{\phi}| \leqq \tilde{C}\tilde{u}_0 \quad \text{in } \tilde{\Omega}\,.$$

Thus

$$|\hat{\phi}| \leqq C\tilde{C}\overline{C}\hat{u}_0 \quad \text{in } \hat{\Omega} ,$$

which yields (8.2). Part (i) of Corollary 2.4 is proved.

In proving part (ii) we may suppose $\lambda = \tilde{\lambda} = 0$. If $\hat{\lambda}_1$ is the principal eigenvalue of $\hat{L}$ in $\hat{\Omega}$ we see from (8.1) — here $\hat{\phi} = \phi \cdot \tilde{\phi} > 0$ — that $\hat{\lambda}_1 \geqq 0$. If $\hat{\lambda}_1 > 0$ we would have a counterexample to the refined maximum principle for $\hat{L}$ — because of Theorem 1.1, since $\hat{\phi} > 0$ in $\hat{\Omega}$, etc. Thus $\hat{\lambda}_1 = 0$ and Corollary 2.4 is proved.

Proof of Theorem 2.3: First we prove part (ii). If $\psi > 0$ in Ω then λ is necessarily real, and $\lambda \leqq \lambda_1$ — which we may take to be zero. If $\lambda < \lambda_1$ then

$$L(-\psi) = \lambda\psi < 0 .$$

Then by Corollary 2.2, ψ is a constant multiple of ϕ_1. Impossible, since $\lambda \neq \lambda_1$.

Proof of Part (i): We will make use of Corollary 2.4 and Theorem 1.1. We may suppose $\lambda_1 = 0$. Let ϕ_1 be the principal eigenfunction of L. Consider the product domain $\hat{\Omega} = \Omega \times \Omega$, with points (x, y), $x \in \Omega$, $y \in \Omega$. Let L_x denote the action of L in the x variables, and L_y its action in the y variable and, as in Corollary 2.4, set $\hat{L} = L_x + L_y$. Let ψ be an eigenfunction of L with eigenvalue $\lambda \neq \lambda_1$. Since $\bar{\psi}$ is an eigenfunction of L with eigenvalue $\bar{\lambda}$ we see from Corollary 2.4 (applied twice) that

$$(\hat{L} + \lambda + \bar{\lambda})(\psi(x)\,\bar{\psi}(y) + \bar{\psi}(x)\,\psi(y)\,) = 0 .$$

The function $\hat{\psi}(x, y) = \psi(x)\bar{\psi}(y) + \bar{\psi}(x)\psi(y)$ is a real eigenfunction for $\hat{L}$. By Corollary 2.4, proved above, $\lambda_1(\hat{L}) = 0$. The principal eigenvalue of $\hat{L} + \lambda + \bar{\lambda}$ is $-(\lambda + \bar{\lambda})$. Thus if $\lambda + \bar{\lambda} < 0$, by Theorem 1.1, we would have a counterexample to the refined maximum principle for $L + \lambda + \bar{\lambda}$. Hence $\lambda + \bar{\lambda} \geqq 0$.

Suppose finally, that $\lambda + \bar{\lambda} = 0$. Applying Theorem 2.2 and Corollary 2.5 (ii), we see that for some real constant $a \neq 0$,

$$\psi(x)\,\bar{\psi}(y) + \bar{\psi}(x)\,\psi(y) = a\,\phi_1(x)\,\phi_1(y) . \tag{8.3}$$

Setting $y = x$ we find $2|\psi(x)|^2 = a\phi_1^2(x)$, so that $a > 0$. Without loss of generality, we may suppose $a = 2$. Then if we set

$$\psi(x) = e^{i\alpha(x)}\,\phi_1(x) ,$$

from (8.3) we find

$$\cos(\alpha(x) - \alpha(y)\,) = \operatorname{Re} e^{i(\alpha(x)-\alpha(y))} \equiv 1 \qquad \forall\, x, y \in \Omega .$$

But then $\alpha(x) \equiv$ constant α, and $e^{-i\alpha}\psi = \phi_1$. This contradicts part (ii) of the theorem, which we have already proved.

Chapter 3. Bounds on λ_1

9. Lower Bounds for λ_1 in Various Circumstances; Proof of Theorem 2.4.

According to Corollary 2.1, if there is a positive function u in Ω satisfying $Lu \leqq 0$, then either $\lambda_1 > 0$, or else $\lambda_1 = 0$ and u is a multiple of the principal eigenvector ϕ_1. In this section we describe various conditions on such functions u which enable us to derive positive lower bounds for λ_1.

We begin with a very simple result.

THEOREM 9.1. *Suppose u is a function in Ω satisfying*

$$Lu \leqq 0$$
$$1 \leqq u \leqq \kappa .$$

Then $\lambda_1(L,\Omega) \geqq \varepsilon/\kappa$ where $\varepsilon > 0$ depends only on Ω, c_0, b.

Proof: We may suppose $\inf_\Omega x_1 = 0$. For $\sigma > 0$ satisfying (3.2), $Le^{\sigma x_1} \geqq 1$ by (3.3)′. If $d = \operatorname{diam}\Omega$ then $\tilde{u} = u - \varepsilon e^{\sigma x_1}$, with $\varepsilon = \frac{1}{2}e^{-\sigma d}$, satisfies

$$\tilde{u} > 0 \quad \text{and} \quad L\tilde{u} \leqq -\varepsilon \quad \text{in } \Omega .$$

Consequently

$$\begin{aligned}\left(L + \frac{\varepsilon}{\kappa}\right)\tilde{u} &\leqq -\varepsilon + \frac{\varepsilon}{\kappa}\left(\kappa - \varepsilon\right) \\ &< 0 .\end{aligned}$$

Therefore $\lambda_1(L + \varepsilon/\kappa) = \lambda_1(L) - \varepsilon/\kappa > 0$.

In subsequent results we shall make use of Lemma 9.2 below, in which

$$q = 1 + \sup|c| . \tag{9.1}$$

The lemma will involve the constant B of Theorem ABP. First, a preliminary result.

LEMMA 9.1. *In the domain Ω, let χ be the characteristic function of a set having positive measure $\leqq (Bq)^{-n}$. Then*

$$\lambda_1(M + q\chi, \Omega) \geqq 0 . \tag{9.2}$$

Proof: Let w be the solution of

$$\begin{aligned} Mw &= -\chi \quad \text{in } \Omega \\ w &\overset{u_0}{=} 0 \quad \text{on } \partial\Omega , \end{aligned}$$

given by Corollary 3.1. According to the corollary,

$$0 < w \leqq B\|\chi\|_{L^n} \leqq \frac{1}{q} .$$

Hence $(M + q\chi)w \leqq 0$, and this implies, by our definition of λ_1, that $\lambda_1(M + q\chi) \geqq 0$.

LEMMA 9.2. *Let K be a fixed closed set in Ω with*

$$|\Omega \setminus K| \leqq \frac{1}{(Bq)^n} .$$

Suppose v is a positive function in Ω such that

$$\lambda_1(L + v, \Omega) \geqq 0 ,$$

and suppose $v \geqq \kappa_1 > 0$ on K. Then

$$\lambda_1(L, \Omega) \geqq \frac{\eta}{1 + \eta} \quad \text{for} \quad \eta = \frac{\kappa_1}{q} .$$

Proof: Let χ be the characteristic function of the set $\{x \in \Omega | v(x) < \kappa_1\}$; this set lies in $\Omega \setminus K$ and hence its measure is $\leqq (Bq)^{-n}$. By Lemma 9.1,

$$\lambda_1(M + q\chi, \Omega) \geqq 0 .$$

Consider the convex combination of $c + v$ and $q\chi$ given by

$$\tilde{c} = \frac{1}{1 + \eta}(c + v) + \frac{\eta}{1 + \eta} q\chi .$$

Then $\tilde{c} \geqq c + \eta/(1 + \eta)$. Indeed where $v(x) \geqq \kappa_1$, there $\chi = 0$, and

$$\begin{aligned} \frac{1}{1 + \eta}\left(c + v\right) - \left(c + \frac{\eta}{1 + \eta}\right) &> \frac{1}{1 + \eta}(\kappa_1 - \eta(1 + c)) \\ &= \frac{\kappa_1}{q(1 + \eta)}(q - (1 + c)) \geqq 0 . \end{aligned}$$

While on the set where $v(x) < \kappa_1$, we have

$$\begin{aligned} \tilde{c} - \left(c + \frac{\eta}{1 + \eta}\right) &= \frac{1}{1 + \eta}(c + v + \kappa_1 - c - \eta(1 + c)) \\ &\geqq \frac{1}{1 + \eta}(\kappa_1 - \eta(1 + c)) \\ &\geqq 0 \qquad \text{as before} . \end{aligned}$$

By Proposition 2.1, in its dependence on c, λ_1 is concave and nonincreasing. Consequently

$$\lambda_1(M + \tilde{c}, \Omega) \geqq 0 ,$$

and so

$$\lambda_1 \left(M + c + \frac{\eta}{1+\eta} , \Omega \right) \geqq 0 .$$

In this and the next section we shall make use of a second result of Krylov and Safonov. It is Theorem 2 on page 118 of [11]. We use it in the following form:

THEOREM 9.2 (K-S). *In a ball D of radius d_0 let u be a positive function satisfying $Mu + cu \leqq 0$ and: on some closed set $\Gamma \subset D$, $Mu + cu \leqq -1$. Assume $|\Gamma|/|D| \geqq \delta > 0$. Then there is a constant $\sigma > 0$ depending only on n, c_0, C_0, b, d_0, and δ, such that on the concentric ball of half the radius of D,*

$$u \geqq \sigma .$$

In addition, we shall make use of a result which may be regarded as half of the Harnack inequality. It is a special case of Theorem 9.22 of [10]:

THEOREM 9.3. *Let v be a positive function in a ball B_{2R} of radius $2R \leqq 1$, satisfying $Lv \leqq 0$ in B_{2R}. Then there are positive constants p, c_1, depending only on n, c_0, C_0, and b, such that in the concentric ball B_R,*

$$\inf_{B_R} v \geqq c_1 \left(\frac{1}{|B_R|} \int_{B_R} v^p \right)^{1/p} .$$

In fact we will use the following corollary:

COROLLARY 9.1. *Assume $v > 0$ and $Lv \leqq 0$ in Ω. Let K be a closed subset of Ω and assume that $v \geqq 1$ in a ball $B_\rho(\bar{x})$ in Ω. Then*

$$v \geqq c_2 > 0 \quad in \ \ K ,$$

where c_2 depends only on K, Ω, ρ, c_0, C_0, and b.

Proof: By decreasing ρ if necessary, we may assume that $\operatorname{dist}(K, \partial\Omega) \geqq 2\rho$ and $B_{2\rho}(\bar{x}) \subset \Omega$. There exist $r > 0$, and a positive integer N, depending only on K, Ω, and ρ, such that for any $x \in K$ there is a sequence of balls $B^1(x^1), \dots, B^N(x^N)$, each of radius r, with the properties: (i) $B^i_{2r}(x^i) \subset \Omega$ for $i = 1, \dots, N$; (ii) $x^1 = \bar{x}$, $x^N = x$; (iii) for $j = 1, \dots, N-1$, $B^j \cap B^{j+1}$ contains a ball of radius $r/9$.

Consider any point x in K and the corresponding sequence of balls. Let $v_j = \inf_{B^j} v$, so that $v_1 \geqq 1$. By Theorem 9.3,

$$\begin{aligned} rlv_j &\geqq c_1\left(\frac{1}{|B_r|}\int_{B^j} v^p\right)^{1/p} \\ &\geqq c_1\left(\frac{1}{|B_r|}\int_{B^j\cap B^{j-1}} v^p\right)^{1/p} \\ &\geqq c_1 v_{j-1} 9^{-n/p} . \end{aligned}$$

Consequently the desired conclusion holds with

$$c_2 = c_1^N 9^{-Nn/p} .$$

We now present our first application.

PROPOSITION 9.1. *Let u be a positive function in Ω satisfying $Lu \leqq 0$. Assume that $Lu \leqq -1$ on a closed set Γ in Ω of positive measure. Let K be a closed subset of Ω, containing Γ, such that*

$$|\Omega \setminus K| \leqq (Bq)^{-n} . \tag{9.3}$$

Set $\kappa = \max_K u$. Then

$$\lambda_1(L,\Omega) \geqq \delta_1 > 0 ,$$

where δ_1 depends only on c_0, C_0, b, Ω, $|\Gamma|$, K, and κ.

Proof of Proposition 9.1: (i) We first reduce consideration to a domain with smooth boundary. In our proof of Theorem 2.1 we considered a sequence Ω_j of expanding subdomains of Ω, with $\partial\Omega_j$ smooth, and with $\Omega = \cup\Omega_j$. There we showed that the principal eigenvalues μ_j of Ω_j were such that

$$\mu_j \searrow \lambda_1 .$$

We may now choose the Ω_j so that they contain K. Then it suffices to prove (9.3) for each such Ω_j, for j sufficiently large (with δ independent of j). Thus we may suppose $\partial\Omega$ is smooth, and that $K \subset \Omega$. With $\rho = \frac{1}{2}\operatorname{dist}(K, \partial\Omega)$, K may be covered by a finite number of open balls with centres in K. It follows easily that there is a constant $\delta > 0$, depending only on $|\Gamma|$ and K, such that for one of these balls, $B_\rho(\bar{x})$,

$$\frac{|\Gamma \cap B_\rho(\bar{x})|}{|B_\rho|} \geqq \delta . \tag{9.4}$$

(ii) With

$$h = \frac{1}{2}\,\frac{1}{1+u} ,$$

in our new Ω, let w be the solution of

$$\begin{aligned} Lw - hw &= Lu \quad \text{in } \Omega\,, \\ w &= u \quad \text{on } \partial\Omega\,. \end{aligned}$$

This is solvable because we have uniqueness. Now

$$(L-h)(w-u) \geqq hu \quad \text{in } \Omega\,,$$

and

$$w - u = 0 \quad \text{on } \partial\Omega\,.$$

It follows from the maximum principle (for $\lambda_1(L-h) > 0$) that $0 < w < u$.

(iii) Note that

$$\left(L + \frac{wh}{u}\right)\left(u - w\right) = -\frac{w^2 h}{u} < 0\,.$$

Thus $\lambda_1(L + wh/u) \geqq 0$. The desired result then follows from Lemma 9.2 once we have established a lower bound for wh/u on K. Since $u \leqq \kappa$ on K we have only to obtain a lower bound for w on K.

To this end we now apply Theorem 9.2, with Γ replaced by $\Gamma \cap B_\rho$, and conclude that

$$w \geqq \sigma > 0 \quad \text{in } B_{\rho/2}(\bar{x})\,,$$

where σ depends only on c_0, C_0, b, ρ, δ. Let v be the solution in $\Omega \setminus B_{\rho/2}(\bar{x})$ of

$$\begin{aligned} (L-h)v &= 0 \quad \text{in } \Omega \setminus B_{\rho/2} \\ v &= 0 \quad \text{on } \partial\Omega \\ v &= \sigma \quad \text{on } \partial B_{\rho/2}\,. \end{aligned}$$

Since $(L-h)(v-w) \geqq 0$, it follows from the maximum principle that $0 < v \leqq w$. Then it suffices to derive a lower bound for v on K. Let us set the origin at $\bar{x}$. With the aid of a standard comparison function we easily obtain a lower bound σ_0 for v on $\{|x| = 3\rho/4\}$. Namely, consider the function

$$g = e^{-\tau|x|^2} - e^{-\tau\rho^2} \quad \text{for} \quad \frac{\rho}{2} \leqq |x| \leqq \rho\,.$$

Then

$$\begin{aligned} e^{\tau|x|^2}(L-h)g &= 4\tau^2 a_{ij}x_i x_j - 2\tau \sum a_{ii} \\ &\quad - 2\tau \sum b_i x_i + (c-h)(1 - e^{-\tau(\rho^2 - |x|^2)}) \\ &> 0 \qquad \text{for } \tau \text{ sufficiently large}\,. \end{aligned}$$

By the maximum principle,

$$v \geqq \frac{\sigma}{e^{-\tau\rho^2/4} - e^{-\tau\rho^2}}\, g \quad \text{in} \quad \rho/2 \leqq |x| \leqq \rho\,.$$

Thus

$$v \geqq \sigma_0 \quad \text{if} \quad |x| = \frac{3}{4}\rho .$$

If we now apply the Krylov-Safonov Harnack inequality to v in $\Omega \setminus \overline{B_{\rho/2}}$ we find

$$v \geqq \kappa_1 \text{ on } K \setminus B_{\rho/2} .$$

Proposition 9.1 is proved.

Next, a related result — in which u_0 is the function constructed in Section 3.

PROPOSITION 9.2. *Suppose u is a positive function in Ω, with $u \leqq 1$, satisfying $Lu \leqq 0$. With q as in* (9.1), *suppose that*

(9.5) $$u(x) \geqq (2q+1)\, u_0(x) \quad \text{in a ball} \quad B(\bar{x}) \text{ in } \Omega .$$

Then

$$\lambda_1 \geqq \delta_2 > 0$$

where δ_2 depends only on c_0, C_0, b, Ω and the ball $B(\bar{x})$.

Proof: As in step (i) in the proof of Proposition 9.1, we may suppose that $\partial\Omega$ is smooth and that $u \in W^{2,n}(\Omega)$. Let v be the solution of

$$\begin{aligned} (M-q)v &= Lu && \text{in } \Omega \\ v &= u && \text{on } \partial\Omega . \end{aligned}$$

Using the maximum principle one sees that $0 < v < u$. Then

$$|(M-q)(u-v)| = (q+c)u \leqq 2q ,$$

for $0 < u \leqq 1$. Since $(M-q)u_0 \leqq -1$ it follows by the maximum principle that

$$u - v \leqq 2q\, u_0(x) \quad \text{in } \Omega .$$

By (9.5), $v \geqq u_0(x)$ in $B(\bar{x})$. In particular,

(9.6) $$v \geqq \kappa = \inf_{B(\bar{x})} u_0 \quad \text{in } B(\bar{x}) .$$

Next we observe that

$$\begin{aligned} (L+v)(u-v) &\leqq v(u-v-c-q) \\ &\leqq 0 , \end{aligned}$$

since $0 \leqq v < u \leqq 1$. The assertion of the theorem now follows with the aid of Lemma 9.2. We have only to obtain a lower bound for v on some closed set K in Ω, containing $B(\bar{x})$ and such that

$$|\Omega \setminus K| \leqq (Bq)^{-n} .$$

In view of (9.6) such a lower bound for v is supplied by Corollary 9.1.

The last result in this section, from which Theorem 2.4 follows easily, is

PROPOSITION 9.3. *Let u be a positive continuous function in Ω, $u < 1$, which belongs to $W^{2,n}_{\mathrm{loc}}$ in $(\Omega \setminus \Gamma)$, where Γ is a closed set in Ω having positive measure. Assume that $Lu \leqq 0$ in $\Omega \setminus \Gamma$, and that for some constant $\kappa > 0$, $u \geqq \kappa$ on Γ. Then*

$$\lambda_1(L, \Omega \setminus \Gamma) \geqq \delta_3 > 0 ,$$

where δ_3 depends only on c_0, C_0, b, Ω, $|\Gamma|$, $\mathrm{dist}(\Gamma, \partial\Omega)$, and κ.

Proof of Theorem 2.4: Fix $\alpha > 0$, $\alpha \leqq |\Gamma|$. For $\eta > 0$, set

$$\Omega_\delta = \{x \in \Omega \mid \mathrm{dist}(x, \partial\Omega) > \eta\} .$$

Fix η so small that $|\Omega \setminus \Omega_\eta| \leqq \alpha/2$. Then

$$\Gamma' = \Gamma \cap \Omega_\eta$$

satisfies $|\Gamma'| \geqq \alpha/2$; the choice of η depends only on Ω and $|\Gamma|$. We will work with Γ' in place of Γ, and for convenience, we denote it by Γ. Now $\Gamma \subset \Omega_\eta$. The principal eigenfunction ϕ_1 constructed in Theorem 2.1 and normalized by $\phi_1(x_0) = 1$ for some fixed point x_0, satisfies $\phi_1 \leqq k$. Set $u = \phi_1/k$; then $u(x_0) = 1/k$. By the Krylov-Safonov Harnack inequality, $u \geqq \kappa > 0$ on Γ, with κ depending only on Ω, c_0, C_0, and b and η. Proposition 9.3 applied to the operator $L + \lambda_1(\Omega)$ then yields the desired conclusion.

Proof of Proposition 9.3: As in the proof of Theorem 2.4, just presented, we first replace Γ by a smaller set, which we continue to call Γ, but which lies in Ω_η, given above. In addition, (i) as in the proof of Proposition 9.1, we need only consider the case that $\partial\Omega$ is smooth and $u \in W^{2,n}(\Omega)$ (i.e., we need only consider the Ω_j in Ω). Let K be a closed set in Ω with $\mathrm{dist}(K, \partial\Omega) \leqq \eta$ and such that

$$|\Omega \setminus K| \leqq (Bq)^{-n} . \tag{9.7}$$

As in the proof of Proposition 9.1, there is a ball $B_\rho(\bar{x})$ with centre $\bar{x}$ in K and radius $\rho = \frac{1}{2}\,\mathrm{dist}(K, \partial\Omega)$, such that

$$\frac{|\Gamma \cap B_\rho(\bar{x})|}{|B_\rho|} \geqq \delta > 0 , \tag{9.8}$$

where δ depends only on $|\Gamma|$ and K.

(ii) Let v be the solution in Ω of

$$\begin{aligned} (M - q)v &= -\chi_\Gamma \quad \text{in } \Omega \\ v &= 0 \quad \text{on } \partial\Omega . \end{aligned}$$

By Theorem ABP, $0 < v \leqq B|\Omega|^{1/n} = \hat{C}$ in Ω. For $t = \kappa/(2\hat{C})$, we have

$$u > tv \quad \text{on} \quad \Gamma .$$

Since $(M-q)u \leqq 0$ on $\Omega \setminus \Gamma$, it follows from the maximum principle that $u - tv > 0$ in $\Omega \setminus \Gamma$. Furthermore, in $\Omega \setminus \Gamma$,

$$\begin{aligned} (L + tv)(u - tv) &\leqq -tMv - tcv + tv(u - tv) \\ &= -t(q + c)v + tv(u - tv) \\ &\leqq tv(-1 + u - tv) \\ &< 0 , \end{aligned}$$

since $u \leqq 1$. Thus

$$\lambda_1(L + tv, \Omega \setminus \Gamma) \geqq 0 .$$

Now we apply Theorem 9.2 in the ball $B_\rho(\bar{x})$ — using (9.8) — and conclude that

$$v \geqq \sigma > 0 \quad \text{in} \quad B_{\rho/2}(\bar{x}) ,$$

where σ depends only on n, c_0, C_0, b, ρ, and δ. Applying, next, Corollary 9.1, we find that

$$v \geqq c_2 \sigma > 0 \quad \text{in} \quad K$$

where c_2 depends only on K, ρ, c_0, C_0, and b. Since $|(\Omega \setminus \Gamma) \setminus K| \leqq (Bq)^{-n}$, the conclusion of the proposition follows from Lemma 9.2.

10. Proof of Theorem 2.5

As in Section 9 we will make use of Theorem 9.2.

Proof of Theorem 2.5: (i) In our proof of Theorem 2.1 we considered a sequence Ω_j, of expanding subdomains of Ω, with $\partial\Omega_j$ smooth, and with $\Omega = \cup\Omega_j$. There we showed that the principal eigenvalues μ_j of Ω_j were such that

$$\mu_j \searrow \lambda_1 .$$

Thus in proving Theorem 2.5 it suffices, as in Section 9, to prove that

$$\lambda_1(M, \Omega_j) \geqq \frac{\tau}{\rho^2} .$$

Clearly each Ω_j satisfies the conditions of Theorem 2.5. In $\overline{\Omega_j}$ the coefficients $a_{ik}(x)$ are continuous. We may extend $a_{ik}(x)$ as continuous functions to all of $\mathbb{R}^n$, satisfying (1.2). In addition we may suppose $b_i = c = 0$ outside Ω_j.

Thus it suffices to prove Theorem 2.5 in case $\partial\Omega$ is smooth and in case L is defined and satisfies (1.2) and (1.3) in all of $\mathbb{R}^n$.

(ii) For a suitable positive constant θ depending only on n, c_0, C_0, b, d_0, and δ, we will prove that the solution in Ω of

$$(10.1)\qquad \begin{aligned} \left(M - \frac{\theta}{\rho^2}\right) u &= 0 \quad \text{in } \Omega \\ u &= 1 \quad \text{on } \partial\Omega\,, \end{aligned}$$

satisfies

$$(10.2)\qquad u \geqq \theta \quad \text{in } \Omega\,.$$

The positive function $1 - u$ then satisfies

$$\begin{aligned} \left(M + \frac{\theta}{\rho^2}\frac{\theta}{1-\theta}\right)\left(1 - u\right) &= \frac{\theta}{\rho^2}\left(-u + \frac{\theta}{1-\theta}(1-u)\right) \\ &\leqq 0\,. \end{aligned}$$

This implies the conclusion of the theorem with

$$\tau = \frac{\theta^2}{1-\theta}\,.$$

(iii) We now derive (10.2) — with θ to be chosen. For any $x \in \Omega$, let v_1 be the solution of

$$(10.3)\qquad \begin{aligned} Mv_1 &= -\chi(B(x,\rho_x)\setminus\Omega_x) \quad &&\text{in } B(x,\rho_x) \\ v_1 &= 0 &&\text{on } \partial B(x,\rho_x)\,. \end{aligned}$$

Applying Theorem ABP of Section 1 we find

$$(10.4)\qquad v_1 \leqq B\rho_x^2$$

where B depends only on n, c_0, b, and d_0. The factor ρ_x^2 is obtained by stretching the ball $B(x, \rho_x)$ to a unit ball and then applying Theorem ABP.

Applying now Theorem 9.2 to (10.3) we find in turn that

$$(10.5)\qquad v_1(x) \geqq \sigma\rho_x^2$$

where σ is a positive number depending only on n, c_0, C_0, b, d_0, and δ. Again the factor ρ_x^2 is obtained by scaling.

With θ still to be chosen, let v_2 be the solution of

$$(10.6)\qquad \begin{aligned} \left(M - \frac{\theta}{\rho^2}\right) v_2 &= -\frac{\theta}{\rho_x^2} v_1 \quad &&\text{in } B(x,\rho_x) \\ v_2 &= 0 &&\text{on } \partial B(x,\rho_x)\,. \end{aligned}$$

Clearly $v = v_1 - v_2$ satisfies

$$\left(M - \frac{\theta}{\rho^2}\right)\left(v_1 - v_2\right) = -\chi(B(x,\rho_x)\setminus\Omega_x)\,.$$

Applying Theorem ABP (with suitable stretching) to (10.6), and using (10.4), we find, with B as above,

$$v_2 \leqq \frac{\theta}{\rho_x^2} B\rho_x^2 B\rho_x^2 = \theta B^2\rho_x^2\,.$$

On $B(x,\rho_x)\cap\partial\Omega_x$, $v \leqq v_1 \leqq B\rho_x^2$. By the maximum principle applied in $B(x,\rho_x)\cap\Omega_x$, it follows that the solution u of (10.1) satisfies

$$\begin{aligned} u(x) &\geqq \frac{1}{B\rho_x^2} v(x) \\ &\geqq \frac{1}{B\rho_x^2}\left(\sigma\rho_x^2 - \theta B^2\rho_x^2\right)\,. \end{aligned}$$

Setting

$$\theta = \frac{\sigma}{2B^2}$$

we obtain (10.2).

We end with a strong estimate for our function u_0 constructed in Section 3. From Theorem ABP we have

$$u_0 \leqq B|\Omega|^{1/n}\,.$$

By a scaling argument, in case diam $\Omega \leqq 1$ one has, in fact, $B = B'$ diam Ω, where B' depends only on n, c_0 and b. Thus

$$u_0 \leqq B' \operatorname{diam} \Omega |\Omega|^{1/n}\,.$$

With the aid of the function u constructed in the proof of Theorem 2.5 we now easily obtain a better estimate:

THEOREM 10.1.

$$u_0 \leqq k|\Omega|^{2/n} \tag{10.7}$$

where k depends only on n, c_0, C_0, and $b|\Omega|^{1/n}$.

For L in semidivergence form, (10.7) was proved in [25].

Proof: As in the proof of Theorem 2.5 we may work with some Ω_j, i.e., we may assume $\partial\Omega$ is smooth. We proceed first as in the proof of Theorem 2.6. The

ball with volume $2|\Omega|$ has radius $\rho \leqq d_0$ with d_0 depending only on n. Then for every $x \in \Omega$, the ball $B(x,\rho)$ satisfies the condition (2.2) in Theorem 2.5, with $\delta = 1/2$:

$$|B(x,\rho) \cap \Omega^c| \geqq \frac{1}{2}|B(x,\rho)| \,.$$

The function u and the constant θ constructed in the proof of Theorem 2.5, satisfy

$$\begin{aligned} \left(M - \frac{\theta}{\rho^2}\right) u &= 0 && \text{in } \Omega \\ u &= 1 && \text{on } \partial\Omega \\ u &\geqq \theta && \text{in } \Omega \qquad \text{(by (10.2))}. \end{aligned}$$

Here θ depends only on n, c_0, C_0, and b. Then

$$M(1-u) = -\frac{\theta}{\rho^2} u \leqq -\frac{\theta^2}{\rho^2} \,.$$

By the maximum principle,

$$u_0 \leqq \frac{\rho^2}{\theta^2}(1-u) \leqq \frac{\rho^2}{\theta^2} \,,$$

which yields (10.7).

Theorem 10.1 suggests the following

QUESTION. May the conclusion (1.20) in Theorem ABP be replaced by

$$\sup_{\Omega} w \leqq \widetilde{B}(n, c_0, C_0, b|\Omega|^{1/n})|\Omega|^{1/n}\|f\|_{L^n} \,?$$

This question has been recently answered in the affirmative by Mr. Xavier Cabre.

Acknowledgements. The first author was supported by ARO-DAAL-003-88-K-0047, the second by NSF grant DMS-8806731, and the third by NSF grant DMS-9201222 and ARO grant DAAL03-92-G-0317. The first and third authors express their gratitude to the Institute for Advanced Study for its hospitality.

Bibliography

[1] Agmon, S., *On positivity and decay of solutions of second order elliptic equations on Riemannian manifolds*, pp. 19–52 in: *Methods of Functional Analysis and Theory of Elliptic Equations*, D. Greco, ed., Liguori Ed. Napoli, 1983.

[2] Bakelman, I. Ya., *Convex Functions and Elliptic Equations*, manuscript of book.

[3] Barta, J., *Sur la vibration fondamentale d'une membrane,* C. R. Acad. Sci. Paris 204, 1937, pp. 472–473.
[4] Berestycki, H., and Nirenberg, L., *Monotonicity, symmetry and antisymmetry of solutions of semilinear elliptic equations,* J. Geom. Phys. 5, 1988, pp. 237–275.
[5] Berestycki, H., and Nirenberg, L., *On the method of moving planes and the sliding method,* Boll. Soc. Brasil Mat. Nova Ser. 22, 1991, pp. 1–37.
[6] Birindelli, I., *Second-Order Elliptic Equations in General Domains: Hopf's Lemma and Anti-Maximum Principle,* Doctoral Dissertation, New York University, 1992.
[7] Bony, J. M., *Principe du maximum dans les espaces de Sobolev,* C. R. Acad. Sci. Paris Série A 265, 1967, pp. 333–336.
[8] Donsker, M. D., and Varadhan, S. R. S., *On a variational formula for the principal eigenvalue for operators with maximum principle,* Proc. Nat. Acad. Sci. USA 72, 1975, pp. 780–783.
[9] Donsker, M. D., and Varadhan, S. R. S., *On the principal eigenvalue of second-order elliptic differential operators,* Comm. Pure Appl. Math. 29, 1976, pp. 595–621.
[10] Gilbarg, D., and Trudinger, N. S., *Elliptic Partial Differential Equations of Second Order,* 2nd ed., Grundlehren der mathematischen Wissenschaften, No. 224, Springer-Verlag, Berlin–New York, 1983.
[11] Krylov, N. V., *Nonlinear elliptic and parabolic equations of the second order,* Math. and its Applications, Reidel, Norwell, Massachusetts, 1987.
[12] Lieb, E. H., *On the lowest eigenvalue of the Laplacian for the intersection of two domains,* Invent. Math. 74, 1983, pp. 441–448.
[13] Lions, P. L., *A remark on Bony's maximum principle,* Proc. AMS 88, 1983, pp. 503–508.
[14] Miller, K., *Barriers on cones for uniformly elliptic operators,* Ann. Mat. Pura Appl. 76, 1967, pp. 93–106.
[15] Nussbaum, R. D., and Pinchover, Y., *On variational principles for the generalized principal eigenvalue of second order elliptic operators and some applications,* J. Analyse Math. 59, 1992, pp. 161–177.
[16] Protter, M. H., and Weinberger, H. F., *On the spectrum of general second order operators,* Bull. AMS 72, 1966, pp. 251–255.
[17] Protter, M. H., and Weinberger, H. F., *Maximum principles in differential equations,* Prentice-Hall, Englewood Cliffs, New Jersey, 1967.
[18] Pucci, C., *Maximum and minimum first eigenvalues for a class of elliptic operators,* Proc. AMS 17, 1966, pp. 788–795.
[19] Rabinowitz, P. H., *Théorie du degré topologique et applications à des problèmes aux limites non linéaires,* Lecture Notes Lab. Analyse Numerique, Université Paris VI, 1975.
[20] Serrin, J., *A remark on the preceding paper of Amann,* Arch. Rat. Mech. Anal. 44, 1972, pp. 183–186.
[21] Stroock, D., and Varadhan, S. R. S., *On degenerate elliptic-parabolic operators of second order and their associated diffusions,* Comm. Pure Appl. Math. 25, 1972, pp. 651-713.
[22] Venturino, M., *Primo antovalore di operatori lineari ellittici in forma non variazionale,* Boll. Un. Math. It. 5, 1978, pp. 576–591.
[23] Walter, W., *A theorem on elliptic differential inequalities with an application to gradient bounds,* Math. Z. 200, 1989, pp. 293–299.
[24] Zhao, Z., *Subcriticality, positivity, and gaugeability of the Schrödinger operator,* Bull. AMS, 23, 1990, pp. 513–517.
[25] Brezis, H., and Lions, P. L., *An estimate related to the strong maximum principle,* Boll. Un. Math. Ital. A 17, 1980, pp. 503–508.

Received June 1992.

Comm. Pure Appl. Math. 50 (1997) 667-706

Diffusion Semigroups and Diffusion Processes Corresponding to Degenerate Divergence Form Operators

J. QUASTEL
University of California, Davis

AND

S. R. S. VARADHAN
Courant Institute

1 Introduction

In this article we study existence and uniqueness problems for semigroups and processes determined by time-dependent, second-order linear partial differential operators of the form

$$L_t u = \nabla \cdot a \nabla u + \sigma c \cdot \nabla u \,. \tag{1.1}$$

For simplicity we will take the space to be the d-dimensional torus $\mathbb{T}^d$. The matrix $a : [0,T] \times \mathbb{T}^d \to \mathbb{R}^d \otimes \mathbb{R}^d$ is assumed to be symmetric and nonnegative definite. We define σ to be the symmetric, nonnegative definite square root of a, and $c : [0,T] \times \mathbb{T}^d \to \mathbb{R}^d$ to be a function about which assumptions will be made later. The focus of our study is on cases where the coefficients a and c have minimal regularity, and a is allowed to degenerate.

From the point of view of partial differential equations, one considers parabolic equations of the form

$$\frac{\partial u}{\partial t} + L_t u = 0 \tag{1.2}$$

on $[0,T] \times \mathbb{T}^d$, with a (final) boundary condition $u(T,x) = f(x)$, and tries to prove existence and uniqueness in some suitable class. The class of potential solutions is to be chosen carefully. Formally one can try to obtain an a priori energy estimate by multiplying the solution by u and integrating. From the maximum principle one expects the solution to be in $L^\infty\big([0,T] \times \mathbb{T}^d\big)$ if we start with an f in $L^\infty\big(\mathbb{T}^d\big)$. It is therefore not unnatural to look for solutions in the class $H_{1,a}$ consisting of u such that

$$\int_0^T dt \int_{\mathbb{T}^d} dx\, \nabla u \cdot a \nabla u < \infty \tag{1.3}$$

provided we impose some control on c. But a serious question that arises is the meaning of a condition like (1.3), especially when a is degenerate. If $a = \sigma^2$ with a symmetric σ, what we have is

$$\int_0^T dt \int_{\mathbb{T}^d} dx\, |\sigma \nabla u|^2 < \infty\,,$$

and we can try to define $\sigma\nabla u$ as a distribution. It is therefore natural to impose on σ the condition

$$\int_0^T dt \int dx \sum_{i,j=1}^{d} |\nabla \sigma_{ij}|^2 < \infty\,.$$

Another issue that has to be faced is the control of large values of a or, equivalently, σ. In any case, we have assumed no regularity of a in t, and some control on a is needed. Our motivation comes from an effort to understand the motion of a tagged particle in an interacting particle system away from equilibrium. Such systems in their hydrodynamic scaling limits can be described approximately by their particle density $\rho(t, x)$, which is a function of time and space. A tagged particle in equilibrium at a constant density ρ will diffuse like a Brownian motion with a diffusion matrix $a(\rho)$. One expects that, to within first-order corrections that are due to density gradients, the leading term in the generator for the tagged particle in nonequilibrium will given by

$$\frac{1}{2}\nabla a(\rho(t, x))\nabla u\,.$$

In other words, one expects the generator to be

$$L_t u = \frac{1}{2}\nabla a(\rho(t, x))\nabla u + \sigma c \nabla u$$

for some decent choice of c. Self-consistency dictate that ρ itself should be a solution of

$$\frac{\partial \rho}{\partial t} = L_t^* \rho\,.$$

We cannot expect $a(t, x) = a(\rho(t, x))$ to have any more regularity than ρ even when $a(\rho)$ is a fairly regular function of ρ. Although $a(\rho)$ is nondegenerate or elliptic, the ellipticity could degenerate at the fringes, especially at high density. For what are known as simple exclusion models, there is a natural upper bound of 1 for the density and $a(\rho) \to 0$ as $\rho \to 1$. If the regularity of ρ is not strong enough to force continuity of ρ, we do not even have local ellipticity. In the context of large deviation theory [6], the only regularity one

has on ρ is the finiteness of some energylike rate function, and so, for wide applicability, it is important to deal with a satisfying minimal regularity. It will be useful to investigate the case with minimal control on the size of a, namely $a \in L^1([0,T] \times \mathbb{T}^d)$. We will therefore assume that

$$\int_0^T dt \int dx \left\{ \sum_{i,j=1}^{d} |\nabla \sigma_{ij}|^2 + \sum_{i,j=1}^{d} |\sigma_{ij}|^2 \right\} < \infty . \tag{A1}$$

If one can establish that $a(\rho)$ is Lipschitz in ρ, then any minimal regularity that one has on $\rho(t,x)$ can be transferred to $a(t,x)$.

From the probabilistic point of view, one is interested in proving a uniqueness theorem for nonnegative weak solutions of the adjoint equation

$$\frac{\partial p}{\partial t} = L_t^* p \,, \tag{1.4}$$

with the initial condition $p(0,x) = p_0(x)$. Section 2 contains some basic approximation tools that are needed, and in Section 3 we establish the existence and uniqueness theorems for parabolic equations under suitable conditions on a and c. The applications that we have in mind have a peculiar twist to the usual martingale problem corresponding to a given generator. We are given L_t as well as a nonnegative weak solution p of the equation (1.4).

We want to show the existence and uniqueness of a measure P on the space of continuous functions that has p for its marginals and is a solution to the martingale problem for L_t. The vector field c can have singularities and, in fact, need not even be defined where $p = 0$, which can be a set of positive Lebesgue measure. In our application c will involve some sort of logarithmic gradient of p. In a much simpler context, this is like taking an arbitrary nonnegative function $p(x)$ and writing it as a solution of

$$0 = \frac{\partial p}{\partial t} = \frac{1}{2}\Delta p - \nabla \cdot \left(\frac{\nabla p}{2p} \right) p$$

and asking if there is a unique process with generator

$$L_t = L = \frac{1}{2}\Delta + \frac{\nabla p}{2p} \cdot \nabla$$

that has p for an invariant measure. The first-order term $\frac{\nabla p}{2p}$ can be quite bad. This is the motivation for Sections 4 and 5. For some of the results in these sections, we need to strengthen the hypothesis A1 by replacing it with A2.

$$\int_0^T dt \int dx \ \mathrm{Tr}\left(\dot{a} a^{-1} \dot{a} \right) < \infty . \tag{A2}$$

Here $\dot{a}$ denotes $\frac{\partial a}{\partial x_i}$ for any $i = 1, \ldots, d$, and

$$\operatorname{Tr}\left(\dot{a}a^{-1}\,\dot{a}\right) = \sum_{i,j,k,\ell=1}^{d} \frac{\partial a_{jk}}{\partial x_i}\left(a^{-1}\right)_{k\ell}\frac{\partial a_{\ell j}}{\partial x_i}\,.$$

The condition (A2) is technical; we need it in order to prove the uniqueness of very weak solutions, which is the type of uniqueness that comes up when one studies the martingale problem for L_t [8]. It implies both (A1) and the classical finite energy condition $\int_0^T dt \int dx (\nabla \cdot a)a^{-1}(\nabla \cdot a) < \infty$. In one dimension it is equivalent to either, while in higher dimensions it is considerably stronger in general. We do not claim that it is optimal. However, when $a(t,x) = a(\rho(t,x))$ for a nice function $a(\rho)$, the two conditions (A1) and (A2) are not too far apart and both demand the finiteness of some weighted quadratic norms on $\nabla\rho$.

2 Some Approximation Results

We start with the bilinear form

$$\langle \phi, \psi \rangle_{1,a} = \int_0^T dt \int dx \; (\nabla\phi \cdot a \nabla\psi + \phi\psi) \tag{2.1}$$

that defines an inner product on $C^\infty\left([0,T] \times \mathbb{T}^d\right)$. The Hilbert space $H_{1,a}$ is defined as the completion of $C^\infty\left([0,T] \times \mathbb{T}^d\right)$ under the norm

$$\|\phi\|_{H_{1,a}} = \langle \phi, \phi \rangle_{1,a}^{1/2}\,.$$

If ϕ_n is a Cauchy sequence of smooth functions in the space $H_{1,a}$ that converges to $\phi \in L_2([0,T] \times \mathbb{T}^d)$, the corresponding sequence $f_n = \sigma\nabla\phi_n$ has a limit f in $L_2([0,T] \times \mathbb{T}^d)$ and it is natural to call $f = \sigma\nabla\phi$. On the other hand, for $\phi \in L_2([0,T], \mathbb{T}^d)$ we can try to define the norm by duality as given by

$$N^2(\phi) = \sup\left\{2\int_0^T dt \int dx \phi \left(\nabla \cdot \sigma g + f\right) - \int_0^T dt \int dx \left(|g|^2 + |f|^2\right)\right\} \tag{2.2}$$

with the supremum being taken over $f, g_1, \ldots, g_d \in C^\infty([0,T] \times \mathbb{T}^d)$. As long as (A1) holds, the above definition makes sense and for $\phi \in H_{1,a}$

$$N(\phi) = \|\phi\|_{H_{1,a}}\,.$$

A key point is that if (A1) holds, the converse is also true. If $N(\phi) < \infty$, then $\phi \in H_{1,a}$ and

$$\|\phi\|_{H_{1,a}} = N(\phi)\,.$$

This requires a certain amount of work, because the way in which the usual mollification process of convolution by a smooth function works on $H_{1,a}$ is far from clear. It will be proved below.

We first prove the easy fact that condition (A2) implies condition (A1).

LEMMA 2.1 *Let $a : [0,T] \times \mathbb{T}^d \to \mathbb{R}^d \otimes \mathbb{R}^d$, and let σ denote the nonnegative definite square root of a. Then*

$$\sum_{i,j} |\dot\sigma_{ij}|^2 \le \operatorname{Tr}\left(\dot a a^{-1}\,\dot a\right).$$

PROOF: Suppose that a is diagonal at some point (t,x). Then

$$\dot\sigma_{ij} = \frac{\dot a_{ij}}{\sqrt{a_{ii}} + \sqrt{a_{jj}}}\,.$$

Therefore at (t,x),

$$\sum_{i,j} |\dot\sigma_{ij}|^2 = \sum_{i,j}\left(\frac{\dot a_{ij}}{\sqrt{a_{ii}} + \sqrt{a_{jj}}}\right)^2 \le \sum_{i,j} \frac{\dot a_{ij}\dot a_{ji}}{a_{jj}} = \operatorname{Tr}\left(\dot a a^{-1}\,\dot a\right).$$

But $\operatorname{Tr}(MM^*) = \sum_{ij} |M_{ij}|^2$ and $\operatorname{Tr}(M)$ are invariant under an orthogonal change of basis, and a is diagonalizable at every point. The lemma follows. ∎

Let $\mathcal{G}_\varepsilon$ be the fundamental solution of the heat equation at time ε^2, and let $\phi_\varepsilon := \phi * \mathcal{G}_\varepsilon$ be the spatial convolution. The following lemma provides a crucial estimate that will be used repeatedly:

LEMMA 2.2 *Let $f \in W^{1,2}$. Then the family of functions*

$$g_\varepsilon(x) = \int dy\, |f(x+y) - f(x)|^2\, \frac{|\nabla \mathcal{G}_\varepsilon|^2}{\mathcal{G}_\varepsilon}(y)$$

for $\varepsilon > 0$ is uniformly integrable.

PROOF: Recall that a family of functions g_γ, $\gamma \in \Gamma$, is uniformly integrable if and only if there exists a function $\Psi : \mathbb{R}_+ \to \mathbb{R}_+$ such that

(a) $\displaystyle\lim_{x\to\infty} \frac{\Psi(x)}{x} = \infty\,.$

(b) $\sup_{\gamma\in\Gamma} \int dx\, \Psi\left(|g_\gamma(x)|\right) < \infty$.

∎

Furthermore, if g is any function in L^1, then there exists an increasing, convex function Ψ satisfying (a) such that $\Psi(|g|) \in L^1$. Choose such a Ψ for $|\nabla f|^2$. For points $x \in \mathbb{T}^d$ and $y \in \mathbb{R}^d$, the point $x+y$ is of course well-defined on $\mathbb{T}^d$, and the convolution can be performed by integrating with respect to y on $\mathbb{R}^d$. For $x \in \mathbb{T}^d$ and $y \in \mathbb{R}^d$, let us define

$$h_y(x) = \frac{|f(x+y) - f(x)|^2}{\|y\|^2}.$$

By Jensen's inequality,

$$\begin{aligned}
\int dx\, \Psi(h_y(x)) &= \int dx\, \Psi\left(\frac{|f(x+y) - f(x)|^2}{\|y\|^2}\right) \\
&= \int dx\, \Psi\left(\left|\int_0^1 dt\, \frac{\langle y \cdot \nabla f(x+ty)\rangle}{\|y\|}\right|^2\right) \\
&\le \int dx \int_0^1 dt\, \Psi\left(|\nabla f(x+ty)|^2\right) \\
&= \int dx\, \Psi\left(|\nabla f|^2\right)
\end{aligned}$$

The last term is finite, and therefore the family h_y, $y \in \mathbb{R}^d$, is uniformly integrable. Let

$$c_\varepsilon = \int_{\mathbb{R}^d} dy \|y\|^2 \frac{|\nabla \mathcal{G}_\varepsilon|^2}{\mathcal{G}_\varepsilon}(y)$$

and note that $c_\varepsilon = c < \infty$ is independent of ε. Therefore

$$\begin{aligned}
\int dx\, \Psi\left(c^{-1} g_\varepsilon(x)\right) &= \int dx\, \Psi\left(\int dy\, |f(x+y) - f(x)|^2\, c^{-1} \frac{|\nabla \mathcal{G}_\varepsilon|^2}{\mathcal{G}_\varepsilon}(y)\right) \\
&\le \int dx \int dy\, \Psi\left(h_y(x)\right) c^{-1} \|y\|^2 \frac{|\nabla \mathcal{G}_\varepsilon|^2}{\mathcal{G}_\varepsilon}(y) \\
&\le \sup_{y\in\mathbb{R}^d} \int dx\, \Psi\left(h_y(x)\right).
\end{aligned}$$

We have just shown that the last term is bounded. Therefore the family $\{c^{-1} g_\varepsilon :\ \varepsilon > 0\}$ is uniformly integrable and so is the family $\{g_\varepsilon :\ \varepsilon > 0\}$.

The basic fact regarding convolutions in $H_{1,a}$ is contained in the next lemma. Although the lemma is an easy consequence of the earlier lemma, its conclusion is far from obvious. When σ is not invertible, control over $|\sigma\nabla u|$ can hide a lot of cancellations. Convolving u while leaving σ alone could royally mess up these cancellations. The assumption (A1) prevents this.

LEMMA 2.3 *Suppose that* (A1) *holds.*

(i) *If* $\phi \in L^2\left([0,T] \times \mathbb{T}^d\right)$, *then* $\|\sigma\nabla\phi_\varepsilon - (\sigma\nabla\phi)_\varepsilon\|_{L^1\left([0,T]\times\mathbb{T}^d\right)} \to 0.$

(ii) *If* $\phi \in L^\infty\left([0,T] \times \mathbb{T}^d\right)$, *then* $\|\sigma\nabla\phi_\varepsilon - (\sigma\nabla\phi)_\varepsilon\|_{L^2\left([0,T]\times\mathbb{T}^d\right)} \to 0.$

In particular, for $\phi \in H_{1,a}$ *with* $\sigma\nabla\phi = f$, $\|\sigma\nabla\phi_\varepsilon - f\|_{L^1\left([0,T]\times\mathbb{T}^d\right)} \to 0$; *if we assume in addition that* $\phi \in L^\infty\left([0,T]\times\mathbb{T}^d\right)$, *then* $\|\sigma\nabla\phi_\varepsilon - f\|_{L^2\left([0,T]\times\mathbb{T}^d\right)} \to 0.$

PROOF: Let us prove (i) first. After an integration by parts we can write

$$
\begin{aligned}
&|\sigma\nabla\phi_\varepsilon - (\sigma\nabla\phi)_\varepsilon|(x) \\
&\quad= \left|\int dy\,(\sigma(x+y) - \sigma(x))\,\nabla\phi(x+y)\mathcal{G}_\varepsilon(y)\right| \\
&\quad= \left|\int dy\left[\nabla\cdot\sigma(x+y) + (\sigma(x+y)-\sigma(x))\frac{\nabla\mathcal{G}_\varepsilon}{\mathcal{G}_\varepsilon}(y)\right]\right. \\
&\qquad\qquad \left.\phi(x+y)\mathcal{G}_\varepsilon(y)\right| \\
&\quad\le \frac{1}{2}\int dy\left|\nabla\cdot\sigma(x+y) + (\sigma(x+y)-\sigma(x))\frac{\nabla\mathcal{G}_\varepsilon}{\mathcal{G}_\varepsilon}(y)\right|^2 \mathcal{G}_\varepsilon(y) \\
&\qquad + \frac{1}{2}\int dy\,\phi^2(x+y)\mathcal{G}_\varepsilon(y)\,.
\end{aligned}
\tag{2.3}
$$

Integrating both sides with respect to x,

$$
\sup_{\varepsilon>0}\|\sigma\nabla\phi_\varepsilon - (\sigma\nabla\phi)_\varepsilon\|_{L^1([0,T]\times\mathbb{T}^d)} \le \frac{1}{2}\left[C^2 + \|\phi\|^2_{L^2([0,T]\times\mathbb{T}^d)}\right],
$$

which in turn implies

$$
\sup_{\varepsilon>0}\|\sigma\nabla\phi_\varepsilon - (\sigma\nabla\phi)_\varepsilon\|_{L^1([0,T]\times\mathbb{T}^d)} \le C\|\phi\|_{L^2([0,T]\times\mathbb{T}^d)}\,. \tag{2.4}
$$

We note that for smooth ϕ the claim is obvious, and given the estimate (2.2), the result follows for ϕ in $L^2([0,T]\times\mathbb{T}^d)$.

We now turn to (ii). Since we already have convergence in $L^1([0,T]\times\mathbb{T}^d)$, we only need to establish the uniform integrability of the functions $|\sigma\nabla\phi_\varepsilon - (\sigma\nabla\phi)_\varepsilon|^2$. They are dominated by

$$\|\phi\|^2_{L^\infty([0,T]\times\mathbb{T}^d)} \int dy \left|\nabla\cdot\sigma(x+y) + (\sigma(x+y)-\sigma(x))\frac{\nabla\mathcal{G}_\varepsilon}{\mathcal{G}_\varepsilon}(y)\right|^2 \mathcal{G}_\varepsilon(y)\,. \tag{2.5}$$

Since we have assumed in (A1) that $|\nabla\cdot\sigma|^2 \in L^1([0,T]\times\mathbb{T}^d)$, the family $\{(|\nabla\cdot\sigma|^2)_\varepsilon : \varepsilon > 0\}$ is uniformly integrable. The remaining piece in (2.3) is uniformly integrable as a consequence of Lemma 2.2. ∎

The following is a useful variant of Lemmas 2.2 and 2.3:

LEMMA 2.4 *Let $\mathcal{H}_\varepsilon$ be any sequence of functions such that $\frac{\|y\|^2}{\varepsilon^2}\mathcal{H}_\varepsilon(y)$ is (a bounded multiple of) an approximate identity on $\mathbb{R}^d$.*

(i) *For any $f \in W^{1,2}$, the family of functions*

$$\tilde{g}_\varepsilon(x) = \varepsilon^{-2}\int dy\, |f(x+y)-f(x)|^2\,\mathcal{H}_\varepsilon(y)$$

for $\varepsilon > 0$ is uniformly integrable.

(ii) *Suppose that* (A1) *holds and $\phi \in L^\infty\left([0,T]\times\mathbb{T}^d\right)$. Then*

$$\int dx \int dz\, |\sigma(x+z)\nabla\phi_\varepsilon(x) - (\sigma\nabla\phi)_\varepsilon(x)|^2\mathcal{H}_\varepsilon(z) \to 0\,.$$

PROOF: To prove (i), we use the notation of the proof of Lemma 2.2. If

$$\tilde{g}_\varepsilon(x) = \int dy\, h_y(x)\|y/\varepsilon\|^2\mathcal{H}_\varepsilon(y)\,,$$

then

$$\begin{aligned}\sup_{\varepsilon>0}\int dx\,\Psi(\tilde{g}_\varepsilon(x)) &\le \sup_{\varepsilon>0}\int dx\int dy\,\Psi(h_y(x))\left\|\frac{y}{\varepsilon}\right\|^2\mathcal{H}_\varepsilon(y)\\ &\le \sup_{y\in\mathbb{R}^d}\int dx\,\Psi(h_y(x))\,.\end{aligned}$$

Since $\{h_y(x) :\ y\in\mathbb{R}^d\}$ is uniformly integrable, so is $\{\tilde{g}_\varepsilon(x),\ \varepsilon > 0\}$. ∎

To prove (ii) from (i), we use exactly the same argument as in the proof of Lemma 2.3 to reduce it to the uniform integrability of

$$\int dz \int dy\, |f(x+y) - f(x+z)|^2 \frac{|\nabla \mathcal{G}_\varepsilon|^2}{\mathcal{G}_\varepsilon}(y)\mathcal{H}_\varepsilon(z)$$

for any $f \in W^{1,2}$. Now expand $|f(x+y) - f(x+z)| \leq |f(x+y) - f(x)| + |f(x+z) - f(x)|$. For the first term, appeal to the previous lemma. For the second term, use (i) and the fact that

$$\int dy \frac{|\nabla \mathcal{G}_\varepsilon|^2}{\mathcal{G}_\varepsilon}(y) \leq C\varepsilon^{-2}.$$

LEMMA 2.5 *Suppose that* (A1) *holds and* $N(\phi) < \infty$ *for some* $\phi \in L_2([0,T] \times \mathbb{T}^d)$. *Then there exists* $f = (f_1, \dots, f_d) \in L^2([0,T] \times \mathbb{T}^d)$ *such that for all* $\psi = (\psi_1, \dots, \psi_d) \in C^\infty$,

$$\int_0^T dt \int dx\, \phi \nabla \cdot (\sigma\psi) = -\int_0^T dt \int dx\, f\psi\,,$$

and $N^2(\phi) = \|f\|^2_{L^2([0,T]\times\mathbb{T}^d)} + \|\phi\|^2_{L^2([0,T]\times\mathbb{T}^d)}$. *Furthermore, we have the following chain rule: Let* $F : \mathbb{R} \to \mathbb{R}$ *satisfy* $|F(x)| \leq C|x|$ *and* $|F'(x)| \leq C$ *for all* $x \in \mathbb{R}$ *and some* $C < \infty$. *Then for all* $\psi \in C^\infty$,

$$\int_0^T dt \int dx\, F(\phi)\nabla \cdot (\sigma\psi) = -\int_0^T dt \int dx\, F'(\phi) f\psi\,. \tag{2.6}$$

PROOF: The first statement is obvious. To prove (2.4), note first that it holds for $\phi \in C^\infty$. In particular,

$$\int_0^T dt \int dx\, F(\phi_\varepsilon)\nabla \cdot (\sigma\psi) = -\int_0^T dt \int dx\, F'(\phi_\varepsilon)\sigma\nabla\phi_\varepsilon\psi\,.$$

Since $\phi \in L^2$, $\phi_\varepsilon \to \phi$ strongly in L^2. Since $|F(x)| \leq C|x|$ and $\nabla\cdot(\sigma\psi) \in L^2$, we conclude

$$\int_0^T dt \int dx\, F(\phi_\varepsilon)\nabla \cdot (\sigma\psi) \to \int_0^T dt \int dx\, F(\phi)\nabla \cdot (\sigma\psi)\,.$$

By Lemma 2.3, we have $\sigma\nabla\phi_\varepsilon \to f$ strongly in L^1. Since F' is bounded, we conclude that

$$\int_0^T dt \int dx\, F'(\phi_\varepsilon)\sigma\nabla\phi_\varepsilon\psi \to \int_0^T dt \int dx\, F'(\phi) f\psi\,.$$

This completes the proof. ∎

THEOREM 2.6 *Suppose that* (A1) *holds, and* $\phi \in L_2([0,T] \times \mathbb{T}^d)$ *with* $N(\phi) < \infty$. *Then* $\phi \in H_{1,a}$.

PROOF: First, let us suppose that ϕ is in $L^\infty([0,T] \times \mathbb{T}^d)$. As before, let ϕ_ε denote the convolution of ϕ with the heat kernel $\mathcal{G}_\varepsilon$. For $\varepsilon_1, \varepsilon_2 > 0$,

$$\begin{aligned} \|\phi_{\varepsilon_1} - \phi_{\varepsilon_2}\|_{H_{1,a}} \le{}& \|\sigma\nabla\phi_{\varepsilon_1} - (\sigma\nabla\phi)_{\varepsilon_1}\|_{L^2([0,T]\times\mathbb{T}^d)} \\ &+ \|\sigma\nabla\phi_{\varepsilon_2} - (\sigma\nabla\phi)_{\varepsilon_2}\|_{L^2([0,T]\times\mathbb{T}^d)} \\ &+ \|f_{\varepsilon_1} - f_{\varepsilon_2}\|_{L^2([0,T]\times\mathbb{T}^d)} + \|\phi_{\varepsilon_1} - \phi_{\varepsilon_2}\|_{L^2([0,T]\times\mathbb{T}^d)} \,. \end{aligned}$$

All terms on the right tend to 0 as $\varepsilon_1, \varepsilon_2 \to 0$. For the first two terms we use Lemma 2.3 and the others are obvious.

If $N(\phi) < \infty$ but ϕ is only in $L^2([0,T] \times \mathbb{T}^d)$, we truncate it at level C and define

$$\phi_C(x) = \begin{cases} \phi & \text{if } |\phi| \le C\,, \\ C & \text{if } |\phi| \ge C\,. \end{cases}$$

By the previous lemma, for all $\psi \in C^\infty$,

$$\int_0^T dt \int dx\, \phi_C\, \nabla\cdot(\sigma\psi) = -\iint_{|\phi|\le C} dx\, dt\, f\psi\,.$$

Then for $0 < C_1 < C_2 < \infty$,

$$\|\phi_{C_1} - \phi_{C_2}\|_{1,a}^2 = \iint_{C_1\le|\phi|\le C_2} dx\, dt\, |f|^2 + \int_0^T\int dt\, dx\, |\phi_{C_1} - \phi_{C_2}|^2\,.$$

In particular, ϕ_C has a limit in $H_{1,a}$ as $C \to \infty$, which is of course ϕ. ∎

3 Existence and Uniqueness for a Class of Degenerate Parabolic Equations

If we are given the basic divergence form operator

$$L_t^0 = \nabla\cdot a\nabla\,,$$

there are several natural problems involving the operator that one can consider. For some suitable c one can study

$$\begin{cases} \dfrac{\partial u}{\partial t} + \nabla\cdot a\nabla u + \sigma c\cdot\nabla u = g & \text{for } 0 \le t \le T\,, \\ u(T,x) = f(x)\,. \end{cases} \tag{3.1}$$

Because of the maximum principle, if $f, g \in L^\infty$ we can look for a solution $u \in L^\infty$.

We could also exploit the energy relation and consider, for suitable c_1, c_2, g, and h, solutions in L^2 of

$$\text{(3.2)} \qquad \begin{cases} \dfrac{\partial u}{\partial t} + \nabla \cdot a\nabla u + \sigma c_1 \cdot \nabla u + \nabla \cdot \sigma c_2 u \\ \quad = g + \nabla \cdot \sigma h & \text{for } 0 \le t \le T\,, \\ u(T,x) = f(x) \end{cases}$$

The adjoint problem

$$\text{(3.3)} \qquad \begin{cases} \dfrac{\partial p}{\partial t} = \nabla \cdot a\nabla p - \nabla \cdot \sigma c p & \text{for } 0 \le t \le T\,, \\ p(0,x) = p_0(x) \end{cases}$$

is important in probability theory as a Fokker-Planck equation, and p is a nonnegative probability density, usually with a total integral equal to 1.

We will have something to say about each one of these problems. The minimal regularity that our solutions satisfy will be that they are in $H_{1,a}$ and

$$\sup_{0 \le t \le T} \int_{\mathbb{T}^d} |u(t,x)|^2 \, dx < \infty\,.$$

All of our equations have the form

$$\text{(3.4)} \qquad \frac{\partial u}{\partial t} = \nabla \cdot \sigma F + G\,,$$

and this is to be interpreted in a suitable sense. We say that (3.4) is satisfied in the sense of distributions if, for any C^∞ test function,

$$\text{(3.5)} \qquad \frac{d}{dt} \int_{\mathbb{T}^d} \phi(t,x) u(t,x)\, dx = \int_{\mathbb{T}^d} \left[-F \cdot \sigma \nabla \phi + \phi G + u \frac{\partial \phi}{\partial t} \right] dx\,.$$

It does not make any difference if we limit ourselves to using C^∞ functions $\phi(x)$ that do not depend on t.

We begin with a lemma that will be useful in dealing with such solutions.

LEMMA 3.1 *Suppose $u_1, u_2, \ldots, u_k$ are functions such that for each $i = 1, \ldots, k$,*

$$\begin{cases} u_i \in L^\infty([0,T] \times \mathbb{T}^d) \cap H_{1,a} \\ \dfrac{\partial u_i}{\partial t} = \nabla \cdot \sigma f_i + g_i \end{cases}$$

in the sense of (3.5), *with* $f_i \in L^2([0,T]\times\mathbb{T}^d)$ *and* $g_i \in L^1([0,T]\times\mathbb{T}^d)$. *Then for any smooth function* $F:\mathbb{R}^k\to\mathbb{R}$, *if* $v = F(u) = F(u_1,\dots,u_k)$,

$$\frac{\partial v}{\partial t} = \sum_{i=1}^{k} \nabla\cdot\frac{\partial F}{\partial x_i}(u)\,\sigma f_i - \sum_{i,j=1}^{k}\frac{\partial^2 F}{\partial x_i\partial x_j}(u) f_i\cdot\sigma\nabla u_j + \sum_{i=1}^{k}\frac{\partial F}{\partial x_i}(u) g_i\,,$$

again in the sense of (3.5). *We can assume without loss of generality that for every* i, f_i *is in the range of* σ *for almost all* (t,x).

PROOF: Let us take the case $k=1$. We convolute u by the heat kernel G_ε at time ε^2 to get u_ε and let $v_\varepsilon = F(u_\varepsilon)$. It is not hard to see that

$$\begin{aligned}\frac{\partial v_\varepsilon}{\partial t} &= F'(u_\varepsilon)\left(\frac{\partial u}{\partial t} * G_\varepsilon\right) = F'(u_\varepsilon)[\nabla(\sigma f)_\varepsilon + g_\varepsilon]\\ &= \nabla F'(u_\varepsilon)(\sigma f)_\varepsilon + F'(u_\varepsilon)g_\varepsilon - F''(u_\varepsilon)\nabla u_\varepsilon\cdot(\sigma f)_\varepsilon\,.\end{aligned}$$

We use a test function ϕ that depends only on x and write

$$\begin{aligned}\frac{d}{dt}\int dx\, v_\varepsilon\phi = &-\int dx\, F'(u_\varepsilon)\nabla\phi\cdot(\sigma f)_\varepsilon + \int dx\, F'(u_\varepsilon)\phi g_\varepsilon\\ &-\int dx\,\phi F''(u_\varepsilon)\nabla u_\varepsilon\cdot(\sigma f)_\varepsilon\,.\end{aligned}$$

We want to let $\varepsilon\to 0$ and conclude that

$$\begin{aligned}\frac{d}{dt}\int dx\, v\phi = &-\int dx\, F'(u)\nabla\phi\cdot(\sigma f) + \int dx\, F'(u)\phi g\\ &-\int dx\,\phi F''(u)\nabla u\cdot(\sigma f)\,,\end{aligned}$$

which establishes the lemma. The only tricky part is the convergence of $\nabla u_\varepsilon\cdot(\sigma f)_\varepsilon$ to $\nabla u\cdot\sigma f$ in $L^1([0,T]\times\mathbb{T}^d)$. To this end we write

$$\nabla u_\varepsilon\cdot(\sigma f)_\varepsilon = \sigma_\varepsilon\nabla u_\varepsilon\cdot(\sigma_\varepsilon)^{-1}(\sigma f)_\varepsilon$$

and Lemma 2.4 assures us that $\sigma_\varepsilon\nabla u_\varepsilon\to\sigma\nabla u$ in $L^2([0,T]\times\mathbb{T}^d)$. It suffices to check that $(\sigma_\varepsilon)^{-1}(\sigma f)_\varepsilon\to f$ in $L^2([0,T]\times\mathbb{T}^d)$. By the convexity of the map $\xi, A\to\langle\xi, A^{-1}\xi\rangle$, it follows that

$$\left\|(\sigma_\varepsilon)^{-1}(\sigma f)_\varepsilon\right\|_{L^2}\le\|f\|_{L^2}\,,$$

and with a little computation one can show the convergence of $(\sigma_\varepsilon)^{-1}(\sigma f)_\varepsilon$ to f in $L^2([0,T]\times\mathbb{T}^d)$. ∎

Remark. In principle, if we assume that the g_i are in L^2, there is no a priori reason why we need u_i to be in L^∞ for the lemma. The final formula makes sense if u_i are just in $H_{1,a}$, but we find that we have to strengthen the definition of (3.4) with a larger class of test functions in order to drop the L^∞ assumption. For bounded functions in $H_{1,a}$ the validity for the larger class follows already from the validity for smooth test functions, whereas for a general function in $H_{1,a}$ one has to assume it. The culprit is Lemma 2.3, where ϕ has to be L^∞ in order to imply convergence in L^2.

The simplest situation is equation (3.1). Here $F = \sigma\nabla u \in L^2$, and if we assume that $c \in L^\infty$, then $G = -\sigma c\cdot\nabla u + g \in L^2$. In this case it will suffice to interpret (3.4) in the sense of (3.5) with C^∞ test functions.

LEMMA 3.2 *If $c \in L^\infty$, equation* (3.1) *has at most one bounded solution.*

PROOF: Since we can take the difference of two solutions to get a new solution, it amounts to proving that any bounded solution of

$$\begin{cases} \dfrac{\partial u}{\partial t} + \nabla\cdot a\nabla u + \sigma c\cdot\nabla u = 0 & \text{for } 0 \le t \le T\,, \\ u(T,x) = 0 \end{cases}$$

is identically zero. From Lemma 3.1

$$\frac{\partial\phi(u)}{\partial t} = -\nabla\cdot\phi'(u)a\nabla u - \phi'(u)\sigma c\cdot\nabla u = \phi''(u)\nabla u\cdot a\nabla u\,.$$

It is not difficult to see that $\phi(u)$ can be replaced by a more general function $\phi(t,u)$, and this produces an extra term, namely, $\frac{\partial\phi}{\partial t}(t,u)$. Since u is bounded, the choice $\phi(t,u) = e^{kt}u^2$ is allowed, and we have

$$\frac{\partial e^{kt}u^2}{\partial t} = -2e^{kt}[\nabla u a\nabla u + 2u\sigma c\nabla u - \nabla u\cdot a\nabla u] + ke^{kt}u^2\,.$$

Integrating over space we get

$$\frac{d}{dt}e^{kt}\int u^2(t,x)\,dx = e^{kt}\int\left[-2u\sigma c\nabla u + 2\langle\nabla u, a\nabla u\rangle + ku^2\right]dx$$
$$\ge 0$$

provided $k > \frac{1}{2}C^2$ where C is an upper bound for $\|c\|$. In particular,

$$\int u^2(t,x)\,dx \le \int u^2(T,x)\,dx = 0$$

and we are done. ∎

For proving existence we have to approximate the equation by more regular ones.

LEMMA 3.3 *Suppose that* (A1) *holds. Then there exists a family of positive definite matrices* $\{a_\varepsilon\}$ *defined for* $\varepsilon > 0$ *with positive definite square roots* $\{\sigma_\varepsilon\}$ *such that for some* $c_\varepsilon > 0$

(i) $c_\varepsilon\,|\xi|^2 \le \xi\cdot a_\varepsilon(t,x)\xi \le c_\varepsilon^{-1}|\xi|^2\,,\quad (t,x)\in[0,T]\times\mathbb{T}^d\,,\ \xi\in\mathbb{R}^d\,,$

(ii) $\|(\sigma_\varepsilon)_{ij}-\sigma_{ij}\|_{L^2([0,T]\times\mathbb{T}^d)}\to 0\,,\quad i,j=1,\dots,d,$

(iii) $\|(\dot\sigma_\varepsilon)_{ij}-\dot\sigma_{ij}\|_{L^2([0,T]\times\mathbb{T}^d)}\to 0\,,\quad i,j=1,\dots,d.$

The dot denotes the partial derivative in any spatial direction. We can assume without loss of generality that σ_ε *is smooth for each* $\varepsilon > 0$.

PROOF: For instance,

$$\sigma_\varepsilon = (I+\varepsilon\sigma)^{-1}(\sigma+\varepsilon I)$$

clearly works. We can always mollify it somewhat to get smoothness. Another choice is to convolute a with $\mathcal{G}_\varepsilon$ to smooth it and take the positive definite square root. Although it is not obvious, it is nevertheless true that $\|\nabla\sigma_\varepsilon - \nabla\sigma\|_{L^2}^2 \to 0$. If we need it, we can always add a small $\delta(\varepsilon)I$ to a or a_ε to make it elliptic. ∎

We now prove existence.

LEMMA 3.4 *Suppose* (A1) *holds and* c, f, *and* $g\in L^\infty$. *Then there is a* $u\in H_{1,a}$ *that is a solution to the equation* (3.1), *in the sense of* (3.5).

PROOF: For each $\varepsilon > 0$, it is known that a regular solution u_ε of

$$\begin{cases} \dfrac{\partial u_\varepsilon}{\partial t} + \nabla\cdot a_\varepsilon\nabla u_\varepsilon + \sigma_\varepsilon c\cdot\nabla u_\varepsilon = g \quad \text{for } 0\le t\le T\,,\\ u_\varepsilon(T,x) = f(x) \end{cases}\tag{3.6}$$

exists satisfying

$$\begin{cases} \sup\limits_\varepsilon \|u_\varepsilon\|_{L^\infty}\le C\\ \sup\limits_\varepsilon \displaystyle\int_0^T dt\int dx\,|\sigma_\varepsilon\nabla u_\varepsilon|^2 \le C\,. \end{cases}\tag{3.7}$$

∎

It is clear from the above estimates that we can choose a subsequence along which $u_\varepsilon \rightharpoonup u$ and $\sigma_\varepsilon \nabla u_\varepsilon \rightharpoonup v$ weakly in $L^2([0,T] \times \mathbb{T}^d)$. We will show that u is a solution of (3.1). Clearly $u \in L^\infty$. From

$$\frac{\partial u_\varepsilon}{\partial t} = \nabla \cdot \sigma_\varepsilon F_\varepsilon + G_\varepsilon \tag{3.8}$$

with $F_\varepsilon = -\sigma_\varepsilon \nabla u_\varepsilon$ and $G_\varepsilon = -c \cdot \sigma_\varepsilon \nabla u_\varepsilon + g$, by taking limits we get

$$\frac{\partial u}{\partial t} = \nabla \cdot \sigma F + G \tag{3.9}$$

with $F = -v$ and $G_\varepsilon = -c \cdot v + g$. It remains only to show that $u \in H_{1,a}$ and $\sigma \nabla u = v$. But we have the relation

$$\int_0^T \int_{\mathbb{T}^d} u_\varepsilon(t,x) \nabla \cdot \sigma_\varepsilon \psi \, dx \, dt = - \int_0^T \int_{\mathbb{T}^d} \sigma_\varepsilon \nabla u_\varepsilon \cdot \psi \, dx \, dt \,. \tag{3.10}$$

The strong convergence of σ_ε and its gradient along with the weak convergence in L^2 of u_ε allows us to pass to the limit to conclude that

$$\int_0^T \int_{\mathbb{T}^d} u(t,x) \nabla \cdot \sigma \psi \, dx \, dt = - \int_0^T \int_{\mathbb{T}^d} v \cdot \psi \, dx \, dt \tag{3.11}$$

holds for all smooth test functions. This proves that $u \in H_{1,a}$ and $\sigma \nabla u = v$. The boundary value is preserved, at least in the sense of a weak limit as $t \to T$, and we are done.

The next step is to establish the energy identity and some other estimates on the solution. The same argument that we used to prove uniqueness in Lemma 3.2 yields the following identities for each $\varepsilon > 0$ as well as for the limit:

$$\begin{aligned}
&\int_{\mathbb{T}^d} e^{kT} u_\varepsilon^2(T,x) dx - \int_{\mathbb{T}^d} e^{kt} u_\varepsilon^2(t,x) dx \\
&= \int_t^T \int_{\mathbb{T}^d} e^{kt} \left[2 \nabla u_\varepsilon \cdot a_\varepsilon \nabla u_\varepsilon - 2 u_\varepsilon \, c \cdot \sigma_\varepsilon \nabla u_\varepsilon + 2 u_\varepsilon g + k u_\varepsilon^2 \right] dx \, ds \\
&= 2 \int_t^T \int_{\mathbb{T}^d} e^{kt} \left\| \sigma_\varepsilon \nabla u_\varepsilon - \frac{1}{2} c u_\varepsilon \right\|^2 dx \, ds \\
&\quad + \int_t^T \int_{\mathbb{T}^d} e^{kt} \left[k - \frac{1}{2} \|c\|^2 \right] u_\varepsilon^2 \, dx \, ds + 2 \int_t^T \int_{\mathbb{T}^d} u_\varepsilon g \, dx \, ds \,.
\end{aligned}$$

We can rewrite this as

$$
\begin{aligned}
&\int_{\mathbb{T}^d} e^{kt} u_\varepsilon^2(t,x)dx + 2\int_t^T \int_{\mathbb{T}^d} e^{ks} \left\| \sigma_\varepsilon \nabla u_\varepsilon - \frac{1}{2} c u_\varepsilon \right\|^2 dx\,ds \\
&\qquad + \int_t^T \int_{\mathbb{T}^d} e^{ks} \left[k - \frac{1}{2}\|c\|^2 \right] u_\varepsilon^2 \, dx\,ds \\
&\qquad = \int_{\mathbb{T}^d} e^{kT} f^2(x) dx + 2\int_t^T \int_{\mathbb{T}^d} e^{ks} u_\varepsilon g\,dx\,ds
\end{aligned} \tag{3.12}
$$

and a similar equation with $\varepsilon = 0$ in the limit. Since the left-hand side is lower semicontinuous under weak convergence and the right-hand side is continuous and equality holds in the limit, we conclude that

$$
\lim_{\varepsilon\to 0} \|u_\varepsilon - u\|_{L^2([0,T]\times\mathbb{T}^d)} = 0
$$

and

$$
\lim_{\varepsilon\to 0} \|\sigma_\varepsilon \nabla u_\varepsilon - \sigma\nabla u\|_{L^2([0,T]\times\mathbb{T}^d)} = 0\,.
$$

The mappings $t \to u_\varepsilon(t,\cdot)$ are uniformly weakly continuous as maps into $L^2(\mathbb{T}^d)$, and the above arguments actually tell us that for each $t \in [0,T]$,

$$
\lim_{\varepsilon\to 0} \|u_\varepsilon(t,\cdot) - u(t,\cdot)\|_{L^2(\mathbb{T}^d)} = 0\,.
$$

Furthermore, we can read off from equation (3.12) the equicontinuity of $\|u_\varepsilon(t,\cdot)\|^2$ as a family of functions of t. We conclude that

$$
\lim_{\varepsilon\to 0} \sup_{0\le t\le T} \|u_\varepsilon(t,\cdot) - u(t,\cdot)\|_{L^2(\mathbb{T}^d)} = 0
$$

We combine these observations in the form of the following theorem:

THEOREM 3.5 *Under the assumptions* (A1), *if* c, f, $g \in L^\infty$, *there exists a unique* $u \in H_{1,a} \cap L^\infty$ *that solves* (3.1) *in the sense of* (3.5). *Moreover, the solution admits the following energy identity and estimate*:

$$
\begin{aligned}
&\int_{\mathbb{T}^d} f^2(x)dx - \int_{\mathbb{T}^d} u^2(t,x)dx \\
&\qquad = \int_t^T \int_{\mathbb{T}^d} [2\nabla u \cdot a\nabla u - 2uc\cdot\sigma\nabla u + 2ug]\,dx\,ds
\end{aligned} \tag{3.13}
$$

and

$$
\begin{aligned}
&\sup_{0\le t\le T} \|u(t,\cdot)\|^2_{L^2(\mathbb{T}^d)} + \int_0^T \int_{\mathbb{T}^d} |\sigma\nabla u|^2\,dx\,dt \\
&\qquad \le C_T \left[\|f\|^2_{L^2(\mathbb{T}^d)} + \|g\|^2_{L^2([0,T]\times\mathbb{T}^d)} \right]
\end{aligned} \tag{3.14}
$$

where the constant C_T *depends on* T *and the* L^∞ *norm of* c.

PROOF: Except for the estimate (3.14), everything else has already been established. From the identity (3.13) and Schwarz's inequality, we get

$$\begin{aligned}\sup_{0\le t\le T}\|u(t,\cdot)\|^2_{L^2(\mathbb{T}^d)}&+\int_0^T\int_{\mathbb{T}^d}|\sigma\nabla u|^2\,dx\,dt\\&\le\|f\|^2_{L^2(\mathbb{T}^d)}+C_T\left[\|f\|^2_{L^2(\mathbb{T}^d)}+\|g\|^2_{L^2([0,T]\times\mathbb{T}^d)}\right]\\&\quad+\frac{1}{2}\sup_{0\le t\le T}\|u(t,\cdot)\|^2_{L^2(\mathbb{T}^d)}\end{aligned}$$

and we are done. ∎

We now turn to the study of equation (3.2). If we assume that everything in sight is smooth and the problem is elliptic, then (3.2) clearly has a smooth solution. If we multiply by u and integrate, we get the energy identity

$$\begin{aligned}\int_{\mathbb{T}^d}u^2(t,x)dx&+\int_t^T\int_{\mathbb{T}^d}|\sigma\nabla u|^2\,dx\,ds\\&=\int_{\mathbb{T}^d}f^2dx+\int_t^T\int_{\mathbb{T}^d}(uc_1-uc_2+h)\sigma\nabla u\,dx\,ds\\&\quad-\int_t^T\int_{\mathbb{T}^d}gu\,dx\,ds\,.\end{aligned}\tag{3.15}$$

It is but one step from this to get an a priori estimate

$$\begin{aligned}\sup_{0\le t\le T}\|u(t,\cdot)\|^2_{L^2(\mathbb{T}^d)}&+\int_0^T\int_{\mathbb{T}^d}|\sigma\nabla u|^2\,dx\,dt\\&\le C_T\left[\|f\|^2_{L^2(\mathbb{T}^d)}+\|g\|^2_{L^2([0,T]\times\mathbb{T}^d)}+\|h\|^2_{L^2([0,T]\times\mathbb{T}^d)}\right]\end{aligned}\tag{3.16}$$

where the constant depends on T and L^∞ bounds for c_1 and c_2.

In the same fashion as before, we can pass to the limit via solutions of a mollified equation involving smooth a_ε, σ_ε, $c_{1,\varepsilon}$, $c_{2,\varepsilon}$, g_ε, h_ε, and f_ε. When the dust settles what we have is the following:

THEOREM 3.6 *For any c_1, c_2 in L^∞ and f, g, h in L^2, there exists a function $u\in H_{1,a}$ satisfying the bounds* (3.16), *that is, a solution of* (3.2) *in the sense of* (3.5).

The difficult question is that of uniqueness. The sticky point is Lemma 3.1, which requires that the functions u_i be in L^∞ in order to prove the energy

identity. We can save the situation by demanding a little bit more from our potential solutions. If we have a function $u \in H_{1,a}$ with $\frac{du}{dt} \in H_{-1,a}$, what we have is a function $u \in H_{1,a}$ that satisfies (3.4) in the sense of (3.5). We could try a smooth ϕ or test against a $\phi \in H_{1,a}$ that satisfies

$$\frac{\partial \phi}{\partial t} = \nabla \cdot \sigma \xi + \eta$$

verifying

$$\frac{d}{dt} \int_{\mathbb{T}^d} u(t,x)\phi(t,x)dx = \int_{\mathbb{T}^d} [-F \cdot \sigma \nabla \phi + G\phi - \xi \cdot \sigma \nabla u + u\eta]\, dx\,.$$

Although it is sufficient to use smooth test functions for $u \in L^\infty$, for $u \in L^2$ we need a larger class. We can get away with using test functions $\phi \in L^\infty$. So let us adopt the convention that we will say that (3.4) holds in the *distributional* sense if we test with smooth functions and in the *weak* sense if we enlarge the class of test functions to include L^∞ functions. We do not know if the distinction is real or only technical. We will prove existence and uniqueness of weak solutions. Just for the record, we note the following:

LEMMA 3.7 *If $u \in L^\infty$ and* (3.4) *holds in the distributional sense, then it holds in the weak sense.*

PROOF: This is basically the same as Lemma 3.1. Start with $u, v \in L^\infty$ satisfying

$$\frac{\partial u}{\partial t} = \nabla \cdot \sigma f + g\,, \qquad \frac{\partial v}{\partial t} = \nabla \cdot \sigma \xi + \eta\,,$$

mollify both by $\mathcal{G}_\varepsilon$, compute

$$\frac{d}{dt} \int_{\mathbb{T}^d} u_\varepsilon(t,x) v_\varepsilon(t,x)\, dx\,,$$

integrate by parts the obvious terms, and pass to the limit as $\varepsilon \to 0$. ∎

We now strengthen Theorem 3.6 to get the slightly stronger version of existence, and then prove uniqueness.

THEOREM 3.8 *Let $c_1, c_2 \in L^\infty$ and $f, g, h \in L^2$. Then equation* (3.2) *admits a solution in the weak sense. The weak solution is unique and satisfies the energy identity and estimate.*

PROOF: We do not have to work any harder to construct a solution in the weak sense. We use the special choice

$$a_\varepsilon = a * \mathcal{G}_\varepsilon$$

to solve our approximate problems. If we mollify our bounded test function v by $\mathcal{G}_\varepsilon$ as well, by Lemma 3.1 we get

$$\frac{\partial v_\varepsilon}{\partial t} = \nabla\sigma_\varepsilon\,\xi_\varepsilon + \eta_\varepsilon\,,$$

and we calculate

$$\begin{aligned} &\frac{d}{dt}\int_{\mathbb{T}^d} u_\varepsilon(t,x)v_\varepsilon(t,x)dx = \\ &\int_{\mathbb{T}^d}\left[-F_\varepsilon\cdot\sigma_\varepsilon\nabla v_\varepsilon + G_\varepsilon v_\varepsilon - \xi_\varepsilon\cdot\sigma_\varepsilon\nabla u_\varepsilon + \eta_\varepsilon u_\varepsilon\right]dx \end{aligned} \tag{3.17}$$

where

$$\begin{gathered} F_\varepsilon = -\sigma_\varepsilon\nabla u_\varepsilon - c_{2,\varepsilon}u_\varepsilon + h_\varepsilon\,, \quad G_\varepsilon = -c_{1,\varepsilon}\cdot\sigma_\varepsilon\nabla u_\varepsilon + g_\varepsilon\,, \\ \xi_\varepsilon = \xi * \mathcal{G}_\varepsilon\,, \quad \text{and} \quad \eta_\varepsilon = \eta * \mathcal{G}_\varepsilon\,. \end{gathered}$$

As $\varepsilon \to 0$, F_ε, G_ε, $\sigma_\varepsilon\nabla u_\varepsilon$, and u_ε converge weakly to $F = \sigma\nabla u - c_2 u + h$, $G = c_1\sigma\nabla u + g$, $\sigma\nabla u$, and u, respectively. On the other hand, $\sigma_\varepsilon\nabla v_\varepsilon$, v_ε, ξ_ε, and η_ε converge strongly to $\sigma\nabla v$, v, ξ, and η, respectively. We can therefore take the limit as $\varepsilon \to 0$ in (3.18), and we are done. ∎

We now turn to the proof of uniqueness as well as the energy identity. Let us look at the special case of (3.2) with $c_1 = c_2 = 0$:

$$\frac{\partial u}{\partial t} + \nabla\cdot a\nabla u = g + \nabla\cdot\sigma h\,, \qquad u(T,x) = f(x)\,, \tag{3.18}$$

with f, g, h in $L^2(\mathbb{T}^d)$. We can now use the solution of

$$\frac{\partial v}{\partial s} = \nabla\cdot a\nabla v \quad \text{in } [t,T]\times\mathbb{T}^d\,, \qquad v(t,x) = \phi(x)\,,$$

in $L^\infty \cap H_{1,a}$ as a test function to conclude that the difference w of two solutions of (3.18) in the weak sense satisfies

$$\int_{\mathbb{T}^d}\phi(x)w(t,x) = \int_t^T\int_{\mathbb{T}^d}[-w\nabla\cdot a\nabla v + v\nabla\cdot a\nabla w]dx\,ds = 0\,,$$

proving the uniqueness of such solutions in the special case. If we approximate f, g in L^2 by smooth functions f_n, g_n, respectively, and take $h = 0$, we can solve (3.18) with $f_n, g_n, 0$ replacing f, g, h. From the estimate (3.16) it follows that the solutions u_n have a limit in a very strong sense, and the limit will be a solution of (3.18) in the weak sense. Because of uniqueness, the limit has to equal u. In particular, u as the limit of u_n will satisfy the energy identity

$$\int_{\mathbb{T}^d} u^2(t_2, x)dx - \int_{\mathbb{T}^d} u^2(t_1, x)dx = 2\int_{t_1}^{t_2}\int_{\mathbb{T}^d} [|\sigma\nabla u|^2 + ug]dx\, dt\,.$$

We can replace g by $g + \nabla\sigma h$ so long as h is smooth. Again, we can approximate $h \in L^2$ by smooth h, and we conclude that any solution in the weak sense of (3.18) will have to satisfy

$$\begin{aligned}(3.19)\qquad &\int_{\mathbb{T}^d} u^2(t_2, x)dx - \int_{\mathbb{T}^d} u^2(t_1, x)dx \\ &\quad = 2\int_{t_1}^{t_2}\int_{\mathbb{T}^d} [|\sigma\nabla u|^2 + ug - h\nabla u]dx\, dt\,.\end{aligned}$$

Now if we take any solution of (3.2) in the weak sense, we can think of it as a solution in the weak sense of

$$(3.20)\qquad \frac{\partial u}{\partial t} + \nabla\cdot a\nabla u = \tilde{g} + \nabla\cdot\sigma\tilde{h}\,, \qquad u(T, x) = f(x)\,,$$

with

$$\tilde{g} = g - c_1\cdot\sigma\nabla u \quad\text{and}\quad \tilde{h} = h - \nabla\cdot\sigma c_2 u\,,$$

and (3.19) will hold with $\tilde{g}$ and $\tilde{h}$ replacing g and h. This is nothing else but the energy identity for (3.2). Once we have the energy identity for all solutions of (3.2) in the weak sense, uniqueness is immediate.

Remark. One of the consequences of uniqueness is that if $u \in H_{1,a}$ and

$$\frac{\partial u}{\partial t} = \nabla\cdot\sigma F + G$$

in the weak sense, with $F, G \in L^2$, then we can rewrite

$$\frac{\partial u}{\partial t} + \nabla\cdot a\nabla u = \nabla\cdot\sigma[F + \sigma\nabla u] + G\,,$$

and as we saw in the earlier argument, u is a limit of bounded solutions. In other words, there are bounded functions u_n such that

$$\frac{\partial u_n}{\partial t} = \nabla\cdot\sigma F_n + G_n$$

with u_n, F_n, and G_n converging in L^2, respectively, to u, F, and G. In particular, Lemma 3.1 is now valid, and we can use test functions ϕ that satisfy

$$\frac{\partial \phi}{\partial t} = \nabla \cdot \sigma \xi + \eta$$

in the weak sense.

Remark. At this point one may have the impression that to get the weak solution one has to be careful in the choice of a_ε used to approximate the problem. In fact, it is not necessary. If we use arbitrary a_ε satisfying Lemma 3.3, we can solve at the same time equations of the form

$$\frac{\partial v_\varepsilon}{\partial t} = \nabla \cdot a_\varepsilon \nabla v_\varepsilon \,, \qquad v_\varepsilon(0,x) = f(x)\,,$$

and, as we saw earlier, $v_\varepsilon \to v$ very nicely as $\varepsilon \to 0$ where v solves

$$\frac{\partial v}{\partial t} = \nabla \cdot a \nabla v \,, \qquad v(0,x) = f(x)\,.$$

It is easy to check that such v's can be used as test functions. But that is enough to establish uniqueness.

If we want to go beyond bounded c in equation (3.1), it is not very natural to replace the condition $c \in L^\infty$ by $c \in L^2$. If we take the simplest case of $L_t = L = \Delta + b(x) \cdot \nabla$ and impose the condition that $b(x) \in L^p$, there is an easy counterexample of a smooth solution to $u_t - Lu \equiv 0$ provided the dimension $d > p$. As for relaxing the condition $\nabla \sigma \in L^2$, even in $d = 1$ the operator $L = D_x |x|^\alpha D_x$ has problems at 0 as soon as $\alpha < 1$. It is easy to see that if we approximate by $D_x a(x) D_x$ with smooth $a(\cdot)$ from below, we have the Green's function $G(x,y) = 0$ for $x < 0,\ y > 0$ or $x > 0,\ y < 0$ for all the approximations. On the other hand, if we approximate by $D_x(\varepsilon + |x|^\alpha) D_x$, then we have $G(x,y) > 0$ for all x, y even in the limit.

We now turn to the study of equation (3.3), which as we said earlier is important in probability theory. In the context that motivated this study, where the operator L_t comes from a specific solution of a nonlinear problem, we are given a particular solution of equation (3.3). In dealing with simple exclusion models, the densities are all bounded by 1, and after renormalization they remain bounded. We are concerned with related existence and uniqueness results that are of interest in probability theory and are implied by the existence of a particular solution. We will deal exclusively with bounded densities, and therefore Lemma 3.1 applies.

We will use relative entropy as our main tool for proving uniqueness. Let us recall the definition. If u_1 and u_2 are nonnegative integrable functions with $\int u_1\,dx = \int u_2\,dx = 1$,

$$h(u_1/u_2) = \int dx\, u_1 \log(u_1/u_2)\,.$$

LEMMA 3.9 *Let* (A1) *hold. Suppose that* u_i, $i = 1, 2$, *are bounded, nonnegative solutions of the forward equations*

$$\frac{\partial u_i}{\partial t} = L_t^* u_i = \nabla \cdot a \nabla u_i - \nabla \cdot \sigma c_i u_i\,, \qquad u_i(0) = u_i^0\,,$$

satisfying

$$\begin{cases} \int_0^T dt \int dx \dfrac{\nabla u_i \cdot a \nabla u_i}{u_i} < \infty\,, & i = 1,2, \\ \int_0^T dt \int dx\, |c_i|^2 u_i < \infty\,, & i = 1,2, \\ \int_0^T dt \int dx\, |c_2 - c_1|^2 u_1 < \infty\,, & \\ h(u_1^0/u_2^0) < \infty\,. & \end{cases}$$

Then the relative entropy at time $T > 0$ *is finite and*

$$\begin{aligned} &h(u_1(T)/u_2(T)) - h(u_1^0/u_2^0) \\ &\quad \le -\frac{1}{2}\int_0^T dt \int dx \left(\frac{\nabla u_1}{u_1} - \frac{\nabla u_2}{u_2}\right) \cdot a \left(\frac{\nabla u_1}{u_1} - \frac{\nabla u_2}{u_2}\right) u_1 \\ &\qquad + \frac{1}{2}\int_0^T dt \int dx\, (c_2 - c_1)^2 u_1\,. \end{aligned}$$

PROOF: We note first that our equations can be cast in the form

$$\frac{\partial u_i}{\partial t} = \nabla \cdot \sigma F_i$$

with F_i in $L^2([0,T] \times \mathbb{T}^d)$. We wish to apply Lemma 3.1 for the function $H(u_1, u_2) = u_1 \log \frac{u_1}{u_2}$. Due the singularity near 0, we have to do it with $H_\delta(u_1, u_2) = u_1 \log \frac{u_1+\delta}{u_2+\delta}$ and then let $\delta \to 0$. This can be done, and after an

integration by parts one obtains

$$\begin{aligned}
&h(u_1(T)/u_2(T)) - h(u_1^0/u_2^0)\\
&\quad = -\int_0^T dt \int dx \left(\frac{\nabla u_1}{u_1} - \frac{\nabla u_2}{u_2}\right) \cdot a \left(\frac{\nabla u_1}{u_1} - \frac{\nabla u_2}{u_2}\right) u_1\\
&\qquad - \int_0^T dt \int dx \left(\frac{\nabla u_1}{u_1} - \frac{\nabla u_2}{u_2}\right) \cdot \sigma(c_2 - c_1) u_1 .
\end{aligned}$$

The lemma follows by Schwarz's inequality. ∎

A consequence of the lemma is the following:

COROLLARY 3.10 *If condition* (A1) *holds, a nonnegative bounded solution of*

$$\frac{\partial u}{\partial t} = \nabla \cdot a \nabla u - \nabla \cdot \sigma c u$$

in the class satisfying

$$\begin{cases} \displaystyle\int_0^T \int_{\mathbb{T}^d} \frac{|\sigma \nabla u|^2}{u}\, dx\, dt < \infty \\ \displaystyle\int_0^T \int_{\mathbb{T}^d} |c|^2 u\, dx\, dt < \infty \end{cases}$$

is unique.

Before we prove the next theorem on the existence and uniqueness of solutions for (3.3) in certain situations, we prove the following lemma that we need:

LEMMA 3.11 *Suppose that* (A1) *holds and* $c \in L^2\left([0,T] \times \mathbb{T}^d\right)$. *Then for each bounded nonnegative* u^0 *there exists a weak solution of*

$$\begin{cases} \dfrac{\partial u}{\partial t} = \nabla \cdot a \nabla u + c \cdot \sigma \nabla u \quad \textit{in } [0,T] \times \mathbb{T}^d \\ u(0) = u^0 \end{cases}$$

satisfying $0 \le u \le \|u^0\|_\infty$ *and*

$$\int_0^T dt \int dx\, \frac{\nabla u \cdot a \nabla u}{u} \le C \tag{3.21}$$

with $C < \infty$ *depending only on* $\|u^0\|_\infty$ *and* $\|c\|_{L^2([0,T]\times\mathbb{T}^d)}$.

PROOF: Let c^n be a truncation of c at level n. For each integer n, by reversing time in Theorem 3.8, we obtain a unique weak solution $u_n \geq 0$ of

$$\frac{\partial u_n}{\partial t} = \nabla \cdot a \nabla u_n + \sigma c^n \cdot \nabla u_n \,, \quad u_n(0) = u^0 \,, \tag{3.22}$$

satisfying $0 \leq u_n \leq \|u^0\|_\infty$. If we take $\phi(u) = u \log u$ in Lemma 3.1 and integrate by parts (we can replace u by $u + \delta$ and let $\delta \to 0$ at the end), we get

$$\int_0^T dt \int dx \, \frac{|\sigma \nabla u_n|^2}{u_n} \leq C \,,$$

and $\|u_n\|_{H_{1,a}} \leq C$ where $C < \infty$ depends only on $\|u^0\|_\infty$ and $\|c\|_{L^2([0,T] \times \mathbb{T}^d)}$. From the definition, for any $\phi \in C_0^\infty\left([0,T) \times \mathbb{T}^d\right)$,

$$\begin{aligned} 0 = &- \int_0^T dt \int dx \, u_n \frac{\partial \phi}{\partial t} - \int dx \, f \phi(0,x) \\ &+ \int_0^T dt \int dx \, \nabla u_n \cdot a \nabla \phi - \int_0^T dt \int dx \, c^n \cdot \sigma \nabla u_n \, \phi \,. \end{aligned} \tag{3.23}$$

We can extract subsequences $u_n \rightharpoonup u$ and $\sigma \nabla u_n \rightharpoonup v$ weakly in $L^2([0,T] \times \mathbb{T}^d)$. We can identify $v = \sigma \nabla u$ as in the proof of Lemma 3.4. By lower semicontinuity we obtain (3.21). Since $c^n \to c$ strongly in L^2, we can take the limit as $n \to \infty$ in (3.23) to verify that u is the desired weak solution. ∎

Let us remark once again that there is in general no uniqueness in this context.

The final theorem in this section is the following:

THEOREM 3.12 *Suppose that* (A1) *holds and in addition that there exists a bounded, nonnegative weak solution ρ of the equation $\partial_t \rho = L_t^* \rho = \nabla \cdot a \nabla \rho - \nabla \cdot \sigma c \rho$, $\rho(0,x) = \rho^0(x)$, satisfying*

$$\begin{cases} \displaystyle\int_0^T dt \int dx \, \frac{\nabla \rho \cdot a \nabla \rho}{\rho} < \infty \,, \\ \displaystyle\int_0^T dt \int dx \, |c|^2 \rho < \infty \,. \end{cases}$$

Then, for any $0 \leq R^0 \leq C\rho^0$ for some $C < \infty$, there exists a unique weak solution R of the equation $\partial_t R = L_t^* R$, $R(0,x) = R^0(x)$, satisfying $0 \leq R \leq C\rho$ on $[0,T] \times \mathbb{T}^d$ and

$$\int_0^T dt \int dx \, \frac{\nabla R \cdot a \nabla R}{R} < \infty \,.$$

PROOF: First of all, note that the uniqueness part of the lemma follows immediately from Corollary 3.10. We now prove the existence part.

Step 1. Suppose that $\rho \geq \delta$ for some $\delta > 0$, and

$$\int_0^T dt \int dx\, |c|^2 < \infty .$$

Let

$$g = 2\sigma \frac{\nabla \rho}{\rho} - c .$$

Since $g \in L^2\left([0,T] \times \mathbb{T}^d\right)$, Theorem 3.5 gives us a bounded, nonnegative, weak solution u of the equation

$$\frac{\partial u}{\partial t} = \nabla \cdot a \nabla u + \sigma g \cdot \nabla u , \qquad u(0,x) = R^0/\rho^0 ,$$

satisfying

$$\int_0^T dt \int dx\, \frac{\nabla u \cdot a \nabla u}{u} < \infty .$$

Now,

$$\begin{aligned}
&\int_0^T dt \int dx \left(\frac{\partial R}{\partial t}\phi + \nabla R \cdot a \nabla \phi + \sigma c \cdot R \nabla \phi \right) \\
&\qquad = \int_0^T dt \int dx \left(\frac{\partial u}{\partial t}\psi + \nabla u \cdot a \nabla \psi - \sigma g \cdot \nabla u\, \psi \right. \\
&\qquad\qquad \left. + \frac{\partial \rho}{\partial t}\xi + \nabla \rho \cdot a \nabla \xi + \sigma c \cdot \rho \nabla \xi \right)
\end{aligned} \tag{3.24}$$

where $\psi = \rho\phi$ and $\xi = u\phi$. If ϕ is a smooth test function, then since both ρ and u are elements of $H_{1,a} \cap L^\infty$, so are ψ and ξ. Therefore the right-hand side, and consequently the left-hand side, of (3.24) vanishes for any smooth ϕ. It follows that $R = \rho u$ is a weak solution of the equation

$$\frac{\partial R}{\partial t} = \nabla \cdot a \nabla R - \nabla \cdot \sigma c R , \qquad R(0) = R^0 .$$

Furthermore, since u and ρ are bounded,

$$\begin{aligned}
&\int_0^T dt \int dx\, \frac{\nabla R \cdot a \nabla R}{R} \\
&\qquad \leq \int_0^T dt \int dx\, \frac{\nabla u \cdot a \nabla u}{u}\rho + \int_0^T dt \int dx\, \frac{\nabla \rho \cdot a \nabla \rho}{\rho} u < \infty .
\end{aligned}$$

This completes the proof for the case $\int dt \int dx\, |c|^2 < \infty$.

Step 2. In the general case, let

$$\rho_\delta = \rho + \delta\,,$$

and note that ρ_δ is a weak solution of the forward equation with c replaced by c_δ, which is given by

$$c_\delta = \frac{c\rho}{\rho_\delta}\,.$$

This ρ_δ satisfies the assumptions of the lemma, and furthermore $\rho_\delta \geq \delta$ and

$$\int_0^T dt \int dx\, c_\delta^2 < \infty\,.$$

Let $R^0 \leq C\rho^0$. By step 1 there exists a weak solution R_δ of the forward equation

$$\frac{\partial R_\delta}{\partial t} = \nabla \cdot a \nabla R_\delta - \nabla \cdot \sigma c_\delta R_\delta\,, \quad R_\delta(0) = R^0\,,$$

satisfying

$$\int_0^T dt \int dx\, \frac{\nabla R_\delta \cdot a \nabla R_\delta}{R_\delta} < \infty\,.$$

From Lemma 3.9,

$$\begin{aligned} \int_0^T dt \int dx \frac{\nabla R_\delta \cdot a \nabla R_\delta}{R_\delta} &\leq 4\left[h(R_\delta(T)/\rho_\delta(T)) - h(R^0/\rho_\delta^0)\right] \\ &\quad + 2\int_0^T dt \int dx \frac{\nabla \rho \cdot a \nabla \rho}{\rho}\,. \end{aligned} \tag{3.25}$$

The right-hand side is bounded independently of δ.

Since R_δ and u_δ are uniformly bounded, we can extract a subsequence along which $R_\delta \rightharpoonup R$ and $u_\delta \rightharpoonup u$ weakly in L^∞. In other words, for any $\phi \in L^1([0,T] \times \mathbb{T}^d)$,

$$\int_0^T dt \int dx\, \phi R_\delta \to \int_0^T dt \int dx\, \phi R\,,$$

and similarly for u_δ. But then since ρ is bounded,

$$\int_0^T dt \int dx\, \phi \rho_\delta u_\delta \to \int_0^T dt \int dx\, \phi \rho u\,.$$

Therefore $R = \rho u$. Also by (3.25), as in the proof of Lemma 3.3, we can find a subsequence $\sigma \nabla R_\delta \rightharpoonup \sigma \nabla R$ in L^2. For any $\phi \in C_0^\infty\left([0,T) \times \mathbb{T}^d\right)$,

$$\begin{aligned} 0 = &- \int_0^T dt \int dx\, R_\delta \frac{\partial \phi}{\partial t} - \int dx\, R^0 \phi(0,x) \\ &+ \int_0^T dt \int dx\, \nabla R_\delta \cdot a \nabla \phi - \int_0^T dt \int dx\, R_\delta c_\delta \cdot \sigma \nabla \phi\,. \end{aligned} \tag{3.26}$$

Write

$$\int_0^T dt \int dx\, u_\delta \rho c \cdot \sigma \nabla \phi\,.$$

Note that $\rho c \cdot \sigma \nabla \phi \in L^1\left([0,T] \times \mathbb{T}^d\right)$. Therefore we can let $\delta \to 0$ to get

$$\int_0^T dt \int dx\, u \rho c \cdot \sigma \nabla \phi = \int_0^T dt \int dx\, R c \cdot \sigma \nabla \phi\,.$$

We can take the limit of the other terms in (3.26) as in the proof of the previous theorems and find that R is the desired weak solution.

∎

Remarks. We make a few comments regarding some earlier work.

1. The special case of $a \equiv I$ has been studied in detail by Eric Carlen in [2].

2. In the time homogeneous case, the energy inequality holds for distributional solutions (using only smooth test functions) by the following lemma, which is a special case of a general theorem of Lions and Magenes [5], and therefore such solutions are unique.

LEMMA 3.13 *Let $V \subset H \subset V'$ be separable Hilbert spaces with V and V' being in duality relative to the inner product of H. Suppose we have a curve $u(t) : 0 \le t \le T$ satisfying $u \in L^2(0,T;V)$ and $u' \in L^2(0,T;V')$. Then u is almost everywhere equal to a function continuous from $[0,T]$ into H, and for $0 \le t_1 \le t_2 \le T$,*

$$\|u(t_2)\|_H^2 - \|u(t_1)\|_H^2 = 2 \int_{t_1}^{t_2} \langle u(t), u'(t) \rangle dt\,.$$

3. There are a few previous results of existence in the time homogeneous case. Let

$$\mathcal{E}(u,v) = \int dx \nabla\phi \cdot a\nabla\psi\,.$$

If $a \in W^{1,2}$, then L_0 with $b = 0$ maps C^∞ into L^2 so $\mathcal{E}$ comes from an operator and therefore it has a minimal self-adjoint extension, the Freidrichs extension. If only $\sigma \in W^{1,2}$, this is no longer the case; however, Albeverio, Hoegh-Krohn, and Streit [1] proved that if $\sigma \in W^{1,2}$, $\mathcal{E}$ with domain C_0^∞ is closable on L^2. The closure $\bar{\mathcal{E}}$ is then the quadratic form of a self-adjoint operator L_0.

In particular, for each $u_0 \in L^2$ there exists a weak solution u of

$$\frac{\partial u}{\partial t} - \nabla\cdot a\nabla u = 0 \quad \text{in } [0,T]\times\mathbb{T}^d\,, \quad u(0,x) = u^0(x)\,,$$

The closure is a regular Dirichlet form, so for almost every $x \in \mathbb{T}^d$ there exists a diffusion process P with generator L_0 and $P(x(0) = x) = 1$. Detailed properties of P can be found in [4].

The argument in [1] helps explain the condition (A1). For

$$\phi \in C_0^\infty(\mathbb{T}^d;\mathbb{R}) \quad \text{and} \quad \psi \in C_0^\infty(\mathbb{T}^d;\mathbb{R}^d)\,,$$

let

$$\lambda(\phi) = \int dx\, (\sigma\nabla\phi)\cdot\psi\,.$$

Integrating by parts and applying Schwarz's inequality,

$$|\lambda| \le (\|\psi\|_\infty + \|\nabla\psi\|_\infty)\int dx\ \left(|\nabla\cdot\sigma|^2 + \|\sigma\|^2\right),$$

so if $\int dx\ \left(|\nabla\cdot\sigma|^2 + \|\sigma\|^2\right) < \infty$, then λ is bounded on L^2. Therefore $(\sigma\nabla)^*$ is densely defined, and consequently $\sigma\nabla$ is closable. Finally, if $\sigma\nabla$ is closable, then the quadratic form $\mathcal{E}(\phi,\psi) = \int dx\, \sigma\nabla\phi\cdot\sigma\nabla\psi$ is closable.

4. The difference in difficulty between existence and uniqueness is made very apparent by the situation in one dimension. As we saw already, uniqueness fails for the operator $L = D_x|x|^\alpha D_x$ if $\alpha < 1$. On the other hand, $\mathcal{E}$ is closable under much more general conditions. A point x is called *regular* for $a(x)$ if a^{-1} is locally integrable at x, and a nonregular point is called *singular*. The following theorem is well-known; a proof can be found in [4]:

5. These results are somewhat similar to those in [8] for first-order equations.

6. See [3] for a discussion of the elliptic case.

THEOREM 3.14 *In one dimension, the form $\mathcal{E}$ is closable if and only if $a(x)$ vanishes almost everywhere on its singular set.*

7. Let $m \in L^1(\mathbb{T}^d)$. One can consider the Dirichlet form

$$\mathcal{E}(u,v) = \int m(x)dx\, \nabla\phi \cdot a\nabla\psi\,.$$

The corresponding operator

$$Lu = \nabla \cdot a\nabla u + \frac{\nabla m}{m} \cdot a\nabla u \tag{3.27}$$

is reversible on $L^2(m(x)dx)$ instead of $L^2(dx)$. Certainly there is no reason to prefer one to the other. However, since our methods are not based on Dirichlet forms, operators of the form (3.27) fit easily into the theory. This is a case in which the conditions of Theorem 3.12 are very easy to check. If m is nonnegative and bounded, and

$$\int dx \frac{\nabla m \cdot a\nabla m}{m} < \infty\,,$$

then m itself is an a priori weak solution of the forward equation $\partial_t m - L^* m = 0$ satisfying the conditions of the theorem. So if $\sigma_{ij} \in W^{1,2}$, $i,j = 1,\dots,d$, then L^* generates a nice semigroup on $L^\infty(m(x)dx)$.

4 Very Weak Solutions

A *very weak* solution of the forward equation

$$\begin{cases} \dfrac{\partial u}{\partial t} = \nabla \cdot a\nabla u - \nabla \cdot \sigma c u & \text{in } [0,T] \times \mathbb{T}^d \\ u(0) = u^0 \end{cases} \tag{4.1}$$

is a function $u \in L^\infty$ satisfying

$$\begin{aligned} 0 = &-\int_0^T dt \int dx\, u\frac{\partial\phi}{\partial t} - \int dx\, u^0\phi(0,x) \\ &- \int_0^T dt \int dx\, u\,(\nabla \cdot a\nabla\phi) + \int_0^T dt \int dx\, u\sigma c \cdot \nabla\phi \end{aligned} \tag{4.2}$$

for all $\phi \in C_0^\infty\left([0,T) \times \mathbb{T}^d\right)$. In order for the last term in (4.2) to make sense, we will assume that

$$\int_0^T dt \int dx\, |c|^2 u < \infty\,. \tag{4.3}$$

If (A1) holds, then the rest of (4.2) makes sense for any smooth ϕ since

$$\nabla \cdot a \nabla \phi = (\nabla \cdot a)\, \nabla \phi + a \nabla \cdot \nabla \phi\,,$$

and (A1) implies in particular that $\nabla \cdot a$ and a are in $L^1\left([0,T] \times \mathbb{T}^d\right)$.

Such questions come up naturally when we study the existence and uniqueness of solutions to the martingale problem corresponding to the operator

$$L_t = \nabla \cdot a \nabla u + \sigma c \cdot \nabla u\,.$$

It is natural to write the operator in nondivergence form and ask questions of uniqueness for weak solutions of the adjoint equation, having defined the weak solution by using only smooth test functions. In this context it is not natural to assume a priori that u has any regularity.

THEOREM 4.1 *Suppose that* (A2) *holds. Assume that* $u \in L^\infty\left([0,T] \times \mathbb{T}^d\right)$ *is a very weak solution of the forward equation* (4.1) *and that* (4.3) *holds. Then we have the following:*

(i) $u \in H_{1,a}$, *and* u *is a weak solution of* (4.1).

(ii) *If, in addition,* u *is nonnegative, then we also have*

$$\int_0^T dt \int dx \frac{\langle \nabla u, a \nabla u\rangle}{u} < \infty\,.$$

PROOF: Suppose we have a very weak solution of the forward equation (4.1). Let $\mathcal{G}_\varepsilon$ be the fundamental solution of the heat equation at time $t = \varepsilon^2$ and $u_\varepsilon = u * \mathcal{G}_\varepsilon$. Then it follows that

$$\frac{\partial u_\varepsilon}{\partial t} = \nabla \cdot (a \nabla u)_\varepsilon - \nabla \cdot (\sigma c u)_\varepsilon\,. \tag{4.4}$$

Multiply equation (4.4) by u_ε, integrate over space and time, and integrate by parts to obtain

$$\begin{aligned} &\int dx\, u_\varepsilon^2(T) - \int dx\, u_\varepsilon^2(0) \\ &\quad = -\int_0^T dt \int dx\, \nabla u_\varepsilon \cdot (a \nabla u)_\varepsilon + \int_0^T dt \int dx\, (\sigma c u)_\varepsilon \cdot \nabla u_\varepsilon\,. \end{aligned} \tag{4.5}$$

∎

Consider the principal term

$$\begin{aligned}
&\int dx\,\nabla u_\varepsilon\cdot(a\nabla u)_\varepsilon \\
&= \int dx\,(\nabla u_\varepsilon)\cdot a(\nabla u_\varepsilon) \\
&\quad + \int dx\,\nabla u_\varepsilon(x)\cdot\int dy\;(a(x+y)-a(x))\,\nabla u(x+y)\mathcal{G}_\varepsilon(y)\,.
\end{aligned}\tag{4.6}$$

An integration by parts gives

$$\begin{aligned}
&\int dx\,\nabla u_\varepsilon(x)\int dy\;(a(x+y)-a(x))\,\nabla u(x+y)\mathcal{G}_\varepsilon(y) \\
&\quad = \int dx\,\nabla u_\varepsilon(x)\int dy\,\nabla\cdot a(x+y)\,u(x+y)\mathcal{G}_\varepsilon(y) \\
&\qquad + \int dx\,\nabla u_\varepsilon(x)\int dy\;(a(x+y)-a(x))\,u(x+y)\nabla\mathcal{G}_\varepsilon(y)\,.
\end{aligned}$$

We use Schwarz's inequality to bound the first term on the right-hand side by

$$\frac{1}{4}\int dx\,\nabla u_\varepsilon\cdot a_\varepsilon\nabla u_\varepsilon+\|u\|_{L^\infty}^2\int dx\,(\nabla\cdot a)a^{-1}(\nabla\cdot a)\,.$$

By assumption, the second term above is bounded by some constant $B_1<\infty$. By Lemma 2.4,

$$\int dx\,\nabla u_\varepsilon\cdot a_\varepsilon\nabla u_\varepsilon=\int dx\;|(\sigma\nabla u)_\varepsilon|^2+o(1)\,.$$

The second term on the right-hand side of (4.6) can be handled similarly. We prefer to do the convolution on $\mathbb{R}^d$ rather than on $\mathbb{T}^d$:

$$\begin{aligned}
&\int dx\,\nabla u_\varepsilon(x)\int_{\mathbb{R}^d} dy\;(a(x+y)-a(x))\,u(x+y)\nabla\mathcal{G}_\varepsilon(y) \\
&\quad = \int dx\,\nabla u_\varepsilon(x)\int_{\mathbb{R}^d} dy\int_0^1 y\cdot\nabla a(x+ty)dt\,u(x+y)\nabla\mathcal{G}_\varepsilon(y)\,.
\end{aligned}$$

We use Schwarz's inequality to dominate this by

$$\begin{aligned}
&\frac{1}{4}\int_0^1 dt\int dx\int_{\mathbb{R}^d} dy\,\nabla u_\varepsilon(x)\cdot a(x+ty)\nabla u_\varepsilon(x)\mathcal{G}_\varepsilon(y) \\
&\quad + \|u\|_{L^\infty}^2\int_0^1 dt\int dx\int_{\mathbb{R}^d} dy\,y\cdot\nabla a(x+yt)\nabla\mathcal{G}_\varepsilon a^{-1}(x+ty)y \\
&\qquad\qquad \cdot\nabla a(x+ty)\nabla\mathcal{G}_\varepsilon(y)\mathcal{G}_\varepsilon^{-1}(y)
\end{aligned}\tag{4.7}$$

By Lemma 2.4,

$$\int_0^1 dt \int dx \int_{\mathbb{R}^d} dy \, \nabla u_\varepsilon(x) \cdot a(x+ty) \nabla u_\varepsilon(x) \mathcal{G}_\varepsilon(y)$$
$$= \int dx \, |(\sigma \nabla u)_\varepsilon|^2 + o(1).$$

The integrand in the last term in (4.11) is dominated by a dimensional constant times

$$\sum_{i,j,k,\ell} \left(a_{jk,i} a_{j\ell}^{-1} a_{\ell k,i} \right) (x+ty) \left(y_i \mathcal{G}_{\varepsilon,k}(y) \right)^2 \mathcal{G}_\varepsilon^{-1}(y) \,.$$

For any $f \in L^1$ it is straightforward to show that

$$\int_0^1 dt \int dx \int dy f(x+ty) \left(y_i \frac{\mathcal{G}_{\varepsilon,k}}{\mathcal{G}_\varepsilon}(y) \right)^2 \mathcal{G}_\varepsilon(y)$$

is bounded uniformly in ε. Therefore by (A2) the last term in (4.7) is bounded by $B_2 < \infty$ independently of ε. For the lower-order part of (4.5),

$$\int dx \, (\sigma c u)_\varepsilon \cdot \nabla u_\varepsilon \leq \frac{1}{4} \int dx \, \nabla u_\varepsilon a_\varepsilon \nabla u_\varepsilon + \int dx \, |c|^2 u$$
$$= \frac{1}{4} \int dx \, |(\sigma \nabla u)_\varepsilon|^2 + o(1) + B_3 \,,$$

where $B_3 < \infty$. From (4.5) we conclude that

$$\int dx \, u_\varepsilon^2(T) - \int dx \, u_\varepsilon^2(0) = -\frac{1}{4} \int dx \, |(\sigma \nabla u)_\varepsilon|^2 + B_1 + B_2 + B_3 + o(1) \,.$$

Since u is uniformly bounded, we conclude that

$$\int dx \, |(\sigma \nabla u)_\varepsilon|^2 \leq C$$

for some C independent of ε. Letting $\varepsilon \to 0$ and using the lower semicontinuity of the norm, the theorem is proved.

Suppose in addition u is nonnegative. Then we want to show that

$$\int dt \int dx \frac{\langle \nabla u, a \nabla u \rangle}{u} < \infty \,.$$

We have

$$\frac{\partial u}{\partial t} = \nabla \cdot \sigma F$$

with

$$F = \sigma\nabla u - cu \in L^2([0,T] \times \mathbb{T}^d)\,.$$

We can apply Lemma 3.1 to conclude that

$$\begin{aligned}
&\int dx\, h(u(T)) - \int dx\, h(u(0)) \\
&\quad = \int_0^T dt \int dx\, h'(u(t,x))\nabla \cdot \sigma F \\
&\quad = -\int_0^T dt \int dx\, h''(u(t,x))\,\langle F, \sigma\nabla u\rangle \\
&\quad = -\int_0^T dt \int dx\, h''(u(t,x))[\langle \nabla u, a\nabla u\rangle - \langle cu, \sigma\nabla u\rangle]
\end{aligned}$$

We can take $h_\delta(u) = (u+\delta)\log(u+\delta)$ to obtain

$$\begin{aligned}
&\int dx\, h_\delta(u(T)) - \int dx\, h_\delta(u(0)) \\
&\quad = -\int_0^T dt \int dx\, \frac{\langle \nabla u, a\nabla u\rangle}{u+\delta} + \int_0^T dt \int dx\, \frac{\langle c, \sigma\nabla u\rangle\, u}{u+\delta}\,.
\end{aligned}$$

We can estimate by Schwarz's inequality

$$|\,\langle c, \sigma\nabla u\rangle\, u| \le |c|^2u^2 + \frac{1}{4}\,\langle \nabla u, a\nabla u\rangle$$

and therefore

$$\left|\int_0^T dt \int dx\, \frac{c\cdot\sigma\nabla u}{u+\delta}\right| \le \int_0^T dt \int dx\, \frac{|c|^2u^2}{u+\delta} + \frac{1}{4}\int_0^T dt \int dx\, \frac{\langle \nabla u, a\nabla u\rangle}{u+\delta}\,.$$

This yields the bound

$$\begin{aligned}
&\int_0^T dt \int dx\, \frac{\langle \nabla u, a\nabla u\rangle}{u+\delta} \\
&\quad \le C\left[\int_0^T dt \int dx\, \frac{|c|^2u^2}{u+\delta} + \int dx\, h_\delta(u(0)) - \int dx\, h_\delta(u(T))\right]
\end{aligned} \tag{4.8}$$

with a constant C not depending on δ. We can let $\delta \to 0$ in (4.8) to establish

$$\int_0^T dt \int dx \frac{\langle \nabla u, a\nabla u\rangle}{u} < \infty\,.$$

We can now state the uniqueness theorem for very weak solutions.

THEOREM 4.2 *Suppose ρ_i, $i = 1,2$, are two bounded, nonnegative, very weak solutions satisfying*

$$\begin{cases} \dfrac{\partial \rho_i}{\partial t} = \nabla \cdot a \nabla \rho_i - \nabla \cdot \sigma c \rho_i \\ \rho_1(0,x) = \rho_2(0,x) \quad \textit{on } \mathbb{T}^d \end{cases}$$

with

$$\int_0^T \int dx |c|^2 \rho_i < \infty \,.$$

Then $\rho_1 \equiv \rho_2$.

PROOF: Since we know now that very weak solutions are really weak solutions, we can use Corollary 3.10. ∎

5 Existence and Uniqueness for a Class of Degenerate Diffusion Processes

In this section we will construct diffusion processes corresponding to the operator L_t with diffusion matrix satisfying (A1) and obtain uniqueness if (A2) holds as well. Now for such wild coefficients, existence of a process for any starting point x is just not in the cards. So we give the following slightly nonstandard definition: Let $\Omega = C\left([0,T];\mathbb{T}^d\right)$, $\mathcal{F}$ the Borel σ-field, and $\mathcal{F}_t = \sigma(x(s) : 0 \le s \le t)$, $0 \le t \le T$. Given a bounded density function ρ^0, a probability measure P on $(\Omega, \mathcal{F})$ is a solution of the martingale problem for L_t with initial distribution f if for every $\phi \in C^\infty(\mathbb{T}^d)$,

1. $E^P\left[\phi(x_0)\right] = \int dx\, \phi f$, and
2. $\phi(x_t) - \int_0^t [L_u \phi](x_u) du$ is a martingale with respect to P and $\mathcal{F}_t$.

LEMMA 5.1 *Let $a^\alpha : [0,T] \times \mathbb{T}^d \to \mathbb{R}^d \otimes \mathbb{R}^d$ be nonnegative, definite, and symmetric-matrix-valued, and let $c^\alpha : [0,T] \times \mathbb{T}^d \to \mathbb{R}^d$ be a family depending on a parameter $\alpha \in \mathcal{A}$. Let σ^α be the symmetric, nonnegative, definite square root of a^α. For each $\alpha \in \mathcal{A}$ let P^α be a solution of the martingale problem with initial distribution π^α and generator*

$$L^\alpha = \nabla \cdot a^\alpha \nabla + \sigma^\alpha c^\alpha \cdot \nabla \,.$$

Let $\pi^{\alpha,t}$ be the marginals of P^α. Suppose that for each $i,j = 1,\dots,d$, $(\sigma^\alpha_{ij})^2$ and $[\nabla \cdot a^\alpha + \sigma^\alpha c^\alpha]_{ij}$ are uniformly integrable with respect to $dt\ \pi^{\alpha,t}(dx)$ on $[0,T] \times \mathbb{T}^d$. Then the family P^α, $\alpha \in \mathcal{A}$, is tight.

PROOF: To show tightness we will prove that for every $\varepsilon > 0$,

$$\limsup_{\delta\to 0}\sup_{\alpha\in\mathcal{A}} P^{\alpha}\left(\sup_{\substack{0\le s<t\le T\\ |t-s|<\delta}} |x(t)-x(s)| > \varepsilon\right) = 0\,. \tag{5.1}$$

Under P^{α} we have

$$x(t)-x(s) = \int_s^t \sigma^{\alpha}d\beta + \int_s^t (\nabla\cdot a^{\alpha} + \sigma^{\alpha}c^{\alpha})\,du \tag{5.2}$$

where β is a standard d-dimensional Brownian motion and the right-hand side is computed on $\mathbb{R}^d$, then projected onto the torus. Let

$$y_{ij}(t) = \int_0^t \sigma_{ij}^{\alpha}d\beta_j\,.$$

For each i, j we can write $y_{ij}(t)$ as a random time change of another Brownian motion $\hat{\beta}(\cdot)$ in $\mathbb{R}^1$,

$$y_{ij}(t) = \hat{\beta}\left(\int_0^t (\sigma_{ij}^{\alpha})^2du\right).$$

Since Brownian motions are almost surely continuous, the desired estimate (5.1) for the first term on the right-hand side of (5.2) will follow if we can show

$$\limsup_{\delta\to 0}\sup_{\alpha\in\mathcal{A}} P^{\alpha}\left(\sup_{\substack{0\le s<t\le T\\ |t-s|<\delta}} \left|\int_s^t (\sigma_{ij}^{\alpha})^2du\right| > \varepsilon\right) = 0\,. \tag{5.3}$$

If we replace $(\sigma_{ij}^{\alpha})^2$ by $(\sigma_{ij}^{\alpha})^2\wedge\ell$ for some $\ell > 0$, then (5.3) is immediate. But then

$$P^{\alpha}\left(\sup_{\substack{0\le s<t\le T\\ |t-s|<\delta}} \left|\int_s^t \left[(\sigma_{ij}^{\alpha})^2 - (\sigma_{ij}^{\alpha})^2\wedge\ell\right]du\right| > \varepsilon\right) \le \varepsilon^{-1}E^{P^{\alpha}}\left[\int_0^T \left[(\sigma_{ij}^{\alpha})^2 - (\sigma_{ij}^{\alpha})^2\wedge\ell\right]du\right].$$

The right-hand side vanishes in the limit of large ℓ, uniformly in α, by the uniform integrability of $(\sigma_{ij}^{\alpha})^2$ with respect to $dt\,\pi^{\alpha,t}(dx)$. The second term on the right-hand side of (5.2) can be handled in a similar fashion. In fact, if we examine the preceding argument, we see that the key point is the uniform integrability of each coordinate of $(\nabla\cdot a^{\alpha} + \sigma^{\alpha}c^{\alpha})$ with respect to $dt\,\pi^{\alpha,t}(dx)$. ∎

LEMMA 5.2 *Suppose that* (A1) *holds. Then there exists a solution P_0 of the martingale problem for*

$$L_t = \nabla \cdot a\nabla$$

with initial distribution $f \equiv 1$.

PROOF: For each $\varepsilon > 0$ let P_ε be the solution of the martingale problem for

$$L_t^\varepsilon = \nabla \cdot a_\varepsilon \nabla$$

with initial distribution $f \equiv 1$ where a_ε is chosen according to Lemma 3.3. Note that all marginals are given by $\rho_\varepsilon \equiv 1$, and the conditions of Lemma 5.1 then follow from (A1). By Lemma 5.1 there is a weak limit P_0. We have to show that P_0 solves the martingale problem for L_t. This amounts to the statement that for any $\phi \in C^\infty(\mathbb{T}^d)$ and any bounded continuous $F \in \mathcal{F}_s$,

$$\begin{aligned} E^{P_\varepsilon}&\left[\left(\phi(x_t) - \phi(x_s) - \int_s^t [L_u^\varepsilon]\phi(x_u)du\right) F\right] \\ &\to E^{P_0}\left[\left(\phi(x_t) - \phi(x_s) - \int_s^t [L_u\phi](x_u)du\right) F\right]. \end{aligned}$$

By the Markov property it suffices to take $F(x(\cdot))$ of the form $f(x_s)$. The boundary terms present no problem and therefore one only has to verify that for bounded continuous f

$$E^{P_\varepsilon}\left[\int_s^t [(\nabla \cdot a_\varepsilon \nabla)\phi](x_u)du\, f(x_s)\right] \to E^{P_0}\left[\int_s^t [\nabla \cdot a\nabla\phi](x_u)du\, f(x_s)\right],$$

which is not difficult. ∎

THEOREM 5.3 *Suppose that* (A1) *holds and that there exists a bounded nonnegative weak solution ρ of the forward equation $\partial_t\rho - L_t^*\rho = 0$, $0 \le t \le T$, satisfying*

$$\begin{cases} \displaystyle\int_0^T dt \int dx \frac{\nabla\rho \cdot a\nabla\rho}{\rho} < \infty\,, \\ \displaystyle\int_0^T dt \int dx\, |c|^2\rho < \infty\,. \end{cases}$$

Then for any $0 \le f(x) \le C\rho(0,x)$ with $\int dx\, f = 1$ there exists a solution P of the martingale problem for L_t with initial density f and having marginal densities $R(t,x)$, the unique weak solution of $\partial_t R - L_t^* R = 0$, $R(0) = f$ satisfying $0 \le R \le C\rho$ and $\int_0^T dt \int dx\, \frac{\nabla R \cdot a\nabla R}{R} < \infty$.

PROOF: By Theorem 3.12, for any $0 \le f(x) \le C\rho(s,x)$ there exists a weak solution $u(t,x) = \left[P^*_{s,t} f\right](x)$ of the forward equation after time s with $\int_s^T dt \int dx \, \frac{\nabla u \cdot a \nabla u}{u} < \infty$ and

$$P^*_{s,t} : L^\infty(\rho(s)) \to L^\infty(\rho(t))$$

is bounded. For any n and any partition $0 \le t_1 < t_2 < \cdots < t_n \le T$, let $P^{t_1,\dots,t_n}$ be the probability measure on $(\mathbb{T}^d)^n$ given as follows: For any bounded functions $f_i : \mathbb{T}^d \to \mathbb{R}$, $i = 1, \dots, n$,

$$\begin{aligned} &\int f_1(x_1) \cdots f_n(x_n) dP^{t_1,\dots,t_n} \\ &\quad = \int dx \, f_n(x) [P^*_{t_{n-1},t_n} f_{n-1} P^*_{t_{n-2},t_{n-1}} \cdots P^*_{t_1,t_2} f_1 P^*_{0,t_1} f](x) . \end{aligned}$$

Let P be the unique probability measure on $(\mathbb{T}^d)^{[0,T]}$ such that

$$E^P \left[f_1(x_{t_1}) \cdots f_n(x_{t_n}) \right] = \int f_1(x_1) \cdots f_n(x_n) dP^{t_1,\dots,t_n}.$$

Let P_0 be a solution of the martingale problem for $\nabla \cdot a \nabla$ with initial density $f \equiv 1$. Such a solution exists by the previous theorem. We will show that the relative entropy satisfies

$$H(P/P_0) \le \int dx \, f \log f + \frac{C}{4} \int_0^T dt \int dx |c|^2 \rho. \tag{5.4}$$

Then P is a probability measure on $C([0,T]; \mathbb{T}^d)$ and by construction P solves the martingale problem. To prove (5.4) we note that, because

$$H(P/P_0) = \sup h \left(P^{t_1,\dots,t_n} / P_0^{t_1,\dots,t_n} \right)$$

where the supremum runs over all partitions of $[0,T]$, it is sufficient to show that

$$h \left(P^{t_1,\dots,t_n} / P_0^{t_1,\dots,t_n} \right) \le \int dx \, f \log f + \frac{C}{4} \int_0^T dt \int dx |c|^2 \rho \tag{5.5}$$

for all partitions. By Lemma 3.9,

$$\begin{aligned} &h \left(P^{t_1,\dots,t_n} / P_0^{t_1,\dots,t_n} \right) \\ &\quad \le h \left(P^{t_1,\dots,t_{n-1}} / P_0^{t_1,\dots,t_{n-1}} \right) + \frac{1}{4} \int_{t_{n-1}}^{t_n} dt \int dx \, |c|^2 R \end{aligned}$$

from which we establish (5.5) by induction. ∎

Remark. If one is interested in the more standard concept of solutions of the martingale problem starting from a point x at time s, then Theorem 5.3 provides us with the result that for each s, there exists a solution starting from x at time s for almost every x with respect to $\rho(s,x)dx$. This is the most that can be expected under our conditions.

Now we consider the problem of uniqueness of our process. Although we have an existence and uniqueness theorem for weak solutions under assumption (A1), this is not of much use in proving uniqueness of solutions to the martingale problem. Even if we start with a solution to the martingale problem with marginals having densities in class $H_{1,a}$, the conditionals are not necessarily in the same regularity class. Assumption (A2) will be important in this context. Suppose we have an operator of the form

$$L_t = \nabla \cdot a\nabla + \sigma c \cdot \nabla$$

with a satisfying (A2). We will limit overselves to the class $\mathcal{P}$ of processes P that satisfy the following conditions:

1. There exists a constant C such that all the one-dimensional marginals of P have densities bounded by C.

2. If $\rho(t,x)$ is the marginal density of P, then

$$\int_0^T \int dx\, |c|^2 \rho(t,x) = E^P\left[\int_0^T |c(t,x(t))|^2 dt\right] < \infty .$$

We can now state and prove our main theorem.

THEOREM 5.4 *Assume that* (A2) *is valid. For any bounded initial density* $\rho_0(x)$, *the solution to the martingale problem for*

$$L_t = \nabla \cdot a\nabla + \sigma c \cdot \nabla$$

is unique in the class $\mathcal{P}$. *If we assume that we have a nonnegative solution* $\rho(t,x)$ *that satisfies*

$$\begin{cases} \dfrac{\partial \rho}{\partial t} = L_t^* \rho = \nabla \cdot a \nabla \rho - \nabla \cdot \sigma c \rho \\ \displaystyle\int_0^T dt \int dx\, \|c\|^2 \rho < \infty , \end{cases}$$

then a solution to the martingale problem for L_t *exists* (*uniquely*) *in the class* $\mathcal{P}$ *and has the solution* $\rho(t,x)$ *given above for its marginal densities.*

PROOF: We need only uniqueness to complete the proof. We will prove by induction the following statement: If P_1 and P_2 are two solutions to the martingale problem in class $\mathcal{P}$ and if $0 = t_1 < t_2 < \cdots < t_n \le T$ are n time points, the distribution of P_1 and P_2 agree on the sigma field $\sigma\{x(t_i),\ i = 1,\dots,n\}$. Assume that the statement is true for $0 = t_1 < t_2 < \cdots < t_n \le T$ and we want to prove it for $0 = t_1 < t_2 < \cdots < t_n < t_{n+1} \le T$. In other words, for any $A \in \sigma\{x(t_i),\ i = 1,\dots,n\}$ and Borel set $B \subset \mathbb{T}^d$, we would like to show that

$$P_1[A \cap \{x(t_{n+1}) \in B\}] = P_2[A \cap \{x(t_{n+1}) \in B\}]. \tag{5.6}$$

Clearly we can assume without loss of generality that $P_i(A) > 0$. Define Q_1 and Q_2 by

$$Q_i(E) = \frac{P_i(A \cap E)}{P_i(A)} \quad \text{for } i = 1, 2.$$

Then one can check that on the interval $[t_n, T]$, Q_1 and Q_2 are in class $\mathcal{P}$ and are again solutions to the martingale problem. From the uniqueness of the marginals we conclude that

$$Q_1[x(t_{n+1}) \in B] = Q_2[x(t_{n+1}) \in B]$$

which is equivalent to (5.6). This establishes the induction. ∎

Acknowledgement. J. Quastel would like to thank S. R. S. Varadhan, H.-T. Yau, and the Courant Institute for their warm hospitality while this work was being completed. J. Quastel was partially supported by U.S. National Science Foundation Grant No. DMS-9504791 and partially supported by the Alfred P. Sloan Foundation. S. R. S. Varadhan was partially supported by U.S. National Science Foundation Grant No. DMS-9503419 and Army Research Office Grant No. ARO-DAAH04-95-1-0666.

Bibliography

[1] Albeverio, S., Hoegh-Krohn, R., and Streit, L., *Energy forms, Hamiltonians, and distorted Brownian paths*, J. Math. Phys. 18, 1977, pp. 907–917.

[2] Carlen, E. A., *Conservative Diffusions*, Comm. Math. Phys. 94, 1984, pp. 293–315.

[3] DiPerna, R. J., and Lions, P.-L., *Ordinary differential equations, transport theory and Sobolev spaces*, Invent. Math. 98, 1989, pp. 511–547.

[4] Fukushima, M., *Dirichlet Forms and Markov Processes*, North-Holland Mathematical Library No. 23, North-Holland, Amsterdam–New York; Kodansha, Tokyo, 1980.

[5] Lions, J.-L., and Magenes, E., *Non-Homogeneous Boundary Value Problems and Applications*, Grundlehren der Mathematischen Wissenschaften Vols. 181–182, Springer, New York–Heidelberg, 1972.

[6] Quastel, J., Rezakhanlou, F., and Varadhan, S. R. S., *Large deviations for tagged particles in the symmetric exclusion process*, to appear.

[7] Stroock, D. W., *Diffusion semigroups corresponding to uniformly elliptic divergence form operators*, pp. 316–347 in: *Séminaire de Probabilités* 22, Lecture Notes in Mathematics No. 1321, Springer, Berlin–New York, 1988.

[8] Stroock, D. W., and Varadhan, S. R. S., *Multidimensional Diffusion Processes*, Grundlehren der Mathematischen Wissenschaften No. 233, Springer, Berlin–New York, 1979.

J. QUASTEL
Department of Mathematics
University of California
Davis, CA 95616
E-mail:
quastel@math.ucdavis.edu

S. R. S. VARADHAN
Courant Institute
251 Mercer Street
New York, NY 10012
E-mail:
varadhan@cims.nyu.edu

Received May 1996.

Random Media

Lecture Notes in Control and Inform. Sci. no. 25, Springer, 1980, 190-206

DIFFUSION IN REGIONS WITH MANY SMALL HOLES

G. C. Papanicolaou* and S. R. S. Varadhan**

Courant Institute of Mathematical Sciences, New York University

1. FORMULATION AND STATEMENT OF MAIN THEOREM

Let D be a bounded open set containing the origin, having C^2 boundary and with diameter less than or equal to one. For each N = 1,2,..., let $y_1^{(N)}, y_2^{(N)} \dots y_N^{(N)}$ be points in R^3 and define sets $D_i^{(N)}$ by

$$D_i^{(N)} = \{x \in R^3 \mid N(x - y_i^{(N)}) \in D\}, \quad i = 1,2,\dots,N. \tag{1.1}$$

We shall call the set $D_i^{(N)}$ the hole centered at $y_i^{(N)}$ with diameter less than or equal to N^{-1}. Let $G^{(N)}$ denote the region

$$G^{(N)} = R^3 - \bigcup_{i=1}^{N} D_i^{(N)} \tag{1.2}$$

which is R^3 with holes of diameter $\leq N^{-1}$ centered at $y_1^{(N)}, \dots, y_N^{(N)}$.

We shall analyze the asymptotic behavior of $u^{(N)}(x,t)$ as $N \to \infty$ which is the solution of

$$\begin{aligned} \frac{\partial u^{(N)}(x,t)}{\partial t} &= \frac{1}{2}\Delta u^{(N)}(x,t), & t > 0,\ x \in G^{(N)} \\ u^{(N)}(x,t) &= 0, & t > 0,\ x \in \partial G^{(N)} = \bigcup_{i=1}^{N} \partial D_i^{(N)}, \\ u^{(N)}(x,0) &= f(x), & x \in G^{(N)}, \end{aligned} \tag{1.3}$$

with f(x) a given bounded continuous function with compact support in R^3. The precise meaning in which problem (1.3) is taken is as follows.

Let P_x be the Brownian motion measure on continuous trajectories on R^3 starting from $x \in R^3$; its transition probability density p(t,x,y) is given by

$$p(t,x,y) = (2\pi t)^{-3/2} \exp\left\{-\frac{|x-y|^2}{2t}\right\}. \tag{1.4}$$

Let $E_x\{\cdot\}$ denote expectation with respect to P_x. Then we set by definition

$$u^{(N)}(x,t) = E_x\left\{f(x(t))\, \chi_{\{\tau^{(N)} > t\}}\right\}. \tag{1.5}$$

---

* Research supported under Grant AFOSR-78-3668 with the U. S. Air Force.
**Research supported under Grant No. NSF-MCS-77-02687 with the National Science Fndn.

Here χ_A is the indicator function of a set A of trajectories and equals one if the trajectory is in A and zero if not. The stopping time $\tau^{(N)}$ is defined by

$$\tau^{(N)} = \min\left\{\tau_1^{(N)}, \tau_2^{(N)}, \ldots, \tau_N^{(N)}\right\} \tag{1.6}$$

where $\tau_j^{(N)}$ is the first time the trajectory reaches $\partial D_j^{(N)}$.

For the theorem that follows we shall need the following two hypotheses regarding the location of the hole centers $y_1^{(N)}, \ldots, y_N^{(N)}$.

First, there exists a bounded continuous function of compact support $V(x) \geq 0$ such that for each $\phi \in C_0(R^3)$ (continuous and with compact support)

$$\lim_{N\uparrow\infty} \frac{1}{N} \sum_{i=1}^{N} \phi(y_i^{(N)}) = \int_{R^3} \phi(x)\, V(x)\, dx\,. \tag{1.7}$$

Second, we assume that

$$\lim_{N\uparrow\infty} \frac{1}{N^2} \sum_{\substack{i,j=1\\ i\neq j}}^{N} \frac{1}{|Y_i^{(N)} - y_j^{(N)}|} = \int_{R^3}\int_{R^3} \frac{V(x)\; V(y)}{|x - y|}\, dx\, dy \tag{1.8}$$

Let α denote the capacitance of the set D. Since ∂D is C^2, α is given by

$$\alpha = \int_{\partial D} \frac{\partial w(x)}{\partial \hat{n}(x)}\, dS(x) \tag{1.9}$$

where $\hat{n}(x)$ is the unit inner normal at $x \in \partial D$ and dS denotes element of surface area. The capacitory potential $w(x)$ is the solution of

$$\begin{aligned} \Delta w(x) &= 0\,, & x &\in R^3 - D, \\ w(x) &= 1\,, & x &\in \partial D\,, \\ w(x) &\to 0 \quad \text{as} & |x| &\to \infty\,. \end{aligned} \tag{1.10}$$

THEOREM 1. <u>Under hypotheses</u> (1.7) <u>and</u> (1.8), <u>given</u> $\varepsilon > 0$ <u>and</u> $T < \infty$ <u>there exists an integer</u> N_0 <u>such that for all</u> $N \geq N_0$ <u>there is a set</u> $G_\varepsilon^{(N)} \subset G^{(N)}$ <u>such that</u>

$$\sup_{0\leq t\leq T}\ \sup_{x\in G_\varepsilon^{(N)}} |u^{(N)}(x,t) - u(x,t)| < \varepsilon \tag{1.11}$$

<u>and</u>

$$\text{vol}(R^3 - G_\varepsilon^{(N)}) < \varepsilon\,. \tag{1.12}$$

Here u(x,t) is given by

$$u(x,t) = E_x\left\{f\big(x(t)\big) \exp\left[-\alpha\int_0^t V\big(x(s)\big)\,ds\right]\right\} . \tag{1.13}$$

Remarks.

1. The limit function u(x,t) is the solution of

$$\begin{aligned} u_t &= \frac{1}{2}\Delta u - \alpha V u , && t > 0 \\ u(0,x) &= f(x) , && x \in R^3, \end{aligned} \tag{1.14}$$

by the Feynman-Kac formula.

2. The above theorem implies that

$$\sup_{0\le t\le T} \int_{G^{(N)}} |u^{(N)}(x,t) - u(x,t)|^2\, dx \to 0 \tag{1.15}$$

as $N \to \infty$ for each $T < \infty$.

3. Theorems valid in bounded regions with various boundary conditions can also be obtained but are not discussed further here.

If we think of $u^{(N)}(x,t)$ as the temperature at x at time t of a solid occupying $G^{(N)}$ of conductivity 1/2 then, the effect of keeping the holes $D_i^{(N)}$ in $G^{(N)}$ at temperature zero appears in the limit $N \to \infty$ as a continuous heat removal rate $\alpha V(x)$ which is the limiting capacitance per unit volume. The scaling is such that this quantity has a finite limit. It is clear that condition (1.7) is a natural one; it gives the limiting density of hole centers V(x). The meaning of condition (1.8) is less immediate although it is clear that it puts a restriction on how close hole centers can be from each other.

We shall consider next two particular cases in which (1.8) follows from simpler conditions.

Suppose first that the holes $D_i^{(N)}$ are spheres with centers at $y_1^{(N)},\ldots,y_N^{(N)}$ and diameters equal to N^{-1}. Suppose that there is a constant $\gamma > 0$ such that for all N = 1,2,...

$$|y_i^{(N)} - y_j^{(N)}| \ge \gamma\, N^{-1/3} , \qquad i,j = 1,2,\ldots,N. \tag{1.16}$$

We assume that((1.7) holds and note that now the capacitance $\alpha = 2\pi$ (the capacitance of a sphere of unit diameter). We shall show that (1.16) implies (1.8).

Let $B_j^{(N)}$ denote the sphere of radius $\gamma N^{-1/3}/2$ with center at $y_i^{(N)}$. For a fixed $y_i^{(N)}$ satisfying (1.16) the function $|y_i^{(N)} - y|^{-1}$ is harmonic for $y \neq y_i^{(N)}$. Thus by the mean value theorem

$$|y_i^{(N)} - y_j^{(N)}|^{-1} = \frac{1}{b^{(N)}} \int_{B_j^{(N)}} \frac{dy}{|y_i^{(N)} - y|}, \qquad j \neq i \tag{1.17}$$

where $b^{(N)} = \pi\gamma^3/6N$ is the volume of $B_j^{(N)}$. With y fixed in $B_j^{(N)}$ we may use (1.17) again for $|y_i^{(N)} - y|^{-1}$. This yields

$$|y_i^{(N)} - y_j^{(N)}|^{-1} = \frac{1}{(b^{(N)})^2} \int_{B_i^{(N)}} \int_{B_j^{(N)}} \frac{dx\ dy}{|x - y|}$$

and hence

$$\frac{1}{N^2} \sum_{\substack{i,j=1 \\ i\neq j}}^{N} \frac{1}{|y_i^{(N)} - y_j^{(N)}|} = \sum_{\substack{i,j=1 \\ i\neq j}}^{N} \frac{1}{(Nb^{(N)})^2} \int_{B_i^{(N)}} \int_{B_j^{(N)}} \frac{dx\ dy}{|x - y|} \tag{1.18}$$

To prove that (1.7) and (1.16) imply (1.8) it follows from (1.18) that we must show that

$$\lim_{N\uparrow\infty} \frac{1}{N^2} \sum_{i,j=1}^{N} \frac{1}{(Nb^{(N)})^2} \int_{B_i^{(N)}} \int_{B_j^{(N)}} \frac{dx\ dy}{|x - y|} = \int\int \frac{V(x)\ V(y)}{|x - y|}\ dx\ dy \tag{1.19}$$

The addition of the diagonal terms on the right side of (1.19) makes a negligible contribution for N large. To show (1.19) define

$$\psi^{(N)}(x) = \frac{1}{N} \sum_{i=1}^{N} \frac{1}{b^{(N)}} \chi_{B_i^{(N)}}(x) \tag{1.20}$$

and note that $\psi^{(N)}(x)$ is uniformly bounded by a constant independent of N and x. In terms of $\psi^{(N)}$ we rewrite (1.19) as

$$\lim_{N\uparrow\infty} \int\int \frac{\psi^{(N)}(x)\ \psi^{(N)}(y)}{|x - y|}\ dx\ dy = \int\int \frac{V(x)\ V(y)}{|x - y|}\ dx\ dy \tag{1.21}$$

which we prove as follows.

Let $\pi^{(N)}$ be a measure defined on the Borel sets of R^3 by

$$\pi^{(N)}(A) = \frac{1}{N}\ [\#\text{ of } i = 1,2,\ldots,N \text{ for which } y_i^{(N)} \in A]\ . \tag{1.22}$$

Hypothesis (1.7) says that $\pi^{(N)}$ converges weakly as $N \to \infty$ to the measure π whose density with respect to Lebesgue measure is $V(x)$. Hence the product measure $\pi^{(N)} \times \pi^{(N)}$ on the Borel sets of R^6 converges weakly to $\pi \times \pi$. Let $\sigma^{(N)}$ be the uniform measure on the ball in R^3 centered at the origin with radius $\gamma N^{-1/3}/2$. Then, if $*$ denotes convolution and $\pi \cdot \phi$ stands for the integral of ϕ with respect to π, we have

$$(1.23) \quad \left(\pi^{(N)} * \sigma^{(N)}\right) \times \left(\pi^{(N)} * \sigma^{(N)}\right) \cdot \phi = \int\int \psi^{(N)}(x)\ \psi^{(N)}(y)\ \phi(x,y)\ dx\ dy$$

for any $\phi \in C_0(R^6)$. It is easily verified that if $\pi^{(N)} \to \pi$ weakly then $\pi^{(N)} * \sigma^{(N)} \to \pi$ weakly also. Hence by (1.23) we have

$$(1.24) \quad \lim_{N\to\infty} \iint \psi^{(N)}(x)\ \psi^{(N)}(y)\ \phi(x,y)\ dx\ dy = \iint V(x)\ V(y)\ \phi(x,y)\ dx\ dy$$

for each $\phi \in C_0(R^6)$. Now we approximate the function $|x - y|^{-1}$ by $(|x-y|+\delta)^{-1}$ with $\delta > 0$. In order that (1.21) be valid (hence (1.19) and hence (1.8)) it suffices to note that the $\psi^{(N)}$ are bounded uniformly in N and x and for any compact set $K \subset R^3$

$$\int_K \int_K \left(\frac{1}{|x - y|} - \frac{1}{|x - y| + \delta}\right) dx\ dy \to 0 \quad \text{as} \quad \delta \downarrow 0 .$$

We have thus shown that (1.7) and (1.16) imply (1.8).

The second case we consider in which (1.8) can be verified from other conditions is when <u>the hole centers are distributed randomly</u> as follows.

For each $N = 1,2,\ldots$, let $y_1^{(N)}, y_2^{(N)}, \ldots, y_N^{(N)}$ be independent identically distributed random variables in R^3 with density $V(x)$ that is,

$$(1.25) \qquad \text{Prob}\left\{y_i^{(N)} \in A\right\} = \int_A V(x)\ dx$$

for any Borel subset A in R^3. We assume that $V(x)$ is bounded, continuous, has compact support and with no loss in generality

$$\int_{R^3} V(x)\ dx = 1$$

We denote expectation with respect to (1.25) by $\langle \cdot \rangle$ to distinguish it from

expectation relative to P_x. Thus, for any $\phi \in C_0(R^3)$

$$\langle \phi(y_i^{(N)}) \rangle = \int \phi(y)\ V(y)\ dy\ , \qquad i = 1,2,\ldots,N. \tag{1.26}$$

and by the strong law of large numbers

$$\lim_{N\to\infty} \frac{1}{N} \sum_{i=1}^{N} \phi(y_i^{(N)}) = \int V(y)\ \phi(y)\ dy \tag{1.27}$$

with probability one. Therefore (1.7) holds with probability one.

Condition (1.8) holds now in probability. To see this we note that for any $1 \leq i < j \leq N$

$$\langle |y_i^{(N)} - y_j^{(N)}|^{-1} \rangle = \int\int \frac{V(x)\ V(y)}{|x - y|}\ dx\ dy \tag{1.28}$$

and

$$\langle\ |\ y_i^{(N)} - y_j^{(N)}|^{-2} \rangle = \iint \frac{V(x)\ V(y)}{|x - y|^2}\ dx\ dy < \infty \tag{1.29}$$

Let

$$S_N = \frac{2}{N(N-1)} \sum_{1 \leq i < j \leq N} \left\{ |y_i^{(N)} - y_j^{(N)}|^{-1} - \int\int \frac{V(x)\ V(y)}{|x - y|}\ dx\ dy \right\} .$$

Using (1.28) and (1.29) we verify that $\langle S_N^{\ 2} \rangle \to 0$ as $N \to \infty$. This implies that for each $\varepsilon > 0$ and $\delta > 0$ there is an N_0 such that for all $N \geq N_0$

$$\text{Prob}\left\{ |S_N| > \varepsilon \right\} < \delta\ , \tag{1.30}$$

that is, (1.8) holds in probability.

Let $(\Omega^{(N)}, F^{(N)}, P^{(N)})$ be a probability space on which $y_1^{(N)},\ldots,y_N^{(N)}$ are defined. If $\omega \in \Omega^{(N)}$ then $u^{(N)}(x,t) = u^{(N)}(x,t,\omega)$ in this case. Theorem 1 takes the the following form.

THEOREM 2. <u>When the hole centers are independent identically distributed random variables with density</u> $V(x)$ <u>then given</u> $\varepsilon > 0$, $\delta > 0$ <u>and</u> $T < \infty$ <u>there is an</u> N_0 <u>such that for all</u> $N \geq N_0$ <u>there is a set</u> $\Omega_{\varepsilon,\delta}^{(N)} \subset \Omega^{(N)}$ <u>and a set</u> $G_\varepsilon^{(N)}(\omega) \subset G^{(N)}(\omega)$, $\omega \in \Omega_{\varepsilon,\delta}^{(N)}$, <u>such that</u>

$$\sup_{0 \leq t \leq T}\ \sup_{x \in G_\varepsilon^{(N)}(\omega)} |u^{(N)}(x,t,\omega) - u(x,t)| < \varepsilon\ , \qquad \omega \in \Omega_{\varepsilon,\delta}^{(N)}\ , \tag{1.31}$$

<u>and</u>

(1.32) $$\mathrm{vol}\left[R^3 - G_\varepsilon^{(N)}(\omega)\right] < \varepsilon , \qquad \omega \in \Omega_{\varepsilon,\delta}^{(N)} ,$$

<u>while</u>

(1.33) $$\mathrm{Prob}\left\{\Omega^{(N)} - \Omega_{\varepsilon,\delta}^{(N)}\right\} < \delta .$$

<u>Here</u> $u(x,t)$ <u>is given by</u> (1.13) <u>as in Theorem 1</u>.

<u>Remarks</u>.

1. This theorem implies that (1.15) holds in probability, that is, for each $\delta > 0$ and $T < \infty$,

(1.34) $$\mathrm{Prob}\left\{\sup_{0\le t\le T}\int_{G^{(N)}} |u^{(N)}(x,t) - u(x,t)|^2\, dx > \delta\right\} \to 0 \quad \text{as} \quad N \uparrow \infty,$$

where $u^{(N)}(x,t) = u^{(N)}(x,t,\omega)$, $\omega \in \Omega^{(N)}$.

2. The proof of Theorem 2 is immediate from Theorem 1 and the validity of (1.8) in probability. The set $\Omega_{\varepsilon,\delta}^{(N)}$ is selected so that $|S_N|$ (cf. (1.30)) is suitably small and thereafter randomness plays no role whatsoever. The estimates are obtained for each $\omega \in \Omega_{\varepsilon,\delta}^{(N)}$ as in the deterministic case.

The content of Theorem 2, in the form (1.34), was obtained by Kac [1] using properties of the Wiener sausage. It was also obtained by Rauch and Taylor in [2].

A comprehensive analysis of problems of the form considered here is given in [3] using analytical methods.

2. PROOF OF THEOREM 1

We shall show the following fact. Given $\varepsilon > 0$ and $t_0 < \infty$ there exists an integer N_0 such that for all $N \ge N_0$ there is an open set $G_\varepsilon^{(N)} \subset G^{(N)}$ such that

(2.1) $$\left|P_x\left\{\tau^{(N)} \le t\right\} - \alpha\int_0^t\int V(y)\, p(s,x,y)\, dy\, ds\right| < \varepsilon + g(t)$$

for all $x \in G_\varepsilon^{(N)}$ and $0 \le t \le t_0$ and moreover

(2.2) $$\mathrm{vol}\left(R^3 - G_\varepsilon^{(N)}\right) < \varepsilon ,$$

and

$$\overline{\lim_{t\downarrow 0}}\, t^{-1} g(t) = 0 \,. \tag{2.3}$$

Before proving this we shall use it to complete the proof of Theorem 1.

We define on $\hat{C}(R^3)$, the bounded continuous functions on R^3 that vanish at infinity, three semigroups as follows.

$$(T_t f)(x) = E_x\left\{f(x(t))\right\} \tag{2.4}$$

$$(T_t^{(N)} f)(x) = E_x\left\{f\big(x(t)\big)\ \chi_{\{\tau^{(N)} > t\}}\right\} \tag{2.5}$$

$$(T_t^V f)(x) = E_x\left\{f\big(x(t)\big)\ \exp\left(-\alpha \int_0^t V\big(x(s)\big)\, ds\right)\right\} . \tag{2.6}$$

Let $\Delta > 0$ be fixed; it will be chosen appropriately later. We rewrite the left side of (1.11).

$$\begin{aligned} I_\varepsilon^{(N)} &\equiv \sup_{0\le t\le t_0}\ \sup_{x\in G_\varepsilon^{(N)}} \left| T_t^{(N)} f(x) - T_t^V f(x) \right| \\ &= \sup_{0\le t\le t_0}\ \sup_{x\in G_\varepsilon^{(N)}} \left\{ \left| \sum_{k=1}^{[t/\Delta]} T^{(N)}_{(k-1)\Delta} [T_\Delta^{(N)} - T_\Delta^V] T^V_{([t/\Delta]-k)\Delta} f(x) \right.\right. \\ &\quad \left.\left. + \left(T_t^{(N)} - T^{(N)}_{\Delta[t/\Delta]}\right) f(x) + \left(T_t^V - T^V_{\Delta[t/\Delta]}\right) f(x) \right| \right\} \end{aligned} \tag{2.7}$$

Here $[t/\Delta]$ denotes the integer part of t/Δ. Since both $T_t^{(N)}$ and T_t^V are contractions, (2.7) can be estimated as follows.

$$\begin{aligned} I_\varepsilon^{(N)} &\le 2\|f\| \sum_{k=2}^{[t_0/\Delta]} \sup_{x\in G_\varepsilon^{(N)}} T_{(k-1)\Delta}\, \chi_{[R^3-G_\varepsilon^{(N)}]}(x) \\ &\quad + \frac{t_0}{\Delta} \sup_{0\le t\le t_0}\ \sup_{x\in G_\varepsilon^{(N)}} \left| \left(T^{(N)} - T^V\right) T_t^V f(x) \right| + 2 \sup_{0\le t\le \Delta}\ \sup_{x\in G_\varepsilon^{(N)}} \left| T_t^{(N)} f(x) - f(x) \right| \\ &\quad + 2\|f\| \sup_{\Delta\le t\le t_0}\ \sup_{x\in G_\varepsilon^{(N)}} T_{\Delta[t/\Delta]}\, \chi_{[R^3-G_\varepsilon^{(N)}]}(x) + \sup_{0\le t\le \Delta}\ \sup_{x\in G} \left| T_t^V f(x) - f(x) \right| . \end{aligned} \tag{2.8}$$

Here $\|f\| = \sup_x |f(x)|$. Using the fact that $T_t^{(N)} g \le T_t g$ for $g \ge 0$ we may estimate (2.8) further by

$$(2.9)\quad I_\varepsilon^{(N)} \le 2\|f\|\left(\frac{t_0}{\Delta}+1\right)\sup_{\Delta\le t\le t_0}\ \sup_{x\in G_\varepsilon^{(N)}} T_t\,\chi_{[R^3-G_\varepsilon^{(N)}]}(x)$$

$$+\frac{t_0}{\Delta}\sup_{0\le t\le t_0}\ \sup_{x\in G_\varepsilon^{(N)}}\left|\left(T_\Delta^{(N)}-T_\Delta^V\right)T_t^V f(x)\right| + 2\sup_{0\le t\le\Delta}\ \sup_{x\in G_\varepsilon^{(N)}}\left|T_t^{(N)}f(x)-f(x)\right|$$

$$+\sup_{0\le t\le\Delta}\ \sup_{x\in G}\left|T_t^V f(x)-f(x)\right| .$$

We estimate next each term on the right side of (2.9) starting with the first term. We have

$$(2.10)\qquad \sup_{\Delta\le t\le t_0}\ \sup_{x\in G_\varepsilon^{(N)}} T_t\,\chi_{[R^3-G_\varepsilon^{(N)}]}(x) \le \theta(\Delta)\ \mathrm{vol}\,(R^3-G_\varepsilon^{(N)})$$

where $\theta(\Delta)\to\infty$ as $\Delta\to\infty$ but is finite for $\Delta>0$ and fixed. Thus the first term on the right of (2.9) can be controlled for $\Delta>0$ fixed by letting N be large and using (2.2).

Let us consider the second term on the right side of (2.9). Let $V(x)\le C$. Then

$$(2.11)\quad \sup_{x\in G_\varepsilon^{(N)}}\left|T_\Delta^{(N)}f(x)-T_\Delta^V f(x)\right| =$$

$$=\sup_{x\in G_\varepsilon^{(N)}}\left|E_x\{f(x(\Delta))\,\chi_{\{\tau^{(N)}\le\Delta\}}\} - E_x\left\{\left(1-\exp\left(-\alpha\int_0^\Delta V(x(s))\,ds\right)\right)f(x(\Delta))\right\}\right|$$

$$\le \sup_{x\in G_\varepsilon^{(N)}}\left|E_x\left\{\left(1-\exp\left(-\alpha\int_0^\Delta V(x(s))\,ds\right)-\alpha\int_0^\Delta V(x(s))\,ds\right)f(x(\Delta))\right\}\right|$$

$$+\sup_{x\in G_\varepsilon^{(N)}}\left|E_x\left\{f(x(\Delta))\,\chi_{\{\tau^{(N)}\le\Delta\}}\right\}-E_x\left\{\alpha\int_0^\Delta V(x(s))\,ds\,f(x(\Delta))\right\}\right|$$

$$\le \alpha C\Delta^2\|f\| + \|f\|\sup_{x\in G_\varepsilon^{(N)}}\left|P_x\left\{\tau^{(N)}\le\Delta\right\}-E_x\left\{\int_0^\Delta \alpha V(x(s))\,ds\right\}\right|$$

$$+\sup_{x\in G_\varepsilon^{(N)}}\left|E_x\left\{\left(f(x(\Delta))-f(x)\right)\chi_{\{\tau^{(N)}\le\Delta\}}\right\}\right| + \alpha C\Delta\sup_x E_x\left\{\left|f(x(\Delta))-f(x)\right|\right\}.$$

Let $\omega(f,\delta)$ be the modulus of continuity of f,

$$(2.12)\qquad \omega(f,\delta) = \sup_{|x-y|\le\delta}|f(x)-f(y)|$$

and note that $\omega(f,\delta)\to 0$ as $\delta\to 0$ since $f\in\hat{C}(R^3)$. For each $\delta>0$ we have

$$(2.13)\quad \sup_{x\in G_\varepsilon^{(N)}} \left| E_x\left\{ \left(f\left(x(\Delta)\right) - f(x)\right) \chi_{\{\tau^{(N)} \leq \Delta\}} \right| \right.$$

$$\leq \omega(f,\delta) \sup_{x\in G_\varepsilon^{(N)}} \left| P_x\left\{\tau^{(N)} \leq \Delta\right\} - E_x\left\{ \int_0^\Delta \alpha V(x(s))\, ds \right\} \right.$$

$$+ \alpha C\Delta\, \omega(f,\delta) + 2\|f\|\, P_x\left\{ \left| x(\Delta) - x \right| > \delta \right\}$$

and similarly

$$(2.14)\qquad E_x\left\{ \left| f\left(x(\Delta)\right) - f(x) \right| \right\} \leq \omega(f,\delta) + 2\|f\|\, P_x\left\{ |x(\Delta) - x| > \delta \right\} .$$

Using (2.13) and (2.14) in (2.11) we see that

$$(2.15)\quad \sup_{x\in G_\varepsilon^{(N)}} \left| T_\Delta^{(N)} f(x) - T_\Delta^V f(x) \right|$$

$$\leq \alpha C\Delta^2 \|f\| + \left(\|f\| + \omega(\delta,f)\right) \sup_{x\in G_\varepsilon^{(N)}} \left| P_x\left\{\tau^{(N)} \leq \Delta\right\} - E_x\left\{\alpha \int_0^\Delta V(x(s))\, ds \right\} \right|$$

$$+ 2\alpha C\Delta\omega(\delta,f) + 2\left(1 + \alpha C\Delta\right)\|f\|\, P_x\left\{ |x(\Delta) - x| > \delta \right\}.$$

Now we note that the third term on the right side of (2.9) is precisely the one estimated by (2.13). We also note that since $f \in \hat{C}(R^3)$ and V is bounded continuous and of compact support in R^3.

$$(2.16)\qquad \sup_{0\leq t\leq t_0} \omega(T_t^V f,\delta) \to 0 \quad \text{as} \quad \delta \to 0$$

and

$$(2.17)\qquad \sup_{0\leq t\leq \Delta}\ \sup_x \left| T_t^V f(x) - f(x) \right| \to 0 \quad \text{as} \quad \Delta \to 0 .$$

In addition, since P_x is the Brownian motion measure

$$(2.18)\qquad \sup_x \Delta^{-1} P_x\left\{ |x(\Delta) - x| > \delta \right\} \to 0 \quad \text{as} \quad \Delta \downarrow 0 \quad \text{for each} \quad \delta > 0.$$

If we replace f by $T_t^V f$ in (2.15) use (2.13), (2.16), (2.17), (2.18) and in particular (2.1) and (2.3) we see that all terms on the right side of (2.9) can be made small by first choosing δ small and fixing it, next choosing Δ small enough and fixing it and finally choosing N large enough.

This completes the proof of Theorem 1 given the validity of (2.1), (2.2), and (2.3).

We turn now to the proof of (2.1), (2.2) and (2.3) which is given in a series of lemmas. The first lemma is a "one hole" estimate, the second gives the upper half of (2.1) and the third lemma gives the lower half of (2.1).

LEMMA 1. <u>Let</u> D <u>be a bounded open set in</u> R^3 <u>containing the origin, having</u> C^2 <u>boundary and capacitance</u> α. <u>Let</u>

$$D^{(N)} = \{x \in R^3 \mid N(x - y) \in D\}, \qquad y \in R^3$$

<u>and let</u> $\sigma^{(N)}$ <u>be the first time the Brownian motion reaches</u> $\partial D^{(N)}$ <u>starting from</u> $x \in R^3 - D^{(N)}$. <u>Then</u>

$$N\, P_x\left\{\sigma^{(N)} \leq t\right\} \to \alpha \int_0^t p(s,x,y)\, ds \quad \text{as} \quad N \to \infty \tag{2.19}$$

<u>uniformly on compact sets of</u> x <u>and</u> y <u>with</u> $x \neq y$ <u>where</u> $p(s,x,y)$ <u>is given by</u> (1.4).

<u>Proof</u>: We may take the hole center y to be the origin. Let

$$v^{(N)}(x,t) = NP_x\left\{\sigma^{(N)} \leq t\right\} \tag{2.20}$$

and

$$\hat{v}^{(N)}(x,\lambda) = \int_0^\infty e^{-\lambda t}\, v^{(N)}(x,t)\, dt, \qquad \lambda > 0. \tag{2.21}$$

The result (2.19) is equivalent to showing that

$$\hat{v}^{(N)}(x,\lambda) \to \frac{\alpha}{\lambda\, 4\pi|x|} \exp\left(-\sqrt{2\lambda}\,|x|\right) \tag{2.22}$$

for each $\lambda > 0$, uniformly on compact sets of x excluding $x = 0$. By change of variables we find that

$$\hat{v}^{(N)}(x,\lambda) = \frac{N}{\lambda} w\left(Nx, \frac{\lambda}{N^2}\right) \tag{2.23}$$

where $w(x,\mu)$ satisfies

$$\begin{aligned} &2\mu w - \Delta w = 0 \quad \text{in} \quad R^3 - D \\ &w(x,\mu) = 1 \quad \text{for} \quad x \in \partial D. \end{aligned} \tag{2.24}$$

If $\hat{n}$ denotes the unit inner normal to D at the boundary, Green's identity gives

$$\begin{aligned} \frac{N}{\lambda} w\left(Nx, \frac{\lambda}{N^2}\right) &= \frac{1}{4\pi\lambda} \int_{\partial D} \frac{\exp(-\sqrt{2\lambda}|x-N^{-1}y|)}{|x-N^{-1}y|} \frac{\partial w(y,N^{-2}\lambda)}{\partial \hat{n}}\, dS \\ &\to \frac{1}{4\pi\lambda} \frac{\exp\{-\sqrt{2\lambda}|x|\}}{|x|} \int_{\partial D} \frac{\partial w(y,0)}{\partial \hat{n}}\, dS = \frac{\alpha}{4\pi\lambda} \frac{\exp(-\sqrt{2\lambda}\,|x|)}{|x|} \end{aligned} \tag{2.25}$$

uniformly on compact x sets excluding $x = 0$. Note that we have used (1.9). The proof of Lemma 1 is complete.

LEMMA 2. *Given $\varepsilon > 0$ and $t_0 < \infty$ there exists an N_0 such that for all $N \geq N_0$ there is a set $G_\varepsilon^{(N)} \subset G^{(N)}$ such that*

$$P_x\{\tau^{(N)} \leq t\} \leq \int_0^t \int \alpha V(y)\, p(s,x,y)\, dy\, ds + \varepsilon \tag{2.26}$$

for all $x \in G_\varepsilon^{(N)}$ and $0 \leq t \leq t_0$ and

$$\text{vol}\,(R^3 - G_\varepsilon^{(N)}) < \varepsilon. \tag{2.27}$$

Proof: Let $\varepsilon > 0$ be given and fix $\delta > 0$ to be chosen later. We write

$$\begin{aligned} P_x\left\{\tau^{(N)} \leq t\right\} &\leq \sum_{i=1}^{N} P_x\left\{\tau_i^{(N)} \leq t\right\} \\ &= \sum_{\substack{i=1 \\ |x-y_i^{(N)}|>\delta}}^{N} P_x\left\{\tau_i^{(N)} \leq t\right\} + \sum_{\substack{i=1 \\ |x-y_i^{(N)}|\leq\delta}}^{N} P_x\left\{\tau_i^{(N)} \leq t\right\}. \end{aligned} \tag{2.28}$$

We estimate first the second term on the right of (2.28).

From Lemma 1 and the estimate

$$\int_0^t p(s,x,y)\, ds \leq c_2\, |x - y|^{-1} \exp\left(-\, |x - y|^2/4t\right)$$

we conclude that there is a constant c_1 independent of x, $y_i^{(N)}$ and N such that

$$P_x\left\{\tau_i^{(N)} \leq t\right\} \leq c_1\, N^{-1} |x - y_i^{(N)}|^{-1} \exp\left(-\, |x - y_i^{(N)}|^2/4t\right) \tag{2.29}$$

for all $t < \infty$. Using (2.29) we obtain for all t

$$\sum_{\substack{i=1 \\ |x-y_i^{(N)}|\leq\delta}}^{N} P_x\left\{\tau^{(N)} \leq t\right\} \leq \frac{c_1}{N} \sum_{\substack{i=1 \\ |x-y_i^{(N)}|\leq\delta}}^{N} |x - y_i^{(N)}|^{-1} \tag{2.30}$$

Now we choose $\delta = \delta(\varepsilon)$ so that

$$\int_{|z|\leq\delta} |z|^{-1}\, dz \leq \frac{\varepsilon^2}{2c_1} \tag{2.31}$$

Then

$$(2.32)\quad \int dx \frac{c_1}{N} \sum_{\substack{i=1 \\ |x-y_i^{(N)}|\le\delta}}^{N} |x-y_i^{(N)}|^{-1} = \frac{c_1}{N} \sum_{i=1}^{N} \int dx\, |x-y_i^{(N)}|^{-1} \chi_{\{|x-y_i^{(N)}|\le\delta\}} \le \frac{\varepsilon^2}{2}$$

We define the set $G_\varepsilon^{(N)} \subset G^{(N)}$ by

$$(2.33)\quad G_\varepsilon^{(N)} = \left\{ x \in R^3 \;\middle|\; \frac{c_1}{N} \sum_{\substack{i=1 \\ |x-y_i^{(N)}|\le\delta(\varepsilon)}}^{N} |x - y_i^{(N)}|^{-1} \le \frac{\varepsilon}{2} \right\}$$

By Tchebychev's inequality

$$(2.34)\qquad \mathrm{vol}\,(R^3 - G_\varepsilon^{(N)}) \le \varepsilon$$

and by definition of $G_\varepsilon^{(N)}$ and (2.28)

$$(2.35)\qquad P_x\left\{\tau^{(N)} \le t\right\} \le \sum_{\substack{i=1 \\ |x-y_i^{(N)}|>\delta(\varepsilon)}}^{N} P_x\left\{\tau_i^{(N)} \le t\right\} + \frac{\varepsilon}{2}$$

for all $x \in G^{(N)}$ and all $t \ge 0$.

We apply Lemma 1 to the first term on the right of (2.35) to obtain

$$(2.36)\qquad \sum_{\substack{i=1 \\ |x-y_i^{(N)}|>\delta(\varepsilon)}}^{N} P_x\left\{\tau_i^{(N)} \le t\right\} = \frac{1}{N} \sum_{i=1}^{N} \alpha \int_0^t p(s,x,y_i^{(N)})\, ds\, \chi_{\{|x-y_i^{(N)}|>\delta\}} + h_N$$

where $h_N \to 0$ as $N \to \infty$ uniformly in $x \in G_\varepsilon^{(N)}$ and $t \in [0,t_0]$. Replacing $\chi_{\{|z|>\delta\}}$ by a smooth bounded function $\phi_\delta(z)$ that vanishes for $|z| \le \delta/2$ and using hypothesis (1.7) we obtain (2.26). The uniformity in t and $x \in G_\varepsilon^{(N)}$ is obtained by approximating the function

$$\alpha \int_0^t p(s,x,y)\, ds\, \phi_\delta(x - y) ,$$

which is uniformly continuous in $[0,t_0] \times R^3 \times R^3$, by finite sums of products of functions of one variable t, x and y at a time and then using (1.7). After this is done we let $\delta \downarrow 0$ and this gives (2.26). The proof is complete.

LEMMA 3. <u>Given</u> $\varepsilon > 0$ <u>and</u> $t_0 < \infty$ <u>there exists an</u> N_0 <u>such that for all</u> $N \ge N_0$

$$(2.37)\qquad \int_0^t \alpha\, V(y)\, p(s,x,y)\, dy\, ds - \varepsilon - g(t) \le P_x\left\{\tau^{(N)} \le t\right\}$$

<u>for all</u> $x \in R^3$ <u>and</u> $0 \le t \le t_0$ <u>where</u> $g(t) \ge 0$ <u>and</u>

$$(2.38)\qquad \overline{\lim_{t\downarrow 0}}\, \frac{1}{t}\, g(t) = 0 .$$

<u>Proof</u>: Let $\hat{\tau}^{(N)}$ be the minimum of $\tau_i^{(N)}$ (cf. (1.6)) over those $i = 1,2,\dots,N$ for which $|y_i^{(N)} - y_j^{(N)}| \geq 3/N$ for all $j \neq i$ and $|x - y_i^{(N)}| \geq \delta$ with $\delta > 0$ to be chosen later. We have

$$(2.39)\quad P_x\left\{\tau^{(N)} \leq t\right\} \geq P_x\left\{\hat{\tau}^{(N)} \leq t\right\}$$

$$\geq \sum_{\substack{i=1\\ |x-y_i^{(N)}|\geq\delta \\ |y_i^{(N)}-y_j^{(N)}|\geq 3/N}}^{N} P_x\left\{\tau_i^{(N)} \leq t\right\} - \frac{1}{2}\sum_{\substack{i\neq j\\ |x-y_i^{(N)}|\geq\delta \\ |y_i^{(N)}-y_j^{(N)}|\geq 3/N}} P_x\left\{\tau_i^{(N)} \leq t,\ \tau_j^{(N)} \leq t\right\}$$

where we have used the principle of inclusion and exclusion for estimating the probability of the union of sets. We look first at the single sum on the right of (2.39). We have the estimate

$$(2.40)\quad \sum_{\substack{i=1\\ |x-y_i^{(N)}|\geq\delta \\ |y_i^{(N)}-y_j^{(N)}|\geq 3/N}}^{N} P_x\left\{\tau_i^{(N)} \leq t\right\} = \sum_{\substack{i=1\\ |x-y_i^{(N)}|\geq\delta}}^{N} P_x\left\{\tau_i^{(N)} \leq t\right\} - \sum_{\substack{i=1\\ |y_i^{(N)}-y_j^{(N)}|\leq 3/N \\ |x-y_i^{(N)}|\geq\delta}}^{N} P_x\left\{\tau_i^{(N)} \leq t\right\}$$

where the second sum on the right is over all $i = 1,2,\dots,N$ for which $|y_i^{(N)}-y_j^{(N)}| \leq \frac{3}{N}$ for some $j \neq i$. The first sum on the right side of (2.40) is estimated exactly as was done in Lemma 2.

We must show that the second sums on the right of (2.39) and of (2.40) are small. We use (2.29) with $t = \infty$ to obtain

$$(2.41)\quad \sum_{\substack{i=1\\ |y_i^{(N)}-y_j^{(N)}|\leq 3/N \\ |x-y_i^{(N)}|\geq\delta}}^{N} P_x\left\{\tau_i^{(N)} \leq t\right\} \leq \frac{c_1}{N}\sum_{\substack{i=1\\ |y_i^{(N)}-y_j^{(N)}|\leq 3/N}}^{N} |x-y_i^{(N)}|^{-1}\,\chi_{\{|x-y_i^{(N)}|\geq\delta\}}$$

We then note that we also have the estimate

$$(2.42)\quad \frac{c_1}{N}\sum_{\substack{i=1\\ |y_i^{(N)}-y_j^{(N)}|\leq\frac{3}{N}}}^{N} |x-y_i^{(N)}|^{-1}\,\chi_{\{|x-y_i^{(N)}|\geq\delta\}} \leq \frac{3c_1}{N^2}\sum_{i\neq j}\frac{\phi_\delta(|x-y_i^{(N)}|)}{|x-y_k^{(N)}|}\,\frac{\psi_\gamma(|y_i^{(N)}-y_j^{(N)}|)}{|y_i^{(N)}-y_j^{(N)}|}$$

for all sufficiently large N. The function $\psi_\gamma(s)$, $s \geq 0$, is defined for each $\gamma > 0$ so that it is smooth and it equals one for $s \leq \gamma$ and zero for $s \geq 2\gamma$. The function $\phi_\delta(s) \geq \chi_{\{s\geq\delta\}}$ is a smooth function that equals one when $s \geq \delta$ and zero when $s \leq \delta/2$.

LEMMA 4. *Let* μ_n *be a sequence of measures on a separable metric space* X *and suppose that the* μ_n *converge to a measure* μ *weakly as* $n \to \infty$. *Let* $f \geq 0$ *be a function on* X *such that*

$$\int_X f(x)\, \mu_n(dx) \to \int_X f(x)\, \mu(dx) \quad \text{as} \quad n \to \infty \tag{2.43}$$

and assume there exists a sequence of bounded continuous functions $f_k(x) \geq 0$ *on* X, $k = 1,2,\ldots$ *such that*

$$f_k(x) \uparrow f(x) \quad \text{as} \quad k \to \infty \tag{2.44}$$

for each $x \in X$. *Then the measures* $f\mu_n$ *converge weakly to* $f\mu$ *as* $n \to \infty$, *i.e., for any bounded continuous function* ϕ *on* X

$$\int_X \phi(x)\, f(x)\, \mu_n(dx) \to \int_X \phi(x)\, f(x)\, \mu(dx) . \tag{2.45}$$

We may apply this lemma to the right side of (2.42) by using hypothesis (1.8). We conclude that the right side of (2.42) tends as $N \to \infty$, uniformly in x, to

$$3c_1 \int_{R^3} \int_{R^3} \frac{\phi_\delta(|x - y|)\, \psi_\gamma(|y - z|)}{|x - y|\, |y - z|} V(y)\, V(z)\, dy\, dz .$$

Since $\gamma > 0$ can be chosen as small as desired we conclude that the second term on the right of (2.40) can be made arbitrarily small for all t and x by choosing N large. Moreover this estimate is independent of $\delta > 0$.

We look next at the double sum on the right of (2.39). We have that since $|y_i^{(N)} - y_j^{(N)}| \geq \frac{3}{N}$,

$$P_x\left\{\tau_i^{(N)} \leq t,\ \tau_j^{(N)} \leq t\right\} = \int_0^t \int_{\partial D_j^{(N)}} P_x\left\{\tau_i^{(N)} \in ds,\ x(\tau_i^{(N)}) \in dz\right\} P_z\left\{\tau_j^{(N)} \leq t-s\right\}$$

$$\leq P_x\left\{\tau_i^{(N)} \leq t\right\} \sup_{z \in D_i^{(N)}} P_z\left\{\tau_j^{(N)} \leq t\right\}$$

$$\leq \frac{c_1 \hat{c}_1}{N^2} \frac{\exp\{-|x-y_i^{(N)}|^2/4t\}}{|x - y_i^{(N)}|} \frac{\exp\{-|y_i^{(N)}-y_j^{(N)}|^2/4t\}}{|y_i^{(N)} - y_j^{(N)}|}$$

In the last inequality we make use of the fact that $|x - y_i^{(N)}| \geq \delta > 0$ and (2.29) with $\hat{c}_1$ a slightly bigger constant to take care of the sup over $z \in D_i^{(N)}$. Thus

$$\sum_{\substack{i\neq j \\ |x-y_i^{(N)}|\geq\delta \\ |y_i^{(N)}-y_j^{(N)}|\geq\frac{3}{N}}} P_x\left\{\tau_i^{(N)} \leq t,\ \tau_j^{(N)} \leq t\right\} \leq \frac{c_1\hat{c}_1}{N^2} \sum_{i\neq j} \frac{\exp(-|x-y_i^{(N)}|^2/4t)}{|x - y_i^{(N)}|}$$

$$\cdot \frac{\exp(-|y_i^{(N)}-y_j^{(N)}|/4t)}{|y_i^{(N)} - y_j^{(N)}|}\ \phi_\delta(|x - y_i^{(N)}|) .$$

Applying Lemma 4 we see that the double sum is bounded independently of N by a constant times

$$c_1\hat{c}_1\iint \frac{\exp(-|x-y|^2/4t)}{|x-y|}\ \frac{\exp(-|y-z|^2/4t)}{|y-z|}\ V(y)\ V(z)\ dy\ dz$$

and this is in turn less than a constant times t^2 uniformly in x.

Returning to the statement of our Lemma 3 we see that we may take for $g(t)$ a constant times t^2 and hence (2.38) holds. This proves Lemma 3.

The proof of Theorem 1 is complete, since Lemmas 2 and 3 give (2.1), (2.2), and (2.3), except for the real variables Lemma 4. We prove this next.

Proof of Lemma 4: Let ϕ be continuous and assume with no loss in generality that $0 \leq \phi \leq 1$.

We have

$$\int \phi f\mu_n(dx) \geq \int \phi f_k\, \mu_n(dx) \quad \text{for} \quad k = 1,2,...$$

and hence by the weak convergence of μ_n

$$\underline{\lim}_{n\uparrow\infty} \int \phi f\mu_n(dx) \geq \int \phi f_k\mu(dx) .$$

Letting $k \to \infty$ we have

$$\underline{\lim}_{n\uparrow\infty} \int \phi f\mu_n(dx) \geq \int \phi f\mu(dx) .$$

Next we repeat the above with ϕ replaced by $1-\phi$. We have

$$\int f\mu_n(dx) - \int \phi f\mu_n(dx) \geq \int f_k\mu_n(dx) - \int \phi f_k\mu_n(dx)$$

and hence for each $k = 1,2,\ldots$, by the weak convergence of μ_n and (2.43),

$$\int f\mu(dx) - \overline{\lim_{n\uparrow\infty}} \int \phi f\mu_n(dx) \geq \int f_k\mu(dx) - \int \phi f_n\mu(dx)$$

By the monotone convergence theorem as $k \to \infty$ this gives

$$\overline{\lim_{n\uparrow\infty}} \int \phi f\mu_n(dx) \leq \int \phi f\mu(dx)$$

and the proof is complete.

REFERENCES

[1] M. Kac, Probabilistic methods in some problems of scattering theory, Rocky Mountain J. Math. 4 (1974) pp. 511-538.

[2] J. Rauch and M. Taylor, Potential and scattering theory on wildly perturbed domains, J. Funct. Anal. 18 (1975) pp. 27-59.

[3] E. I. Khruslov and V. A. Marchenko, Boundary value problems in regions with fine-grained boundaries, Naukova Dumka, Kiev, 1974.

COLLOQUIA MATHEMATICA SOCIETATIS JÁNOS BOLYAI
27. RANDOM FIELDS, ESZTERGOM (HUNGARY), 1979.

BOUNDARY VALUE PROBLEMS WITH RAPIDLY OSCILLATING RANDOM COEFFICIENTS

G.C. PAPANICOLAOU* – S.R.S. VARADHAN**

1. INTRODUCTION AND STATEMENT OF THE PROBLEM

We shall introduce the problem under consideration and its analysis in an informal way at first. The detailed mathematical formulation begins in the next section and the main theorem is stated in Section 3. Another version of this theorem is given in Section 5 and a probabilistic one in Section 8.

Consider a conductor occupying a region $\mathcal{O}$ in R^3 and suppose that the conducting material is inhomogeneous as for example in the case of a composite or a mixture of several materials with different conductivities. We shall model the material with a conductivity that is a random function of position changing rapidly as x varies over lengths comparable to the size of the region $\mathcal{O}$. To articulate this last feature we introduce a parameter $\epsilon > 0$ which is the ratio of a typical length scale associated

*Research supported by Contract AFOSR-78-3668.

**Research supported by Grant NSF-MCS 77-02687, and by a grant from IBM Corporation.

with the region $\mathcal{O}$ to a typical length scale associated with the variations in conductivity. We take then the conductivity to be a random function of the form $a\left(\frac{x}{\epsilon}\right)$ where $a(x)$ is a given stationary random function that is strictly positive and bounded. The temperature $u^{\epsilon}(x)$ satisfies the stochastic heat equation

$$(1.1) \qquad -\nabla \cdot \left(a\left(\frac{x}{\epsilon}\right) \nabla u^{\epsilon}(x)\right) = f(x) \qquad (x \in \mathcal{O}),$$

and we assume for simplicity here that $u^{\epsilon}(x) = 0$ for $x \in \partial\mathcal{O}$, the boundary of $\mathcal{O}$. The function $f(x)$ is a given deterministic (for simplicity) function which represents the density of heat sources in $\mathcal{O}$.

The problem now is to analyze the behaviour of the random temperature distribution $u^{\epsilon}(x)$ as $\epsilon \to 0$. It is found that when the random conductivity is strictly stationary and ergodic (cf. Section 2), there exist constants q_{ij} $(i, j = 1, 2, 3)$ such that if $u(x)$ is the solution of the deterministic heat equation

$$(1.2) \qquad -\sum_{i,j=1}^{3} q_{ij} \frac{\partial^2 u(x)}{\partial x_i \partial x_j} = f(x) \qquad (x \in \mathcal{O},\ u(x) = 0,\ x \in \partial\mathcal{O}),$$

then

$$(1.3) \qquad \int_{\mathcal{O}} E\{|u^{\epsilon}(x) - u(x)|^2\}\, dx \to 0 \quad \text{as} \quad \epsilon \to 0,$$

where $E\{\cdot\}$ denotes expectation value. The tensor $q = (q_{ij})$ is called the effective conductivity tensor and can be computed by solving certain auxiliary problems (cf. Section 3).

This result shows that the randomly inhomogeneous conducting medium will behave like a homogeneous deterministic medium with conductivity tensor q when ϵ is small. When the conductivity $a(x)$ is a periodic function the above convergence result, considered from a variety of viewpoints and in several more general or refined forms, has been obtained and discussed by several authors [1]-[10]. When $a(x)$ is almost periodic the result was obtained by Kozlov [11]. In the stochastic case (1.3) was obtained by Jurinskii [12] assuming that the

random conductivity $a(x)$ satisfies a mixing condition, and in general by Kozlov [13].*

Our purpose here is to review the problem and formulate it in a suitable analytical framework, which is described in Section 2, in which one can obtain (1.3) and related facts with just stationarity, ergodicity and uniform ellipticity assumptions on $a(x)$. Thus the periodic, almost periodic and random cases come out of the same theorem in a unified way.

In the remainder of this section we shall describe briefly the calculations that enter in obtaining (1.3) and in determining the effective conductivity tensor (q_{ij}).

Let $(\Omega, \mathscr{F}, P)$ be a probability space with $\omega \in \Omega$ labeling the realization of the medium $(a(x) = a(x, \omega))$. It is natural to attempt to expand the temperature $u^\epsilon(x) = u^\epsilon(x, \omega)$ in the form

$$(1.4) \qquad u^\epsilon(x, \omega) = u(x) + \epsilon u_1\left(x, \frac{x}{\epsilon}, \omega\right) + \epsilon^2 u_2\left(x, \frac{x}{\epsilon}, \omega\right) + \dots$$

which is the usual formalism of multiple scales used in [10] and in many other situations. Nothing that if $v^\epsilon(x) = v\left(x, \frac{x}{\epsilon}\right)$, then

$$(1.5) \qquad \nabla v^\epsilon(x) = \left[\nabla_x v(x, y) + \frac{1}{\epsilon} \nabla_y v(x, y)\right]_{y = \frac{x}{\epsilon}},$$

we insert (1.4) into (1.1), use (1.5) and equate coefficients of powers of ϵ. This gives the following sequence of problems in which $y = \frac{x}{\epsilon}$.

$$(1.6) \qquad -\nabla_y \cdot (a(y, \omega) \nabla_y u(x)) = 0$$

$$(1.7) \qquad \begin{aligned} &-\nabla_y \cdot (a(y, \omega) \nabla_y u_1(x, y, \omega)) - \nabla_y \cdot (a(y, \omega) \nabla_x u(x)) - \\ &-\nabla_x \cdot (a(y, \omega) \nabla_y u(x)) = 0 \end{aligned}$$

*The authors obtained the results given here after seeing that Jurinskii's analysis did not require mixing. Kozlov's work than came to our attention wherein similar results and methods of analysis as here are given.

$$
(1.8)\qquad \begin{aligned} &-\nabla_y\cdot(a(y,\omega)\nabla_y u_2(x,y,\omega)) - \nabla_y\cdot(a(y,\omega)\nabla_x u_1(x,y,\omega)) - \\ &-\nabla_x\cdot(a(y,\omega)\nabla_y u_1(x,y,\omega)) - \nabla_x\cdot(a(y,\omega)\nabla_x u(x)) = \\ &= f(x), \end{aligned}
$$

and similarly for the higher order terms.

Clearly (1.6) is satisfied since u was taken as a function of x only. The function $u(x)$ is usually determined from solvability conditions in the equations for u_1 or u_2, etc. We consider (1.7) and note that if we let

$$
(1.9)\qquad u_1(x,y,\omega) = \sum_{k=1}^{d} \chi^k(y,\omega)\,\frac{\partial u(x)}{\partial x_k},
$$

then (1.7) will be satisfied if $\chi^k(y,\omega)$ solves the problem

$$
(1.10)\qquad -\nabla_y\cdot(a(y,\omega)\nabla_y\chi^k(y,\omega)) = \nabla_y\cdot(a(y,\omega)e_k) \qquad (k=1,2,3),
$$

where e_k is the column vector with components (δ_{ik}). Once (1.10) has been solved in a suitable sense, (1.9) is inserted into (1.8) which becomes another equation of the form (1.10) with a different right-hand side. The condition that this form of (1.8) have a suitable solution leads to the equation (1.2) that determines $u(x)$.

The problem then reduces effectively to the analysis of the stochastic partial differential equation (1.10) which we rewrite as

$$
(1.11)\qquad -\nabla_y\cdot(a(y,\omega)\nabla_y\chi(y,\omega)) = h(y,\omega) \qquad (y\in R^3),
$$

with $a(y,\omega)$ and $h(y,\omega)$ given stationary stochastic processes. In the periodic case, that is when for each $\omega\in\Omega$, $a(y,\omega)$ and $h(y,\omega)$ are periodic of period one in y_1, y_2 and y_3 with $y=(y_1,y_2,y_3)$, problem (1.11) is elementary. If $a(y,\omega)$ is strictly positive and bounded, uniformly in y and ω, (1.11) interpreted in a weak or variational form has a unique periodic solution $\chi(y,\omega)$ if and only if the integral of h over the period cell vanishes. For uniqueness we require that the integral of χ over the period cell also vanishes. In the almost periodic and stationary random case it is not possible in general to find almost periodic or stationary solutions $\chi(y,\omega)$ even when h has mean zero. We show in Theorem 2 below that

(1.11) always has a solution $\chi(y, \omega)$ that is not stationary but its gradient $\nabla\chi(y, \omega)$ is stationary. Furthermore when a and h are ergodic and h has mean zero then $\chi(y, \omega)$ increases slower than $|y|$ in mean square as $|y| \to \infty$.

Returning to (1.10) and (1.9) we see that this is sufficient to make some sense out of (1.4) because then $\epsilon u_1\left(x, \frac{x}{\epsilon}, \omega\right)$ is truly small in mean square as $\epsilon \to 0$. We also see that the solvability condition for (1.8) leads to (1.2) where, if we let $a_{ij}(x, \omega) = a(x, \omega)\delta_{ij}$, we have

$$q_{ij} = E\left\{\sum_{k=1}^{3} a_{ik}(x)\left(\delta_{kj} + \frac{\partial \chi^j(x)}{\partial x_k}\right)\right\} \qquad (i, j = 1, 2, 3). \tag{1.12}$$

Note that the effective conductivity $q = (q_{ij})$ is indeed constant, independent of x, since $\nabla\chi^k(x, \omega)$ is stationary. Formula (1.12) is exactly the one that holds in the periodic case and had been obtained previously. We show here that it holds also in the almost periodic and stationary random case when the solution χ^k of (1.10) is interpreted suitably.

In the method of averaging for ordinary differential equations one encounters a similar but simpler problem with the one encountered here namely that integrals of almost periodic functions with mean zero are not in general almost periodic but grow more slowly than linearly as their argument tends to infinity. It is also encountered in the analysis of stochastic ordinary differential equations by multiscaling which is frequently similar to averaging (cf. [14] and references therein).

Let us give a physical interpretation for the formula (1.12) defining the effective conductivity (q_{ij}). Conductivity is by Fourier's law heat flow per unit temperature gradient. This is, of course, already a macroscopic concept but we take it to hold microscopically here while allowing for very rapid irregular fluctuations. If the temperature distribution $u^\epsilon(x)$ stabilizes when ϵ, the scale of fluctuations, is small, then locally the conductor sees a uniform temperature gradient impressed on it. Consider now the conductor in this local picture and suppose unit temperature gradient is impressed in the y_j direction. Compute the resulting temperature distribution $y_j + \chi^j(y, \omega)$ (this is the meaning of (1.10)) and the heat flux

$J_{ij}(y, \omega)$ in the y_i direction due to the impressed temperature gradient in the y_j direction

$$(1.13) \qquad J_{ij}(y, \omega) = \sum_{k=1}^{3} a_{ik}(y, \omega)\left(\delta_{kj} + \frac{\partial \chi^j(y, \omega)}{\partial y_k}\right).$$

Then average the result over realizations. This gives the effective conductivity (q_{ij}) in (1.12) as average local heat flux per unit local temperature gradient. Note finally that if $w^k(y, \omega) = y_k + \chi^k(y, \omega)$ is the temperature distribution in the local picture so that $\nabla \cdot (a \nabla w^k) = 0$, which is (1.10), then the statement that unit temperature gradient is impressed in the y_k direction is interpreted properly by the statement that $w^k(y, \omega) - y_k = = \chi^k(y, \omega) = o(|y|)$ (in mean square) as $|y| \to \infty$ which is what is shown below.

2. THE ABSTRACT FRAMEWORK

We begin with a precise formulation of (1.1). Let $(\Omega, \mathscr{F}, P)$ be a probability space and let $(a_{ij}(y, \omega))$ $(i, j = 1, 2, \ldots, d)$ be a strictly stationary matrix-valued random field with $y \in R^d$. It is assumed throughout that there is a positive constant a_0 such that

$$(2.1) \qquad a_0 |\xi|^2 \leqslant \sum_{i,j=1}^{d} a_{ij}(y, \omega)\xi_i \xi_j \leqslant a_0^{-1} |\xi|^2$$

for all $y \in R^d$, $\omega \in \Omega$ and $\xi = (\xi_1, \ldots, \xi_d)$. Strict stationarity means that for any $h \in R^d$, any integer $n = 1, 2, \ldots$ and any points $y_1, y_2, \ldots, y_n$ in R^d the joint distribution of $a_{ij}(y_1, \omega), a_{ij}(y_2, \omega), \ldots$ $\ldots, a_{ij}(y_n, \omega)$ is the same as that of $a_{ij}(y_1 + h, \omega), \ldots, a_{ij}(y_n + h, \omega)$ $(i, j = 1, 2, \ldots, d)$. The processes $a_{ij}(y, \omega)$ are assumed to be stochastically continuous.

$$(2.2) \qquad \lim_{|h| \downarrow 0} P\Big\{\sum_{i,j=1}^{d} |a_{ij}(y + h, \omega) - a_{ij}(y, \omega)| > \delta\Big\} = 0,$$

for all $\delta > 0$ and $y \in R^d$. On account of this assumption the process can be taken to be separable (with separating set the rationals in R^d) and jointly measurable in $(y, \omega) \in R^d \times \Omega$ (R^d taken with its Lebesgue

measurable sets and the Lebesgue measure). We take $\mathscr{F}$ to be countably generated.

We shall now introduce some notation that will be used throughout. Let $\mathscr{H} = L^2(\Omega, \mathscr{F}, P)$ be the Hilbert space of square integrable functions on Ω with inner product

$$E\{gh\} = \int_{\Omega} P(d\omega) g(\omega) h(\omega) \tag{2.3}$$

and norm $E^{\frac{1}{2}}\{|g|^2\}$. $\mathscr{H}$ is separable since $\mathscr{F}$ is countably generated. Let $\mathscr{O} \subset R^d$ be an open set. Denote by $H = L^2(\mathscr{O}; \mathscr{H}) = L^2(\mathscr{O} \times \Omega)$ the Hilbert space of square integrable functions on $\mathscr{O}$ with values in $\mathscr{H}$ and with inner product

$$(u, v) = \int_{\mathscr{O}} dx E\{u(x) v(x)\} = \int_{\mathscr{O}} dx \int_{\Omega} P(d\omega)\, u(x, \omega) v(x, \omega). \tag{2.4}$$

We denote by $H^1 = H^1(\mathscr{O}; \mathscr{H})$ the Hilbert space of $\mathscr{H}$ valued functions whose distribution derivatives are square integrable over $\mathscr{O}$ and with inner product

$$\begin{aligned} (u, v)_1 &= (u, v) + \sum_{i=1}^{d} \left(\frac{\partial u}{\partial x_i}, \frac{\partial v}{\partial x_i} \right) = \\ &= \int_{\mathscr{O}} dx \int_{\Omega} P(d\omega) + \sum_{i=1}^{d} \int_{\mathscr{O}} dx \int_{\Omega} P(d\omega) \frac{\partial u(x, \omega)}{\partial x_i} \frac{\partial v(x, \omega)}{\partial x_i}. \end{aligned} \tag{2.5}$$

We denote by $H_0^1 = H_0^1(\mathscr{O}; \mathscr{H})$ the Hilbert space of $\mathscr{H}$ valued functions u with square integrable distribution derivatives and such that $u = 0$ on $\partial\mathscr{O}$.

Let $\{e_n(\omega)\}$ be an orthonormal basis in $\mathscr{H}$. Then $u \in H^1(\mathscr{O}; \mathscr{H})$ if and only if $u_n(x) = E\{u(x) e_n\}$ is in $H^1(\mathscr{O}; R^1)$ and

$$\sum_{n=1}^{\infty} |u_n|^2_{H^1(\mathscr{O}; R^1)} < \infty.$$

We shall denote by $\mathscr{D}(\mathscr{O}; \mathscr{H})$ or $C_0^\infty(\mathscr{O}; \mathscr{H})$ the space of infinitely differentiable $\mathscr{H}$-valued functions that vanish outside a compact subset of $\mathscr{O}$.

If $u_n(x)$ are the coordinates of u relative to the basis $\{e_n\}$, then $u \in \in \mathscr{D}(\mathcal{O}; \mathscr{H})$ if and only if the u_n are infinitely differentiable, vanish outside a fixed compact subset of $\mathcal{O}$ and

$$\sum_{n=1}^{\infty} \int_{\mathcal{O}} dx \left| \left(\frac{\partial}{\partial x_1}\right)^{k_1} \cdots \left(\frac{\partial}{\partial x_d}\right)^{k_d} u_n(x) \right|^2 < \infty$$

for all multiindices $k = (k_1, k_2, \ldots, k_d)$ with $|k| = 1, 2, \ldots$.

Let V be a closed subset of $H^1(\mathcal{O}; \mathscr{H})$ containing $H_0^1(\mathcal{O}; \mathscr{H})$ i.e., $H_0^1 \subseteq V \subseteq H^1$. The precise meaning of the problem under considerating (1.1) with general homogeneous variational boundary conditions is this. Find $u^\epsilon(x, \omega) \in V$ such that

$$\int_{\mathcal{O}} dx \int_{\Omega} P(d\omega) \sum_{i,j=1}^{d} a_{ij}\left(\frac{x}{\epsilon}, \omega\right) \frac{\partial u^\epsilon(x, \omega)}{\partial x_j} \frac{\partial \varphi(x, \omega)}{\partial x_i} +$$

$$+ \alpha \int_{\mathcal{O}} dx \int_{\Omega} P(d\omega) u^\epsilon(x, \omega) \varphi(x, \omega) = \tag{2.6}$$

$$= \int_{\mathcal{O}} dx \int_{\Omega} P(d\omega) f(x) \varphi(x, \omega),$$

for all $\varphi(x, \omega) \in V$. Here $\alpha > 0$ is a fixed constant and $f(x) \in L^2(\mathcal{O}; R^1)$ is given. In view of (2.1) this problem has a unique solution for each $\epsilon > 0$ by the Lax – Milgram lemma [15]. By letting $\varphi = u^\epsilon$ in (2.6) and using (2.1) we obtain the bound

$$(u^\epsilon, u^\epsilon)_1 \leqslant C \int_{\mathcal{O}} dx \, |f(x)|^2 \tag{2.7}$$

which C a constant independent of ϵ. The case $V = H_0^1$ corresponds to the Dirichlet problem while $V = H^1$ to the generalized Neumann problem. For additional information regarding boundary value problems in variational form we refer to [15].

Before going to the theorem regarding the behavior of u^ϵ as $\epsilon \to 0$ in the following sections we shall consider more closely the implications of stationarity and stochastic continuity of the coefficients (a_{ij}).

We may take the probability space $(\Omega, \mathscr{F}, P)$ to be as follows. The

set Ω is the set of Lebesgue measurable $d \times d$ matrix-valued functions on R^d. The value of $\omega \in \Omega$ at $y \in R^d$ is defined almost everywhere and is denoted by $\omega_{ij}(y, \omega)$. Thus Ω is the set of all coefficients for (2.6). We take for the σ-algebra $\mathscr{F}$ the one generated by cylinder sets with base points that have rational coordinated in R^d and range sets that are spheres in R^{d^2} with rational centers and rational radii so that $\mathscr{F}$ is countably generated. The probability measure P is defined on $(\Omega, \mathscr{F})$ and invariant with respect to the translation group $\tau_x: \Omega \to \Omega$ defined by

$$(2.8) \qquad (\tau_x \omega)(y) = \omega(y - x) \qquad (x, y \in R^d).$$

The translation group τ_x will also be assumed to be ergodic: the only sets A in $\mathscr{F}$ that are invariant, i.e. $\tau_x A \subset A$, have $P(A) = 0$ or 1. We note that the mapping $(x, \omega) \to \tau_x \omega$ is jointly measurable in (x, ω) with respect to $\mathscr{L}(R^d) \times \mathscr{F}$ with $\mathscr{L}(R^d)$ the σ-algebra of Lebesgue measurable sets in R^d.

Let $\tilde{f} \in \mathscr{H} = L^2(\Omega, \mathscr{F}, P)$. For almost all ω we let

$$(2.9) \qquad (T_x \tilde{f})(\omega) = \tilde{f}(\tau_{-x} \omega) \qquad (x \in R^d)$$

and note that the operators T_x form a unitary group on $\mathscr{H}$. This group is strongly continuous. To see this we note that any $\tilde{f}$ in $\mathscr{H}$ can be approximated by a function in $\mathscr{H}$ that is measurable with respect to the algebra of cylinder sets and therefore depends on the value of ω at a finite number of points $y \in R^d$. But for such functions the strong continuity of T_x follows from the stochastic continuity (2.2) (the $a_{ij}(x, \omega)$ is identified with $\omega_{ij}(x, \omega)$ below).

With any $\tilde{f}$ in $\mathscr{H}$ we may associate the stationary process

$$(2.10) \qquad f(x, \omega) = (T_x \tilde{f})(\omega) = f(\tau_{-x} \omega).$$

We shall use the tilde notation exclusively for associating a function with its translates that form the stationary process. In particular we define

$$(2.11) \qquad \tilde{a}_{ij}(\omega) = \omega_{ij}(0, \omega) \qquad (i, j = 1, 2, \ldots, d)$$

for P almost all ω so that

$$(2.12) \qquad a_{ij}(x,\omega) = (T_x \tilde{a}_{ij})(\omega) = \tilde{a}_{ij}(\tau_{-x}\omega) = \omega_{ij}(x,\omega).$$

Thus the stationary random coefficients in (2.6) are the sample points themselves. Note that the functions $\tilde{a}_{ij}(\omega)$ satisfy (2.1) which can also be expressed by saying that the support of P is all $d \times d$ matrix functions that satisfy (2.1).

When one coordinate variable of $x = (x_1, \ldots, x_d)$ varies at a time in the group T_x, we obtain d one-parameter strongly continuous unitary groups in $\mathscr{H}$ that commute with each other. Let $D_1, D_2, \ldots, D_d$ denote the infinitesimal generators of these groups. They are closed and densely defined linear operators with domains $\mathscr{D}(D_i)$ in $\mathscr{H}$. For $f \in \mathscr{D}(D_i)$ we have

$$(2.13) \qquad (D_i\tilde{f})(\omega) = \frac{\partial}{\partial x_i}(T_x\tilde{f})(\omega)\Big|_{x=0}$$

where the differentiation is in the $\mathscr{H} = L^2(\Omega, \mathscr{F}, P)$ sense. Since T_x is a unitary group, the generators are skew adjoint

$$(2.14) \qquad E\{\tilde{g}D_i\tilde{f}\} = -E\{\tilde{f}D_i\tilde{g}\}$$

for all $\tilde{f}, \tilde{g}$ in $\mathscr{D}(D_i)$.

The hypothesis that the action of the translation group τ_x is ergodic on Ω takes the following form in $\mathscr{H}$: the only functions in $\mathscr{H}$ that are invariant under T_x are the constant functions.

The linear subspace

$$(2.15) \qquad \mathscr{H}^1 = \bigcap_{i=1}^{d} \mathscr{D}(D_i)$$

is a dense subset of $\mathscr{H}$. Equipped with the inner product

$$E\{\tilde{f}\tilde{g}\} + \sum_{i=1}^{d} E\{D_i\tilde{f}D_i\tilde{g}\},$$

$\mathscr{H}^1$ is a Hilbert space since the D_i are closed. Let $H_{\text{loc}}(R^d; \mathscr{H})$ be the space of functions from R^d to $\mathscr{H}$ with inner product

$$\int_{\mathscr{O}} dx E\{f(x)g(x)\} = \text{vol}(\mathscr{O})E\{\tilde{f}\tilde{g}\}$$

and finite norm for every bounded open set $\mathcal{O}$. Here $f(x, \omega) = \tilde{f}(\tau_{-x}\omega)$ and $g(x, \omega) = \tilde{g}(\tau_{-x}\omega)$ and we have used the translation invariance of P. Let $H_S(R^d; \mathcal{H})$ be the space of all stationary random processes on R^d Clearly H_S is in one-to-one correspondence with $\mathcal{H}$ since it is simply the space of all translates of $\mathcal{H}$. H_S is also a closed subset of $H_{\text{loc}}(R^d; \mathcal{H})$ that is invariant under T_x. The group T_x acts on $H_{\text{loc}}(R^d; \mathcal{H})$ in the manner $f(x, \omega) \to f(x, \tau_{-x}\omega)$.

We may similarly identify $\mathcal{H}^1$ with the set of mean square differentiable, stationary processes $H_S^1(R^d; \mathcal{H})$. Clearly $H_S^1(R^d; \mathcal{H})$ is a closed subset of $H_{\text{loc}}^1(R^d; \mathcal{H})$ i.e. the set of all $\mathcal{H}$ valued functions on R^d that have square integrable distribution derivatives over every bounded open set $\mathcal{O}$. We note that if $f \in H_S^1$, then its x derivatives form stationary processes and

$$\frac{\partial f(x, \omega)}{\partial x_i} = D_i f(x, \omega)$$

with equality holding $\mu \times P$ almost everywhere (μ = Lebesgue measure on R^d).

The unitary group of operators T_x is ergodic on $\mathcal{H}$ since the action τ_x is ergodic on Ω. Let B_N be the cube of side $2N$ in R^d centered at the origin. By the mean ergodic theorem, for any $\tilde{f} \in \mathcal{H}$,

$$(2.16) \qquad \frac{1}{(2N)^d} \int_{B_N} (T_x \tilde{f})(\omega) dx \to E\{\tilde{f}\}$$

in $\mathcal{H}$ as $N \to \infty$. We also have the individual ergodic theorem [16] which says that (2.16) is valid for P almost all ω when $\tilde{f} \in L^1(\Omega, \mathcal{F}, P)$. We note finally thet the spectral resolution of T_x is

$$(2.17) \qquad T_x = \int_{R^d} e^{i\lambda x} U(d\lambda)$$

where $U(d\lambda)$ is the associated projection valued measure.

Periodic and almost periodic coefficients (a_{ij}) can be considered within the above framework in the following manner.

Consider the periodic case first. Let $(\widetilde{a}_{ij}(x))$ be the given Lebesgue measurable periodic functions of period one in each coordinate and satisfying (2.1). Let T^d be the unit d-dimensional torus. We take $\Omega = \{\widetilde{a}_{ij}(\cdot + \omega)$ with $\omega \in T^d\}$ so that in fact Ω is identical with T^d in this case and is a much smaller space than it can be in general. Let $\mathscr{F}$ be the σ-algebra of Lebesgue measurable sets and P Lebesgue measure on T^d which is invariant under translation $\tau_x \omega = \omega - x \pmod 1$. The action of τ_x on Ω is ergodic and the infinitesimal generators of the unitary group T_x are now the usual partial derivatives

$$D_i = \frac{\partial}{\partial \omega_i} \qquad (\omega = (\omega_1, \ldots, \omega_d)).$$

The coefficients a_{ij} are given by $a_{ij}(x, \omega) = \widetilde{a}_{ij}(x + \omega)$.

The periodic case is thus put into the present abstract framework by the essentially trivial process of letting the center of the period cell be a random variable that is uniformly distributed over the unit torus.

In the almost periodic case we shall only consider coefficients (a_{ij}) that are continuous as follows.

Consider R^d as a locally compact abelian group. There exists [17] a compact abelian group G, the Bohr compactification of R^d, containing R^d as a dense subgroup and such that the continuous almost periodic functions on R^d are precisely the continuous functions on G, $C(G)$, restricted to R^d. Let $(\widetilde{a}_{ij}(x))$ be the given continuous almost periodic coefficients satisfying (2.1). These functions are restrictions to R^d of functions $(\widetilde{a}_{ij}(g))$ in $C(G)$ which are uniquely defined by continuity and density of R^d in G. Let τ_g be the group action on G which on R^d reduces to the usual translation. We take for Ω the subset of all continuous functions on G with values $d \times d$ matrices that consists of the given functions $(\widetilde{a}_{ij}(x))$ and all their translates by τ_g, that is

$$\Omega - \{\widetilde{a}_{ij}(\tau_{-g} x), \ g \in G\}.$$

Thus the set Ω may be identified with G. For $\mathscr{F}$ we take the σ-algebra generated by "rational" cylinder sets in Ω as above. This $\mathscr{F}$ is countably generated and it is a sub σ-algebra of the one relative to which

all continuous $d \times d$ matrix functions on G are measurable (this one is not countably generated). We take for P the Haar measure which is invariant under τ_g. Moreover, the action τ_g is ergodic on Ω.

3. WEAK CONVERGENCE THEOREM

Let the coefficients $a_{ij}(x, \omega) = \tilde{a}_{ij}(\tau_{-x}\omega)$ be given satisfying (2.1) with $x \in R^d$, $\omega \in \Omega$, $(\Omega, \mathscr{F}, P)$ the probability space introduced in the previous section with $\tau_x \colon \Omega \to \Omega$ the measure preserving, ergodic translation group. Let $u^\epsilon \in V$ be the solution of (2.6) for fixed $f \in L^2(\mathcal{O}; R^1)$, $\alpha > 0$ and each $\epsilon > 0$.

Let $\{e_n(\omega)\}$ be an orthonormal basis in $\mathscr{H} = L^2(\Omega, \mathscr{F}, P)$ with $e_1(\omega) \equiv 1$ and let $V_1 = E\{Ve_1\}$ by which we mean the set of all first coordinates of elements in V. V_1 is a closed subspace of $H^1(\mathcal{O}; R^1)$ and contains $H_0^1(\mathcal{O}; R^1)$. V_1 is also a closed subspace of V.

Theorem 1. *The solution $u^\epsilon \in V$ of (2.6) converges weakly in V to the solution $u(x) \in V_1$ of the deterministic variational problem*

$$\int_{\mathcal{O}} dx \sum_{i,j=1}^{d} q_{ij} \frac{\partial u(x)}{\partial x_j} \frac{\partial \varphi(x)}{\partial x_i} + \alpha \int_{\mathcal{O}} dx u(x)\varphi(x) = \int_{\mathcal{O}} dx f(x)\varphi(x), \tag{3.1}$$

for all $\varphi(x) \in V_1$. Here (q_{ij}) is a constant matrix that satisfies (2.1) *and is defined in Theorem* 2 *below. It is symmetric when (a_{ij}) is symmetric.*

Remark 1. By weak convergence we mean that for every $\varphi(x, \omega) \in \in V$,

$$(u^\epsilon, \varphi)_1 \to (u, \varphi)_1 \quad \text{as} \quad \epsilon \to 0 \tag{3.2}$$

where the inner product is defined by (2.5).

Remark 2. Another convergence result is given in Section 5 and still another in Section 8.

Theorem 2. *There exist uniquely defined functions $\tilde{\psi}_i^k(\omega)$ in $\mathscr{H}$, $(i, k = 1, 2, \ldots, d)$, such that*

$$\sum_{i,j=1}^{d} E\{\tilde{a}_{ij}(\delta_{jk} + \tilde{\psi}_j^k) D_i \tilde{\varphi}\} = 0, \quad \forall \tilde{\varphi} \in \mathscr{H}^1 \quad (k = 1, 2, \ldots, d), \tag{3.3}$$

$$E\{\widetilde{\psi}_i^k\} = 0, \tag{3.4}$$

$$E\{\widetilde{\psi}_i^k D_i \widetilde{\varphi}\} = E\{\widetilde{\psi}_i^k D_j \widetilde{\varphi}\}, \quad \forall \widetilde{\varphi} \in \mathscr{H}^1 \qquad (i, j, k = 1, 2, \ldots, d). \tag{3.5}$$

The coefficients (q_{ij}) *in* (3.1) *are defined by*

$$q_{ij} = E\Big\{\sum_{k=1}^{d} \widetilde{a}_{ik}(\delta_{kj} + \widetilde{\psi}_k^j)\Big\} \qquad (i, j = 1, 2, \ldots, d). \tag{3.6}$$

There exist furthermore uniquely defined processes $\chi^k(x, \omega)$ $(k = 1, 2, \ldots, d)$ *which are in* $H^1_{\mathrm{loc}}(R^d; \mathscr{H})$, *they are not stationary,* $\chi^k(0, \omega) = 0$ *and*

$$\frac{\partial \chi^k(x, \omega)}{\partial x_i} = \psi_i^k(x, \omega) = \widetilde{\psi}_i^k(\tau_{-x}\omega) \qquad (i, k = 1, 2, \ldots, d) \tag{3.7}$$

so that their gradients are stationary. For any compact subset K *of* R^d *they satisfy the estimate*

$$\lim_{e \downarrow 0} \sup_{x \in K} E\Big\{\Big(\epsilon \chi^k\Big(\frac{x}{\epsilon}\Big)\Big)^2\Big\} = 0 \qquad (k = 1, 2, \ldots, d). \tag{3.8}$$

Remark 1. In the periodic case there exist functions $\widetilde{\chi}^k(\omega)$ in $\mathscr{H}^1$ which satisfy

$$\int_{T^d} d\omega \sum_{i,j=1}^{d} \widetilde{a}_{ij}(\omega)\Big(\delta_{jk} + \frac{\partial \widetilde{\chi}^k(\omega)}{\partial \omega_j}\Big) \frac{\partial \widetilde{\varphi}(\omega)}{\partial \omega_i} = 0, \tag{3.9}$$

for all $\widetilde{\varphi} \in \mathscr{H}^1$ $(k = 1, 2, \ldots, d)$. This is the usual cell problem in homogenization (cf. for example [10], Chapter 1, Sections 2 and 3). In the general case one cannot obtain stationary $\chi^k(x, \omega)$ $(= \widetilde{\chi}^k(\tau_{-x}\omega))$ but it turns out that Theorem 2 is sufficient.

Remark 2. Theorem 2 also be stated in the following way. There are functions $\chi^k(x, \omega)$ in $H^1_{\mathrm{loc}}(R^d; \mathscr{H})$ whose gradients are in $H_S(R^d; \mathscr{H})$ (i.e. stationary) such that if B_N is as in (2.16),

$$\begin{aligned} &\lim_{N \uparrow \infty} \frac{1}{(2N)^d} \int_{B_N} dx \sum_{i,j=1}^{d} a_{ij}(a, \omega) \times \\ &\times \Big(\delta_{jk} + \frac{\partial \chi^k(x, \omega)}{\partial x_j}\Big) \frac{\partial \varphi(x, \omega)}{\partial x_i} = 0 \end{aligned} \tag{3.10}$$

for all $\varphi \in H_S^1(R^d; \mathscr{H})$ $(k = 1, 2, \ldots, d)$ and with the limit in (3.10) being in $\mathscr{H}$ or P almost everywhere. The gradients of χ^k have mean zero and

$$(3.11) \qquad q_{ij} = \lim_{N\uparrow\infty} \frac{1}{(2N)^d} \int_{B_N} dx \sum_{k=1}^{d} a_{ik}(x, \omega)\Big(\delta_{kj} + \frac{\partial \chi^k(x, \omega)}{\partial x_j}\Big),$$

with the limit again in $\mathscr{H}$ or P almost everywhere.

The translation of Theorem 2 to the above statement is immediate in view of the ergodic theorem (2.16). The statement (3.10), (3.11) is perhaps a bit more physical and in line with the discussion in Section 1. However, we have found the abstract formulation very convenient analytically.

Remark 3. In the next section we show that formula (3.6) for q_{ij} can also be written in the form

$$(3.12) \qquad q_{ij} = E\Big\{\sum_{k,l=1}^{d} \tilde{a}_{lk}(\delta_{kj} + \tilde{\psi}_k^j)(\delta_{li} + \tilde{\psi}_l^i)\Big\}.$$

From this formula we see that (q_{ij}) is symmetric when a_{ik} is symmetric. Using (2.1) and (3.4) we also see that (2.1) holds for the (q_{ij}). For example

$$\sum_{i,j=1} q_{ij}\xi_i\xi_j \geqslant a_0 E\Big\{\sum_{i,i,j=1}^{d} (\delta_{kj} + \tilde{\psi}_k^j)\xi_j(\delta_{ki} + \tilde{\psi}_k^i)\xi_i\Big\} =$$

$$= a_0 |\xi|^2 + a_0 E\Big\{\sum_{k=1}^{d}\Big(\sum_{j=1}^{d} \tilde{\psi}_k^j \xi_j\Big)^2\Big\} \geqslant a_0 |\xi|^2.$$

4. PROOF OF WEAK CONVERGENCE THEOREM

The proof of Theorem 1 is a modification of Tartar's proof for the periodic case (cf. [10], Chapter 1, Section 3). Of course we use in it Theorem 2 so we shall prove it first.

Proof of Theorem 2. For each $\beta > 0$ consider the problem: Find $\tilde{\chi}^{k,\beta}$ in $\mathscr{H}^1$ such that

$$\int_\Omega P(d\omega) \sum_{i,j=1}^{d} \tilde{a}_{ij}(\omega)(\delta_{jk} + D_j\tilde{\chi}^{k,\beta}(\omega))D_i\tilde{\varphi}(\omega) + \tag{4.1}$$
$$+ \beta \int_\Omega P(d\omega)\tilde{\chi}^{k,\beta}(\omega)\tilde{\varphi}(\omega) = 0$$

for all $\tilde{\varphi} \in \mathscr{H}^1$ $(k = 1, 2, \ldots, d)$. This problem has a unique solution by the Lax – Milgram lemma [15]. Letting $\tilde{\varphi} = \tilde{\chi}^{k,\beta}$ in (4.1) and unisg (2.1) yields

$$\int_\Omega P(d\omega) \sum_{j=1}^{d} (D_j\tilde{\chi}^{k,\beta}(\omega))^2 \leqslant C_1 \tag{4.2}$$

$$\beta \int_\Omega P(d\omega)(\tilde{\chi}^{k,\beta}(\omega))^2 \leqslant C_2 \tag{4.3}$$

for $k = 1, 2, \ldots, d$ with C_1 and C_2 constants independent of β.

Because of (4.2) there is a subsequence $\beta' \to 0$ such that $D_j\tilde{\chi}^{k,\beta'} \to$ $\to \tilde{\psi}_j^k$ (some limit) in $\mathscr{H}$ weakly. Using (4.3) we can pass to the limit in (4.1) along this subsequence to obtain

$$\int_\Omega P(d\omega) \sum_{i,j=1}^{d} \tilde{a}_{ij}(\omega)(\delta_{jk} + \tilde{\psi}_j^k(\omega))D_i\tilde{\varphi}(\omega) = 0, \quad \forall \tilde{\varphi} \in \mathscr{H}^1 \quad (k = 1, 2, \ldots, d). \tag{4.4}$$

By passage to the limit we obtain in addition the following:

$$\int_\Omega P(d\omega)\tilde{\psi}_j^k(\omega)D_i\tilde{\varphi}(\omega) = \int_\Omega P(d\omega)\tilde{\psi}_i^k(\omega)D_j\tilde{\varphi}(\omega), \quad \forall \tilde{\varphi} \in \mathscr{H}^1 \quad (i, j, k = 1, 2, \ldots, d) \tag{4.5}$$

and

$$\int_\Omega P(d\omega)\tilde{\psi}_j^k(\omega) = 0. \tag{4.6}$$

We shall show next that (4.4)-(4.6) has a unique solution which proves the first part of Theorem 2. For this purpose we use the spectral resolution of the unitary group T_x,

$$(4.7) \qquad T_x = \int_{R^d} e^{ix\lambda} U(d\lambda) \qquad (x \in R^d),$$

with $U(d\lambda)$ the spectral projections.

For each $\gamma > 0$ we define

$$(4.8) \qquad \widetilde{g}^{k,\beta}(\omega) = \int_{R^d} \sum_{j=1}^{d} \frac{(-i\lambda_j - \gamma)}{|i\lambda - \gamma|^2} U(d\lambda) \widetilde{\psi}_j^k(\omega) \qquad (k = 1, 2, \ldots, d)$$

where $|i\lambda - \gamma|^2 = \sum_{j=1}^{d} (i\lambda_j - \gamma)(-i\lambda_j - \gamma)$. On using (4.5) we conclude that $\widetilde{g}^{k,\gamma} \in \mathscr{H}^1$ and

$$(4.9) \qquad (D_j - \gamma)\widetilde{g}^{k,\gamma} = \widetilde{\psi}_i^k \qquad (k, j = 1, 2, \ldots, d).$$

Moreover,

$$|\gamma\widetilde{g}^{k,\gamma}|^2 = \int_{\Omega} P(d\omega)(\gamma\widetilde{g}^{k,\gamma}(\omega))^2 =$$

$$(4.10) \qquad = \int_{R^d} \gamma^2 \sum_{j,l=1}^{d} \frac{(-i\lambda_j - \gamma)(i\lambda_l - \gamma)}{|i\lambda - \gamma|^4} (U(d\lambda)\widetilde{\psi}_j^k, \widetilde{\psi}_l^k) \to$$

$$\to (U(\{0\})\widetilde{\psi}_j^k, \widetilde{\psi}_l^k) \quad \text{as} \quad \gamma \to 0,$$

by the Lebesgue dominated convergence theorem. But $U(\{0\})$ is the projection operator into the functions invariant by T_x. The only such functions are the constants, by ergodicity, and because of (4.6) we conclude that

$$(4.11) \qquad \lim_{\gamma \downarrow 0} |\gamma\widetilde{g}^{k,\gamma}|^2 = 0 \qquad (k = 1, 2, \ldots, d).$$

Now suppose $\widetilde{\psi}_j^k$ is a solution of the homogeneous version of (4.4)-(4.6). From (4.4) we have

$$\int_{\Omega} P(d\omega) \sum_{i,j=1}^{d} \widetilde{a}_{ij}(\omega)\widetilde{\psi}_j^k(\omega) D_i\widetilde{\varphi}(\omega) \qquad (k = 1, 2, \ldots, d).$$

Substitute for $\widetilde{\varphi}$ the function $\widetilde{g}^{k,\gamma}$. Using (4.9), passing to the limit $\gamma \to 0$ and using (4.11) we find that

$$\int_{\Omega} dP(\omega) \sum_{i,j=1}^{d} \widetilde{a}_{ij}(\omega)\widetilde{\psi}_j^k(\omega)\widetilde{\psi}_i^k(\omega) = 0 \qquad (k = 1, 2, \ldots, d).$$

From the uniform ellipticity we conclude that $\widetilde{\psi}_j^k = 0$ and hence uniqueness has been shown.

We continue with the construction of $\chi^k(x, \omega)$ satisfying (3.7) and (3.8).

We define $\chi^k(x, \omega)$ by

$$\chi^k(x, \omega) = \int_{R^d} (e^{i\lambda x} - 1) \frac{1}{|\lambda|^2} \sum_{j=1}^{d} (-i\lambda_j) U(d\lambda)\widetilde{\psi}_j^k(\omega) \qquad (k = 1, 2, \ldots, d), \tag{4.12}$$

which is nonstationary because it is not of the form $\chi^k = T_x \widetilde{\chi}^k$. However, $\chi^k(x, \omega)$ is in $H^1_{\mathrm{loc}}(R^d; \mathscr{H})$ as can be verified directly, $\chi^k(0, \omega) = 0$ and (3.7) holds.

It remains to show (3.8). We have

$$\int_{\Omega} P(d\omega)\left(\epsilon\chi^k\left(\frac{x}{\epsilon}, \omega\right)\right)^2 = \int_{R^d} \left| \frac{e^{ix\frac{\lambda}{\epsilon}} - 1}{\left|\frac{\lambda}{\epsilon}\right|} \right|^2 \sum_{j,l=1}^{d} \frac{\lambda_j\lambda_l}{|\lambda|^2} \hat{R}_{jl}^k(d\lambda) \tag{4.13}$$

where

$$\hat{R}_{jl}^k(d\lambda) = \int_{\Omega} P(d\omega) U(d\lambda)\widetilde{\psi}_j^k(\omega)\widetilde{\psi}_l^k(\omega), \tag{4.14}$$

is the power spectral matrix measure of the stationary process $\psi_j^k(x, \omega) = \widetilde{\psi}_j^k(\tau_{-x}\omega)$. From the estimate

$$\frac{1}{|\lambda|^2} \sum_{j,l=1}^{d} \lambda_j\lambda_l \hat{R}_{jl}^k(d\lambda) \leqslant \sum_{j=1}^{d} \hat{R}_{jj}^k(d\lambda),$$

which follows from Schwartz's inequality, we obtain

$$\int_{\Omega} P(d\omega)\left(\epsilon\chi^k\left(\frac{x}{\epsilon}, \omega\right)\right)^2 \leqslant \int_{R^d} \left| \frac{e^{ix\frac{\lambda}{\epsilon}} - 1}{\left|\frac{\lambda}{\epsilon}\right|} \right|^2 \sum_{j=1}^{d} \hat{R}_{jj}(d\lambda). \tag{4.15}$$

By the ergodicity hypothesis and (4.6) it follows that $\hat{R}_{jj}(\{0\}) = 0$. Application of the Lebesgue convergence theorem to (4.15) yields the result (3.8). The proof of Theorem 2 is complete.

Before continuing with the proof of Theorem 1 we observe that formula (3.12) for (q_{ij}) is obtained by replacing $D_i\widetilde{\varphi}$ in (3.3) by $\widetilde{\psi}_i^l$ and adding the result to (3.6). That this substitution is permissible follows from the uniqueness argument for $\widetilde{\psi}_j^k$ given above.

Proof of Theorem 1. We follow Tartar's elegant argument ([10], Chapter 1, Section 3). Define

$$(4.16)\qquad \xi_i^\epsilon(x, \omega) = \sum_{j=1}^{d} a_{ij}\left(\frac{x}{\epsilon}, \omega\right) \frac{\partial u^\epsilon(x, \omega)}{\partial x_j} \qquad (i = 1, 2, \ldots, d),$$

which is in $L^2(\mathscr{O}; \mathscr{H})$. In view of (2.1) and (2.7) there is a subsequence, denoted again by $\xi_i^\epsilon(x, \omega), u^\epsilon(x, \omega)$ which converges weakly in $(L^2(\mathscr{O}; \mathscr{H}))^{d+1}$ to some limit $\xi_i(x, \omega), u(x, \omega)$. Passing to the limit in (2.6) we obtain

$$(4.17)\qquad \begin{aligned} &\int_{\mathscr{O}} dx \int_{\Omega} P(d\omega) \sum_{i=1}^{d} \xi_i(x, \omega) \frac{\partial \varphi(x, \omega)}{\partial x_i} + \\ &+ \int_{\mathscr{O}} dx \int P(d\omega) u(x, \omega) \varphi(x, \omega) = \int_{\mathscr{O}} dx \int_{\Omega} P(d\omega) f(x) \varphi(x, \omega) \end{aligned}$$

for all $\varphi(x, \omega) \in V$.

Let $\chi^k(x, \omega)$ be the function defined in Theorem 2 with $(\widetilde{a}_{ij})$ replaced by $(\widetilde{a}_{ji})$. Define $w_k(x, \omega)$ by

$$w_k(x, \omega) = x_k + \chi^k(x, \omega).$$

We may rewrite (3.3) in the form

$$(4.18)\qquad \int_{\Omega} P(d\omega) \sum_{i,j=1}^{d} a_{ij}(x, \omega) \frac{\partial w_k(x, \omega)}{\partial x_i} \frac{\partial \varphi(x, \omega)}{\partial x_j} = 0$$

for all $\varphi(x, \omega) = \widetilde{\varphi}(\tau_{-x}\omega)$ with $\widetilde{\varphi} \in \mathscr{H}^1$ and $k = 1, 2, \ldots, d$. Since $D_j\varphi = \dfrac{\partial \varphi}{\partial x_j}$ we may rewrite (4.18) in the form

$$\sum_{i,j=1}^{d} \frac{\partial}{\partial x_j} \left(a_{ij}(x, \omega) \frac{\partial w_k(x, \omega)}{\partial x_j} \right) = 0 \tag{4.19}$$

which makes sense as an equality of distribution valued stationary processes.

Define

$$w_k^\epsilon(x, \omega) = \epsilon w_k^\epsilon\left(\frac{x}{\epsilon}, \omega\right) = x_k + \epsilon \chi^k\left(\frac{x}{\epsilon}, \omega\right). \tag{4.20}$$

Clearly (4.19) may also be written in scaled form

$$\sum_{i,j=1}^{d} \frac{\partial}{\partial x_i} \left(a_{ij}\left(\frac{x}{\epsilon}, \omega\right) \frac{\partial w_k^\epsilon(x, \omega)}{\partial x_i} \right) = 0. \tag{4.21}$$

Let $\theta(x, \omega)$ be any function in $\mathscr{D}(\mathscr{O}; L^\infty(\Omega, \mathscr{F}, P))$. Since $u^\epsilon(x, \omega) \in$ $\in V \subset H^1(\mathscr{O}; \mathscr{H})$ it follows that $\theta(x, \omega) u^\epsilon(x, \omega) \in H_0^1(\mathscr{O}; \mathscr{H})$. Multiplying (4.21) by θu^ϵ, integrating and integrating by parts gives the identity

$$\begin{aligned} &\int_{\mathscr{O}} dx \int P(d\omega) \sum_{i,j=1}^{d} a_{ij}\left(\frac{x}{\epsilon}, \omega\right) \frac{\partial}{\partial x_j} (\theta(x, \omega) u^\epsilon(x, \omega)) \times \\ &\times \frac{\partial w_k^\epsilon(x, \omega)}{\partial x_i} = 0. \end{aligned} \tag{4.22}$$

The functions $w_k^\epsilon(x, \omega)$ are in $H_{\text{loc}}^1(R^d; \mathscr{H})$ and hence $\theta(x, \omega) w_k^\epsilon(x, \omega)$ is in $H_0^1(\mathscr{O}; \mathscr{H}) \subseteq V$. Thus, replacing φ by θw_k^ϵ in (2.6) gives

$$\begin{aligned} &\int_{\mathscr{O}} dx \int_{\Omega} P(d\omega) \sum_{i=1}^{d} \xi_i^\epsilon(x, \omega) \frac{\partial}{\partial x_i} (\theta(x, \omega) w_k^\epsilon(x, \omega)) + \\ &+ \int_{\mathscr{O}} dx \int_{\Omega} P(d\omega) u^\epsilon(x, \omega) \theta(x, \omega) w_k^\epsilon(x, \omega) = \\ &= \int_{\mathscr{O}} dx \int_{\Omega} P(d\omega) f(x) \theta(x, \omega) w_k^\epsilon(x, \omega). \end{aligned} \tag{4.23}$$

Subtracting (4.22) from (4.23) and cancelling some terms yields the identity

$$\begin{aligned}
&\int_{\mathcal{O}} dx \int_{\Omega} P(d\omega) \sum_{i=1}^{d} \xi_i^{\epsilon}(x,\omega) w_k^{\epsilon}(x,\omega) \frac{\partial \theta(x,\omega)}{\partial x_i} - \\
&\quad - \int_{\mathcal{O}} dx \int_{\Omega} P(d\omega) \sum_{i,j=1}^{d} a_{ij}\left(\frac{x}{\epsilon},\omega\right) \frac{\partial w_k^{\epsilon}(x,\omega)}{\partial x_i} u^{\epsilon}(x,\omega) \times \\
&\quad \times \frac{\partial \theta(x,\omega)}{\partial x_j} + \alpha \int_{\mathcal{O}} dx \int_{\Omega} P(d\omega) u^{\epsilon}(x,\omega)\theta(x,\omega) w_k^{\epsilon}(x,\omega) = \\
&\quad = \int_{\mathcal{O}} dx \int_{\Omega} P(d\omega) f(x)\theta(x,\omega) w_k^{\epsilon}(x,\omega).
\end{aligned} \tag{4.24}$$

We may now pass to the limit $\epsilon \to 0$ in (4.24) along the subsequence. From (4.20) and (3.8) we conclude that

$$\int_{\mathcal{O}} dx \int_{\Omega} P(d\omega)(w_k^{\epsilon}(x,\omega) - x_k)^2 \to 0 \quad \text{as} \quad \epsilon \to 0. \tag{4.25}$$

Using (4.25) to pass to the limit in (4.24) and then using (4.17) gives us the relation

$$\begin{aligned}
&\int_{\mathcal{O}} dx \int_{\Omega} P(d\omega) \xi_k(x,\omega)\theta(x,\omega) = \\
&\quad = - \lim_{\epsilon \downarrow 0} \int_{\mathcal{O}} dx \int_{\Omega} P(d\omega) \sum_{i,j=1}^{d} a_{ij}\left(\frac{x}{\epsilon},\omega\right) \frac{\partial w_k^{\epsilon}(x,\omega)}{\partial x_i} \times \\
&\quad \times u^{\epsilon}(x,\omega) \frac{\partial \theta(x,\omega)}{\partial x_j}
\end{aligned} \tag{4.26}$$

for $k = 1, 2, \ldots, d$. We shall show that the right side in (4.26) equals

$$\int_{\mathcal{O}} dx \int_{\Omega} P(d\omega) \sum_{j=1}^{d} q_{kj} u(x,\omega) \frac{\partial \theta(x,\omega)}{\partial x_j} \tag{4.27}$$

with (q_{ij}) defined by (3.6) (after noting that $(\tilde{a}_{ij})$ has been replaced by $(\tilde{a}_{ji})$). Assuming this fact we shall complete the proof.

Since $\theta(x,\omega) \in \mathscr{D}(\mathcal{O}; L^{\infty}(\Omega, \mathscr{F}, P))$ is arbitrary and $\mathscr{D}(\mathcal{O}; L^{\infty}(\Omega, \mathscr{F}, P))$ is dense in $L^2(\mathcal{O}; \mathscr{H})$ we conclude that

$$\xi_k(x,\omega) = \sum_{j=1}^{d} q_{kj} \frac{\partial u(x,\omega)}{\partial x_j} \qquad (k = 1, 2, \ldots, d). \tag{4.28}$$

Inserting this expression into (4.17) gives

$$\int_{\mathcal{O}} dx \int_{\Omega} P(d\omega) \sum_{i,j=1}^{d} q_{ij} \frac{\partial u(x,\omega)}{\partial x_j} \frac{\partial \varphi(x,\omega)}{\partial x_i} + \tag{4.29}$$

$$+ \alpha \int_{\mathcal{O}} dx \int_{\Omega} P(d\omega) u(x,\omega)\varphi(x,\omega) = \int_{\mathcal{O}} dx \int_{\Omega} P(d\omega) f(x) \varphi(x,\omega)$$

for all $\varphi(x,\omega) \in V$. This problem has a unique solution in V since (q_{ij}) is positive definite and α is positive. But the solution $u(x) \in V_1$ of (3.1) satisfies (4.29) since the (q_{ij}) are constants. Thus $u(x) = u(x,\omega)$, i.e. the limit is deterministic and we have the result stated in Theorem 1.

It remains to show that the right side of (4.26) equals (4.27). Define g^{kj} by

$$g^{kj}(x,\omega) = \tilde{g}^{kj}(\tau_{-x}\omega) = \sum_{i=1}^{d} a_{ij}(x,\omega) \frac{\partial w_k(x,\omega)}{\partial x_i} - q_{kj}. \tag{4.30}$$

In terms of g^{kj} what we must show is that

$$\lim_{\epsilon \downarrow 0} \int_{\mathcal{O}} dx \int_{\Omega} P(d\omega) \sum_{j=1}^{d} g^{kj}\left(\frac{x}{\epsilon},\omega\right) u^{\epsilon}(x,\omega) \frac{\partial \theta(x,\omega)}{\partial x_j} = 0 \tag{4.31}$$

$$(k = 1, 2, \ldots, d).$$

Define $h_l^{kj}(x,\omega)$ by

$$h_l^{kj}(x,\omega) = \int_{R^d} (e^{ix\lambda} - 1) \frac{(-i\lambda_l)}{|\lambda|^2} U(d\lambda) \tilde{g}^{kj}(\omega). \tag{4.32}$$

Since $E\{\tilde{g}^{kj}\} = 0$, a calculation similar to the one used for χ^k below (4.12) gives

$$\lim_{\epsilon \downarrow 0} \int_{\mathcal{O}} dx \int_{\Omega} P(d\omega) \left(\epsilon h_l^{kj}\left(\frac{x}{\epsilon},\omega\right)\right)^2 = 0 \qquad (k,j,l = 1, 2, \ldots, d). \tag{4.33}$$

We also have that

$$\sum_{l=1}^{d} \frac{\partial h_l^{kj}(x,\omega)}{\partial x_l} = \int_{R^d} e^{ix\lambda} U(d\lambda) \tilde{g}^{kj} = g^{kj}(x,\omega). \tag{4.34}$$

Now we use (4.34) suitably scaled in (4.31) and integrate by parts.

$$\int_{\mathcal{O}} dx \int_{\Omega} P(d\omega) \sum_{j,l=1}^{d} \left(\epsilon \frac{\partial}{\partial x_l} h_l^{kj}\left(\frac{x}{\epsilon}, \omega\right)\right) u^{\epsilon}(x,\omega) \frac{\partial \theta(x,\omega)}{\partial x_j} =$$

$$(4.35) \qquad = - \int_{\mathcal{O}} dx \int_{\Omega} P(d\omega) \sum_{j,l=1}^{d} \epsilon h_l^{kj}\left(\frac{x}{\epsilon}, \omega\right) \times$$

$$\times \left[\frac{\partial u^{\epsilon}(x,\omega)}{\partial x_l} \frac{\partial \theta(x,\omega)}{\partial x_j} + u^{\epsilon}(x,\omega) \frac{\partial^2 \theta(x,\omega)}{\partial x_j \partial x_l}\right].$$

The result (4.31) follows from this by using (4.34) and the estimate (2.7) which is uniform in ϵ. The proof of Theorem 1 is complete.

5. STRONG CONVERGENCE THEOREM

We shall state and prove the strong convergence theorem for the whole of R^d since the result is local in character. The cutoff arguments of [10], Chapter 1, Section 5 apply here also so that the same problem over a subset $\mathcal{O}$ with Dirichlet conditions can be analyzed without difficulty.

We assume that all hypotheses about (a_{ij}) stated in Section 2 and at the beginning of Section 3 hold here also. With $f \in L^2(R^d; R^1)$ and $\alpha > 0$ fixed we consider the problem: find $u^{\epsilon}(x,\omega)$ in $H^1(R^d; \mathcal{H})$ such that for each $\epsilon > 0$

$$\int_{R^d} dx \int_{\Omega} P(d\omega) \sum_{i,j=1}^{d} a_{ij}\left(\frac{x}{\epsilon}, \omega\right) \frac{\partial u^{\epsilon}(x,\omega)}{\partial x_j} \frac{\partial \varphi(x,\omega)}{\partial x_i} +$$

$$(5.1) \qquad + \alpha \int_{R^d} dx \int_{\Omega} P(d\omega) u^{\epsilon}(x,\omega)\varphi(x,\omega) =$$

$$= \int_{R^d} dx \int_{\Omega} P(d\omega) f(x) \varphi(x,\omega)$$

for all $\varphi(x,\omega)$ in $H^1(R^d; \mathcal{H})$. This problem has a unique solution satisfying (2.7) with $\mathcal{O}$ replaced by R^d.

Theorem 3. *Let $u^{\epsilon}(x,\omega) \in H^1(R^d; \mathcal{H})$ be the solution of* (5.1) *and let $u(x) \in H^1(R^d; R^1)$ be the solution of*

$$(5.2)\qquad \int_{R^d} dx \sum_{i,j=1}^{d} q_{ij} \frac{\partial u(x)}{\partial x_j} \frac{\partial \varphi(x)}{\partial x_i} + \alpha \int_{R^d} dx u(x)\varphi(x) = \int_{R^d} dx f(x)\varphi(x)$$

for all $\varphi \in H^1(R^d; R^1)$ *and with* (q_{ij}) *given by* (3.6). *Then*

$$(5.3)\qquad \lim_{\epsilon \downarrow 0} \int_{R^d} dx \int_{\Omega} P(d\omega) | u^{\epsilon}(x, \omega) - u(x)|^2 = 0$$

and

$$(5.4)\qquad \lim_{\epsilon \downarrow 0} \int_{R^d} dx \int_{\Omega} P(d\omega) \sum_{i=1}^{d} \left| \frac{\partial u^{\epsilon}(x, \omega)}{\partial x_i} - \frac{\partial u(x)}{\partial x_i} - \sum_{k=1}^{d} \psi_i^k \left(\frac{x}{\epsilon}, \omega\right) \frac{\partial u(x)}{\partial x_k} \right|^2 = 0$$

where $\psi_i^k(x, \omega) = \widetilde{\psi}_i^k(\tau_{-x}\omega)$ *and* $\widetilde{\psi}_i^k(\omega)$ *is defined in Theorem* 2.

Remark 1. This theorem complements Theorem 1 since it gives stronger convergence. Note in particular that u^{ϵ} does not tend to u in $H^1(R^d; \mathscr{H})$ strongly as can be seen from (5.4). The principal advantage of Theorem 1 is that it works for arbitrary homogeneous boundary value problems in variational form while the present theorem is of a local character.

Remark 2. We shall give two proofs of Theorem 3 both of which are more direct and hence more intuitive than Tartar's proof of Theorem 1. The first proof is exactly analogous to the one given in the periodic case and uses Theorem 2. The second uses Theorem 2 without taking advantage of the nonstationary χ^k and (3.8). As a result more information is needed about problem (4.1) that approximates (3.3)-(3.5). This being of independent interest it is stated as Theorem 4 in the next section. In the second proof and Theorem 4 we require that the (a_{ij}) be symmetric since we use the spectral theorem.

6. PROOF OF THE STRONG CONVERGENCE THEOREM

In view of (2.7) it suffices to prove (5.3) and (5.4) for each $f(x) \in C_0^{\infty}(R^d)$. In that case (5.2) has a C^{∞} solution $u(x)$ that goes to zero exponentially fast $(\alpha > 0)$ as $|x| \to \infty$.

Let

$$(6.1)\qquad z^{\epsilon}(x,\omega)=u^{\epsilon}(x,\omega)-u(x)-\sum_{k=1}^{d}\epsilon\chi^{k}\left(\frac{x}{\epsilon},\omega\right)\frac{\partial u(x)}{\partial x_k}$$

where $\chi^k(x,\omega)$ is the process defined in Theorem 2. We note first that because of the estimate

$$(6.2)\qquad \int_{\Omega} P(d\omega)(\chi^k(x,\omega))^2 \leqslant C_2(1+|x|^2) \quad \text{for all } x,$$

which is a cruder version of (3.8) and follows from (4.13), and because of (2.7) there is a constant C_1 independent of ϵ such that

$$(6.3)\qquad \int_{R^d} dx \int_{\Omega} P(d\omega) \sum_{i=1}^{d} \left|\frac{\partial z^{\epsilon}(x,\omega)}{\partial x_i}\right|^2 \leqslant C_1 < \infty .$$

We note further that

$$(6.4)\qquad \lim_{\epsilon\downarrow 0} \int_{R^d} dx \int_{\Omega} P(d\omega)\left(\sum_{k=1}^{d}\epsilon\chi^k\left(\frac{x}{\epsilon},\omega\right)\frac{\partial u(x)}{\partial x_k}\right)^2 = 0$$

which follows from (3.8), the rapid decay of $u(x)$ and its derivatives as $|x|\to\infty$ and from (6.2).

We write (5.1) in the form

$$(6.5)\qquad \mathscr{L}_{\epsilon}u^{\epsilon}=f$$

with

$$(6.6)\qquad \mathscr{L}_{\epsilon}=-\sum_{i,j=1}^{d}\frac{\partial}{\partial x_i}\left(a_{ij}\left(\frac{x}{\epsilon},\omega\right)\frac{\partial}{\partial x_j}\right)+\alpha .$$

By direct computation and by using Theorem 2 we find that

$$(6.7)\qquad \mathscr{L}_{\epsilon}z^{\epsilon}=\sum_{i,j,k=1}^{d}\left[a_{ik}\left(\frac{x}{\epsilon},\omega\right)\left(\delta_{kj}+\frac{\partial\chi^j\left(\frac{x}{\epsilon},\omega\right)}{\partial x_u}\right)-q_{ij}\right]\times$$

$$\times\frac{\partial^2 u(x)}{\partial x_i\,\partial x_j}-\epsilon\sum_{i,j,k=1}^{d}\frac{\partial}{\partial x_i}\left[a_{ij}\left(\frac{x}{\epsilon},\omega\right)\chi^k\left(\frac{x}{\epsilon},\omega\right)\right]\frac{\partial^2 u(x)}{\partial x_k\,\partial x_j}+$$

$$+ \epsilon \sum_{i,j,k=1}^{d} a_{ij}\left(\frac{x}{\epsilon}, \omega\right) \chi^k\left(\frac{x}{\epsilon}, \omega\right) \frac{\partial^3 u(x)}{\partial x_k \partial x_j \partial x_i} -$$

$$- \alpha \sum_{k=1}^{d} \chi^k\left(\frac{x}{\epsilon}, \omega\right) \frac{\partial u(x)}{\partial x_k}.$$

The precise meaning of this identity is that if both sides are multiplied by $\varphi(x, \omega) \in H^1(R^d; \mathscr{H})$, integrated with respect to x and Ω and $\mathscr{L}^\epsilon$ is written in variational form, then a correct equation results. In this equation we may replace φ by z^ϵ in view of (6.3). Using then the uniform ellipticity (2.1) and (6.4) (along with some obvious generalizations of it) we obtain the estimate

$$\overline{\lim_{\epsilon \downarrow 0}}\, a_0 \int_{R^d} dx \int_{\Omega} P(d\omega) \sum_{i=1}^{d} \left| \frac{\partial z^\epsilon(x, \omega)}{\partial x_i} \right|^2 +$$

$$+ \overline{\lim_{\epsilon \downarrow 0}}\, \alpha \int_{R^d} dx \int_{\Omega} P(d\omega) |z^\epsilon(x, \omega)|^2 \leqslant \tag{6.8}$$

$$\leqslant \lim_{\epsilon \downarrow 0} \left| \int_{R^d} dx \int_{\Omega} P(d\omega) \sum_{i,j,k=1}^{d} \left[a_{ik}\left(\frac{x}{\epsilon}, \omega\right) \times \right.\right.$$

$$\left.\left. \times\ \delta_{kj} + \psi_k^j\left(\frac{x}{\epsilon}, \omega\right)\right) - q_{ij} \right] \frac{\partial^2 u(x)}{\partial x_j \partial x_i} z^\epsilon(x, \omega) \Bigg|.$$

We observe now that the right side of (6.8) is precisely of the same form as the limit in (4.27) i.e. just like (4.32). Since the uniform in ϵ estimate (6.3) holds the arguments of Section 4 proving (4.31) applies again and the proof of Theorem 3 is complete.

The second proof of Theorem 3 goes as follows. Instead of defining the error z^ϵ with the term

$$\sum_{k=1}^{d} \epsilon \chi^k\left(\frac{x}{\epsilon}, \omega\right) \frac{\partial u(x)}{\partial x_k}$$

we define it with $\chi^k\left(\frac{x}{\epsilon}, \omega\right)$ replaced by $\chi^{k,\epsilon^2}\left(\frac{x}{\epsilon}, \omega\right)$ where $\chi^{k,\beta}(x, \omega) = \tilde{\chi}^{k,\beta}(\tau_{-x}\omega)$ is stationary and $\tilde{\chi}^{k,\beta}(\omega)$ solves (4.1). A computation nearly identical to the one that gave (6.7) gives an equation for the new error z^ϵ. Inspection of this equation reveals that Theorem 3 will follow if we can prove that

$$(6.9)\qquad \lim_{\beta\downarrow 0} \beta \int_\Omega P(d\omega)(\widetilde{\chi}^{k,\beta}(\omega))^2 = 0$$

and

$$(6.10)\qquad \lim_{\beta\downarrow 0} \int_\Omega P(d\omega)|D_j\widetilde{\chi}^{k,\beta}(\omega) - \widetilde{\psi}_j^k(\omega)|^2 = 0 \qquad (j,k = 1,2,\ldots,d).$$

The two estimates (6.9) and (6.10) say that the approximating solutions $\widetilde{\chi}^{k,\beta}$ of (3.3)-(3.5), which satisfy (4.1), not only converge weakly in $\mathscr{H}^1$ but also strongly. We state this as a separate theorem.

Theorem 4. *Estimates* (6.9) *and* (6.10) *hold for* $\widetilde{\chi}^{k,\beta}$, *the solution of* (4.1), *provided the* $(\widetilde{a}_{ij})$ *are symmetric.*

Proof of Theorem 4. Suppose first that (6.9) holds. Subtracting (3.3) from (4.1), replacing $\widetilde{\varphi}$ by $\widetilde{\chi}^{k,\beta}$ and using (6.9) we obtain

$$(6.11)\qquad \lim_{\beta\downarrow 0} \int_\Omega P(d\omega) \sum_{i,j=1}^d \widetilde{a}_{ij}(\omega)(D_j\widetilde{\chi}^{k,\beta}(\omega) - \widetilde{\psi}_j^k(\omega))D_i\widetilde{\chi}^{k,\beta}(\omega) = 0.$$

By the replacement used in Section 4 to prove uniqueness of the solution of (3.3)-(3.5) we conclude that

$$(6.12)\qquad \int_\Omega P(d\omega) \sum_{i,j=1}^d (D_j\widetilde{\chi}^{k,\beta}(\omega) - \widetilde{\psi}_j^k(\omega))\widetilde{\psi}_j^k(\omega) = 0.$$

Subtracting (6.12) from (6.11) and using the uniform ellipticity gives (6.10).

So it remains to prove (6.9). Let A be the operator

$$(6.13)\qquad A = -\sum_{i,j=1}^d D_i(\widetilde{a}_{ij}D_j\ \cdot)$$

defined on $\mathscr{H}^1$ as a quadratic form. We denote by A also its Friedrichs extension [19], p. 372 in $\mathscr{H}$. This operator is a nonnegative selfadjoint operator (using symmetry of $(\widetilde{a}_{ij})$) with spectral resolution

$$(6.14)\qquad A = \int_0^\infty \lambda G(d\lambda).$$

Write

$$(6.15)\qquad \tilde{f}^k(\omega) = \sum_{i=1}^{a} D_i(\tilde{a}_{ij}(\omega)) \qquad (k = 1, 2, \ldots, d),$$

with the right side interpreted in the weak sense

$$(6.16)\qquad \int_\Omega P(d\omega)\tilde{g}D_i\tilde{a}_{ik} = -\int_\Omega P(d\omega)\tilde{a}_{ik}D_i\tilde{g}, \quad \forall \tilde{g} \in \mathscr{H}^1 .$$

We shall show below that $\tilde{f}^k$ is in the range of $A^{\frac{1}{2}}$. Assuming this we shall complete the proof of (6.9).

In terms of the spectral resolution we may write $\tilde{\chi}^{k,\beta}$ in the form

$$\sqrt{\beta}\tilde{\chi}^{k,\beta} = \sqrt{\beta}(\beta + A)^{-1}\tilde{f}^k = \int_0^\infty \frac{\sqrt{\beta}}{\beta + \lambda} G(d\lambda)\tilde{f}^k(\omega)$$

so that

$$(6.17)\qquad \beta \int_\Omega P(d\omega)(\tilde{\chi}^{k,\beta}(\omega))^2 = \int_0^\infty \frac{\beta}{(\beta + \lambda)^2} E\{\tilde{f}^k G(d\lambda)\tilde{f}^k\}.$$

Since we have assumed that $\tilde{f}^k$ is in the range of $A^{\frac{1}{2}}$, there exists an $\tilde{h}^k$ in $\mathscr{D}(A^{\frac{1}{2}})$ such that $\tilde{f}^k = A^{\frac{1}{2}}h^{\frac{1}{2}}$. But then

$$(6.18)\qquad E\{\tilde{f}^k G(d\lambda)\tilde{f}^k\} = E\{A^{\frac{1}{2}}\tilde{h}^k G(d\lambda)A^{\frac{1}{2}}\tilde{h}^k\} = Eh^k G(d\lambda)h^k$$

and (6.17) becomes

$$(6.19)\qquad \beta \int_\Omega P(d\omega)(\tilde{\chi}^{k,\beta}(\omega))^2 = \int_0^\infty \frac{\lambda^\beta}{(\lambda + \beta)^2} E\{h^k G(d\lambda)\tilde{h}^k\}.$$

Since $\lambda^\beta(\lambda + \beta)^{-2}$ is bounded by $\frac{1}{4}$ independently of $\beta > 0$ the Lebesgue convergence theorem applied to (6.19) gives the desired result (6.9).

We must now show that $\tilde{f}^k$ of (6.15) is in the range of $A^{\frac{1}{2}}$.

Lemma. *If there is a constant* $C < \infty$ *such that*

$$(6.20)\qquad |E\{\tilde{f}\tilde{g}\}|^2 \leqslant CE\{\tilde{g}A\tilde{g}\} \quad \forall \tilde{g} \in \mathscr{D}(A)$$

then f *is in the range of* $A^{\frac{1}{2}}$ *and* $E\{|A^{-\frac{1}{2}}\tilde{f}|^2\} \leqslant C$.

We prove the lemma after verifying that the f^k of (6.15) satisfy (6.20). Since $\mathscr{D}(A) \subset \mathscr{H}^1$; for all $\tilde{g} \in \mathscr{D}(A)$ we have

$$
\begin{aligned}
|E\{\tilde{f}^k \tilde{g}\}|^2 &= \Big| E\Big\{\tilde{g} \sum_{i=1}^{d} D_i \tilde{a}_{ik}\Big\}\Big|^2 = \\
&= \Big|\sum_{i=1}^{d} E\{\tilde{a}_{ik} D_i \tilde{g}\}\Big|^2 \leqslant \\
&\leqslant \Big(\frac{1}{a_0}\Big)^2 \sum_{i=1}^{d} E\{|D_i \tilde{g}|^2\} \leqslant && \text{(using (2.1))} \\
&\leqslant \Big(\frac{1}{a_0}\Big)^3 \sum_{i,j=1}^{d} E\{D_j \tilde{g} \tilde{a}_{ij} D_j \tilde{g}\} = && \text{(using (2.1))} \\
&= \Big(\frac{1}{a_0}\Big)^3 E\{\tilde{g} A \tilde{g}\}.
\end{aligned}
$$

Then (6.2) holds and the proof of Theorem 4 is complete.

Proof of Lemma. Put

$$
(6.21) \qquad \tilde{h}_\beta = (\beta + A)^{-\frac{1}{2}} \tilde{f} \qquad (\beta > 0).
$$

Then for all $\tilde{g} \in \mathscr{D}(A)$

$$
\begin{aligned}
|E\{\tilde{g}\tilde{h}_\beta\}| &= |E\{\tilde{g}(\beta + A)^{-\frac{1}{2}} \tilde{f}\}|^2 = \\
&= |E\{\tilde{f}(\beta + A^{-\frac{1}{2}} \tilde{g}\}|^2 \leqslant \\
&\leqslant CE\{(\beta + A)^{-\frac{1}{2}} \tilde{g} A (\beta + A)^{-\frac{1}{2}} \tilde{g}\} = && \text{(by (6.20))} \\
&= CE\{A(\beta + A)^{-1} \tilde{g}\tilde{g}\} \leqslant \\
&\leqslant CE\{|\tilde{g}|^2\}.
\end{aligned}
$$

Since $\mathscr{D}(A)$ is dense in $\mathscr{H}$ this means that $E\{|h_\beta|^2\} \leqslant C$ and hence there exists a subsequence, also denoted by $\tilde{h}_\beta$, that converges weakly in $\mathscr{H}$ to some limit $\tilde{h}$ as $\beta \to 0$.

By (6.21), $(\beta + A)^{\frac{1}{2}} \widetilde{h}_\beta = \widetilde{f}$ and hence

$$E\{\widetilde{h}_\beta (\beta + A)^{\frac{1}{2}} \widetilde{g}\} = E\{\widetilde{f}\,\widetilde{g}\}, \quad \forall \widetilde{g} \in \mathscr{D}(A^{\frac{1}{2}}).$$

Passing to the limit $\beta \to 0$ gives

$$E\{\widetilde{h} A^{\frac{1}{2}} \widetilde{g}\} = E\{\widetilde{f}\,\widetilde{g}\}, \quad \forall \widetilde{g} \in \mathscr{D}(A^{\frac{1}{2}}).$$

This implies that $\widetilde{h}$ is in the domain of $A^{\frac{1}{2}}$ and that $A^{\frac{1}{2}} \widetilde{h} = \widetilde{f}$, (cf. [19], p. 322) as was to be shown.

7. THE BLOCH REPRESENTATION

Problem (5.1) can be solved explicitly in terms of Fourier transforms and the solution of a problem in the abstract space $(\Omega, \mathscr{F}, P)$. We call this the Bloch representation, or the Bloch expansion, because in the periodic case the problem in the abstract space is a boundary value problem over a period cell that can be solved by Bloch waves (cf. [10], Chapter 4, Section 3). Although we do not make use of this representation in our asymptotic analysis we give it here since it may be useful in other contexts.

We write (5.1) with $\epsilon = 1$ in the form

$$-\sum_{j,l=1}^{d} \frac{\partial}{\partial x_j} \left[a_{jl}(x, \omega) \frac{\partial u(x, \omega)}{\partial x_l} \right] + \alpha u(x, \omega) = f(x) \qquad (x \in R^d). \tag{7.1}$$

Since the coefficients $a_{jl}(x, \omega) = \widetilde{a}_{jl}(\tau_{-x} \omega)$ depend on ω through $\tau_{-x}\omega$ we may look for u in the form

$$u(x, \omega) = v(x, \tau_{-x} \omega). \tag{7.2}$$

We then have $\dfrac{\partial u}{\partial x_i} = \dfrac{\partial v}{\partial x_i} + D_i v$, and hence (7.1) becomes

$$-\sum_{j,l=1}^{d} \left(D_j + \frac{\partial}{\partial x_j}\right) \left[\widetilde{a}_{jl}(\omega) \left(D_l + \frac{\partial}{\partial x_l}\right) v(x, \omega) \right] + \alpha v(x, \omega) = f(x) \qquad (x \in R^d). \tag{7.3}$$

Suppose $f(x)$ is a smooth function decaying rapidly as $|x| \to \infty$ and let $\tilde{f}(k)$ be its Fourier transform

$$\hat{f}(k) = \frac{1}{(2\pi)^{\frac{d}{2}}} \int_{R^d} dx e^{-ikx} f(x).$$

We write formally at the moment

$$v(x, \omega) = \frac{1}{(2\pi)^{\frac{d}{2}}} \int_{R^d} dk e^{ikx} \hat{v}(k, \omega). \tag{7.4}$$

Then $\hat{v}(k,\omega)$ satisfies

$$- \sum_{j,l=1}^{d} (D_j + ik_j)[\tilde{a}_{jl}(\omega)(D_l + ik_l)\hat{v}(k, \omega)] + \alpha\hat{v}(k, \omega) = \hat{f}(k) \tag{7.3'}$$

which is an equation in the abstract space $(\Omega, \mathscr{F}, P)$ with $k \in R^d$ a parameter.

Define formally for each $k \in R^d$

$$A(k) = - \sum_{j,l=1}^{d} (D_j + ik_j)[\tilde{a}_{jl}(\omega)(D_l + ik_l)]. \tag{7.5}$$

The precise definition of $A(k)$ is as a quadratic form over $\mathscr{H}^1$ and then by the Friedrichs extension of this form [19]. It is a nonnegative selfadjoint operator with domain $\mathscr{D}(A(k))$ dense in $\mathscr{H}^1$ (we always assume uniform ellipticity (2.1)). It can be verified easily that for each $\alpha > 0$ the operator $(A(k) + \alpha)^{-1}$ is bounded in $\mathscr{H}$ uniformly in $k \in R^d$ and is strongly continuous as a function of k.

Let

$$\hat{v}(k, \omega) = (A(k) + \alpha)^{-1} \hat{f}(k)(\omega) \tag{7.6}$$

with $\hat{f}(k)$ smooth and rapidly decaying as $|k| \to \infty$. Then the integral (7.4) makes sense for almost all ω and we have Parseval's identity,

$$\int_{R^d} dx \int_{\Omega} P(d\omega) |v(x, \omega)|^2 = \int_{R^d} dx \int_{\Omega} P(d\omega) |\hat{v}(k, \omega)|^2 .$$

This and the boundedness of $(A(k) + \alpha)^{-1}$ uniformly in k imply that

the mapping $f(x) \to v(x, \omega)$ is bounded from $L^2(R^d)$ to $L^2(R^d; \mathscr{H})$ and the formula

$$(7.7) \qquad u(x, \omega) = \frac{1}{(2\pi)^{\frac{d}{2}}} \int_{R^d} dk e^{ikx} (A(k) + \alpha)^{-1} \hat{f}(k)(\tau_{-x}\omega)$$

extends by continuity to all $f \in L^2(R^d)$.

We summarize the discussion of this section as follows: The solution $u(x, \omega) \in H^1(R^d; \mathscr{H})$ of (5.1) with $\epsilon = 1$ (and with $\alpha > 0$ and (2.1)) for $f \in L^2(R^d)$, admits the representation (7.7) with the integral understood in the sense of $L^2(R^d; \mathscr{H})$. The operator $A(k)$ is the Friedrichs extension of (7.5) defined as a quadratic form on $\mathscr{H}^1$ for each $k \in R^d$.

Thus the problem of solving (7.1) is reduced to solving the abstract cell problems (7.3′) (the shifted cell problems in the terminology of [10]) for each $k \in R^d$ and then constructing the Fourier integral (7.7).

8. PROBABILISTIC CONVERGENCE THEOREM

We shall assume that the coefficients $a_{ij}(x, \omega) = \tilde{a}_{ij}(\tau_{-x}\omega)$ are symmetric and satisfy (2.1) on $(\Omega, \mathscr{F}, P)$ with the translation group τ_x ergodic on M as described in Section 2. We shall assume in addition that

(8.1) for almost all ω the coefficients $(a_{ij}(x, \omega)$ are continuous functions of $x \in R^d$,

(8.2) the functions $b_j(x, \omega) = \frac{1}{2} \sum_{i=1}^{d} \frac{\partial}{\partial x_i} (a_{ij}(x, \omega))$ are bounded and measurable in (x, ω).

For each $\epsilon > 0$ define the operator

$$(8.3) \qquad \mathscr{L}^{\epsilon}_{\omega} = \frac{1}{2} \sum_{i,j=1}^{d} a_{ij}\left(\frac{x}{\epsilon}, \omega\right) \frac{\partial^2}{\partial x_i \partial x_j} + \frac{1}{\epsilon} \sum_{j=1}^{d} b_j\left(\frac{x}{\epsilon}, \omega\right) \frac{\partial}{\partial x_j}.$$

Let $X = \mathscr{C}([0, \infty); R^d)$ be the space of continuous trajectories in R^d with the topology of uniform convergence on compact sets. Let Σ denote the σ-algebra of Borel sets and let Σ_t $(t \geqslant 0)$, denote the σ-algebra generated by events depending on paths up to time t. A point in X is

denoted by ζ and the point in R^d at time t through which ζ passes is denoted by $x(t) = x(t, \zeta)$.

Under assumptions (8.1), (8.2) and (2.1) there is a uniquely defined diffusion process on R^d associated with $\mathscr{L}^\epsilon_\omega$. That is, for each $\epsilon > 0$, $x \in R^d$ and almost all $\omega \in \Omega$ there is a probability measure $Q_x(\cdot, \omega)$ on (X, Σ) which solves the martingale problem [20] associated with $\mathscr{L}^\epsilon_\omega$:

(8.4)

(i) $Q^\epsilon_x(x(0) = x, \omega) = 1$,

(ii) for each $f(x)$ in $C^2(R^d)$ that is bounded and has bounded derivatives the expression

$$f(x(t)) - \int_0^t (\mathscr{L}^\epsilon_\omega f)(x(s))\, ds$$

is a $(\Sigma_t, Q^\epsilon_x(\cdot, \omega))$ martingale.

Theorem 5. *Let* Q_x *be the diffusion process (Brownian motion) associated with*

$$\mathscr{L} = \frac{1}{2} \sum_{i,j=1}^{d} q_{ij} \frac{\partial^2}{\partial x_i \partial x_j} \tag{8.5}$$

with the constant coefficients (q_{ij}) *defined in Theorem 2. Let* $\varphi(x)$ *be any nonnegative square integrable function on* R^d *and let* $F(\zeta)$ *be any bounded continuous function on* X. *Then*

$$\lim_{\epsilon \downarrow 0} \int_\Omega dP(\omega) \Big| \int_{R^d} dx \varphi(x) \int_X F(\zeta) Q^\epsilon_x(d\zeta, \omega) - \int_{R^d} dx \varphi(x) \int_X F(\zeta) Q_x(d\zeta) \Big|^2 = 0. \tag{8.6}$$

Remark 1. In the periodic case it is possible to analyze processes with generators of the form (8.3) that do not satisfy (8.2) i.e. are not in divergence form. However, the large drift b_j must satisfy certain other conditions (cf. [10], Chapter 3 and [9]).

Remark 2. The asymptotic analysis of the process corresponding to (8.3) with $b_j \equiv 0$ can also be carried out but requires additional consideration. It is given in [22].

Remark 3. The main advantage of the probabilistic convergence Theorem 5 is that it can be used to obtain convergence results for functionals F that are associated with solutions of boundary value problems. Thus, it provides a convenient tool for localization i.e. for separating what happens away from boundaries from what happens near them.

Remark 4. Let us assume that the matrix (a_{ij}) has a symmetric square root

$$(8.7)\qquad a_{ij}(x,\omega) = \sum_{k=1}^{d} \sigma_{ik}(x,\omega)\sigma_{jk}(x,\omega) \qquad (i,j = 1,2,\ldots,d),$$

where $\sigma_{ij}(x,\omega) = \widetilde{\sigma}_{ij}(\tau_{-x}\omega)$. Let us also assume that b_j and σ_{ij} satisfy a Lipschitz condition with Lipschitz constant independent of ω. Let $\beta(t) = (\beta_j(t))$ denote the standard Brownian motion processs on R^d. Then the Ito stochastic differential equations

$$(8.8)\qquad \begin{gathered} dx_j(t) = b_j(x(t),\omega)\,dt + \sum_{k=1}^{d} \sigma_{jk}(x(t),\omega)d\beta_j(t) \\ (x(0) = x \in R^d,\ j = 1,2,\ldots,d), \end{gathered}$$

have for almost all $\omega \in \Omega$ a solution which is a diffusion process (with path measure $Q_x^\epsilon(\cdot,\omega)$ as above) expressed explicitly as a functional of a standard Brownian motion. We have what may be called a diffusion process in a random medium or a random environment with the realization of the medium being labeled by $\omega \in \Omega$. A natural question to ask is how does $x(t) - x$ behave for t large? Does $t^{-\frac{1}{2}}(x(t) - x)$ tend in some sense to a Gaussian as $t \uparrow \infty$? In fact if we let

$$(8.9)\qquad x^\epsilon(t) = \epsilon x\left(\frac{t}{\epsilon^2}\right) \qquad (x^\epsilon(0) = x),$$

we find that (8.8) becomes

$$(8.10)\qquad \begin{gathered} dx_j^\epsilon(t) = \frac{1}{\epsilon}\, b_j\left(\frac{x^\epsilon(t)}{\epsilon},\omega\right) dt + \sum_{k=1}^{d} \sigma_{jk}\left(\frac{x^\epsilon(t)}{\epsilon},\omega\right) d\beta_j(t) \\ (x^\epsilon(0) = x,\ j = 1,2,\ldots,d), \end{gathered}$$

which is precisely the Ǐto equation that goes with (8.3). Thus, Theorem 5 says that indeed $x^\epsilon(t)$ tends to Brownian motion (generator (8.5)) as $\epsilon \to 0$ weakly (as a measure in X) in $L^2(\Omega, \mathscr{F}, P)$, i.e. in the sense (8.6).

We note again that Theorem 5 has been shown only under condition (8.2) which says that (8.3) is in divergence form (cf. Remark 2 above).

9. PROOF OF PROBABILISTIC CONVERGENCE THEOREM

We shall first show that the measures $Q_x^\epsilon(\cdot, \omega)$ $(\epsilon > 0,\ \omega \in \Omega,\ x \in R^d)$ are relatively weakly compact. For this purpose it suffices to show that

$$
\begin{aligned}
&E^{Q_x^\epsilon}\{|x(t) - x(s)|^4\} = \\
(9.1)\qquad &= \int_X |x(t,\zeta) - x(s,\zeta)|^4 Q_x^\epsilon(d\zeta, \omega) \leqslant C|t-s|^2,
\end{aligned}
$$

where C is a constant independent of t, s, ϵ, ω and x. To prove (9.1) we use the divergence form of (8.3) and Nash's estimates [21].

Specifically, suppose for the moment that the coefficients (a_{ij}) are smooth and satisfy (2.1). Then the diffusion equation

$$\frac{\partial u}{\partial t} = \mathscr{L}_\omega^\epsilon u$$

has a fundamental solution $\varphi^\epsilon(t, x, y; \omega)$. The estimate of Nash says that there is a constant $\gamma > 0$ that depends only on the uniform ellipticity constants in (2.1) such that

$$
(9.2)\qquad \varphi^\epsilon(t, x, y; \omega) \leqslant \frac{\gamma}{t^{\frac{d}{2}}}\, e^{-\gamma \frac{|x-y|}{\sqrt{t}} \log \gamma \frac{|x-y|}{\sqrt{t}}},
$$

for all $t \geqslant 0$, $x, y \in R^d$ and $\omega \in \Omega$. Estimate (9.2) immediately yields (9.1) (after the preliminary smoothing is removed).

Compactness of Q_x^ϵ, with $|x| \leqslant M$, means that for any $\delta > 0$ there is a compact set K_δ in $X = C([0,\infty); R^d)$ such that

(9.3) $\quad Q_x^\epsilon(K_\delta, \omega) \geqslant 1 - \delta$

for all $\epsilon > 0$, $\omega \in \Omega$ and $|x| \leqslant M$. Thus, to prove (8.6) it suffices to prove

$$\lim_{\epsilon \downarrow 0} \int_\Omega dP(\omega) \Big| \int_{|x| \leqslant M} dx \varphi(x) \int_{K_\delta} F(\zeta) Q_x^\epsilon(d\zeta, \omega) - \int_{|x| \leqslant M} dx \varphi(x) \int_{K_\delta} F(\zeta) Q_x(d\zeta) \Big|^2 = 0 \tag{9.4}$$

or, since $\varphi \in L^2(R^d)$,

$$\lim_{\epsilon \downarrow 0} \int_\Omega dP(\omega) \int_{|x| \leqslant M} dx \Big| \int_{K_\delta} F(\zeta) Q_x^\epsilon(d\zeta, \omega) - \int_{K_\delta} F(\zeta) Q_x(d\zeta) \Big|^2 = 0. \tag{9.5}$$

Now $F(\zeta)$ is a bounded continuous function on X so in particular $F \in C(K_\delta)$ (the space of bounded continuous functions on the compact set K_δ). Finite linear combinations of functions in $C(K_\delta)$ of the form

(9.6) $\quad f_1(x(t_1)) \ldots f_N(x(t_N)) \qquad (0 \leqslant t_1 < t_2 < \ldots < t_N < \infty,$

with $f_j(x)$ in $C_0^\infty(R^d)$ form an algebra in $C(K_\delta)$ that contains the constant functions and separates points. This algebra is thus dense in $C(K_\delta)$ by the Stone – Weierstrass theorem. It suffices then to prove (9.5) with F of the form (9.6). But then (9.5) will follow if we can show that for each $T < \infty$ and for each $f(x) \in C^\infty(R^d)$ that decays to zero as $|x| \to \infty$, say, exponentially fast we have

$$\lim_{\epsilon \downarrow 0} \sup_{0 \leqslant t \leqslant T} \int_\Omega dP(\omega) \int_{R^d} dx \, |E^{Q_x^\epsilon}(f(x(t))) - E^{Q_x}(f(x(t)))|^2 = 0. \tag{9.7}$$

Here we use the fact that $u(t, x) = E^{Q_x}\{f(x(t))\}$ satisfies the heat equation

$$\frac{\partial u}{\partial t} = \mathscr{L} u \qquad (t > 0, \; u(0, x) = f(x)) \tag{9.8}$$

and has smooth and decaying solution as $|x| \to \infty$ when $f(x)$ is also smooth and decaying as $|x| \to \infty$.

Let

$$(9.9)\qquad u^{\epsilon}(t,x,\omega)=E^{Q^{\epsilon}_{x}}\{f(x(t))\}.$$

This function is a suitable weak solution of the diffusion equation

$$(9.10)\qquad \frac{\partial u^{\epsilon}}{\partial t}=\mathscr{L}^{\epsilon}_{\omega}u^{\epsilon}\qquad (t>0,\ u^{\epsilon}(0,x,\omega)=f(x)),$$

with $\mathscr{L}^{\epsilon}_{\omega}$ given by (8.3). Since $\mathscr{L}^{\epsilon}_{\omega}$ is in divergence form we consider (9.10) in the following manner.

Let $\mathscr{H}=L^2(\Omega,\mathscr{F},P)$, $H=L^2(R^d;\mathscr{H})$ and $H^1=H^1(R^d;\mathscr{H})$ with the usual inner products (cf. Section 2). On H^1 which is dense in H the operator $A^{\epsilon}=-\mathscr{L}^{\epsilon}_{\omega}$ defines a nonnegative, symmetric bilinear form. We denote the Friedrichs extension of A also by A^{ϵ} and note that it is a densely defined, selfadjoint nonnegative operator on H. Let A be the operator defined by $-\mathscr{L}$ on H (which is essentially the Laplacian). From Theorem 3 we know that for each $\alpha>0$

$$(9.11)\qquad (A^{\epsilon}+\alpha)^{-1}\to(A+\alpha)^{-1}$$

strongly in H, as $\epsilon\downarrow 0$. Let T^{ϵ}_t and T_t be, respectively, the semigroup defined by $\exp(-A^{\epsilon}t)$ and $\exp(At)$ via the spectral theorem for the selfadjoint operators A^{ϵ} and A. From the strong resolvent convergence (9.11) it follows that the semigroups converge

$$(9.12)\qquad T^{\epsilon}_t\to T_t,$$

strongly in H, as $\epsilon\downarrow 0$, uniformly in finite t intervals. But (9.12) is precisely the statement (9.7) which was to be shown. The proof Theorem 5 is complete.

REFERENCES

[1] E. di Giorgi – S. Spagnolo, Sulla convergenza degli integrali dell'energia per operatori ellittici del secundo ordine, *Boll. U.M.I.*, 8 (1973), 391-411.

[2] I. Babuska, Solution of interface problems by homogenization I, II, III, *SIAM J. Math. Analysis,* 7 (1976), 603-634, 635-645 and 8 (1977), 923-937.

[3] S. Spagnolo, Convergence in energy, in: *Numerical methods of partial differential equations,* III, Ed. B. Hubbard, Academic Press, New York, 1976, 469-498.

[4] N.S. Bakhbalov, *Doklady Akad. Nauk,* 218 (# 5), 1974.

[5] V.L. Berdicevskii, *Doklady Akad. Nauk,* 222 (# 3), 1975.

[6] E. Sanchez Palencia, Compartements local et macroscopique d'un type de milieux physique heterogenes, *Int. J. Eng. Sci.,* 12 (1974), 331-351.

[7] E. Larsen, Neutron transport and diffusion in inhomogeneous media, *J. Math. Phys.,* 16 (1975), 1421-1427.

[8] J.B. Keller, Effective behavior of heterogeneous media, in: *Statistical mechanics and statistical methods in theory and application,* Ed. U. Landman, Plenum Press, 1977.

[9] G. Papanicolaou – D. Stroock – S.R.S. Varadhan, Martingale approach to some limit theorems, in: *Statistical mechanics, dynamical systems and the Duke Turbulance Conference,* Ed. D. Ruelle, Duke Univ. Math. Series, Vol. 3, Durham, North Carolina, 1977.

[10] A. Bensoussan – J.L. Lions – G.C. Papanicolaou, *Asymptotic Analysis for Periodic Structures* North-Holland, Amsterdam, 1978.

[11] C.M. Kozlov, *Doklady Akad. Nauk,* 236 (# 5), 1977.

[12] V.V. Jurinskii, *Vilnius Conference Abstracts,* Sept. 1978, 54.

[13] C.M. Kozlov, *Doklady Akad. Nauk,* 241 (# 5), 1978).

[14] G.C. Papanicolaou, Introduction to the asymptotic analysis of stochastic equations, in: *AMS book series in applied mathematics,* 16, Providence, R.I., 1977.

[15] J.L. Lions, *Equations differentielles operationnelles et problemes aux limites,* Berlin, Springer, 1961.

[16] N. Wiener, The ergodic theorem, *Duke Math. Journal,* 5 (1939), 1-18.

[17] A. Guichardet, *Analyse harmonique commutative,* Paris, Dunod, 1968).

[18] H. Bohr, *Almost Periodic Functions,* Chelsea, New York, 1947.

[19] T. Kato, *Perturbation Theory of Linear Operators,* Springer, Berlin, 1966.

[20] D. Stroock – S.R.S. Varadhan, *Multidimensional Diffusion Processes,* Springer, Berlin – Heidelberg – New York, 1979.

[21] J. Nash, Continuity of solutions of parabolic and elliptic equations, *Am. J. Math.,* 80 (1958), 931-954.

[22] G.C. Papanicolaou – S.R.S. Varadhan, (to appear).

G.C. Papanicolaou – S.R.S. Varadhan
Courant Institute, 251 Mercer Str., New York, N.Y., 10012, USA.

Statistics and Probability: Essays in honor of C. R. Rao, North-Holland, 1982, 547-552

DIFFUSIONS WITH RANDOM COEFFICIENTS

George C. PAPANICOLAOU and S.R.S. VARADHAN
Courant Institute of Mathematical Sciences, New York University, NY, U.S.A.

1. Introduction

Here we will be studying diffusion processes in R^d whose coefficients are random, stationary, stochastic processes. We will be mainly concerned with the behavior of such diffusion processes for large times for which we will establish a form of the ergodic theorem and deduce a central limit theorem from it.

Our diffusion processes will always start at $t=0$ from the point $x=0$ in R^d and will be governed by the diffusion generator:

$$L_\omega = \tfrac{1}{2}\sum a_{ij}(x,\omega)\frac{\partial^2}{\partial x_i \partial x_j}.$$

Here $\omega \in \Omega$ where $(\Omega, \mathcal{F}, \mu)$ is a probability space and $\{a_{ij}(x,\omega)\}$ constitutes a stationary stochastic process with $x \in R^d$ as the parameter set. For each x, the random variables $\{a_{ij}(x,\omega)\}$ constituting the diffusion coefficients at x, can be viewed as taking values in the space S_d of symmetric positive semi-definite matrices of size d. We will assume, throughout this paper, that the coefficients $\{a_{ij}(x,\omega)]$ are all almost surely continuous in x and that there exist constants $0 < C_1 \leqslant C_2 < \infty$ which are non-random such that,

$$C_1 \sum \xi_j^2 \leqslant \sum a_{ij}(x,\omega)\xi_i\xi_j \leqslant C_2 \sum \xi_j^2, \quad \forall x \in R^d, \quad \omega \in \Omega.$$

The last assumption we need is the ergodicity of the coefficient process $\{a(x,\omega)\}$ with respect to translations in space.

For each $\omega \in \Omega$, L_ω governs a diffusion and corresponding to the initial conditions of starting from $x=0$ at $t=0$ we have a measure P_ω on the space $C[[0,\infty); R^d]$ of continuous trajectories on $[0,\infty)$ with values in R^d. We want to look at the behavior of the distribution of $x_t/\sqrt{t}$ under P_ω as $t \to \infty$. Some form of the ergodic theorem should produce an averaging effect and the limiting distribution of $x_t/\sqrt{t}$ under P_ω should be independent of ω and should be a normal distribution with mean zero and covariance a. The identification of $\bar{a}$ should be made as a precise form of averaging. This is the goal of this article.

Instead of random coefficients, one can take periodic coefficients and this case has been studied extensively in refs. [2, 1]. The case of random coefficients when L_ω is in divergence form has been studied in refs. [3, 4].

When the dimension, d, is one, we can use the method of random time change to study the problem explicitly and an easy computation establishes the result and calculates the limiting variance $\bar{a}$ to be the harmonic mean:

$$\bar{a}=\left[E\frac{1}{a(0;\omega)}\right]^{-1}.$$

2. An ergodic theorem

Although we formulated the theorem as a central limit theorem it is in reality an ergodic theorem that underlies the phenomenon. To see this, from the fact that P_ω is the diffusion corresponding to L_ω it follows that,

$$E^{P_\omega}\left[\exp\left\{\mathrm{i}\langle\theta, x(t)\rangle+\tfrac{1}{2}\left\langle\theta, \int_0^t a[x(s),\omega]\,\mathrm{d}s\,\theta\right\rangle\right\}\right]=1 \quad \text{a.e. } \omega.$$

By replacing θ by $\theta/\sqrt{t}$ we can rewrite this as:

$$E^{P_\omega}\left[\exp\left\{\mathrm{i}\left\langle\theta, \frac{x(t)}{\sqrt{t}}\right\rangle+\tfrac{1}{2}\left\langle\theta, \frac{1}{t}\int_0^t a[x(s),\omega]\,\mathrm{d}s\,\theta\right\rangle\right\}\right]=1.$$

If we had an ergodic theorem asserting,

$$\lim_{t\to\infty}\frac{1}{t}\int_0^t a(x(s),\omega)\,\mathrm{d}s=\bar{a}, \quad \text{a.e. } P_\omega, \text{a.e. } \mu, \tag{2.1}$$

it then follows that,

$$\lim_{t\to\infty} E^{P_\omega}\left[\exp\left\{\mathrm{i}\left\langle\theta, \frac{x(t)}{\sqrt{t}}\right\rangle\right\}\right]=\exp\{-\tfrac{1}{2}\langle\theta, \bar{a}\theta\rangle\},$$

which is the central limit theorem that we are after.

Now we will set up the basic notation that we will be using. Let Ω denote the space of all continuous maps $\omega: R^d\to S_d$ satisfying

$$C_1\sum\xi_j^2\leqslant\sum\omega_{ij}(x)\xi_i\xi_j\leqslant C_2\sum\xi_j^2,$$

for all $\xi\in R^d$ and $x\in R^d$. We will use also the customary notation $a_{ij}(x,\omega)$ to denote $\omega_{ij}(x)$ so we may conveniently view $a(x,\omega)=\{a_{ij}(x,\omega)\}$ either as a function of x for each fixed $\omega\in\Omega$ or as a function on Ω for each fixed x. We also have the translation maps $\tau_x:\Omega\to\Omega$ defined by:

$$a(y,\tau_x\omega)=a(x+y,\omega), \quad \forall x\in R^d, \quad y\in R^d \quad \text{and} \quad \omega\in\Omega.$$

There is a canonical Markov process with Ω as state space. Let us fix $\omega\in\Omega$. Then we have the diffusion operator,

$$L_\omega=\tfrac{1}{2}\sum a_{ij}(x,\omega)\frac{\partial^2}{\partial x_i\partial x_j},$$

on R^d and this generates a diffusion measure P_ω corresponding to these coefficients, with initial conditions corresponding to starting from the origin at time 0. P_ω is a measure on the space $C[[0,\infty); R^d]$ of trajectories starting from 0 at time 0.

Given a trajectory $x(t)$ with $x(0)=0$ we can map it onto a trajectory in the Ω space by setting

$$\omega_t = \tau_{x(t)}\omega.$$

This induces a measure Q_ω on the space of trajectories in Ω starting from ω at time 0 and a moment's reflection tells us that this is in fact a Markov process with Ω as state space. Let us denote by $q(t,\omega,\mathrm{d}\omega')$ the transition probability of this Markov process. If we think of the topology of Ω as coming from uniform convergence on compact sets it is easy to see that $q(t,\omega,\mathrm{d}\omega')$ has the Feller property. If we denote by $A(\omega)$ the function $a(0,\omega)$ then the property (2.1) that we want is really,

$$\lim_{t\to\infty}\frac{1}{t}\int_0^t A(\omega_s)\,\mathrm{d}s=\bar{a} \quad \text{a.e. } Q_\omega \quad \text{a.e. } \mu, \tag{2.2}$$

where μ is a stationary ergodic measure on Ω. In other words, we need an ergodic theory for the canonical Markov process $\{Q_\omega\}$ on Ω. In the next section we will establish the following theorem.

Theorem 2.1. *Given any ergodic stationary measure μ on Ω, there exists a unique λ on Ω which is mutually absolutely continuous with respect to μ, and which is an ergodic invariant measure with respect to our canonical Markov process on Ω.*

As an immediate consequence we have:

Theorem 2.2. *For any ergodic stationary measure μ on Ω,*

$$\lim_{t\to\infty}\frac{1}{t}\int_0^t A(\omega_s)\,\mathrm{d}s=\int_\Omega A(\omega)\lambda(\mathrm{d}\omega), \quad a.e.\ Q_\omega, \quad a.e.\ \mu,$$

where λ is determined by μ according to Theorem 2.1.

3. Proof

In order to prove our main theorem we need two results from the theory of partial differential equations which we shall state as lemmas without proofs.

Lemma 3.1. *Let $a(x)$ be a set of coefficients on R^d which is periodic in each variable of period 1. We will think of*

$$L=\tfrac{1}{2}\sum a_{ij}\frac{\partial^2}{\partial x_i\,\partial x_j}$$

as a diffusion process on the d-dimensional torus, T_d. Then the resolvent,

$$R=(I-L)^{-1},$$

maps $L_d(T_d)$ into $L_\infty(T_d)$ with a bound K that depends only on the upper and lower bounds for eigenvalues of the coefficient matrix $\{a(x)\}$ and not on the smoothness of the coefficients:

Remark. This result can be found in ref. [6] in a slightly different form. The modifications needed to reduce it to Lemma 3.1 are routine.

Lemma 3.2. *Let $a(x)$ be a set of coefficients in our space Ω. Let $p(t,x,\mathrm{d}y)$ be the transition probability for the diffusion on R^d governed by a. Then for each $x\in R^d$ and $t>0$, the measure $p(t,x,\mathrm{d}y)$ is equivalent to the Lebesgue measure on R^d.*

Remark. The existence of a density for $p(t,x,\mathrm{d}y)$ can be found in ref. [7] and the fact that the density cannot vanish on a set of positive measure is essentially the strong maximum principle.

The next ingredient is the fact that any stationary process is a limit of periodic processes:

Lemma 3.3. *Let μ be a stationary process on Ω. Then we can find for a sequence $l_n\to\infty$, periodic functions $\omega_n\in\Omega$ of period l_n in each variable, such that, the measure μ_n on Ω obtained as the distribution of $\tau_x\omega_n$ where x is random and distributed uniformly on the cube of size l_n, converges weakly to μ as $n\to\infty$.*

Remark. This is a well-known fact from the theory of stationary process. See for instance reference [5] for a proof.

Proof of Theorem 2.1. Let us start with the coefficients $\omega_n\in\Omega$ given by Lemma 3.3 which is periodic of period l_n. By the map $x\to x/l_n$ we can shrink the torus of size l_n to one of unit size. If ω_n represents the coefficients $a^{(n)}(x)$ then we consider the coefficients $a^{(n)}(l_nx)$ on the unit torus. The diffusion generated by this on the torus will have an invariant density $\phi_n(x)$ representing an invariant probability distribution for this diffusion. Although the coefficients are rougher the upper and lower bounds C_1, C_2 stay the same, i.e.

$$C_1\sum\xi_j^2\leqslant a_{ij}^{(n)}(l_nx)\xi_i\xi_j\leqslant C_2\sum\xi_j^2.$$

From Lemma 3.1 we know that the resolvent R maps L_d into L_∞ boundedly with a bound K independent of n. The adjoint R^* therefore maps L_1 boundedly into $L_{d/(d-1)}$ with the same bound K. Since $R^*\phi_n=\phi_n$ and ϕ_n has L_1 norm one we obtain,

$$\|\phi_n\|_{d/(d-1)}\leqslant K \quad \text{for all } n.$$

If we now go back to the torus of size l_n and denote by $\tilde{\phi}_n(x)$, the function $\phi_n(x/l_n)$ then,

$$\|\tilde{\phi}_n\|_{d/(d-1)}\leqslant K,$$

provided we compute the norms with respect to the normalized Lebesgue measure

on the torus of size l_n. We denote the probability measure represented by $\tilde{\phi}_n$ by $\tilde{\lambda}_n$ and map it into Ω by the map $x \to \tau_x \omega_n$. The measure induced on Ω is denoted by λ_n. It is easily verified that λ_n is an invariant measure for the canonical Markov process, $\lambda_n \ll \mu_n$ and that

$$\int_\Omega \left| \frac{d\lambda_n}{d\mu_n} \right|^{d/(d-1)} d\mu_n \leqslant K^{d/(d-1)}.$$

Since μ_n is converging weakly to μ and K is independent of n, it is routine to check that λ_n has a weakly convergent subsequence, and the limit λ so obtained is again an invariant measure for our canonical Markov process. In addition, $\lambda \ll \mu$ and:

$$\int_\Omega \left| \frac{d\lambda}{d\mu} \right|^{d/(d-1)} d\mu \leqslant K^{d/(d-1)}.$$

We now assume that μ is ergodic and show that $\mu \ll \lambda$ and λ is ergodic for the canonical process. If μ is not absolutely continuous with respect to λ then since $\lambda \ll \mu$ we can use the Lebesgue decomposition to find a set $E \subset \Omega$ with the properties $\lambda(E)=0$, $0<\mu(E)<1$ and $\mu \ll \lambda$ on E^c. From $\lambda(E^c)=1$ we conclude,

$$q(t, \omega, E^c)=1 \quad \text{a.e. } \omega \quad \text{w.r.t. } \lambda.$$

Since $\mu \ll \lambda$ on E^c this implies

$$q(t, \omega, E^c)=1 \quad \text{a.e. } \omega \text{ on } E^c \quad \text{w.r.t. } \mu.$$

From the definition of q and Lemma 3.2, it now follows that

$$\tau_x \omega \in E^c \quad \text{a.e. } x \quad \text{a.e. } \omega \text{ on } E^c \quad \text{w.r.t. } \mu.$$

This in turn implies by Fubini's theorem that

$$\tau_x E^c = E^c \quad \text{for almost all } x,$$

as elements in the measure algebra of μ. Since $x \to \tau_x$ is continuous we conclude that E^c is an invariant set. By ergodicity of μ we must have $\mu(E^c)=0$ or 1 and we have a contradiction.

Let us now prove that as an invariant measure for our canonical process λ is ergodic; if not, we have a set $E \subset \Omega$ with $0<\lambda(E)<1$ such that

$$q(t, \omega, E)=1 \quad \text{a.e. } \omega \text{ in } E \quad \text{w.r.t. } \lambda.$$

Since $\mu \ll \lambda$, we conclude

$$q(t, \omega, E)=1 \quad \text{a.e. } \omega \text{ in } E \quad \text{w.r.t. } \mu.$$

Just as in the preceding argument we conclude now by Lemma 3.2 that E is invariant and by ergodicity of μ that $\mu(E)=0$ or 1. Since $\lambda \ll \mu$ we must also have $\lambda(E)=0$ or 1.

This completes the proof of Theorem 2.1. Theorem 2.2 of course follows immediately. Note that the ergodicity of any arbitrary λ which is absolutely continuous with respect to μ implies the uniqueness of λ.

4. Remarks

We remark that without any work we can obtain the weak convergence of the distribution of the process $(x(kt))/\sqrt{k}$ as $k \to \infty$ under Q_ω to a suitable Brownian motion.

If $d\lambda/d\mu = \psi(\omega)$ then $\psi(\tau_x\omega)$ defines a stochastic process $\psi(x,\omega)$ with the following properties: For a.e. ω $\psi(x,\omega)$ is a σ-finite non-negative invariant density for L_ω on R^d. It is a stationary stochastic process in x and is normalized to have $E\psi(x,\omega) = 1$.

References

[1] Freidlin, M.I. (1964). Dirichlet's problem for an equation with periodic coefficients depending on a small parameter. *Th. Prob. Appl.* **9**, 121–125.

[2] Bensoussan, A., Lions, J.L. and Papanicolaou, G. (1978). *Asymptotic Methods for Periodic Structures*. North-Holland, Amsterdam.

[3] Kozlov, C.M. (1978). *Dokl. Akad. Nauk.* **5**, 236–239.

[4] Papanicolaou, G. and Varadhan, S. Boundary value problems with rapidly oscillating random coefficients. *Seria Coll. Math. Soc. János Bolyai*, North-Holland, Amsterdam, to appear.

[5] Parthasarathy, K.R. (1961). On the category of ergodic measures. *Ill. J. Math.* **5**, 648–655.

[6] Pucci, C. (1966). Limitazioni per soluzioni di equazioni ellitiche. *Ann. Matematica Pura Applic.*, Series IV, 15–30.

[7] Stroock, D. and Varadhan, S.R.S. (1978). *Multidimensional Diffusion Processes*. Springer, New York.

Comm. Pure Appl. Math. 38 (1985) 819-834

Ohrnstein–Uhlenbeck Process in a Random Potential

G. PAPANICOLAOU AND S. R. S. VARADHAN
Courant Institute

1. Introduction

Let $(\Omega, \mathscr{F}, P)$ be a probability space on which a group τ_x of measure preserving transformations indexed by $x \in \mathbb{R}^d$ acts ergodically. The action is assumed stochastically continuous. Let $Q(\omega)$ be a function from Ω to $\mathbb{R}$ that is bounded along with $D_iQ(\omega)$ and $D_iD_jQ(\omega)$, $i, j = 1, 2, \cdots, d$. Here the operators D_i are defined by

$$D_iF(\omega) = \frac{\partial}{\partial x_i} F(\tau_{-x}\omega)\bigg|_{x=0} \tag{1.1}$$

for functions F for which the derivative on the right exists in $L^2(\Omega, \mathscr{F}, P)$.

For each $\omega \in \Omega$ we let

$$\tilde{Q}(x, \omega) = Q(\tau_{-x}\omega) \tag{1.2}$$

define a stationary random function on $\mathbb{R}^d$. If $q_i(\omega) = D_iQ(\omega)$ and

$$\tilde{q}_i(x, \omega) = q_i(\tau_{-x}\omega), \tag{1.3}$$

then $\tilde{q}_i(x, \omega)$, $i = 1, 2, \cdots, d$, are also stationary random functions. We shall consider the Ohrnstein-Uhlenbeck process on $\mathbb{R}^d \times \mathbb{R}^d$, in the random potential $-\tilde{Q}(x, \omega)$. This means that the infinitesimal generator of this process on $\mathbb{R}^d \times \mathbb{R}^d$ has for each $\omega \in \Omega$ the form

$$\tilde{L}_\omega = \sum_{i=1}^{d} \left(\frac{\partial^2}{\partial v_i^2} - v_i \frac{\partial}{\partial v_i} + \tilde{q}_i(x, \omega) \frac{\partial}{\partial v_i} + v_i \frac{\partial}{\partial x_i} \right). \tag{1.4}$$

If we start a particle at $x = 0$ with initial velocity distribution $G(dv)$ where

$$G(dv) = \left(\frac{1}{2\pi} \right)^{d/2} \exp\{-\tfrac{1}{2}|v|^2\}\, dv, \tag{1.5}$$

then we shall show that the scaled process

$$x^{\varepsilon}(t) = \varepsilon x(t/\varepsilon^2) \tag{1.6}$$

obeys the central limit theorem in probability with respect to ω. This means that for small ε, for most initial velocities and most realizations of the random potential $-\tilde{Q}(x, \omega)$, the distribution of the scaled process $x^{\varepsilon}(\cdot)$ is nearly that of a d-dimensional Brownian motion with some nonsingular diffusion matrix (γ_{ij}).

The problem can be formulated equivalently for the diffusion with generator L on $\Omega \times \mathbb{R}^d$ given by

$$L = \sum_{i=1}^{d} \left(\frac{\partial^2}{\partial v_i^2} - v_i \frac{\partial}{\partial v_i} + q_i(\omega) \frac{\partial}{\partial v_i} + v_i D_i \right). \tag{1.7}$$

What we want to show now is that the central limit theorem holds for the position coordinates which are the additive functionals

$$x(t) = \int_0^t v(s)\, ds \tag{1.8}$$

with the scaling $x^{\varepsilon}(t) = \varepsilon x(t/\varepsilon^2)$ as $\varepsilon \to 0$.

Let $d\tilde{P} = ce^{Q(\omega)}\, dP$, where c is the normalization constant, and let $\tilde{\pi} = \tilde{P} \times G$. Then $\tilde{\pi}$ is an invariant measure for the Markov process on $\Omega \times \mathbb{R}^d$ generated by L of (1.7). Assume that the starting point (ω, v) is distributed by $\tilde{\pi}$.

THEOREM 1.1. *The distribution of $x^{\varepsilon}(\cdot)$ converges weakly as $\varepsilon \to 0$ to that of Brownian motion in $\mathbb{R}^d$ with diffusion matrix (γ_{ij}). The diffusion matrix is nonsingular and is characterized in Section* 3 (*cf*. (3.2)).

The proof is given in three sections. In the next section we study the solution of the resolvent equation

$$\lambda u_\lambda - L u_\lambda = l(v), \qquad \lambda > 0, \tag{1.9}$$

where $l(v)$ is a linear functional $\sum_{i=1}^{d} l_i v_i$. In Section 3 we show that the finite-dimensional distributions of $x^{\varepsilon}(\cdot)$ converge as $\varepsilon \to 0$ to those of a Brownian motion. In Section 4 we establish compactness of the distribution of $x^{\varepsilon}(\cdot)$ as $\varepsilon \to 0$ in the space of continuous trajectories which implies the weak convergence of the processes.

The method followed here is similar to the one in [1]. For periodic potentials convergence to Brownian motion was shown by Rodenhausen [2]. Using the solution u_λ of (1.9) one first converts the functional (1.8) into an approximate martingale and verifies the conditions for the martingale central limit theorem (see [3]).

2. Analysis of the Resolvent Equation

Let us introduce the notation

$$A = \sum_{i=1}^{d} \left(\frac{\partial^2}{\partial v_i^2} - v_i \frac{\partial}{\partial v_i} \right),$$

(2.1)

$$J = \sum_{i=1}^{d} \left(q_i(\omega) \frac{\partial}{\partial v_i} + v_i D_i \right),$$

so that (1.9) becomes

(2.2) $$\lambda u_\lambda - A u_\lambda - J u_\lambda = l(v).$$

We must first make precise the meaning of this equation and the kind of function the solution u_λ is.

Let $\mathscr{H}_0$ be the Hilbert space of functions $f(\cdot)$ from $\Omega \times \mathbf{R}^d$ to R with norm

(2.3) $$\|f\|_0^2 = \int_\Omega |f(\omega, \cdot)|_0^2 \tilde{P}(d\omega) = \int_\Omega \int_{\mathbf{R}^d} |f(\omega, v)|^2 G(dv) \tilde{P}(d\omega).$$

For functions g from $\mathbf{R}^d \to \mathbf{R}$ we introduce the norms

(2.4) $$|g|_0^2 = \int g^2(v) G(dv),$$

$$|g|_1^2 = \int \left\{ g^2(v) + \sum_{i=1}^{d} \left(\frac{\partial g(v)}{\partial v_i} \right)^2 \right\} G(dv)$$

(2.5)

$$= \int g(v)(I - A) g(v) G(dv),$$

(2.6) $$|g|_{-1}^2 = \int g(v)(I - A)^{-1} g(v) G(dv).$$

In addition to the space $\mathscr{H}_0$ we also introduce $\mathscr{H}_1$ and $\mathscr{H}_{-1}$ which are Hilbert spaces of functions $f(\omega, v)$ with norms

(2.7) $$\|f\|_1^2 = \int_\Omega |f(\omega, \cdot)|_1^2 \tilde{P}(d\omega),$$

(2.8) $$\|f\|_{-1}^2 = \int_\Omega |f(\omega, \cdot)|_{-1}^2 \tilde{P}(d\omega).$$

We know that $\mathscr{H}_1 \subset \mathscr{H}_0 \subset \mathscr{H}_{-1}$ and that $\mathscr{H}_{-1}$ is the dual of $\mathscr{H}_1$ relative to $\mathscr{H}_0$.

For functions f in $\mathscr{H}_1$ we define the Dirichlet form

$$[f,f]=\int_\Omega\int_{\mathbf{R}^d}\sum_{i=1}^{d}\left(\frac{\partial f(\omega,v)}{\partial v_i}\right)^2 G(dv)\tilde{P}(d\omega) \tag{2.9}$$
$$= -(Af,f),$$

where (,) is the inner product in $\mathscr{H}_0$. The null space of the Dirichlet form (2.9) consists of functions $f(\omega)$ that are constant in v. On functions $f(\omega,v)$ such that $\int f(\omega,v)\,dG(v)\equiv 0$ in ω, the Dirichlet form [,] is equivalent to the inner product $(\ ,\ )_1$ in $\mathscr{H}_1$.

In the next lemma we clarify the sense in which the resolvent equation (2.2) is to be taken.

LEMMA 2.1. *For each $\lambda > 0$ equation (2.2) has a solution u_λ in $\mathscr{H}_1$ with Au_λ in $\mathscr{H}_{-1}$ and Ju_λ in $\mathscr{H}_{-1}$. The equation itself is satisfied in the space $\mathscr{H}_{-1}$.*

Proof: The proof is routine so we shall merely outline it. Equation (2.2) is regularized first by considering

$$\lambda u_\lambda^\beta-\beta e^{-Q}\sum_{j=1}^{d}D_j\left(e^{Q}D_ju_\lambda^\beta\right)-Au_\lambda^\beta-Ju_\lambda^\beta=l(v) \tag{2.10}$$

or as a bilinear form

$$\lambda\left(u_\lambda^\beta,\phi\right)+\beta\sum_{j=1}^{d}\left(D_ju_\lambda^\beta,D_j\phi\right)+\left[u_\lambda^\beta,\phi\right]-\left(Ju_\lambda^\beta,\phi\right)=\left(l(v),\phi\right) \tag{2.11}$$

with $\phi\in\mathscr{H}_1$ such that $D_j\phi$ is in $\mathscr{H}_0$, $j=1,\cdots,d$. From (2.1) we see that

$$\left(Ju_\lambda^\beta,\phi\right)=\sum_{j=1}^{d}\left\{\left(D_ju_\lambda^\beta,\frac{\partial\phi}{\partial v_j}\right)-\left(\frac{\partial u_\lambda^\beta}{\partial v_j},D_j\phi\right)\right\} \tag{2.12}$$

and hence, for each $\lambda>0$ and $\beta>0$, (2.11) has a unique solution u_λ^β in $\mathscr{H}_1$ with $D_ju_\lambda^\beta$ in $\mathscr{H}_0$, $j=1,\cdots,d$.

Letting $\phi=u_\lambda^\beta$ in (2.11) gives the estimates

$$\lambda\left(u_\lambda^\beta,u_\lambda^\beta\right)\leqq C,\qquad \beta\sum_{j=1}^{d}\left(D_ju_\lambda^\beta,D_ju_\lambda^\beta\right)\leqq C, \tag{2.13}$$
$$\left[u_\lambda^\beta,u_\lambda^\beta\right]\leqq C,$$

with C a constant independent of β and λ. From these estimates it follows that

$u_\lambda^\beta \to u_\lambda$ as $\beta \to 0$ weakly in $\mathscr{H}_1$ and noting that J is skew adjoint we can pass to the limit in (2.11) to get

$$\lambda(u_\lambda, \phi) + [u_\lambda, \phi] + (u_\lambda, J\phi) = (l(v), \phi). \tag{2.14}$$

From (2.14) we conclude that

$$|(u_\lambda, J\phi)| \leqq C\|\phi\|_1, \tag{2.15}$$

where C is independent of ϕ. This means that $(u_\lambda, J\phi)$ is a continuous linear functional on $\mathscr{H}_1$ and so $(u_\lambda, J\phi) = (\psi, \phi)$ for some ψ in $\mathscr{H}_{-1}$ which we denote by $-Ju_\lambda$. This solves (2.2).

In order to analyze the resolvent equation (2.2) as $\lambda \to 0$ we need the following lemma.

LEMMA 2.2. *Let*

$$a_\lambda(\omega) = \int u_\lambda(\omega, v) G(dv) \tag{2.16}$$

be the mean of the solution u_λ of (2.2). *Write*

$$u_\lambda = a_\lambda + f_\lambda. \tag{2.17}$$

Then

$$\|f_\lambda\|_1 \leqq C \tag{2.18}$$

and there exist functions $b_\lambda(\omega)$ and $c_\lambda(\omega)$ such that $a_\lambda = b_\lambda + c_\lambda$ and

$$\sup_{\lambda \leqq 1} \int \left\{ |a_\lambda(\omega)|^2 + \sum_{i=1}^d |D_i c_\lambda(\omega)|^2 \right\} \tilde{P}(d\omega) < \infty. \tag{2.19}$$

Proof: The bound (2.18) follows immediately from the Poincaré inequality for functions in $\mathscr{H}_1$ that have mean identically zero in ω. To get (2.19) we note from (2.13) that there is a constant C independent of λ such that

$$\lambda\|u_\lambda\|_0^2 \leqq C, \qquad [u_\lambda, u_\lambda] \leqq C. \tag{2.20}$$

In fact, $C = \|l\|^2$, the norm of the linear functional $l(v)$ on $\mathbb{R}^d$. We also note that $(Ju_\lambda, u_\lambda) = 0$, since J is skew adjoint, and that by (2.15)

$$\|Ju_\lambda\|_{-1} \leqq C. \tag{2.21}$$

Let $\phi(v)$ be any function of v such that $\|\phi\|_1 < \infty$. Then from the boundedness of D_jQ and (2.1) we conclude that $\Sigma_j v_j D_j u_j$ is bounded in $\mathscr{H}_{-1}$ and

$$\int \left(\sum_{i=1}^d v_i D_i u_\lambda \right) \phi(v) G(dv) = F_{\lambda, \phi}(\omega)$$

are bounded in $L^2(\Omega, \mathscr{F}, \tilde{P})$ as $\lambda \to 0$ for each ϕ. Note that, for any $H(\omega) \in L^2(\Omega, \mathscr{F}, P)$,

$$\left|\int F_{\lambda,\phi}(\omega) H(\omega) \tilde{P}(d\omega)\right| = \left|\int\left(\sum_i v_i D_i u_\lambda\right)\phi(v) H(\omega)\, d\tilde{\pi}\right|$$

$$\leq \left\|\sum_i v_i D_i u_\lambda\right\|_{-1} \|\phi H\|_1$$

$$= \left\|\sum_i v_i D_i u_\lambda\right\|_{-1} \|\phi(v)\|_1 \|H(\omega)\|_0.$$

Now choose $\phi = v_j$, $j = 1, \cdots, d$, and write

$$F_\lambda^j(\omega) = \sum_{i=1}^{d} D_i \int v_i v_j u_\lambda G(dv)$$

$$= \sum_{i=1}^{d} D_i \int (v_i v_j - \delta_{ij}) u_\lambda(\omega, v) G(dv) + D_j \int u_\lambda(\omega, v) G(dv) \tag{2.22}$$

$$= \sum_{i=1}^{d} D_i W_{ij}^\lambda(\omega) + D_j a_\lambda(\omega).$$

Since $v_i v_j - \delta_{ij}$ has mean zero,

$$W_{ij}^\lambda(\omega) = \int (v_i v_j - \delta_{ij}) u_\lambda(\omega, v) G(dv)$$

$$= \int (v_i v_j - \delta_{ij}) f_\lambda(\omega, v) G(dv)$$

and hence $W_{ij}^\lambda(\omega)$ are bounded in $L^2(\Omega, \mathscr{F}, \tilde{P})$ as $\lambda \to 0$ because of (2.18).

Therefore, for each $j = 1, 2, \cdots, d$,

$$D_j a_\lambda(\omega) = F_\lambda^j(\omega) + \sum_{i=1}^{d} D_i W_{ij}^\lambda(\omega) \tag{2.23}$$

with F_λ^j and W_{ij}^λ both bounded in $L^2(\Omega, \mathscr{F}, \tilde{P})$ as $\lambda \to 0$. Now P and $\tilde{P}$ are equivalent measures with bounded Radon Nikodym derivative so we may replace $\tilde{P}$ by P in our estimates. Equation (2.23) can be solved for a_λ in the following formal way which can be justified easily using the spectral resolution of the unitary group induced by the translations τ_x on $L^2(\Omega, \mathscr{F}, P)$. From (2.23) we

have

$$\sum_j D_j^2 a_\lambda = \sum_j D_j F_\lambda^j + \sum_{i,j=1}^{d} D_i D_j W_{ij}^\lambda. \tag{2.24}$$

Dividing by $\Sigma_j D_j^2$ we have

$$\begin{aligned} a_\lambda(\omega) &= \Big(\sum_j D_j^2\Big)^{-1} \sum_i D_i F_\lambda^i(\omega) + \Big(\sum_j D_j^2\Big)^{-1} \sum_{i,k} D_i D_k W_{ik}^\lambda(\omega) \\ &= c_\lambda(\omega) + b_\lambda(\omega). \end{aligned} \tag{2.25}$$

It is easily seen that $\|b_\lambda\|_0$ is controlled by $\|W_{ij}^\lambda\|$, $i, j = 1,\cdots,d$, and that $\|D_j c_\lambda\|_0$ is controlled by $\|F_\lambda^j\|_0$ as $\lambda \to 0$ since the operators $(\Sigma D_j^2)^{-1} D_i D_j$ are bounded in $L^2(\Omega, \mathcal{F}, P)$. This completes the proof of the lemma.

3. Convergence of Finite-Dimensional Distributions

In this section we prove that the finite-dimensional distributions of the scaled process $x^\varepsilon(t)$ defined by (1.8) converge to those of a nondegenerate Brownian motion on $\mathbb{R}^d$. The proof is based on the following theorem concerning the behavior as $\lambda \to 0$ of the solution u_λ of the resolvent equation (2.2).

THEOREM 3.1. *We have that*

$$\lim_{\lambda \downarrow 0} \lambda \|u_\lambda\|_0^2 = 0 \tag{3.1}$$

and

$$\lim_{\substack{\lambda \downarrow 0 \\ \mu \downarrow 0}} [u_\lambda, u_\mu] = \sum_{i,j=1}^{d} l_i l_j \gamma_{ij}, \tag{3.2}$$

where (γ_{ij}) *is a nondegenerate, positive, symmetric matrix and* l_i *are the components of the linear functional* $l(v) = \Sigma l_i v_i$ *on* $\mathbb{R}^d$.

Proof: From Lemma 2.2, (2.17), (2.18) and (2.19), we know that we can write u_λ as

$$u_\lambda(\omega, v) = g_\lambda(\omega, v) + c_\lambda(\omega), \tag{3.3}$$

where

$$\|g_\lambda\|_1 \leq C \quad \text{and} \quad \|D_i c_\lambda\|_0 \leq C, \qquad i = 1,\cdots,d. \tag{3.4}$$

Taking weakly convergent subsequences we can assume that along the subsequence

$$u_{\lambda_n}(\omega, v) = u^{(n)}(\omega, v) = g^{(n)}(\omega, v) + c^n(\omega) \tag{3.5}$$

with

$$\lim_{n \to \infty} g^{(n)} = g \quad \text{weakly in} \quad \mathscr{H}_1 \tag{3.6}$$

and

$$\lim_{n \to \infty} D_i c^{(n)} = c_i \quad \text{weakly in} \quad L^2(\Omega, \mathscr{F}, \tilde{P}). \tag{3.7}$$

In addition, from the equation (2.2) we know that Ju^n converges weakly in $\mathscr{H}_{-1}$ and

$$\begin{aligned} \lim_{n \to \infty} Ju^{(n)} &= \lim_{n \to \infty} \left[Jg^{(n)} + Jc^{(n)} \right] \\ &= Jg + \sum_i v_i c_i(\omega). \end{aligned} \tag{3.8}$$

Multiplying (2.2) for $\lambda = \lambda_n$ with $u^{(m)}$ and integrating, we get

$$\lambda_n(u^{(n)}, u^{(m)}) + [u^{(n)}, u^{(m)}] - (Ju^{(n)}, u^{(m)}) = (l, u^{(m)}). \tag{3.9}$$

We note that

$$\lim_{n \to \infty} \lambda_n u^{(n)} = 0 \quad \text{in} \quad L^2(\Omega \times \mathbb{R}, \tilde{\pi}) \tag{3.10}$$

and

$$\begin{aligned} \lim_{n \to \infty} [u^{(n)}, u^{(m)}] &= \lim_{n \to \infty} [g^{(n)}, u^{(m)}] \\ &= [g, u^{(m)}]. \end{aligned}$$

Letting $m \to \infty$ we then get

$$\lim_{m \to \infty} \lim_{n \to \infty} [u^{(n)}, u^{(m)}] = [g, g]. \tag{3.11}$$

Consider

$$\begin{aligned} (Ju^{(n)}, u^{(m)}) &= \left(Jg^{(n)} + Jc^{(n)}, g^{(m)} + c^{(m)} \right) \\ &= \left(Jg^{(n)}, g^{(m)} \right) + \left(Jg^{(n)}, c^{(m)} \right) \\ &\quad + \left(Jc^{(n)}, g^{(m)} \right) + \left(Jc^{(n)}, c^{(m)} \right) \\ &= \left(Jg^{(n)}, g^{(m)} \right) + \left(Jc^{(n)}, g^{(m)} \right) - \left(g^{(n)}, Jc^{(m)} \right). \end{aligned} \tag{3.12}$$

Now

$$(3.13) \qquad Jc^{(n)} = \sum_{i=1}^{d} v_i D_i c^{(n)} \to \sum_{i=1}^{d} v_i c_i(\omega)$$

weakly in $L^2(\Omega \times \mathbb{R}^d, \tilde{\pi})$. Therefore, from (3.12)

$$\begin{aligned} \lim_{n\to\infty} (Ju^{(n)}, u^{(m)}) &= (Jg, g^{(m)}) - \left(g, \textstyle\sum v_i c_i\right) + \left(\textstyle\sum v_i c_i, g\right) \\ &= (Jg, g^{(m)}) \end{aligned}$$

and hence

$$(3.14) \qquad \lim_{m\to\infty} \lim_{n\to\infty} (Ju^{(n)}, u^{(m)}) = (Jg, g) = 0,$$

since J is skew symmetric and $g \in \mathscr{H}_1$.

Letting $n \to \infty$ and then $m \to \infty$ in (3.9) and using (3.10), (3.11) and (3.14) gives

$$(3.15) \qquad \begin{aligned} [g, g] &= \lim_{m\to\infty} (l(v), u^{(m)}) \\ &= \lim_{m\to\infty} (l(v), g^{(m)}) \\ &= (l(v), g). \end{aligned}$$

Setting $n = m$ in (3.9) and letting $n \to \infty$ we get

$$(3.16) \qquad \lim_{n\to\infty} \left\{\lambda_n \|u^{(n)}\|_0^2 + [u^{(n)}, u^{(n)}]\right\} = (l(v), g).$$

By lower semicontinuity,

$$\varliminf_{n\to\infty} \left\{\lambda_n \|u^{(n)}\|_0^2 + [u^{(n)}, u^{(n)}]\right\} \geqq [g, g] + \varliminf \lambda_n \|u^{(n)}\|_0^2.$$

Thus,

$$\varlimsup \lambda_n \|u^{(n)}\|_0^2 + [g, g] \leqq (l(v), g)$$

which in view of (3.15) implies our first conclusion (3.1). Moreover, (3.1), (3.15) and (3.16) imply that

$$\lim_{n\to\infty} [g^{(n)}, g^{(n)}] = [g, g]$$

and hence $g^{(n)}$ converges strongly to g in $\mathscr{H}_1$, not only weakly as we had in (3.6). This demonstrates (3.2) for subsequences.

We must show now that the limits are unique so that arbitrary sequences converge. Suppose two subsequences converge to the limits g and g' in $\mathscr{H}_1$ strongly. Then one has

$$[g, g'] - (Jg, g') = (l(v), g'),$$

$$[g', g] - (Jg', g) = (l(v), g),$$

$$[g, g] = (l(v), g),$$

$$[g', g'] = (l(v), g').$$

These relations imply that

$$[g - g', g - g'] = 0$$

and this establishes uniqueness.

The nondegeneracy of the matrix (γ_{ij}) is obtained by the following considerations. We multiply (2.2) by $l(v)$ and integrate with respect to the measure $\pi(d\omega, dv) = P(d\omega)G(dv)$. If we denote inner products with respect to this measure with a prime and take the limit as $\lambda \to 0$ we get

$$[l(v), g]' - \left(\sum_i q_i \frac{\partial g}{\partial v_i}, l(v)\right)' = (l(v), l(v))' = \|l\|^2.$$

But

$$\left(\sum q_i(\omega) \frac{\partial g}{\partial v_i}, l(v)\right)' \leqq C_1 \|l\| ([g, g]')^{1/2}$$

$$\leqq C_2 \|l\| [g, g]^{1/2}$$

and

$$[l(v), g]' \leqq [l(v), l(v)]^{1/2} ([g, g]')^{1/2}$$

$$\leqq C_2 \|l\| [g, g]^{1/2},$$

where C_1, C_2 and C_3 are constants. Therefore,

$$\|l\|^2 \leqq C_4 \|l\| [g, g]^{1/2}$$

and hence

$$[g, g] \geqq \frac{1}{C_4} \|l\|^2,$$

where C_4 is another constant. The proof of Theorem 3.1 is complete.

With the help of Theorem 3.1 the proof of convergence of the finite-dimensional distributions of $x^\varepsilon(\cdot)$ follows (see [1]). First we note that

$$l(x^\varepsilon(t)) = \frac{1}{\varepsilon}\int_0^t l(v^\varepsilon(s))\,ds, \qquad v^\varepsilon(t) = v(t/\varepsilon^2).$$

If we let $\omega^\varepsilon(t) = \omega(t/\varepsilon^2)$ and take into account the resolvent equation (2.2) we see that

$$l(x^\varepsilon(t)) - M^\varepsilon_{l,\lambda}(t)$$
$$= \varepsilon u_\lambda(\omega, v) - \varepsilon u_\lambda(\omega^\varepsilon(t), v^\varepsilon(t)) + \frac{\lambda}{\varepsilon}\int_0^t u_\lambda(\omega^\varepsilon(s), v^\varepsilon(s))\,ds,$$

where $M^\varepsilon_{l,\lambda}(t)$ is a zero mean, square integrable martingale. Choosing $\lambda = \varepsilon^2$ and writing $M^\varepsilon_l(t) = M^\varepsilon_{l,\varepsilon^2}(t)$, we see that

$$E_{\tilde{\pi}}\left\{\left(l(x^\varepsilon(t)) - M^\varepsilon_l(t)\right)^2\right\} \to 0 \quad \text{as} \quad \varepsilon \to 0, \tag{3.17}$$

by Theorem 3.1, estimate (3.1). The expectation in (3.17) is with respect to the stationary Markov process $(\omega^\varepsilon(t), v^\varepsilon(t))$ on $\Omega \times \mathbb{R}^d$ generated by $\varepsilon^{-2}L$, where L is defined by (1.7). This means that $\omega^\varepsilon(0) = \omega$, $v^\varepsilon(0) = v$ are distributed by $\tilde{\pi}(d\omega, dv) = \tilde{P}(d\omega) \times G(dv)$ as defined in Section 1.

It remains then in view of (3.17) to show that the square integrable martingales $M^\varepsilon_l(t)$ converge weakly to Brownian motion. This is the martingale central limit theorem (see [3]) for which it is enough to check that the quadratic variations converge in mean square. But

$$E_{\tilde{\pi}}|M^\varepsilon_\lambda(t)|^2 = t[u_{\varepsilon^2}, u_{\varepsilon^2}] \to t[g, g]$$

as $\varepsilon \to 0$ by Theorem 3.1. So the martingale central limit theorem can be applied and we conclude that the finite-dimensional distributions of $x^\varepsilon(t)$ converge weakly as $\varepsilon \to 0$ to those of a Brownian motion with a nondegenerate covariance matrix (γ_{ij}) as described in Theorem 3.1.

4. Compactness for the Processes

In this section we prove that the family $x^\varepsilon(\cdot)$ of stochastic processes satisfies the condition

$$\lim_{h\downarrow 0}\overline{\lim_{\varepsilon\downarrow 0}} Q\left\{\sup_{\substack{0<s<t<T\\|s-t|\leq h}} |x^\varepsilon(s) - x^\varepsilon(t)| > \delta\right\} = 0 \tag{4.1}$$

for each $\delta > 0$ and $T < \infty$. Here Q is the law of the process $(x^\varepsilon(\cdot), v^\varepsilon(\cdot))$

generated by $\varepsilon^{-2}L$ with L defined in (1.7) and with $\tilde{\pi}$ as initial distribution. It is enough for (4.1) to obtain the estimate

$$\lim_{h\downarrow 0}\frac{1}{h}\overline{\lim_{\varepsilon\downarrow 0}}\, Q\left\{\sup_{0<s<h}|x^{\varepsilon}(s)-x^{\varepsilon}(0)|>\delta\right\}=0 \tag{4.2}$$

for each $\delta > 0$. To prove (4.2) it suffices to obtain for each component $x_i(s)$ an estimate of the form

$$\lim_{h\downarrow 0}\frac{1}{h}\overline{\lim_{\varepsilon\downarrow 0}}\, Q\left\{\sup_{0<s<h/\varepsilon^2}\{x_i(s)-x_i(0)\}\geqq\frac{\delta}{\varepsilon}\right\}=0 \tag{4.3}$$

for each $\delta > 0$.

Let τ_l be the time it takes for $x_i(t)$ to reach level l assuming $x_i(0) = 0$. Then (4.3) can be reduced to

$$\lim_{h\downarrow 0}\frac{1}{h}\overline{\lim_{\varepsilon\downarrow 0}}\, Q\left\{\tau_{\delta/\varepsilon}\leqq h/\varepsilon^2\right\}=0 \tag{4.4}$$

for each $\delta > 0$. From the Tchebyshev inequality,

$$Q\left\{\tau_{\delta/2}\leqq h/\varepsilon^2\right\}\leqq\exp\{\lambda h\}E\left\{\exp\left\{-\varepsilon^2\lambda\tau_{\delta/\varepsilon}\right\}\right\}.$$

Therefore, (4.4) can be deduced from

$$E\left\{\exp\left\{-\varepsilon^2\lambda\tau_{\delta/2}\right\}\right\}\leqq F_{\varepsilon}(\lambda,\delta), \tag{4.5}$$

where for each $\lambda > 0$

$$\overline{\lim_{\varepsilon\downarrow 0}}\, F_{\varepsilon}(\lambda,\delta)\leqq F(\lambda,\delta)$$

and

$$\inf_{\lambda>0} e^{\lambda h}F(\lambda,\delta)=o(h) \tag{4.6}$$

as $h \to 0$ for each $\delta > 0$.

Consider therefore the solution $u_\lambda(x, v, L, \omega)$ of

$$\lambda u_\lambda - Au_\lambda - \sum_{i=1}^{d}\tilde{q}_i(x,\omega)\frac{\partial u_\lambda}{\partial v_i}-\sum_{i=1}^{d}v_i\frac{\partial u_\lambda}{\partial x_i}=0$$

for $x_1 < L$ and

$$u_\lambda = 1 \quad\text{for}\quad x_1 = L,\ v_1 > 0.$$

We look at the solution on the plane $x_1 = 0$. If we write $x = (x_1, y)$ with $y = (x_2, x_3, \cdots, x_d)$, then if we take expectations with respect to π rather than $\tilde{\pi}$ we make errors which affect estimates by at most a constant. On the other hand, we gain translation invariance. Consider

$$\int u_\lambda((0, y), v, L, \omega)\pi(d\omega, dv) = \theta_\lambda(L)$$

which does not depend on y by stationarity. This function is monotone decreasing in L as can be seen by comparing stopping times. To get (4.6) we need an estimate for $\theta_{\varepsilon^2\lambda}(\delta/\varepsilon)$ as $\varepsilon \to 0$. For simplicity we shall assume that $d = 1$ and then explain at the end of the section the necessary changes to pass to $d > 1$. In $d = 1$ we suppress the $\omega \in \Omega$ and the $\tilde{}$ in q.

Let $u_\lambda(x, v)$ be the solution of

$$\text{(4.7)} \qquad \lambda u_\lambda - A u_\lambda - q(x)\frac{\partial u_\lambda}{\partial v} - v\frac{\partial u_\lambda}{\partial x} = 0, \qquad x < 0,$$

$$u_\lambda = 1 \quad \text{for} \quad x = 0, \quad v > 0, \quad q(x) = \frac{d}{dx}Q(x).$$

Define the following quantities:

$$\text{(4.8)} \qquad \begin{aligned} a_\lambda(x) &= e^{Q(x)}\int u_\lambda^2(x, v)G(dv), \\ b_\lambda(x) &= e^{Q(x)}\int\left(\frac{\partial u_\lambda}{\partial v}\right)^2(x, v)G(dv), \\ c_\lambda(x) &= e^{Q(x)}\int u_\lambda\frac{\partial u_\lambda}{\partial v}G(dv) \\ &= \tfrac{1}{2}e^{Q(x)}\int v u_\lambda^2(x, v)G(dv). \end{aligned}$$

Multiplying (4.7) by u_λ and integrating over v with respect to $G(dv)$, one obtains

$$\text{(4.9)} \qquad \lambda a_\lambda(x) + \tfrac{1}{2}b_\lambda(x) = \frac{d}{dx}c_\lambda(x).$$

From (4.9) we see that $c_\lambda(x)$ is nondecreasing in x and $\lim c_\lambda(x)$ as $x \to -\infty$ is zero. Since $|u_\lambda| \leqq 1$, (4.8) implies that

$$\text{(4.10)} \qquad |c_\lambda(x)| \leqq k\sqrt{b_\lambda(x)}$$

with $k = \sup_x e^{Q(x)}$. For $x = 0$ we get from (4.8) again $|c_\lambda(0)| \leqq \frac{1}{2}k$.

LEMMA 4.1. *For every $x < 0$ and $\lambda > 0$,*

$$c_\lambda(x) \leqq \frac{2k^2}{|x|}.$$

Proof: Relations (4.9) and (4.10) yield the inequality

$$\frac{dc_\lambda(x)}{dx} \geqq \tfrac{1}{2}b_\lambda(x) \geqq \frac{1}{2k^2}c_\lambda^2(x).$$

Integrating and using the estimate $|c_\lambda(0)| \leqq \frac{1}{2}k$ gives the lemma.

LEMMA 4.2. *We also have, for every $x < 0$ and $\lambda > 0$,*

$$c_\lambda(x) \leqq \frac{4k^2}{|x|}\exp\{-\sqrt{2\lambda}\,\tfrac{1}{2}|x|\}.$$

Proof: From (4.9) again we have

$$\begin{aligned}\frac{dc_\lambda(x)}{dx} &= \lambda a_\lambda(x) + \tfrac{1}{2}b_\lambda(x)\\ &\geqq \sqrt{2\lambda}\sqrt{a_\lambda(x)b_\lambda(x)}\\ &\geqq \sqrt{2\lambda}\,c_\lambda(x)\end{aligned}$$

by the Schwartz inequality. Integrating we get

$$c_\lambda(x) \leqq \exp\{-\sqrt{2\lambda}\,|\tfrac{1}{2}x|\}$$

and use now the estimate of Lemma 4.1 for $c_\lambda(\frac{1}{2}x)$.

Since averaging with respect to ω allows us to use stationarity, starting from zero and going to L is equivalent to starting from $-L$ and getting to level 0. Since the hitting times τ_L for successive levels are monotonically increasing in L, the averages $\bar{a}_\lambda(x)$ of $a_\lambda(x)$ with respect to the random $Q(\omega)$ will be monotone in x. We also have, from (4.8),

$$\theta_\lambda(x) \leqq \sqrt{a_\lambda(x)}\,.$$

We can now complete the proof of the necessary estimate for (4.6).

LEMMA 4.3. *For each $\delta > 0$ and $\lambda > 0$,*

$$\overline{\lim_{\varepsilon\downarrow 0}}\,\bar{a}_{\varepsilon^2\lambda}(\delta/\varepsilon) \leqq \frac{c}{\lambda\varepsilon^2}\exp\{-\tfrac{1}{4}\delta\sqrt{2\lambda}\}.$$

Proof: From equation (4.9) and the estimate of Lemma 4.2 we obtain

$$\lambda \int_{-\infty}^{y} a_\lambda(x)\,dx \leqq c_\lambda(y) \leqq \frac{4k^2}{|y|} \exp\{-\sqrt{2\lambda}\,\tfrac{1}{2}|y|\}.$$

Now we can average with respect to ω in Q and this estimate will still hold for the average $\bar{a}_\lambda(x)$ which is monotone in x. By the monotonicity,

$$\int_{-\infty}^{y} \bar{a}_\lambda(x)\,dx \geqq \int_{2y}^{y} \bar{a}_\lambda(x)\,dx$$

$$\geqq |y|\bar{a}_\lambda(2y), \qquad y < 0.$$

Therefore,

$$\lambda \bar{a}_\lambda(2y) \leqq \frac{4k^2}{\lambda |y|^2} \exp\{-\sqrt{2\lambda}\,\tfrac{1}{2}|y|\}.$$

Replace λ by $\varepsilon^2\lambda$ and y by $\delta/2\varepsilon$. Then

$$\bar{a}_{\varepsilon^2\lambda}(\delta/\varepsilon) \leqq \frac{16k^2}{\lambda\delta^2} \exp\{-\tfrac{1}{4}\delta\sqrt{2\lambda}\}$$

which completes the proof of the lemma.

To see how Lemma 4.3 gives the estimate (4.6) we note that

$$E\left\{\exp\{-\varepsilon^2\lambda\tau_{\delta/\varepsilon}\}\right\} = \theta_{\varepsilon^2\lambda}(\delta/\varepsilon) \leqq \sqrt{\bar{a}_{\varepsilon^2\lambda}(\delta/\varepsilon)}$$

$$\leqq \frac{4k}{\sqrt{\lambda}\,\delta} \exp\{-\tfrac{1}{8}\delta\sqrt{2\lambda}\}$$

which is an estimate independent of ε. To verify that

$$\inf_{\lambda>0} e^{\lambda h} \frac{4k}{\sqrt{\lambda}\,\delta} \exp\{-\tfrac{1}{8}\delta\sqrt{2\lambda}\} = o(\delta)$$

as $\delta \to 0$ we choose $\lambda = \rho^2/h^2$ for a suitable ρ. In fact we get exponential decay in $1/\delta$ provided $\rho < \delta\sqrt{2}/8$. We have therefore proven

THEOREM 4.4. *The family $x^\varepsilon(\cdot)$ of stochastic processes is uniformly tight and therefore we have weak convergence to Brownian motion in the space of continuous functions in $\mathbb{R}^d$.*

It is necessary to explain the main steps in modifying the proof for $d > 1$. Equation (4.9) takes the form

$$\lambda a_\lambda(x) + \tfrac{1}{2} b_\lambda(x) = \nabla \cdot c_\lambda(x),$$

where $c_\lambda(x)$ is now vector-valued and $x = (x_1, y)$. We average over large boxes in the y coordinates. Since all the calculations are done with respect to P which is translation invariant and ergodic, the ergodic theorem can be used. If we denote averages by an overbar the equation becomes

$$\lambda \bar{a}_\lambda(x) + \tfrac{1}{2}\bar{b}_\lambda(x) = \nabla \cdot \bar{c}_\lambda(x)$$

with $\bar{a}_\lambda$, $\bar{b}_\lambda$ and $\bar{c}_\lambda$ depending only on x_1 now. Therefore, we actually have

$$\lambda \bar{a}_\lambda(x_1) + \tfrac{1}{2}\bar{b}_\lambda(x_1) = \frac{d}{dx_1}\bar{c}_\lambda(x_1)$$

which is the situation we started with in (4.9) above.

Acknowledgment. The work of the first author was supported by NSF Grant DMS-83-12229 and Grant AFOSR-85-0017. That of the second author was supported by NSF Grant MCS-8301364, ONR Contract N00014-81-K-0012 and a John S. Guggenheim Memorial Foundation Fellowship.

Bibliography

[1] Kipnis, C., and Varadhan, S. R. S., *Central limit theorem for additive functionals of reversible Markov processes and application to simple exclusions*, to appear in Comm. in Math. Physics.

[2] Rodenhausen, H., *A central limit theorem for Brownian motion in an external periodic force field*, preprint, Heidelberg, 1985.

[3] Hall, P., and Heyde, C. C., *Martingale Limit Theorems and Applications*, Academic Press, New York, 1980.

Received March, 1985.

Comm. Pure Appl. Math. 56 (2003) 1222-1245

Large Deviations for Random Walks in a Random Environment

S. R. S. VARADHAN
Courant Institute

To the memory of Jürgen K. Moser

Abstract

We prove two large deviation principles for random walks in a random environment: one for quenched walks in general ergodic environments and the other for averaged walks in a product environment.

1 Introduction

We will be concerned with establishing a large deviation principle for random walks in a random environment. The walk will take place in $\mathbb{Z}^d$, $d \geq 1$. We denote, by $\mathcal{P}$, the space of all probability distributions $p(z) : z \in \mathbb{Z}^d$, with $p(z) \geq 0$ and $\sum_z p(z) = 1$. The environment will be specified by a probability space $(\Omega, \Sigma, \mathbf{P})$ on which $\mathbb{Z}^d$ acts ergodically as measure-preserving transformations and a map $p(\omega, z)$ of $\Omega \to \mathcal{P}$. The random environment will consist, for each ω, of transition probabilities $\pi(\omega, x, y) = p(T_x\omega, y - x)$ that define a Markov chain on $\mathbb{Z}^d$. One can always think of Ω as the canonical product space of maps $x \to p_x(\cdot)$ of $\mathbb{Z}^d \to \mathcal{P}$. A special random environment is when $\mathbf{P}$ is the product measure determined by its one-dimensional marginal β on $\mathcal{P}$. The support of β will be denoted by $\mathcal{K}$ and the convex hull of $\mathcal{K}$ by $\hat{\mathcal{K}}$. For each $\omega \in \Omega$ we have a Markov chain with this transition probability starting at time 0 from zero, and this will be denoted by P^ω. This will be referred to as the "quenched" walk. We define the annealed or averaged walk P_0 by

$$P_0 = \int_\Omega P^\omega \mathbf{P}(d\omega) .$$

We will be interested in two types of large deviations. The quenched large deviation principle is a large deviation principle for the distribution

$$P^\omega\left[\frac{x_n}{n} \in A\right]$$

on $\mathbb{R}^d$. Here $x_n = z_1 + z_2 + \cdots + z_n$ is the position of the Markov chain after n steps, and $\mathbf{z} = (z_1, z_2, \ldots, z_n, \ldots)$ are the successive steps of the Markov chain.

Although P^ω is random, under suitable hypotheses there exists a large deviation principle with a rate function $h(\xi)$ that is nonrandom. We expect similarly

$$P_0\left[\frac{x_n}{n} \in A\right] = \int_\Omega P^\omega\left[\frac{x_n}{n} \in A\right]\mathbf{P}(d\omega)$$

to have a large deviation principle with a rate function $H(\xi)$. By Jensen's inequality we expect $h(\xi) \geq H(\xi)$. There is a statistical translation invariance in the problem, and that can be used through the subadditive ergodic theorem to establish the large deviation principle for P^ω directly by a soft argument. We need only that the environment $\mathbf{P}$ be translation invariant and ergodic and satisfy some mild regularity conditions. Unfortunately, this says very little about $h(\xi)$ except that it exists and is convex. The existence of a large deviation principle for P_0 is slightly harder because of the lack of the Markov property. In fact, P_0 has a long-term memory that has to be tackled. However, we end up with a "formula" for the rate function $H(\xi)$.

The problem has been considered by others recently, in particular by Zerner [8] and Comets, Gantert, and Zeitouni [1]. See Zeitouni [7] or Sznitman [5] for a survey. Our proof, while using some of the ideas to be found in their work, is quite different.

2 The "Quenched" Large Deviations

Let a stationary and ergodic random environment $\{\pi(\omega, x, y)\}$ be given. In other words, we have a probability space $(\Omega, \Sigma, \mathbf{P})$ on which $\{T_z : z \in \mathbb{Z}^d\}$ acts ergodically as measure-preserving transformations and $\pi(\omega, x, y) = p(T_x\omega, y-x)$.

Hypothesis A. $p(\omega, \pm\mathbf{e}_i) \geq \rho$ for some nonrandom $\rho > 0$, where $\{\pm\mathbf{e}_i\}$ are the nearest neighbors of 0 along the coordinate axes. There is a nonrandom constant C such that $p(\omega, x, y) = 0$ if $|x - y| > C$.

Let $\{\pi^{(n)}(\omega, x, y)\}$ be the n-step transition probability derived from $\pi(\omega, x, y)$. Let $c > 0$ be given. Define for each ω, t, x, and y,

$$q_c(\omega, t, x, y) = \sup_{n\geq 0}[\pi^{(n)}(\omega, x, y)e^{-c|n-t|}] .$$

For each c and t, $q_c(\omega, t, x, y) = q_c(T_x\omega, t, 0, y - x)$ is a stationary process with respect to translations $\{T_z : z \in \mathbb{Z}^d\}$. From the lower bound on $\pi(\omega, x, x \pm \mathbf{e}_i)$, it follows that $q_c(\omega, t, x, y) > 0$ and

$$|\log q_c(\omega, t, x, y) - \log q_c(\omega, s, x, y)| \leq c|t - s| .$$

Since we can go from $x + z_1$ to x in $|z_1|$ steps and from y to $y + z_2$ in $|z_2|$ steps with probabilities at least $\rho^{|z_1|}$ and $\rho^{|z_2|}$, respectively, we obtain

$$\begin{aligned} q_c(\omega, t, x + z_1, y + z_2) &= \sup_{n\geq 0} \left[\pi^{(n)}(\omega, x + z_1, y + z_2)e^{-c|n-t|}\right] \\ &= \sup_{n,n_1,n_2\geq 0} \left[\pi^{(n+n_1+n_2)}(\omega, x + z_1, y + z_2)e^{-c|n+n_1+n_2-t|}\right] \\ &\geq \rho^{|z_1|+|z_2|}e^{-c[|z_1|+|z_2|]} \sup_{n\geq 0} \left[\pi^{(n)}(\omega, x, y)e^{-c|n-t|}\right] \\ &= \rho^{|z_1|+|z_2|}e^{-c[|z_1|+|z_2|]}q_c(\omega, t, x, y)\,. \end{aligned}$$

Therefore, for each $c < \infty$, there is a constant c' such that the function

$$g_c(\omega, t, x, y) = -\log q_c(\omega, t, x, y)$$

satisfies

$$|g_c(\omega, t, x, y) - g_c(\omega, t + s, x + z_1, y + z_2)| \leq c'[s + |z_1| + |z_2|]\,.$$

In addition,

$$\begin{aligned} q_c(\omega, t + s, 0, x + y) &= \sup_{n\geq 0}[\pi^{(n)}(\omega, 0, x + y)e^{-c|n-(t+s)|}] \\ &= \sup_{n,m\geq 0} [\pi^{(n+m)}(\omega, 0, x + y)e^{-c|(n+m)-(t+s)|}] \\ &= \sup_{n,m\geq 0} \left[e^{-c|(n+m)-(t+s)|} \sum_z \pi^{(n)}(\omega, 0, z)\pi^{(m)}(\omega, z, x + y)\right] \\ &\geq \sup_{n,m\geq 0} [e^{-c|n-t|-c|m-s|}\pi^{(n)}(\omega, 0, x)\pi^{(m)}(\omega, x, x + y)] \\ &= q_c(\omega, t, 0, x)\, q_c(\omega, s, x, x + y)\,, \end{aligned}$$

i.e.,

$$g_c(\omega, t + s, 0, x + y) \leq g_c(\omega, t, 0, x) + g_c(\omega, s, x, x + y)\,,$$

and g_c is stationary and ergodic relative to $x \to x + z$, $y \to y + z$.

We can use the subadditive ergodic theorem to prove the following theorem.

THEOREM 2.1 *Let $(\Omega, \mathcal{F}, P)$ be a probability space on which $\{T_z : z \in \mathbb{Z}^d\}$ acts ergodically as measure-preserving transformations. Let a nonnegative function $V(\omega, t, z_1, z_2)$ on $\Omega \times [0, \infty) \times \mathbb{Z}^d \times \mathbb{Z}^d$ be given, with the properties*

- $|V(\omega, t, z_1, z_2) - V(\omega, t', z_1', z_2')| \leq C[|t - t'| + |z_1 - z_1'| + |z_2 - z_2'|]$ *for all $\omega, t, t', z_1, z_2, z_1'$, and z_2' with a uniform constant C,*
- $V(\omega, t+s, z_1, z_3) \leq V(\omega, t, z_1, z_2)+V(\omega, s, z_2, z_3)$ *for all $\omega, z_1, z_2, z_3 \in \mathbb{Z}^d$ and $t, s \geq 0$, and*

- $V(\omega, t, z+z_1, z+z_2) = V(T_z\omega, t, z_1, z_2)$ *for all* ω, t, z, z_1, *and* z_2.

Then there exists a deterministic convex function $v(\xi)$ *on* $\mathbb{R}^d$ *such that for a.a.* ω

$$\lim_{\substack{n\to\infty \\ y_n/n\to\eta}} \frac{1}{n} V(\omega, nt, 0, y_n) = tv\left(\frac{\eta}{t}\right). \tag{2.1}$$

Remark 2.2. Subadditive ergodic theorems have been used before to prove similar results. Some smoothing out of the subadditive functional is needed for it to be effective; see Rezakhanlou and Tarver [4] for an example. In Zerner [8] the subadditivity is used for passage times, which gives slightly weaker results. Our improvement here comes from applying the subadditivity argument directly to the transition probability.

PROOF:

$$V^{(n)}(t, \xi, \eta) = \frac{1}{n} V(\omega, nt, n\xi, n\eta)$$

are uniformly Lipschitz in t, ξ, and η and are defined on $[0,\infty)\times\frac{1}{n}\mathbb{Z}^d\times\frac{1}{n}\mathbb{Z}^d$. They have subsequences that converge in distribution to an almost surely continuous stochastic process $V^\infty(t,\xi,\eta)$ defined on $[0,\infty)\times\mathbb{R}^d\times\mathbb{R}^d$ that is stationary under the transformations $\{T_\zeta : \zeta\in\mathbb{R}^d\}$, $\xi\to\xi+\zeta$, and $\eta\to\eta+\zeta$ and satisfies the subadditivity property

$$V^\infty(t+s,\xi,\zeta) \le V^\infty(t,\xi,\eta) + V^\infty(s,\eta,\zeta)$$

with probability 1.

If $\xi\in\mathbb{Z}^d$, then

$$\lim_{n\to\infty} V^{(n)}(t, 0, \xi) = \hat{v}(t,\xi) \tag{2.2}$$

exists almost surely and is a deterministic function of t and ξ by the subadditive ergodic theorem (see [2] for a proof) and $\hat{v}(kt, k\xi) = k\hat{v}(t,\xi)$ for all $k\in\mathbb{Z}^+$, $t\in\mathbb{R}$, and $\xi\in\mathbb{Z}^d$. One can therefore extend $\hat{v}(t,\xi)$ for ξ with rational coordinates by the relation $\hat{v}(t,\xi) = \frac{1}{k}\hat{v}(kt,k\xi)$, and $\hat{v}$ will satisfy $\hat{v}(rt, r\xi) = r\hat{v}(t,\xi)$ for all positive rational r and $\xi\in\mathbb{R}^d$ with rational coordinates. Since $\hat{v}$ is Lipschitz in t and ξ, it has a natural extension as a function of the form $tv(\xi/t)$.

The next step is to deduce (2.1) from (2.2). Let $\xi\in\mathbb{R}^d$ have rational coordinates. Then $q\xi\in\mathbb{Z}^d$ for some integer q. Clearly

$$\lim_{n\to\infty}\frac{1}{n} V(\omega, nqt, 0, nq\xi) = \hat{v}(qt, q\xi) = q\hat{v}(t,\xi).$$

By writing $n = n'q + r$ with a remainder r, $0\le r\le q-1$, and using the Lipschitz property, we can control the error and

$$\frac{1}{n}|V(\omega, nt, 0, y_n) - V(\omega, n'qt, 0, n'q\,\xi)| \le \frac{C}{n}[|n-n'q|t + |y_n - n'q\,\xi|] \to 0.$$

Therefore

$$\frac{1}{n}V(\omega, nt, 0, y_n) \to \hat{v}(t,\xi) = tv\left(\frac{\xi}{t}\right).$$

It is routine now to approximate arbitrary $\xi \in \mathbb{R}^d$ by rational ξ and conclude that (2.1) holds for all $\xi \in \mathbb{R}^d$. In particular, $V^\infty(t, 0, \xi)$ is almost surely equal to $tv(\xi/t)$. From stationarity

$$V^\infty(t,\xi,\eta) = tv\left(\frac{\eta-\xi}{t}\right)$$

almost surely. The subadditivity of V^∞ tells us that

$$(t+s)v\left(\frac{\eta}{t+s}\right) \le tv\left(\frac{\xi}{t}\right) + sv\left(\frac{\eta-\xi}{s}\right),$$

which is exactly the convexity of v. □

In order to prove the large deviation principle we apply Theorem 2.1 to g_c to get a limit h_c and then let $c \to \infty$. The functions h_c are nondecreasing, and the limit h, which is also convex, is easily seen to be the rate function.

THEOREM 2.3 *There exists a nonrandom convex rate function $h(\xi)$ such that, for almost all ω with respect to P,*

$$\limsup_{n\to\infty} \frac{1}{n}\log P^\omega\left[\frac{x_n}{n} \in C\right] \le -\inf_{\xi\in C} h(\xi) \quad \textit{for closed } C \subset \mathbb{R}^d,$$

$$\liminf_{n\to\infty} \frac{1}{n}\log P^\omega\left[\frac{x_n}{n} \in G\right] \ge -\inf_{\xi\in G} h(\xi) \quad \textit{for open } G \subset \mathbb{R}^d.$$

PROOF: For every c and any $y \in \mathbb{Z}^d$

$$\pi^{(n)}(\omega, 0, y) \le q_c(\omega, n, 0, y).$$

Since x_n must be one of at most An^d lattice points,

$$\begin{aligned}\limsup_{n\to\infty} \frac{1}{n}\log P^\omega\left[\frac{x_n}{n} \in C\right] &= \limsup_{n\to\infty} \frac{1}{n}\log \sum_{y\in nC} \pi^{(n)}(\omega,0,y)\\ &\le \limsup_{n\to\infty} \sup_{y\in nC} \frac{1}{n}\log \pi^{(n)}(\omega,0,y)\\ &\le \limsup_{n\to\infty} \sup_{y\in nC} \frac{1}{n}\log q_c(\omega,n,0,y)\\ &= -\inf_{\xi\in C} h_c(\xi).\end{aligned}$$

The above upper bound holds with any c, and so we can let $c \to \infty$. For the lower bound, if $h(\xi) = \infty$ there is nothing to prove. Assume $h(\xi) = \ell < \infty$. Let $\epsilon > 0$ be given. Choose c large enough such that $(\ell+1)C < c\epsilon$, where C is an upper

bound for the size of the jumps. Clearly $h_c(\xi) \le \ell$. Then, if $x_n = n\xi + o(n)$, for almost all ω,

$$\sup_{k\ge 0}\left[\pi^{(k)}(\omega, 0, x_n)e^{-c|k-n|}\right] \ge e^{-nh(\xi)+o(n)} \ge e^{-n\ell+o(n)}.$$

For sufficiently large n, the supremum is clearly attained in the set $\{k : |k-n| \le (\ell+1)n/c\}$; hence

$$\sup_{k:c|k-n|\le(\ell+1)n} \pi^{(k)}(\omega, 0, x_n) \ge e^{-n\ell+o(n)}.$$

Since the chain moves with a maximum speed of C, if it is at x_n at some time within $\frac{\ell+1}{c}n$ of n, then at time n, it is within $\frac{\ell+1}{c}nC$ or ϵn of x_n, and we are done. □

While we cannot calculate $h(\cdot)$, we will be able to say later on something about where $h(\cdot)$ is 0.

3 Assumptions and Statement of Results for the "Averaged" Case

We will now state our assumptions on β, the probability distribution on $\mathcal{P}$. The environment is the product measure $\mathbf{P}$ with marginal β.

Hypothesis B. There are constants $\rho > 0$, $c \ge 1$, $C < \infty$, and a deterministic function $p_0(z) \ge 0$ with $p_0(\pm\mathbf{e}_j) \ge \rho > 0$ and $p_0(z) = 0$ for $|z| > C$ such that the distribution β is supported on $p(\cdot)$ with the following property:

$$cp_0(z) \ge p(z) \ge p_0(z) \quad \text{for all } z \in \mathbb{Z}^d.$$

Remark 3.1. These conditions are rather strong. We are demanding that all transition probabilities, although they can be random, have to be either zero or uniformly positive. We are imposing a uniform bound on the size of the jumps and insisting that the probabilities of nearest-neighbor jumps be nonzero and thus have a uniform lower bound. These conditions help considerably in simplifying the proof. One expects that with the use of standard tools one can replace them with assumptions on the behavior of the corresponding tail distributions.

Before we can state the large deviation principle for the "averaged" walk P_0, we have to describe P_0 in a special manner. Let us start with a walk $\{0, x_1, x_2, \dots, x_n\}$ of n steps. We view the walk relative to the current position of the walker; i.e., we represent the walk by w_n, which is the walk

$$w_n = (-x_n, x_1 - x_n, \dots, x_{n-1} - x_n, 0).$$

This walk might have visited the site x, n_x times and made jumps of size z to the site $x+z$, $n_{x,z}$ times. The last visit to 0 is excluded so that $\sum_z n_{x,z} = n_x$ and $\sum_x n_x = n$. It is not difficult to see that

$$P_0[w_n] = \Pi_x \int_{\mathcal{P}} \Pi_z [p(z)]^{n_{x,z}} \beta(dp). \tag{3.1}$$

We only have to write down the probability under P^ω and integrate with respect to $\mathbf{P}$, noting that $\{\pi(\omega, x, x+\cdot)\} = \{p_x(\cdot)\}$ are independent for distinct values of x. Given w_n and a new jump of size z, a new walk w_{n+1} of length $n+1$ is generated by translating the walk w_n by $-z$ so that it now ends at $-z$ instead of 0, and then making the new jump of size z in order to end up once again at 0. We can write this as a transformation $w_{n+1} = T_z w_n$. To undo this transformation is easy. We need to erase the last jump, already determined by the walk, which we will denote by z_0, and shift the whole path by z_0 to end up again at 0. We can call this deterministic map T^*. It is easy to see that $T^* T_z w = w$ for all w and z.

One can calculate the conditional probability

$$\text{(3.2)}\quad \text{Prob}[w_{n+1} = T_{z'} w_n \mid w_n] = \frac{\text{Prob}[w_{n+1}]}{\text{Prob}[w_n]} = \frac{\int_{\mathcal{P}} p(z')\Pi_z[p(z)]^{n_{0,z}}\beta(dp)}{\int_{\mathcal{P}} \Pi_z[p(z)]^{n_{0,z}}\beta(dp)},$$

which has the following natural Bayesian interpretation. We think of $\beta(dp)$ as a "prior" on multinomials $p(\cdot)$, and given n_0 past trials with $n_{0,z}$ occurrences of type z, the "posterior" has distribution

$$c\Pi_z[p(z)]^{n_{0,z}}\, d\beta$$

with normalization

$$c^{-1} = \int_{\mathcal{P}} \Pi_z[p(z)]^{n_{0,z}}\beta(dp).$$

Then $P_0[w_{n+1} = T_{z'} w_n \mid w_n]$ is just the expectation of $p(z')$ under the "posterior" $\beta(w_n, dp)$. Because the environment is a *product* measure, the conditional probability of a new jump of size z depends only on the number of prior visits to the current site and the types of jumps that have occurred in the past from that site. Notice that in our notation, z_0 is always the last jump and z_1 the new jump. After each jump, all the previous jumps are renamed by a shift and z_1 becomes the new z_0.

Let us denote by $\mathbf{W}_n$ the space of walks w_n of length n that end at 0. We have a Markov chain $\{w_n\}$, $w_n \in \mathbf{W}_n$, with transition probability $\mathbf{q}(w, T_{z'}w) = q(w, z')$ given by (3.2). We note that q does not explicitly depend on n. The information needed from w is the number of prior visits to 0 and the frequency count of the jumps of different sizes that occurred after those visits. We can define a "stationary" Markov chain on $\mathbf{W} = \bigcup_n \mathbf{W}_n$ with transition probability $\mathbf{q}(w, w')$, which is zero unless $w' = T_z w$ for some z in the support of $p_0(\cdot)$, in which case it is $q(w, z)$. The natural starting point for this Markov chain is from $\mathbf{W}_0$ consisting of a single point ϕ that has no prior history, starting with a clean slate from 0. But we can start from any $w \in \mathbf{W}_n \subset \mathbf{W}$, and we will have the corresponding measure Q^w for the chain in $\mathbf{W}$.

The measure Q^ϕ on $\mathbf{W}$ corresponds naturally to P_0 except that the origin is at the wrong end. Since the chain marches from $\mathbf{W}_n \to \mathbf{W}_{n+1}$, it is clearly transient and one natural way to compactify the state space is to add the set $\mathbf{W}_\infty$ of walks of infinite length $w = \{\{x_j\} : j \le 0\}$ with $x_0 = 0$. Unless the walk w is transient, i.e., $\lim_{n\to-\infty} |x_n| = \infty$, $q(w, z)$ is not well-defined. The set of transient walks $\mathbf{W}_\infty^{\text{tr}}$ is

a subset of $\mathbf{W}_\infty$, which is in fact dense in $\mathbf{W}_\infty$. There is a natural Markov chain with transition probability $\mathbf{q}(w, w')$ on $\mathbf{W}_\infty^{\text{tr}}$ with $\mathbf{q}(w, T_z w)$ still given by the same formula by $q(w, z)$. Both T^* and T_z continue to act the same way as before, on $\mathbf{W}_\infty$ as well as $\mathbf{W}_\infty^{\text{tr}}$.

We denote by $\mathcal{I}$ the set of probability measures on $\mathbf{W}_\infty^{\text{tr}}$ that are invariant under T^*. These correspond to transient processes with stationary increments. We get a shift-invariant measure on $\mathbf{z} = \{z_j\}$, initially for $j \le 0$, which is easily extended to $-\infty < j < \infty$ to be shift invariant in both directions. The ergodic measures are the extremals in $\mathcal{I}$, and these are denoted by $\mathcal{E}$. If $\mu \in \mathcal{I}$, the conditional probability $q_\mu(w, z)$ of the next jump being equal to z given the past history is well-defined. Standard ergodic theory even yields a version $\hat{q}$ of this conditional probability that is universal, i.e., independent of μ. But this is cheating because the ergodic or extremal μ's are supported on mutually disjoint sets and one defines $\hat{q}(w, z)$ to be equal to $q_\mu(w, z)$ on the support of μ. There is a natural large deviation rate functional in this context given by

$$I(\mu) = \int_{\mathbf{W}_\infty^{\text{tr}}} \left[\sum_z \hat{q}(w, z) \log \frac{\hat{q}(w, z)}{q(w, z)} \right] \mu(dw) .$$

Note that one consequence of the universality of $\hat{q}$ is the affine linearity of $I(\cdot)$ on $\mathcal{I}$. Of course, each $\mu \in \mathcal{I}$ being a process with stationary increments has a mean "drift" $m(\mu) = E^\mu[z_0]$. It is natural to define by contraction, especially for $\xi \neq 0$, the rate function

$$H(\xi) = \inf_{\substack{\mu \in \mathcal{E} \\ m(\mu) = \xi}} I(\mu) . \tag{3.3}$$

We can now state the main theorem regarding the "averaged" measure P_0.

THEOREM 3.2 *The function $H(\xi)$ defined above for $\xi \neq 0$ extends to $\mathbb{R}^d$ as a convex function and is the rate function for large deviations of x_n/n under "averaged" measure P_0:*

$$\limsup_{n\to\infty} \frac{1}{n} \log P_0\left[\frac{x_n}{n} \in C\right] \le -\inf_{\xi \in C} H(\xi) \quad \textit{for closed } C \subset \mathbb{R}^d,$$

$$\liminf_{n\to\infty} \frac{1}{n} \log P_0\left[\frac{x_n}{n} \in G\right] \ge -\inf_{\xi \in G} H(\xi) \quad \textit{for open } G \subset \mathbb{R}^d.$$

Remark 3.3. Although the result is not surprising and the lower bound is in fact an elementary consequence of the ergodic theorem, the proof of the upper bound is a bit technical. The lack of compactness as well as the lack of continuity of $q(w, z)$ in w as we move away from $\mathbf{W}_\infty^{\text{tr}}$ or even when restricted to it, are the consequences of real long-term memory in the model, and this has to be understood and handled properly in order to prove the upper bound.

4 Some Mixing Properties

In this section we will consider the mixing properties of our Markov chain on the space $\mathbf{W} \cup \mathbf{W}_\infty^{\text{tr}}$. The probabilities $\{q(w, z)\}$ of transition from w to $T_z w$ are given by formula (3.2). In fact, we only use the fact that $q(w, z)$ depends on local information, i.e., on $\{n_{x,z} : x \in F\}$ for some finite F. We assume bounded step size as well as ellipticity, i.e., for some fixed function $p_0(\cdot) \geq 0$ with $p_0(z) = 0$ for $|z| > C$ and $p_0(\pm\mathbf{e}_i) > 0$ (where $\pm\mathbf{e}_i$ are the nearest neighbors of 0) and a constant $c > 0$, we have

$$cp_0(z) \geq q(w, z) \geq p_0(z) .$$

We want to consider measures on $\mathbf{W} \cup \mathbf{W}_\infty^{\text{tr}}$ that are invariant for the Markov chain. Such measures are clearly supported on $\mathbf{W}_\infty^{\text{tr}}$ and are invariant under T^* as well. We will be interested in ergodic measures μ with mean $m(\mu) \neq 0$.

Given a starting point w, we have the measure Q^w corresponding to the Markov chain $\{w_j\}$ starting with this initial walk, i.e., $w_0 = w$. The measure in particular will generate a random walk $\{x_j : j \geq 1\}$ in the future with steps $\mathbf{z} = \{z_j\}$ where $z_j = z_0(w_j)$ and $x_k = z_1 + z_2 + \cdots + z_k$. Remember that $z_0(w)$ is the last jump of w, and z_j will be the last jump after j steps. Given a set $A \subset \mathbb{Z}^d$, let us denote by

$$H(n, A, \mathbf{z}) = \#\big\{k : 0 \leq k \leq n \text{ and } x_k \in A\big\}$$

the number of visits to A by $x_k = z_1 + z_2 + \cdots + z_k$ during $0 \leq k \leq n$. Clearly, by our ellipticity condition, if we define the range $S(w)$ of the past as

$$S(w) = \{x_j(w) : j \leq 0\} ,$$

then

$$\left| \log \frac{dQ^{w_1}}{dQ^{w_2}}(\mathbf{z}) \bigg|_{\mathcal{F}_n} \right| \leq CH(n, S(w_1) \cup S(w_2), \mathbf{z}) .$$

Here $\mathcal{F}_n$ is the σ-field generated by $z_1, z_2, \ldots, z_n$ and $\frac{d\alpha}{d\beta}|_{\mathcal{F}}$ is the Radon-Nikodym derivative $\frac{d\alpha}{d\beta}$ on the σ-field $\mathcal{F}$.

Let us fix a direction $s \in \mathbb{S}^{d-1}$ and define the events $E_s \subset F_s$,

$$E_s = \big\{\mathbf{z} : \lim_{n\to\infty} \langle s, x_n\rangle = \infty\big\} , \qquad F_s = \big\{\mathbf{z} : \limsup_{n\to\infty} \langle s, x_n\rangle = \infty\big\} ,$$

and their probabilities $U(w) = Q^w[E_s]$ and $V(w) = Q^w[F_s]$ with $U(w) \leq V(w)$. Let

$$\ell(w) = \sup_{y \in S(w)} \langle s, y\rangle .$$

LEMMA 4.1 *If for some $\bar{w}$ with $\ell(\bar{w}) < \infty$, we have $U(\bar{w}) > 0$, then for every w with $\ell(w) < \infty$ we have $U(w) = V(w) > 0$ and*

$$1 - U(w) = Q^w\big[\lim_{n\to\infty} \langle s, x_n\rangle = -\infty\big] .$$

PROOF: We will first prove an estimate of the form

$$U(w) \geq u(\ell(w))$$

for some function $u(\ell)$ that is nonincreasing in ℓ and positive for each $\ell > 0$. Since $U(\bar{w}) > 0$, for every ℓ there is a $k = k(\ell)$ such that

$$Q^{\bar{w}}\big[\langle s, x_n\rangle \geq \max\{\ell, \ell(\bar{w})\} + 1 \text{ for } n \geq k(\ell)\big] \geq \frac{1}{2}U(\bar{w}) .$$

Then for any w and $\mathbf{z}$ satisfying $\langle s, x_n\rangle \geq \max\{\ell(w), \ell(\bar{w})\}+1$ for all $n \geq k(\ell(w))$,

$$\sup_n H(n, S(w) \cup S(\bar{w}), \mathbf{z}) \leq k(\ell(w))$$

providing a lower bound

$$U(w) \geq \frac{1}{2}U(\bar{w})e^{-Ck(\ell(w))} .$$

Since

$$U(w_n) = Q^w[E_s \mid \mathcal{F}_n]$$

is a Q^w martingale, it must converge to 0 almost surely on E_s^c. This means that

$$\lim_{n\to\infty} \ell(w_n) = \infty \quad \text{a.e. } Q^w \text{ on } E_s^c .$$

But clearly $\ell(w_n) = 0$ infinitely often on F_s, proving that F_s and E_s^c are almost surely disjoint. In fact, because of ellipticity, $\limsup_{n\to\infty}\langle s, x_n\rangle$ can only be $\pm\infty$, proving the lemma. □

THEOREM 4.2 *Let* $\{\alpha_i : i = 1, 2\}$ *be two invariant ergodic measures for* $q(\,\cdot\,,\cdot\,)$ *such that* $m(\alpha_i) \neq 0$ *for* $i = 1, 2$. *Then* $m(\alpha_1) = cm(\alpha_2)$ *for some* $c < 0$.

PROOF: Assume otherwise. Then there is an $s \in \mathbb{S}^{d-1}$ such that $\langle s, m(\alpha_i)\rangle > 0$ for $i = 1$ and $i = 2$. From our assumptions

$$\sup_n H(n, S(w), \mathbf{z}) < \infty$$

a.e. $\alpha_i(dw)Q^w(d\mathbf{z})$ for both $i = 1$ and $i = 2$. This proves the mutual absolute continuity of $Q^{w_1}(d\mathbf{z})$ and $Q^{w_2}(d\mathbf{z})$ for almost all w_1 and w_2 with respect to $\alpha_1 \times \alpha_2$. This implies, by the ergodic theorem, that $\alpha_1 = \alpha_2$. □

Remark 4.3. A zero-one law of Zerner and Merkl [9] proves in $d = 2$ ($d = 1$ is trivial) that for any $w \in \mathbf{W}$, in particular for $w = \phi$, $U(w) = 0$ or 1. This shows that in Theorem 4.2 there is at most one invariant measure.

5 Compactification

The compactification of $\mathbf{W} = \bigcup_n \mathbf{W}_n$ should be done with some care. Because the jump sizes are all bounded by C, the space $\mathbf{W}_\infty$ is already compact if we only demand that $x_j(w)$, the position of the walk at time j, be continuous as maps into the discrete space $\mathbb{Z}^d$ for each $j \le 0$. The transition probabilities $q(w, z)$ are only defined on the transient piece $\mathbf{W}_\infty^{\text{tr}}$, and even when restricted to it they are not continuous since some of the visits to 0 from the remote past could disappear in the limit. The compactification $\overline{\mathbf{W}}$ will naturally be a ramification of $\mathbf{W}_\infty$. Here is the list of things we want of the compactification $\overline{\mathbf{W}}$.

(1) $\mathbf{W}$ is dense in $\overline{\mathbf{W}}$ and is unramified.
(2) The functions $\{x_j(w)\}$, defined on $\mathbf{W}_n$ for $n \ge -j$, have continuous extensions, as maps into $\mathbb{Z}^d$, to the closure of $\bigcup_{n \ge -j} \mathbf{W}_n$ in $\overline{\mathbf{W}}$.
(3) The number of visits and types of jumps $\{n_{x,z}\}$ are continuous as maps into the space of integers with ∞ added, i.e., into $\mathbb{Z}^+ = \{0, 1, 2, \dots, \infty\}$.
(4) The transition probabilities $\{q((T^*)^k w, z)\}$ and $\{q(T_{z_k} \cdots T_{z_2} T_{z_1} w, z)\}$ are continuous maps for every $k \ge 0$ and every choice of $z_1, z_2, \dots, z_k$ and z from $\mathbb{Z}^d$.
(5) The set $\mathbf{W}_\infty^{\text{tr}}$, while ramified, is embedded canonically as a subset of $\overline{\mathbf{W}}$ characterized by the property that for each $x \in \mathbb{Z}^d$,

$$n_x(w) = \sum_{j \le -1} \mathbf{1}_{\{x_j(w) = x\}} \,.$$

In general, only the inequality

$$n_x(w) \ge \sum_{j \le -1} \mathbf{1}_{\{x_j(w) = x\}}$$

is true. Of course, if for some x, equality holds, then for that x and any z,

$$n_{x,z}(w) = \sum_{j \le -1} \mathbf{1}_{\{x_j(w) = x, x_{j+1} = x+z\}} \,.$$

We continue to call it $\mathbf{W}_\infty^{\text{tr}}$.
(6) The maps T^* and $\{T_z\}$ are continuous maps of $\overline{\mathbf{W}}$ into itself. (With T^* we need to leave out $\mathbf{W}_0 = \{\phi\}$.)

The existence of such a compactification, unique if it is minimal, is a routine exercise in real analysis. If we want to compactify a separable metric space $\mathbf{X}$ in such a way that a given countable collection of functions $\{\phi_j(x)\}$ (assumed without loss of generality to satisfy $0 \le \phi_j(x) \le 1$) remain continuous, then we embed X into the countable product of the unit interval, $[0, 1]$, using $x \to \{\phi_j(x)\}$, and take the closure $\hat{Y}$ of the range Y for the compactification. It is clear that as we enlarge the collection, the compactification gets ramified.

The compactification is only for convenience and will not solve any real problems. All it does is sweep under the rug some of the natural difficulties of the

long-range dependence in our model. We have to prove later that we do not have to look too closely underneath the rug. This is what we do in the next section. We end this section with some discussion about invariant measures and a lemma.

For any $w \in \overline{\mathbf{W}}$ we denote its range $S(w)$ as the set

$$S(w) = \{x : n_x(w) > 0\} \cup \{0\}.$$

Although $S(w)$ contains the actual range $\{x_j(w) : j \leq 0\}$, it may in general be larger. However, for $w \in \mathbf{W}^{\mathrm{tr}}_\infty$ they are the same.

We denote by $\overline{\mathcal{I}}$ the convex set of T^* invariant probability measures on $\overline{\mathbf{W}}$ and by $\overline{\mathcal{E}}$ its extremals or the ergodic ones. For each $\mu \in \overline{\mathcal{I}}$, there is a natural process with stationary increments $z_j(w) = x_j(w) - x_{j-1}(w)$, and we can define its mean drift by

$$m(\mu) = E^\mu[z_0] = E^\mu[-x_{-1}].$$

LEMMA 5.1 *Suppose $\alpha \in \overline{\mathcal{E}}$ and $m(\alpha) \neq 0$. Let $s \in \mathbb{S}^{d-1}$ be such that $\langle s, m(\alpha)\rangle > 0$. Let $S(w)$ denote the support of w defined earlier. If*

$$\alpha\Big[w : \sup_{x\in S(w)} \langle s, x\rangle < \infty\Big] > 0,$$

then α is supported on $\mathbf{W}^{\mathrm{tr}}_\infty$.

PROOF: The proof is elementary. By ergodicity, the above probability, if it is not 0, is actually 1. Because $\langle s, m(\alpha)\rangle > 0$, the whole scene is drifting away in the opposite direction, and all the new visits are eventually to sites that are not in the current $S(w)$. That means eventually there are no excess visits other than the ones generated from the walk. Since the scene is invariant, this was always true. In other words, $\alpha[\mathbf{W}^{\mathrm{tr}}_\infty] = 1$. □

6 A Combinatorial Result

We start with n steps $z_1, z_2, \ldots, z_n$, and let $x_n = z_1 + z_2 + \cdots + z_n$. This generates a history of n walks, $\{w_j : 1 \leq j \leq n\}$ where $w_j = T_{z_j} \cdots T_{z_2} T_{z_1}\phi$ and ϕ is the starting point with no history. An empirical measure on $\mathbf{W}$ can be defined by

$$\mathcal{R}_n = \frac{1}{n}\Bigg[\sum_{j=1}^{n} \delta_{T_{z_j}\cdots T_{z_1}\phi}\Bigg].$$

It is viewed as a measure on the compactification $\overline{\mathbf{W}}$. We assume that $x_n/n \to \xi \in \mathbb{R}^d$. Let us denote by $\mathcal{C}$ the set of limit points of $\mathcal{R}_n$ in the space of probability measures $\mathcal{M}(\overline{\mathbf{W}})$ on the compactified space $\overline{\mathbf{W}}$. We recall that $\overline{\mathcal{I}}$ is the set of T^* invariant measures on $\overline{\mathbf{W}}$, and $\overline{\mathcal{E}}$ is the set of extremals in $\overline{\mathcal{I}}$. $\mathcal{E}$ was defined as the set of ergodic measures supported on $\mathbf{W}^{\mathrm{tr}}_\infty$, and we denote by $\mathcal{E}_0$ measures $\alpha \in \mathcal{E}$ with $m(\alpha) \neq 0$. Clearly, $\mathcal{E}_0 \subset \mathcal{E} \subset \overline{\mathcal{E}}$.

THEOREM 6.1 *Every $\mu \in \mathcal{C}$ is invariant under T^*. As such, it has an integral representation over the ergodic measures (extremals)*

$$\mu = \int_{\mathcal{E}} \alpha \hat{\mu}(d\alpha), \qquad \xi = \int_{\mathcal{E}} m(\alpha)\hat{\mu}(d\alpha).$$

We can write $\hat{\mu} = \hat{\mu}_1 + \hat{\mu}_2$ in such a way that $\hat{\mu}_2$ is supported on $\mathcal{E}_0$ and

$$\xi = \int_{\mathcal{E}_0} m(\alpha)\hat{\mu}_2(d\alpha), \qquad 0 = \int_{\mathcal{E}} m(\alpha)\hat{\mu}_1(d\alpha).$$

(It is not claimed that $\hat{\mu}_1(\mathcal{E}_0) = 0$).

PROOF: The theorem is essentially combinatorial in character. All the probabilities come from the empirical measures. The first part of the theorem is routine ergodic theory. Let us concentrate on the second part. If $\xi = 0$, there is nothing to prove, and we can take $\hat{\mu}_2 = 0$. Let us therefore assume that $\xi \neq 0$. Let us define B_μ as

$$B_\mu = \left\{ \int_{\mathcal{E}_0} \varphi(\alpha) m(\alpha)\hat{\mu}(d\alpha) : 0 \leq \varphi \leq 1 \right\}.$$

$B_\mu \subset \mathbb{R}^d$ is obviously a compact convex set in $\mathbb{R}^d$ containing the origin. The problem is to show that $\xi \in B_\mu$. We can, in that case, define $\hat{\mu}_1$ and $\hat{\mu}_2$ by $d\hat{\mu}_1 = (1 - \varphi(\alpha))d\hat{\mu}$ and $d\hat{\mu}_2 = \varphi(\alpha)d\hat{\mu}$. Assume that $\xi \notin B_\mu$. Then there is $s \in \mathbb{S}^{d-1}$ and $a > a' > 0$ such that

$$\langle s, \xi \rangle \geq a > a' \geq \sup_{\eta \in B_\mu} \langle s, \eta \rangle. \tag{6.1}$$

We now consider the sequence of real numbers $y_j = \langle s, z_1 + z_2 + \cdots + z_j \rangle$, and define successive times $\tau_1 \leq \tau_2 \cdots \leq \tau_j \leq \tau_{j+1} \leq \cdots$ as

$$\tau_j = n \wedge \inf\{i : y_i \geq j\}.$$

These are times at which y_i reaches successive predetermined levels that increase in unit steps. Notice that $\{y_j\}$ can jump over more than one level in a single step, in which case $\tau_j = \tau_{j+1}$. There will be a last $\tau_r < n$ with $y_{\tau_r} \geq r$, and we will then have $y_n \leq (r+1)$. We pick a (large) integer k, fix it, and call the interval $[\tau_j + 1, \tau_{j+1}]$ a *good run* if $y_i \geq y_{\tau_j} - k$ for all $i \in [\tau_j + 1, \tau_{j+1}]$, and a *bad run* otherwise. If $\tau_j = \tau_{j+1}$, the run is empty and it is natural to call it good. Each bad run uses up at least k/C steps, and therefore there can be at most nC/k bad runs. Each bad run gains a distance that is at most $C + 1$, and therefore the sum total of gains during bad runs is at most $\frac{nC(C+1)}{k}$. The last incomplete run may be negative, but in any case, is bounded above by 1. Therefore

$$\sum_{\text{good runs}} (y_{\tau_{j+1}} - y_{\tau_j}) \geq n\left[a - \frac{C(C+1)}{k}\right] - 1. \tag{6.2}$$

Note that during a good run, the walker currently at site x has never before visited any site x' with $\langle s, x'-x\rangle > k+1$. The number of steps j that are part of a good run is at least

$$n\left(\frac{a}{C}-\frac{(C+1)}{k}\right)-\frac{1}{C}.$$

Let us denote by

$$\nu_{n,k}=\frac{1}{n}\sum_{j\in\text{good runs}}\delta_{w_j},$$

the empirical distribution of w_j over all the good runs, still normalized by n. The total mass of $\nu_{n,k}$ can be less than 1. It is at least $\frac{a}{C}-\frac{(C+1)}{k}+o(1)$. We now let $n\to\infty$, along a subsequence if necessary, and get a limit ν_k. We can assume (by the diagonalization procedure) that we have taken one subsequence that works for all k. The limiting measure ν_k may not be invariant because it is the limit of orbital averages over a large number of blocks with gaps between them and the boundaries of the blocks, i.e., two sites per block, constitute a nonnegligible fraction of the total length of all the blocks.

As we let $k\to\infty$, ν_k is monotone and the limit ν, which exists, is invariant. This is because although a proportion of j among $[1,n]$ may be omitted as part of a bad run, the number of such runs is at most $n\frac{C(C+1)}{k}$. The mass contributed by the boundary is at most $\frac{2C(C+1)}{k}$ and is negligible for large k. We get in the limit a measure $\nu\le\mu$, which may still have total mass less than 1 but will now be invariant. There is then a measure $\hat{\nu}$ on $\overline{\mathcal{E}}$ that represents it. Note that $\hat{\nu}\le\hat{\mu}$. Since $\nu_{n,k}$ was constructed during good runs, we have for every n,

$$\nu_{n,k}\big[w: S(w)\cap\{x:\langle s,x\rangle\ge k+1\}\ne\varnothing\big]=0,$$

which leads to

$$\nu_k\big[w: S(w)\cap\{x:\langle s,x\rangle\ge k+1\}\ne\varnothing\big]=0.$$

Since the convergence of ν_k to ν is monotone, if $\delta(k)$ is defined by

$$\delta(k)=\|\nu_k-\nu\|=|\nu(\overline{\mathbf{W}})-\nu_k(\overline{\mathbf{W}})|,$$

then $\delta(k)\to 0$ as $k\to\infty$. Therefore,

$$\lim_{k\to\infty}\nu\big[w: S(w)\cap\{x:\langle s,x\rangle\ge k+1\}\ne\varnothing\big]=0$$

or

$$\sup_{x\in S(w)}\langle s,x\rangle<\infty$$

a.e. ν. This forces

$$\hat{\nu}\big[\alpha:\alpha[\sup_{x\in S(w)}\langle s,x\rangle<\infty]<1\big]=0,$$

and in view of Lemma 5.1, we can conclude that $\hat{\nu}$ is supported on $\mathcal{E}$. Relation (6.2), on the other hand, translates into

$$\int \langle s, m(\alpha)\rangle \hat{\nu}(d\alpha) \geq a$$

after the limit on n and k are taken. Finally,

$$\int_{\mathcal{E}_0} \langle s, m(\alpha)\rangle \hat{\nu}(d\alpha) = \int_{\mathcal{E}} \langle s, m(\alpha)\rangle \hat{\nu}(d\alpha) \geq a\,.$$

Because $\nu \leq \mu$, it follows that $\hat{\nu} \leq \hat{\mu}$ and $\hat{\nu}(d\alpha) = \varphi(\alpha)\hat{\mu}(d\alpha)$ for some $0 \leq \varphi \leq 1$. Hence

$$\sup_{0\leq\phi\leq 1} \int_{\mathcal{E}_0} \langle s, m(\alpha)\rangle \phi(\alpha)\hat{\mu}(d\alpha) \geq a\,,$$

which contradicts (6.1). □

COROLLARY 6.2 *Suppose that μ_n is a sequence of ergodic invariant measures on $\overline{W}$ that are supported on W^{tr}_{∞} and μ is its weak limit. Assume that $m(\mu) = \xi \neq 0$. Then, while μ may not be ergodic, it has a representation*

$$\mu = \int_{\overline{\mathcal{E}}} \alpha\, d\hat{\mu}_1(d\alpha) + \int_{\mathcal{E}_0} \alpha\, d\hat{\mu}_2(d\alpha)$$

with

$$\int_{\mathcal{E}_0} m(\alpha) d\hat{\mu}_2 = \xi \quad \textit{and} \quad \int_{\hat{\mathcal{E}}} m(\alpha) d\hat{\mu}_1 = 0\,.$$

7 Proof of the "Averaged" Large Deviations

We will provide a proof of Theorem 3.2. We will begin with some general facts regarding the local decay rate for Markov chains on a countable state space.

Let $\mathbf{X}$ be a countable state space and $\pi(x, y)$ the transition probability of a Markov chain on $\mathbf{X}$. We are interested in the local decay rate of the n-step transition probabilities $\pi^{(n)}(x, y)$. We will assume that the chain is irreducible, with some period p. In addition, we will assume that for each x, $\pi(x, \cdot)$ is supported on a finite set $F_x \subset \mathbf{X}$. From the subadditivity of $-\log \pi^{(n)}(x, x)$ it follows that there exists $0 \leq \lambda < \infty$ such that

$$\limsup_{\substack{n\to\infty \\ n=0\,(\mathrm{mod}\, p)}} \frac{1}{n} \log \pi^{(n)}(x, x) = -\lambda\,,$$

and by irreducibility is independent of x. For any two $x, y \in \mathbf{X}$, if we let $n \to \infty$ along the subsequence dictated by the periodicity, we have, with the same λ,

$$\lim_{n\to\infty} \frac{1}{n} \log \pi^{(n)}(x, y) = -\lambda\,.$$

For each $0 \le \sigma < \infty$ we consider the convex set $\mathcal{U}_\sigma$ of positive solutions of

$$\sum_y \pi(x, y)u(y) = e^{-\sigma} u(x) ,$$

normalized so that $u(x_0) = 1$ for some fixed point $x_0 \in \mathbf{X}$. In general, $\mathcal{U}_\sigma$ may be empty. Note that the function $1 \in \mathcal{U}_0$. $\mathcal{U}_\sigma$ is nonempty if and only if $\sigma \le \lambda$. For any finite set F we define the exit time $\tau_F = \inf\{n : X_n \notin F\}$ and compute for σ and $x \in F$ the expected value

$$u_F(\sigma, x) = E\big[e^{\sigma\tau_F} \mid X_0 = x\big],$$

which may or may not be finite. Let us define

$$\lambda_F = \Big\{ \sup \sigma : u_F(\sigma, x) < \infty \text{ for all } x \in F \Big\} .$$

Then $\lambda = \inf_F \lambda_F$. The proofs of these assertions are elementary and depend on the fact that $u_F(\sigma, x)$ is the solution of

$$\begin{aligned} \sum_y \pi(x, y)u(y) = e^{-\sigma} u(x) \quad & \text{for } x \in F , \\ u(y) = 1 \quad & \text{for } x \in F^c . \end{aligned}$$

For any finite set F, the value λ_F can be characterized as $-\log \operatorname{spect} \pi_F$ where π_F is the substochastic matrix $\{\{\pi(x, y)\} : x, y \in F\}$, and spect refers to its spectral radius. See [3] for a discussion of these ideas.

We next consider a simple dynamic programming problem. At each site x, a choice of $\pi(x, \cdot)$ can be made from a set $\mathcal{P}_x$ of possible distributions supported on some fixed finite set N_x. The goal is to maximize $u_F(\sigma, x)$. The Bellman equation for this problem is

$$\begin{aligned} \sup_{\pi(x,\cdot)\in\mathcal{P}_x} \sum_y \pi(x, y)u(y) = e^{-\sigma} u(x) \quad & \text{for } x \in F , \\ u(y) = 1 \quad & \text{for } x \in F^c , \end{aligned}$$

which may or may not have a solution for a given σ. There will be a largest value $\hat{\lambda}_F$ such that $u = \hat{u}_F$ exists if $\sigma < \hat{\lambda}_F$. It is clear from the Bellman equation that $\hat{u}_F$ and $\hat{\lambda}_F$ are not changed if we replace each $\mathcal{P}_x$ by its convex hull. Moreover,

$$\hat{\lambda}_F = \inf_{\pi:\pi_x\in\mathcal{P}_x, x\in F} \lambda_F(\pi) .$$

Although the quenched rate function $h(\xi)$ is not in general explicit, $h(0)$ has a natural and important interpretation. Consider for each environment ω the n-step transition probability $\pi^{(n)}(\omega, 0, 0)$. From the Chapman-Kolmogorov equations, we have

$$\begin{aligned} \pi^{(n+m)}(\omega, x, x) &= \sum_{y\in z^d} \pi^{(n)}(\omega, x, y)\pi^{(m)}(\omega, y, x) \\ &\ge \pi^{(n)}(\omega, x, x)\pi^{(m)}(\omega, x, x) , \end{aligned}$$

making $-\log \pi^{(n)}(\omega, x, x)$ subadditive. Therefore, if p is the period, then

$$k(x,\omega) = -\lim_{n\to\infty} \frac{1}{np} \log \pi^{(np)}(\omega, x, x) = \inf_n \left[-\frac{1}{n} \log \pi^{(n)}(\omega, x, x) \right]$$

exists for every x for almost all ω. Moreover, by ellipticity $k(x,\omega) = k(\omega)$ is independent of x, and, by ergodicity, it is then equal to a constant $k = k(\beta)$. In particular,

$$\pi^{(n)}(\omega, 0, 0) \leq e^{-k(\beta)n}$$

for every n and for every ω in the support of $\mathbf{P}$. The value $k(\beta)$ is the largest constant for which this is true. According to our earlier discussion, we can define $k(\beta)$ alternately in the following way. For each x, let us choose $\mathcal{P}_x$ as $\{p(\cdot - x)\}$ as $p(\cdot)$ varies over $\mathcal{K}$. Then, for a finite set F, the largest spectral radius over all possible choices of π with $\pi(x,\cdot) \in \mathcal{P}_x$ is $e^{-\hat{\lambda}_F}$, and

$$k(\beta) = \inf_F \hat{\lambda}_F$$

depends only on the convex hull $\hat{\mathcal{K}}$ of the support $\mathcal{K}$ of the distribution β on $\mathcal{P}$.

Now we can integrate with respect to $\mathbf{P}$ to obtain

$$P_0[x_n = 0] \leq e^{-k(\beta)n}\,.$$

Of course, $k(\beta)$ can and often is 0. The case $k(\beta) = 0$ is called the "nestling" case, and $k(\beta) > 0$ the "nonnestling" case. There is a simple formula for $k(\beta)$.

LEMMA 7.1 *The value of $k(\beta)$ is given by*

$$k(\beta) = -\log\left[\inf_{\theta\in\mathbb{R}^d} \sup_{p\in\hat{\mathcal{K}}} \left[\sum_z e^{\langle\theta,z\rangle} p(z) \right] \right],$$

where $\hat{\mathcal{K}}$ is the convex hull of the support $\mathcal{K}$ of β. In order that $k(\beta) > 0$, it is necessary and sufficient that 0 *not be in the closed convex hull of the support of the distribution of $m(p) = \sum_z zp(z)$ induced by the distribution β on $\mathcal{P}$.*

PROOF: Suppose there is θ such that

$$\sup_{p\in\hat{\mathcal{K}}} \left[\sum_z e^{\langle\theta,z\rangle} p(z) \right] \leq e^{-\lambda}\,.$$

Then, for any environment ω, one has by induction

$$E^{P^\omega}\left[\exp[\langle\theta, x_n\rangle]\right] \leq e^{-n\lambda}\,,$$

and this in turn implies immediately that

$$\pi^{(n)}(\omega, 0, 0) \leq e^{-n\lambda}$$

or $k(\beta) \geq \lambda$. On the other hand, because

$$\Psi(p(\cdot), \theta) = \log\left[\sum_z e^{\langle\theta,z\rangle} p(z) \right]$$

is convex in θ and concave in $p(\cdot)$, by standard minimax argument

$$\inf_{\theta\in\mathbb{R}^d}\sup_{p\in\hat{\mathcal{K}}}\left[\sum_z e^{\langle\theta,z\rangle}p(z)\right] = \sup_{p\in\hat{\mathcal{K}}}\inf_{\theta\in\mathbb{R}^d}\left[\sum_z e^{\langle\theta,z\rangle}p(z)\right].$$

If there is a $p(\cdot)\in\hat{\mathcal{K}}$ such that

$$c = \inf_\theta \sum_z e^{\langle\theta,z\rangle}p(z) \geq e^{-\lambda},$$

then the uniform environment of the simple random walk with jump distributed according to $p(\cdot)$ will have the property

$$\lim_{n\to\infty}\frac{1}{n}\log\pi^{(n)}(0,0) = \log c \geq -\lambda,$$

thus establishing $k(\beta)\leq\lambda$. This proves the first part of the lemma.

Now, $k(\beta)=0$ if and only if there is a $p(\cdot)\in\hat{\mathcal{K}}$ such that

$$\inf_{\theta\in\mathbb{R}^d}\left[\sum_z e^{\langle\theta,z\rangle}p(z)\right] = 1,$$

i.e., $\sum_z zp(z)=0$. □

Since we have a Feller process with transition probabilities $q(w,w')$ on the compact space $\overline{\mathbf{W}}$, the standard theory of large deviation (see, e.g., [6]) applies, and we get a lower-semicontinuous rate function $\mathcal{J}(\mu)$ defined by

$$\mathcal{J}(\mu) = \sup_{u\in C^+(\overline{\mathbf{W}})}\int \log\frac{u(w)}{(qu)(w)}\mu(dw),$$

which is infinite unless $\mu\in\overline{\mathcal{I}}$, the set of invariant measures on $\overline{\mathbf{W}}$. For the empirical measure $\mathcal{R}_n$, we get the estimates

$$\limsup_{n\to\infty}\frac{1}{n}\sup_w \log Q^w\big[\mathcal{R}_n\in C\big] \leq -\inf_{\mu\in C}\mathcal{J}(\mu)$$

for closed sets C, and for open sets G we have

$$\liminf_{n\to\infty}\frac{1}{n}\sup_w \log Q^w\big[\mathcal{R}_n\in G_\mu\big] \geq -\inf_{\mu\in G\cap\hat{\mathcal{E}}}\mathcal{J}(\mu).$$

The lower bound is rather weak (since it may not hold for a given w) and can only be improved under some sort of transitivity condition.

LEMMA 7.2 *With $k=k(\beta)$ as defined before Lemma* 7.1,

$$\inf_{\substack{\mu\in\hat{\mathcal{I}}\\ m(\mu)=0}}\mathcal{J}(\mu)\geq k(\beta).$$

PROOF: We need to find a u and use the formula

$$\mathcal{J}(\mu) \geq \int \log \frac{u(w)}{(qu)(w)} \mu(dw) .$$

We will try $u(w) = e^{\langle \theta, z_0(w) \rangle}$. Since $m(\mu) = 0$, we get the lower bound

$$\mathcal{J}(\mu) \geq - \inf_{\theta \in \mathbb{R}^d} \int \log \left[\sum_z q(w, z) e^{\langle \theta, z \rangle} \right] \mu(dw) .$$

Since $q(w, \cdot) \in \mathcal{K}$, we are done. □

LEMMA 7.3 *For* $0 \neq \xi \in \mathbb{R}^d$ *and any open set* $G \ni \xi$

$$\liminf_{n \to \infty} P_0 \left[\frac{x_n}{n} \in G \right] \geq -H(\xi) .$$

PROOF: Let $\alpha \in \mathcal{E}$ with $m(\alpha) = \xi$ be given. Let us denote by

$$\hat{\alpha}(\omega, z) = \alpha[z_1 = z \mid \omega] ,$$

the conditional probability of the next step being z given the past. The rate function is given by

$$I(\alpha) = E^{\alpha} \left[\sum_z \hat{\alpha}(\omega, z) \log \frac{\hat{\alpha}(w, z)}{q(w, z)} \right] .$$

By the ergodic theorem, for any $\epsilon > 0$, as $n \to \infty$

$$\alpha \left[|\bar{z}(n) - \xi| \leq \epsilon, |\bar{I}(n) - I(\alpha)| \leq \epsilon \right] \to 1 ,$$

where

$$\bar{z}(n) = \frac{1}{n} \sum_{i=1}^{n} z_i , \quad \bar{I}(n) = \frac{1}{n} \sum_{i=1}^{n} \log \frac{\hat{\alpha}(w_{i-1}, z_i)}{q(w_{i-1}, z_i)} .$$

Moreover, we can assume that ℓ is large enough that for all n,

$$\alpha[H(n, S(w), \mathbf{z}) \leq \ell] \geq \frac{1}{2}$$

for some $s \in \mathbb{S}^{d-1}$ with $\langle s, \xi \rangle > 0$. We wish to estimate

$$
\begin{aligned}
&Q^{\phi}[|\bar{z}(n) - \xi| \le \epsilon] \\
&\quad \ge Q^{\phi}[|\bar{z}(n) - \xi| \le \epsilon, H(n, S(w), \mathbf{z}) \le \ell] \\
&\quad \ge e^{-C\ell} \sup_{w} P^{w}[|\bar{z}(n) - \xi| \le \epsilon, H(n, S(w), \mathbf{z}) \le \ell] \\
&\quad \ge e^{-C\ell} \int P^{w}[|\bar{z}(n) - \xi| \le \epsilon, H(n, S(w), \mathbf{z}) \le \ell]\alpha(dw) \\
&\quad = e^{-C\ell} \int_{\{|\bar{z}(n)-\xi|\le\epsilon\}\cap\{H(n,S(w),\mathbf{z})\le\ell\}} \mathcal{R}_n(w)\alpha(dw) \\
&\quad \ge e^{-C\ell} e^{-n(I(\alpha)+\epsilon)} \alpha\big[|\bar{z}(n) - \xi| \le \epsilon, |\bar{I}(n) - I(\alpha)| \le \epsilon, H(n, S(w), \mathbf{z}) \le \ell\big].
\end{aligned}
$$

□

THEOREM 7.4 *The function*

$$H(\xi) = \inf_{\substack{\mu \in \mathcal{E} \\ m(\mu)=\xi}} \mathcal{J}(\mu)$$

is a convex function in every half-space $\{\xi : \langle s, \xi \rangle > 0\} \subset \mathbb{R}^d$. The limit $\lim_{\xi \to 0} H(\xi)$ exists and equals $k(\beta)$. If we define $H(0)$ as this limit, then $H(\cdot)$ is a convex function on $\mathbb{R}^d$.

PROOF: First, assume that $\xi = a\xi_1 + (1-a)\xi_2$, with both ξ_1 and ξ_2 belonging to the same half-space; i.e., there exists $s \in \mathbb{S}^{d-1}$ such that $\langle s, \xi_i \rangle > 0$ for $i = 1, 2$. For any $\epsilon > 0$, there are processes with stationary increments μ_i that are ergodic with means ξ_i and $\mathcal{J}(\mu_i) \le H(\xi_i) + \epsilon$. The measure $\mu = a\mu_1 + (1-a)\mu_2$ is not ergodic but has stationary increments and $I(\mu) \le aH(\xi_1) + (1-a)H(\xi_2) + \epsilon$. It can be approximated by ergodic ones by the usual method of taking a block of length ℓ and constructing (for the increments) a product measure over distinct blocks and then averaging over ℓ translates to produce something that is ergodic and stationary. If we call this approximation μ_ℓ, then it is ergodic and $m(\mu_\ell) = m(\mu)$ for every ℓ.

One needs only to prove that

$$\lim_{\ell \to \infty} \mathcal{J}(\mu_\ell) = \mathcal{J}(\mu).$$

From the lower semicontinuity of $\mathcal{J}(\cdot)$, it is clearly enough to prove that $\mu_\ell \to \mu$ weakly as measures on $\overline{\mathbf{W}}$. From Corollary 6.2, it follows that any limit μ' of μ_ℓ

along a subsequence has a representation of the form

$$\mu' = \theta \int_{\mathcal{E}_0} \alpha \nu(d\alpha) + (1-\theta) \int_{\overline{\mathcal{E}}} \alpha \nu'(d\alpha)$$

with

$$\int_{\overline{\mathcal{E}}} m(\alpha) \nu'(d\alpha) = 0 .$$

However, μ' on $\overline{\mathbf{W}}$ has the same projection as μ on $\mathbf{W}_\infty$, which is supported on $\mathbf{W}_\infty^{\mathrm{tr}}$. Therefore both ν and ν' can only be supported on measures α with $m(\alpha) = \xi_1$ or ξ_2. Since 0 is not in the convex hull of ξ_1 and ξ_2, we must have $\nu' = 0$. Therefore $\mu' = \mu$ and we are done.

Now we turn to the proof of $\lim_{\xi \to 0} H(\xi) = k(\beta)$. If $m(\mu_n) = \xi_n$ and $\xi_n \to \xi$, then along a subsequence $\mu_n \to \mu$ and $m(\mu) = 0$ even if μ is not ergodic. By the semicontinuity of $\mathcal{J}(\cdot)$

$$\mathcal{J}(\mu) \leq \liminf_{n \to \infty} \mathcal{J}(\mu_n) \leq \liminf_{n \to \infty} H(\xi_n) .$$

But from Lemma 7.2

$$k(\beta) \leq \inf_{m(\mu)=0} \mathcal{J}(\mu) .$$

Hence

$$k(\beta) \leq \liminf_{\xi \to 0} H(\xi) .$$

To prove the inequality in the reverse direction, given $\xi \in \mathbb{S}^{d-1}$, for sufficiently small δ, we have to construct an ergodic $\alpha \in \mathcal{E}$ with $m(\alpha) = \delta\xi$ and $\mathcal{J}(\alpha) \leq k(\beta) + \epsilon(\delta)$, where $\epsilon(\delta) \to 0$ with δ. This can be accomplished as follows. Let $\xi \in \mathbb{S}^{d-1}$ and integers L, k, and $\delta > 0$ be given. Let us select a deterministic path connecting the origin and a lattice site close to $k\delta\xi$ in $Ck\delta$ steps, where each step has conditional probability at least ρ. Assume that the path is such that the endpoint is a boundary point of a cube C_L of size L and the path has no other common points with the cube C_L. We will construct a probability distribution $\mu_{L,k}$ of a walk that will start where the deterministic path ends, remain inside the cube C_L for $k(1 - C\delta)$ steps, and end up at the boundary point where it entered, essentially making a loop. The two walks together will produce a probability distribution, which we will denote by $\mu_{L,\delta,k,\xi}$, which in k steps goes from 0 to $k\delta\xi$. We can take a countable product of this and get a process whose increments will be stationary under the k shift. We can do the averaging over k steps and get an ergodic process with stationary increments, which we call $\alpha_{L,\delta,k,\xi}$. It is not difficult to estimate $\mathcal{J}(\alpha_{L,\delta,k,\xi})$. The deterministic part contributes at most $Ck\delta \log(1/\rho)$, and the part confining the walk to a box of size L depends on the choice of $\mu_{L,k}$; it can be chosen so that it contributes at most $\sigma_L k(1 - C\delta) + o(k)$, where σ_L is given by

$$\sigma_L = -\liminf_{n \to \infty} \frac{1}{n} \log Q^{\phi}[z_1 + z_2 + \cdots + z_j \in C_L \text{ for all } 1 \leq j \leq n]$$

and $\sigma_L \to k(\beta)$ as $L \to \infty$. This leads to an estimate of the form

$$\mathcal{J}(\alpha_{L,\delta,k,\xi}) \le \sigma_L(1 - C\delta) + C\delta \log \frac{1}{\rho} + o(1)$$

as $k \to \infty$. We let $k \to \infty$ and then $L \to \infty$ to obtain the estimate

$$H(\delta\xi) \le k(\beta)(1 - C\delta) + C\delta \log \frac{1}{\rho}$$

for arbitrary $\xi \in \mathbb{S}^{d-1}$ and δ small enough.

Finally, to prove convexity at 0, we need to show that for any $\{\xi_j\}$ with convex combination $\sum_j a_j\xi_j = 0$ we have $\sum_j a_j k(\xi_j) \ge k(\beta)$. But $\mu = \sum_j a_j\mu_j$ with $m(\mu_j) = \xi_j$ has $m(\mu) = 0$, and therefore

$$\sum_j a_j \mathcal{J}(\mu_j) = \mathcal{J}(\mu) \ge k(\beta)\,,$$

and we are done. □

We now prove Theorem 3.1.

PROOF OF THEOREM 3.1: Lemma 7.4 provides the upper bound at 0, and Theorem 7.6 and Lemma 7.5 establish lower bounds. We need to prove only the upper bound for $\xi \neq 0$. In view of Theorem 5.1, we have the following inequalities valid for closed sets C not containing 0:

$$\begin{aligned}
\limsup_{n\to\infty} \frac{1}{n} \log Q^{\phi}\left[\frac{x_n}{n} \in C\right] &\le -\inf_{\xi\in C}\inf_{\theta}\left[(1-\theta)k(\beta) + \theta H\left(\frac{\xi}{\theta}\right)\right] \\
&\le -\inf_{\xi\in C}\inf_{\theta}\left[(1-\theta)H(0) + \theta H\left(\frac{\xi}{\theta}\right)\right] \\
&\le -\inf_{\xi\in C} H(\xi)\,.
\end{aligned}$$

□

8 Concluding Remarks

One of the interesting issues in the large-deviation estimate for the averaged process is where the major contribution to the integral

$$P_0\left[\frac{x_n}{n} \simeq \xi\right] = \int P^{\omega}\left[\frac{x_n}{n} \simeq \xi\right]\mathbf{P}(d\omega) = e^{-nH(\xi)+o(n)}$$

comes from. The answer is easily provided by the properties of entropy. Let λ_1 and λ_2 be two measures on a product space $X \times Y$. We can disintegrate them as

$\alpha_i(dx)p_i(x,dy)$, $i = 1,2$, or as $\beta_i(dy)q_i(y,dx)$, $i = 1,2$. The relative entropy $h(\lambda_1 \mid \lambda_2)$ can be computed in two different ways:

$$\begin{aligned} h_{X\times Y}(\lambda_1 \mid \lambda_2) &= h_X(\alpha_1 \mid \alpha_2) + \int_X h_Y((p_1(x,\cdot) \mid p_2(x,\cdot))\alpha_1(dx) \\ &= h_Y(\beta_1 \mid \beta_2) + \int_Y h_X((q_1(y,\cdot) \mid q_2(y,\cdot))\beta_1(dy)\,. \end{aligned}$$

If we identify X with the environment and Y with the walk, then $\alpha_2 = \mathbf{P}$, $p_2(x,\cdot) = P^{\omega}$, $\beta_2 = P_0$. We want to choose β_1 to be the measure μ that provides the infimum in the definition of $H(\xi)$. It does not pay to have $h_{X\times Y}(\lambda_1 \mid \lambda_2)$ any larger than $h_Y(\beta_1 \mid \beta_2) \simeq nH(\xi)$. This is, of course, achieved when $q_1 = q_2$. If μ is the optimal process with stationary increments that produces the large deviation, i.e., the μ that achieves the minimum in the definition of $H(\xi)$, then the strange environment that produces the best large deviation for the averaged walk is $\int P_0^w \mu(dw)$, the average with respect to μ of the a posteriori distribution P_0^w of the environment given the walk w. In $d \ge 2$ this is not very informative because the averaged environment is not translation invariant, whereas in $d = 1$, it is essentially translation invariant over a long segment.

Another question that we can answer to some degree is about the set where $H(\xi)$ and $h(\xi)$ are zero.

THEOREM 8.1 *Although we can only assert in general that $H(\xi) \le h(\xi)$, their zero set is the same. In other words, $H(\xi) = 0$ implies that $h(\xi) = 0$.*

PROOF: Before we consider the possibilities for the zero set $N = \{\xi : H(\xi) = 0\}$, let us note that there can be at most two invariant measures in $\hat{E}$ with nonzero mean. If there are two with means $\{m_i : i = 1,2\}$, then $m_1 = -cm_2$ for some $c > 0$. It follows from Theorem 7.7 that if $H(\xi) = 0$ for some $\xi \ne 0$ and there is no ergodic invariant measure with mean ξ, then $H(0) = 0$ and there is an invariant measure in $\hat{E}$ with mean $c\xi$ for some $c > 1$. If there is an invariant measure in $\hat{E}$ with nonzero mean ξ, then

$$Q^{\phi}\left[\lim_{n\to\infty} \frac{x_n}{n} = \xi\right] > 0\,,$$

so that $h(\xi) = 0$. This means that the set N consists of either a single point or a line segment through the origin that can extend in one or both directions. In any direction that it extends there is an ergodic invariant measure with mean equal to the endpoint. If N consists only of a single point $\xi \ne 0$ and this then there is an invariant measure with mean ξ, and there is a law of large numbers asserting that the mean speed is ξ. Since $H(0) = h(0)$, this proves the assertion of the theorem. □

Remark 8.2. The possibility of the zero set N extending in both directions is ruled out in dimensions 1 and 2. It is easily checked in $d = 1$, and for $d = 2$ it is a consequence of a recent result of Zerner and Merkl [9].

Remark 8.3. A related question is the validity of the law of large numbers. The general results will show that $\lim_{n\to\infty} x_n/n = U$ exists and U can take at most three values, the two extremes of the zero set of $H(\cdot)$ and the value 0. In $d = 1$, U is always a constant, and Zerner and Merkl in [9] show the same result for $d = 2$ as well. But for $d \geq 3$ the question remains unresolved.

Acknowledgment. The author wishes to thank Dr. Firas Rassoul-Agha for valuable discussions. This research was supported by National Science Foundation Grant DMS-0104343.

Bibliography

[1] Comets, F.; Gantert, N.; Zeitouni, O. Quenched, annealed and functional large deviations for one-dimensional random walk in random environment. *Probab. Theory Related Fields* **118** (2000), no. 1, 65–114.

[2] Liggett, T. M. An improved subadditive ergodic theorem. *Ann. Probab.* **13** (1985), no. 4, 1279–1285.

[3] Ney, P.; Nummelin, E. Markov additive processes. I. Eigenvalue properties and limit theorems. *Ann. Probab.* 15 (1987), no. 2, 561–592.

[4] Rezakhanlou, F.; Tarver, J. E. Homogenization for stochastic Hamilton-Jacobi equations. *Arch. Ration. Mech. Anal.* **151** (2000), no. 4, 277–309.

[5] Sznitman, A. S. Topics in random walks in random environment. Preprint. Available at: `http://www.math.ethz.ch/~sznitman/preprint.shtml`.

[6] Varadhan, S. R. S. Large deviations and applications. CBMS-NSF Regional Conference Series in Applied Mathematics, 46. Society for Industrial and Applied Mathematics (SIAM), Philadelphia, 1984.

[7] Zeitouni, O. Random walks in random environments. *Proceedings of the International Congress of Mathematicians, Vol. III (Beijing, 2002)*, 117–127. Higher Education Press, Beijing, 2002.

[8] Zerner, M. P. W. Lyapounov exponents and quenched large deviations for multidimensional random walk in random environment. *Ann. Probab.* **26** (1998), no. 4, 1446–1476.

[9] Zerner, M. P. W.; Merkl, F. A zero-one law for planar random walks in random environment. *Ann. Probab.* **29** (2001), no. 4, 1716–1732.

S. R. S. VARADHAN
Courant Institute
251 Mercer Street
New York, NY 10012-1185
E-mail: `varadhan@cims.nyu.edu`

Received September 2002.

Proc. Indian Acad. Sci. Math. Sci. 114 (2004) 309-318

Random walks in a random environment*

S R S VARADHAN

Department of Mathematics, Courant Institute of Mathematical Sciences, New York University, NY 10012, USA
E-mail: varadhan@cims.nyu.edu

MS received 16 June 2004

Abstract. Random walks as well as diffusions in random media are considered. Methods are developed that allow one to establish large deviation results for both the 'quenched' and the 'averaged' case.

Keywords. Large deviations; random walks in a random environment.

1. Introduction

A random walk on $\mathbb{Z}^d$ is a stochastic process $\{S_n: n \geqslant 0\}$ defined as

$$S_0 = 0, \quad S_n = X_1 + \cdots + X_n \quad \text{for } n \geqslant 1,$$

where $\{X_i\}$ are independent and identically distributed random variables with a common distribution $\pi(z) = P[X_i = z]$ for $z \in \mathbb{Z}^d$. The relationship between the behavior of S_n as $n \to \infty$ and properties of $\pi(\cdot)$ is quite well-understood and is easy to analyse, using often Fourier analysis as the main tool.

For instance if $\sum_{z\in\mathbb{Z}^d} |z|\, \pi(z) < \infty$, the law of large numbers states that

$$\lim_{n\to\infty} \frac{S_n}{n} = \sum_z z\, \pi(z).$$

If $\sum_{z\in\mathbb{Z}^d} |z|^2\, \pi(z) < \infty$, and $\sum_{z\in\mathbb{Z}^d} z\, \pi(z) = 0$, then the central limit theorem asserts that the distribution of $S_n/\sqrt{n}$ is asymptotically Gaussian on $\mathbb{R}^d$ with mean 0 and covariance $C = \{C_{r,s}\}$ given by

$$C_{r,s} = \sum_{z\in\mathbb{Z}^d} z_r z_s\, \pi(z).$$

Cramér's theorem on large deviations provides exponential rates of convergence of

$$P\left[\frac{S_n}{n} \in A\right]$$

*This is essentially a transcript of the plenary talk given at the Joint India–AMS Mathematics Meeting held in December 2003 in Bangalore, India.

for sets A excluding the mean $m = \sum_{z \in \mathbb{Z}^d} z\,\pi(z)$. Assuming that

$$M(\theta) = \sum_{z \in \mathbb{Z}^d} e^{\langle z, \theta \rangle} \pi(z) < \infty$$

for $\theta \in \mathbb{R}^d$, for a wide class of sets A,

$$\lim_{n \to \infty} \frac{1}{n} \log P\left[\frac{S_n}{n} \in A\right]$$

exists and is given by $-\inf_{x \in A} I(x)$ where

$$I(x) = \sup_{\theta \in \mathbb{R}^d} [\langle \theta, x \rangle - \log M(\theta)].$$

The situation is quite different if we drop the hypothesis of independence of increments of S_n and replace it with the assumption that S_n is a Markov process. In other words, there exists $p(z', z)$ such that

$$P[S_{n+1} = z | S_1, S_2, \ldots, S_n] = p(S_n, z).$$

The random walk case considered earlier is a special case where $p(z', z)$ takes the form $\pi(z - z')$. Sometimes it is more convenient to denote $p(z', z)$ by $\pi(z', z - z')$ so that the independent case is when $\pi(z', \cdot)$ is independent of z'. In this generality not much can be said. However if $\pi(z', z)$ is periodic in z', the phase gets averaged out and some homogenization takes place with $\pi(z', z)$ getting effectively replaced by an averaged $\hat{\pi}(\cdot)$. Results on law of large numbers, central limit theorem and large deviation estimates can be established, with some extra work, quite similar to the independent case.

We will investigate the case when $\pi(z', z)$ are random but spatially homogeneous in z'. This can viewed as a limiting version of the periodic case when the period becomes large.

2. Formulation

We will start with a probability space (Ω, Σ, P) on which $\mathbb{Z}^d$ acts ergodically as a family τ_z of measure preserving transformations. We are given $\pi(\omega, z)$ which is a probability distribution on $\mathbb{Z}^d$ for each ω, that are measurable functions of ω. One can then generate random transition probabilities $\pi(\omega, z', z)$ by defining

$$\pi(\omega, z', z) = \pi(\tau_{z'}\omega, z).$$

For each ω, $\pi(\omega, z', z)$ can serve as the transition probability of a Markov process on $\mathbb{Z}^d$ and the measure corresponding to this process, starting from 0, is denoted by Q^ω. This is of course random (depends on ω) and is called the random walk in the random environment ω. One can ask the same questions about this random walk and in some form they may be true for almost all omega with respect to P. The law of large numbers, if valid, will take the form

$$P\left[\omega\colon \lim_{n \to \infty} \frac{S_n}{n} = m(P) \quad \text{a.e. } Q^\omega\right] = 1. \tag{1}$$

Such statement concerning the almost sure behavior of Q^ω for almost all ω with respect to P are said to deal with the 'quenched' version. Sometimes one wishes to study the behavior of the 'averaged' measure

$$\bar{Q} = \int Q^\omega P(\mathrm{d}\omega). \tag{2}$$

The law of large numbers is the same because (1) is equivalent to

$$\bar{Q}\left[\omega\colon \lim_{n\to\infty} \frac{S_n}{n} = m(P)\right] = 1.$$

On the other hand, questions on the asymptotic behavior of probabilities, like for instance, the central limit theorem or large deviations could be different for the quenched and the averaged cases.

A special environment called the 'product environment' is one in which $\pi(\omega, z', z)$ are independent for different z' and have a common distribution β which is a probability measure on the space $\mathcal{M}$ of all probability measures on $\mathbb{Z}^d$. In this case the canonical choice for Ω is the countable product of $\mathcal{M}$. The product measure having marginals β is of course the choice for P.

One can consider the continuous versions of these problems. For instance instead of the action of $\mathbb{Z}^d$ we can have $\mathbb{R}^d$ acting on (ω, Σ, P) ergodically and consider a diffusion on $\mathbb{R}^d$ with a random infinitesimal generator

$$(\mathcal{L}^\omega u)(x) = \frac{1}{2}(\Delta u)(x) + \langle b(\omega, x), (\nabla u)(x)\rangle \tag{3}$$

acting on smooth function on $\mathbb{R}^d$. Here $b(\omega, x)$ is generated from a map $b(\omega)\colon \Omega \to \mathbb{R}^d$ by the action $\{\tau_x\}$ of $\mathbb{R}^d$,

$$b(\omega, x) = b(\tau_x \omega).$$

Again there is the quenched measure Q^ω that corresponds to the diffusion with generator $\mathcal{L}^\omega$ that starts from 0 at time 0 and the averaged measure that is given by the same formula (2). This model is referred to as diffusion with a random drift. Exactly the same questions can be asked in this context.

3. Results: One-dimensional case

We shall describe the results in the one-dimensional case, which is quite well-understood. A very good reference with a full bibliography is a survey by Zeitouni [9] at the Beijing congress as well as his notes from St. Flour [10]. For simplicity we will take the special case of the product environment in which only $\pi(\omega, z, z \pm 1)$ are non-zero and are random variables $p(z), q(z) = 1 - q(z)$ that are independent for different values of z, β being the distribution on [0, 1] of $p(0)$. To avoid some technical details we shall assume that β has no mass near the edges i.e. is supported on a closed subinterval of (0, 1).

The question of when S_n is transient was answered in [6].

Theorem 3.1. *In order that*

$$Q^\omega\left[\lim_{n\to\infty} S_n = +\infty\right] = 1$$

a.e. P it is necessary and sufficient that

$$E\left[\log\frac{p(0)}{q(0)}\right]=\int\log\frac{p}{1-p}\beta(\mathrm{d}p)>0. \tag{4}$$

In the original random walk where $p(z)\equiv p$ is a constant this is equivalent to $p>\frac{1}{2}$ or $p-q>0$. In such a case the law of large numbers asserts that

$$\lim_{n\to\infty}\frac{S_n}{n}=p-q>0.$$

In our context the law of large numbers, also proved in [6], states the following theorem.

Theorem 3.2.

$$\lim_{n\to\infty}\frac{S_n}{n}=m$$

a.e. Q^ω for almost all ω with respect to P where m is given by

$$m=\begin{cases}\frac{1-E[\frac{q}{p}]}{1+E[\frac{q}{p}]}, & if\ E[\frac{q}{p}]<1,\\ 0, & otherwise.\end{cases} \tag{5}$$

It is therefore possible that $E[\log\frac{p}{q}]>0$ while $E[\frac{q}{p}]\geqslant 1$, in which case $S_n\to\infty$ but $\frac{S_n}{n}\to 0$. In the original random walk this could not happen. In the one-dimensional case further detailed analysis can be carried out in this case. The problem comes from 'traps' where the environment conspires to hold the particle for a long time. As the particle moves to ∞, it encounters increasingly deeper traps that hold it even longer producing a critical slowing down. This phenomenon discovered by Sinai [5] is well-understood. In one-dimension the traps can not be avoided. Therefore the one-dimensional case is strikingly different from the higher dimensional ones.

Assuming that $m>0$, one can study the behavior of $\xi_n=(S_n-nm)/\sqrt{n}$ and ask for a central limit theorem. It turns out that the fluctuations of ξ_n have two sources; from the environment $\{p(x)\colon 0\leqslant x\leqslant nm\}$ and fluctuations of the walk in this environment. Under Q^ω, the distribution of ξ_n is Gaussian with a random mean which depends on the environment and has an asymptotic Gaussian distribution by itself. Hence while there is a central limit theorem for the averaged $\bar{Q}$ there is none for Q^ω. This phenomenon is not expected to persist in higher dimensions.

There are large deviation results regarding the limits

$$\lim_{n\to\infty}\frac{1}{n}\log Q^\omega\left[\frac{S_n}{n}\simeq a\right]=I(a)$$

and

$$\lim_{n\to\infty}\frac{1}{n}\log \bar{Q}\left[\frac{S_n}{n}\simeq a\right]=\bar{I}(a).$$

The difference between I and $\bar{I}$ has a natural explanation in terms of the large deviation behavior of the environment. It is related to the following question. If at time n, a particle is near na instead of being around nm which is what we should expect, what does it say about the environment in $[0,na]$? Did the particle behave strangely in a normal environment or did it encounter a strange environment? It is in fact a combination of both as is seen in the relation of $\bar{I}$ to I which is essentially Bayes' rule. See for instance [2] and [3].

4. Higher dimension

If $d \geqslant 2$ the law of large numbers and central limit theorems are not that well-understood. For instance the questions of recurrence, transience and the existence of a non-zero limit $m = \lim_{n\to\infty} \frac{S_n}{n}$ have only partial answers with sufficient conditions that are not that easy to check. Even when we limit ourselves to a product environment there is no clear analog of the condition (4).

However if $d \geqslant 3$, under some symmetry conditions that ensure that $m = 0$, a central limit theorem was proved by Bricmont and Kupiainen [1], provided the randomness in the environment is small. A recent preprint by Sznitman and Zeitouni [7] provides an alternate proof.

Large deviation results however exist in much wider generality, both in the quenched and the averaged cases. The large deviation principle is essentially the existence of the limits

$$\lim_{n\to\infty} \frac{1}{n} \log E[\exp[\langle\theta, S_n\rangle] = \Psi(\theta).$$

The expectation is with respect to Q^ω or $\bar{Q}$, which could produce different limits for Ψ. The law of large numbers and central limit theorem involve the differentiability of Ψ at $\theta = 0$, which is harder. In fact large deviation results have been proved by general sub-additivity arguments for the quenched case. Roughly speaking fixing ω,

$$Q^\omega[S_{k+\ell} \simeq (k+\ell)a] \geqslant Q^\omega[S_k \simeq ka] \times Q^{\tau_{ka}\omega}[S_\ell \simeq \ell a].$$

One can then use some version of the sub-additive ergodic theorem. This idea has to be cleaned up a bit, but is not hard. See for instance [11] or [8]. We will however explore, in the next two sections a proof of the large deviation principle for the averaged case and an alternate proof for the quenched case.

5. Large deviations: The averaged case

We want to prove that the limits

$$\lim_{n\to\infty} \frac{1}{n} \log \bar{Q}\left[\frac{S_n}{n} \simeq a\right] = -\bar{I}(a) \tag{6}$$

or equivalently

$$\lim_{n\to\infty} \frac{1}{n} \log E^{\bar{Q}}[\exp[\langle\theta, S_n\rangle]] = \bar{\Psi}(\theta)$$

exist. The problem is that the measure $\bar{Q}$ is not very nice. As the random walk explores $\mathbb{Z}^d$ it learns about the environment and in the case of the product environment, when it returns to a site that it has visited before, the experience has not been forgotten and leads to long term correlations. However, if we are interested in the behavior $S_n \simeq na$ with $a \neq 0$, the same site is not visited too often and the correlations should decay fast. This can be exploited to provide a proof of (6).

One can use Bayes' rule to calculate the conditional distribution

$$\bar{Q}[S_{n+1} = S_n + z | S_1, S_2, \ldots, S_n] = q(z|w),$$

where w is the past history of the walk. Before we do that it is more convenient to shift the origin as we go along so that the current position of the random walk is always the origin and the current time is always 0. Then an n step walk looks like $w = \{S_0 = 0, S_{-1}, \ldots, S_{-n}\}$. We pick a z with probability $q(z|w)$. We get a new walk of $n+1$ steps $w' = \{S'_0 = 0, S'_{-1}, \ldots, S'_{-(n+1)}\}$ given by $S'_{-(k+1)} = S_{-k} - z$ for $k \geqslant 0$. We can now calculate $q(z|w)$. We need to know all the numbers $\{k(w, x, z)\}$ of the number of times the walk has visited x in the past and jumped from x to $x + z$. It is not hard to see that the *a posteriori* probability can be calculated as

$$q(z|w) = \frac{\int \pi(z)\Pi_{z'}\pi(z')^{k(w,0,z')}\beta(\mathrm{d}\pi)}{\int \Pi_{z'}\pi(z')^{k(w,0,z')}\beta(\mathrm{d}\pi)}.$$

While this makes sense initially only for walks of finite length it can clearly be extended to all transient paths. Note that although we only use $k(w, 0, z')$, in order to obtain the new $k(w', 0, z')$ we would need to know $k(w, z, z')$.

Let us suppose that R is a process with stationary increments $\{z_j\}$. If the increments process were ergodic and had a non-zero mean a we can again make the current position the origin and the process will be transient. $q(z|w)$ would exist a.e. R and can be compared to the corresponding conditional probabilities $r(z|w)$ under R. The relative entropy

$$H(R) = E^R\left[\sum_z r(z|w)\log\frac{r(z|w)}{q(z|w)}\right]$$

is then well-defined.

Theorem 5.1. *The function*

$$\bar{I}(a) = \inf_{\substack{R:\ \int z_1\,\mathrm{d}R = a\\ R\text{ ergodic}}} H(R)$$

defined for $a \neq 0$ extends as a convex function to all of R^d, and with this $\bar{I}$, (6) *is satisfied.*

The proof, which uses basic large deviation techniques, can be found in [8].

Remark 5.1. It is interesting to note that the rate functions $I(0)$ and $\bar{I}(0)$ coincide at 0 and can be calculated explicitly. Let $\mathcal{C}$ be the convex hull of the support of β in $\mathcal{M}(\mathbb{Z}^d)$. Then

$$I(0) = \bar{I}(0) = -\inf_{\pi\in\mathcal{C}}\inf_{\theta\in\mathbb{R}^d}\log\sum_z \mathrm{e}^{\langle\theta,z\rangle}\pi(z).$$

In particular $I(0) > 0$ if and only if 0 is not in the range of $\sum z\pi(z)$ as π varies over $\mathcal{C}$. This is referred to in the literature as the 'non-nestling' case.

6. Large deviations: The quenched case

Although a proof using the subadditive ergodic theorem exists, we will provide an alternate approach that is more appealing. We will illustrate this in the context of Brownian motion with a random drift (3).

We can define a diffusion on Ω with generator

$$\mathcal{L} = \frac{1}{2}\Delta + \langle b(\omega), \nabla\rangle,$$

where $\nabla = \{D_i\}$ are the generators of the translation group $\{\tau_x : x \in \mathbb{R}^d\}$. This is essentially the image of lifting the paths $x(t)$ of the diffusion on $\mathbb{R}^d$ corresponding to $\mathcal{L}^\omega$ to Ω by

$$\omega(t) = \tau_{x(t)}\omega.$$

While there is no possibility of having an invariant measure on $\mathbb{R}^d$, on Ω one can hope to find an invariant density $\varphi(\omega)$ provided we can find $\varphi(\omega) \geqslant 0$ in $L_1(P)$, that solves

$$\frac{1}{2}\Delta\varphi = \nabla \cdot (b\varphi).$$

If such a φ exists, then we have an ergodic theorem for the diffusion process Q^ω corresponding $\mathcal{L}$ on Ω,

$$\lim_{t\to\infty} \frac{1}{t}\int_0^t f(\omega(s))\mathrm{d}s = \int f(\omega)\varphi(\omega)\mathrm{d}P \quad \text{a.e. } Q^\omega \quad \text{a.e. } P. \tag{7}$$

This translates to an ergodic theorem on R^d as well

$$\lim_{t\to\infty} \frac{1}{t}\int_0^t f(\omega, x(s))\mathrm{d}s = \int f(\omega)\varphi(\omega)\mathrm{d}P \quad \text{a.e. } Q^\omega \quad \text{a.e. } P$$

where now Q^ω is the quenched process in the random environment. Since

$$x(t) = \int_0^t b(\omega, x(s)\mathrm{d}s + \beta(t),$$

it is clear that

$$\lim_{t\to\infty} \frac{x(t)}{t} = \int b(\omega)\varphi(\omega)\mathrm{d}P \quad \text{a.e. } Q^\omega \quad \text{a.e. } P$$

providing a law of large numbers for $x(t)$. While we can not be sure of finding φ for a given b it is easy to find a b for a given φ. For instance, we could take $b = \nabla\varphi/2\varphi$. Or more generally $b = (\nabla\varphi/2\varphi) + (c/\varphi)$ with $\nabla \cdot c = 0$. If we change b to $b' = (\nabla\varphi/2\varphi) + c$ with $\nabla \cdot c = 0$, the new process will have relative entropy

$$E^{Q^{b',\omega}}\left[\frac{1}{2}\int_0^t \left\| b(\omega(s)) - \frac{\nabla\varphi(\omega(s))}{2\varphi(\omega(s))} - \frac{c(\omega(s))}{\varphi(\omega(s))}\right\|^2 \mathrm{d}s\right].$$

Moreover, for almost all ω with respect to P, almost surely with respect to $Q^{b',\omega}$,

$$\lim_{t\to\infty} \frac{x(t)}{t} = \int \left[\frac{\nabla\varphi}{2\varphi} + \frac{c}{\varphi}\right]\varphi\mathrm{d}p = \int c\mathrm{d}P.$$

If we fix $\int c\mathrm{d}P = a$, the bound

$$\liminf_{t\to\infty} \frac{1}{t}\log Q^\omega\left[\frac{x(t)}{t} \simeq a\right] \geqslant -\frac{1}{2}\int \left\| b - \frac{\nabla\varphi}{2\varphi} - \frac{c}{\varphi}\right\|^2 \varphi\mathrm{d}P$$

is easily obtained. If we define

$$I(a) = \inf_{\substack{\nabla\cdot c=0\\ \int c\,\mathrm{d}P=a}} \frac{1}{2}\int \left\| b - \frac{\nabla\varphi}{2\varphi} - \frac{c}{\varphi}\right\|^2 \varphi\,\mathrm{d}P,$$

then

$$\liminf_{t\to\infty} \frac{1}{t}\log Q^{\omega}\left[\frac{x(t)}{t} \simeq a\right] \geqslant -I(a).$$

Of course these statements are valid a.e. Q^{ω} a.e. P. One can check that I is convex and the upper bound amounts to proving the dual estimate

$$\lim_{t\to\infty} \frac{1}{t}\log E^{Q^{\omega}}[\mathrm{e}^{\langle\theta, x(t)\rangle}] \leqslant \Psi(\theta),$$

where

$$\psi(\theta) = \sup_a [\langle a, \theta\rangle - I(a)].$$

We need a bound on the solution of

$$u_t = \frac{1}{2}\Delta u + \langle b, \nabla u\rangle$$

with $u(0) = \exp[\langle\theta, x\rangle]$. By Hopf-Cole transformation $v = \log u$ this reduces to estimating

$$v_t = \frac{1}{2}\Delta v + \frac{1}{2}\|\nabla v\|^2 + \langle b, \nabla v\rangle$$

with $v(0) = \langle\theta, x\rangle$. This can be done if we can construct a subsolution

$$\frac{1}{2}\nabla\cdot w + \frac{1}{2}\|\nabla w\|^2 + \langle b, w\rangle \leqslant \psi(\theta)$$

on Ω, where $w\colon \Omega \to \mathbb{R}^d$ satisfies $\int w\,\mathrm{d}P = \theta$ and w is closed in the sense that $D_i w_j = D_j w_i$. The existence of the subsolution comes from convex analysis.

$$\begin{aligned}
\psi(\theta) &= \sup_{\substack{\varphi\\ \nabla\cdot c=0}} \left[\int \langle c, \theta\rangle \mathrm{d}P - \frac{1}{2}\int \left\| b - \frac{\nabla\varphi}{2\varphi} - \frac{c}{\varphi}\right\|^2 \varphi\,\mathrm{d}P\right]\\
&= \sup_{\varphi}\sup_{c}\inf_{u} \left[\int \langle c, \theta + \nabla u\rangle \mathrm{d}P - \frac{1}{2}\int \left\| b - \frac{\nabla\varphi}{2\varphi} - \frac{c}{\varphi}\right\|^2 \varphi\,\mathrm{d}P\right]\\
&= \sup_{\varphi}\inf_{u}\sup_{c} \left[\int \langle c, \theta + \nabla u\rangle \mathrm{d}P - \frac{1}{2}\int \left\| b - \frac{\nabla\varphi}{2\varphi} - \frac{c}{\varphi}\right\|^2 \varphi\,\mathrm{d}P\right]\\
&= \sup_{\varphi}\inf_{u} \int \sup_{c} \left[\langle c, \theta + \nabla u\rangle - \frac{1}{2}\left\| b - \frac{\nabla\varphi}{2\varphi} - \frac{c}{\varphi}\right\|^2 \varphi\right] \mathrm{d}P
\end{aligned}$$

$$
\begin{aligned}
&= \sup_{\varphi} \inf_{u} \int \left[\left\langle b - \frac{\nabla \varphi}{2\varphi}, \theta + \nabla u \right\rangle + \frac{1}{2} \|\theta + \nabla u\|^2 \right] \varphi \mathrm{d}P \\
&= \sup_{\varphi} \inf_{u} \int \left[\left[\langle b, \theta + \nabla u \rangle + \frac{1}{2} \|\theta + \nabla u\|^2 \right] \varphi - \frac{1}{2} \langle \nabla u, \nabla \varphi \rangle \right] \mathrm{d}P \\
&= \sup_{\varphi} \inf_{u} \int \left[\frac{1}{2} \Delta u + \langle b, \theta + \nabla u \rangle + \frac{1}{2} \|\theta + \nabla u\|^2 \right] \varphi \mathrm{d}P \\
&= \sup_{\varphi} \inf_{\substack{w \text{ closed} \\ \int w \mathrm{d}P = \theta}} \int \left[\frac{1}{2} \nabla \cdot w + \langle b, w \rangle + \frac{1}{2} \|w\|^2 \right] \varphi \mathrm{d}P \\
&= \inf_{\substack{w \text{ closed} \\ \int w \mathrm{d}P = \theta}} \sup_{\varphi} \int \left[\frac{1}{2} \nabla \cdot w + \langle b, w \rangle + \frac{1}{2} \|w\|^2 \right] \varphi \mathrm{d}P \\
&= \inf_{\substack{w \text{ closed} \\ \int w \mathrm{d}P = \theta}} \sup_{\omega} \left[\frac{1}{2} \nabla \cdot w + \langle b, w \rangle + \frac{1}{2} \|w\|^2 \right]
\end{aligned}
$$

which proves the existence of the subsolution.

Remark 6.1. This can be viewed as showing the existence of a limit as $\varepsilon \to 0$ (homogenization) of the solution of

$$u_t^\varepsilon = \frac{\varepsilon}{2} \Delta u^\varepsilon + \frac{1}{2} \|\nabla u^\varepsilon\|^2 + \left\langle b\left(\frac{x}{\varepsilon}, \omega\right), \nabla u^\varepsilon \right\rangle$$

with $u^\varepsilon(0, x) = f(x)$. The limit satisfies

$$u_t = \Psi(\nabla u)$$

with $u(0, x) = f(x)$.

This can be generalized to equations of the form

$$u_t^\varepsilon = \frac{\varepsilon}{2} \Delta u^\varepsilon + H\left(\frac{x}{\varepsilon}, \nabla u^\varepsilon, \omega\right).$$

We have left out the details as well as the hypothesis needed to prove the results. They will appear in [4].

Acknowledgments

This research was supported by a grant from the National Science Foundation DMS-0104343.

References

[1] Bricmont J and Kupiainen A, Random walks in asymmetric random environments, *Comm. Math. Phys.* **142(2)** (1991) 345–420

[2] Comets Francis, Gantert Nina and Zeitouni Ofer, Quenched, annealed and functional large deviations for one-dimensional random walk in random environment, *Probab. Theory Related Fields* **118(1)** (2000) 65–114

[3] Greven Andreas and den Hollander Frank, Large deviations for a random walk in random environment, *Ann. Probab.* **22(3)** (1994) 1381–1428

[4] Kosygina E, Rezakhnalou F and Varadhan S R S, Homogenization of random Hamilton–Jacobi–Bellman equations (in preparation)

[5] Sinai Ya G, The limit behavior of a one-dimensional random walk in a random environment, *Theor. Probab. Appl.* **27(2)** (1982) 256–268

[6] Solomon Fred, Random walks in a random environment, *Ann. Probab.* **3** (1975) 1–31

[7] Sznitman A S and Zeitouni O, On the diffusive behavior of isotropic diffusions in a random environment, *C. R. Acad. Sci. Paris, Ser. I* **339** (2004) 429–434

[8] Varadhan S R S, Large deviations for random walks in a random environment. Dedicated to the memory of Jürgen K Moser, *Comm. Pure Appl. Math.* **56(8)** (2003) 1222–1245

[9] Zeitouni Ofer, Random walks in random environments, Proceedings of ICM 2002, vol. III, pp. 117–127

[10] Zeitouni Ofer, Lecture Notes on RWRE, Notes from the St. Flour Summer School in Probability (2001), available at http://www.ee.technion.ac.il/zeitouni/ps/notes1.ps

[11] Zerner Martin P W, Lyapunov exponents and quenched large deviations for multidimensional random walk in random environment, *Ann. Probab.* **26(4)** (1998) 1446–1476

Comm. Pure Appl. Math. 59 (2006) 1489-1521

Stochastic Homogenization of Hamilton-Jacobi-Bellman Equations

ELENA KOSYGINA
Baruch College

FRAYDOUN REZAKHANLOU
University of California, Berkeley

AND

S. R. S. VARADHAN
Courant Institute

Abstract

We study the homogenization of some Hamilton-Jacobi-Bellman equations with a vanishing second-order term in a stationary ergodic random medium under the hyperbolic scaling of time and space. Imposing certain convexity, growth, and regularity assumptions on the Hamiltonian, we show the locally uniform convergence of solutions of such equations to the solution of a deterministic "effective" first-order Hamilton-Jacobi equation. The effective Hamiltonian is obtained from the original stochastic Hamiltonian by a minimax formula. Our homogenization results have a large-deviations interpretation for a diffusion in a random environment.

1 Introduction

Homogenization of a differential equation is the process of replacing rapidly varying coefficients (or functions) by new ones such that the solutions are close. A simple illustration is provided by the following example: Suppose that $a(x)$ is a periodic or stationary random function on $\mathbb{R}$ satisfying $a \leqslant a(x) \leqslant b$ for some constants $0 < a < b < \infty$. For $\varepsilon > 0$ one can consider the elliptic operator

$$\mathcal{L}_\varepsilon = \frac{d}{dx} a\left(\frac{x}{\varepsilon}\right) \frac{d}{dx}.$$

For small ε, $\mathcal{L}_\varepsilon$ can be replaced by

$$\mathcal{L} = \frac{d}{dx} \bar{a} \frac{d}{dx},$$

where $\bar{a} = [\mathbb{E}[1/a]]^{-1}$ is the "harmonic mean" of $a(\cdot)$. $\mathbb{E}$ is the expectation in the random case or the average over a period in the periodic case. As $\varepsilon \to 0$, solutions

of parabolic or elliptic equations of the form

$$\frac{du}{dt} = \mathcal{L}_\varepsilon u, \quad u(0,x) = f(x),$$

or

$$\lambda u - \mathcal{L}_\varepsilon u = f$$

are close to the corresponding solutions with $\mathcal{L}_\varepsilon$ replaced by $\mathcal{L}$.

There is a vast literature on periodic and quasi-periodic homogenization of linear and nonlinear partial differential equations (see, for example, monographs [1, 2, 5, 7, 10, 15] and references therein). Many results obtained in the periodic setting are based on the existence of "correctors," i.e., periodic solutions to an auxiliary problem, which arise from a formal expansion in a scaling parameter ε. In connection with this approach for Hamilton-Jacobi type and more general equations, let us mention just several representative works [4, 6, 8, 11], where one can find further references. It turns out [12] that for Hamilton-Jacobi equations with stationary ergodic Hamiltonians, correctors may not exist in general. Homogenization results for stationary ergodic environments are usually obtained by an application of a version of the ergodic theorem (see, for instance, [3, 10, 16, 17, 19]).

In the current work we will be interested in solutions to certain Hamilton-Jacobi-Bellman equations. These are solutions u_ε to equations of the type

$$\frac{\partial u_\varepsilon}{\partial t} = \frac{\varepsilon}{2}\Delta u_\varepsilon + H\left(\nabla u_\varepsilon, \frac{x}{\varepsilon}, \omega\right), \quad (t,x) \in [0,\infty) \times \mathbb{R}^d, \tag{1.1}$$

with initial condition $u_\varepsilon(0,x) = f(x)$. Here $H(p,x,\omega)$ is a convex function of p that is a stationary random process in x. We are interested in establishing a result of the form $u_\varepsilon \to u$ as $\varepsilon \to 0$ where u is the solution of

$$\frac{\partial u}{\partial t} = \overline{H}(\nabla u), \quad u(0,x) = f(x). \tag{1.2}$$

The solutions u_ε and u have natural variational representations and we will consider these problems from this perspective. Therefore, it will not be necessary for us to characterize these solutions as viscosity solutions.

These types of problems arise naturally if we consider questions of large deviations for diffusions in a random environment. More specifically, consider a diffusion process on $\mathbb{R}^d$ with generator

$$\mathcal{A}_\omega = \frac{1}{2}\Delta + \langle b(x,\omega), \nabla\rangle,$$

i.e., Brownian motion with a random drift $b(x,\omega)$ that is assumed to be an ergodic stationary process in $x \in \mathbb{R}^d$. We are interested in considering the probabilities of large deviations of $x(t)/t$ as $t \to \infty$. This amounts to establishing the limiting behavior

$$\lim_{t\to\infty} \frac{1}{t} \log E^{Q_0^{b,\omega}}[\exp[\langle\theta, x(t)\rangle]] = \overline{H}(\theta).$$

Here $Q_x^{b,\omega}$ is the diffusion process corresponding to $\mathcal{A}_\omega$ starting from $x \in \mathbb{R}^d$ at time 0. The above limit is supposed to exist almost surely in ω. The probabilities of large deviations are now determined in the standard manner, the rate function being the conjugate function $\mathcal{I}(y)$ defined for $y \in \mathbb{R}^d$, by

$$\mathcal{I}(y) = \sup_{\theta \in \mathbb{R}^d} [\langle \theta, y \rangle - \overline{H}(\theta)]. \tag{1.3}$$

The quantity

$$v(t, x) = E^{Q_x^{b,\omega}} [\exp\langle \theta, x(t) \rangle]$$

solves the equation

$$\frac{\partial v}{\partial t} = \frac{1}{2} \Delta v + \langle b(x, \omega), \nabla v \rangle, \qquad v(0, x) = \exp\langle \theta, x \rangle.$$

We are interested in the behavior of $\frac{1}{t} \log v(t, 0)$ as $t \to \infty$. If we define

$$u_\varepsilon(t, x) = \varepsilon \log v\left(\frac{t}{\varepsilon}, \frac{x}{\varepsilon}\right),$$

then the problem reduces to the study of the behavior of $u_\varepsilon(1, 0)$ as $\varepsilon \to 0$. The equation satisfied by u_ε is

$$\frac{\partial u_\varepsilon}{\partial t} = \frac{\varepsilon}{2} \Delta u_\varepsilon + \frac{1}{2} \|\nabla u_\varepsilon\|^2 + \left\langle b\left(\frac{x}{\varepsilon}, \omega\right), \nabla u_\varepsilon \right\rangle, \qquad u(0, x) = \langle \theta, x \rangle.$$

This is, of course, a special case of (1.1) with

$$H(p, x, \omega) = \frac{1}{2} \|p\|^2 + \langle b(x, \omega), p \rangle.$$

Remark 1.1. After our work was completed, we received a preprint [13] that deals with more general "viscous" Hamilton-Jacobi equations and in the special case that we consider overlaps considerably with our results. The main tools in [13] are the uniform gradient estimate on solutions u_ε and the subadditive ergodic theorem. The approach presented in our paper is different. It is based on a direct application of the standard ergodic theorem to obtain a uniform lower bound on solutions and on a straightforward construction of "supercorrectors" for an upper bound. The minimax formula for the effective Hamiltonian arises naturally in the proof. Another characteristic feature of our method is that it does not require a uniform gradient estimate of u_ε even for subquadratic Hamiltonians. The tradeoff is that we need a stronger regularity assumption on the Hamiltonian (see (H3)) in the next section.

In the special case of

$$H(p, \omega) = a(\omega) \|p\|^\alpha$$

with $\alpha > 1$, in addition to uniform upper and lower bounds on a, we would need uniform continuity of $a(\tau_y \omega)$, whereas [13] would assume a uniform bound on the gradient in y of $a(\tau_y \omega)$. Moreover, if there is a way of getting uniform gradient bounds for perturbations of the form $H(p, \omega) + \delta \|p\|^k$ that are uniform in δ, we

can easily choose $k > d$ and put ourselves in a situation where (H4) holds. It is not hard to let $\delta \to 0$ and recover Theorem 2.2.

2 Main Results

We consider a probability space $(\Omega, \mathcal{F}, \mathbb{P})$ on which $\mathbb{R}^d$ acts as a group $\{\tau_x : x \in \mathbb{R}^d\}$ of measure-preserving transformations. $\mathbb{P}$ is assumed to be ergodic under this action. Let the function $H(p, \omega) : \mathbb{R}^d \times \Omega \to \mathbb{R}$ be convex in p and have the following properties:

(H1) For all $p \in \mathbb{R}^d$, $\omega \in \Omega$,

$$c_1\|p\|^\alpha - c_2 \leqslant H(p, \omega) \leqslant c_3\|p\|^\beta + c_4 \tag{2.1}$$

for some positive constants c_1, c_2, c_3, and c_4, $1 < \alpha \leqslant \beta < \infty$. Equivalently, its convex conjugate $L(q, \omega)$, defined by

$$L(q, \omega) = \sup_p[\langle p, q\rangle - H(p, \omega)], \tag{2.2}$$

satisfies

$$c_1\|q\|^{\beta'} - c_2 \leqslant L(q, \omega) \leqslant c_3\|q\|^{\alpha'} + c_4 \tag{2.3}$$

with $\alpha' = \alpha/(\alpha - 1)$, $\beta' = \beta/(\beta - 1)$, and possibly a different set of constants.

(H2) $H(p, \tau_x\omega)$ is uniformly continuous in x; i.e., for any $\ell < \infty$,

$$\lim_{\delta\to 0} \sup_{\|x\|\leqslant\delta} \sup_{\|p\|\leqslant\ell} \sup_{\omega} |H(p, \tau_x\omega) - H(p, \omega)| = 0. \tag{2.4}$$

It is not difficult to see, by using relation (2.2) and the bounds (2.1), that $L(q, \omega)$ satisfies a similar estimate

$$\lim_{\delta\to 0} \sup_{\|x\|\leqslant\delta} \sup_{\|q\|\leqslant\ell} \sup_{\omega} |L(q, \tau_x\omega) - L(q, \omega)| = 0. \tag{2.5}$$

For some of the results we will need stronger regularity or convexity assumptions.

(H3) There exists a function $\nu(\delta) \to 0$ as $\delta \to 0$ and $C > 0$ such that for $\|x\| \leqslant \delta$ and $\omega \in \Omega$

$$L(q, \tau_x\omega) \leqslant (1 + \nu(\delta))L(q, \omega) + C\nu(\delta). \tag{2.6}$$

Equivalently, there exists a function $\nu(\delta) \to 0$ as $\delta \to 0$ and $C > 0$ such that for $\|x\| \leqslant \delta$ and $\omega \in \Omega$

$$H(p, \tau_x\omega) \geqslant (1 + \nu(\delta))H((1 + \nu(\delta))^{-1}p, \omega) - C\nu(\delta). \tag{2.7}$$

(H4) Assumption (H1) holds with $\alpha > d$.

(H5) Assumption (H1) holds with $\alpha > 2$ and there exist functions $\gamma(\delta) > 0$ and $C(\delta) \to 0$ as $\delta \to 0$ such that for all $\omega \in \Omega$

$$(1-\delta)H((1-\delta)^{-1}p, \omega) \geqslant H(p, \omega) + \gamma(\delta)\|p\|^2 - C(\delta). \tag{2.8}$$

For (2.8) to hold, it is sufficient that (uniformly in ω) $D^2 H(p, \omega) \geqslant cI$ on the set $\{p \in \mathbb{R}^d : \|p\| \geqslant k\}$ for some $c, k > 0$. Here I is the identity matrix.

For any given $\varepsilon > 0$ and $\omega \in \Omega$, we consider the solution $u_\varepsilon = u_\varepsilon(t, x, \omega)$ of equation (1.1), which we write as

$$\frac{\partial u_\varepsilon}{\partial t} = \frac{\varepsilon}{2}\Delta u_\varepsilon + H(\nabla u_\varepsilon, \tau_{x/\varepsilon}\omega), \quad (t, x) \in [0, \infty) \times \mathbb{R}^d, \tag{2.9}$$

with the initial condition $u_\varepsilon(0, x) = f(x)$. We assume that f satisfies the following condition:

(H6) The function f is uniformly continuous on $\mathbb{R}^d$. This implies that for every $\delta > 0$ there is a constant $K_\delta > 0$ such that for all $x, y \in \mathbb{R}^d$

$$|f(x) - f(y)| \leqslant K_\delta \|x - y\| + \delta. \tag{2.10}$$

We note that the solution $u_\varepsilon(t, x, \omega)$ of (2.9) is equal to $\varepsilon v_\varepsilon(t/\varepsilon, x/\varepsilon, \omega)$, the rescaled version of v_ε that solves

$$\frac{\partial v_\varepsilon}{\partial t} = \frac{1}{2}\Delta v_\varepsilon + H(\nabla v_\varepsilon, \tau_x\omega), \quad (t, x) \in [0, \infty) \times \mathbb{R}^d, \tag{2.11}$$

with $v_\varepsilon(0, x) = \varepsilon^{-1} f(\varepsilon x)$.

We now construct the convex function $\overline{H}(p)$ that appears in (1.2). The translation group $\{\tau_x : x \in \mathbb{R}^d\}$ acting on $L^2(\Omega, \mathcal{F}, \mathbb{P})$ will have infinitesimal generators $\{\nabla_i : 1 \leqslant i \leqslant d\}$ in the coordinate directions and the corresponding Laplace operator $\Delta = \sum_i \nabla_i^2$. For reasonable choices of $b(\omega) : \Omega \to \mathbb{R}^d$, the operator

$$\mathcal{A}_b = \frac{1}{2}\Delta + \langle b(\omega), \nabla \rangle$$

will define a Markov process on Ω. Construction of this Markov process is not difficult. Given a starting point $\omega \in \Omega$, we define $b(x, \omega) : \mathbb{R}^d \to \mathbb{R}^d$ by $b(x, \omega) = b(\tau_x\omega)$. This allows us to define the diffusion $Q_0^{b,\omega}$, starting from 0 at time 0, in the random environment that corresponds to the generator

$$\frac{1}{2}\Delta + \langle b(x, \omega), \nabla \rangle.$$

The diffusion is then lifted to Ω by evolving ω randomly in time by the rule $\omega(t) = \tau_{x(t)}\omega$. The induced measure $P^{b,\omega}$ defines the Markov process on Ω that corresponds to $\mathcal{A}_b$. The problem of finding the invariant measures for the process $P^{b,\omega}$ with generator $\mathcal{A}_b$ on Ω is very hard and nearly impossible to solve. However,

if we can find a density $\varphi > 0$ such that $\varphi\, d\mathbb{P}$ is an invariant ergodic probability measure for $\mathcal{A}_b$, then we have by the ergodic theorem

$$\lim_{t\to\infty} \frac{1}{t}\int_0^t F(\omega(s))ds = \int_\Omega F(\omega)\varphi(\omega)d\mathbb{P}$$

a.e. $P^{b,\omega}$ or in $L^1(P^{b,\omega})$ for almost all ω with respect to $\mathbb{P}$.

Let us denote by **B** the space of essentially bounded maps from $\Omega \to \mathbb{R}^d$ and by **D** the space of probability densities $\varphi : \Omega \to \mathbb{R}$ relative to $\mathbb{P}$, with φ, $\nabla\varphi$, and $\nabla^2\varphi$ essentially bounded and φ in addition having a positive essential lower bound. Let us denote by $\mathcal{E}$ the following subset of $\mathbf{B}\times\mathbf{D}$:

$$\mathcal{E} = \left\{(b,\varphi)\in \mathbf{B}\times\mathbf{D} : \tfrac{1}{2}\Delta\varphi = \nabla\cdot(b\varphi)\right\}. \tag{2.12}$$

Here we assume that the equation $\frac{1}{2}\Delta\varphi = \nabla\cdot(b\varphi)$ is satisfied in the weak sense. We define the convex function $\overline{H}$ on $\mathbb{R}^d$ by

$$\overline{H}(p) = \sup_{(b,\varphi)\in\mathcal{E}} \left[\langle p, \mathbb{E}[b(\omega)\varphi(\omega)]\rangle - \mathbb{E}[L(b(\omega),\omega)\varphi(\omega)]\right], \tag{2.13}$$

where $\mathbb{E}$ denotes the expectation with respect to $\mathbb{P}$. The corresponding variational solution of (1.2) is given by

$$u(t,x) = \sup_y \left(f(y) - t\mathcal{I}\left(\frac{y-x}{t}\right)\right), \tag{2.14}$$

where $\mathcal{I}$ is related to $\overline{H}$ by the duality relation (1.3). Our main result is stated below.

THEOREM 2.1 (Lower Bound) *Assume* (H1), (H2), *and* (H6). *Let* $u_\varepsilon(t,x,\omega)$ *be the solution of* (2.9) *and* $u(t,x)$ *be given by* (2.14). *Then with probability* 1 *for any* $\ell, T > 0$, *we have*

$$\liminf_{\varepsilon\to 0} \inf_{0\leqslant t\leqslant T} \inf_{\|x\|\leqslant \ell} (u_\varepsilon(t,x,\omega) - u(t,x)) \geqslant 0.$$

THEOREM 2.2 (Upper Bound) *Assume* (H1), (H2), (H6), *and either* (H3), (H4), *or* (H5). *Let* $u_\varepsilon(t,x,\omega)$ *be the solution of* (2.9) *and* $u(t,x)$ *be given by* (2.14). *Then with probability* 1 *for any* $\ell, T > 0$, *we have*

$$\limsup_{\varepsilon\to 0} \sup_{0\leqslant t\leqslant T} \sup_{\|x\|\leqslant \ell} (u_\varepsilon(t,x,\omega) - u(t,x)) \leqslant 0.$$

Theorems 2.1 and 2.2 immediately imply the homogenization result under appropriate conditions.

THEOREM 2.3 *Assume* (H1), (H2), (H6), *and either* (H3), (H4), *or* (H5). *Then with probability* 1

$$\lim_{\varepsilon\to 0} |u_\varepsilon(t,x,\omega) - u(t,x)| = 0 \tag{2.15}$$

locally uniformly in $(t,x)\in[0,\infty)\times\mathbb{R}^d$.

Remark 2.4. It is easy to see from the definition of $\overline{H}(p)$ that it satisfies the same upper and lower bounds given in (2.1). We can always choose $b(\omega) = b$ a constant vector $b \in \mathbb{R}^d$, and $\mathbb{P}$ is still an ergodic invariant measure for the Markov process with generator $\mathcal{A}_b$ on Ω. Therefore, using the dual upper bound on L, which is uniform in ω, we get

$$\overline{H}(p) \geqslant \sup_{b \in \mathbb{R}^d} [\langle p, b\rangle - \mathbb{E}[L(b, \omega)]] \geqslant c_1 \|p\|^\alpha - c_2. \tag{2.16}$$

It is just as easy to note that

$$\overline{H}(p) \leqslant \sup_{b \in \mathbb{R}^d} [\langle p, b\rangle - \inf_\omega L(b, \omega)] \leqslant c_3 \|p\|^\beta + c_4. \tag{2.17}$$

3 Outline of Proof

The first step in establishing the lower bound is the variational representation of solutions of Hamilton-Jacobi-Bellman equations (see [9]). Let $\mathcal{C}$ be the set of all bounded maps $c(s, x)$ from $[0, T] \times \mathbb{R}^d$ to $\mathbb{R}^d$ such that $\sup_{s,x} \|c(s, x)\| < \infty$. Consider the diffusion Q_x^c on $\mathbb{R}^d$ starting from $x \in \mathbb{R}^d$ at time 0 with time-dependent generator

$$\frac{1}{2}\Delta + c(s, x) \cdot \nabla$$

in the time interval $[0, t]$. For each $c \in \mathcal{C}$ and $\omega \in \Omega$ we consider

$$v_c(t, x, \omega) = E^{Q_x^c}\left(f(x(t)) - \int_0^t L\big(c(s, x(s)), \tau_{x(s)}\omega\big) ds \right),$$

where $L(c, \omega)$ is as in (2.2). If

$$v(t, x, \omega) = \sup_{c \in \mathcal{C}} v_c(t, x, \omega),$$

then v is the solution of

$$\frac{\partial v}{\partial t} = \frac{1}{2}\Delta v + H(\nabla v, \tau_x \omega) \tag{3.1}$$

with $v(0, x) = f(x)$.

There is a simple relation between $v(t, y, \cdot)$ and $v(t, 0, \cdot)$. If we define $f^y(x) = f(x + y)$, then the solution of (3.1) with initial data $v(0, x) = f^y(x)$ and $\omega' = \tau_y \omega$ is given by

$$v^y(t, x, \omega') = v^y(t, x, \tau_y \omega) = v(t, x + y, \omega). \tag{3.2}$$

In particular,

$$v(t, y, \omega) = v^y(t, 0, \tau_y \omega).$$

The solution u_ε of (2.9) with initial data $f(x)$ is related to the solution v_ε of (3.1) with initial data $\varepsilon^{-1} f(\varepsilon x)$ by

$$u_\varepsilon(t, x, \omega) = \varepsilon v_\varepsilon(\varepsilon^{-1} t, \varepsilon^{-1} x, \omega). \tag{3.3}$$

Relation (3.2) translates to

$$u_\varepsilon^y(t,x,\omega) = u_\varepsilon(t, x+y, \tau_{-y/\varepsilon}\omega).$$

We therefore obtain the following variational expression for $u_\varepsilon(t,x)$:

$$\begin{aligned} u_\varepsilon(t,x,\omega) &= \sup_{c\in\mathcal{C}} E^{Q^c_{x/\varepsilon}}\left(f(\varepsilon x(\varepsilon^{-1}t)) - \varepsilon \int_0^{t/\varepsilon} L\big(c(s,x(s)), \tau_{x(s)}\omega\big) ds\right) \\ &= \sup_{c\in\mathcal{C}} E^{Q^{\varepsilon,c}_x}[f(x(t)) - \xi_\varepsilon(t)] \end{aligned} \tag{3.4}$$

where $Q^{\varepsilon,c}_x$ is the diffusion on $\mathbb{R}^d$ starting from x corresponding to the generator

$$\frac{\varepsilon}{2}\Delta + c(\varepsilon^{-1}s, \varepsilon^{-1}x)\cdot\nabla;$$

i.e., almost surely with respect to $Q^{\varepsilon,c}_x$, $x(t)$ satisfies

$$x(t) = x + \int_0^t c(\varepsilon^{-1}s, \varepsilon^{-1}x(s))ds + \sqrt{\varepsilon}\beta(t) \tag{3.5}$$

and

$$\xi_\varepsilon(t) = \int_0^t L\big(c(\varepsilon^{-1}s, \varepsilon^{-1}x(s)), \tau_{\varepsilon^{-1}x(s)}\omega\big)ds. \tag{3.6}$$

Since the supremum over $c\in\mathcal{C}$ is taken for each ω, one can choose c to depend on ω. A special choice for $c(t,x)$, one that depends on $\omega\in\Omega$ but not on t, is the choice $c(t,x) = c(t,x,\omega) = c(x,\omega) = b(\tau_x\omega)$ with $(b,\varphi)\in\mathcal{E}$. With that choice we can consider either the process $\{Q^{b,\omega}_x\}$ on $\mathbb{R}^d$ or the process $\{P^{b,\omega}\}$ with values in Ω. It is easy to see that for any $y\in\mathbb{R}^d$, the translation map $\hat{\tau}_y$ on $C([0,T];\mathbb{R}^d)$ defined by $x(\cdot)\to x(\cdot)+y$ has the property

$$Q^{b,\omega}_y = Q^{b,\tau_y\omega}_0 \hat{\tau}_y^{-1},$$

which is essentially a restatement of (3.2). Since $(b,\varphi)\in\mathcal{E}$, by the ergodic theorem we have

$$\lim_{\varepsilon\to 0} \varepsilon \int_0^{t/\varepsilon} b(\omega(s))ds = t\int b(\omega)\varphi(\omega)d\mathbb{P} \overset{\text{def}}{=} m(b,\varphi)t \tag{3.7}$$

and

$$\lim_{\varepsilon\to 0} \varepsilon \int_0^{t/\varepsilon} L\big(b(\omega(s)), \omega(s)\big)ds = t\int L(b(\omega),\omega)\varphi(\omega)d\mathbb{P} \overset{\text{def}}{=} h(b,\varphi)\,t. \tag{3.8}$$

Both limits are valid in $L^1(P^{b,\omega})$ for $\mathbb{P}$ almost all ω. If we define $\mathbf{A}\subset\mathbb{R}^d\times\mathbb{R}$ as

$$\mathbf{A} = \{(m(b,\varphi), h(b,\varphi)) : (b,\varphi)\in\mathcal{E}\}, \tag{3.9}$$

then

$$\liminf_{\varepsilon\to 0} u_\varepsilon(t,0,\omega) \geqslant [f(mt) - ht]$$

for every $(m, h) \in \mathbf{A}$. Therefore, for almost all ω with respect to $\mathbb{P}$

$$\begin{aligned}\liminf_{\varepsilon \to 0} u_\varepsilon(t, 0, \omega) &\geqslant \sup_{(m,h)\in\mathbf{A}} [f(mt) - ht] \\ &= \sup_{y\in\mathbb{R}^d} (f(y) - t\mathcal{I}(y/t)) = u(t, 0).\end{aligned} \tag{3.10}$$

This is a very weak form of convergence, and work has to be done in order to strengthen it to locally uniform convergence.

The upper bound is first obtained for linear f and then extended to general f. By using the convex duality and the minimax theorem, the right-hand side of (3.10) is rewritten in terms of the dual problem. If we take $f(x) = \langle p, x\rangle$, we have established an asymptotic lower bound for u_ε, which is the solution

$$u(t, x) = \langle p, x\rangle + t\overline{H}(p)$$

of (1.2) with $u(0, x) = \langle p, x\rangle$. Here

$$\begin{aligned}\overline{H}(p) &= \sup_{(b,\varphi)\in\mathcal{E}} \mathbb{E}\big[[\langle p, b(\omega)\rangle - L(b(\omega), \omega)]\varphi(\omega)\big] \\ &= \sup_{\varphi}\sup_{b}\inf_{\psi} \mathbb{E}\big[[\langle p, b(\omega)\rangle - L(b(\omega), \omega) + \mathcal{A}_b\psi(\omega)]\varphi(\omega)\big] \\ &= \sup_{\varphi}\inf_{\psi}\sup_{b} \mathbb{E}\big[[\langle p, b(\omega)\rangle - L(b(\omega), \omega) + \mathcal{A}_b\psi(\omega)]\varphi(\omega)\big] \\ &= \sup_{\varphi}\inf_{\psi}\sup_{b} \mathbb{E}\big[[\langle p + \nabla\psi(\omega), b(\omega)\rangle - L(b(\omega), \omega) + \tfrac{1}{2}\Delta\psi(\omega)]\varphi(\omega)\big] \\ &= \sup_{\varphi}\inf_{\psi} \mathbb{E}\big[[H(p + \nabla\psi(\omega), \omega) + \tfrac{1}{2}\Delta\psi(\omega)]\varphi(\omega)\big] \\ &= \inf_{\psi}\sup_{\varphi} \mathbb{E}\big[[H(p + \nabla\psi(\omega), \omega) + \tfrac{1}{2}\Delta\psi(\omega)]\varphi(\omega)\big] \\ &= \inf_{\psi}\operatorname*{ess\,sup}_{\omega} \big[H(p + \nabla\psi(\omega), \omega) + \tfrac{1}{2}\Delta\psi(\omega)\big].\end{aligned}$$

We have used the fact that

$$\inf_{\psi} \mathbb{E}[\varphi\mathcal{A}_b\psi(\omega)] = -\infty$$

unless $\varphi\, d\mathbb{P}$ is an invariant measure for $\mathcal{A}_b$, in which case it is 0. It follows that for any $\delta > 0$, there exists a "ψ" such that

$$\frac{1}{2}\Delta\psi(\omega) + H(\theta + \nabla\psi(\omega), \omega) \leqslant \overline{H}(\theta) + \delta.$$

The "ψ" is a weak object, and one has to do some work before we can use it as a test function and obtain the upper bound by comparison. The interchange of inf and sup that we have done freely needs justification.

4 Lower Bounds

We begin with some estimates on admissible controls and $u_\varepsilon(t,x,\omega)$. Recall the notation introduced in (3.5) and (3.6).

LEMMA 4.1 *Assume* (2.3) *and* (2.10). *Then in the variational formula* (3.4), *the supremum over $\mathcal{C}$ can be replaced with the supremum over the subset $\mathcal{C}^* \subset \mathcal{C}$ of controls that satisfy the following condition: for each $\delta > 0$ there is $C_\delta > 0$, which depends only on δ and the constants in* (2.3) *and* (2.10), *such that*

$$\sup_{x,\omega} E^{Q_x^{\varepsilon,c}} |\xi_\varepsilon(t)| \leqslant C_\delta(t+\sqrt{\varepsilon t}) + 2\beta\delta. \tag{4.1}$$

In particular, for all $c \in \mathcal{C}^*$,

$$\sup_{x,\omega} E^{Q_x^{\varepsilon,c}}\left[\int_0^t \|c(\varepsilon^{-1}s, \varepsilon^{-1}x(s))\|^{\beta'} ds\right] \leqslant C_\delta(t+\sqrt{\varepsilon t}) + 2\beta\delta. \tag{4.2}$$

PROOF: There is a trivial lower bound with $c \equiv 0$ that can be obtained from (2.3), (2.10), and (3.5): for every $\delta > 0$

$$\begin{aligned} u_\varepsilon(t,x,\omega) - f(x) &\geqslant E^{Q_x^{\varepsilon,0}}[f(x(t)) - f(x) - \xi_\varepsilon(t)] \\ &\geqslant -K_\delta E^{Q_x^{\varepsilon,0}}[\|x(t)-x\|] - c_4 t - \delta \\ &\geqslant -K_\delta\sqrt{\varepsilon t} - c_4 t - \delta. \end{aligned} \tag{4.3}$$

So we need to consider only those $c \in \mathcal{C}$ for which

$$E^{Q_x^{\varepsilon,c}}[f(x(t)) - f(x) - \xi_\varepsilon(t)] \geqslant -K_\delta\sqrt{\varepsilon t} - c_4 t - \delta.$$

In view of (2.10), such a c has to satisfy

$$E^{Q_x^{\varepsilon,c}}[K_\delta\|x(t)-x\| - \xi_\varepsilon(t)] \geqslant -K_\delta\sqrt{\varepsilon t} - c_4 t - 2\delta. \tag{4.4}$$

The following simple inequalities allow us to estimate $E^{Q_x^{\varepsilon,c}}\|x(t)-x\|$ in terms of $E^{Q_x^{\varepsilon,c}}\xi_\varepsilon(t)$:

$$E^{Q_x^{\varepsilon,c}}\|x(t)-x\| \leqslant E^{Q_x^{\varepsilon,c}}\left[\int_0^t \|c(\varepsilon^{-1}s, \varepsilon^{-1}x(s))\|ds + \sqrt{\varepsilon t}\right], \tag{4.5}$$

$$\begin{aligned} &E^{Q_x^{\varepsilon,c}}\int_0^t \|c(\varepsilon^{-1}s, \varepsilon^{-1}x(s))\|ds \\ &\qquad \leqslant t^{1/\beta}\left(E^{Q_x^{\varepsilon,c}}\int_0^t \|c(\varepsilon^{-1}s, \varepsilon^{-1}x(s))\|^{\beta'} ds\right)^{1/\beta'}, \end{aligned} \tag{4.6}$$

$$E^{Q_x^{\varepsilon,c}}\int_0^t \|c(\varepsilon^{-1}s, \varepsilon^{-1}x(s))\|^{\beta'} ds \leqslant c_1^{-1}\big(E^{Q_x^{\varepsilon,c}}\xi_\varepsilon(t) + c_2 t\big). \tag{4.7}$$

The last inequality follows from (2.3). Notice also that $\xi_\varepsilon(t) + c_2 t \geqslant 0$ and is nondecreasing for all t. Denoting $E^{Q_x^{\varepsilon,c}}(\xi_\varepsilon(t) + c_2 t)$ by $\Theta(t)$, we obtain from (4.4) that the set of controls $\mathcal{C}$ can be reduced to those for which

$$K_\delta\big[c_1^{-1/\beta'} t^{1/\beta}\Theta(t)^{1/\beta'} + \sqrt{\varepsilon t}\big] - \Theta(t) + K_\delta\sqrt{\varepsilon t} + (c_4+c_2)t + 2\delta \geqslant 0. \tag{4.8}$$

The above relation is of the form $A\Theta^{1/\beta'} - \Theta + B \geqslant 0$. By Young's inequality this implies that

$$\frac{A^\beta}{\beta} + \frac{\Theta}{\beta'} - \Theta + B \geqslant 0.$$

Since $1/\beta + 1/\beta' = 1$, we conclude that

$$\Theta \leqslant A^\beta + \beta B.$$

This immediately gives (4.1). Inequality (4.2) follows from (4.1) and (4.7). □

LEMMA 4.2 *Assuming the bounds* (2.1) *and* (2.3), *and* (2.10) *on* f, *there is a function* $c(t) \to 0$ *as* $t \to 0$ *such that for all* $0 < \varepsilon \leqslant 1$,

$$\sup_\omega |u_\varepsilon(t, x, \omega) - f(x)| \leqslant c(t). \tag{4.9}$$

PROOF: We start with the variational formula (3.4)

$$u_\varepsilon(t, x, \omega) - f(x) = \sup_{c \in \mathcal{C}^*} E^{Q_x^{\varepsilon,c}}[f(x(t)) - f(x) - \xi_\varepsilon(t)]$$

where $\mathcal{C}^*$ is as in Lemma 4.1. From (3.5), (2.10), and (4.5) we get for every $\delta > 0$

$$\begin{aligned} &E^{Q_x^{\varepsilon,c}}[f(x(t)) - f(x) - \xi_\varepsilon(t)] \\ &\leqslant E^{Q_x^{\varepsilon,c}}[K_\delta \|x(t) - x\| - \xi_\varepsilon(t)] + \delta \\ &\leqslant K_\delta \sqrt{\varepsilon t} + E^{Q_x^{\varepsilon,c}}\left[K_\delta \int_0^t \|c(\varepsilon^{-1}s, \varepsilon^{-1}x(s))\| ds - \xi_\varepsilon(t)\right] + \delta. \end{aligned}$$

Combining this with (4.1), we get an upper bound. Inequality (4.3) provides a lower bound. □

The starting point for lower bounds is the ergodic theorem. Let us begin with $(b, \varphi) \in \mathcal{E}$ and recall the definitions in (3.7) and (3.8):

$$\begin{aligned} m(b, \varphi) &= \int b(\omega)\varphi(\omega) d\mathbb{P}, \\ h(b, \varphi) &= \int L(b(\omega), \omega)\varphi(\omega) d\mathbb{P}. \end{aligned}$$

For the Markov process on Ω with generator

$$\mathcal{A}_b = \frac{1}{2}\Delta + \langle b(\omega), \nabla \rangle$$

$\varphi \, d\mathbb{P}$ is an ergodic invariant probability measure. By the ergodic theorem for almost all ω with respect to $\mathbb{P}$, the corresponding diffusion on Ω starting from ω, i.e., $P^{b,\omega}$, satisfies

$$\lim_{\varepsilon \to 0} \varepsilon \int_0^{t/\varepsilon} b(\omega(s)) ds = t \int b(\omega)\varphi(\omega) d\mathbb{P} = m(b, \varphi)t,$$

$$\lim_{\varepsilon \to 0} \varepsilon \int_0^{t/\varepsilon} L(b(\omega(s)), \omega(s)) ds = t \int L(b(\omega), \omega)\varphi(\omega) d\mathbb{P} = h(b, \varphi)t,$$

in $L^1(P^{b,\omega})$, uniformly in any finite t-interval. We can assume that this holds for every $\omega \in N$ where $\mathbb{P}(N) = 1$. In fact, by Egoroff's theorem, given any $\eta > 0$, there exists N_η with $\mathbb{P}(N_\eta) \geqslant 1 - \eta$, such that

$$\lim_{\varepsilon\to 0} \sup_{\omega\in N_\eta} \sup_{0\leqslant t\leqslant T} E^{P^{b,\omega}}\left(\left|\varepsilon \int_0^{t/\varepsilon} b(\omega(s))ds - m(b,\varphi)t\right|\right) = 0 \tag{4.10}$$

and

$$\lim_{\varepsilon\to 0} \sup_{\omega\in N_\eta} \sup_{0\leqslant t\leqslant T} E^{P^{b,\omega}}\left(\left|\varepsilon \int_0^{t/\varepsilon} L(b(\omega(s)), \omega(s))ds - h(b,\varphi)t\right|\right) = 0. \tag{4.11}$$

These properties can be expressed in terms of $Q_0^{b,\omega}$ as well:

$$\lim_{\varepsilon\to 0} \sup_{\omega\in N_\eta} \sup_{0\leqslant t\leqslant T} E^{Q_0^{b,\omega}}\left(\left|\varepsilon \int_0^{t/\varepsilon} b(\tau_{x(s)}\omega)ds - m(b,\varphi)t\right|\right) = 0, \tag{4.12}$$

$$\lim_{\varepsilon\to 0} \sup_{\omega\in N_\eta} \sup_{0\leqslant t\leqslant T} E^{Q_0^{b,\omega}}\left(\left|\varepsilon \int_0^{t/\varepsilon} L(b(\tau_{x(s)}\omega), \tau_{x(s)}\omega)ds - h(b,\varphi)t\right|\right) = 0. \tag{4.13}$$

LEMMA 4.3 *Assume* (H1) *and* (H6). *Let* $(b,\varphi) \in \mathcal{E}$ *and* $m = m(b,\varphi)$, $h = h(b,\varphi)$. *Then*

$$\liminf_{\varepsilon\to 0} \inf_{\omega\in N_\eta} \inf_{0\leqslant t\leqslant T} [u_\varepsilon(t,0,\omega) - f(mt) + ht] \geqslant 0. \tag{4.14}$$

PROOF: From the variational formula (3.4), we have for any $c \in \mathcal{C}$,

$$u_\varepsilon(t,0,\omega) \geqslant E^{Q_0^{c,\omega}}\left(f(\varepsilon x(\varepsilon^{-1}t)) - \varepsilon \int_0^{t/\varepsilon} L(c(s,x(s)), \tau_{x(s)}\omega)ds\right).$$

Choose $c(s,x) = c(x) = b(\tau_x\omega)$. Since f satisfies (2.10) and

$$x(t) = \int_0^t b(\tau_{x(s)}\omega)ds + \beta(t) = \int_0^t b(\omega(s))ds + \beta(t),$$

(4.14) follows from the ergodic theorem just as (4.10) and (4.11). □

LEMMA 4.4 *Assume* (H1), (H2), *and* (H6). *Let* $(b,\varphi) \in \mathcal{E}$, $m = m(b,\varphi)$, *and* $h = h(b,\varphi)$. *Then*

$$\lim_{r\to 0} \liminf_{\varepsilon\to 0} \inf_{\omega\in N_\eta} \inf_{0\leqslant t\leqslant T} \inf_{\|y\|\leqslant r} [u_\varepsilon(t,y,\omega) - f(mt) + ht] \geqslant 0. \tag{4.15}$$

PROOF: By Lemma 4.2, we can assume without loss of generality that $t \geqslant r$. We have a process $x(s)$ that starts from 0 having for its distribution $Q_0^c = Q_0^{b,\omega}$ with $c(s,x) = b(\tau_x\omega)$. It is enough to construct a process $y(s)$ with generator $\frac{1}{2}\Delta + \langle \hat{c}(s,\cdot), \nabla\rangle$ that starts from $\varepsilon^{-1}y$, which is close to $x(s)$ in the sense that

$$\varepsilon\|y(\varepsilon^{-1}t) - x(\varepsilon^{-1}t)\|$$

and

$$\varepsilon\left|\int_0^{t/\varepsilon} L(c(s,x(s)), \tau_{x(s)}\omega)ds - \int_0^{t/\varepsilon} L(\hat{c}(s,y(s)), \tau_{y(s)}\omega)ds\right|$$

are small. We will construct such a $y(\cdot)$ that is coupled to $x(\cdot)$.

Define

$$y(s) = \begin{cases} x(s) + y/\varepsilon - sy/\|y\| & \text{if } 0 \leqslant s \leqslant \varepsilon^{-1}\|y\| \\ x(s) & \text{if } s \geqslant \varepsilon^{-1}\|y\|. \end{cases}$$

Clearly,

$$\varepsilon\|y(\varepsilon^{-1}t) - x(\varepsilon^{-1}t)\| \leqslant \|y\| \leqslant r.$$

A simple calculation reveals

$$\hat{c}(s,z) = \begin{cases} c(s, z - y/\varepsilon + sy/\|y\|) - y/\|y\| & \text{if } 0 \leqslant s \leqslant \varepsilon^{-1}\|y\| \\ c(s,z) & \text{if } s \geqslant \varepsilon^{-1}\|y\|, \end{cases}$$

and $\|\hat{c}(\cdot,\cdot)\|_\infty \leqslant \|c(\cdot,\cdot)\|_\infty + 1$. We have

$$\begin{aligned} &\varepsilon\left|\int_0^{t/\varepsilon} L(c(s,x(s)), \tau_{x(s)}\omega)ds - \int_0^{t/\varepsilon} L(\hat{c}(s,y(s)), \tau_{y(s)}\omega)ds\right| \\ &= \varepsilon\left|\int_0^{\|y\|/\varepsilon} L(c(s,x(s)), \tau_{x(s)}\omega)ds - \int_0^{\|y\|/\varepsilon} L(\hat{c}(s,y(s)), \tau_{y(s)}\omega)ds\right| \\ &\leqslant 2\|y\| \sup_{|p|\leqslant\|c(\cdot,\cdot)\|_\infty+1} \sup_\omega |L(p,\omega)| \end{aligned}$$

The statement of the lemma follows from the above estimates and (4.14). □

We now turn to the proof of Theorem 2.1.

PROOF: If we want to show that $u_\varepsilon(t,y,\omega) \to u(t,y)$ in probability, we could translate by $\tau_{\varepsilon^{-1}y}$ and the initial condition would be $f_y(x) = f(x+y)$. Since $\mathbb{P}$ is invariant, statistically nothing would have changed. However, it is not obvious that the convergence is almost sure. Nor is it obvious that the convergence is locally uniform in x.

Let us remark that our proof of (4.14) depended only on the constant K_δ in (2.10). Also note that the constant K_δ corresponding to f also works for any translate $f^y(x) = f(x+y)$. As a result, if we decide to start from 0, but consider for some $\ell \geqslant 1$ the family $\{f^y(x) = f(x+y) : \|y\| \leqslant \ell\}$ of initial functions and their corresponding solutions $u_\varepsilon^y(t,x,\omega)$, then as we saw in (4.14), for any given $\eta > 0$, there is a set N_η, with $\mathbb{P}(N_\eta) \geqslant 1-\eta$, such that

$$\lim_{\varepsilon\to0} \sup_{0\leqslant t\leqslant T} \sup_{\|y\|\leqslant\ell} \sup_{\omega\in N_\eta} [f(y+mt) - th - u_\varepsilon^y(t,0,\omega)]^+ = 0.$$

By the ergodic theorem, there is a set N with $\mathbb{P}(N) = 1$ and for $\omega \in N$,

$$\lim_{\varepsilon\to0} \frac{|\{x : \tau_x\omega \in N_\eta, \|x\| \leqslant \ell\varepsilon^{-1}\}|}{|\{x : \|x\| \leqslant \ell\varepsilon^{-1}\}|} = \mathbb{P}(N_\eta) \geqslant 1-\eta.$$

Let $\omega \in N$. Then for $\varepsilon \leqslant \varepsilon_0(\eta)$,

$$\left|\{x : \tau_x\omega \in N_\eta, \|x\| \leqslant \ell\varepsilon^{-1}\}\right| \geqslant (1-2\eta)\left|\{x : \|x\| \leqslant \ell\varepsilon^{-1}\}\right|.$$

In particular, if $\delta(\eta) = \ell(3\eta)^{1/d}$, then every x satisfying $\|x\| \leqslant \varepsilon^{-1}\ell$ is within a distance $\delta(\eta)\varepsilon^{-1}$ of an x' such that $\tau_{x'}\omega \in N_\eta$. Now from (4.15) and the relation (3.2) it follows easily that

$$\lim_{\varepsilon\to 0} \sup_{0\leqslant t\leqslant T} \sup_{\|x\|\leqslant \ell} [f(x+mt) - ht - u_\varepsilon(t,x,\omega)]^+ = 0$$

for $\omega \in \bigcup_{\eta>0} N_\eta$. If we denote by $\mathbf{A} \in \mathbb{R}^d \times \mathbb{R}$ the range of $(m(b,\varphi), h(b,\varphi))$ as (b,φ) varies over $\mathcal{E}$, then it is routine to show that with

$$u(t,x) = \sup_{(m,h)\in\mathbf{A}} [f(x+mt) - ht],$$

$$\lim_{\varepsilon\to 0} \sup_{0\leqslant t\leqslant T} \sup_{\|x\|\leqslant \ell} [u(t,x) - u_\varepsilon(t,x,\omega)]^+ = 0$$

$\mathbb{P}$ a.s., which proves Theorem 2.1. □

5 Convex Analysis

We start with a simple estimate. For any $\varepsilon > 0$, there exists a constant C_ε such that

$$\|m\| \leqslant C_\varepsilon + \varepsilon h$$

for all $(m,h) \in \mathbf{A}$, which is a consequence of (2.3). Therefore, for any $\theta \in \mathbb{R}^d$,

$$\overline{H}(\theta) = \sup_{(m,h)\in\mathbf{A}} [\langle\theta, m\rangle - h] \tag{5.1}$$

is finite. The aim of this section is to show that, for each $\theta \in \mathbb{R}^d$, there exists a "function" $\psi(\omega) = \psi_\theta(\omega)$ such that

$$\frac{1}{2}\Delta\psi(\omega) + H(\nabla\psi(\omega), \omega) \leqslant \overline{H}(\theta)$$

for almost all ω. The function ψ does not really exist, but $\mathbf{W} = \nabla\psi$ exists as a function in some $L^p(\Omega)$ with values in $\mathbb{R}^d$. Moreover, $\nabla \times \mathbf{W} = 0$ and

$$\frac{1}{2}\nabla\cdot\mathbf{W} + H(\mathbf{W},\omega) \leqslant \overline{H}(\theta)$$

in the sense of distributions.

We now represent $\mathcal{E}$ as a limit of an increasing sequence of compact sets. We can assume without loss of generality that $L^p(\Omega,\mathbb{P})$ are all separable spaces for $1 \leqslant p < \infty$. Let L^1_+ be the convex subset of $\varphi \in L^1(\Omega,\mathbb{P})$ consisting of $\varphi(\cdot)$ that satisfy $\varphi(\omega) \geqslant 0$ a.e. $\mathbb{P}$ and $\int_\Omega \varphi(\omega)\,d\mathbb{P} = 1$. Let $\{\varphi_j(\omega)\}$ be a countable dense subset in L^1_+ in the strong L^1 topology. We will also assume that each $\varphi_j(\omega)$ is smooth in the sense that $\varphi_j(\tau_x\omega) = \varphi_j(x,\omega)$ is uniformly bounded above and below (away from 0) on $\mathbb{R}^d$ and has two bounded continuous derivatives in x. The bounds will be independent of ω but could and in fact will depend on j. We denote by $\mathbf{D}_k$ the convex hull of $\{\varphi_j : 1 \leqslant j \leqslant k\}$. In particular, for each k, $\mathbf{D}_k$ is a compact subset of $L^1_+ \subset L^1(\Omega,\mathbb{P})$. Each $\varphi \in \mathbf{D}_k$ has uniform bounds

$$\varphi(\omega) \geqslant c_k, \quad \|\nabla\varphi\| \leqslant C_k, \quad \|\nabla^2\varphi\| \leqslant C_k.$$

Now we consider maps $b : \Omega \to \mathbb{R}^d$. Set

$$\mathbf{B}_r = \{b : \sup_{\omega} \|b(\omega)\| \leqslant r\},$$

$$\mathcal{E}_{r,k} = \left\{(b, \varphi) : b \in \mathbf{B}_r,\ \varphi \in \mathbf{D}_k,\ \tfrac{1}{2}\Delta\varphi = \nabla \cdot (b\,\varphi)\right\},$$

$$\mathcal{E}_k = \bigcup_r \mathcal{E}_{r,k},$$

$$\mathcal{E} = \bigcup_k \mathcal{E}_k.$$

In the definition of $\mathcal{E}_{r,k}$ the equation $\frac{1}{2}\Delta\varphi = \nabla \cdot (b\,\varphi)$ is interpreted in the weak sense. This is sufficient to show that the Markov process $P^{b,\omega}$, which is well-defined, has $\varphi\, d\mathbb{P}$ as an ergodic invariant measure. Let us note that while each $\mathbf{D}_k$ is strongly compact, each $\mathbf{B}_r$ is weakly compact.

For each $\theta \in \mathbb{R}^d$ we have

$$\overline{H}(\theta) \geqslant \sup_{(b,\varphi)\in\mathcal{E}_{r,k}} \left[\int [\langle \theta, b(\omega)\rangle - L(b(\omega), \omega)]\, \varphi(\omega) d\mathbb{P} \right] \tag{5.2}$$

$$= \sup_{\varphi\in\mathbf{D}_k} \sup_{b\in\mathbf{B}_r} \inf_{u\in\mathcal{U}} \left[\int [\langle \theta, b(\omega)\rangle - L(b(\omega), \omega)]\varphi(\omega)\, d\mathbb{P} + \int (\mathcal{A}_b u)\varphi(\omega) d\mathbb{P} \right] \tag{5.3}$$

$$= \sup_{\varphi\in\mathbf{D}_k} \inf_{u\in\mathcal{U}} \sup_{b\in\mathbf{B}_r} \left[\int [\langle \theta, b(\omega)\rangle - L(b(\omega), \omega)]\varphi(\omega)\, d\mathbb{P} + \int (\mathcal{A}_b u)\, \varphi(\omega)\, d\mathbb{P} \right] \tag{5.4}$$

$$= \sup_{\varphi\in\mathbf{D}_k} \inf_{u\in\mathcal{U}} \left[\int \left(\frac{1}{2}\Delta u + H_r(\theta + \nabla u, \omega) \right) \varphi\, d\mathbb{P} \right] \tag{5.5}$$

$$= \inf_{u\in\mathcal{U}} \sup_{\varphi\in\mathbf{D}_k} \left[\int \left(\frac{1}{2}\Delta u + H_r(\theta + \nabla u, \omega) \right) \varphi\, d\mathbb{P} \right] \tag{5.6}$$

Since $\mathcal{E} \supset \mathcal{E}_{r,k}$, (5.2) is obvious. On the other hand, in (5.2), b and φ have to be related by $\frac{1}{2}\Delta\varphi = \nabla\cdot(b\,\varphi)$. This is taken care of by the term $\mathcal{A}_b u = \frac{1}{2}\Delta u + \langle b, \nabla u\rangle$ and

$$\inf_{u\in\mathcal{U}} \int (\mathcal{A}_b u)\varphi\, d\mathbb{P} = -\infty$$

unless $\frac{1}{2}\Delta\varphi = \nabla \cdot (b\,\varphi)$ in the weak sense, in which case it is 0. For this we can take any reasonable linear space $\mathcal{U}$ of test functions. This establishes (5.3). The functional on the right is clearly a concave function of $b \in \mathbf{B}_r$ and is upper

semicontinuous in the weak topology in which $\mathbf{B}_r$ is compact. It is linear and continuous in $u \in \mathcal{U}$. We can apply the minimax theorem (see [18]) and interchange the order of $\inf_u$ and $\sup_b$ for each fixed φ. We thus arrive at (5.4).

Now, the expression inside the integral is a local expression in $b(\omega)$ and the supremum can, therefore, be taken inside the integral. Setting

$$H_r(\xi, \omega) = \sup_{\|q\| \leqslant r} [\langle \xi, q \rangle - L(q, \omega)],$$

we get (5.5). The expression in (5.5) is convex and continuous in $u \in \mathcal{U}$ and linear and continuous in $\varphi \in \mathbf{D}_k$. Moreover, $\mathbf{D}_k$ is compact. We can use the minimax theorem once more to arrive at (5.6).

We have therefore proved the following:

LEMMA 5.1 *Let $\theta \in \mathbb{R}^d$ be given. For each integer $k \geqslant 1$ there exists a function $u_k(\omega) \in \mathcal{U}$ such that*

$$\sup_{\varphi \in \mathbf{D}_k} \int_\Omega \left(\frac{1}{2} \Delta u_k + H_k(\theta + \nabla u_k, \omega) \right) \varphi \, d\mathbb{P} \leqslant \overline{H}(\theta) + \frac{1}{k}.$$

The next step is to obtain some sort of a "weak limit" u from $\{u_k\}$ such that

$$\sup_k \sup_{\varphi \in \mathbf{D}_k} \int_\Omega \left(\frac{1}{2} \Delta u + H(\theta + \nabla u, \omega) \right) \varphi \, d\mathbb{P} \leqslant \overline{H}(\theta).$$

The function u may not exist, but we will show that $\nabla u = v$ exists.

Let $v_k = \nabla u_k$. Notice that the constant function $1 \in \mathbf{D}_k$ for every k, so we can choose $\varphi \equiv 1$. Since $\int \Delta u \, d\mathbb{P} = 0$ for any $u \in \mathcal{U}$, we get the bound

$$\int H_k(\theta + v_k(\omega), \omega) d\mathbb{P} \leqslant \overline{H}(\theta) + \frac{1}{k}. \tag{5.7}$$

THEOREM 5.2 *The sequence $\{v_k\}$ is uniformly integrable in $L^1(\Omega, \mathbb{P})$, and any weak limit v of v_k satisfies*

$$\int v \, d\mathbb{P} = 0, \quad \nabla \times v = 0,$$

$$\sup_{\varphi \in \bigcup_k \mathbf{D}_k} \int_\Omega \left(\frac{1}{2} \nabla \cdot v + H(\theta + v, \omega) \right) \varphi \, d\mathbb{P} \leqslant \overline{H}(\theta),$$

and hence also satisfies

$$\frac{1}{2} (\nabla \cdot v)(\omega) + H(\theta + v(\omega), \omega) \leqslant \overline{H}(\theta) \tag{5.8}$$

as a distribution.

PROOF: Once we establish uniform integrability, the rest is routine. If v is a weak limit point of $\{v_k\}$, convexity of H_k guarantees that

$$\sup_{\varphi\in\mathbf{D}_k}\int_\Omega\left(\frac{1}{2}\nabla\cdot v+H_k(\theta+v,\omega)\right)\varphi\,d\mathbb{P}\leqslant\overline{H}(\theta).$$

Since the $\mathbf{D}_k$ are increasing and $H_k\uparrow H$, we can replace H_k by H. Since $\bigcup_k\mathbf{D}_k$ is dense in L^1_+, the uniform bound on the integrals turns into an almost sure bound on the integrand.

Let us now establish the uniform integrability of $\{v_k\}$. We can get a lower bound on H_k from (2.3),

$$H_k(\xi,\omega)\geqslant\sup_{\|y\|\leqslant k}[\langle y,\xi\rangle-c_3-c_4\|y\|^{\alpha'}]=\sup_{0\leqslant y\leqslant k}[y\|\xi\|-c_3-c_4y^{\alpha'}].$$

Let c be a small positive number. If $\|\xi\|\leqslant(k/c)^{\alpha'-1}$, we pick $y=c\|\xi\|^{1/(\alpha'-1)}$ and get the lower bound

$$H_k(\xi,\omega)\geqslant c\|\xi\|^\alpha-c_3-c_4c^{\alpha'}\|\xi\|^\alpha\geqslant c_5\|\xi\|^\alpha-c_3,$$

where $1/\alpha'+1/\alpha=1$ and $c_5>0$ provided that c is small enough. If $\|\xi\|>(k/c)^{\alpha'-1}$, then we pick $y=k$ and obtain for all sufficiently small c

$$H_k(\xi,\omega)\geqslant k\|\xi\|-c_3-c_4k^{\alpha'}\geqslant k\|\xi\|-c_3-c_4kc^{\alpha'-1}\|\xi\|\geqslant c_6k\|\xi\|-c_3.$$

From the above lower bounds we conclude that

$$\begin{aligned}\int_{\|\xi\|\geqslant A}\|\xi\|d\mathbb{P}&=\int_{\substack{\|\xi\|\geqslant A\\ \|\xi\|\leqslant(k/c)^{\alpha'-1}}}\|\xi\|d\mathbb{P}+\int_{\substack{\|\xi\|\geqslant A\\ \|\xi\|\geqslant(k/c)^{\alpha'-1}}}\|\xi\|d\mathbb{P}\\ &\leqslant\int_{\substack{\|\xi\|\geqslant A\\ \|\xi\|\leqslant(k/c)^{\alpha'-1}}}\frac{\|\xi\|^\alpha}{\|\xi\|^{\alpha-1}}d\mathbb{P}+\int_{\|\xi\|\geqslant(k/c)^{\alpha'-1}}\|\xi\|d\mathbb{P}\\ &\leqslant\frac{1}{c_5A^{\alpha-1}}\int[H_k(\xi,\omega)+c_3]d\mathbb{P}+\frac{1}{kc_6}\int[H_k(\xi,\omega)+c_3]d\mathbb{P}.\end{aligned}$$

Setting $\xi=\theta+v_k$ establishes the uniform integrability of $\{v_k\}$. □

6 Upper Bounds: Preliminaries

Now we use the function v constructed in (5.8) to provide us with an upper bound with the help of the maximum principle. The idea is, roughly speaking, to use $v(x,\omega)=v(\tau_x\omega)$ as a gradient to obtain $V(x,\omega)$ normalized by the condition $V(0,\omega)=0$ for $\mathbb{P}$ a.e. ω. Observe that, since $v\subset L^\alpha(\Omega)$ is irrotational in the weak sense, it has a potential $V(\cdot,\omega)\in W^{1,\alpha}_{\text{loc}}(\mathbb{R}^d)$ for a.e. ω. In fact,

$$V(x,\omega)=\int_{0\to x}\langle v,dz(s)\rangle,\tag{6.1}$$

where $z(s)$ is a path connecting 0 to x. It is now easy to estimate

$$\|V(x,\cdot) - V(y,\cdot)\|_\alpha \leqslant C\|x-y\|.$$

By our choice of the normalization,

$$\|V(x,\cdot)\|_\alpha \leqslant C\|x\|. \tag{6.2}$$

If v were nice, the ergodic theorem would tell us that $V(x,\omega) = o(\|x\|)$ as $\|x\| \to \infty$ a.s.

The function $\widehat{V}_\varepsilon(t,x,\omega) = \langle\theta,x\rangle + t\overline{H}(\theta) + \varepsilon V(x/\varepsilon,\omega)$ would be a supersolution of (1.1). Its initial value $\widehat{V}_\varepsilon(0,x,\omega) = \langle\theta,x\rangle + \varepsilon V(x/\varepsilon,\omega)$ would differ from the affine function $\langle\theta,x\rangle$ by a vanishing term $\varepsilon V(x/\varepsilon,\omega) = o(1)$ as $\varepsilon \to 0$. Moreover, setting $x = 0$ and using our normalization $V(0,\omega) = 0$, we would get by comparison that with probability 1

$$\limsup_{\varepsilon\to 0} u_\varepsilon(t,0,\omega) \leqslant t\overline{H}(\theta) = u(t,0),$$

establishing the upper bound for the affine initial data at $x = 0$. We shall see in Corollary 6.8 at the end of this section that in this case the homogenization takes place under weak hypotheses (H1) and (H2) without assuming the regularity of v. Additional conditions will be needed to strengthen this result to the locally uniform bound and general initial data.

At first, we need to understand the nature of the solution v. It is clear from the lower bound on H that $v \in L^\alpha(\Omega,\mathbb{P})$ where $\alpha > 1$ is as in (2.1). Then for almost all ω with respect to $\mathbb{P}$, $v(x,\omega) = v(\tau_x\omega) \in L^\alpha_{\text{loc}}(\mathbb{R}^d)$ and satisfies as a distribution on $\mathbb{R}^d$,

$$\frac{1}{2}(\nabla\cdot v)(\tau_x\omega) + H(\theta + v(\tau_x\omega),\tau_x\omega) \leqslant \overline{H}(\theta). \tag{6.3}$$

LEMMA 6.1 *For any $r > 0$, there is a constant $C(r)$ independent of ω such that*

$$\sup_x \int_{\|y-x\|\leqslant r} \|v(y,\omega)\|^\alpha\,dy = C(r) < \infty$$

for almost all ω with respect to $\mathbb{P}$.

PROOF: Since the proof is carried out for a fixed ω, we shall drop ω from the notation. Let g be any compactly supported nonnegative test function such that $\int g(x)dx = 1$. From (6.3) and the lower bound in (2.1) we have

$$\int(\nabla\cdot v(y) + c_1\|\theta + v(y)\|^\alpha - c_2)g(y)dy \leqslant \overline{H}(\theta).$$

From this inequality we get

$$\int \|v(y)\|^\alpha g(y)dy \leqslant C_1\int\langle(\nabla g)(y), v(y)\rangle dy + C_2$$

with constants C_1 and C_2 independent of g and ω. Let us take a g supported on $B(0, 2r)$ with

$$\int \left(\frac{\|\nabla g\|}{g}\right)^{\alpha'} g(y)dy = C_3(r) < \infty \quad \text{and} \quad g(y) \geqslant c(r) > 0 \text{ on } B(0, r).$$

Then an application of Young's inequality yields

$$\begin{aligned}\int \|v(y)\|^{\alpha} g(y)dy &\leqslant C_1 \int \left\langle \frac{(\nabla g)(y)}{g(y)}, v(y) \right\rangle g(y)dy + C_2 \\ &\leqslant \frac{1}{2} \int \|v(y)\|^{\alpha} g(y)dy + C_4(r),\end{aligned}$$

providing us with a uniform estimate

$$\int_{B(0,r)} \|v(y)\|^{\alpha} dy \leqslant C(r).$$

Since replacing $g(y)$ by $g(x - y)$ does not change anything, the estimate is uniform in x (which is the same as being uniform in ω). □

We now try to improve the $L^{\alpha}(\Omega, \mathbb{P})$ bound on v, replacing it with a convolution $\hat{v}_{\delta} \in L^{\infty}(\Omega, \mathbb{P})$. The proof of Lemma 6.2 below is straightforward and omitted.

LEMMA 6.2 *Let us define the convolution*

$$v_{\delta}(\omega) = \int v(\tau_{\delta y}\, \omega)\rho(y)dy \tag{6.4}$$

where $\rho(y)$ is a mollifier supported on $\|y\| \leqslant 1$. Then v_{δ} satisfies

$$\|v_{\delta}\|_{\infty} \leqslant C_{\delta}$$

for some constant C_{δ}. If we now define

$$V_{\delta}(x) = \int V(x + \delta y)\rho(y)dy \tag{6.5}$$

as the corresponding convolution of V, then $\nabla V_{\delta} = v_{\delta}$ and

$$\|V_{\delta}(0, \omega)\|_{\infty} \leqslant C_{\delta}$$

with a possibly different C_{δ}. Moreover, by the ergodic theorem for any $\delta > 0$ and $\ell < \infty$,

$$\lim_{\varepsilon \to 0} \varepsilon \sup_{|x| \leqslant \ell} |V_{\delta}(\varepsilon^{-1} x, \omega)| = 0 \quad a.e.\ \mathbb{P}.$$

Under (H3) one can use the mollified functions V_{δ} for comparison.

LEMMA 6.3 *Assume the strong hypothesis* (H3). *Then for each $\delta > 0$, there exists a smooth $\hat{v}_{\delta}$ that is uniformly bounded in ω, satisfies $\nabla \times \hat{v}_{\delta} = 0$, $\mathbb{E}[\hat{v}_{\delta}] = 0$, and*

$$\frac{1}{2} \nabla \cdot \hat{v}_{\delta} + H(\theta + \hat{v}_{\delta}, \omega) \leqslant \overline{H}(\theta) + \hat{\nu}(\delta),$$

where $\hat{\nu}(\delta) \to 0$ as $\delta \to 0$.

PROOF: We start with the convolution v_δ as in (6.4). Clearly,

$$\frac{1}{2}\nabla\cdot v_\delta + \int H(\theta + v(\tau_{\delta y}\omega), \tau_{\delta y}\omega)\rho(y)dy \leqslant \overline{H}(\theta),$$

and (2.7) allows us to conclude that

$$\frac{1}{2}\nabla\cdot v_\delta + (1+\nu(\delta))\int H((1+\nu(\delta))^{-1}(\theta + v(\tau_{\delta y}\omega)), \omega)\rho(y)dy \leqslant \overline{H}(\theta) + C\nu(\delta).$$

The convexity of H can now be used to infer that $\hat{v}_\delta = v_\delta/(1+\nu(\delta))$ satisfies

$$\frac{1}{2}\nabla\cdot \hat{v}_\delta + H((1+\nu(\delta))^{-1}\theta + \hat{v}_\delta, \omega) \leqslant (1+\nu(\delta))^{-1}(\overline{H}(\theta) + C\nu(\delta)).$$

We can replace θ by $(1+\nu(\delta))\theta$ and use the continuity of $\overline{H}$, which is a convex function of θ. □

Next we prove a technical lemma that will allow us to use v_δ and V_δ and avoid dealing with weak supersolutions in the comparison arguments that follow.

LEMMA 6.4 *Assume* (H1) *and* (H2). *Let* $\theta \in \mathbb{R}^d$ *and* $\lambda \in \mathbb{R}$ *be such that there exists a mean zero irrotational vector field* $v \in L^\alpha(\Omega)$, *which is a weak solution of*

$$\frac{1}{2}(\nabla\cdot v)(\omega) + H(\theta + v(\omega), \omega) \leqslant \lambda.$$

Denote by V *its integral normalized by the condition* $V(0,\omega) = 0$ ω*-a.s. Let* $c \in \mathcal{C}$,

$$x(t) = x + \int_0^t c(s, x(s))ds + \beta(t), \quad \xi(t) = \int_0^t L(c(s, x(s)), \tau_{x(s)}\omega)ds.$$

Then there exists $r_c(\delta) \to 0$ *as* $\delta \to 0$, *which depends only on the bound of* $c \in \mathcal{C}$, *such that*

$$\langle\theta, x(t) - x\rangle - t\lambda - \xi(t) \leqslant V_\delta(x,\omega) - V_\delta(x(t),\omega) + \int_0^t \langle\theta + v_\delta(\tau_{x(s)}\omega), d\beta(s)\rangle + r_c(\delta)t.$$

PROOF: Applying Itô's formula to $V_\delta(x,\omega) + \langle\theta, x\rangle$, we get

$$\begin{aligned} V_\delta(x(t),\omega) &- V_\delta(x,\omega) + \langle\theta, x(t) - x\rangle \\ &= \int_0^t \langle\theta + v_\delta(\tau_{x(s)}\omega), c(s, x(s))\rangle + \frac{1}{2}\nabla\cdot v_\delta(\tau_{x(s)}\omega)ds \\ &\quad + \int_0^t \langle\theta + v_\delta(\tau_{x(s)}\omega), d\beta(s)\rangle. \end{aligned} \tag{6.6}$$

From the relationship between H and L it is obvious that for all c and p

$$\langle p, c\rangle \leqslant H(p,\omega) + L(c,\omega).$$

Thus, for every $x \in \mathbb{R}^d$,

$$\begin{aligned} &\langle \theta + v_\delta(\tau_x\omega), c(s,x)\rangle \\ &= \int_{\mathbb{R}^d} \langle \theta + v(\tau_y\omega), c(s,x)\rangle \rho_\delta(x-y)dy \\ &\leqslant \int_{\mathbb{R}^d} (L(c(s,x),\tau_y\omega) + H(\theta + v(\tau_y\omega),\tau_y\omega))\rho_\delta(x-y)dy. \end{aligned} \tag{6.7}$$

Substituting this into (6.6) and using (6.3), we obtain

$$\begin{aligned} &V_\delta(x(t),\omega) - V_\delta(x,\omega) + \langle \theta, x(t)-x\rangle \\ &\leqslant t\lambda + \int_0^t \int_{\mathbb{R}^d} L(c(s,x(s)),\tau_y\omega)\rho_\delta(x(s)-y)dy\,ds \\ &\quad + \int_0^t \langle \theta + v_\delta(\tau_{x(s)}\omega), d\beta(s)\rangle. \end{aligned}$$

Rearranging the terms and subtracting $\xi(t)$ from both sides, we arrive at the inequality

$$\begin{aligned} &\langle \theta, x(t)-x\rangle - t\lambda - \xi(t) \\ &\leqslant V_\delta(x,\omega) - V_\delta(x(t),\omega) + \int_0^t \langle \theta + v_\delta(\tau_{x(s)}\omega), d\beta(s)\rangle \\ &\quad + \int_0^t \int_{\mathbb{R}^d} \big(L(c(s,x(s)),\tau_y\omega) - L(c(s,x(s)),\tau_{x(s)}\omega)\big)\rho_\delta(x(s)-y)dy\,ds. \end{aligned}$$

Since the drift c is bounded and L satisfies (H2), the last term is bounded in absolute value by $r_c(\delta)t$ where

$$r_c(\delta) = \sup_{\|q\|\leqslant\|c(\cdot,\cdot)\|_\infty} \sup_\omega \sup_{\|y\|\leqslant\delta} |L(q,\tau_y\omega) - L(q,\omega)|$$

tends to 0 as $\delta \to 0$. □

As the first application, we obtain an upper bound at $x = 0$ for affine initial data under weak assumptions (H1) and (H2). We will need the following elementary fact, which we state as a lemma:

LEMMA 6.5 *Let $g \in L^1_{\text{loc}}(\mathbb{R}^d)$ and $\frac{1}{2}\Delta g \leqslant C$ in the sense of distributions. Then there exists g_*, a lower semicontinuous modification of g, such that for every $\delta > 0$*

$$g_*(x) \geqslant g_\delta(x) - \frac{C}{d}\delta^2 \quad \textit{for all } x \in \mathbb{R}^d.$$

PROOF: It is a standard fact for $C = 0$. Subtracting a quadratic function from g leads to the above result. □

Remark 6.6. Below when we refer to V or V_ε, we shall always mean a lower semicontinuous modification given by Lemma 6.5 and will not use the subscript $*$.

THEOREM 6.7 *Assume* (H1) *and* (H2). *Let* $\theta \in \mathbb{R}^d$ *and* $\lambda \in \mathbb{R}$ *be as in Lemma* 6.4. *Let* $u_\varepsilon(t, x, \omega)$ *be the solution of* (1.1) *with the initial condition* $f(x) = \langle \theta, x\rangle$. *Then with probability* 1

$$\limsup_{\varepsilon\to 0} u_\varepsilon(t, 0, \omega) \leqslant t\lambda.$$

PROOF: Let V be as in Lemma 6.4 and V_δ be given by (6.5). Set $V_{\delta,\varepsilon}(x, \omega) = \varepsilon V_\delta(x/\varepsilon, \omega)$. Using Lemma 6.4 with $x = 0$, rescaling, and taking the expectation with respect to $Q_0^{\varepsilon,c}$, we get

$$E^{Q_0^{\varepsilon,c}}(\langle\theta, x(t)\rangle - \xi_\varepsilon(t)) \leqslant t\lambda - E^{Q_0^{\varepsilon,c}} V_{\delta,\varepsilon}(x(t), \omega) + V_{\delta,\varepsilon}(0, \omega) + r_c(\delta)t \tag{6.8}$$

where $r_c(\delta) \to 0$ as $\delta \to 0$ for each $c \in \mathcal{C}$.

We claim that with $\mathbb{P}$ probability 1

$$\lim_{\delta\to 0} E^{Q_0^{\varepsilon,c}} |V_{\delta,\varepsilon}(x(t), \omega) - V_\varepsilon(x(t), \omega)| = 0. \tag{6.9}$$

Let $p_{\varepsilon,c}(t, y)$ be the probability density of $x(t)$ under $Q_0^{\varepsilon,c}$. Then

$$E^{Q_0^{\varepsilon,c}} |V_{\delta,\varepsilon}(x(t), \omega) - V_\varepsilon(x(t), \omega)| = \int_{\mathbb{R}^d} |V_{\delta,\varepsilon}(y, \omega) - V_\varepsilon(y, \omega)| p_{\varepsilon,c}(t, y) dy.$$

Since c is bounded and $\varepsilon > 0$, $p_{\varepsilon,c}(t, x, y)$ has sufficient regularity and decay to imply (6.9) in view of (6.2). This allows us to pass to the limit as $\delta \to 0$ in (6.8) for each $\varepsilon > 0$ and obtain

$$E^{Q_0^{\varepsilon,c}}(\langle\theta, x(t)\rangle - \xi_\varepsilon(t)) \leqslant t\lambda - E^{Q_0^{\varepsilon,c}} V_\varepsilon(x(t), \omega) + V_\varepsilon(0, \omega).$$

Recalling that $V_\varepsilon(0, \omega) = 0$ a.s. and applying the lower bound of Lemma 6.5 to V_ε, we see that for every $\delta > 0$

$$E^{Q_0^{\varepsilon,c}}(\langle\theta, x(t)\rangle - \xi_\varepsilon(t)) \leqslant t\lambda - E^{Q_0^{\varepsilon,c}} V_{\delta,\varepsilon}(x(t), \omega) + \frac{C}{d}\varepsilon\delta^2.$$

From the variational formula and the above inequality, we get

$$u_\varepsilon(t, 0, \omega) \leqslant t\lambda - \inf_{c\in\mathcal{C}^*} E^{Q_0^{\varepsilon,c}} V_{\delta,\varepsilon}(x(t), \omega) + \frac{C}{d}\varepsilon\delta^2.$$

By Lemma 6.2, $V_\delta(x, \omega)$ has a bounded gradient. Then by the ergodic theorem $V_\delta(x, \omega)$ is sublinear: for every $\eta > 0$ there is a constant $C_{\eta,\delta}$ such that $|V_\delta(x, \omega)| \leqslant \eta\|x\| + C_{\eta,\delta}$ for all x and a.e. ω. After the rescaling we get

$$|V_{\delta,\varepsilon}(x, \omega)| \leqslant \eta\|x\| + \varepsilon C_{\eta,\delta}.$$

This inequality and Lemma 4.1 (see also (7.1) below) imply the desired statement when we let $\varepsilon \to 0$, then $\delta \to 0$, and finally $\eta \to 0$. □

By (6.3) we can choose $\lambda = \overline{H}(\theta)$. Using Theorem 6.7 and the already proved general lower bound, we obtain the homogenization result for affine initial data at $x = 0$.

COROLLARY 6.8 *Assume* (H1) *and* (H2). *Let* $u_\varepsilon(t, x, \omega)$ *be the solution of* (1.1) *with the initial condition* $f(x) = \langle \theta, x \rangle$. *Then with probability* 1

$$\limsup_{\varepsilon \to 0} u_\varepsilon(t, 0, \omega) = t\overline{H}(\theta).$$

The following obvious corollary will be used in the next section:

COROLLARY 6.9 *Let* $\theta \in \mathbb{R}^d$ *and* $\lambda \in \mathbb{R}$ *admit a* v *as in Lemma* 6.4. *Then* $\overline{H}(\theta) \leqslant \lambda$.

7 Proof of Theorem 2.2

We are now ready to proceed with the proof of Theorem 2.2. Let us recall the formula

$$u_\varepsilon(t, x, \omega) = \sup_{c \in \mathcal{C}} E^{Q_x^{\varepsilon,c}}[f(x(t)) - \xi_\varepsilon(t)],$$

where

$$\xi_\varepsilon(t) = \int_0^t L\big(c(\varepsilon^{-1}s, \varepsilon^{-1}x(s)), \tau_{\varepsilon^{-1}x(s)}\omega\big)ds$$

and $Q_x^{\varepsilon,c}$ is the diffusion process on $C([0, T]; \mathbb{R}^d)$ corresponding to the generator

$$\frac{\varepsilon}{2}\Delta + c(\varepsilon^{-1}s, \varepsilon^{-1}x) \cdot \nabla$$

starting at time 0 from $x \in \mathbb{R}^d$ so that almost surely with respect to $Q_x^{\varepsilon,c}$

$$x_\varepsilon(t) = x(t) = x + \int_0^t c(\varepsilon^{-1}s, \varepsilon^{-1}x(s))ds + \sqrt{\varepsilon}\beta(t),$$

$\beta(\cdot)$ being the standard Brownian motion on $\mathbb{R}^d$.

The following steps require only assumptions (H1), (H2), and (H6).

Step 1. As we saw in Lemma 4.1, the supremum in the variational formula (3.4) can be restricted to the set $\mathcal{C}^*$ that consists of all controls $c \in \mathcal{C}$ that satisfy condition (4.1). Observe that $\|x_\varepsilon(t)\|$ and therefore $f(x_\varepsilon(t))$ is uniformly integrable with respect to $\{Q_x^{\varepsilon,c} \; ; \; \|x\| \leqslant \ell, \; 0 \leqslant t \leqslant T, \; 0 < \varepsilon \leqslant 1, \; c \in \mathcal{C}^*\}$, and

$$\sup_{\|x\| \leqslant \ell} \sup_{0 \leqslant t \leqslant T} \sup_{c \in \mathcal{C}^*} \sup_{0 < \varepsilon \leqslant 1} E^{Q_x^{\varepsilon,c}}[\|x_\varepsilon(t)\| + |\xi_\varepsilon(t)|] < \infty. \tag{7.1}$$

This is a simple consequence of Lemma 4.1 and inequalities (4.5) and (4.6).

Step 2. For any $\delta > 0$, we can find M_δ such that

$$\begin{aligned}
&u_\varepsilon(t,x,\omega) - u(t,x)\\
&\quad = \sup_{c\in\mathcal{C}^*} E^{Q_x^{\varepsilon,c}}(f(x_\varepsilon(t)) - \xi_\varepsilon(t)) - \sup_{y\in\mathbb{R}^d}\left[f(y) - t\mathcal{I}\left(\frac{y-x}{t}\right)\right]\\
&\quad \leqslant \sup_{c\in\mathcal{C}^*} E^{Q_x^{\varepsilon,c}}\left[\left[t\mathcal{I}\left(\frac{x_\varepsilon(t)-x}{t}\right) - \xi_\varepsilon(t)\right]\mathbf{1}_{\|x_\varepsilon(t)\|\leqslant M_\delta}\right] + \delta\\
&\quad = \sup_{c\in\mathcal{C}^*} E^{Q_x^{\varepsilon,c}}\left[\left(t\sup_{\theta\in\mathbb{R}^d}\left(\left\langle\theta, \frac{x_\varepsilon(t)-x}{t}\right\rangle - \overline{H}(\theta)\right) - \xi_\varepsilon(t)\right)\mathbf{1}_{\|x_\varepsilon(t)\|\leqslant M_\delta}\right] + \delta.
\end{aligned}$$

Note that according to (2.16),

$$\overline{H}(\theta) \geqslant c_1\|\theta\|^\alpha - c_2.$$

The above estimate together with estimate (7.1) ensures that large values of θ will not matter. In view of the continuity of $\overline{H}$, it suffices to show that for each fixed $\theta \in \mathbb{R}^d$, $\ell, T, \eta > 0$,

$$\lim_{\varepsilon\to 0}\sup_{c\in\mathcal{C}^*}\sup_{\|x\|\leqslant\ell}\sup_{0\leqslant t\leqslant T} Q_x^{\varepsilon,c}\big\{\langle\theta, x_\varepsilon(t) - x\rangle - t\overline{H}(\theta) - \xi_\varepsilon(t) \geqslant \eta\big\} = 0. \tag{7.2}$$

Here we are using the facts that $\xi_\varepsilon(t)$ is bounded below uniformly in ε and that the family $\{\|x_\varepsilon\|\}$ is uniformly integrable.

PROOF OF THEOREM 2.2 ASSUMING (H3): According to Lemma 6.3, for any $\theta \in \mathbb{R}^d$ and $\delta > 0$, there is a uniformly bounded vector field $\hat{v}_\delta$ that is a potential gradient. Let $V_\delta(x,\omega)$ be its integral as in Lemma 6.2. Then there exists a set N with $\mathbb{P}(N) = 1$ such that for $\omega \in N$, $V_{\delta,\varepsilon}(x,\omega) = \varepsilon V_\delta(\varepsilon^{-1}x,\omega)$ satisfies

$$\lim_{\varepsilon\to 0}\sup_{\|x\|\leqslant\ell}|V_{\delta,\varepsilon}(x,\omega)| = 0, \tag{7.3}$$

$$\sup_{0<\varepsilon\leqslant 1}|V_{\delta,\varepsilon}(x,\omega)| \leqslant M_\delta\|x\|, \tag{7.4}$$

for some constant M_δ. Applying Itô's formula to $V_{\delta,\varepsilon}$, a.e. with respect to $Q_x^{\varepsilon,c}$, we have

$$\begin{aligned}
&V_{\delta,\varepsilon}(x(t),\omega) - V_{\delta,\varepsilon}(x(0),\omega)\\
&\quad = \int_0^t \frac{1}{2}(\nabla\cdot\hat{v}_\delta)(\varepsilon^{-1}x(s),\omega)\,ds + \int_0^t \langle c(\varepsilon^{-1}s, \varepsilon^{-1}x(s)), \hat{v}_\delta(\varepsilon^{-1}x(s),\omega)\rangle\,ds\\
&\qquad + \sqrt{\varepsilon}\int_0^t \langle\hat{v}_\delta(\varepsilon^{-1}x(s),\omega), d\beta(s)\rangle.
\end{aligned}$$

Here and below, we simply write $x(t)$ for $x_\varepsilon(t)$. From (3.5) we also have

$$\langle\theta, x(t) - x\rangle = \int_0^t \langle\theta, c(\varepsilon^{-1}s, \varepsilon^{-1}x(s))\rangle\,ds + \sqrt{\varepsilon}\langle\theta, \beta(t)\rangle.$$

Adding the two, noticing that $\langle c, p\rangle \leqslant H(p,\omega)+L(c,\omega)$, and using the inequality of Lemma 6.3, we obtain

$$V_{\delta,\varepsilon}(x(t),\omega) - V_{\delta,\varepsilon}(x(0),\omega) + \langle\theta, x(t)-x\rangle$$
$$\leqslant [\overline{H}(\theta)+\hat{\nu}(\delta)]t + \xi_\varepsilon(t) + \sqrt{\varepsilon}\int_0^t \langle\theta + \hat{v}_\delta(\varepsilon^{-1}x(s),\omega), d\beta(s)\rangle.$$

Since we consider only $c \in \mathcal{C}^*$, the family $\{\|x_\varepsilon(t)\|\}$ is uniformly integrable. This together with (7.3) and (7.4) imply

$$\lim_{\varepsilon\to 0}\sup_{c\in\mathcal{C}^*}\sup_{\|x\|\leqslant\ell}\sup_{0\leqslant t\leqslant T} Q_x^{\varepsilon,c}\big[|V_{\delta,\varepsilon}(x(t)) - V_{\delta,\varepsilon}(x)| \geqslant \eta\big] = 0.$$

Since $\hat{v}_\delta \in L^\infty(\Omega,\mathbb{P})$,

$$\text{(7.5)}\quad \lim_{\varepsilon\to 0}\sup_{c\in\mathcal{C}^*}\sup_{\|x\|\leqslant\ell}\sup_{0\leqslant t\leqslant T} Q_x^{\varepsilon,c}\left[\sqrt{\varepsilon}\left|\int_0^t \langle\theta+\hat{v}_\delta(\varepsilon^{-1}x(s),\omega), d\beta(s)\rangle\right| \geqslant \eta\right] = 0.$$

From $\lim_{\delta\to 0}\hat{\nu}(\delta) = 0$ and the last three inequalities, we obtain

$$\lim_{\varepsilon\to 0}\sup_{c\in\mathcal{C}^*}\sup_{\|x\|\leqslant\ell}\sup_{0\leqslant t\leqslant T} Q_x^{\varepsilon,c}\big[\langle\theta, x(t)-x\rangle - \xi_\varepsilon(t) - t\overline{H}(\theta) \geqslant \eta\big] = 0,$$

thereby completing the proof of Theorem 2.2 assuming (H3). □

The main problem in the absence of assumption (H3) is that we cannot regularize v to get a bounded v_δ that still satisfies the inequality of Lemma 6.3. In the absence of regularization, one cannot get a bound on $\|\nabla v\|_\infty$, which is needed to control the stochastic integral term (7.5) as well as to obtain the uniformity in the ergodic theorem. This is handled differently in the two cases depending on whether (H4) or (H5) is assumed.

PROOF OF THEOREM 2.2 ASSUMING (H4): In this case since $v \in L^\alpha(\mathbb{P})$ with $\alpha > d$, one uses Sobolev's imbedding to control the $\|V\|_\infty$ locally and then martingale theory to control, also locally, the quadratic variation

$$E^{Q_x^{\varepsilon,c}}\left[\int_0^\tau \|v(\varepsilon^{-1}x(s))\|^2\,ds\right]$$

for a suitable stopping time τ.

Recall $v(x,\omega) = v(\tau_x\omega)$. Since $v(\cdot,\omega) \in L^\alpha_{\mathrm{loc}}(\mathbb{R}^d)$ for almost all ω, Sobolev's imbedding theorem gives us an almost surely continuous choice of $V(x,\omega)$ that can be normalized so that $V(0,\omega) = 0$ for $\mathbb{P}$-a.e. ω.

Step 1. Let $\tau_{\varepsilon,a} = \inf\{t>0: \xi_\varepsilon(t)\geqslant a\}$, $\hat{\tau}_{\varepsilon,b} = \inf\{t>0: \|x_\varepsilon(t)-x\|\geqslant b\}$, and $\sigma = \sigma(\varepsilon,a,b) = \tau_{\varepsilon,a}\wedge\hat{\tau}_{\varepsilon,b}$. Then an argument similar to the proof of Lemma 4.1 shows that for sufficiently large a and b

$$\lim_{\varepsilon\to 0}\sup_{c\in\mathcal{C}^*}\sup_{\|x\|\leqslant\ell}\sup_{0\leqslant t\leqslant T} Q_x^{\varepsilon,c}\{\langle\theta, x(t)-x\rangle - \overline{H}(\theta)t - \xi_\varepsilon(t)\geqslant\eta\}$$
$$= \lim_{\varepsilon\to 0}\sup_{c\in\mathcal{C}^*}\sup_{\|x\|\leqslant\ell}\sup_{0\leqslant t\leqslant T} Q_x^{\varepsilon,c}\{\langle\theta, x(t)-x\rangle - \overline{H}(\theta)t - \xi_\varepsilon(t)\geqslant\eta;\ \sigma\geqslant t\}$$

Let $V_{\delta,\varepsilon}(x,\omega) = \varepsilon V_\delta(x/\varepsilon,\omega)$. Recalling definitions (3.5) and (3.6) and applying the rescaled version of Lemma 6.4, we get for each $c \in \mathcal{C}$ and $\varepsilon > 0$

$$\begin{aligned}
&\langle \theta, x(t) - x\rangle - t\overline{H}(\theta) - \xi_\varepsilon(t) \\
&\quad \leqslant V_{\delta,\varepsilon}(x,\omega) - V_{\delta,\varepsilon}(x(t),\omega) \\
&\qquad + \sqrt{\varepsilon}\int_0^t \langle \theta + v_\delta(\tau_{\varepsilon^{-1}x(s)}\omega), d\beta(s)\rangle + r_c(\delta)t
\end{aligned} \tag{7.6}$$

where $r_c(\delta) \to 0$ as $\delta \to 0$ for each $c \in \mathcal{C}$. An application of Itô's formula gives

$$\begin{aligned}
&\sqrt{\varepsilon}\int_0^t \langle v_\delta(\tau_{x(s)}\omega), d\beta(s)\rangle \\
&\quad = V_{\delta,\varepsilon}(x(t),\omega) - V_{\delta,\varepsilon}(x,\omega) \\
&\qquad - \int_0^t \frac{\varepsilon}{2}\nabla\cdot v_\delta(\tau_{\varepsilon^{-1}x(s)}\omega) + \langle c(s,x(s)), v_\delta(\tau_{\varepsilon^{-1}x(s)}\omega)\rangle\, ds.
\end{aligned} \tag{7.7}$$

Substituting this into (7.6) we obtain, almost surely with respect to $Q_x^{\varepsilon,c}$,

$$\begin{aligned}
&\langle \theta, x(t) - x\rangle - t\overline{H}(\theta) - \xi_\varepsilon(t) \\
&\quad \leqslant \sqrt{\varepsilon}\langle \theta, \beta(t)\rangle \\
&\qquad - \int_0^t \frac{\varepsilon}{2}\nabla\cdot v_\delta(\tau_{\varepsilon^{-1}x(s)}\omega) + \langle c(s,x(s)), v_\delta(\tau_{\varepsilon^{-1}x(s)}\omega)\rangle ds + r_c(\delta)t.
\end{aligned}$$

Setting

$$F_{\delta,\varepsilon}(s,x,\omega) = \frac{\varepsilon}{2}\nabla\cdot v_\delta(\tau_{\varepsilon^{-1}x}\omega) + \langle c(s,x), v_\delta(\tau_{\varepsilon^{-1}x}\omega)\rangle,$$

we observe that it is enough to show that for almost all ω

$$\limsup_{\varepsilon\to 0}\sup_{c\in\mathcal{C}^*}\sup_{\|x\|\leqslant\ell}\sup_{0\leqslant t\leqslant T}\limsup_{\delta\to 0} E^{Q_x^{\varepsilon,c}}\left(\int_0^{t\wedge\sigma} F_{\delta,\varepsilon}(s,x(s),\omega)ds\right)^2 = 0. \tag{7.8}$$

Step 2. Below we shall write E_x instead of $E^{Q_x^{\varepsilon,c}}$ and also drop ω from the notation. Rewriting the square of the integral as twice the time-ordered double integral, we get

$$\begin{aligned}
&E_x\left(\int_0^{t\wedge\sigma} F_{\delta,\varepsilon}(s,x(s))ds\right)^2 \\
&\qquad = 2E_x\int_0^{t\wedge\sigma} F_{\delta,\varepsilon}(s,x(s))\left(\int_s^{t\wedge\sigma} F_{\delta,\varepsilon}(u,x(u))du\right)ds.
\end{aligned}$$

After conditioning on the σ-algebra up to time s and using (7.7), we see that the last integral is equal to

$$2E_x\int_0^{t\wedge\sigma} F_{\delta,\varepsilon}(s,x(s))E_{x(s)}\big(V_{\delta,\varepsilon}(x(t\wedge\sigma)) - V_{\delta,\varepsilon}(x(s))\big)ds.$$

Thus, for all starting points x with $\|x\| \leqslant \ell$

$$
\begin{aligned}
E_x\left[\int_0^{t\wedge\sigma} F_{\delta,\varepsilon}(s,x(s))ds\right]^2
&\leqslant 4 \sup_{\|x\|\leqslant b+\ell} |V_{\delta,\varepsilon}(x)| E_x \int_0^{t\wedge\sigma} |F_{\delta,\varepsilon}(s,x(s))| ds \\
&= 4 \sup_{\|x\|\leqslant b+\ell} |V_{\delta,\varepsilon}(x)| \left[2E_x \int_0^{t\wedge\sigma} [F_{\delta,\varepsilon}(s,x(s))]_+ ds \right. \\
&\qquad \left. - E_x \int_0^{t\wedge\sigma} F_{\delta,\varepsilon}(s,x(s)) ds \right] \\
&\leqslant 8 \sup_{\|x\|\leqslant b+\ell} |V_{\delta,\varepsilon}(x)| \left[E_x \int_0^{t\wedge\sigma} [F_{\delta,\varepsilon}(s,x(s))]_+ ds + \sup_{\|x\|\leqslant b+\ell} |V_{\delta,\varepsilon}(x)| \right].
\end{aligned}
$$

Step 3. Only now we use the assumption $\alpha > d$. Sobolev's imbedding theorem and Lemma 6.1 imply that for almost all ω, the function V satisfies

$$|V(x,\omega) - V(y,\omega)| \leqslant C(\|x-y\|),$$

where $C(r) \to 0$ as $r \to 0$ and $C(r) \leqslant Cr$ for $r \geqslant 1$. By the uniform continuity and the ergodic theorem there is a set $N \subset \Omega$ with $\mathbb{P}(N) = 1$ such that for all $\omega \in N$

$$\lim_{\varepsilon\to 0}\lim_{\delta\to 0} V_{\delta,\varepsilon}(x,\omega) = 0$$

uniformly on bounded subsets of $\mathbb{R}^d$. Thus, a bound on

$$E_x \int_0^{t\wedge\sigma} [F_{\delta,\varepsilon}(s,x(s))]_+ ds$$

will complete the proof. From the definition of $F_{\delta,\varepsilon}$, the rescaled versions of (6.7) and (6.3), and the weak regularity of L, we obtain

$$
\begin{aligned}
E_x \int_0^{t\wedge\sigma} [F_{\delta,\varepsilon}(s,x)]_+ ds \leqslant E_x \int_0^{t\wedge\sigma} \big[&\overline{H}(\theta) + L(\varepsilon^{-1}s, \varepsilon^{-1}x(s)), \tau_{\varepsilon^{-1}x(s)}\omega) \\
&- \langle \theta, c(\varepsilon^{-1}s, \varepsilon^{-1}x(s))\rangle + r_c(\delta)\big]_+ ds
\end{aligned}
$$

as $\delta \to 0$. Lemma 4.1 implies the desired bound (7.8) and finishes the proof. □

Finally, we turn to the proof of Theorem 2.2 under assumptions (H1), (H2), and (H5). First we prove almost sure weak convergence of $u_\varepsilon(t,x,\omega)$ to $u(t,x)$. Looking at (7.6) we observe that the main problem is getting a control on the martingale term. The idea of the proof is to perturb the Hamiltonian H by replacing it with $H^\gamma(p,\omega) = H(p,\omega) + \gamma\|p\|^2$ with small $\gamma > 0$ and then using the "leftover" quadratic term to control the martingale by means of the following lemma (see, for example, section 2.3 of [14]).

LEMMA 7.1 *Let z be a nonanticipating Brownian functional such that*

$$P\left[\int_0^t z^2\,dt < \infty, t \geqslant 0\right] = 1.$$

Then

$$P\left[\sup_{t\geqslant 0}\left(\int_0^t z\,d\beta - \frac{a}{2}\int_0^t z^2\,ds\right) > b\right] \leqslant e^{-ab}.$$

After passing to the limit as $\varepsilon \to 0$ we would let $\gamma \to 0$. Assumption (H5) will assure that the new effective Hamiltonian $\overline{H}^\gamma$ decreases to $\overline{H}^0 = \overline{H}$ as $\gamma \downarrow 0$.

LEMMA 7.2 *For any $\theta \in \mathbb{R}^d$*

$$\lim_{\gamma\to 0} \overline{H}^\gamma(\theta) = \overline{H}(\theta).$$

PROOF: Obviously, $\overline{H}^{\gamma_1} \geqslant \overline{H}^{\gamma_2}$ for $\gamma_1 \geqslant \gamma_2 \geqslant 0$ and $\lim_{\gamma\to 0}\overline{H}^\gamma(\theta) \geqslant \overline{H}(\theta)$. We need to prove that $\lim_{\gamma\to 0}\overline{H}^\gamma(\theta) \leqslant \overline{H}(\theta)$.

Fix $\theta \in \mathbb{R}^d$ and apply Theorem 5.2 to get the gradient v of a supersolution corresponding to $\theta/(1-\delta)$. Then the function $\psi = (1-\delta)v$ satisfies

$$\frac{1}{2}(\nabla\cdot\psi)(\omega) + (1-\delta)H\left(\frac{\theta+\psi(\omega)}{1-\delta}, \omega\right) \leqslant (1-\delta)\overline{H}\left(\frac{\theta}{1-\delta}\right).$$

From (H5) we get

$$(1-\delta)H\left(\frac{\theta+p}{1-\delta}, \omega\right) \geqslant H(\theta+p,\omega) + \gamma(\delta)\|\theta+p\|^2 - C(\delta).$$

Then for all $0 \leqslant \gamma \leqslant \gamma(\delta)$

$$\frac{1}{2}(\nabla\cdot\psi)(\omega) + H^\gamma(\theta+\psi(\omega),\omega) \leqslant (1-\delta)\overline{H}\left(\frac{\theta}{1-\delta}\right) + C(\delta). \tag{7.9}$$

Using Corollary 6.9 with

$$\lambda = (1-\delta)\overline{H}\left(\frac{\theta}{1-\delta}\right) + C(\delta)$$

and $\overline{H}(\theta)$ replaced by $\overline{H}^\gamma(\theta)$, we obtain the estimate

$$\overline{H}^\gamma(\theta) \leqslant (1-\delta)\overline{H}\left(\frac{\theta}{1-\delta}\right) + C(\delta).$$

Finally, we let $\gamma \to 0$ and then $\delta \to 0$. □

Let v_δ and V_δ be convolutions with a mollification kernel $\rho_\delta(x) = \delta^{-d}\rho(x/\delta)$ as in Lemma 6.2.

PROOF OF OF THEOREM 2.2 UNDER (H5): The proof in this case involves several steps.

Step 1. Recall that Q_x^c denotes the law of a diffusion with drift c that starts from a point x. Let us write Q_φ^c for the law of the same diffusion when the initial position is distributed according to a probability measure with a density φ. We start with $H^\gamma(p,\omega) = H(p,\omega)+\gamma\|p\|^2$ and construct the corresponding v^γ using Theorem 5.2. We have that $\mathbb{E}\, v^\gamma = 0$, $\nabla \times v^\gamma = 0$, and

$$\frac{1}{2}\nabla \cdot v^\gamma + H(\theta + v^\gamma, \omega) + \gamma\|v^\gamma\|^2 \leqslant \overline{H}^\gamma(\theta).$$

Then we define V^γ as in (6.1) so that $V^\gamma \in L^\alpha_{\text{loc}}(\mathbb{R}^d)$, $\nabla V^\gamma = v^\gamma$, and $V^\gamma(0,\omega) = 0$ for $\mathbb{P}$-a.e. ω. They are then smoothed to obtain V_δ^γ satisfying $\nabla V_\delta^\gamma = v_\delta^\gamma$ and $V_\delta^\gamma(x) = \int V^\gamma(x+\delta y,\omega)\rho(y)dy$. Let $c \in \mathcal{C}$. Then by Itô's formula,for $\mathbb{P}$-a.e. ω and Q_φ^c-a.e. path $x(\cdot)$

$$\begin{aligned} &V_\delta^\gamma((x(t)) - V_\delta^\gamma(x(0)) + \langle\theta, x(t) - x(0)\rangle \\ &= \int_0^t \left(\frac{1}{2}\nabla \cdot v_\delta^\gamma + c\cdot(\theta + v_\delta^\gamma)\right)(s, x(s))ds + \int_0^t \langle(v_\delta^\gamma((x(s)) + \theta), d\beta(s)\rangle. \end{aligned}$$

We add $\gamma\int_0^t \|\theta + v_\delta^\gamma\|^2\,ds$ to both sides and use the inequality

$$\langle c, p\rangle \leqslant L(c,\omega) + H^\gamma(p,\omega) - \gamma\|p\|^2$$

to obtain

$$\begin{aligned} &\langle\theta, x(t) - x(0)\rangle + \gamma\int_0^t \|\theta + v_\delta^\gamma(x(s))\|^2\,ds \\ &\quad - \int_0^t L(c(s,x(s)), \tau_{x(s)}\omega)ds - t\overline{H}^\gamma(\theta) \\ &\quad \leqslant -V_\delta^\gamma((x(t)) + V_\delta^\gamma(x(0)) + \int_0^t \langle v_\delta^\gamma(x(s)) + \theta, d\beta(s)\rangle + r_c(\delta)t, \end{aligned}$$

Q^c_φ-a.e. as in Lemma 6.4. If the initial distribution of $x(0)$ were given by a nice density $\varphi(x)dx$, then since t and $c \in \mathcal{C}$ are fixed, we could let $\delta \to 0$ to arrive at

$$\begin{aligned}\langle \theta, x(t)-x(0)\rangle &- \int_0^t L(c(s,x(s)), \tau_{x(s)}\omega)ds - t\overline{H}^\gamma(\theta)\\ &\leqslant -V^\gamma((x(t)) + V^\gamma(x(0)) + \int_0^t \langle v^\gamma(x(s))+\theta, d\beta(s)\rangle\\ &\quad - \gamma\int_0^t \|\theta + v^\gamma(x(s))\|^2\, ds\end{aligned}$$

a.e. Q^c_φ.

We can now use Lemma 6.5 to replace $V^\gamma(x(t))$ by $V^\gamma_\delta(x(t))$:

$$\begin{aligned}\langle \theta, x(t)-x(0)\rangle &- \int_0^t L(c(s,x(s)), \tau_{x(s)}\omega)ds - t\overline{H}^\gamma(\theta)\\ &\leqslant -V^\gamma_\delta((x(t)) + V^\gamma(x(0)) + \int_0^t \langle v^\gamma(x(s))+\theta, d\beta(s)\rangle\\ &\quad - \gamma\int_0^t \|\theta + v^\gamma(x(s))\|^2\, ds + C\delta^2\end{aligned}$$

a.e. Q^c_φ. Now we rescale and let $\varepsilon \to 0$, allowing arbitrary choices for $c = c_\varepsilon \in \mathcal{C}^*$ and suitable choices of $\varphi = \varphi_\varepsilon$. In fact, the choices of c_ε will be limited to those that satisfy

$$E^{Q^{c_\varepsilon}_\varphi}\left[\gamma\int_0^t \|\theta + v^\gamma(x(s))\|^2\, ds\right] \leqslant Ct$$

for some C independent of ε. By the ergodic theorem, $\varepsilon V^\gamma_\delta(x(\varepsilon^{-1}t)) \to 0$. By Lemma 7.1 the stochastic integral term goes to 0. If the choice of $\varphi_\varepsilon(y)$ is of the form $\varphi(x_\varepsilon - y)$, with fixed φ and $\varepsilon x_\varepsilon \to x$, then the ergodic theorem again can be used to show that

$$\lim_{\varepsilon\to 0} E^{Q^{c_\varepsilon}_{\varphi_\varepsilon}}[\varepsilon|V^\gamma(x(0))|] = 0.$$

Now one can see that with probability 1

$$\lim_{\varepsilon\to 0}\sup_{0\leqslant t\leqslant T}\sup_{|x|\leqslant \ell}\left|\int u_\varepsilon(t, x+\varepsilon y, \omega)\varphi(y)dy - u(t,x)\right| = 0. \tag{7.10}$$

Step 2. By (H5) the function L satisfies

$$L(q,\omega) \leqslant c_3\|q\|^{\alpha'} + c_4$$

with $\alpha' < 2$. Then it is possible to pick $c(\cdot,\cdot) : [0,t]\times\mathbb{R}^d \to \mathbb{R}^d$ such that the process Q^c_x is the Brownian bridge from x to y with $Q^c_x[x(t) = y] = 1$. For a Brownian bridge the choice of c is

$$c(s,x) = \frac{y-x}{t-s}.$$

Since

$$x(s) = \left(1 - \frac{s}{t}\right)x + \frac{s}{t}y + \beta(s) - \frac{s}{t}\beta(t), \quad 0 \leqslant s \leqslant t,$$

we have that

$$c(s, x(s)) = \frac{y - x(s)}{t - s} = \frac{y - x}{t} + \frac{\beta(t) - \beta(s)}{t - s} - \frac{\beta(t)}{t}$$

and

$$\|c(s, x(s))\|^{\alpha'} \leqslant A\left\|\frac{y - x}{t}\right\|^{\alpha'} + A\left\|\frac{\beta(t) - \beta(s)}{t - s}\right\|^{\alpha'} + A\left\|\frac{\beta(t)}{t}\right\|^{\alpha'}$$

with $A = 3^{\alpha'-1}$. Computing the expected value, we get

$$E^{Q_x^c}\|c(s, x(s))\|^{\alpha'} \leqslant A\left\|\frac{y - x}{t}\right\|^{\alpha'} + \frac{B}{(t - s)^{\alpha'/2}} + \frac{B}{t^{\alpha'/2}}.$$

For $\alpha' < 2$ the integration from 0 to t gives

$$E^{Q_x^c}\left[\int_0^t \|c(s, x)\|^{\alpha'} ds\right] \leqslant At\left\|\frac{x - y}{t}\right\|^{\alpha'} + Ct^{1-\alpha'/2} < \infty,$$

providing the following lower bound for $v_\varepsilon(t, x, \omega)$, the solution of (3.1):

$$v_\varepsilon(t, x, \omega) \geqslant \sup_y \left[\frac{1}{\varepsilon} f(\varepsilon y) - At\left\|\frac{x - y}{t}\right\|^{\alpha'} - Ct^{1-\alpha'/2} - c_4 t\right].$$

Recalling that $u_\varepsilon(t, x, \omega) = \varepsilon v_\varepsilon(t/\varepsilon, x/\varepsilon, \omega)$, we obtain

$$\begin{aligned} u_\varepsilon(t, x, \omega) &\geqslant \sup_y \left[f(\varepsilon y) - At\left\|\frac{x - \varepsilon y}{t}\right\|^{\alpha'} - \varepsilon C\left(\frac{t}{\varepsilon}\right)^{1-\alpha'/2} - c_4 t\right] \\ &= \sup_y \left[f(y) - At\left\|\frac{x - y}{t}\right\|^{\alpha'} - C\varepsilon^{\alpha'/2} t^{1-\alpha'/2} - c_4 t\right]. \end{aligned}$$

Combining the above estimate with the semigroup property, we arrive at

$$u_\varepsilon(t+h, x, \omega) \geqslant \sup_y \left[u_\varepsilon(t, y, \omega) - Ah^{1-\alpha'}\|x-y\|^{\alpha'} - C\varepsilon^{\alpha'/2}h^{1-\alpha'/2} - c_4 h\right]. \tag{7.11}$$

Step 3. It follows from (7.10) that uniformly for $(t, x) \in [0, T]\times\{x : \|x\| \leqslant \ell\}$ and $0 \leqslant h \leqslant 1$

$$\limsup_{\varepsilon\to 0} \inf_{y:\|y-x\|\leqslant\varepsilon} u_\varepsilon(t + h, y, \omega) \leqslant u(t + h, x).$$

Then from (7.11) we obtain

$$\inf_{y:\|y-x\|\leqslant\varepsilon} u_\varepsilon(t + h, y, \omega) \geqslant u_\varepsilon(t, x, \omega) - Ah^{1-\alpha'}\varepsilon^{\alpha'} - C\varepsilon^{\alpha'/2} - c_4 h.$$

The above two inequalities imply the estimate

$$\limsup_{\varepsilon\to 0} u_\varepsilon(t, x, \omega) \leqslant u(t + h, x) + c_4 h$$

uniformly for $(t, x) \in [0, t) \times \{x : \|x\| \leqslant \ell\}$. Since $u(t, x)$ is given explicitly by (2.14), one can now let $h \to 0$ and obtain the upper bound. □

Remark 7.3. If (H5) holds with $\alpha = 2$, then we only get (7.10). This, of course, implies local L^p convergence of u_ε to u.

Remark 7.4. In the absence of (H3), (H4), and (H5), one cannot control the stochastic integral, and one can only prove (7.10) for linear f. If $\alpha > 2$, this convergence can be improved to locally uniform convergence, as we have shown above.

Acknowledgment. Kosygina was partially supported by an NSF-AWM Mentoring Travel Grant and PSC-CUNY Awards #60079-33-34 and #66512-00-35. Rezakhanlou and Varadhan were partially supported, respectively, by NSF grants DMS-0307021 and DMS-0104343.

Bibliography

[1] Bensoussan, A.; Lions, J.-L.; Papanicolaou, G. *Asymptotic analysis for periodic structures.* Studies in Mathematics and Its Applications, 5. North-Holland, Amsterdam-New York, 1978.

[2] Braides, A.; Defranceschi, A. *Homogenization of multiple integrals.* Oxford Lecture Series in Mathematics and Its Applications, 12. The Clarendon Press, New York, 1998.

[3] Caffarelli, L. A.; Souganidis, P. E.; Wang, L. Stochastic homogenization for fully nonlinear, second order partial differential equations. *Comm. Pure Appl. Math.* **58** (2005), no. 3, 319–361.

[4] Capuzzo-Dolcetta, I.; Ishii, H. On the rate of convergence in homogenization of Hamilton-Jacobi equations. *Indiana Univ. Math. J.* **50** (2001), no. 3, 1113–1129.

[5] Cioranescu, D.; Donato, P. *An introduction to homogenization.* Oxford Lecture Series in Mathematics and Its Applications, 17. The Clarendon Press, New York, 1999.

[6] Concordel, M. C. Periodic homogenization of Hamilton-Jacobi equations: additive eigenvalues and variational formula. *Indiana Univ. Math. J.* **45** (1996), no. 4, 1095–1117.

[7] Dal Maso, G. *An introduction to* Γ*-convergence.* Progress in Nonlinear Differential Equations and Their Applications, 8. Birkhäuser-Boston, Boston, 1993.

[8] Evans, L. Periodic homogenisation of certain fully nonlinear partial differential equations. *Proc. Roy. Soc. Edinburgh Sect. A* **120** (1992), no. 3-4, 245–265.

[9] Fleming, W. H.; Soner, H. M. *Controlled Markov processes and viscosity solutions.* Applications of Mathematics (New York), 25. Springer, New York, 1993.

[10] Jikov, V. V.; Kozlov, S. M.; Oleĭnik, O. A. *Homogenization of differential operators and integral functionals.* Translated from the Russian by G. A. Yosifian. Springer, Berlin, 1994.

[11] Lions, P. -L.; Papanicolaou, G.; Varadhan, S. R. S. Homogenization of Hamilton-Jacobi equations. Unpublished manuscript, 1986.

[12] Lions, P. -L.; Souganidis, P. E. Correctors for the homogenization of Hamilton-Jacobi equations in the stationary ergodic setting. *Comm. Pure Appl. Math.* **56** (2003), no. 10, 1501–1524.

[13] Lions, P. -L.; Souganidis, P. E. Homogenization of "viscous" Hamilton-Jacobi equations in stationary ergodic media. Preprint, 2004.

[14] McKean, H. P. Jr. *Stochastic integrals.* Probability and Mathematical Statistics, 5. Academic Press, New York, 1969.

[15] Pankov, A. *G-convergence and homogenization of nonlinear partial differential operators.* Mathematics and Its Applications, 422. Kluwer, Dordrecht, 1997.

[16] Papanicolaou, G. C.; Varadhan, S. R. S. *Diffusions with random coefficients.* North-Holland, Amsterdam, 1982.

[17] Rezakhanlou, F.; Tarver, J. E. Homogenization for stochastic Hamilton-Jacobi equations. *Arch. Ration. Mech. Anal.* **151** (2000), no. 4, 277–309.

[18] Sion, M. On general minimax theorems. *Pacific J. Math.* **8** (1958), 171–176.

[19] Souganidis, P. Stochastic homogenization of Hamilton-Jacobi equations and some applications. *Asymptot. Anal.* **20** (1999), no. 1, 1–11.

ELENA KOSYGINA
Department of Mathematics
Baruch College
One Bernard Baruch Way
New York, NY 10010
E-mail: `elena_kosygina@baruch.cuny.edu`

FRAYDOUN REZAKHANLOU
Department of Mathematics
University of California, Berkeley
815 Evans Hall
Berkeley, CA 94720
E-mail: `rezakhan@math.berkeley.edu`

S. R. S. VARADHAN
Courant Institute
251 Mercer St.
New York, NY 10012
E-mail: `varadhan@cims.nyu.edu`

Received January 2005.

Comm. Pure Appl. Math. 61 (2008) 816-847

Homogenization of Hamilton-Jacobi-Bellman Equations with Respect to Time-Space Shifts in a Stationary Ergodic Medium

ELENA KOSYGINA
Baruch College
CUNY Graduate Center

AND

S. R. S. VARADHAN
Courant Institute

Abstract

We consider a family $\{u_\varepsilon(t,x,\omega)\}$, $\varepsilon > 0$, of solutions to the equation $\partial u_\varepsilon/\partial t + \varepsilon\Delta u_\varepsilon/2 + H(t/\varepsilon, x/\varepsilon, \nabla u_\varepsilon, \omega) = 0$ with the terminal data $u_\varepsilon(T,x,\omega) = U(x)$. Assuming that the dependence of the Hamiltonian $H(t,x,p,\omega)$ on time and space is realized through shifts in a stationary ergodic random medium, and that H is convex in p and satisfies certain growth and regularity conditions, we show the almost sure locally uniform convergence, in time and space, of $u_\varepsilon(t,x,\omega)$ as $\varepsilon \to 0$ to the solution $u(t,x)$ of a deterministic averaged equation $\partial u/\partial t + \overline{H}(\nabla u) = 0$, $u(T,x) = U(x)$. The "effective" Hamiltonian $\overline{H}$ is given by a variational formula.

1 Introduction

Let $(\Omega, \mathcal{F}, \mathbb{P})$ be a probability space and $\omega \in \Omega$. We are interested in the behavior, as $\varepsilon \to 0$, of a family $\{u_\varepsilon(t,x,\omega)\}$, $\varepsilon > 0$, of solutions to the following terminal value problem:

$$\frac{\partial u_\varepsilon}{\partial t} + \frac{\varepsilon}{2}\Delta u_\varepsilon + H\left(\frac{t}{\varepsilon}, \frac{x}{\varepsilon}, \nabla u_\varepsilon, \omega\right) = 0, \quad (t,x) \in [0,T) \times \mathbb{R}^d, \tag{1.1}$$

$$u_\varepsilon(T,x,\omega) = U(x). \tag{1.2}$$

The Hamiltonian $H(t,x,p,\omega)$ is assumed to be convex in p. The dependence of H on (t,x) is realized through the shifts in a stationary ergodic random medium (see the next section for a detailed description). The terminal condition U is a uniformly continuous nonrandom function. Under some additional assumptions on H, we obtain a homogenization result; i.e., we show that, with probability 1, $u_\varepsilon(t,x,\omega)$ converges locally uniformly in t and x to a nonrandom limit $u(t,x)$.

Function $u(t,x)$ is the unique solution of the Hamilton-Jacobi equation

$$\frac{\partial u}{\partial t} + \overline{H}(\nabla u) = 0, \quad (t,x) \in [0,T) \times \mathbb{R}^d, \tag{1.3}$$

with the same terminal value $U(x)$. We also provide a variational representation for the effective Hamiltonian $\overline{H}$.

Homogenization problems for this type of equation with or without a "viscous" term were extensively studied in the periodic, quasi, and almost periodic settings. We mention only several relatively recent papers where one can find further references [1, 2, 6, 8, 10, 17]. For a broader view of homogenization methods, see monographs [3, 4, 5, 9, 13, 19].

One of the standard approaches is based on the construction of so-called correctors. Roughly speaking, the corrector is the first nontrivial error term in a formal asymptotic expansion of u_ε around u. It is obtained as a solution of an auxiliary "cell" problem. The word *cell* refers to the fact that in the periodic case the space variable changes over a torus. The existence of correctors and the uniform convergence of u_ε to a limit were established for a wide range of first- and second-order partial differential equations. This method is very robust and does not require the Hamiltonian to be convex. The most essential assumption on the Hamiltonian is the coercivity in p:

$$H(t,x,p,\omega) \to \infty \quad \text{as } |p| \to \infty,$$

uniformly in t, x, and ω.

The extension of homogenization results to the stationary ergodic setting presented a number of difficulties due to the lack of compactness. It was shown that correctors (or even approximate correctors) need not exist in general [16]. In the case when H is convex in p and independent of t, the homogenization can be proved by using variational methods in combination with some version of the ergodic theorem (see [14, 18]). A further step was taken in [7], which proves homogenization for fully nonlinear uniformly elliptic equations whose solutions do not have a representation formula.

It turns out that the time-space averaging (versus just the space averaging) requires an additional control on u_ε (see, for example, [12] for periodic Hamilton-Jacobi equations and [15] for linear equations in stationary ergodic random media).

In this paper we present a method that allows us to obtain time-space homogenization results for Hamilton-Jacobi-Bellman equations in a stationary ergodic setting. It is essential for our approach that H be convex in p.

In the case when

$$H(t,x,p,\omega) = \frac{1}{2}\sum_{i=1}^{d} p_i^2 - \sum_{i=1}^{d} b_i(t,x,\omega)p_i - W(t,x,\omega),$$

the homogenization problem for (1.1) is closely related to the quenched large-deviations principle for a Brownian motion with a random drift $b = (b_1, b_2,$

$\ldots, b_d)$ in a random potential W (see [14, 18, 21]). Both the drift and the potential are assumed to be stationary ergodic processes (see Remark 2.1).

The paper is organized as follows: In Section 2 we state the assumptions, formulate the main result, and discuss the idea of the proof. In Section 3 we prove the lower bound on u_ε. The construction of approximate supersolutions of (1.1) needed for the upper bound is carried out in Section 4. The proof of the upper bound on u_ε is given in Section 5. A generalization of Wiener's ergodic theorem used for the upper bound is derived in Section 6. The appendix contains proofs of two technical lemmas.

Notation. For $x, y \in \mathbb{R}^k$ we denote by $\langle x, y\rangle$ the standard scalar product of x and y and by $|y|$ the Euclidean norm of y. Borel σ-algebra on $\mathbb{R}^{d+1}$ is denoted by $\mathcal{B}$. If A is a set in $\mathbb{R}^k$, then $|A|$ denotes the Lebesgue measure of A. $B(x,r) \subset \mathbb{R}^k$ is a ball of radius r centered at x. If $f \in L^\alpha$, then $\|f\|_\alpha$ stands for its norm. $\mathbb{E}$ is the expectation with respect to $\mathbb{P}$. $B(t)$, $t \geqslant 0$, is the path of the canonical Brownian motion.

2 Main Result and Preliminary Discussion

Let $\{\tau_{(t,x)} : (t,x) \in \mathbb{R}^{d+1}\}$ be a group of measure-preserving transformations acting ergodically on $(\Omega, \mathcal{F}, \mathbb{P})$. Assume that the map $(t,x,\omega) \mapsto \tau_{(t,x)}\omega$ from $\mathbb{R}^{d+1} \times \Omega$ to Ω is $\mathcal{B} \times \mathcal{F}$ measurable.

Let $H(p,\omega) : \mathbb{R}^d \times \Omega \to \mathbb{R}$ be convex in p and satisfy the following assumptions:

(H1) For all $(p,\omega) \in \mathbb{R}^d \times \Omega$

$$c_1|p|^\alpha - c_2 \leqslant H(p,\omega) \leqslant c_3(|p|^\beta + 1)$$

for some positive constants c_1, c_2, c_3 and $1 < \alpha \leqslant \beta < \infty$.

(H2) $H(p, \tau_{(t,x)}\omega)$ is uniformly continuous in (t,x), uniformly in ω, and locally uniformly in p, i.e., for every $l > 0$

$$\lim_{\delta\to 0} \sup_{|(t,x)|\leqslant\delta} \sup_{|p|\leqslant l} \sup_{\omega\in\Omega} |H(p,\tau_{(t,x)}\omega) - H(p,\omega)| = 0.$$

(H3) There exists a positive function $\nu(\delta) \to 0$ as $\delta \to 0$ and a constant $C > 0$ such that for $|(t,x)| \leqslant \delta$ and $\omega \in \Omega$

$$H(p,\tau_{(t,x)}\omega) \geqslant (1+\nu(\delta))H((1+\nu(\delta))^{-1}p,\omega) - C\nu(\delta).$$

These assumptions can be equivalently stated in terms of the Lagrangian

$$L(q,\omega) = \sup_{p\in\mathbb{R}^d} (\langle p,q\rangle - H(p,\omega)). \tag{2.1}$$

(L1) For all $(p,\omega) \in \mathbb{R}^d \times \Omega$

$$c_4|q|^{\beta'} - c_5 \leqslant L(q,\omega) \leqslant c_6(|q|^{\alpha'} + 1)$$

for some positive of constants c_4, c_5, c_6 and

$$\alpha' = \frac{\alpha}{\alpha - 1}, \quad \beta' = \frac{\beta}{\beta - 1}.$$

(L2) $L(q, \tau_{(t,x)}\omega)$ is uniformly continuous in x, uniformly in ω, and locally uniformly in q.

(L3) There exists a positive function $\nu(\delta) \to 0$ as $\delta \to 0$ and a constant $C > 0$ such that for $|(t, x)| \leqslant \delta$ and $\omega \in \Omega$

$$L(q, \tau_{(t,x)}\omega) \leqslant (1 + \nu(\delta))L(q, \omega) + C\nu(\delta).$$

We assume that the terminal data $U(x)$ is uniformly continuous. Note that such U can be uniformly approximated by functions that are Lipschitz-continuous, and by the maximum principle the u_ε are uniformly approximated as well. So there is no loss of generality in assuming that there is a constant K such that for all $x, y, \in \mathbb{R}^d$

$$|U(x) - U(y)| \leqslant K|x - y|. \tag{2.2}$$

This makes the proof somewhat simpler.

Remark 2.1. Functions $H(t, x, p, \omega)$, $b(t, x, \omega)$, and $W(t, x, \omega)$, from the introduction, should be understood as $H(p, \tau_{(t,x)}\omega)$, $b(\tau_{(t,x)}\omega)$, and $W(\tau_{(t,x)}\omega)$, respectively.

2.1 The Effective Hamiltonian $\overline{H}$

The translation group $\{\tau_{(t,x)} : (t, x) \in \mathbb{R}^{d+1}\}$ acting on $L^2(\Omega, \mathcal{F}, \mathbb{P})$ will have infinitesimal generators D_t, ∇_i, $i = 1, 2, \ldots, d$, in the coordinate directions (see [13, sec. 7.1 and pp. 231–232]). The space gradient $\nabla = (\nabla_1, \nabla_2, \ldots, \nabla_d)$, the divergence $\operatorname{div}(U_1, U_2, \ldots, U_d) = \nabla_1 U_1 + \nabla_2 U_2 + \cdots \nabla_d U_d$, and the Laplace operator $\Delta = \sum_{i=1}^d \nabla_i^2$ are defined in the usual way.

Let us denote by $\mathbf{B}$ the space of measurable, essentially bounded maps from Ω to $\mathbb{R}^d$ and by $\mathbf{D}$ the space of bounded probability densities $\Phi : \Omega \to \mathbb{R}$ relative to $\mathbb{P}$ that are bounded away from 0 and have essentially bounded time-space gradients. Define

$$\mathcal{E} = \{(b, \Phi) \in \mathbf{B} \times \mathbf{D} : D_t\Phi + \operatorname{div}(b\Phi) = \tfrac{1}{2}\Delta\Phi\}. \tag{2.3}$$

We shall always assume that the equation in (2.3) is satisfied in the weak sense: with probability 1 for every $G \in C_0^\infty(\mathbb{R}^{d+1})$

$$\iint \Phi(\tau_{(t,x)}\omega)\partial_t G(t, x) + \big\langle (b\Phi - \tfrac{1}{2}\nabla\Phi)(\tau_{(t,x)}\omega), \nabla G(t, x)\big\rangle dt\, dx = 0. \tag{2.4}$$

For $(b, \Phi) \in \mathcal{E}$ set

$$m(b, \Phi) = \mathbb{E}[b(\omega)\Phi(\omega)], \tag{2.5}$$

$$h(b, \Phi) = \mathbb{E}[L(b(\omega), \omega)\Phi(\omega)]. \tag{2.6}$$

Define a function $\overline{L}$ on $\mathbb{R}^d$ by

$$\overline{L}(q) = \inf_{\substack{b:(b,\Phi)\in\mathcal{E} \\ \mathbb{E}(b\Phi)=q}} h(b,\Phi). \tag{2.7}$$

Observe that $\overline{L}$ is convex. This is a simple consequence of (2.7) and the fact that if $(b_i, \Phi_i) \in \mathcal{E}$, $i = 1, 2$, then for every $\lambda \in [0, 1]$,

$$\left(\frac{\lambda b_1 \Phi_1 + (1-\lambda) b_2 \Phi_2}{\lambda \Phi_1 + (1-\lambda)\Phi_2}, \lambda\Phi_1 + (1-\lambda)\Phi_2\right) \in \mathcal{E}.$$

Let $\overline{H}$ be the convex conjugate of $\overline{L}$:

$$\begin{aligned}\overline{H}(p) &= \sup_{q\in\mathbb{R}^d} [\langle p, q\rangle - \overline{L}(q)] \\ &= \sup_{(b,\Phi)\in\mathcal{E}} [\langle p, m(b,\Phi)\rangle - h(b,\Phi)].\end{aligned} \tag{2.8}$$

The solution $u(t, x)$ of (1.3), with $u(T, x) = U(x)$, is given by the Hopf-Lax-Oleĭnik formula

$$\begin{aligned}u(t,x) &= \sup_{q\in\mathbb{R}^d} \left[U(q) - (T-t)\overline{L}\left(\frac{q-x}{T-t}\right)\right] \\ &= \sup_{(b,\Phi)\in\mathcal{E}} [U(x + (T-t)m(b,\Phi)) - (T-t)h(b,\Phi)].\end{aligned} \tag{2.9}$$

The main result of this paper is the following theorem:

THEOREM 2.2 *Assume that* $H(p, \omega)$ *satisfies* (H1)–(H3) *and that the terminal value* $U(x)$ *satisfies* (2.2) *on* $\mathbb{R}^d$. *Let* $u(t, x)$ *be the unique solution of* (1.2)–(1.3) *with* $\overline{H}$ *given by* (2.8). *Then with probability* 1, *for every* $l > 0$,

$$\lim_{\varepsilon\to 0} \sup_{0\le t\le T} \sup_{|x|\le l} |u_\varepsilon(t,x,\omega) - u(t,x)| = 0.$$

Remark 2.3. To identify the effective Hamiltonian it is enough to consider the set of linear terminal data $\{\langle p, x\rangle,\ p \in \mathbb{R}^d\}$. For each $p \in \mathbb{R}^d$ the terminal value problem for the averaged equation (1.3) with $U(x) = \langle p, x\rangle$ has an obvious solution $u(t, x) = \langle p, x\rangle - (T - t)\overline{H}(p)$. In particular, if we let $(t, x) = (0, 0)$ and $T = 1$, then we get $\overline{H}(p) = u(0, 0)$. Therefore if the homogenization result holds, then

$$\overline{H}(p) = \lim_{\varepsilon\to 0} u_\varepsilon^p(0,0,\omega) \quad \text{a.s.},$$

where u_ε^p solves (1.1)–(1.2) with $U(x) = \langle p, x\rangle$.

2.2 Shifts, Rescaling, and Variational Formulae

Our assumptions on the Hamiltonian allow us to use a variational representation of $u_\varepsilon(t, x, \omega)$, which can be considered the starting point of our analysis.

Denote by $\mathcal{C}$ the space of essentially bounded controls $c = c(s, x)$. Consider the diffusion on $\mathbb{R}^d$, with generator

$$\mathbf{A}_s = \frac{1}{2}\Delta + c(s, x) \cdot \nabla.$$

For $t < T$ and $x \in \mathbb{R}^d$, let $Q^c_{t,x}$ be the measure on $C([t, T); \mathbb{R}^d)$ that corresponds to the diffusion process with the above generator that starts from x at time t. For each $c \in \mathcal{C}$ and $\omega \in \Omega$ we set

$$v^c(t, x, \omega) = E^{Q^c_{t,x}}\left[U(x(T)) - \int_t^T L(c(s, x(s)), \tau_{(s,x(s))}\omega)ds\right].$$

Then

$$v(t, x, \omega) = \sup_{c \in \mathcal{C}} v^c(t, x, \omega)$$

is the unique solution on $(-\infty, T] \times \mathbb{R}^d$ of

$$\frac{\partial v}{\partial t} + \frac{1}{2}\Delta v + H(\nabla v, \tau_{(t,x)}\omega) = 0 \tag{2.10}$$

with $v(T, x) = U(x)$ (see, for example, [11]).

An easy consequence of uniqueness is the covariant nature of the solution. If $v(\cdot, \cdot, \omega)$ is a solution corresponding to ω on $\mathbb{R}^d \times (-\infty, T]$, then $v^{s,y}(\cdot, \cdot, \omega) = v(\cdot + s, \cdot + y, \omega)$ is a solution on $(-\infty, T - s] \times \mathbb{R}^d$ corresponding to $\omega' = \tau_{(s,y)}\omega$ with terminal data $v^{s,y}(T - s, \cdot, \omega) = U(\cdot + y)$. In particular,

$$v(t, x, \omega) = v^{t,x}(0, 0, \tau_{(t,x)}\omega) \tag{2.11}$$

if we match the terminal time and data.

A simple calculation shows $u_\varepsilon(t, x, \omega) = \varepsilon v_\varepsilon(t/\varepsilon, x/\varepsilon, \omega)$, where v_ε solves the unscaled equation (2.10) with the terminal condition

$$v_\varepsilon(T/\varepsilon, x, \omega) = U(\varepsilon x)/\varepsilon.$$

This leads to the following variational formula for the solution to the terminal value problem (1.1)–(1.2):

$$u_\varepsilon(t, x, \omega) = \sup_{c \in \mathcal{C}} E^{Q^c_{t/\varepsilon, x/\varepsilon}}\left[U(\varepsilon x(T/\varepsilon)) - \varepsilon \int_{t/\varepsilon}^{T/\varepsilon} L(c(s, x(s)), \tau_{(s,x(s))}\omega)ds\right]. \tag{2.12}$$

It is sometimes more convenient to rescale the diffusion by considering the process $y_\varepsilon(s) = \varepsilon x(s/\varepsilon)$, $s \in [t, T]$, with the drift $c_\varepsilon(s, x) = c(s/\varepsilon, x/\varepsilon) \in \mathcal{C}$. Below we

drop the subscript ε, since we take a supremum over all drifts in $\mathcal{C}$. After rescaling the generators become

$$\mathbf{A}_s^\varepsilon = \frac{\varepsilon}{2}\Delta + c(s,x)\cdot\nabla \tag{2.13}$$

with controls $c \in \mathcal{C}$. Writing $Q_{t,x}^{\varepsilon,c}$ for the measure associated with $\mathbf{A}_s^\varepsilon$, we get

$$u_\varepsilon(t,x,\omega) = \sup_{c\in\mathcal{C}} E^{Q_{t,x}^{\varepsilon,c}}\left(U(y(T)) - \int_t^T L(c(s,y(s)), \tau_{(s/\varepsilon, y(s)/\varepsilon)}\omega)ds\right). \tag{2.14}$$

Our goal is to show, locally uniform in t and x, the almost sure convergence of $u_\varepsilon(t,x,\omega)$ to $u(t,x)$ given by (2.9).

2.3 Ergodic Theorem and a Lower Bound

Let $b \in \mathbf{B} \equiv L^\infty(\Omega;\mathbb{R}^d)$. Consider a Brownian motion $x(t)$ on $\mathbb{R}^d$ starting from 0 at time 0 with a random drift $b(\tau_{(t,x)}\omega)$, and denote the corresponding measure on $C([0,\infty);\mathbb{R}^d)$ by $Q_{0,0}^{b,\omega}$. Then this diffusion can be "lifted" to Ω as follows. Pick a starting point ω. Define a path $\omega(s)$, which starts from ω at time 0, by $\omega(s) = \tau_{(s,x(s))}\omega$, $s \geqslant 0$.

The measure $P^{b,\omega}$ induced on paths in Ω corresponds to a Markov process on Ω with the generator

$$\mathcal{A}_b = D_t + \frac{1}{2}\Delta + b(\omega)\cdot\nabla.$$

If we can find a positive density Φ on Ω such that $\Phi\, d\mathbb{P}$ is an invariant ergodic probability measure for $\mathcal{A}_b$, then by the ergodic theorem for every $F \in L^1(\Omega, \Phi\, d\mathbb{P})$

$$\lim_{t\to\infty}\frac{1}{t}\int_0^t F(\omega(s))ds = \int_\Omega F(\omega)\Phi(\omega)d\mathbb{P}$$

$P^{b,\omega}$-a.s. or in $L^1(P^{b,\omega})$ for $\mathbb{P}$-a.e. ω.

Finding an invariant density for a given drift is a hard problem. We work in Ω, which is infinite dimensional, while our diffusion is finite dimensional and is very degenerate when viewed on Ω. It is clear, though, that $\mathcal{E}$ is not empty, since it obviously contains pairs $(b, 1)$, where b is a constant. Moreover, it is easy to show that if $(b,\Phi) \in \mathcal{E}$, then $\Phi\, d\mathbb{P}$ is an *ergodic* invariant measure for $\mathcal{A}_b$, and we can use the ergodic theorem.

If we view ω in the formula (2.12) for u_ε as a parameter, we may allow the controls c to be dependent on ω as well. Let us consider only stationary controls, i.e., $c(t,x,\omega) = b(\tau_{(t,x)}\omega)$, where b is such that $(b,\Phi) \in \mathcal{E}$ for some Φ. Setting $(t,x) = 0$, for a.e. ω with respect to $\mathbb{P}$, we obtain almost surely with respect to

$P^{b,\omega}$ and in $L^1(P^{b,\omega})$

$$\lim_{\varepsilon\to 0} \varepsilon \int_0^{T/\varepsilon} b(\omega(s))ds = T \int_\Omega b(\omega)\Phi(\omega)d\mathbb{P} = Tm(b,\Phi),$$

$$\lim_{\varepsilon\to 0} \varepsilon \int_0^{T/\varepsilon} L(b(\omega(s)),\omega(s))ds = T \int_\Omega L(b(\omega),\omega)\Phi\, d\mathbb{P} = Th(b,\Phi).$$

Therefore, for the diffusions $Q_{0,0}^{b,\omega}$ on $\mathbb{R}^d$ we obtain, for almost all ω with respect to $\mathbb{P}$,

$$\lim_{\varepsilon\to 0} E^{Q_{0,0}^{b,\omega}}[\varepsilon|x(T/\varepsilon) - Tm(b,\Phi)|] = 0, \tag{2.15}$$

$$\lim_{\varepsilon\to 0} E^{Q_{0,0}^{b,\omega}}\left[\varepsilon\left|\int_0^{T/\varepsilon} L(b(\tau_{(t,x(t))}\omega,\omega)dt - Th(b,\Phi)\right|\right] = 0. \tag{2.16}$$

It now follows that for each $(b,\Phi)\in\mathcal{E}$,

$$\liminf_{\varepsilon\to 0} u_\varepsilon(0,0,\omega) \geqslant U(Tm(b,\Phi)) - Th(b,\Phi) \quad \mathbb{P}\text{-a.s.} \tag{2.17}$$

Notice that, because of (2.9), this establishes an almost sure lower bound at $(0,0)$. The relation (2.11) and the translation invariance of $\mathbb{P}$ imply that the lower bound holds for arbitrary (t,x) but in probability. More work needs to be done to obtain an almost sure locally uniform lower bound. The proof is similar to the one given in Section 4 of [14], and the details are presented in the next section.

Remark (A few words about an upper bound). An upper bound is essentially obtained by comparison with a family of supersolutions of (1.1). The starting point of the construction is formula (2.8) for the effective Hamiltonian. The main idea is the same as in [14]. There are some difficulties in the construction due to the lack of control on the time derivatives. This problem was not present in the case of the time-independent Hamiltonian.

3 Lower Bound

We begin with an easy estimate on the solution $u_\varepsilon(t,x,\omega)$ of (1.1) and its approximations

$$u_\varepsilon^k(t,x,\omega) \stackrel{\text{def}}{=} \sup_{c\in\mathcal{C}_k} E^{Q_{t,x}^{\varepsilon,c}}\Big(U(y(T)) - \int_t^T L(c(s,y(s)),\tau_{(s/\varepsilon,y(s)/\varepsilon)}\omega)ds\Big), \tag{3.1}$$

where $y(\cdot)$ is a diffusion with the generator (2.13) and $\mathcal{C}_k$ consists of $c\in\mathcal{C}$ that are bounded in norm by k.

LEMMA 3.1 *Assume* (L1) *and* (2.2). *Then there are positive constants* C_1 *and* C_2 *such that*

$$|u_\varepsilon(t,x,\omega) - U(x)| \leqslant C_1(T-t)^{1/2} + C_2|T-t|,$$
$$|u_\varepsilon^k(t,x,\omega) - U(x)| \leqslant C_1(T-t)^{1/2} + C_2|T-t|.$$

PROOF: From the variational formula (2.14), the bounds (L1), and the convexity of $|q|^\alpha$ for $\alpha \geqslant 1$, it follows that, uniformly in ω,

$$\begin{aligned} u_\varepsilon(t,x,\omega) - U(x) &\leqslant \sup_{c\in\mathcal{C}} \Bigg[K\Bigg[E^{Q^{\varepsilon,c}_{t,x}} \int_t^T |c(s,x(s))|ds \Bigg] + CK\varepsilon^{1/2}(T-t)^{1/2} \\ &\qquad - E^{Q^{\varepsilon,c}_{t,x}} \int_t^T (c_4|c(s,x(s))|^{\beta'} - c_5)ds \Bigg] \\ &\leqslant CK\varepsilon^{1/2}(T-t)^{1/2} + (T-t) \sup_{q\in\mathbb{R}^d} [K|q| - c_4|q|^{\beta'} + c_5]. \end{aligned}$$

For the lower bound we can take $c = 0$ to get

$$u_\varepsilon(t,x,\omega) - U(x) \geqslant -C\varepsilon^{1/2}(T-t)^{1/2} - (T-t) \sup_\omega L(0,\omega).$$

The same proof works for u_ε^k as well. □

In the previous section we established a weak version of the lower bound. We state it now as a lemma.

LEMMA 3.2 *Assuming the regularity condition* (2.2) *on* U, *for almost all* ω *with respect to* $\mathbb{P}$,

$$\liminf_{\varepsilon\to 0} [u_\varepsilon(0,0,\omega) - u(0,0)] \geqslant 0. \tag{3.2}$$

PROOF: Relation (2.17) states that for each $(b,\Phi) \in \mathcal{E}$

$$\liminf_{\varepsilon\to 0} [u_\varepsilon(0,0,\omega) - U(Tm(b,\varphi)) + Th(b,\Phi)] \geqslant 0,$$

$\mathbb{P}$ almost surely. Since the supremum over $\mathcal{E}$ is easily reduced to the supremum over a countable subset of $\mathcal{E}$ in the formula (2.9), the assertion (3.2) of the lemma follows. □

The next lemma strengthens the estimate (3.2).

LEMMA 3.3 *Under the conditions of Lemma* 3.2, *for each* $\eta > 0$, *there is a subset* $N_\eta \subset \Omega$ *such that* $\mathbb{P}(N_\eta) \geqslant 1 - \eta$ *and*

$$\liminf_{r\to 0} \liminf_{\varepsilon\to 0} \inf_{\omega\in N_\eta} \inf_{\substack{|y|\leqslant r \\ |t|\leqslant r}} [u_\varepsilon(t,y,\omega) - u(0,0)] \geqslant 0. \tag{3.3}$$

PROOF: It is enough to prove that for almost all ω with respect to $\mathbb{P}$,

$$\liminf_{r\to 0}\liminf_{\varepsilon\to 0}\inf_{\substack{|y|\leqslant r\\ |t|\leqslant r}}[u_\varepsilon(t,y,\omega)-u(0,0)]\geqslant 0, \tag{3.4}$$

and (3.3) will follow by an application of Egoroff's theorem.

Consider functions u_ε^k defined by (3.1). Clearly, $u_\varepsilon^k(t,x,\omega)\nearrow u_\varepsilon(t,x,\omega)$ as $k\uparrow\infty$. Therefore, by Lemma 3.2 it is sufficient to prove that for each $k<\infty$,

$$\liminf_{r\to 0}\liminf_{\varepsilon\to 0}\inf_{\omega}\inf_{\substack{|y|\leqslant r\\ |t|\leqslant r}}[u_\varepsilon(t,y,\omega)-u_\varepsilon^k(0,0,\omega)]\geqslant 0. \tag{3.5}$$

We break the proof into two parts. For fixed k,

$$\liminf_{r\to 0}\liminf_{\varepsilon\to 0}\inf_{\omega}\inf_{|y|\leqslant r}[u_\varepsilon^{k+1}(0,y,\omega)-u_\varepsilon^k(0,0,\omega)]\geqslant 0, \tag{3.6}$$

$$\liminf_{r\to 0}\liminf_{\varepsilon\to 0}\inf_{\omega}\inf_{\substack{|y|\leqslant r\\ |t|\leqslant r}}[u_\varepsilon^{k+1}(t,y,\omega)-u_\varepsilon^k(0,y,\omega)]\geqslant 0. \tag{3.7}$$

Step 1. To prove (3.6) we let $x(\cdot)$ be a path of a diffusion with a drift $b\in\mathcal{C}_k$ starting from 0 at time 0. We shall couple it to $y(\cdot)$, which starts from y at time 0 as follows: Assume that $r<T$. Define

$$y(s)=\begin{cases} y+x(s)-\frac{sy}{|y|}, & 0\leqslant s<|y|,\\ x(s), & |y|\leqslant s\leqslant T.\end{cases}$$

Then $y(\cdot)$ has the drift

$$c(s,z)=\begin{cases} b\left(s,z-y+\frac{sy}{|y|}\right)-\frac{y}{|y|} & \text{if } 0\leqslant s<|y|,\\ b(s,z) & \text{if } |y|\leqslant s\leqslant T,\end{cases}$$

and $c\in\mathcal{C}_{k+1}$. Let us denote the coupled measure by $Q_{0,0,y}^{\varepsilon,b,c}$.

Since the two paths are identical after time $|y|$, they end up at the same place (if $|y|\leqslant r$) and

$$\begin{aligned}
&u_\varepsilon^k(0,0,\omega)-u_\varepsilon^{k+1}(0,y,\omega)\\
&\leqslant \sup_{b\in\mathcal{C}_k}\left[E^{Q_{0,0,y}^{\varepsilon,b,c}}\int_0^{|y|}|L(b(s,x(s)),\tau_{(s/\varepsilon,x(s)/\varepsilon)}\omega)|ds\right.\\
&\qquad\left.+E^{Q_{0,0,y}^{\varepsilon,b,c}}\int_0^{|y|}|L(c(s,y(s)),\tau_{(s/\varepsilon,y(s)/\varepsilon)}\omega)|ds\right]\\
&\leqslant 2|y|\sup_{\substack{\omega\\ |q|\leqslant k+1}}|L(q,\omega)|.
\end{aligned}$$

This proves (3.6).

Remark 3.4. Observe that by using the same argument as in Step 1 we can show that for a fixed k

$$\liminf_{r\to 0} \liminf_{\varepsilon\to 0} \inf_{\omega} \inf_{t\leqslant T} \inf_{|y-z|\leqslant r} [u_\varepsilon^{k+1}(t,y,\omega) - u_\varepsilon^k(t,z,\omega)] \geqslant 0. \tag{3.8}$$

Step 2. Now we turn to (3.7). Let us take the case $t < 0$. Then we can choose $c(s,x) = 0$ until time 0 and then make the optimal choice from $\mathcal{C}_{k+1}$. Therefore

$$u_\varepsilon^{k+1}(t,y,\omega) \geqslant \int_{\mathbb{R}^d} u_\varepsilon^{k+1}(0,z,\omega) p_\varepsilon(t,y;0,z)dz - |t| \sup_{\omega} L(0,\omega) \tag{3.9}$$

where p_ε is the fundamental solution of $\frac{\partial}{\partial t} + \frac{\varepsilon}{2}\Delta$. From Remark 3.4 we can obtain a uniform lower bound on $u_\varepsilon^{k+1}(0,z,\omega) - u_\varepsilon^k(0,y,\omega)$ provided $|z|$ is small. In addition, we have for fixed $r' > 0$,

$$\lim_{r\to 0} \limsup_{\varepsilon\to 0} \sup_{\substack{|y|\leqslant r\\ |t|\leqslant r}} \int_{|z|>r'} p_\varepsilon(t,y;0,z)dz = 0.$$

From the bounds on u_ε^{k+1} of Lemma 3.1 and from (2.2), it is clear that

$$\lim_{r\to 0} \limsup_{\varepsilon\to 0} \sup_{\substack{|y|\leqslant r\\ |t|\leqslant r}} \int_{|z|>r'} |u_\varepsilon^{k+1}(0,z,\omega)| p_\varepsilon(t,y;0,z)dz = 0,$$

which together with (3.9) implies (3.7).

The case when $t > 0$ is similar. We represent $u_\varepsilon^k(0,y,\omega)$ by a variational formula, make the optimal choice of $c \in \mathcal{C}_k$, and get

$$u_\varepsilon^{k+1}(t,y,\omega) - u_\varepsilon^k(0,y,\omega) \geqslant$$
$$\int_{\mathbb{R}^d} \left(u_\varepsilon^{k+1}(t,y,\omega) - u_\varepsilon^k(t,z,\omega)\right) p_\varepsilon^c(0,y;t,z)dz - t \sup_{|q|\leqslant k} |L(q,\omega)|,$$

where p_ε^c is the fundamental solution of $\frac{\partial}{\partial t} + c(t,x) \cdot \nabla + \frac{\varepsilon}{2}\Delta$. Using (3.8), Lemma 3.1, condition (2.2), and properties of p_ε^c with $c \in \mathcal{C}_k$, we can derive (3.7). □

Combining the relation (2.11), Lemma 3.3, and the ergodic theorem, we obtain the almost sure locally uniform lower bound.

THEOREM 3.5 *Assume that $H(p,\omega)$ satisfies* (H1) *and* (H2) *and that the terminal condition $U(x)$ is uniformly continuous on $\mathbb{R}^d$. Let $u(t,x)$ be the unique solution of* (1.2)–(1.3) *with $\overline{H}$ given by* (2.8). *Then with probability* 1, *for every $l > 0$,*

$$\liminf_{\varepsilon\to 0} \inf_{0\leqslant t\leqslant T} \inf_{|x|\leqslant l} [u_\varepsilon(t,x,\omega) - u(t,x)] \geqslant 0.$$

PROOF: We omit the proof, since it is essentially the same as that for theorem 2.1 in [14]. □

4 Construction of Approximate Supersolutions

Let $H(p,\omega)$ satisfy (H1)–(H3). Fix an arbitrary $\theta \in \mathbb{R}^d$. Our goal is ideally to construct, for each $\theta \in \mathbb{R}^d$, a supersolution V such that

$$\partial_t V + \frac{1}{2}\Delta V + H(\theta + \nabla V, \tau_{(t,x)}\omega) \leqslant \overline{H}(\theta), \quad (t,x) \in \mathbb{R}^{d+1}.$$

Actually, it is enough to construct approximate supersolutions $V_\delta(t,x,\omega)$, $\delta \in (0,\delta_0)$, or even just their space-time gradients with mean 0, such that, for a.e. ω,

$$\text{(4.1)} \quad \partial_t V_\delta + \frac{1}{2}\Delta V_\delta + H(\theta + \nabla V_\delta, \tau_{(t,x)}\omega) \leqslant \overline{H}(\theta) + \nu(\delta), \quad (t,x) \in \mathbb{R}^{d+1},$$

where $\nu(\delta) > 0$, $\nu(\delta) \to 0$ as $\delta \to 0$.

The supersolution V is constructed by duality very much like in the time-independent case. At a purely formal level there is no difference. However, in terms of the regularity of the supersolution there is considerable difference. One can only establish the minimal regularity for the supersolution. The inequality (4.1) and the lower bound of (H1) allow us to control $E[|\nabla V|^\alpha]$. The bounds one gets are local L^α bounds on ∇V and an L^∞ upper bound on $\partial_t V + \frac{1}{2}\Delta V$. The fact that $\partial_t V + \frac{1}{2}\Delta V$ has mean 0 provides a local L^1 bound on $\partial_t V + \frac{1}{2}\Delta V$. Convolution in space by a smoothing kernel will get an estimate on ΔV, and this will finally allow us to control $V_t * \varphi$ in L^1_{loc} for smooth φ. For this one has to work with weak compactness in L^1, which is always delicate.

An upper bound on u_ε will then be established in the next section using a probabilistic comparison argument and the ergodic theorem, Theorem 5.3.

At first, we introduce some notation. Let us recall that $\mathbf{B} \equiv L^\infty(\Omega;\mathbb{R}^d)$. We define $\mathbf{B}_r = \{b \in \mathbf{B} : \|b\|_\infty \leqslant r\}$ and

$$H_r(p,\omega) = \sup_{|q|\leqslant r} (\langle p,q\rangle - L(q,\omega)).$$

Clearly $\mathbf{B}_r$ is compact and convex in $L^p(\Omega,\mathbb{P})$ in the weak topology for $1 \leqslant p < \infty$. For each $k > 1$ we define

$$\mathbf{E}_k = \left\{\Phi : k^{-1} \leqslant \Phi(\omega) \leqslant k, \int \Phi(\omega)d\,\mathbb{P} = 1\right\},$$

$$\mathbf{D}_k = \left\{\Phi : k^{-1} \leqslant \Phi(\omega) \leqslant k, |\nabla\Phi| \leqslant k^2, \int \Phi(\omega)d\,\mathbb{P} = 1\right\} \subset \mathbf{E}_k,$$

which are again compact convex sets if appropriate weak topologies are imposed. Observe that $\mathcal{E} = \bigcup_{k>0}\bigcup_{r>0}\mathcal{E}_{r,k}$, where

$$\mathcal{E}_{r,k} = \{(b,\Phi) \in \mathbf{B}_r \times \mathbf{D}_k : D_t\Phi + \operatorname{div}(b\Phi) = \tfrac{1}{2}\Delta\Phi\}.$$

If φ is a mollifier on $\mathbb{R}^d$ and $F \in L^1(\Omega,\mathbb{P})$, we can define

$$\text{(4.2)} \quad F^\varphi(\omega) = \int F(\tau_{(0,x)}\omega)\varphi(x)dx.$$

Then $F^\varphi \in L^1(\Omega, \mathbb{P})$ and is smooth under space shifts. Notice that if $\Phi \in \mathbf{E}_k$, then $\Phi^\varphi(\omega) \in \mathbf{D}_k$ provided k is large enough so that $\|\nabla\varphi\|_\infty \leqslant k$.

We start with formula (2.8) for the effective Hamiltonian and use the minimax theorem to obtain a dual formula.

LEMMA 4.1 *There is a sequence of functions* $\{F_n(\omega)\}$, $n \geqslant 1$, $F_n \in W^{1,\infty}(\Omega)$, $F_n(\tau_{(\cdot,\cdot)}\omega) \in C^\infty(\mathbb{R}^{d+1})$ *a.s. such that*

$$\sup_{\Phi\in\mathbf{D}_n} \mathbb{E}[[D_tF_n + \tfrac{1}{2}\Delta F_n + H_n(\theta + \nabla F_n, \omega)]\Phi] \leqslant \overline{H}(\theta) + \tfrac{1}{n}. \tag{4.3}$$

PROOF: From (2.8) we have

$$\overline{H}(\theta) \geqslant \sup_{(b,\Phi)\in\mathcal{E}_{r,k}} \mathbb{E}[(\langle\theta, b(\omega)\rangle - L(b(\omega),\omega))\Phi(\omega)].$$

The definition of $\mathcal{E}_{r,k}$ contains a constraint. Let $\mathcal{Y}$ be a space of nice test functions (for example, $\mathcal{Y}$ could be a class of functions $F \in W^{1,\infty}(\Omega)$ convoluted with a mollifier from $C_0(\mathbb{R}^{d+1})$ to ensure the smoothness under the shifts). Using the fact that for $\Phi \in \mathbf{D}_k$

$$\begin{aligned}
&\inf_{F\in\mathcal{Y}} \mathbb{E}([D_tF + \langle b, \nabla F\rangle + \tfrac{1}{2}\Delta F]\Phi)\\
&\quad= \inf_{F\in\mathcal{Y}} \mathbb{E}([D_tF + \langle b, \nabla F\rangle]\Phi - \tfrac{1}{2}\langle\nabla F, \nabla\Phi\rangle)\\
&\quad= \begin{cases} 0 & \text{if } D_t\Phi + \operatorname{div}(b\Phi) = \tfrac{1}{2}\Delta\Phi \text{ in the weak sense,}\\ -\infty & \text{otherwise,}\end{cases}
\end{aligned}$$

we can remove the constraint and get the inequality

$$\begin{aligned}
\overline{H}(\theta) &\geqslant \sup_{\Phi\in\mathbf{D}_k}\sup_{b\in\mathbf{B}_r}\inf_{F\in\mathcal{Y}} \mathbb{E}[(D_tF + \langle b, \theta + \nabla F\rangle - L(b,\omega) + \tfrac{1}{2}\Delta F)\Phi]\\
&= \sup_{\Phi\in\mathbf{D}_k}\sup_{b\in\mathbf{B}_r}\inf_{F\in\mathcal{Y}} \mathbb{E}[(D_tF + \langle b, \theta + \nabla F\rangle - L(b,\omega))\Phi - \tfrac{1}{2}\langle\nabla F, \nabla\Phi\rangle].
\end{aligned}$$

The set $\mathbf{B}_r$ is convex and compact in the weak topology while $\mathcal{Y}$ is convex. For fixed Φ, the functional

$$\mathbb{E}[(D_tF + \langle b, \theta + \nabla F\rangle - L(b,\omega))\Phi - \tfrac{1}{2}\langle\nabla F, \nabla\Phi\rangle]$$

is concave and upper semicontinuous in b in the weak topology and is linear in F. We can therefore apply Sion's minimax theorem [20] and interchange the infimum and the supremum to obtain

$$\overline{H}(\theta) \geqslant \sup_{\Phi\in\mathcal{D}_k}\inf_{F\in\mathcal{Y}}\sup_{b\in\mathbf{B}_r} \mathbb{E}[(D_tF + \langle b, \theta + \nabla F\rangle - L(b,\omega))\Phi - \tfrac{1}{2}\langle\nabla F, \nabla\Phi\rangle].$$

For fixed Φ and F the supremum over $b \in \mathbf{B}_r$ can be taken inside the expectation. Recalling the definition of $H_r(p,\omega)$ we get

$$\overline{H}(\theta) \geqslant \sup_{\Phi\in\mathbf{D}_k}\inf_{F\in\mathcal{Y}} \mathbb{E}[(D_tF + H_r(\theta + \nabla F, \omega))\Phi - \tfrac{1}{2}\langle\nabla F, \nabla\Phi\rangle].$$

Now we would like to use the minimax theorem one more time. The functional

$$\mathbb{E}[(D_t F + H_r(\theta + \nabla F, \omega))\Phi - \tfrac{1}{2}\langle \nabla F, \nabla \Phi \rangle]$$

is convex in F and is a continuous linear functional of Φ in the weak topology (requiring the weak convergence of Φ and $\nabla\Phi$). Therefore, for each k and r

$$\overline{H}(\theta) \geqslant \inf_{F\in\mathcal{Y}} \sup_{\Phi\in\mathbf{D}_k} \mathbb{E}[(D_t F + \tfrac{1}{2}\Delta F + H_r(\theta + \nabla F, \omega))\Phi].$$

This immediately implies the statement of the lemma. □

Let us briefly describe the idea of the next step. We would like to take some sort of weak limit of F_n and produce an F such that for almost all ω with respect to $\mathbb{P}$,

$$D_t F(\omega) + \frac{1}{2}\operatorname{div}(\nabla F(\omega)) + H(\theta + \nabla F(\omega), \omega) \leqslant \overline{H}(\theta).$$

Unfortunately, F may not exist. Let us first see what estimates one might expect on F. Taking expectations, one can see that $\mathbb{E}[H(\theta + \nabla F(\omega), \omega)]$ and therefore $\mathbb{E}[|\nabla F|^\alpha]$ can be controlled. With some spatial (in x) mollification, for the convolution $F^\varphi = F * \varphi$, one can get an upper bound on $\mathbb{E}[(D_t F^\varphi)^+]$. Since $D_t F$ should have mean 0, this means one can expect to control $\mathbb{E}[|D_t F^\varphi|]$. At best, we will be able to produce a limiting family of objects f_φ, g such that $\mathbb{P}$-a.s.

$$f_\varphi(\omega) + \frac{1}{2}\operatorname{div} g^\varphi(\omega) + \int H(\theta + g(\tau_{(0,x)}\omega), \tau_{(0,x)}\omega)\varphi(x)dx \leqslant \overline{H}(\theta), \tag{4.4}$$

where

$$g^\varphi = \int g(\tau_{(0,x)}\omega)\varphi(x)dx, \quad \operatorname{div} g^\varphi = -\int \langle g(\tau_{(0,x)}\omega), \nabla\varphi(x)\rangle dx. \tag{4.5}$$

We shall end up with $g \in L^\alpha(\Omega, \mathbb{P})$ and a collection $f_\varphi \in L^1(\Omega, \mathbb{P})$, not necessarily smooth with respect to the shifts, which satisfies for almost all ω and each φ the compatibility condition

$$\int f_\varphi(\tau_{(t,x)}\omega)(\operatorname{div} G(t,x))dt\,dx = \int \langle g^\varphi(\tau_{(t,x)}\omega), G_t(t,x)\rangle dt\,dx \tag{4.6}$$

for $G \in C_0^\infty(\mathbb{R}\times\mathbb{R}^d; \mathbb{R}^d)$. We shall lose control of $\|f_\varphi\|_1$ as $\varphi \to \delta_0$ in $\mathcal{D}'(\mathbb{R}^d)$.

The details of the above construction are presented in the next theorem.

THEOREM 4.2 *Let H satisfy* (H1) *and* (H2). *For every $\theta \in \mathbb{R}^d$ there is $g : \Omega \to \mathbb{R}^d$, $g \in L^\alpha(\Omega, \mathbb{P})$, and for every $r > 0$ there is a nonnegative mollifier $\varphi \in C_0^\infty(\mathbb{R}^d)$, $\varphi(x) = 0$ outside of $B(0,r)$, and a function $f_\varphi \in L^1(\Omega, \mathbb{P})$ such that*

(i) $\mathbb{E}[f_\varphi] = 0$, $\mathbb{E}[g] = \theta$;
(ii) $\nabla \times g^\varphi = 0$ *and* (4.6) *holds, where g^φ is defined in* (4.5);
(iii) *inequality* (4.4) *holds $\mathbb{P}$-a.s.*;

(iv) *there is a constant* $K(\varphi) > 0$, *which depends only on* φ, c_1, α, *and* $\overline{H}(\theta)$, *such that*

$$f_\varphi(\omega) + \frac{c_1}{2}|g^\varphi(\omega)|^\alpha \leqslant K(\varphi) \quad \mathbb{P}\text{-}a.s.$$

PROOF: Let $f_n = D_t F_n$ and $g_n = \theta + \nabla F_n$, where $\{F_n\}$, $n \geqslant 1$, is constructed in Lemma 4.1.

Step 1. The sequence $\{g_n\}$ is uniformly integrable (see theorem 5.2 of [14]). Take a weakly convergent subsequence and call it again $\{g_n\}$. We have $g_n \to g$ as $n \to \infty$ weakly in $L^1(\Omega, \mathbb{P})$ and

$$\mathbb{E}[H_n(g_n(\omega), \omega)] \leqslant \overline{H}(\theta) + \frac{1}{n}.$$

Clearly, $H_k \leqslant H_n$ for $k \leqslant n$. Then

$$\mathbb{E}[H_k(g_n(\omega), \omega)] \leqslant \overline{H}(\theta) + \frac{1}{n}.$$

Let $n \to \infty$ and use the weak convergence of g_n to g and the convexity of H_k to get $\mathbb{E}[H_k(g(\omega), \omega)] \leqslant \overline{H}(\theta)$. Finally, we let $k \to \infty$ to obtain

$$\mathbb{E}[H(g(\omega), \omega)] \leqslant \overline{H}(\theta).$$

Using the lower bound on H, we conclude that $g \in L^\alpha(\Omega, \mathbb{P})$.

Step 2. The only control we have on $\{f_n\}$, $n \geqslant 1$, is inequality (4.3) of Lemma 4.1:

$$\sup_{\Phi \in \mathbf{D}_n} \mathbb{E}\left[\left(f_n + \frac{1}{2}\operatorname{div} g_n + H_n(g_n, \omega)\right)\Phi\right] \leqslant \overline{H}(\theta) + \frac{1}{n}. \tag{4.7}$$

This control is very weak. We mollify f_n by convoluting with φ and obtain some estimates on $f_n^\varphi = f_n * \varphi$, $n \geqslant 1$.

Take a mollifier $\varphi \in C_0^\infty(\mathbb{R}^d)$, $\varphi(x) \geqslant 0$, and $\int \varphi(x)\,dx = 1$ such that

$$\int |\nabla\varphi(x)|\,dx \leqslant \left(\int \left|\frac{\nabla\varphi(x)}{\varphi(x)}\right|^{\alpha'} \varphi(x)\,dx\right)^{1/\alpha'} = C(\varphi) < \infty \tag{4.8}$$

where $(\alpha')^{-1} + \alpha^{-1} = 1$ and $C(\varphi)$ is a constant that depends on φ. We will continue to denote any constant that depends only on φ by $C(\varphi)$, even if it changes during a step. Our first goal is to establish a uniform bound on $\mathbb{E}[|f_n^\varphi|]$. We rewrite (4.7) in the form

$$\mathbb{E}[f_n\Phi] \leqslant -\mathbb{E}\left[\left(\frac{1}{2}\operatorname{div} g_n + H_n(g_n, \omega)\right)\Phi\right] + \overline{H}(\theta) + \frac{1}{n}$$

valid for all $\Phi \in \mathbf{D}_n$, or

$$\mathbb{E}[f_n\Phi^\varphi] \leqslant -\mathbb{E}\left[\left(\frac{1}{2}\operatorname{div} g_n + H_n(g_n, \omega)\right)\Phi^\varphi\right] + \overline{H}(\theta) + \frac{1}{n} \tag{4.9}$$

valid for $\Phi \in \mathbf{E}_n$. We note that H_n has a uniform lower bound $H_n \geqslant -C$. Moving the derivatives to φ as in (4.5), we get

$$\left|\mathbb{E}[\tfrac{1}{2}(\operatorname{div} g_n)\Phi^\varphi]\right| \leqslant C(\varphi)\mathbb{E}[|g_n|]\|\Phi\|_\infty \leqslant cC(\varphi)\|\Phi\|_\infty,$$

where c is a uniform bound on $\mathbb{E}[|g_n|]$. Therefore, for all $\Phi \in \mathbf{E}_n$ there are constants c, C, and $C(\varphi)$ such that

$$\mathbb{E}[f_n^\varphi \Phi] = \mathbb{E}[f\Phi_n^\varphi] \leqslant cC(\varphi)\|\Phi\|_\infty + C + \overline{H}(\theta).$$

If $\Phi > 0$ is not normalized but $\Phi/\mathbb{E}[\Phi] \in \mathbf{E}_n$, then

$$\mathbb{E}[f_n^\varphi \Phi] \leqslant cC(\varphi)\|\Phi\|_\infty + (C + \overline{H}(\theta))\mathbb{E}[\Phi].$$

Consider functions Φ_n, $n = 1, 2, \ldots$, defined by

$$\Phi_n(\omega) = \begin{cases} \frac{1}{2} & \text{if } f_n^\varphi(\omega) \leqslant 0, \\ 2 & \text{if } f_n^\varphi(\omega) > 0. \end{cases}$$

Since $\mathbb{E}[f_n^\varphi] = 0$, we have

$$\mathbb{E}[f_n^\varphi : f_n^\varphi > 0] = -\mathbb{E}[f_n^\varphi : f_n^\varphi \leqslant 0] = \frac{1}{2}\,\mathbb{E}[|f_n^\varphi|].$$

Also, $\Phi_n/\mathbb{E}[\Phi_n] \in \mathbf{E}_4$, $\mathbb{E}[\Phi] \leqslant \|\Phi\|_\infty \leqslant 2$, and

$$\mathbb{E}[f_n^\varphi \Phi_n] = 2\,\mathbb{E}[f_n^\varphi : f_n^\varphi > 0] + \frac{1}{2}\,\mathbb{E}[f_n^\varphi : f_n^\varphi \leqslant 0] = \frac{3}{4}\,\mathbb{E}[|f_n^\varphi|]$$

giving us the bound

$$\sup_{n \geqslant n_0(\varphi)} \mathbb{E}[|f_n^\varphi|] \leqslant C(\theta, \varphi). \tag{4.10}$$

Step 3. We would like to prove the uniform integrability of $\{f_n^\varphi\}$, but at this point we can only establish the uniform integrability of the positive parts of $\{f_n^\varphi\}$, i.e., the uniform integrability of $\{f_n^{\varphi,+}\}$ where $f_n^{\varphi,+} = \max\{f_n^\varphi, 0\}$. This depends on the uniform integrability of g_n established in Step 1. If we truncate $|g_n|$ at level N and write $g_n = g_{n,N} + (g_n - g_{n,N})$, we have for all $\Phi \in \mathbf{E}_k$

$$-\frac{1}{2}\,\mathbb{E}\langle g_n, \nabla\Phi^\varphi\rangle \leqslant \mathbb{E}[|g_n - g_{n,N}|]\|\nabla\Phi^\varphi\|_\infty - \frac{1}{2}\,\mathbb{E}\langle g_{n,N}, \nabla\Phi^\varphi\rangle, \tag{4.11}$$

and $\sup_n \mathbb{E}[|g_n - g_{n,N}|] = \delta(N) \to 0$ as $N \to \infty$ by uniform integrability. Moreover,

$$-\frac{1}{2}\,\mathbb{E}\langle g_{n,N}, \nabla\Phi^\varphi\rangle = \frac{1}{2}\,\mathbb{E}[(\operatorname{div}\, g_{n,N}^\varphi)\Phi].$$

We can estimate

$$\begin{aligned} |\operatorname{div} g_{n,N}^\varphi| &= \left|\int \langle g_{n,N}(\tau_{(0,x)}\omega), \nabla\varphi(x)\rangle\, dx\right| \\ &\leqslant C(\varphi)\left(\int |g_{n,N}(\tau_{(0,x)}\omega)|^\alpha \varphi(x)dx\right)^{1/\alpha}. \end{aligned}$$

On the other hand, for $n \geqslant n_0(N)$ large enough,

$$H_n(g_{n,N}) = H(g_{n,N}) \geqslant c_1|g_{n,N}|^\alpha - c_2$$

and

$$\begin{aligned}\mathbb{E}[H_n(g_n(\omega),\omega))\Phi^\varphi(\omega)] &\geqslant \mathbb{E}[(c_1|g_{n,N}(\omega)|^\alpha - c_2)\Phi^\varphi(\omega)]\\ &= c_1\mathbb{E}\left[\Phi(\omega)\int |g_{n,N}(\tau_{(0,x)}\omega)|^\alpha\varphi(x)dx\right] - c_2.\end{aligned}$$

Therefore,

$$\begin{aligned}&-\frac{1}{2}\mathbb{E}\langle g_{n,N},\nabla\Phi^\varphi\rangle - \mathbb{E}[H_n(g_n(\omega),\omega))\Phi^\varphi(\omega)]\\ &\leqslant \frac{1}{2}C(\varphi)\mathbb{E}\left[\Phi(\omega)\left(\int |g_{n,N}(\tau_{(0,x)}\omega)|^\alpha\varphi(x)dx\right)^{1/\alpha}\right]\\ &\quad - c_1\mathbb{E}\left[\Phi(\omega)\int |g_{n,N}(\tau_{(0,x)}\omega)|^\alpha\varphi(x)dx\right] + c_2\\ &\leqslant \max_{x\geqslant 0}\left(\frac{1}{2}C(\varphi)x - c_1x^\alpha\right) + c_2 \leqslant C(\varphi).\end{aligned}$$

Combining the above inequality with (4.9) and (4.11), we conclude that for all $n \geqslant n_0(N)$

$$\mathbb{E}[f_n^\varphi\Phi] \leqslant \delta(N)\|\nabla\Phi^\varphi\|_\infty + C(\theta,\varphi).$$

Finally, letting $n \to \infty$ and then $N \to \infty$, we obtain for every k

$$\limsup_{n\to\infty}\sup_{\Phi\in\mathbf{E}_k}\mathbb{E}[f_n^\varphi\Phi] \leqslant C(\theta,\varphi). \tag{4.12}$$

Consider functions $\Phi_{n,k,\ell}$ defined by

$$\Phi_{n,k,\ell}(\omega) = \begin{cases}\frac{1}{k} & \text{if } f_n^\varphi(\omega) \leqslant \ell,\\ k & \text{if } f_n^\varphi(\omega) > \ell.\end{cases}$$

Then $\Phi_{n,k,\ell}/\mathbb{E}[\Phi_{n,k,\ell}] \in \mathbf{E}_{k^2}$. Recalling that

$$\mathbb{E}[f_n^\varphi : f_n^\varphi > \ell] + \mathbb{E}[f_n^\varphi : f_n^\varphi \leqslant \ell] = 0,$$

we have

$$\begin{aligned}\mathbb{E}[f_n^\varphi\Phi_{k,n,\ell}] &= k\mathbb{E}[f_n^\varphi : f_n^\varphi > \ell] + \frac{1}{k}\mathbb{E}[f_n^\varphi : f_n^\varphi \leqslant \ell]\\ &= \left(k - \frac{1}{k}\right)\mathbb{E}[f_n^\varphi : f_n^\varphi > \ell],\end{aligned}$$

and by (4.12)

$$\begin{aligned}\limsup_{n\to\infty}\mathbb{E}[f_n^\varphi\Phi_{k,n,\ell}] &\leqslant C(\theta,\varphi)\limsup_{n\to\infty}\left[k\mathbb{P}[f_n^\varphi > \ell] + \frac{1}{k}\mathbb{P}[f_n^\varphi \leqslant \ell]\right]\\ &\leqslant C(\theta,\varphi)\limsup_{n\to\infty}\left[\frac{k}{\ell}\mathbb{E}[|f_n^\varphi|] + \frac{1}{k}\right].\end{aligned}$$

From this and (4.10) we obtain

$$\limsup_{n\to\infty} \mathbb{E}[f_n^\varphi : f_n^\varphi > \ell] \leqslant C(\theta,\varphi)\frac{k/\ell + 1/k}{k - 1/k}.$$

Since k is arbitrary it is clear that $\{f_n^{\varphi,+}\}$ is uniformly integrable.

Step 5. Now we have to deal with the negative parts of $\{f_n^\varphi\}$. The difficulty is that they may not be uniformly integrable, and we have to find a way to split $f_n^{\varphi,-} = -(f_n^\varphi \wedge 0)$ (possibly just along a subsequence) into uniformly integrable and "bad" parts, so that the bad part plays no role in the limit. This is done using the following lemmas.

LEMMA 4.3 *Let $\{h_n\}$, $n \geqslant 1$, be a sequence of nonnegative functions and*

$$\sup_n \mathbb{E}[h_n] \leqslant C.$$

Then there is a subsequence $\{n_j\}$, $j \geqslant 1$, such that $h_{n_j} = \widehat{h}_{n_j} + r(h_{n_j})$, where

(i) $\widehat{h}_{n_j} = h_{n_j}\mathbb{1}_{\{h_{n_j}\leqslant \ell_j\}}$, $\ell_j \to \infty$ *as* $j \to \infty$, *and* $\{\widehat{h}_{n_j}\}$, $j \geqslant 1$, *is uniformly integrable;*

(ii) *remainder terms $r(h_{n_j})$ converge to 0 in probability as $j \to \infty$.*

Properties (i) *and* (ii) *continue to hold if we replace ℓ_j by any slower-growing sequence $\ell'_j \leqslant \ell_j$ such that $\ell'_j \to \infty$ as $j \to \infty$.*

LEMMA 4.4 *Let $\{h_n\}$, $n \geqslant 1$, be a sequence of nonnegative functions such that*

$$\sup_n \mathbb{E}[h_n] \leqslant C,$$

and let $\psi \in C_0^\infty(\mathbb{R}^d)$ be a nonnegative mollifier. Apply Lemma 4.3 to $\{h_n^\psi\}$. Then the convergence of the remainder terms $r(h_{n_j}^\psi)$ to 0 is locally uniform in x, i.e., for every $R > 0$ and $\varepsilon > 0$

$$\mathbb{P}\Big\{\omega : \sup_{|x|\leqslant R} r(h_{n_j}^\psi)(\tau_{(0,x)}\omega) \geqslant \varepsilon\Big\} \to 0 \quad \textit{as } j \to \infty.$$

The proofs of Lemma 4.3 and Lemma 4.4 are elementary and are provided in the appendix.

Return now to f_n^φ. Let $\varphi = \varphi_1 * \varphi_2$, where φ_1 satisfies (4.8). By Lemma 4.3 we can write (considering every subsequence as a whole sequence)

$$f_n^{\varphi_1} - m_n - b_n,$$

where $m_n = f_n^{\varphi_1,+} - \widehat{f_n^{\varphi_1,-}}$ are uniformly integrable, $b_n = r(f_n^{\varphi_1,-})$, and $b_n \to 0$ in probability. Taking a convolution with φ_2 in this decomposition, we get

$$f_n^\varphi = m_n^{\varphi_2} - b_n^{\varphi_2}.$$

It is not hard to show that $m_n^{\varphi_2}$, $n \geqslant 1$, are still uniformly integrable. Apply Lemma 4.4 to $b_n^{\varphi_2}$ and get

$$b_n^{\varphi_2} = \widehat{b_n^{\varphi_2}} + r(b_n^{\varphi_2}),$$

where $\widehat{b_n^{\varphi_2}}$ are uniformly integrable and $r(b_n^{\varphi_2}) \to 0$ in probability locally uniformly with respect to the spatial shifts, i.e.,

$$\sup_{|x| \leqslant R} r(b_n^{\varphi_2})(\tau_{(0,x)}\omega) \to 0$$

in probability as $n \to \infty$. Putting the two decompositions together, we obtain

$$f_n^{\varphi} = k_n - r_n,$$

where $k_n = m_n^{\varphi_2} - \widehat{b_n^{\varphi_2}}$ and $r_n = r(b_n^{\varphi_2})$. This decomposition is "stable" in the following sense: further mollification and decomposition of the remainder terms r_n will not contribute any additional uniformly integrable piece.

Step 6. Let $g \in L^{\alpha}(\Omega; \mathbb{R}^d)$ be a weak limit point of $\{g_n\}$, $n \geqslant 1$. Fix a small $r > 0$. Take $\varphi = \varphi_1 * \varphi_2$, where φ_i, $i = 1, 2$, are mollifiers, $\varphi \equiv 0$ outside of $B(0, r)$, and φ_1 satisfies (4.8). According to Step 4, the positive part of f_n^{φ} is uniformly integrable. Decompose the negative part of f_n^{φ} as in Step 5 and let $\tilde{f}^{\varphi} \in L^1(\Omega, \mathbb{P})$ be a weak limit point of k_n. Then $\mathbb{E}[\tilde{f}^{\varphi}] \geqslant 0$. It is convenient to write $\tilde{f}^{\varphi}$ as $f_{\varphi} + c$ where $c \geqslant 0$ is a constant and $\mathbb{E}[f_{\varphi}] = 0$. We can drop c and inequality (4.4) will still hold. This gives (iii). It is clear also that f_{φ} and g satisfy (i) and (ii).

To show part (iv) we use (iii) and the lower bound on H to get

$$f_{\varphi}(\omega) + \frac{1}{2} \operatorname{div} g^{\varphi}(\omega) + c_1 \int |g(\tau_{(0,x)}\omega)|^{\alpha} \varphi(x)\,dx \leqslant \overline{H}(\theta) + c_2. \tag{4.13}$$

Applying the inequality

$$|\langle p, q\rangle| \leqslant \varepsilon |p|^{\alpha} + C_{\varepsilon} |q|^{\alpha'}$$

with $p = g(\tau_{(0,x)}\omega)$, $q = (\nabla\varphi_1)/\varphi_1$, and $\varepsilon = c_1$ and recalling that φ_1 satisfies (4.8), we obtain

$$\begin{aligned} &\frac{1}{2} \operatorname{div} g^{\varphi_1}(\omega) + c_1 \int |g(\tau_{(0,x)}\omega)|^{\alpha} \varphi_1(x)\,dx \\ &\quad = c_1 \int |g(\tau_{(0,x)}\omega)|^{\alpha} \varphi_1(x)\,dx - \frac{1}{2} \int \langle g(\tau_{(0,x)}\omega), \nabla\varphi_1(x)\rangle\,dx \\ &\quad \geqslant \frac{1}{2}\left(c_1 \int |g(\tau_{(0,x)}\omega)|^{\alpha} \varphi_1(x)\,dx - C_{\varepsilon}[C(\varphi_1)]^{\alpha'}\right). \end{aligned}$$

Convoluting every term of the above inequality with φ_2, substituting in (4.13), and applying Hölder's inequality, we get (iv). □

COROLLARY 4.5 *Let H satisfy* (H1)–(H3). *For each $\theta \in \mathbb{R}^d$ there is a positive function $\nu(\delta) \to 0$ as $\delta \to 0$ and a family of functions $(f_\delta, g_\delta) : \Omega \to \mathbb{R} \times \mathbb{R}^d$, $f_\delta \in L^1(\Omega, \mathbb{P})$, $g_\delta \in L^\alpha(\Omega, \mathbb{P})$, $\delta \in (0, \delta_0)$, such that*

(i) $\mathbb{E}(f_\delta, g_\delta) = (0,0)$, $\nabla \times g_\delta = 0$, *and* $(D_t, \nabla) \times (f_\delta, g_\delta) = 0$;
(ii) $f_\delta(\tau_{(\cdot,\cdot)}\omega), g_\delta(\tau_{(\cdot,\cdot)}\omega) \in C^\infty(\mathbb{R}^{d+1}; \mathbb{R}^{d+1})$ *for* $\mathbb{P}$*-a.e.* ω;
(iii) *there is a constant $K_\delta > 0$, which depends only on δ, c_1, 4α, and $\overline{H}(\theta)$, such that for almost every ω with respect to $\mathbb{P}$*

$$f_\delta(\tau_{(t,x)}\omega) + \frac{c_1}{2}|g_\delta(\tau_{(t,x)}\omega)|^\alpha \leqslant K_\delta \quad \textit{for all } (t,x) \in \mathbb{R}^{d+1};$$

(iv) *for almost every ω with respect to $\mathbb{P}$*

$$f_\delta(\tau_{(t,x)}\omega) + \operatorname{div} g_\delta(\tau_{(t,x)}\omega) + H(\theta + g_\delta(\tau_{(t,x)}\omega), \tau_{(t,x)}\omega) \leqslant \overline{H}(\theta) + \nu(\delta)$$

on $\mathbb{R}^{d+1}$.

PROOF: Assumption (H3) allows us to replace

$$\int H(\theta + g(\tau_{(0,x)}\omega), \tau_{(0,x)}\omega)\varphi(x)dx$$

in (4.4) with $H(\theta + g^\varphi(\omega), \omega)$ modulo a small error. See lemma 6.3 of [14] for details. To gain the regularity with respect to space-time shifts, we can mollify one more time in both t and x and again bring the mollification inside H. □

Now we are ready to construct approximate supersolutions V_δ. Set

$$V_\delta(t,x,\omega) = \int_{(0,0)\to(t,x)} f_\delta(\tau_{(s,x(s))}\omega)ds + \langle g_\delta(\tau_{(s,x(s))}\omega), dx(s)\rangle, \tag{4.14}$$

where $x(s)$ is any smooth path such that $x(0) = 0$ and $x(t) = x$. Since $(D_t, \nabla) \times (f_\delta, g_\delta) = 0$, the integral does not depend on the path. Main properties of V_δ are summarized in the next lemma.

LEMMA 4.6 *Let $V_\delta(t,x,\omega)$ be given by* (4.14). *Then*

(i) *for almost every ω with respect to $\mathbb{P}$ and all $t, s \in \mathbb{R}$, $x, y \in \mathbb{R}^d$,*

$$V_\delta(0,0,\omega) = 0, \qquad V_\delta(s,y,\tau_{(t,x)}\omega) = V_\delta(t+s, x+y, \omega) - V_\delta(t,x,\omega);$$

(ii) $\partial_t V_\delta(t,x,\omega) = f_\delta(\tau_{(t,x)}\omega)$, $\nabla V_\delta(t,x,\omega) = g_\delta(\tau_{(t,x)}\omega)$, *and $V_\delta(t,x,\omega)$ satisfies* (4.1);
(iii) *for all $t, s \in \mathbb{R}$ and $x, y \in \mathbb{R}^d$*

$$\|V_\delta(t,x,\cdot) - V_\delta(s,y,\cdot)\|_1 \leqslant C(\delta)(|t-s| + |x-y|). \tag{4.15}$$

(iv) *There are positive constants* $C_1 = C_1(d)$ *and* $C_2 = C_2(\delta, c_1, \alpha, \overline{H}(\theta))$ *such that with probability* 1 *for all* $(t,x) \in \mathbb{R}^{d+1}$ *and* $\eta > 0$,

$$V_\delta(t,x,\omega) \leqslant C_1 \eta^{-(d+1)} \int_0^\eta \int_{|y| \leqslant s} V_\delta(t-s, x-y, \omega)\,dy\,ds + C_2\eta,$$

$$V_\delta(t,x,\omega) \geqslant C_1 \eta^{-(d+1)} \int_0^\eta \int_{|y| \leqslant s} V_\delta(t+s, x+y, \omega)\,dy\,ds - C_2\eta.$$

PROOF: Parts (i)–(iii) are immediate from (4.14) and Corollary 4.5. We only need to prove (iv).

Let δ, t, x, ω be fixed and (s,y) be an arbitrary vector in $\mathbb{R} \times \mathbb{R}^d$. Consider

$$W(s,y) := V_\delta(t+s, x+sy, \omega).$$

Then by part (iii) of Corollary 4.5

$$\partial_s W(s,y) = f_\delta + \langle g_\delta, y\rangle \leqslant K_\delta - \frac{c_1}{2}|g_\delta|^\alpha + \langle g_\delta, y\rangle \leqslant K_1 + K_2|y|^{\alpha'}.$$

Integrating we get for $s \geqslant 0$

$$V_\delta(t+s, x+sy, \omega) - V_\delta(t,x,\omega) \leqslant [K_1 + K_2|y|^{\alpha'}]s. \tag{4.16}$$

In particular, for $s > 0$

$$V_\delta(t+s, x+y, \omega) - V_\delta(t,x,\omega) \leqslant \left[K_1 + K_2\left(\frac{|y|}{s}\right)^{\alpha'}\right]s.$$

Notice that (4.16) leads to the following a.s. estimates for fixed (t,x) and $\eta > 0$:

$$\begin{aligned} V_\delta(t,x,\omega) &\leqslant \min_{\substack{0 \leqslant s \leqslant \eta \\ |y| \leqslant 1}} V_\delta(t-s, x-sy, \omega) + (K_1+K_2)\eta \\ &\leqslant C_1\eta^{-(d+1)} \int_0^\eta \int_{|y| \leqslant s} V_{\delta,\varepsilon}(t-s, x-y, \omega)\,dy\,ds + C_2\eta. \end{aligned}$$

The other inequality is obtained in a similar way. □

5 Upper Bound

We start with a lemma (see [14, lemma 4.1] for a proof). Let $c \in \mathcal{C}$ and $y_\varepsilon(s)$, $s \in [t,T]$, be a sample path of a diffusion with generator $\mathbf{A}_s^\varepsilon$ (see (2.13)), and

$$\xi_\varepsilon(t) = \int_t^T L(c(s, y_\varepsilon(s)), \tau_{(s/\varepsilon, y_\varepsilon(s)/\varepsilon)}\omega)\,ds. \tag{5.1}$$

LEMMA 5.1 *Assume* (L1) *and* (2.2). *Then in the variational formula* (2.14), *the supremum over* $\mathcal{C}$ *can be replaced with the supremum over the subset* $\mathcal{C}^* \subset \mathcal{C}$ *of controls that satisfy the following condition: there is a* $C > 0$ *that depends only on the constants in* (L1) *and* (2.2) *such that*

$$\sup_{x,\omega} E^{Q^{c,\varepsilon}_{t,x}} |\xi_\varepsilon(t)| \leqslant C\big(T - t + \sqrt{\varepsilon(T-t)}\big). \tag{5.2}$$

In particular, for all $c \in \mathcal{C}^*$

$$\sup_{x,\omega} E^{Q^{c,\varepsilon}_{t,x}} \left[\int_t^T |c(s, y_\varepsilon(s))|^{\beta'} ds \right] \leqslant C\big(T - t + \sqrt{\varepsilon(T-t)}\big). \tag{5.3}$$

The following theorem is the main result of this section:

THEOREM 5.2 *Assume that* $H(p,\omega)$ *satisfies* (H1)–(H3) *and that the terminal condition* $U(x)$ *satisfies* (2.2). *Let* $u(t,x)$ *be the unique solution of* (1.2)–(1.3) *with* $\overline{H}$ *given by* (2.8). *Then with probability* 1 *for every* $\ell > 0$

$$\limsup_{\varepsilon \to 0} \inf_{0 \leqslant t \leqslant T} \inf_{|x| \leqslant \ell} [u_\varepsilon(t,x,\omega) - u(t,x)] \geqslant 0.$$

PROOF: By Lemma 5.1 the supremum in (2.14) can be restricted to controls $c \in \mathcal{C}^*$. Estimate (5.3) implies that $y_\varepsilon(T)$ is uniformly integrable with respect to $\{Q^{\varepsilon,c}_{t,x} : |x| \leqslant \ell,\ 0 \leqslant t \leqslant T,\ 0 < \varepsilon \leqslant 1,\ c \in \mathcal{C}^*\}$. Combining this with (5.2) we get

$$\sup_{|x| \leqslant \ell} \sup_{0 \leqslant t \leqslant T} \sup_{c \in \mathcal{C}^*} \sup_{0 < \varepsilon \leqslant 1} E^{Q^{\varepsilon,c}_x} (|y_\varepsilon(t)| + |\xi_\varepsilon(t)|) < \infty. \tag{5.4}$$

Step 1. Without loss of generality we can assume that $(T - t) \geqslant r$ (see Lemma 3.1). Then using the uniform integrability we get that for every $\delta > 0$ there are $M = M(\delta)$ and $A = A(\delta, r)$ such that

$$\begin{aligned}
&u_\varepsilon(t,x,\omega) - u(t,x) \\
&\quad = \sup_{c \in \mathcal{C}^*} E^{Q^{c,\varepsilon}_{t,x}} (f(y_\varepsilon(T)) - \xi_\varepsilon(t)) - \sup_{y \in \mathbb{R}^d} \left[f(y) - (T-t)\overline{L}\left(\frac{y-x}{T-t}\right)\right] \\
&\quad \leqslant \sup_{c \in \mathcal{C}^*} E^{Q^{c,\varepsilon}_{t,x}} \left[(f(y_\varepsilon(T)) - \xi_\varepsilon(t)) \mathbb{1}_{|y_\varepsilon(T) - x| \leqslant M} \right] + \delta \\
&\qquad - \sup_{y \in \mathbb{R}^d} \left[f(y) - (T-t)\overline{L}\left(\frac{y-x}{T-t}\right)\right] \\
&\quad \leqslant \sup_{c \in \mathcal{C}^*} E^{Q^{c,\varepsilon}_{t,x}} \left[\left((T-t)\overline{L}\left(\frac{y_\varepsilon(T) - x}{T-t}\right) - \xi_\varepsilon(t) \right) \mathbb{1}_{|y_\varepsilon(T) - x| \leqslant M} \right] + \delta \\
&\quad = \sup_{c \in \mathcal{C}^*} E^{Q^{c,\varepsilon}_{t,x}} \Big[\sup_{|\theta| \leqslant A} \big(\langle \theta, (y_\varepsilon(T) - x) \rangle - (T-t)\overline{H}(\theta) - \xi_\varepsilon(t) \big) \mathbb{1}_{|y_\varepsilon(T) - x| \leqslant M} \Big] + \delta \\
&\quad = \sup_{c \in \mathcal{C}^*} \int_{\{|y - x| \leqslant M\} \times \mathbb{R}} \sup_{|\theta| \leqslant A} \big(\langle \theta, y - x \rangle - (T-t)\overline{H}(\theta) - \xi \big) \mu_\varepsilon(dy, d\xi) + \delta,
\end{aligned}$$

where μ_ε is the distribution of $(y_\varepsilon(T), \xi_\varepsilon(t))$ on $\mathbb{R}^d \times \mathbb{R}$ under $Q^{c,\varepsilon}_{t,x}$. Since $\overline{H}$ is continuous, the supremum over θ can be restricted to the set $\{|\theta| \leqslant A, \theta \in \mathbb{Q}\}$, where $\mathbb{Q}$ is the countable set of points in $\mathbb{R}^d$ with rational coordinates. Since $L(\cdot,\omega)$ is superlinear, the integration can be further restricted to the set $\{(y,\xi) : |y-x| \leqslant M,\ \xi \leqslant B\}$ for a sufficiently large B (the proof is similar to that for lemma 4.1 in [14]). Therefore, to obtain an almost sure, locally uniform upper bound, it is enough to show that for each $\theta \in \mathbb{R}^d$, $\ell, \eta > 0$, and all large enough M and B,

$$\lim_{\varepsilon\to 0} \sup_{c\in C^*} \sup_{|x|\leqslant \ell} \sup_{0\leqslant t\leqslant T} Q^{\varepsilon,c}_{t,x}\{\langle\theta, y_\varepsilon(T)-x\rangle - (T-t)\overline{H}(\theta) - \xi_\varepsilon(t) \geqslant \eta; A_{t,x}\} = 0, \tag{5.5}$$

ω-a.s. Here $A_{t,x} = \{|y_\varepsilon(T) - x| \leqslant M, \xi_\varepsilon(t) \leqslant B\}$.

Step 2. Since all computations below are valid for all $\omega \in \Omega'$, $\mathbb{P}(\Omega') = 1$, we fix $\omega \in \Omega'$ and drop it from the notation. Define

$$V_{\delta,\varepsilon}(t,x) = \varepsilon V_\delta(t/\varepsilon, x/\varepsilon). \tag{5.6}$$

By part (ii) of Lemma 4.6, $V_{\delta,\varepsilon}(t,x)$ satisfies

$$\partial_t V_{\delta,\varepsilon} + \frac{\varepsilon}{2}\Delta V_{\delta,\varepsilon} + H(\theta + \nabla V_{\delta,\varepsilon}, \tau_{(t/\varepsilon,x/\varepsilon)}\omega) \leqslant \overline{H}(\theta) + \nu(\delta), \quad (t,x) \in \mathbb{R}\times\mathbb{R}^d.$$

From Itô's formula for $V_{\delta,\varepsilon}(t,x) + \langle\theta, x\rangle$ and the above inequality we get

$$\begin{aligned} &\langle\theta, y_\varepsilon(T)-x\rangle - (T-t)\overline{H}(\theta) - \xi_\varepsilon(t) \\ &\quad\leqslant V_{\delta,\varepsilon}(t,x) - V_{\delta,\varepsilon}(T, y_\varepsilon(T)) \\ &\qquad + \sqrt{\varepsilon}\int_t^T \langle\nabla V_{\delta,\varepsilon}(s, y_\varepsilon(s)) + \theta, dB(s)\rangle + (T-t)\nu(\delta). \end{aligned} \tag{5.7}$$

We also used the fact that $\langle c, p\rangle \leqslant H(p,\omega) + L(c,\omega)$. Therefore, it is enough to show that for each $\theta \in \mathbb{R}^d$, $\ell, \eta > 0$, all sufficiently large M and B, and small $\delta > 0$,

$$\begin{aligned} \lim_{\varepsilon\to 0} \sup_{c\in C^*} \sup_{|x|\leqslant \ell} \sup_{0\leqslant t\leqslant T} Q^{\varepsilon,c}_{t,x}\bigg\{&\sqrt{\varepsilon}\int_t^T \langle\nabla V_{\delta,\varepsilon}(s, y_\varepsilon(s)), dB(s)\rangle \\ &+ V_{\delta,\varepsilon}(t,x) - V_{\delta,\varepsilon}(T, y_\varepsilon(T)) \geqslant \eta; A_{t,x}\bigg\} = 0. \end{aligned} \tag{5.8}$$

Observe that again by Itô's formula

$$
\begin{aligned}
&V_{\delta,\varepsilon}(T, y_\varepsilon(T)) - V_{\delta,\varepsilon}(t,x) - \sqrt{\varepsilon}\int_t^T \langle \nabla V_{\delta,\varepsilon}(s, y_\varepsilon(s)), dB(s)\rangle \\
&\quad = \int_t^T \partial_s V_{\delta,\varepsilon}(s, y_\varepsilon(s)) + \langle \nabla V_{\delta,\varepsilon}(s, y_\varepsilon(s)), c(s, y_\varepsilon(s))\rangle \\
&\qquad + \frac{\varepsilon}{2}\Delta V_{\delta,\varepsilon}(s, y_\varepsilon(s))\, ds.
\end{aligned}
\tag{5.9}
$$

Setting

$$F_{\delta,\varepsilon}(s,y) = \partial_s V_{\delta,\varepsilon}(s,y) + \langle \nabla V_{\delta,\varepsilon}(s,y), c(s,y)\rangle + \frac{\varepsilon}{2}\Delta V_{\delta,\varepsilon}(s,y),$$

we conclude from (5.8) and (5.9) that the relation

$$\lim_{\varepsilon\to 0}\sup_{c\in\mathcal{C}^*}\sup_{|x|\leqslant\ell}\sup_{0\leqslant t\leqslant T} E^{Q^{\varepsilon,c}_{t,x}}\left(\left[\int_t^T F_{\delta,\varepsilon}(s, y_\varepsilon(s))ds\right]^2; A_{t,x}\right) = 0$$

will imply the desired upper bound. The proof follows step 2 of the proof assuming H4 in [14, theorem 2.2].

To complete the argument, we only need to show that for every $\delta, a > 0$

$$\lim_{\varepsilon\to 0}\sup_{\substack{0\leqslant t\leqslant T\\ |x|\leqslant a}} |V_{\delta,\varepsilon}(t,x,\omega)| = 0 \quad \mathbb{P}\text{-a.s.} \tag{5.10}$$

Choosing $\eta = \delta/\varepsilon$ and rescaling the integrals in part (iv) of Lemma 4.6, we get

$$|V_{\delta,\varepsilon}(t,x,\omega)| \leqslant C_1\delta^{-(d+1)}\int_{|s|\leqslant\delta}\int_{|y|\leqslant|s|} |V_{\delta,\varepsilon}(t+s, x+y, \omega)|dy\, ds + C_2\delta.$$

Therefore,

$$\sup_{\substack{0\leqslant t\leqslant T\\ |x|\leqslant a}} |V_{\delta,\varepsilon}(t,x,\omega)| \leqslant C_1\delta^{-(d+1)}\int_{|t|\leqslant T+\delta}\int_{|x|\leqslant a+\delta} |V_{\delta,\varepsilon}(t,x,\omega)|dx\, dt + C_2\delta,$$

and (5.10) is an immediate consequence of the scaling relation (5.6) and the following ergodic theorem. Note that, in this context, the variable t does not play a special role, and we only need to apply the ergodic theorem in $\mathbb{R}^{d+1}$. □

THEOREM 5.3 (Ergodic Theorem) *Assume that* $\{\tau_x : x \in \mathbb{R}^d\}$ *is an ergodic-measure-preserving action on* $(\Omega, \Sigma, \mathbb{P})$. *Let* $f : \Omega \to \mathbb{R}^d$ *satisfy the following conditions*:

$$f \in L^1(\Omega, \mathbb{P}), \quad \mathbb{E}[f] = 0, \quad \textit{and} \quad \nabla\times f = 0,$$

in the sense that for smooth test functions φ

$$\int [f_i(\tau_x\omega)\partial_j\varphi(x) - f_j(\tau_x\omega)\partial_i\varphi(x)]dx = 0 \quad \textit{a.e. } \mathbb{P}; \tag{5.11}$$

i.e., f is formally a gradient. Define the normalized integral $F : \mathbb{R}^d \to L^1(\Omega, \mathbb{P})$ by

$$F(x, \omega) = \int_0^t \langle f(\tau_{x(s)}\omega), dx(s)\rangle,$$

where $x(s)$ is a smooth path in $\mathbb{R}^d$ such that $x(0) = 0$ and $x(t) = x$. Then for every bounded set $D \subset \mathbb{R}^d$

$$\lim_{\varepsilon \to 0} \varepsilon \int_D |F(x/\varepsilon, \omega)| dx = 0 \quad \text{a.e. } \mathbb{P}.$$

This theorem, which can be viewed as one possible generalization of the usual Birkhoff ergodic theorem, does not seem to exist in the literature. We will outline a standard proof in the next section that reduces it to Wiener's multidimensional version of Birkhoff's theorem through a series of steps.

6 Proof of Theorem 5.3

Wiener's multidimensional ergodic theorem states that if $f \in L^1(\Omega, \mathbb{P})$ and $\mathbb{E}[f] = 0$, then

$$\lim_{\varepsilon \to 0} \int_D f(\tau_{x/\varepsilon}\omega) dx = 0 \quad \text{a.e. } \mathbb{P}$$

for a large class of sets that includes all bounded boxes with sides parallel to the axes. It is not hard to strengthen it and show that for any continuous function g on $\mathbb{R}^d$

$$\lim_{\varepsilon \to 0} \int_D f(\tau_{x/\varepsilon}\omega) g(x) dx = 0 \quad \text{a.e. } \mathbb{P}.$$

LEMMA 6.1 *If f is as in Theorem 5.3, then*

$$\lim_{\varepsilon \to 0} \varepsilon \int_D F(x/\varepsilon, \omega) dx = 0 \quad \text{a.e. } \mathbb{P},$$

where D is the unit cube $[0, 1]^d$.

PROOF: We will prove by induction on k that

$$\lim_{\varepsilon \to 0} \varepsilon \int_{D_k} F(x/\varepsilon, \omega) dx_1 \cdots dx_k = 0 \quad \text{a.e. } \mathbb{P}$$

where $D_k = \{x : x \in D, x_j = 0 \text{ for } j \geqslant k + 1\}$. For $k = 1$, this means proving the almost sure convergence

$$\lim_{\varepsilon \to 0} \varepsilon \sup_{0 \leqslant a \leqslant 1} \left| \int_0^{a\varepsilon^{-1}} f_1(\tau_{ye_1}\omega) dy \right| = 0 \quad \text{a.e. } \mathbb{P}.$$

The ergodic theorem will almost yield this except that

$$\lim_{\ell\to\infty}\frac{1}{\ell}\int_0^\ell f_1(\tau_{ye_1}\omega)dy = c(\omega)$$

may not be 0. The action of translations in just one coordinate direction may not be ergodic. The limit $c(\omega)$ may not be invariant under translations in other coordinate directions. If it is, then it is a constant and hence 0. Since $\{f_i\}$ is a gradient, an elementary calculation shows that

$$\int_0^\ell [f_1(\tau_{ye_1}\tau_x\omega) - f_1(\tau_{ye_1}\omega)]dy = \int_{0\to x} \langle f(\tau_{z(t)+\ell e_1}\omega) - f(\tau_{z(t)}\omega), dz(t)\rangle,$$

which goes to 0 if divided by ℓ, proving thereby that $c(\tau_x\omega) = c(\omega)$ for all x. The same argument shows that for $1 \leqslant i \leqslant k \leqslant d$,

$$\lim_{\varepsilon\to 0}\int_{D_k} f_i(\tau_{(x_1e_1+\cdots+x_ke_k)/\varepsilon}\,\omega)dx_1\cdots dx_k = 0,$$

and with a little more work (Riemann sum approximation), for any choice of $1 \leqslant i \leqslant k \leqslant d$, continuous g on $\mathbb{R}^k$, and box $D_k \subset \mathbb{R}^k$,

$$\lim_{\varepsilon\to 0}\int_{D_k} f_i(\tau_{(x_1e_1+\cdots+x_ke_k)/\varepsilon}\,\omega)g(x_1,\ldots,x_k)dx_1\cdots dx_k = 0.$$

Now we do the induction on k. Assume that for any box $D_k \subset \mathbb{R}^k$,

$$\lim_{\varepsilon\to 0}\varepsilon\int_{D_k} F(x_1/\varepsilon,\ldots,x_k/\varepsilon,0,\ldots,0,\omega)dx_1\cdots dx_k = 0.$$

We can integrate by parts and rewrite

$$\varepsilon\int_{D_{k+1}} F(x_1/\varepsilon,\ldots,x_k/\varepsilon,x_{k+1}/\varepsilon,0,\ldots,0,\omega)dx_1\cdots dx_k\,dx_{k+1}$$

as

$$\int_{D_{k+1}} f_{k+1}(\tau_{(x_1e_1+\cdots+x_{k+1}e_{k+1})/\varepsilon}\,\omega)(1-x_{k+1})dx_1\cdots dx_k\,dx_{k+1}$$

$$+\,\varepsilon\int_{D_k} F(x_1/\varepsilon,\ldots,x_k/\varepsilon,0,\ldots,0)dx_1\cdots dx_k.$$

This completes the induction. □

LEMMA 6.2 *Assume that* $f \in L^\infty(\Omega,\mathbb{P})$. *Then*

$$\lim_{\varepsilon\to 0}\varepsilon\int_D |F(x/\varepsilon,\omega)|dx = 0.$$

PROOF: Clearly,

$$\varepsilon|F(x/\varepsilon,\omega)-F(y/\varepsilon,\omega)|\leqslant C|x-y|$$

and $F(0,\omega)\equiv 0$. For fixed ω, if $F^*(x)$ is any weak limit point of $\varepsilon F(x/\varepsilon,\omega)$, it is enough to show that $F^*(x)$ is a constant. Therefore it suffices to show that, for every smooth $\psi:\mathbb{R}^d\to\mathbb{R}^d$ with compact support,

$$\lim_{\varepsilon\to 0}\int \varepsilon F(x/\varepsilon,\omega)\operatorname{div}\psi(x)dx=0$$

a.e. $\mathbb{P}$. We can integrate by parts and reduce it to

$$\lim_{\varepsilon\to 0}\int\langle f(\tau_{x/\varepsilon}\omega),\psi(x)\rangle dx=0,$$

which is an easy consequence of Wiener's multidimensional ergodic theorem. □

LEMMA 6.3 *Let* $f\in L^1(\Omega,\mathbb{P})$ *with* $\mathbb{E}[f]=0$ *satisfying* (5.11). *Then*

$$\limsup_{\varepsilon\to 0}\varepsilon\int_D|F(x/\varepsilon,\omega)|dx\leqslant\frac{1}{2}\mathbb{E}\Big[\sum_{i=1}^d|f_i|\Big]\quad a.e.\ \mathbb{P}. \tag{6.1}$$

PROOF: Let $D=[0,1]^d$, and F be an integrable function on D whose weak gradient $\nabla F=f\in L^1(D;\mathbb{R}^d)$. Then

$$\begin{aligned}
&\int_{D\times D}|F(x)-F(y)|dx\,dy\\
&\leqslant\sum_{i=1}^d\int_{D\times D}|F(\ldots,y_{i-1},x_i,x_{i+1},\ldots)-F(\ldots,y_{i-1},y_i,x_{i+1},\ldots)|dx\,dy\\
&=\sum_{i=1}^d\int_{D\times D}\left|\int_{y_i}^{x_i}f_i(\ldots,y_{i-1},s,x_{i+1},\ldots)ds\right|dx\,dy\\
&\leqslant 2\sum_{i=1}^d\int_D s(1-s)|f_i(\ldots,y_{i-1},s,x_{i+1},\ldots)|dy_1\cdots dy_{i-1}\,ds\,dx_{i+1}\cdots dx_d\\
&\leqslant\frac{1}{2}\int_D\sum_{i=1}^d|f_i(x)|dx.
\end{aligned}$$

For a.e. ω

$$\int_D \varepsilon|F(x/\varepsilon,\omega)|dx$$

$$\leqslant \varepsilon \int_D \left| F(x/\varepsilon,\omega) - \int_D F(y/\varepsilon,\omega)dy \right| dx + \varepsilon \left| \int_D F(y/\varepsilon,\omega)dy \right|$$

$$\leqslant \varepsilon \int_{D\times D} |F(x/\varepsilon,\omega) - F(y/\varepsilon,\omega)|dx\,dy + \varepsilon \left| \int_D F(y/\varepsilon,\omega)dy \right|$$

$$\leqslant \frac{1}{2}\sum_{i=1}^{d} \int_D |f_i(x/\varepsilon,\omega)|dx + \varepsilon \left| \int_D F(y/\varepsilon,\omega)dy \right|.$$

Applying Wiener's ergodic theorem and Lemma 6.1, we get the result. □

Now we return to the proof of Theorem 5.3.

PROOF: In view of Lemmas 6.1 and 6.2, we only need to approximate gradients $f \in L^1(\Omega,\mathbb{P})$ by those in $L^\infty(\Omega,\mathbb{P})$. The approximation itself has to be done with some care. We begin with the relation

$$\frac{\partial}{\partial\lambda}[\lambda^d\varphi(\lambda x)] = d\lambda^{d-1}\varphi(\lambda x) + \lambda^d \sum_j x_j \left(\frac{\partial\varphi}{\partial x_j}\right)(\lambda x) = \frac{1}{\lambda}\sum_{j=1}^{d} \frac{\partial\psi_j^\lambda}{\partial x_j}(x),$$

where $\psi_j^\lambda(x) = x_j\lambda^d\varphi(\lambda x)$ and φ is a mollifier with compact support. This implies that for any $\lambda_1,\lambda_2 > 0$ the function $\lambda_1^d\varphi(\lambda_1 x) - \lambda_2^d\varphi(\lambda_2 x)$ is the divergence of a nice object. Take $\lambda_1 = n$ and $\lambda_2 = n^{-1}$ and define

$$g_n(x) = n^d\varphi(nx) - n^{-d}\varphi(n^{-1}x).$$

Then g_n can be written as $\operatorname{div}\psi^{(n)}$ for some smooth function $\psi^{(n)} : \mathbb{R}^d \to \mathbb{R}^d$. Let $f_j^{(n)}(\omega) = (f_j * g_n)(\omega)$. Using (5.11) it is easy to check that $f^{(n)} = \{f_j^{(n)}\}$, $j = 1,\dots,d$, is a gradient, namely,

$$f^{(n)}(\omega) = \nabla h_n(\omega) \quad \text{where } h_n(\omega) = -\sum_{j=1}^{d}(f_j * \psi_j^{(n)})(\omega) \in L^1(\Omega,\mathbb{P}).$$

Moreover, as $n \to \infty$ each $f_j^{(n)}$ converges in $L^1(\Omega,\mathbb{P})$ to f_j because, by the ergodic theorem, the large-scale average corresponding to λ_2 tends to 0. Now it is routine to truncate h_n, mollify the truncation, and take gradients to get nice bounded gradients that approximate f in $L^1(\Omega,\mathbb{P})$. □

Appendix: Proofs of Technical Lemmas

PROOF OF LEMMA 4.3: For $\ell = 1, 2, \dots$, define $h_n^\ell = h_n \mathbb{1}_{\{h_n \leqslant \ell\}}$. Then $0 \leqslant h_n^\ell \leqslant \ell$, and we can choose a subsequence $\{n_j\}$ such that, for each ℓ, functions $h_{n_j}^\ell \in L^\infty(\Omega)$ converge weakly as $j \to \infty$ to some h^ℓ, $0 \leqslant h^\ell \leqslant \ell$. Notice that

$$\mathbb{E}\, h^\ell = \lim_{j\to\infty} \mathbb{E}\, h_{n_j}^\ell \leqslant \liminf_{j\to\infty} \mathbb{E}\, h_{n_j} \leqslant C.$$

Since the sequence $\{h^\ell\}$, $\ell = 1, 2, \dots$, is nondecreasing, we can define $h(\omega) = \lim_{\ell\to\infty} h^\ell(\omega)$ a.s. Then $\lim_{\ell\to\infty} \mathbb{E}\, h^\ell = \mathbb{E}\, h$.

Set

$$a_j^\ell = \mathbb{E} h_{n_j}^\ell, \quad a^\ell = \mathbb{E} h^\ell, \quad \text{and} \quad a = \mathbb{E} h.$$

Then

$$a_j^\ell \to a^l \text{ as } j \to \infty \quad \text{and} \quad a^\ell \nearrow a \text{ as } \ell \to \infty,$$

and we can choose a subsequence $\{\ell_j\}$, $\ell_j \to \infty$ such that $a_j^{\ell_j} \to a$ as $j \to \infty$. It is clear that for any slower-growing sequence $\ell_j' \leqslant \ell_j$, $\ell_j' \to \infty$,

$$a_j^{\ell_j'} \to a \quad \text{as } j \to \infty.$$

We claim that $\widehat{h}_{n_j} = h_{n_j}^{\ell_j}$ and $r(h_{n_j}) = h_{n_j} - h_{n_j}^{\ell_j}$ satisfy conditions (i) and (ii).

Indeed, for each ℓ and all j such that $\ell_j \geqslant \ell$

$$\begin{aligned}\int\limits_{\widehat{h}_{n_j} > \ell} \widehat{h}_{n_j}\, d\mathbb{P} &= \int\limits_{h_{n_j}^{\ell_j} > l} h_{n_j}^{\ell_j}\, d\mathbb{P} \\ &= a_j^{\ell_j} - \int\limits_{h_{n_j}^{\ell_j} \leqslant l} h_{n_j}^{\ell_j}\, d\mathbb{P} = a_j^{\ell_j} - \int h_{n_j}^\ell\, d\mathbb{P} = a_j^{\ell_j} - a_j^\ell.\end{aligned}$$

Thus,

$$\lim_{\ell\to\infty} \lim_{j\to\infty} \int\limits_{\widehat{h}_{n_j} > \ell} \widehat{h}_{n_j}\, d\mathbb{P} = 0.$$

Also, for every $\varepsilon \in (0, 1)$,

$$\mathbb{P}(r(h_{n_j}) > \varepsilon) = \mathbb{P}(h_{n_j} - h_{n_j}^{\ell_j} > \varepsilon) \leqslant \mathbb{P}(h_{n_j} \geqslant \ell_j) \leqslant \frac{\mathbb{E}\, h_{n_j}}{\ell_j} \leqslant \frac{C}{\ell_j} \to 0$$

as $j \to \infty$. This completes the proof of Lemma 4.3. □

The proof of Lemma 4.4 is based on the following observation:

LEMMA A.1 *Let $h \geqslant 0$ be a measurable function and $\psi \geqslant 0$ be a nice mollifier on $\mathbb{R}^d$ supported on the cube of side a around the origin. Then there is a $c > 0$, which depends only on ψ, such that, for any $l > 0$ and ω,*

$$h^{\psi}(\omega) > \ell \Rightarrow |\{x \in \mathbb{R}^d : |x| \leqslant a\sqrt{d},\ h^{\psi}(\tau_{(0,x)}\omega) \geqslant cl\}| \geqslant c.$$

PROOF OF LEMMA 4.4 ASSUMING LEMMA A.1: For notational convenience we assume that $\{n_j\}$ is the whole sequence: $n_j = j$. We have

$$h_j^{\psi} = \widehat{(h^{\psi})}_j + r(h_j^{\psi}),$$

where $\widehat{(h^{\psi})}_j = h_j^{\psi} \mathbb{1}_{\{h_j^{\psi} \leqslant l_j\}}$ are uniformly integrable, $l_j \to \infty$, and $r(h_j^{\psi}) \to 0$ in probability. We need to show that the last convergence is locally uniform in x.

Fix $R > 0$ and $\varepsilon > 0$ and suppose that $r(h_j^{\psi})(\tau_{(0,x)}\omega) \geqslant \varepsilon$ for some x in the ball of radius R around the origin. This, in fact, means that $h_j^{\psi}(\tau_{(0,x)}\omega) > \ell_j$. From Lemma A.1 it follows that $h_j^{\psi}(\tau_{0,y}\omega) \geqslant c\ell_j$ for y on a set of measure at least c in the ball $|y - x| \leqslant a\sqrt{d}$, where a, c depend only on ψ. Hence, for a ball B of radius $R + a\sqrt{d}$ around the origin

$$\int_B h_j^{\psi}(\tau_{(0,x)}\omega)dx \geqslant c^2\ell_j.$$

Summarizing the above, we have

$$\begin{aligned}\mathbb{P}\big(\omega : \sup_{|x|\leqslant R} r(h_j^{\psi})(\tau_{(0,x)}\omega) \geqslant \varepsilon\big) &\leqslant \mathbb{P}\left(\omega : \int_B h_j^{\psi}(\tau_{(0,x)}\omega)dx \geqslant c^2\ell_j\right) \\ &\leqslant \frac{1}{c^2\ell_j}\mathbb{E}\int_B h_j^{\psi}(\tau_{(0,x)}\omega)dx \\ &\leqslant \frac{|B|}{c^2\ell_j}\mathbb{E}\, h_j^{\psi} \leqslant \frac{C|B|}{c^2\ell_j} \to 0 \quad \text{as } j \to \infty.\end{aligned}$$

□

PROOF OF LEMMA A.1: We have $h^{\psi}(\omega) = \int h(\tau_{(0,x)}\omega)\psi(x)dx$, where ψ is a nice nonnegative mollifier, bounded by A and supported on the cube of side a around the origin. We consider the cube K_b of size $b = a/N$ such that the cube of side a can be partitioned into a finite number N^d of cubes of side b. We have

$$\psi(x) \leqslant A\sum_{j=1}^{N^d} \mathbb{1}_{K_b^j}(x)$$

with $\{K_b^j\}$ being suitable translates of K_b. If $h^\psi(\omega) \geqslant \ell$, then for some j

$$\int h(\tau_{(0,y)}\omega)\mathbb{1}_{K_b^j}(y)dy \geqslant \frac{\ell}{AN^d}.$$

Since ψ is nice and $\int \psi(x)dx = 1$, it is easy to find c and N such that for any j

$$|\{x \in \mathbb{R}^d : |x| \leqslant a\sqrt{d},\ c\mathbb{1}_{K_b^j+x}(\cdot) \leqslant \psi(\cdot)\}| \geqslant c.$$

We now have a lower bound

$$h^\psi(\tau_{(0,x)}\omega) = \int h(\tau_{(0,y)}\omega)\psi(y-x)dy \geqslant \frac{c\ell}{AN^d}$$

provided $\psi(y-x) \geqslant c\mathbb{1}_{K_b^j}(y)$ or $\psi(y) \geqslant c\mathbb{1}_{K_b^j+x}(y)$.

We have shown that there is a positive constant (call it again c) depending only on ψ such that, for any $\ell > 0$, if $h^\psi(\omega) \geqslant \ell$, then

$$|\{x \in \mathbb{R}^d : |x| \leqslant a\sqrt{d},\ h^\psi(\tau_{(0,x)}\omega) \geqslant c\ell\}| \geqslant c.$$

□

Acknowledgment. E. Kosygina was partially supported by the CUNY Research Foundation under PSC-CUNY awards #66512-00-35 and #67332-00-36, and by a Eugene M. Lang Junior Faculty Research Fellowship. S. R. S. Varadhan was supported by the National Science Foundation through grants DMS-0104343 and DMS-0604380.

Bibliography

[1] Alvarez, O.; Bardi, M. Singular perturbations of nonlinear degenerate parabolic PDEs: a general convergence result. *Arch. Ration. Mech. Anal.* **170** (2003), no. 1, 17–61.

[2] Arisawa, M. Quasi-periodic homogenizations for second-order Hamilton-Jacobi-Bellman equations. *Adv. Math. Sci. Appl.* **11** (2001), no. 1, 465–480.

[3] Bakhvalov, N. S.; Panasenko, G. P. *Homogenisation: averaging processes in periodic media. Mathematical problems in the mechanics of composite materials.* Mathematics and Its Applications (Soviet Series), 36. Kluwer, Dordrecht, 1989.

[4] Bensoussan, A.; Lions, J. L.; Papanicolaou, G. *Asymptotic analysis for periodic structures.* Studies in Mathematics and Its Applications, 5. North-Holland, Amsterdam, 1978.

[5] Braides, A.; Defranceschi, A. *Homogenization of multiple integrals.* Oxford Lecture Series in Mathematics and Its Applications, 12. Clarendon, Oxford University Press, New York, 1998.

[6] Buckdahn, R.; Ichihara, N. Limit theorem for controlled backward SDEs and homogenization of Hamilton-Jacobi-Bellman equations. *Appl. Math. Optim.* **51** (2005), no. 1, 1–33.

[7] Caffarelli, L. A.; Souganidis, P. E.; Wang, L. Homogenization of fully nonlinear, uniformly elliptic and parabolic partial differential equations in stationary ergodic media. *Comm. Pure Appl. Math.* **58** (2005), no. 3, 319–361.

[8] Capuzzo-Dolcetta, I.; Ishii, H. On the rate of convergence in homogenization of Hamilton-Jacobi equations. *Indiana Univ. Math. J.* **50** (2001), no. 3, 1113–1129.

[9] Cioranescu, D.; Donato, P. *An introduction to homogenization.* Oxford Lecture Series in Mathematics and Its Applications, 17. Clarendon, Oxford University Press, New York, 1999.

[10] Evans, L. C. A survey of partial differential equations methods in weak KAM theory. *Comm. Pure Appl. Math.* **57** (2004), no. 4, 445–480.

[11] Fleming, W. H.; Soner, H. M. *Controlled Markov processes and viscosity solutions*. Applications of Mathematics, 25. Springer, New York, 1993.

[12] Ishii, H. Homogenization of the Cauchy problem for Hamilton-Jacobi equations. *Stochastic analysis, control, optimization and applications*, 305–324. Systems & Control: Foundations & Applications. Birkhäuser, Boston, 1999.

[13] Jikov, V. V.; Kozlov, S. M.; Oleĭnik, O. A. *Homogenization of differential operators and integral functionals*. Springer, Berlin, 1994.

[14] Kosygina, E.; Rezakhanlou, F.; Varadhan, S. R. S. Stochastic homogenization of Hamilton-Jacobi-Bellman equations. *Comm. Pure Appl. Math.* **59** (2006), no. 10, 1489–1521.

[15] Landim, C.; Olla, S.; Yau, H. T. Convection-diffusion equation with space-time ergodic random flow. *Probab. Theory Related Fields* **112** (1998), no. 2, 203–220.

[16] Lions, P.-L.; Souganidis, P. E. Correctors for the homogenization of Hamilton-Jacobi equations in the stationary ergodic setting. *Comm. Pure Appl. Math.* **56** (2003), no. 10, 1501–1524.

[17] Lions, P.-L.; Souganidis, P. E. Homogenization of degenerate second-order PDE in periodic and almost periodic environments and applications. *Ann. Inst. H. Poincaré Anal. Non Linéaire* **22** (2005), no. 5, 667–677.

[18] Lions, P.-L.; Souganidis, P. E. Homogenization of "viscous" Hamilton-Jacobi equations in stationary ergodic media. *Comm. Partial Differential Equations* **30** (2005), no. 1-3, 335–375.

[19] Marchenko, V. A.; Khruslov, E. Ya. *Homogenization of partial differential equations*. Progress in Mathematical Physics, 46. Birkhäuser, Boston, 2006.

[20] Sion, M. On general minimax theorems. *Pacific J. Math.* **8** (1958), 171–176.

[21] Sznitman, A.-S. *Brownian motion, obstacles and random media*. Springer Monographs in Mathematics. Springer, Berlin, 1998.

ELENA KOSYGINA
Baruch College
Department of Mathematics
One Bernard Baruch Way
Box B6-230
New York, NY 10010
E-mail: elena_kosygina@baruch.cuny.edu

S. R. S. VARADHAN
Courant Institute
251 Mercer Street
New York, NY 10012
E-mail: varadhan@cims.nyu.edu

Received September 2006.

Comm. Pure Appl. Math. 61 (2008) 1298-1329

Behavior of the Solution of a Random Semilinear Heat Equation

S. R. S. VARADHAN
Courant Institute

AND

NIKOS ZYGOURAS
University of Southern California

Abstract

We consider a semilinear heat equation in one space dimension, with a random source at the origin. We study the solution, which describes the equilibrium of this system, and prove that, as the space variable tends to infinity, the solution becomes a.s. asymptotic to a steady state. We also study the fluctuations of the solution around the steady state.

1 Introduction

One of the natural questions to ask in the study of random partial differential equations is the effect of randomness on the long term or asymptotic behavior. For linear equations, diffusions in a random environment is a typical case, and the large-time behavior has a long history. The case when the motion in the environment is either reversible or balanced has been investigated in [5, 6] and the more difficult general case has been investigated in [1, 8].

On the other hand, even for the simplest of nonlinear equations the effect of random initial or boundary data, at locations far away from the random input is hard to analyze. Turbulence, where "far away" is in the frequency domain, has been a particularly challenging problem, with some recent progress. In [2], for the Navier-Stokes equation with random forcing in two space dimensions, the existence of a unique equilibrium is proved, though very little information is provided about its properties.

In this article we provide a detailed description of the equilibrium state of a randomly perturbed, nonlinear system. In particular, we consider a nonlinear heat equation in one space dimension, with a source at the origin that is random in time and with quadratic dissipation. There is a unique stochastic stationary state where there is equilibrium between the source and dissipation. The solution, which describes this state, decays far away from 0 in space, and we are concerned here with the effect of randomness in the source at the origin on the decay at infinity of the solution.

We will consider the heat equation on $\mathbb{R} \times \mathbb{R}$, with a quadratic dissipative term and a stationary random source at the origin.

$$u_t + u_{xx} - u^2 + \lambda(t)\delta_0(x) = 0, \quad -\infty < x < \infty, \ -\infty < t < \infty, \tag{1.1}$$

where $\lambda(t) = \lambda(t, \omega) = \lambda(\theta_t \omega)$ is a stationary process represented through an ergodic measure-preserving action of the translation group $\omega \to \theta_t \omega$ acting on a probability space $(\Omega, \mathcal{F}, P)$. We assume, in order to simplify the analysis, that $\lambda(t)$ satisfies $0 < \lambda_1 \le \lambda(t) \le \lambda_2 < \infty$.

We can think of the solution of this equation as describing, under suitable rescaling, the density of a system of annihilating Brownian particles. The particles appear at the origin with an intensity determined by the source term $\lambda(t)$, performing subsequently independent Brownian motions, and they are killed as a result of binary collisions resulting in the term $-u^2$. This fact has been made rigorous in the case of dimension greater or equal to 2 and in the absence of the source term in [7].

It is worth pointing out that similar methods apply to nonlinear equations of the form

$$u_t + u_{xx} - u^p + \lambda(t)\delta_0(x) = 0, \quad -\infty < x < \infty, \ -\infty < t < \infty,$$

or even more generally to equations of the form

$$u_t + u_{xx} - f(u) + \lambda(t)\delta_0(x) = 0, \quad -\infty < x < \infty, \ -\infty < t < \infty,$$

with $f(u) \simeq u^p$ as $u \to 0$.

It is not hard to show that there is a unique positive and bounded solution $u(t, x, \omega)$ of the above equation on $\mathbb{R} \times \mathbb{R} \times \Omega$, and this solution is covariant in the sense that $u(t + \tau, x, \omega) = u(t, x, \theta_\tau \omega)$. Since the unique solution will be symmetric around $x = 0$, it is more convenient to consider it as a solution of

$$u_t + u_{xx} - u^2 = 0, \quad 0 < x < \infty, \ -\infty < t < \infty, \tag{1.2}$$

with Neumann boundary data, i.e., $u_x(t, 0) = -\frac{1}{2}\lambda(t)$. The solution is smooth for $x > 0$. To prove its existence we only need to notice that

(1) If we prescribe terminal data, say $u(T, x) = 0$ at $t = T$, then the solutions $u^T(t, x) \uparrow$ as $T \uparrow \infty$ by the maximum principle.
(2) By the maximum principle, the solutions $u^T(t, x)$ are all dominated by any positive solution of

$$u_t + u_{xx} - u^2 = 0; \quad u_x(t, 0) = -\frac{1}{2}\lambda_2. \tag{1.3}$$

(3) There are positive solutions of equation (1.3) that do not depend on t, given by

$$u_a(x) = \frac{6}{(x + a)^2}, \quad a > 0,$$

satisfying

$$u_{xx} - u^2 = 0, \quad u_x(t,0) = -\frac{12}{a^3}.$$

Therefore $u_a(x)$, with $a = (24/\lambda_2)^{1/3}$ is a positive solution of equation (1.3) that provides the necessary bound to prove existence. Uniqueness within the class of bounded solutions is easily established. Again by the maximum principle it is not hard to see that if $\lambda(t)$ satisfies the bounds $0 < \lambda_1 \leq \lambda(t) \leq \lambda_2 < \infty$, then

$$\frac{6}{(x+a_1)^2} \leq u(t,x) \leq \frac{6}{(x+a_2)^2}$$

where $a_i = (24/\lambda_i)^{1/3}$. We write

$$u(t,x) = \frac{6}{(x+a(t,x))^2}.$$

The goal of this article is to establish the following two theorems:

THEOREM 1.1 *If $\lambda(\cdot)$ is ergodic, then there exists a constant $\bar{a}$ such that*

$$\lim_{x\to\infty} a(0,x) = \bar{a}$$

exists for almost all a with respect to P.

For the second theorem we will assume that there exists $0 < A < \infty$ such that $\lambda(\cdot)$ has the property $\{\lambda(s) : s \leq a\}$ and $\{\lambda(s) : s \geq b\}$ are independent if $b - a \geq A$. We will use the standard notation $E^P[\cdot]$ to denote expectation with respect to P. This condition is stronger than what we need, but we assume it in order to simplify the proofs.

THEOREM 1.2 *If, in addition, $\lambda(\cdot)$ satisfies the mixing condition stated above, then as $x \to \infty$, $x(a(0,x) - \bar{a})$ has a limiting normal distribution with mean* 0 *and variance*

$$\sigma^2 = \lim_{x\to\infty} E^P[x^2(a(0,x) - \bar{a})^2].$$

We can show that $\sigma^2 > 0$ for a large class of examples.

Notice that Theorems 1.1 and 1.2 imply that the solution $u(t,x)$ has, for every t, the following asymptotic expansion in x for x tending to infinity:

$$u(t,x) \sim \frac{6}{x^2} - \frac{12\bar{a}}{x^3} + \frac{12\,G(t)}{x^4}$$

with $G(t)$ a Gaussian random variable.

Idea of proof. We first write the equation satisfied by a. After some calculation

$$u_t + u_{xx} - u^2 = 0$$

becomes

$$a_t + a_{xx} - \frac{6a_x}{(x+a)} - \frac{3a_x^2}{(x+a)} = 0. \tag{1.4}$$

The operator

$$a_{xx} - \frac{6a_x}{(x+a)} - \frac{3a_x^2}{(x+a)}$$

is viewed as a perturbation of the Bessel operator

$$a_{xx} - \frac{6}{x}a_x$$

by a Feynman-Kac term $c(t,x)$, and we write equation (1.4) as

$$a_t + a_{xx} - \frac{6}{x}a_x - c(t,x)a = 0 \tag{1.5}$$

where c is given by

$$c(t,x) = \frac{3a_x^2}{a(x+a)} - \frac{6a_x}{x(x+a)}. \tag{1.6}$$

We will consider the equation (1.5) in a domain $(t,x) \in (-\infty,\infty) \times [\ell,\infty)$. If $\ell > 0$, there are no singularities. We will have to choose ℓ to be sufficiently large, and this choice will be made later on.

Let us consider the space-time Bessel process $Q_{t,x}$ with generator

$$\frac{\partial}{\partial t} + \frac{\partial^2}{\partial x^2} - \frac{6}{x}\frac{\partial}{\partial x}$$

starting from (t,x) and the exit time τ_ℓ from (ℓ,∞),

$$\tau_\ell = \inf\{s : x(s) \le \ell\}.$$

Then

$$a(t,x,\omega) = E^{Q_{t,x}}[a(\tau_\ell,\ell,\omega)R(t,\tau_\ell,\omega)]$$

where

$$R(t,\sigma,\omega) = \exp[-r(t,\sigma,\omega)]$$

is the Feynman-Kac term with

$$r(t,\sigma,\omega) = \int_t^\sigma c(s,x(s),\omega)ds.$$

We will in Theorem A.2 prove a uniform bound of the form $\sup_{x,t,\omega}|xa_x| \le C$. This in turn will imply a uniform bound of the form $\sup_{x,t,\omega} x^3|c(t,x)| \le C$. We can then control $E^{Q_{t,x}}[R(t,\tau_\ell)]$ in terms of $E^{Q_{t,x}}[\exp[\int_t^{\tau_\ell} V(x(s))ds]]$ with $V(x) = C/x^3$, which in turn can be controlled if $\ell = \ell(C)$ is large enough. The expectation is with respect to the Bessel process $Q_{t,x}$, and $R(t,\tau_\ell) = R(t,\tau_\ell,\omega)$

depends on the random boundary conditions $\lambda(t,\omega)$ at $x=0$. If we define, for a fixed suitably large ℓ, the conditional expectation

$$g(t,x,\tau,\ell,\omega)=E^{Q_{t,x}}[R(t,\tau_\ell,\omega)\mid\tau_\ell=\tau]=E^{Q^{\tau,\ell}_{t,x}}[R(t,\tau,\omega)] \tag{1.7}$$

where $Q^{\tau,\ell}_{t,x}$ is the Bessel Bridge conditioned to exit from $[\ell,\infty)$ at time τ, then we can represent

$$a(t,x,\omega)=E^{Q_{t,x}}[a(\tau_\ell,\ell,\omega)g(t,x,\ell,\tau_\ell,\omega)].$$

We will show that, for fixed ℓ, as $x\to\infty$ and $t\to-\infty$ so long as $|t|\sim x^2$, $g(t,x,\tau,\ell,\omega)$ is nearly independent of t and x and can be approximated by a function $h(\tau,\ell,\omega)$ of τ that is covariant, i.e., $h(\tau,\ell,\omega)=h(0,\ell,\theta_\tau\omega)$. It will then follow that

$$\begin{aligned}a(0,x,\omega)&\simeq E^{Q_{0,x}}[a(\tau_\ell,\ell,\omega)h(\tau_\ell,\ell,\omega)]\\&=\int a(\tau,\ell,\omega)h(\tau,\ell,\omega)p(x,\ell,\tau)d\tau\end{aligned} \tag{1.8}$$

where $p(x,\ell,\tau)$ is the density of the hitting time τ_ℓ of the Bessel process $Q_{0,x}$. This will give us a law of large numbers with

$$\lim_{x\to\infty}a(0,x,\omega)=\bar a=\int a(0,\ell,\omega)h(0,\ell,\omega)dP,$$

and if the approximation in (1.8) is good enough, then

$$x(a(0,x)-\bar a)\simeq x\int_0^\infty[(a(\tau,\ell,\omega)h(\tau,\ell,\omega)-\bar a)]p(x,\ell,\tau)d\tau.$$

Since $p(x,\ell,\tau)$, the density of the distribution of the exit time τ_ℓ of the Bessel process $Q_{0,x}$, admits a diffusive scaling limit

$$p(x,\ell,\tau)\simeq\frac{1}{x^2}\,p\Big(1,0,\frac{\tau}{x^2}\Big)$$

if $a(t,\ell,\omega)h(t,\omega)$ is sufficiently mixing in t, then a CLT can be proved with the required scaling.

The fixed constant $\ell>0$ will be chosen later. We consider the space-time Bessel process starting from $x>\ell$ at time $t<\tau$, and condition it so that the exit time from the interval (ℓ,∞) is τ, i.e., $\tau_\ell=\tau$, and denote the conditioned process (Bessel bridge) by $Q^{\tau,\ell}_{t,x}$. Since the Bessel process is not explicitly time dependent, $Q^{\tau,\ell}_{t,x}$ is covariant with respect to time translations, i.e., $Q^{\tau,\ell}_{t,x}(A)=Q^{\tau-t,\ell}_{0,x}(\theta_t A)$. As t and x go to $-\infty$ and ∞ appropriately, i.e., $t\sim x^2$, this process will have a limit on the space $C[(-\infty,0],[\ell,\infty)]$ that we will call $Q^{\tau,\ell}_*$. Then with

$$R(-\infty,\sigma,\omega,x(\cdot))=\exp\Big[-\int_{-\infty}^{\sigma}c(s,x(s),\omega)ds\Big]$$

and

$$h(\tau,\ell,\omega)=E^{Q^{\tau,\ell}_*}[R(-\infty,\tau,\omega,x(\cdot))]=h(0,\ell,\theta_\tau\omega), \tag{1.9}$$

$g(t,x,\tau,\ell,\omega)$ will be approximated by $h(\tau,\ell,\omega) = h(0,\ell,\theta_\tau\omega)$. Therefore the solution $a(0,x,\omega)$ will be approximated by $\widehat{a}(0,x)$, defined by

$$\widehat{a}(0,x,\omega) = \int a(\tau,\ell,\omega)h(\tau,\ell,\omega)p(x,\ell,\tau)d\tau.$$

For fixed ℓ, $a(\tau,\ell,\omega)$ and $h(\tau,\ell,\omega)$ are stationary processes in τ. While the density $p(x,\ell,\tau)$ of the distribution of τ_ℓ under $Q_{0,x}$ is not explicit, its Fourier transform is in closed form and in addition, it has the scaling property

$$x^2 p(x,\ell,x^2 t) = p\left(1,\frac{\ell}{x},t\right).$$

Its behavior as $x \to \infty$ is therefore easy to analyze. This asymptotic behavior is combined with the standard ergodic theorem to prove that

$$\lim_{x\to\infty} \widehat{a}(0,x,\omega) = \bar{a} = E^P[a(0,\ell,\omega)h(0,\ell,\omega)] \quad \text{a.s.} \tag{1.10}$$

The estimate that plays the central role in our analysis is

$$\lim_{x\to\infty} x^2 E^P[|\widehat{a}(0,x,\omega) - a(0,x,\omega)|^2] = 0.$$

This estimate reduces the CLT to proving a CLT for $x(\widehat{a}(0,x,\omega) - \bar{a})$. This will further require us to prove some mixing properties for $h(\tau,\ell,\omega)$ and $a(\tau,\ell,\omega)$ as processes in τ. We will show that they inherit these from similar properties of the source process $\lambda(\cdot,\omega)$.

2 Bessel Process

We will now define the entrance process $Q_*^{\tau,\ell}$. Because of its covariant nature, we need only define $Q_*^{0,\ell}$.

Let us recall that $Q_{t,x}^{0,\ell}$ is the distribution of the Bessel process starting at time $t < 0$ from $x > \ell$ conditioned to exit from (ℓ,∞) at time 0, i.e., $\tau_\ell = 0$. This defines a Markov process on $C[[t,0];[\ell,\infty)]$ with transition probability densities $s < \sigma < 0$ and $y,z > \ell$

$$q^{0,\ell}(s,y,\sigma,z) = \frac{q^D(\sigma - s,y,z)p(z,\ell,-\sigma)}{p(y,\ell,-s)}$$

where q^D is the transition density of the Bessel process with Dirichlet boundary condition at $x = \ell$.

The entrance process $Q_*^{0,\ell}$ is characterized by two properties. For every $x > \ell$, the distribution of the exit time τ_x from (x,∞) under $Q_*^{0,\ell}$ is given by the density $f^{0,\ell}(x,t) = p(x,\ell,-t)$ and the conditional distribution of $Q_*^{0,\ell}$ after time τ_x, given the past is given by $Q_{\tau_x,x}^{0,\ell}$. Uniqueness within the class of measures supported on paths $\{x(\cdot) : x(t) \to \infty \text{ as } t \to -\infty\}$ is an easy consequence of $\tau_x \to -\infty$ as $x \to \infty$.

Existence will follow from Kolmogorov's consistency theorem provided that $\{f^{0,\ell}(x,t)\}$ is consistent as a family of distributions of exit times. Under any $Q^{0,\ell}_{t,x}$ the successive hitting times $\{\tau_y\}$ of levels $x > y > \ell$ are Markovian with the transition density given by

$$p^{0,\ell}(y_1,\tau_1,y_2,\tau_2) = \frac{p(y_1,y_2,\tau_2-\tau_1)p(y_2,\ell,-\tau_2)}{p(y_1,\ell,-\tau_1)} \tag{2.1}$$

and

$$\begin{aligned}\int f^{0,\ell}(x,s)\,p^{0,\ell}(s,x,t,y)ds &\int p(x,\ell,-s)\frac{p(x,y,t-s)p(y,\ell,-t)}{p(x,\ell,-s)}ds\\ &= \int p(x,y,t-s)p(y,\ell,-t)ds\\ &= p(y,\ell,-t) = f^{0,\ell}(y,t),\end{aligned}$$

which proves the necessary consistency.

If we condition the hitting time of level ℓ to be any arbitrary time τ, then the corresponding $Q^{\tau,\ell}_*$ is just the time shift by τ of $Q^{0,\ell}_*$.

One of the quantities we will need to estimate is

$$Q^{0,\ell}_*[x(-s) \le y]$$

THEOREM 2.1 *For $y \ge \ell, s > 0$*

$$Q^{0,\ell}_*[x(-s) \le y] \le c\int_{s/y^2}^{\infty} e^{-\frac{1}{4t}}t^{-\frac{9}{2}}\,dt$$

where

$$c^{-1} = \int_0^\infty e^{-\frac{1}{4t}}t^{-\frac{9}{2}}\,dt.$$

In particular, for any $r > 0$ there is a constant C_r such that

$$E^{Q^{0,\ell}_*}[x(-s)^{-r}] \le C_r s^{-\frac{r}{2}}.$$

For small s we can use the bound $x(-s) \ge \ell$. In addition, we have the moment estimates

$$\int \tau^k p(x,\ell,\tau)\,d\tau \le C_k x^{2k} \tag{2.2}$$

for $k = 1,2,3$.

PROOF: Clearly

$$\begin{aligned}Q^{0,\ell}_*[x(-s) \le y] \le Q^{0,\ell}_*[\tau_y \le -s] &= \int_s^\infty p(y,\ell,t)dt\\ &= Q_{0,y}[\tau_\ell \ge s] \le Q_{0,y}[\tau_0 \ge s]\end{aligned}$$

and from the scaling properties of the Bessel process,

$$Q_{0,y}[\tau_0 \geq s] = Q_{0,1}\left[\tau_0 \geq \frac{s}{y^2}\right] = c\int_{s/y^2}^{\infty} e^{-\frac{1}{4t}} t^{-\frac{9}{2}}\,dt.$$

Since $\tau_\ell \leq \tau_0$, the estimates (2.2) on the moments of τ_ℓ follow from the scaling property of the distribution of τ_0 and the tail behavior of $p(1,0,t) = ce^{-1/4t}t^{-9/2}$. We can also bound

$$\begin{aligned}
E^{Q_*^{0,\ell}}[x(-s)^{-r}] &= \int_\ell^\infty y^{-r}\,dQ_*^{0,\ell}[x(-s) \leq y] \\
&= r\int_\ell^\infty y^{-(r+1)} Q_*^{0,\ell}[x(-s) \leq y]dy \\
&\leq Cr\int_\ell^\infty y^{-(r+1)}dy \int_{s/y^2}^\infty e^{-\frac{1}{4t}} t^{-\frac{9}{2}}\,dt \\
&\leq Cr\int_\ell^\infty y^{-(r+1)}\left(1+\frac{s}{y^2}\right)^{-\frac{7}{2}} dy \\
&= Crs^{-\frac{r}{2}}\int_\ell^\infty \frac{y^{6-r}}{(y^2+1)^{7/2}}\,dy \\
&= C_r s^{-\frac{r}{2}}.
\end{aligned}$$

□

We will need the following estimates on $p(x,\ell,\tau)$, the probability density of the exit time τ_ℓ from (ℓ,∞) of the Bessel process starting from level $x > \ell$ at time 0.

THEOREM 2.2 *There is a constant C such that for all $x \geq 2\ell$ and $\sigma \geq 0$,*

$$\int_0^\infty |p(x,\ell,\tau+\sigma) - p(x,\ell,\tau)|d\tau \leq \frac{C\sigma}{x^2} \tag{2.3}$$

and for $x \geq 2y$, $y \geq \ell$,

$$\int_0^\infty |p(x,\ell,\tau) - p(x,y,\tau)|d\tau \leq \frac{Cy^2}{x^2}. \tag{2.4}$$

PROOF: We note that the Laplace transform

$$\begin{aligned}
\psi(x,0,\lambda) &= \int e^{-\lambda\tau} p(x,0,\tau)d\tau \\
&= e^{-x\sqrt{\lambda}}\left[1 + x\sqrt{\lambda} + \frac{6x^2}{15}\lambda + \frac{x^3}{15}\lambda^{\frac{3}{2}}\right]
\end{aligned}$$

can be explicitly calculated as the solution of

$$\psi_{xx} - \frac{6}{x}\psi_x = \lambda\psi$$

with $\psi(0) = 1, \psi(\infty) = 0$. It is easy to see that for $0 \le a < b$,

$$\psi(b, a, \lambda) = \int e^{-\lambda\tau} p(b, a, \tau) d\tau = \frac{\psi(b, 0, \lambda)}{\psi(a, 0, \lambda)}.$$

By analytic continuation, setting $\lambda = -i\xi$, the Fourier transform

$$\widehat{p}(1, a, \xi) = \int_0^\infty e^{i\xi\tau} p(1, a, \tau) d\tau = \frac{\psi(1, 0, -i\xi)}{\psi(a, 0, -i\xi)}$$

is seen to satisfy

$$\sup_{0 \le a \le \frac{1}{2}} \int_{-\infty}^\infty [1 + \xi^2] |D_\xi^r \widehat{p}(1, a, \xi)|^2 \, d\xi \le C$$

for $r = 1, 2, 3$. The only subtle point here is the differentiability of $\psi(1, 0, \lambda)$ at $\lambda = 0$, which follows from an explicit expansion of the form $\psi(1, 0, \lambda) = 1 + c_1\lambda + c_2\lambda^2 + c_3\lambda^3 + O(\lambda^{7/2})$ in powers of $\sqrt{\lambda}$. The Fourier transform $\widehat{p}$ therefore has three continuous derivatives at $\xi = 0$. This implies by the Plancherel identity

$$\sup_{0 \le a \le \frac{1}{2}} \int_0^\infty (1 + \tau^2)^3 |p_\tau(1, a, \tau)|^2 \, d\tau \le C. \tag{2.5}$$

Therefore

$$\sup_{0 \le a \le \frac{1}{2}} \int_0^\infty (1 + |\tau|) |p_\tau(1, a, \tau)| d\tau \le C, \tag{2.6}$$

which in turn implies that for $0 \le a \le \frac{1}{2}$,

$$\int_0^\infty |p(1, a, \tau) - p(1, a, \tau + \sigma)| d\tau \le C\sigma.$$

The scaling relation $p(x, \ell, \tau) = \frac{1}{x^2} p(1, \frac{\ell}{x}, \frac{\tau}{x^2})$ immediately establishes the first part of the theorem. As for the second part,

$$p(x, \ell, \tau) - p(x, y, \tau) = \int_0^\infty [p(x, y, \tau - \sigma) - p(x, y, \tau)] p(y, \ell, \sigma) d\sigma,$$

and using (2.2)

$$\begin{aligned}
&\int_0^\infty |p(x, \ell, \tau) - p(x, y, \tau)| d\tau \\
&\quad \le \int_0^\infty \int_0^\infty |p(x, y, \tau - \sigma) - p(x, y, \tau)| p(y, \ell, \sigma) d\sigma \, d\tau \\
&\quad = \frac{C}{x^2} \int \sigma p(y, \ell, \sigma) d\sigma \\
&\quad \le \frac{Cy^2}{x^2}. \qquad (2.7)
\end{aligned}$$

□

3 The Main Estimate

We will now estimate the difference between $g(t,x,\tau,\ell,\omega)$ and $h(\tau,\ell,\omega)$. But before that we will need some preliminary estimates.

We start with Khasminski's lemma, which we state and prove for completeness.

LEMMA 3.1 *Let $\{P_x\}$ be a Markov family on a state space X with $x(t)$ as its trajectory. τ is the exit time from a set G, and $V(x) \geq 0$ is a nonnegative function on X. If*

$$\sup_{x\in G} E^{P_x}\left[\int_0^\tau V(x(s))ds\right] \leq \theta,$$

then for $n \geq 2$

$$E^{P_x}\left[\left[\int_0^\tau V(x(s))ds\right]^n\right] \leq n\theta E^{P_x}\left[\left[\int_0^\tau V(x(s))ds\right]^{n-1}\right]. \tag{3.1}$$

Moreover, if $\alpha > 0$ satisfies $\alpha\theta < 1$,

$$E^{P_x}\left[\exp\left[\alpha\int_0^\tau V(x(s))ds\right]\right] \leq \frac{1}{1-\alpha\theta}. \tag{3.2}$$

PROOF:

$$\begin{aligned} &E_x\left[\left[\int_0^\tau V(x(s))ds\right]^n\right] \\ &= n!E_x\left[\int\cdots\int_{0\leq s_1\cdots<s_n<\tau} V(x(s_1))\cdots V(x(s_n))ds_1\cdots ds_n\right] \\ &\leq n\theta E_x\left[\left[\int_0^\tau V(x(s))ds\right]^{n-1}\right]. \end{aligned}$$

By induction

$$E_x\left[\left[\int_0^\tau V(x(s))ds\right]^n\right] \leq n!\theta^{n-1}E_x\left[\left[\int_0^\tau V(x(s))ds\right]\right] \leq n!\theta^n. \tag{3.3}$$

If $\alpha\theta < 1$, then summing the exponential series we get

$$E^{P_x}\left[\exp\left[\alpha\int_0^\tau V(x(s))ds\right]\right] \leq \sum_{n=0}^\infty (\alpha\theta)^n = \frac{1}{1-\alpha\theta}.$$

□

We now look at the space-time Bessel process with generator

$$\mathcal{A} = \frac{\partial}{\partial t} + \frac{\partial}{\partial x^2} - \frac{6}{x}\frac{\partial}{\partial x}$$

in the domain $(-\infty,\infty)\times[\ell,\infty)$ stopped when it exits at $x=\ell$. For the function $u(x)=C/4\ell - C/4x$, we see that

$$(\mathcal{A}u)(x) = -\frac{C}{x^3}, \quad u(\ell)=0,$$

and therefore with $V(x)=C/x^3$ and $\theta = C/4\ell$

$$\sup_{t,x\geq\ell} E^{Q_{t,x}}\Big[\int_t^{\tau_\ell} V(x(s))ds\Big] \leq \theta. \tag{3.4}$$

LEMMA 3.2 *There exists ℓ such that for all $x>y\geq\ell$, and $n\geq 1$*

$$\begin{aligned}\sup_{\omega,x,t} y^n E^{Q_{t,x}}\left[\left[\int_t^{\tau_y} |c(s,x(s),\omega)|ds\right]^n\right] &= \sup_{\omega,x,t} y^n \int E^{Q_{t,x}^{\tau,\ell}}\left[\left[\int_t^{\tau_y} |c(s,x(s),\omega)|ds\right]^n\right] p(x,\ell,\tau-t)d\tau \\ &\leq C_n.\end{aligned} \tag{3.5}$$

If we replace the supremum over ω by the mean value, then

$$\begin{aligned}\limsup_{y\to\infty}\sup_{x,t} y^n E^P\left[E^{Q_{t,x}}\left[\int_t^{\tau_y} |c(s,x(s),\omega)|ds\right]^n\right] &= \limsup_{y\to\infty}\sup_{x,t} y^n E^P\left[\int E^{Q_{t,x}^{\tau,\ell}}\left[\left[\int_t^{\tau_y} |c(s,x(s),\omega)|ds\right]^n\right] p(x,\ell,\tau-t)d\tau\right] \\ &= 0.\end{aligned} \tag{3.6}$$

The quantities really do not depend on t, due to stationarity.

Moreover, given $\alpha>0$, there exist $\ell(\alpha)$ and a constant $C=C(\alpha)$ such that if $x>\ell\geq\ell(\alpha)$, then

$$\begin{aligned}\sup_{\omega,x,t} E^{Q_{t,x}}\left[\exp\left[\alpha\int_t^{\tau_\ell} |c(s,x(s),\omega)|ds\right]\right] &= \sup_{\omega,x,t}\int E^{Q_{t,x}^{\tau,\ell}}\left[\exp\left[\alpha\int_t^{\tau} |c(s,x(s),\omega)|ds\right]\right] p(x,\ell,\tau-t)d\tau \\ &\leq C(\alpha)\end{aligned} \tag{3.7}$$

PROOF: Since from (A.7) we have the uniform bound $|c(s,x,\omega)|\leq C/x^3$, (3.3), (3.2) and (3.4) can be applied; this proves (3.5) and (3.7). Once we have (3.5) to prove (3.6), it is enough to prove it for $n=1$. For $n=1$, if we are allowed to average with respect to ω, then by stationarity, $E^P[|c(s,x)|]$ is independent of s. We use the uniform bound $\sup_x G(x,z)\leq Cz$ on the Green's function

$$G(x,z) = \frac{1}{7} z^{-6}((\min\{x,z\})^7 - y^7)$$

of the Bessel operator $D_{xx} - (6/x)D_x$ in the domain $(t,\infty)\times(y,\infty)$ to conclude that

$$E^P\left[E^{Q_{t,x}}\left[\int_t^{\tau_y} |c(s,x(s),\omega)|ds\right]\right] = E^{Q_{t,x}}\left[\int_t^{\tau_y} E^P[|c(s,x(s),\omega))|]ds\right]$$
$$\le C\int_y^\infty zE^P|c(s,z,\omega))|dz = o\left(\frac{1}{y}\right)$$

in view of (A.9), and so (3.6) follows. We will need estimates (3.7) for $\alpha \le 4$, and this will dictate the choice of ℓ. □

LEMMA 3.3 *Estimates similar to Lemma* 3.2 *are valid for the entrance processes* $Q_*^{\tau,\ell}$. *There exists* ℓ *such that for* $y > \ell$,

$$\sup_{\omega,\tau} y^n E^{Q_*^{\tau,\ell}}\left[\left[\int_{-\infty}^{\tau_y} |c(s,x(s),\omega)|ds\right]^n\right] \le C_n. \tag{3.8}$$

If we replace the supremum over ω *by the mean value, then*

$$\limsup_{y\to\infty}\sup_\tau\ y^n E^P\left[E^{Q_*^{\tau,\ell}}\left[\int_{-\infty}^{\tau_y} |c(s,x(s),\omega)|ds\right]^n\right] = 0 \tag{3.9}$$

The quantities are again independent of τ.

Moreover, given $\alpha > 0$, *there exist* $\ell = \ell(\alpha)$ *and a constant* $C = C(\alpha)$ *such that if* $x > \ell > \ell(\alpha)$, *then*

$$\sup_{\omega,\tau} E^{Q_*^{\tau,\ell}}\left[\exp\left[\alpha\int_{-\infty}^{\tau} |c(s,x(s),\omega)|ds\right]\right] \le C(\alpha) \tag{3.10}$$

PROOF: Since the entrance process is covariant with respect to translations, we can assume that $\tau = 0$. Let us estimate

$$y^n E^{Q_*^{0,\ell}}\left[\left[\int_{\tau_x}^{\tau_y} |c(s,x(s),\omega)|ds\right]^n\right]$$

uniformly in $x > y$. If we condition with respect to $\tau_x = t$, then using the bound of $V(x) = C/x^3$ for $|c(s,x,\omega)|$

$$\begin{aligned}
&E^{Q_*^{0,\ell}}\left[\left[\int_{\tau_x}^{\tau_y} |c(s,x(s),\omega)|ds\right]^n\right]\\
&\le \int E^{Q_{x,t}^{\ell,0}}\left[\left[\int_t^{\tau_y} V(x(s))ds\right]^n\right] f^{0,\ell}(x,t)dt\\
&= \int E^{Q_{x,0}^{\ell,-t}}\left[\left[\int_0^{\tau_y} V(x(s))ds\right]^n\right] f^{0,\ell}(x,t)dt\\
&= \int E^{Q_{x,0}^{\ell,t}}\left[\left[\int_0^{\tau_y} V(x(s))ds\right]^n\right] p(x,\ell,t)dt\\
&= E^{Q_{x,0}}\left[\left[\int_0^{\tau_y} V(x(s))ds\right]^n\right].
\end{aligned}$$

This proves (3.8) and (3.9). The proof of (3.10) is similar. □

We now turn to the proof of the main estimate of this section. For functions g of ω (and possibly other variables) we define the norm $\|g\|_{2,P}$ by

$$\|g\|_{2,P} = [E^P|g(\omega)|^2]^{\frac{1}{2}}.$$

THEOREM 3.4 *Let* $g(0,x,\tau,\ell,\omega)$ *and* $h(\tau,\ell,\omega)$ *be as defined earlier in* (1.7) *and* (1.9). *Then*

$$\lim_{x\to\infty} x\int_0^\infty \left[E^P[|g(0,x,\tau,\ell,\omega)-h(\tau,\ell,\omega)|^2]\right]^{\frac{1}{2}} p(x,\ell,\tau)d\tau = 0.$$

Therefore

$$\lim_{x\to\infty} x\|\widehat{a}(0,x,\omega)-a(0,x,\omega)\|_{2,P} = 0.$$

PROOF: The proof consists of several steps. First we truncate $r(-\infty,\tau_\ell,\omega)$ and replace it with $r(\tau_x,\tau_\ell,\omega)$. For $j\ge 1$, we define

$$\sigma_j = \tau_{2^j\ell} = \inf\{t : x(t)\le 2^j\ell\}$$

to be the hitting time of the level $y_j = 2^j\ell$. Define $R(\tau_a,\tau_\ell,\omega) = e^{-r(\tau_a,\tau_\ell,\omega)}$. Let n be the largest integer such that $2^n\ell\le x$. $R(\tau_x,\tau_\ell,\omega) = e^{-r(\tau_x,\tau_\ell,\omega)}$ will be successively replaced by $R(\sigma_j,\tau_\ell,\omega) = e^{-r(\sigma_j,\tau_\ell,\omega)}$ with $j = 1,\dots,n$.

Step A. In order to estimate the successive differences, we will need to estimate the size of the difference between $R(\tau_a,\tau_\ell,\omega) = e^{-r(\tau_a,\tau_\ell,\omega)}$ and $R(\tau_b,\tau_\ell,\omega) = e^{-r(\tau_b,\tau_\ell,\omega)}$ where $a>b>\ell$. More precisely, if $Q_{0,x}^{\tau,\ell}$ is the Bessel bridge, then we will need an estimate of

$$H(x,a,b) = \int \left\|E^{Q_{0,x}^{\tau,\ell}}[R(\tau_a,\tau_\ell)-R(\tau_b,\tau_\ell)]\right\|_{2,P}^2 p(x,\ell,\tau)\,d\tau.$$

First, we use the bound $|e^x-e^y|\le e^{\max\{x,y\}}|x-y|$ to estimate

$$|R(\tau_a,\tau_\ell)-R(\tau_b,\tau_\ell)|\le e^{\xi}|r(\tau_a,\tau_b)|$$

where

$$\xi\le\int_0^{\tau_\ell}|c(s,x(s),\omega)|ds\le\int_0^{\tau_\ell}V(x(s))ds$$

and $V(x) = C/x^3$:

$$\begin{aligned} E^{Q_{0,x}^{\tau,\ell}}\left[|R(\tau_a,\tau_\ell)-R(\tau_b,\tau_\ell)|^2\right] &\le E^{Q_{0,x}^{\tau,\ell}}[e^{2\xi}|r(\tau_a,\tau_b)|^2] \\ &\le \left[E^{Q_{0,x}^{\tau,\ell}}[e^{4\xi}]\right]^{\frac{1}{2}}\left[E^{Q_{0,x}^{\tau,\ell}}[|r(\tau_a,\tau_b)|^4]\right]^{\frac{1}{2}}. \end{aligned}$$

Therefore

$$H(x,a,b) \le \left[E^P\left[\int [E^{Q_{0,x}^{\tau,\ell}}[e^{4\xi}]]p(x,\ell,\tau)\,d\tau\right]\right]^{\frac{1}{2}}$$

$$\times\left[E^P\left[\int [E^{Q_{0,x}^{\tau,\ell}}[|r(\tau_a,\tau_b)|^4]p(x,\ell,\tau)\,d\tau\right]\right]^{\frac{1}{2}}$$

$$= [E^P[[E^{Q_{0,x}}[e^{4\xi}]]]]^{\frac{1}{2}} \times [E^P[[E^{Q_{0,x}}[|r(\tau_a,\tau_b)|^4]]]^{\frac{1}{2}}$$

$$\le \frac{\delta(b)}{b^2} \tag{3.11}$$

where $\delta(b) \to 0$ as $b \to \infty$ in view of the estimate (3.6). In what follows we will use $\delta(b)$ to denote quantities that are $o(1)$ as $b \to \infty$. Since the estimate is uniform in x and a, we have the same estimate on

$$H(b) = \sup_\tau \left\| E^{Q_*^{\tau,\ell}}[R(-\infty,\tau_\ell) - R(\tau_b,\tau_\ell)]\right\|_{2,P}^2 \le \frac{\delta(b)}{b^2}. \tag{3.12}$$

We now replace $R(-\infty,\tau_\ell)$ in the definition of

$$h(\tau,\ell,\omega) = E^{Q_*^{\tau,\ell}}[R(-\infty,\tau_\ell)]$$

by $R(\tau_x,\tau_\ell)$ and define

$$h_x(\tau,\ell,\omega) = E^{Q_*^{\tau,\ell}}[R(\tau_x,\tau)] = h_x(0,\ell,\theta_\tau\omega).$$

Then, from (3.12),

$$\sup_\tau \|h(\tau,\ell,\omega) - h_x(\tau,\ell,\omega)\|_{2,P} \le \frac{\delta(x)}{x}.$$

Step B. Now we take up the difference between $h_x(\tau,\ell,\omega)$ and $g(t,x,\tau,\ell,\omega)$:

$$g(t,x,\tau,\ell,\omega) - h_x(\tau,\ell,\omega) = E^{Q_{0,x}^{\tau,\ell}}[R(\tau_x,\tau_\ell)] - E^{Q_*^{\tau,\ell}}[R(\tau_x,\tau_\ell)].$$

We represent

$$R(\tau_x,\tau_\ell) = 1 + \sum_{j=1}^{n}[R(\sigma_j,\tau_\ell) - R(\sigma_{j-1},\tau_\ell)] + [R(\tau_x,\tau_\ell) - R(\sigma_n,\tau_\ell)]$$

to obtain

$$g(0,x,\tau,\ell,\omega) - h_x(\tau,\ell,\omega) = \sum_{j=1}^{n}[g_j(0,x,\tau,\ell,\omega) - h_j(\tau,\ell,\omega)]$$
$$+ [g_{n+1}(0,x,\tau,\ell,\omega) - h_{x,n+1}(\tau,\ell,\omega)],$$

where for $1 \le j \le n$

$$g_j(0,x,\tau,\ell,\omega) = E^{Q^{\tau,\ell}_{0,x}}[R(\sigma_j,\tau_\ell) - R(\sigma_{j-1},\tau_\ell)]$$
$$h_j(\tau,\ell,\omega) = E^{Q^{\tau,\ell}_{*}}[R(\sigma_j,\tau_\ell) - R(\sigma_{j-1},\tau_\ell)]$$

and

$$g_{n+1}(0,x,\tau,\ell,\omega) = E^{Q^{\tau,\ell}_{0,x}}[R(0,\tau_\ell) - R(\sigma_n,\tau_\ell)]$$
$$h_{x,n+1}(\tau,\ell,\omega) = E^{Q^{\tau,\ell}_{*}}[R(\tau_x,\tau_\ell) - R(\sigma_n,\tau_\ell)].$$

Step B1. First we estimate $\Delta(x,n+1)$, which is defined as

$$\begin{aligned}
&\Delta(x,n+1)\\
&= \int \|g_{n+1}(0,x,\tau,\ell,\omega) - h_{x,n+1}(x,\tau,\ell,\omega)\|_{2,P}\, p(x,\ell,\tau)\,d\tau\\
&\le \int [\|g_{n+1}(0,x,\tau,\ell,\omega)\|_{2,P} + \|h_{x,n+1}(\tau,\ell,\omega)\|_{2,P}]\, p(x,\ell,\tau)\,d\tau.
\end{aligned}$$

Recalling that $y_j = 2^j\ell$ and that n is defined by $y_n \le x < 2y_n$,

$$\int \|g_{n+1}(0,x,\tau,\ell,\omega)\|_{2,P}\, p(x,\ell,\tau)\,d\tau = \sqrt{H(x,x,y_n)}$$

and

$$\int \|h_{x,n+1}(\tau,\ell,\omega)\|_{2,P}\, p(x,\ell,\tau)\,d\tau \le \sqrt{H(x)} + \sqrt{H(y_n)}.$$

Therefore

$$\begin{aligned}
\Delta(x,n+1) &\le \sqrt{H(x,x,y_n)} + \sqrt{H(x)} + \sqrt{H(y_n)}\\
&\le 2\,\frac{\delta(y_n)}{y_n} + \frac{\delta(x)}{x} \le 5\,\frac{\delta(x/2)}{x}.
\end{aligned}$$

Step B2. We will now estimate for $1 \le j \le n$ the quantity

$$\Delta(x,j) = \int \|g_j(0,x,\tau,\ell,\omega) - h_j(\tau,\ell,\omega)\|_{2,P}\, p(x,\ell,\tau)\,d\tau.$$

By conditioning we can write

$$\begin{aligned}
g_j(0,x,\tau,\ell,\omega) - h_j(\tau,\ell,\omega) \int \big[E^{Q^{\tau,\ell}_{\sigma_j,y_j}}[R(\sigma_j,\tau_\ell) - R(\sigma_{j-1},\tau_\ell)]\big]\\
\times [p^{\tau,\ell}(0,x,y_j,\sigma_j) - p(y_j,\ell,\tau-\sigma_j)]\,d\sigma_j
\end{aligned}$$

and

$$\begin{aligned}\|g_j(0,x,\tau,\ell,\omega) &- h_j(\tau,\ell,\omega)\|_{2,P}\\ &\le \int \left\| E^{Q^{\tau,\ell}_{\sigma_j,y_j}}[R(\sigma_j,\tau_\ell)-R(\sigma_{j-1},\tau_\ell)]\right\|_{2,P}\\ &\qquad \times |p^{\tau,\ell}(0,x,y_j,\sigma_j)-p(y_j,\ell,\tau-\sigma_j)|\,d\sigma_j\end{aligned}$$

$$\begin{aligned}\Delta(x,j) \le \iint &\left\| E^{Q^{\tau,\ell}_{\sigma_j,y_j}}[R(\sigma_j,\tau_\ell)-R(\sigma_{j-1},\tau_\ell)]\right\|_{2,P}\\ &\times |p^{\tau,\ell}(0,x,y_j,\sigma_j)-p(y_j,\ell,\tau-\sigma_j)|d\sigma_j\, p(x,\ell,\tau)d\tau\end{aligned}$$

Recall from (2.1) that

$$p^{\tau,\ell}(0,x,y_j,\sigma_j)=\frac{p(x,y_j,\sigma_j)p(y_j,\ell,\tau-\sigma_j)}{p(x,\ell,\tau)}.$$

If we define for $a>b$

$$F(a,b,\tau)=\left\| E^{Q^{\tau,\ell}_{0,a}}[R(0,\tau_\ell)-R(\tau_b,\tau_\ell)]\right\|_{2,P},$$

then by stationarity

$$\begin{aligned}&\Delta(x,j)\\ &\le \iint F(y_j,y_{j-1},\tau-\sigma_j)|p^{\tau,\ell}(0,x,y_j,\sigma_j)-p(y_j,\ell,\tau-\sigma_j)|p(x,\ell,\tau)d\sigma_j\,d\tau\\ &=\int F(y_j,y_{j-1},\tau-\sigma_j)|p(x,y_j,\sigma_j)-p(x,\ell,\tau)|\,p(y_j,\ell,\tau-\sigma_j)d\sigma_j\,d\tau\\ &=\int F(y_j,y_{j-1},\tau)|p(x,y_j,\sigma_j)-p(x,\ell,\tau+\sigma_j)|p(y_j,\ell,\tau)d\sigma_j\,d\tau.\end{aligned}$$

We can use estimates (2.3) and (2.4) to obtain

$$\int |p(x,y_j,\sigma_j)-p(x,\ell,\sigma_j+\tau)|d\sigma_j \le C\frac{(y_j^2+\tau)}{x^2}$$

and arrive at

$$\Delta(x,j)\le \frac{C}{x^2}\int F(y_j,y_{j-1},\tau)(y_j^2+\tau)p\ \ y_j,\ell,\tau)d\tau.$$

Moreover, in view of (3.11),

$$\begin{aligned}&\int |F(y_j,y_{j-1},\tau)|^2 p(y_j,\ell,\tau)d\tau\\ &\quad=\int E^P\left[\left|E^{Q^{\tau,\ell}_{0,y_j}}[R(0,\tau_\ell)-R(\tau_{y_{j-1}},\tau_\ell)]\right|^2\right]p(y_j,\ell,\tau)d\tau\\ &\quad\le E^P\int E^{Q^{\tau,\ell}_{0,y_j}}[[R(0,\tau_\ell)-R(\tau_{y_{j-1}},\tau_\ell)]^2]p(y_j,\ell,\tau)d\tau\\ &\quad=E^P E^{Q_{0,y_j}}[[R(0,\tau_\ell)-R(\tau_{y_{j-1}},\tau_\ell)]^2]=\end{aligned}$$

$$
\begin{aligned}
&= H(y_j, y_j, y_{j-1}) \\
&\le \frac{\delta(y_{j-1})}{y_{j-1}^2}
\end{aligned}
$$

and from (2.2)

$$\int (y_j^2 + \tau)^2 p(y_j, \ell, \tau) d\tau \le C y_j^4.$$

Applying now Schwartz's inequality,

$$\Delta(x, j) \le C \,\frac{y_j^2}{x^2} \frac{\delta(y_{j-1})}{y_{j-1}} \le 2C \,\frac{y_j \delta(y_{j-1})}{x^2},$$

we finally have

$$
\begin{aligned}
&\int \|g(0, x, \tau, \ell, \omega) - h(\tau, \ell, \omega)\|_{2,P}\, p(x, \ell, \tau) d\tau \\
&\le \frac{C}{x^2} \sum_{j=1}^{n} \delta(y_{j-1})\, y_j + \frac{\delta(x/2)}{x} \\
&= o\left(\frac{1}{x}\right).
\end{aligned}
$$

□

4 Law of Large Numbers

We will now provide a proof of the law of large numbers stated in Theorem 1.1. Our proof will rely on the main estimate of Theorem 3.4. Though, let us mention that the law of large numbers can be derived by much easier estimates, bypassing the much stronger one of Theorem 3.4. We prefer to give the proof by means of Theorem 3.4, because, on the one hand, it provides an expression for the limit $\bar{a}$, (1.10), and on the other hand, the representation that Theorem 3.4 provides will be crucial for the proof of the central limit theorem later on.

In view of the bound $|x a_x(t, x, \omega)| \le C$, it is sufficient to prove

$$\lim_{n\to\infty} a(0, n, \omega) = \int a(\tau, \ell, \omega) g(0, n, \tau, \ell, \omega) p(\ell, n, \tau) d\tau = \bar{a}$$

with probability 1. Since

$$\widehat{a}(0, n, \omega) = \int a(\tau, \ell, \omega) h(\tau, \ell, \omega) p(\ell, n, \tau) d\tau$$

from the estimate of Theorem 3.4, it is clear that

$$E^P[|a(0, n, \omega) - \widehat{a}(0, n, \omega)|^2] = o(n^{-2}),$$

and hence by an application of the Borel-Cantelli lemma,

$$|a(0, n, \omega) - \widehat{a}(0, n, \omega)| \to 0$$

with probability 1. It is therefore sufficient to prove

$$\widehat{a}(0,n,\omega) = \int a(\tau,\ell,\omega)p(n,\ell,\tau)d\tau \to \bar{a}$$

a.e. P. If $x(t)$ is an arbitrary stationary stochastic process with finite first moment, and we wish to prove almost sure convergence to $E[x(t)]$ of averages of the form

$$X_n = \int x(t)f(n,t)dt$$

the following conditions on $\{f(n,t)\}$ are sufficient.

(1) $f(n,t) \geq 0$ and $\int f(n,t)dt = 1$.
(2) For every $h > 0$, $\int |f(n,t) - f(n,t+h)|dt \to 0$ as $n \to \infty$.
(3)

$$\sup_n \int (1+|t|) \left|\frac{\partial f(n,t)}{\partial t}\right| dt \leq C.$$

The standard proof of the ergodic theorem proceeds by proving it for $x(t+h) - x(t)$ and then approximating $x(t)$, when $E[x(t)] = 0$ by linear combinations of these in L_1. Finally, it all comes down to the maximal lemma, [9, section 6.1]. Since we know that the maximal lemma is valid for the standard averages, the problem reduces to estimating, for $x(t) \geq 0$, $\sup_n \int x(t)f(n,t)dt$ in terms of $\sup_\tau z(\tau)$ where

$$z(\tau) = \frac{1}{\tau}\int_0^\tau x(t)\,dt$$

$$\begin{aligned}\int x(t)f(n,t)dt &= \int \frac{d}{dt}[tz(t)]f(n,t)dt \\ &= -\int tz(t)\frac{df(n,t)}{dt}\,dt \\ &\leq \sup_\tau z(\tau)\int (1+|t|)\left|\frac{\partial f(n,t)}{\partial t}\right| dt.\end{aligned}$$

We need a uniform bound

$$\int (1+|\tau|)\left|\frac{\partial p(n,\ell,\tau)}{\partial \tau}\right| d\tau \leq C,$$

which follows from (2.6) by rescaling.

5 Central Limit Theorem

Our goal is to prove a central limit theorem for the random variable

$$x\int [a(s,\ell,\omega)h(s,\ell,\omega) - \bar{a}]p(x,\ell,s)ds = x\int f(\theta_s\omega)p(x,\ell,s)ds$$

as $x \to \infty$ where $f(\omega) = [a(0, \ell, \omega)h(0, \ell, \omega) - \bar{a}]$. First we replace $p(x, \ell, s)$ by $p(x, 0, s)$, which has the scaling property $x^2 p(x, 0, x^2 s) = p(1, 0, s)$, and the difference

$$\Delta_x = x \int [p(x, \ell, s) - p(x, 0, s)] f(\theta_s \omega) ds$$

can be estimated easily with the help of (2.7):

$$\begin{aligned} \|\Delta_x\|_{2,P} &\le x \|f(\omega)\|_{2,P} \int |p(x, \ell, s) - p(x, 0, s)| ds \\ &\le \frac{C\ell^2}{x} \|f(\omega)\|_{2,P}. \end{aligned}$$

The next step is to reduce it to the standard type of central limit theorem. If

$$\xi(t) = \int_0^t f(\theta_s \omega) ds,$$

we have the identity

$$\begin{aligned} x \int_0^\infty f(\theta_s \omega) p(x, 0, s) ds &= x \int p(x, 0, s) d\xi(s) \\ &= -x \int_0^\infty \xi(s) p_s(x, 0, s) ds \\ &= -x^3 \int_0^\infty \xi(x^2 s) p_s(x, 0, x^2 s) ds \\ &= -\int_0^\infty x^{-1} \xi(x^2 s) p_s(1, 0, s) ds. \end{aligned}$$

If $x^{-1}\xi(x^2 t)$ converges to a Brownian motion with variance σ^2 and we have an estimate

$$E^P[|\xi(t)|^2] \le Ct,$$

then the tails

$$e_T(\omega) = \int_T^\infty x^{-1} \xi(x^2 s) p_s(1, 0, s) ds$$

can be controlled in $L_2(P)$ uniformly in x. It is now easy to derive from a central limit theorem for $x^{-1}\xi(x^2 t)$ similar theorems for $x \int f(\theta_s \omega) p(x, 0, s) ds$ and $x \int f(\theta_s \omega) p(x, \ell, s) ds$. The limiting variance will now be equal to

$$\sigma^2 \int_0^\infty |p(1, 0, s)|^2 ds.$$

We summarize all of this as follows:

THEOREM 5.1 *Let $f(\omega)$ be a function in $L_2(P)$ with mean 0 such that for some constant C,*

$$E^P\left[\left|\int_0^t f(\theta_s \omega) ds\right|^2\right] \le Ct, \tag{5.1}$$

and

$$x^{-1}\int_0^{x^2t} f(\theta_s\omega)ds$$

satisfies a central limit theorem with variance σ^2*. Then a central limit theorem holds for*

$$x\int_0^{\infty} f(\theta_s\omega)p(x,\ell,s)ds$$

with limiting variance

$$\sigma^2\int_0^{\infty} |p(1,0,s)|^2\,ds.$$

We now assume that that our source process has the property that $\{\lambda(s,\omega): s\le a\}$ and $\{\lambda(s,\omega): s\ge b\}$ are independent if $b-a\ge A$ for some $A<\infty$. We will prove the CLT under this condition. First we establish a general theorem.

Let $(\Omega,\mathcal{F},P)$ be a probability space and θ_t a one-parameter family of measure-preserving transformations. Assume that $\{\mathcal{F}_t^s; s\le t\}$ is a family of sub σ-fields (corresponding to events observable during time $[s,t]$) satisfying $\mathcal{F}_t^{s'}\subset\mathcal{F}_t^s$ if $t\ge s'\ge s$, $\mathcal{F}_{t'}^s\subset\mathcal{F}_t^s$ if $s\le t'\le t$ and $\theta_\tau\mathcal{F}_t^s=\mathcal{F}_{t+\tau}^{s+\tau}$. Let $\mathcal{F}_t=\vee_{s\le t}\mathcal{F}_t^s$ and $\mathcal{F}^s=\vee_{t\ge s}\mathcal{F}_t^s$. Finally, $\mathcal{F}=\vee_s\mathcal{F}^s=\vee_t\mathcal{F}_t$. Assume that for some finite A, $\mathcal{F}^t$ and $\mathcal{F}_s$ are independent under P provided $t>s+A$. It is not very hard to show that if $f(\omega)$ is measurable with respect to some $\mathcal{F}_b^a$ with $\int f(\omega)dP=0$ and $\int f^2(\omega)dP<\infty$, then

$$x^{-1}\xi(x^2t)=x^{-1}\int_0^{tx^2} f(\theta_s\omega)ds$$

satisfies a central limit theorem as $x\to\infty$ and converges to a Brownian motion with a limiting covariance variance $\sigma^2 s\wedge t$ where σ^2 is given by

$$\begin{aligned}\sigma^2&=\int_{-\infty}^{\infty}\left[\int f(\omega)f(\theta_t\omega)dP\right]dt\\&=\int_{-A-(b-a)}^{A+(b-a)}\left[\int f(\omega)f(\theta_t\omega)dP\right]dt\le 2(A+b-a)\int f^2(\omega)dP.\end{aligned}$$

If f is only $\mathcal{F}$ measurable, then it needs to be well approximated by functions measurable with respect to $\mathcal{F}_b^a$.

First we obtain an estimate on the variance of $\xi(t)=\int_0^t f(\theta_s\omega)ds$.

THEOREM 5.2 *Assume that* $f_k=E^P[f(\omega)|\mathcal{F}_k^{-k}]$ *satisfies* $E^P[|f-f_k|^2]\le Ck^{-\alpha}$ *for* $k\ge 1$ *with* $\alpha>0$. *Then*

$$\|\xi(t)\|_{2,P}\le\begin{cases}Ct^{1-\frac{\alpha}{2}} & \text{if } 0<\alpha<1,\\ C\sqrt{t}\log t & \text{if } \alpha=1,\\ C\sqrt{t} & \text{if } \alpha>1.\end{cases}$$

PROOF: Let us write

$$f = f_1 + \sum_{j=1}^{\infty}[f_{2^j} - f_{2^{j-1}}]$$

with the corresponding integrals

$$\xi(t) = \sum_{j=0}^{\infty} \eta_j(t)$$

with

$$\eta_j(t) = \int_0^t [f_{2^j}(\theta_s\omega) - f_{2^{j-1}}(\theta_s\omega)]ds \tag{5.2}$$

for $j \geq 1$, and

$$\eta_0(t) = \int_0^t f_1(\theta_s\omega)ds.$$

Then

$$\|\xi(t)\|_{2,P} \leq \sum_{j=0}^{\infty} \|\eta_j(t)\|_{2,P}.$$

Since the correlations $E^P[f_{2^j}(s)f_{2^j}(s+t)]$ vanish if $|t| \geq A + 2^j$, it is easy to obtain the bound

$$\begin{aligned}\|\eta_j(t)\|_{2,P}^2 &\leq Ct\min\{t, A+2^j\}\|f_{2^j} - f_{2^{j-1}}\|_{2,P}^2\\ &\leq Ct\min\{t, A+2^j\}2^{-j\alpha}\\ &\leq Ct\min\{t, 2^j\}2^{-j\alpha}\end{aligned}$$

where C is now a constant that depends on A. An easy estimation of the sum

$$\sum_j [\min\{t, 2^j\}2^{-j\alpha}]^{\frac{1}{2}}$$

proves the the thorem. □

We note that if $\alpha > 1$, we have the estimate

$$\min\{t, 2^j\}2^{-j\alpha} \leq C2^{-j(\alpha-1)}$$

and can use it to prove the central limit theorem for $\xi(t)$.

THEOREM 5.3 *Let $f \in L_2(P)$. Assume that $f_k(\omega) = E^P[f(\omega)|\mathcal{F}_k^{-k}]$ satisfies*

$$E^P[|f - f_k|^2] \leq Ck^{-\alpha}$$

for some $\alpha > 1$. Then the central limit theorem is valid for

$$x^{-1}\int_0^{tx^2} f(\theta_s\omega)ds$$

with the limiting Brownian motion having covariance $\sigma^2 s \wedge t$*, where*

$$\sigma^2 = \lim_{k\to\infty} \int_{-\infty}^{\infty} E^P[f_{2^k}(\omega) f_{2^k}(\theta_s\omega)]ds = \int_{-\infty}^{\infty} E^P[f(\omega)f(\theta_s\omega)]ds.$$

PROOF: If we write

$$f = f_1 + \sum_{j\geq 1}[f_{2^j} - f_{2^{j-1}}],$$

since each finite sum satisfies the central limit theorem ([3]), it is clearly sufficient to show that $\eta_j(t)$ defined in equation (5.2) satisfies

$$\|\eta_j(t)\|_{2,P} \leq \delta_j \sqrt{t}$$

for some δ_j satisfying $\sum_j \delta_j < \infty$. Clearly $\delta_j = 2^{-j\frac{1-\alpha}{2}}$ works. □

Remark 5.4. Because of Theorem 5.1 the CLT holds for

$$x \int [a(s,\ell,\omega)h(s,\ell,\omega) - \bar{a}]p(x,\ell,s)ds$$

provided we get an estimate of the type used in Lemma 5.1 with $\alpha > 1$ for $f(\omega) = a(0,\ell,\omega)h(0,\ell,\omega)$.

Since a and h are bounded functions of ω, it is sufficient to prove the estimate separately on a and h.

THEOREM 5.5 *Let* $f = a(0,\ell,\omega)$ *and* $f_k = E^P[f \mid \mathcal{F}_k^{-k}]$*. Then*

$$E^P[|f - f_k|^2] \leq Ck^{-4}.$$

PROOF: Since $u(t,\ell) = 6/(\ell + a(t,\ell))^2$ and we have uniform bounds on u, $a(t,\ell)$ depends in a Lipschitz manner on $u(t,\ell)$ with a Lipschitz constant that depends on ℓ. For a fixed ℓ, we can therefore take $f = u(0,\ell,\omega)$ instead of $a(0,\ell,\omega)$. According to Theorem A.4, the total variation of $u(0,\ell,\omega)$ if we vary $\lambda(\cdot)$ arbitrarily outside $[-k,k]$ is dominated by $C\ell k^{-2}$. The theorem is now an easy consequence. □

A corresponding estimate for $f(\omega) = h(0,\ell,\omega)$ is more complicated. We do it in several steps, with each step formulated as a lemma.

Starting from the measure P on the space Ω of functions $\lambda(t)$ on $(-\infty,\infty)$, we define a measure P^k on the subset of $\Omega\times\Omega$ consisting of functions $(\omega_1,\omega_2) = (\lambda_1(t),\lambda_2(t))$ such that $\lambda_1(t) = \lambda_2(t)$ if $|t| \leq k$. The measure P^k is uniquely defined by the property that the distribution of the process $\lambda(t) = \lambda_1(t) = \lambda_2(t)$ in the interval $[-k,k]$ is the same as under P and for $|t| \geq k$, $\lambda_1(t)$ and $\lambda_2(t)$ are conditionally independent given $\{\lambda(t) : |t| \leq k\}$, each component having the same conditional distribution as that of $\{\lambda(t) : |t| \geq k\}$ given $\{\lambda(t) : |t| \leq k\}$ under P. This guarantees that under P^k the marginal distributions of both $\lambda_1(\cdot)$ and $\lambda_2(\cdot)$ are P, the two components are almost surely identical for $|t| \leq k$ and are conditionally independent given their common values on $[k,k]$.

Then it is easy to see that

$$E^P\big[\big|f(\omega)-E[f\mid\mathcal{F}_k^{-k}]\big|^2\big]=\frac{1}{2}E^{P^k}[|f(\omega^1)-f(\omega^2)|^2].$$

LEMMA 5.6 *Let $c(t,x,\omega)$ be as in equation* (1.6) *and $f(\omega)=h(0,\ell,\omega)$. Then for any $\beta>0$,*

$$\begin{aligned}&E^{P^k}[|f(\omega^1)-f(\omega^2)|^2]\\&\quad\le CE^{P^k}E^{Q_*^{0,\ell}}\left[\left|\int_{-\infty}^0[c(t,x(t),\omega^1)-c(t,x(t),\omega^2)]dt\right|^2\right]\\&\quad\le C_\beta E^{P^k}E^{Q_*^{0,\ell}}\left[\int_{-\infty}^0(1+|t|)^{1+\beta}|c(t,x(t),\omega^1)-c(t,x(t),\omega^2)|^2dt\right].\end{aligned}$$

PROOF: We start with the formula

$$f(\omega^i)=E^{Q_*^{0,\ell}}\left[\exp\left[-\int_{-\infty}^0c(t,x(t),\omega^i)dt\right]\right]$$

and use the inequality

$$|e^x-e^y|\le e^{\max\{x,y\}}|x-y|.$$

From (3.10) we have the bounds

$$\sup_\omega E^{Q_*^{0,\ell}}\left[\exp\left[2\int_{-\infty}^0|c(t,x(t),\omega^i)|dt\right]\right]\le C.$$

Hence we obtain

$$\begin{aligned}&|f(\omega^1)-f(\omega^2)|^2\\&\quad\le CE^{Q_*^{0,\ell}}\left[\left|\int_{-\infty}^0[c(t,x(t),\omega^1)-c(t,x(t),\omega^2)]dt\right|^2\right]\\&\quad\le C_\beta E^{Q_*^{0,\ell}}\left[\int_{-\infty}^0(1+|t|)^{1+\beta}|c(t,x(t),\omega^1)-c(t,x(t),\omega^2)|^2dt\right].\end{aligned}$$

□

We use the estimates on the difference of two solutions provided in Theorem A.5 to continue with our estimation. We get a preliminary estimate that is not good enough, but will bootstrap from it to a better one.

LEMMA 5.7 *For any $\beta>0$, there exists a constant C_β such that*

$$E^{P^k}[|f(\omega^1)-f(\omega^2)|^2]\le C_\beta k^{-1+\beta}.$$

PROOF: On the difference $|c(t,x,\omega^1)-c(t,x,\omega^2)|$ we have two different estimates. For $|t|\le k/2$, from Theorem A.5

$$|c(t,x,\omega^1)-c(t,x,\omega^2)|\le\frac{C}{kx},$$

and, from the bounds (A.7) on $c(t,x,\omega)$,

$$|c(t,x,\omega^1) - c(t,x,\omega^2)| \le \frac{C}{x^3}.$$

We now use the estimates from Theorem 2.1 and the fact that $x \ge \ell$ to conclude that, for $|t| \le k/2$

$$E^{Q_*^{0,\ell}}\left[|c(t,x(t),\omega^1) - c(t,x(t),\omega^2)|^2\right] \le \frac{C}{k^2(1+|t|)} \tag{5.3}$$

and for all t,

$$E^{Q_*^{0,\ell}}\left[|c(t,x(t),\omega^1) - c(t,x(t),\omega^2)|^2\right] \le \frac{C}{(1+|t|)^3}. \tag{5.4}$$

If we now use the estimates (5.3) for $t \le k$ and (5.4) for $t \ge k$, it is easy to see that

$$E^{Q_*^{0,\ell}}\left[\int_{-\infty}^0 (1+|t|)^{1+\beta} \left|c(t,x(t),\omega^1) - c(t,x(t),\omega^2)\right|^2 dt\right] \le C_\beta k^{-1+\beta}.$$

□

LEMMA 5.8 *Given any $\gamma < 2$, there is a constant C_γ such that*

$$E^P[|a(0,x,\omega) - \bar{a}|^2] \le C_\gamma x^{-\gamma}.$$

PROOF: Applying the second part of Theorem 5.2 to

$$f(\omega) = a(0,\ell,\omega)h(0,\ell,\omega),$$

we see that the lemma is valid with $\gamma = 2(1-\beta)$. □

LEMMA 5.9 *There exists $\alpha > 1$ and a constant C such that*

$$E^{P^k}[|f(\omega^1) - f(\omega^2)|^2] \le Ck^{-\alpha}.$$

PROOF: Now we go back and improve on the basic estimate

$$\sup_{t,\omega} |c(t,x,\omega)| \le \frac{C}{x^3}$$

in the mean by getting a better estimate in the mean on $a_x(t,x)$. From

$$E^P[|a(t,x,\omega) - \bar{a}|^2] \le C_\beta x^{-2(1-\beta)}$$

one sees, using Theorem A.1, that

$$E^P[|a_x(t,x,\omega)|^2] \le C_\beta x^{-4+2\beta},$$

which in turn improves the estimate on $c(t,x,\omega)$ to

$$E^P[|c(t,x,\omega)|^2] \le x^{-8+2\beta}.$$

Now we return to the proof of Lemma 5.6 and use (5.3) for $|t| \le k^\delta$ where $\delta < 1$. The new estimate is

$$E^{P^k}[|f(\omega^1) - f(\omega^2)|^2] \le C_\beta k^{\delta(1+\beta)-2} + C_\beta k^{-\delta(4-2\beta)}.$$

One can pick $\delta < 1$ and $\beta > 0$ such that this works out to

$$E^{P^k}[|f(\omega^1) - f(\omega^2)|^2] \le Ck^{-\alpha}$$

for some $\alpha > 1$. □

This completes the proof of the central limit theorem. We finally establish the positivity of the variance in a special class of examples.

Let $y(t)$ be a stationary Gaussian process with mean 0 and covariance $\rho(t-s) = E^P[y(s)y(t)]$. We assume that $\rho(\cdot) \ge 0$, has compact support and therefore satisfies our mixing condition. Let $\phi(y)$ be a smooth function on $\mathbb{R}$ satisfying $\lambda_1 \le \phi(y) \le \lambda_2$. Assume further that $\phi(y)$ is monotone and $\phi'(y) > 0$. Use the function ϕ to construct the random source as $\lambda(t) = \phi(y(t))$. Then the limiting variance

$$\sigma^2 = \lim_{x\to\infty} x^2 E^P[|a(0,x,\omega) - \bar{a}|^2] > 0.$$

The proof proceeds by a perturbation argument. Suppose for each $x > 0$, $f(x,t) \ge 0$ is a function with compact support in t. We denote by P^x the Gaussian process with mean

$$m(x,t) = \int \rho(t-s) f(x,s)\,ds$$

and covariance $\rho(t-s)$. Then the Radon-Nikodym derivative $R(x,\omega)$ of P^x with respect to P is given by

$$\frac{dP^x}{dP} = \exp\left[\int f(x,t)y(t)\,dt - \frac{1}{2}\iint \rho(t-s) f(x,s) f(x,t)\,ds\,dt\right]$$

and

$$\int [R(x,\omega)]^2 dP = e^{H(x)}$$

where

$$H(x) = \iint \rho(t-s) f(x,s) f(x,t)\,ds\,dt.$$

The resulting source process can also be represented as

$$\lambda^x(t) = \phi(y(t) + m(x,t))$$

where $y(\cdot)$ has P for its distribution. Because

$$E^{P^x}[x|a(0,x,\omega) - \bar{a}|] \le e^{\frac{1}{2}H(x)}\big[E^P[x^2|a(0,x,\omega) - \bar{a}|^2]\big]^{\frac{1}{2}}$$

if $\sigma^2 = 0$ and $f(x,t)$ is chosen so that $H(x)$ remains bounded as $x \to \infty$, then

$$E^{P^x}[x|a(x,0,\omega) - \bar{a}|] \to 0.$$

On the other hand, we can solve

$$u_t + u_{xx} - u^2 = 0$$

with different Neumann boundary data corresponding to $\lambda(t) = \phi(y(t))$ and $\lambda^x(t) = \phi(y(t) + m(x,t))$ and get two solutions u_1 and u_2. Since $m(x,t) \geq 0$ and $\phi(\cdot)$ is monotone, it follows that $u_2 \geq u_1$ and the difference v will satisfy

$$v_t + v_{xx} - (u_1 + u_2)v = 0, \quad v_x(t,0) = -\frac{1}{2}[\phi(y(t) + m(x,t)) - \phi(y(t))].$$

From the relation

$$u(t,x) = \frac{6}{(x + a(t,x))^2}$$

and the lower bounds

$$u(t,x) \geq \frac{6}{(x + a_1)^2},$$

it is not hard to obtain an estimate of the form (see (A.13))

$$a_1(t,x) - a_2(t,x) \geq c_2 x^3 (u_2(x) - u_1(x)) = c_2 x^3 v(t,x).$$

If $\sigma^2 = 0$, both $E^{P^x}[x|a(x,0,\omega) - \bar{a}|]$ and $E^P[x|a(x,0,\omega) - \bar{a}|] \to 0$, and this implies

$$E^P[x^4 v(0,x,\omega)] \leq C E^P[x[a_1(0,x,\omega) - a_2(0,x,\omega)]] \to 0$$

as $x \to \infty$, and we will show that does not happen for a suitable choice of $f(t,x)$.

We have an upper bound of $g = 12/(x + a_2)^2$ on $u_1 + u_2$. This will provide a lower bound on v,

$$v(0,x,\omega) \geq \int [\phi(y(t) + m(x,t)) - \phi(y(t))] q_g(0,x,t)\,dt$$

where q_g is as in the proof of Theorem A.4. We pick

$$f(x,t) = \frac{1}{x} f\left(\frac{t}{x^2}\right)$$

where $f > 0$ is compactly supported and equal to 1 on $[-T,T]$. It is easy to verify that

$$\sup_{x \geq 1} H(x) = \sup_{x \geq 1} x^{-2} \iint \rho(t-s) f\left(\frac{t}{x^2}\right) f\left(\frac{s}{x^2}\right) dt\,ds < \infty.$$

On the other hand, we have

$$\begin{aligned} &E^P[x^4 v(0,x,\omega)] \\ &\quad \geq x^4 \int \left[E^P\left[\phi\left(y(t) + \frac{1}{x} \int \rho(t-s) f\left(\frac{s}{x^2}\right) ds \right) - \phi(y(t)) \right] \right] q_g(0,x,t)\,dt \\ &\quad \simeq x^3 \int E^P[\phi'(y(t))] q_g(0,x,t)\,dt \int \rho(t-s) f\left(\frac{s}{x^2}\right) ds. \end{aligned}$$

From (A.11) it follows that for large x,

$$\int q_g(0,x,t)\,dt \geq \frac{C_1}{x^3},$$

$$\int t^2 q_g(0,x,t)\,dt \leq C_2 x.$$

This implies that if T is chosen large enough, then for all $x \geq 1$,

$$\int_0^{x^2T} q_g(0,x,t)dt \geq \frac{C_1}{x^3} - \frac{C_2 x}{x^4T^2} \geq \frac{C_3}{x^3}.$$

By stationarity, $E^P[\phi'(y(t))]$ is a positive constant and

$$\inf_{x\geq 1} \inf_{0\leq t\leq x^2T} \int \rho(t-s) f\left(\frac{s}{x^2}\right) ds \geq \int_0^{x^2T} \rho(s)ds > 0,$$

establishing the required contradiction.

Appendix

Since our goal is to control the Feynman-Kac term, we will need some estimates on $c(t,x,\omega)$. Some of them will be uniform in ω and some will be only in the mean. In addition, we will need some estimates on the transition probabilities of the Bessel process. We will collect all of these estimates in this section.

We will need the following estimate from the theory of parabolic partial differential equation:

THEOREM A.1 *For any t_0 and $x_0 > 0$, let $D^r_{t_0,x_0} = \{(t,x) : |x-x_0| \leq r,\ t_0 \leq t \leq t_0 + r^2\}$. If $v(t,x)$ is a solution of an equation of the form*

$$v_t + v_{xx} + b(t,x)v_x + q(t,x)v = 0 \tag{A.1}$$

in $G \subset (-\infty,\infty)\times[0,\infty)$, and $D^1_{t_0,x_0} \subset D^2_{t_0,x_0} \subset G$. Assume that b and q are bounded by a constant B. Then

$$\sup_{(s,y)\in D^1_{t_0,x_0}} |v_x(s,y)| \leq C\|v(t,x)\|_{L_2[D^2_{t_0,x_0}]} \leq 4C \sup_{(s,y)\in D^2_{t_0,x_0}} |v(s,y)|$$

where C depends only on B. In particular, if $v(t,x)$ is a solution of an equation of the form (A.1), *with $b(t,x)$ and $q(t,x)$ satisfying*

$$\sup_{t,x} x|b(t,x)| \leq C,$$

$$\sup_{t,x} x^2|q(t,x)| \leq C,$$

$$\sup_{t,x} x^\alpha|v(t,x)| \leq C,$$

then

$$\sup_{t,x} x^{\alpha+1}|v_x(t,x)| \leq C.$$

PROOF: A proof of the first half can be found in [4]. To see the second half, let v be a solution of (A.1) in $(-\infty,\infty)\times[0,\infty)$. If we define $v^k(t,x) = \frac{1}{k}v(k^2t,kx)$, then $v^k(t,x)$ satisfies

$$v^k_t(t,x) + v^k_{xx} + b^k(t,x)v^k_x(t,x) + q^k(t,x)v^k(t,x) = 0$$

where $b^k(t,x) = kb(k^2t,kx)$ and $q^k(t,x) = k^2q(k^2t,kx)$. In particular, if we assume that $x|b(t,x)| \le C$ and $x^2|q(t,x)| \le C$ for $x \ge 4$, then $b^k(t,x)$ and $q^k(t,x)$ are uniformly bounded on $[1,\infty) \times (-\infty,\infty)$ for $k \ge 1$ and $D^r_{t,x} \subset (-\infty,\infty) \times (4,\infty)$ for $r = 1, 2$. We can now get a bound on v_x:

$$|v_x(k^2t,kx)| \le \frac{C}{k}\|v(k^2s,ky)\|_{2,D^2_{t,x}} \le \frac{C}{k} \sup_{(s,y)\in D^2_{t,x}} |v(k^2s,ky)|.$$

□

THEOREM A.2 *Consider a smooth solution u of*

$$u_t + u_{xx} - u^2 = 0, \quad x > 0, \ -\infty < t < \infty, \tag{A.2}$$

with $u_x(t,0) = -\frac{1}{2}\lambda(t)$. Let $a(t,x)$ be the corresponding solution

$$a(t,x) = \sqrt{\frac{6}{u(t,x)}} - x \tag{A.3}$$

of

$$a_t + a_{xx} - \frac{6a_x}{(x+a)} - \frac{3a_x^2}{(x+a)} = 0 \tag{A.4}$$

on $G = [-\infty,\infty) \times (0,\infty)$. The following bounds are valid for all t, x, and ω:

$$0 < a_2 \le a(t,x) \le a_1 < \infty \tag{A.5}$$

and

$$\sup_{t,x,\omega} |xa_x(t,x,\omega)| \le C < \infty. \tag{A.6}$$

In particular, the function $c(t,x,\omega)$ defined in (1.6) *satisfies the uniform bound*

$$\sup_{t,\omega} |c(t,x,\omega)| \le \frac{C}{x^3}. \tag{A.7}$$

PROOF: We saw already that (A.5) is valid by the maximum principle. We will establish (A.6). We start with equation (A.2) and differentiate it with respect to x. Then for the derivative $v = u_x$, the equation satisfies

$$v_t + v_{xx} - 2uv = 0, \quad |v(t,0)| = \frac{1}{2}\lambda(t) \le \frac{\lambda_2}{2}.$$

We also have an estimate on u of the form

$$u(t,x) \ge \frac{6}{(x+a_2)^2},$$

and $w = 1/(x+a_2)^3$ solves the equation

$$w_t + w_{xx} - \frac{12}{(x+a_2)^2}w = 0, \quad w(t,0) = \frac{1}{a_2^3}.$$

Standard comparison using the maximum principle gives the bound.

To derive (A.6) we rewrite the equation for a in the form

$$a_t + a_{xx} - b(t,x)a_x = 0 \tag{A.8}$$

where

$$b(t,x) = \frac{6 + 3a_x(t,x)}{(x+a)}$$

satisfies $|xb(t,x)| \le C$. It follows from Theorem A.1 that

$$\sup_{t,x,\omega} |xa_x(t,x,\omega)| \le C,$$

which proves (A.6). (A.7) is now an easy consequence of the formula (1.6) for $c(t,x,\omega)$. □

We now improve slightly on this estimate but in the mean with respect to ω.

THEOREM A.3 *There is a finite constant C such that*

$$E^P\left[\int_0^\infty x|a_x(0,x,\omega)|^2\,dx\right] \le C < \infty. \tag{A.9}$$

PROOF: Multiply the equation (A.4) for a by xa and take expectations:

$$E^P[xaa_{xx}] = E^P\left[\frac{6xaa_x}{(x+a)}\right] + E^P\left[\frac{3xaa_x^2}{(x+a)}\right].$$

Integrate with respect to x from y_1 to y_2 and integrate by parts the term on the left:

$$E^P[xaa_x]|_{y_1}^{y_2} - \int_{y_1}^{y_2} E^P[xa_x^2]dx - \int_{y_1}^{y_2} E^P[aa_x]dx =$$

$$\int_{y_1}^{y_2} E^P\left[\frac{6xaa_x}{(x+a)}\right]dx + \int_{y_1}^{y_2} E^P\left[\frac{3xaa_x^2}{(x+a)}\right]dx.$$

Let us split the term

$$\frac{6xaa_x}{(x+a)} = 6aa_x - \frac{6a^2a_x}{(x+a)}$$

and note that

$$\int_{y_1}^{y_2} aa_x\,dx = \frac{1}{2}[a^2(y_2) - a^2(y_1)].$$

All the terms except $\int_{y_1}^{y_2} E^P[xa_x^2]dx$ are seen to remain bounded as $y_2 \to \infty$. We finally get

$$\int_0^\infty E^P[xa_x^2]dx \le C.$$

□

We now begin estimating the difference between two solutions of (A.2).

THEOREM A.4 *Let $v(t,x) = u^1(t,x) - u^2(t,x)$ be the difference of two solutions corresponding to two different initial values $\lambda^1(\cdot)$ and $\lambda^2(\cdot)$ that agree on $\{t : |t| \le k\}$. Then, for $x > 0$ and $|t| \le \frac{k}{2}$, v satisfies*

$$|v(t,x)| \le \frac{Cx}{k^2}, \qquad |v(t,x)| \le \frac{C}{kx}, \qquad |v_x(t,x)| \le \frac{C}{kx^2}.$$

PROOF: Let $v = u^1 - u^2$ be the difference of two solutions with boundary data $-\frac{1}{2}\lambda^1(\cdot)$ and $-\frac{1}{2}\lambda^2(\cdot)$ that agree on $\{t : |t| \le k\}$. Then with $\sigma(t) = \lambda_1(t) - \lambda_2(t)$

$$v_t + v_{xx} - (u^1 + u^2)v = 0 : v_x(t,0) = -\frac{1}{2}\sigma(t).$$

Any bounded solution $w(t,x)$ of an equation of the form

$$w_t + w_{xx} - gw = 0 : w_x(t,0) = -\frac{1}{2}\sigma(t)$$

with a positive g has, by the maximum principle, a representation

$$w(t,x) = \int_t^\infty \sigma(\tau) q_g(t,x,\tau) d\tau \tag{A.10}$$

with a nonnegative q_g. A formula for q_g can be written down in terms of a Feynman-Kac formula

$$q_g(t,x,\tau) = \frac{1}{2\sqrt{\pi(\tau - t)}} e^{-\frac{x^2}{4(\tau-t)}} E_{x,t,\tau}\left[\exp\left[-\int_t^\tau g(s,x(s))ds\right]\right]$$

where $E_{x,t,\tau}$ is expectation with respect to Brownian motion (with variance 2) starting from $x > 0$ at time t and conditioned to exit $(0,\infty)$ at time τ. In particular, our solution v is given by

$$v(t,x) = \int_t^\infty \sigma(\tau) q_{u_1+u_2}(t,x,\tau) d\tau.$$

Since we have a lower bound of the form $u_i \ge \frac{6}{(x+a_1)^2}$, we can bound $q_{u_1+u_2}$ by q_g, corresponding to $g = \frac{12}{(x+a_1)^2}$. We estimate q_g by exhibiting special solutions of the form

$$w(t,x) = \frac{a(t)}{(x+a_1)^3} + \frac{b(t)}{(x+a_1)} + c(t)(x+a_1). \tag{A.11}$$

w will satisfy

$$w_t + w_{xx} - \frac{12}{(x+a_1)^2} w = 0$$

provided $a'(t) = 10b(t)$, $b'(t) = 12c(t)$, and $c'(t) = 0$. This yields the estimates

$$\int_t^\infty (\tau - t)^2 q_{u_1+u_2}(t,x,\tau) d\tau \le Cx,$$

$$\int_t^\infty q_{u_1+u_2}(t,x,\tau) d\tau \le Cx^{-3},$$

for $x \geq \ell$. By interpolation we also have

$$\int_t^\infty |\tau - t| q_{u_1+u_2}(t, x, \tau) d\tau \leq C x^{-1}.$$

This takes care of the estimates. □

We now translate these into estimates on $a^1(t,x) - a^2(t,x)$.

Let $\lambda^i(\cdot)$ be two different initial conditions at $x = 0$ that agree in $|t| \leq k$. Let $u^i(t,x)$ be the corresponding solutions

$$u_t^i(t,x) + u_{xx}^i(t,x) - u^i(t,x)^2 = 0, \quad u^i(t,0) = \lambda^i(t),$$

with

$$a^i(t,x) = \sqrt{\frac{6}{u^i(t,x)}} - x$$

and

$$c^i(t,x) = \frac{3 a_x^i(t,x)^2}{a^i(t,x)(x + a^i(t,x))} - \frac{6 a_x^i(t,x)}{x(x + a^i(t,x))}. \tag{A.12}$$

We have the following estimates:

THEOREM A.5 *For $x \geq \ell$ and $|t| \leq \frac{k}{2}$*

$$|a^1(t,x) - a^2(t,x)| \leq \frac{Cx^2}{k},$$
$$|a_x^1(t,x) - a_x^2(t,x)| \leq \frac{Cx}{k},$$
$$|c^1(t,x) - c^2(t,x)| \leq \frac{C}{kx}.$$

PROOF: Since

$$a^1(t,x) - a^2(t,x) = \frac{\sqrt{6}(u^2(t,x) - u^1(t,x))}{\sqrt{u^1(t,x)}\sqrt{u^2(t,x)}(\sqrt{u^1(t,x)} + \sqrt{u^2(t,x)})} \tag{A.13}$$

and from Theorem A.4, for $v = u_1 - u_2$ we have the estimates

$$|v(t,x)| \leq \frac{C}{xk},$$

and from equation (A.13)

$$|a^1(t,x) - a^2(t,x)| \leq Cx^3 |u^1(t,x) - u^2(t,x)|,$$

which proves the first estimate. From Theorem A.1 it follows that

$$|u_x^1(t,x) - u_x^2(t,x)| \leq \frac{C}{x^2 k},$$

and this in turn implies the second estimate. The last estimate follows from equation (A.12) and the first two estimates of the theorem. □

Bibliography

[1] Bricmont, J.; Kupiainen, A. Random walks in asymmetric random environments. *Comm. Math. Phys.* **142** (1991), no. 2, 345–420.

[2] Hairer, M.; Mattingly, J. C. Ergodicity of the 2D Navier-Stokes equations with degenerate stochastic forcing. *Ann. of Math. (2)* **164** (2006), no. 3, 993–1032.

[3] Hall, P.; Heyde, C. C. *Martingale limit theory and its application*. Probability and Mathematical Statistics. Academic Press [Harcourt Brace Jovanovich], New York–London, 1980.

[4] Lieberman, G. M. *Second order parabolic differential equations*. World Scientific, River Edge, N.J., 1996.

[5] Papanicolaou, G. C. Diffusions and random walks in random media. *The mathematics and physics of disordered media (Minneapolis, Minn., 1983)*, 391–399. Lecture Notes in Mathematics, 1035. Springer, Berlin, 1983.

[6] Papanicolaou, G. C., Varadhan, S. R. S. Diffusions with random coefficients. *Statistics and probability: essays in honor of C. R. Rao*, 547–552. North-Holland, Amsterdam, 1982.

[7] Sznitman, A.-S. Propagation of chaos for a system of annihilating Brownian spheres. *Comm. Pure Appl. Math.* **40** (1987), no. 6, 663–690.

[8] Sznitman, A.-S.; Zeitouni, O. An invariance principle for isotropic diffusions in random environment. *Invent. Math.* **164** (2006), no. 3, 455–567.

[9] Varadhan, S. R. S. *Probability theory*. Courant Lecture Notes in Mathematics, 7. New York University, Courant Institute of Mathematical Sciences, New York; American Mathematical Society, Providence, R.I., 2001.

S. R. S. VARADHAN
Courant Institute
251 Mercer Street
New York, NY 10012-1185
E-mail: varadhan@cims.nyu.edu

NIKOS ZYGOURAS
University of Southern California
Department of Mathematics
3620 Vermont Avenue, KAP 108
Los Angeles, CA 90089-2532
E-mail: zygouras@usc.edu

Received January 2007.

Volume 4

Particle Systems and their Large Deviations

Volume 3

Large Deviations

Random Media

Review Articles

Volume 2

PDE, SDE, Diffusions

Contents

Volume 1

Limit Theorems

Published by:

Hindustan Book Agency (India)
P 19 Green Park Extension
New Delhi 110 016
India

email: info@hindbook.com
www.hindbook.com

ISBN 978-93-80250-39-7 (4 volume set)

Printed and bound by Gopsons Papers Ltd., Noida

Collected Papers of S. R. S. Varadhan

Volume 2
PDE, SDE, Diffusions, Random Media

Editors

Rajendra Bhatia
Abhay G. Bhatt
K. R. Parthasarathy

HINDUSTAN BOOK AGENCY

Collected Papers of S. R. S. Varadhan

Volume 1: Limit Theorems,
Review Articles

Volume 2: PDE, SDE, Diffusions,
Random Media

Volume 3: Large Deviations

Volume 4: Particle Systems and their Large Deviations

Collected Papers of
S. R. S. Varadhan

Volume 2
PDE, SDE, Diffusions, Random Media

Collected Papers of
S. R. S. Varadhan

Volume 4
Particle Systems and their Large Deviations

Collected Papers of S. R. S. Varadhan

Volume 1: Limit Theorems,
Review Articles

Volume 2: PDE, SDE, Diffusions,
Random Media

Volume 3: Large Deviations

Volume 4: Particle Systems and their Large Deviations

Collected Papers of
S. R. S. Varadhan

Volume 4
Particle Systems and their Large Deviations

Editors

Rajendra Bhatia
Abhay G. Bhatt
K. R. Parthasarathy

HINDUSTAN
BOOK AGENCY

Published by:

Hindustan Book Agency (India)
P 19 Green Park Extension
New Delhi 110 016
India

email: info@hindbook.com
www.hindbook.com

ISBN 978-93-80250-39-7 (4 volume set)

Printed and bound by Gopsons Papers Ltd., Noida

Contents

Volume 1

Limit Theorems

Review Articles

Volume 2

PDE, SDE, Diffusions

Random Media

Volume 3

Large Deviations

Volume 4

Particle Systems and their Large Deviations

Particle Systems and their Large Deviations

Commun. Math. Phys. 118, 31–59 (1988)

Communications in Mathematical Physics

Nonlinear Diffusion Limit for a System with Nearest Neighbor Interactions

M. Z. Guo[1], G. C. Papanicolaou[2] and S. R. S. Varadhan[2]*

[1] Department of Mathematics, Beijing University, Beijing, People's Republic of China

[2] Courant Institute, New York University, 251 Mercer Street, New York, NY 10012, USA

Abstract. We consider a system of interacting diffusions. The variables are to be thought of as charges at sites indexed by a periodic one-dimensional lattice. The diffusion preserves the total charge and the interaction is of nearest neighbor type. With the appropriate scaling of lattice spacing and time, a nonlinear diffusion equation is derived for the time evolution of the macroscopic charge density.

1. Notation and Summary

We will study the problem of passage to hydrodynamic limit under diffusion type scaling for a system of charges that are located at various sites of a periodic one-dimensional lattice. The charges migrate between adjacent sites randomly according to a well defined diffusion law. The algebraic sum of the charges is always conserved. The charges themselves are of indeterminate sign. Under diffusion type scaling of lattice width and time, a deterministic limit is obtained for the macroscopic charge density and the limit is characterized as the solution of a nonlinear heat equation. We will now develop the notation and end this section with a precise statement of the results as well as a sketch of the proof.

S is the unit circle represented as the interval $0 \leqq \theta \leqq 1$ with end points identified. The scaling parameter is N and the scaled lattice consists of sites j/N in S for $j = 1, 2, \ldots, N$. The sites adjacent to j/N are $j \pm 1/N$ with addition being modulo N. The charge at site j/N is represented by the variable x_j. The charge configuration $(x_1, \ldots, x_N)$ is represented as a vector $\vec{x}$ in R^N. The configuration $\vec{x}$ changes with time and as a function of time $\vec{x}(t)$ undergoes a diffusion in R^N. The infinitesimal generator of the diffusion is given by

$$L_N = \frac{N^2}{2} \Sigma \left(\frac{\partial}{\partial x_i} - \frac{\partial}{\partial x_{i+1}} \right)^2 - \frac{N^2}{2} \Sigma (\phi'(x_i) - \phi'(x_{i+1})) \left(\frac{\partial}{\partial x_i} - \frac{\partial}{\partial x_{i+1}} \right). \tag{1.1}$$

* Work supported by the National Science Foundation under grants no. DMS 8600233 and DMS 8701895

Here ϕ is a continuously differentiable function from $R \to R$ satisfying the properties to be listed below and $\phi'(x) = d\phi(x)/dx$. The factor N^2 in (1.1) with lattice spacing of $1/N$ represents the effect of diffusion scaling,

$$\int e^{-\phi(x)} dx = 1, \tag{1.2}$$

$$\int e^{\lambda x - \phi(x)} dx = M(\lambda) < \infty \quad \forall \lambda \in R, \tag{1.3}$$

or equivalently there exists $\omega(x)$ which is symmetric and convex on R satisfying

$$\lim_{x \to \infty} \frac{|x|}{\omega(x)} = 0, \quad \omega(x) \geqq |x| \quad \text{for all } x \tag{1.4}$$

and

$$\int e^{\omega(x) - \phi(x)} dx < \infty. \tag{1.5}$$

In addition

$$\int e^{\sigma|\phi'(x)| - \phi(x)} dx < \infty \quad \forall \sigma > 0. \tag{1.6}$$

Given such a ϕ we can define a density $\Phi_N(x) = \exp[-\Sigma \phi(x_j)]$ on R^N, and this will be the density relative to Lebesgue measure of a probability measure on R^N. The generator L_N is formally symmetric relative to the density Φ_N and defines a diffusion process with invariant density Φ_N with respect to which the process is reversible. The Dirichlet form for the diffusion is given by

$$N^2 \hat{D}_N(u) = \frac{N^2}{2} \int \left\{ \Sigma \left(\frac{\partial u}{\partial x_i} - \frac{\partial u}{\partial x_{i+1}} \right)^2 \right\} \Phi_N d\vec{x}. \tag{1.7}$$

The diffusion is not ergodic for the invariant measure Φ_N. The sum $x_1 + \cdots + x_N$ is conserved by the diffusion and the hyperplanes $x_1 + \cdots + x_N = Ny$ of average charge y are invariant sets. For every y the diffusion restricted to such a hyperplane is elliptic and ergodic. The invariant measure on the hyperplane is the conditional distribution of Φ_N given $x_1 + \cdots + x_N/N = y$.

We start with an initial distribution for $\vec{x}(0)$ given by some density $f_N^0(x_1, \ldots, x_N)$ relative to Φ_N. Then the evolution will give us a density $f_N^t(x_1, \ldots, x_N)$ for $t \geqq 0$. To obtain f_N^t we have to solve the heat equation

$$\frac{\partial f_N^t}{\partial t} = L_N f_N^t \quad \text{for} \quad t > 0, \quad f_N^t|_{t=0} = f_N^0. \tag{1.8}$$

Associated with the charge configuration $\vec{x}$ we have the measure,

$$\mu_N = \frac{1}{N} \Sigma x_j \delta_{j/N},$$

viewed as a random signed measure on S. If for every smooth function $J(\theta)$ on S,

$$\lim_{N \to \infty} \langle J, \mu_N \rangle = \lim_{N \to \infty} \frac{1}{N} \Sigma J\left(\frac{i}{N}\right) x_i = \int J(\theta) m(\theta) d\theta$$

exists, then we say that m is the asymptotic macroscopic charge density. Of our initial density f_N^0 we assume that for some nice $m_0(\theta)$, every smooth $J(\cdot)$ and each $\delta > 0$,

$$\lim_{N\to\infty} \int_{E_N} f_N^0 \Phi_N dx = 0, \tag{1.9}$$

where

$$E_N = \left[\vec{x} : \left| \frac{1}{N} \Sigma J\left(\frac{i}{N}\right) x_i - \int m_0(\theta) J(\theta) d\theta \right| \geqq \delta \right].$$

In addition we assume that the entropy

$$H_N(f_N^0) = \int f_N^0 \log f_N^0 \Phi_N d\vec{x} \tag{1.10}$$

satisfies a bound

$$H_N(f_N^0) \leqq CN, \tag{1.11}$$

for some constant C.

Under these conditions the randomly changing charge configurations at time $t \geqq 0$ are shown to have an asymptotic deterministic charge density $m(t, \theta)$ which is characterized as the unique solution of a certain nonlinear heat equation with initial condition $m_0(\theta)$. But to describe this equation we have to develop some more notation. We look at the function $M(\lambda)$ defined in (1.3) and define

$$\rho(\lambda) = \log M(\lambda), \quad h(y) = \sup_{\lambda} [\lambda y - \rho(\lambda)]. \tag{1.12}$$

Then $h(\cdot)$ and $\rho(\cdot)$ are a pair of conjugate convex functions and

$$\lambda = h'(y) \quad \text{if and only if} \quad y = \rho'(\lambda). \tag{1.13}$$

i.e. h' and ρ' are inverses of each other. By elementary calculation one can check that

$$\frac{1}{M(\lambda)} \int e^{\lambda x - \phi(x)} x\, dx = \rho'(\lambda), \tag{1.14}$$

so that the value of λ that makes (1.14) equal to y is $\lambda = h'(y)$. One knows also that ρ' and h' are smooth strictly increasing functions. We can now state our main result.

Theorem 1.1. *If the initial distribution of charges satisfies (1.9) and (1.11), then for every $t > 0$, arbitrary smooth $J(\cdot)$ and each $\delta > 0$,*

$$\lim_{N\to\infty} \int_{E_N^t} f_N^t \Phi_N d\vec{x} = 0 \tag{1.15}$$

with

$$E_N^t = \left\{ (x_1, \ldots, x_N) : \left| \frac{1}{N} \Sigma J\left(\frac{1}{N}\right) x_i - \int m(\theta, t) J(\theta) d\theta \right| \geqq \delta \right\},$$

where $m(\theta, t)$ is the unique weak solution of the nonlinear diffusion equation

$$\frac{\partial m}{\partial t} = \tfrac{1}{2} [h'(m(\theta, t))]_{\theta\theta}, \quad m(t, \theta)|_{t=0} = m_0(\theta). \tag{1.16}$$

We will now give a sketch of the proof.

Let us fix a smooth J and consider the stochastic process

$$\xi_N(t)=\frac{1}{N}\Sigma J\left(\frac{i}{N}\right)x_i(t).$$

It is easy to write down a stochastic differential equation satisfied by $\xi_N(t)$ in the form

$$d\xi_N(t)=a_N(t)d\beta+b_N(t)dt. \tag{1.17}$$

The way the problem is scaled implies that $a_N(t)\to 0$ as $N\to\infty$. An explicit calculation of $b_N(t)$ can be made and we obtain

$$b_N(t)=\frac{N^2}{2}\Sigma J\left(\frac{i}{N}\right)[\phi'(x_{i-1}(t))-2\phi'(x_i(t))+\phi'(x_{i+1}(t))]. \tag{1.18}$$

Using summation by parts and the smoothness of $J(\theta)$ we can pass to

$$b_N(t)\simeq\tfrac{1}{2}\Sigma J''\left(\frac{i}{N}\right)\phi'(x_i(t)). \tag{1.19}$$

At this point one has to justify the ansatz that the density $f_N^t(x_1,\dots,x_N)$ looks like a slowly varying family of local Gibbs states. This means that averages of $\phi'(x_i)$ can be replaced by their mean values $(1/M(\lambda))\int e^{\lambda x-\phi(x)}\phi'(x)dx$, where $\lambda=h'(m)$ and m is the local macroscopic density. One calculates easily that

$$\frac{1}{M(\lambda)}\int e^{\lambda x-\phi(x)}\phi'(x)dx=\lambda=h'(m), \tag{1.20}$$

and therefore

$$\lim_{N\to\infty} b_N(t)=\tfrac{1}{2}\int J''(\theta)h'(m(\theta,t))d\theta. \tag{1.21}$$

The crucial step in the whole proof is the replacement of (1.19) by (1.21). Since the ansatz is too vague to be formulated rigorously we obtain certain elementary bounds on the average $1/T\int_0^T f_N^t(x_1,\dots,x_N)dt$ as a consequence of (1.11). These are then shown to imply certain inequalities, which in turn justify the passage from (1.19) to (1.21).

2. Entropy and its Rate of Change

Since we will have to consider probability measures on R^N as well as many of its projections we now introduce some uniform notation. A probability measure α on R^N, if it is absolutely continuous will be represented by its density f with respect to the density Φ_N so that its actual Lebesgue density is $f\Phi_N$. If $F\subset\{1,2,\dots,N\}$ is a subset of cardinality k we can project α from R^N to R^k by looking at the distribution of $\{x_j\}$ for $j\in F$. We will denote the projection by α^F and the corresponding density on R^k, relative to Φ_k by f^F. We have the obvious formula

$$f^F(x_F)=\int f(x)\Phi_{F^c}(x_{F^c})dx_{F^c}.$$

We have used the obvious notation

$$\Phi_A(x) = \exp\Big[-\sum_{j\in A}\phi(x_j)\Big], \quad x_A = \{x_j : j\in A\}, \quad dx_A = \prod_{j\in A} dx_j.$$

We consider the following diffusion generators on R^N defined for a pair $(i,j)\in [1,2,\ldots,N]$,

$$\hat{L}_{i,j} = \frac{1}{2}\left(\frac{\partial}{\partial x_i} - \frac{\partial}{\partial x_j}\right)^2 - \tfrac{1}{2}(\phi'(x_i) - \phi'(x_j))\left(\frac{\partial}{\partial x_i} - \frac{\partial}{\partial x_j}\right). \tag{2.1}$$

$\hat{L}_{i,j}$ can be viewed as an operator acting on functions depending on the variables $x_F = \{x_j : j\in F\}$, provided $i,j\in F$. One should think of $\hat{L}_{i,j}$ as representing the exchange of charge between sites i and j along a "direct bond" linking them. Our generator

$$\underline{L}_N = N^2 \sum_{i=0}^{N-1} \hat{L}_{i,i+1} \tag{2.2}$$

is a sum over nearest neighbor bonds. But we can consider sums that involve other bonds as well. Each $\hat{L}_{i,j}$ is reversible relative to the weight Φ_N.

There are two functionals on probability measures on R^N that are relevant for our purpose. The first one is the entropy. If α is a probability distribution with density f then

$$H_N(\alpha) = \int_{R^N} f \log f\, \Phi_N d\vec{x}.$$

It is known that entropy is a lower semicontinuous convex functional of α and can be defined equivalently by the basic entropy inequality as the smallest constant H for which

$$\int U f \Phi_N dx \leqq \log \int e^U \Phi_N dx + H \tag{2.3}$$

for all bounded measurable functions U on R^N.

If we let the initial distribution evolve according to the forward equation or equivalently the density f evolve according to (1.8), then the function

$$H_N(t) = H(\alpha_N^t) = \int f_N^t \log f_N^t \Phi_N dx$$

is nonincreasing in t and one can compute

$$\frac{dH_N(t)}{dt} = -\frac{N^2}{2}\int\left\{\sum_{i=0}^{N-1}\frac{1}{f_N^t}\left(\frac{\partial f_N^t}{\partial x_i} - \frac{\partial f_N^t}{\partial x_{i+1}}\right)^2\right\}\Phi_N dx. \tag{2.4}$$

Corresponding to each operator $\hat{L}_{i,j}$ we can define the Dirichlet form

$$\hat{D}_{i,j}(u) = \frac{1}{2}\int_{R^N}\left(\frac{\partial u}{\partial x_i} - \frac{\partial u}{\partial x_j}\right)^2 \Phi_N dx, \tag{2.5}$$

and the corresponding $\hat{I}_{i,j}(f)$ by

$$\hat{I}_{i,j}(f) = \hat{D}_{i,j}(\sqrt{f}) = \frac{1}{8}\int\frac{1}{f}\left(\frac{\partial f}{\partial x_i} - \frac{\partial f}{\partial x_j}\right)^2 \Phi_N dx. \tag{2.6}$$

We can rewrite (2.4) in the form

$$\frac{dH_N(t)}{dt} = -4N^2 \sum_{j=0}^{N-1} \hat{I}_{j,j+1}(f_N^t). \tag{2.7}$$

The variational characterization (2.3) of entropy defines $H_N(\alpha)$ as a natural lower semicontinuous convex functional of α which is finite only when α has a density f for which $\int f \log f \,\Phi_N dx$ is finite. Similarly there is also a variational characterization of $I(\alpha)$,

$$I(\alpha) = \sup_{u>0}\left[-\int\left(\frac{Lu}{u}\right)(x)\alpha(dx)\right]. \tag{2.8}$$

If α should have a nice density f then

$$I(\alpha) = D(\sqrt{f}), \tag{2.9}$$

where L is any self adjoint operator with respect to Φ_N and D is the Dirichlet form of L. Typically $L = L_N$ or $L_{i,j}$. Since L is not elliptic $\hat{I}(\alpha)$ can be finite without α having a density f. However the restriction of α to any invariant hyperplane will have a density relative to the restriction of Φ_N to the hyperplane and the Dirichlet form is only computed from tangential derivatives on the hyperplane.

We want to establish some simple inequalities relating to the behavior of $I(f)$ under projections. Let α be a probability measure on R^N and α^F its projection onto R^k corresponding to $\{x_j: j \in F\}$, where k is the cardinality of F. We then have

Lemma 2.1. *For any pair $i, j \in F$,*

$$\hat{I}_{i,j}(\alpha^F) \leqq \hat{I}_{i,j}(\alpha). \tag{2.10}$$

Proof.

$$\hat{I}_{i,j}(\alpha^F) = \sup_{u>0}\left[-\int\left(\frac{L_{ij}u}{u}\right)(x_F)\alpha^F(dx_F)\right] \tag{2.11}$$

over all functions u of x_F. Since α^F is the projection of α,

$$\hat{I}_{i,j}(\alpha^F) \leqq \sup_{u>0}\left[-\int\left(\frac{L_{ij}u}{u}\right)(x)\alpha(dx)\right] = \hat{I}_{i,j}(\alpha). \tag{2.12}$$

The supremum in (2.11) is only over functions of x_F, and one gets the inequality in (2.12) because the supremum there is over a larger class, namely all functions of x.

Let f be a probability density on R^N. Let $l \geqq 1$ be given. We consider the set $F_{i,l} = \{j: |j-i| \leqq l\}$ and $f^{i,l} = f^{F_{i,l}}$ the projection of f onto R^{2l+1}. We define $f^l = (1/N)\sum_{i=1}^{N} f^{i,l}$ as a measure on R^{2l+1}.

Lemma 2.2. *For any $l \geqq 1$ and α,*

$$\sum_{i=-l}^{l-1} \hat{I}_{i,i+1}(f^l) \leqq \frac{2l}{N}\sum_{i=0}^{N-1} \hat{I}_{i,i+1}(f). \tag{2.13}$$

Remark. We have abused the notation somewhat here. On the left f^l is a measure on R^{2l+1} with coordinates indexed by variables $\{x_j:|j|\leqq l\}$ and $\hat{I}_{i,i+1}$ are for nearest neighbor bonds between them. But the sites $|j|\leqq l$ are arranged in a chain with two ends. On the right however the N sites are arranged cyclically.

Proof. By convexity

$$\begin{aligned}\hat{I}_{i,i+1}(f^l)&=\hat{I}_{i,i+1}\left(\frac{1}{N}\sum_{r=1}^{N}f^{r,l}\right)\leqq\frac{1}{N}\sum_{r=1}^{N}\hat{I}_{i,i+1}(f^{r,l})\\&=\frac{1}{N}\sum_{r=1}^{N}\hat{I}_{r+i,r+i+1}(f^{r,l})\quad\text{[by abuse of notation]}\\&\leqq\frac{1}{N}\sum_{r=1}^{N}\hat{I}_{r+i,r+i+1}(f)\quad\text{[by Lemma 2.1]}\\&=\frac{1}{N}\sum_{r=1}^{N}\hat{I}_{r,r+1}(f).\end{aligned}$$

This proves the lemma.

Similarly let f on R^N be given along with $l\geqq 1$ and m such that $2l+1\leqq m\leqq N-(2l+1)$. Consider the set $F_{m,l,i}=\{j:|j-i|\leqq l \text{ or } |j-m-i|\leqq l\}$. We project f onto the coordinates in $F_{m,l,i}$ and the resulting density on $R^{2l+1}\times R^{2l+1}$ is denoted by $f^{m,l,i}$. We average this over location i and denote $f^{m,l}=(1/N)\sum_{i=1}^{N}f^{m,l,i}$. For clarity we think of the variables in $R^{2l+1}\times R^{2l+1}$ as $\{y_j:|j|\leqq l\}$ and $\{z_j:|j|\leqq l\}$. We define three forms

$$\hat{I}^Y_{i,i+1}(f)=\frac{1}{8}\int\frac{1}{f}\left(\frac{\partial f}{\partial y_i}-\frac{\partial f}{\partial y_{i+1}}\right)^2\Phi_{2l+1}(y)\Phi_{2l+1}(z)dy\,dz,$$

$$\hat{I}^Z_{i,i+1}(f)=\frac{1}{8}\int\frac{1}{f}\left(\frac{\partial f}{\partial z_i}-\frac{\partial f}{\partial z_{i+1}}\right)^2\Phi_{2l+1}(y)\Phi_{2l+1}(z)dy\,dz,$$

$$\hat{I}^{Y,Z}(f)=\frac{1}{8}\int\frac{1}{f}\left(\frac{\partial f}{\partial z_0}-\frac{\partial f}{\partial y_0}\right)^2\Phi_{2l+1}(y)\Phi_{2l+1}(z)dy\,dz.$$

We then have the analog of Lemma 2.2.

Lemma 2.3. *For any m such that $2l+1\leqq m\leqq N-(2l+1)$,*

$$\sum_{i=-l}^{l-1}\hat{I}^Y_{i,i+1}(f^{m,l})\leqq\frac{2l}{N}\sum_{i=0}^{N-1}\hat{I}_{i,i+1}(f),\tag{2.14}$$

$$\sum_{i=-l}^{l-1}\hat{I}^Z_{i,i+1}(f^{m,l})\leqq\frac{2l}{N}\sum_{i=0}^{N-1}\hat{I}_{i,i+1}(f),\tag{2.15}$$

$$I^{Y,Z}(f^{m,l})\leqq\frac{m^2}{N}\sum_{i=0}^{N-1}\hat{I}_{i,i+1}(f).\tag{2.16}$$

Proof. Inequalities (2.12) and (2.15) are analogs of inequality of (2.13). We therefore only have to prove (2.16). By convexity and (2.10),

$$\hat{I}^{Y,Z}(f^{m,l}) \leqq \frac{1}{N}\sum_{i=0}^{N-1} \hat{I}^{Y,Z}(f^{m,l,i}) \leqq \frac{1}{N}\sum_{i=0}^{N-1} \hat{I}_{i,i+m}(f).$$

By Schwartz's inequality,

$$\hat{I}_{i,i+l}(f) \leqq l \sum_{j=i}^{i+l-1} \hat{I}_{j,j+1}(f).$$

Summing with respect to i from $i=0$ to $N-1$ we get (2.16).

Let us start our evolution with an initial distribution satisfying (1.1). Then according to formula (2.4) we obtain the following estimates:

$$\int f_N^t \log f_N^t \Phi_N dx \leqq CN \quad \text{for all} \quad t \geqq 0, \tag{2.17}$$

and

$$\frac{1}{2}\int_0^T dt \int \left\{ \sum_{i=0}^{N-1} \frac{1}{f_N^t} \left(\frac{\partial f_N^t}{\partial x_i} - \frac{\partial f_N^t}{\partial x_{i+1}} \right)^2 \right\} \Phi_N dx \leqq \frac{C}{N} \quad \text{for all } T. \tag{2.18}$$

Now if we use the fact that both the expressions $\int f \log f \Phi_N dx$ and

$$\tfrac{1}{2}\int \left\{ \sum_{i=0}^{N-1} \frac{1}{f} \left(\frac{\partial f}{\partial x_i} - \frac{\partial f}{\partial x_{i+1}} \right)^2 \right\} \Phi_N dx$$

are convex functionals of f, we obtain for the density $(1/T)\int_0^T f_N^t dt$ or taking $T=1$ and setting

$$\bar{f}_N = \int_0^1 f_N^t dt.$$

Theorem 2.4.

$$\int \bar{f}_N \log \bar{f}_N \Phi_N dx \leqq CN, \tag{2.19}$$

$$\frac{1}{2}\int \left\{ \sum_{i=1}^{N-1} \frac{1}{\bar{f}_N} \left(\frac{\partial \bar{f}_N}{\partial x_i} - \frac{\partial \bar{f}_N}{\partial x_{i+1}} \right)^2 \right\} \Phi_N dx \leqq \frac{C}{N}. \tag{2.20}$$

In addition

$$\sup_N \int \left\{ \frac{1}{N} \sum \omega(x_i) \right\} \bar{f}_N \Phi_N dx \leqq C'. \tag{2.21}$$

Proof. Only (2.21) needs an explanation. By estimate (2.3) and (2.19),

$$\int \{ \sum \omega(x_i) \} \bar{f}_N \phi_N dx \leqq \log \int e^{\Sigma \omega(x_i)} \Phi_N dx + CN \leqq C_1 N + CN.$$

3. Limit Theorems for Densities

The aim of this section is to prove two theorems concerning the probability densities of sums of independent random variables. Although the results are essentially well known we need special forms of these results for our use. We will give a quick sketch of the proofs as we go along.

Lemma 3.1. *Let $f(x)$ be a probability density on R such that $\int x f(x) dx = 0$,*

$\int x^2 f(x)dx = \sigma^2$, $\int x^4 f(x)dx \leqq M$ *and* $\int |f'(x)|dx \leqq M$. *Then there is a number* $\delta = \delta(M)$ *depending only on* M *such that*

$$|\int e^{i\xi x} f(x)dx| \leqq (1+|\xi|^2)^{-\delta} \quad \text{for all} \quad \xi \in R. \tag{3.1}$$

Proof. If $\phi(\xi) = \int e^{i\xi x} f(x)dx$, then $|\phi(\xi)| \leqq 1$ and moreover

$$|\phi(\xi) - 1 + \frac{\sigma^2 \xi^2}{2}| \leqq C_M |\xi|^4, \tag{3.2}$$

$$|\phi(\xi)| \leqq \frac{C_M}{|\xi|}, \tag{3.3}$$

and for every $\varepsilon > 0$,

$$\sup_{|\xi| \geqq \varepsilon} |\phi(\xi)| \leqq \theta_M(\varepsilon) < 1, \tag{3.4}$$

where C_M and $\theta_M(\varepsilon)$ are constants depending only on M and $\varepsilon > 0$ as indicated. Moreover σ^2 is bounded above as well as below away from zero in terms of M. To establish (3.1) for all ξ we use (3.2) for small $|\xi|$, (3.3) for large $|\xi|$ and (3.4) for intermediate $|\xi|$.

Lemma 3.2. *Let* $f_n(x)$ *be the density of* $x_1 + \cdots + x_n/\sqrt{n}$, *where* $x_1, \ldots, x_n$ *are independent and identically distributed with a common density* $f(x)$ *satisfying the assumptions of Lemma 3.1. Then*

$$\lim_{n\to\infty} f_n(x) = \frac{1}{\sqrt{2\pi}\sigma} e^{-x^2/2\sigma^2}, \tag{3.5}$$

uniformly on R *and in fact the derivatives of* $f_n(x)$ *of all orders converge uniformly to the corresponding derivatives.*

Proof. We use the formula

$$f_n(x) = \frac{1}{2\pi} \int e^{-ix\xi} \left[\phi\left(\frac{\xi}{\sqrt{n}}\right) \right]^n d\xi. \tag{3.6}$$

For sufficiently large n, the above formula is well defined because of inequality (3.1). Clearly $[\phi(\xi/\sqrt{n})]^n \to e^{-(\xi^2/2)\sigma^2}$ as $n \to \infty$. It is then a question of verifying that we can apply the dominated convergence theorem. From the monotonicity of the function $(1 + A/t)^t$ in t for $t \geqq 0$ we can estimate

$$\sup_{n \geqq l} |\xi|^k \left| \phi\left(\frac{\xi}{\sqrt{n}}\right) \right|^n \leqq \sup_{n \geqq l} |\xi|^k \left(1 + \frac{|\xi|^2}{n}\right)^{-n\delta} \leqq |\xi|^k \left(1 + \frac{|\xi|^2}{l}\right)^{-l\delta},$$

which is integrable for l large enough. We can now differentiate (3.6) with respect to x as often as we want and pass to the limit.

Let $e^{-\phi(x)}$ be our basic probability density and $M(\lambda) = \int e^{\lambda x - \phi(x)} dx$. We denote $\rho(\lambda) = \log M(\lambda)$ and

$$\rho'(\lambda) = \frac{M'(\lambda)}{M(\lambda)} = \int \frac{1}{M(\lambda)} e^{\lambda x - \phi(x)} x \, dx. \tag{3.7}$$

Let us define

$$f(x,\lambda)=\frac{1}{M(\lambda)}e^{\lambda(x+\rho'(\lambda))}e^{-\phi(x+\rho'(\lambda))}. \tag{3.8}$$

Then

$$\int f(x,\lambda)dx=1, \tag{3.9}$$

$$\int xf(x,\lambda)dx=0, \tag{3.10}$$

$$\int x^2 f(x,\lambda)dx=\sigma^2(\lambda)=\frac{M''(\lambda)}{M(\lambda)}-\left[\frac{M'(\lambda)}{M(\lambda)}\right]^2, \tag{3.11}$$

and

$$\int\left|\frac{\partial f}{\partial x}(x,\lambda)\right|dx \leqq C_\lambda \tag{3.12}$$

is bounded if λ is bounded.

Lemma 3.3. *Let $f_n(x,\lambda)$ be the probability density of $x_1+\cdots+x_n/\sqrt{n}$, where $x_1,\ldots,x_n$ are independent and identically distributed having a common density $f(x,\lambda)$. Then*

$$\lim_{n\to\infty} f_n(x,\lambda)=\frac{1}{\sqrt{2\pi}\,\sigma(\lambda)}\exp\left[-\frac{1}{2\sigma^2(\lambda)}x^2\right] \tag{3.13}$$

uniformly in x and bounded λ intervals. In addition the first partial derivative of f_n with respect to λ converges to the corresponding partial derivative on the right uniformly on bounded λ sets and uniformly in x.

Proof. Let

$$\psi(\xi,\lambda)=\int e^{i\xi x}f(x,\lambda)dx.$$

Then

$$f_n(x,\lambda)=\frac{1}{2\pi}\int e^{-i\xi x}\left[\psi\left(\frac{\xi}{\sqrt{n}},\lambda\right)\right]^n d\xi.$$

Let us compute $\partial f_n/\partial\lambda$. Computing the derivative inside we note first

$$\frac{\partial\psi}{\partial\lambda}=\int e^{i\xi x}\frac{\partial f}{\partial\lambda}(x,\lambda)dx=\int(e^{i\xi x}-1-i\xi x)\frac{\partial f}{\partial\lambda}(x,\lambda)dx,$$

because $\int(\partial f/\partial\lambda)(x,\lambda)dx=\int x(\partial f/\partial\lambda)(x,\lambda)dx=0$. Therefore

$$\left|\frac{\partial\phi}{\partial\lambda}\right|\leqq|\xi|^2\int x^2\left|\frac{\partial f}{\partial\lambda}(x,\lambda)\right|dx\leqq C_\lambda|\xi|^2.$$

Therefore

$$\left|n\frac{\partial\phi}{\partial\lambda}\left(\frac{\xi}{\sqrt{n}},\lambda\right)\right|\leqq n\cdot C_\lambda\left(\frac{|\xi|}{\sqrt{n}}\right)^2\leqq C_\lambda|\xi|^2. \quad\square$$

We are now ready to prove our first main theorem of the section:

Theorem 3.4. *Let $\phi_n(x)$ be the density of $x_1+\cdots+x_n/n$, where $x_1,\ldots,x_n$ are independent and identically distributed with a common density $e^{-\phi(x)}$. Then*

$$\lim_{n\to\infty}\frac{1}{n}\log\phi_n(x) = -h(x), \tag{3.14}$$

$$\lim_{n\to\infty}\frac{1}{n}\frac{\phi_n'(x)}{\phi_n(x)} = -h'(x), \tag{3.15}$$

uniformly on compact x-intervals.

Proof. We use Cramer's trick. Let $y \in R$, $\lambda = h'(y)$ and $y = \rho'(\lambda)$. Consider

$$f(x, y) = \frac{1}{M(\lambda)} e^{\lambda(x+y) - \phi(x+y)}.$$

We want to apply Lemma 3.3 to the above density treating y as a parameter. Then density $f_n(x, y)$ of $x_1 + \cdots + x_n/\sqrt{n}$ at $x = 0$ is related to the density $\hat{f}_n(x, y)$ of $x_1 + \cdots + x_n/\sqrt{n}$ at $x = 0$ by a simple factor of $\sqrt{n}$ and a calculation yields

$$\phi_n(y) = [M(\lambda)]^n e^{-n\lambda y}\sqrt{n}\, f_n(0, y). \tag{3.16}$$

We also know that $\log M(\lambda) - \lambda y = \rho(\lambda) - \lambda y = -h(y)$. Equations (3.14) and (3.15) are now easy consequences of Lemma 3.3.

We are now interested in studying the conditional distribution of $x_1, \ldots, x_n$ on the hyperplane $x_1 + \cdots + x_n/n = y$, where $x_1, \ldots, x_n$ are independent and identically distributed random variables with a common density $e^{-\phi(x)}$. We have clearly a smooth density $\phi_n(x)$ for $x_1 + \cdots + x_n/n$, and therefore the conditional density is well defined on each hyperplane. Let us denote by $v_y^{(n)}(dx_1, \ldots, dx_n)$ the measure on R^n concentrated on the hyperplane $x_1 + \cdots + x_n/n = y$. We want to establish the following main theorem.

Theorem 3.5. *Let F be a bounded continuous function on R^k for some $k \geqq 1$. Let for each $y \in R$, $\alpha_y(dx)$ denote the probability measure on R with density $(1/M(\lambda))e^{\lambda x - \phi(x)}$, where $\lambda = h'(y)$. Then for every $\varepsilon > 0$,*

$$\lim_{n\to\infty} v_y^{(n)}\left[(x_1, \ldots, x_n): \left|\frac{1}{n-k+1}\sum_{i=1}^{n-k+1} F(x_i, \ldots, x_{i+k-1}) \right.\right.$$
$$\left.\left. - \int F(z_1, \ldots, z_k)\alpha_y(dz_1)\alpha_y(dz_2)\cdots\alpha_y(dz_k)dy\right| \geqq \varepsilon\right] = 0 \tag{3.17}$$

locally uniformly in $y \in R$.

Proof. Let us remark that $v_y^{(n)}(dx_1, \ldots, dx_n)$ is symmetric with respect to the permutation group acting on the variables $x_1, \ldots, x_n$. Therefore by Hewitt–Savage 0–1 law and its implication we need to check the theorem only when $k = 1$ and $F(x)$ is a bounded continuous function on R with a uniformly bounded first derivative. This will follow from the following lemma.

Lemma 3.6. *For every $\theta \in R$*

$$\lim_{n\to\infty}\frac{1}{n}\log\int e^{\theta(F(x_1) + \cdots + F(x_n))} dv_y^{(n)}(dx_1, \ldots, dx_n) = G(\theta, y) \tag{3.18}$$

exists and the limit $G(\theta)$ *is differentiable at* $\theta = 0$ *with*

$$\left.\frac{dG(\theta, y)}{d\theta}\right|_{\theta=0} = \int F(x)\alpha_y(dx). \tag{3.19}$$

Proof. Let us denote by $H_n(y)$ the function

$$H_n(y, \theta) = \int_{x_1 + \cdots + x_n = ny} e^{\theta(F(x_1) + \cdots + F(x_n)) - \phi(x_1) - \phi(x_2) - \cdots - \phi(x_n)} dx_1 \cdots dx_n,$$

$$\phi_n(y) = \int_{x_1 + \cdots + x_n = ny} e^{-\Sigma\phi(x_j)} dx_1 \cdots dx_n.$$

Proving (3.18) reduces to showing that

$$\lim_{n\to\infty} \frac{1}{n} \log H_n(y, \theta) - \lim_{n\to\infty} \frac{1}{n} \log \phi_n(y) = G(\theta, y)$$

exists locally uniformly in y. Let us denote

$$a(\theta) = \int e^{\theta F(x) - \phi(x)} dx, \quad M(\lambda, \theta) = \frac{1}{a(\theta)} \int e^{\lambda x + \theta F(x) - \phi(x)} dx.$$

Then essentially by replacing $\phi(x)$ with $\phi(x) - \theta F(x)$ we can apply Theorem 3.4 and obtain

$$\lim_{n\to\infty} \frac{1}{n} \log H_n(y, \theta) = \log a(\theta) - h(\theta, y),$$

where

$$h(\theta, y) = \sup_{\lambda} [\lambda y - \log M(\lambda, \theta)].$$

Equation (3.18) now follows with

$$G(\theta, y) = \log a(\theta) - h(\theta, y) + h(y).$$

It is now elementary to conclude that

$$\left.\frac{dG(\theta, y)}{d\theta}\right|_{\theta=0} = \frac{a'(0)}{a(0)} + \left.\frac{1}{M(\lambda^*, 0)} \frac{\partial M}{\partial\theta}\right|_{(\lambda^*, 0)}, \tag{3.20}$$

where $\lambda^* = h'(y)$. The right-hand side of (3.20) is easily computed to be $(1/M(\lambda)) \int F(x) e^{\lambda^* x - \phi(x)} dx$. This proves (3.19).

Theorem 3.5 follows from Lemma 3.6 by exponential Tchebychev bounds.

4. Local Gibbs States

A local Gibbs state is a vague term which refers to a probability distribution on R^N with a density relative to Φ_N which looks somewhat like $c \exp[\Sigma \lambda_i x_i]$. Here λ_i are constants that are slowly varying so that $\lambda_i = \bar{\lambda}(i/N)$ for some smooth function $\bar{\lambda}$ on S. If we define $\bar{m}(\theta) = \rho'(\bar{\lambda}(\theta))$, then the above local Gibbs state corresponds to a macroscopic charge density of $\bar{m}(\theta)$ on S. This refers to the fact that if $J(\theta)$ is a smooth function on S then $(1/N)\Sigma J(i/N)x_i$ as a random variable on R^N is almost a constant for large N, being nearly equal to $\int J(\theta)\bar{m}(\theta)d\theta$ under our local

Gibbs density. For such a density $f=c\exp[\Sigma\lambda_i x_i]$ one can calculate

$$\sum_{i=1}^{N}\hat{I}_{i,i+1}(f)=\frac{1}{8}\int\frac{1}{f}\sum_{i=1}^{N}\left(\frac{\partial f}{\partial x_i}-\frac{\partial f}{\partial x_{i+1}}\right)^2\Phi_N dx$$
$$=\frac{1}{8}\sum_{i=1}^{N}(\lambda_i-\lambda_{i+1})^2=\frac{1}{N}\int\left(\frac{\partial\bar{\lambda}}{\partial\theta}\right)^2 d\theta\leq\frac{C}{8N}. \tag{4.1}$$

On the other hand if $f=c\exp\Sigma\lambda_i x_i$, one can then calculate explicitly and show that for any function $F(x_{-k},\ldots,x_k)$ depending on $(2k+1)$ coordinates $(1/N)\Sigma J(i/N)F(x_{-k+i},\ldots,x_{k+i})$ is also almost a constant relative to our Gibbs state and compute the constant explicitly in terms of $\bar{m}(\theta)$. The aim of this section is to show that the above consequence for local averages follows from entropy bounds of the type (4.1). This will be the key step of our main theorem.

We will be dealing with sets of probability measures or densities $\underline{A}_{N,C_1,C_2}$ defined in terms of two constants C_1,C_2 but in dimensions N, that vary. The purpose will be to obtain for these classes certain results that are valid uniformly in N so long as C_1 and C_2 are held fixed.

Let $\underline{A}_{N,C_1,C_2}$ be the class of densities f on R^N satisfying the following bounds:

$$\frac{1}{N}\int[\Sigma\omega(x_i)]f\,\Phi_N dx\leq C_1, \tag{4.2}$$

$$\sum_{j=1}^{N}\hat{I}_{j,j+1}(f)\leq\frac{C_2}{N}. \tag{4.3}$$

Here $\omega(\cdot)$ is the function defined in (1.4).

Let F be a function depending on the $2k+1$ variables $\{x_j:|j|\leq k\}$ on R^{2k+1} which is bounded and continuous on R^{2k+1}. Given such a function let us define $\tilde{F}(y)$ for $y\in R$ by

$$\tilde{F}(y)=\frac{1}{[M(\lambda)]^{2k+1}}\int\exp[\lambda\Sigma x_j]F(x_{-k},\ldots,x_k)\Phi_{2k+1}dx, \tag{4.4}$$

with $\lambda=h'(y)$. Let us define for any integer $i\in[1,2,\ldots,N]$ and l the quantities $\xi_i=F(x_{i-k},\ldots,x_{i+k})$ and $\bar{\xi}_{i,l}=\tilde{F}((1/2l+1)(x_{i-l}+\cdots+x_{i+l}))$. Finally let $J(\cdot)$ be a smooth function of $\theta\in S$. The main result of this section is the following theorem.

Theorem 4.1. *For any given constant C_1 and C_2, integer k and choices of $F(\cdot)$ and $J(\cdot)$,*

$$\lim_{\varepsilon\to 0}\limsup_{N\to\infty}\sup_{f\in\underline{A}_{N,C_1,C_2}}\int\frac{1}{N}\left|\Sigma\left[J\left(\frac{i}{N}\right)(\xi_i-\bar{\xi}_{i,[N\varepsilon]}\right)\right|f\,\Phi_N dx=0. \tag{4.5}$$

We will first prove some auxiliary lemmas and theorems before finally returning to the proof of Theorem 4.1. The first step is Theorem 4.2 which we now state and prove.

Theorem 4.2. *For any C_1,C_2 and $F(\cdot)$,*

$$\lim_{l\to\infty}\limsup_{N\to\infty}\sup_{f\in\underline{A}_{N,C_1,C_2}}\int\frac{1}{N}\sum_{j=1}^{N}\left|\frac{1}{2l+1}\sum_{i=j-l}^{j+l}\xi_i-\bar{\xi}_{j,l}\right|f\,\Phi_N dx=0. \tag{4.6}$$

Proof. At first glance the above theorem looks hard because N is becoming large and things can get out of control. We will show however that the hard work has all been done already and we can obtain (4.6) from our earlier results in Sects. 2 and 3.

First Step. Since F is a bounded function and l is becoming large while k remains fixed we can replace $(1/2l+1)\sum_{i=j-l}^{j+l}\xi_i$ by $(1/2l-2k+1)\sum_{i=j-l+k}^{j+l-k}\xi_i$, and the error will be uniformly small.

Second Step. If we denote by $\psi_l(y_{-l},\dots,y_l)$ the function

$$\psi_l(y_{-l},\dots,y_l)=\frac{1}{2(l-k)+1}\sum_{i=-(l-k)}^{(l-k)}F(y_{i-k},\dots,y_{i+k})-\tilde{F}\left(\frac{y_{-l}+\dots+y_l}{2l+1}\right),$$

then

$$\frac{1}{N}\int\sum_{j=1}^{N}\left|\frac{1}{2l+1}\sum_{i=j-l}^{j+l}\xi_i-\xi_{j,l}\right|f\Phi_N dx=\int|\psi_l(y_{-l},\dots,y_l)|f^l\Phi_{2l+1}dy$$

[see Lemma 2.2 for the definition of f^l]. If we denote by $\underline{B}^l_{N,C_1,C_2}$ the range of f^l as f varies over $\underline{A}_{N,C_1,C_2}$, then

$$\sup_{f\in\underline{A}_{N,C_1,C_2}}\int\frac{1}{N}\sum_{j=1}^{N}\left|\frac{1}{2l+1}\sum_{i=j+l}^{j-l}\xi_i-\xi_{i,l}\right|f\Phi_N dx=\sup_{g\in\underline{B}^l_{N,C_1,C_2}}\int|\psi_l|g\Phi_{2l+1}dy.$$

Now $\underline{B}^l_{N,C_1,C_2}$ are all distributions on R^{2l+1}. We will verify that $\underline{B}^l_{N,C_1,C_2}$ is a tight family on R^{2l+1} as $N\to\infty$ and obtain information about its limit points $\underline{B}^l_{C_1,C_2}$.

From (4.2) we obtain

$$\frac{1}{2l+1}\int\left[\sum_{i=-l}^{l}\omega(y_i)\right]g\Phi_{2l+1}(y)dy\leqq C_1\quad\text{for all}\quad g\in\underline{B}^l_{N,C_1,C_2}.\qquad(4.7)$$

This implies tightness immediately. If β is any element of $\underline{B}^l_{C_1,C_2}$, by lower semicontinuity and (4.7) we obtain

$$\sup_{\beta\in\underline{B}^l_{C_1,C_2}}\frac{1}{2l+1}\int\Sigma\omega(y_i)d\beta\leqq C_1.$$

Further the function

$$I_l(\beta)=\sum_{j=-l}^{l-1}\hat{I}_{j,j+1}(\beta)$$

is lower semicontinuous. From (4.3), Lemma 2.2 and lower semicontinuity we find that $I_l(\beta)=0$ for $\beta\in\underline{B}^l_{C_1,C_2}$.

Third Step. We can now replace the $\limsup_{N\to\infty}\sup_{f\in\underline{A}_{N,C_1,C_2}}$ in (4.6) by $\sup_{\beta\in\underline{B}^l_{C_1,C_2}}$. Moreover if $\beta\in\underline{B}^l_{C_1,C_2}$,

$$\frac{1}{2l+1}\int[\Sigma\omega(y_i)]d\beta\geqq\frac{1}{2l+1}\int[\Sigma|y_i|]d\beta.$$

It is therefore sufficient to prove that

$$\lim_{l\to\infty} \sup_{\beta\in\Gamma_{l,C_1}} \int |\psi_l| d\beta = 0, \tag{4.8}$$

where

$$\Gamma_{l,C_1} = \left\{\beta: \frac{1}{2l+1}\int [\Sigma |y_i|] d\beta \leqq C_1, I_l(\beta) = 0\right\}. \tag{4.9}$$

Final Step. If we define the generator

$$L^{[l]} = \frac{1}{2}\sum_{i=-l}^{l-1}\left(\frac{\partial}{\partial y_i} - \frac{\partial}{\partial y_{i+1}}\right)^2 - \frac{1}{2}\sum_{i=-l}^{l-1}(\phi'(y_i) - \phi'(y_{i+1}))\left(\frac{\partial}{\partial y_i} - \frac{\partial}{\partial y_{i+1}}\right),$$

then $I_l(\beta) = 0$ implies that β is an invariant measure for the diffusion on R^{2l+1} with generator $L^{[l]}$. We know that these are convex combinations of the ergodic ones concentrated on the various hyperplanes $(1/2l+1)\sum_{|j|\leqq l} y_j = y$. Therefore

$$\beta = \int \nu_y^{(2l+1)} d\hat{\beta},$$

where $\hat{\beta}$ is the distribution of y under β. From (4.9) it follows that $\int |y| d\hat{\beta} \leqq (1/2l+1)\int (\Sigma |y_i|) d\beta \leqq C_1$. One can now obtain (4.8) from Theorem 3.5 and the above bound.

If we take a look at what has been achieved in Theorem 4.2 for a "Local Gibbs state," i.e. a state satisfying (4.2) and (4.3) for some constants C_1 and C_2, then relative to such a probability distribution one can always replace $(1/N)\Sigma J(i/N)\xi_i$ by $(1/N)\Sigma J(i/N)\left((1/2l+1)\sum_{j=i-l}^{i+l}\xi_i\right)$, and according to Theorem 4.2 this can be replaced asymptotically by $(1/N)\Sigma J(i/N)\tilde{\xi}_{i,l}$, provided l is large and fixed. However Theorem 4.1 demands that l be chosen as $[N\varepsilon]$ for arbitrarily small but fixed $\varepsilon > 0$. The next two theorems are to make the replacement possible.

Theorem 4.3. *For any $\delta > 0$ and C_1 and C_2 finite*

$$\lim_{\varepsilon\to 0}\lim_{l\to\infty}\sup\limsup_{N\to\infty}\ \sup_{2l+1\leqq|m|\leqq\varepsilon N}\ \sup_{f\in\underline{A}_{N,C_1,C_2}}$$

$$\frac{1}{N}\int\left[\#i: \left|\frac{x_{i-l}+x_{i+l}}{2l+1} - \frac{x_{i+m-l}+x_{i+m+l}}{2l+1}\right| \geqq \delta\right] f\,\Phi_N dx = 0. \tag{4.10}$$

Proof. The proof is very much like the proof of Theorem 4.2. We first rewrite the integral through the projections $\alpha^{m,l}$ introduced before Lemma 2.3. Let us note that the measure α and the density f will be used interchangeably. The integral then is

$$\alpha^{m,l}\left[(y_j, z_j): \left|\frac{1}{2l+1}\Sigma y_j - \frac{1}{2l+1}\Sigma z_j\right| \geqq \delta\right]. \tag{4.11}$$

First Step. Let us denote by $\underline{B}^{m,l}_{N,C_1,C_2}$ the range of $f^{m,l}$ as α or f varies over $\underline{A}_{N,C_1,C_2}$. These are probability measures on $R^{2l+1}\times R^{2l+1}$ with coordinates $\{y_j: |j|\leqq l\}$ and $\{z_j: |j|\leqq l\}$. We denote by $\underline{B}^{l,\varepsilon}_{C_1,C_2}$ the set of limit points of $\sum_{2l+1\leqq|m|\leqq N\varepsilon}\underline{B}^{m,l}_{N,C_1,C_2}$. The basic estimate (4.2) yields for any $g\in\underline{B}^{m,l}_{N,C_1,C_2}$,

$$\frac{1}{4l+2}\int[\Sigma\omega(y_i)+\Sigma\omega(z_i)]g\Phi_{4l+2}dydz \leqq C_1, \tag{4.12}$$

and thereby tightness. The lower bound on the gap $|m|\geqq 2l+1$ is to make sure that the two sets of blocks of size $2l+1$ with their centers separated by m have no common sites. If β is any limit point in $\underline{B}^{l,\varepsilon}_{C_1,C_2}$ we obtain the following estimates:

$$\frac{1}{4l+2}\int[\Sigma\omega(y_i)+\Sigma\omega(z_i)]d\beta \leqq C_1, \tag{4.13}$$

$$\sum_{i=-l}^{l-1}\hat{I}^Y_{i,i+1}(\beta)=0, \tag{4.14}$$

$$\sum_{i=-l}^{l}\hat{I}^Z_{i,i+1}(\beta)=0, \tag{4.15}$$

$$\hat{I}^{Y,Z}(\beta)\leqq C_2\varepsilon^2. \tag{4.16}$$

These are obtained from lower semicontinuity and Lemma 2.3.

Second Step. (4.14) and (4.15) imply that the measure β on $R^{2l+1}\times R^{2l+1}$ is invariant with respect to the diffusions

$$L^Y=\frac{1}{2}\sum_{i=-l}^{l-1}\left(\frac{\partial}{\partial y_i}-\frac{\partial}{\partial y_{i+1}}\right)^2-\frac{1}{2}\sum_{i=-l}^{l-1}(\phi'(y_i)-\phi'(y_{i+1}))\left(\frac{\partial}{\partial y_i}-\frac{\partial}{\partial y_{i+1}}\right),$$

and a similar one involving the second set of coordinates. This implies that

$$\beta(dy,dz)=\int v_a^{(2l+1)}(dy)v_b^{(2l+1)}(dz)\hat{\beta}(da,db), \tag{4.17}$$

where $\hat{\beta}$ is the distribution of $a=(1/2l+1)\Sigma y_i$ and $b=(1/2l+1)\Sigma z_i$ on R^2. Here v is as defined in Theorem 3.5.

Third Step. From (4.3) for β one can clearly obtain

$$\int[|a|+|b|]\hat{\beta}(da,db)\leqq 2C_1. \tag{4.18}$$

According to Lemma 4.4 the inequality (4.16) provides an estimate

$$\bar{I}^{(l)}(\hat{\beta})\leqq C_2\varepsilon^2(2l+1)^2, \tag{4.19}$$

where

$$\bar{I}^{(l)}(\hat{\beta})=\sup_{u>0}\int\left[-\frac{G^{(l)}u}{u}\right]d\hat{\beta}(a,b), \tag{4.20}$$

and $G^{(l)}$ is the generator of the diffusion

$$G^{(l)}u=\frac{1}{2}\left(\frac{\partial}{\partial a}-\frac{\partial}{\partial b}\right)^2u-\tfrac{1}{2}(h'_{2l+1}(a)-h'_{2l+1}(b))\left(\frac{\partial u}{\partial a}-\frac{\partial u}{\partial b}\right) \tag{4.21}$$

on R^2. Here $h_{2l+1}(a)=-\log\psi_{2l+1}(a)$ and $\psi_k(a)$ is the density of $x_1+\dots+x_k/k$, where $x_1,\dots,x_k$ are independent with common density $e^{-\phi(x)}$.

Let us consider the range $\Gamma^{l,\varepsilon}_{C_1,C_2}$ of $\hat{\beta}$ on R^2 as β varies over $\underline{B}^{l,\varepsilon}_{C_1,C_2}$. Then $\hat{\beta}$ satisfies (4.18) and (4.19). We now consider the set of limit points of $\Gamma^{l,\varepsilon}_{C}$ as $l\to\infty$ and denote it by $\Gamma^{\varepsilon}_{C_1,C_2}$.

We have tightness from (4.18) and Lemma 4.5 gives us the estimate

$$\int [h'(a)-h'(b)]^2 d\hat{\beta}(a,b) \leq 8C_2\varepsilon^2 \tag{4.22}$$

for every $\hat{\beta}\in\Gamma^{\varepsilon}_{C_1,C_2}$. We also have the companion estimate: for every $\hat{\beta}\in\Gamma^{\varepsilon}_{C_1,C_2}$,

$$\tfrac{1}{2}\int [|a|+|b|]d\hat{\beta}(a,b) \leq C_1. \tag{4.23}$$

Finally if we let $\varepsilon\to 0$ and again look at the limit points, denoting the set by $\Gamma^{0}_{C_1,C_2}$, then

$$\int [h'(a)-h'(b)]^2 d\hat{\beta}(a,b) = 0 \tag{4.24}$$

for every $\hat{\beta}\in\Gamma^{0}_{C_1,C_2}$. From the strict convexity of h, h' is a strictly monotone map and β is then concentrated on the diagonal of R^2. But this is precisely the content of the theorem because of (4.11).

We now have two lemmas to prove in order to complete the proof of Theorem 4.3:

Lemma 4.4. *Let β be a probability measure on $R^{2l+1}\times R^{2l+1}$ of the form*

$$\beta(dy,dz) = \int v_a^{(2l+1)}(dy)v_b^{(2l+1)}(dz)\hat{\beta}(da,db),$$

where $v_a^{(2l+1)}$ is the restriction of Φ_{2l+1} to the hyperplane of average charge a. Then

$$\bar{I}^{(2l+1)}(\hat{\beta}) \leq (2l+1)^2 I^{Y,Z}(\beta),$$

where $\bar{I}$ is defined in (4.20) and $I^{Y,Z}$ in Lemma 2.3.

Proof. We recall that $e^{-h_{2l+1}(a)}$ is the density at a of the mean $(x_1+\cdots+x_{2l+1})/(2l+1)$, where $x_1,\ldots,x_{2l+1}$ are independent, all having $e^{-\phi(x)}$ as density. We notice first that

$$\begin{aligned}
&\int g(a)e^{-h_{2l+1}(a)}\int \phi'(y_0)v_a^{(2l+1)}(dy)da\\
&=\int g\left(\frac{1}{2l+1}\Sigma y_j\right)\phi'(y_0)e^{-\Sigma\phi(y_j)}dy\,da\\
&=\frac{1}{2l+1}\int g'\left(\frac{1}{2l+1}\Sigma y_j\right)e^{-\Sigma\phi(y_j)}dy\,da=\frac{1}{2l+1}\int g'(a)e^{-h_{2l+1}(a)}da\\
&=\frac{1}{2l+1}\int g(a)h'_{2l+1}(a)e^{-h_{2l+1}(a)}da.
\end{aligned}$$

Since g is arbitrary and v_a and $h'_{2l+1}(a)$ are continuous in a, we obtain

$$\int \phi'(y_0)v_a^{(2l+1)}(dy) = \frac{1}{2l+1}h'_{2l+1}(a),$$

$$I^{Y,Z}(\beta) = \sup_{u>0}\left[-\int\left(\frac{Gu}{u}\right)\beta(du,dz)\right], \tag{4.25}$$

where

$$G = \frac{1}{2}\left(\frac{\partial}{\partial y_0}-\frac{\partial}{\partial z_0}\right)^2 - \tfrac{1}{2}[\phi'(y_0)-\phi'(z_0)]\left(\frac{\partial}{\partial y_0}-\frac{\partial}{\partial z_0}\right).$$

If $\beta = \int v_a^{(2l+1)}(dy) v_b^{(2l+1)}(dz)\hat{\beta}(da, db)$, we can take u to be a function of a and b of the form $u((1/2l+1)\Sigma y_j, (1/2l+1)\Sigma z_j)$. Then

$$I^{Y,Z}(\beta) \geqq \sup_u \left[-\int \frac{1}{u} \left[\frac{1}{2} \cdot \left(\frac{1}{2l+1} \right)^2 \left(\frac{\partial}{\partial a} - \frac{\partial}{\partial b} \right)^2 u \right.\right.$$

$$\left.\left. - \frac{1}{2(2l+1)} (\phi'(y_0) - \phi'(z_0))(u_a - u_b) \right] \beta(dy\, dz) \right].$$

If we now use (4.25),

$$I^{Y,Z}(\beta) \geqq \frac{1}{(2l+1)^2} \left[\sup_u \left\{ -\int \frac{1}{2u} \left[\left(\frac{\partial}{\partial a} - \frac{\partial}{\partial b} \right)^2 u \right.\right.\right.$$

$$\left.\left.\left. - (h_{2l+1}(a) - h_{2l+1}(b))(u_a - u_b) \right] \hat{\beta}\, da\, db \right.\right.$$

$$= \frac{1}{(2l+1)^2} \bar{I}^l(\hat{\beta}).$$

Lemma 4.5. *If $\{\hat{\beta}_l\}$ is a sequence of probability measures on R^2 satisfying*

$$\int [|a| + |b|] \hat{\beta}_l(da, db) \leqq 2C_1 \tag{4.26}$$

and

$$I^{(l)}(\hat{\beta}_l) \leqq C_2(2l+1)^2 \varepsilon^2, \tag{4.27}$$

and if $\hat{\beta}$ is any limit point of $\hat{\beta}_l$, then

$$\int [h'(a) - h'(b)]^2 \hat{\beta}(da, db) \leqq 8C_2 \varepsilon^2.$$

Proof. (4.27) implies for every u

$$\frac{1}{2} \left[\int \frac{1}{u} \left[\left(\frac{\partial}{\partial a} - \frac{\partial}{\partial b} \right)^2 u - (h'_{2l+1}(a) - h'_{2l+1}(b))(u_a - u_b) \right] \hat{\beta}_l(da, db) \right]$$

$$\geqq -C\varepsilon^2(2l+1)^2.$$

Taking $u = \exp[(2l+1)v(a,b)]$ for some v,

$$\left[\int \left\{ (2l+1) \left(\frac{\partial}{\partial a} - \frac{\partial}{\partial b} \right)^2 v + (2l+1)^2 \left(\frac{\partial}{\partial a} - \frac{\partial}{\partial b} \right)^2 \right.\right.$$

$$\left.\left. - (2l+1)(h'_{2l+1}(a) - h'_{2l+1}(b)) \left(\frac{\partial v}{\partial a} - \frac{\partial v}{\partial b} \right) \hat{\beta}_l(da, db) \right\} \right] \geqq -2C\varepsilon^2(2l+1)^2.$$

If v is a constant outside a compact set one can divide (4.27) by $(2l+1)$ and let $l \to \infty$ to obtain

$$\int \left[\left(\frac{\partial v}{\partial a} - \frac{\partial v}{\partial b} \right)^2 - (h'(a) - h'(b)) \left(\frac{\partial v}{\partial a} - \frac{\partial v}{\partial b} \right) \right] \beta(da, db) \geqq -2C\varepsilon^2. \tag{4.28}$$

We are done if we let

$$v(a,b) = \tfrac{1}{2}[(h(a) + h(b)]$$

except, then v is not constant outside a compact set. But we can take

$$v_\tau(a,b)=\min\left[\tfrac{1}{2}[h(a)+h(b)],\tau\right],$$

and let $\tau\to\infty$. v_τ is not quite smooth enough. But one can mollify v_τ somewhat.

We now prove a slightly strengthened version of Theorem 4.3.

Theorem 4.6. *For any given finite constants* C_1, C_2,

$$\lim_{\varepsilon\to 0}\limsup_{l\to\infty}\limsup_{N\to\infty}\sup_{2l+1\leqq|m|\leqq\varepsilon N}\sup_{f\in \underline{A}_{N,C_1,C_2}}$$

$$\int\left\{\frac{1}{N}\Sigma\left|\frac{x_{i+l}+\cdots+x_{i+l}}{2l+1}-\frac{x_{i+m-k}+\cdots+x_{i+m+k}}{2l+1}\right|\right\}f\,\Phi_N\,dx=0. \quad (4.29)$$

Proof. Given Theorem 4.5 all we need to derive Theorem 4.6 is some uniform integrability of $(1/2l+1)\sum_{|j|\leqq l}|y_j|$ relative to all the distributions α^l with α varying over $\underline{A}_{N,C_1,C_2}$. But this is precisely the role of condition (4.2) especially with $\omega(x)$ satisfying (1.4).

Finally we prove

Theorem 4.7. *For any* C_1, C_2,

$$\lim_{\varepsilon\to 0}\limsup_{l\to\infty}\limsup_{N\to\infty}\sup_{f\in \underline{A}_{N,C_1,C_2}}$$

$$\int\frac{1}{N}\Sigma\left|\frac{x_{i-l}+\cdots+x_{i+l}}{2l+1}-\frac{x_{i-[N\varepsilon]}+\cdots+x_{i+[N\varepsilon]}}{2[N\varepsilon]+1}\right|f\,\Phi_N\,dx=0.$$

Proof. Theorem 4.7 follows immediately from Theorem 4.6. An average over a long block of length $2[N\varepsilon]+1$ is just an average of averages over short blocks of length $(2l+1)$. According to Theorem 4.6 they do not differ by much among themselves so the average is close to any one of them.

Theorem 4.7 allows us to deduce Theorem 4.1 from Theorem 4.2. One uses the continuity of the function $\tilde{F}(y)$. Any trouble arising from large values of $(x_{i-l}+\cdots+x_{i+l}/l)[\tilde{F}(y)$ need not be uniformly continuous] is taken care of by (4.2) which tells us that there cannot be too many such blocks.

5. Hydrodynamic Limit

Let us start our evolution with an initial distribution f_N^0 satisfying (1.9) and (1.11). We will study the evolution in a fixed interval $[0,T]$. In fact without loss of generality we will assume that $T=1$. With f_N^0 as initial distribution we have a measure P_N on the space $C\{[0,1]:R^N\}$ of trajectories $\{x_j(t): 0\leqq t\leqq 1, 1\leqq j\leqq N\}$. Let us denote by $\underline{M}_l$ the compact metric space (under weak convergence) of all signed measures on S of total variation at most l. The space $\underline{M}=U_l\underline{M}_l$ is viewed mainly as a measurable space. We denote by Ω_l the space $C\{[0,1];\underline{M}_l\}$ of all weakly continuous maps of $[0,1]$ into $\underline{M}_l$. Ω_l is a complete separable metric space with uniform convergence in weak topology. The space $\Omega=U_l\Omega_l$ is again viewed only as a measurable space. Given a collection of trajectories $\{x_j(t)\}$ we can associate a point in Ω by

$$\mu_N(t) = \frac{1}{N} \Sigma x_j(t) \delta_{j/N}. \tag{5.1}$$

This introduces a measure Q_N on Ω as the distribution of $\mu_N(\cdot)$, where $\{x_j(\cdot)\}$ have P_N for their distribution. According to Lemmas 6.1 and 6.2, from any subsequence of Q_N we can choose a further subsequence that converges weakly on Ω. What this means is that for any bounded function F whose restriction to Ω_l is continuous for every l, we have convergence of expectations. Let Q be any limit point along any subsequence. The main theorem of the section is the following:

Theorem 5.1. *Let Q be any limit point. Then*

a) $Q[\mu(\cdot): \mu(t) = m(t, \theta) d\theta$ a.e. $t] = 1$,

b) $Q[\mu(\cdot): \mu(0) = m_0(\theta) d\theta] = 1$,

c) $Q[\mu(\cdot): \int_0^1 \int_S h(m(t, \theta)) d\theta dt < \infty] = 1$,

d) $Q[\mu(\cdot): \int_0^1 \int_S \{[h'(m(t, \theta))]_\theta\}^2 d\theta dt < \infty] = 1$,

e) $Q[\mu(\cdot): \int_S J(\theta) \mu(t, d\theta) - \int_S J(\theta) \mu(0, d\theta) = \frac{1}{2} \int_0^t \int_S J''(\theta) h'(m(s, \theta)) d\theta ds] = 1.$

Proof. From the monotonicity of the entropy (see Eq. 2.4) and Lemma 6.3 we conclude that for each $t > 0$, $Q[\mu: \mu(t, d\theta) = m(t, \theta) d\theta] = 1$ and $\int_S h(m(t, \theta)) d\theta \leqq C$ for all $t \geqq 0$. Now Fubini's theorem will yield a) as well as c). b) is just a restatement of condition (1.9). To prove d) we use Fubini's theorem and Lemma 6.6. We note that

$$E^Q\left[\int_0^1 \int_S \{[h'(m(\theta, t))]_\theta\}^2 d\theta dt \right] = E^{\tilde{Q}}\left[\int_S \{[h'(m(\theta))]_\theta\}^2 d\theta \right]$$

where $\tilde{Q} = \int_0^1 Q_t dt$ and Q_t is the marginal at time t of Q. Lemma 6.6 of course applies to $\tilde{Q}$.

We now turn to the proof of e). Let us consider the cutoff function ψ_l of Lemma 6.4 and pick $\varepsilon > 0$. We then consider for some $0 < t \leqq 1$,

$$F_{l,\varepsilon}^{N,t}(\mu(\cdot)) = \int_S J(\theta) \mu(t, d\theta) - \int_S J(\theta) \mu(0, d\theta)$$
$$- \frac{1}{2N} \int_0^t \sum_{j=1}^N J''\left(\frac{j}{N}\right) \tilde{\psi}_l \left(\frac{x_{j-[N\varepsilon]}(s) + \dots + x_{j+[N\varepsilon]}(s)}{1 + [2N\varepsilon]} \right) ds.$$

As functionals $F_{l,\varepsilon}^{N,t}(\mu(\cdot))$ converge as $N \to \infty$ to a limit given by

$$F_{l,\varepsilon}^t(\mu(\cdot)) = \int_S J(\theta) \mu(t, d\theta) - \int_S J(\theta) \mu(Om, d\theta)$$
$$- \int_0^t \int_s J'(\theta) \tilde{\psi}_l \left(\frac{\mu(s, [\theta - \varepsilon, \theta + \varepsilon])}{2\varepsilon} \right) d\theta ds,$$

and the convergence is uniform on compact subsets. Therefore

$$E^Q[|F^t_{l,\varepsilon}(\mu(\cdot))|] \leqq \overline{\lim_{N\to\infty}}\, E^{Q_N}[|F^{N,t}_{l,\varepsilon}(\mu(\cdot))|]. \tag{5.2}$$

Let us look at the right-hand side first. By Itô's formula

$$\begin{aligned}
&\frac{1}{N}\Sigma J\left(\frac{i}{N}\right)x_i(t) - \frac{1}{N}\Sigma J\left(\frac{i}{N}\right)x_i(0)\\
&= \frac{N^2}{2}\int_0^t \frac{1}{N}\Sigma J\left(\frac{i}{N}\right)\{\phi'(x_{i-1}(s)) - 2\phi'(x_i(s)) + \phi'(x_{i+1}(s))\}\,ds + M_N(t)\\
&= \frac{N^2}{2N}\int_0^t \Sigma\left(J\left(\frac{i+1}{N}\right) - 2J\left(\frac{i}{N}\right) + J\left(\frac{i-1}{N}\right)\right)\phi'(x_i(s))\,ds + M_N(t),
\end{aligned}$$

where $M_N(t)$ is a Martingale and an explicit calculation yields

$$E^{P_N}[|M_N(t)|^2] = \frac{1}{N^2}N^2 t\Sigma\left[J\left(\frac{i+1}{N}\right) - J\left(\frac{i}{N}\right)\right]^2 = \frac{Ct}{N}, \tag{5.3}$$

and tends to zero with N. Moreover because of Lemmas 6.4 and (5.3),

$$E^{P_N}\left[\left|\frac{1}{N}\Sigma J\left(\frac{i}{N}\right)x_i(t) - \frac{1}{N}\Sigma J\left(\frac{i}{N}\right)x_i(0) - \frac{1}{2N}\int_0^t \Sigma J''\left(\frac{i}{N}\right)\phi'(x_i(s))\,ds\right|\right] \to 0. \tag{5.4}$$

Let us consider

$$E^{P_N}\left|\frac{1}{2N}\int_0^t \Sigma J''\left(\frac{i}{N}\right)\phi'(x_i(s))\,ds - \frac{1}{2N}\int_0^t \Sigma J''\left(\frac{i}{N}\right)\psi_l(x_i(s))\,ds\right| = \Delta_N(l). \tag{5.5}$$

According to Lemma 6.4,

$$\lim_{l\to\infty}\limsup_{N\to\infty} \Delta_N(l) = 0. \tag{5.6}$$

Let us consider

$$\begin{aligned}
E^{P_N}\bigg|&\frac{1}{2N}\int_0^t \Sigma J''\left(\frac{i}{N}\right)\psi_l(x_i(s)\,ds\\
&- \frac{1}{2N}\int_0^t \Sigma J''\left(\frac{i}{N}\right)\tilde{\psi}_l\left(\frac{x_{i-N\varepsilon}(s) + \cdots + x_{i+N\varepsilon}(s)}{2N\varepsilon+1}\right)ds\bigg| = \bar{\Delta}_N(l,\varepsilon).
\end{aligned} \tag{5.7}$$

According to Theorem 4.1 and then (2.4),

$$\lim_{\varepsilon\to 0}\overline{\lim_{N\to\infty}}\, \bar{\Delta}_N(l,\varepsilon) = 0 \text{ for every } l. \tag{5.8}$$

If we combine (5.4), (5.5) and (5.6) we obtain

$$\lim_{l\to\infty}\limsup_{\varepsilon\to 0}\limsup_{N\to\infty} E^{Q_N}[|F^{N,t}_{l,\varepsilon}(\mu(\cdot))|] = 0,$$

and combined with (5.2) this yields

$$\lim_{l\to\infty}\lim_{\varepsilon\to 0} E^Q[|F^t_{l,\varepsilon}(\mu(\cdot))|]=0.$$

We can let $\varepsilon\to 0$ inside because $F^t_{l,\varepsilon}$ is uniformly bounded and obtain using a)

$$\lim_{l\to\infty} E^Q[|F^t_l(\mu(\cdot))|]=0,$$

where

$$F^t_l(\mu(\cdot))=\langle J,\mu(t)\rangle-\langle J,\mu(0)\rangle-\tfrac{1}{2}\int_0^t\int_s J''(\theta)\tilde{\psi}_l(m(s,\theta))\,ds\,d\theta.$$

Now we let $l\to\infty$. We can use Lemmas 6.5 and 6.3 and the dominated convergence theorem to conclude that

$$E^Q[|F^t(\mu(\cdot))|]=0,$$

where

$$F^t(\mu(\cdot))=\langle J,\mu(t)\rangle-\langle J,\mu(0)\rangle-\tfrac{1}{2}\int_0^t\int_s J\cdot(\theta)h'(m(s,\theta))\,ds\,d\theta.$$

This proves the theorem.

Now to complete the proof of Theorem 1.1 all we need is to appeal to a uniqueness theorem for weak solutions of

$$\frac{\partial m}{\partial t}=\tfrac{1}{2}[h'(m(t,\theta))]_{\theta\theta},\quad m(t,\theta)|_{t=0}=m_0(\theta).$$

6. Auxiliary Lemmas

In this section we shall prove some lemmas that were used in Sect. 5 where the hydrodynamic limit was identified. The first two lemmas show compactness of the process $\mu_N(t)$ defined by (5.1), in the sense described in that section. Lemma 6.3 is essentially (a) of Theorem 5.1 stated in a general way. Lemmas 6.4 and 6.5 are used in implementing the truncation in (c) of Theorem 5.1. Lemma 6.6 is needed in the proof of uniqueness in the next section.

Lemma 6.1. *Let P^{Φ_N} be the law of the process $\xi_N(t)=(x_1(t),\ldots,x_N(t))$ with Φ_N as initial distribution and let P_N be its law with initial density a_N (relative to Φ_N) such that*

$$\int_{R^N} a_N\log a_N\cdot\Phi_N dx\leqq CN$$

with C a constant. Then

$$\lim_{l\uparrow\infty}\overline{\lim_{N\to\infty}} P_N\left\{\sup_{0\leqq t\leqq 1}\frac{1}{N}\sum_{j=1}^N |x_j(t)|\geqq l\right\}=0.$$

Proof. By Lemma 1.12 of [K–V] we have that for any symmetric function g on $\mathbb{R}^N$

$$P^{\Phi_N}\left\{\sup_{0\leqq t\leqq 1} g(\xi_N(t))\geqq l\right\}\leqq\frac{3}{l}\sqrt{A+B},$$

where

$$A = \int_{R^N} g^2 \Phi_N dx,$$

$$B = \frac{N^2}{2} \int_{R^N} \left\{ \sum_{i=0}^{N-1} \frac{1}{g} \left(\frac{\partial g}{\partial x_i} - \frac{\partial g}{\partial x_{i+1}} \right)^2 \right\} \Phi_N dx$$

are assumed finite. Let

$$g = \exp\left\{ \sum_{i=0}^{N-1} |x_i| \right\}.$$

Then there are constants C_1 and C_2 independent of N such that

$$A \leqq C_1^N \quad \text{and} \quad B \leqq N^2 C_2 C_1^N.$$

Therefore

$$P^{\Phi_N}\left\{ \sup_{0 \leqq t \leqq 1} \frac{1}{N} \sum_{i=0}^{N-1} |x_i(t)| \geqq l \right\} \leqq \frac{3}{e^{Nl}} (C_1^N + N^2 C_2 C_1^N),$$

which is equivalent to

$$P^{\Phi_N}\left\{ \sup_{0 \leqq t \leqq 1} \frac{1}{N} \sum_{i=0}^{N-1} |x_i(t)| \geqq l \right\} \leqq C_3 e^{-C_4 N l}$$

for some constants C_3 and C_4.

We now use the basic inequality (2.3) which is

$$\int g f_N \Phi_N dx \leqq \log \int e^g \Phi_N dx + \int f_N \log f_N \cdot \Phi_N dx,$$

to obtain the inequality

$$P_N(A) \leqq \frac{\log 2 + H_N}{\log(1 + 1/P^{\Phi_N}(A))}$$

for any Borel set A in $\mathbb{R}^N$. Here H_N is the entropy of the law P_N relative to P^{Φ_N} and we know by hypothesis that $H_N \leqq CN$. This inequality and the estimate above give

$$P_N\left\{ \sup_{0 \leqq t \leqq 1} \frac{1}{N} \sum_{i=0}^{N-1} |x_i(t)| \geqq l \right\} \leqq \frac{\log 2 + CN}{\log(1 + 1/C_3 e^{-C_4 N l})}.$$

This implies that

$$\overline{\lim_{N \to \infty}} P_N\left\{ \sup_{0 \leqq t \leqq 1} \frac{1}{N} \sum_{i=1}^{N-1} |x_i(t)| \geqq l \right\} \leqq \frac{C_5}{l},$$

for some constant C_5 and hence the lemma is proved.

Lemma 6.2. *For every test function J on S,*

$$\lim_{\delta \downarrow 0} \overline{\lim_{N \to \infty}} P_N\left\{ \sup_{\substack{0 \leqq t,s \leqq 1 \\ |t-s| \leqq \delta}} |\langle J, \mu_N(t) \rangle - \langle J, \mu_N(s) \rangle| > \varepsilon \right\} = 0$$

for all $\varepsilon > 0$.

Proof. By Itô's formula and a calculation similar to the one in Sect. 5 for the identification we see that we must show that

$$\overline{\lim_{N\to\infty}} P_N\left\{\sup_{\substack{0\leqq s,t\leqq 1\\ |s-t|\leqq\delta}}\left|\int_s^t \frac{1}{N}\sum_{j=1}^N J''\left(\frac{j}{N}\right)\phi'(x_j(\sigma))d\sigma\right|>\varepsilon\right\}\to 0$$

as $\delta\to 0$ for all $\varepsilon>0$. In fact since $|J''|$ is bounded it suffices to show that

$$\lim_{\delta\downarrow 0}\overline{\lim_{N\to\infty}} P_N\left\{\sup_{\substack{0\leqq s,t\leqq 1\\ |s-t|\leqq\delta}}\left|\int_s^t \frac{1}{N}\sum_{j=1}^N \phi'(x_j(\sigma))d\sigma\right|>\varepsilon\right\}=0.$$

To prove this we use a truncation ψ_l of ϕ' defined by

$$\phi' \quad \text{for} \quad |\phi'|\leqq l, \quad \psi_l=l \quad \text{for} \quad \phi'>l, \quad -l \quad \text{for} \quad \phi'<-l.$$

It then suffices to show that the estimate is valid for ϕ' replaced by ψ_l, which is obvious since ψ_l is bounded, and then that

$$\lim_{l\uparrow\infty}\overline{\lim_{N\to\infty}} P_N\left\{\int_0^1\frac{1}{N}\sum_{j=1}^N|\psi_l(x_j(\sigma))-\phi'(x_j(\sigma))|>\varepsilon\right\}=0$$

for all $\varepsilon>0$. Now

$$P_N\left\{\int_0^1\frac{1}{N}\sum_{j=1}^N|\psi_l(x_j(\sigma))-\phi'(x_j(\sigma))|>\varepsilon\right\}\leqq\frac{1}{\varepsilon}\int_0^1 E^{P_N}\left\{\frac{1}{N}\sum_{j=1}^N|\psi_l(x_j(\sigma))-\phi'(x_j(\sigma))|\right\}d\sigma,$$

and for the expectation on the right we can use the basic inequality (2.3). This gives for any $0\leqq t\leqq 1$ and any $\gamma>0$,

$$\int_{R^N}\frac{1}{N}\sum_{j=1}^N|\psi_l(x_j)-\phi'(x_j)|f_N\Phi_N dx\leqq\frac{1}{N\gamma}\log\left(\int_R e^{\gamma|\psi_l-\phi'|-\phi}dx\right)^N+\frac{1}{N\gamma}\int f_N\log f_N D_N dx$$

$$\leqq\frac{c}{\gamma}+\frac{1}{\gamma}\log\left(\int_{|\phi'|>l}e^{\gamma|\phi'|-\phi}dx\right).$$

Thus

$$\lim_{l\to\infty}\overline{\lim_{N\to\infty}} P_N\left\{\int_0^1\frac{1}{N}\sum_{j=1}^N|\psi_l(x_j(\sigma))-\phi'(x_j(\sigma))|d\sigma>\varepsilon\right\}\leqq\frac{c}{\gamma},$$

where we use the hypothesis that

$$\int e^{\gamma|\phi'|-\phi}dx<\infty$$

for all $\gamma>0$. Letting $\gamma\to\infty$ gives the result.

Lemma 6.3. *Let f_N be the density relative to Φ_N of a random variable $\xi_N=(x_1,x_2,\dots,x_N)$ on $\mathbb{R}^N$ such that*

$$\int_{R^N} f_N\log f_N\Phi_N dx\leqq CN \tag{6.1}$$

with C independent of N. Let Q_N be the law of the empirical measure

$$\mu=\frac{1}{N}\sum_{j=1}^N\delta_{j/N}x_j.$$

Then for any weak limit Q of Q_N

$$Q\{\mu:\mu \ll d\theta\}=1,$$

and if $m(\theta)=d\mu/d\theta$ then

$$E^Q\left\{\int_S h(m(\theta))d\theta\right\}\leqq C,$$

where C is the constant in (6.1) and $h(x)$ is the conjugate convex function of $\rho(\lambda)=\log M(\lambda)$, $M(\lambda)=\int e^{\lambda y-\phi(y)}dy$.

Proof. Let W^N be the law of μ induced by Φ_N itself. Under W^N, μ is the empirical measure of N independent, identically distributed random variables with mean

$$m=\int xe^{-\phi(x)}dx,$$

and finite moments of all orders. The law of large numbers tells us that W^N converges weakly to the deterministic law concentrated on the measure $md\theta$ (Lebesgue measure multiplied by m). Moreover W^N has the large deviations property [V] with I function,

$$I(\mu)=\begin{cases}\int_S h(g(\theta))d\theta & \text{if } \mu\ll d\theta,\ \dfrac{d\mu}{d\theta}=g\\ +\infty & \text{otherwise}\end{cases}.$$

The large deviations property leads also to the result

$$\lim_{N\to\infty}\frac{1}{N}\log\int e^{Nu}\Phi_N=\sup\left\{u(\mu)-\int_S h(g(\theta))d\theta\right\}$$

for any bounded, continuous function u of μ.

Let $\{f_j\}$, $j=1,2,\ldots$ be a sequence of functions on S that are dense in $C(S)$ and let

$$u_k(\mu)=\sup_{1\leqq j\leqq k}\left\{\int_S f_j(x)\mu(d\theta)-\int_S \rho(f_j(\theta))d\theta\right\}.$$

We can pass to the limit in the basic inequality (2.3) in the form

$$\int_{R^N} u_k f_N\Phi_N dx\leqq\frac{1}{N}\log\int e^{Nu_k}\Phi_N dx+\frac{1}{N}\int f_N\log f_N\cdot\Phi_N dx.$$

Since the entropy per variable is bounded by C we conclude that

$$E^Q\{u_k(\cdot)\}\leqq\sup_{\{\mu:d\mu/d\theta=g\in L_1\}}\left\{u_k(\mu)-\int_S h(g(\theta))d\theta\right\}+C.$$

By convex duality

$$I(M)=\int_S h(g(\theta))d\theta=\sup_{f\in C(S)}\left\{\int_S f(\theta)\mu(d\theta)-\int_S\rho(f(\theta))d\theta\right\}\geqq u_k(\mu).$$

Using this we get the estimate

$$E^Q\{u_k(\cdot)\}\leqq C.$$

Since $u_k(\mu)\to I(\mu)$ as $k\to\infty$, by Fatou's lemma

$$E^Q\{I(\cdot)\}\leqq C.$$

This implies that

$$Q\{\mu:\mu\ll d\theta\}=1$$

and that

$$E^Q\left\{\int_S h(m(\theta))d\theta\right\}\leqq C$$

where $d\mu/d\theta=m(\theta)$.

Lemma 6.4. *Let $\psi_l(x)$ be a truncation of $\phi'(x)$.*

$$\psi_l(x)=\begin{matrix}\phi'(x), & |\phi'(x)|\leqq l\\ l\quad, & \phi'(x)>l\\ -l\ , & \phi'(x)<-l\end{matrix}\ .$$

Let

$$\tilde\psi_l(x)=\frac{1}{M(\lambda)}\int e^{\lambda y-\phi(y)}\psi_l(y)dy,$$

with $\lambda=h'(x)$. Then for each x,

$$\tilde\psi_l(x)\to h'(x)\quad\text{as}\quad l\to\infty.$$

Proof. We will use our hypotheses

$$\int e^{\lambda x-\phi(x)}dx<\infty\qquad\text{for all }\lambda,$$
$$\int e^{\sigma|\phi'(x)|-\phi(x)}dx<\infty\quad\text{for all }\sigma.$$

Clearly

$$\int e^{\sigma|\psi_l-\phi'|-\phi}dx\to 1\qquad\text{on}\quad l\to\infty,$$

for all $\sigma\geqq 0$. Now

$$\frac{1}{M(\lambda)}\int e^{\lambda y-\phi(y)}\phi'(y)dy=\lambda=h'(x).$$

Hence

$$|\tilde\psi_l(x)-h'(x)|=\left|\frac{1}{M(\lambda)}\int e^{\lambda y-\phi(y)}(\psi_l(y)-\phi'(y))dy\right|\leqq\int\frac{e^{\lambda y-\phi(y)}}{M(\lambda)}|\psi_l-\phi'|dy.$$

By the basic inequality (2.3), for any $\sigma>0$,

$$\int\frac{e^{\lambda y-\phi(y)}}{M(\lambda)}|\psi_l-\phi'|dy\leqq\frac{1}{\sigma}\log\int e^{\sigma|\psi_l-\phi'|-\phi}dy+\frac{1}{\sigma}\int\frac{e^{\lambda y}}{M(\lambda)}[\lambda y-\rho(\lambda)]e^{-\phi(y)}dy$$
$$=\frac{1}{\sigma}\log\int e^{\sigma|\psi_l-\phi'|-\phi}dy+\frac{h(x)}{\sigma}.$$

Here we use the fact that at $\lambda=h'(x)$,

$$\frac{1}{M(\lambda)}\int ye^{\lambda y-\phi(y)}dy=x.$$

Passing to the limit for x fixed we see that

$$\overline{\lim_{l\to\infty}}\ |\tilde{\psi}_l(x) - h'(x)| \leqq \frac{h(x)}{\sigma}$$

because of our basic hypotheses on ϕ and ϕ'. Letting now $\sigma \to \infty$ we obtain the desired result.

Lemma 6.5. *There is a constant C_ε such that for any $\varepsilon > 0$,*

$$|h'(x)| \leqq C_\varepsilon + \varepsilon h(x).$$

Proof. We will actually show that

$$\overline{\lim_{l\to\infty}}\ |\tilde{\psi}_l(x)| \leqq C_\varepsilon + \varepsilon h(x),$$

which combined with Lemma 6.4 gives the result.

Now

$$\tilde{\psi}_l(x) = \frac{1}{M(\lambda)} \int e^{\lambda y - \phi(y)} \psi_l(y)\, dy,$$

and as in the proof of Lemma 6.4

$$|\tilde{\psi}_l(x)| \leqq \frac{1}{\sigma} \log \int e^{\sigma|\psi_l(y)| - \phi(y)} dy + \frac{h(x)}{\sigma},$$

for any $\sigma > 0$. With

$$\tilde{C}_\sigma = \log \int e^{\sigma|\phi'| - \phi} dy,$$

we then get

$$\overline{\lim_{l\to\infty}}\ |\tilde{\psi}_l(x)| \leqq \frac{\tilde{C}_\sigma}{\sigma} + \frac{1}{\sigma} h(x).$$

If we let $\varepsilon = 1/\sigma$ and $C_\varepsilon = \varepsilon \tilde{C}_{1/\varepsilon}$ this is the same as the statement above.

Lemma 6.6. *Let us assume that the conditions of Lemma 6.3 are fulfilled and in addition*

$$\frac{1}{8} \int_S \frac{1}{f_N} \sum_{i=1}^{N} \left(\frac{\partial f_N}{\partial x_i} - \frac{\partial f_N}{\partial x_{i+1}} \right)^2 \Phi_N dx \leqq \frac{C'}{N}, \tag{6.2}$$

where C' is some nonnegative constant. Then any limit point Q obtained in that lemma satisfies in addition

$$E^Q \left[\int_S \left[\frac{\partial}{\partial \theta} h'(m(\theta)) \right]^2 d\theta \right] \leqq 8C' \tag{6.3}$$

with the same constant C' as in (6.2).

Proof. This follows immediately from Eq. (4.22).

7. Uniqueness

We will show in this section that a weak solution of the equation

$$\frac{\partial u}{\partial t} = \tfrac{1}{2}(h'(u))_{\theta\theta}, \quad u(0,\theta) = f(\theta) \tag{7.1}$$

is unique within the class of measures $\mu(t, d\theta)$ satisfying

$$\sup_{0 \leq t \leq T} \|\mu(t,\cdot)\| < \infty \quad \text{and} \quad \mu(t,\cdot) \text{ is weakly continuous in } t, \tag{7.2}$$

$\mu(t, d\theta) = u(t,\theta)\, d\theta$ for almost all t and

$$\int_0^T \int_S h(u(t,\theta))d\theta < \infty, \tag{7.3}$$

$$\int_0^T \int_S \{[h'(u(t,\theta))]_\theta\}^2 \leq C < \infty. \tag{7.4}$$

A weak solution is of course given by

$$\int_S J(\theta)\mu(t,d\theta) - \int_S J(\theta)\mu(0,d\theta) = \int_0^t \int_S \tfrac{1}{2} J''(\theta) h'(u(s,\theta)) ds\, d\theta, \tag{7.5}$$

for all smooth test functions $J(\cdot)$.

We first take $J = 1$ and check that $\int u(t,\theta)\, d\theta = \int f(\theta)\, d\theta = a$ for all t and all solutions. The function $u(t,\theta) - a$ is denoted by $\tilde{u}(t,\theta)$ and has mean zero on S. Therefore there is a function $\tilde{v}(t,\theta)$ of mean zero, such that

$$\frac{\partial \tilde{v}(t,\theta)}{\partial \theta} = \tilde{u}(t,\theta), \tag{7.6}$$

and (7.5) can be rewritten in a suitable sense as

$$\frac{\partial}{\partial t} \int_S \tilde{u}(t,\theta) J(\theta) d\theta = \frac{1}{2} \int_S J''(\theta)[h'(u(t,\theta))] d\theta \tag{7.7}$$

or

$$\frac{\partial}{\partial t} \int_S J'(\theta) \tilde{v}(t,\theta) d\theta = \frac{1}{2} \int_S J'(\theta)[h'(u(t,\theta))]_\theta d\theta. \tag{7.8}$$

Because of (7.4) this means that $\tilde{v}(t,\cdot)$ is differentiable in t as a map into $L_2(S)$ and

$$\frac{d\tilde{v}}{dt} = \tfrac{1}{2}[h'(u(t,\theta))]_\theta. \tag{7.9}$$

Now if we denote the difference between two solutions by w, then $w = u - v = \tilde{u} - \tilde{v}$ and

$$\frac{dw}{dt} = \tfrac{1}{2}[h'(u(t,\theta))]_\theta - \tfrac{1}{2}[h'(v(t,\theta))]_\theta. \tag{7.10}$$

If we now calculate

$$\begin{aligned}
\frac{d}{dt}\|w\|^2 &= \int_S w(t,\theta)[h'(u(t,\theta)) - h'(v(t,\theta))]_\theta d\theta \\
&= -\int_S (h'(u(t,\theta)) - h'(v(t,\theta)))(u(t,\theta) - v(t,\theta)) d\theta \\
&\leq 0.
\end{aligned}$$

Since $\| w(0) \| = 0$ we are done. The steps need a little bit of justification but it is routine.

8. Concluding Remarks

There have been several examples where hydrodynamic limits have been established for interacting systems. For example De Masi et al. [DE] and Dobrusin [D]. See also Spohn [S] for the closely related problem of fluctuation theory. Our work is related to and was motivated by the work of Fritz [FR]. Funaki [FU] has some results extending Fritz's work. But our methods are very different from all of the earlier methods. We use basically entropy estimates. They have the advantage of universality and with modifications the method should be applicable to other models that are reversible and when the scaling is of diffusion type. Moreover the method allows us to proceed farther and do large deviation theory. See for instance Donsker and Varadhan [DV]. Although we treat the case of dimension one it is not an essential condition. With only a change in notation the method works in any number of dimensions.

Replacing the circle or torus by R^d poses more of a challenge. The problem in some sense is essentially local, and it should be possible to localize. Fritz has conveyed to us some ideas in private communication, but it still remains to be done.

References

[KV] Kipnis, C., Varadhan, S. R. S.: Central limit theorem for additive functionals of reversible Markov processes and applications to simple exclusions. Commun. Math. Phys. **104**, 1–19 (1986)

[DE] De Masi, A., Ianiro, N., Pellegrinotti, A., Presutti, E.: A survey of the hydrodynamical behavior of many-particle systems. In: Nonequilibrium phenomena II. From stochastics to hydrodynamics. Lebowitz, J. L., Montroll, E. N. (eds.) pp. 123–294. Amsterdam: North-Holland 1984

[D] Dobrushin, R., Seigmund-Schultze, R.: The hydrodynamic limit for systems of particles with independent evolution, Akademie der Wissenschaften der DDR, Berlin, 1981

[S] Spohn, H.: Equilibrium fluctuations for interacting Brownian particles. Commun. Math. Phys. **103**, 1–33 (1986)

[FR] Fritz, J.: On the hydrodynamic limit of a scalar Ginzburg–Landau lattice model, IMA Vol. **9**. Berlin, Heidelberg, New York: Springer 1987

[F] Funaki, T.: preprint 1987

[DV] Donsker, M., Varadhan, S. R. S.: preprint 1987

[V] Varadhan, S. R. S.: Asymptotic probabilities and differential equations. Commun. Pure Appl. Math. **19**, 261–286 (1966)

Communicated by J. L. Lebowitz

Received October 29, 1987

Comm. Pure Appl. Math. 42 (1989) 115-137

Hydrodynamics and Large Deviation for Simple Exclusion Processes

C. KIPNIS
Centre de Mathematiques Appliquées, Ecole Polytechnique

S. OLLA AND S. R. S. VARADHAN
Courant Institute

Abstract

We prove the hydrodynamical limit for weakly asymmetric simple exclusion processes. A large deviation property with respect to this limit is established for the symmetric case. We treat also the situation where a slow reaction (creation and annihilation of particles) is present.

0. Introduction

When we study the hydrodynamical limit (reduced description) of an infinite particle system we can obtain the same limiting equation for different systems. For instance, both symmetric simple exclusion and a system of independent symmetric random walks give rise to the heat equation. However, in the study of the large deviations, the rate functions are different for these two systems.

In this spirit the large deviations from the hydrodynamical limit for independent Brownian motions were studied in [7], using an approach that can be also used for strongly interacting systems, like simple exclusion, as we show in this paper.

The case of independent particles is simple because only the empirical density field appears in the martingales considered. In the case of strongly interacting systems the density field is no longer autonomous and we need to control the deviations of the empirical correlation fields. In Theorem 2.1 we prove that the probability that any correlation field deviates from an appropriate function of the density field is superexponentially small, whereas the large deviations of the density fields are only exponentially small.

The other ingredient needed for the lower bound is a large number of hydrodynamical limit theorems for several small perturbations of the original system. The perturbations arise naturally and all the necessary hydrodynamical limits follow easily from the superexponential estimate (2.1) (see Section 3). For the symmetric simple exclusion, these small perturbations are the weakly asymmetric exclusion processes with an asymmetry that depends on space and time (see also [4] for a very different proof of the hydrodynamical limit for these systems).

The method used here for the proof of Theorem 2.1 is the one used in [8] and [5] to obtain the hydrodynamical equation and large deviation principle for the Ginzburg-Landau model on the lattice. The method is quite general and could be used to study the zero range process. It is also powerful enough to allow one to treat small perturbation-like reaction diffusion (see Section 6). However it is very heavily dependent on the reversibility of the system in equilibrium and so one still has to understand how to deal with genuinely nonreversible situations, like the asymmetric simple exclusion.

1. Notation and Results

We consider, for each integer N, simple exclusion processes on the configuration space $X_N = \{0,1\}^{T_N}$, where $T_N = \{i/N \in T,\ i = 1,\cdots, N\}$ and T is the unit circle ($T = \mathbf{R}/\mathbf{Z}$). A configuration in X_N is denoted by $\eta = \{\eta(j);\ j = 1,\cdots, N\}$, $\eta(j) = 0$ or 1.

Given a function $\gamma(\theta)$ on T such that $0 \leq \gamma(\cdot) \leq 1$, we denote by ν_γ^N the Bernoulli distribution on X_N with profile, γ, i.e., the product measure on X_N with marginals given by $\nu_\gamma^N(\eta(j)) = \gamma(j/N)$. Define also, for $\alpha \in [0,1]$ and $\phi \in \mathbf{C}(X_N)$,

$$\tilde{\phi}(\alpha) = \nu_\alpha^N(\phi). \tag{1.1}$$

We have used α to denote the constant function α as well. The (speeded up) symmetric simple exclusion process is the Markov process on X_N generated by

$$(L_N f)(\eta) = \tfrac{1}{2}N^2 \sum_{i=1}^{N} \left[f(\eta^{i,i+1}) - f(\eta)\right], \tag{1.2}$$

where

$$\eta^{i,i+1}(j) = \begin{cases} \eta(j), & j \neq i, i+1, \\ \eta(i), & j = i+1, \\ \eta(i+1), & j = i. \end{cases}$$

For a fixed interval of time $[0, t]$, we denote by $\mathbf{P}_N^\gamma$ the corresponding probability measure on the path space $E_N = D([0, t]; X_N)$ (right continuous trajectories on X_N with left limits), the initial measure being the Bernoulli distribution ν_γ^N. We shall also consider the weakly asymmetric simple exclusion processes (WASEP) generated by

$$(1.3)\quad \left(L_N^H f\right)(\eta) = \tfrac{1}{2}N^2 \sum_{i=1}^{N} \left(c_{i,i+1}^H(\eta, s) + c_{i+1,i}^N(\eta, s)\right)\left(f(\eta^{i,i+1}) - f(\eta)\right),$$

where

$$c_{i,i+1}^H(\eta, s) = \exp\left\{H\left(\frac{i+1}{N}, s\right) - H\left(\frac{i}{N}, s\right)\right\}\eta(i)(1 - \eta(i+1)),$$

$$c_{i+1,i}^H(\eta, s) = \exp\left\{H\left(\frac{i}{N}, s\right) - H\left(\frac{i+1}{N}, s\right)\right\}\eta(i+1)(1 - \eta(i)),$$

and $H(\theta, s)$ is a continuous function on $T \times [0, t]$. The corresponding measure on E_N will be denoted by $\mathbf{P}_N^{H,\gamma}$, with ν_N^γ as initial distribution. The number of jumps occurring from the site j to the site $j + 1$ up to time s will be denoted by $J_s^{j,j+1}$. Thus, under $\mathbf{P}_N^{H,\gamma}$-probability, $J_s^{j,j+1}$ is a Poisson process with (varying) intensity $c_{j,j+1}^H(\eta_s, s)$. $J_s^{j+1,j}$ is defined similarly.

We denote by M_1 the set of all measurable functions $\rho(\theta)$ on T such that $0 \leqq \rho(\theta) \leqq 1$. We shall use the notation

$$\langle \rho, G\rangle = \int_T G(\theta)\rho(\theta)\, d\theta, \qquad \rho \in M_1,\ G \in \mathbf{C}(T),$$

and consider on M_1 the weak topology induced by $\mathbf{C}(T)$ with the duality $\langle \cdot, \cdot\rangle$. Given any path $\{\eta_s,\ s \in [0, t]\} = \eta \in E_N$, the empirical density is defined by

$$\mu^N(s, \theta, \eta) = \sum_{i=1}^{N} \eta_s(i)\chi_{(i/N,(i+1)/N)}(\theta)$$

so that $\mu^N(\cdot; \eta) \in D([0, t]; M_1)$.

In Section 5 we shall consider the Hilbert space $H^1(\rho)$ for $\rho \in D([0, t], M_1)$ defined in the following way: Consider in $\mathbf{C}^{2,1}(T \times [0, t])$ the equivalence relation $G(\theta, s) \sim G'(\theta, s)$ if $G(\theta, s) - G'(\theta, s)$ is constant in θ for every $s \in [0, t]$. On $\mathbf{C}^{2,1}/\sim$ consider the scalar product

$$[G, G'] = \int_0^t \left\langle \rho_s, \frac{\partial G}{\partial \theta} \cdot \frac{\partial G'}{\partial \theta}(\cdot, s) \right\rangle ds;$$

then $\mathbf{H}^1(\rho)$ is the completion of $\mathbf{C}^{2,1}/\sim$ with respect to this scalar product.

In Section 3, we prove the convergence to a deterministic density of μ^N in $\mathbf{P}_N^{H,\gamma}$-probability as $N \to \infty$ (law of large numbers). In Sections 4 and 5, we establish a large deviation principle for the $\mathbf{P}_N^{\alpha}$-distribution of μ^N. The results on $\mathbf{P}_N^{H,\gamma}$ of Section 3 are needed in Section 5 in order to obtain lower bounds for the probability of large deviations.

2. Superexponential Estimates

In this section, ϕ will be a function defined in X_N for any N, depending only on a fixed number of sites. In Sections 4 and 5, we shall use only the case $\phi(\eta) = \eta(0)\eta(1)$.

THEOREM 2.1. *For any $\delta > 0$,*

$$\overline{\lim_{\varepsilon \downarrow 0}}\ \overline{\lim_{N \to \infty}} \frac{1}{N} \log \mathbf{P}_N^{1/2}\left(\frac{1}{N} \int_0^t V_{N,\varepsilon}(\eta_s)\, ds \geq \delta \right) = -\infty, \tag{2.1}$$

where

$$V_{N,\varepsilon}(\eta) = \sum_{i=1}^{N} \left| \frac{1}{2\varepsilon N + 1} \sum_{|i-j| \leq N\varepsilon} \tau_j \phi(\eta) - \tilde{\phi}\left(\frac{1}{2\varepsilon N + 1} \sum_{|i-j| \leq N\varepsilon} \eta(j) \right) \right|,$$

and τ_j is the space shift on X_N.

Proof: For any $N \in \mathbb{N}$, $\varepsilon > 0$, $a < \infty$, consider the operator $L_N + aV_{N,\varepsilon}$ on $\mathbf{L}^2(X_N, \nu^{1/2})$. It is selfadjoint. We denote by $\lambda_{N,\varepsilon}(a)$ its largest eigenvalue.

By the Feynman-Kac formula and the spectral theorem, we have

$$\mathbf{E}^{\mathbf{P}_N^{1/2}}\left(\exp\left\{ a \int_0^t V_{N,\varepsilon}(\eta_s)\, ds \right\} \right) \leq \exp\{ t\lambda_{N,\varepsilon}(a) \}.$$

Thus by Chebychev inequality,

$$\mathbf{P}_N^{1/2}\left(\frac{1}{N}\int_0^t V_{N,\varepsilon}(\eta_s)\,ds \geqq \delta\right)$$
$$\leqq \exp\{t\lambda_{N,\varepsilon}(a) - N\delta a\} = \exp\left\{N\left(\frac{t}{N}\lambda_{N,\varepsilon}(a) - \delta a\right)\right\}.$$

Then it is enough to show that, for any $a > 0$,

$$\overline{\lim_{\varepsilon\downarrow 0}}\ \overline{\lim_{N\to\infty}}\ \frac{1}{N}\lambda_{N,\varepsilon}(a) = 0$$

in order to see that

$$\overline{\lim_{\varepsilon\downarrow 0}}\ \overline{\lim_{N\to\infty}}\ \frac{1}{N}\log\mathbf{P}_N^{1/2}\left(\frac{1}{N}\int_0^t V_{N,\varepsilon}(\eta_s)\,ds \geqq \delta\right) \leqq -a\delta,$$

and then taking the limit as a goes to infinity we obtain (2.1).

By the classical variational formula for the largest eigenvalue of a symmetric matrix,

$$\lambda_{N,\varepsilon}(a) = \sup_{f_N}\left\{a\sum_{\eta\in X_N} V_{N,\varepsilon}(\eta)f_N(\eta)2^{-N} - N^2\mathbf{D}_N(f_N)\right\},$$

where the sup is taken over all the distributions on X_N (i.e., $\sum_{\eta\in X_N} f_N(\eta)2^{-N} = 1$) and $\mathbf{D}_N$ is the Dirichlet form defined by

$$\mathbf{D}_N(f_N) = \frac{1}{4}\sum_{\eta\in X_N} 2^{-N}\sum_{i=1}^{N}\left(\sqrt{f(\eta^{i,i+1})} - \sqrt{f(\eta)}\right)^2.$$

Observe that $V_{N,\varepsilon}$ and $\mathbf{D}_N$ are invariant under space translations; thus, by convexity, the sup in the variational formula can be restricted to translation invariant f. Since $V_{N,\varepsilon}(\eta)$ is uniformly bounded by CN, it is enough to show that, for any $c > 0$,

$$\overline{\lim_{\varepsilon\downarrow 0}}\ \overline{\lim_{N\to\infty}}\ \sup_{f_N:\,\mathbf{D}_N(f_N)\leqq c/N}\frac{1}{N}\sum_{\eta\in X_N} V_{N,\varepsilon}(\eta)f_N(\eta)2^{-N} = 0.$$

By the translation invariance of the f it suffices to show that

$$\overline{\lim_{\varepsilon\downarrow 0}}\ \overline{\lim_{N\to\infty}}\ \sup_{f_N:\,\mathbf{D}(f_N)\leqq c/N}\sum_{\eta\in X_N}\left|\frac{1}{\varepsilon N+1}\sum_{|j|\leqq\varepsilon N}\tau_j\phi(\eta) - \tilde{\phi}\left(\frac{1}{2\varepsilon N+1}\sum_{|j|\leqq\varepsilon N}\eta(j)\right)\right| f_N(\eta)2^{-N} = 0. \tag{2.2}$$

Observe that, for any $\eta \in X_N$,

$$\frac{1}{2\varepsilon N+1}\sum_{|j|\leq \varepsilon N}\tau_j\phi(\eta)=\frac{1}{2\varepsilon N+1}\sum_{|j|\leq \varepsilon N}\left(\frac{1}{2k+1}\sum_{|j-l|\leq k}\tau_l\phi(\eta)\right)+O\left(\frac{k}{N}\right),$$

where $O(k/N)\to 0$ as $N\to\infty$. Therefore,

$$\left|\frac{1}{2\varepsilon N+1}\sum_{|j|\leq \varepsilon N}\tau_j\phi(\eta)-\tilde{\phi}\left(\frac{1}{2\varepsilon N+1}\sum_{|j'|\leq \varepsilon N}\eta(j')\right)\right|$$

$$\leqq \frac{1}{2\varepsilon N+1}\sum_{|j|\leq \varepsilon N}\left|\frac{1}{2k+1}\sum_{|j-l|\leq k}\tau_l\phi(\eta)-\tilde{\phi}\left(\frac{1}{2\varepsilon N+1}\sum_{|j'|\leq \varepsilon N}\eta(j')\right)\right|$$

$$+O\left(\frac{k}{N}\right)$$

$$\leqq \frac{1}{2\varepsilon N+1}\sum_{|j|\leq \varepsilon N}\left|\frac{1}{2k+1}\sum_{|j-l|\leq k}\tau_l\phi(\eta)-\tilde{\phi}\left(\frac{1}{2k+1}\sum_{|j-l|\leq k}\eta(l)\right)\right|$$

$$+\frac{1}{2\varepsilon N+1}\sum_{|j|\leq \varepsilon N}\left|\tilde{\phi}\left(\frac{1}{2\varepsilon N+1}\sum_{|j'|\leq N}\eta(j')\right)-\tilde{\phi}\left(\frac{1}{2k+1}\sum_{|j-l|\leq k}\eta(l)\right)\right|$$

$$+O\left(\frac{k}{N}\right)$$

$$\leqq \frac{1}{2\varepsilon N+1}\sum_{|j|\leq \varepsilon N}\left|\frac{1}{2k+1}\sum_{|j-l|\leq k}\tau_l\phi(\eta)-\tilde{\phi}\left(\frac{1}{2k+1}\sum_{|j-l|\leq k}\eta(l)\right)\right|$$

$$+\frac{\|\tilde{\phi}'\|_\infty}{2\varepsilon N+1}\sum_{|j|\leq \varepsilon N}\left|\frac{1}{2\varepsilon N+1}\sum_{|j'|\leq N}\eta(j')-\frac{1}{2k+1}\sum_{|j-l|\leq k}\eta(l)\right|+O\left(\frac{k}{N}\right)$$

$$\leqq \frac{1}{2\varepsilon N+1}\sum_{|j|\leq \varepsilon N}\left|\frac{1}{2k+1}\sum_{|j-l|\leq k}\tau_l\phi(\eta)-\tilde{\phi}\left(\frac{1}{2k+1}\sum_{|j-l|\leq k}\eta(l)\right)\right|$$

$$+\|\tilde{\phi}'\|_\infty\frac{1}{(2\varepsilon N+1)^2}\sum_{|j|\leq \varepsilon N}\sum_{|j'|\leq \varepsilon N}\left|\frac{1}{2k+1}\sum_{|j'-l|\leq k}\eta(l)-\frac{1}{2k+1}\sum_{|j-l\leq k}\eta(l)\right|$$

$$+O\left(\frac{k}{N}\right).$$

Then, using again the translation invariance of the f's considered, equation (2.2) is a consequence of the following two lemmas.

LEMMA 2.2. *For any* $c > 0$,

$$\varlimsup_{k\to\infty} \varlimsup_{N\to\infty} \sup_{f_N:\, \mathbf{D}_N(f_N)\le c/N} \sum_{\eta\in X_N} \left| \frac{1}{2k+1} \sum_{|j|\le k} \tau_j \phi(\eta) - \tilde{\phi}\left(\frac{1}{2k+1} \sum_{|j|\le k} \eta(j) \right) \right| \cdot f_N(\eta) 2^{-N} = 0.$$

LEMMA 2.3. *For any* $c > 0$,

$$\varlimsup_{k\to\infty} \varlimsup_{\varepsilon\downarrow 0} \varlimsup_{N\to\infty} \sup_{|r|\le N\varepsilon} \sup_{f_N:\, \mathbf{D}_N(f_N)\le c/N} \sum_{\eta\in X_N} \left| \frac{1}{2k+1} \sum_{|j|\le k} \eta(j) - \frac{1}{2k+1} \sum_{|j+r|\le k} \eta(j) \right| f_N(\eta) 2^{-N} = 0.$$

We first prove these two lemmas, and then show how they imply (2.2).

Proof of Lemma 2.2: For a function g on $\{0,1\}^{2k+1}$, define

$$\mathbf{D}^*_{2k+1}(g) = \frac{1}{4} \sum_{\eta\in\{0,1\}^{2k+1}} 2^{-(2k+1)} \sum_{j=-k}^{k-1} \left(\sqrt{g(\eta^{j,j+1})} - \sqrt{g(\eta)} \right)^2.$$

Let $f_N^k(\eta) = 2^{-N+2k+1} \Sigma_{\eta(j):\, |j|>k} f_N(\eta)$ so that $2^{-(2k+1)} f_N^k$ is the marginal probability distribution on $\{0,1\}^{2k+1}$. Then

$$\mathbf{D}^*_{2k+1}(f_N^k) = \frac{1}{4} \sum_{\eta(j),\, |j|\le k} 2^{-N} \sum_{j=-k}^{k-1} \left(\sqrt{\sum_{\eta(j),\, |j|>k} f_N(\eta^{j,j+1})} - \sqrt{\sum_{\eta(j),\, |j|>k} f(\eta)} \right)^2$$

$$\le \frac{1}{4} \sum_{\eta\in X_N} 2^{-N} \sum_{j=-k}^{k-1} \left(\sqrt{f_N(\eta^{j,j+1})} - \sqrt{f_N(\eta)} \right)^2 \tag{2.3}$$

$$= \frac{1}{4} \sum_{j=-k}^{k-1} \sum_{\eta\in X_N} 2^{-N} \left(\sqrt{f_N(\eta^{j,j+1})} - \sqrt{f_N(\eta)} \right)^2$$

$$= \frac{2k}{N} \mathbf{D}_N(f_N)$$

where we have used the inequality

$$\left(\sqrt{\sum_j a_j} - \sqrt{\sum_j b_j}\right)^2 \leqq \sum_j \left(\sqrt{a_j} - \sqrt{b_j}\right)^2, \qquad a_j, b_j \geqq 0,$$

(which follows directly from Schwartz' inequality) and the translation invariance of f_N. Using inequality (2.3) we have

$$\begin{aligned}
&\sup_{f_N:\, \mathbf{D}_N(f_N) \leq c/N} \sum_{\eta \in X_N} \left| \frac{1}{2k+1} \sum_{|j| \leq k} \tau_j \phi(\eta) - \tilde{\phi}\left(\frac{1}{2k+1} \sum_{|j| \leq k} \eta(j) \right) \right| f_N(\eta) 2^{-N} \\
&\leqq \sup_{g_k:\, \mathbf{D}^*_{2k+1}(g_k) \leq 2ck/N^2} \sum_{\eta \in X_{2k+1}} \left| \frac{1}{2k+1} \sum_{|j| \leq k} \tau_j \phi(\eta) - \tilde{\phi}\left(\frac{1}{2k+1} \sum_{|j| \leq k} \eta(j) \right) \right| g_k(\eta) 2^{-(2k+1)} + O\left(\frac{1}{k}\right).
\end{aligned}$$

On the right-hand side of the last equation a small error $O(1/k) \xrightarrow{k \to \infty} 0$ (uniformly in η) is made by the finite range of ϕ. By the compactness of the level sets of $\mathbf{D}^*_{2k+1}$ we have

$$\begin{aligned}
&\overline{\lim_{N \to \infty}} \sup_{g_k:\, \mathbf{D}^*_{2k+1}(g_k) \leq 2ck/N^2} \sum_{\eta \in X_{2k+1}} \left| \frac{1}{2k+1} \sum_{|j| \leq k} \tau_j \phi(\eta) - \tilde{\phi}\left(\frac{1}{2k+1} \sum_{|j| \leq k} \eta(j) \right) \right| g_k(\eta) 2^{-(2k+1)} \\
&\leqq \sup_{g_k:\, \mathbf{D}^*_{2k+1}(g_k) = 0} \sum_{\eta \in X_{2n+1}} \left| \frac{1}{2k+1} \sum_{|j| \leq k} \tau_j \phi(\eta) - \tilde{\phi}\left(\frac{1}{2k+1} \sum_{|j| \leq k} \eta(j) \right) \right| g_k(\eta) 2^{-(2k+1)}.
\end{aligned}$$

The set of probability distributions $\{g_k(\eta)2^{-(2k+1)}: \mathbf{D}_{2k+1}(g_k) = 0\}$ on X_{2k+1} is the convex set of all the symmetric distributions whose extremals are the uniform distributions with fixed number of particles. Then the last expression equals

$$\sup_{l=1,\cdots,2k+1} \sum_{\eta:\Sigma_j\eta(j)=l} \binom{2k+1}{l}^{-1} \left| \frac{1}{2k+1} \sum_{j=1}^{2k+1} \tau_j\phi(\eta) - \tilde{\phi}\left(\frac{l}{2k+1}\right) \right|,$$

which goes to zero as $k \to \infty$ by the uniform convergence of the finite-dimensional distribution of the uniform measure to the corresponding Bernoulli measure with the same mean.

Proof of Lemma 3.3: Let $f_N(\eta)$ be as in the proof of Lemma 3.2 and define

$$f_N^{r,k}(\eta) = 2^{-N+2(2k+1)} \sum_{\substack{\eta(j),|j|>k \\ |r-j|>k}} f_N(\eta).$$

For a function $g(\eta_1, \eta_2)$ on $X_{2k+1} \times X_{2k+1}$ define

$$\mathbf{D}_k^1(g) = \frac{1}{4} \sum_{\eta_1,\eta_2} 2^{-2(2k+1)} \sum_{i=-k}^{k-1} \left(\sqrt{g\left(\eta_1^{i,i+1}, \eta_2\right)} - \sqrt{g(\eta_1, \eta_2)}\right)^2,$$

$$\mathbf{D}_k^2(g) = \frac{1}{4} \sum_{\eta_1,\eta_2} 2^{-2(2k+1)} \sum_{i=-k}^{k-1} \left(\sqrt{g\left(\eta_1, \eta_2^{i,i+1}\right)} - \sqrt{g(\eta_1, \eta_2)}\right)^2,$$

$$\Delta_k(g) = \frac{1}{4} \sum_{\eta_1,\eta_2} 2^{-2(2k+1)} \left(\sqrt{g\left((\eta_1, \eta_2)^0\right)} - \sqrt{g(\eta_1, \eta_2)}\right)^2,$$

where $(\eta_1, \eta_2)^0$ is obtained from (η_1, η_2) by exchanging the values of $\eta_1(0)$ and $\eta_2(0)$.

By the same argument as in the proof of inequality (2.3) we have

$$\mathbf{D}_k^1\left(f_N^{r,k}\right) \leq \frac{2k}{N} \mathbf{D}_N(f_N)$$

and

$$\mathbf{D}_k^2\left(f_N^{r,k}\right) \leq \frac{2k}{N} \mathbf{D}_N(f_N),$$

and by Schwartz's inequality

$$\Delta_k\left(f_N^{r,k}\right) \leq \sum_{\eta \in X_N} 2^{-N}\left(\sqrt{f_N(\eta^{0,r})} - \sqrt{f_N(\eta)}\right)^2.$$

Because

$$\eta^{0,r} = \left(\cdots\left(\left(\left(\cdots\left(\left(\eta^{0,1}\right)^{1,2}\right)\cdots\right)^{r-1,r}\right)^{r-1,r-2}\right)\cdots\right)^{1,0},$$

using telescoping sums, and Schwartz's inequality, we obtain

$$\Delta_k\left(f_N^{r,k}\right) \leqq (2r-1)^2 \sum_{\eta \in X_N} 2^{-N}\left(\sqrt{f_N(\eta^{0,1})} - \sqrt{f_N(\eta)}\right)^2$$

$$= \frac{(2r-1)^2}{N} \mathbf{D}_N(f_N),$$

where we have used also the translation invariance of f_N.

For any $c > 0$, $\varepsilon > 0$, $N \in \mathbf{N}$, define the set of distributions on $\{0,1\}^{2k+1} \times \{0,1\}^{2k+1}$:

$$A_{N,\varepsilon} = \left\{ g(\eta_1, \eta_2) \colon \mathbf{D}_k^1(g) \leqq \frac{2kc}{N^2}; \quad \mathbf{D}_k^2(g) \leqq \frac{2kc}{N^2}; \quad \Delta_k(g) \leqq \varepsilon^2 c \right\}.$$

Then we have

$$\sup_{|r| \leqq \varepsilon N} \sup_{\mathbf{D}_N(f_N) \leqq c/N} \sum_{\eta \in X_N} \left| \frac{1}{2k+1} \sum_{|j| \leqq k} \eta(j) - \frac{1}{2k+1} \sum_{|j+r| \leqq k} \eta(j) \right| f_N(\eta)^{2^{-N}}$$

$$\leqq \sup_{g \in A_{N,\varepsilon}} \sum_{\eta_1, \eta_2} \left| \frac{1}{2k+1} \sum_{j=-k}^{k} \eta_1(j) - \frac{1}{2k+1} \sum_{j=-k}^{k} \eta_2(j) \right| g(\eta_1, \eta_2) 2^{-2(2k+1)}.$$

As in the proof of Lemma 2.2, by compactness of the sets $A_{N,\varepsilon}$ we only have to show that

$$\overline{\lim_{k \to \infty}} \sup_{\substack{\mathbf{D}_u^1(g) = 0 \\ \mathbf{D}_u^2(g) = 0 \\ \Delta_k(g) = 0}} \sum_{\eta_1, \eta_2} \left| \frac{1}{2k+1} \sum_{j=-k}^{k} \eta_1(j) - \frac{1}{2k+1} \sum_{j=-k}^{k} \eta_2(j) \right| \cdot g(\eta_1, \eta_2) 2^{-2(2k+1)} = 0. \tag{2.4}$$

The three conditions $\mathbf{D}_k^1(g) = 0$, $\mathbf{D}_k^2(g) = 0$ and $\Delta_k(g) = 0$ imply that g is symmetric in $X_{2k+1} \times X_{2k+1}$. The supremum on $X_{2k+1} \times X_{2k+1}$ is obtained at the uniform distribution on $X_{2k+1} \times X_{2k+1}$ with a fixed number of particles. The same argument as used at the end of Lemma 2.2 yields (2.4).

Remark. Theorem 2.1 has been proved for $\mathbf{P}_N^{1/2}$, i.e., in equilibrium. It extends immediately to $\mathbf{P}_N^{\gamma}$ for any initial profile $\gamma(\theta)$ observing that, for $A \subset E_N$,

$$\mathbf{P}_N^{\gamma}(A) = \sum_{\eta \in X_N} \frac{d\nu_N^{\gamma}}{d\nu_N^{1/2}}(\eta) \mathbf{P}_N^{\eta}(A) \nu_N^{1/2}(\eta) \leqq 2^N \mathbf{P}_N^{1/2}(A).$$

By a similar argument it can be extended also to the weakly asymmetric exclusion $\mathbf{P}_N^{H,\gamma}$, for H regular, as will be shown in the next section.

3. Hydrodynamical Limits

We derive here several hydrodynamical limits for the WASEP $\mathbf{P}_N^{H,\gamma}$ defined in Section 1. $H(\theta, s)$ will be a smooth function here.

THEOREM 3.1. *As N tends to infinity, the empirical density $\mu^N(s, \theta, \eta)$ converges in $\mathbf{P}_N^{H,\gamma}$-probability to the unique weak solution $\rho(s, \theta)$ of the equation*

$$\frac{\partial}{\partial t}\rho = \frac{1}{2}\frac{\partial^2}{\partial\theta^2}\rho - \frac{\partial}{\partial\theta}\left(\frac{\partial H}{\partial\theta}\rho(1-\rho)\right), \tag{3.1}$$

$$\rho(0,\theta) = \gamma(\theta).$$

The uniqueness of the weak solution of (3.1) in the class of bounded functions follows from the boundedness of $\gamma(\theta)$.

Before proving Theorem 3.1 we extend the superexponential estimate of Section 2 to $\mathbf{P}_N^{H,\gamma}$. This will be a simple consequence of the Girsanov formula.

THEOREM 3.2. *For any $\delta > 0$ the superexponential estimate* (2.1) *also holds for* $\mathbf{P}_N^{H,\gamma}$.

Proof of Theorem 3.2: The Radon-Nykodym derivative of $\mathbf{P}_N^{H,\gamma}$ with respect to $\mathbf{P}_N^{\gamma}$ is given by the following Girsanov formula:

$$\begin{aligned} Z_N^H = \exp\Bigg\{ \sum_{i=1}^N \Bigg[& \int_0^t \left(H\left(\frac{i+1}{N}, s\right) - H\left(\frac{i}{N}, s\right)\right) dJ_s^{i,i+1} \\ & + \int_0^t \left(H\left(\frac{i}{N}, s\right) - H\left(\frac{i+1}{N}, s\right)\right) dJ_s^{i+1,i} \\ & - \tfrac{1}{2}N^2 \int_0^t \left(\exp\left\{ H\left(\frac{i+1}{N}, s\right) - H\left(\frac{i}{N}, s\right)\right\} - 1\right) \eta_s(i)(1-\eta_s(i+1))\, ds \\ & - \tfrac{1}{2}N^2 \int_0^t \left(\exp\left\{ H\left(\frac{i}{N}, s\right) - H\left(\frac{i+1}{N}, s\right)\right\} - 1\right) \eta_s(i+1)(1-\eta_s(i))\, ds \Bigg]\Bigg\}. \end{aligned}$$

We integrate by parts the first two terms in the exponential above, reorder the sums and write Z_N^H as

$$Z_N^H = \exp\Bigg\{ \sum_{i=1}^{N} \Bigg[H\left(\frac{i}{N}, t\right)\eta_t(i) - H\left(\frac{i}{N}, 0\right)\eta_0(i) - \int_0^t \frac{\partial H}{\partial t}\left(\frac{i}{N}, s\right)\eta_s(i)\, ds$$

$$- \tfrac{1}{2}N^2 \int_0^t \Bigg(\exp\left\{ H\left(\frac{i+1}{N}, s\right) - H\left(\frac{i}{N}, s\right)\right\}$$

$$+ \exp\left\{ H\left(\frac{i-1}{N}, s\right) - H\left(\frac{i}{N}, s\right)\right\} - 2\Bigg)\eta_s(i)\, ds$$

$$+ \tfrac{1}{2}N^2 \int_0^t \Bigg(\exp\left\{ H\left(\frac{i+1}{N}, s\right) - H\left(\frac{i}{N}, s\right)\right\}$$

$$+ \exp\left\{ H\left(\frac{i}{N}, s\right) - H\left(\frac{i+1}{N}, s\right)\right\} - 2\Bigg)\eta_s(i)\eta_s(i+1)\, ds \Bigg]\Bigg\}.$$

It follows easily then that there exists a constant c such that, for all N,

$$Z_B^H \leqq \exp\{cN\}.$$

Therefore, the superexponential inequality is also satisfied by $\mathbf{P}_N^{N,\gamma}$.

Proof of Theorem 3.1: For any $G(\theta, s) \in \mathbf{C}^{2,1}(T \times [0, t])$ consider the martingale

$$M_t^G = \frac{1}{N} \sum_{i=1}^{N} \Bigg\{ \int_0^t \left(G\left(\frac{i+1}{N}, s\right) - G\left(\frac{i}{N}, s\right)\right)\left(dJ_s^{i,i+1} - N^2 c_{i,i+1}^H(\eta_s, s)\, ds\right)$$

$$+ \int_0^t \left(G\left(\frac{i}{N}, s\right) - G\left(\frac{i+1}{N}, s\right)\right)\left(dJ_s^{i+1,i} - N^2 c_{i+1,i}^H(\eta_s, s)\, ds\right)\Bigg\}$$

(3.2)

$$= \frac{1}{N} \sum_{i=1}^{N} \Bigg\{ \eta_t(i) G\left(\frac{i}{N}, t\right) - \eta_0(i) G\left(\frac{i}{N}, 0\right)$$

$$- \int_0^t \eta_s(i) \frac{\partial G}{\partial s}\left(\frac{i}{N}, s\right) ds - \int_0^t G\left(\frac{i}{N}, s\right) L_N^H(\eta_s(i))\, ds \Bigg\}.$$

The quadratic variation of M_t^G is given by

(3.3)
$$V_t^G = \sum_{i=1}^{N} \int_0^t \left[\left(G\left(\frac{i+1}{N}, s\right) - G\left(\frac{i}{N}, s\right) \right)^2 c_{i,i+1}^H(\eta_s, s) + \left(G\left(\frac{i}{N}, s\right) - G\left(\frac{i+1}{N}, s\right) \right)^2 c_{i+1,i}^H(\eta_s, s) \right] ds$$

which goes to zero uniformly in η as N tends to infinity. Observe that the first three terms in the expression of M_t^G are functions of the density field, whereas the last term can be written as

$$\frac{1}{N} \int_0^t \sum_{i=1}^{N} G\left(\frac{i}{N}, s\right) L_{N,s}^H[\eta_s(i)]\, ds$$

$$= \tfrac{1}{2} N \int_0^t \sum_{i=1}^{N} \Bigg\{ \Bigg[G\left(\frac{i-1}{N}, s\right) \exp\left\{ H\left(\frac{i-1}{N}, s\right) - H\left(\frac{i}{N}, s\right) \right\}$$
$$+ G\left(\frac{i+1}{N}, s\right) \exp\left\{ H\left(\frac{i+1}{N}, s\right) - H\left(\frac{i}{N}, s\right) \right\}$$
$$- G\left(\frac{i}{N}, s\right) \left(\exp\left\{ H\left(\frac{i+1}{N}, s\right) - H\left(\frac{i}{N}, s\right) \right\} \right.$$
$$\left. + \exp\left\{ H\left(\frac{i-1}{N}, s\right) - H\left(\frac{i}{N}, s\right) \right\} \right) \Bigg] \eta_s(i)$$
$$- \Bigg[G\left(\frac{i+1}{N}, s\right) \left(\exp\left\{ H\left(\frac{i+1}{N}, s\right) - H\left(\frac{i}{N}, s\right) \right\} \right.$$
$$\left. - \exp\left\{ H\left(\frac{i}{N}, s\right) - H\left(\frac{i+1}{N}, s\right) \right\} \right)$$
$$+ G\left(\frac{i}{N}, s\right) \left(\exp\left\{ H\left(\frac{i}{N}, s\right) - H\left(\frac{i+1}{N}, s\right) \right\} \right.$$
$$\left. - \exp\left\{ H\left(\frac{i+1}{N}, s\right) - H\left(\frac{i}{N}, s\right) \right\} \right) \Bigg] \eta_s(i)\eta_s(i+1) \Bigg\}\, ds$$

$$= \frac{1}{N} \int_0^t \sum_{i=1}^{N} \left[\frac{1}{2} \frac{\partial^2 G}{\partial \theta^2}\left(\frac{i}{N}, s\right) + \left(\frac{\partial G}{\partial \theta} \cdot \frac{\partial H}{\partial \theta} \right)\left(\frac{i}{N}, s\right) \right] \eta_s(i)\, ds$$
$$- \frac{1}{N} \int_0^t \sum_{i=1}^{N} \left(\frac{\partial G}{\partial \theta} \cdot \frac{\partial H}{\partial \theta} \right)\left(\frac{i}{N}, s\right) \eta_s(i)\eta_s(i+1)\, ds + O\left(\frac{1}{N}\right).$$

Both functions G and H being regular, there exists, for every $a > 0$, an ε_0 such that, for all $\varepsilon < \varepsilon_0$ and for all $\eta(\cdot)$,

$$\left| \frac{1}{N} \int_0^t \sum_{i=1}^N \frac{\partial G}{\partial \theta}\left(\frac{i}{N}, s\right) \frac{\partial H}{\partial \theta}\left(\frac{i}{N}, s\right) \eta_s(i)\eta_s(i+1)\, ds \right.$$

$$\left. - \frac{1}{N} \int_0^t \sum_{i=1}^N \frac{\partial G}{\partial \theta}\left(\frac{i}{N}, s\right) \frac{\partial H}{\partial \theta}\left(\frac{i}{N}, s\right) \left(\frac{1}{2\varepsilon N + 1} \sum_{|i-j| \leqq N\varepsilon} \eta_s(j)\eta_s(j+1) \right) ds \right| \leqq a;$$

using the superexponential estimate (Theorem 3.2) we obtain, for any $a > 0$,

$$\varlimsup_{\varepsilon \downarrow 0} \varlimsup_{N \to \infty} \frac{1}{N} \log \mathbf{P}_N^{H, \rho_0} \left(\frac{1}{N} \int_0^t \left| \sum_{i=1}^N \frac{\partial G}{\partial \theta}\left(\frac{i}{N}, s\right) \frac{\partial H}{\partial \theta}\left(\frac{i}{N}, s\right) \eta_s(i)\eta_s(i+1) \right.\right.$$

$$\left.\left. - \sum_{i=1}^N \frac{\partial G}{\partial \theta}\left(\frac{i}{N}, s\right) \frac{\partial H}{\partial \theta}\left(\frac{i}{N}, s\right) \left(\frac{1}{2\varepsilon N + 1} \sum_{|i-j| \leqq N\varepsilon} \eta_s(j) \right)^2 \right| ds \geqq a \right) = -\infty.$$

Hence, we have proven for all $\delta > 0$, that, for ε small enough,

$$\begin{aligned} \lim_{N \to \infty} \mathbf{P}_N^{H, \gamma} \Bigg(\sup_{0 \leqq \tau \leqq t} \Bigg| & \langle \mu_\tau^N(\eta), G(\cdot, \tau) \rangle - \langle \mu_0^N(\eta), G(\cdot, 0) \rangle \\ & - \int_0^\tau \left\langle \mu_s^N(\eta), \left(\frac{\partial}{\partial s} + \frac{1}{2} \frac{\partial^2}{\partial \theta^2} \right) G(\cdot, s) \right\rangle ds \\ & - \int_0^\tau \left\langle \mu_s^N(\eta), \left(\frac{\partial G}{\partial \theta} \cdot \frac{\partial H}{\partial \theta} \right)(\cdot, s) \right\rangle ds \\ & + \int_0^\tau \left\langle (\mu_s^N(\eta) * \alpha_\varepsilon)^2, \left(\frac{\partial G}{\partial \theta} \cdot \frac{\partial H}{\partial \theta} \right)(\cdot, s) \right\rangle ds \Bigg| \geqq \delta \Bigg) = 0, \end{aligned} \tag{3.4}$$

where $\alpha_\varepsilon(\theta)$ is defined as $\alpha_\varepsilon(\theta) = (1/2\varepsilon)\chi_{[-\varepsilon, \varepsilon]}(\theta)$ and $*$ is the convolution.

Now observe that the sequence of distribution μ^N is weakly compact by (3.2) and (3.3). The above estimate (3.4) shows that any weak limit point of the distribution of μ^N is concentrated on the set of trajectories ρ_t such that, for all $\delta > 0$, there exists an $\bar{\varepsilon}$ such that, for all $\varepsilon < \bar{\varepsilon}$ and all $\tau \in [0, t]$,

$$\left| \langle \rho_\tau, G_\tau \rangle - \langle \rho_0, G_0 \rangle - \int_0^\tau \left\langle \rho_s, \left(\frac{\partial}{\partial s} + \frac{1}{2} \frac{\partial^2}{\partial \theta^2} \right) G_s \right\rangle ds \right.$$

$$\left. - \int_0^\tau \left\langle \rho_s, \frac{\partial G_s}{\partial \theta} \frac{\partial H_s}{\partial \theta} \right\rangle ds + \int_0^\tau \left\langle (\rho_s * \alpha_\varepsilon)^2, \frac{\partial G_s}{\partial \theta} \frac{\partial H_s}{\partial \theta} \right\rangle ds \right| \leqq \delta.$$

Letting ε tend to zero we have then that ρ_s satisfies (3.1) in a weak sense.

4. Large Deviations; Upper Bounds

In this and the next section we prove a large deviation principle for the empirical density μ^N for the symmetric simple exclusion process. For any $\rho \in D([0, t]: M_1)$ define the linear functional on $\mathbf{C}^{2,1}(T \times [0, t])$:

$$(4.1)\qquad \begin{aligned} l(\rho; G) &= \langle \rho_t, G(\cdot, t)\rangle - \langle \rho_0, G(\cdot, 0)\rangle \\ &\quad - \int_0^t \left\langle \rho_s, \left(\frac{\partial}{\partial s} + \frac{1}{2}\frac{\partial^2}{\partial \theta^2}\right) G(\cdot, s)\right\rangle ds. \end{aligned}$$

Then introduce the following rate functions:

$$(4.2)\quad I_0(\rho) = \sup_{G \in C^{2,1}} \left\{ l(\rho; G) - \frac{1}{2}\int_0^t \left\langle \rho_s(1 - \rho_s), \left(\frac{\partial G}{\partial \theta}(\cdot, s)\right)^2 \right\rangle ds \right\},$$

$$(4.3)\quad \begin{aligned} &h(\rho_0; \gamma) = \\ &\sup_{\phi_1, \phi_2 \in C(T)} \left\{ \langle \rho_0, \phi_0\rangle + \langle 1 - \rho_0, \phi_1\rangle - \left\langle 1, \log\left(\gamma_0 e^{\phi_0} + (1 - \gamma_0) e^{\phi_1}\right)\right\rangle \right., \end{aligned}$$

$$(4.4)\qquad I_\gamma(\rho) = I_0(\rho) + h(\rho_0; \gamma).$$

THEOREM 4.1. *For any compact set $C \subset D([0, t], M_1)$,*

$$(4.5)\qquad \overline{\lim_{N \to \infty}} \frac{1}{N} \log \mathbf{P}_N^\gamma(\mu^N \in C) \leqq - \inf_{\rho \in C} I_\gamma(\rho).$$

Proof: For a given $G \in \mathbf{C}^{2,1}(T \times [0, t])$, $\varepsilon > 0$, $\delta > 0$, consider the set of paths in the configuration space

$$B^\delta_{\varepsilon, N, G} = \left\{ \eta_s : \left| \frac{1}{N} \sum_{i=1}^N \int_0^t \left(\frac{\partial G}{\partial \theta}\right)^2 \left(\frac{i}{N}, s\right) \left(\eta_s(i)\eta_s(i+1) \right.\right.\right.$$
$$\left.\left.\left. - \left[\frac{1}{2N\varepsilon + 1} \sum_{|i-j| \leqq N\varepsilon} \eta(j)\right]^2 \right) ds \right| \geqq \delta \right\}.$$

By the superexponential estimate for any $\delta > 0$,

$$
\begin{aligned}
&\overline{\lim_{N\to\infty}} \frac{1}{N} \log \mathbf{P}_N^\gamma\left(\mu^N(\eta) \in C\right) \\
(4.6)\qquad &= \overline{\lim_{\varepsilon\to 0}}\ \overline{\lim_{N\to\infty}} \frac{1}{N} \log \mathbf{P}_N^\gamma\left(\mu^N(\eta) \in C;\ \eta \in B^\delta_{\varepsilon, N, G}\right).
\end{aligned}
$$

Consider the exponential martingale

$$
\begin{aligned}
Z_N^G = \exp\Bigg\{ &\sum_{i=1}^N G\left(\frac{i}{N}, t\right)\eta_t(i) - \sum_{i=1}^N G\left(\frac{i}{N}, 0\right)\eta_0(i) \\
&- \sum_{i=1}^N \int_0^t \frac{\partial G}{\partial s}\left(\frac{i}{N}, s\right)\eta_s(i)\, ds \\
(4.7)\qquad &- \tfrac{1}{2}N^2 \sum_{i=1}^N \int_0^t \Bigg(\exp\left\{G\left(\frac{i+1}{N}, s\right) - G\left(\frac{i}{N}, s\right)\right\} \\
&\qquad + \exp\left\{G\left(\frac{i-1}{N}, s\right) - G\left(\frac{i}{N}, s\right)\right\} - 2\Bigg)\eta_s(i)\, ds \\
&+ \tfrac{1}{2}N^2 \sum_{i=1}^N \int_0^t \Bigg(\exp\left\{G\left(\frac{i+1}{N}, s\right) - G\left(\frac{i}{N}, s\right)\right\} \\
&\qquad + \exp\left\{G\left(\frac{i}{N}, s\right) - G\left(\frac{i+1}{N}, s\right)\right\} - 2\Bigg)\eta_s(i)\eta_s(i+1)\, ds \Bigg\} \\
= \exp\Bigg\{ &\sum_{i=1}^N G\left(\frac{i}{N}, t\right)\eta_s(i) - \sum_{i=1}^N G\left(\frac{i}{n}, 0\right)\eta_0(i) \\
&- \sum_{i=1}^N \int_0^t \left(\frac{\partial G}{\partial s}\left(\frac{i}{N}, s\right) + \frac{1}{2}\frac{\partial^2 G}{\partial \theta^2}\left(\frac{i}{N}, s\right) + \frac{1}{2}\left(\frac{\partial G}{\partial \theta}\right)^2\left(\frac{i}{N}, s\right)\right)\eta_s(i)\, ds \\
&+ \sum_{i=1}^N \int_0^t \frac{1}{2}\left(\frac{\partial G}{\partial \theta}\right)^2\left(\frac{i}{N}, s\right)\eta_s(i)\eta_s(i+1)\, ds + O(N)\Bigg\},
\end{aligned}
$$

where $O(N)$ is uniformly bounded in N. There exists a constant $c(\varepsilon)$ such that, for any N and any $\eta \in B^{\delta}_{\varepsilon, N, G}$,

$$\exp\left\{ N\left\{ l(\mu^N * \alpha_\varepsilon, G) - \frac{1}{2}\int_0^t \left\langle (\mu^N * \alpha_\varepsilon)(1 - (\mu^N * \alpha_\varepsilon)), \left(\frac{\partial G}{\partial \theta}\right)^2 (\cdot, s) \right\rangle ds \right\}\right\}$$

(4.8)

$$\leqq Z^G_N \exp\{ N(c(\varepsilon) + \delta)\}.$$

Then, in view of (4.6), (4.7), (4.8), using Chebychev's inequality, for a given $G \in \mathbf{C}^{2,1}(T \times [0, t])$ and $\phi_0, \phi_1 \in \mathbf{C}(T)$, we have

$$\overline{\lim_{N \to \infty}} \frac{1}{N} \log \mathbf{P}^{\gamma}_N(\mu^N \in C)$$

$$\leqq - \inf_{\rho \in C} \left\{ \langle \rho_0, \phi_0 \rangle + \langle 1 - \rho_0, \phi_1 \rangle - \left\langle 1, \log(\gamma e^{\phi_0} + (1 - \gamma) e^{\phi_1}) \right\rangle \right.$$

$$\left. + l(\rho * \alpha^\varepsilon, G) - \frac{1}{2}\int_0^t \left\langle (\rho * \alpha_\varepsilon)(1 - (\rho * \alpha_\varepsilon)), \left(\frac{\partial G}{\partial \theta}\right)^2 (\cdot, s) \right\rangle ds \right\} + c(\varepsilon) + \delta,$$

where α_ε is defined as in Section 3. Taking the lim sup as ε tends to zero in the right-hand side of the last expression and optimizing with respect to G, ϕ_0, ϕ_1, we obtain

$$\overline{\lim_{N \to \infty}} \frac{1}{N} \log \mathbf{P}^{\gamma}_N(\mu^N \in C) \leqq - \sup_{G, \phi_1, \phi_0} \overline{\lim_{\varepsilon \downarrow 0}} \inf_{\rho \in C} \left\{ \langle \rho_0, \phi_0 \rangle + \langle 1 - \rho_0, \phi_1 \rangle \right.$$

$$- \left\langle 1, \log(\gamma e^{\phi_0} + (1 - \gamma) e^{\phi_1}) \right\rangle + l(\rho * \alpha^\varepsilon, G)$$

$$\left. - \frac{1}{2}\int_0^t \left\langle (\rho_s * \alpha_\varepsilon)(1 - (\rho_s * \alpha_\varepsilon)), \left(\frac{\partial G}{\partial \theta}\right)^2 (\cdot, s) \right\rangle ds \right\}.$$

Since I_γ is lower semicontinuous we can let $\varepsilon \to 0$ and obtain (4.5) by the same argument as used in [9], Lemma 11.3.

THEOREM 4.1. *For any closed set $C \subset D([0, t], M_1)$,*

$$\overline{\lim_{N \to \infty}} \frac{1}{N} \log \mathbf{P}^{\gamma}_N(\mu^N \in C) \leqq - \inf_{\rho \in C} I_\gamma(\rho). \tag{4.9}$$

Proof: It is enough to show that, for any $\varepsilon > 0$, $G \in \mathbf{C}^2(T)$,

$$(4.10) \quad \lim_{\delta \to 0} \lim_{N \to \infty} \frac{1}{N} \log \mathbf{P}_N^\gamma \left(\sup_{|s-t|<\delta} |\langle \mu_N(t), G\rangle - \langle \mu_N(s), G\rangle| \geqq \varepsilon \right) = -\infty.$$

Consider the martingale

$$M_N^G(t) = \frac{1}{N} \sum_{i=1}^{N} \left[G\left(\frac{i}{N}\right)(\eta_t(i) - \eta_0(i)) - \int_0^t G\left(\frac{i}{N}\right) L\eta_s(i)\, ds \right].$$

By Chebychev's inequality and Doob's inequality, for any $\lambda > 0$:

$$\mathbf{P}_N^\gamma \left(\sup_{0 \leqq t \leqq \delta} |M_N^G(t)| \leqq \varepsilon \right) \leqq \exp\{-\varepsilon\lambda N\} E^{P_N^\gamma}(\exp\{\lambda N M_N^T(\varepsilon)\})$$

$$\leqq \exp\{-\varepsilon\lambda N + \delta\lambda^2 C N + o(N)\},$$

where C is a constant not depending on δ and λ (see formula (4.7)). Then, for any $\lambda > 0$,

$$\overline{\lim_{\delta \to 0}}\ \overline{\lim_{N \to \infty}} \frac{1}{N} \log \mathbf{P}_N^\gamma \left(\sup_{0 \leqq t \leqq \delta} |M_N^G(t)| > \varepsilon \right) \leqq -\varepsilon\lambda,$$

and λ can be chosen as large as one wishes. By a 3 epsilon argument this implies (4.10).

5. Large Deviations; Lower Bound

By definition (4.3) it follows that if ρ_0 and γ are such that $h(\rho_0, \gamma) < +\infty$, we have

$$(5.1) \quad h(\rho_0, \gamma) = \left\langle \rho_0, \log \frac{\rho_0}{\gamma} \right\rangle + \left\langle 1 - \rho_0, \log \frac{1 - \rho_0}{1 - \gamma} \right\rangle.$$

Furthermore, we have the following:

LEMMA 5.1. *If $I_0(\rho) < +\infty$, there exists $H \in \mathbf{H}'(\rho(1-\rho))$ (defined in Section 1) such that*

$$(5.2) \quad I_0(\rho) = \frac{1}{2} \int_0^t \left\langle \rho_s(1-\rho_s), \left(\frac{\partial H}{\partial \theta}\right)^2 \right\rangle ds$$

and that ρ *satisfies in the weak sense the equation*

$$\frac{\partial \rho}{\partial s} = \frac{1}{2}\frac{\partial^2 \rho}{\partial \theta^2} - \frac{\partial}{\partial \theta}\left(\frac{\partial H}{\partial \theta}\rho(1-\rho)\right). \tag{5.3}$$

Proof: Definition (4.2) implies that, for any $G \in \mathbf{C}^{2,1}(T \times [0, t])$,

$$|l(\rho; G)|^2 \leqq 2I_0(\rho)\int_0^t \left\langle \rho_s(1-\rho_s), \left|\frac{\partial G}{\partial \theta}\right|^2 \right\rangle ds.$$

Then $l(\rho; \cdot)$ can be extended to a bounded linear function on $\mathbf{H}^1(\rho(1-\rho))$ and by Riesz representation there exists $H \in \mathbf{H}^1(\rho(1-\rho))$ such that, for any $G \in \mathbf{H}^1(\rho(1-\rho))$,

$$l(\rho; G) = \int_0^t \left\langle \rho_s(1-\rho_s), \frac{\partial G}{\partial \theta}\frac{\partial H}{\partial \theta} \right\rangle ds,$$

that is (5.3) is satisfied in the weak sense. Furthermore,

$$\begin{aligned} I_0(\rho) &= \sup_G \left\{ \int_0^t \left\langle \rho_s(1-\rho_s), \frac{\partial G}{\partial \theta}\frac{\partial H}{\partial \theta} \right\rangle ds - \frac{1}{2}\int_0^t \left\langle \rho_s(1-\rho_s), \left|\frac{\partial G}{\partial \theta}\right|^2 \right\rangle ds \right\} \\ &= \frac{1}{2}\int_0^t \left\langle \rho_s(1-\rho_s), \left|\frac{\partial H}{\partial \theta}\right|^2 \right\rangle ds - \inf_G \frac{1}{2}\int_0^t \left\langle \rho_s(1-\rho_s), \left|\frac{\partial H}{\partial \theta} - \frac{\partial G}{\partial \theta}\right|^2 \right\rangle ds \\ &= \frac{1}{2}\int_0^t \left\langle \rho_s(1-\rho_s), \left|\frac{\partial H}{\partial \theta}\right|^2 \right\rangle ds. \end{aligned}$$

THEOREM 5.2. *Let* $0 < \gamma(\theta) < 1$ *for any* $\theta \in T$. *Thus, for any open neighborhood* A *of a given* $\rho \in \mathbf{D}([0, t], M_1)$,

$$\varliminf_{N \to \infty} \frac{1}{N} \log \mathbf{P}_N^\gamma(\mu^N \in A) \geqq -I_\gamma(\rho). \tag{5.4}$$

Proof: If $I_\gamma(\rho) = +\infty$, (5.4) is trivial. Suppose then $I_\gamma(\rho) < \infty$. Since $I_\gamma(\rho)$ is lower semicontinuous, convex, and invariant under spatial translations, ρ can be approximated by spatial convolutions ρ^ε such that $\lim_{\varepsilon \downarrow 0} I_\gamma(\rho^\varepsilon) = I_\gamma(\rho)$. So, for ε small enough we have $\rho^\varepsilon \in A$ and $I_\gamma(\rho^\varepsilon) < +\infty$. Choosing a suitable convolution, the H identified by Lemma 5.1 can be assumed to be regular without any loss of generality.

Consider the weakly asymmetric exclusion process, $\tilde{\mathbf{P}}_N$ being the corresponding probability measure on Ω_N with initial distribution a Bernoulli distribution with profile ρ_0. Then

$$\mathbf{P}_N^\gamma(\mu^N \in A) = \mathbf{E}^{\tilde{P}_N}\left(I_{[\mu^N \in A]} \frac{d\nu_\gamma^N}{d\nu_{\rho_0}^N}(\eta_0) \frac{d\mathbf{P}_N^{\rho_0}}{d\tilde{\mathbf{P}}_N} \right) \tag{5.5}$$

and, by Girsanov's formula,

$$\begin{aligned}
\log \frac{d\mathbf{P}_N^{\rho_0}}{d\tilde{\mathbf{P}}_N} &= \sum_{i=1}^{N} \Bigg\{ \int_0^t \Big(H\Big(\frac{i}{N}, s\Big) - H\Big(\frac{i+1}{N}, s\Big)\Big)\, dJ_1^{i,i+1} \\
&\quad + \int_0^t \Big(H\Big(\frac{i+1}{N}, s\Big) - H\Big(\frac{i}{N}, s\Big)\Big)\, dJ_s^{i+1,i} \\
&\quad + \tfrac{1}{2} N^2 \int_0^t \Big(\exp\Big\{ H\Big(\frac{i+1}{N}, s\Big) - H\Big(\frac{i}{N}, s\Big)\Big\} - 1\Big) \eta_s(i)(1-\eta_s(i+1))\, ds \\
&\quad + \tfrac{1}{2} N^2 \int_0^t \Big(\exp\Big\{ H\Big(\frac{i+1}{N}, s\Big) - H\Big(\frac{i}{N}, s\Big)\Big\} - 1\Big) \eta_s(i+1)(1-\eta_s(i))\, ds \Bigg\} \\
&= \sum_{i=1}^{N} \Bigg\{ H\Big(\frac{i}{N}, 0\Big)\eta_0(i) - H\Big(\frac{i}{N}, t\Big)\eta_t(i) + \int_0^t \frac{\partial H}{\partial s}\Big(\frac{i}{N}, s\Big)\eta_s(i)\, ds \\
&\quad + \tfrac{1}{2} N^2 \int_0^t \Big(\exp\Big\{ H\Big(\frac{i+1}{N}, s\Big) - H\Big(\frac{i}{N}, s\Big)\Big\} \\
&\qquad\qquad + \exp\Big\{ H\Big(\frac{i-1}{N}, s\Big) - H\Big(\frac{i}{N}, s\Big)\Big\} - 2\Big) \eta_s(i)\, ds \\
&\quad - \tfrac{1}{2} N^2 \int_0^t \Big(\exp\Big\{ H\Big(\frac{i+1}{N}, s\Big) - H\Big(\frac{i}{N}, s\Big)\Big\} \\
&\qquad\qquad + \exp\Big\{ H\Big(\frac{i}{N}, s\Big) - H\Big(\frac{i+1}{N}, s\Big)\Big\} - 2\Big) \eta_s(i)\eta_s(i+1)\, ds \Bigg\} \\
&= \sum_{i=1}^{N} \Bigg\{ H\Big(\frac{i}{N}, 0\Big)\eta_0(i) - H\Big(\frac{i}{N}, t\Big)\eta_t(i) + \int_0^t \frac{\partial H}{\partial s}\Big(\frac{i}{N}, s\Big)\eta_s(i)\, ds \\
&\quad + \frac{1}{2} \int_0^t \Big(\frac{\partial^2 H}{\partial \theta^2}\Big(\frac{i}{N}, s\Big) + \Big(\frac{\partial H}{\partial \theta}\Big)^2\Big(\frac{i}{N}, s\Big) + O\Big(\frac{1}{N}\Big)\Big) \eta_s(i)\, ds \\
&\quad - \frac{1}{2} \int_0^t \Big(\Big(\frac{\partial H}{\partial \theta}\Big)^2\Big(\frac{i}{N}, s\Big) + O\Big(\frac{1}{N}\Big)\Big) \eta_s(i)\eta_s(i+1)\, ds \Bigg\} \\
&= \sum_{i=1}^{N} \Bigg\{ H\Big(\frac{i}{N}, 0\Big)\eta_0(i) - H\Big(\frac{i}{N}, t\Big)\eta_t(i) \\
&\quad + \int_0^t \Big(\frac{\partial}{\partial s} + \frac{1}{2}\frac{\partial^2}{\partial \theta^2}\Big) H\Big(\frac{i}{N}, s\Big)\eta_s(i)\, ds \\
&\quad + \frac{1}{2} \int_0^t \Big(\frac{\partial H}{\partial \theta}\Big)^2\Big(\frac{i}{N}, s\Big)\eta_s(i)(1-\eta_s(i+1))\, ds + O\Big(\frac{1}{N}\Big)\Bigg\}.
\end{aligned}$$

By Theorem 3.1 (hydrodynamical limit),

$$\lim_{N\to\infty} \tilde{\mathbf{P}}_N(\mu^N \in A) = 1.$$

Then, by the Jensen inequality, (5.5), (5.6), and Theorem 3.1,

$$\varliminf_{N\to\infty} \frac{1}{N} \log \mathbf{P}_N^{\gamma}(\mu^N \in A)$$

$$\geqq -\left[\langle \rho_t, H(\cdot, t)\rangle - \langle \rho_s, H(\cdot, 0)\rangle - \int_0^t \left\langle \rho_s, \left(\frac{\partial}{\partial s} + \frac{1}{2}\frac{\partial^2}{\partial \theta^2}\right) H(\cdot, s)\right\rangle ds\right]$$

$$+\frac{1}{2}\int_0^t \left\langle \rho_s(1-\rho_s), \left(\frac{\partial H}{\partial \theta}\right)^2\right\rangle ds - h(\rho_0, \gamma)$$

$$= -\int_0^t \left\langle \rho_s(1-\rho_s), \left(\frac{\partial H}{\partial \theta}\right)^2\right\rangle ds + \frac{1}{2}\int_0^t \left\langle \rho_s(1-\rho_s), \left(\frac{\partial H}{\partial \theta}\right)^2\right\rangle ds - h(\rho_0, \gamma)$$

$$= -\frac{1}{2}\int_0^t \left\langle \rho_s(1-\rho_s), \frac{\partial H}{\partial \theta}\Big)^2\right\rangle ds - h(\rho_0, \gamma)$$

$$= -I_\gamma(\rho)$$

THEOREM 5.3. *For any open set $A \subset \mathbf{D}([0, t], M_1)$,*

$$\varliminf_{N\to\infty} \frac{1}{N} \log \mathbf{P}_N^{\gamma}(\mu^N \in A) \geqq -\inf_{\rho \in A} I_\gamma(\rho).$$

Appendix
Reaction-Diffusion Equations

Consider the Markov process on X_N generated by

$$L_N^R = L_N + L_I,$$

where L_N is the generator of the speeded up symmetric simple exclusion process as above, i.e.,

$$(L_N f)(\eta) = \tfrac{1}{2}N^2 \sum_{i=1}^{N} \left[f(\eta^{i,i+1}) - f(\eta)\right]$$

and L_I is a process that creates and annihilates particles independently at each site, that is

$$(L_I f)(\eta) = \sum_{i=1}^{N} \left(f(\eta^i) - f(\eta)\right)$$

with

$$\eta^i(j) = \begin{cases} 1 - \eta(i), & j = i, \\ \eta(j), & j \neq i. \end{cases}$$

Denote by $\mathbf{P}_N^R$ the corresponding probability measure on the path space E_N.

The hydrodynamical limit for the process can be easily proved (as for the symmetric simple exclusion) and the limit equation is

$$\frac{\partial \rho}{\partial t} = \frac{1}{2}\frac{\partial^2 \rho}{\partial \theta^2} + (1 - 2\rho).$$

The invariant reversible measure for this process is the Bernoulli distribution $\nu_N^{1/2}$.

Theorem 2.1 extends immediately to $\mathbf{P}_N^R$: in the notation of Section 2, let $\lambda_{N,\varepsilon}^R(a)$ be the largest eigenvalue of $L_N^R + aV_{N,\varepsilon}$, then $\lambda_{N,\varepsilon}^R(a) \leqq \lambda_{N,\varepsilon}(a)$ and the argument of the proof of Theorem 2.1 can be repeated identically. As in Section 2, one can extend Theorem 2.1 to the new equilibrium process $\mathbf{P}_N^R$ starting with a product distribution with profile $\gamma(\theta)$.

Consider now the Markov process generated by

$$L^{RD} = L_N + L_c,$$

where L_N is defined by (1.2) and

$$(L_c f)(\eta) = \sum_{i=1}^{N} c(\tau_i \eta)\left(f(\eta^i) - f(\eta)\right),$$

where $c(\eta)$ is a strictly positive local function, and hence bounded below and above.

Let $\mathbf{P}_N^{RD,\gamma}$ be the corresponding probability measure on E_N. Now $\mathbf{P}_N^{RD,\gamma}$ is absolutely continuous with respect to $\mathbf{P}_N^{R,\gamma}$ and by the same argument as used in Theorem 3.2, Theorem 2.1 extends to $\mathbf{P}_N^{RD,\gamma}$. This permits us, as in Theorem 3.1, to deduce the hydrodynamical limit for this process. The limit equation is given by

$$\frac{\partial \rho(\theta, t)}{\partial t} = \frac{1}{2}\frac{\partial^2 \rho}{\partial \theta^2} + F(\rho(\theta, t)),$$

$$\rho(0, \theta) = \gamma(\theta),$$

where, for any constant ρ, $F(\rho) = \nu^{\rho}(c(\eta)(1 - 2\eta(0)))$.

A similar result in the infinite-volume case was obtained using correlation methods in [2].

Acknowledgment. The work for this paper was supported in part by NSF grants DMS-8504683 and DMS-8600233.

Bibliography

[1] Dawson, D. A., and Gartner, J., *Large deviations from the McKean-Vlasov limit for weakly interacting diffusions*, Stochastics 20, 1987, pp. 247–308.

[2] DeMasi, A., Ferrari, P., and Lebowitz, J. L., *Reaction-diffusion equations for interacting particle systems*, J. Stat. Phys. 44, 1986, p. 589.

[3] DeMasi, A., Ianiro, N., Pellegrinotti, A., and Presutti, E., *A survey of the hydrodynamical behavior of many particle systems*, Studies in Statistical Mechanics 11, J. L. Lebowitz and E. W. Montroll, eds., North-Holland, 1984, pp. 123–294.

[4] DeMasi, A., Presutti, E., and Scacciatelli, E., *The Weakly Asymmetric Simple Exclusion Processes*, CARR Reports in Mathematical Physics, 1987.

[5] Donsker, M. D., and Varadhan, S. R. S., *Large deviations from a hydrodynamic scaling limit*, Comm. Pure Appl. Math, Vol. 42, issue 3, 1989, to be published.

[6] Gartner, J., *Convergence towards Burger's equation and propagation of chaos for weakly asymmetric exclusion processes*, Stochastic Processes and their Applications 27, 1988, pp. 233–260.

[7] Kipnis, C., and Olla, S., *Large deviations from the hydrodynamical limit for a system of independent Brownian particles*, to appear in Stochastics.

[8] Papanicolaou, F., and Varadhan, S. R. S., *Nonlinear diffusion limit for a system with nearest neighbor interactions*, Comm. Math. Phys. 118, 1988, pp. 31–59.

[9] Varadhan, S. R. S., *Large Deviations and Applications*, CBMS-NSF Regional Conference Series in Applied Mathematics 46, Philadelphia, Society for Industrial and Applied Mathematics, 1984.

Received March, 1988

Comm. Pure Appl. Math. 42 (1989) 243-270

Large Deviations from a Hydrodynamic Scaling Limit

M. D. DONSKER AND S. R. S. VARADHAN
Courant Institute

1. Introduction

The problem of describing how a large system evolves towards it equilibrium in terms of the evolution of certain macroscopic quantities can be formulated and studied in widely different contexts. Recently in [1] one such model has been studied using estimates based on entropy and its rate of change. The main result in [1] is a law of large numbers type result which asserts that the macroscopic functions evolve according to a specific deterministic motion (to be more specific a nonlinear diffusion equation) as the size gets large and the fluctuations disappear in the scaling. The present article is a complement to [1] and we are interested here in estimating precisely the probabilities of significant (large) deviations from the deterministic limit. A similar result in a different context has been derived in [2]. We now proceed to formulate the problem in precise terms.

Let S denote the unit circle or a unit interval $0 \leqq \theta \leqq 1$ with $\theta = 0$ and $\theta = 1$ identified. For any integer $N \geqq 1$ let S_N denote the periodic lattice $\{j/N,\ j = 1, \cdots, N\}$ viewed as a subset of S. At each site j of S_N there is a real-valued random variable $x_j = x_j(t)$ which varies in time according to the following system of stochastic differential equations:

$$(1.1)\qquad \begin{aligned} dx_j(t) &= dz_{j-1,j}(t) - dz_{j,j+1}(t), \\ dz_{j,j+1}(t) &= \tfrac{1}{2}N^2\big[\phi'(x_j(t)) - \phi'(x_{j+1}(t))\big]\,dt + N d\beta_{j,j+1}(t), \end{aligned}$$

where ϕ is a fixed function from $\mathbb{R}$ into $\mathbb{R}$ satisfying the hypotheses to be listed shortly. We can visualize $x_j(t)$ as the charge at time t located at site j of S_N. The first equation of (1.1) describes the net change at site j as the inflow minus the outflow. Although all charges are expressed as flowing from j to $j+1$ the sign of the charge can be either positive or negative. The flow along the bond connecting j and $j+1$ is proportional to $[\phi'(x_j(t)) - \phi'(x_{j+1}(t)]$ with an additional Brownian noise. The factor N^2 in front of the dt term and N in front of the Brownian term are the result of speeding up time by a factor of N^2. The Brownian motions $\beta_{j,j+1}(t)$ for the different bonds are assumed to be mutually independent. The addition of indices $j \pm 1$ are always modulo N and time has been speeded up by a factor of N^2 to match with the spatial scaling by the factor

$1/N$ represented in the lattice width. We note that the total charge $x_1 + \cdots + x_N$ is conserved over time.

We assume that the function $\phi: \mathbb{R} \to \mathbb{R}$ is continuously differentiable and satisfies

$$\int_{-\infty}^{\infty} \exp\{-\phi(x)\}\, dx = 1,$$

$$\int_{-\infty}^{\infty} \exp\{\lambda x - \phi(x)\}\, dx = M(\lambda) < \infty \quad \text{for all} \quad \lambda \in \mathbb{R}, \tag{1.2}$$

$$\int_{-\infty}^{\infty} \exp\{\sigma|\phi'(x)| - \phi(x)\}\, dx < \infty \quad \text{for all} \quad \sigma \geqq 0.$$

We can assume without loss of generality the existence of a function $w(x)$ which is convex and satisfies $w(x) \geqq |x|$, $w(x)/|x| \to \infty$ as $|x| \to \infty$ and

$$\int_{-\infty}^{\infty} \exp\{w(x) - \phi(x)\}\, dx < \infty. \tag{1.3}$$

If we let $\rho(\lambda) = \log M(\lambda)$ and define $h(x) = \sup_{\lambda \in \mathbb{R}}[\lambda x - \rho(\lambda)]$, then $h(\cdot)$ and $\rho(\cdot)$ are a pair of conjugate convex functions: $h'(\rho'(\lambda)) = \lambda$ and $\rho'(h'(x)) = x$. Now the prescription (1.1) leads to a diffusion on $\mathbb{R}^N$ with coordinates $(x_1, \cdots, x_N)$ which has an infinitesimal generator L_N given by

$$\begin{aligned} L_N = \frac{1}{2} N^2 \Bigg\{ &\sum_{j=1}^{N} \left[\frac{\partial}{\partial x_j} - \frac{\partial}{\partial x_{j+1}} \right]^2 \\ &+ \sum_{j=1}^{N} \left[\phi'(x_{j-1}) - 2\phi'(x_j) + \phi'(x_{j+1}) \right] \frac{\partial}{\partial x_j} \Bigg\}. \end{aligned} \tag{1.4}$$

If $\omega_N(x_1, \cdots, x_N) = \exp\{-\sum_{j=1}^{N} \phi(x_j)\}$, then one can verify easily that the operator L_N is symmetric on $\mathbb{R}^N$ with respect to the weight ω_N. In particular ω_N is an invariant probability density for the diffusion with generator L_N. However, the diffusion is not ergodic on $\mathbb{R}^N$. If the process is restricted to a hyperplane $x_1 + \cdots + x_N = \text{const}$, then it is ergodic and the invariant distribution is the restriction or the conditional of ω_N given the sum $x_1 + x_2 + \cdots + x_N$.

Let $\Omega^{(N)} = C[[0, T]; \mathbb{R}^N]$ be the space of continuous functions $\xi_N(t)$ on $0 \leqq t \leqq T$ for some $T > 0$ with values in $\mathbb{R}^N$. Corresponding to the generator L_N and initial density ω_N we have a measure P_N on $\Omega^{(N)}$. We denote by M_S the space of all signed measures on S and by $\bar{\Omega}$ the space of continuous maps of

$[0, T]$ into M_S. We can map $\mathbb{R}^N$ into M_S by sending $\xi = (x_1, \cdots, x_N)$ into

$$\pi_N \xi = \frac{1}{N} \sum_{j=1}^{N} x_j \delta_{j/N}, \tag{1.5}$$

where $\delta_{j/N}$ is the Dirac measure at the site j of S_N. There is the associated map $\hat{\pi}_N$ of $\Omega^{(N)}$ into $\overline{\Omega}$ defined by

$$(\hat{\pi}_N \xi_N)(t) = \pi_N \xi_N(t). \tag{1.6}$$

This of course induces a measure Q_N on $\overline{\Omega}$:

$$Q_N = P_N \hat{\pi}_N^{-1}. \tag{1.7}$$

Let us denote $a_0 = \int x \exp\{-\phi(x)\}\, dx$ and $\alpha_0(d\theta) = a_0\, d\theta \in M_S$. We let $\alpha_0(\cdot)$ be the element of $\overline{\Omega}$ which is equal to α_0 for all $t \in [0, T]$. From the law of large numbers, Q_N on $\overline{\Omega}$ converges weakly to $\delta_{\alpha_0(\cdot)}$ as $N \to \infty$. The aim of this article is to obtain a large deviation result for Q_N on $\overline{\Omega}$. We describe the results now. An element of $\overline{\Omega}$ will be denoted by $\mu(\cdot, d\theta)$. To describe a large deviation result for Q_N we need a rate function I: $\overline{\Omega} \to [0, \infty]$ such that, for any closed set C and open set G in $\overline{\Omega}$ (in a suitable topology),

$$\limsup_{N \to \infty} \frac{1}{N} \log Q_N(C) \leqq -\inf_{\mu(\cdot,\cdot) \in C} I(\mu(\cdot, \cdot)) \tag{1.8}$$

and

$$\liminf_{N \to \infty} \frac{1}{N} \log Q_N(G) \geqq -\inf_{\mu(\cdot,\cdot) \in G} I(\mu(\cdot, \cdot)). \tag{1.9}$$

We shall establish such a result with the following rate function:

$$I(\mu(\cdot, \cdot)) = \infty \quad \text{unless} \quad \mu(t, d\theta) = m(t, \theta)\, d\theta$$

for every t, and in such a case

$$\begin{aligned} I(\mu(\cdot, \cdot)) &= \int_S h(m(0, \theta))\, d\theta \\ &\quad + \frac{1}{2} \int_0^T \left\| \frac{\partial m(t, \theta)}{\partial t} - \frac{1}{2}[h'(m(t, \theta))]_{\theta\theta} \right\|_{-1}^2 dt. \end{aligned} \tag{1.10}$$

Here $h(\cdot)$ is the function defined by

$$h(x) = \sup_{\lambda \in \mathbb{R}} [\lambda x - \rho(\lambda)]$$

and $\| \ \|_{-1}$ denotes the H_{-1} norm.

In Section 2 we prove the upper bound (1.8) and in Section 4 the lower bound (1.9).

2. The Upper Bound

In this section we prove the upper bound (1.8) but we need several preliminary results. Let F: $\mathbb{R} \to \mathbb{R}$ be a suitable continuous function on $\mathbb{R}$. With $M(\lambda)$ and $h(x)$ as defined in the Introduction we define

$$\tilde{F}(y) = \frac{1}{M(\lambda)} \int \exp\{\lambda x - \phi(x)\} F(x)\, dx, \tag{2.1}$$

where $\lambda = h'(y)$.

We will be interested in $F(x) = \phi'(x)$. But since ϕ' may be unbounded we shall approximate it by bounded functions $F(\cdot)$. We note now that if $F(x) = \phi'(x)$, then from (2.1) we obtain that $\tilde{F}(y) = h'(y)$.

Let $\alpha > 0$, $\varepsilon > 0$, and let k be a positive integer. Define $V = V^{(N)}(\delta, \varepsilon, k, F, \alpha, \xi)$ by

$$V^{(N)}(\delta, \varepsilon, k, F, \alpha, \xi)$$

$$= \alpha \sum_{j=1}^{N} \left| \tilde{F}\left(\frac{1}{1 + 2[N\varepsilon]} \sum_{|i-j| \leqq N\varepsilon} x_i \right) - \frac{1}{2k+1} \sum_{|i-j| \leqq k} F(x_i) \right| - \delta \sum_{j=1}^{N} w(x_j).$$

We denote by $\Lambda^{(N)}(\delta, \varepsilon, k, F, \alpha)$ the largest eigenvalue of the operator $L_N + V$. We then have by the spectral theorem the bound

$$E^{P_N}\left\{ \exp\left\{ \int_0^T V(\xi(s))\, ds \right\} \right\} \leqq \exp\{ T \Lambda^{(N)}(\delta, \varepsilon, k, F, \alpha) \}. \tag{2.2}$$

Essential for our purposes will be Lemma 2.2 to be proved below which asserts that, for each fixed bounded continuous F, $\alpha > 0$ and $\delta > 0$,

$$\overline{\lim_{\varepsilon \to 0}}\ \overline{\lim_{k \to \infty}}\ \overline{\lim_{N \to \infty}}\ \frac{1}{N} \Lambda^{(N)}(\delta, \varepsilon, k, F, \alpha) \leqq 0. \tag{2.3}$$

Let us now take smooth functions G: $S \to \mathbb{R}$ and J: $[0, T] \times S \to \mathbb{R}$. Consider

$$
\begin{aligned}
K_N(J, G, F, \varepsilon, \xi(\cdot)) = {} & \sum_{j=1}^{N} J\left(T, \frac{j}{N}\right)x_j(T) - \sum_{j=1}^{N} G\left(\frac{j}{N}\right)x_j(0) \\
& - \int_0^T \left\{ \sum_{j=1}^{N} J_t\left(t, \frac{j}{N}\right)x_j(t) \right\} dt \\
& - \frac{1}{2}\int_0^T \left\{ \sum_{j=1}^{N} J_{\theta\theta}\left(t, \frac{j}{N}\right)\tilde{F}\left(\frac{1}{2[N\varepsilon]+1}\sum_{|i-j|\leq N\varepsilon} x_i(t)\right)\right\} dt \\
& - \frac{1}{2}N\int_0^T\int_s [J_\theta(t,\theta)]^2\, dt\, d\theta
\end{aligned}
\tag{2.4}
$$

and denote

$$
\Lambda(\delta, \varepsilon, k, F, \alpha) = \varlimsup_{N\to\infty} \frac{1}{N}\Lambda^{(N)}(\delta, \varepsilon, k, F, \alpha). \tag{2.5}
$$

We let C be a bound for $|J_{\theta\theta}(\cdot, \cdot)|$.

LEMMA 2.1. *Let $p > 1, q > 1$ be real numbers satisfying $1/p + 1/q = 1$. Define*

$$
\begin{aligned}
\psi(J, G, F, \varepsilon, k, q) = {} & \frac{1}{p}\int_S \log[M(pJ(0,\theta) - pG(\theta))]\, d\theta \\
& + \frac{1}{2}p\int_0^T\int_S J_\theta^2(t,\theta)\, dt\, d\theta \\
& + \frac{T}{8q}\Lambda\left(\frac{1}{T}, \varepsilon, k, F, 4qC\right) \\
& + \frac{1}{4q}\log\left[\int_{-\infty}^{\infty} \exp\{2qCT|\phi'(x) - F(x)| - \phi(x)\}\, dx\right] \\
& + \frac{1}{8q}\log\left[\int_{-\infty}^{\infty} \exp\{w(x) - \phi(x)\}\, dx\right] \\
& - \frac{1}{2}\int_0^T\int_S J_\theta^2(t,\theta)\, dt\, d\theta.
\end{aligned}
$$

Then

$$
\begin{aligned}
& \varlimsup_{N\to\infty} \frac{1}{N}\log E^{P_N}\{\exp\{K_N(J, G, F, \varepsilon, \xi(\cdot))\}\} \\
& \leq \psi(J, G, F, \varepsilon, k, q).
\end{aligned}
\tag{2.6}
$$

Proof: Let

$$K_N^*(J,G;\xi(\cdot)) = \sum_{j=1}^{N} J\left(T,\frac{j}{N}\right)x_j(T) - \sum_{j=1}^{N} G\left(\frac{j}{N}\right)x_j(0)$$

$$-\int_0^T \sum_{j=1}^{N} J_t\left(t,\frac{j}{N}\right)x_j(t)\,dt$$

$$-\frac{1}{2}N^2\int_0^T \sum_{j=1}^{N}\left[J\left(t,\frac{j+1}{N}\right) - 2J\left(t,\frac{j}{N}\right)\right.$$

$$\left.+J\left(t,\frac{j-1}{N}\right)\right]\phi'(x_j(t))\,dt$$

$$-\frac{1}{2}N^2\int_0^T \sum_{j=1}^{N}\left[J\left(t,\frac{j}{N}\right) - J\left(t,\frac{j+1}{N}\right)\right]^2 dt.$$

The functional $\exp\{K_N^*\}$ is a martingale in T relative to the measure P_N and therefore

$$E^{P_N}\{\exp\{K_N^*(J,G;\xi(\cdot))\}\} = E^{P_N}\left\{\exp\left\{\sum_{j=1}^{N}\left[J\left(0,\frac{j}{N}\right) - G\left(\frac{j}{N}\right)\right]x_j(0)\right\}\right\}$$

$$= \int_{-\infty}^{\infty}\cdots\int_{-\infty}^{\infty}\exp\left\{\sum_{j=1}^{N}\left[J\left(0,\frac{j}{N}\right) - G\left(\frac{j}{N}\right)\right]x_j\right\}$$

(2.7)

$$\cdot\exp\left\{-\sum_{j=1}^{N}\phi(x_j)\right\}dx_1\cdots dx_N$$

$$= \exp\left\{\sum_{j=1}^{N}\log\left[M\left(J\left(0,\frac{j}{N}\right) - G\left(\frac{j}{N}\right)\right)\right]\right\}.$$

The last term in the exponent K_N^* is independent of $\xi(\cdot)$ and can be moved outside the integration with respect to P_N. So we rewrite (2.7) as

$$E^{P_N}\{\exp\{S_1 + S_2 + S_3 + S_4\}\}$$

(2.8)
$$= \exp\left\{\sum_{j=1}^{N} \log\left[M\left(J\left(0, \frac{j}{N}\right) - G\left(\frac{j}{N}\right)\right)\right] + \frac{1}{2}N^2\int_0^T \sum_{j=1}^{N}\left[\left(J\left(t, \frac{j+1}{N}\right) - J\left(t, \frac{j}{N}\right)\right)\right]^2 dt\right\},$$

where

$$S_1 = \sum_{j=1}^{N} J\left(T, \frac{j}{N}\right)x_j(T),$$

$$S_2 = -\sum_{j=1}^{N} G\left(\frac{j}{N}\right)x_j(0),$$

$$S_3 = -\int_0^T \sum_{j=1}^{N} J_t\left(t, \frac{j}{N}\right)x_j(t)\,dt,$$

and

$$S_4 = -\frac{1}{2}N^2\int_0^T \sum_{j=1}^{N}\left[J\left(t, \frac{j+1}{N}\right) - 2J\left(t, \frac{j}{N}\right) + J\left(t, \frac{j-1}{N}\right)\right]\phi'(x_j(t))\,dt.$$

Let us denote

(2.9)
$$F_1 = S_1 + S_2 + S_3 + S_4$$

and

(2.10)
$$F_2 = S_1 + S_2 + S_3 + S_4',$$

where

$$S_4' = -\frac{1}{2}\int_0^T \sum_{j=1}^{N} J_{\theta\theta}\left(t, \frac{j}{N}\right)\tilde{F}\left[\frac{1}{1+2[N\varepsilon]}\sum_{|i-j|\leq N\varepsilon} x_i(t)\right]dt.$$

We can write

$$E^{P_N}[\exp\{F_2\}] = E^{P_N}[\exp\{F_1 + (F_2 - F_1)\}].$$

From Hölder's inequality we obtain

$$E^{P_N}[\exp\{F_2\}] \leqq \{E^{P_N}[\exp\{pF_1\}]\}^{1/p}\{E^{P_N}[\exp\{q(F_2-F_1)\}]\}^{1/q}$$

$$\tag{2.11} \begin{aligned} &\leqq \exp\left\{\frac{1}{p}\sum_{j=1}^{N}\log\left(M\left(pJ\left(0,\frac{j}{N}\right)-pG\left(\frac{j}{N}\right)\right)\right)\right. \\ &\qquad \left.+p\frac{1}{2}N^2\int_0^T\sum_{j=1}^{N}\left[J\left(t,\frac{j+1}{N}\right)-J\left(t,\frac{j}{N}\right)\right]^2 dt\right\} \\ &\qquad \cdot\{E^{P_N}[\exp\{q(F_2-F_1)\}]\}^{1/q}. \end{aligned}$$

We write

$$F_2 - F_1 = S_4' - S_4 = F_3 + F_4,$$

where

$$\begin{aligned} F_3 = F_3(J,\xi(\cdot)) = \frac{1}{2}N^2\int_0^T\sum_{j=1}^{N}&\left[J\left(t,\frac{j+1}{N}\right)\right. \\ &\left.-2J\left(t,\frac{j}{N}\right)+J\left(t,\frac{j-1}{N}\right)\right]\phi'(x_j(t))\,dt \\ -\frac{1}{2}\int_0^T\sum_{j=1}^{N}&J_{\theta\theta}\left(t,\frac{j}{N}\right)\phi'(x_j(t))\,dt, \end{aligned}$$

and

$$\begin{aligned} F_4 = F_4(J,F,\varepsilon,\xi(\cdot)) &= \frac{1}{2}\int_0^T\sum_{j=1}^{N}J_{\theta\theta}\left(t,\frac{j}{N}\right)\phi'(x_j(t))\,dt \\ &\quad -\frac{1}{2}\int_0^T\sum_{j=1}^{N}J_{\theta\theta}\left(t,\frac{j}{N}\right)\tilde{F}\left(\frac{1}{1+2[N\varepsilon]}\sum_{|i-j|\leqq N\varepsilon}x_i(t)\right)dt. \end{aligned}$$

We can estimate the second term of (2.11) with the help of Schwartz's inequality by

$$\tag{2.12} \begin{aligned} &[E^{P_N}\{\exp\{q(F_2-F_1)\}\}]^{1/q} \\ &\quad = [E^{P_N}\{\exp\{q(F_3+F_4)\}\}]^{1/q} \\ &\quad \leqq [E^{P_N}\{\exp\{2qF_3\}\}]^{1/2q}[E^{P_N}\{\exp\{2qF_4\}\}]^{1/2q}. \end{aligned}$$

Since J is twice continuously differentiable in θ, there is a function $\eta(N)$ that goes to 0 as $N \to \infty$ such that

$$|F_3| \leqq \eta(N) \int_0^T \sum_{j=1}^N |\phi'(x_j(t))| \, dt. \tag{2.13}$$

Thus by Jensen's inequality, and (2.13),

$$\begin{aligned}
E^{P_N}\{\exp\{2qF_3\}\} &\leqq E^{P_N}\left\{\exp\left\{2q\eta(N)\int_0^T \sum_{j=1}^N |\phi'(x_j(t))| \, dt\right\}\right\} \\
&= E^{P_N}\left\{\exp\left\{2Tq\eta(N) \cdot \frac{1}{T}\int_0^T \sum_{j=1}^N |\phi'(x_j(t))| dt\right\}\right\} \\
&\leqq E^{P_N}\left\{\frac{1}{T}\int_0^T \exp\left\{2Tq\eta(N) \sum_{j=1}^N |\phi'(x_j(t)|\right\} dt\right\} \\
&= E^{P_N}\left\{\exp\left\{2Tq\eta(N) \sum_{j=1}^N |\phi'(x_j(0)|\right\}\right\} \\
&= \left[\int_{-\infty}^{\infty} \exp\{2Tq\eta(N)|\phi'(x)| - \phi(x)\} \, dx\right]^N.
\end{aligned} \tag{2.14}$$

The term $E^{P_N}\{\exp\{2qF_4\}\}$ is handled in the same manner. We write $F_4 = F_5 + F_6$, where

$$\begin{aligned}
F_5 &= F_5(J, F, \varepsilon, \xi(\cdot)) \\
&= \frac{1}{2}\int_0^T \sum_{j=1}^N J_{\theta\theta}\left(t, \frac{j}{N}\right)\left[F(x_j(t)) - \tilde{F}\left(\frac{1}{[2N\varepsilon]+1} \sum_{|i-j| \leqq N\varepsilon} x_i(t)\right)\right] dt
\end{aligned} \tag{2.15}$$

and

$$\begin{aligned}
F_6 &= F_6(J, F, \xi(\cdot)) \\
&= \frac{1}{2}\int_0^T \sum_{j=1}^N J_{\theta\theta}\left(t, \frac{j}{N}\right)\left[\phi'(x_j(t)) - F(x_j(t))\right] dt.
\end{aligned} \tag{2.16}$$

Thus,

$$E^{P_N}\{\exp\{2qF_4\}\} \leqq \left[E^{P_N}\{\exp\{4qF_5\}\}^{1/2}\left[E^{P_N}\{\exp\{4qF_6\}\}\right]^{1/2}\right]. \tag{2.17}$$

Since $J_{\theta\theta}$ is bounded by a constant C we obtain, by Jensen's inequality, in a similar fashion

$$\begin{aligned}
&E^{P_N}\{\exp\{4qF_6\}\} \\
&\leqq \left(\int_{-\infty}^{\infty} \exp\{2qCT|\phi'(x) - F(x)| - \phi(x)\} \, dx\right)^N.
\end{aligned} \tag{2.18}$$

We write $F_5 = F_7 + F_8$, where

$$F_7 = \frac{1}{2}\int_0^T \sum_{j=1}^N J_{\theta\theta}\left(t, \frac{j}{N}\right)\left[F(x_j(t)) - \tilde{F}\left(\frac{1}{1+2[N\varepsilon]}\sum_{|i-j|\leqq N\varepsilon} x_i(t)\right)\right] dt$$

$$- \frac{1}{8qT}\int_0^T \sum_{j=1}^N w(x_j(t))\, dt$$

and

$$F_8 = \frac{1}{8qT}\int_0^T \sum_{j=1}^N w(x_j(t))\, dt,$$

where $w(\cdot)$ is the function introduced in equation (1.3). Exactly as before we have

$$E^{P_N}\{\exp\{4qF_5\}\} \leqq E^{P_N}\{\exp\{8qF_7\}\}^{1/2} E^{P_N}\{\exp\{8qF_8\}\}^{1/2} \tag{2.19}$$

and

$$E^{P_N}\{\exp\{8qF_8\}\} \leqq \left(\int_{-\infty}^{\infty} \exp\{w(x) - \phi(x)\}\, dx\right)^N. \tag{2.20}$$

Let k be a positive integer; we write, for each k, $F_7 = F_9 + F_{10}$ where

$$F_9 = \int_0^T \left\{\frac{1}{2}\sum_{j=1}^N J_{\theta\theta}\left(t, \frac{j}{N}\right)\left[\frac{1}{2k+1}\sum_{|i-j|\leqq k} F(x_i(t)) - \tilde{F}\left(\frac{1}{1+2[N\varepsilon]}\sum_{|i-j|\leqq N\varepsilon} x_i(t)\right)\right] - \frac{1}{8qT}\sum_{j=1}^N w(x_j(t))\right\} dt$$

and

$$F_{10} = \frac{1}{2}\int_0^T \sum_{j=1}^N J_{\theta\theta}\left(t, \frac{j}{N}\right)\left[F(x_j(t)) - \frac{1}{2k+1}\sum_{|i-j|\leqq k} F(x_i(t))\right] dt$$

$$= \frac{1}{2}\int_0^T \sum_{j=1}^N F(x_j(t))\left[J_{\theta\theta}\left(t, \frac{j}{N}\right) - \frac{1}{2k+1}\sum_{|i-j|\leqq k} J_{\theta\theta}\left(t, \frac{i}{N}\right)\right] dt.$$

If we let

$$\Delta(h) = \sup_{|\theta'| \leq h, 0 \leq t \leq T} |J_{\theta\theta}(t, \theta) - J_{\theta\theta}(t, \theta + \theta')|,$$

then

$$|F_{10}| \leq \frac{1}{2} T \cdot N \|F\|_{\infty} \Delta\left(\frac{k}{N}\right). \tag{2.21}$$

If we now start with K_N as defined by (2.4) we see from (2.10) that

$$K_N = F_2 - \frac{1}{2} N \int_0^T \int_S [J_\theta(t, \theta)]^2 \, dt \, d\theta$$

and similarly

$$K_N^* = F_1 - \frac{1}{2} N^2 \int_0^T \left\{ \sum_{j=1}^N \left[J\left(t, \frac{j+1}{N}\right) - J\left(t, \frac{j}{N}\right) \right]^2 \right\} dt.$$

If we want to estimate $E^{P_N}\{\exp\{K_N\}\}$ we can write

$$\begin{aligned}
\frac{1}{N} \log E^{P_N}\{\exp\{K_N\}\} &= \frac{1}{N} \log E^{P_N}\{\exp\{K_N(J, G, F, \varepsilon, \xi(\cdot))\}\} \\
&= \frac{1}{N} \log E^{P_N}\{\exp\{F_2\}\} - \frac{1}{2} \int_0^T \int_S [J_\theta(t, \theta)]^2 \, dt \, d\theta \\
&\leq \frac{1}{Np} \sum_{j=1}^N \log\left(M\left(pJ\left(0, \frac{j}{N}\right) - pG\left(\frac{j}{N}\right)\right)\right) \\
&\quad + \frac{1}{2} Np \int_0^T \sum_{j=1}^N \left[J\left(t, \frac{j+1}{N}\right) - J\left(t, \frac{j}{N}\right) \right]^2 dt \\
&\quad + \frac{1}{Nq} \log E^{P_N}\{\exp\{q(F_2 - F_1)\}\} \\
&\quad - \frac{1}{2} \int_0^T \int_S [J_\theta(t, \theta)]^2 \, dt \, d\theta
\end{aligned}$$

by using (2.11). We can proceed through (2.12), (2.14), (2.17)–(2.21) and finally (2.2). We then obtain

$$\begin{aligned}
\frac{1}{N}\log E^{P_N}\{\exp\{K_N\}\} \leqq \frac{1}{Np}\sum_{j=1}^{N}\log\left(M\left(pJ\left(0,\frac{j}{N}\right)-pG\left(\frac{j}{N}\right)\right)\right) \\
+\frac{1}{2}Np\int_0^T\sum_{j=1}^{N}\left[J\left(t,\frac{j+1}{N}\right)-J\left(t,\frac{j}{N}\right)\right]^2 dt \\
-\frac{1}{2}\int_0^T\int_S[J_\theta(t,\theta)]^2\,dt\,d\theta \\
+\frac{1}{2q}\log\left[\int_{-\infty}^{\infty}\exp\{2Tq\eta(N)|\phi'(x)|-\phi(x)\}\,dx\right] \\
+\frac{1}{4q}\log\left[\int_{-\infty}^{\infty}\exp\{2qCT|\phi'(x)-F(x)|-\phi(x)\}\,dx\right] \\
+\frac{1}{8q}\log\left[\int_{-\infty}^{\infty}\exp\{w(x)-\phi(x)\}\,dx\right] \\
+\frac{1}{2}T\|F\|_\infty\Delta\left(\frac{k}{N}\right)+\frac{T}{8Nq}\Lambda^{(N)}\left(\frac{1}{T},\varepsilon,k,F,4qC\right).
\end{aligned}$$

Now we easily complete Lemma 2.1 by letting $N \to \infty$.

We next establish Lemma 2.2.

LEMMA 2.2. *For any fixed $\delta > 0$, $\alpha > 0$ and bounded continuous function $F(x)$,*

$$\overline{\lim_{\varepsilon\to 0}}\ \overline{\lim_{k\to\infty}}\ \Lambda(\delta,\varepsilon,k,F,\alpha) \leqq 0. \tag{2.22}$$

Proof: For any positive constants C_1 and C_2 let $\mathbf{F}_{N,C_1,C_2}$ denote the space of densities $f(x_1,\cdots,x_N)$ relative to

$$\omega_N(x_1,\cdots,x_N)\,dx_1\cdots dx_N = \exp\left\{-\sum_{j=1}^{N}\phi(x_j)\right\}dx_1\cdots dx_N$$

satisfying

$$\int_{\mathbb{R}^N} f(x_1,\cdots,x_N)\,\omega_N(x_1,\cdots,x_N)\,dx_1\cdots dx_N = 1, \tag{2.23}$$

$$\begin{aligned}\int_{\mathbb{R}^N}[w(x_1)+\cdots+w(x_N)]\,f(x_1,\cdots,x_N)\\ \cdot\,\omega_N(x_1,\cdots,x_N)\,dx_1\cdots dx_N \leqq C_1 N,\end{aligned} \tag{2.24}$$

$$\begin{aligned}I_N(f) = \frac{1}{8}\int_{\mathbb{R}^N}\left\{\sum_{j=1}^{N}\left(\frac{\partial f}{\partial x_j}-\frac{\partial f}{\partial x_{j+1}}\right)^2\frac{1}{f}\right\}\\ \cdot\,\omega_N(x_1,\cdots,x_N)\,dx_1\cdots dx_n \leqq \frac{C_2}{N}.\end{aligned} \tag{2.25}$$

Let F be a bounded and continuous function on $\mathbb{R}$ and let $\tilde{F}$ be defined by (2.1). In [1] (see Theorems 4.2 and 4.7 and the comments at the end of the proof of Theorem 4.7) it is shown that

$$\overline{\lim_{\varepsilon\to 0}}\ \overline{\lim_{k\to\infty}}\ \overline{\lim_{N\to\infty}}\ \sup_{f\in \mathbf{F}_{N,C_1,C_2}} \int \frac{1}{N} V^{(N)}(0, \varepsilon, k, F, 1, \xi) \cdot f\omega_N \, dx_1 \cdots dx_N = 0. \tag{2.26}$$

By the variational formula for the largest eigenvalue,

$$\begin{aligned}\Lambda^N(\delta, \varepsilon, k, F, \alpha) &= \sup_{f:\ \int f\omega_N\, dx_1\cdots dx_N = 1} \left\{ \int f V^{(N)}(\delta, \varepsilon, k, F, \alpha, \xi)\omega_N \, dx_1 \cdots dx_N - N^2 I_N(f) \right\} \\ &= \sup_{f:\ \int f\omega_N\, dx_1\cdots dx_N = 1} \left\{ \int \left[\alpha V^{(N)}(0, \varepsilon, k, F, 1, \xi) - \delta \sum_{j=1}^{N} w(x_j) \right] \cdot f w_N \, dx_1 \cdots dx_N - N^2 I_N(f) \right\}.\end{aligned} \tag{2.27}$$

Since $V^{(N)}$ is bounded by AN for some constant A depending only on F there exist constants C_1 and C_2 depending only on α, δ and A such that

$$\frac{1}{N}\Lambda^N(\delta, \varepsilon, k, F, \alpha) \leqq \sup_{f\in \mathbf{F}_{N,C_1,C_2}} \int \frac{1}{N} V^{(N)}(0, \varepsilon, F, 1, \xi) f\omega_N \, dx_1 \cdots dx_N. \tag{2.28}$$

The lemma follows from (2.26).

Before we proceed to establish the large deviation estimates for the measures Q_N on the space $\overline{\Omega} = C[[0, T];\ M_S]$, we need to say a few words about the topology on $\overline{\Omega}$. The space M_S with weak convergence is not a good topological space. We shall, therefore, consider only the spaces M_S^l of measures on S of total variation at most l. Each M_S^l is a compact metric space under weak convergence. $\overline{\Omega}$ will be thought of as $\overline{\Omega} = \bigcup_l \overline{\Omega}^{(l)}$, where $\overline{\Omega}^l = C[[0, T];\ M_S^l]$. We shall establish estimates that imply

$$\lim_{l\to\infty}\ \overline{\lim_{N\to\infty}}\ \frac{1}{N}\log Q_N[\overline{\Omega} - \overline{\Omega}^l] = -\infty, \tag{2.29}$$

and therefore the large deviation principles take place on $\overline{\Omega}^{(l)}$ for fixed l which are nice complete separable metric spaces. This means that the upper bound will be valid for sets C such that $C \cap \overline{\Omega}^{(l)}$ is closed in $\overline{\Omega}^{(l)}$ for each l and the lower bound will be valid for sets G such that $G \cap \overline{\Omega}^{(l)}$ is open in $\overline{\Omega}^{(l)}$ for each l.

The next lemma which is Lemma 2.3 is to show that the large deviation principle has to be focused only on those paths $\mu(t, d\theta)$ in $\overline{\Omega}$ of the form $\mu(t, d\theta) = m(t, \theta)\, d\theta$ for every t in $[0, T]$.

LEMMA 2.3. *Let $\mu_0(\cdot, d\theta)$ be an element in $\overline{\Omega}$. Let $N^t_{J,\mu_0,\varepsilon}$ be the set defined by*

$$N^t_{J,\mu_0,\varepsilon} = \left\{\mu : \left| \int_S J(\theta)\mu(t, d\theta) - \int_S J(\theta)\mu_0(t, d\theta) \right| < \varepsilon \right\}$$

and assume μ_0 is such that

$$\varlimsup_{N\to\infty} \frac{1}{N} \log Q_N\left(N^t_{J,\mu_0,\varepsilon}\right) \geqq -\rho \tag{2.30}$$

for all t, $\varepsilon > 0$ and nice J. Then

$$\mu_0(t, d\theta) = m_0(t, \theta)\, d\theta \quad \textit{for every} \quad t \in [0, T]$$

and

$$\sup_{0 \leqq t \leqq T} \int_S h(m_0(t, \theta))\, d\theta \leqq \rho.$$

Proof:

$$\begin{aligned} E^{Q_N}\left[\exp\left\{ N\int_S J(\theta)\mu(t, d\theta)\right\}\right] &= E^{P_N}\left[\exp\left\{ \sum_{j=1}^N J\left(\frac{j}{N}\right) x_j(t)\right\}\right] \\ &= \int_{\mathbb{R}^N} \exp\left\{ \sum_{j=1}^N J\left(\frac{j}{N}\right) x_j \right\} \omega_N\, dx_1 \cdots dx_N \\ &= \exp\left\{ \sum_{j=1}^N \log M\left(J\left(\frac{j}{N}\right)\right)\right\}. \end{aligned}$$

From this we infer by Tchebechev's inequality that

$$\varlimsup_{N\to\infty} \frac{1}{N} \log Q_N\left[N^t_{J,\mu_0,\varepsilon}\right] \leqq \int_S \log M(J(\theta))\, d\theta - \int_S J(\theta)\mu_0(t, d\theta) + \varepsilon.$$

Thus, from (2.30),

$$\rho \geqq \sup_{J,t,\varepsilon}\left[\int_S J(\theta)\mu_0(t,d\theta) - \int_S \log M(J(\theta))\,d\theta - \varepsilon\right]$$

(2.31)

$$= \sup_t \sup_J \left[\int_S J(\theta)\mu_0(t,d\theta) - \int_S \log M(J(\theta))\,d\theta\right].$$

The lemma follows from the fact that h and $\log M$ are conjugate convex functions.

The impact of the lemma is that for large deviation purposes we only need to look at $\mu(\cdot, d\theta)$ of the form $m(\cdot,\theta)\,d\theta$. Since the total charge is conserved, $\int m(t,\theta)\,d\theta$ is a constant. At this stage we cannot make any statements concerning the regularity of $m(t,\theta)$ as a function on $[0,T]\times S$. The following lemma will play a role in the calculation of the rate function for our large deviation principle.

LEMMA 2.4. *Let $f(t,\theta)$ be a weakly continuous map $t \to f(t,\cdot)$ of $[0,T]$ into $L_1(S)$ such that $\int_S f(t,\theta)\,d\theta$ is a constant. Let $\sigma(t,\theta)$ be a function on $[0,T)\times S$ which is integrable. Let* **A** *be the class of smooth functions on $[0,T]\times S$. Then*

$$\sup_{J\in\mathbf{A}}\left[\int_0^T\int_S J(t,\theta)[f_t(t,\theta) - \sigma_{\theta\theta}(t,\theta)]\,dt\,d\theta - \frac{1}{2}\int_0^T\int_S J_\theta^2(t,\theta)\,dt\,d\theta\right]$$

$$= \sup_{J\in\mathbf{A}}\left[\int_S J(T,\theta)f(T,\theta)\,d\theta - \int_S J(0,\theta)f(0,\theta)\,d\theta\right.$$

$$-\int_0^T\int_S J_t(t,\theta)f(t,\theta)\,dt\,d\theta$$

$$\left.-\int_0^T\int_S J_{\theta\theta}(t,\theta)\sigma(t,\theta)\,dt\,d\theta - \frac{1}{2}\int_0^T\int_S J_\theta^2(t,\theta)\,dt\,d\theta\right]$$

$$= \frac{1}{2}\int_0^T \|f_t(t,\theta) - \sigma_{\theta\theta}(t,\theta)\|_{-1}^2\,dt.$$

Proof: The first relation is only a formal one and f_t and $\sigma_{\theta\theta}$ make sense as distributions only. The rigorous statement is only the equality between the second and third expressions. This is carried out by routine methods using the duality between $\int_S(\partial J/\partial\theta)^2\,d\theta$ and $\|J\|_{-1}^2$.

Our main theorem on upper bounds is the following.

THEOREM 2.5. *Let C be a closed set in $\bar{\Omega}$. Then*

$$\overline{\lim_{N\to\infty}} \frac{1}{N}\log Q_N(C) \leqq -\inf_{\mu(\cdot,\cdot)\in C} I(\mu(\cdot,\cdot)), \tag{2.33}$$

where $I(\mu(\cdot,\cdot) = \infty$ unless $\mu(t,d\theta) = m(t,\theta)\,d\theta$ and

(2.34)

$$I(\mu(\cdot,\cdot)) = \int_S h(m(0,\theta))\,d\theta + \frac{1}{2}\int_0^T \left\| \frac{\partial m(t,\theta)}{\partial t} - \frac{1}{2}\left[h'(m(t,\theta))\right]_{\theta\theta} \right\|_{-1}^2 dt.$$

Proof: From the definition of Q_N measure on $\bar{\Omega}$ we observe that

$$E^{P_N}\left[\exp\{K_N(J,G,F,\varepsilon,\xi(\cdot))\}\right] = E^{Q_N}\left[\exp\{N\bar{K}(J,G,F,\varepsilon,\mu(\cdot,\cdot))\}\right],$$

where

$$\begin{aligned}\bar{K}(J,G,F,\varepsilon,\mu(\cdot,\cdot)) = {} & \int_S J(T,\theta)\mu(T,d\theta) - \int_S G(\theta)\mu(0,d\theta) \\ & - \int_0^T\int_S J_t(t,\theta)\mu(t,d\theta)\,dt \\ & - \frac{1}{2}\int_0^T\int_S J_{\theta\theta}(t,\theta)\tilde{F}\left(\frac{1}{2\varepsilon}\int_{\theta-\varepsilon}^{\theta+\varepsilon}\mu(t,d\theta')\right)d\theta\,dt \\ & - \frac{1}{2}\int_0^T\int_S\left[J_\theta(t,\theta)\right]^2 dt\,d\theta.\end{aligned}$$

Let U be a neighborhood of a point $\mu(\cdot,\cdot)$ in $\bar{\Omega}$. From Lemma 2.1 and a Tchebechev type argument we deduce that

$$\begin{aligned}\overline{\lim_{N\to\infty}} \frac{1}{N}\log Q_N(U) & \leqq \psi(J,G,F,\varepsilon,k,q) \\ & \quad - \inf_{\mu(\cdot,\cdot)\in U} \bar{K}(J,G,F,\varepsilon,\mu(\cdot,\cdot)) \\ & = \sup_{\mu(\cdot,\cdot)\in U}\left\{\psi(J,G,F,\varepsilon,k,q) - \bar{K}(J,G,F,\varepsilon,\mu(\cdot,\cdot))\right\}.\end{aligned} \tag{2.35}$$

Since the estimate in (2.35) holds for all permissible $J, G, F, \varepsilon, k, q$ we can have

$$\overline{\lim_{N\to\infty}} \frac{1}{N}\log Q_N(U) \leq \inf_{J,G,F,\varepsilon,k,q}\ \sup_{\mu(\cdot,\cdot)\in U} [\psi - \bar{K}]. \tag{2.36}$$

By standard arguments in large deviations (see [3]) for compact sets $C \subset \bar{\Omega}$ this will produce an estimate of the form

$$\overline{\lim_{N\to\infty}} \frac{1}{N} Q_N(C) \leq \sup_{\mu(\cdot,\cdot)\in C}\ \inf_{J,G,F,\varepsilon,k,q} [\psi - \bar{K}]. \tag{2.37}$$

Some auxiliary "exponential tightness" estimates are needed to go from (2.37) for compact sets to similar estimates for arbitrary closed sets. This is carried out in Lemmas 2.6–2.8 below. In order to complete the argument, because of Lemma 2.3, it suffices to show that

$$\inf_{J,G,F,\varepsilon,k,q} [\psi - \bar{K}] \leq -I(\mu(\cdot,\cdot)), \tag{2.38}$$

where $I(\mu(\cdot,\cdot))$ is given by (2.34) and $\mu(\cdot, d\theta) = m(t,\theta)\,d\theta$ for every $t \in [0,T]$ with $\sup_{0\leq t\leq T}\int_S h(m(t,\theta))\,d\theta < \infty$. We shall estimate (2.38) by first letting $k \to \infty$, then letting $\varepsilon \to 0$, and then taking infima over J and G. We next let F tend to ϕ' and finally let $q \to \infty$. Let us denote by l_0 the infinum in (2.38). By letting $k \to \infty$ and $\varepsilon \to 0$, we can use Lemma 2.2 to conclude that

$$\begin{aligned} l \leq\ & \frac{1}{p}\int_S \log[M(pJ(0,\theta) - pG(\theta))]\,d\theta + \frac{1}{2}p\int_0^T\int_S [J_\theta(t,\theta)]^2\,dt\,d\theta \\ & + \frac{1}{4q}\log\left(\int_{-\infty}^{\infty} \exp\{2qCT|\phi'(x) - F(x)| - \phi(x)\}\,dx\right) \\ & + \frac{1}{8q}\log\left(\int_{-\infty}^{\infty} \exp\{w(x) - \phi(x)\}\,dx\right) \\ & + \frac{1}{2}\int_0^T\int_S J_{\theta\theta}(t,\theta)\tilde{F}(m(t,\theta))\,d\theta\,dt \\ & - \int_S J(T,\theta)m(T,\theta)\,d\theta + \int_S G(\theta)m(0,\theta)\,d\theta \\ & + \int_0^T\int_S J_t(t,\theta)m(t,\theta)\,dt\,d\theta. \end{aligned} \tag{2.39}$$

Now we let $G(\theta) = J(0,\theta) - H(\theta)$ and take the infinum over J and H separately. Since $h(x)$ and $\rho(\lambda) = \log M(\lambda)$ are conjugate convex functions,

$$\inf_H \left[\frac{1}{p} \int_S \log M(pH(\theta)) - \int_S H(\theta) m(0,\theta)\, d\theta \right] \tag{2.40}$$
$$= -\frac{1}{p} \int_S h(m(0,\theta))\, d\theta.$$

Also, from Lemma 2.4,

$$\inf_J \Bigg[\frac{1}{2} p \int_0^T \int_S J_\theta^2(t,\theta)\, dt\, d\theta + \frac{1}{2} \int_0^T \int_S J_{\theta\theta}(t,\theta) \tilde{F}(m(t,\theta))\, dt\, d\theta$$
$$- \int_S J(T,\theta) m(T,\theta)\, d\theta + \int_S J(0,\theta) m(0,\theta)\, d\theta$$
$$+ \int_0^T \int_S J_t(t,\theta) m(t,\theta)\, dt\, d\theta \Bigg] \tag{2.41}$$
$$= -\sup_J \Bigg[\int_0^T \int_S J(t,\theta) \left\{ \frac{\partial m}{\partial t} - \frac{1}{2} \left[\tilde{F}(m(t,\theta)) \right]_{\theta\theta} \right\} d\theta\, dt$$
$$- \frac{1}{2} p \int_0^T \int_S J_\theta^2(t,\theta)\, d\theta\, dt \Bigg]$$
$$= -\frac{1}{2p} \int_0^T \left\| \frac{\partial m(t,\theta)}{\partial t} - \frac{1}{2} \left(\tilde{F}(m(t,\theta)) \right)_{\theta\theta} \right\|_{-1}^2 dt.$$

From (2.39), (2.40) and (2.41) we obtain

$$l \leqq -\frac{1}{p} \int_S h(m(0,\theta))\, d\theta - \frac{1}{2p} \int_0^T \left\| \frac{\partial m}{\partial t}(t,\theta) - \frac{1}{2} \left[\tilde{F}(m(t,\theta)) \right]_{\theta\theta} \right\|_{-1}^2 dt$$
$$+ \frac{1}{4q} \log \left(\int_{-\infty}^{\infty} \exp\{ 2qCT |\phi'(x) - F(x)| - \phi(x) \}\, dx \right) \tag{2.42}$$
$$+ \frac{1}{8q} \log \left(\int_{-\infty}^{\infty} \exp\{ w(x) - \phi(x) \}\, dx \right).$$

We let $F \to \phi'$ and then finally let $q \to \infty$. Because of our hypotheses (1.2) we can take F to be a standard truncation of ϕ' such that

$$\int_{-\infty}^{\infty} \exp\{ C |\phi'(x) - F(x)| - \phi(x) \}\, dx \to 1,$$

as we remove the truncation and let $F \to \phi'$. By Lemma 6.4 of [1] this implies that $\tilde{F}(x) \to h'(x)$ as $F \to \phi'$. Moreover, according to Lemma 6.5 of [1], for each $\varepsilon > 0$ we can get a uniform bound of the form $|\tilde{F}(x)| \leqq C_\varepsilon + \varepsilon h(x)$ valid for the sequence F approximating ϕ'. Therefore, $\tilde{F}(m(t, \theta)) \to h'(m(t, \theta))$ and because $\sup_{0 \leqq t \leqq T} \int_S h(m(t, \theta))\, d\theta < \infty$, by uniform integrability $\tilde{F}(m(t, \theta))$ will converge to $h'(m(t, \theta))$ weakly. By lower semicontinuity,

$$l \leqq -\frac{1}{p}\int_S h(m(0, \theta))\, d\theta$$

$$-\frac{1}{2p}\int_0^T \left\| \frac{\partial m}{\partial t}(t, \theta) - \frac{1}{2}\left[h'(m(t, \theta))\right]_{\theta\theta} \right\|_{-1}^2 dt \tag{2.43}$$

$$+\frac{1}{8q}\log\left(\int_{-\infty}^{\infty} \exp\{w(x) - \phi(x)\}\, dx\right).$$

We let $p \to 1$ $(q \to \infty)$ and we are done.

We must prove three lemmas, 2.6, 2.7 and 2.8, to obtain the upper bound for closed sets when we already have it for compact sets. Lemmas 2.6 and 2.8 serve the purpose and Lemma 2.7 is needed to derive Lemma 2.8.

LEMMA 2.6.

$$\overline{\lim_{l \to \infty}}\ \overline{\lim_{N \to \infty}}\ \frac{1}{N}\log Q_N\left[\sup_{0 \leqq t \leqq T} \|\mu(t, \cdot)\| \geqq l\right] = -\infty.$$

Proof: This is basically a step in the proof of Lemma 6.1 of [1] where it is derived for $T = 1$. In fact a bound of the form

$$Q_N\left[\sup_{0 \leqq t \leqq 1} \|\mu(t, \cdot)\| \geqq l\right] \leqq C_1 \exp\{-C_2 N l\}$$

is established there.

LEMMA 2.7. *For every* $\varepsilon > 0$,

$$\overline{\lim_{\delta \to 0}}\ \overline{\lim_{N \to \infty}}\ \frac{1}{N}\log P_N\left\{\sup_{0 \leqq \sigma \leqq t_1 < t_2 \leqq T;\ |t_1 - t_2| \leqq \delta} \int_{t_1}^{t_2} \frac{1}{N}\sum_{j=1}^N \left|\phi'(x_j(t))\right| dt \geqq \varepsilon\right\} = -\infty.$$

Proof: For any $\sigma > 0$ we obtain by Jensen's inequality

$$E^{P_N}\left[\exp\left\{\sigma\int_0^T \sum_{j=1}^N \left|\phi'(x_j(t))\right|\right\}\right] = E^{P_N}\left[\exp\left\{\sigma T\frac{1}{T}\int_0^T \sum_{j=1}^N \left|\phi'(x_j(t))\right| dt\right\}\right]$$

$$\leqq \frac{1}{T}E^{P_N}\left\{\int_0^T \exp\left\{\sigma T\sum_{j=1}^N \left|\phi'(x_j(t))\right|\right\} dt\right\}$$

$$= \left(\int_{-\infty}^{\infty} \exp\{\sigma T|\phi'(x)| - \phi(x)\}\, dx\right)^N$$

$$= \exp\{NG(\sigma T)\},$$

where $G(\sigma) = \log(\int_{-\infty}^{\infty}\exp\{\sigma|\phi'(x)| - \phi(x)\}\, dx)$. Therefore, for any $l > 0$,

$$P_N\left\{\frac{1}{N}\int_0^T \sum \left|\phi'(x_j(t))\right| dt \geqq l\right\} \leqq \exp\{-\sigma l N + NG(\sigma T)\}.$$

We obtain immediately

$$\text{(2.44)}\quad \overline{\lim_{N\to\infty}}\, \frac{1}{N}\log P_N\left\{\frac{1}{N}\int_0^T \sum_{j=1}^N \left|\phi'(x_j(t))\right| dt \geq l\right\} \leqq -\sigma l + G(\sigma T).$$

In view of stationarity,

$$P_N\left\{\sup_{0<t_1<t_2\leqq T} \frac{1}{N}\int_{t_1}^{t_2} \sum_{j=1}^N \left|\phi'(x_j(t))\right| dt \geqq \varepsilon\right\}$$

$$\leqq \left(\left[\frac{T}{\delta}\right] + 1\right)P_N\left\{\frac{1}{N}\int_0^{2\delta} \sum_{j=1}^N \left|\phi'(x_j(t))\right| dt \geqq \varepsilon\right\}.$$

If we use (2.44) with $T = 2\delta$ and $l = \varepsilon$, we get

$$\limsup_{N\to\infty} \frac{1}{N}\log P_N\left\{\sup_{0<t_1<t_2\leqq T} \frac{1}{N}\int_{t_1}^{t_2} \sum_{j=1}^N \left|\phi'(x_j(t))\right| dt \geqq \varepsilon\right\}$$

$$\leqq -\sigma\varepsilon + G(2\sigma\delta).$$

Letting $\delta \to 0$, we see that $G(2\sigma\delta) \to 0$ and so we obtain Lemma 2.7 for $\varepsilon > 0$ since $\sigma > 0$ is arbitrary.

LEMMA 2.8. *Let f be a smooth function on S. Then, for any $\varepsilon > 0$,*

$$\overline{\lim_{\delta\to 0}}\ \overline{\lim_{N\to\infty}} \frac{1}{N}\log P_N\left\{ \sup_{\substack{0\le t_1<t_2\le T\\ |t_1-t_2|\le\delta}} \left| \frac{1}{N}\sum_{j=1}^{N} f\left(\frac{j}{N}\right)\left(x_j(t_2)-x_j(t_1)\right)\right| \ge \varepsilon \right\} = -\infty.$$

Proof: If we define

$$\eta(t) = \frac{1}{N}\sum_{j=1}^{N} f\left(\frac{j}{N}\right)x_j(t), \tag{2.45}$$

then from Itô's formula

$$d\eta(t) = A(t)\,dt + dM(t), \tag{2.46}$$

where

$$\begin{aligned} A(t) &= \frac{1}{2}N\sum_{j=1}^{N} f\left(\frac{j}{N}\right)\left[\phi'(x_{j-1}(t)) - 2\phi'(x_j(t)) + \phi'(x_{j+1}(t))\right] \\ &= \frac{1}{2}N\sum_{j=1}^{N}\left[f\left(\frac{j-1}{N}\right) - 2f\left(\frac{j}{N}\right) + f\left(\frac{j+1}{N}\right)\right]\phi'(x_j(t)) \end{aligned} \tag{2.47}$$

and the martingale $M(t)$ has the quadratic variation:

$$\begin{aligned} \langle M(t), M(t)\rangle &= t\sum_{j=1}^{N}\left[f\left(\frac{j+1}{N}\right) - f\left(\frac{j}{N}\right)\right]^2 \\ &= \frac{C_N t}{N}, \end{aligned} \tag{2.48}$$

where C_N is a bounded sequence. From the smoothness of f it follows that

$$A(t) \le \frac{C}{N}\sum_{j=1}^{N}\left|\phi'(x_j(t))\right|. \tag{2.49}$$

From Lemma 2.7 and the estimate (2.49) we obtain, for any $\varepsilon > 0$,

$$\begin{aligned} &\overline{\lim_{\delta\to 0}}\ \overline{\lim_{N\to\infty}} \frac{1}{N}\log P_N\left\{ \sup_{0\le t_1<t_2\le T;\ |t_1-t_2|\le\delta} \left|A(t_2) - A(t_1)\right| \ge \tfrac{1}{2}\varepsilon\right\} \\ &= -\infty. \end{aligned} \tag{2.50}$$

For the martingale part we have, because of (2.48),

$$\begin{aligned}
&\overline{\lim_{\delta\to 0}}\ \overline{\lim_{N\to\infty}}\ \frac{1}{N}\log P_N\left\{\sup_{0\leqq t_1<t_2\leqq T;\ |t_1-t_2|\leqq\delta}|M(t_2)-M(t_1)|\geqq\frac{1}{2}\varepsilon\right\}\\
&\leqq \overline{\lim_{\delta\to 0}}\ \overline{\lim_{N\to\infty}}\ \frac{1}{N}\log P_N\left\{\sup_{0\leqq t_1<t_2\leqq T;\ |t_1-t_2|\leqq\delta}|\beta(t_2)-\beta(t_1)|\geqq C\varepsilon\sqrt{N}\right\}\\
&\leqq \overline{\lim_{\delta\to 0}}\ \overline{\lim_{N\to\infty}}\ \frac{1}{N}\log P_N\left\{\sup_{0\leqq t\leqq 2\delta}|\beta(t)|\geqq C\varepsilon\sqrt{N}\right\}\\
&\leqq \overline{\lim_{\delta\to 0}}\left[-\frac{C^2\varepsilon^2}{4\delta}\right]\\
&= -\infty.
\end{aligned}$$

With the help of Lemmas 2.6 and 2.8 the upper bound can be derived for arbitrary closed sets.

It is worthwhile to point out at this time that the basic ingredient of the method is a super exponential estimate that replaces

$$\frac{1}{N}\int_0^t \sum J\left(\frac{i}{N}\right)\phi'(x_i(s))\,ds$$

by

$$\int_0^t\int_S J(\theta)h'(m(\theta,s))\,ds\,d\theta$$

without any change in the large deviation rates. This is possible because of an estimate of the type

$$\lim_{\varepsilon\to 0}\ \overline{\lim_{N\to\infty}}\ \frac{1}{N}\log P_N\left[\frac{1}{N}\int_0^T W_{N,\varepsilon}(x_1(s),\cdots,x_N(s))\,ds\geqq\delta\right]=-\infty,$$

where

$$W_{N,\varepsilon}(x_1,\cdots,x_N)=\sum_{i=1}^{N}|G_i^{N,\varepsilon}(x_1,\cdots,x_N)|$$

and

$$G_i^{N,\varepsilon}(x_1,\cdots,x_N)=\left|\frac{1}{2[N\varepsilon]+1}\sum_{|j-i|\leqq N\varepsilon}F(x_{j-k},\cdots,x_{j+k})\right.$$
$$\left.-\tilde{F}\left(\frac{x_{i-[N\varepsilon]}+\cdots+x_{i+[N\varepsilon]}}{1+2[N\varepsilon]}\right)\right|.$$

Here F is an arbitrary bounded continuous function of $2k+1$ variables $(y_{-k},\cdots,y_k)$ and $\tilde{F}(m)$ is given by

$$\tilde{F}(m)=\frac{1}{[M(\lambda)]^{2k+1}}\int\exp\left\{\lambda\sum y_i-\sum\phi(y_i)\right\}F(y_{-k},\cdots,y_k)\,dy_{-k}\cdots dy_k$$

with $\lambda=h'(m)$.

The super exponential estimate follows easily from Lemma 2.2 by using Chebechev's inequality. The fact that we have a more general function $F(y_{-k},\cdots y_k)$ rather than a function of y_0 alone is of no consequence because the more general case is covered by the results in [1], which are used in the proof of Lemma 2.2.

3. Hydrodynamics of Mildly Perturbed Systems

In [1] for the evolution of the Markov process with generator

$$L_N=\frac{1}{2}N^2\left[\sum_{i=1}^{N}\left(\frac{\partial}{\partial x_i}-\frac{\partial}{\partial x_{i+1}}\right)^2-\sum_{i=1}^{N}(\phi'(x_i)-\phi'(x_{i+1}))\left(\frac{\partial}{\partial x_i}-\frac{\partial}{\partial x_{i+1}}\right)\right]$$

it is shown that if we start with density $f_N^0(x_1,\cdots,x_N)$ [relative to the weight $\omega_N(x_1,\cdots,x_N)$] satisfying

$$\int\left(f_N^0\log f_N^0\right)\omega_N\,dx_1\cdots dx_N\leqq CN, \tag{3.1}$$

$$\lim_{N\to\infty}\frac{1}{N}\sum_{i=1}^{N}J\left(\frac{i}{N}\right)x_i=\int J(\theta)m_0(\theta)\,d\theta \tag{3.2}$$

in probability relative to the distributions determined by the densities $\{f_N^0\}$, then for the evolved density $f_N^t(x_1,\cdots,x_N)$ one has, for $t>0$, the analogue of (3.2)

$$\lim_{N\to\infty}\frac{1}{N}\sum_{i=1}^{N}J\left(\frac{i}{N}\right)x_i=\int J(\theta)m(t,\theta)\,d\theta \tag{3.3}$$

in probability relative to the distributions determined by the densities $\{f_N^t\}$. Here $m(t,\theta)$ is obtained by solving the "fluid" equation

$$\begin{aligned}\frac{\partial m(t,\theta)}{\partial t}&=\frac{1}{2}[h'(m(t,\theta))]_{\theta\theta}, \qquad t>0,\\ m(0,\theta)&=m_0(\theta).\end{aligned} \tag{3.4}$$

The proof is based on a set of estimates that establish for the distributions Q_N of the measure-valued process $(1/N)\sum_{j=1}^{N}\delta_{j/N}x_j(t)$ first compactness and then the fact that the limiting measure is concentrated on the weak solutions of (3.4).

The crucial step in showing that any limit point of Q_N is concentrated on the weak solutions of (3.4) is to show that the term

$$\int_0^T \frac{1}{N} \sum_{i=1}^N J\left(\frac{i}{N}\right) \phi'(x_i(t))\, dt = A_N$$

can be replaced by

$$\int_0^T \frac{1}{N} \sum_{i=1}^N J\left(\frac{i}{N}\right) h'\left(\frac{1}{2[N\varepsilon]+1} \sum_{|j-i| \leqq N\varepsilon} x_j(t)\right) dt = B_N(\varepsilon)$$

for the measure P_N^0 of the process with f_N^0 as initial density. More precisely, one needs an estimate of the form

$$\lim_{\varepsilon \to 0} \overline{\lim_{N \to \infty}} P_N^0\left[|A_N - B_N(\varepsilon)| \geqq \delta\right] = 0. \tag{3.5}$$

This particular estimate is derived from two basic estimates. If we denote by f_N^t the density relative to ω_N of the marginal at time t of the process P_N^0, then, for the average marginal

$$\bar{f}_N = \frac{1}{T} \int_0^T f_N^t\, dt,$$

we have

$$\int \left[\sum_{i=1}^N \left(\frac{\partial \bar{f}_N}{\partial x_i} - \frac{\partial \bar{f}_N}{\partial x_{i+1}}\right)^2\right] \frac{1}{\bar{f}_N} \omega_N\, dx_1 \cdots dx_N \leqq \frac{C}{N} \tag{3.6}$$

and for the marginals f_N^t themselves

$$\int f_N^t \log f_N^t \omega_N\, dx_1 \cdots dx_N \leqq CN. \tag{3.7}$$

In [1], (3.5) is derived purely as a consequence of (3.6) and (3.7). If we change the evolution to one of the form

$$\begin{aligned} L_{N,t} = \frac{1}{2} N^2 \Bigg[& \sum_{j=1}^N \left(\frac{\partial}{\partial x_j} - \frac{\partial}{\partial x_{j+1}}\right)^2 \\ & - \sum_{j=1}^N (\phi'(x_j) - \phi'(x_{j+1})) \left(\frac{\partial}{\partial x_j} - \frac{\partial}{\partial x_{j+1}}\right) \Bigg] \\ & + N \sum_{j=1}^N g\left(t, \frac{j}{N}\right) \left(\frac{\partial}{\partial x_j} - \frac{\partial}{x_{j+1}}\right), \end{aligned} \tag{3.8}$$

where g is a smooth function on $[0, T] \times S$, then starting from an initial density f_N^0 we look at the evolving density g_N^t. The entropy

$$H_N(t) = \int g_N^t \log g_N^t \omega_N \, dx_1 \cdots dx_N$$

while non-negative is not necessarily monotone under the evolution (3.8). However an elementary calculation yields

$$\begin{aligned} \frac{dH_N(t)}{dt} &= -\frac{1}{2}N^2 \int \left[\sum_{i=1}^N \left(\frac{\partial g_N^t}{\partial x_i} - \frac{\partial g_N^t}{\partial x_{i+1}} \right)^2 \right] \frac{1}{g_N^t} \omega_N \, dx_1 \cdots dx_N \\ &\quad - N \int \sum_{i=1}^N g\left(t, \frac{i}{N}\right) \left[\frac{\partial g_N^t}{\partial x_i} - \frac{\partial g_N^t}{\partial x_{i+1}} \right] \omega_N \, dx_1 \cdots dx_N. \end{aligned} \tag{3.9}$$

Denoting

$$I_N(t) = \int \left[\sum_{i=1}^N \left(\frac{\partial g_N^t}{\partial x_i} - \frac{\partial g_N^t}{\partial x_{i+1}} \right)^2 \frac{1}{g_N^t} \right] \omega_N \, dx_1 \cdots dx_N, \tag{3.10}$$

we obtain, by use of Schwartz's inequality,

$$\begin{aligned} &\left| \int \sum_{i=1}^N g\left(t, \frac{i}{N}\right) \left[\frac{\partial g_N^t}{\partial x_i} - \frac{\partial g_N^t}{\partial x_{i+1}} \right] \omega_N \, dx_1 \cdots dx_N \right| \\ &\quad \leqq \left\{ \sum_{j=1}^N \left[g\left(t, \frac{j}{N}\right) \right]^2 \right\}^{1/2} \{ I_N(t) \}^{1/2} \leqq C N^{1/2} [I_N(t)]^{1/2}; \end{aligned} \tag{3.11}$$

we have used here $\int g_N^t \omega_N \, dx_1 \cdots dx_N = 1$. Combining (3.9), (3.10) and (3.11), we have

$$\begin{aligned} \frac{dH_N(t)}{dt} &\leqq -\frac{1}{2} N^2 I_N(t) + C N^{3/2} [I_N(t)]^{1/2}, \\ H_N(0) &\leqq CN, \\ H_N(t) &\geqq 0. \end{aligned} \tag{3.12}$$

These inequalities are sufficient to yield

$$H_N(T) \leqq C_T N \tag{3.13}$$

and

$$\int_0^T I_N(t) \, dt \leqq \frac{C_T}{N}. \tag{3.14}$$

By convexity one obtains, for

$$\bar{g}_N = \frac{1}{T}\int_0^T g_N^t \, dt,$$

the inequality

$$I_N(\bar{g}_N) \leqq \frac{C_T}{N}. \tag{3.15}$$

The rest of the proof is exactly like in [1] and one obtains a hydrodynamic limit to the unique weak solution of

$$\begin{aligned} \frac{\partial m(t,\theta)}{\partial t} &= \frac{1}{2}[h'(m(t,\theta))]_{\theta\theta} - \frac{\partial g(t,\theta)}{\partial \theta} \quad \text{for} \quad t > 0, \\ m(t,\theta)|_{t=0} &= m_0(\theta). \end{aligned} \tag{3.16}$$

4. The Lower Bound

In this section we shall prove the lower bound (1.9) of the introduction. Let us consider the evolution P_N^0 obtained by a time dependent generator

$$\begin{aligned} L_{N,t} = \frac{1}{2}N^2 \Bigg[\sum_{j=1}^N \left(\frac{\partial}{\partial x_j} - \frac{\partial}{\partial x_{j+1}} \right)^2 \\ - \sum_{j=1}^N (\phi'(x_j) - \phi'(x_{j+1}))\left(\frac{\partial}{\partial x_j} - \frac{\partial}{\partial x_{j+1}} \right) \Bigg] \\ + N \sum_{j=1}^N g\left(t, \frac{j}{N}\right)\left(\frac{\partial}{\partial x_j} - \frac{\partial}{\partial x_{j+1}} \right) \end{aligned} \tag{4.1}$$

with initial density

$$f_N^0 = \prod_{j=1}^N \frac{1}{M(\lambda_{N,j})} \exp\{\lambda_{N,j} x_j\}, \tag{4.2}$$

where

$$\lambda_{N,j} = h'(m_0(j/N)). \tag{4.3}$$

Here $m_0(\theta)$ on S and $g(t,\theta)$ on $[0,T] \times S$ are smooth functions.

First we note that according to the discussion in Section 3, if U is any weak neighborhood of the unique weak solution of

$$\frac{\partial m(t,\theta)}{\partial t} = \tfrac{1}{2}[h'(m,t,\theta)]_{\theta\theta} - \frac{\partial}{\partial\theta}g(t,\theta), \qquad m(0,\theta) = m_0(\theta), \tag{4.4}$$

then

$$P_N^0(U) \to 1 \tag{4.5}$$

as $N \to \infty$. We are interested in estimating $P_N(U)$ and what we have is an explicit formula for the Radon-Nikodym derivative dP_N^0/dP_N by combining f_N^0 with the Cameron-Martin-Girsanov formula,

$$\frac{dP_N}{dP_N^0} = \exp\left\{\sum_{j=1}^{N}\log M(\lambda_{N,j}) - \sum_{j=1}^{N}\lambda_{N,j}x_j(0)\right\}$$
$$+\exp\left\{-\sum_{j=1}^{N}\int_0^T g\left(t,\frac{j}{N}\right)d\hat{x}_j(t) - \frac{1}{2}\sum_{j=1}^{N}\int_0^T\left[g\left(t,\frac{j}{N}\right)\right]^2 dt\right\},$$

where $\{\hat{x}_j(t)\}$ are independent Brownian motions. An elementary calculation yields

$$\begin{aligned}\lim_{N\to\infty}\frac{1}{N}\log\frac{dP_N}{dP_N^0} &= \int_S \rho(h'(m_0(\theta)))\,d\theta - \int_S h'(m_0(\theta))m_0(\theta)\,d\theta \\ &\quad -\frac{1}{2}\int_0^T\int_S[g(t,\theta)]^2\,d\theta\,dt \\ &= -\int_S h(m(0,\theta))\,d\theta - \frac{1}{2}\int_0^T\int_S[g(t,\theta)]^2\,dt\,d\theta \\ &= -I(m(\cdot,\cdot))\end{aligned} \tag{4.6}$$

in probability with respect to P_N^0. If we combine (4.6) and (4.5) we get the lower bound (1.9). We make the following remark to complete the argument.

Remark. Given any $m(t,\theta)$ which is weakly continuous in t and satisfies

$$I(m(\cdot,\cdot)) = \int_0^T\left\|\frac{\partial m}{\partial t} - \frac{1}{2}[h'(m(t,\theta))]_{\theta\theta}\right\|_{-1}^2 dt = l,$$

it can be approximated by a sequence m_k of smooth functions $m_k(t, \theta)$ such that

$$I(m_k(\cdot, \cdot)) \to l.$$

Acknowledgment. The work for this paper was supported in part by the National Science Foundation under grant DMS 8600233.

Bibliography

[1] M. Z. Guo, G. C. Papanicolaou, and S. R. S. Varadhan, *Nonlinear diffusion limit for a system with nearest neighbor interactions*, to appear in Comm. Math. Physics.

[2] C. Kipnis, S. Olla, and S. R. S. Varadhan, *Hydrodynamics and large deviations for simple exclusion process*, Comm. Pure Appl. Math. Vol 42, 1989, pp. 115–137, this issue.

[3] S. R. S. Varadhan, *Large deviations and applications*, CBMS-NSF Regional Conference Series in Applied Mathematics, 46, Society for Industrial and Applied Mathematics, 1984.

Received April, 1988.

Analysis, Academic Press, 1990, 677-694

On the Derivation of Conservation Laws for Stochastic Dynamics

S. R. S. Varadhan

Courant Institute, New York University

Dedicated with respect and affection to Professor Jurgen K. Moser on his sixtieth birth day.

1. Introduction.

The problem of deriving the macroscopic conservation laws, or the Euler equations, in a rigorous fashion from the microscopic equations of classical mechanics is still largely open, even though the physics is very well understood. With suitable scaling of space and time the following general picture emerges. At the microscopic or grain level particles are moving very rapidly. In fact the motion is so rapid that it is impossible to follow the motion of individual particles in the prevailing time scale. But the conservation laws deal with the evolution of the spatial distribution of quantities that are conserved in encounters between particles at the microscopic level. They involve exchange of sorts between particles and as a result the macroscopic distribution of these quantities undergoes only changes of order one in the prevailing time scale. Much of the motion of the particles appear as thermal motion and only a small part acts coherently to produce macroscopic changes in the distribution of these conserved quantities.

Supported by the National Science Foundation under grant number DMS 8600233.

In order to derive the laws of evolution of these conserved quantities one has to compute its rate of change in time. This is often easy enough. But the rate of change at a given instant of time and a point in macroscopic space depends on the precise location of the particles at that time and around that point in space. In other words the rate of change of macroscopic quantities depends on microscopic information which changes rather rapidly in space and time. While this is a nuisance from the mathematical point of view, it is at this stage that physics enters. The rapid change allows one to average out the microscopic information and replace it by its average value for given macroscopic data. By some form of "Ergodic" phenomenon the space and time averages can be replaced by an "ensemble" average. What we need then is the correct family of "ensembles" indexed by a parameter set that corresponds to the individual values of the macroscopic quantities. Such a family is provided by the Gibbs formalism of equilibrium statistical mechanics. If we can avoid some of its problems like phase transition etc. one can formally derive an Euler equation or conservation law. There is however no well defined separation of scales. In a sense one has to demonstrate the irrelevancy of all the intermediate scales between the macroscopic and the microscopic ones. Another complication is that ergodic theorems require noise, for in a sense they are almost sure theorems. If the system is a classical mechanical one the only possible source of noise is in the initial position in phase space. The Liouville measure is all that we have got. Since the prevailing time scale is an eternity in the original dynamical time we have then the unenviable task of following the dynamics very closely for a very long time.

At this point an Applied Mathematician should go on to find an approximate solution to the exact problem he faces. We will instead find an exact solution to what is at best an approximate problem. One way to make the problem tractable is to introduce some noise in the dynamics, making it a real stochastic process. If there is enough noise, the ergodic theory can be made to depend on the new noise that comes in continuously in time and the need for eternal memory can be dispensed with.

Several such examples have been worked out in recent times. Depending on the circumstances we can have either parabolic or hyperbolic scaling of space and time. In the former case one ends up with a nonlinear heat equation for the conservation law which is diffusive whereas in the latter case we end up with a nonlinear hyperbolic equation of first order. The first situation is a lot easier to handle because the solutions are always smooth and we do not have to face shocks.

There are several examples of both types that have been studied before. There is one example of classical Hamiltonian dynamics, i.e. hard rods under elastic collision in one dimension that has been studied in [2]. The asymmetric random walk with simple exclusion in one dimension has been studied in [1]. The weakly asymmetric random walk with simple exclusion in one dimension and interacting Brownian motion with hard rod potential in one dimension where diffusive scaling holds is established in [4] and [9] respectively. See [3] for an exhaustive survey.

These examples are mostly one dimensional and they depend to a very large extent on the specifics of the model. We have in last few years developed a method that works for a large class of models. The method appears to be flexible and general. But however it is a method that can only work in situations that call for parabolic scaling. The basic assumption is that the Stochastics Dynamics is reversible with respect to a suitable probability measure. This allows us to use the Dirichlet form rather effectively.

We will illustrate the method by outlining the steps for one particular model. We will then describe some of the other models where the method has worked.

2. The Model

We call our basic variable quantity charge which can be any real number. This variable is denoted by x, $-\infty < x < \infty$. The size of the system is a large integer N which is our scaling parameter and goes to ∞. For each N , we have N charges $x_1, x_2, \cdots, x_N$. We visualize these charges as being located at sites $1, 2, \cdots, N$. These sites are placed at a distance $1/N$ apart on the unit interval with end points identified. In other words the N sites form a one dimensional periodic lattice with spacing $\frac{1}{N}$. We denote by S the unit interval with the end points identified and by θ its typical point. The charges $x_1, x_2, \cdots, x_N$ vary over time and we describe them by functions $\{x_1(t), x_2(t), \cdots, x_N(t)\}$, $0 \leq t \leq T$. We speak of the charge configuration $\{x_1(t), x_2(t), \cdots, x_N(t)\}$ at time t. The way the charges change over time is described below.

$$dx_i(t) = dz_{i-1,i} - dz_{i,i+1} \tag{2.1}$$

One thinks of $dz_{i,i+1}$ as the flow of charge from site i to site i+1 along a "bond" connecting neighboring sites. Equation (2.1) then expresses the conservation law that the net change of charge at any site is the difference between the inflow and outflow at that site. It is clear that $\sum_{j=1}^{N} x_j(t)$ is conserved. We now have to specify the law that governs the flow of charge along a bond. We could for instance say that for some $c > 0$

$$dz_{i,i+1} = c[x_i(t) - x_{i+1}(t)]dt \tag{2.2}$$

This will mean that the charges tend to equalize at the different sites. We can make the model nonlinear and say that

$$dz_{i,i+1} = [\psi(x_i(t)) - \psi(x_{i+1}(t))]\ dt \tag{2.3}$$

for some suitable function ψ. We can also introduce some noise into the system by defining

$$dz_{i,i+1} = [\psi(x_i(t)) - \psi(x_{i+1}(t))]dt + d\beta_{i,i+1}(t) \tag{2.4}$$

where $\beta_{i,i+1}(t)$ are N independent Brownian motions. Our model is specified by (2.4). The charge configuration performs a diffusion on R^N with an infinitesimal generator given by

$$\frac{1}{2}\sum_{i=1}^{N}(\frac{\partial}{\partial x_i} - \frac{\partial}{\partial x_{i+1}})^2 + \sum_{i=1}^{N}[\psi(x_{i-1}) - 2\psi(x_i) + \psi(x_{i+1})]\frac{\partial}{\partial x_i} \tag{2.5}$$

Since the space has already been scaled by a factor of $\frac{1}{N}$ we should speed up time by a factor of N^2 and we redefine our generator as

$$\begin{aligned} L_N = {} & \frac{N^2}{2}\sum_{i=1}^{N}(\frac{\partial}{\partial x_i} - \frac{\partial}{\partial x_{i+1}})^2 \\ & + N^2\sum_{i=1}^{N}[(\psi(x_{i-1}) - 2\psi(x_i) + \psi(x_{i+1})]\frac{\partial}{\partial x_i} \end{aligned} \tag{2.6}$$

(The summations are all cyclic sums)

3. The Problem

Suppose we start with a random configuration of charges at time t=0. For example we may choose a smooth function or "profile" $a_0(\theta)$ defined on S and define $a_i = a_0(\frac{i}{N})$. We may make the configuration $(x_1, x_2, \cdots, x_N)$ at time t=0, to be independent random variables with x_i having a uniform distribution in the interval $[a_i - \frac{1}{2}, a_i + \frac{1}{2}]$. By the ergodic theorems one can expect the noises to have no effect globally. We will therefore limit ourselves to calculations of mean value only. It is easy to see that for any smooth function $J(\theta)$ on S

$$\begin{aligned}\lim_{N\to\infty} E[\frac{1}{N}\sum_{i=1}^{N} x_i J(\frac{i}{N})] &= \lim_{N\to\infty}\frac{1}{N}\sum_{i=1}^{N} a_i J(\frac{i}{N}),\\ &= \lim_{N\to\infty}\frac{1}{N}\sum_{i=1}^{N} a_0(\frac{i}{N})J(\frac{i}{N}),\\ &= \int_S a_0(\theta)d\theta \end{aligned} \tag{3.1}$$

One can then think of $a_0(\theta)$ as the "profile" of the macroscopic charge density at time t=0. If we think of $a(t,\theta)$ as the profile at time t we can calculate using the infinitesimal generator L_N

$$\begin{aligned}\frac{\partial}{\partial t}\int a(t,\theta)d\theta &= \lim_{N\to\infty}\frac{d}{dt}\frac{1}{N}E[\sum_{i=1}^{N} J(\frac{i}{N})x_i(t)]\\ &= N^2 E[\frac{1}{N}\sum_{i=1}^{N}\{\psi(x_{i-1}(t)) - 2\psi(x_i(t)) + \psi(x_{i+1}(t))\}J(\frac{i}{N})]\\ &= N^2 E[\frac{1}{N}\sum_{i=1}^{N}\{J(\frac{i-1}{N}) - 2J(\frac{i}{N}) + J(\frac{i+1}{N})\}\psi(x_i(t))]\\ &\approx E[\frac{1}{N}\sum_{i=1}^{N} J''(\frac{i}{n})\psi(x_i(t))] \end{aligned} \tag{3.2}$$

Unfortunately knowing that $\frac{1}{N}\sum_{i=1}^{N} J(\frac{1}{N})x_i(t)$ tends to $\int_S J(\theta)a(t,\theta)d\theta$ does not allow us to compute the limiting value of

$\frac{1}{N}\sum_{i=1}^{N} J''(\frac{i}{N})\psi(x_i(t))$. We require more detailed information about the distribution of the values $\{x_1(t), x_2(t), \cdots, x_N(t)\}$. Since $\psi(x_i(t))$ changes very rapidly one should be able to replace it with its average value for a given value of the macroscopic charge density. If we denote this average value for a given charge density "a" by $\lambda(a)$ one precise way of stating this is equation (3.3)

$$\lim_{\epsilon\to 0}\limsup_{N\to\infty} E[\int_0^T |\frac{1}{N}\sum_{i=1}^{N} J''(\frac{i}{N})\psi(x_i(s)) - \frac{1}{N}\sum_{i=1}^{N} J''(\frac{i}{N})\lambda(\frac{x_{i-N\epsilon}(s)+\cdots+x_{i+N\epsilon}(s)}{1+2N\epsilon})|\, ds] = 0 \tag{3.3}$$

If we can do this for some function $\lambda(a)$ then it is straight forward to establish

$$\frac{d}{dt}\int J(\theta)a(t,\theta)d\theta = \int J''(\theta)\lambda(a(t,\theta))d\theta \tag{3.4}$$

or

$$\frac{\partial}{\partial t}a(t,\theta) = \frac{\partial^2}{\partial\theta^2}[\lambda(a(t,\theta))] \tag{3.5}$$

with

$$a(0,\theta) = a_0(\theta) \tag{3.6}$$

Our problem is to determine $\lambda(a)$ in terms of the function $\psi(x)$ and establish (3.3) for the function $\lambda(.)$ so determined.

4. The Solution

We assume that $\psi(x) = \frac{1}{2}\frac{d\phi}{dx}$ for some ϕ and that $\int e^{-\phi(x)}dx < \infty$. We can then normalize ϕ so that $\int e^{-\phi(x)}dx = 1$. It is then easy to check that $\Phi_N(x_1, x_2, \cdots, x_N) = \exp[-\sum_{i=1}^{N}\phi(x_i)]$ is an invariant density and that L_N is self adjoint with respect to Φ_N. However L_N is not elliptic on R^N. It is elliptic only when restricted to the hyperplanes $\frac{1}{N}\sum_{i=1}^{N} x_i = a$ for each fixed a. So the restrictions of Φ_N to these hyperplanes are invariant densities when they are suitably normalized. The diffusion is reversible and ergodic on each one of these hyperplanes.We assume further that

$$M(\sigma) = \int e^{\sigma x - \phi(x)} dx < \infty \ forall - \infty < \sigma < \infty \tag{4.1}$$

We also make the technical hypothesis that $\psi = \frac{1}{2}\frac{d\phi}{dx}$ is continuous as a function of x and that

$$\int e^{\sigma|\psi(x)| - \phi(x)} dx < \infty \ forall \ \sigma \geq 0. \tag{4.2}$$

The parameter $a = \frac{1}{N}\sum_{i=1}^{N} x_i$ defining the hyperplane is thought of as the macroscopic parameter defining the average charge density. We take a large block $[1, 2, \cdots, l]$ of sites and try to calculate

$$\lim_{\ell \to \infty} E[\psi(x_i) \mid \frac{x_1 + x_2 + \cdots + x_\ell}{\ell} = a]$$

assuming that $x_1, x_2, \cdots, x_l$ have a distibution specified by the density $\Phi_l(x_1, x_2, \cdots, x_l) = exp[-\sum_{i=1}^{\ell} \phi(x_i)]$. Using the Gibbs formalism which reduces in this case to the theory of large deviations developed by Cramér in the thirties, we find the following answer:

$$\lim_{l \to \infty} E[\psi(x_i) \mid \frac{x_1 + x_2 + \cdots + x_l}{l} = a]$$

$$= \frac{1}{M(\sigma^*(a))} \int \psi(x) e^{\sigma^*(a)x - \phi(x)} dx \tag{4.3}$$

where $\sigma^*(a)$ is determined so that

$$a = \frac{1}{M(\sigma^*(a))} \int x e^{\sigma^*(a)x - \phi(x)} dx. \tag{4.4}$$

From the definition (4.1) we see that (4.4) reduces to

$$\frac{M'(\sigma^*(a))}{M(\sigma^*(a)} = a \tag{4.5}$$

or $\sigma = \sigma^*(a)$ is the solution of

$$\frac{M'(\sigma)}{M(\sigma)} = a \tag{4.6}$$

The function $\rho(\sigma) = log M(\sigma)$ is a convex function of σ and if $h(a)$ is its conjugate convex function

$$h(a) = \sup_{\sigma}[a\sigma - \rho(\sigma)] \tag{4.7}$$

then the solution of (4.6) is

$$\sigma^*(a) = h'(a) \tag{4.8}$$

Since $\psi(x) = \frac{1}{2}\frac{d\phi}{dx}$, (4.3) can be calculated by integrating by parts and we get

$$\frac{1}{M(\sigma)}\int \psi(x)e^{\sigma x - \phi(x)}dx = \frac{\sigma}{2} \tag{4.9}$$

This provides us with the formula

$$\lambda(a) = \frac{1}{2}h'(a)$$

and finally the equation for our macroscopic evolution

$$\frac{\partial a(t,\theta)}{\partial\theta} = \frac{1}{2}\frac{\partial^2}{\partial\theta^2}[h'(a(t,\theta))] \tag{4.10}$$

with

$$a(0,\theta) = a_0(\theta)$$

5. The Method

The problem is to establish (3.3). It is more convenient to break it up into two parts namely (5.1) and (5.2) below.

$$\begin{aligned}\lim_{k\to\infty}\limsup_{N\to\infty} E[\frac{1}{N}\int_0^T \sum_{i=1}^{N} | \frac{1}{2k+1}\sum_{j=i-k}^{i+k} G(x_j(t)) \\ - \hat{G}(\frac{1}{2k+1}\sum_{j=i-k}^{i+k} x_j(t)) \mid dt] = 0\end{aligned} \tag{5.1}$$

and

$$\lim_{\substack{\epsilon\to 0\\ k\to\infty}} \limsup_{N\to\infty} \sup_{0\le l\le \epsilon N} E[\frac{1}{N}\int_0^T \sum_{i=1}^{N} | \frac{1}{2k+1}\sum_{j=i-k}^{i+k} x_j(t) - \frac{1}{2k+1}\sum_{j=i-k+l}^{i+k+l} x_j(t) \mid dt] = 0 \tag{5.2}$$

In formula (5.1) $G(x)$ is a bounded continuous function on R and $\hat{G}(a)$ is defined by

$$\hat{G}(a) = \frac{1}{M(\sigma)}\int e^{\sigma x-\phi(x)}\, G(x)\, dx \tag{5.3}$$

with $\sigma = h'(a)$.

We want really to use (5.1) with $G(x) = \psi(x)$ which is unbounded and get $\hat{G}(a) = \frac{1}{2}h'(a)$ in that case. However (4.2) allows us to negotiate this successfully by truncation. To see why (5.1) and (5.2) imply (3.4), since $J(\theta)$ is a smooth function we can replace $\frac{1}{N}\sum J(\frac{i}{N})\psi(x_i(t))$ by $\frac{1}{N}\sum J(\frac{i}{N})\frac{1}{2k+1}\sum_{j=i-k}^{j=i+k}\psi(x_j(t))$ with large k, and by (5.1) this can be replaced by

$$\frac{1}{2N}\sum J(\frac{i}{N})h'(\frac{x_{i-k}(t)+\cdots+x_{i+k}(t)}{2k+1}).$$

According to (5.2) the averages over blocks of length k,for large enough k, do not vary by much when a block is moved by a distance of $N\epsilon$ sites, ϵ being a small number. Therefore $\dfrac{x_{i-k}(t)+\cdots+x_{i+k}(t)}{2k+1}$ with large k, can be replaced by $\dfrac{x_{i-N\epsilon}(t)+\cdots+x_{i+N\epsilon}(t)}{1+2N\epsilon}$ with small ϵ and this is precisely (3.3). If α_0^N is the initial distribution on R^N, and α_t^N is the distribution at time t, obtained by the evolution, (5.1) and (5.2) can be reexpressed in terms of $\bar{\alpha}^N = \frac{1}{T}\int_0^T \alpha_t^N dt$.

$$\lim_{k\to\infty} \limsup_{N\to\infty} E^{\bar{\alpha}^N}[\frac{1}{N}\sum_{i=1}^{N} | \frac{1}{2k+1}\sum_{j=i-k}^{i+k} G(x_j) - \hat{G}(\frac{x_{i-k}+\cdots+x_{i+k}}{2k+1} |] = 0 \tag{5.1a}$$

and

$$\lim_{\substack{k\to\infty\\ \epsilon\to 0}} \limsup_{N\to\infty} \sup_{l\le \epsilon N} E^{\bar{\alpha}^N}\Big[\frac{1}{N}\sum_{i=1}^{N} \Big| \frac{1}{2k+1}\sum_{j=i-k}^{i+k} x_j - \frac{1}{2k+1}\sum_{j=i+l-k}^{i+l+k} x_j \Big|\Big] = 0 \tag{5.2a}$$

In fact we can project the distribution $\bar{\alpha}^N$ onto R^{2k+1} corresponding to the coordinates $\{x_{i-k},\cdots,x_{i+k}\}$ and average the resulting distibution over $i = 1,2\cdots,N$. If we denote the average by $\bar{\alpha}^{N,k}$ (5.1a) can be expressed as

$$\lim_{k\to\infty} \limsup_{N\to\infty} E^{\bar{\alpha}^{N,k}}\Big[\Big| \frac{1}{2k+1}\sum_{i=-k}^{k} G(y_i) - \hat{G}\Big(\frac{y_{-k}+\cdots+y_k}{2k+1}\Big) \Big|\Big] = 0 \tag{5.1b}$$

Similarly we can project $\bar{\alpha}^N$ onto $R^{2k+1}\times R^{2k+1}$ corresponding to the coordinates $[x_{i-k},\cdots,x_{i+k};x_{i+l-k},\cdots,x_{i+l+k}]$, average the resulting distribution over $i = 1,2,\cdots,N$ and call it $\bar{\alpha}^{N,k,l}$. Then (5.2a) reduces to

$$\lim_{\substack{k\to\infty\\ \epsilon\to 0}} \limsup_{N\to\infty} \sup_{l\le N\epsilon} E^{\bar{\alpha}^{N,k,l}}\Big[\Big| \frac{1}{2k+1}\sum_{i=-k}^{k} y_i - \frac{1}{2k+1}\sum_{i=-k}^{k} z_i \Big|\Big] = 0 \tag{5.2b}$$

We have used the dummy variables $\{y_j\}$ for R^{2k+1} in (5.1b) and variables $\{y_j\}$ and $\{z_j\}$ for $R^{2k+1}\times R^{2k+1}$ in (5.2b).

The advantage in (5.2a) and (5.2b) is that N goes to infinity first with k and ϵ fixed. Modulo some compactness which can be taken care of, we can denote by $\mathcal{A}_k$ the set of limit points of $\bar{\alpha}^{N,k}$ as $N\to\infty$ and reduce (5.1b) to (5.1c) below:

$$\lim_{k\to\infty} \sup_{\alpha\in\mathcal{A}_k} E^{\alpha}\Big[\Big| \frac{1}{2k+1}\sum_{i=-k}^{k} G(y_i) - \hat{G}\Big(\frac{y_{-k}+\cdots+y_k}{2k+1}\Big) \Big|\Big] = 0 \tag{5.1c}$$

Similarly if we denote by $\mathcal{A}_{k,\epsilon}$ the limit points of $\bar{\alpha}^{N,k,l}$ as $N \to \infty$ $l \to \infty$ with $l \leq N\epsilon$ we need to show

$$\lim_{\substack{k\to\infty \\ \epsilon\to 0}} \sup_{\alpha\in\mathcal{A}_{k,\epsilon}} E^{\alpha}[| \frac{1}{2k+1}\sum y_i - \frac{1}{2k+1}\sum z_i |] = 0 \tag{5.2c}$$

It is now clear that we need to obtain information about the sets $\mathcal{A}_k$ and $\mathcal{A}_{k,\epsilon}$ in order to carry out our program.

In order to simplify our presentation we will pretend that all the probability distributions that we meet with on various R^d with $d = 2k+1$, $4k+2$ or N all have densities with respect to the corresponding Lebesgue measure on R^d . More over we will compute the densities relative to $\Phi_d = exp[-\sum \phi(x_i)]$. So we think of $\alpha(dx) = f\Phi_d dx$ with f representing the distribution on R^d.

We need the following functionals:

The Entropy Functional.

$$H_N(\alpha) = H_N(f) = \int_{R^N} f log f \ \Phi_N \ dx \tag{5.4}$$

The Dirichlet Functionals:

$$I_N(\alpha) = I_N(f) = \frac{1}{2}\int_{R^N} \frac{1}{f}\sum_{i=1}^{N}(\frac{\partial f}{\partial x_i} - \frac{\partial f}{\partial x_{i+1}})^2 \ \Phi_N dx \tag{5.5}$$

$$J_k(\alpha) = J_k(f) = \frac{1}{2}\int_{R^{2k+1}} \frac{1}{f}\sum_{i=-k}^{k-1}(\frac{\partial f}{\partial x_i} - \frac{\partial f}{\partial x_{i+1}})^2 \ \Phi_{2k+1} dx \tag{5.6}$$

$$\begin{aligned} J_{k,k}(\alpha) &= J_{k,k}(f) = \\ &= \frac{1}{2}\int_{R^{2k+1}\times R^{2k+1}} \frac{1}{f}\sum_{i=-k}^{k-1}[(\frac{\partial f}{\partial y_i} - \frac{\partial f}{\partial y_{i+1}})^2 \\ &+ (\frac{\partial f}{\partial z_i} - \frac{\partial f}{\partial z_{i+1}})^2] \ \Phi_{2k+1}(y)\Phi_{2k+1}(z)dy \ dz \end{aligned} \tag{5.7}$$

and finally,

$$\bar{J}_{k,k}(\alpha) = \bar{J}_{k,k}(f)$$
$$= \frac{1}{2}\int_{R^{2k+1}\times R^{2k+1}} \frac{1}{f}(\frac{\partial f}{\partial y_0} - \frac{\partial f}{\partial z_0})^2 \Phi_{2k+1}(y)\ \Phi_{2k+1}(z)dy\ dz \tag{5.8}$$

It turns out that all these functionals are nonnegative, lower semicontinuous and convex. Furthermore if α_t^N evolves according to our evolution then

$$\frac{d}{dt}H_N(\alpha_t^N) = \ -N^2 I_N(\alpha_t^N) \tag{5.9}$$

If we now impose the initial assumption that $H_N(\alpha_0^N) \leq CN$, where C is a constant independent of N, then it follows that for some other constant,

$$\int_0^T I_N(\alpha_t^N)dt \ \leq \frac{C}{N} \tag{5.10}$$

and by convexity for each fixed T,

$$I_N(\bar{\alpha}^N) \leq \frac{C}{N} \tag{5.11}$$

There are other inequalities that result from (5.11) because of convexity.

$$J_k(\bar{\alpha}^{N,k}) \leq \frac{2k}{N} I_N(\bar{\alpha}^N) \tag{5.12}$$

$$J_{k,k}(\bar{\alpha}^{N,k,l}) \leq \frac{4k}{N} I_N(\bar{\alpha}^N) \tag{5.13}$$

and

$$\bar{J}_{k,k}(\bar{\alpha}^{N,k,l}) \leq \frac{l^2 N^2}{N} I_N(\bar{\alpha}^N) \tag{5.14}$$

It now follows immediately that on $\mathcal{A}_k$ and $\mathcal{A}_{k,\epsilon}$ we have

$$J_k(\alpha) = 0 \text{ for all } \alpha \in \mathcal{A}_k \tag{5.15}$$

$$J_{k,k}(\alpha) = 0 \text{ for all } \quad \alpha \in \mathcal{A}_{k,\epsilon} \tag{5.16}$$

and

$$\bar{J}_{k,k}(\alpha) \leq C\epsilon^2 \text{ for all } \quad \alpha \in \mathcal{A}_{k,\epsilon} \tag{5.17}$$

From $J_k(\alpha) = 0$ it follows that α resricted to any hyperplane is the same as Φ_{2k+1} restricted to the same hyperplane. This reduces (5.1c) to Cramér's theory. To go from (5.16) and (5.17) to (5.2c) is similar but slightly more complex. The main advantage is that in some sense the calculations are reduced to the equilibrium case.

7. The Ubiquitous Large Deviations.

The concept of Large Deviations has been mentioned in the previous sections as a tool for computing certain quantities in equilibrium. This is at the level of calculating thermodynamical quantities in equilibrium statistical mechanics. The relevance of large deviations in this context is developed in the article [5]. However the "Dirichlet" functionals (5.4)-(5.8) introduced in section 5 are those that play a role in the theory of large deviations in a "dynamical" context rather than an "equilibrium context". This of course is not just a coincidence and we will in this section try to explain a direct connection and some of its consequences.

The basic Probability measure P_N is obtained on the space of trajectories $\{x_1(t), \cdots, x_N(t);\ \ 0 \leq t \leq T\}$ starting from the initial measure α_0^N and the evolution governed by L_N. In addition to P_N we have the measure Q_N which is obtained for the same evolution if we start instead with the invariant distribution at time 0 namely $\Phi_N(x)dx$. The measure Q_N corresponds to a stationary Markov process which is reversible. The initial condition on α_0^N can be thought of as a bound on the relative entropy $H(P_N; Q_N)$ of P_N with respect to Q_N. We have certain sets $E_N^\epsilon(T)$ defined by (3.4) and the basic problem in (3.4) is to show that

$$\lim_{\epsilon \to 0} \limsup_{N \to \infty} P_N[E_N^\epsilon(T)] = 0 \tag{6.1}$$

There is a very elementary inequality

$$P_N(A) \leq \frac{1 + H(P_N; Q_N)}{log \frac{1}{Q_N(A)}} \tag{6.2}$$

valid for all sets A. Therefore to show (6.1) when we have a bound of the form $H(P_N; Q_N) \leq CN$, it is sufficient to prove

$$\lim_{\epsilon \to 0} \limsup_{N \to \infty} \frac{1}{N} log Q_N[E_N^\epsilon] = -\infty \tag{6.3}$$

In other words superexponential estimates for Q_N will yield the result that probabilities go to zero under P_N.

Since the event $E_N^\epsilon(T)$ can be expressed as

$$E_N^\epsilon(T) = \{\omega : \frac{1}{N} \int_0^T W_{N,\epsilon}(x_1(s), \cdots, x_N(s)) ds \geq \delta\}$$

where $\delta > 0$ is fixed, we can get a bound on $Q_N(E_N^\epsilon(T))$ by Tchbechev's inequality if we can estimate

$$E^{Q_N}\{exp[\int_0^T W_{N,\epsilon}(x_1(s), \cdots, x_N(s)) ds]\} \tag{6.4}$$

Since Q_N is the stationary process with reversible initial distribution, an application of Feynman-Kac formula and the spectral theorem allows us to estimate (6.4) by $exp[\Lambda_{N,\epsilon} T]$ where $\Lambda_{N,\epsilon}$ is the principal eigenvalue of $L_N + W_{N,\epsilon}$.Our goal is to establish

$$\lim_{\epsilon \to 0} \limsup_{N \to \infty} \frac{1}{N} \Lambda_{N,\epsilon} = 0 \tag{6.5}$$

Since $\Lambda_{N,\epsilon}$ is given by the variational formula

$$\Lambda_{N,\epsilon} = \sup_{\alpha} [E^\alpha[W_{N,\epsilon}] - N^2 I_N(\alpha)] \tag{6.6}$$

the problem is one of controlling $E^\alpha[W_{N,\epsilon}]$ in terms of $I_N(\alpha)$. Since $W_{N,\epsilon}$ is of order N, it is a question of showing that if $I_N(\alpha) \leq CN$ then

$$\lim_{\epsilon \to 0} \limsup_{N \to \infty} \frac{1}{N} E^\alpha[W_{N,\epsilon}] = 0 \tag{6.7}$$

At the end of section 5 we did in fact establish (6.7).

8. Other Models

We shall discuss very briefly some of the other models where this general method works after some modifications necessitated by the circumstances.

A. The Weakly Asymmetric case.

We can modify our infinitesimal generator L_N by adding a weaker lower order term.

$$\hat{L}_N = L_N + N \sum_{i=1}^{N} b(x_i, x_{i+1})(\frac{\partial}{\partial x_i} - \frac{\partial}{\partial x_{i+1}})$$

The final equation gets modified by

$$\frac{\partial a(t,\theta)}{\partial t} = \frac{1}{2}\frac{\partial^2}{\partial \theta^2}(h'(a(t,\theta)) + \frac{\partial}{\partial \theta}(\hat{b}(a(t,\theta)))$$

where

$$\hat{b}(a) = \frac{1}{M(\theta)^2} \int\int e^{\sigma x_1 + \sigma x_2} b(x_1, x_2) e^{-\phi(x_1)-\phi(x_2)} dx_1 dx_2$$

with $\sigma = h'(a)$.To establish this result we follow basically the same procedure as before, except we now have to deal with the measure $\hat{P}_N$ corresponding to the new evolution. By Girsanov's formula we can compute the Radon -Nikodym derivative $\frac{d\hat{P}_N}{dQ_N}$ and show that $H(\hat{P}_N; Q_N)$ has a bound CN. We are now back in the same situation as before.

B. The Weakly Asymmetric Random Walk with Simple Exclusion.

In this model we have a certain number ℓ of particles located on some of the sites of our basic periodic lattice. There can be at most one particle per site and each particle tries to move to the site on the left or on the right after waiting for an exponential time with mean $\frac{1}{N^2}$. If the site it wants to move to is occupied it gives up the turn and waits again. Although the conservation law for this evolution is linear under the assumption that the probabilities of opting to move left or right are equal to $\frac{1}{2}$ and in fact is rather trivial to establish, if the probabilities were perturbed by an amount of order $\frac{1}{N}$ then our method can handle this case with ease, even when these probabilities are allowed to depend on the immediate environment.

C. The Non-gradient Case of Our Model.

If we alter our model in section 2 so that the "Dirichlet" form is

$$\frac{N^2}{2}\int_{R^N}\sum_{i=1}^{N} c(x_i, x_{i+1})(\frac{\partial u}{\partial x_i} - \frac{\partial u}{\partial x_{i+1}})^2 \Phi_N(x) dx_1 \cdots dx_N,$$

then the generator L_N can be computed to be

$$L_N = \frac{N^2}{2}\sum_{i=1}^{N}(\frac{\partial}{\partial x_i} - \frac{\partial}{\partial x_{i+1}})c(x_i, x_{i+1})(\frac{\partial}{\partial x_i} - \frac{\partial}{\partial x_{i+1}})$$
$$-\frac{N^2}{2}\sum_{i=1}^{N} c(x_i, x_{i+1})(\phi'(x_i) - \phi'(x_{i+1}))(\frac{\partial}{\partial x_i} - \frac{\partial}{\partial x_{i+1}})$$

The situation studied earlier correspond to $c \equiv 1$. The essential difference is that in arriving at equation (3.3) we performed summation by parts twice. If $c(x_1, x_2)$ is not a constant then we have to stop at

$$\frac{\partial}{\partial t} E \sum_{i=1}^{N} J(\frac{i}{N}) x_i(t) \simeq \sum_{i=1}^{N} J'(\frac{i}{N}) W(x_i(t), x_{i+1}(t))$$

for some W determined in terms of "c" and"ϕ".We cannot do summation by parts once more. This leads to additional complications. But still one can arrive at an equation

$$\frac{\partial a(t,\theta)}{\partial t} = \frac{1}{2}\frac{\partial}{\partial\theta}(\hat{c}(a(t,\theta))(\frac{\partial}{\partial\theta}h'(a(t,\theta))))$$

The main problem is to derive the formula for expressing $\hat{c}(a)$ in terms of $c(x_1, x_2)$.Such a formula exists in the literature under the name of Green -Kubo formula. We provide a different but equivalent formula, one we can rigorously establish.

For a given value of "a" ,we pick $\sigma = h'(a)$ and consider P_a the infinite product measure on R^∞, $\prod \frac{1}{M(\sigma)} e^{\sigma x_i - \phi(x_i)} dx_i$. For any nice function f depending on a finite set $(x_1, \cdots, x_k)$ of coordinates we define the formal function

$$Z = Z_f = \sum_{j=-\infty}^{\infty} f(\tau^j x)$$

where τ is the shift on the infinite product space. Although Z does not make any sense

$$\xi = \xi_f = \frac{\partial Z}{\partial x_1} - \frac{\partial Z}{\partial x_2}$$

makes sense and

$$\hat{c}(a) = \inf_f E^{P_a}[(1 - \xi_f)^2]$$

where the infimum is taken over all nice functions and k as well.

D. The Interacting Diffusion Model.

In this last model that we consider we have a system (of N particles) undergoing a diffusion on the N-torus T^N. The N coordinates represent the position on the 1-torus of each of the particles. The nature of the interacting evolution is specified by the N-partcle generator

$$L_N = \frac{1}{2} \sum_{i=1}^{N} \frac{\partial}{\partial x_i{}^2} - \frac{N}{2} \sum_{i,j=1}^{N} V'(N(x_i - x_j)) \frac{\partial}{\partial x_i}$$

V(x-y) is a binary repulsive potential scaled to adjust to the mean interparticle distance. As $N \to \infty$, one expects the macroscopic density of the particles to undergo a non linear evolution given by

$$\frac{\partial \rho(t,\theta)}{\partial t} = \frac{1}{2} \frac{\partial}{\partial \theta} (\hat{c}(\rho(t,\theta)) \frac{\partial \rho(t,\theta)}{\partial \theta})$$

where $\hat{c}$ is a thermodynamic function to be computed for the model in terms of the function V.

9. Notes and Comments

The basic model that we considered in section 2, was considered by J. Fritz in [7] under the assumption that the function ϕ is convex.

The method used was quite different. The method outlined here has appeared in [6]. The large deviation aspects of the model can be found in [5], where the weakly asymmetric case is treated as well. For the large deviation aspects of the symmetric random walk with simple exclusions as well as an analysis of the weakly asymmetric case in that context see [8]. The work on the nongradient case and the general interacting diffusion model is still to be completed.

Bibliography

[1] A. Benassi, J. P. Fouque, Hydrodynamic limit for the simple exclusion process, Annals of Probability, Vol. 15 (1987) pp. 546-566.

[2] C. Boldrighini, R. L. Dobrushin and Yu. M. Sukhov, One dimensional hard rod caricature of hydrodynamics, J. Stat. Phys. 31 (1983) pp. 577-616.

[3] A. DeMasi, N. Ianiro, A. Pellegrinotti, and E. Presutti, A survey of the hydrodynamical behavior of many-particle systems, Non Equilibrium Phenomena II, from Stochastics to Hydrodynanmics, J. L. Lebowitz and E. W. Montroll, Editors: North-Holland Physics Publishing, Amsterdam, 1984.

[4] A. DeMasi, E. Presutti and E. Scacciatelli, The weakly asymmetric simple exclusion process, Carr Reports in Mathematical Physics, No. 4/87, February 1987.

[5] M. D. Donsker and S. R. S. Varadhan, Large deviations from a hydrodynamical scaling limit, to appear in Comm. Pure Appl. Math.

[6] M. Z. Guo, G. C. Papanicolaou and S. R. S. Varadhan, Nonlinear diffusion limit for a system of nearest neighbor interactions, Comm. Math. Phys., Vol. 118 (1988) pp. 31-59.

[7] J. Fritz, On the hydrodynamic limit of a scalar Ginzburg-Landau lattice model, IMA, Vol. 9, Hydrodynamic behavior and interacting particle systems: G. C. Papanicolaou, Editor, Berlin - Heidelberg - New York, Springer, 1987.

[8] C. Kipnis, S. Olla and S. R. S. Varadhan, Hydrodynamics and large deviation for simple exclusion processes, to appear in Comm. Pure Appl. Math.

[9] H. Rost, Diffusion de sphères dures dans la droite réelle: comportment macroscopique et équilibra local, Lecture Notes in Mathematics No. 1059, pp. 127-144, Springer-Verlag, Heidelberg, 1984.

Commun. Math. Phys. 135, 313–353 (1991)

Communications in Mathematical Physics

Scaling Limits for Interacting Diffusions*

S. R. S. Varadhan

Courant Institute of Mathematical Sciences, New York University, 251 Mercer Street, New York, NY 10012, USA

Received February 26, 1990; in revised form June 22, 1990

Abstract. We consider a large number of particles diffusing on a circle interacting through a drift resulting from the gradient of a pair potential whose support is of the order of the interparticle distance. We derive a nonlinear bulk diffusion equation for the density of the particle distribution on the circle. The diffusion coefficient is determined as a function of density in terms of standard thermodynamical objects.

1. Introduction

In this article we study the hydrodynamic limit for interacting Brownian motions on the one-dimensional circle. The interaction is between pairs of particles and is repulsive in nature. The scaling is such that the range of the interaction is of the same order as interparticle distance and therefore each particle interacts with only a finite number of nearby particles at any given time.

We obtain a nonlinear bulk diffusion equation and the diffusion coefficient is naturally expressed in terms of the thermodynamic functions of our one dimensional system. The main limitations are the finiteness of volume that is forced because our basic space is the circle, the repulsive nature of the interactions that is assumed and that we are in one space dimension.

In this context the fluctuations around equilibrium have been studied earlier by H. Spohn [5] and the self diffusion in equilibrium by Guo and Papanicolaou [1]. Their results have been derived for infinite volume and arbitrary space dimension. Of course these results deal essentially with equilibrium problems where a lot of control through estimates is available.

Our method is similar to the one used in Guo, Papanicolaou and Varadhan [2] and uses estimates based on entropy and its rate of change. Section 2 describes

* This research was supported by a grant from the National Science Foundation DMS-89-01682

the results precisely and Sects. 3–9 contain detailed proofs. Section 10 contains some results concerning the Gibbs measures of one dimensional systems. These results are all well known to the experts although it is difficult to find precise references to the results in the form we need. We have therefore included a quick exposition with sketches of proofs.

Along with S. Olla in [3] we have modified the methods of this article to extend these results to the case of interacting Ornstein-Uhlenbeck processes and they will appear in the next article.

2. Summary

We denote by S the circle of unit circumference and consider a system of N interacting Brownian motions with S as a state space satisfying the following system of stochastic differential equations:

$$dx_i(t) = -N \sum_{j:j\neq i} V'(N(x_i(t)-x_j(t))dt + d\beta_i(t) , \quad i=1,2,\ldots,N . \tag{2.1}$$

Here, $\beta_1,\ldots,\beta_N$ are N independent Brownian motions and $V(\cdot)$ is an even function on R satisfying the following assumptions:

(i) $V\geqq 0$, $V(0)>0$ and V has compact support.
(ii) V is once continuously differentiable.
(iii) $V(\cdot)$ is repulsive in the sense that

$$\psi(z) = -zV'(z)\geqq 0 . \tag{2.2}$$

Then the process $[x_1(t),\ldots,x_N(t)]$ is a Markov process of diffusion type on S^N, the N-fold copy of S, with an infinitesimal generator given by

$$L_N = \frac{1}{2}\sum_{i=1}^{N}\frac{\partial^2}{\partial x_i^2} - N\sum_{j\neq i} V'(N(x_i-x_j))\frac{\partial}{\partial x_j} . \tag{2.3}$$

It is easily verified that

$$L_N = \frac{1}{2}\sum_{i=1}^{N}\left(e^{\sum_{j,k} V(N(x_j-x_k))}\frac{\partial}{\partial x_i} e^{-\sum_{j,k} V(N(x_j-x_k))}\frac{\partial}{\partial x_i}\right) , \tag{2.4}$$

so that L_N is formally symmetric with respect to the measure

$$d\mu_N(x) = \frac{1}{Z_N}\exp\left[-\sum_{i,j} V(N(x_i-x_j))\right] dx_1\ldots dx_N . \tag{2.5}$$

It is straightforward to verify that the Markov process $x(t)$ with L_N as generator is in fact reversible with respect to the invariant measure μ_N given by (2.5). The constant Z_N is for normalization and is so defined to make μ_N into a probability measure.

We start $x(0)$ with an initial distribution with density $f_N^0(x_1,\ldots,x_N)$ with respect to the invariant density μ_N. The density at time t with respect to μ_N is denoted by $f_N^t(x_1,\ldots,x_N)$ and is given as a solution of the forward equation

$$\frac{\partial f_N^t}{dt} = L_N f_N^t \quad \text{with} \quad f_N^t|_{t=0} = f_N^0 . \tag{2.6}$$

The empirical distribution of the process at time t is defined by

$$\xi_N(t, A) = \frac{1}{N} \sum_{i=1}^{N} \chi_A(x_i(t)) \quad \text{for} \quad A \subset S , \tag{2.7}$$

and $\xi_N(t)$ is viewed as a random measure on S. If we denote by $M_1(S)$ the space of probability measures on S, one can view $\xi_N(\cdot)$ as a stochastic process with values in $M_1(S)$. In view of the continuity of the trajectories, we can pick a time $T < \infty$ and consider the space $C[[0,T], M_1(S)]$ of measure valued continuous functions on $[0, T]$, and our basic markov process with initial density f_N^0 will induce a measure Q_N on $C[[0, T], M_1(S)]$. Our main result is to show that under suitable assumptions on f_N^0, the measures Q_N as $N \to \infty$ will concentrate on a single measure valued trajectory which is the solution of a certain nonlinear diffusion equation.

In order to describe this nonlinear diffusion equation we have to introduce some thermodynamic functions of one dimensional systems with pair interaction given by $V(\cdot)$. The partition function in a finite region $[0, l]$ with activity λ is given by

$$\hat{Z}(l, \lambda) = \sum_{n=0}^{\infty} \frac{e^{n\lambda}}{n!} \int_0^l \cdots \int_0^l e^{-\sum_{i,j} V(x_i - x_j)} dx_1 \ldots dx_n . \tag{2.8}$$

It is known [4] that the free energy defined by

$$F(\lambda) = \lim_{l \to \infty} \frac{\log \hat{Z}(l, \lambda)}{l} \tag{2.9}$$

exists and is a convex function of λ for all λ,

$$\rho(\lambda) = \frac{dF}{d\lambda} , \tag{2.10}$$

is the "density" corresponding to the activity λ and is a continuous strictly monotone function of λ. This function of course can be inverted to yield $\lambda = \lambda(\rho)$ as a function of ρ. The free energy expressed as a function of ρ, i.e.

$$P(\rho) = F(\lambda(\rho)) \tag{2.11}$$

is called the pressure and is again a continuous strictly monotone function of the density ρ.

Our main result can now be stated. Let us assume that the initial densities $\{f_N^0\}$ have the following behavior as $N \to \infty$:

$$\int f_N^0 \log f_N^0 d\mu_N \leq CN \quad \text{for some} \quad C < \infty \text{ and all } N ; \tag{2.12}$$

$$\int_{A_{N,J,\varepsilon}} f_N^0 d\mu_N \to 0 \quad \text{as} \quad N \to \infty \tag{2.13}$$

for every smooth test function J on S. Here $A_{N,J,\varepsilon}$ is the set of configurations $(x_1, \ldots, x_N)$ such that

$$A_{N,J,\varepsilon} = \left\{ x : \left| \frac{1}{N} \sum J(x_i) - \int J(\theta) \rho_0(\theta) d\theta \right| > \varepsilon \right\} ,$$

and $\rho_0(\theta)$ is some fixed density on S such that $\rho_0(\theta) \geq 0$ and $\int \rho_0(\theta) d\theta = 1$. In order for (2.12) and (2.13) to be compatible $\rho_0(\theta)$ will have to satisfy some minimal regularity conditions. In any case (2.13) asserts that

$$\frac{\delta_{x_1}+\ldots+\delta_{x_N}}{N} \Rightarrow \rho_0(\theta)d\theta$$

in the sense of weak convergence, in probability with respect to the initial distribution $f_N^0 d\mu_N$.

Under these assumptions we establish that Q_N converges weakly to a limit which is the degenerate distribution on a single trajectory in $M_1(S)$ described by $\rho(\theta,t)d\theta$, where $\rho(\theta,t)$ is defined as the unique solution (satisfying mild regularity conditions) of

$$\frac{\partial\rho(\theta,t)}{\partial t}=\frac{1}{2}[P(\rho(\theta,t)]_{\theta\theta} \tag{2.14}$$

with

$$\rho(\theta,t)|_{t=0}=\rho_0(\theta)\ . \tag{2.15}$$

3. Outline of Proof

The proof of our result is based on the following ideas. For the sequence $\{Q_N\}$ of probability measures on the space $\Omega_T=C[[0,T],M_1(S)]$ of measure valued trajectories we first establish compactness under weak convergence.

Theorem 3.1. *The sequence $\{Q_N\}$ is tight on the space Ω_T.*

In order to establish that the limit Q_N exists and is concentrated on a single trajectory we establish properties satisfied by the support of any limit point Q, enough properties so that one is left with a single trajectory satisfying all of them.

Let us consider a smooth function $u(\theta)$ on S and the corresponding functional

$$\xi_N(t)=\frac{1}{N}\sum u(x_i(t))\ .$$

By Itô's formula we see that

$$\begin{aligned} d\xi_N(t)=&\frac{1}{2N}\sum_{i=1}^{N}u''(x_i(t))dt\\ &-\sum_{i,j=1}^{N}V'(N(x_i(t))-x_j(t))u'(x_i(t))dt\\ &+\frac{1}{N}\sum_{i=1}^{N}u'(x_i(t))d\beta_i(t)\ . \end{aligned} \tag{3.1}$$

By the law of large numbers, as $N\rightarrow\infty$, the third term goes to zero. As for the second term

$$\sum_{i,j=1}^{N} V'(N(x_i(t)-x_j(t)))u'(x_i(t))$$

$$=\frac{1}{2}\sum V'(N(x_i(t)-x_j(t)))(u'(x_i(t))-u'(x_j(t)))$$

(by skew symmetry of V')

$$\simeq\frac{1}{2}\sum V'(N(x_i(t)-x_j(t)))u''(x_i(t))(x_i(t)-x_j(t))$$

by (mean value theorem)

$$\simeq -\frac{1}{2N}\sum \psi(N(x_i(t)-x_j(t)))u'(x_i(t)) \ .$$

If we are at some point (θ, t) in space time and the local density of particles is $\rho(\theta, t)$ then the system should be in local equilibrium there in a Gibbs state with density ρ and it is not a difficult calculation to show that in an exact Gibbs state of density ρ

$$\lim_{l\to\infty}\frac{1}{l}\sum_{x_i, x_j\in[0,l]}\psi(x_i-x_j)=P(\rho)-\rho \ . \tag{3.2}$$

One therefore hopes to estabish that if at a typical time t, $\frac{1}{N}[\delta_{x_1(t)}+\ldots+\delta_{x_N(t)}]$ is close to $\rho(t,\theta)d\theta$, then $\frac{1}{2N}\sum\psi(N(x_i(t)-x_j(t))u''(x_i(t))+\frac{1}{2N}\sum u''(x_i(t))$ should be close to $\frac{1}{2}\int P(\rho(t,\theta))u''(\theta)d\theta$.

In this manner we establish

Theorem 3.2. *Any limit point Q of Q_N satisfies*

(i) $Q[\alpha(t,\cdot):\alpha(t,d\theta)=\rho(t,\theta)d\theta$ *for* a.e. $t]=1$.

(ii) $E^Q[\rho:\rho(0,\theta)=\rho_0(\theta)$ a.e. $\theta]=1$.

(iii) $Q[P:\int u(\theta)\rho(t,\theta)d\theta-\int u(\theta)\rho_0(\theta)d\theta=\frac{1}{2}\int_0^t\int u''(\theta)P(\rho(t,\theta))d\theta]=1$.

Given Theorem 3.2, Q lives on the set of weak solutions of

$$\begin{cases}\dfrac{\partial\rho}{\partial t}=\dfrac{1}{2}(P(\rho))_{\theta\theta} \ \text{with}\\ \rho(t,\theta)|_{t=0}=\rho_0(\theta) \ .\end{cases} \tag{3.3}$$

One can establish the uniqueness of weak solutions of (3.3) under the assumptions

$$\int_0^T\int_S \rho^3(t,\theta)dtd\theta<\infty \tag{3.4}$$

and

$$\int_0^T \int_S \frac{1}{\rho(t,\theta)} \left[\frac{\partial}{\partial\theta} P(\rho(t,\theta))\right]^2 d\theta < \infty . \tag{3.5}$$

In order then to complete the proof we only need to establish

Theorem 3.3. *Any limit point Q satisfies*

$$E^Q\left[\int_0^T \int_S \rho^3(t,\theta)\,dt\,d\theta\right] < \infty$$

and

$$E^Q\left[\int_0^T \int_S \frac{1}{\rho(t,\theta)} \left[\frac{d}{d\theta} P(\rho(t,\theta))\right]^2 dt\,d\theta\right] < \infty .$$

Finally a word about the strategy to be used in establishing Theorems 3.1., 3.2 and 3.3. Since $N\to\infty$ we need to develop estimates that hold uniformly in N. We will need properties of the solutions f_N^t, and $\bar{f}_N = \frac{1}{T}\int_0^T f_N^s\,ds$, where f_N^t solves

$$\frac{\partial f_N^t}{\partial t} = L_N f_N^t .$$

Our assumptions is that $\int f_N^0 \log f_N^0\,d\mu_N \leqq CN$. One knows that

$$\int f_N^t \log f_N^t\,d\mu_N = H_N(t)$$

is nonincreasing in t and

$$\frac{d}{dt} H_N(t) = -I_N(t) = -\frac{1}{2}\int \frac{|\nabla f_N^t|^2}{f_N^t}\,d\mu_N .$$

The functionals

$$H_N(t) = \int f \log f\,d\mu_N$$

and

$$I_N(t) = \frac{1}{2}\int \frac{|\nabla f|^2}{f}\,d\mu_N$$

are lower semicontinuous, non-negative and convex. Therefore one gets easily

$$\begin{aligned} H_N(\bar{f}_N) &= H_N\left(\frac{1}{T}\int_0^T f_N^s\,ds\right) \\ &\leqq \frac{1}{T}\int_0^T H(f_N^s)\,ds \\ &\leqq H(f_N^0) \leqq CN . \end{aligned}$$

Similarly,

$$
\begin{aligned}
I_N(\bar{f}_N) &\leqq \frac{1}{T}\int_0^T I_N(f_N^s)\,ds \\
&= \frac{1}{T}[H_N(f_N^0) - H_N(f_N^T)] \\
&\leqq \frac{1}{T} H_N(f_N^0) \\
&\leqq \frac{C}{T} N .
\end{aligned}
$$

In some sense our procedures are based solely on the two estimates derived above.

4. Some Estimates Based on Entropy

Given a density f_N relative to the invariant measure μ_N with entropy $H_N(f_N) = \int f_N \log f_N \, d\mu_N$, we shall obtain in this section some preliminary estimates based on a bound of the type

$$
H_N(f_N) \leqq AN . \tag{4.1}
$$

Lemma 4.1. *There exists a constant C such that for every N and every f_N satisfying* (4.1) *we have*

$$
E^{f_N}\left\{\frac{1}{N}\sum_{i,j} V(N(x_i - x_j))\right\} \leqq A + C ,
$$

where A is the same constant as in (4.1).

Proof.

$$
\begin{aligned}
&E^{f_N}\left\{\frac{1}{N}\sum_{i,j} V(N(x_i - x_j))\right\} \\
&\leqq \frac{1}{N}\log E^{\mu_N}\left\{\exp\left[\sum V(N(x_i - x_j))\right]\right\} + A \\
&\leqq -\frac{1}{N}\log Z_N + A = -\frac{1}{N}\log \int e^{-\sum V(N(x_i - x_j))}\,dx_1 \ldots dx_N + A \\
&\leqq \frac{1}{N}\int\left(\sum V(N(x_i - x_j))\right) dx_1 \ldots dx_N + A \\
&\leqq C + A .
\end{aligned}
$$

From now on for the rest of the section we want to derive some consequences of an estimate of the form

$$
E^{f_N}\left\{\frac{1}{N}\sum V(N(x_i - x_j))\right\} \leqq B \quad \text{for all } N . \tag{4.2}
$$

Let us divide the circle S into a large number l of small arcs length $\frac{1}{l}$. Let us denote them by $S_1, S_2, \ldots, S_l$. If we are given a configuration of N points $x_1, \ldots, x_N$ we can form the frequency counts

$$g_i = \sum_{j=1}^{N} \chi_{s_i}(x_j)$$

of the number of points among $x_1, \ldots, x_N$ that fall in a given arc S_i for $1 \leqq i \leqq l$. We wish to estimate $\sum_{i=1}^{l} g_i^2$ in terms of $\sum_{i,j} V(N(x_i - x_j))$. Let us subdivide each arc S_i into $2k$ arcs of length $\frac{1}{2lk}$ and we assume that the size $\frac{1}{2lk}$ is of the order $\frac{\delta}{N}$, where δ is small, independent of N, and is so small that $V(x) \geqq \eta$ for $|x| \leqq \delta$ for some positive constant $\eta > 0$. For $1 \leqq i \leqq 2lk$, if we denote by h_i the frequency counts for the smaller arcs, then

$$\sum_1^l g_i^2 \leqq 2k \sum_1^{2lk} h^2 i .$$

If we translate the basic arcs S_i by half their lengths and denote them by $S_{i+1/2}$ with their frequency counts $g_{i+1/2}$, then we have just as well

$$\sum_1^l g_{i+1/2}^2 \leqq 2k \sum_1^{2lk} h_i^2 .$$

If from among $x_1, \ldots, x_N$, we have a pair x_r, x_s with $|x_r - x_s| \leqq \frac{1}{2l}$, then they must belong either to the same S_i or $S_{i+1/2}$ for some i. Therefore

$$\begin{aligned} \sum_{i,j} \chi_{|x| \leqq 1/2}(|(x_i - x_j)|) &\leqq \sum g_i^2 + \sum g_{i+1/2}^2 \\ &\leqq (4k) \sum_1^{2lk} h_i^2 \\ &\leqq \frac{4k}{\eta} \sum V(N(x_i - x_j)) . \end{aligned} \tag{4.3}$$

If $W(x)$ is a function that is supported on $\left[-\frac{1}{2l}, \frac{1}{2l}\right]$ and is bounded there by a constant $\|W\|$ then

$$\sum_{i,j} W(x_i - x_j) \leqq \|W\| \cdot \frac{4k}{\eta} \sum V(N(x_i - x_j)) , \tag{4.4}$$

provided only that $\frac{N}{2lk} = \delta$ is sufficiently small.

Lemma 4.2. *If W is a fixed compactly supported function then we can estimate*

$$E^{f_N} \frac{\lambda}{N^2} \sum W(\lambda(x_i - x_j)) \leqq \|W\| \cdot B \cdot C ,$$

where $\|W\|$ is a bound for W, B is any constant satisfying (4.2) *and C is a universal constant depending only on V and the size of the support of $W(\cdot)$. This estimate holds uniformly for all λ in the range $\varepsilon_0 \leqq \lambda \leqq N$.*

Proof. We need basically l to be $c\lambda$ for fixed constant c in (4.4) and that forces on us a choice of $k = CN/\lambda$ for some other constant C. Now Lemma 4.2 is an immediate consequence of (4.2) and (4.4).

5. Some Estimates Based on the Dirichlet Form

In this section we want to explore some of the consequences of assuming that the density f_N satisfies

$$E^{f_N}\left[\frac{1}{N}\sum V(N(x_i - x_j))\right] \leqq B \tag{5.1}$$

and

$$\frac{1}{2}\int |\nabla f_N|^2 \frac{1}{f_N} d\mu_N = I_N(f_N) \leqq DN . \tag{5.2}$$

The basic estimate uses integration by parts for any "suitable" test function $u = u(x_1, \ldots, x_N)$

$$\begin{aligned} |E^{f_N}[L_N u]| &= |\int (L_N u) f_N d\mu_N| \\ &= |\frac{1}{2}\int \langle \nabla u, \nabla f_N \rangle d\mu_N| \\ &\leqq \left(\frac{1}{2}\int |\nabla u|^2 f_N d\mu_N\right)^{1/2} \left(\frac{1}{2}\int |\nabla f_N|^2 \frac{1}{f_N} d\mu_N\right)^{1/2} \\ &\leqq \left(\frac{1}{2}\int |\nabla u|^2 f_N d\mu_N\right)^{1/2} (DN)^{1/2} . \end{aligned} \tag{5.3}$$

Of course such an estimate is only as good as the test functions u that we can find to use in the estimate. The following class of test functions will be particularly useful:

$$u_N(x_1, \ldots, x_N) = \sum_{i,j} \mathscr{G}_{N,\varepsilon,\lambda}(x_i - x_j) , \tag{5.4}$$

where $\mathscr{G}_{N,\varepsilon,\lambda}(x)$ is a function on R that has sufficiently small support that it can be viewed as a function on S. For each fixed λ and ε we will have a choice of u_N made for all sufficiently large N. The function $\mathscr{G}_{N,\varepsilon,\lambda}(x)$ is given by

$$\mathscr{G}_{N,\varepsilon,\lambda}(x) = \int_{-\infty}^{x} G_{N,\varepsilon,\lambda}(y) dy \tag{5.5}$$

and

$$G_{N,\varepsilon,\lambda}(x) = \int_{-\infty}^{x} g_{N,\varepsilon,\lambda}(y) dy , \tag{5.6}$$

where $g_{N,\varepsilon,\lambda}(\cdot)$ is a function R with support in a small interval around the origin, satisfying in addition

$$\int_{-\infty}^{\infty} g_{N,\varepsilon,\lambda}(y)dy=0 \tag{5.7}$$

and

$$\int_{-\infty}^{\infty} y g_{N,\varepsilon,\lambda}(y)dy=0 \ , \tag{5.8}$$

so that G and $\mathscr{G}$ have the same support as g. The function $g_{N,\varepsilon,\lambda}(y)$ will be of the form

$$g_{N,\varepsilon,\lambda}(y)=(N\varepsilon)g(N\varepsilon y)-\lambda g(\lambda y) \ , \tag{5.9}$$

where g is a nonnegative, smooth, symmetric function with compact support with total integral 1,

$$\int g(y)dy=1 \ . \tag{5.10}$$

Clearly if λ is big enough and $\varepsilon>0$ is arbitrary then for sufficiently large $N, u_N(x_1,\ldots,x_N)$ is well defined.

In order to use the integration by parts estimate (5.3) we have to compute $\mathscr{L}_N u_N$ and $|\nabla u_N|^2$,

$$\begin{aligned}\mathscr{L}_N u_N= & N\varepsilon \sum g(N\varepsilon\,(x_i-x_j))-\lambda \sum g(\lambda(x_i-x_j)) \\ & -2N\sum_i\Big(\sum_j [G(N\varepsilon(x_i x_j))-G(\lambda\,(x_i-x_j))]\Big) \\ & \cdot\Big(\sum_k V'\,(N(x_i-x_k))\Big) \ ,\end{aligned} \tag{5.11}$$

$$|\nabla u_N|^2=4\sum_i|\sum_j [G(N\varepsilon(x_i-x_j))-G(\lambda\,(x_i-x_j))]|^2 \ . \tag{5.12}$$

One can use the fact that V' is an odd function to rewrite (5.11) in the form

$$\begin{aligned}\mathscr{L}_N u_N= & N\varepsilon \sum g(N\varepsilon(x_i-x_j))-\lambda \sum g(\lambda\,(x_i-x_j)) \\ & -N\sum_{i,j,k} [G(N\varepsilon(x_i-x_j))-G(N\varepsilon(x_k-x_j)) \\ & -G(\lambda\,(x_i-x_j))+G(\lambda\,(x_k-x_j))]\,V'\,(N(x_i-x_k)) \ .\end{aligned} \tag{5.12}$$

Let us take the special case of $\varepsilon=1$ and $\lambda=1$. We assume that the support of g is small enough that $\lambda=1$ works. Otherwise we can take any value λ_0 for λ.

We can write inequality (5.3) in the form

$$\left|E^{f_N}\frac{1}{N^2}\mathscr{L}_N\mathscr{U}_N\right|\leqq\frac{D^{1/2}}{N^{3/2}}\left[E^{f_N}\left\{\frac{1}{2}|\nabla u_N|^2\right\}\right]^{1/2} \ , \tag{5.13}$$

with $\varepsilon=1$, the first term on the right-hand side of (5.11) is

$$E^{f_N}\left\{\frac{1}{N}\sum g(N(x_i-x_j))\right\} \ ,$$

which is uniformly bounded because of (5.1) and Lemma 4.2. The second term is

$$E^{f_N}\left\{\frac{1}{N^2}\sum g(x_i-x_j)\right\} ,$$

which is bounded by a bound for g. The term

$$\frac{1}{N}\sum_{i,j,k}[G(x_i-x_j)-G(x_k-x_j)]V'(N(x_i-x_k))$$

$$\leq\frac{C}{N}\sum_{i,j,k}|x_i-x_k|\,|V'(N(x_i-x_k))|$$

by using a bound on the derivative of G. We can use the function $\psi(x)=-xV'(x)$ and write this term as

$$\frac{C}{N^2}\sum_{i,j,k}\psi(N(x_i-x_k))\leq\frac{C}{N}\sum_{i,k}\psi(N(x_i-x_k)) ,$$

which is bounded by (5.1) and Lemma 4.2. It is easy to estimate

$$|\nabla u_N|^2\leq CN^3$$

by using a bound for G. Since in the inequality (5.3) every term has been bounded except for a single term we obtain

$$E^{f_N}\left\{\frac{1}{N}\sum_{i,j,k}[G(N(x_k-x_j))-G(N(x_i-x_j))]V'(N(x_i-x_k))\right\}\leq C , \tag{5.14}$$

where C is a bound depending only on B and D. The function $F(x-y,y-z)=V'(x-z)[G(z-y)-G(x-y)]$ is nonnegative and one can check easily that it is possible to pick g in such a manner that

$$F(x-y,y-z)\geq\psi(x-z)\psi(x-y) . \tag{5.15}$$

From (5.14) and (5.15) we obtain

$$E^{f_N}\left\{\left(\frac{1}{N}\sum_{i,j}\psi(N(x_i-x_j))\right)^2\right\}\leq E^{f_N}\left\{\frac{1}{N}\sum_i\left(\sum_j\psi(N(x_i-x_j))\right)^2\right\}$$

$$=E^{f_N}\left\{\frac{1}{N}\sum_{i,j,k}\psi(N(x_i-x_j))\psi(N(x_i-x_k))\right\}$$

$$\leq E^{f_N}\left\{\frac{1}{N}\sum_{i,j,k}F(N(x_i-x_j),N(x_j-x_k))\right\}$$

$$\leq C .$$

In other words (5.1) alone is sufficient to provide a bound for

$$E^{f_N}\left\{\frac{1}{N}\sum\psi(N(x_i-x_j))\right\}$$

while (5.1) and (5.2) provide a bound for

$$E^{f_N}\left\{\frac{1}{N}\sum \psi(N(x_i-x_j))\right\}^2 .$$

We have therefore proved Lemma 5.1.

Lemma 5.1. *There is a constant C depending only on B and D such that*

$$E^{f_N}\left\{\frac{1}{N}\sum \psi(N(x_i-x_j))\right\}^2 \leqq C ,$$

$$E^{f_N}\left\{\frac{1}{N}\sum_i \left[\sum_j \psi(N(x_i-x_j))\right]^2\right\} \leqq C .$$

6. Compactness

In this section we will establish Theorem 3.1. Since the state space $M_1(S)$ is compact we need only establish estimates of modulus of continuity to appeal to Prohorov's theorem. It is therefore enough to prove

Lemma 6.1. *For any smooth function J on S, and* $\varepsilon > 0$,

$$\lim_{\delta\to 0}\limsup_{N\to\infty} Q_N\left[\sup_{\substack{0\leqq s<t\leqq T\\ |s-t|\leqq\delta}}\left|\frac{1}{N}\sum J(x_i(s))-\frac{1}{N}\sum J(x_i(t))\right|\geqq\varepsilon\right]=0 . \tag{6.1}$$

Proof. Let us take $s<t$. Then

$$\begin{aligned}
&\frac{1}{N}\sum J(x_i(t))-\frac{1}{N}\sum J(x_i(s))\\
&=\frac{1}{2N}\sum\int_s^t J''(x_i(\sigma))d\sigma+\frac{1}{N}\int_s^t\sum J'(x_i(\sigma))d\beta_i(\sigma)\\
&\quad-\int_s^t\sum V'(N(x_i(\sigma))-x_j(\sigma)))J'(x_i(\sigma))d\sigma\\
&=\eta_1+\eta_2+\eta_3 .
\end{aligned} \tag{6.2}$$

The first term is clearly bounded by

$$|\eta_1|\leqq\tfrac{1}{2}C(t-s) , \tag{6.3}$$

where C is a bound for J''.

The second term is a Martingale term and by Doob's inequality

$$\begin{aligned}
&E\sup_{0\leqq t\leqq T}\left|\frac{1}{N}\sum\int_0^t J'(x_i(\sigma))d\beta_i(\sigma)\right|^2\\
&\leqq C\exp_{0\leqq t\leqq T} E\left|\frac{1}{N}\sum\int_0^t J'(x_i(\sigma))d\beta_i(\sigma)\right|^2
\end{aligned}$$

$$= C\frac{1}{N^2}\int_0^T \sum |J'(x_i(\sigma))|^2 d\sigma$$

$$\leq \frac{C \cdot C_1^2 T}{N} , \tag{6.4}$$

where C_1 is a bound for J'.

The third term can be estimated by estimating the integrand

$$\sum_{ij} V'(N(x_i(\sigma) - x_j(\sigma)))J'(x_i(\sigma))$$

$$= \sum_{ij} V'(N(x_i(\sigma) - x_j(\sigma)))(J'(x_i(\sigma)) - J'(x_j(\sigma)))$$

$$\leq C \sum_{i,j} |V'(N(x_i(\sigma) - x_j(\sigma)))| \, |x_i(\sigma) - x_j(\sigma)| ,$$

where C is a bound for J''. Recalling that $\psi(z) = -zV'(z)$ we can estimate

$$|\eta_3| \leq C\int_s^t \frac{1}{N}\sum |\psi(N(x_i(\sigma) - x_j(\sigma))| d\sigma .$$

One can then complete the proof of Lemma 4.1 if we show the following estimate:

$$\sup_N E^{Q_N}\int_0^T \left[\frac{1}{N}\sum \psi(N(x_i(\sigma) - x_j(\sigma)))\right]^2 d\sigma < \infty .$$

This follows from Lemma 5.1.

7. Local Gibbs States

In establishing the hydrodynamic scaling limit one of the problems we have to deal with is expressing average microscopic quantities like $\frac{1}{N}\sum \psi(N(x_i - y_j))U(x_i)$ purely in terms of the macroscopic density function. This is of course possible only if the configuration $x_1, \ldots, x_n$ is suitably organized and we have to establish that during our stochastic dynamics most of the time the configurations are suitably organized with a very high probability.

To make this precise let us fix a value of l and consider a functional $H(\omega)$ which is a bounded and continuous functional of the configuration of points in the interval $(-l, l)$. We can view $H(\omega)$ as a functional of a point process on R, which depends however only on the configuration of points in $(-l, l)$. If we have a set $(x_1, \ldots, x_N)$ of N points in S, for any given x, we can look at the points x_i: $|x_i - x| < \frac{1}{4}$ and consider the points $N(x_i - x)$ on the line $(-\infty, \infty)$ obtained from the original $(x_1, \ldots, x_N)$ on S. The resulting point process on R, while it is somewhat arbitrary due to the cutoff at the edges namely around $\pm\frac{N}{4}$, for every l the configuration is natural and well defined on $[-l, l]$, provided N is large enough. We can evaluate our functional $H(\omega)$ at this configuration, which we denote by ω_N^x and obtain a value $H(\omega_N^x)$.

We are interested in the quantity

$$\xi_N(x_1,\dots,x_N)=\int_S H(\omega_N^x)\,U(x)\,dx\ . \tag{7.1}$$

Corresponding to the set $x_1,\dots,x_N$ we have the empirical distribution

$$\alpha_N(dx)=\frac{1}{N}(\delta_{x_1}+\dots+\delta_{x_N})\ .$$

We pick a mollifier $h(x)$ with compact support, i.e. a nonnegative function $h(x)$ with $\int h(x)dx=1$, and assuming that the support of h is contained in $[-\frac{1}{2},\frac{1}{2}]$, we can lift h as a function on S, and for $\lambda\geqq 1$, $\lambda h(\lambda x)$ is a mollifier on S as well. We can then pick for $\lambda\geqq 1$, a natural version of local density for the configuration $(x_1,\dots,x_N)$, by

$$\rho_\lambda(x)=\frac{\lambda}{N}\sum h(\lambda(x_i-x))$$

$$=(h_\lambda*\alpha_N)(x)\ ,\qquad\text{where}$$

$$h_\lambda=\lambda h(\lambda x)\ . \tag{7.2}$$

For a one dimensional system, there are no phase transitions and we have a unique Gibbs state P_ρ viewed as a stationary point process with density ρ that corresponds to the given pair interaction $V(x)$. Since we expect the arrangement of points near x to be a random arrangement from P_ρ with $\rho=\rho_\lambda$, by (7.1) we expect to replace $\xi_N(x_1,\dots,x_N)$ by $\eta_{N,\lambda}(x_1,\dots,x_N)$, where

$$\eta_{N,\lambda}(x_1,\dots,x_N)=\int_S \hat{H}(\rho_\lambda(x))\,U(x)\,dx \tag{7.3}$$

with

$$\hat{H}(\rho)=E^{P_\rho}[H(\omega)]\ . \tag{7.4}$$

The error between ξ_N and $\eta_{N,\lambda}$ is expected to be small provided λ is large and N is large and $(x_1,\dots,x_N)$ is a typical configuration from a nice distribution $f_N d\mu_N$.

Let us denote by $\mathscr{A}_{N,B,D}$ the set of probability densities f_N on S^N that satisfy

$$F^{f_N}\left\{\frac{1}{N}\sum V(N(x_i-x_j))\right\}\leqq B\ , \tag{7.5}$$

$$\frac{1}{2}\int\frac{|\nabla f_N|^2}{f_N}\,d\mu_N\leqq DN\ . \tag{7.6}$$

Because of translation invariance of the expressions (7.5) and (7.6), if we replace f_N by

$$\hat{f}_N(x_1,\dots,x_N)=\int_S f_N(x_1+a,\dots,x_N+a)\,da$$

due to linearity in the case of (7.5) and convexity in the case of (7.6), $\hat{f}_N$ will satisfy the inequalities (7.5) and (7.6). In other words, if $f_N\in\mathscr{A}_{N,B,D}$ so does $\hat{f}_N$. Let us denote by $\mathscr{A}^0_{N,B,D}$, the translation invariant densities in $\mathscr{A}_{N,B,D}$. Our goal is to establish that for each fixed l and $H(\cdot)$

Theorem 7.1.

$$\lim_{\lambda\to\infty}\limsup_{N\to\infty}\sup_{f_N\in\mathscr{A}_{N,B,D}} E^{f_N}|\xi_N-\eta_{N,\lambda}|=0\ .$$

Since P_ρ depends continuously on ρ, $\hat{H}(\rho)$ is continuous in ρ and we can establish Theorem 7.1 in two steps.

Theorem 7.2.

$$\lim_{\varepsilon\to 0}\limsup_{N\to\infty}\sup_{f_N\in\mathscr{A}_{N,B,D}} E^{f_N}|\xi_N-\eta_{N,N_\varepsilon}|=0\ ,$$

and

Theorem 7.3.

$$\lim_{\substack{\varepsilon\to 0\\ \lambda\to\infty}}\limsup_{N\to\infty}\sup_{f_N\mathscr{A}_{N,B,D}} E^{f_N}\int_S|\rho_\lambda(x)-\rho_{\varepsilon N}(x)|\,dx=0\ .$$

Remark. In view of the comment made earlier about translation invariance in Theorems 7.1, 7.2 and 7.3 we can in the proof replace $\mathscr{A}_{N,B,D}$ by $\mathscr{A}^0_{N,B,D}$.

We can write

$$\begin{aligned}
&E^{f_N}|\int H(\omega_N^x)U(x)dx-\int\hat{H}(\rho_\lambda(x))U(x)dx|\\
&\leqq E^{f_N}|\int\left[\frac{N}{2a}\int_{|y-x|\leqq a/N}H_0(\omega_N^x)dx\right]U(y)dy-\int\hat{H}(\rho_\lambda(x))U(x)dx|\\
&\quad+\|H\|E^{f_N}\int\frac{N}{2a}\int_{|U-x|\leqq a/N}|U(y)-u(x)|\,dy\,dx\ .
\end{aligned}$$

To fixed a, H and any sequence $\lambda=\lambda_N$,

$$\begin{aligned}
&\limsup_{N\to\infty}\sup_{f_N\in\mathscr{A}_{N,B,D}} E^{f_N}|\int H(\omega_N^x)U(x)dx-\int\hat{f}(\rho_{\lambda_N}(x))U(x)dx|\\
&\leqq\|U\|\limsup_{N\to\infty}\sup_{f_N\in\mathscr{A}_{N,B,D}} E^{f_N}\left[\int dy\left|\frac{N}{2a}\int_{|y-x|\leqq a/N}H(\omega_N^x)dx-\hat{H}(\rho_{\lambda_N}(y))\right|\right]
\end{aligned}$$

by assuming continuity of the function $U(x)$.

If we now replace f_N by $\hat{f}_N$ then

$$\begin{aligned}
&E^{f_N}\left[\int dy\left|\frac{N}{2a}\int_{|y-x|\leqq a/N}H(\omega_N^x)dx-\hat{H}(\rho_{\lambda_N}(y)\right|\right]\\
&=E^{\hat{f}_N}\left[\int dy\left|\frac{N}{2a}\int_{|y-x|\leqq a/N}H(\omega_N^x)dx-\hat{H}(\rho_{\lambda_N}(y)\right|\right]\\
&=E^{\hat{f}_N}\left[\left|\frac{N}{2a}\int_{|x|\leqq a/N}H(\omega_N^x)\,dx-\hat{H}(\rho_{\lambda_N}(0))\right|\right]\ .
\end{aligned}$$

The proof of Theorem 7.2 can therefore be reduced to the proof of the following theorem.

Theorem 7.4.

$$\overline{\lim_{\varepsilon\to 0}}\ \limsup_{N\to\infty}\ \sup_{f_N\in\mathscr{A}^0_{N,B,D}} E^{f_N}\left|\frac{N\varepsilon}{2}\int_{|x|\leqq 1/N\varepsilon} H(\omega_N^x)\,dx-\hat{H}(\rho_{N_\varepsilon}(0))\right|=0\ .$$

The proof of Theorem 7.4 proceeds as follows.

The computation of $\int_{|x|\leqq 1/N\varepsilon} H(\omega_N^x)dx$ and $\rho_{N_\varepsilon}(0)$ requires only a knowledge of the location of the particles from among $x_1,\dots,x_N$ that belong to some interval of the form $|x|\leqq\frac{1}{N\varepsilon}+\frac{l_0}{N}$, where $[-l_0,l_0]$ is such that $H(\omega)$ depends only on the configuration there. So if we project $f_N d\mu_N$ onto the configurations in such an interval and expand the interval by a factor of N, we will get a set $\mathscr{B}^0_{N,B,D,\varepsilon}$ of point processes on the interval $|x|\leqq\frac{1}{\varepsilon}+l_0$. Theorem 7.4 reduces to proving that

$$\lim_{\varepsilon\to 0}\ \limsup_{N\to\infty}\ \sup_{\gamma\in\mathscr{B}^0_{N,B,D,\varepsilon}} E^{\gamma}\left|\frac{\varepsilon}{2}\int_{-1/\varepsilon}^{1/\varepsilon} H(\omega^x)dx-\hat{H}(\rho_\varepsilon(0))\right|=0\ . \tag{7.7}$$

Let us note that because of translation invariance and blowup by a factor of N, the local density of particles is normalized at 1. Therefore the collection $\mathscr{B}^0_{N,B,D,\varepsilon}$ of point processes on $\left[-\left(\frac{1}{\varepsilon}+l_0\right),\left(\frac{1}{\varepsilon}+l_0\right)\right]$ is compact as $N\to\infty$ and we denote by $\hat{\mathscr{B}}_{B,D,\varepsilon}$ the set of limit points. Verifying (7.7) reduces to verifying

$$\lim_{\varepsilon\to 0}\ \sup_{\nu\in\hat{\mathscr{B}}_{B,D,\varepsilon}} E^{\nu}\left[\left|\frac{\varepsilon}{2}\int_{-1/\varepsilon}^{1/\varepsilon} H(\omega^x)dx-\hat{H}(\rho_\varepsilon(0))\right|\right]=0\ . \tag{7.8}$$

In the appendix we have introduced the collection of Gibbs measure with boundary condition ω and particle number n on the interval $[-l,l]$. Let us denote the convex hull of these measures with n and ω varying by Γ. The subset with the average particle density less than or equal to 1 will be denoted by $\Gamma_l^{(1)}$. According to Theorem 10.3 of the appendix,

$$\lim_{l\to\infty}\ \sup_{\nu\in\Gamma^{(1)}_{l+l_0}} E^{\gamma}\left[\left|\frac{1}{2l}\int_{-l}^{l} H(\omega^x)dx-\hat{H}(\rho_l(0))\right|\right]=0\ ,$$

and we will complete the proof of (7.8) by establishing

Lemma 7.5. *For any B, D and $\varepsilon>0$*

$$\hat{\mathscr{B}}_{B,D,\varepsilon}\subset\Gamma^{(1)}_{1/\varepsilon+l_0}\ .$$

Proof. Since the density can only drop under the weak limit it is clear that it is sufficient to prove $\mathscr{B}_{B,D,\varepsilon}\subset\Gamma_{l_0+1/\varepsilon}$.

Let us assume that V is supported on some $[-c_0,c_0]$ and look at the set of limit points of the point process on a slightly larger interval of the form $\left[-\left(\frac{1}{\varepsilon}+l_0+c_0\right),\frac{1}{\varepsilon}+l_0+c_0\right]$ in the expanded scale. Let γ be any limit point. We want to characterize the restriction of ν as a point process on

$\left[-\left(\frac{1}{\varepsilon}+l_0\right),\left(\frac{1}{\varepsilon}+l_0\right)\right]$ which is a member of $\Gamma_{l_0+1/\varepsilon}$. Let us consider a diffusion on the configuration space of point processes on the interval $\left[-\left(\frac{1}{\varepsilon}+l_0+c_0\right),\frac{1}{\varepsilon}+l_0+c_0\right]$ under which the particles in $\left[-\left(\frac{1}{\varepsilon}+l_0\right),\left(\frac{1}{\varepsilon}+l_0\right)\right]$ diffuse with reflecting boundary conditions at the end point and those outside $\left[-\left(\frac{1}{\varepsilon}+l_0\right),\left(\frac{1}{\varepsilon}+l_0\right)\right]$ stay put. The generator is given by

$$L=\frac{1}{2}\sum\frac{\partial^2}{\partial x_i^2}-\sum_{i,j}V'(x_i-x_j)\frac{\partial}{\partial x_i}-\frac{1}{2}\sum_{i,\alpha}V'(x_i-y_\alpha)\frac{\partial}{\partial x_i},$$

where $x_1,\dots,x_n$ are the particles inside and $\{y_\alpha\},\dots$ are the particles outside. The outside configuration ω and the number n of particles inside do not change and for this diffusion L the family $\mu^{\omega}_{n,l'}$ with $l'=\left(\frac{1}{\varepsilon}+l_0\right)$ is precisely the set of extremal invariant measures. Therefore all we need to show is that ν is invariant for the diffusion L.

Let us take a function $u(x_1,\dots,x_n,b)$ of the form $u(x_1,\dots,x_n)\phi(b)$ with a smooth u, satisfying Neumann boundary conditions. Strictly speaking $u(x_1,\dots,x_n,b)$ is not a continuous function viewed as a function on configurations in $\left[-\left(\frac{1}{\varepsilon}+l+k_0\right),\frac{1}{\varepsilon}+l+k_0\right]$ because of the distinction between outside and inside variables. Its discontinuity points are on the set where a particle sits exactly on the boundary. These have probability zero because the homogeneity in space makes any local density at most one in the limit. So we can calculate

$$\begin{aligned}\int_{A_n}Lud\nu&=\lim_{N\to\infty}\int_{A_n}Lud\nu_N\\&=\lim_{N\to\infty}\int_{A_{n,N}}\hat{L}_N\hat{u}_Nf_Nd\mu_N,\end{aligned}$$

where $f_N d\mu_N$ is the measure on S^N which produced the ν_N by restricting in the set $\mathscr{B}^0_{N,B,D,\varepsilon'}$ for some choice of ε'. A_n is the set with n points in $\left[-\left(\frac{1}{\varepsilon}+l_0\right),\left[\frac{1}{\varepsilon}+l_0\right]\right]$ and $A_{n,N}$ is the set with n points in $\left[-\frac{\left(\frac{1}{\varepsilon}+l_0\right)}{N},\frac{\left(\frac{1}{\varepsilon}+l_0\right)}{N}\right]$. An elementary calculation using integration by parts yields

$$\int_{A_{n,N}}\hat{L}_N\hat{u}_Nf_Nd\mu_N=\frac{1}{2N^2}\int_{A_{n,N}}\sum{}'\frac{\partial\hat{u}_n}{\partial x_i}\cdot\frac{\partial f_N}{\partial x_i}d\mu_N,$$

where $\sum'$ is the summation over those variables $x_1,\dots,x_n$ inside $\left[-\frac{\left(\frac{1}{\varepsilon}+l_0\right)}{N},\frac{\left(\frac{1}{\varepsilon}+l_0\right)}{N}\right]$. The factor $\frac{1}{N^2}$ comes from the stretching. $\frac{\partial\hat{u}_n}{\partial x_i}$ are of order N and therefore we only have to prove

$$\sup_N \frac{1}{2} \int_{A_{n,N}} \sum{}' \left(\frac{\partial f_N}{\partial x_i}\right)^2 \cdot \frac{1}{f_N} d\mu_N < \infty \ .$$

We can calculate by translation invariance

$$\int \frac{1}{2} \sum{}' \left(\frac{\partial f_N}{\partial x_i}\right)^2 \frac{1}{f_N} d\mu_N = \frac{2\left(\frac{1}{\varepsilon}+l_0\right)}{N} \int \frac{1}{2} \sum \left(\frac{\partial f_N}{\partial x_i}\right)^2 \frac{1}{f_N} d\mu_N$$

$$\leqq \frac{2\left(\frac{1}{\varepsilon}+l_0\right)}{N} \cdot ND$$

$$= 2\left(\frac{1}{\varepsilon}+l_0\right) D \ ,$$

and we are done.

This proves Theorem 7.2.

We now turn to the proof of Theorem 7.3. We have to carry out some construction. Given any set $(x_1, \ldots, x_N)$ of N points in S, we have earlier associated the empirical distribution

$$\alpha_N(dx) = \frac{1}{N} [\delta_{x_1} + \ldots + \delta_{x_N}]$$

and a mollified density

$$\rho_\lambda(x) = \frac{\lambda}{N} \sum h(\lambda(x - x_i)) \ .$$

For most configurations the density $\rho_{\varepsilon N}(x)$ can be highly oscillatory and $\int \hat{H}(\rho_{\varepsilon N}(x))dx$ and $\int \hat{H}(\rho_\lambda(x))dx$ for fixed ε and λ can be very far apart. To rule this out we plan to calculate the Young measure associated with the functions $\rho_{\varepsilon N}(\cdot)$ and show that for N large and ε small, these are nearly degenerate. Corresponding to a mollified density $\rho_\lambda(x)$ we can define a probability measure π on $S \times R^+$ by the relation

$$\int F(x, \rho) \pi(dx, d\rho) = \int_S F(x, \rho_\lambda(x)) dx \ .$$

π clearly enjoys the properties

$$\int \rho \pi(dx, d\rho) = \int \rho_\lambda(x) dx = 1 \ , \tag{7.9}$$

$$\int F(x) \pi(dx, d\rho) = \int F(x) dx \ , \tag{7.10}$$

or π projects to Lebesgue measure on S. If we denote by $\mathscr{M}$ the space of probability measures on $S \times R^1$, then through π and ρ_λ we map $x_1, \ldots, x_N$ into $\mathscr{M}$. We already have a map $\mathscr{L}_n$ that maps $(x_1, \ldots, x_n)$ into $M_1(S)$ by

$$\alpha_N(dx) = \frac{1}{N} [\delta_{x_1} + \ldots + \delta_{\alpha_N}] \ .$$

We now take both the maps and map $(x_1,\ldots,x_N)$ into the pair α_N,π, a point in $M_1(S)\times\mathscr{M}$. With the choice of $\lambda=N\varepsilon$, this produces an induced probability measure $\hat{Q}_{N,\varepsilon}$ on $M_1(S)\times\mathscr{M}$ for each f_N from $\mathscr{A}_{N,B,D}$ and choice of ε. The range of such $\hat{Q}_{N,\varepsilon}$ will be denoted by $\hat{\mathscr{A}}_{N,B,D,\varepsilon}$, their limit points as $N\to\infty$ by $\hat{\mathscr{A}}_{B,D,\varepsilon}$ and the limit points of these as $\varepsilon\to0$ by $\hat{\mathscr{A}}_{B,D}$. Our goal is to prove Theorem 7.3 and the main step is

Theorem 7.6. *Let $\hat{Q}\in\hat{\mathscr{A}}_{B,D}$. Then for almost all (α,π) with respect to $\hat{Q}$,*

$$\alpha(dx)=\rho(x)dx$$

and

$$\pi(dx,d\rho)=dx\,\delta_{\rho(x)}(d\rho)$$

for some function $\rho(x)$ in $L_1(S)$, i.e. *the young measures are trivial.*

We will reach Theorem 7.6 by a series of lemmas:

Lemma 7.7. *Let $\hat{Q}\in\hat{\mathscr{A}}_{B,D}$. Then $\hat{Q}$ has the property*

$$\hat{Q}[\alpha:\alpha(dx)=\rho(x)dx \text{ for some } \rho]=1\ ,$$

and in fact

$$E^{\hat{Q}}\left[\int_S\rho^2(x)dx\right]\leqq C\cdot B\ ,$$

where C is a universal constant.

Proof. From Lemma 4.2 we can conclude that

$$E^{\hat{Q}}\int_S\lambda\,W(\lambda(x-y))\alpha(d\alpha)\alpha(dy)\leqq\|W\|\,CB\ ,$$

where W is a nonnegative function of integral 1. Since the bound is uniform in λ we can let $\lambda\to\infty$ and establish Lemma 7.7 by Fatou's lemma.

Lemma 7.8. *For any Q_ε in $\hat{\mathscr{A}}_{B,D,\varepsilon}$,*

$$Q_\varepsilon[(\alpha,\pi):\int\rho\,\pi(dx,d\rho)=1]=1\ , \tag{7.11}$$

$$Q_\varepsilon[(\alpha,\pi):\pi(dx,d\rho)=dx\,\pi_x(d\rho)]=1\ , \tag{7.12}$$

$$Q_\varepsilon[(\alpha,\pi):\alpha(dx)=\rho(x)dx \text{ with } \rho(x)=\int\rho\,\pi_x(d\rho)]=1\ . \tag{7.13}$$

Proof.

$$\begin{aligned}E^{\hat{Q}_{N,\varepsilon}}\int\rho^2\pi(dx,d\rho)&=E^{f_N}\int_S\rho_{\varepsilon N}^2(x)dx\\&=E^{f_N}\int_S\left|\varepsilon\sum h(\varepsilon N(x_i-x))\right|^2dx\\&=\frac{\varepsilon}{N}E^{f_N}\sum W(\varepsilon N(x_i-x_j))\\&\leqq\|W\|\,BC\quad\text{(by Lemma 4.2)}\ ,\end{aligned}$$

Now (7.11) is a consequence of the unitform integrability (7.9). Relation (7.12) follows from (7.10). (7.13) is just the routine fact that the mean of the Young measures defines the function which is the weak limit.

We now try to use the bound on the Dirichlet form and use inequality (5.3). The expression for $\mathscr{L}_N U_N$ given by (5.11) can be written as

$$\frac{1}{N^2}\mathscr{L}_N U_N = T_1 - T_2 + T_3 - T_4 \ , \tag{7.14}$$

where

$$T_1 = \frac{\delta}{N}\sum g(N\delta(x_i - x_j)) \ ,$$

$$T_2 = \frac{\lambda}{N^2}\sum g(\lambda(x_i - x_j))$$

$$T_3 = \frac{1}{N}\sum_{i,j,k}[G(N\delta(x_k - x_j)) - G(N\delta(x_i - x_j))]V'(N(x_i - x_k)) \ ,$$

$$T_4 = \frac{1}{N}\sum_{i,j,k}[G(\lambda(x_k - x_j)) - G)\lambda(x_i - x_j))]V'(N(x_i - x_k)) \ .$$

The inequality (5.3) takes the form

$$E^{f_N}[T_1 + T_3] \leqq E^{f_N}[T_2 + T_4] + D^{1/2}(E^{f_N}T_5)^{1/2} \ , \tag{7.15}$$

where

$$T_5 = \frac{4}{N^3}\sum_i \Big| \sum_j G(N\delta(x_i - x_j)) - G(\lambda(x_i - x_j))|^2 \ .$$

Because $|G(N\delta(x_i - x_j)) - G(\lambda(x_i - x_j))|$ is bounded by 1 and $G(N\delta x) - G(\lambda x)$ is supported on some interval of size λ^{-1} we can bound

$$|G(N\delta(x_i - x_j)) - G(\lambda(x_i - x_j))| \leqq \chi(\lambda(x_i - x_j)) \ ,$$

where $\chi(x)$ is a bounded nonnegative continuous function with compact support. Therefore in (7.15), T_5 can be replaced by

$$T_6 = \frac{4}{N^2}\sum_i\sum_j \chi(\lambda(x_i - x_j)) \ .$$

We next want to chase the inequality (7.15) through subsequences to get a limit point of $\hat{Q}_{N,\varepsilon}$ in $\hat{\mathscr{A}}_{B,D,\varepsilon}$ and then again through limits finally to a limit $\hat{Q}$ in $\hat{\mathscr{A}}_{B,D}$.

We have

$$\begin{aligned} T_1 &= \frac{\delta}{N}\sum g(N\delta(x_i - x_j)) \\ &= \delta \int_S \sum g(N\delta(x_i - x_j))h(N(x_i - x))dx \\ &= \delta \int_S H(\omega_N^x)dx \ , \end{aligned}$$

where

$$H(\omega) = \sum g(\delta(x_i - x_j))h(x_i) \ .$$

If it were not for the fact that $H(\omega)$ is unbounded we could apply Theorem 7.2

and we would have

$$\lim_{N\to\infty} E^{f_N} T_1 = E^Q \int dx[\int \hat{H}_\delta(\rho)\pi_x(d\rho)]$$

with

$$\hat{H}_\delta(\rho) = E_\rho[\delta \sum g(\delta(x_i - x_j))h(x_i)]$$
$$= \int \delta g(\delta(x-y))h(x)R_\rho^{(2)}(dx, dy) ,$$

where $R_\rho^{(2)}(dx, dy)$ is the two point correlation function relative to the Gibbs measure with density ρ. However since $H(\omega)$ is nonnegtive and continuous Fatou's lemma can be used to conclude

$$E^Q \int_S dx[\int \hat{H}_\delta(\rho)\pi_x(d\rho)] \leqq \lim_{N\to\infty} E^{f_N} T_1 . \tag{7.16}$$

We next look at T_3,

$$T_3 = \frac{1}{N} \sum_{i,j,k} [G(N\delta(x_k - x_j)) - G(N\delta(x_i - x_j))] V'(N(x_i - x_k))$$
$$= \int_S dx \sum [G(N\delta(x_k - x_j) - G(N\delta(x_i - x_j)] V'(N(x_i - x_k))h(N(x_i)$$
$$= \int_S H_1(\omega_N^x)dx ,$$

where

$$H_1(\omega) = \sum [G(\delta(x_k - x_j)) - G(\delta(x_i - x_j))] V'(x_i - x_k)h(x_i) .$$

One can check that because of the monotonicity of G and the fact that $xV'(x) \leqq 0$ for all x, $H_1(\omega)$ is nonnegative. One can then obtain by exactly the same reasoning as above

$$E^Q \int_S dx[\int \hat{H}_{1,\delta}(\rho)\pi_x(d\rho)]] \leqq \lim_{N\to\infty} E^{f_N} T_3 , \tag{7.17}$$

where

$$\hat{H}_{1,\delta}(\rho) = \iint [G(\delta(z-y)) - G(\delta(x-y))] V'(x-y)h(x)R_\rho^{(3)}(dx, dy, dz) ,$$

where $R_\rho^{(3)}(dx, dy, dz)$ is the three point correlation.

We now turn to T_2:

$$E^Q \iint \lambda g(\lambda(x-y))d\alpha(x)d\alpha(y) = E^Q \iint \lambda g(\lambda(x-y))\rho(x)\rho(y)dxdy$$
$$= \lim_{N\to\infty} E^{f_N} T_2 . \tag{7.18}$$

As for T_4

$$T_4 = \frac{1}{N} \sum_{i,j,k} [G(\lambda(x_k - x_j)) - G(\lambda(x_i - x_j))] V'(N(x_i - x_k)) .$$

We know that $V'(N(x_i - x_k)) = 0$ unless $|x_i - x_k| \leqq C/N$. Therefore we can expand $G(\lambda(x_k - x_j)) - G(\lambda(x_i - x_j))$ by Taylor's formula with remainder and

$$|G(\lambda(x_k-x_j))-G(\lambda(x_i-x_j))|\leqq\lambda g(\lambda(x_k-x_j))(x_k-x_i)+\frac{C}{N}|x_k-x_i| ,$$

where the error term depends only on λ.

We can now write

$$T_4=\frac{\lambda}{N^2}\sum g(\lambda(x_i-x_j))\psi(N(x_i-x_k))+\text{Error} ,$$

$$E^{f_N}[|\text{Error}|]\leqq E^{f_N}\frac{C}{N^2}\sum_{i,j,k}\psi(N(x_i-x_j))$$

$$\leqq\frac{C}{N}\text{ by Lemma 4.2.}$$

We now try to evaluate

$$\lim_{n\to\infty}E^{f_N}\left[\frac{\lambda}{N^2}\sum g(\lambda(x_k-x_j))\psi(N(x_i-x_k))\right] .$$

The function $\lambda g(\lambda(x-y))$ on $S\times S$ can be approximated by functions of the form

$$\sum_{r=1}^{l}a_r(x)b_r(y)$$

uniformly on $S\times S$ and if we denote by

$$\lambda g(\lambda(x-y))-\sum_{r=1}^{l}a_r(x)b_r(y)=c(x,y) ,$$

then

$$E^{f_N}\left|\frac{1}{N^2}\sum c(x_k,x_j)\psi(N(x_i-x_k))\right|\leqq\frac{1}{N^2}\|C\|NE^{f_N}\left[\sum\psi(N(x_i-x_j))\right]$$

$$\leqq\|C\|\quad\text{by Lemma 4.2.}$$

So the error is controlled uniformly in N. We can compute

$$\lim_{N\to\infty}E^{f_N}\frac{1}{N^2}\sum a(x_k)b(x_j)\psi(N(x_i-x_k))$$

$$=\lim_{N\to\infty}E^{f_N}\left[\frac{1}{N}\sum b(x_j)\cdot\frac{1}{N}\sum a(x_k)\psi(N(x_i-x_k))\right.$$

$$\left|\frac{1}{N}\sum a(x_k)\psi(N(x_i-x_k))-\int_S a(y)\sum_{i,k}\psi(N(x_i-x_k))h(N(x_k-y))\right|$$

$$\leqq\left|\frac{1}{N}\sum\psi(N(x_i-x_k))\int[a(y)-(x_k)]Nh(N(x_k-y))dy\right|$$

$$\leqq\frac{1}{N}\sum\psi(N(x_i-x_k)\cdot\varepsilon_N ,$$

where ε_N depends only on the modulus of continuity of a and goes to zero with N. Again by Lemma 4.2 we can control the error uniformly in N.

Now we can consider

$$\lim_{N\to\infty} E^{f_N}\left\{\frac{1}{N}\sum b(x_j)\cdot \int_S a(y)\sum \psi(N(x_i-x_k))h(N(x_k-y))dy\right\} ,$$

and working as before we should end up with

$$E^Q\left\{\int_S b(x)\rho(x)dx \int_S a(y)\int_R \hat{H}_3(\rho)\pi_y(d\rho)\right\}$$

$$=E^Q\left\{\iint b(x)a(y)\rho(x)\int_R \hat{H}_3(\rho)\pi_y(d\rho)\right\} ,$$

where

$$\hat{H}_3(\rho)=\iint \psi(x-y)h(y)R_\rho^{(2)}(dx,dy) .$$

However a serious problem here is that we are operating on the right side of the inequality and cannot afford the luxury of Fatou's lemma. We have to have an actual identification of limit and therefore have to prove that the functionals

$$H_3(\omega)=\sum \psi(x_i-x_k)h(N(x_k-y))$$

are uniformly integrable and can be truncated to yield uniformly bounded versions with uniformly small errors. What we can use for this is an estimate of the form

$$E^{f_N}\int \left|\sum_{i,k} \psi(N(x_i-x_k))h(N(x_k-y))\right|^{4/3} dy \leqq C .$$

a and b are uniformly bounded and cause no problems.

Lemma 7.9. *If $f_N \in \mathscr{A}_{N,B,D}$, then*

$$E^{f_N}\int \left|\sum_{i,k} \psi(N(x_i-x_k))\hat{H}(N(x_k-y))\right|^{4/3} dy \leqq C ,$$

where C depends only on B and D.

Proof. Let us denote $\psi(N(x_i-x_k))$ by ψ_{jk}, $h(N(x_k-y))$ by h_k and $\sum_i \psi_{ik}$ by T_k. Then

$$\left|\sum_{i,k} \psi_{ik}h_k\right| = |\sum h_k T_k| \leqq (\sum h_k)^{1/2}(\sum h_k T_k^2)^{1/2}$$

$$\left|\sum_{ik} \psi_{ik}h_k\right|^{4/3} \leqq (\sum h_k)^{2/3}(\sum h_k T_k^2)^{2/3}$$

$$E^{f_N}\int \left|\sum_{i,k} \psi_{ik}h_k\right|^{4/3} dy \leqq E^{f_N}\{\int (\sum h_k)^{4/3}(\sum h_k T_k^2)^{2/3}dy\}$$

$$\leqq (E^{f_N}\int (\sum h_k)^2 dy)^{1/3}(E^{f_N}\int (\sum h_k T_k^2)dy)^{2/3} .$$

But

$$E^{f_N} \int |\sum h(N(x_k - y))|^2 dy = \frac{1}{N} E^{f_N} \sum \chi(N(x_i - x_j))$$

is bounded. Moreover

$$E^{f_N} \int \sum h(N(x_k - y)) T_k^2 dy = E^{f_N} \frac{1}{N} \sum_i \Big| \sum_j \psi(N(x_i - x_j)) |^2 ,$$

and this was shown to us uniformly bounded before in Lemma 5.1.

Turning to our original estimate now we have

$$\lim_{N\to\infty} E^{f_N} T_4 = E^Q \left[\int \int \lambda g(\lambda(x-y)) \rho(x) dx dy \int \hat{H}_3(\rho) \pi_y(d\rho) \right] . \tag{7.19}$$

We now look at the final term

$$\lim_{N\to\infty} E^{f_N} T_6 = 4 E^Q \left[\int \int \chi(\lambda(x-y)) \rho(x) \rho(y) dx dy \right] .$$

Combining (7.15) with (7.16) through (7.19) we obtain for any measure Q in $\hat{\mathscr{A}}_{b,D}$,

$$\begin{aligned} E^Q \Bigg[& \int_S dx \left[\int \hat{H}_\delta(\rho) \pi_x(d\rho) \right] \\ & + \int_S dx \left[\int \hat{H}_{1,\delta}(\rho) \pi_x(d\rho) \right] \Bigg] \\ & \leqq E^Q \Bigg[\int\int \lambda g(\lambda(x-y)) \rho(x) \rho(y) dx dy \\ & + \int \int \lambda g(\lambda(x-y)) \rho(x) dx dy \int \hat{H}_3(\rho) \pi_y(d\rho) \Bigg] \\ & + \sqrt{D} \left[4 E^Q \left[\int \int \chi(\lambda(x-y)) \rho(x) \rho(y) dx dy \right] \right]^{1/2} . \end{aligned} \tag{7.20}$$

In the relation (7.20) the right-hand side is clearly bounded uniformly in terms of our constants B and D for each fixed λ. We first let $\delta \to 0$ in the left. From the factoring properties of the correlation functions $R_\rho^{(2)}(dx, dy)$ and $R_\rho^{(3)}(dx, dy, dz)$ we can conclude by an application of Fatou's lemma

$$\lim_{\delta\to 0} \int_S dz \int [\hat{H}_\delta(\rho) + \hat{H}_{1,\delta}(\rho)] \pi_x(d\rho) \geqq \int_S dx \int \rho P(\rho) \pi_x(d\rho) ,$$

since by Theorem 10.4 $P(\rho) \sim \rho^2$ for large ρ, it follows that

Lemma 7.10. *For* $Q \in \hat{\mathscr{A}}_{B,D}$,

$$\sup_{Q \in \mathscr{A}_{B,\nu}} E^Q \left[\int \rho^3(x) dx \right] \leqq E^Q \left[\int_S dx \int \rho^3 \pi_x(d\rho) \right] < \infty .$$

We are interested in letting $\lambda\to\infty$ on the right-hand side. Let us look at each of the three terms on the right. Clearly $\int\int\lambda g(\lambda(x-y))\rho(x)\rho(y)dxdy \leqq \int\rho^2(x)dx$, and since $E^Q\int\rho^2(x)dx<\infty$ we can use the dominated convergence theorem to conclude that

$$\lim_{\lambda\to\infty} E^Q\int\int\lambda g(\lambda(x-y))\rho(x)\rho(y)dxdy = E^Q\int\rho^2(x)dx\ . \tag{7.21}$$

The second term involves

$$\int\int\lambda g(\lambda(x-y))\rho(x)dxdy\int\hat{H}_3(\rho)\pi_y(d\rho)$$

and

$$\lim_{\lambda\to\infty}\int\int\lambda g(\lambda(x-y))\rho(x)dy\int\hat{H}_3(\rho)\pi_y(d\rho)=\int\rho(y)\int\hat{H}_3(\rho)\pi_y(d\rho)\ .$$

We can bound

$$\int\int\lambda g(x-y))\rho(x)dxdy\int P(\rho)\pi_y(d\rho)$$

$$\left[\int\left(\int\lambda g(\lambda(x-y))\rho(x)dx\right)^3\right]^{1/3}\left[\int\left[\int P(\rho)\pi_y(d\rho)\right]^{3/2}dy\right]^{2/3}$$

$$\leqq\tfrac{1}{3}\int\lambda g(\lambda(x-y))\rho^3(x)dxdy+\tfrac{2}{3}\int[P(\rho)]^{3/2}\pi_y(d\rho)dy$$

$\leqq A$ by Lemma 7.10.

The third term clearly goes to zero as $\lambda\to\infty$. Therefore for every $Q\in\hat{\mathscr{A}}_{B,D}$,

$$E^Q\left[\int_S dx\int\rho P(\rho)\pi_x(d\rho)\right]\leqq E^Q\left[\int_S dx\left[\int\rho\pi_x(d\rho)\int P(\rho)\pi_x(d\rho)\right]\right]\ . \tag{7.22}$$

Proof of Theorem 7.6. We can think of Eq. (7.22) as

$$E^Q\int_S dx\left\{\int\rho P(\rho)\pi_x(d\rho)-\int\rho\pi_x(d\rho)\int P(\rho)\pi_x(d\rho)\right\}\leqq 0\ . \tag{7.23}$$

This means that on the average the covariance

$$\int\rho P(\rho)\pi_x(d\rho)-\int\rho\pi_x(d\rho)\int P(\rho)\pi_x(d\rho)$$

is nonnegative. However

$$\int\rho P(\rho)\pi_x(d\rho)-\int\rho\pi_x(d\rho)\int P(\rho)\pi_x(d\rho)$$

$$=\int\left(\rho-\int\rho\pi_x(d\rho)\right)\left(P(\rho)-P\left(\int\rho\pi_x d\rho\right)\right)\pi_x(d\rho)$$

$$\geqq 0\ ,$$

because $P(\rho)$ is a nondecreasing function of ρ. The strictly increasing nature of $P(\rho)$ dictates that we have strict positivity unless $\pi_x(d\rho)$ is degenerate. This completes the proof of Theorem 7.6.

Having proved Theorem 7.6 the first application is an estimate on the regularity of densities in the support of any $Q\in\hat{\mathscr{A}}_{B,D}$.

Theorem 7.11. *For any* $Q \in \hat{\mathscr{A}}_{B,D}$,

$$E^Q\left[\int_S \{[P(\rho(\theta))]_\theta\}^2 \frac{1}{\rho(\theta)}\right] \leqq 2D .$$

Proof. Let us consider k functions $u_1(x), \ldots, u_k(x)$ from $C^\infty(S)$ and l functions $F_1(x), \ldots, F_l(x)$ also from $C^\infty(S)$. We consider as well the functions $g_1, \ldots, g_k$ from $C_0^\infty(R^l)$. Let us denote by $\hat{x} = (x_1, \ldots, x_N)$ a point of S^N. Then we define on S^N

$$U_n(\hat{x}) = \frac{1}{N} \sum_{j=1}^N u_r(x_j) ,$$

$$G_r(\hat{x}) = g_r\left(\frac{1}{N} \sum_{j=1}^N F_1(x_j), \ldots, \frac{1}{N} \sum_{j=1} F_l(x_j)\right) .$$

We can perform integration by parts and write

$$\int \left\{\sum_{r=1}^k [L_N U_r(\hat{x})][G_r(\hat{x})]\right\} f_N(\hat{x}) d\mu_N$$

$$= -\frac{1}{2} \int \left\{\sum_{r=1}^k \langle \nabla U_r, \nabla G_r \rangle + \sum_{r=1}^k \left\langle U_r, \frac{\nabla f_n}{f_N} \right\rangle G_r\right\} f_N d\mu_N$$

$$= T_1 + T_2 . \tag{7.24}$$

Since $\left|\frac{\partial G_r}{\partial x_j}\right| \leqq \frac{C}{N}$ it follows by an elementary calculation that $|T_1| \to 0$ as $N \to \infty$. We can estimate T_2 by Schwartz's inequality to obtain

$$|T_2| \leqq \frac{1}{2}\sqrt{2D}\left(\int \frac{1}{N} \sum_{j=1}^N \left(\sum_{r=1}^k u_r'(x_j) G_r(\hat{x})\right)^2 \cdot f_N d\mu_N\right)^{1/2} .$$

The left-hand side of (7.24) can be calculated to be approximately

$$\frac{1}{2} \int \sum_{r=1}^k \left\{\frac{1}{N} \sum_{j=1}^N u_r''(x_j) G_r(\hat{x}) \left(1 + \sum_{i=1}^N \psi(N(x_i - x_j))\right)\right\} f_N d\mu_N .$$

For any $Q \in \hat{\mathscr{A}}_{B,D}$, we can therefore obtain the inequality

$$\frac{1}{2}\left|E^Q \int \sum_{r=1}^k u_r''(x) G_r(\langle\rho, F_1\rangle, \ldots, \langle\rho, F_l\rangle) P(\rho(x)) dx\right|$$

$$\leqq \tfrac{1}{2}\sqrt{2D}\left[E^Q \int \left|\sum u_r'(x) G_r(\langle\rho, F_1\rangle, \ldots, \langle\rho, F_l\rangle)\right|^2 \phi(x) dx\right]^{1/2} .$$

It is not difficult to pass from this estimate to

$$\left|E^Q \int \psi_{xx}(x, \omega) P(\rho(x, \omega)) dx\right| \leqq \sqrt{2D}\left\{E^Q \int \psi_x^2(x, \omega) \rho(x, \omega) dx\right\} . \tag{7.25}$$

Here ω is the generic point and Q is the measure on random densities $\rho(\cdot)$.

The estimate (7.25) will hold if the test function $\psi(x, \omega)$ is uniformly bounded along with ψ_x and ψ_{xx}. Since we can replace ψ_x by ψ we can assume

$$\left| E^{Q} \int \psi_x(x,\omega) P(\rho(x,\omega)) dx \right| \leqq \sqrt{2D} \left\{ E^{Q} \int \psi^2(x,\omega) \rho(x,\omega) dx \right\}^{1/2} \quad (7.26)$$

for all bounded ψ with ψ_x bounded and having $\int \psi dx \equiv 0$ for almost all ω.

Since $E^{Q} \int \rho^2(x,\omega) dx$ has a uniform bound in $\hat{\mathscr{A}}_{B,D}$ we will have a bound of the form

$$\left| E^{Q} \int \psi_x(x,\omega) P(\rho(x,\omega)) dx \right| \leqq C \left(E^{Q} \int \psi^4(x,\omega) dx \right)^{1/4}, \quad (7.27)$$

provided $\int \psi(x,\omega) dx \equiv 0$. This is enough to yield the estimate

$$E^{Q} \int |[P(\rho(x,\omega))]_x|^{4/3} dx \leqq C .$$

We can now rewrite (7.26) in the form

$$\left| E^{Q} \int \psi(x,\omega) (P(\rho(x,\omega)))_x dx \right| \leqq \sqrt{2D} \left\{ E^{Q} \int \psi^2(x,\omega) \rho(x,\omega) dx \right\}^{1/2} \quad (7.28)$$

for all $\psi(x,\omega)$ such that $E^{Q} \int \psi^4(x,\omega) dx < \infty$ and satisfy $\int \psi(x,\omega) dx \equiv 0$. Let $\psi(x,\omega)$ be any function with $E^{Q} \int \psi^4(x,\omega) dx < \infty$. We write $\psi(x,\omega) = \psi_1(x,\omega) + h(\omega)$, where $\psi_1(x,\omega)$ has the property $\int \psi_1(x,\omega) dx \equiv 0$. Then because $(P(\rho(x,\omega)))_x$ is orthogonal to constants,

$$\begin{aligned} &\left| E^{Q} \int \psi(x,\omega) P(\rho(x,\omega))_x dx \right| \\ &= \left| E^{Q} \int \psi_1(x,\omega) (P(\rho(x,\omega)))_x dx \right| \\ &\leqq \sqrt{2D} \left\{ E^{Q} \int [\psi(x,\omega) - h(\omega)]^2 P(x,\omega) dx \right\}^{1/2} \\ &\leqq \sqrt{2D} \left\{ E^{Q} \int [\psi(x,\omega) - h(\omega)]^2 \right. \\ &\quad \left. \cdot [1+\rho(x,\omega)] dx \right\}^{1/2} . \end{aligned} \quad (7.29)$$

Since $h(\omega) = 0 \int \psi(x,\omega) dx$ we can bound

$$E^{Q} h^2(\omega) \leqq E^{Q} \int \psi^2(x,\omega)[1+\rho(x,\omega)] dx .$$

Therefore

$$\left| E^{Q} \int \psi(x,\omega) (P(\rho(x,\omega)))_x dx \right| \leqq C \left[E^{Q} \int \psi^2(x,\omega)[1+\rho(x,\omega)] dx \right]^{1/2}$$

for all functions $\psi(x,\omega)$ with $E^{Q} \int \psi^4(x,\omega) dx < \infty$. This is enough to conclude that

$$E^{Q} \int \frac{\{P(\rho(x,\omega))_x\}^2}{1+\rho(x,\omega)} dx < \infty . \quad (7.30)$$

We now go back to (7.29) and use for $\psi(x,\omega)$ the function $\frac{[P(\rho(x,\omega))]_x}{\varepsilon+\rho(x,\omega)}$. (This has to be justified.) Then $h(\omega)=0$.

We now obtain

$$E^Q \int \frac{[P(\rho(x,\omega))_x]^2}{\varepsilon+\rho(x,\omega)}\,dx$$
$$\leqq \sqrt{2D}\left(E^Q \int \frac{(P(\rho(x,\omega)))_x^2}{[\varepsilon+\rho(x,\omega)]^2}\,\rho(x,\omega)\,dx\right)^{1/2}$$
$$\leqq \sqrt{2D}\left(E^Q \int \frac{(P(\rho(x,\omega)))_x^2}{\varepsilon+\rho(x,\omega)}\,dx\right)^{1/2}.$$

Because of (7.30) we can let $\varepsilon \to 0$ and obtain our theorem.

Now we have all the ingredients necessary to prove Theorem 7.3. If we calculate

$$\lim_{\lambda\to 0}\ \limsup_{\varepsilon\to 0}\ \limsup_{N\to\infty}\ \sup_{Q\in\mathscr{A}_{N,B,D}} E^Q\left[\int |\rho_\lambda(x)-\rho_{N\varepsilon}(x)|\,dx\right]$$
$$=\lim_{\lambda\to 0}\ \sup_{Q\in\mathscr{A}_{B,D}} E^Q\left[\int\left|\int h_\lambda(x-y)\rho(y)\,dy-\rho(x)\right|dx\right]$$
$$=0$$

because of the estimate in Theorem 7.11. Because $P(\rho)$ is strictly monotone and continuous in ρ any control on $\int \frac{(P(\rho))_\theta^2}{\rho}\,d\theta$ will provide control on the L_1 modulus of continuity of $\rho(\cdot)$.

8. Hydrodynamic Limit

Let Q_N be the family of probability measures induced on $C[[0,T];M_1(S)]$ by looking at the trajectory $\frac{\delta_{x_1(\cdot)}+\cdots\delta_{x_n(\cdot)}}{\varepsilon}$ in the space of probability measures. We have already established Theorem 3.1 asserting the compactness of Q_N. Let Q be any weak limit point. The results of Sect. 7 establish Theorems 3.2 and 3.3. All we need is the following lemma.

Lemma 8.1. *If we have the bounds*

$$E^Q \int \rho^2(\theta,\omega)\,d\theta \leqq C\,,$$
$$E^Q \int \frac{(P(\rho))_\theta^2}{\rho}\,d\theta \leqq C\,,$$

then for some other constant C,

$$E^Q \int \rho^3(\theta)\,d\theta \leqq C\,.$$

Proof. Let us consider

$$S(\rho)=[1+P(\rho)]^{3/4},$$

$$(S(\rho))_\theta=\tfrac{3}{4}[1+P(\rho)]^{-1/4}[P(\rho)]_\theta,$$

$$\int (S(\rho))_\theta^2 d\theta \leqq \frac{9}{16}\int \frac{(P(\rho))_\theta^2}{[1+P(\rho)]^{1/2}}d\theta,$$

$$\leqq C$$

because of the bound $P(\rho)\geqq c\rho^2$ for large ρ. Since $P(\rho)\geqq C\rho^2$ we have $S(\rho)\geqq C\rho^{3/2}$ for large ρ and we have $S(\rho)\leqq C\rho^{3/2}$ with another constant. In other words we can control $\int S(\rho)d\theta$ and $\int (S(\rho))_\theta^2 d\theta$, and hence control $\int S^2(\rho)d\theta$.

Remark. Although the conclusion of the lemma is already implied by Lemma 7.10, it follows from the first two estimates assumed in the lemma.

9. Uniqueness

Let $\rho_0(\theta)$ be a function on $L_2(S)$. Assume $\rho(t,\theta)$ is a weak solution of

$$\frac{\partial\rho(t,\theta)}{\partial t}=\frac{1}{2}(P(\rho(t,\theta)))_{\theta\theta}$$

with

$$\rho(0,\theta)=\rho_0(\theta).$$

Assume that the solution $\rho(t,\theta)$ has the property

$$\int_S\int_0^T \rho^3(t,\theta)d\theta dt<\infty \tag{9.1}$$

and

$$\int_S\int_0^T \frac{[P(\rho(t,\theta))]_\theta^2}{\rho(t,\theta)}dt d\theta<\infty. \tag{9.2}$$

Theorem 9.1. *A weak solution satisfying* (9.1) *and* (9.2) *is unique.*

Let if possible $u(t,\theta)$ and $v(t,\theta)$ be two solutions corresponding to the same initial data $\rho_0(\theta)$. Let us define the function

$$\omega(\theta,t)=\tfrac{1}{2}\int_0^t\{[P(u(s,\theta))]_\theta-[P(v(s,\theta))]_\theta\}ds. \tag{9.3}$$

Using the fact that $\int\int_0^t \rho(s,\theta)ds d\theta<\infty$, it follows from (9.2) that

$$[P(u)]_\theta-[P(v)]_\theta\in L_1(S\times[0,t]).$$

Hence $\omega(\theta,t)$ is well defined as a continuous function of t for almost all θ. An elementary computation yields

$$\frac{\partial\omega}{\partial\theta}=u-v \tag{9.4}$$

in the sense of distributions. Now

$$\frac{\partial}{\partial t}[\omega(\theta,t)]^2 = 2\omega(\theta,t)\frac{\partial w}{\partial t}$$
$$= \omega(\theta,t)[P(u)-P(v)]_\theta \ .$$

We want to show that

$$\int \int_0^t |\omega(\theta,t)| \{|[P(u)]_\theta| + |[P(v)]_\theta\} d\theta\, dt < \infty \ . \qquad (9.5)$$

Since $\int\int \frac{|[P(u)]_\theta|^2}{u} d\theta\, dt$ and $\int\int \frac{|[P(v)]_\theta|^2}{v} d\theta\, dt$ are finite, it is sufficient to prove that

$$\int \int_0^t |\omega(\theta,t)|^2 u(\theta,t) d\theta\, dt$$

and

$$\int \int_0^t |\omega(\theta,t)|^2 v(\theta,t) d\theta\, dt$$

are finite. (9.1) and (9.4) are sufficient now to conclude (9.5). We therefore conclude that

$$\chi(\theta,t) = [\omega(\theta,t)]^2$$

is a continuously differentiable map of $[0,t]$ into $L_1(S)$. Moreover $\omega(\theta,t)\to 0$ a.e. θ as $t\to 0$,

$$\int \frac{\partial \chi}{\partial t} d\theta = \int \omega(\theta,t)[P(u(\theta,t)) - P(v(\theta,t))]_\theta d\theta$$
$$= -\int \left(\frac{\partial \omega}{\partial \theta}\right)(P(u)-P(v))d\theta$$
$$= -\int (u-v)(P(u)-P(v))d\theta \leqq 0 \ .$$

Therefore $\int \chi(\theta,t)d\theta \equiv 0$. This proves uniqueness.

10. Appendix

In this section we will establish some results concerning finite and infinite volume Gibbs measures for our continuous system with finite range interaction. Although these results are not new and are part of the folklore in the field, there does not appear to be any source that we can refer to for a precise statement of the results in the form we need them. We will therefore provide a proof.

Our main assumption is that we have a repulsive pair potential, i.e. a function $V(x)$, which is nonnegative, even, continuously differentiable and has compact support. We denote by $[-c_0, c_0]$ the interval of support of the function $V(\cdot)$ and we assume further that $\psi(x) = -xV'(x) \geqq 0$ and that $V(0) > 0$.

For any $l > 0$, we consider the exterior $[-l,l]^c$ of the interval $[-l,l]$ relative to the real line. A configuration ω is the realization of a point process and represents a locally finite collection of points. Since our interaction has range c_0 we are only interested in the part of the realization that has to do with the border $[-l-c_0, -l]\cup[l, l+c_0]$ and therefore we can think of ω as an arbitrary but finite configuration of points in the border.

We shall think of $\omega=\{y_a\}$, as a finite set of points located in $[-l-c_0, -l]\cup[l, l+c_0]$. The canonical partition function $Z(n,l,\omega)$ is defined as follows for $n\geq 0$, $l>0$ and all ω,

$$Z(n,l,\omega)=1 \quad \text{if} \quad n=0 \quad \text{for all } l \text{ and } \omega \ , \tag{10.1}$$

$$Z(n,l,\omega)=\frac{1}{n!}\int_{-l}^{l}\cdots\int_{-l}^{l}\exp\left[-\sum_{i\neq j=1}^{n}V(x_i-x_j)\right.$$
$$\left.-2\sum_{i,\alpha}V(x_i-y_\alpha)\right]dx_1\dots dx_n \ . \tag{10.2}$$

The canonical Gibbs measure $\mu_{n,l,\omega}$ is a point process on $[-l,l]$ with exactly n points or equivalently a measure on $[-l,l]^n$ with a density given by

$$f_{n,l,\omega}(x_1,\dots,x_n)=c_n\exp\left[-\sum_{i\neq j=1}^{n}V(x_i-x_j)-2\sum_{i,\alpha}V(x_i-y_\alpha)\right] \ .$$

The normalizing constant c_n is of course given by

$$c_n^{-1}=n!Z(n,l,\omega) \ .$$

When ω is the empty set we denote $Z(n,l,\omega)$ by $Z^0_{n,l}$ and $\mu_{n,l,\omega}$ by $\mu^0_{n,l}$.

The grand canonical partition function with activity λ is defined for every real λ and

$$\hat{Z}(\lambda,l,\omega)=\sum_{n=0}^{\infty}e^{\lambda n}Z(n,l,\omega) \ ,$$

and the grand canonical Gibbs measure is a point process $\hat{\mu}_{\lambda,l,\omega}$ on $[-l,l]$ which is thought of as a probability measure on

$$\bigcup_{n=0}^{\infty}[-l,l]^n \ .$$

It is given by

$$\hat{\mu}_{\lambda,l,\omega}=[\hat{Z}(\lambda,l,\omega)]^{-1}\sum e^{\lambda n}Z(n,l,\omega)\mu_{n,l,\omega} \ , \tag{10.3}$$

i.e. it is a convex combination of canonical Gibbs measures and the weights are proportional to $e^{\lambda n}Z(n,l,\omega)$.

Although we have only defined these Gibbs measures on intervals of the form $[-l,l]$ it is clear that they enjoy translation invariance of sorts and can be defined on any interval $[a,b]$ and depend only on the configuration in the appropriate border. Moreover, there is an internal consistency in the following two senses.

First, by definition, the canonical Gibbs measure is the conditional distribution of the grand canonical Gibbs measure when it is conditioned by the number of particles in the set.

Moreover if we take the grand canonical Gibbs distribution on $[-l,l]$ and condition it with respect to the configuration in $[a,b]^c$, where $-l\leq a<b\leq l$, then

the conditional distribution is the grand canonical Gibbs measure on $[a,b]$ with the same activity λ.

The infinite volume Gibbs measure μ_λ corresponding to activity λ, is defined as a stationary point process such that for every interval $[a,b]$ the conditional distribution of the configuration in $[a,b]$ given the configuration in $[a,b]^c$ is the grand canonical Gibbs measure $\hat{\mu}_{\lambda,l,\omega}$ with $l=\frac{1}{2}(b-a)$ and ω representing the exterior configuration. We shall state the main theorems we will use in our article and provide a sketch of the proof at the end, after establishing a few key estimates as lemmas.

Theorem 10.1. *The following thermodynamic functions exists:*

$$F(\lambda)=\lim_{l\to\infty}\frac{1}{2l}\log\hat{Z}^0_{\lambda,l}\quad \text{for } \lambda \text{ real},$$

$$\Psi(\rho)=\lim_{\substack{l\to\infty\\ n\to\infty\\ n/2k\to\rho}}\frac{1}{2l}\log Z^0_{n,k}\quad \text{for } \rho\geqq 0\,.$$

The function $F(\lambda)$ is convex in λ and the function $\Psi(\rho)$ is concave in ρ. Moreover they are related by the equation

$$F(\lambda)=\sup_\rho[\lambda\rho+\Psi(\rho)]\,. \tag{10.4}$$

Theorem 10.2. *For each value of the activity parameter λ, there exists exactly one Gibbs measure μ_λ with activity λ. The point process μ_λ depends continuously on the parameter λ. As a point process its density is given by*

$$\rho(\lambda)=F'(\lambda)=\frac{dF(\lambda)}{d\lambda}\,.$$

In particular $F(\lambda)$ is once continuously differentiable. The function $\rho(\lambda)$ is continuous and nondecreasing. Its inverse is

$$\lambda=-\Psi'(\rho)=\frac{d\Psi(\rho)}{d\rho}\,.$$

We have in addition

$$\lim_{\lambda\to\infty}\rho(\lambda)=\infty\,,\qquad \lim_{\lambda\to-\infty}\rho(\lambda)=0\,.$$

If I is a unit interval and $N(I)$ denotes the number of points in the interval I, then

$$E^{\mu_\lambda}[\{N(I)\}^r]<\infty\quad \textit{for every}\quad r\geqq 1\,.$$

Theorem 10.3. *Let $n\to\infty$ and $l\to\infty$ in such a manner that $\frac{n}{l}\to\rho$. Let $H(\omega)$ be a bounded continuous local function depending on the configuration in some fixed finite interval $[-l_0,l_0]$. Let us denote by $H_x(\omega)$ the same functional evaluated at the configuration in $[-l_0+x,l_0+x]$. We denote by $\hat{H}(\lambda)$ the expected value $E^{\mu_\lambda}[H(\omega)]$. Then for every $\varepsilon>0$,*

$$\lim\mu^0_{n,l}\left[\left|\frac{1}{2l}\int_{-(l-l_0)}^{(l-l_0)}H_x(\omega)dx-\hat{H}(\lambda)\right|\geqq\varepsilon\right]=0\,, \tag{10.5}$$

where $\rho=F'(\lambda)$.

Moreover for every function $H(x)$ *on* $[-1,1]$ *which is nonnegative, has compact support with* $\int \phi(x)dx=1$,

$$\lim \mu^0_{n,l}\left[\left|\frac{1}{l}\sum\phi\left(\frac{x_i}{l}\right)-\rho\right|\geqq\varepsilon\right]=0 . \tag{10.6}$$

Remark 1. It follows from (10.5) that if λ_1,λ_2 are two different activities such that $F'(\lambda_1)=F'(\lambda_2)=\rho$, then $\hat{H}(\lambda_1)=\hat{H}(\lambda_2)$ for all functions $H(\omega)$. This in turn implies $\mu_1=\mu_{\lambda_1}$ and $\lambda_1=\lambda_2$. Therefore $\rho(\lambda)$ is a strictly increasing function of λ and its inverse $-\Psi'(\rho)$ is a strictly increasing continuous function of ρ.

We define the function

$$P(\rho)=F(\lambda(\rho)) . \tag{10.7}$$

Theorem 10.4. *There exists constants* $0<c_1<c_2<\infty$ *such that*

$$c_1\rho^2\leqq P(\rho)\leqq c_2\rho^2 \quad \textit{for } \rho \textit{ large} .$$

Since $E^{\mu_\lambda}[\{N(I)\}^r\}<\infty$ for every μ_λ, we consider for every function $f(x_1,\dots,x_k)$ of k-variables which is smooth and has compact support on R^k, the variable

$$\xi_f=\sum f(x_1,\dots,x_k) ,$$

where the summation is over all k-tuples of k distinct points in the configuration. ξ_f has a finite expectation and

$$E^{\mu_\lambda}\xi_f=\int f(x_1,\dots,x_k)R^\lambda_k(dx_1,\dots,dx_k) .$$

The measure $R^\lambda_k(dx_1,\dots,dx_k)$ is the k-point correlation measure. It has translation invariance relative to $x_1,\dots,x_k\to x_1+a,\dots,x_k+a$. The one point measure $R^\lambda_k(dx_1)=\rho(\lambda)dx_1$, where $\rho(\lambda)$ is the density.

We need the following identity between $P(\rho)$ and $R^\lambda_2(dx_1,dx_2)$.

Theorem 10.5. *If* λ *and* ρ *are related by* $\lambda=\lambda(\rho)$, *then*

$$P(\rho)=\rho+\int\psi(x_1-x_2)\phi(x_2)R^\lambda_1(dx_1,dx_2) , \tag{10.8}$$

where $\phi(x)$ *is any function on R that is smooth, has compact support, is nonnegative and has* $\int\phi(x)dx=1$.

Finally the ergodicity of μ^λ implies some asymptotic factorization of the correlation measures.

Theorem 10.6. *Let* g_1,g_2 *be functions on R that are smooth and have compact support. Let* g_3 *be a smooth function on* R^2 *having compact support. Then for any* λ,

$$\lim_{k\to 0} k\int g_1(k(x_1-x_2))g_2(x_2)R^\lambda_2(dx_1,dx_2)=\rho^2\left(\int g_1(x)dx\right)\left(\int g_2(x)dx\right) , \tag{10.9}$$

$$\lim_{k\to 0} k\int g(k(x_1-x_2))g_2(x_2,x_3)R^\lambda_3(dx_1,dx_2,dx_3)$$
$$=\rho\left(\int g_1(x)dx\right)\left(\int g_2(x_2,x_3)R^\lambda_2(dx_1,dx_2)\right) . \tag{10.10}$$

We adopt the following conventions regarding constants. Constants will always be independent of n, l and ω. A, B, C will denote constants and will be used at different times of denote different constants. We will use γ to denote constants that are strictly positive. The constant c_0 is the range of the interaction $V(\cdot)$. The quantity $\frac{n}{2l}$ will be denoted by ρ.

Lemma 10.7. *For every $l > c_0$ and ω we have*

$$Z^0_{n,l-c_0} \leqq Z(n,l,\omega) \leqq Z^0_{n,l} \, . \tag{10.11}$$

Proof. For $n=0$, there is nothing to prove. The upper bound is obtained by dropping the interaction terms involving the points in the border. For the lower bound we limit the integration to the situation where all the points $x_1, \ldots, x_n$ lie inside $[-(l-c_0), (l-c_0)]$.

Lemma 10.8. *There is a constant C such that*

$$\frac{d \log Z^0_{n,l}}{dl} \leqq \frac{n}{l} + C\frac{n^2}{l^2} \, . \tag{10.12}$$

Proof. By differentiation we can write

$$\frac{d \log Z^0_{n,l}}{dl} = \frac{n}{l} + \frac{1}{l} \int \sum \psi(x_i - x_j) d\mu^0_{n,l} \, .$$

[Change variables $x_i = ly_i$, differentiate and change back.] From formula (4.4) we have an estimate

$$\sum \psi(x_i - x_j) \leqq C_1 \sum V(x_i - x_j)$$

for some C_1. Since

$$d\mu^0_{n,l} = c_n \exp[-\sum V(x_i - x_j)] d\theta \, ,$$

where $d\theta$ is volume on $[-l, l]^N$, it follows that

$$\int \sum V(x_i - x_j) d\mu^0_{n,l} \leqq \int \sum V(x_i - x_j) d\hat{\theta} \, ,$$

where $d\hat{\theta}$ is normalized volume. We can now calculate $\int \sum V(x_i - x_j) d\hat{\theta}$ and obtain (10.12).

Lemma 10.9. *For every n and $l > 0$,*

$$z^0_{n+1,l} \leqq \frac{2l}{n+1} Z^0_{n,l} \, . \tag{10.13}$$

Proof. If we ignore any interaction involving x_{n+1} and integrate out x_{n+1} we obtain (10.13).

Lemma 10.10. *Let $l = l_1 + l_2 + \frac{1}{2}c_0$ with $l_1 > 0$, $l_2 > 0$. Then*

$$Z^0_{n,l} \geqq \sum_{\substack{n_1+n_2=n \\ n_1 \geqq 0, n_2 \geqq 0}} Z^0_{n,l_1} Z^0_{n_2,l_2} \, .$$

In particular for any n_1, n_2 with $n_1 + n_2 = n$,

$$Z^0_{n,l} \geqq Z^0_{n_1,l_1} Z^0_{n_2,l_2} \, . \tag{10.14}$$

Proof. Let us divide the interval $[-l, l]$ into three parts I_1, I and I_2 with lengths $2l_1$, c_0 and $2l_2$ arranged in order. We limit the integration to the domain where I_1 and I_2 contain all the points $(x_1, \ldots, x_n)$. Here n_1 and n_2 represent the number of particles in I_1 and I_2 respectively. The combinatorial factors adjust themselves and we get Lemma 10.10.

Lemma 10.11. *There exist constants A and B such that*

$$\mu_{n,l,\omega} \leqq A e^{B\rho^2} \mu_{n,l}^0, \qquad \text{provided } l \geqq 2c_0 .$$

Proof. The density $f_{n,l,\omega}$ would be dominated by $f_{n,l}^0$ if it were not for the normalizing constant. We therefore need only estimate the ratio

$$Z_{n,l}^0 [Z(n, l, \omega)]^{-1} .$$

Lemmas 10.7 and 10.8 will establish a bound for this if we can keep l away from c_0.

Lemma 10.12. *There are positive constants γ, c_1 and c_2 such that*

$$\frac{(2l)^n}{n!} \exp\left[-c_2 \frac{n^2}{l}\right] \leqq Z_{n,l}^0 \leqq \frac{(2l)^n}{n!} \exp\left[-\gamma \frac{n^2}{l} + c_1 n\right] . \tag{10.15}$$

Proof.

$$\frac{n! Z_{n,l}^0}{(2l)^n} = \frac{1}{(2l)^n} \int_{-l}^{l} \cdots \int_{-l}^{l} \exp\left[-\sum_{i \neq j = 1}^{n} V(x_i - x_j)\right] dx_1 \ldots dx_n .$$

The lower bound follows from Jensen's inequality. On the other hand there are constants $\gamma > 0$ and c_1 such that

$$\sum_{i \neq j = 1}^{n} V(x_i - x_j) \geqq \gamma \frac{n^2}{l} - c_1 n$$

for every configuration $(x_1, \ldots, x_n)$.

Lemma 10.13. *Let $I \subset [-l, l]$ be an arbitrary interval of length 1. Let E_k denote the event that the set I contains exactly k of the points from $(x_1, \ldots, x_n)$. Then there are constants A, B, C and $\nu > 0$ such that*

$$\mu_{n,l}^0 (E_k) \leqq A \exp[-\nu k^2 + Bk\rho + c\rho^2] \tag{10.16}$$

for all n and $l > \frac{1}{2}$.

Proof. Let $i = [a, a+1]$ for some $-l \leqq a \leqq l - 1$. Let $I_1 = [-l, a]$ and $I_2 = [a+1, l]$. We let $l_1 = \dfrac{l+a}{2}$ and $l_2 = \frac{1}{2}(l - a - 1)$. Suppose that I_1, I_2 contain n_1, n_2 points respectively and $n_1 + n_2 + k = n$. If we ignore all interactions between points in different intervals and use the bound

$$\sum_{\substack{x_i, x_j \in I \\ i \neq j}} V(x_i - x_j) \geqq \nu_1 k^2 - c_1 k ,$$

we obtain

$$\mu_{n,l}^0 (E_k) \leqq \frac{1}{k!} \exp[-\nu_1 k^2 + c_1 k] \cdot \frac{1}{Z_{n,l}^0} \sum_{n_1 + n_2 = n-k} Z_{n_1,l_1}^0 Z_{n_2,l_2}^0 . \tag{10.17}$$

From Lemma 10.10,

$$\sum_{n_1+n_2=n-k} Z^0_{n_1,l_1} Z^0_{n_2,l_2} \leqq Z^0_{n-k,l_1+l_2+c_0/2}$$
$$= Z^0_{n-k,l-(1/2)+(c_0/2)} .$$

We can now apply (10.14) to get

$$Z^0_{n,l+c_0+(kl/n)} \geqq Z^0_{n-k,l-(1/2)+(c_0/2)} \cdot Z^0_{k,(1/2)+(kl/n)} .$$

(10.17) then becomes

$$\mu^0_{n,l}(E_k) \leqq \frac{1}{k!} \exp[-\nu_1 k^2 + c_1 k] \frac{Z^0_{n,l+c_0+(kl/n)}}{Z^0_{n,l}} \cdot \frac{1}{Z^0_{k,(1/2)+(kl/n)}} . \tag{10.18}$$

We can use bounds from Lemma 10.8 and the lower bound of Lemma 10.12 to complete the proof.

Lemma 10.14. *If N_I is the number of particles in the interval I then for some $\nu_1 > 0$ and $l > \frac{1}{2}$,*

$$E^{\mu^0_{n,l}}[\exp\{\nu_1 N_I^2\}] \leqq A \exp[B\rho^2] ,$$

where A and B are some constants.

Proof. We pick $\nu_1 < \nu$ and sum

$$\sum \mu^0_{n,l}(E_k) e^{\nu_1 k^2} ,$$

using the bounds in Lemma 10.13.

Lemma 10.15. *For every m, there exists a constant c_m such that*

$$E^{\mu^0_{n,l}}[N_I^m] \leqq C_m(1+\rho)^m ,$$

for all n and l.

Proof. For any random variable X there is a constant c_m independent of X such that

$$E|X|^m \leqq c_m(1+\log Ee^{X^2})^{m/2} .$$

If we now combine Lemma 10.11 with Lemmas 10.14 and 10.15 we get

Lemma 10.16. *For all $l > \frac{1}{2}$ and n,*

$$\sup_\omega \mathrm{E}^{\mu^\omega_{n,l}}[\exp[\nu N_I^2]] \leqq A \exp[B\rho^2] , \tag{10.19}$$

$$\sup_\omega E^{\mu^{\omega}_{n,l}}[[N_I]^m] \leqq C_m(1+\rho)^m . \tag{10.20}$$

We now turn to the grand canonical Gibbs distribution and estimates on them.

Lemma 10.17. *Let $\hat{\mu}_{\lambda,l,\omega}$ be the grand canonical Gibbs measure. Then for any given $\lambda_0 > 0$ for all $\lambda \leqq \lambda_0$, $l > \frac{1}{2}$ and ω we have the following bounds:*

$$E^{\hat{\mu}_{\lambda,l,\omega}}[\exp\{\nu N_I^2\}] \leqq A \exp[B(1+\lambda_0)^2] , \tag{10.21}$$

$$E^{\hat{\mu}_{\lambda,l,\omega}}[[N_I]^m] \leqq C_m(1+\lambda_0)^m . \tag{10.22}$$

Proof. According to Lemma 10.17 there exists $\alpha > 0$ such that

$$E^{\mu^{\omega_{n,l}}}[\exp[\alpha N_I^2]] \leqq A \exp[B\rho^2] \ .$$

We can replace α by $\varepsilon\alpha$ and then B can be replaced by $\varepsilon\beta$ by Hölder's inequality.

We pick ε so small that εB is less than $\frac{\nu}{2}$, where ν is as in Lemma 10.13. We will continue to denote $\varepsilon\alpha$ by α. We now have

$$E^{\mu_{n,l,\omega}}[\exp[\alpha N_I^2]] \leqq A \exp\left[\frac{\nu}{2}\rho^2\right] .$$

To estimate $E^{\hat{\mu}_{\lambda,l,\omega}}[\exp[N_I^2]$ we must bound

$$\frac{A}{\hat{Z}(\lambda, l, \omega)} \sum e^{\lambda n} Z(n, l, \omega) \exp\left[\frac{\nu}{2}\rho^2\right] = T ,$$

$$\mathrm{T}^{2l} \leqq \frac{a^{2l}}{\hat{Z}(\lambda, l, \omega)} \sum e^{\lambda n} Z(n, l, \omega) \exp\left[\frac{\nu}{2}(2l)\rho^2\right] .$$

We use the trivial bound $\hat{Z}(\lambda, l, \omega) \geqq 1$. Then

$$T^{2l} \leqq A^{2l} \sum_{n=0}^{\infty} e^{\lambda n} \frac{(2l)^n}{n!} e^{-\nu(n^2/l) + cn + (r/4)\cdot(n^2/l)} .$$

One can estimate $\frac{(2l)^n}{n!}$ by e^{2l} and obtain (10.21) by direct estimation of the sum (more or less Gaussian sum). The inequality (10.22) follows from (10.21). The fact that the summation is over nonnegative n makes the estimates uniform for λ in $(-\infty, \lambda_0]$.

Now we turn our attention to a proof of our theorems.

Proof of Theorem 10.1. The existence of the free energy $F(\lambda)$ and the specific energy or entropy functional $\Psi(\rho)$ follows from standard subadditivity arguments. Lemma 10.10 provides the required subadditivity as well as the concavity. The formula for $\hat{Z}$ in terms of Z can be used with a standard Laplace asymptotic formula to calculate the free energy in terms of the function $\Psi(\rho)$ i.e. to prove formula (10.4).

Proof of Theorem 10.2. For any real λ, we consider the family $\hat{\mu}^0_{\lambda,l}$ of grand canonical Gibbs measures as point processes on $[-l, l]$. We let $l \to \infty$. Lemma 10.17 allows us to take a subsequence and prove the existence of a Gibbs measure which may perhaps be nonstationary. A standard averaging argument over translations and another limit produces a stationary candidate.

For uniqueness let P_1, P_2 be two grand canonical Gibbs measures for the same activity λ. Let $[-a, a]$ be any arbitrary finite interval and we denote by $\| P_1 - P_2 \|_a$ the variation norm on the σ field configurations on $[-a, a]$.

We have to show that $\| P_1 - P_2 \|_a = 0$ for every a. Let us take $J_k = [x : (k-1)c_0 \leqq |x| \leqq kc_0]$ and define the conditional expectation operator,

$$(T_k F)(\omega) = \int F(\omega')\hat{\mu}_{\lambda, kc_0, \omega}(d\omega') ,$$

taking functions on the configuration space of J_k into those of J_{k+1}. Proving uniqueness can easily be reduced to proving

$$|||T_k F||| \leqq \theta\, |||F|||$$

for some $\theta < 1$ for all k and F. Here

$$|||F||| = \tfrac{1}{2} \sup_{\omega_1,\omega_2} |F(\omega_1) - F(\omega_2)|$$

$$= \inf_c \sup_\omega |F(\omega) - c| \ ,$$

where the infimum is taken over constants. This however needs only the following lemmas.

Lemma 10.18. *Let E_k be the event that the interval J_k contains no points of the configuration. Then there is a $\delta > 0$ depending only on λ such that*

$$\hat{\mu}_{\lambda,kc_0,\omega}(E_k) \geqq \delta$$

for all $k \geqq 1$ and ω.

Proof. All we need is to estimate the ratio

$$\tau = \frac{1}{\hat{Z}(\lambda, kc_0, \omega)} \left(\sum_{n=0}^{\infty} e^{\lambda n} Z(n, kc_0, \omega) \mu_{n,kc_0,\omega}(E_k) \right)$$

$$\tau \geqq \frac{1}{\hat{Z}(\lambda, kc_0, \omega)} \sum_{n=0}^{\infty} e^{\lambda n} Z^0_{n,(k-1)c_0}$$

$$\geqq \frac{\sum_{n=0}^{\infty} e^{\lambda n} Z^0_{n,(k-1)c_0}}{\sum_{n=0}^{\infty} e_{\lambda n} Z^0_{n,kc_0}} \ .$$

We use the estimate

$$\frac{Z^0_{n,(k-1)c_0}}{Z_{n,kc_0}} \geqq e^{-A\rho^2 - B}$$

from Lemma 10.8 and we write

$$\tau \geqq \frac{\sum_{n=0}^{\infty} e^{\lambda n} Z^0_{n,kc_0} e^{-A(n/2l)^2}}{\sum_{n=0}^{\infty} e^{\lambda n} Z^0_{n,kc_0}} \cdot e^{-B}$$

$$\geqq e^{-B} e^{-AR^2} \left(1 - \sum_{n \geqq 2lR} e^{\lambda n} Z^0_{n,kc_0} e^{-A(n/2l)^2} \right) .$$

We have used $\sum_{n=0}^{\infty} e^{\lambda n} Z^0_{n,kc_0} \geqq 1$.

Now we use the estimates of Lemma 10.12 to complete the proof.

Continuous dependence is an easy consequence of uniform bounds and uniqueness.

Uniqueness of the Gibbs measure yields the differentiability of $F(\lambda)$ and the continuity of $\rho(\lambda)$. The bounds on the moments of $N(I)$ are contained in Lemma 10.17. The relation between $F'(\lambda)$ and $\Psi'(\rho)$ is a consequence of the Legendre transform (10.4). To see the limits of $\rho(\lambda)$ as $\lambda \to \pm\infty$, the behavior as $\lambda \to -\infty$ will follow from the easily established relation $\lim\limits_{\lambda \to -\infty} F(\lambda) = 0$. The behavior for $\lambda \to \infty$ is obtained by proving $c_1\lambda^2 \leqq F(\lambda) \leqq c_2\lambda^2$ as $\lambda \to \infty$. These are deduced from Lemma 10.12.

Proof of Theorem 10.3. This theorem, known often as the theorem on the equivalence of ensembles, is an easy consequence of large deviation theory. We note first that the uniqueness of Gibbs measures implies that each μ_λ is ergodic, and in fact, for nice local functions $H(\omega)$,

$$\overline{\lim_{l \to \infty}} \frac{1}{2l} \log \mu_\lambda \left\{ \left| \frac{1}{2l} \int_{-l}^{l} H_x(\omega)dx - \hat{H}(\lambda) \right| \geqq \varepsilon \right\} \leqq -\eta(\varepsilon) < 0 . \quad (10.23)$$

If E_l is the event that the set $[-l-1, -l] \cup [l, l+1]$ has no points in it, then from Lemma 10.18 it follows easily that

$$\mu_\lambda[E_l] \geqq \delta > 0 \quad \text{uniformly as} \quad l \to \infty . \quad (10.24)$$

Combining (10.23) and (10.24)

$$\overline{\lim_{l \to \infty}} \frac{1}{2l} \log \hat{\mu}^0_{\lambda,l} \left\{ \left| \frac{1}{2l} \int_{-l}^{l} H_x(\omega)dx - yZH(\lambda) \right| \geqq \varepsilon \right\} \leqq -\eta(\varepsilon) < 0 . \quad (10.25)$$

One has a similar result for the number of points $N(l)$ in the interval $[-l, l]$,

$$\overline{\lim_{l \to \infty}} \frac{1}{2l} \log \mu^0_{\lambda,l} \left\{ \left| \frac{N(l)}{2l} - \rho(\lambda) \right| \geqq \varepsilon \right\} \leqq -\eta(\varepsilon) < 0 . \quad (10.26)$$

In particular

$$\underline{\lim_{l \to \infty}} \mu^0_{\lambda,l} \left\{ \left| \frac{N(l)}{2l} - \rho(\lambda) \right| \leqq \varepsilon \right\} = 1 . \quad (10.27)$$

If we look at the ratios

$$\frac{\mu^0_{\lambda,l}[N(l) = k+1]}{\mu^0_{\lambda,l}[N(l) = k]} = \frac{Z^0_{k+1,l}}{Z^0_{k,l}}$$

in the range where $\left| \frac{k}{2l} - \rho(\lambda) \right| \leqq \varepsilon$, then Lemma 10.9 provides an upper bound for the ratio and Lemmas 10.10 and 10.8 provide a lower bound for it. Since $\rho(\lambda) > 0$ we can make the bounds uniform over k in the range we are interested in. In turn this means

$$\mu^0_{\lambda,l} \left[\left| \frac{N(l)}{2l} - \rho(\lambda) \right| \leqq \varepsilon \right] \leqq e^{cl\varepsilon} \cdot 2l\varepsilon \cdot \inf_{|(k/2l) - \rho(\lambda)| \leqq \varepsilon} \mu^0_{\lambda,l}[N(l) = k] . \quad (10.28)$$

It follows now from (10.28) that

$$\lim_{\substack{l\to\infty\\ k\to\infty\\ k/2l\to\rho(\lambda)}} \frac{1}{2l}\log\mu^0_{\lambda,l}[N(l)=k]=0\ . \tag{10.29}$$

If we combine (10.23) with (10.29) we obtain the first half of Theorem 10.3. The second half is similar. We start with the estimate from large deviations.

$$\overline{\lim_{l\to\infty}}\ \frac{1}{2l}\log\hat{\mu}_\lambda\left\{\left|\sum\phi(lx_i)-\rho(\lambda)\right|\geqq\varepsilon\right\}\leqq-\eta(\varepsilon)<0 \tag{10.30}$$

if ϕ is nonnegative, is supported on $[-1,1]$, and has $\int\phi(x)dx=1$. This covers Theorem 10.3 except when $n/2l\to 0$. But this case is trivial anyway since most intervals in $[-l,l]$ would have to be empty.

Proof of Theorem 10.4. It follows from the bounds in Lemma 10.2 that for $\lambda\to+\infty$ there exist constants $0<c_1<c_2<\infty$ such that

$$c_1\lambda^2\leqq F(\lambda)\leqq c_2\lambda^2\ . \tag{10.31}$$

Since $F(\lambda)$ is convex, this implies for some $0<c_1<c_2<\infty$,

$$c_1\lambda\leqq\rho(\lambda)=F'(\lambda)\leqq c_2\lambda\ . \tag{10.32}$$

Now because $P(\rho)=F(\lambda(\rho))$, (10.31) and (10.32) imply our theorem.

Proof of Theorem 10.5. We have the identity from Lemma 10.8,

$$\begin{aligned}\frac{d\log Z^0_{m,l}}{dl}&=\frac{n}{l}+\frac{1}{l}\int\sum\psi(x_i-x_j)d\mu^0_{nl}\\ &=\frac{n}{l}+\frac{1}{l}\int\int_{-l}^{l}\sum\psi(x_i-x_j)\phi(x_j-x)dx\,d\mu^0_{nl}\\ &\quad+\text{a small error}\ .\end{aligned}$$

Clearly

$$\lim_{l\to\infty}\frac{dZ^0_{n,l}}{dl}=2\rho+2\int\Psi(x_1-x_2)\phi(x_2)R^\lambda_2(dx_1,dx_2)\ . \tag{10.33}$$

Since $Z^0_{n,l}\sim 2l\Psi(\rho)=2l\Psi\left(\frac{n}{2l}\right)$,

$$\frac{dZ^0_{n,l}}{dl}\simeq 2\Psi\left(\frac{n}{2l}\right)-2\cdot l\cdot\frac{n}{2l^2}\Psi'\left(\frac{n}{2l}\right)\ .$$

Therefore

$$\begin{aligned}\lim_{l\to\infty}\frac{dZ^0_{n,l}}{dl}&=2\Psi(\rho)-2\rho\Psi'(\rho)\\ &=2\Psi(\rho)+2\lambda\rho\\ &=2F(\lambda)\ .\end{aligned} \tag{10.34}$$

Combining (10.33) and (10.34)

$$P(\rho) = F(\lambda) = \rho + \int \psi(x_1 - x_2)\phi(x_2) R_2^\lambda(dx_1, dx_2) \ .$$

Proof of Theorem 10.6. Consider

$$\xi_{k,f} = \sum f_k(x_1, x_2) \ ,$$

where $f_k(x_1, x_2) = kg_1(k(x_1 - x_2))g_2(x_2)$,

$$\lim_{k \to \infty} \xi_{k,f} = \rho \sum g_2(x_2) \quad \text{a.e. by the ergodic theorem} \ .$$

If we take expectations with respect to μ_λ we obtain (10.9). Relation (10.10) is similar. Lemma 10.17 provides enough uniform integrability to justify interchanging limits and integration.

References

1. Guo, M.Z., Papanicolaou, G.C.: Self diffusion of interacting Brownian particles, Probabilistic Methods in Mathematical Physics, pp. 113–151. Ito, K, Ikeda N. (eds.). Tokyo: Kinokuniya 1987
2. Guo, M.Z., Papanicolaou, G.C., Varadhan, S.R.S.: Nonlinear diffusion limit for a system with nearest neighbor interaction. Commun. Math. Phys. **118**, 31–59 (1988)
3. Spohn, H.: Equilibrium fluctuations for interacting Brownian particles. Commun. Math. Phys. **103**, 1–33 (1986)
4. Olla, S., Varadhan, S.R.S.: Scaling limit for interacting Ornstein-Uhlenbeck Processes. Commun. Math. Phys. **135**, 355–378 (1991)
5. Ruelle, David: Statistical mechanics; rigorous results. New York: Benjamin 1969

Communicated by J.L. Lebowitz

Commun. Math. Phys. 135, 355–378 (1991)

Communications in Mathematical Physics

Scaling Limit for Interacting Ornstein-Uhlenbeck Processes *

Stefano Olla and S. R. S. Varadhan

Courant Institute of Mathematical Sciences, New York University, 251 Mercer Street, New York, NY 10012, USA

Received February 26, 1990; in revised form July 26, 1990

Abstract. The problem of describing the bulk behavior of an interacting system consisting of a large number of particles comes up in different contexts. See for example [1] for a recent exposition. In [4] one of the authors considered the case of interacting diffusions on a circle and proved that the density of particles evolves according to a nonlinear diffusion equation. The interacting particles evolved according to a generator that was symmetric in equilibrium. In this article we consider interacting Ornstein-Uhlenbeck processes. Here the diffusion generator is not symmetric relative to the equilibrium and the earlier methods have to be modified considerably. We use some ideas that were employed in [3] to extend the central limit theorem from the symmetric to nonsymmetric cases.

1. The Model and Its Macroscopic Equation

Let S be the circle of circumference 1. For each positive integer N we consider a system of N interacting particles with positions on S and velocities in R. The system is described by the following stochastic differential equations in phase space $(\underline{x}, \underline{v}) = \{(x_1, v_1), (x_2, v_2), \ldots, (x_N, v_N)\}$,

$$\begin{aligned} dx_i(t) &= N v_i(t)\,dt \\ dv_i(t) &= -N^2 \sum_{j \neq i} 2V'(N(x_i(t) - x_j(t)))\,dt \\ &\quad - \frac{N^2}{2} v_i(t)\,dt + N\,dw_i(t) \end{aligned} \tag{1.1}$$

for $i = 1, 2, \ldots, N$; $0 \leqq t \leqq T$. Here $\{w_i(t), i = 1, 2, \ldots, N\}$ are N independent Wiener processes and V is an even function on R with compact support describing a pair interaction.

* This research is supported in part by the National Science Foundation, grant nos. DMS 89-01682 and DMS-88-06727

We make the following assumptions on V:

(i) $V \geqq 0$, $V(0) > 0$ and V has compact support (super stability).
(ii) V is once continuously differentiable.
(iii) $\psi(z) = -zV'(z) \geqq 0$ (Repulsive Interaction).

The generator of the Markov Process $(\underline{x}(t), \underline{v}(t))$ on $(S \times R)^N$ is given by

$$L_N = S_N + J_N,$$

where

$$\begin{aligned} S_N &= \frac{N^2}{2} \sum_{i=1}^{N} \left(\frac{\partial^2}{\partial v_i^2} - v_i \frac{\partial}{\partial v_i} \right), \\ J_N &= N \sum_{i=1}^{N} \left\{ v_i \frac{\partial}{\partial x_i} - N \sum_{j \neq i} 2V'(N(x_i - x_j)) \frac{\partial}{\partial v_i} \right\}. \end{aligned} \tag{1.2}$$

The "equilibrium" measure on $(S \times R)^N$ for L_N is given by

$$d\mu_N(\underline{x}, \underline{v}) = \frac{1}{Z_N} e^{-\frac{1}{2}\sum_{i=1}^{N} v^2 - \sum_{i \neq j} V(N(x_i - x_j))} d^N\underline{x}\, d^N\underline{v}, \tag{1.3}$$

where Z_N is the normalizing constant. We note that S_N and J_N are respectively the symmetric and antisymmetric part of L_N with respect to μ_N.

We will assume that the initial distribution of the process has a density $f_N^0(\underline{x}, \underline{v})$ with respect to μ_N and it satisfies the following entropy bound:

$$H_N(f_N^0) = \int f_N^0(\underline{x}, \underline{v}) \log f_N^0(\underline{x}, \underline{v})\, d\mu_N(\underline{x}, \underline{v}) \leqq CN \tag{1.4}$$

for some suitable constant C.

The empirical distribution of the process at time t is the probability measure on S defined by

$$\alpha_N(t, d\theta) = \frac{1}{N} \sum_{i=1}^{N} \delta_{x_i(t)}(d\theta). \tag{1.5}$$

We view α_N as an element of $C[[0,T], M_1(S)]$, the space of continuous functions on $[0,T]$ with values in the space $M_1(S)$ of probability measures on S, endowed with the weak topology. We denote by Q_N the distribution of α_N on $C[[0,T], M_1(S)]$. The aim of this article is to study the limiting behavior of Q_N. The main result is that as $N \to \infty$, Q_N will concentrate around a single trajectory which is the solution of a nonlinear diffusion equation.

Theorem 1.1 (Hydrodynamic Limit). *Assume that $H_N(f_N^0) \leqq CN$ and that there exists a function $m_0(\theta)$ such that $m_0(\theta) \geqq 0$, $\int m_0(\theta)\, d\theta = 1$ and for every $J : S \to R$ which is smooth and every $\delta > 0$,*

$$\lim_{N \to \infty} \int_{A_{N,\delta,J}} f_N^0(\underline{x}, \underline{v})\, d\mu_N(\underline{x}, \underline{v}) = 0,$$

where

$$A_{N,\delta,J} = \left\{ (\underline{x}, \underline{v}) \in (S \times R)^N : \left| \frac{1}{N} \sum_{i=1}^{N} J(x_i) - \int J(\theta) m_0(\theta)\, d\theta \right| \geqq \delta \right\},$$

then $Q = \lim_{N\to\infty} Q_N$ *exists and it is concentrated on the trajectory of the form* $\mu(t, d\theta)$ *in* $C[[0,T], M_1(S)]$ *satisfying*

(i) $\mu(t, d\theta) = m(t,\theta)d\theta$ *for all* t.

(ii) $\int_0^T dt \int_S d\theta [m(t,\theta)]^3 < \infty$.

(iii) $\int_0^T dt \int_S \frac{1}{m(t,\theta)} \left[\frac{\partial}{\partial\theta} P(m(t,\theta))\right]^2 d\theta < \infty$.

(iv) $m(t,\theta)$ *is a weak solution of the equation*

$$\frac{\partial m}{\partial t} = 2\frac{\partial^2 P(m(\theta,t))}{\partial\theta^2} \tag{1.6}$$

with initial condition $m(t,\theta)$ *at* $t=0$ *given by* $m_0(\theta)$. *Furthermore the weak solution* (1.6) *satisfying* (ii) *and* (iii) *above is unique. The function* $P(\cdot)$ *is the thermodynamic function "pressure" which is defined in Sect. 2.*

This theorem will be proved in Sect. 5 using the basic results of Sects. 3 and 4. The main idea is the following.

Let $f_N^t(\underline{x}, \underline{v})$ be the solution of the forward equation

$$\frac{\partial f_N^t}{\partial t} = L_N^* f_N^t \tag{1.7}$$

with initial condition f_N^0, where L_N^* is adjoint of L_N with respect to the measure μ_N. The core of the problem is to prove certain local ergodic properties of the time average

$$\bar{f}_N(\underline{x}, \underline{v}) = \frac{1}{T}\int_0^T f_N^t(\underline{x}, \underline{v})dt. \tag{1.8}$$

These local ergodic properties will be established in Sect. 4 and are based on certain bounds on entropy and its rate of change. An elementary computation yields

$$\frac{d}{dt} H_N(f_N^t) = -\frac{N^2}{2}\int \frac{1}{f_N^t}\sum_{i=1}^N \left(\frac{\partial f_N^t}{\partial v_i}\right)^2 d\mu_N \leqq 0. \tag{1.9}$$

This implies immediately that $H_N(f_N^t) \leqq CN$ for all $t \geqq 0$. Defining for any density f relative to μ_N

$$I_N(f) = \frac{1}{2}\int \frac{1}{f}\sum_{i=1}^N \left(\frac{\partial f}{\partial v_i}\right)^2 d\mu_N,$$

we have

$$\int_0^T I_N(f_N^t)dt = -\frac{1}{N^2}(H_N(f_N^T) - H_N(f_N^0)) \leqq \frac{C}{N}.$$

Since both $I_N(f)$ and $H_N(f)$ are convex functionals of f we have

$$I_N(\bar{f}_N) \leqq \frac{C}{TN}, \tag{1.10}$$

$$H_N(\bar{f}_N) \leqq CN. \tag{1.11}$$

2. Canonical and Grand Canonical Gibbs Measures

Let $\mathscr{X}$ denote the set of all locally finite configurations of particles on the real line R. By a locally finite configuration we mean a countable subset of R with no accumulation points in R. For each value of a real parameter λ called the chemical activity the Grand Canonical Partition function in a finite interval Λ is defined as

$$Z(\lambda, \Lambda)=\sum_{n\geq 0}\frac{e^{n\lambda}}{n!}\int_{\Lambda^n}\exp\left[-\sum_{i\neq j}V(q_i-q_j)\right]dq_1\ldots dq_n. \tag{2.1}$$

The free energy as a function $F(\lambda)$ of λ is defined by

$$F(\lambda)=\lim_{\Lambda\uparrow R}\frac{1}{|\Lambda|}\log Z(\lambda,\Lambda). \tag{2.2}$$

By the assumptions made on the interaction V, the limit (2.2) exists and defines a convex function $F(\lambda)$ of λ. In our case since the dimension is 1, the function $F(\lambda)$ is continuously differentiable with a derivative

$$\varrho(\lambda)=\frac{dF(\lambda)}{d\lambda}. \tag{2.3}$$

The function $\varrho(\lambda)$ is continuous and strictly increasing with $\varrho(\lambda)\to 0$ as $\lambda\to-\infty$ and $\varrho(\lambda)\to+\infty$ as $\lambda\to+\infty$. The inverse $\lambda=\lambda(\varrho)$ exists as a continuous increasing function and

$$P(\varrho)=F(\lambda(\varrho)) \tag{2.4}$$

defines the pressure $P(\varrho)$ as a continuous, strictly increasing function of ϱ.

The canonical Gibbs measure in a finite interval Λ with particle number n and external boundary condition ω is a probability measure $\mu^{\omega}_{n,\Lambda}$ on configurations in Λ with n points or equivalently, a symmetric probability measure on Λ^n. The external boundary condition is a configuration on Λ^c. We define

$$d\mu^{\omega}_{n,\Lambda}=\frac{1}{n!}[Z^{\omega}_{n,\Lambda}]^{-1}\exp[-\sum V(q_i-q_j)-2\sum V(q_i-y_\alpha)]dq_1\ldots dq_n, \tag{2.5}$$

where $Z^{\omega}_{n,\Lambda}$ is the normalization constant,

$$Z^{\omega}_{n,\Lambda}=\frac{1}{n!}\int_{\Lambda^n}\exp[-\sum V(q_i-q_j)-2\sum V(q_i-y_\alpha)]dq_1\ldots dq_N. \tag{2.6}$$

The Grand Canonical Gibbs measure on a finite interval Λ with activity λ is a convex combination of $\mu^{\omega}_{n,\Lambda}$ with special weights. We define

$$\hat{\mu}^{\omega}_{\lambda,\Lambda}=[\hat{Z}^{\omega}_{\lambda,\Lambda}]^{-1}\sum_{n=0}^{\infty}e^{\lambda n}Z^{\omega}_{n,\Lambda}\mu^{\omega}_{n,\Lambda},$$

where

$$\hat{Z}^{\omega}_{\lambda,\Lambda}=\sum_{n=0}^{\infty}e^{\lambda n}Z^{\omega}_{n,\lambda}.$$

The Grand Canonical Gibbs measure on the infinite line with activity λ is a point process μ_λ on the configuration space $\mathscr{X}$ such that for every finite interval Λ the conditional distribution of the configuration in Λ given the configuration ω in the exterior Λ^c, is given by $\hat{\mu}^{\omega}_{\lambda,\Lambda}$ for almost all ω.

One of the important aspects of the absence of phase transition in our one dimensional system is the following theorem. It is stated here without proof. A sketch of the proof can be found in the appendix of [4].

Theorem 2.1. *For each λ, there is exactly one Grand Canonical Gibbs measure μ_λ on $\mathscr{X}$ corresponding to the activity λ. It is a stationary point process with density $\varrho=\varrho(\lambda)$ given by* (2.3). *If $\mu^{\omega}_{n,\Lambda}$ are canonical Gibbs measures such that $\Lambda\uparrow R$ and $\frac{n}{|\Lambda|}\to\varrho$, then for any continuous bounded local function $H(\omega')$ on the configuration space,*

$$\lim_{\substack{n/|\Lambda|\to\varrho\\ \Lambda\uparrow R}}\ \sup_{\omega}\mu^{\omega}_{n,\Lambda}\left[\left|\frac{1}{|\Lambda|}\int_{\Lambda}H(\omega'_x)dx-\hat{H}(\varrho)\right|\geqq\delta\right]=0. \tag{2.7}$$

Here $\hat{H}(\varrho)=E^{\mu_\lambda}[H(\omega')]$ with λ chosen so that $\varrho(\lambda)=\varrho$. The configuration ω' translated in space by x is denoted by ω'_x.

The correlation measures $R^{(k)}_\lambda(dz_1,\dots,dz_k)$ are defined so that

$$E^{\mu_\lambda}\left[\sum_{(z_1,\dots,z_k)}f(z_1,\dots,z_k)\right]=\int_{R^k}f(z_1,\dots,z_k)R^{(k)}_\lambda(dz_1,\dots,dz_k).$$

On the left-hand side the summation is over all k-tuples $(z_1,\dots,z_k)$ that belong to the configuration. $f(z_1,\dots,z_k)$ is assumed to be a bounded continuous function with compact support. $R^{(k)}_\lambda(dz_1,\dots,dz_k)$ is a σ-finite measure on R^k invariant with respect to diagonal translations $z_1,\dots,z_k\to z_1+a,\dots,z_k+a$. In particular

$$R^{(1)}_\lambda(dz)=\varrho(\lambda)dz.$$

We have the identity proved in the appendix of [4].

Theorem 2.2. *For every λ and ϱ related by $\varrho=\varrho(\lambda)$,*

$$\varrho+\iint\varphi(x-y)h(y)R^{(2)}_\lambda(dx,dy)=P(\varrho),$$

where $h(y)$ is any function which has compact support with $\int_{-\infty}^{\infty}h(y)dy=1$.

3. Some Estimates Based on Entropy and the Dirichlet Form

In this section our aim is to obtain some elementary bounds based on the inequalities

$$H_N(f_N)\leqq AN \tag{3.1}$$

and

$$I_N(f_N)\leqq\frac{C}{N} \tag{3.2}$$

for an arbitrary density $f_N(\underline{x},\underline{v})$ relative to μ_N.

Theorem 3.1. *Let f_N satisfy* (3.1). *Then*

(i)
$$\int|\underline{v}|^2f_Nd\mu_N\leqq N(\log\sqrt{2}+A). \tag{3.3}$$

(ii) *For any function ϕ on $C(R)$ with compact support there exists a constant C' such that for all N and $\lambda \leqq 1$,*

$$\int \lambda \sum_{i,j} \phi(\lambda N(x_i - x_j)) f_N d\mu_N \leqq C'N. \tag{3.4}$$

Proof.

$$\text{(i)}\ \tfrac{1}{4}\int \left[\sum_{i=1}^{N} v_i^2\right] f_N d\mu_N \leqq \log \int \exp[\tfrac{1}{4}\sum v_i^2] d\mu_N + H_N(f_N)$$

$$= N(\log\sqrt{2} + A).$$

(ii) The proof can be found in Lemma 4.2 of [4].

In the following we will denote the Gaussian product distribution on R^N by

$$G_N(d\underline{v}) = \frac{1}{(2\pi)^{N/2}} e^{-(1/2)|\underline{v}|^2} d^N\underline{v}, \tag{3.5}$$

and the equilibrium distribution on S^N by

$$d\mu_N^*(\underline{x}) = \frac{1}{Z_N^*} \exp\left(-\sum_{i \neq j} V(N(x_i - x_j))\right) d^N\underline{x} \tag{3.6}$$

so that $d\mu_N(\underline{x}, \underline{v}) = dG_N(\underline{v}) \cdot \mu_N^*(\underline{x})$. Given a probability density $f_N(x, v)$ relative to μ_N, we denote by $f_N(\underline{v}|\underline{x})$ the density of the conditional distribution of $\underline{v}$ given $\underline{x}$, relative to $G_N(d\underline{v})$. By the logarithmic Sobolev inequality (see [1] for a proof) for any configuration $\underline{x}$ we have

$$\int f_N(\underline{v}|\underline{x}) \log f_N(\underline{v}|\underline{x}) G_N(d\underline{v}) \leqq 2\int \frac{1}{f_N(\underline{v}|\underline{x})} |\nabla_v f_N(\underline{v}|\underline{x})|^2 G_N(d\underline{v}).$$

If we denote the marginal of $f_N(\underline{x}, \underline{v})$ by

$$f_N^*(\underline{x}) = \int f_N(\underline{x}, \underline{v}) G_N(d\underline{v}),$$

then by a simple calculation

$$\int_{S^N} f_N^*(\underline{x}) d\mu_N^*(\underline{x}) \int_{R^N} f_N(\underline{v}|\underline{x}) \log f_N(\underline{v}|\underline{x}) G_N(d\underline{v})$$

$$\leqq 2\int \frac{1}{f_N(\underline{x}, \underline{v})} |\nabla_v f_N(\underline{x}, \underline{v})|^2 d\mu_N(\underline{x}, \underline{v}) = 4I_N(f_N). \tag{3.7}$$

We have therefore proved

Lemma 3.2. *Let f_N be such that $I_N(f_N) \leqq C/N$. Then the conditional density $f_N(\underline{v}|\underline{x})$ satisfies*

$$\int f_N^*(\underline{x}) d\mu_N^*(\underline{x}) \int G_N(d\underline{v}) f_N(\underline{v}|\underline{x}) \log f_N(\underline{v}|\underline{x}) \leqq \frac{4C}{N}. \tag{3.8}$$

4. Local Equilibrium Distributions

We would like to show that, relative to the probability distribution $\bar{f}_N d\mu_N$, "average microscopic quantities" like $\frac{1}{N}\sum \psi(N(x_i - x_j)) J(x_i)$ are close, for large N, to the expression involving only the macroscopic density function.

The approach we use here is similar to the one used in [4], but the information we have on $\bar{f}_N$ here are different. In [4] two basic estimates available on $\bar{f}_N$ are used,

$$\int \bar{f}_N \log \bar{f}_N d\mu_N \leqq BN \tag{4.1}$$

and

$$\frac{1}{2}\int \frac{1}{\bar{f}_N} \sum_{i=1}^{N} \left(\frac{\partial \bar{f}_N}{\partial x_i}\right)^2 d\mu_N \leqq DN. \tag{4.2}$$

In our case we do not have (4.2) but instead have

$$\frac{1}{2}\int \frac{1}{\bar{f}_N} \sum_{i=1}^{N} \left(\frac{\partial \bar{f}_N}{\partial v_i}\right)^2 d\mu_N \leqq \frac{D}{N}. \tag{4.3}$$

As we saw in the previous section, by the logarithmic Sobolev inequality this will tell us that the distribution of the velocities conditioned on the positions are all very close to the equilibrium. But this says nothing about the distribution of positions. To infer from this that the local position distributions are close to equilibrium distributions and that the macroscopic density has no unnecessary fluctuations, we have to actually use information that $\bar{f}_N$ are time averages of the solution to the forward equation.

Consider a function $\Phi(\omega)$ on the configuration of points in R which is bounded, continuous and localized in some finite interval $[-l, l]$. For any given $x \in S$ and any $\underline{x} \in S^N$ we can consider the configuration ω_N^x on the line

$$\omega_N^x = \{N(x_i - x) : |x_i - x| < \tfrac{1}{4}\}.$$

If N is sufficiently large $\Phi(\omega_N^x)$ makes sense and we are interested in the quantity

$$\xi_N(\underline{x}) = \int_S \Phi(\omega_N^x) J(x) dx,$$

let $h(x)$ be a nonnegative smooth function with $\int h(x)dx = 1$, supported on $[-\frac{1}{2}, \frac{1}{2}]$. For $\lambda > 1$, we define

$$\begin{aligned} \varrho_\lambda(x) &= \frac{\lambda}{N} \sum_{i=1}^{N} h(\lambda(x_i - x)) \\ &= (h_\lambda * \alpha_N)(x), \end{aligned}$$

where $h_\lambda(x) = \lambda h(\lambda x)$ and $\alpha_N(dx) = \frac{1}{N}[\delta_{x_1} + \ldots \delta_{x_N}]$ is the empirical distribution of the configuration $(x_1, \ldots, x_N)$ on S. We denote by $\hat{\Phi}(\varrho)$ the expected value

$$E^{\mu_\varrho}[\Phi(\omega)] = \hat{\Phi}(\varrho),$$

where μ_ϱ is the Gibbs measure with density ϱ or activity $\lambda = \lambda(\varrho)$. If we define

$$\eta_{N,\lambda}(\underline{x}) = \int_S \hat{\Phi}(\varrho_\lambda(x)) J(x) dx,$$

then we want to prove

Theorem 1.

$$\lim_{\lambda \to \infty} \limsup_{N \to \infty} E^{\bar{f}_N} |\xi_N(\underline{x}) - \eta_{N,\lambda}(\underline{x})| = 0.$$

As in [4] we establish Theorem 1.1 in two steps.

Theorem 4.2.

$$\lim_{\varepsilon\to 0}\limsup_{N\to\infty} E^{\bar f_N}|\xi_N(\underline{x})-\eta_{N,N_\varepsilon}(x)|=0.$$

Theorem 4.3.

$$\lim_{\substack{\varepsilon\to 0\\ \lambda\to\infty}}\limsup_{N\to\infty} E^{\bar f_N}\int_S|\varrho_\lambda(x)-\varrho_{\varepsilon N}(x)|\,dx=0.$$

Let $\hat f_N(\underline{x},\underline{v})=\int_S \bar f_N(\underline{x}+a,v)\,da$. Then by convexity we have that $\hat f_N$ satisfies

$$H_N(\hat f_N)\leqq AN,$$
$$I_N(\hat f_N)\leqq \frac{C}{N}.$$

In exactly the same way as in [4] the proof of Theorem 4.2 can be reduced to

Theorem 4.4.

$$\lim_{\varepsilon\to 0}\limsup_{N\to\infty} E^{\hat f_N}\left|\frac{N\varepsilon}{2}\int_{|x|\leqq 1/N\varepsilon}\Phi(\omega_N^x)\,dx-\hat\Phi(\varrho_{N\varepsilon}(0))\right|=0.$$

Proof of Theorem 4.4. Both $\int_{|y|\leqq 1/N\varepsilon}\Phi(\omega_y^N)\,dy$ and $\varrho_{N\varepsilon}(0)$ depend only on the configuration $\underline{x}$ in the interval $\left[-\frac{1}{\varepsilon N}-\frac{l}{N},\frac{1}{\varepsilon N}+\frac{l}{N}\right]$ centered around the origin of S. Here the function Φ is assumed to depend only on the configuration in $[-l,l]$. If we project $\hat f_N d\mu_N$ onto configurations on this interval and expand the interval by a factor N, we will get a point process $\hat v_{N,\varepsilon}$ on the interval $\left[-\frac{1}{\varepsilon}-l,\frac{1}{\varepsilon}+l\right]$. All we have to prove is that

$$\lim_{\varepsilon\to 0}\limsup_{N\to\infty} E^{\hat v_{N,\varepsilon}}\left|\frac{\varepsilon}{2}\int_{-1/\varepsilon}^{1/\varepsilon}\Phi(\omega^x)\,dx-\hat\Phi(\varrho_\varepsilon(0))\right|=0,$$

where ω^x is the configuration ω translated by x in space. Since the density of particles is 1 under $\hat f_N d\mu_N$, $\hat v_{N,\varepsilon}$ is compact and we denote by $\hat{\mathscr{B}}_\varepsilon$ the set of all limit points of $\hat v_{N,\varepsilon}$ as $N\to\infty$. We have to prove that

$$\lim_{\varepsilon\to 0}\sup_{v\in\hat{\mathscr{B}}_\varepsilon} E^v\left[\left|\frac{\varepsilon}{2}\int_{-1/\varepsilon}^{1/\varepsilon}\Phi(\omega^x)\,dx-\hat\Phi(\varrho_\varepsilon(0))\right|\right]=0.$$

Let $\mu_{n,l}^\omega$ be the Gibbs measure on $[-l,l]$ with exterior boundary condition ω and n particles inside $[-l,l]$. We denote by Γ_l the convex hull of these measures, and by $\Gamma_l^{(1)}$ those for which expected particle density in $[-l,l]$ is at most 1. Then according to Theorem 2.1,

$$\lim_{k\to\infty}\sup_{v\in\Gamma_{l+k}^{(1)}} E^v\left[\left|\frac{1}{2k}\int_{-k}^{k}\Phi(\omega^x)\,dx-\hat\Phi(\varrho_{1/k}(0))\right|\right]=0,$$

and we therefore need only establish the lemma

Lemma 4.5. *For every l and ε*

$$\hat{\mathscr{B}}_\varepsilon\subset\Gamma_{(1/\varepsilon)+l}^{(1)}.$$

Proof. Let c_0 be the range of the interaction V, i.e. $V(x)=0$ if $|x|\geqq c_0$. Let us consider the limiting point process on the interval $\left[-\frac{1}{\varepsilon}-l-c_0, \frac{1}{\varepsilon}+l+c_0\right]$, i.e. any distribution $\nu\in\hat{\mathscr{B}}_{\varepsilon'}$, where $1/\varepsilon'+l=1/\varepsilon+l+c_0$. We want to show that $\nu\in\Gamma^{(1)}_{l+(1/\varepsilon)}$.

Consider the function

$$u_N(\underset{\sim}{x},\underset{\sim}{v})=\sum_{i=1}^{N} v_i\Phi(\omega_N^0)g(Nx_i),$$

where Φ is any function of the configuration of points in $[-l-c_0, l+c_0]$ which is bounded continuous and differentiable, and g is any positive continuous function with compact support contained in $[-l,l]$. By direct computation we have

$$\sum_{i,j}2V'(N(x_i-x_j))\Phi(\omega_N^0)g(N(x_i))-\sum_{i,j}\frac{1}{N}v_iv_j\frac{\partial}{\partial x_i}(\Phi(\omega_N^0)g(N(x_j)))$$
$$=u_N(\underset{\sim}{x},\underset{\sim}{v})-\frac{1}{N^2}(L_Nu_N)(\underset{\sim}{x},\underset{\sim}{v}).$$

On integrating by parts:

$$\begin{aligned}|\textstyle\int u_N(\underset{\sim}{x},\underset{\sim}{v})\hat{f}_N(\underset{\sim}{x},\underset{\sim}{v})d\mu_N|&=\left|\int_S da\int u_N(\underset{\sim}{x}+a,\underset{\sim}{v})\bar{f}_N(\underset{\sim}{x},\underset{\sim}{v})d\mu_N\right|\\ &=\left|\int_S da\int\Phi(\omega_N^a)\sum_{i=1}^{N}g(N(x_i-a))\frac{\partial\bar{f}_N}{\partial v_i}d\mu_N\right|\\ &\leqq\|\Phi\|_\infty\left(\int\sum_i g(Nx_i)\hat{f}_Nd\mu_N\right)^{1/2}\\ &\quad\times\left(\int_S da\int\frac{1}{\bar{f}_N}\sum\left(\frac{\partial\bar{f}_N}{\partial v_i}\right)^2g(N(x_i-a))d\mu_N\right)^{1/2}\\ &\leqq\|\Phi\|_\infty\left(\frac{C}{N}\cdot\int\frac{1}{\bar{f}_N}\sum\left(\frac{\partial\bar{f}_N}{\partial v_i}\right)^2d\mu_N\right)^{1/2}\leqq\frac{C'}{N}.\end{aligned}$$

If g_N is any probability density with respect to μ_N and $\hat{g}_N$ is its average

$$\int_S g_N(\underset{\sim}{x}+a,\underset{\sim}{v})da=\hat{g}_N(\underset{\sim}{x},\underset{\sim}{v}),$$

$$\begin{aligned}\textstyle\int u_N(\underset{\sim}{x},\underset{\sim}{v})\hat{g}_Nd\mu_N&=\int_S\int u_N(\underset{\sim}{x}+a,\underset{\sim}{v})g_N(\underset{\sim}{x},\underset{\sim}{v})d\mu_Nda\\ &\leqq\|\Phi\|_\infty\left(\int\frac{1}{N}\sum|v_i|g_N(\underset{\sim}{x},\underset{\sim}{v})d\mu_N\right)\\ &\leqq\|\Phi\|_\infty\left(\int\frac{1}{N}\sum|v_i|^2g_N(\underset{\sim}{x},\underset{\sim}{v})d\mu_N\right)^{1/2}\\ &\leqq\|\Phi\|_\infty\left[\frac{4}{N}\log\int\exp\left[\frac{1}{4}\sum v_i^2\right]d\mu_N+\frac{4}{N}H(g_N)\right]^{1/2}\\ &\leqq C'\|\Phi\|_\infty,\end{aligned}$$

provided $H(g_N) \leqq CN$. If we now use the explicit formula $\hat{f}_N = \frac{1}{T}\int_0^T \hat{f}_N^t dt$ and calculate

$$\int L_N u_N \hat{f}_N d\mu_N = \frac{1}{T}\int_0^T \int u_N \frac{\partial \hat{f}_N^t}{\partial t} dt d\mu_N$$
$$= \frac{1}{T}\left[\int u_N \hat{f}_N^T d\mu_N - \int u_N \hat{f}_N^0 d\mu_N\right],$$

then

$$\left|\int L_N u_N \hat{f}_N d\mu_N\right| \leqq 2C' \|\Phi\|_\infty$$

because $H(\hat{f}_N^T) \leqq H(\hat{f}_N^0) \leqq CN$.

We have therefore established

$$\lim_{N\to\infty} E^{\hat{f}_N}\left[\sum_{i,j} 2V'(N(x_i - x_j))\Phi(\omega_N^0)g(Nx_i) - \frac{1}{N}\sum_{i,j} v_i v_j \frac{\partial}{\partial x_i}(\Phi(\omega_N^0)g(Nx_j))\right] = 0. \tag{4.4}$$

Let us define the following $N \times N$ matrix

$$F_{ij}(\underline{x}) = \frac{1}{N}\frac{\partial}{\partial x_i}(\Phi(\omega_N^0)g(Nx_j)).$$

Because of the scaling $\{F_{ij}(\underline{x})\}$ are uniformly bounded. Then by the entropy estimates of Sect. 3,

$$\int \frac{1}{N}\sum_{i=1}^N (v_i^2 - 1)F_{ii}(\underline{x})\hat{f}_N(\underline{x}, \underline{v})d\mu_N$$
$$\leqq \int \hat{f}_N^*(\underline{x})d\mu_N^*(\underline{x})\left\{\log\int \exp\left[\frac{1}{N}\sum(v_i^2 - 1)F_{ii}(\underline{x})\right]G_N(d\underline{v}) + \int \hat{f}_N(\underline{v}|\underline{x})\log \hat{f}_N(\underline{v}|\underline{x})G_N(d\underline{v})\right\}$$
$$\leqq \int \hat{f}_N^*(\underline{x})d\mu_N^*(\underline{x})\left\{\log\int \exp\left[\frac{1}{N}\sum(v_i^2 - 1)F_{ii}(\underline{x})\right]G_N(d\underline{v})\right\} + \frac{2C}{TN}.$$

By an elementary calculation, essentially because of the law of large numbers we can conclude that the last term goes to zero with N. Therefore

$$\limsup_{N\to\infty} \frac{1}{N}\sum[v_i^2 - 1]F_{ii}(\underline{x})\hat{f}_N(\underline{x}, \underline{v})d\mu_N \leqq 0.$$

Since the same argument works with change of sign we actually have

$$\lim_{N\to\infty}\left|\frac{1}{N}\int \sum(v_i^2 - 1)F_{ii}(\underline{x})\hat{f}_N(\underline{x}, \underline{v})d\mu_N\right| = 0. \tag{4.5}$$

Using an exactly similar argument on the off diagonal elements

$$\frac{1}{N}\int\sum_{i\neq j} v_i v_j F_{ij}(\underline{x})\hat{f}_N(\underline{x},\underline{v})d\mu_N$$

$$\leq \int \hat{f}_N^*(\underline{x})d\mu_N^*\left\{\frac{1}{\sigma}\log\int\exp\left[\frac{\sigma}{N}\sum_{i\neq j} v_i v_j F_{ij}(\underline{x})\right]G_N(d\underline{v})+\frac{2C}{\sigma NT}\right\}.$$

We can estimate by direct calculation

$$\int\exp\left[\frac{\sigma}{N}\sum_{i\neq j} v_i v_j F_{ij}\right]G_N(d\underline{v})=\int\exp\left[\frac{\sigma}{N}\sum_{i\neq j} v_i v_j \bar{F}_{ij}\right]G_N(d\underline{v})$$

$$=\exp\left[\frac{1}{2}\log\det\left|I-\frac{2\sigma}{N}\bar{F}\right|\right],$$

where $\bar{F}$ is the matrix $\bar{F}_{ij}$ for $i\neq j$ and $\bar{F}_{ii}=0$ with $\bar{F}_{ij}=\frac{1}{2}(F_{ij}+F_{ji})$. Since $\bar{F}_{ij}$ is bounded it is easy to obtain the estimate (note $\operatorname{tr}\bar{F}=0$)

$$\log\det\left|I-\frac{2\sigma}{N}\bar{F}\right|\leq C\frac{\sigma^2}{N^2}\operatorname{Tr}(\bar{F})^2\leq C\cdot\sigma^2,$$

provided σ is suitably small. We therefore obtain

$$\limsup_{N\to\infty}\frac{1}{N}\int\sum_{i\neq j} v_i v_j F_{ij}(\underline{x})\hat{f}_N(\underline{x},\underline{v})d\mu_N$$

$$\leq C\sigma \quad \text{for all small } \sigma>0$$

$$=0.$$

Again since the argument is insensitive to sign we have

$$\limsup_{N\to\infty}\left|\frac{1}{N}\int\sum_{i\neq j} v_i v_j F_{ij}(\underline{x})\hat{f}_N(\underline{x},\underline{v})d\mu_N\right|=0. \tag{4.6}$$

Combining (4.4) with (4.5) and (4.6) we get

$$\lim_{N\to\infty}E^{\hat{f}_N}\left[\sum_{i,j}2V'(N(x_i-x_j))\Phi(\omega_N^0)g(Nx_i)\right.$$
$$\left.-\frac{1}{N}\sum\frac{\partial}{\partial x_j}(\Phi(\omega_N^0)g(Nx_j))\right]=0. \tag{4.7}$$

Since the function $\Phi(\omega)$ can be used just as well to cut off configurations having too many particles, we get from (4.7)

$$E^{\nu}\left[\sum_{i,j}2V'(y_i-y_j)\Phi(\omega^0)g(y_i)-\sum\frac{\partial}{\partial y_i}(\Phi(\omega^0)g(y_i))\right]=0 \tag{4.8}$$

for any ν which is a limit point. But (4.8) is enough to guarantee that $\nu\in\Gamma^{(1)}_{1+1/\varepsilon}$.

To see this we choose the function $\Phi(\omega)$ to be of the form $\psi_1(\omega_{\Lambda_l^c})\psi_2(\omega_{\Lambda_l})$, where ψ_1 and ψ_2 are functions of the configurations in the interior of $\Lambda_l=[-l,l]$ and its exterior Λ_l^c. We can choose $\psi_2(\omega_{\Lambda_l})$ to be of the form

$$\psi_2(\omega_{\Lambda_l})=\phi(y_1,\ldots,y_n) \quad \text{if } |\omega_{\Lambda_l}|=n$$
$$=0 \quad \text{otherwise},$$

where $|\omega_{A_l}|$ is the cardinality of the configuration inside $[-l,l]$ and $y_1,\dots,y_n$ are the locations of the n particles inside $[-l,l]$. Here $\phi(y_1,\dots,y_n)$ is a smooth symmetric function of $y_1,\dots,y_n$. Although such a $\psi_2(\omega_{A_l})$ is not smooth on configuration space it is easily approximated by smooth ones. There is no difficulty because the probability of finding a particle on the boundary is zero under v. If we denote by $v=v_{A_l}(dy_1,\dots,dy_n|\omega_{A_l^c},n)$ the conditional distribution of the configuration on $[-l,l]$ given the configuration on $\omega^c_{A_l}$ and the cardinality $|\omega_A|$ of ω_A, then

$$\int\left\{\sum_{i=1}^{n}\sum_{y'_j\in\omega^c_{A_l}} 2V'(y_i-y'_j)\phi(y_1,\dots,y_n)g(y_i) - \sum_{i=1}^{n}\frac{\partial}{\partial y_i}(\phi(y_1,\dots,y_n)g(y_i))\right\}dv'=0.$$

This identifies v' as the canonical Gibbs measure.

Proof of Theorem 4.3. For any configuration $\underline{x}\in S^N$ we have associated the empirical measure

$$\alpha_N(dx)=\frac{1}{N}\sum\delta_{x_i}(dx)$$

and smoothed version

$$\varrho_\lambda(x)=\frac{\lambda}{N}\sum_{i=1}^{N}h(\lambda(x-x_i)).$$

We have to prove that $\varrho_\lambda(x)$ for λ large but fixed and $\varrho_{\varepsilon N}(x)$ with ε small but fixed are close. This amounts to proving that $\varrho_{\varepsilon N}(x)$ does not have any oscillations. Just as in [4] we calculate the Young measures associated with $\varrho_{\varepsilon N}(\cdot)$ and show that as $N\to\infty$ and $\varepsilon\to 0$, these converge to a degenerate measure for almost every configuration. Corresponding to any density $\varrho(x)$ we define a measure π on $S\times R^+$ by

$$\int F(x,\varrho)\pi(dx,d\varrho)=\int_S F(x,\varrho(x))dx.$$

Clearly

$$\int\varrho\pi(dx,d\varrho)=\int\varrho(x)dx=1,$$
$$\int F(x)\pi(dx,d\varrho)=\int F(x)dx.$$

If we denote by $\mathcal{M}$ the space of probability measures on $S\times R^+$, then through the map $\varrho(\cdot)\to\pi$ we can map the random density

$$\varrho_{\varepsilon N}(x)=(h_{\varepsilon N}*\alpha_N)(x)$$

into $\mathcal{M}$. If we consider the pair α_N and π we get a map of S^N into $M_1(S)\times\mathcal{M}$. We start with our basic $\hat{f}_N d\mu_N$ on S^N and denote by $\hat{Q}_{N,\varepsilon}$ the induced distribution on $M_1(S)\times\mathcal{M}$. In order to establish Theorem 4.3, the main step, after establishing compactness of $\hat{Q}_{N,\varepsilon}$ as $N\to\infty$ and then as $\varepsilon\to 0$, is to show

Theorem 4.6. *Let $\hat{Q}$ be any limit point of $\hat{Q}_{N,\varepsilon}$ as $N\to\infty$ and then as $\varepsilon\to 0$. Then for almost all (α,π) with respect to $\hat{Q}$*

$$\alpha(dx)=\varrho(x)dx,$$
$$\pi(dx,d\varrho)=dx\delta_{\varrho(x)}(d\varrho)$$

for some function $\varrho(x)\in L_1(S)$.

We will prove Theorem 4.6 by means of the following lemmas.

Lemma 4.7. *Let $\hat{Q}$ be the limit measure as above. Then*

$$\hat{Q}[\alpha:\alpha(dx)=\varrho(x)dx \text{ for some } \varrho\in L_1(S)]=1\,,$$
$$E^{\hat{Q}}\left[\int_S \varrho^2(x)dx\right]\leqq C\cdot B\,,$$

where C is a universal constant and B is the bound on entropy.

Proof. Identical to Lemma 7.7 in [4].

Lemma 4.8. *Let $\hat{Q}_\varepsilon$ be any limit measure of $\hat{Q}_{N,\varepsilon}$ as $N\to\infty$. Then*

(i) $$\hat{Q}_\varepsilon[(\alpha,\pi):\int\varrho\pi(dx,d\varrho)=1]=1\,,$$

(ii) $$\hat{Q}_\varepsilon[(\alpha,\pi):\pi(dx,d\varrho)=dx\pi_x(d\varrho)]=1\,,$$

(iii) $$\hat{Q}_\varepsilon[(\alpha,\pi):\alpha(dx)=\varrho(x)dx\,,\quad \text{with}\quad \varrho(x)=\int\varrho\pi_x(d\varrho)]=1\,.$$

Proof. Since this lemma is based only on entropy bounds, like Lemma 4.7 this proof is identical to Lemma 7.8 of [4].

It remains to prove that as $\varepsilon\to 0$ the limit points of $\hat{Q}_\varepsilon$ are supported by degenerate Young measures. Consider a function g which is nonnegative, smooth, symmetric with compact support in $[-\frac{1}{4},\frac{1}{4}]$ and having $\int g(x)dx=1$. For every N, δ, and λ we define

$$G_{\delta,N,\lambda}(x)=\int_{-\infty}^{x} g_{\delta,N,\lambda}(y)dy\,,$$

where

$$g_{\delta,N,\lambda}(x)=N\delta g(N\delta x)-\lambda g(\lambda x)\,.$$

We note that $G_{\delta,N,\lambda}$ has small compact support on R and is therefore well defined as a function on S provided $N\delta\geqq 1$ and $\lambda\geqq 1$. We consider the following test function on $S^N\times R^N$:

$$W_{\delta,N,\lambda}(\underline{x},\underline{v})=\frac{1}{N}\sum_{i=1}^{N} v_i\sum_{j\neq i} G_{\delta,N,\lambda}(x_i-x_j)\,.$$

We then have by calculation

$$\begin{aligned}&\frac{1}{N^2}\sum_{i,j}\Big\{2\sum_{k\neq i} NV'(N(x_i-x_k))\sum_{j\neq i} G_{\delta,N,\lambda}(x_i-x_j)\\&\quad-\sum_l v_i v_l\frac{\partial}{\partial x_l}\Big(\sum_{j\neq i} G_{\delta,N,\lambda}(x_i-x_j)\Big)\Big\}\\&=-W_{\delta,N,\lambda}(\underline{x},\underline{v})+\frac{1}{N^2}L_N W_{\delta,N,\lambda}\,.\end{aligned}\tag{4.9}$$

We integrate by parts and use Schwartz' inequality to obtain

$$
\begin{aligned}
|\int W_{\delta,N,\lambda}(\underline{x},\underline{v})\hat{f}_N(x,v)d\mu_N| &= \left|\frac{1}{N}\int\sum v_i\sum_{j\neq i}G_{\delta,N,\lambda}(x_i-x_j)\hat{f}_N(\underline{x},\underline{v})d\mu_N\right| \\
&= \left|\int\frac{1}{N}\sum_{i=1}^{N}\frac{\partial\hat{f}_N}{\partial v_i}\sum_{j\neq i}G_{\delta,N,\lambda}(x_i-x_j)d\mu_N\right| \\
&\leqq \left(\int\frac{1}{N}\sum_{i=1}^{N}\Big(\sum_{j\neq i}G_{\delta,N,\lambda}(x_i-x_j)\Big)^2\hat{f}_N d\mu_N\right)^{1/2} \\
&\quad\times\left(\int\frac{1}{N}\sum_{i=1}^{N}\frac{1}{\hat{f}_N}\left(\frac{\partial\hat{f}_N}{\partial v_i}\right)^2 d\mu_N\right)^{1/2} \\
&\leqq \sqrt{\frac{C}{T}}\left(\int\frac{1}{N^3}\sum_{i=1}^{N}\Big(\sum_{j\neq i}G_{\delta,N,\lambda}(x_i-x_j)\Big)^2\hat{f}_N d\mu_N\right)^{1/2} \\
&\leqq \sqrt{\frac{C}{T}}\left(\int\frac{1}{N^2}\sum_{i,j}G^2_{\delta,N,\lambda}(x_i-x_j)\hat{f}_N d\mu_N\right)^{1/2}.
\end{aligned}
\tag{4.10}
$$

From the definition of $\hat{f}_N$,

$$
\frac{1}{N^2}\int(L_N W_{\delta,N,\lambda})\hat{f}_N d\mu_N=\frac{1}{N^2T}\left(\int W_{\delta,N,\lambda}\hat{f}^T_N d\mu_N-\int W_{\delta,N,\lambda}f^0_N d\mu_N\right),
$$

since $H(\hat{f}^T_N)\leqq H(\hat{f}^0_N)\leqq CN$, by Lemma 3.1 and Schwartz's inequality,

$$
\begin{aligned}
&|\int W_{\delta,N,\lambda}\hat{f}^t_N d\mu_N| \\
&\leqq\left(\int\frac{1}{N}\sum_{i=1}^{N}v_i^2\hat{f}^t_N d\mu\right)^{1/2}\left(\int\frac{1}{N}\sum_{i=1}^{N}\Big(\sum_{j\neq i}G_{\delta,N,\lambda}(x_i-x_j)\Big)^2\hat{f}^t_N d\mu_N\right)^{1/2} \\
&\leqq C_1\left(\int\frac{1}{N}\cdot N\cdot N^2\cdot C_2^2\cdot f^t_N d\mu_N\right)^{1/2} \\
&\leqq C_3 N.
\end{aligned}
\tag{4.11}
$$

We have used here the fact that $G_{\delta,N,\lambda}$ is bounded by some C_2. Using (4.9), (4.10), and (4.11) it follows that

$$
\begin{aligned}
\lim_{\lambda\to\infty}\lim_{N\to\infty}\int\frac{1}{N^2}\sum_{i=1}^{N}\Bigg\{&2\sum_{k\neq i}V'(N(x_i-x_k))\Big(\sum_{j\neq i}G_{\delta,N,\lambda}(x_i-x_j)\Big) \\
&-\sum_{l=1}^{N}v_iv_l\frac{\partial}{\partial x_l}\Big(\sum_{j\neq i}G_{\delta,N,\lambda}(x_i-x_j)\Big)\Bigg\}\hat{f}_N(\underline{x},\underline{v})d\mu_N=0.
\end{aligned}
\tag{4.12}
$$

Let $A^{ij}_N(\underline{x})$ be the symmetric matrix

$$
\begin{aligned}
[A_N(\underline{x})]_{ij}&=\frac{1}{N}g_{\delta,N,\lambda}(x_i-x_j) \quad \text{for} \quad i\neq j \\
&=0 \qquad\qquad\qquad\quad\ \text{for} \quad i=j.
\end{aligned}
$$

We can then use an argument similar to one used in Theorem 4.1 [Eqs. (4.7) and (4.8)] to get

$$\int \frac{1}{N^2} \sum_{i \neq j} v_i v_j g_{\delta, N, \lambda}(x_i - x_j) \hat{f}_N d\mu_N$$
$$\leq \frac{1}{\sigma} \int \hat{f}_N^*(\underline{x}) d\mu_N^* \left\{ \frac{1}{2} \log \det \left[I - \frac{2\sigma}{N} A_N(\underline{x}) \right] \right\} + \frac{C}{2TN\sigma}.$$

Just as before we conclude that

$$\lim_{N \to \infty} \left| \int \frac{1}{N^2} \sum_{i \neq j} v_i v_j g_{\delta, N, \lambda}(x_i - x_j) \hat{f}_N d\mu_N \right| = 0. \tag{4.13}$$

We treat the remaining term in the same manner:

$$\int \frac{1}{N^2} \sum_{i=1}^{N} (v_i^2 - 1) \sum_{j \neq i} g_{\delta, N, \lambda}(x_i - x_j) \hat{f}_N d\mu_N$$
$$\leq \frac{1}{\sigma} \int \hat{f}_N^*(x) d\mu_N^* \left\{ \log \int \exp \left[\frac{\sigma}{N^2} \sum_{i=1}^{N} (v_i^2 - 1) \right. \right.$$
$$\left. \left. \times \left(\sum_{j \neq i} g_{\delta, N, \lambda}(x_i - x_j) \right) \right] G_N(dv) \right\} + \frac{C}{TN\sigma}.$$

If we denote $\sum_{j \neq i} g_{\delta, N, \lambda}(x_i - x_j)$ by ξ_i then $|\xi_i| \leq CN^2$. Moreover

$$\int \exp \left[\frac{\sigma \xi_i}{N^2} (v_i^2 - 1) \right] e^{-v_i^2/2} \frac{1}{\sqrt{2\pi}} dv_i = e^{-\frac{\sigma \xi_i}{N^2} - \frac{1}{2} \log \left(1 - \frac{2\xi_i \sigma}{N^2} \right)}.$$

Therefore for small σ,

$$\log \int \exp \left[\frac{\sigma}{N^2} \sum_{i=1}^{N} (v_i^2 - 1) \xi_i \right] G_N(dv) = -\frac{\sigma}{N^2} \sum \left[\xi_i - \frac{1}{2} \log \left(1 - \frac{2\xi_i \sigma}{N^2} \right) \right]$$
$$\leq -\frac{\sigma}{N^2} \sum \xi_i + \frac{\sigma}{N^2} \sum \xi_i + c_1 \frac{\sigma^2}{2N^4} \sum \xi_i^2$$
$$\leq c_2 \sigma^2.$$

We have now as before

$$\lim_{N \to \infty} \left| \frac{1}{N^2} \int \sum_{i=1}^{N} (v_i^2 - 1) \sum_{j \neq i} g_{\delta, N, \lambda}(x_i - x_j) \hat{f}_N d\mu_N \right| = 0. \tag{4.14}$$

If we combine (4.12), (4.13), and (4.14) we get

$$\lim_{\lambda \to \infty} \lim_{N \to \infty} \left| \int \frac{1}{N^2} \sum_{i=1}^{N} \left\{ \sum_{j, k \neq i} NV'(N(x_i - x_j)) G_{\delta, N, \lambda}(x_i - x_j) \right. \right.$$
$$\left. \left. - \sum_{j \neq i} g_{\delta, N, \lambda}(x_i - x_j) \right\} \hat{f}_N d\mu_N \right| = 0. \tag{4.15}$$

From this point on the rest of the proof proceeds exactly like in [4]. First we note that with $\delta = 1$ and $\lambda = 1$,

$$\sup_N \left| \int \frac{1}{N^2} \sum_{i=1}^{N} \{ NV'(N(x_i - x_j)) G_{1, N, 1}(x_i - x_j) \} \hat{f}_N d\mu_N \right| \leq C. \tag{4.16}$$

This provides us with the estimate

$$\sup_N \left| \int \frac{1}{N} \sum_{i=1}^{N} \left\{ \sum_{j=1}^{N} \psi(N(x_i - x_j)) \right\}^2 \hat{f}_N d\mu_N \right| \leqq C. \tag{4.17}$$

According to Lemma 7.9 of [4] this will give the required uniform integrability to take the limits as $N \to \infty$ and then $\lambda \to \infty$ and $\delta \to 0$. We remark that once $N \to \infty$, according to Lemma 4.5 we are in the local Gibbs situation and that puts us in the circumstances of [4]. In particular we get Lemma 7.10 of [4] establishing for any possible limit point the bound

$$E^Q \int_S dx \int \varrho^3 \pi_x(d\varrho) \leqq C. \tag{4.18}$$

5. Hydrodynamic Limit

We will prove in this section the main part namely (iv) for Theorem 1.1. Parts (i) and (ii) were essentially established in the previous section. We will prove (iii) in the next section. Of course we need some estimates to prove compactness. These will also be deferred to the next section.

For any function J on S which is smooth by Eq. (1.1) we have

$$\frac{1}{N} \sum_{i=1}^{N} J(x_i(T)) - \frac{1}{N} \sum_{i=1}^{N} J(x_i(0)) = \int_0^T \sum_{i=1}^{N} J'(x_i(t)) v_i(t) dt. \tag{5.1}$$

On the other hand

$$\begin{aligned} d\left(\sum_{i=1}^{N} J'(x_i(t)) v_i(t) \right) = {} & N \sum_{i=1}^{N} J''(x_i(t)) v_i^2(t) dt \\ & - N \sum_{i=1}^{N} \left(2 \sum_{j \neq i} N V'(N(x_i(t) - x_j(t))) \right) J'(x_i(t)) dt \\ & - \frac{N^2}{2} \sum_{i=1}^{N} J'(x_i(t)) v_i(t) dt \\ & + N \sum_{i=1}^{N} J'(x_i(t)) dw_i(t), \end{aligned} \tag{5.2}$$

we can rewrite this in the form

$$\begin{aligned} \int_0^T \sum_{i=1}^{N} J'(x_i(t)) v_i(t) dt = {} & \frac{2}{N} \int_0^T \sum_{i=1}^{N} J''(x_i(t)) v_i^2(t) dt \\ & - \frac{2}{N} \int_0^T \sum_{i=1}^{N} \left(2 \sum_{j \neq i} N V'(N(x_i(t) - x_j(t))) \right) J'(x_i(t)) ct \\ & + \frac{2}{N} \int_0^T \sum_{i=1}^{N} J'(x_i(t)) dw_i(t) \\ & - \frac{2}{N^2} \left(\sum_{i=1}^{N} J'(x_i(T)) v_i(T) - \sum_{i=1}^{N} J'(x_i(0)) v_i(0) \right). \end{aligned} \tag{5.3}$$

Combining it with (5.1)

$$\begin{aligned}\frac{1}{N}\sum J(x_i(T))-\frac{1}{N}\sum J(x_i(0))\\ =2\int_0^T F_N(t)dt+\frac{2}{N}\int_0^T\sum_{i=1}^N J'(x_i(t))dw_i(t)\\ -\frac{2}{N^2}\left(\sum_{i=1}^N J'(x_i(T))v_i(T)-\sum_{i=1}^N J'(x_i(0))v_i(0)\right),\end{aligned} \tag{5.4}$$

where

$$F_N(t)=\frac{1}{N}\sum_{i=1}^N\left\{v_i^2(t)J''(x_i(t))-\sum_{j\neq i}2NV'(N(x_i(t)-x_j(t)))J'(x_i(t))\right\}. \tag{5.5}$$

From (i) of Theorem 3.1 it is easy to show that

$$\lim_{N\to\infty}E^{f_N^0}\left(\frac{2}{N^2}\left|\sum_{i=1}^N J'(x_i(T))v_i(T)-\sum_{i=1}^N J'(x_i(0))v_i(0)\right|\right)=0, \tag{5.6}$$

and because $w_1(t),\dots,w_N(t)$ are independent Brownian motions

$$\lim_{N\to\infty}E^{f_N^0}\left|\frac{2}{N}\int_0^T\sum_{i=1}^N J'(x_i(t))dw_i(t)\right|=0. \tag{5.7}$$

The next step is to replace v_i^2 by 1 in (5.5).

Lemma 5.1.

$$\lim_{N\to\infty}E^{f_N^0}\left(\int_0^T\left|\frac{1}{N}\sum J''(x_i(t))(v_i^2(t)-1)\right|dt\right)=0.$$

Proof. By entropy inequality and logarithmic Sobolev inequality,

$$\begin{aligned}\int\left|\frac{1}{4}\sum J''(x_i)(v_i^2-1)\right|\bar{f}_N d\mu_N\\ \leqq\int f_N^*(\underline{x})d\mu_N^*\log\left\{\int\exp\left[\left|\frac{1}{N}\sum_{i=1}^N(v_i^2-1)J''(x_i)\right|\right]G_N(d\underline{v})\right\}+\frac{C}{2NT}.\end{aligned}$$

By the law of large numbers

$$\int\exp\left[\left|\frac{1}{N}\sum_{i=1}^N(v_i^2-1)J''(x_i)\right|\right]G_N(d\underline{v})\to 1$$

as $N\to\infty$, uniformly over $\underline{x}$. This proves the lemma.

Now the velocities are completely out of the picture and we only have to look at

$$\int_0^T\frac{2}{N}\sum_{i=1}^N\left\{J''(x_i(t))-\sum_{j\neq i}2NV'(N(x_i(t)-x_j(t)))J'(x_i(t))\right\}dt.$$

We replace $J'(x_i(t))$ by $\frac{1}{2}[J'(x_i(t))-J'(x_j(t))]$, which we can, because of the skew symmetry of V'. As in [4], we proceed to replace

$$J'(x_i(t))-J'(x_j(t))\quad\text{by}\quad(x_i(t)-x_j(t))J''(x_i(t))$$

to get

$$\int_0^T \frac{2}{N} \sum_{i=1}^N J''(x_i(t)) \left(1 + \sum_{j=1}^N \psi(N(x_i(t) - x_j(t)) \right) dt,$$

where

$$\psi(z) = -zV'(z).$$

Now we can use Theorem 4.1 as in [4] and show that any weak limit will satisfy (iv) of Theorem 1.1.

6. Some Auxiliary Lemmas

Lemma 6.1. *There exists a constant C such that for all N,*

$$\int \frac{1}{N} \sum_{i=1}^N \left| \sum_{j=1}^N \psi(N(x_i - x_j)) \right|^2 \bar{f}_N d\mu_N \leqq C.$$

Proof. This is precisely (4.17).

Lemma 6.2 (Compactness). *For every smooth $J(\cdot)$ on S, and $\varepsilon > 0$,*

$$\lim_{\delta \to 0} \limsup_{N \to \infty} P^{f_N^0} \left[\sup_{\substack{0 \leqq s < t \leqq T \\ |s-t| \leqq \delta}} |\langle J, \xi_N(t) \rangle - \langle J, \xi_N(s) \rangle| \geqq \varepsilon \right] = 0.$$

Proof. According to Eq. (5.1),

$$\begin{aligned} \langle J, \xi_N(t) \rangle - \langle J, \xi_N(s) \rangle &= \frac{1}{N} \sum J(x_i(t)) - \frac{1}{N} \sum J(x_i(s)) \\ &= \int_s^t \sum_{i=1}^N J'(x_i(\sigma)) v_i(\sigma) d\sigma. \end{aligned}$$

We can rewrite this using (5.3) as

$$\begin{aligned} \int_s^t \sum_{i=1}^N J'(x_i(\sigma)) v_i(\sigma) d\sigma &= \frac{2}{N} \int_s^t \sum_{i=1}^N J''(x_i(\sigma)) v_i^2(\sigma) d\sigma \\ &- \frac{2}{N} \int_s^t \sum_{i=1}^N 2 \sum_{j \neq i} NV'(N(x_i(\sigma) - x_j(\sigma))) J'(x_i(\sigma)) d\sigma \\ &+ \frac{2}{N} \int_s^t \sum_{i=1}^N J'(x_i(\sigma)) d\omega_i(\sigma) \\ &- \frac{2}{N^2} \left[\sum_{i=1}^N J'(x_i(t)) v_i(t) - \sum_{i=1}^N J'(x_i(s)) v_i(s) \right] \\ &= A_1(s,t) + A_2(s,t) + A_3(s,t) + A_4(s,t), \end{aligned}$$

$$A_1(s,t) = \frac{2}{N} \int_s^t \sum_{i=1}^N J''(x_i(\sigma)) [v_i^2(\sigma) - 1] d\sigma + \frac{2}{N} \int_s^t \sum J''(x_i(\sigma)) d\sigma,$$

$$|A_1(s,t)| \leqq 2 \int_0^T \left| \frac{1}{N} \sum J''(x_i(\sigma)) (v_i^2(\sigma) - 1) \right| d\sigma + 2|t-s| \, \|J''\|_\infty.$$

Since $E^{f_N^0}\left[\int_0^T\left|\frac{1}{N}\sum J''(x_i(\sigma))(v_i^2(\sigma)-1)\right|d\sigma\right]\to 0$ as $N\to\infty$ it follows that

$$\lim_{\delta\downarrow 0}\overline{\lim_{N\to\infty}}P^{f^0}\left[\sup_{\substack{0\leq s\leq t\leq T\\|t-s|\leq\delta}}|A_1(s,t)|\geqq\varepsilon\right]=0,$$

$$A_2(s,t)=-\frac{2}{N}\int_s^t\sum_{i=1}^N NV'(N(x_i(\sigma)-x_j(\sigma)))(J'(x_i(\sigma)))d\sigma,$$

and

$$\begin{aligned}|A_2(s,t)|&\leqq\frac{2}{N}\|J''\|_\infty\int_s^t\sum_{i,j}\psi(N(x_i(\sigma)-x_j(\sigma)))d\sigma\\&\leqq(t-s)^{1/2}\cdot\frac{2}{N}\|J''\|_\infty\left[\int_s^t\Big(\sum_{i,j}\psi(N(x_i(\sigma)-x_j(\sigma)))\Big)^2d\sigma\right]^{1/2}\\&\leqq(t-s)^{1/2}\,2\|J''\|_\infty\left(\int_0^T\frac{1}{N}\sum_{i=1}^N\Big[\sum_{j\neq i}\psi(N(x_i(\sigma)-x_j(\sigma)))\Big]^2d\sigma\right)^{1/2}.\end{aligned}$$

From Lemma 6.1

$$\lim_{\delta\to 0}\limsup_{N\to\infty}P^{f_N^0}\left[\sup_{\substack{0\leq s\leq t\leq T\\|t-s|\leq\delta}}|A_2(s,t)|\geqq\varepsilon\right]=0.$$

From Doob's inequality

$$P^{f_N^0}\left[\sup_{0\leqq t\leqq T}\left|\frac{2}{N}\int_0^t\sum_{i=1}^N J'(x_i(\sigma))d\omega_i(\sigma)\right|\geqq\varepsilon\right]\leqq\frac{2}{\varepsilon^2}\cdot\frac{4}{N^2}\cdot N\cdot\|J'\|_\infty^2.$$

Therefore

$$\lim_{N\to\infty}P^{f_N^0}\left[\sup_{0\leqq s\leqq t\leqq T}|A_3(s,t)|\geqq\varepsilon\right]=0.$$

In order to handle $A_4(s,t)$ it is sufficient to prove

$$\lim_{N\to\infty}P^{f_N^0}\left[\sup_{0\leqq t\leqq T}\frac{1}{N^2}\left|\sum_{i=1}^N J'(x_i(t))v_i(t)\right|\geqq\varepsilon\right]=0. \tag{6.1}$$

Let us denote by $\eta_N(t)=\sum_{i=1}^N J'(x_i(t))v_i(t)$. We can rewrite (5.2) in the form

$$\begin{aligned}d\eta_N(t)=&-\frac{N^2}{2}\eta_N(t)dt\\&+N\sum_{i=1}^N J''(x_i(t))v_i^2(t)dt\\&-N\sum_{i=1}^N\Big(2\sum_{j\neq i}NV'(N(x_i(t)-x_j(t)))J'(x_i(t))\Big),dt\\&+N\sum_{i=1}^N J'(x_i(t))d\omega_i(t).\end{aligned} \tag{6.2}$$

Integrating this we get

$$\begin{aligned}\eta_N(t) = \eta_N(0)\, e^{-N^2 t/2} \\ + N^2 \int_0^t A_N(s) e^{-(N^2/2)(t-s)} ds \\ + N^2 \int_0^t B_N(s) e^{-(N^2/2)(t-s)} ds \\ + N^2 \int_0^t e^{-(N^2/2)(t-s)} dM_N(s),\end{aligned} \tag{6.3}$$

where

$$A_N(s) = \frac{1}{N} \sum J''(x_i(s)) v_i^2,$$

$$|B_N(s)| \leqq \frac{1}{N} \|J''\|_\infty \sum_{i,j} \psi(N(x_i(s) - x_j(s))),$$

$$M_N(s) = \frac{1}{N} \int_0^s \sum J'(x_i(\sigma)) d\omega_i(\sigma).$$

Since we have a uniform bound on

$$\begin{aligned} E^{f_N^0}[|\eta_N(0)|] &\leqq \log \int \exp[|\sum J'(x_i) v_i|] d\mu_N + CN \\ &\leqq C'N,\end{aligned}$$

$$P^{f_N^0}\left[\frac{1}{N^2} \sup_{0 \leqq t \leqq T} |\eta_N(0) e^{-N^2 t/2}| \geqq \delta\right] \to 0 \quad \text{as} \quad N \to \infty.$$

In order to establish (6.1) we need only prove

$$P^{f_N^0}\left[\sup_{0 \leqq t < T} \int_0^t e^{-(N^2/2)(s-t)} |A_N(s)| ds \geqq \varepsilon\right] \to 0 \quad \text{as} \quad N \to \infty, \tag{6.4}$$

$$P^{f_N^0}\left[\sup_{0 \leqq t < T} \int_0^t e^{-(N^2/2)(s-t)} |B_N(s)| ds \geqq \varepsilon\right] \to 0 \quad \text{as} \quad N \to \infty, \tag{6.5}$$

and

$$P^{f_N^0}\left[\sup_{0 \leqq t < T} \left|\int_0^t e^{-(N^2/2)(s-t)} dM_N(s)\right| \geqq \varepsilon\right] \to 0 \quad \text{as} \quad N \to \infty,$$

$$\begin{aligned} |A_N(s)| &\leqq \left|\frac{1}{N} \sum J''(x_i(s))\right| + \left|\frac{1}{N} \sum J''(x_i(s))(v_i^2(s) - 1)\right| \\ &\leqq \|J''\|_\infty + \left|\frac{1}{N} \sum J''(x_i(s))(v_i^2(s) - 1)\right|. \end{aligned} \tag{6.6}$$

Clearly,

$$\sup_{0\leq t\leq T}\int_0^T e^{-(N^2/2)(t-s)}ds\to 0 \quad \text{as} \quad N\to\infty,$$

$$\sup_{0\leq t\leq T}\int_0^T e^{-(N^2/2)(t-s)}\left|\frac{1}{N}\sum J''(x_i(s))(v_i^2(s)-1)\right|ds$$

$$\leq \int_0^T\left|\frac{1}{N}\sum J''(x_i(s))(v_i^2(s)-1)\right|ds,$$

and since $E^{f_N^0}\left[\int_0^T\left|\frac{1}{N}\sum J''(x_i(s))(v_i^2(s)-1)\right|ds\right]\to 0$ as $N\to\infty$ we have (6.4).

We estimate

$$\int_0^t e^{-(N^2/2)(t-s)}|B_N(s)|ds\leq\left(\int_0^t e^{N^2(t-s)}ds\right)^{1/2}\left(\int_0^t |B_N(s)|^2ds\right)^{1/2}$$

$$\leq\frac{1}{N}\left(\int_0^T |B_N(s)|^2ds\right)^{1/2}$$

$$\int_0^T |B_N(s)|^2ds\leq\frac{1}{N}\int_0^T\sum_{i=1}^N\left(\sum_{j=1}^N\psi(N(x_i(s)-x_j(s)))\right)^2ds.$$

Since we have a uniform bound on the expectation of the right-hand side from Lemma 6.1 we have (6.5).

Finally we turn to (6.6). Let us pretend N^2T is an integer. Then

$$\sup_{0\leq t\leq T}\left|e^{-(N^2/2)t}\int_0^t e^{(N^2/2)s}dM_N(s)\right|$$

$$\leq\sup_{1\leq k\leq N^2T}e^{-(N^2/2)(k-1)/N^2}\sup_{0\leq t\leq k/N^2}\left|\int_0^t e^{N^2s/2}dM_N(s)\right|.$$

We can estimate

$$P^{f_N^0}\left[\sup_{0\leq t\leq T}\left|e^{-N^2t/2}\int_0^t e^{(N^2/2)s}dM_N(s)\right|\geq\varepsilon\right]$$

$$\leq\sum_{k=1}^{N^2T}P^{f_N^0}\left[\sup_{0\leq t\leq k/N^2}\left|\int_0^t e^{-(N^2/2)s}dM_N(\varepsilon)\right|\geq\varepsilon e^{(k-1)/2}\right]$$

$$\leq\sum_{k=1}^{N^2T}\frac{e^{-(k-1)}}{\varepsilon^2}E^{f_N^0}\left[\int_0^{k/N^2}e^{(N^2/2)s}dM_N(s)\right]^2$$

$$\leq\sum_{k=1}^{N^2T}\frac{1}{\varepsilon^2}\cdot e^{-(k-1)}e^k\|J'\|_\infty^2\cdot\frac{1}{N^3}$$

$$=\frac{T}{N}\cdot e\cdot\|J'\|_\infty^2\cdot\frac{1}{\varepsilon^2}$$

$\to 0$ as $N\to\infty$ and we are done.

Lemma 6.3. *Let Q be any limit point of the distributions of $\xi_N(\cdot)$ under $P^{f_N^0}$ as $N\to\infty$, where f_N^0 satisfies $H_N(f_N^0)\leq CN$. Then*

$$E^Q\left(\int_0^T dt\int_S d\theta\frac{1}{q(\theta,t)}\left[\frac{\partial}{\partial\theta}P(q(\theta,t))\right]^2\right)\leq\frac{C}{2},$$

with the same constant.

Proof. Given two integers n and l we consider a family of functions $u_1(x), \ldots, u_n(x)$, $F_1(x), \ldots, F_l(x)$ which are smooth functions on S. We consider also a family $g_1(y_1, \ldots, y_l), \ldots, g_n(y_1, \ldots, y_l)$ of smooth functions with compact support on R^l. We use these functions to define a family $U_1, \ldots, U_n$ of functions on $S^N \times R^N$ and $G_1, \ldots, G_n$ on S^N,

$$U_r(\underline{x}, \underline{v}) = \sum_{j=1}^{N} v_j u_r(x_j),$$

$$G_r(\underline{x}) = g_r\left(\frac{1}{N}\sum_{j=1}^{N} F_1(x_j), \ldots, \frac{1}{N}\sum_{j=1}^{N} F_l(x_j)\right).$$

By a direct calculation for each r,

$$\frac{1}{N}\sum_{i=1}^{N}\left\{2\sum_{j=1}^{N} NV'(N(x_i-x_j))u_r(x_i) - v_i^2 u_r'(x_i)\right\}$$
$$= -\frac{1}{2}U_r(\underline{x}, \underline{v}) - \frac{1}{N^2}L_N U_r(x, v). \tag{6.7}$$

We want to prove first that

$$\lim_{N\to\infty}\frac{1}{N^2}\int (L_N U_r)(\underline{x}, \underline{v}) \cdot G_r(\underline{x})\bar{f}_N d\mu_N = 0. \tag{6.8}$$

In fact since $G_r(\underline{x})$ is only a function of $\underline{x}$ and L_N is first order in x derivatives,

$$(L_N U_r)G_r = L_N(U_r G_r) - U_r(L_N G_r)$$

and

$$\frac{1}{N^2}\int L_N(U_r G_r)\bar{f}_N d\mu_N = \frac{1}{N^2 T}\left[\int U_r G_r f_N^T d\mu_N - \int U_r G_r f_N^0 d\mu_N\right].$$

Since both f_N^T and f_N^0 satisfy entropy bounds,

$$H_N(f_N) \leqq CN,$$

$$\left|\int U_r G_r f_N d\mu_N\right| \leqq \|G_r\|\,\|U_r\|\sqrt{N}\left(\int \sum v_i^2 f_N d\mu_N\right)^{1/2}$$
$$\leqq C'N,$$

where f_N can be either f_N^T or f_N^0. Therefore

$$\lim_{N\to\infty}\frac{1}{N^2}\int L_N(U_r G_r)\bar{f}_N d\mu_N = 0. \tag{6.9}$$

Moreover

$$\frac{1}{N^2}\int U_r(L_N G_r)\bar{f}_N d\mu_N = \frac{1}{N^2}\int \sum_{i,j} v_i v_j u_r(x_i)\frac{\partial G_r}{\partial x_j}(\underline{x})\bar{f}_N d\mu_N.$$

By the arguments we have used in the proof of Theorem 4.6 we can replace $v_i v_j$ by δ_{ij} and because it is obvious that

$$\lim_{N\to\infty}\frac{1}{N^2}\int \sum_{r=1}^{N} u_r(x_i)\frac{\partial G_r}{\partial x_i}(\underline{x})\bar{f}_N d\mu_N = 0,$$

we conclude that

$$\lim_{N\to\infty} \frac{1}{N^2} \int U_r(L_N G_r)\bar{f}_N d\mu_N = 0. \tag{6.10}$$

(6.8) follows immediately from (6.9) and (6.10).

We now turn to the relation (6.7). Let us use $\bar{f}_N$ to construct a measure $\bar{Q}_N$ on $M_1(S)$ by the map $(x_1,\dots,x_N)\to\frac{1}{N}(\delta_{x_1}+\dots\delta_{x_N})$ and let $\bar{Q}$ be any weak limit. According to the results of Sects. 4 and 5,

$$\lim_{N\to\infty} \int \frac{1}{N}\sum_{i=1}^{N}\left\{2\sum_{j=1}^{N} NV'(N(x_i-x_j))(u_r(x_i)-u_r'(x_i))\right\} G_r(\underline{x})\bar{f}_N d\mu_N$$

along the subsequence giving the weak limit $\bar{Q}$ can be represented as

$$-E^{\bar{Q}}\{\int P(q(x))u_r'(x)G_r(\int F_1(x)q(x)dx,\dots,\int F_l(x)q(x)dx)\}.$$

If we now use (6.7) and (6.8) and sum over $r=1,2,\dots$, we obtain

$$\begin{aligned}
&\left|E^{\bar{Q}}\left\{\int P(q(x))\sum_{r=1}^{n} u_r'(x)G_r(\int F_1(x)q(x)dx,\dots,\int F_l(x)q(x)dx)dx\right\}\right| \\
&\leqq \limsup_{N\to\infty}\left|\frac{1}{2}\int\sum_{r=1}^{n} U_r(\underline{x},\underline{v})G_r(\underline{x})\bar{f}_N d\mu_N\right| \\
&= \limsup_{N\to\infty}\left|\frac{1}{2}\int\sum_{r=1}^{n}\sum_{i=1}^{N} u_r(x_i)G_r(\underline{x})v_i\bar{f}_N d\mu_N\right| \\
&= \limsup_{N\to\infty}\left|\frac{1}{2}\int\sum_{r=1}^{n}\sum_{i=1}^{N} u_r(x_i)G_r(\underline{x})\frac{\partial\bar{f}_N}{\partial v_i}d\mu_N\right| \\
&\leqq \limsup_{N\to\infty}\frac{1}{2}\left(\int\frac{1}{\bar{f}_N}\sum_{i=1}^{N}\left(\frac{\partial\bar{f}_N}{\partial v_i}\right)^2 d\mu_N\right)^{1/2} \\
&\quad\times\left(\int\sum_{i=1}^{N}\left(\sum_{r=1}^{n} u_r(x_i)G_r(\underline{x})\right)^2\bar{f}_N d\mu_N\right)^{1/2} \\
&\leqq \limsup_{N\to\infty}\frac{1}{2}\left(\frac{2C}{NT}\right)^{1/2} N^{1/2}\left(\int\frac{1}{N}\sum_{i=1}^{N}\left(\sum_{r=1}^{n} u_r(x_i)G_r(\underline{x})\right)^2\bar{f}_N d\mu_N\right)^{1/2} \\
&= \frac{1}{2}\left(\frac{2C}{T}\right)^{1/2}\left(E^{\bar{Q}}\int q(x)dx\left(\sum_{r=1}^{n} u_r(x)G_r(\int F_1(x)q(x)dx,\dots,\int F_l((x)q(x)dx)\right)\right)^{1/2}.
\end{aligned} \tag{6.11}$$

Denoting by ω the typical point in $M_1(S)$ we can rewrite (6.11) as

$$\left|E^{\bar{Q}}\int P(q(x))\frac{\partial\hat{G}}{\partial x}(x,\omega)dx\right| \leqq \left(\frac{C}{2T}\right)^{1/2}(E^{\bar{Q}}\int\hat{G}^2(x,\omega)q(x)dx)^{1/2} \tag{6.12}$$

for test functions $\hat{G}$ from a suitable class. First we have it for $\hat{G}$ of the form

$$\sum u_r(x)G_r(\langle F_1,q\rangle,\dots,\langle F_l,q\rangle).$$

Then we can obtain the inequality for $\hat{G}$ of the form

$$\sum u_r(x)\hat{G}_r(\omega).$$

The next step is to use test functions of general form $\hat{G}(x,\omega)$ but with reasonable bounds on $\partial\hat{G}/\partial x$. Then by standard regularization techniques one can prove

$$E^Q \int \frac{1}{q(x)} \left(\frac{\partial P(q(x))}{\partial x}\right)^2 dx \leqq \frac{C}{2T}. \tag{6.13}$$

It is easy to see that if Q is a weak limit on $C[[0,T]; M_1(S)]$ then $\bar{Q}$ is just the average marginal distribution over the time interval $[0,T]$. Therefore

$$\begin{aligned} E^Q \int_0^T \int \frac{1}{q(x,t)} \left(\frac{\partial}{\partial x} P(q(x,t))\right)^2 dx\,dt &\leqq \frac{C}{2T} \cdot T \\ &= \frac{C}{2}. \end{aligned} \tag{6.14}$$

Lemma 6.4. *There is at most one weak solution of the equation*

$$\frac{\partial \varrho(t,x)}{\partial t} = 2(P(\varrho(t,x)))_{xx}$$

with initial condition $\varrho(0,x)=\varrho_0(x)$ *among the class of nonnegative solutions satisfying*

$$\int_0^T \int \varrho^3(t,x)\,dt\,dx < \infty$$

$$\int_0^T \int \left(\frac{\partial P(\varrho(t,x))}{\partial x}\right)^2 \frac{1}{\varrho(t,x)}\,dt\,dx < \infty.$$

Proof. This has been carried out in [4].

References

1. Deuschel, J.D., Stroock, D.W.: Large deviations. New York: Academic Press 1989
2. De Masi, A., Presutti, E.: Lectures on the collective behavior of particle systems. C.A.R.R. Reports in Mathematical Physics, 5/89, 1989
3. Papanicolaou, G., Varadhan, S.R.S.: Ornstein-Uhlenbeck process in a random potential. Commun. Pure Appl. Math. **35**, 819–834 (1985)
4. Varadhan, S.R.S.: Scaling limits for interacting diffusions. Commun. Math. Phys. **135**, 313–353 (1991)

Communicated by J. L. Lebowitz

Mathematical Physics, X, Springer, 1992, 103-112

Entropy Methods in Hydrodynamical Scaling

S. R. S. Varadhan

Courant Institute, NYU, New York 10012

We shall describe these methods by examining very closely an explicit model. Consider N lattice sites arranged periodically in one dimension with a lattice spacing of $1/N$. We have spin variables x_j attached to each site j/N, the sites being viewed as equally spaced points on the circle of unit circumference. The spins x_j vary in time in such a manner that they undergo a diffusion on $\mathbb{R}^N$ denoted by $\{x_1(t), \ldots, x_N(t)\}$. The diffusion process is described by

$$dx_i(t) = dz_{i-1,i}(t) - dz_{i,i+1}(t)\,,$$

$$dz_{i,i+1}(t) = \frac{N^2}{2}\left[\phi'(x_i(t)) - \phi'(x_{i+1}(t))\right] dt + N\, d\beta_{i,i+1}(t)\,.$$

Here $\phi'(x) = d\phi/dx$ for a suitable function $\phi(x)$. One can think of $\phi(x)$ as somewhat like $x^2/2$. If ϕ where $x^2/2$ we would have a linear model and our model should be thought of, as a nonlinear version. The random process $\beta_{i,i+1}(t)$ are N, independent Brownian motions. Equivalently one could write down the generator of the process

$$L_N = \frac{N^2}{2}\sum\left(\frac{\partial}{\partial x_i} - \frac{\partial}{\partial x_{i+1}}\right)^2 - \frac{N^2}{2}\sum\left[\phi'(x_i) - \phi'(x_{i+1})\right]\left[\frac{\partial}{\partial x_i} - \frac{\partial}{\partial x_{i+1}}\right].$$

One can also describe the process through its Dirichlet form

$$\frac{N^2}{2}\int\sum\left(\frac{\partial u}{\partial x_i} - \frac{\partial u}{\partial x_{i+1}}\right)^2 e^{-\Sigma\phi(x_i)}\prod dx_i\,.$$

We normalize ϕ so that $\int \exp[-\phi(x)]\,dx = 1$. We assume

$$M(\theta) = \int \exp\left[\theta x - \phi(x)\right] dx$$

is finite for all $\theta \in \mathbb{R}$. Then

$$h(x) = \sup_\theta[\theta x - \log M(\theta)]$$

is strictly convex and

$$\lambda = h'(x)$$
$$x = \frac{M'(\lambda)}{M(\lambda)} = [\log M(\lambda)]'$$

are inverse functions. The measure $\prod e^{-\phi(x_i)}\,dx_i$ is invariant for the diffusion. It is however not ergodic. $x_1 + \cdots + x_N$ is a conserved quantity. The ergodic pieces are restrictions of $\prod e^{-\phi(x_i)}\,dx_i$ to various hyperplanes $\frac{1}{N}\sum x_i = a$. The factor N^2 corresponds to speeding up of time. Since space has already been scaled by a factor of N, we are clearly anticipating a diffusive limit in our choice of parabolic scaling of space and time. We would like to think of

$$\nu_N(t) = \frac{1}{N}\sum x_j(t)\,\delta_{j/N}$$

as a randomly changing signed measure on the circle S. At time 0, we have a random initial configuration with density

$$f_N^0(x_1,\ldots,x_N)\exp\left[-\sum \phi(x_i)\right]\prod dx_i$$

on $\mathbb{R}^N$. The distribution of the configuration at time t will be given by a density

$$f_N^t(x_1,\ldots,x_N)\exp\left[-\sum \phi(x_i)\right]\prod dx_i\,,$$

where f_N^t satisfies the Kolmogorov equation

$$\frac{\partial f_N^t}{\partial t} = L_N f_N^t$$
$$f_N^t\big|_{t=0} = f_N^0\,.$$

Although in principle we should use L_N^*, because L_N is symmetric with respect to the weight $\exp[-\sum\phi(x_i)]$ we do not have to take the adjoint. Suppose at time 0 for some smooth deterministic profile $m_0(\theta)$

$$\nu_N(0) \sim m_0(\theta)\,d\theta\ ,$$

or more precisely

$$\frac{1}{N}\sum J\left(\frac{i}{N}\right)x_i \to \int J(\theta)\,m_0(\theta)\,d\theta$$

in probability with respect to $f_N^0(x_1,\ldots,x_N)\exp[-\sum\phi(x_i)]\,\Pi\,dx_i$ for every nice test function $J(\theta)$ on S. Under suitable regularity conditions of a simple nature on f_N^0 we want to conclude that for some profile $m(\theta,t)$ we have

$$\frac{1}{N}\sum J\left(\frac{i}{N}\right)x_i \to \int J(\theta)\,m(\theta,t)\,d\theta$$

in probability with respect to $f_N^t(x_1, \dots, x_N)\exp[-\sum \phi(x_i)] \prod dx_i$. In addition we should be able to describe how to compute $m(\theta, t)$ from the initial profile $m_0(\theta)$. Hopefully $m(\theta, t)$ does not depend on f_N^0 in any manner other than through the initial profile $m_0(\theta)$ determined by it. The main step is to compute

$$\begin{aligned} d\frac{1}{N}\sum J(\frac{i}{N})x_i(t) &= -\frac{N^2}{2}\cdot\frac{1}{N}\sum\left[J(\frac{i+1}{N}) - J(\frac{i}{N})\right]\left[\phi'(x_{i+1}(t))\right.\\ &\quad \left. - \phi'(x_i(t))\right] dt \\ &\quad + N\cdot\frac{1}{N}\sum\left[J(\frac{i+1}{N}) - J(\frac{i}{N})\right] d\beta_{i,i+1}(t) \\ &= \frac{1}{2N}\cdot N^2\sum\left[J(\frac{i+1}{N}) - 2J(\frac{i}{N})\right. \\ &\quad \left. + J(\frac{i-1}{N})\right]\phi'(x_i(t)) + d(\text{``noise''}). \end{aligned}$$

Approximately "noise" $= \frac{1}{N}\sum J'(\frac{i}{N})\, d\beta_{i,i+1}(t)$ and goes to zero by the law of large numbers. We therefore have roughly

$$d(\frac{1}{N}\sum J(\frac{i}{N})\, x_i(t)) \simeq \frac{1}{2N}\sum J''(\frac{i}{N})\, \phi'(x_i(t))\, dt.$$

The problem is that the measure

$$\frac{1}{2N}\sum \phi'(x_i)\delta_{i/N}$$

is not so easily expressed in terms of $\frac{1}{N}\sum x_i\delta_{i/N}$. We need to know how the spins arrange themselves so that a knowledge of

$$\frac{1}{N\epsilon}\sum_B x_j = m$$

determines

$$\frac{1}{N\epsilon}\sum_B \phi'(x_j)$$

over a block B of size $N\epsilon$ (ϵ is a small positive number).

On the infinite lattice the invariant measures for the diffusion are

$$\prod \frac{1}{M(\lambda)}\, e^{\lambda x_i - \phi(x_i)}\, dx_i$$

for various values of λ. These are the "Gibbs" measures in this context.

$$\begin{aligned} \text{The average spin } &= \int x\cdot\frac{1}{M(\lambda)}\, e^{\lambda x - \phi(x)}\, dx \\ &= \frac{M'(\lambda)}{M(\lambda)} = m\,. \end{aligned}$$

Therefore we know that λ has to be chosen so that

$$\lambda = h'(m)\,.$$

Moreover

$$\int \phi'(x) \frac{1}{M(\lambda)}\, e^{\lambda x - \phi(x)}\, dx = \lambda$$

so that, if the average spin is m and the spins are organized according to a Gibbs ensemble then the average of $\phi'(x)$ is $h'(m)$.

If we assume that locally the system is close to Gibbs ensembles then we get

$$\frac{\partial}{\partial t} \int J(\theta)\, m(\theta,t)\, d\theta = \frac{1}{2} \int J''(\theta)\, h'(m(\theta,t)\, d\theta\,,$$

or equivalently the nonlinear heat equation

$$\frac{\partial m(\theta,t)}{\partial t} = \frac{1}{2} \big[h'(m(\theta,t)\,)\big]_{\theta\theta}$$

with initial condition

$$m(\theta,0) = m_0(\theta)\,.$$

This is the hydrodynamic equation. The problem is to prove that the density $f_N^t(x_1,\ldots,x_N)\exp\big[-\sum\phi(x_i)\big]\prod dx_i$ has the correct profile.

A version of the model on the infinite lattice was originally considered by J. Fritz [2]. The method we will outline now was developed by Guo-Papanicolaou-Varadhan in [3]. The basic step is to prove

$$\frac{1}{N}\sum_i \Big|\frac{1}{2N\epsilon}\sum_{|j-i|\le N\epsilon} \phi'(x_j) - h'\Big(\frac{1}{2N\epsilon}\sum_{|j-i|\le N\epsilon} x_j\Big)\Big| = \psi_{N,\epsilon}$$

is small most of the time. If we take a finite interval $0\le t\le T$ for time and define

$$\bar{f}_N(x_1,\ldots,x_n) = \frac{1}{T}\int_0^T f_N^t(x_1,\ldots,x_N)\, dt$$

then we need

$$\lim_{\epsilon\to 0}\ \overline{\lim}_{N\to\infty} \int \psi_{N,\epsilon} \bar{f}_N\, e^{-\sum\phi(x_i)} \prod dx_i = 0\,.$$

However, ϕ' is an unbounded function. If we assume

$$\int e^{\theta|\phi'(x)|-\phi(x)}\, dx < \infty$$

for all $\theta > 0$, then we can do effective truncation. For any bounded continuous function $g(x)$ on the line let us define

$$\hat{g}(m) = \frac{1}{M(\lambda)}\int e^{\lambda x - \phi(x)}\, dx$$

with $\lambda = h'(m)$. Then we need to prove that for

$$\tilde{\psi}_{N,\epsilon} = \frac{1}{N}\sum_i \left|\frac{1}{2N\epsilon}\sum_{|j-i|\le N\epsilon} g(x_i) - \hat{g}\left(\frac{1}{2N\epsilon}\sum_{|j-i|\le N\epsilon} x_j\right)\right|$$

$$\lim_{\epsilon\to 0}\limsup_{N\to\infty}\int \tilde{\psi}_{N,\epsilon}\bar{f}_N\, e^{-\sum\phi(x_i)}\prod dx_i = 0\,.$$

The key idea in [3] is the following. If we assume initially that

$$\int f_N^0 \log f_N^0\, e^{-\sum\phi(x_i)}\prod dx_i \le C\,N$$

for some constant C independent of N, then $H_N(t)$ defined by

$$H_N(t) = \int f_N^t \log f_N^t\, e^{-\sum\phi(x_i)}\prod dx_i$$

is decreasing in t and

$$\begin{aligned}\frac{dH_N(t)}{dt} &= -\frac{N^2}{2}\int\sum\left(\frac{\partial f_N^t}{\partial x_i} - \frac{\partial f_N^t}{\partial x_{i+1}}\right)^2 \frac{1}{f_N^t} e^{-\sum\phi(x_i)}\prod dx_i\\ &= -N^2\, I_N(t) \qquad \text{(by definition)}\,.\end{aligned}$$

Clearly

$$\int_0^T I_N(t)\,dt \le \frac{C}{N}\,.$$

If we define

$$I_N(f) = \frac{1}{2}\int\sum\left(\frac{\partial f}{\partial x_i} - \frac{\partial f}{\partial x_{i+1}}\right)^2\cdot\frac{1}{f}e^{-\sum\phi(x_i)}\prod x_i$$

then $I_N(t) = I_N(f_N^t)$. By convexity

$$\begin{aligned}I_N(\bar{f}_N) &\le \frac{1}{T}\int_0^T I_N(f_N^t)\,dt\\ &\le \frac{C}{TN}\,.\end{aligned}$$

Moreover, again by convexity and the monotonicity of $H_N(t)$,

$$\begin{aligned}H_N(\bar{f}_N) &= \int \bar{f}_N \log \bar{f}_N\, e^{-\sum\phi(x_i)}\prod dx_i\\ &\le \frac{1}{T}\int_0^T H_N(f_N^t)\,dt\\ &\le C\,N\,.\end{aligned}$$

The basic step in [3] is to show that for any densities f_N relative to $\exp[-\sum\phi(x_i)]\prod dx_i$ satisfying

$$H_N(f_N) \le C\,N$$

and

$$I_N(f_N) \le \frac{C}{N}$$

$$\lim_{\epsilon\to 0}\limsup_{N\to\infty}\int \tilde{\psi}_{N,\epsilon}\, f_N\, e^{-\sum\phi(x_i)} \prod dx_i = 0\,.$$

This is carried out in two steps. First we replace $N\epsilon$ by k and define

$$\tilde{\psi}_{N,k} = \frac{1}{N}\sum_i \Big|\frac{1}{2k+1}\sum_{|j-i|\le k} g(x_j) - \hat{g}\big(\frac{1}{2k+1}\sum_{|j-i|\le k} x_j\big)\Big|\,.$$

Then

$$\lim_{k\to\infty}\limsup_{N\to\infty}\int f_N\tilde{\psi}_{N,k}\, e^{-\sum\phi(x_i)}\, c\prod dx_i = 0\,. \tag{1}$$

This is relatively easy and is proved in fair generality. If we define

$$f_k^N = \frac{1}{N}\sum_i (\text{marginal density of } x_{i-k},\ldots,x_{i+k})$$

then (1) is the same as

$$\lim_{k\to\infty}\limsup_{N\to\infty}\int \Big|\frac{1}{2k+1}\sum g(x_i) - \hat{g}\left(\frac{1}{2k+1}\sum x_i\right)\Big| f_N^k e^{-\sum\phi(x_i)}\prod dx_i$$
$$= 0\,.$$

We can let $N\to\infty$, keeping k fixed and see what the limit points of f_N^k are as $N\to\infty$. If we consider

$$I_k(f) = \frac{1}{2}\int\sum_{i=-k}^{k-1}\left(\frac{\partial f}{\partial x_i} - \frac{\partial f}{\partial x_{i+1}}\right)^2 \frac{1}{f}\, e^{-\sum\phi(x_i)}\prod dx_i$$

then by convexity

$$I_k(f_N^k) \le \frac{2k}{N^2}\cdot C\,.$$

By passing to the limit we conclude that for any limit point f^k, $I_k(f^k) = 0$. This forces f^k to be essentially a Gibbs measure and what we need is an easy consequence that is generally known as equivalence of ensembles.

To go from k large to $N\epsilon$ with ϵ small is harder. It can be done somewhat easily in our model, but can sometimes be very difficult. In the case of hyperbolic scaling one can see a priori that there have to be difficulties.

This method has worked successfully in various models, see for instance [3], [5], [7] and [8].

We will now discuss another method based again on entropy developed in [9] by H. T. Yau for this model. This method avoids the passage from k to $N\epsilon$ and therefore has wider applicability. See for instance [6], [10] where application is made to classical Hamiltonian systems with small noise.

Assume that $m(\theta,t)$ is given as a smooth solution of

$$\frac{\partial m}{\partial t} = \frac{1}{2}(h'(m(\theta,t)\,)_{\theta\theta}\,.$$

We go from $m(\theta,t)$ to $\lambda(\theta,t)$ where

$$\lambda(\theta,t) = h'(m(\theta,t)\,)\,.$$

λ is also a smooth function. We expect f_N^t to be approximately of the form

$$g_N^t(x_1,\ldots,x_N) = \prod_{j=1}^{N} \frac{1}{M(\lambda(j/N,t)\,)}\, e^{\lambda(j/N,t)x_j}\,.$$

Let us assume $f_N^0 = g_N^0$, a special locally Gibbsian ensemble. We want to see how close f_N^t is to g_N^t for $t > 0$. As a measure of closeness we take relative entropy

$$\overline{H}_N(t) = \int \log \frac{f_N^t}{g_N^t} \cdot f_N^t\, e^{-\sum \phi(x_i)} \prod dx_i\,,$$

$$\overline{H}_N(0) = 0\,.$$

One can show that

$$\lim_{N\to\infty} \frac{1}{N}\overline{H}_N(t) = 0 \quad \text{for} \quad t > 0\,.$$

This proves hydrodynamic scaling. The profile is correct for g_N^t with exponentially small error probabilities using the large deviation theory. The entropy inequality

$$Q(A) \le \frac{H(Q,P)+1}{\log 1/P(A)}$$

takes care of the rest. We need to compute $\frac{d\overline{H}_N(t)}{dt}$. Although it is somewhat of a mess

$$
\begin{aligned}
\frac{\partial}{\partial t}\overline{H}_N(t) &= -\frac{\partial}{\partial t}\int f_N^t \log\frac{g_N^t}{f_N^t}\, e^{-\sum\phi(x_i)}\prod dx_i \\
&= -\int \frac{\dot{g}_N^t}{g_N^t} f_N^t\, e^{-\sum\phi(x_i)}\prod dx_i \\
&\quad -\int \dot{f}_N^t \log\frac{g_N^t}{f_N^t}\, e^{-\sum\phi(x_i)}\prod dx_i \\
&= -\int \frac{\dot{g}_N^t}{g_N^t} f_N^t\, e^{-\sum\phi(x_i)}\prod dx_i \\
&\quad +\int f_N^t (L_N \log\frac{f_N^t}{g_N^t})\, e^{-\sum\phi(x_i)}\prod dx_i \\
&\le \int \frac{(-\partial/\partial t + L_N^*)g_N^t}{g_N^t} f_N^t\, e^{-\sum\phi(x_i)}\prod dx_i\,.
\end{aligned}
$$

Based on the above general calculation in our special case, since we have

$$g_N^t = \exp\left[-\sum \log M(\lambda(\tfrac{j}{N},t)) + \sum \lambda(\tfrac{j}{N},t)x_j\right].$$

We can compute

$$\frac{1}{N}\frac{\partial}{\partial t}\overline{H}_N(t) = E^{f_N^t}\{\psi_N\},$$

where

$$
\begin{aligned}
\psi_N &= \frac{1}{N}\frac{(-(\partial/\partial t) + L_N^*)g_N^t}{g_N^t} \\
&= \frac{1}{N}\sum m(\tfrac{j}{N},t)\,\dot{\lambda}(\tfrac{j}{N},t) + \frac{1}{2N}\sum \lambda_{\theta\theta}(\tfrac{j}{N},t)\,\phi'(x_j) \\
&\quad -\frac{1}{N}\sum \dot{\lambda}(\tfrac{j}{N},t)x_j + \frac{1}{2N}\sum\left[\lambda_0(\tfrac{j}{N},t)\right]^2 \\
&\simeq \frac{1}{N}\sum m(\tfrac{j}{N},t)\,\dot{\lambda}(\tfrac{j}{N},t) + \frac{1}{2N}\sum \lambda_{\theta\theta}(\tfrac{j}{N},t)\,h'(\xi_{j,k}) \\
&\quad -\frac{1}{N}\sum \dot{\lambda}(\tfrac{j}{N},t)x_j + \frac{1}{2N}\sum\left[\lambda_0(\tfrac{j}{N},t)\right]^2 \\
&\simeq \frac{1}{N}\sum m(\tfrac{j}{N},t)\,\dot{\lambda}(\tfrac{j}{N},t) + \frac{1}{2N}\sum \lambda_{\theta\theta} h'(m(\tfrac{j}{N},t)) \\
&\quad +\frac{1}{2N}\sum \lambda_{\theta\theta}(\tfrac{j}{N},t)\,h''(m(\tfrac{j}{N},t))\,(\xi_{j,k} - m(\tfrac{j}{N},t)) \\
&\quad -\frac{1}{N}\sum \dot{\lambda}(\tfrac{j}{N},t)x_j + \frac{1}{2N}\sum\left[\lambda_\theta(\tfrac{j}{N},t)\right]^2 + \frac{1}{N}(\text{Error})\,.
\end{aligned}
$$

The equation $\partial m/\partial t = \frac{1}{2}(h'(m))_{\theta\theta}$ is equivalent to

$$\frac{\partial\lambda}{\partial t} = h''(m)\frac{\partial m}{\partial t} = \frac{1}{2}h''(m)\,\lambda_{\theta\theta}\,.$$

So only the error term survives.

$$\begin{aligned}\frac{d\frac{1}{N}\overline{H}_N(t)}{dt} &\le E^{f_N^t}\{\frac{1}{N}(\text{error})\}\\ &\le \frac{1}{\alpha N}\overline{H}_N(t) + \frac{1}{\alpha N}\log E^{g_N^t}\left\{e^{\alpha\{\text{error}\}}\right\}\end{aligned}$$

One shows that

$$\frac{1}{N}\log E^{g_N^t}\left\{e^{\alpha(\text{error})}\right\} \to 0 \quad \text{as} \quad N \to \infty .$$

Then Gronwall's inequality will take care of the rest.

Large Deviation Theory

The method yields superexponential estimates in equilibrium for deviations from the Boltzmann-Gibbs principle and this allows one to calculate the large deviation rates for deviations from hydrodynamical limit rather explicitly because fluctuations can be averaged with respect to local Gibbs distributions without affecting the expnential rate for the probabilities of large deviation. See [1], [4] for examples.

Nongradient Systems

We can use ideas from large deviation theory to handle the following modification of the model that we just considered. We change the Dirichlet form to

$$\frac{N^2}{2}\int \sum a(x_i, x_{i+1})\left(\frac{\partial u}{\partial x_i} - \frac{\partial u}{\partial x_{i+1}}\right)^2 e^{-\sum \phi(x_i)}\prod dx_i .$$

The corresponding operator is

$$\begin{aligned}L_N = &\frac{N^2}{2}\sum a(x_i, x_{i+1})\left(\frac{\partial}{\partial x_i} - \frac{\partial}{\partial x_{i+1}}\right)^2\\ &- \frac{N^2}{2}\sum W(x_i, x_{i+1})\left(\frac{\partial}{\partial x_i} - \frac{\partial}{\partial x_{i+1}}\right),\end{aligned}$$

where

$$W(x,y) = a(x,y)\left[\phi'(x) - \phi'(y)\right] - \left[a_x(x,y) - a_y(x,y)\right].$$

If we compute

$$d\left(\frac{1}{N}\sum J\left(\frac{i}{N}\right)x_i(t)\right)$$

and ignore the noise terms that are negligible, we now get

$$d\left(\frac{1}{N}\sum J\left(\frac{i}{N}\right)x_i(t)\right) \simeq \sum J'\left(\frac{i}{N}\right)W(x_i(t), x_{i+1}(t))\,dt .$$

There is a factor of $1/N$ missing because we could do summation by parts only once. On the other hand $E\{W(x_i, x_{i+1})\} = 0$ under every Gibbs measure so we need to do more analysis. By using ideas from large deviations one can establish hydrodynamical limits to an equation

$$\frac{\partial m}{\partial t} = \frac{1}{2} \frac{\partial}{\partial \theta} \left(\hat{a}(m(\theta, t)) \right) \left(h'(m(\theta, t))_\theta \right),$$

where $\hat{a}(m)$ is the Green-Kubo diffusion coefficient. It turns out one can actually give a variational characterization of $\hat{a}(m)$.

References

1. Donsker, M.D., Varadhan, S.R.S.: Comm.Pure Appl.Math. **42**, 243 (1989)
2. Fritz, J.: IMA **9** Springer-Verlag, Berlin – Heidelberg – New York (1987)
3. Guo, M.Z., Papanicolaou, G.C., Varadhan, S.R.S.: Comm.Math.Phys. **118**, 31 (1988)
4. Kipnis, C., Olla, S., Varadhan, S.R.S.: Comm.Pure Appl.Math. **42**, 115 (1989)
5. Olla, S., Varadhan, S.R.S.: Comm.Math.Phys. **135**, 355 (1991)
6. Olla, S., Varadhan, S.R.S., Yau, H.T.: In preparation
7. Freydoun Rezhakhanlou: Comm.Math.Phys. **129**, 445 (1990)
8. Varadhan, S.R.S.: Comm.Math.Phys. **135**, 313 (1991)
9. Yau, H.T.: Lett.Math.Phys. **22**, 63 (1991)
10. Yau, H.T.: These proceedings

Commun. Math. Phys. 155, 523–560 (1993)

Communications in
Mathematical
Physics

Hydrodynamical Limit for a Hamiltonian System with Weak Noise

S. Olla*, S. R. S. Varadhan, and H. T. Yau*****

Courant Institute of Mathematical Sciences, New York University, 251 Mercer Street, New York, NY 10012, USA

Received July 24, 1992; in revised form November 16, 1992

Abstract. Starting from a general Hamiltonian system with superstable pairwise potential, we construct a stochastic dynamics by adding a noise term which exchanges the momenta of nearby particles. We prove that, in the scaling limit, the time conserved quantities, energy, momenta and density, satisfy the Euler equation of conservation laws up to a fixed time t provided that the Euler equation has a smooth solution with a given initial data up to time t. The strength of the noise term is chosen to be very small (but nonvanishing) so that it disappears in the scaling limit.

1. Introduction

Let us begin by considering N particles on R^3, evolving according to a system of Hamiltonian equations. If (x_α, p_α), $\alpha = 1, 2, \dots, N$ are respectively the positions and momenta of the N indivdual particles with components $\{x^i_\alpha, p^i_\alpha\}$, $i = 1, 2, 3$ then the Hamiltonian equations of motion in the phase space $(R^3 \times R^3)^N$ are given by

$$\begin{cases} \dfrac{dx^i_\alpha}{dt} = \dfrac{\partial \mathscr{H}}{\partial p^i_\alpha} \\ \dfrac{dp^i_\alpha}{dt} = -\dfrac{\partial \mathscr{H}}{\partial x^i_\alpha} \end{cases} . \tag{1.1}$$

Here $\mathscr{H}(x,p)$ is the Hamiltonian

$$\mathscr{H}(x,p) = \frac{1}{2}\sum_\alpha \sum_{i=1}^{3} (p^i_\alpha)^2 + \frac{1}{2}\sum_{\alpha \neq \beta} V(x_\alpha - x_\beta).$$

---

* Permanent address: Politecnico di Torino, Dipartimento di Matematica, Consa Duca Degli Abruzzi 24, I-10129 Torino, Italy

** Research partially supported by U.S. National Science Foundation grants DMS 89001682, DMS 920-1222 and a grant from ARO, DAAL03-92-G-0317

*** Research partially supported by U.S. National Science Foundation grants DMS-9101196, DMS-9100383, and PHY-9019433-A01, Sloan Foundation Fellowship and David and Lucile Packard Foundation Fellowship

Here $V(x)$ is a radial function defining the pair interaction which is assumed to be short range, i.e. $V(x)$ has compact support on R^3. In addition to the total number of particles which is clearly conserved there are four additional conserved quantities of interest. They are the three components of the total momenta and the total energy given by the Hamiltonian itself. We have of course assumed tacitly the masses are all equal to unity.

Let $\varepsilon > 0$ be a small scale parameter. It enters in several places. The interaction $V(x)$ is assumed to have an effective range of order ε, so that $V(x)$ is taken to be of the form $V(x/\varepsilon)$. The number of particles N is assumed to be of order ε^{-3} so that even as $\varepsilon \to 0$ each particle typically interacts with a finite number of particles. We consider the following signed measures representing the spatial distribution of the conserved quantities, as a function of time:

$$
\begin{aligned}
\xi^0_\varepsilon(dx,t) &= \frac{1}{N}\sum_\alpha \delta(x - x_\alpha(t))dx\,, \\
\xi^i_\varepsilon(dx,t) &= \frac{1}{N}\sum_\alpha \delta(x - x_\alpha(t))p^i_\alpha(t)dx\,, \qquad i = 1,2,3 \\
\xi^4_\varepsilon(dx,t) &= -\frac{1}{N}\sum_\alpha \delta(x - x_\alpha(t))h_\alpha(t)dx\,, \\
h_\alpha(t) &= \frac{1}{2}\sum_{i=1}^{3}(p^i_\alpha)^2 + \frac{1}{2}\sum_{\beta\neq\alpha} V(x_\alpha - x_\beta)\,.
\end{aligned}
\tag{1.2}
$$

Note that ξ^4_ε has a negative sign and is the negative of energy instead of energy. We choose this convention for notational convenience later. We shall disregard this negative sign and use energy to refer to ξ^4 freely without further comment. As $\varepsilon \to 0$, $\xi^\mu_\varepsilon(dx,t)$ is expected to have limits $\xi^\mu(x,t)dx$ and these spatial densities of mass, momenta and energy in the limit are expected to satisfy the system of Euler equations for our model,

$$
\begin{cases}
\dfrac{d\varrho(x)}{dt} + \sum\limits_{j=1}^{3} \dfrac{\partial}{\partial x_j}[\varrho(x)\pi_j(x)] = 0 \\
\dfrac{d}{dt}(\varrho(x)\pi^i(x)) + \sum\limits_{j=1}^{3} \dfrac{\partial}{\partial x_j}[\varrho(x)\pi^i(x)\pi^j(x) + \delta_{ij}P] = 0 \\
\dfrac{d}{dt}(\varrho(x)e(x)) + \sum\limits_{j=1}^{3} \dfrac{\partial}{\partial x_j}[\varrho(x)e(x)\pi^j(x) - \pi^j(x)P] = 0\,.
\end{cases}
\tag{1.3}
$$

Here ϱ is the density; $\pi(x)$ is the velocity per particle; $e(x)$ is the energy per particle. $P(x)$ is the pressure which is a function of ϱ, π and e. This is a symmetric hyperbolic system of conservation laws.

Although one expects such a result on physical grounds there is very little in the way of rigorous proof that is available (cf. [DeM, Sp, Si]). For the special case of one dimensional hard rods with elastic collision the result is known [BDS]. This case is however unusual because there are infinitely many conserved quantities.

To get back to our general 3-dimensional model we have to modify the Hamiltonian in three ways before we can establish a result of the type described earlier.

The first modification is technical in nature and we replace the physical space R^3 by a finite cube with periodic boundaries i.e. the 3-torus T^3 (with unit volume). The phase space then becomes $T^3 \times R^3$ for each particle. Since $V(x/\varepsilon)$ has very small support it is easily extended as a periodic function on T^3 for all sufficiently small $\varepsilon > 0$.

The second modification is needed because we cannot handle the large velocities that might arise in our problem. We do not have effective truncation techniques. This forces us to modify the kinetic energy part $\frac{1}{2}\sum(p^i)^2$ to a function $\phi(p)$ which has a bounded gradient, thereby making the velocities uniformly bounded. A good example to keep in mind is the relativistic kinetic energy

$$\phi(p) = \left(c^2 + \sum_{i=1}^{3}(p^i)^2\right)^{1/2} - c^2 .$$

In the classical limit as $c \to \infty$, $\phi(p)$ becomes the classical kinetic energy.

The final modification is intrinsically more serious. This is the addition of a small amount of noise to the Hamiltonian system. This is needed because the Hamiltonian system in infinite volume has very poor ergodicity properties. One can write a family of spatially homogeneous invariant Gibbs measures on the infinite volume configuration space of positions and momenta. These are indexed by five parameters that correspond to density, the three components of average momenta and the average energy or temperature. Excluding complications of phase transition one would like to know that any spatially homogeneous invariant probability measure for the infinite volume evolution satisfying some mild regularity conditions is a superposition of these Gibbs distributions. This fact is essentially a form of a strong ergodic theorem. Since such a result is needed, but is unavailable, we modify the dynamics with noise so that the ergodic theorem we seek is available in part due to the noise. The noise itself is added only to the velocities when two particles are close, i.e. within a distance of order ε. They exchange velocities randomly and continuously but in such a way as to conserve the combined momenta and energy of the pair of particles. The strength of this noisy exchange is carefully regulated. It must be strong enough to provide us with the ergodicity that we need but at the same time should be weak enough that it does not alter the final Euler equations that one wants to derive. A precise formulation of the model and a statement of the results can be found in the next section.

There is a further restriction in our derivation: even if the initial data for the Euler equations are smooth and in the one-phase region, the solution will develop shocks or may enter in the region of phase coexistence. Our derivation is valid only in the smooth regime of the Euler equations.

Our approach in this paper is based on the relative entropy method used in [Y] for diffusion scaling limits (which simplifies the previous approach of [GPV] in that context).

2. Summary

2.1. The Model

The phase space in our model is $(T^3 \times R^3)^N$. The Hamiltonian is

$$\mathscr{H}(x,p) = \mathscr{H}_{\varepsilon,N}(x,p) = \sum_{\alpha} \phi(p_\alpha) + \frac{1}{2}\sum_{\alpha}\sum_{\beta\neq\alpha} V\left(\frac{x_\alpha - x_\beta}{\varepsilon}\right). \tag{2.1}$$

In this paper, we shall assume that $N = \varepsilon^{-3}$ for simplicity of notation. Since $V(\cdot)$ has compact support we can assume that for all sufficiently small ε, $\mathscr{H}_\varepsilon(x,p)$ is well defined on our phase space. Of the functions $\phi(p)$ on R^3 defining kinetic energy and $V(x)$ on R^3 defining the potential energy we assume the following:
(i) $V \in C^1(\mathbb{R}^3)$ is radial with compact support.
(ii) V can be written as $V(x) = V_0(x) + V_1(x)$ where $V_0 \geq 0$, $V_0(0) > 0$ and V_1 is positive definite.
(iii) $\phi \in C^2(\mathbb{R}^3)$, strictly convex and

$$\left|\frac{\partial \phi}{\partial p^j}\right| < c', \qquad \left|\frac{\partial^2}{\partial p^i \partial p^j}\right| < c'' \ \forall p. \tag{2.2}$$

The assumption that the potential V is radial is not essential, but simplifies notations.

The Hamiltonian defines a vector field on the phase space which is the Liouville operator

$$L_\varepsilon = \sum \frac{\partial \mathscr{H}_\varepsilon}{\partial p^i_\alpha} \frac{\partial}{\partial x^i_\alpha} - \sum \frac{\partial \mathscr{H}_\varepsilon}{\partial x^i_\alpha} \frac{\partial}{\partial p^i_\alpha}. \tag{2.3}$$

If it were classical evolution then we will have just L_ε. Our perturbation of L_ε is by a second order operator involving second derivatives in $\partial/\partial p^i_\alpha$. The precise form is not important. Let us just consider two sets of momenta (p^1_1, p^2_1, p^3_1) and (p^1_2, p^2_2, p^3_2) constituting $R^3 \times R^3 = R^6$. The four laws of conservation are

$$\begin{aligned} p^i_1 + p^i_2 &= c_i, \qquad i = 1,2,3, \\ \phi(p_1) + \phi(p_2) &= c_4. \end{aligned} \tag{2.4}$$

For each c_1, c_2, c_3, c_4 in a certain open set we have a hypersurface Γ_c in $\mathbb{R}^6$ of dimension 2. Observe that the assumption (iii) implies that almost all points in $\mathbb{R}^6$ are smooth points of the foliation Γ_c.

Let $\gamma(c)$ be a strictly positive function of c_1, c_2, c_3, c_4 that decays fast enough at infinity (for example exponentially). Let us consider the operator

$$\hat{L}_{1,2} = -\tfrac{1}{2}\gamma(c)\Delta_{c;1,2}, \tag{2.5}$$

where $\Delta_{c;1,2}$ is the Laplace-Beltrami operator on the hypersurface Γ_c. This is just to make a canonical choice of the noise we introduced in the dynamics. We can choose in fact a more general operator $\hat{L}_{1,2}$, as long as it is elliptic of second order acting only on tangential directions of Γ_c. We note that the operator $\hat{L}_{1,2}$ is clearly selfadjoint with respect to Lebesgue measure on R^6. The subscript $1,2$ in $\hat{L}_{1,2}$ refers to the fact that we have used particle labels $1,2$ for our momenta p^i_1 and p^i_2. Clearly there are similar operators $\hat{L}_{\alpha,\beta}$ for every pair α, β and these can all be viewed as acting on functions defined on our phase space. Our noisy generator then is of the form

$$\begin{aligned} \tilde{L}_\varepsilon &= L_\varepsilon + \hat{L}_\varepsilon, \\ \hat{L}_\varepsilon &= \theta(\varepsilon) \sum_{\alpha \neq \beta} \psi(\varepsilon, x_\alpha - x_\beta) \hat{L}_{\alpha,\beta}, \end{aligned} \tag{2.6}$$

where $\psi(\varepsilon, x_\alpha - x_\beta)$ provides control of the intensity of the noise and spatial cutoff for the exchange of velocities by limiting it to particles that are close in position space. The choice of the function $\psi(\varepsilon, x)$ is made as follows. We pick a function

$\psi(x)$ which is smooth, strictly positive for all x in R^3 and decays rapidly at ∞. Then define

$$\psi(\varepsilon, x) = \sum_n \psi\left(\frac{x-n}{\varepsilon}\right). \tag{2.7}$$

The summation is over all the lattice points in $\mathbb{R}^3$ so that $\psi(\varepsilon, x)$ is a function on T^3. This essentially turns the interaction on at all distances in the scale ε but with the intensity decaying very rapidly with the distance. Finally $\theta(\varepsilon)$ controls the intensity as a function of ε. It will go to infinity as $\varepsilon \to 0$ but such that $\varepsilon\theta(\varepsilon) \to 0$.

2.2. *The Formal Derivation of the Euler Equation*

Let us pretend for the purpose of our derivation that no noise term is present. We will associate with each configuration $(x_1, p_1), \ldots, (x_N, p_N)$ a group of five signed measures $\xi_\varepsilon^\mu(dx)$, $\mu = 0, 1, 2, 3, 4$ on T^3:

$$\xi_\varepsilon^\mu(dx) = \frac{1}{N}\sum_\alpha \delta(x - x_\alpha)\zeta_{\alpha,\varepsilon}^\mu dx\,, \tag{2.9}$$

where $\zeta_{\alpha,\varepsilon}^0 = 1$, $\zeta_{\alpha,\varepsilon}^i = p_\alpha^i$, $i = 1, 2, 3$ and

$$\zeta_{\alpha,\varepsilon}^4 = -\left\{\phi(p_\alpha) + \frac{1}{2}\sum_{\beta\neq\alpha} V\left(\frac{x_\alpha - x_\beta}{\varepsilon}\right)\right\}. \tag{2.10}$$

We shall adopt the convention of using μ, ν for the indices for conserved quantities; i for the three dimensions in $\mathbb{R}^3$ and α, β for the particles from 1 to $N = \varepsilon^{-3}$. For any test function $J(x)$ on T^3 clearly

$$\langle J, \xi_\varepsilon^\mu\rangle = \frac{1}{N}\sum_\alpha J(x_\alpha)\zeta_{\alpha,\varepsilon}^\mu\,.$$

Because of the evolution, (x_α, p_α) are functions $\{x_\alpha(t), p_\alpha(t)\}$ of t, and this makes $\xi_\varepsilon^\mu(dx)$ depend on t as well. We denote these by $\xi_\varepsilon^\mu(dx, t)$. We can use the Hamiltonian equations to compute $d\langle J, \xi_\varepsilon^\mu(t)\rangle$ and write it in the following form:

$$d\langle J, \xi_\varepsilon^\mu(t)\rangle = \sum_{i=1}^3 \langle J_i, \eta_{i,\varepsilon}^\mu(t)\rangle dt + \Omega_\varepsilon^\mu(J)\,. \tag{2.11}$$

Here

$$\eta_{i,\varepsilon}^\mu(t) = \frac{1}{N}\sum_\alpha \delta(x - x_\alpha(t))\theta_{i,\alpha,\varepsilon}^\mu(t) \tag{2.12}$$

with $\theta^{\mu}_{i,\alpha,\varepsilon}$ given by

$$\begin{aligned}\theta^0_{i,\alpha,\varepsilon}(t) &= \phi_i(p_\alpha(t)),\\ \theta^j_{i,\alpha,\varepsilon}(t) &= p^j_\alpha(t)\phi_i(p_\alpha(t))\\ &\quad -\frac{1}{2}\sum_{\beta\neq\alpha}\varepsilon^{-1}(x^i_\alpha - x^i_\beta)V_j((x_\alpha(t)-x_\beta(t))\varepsilon^{-1}),\\ \theta^4_{i,\alpha,\varepsilon}(t) &= \Bigg[\zeta^4_\alpha(t)\phi_i(p_\alpha(t)) + \frac{1}{2}\sum_{\beta\neq\alpha}\varepsilon^{-1}(x^i_\alpha(t)-x^i_\beta(t))\\ &\quad \times\left\{\sum_{j=1}^{3} V_j(\varepsilon^{-1}(x_\alpha(t)-x_\beta(t))\frac{1}{2}(\phi_j(p_\alpha(t))+\phi_j(p_\beta(t))\right\}\Bigg].\end{aligned} \tag{2.13}$$

We have used the notation $\phi_j = \partial\phi/\partial p^j$ and $V_j = \partial V/\partial x_j$.

The $\Omega^{\mu}_{\varepsilon}$ terms are error terms and they are explicitly given by

$$\begin{aligned}\Omega^0_\varepsilon &= 0,\\ \Omega^i_\varepsilon &= -\frac{\varepsilon^2}{2}\sum_{\alpha\neq\beta}\left[J(x_\alpha(t)) - J(x_\beta(t)) - \sum_{r=1}^{3}(x^r_\alpha(t)-x^r_\beta(t))J_r(x_\alpha(t))\right]\\ &\quad\times V_i\left(\frac{1}{\varepsilon}(x_\alpha(t)-x_\beta(t))\right),\\ \Omega^4_\varepsilon &= \frac{\varepsilon^2}{2}\sum_{\alpha\neq\beta}\left[J(x_\alpha(t)) - J(x_\beta(t)) - \sum_{r=1}^{3}(x^r_\alpha(t)-x^r_\beta(t))J_r(x_\alpha(t))\right]\\ &\quad\times\left[\sum_{s=1}^{3}V_s\left(\frac{1}{\varepsilon}(x_\alpha(t)-x_\beta(t))\right)\frac{\phi_s(p_\alpha(t))+\phi_s(p_\beta(t))}{2}\right].\end{aligned} \tag{2.14}$$

We will show that as $\varepsilon\to 0$ the $\Omega^{\mu}_{\varepsilon}$ terms become negligible.

In order to close the equations specified by (2.11) we need to express the quantities $\eta^{\mu}_{i,\varepsilon}$ as $\varepsilon\to 0$ back in terms of ξ^{μ}_{ε}. This is done using Gibbs distributions (cf. Sect. 5). Given the parameters of density, the three components of momenta and average energy one expects to have a Gibbs distribution uniquely specified by these parameters. The Gibbs distribution describes the statistics of local configurations as functions of these five parameters. If one takes averages with respect to the Gibbs distribution then $\eta^{\mu}_{i,\varepsilon}$ can be replaced by $A^{\mu}_i(q)$ as ξ^{μ}_{ε} gets replaced by q^{μ} in the weak limit as $\varepsilon\to 0$. The Euler equations then take the form

$$\frac{\partial q^\mu}{\partial t} + \sum_{r=1}^{3}[A^{\mu}_i(q)] = 0, \qquad \mu = 0,1,2,3,4. \tag{2.15}$$

The matrix A can be computed explicitly. If $\phi(p)=p^2$, then by comparing with (1.3) A is given by

$$\begin{aligned}A^0_j &= q^j,\\ A^i_j &= \delta_{ij}P + q_iq_j/q_0,\\ A^4_j &= q^j(q_4-P)/q_0.\end{aligned} \tag{2.16}$$

Note that in order to establish a connection with the Euler equation in (1.3), q^μ is related to density, momenta and energy by

$$q^0 = \varrho\,, \qquad q^i = \varrho\pi^i\,; \qquad q^4 = \varrho e\,, \tag{2.17}$$

in other words q^1, q^2, q^3, and q^4 are momenta and energy per *volume* instead of *per particle* as in the usual Euler equation (1.3). In the general case A is still explicitly computable, as seen in the next section.

2.3. A Precise Statement of the Results

We start with our Hamiltonian in the microscopic scale on configurations of N points in the phase space $(\Lambda \times R^3)^N$ where Λ is some domain in the physical space R^3. If we denote the configuration by (x_α, p_α) then the Hamiltonian is

$$\mathscr{H}_N(p,x) = \sum_\alpha \phi(p_\alpha) + \frac{1}{2}\sum_{\alpha\neq\beta} V(x_\alpha - x_\beta)\,. \tag{2.18}$$

Taking Λ to be a bounded cube we can define a partition function

$$Z_\Lambda(\lambda^0,\lambda^1,\lambda^2,\lambda^3,\lambda^4) = \sum_{N=0}^\infty \frac{1}{N!}\int\cdots\int_{(\Lambda\times\mathbb{R}^3)^N} \exp\left[\sum_{\alpha=1}^N\sum_{\mu=0}^4 \lambda^\mu\zeta^\mu_\alpha\right] \times dx_1\ldots dx_N\, dp_1\ldots dp_N\,, \tag{2.19}$$

and the pressure $\left(\text{note the definition of pressure where the prefactor } T = \frac{1}{\lambda^4} \text{ has been omitted}\right)$

$$\psi(\lambda^0,\ldots,\lambda^4) = \lim_{\Lambda\Uparrow R^3} \frac{1}{|\Lambda|}\log Z_\Lambda\,. \tag{2.20}$$

Here $\zeta^\mu_\alpha = \zeta^\mu_{\alpha,\varepsilon=1}$ with $\zeta^\mu_{\alpha,\varepsilon}$ defined in (2.10). See Sect. 5 or [R] for the existence of the limit (2.20).

The Gibbs measure corresponding to $\lambda^0,\lambda^1,\lambda^2,\lambda^3,\lambda^4$ is a point process on $R^3 \times R^3$ which is invariant under space translations and has the DLR property, namely the conditional distribution of the possible configurations in a box Λ, given the configuration outside has density proportional to

$$\frac{e^{\lambda^0 n}}{n!}\exp\left[-\lambda^4\tilde{\mathscr{H}}_n + \sum_{i=1}^3\sum_{\alpha=1}^n \lambda^i p^i_\alpha\right] \tag{2.21}$$

on $\bigcup_n(\Lambda\times\mathbb{R}^3)^n$, where $\tilde{\mathscr{H}}_n$ is the Hamiltonian of the n-point configuration we introduced above to which has been added the interaction energy due to the configuration outside the box. This is the term

$$\sum_\alpha\sum_y V(x_\alpha - y)\,,$$

where α runs over the particles inside Λ and y runs over the locations of all particles outside Λ. The general theory of equilibrium statistical mechanics provides with a set

W of possible values of $\lambda = (\lambda^0, \lambda^1, \lambda^2, \lambda^3, \lambda^4)$ in R^5 such that W is an open set in R^5 and for each λ in W we have a unique Gibbs measure and the Gibbs measure is ergodic (with respect to space translations). The five quantities which are averages of density, momenta and energy are given as functions of λ and they can be computed from the free energy. As λ varies over W, these averages $q = (q^0, \dots, q^4)$ vary over a set U and the correspondence between λ and q is one to one and smooth as long as λ and q are restricted respectively to W and U. To make the correspondence between λ and q transparent, let us define the thermodynamic entropy function

$$\hat{s}(q) = \sup_{\lambda} \left[\sum_{\mu=0}^{4} q^\mu \cdot \lambda^\mu - \psi(\lambda) \right]. \tag{2.22}$$

Then λ and q are related by the formulae

$$\lambda^\mu = \partial \hat{s} / \partial q^\mu, \qquad q^\mu = \partial \psi / \partial \lambda^\mu. \tag{2.23}$$

Let us denote the Gibbs measure with chemical potentials λ by $d\mu_\lambda$, or $d\mu_q$ if one specifies the parameter q. The matrix A can now be computed explicitly by

$$A_j^\mu(q) = \lim_{k \to \infty} (2k)^{-3} \int \sum_\alpha \theta_{j,\alpha}^\mu I_k(x_\alpha) d\mu_q. \tag{2.24}$$

Here $\theta_{j,\alpha}^\mu = \theta_{j,\alpha,\varepsilon=1}^\mu$ and I_k is the indicator function of $D_k = \{x \in \mathbb{R}^3 | \, |x_i| \le k, i = 1, 2, 3\}$. Explicitly, A is given by

$$\begin{aligned} A_j^0 &= -\frac{\lambda^j}{\lambda^4} q^0, \\ A_j^i &= -\frac{\lambda^i q^j}{\lambda^4} - \frac{P}{\lambda^4} \delta_{i,j}, \\ A_j^4 &= -\frac{\lambda^j}{\lambda_4} q^4 + \frac{\lambda^j}{\lambda^4} P. \end{aligned} \tag{2.25}$$

Here $\lambda = \partial \hat{s} / \partial q$ (2.23) and P is the pressure as a function of q. By the virial theorem for dilations we have a diect relation between P and the interaction V (cf. [V]):

$$\frac{1}{2} E^{\mu_q} \left[\sum_{\alpha \ne \beta} V_i(x_\alpha - x_\beta) \chi(x_\alpha)(x_\alpha - x_\beta)^j \right] = \delta_{i,j} \left(\frac{P}{\lambda^4} - q^4 \right), \tag{2.26}$$

where χ is a non-negative function with compact support and total integral 1.

In order to see where (2.25) come from, let us compute the expectation of $p^j \phi_i(p)$ with respect to the density $d\omega_\lambda = \exp[\lambda^4 \phi(p) + \boldsymbol{\lambda} \cdot \mathbf{p}]$/Normalization,

$$\int p^j \phi_i d\omega_\lambda(p) = -q^j \frac{\lambda^i}{\lambda^4} - \frac{\lambda_{ij}}{\lambda^4}.$$

The second term in the above expression is the contribution of ideal gas to the pressure tensor. This phenomena occurs also in computing A_j^i and A_j^4.

If we give C^∞ initial data to our hyperbolic system, up to a finite positive time, there exists a C^∞ solution. Then we can assume that we are given a C^∞ solution of our hyperbolic system of conservation laws in some time interval $0 \le t \le T_0$, such that $q(t, x)$ for $t \in [0, T_0]$ and $x \in T^3$ lies in a compact set $U_1 \subset U$ with $\text{dist}(U_1, \mathbb{R}^5 \backslash U) \ge 2\delta_1$, for some constant δ_1. Then the corresponding λ is a C^∞

function $\lambda(t, x)$ that lies in a compact set $W_1 \subset W$ with dist$(W_1, \mathbb{R}^5 - W) \geq 2\delta_2$ for some constant δ_2.

Let us consider the following time dependent density function $\hat{g}_N(t)$ on the phase space of N particles in $(T^3 \times \mathbb{R}^3)^N$,

$$\hat{g}_N(t, x_1, \ldots, x_N, p_1, \ldots, p_N) = \frac{1}{c_N(t)} \exp\left[\sum_{\alpha=1}^{N} \sum_{\mu=0}^{4} \lambda^\mu(t, x_\alpha)\zeta^\mu_{\alpha,\varepsilon}(\mathbf{x}, \mathbf{p})\right], \quad (2.27)$$

where $c_N(t)$ is the normalizing constant.

We can take $\hat{g}_N(0, x_1, \ldots, x_N, p_1, \ldots, p_N)$ as the initial value at $t = 0$ of the Kolmogorov or Fokker-Planck equation

$$\frac{\partial g}{\partial t} = \tilde{L}^*_\varepsilon g, \qquad g|_{t=0} = \hat{g}|_{t=0}$$

and the solution $g_N(t, x_1, \ldots, x_N, p_1, \ldots, p_N)$ is the actual distribution at time t of our noisy evolution.

Our main theorem is the following:

Theorem 2.1. *Under the hypotheses listed above, for any* $t \in [0, T_0]$,

$$\lim_{N\to\infty} \frac{1}{N} \int \underset{(T^3\times R^3)^N}{\int} g_N \log \frac{g_N}{\hat{g}_N}\, dx_1 \ldots dx_N dp_1 \ldots dp_N = 0\,.$$

Remark. The above theorem implies hydrodynamic behavior. For instance it is elementary to check that under $\hat{g}_N$-probability $\frac{1}{N}\sum J(x_\alpha)$ is nearly equal to $\int J(x)q^0(t, x)dx$ and in fact by the theory of large deviation the probability that $\frac{1}{N}\sum J(x_\alpha)$ deviates significantly from $\int J(x)q^0(t, x)dx$ is exponentially small under $\hat{g}_N$ (Corollary 5.8). By entropy estimates (2.29) the probability is small under g_N and this establishes the correct hydrodynamic behavior.

More precisely, for any two probability measures α and β the well known entropy inequality (in fact, a special case of Jensen's inequality) states that for any measurable function F,

$$E^\alpha[F] \leq \log E^\beta[\exp F] + H(\alpha/\beta)\,. \quad (2.28)$$

Here $H(\alpha/\beta)$ denotes the relative entropy of α with respect to β [indeed H is defined to be the smallest constant for which (2.28) holds]. A special case of (2.28) gives

$$E^\alpha(A) \leq \frac{\log 2 + H(\alpha/\beta)}{\log(1 + 1/E^\beta(A))}\,. \quad (2.29)$$

Hence if $N^{-1}H(\alpha/\beta) \to 0$ as $N \to \infty$ then any set A which has exponentially small probability with respect to β will have small probability with respect to α. In our case β is local Gibbs state $\hat{g}_N$ and A is the set

$$A_\delta = \{|\langle J, \xi^\mu_\varepsilon\rangle - \langle J, q^\mu\rangle| > \delta\}\,.$$

For arbitrary $\delta > 0$, $E^{\hat{g}}[A_\delta] \leq \exp[-const(\delta)N]$ for some const$(\delta) > 0$ by large deviation theory in the appendix (see Corollary 5.8). Hence we conclude the following

Corollary 2.2. *Under the assumption of Theorem 2.1,*

$$\overline{\lim_{\varepsilon\to 0}}\, E^{g_N}[A_\delta] = 0\,.$$

This corollary establishes the hydrodynamical limit under the usual definition.

3. An Outline of the Proof

On the phase space $(T^3 \times \mathbb{R}^3)^N$ we have two time dependent families of probability densities defined for $0 \le t \le T_0$. One of them $\hat{g}_N(t, \mathbf{x}, \mathbf{p})$ is a family of local Gibbs measures constructed from a smooth solution of the Euler equations, i.e.

$$\log \hat{g}_N = \sum_{\alpha=1}^{N} \sum_{\mu=0}^{4} \lambda^\mu(t, x_\alpha)\zeta^\mu_{\alpha,\varepsilon} - \log c_N(t), \tag{3.1}$$

where ζ was defined by (2.10) and $c_N(t)$ is the normalization constant

$$c_N(t) = \int \hat{g}_N(t, \mathbf{x}, \mathbf{p}) d\mathbf{x}\, d\mathbf{p}. \tag{3.2}$$

The other family $g_N(t, \mathbf{x}, \mathbf{p})$ is the actual probability density of the noisy dynamics assuming that at time $t = 0$ the initial distribution is the same as $\hat{g}_N(t, \mathbf{x}, \mathbf{p})$ at $t = 0$. In other words g_N is the solution of the Fokker-Planck or Kolmogorov equation

$$\frac{\partial g_N}{\partial t} = \tilde{L}^*_\varepsilon g_N, \qquad g_N|_{t=0} = \hat{g}_N(0, \mathbf{x}, \mathbf{p}). \tag{3.3}$$

Our goal is to show that g_N is close to $\hat{g}_N$ for $0 \le t \le T_0$ by establishing

$$\lim_{N\to\infty} \frac{1}{N} \int \log \frac{g_N}{\hat{g}_N} \cdot g_N d\mathbf{x}\, d\mathbf{p} = 0.$$

Let us define

$$H_N(t) = \int \log \frac{g_N}{\hat{g}_N} g_N d\mathbf{x}\, d\mathbf{p}. \tag{3.4}$$

We know by construction that $N^{-1}H_N(0) = 0$. The first step is to establish a differential inequality for $H_N(t)$.

Lemma 3.1. *Let $f(t, x)$ be a solution of the forward equation of a Markov process with generator L, being represented as densities relative to some fixed reference measure μ. Let $u(t, x)$ be any arbitrary family of densities relative to μ. We shall assume enough smoothness to make our computations valid. Then for every $t > 0$,*

$$\frac{\partial}{\partial t} \int \log \frac{f(t,x)}{u(t,x)} f(t,x) d\mu \le \int \left(L^* u - \frac{\partial u}{\partial t} \right) \cdot \frac{1}{u} f\, d\mu(x).$$

Proof. Let us differentiate $\int \log \frac{f(t,x)}{u(t,x)} \cdot f(t,x) d\mu$ with respect to t. An elementary calculation yields

$$\frac{\partial}{\partial t} \int \log \frac{f(t,x)}{u(t,x)} f(t,x) d\mu = -\int \frac{1}{u} \frac{\partial u}{\partial t} \cdot f\, d\mu + \int L^* f \log \frac{f}{u}\, d\mu + \int \frac{\partial f}{\partial t}\, d\mu.$$

The third term $\int \frac{\partial f}{\partial t}\, d\mu$ integrates to zero because $\int f\, d\mu = 1$ for all $t > 0$. Integrating by parts,

$$\begin{aligned} \int L^* f \log \frac{f}{u}\, d\mu &= \int f L \log \frac{f}{u}\, d\mu \le \int f \frac{L(f/u)}{f/u}\, d\mu \\ &= \int u \left(L \frac{f}{u} \right) d\mu = \int (L^* u) \frac{f}{u}\, d\mu = \int \left(\frac{L^* u}{u} \right) f \cdot d\mu. \end{aligned}$$

The inequality $L \log u \le Lu/u$ for any nonegative u is a consequence of the maximum principle for L that dictates that for any convex function ϕ,

$$L\phi(u) \ge \phi'(u)Lu\,. \quad \square$$

Remark. For the case we are interested in with L given by (2.6) the error term in Lemma 3.1 can be computed explicitly,

$$\begin{aligned}\frac{\partial}{\partial t}\int g_N \log\frac{g_N}{u}\,d\mathbf{x}\,d\mathbf{p} &= \int u^{-1}\Big(L_\varepsilon^* u - \frac{\partial u}{\partial t}\Big) g_N d\mathbf{x}\,d\mathbf{p}\\ &\quad + \theta(\varepsilon)\int \sum_{\alpha\neq\beta}\psi(\varepsilon, x_\alpha - x_\beta)\gamma(c)\\ &\quad \times |\nabla_{c;\alpha,\beta} h_N|^2 h_N^{-1} u(t,x) d\mathbf{x}\,d\mathbf{p}\,,\end{aligned}$$

where $h = g_N/u$. The last term is traditionally called the Dirichlet form. In particular if we let $u = \hat{g}_{\lambda,N}$ with λ constant, then the first term on the right side vanishes. Let $\bar{h} = T_0^{-1}\int_0^{T_0} h(t)dt$ be the time average and integrate from 0 to T_0,

$$E^{\hat{g}_{\lambda,N}}\Big[\sum_{\alpha\neq\beta}\psi(\varepsilon, x_\alpha - x_\beta)\gamma(c)\,|\nabla_{c;\alpha,\beta}\bar{h}_N|^2 \bar{h}_N^{-1}\Big] \le \frac{H_N(0)}{T_0}\,. \tag{3.5}$$

Here we have used the convexity of the Dirichlet form with respect to h. For a more detailed explanation of this, see [GPV].

Lemma 3.1 was first proved in [Y] for the special case of the Ginzburg-Landau model. Our proof here generalizes it to include all stochastic processes.

Returning to our problem at hand let us write

$$\log\hat{g}_N(t,\mathbf{x},\mathbf{p}) = u_N(t,\mathbf{x},\mathbf{p})\,. \tag{3.6}$$

We want to calculate the function

$$\psi_N(t,\mathbf{x},\mathbf{p}) = \hat{g}_N^{-1}\Big[\tilde{L}^*\hat{g}_N - \frac{\partial}{\partial t}\hat{g}_N\Big].$$

Lemma 3.2. *Recall that $\tilde{L}_\varepsilon = L_\varepsilon + \hat{L}_\varepsilon$, where $\hat{L}_\varepsilon$ is the contribution from the noise terms and L_ε is the pure Hamiltonian term or the Liouville operator. Then*

$$\limsup_{N\to\infty}\sup_{0\le t\le T_0}\frac{1}{N}\int(\hat{L}_\varepsilon\hat{g}_N)\cdot\frac{1}{\hat{g}_N}\cdot g_N d\mathbf{x}\,d\mathbf{p} = 0\,.$$

Proof. Let us note that $\hat{L}_\varepsilon$ has the form

$$\hat{L}_\varepsilon = \theta(\varepsilon)\sum\psi(\varepsilon, x_\alpha - x_\beta)\hat{L}_{\alpha,\beta}$$

and that $L_{\alpha,\beta}$ acts only on the momentum variables. It is easy to estimate, using the conservation properties of $\hat{L}_{\alpha\beta}$,

$$\hat{g}_N^{-1}\hat{L}_\varepsilon\hat{g}_N \le C\varepsilon\theta(\varepsilon)\sum\psi(\varepsilon, x_\alpha - x_\beta)\,.$$

We therefore have only to prove

$$\limsup_{N\to\infty} \frac{1}{N} \int \sum_{\alpha,\beta} \psi(\varepsilon, x_\alpha - x_\beta) g_n d\mu \le C$$

in order to establish the lemma, because we have assumed that $\varepsilon\theta(\varepsilon) \to 0$. This is a consequence of the fast decay of ψ and Lemma 5.9. □

We can now concentrate on the Liouville term $\left(L_\varepsilon^* - \frac{\partial}{\partial t}\right) u_N$.

Lemma 3.3. *We can write*

$$\left(L_\varepsilon^* - \frac{\partial}{\partial t}\right) u_N = \psi_N^{(1)}(t,x,p) + \varepsilon_N(t,x,p)\,,$$

where

$$\psi_N^{(1)}(t,x,p) = -\sum_{\mu=0}^{4} \sum_{j=0}^{3} \sum_{\alpha} \lambda_j^\mu(t,x_\alpha)\theta_{j,\alpha,\varepsilon}^\mu + a_N(t)$$

and

$$\limsup_{N\to\infty} \sup_{0\le t\le T_0} \frac{1}{N} \int \varepsilon_N(t,x,p) g_N d\mu = 0\,.$$

Here $\theta_{j,\alpha,\varepsilon}^\mu$, $j = 1,2,3$ was defined in (2.13), $\theta_{0,\alpha,\varepsilon}^\mu = \zeta_{\alpha,\varepsilon}^\mu$ *(see* (2.10)*)*, $\lambda_0^\mu = \partial\lambda^\mu/\partial t$, $\lambda_j^\mu = \partial\lambda^\mu/\partial x_j$ *and a_N is given by*

$$a_N(t) = \frac{d}{dt} - [\log c_N(t)] = -\int \sum_{\mu=0}^{4} \sum_{\alpha=1}^{N} \lambda_0^\mu(t,x_\alpha)\zeta_{\alpha,\varepsilon}^\mu(\mathbf{x},\mathbf{p})\hat{g}_N(t,\mathbf{x},\mathbf{p}) d\mathbf{x}\, d\mathbf{p}\,.$$

Proof. Proof is just by direct computation (see computation in Sect. 2.2). The error term arises because of terms of the form [cf. (2.14)]

$$\varepsilon^{-1} \sum_{\alpha,\beta} V_i\left(\frac{x_\alpha - x_\beta}{\varepsilon}\right) \lambda^i(t,x_\alpha) - \sum_{\alpha,\beta} \sum_{j=1}^{3} \frac{\partial\lambda^i}{\partial x_j}(t,x_\alpha)(x_\alpha^j - x_\beta^j) V_i\left(\frac{x_\alpha - x_\beta}{\varepsilon}\right).$$

Since V_i is antisymmetric, we can rewrite it as

$$\varepsilon^{-1} \sum_{\alpha,\beta} \left[\lambda^i(t,x_\alpha) - \lambda^i(t,x_\beta) - \sum_{j=1}^{3} \frac{\partial\lambda^i}{\partial x_j}(t,x_\alpha)(x_\alpha^j - x_\beta^j)\right] V((x_\alpha - x_\beta)/\varepsilon)\,.$$

By Taylor's theorem, it is bounded by

$$(2\varepsilon)^{-1} \sum_{\alpha,\beta} |V_i(x_\alpha - x_\beta)/\varepsilon)| \cdot O(\varepsilon^2)\,.$$

By (3.5) and Lemma 5.9, $E^g\left[\left|\sum_{\alpha,\beta} V_j((x_\alpha - x_\beta)/\varepsilon)\right|\right] \le \text{const}\, N$. Thus we have proved Lemma 3.3. □

Another easy consequence of Lemma 5.9 is

Lemma 3.4. *For some constant $C < \infty$,*

$$\int \sum_{\alpha} |\theta^{\mu}_{j,\alpha}| g_N dx\, dp \leq CN\,, \qquad \int \sum_{\alpha} |\theta^{\mu}_{j,\alpha}| \hat{g}_N dx\, dp \leq CN\,.$$

For any configuration (x_α, p_α) on $(T^3 \times R^3)^N$ and for any $x \in T^3$ we shall expand a neighborhood of x by $1/\varepsilon$ and let $y_\alpha = \frac{1}{\varepsilon}(x_\alpha - x)$. The resulting configuration can be thought of as a configuration in R^3 although it is a little fuzzy at the edges. However the configuration $\{y_\alpha\}$ is well-defined on compact domains in R^3 provided ε is sufficiently small. Of course ε has to be smaller if the compact domain is larger. The configuration so obtained will be denoted by $\omega_{\varepsilon,x}$. If $F(\omega)$ is a functional on configuration space in R^3 that depends only on particles in a compact domain, then $F(\omega_{x,\varepsilon})$ is well defined if ε is small enough. Moreover we can attach the momenta p_α to each x_α so the functional $F(\omega)$ can be a functional of configurations in the phase space.

Recall $\theta^{\mu}_{i,x,\varepsilon}$ defined in (2.13). We shall denote $\theta^{\mu}_{i,\alpha,\varepsilon=1}$ by $\theta^{\mu}_{i,\alpha}$. Let $\chi(x)$ be a positive smooth function on $\mathbb{R}^3$ with compact support and total integral 1. Define the local functional on the configurations of phase space by

$$\theta^{\mu}_{i}(\omega) = \sum_{\alpha} \chi(x_\alpha)\theta^{\mu}_{i,\alpha}(\omega)\,. \tag{3.7}$$

Let $I_k(x)$ be the indicator function of $D_k = \{x \in \mathbb{R}^3 | \, |x_i| \leq k\}$ and let $\chi_k = (2k)^{-3}\chi * I_k$. For any local functional F, define F_k to be its average over D_k,

$$F_k(\omega) = (2k)^{-3} \int I_k(y) F(\tau^y \omega) dy\,. \tag{3.8}$$

Here τ^y is the space translation by y. With these conventions, $\theta^{\mu}_{i,k} = (\theta^{\mu}_{i})_k$ is well defined.

Lemma 3.5. *For any smooth function J and $\mu = 0, \ldots, 4$, $j = 1, 2, 3$,*

$$\overline{\lim_{k\to\infty}}\ \overline{\lim_{N\to\infty}}\ E^{g}\left[N^{-1} \left| \sum_{\alpha} J(x_\alpha)\theta^{\mu}_{j,\alpha,\varepsilon}(\omega) - \int_{T^3} J(x)\theta^{\mu}_{j,k}(\omega_{\varepsilon,x}) dx \right| \right] = 0\,. \tag{3.9}$$

Remark. Similarly we can define $\zeta^{\mu}_{k} = (\zeta^{\mu})_k$ with $\zeta^{\mu}_{\alpha} = \zeta^{\mu}_{\alpha,\varepsilon=1}$ defined in (2.10). A parallel identity holds for ζ. Intuitively, $(\zeta^{\mu})_k(\omega_{e,x})$ is the local empirical density, momenta and energy.

Proof. By definition

$$\begin{aligned} \int_{T_3} J(x)\theta^{\mu}_{i,k}(\omega_{\varepsilon,x}) dx &= \sum_{\alpha} \int_{T_3} dx\, J(x)\chi_k(\varepsilon^{-1}(x + x_\alpha))\theta^{\mu}_{i,\alpha,\varepsilon} \\ &= \frac{1}{N} \sum_{\alpha} J_{k,\varepsilon}(x_\alpha)\theta^{\mu}_{i,\alpha,\varepsilon}\,. \end{aligned}$$

Here $J_{k,\varepsilon}$ is defined by

$$J_{k,\varepsilon}(z) = \int_{T_3} J(x) N \chi_k(\varepsilon^{-1}(x + z)) dx\,.$$

We can bound the expectation in (3.9) by

$$E^g\left[\sum_\alpha |J - J_{k,\varepsilon}|(x_\alpha)\,|\theta^\mu_{j,\alpha,\varepsilon}(\omega)|\right] \le \|J - J_{k,\varepsilon}\|_\infty E^g\left[\sum_\alpha |\theta^\mu_{j,\alpha,\varepsilon}|\right].$$

By Lemma 3.4, $N^{-1}E^g\left[\sum_\alpha |\theta^\mu_{j,\alpha,\varepsilon}|\right]$ is bounded. Lemma 3.5 thus follows from the smoothness assumption of J. □

Combining Lemma 3.1–3.5 we have

Lemma 3.6. *Assume that λ^μ is smooth up to time T. Then*

$$\overline{\lim_{k\to\infty}}\ \overline{\lim_{N\to\infty}}\ \sup_{0\le t\le T} N^{-1}\frac{dH_N(t)}{dt}$$

$$+ E^g\left[\int_{T^3}\sum_{\mu=0}^{4}\sum_{j=0}^{3}\lambda^\mu_j(t,x)\theta^\mu_{j,k}(\omega_{x,\varepsilon})dx\right] + N^{-1}a_N(t) \le 0\,.$$

Furthermore, we can compute a_N as

$$\lim_{N\to\infty} N^{-1}a_N(t) = \int\sum_{\mu=0}^{4}\lambda^\mu_0(x,t)q^\mu(x,t)dx\,.$$

Proof. The only thing that needs to be proved is the formula for a_N. But it is a consequence of Lemma 3.3 and large deviations in Sect. 5 [cf. (5.20), (5.21)]. □

We now introduce some cutoffs. There are two sources of unboundedness. The first kind appears due to the unboundedness of $\phi(p_\alpha)$ and p^i_α. Let us introduce a cutoff l and replace $\phi(p_\alpha)$ by $\min\{\phi(p_\alpha), l\}$ (similarly for p^i_α). Let us denote the resulting θ with these replacements by $\theta^\mu_{j,\alpha,\varepsilon,l}, \theta^\mu_{j,k,l}$, etc. The next lemma shows that the error we made is negligible. Its proof is again an easy application of the entropy inequality (2.28).

Lemma 3.7. *For $\mu = 0,\dots,4$, $j = 1,2,3$,*

$$\overline{\lim_{l\to\infty}}\ \overline{\lim_{N\to\infty}}\ E^{\tilde g_N}\left[\int_{T^3}|\theta^\mu_{j,k,l}(\omega_{x,\varepsilon}) - \theta^\mu_{j,k}(\theta_{x,\varepsilon})|dx\right] = 0\,.$$

Here k is arbitrary positive and $\tilde g_N$ can either be g_N (3.3) or $\hat g_N$ (3.1).

The next step we have to be concerned about is that some of the empirical quantities ζ^μ_k characterizing the local equilibrium states may not lie in the admissible region U defined before (2.22). Let U_2 be a compact set satisfying [with δ_2 defined before (2.22)]

$$\mathrm{dist}(U_1, \mathbb{R}^3\backslash U_2) \ge \delta_2\,,\qquad \mathrm{dist}(U_2, \mathbb{R}^3\backslash U) \ge \delta_2\,.$$

Define an indicator function σ_k by

$$\sigma_k(\omega) = 1_{U_2}\{\zeta^\mu_k(\omega)\}\,. \tag{3.10}$$

The next lemma states that the error we made from introducing this cutoff is again negligible. Its proof relies on large deviation and will be proved in Sect. 5 after (5.22).

Lemma 3.8. *There exist $\delta_0 > 0$ such that for any local functional A of the type* (5.18),

$$\overline{\lim_{k\to\infty}}\ \overline{\lim_{N\to\infty}}\ E^{g_N}\left[\int_{T^3}(1-\sigma_k)(\omega_{\varepsilon,x})A(\omega_{\varepsilon,x})dx\right] - \delta_0^{-1}N^{-1}H(g_N|\hat{g}_N) \le 0.$$

Now we state the local ergodic theorem to be proved in the next section.

Definition 3.9. For any local functional $F(\omega)$ define

$$\hat{F}(q) = E^{\mu_q}[F].$$

Here μ_q is the unique Gibbs measure with average density, momenta and energy given by q. This is well defined if $q \in U$.

The next theorem will be proved in Sect. 4.

Theorem 3.10 (Local ergodic theorem.) *For any bounded local functional F we have*

$$\overline{\lim_{k\to\infty}}\ \overline{\lim_{N\to\infty}}\int_0^T dt\, E_N^g\left[\int_{T^3}|F_k(\omega_{\varepsilon,x}) - \hat{F}(\zeta_k(\omega_{\varepsilon,x}))|\sigma_k(\theta_{\varepsilon,x})dx\right] = 0.$$

Let us pause to see what we have so far. By Lemma 3.6 and cutoffs in Lemmas 3.7 and 3.8 we have

$$\begin{aligned}
&\overline{\lim_{l\to\infty}}\ \overline{\lim_{k\to\infty}}\ \overline{\lim_{N\to\infty}}\ \sup_{0\le t\le T} N^{-1}\frac{dH_N(t)}{dt}\\
&+E^g\left[\int_{T^3}\sum_{\mu=0}^{4}\sum_{j=0}^{3}\lambda_j^\mu(t,x)\theta_{j,k,l}^\mu(\omega_{x,\varepsilon})\sigma_k(\omega_{x,\varepsilon})dx\right]\\
&+N^{-1}a_N(t) - \delta_0^{-1}N^{-1}H_N(t) \le 0.
\end{aligned} \tag{3.11}$$

By Theorem 3.10 we can replace $\theta_{j,k,l}^\mu(\omega_{x,\varepsilon})$ by $\hat{\theta}_{j,l}^\mu(\zeta_k(\omega_{\varepsilon,x}))$. By applying Lemma 3.7 to $\hat{g}_N$ we can replace $\hat{\theta}_{j,l}^\mu$ by $\hat{\theta}_j^\mu$ in the limit $l \to \infty$ and thus we can rewrite (3.11) as

$$\begin{aligned}
&\overline{\lim_{l\to\infty}}\ \overline{\lim_{k\to\infty}}\ \overline{\lim_{N\to\infty}}\ \sup_{0\le t\le T} N^{-1}\frac{dH_N(t)}{dt}\\
&+E^{g_N}\left[\int_{T^3}\sum_{\mu=0}^{4}\sum_{j=0}^{3}\lambda_j^\mu(t,x)\hat{\theta}_j^\mu\cdot I_{U_2})(\zeta_k(\omega_{x,\varepsilon})\right]dx\\
&+N^{-1}a_N(t) - \delta_0^{-1}N^{-1}H_N(t) \le 0.
\end{aligned} \tag{3.12}$$

The function $\hat{\theta}_j^\mu$ can be easily computed. By definition of θ in Lemma 3.3, (2.24) and (2.25),

$$\hat{\theta}_j^\mu(q) = A_j^\mu(q), \qquad j = 1,2,3,\ \mu = 0,1,2,3,4. \tag{3.13}$$

Clearly, by definition $\theta_{0,\alpha}^\mu = \zeta_\alpha^\mu$ and thus

$$\hat{\theta}_0^\mu(q) = q^\mu. \tag{3.14}$$

Observe that, since $q(x)$ is smooth, integrating by parts we have by (2.25)

$$\int \sum_{\mu=0}^{4} \sum_{j=1}^{3} A_j^{\mu}(q(x))\lambda_j^{\mu}(q(x))dx = 0\,.$$

Hence we can rewrite the expectation in (3.12) as

$$E^{g_N}\left[\int_{T^3} \Omega(\lambda(t,x), \zeta_k(\omega_{\varepsilon,x}))dx\right], \tag{3.15}$$

where

$$\Omega(\boldsymbol{\lambda}, \boldsymbol{\zeta}) = \left\{\sum_{\mu=0}^{4}\sum_{j=1}^{3} \lambda_j^{\mu}[A_j^{\mu}(\boldsymbol{\zeta}) - A_j^{\mu}(\mathbf{q})] + \sum_{\mu=0}^{4} \lambda_0^{\mu}(\zeta^{\mu} - q^{\mu})\right\} I_{U_2}(\boldsymbol{\zeta}) \tag{3.16}$$

with $q = \partial\psi/\partial\lambda$ (2.23). By entropy bound (2.28) for any constant $\delta > 0$,

$$E^{g_N}\left[\int_{T^3} \Omega\, dx\right] \geq -(\delta \cdot N)^{-1} \log E^{\hat{g}_N}\left[\exp\left\{ - \delta \cdot N \int_{T^3} \Omega\, dx\right\}\right] - \delta^{-1} H_N/N\,.$$

By the large deviation theory in Sect. 5 (5.21) the expectation with respect to $\hat{g}_N$ can be computed explicitly as

$$\begin{aligned} &\overline{\lim_{k\to\infty}}\ \overline{\lim_{N\to\infty}}\ N^{-1} \log E^{\hat{g}_N}\left\{\exp\left[- \delta \cdot N \int_{T^3} \Omega\, dx\right]\right\} \\ &= \sup_{\zeta(x)}\left[- \delta \int \Omega(\lambda(x), \zeta(x))dx - I(\zeta(x), \lambda(x))\right]. \end{aligned} \tag{3.17}$$

Here the sup is over all integrable functions $\zeta(x)$ and

$$I(\zeta, \lambda) = \hat{s}(\zeta) + \psi(\lambda) - \lambda\zeta$$

with $\hat{s}$ and ψ defined in (2.21) and (2.20). The function $I(\zeta, \lambda)$ is uniformly strictly convex and vanishing at $\zeta = q = \partial\psi/\partial\lambda$. Also the function $\Omega(\lambda, \zeta)$ vanishes if $\zeta = q = \partial\psi/\partial\lambda$. If we have the first derivatives of Ω also vanish at $\zeta = q$, then since Ω is bounded (thanks to the cutoff I_{U_2}) the sup is zero. To summarize, (3.17) is zero provided

$$\left.\frac{\partial\Omega}{\partial\zeta^{\mu}}\right|_{\zeta=0} = 0\,, \qquad \mu = 0, \ldots, 4\,. \tag{3.18}$$

A direct computation shows that (3.18) is equivalent to

$$\lambda_0^{\mu} + \sum_{\nu,j} \lambda_j^{\nu} \frac{\partial A^{\nu}}{\partial q^{\mu}} = 0\,. \tag{3.19}$$

By (2.15)

$$\lambda_0^{\mu} = \sum_{\alpha} \frac{\partial\lambda^{\mu}}{\partial q^{\alpha}} \frac{\partial q^{\alpha}}{\partial t} = \sum_{\alpha}\sum_{\beta}\sum_{j} \frac{\partial\lambda^{\mu}}{\partial q^{\alpha}} \frac{\partial A_j^{\alpha}}{\partial q^{\beta}} \frac{\partial q^{\beta}}{\partial x_j}\,.$$

Hence (3.19) is satisfied provided the Euler equation (2.15) is satisfied and

$$\sum_{\alpha} \frac{\partial \lambda^{\mu}}{\partial q^{\alpha}} \frac{\partial A_j^{\alpha}}{\partial q^{\beta}} = \sum_{\nu} \frac{\partial \lambda^{\nu}}{\partial q^{\beta}} \frac{\partial A_j^{\nu}}{\partial q^{\mu}}, \qquad j = 1, 2, 3. \tag{3.20}$$

Since $\lambda = \partial \hat{s}/\partial \zeta$, we can write (3.20) as

$$\sum_{\alpha} \frac{\partial^2 \hat{s}}{\partial q^{\alpha} \partial q^{\mu}} \frac{\partial A_j^{\alpha}}{\partial q^{\beta}} = \sum_{\nu} \frac{\partial^2 \hat{s}}{\partial q^{\beta} \partial q^{\nu}} \frac{\partial A_j^{\nu}}{\partial q^{\mu}}, \qquad j = 1, 2, 3. \tag{3.21}$$

This means that the Euler equation is symmetric with symmetrizer $\partial^2 \hat{s}/\partial \zeta \partial \zeta$. One can check by explicit computation that (3.21) holds and thus conclude our main Theorem 2.1. In fact consider the rate of change of local entropy and using the Euler equation we have [by (2.25)]

$$\frac{d}{dt} \int_{T^3} \hat{s}(q) dx = \int_{T^3} \sum_{\alpha} \frac{\partial \hat{s}}{\partial q^{\alpha}} \frac{\partial A_j^{\alpha}}{\partial q^{\beta}} \frac{\partial q^{\beta}}{\partial x^j} dx = 0. \tag{3.22}$$

Since (3.22) is zero with respect to all choices of initial conditions, (3.22) is identical to zero even if we take q^{α} as independent variables. Hence the compatibility conditions require

$$\frac{\partial}{\partial q^{\beta}} \left[\frac{\partial \hat{s}}{\partial q^{\alpha}} \frac{\partial A_j^{\alpha}}{\partial q^{\beta}} \right] = \frac{\partial}{\partial q^{\alpha}} \left[\frac{\partial \hat{s}}{\partial q^{\beta}} \frac{\partial A_j^{\beta}}{\partial q^{\alpha}} \right], \qquad j = 1, 2, 3. \tag{3.23}$$

Doing the differentiation and using $\partial \hat{s}/\partial q^{\beta} \partial q^{\alpha} = \partial^2 \hat{s}/\partial q^{\alpha} \partial q^{\beta}$ we also conclude (3.21).

4. Local Ergodic Theorem

The aim of this section is to prove Theorem 3.10. To achieve this we proceed in several steps.

Let $\varepsilon > 0$ be given. Then we defined earlier (paragraph after Lemma 3.4) a local map $\omega_{\varepsilon,x}$ which maps a local configuration in the phase space around x to a configuration in $\mathbb{R}^3$. As $\varepsilon \to 0$ the resulting configurations are getting well defined in larger and larger cubes around the origin in $\mathbb{R}^3$. Let us denote the measure obtained from the density $g(t, \mathbf{x}, \mathbf{p})$ through the map by $Q_{t,x,\varepsilon}$. We define

$$\bar{Q}_{\varepsilon} = \frac{1}{T_0} \int_0^{T_0} dt \int_{T^3} dx \, Q_{t,x,\varepsilon} \,. \tag{4.1}$$

Roughly speaking, if we pick a random time "t" in $[0, T_0]$ and a random point "x" in T^3 and blow up the space by a factor of ε^{-1}, the statistics of the configuration in phase space in a large ε-vicinity of this point is given by $\bar{Q}_{\varepsilon}$. As $\varepsilon \to 0$, $\bar{Q}_{\varepsilon}$ gets to be well defined as a point process on all of $\mathbb{R}^3$:

The following list contains the main steps in proving local ergodic theorem (3.10).
(A) The family $\{\bar{Q}_{\varepsilon}\}$ is compact. It also follows that any limiting point process has finite entropy.

(B) Conditioned on the positions $\mathbf{x}$, the distribution of the momenta is a mixture of product states $\prod_i \exp\{-a\phi(p_i) - bp_i\}$/Normalization. This can be proven with the help of the noise introduced in the dynamics.
(C) We shall define the energy, momenta and density associated with a configuration and prove that they are constants of the motion determined by the generator L (in the sense specified later).
(D) By conditioning on energy, momenta and density, any limiting process is invariant under the Liouville operator L and hence is Gibbs with potential V.
(E) The local ergodic Theorem 3.10 is just the law of large numbers for Gibbs states.

Lemma 4.1. *As $\varepsilon \to 0$ the family $\{\bar{Q}_\varepsilon\}$ of point processes is compact and any limit point is a stationary point process in $\mathbb{R}^3$ with values in momentum space.*

Proof. To prove compactness we have to prove that:

$$E^{\bar{Q}_\varepsilon}\left[\sum_\alpha \mathbf{1}_\Lambda(x_\alpha)\right] \le |\Lambda|\,, \tag{4.2}$$

$$\lim_{l\to\infty}\limsup_{\varepsilon\to 0} E^{\bar{Q}_\varepsilon}\left[\sum_\alpha \mathbf{1}_\Lambda(x_\alpha)(1-\mathbf{1}_l(p_\alpha))\right] = 0 \tag{3.3}$$

for any bounded set Λ in $\mathbb{R}^3$. Here $\mathbf{1}_l(p)$ is the characteristic function of the set $\{p\colon \|p\| \le l\}$. In (4.2) actually equality holds. In fact an elementary computation yields:

$$E^{\bar{Q}_\varepsilon}\left[\sum_\alpha \mathbf{1}_\Lambda(x_\alpha)\right] = \frac{1}{T_0}\int_0^{T_0} dt \int_{T^3} dx \sum_\alpha \mathbf{1}_\Lambda(\varepsilon^{-1}(x_\alpha - x))g(t,\mathbf{x},\mathbf{p})d\mathbf{x}\,d\mathbf{p}\,.$$

Since

$$\int_{T^3} \mathbf{1}_\Lambda(\varepsilon^{-1}(x_\alpha - x))dx = \varepsilon^3|\Lambda| = N^{-1}|\Lambda|\,,$$

(4.2) follows.

Let us denote by $\bar{g}_N(\mathbf{x},\mathbf{p}) = \frac{1}{T_0}\int_0^{T_0} g(t,\mathbf{x},\mathbf{p})dt$. Then

$$\begin{aligned}
&E^{\bar{Q}_\varepsilon}\left(\sum_\alpha \mathbf{1}_\Lambda(x_\alpha)(1-\mathbf{1}_l(p_\alpha))\right)\\
&= \int\int_{T^3} dx \sum_\alpha \mathbf{1}_\Lambda(\varepsilon^{-1}(x_k - x))(1-\mathbf{1}_l(p_\alpha))\bar{g}_N(\mathbf{x},\mathbf{p})d\mathbf{x}\,d\mathbf{p}\\
&= |\Lambda|\varepsilon^3 \int \sum_\alpha (1-\mathbf{1}_l(p_\alpha))\bar{g}_N(\mathbf{x},\mathbf{p})d\mathbf{x}\,d\mathbf{p}\,.
\end{aligned}$$

We can use as a reference measure a Gibbs measure $\hat{g}_{\lambda,N} = \exp\left\{ \sum_{\alpha=1} \sum_{\mu=0}^{4} \lambda^\mu \zeta^\mu_{\alpha,\varepsilon} + c_{N,\lambda} \right\}$. Then by the entropy bound (2.28):

$$\frac{|\Lambda|}{N} \int \sum_\alpha [1 - \mathbf{1}_l(p_\alpha)] \bar{g}_N(\mathbf{x}, \mathbf{p}) d\mathbf{x}\, d\mathbf{p} \le \frac{|\Lambda|}{\delta N} \left\{ \log \int e^{\delta \sum_\alpha (1-\mathbf{1}_l(p_\alpha))} \bar{g}_{\lambda,N}(\mathbf{x}, \mathbf{p}) d\mathbf{x}\, d\mathbf{p} \right\} + \frac{|\Lambda|}{\delta N} H_N(\bar{g}_N | \hat{g}_{\lambda,N}).$$

Since $H_N(g(0, \mathbf{x}, \mathbf{p})|\hat{g}\lambda) \le CN$ for some $C > 0$, and decreases in time, by convexity we have

$$H_N(\bar{g}_N | \hat{g}_{\lambda,\mu}) \le \frac{CN}{T_0}. \tag{4.4}$$

We let $N \to \infty$ and $l \to \infty$. Note that $\{p_\alpha\}_{\alpha=1}^N$ are independent under $\hat{g}_\lambda$ and the first term can be calculated explicitly and goes to zero as $l \to \infty$. Finally let $\delta \to \infty$ and (4.3) follows. □

Lemma 4.2 *Let Q be any limit point of $\{\bar{Q}_\varepsilon\}$. For any $\delta > 0$ let P_δ the Poisson process on $\mathbb{R}^3 \times \mathbb{R}^3$ with intensity $dx \times \frac{e^{-\delta\phi(p)}}{c_\delta} dp$ (cf. Sect. 5, c_δ is a normalization constant). Then there exists a constant $C < \infty$ independent of Q such that the density of relative entropy*

$$h(Q|P_\delta) \le C \tag{4.5}$$

(h is defined by 5.1).

The proof of Lemma 4.2 is a simple consequence of 4.4 and the results contained in Sect. 5. We will postpone the proof to the end of Sect. 5.

Lemma 4.3. *Let Q be any limit point. Then for any bounded function W with compact support contained in $D_r = \{|x| \le r\}$ we have*

$$E^Q\left(\sum_{\alpha \ne \beta} W(x_\alpha - x_\beta) I_{D_r}(x_\alpha) \right) \le \text{const}\, r^6. \tag{4.6}$$

In particular, let $N_r = \sum_\alpha I_{D_r}(x_\alpha)$. Then

$$E^Q(N_r^2) \le \text{const}\, r^6. \tag{4.7}$$

Lemma 4.3 is the "local version" of Lemma 5.9 and can be proved in the same way.

Lemma 4.4. *If Q is any limit point and we restrict Q to any region W then Q is a measure on the union of particle spaces Ω_n. Ω_n consists of n copies of the phase-space over W, i.e. $(W \times R^3)^n$. The restriction of Q of Ω_n is invariant with respect to $\hat{L}_{\alpha,\beta}$ (2.6) for every $\alpha, \beta \in \{1, 2, \ldots, n\}$ provided $n \ge 2$.*

Proof. We will use here the assumption made on the random interaction, i.e.

$$\psi(\varepsilon, x) > \delta_{l_0} \quad \text{if } |\varepsilon^{-1} x| < k.$$

Using again $\hat{g}_{\lambda,N}$ as reference measure (for any $\lambda \in V$), we have from (3.5),

$$E^{\hat{g}_{\lambda,N}} \left[\theta(\varepsilon) \sum_{\alpha \ne \beta} \psi(\varepsilon, x_\alpha - x_\beta) \gamma(c) \frac{|\nabla_{c;\alpha,\beta} \bar{g}_N|^2}{\bar{g}_N} \right] \le c' N.$$

Since $\theta(\varepsilon) \to \infty$ as $\varepsilon \to 0$ and ψ is uniformly positive in the sense described above, the assertion follows by elementary considerations by taking the limit as $N \to \infty$. □

Lemma 4.5. *Let Q be any limit point of $\{\bar{Q}_\varepsilon\}$. Then Q, when restricted to Ω_n for $n \geq 2$ and given the configuration of positions, is uniform on every energy-momentum hypersurface*

$$\sum_{\alpha=1}^{n} \phi(p_\alpha) = C_0,$$

$$\sum_{\alpha=1}^{n} p_\alpha^i = C_i$$

(uniform in the sense of the restriction of the volume element dp).

Proof. This requires that $\sum_{\alpha,\beta} \hat{L}_{\alpha,\beta}$ be elliptic on the above hypersurface and that the hypersurface be connected. Our assumptions on ϕ ensure this except for special values of c_0, c_1, c_2, c_3 when the hypersurface collapses to a point. □

Lemma 4.6. *Given the configuration of positions then the configuration of momenta is a mixture of product measures of the form*

$$\pi(dp|\lambda) = \exp\left[\sum_{i=1}^{3}\sum_{\alpha} \lambda^i p_\alpha^i - \lambda^4 \sum_\alpha \phi(p_\alpha)\right] /\text{Normalization}, \tag{4.8}$$

where λ^i are constants restricted to the region where the exponent becomes negative as $|p_\alpha| \to \infty$.

Proof. Lemma 4.6 is just a consequence of Lemma 4.5 and the equivalence of ensembles theorem. □

In the following L will denote the Liouville operator defined on Ω, the space of configurations in the infinite volume.

Lemma 4.7. *If Q is any limit point of $\{\bar{Q}_\varepsilon\}$ then*

$$\int LF(\omega)dQ(\omega) = 0 \tag{4.9}$$

for all bounded smooth local functionals F.

Proof. By definition we can choose a subsequence such that for any bounded smooth local function F

$$E^Q[LF] = \lim_{\varepsilon\to 0} E^{\bar{Q}_\varepsilon}[LF].$$

It is therefore enough to prove that $E^{\bar{Q}_\varepsilon}[LF] \to 0$ as $\varepsilon \to 0$. Define $G(\omega)$ by

$$G(\omega) = (LF)(\omega) = \sum_\alpha \sum_{i=1}^{3} \phi_i(p_\alpha)\frac{\partial F}{\partial x_\alpha^i} - \sum_{\alpha\neq\beta}\sum_{i=1}^{3} V_i(x_\alpha - x_\beta)\frac{\partial F}{\partial p_\alpha^i}.$$

Define $\bar{g}_N$ to be the average of g_N

$$\bar{g}_N(\mathbf{x},\mathbf{p}) = \frac{1}{T}\int_0^T g_N(t,\mathbf{x},\mathbf{p})dt.$$

Then we can express $E^{\bar{Q}_\varepsilon}[G(\omega)]$ by

$$E^{\bar{Q}_\varepsilon}[G(\omega)] = E^{\bar{g}}\left\{\int_{T^3} G(\omega_{\varepsilon,x})dx\right\}.$$

Define $\bar{F}_\varepsilon$ to be the space average of $F(\omega_{\varepsilon,x})$,

$$\bar{F}_\varepsilon(\mathbf{x},\mathbf{p}) = \int_{T^3} F(\omega_{\varepsilon,x})dx\,.$$

Recall L_ε (2.4) is the Liouville operator with scale ε. We can compute $L_\varepsilon \bar{F}_\varepsilon$ as

$$L_\varepsilon \bar{F}_\varepsilon = \varepsilon^{-1}\int_{T^3} G(\omega_{\varepsilon,x})dx\,. \tag{4.10}$$

Recall by Ito's formula

$$E^{g_N}\left[\int_0^T \varepsilon\tilde{L}_\varepsilon \bar{F}_\varepsilon(\mathbf{x}(t),\mathbf{p}(t))dt\right] = \varepsilon E^{g_N}[\bar{F}_\varepsilon(\mathbf{x}(T),\mathbf{p}(T))] - \varepsilon E^{g_N}[\bar{F}_\varepsilon(\mathbf{x}(0),\mathbf{p}(0))]\,. \tag{4.11}$$

Since $\bar{F}_\varepsilon$ is bounded, the right side of (4.11) vanishes as $\varepsilon \to 0$. By definition of $\tilde{L}_\varepsilon$ and (4.10) we have

$$\lim_{\varepsilon\to 0} E^{\bar{Q}_\varepsilon}[G(\omega)] = \lim_{\varepsilon\to 0} E^{\bar{g}_N}[\varepsilon\hat{L}_\varepsilon \bar{F}_\varepsilon(\mathbf{x},\mathbf{p})]\,. \tag{4.12}$$

By assumption that F is bounded smooth function we can bound $\hat{L}_\varepsilon \bar{F}_\varepsilon$ by

$$|E^{\hat{g}_N}[\hat{L}_\varepsilon \bar{F}_\varepsilon(\mathbf{x},\mathbf{p})]| \le \varepsilon^4 \,\mathrm{const}\,\theta(\varepsilon) E^{\bar{g}_N}\left[\sum_{\alpha\neq\beta}\psi(\varepsilon, x_\alpha - x_\beta)\right].$$

By Lemma 5.9 and the fast decay of ψ, $\varepsilon^3 E^{\bar{g}_N}\left[\sum_{\alpha\neq\beta}\psi(\varepsilon,x_\alpha - x_\beta)\right] \le$ const. Since $\varepsilon\theta(\varepsilon) \to 0$, $\lim_{\varepsilon\to 0} E^{\bar{Q}_\varepsilon}[G(\omega)] = 0$. This proves Lemma 4.7. □

Lemma 4.8. *For any configuration $\omega = \{(x_\alpha, p_\alpha)\}$, let $\mathbf{z}$ be the density, momenta and kinetic energy associated with the configuration defined by*

$$\begin{aligned} z^\mu(\omega) &= \lim_{\delta\to 0} z^\mu_{\chi,\delta}(\omega) = \lim_{\delta\to 0}\delta^3\sum_{\alpha=1}^N \chi(\delta x_\alpha)\zeta^\mu_\alpha(\omega)\,, \qquad \mu = 0,1,2,3\,, \\ z^4(\omega) &= \lim_{\delta\to 0} z^\mu_{\chi,\delta}(\omega) = \lim_{\delta\to 0}\delta^3\sum_{\alpha=1}^N \chi(\delta x_\alpha)\phi(p_\alpha)\,. \end{aligned} \tag{4.13}$$

Here ζ^μ_α was defined in (2.10) and χ is a cutoff function of total integral one (cf. (3.7)). For any limit point Q of $\{\bar{Q}_\varepsilon\}$, $\mathbf{z}^\mu(\omega)$ exist almost everywhere and are independent of the cutoff χ. Furthermore, $\mathbf{z}(\omega)$ are constants of the motion defined by L in the sense that

$$\int h(\mathbf{z}(\omega))LF(\omega)dQ = 0 \tag{4.14}$$

for all local smooth functionals F and all smooth functions h with compact support.

Proof. Because of the stationarity of $\{x_\alpha\}$ as a point process and the conditionals of p being independent (4.8), these limits clearly exist if the moments are finite: essentially one needs finiteness of the kinetic energy density or

$$\int z^4(\omega)dQ < \infty .$$

This is implied by (4.5) and (2.28). It also follows from (4.7) that $\mathbf{z}$ is independent of the cutoff χ.

We now turn to the proof of (4.14). By (4.9),

$$\begin{aligned} 0 &= \int L(Fh(z^\mu_{\chi,\delta}(\omega))dQ \\ &= \int (LF)h(z^\mu_{\chi,\delta}(\omega))dQ + \int FLh(z^\mu_{\chi,\delta}(\omega))dQ . \end{aligned}$$

The first term converges to $\int LFh(z^\mu_\chi(\omega))dQ$ as $\delta \to 0$. We only have to show that the second term converges to zero as $\delta \to 0$. Clearly, it suffices to show that as $\delta \to 0$,

$$\int |Lz^\mu_{\chi,\delta}|dQ \to 0 , \qquad \mu = 0, \dots, 4 . \tag{4.15}$$

By definition,

$$Lz^0_{\chi,\delta} = \delta\delta^3 \sum_{\alpha,i} \chi_i(\delta x_\alpha)p^i_\alpha .$$

Hence $\int |Lz^0_{\chi,\delta}|dQ \to 0$ as $\delta \to 0$. Similarly,

$$\begin{aligned} Lz^i_{\chi,\delta} &= \delta\delta^3 \sum_{\alpha,j} \chi_j(\delta x_\alpha)p^i_\alpha \phi_j(p_\alpha) \\ &\quad - \delta^3 \sum_{\alpha\neq\beta} \chi(\delta x_\alpha)V_i(x_\alpha - x_\beta) , \qquad i = 1, 2, 3 . \end{aligned} \tag{4.16}$$

The first term clearly goes to zero as $\delta \to 0$. As for the second term we use the antisymmetry of V_i to write it as

$$\frac{1}{2}\delta^3 \sum_{\alpha\neq\beta} V_i(x_\alpha - x_\beta)\,[\chi(\delta x_\alpha) - \chi(\delta x_\beta)] . \tag{4.17}$$

Since χ is smooth with compact support

$$|\chi(\delta x_\alpha) - \chi(\delta x_\beta)| \le \text{const}\,\delta[1(|x_\alpha| \le \delta^{-1}C) + 1(|x_\beta| \le \delta^{-1}C)| ,$$

for some constant C. So (4.19) is bounded by

$$\text{const}\,\delta \cdot \delta^3 \sum_{\alpha\neq\beta} W(x_\alpha - x_\beta)1(|x_\alpha| \le \delta^{-1}C) , \tag{4.18}$$

where W is a smooth positive function with compact support. By (4.6) we can bound the expectations of (4.18) by const δ. Hence

$$\lim_{\delta\to 0} \int |Lz^i_{\chi,\delta}|dQ = 0 , \qquad i = 1, 2, 3 .$$

Finally we have to bound $\int |Lz^0_{\chi,\delta}|dQ$,

$$E^Q[|Lz^0_{\chi,\delta}|] = E^Q\left[\left|\delta\delta^3 \sum_{i,\alpha} \chi_i(\delta x_\alpha)\phi_i(p_\alpha)\phi(p_\alpha)\right|\right] + E^Q\left[\left|\delta^3 \sum_{\alpha\neq\beta}\sum_i \chi(\delta x_\alpha)\phi_i(p_\alpha)V_i(x_\alpha - x_\beta)\right|\right]. \tag{4.19}$$

The first term on the right side of (4.19) is easy as before. The second term is trickier. Let $w_i(\mathbf{z})$ and $\sigma_i(\mathbf{t})$ denote the expectation and variance of $\phi_i(p_\alpha)$ with respect to Q conditioned on $\mathbf{z}$. Note that since the extremal measures of momenta distribution are characterized by (4.8), w_i and σ_i depend on $\mathbf{z}$ only and are independent of the other parameter or structure of Q. We can now bound the second term of (4.19) by

$$\begin{aligned} &E^Q\left[\left|\delta^3 \sum_i \sum_{\alpha\neq\beta} \chi(\delta x_\alpha)\phi_i(p_\alpha)V_i(x_\alpha - x_\beta)\right|\right] \\ &= E^Q\left[\left|\delta^3 \sum_i \chi(\delta x_\alpha)\left[\phi_i(p_\alpha) - w_i\right]\sum_{\beta\neq\alpha} V_i(x_\alpha - x_\beta)\right|\right] \\ &+ E^Q\left[\left|\delta^3 \sum_{\alpha\neq\beta}\sum_i \chi(\delta x_\alpha)V_i(x_\alpha - x_\beta)w_i\right|\right]. \end{aligned} \tag{4.20}$$

The second term of (4.20) can be bounded as before. Using the Schwartz inequality the first term can be bounded by

$$\begin{aligned} &E^Q\left\{\sum_{i=1}^3 \sigma_i(\mathbf{z})\delta^3 E^Q\left[\sum_\alpha \chi(\delta x_\alpha)^2\left(\sum_\beta V_i(x_\alpha - x_\beta)\right)^2 \Big| \mathbf{z}\right]^{1/2}\right\} \\ &\leq \operatorname{const}\delta^3 E^Q\left[\sum_{i=1}^3\sum_\alpha \chi(\delta x_\alpha)^2\left(\sum_{\beta\neq\alpha} V_i(x_\alpha - x_\beta)\right)^2\right]^{1/2}, \end{aligned} \tag{4.21}$$

where we use the boundedness of $\sigma_i(z)$. To bound the second expectation, let us divide the set $\{x|\,|x| \leq 2\delta^{-1}\}$ into boxes of size $2R$ with R denoting the range of V. Let σ index the boxes and let N_σ be the number of particles in the σ box.

$$\delta^3 E^Q\left[\sum_i\sum_\alpha \chi(\delta x_\alpha)^2\left(\sum_\beta V_i(x_\alpha - x_\beta)\right)^2\right]^{1/2} \leq \operatorname{const}\delta^3 E^Q\left[\sum_\sigma N_\sigma^3\right]^{1/2}:$$

By convexity and the inequality $\left(\sum_\sigma N_\sigma^3\right)^{1/3} \leq \left(\sum_\sigma N_\sigma^2\right)^{1/2}$ we see that (4.21) is bounded by

$$\operatorname{const}\delta^3 E^Q\left[\left(\sum_\sigma N_\sigma^2\right)^{3/4}\right] \leq \delta^3\left\{E^Q\left[\sum_\sigma N_\sigma^2\right]\right\}^{3/4}. \tag{4.22}$$

By (4.7), $E^Q\left[\sum_\upsilon N_\sigma^2\right]$ is bounded by δ^{-3} and (4.22) is bounded by $\operatorname{const}\delta^{3/4}$. We have thus concluded that (4.17) vanishes as $\delta \to 0$ and thus Lemma 4.9. □

By Lemma 4.6 we can represent the limit point Q as

$$Q = \int \beta(d\mathbf{z})\nu(d\mathbf{x}|\mathbf{z})\pi(d\mathbf{p}|\mathbf{z}), \tag{4.23}$$

where π is defined in (4.3) with λ chosen in such a way as to have average momenta and kinetic energy given by $\mathbf{z}$. Furthermore Lemmas 4.7 and 4.8 imply that

$$\int LF(\omega)\nu(d\mathbf{x}|\mathbf{z})\pi(d\mathbf{p}|\mathbf{z}) = 0 \tag{4.24}$$

for almost all $\mathbf{z}$. From (4.24) and the translation invariance of Q, $\nu \otimes \pi$ is a Gibbs state satisfying the DLR equation. We state this as the following.

Lemma 4.10. *Any extremal limit point of the family $\{\bar{Q}_\varepsilon\}$ defined in* (4.1) *is a Gibbs state. It is characterized by the five parameters z^μ, $\mu = 0, \dots, 4$.*

Proof. By the translation invariance of $\nu(d\mathbf{x}|\mathbf{z})$ we can rewrite (4.24) as

$$\int \left[\sum_\alpha \sum_i (\phi_i(p_\alpha) - w_i(\mathbf{z})) \frac{\partial F}{\partial x^i_\alpha} - \sum_{\beta\neq\alpha} \sum_i V_i(x_\alpha - x_\beta) \frac{\partial F}{\partial p^i_\alpha}\right] \nu(d\mathbf{x}|\mathbf{z})\pi(d\mathbf{p}|\mathbf{z}) = 0,$$

where $w_i(\mathbf{z})$ is the expectation of $\phi_i(p_\alpha)$ with respect to $\pi(d\mathbf{p}|\mathbf{z})$ (as in the proof of Lemma 4.9). Then the proof follows easily along the line of the proof of Lemma 4.5 in [OV]. In fact the particular form of ϕ does not play any role here. □

Now the local ergodic Theorem 3.10 is just a simple corollary of Lemma 4.10. By taking the limit $\varepsilon \to 0$ we only have to check that

$$\overline{\lim_{k\to\infty}} \sup_{Q\in A} E^Q[|F_k(\omega) - \hat{F}(\zeta_k(\omega))|\sigma_k(\omega)] = 0, \tag{4.25}$$

where A denotes the set of limit points of $\bar{P}_\varepsilon$. The characteristic function σ_k reduces the problem to verification of (4.25) with conserved quantities in the single phase region of Gibbs states with potential V and kinetic energy $\phi(p)$. Combining with Lemma 4.10 we only have to check (4.25) for Gibbs states in the single phase region. This is simply the law of large numbers for Gibbs states.

5. Large Deviations for Gibbs Point Processes

We will develop here a large deviation theory for Gibbs point processes, needed for the proof of Theorem 2.1 and Corollary 2.2. For Gibbs processes on a lattice this theory has already been developed in [O] and [Re]. For an account of general theory of large deviation, see [V1].

Initially we want to develop a large deviation theory for a Poisson point process P on $\mathbb{R}^3 \times \mathbb{R}^3$. The intensity of the Poisson process will be the measure $dx \times f(p)dp$. The density $f(p)$ will be nonnegative and $\int_{\mathbb{R}^3} f(p)dp = 1$. A point of $\mathbb{R}^3 \times \mathbb{R}^3$ will be denoted by (x, p) and the sample space Ω will consist of points ω, denoting a configuration $\{(x_\alpha, p_\alpha)\}$. Any bounded region B in $\mathbb{R}^3$ will have only a finite number of x_a in it and one can think of p_α as tags and consider the corresponding finite configuration in $B \times \mathbb{R}^3$. The configuration x_α alone will constitute a Poisson point process on $\mathbb{R}^3$ with intensity given by the Lebesgue measure. The Poisson measure P on Ω is invariant with respect to the group of x-translation τ_x. Suppose Q is any other τ_x-invariant point process measure on Ω. Let D_l be a cube of side l in $\mathbb{R}^3$; we can define the relative entropy $H_l(Q|P)$ of the restriction of Q to $D_l \times \mathbb{R}^3$, with respect to the same restriction of P. By restriction of course we mean restriction to the σ-field $\mathscr{F}_l$ generated by configurations in that domain. We define then

$$H_l(Q|P) = \sup_{F_l}\{E^Q(F_l) - \log E^P(e^{F_l})\}, \tag{5.1}$$

where the sup is taken over all the bounded continuous $\mathscr{F}_l$-measurable functions F_l on Ω. Because of independence properties and stationarity of P, H_l is superadditive and one can define the entropy density as

$$h(Q/P) = \lim_{\delta\to\infty} \frac{1}{l^3} H_l(Q|P) = \sup_l \frac{1}{l^3} H_l(Q|P). \tag{5.2}$$

To any configuratioin $\omega \in \Omega$, we can associate a σ-finite measure on $\mathbb{R}^3 \times \mathbb{R}^3$ by counting the number of points (x_α, p_α) in any given set. Then the space Ω can be made into a topological space under vague convergence. That is to say that for any test function $\phi(x,p)$ which is continuous and has compact support in $\mathbb{R}^3 \times \mathbb{R}^3$, the function on Ω

$$\sum_\alpha \phi(x_\alpha, p_\alpha)$$

is continuous on Ω. This in turn provides a weak convergence definition for the space $\mathscr{M}$ of stochastic point processes on Ω and we shall limit ourselves mainly to the τ_x-invariant ones $\mathscr{M}_0$.

Let $N_B(\omega)$ be the number of points of the configuration ω in $B \times \mathbb{R}^3$, where B is a bounded set in $\mathbb{R}^3$. For any stationary point process Q, if for any bounded $B \subset \mathbb{R}^3$

$$E^Q[N_B] < +\infty,$$

then there exists a positive σ-finite measure $q(dx, dp)$ on $\mathbb{R} \times \mathbb{R}^3$ such that

$$E^Q[N_B] = \int_{B\times\mathbb{R}^3} q(dx, dp).$$

By stationarity we have

$$q(dx, dp) = \varrho\, dx \cdot \pi(dp),$$

where ϱ is a constant density of the x-configurations and π is the distribution of the "p_α."

Theorem 5.1. (i) *$h(Q|P)$ is lower semicontinuous.*
(ii) *for any finite number l, the set $\{Q\colon h(Q|P) \le l\}$ is compact.*
(iii) *if $Q_n \rightharpoonup Q$ with $h(Q_n|P) \le c$ for any n, then $q_n(dx, dp) = \varrho_n dx \cdot \pi_n(dp)$ converges to $q(dx, dp) = \varrho\, dx \cdot \pi(dp)$ of the limiting Q.*

Proof. (i): The lower semicontinuity of H_l follows easily from this definition and, by (5.1), $h(Q)$ is also lower semicontinuous.

(ii) and (iii): By the entropy inequality it is easy to show that if $Q \in \mathrm{M}_0$ and $h(Q|P) \le c$, then for any bounded $B \subset \mathbb{R}^3$ there exists a constant c_B,

$$E^Q(N_B \log N_B) \le c_B.$$

Then (ii) and (iii) follow by standard arguments. □

The following Theorems 5.2 and 5.3 are the large deviation theory relative to a Poisson point process. For the upper bound contained in Theorem 5.2 let us assume that $F(\omega)$ is a local function measurable with respect to configurations on some $B \times \mathbb{R}^3$ and satisfies a bound

$$F(\omega) \le c_1 + c_2 N_{B_0}(\omega)$$

for some constant c_1, c_2.

Theorem 5.2. *Under the above assumption*

$$\limsup_{l\to\infty} \frac{1}{l^3} \log E^P \left[\exp \int_{D_l} F(\tau_x \omega) dx \right] \le \sup_{Q \in \mathscr{M}_0} [E^Q[F] - h(Q|P)]$$

and the right-hand side is finite.

Proof. Let us first assume that $F(\omega)$ is bounded in absolute value. Then by the entropy inequality we have

$$\log E^P \left[\exp \int_{D_l} F(\tau_x \omega) dx \right] = \sup_{Q'} \left\{ E^{Q'} \left[\int_{D_l} F(\tau_x \omega) dx \right] - H_{l+l_0}(Q'|P) \right\},$$

where the supremum is taken over all point processes Q' on $D_{l+l_0} \times \mathbb{R}^3$. Here we are assuming that $B_0 \subset D_{l_0}$. We extend the measure Q' where the supremum is attained to a point process on $\mathbb{R}^3 \times \mathbb{R}^3$ by taking independent copies on all disjoint translated cubes of $\mathbb{R}^3 \times \mathbb{R}^3$. The measure Q'' so obtained is not stationary but can be made stationary by defining

$$\hat{Q}_{l+l_0} = \frac{1}{(l+l_0)^3} \int_{D_{l+l_0}} \tau_x Q'' dx .$$

By (5.2), convexity of H_l and independence properties of the Poisson measure P we have

$$\begin{aligned} h(\hat{Q}_{l+l_0}|P) &= \lim_{n\to\infty} \frac{1}{(n(l+l_0))^3} H_{n(l+l_0)}(\hat{Q}_{l+l_0}|P) \\ &\le \lim_{n\to\infty} \frac{1}{(n(l+l_0))^3} \frac{1}{(l+l_0)^3} \int_{D_{l+l_0}} H_{n(l+l_0)}(\tau_x Q''|P) dx \\ &\le \frac{1}{(l+l_0)^3} H_{l+l_0}(Q''|P) \\ &= \frac{1}{(l+l_0)^3} H_{l+l_0}(Q'|P) . \end{aligned}$$

Furthermore we have

$$\left| E^{\hat{Q}_l}(F(\omega)) - \frac{1}{l^d} E^{Q'} \left(\int_{D_l} F(\tau_x \omega) dx \right) \right| \le \|F\|_\infty \cdot o(1)$$

as $l \to \infty$. This proves the upper bound for bounded F.

Now assume F only bounded above and define $F_k = \max\{F, -k\}$. Then we have

$$\limsup_{l\to\infty} \frac{1}{l^3} \log E^P \left(\exp \int_{D_e} F(\tau_x \omega) dx \right) \le \sup_{Q \in \mathscr{M}_0} \{E^Q[F_k(\omega)] - h(Q|P)\} = \sigma_k .$$

We need only to establish $\lim_{k\to\infty} \sigma_k \leq \sup_{Q\in M_0} \{E^Q(F) - h(Q|P)\}$ (the limit exists since σ_k is decreasing.) If $\lim_{k\to\infty} \sigma_k = -\infty$ there is nothing to prove. Let us assume $\lim_{k\to\infty} \sigma_k = \sigma > -\infty$. Since F_k is bounded above we can find Q_k such that $h(Q_k|P)$ is bounded and $E^{Q_k}[F_k] - h(Q_k|P) \to \sigma$. The boundedness of $h(Q_k|P)$ means that along some subsequence we will have a weak limit Q and moreover $Q_k \to Q$ strongly, i.e. for each Borel set in the σ-field corresponding to every $B \times \mathbb{R}^d$. In particular

$$E^{Q_k}(F) \to E^Q(F)$$

for every bounded measurable F. Therefore

$$\lim_{k\to\infty} E^{Q_k}(F_k) \leq E^Q(F)\,.$$

The lower semicontinuity of $h(Q|P)$ completes the argument.

For removing the upper bound on F we replace F by $F_k = \min(F, k)$ and we will use the fact that $E^P e^{\lambda N_{B_0}} < +\infty$ for every $\lambda < +\infty$. By Hölder's inequality if $\frac{1}{p} + \frac{1}{q} = 1$ we can bound

$$\begin{aligned} E^P\left\{ \exp\left[\int_{D_e} F(\tau_x\omega)dx \right] \right\} &\leq \left(E^P\left\{ \exp\left[p \int_{D_e} F_k(\tau_x\omega)dk \right] \right\} \right)^{1/p} \\ &\quad \times \left(E^P\left\{ \exp\left[q \int_{D_e} (F - F_k)(\tau_x\omega)dx \right] \right\} \right)^{1/q} \\ &= T_1 \cdot T_2\,. \end{aligned}$$

If we take log and divide by $1/l^3$ we get for the first term

$$\limsup_{l\to\infty} \frac{1}{l^3} \log T_1 \leq \frac{1}{p} \sup_{Q\in\mathscr{M}_0} \{E^Q[pF] - h(Q|P)\}$$

while the second term is estimated by

$$T_2 \leq E^P\left\{ \exp\left[q \int_{D_e} [c_2 N_{B_0}(\tau_x\omega) + c_1 - k]^+ dx \right] \right\}^{1/q}.$$

We can think of D_l of disjoint union of translates of B_0 and if we denote by

$$\xi_x(\omega) = \sum_{\alpha}^{l^3/l_0^3} [c_2 N_{D_{l_0}}(\tau_{x+\alpha}\omega) - (k - c_1)]^+,$$

then we can bound T_2 by (using Jensen's inequality)

$$T_2 \le E^P\left\{\exp\left[q\int_{D_{l_0}} \xi_x\, dx\right]\right\}^{1/q}$$
$$\le \left(\frac{1}{l_0^3}\int_{D_{l_0}} E^P\{\exp[ql_0^3\xi_x]\}\right)^{1/q}$$
$$= (E^P\{\exp(ql_0^3[c_2 N_{D_{l_0}}(\omega) - k - c_1)]^+)\})^{(l^3/l_0^3)\cdot(1/q)}.$$

We obtain then

$$\limsup_{l\to\infty}\frac{1}{l^3}\log T_2 \le \frac{1}{l_0^3}\frac{1}{q}\log E^P[\exp(ql_0^3[c_2 N_{D_{l_0}}(\omega) - (k-c_1)]^+)].$$

Let $k \to \infty$, then the right-hand side goes to zero. Now we let $p \to 1$ and we get the right bound. □

For the following lower bound we can assume $F(\omega)$ to be an arbitrary function measurable with respect to the configurations on some $D_{l_0} \times \mathbb{R}^3$. Then we can prove

Theorem 5.3.

$$\liminf_{l\to\infty}\frac{1}{l^3}\log E^P\left(\exp\left[\int_{D_l} F(\tau_x\omega)dx\right]\right) \ge \sup_{Q\in\mathscr{M}_0}[E^Q(F) - h(Q|P)],$$

where the sup *is taken over all* Q *such that* $h(Q|P) < +\infty$ *and* $E^Q(F^-) < +\infty$.

Proof. We can take any permissible Q and since $h(Q|P) < \infty$ we know that

$$\left.\frac{dQ}{dP}\right|_{F_{D_{l+l_0}\times\mathbb{R}^3}} = \psi_{l+l_0}$$

exists. Then by Jensen's inequality

$$E^P\left[\exp\left(\int_{D_l} F(\tau_x\omega)dx\right)\right] = E^Q\left[\exp\left(\int_{D_l} F(\tau_x\omega)dx - \log\psi_{l+l_0}\right)\right]$$
$$\ge \exp\left[E^Q\left(\int_{D_l} F(\tau_x\omega)dx\right) - H_{l+l_0}(Q|P)\right]$$
$$\ge \exp[l^3 E^Q(F) - (l+l_0)^3 h(Q|P)].$$

Taking log, dividing by l^3 and letting $l \to \infty$ we obtain the lower bound. □

Remark. The above proof can be easily modified to yield a lower bound for

$$\liminf_{l\to\infty}\frac{1}{l^3}\log E^P\left(\chi_l \exp\left[\int_{D_l} F(\tau_x\omega)dx\right]\right) \ge \sup_{Q\in A_e}[E^Q(F) - h(Q|P)],$$

where χ_l is defined by

$$\chi_l = \left\{ \omega \colon \frac{1}{D_l} \int_{D_l} G(\tau_x \omega) dx \in (a,b) \right\} \quad \text{for some} \quad -\infty < a < b < \infty$$

and

$$A_e = \left\{ Q \colon \int G(\omega) dQ \in (a,b), Q \text{ ergodic} \right\},$$

and by arguments standard in large deviation theory we can replace A_e by

$$A = \left\{ Q \colon \int G(\omega) dQ \in (a,b) \right\}.$$

We now turn our attention to a similar problem for a Poisson point process P_γ with intensity $\gamma\, dx \times f(p)dp$ on $T^3 \times \mathbb{R}^3$ as $\gamma \to \infty$. We will have a family $F(x,\omega)$ of functions depending on a fixed domain $D_{l_0} \times \mathbb{R}^3$ in $\mathbb{R}^3 \times \mathbb{R}^3$ and depending in some continuous manner on $x \in T^3$ as a parameter. We will think of $\gamma = \varepsilon^{-3}$, and $\varepsilon \to 0$ as $\gamma \to \infty$. If x is any point in T^3 we can define $\omega_{x,\varepsilon}$, a configuration in $\mathbb{R}^3 \times \mathbb{R}^3$ in the same fashion as in Sect. 3. Then if ε is small enough $F(x, \omega_{x,\varepsilon})$ is well defined. We will be interested in the behavior of

$$\frac{1}{\gamma} \log E^{P_\gamma} \left[\exp \left\{ \gamma \int_{T^3} F(x, \omega_{x,\varepsilon}) dx \right\} \right]$$

as $\gamma \to \infty$. Remember that $\gamma = \varepsilon^{-3}$.

Theorem 5.4. *Assume that $F(\omega)$ is a bounded function measurable on $D_{l_0} \times \mathbb{R}^3$. Then*

$$\lim_{\gamma \to \infty} \frac{1}{\gamma} \log E^{P_\gamma} \left[\exp \left\{ \gamma \int_{T^3} F(\omega_{x,\varepsilon}) dx \right\} \right] = \sup_{Q \in \mathscr{M}_0} [E^Q[F] - h(Q|P)].$$

Proof. It is no different from that of Theorems 5.2 and 5.3. If we ignore strips of size ε, adding up to a volume $O(\varepsilon^{-2})$ and expand the rest out to a cube of side nearly $1/\varepsilon$ we are back to the earlier case. □

We now consider integrals of the form

$$Z(\gamma) = E^{P_\gamma} \{ \exp F_\gamma(\omega) \}, \tag{5.3}$$

where $(\varepsilon^{-3} = \gamma)$

$$\begin{aligned} F_\gamma(\omega) = {} & \sum_\alpha A_0(x_\alpha) + \sum_\alpha \sum_{i=1}^{3} A_i(x_\alpha) p_\alpha^i - \frac{1}{2} \sum_\alpha A_4(x_\alpha) \sum_{\beta \neq \alpha} V\left(\frac{x_\alpha - x_\beta}{\varepsilon} \right) \\ & - \sum_\alpha A_5(x_\alpha) \phi(p_\alpha) + \gamma \int_{T^3} G(x, \omega_{x,\varepsilon}) dx . \end{aligned} \tag{5.4}$$

We make the following assumptions:

1) A_j, $j = 0, \ldots, 5$ are continuous functions of $x \in T^3$.

2) $V(x)$ is a continuous function with compact support in $\mathbb{R}^3$ such that $V = V_1 + V_2$ with $V_1(x) \geq 0$, $V_1(0) > 0$, and V_2 being a positive definite potential.

3) $\sum_{i=1}^{3} A_i(x)p^i \leq A_5(x)\phi(p)$ for all x and p.
4) $G(x,\omega) = \sum_s g_s(x)G_s(\omega)$, where $g_s(x)$ are bounded measurable functions depending on configurations in some fixed domain $D_{l_0} \times \mathbb{R}^3$.
5) $A_4(x) \geq a_0 > 0$.

Theorem 5.5. *Under the above assumptions*

$$\lim_{\gamma\to\infty} \frac{1}{\gamma} \log Z(\gamma) = \int_{T^3} \sup_{Q\in\mathscr{M}_0} [E^Q(\hat{F}_x(\omega)) - h(Q|P)]dx\,,$$

where

$$\hat{F}_x(\omega) = \sum_{i=0}^{5} A_i(x)\Psi_i(\omega) + \sum_s g_s(x)G_s(\omega)$$

and

$$\Psi_0(\omega) = \sum_\alpha u(x_\alpha)\,,$$

$$\Psi_i(\omega) = \sum_\alpha u(x_\alpha)p^i_\alpha\,, \qquad i = 1,2,3\,,$$

$$\Psi_4(\omega) = -\frac{1}{2}\sum_\alpha u(x_\alpha)V(x_\alpha - x_\beta)\,,$$

$$\Psi_5(\omega) = -\sum_\alpha u(x_\alpha)\phi(p_\alpha)\,.$$

Here u is a positive function of compact support supported in D_{l_0} with $\int_{\mathbb{R}^3} u(x)dx = 1$.

Remark. The assumption that V is bounded and is of finite range is not essential. The proof we give in the following is taken from [CLY] where one needs to control Coulomb potential. We prefer to keep our proof in a simple form.

Proof. The proof is just an easy step away from Theorems 2, 3, and 4. First let us remark that we can find approximations for $A_i(x)$ and $g_i(x)$ that are constant on some cubic partition of T^3, both from above and below, and we can assume without loss of generality that all these functions are piecewise constant.

The only restriction on the function u is that it has compact support, is non-negative and is of total integral one. It is not hard to see that by the stationarity of Q the variational formula gives the same value independent of the choice of u (subjected to the previous conditions, of course). In the following we prove Theorem 5.5 with u being a characteristic function of a cube.

Divide T^3 into disjoint boxes B_σ of size εl. We can translate the boxes by τ with $0 \leq \tau_i \leq \varepsilon l$, $i = 1,2,3$. Let $F_{\sigma+\tau}(\omega)$ denote the contributions of F_γ in the box B_σ translated by τ. There are ambiguities in this definition which we shall clarify as follows. First of all we can assume A_j is constant in the box $\alpha + \tau$ and replace $A_j(x_\alpha)$ by $A_j(\sigma+\tau)$. Next $V(x_\alpha - x_\beta/\varepsilon)$ is replaced by

$$V(x_\alpha - x_\beta/\varepsilon)1_{\sigma+\tau}(x_\alpha)1_{\sigma+\tau}(x_\beta)\,,$$

namely, we neglect the interactions between neighboring boxes. Certainly we made certain errors. It's not hard to see that there is a bounded function with compact support W such that

$$F(\omega) \le (\varepsilon l)^{-3} \int d\tau \sum_{\sigma} F_{\sigma+\tau}(\omega) + l^{-1} \sum_{\alpha \ne \beta} W(x_\alpha - x_\beta/\varepsilon).$$

The last term is in general of order N^2 but one can control it in the following way. Let

$$F^{(\theta)} = F + \theta \sum_{\beta \ne \alpha} V(x_\alpha - x_\beta/\varepsilon).$$

Apply previous bound to $F^{(\theta)}$,

$$F(\omega) \le (\varepsilon l)^{-3} \int d\tau \sum_{\sigma} (F^{(\theta)})_{\sigma+\tau}(\omega) + \Omega,$$

$$\Omega = \sum_{\alpha \ne \beta} l^{-1}(1+\theta) W(x_\alpha - x_\beta/\varepsilon) - \theta V(x_\alpha - x_\beta/\varepsilon).$$

By assumption $V = V_1 + V_2$ with V_2 positive definite and V_1 nonnegative with $V_1(0) > 0$. By Lemma 5.7 below, for ℓ large enough there exists a $\delta > 0$ such that:

$$\Omega \le \delta^{-1} l^{-1} N - \theta \sum_{\alpha \ne \beta} V_2(x_\alpha - x_\beta/\varepsilon) \le (\delta^{-1} l^{-1} + \mathrm{const}\,\theta) N.$$

Hence

$$\begin{aligned} \log Z(\gamma) &\le (\delta^{-1} l^{-1} + \mathrm{const}\,\theta) N \\ &\quad + \log E^{P_\gamma}\Big[\exp A_{v_\tau} \sum (F^{(\theta)})_{\sigma+\tau}(\theta)\Big]. \end{aligned}$$

By Jensen's inequality the last expectation is bounded by

$$Av_\tau \log E^{P_\gamma}\Big[\exp\Big(\sum (F^{(\theta)})_{\sigma+\tau}(\omega)\Big)\Big].$$

Note that for τ fixed there is no interaction between neighboring boxes. So the last expectation is nothing but

$$Av_\tau \sum_{\sigma+\tau} E^{P_\gamma}[\exp((F^{(\theta)})_{\sigma+\tau}(\omega))].$$

We can now apply Theorem 5.2 and let $l \to \infty$ and then $\theta \to 0$ to conclude our upper bound.

The lower bound is much easier. First let us divide T^3 into disjoint cubes of size εl. Let A_α denote such a typical cube and let $B_\alpha \subset A_\alpha$ be defined by

$$B_\alpha = \{x \mid \mathrm{dist}(x, A_\alpha^c) \ge l_0\}.$$

Since we need only lower bound, we can restrict to the set that no particle exists in the corridors $T^3 \setminus \bigcup_\alpha B_\alpha$. The probability of such an event under P_γ is $\exp[-C(l)N]$, where $C(l) \to 0$ as $l \to \infty$. Moreover, the integrand factors into product over cubes as no interaction presents in this event. We now use Theorem 5.3 for each cube and sum them up to the right side of Theorem 5.3. Note that the errors caused by restricting to the specific event mentioned above is negligible in the limit $\gamma \to \infty$ and $l \to \infty$. $\square$

Consider now the canonical partition function with the number of particles fixed at N that goes to ∞ as ε^{-3}:

$$Z_N^c = E^{P_N}\{\exp[F_\gamma(\omega)]\}, \tag{5.5}$$

where P_N is the measure on $(T^3 \times \mathbb{R}^3)^N$ defined by

$$dP_N = \prod_{\alpha=1}^N dx_\alpha \times f(p_\alpha)dp_\alpha .$$

Then the previously defined [see (5.3)] grand canonical partition function is related to the canonical one by

$$Z_\gamma = e^{-\gamma} \sum_{N=0}^{\infty} \frac{\gamma^N}{N!} Z_N^c . \tag{5.6}$$

Theorem 5.6. *Suppose that* $\lim_{N\to\infty} N\gamma^{-1} = 1$. *Then*

$$\lim_{N\to\infty} \frac{1}{N} \log Z_N^c = \sup_{\substack{\{Q_x\} \\ \int_{T^3} \varrho(Q_x)dx=1}} \left\{ \int dx[E^{Q_x}\{F_x(\omega)\} - h(Q_x|P)] \right\},$$

(here $\varrho(Q)$ is the density of Q).

Proof. If we consider

$$Z_\gamma^{a,b} = \sum_{\gamma a < N < \gamma b} \frac{e^{-\gamma^N}}{N!} Z_N^c ,$$

then the remark after Theorem 5.3 yields

$$\liminf_{\gamma\to\infty} \frac{1}{\gamma} \log Z_\gamma^{a,b} \geq \sup \left\{ \int_{T^3} dx[E^{Q_x}\{F_x(\omega)\} - h(Q_x|P)]; \quad Q_x \colon a < \int_{T^3} \varrho(Q_x)dx < b \right\}.$$

We note that

$$Z_{N-K}^c \leq (\text{const})^K Z_N^c .$$

To see this, let us integrate out one particle, say x_1, in Z_N^c. For any distribution of $x_2, \ldots, x_N$ we can always find a set A of measure at least α with $\alpha > 0$ such that

$$\sup_{x\in A} \sum_{j=2}^N |V(x - x_j)| \leq \text{const} .$$

for some constant independent of N. By restricting x_1 integration to A we get the desired bound for $k = 1$. The general case follows by induction.

Now let us choose $a = 1 - \delta$ and $b = 1$ to have a lower bound for $Z_\gamma^{a,b}$. We obtain a lower bound for Z_N^c by using the upper bound on Z_{N-k}^c. Finally let $\delta \to 0$ at the end to conclude the lower bound of Theorem 5.6.

To obtain the upper bound we define

$$Z_\gamma^\lambda = \sum_{N=0}^{\infty} e^{-\gamma} \frac{\gamma^N}{N!} e^{\lambda N} Z_N^c .$$

Our methods can evaluate

$$\limsup_{\gamma\to\infty} \frac{1}{\gamma} \log Z_\gamma^\lambda$$

for any real λ. We then note that

$$\lim_{N\to\infty} \frac{1}{N} \log \frac{e^{-N} N^N}{N!} = 0 .$$

Therefore

$$\begin{aligned} \limsup_{N\to\infty} \frac{1}{N} \log Z_N^c &\le \limsup_{N\to\infty} \left[\frac{1}{N} \log Z_N^\lambda - \lambda\right] \\ &\le \inf_\lambda \limsup_{N\to\infty} \left[\frac{1}{N} \log Z_N^\lambda - \lambda\right], \end{aligned}$$

which yields the right-hand side of our theorem. □

Lemma 5.7. *Suppose W is a positive bounded function with compact support on $\mathbb{R}^3$. Then there is a $\delta > 0$ such that*

$$\delta \sum_{\alpha\neq\beta}^{N} W(x_\alpha - x_\beta) \le \sum_{\alpha\neq\beta}^{N} V(x_\alpha - x_\beta) + N .$$

Proof. Proof of Lemma 4.2 in [V] generalized to more dimensions. □

We will review now certain facts from thermodynamics of continuous systems, based partly on the large deviation theory developed earlier.

Suppose we are given a kinetic energy function $\phi(p)$ for $p \in \mathbb{R}^3$, which is continuous nonnegative, and growing at least linearly at ∞. We consider the set L of $(\lambda_1, \lambda_2, \lambda_3, \lambda_4)$ in $\mathbb{R}^3 \times \mathbb{R}_+$ satisfying

$$\int \exp\left[\sum_{i=1}^{3} \lambda_i p_i - \lambda_4 \phi(p)\right] dp < +\infty .$$

Let L^0 be the interior of L. Suppose V is a pair potential of the class described earlier, then for any configuration (x_α, p_α) in a bounded cube D the grand canonical partition function is defined, for $\lambda = (\lambda_0, \lambda_1, \dots, \lambda_4)$, $\lambda_0 \in \mathbb{R}$ and $(\lambda_1, \dots, \lambda_4) \in L_0$, by:

$$Z_D(\lambda) = \sum_{N=0}^{\infty} e^{-|D|} \frac{e^{\lambda_0 N}}{N!} \int \cdots \int_{(D\times\mathbb{R}^3)^N} \exp\left[\sum_\alpha \sum_{i=1}^{3} \lambda_i p_\alpha^i - \lambda_4 \mathscr{H}_N\right] \prod_{\alpha=1}^{N} dx_\alpha dp_\alpha , \quad (5.7)$$

where

$$\mathscr{H}_N = \frac{1}{2} \sum_{\alpha\neq\beta} V(x_\alpha - x_\beta) + \sum_\alpha \phi(p_\alpha) .$$

The pressure is defined by

$$\psi(\lambda) = \lim_{D \nearrow \mathbb{R}^3} \frac{1}{|D|} \log Z_D(\lambda) . \tag{5.8}$$

Although the existence of this limit can be proved in many ways we will use large deviation theory to see it. We can pick δ such that $(\lambda_1, \lambda_2, \lambda_3, \lambda_4 - \delta) \in L^0$ and normalize $e^{-\delta\phi(p)}$ to be a probability density. We then see by large deviation theory that $\psi(\lambda)$ exists (Theorems 5.2 and 5.3, observe that the integrand is bounded above) and

$$\psi(\lambda) = \sup_{Q \in \mathscr{M}_0} \{E^Q(F_{\lambda,\delta}) - h(Q|P_\delta)\} , \tag{5.9}$$

where (recall u from Theorem 5.5)

$$\begin{aligned} F_{\lambda,\delta} = \lambda_0 \sum_\alpha u(x_\alpha) + \sum_{i=1}^3 \sum_\alpha \lambda_i u(x_\alpha) p_\alpha^i - \frac{\lambda_4}{2} \sum_{(\alpha \neq \beta)} u(x_\alpha) V(x_\alpha - x_\beta) \\ - (\lambda_4 - \delta) \sum_\alpha u(x_\alpha)\phi(p_\alpha) + (\log c_\delta) \sum_\alpha u(x_\alpha) . \end{aligned}$$

Here $c_\delta = \int_{\mathbb{R}^3} e^{-\delta\phi(p)} dp$ and $u \geq 0$ is a compactly supported function with $\int_{\mathbb{R}^3} u(x)dx = 1$. P_δ is the Poisson process with intensity $dx \times \exp(-\delta\phi(p))dp$/Normalization. It is easy to check that

$$h(Q|P_\delta) - \delta E^Q \left[\sum_\alpha u(x_\alpha)\phi(p_\alpha) \right] - (\log c_\delta) E^Q \left(\sum_\alpha u(x_\alpha) \right)$$

is independent of δ and defines a functional $s(Q)$. Then

$$\psi(\lambda) = \sup_{Q \in \mathscr{M}_0} [E^Q[F_\lambda] - s(q)] . \tag{5.10}$$

Here

$$\begin{aligned} F_\lambda = \sum_{i=0}^4 \lambda_i F^i = \lambda_0 \sum_\alpha u(x_\alpha) + \sum_{i=1}^3 \lambda_i \sum_\alpha u(x_\alpha) p_\alpha^i \\ - \lambda_4 \sum_\alpha \left[\phi(p_\alpha) + \frac{1}{2} \sum_{\beta \neq \alpha} V(x_\alpha - x_\beta) \right] u(x_\alpha) . \end{aligned}$$

We can fix $E^Q(F^i) = q_i$ for $i = 0, 1, \ldots, 4$ and define the thermodynamic entropy function

$$\hat{s}(q) = \inf_{Q : E^Q[F_i] = q_i} s(Q) . \tag{5.11}$$

Then $\hat{s}(q)$ is a convex function of $q \in \mathbb{R}^5$ and ψ and $\hat{s}$ are related by

$$\psi(\lambda) = \sup_q \left[\sum_{i=0}^4 \lambda_i q_i - \hat{s}(q) \right] . \tag{5.12}$$

There is an open subset of U of $L_0 \times \mathbb{R}$ and an open subset $\hat{U}$ of $\mathbb{R}^5$ such that $\nabla\psi$ and $\nabla\hat{s}$ map each onto the other in a smooth one-to-one manner (cf. [R]). Moreover by restricting to a smaller nonempty subset we can assume that the variation in (5.10) is attained at a unique measure Q_λ and Q_λ is the "Gibbs" measure corresponding to the activity λ. From the uniqueness it follows that Q_λ is ergodic as a stationary point process.

Now we want to study local "Gibbs" families of measures. Let $\lambda(x)$ be a smooth map of $T^3 \times U$. Let $q(x)$ be the corresponding map into $\hat{U}$. Assume $\int_{T^3} q_0(x)dx = 1$. Let us define on $(T^3 \times \mathbb{R}^3)^N$

$$G(\underline{x}, \underline{p}) = \sum_\alpha \lambda_0(x_\alpha) + \sum_\alpha \sum_{i=1}^3 \lambda_i(x_\alpha)p_\alpha^i - \sum_\alpha \lambda_4(x_\alpha)h_\alpha\,, \tag{5.13}$$

where

$$h_\alpha = \phi(p_\alpha) + \frac{1}{2}\sum_{\beta\neq\alpha} V\left(\frac{x_\alpha - x_\beta}{\varepsilon}\right) \tag{5.14}$$

and $N \to \infty$ as $\varepsilon \to 0$ with $N\varepsilon^3 \to 1$.

We define on $(T^3 \times \mathbb{R}^3)^N$ a local Gibbs measure $\hat{\mu}_N = \hat{g}_N d\underline{x}d\underline{p}$ with density given by

$$\hat{g}_N(\underline{x}, \underline{p}) = \frac{1}{Z_N^c}\exp[G(\underline{x}, \underline{p})]\,. \tag{5.15}$$

Note that $\hat{g}_N$ is the same function as defined in 2.27. According to Theorem 5.6

$$\lim_{\varepsilon\to 0}\frac{1}{N}\log Z_N^c = \sup\left\{\int_{T^3} dx[E^{Q_x}[F_x(\omega)] - s(Q_x)]; \{Q_x\} : \int \varrho(Q_x)dx = 1\right\}$$

$$= \sup_{\zeta :\, \int \zeta_0(x)dx = 1}\left\{\int\left[\sum_i \lambda_i(x)\zeta_i(x) - \hat{s}(\zeta(x))\right]dx\right\}. \tag{5.16}$$

Since the supremum is attained anyway at a $\zeta_0(x)$ with $\int \zeta_0(dx) = 1$, we can drop the restriction and

$$\lim_{N\to\infty}\frac{1}{N}\log Z_{N,\lambda}^c = \int \psi(\lambda(x))dx\,. \tag{5.17}$$

Let us take a function [h_α defined in (5.14)]

$$\Omega(\underline{x}, \underline{p}) = N^{-1}\left\{b_0(x) + \sum_{i=1}^3\sum_\alpha b_i(x_\alpha)p_\alpha^i + \sum_\alpha b_4(x_\alpha)h_\alpha\right\}$$

$$+ a\int K(x, \omega_{x,\varepsilon})dx\,, \tag{5.18}$$

where $b_i(x)$ are continuous, $b_4 < \lambda_4$ and $K(x, \omega)$ is bounded and local. Then if we define the partition function

$$\bar{Z}_N = \int \exp[\Omega(\underline{x}, \underline{p})]d\hat{\mu}_N\,, \tag{5.19}$$

then

$$\lim_{N\to\infty}\frac{1}{N}\log\bar{Z}_N = \sup_{\{Q_x\};\int\zeta_0(x)dx=1}\int\left[E^{Q_x}(\Omega_x(\omega)) + \sum_i \lambda_i(x)\zeta_i(x) - s(Q_x)\right]dx$$

$$- \int \psi(\lambda(x))dx\,, \tag{5.20}$$

where $\zeta_i(x) = E^{Q_x}[F_i]$. Clearly one can move the sup inside and drop the constraint $\int \zeta_0(x)dx = 1$ to have an upper bound. Together with (5.11) we have

$$\lim_{N\to\infty} N^{-1} \log \bar{Z}_N \\ \leq \int \sup_Q \left\{ E^Q(K(x, \omega_{x,\varepsilon})) + \sum_\mu b_\mu(x)\zeta_\mu - I(\zeta, \lambda(x)) \right\} dx . \tag{5.21}$$

Here

$$I(\zeta, \lambda) = \hat{s}(\zeta) + \psi(\lambda) - \lambda\zeta .$$

These estimates provide an easy corollary.

Corollary 5.8. *The measures $\hat{\mu}_N$ have a law of large numbers. If*

$$m = \int E^{\hat{\mu}_{\lambda(x)}}(\Omega_x(\omega))dx ,$$

where $\hat{\mu}_{\lambda(x)}$ is the Gibbs point process corresponding to $\lambda(x)$ then

$$\hat{\mu}_N \left\{ (\underline{x}, \underline{p}) \colon \left| \frac{1}{N} \Omega(\underline{x}, \underline{p}) - m \right| \geq \delta \right\}_{N\to\infty} \to 0$$

exponentially fast in N for each $\delta > 0$.

Let us map the configuration $(\underline{x}, \underline{p})$ into the set of maps from T^3 into the stationary point processes on $\mathbb{R}^3 \times \mathbb{R}^3$. Let K be an arbitrary number. We map the configuration in a box of size εK around $x \in T^3$ to a configuration in a box of size K in $\mathbb{R}^3$ by $y_\alpha = (x_\alpha - x)\varepsilon^{-1}$. Then $\{(y_\alpha, p_\alpha)\} = \omega_k^x$ is a configuration in $B_k \times \mathbb{R}^3$. We can extend it periodically to a configuration $\tilde{\omega}_k^x$ in $\mathbb{R}^3 \times \mathbb{R}^3$, and averaging over translations on B_k we obtain a stationary random point process $R_{x,\varepsilon,k,\omega}$ defined by

$$E^{R_{x,\varepsilon,k,\omega}}[A(\omega')] = \frac{1}{|B_k|} \int_{B_k} dy\, A(\tau_y \tilde{\omega}_k^x) .$$

Let $\hat{\pi}_{N,k}$ be the measures induced by this map by $d\hat{\mu}_N = \hat{g}_N d\underline{x}d\underline{p}$ on the space of maps

$$x \in T^3 \to R_x \in M_0 .$$

Our basic result can be thought of as a large deviation result for $\hat{\pi}_{N,k}$. Then the estimate (5.20) can be rewritten as $(\zeta_\mu(x) = E^{Q_x}(F_\mu))$

$$\lim_{k\to\infty} \lim_{\varepsilon\downarrow 0} \frac{1}{N} \log E^{\hat{\pi}_{N,k}} \left(\exp N \int_{T^3} dx\, E^{R_{x,w}}(\Omega_x) \right) \\ = \sup_{\{Q_x\} \colon \int \zeta_0(x)dx = 1} \int dx \left[E^{Q_x}(\Omega_x) + \sum_\mu \lambda_\mu(x)\zeta_\mu(x) \right. \\ \left. - s(Q_x) - \psi(\lambda(x)) \right] . \tag{5.22}$$

Proof of Lemma 3.8. By entropy bound (2.28) the expectation with respect to g_N is bounded by

$$N^{-1}\delta_0^{-1} \log E^{\hat{g}_N}\left[\exp\left\{\delta_0 N \int_{T^3} (1-\sigma_k) A \, dx\right\}\right] + \delta^{-1}N^{-1}H\,.$$

The first term can be bounded using (5.21) in the limit $N \to \infty$ and $k \to \infty$ by

$$\delta_0^{-1} \int \sup_{\zeta(x)} \left\{\delta_0 \sum_\mu b_\mu(x)\zeta_\mu(x)(1 - 1_{u_2}(\zeta(x)) - I(\zeta(x), \lambda(x))\right\} dx\,. \tag{5.23}$$

Since $I(\zeta(x), \lambda(x)) > \varepsilon > 0$ in the region $1 - 1_{u_2}(\zeta(x)) \neq 0$ and I is convex in $\zeta(x)$, the sup in (5.23) is zero if δ_0 is small enough. We have thus proved Lemma 3.8. □

We can now prove Lemma 4.2.

Proof of Lemma 4.2. Let Q be any limit of $\bar{Q}_\varepsilon$. By (5.9)

$$h(Q|P_\delta) \leq E^Q(F_{\lambda,\delta}) - \psi(\lambda)\,.$$

By the entropy inequality and 4.4

$$E^Q(F_{\lambda,\delta}) \leq \lim_{N\to\infty} \frac{1}{N} \log E^{P_N}\left(e^{N\int F_{\lambda,\delta}(\omega_\varepsilon, x)dx}\right) + C \leq \psi(\lambda) + C\,,$$

where the last inequality is due to 5.17. □

Finally we state:

Lemma 5.9. *If $g_N(\underline{x}, \underline{p})$ is a symmetric probability density on $(T^3 \times \mathbb{R}^3)^N$ such that the relative entropy with respect to $\hat{g}_{N,\lambda}$*

$$H(g_N|\hat{g}_{N,\lambda}) \leq cN\,.$$

Then

$$E^{g_N}\left(\frac{1}{N} \sum_{(\alpha\neq\beta)} V\left(\frac{x_\alpha - x_\beta}{\varepsilon}\right)\right) \leq c'\,,$$

and therefore by Lemma 5.7

$$E^{g_N}\left(\frac{1}{N} \sum_{(\alpha\neq\beta)} W\left(\frac{x_\alpha - x_\beta}{\varepsilon}\right)\right) \leq C_W$$

for all W continuous and having compact support.

Proof. It is an easy consequence of the entropy inequality and the theory developed above. □

References

[CLY] Conlon, J.G., Lieb, E.H., Yau, H.T.: The Coulomb gas at low temperature and low density. Commun. Math. Phys. **125**, 153–218 (1989)

[DeM] DeMasi, A., Ianiro, N., Pellegrinotti, A., Presutti, E.: A survey of the hydrodynamical behavior of many-particle systems. In: Nonequilibrium phenomena. II. From stochastic to hydrodynamics. Lebowitz, J.L., Montroll, E.W. (eds.), pp. 123–234, Amsterdam: North-Holland, 1984

[Sp] Spohn, H.: Large Dynamics of interacting particles. Berlin, Heidelberg, New York: Springer 1991

[Si] Sinai, Ya.G.: Dynamics of local equilibrium Gibbs distributions and Euler equations. The one-dimensional case. Selecta Math. Sov. **7**(3), 279–289 (1988)

[BDS] Boldrighini, C., Dobrushin, R.L., Suhov, Yu.M.: One-dimensional hard rod caricatures of hydrodynamics. J. Stat. Phys. **31**, 577–616 (1983)

[Y] Yau, H.T.: Relative entropy and the hydrodynamics of Ginzburg-Landau models. Lett. Math. Phys. **22**, 63-80 (1991)

[GPV] Guo, M.Z., Papanicolaou, G.C., Varadhan, S.R.S.: Nonlinear diffusion limit for a system with nearest neighbor interactions. Commun. Math. Phys. **118**, 31–53 (1988)

[V] Varadhan, S.R.S.: Scaling limits for interacting diffusions. Commun. Math. Phys. **135**, 313–353 (1991)

[O] Olla, S.: Large deviations for Gibbs random fields. Prob. Theory Related Fields **77**, 343–357 (1988)

[OV] Olla, S., Varadhan, S.R.S.: Scaling limit for interacting Ornstein-Uhlenbeck processes. Commun. Math. Phys. **135**, 355–378 (1991)

[Re] Rezakhanlou, F.: Hydrodynamic limit for a system with finite range interactions. Commun. Math. Phys. **129**, 445–480 (1990)

[R] Ruelle, D.: Statistical Mechanics. Reading, MA: Benjamin 1969

[V1] Varadhan, S.R.S.: Large deviation and applications. Philadelphia: SIAM 1984

Communicated by J. L. Lebowitz

Pitman Res. Notes Math. Ser., 283, Longman Sci. Tech., 1993, 75-128

S R S VARADHAN

Nonlinear diffusion limit for a system with nearest neighbor interactions – II

1. Introduction.

This paper is an extension of Guo-Papanicolaou and Varadhan [2] as well as Donsker and Varadhan. In [1] we considered the following situation: let S denote the unit circle viewed as the interval $0 \leq \theta \leq 1$ with the end points $\theta = 0$ and $\theta = 1$ identified. For any integer $N \geq 1$, let S_N denote the periodic lattice $\{j/N\}$ with $j = 1, 2, \ldots, N$. The sites adjacent to j/N are $(j \pm 1)/N$, addition being modulo N. The "charge" at site j is represented by the variable x_j. The charge configuration $(x_1, \ldots, x_N)$ is represented as a vector x in R^N. The colnfiguration x changes with time and as a function of time $x(t)$ undergoes a diffusion in R^N. The infinitesimal generator of the diffusion is given by

$$\begin{aligned}(1.1)\qquad \mathcal{L}_N = &\frac{N^2}{2}\sum_{i=1}^{N}\Big(\frac{\partial}{\partial x_i} - \frac{\partial}{\partial x_{i+1}}\Big)^2 \\ &- \frac{N^2}{2}\sum_{i=1}^{N}\big(\phi'(x_i) - \phi'(x_{i+1})\big)\Big(\frac{\partial}{\partial x_i} - \frac{\partial}{\partial x_{i+1}}\Big)\end{aligned}$$

Here ϕ is a continuously differentiable fucntion from R into R satisfying the properties (1.2), (1.3) and (1.4) to be listed below and $\phi'(x) = \dfrac{d\phi(x)}{dx}$. The factor N^2 in (1.1) with the lattice spacing of $1/N$ represents the diffusion scaling of space and time.

$$(1.2)\qquad \int_{-\infty}^{\infty} e^{-\phi(x)}\,dx = 1$$

$$(1.3)\qquad \int_{-\infty}^{\infty} e^{\lambda x - \phi(x)} = M(\lambda) < \infty \quad \text{for all } \lambda \in R$$

$$(1.4)\qquad \int_{-\infty}^{\infty} e^{\sigma|\phi'(x)| - \phi(x)} < \infty \quad \text{for all } \sigma > 0\,.$$

We can reexpress (1.3) and (1.4) in an equivalent form

$$\int_{-\infty}^{\infty} e^{\omega(x)-\phi(x)}\,dx < \infty \tag{1.5}$$

where $\omega(x)$ is a symmetric function of x satisfying $\omega(x) \geq |x| + 1$ and

$$\lim_{|x|\to\infty} \frac{|x|}{\omega(x)} = 0\,, \qquad \lim_{|x|\to\infty} \frac{|\phi'(x)|}{\omega(x)} = 0 \tag{1.6}$$

Given such a ϕ we can define a density $\Phi_N(x) = \exp\left[-\sum \phi(x_j)\right]$ on R^N and this will be the density relative to Lebesgue measure of a probability measure on R^N. The generator $\mathcal{L}_N$ is formally symmetric relative to the density Φ_N and defines a diffusion process with invariant density Φ_N with respect to which it is reversible. The Dirichlet form of the diffusion is given by

$$N^2\widehat{\mathcal{D}}_N(u) = \frac{N^2}{2}\int \Big\{\sum_{i=1}^{N}\Big(\frac{\partial u}{\partial x_i} - \frac{\partial u}{\partial x_{i+1}}\Big)^2\Big\}\,\Phi_N\,dx \tag{1.7}$$

The diffusion with invariant density Φ_N is not ergodic. The sum $x_1 + \cdots + x_N$ is conserved by the diffusion and the hyperplanes $x_1 + \cdots + x_N = Ny$ of average charge y are invariant sets. For each fixed y, the diffusion restricted to the hyperplane is nondegenerate and ergodic. The invariant measure on the hyperplane is the conditional distribution of Φ_N given $(x_1 + \cdots + x_N)/N = y$.

We start with an initial distribution for $x(0)$ given by a density $f_N^0(x_1, \ldots, x_N)$ relative to Φ_N. Then the evolution will give us a density for $t \geq 0$ to be denoted by $f_N^t(x_1, \ldots, x_N)$. To obtain f_N^t we have to solve the equation

$$\begin{aligned}\frac{\partial f_N^t}{\partial t} &= \mathcal{L}_N f_N^t \\ f_N^t\big|_{t=0} &= f_N^0\end{aligned} \tag{1.8}$$

For each charge configuration x we have an associated random measure

$$\mu_N = \frac{1}{N}\sum x_i \delta_{i/N}$$

viewed as a signed measure on S. If we have a sequence of configurations for which the limits

$$\lim_{N\to\infty}\langle J,\mu_N\rangle = \lim_{N\to\infty}\frac{1}{N}\sum J(\frac{i}{N})x_i$$
$$= \int J(\theta)\,m(\theta)\,d\theta$$

exists for all bounded continuous test functions $J(\cdot)$ on S, with some choice of a function $m(\theta)$ in $L_1(S)$, we speak of $m(\theta)$ as the macroscopic charge density. For our initial density f_N^0 we assume that for some nice $m_0(\theta)$ for every smooth $J(\cdot)$ and each $\delta > 0$,

$$\lim_{N\to\infty}\int_{E_N} f_N^0\Phi_N\,dx = 0 \tag{1.9}$$

where

$$E_N = \{x : |\frac{1}{N}\sum J(\frac{i}{N})x_i - \int m_0(\theta)\,J(\theta)\,d\theta| \geq \delta\}$$

In addition we assume that the entropy

$$H_N(f_N^0) = \int f_N^0 \log f_N^0 \Phi_N\,dx \tag{1.10}$$

satisfies a bound

$$H_N(f_N^0) \leq CN \tag{1.11}$$

for some constant C independent of N.

Under these conditions the randomly changing charge configurations at times $t > 0$ are shown to have an asymptotic charge density $m(t,\theta)$ which is characterized as the unique solution of a certain nonlinear heat equation with initial condition $m_0(\theta)$. Before we can describe this equation we have to introduce some notation. We look at the function $M(\lambda)$ defined in equation (1.3) and define

$$\begin{aligned}\rho(\lambda) &= \log M(\lambda)\\ h(y) &= \sup_{\lambda}\left[\lambda y - \rho(\lambda)\right]\end{aligned} \tag{1.12}$$

Then $h(\cdot)$ and $\rho(\cdot)$ are a pair of conjugate convex functions and we have the dual variables λ and y related by

$$\lambda = h'(y) \quad \text{or} \quad y = \rho'(\lambda) \tag{1.13}$$

i.e. h' and ρ' are inverse functions of each other. By elementary calculation one can check that

$$\frac{1}{M(\lambda)} \int e^{\lambda x - \phi(x)}\, x\, dx = \rho'(\lambda) \tag{1.14}$$

so that the value of λ for which (1.14) equals a given value y is given by $h'(y)$. It is known that h' and ρ' are strictly increasing functions. The main result of [2] is the following theorem.

Theorem 1.1. *If the initial distribution of charges satisfies* (1.9) *and* (1.11) *then for every* $t > 0$ *and arbitrary continuous* $J(\cdot)$ *and each* $\delta > 0$

$$\lim_{N\to\infty} \int_{E_N^t} f_N^t \Phi_N\, dx = 0 \tag{1.15}$$

where

$$E_N^t = \{x : |\frac{1}{N} \sum J(\frac{i}{N}) x_i - \int J(\theta)\, m(\theta, t) d\theta| \geq \delta\}$$

and $m(\theta, t)$ *is the unique weak solution of the nonlinear diffusion equation*

$$\frac{\partial m}{\partial t} = \frac{1}{2}\left[h'(m(\theta, t))\right]_{\theta\theta} \tag{1.16}$$

The results of [2] are modified in this paper. The basic diffusion operator $\mathcal{L}_N$ is given by

$$\begin{aligned} \mathcal{L}_N = {} & \frac{N^2}{2} \sum a(x_i, x_{i+1}) \left(\frac{\partial}{\partial x_i} - \frac{\partial}{\partial x_{i+1}}\right)^2 \\ & - \frac{N^2}{2} \sum W(x_i, x_{i+1}) \left(\frac{\partial}{\partial x_i} - \frac{\partial}{\partial x_{i+1}}\right) \end{aligned} \tag{1.17}$$

where

$$W(x,y) = -a_1(x,y) + a_2(x,y) + a(x,y)(\phi'(x) - \phi'(y)) \tag{1.18}$$

Here $a(x,y)$ is a function of x, y with bounded continuous first derivatives satisfying

$$0 < c_1 \leq a(x,y) \leq c_2 < \infty \tag{1.19}$$

The functions a_1 and a_2 are the partial derivatives

$$a_1 = \frac{\partial a}{\partial x} \quad \text{and} \quad a_2 = \frac{\partial a}{\partial y}$$

The new operation $\mathcal{L}_N$ is again symmetric with respect to the weight Φ_N, but now the Dirichlet form is given by

$$N^2 D_N(u) = \frac{N^2}{2} \int \sum a(x_i, x_{i+1}) \left(\frac{\partial u}{\partial x_i} - \frac{\partial u}{\partial x_{i+1}}\right)^2 e^{-\Sigma\phi(x_i)}\, dx \tag{1.20}$$

Of the initial density f_N^0 we assume that (1.9) holds.

We have to make an additional hypothesis on ϕ, one of strong convexity. For simplicity we assume that

$$0 < k_1 \leq \phi''(x) \leq k_2 < \infty \tag{1.21}$$

This is used at the end to provide an estimate on a spectral gap. This condition can be weakened considerably. We will say more about this at the end.

The main result of our paper is the following theorem in our modified circumstances.

Theorem 1.2. *If the initial distribution of charges satisfies* (1.9) *and* (1.11) *then the evolution determined by the operator* (1.17) *satisfies for every* $t > 0$, $\delta > 0$ *and smooth* $J(\cdot)$

$$\lim_{N\to\infty} \int_{E_N^t} f_N^t \, \Phi_N \, dx = 0 \tag{1.22}$$

with

$$E_N^t = \{x : |\frac{1}{N}\sum J(\frac{i}{N})x_i - \int J(\theta)\, m(\theta,t)\, d\theta| \geq \delta\}$$

and $m(\theta,t)$ is the unique weak solution of the nonlinear diffusion equation

$$\begin{aligned} \frac{\partial m(\theta,t)}{\partial t} &= \frac{1}{2}\left(\hat{a}(m(\theta,t))\, h'(m(\theta,t))_\theta\right)_\theta \\ m(\theta,t)\big|_{t=0} &= m_0(\theta) \end{aligned} \tag{1.23}$$

where the function $\hat{a}(m)$ is determined in the manner described below.

For each $m \in R$, we have $\lambda = h'(m)$ and with this choice of λ, we have a product measure Q_m on $\prod_{-\infty}^{\infty} R = \Omega$ with each coordinate having the distribution $\frac{1}{M(\lambda)}\, e^{\lambda x - \phi(x)}\, dx$. Let $g(x_{-\ell}, \ldots, x_\ell)$ be a smooth function of $(2\ell + 1)$ variables. We will view this as a function $g(w)$ defined on Ω. If we denote by T the shift operator $(T\omega)(i) = x_{i+1}$ if $\omega(i) = x_i$, we can form the formal infinite sum

$$\xi = \sum_{k=-\infty}^{\infty} g(T^k \omega) \tag{1.27}$$

Although ξ does not really make sense the partial derivatives $\frac{\partial \xi}{\partial x_i}$ are all well defined. We can now write down the formula for $\hat{a}(m)$:

$$\hat{a}(m) = \inf_g E^{Q_m}\left[a(x_1, x_2)\left[1 - \left(\frac{\partial \xi}{\partial x_1} - \frac{\partial \xi}{\partial x_2}\right)\right]^2\right] \tag{1.24}$$

where the infimum is taken over all functions g varying ℓ as well as the function of the $(2\ell + 1)$ variables.

2. A sketch of the proof

We can write a system of stochastic differential equations to describe our diffusion

$$(2.1)\quad \begin{cases} dx_i(t) = \frac{N^2}{2}\left[W(x_{i-1}(t), x_i(t)) - W(x_i(t), x_{i+1}(t))\right] dt \\ \qquad + N\left[\sigma(x_{i-1}(t), x_i(t))\, d\beta_{i-1,i}(t) - \sigma(x_i(t), x_{i+1}(t))\, d\beta_{i,i+1}(t)\right] \end{cases}$$

where $\{\beta_{i,i+1}(t)\}$ are N-independent Brownian motions and $\sigma(x_1, x_2)$ is a smooth square root of $a(x_1, x_2)$. We compute for an arbitrary smooth function $J(\theta)$ on S,

$$(2.2)\quad \begin{aligned} d[\frac{1}{N}\sum J(\frac{i}{N})x_i(t)] &= \frac{N}{2}\sum J(\frac{i}{N})\{W(x_{i-1}(t), x_i(t)) - W(x_i(t), x_{i+1}(t))\}\, dt \\ &\quad + \sum J(\frac{i}{N})\{\sigma(x_{i-1}(t), x_i(t))\, d\beta_{i-1,i}(t) \\ &\qquad - \sigma(x_i(t), x_{i+1}(t))\, d\beta_{i,i+1}(t)\} \\ &= \frac{N}{2}\sum W(x_i(t), x_{i+1}(t))[J(\frac{i}{N}) - J(\frac{i+1}{N})]\, dt \\ &\quad + \sum \sigma(x_i(t), x_{i+1}(t))[J(\frac{i}{N}) - J(\frac{i+1}{N})]\, d\beta_{i,i+1}(t) \end{aligned}$$

Let us denote by

$$(2.3)\quad \lambda_N(t) = \frac{1}{N}\sum x_i(t)\, \delta_{i/N}$$

$$(2.4)\quad \mu_N(t) = \sum_{i=1}^{N} [\int_0^t W(x_i(s), x_{i+1}(s))\, ds]\, \delta_{i/N}$$

$$(2.5)\quad \nu_N(t) = \frac{1}{N}\sum_{i=1}^{N} [\int_0^t \sigma(x_i(s), x_{i+1}(s))\, d\beta_{i,i+1}(s)]\, \delta_{i/N}$$

Using these signed measures we can rewrite our stochastic differential equation (2.2) in the form

$$(2.6)\quad \langle J, \lambda_N(t)\rangle - \langle J, \lambda_N(0)\rangle = -\frac{1}{2}\langle J'_N, \mu_N(t)\rangle - \frac{1}{2}\langle J'_N, \nu_N(t)\rangle$$

where

$$J'_N(\theta) = N\big[J(\theta + \frac{1}{N}) - J(\theta)\big] \tag{2.7}$$

and $\langle J, \alpha\rangle$ for a function J on S and a measure α on S is the usual integral $\int J(\theta)\, d\alpha(\theta)$.

We have our basic diffusion proces Q_N, a measure on $C[[0,\infty), R^N]$ corresponding to our diffusion with initial density $f_N^0 \Phi_N$. We can start from equilibrium and then our initial density will be Φ_N and the corresponding measure on $C[[0,\infty\ , R^N]$ will be denoted by P_N.

Our first goal is to establish some form of compactness for the triplet $\{\lambda_N(\cdot), \mu_N(\cdot), \nu_N(\cdot)\}$ as random functions on $[0,\infty)$ with values in the space of signed measures on S. The compactness of course is for the distribution of these objects induced from Q_N. Our aim is to prove compactness for the sequence $\widehat{Q}_N$ of probability measures so induced and if $\widehat{Q}$ is any limit point representing the distribution of a triplet $\lambda(\cdot)\ \mu(\cdot)\ \nu(\cdot)$ of signed measure valued random functions on $[0,\infty)$, we will establish various things for the measure $\widehat{Q}$. We prove that for every t, for almost all $\lambda(\cdot)$ with respect to $\widehat{Q}$, $\lambda(t)$ is absolutely continuous in θ and has a density given by $m(\theta,t)$. We prove that $\mu(t)$ has a representation almost surely of the form

$$\mu_t = [\int_0^t \eta(\theta,s)\, ds]\, d\theta$$

It is rather easy to check that $\nu(t) \equiv 0$ almost surely. At this point we clearly have established in a weak sense

$$\frac{\partial m}{\partial t} = \frac{1}{2}\eta_\theta \tag{2.8}$$

which is an easy consequence of (2.6). We now have to identify η in terms of m. We have to show that

$$\eta(\theta,t) = \hat{a}(m(\theta,t))\,[h'(m(\theta,t))]_\theta \quad \text{a.e. } \widehat{Q} \tag{2.9}$$

From the earlier work we know that

(2.10) $$\frac{1}{N}\sum J(\frac{i}{N})\,\phi'(x_i(t)) \sim \int_S J(\theta)\,h'(m(t,\theta))\,d\theta$$

So roughly speaking

(2.11) $$N[\phi'(x_{i+1}(t)) - \phi'(x_i(t))] \sim [h'(m(t,\theta))]_\theta$$

On the other hand

(2.12) $$NW(x_i(t)\ ,\ x_{i+1}(t)) \sim \eta(t,\theta)$$

We therefore have to prove

(2.13) $$NW(x_i(t)\ ,\ x_{i+1}(t)) \sim N\hat{a}(m(t,\theta))\,[\phi'(x_{i+1}(t) - \phi'(x_i(t))]$$

By some large deviation estimates this is reduced to a problem concerning P_N or the process in equilibrium. After additional calculations this can be reduced, in a very rough sense, to showing the following property: for a given m let us pick $\lambda = h'(m)$ and consider the product measure Q_m on R^∞ corresponding to

$$\prod \frac{1}{M(\lambda)} e^{\lambda x_j - \phi(x_j)}\,dx_j\ .$$

With Q_m as invariant measure we can construct a diffusion which is symmetric and has a Dirichlet form

$$\frac{1}{2}\int \sum a(x_i, x_{i+1})(\frac{\partial u}{\partial x_i} - \frac{\partial u}{\partial x_{i+1}})^2\,dP_m$$

Relative to the process we consider

$$B_N(t) = \sum_{i=-N}^{N} W(x_i(t), x_{i+1}(t))$$

and

$$A_N(t) = \sum_{i=-N}^{N} [\phi'(x_{i+1}(t)) - \phi'(x_i(t))]$$

which are both mean zero functions and satisfy a central limit theorem. The constant $\hat{a}(m)$ will have the property

$$\limsup_{N\to\infty} \frac{1}{2N} \limsup_{t\to\infty} \frac{1}{t} E\left| \int_0^t [B_N(s) - \hat{a}(m)\, A_N(s)]\, ds\right|^2 = 0$$

Roughly speaking $W(x_i, x_{i+1})$ has fluctuation in space and time and as one averages over space and time only the component proportional to the fluctuations of $\phi'(x_{i+1}) - \phi'(x_i)$ survives. The diffusion coefficient turns out to be the coefficient of this component.

3. Compactness Estimates:

The following estimates on the measure Q_N will establish the compactness of the induced sequence $\widehat{Q}_N$.

Theorem 3.1. *For each $T < \infty$, $\epsilon > 0$ and smooth function J on S*

$$\lim_{\ell\to\infty} \limsup_{N\to\infty} Q_N\{ \sup_{0\le t\le T} \|\lambda_N(t)\| \ge \ell\} = 0 \tag{3.1}$$

and

$$\lim_{\delta\to 0} \limsup_{N\to\infty} Q_N\{ \sup_{\substack{0\le s<t\le T\\ |s-t|\le\delta}} |\langle J, \lambda_N(t) - \lambda_N(s)\rangle| \ge \epsilon\} = 0 \tag{3.2}$$

Here we denote the variation norm of λ by $\|\lambda\|$.

We also have the identical estimates when λ_N is replaced by μ_N or ν_N. Because of equation (2.6) the estimates rub off on each other and we do not have to establish all six of them. We will develop these estimates one at a time.

Lemma 3.2.

$$\lim_{\ell\to\infty} \limsup_{N\to\infty} Q_N\{ \sup_{0\le t\le T} \|\lambda_N(t)\| \ge \ell\} = 0$$

Proof: The proof is very similar to the one of Lemma 6.1 in [2]. Because of the inequality $h(Q_N, P_N) \le CN$, for the relative entropy of Q_N with respect to P_N on $C([0,\infty), R^N)$ it is sufficient to show

$$\lim_{\ell\to\infty} \limsup_{N\to\infty} \frac{1}{N} \log P_N\{ \sup_{0\le t\le T} \|\lambda_N(t)\| \ge \ell\} = -\infty \tag{3.3}$$

This was established in [2] for the case when $a(x_1, x_2) \equiv 1$, using properties of the Dirichlet form. Since the Dirichlet form with $a(x_1, x_2)$ is bounded above and below by a constant multiple of the Dirichlet form with $a = 1$, for this part of the argument it makes no difference if $a(x_1, x_2)$ is not identically one so long as it is bounded above and below, away from zero.

Lemma 3.3.

$$\lim_{\ell\to\infty} \limsup_{N\to\infty} Q_N\{ \sup_{0\le t\le T} \|\nu_N(t)\| \ge \ell\} = 0 \tag{3.4}$$

Proof: Looking at the definition of $\nu_N(t)$, clearly

$$E^{Q_N}\{ \sup_{0\le t\le T} \|\nu_N(t)\|\} \le \frac{1}{N} E^{Q_N}\{\sum_{j=1} \sup_{0\le t\le T} |\int_0^t \sigma(x_i(s), x_{i+1}(s))d\beta_{i,i+1}(s)|\}$$

In order to establish a uniform bound in N, we need only to note that by Doob's inequality there is a simple bound for

$$E^{Q_N}[\sup_{0\le t\le T} |\int_0^t \sigma(x_i(s), x_{i+1}(s))\, d\beta_{i,i+1}(s)|]$$

depending only on T and a bound for σ.

Lemma 3.4. *For every smooth J,*

$$\limsup_{N\to\infty} Q_N\{ \sup_{0\le t\le T} |\langle J, \nu_N(t)\rangle| \ge \epsilon\} = 0 \tag{3.5}$$

Proof: Since $\langle J, \nu_N(t)\rangle$ is a Q_N-martingale we only need to compute $E^{Q_N}\{\langle J, \nu_N(T)\rangle^2\}$ and verify that it goes to zero. Since $\beta_{i,i+1}$ are independent Brownian motions

$E^{Q_N}\{\langle J, \nu_N(T)\rangle^2\}$ is clearly of order $O(\frac{1}{N})$ and goes to zero as $N \to \infty$.

Lemma 3.5. *For every smooth J on S,*

$$\lim_{\delta\to 0} \limsup_{N\to\infty} Q_N\{ \sup_{\substack{0\le s<t\le T\\ |s-t|\le\delta}} |\langle J, \mu_N(t) - \mu_N(s)\rangle| \ge \epsilon\} = 0 \tag{3.6}$$

Proof: Because of the entropy estimate $h(Q_N, P_N) \le CN$ it is sufficient to establish

$$\lim_{\delta\to 0} \limsup_{N\to\infty} \frac{1}{N} \log P_N\{ \sup_{\substack{0\le s<t\le T\\ |s-t|\le\delta}} |\langle J, \mu_N(t) - \mu_N(s)\rangle| \ge \epsilon\} = -\infty \tag{3.7}$$

Our first step in deriving (3.7) is a bound of the form

$$E^{P_N}\{\exp[N\rho\langle J,\mu_N(t)-\mu_N(s)\rangle]\}\le\exp[\frac{N}{2}(t-s)\rho^2C_J] \tag{3.8}$$

where C_J is a constant depending only on J. We note

$$\begin{aligned}&E^{P_N}\{\exp[N\rho\langle J,\mu_N(t)-\mu_N(s)\rangle]\}\\&=E^{P_N}\{\exp[N\rho\int_s^t\sum J(\frac{i}{N})W(x_i(\sigma),x_{i+1}(\sigma))\,d\sigma]\}\\&=E^{P_N}\{\exp[N\rho\int_0^{t-s}\sum J(\frac{i}{N})W(x_i(\sigma),x_{i+1}(\sigma))\,d\sigma]\}\quad\text{(by stationarity)}\\&\le\exp[(t-s)\Lambda_N(J)]\quad\text{(by spectral theorem)}\end{aligned}$$

where $\Lambda_N(J)$ is the eigenvalue

$$\Lambda_N(J)=\sup[N\rho\sum J(\frac{i}{N})\int W(x_i,x_{i+1})\,f\,\Phi_N\,dx-N^2D_N(\sqrt{f})]$$

If we consider the terms

$$A_{N,i}=N\rho J(\frac{i}{N})\int W(x_i,x_{i+1})\,f\,\Phi_N\,dx-\frac{N^2}{8}\int a(x_i,x_{i+1})\,(\frac{\partial f}{\partial x_i}-\frac{\partial f}{\partial x_{i+1}})^2\frac{1}{f}\,\Phi_N\,dx$$

for a fixed i, after performing an integration by parts

$$\begin{aligned}A_{N,i}&=N\rho J(\frac{i}{N})\int a(x_i,x_{i+1})\,(\frac{\partial f}{\partial x_i}-\frac{\partial f}{\partial x_{i+1}})\,\Phi_N\,dx\\&\quad-\frac{N^2}{8}\int a(x_i,x_{i+1})\,(\frac{\partial f}{\partial x_i}-\frac{\partial f}{\partial x_{i+1}})^2\frac{1}{f}\,\Phi_N\,dx\\&\le\rho^2J^2(\frac{i}{N})C\end{aligned}$$

by using Schwartz's inequality as well as the normalization $\int f\Phi_N\,dx=1$. C can be a bound for $a(x_1,x_2)$. We see easily that

$$\Lambda_N(J)\le\frac{N\rho^2}{2}C_J$$

establishing (3.8). To go from (3.8) to (3.7) we use a theorem of Garsia Rodemick and Rumsey [4], which states that for any function $f(t)$ continuous on $[0,T]$ if $\int_0^T\int_0^T\psi(\frac{|f(s)-f(t)|}{\sqrt{|t-s|}})\,dt\,ds\le B$ then

(3.9) $$\sup_{\substack{0<s<t\le T\\ |s-t|\le\delta}} |f(s)-f(t)| \le 4\int_0^\delta \psi^{-1}\Big(\frac{4B}{u^2}\Big)\frac{du}{\sqrt{u}}\,.$$

In particular we can take $f(s) = \langle J, \mu_N(s)\rangle$ and $\psi(x) = \cosh N\rho x - 1$. We then get from (3.8) some estimate for $E^{P_N}\{B\}$. On the other hand we can use (3.9) to turn $P_N\{\sup_{\substack{0<s<t\le T\\ |s-t|\le\delta}} |f(s)-f(t)| \ge \epsilon\}$ into estimating the tail properties of B. Since we have control on $E^{P_N}(B)$ we can use Tchebycheff's inequality. However because ψ is exponential ψ^{-1} is like a logarithm for large values and the Tchebycheff inequality has an exponential character to it. It is somewhat messy to carry out. But a few simple calculations will yield the lemma.

Lemma 3.6.

$$\lim_{\ell\to\infty}\limsup_{N\to\infty} Q_N\{\sup_{0\le t\le T} |\langle 1, \mu_N(t)\rangle| \ge \ell\} = 0\,.$$

Since $\mu_N(0) = 0$ and $J \equiv 1$ is allowed in Lemma 3.5 we can control $\sup_{0\le t\le T} |\langle 1, \mu_N(t)\rangle|$ in terms of the modulus continuity estimates of Lemma 3.5.

We now claim that we have all the individual pieces needed to complete the compactness proof in Theorem 3.1. First we look at estimates for the variation. For λ_N and ν_N it was established in Lemmas 3.2 and 3.3. We cannot say immediately that because of (2.6) we have variation bounds on μ_N. The problem is that the only test functions we see are those of the form J' that have mean zero. However we can use Lemma 3.6 to remedy the situation. As for the modulus of continuity estimates once the variation is controlled the relation (2.6) is enough to pass to λ_N from controlling the modulus of continuity of μ_N and ν_N.

Remark 1. The space of all continuous signed measure valued functions on an interval $[0, T]$ is not a good space as a topological space. However the set of all continuous measure valued functions $\alpha(t)$ on $[0, T]$ with $\sup_{0\le t\le T} \|\alpha(t)\| \le \ell$, is a nice complete separable metric space. By obtaining estimates on the complement we avoid having to look at the space of all signed measures straight in its face.

Remark 2. A more convenient norm may be some negative Sobolev norm rather than bounded variation norm. This may yield slightly weaker results. But this generalizes well to higher dimensions.

In Theorem 3.1 we have obtained the compactness of the sequence $\{\widehat{Q}_N\}$. Suppose $\widehat{Q}$ is any limit point of the sequence, it is our aim to show that $\widehat{Q}$ is concentrated on a single trajectory, i.e. the solution of the nonlinear diffusion equation with the correct initial condition. But in order to do that we will need some preliminary estimates to be satisfied by the measure $\widehat{Q}$. The estimates will establish some mimimal smoothness of the measure valued functions $\lambda(t, d\theta)$ and $\mu(t, d\theta)$. We note that $\nu(t, d\theta)$ has been shown to be identically zero.

We want to show that almost surely with respect to $\widehat{Q}$

$$\lambda(t, d\theta) = m(t, \theta)\, d\theta$$
$$\mu(t, d\theta) = \Big(\int_0^t \eta(s, \theta)\, ds \Big)\, d\theta$$

with $m(t, \theta)$ being absolutely continuous in θ. We will show all of this by establishing a priori estimates for

$$E^{\widehat{Q}}\{ \sup_{0 \le t \le T} \int_S h(m(t, \theta)\,)\, d\theta \}$$
$$E^{\widehat{Q}}\{ \int_0^T \int_S \{ \frac{\partial}{\partial \theta} [h'(m(t, \theta)\,)] \}^2\, d\theta\, ds \}$$

and

$$E^{\widehat{Q}}\{ \int_0^T \int_S \eta^2(s, \theta)\, d\theta\, ds \}$$

in terms of the constant H appearing in the entropy bound (1.11) i.e.

$$\limsup_{N \to \infty} \frac{1}{N} H_N(f_N^0) \le H\ . \tag{3.10}$$

The basic tool is the following lemma from the theory of large deviations.

Lemma 3.7. *Let α_N and β_N be two probability measures on a topological space with α_N converging weakly to a limit α. We assume that for some continuous function $F(\cdot)$,*

$$\limsup_{N\to\infty} \frac{1}{N} \log \int \exp\left[NF(\cdot)\right] d\beta_N \leq C \tag{3.11}$$

and that

$$\limsup_{N\to\infty} \frac{1}{N} h(\alpha_N;\beta_N) \leq H \ . \tag{3.12}$$

Then $\max(F,0)$ *is integrable with respect to* α *and*

$$\int F(\cdot)\, d\alpha \leq C + H \ .$$

Proof: If we assume that F is bounded below then by the entropy inequality

$$\int F(\cdot)\, d\alpha_N \leq \frac{1}{N} \log \int \exp\left[NF(\cdot)\right] d\beta_N + \frac{1}{N} h(\alpha_N;\beta_N)$$

and we can let $N \to \infty$. We can conclude that

$$\int F\, d\alpha \leq C + H \ .$$

If F is not bounded below we take the truncation $F_\ell(\cdot) = \max\left[F(\cdot), -\ell\right]$. Then

$$\int \exp\left[NF_\ell(\cdot)\right] d\beta_N \leq \int \exp\left[NF(\cdot)\right] d\beta_N + e^{-N\ell}$$

It follows that

$$\int F_\ell(\cdot)\, d\alpha \leq \max\left[C, -\ell\right] + H \ .$$

This implies the integrability of $\max\{F, 0\}$ and

$$\int F(\cdot)\, d\alpha \leq \int F_\ell(\cdot)\, d\alpha \leq \max[C, -\ell] + H \ .$$

We let $\ell \to \infty$.

This lemma by itself is not a big tool. However assume that we have a family $\mathcal{F}$ of functions F for each of which the assumptions of Lemma 3.7 holds with the constant $C = C(F)$ depending on $F \in \mathcal{F}$. Then it is true gratuitously that

$$\sup_{F\in\mathcal{F}} E^{\alpha}\{F(\cdot)-C(F)\} \le H \ .$$

However the following lemma, although an easy consequence is a big step.

Lemma 3.8. *Assume that* (3.11) *holds for each $F \in \mathcal{F}$ with a constant C depending on F Then*

$$E^{\alpha}\{ \sup_{F\in\mathcal{F}} [F(\cdot)-C(F)]\} \le H \ . \tag{3.13}$$

Proof: We can assume without loss of geneality that $\mathcal{F}$ is a finite set. If we can establish (3.13) for finite sets that we can pick finite subsets that increase to cover all of $\mathcal{F}$ and use monotone convergence theorem to complete the proof. Any technical problems are of a routine nature. The idea is that for a finite set $F_1,\ldots,F_k$ with $C_k = C(F_k)$ we can define $F = \sup_{1\le i\le k}\{F_i - C(F_i)\}$ and a routine estimate in (3.11) shows that $C(F) \le 0$. Therefore by Lemma 3.7

$$\int F\,d\alpha \le H \ ,$$

and we are done.

Theorem 3.9. *There exists a constant C such that*

$$E^{\widehat{Q}}\{ \sup_{0\le t\le T} \int_S h(m(t,\theta))\,d\theta\} \le H \tag{3.14}$$

$$E^{\widehat{Q}}\{ \int_0^T \int_S \{\frac{\partial}{\partial\theta}[h'(m(t,\theta))]\}^2\,d\theta\,ds\} \le CH \tag{3.15}$$

$$E^{\widehat{Q}}\{ \int_0^T \int_S \eta^2(s,\theta)\,ds\,d\theta\} \le CH \ . \tag{3.16}$$

Proof: Let us consider the equilibrium process P_N with an entropy bound $h(Q_N,P_N) \le HN$ and the two induced processes $\widehat{Q}_N$ and $\widehat{P}_N$. Because of Lemma 3.8 we only need to prove exponential bounds under $\widehat{P}_N$ or P_N. To establish (3.14) we start with

$$E^{P_N}\{\exp[\sum U(\frac{j}{N})x_j(t)]\} = \exp\{\sum \rho(U(\frac{j}{N}))\}$$

This leads to the functional

$$F = \int U(\theta)\,\lambda(t, d\theta)$$

and

$$C(F) = \int \rho(U(\theta))\,d\theta$$

According to Lemma 3.8 we should have a bound

$$E^{\widehat{Q}}\{ \sup_{t,U(\cdot)} [\int U(\theta)\,\lambda(t,d\theta) - \int \rho(U(\theta))\,d\theta] \} \leq H \,.$$

This is exactly (3.14).

To establish (3.15) we know from results of [2] that $\sum U(\frac{i}{N})[\phi'(x_i(t)) - \phi'(x_{i+1}(t))]$ is nearly the same as $\frac{1}{N}\sum U'(\frac{i}{N})\,\phi'(x_i(t))$ and this is close to $\int U'(\theta)\,h'(m(\theta,t))\,d\theta$ in the limit.

We can make U depend on time as well so that $\int_0^T \sum U(\frac{i}{N},t)[\phi'(x_i(t)-\phi'(x_{i+1}(t))]\,dt$ approximates $\int_0^T U_\theta(\theta,t)\,h'(m(\theta,t))\,dt$. We can estimate

$$E^{P_N}\{\exp[N\int_0^T \sum U(\frac{i}{N},t)[(\phi'(x_i(t)) - \phi'(x_{i+1}(t))]\,dt\}$$

by Feynman-Kac formula to obtain the bound

$$E^{P_N}\{\exp[N\int_0^T U(\frac{i}{N},t)[\phi'(x_i(t)) - \phi'(x_{i+1}(t))]\,dt\} \leq \exp[\int_0^T \Lambda_N(t)\,dt]$$

where $\Lambda_N(t)$ is the eigenvalue

$$\begin{aligned}
\Lambda_N(t) &= \sup\Big[E^f\{N\sum U(\frac{i}{N},t)[\phi(x_i) - \phi'(x_{i+1})]\} - N^2 D_N(\sqrt{f})\Big] \\
&= \sup\Big[E^f\{N\sum U(\frac{i}{N},t)(\frac{\partial f}{\partial x_i} - \frac{\partial f}{\partial x_{i+1}})\} - N^2 D_N(\sqrt{f})\Big] \\
&\leq \sup\Big[E\{N\sum U(\frac{i}{N},t)(\frac{\partial f}{\partial x_i} - \frac{\partial f}{\partial x_{i+1}}) - C\frac{N^2}{2}\sum(\frac{\partial f}{\partial x_i} - \frac{\partial f}{\partial x_{i+1}})^2\frac{1}{f}\}\Big] \\
&\leq \frac{1}{2C}\sum U^2(\frac{i}{N},t)
\end{aligned}$$

We can use Lemmma 3.8 again with

$$F = \int\int U_\theta(\theta,t)\,h'(m(t,\theta))\,d\theta\,dt$$

and

$$C(F) = \frac{1}{2C} \int\int U^2(\theta, s)\, d\theta\, ds$$

This yields (3.15) and (3.16) is similar if we start with a bound for the expected value of

$$\exp\{N \int_0^T \sum U(\frac{i}{N}, t)\, W(x_i(t), x_{i+1}(t))\, dt\}$$

in equilibrium. The eigenvalue is easily estimated because

$$\begin{aligned} E^f W(x_i, x_{i+1}) &= E\{fW(x_i, x_{i+1})\} \\ &= E\{a(x_i, x_{i+1})(\frac{\partial f}{\partial x_i} - \frac{\partial f}{\partial x_{i+1}})\} \end{aligned}$$

This leads to an application of Lemma 3.8 with

$$F(\mu) = \int_0^T \int_S U(t, \theta)\, \mu(dt, d\theta)$$

and

$$C(F) = C \int_0^T \int_S U^2(t, \theta)\, dt\, d\theta\ .$$

4. Some bounds for the principal eigenvalue.

We will provide a proof for a general fact concerning the behavior of the principal eigenvalue of $L + \epsilon V$ where L is an elliptic operator in a Euclidean space R^k and V is a suitable potential and ϵ is a small parameter. Although the result is very well known, we need the results under very special circumstances and we will provide a straightforward proof.

We assume that L is a symmetric operator with a Dirichlet form

$$\mathcal{D}(u) = \frac{1}{2} \int_{R^k} \langle a(x)\, \nabla u, \nabla u \rangle\, e^{-\psi(x)}\, dx \tag{4.1}$$

Here $e^{-\psi(x)}$ is a probability density and the diffusion coefficients $a(x) = \{a_{ij}(x)\}$ is assumed to be uniformly elliptic, i.e.

$$m \sum \xi_i^2 \le \sum a_{ij} \xi_i \xi_j \le M \sum \xi_i^2\ ,$$

for some $0 < m \le M < \infty$. We further assume that there exists a nonnegative function $\omega(x)$ such that $|V(x)| = o(\omega(x))$ as $|x| \to \infty$, $\omega(x) \to \infty$ as $|x| \to \infty$ and $\int \omega\, e^{-\psi}\, dx \le C$. We further assume the existence of a uniformly bounded vector valued function $H(x)$ with

$\|H(x)\| \le C$, such that for all test functions $u(x)$,

$$\int V(x)\,u(x)\,e^{-\psi(x)}\,dx = \int \langle \nabla u(x), H(x)\rangle\, e^{-\psi(x)}\,dx \tag{4.2}$$

i.e. V is the divergence of a bounded function. Given L, V and ω for each real ϵ, we consider the operation

$$L_\epsilon = L + \epsilon V - \epsilon^2\omega$$

and denote by $\sigma(\epsilon)$ the principal eigenvalue of L_ϵ. Since $\epsilon V - \epsilon^2\omega \to -\infty$ at $|x| \to \infty$, for each $\epsilon \ne 0$ the principal eigenvalue is a proper eigenvalue with a corresponding eigenfunction $f_\epsilon(\cdot)$, that satisfies

$$\begin{aligned}\sigma(\epsilon) &= \sup_{g:|g|_2=1} \Big\{\epsilon\int V(x)\,g^2(x)\,e^{-\psi(x)}\,dx - \epsilon^2\int \omega(x)\,g^2(x)\,e^{-\psi(x)}\,dx - \mathcal{D}(g)\Big\}\\ &= \Big\{\epsilon\int V(x)\,f_\epsilon^2(x)\,e^{-\psi(x)}\,dx - \epsilon^2\int \omega(x)\,f_\epsilon^2(x)\,e^{-\psi(x)}\,dx - \mathcal{D}(f_\epsilon)\Big\}\end{aligned} \tag{4.3}$$

where

$$L_\epsilon f_\epsilon = \sigma(\epsilon)\,f_\epsilon\ .$$

The following bounds are easy to establish for $\sigma(\epsilon)$.

$$\begin{aligned}\sigma(\epsilon) &\le \sup_{g:|g|_2=1}\Big\{\epsilon\int V(x)\,g^2(x)\,e^{-\psi}\,dx - \mathcal{D}(g)\Big\}\\ &= \sup_{g:|g|_2=1}\Big\{\epsilon\int 2g\langle\nabla g, H\rangle\,e^{-\psi}\,dx - \mathcal{D}(g)\Big\}\\ &\le \sup_{g:|g|_2=1}\Big\{2\epsilon\Big(\int \langle\nabla g, H\rangle^2\,e^{-\psi}\,dx\Big)^{1/2} - \mathcal{D}(g)\Big\}\\ &\le \sup\Big\{2\epsilon C_1\sqrt{\mathcal{D}(g)} - \mathcal{D}(g)\Big\}\\ &= \epsilon^2 C_1^2\end{aligned} \tag{4.4}$$

On the other hand because $\mathcal{D}(g) = 0$ if $g \equiv 1$,

$$\begin{aligned}\sigma(\epsilon) &\ge \epsilon\int V(x)\,e^{-\psi(x)}\,dx - \epsilon^2\int \omega(x)\,e^{-\psi}\,dx\\ &= -C_2\epsilon^2\ .\end{aligned} \tag{4.5}$$

Moreover the relation (4.2) implies a bound

$$|\int V(x)\,u(x)\,e^{-\psi(x)}\,dx| = |\int \langle\nabla u(x), H(x)\rangle\, e^{-\psi}\,dx| \le A\sqrt{\mathcal{D}(u)} \tag{4.6}$$

for some constant A. The best possible value for A is related to the variance in the central limit theorem by

$$\lim_{t\to\infty} E\Big(\frac{1}{\sqrt{t}}\int_0^t V(x(\sigma))\,d\sigma\Big)^2 = 2A^2\ , \tag{4.7}$$

where $x(t)$ is the stationary diffusion corresponding to L.

Our main result concerning the behavior of $\sigma(\epsilon)$ is

Theorem 4.1. *With the value of A given by* (4.7)

$$\limsup_{\epsilon\to 0} \frac{\sigma(\epsilon)}{\epsilon^2} \le A^2\ .$$

Proof: We have the eigenfunction $f_\epsilon(x)$ satisfying

$$Lf_\epsilon + \epsilon V f_\epsilon - \epsilon^2 \omega f_\epsilon = \sigma_\epsilon f_\epsilon$$

and

$$\sigma(\epsilon) = -\frac{1}{2}\int \langle a\nabla f_\epsilon, \nabla f_\epsilon\rangle\, e^{-\psi}\,dx + \epsilon\int V f_\epsilon^2\, e^{-\psi}\,dx - \epsilon^2\int \omega f_\epsilon^2\, e^{-\psi}\,dx$$

We have the bound

$$|\int V f_\epsilon^2\, e^{-\psi}\,dx| = |\int 2 f_\epsilon \langle H, \nabla f_\epsilon\rangle\, e^{-\psi}\,dx| \le C_3\sqrt{\mathcal{D}(f_\epsilon)} \tag{4.8}$$

The lower bound (4.5) and (4.8) yield bounds of the form

$$\mathcal{D}(f_\epsilon) \le C\epsilon^2 \tag{4.9}$$

$$\int \omega f_\epsilon^2\, e^{-\psi}\,dx \le C \tag{4.10}$$

Equations (4.9) and (4.10) imply that

$$\lim_{\epsilon \to 0} \int |f_\epsilon - 1|^2 e^{-\psi} \, dx = 0 \tag{4.11}$$

Let us consider the function

$$g_\epsilon = \frac{f_\epsilon - 1}{\epsilon} \tag{4.12}$$

Then $\mathcal{D}(g_\epsilon) = \frac{1}{\epsilon^2}\mathcal{D}(f_\epsilon) \leq C$. Therefore ∇g_ϵ has a weak limit in $L_2(loc)$ and it is the gradient of a function g in $L_2(loc)$. From the formula for $\sigma(\epsilon)$

$$\epsilon \int V f_\epsilon^2 \, e^{-\psi} \, dx \geq \sigma(\epsilon) + \mathcal{D}(f_\epsilon)$$

or

$$\frac{1}{\epsilon} \int V(1 + \epsilon g_\epsilon)^2 \, e^{-\psi} \, dx \geq \frac{\sigma(\epsilon)}{\epsilon^2} + \mathcal{D}(g_\epsilon)$$

We can calculate

$$\begin{aligned}
\frac{1}{\epsilon} \int V(1 + \epsilon g_\epsilon)^2 \, e^{-\psi} \, dx &= \frac{1}{\epsilon} \int 2(1 + \epsilon g_\epsilon)\langle H, \epsilon \nabla g_\epsilon \rangle \, e^{-\psi} \, dx \\
&= 2 \int (1 + \epsilon g_\epsilon)\langle H, \nabla g_\epsilon \rangle \, e^{-\psi} \, dx \\
&= 2 \int \langle H, \nabla g_\epsilon \rangle \, e^{-\psi} \, dx + 2\epsilon \int g_\epsilon \langle H, \nabla g_\epsilon \rangle \, e^{-\psi} \, dx
\end{aligned}$$

Because $\int |\epsilon g_\epsilon|^2 \, e^{-\psi} \, dx = \int |f_\epsilon - 1|^2 \, e^{-\psi} \, dx \to 0$ as $\epsilon \to 0$, $|H| \leq C$ and $\int |\nabla g_\epsilon|^2 \, e^{-\psi} \, dx$ is uniformly bounded, for the weak limit g of g_ϵ in $L_2(loc)$ we end up with,

$$2 \int \langle H, \nabla g \rangle \, e^{-\psi} \, dx \geq \limsup_{\epsilon \to 0} \frac{\sigma(\epsilon)}{\epsilon^2} + \mathcal{D}(g)$$

If we have in addition the bound (4.6) then

$$\limsup_{\epsilon \to 0} \frac{\sigma(\epsilon)}{\epsilon^2} \leq 2A\sqrt{\mathcal{D}(g)} - \mathcal{D}(g)$$

for some g. Therefore

$$\limsup_{\epsilon\to 0}\frac{\sigma(\epsilon)}{\epsilon^2}\le \sup_d\left[2Ad-d^2\right] \le A^2 .$$

Remark 1. The bound (4.6) is strictly speaking valid only for functions u in $C_0^\infty(R^k)$. But it is very easy to pass to the limit and extend it to functions u with ∇u in L_2.

Remark 2. We can let V, ψ, H depend on the small parameters ϵ and behave regularly as $\epsilon\to 0$. We can then get the limits to hold uniformly as well.

5. The diffusion coefficient

Let $\Omega=\prod_{-\infty}^{\infty}R$ be the countable product of the real line. A point ω in Ω is a doubly infinite sequence $\omega=\{x_j:-\infty<j<\infty\}$. We denote by T the shift operator defined by $\omega'=T\omega=\{x'_j\}$ where $x'_j=x_{j+1}$. We will be looking at a class of product measures $\{Q_m\}$ on Ω of the form $dQ_m=\prod e^{-\psi(x_i)}\,dx_i$ where $\psi_\lambda(x)=\psi(x)=\phi(x)-\lambda x+\log M(\lambda)$. The value of λ is determined so that

$$\int x\,e^{-\psi_\lambda(x)}\,dx=m$$

We know that the choice of λ is $\lambda=h'(m)$. For the moment we think of m and λ as having been chosen. The corresponding ψ leads to a Q. We make the additional assumption that for some $0<c_1\le c_2<\infty$, $c_1\le\psi''(x)\le c_2$ for all x. Of course if ϕ satisfies the assumption then so do $\psi_\lambda(x)$ for all values of λ. The assumption on $\phi(\cdot)$ is far more stringent than what we really need. See the comments at the end of the article for improvement of conditions.

Let k be a positive integer. We denote by $\mathcal{A}_k$ the space of smooth functions $F(x_{-k},\dots,x_k)$ of $2k+1$ variables. We denote by $\mathcal{A}=\cup_k\mathcal{A}_k$. If we take a function F in $\mathcal{A}_k$ we can consider the formal sum

$$\widetilde{F}(\omega)=\sum_{-\infty<j<\infty}F(x_{j-k},\dots,x_{j+k}) \tag{5.1}$$

$\widetilde{F}(\omega)$ is not defined properly. However

$$\frac{\partial\widetilde{F}(\omega)}{\partial x_i} = \sum_{-\infty<j<\infty} \frac{\partial}{\partial x_i} F(x_{j-k},\ldots,x_{j+k})$$
$$= \sum_{i-k\le j\le i+k} \frac{\partial}{\partial x_i} F(x_{j-k},\ldots,x_{j+k})$$

is well defined. The formal invariance

$$\widetilde{F}(T\omega) = \widetilde{F}(\omega)$$

leads to the precise covariance

$$\frac{\partial\widetilde{F}(\omega)}{\partial x_i} = \left(\frac{\partial\widetilde{F}}{\partial x_0}\right)(T^i\omega) \tag{5.2}$$

We are particularly interested in

$$\xi = \frac{\partial\widetilde{F}(\omega)}{\partial x_1} - \frac{\partial\widetilde{F}(\omega)}{\partial x_0} = \left(\frac{\partial\widetilde{F}}{\partial x_0}\right)(T\omega) - \left(\frac{\partial\widetilde{F}}{\partial x_0}\right)(\omega)\,. \tag{5.3}$$

It is clear that for each $F \in \mathcal{A}$, ξ is well defined as an element of $L_2(Q)$. We are interested in the quantity

$$\hat{a} = \inf_{F\in\mathcal{A}} E^Q\{a(x_0,x_1)[1-\xi]^2\} \tag{5.4}$$

If we pick $\lambda = h'(m)$ and think of Q as Q_m then $\hat{a} = \hat{a}(m)$ is a function of m. It will turn out that $\hat{a}(m)$ is the diffusion coefficient. The following bounds are elementary.

Lemma 5.1. *For every m*

$$\left[E^{Q_m}\{\frac{1}{a(x_0,x_1)}\}\right]^{-1} \le \hat{a}(m) \le E^{Q_m}\{a(x_0,x_1)\}$$

Proof: In (5.4) we can take $F = 0$ and this gvies the bound on the right. To see the lower bound we note that because of (5.3), $E^Q\xi = 0$ or $E^Q(1-\xi) = 1$. By Schwartz's inequality we can get the left half of the inequality.

Let us denote by $\mathcal{H}_{0,1} \subset L_2(Q)$ the closure of the set of functions ξ obtained by varying F over $\mathcal{A}$. The closure is of course taken in $L_2(Q)$. Clearly

$$\hat{a} = \inf_{\xi \in \mathcal{H}_{0,1}} E^Q\big[a(x_0, x_1)(1-\xi)^2\big]$$

and this infimum is attained at a unique $\xi \in \mathcal{H}_{0,1}$ by the standard projection theorem. For the purposes of the next section we would like to have an intrinsic characterization of the space $\mathcal{H}_{0,1}$. Clearly

$$E^Q \xi = 0 \tag{5.5}$$

If we define $\xi_{j,j+1}(\omega) = \xi(T^j \omega)$, then in a weak sense

$$\Big(\frac{\partial}{\partial x_{i+1}} - \frac{\partial}{\partial x_i}\Big)\xi_{j,j+1} = \Big(\frac{\partial}{\partial x_{j+1}} - \frac{\partial}{\partial x_j}\Big)\,\xi_{i,i+1} \tag{5.6}$$

i.e.

$$\begin{aligned}\int \xi_{j,j+1}\Big[\frac{\partial G}{\partial x_{i+1}} - \frac{\partial G}{\partial x_i} - G(\psi'(x_{i+1}) - \psi'(x_i))\Big]\,dQ \\ = \int \xi_{i,i+1}\Big[\frac{\partial G}{\partial x_{j+1}} - \frac{\partial G}{\partial x_j} - G(\psi'(x_{j+1}) - \psi'(x_j))\Big]\,dQ\end{aligned} \tag{5.7}$$

for suitable test functions G. We note that $\psi'_\lambda(x) - \psi'_\lambda(y) = \phi'(x) - \phi'(y)$ for all λ. The main theorem of the section is the following:

Theorem 5.2. *If $\xi \in L_2(Q)$ satisfies* (5.5) *and* (5.7) *then $\xi \in \mathcal{H}_{0,1}$.*

Proof of Theorem 5.2: We have already defined $\xi_{i,i+1}(\omega)$ as $\xi(T^i\omega)$. Formally we expect

$$\xi_{i,i+1}(\omega) = \frac{\partial \widetilde{F}}{\partial x_{i+1}} - \frac{\partial \widetilde{F}}{\partial x_i} \tag{5.6$'$}$$

However since $\widetilde{F}$ is not expected to exist and in fact even F may not exist it is impossible to integrate (5.6). We first come down to finite dimensions. We define

$$\xi^{(N)}_{i,i+1} = E^Q\big[\xi_{i,i+1} \,|\, \mathcal{F}^N_{-N}\big] \tag{5.7$'$}$$

where $\mathcal{F}_{-N}^{N}$ is the sub σ-field of Ω generated by $x_{-N}, \ldots, x_N$. Then $\xi_{i,i+1}^{(N)}$ are functions of $x_{-N}, \ldots, x_N$ only. Because Q is a product measure it is easy to check that (5.4) is satisfied with $\xi_{i,i+1}^{(N)}$ replacing $\xi_{i,i+1}$ provided $-N \le i \le i+1 \le N$ and $-N \le j \le j+1 \le N$. We can then integrate these to obtain functions $g^{(N)}(x_{-N}, \ldots, x_N)$ such that

$$(\frac{\partial}{\partial x_{i+1}} - \frac{\partial}{\partial x_i})\, g^{(N)} = \xi_{i,i+1}^{(N)} \quad \text{if} \quad -N \le i \le i+1 \le N \tag{5.8}$$

Moreover we can normalize $g^{(N)}$ so that

$$E^Q[g^{(N)} \,|\, x_{-N} + \cdots + x_N = y] = 0 \quad \text{for every} \quad y\ . \tag{5.9}$$

On the other hand given $\epsilon > 0$ there is an integer N_0 such that

$$E^Q[\xi_{0,1} - \xi_{0,1}^{(N)}]^2 \le \epsilon \quad \text{if} \quad N \ge N_0$$

and by translation invariance

$$E^Q[\xi_{i,i+1} - \xi_{i,i+1}^{(N)}]^2 \le \epsilon \quad \text{if} \quad \{i - N_0, i + N_0\} \subset \{-N, N\} \tag{5.10}$$

or if i is not near the boundary of $\{-N, N\}$.

In fact for any given ϵ for (5.10) to hold i need only be at least a distance N_0 from the two edges. In any case $E^Q[\xi_{i,i+1}^{(N)}]^2 \le E^Q \xi_{i,i+1}^2 = E^Q \xi^2$. Let us try to take

$$F_N = \frac{1}{2N+1} g^{(N)}(x_{-N}, \ldots, x_N)\ .$$

Then

$$\widetilde{F}_N = \sum_j \frac{1}{2N+1} g^{(N)}(x_{j-N}, \ldots, x_{j+N})$$

and

$$\begin{aligned}
\frac{\partial \widetilde{F}_N}{\partial x_1} - \frac{\partial \widetilde{F}_N}{\partial x_0} &= \frac{1}{2N+1} \sum_{-N \le r \le r+1 \le N} (\frac{\partial g^{(N)}}{\partial x_{r+1}} - \frac{\partial g^{(N)}}{\partial x_r})(T^{-r}\omega) \\
&\quad + \frac{1}{2N+1} (\frac{\partial g^{(N)}}{\partial x_{-N}})(T^{(N+1)}\omega) \\
&\quad - \frac{1}{2N+1} (\frac{\partial g^{(N)}}{\partial x_N})(T^{-N}\omega) \\
&= \frac{1}{2N+1} \sum_{-N \le r \le r+1 \le N} \xi_{r,r+1}^{(N)}(T^{-r}\omega) \\
&\quad + \text{ Errors.}
\end{aligned}$$

It is clear that if $\xi^{(N)}_{r,r+1}$ is close to $\xi_{r,r+1}$ then $\xi^{(N)}_{r,r+1}(T^{-r}\omega)$ is close to $\xi(\omega)$ by translation invariance of Q. Therefore if N is large then

$$\frac{1}{2N+1}\sum_{-N\le r\le r+1\le N}\xi^{(N)}_{r,r+1}(T^{-r}\omega)$$

is close to ξ and

$$\lim_{N\to\infty}E^Q\Big[\frac{1}{2N+1}\sum_{-N\le r\le r+1\le N}\xi^{(N)}_{r,r+1}(T^{-r}\omega)-\xi\Big]^2=0$$

We only have to worry about the two single terms and by translation invariance we need

$$(5.11)\qquad \begin{cases}\lim\limits_{N\to\infty}\dfrac{1}{N^2}E^Q\big(\dfrac{\partial g^{(N)}}{\partial x_N}\big)^2=0\\[2mm] \lim\limits_{N\to\infty}\dfrac{1}{N^2}E^Q\big(\dfrac{\partial g^{(N)}}{\partial x_{-N}}\big)^2=0\,.\end{cases}$$

This is hard because the way $g^{(N)}$ was constructed was integrating its "gradient" $\frac{\partial}{\partial x_{i+1}}-\frac{\partial}{\partial x_r}$ on each hyperplane $\frac{1}{2N+1}\sum_{-N}^{N}x_r=y$. There is no way therefore to estimate $\frac{\partial g^{(N)}}{\partial x_N}$ which invovles a gradient in a nontangential direction. The way we avoid the difficulty is to construct $g^{(2N)}$ corresponding to a much larger interval and then take

$$(5.12)\qquad \tilde g^{(N)}(x_{-N},\dots,x_N)=E^Q\big[g^{(2N)}(x_{-2N},\dots,x_{2N})\,|\,\mathcal{F}^N_{-N}\big]$$

It is not difficult to see that as far as $\frac{\partial\tilde g_N}{\partial x_{r+1}}-\frac{\partial\tilde g_N}{\partial x_r}$, for $-N\le r\le r+1\le N$, are concerned these are all close to $\xi(T^r\omega)$ which is almost $\mathcal{F}^N_{-N}$ measurable if r is not close to the edges. So they work just as before.

We only need to estiamte $\frac{\partial\tilde g^{(N)}}{\partial x_N}$ and $\frac{\partial\tilde g^{(N)}}{\partial x_{-N}}$.

$$\begin{aligned}\frac{\partial\tilde g^{(N)}}{\partial x_{-N}}&=E^Q\Big[\frac{\partial g^{(2N)}}{\partial x_{-N}}\,|\,\mathcal{F}^N_{-N}\Big]\\&=E^Q\Big[\frac{\partial g^{(2N)}}{\partial x_{-N}}-\psi'(x_r)\,g^{(2N)}+\psi'(x_r)\,g^{(2N)}\,|\,\mathcal{F}^N_{-N}\Big]\\&=E^Q\Big[\frac{\partial g^{(2N)}}{\partial x_{-N}}-\frac{\partial g^{(2N)}}{\partial x_r}\,|\,\mathcal{F}^N_{-N}\Big]\\&\quad+E^Q\big[\psi'(x_r)\,g^{(2N)}\,|\,\mathcal{F}^N_{-N}\big]=T_1+T_2\,.\end{aligned}$$

This is valid for any $-N\le r\le N$. Suppose we average with respect to r over the range.

Then

$$\frac{1}{2N+1}\sum_{-N}^{N}(\frac{\partial g^{(2N)}}{\partial x_{-N}}-\frac{\partial g^{(2N)}}{\partial x_r})$$

is close to

$$\frac{1}{2N+1}\sum_{-N}^{N}\{\xi(T^{-(N)}\omega)+\cdots+\xi(T(r-1)\omega)\}$$

From the ergodic theorem we see that

$$E^Q|T_1|^2 \leq o(N^2) \quad \text{as } N\to\infty .$$

We have on the other hand

$$E^Q[\frac{1}{2N+1}\sum_{r=-N}^{N}\psi'(x_r)\cdot g^{(2N)}\,|\,\mathcal{F}_{-N}^{N}]=T_2$$

and

$$\begin{aligned}E^Q|T_2|^2 &\leq (E^Q|\frac{1}{2N+1}\sum\psi'(x_r)|^2)\cdot(E^Q|g^{(2N)}|^2)\\ &\leq \frac{1}{N}\cdot(E^Q|g^{(2N)}|^2)\end{aligned}$$

Because $E^Q\psi'(x_0)=0$ and $E^Q|\psi'(x_0)|^2<\infty$.

In order to complete the proof we need only establish

Lemma 5.3. *As* $N\to\infty$

$$\lim_{N\to\infty}\frac{1}{N^3}E^Q[g^{(2N)}]^2=0 .$$

Proof: Since $G^{(2N)}$ is constructed by integrating its gradient on each hyperplane what we need is an estimate on the gradient and a Poincaré estimate on the mean square norm of the function in terms of its gradient. On the hyperplane $\frac{1}{4N+1}\sum_{|r|\leq 2n}x_n=y$ we have the weight λ given by $\exp(-\sum\psi(x_r))$. We need to look at the gradients $\frac{\partial u}{\partial x_{r+1}}-\frac{\partial u}{\partial x_r}$ and estimate $\int u^2\,d\lambda$ in terms of the gradients uniformly in y. From the convexity of ψ it is easy to see that if u is normalized so that $\int u\,d\lambda=0$, we have

$$\int u^2\,d\lambda\leq C\frac{1}{N}\int\sum_{-N}^{N}\sum_{-N}^{N}(\frac{\partial u}{\partial x_i}-\frac{\partial u}{\partial x_j})^2\,d\lambda$$

with a constant C independent of N and y. From the law of large numbers one can see as before that

$$\frac{1}{N^4} E^Q \Big[\sum_{-N}^{N} \sum_{-N}^{N} \Big(\frac{\partial g^{(2N)}}{\partial x_i} - \frac{\partial g^{(2N)}}{\partial x_j} \Big)^2 \Big] \to 0$$

as $N \to \infty$. The lemma is therefore true.

Remark. The convexity is only used in this lemma to establish easily a spectral gap. One can weaken the hypothesis considerably. See a forthcoming note by Varadhan and Yau for the best possible conditions on $\phi(x)$ for the spectral gap to be of the required form.

6. Some estimates on central limit theorem variances.

Let us suppose that L is the infinitesimal generator of a Markov process which is reversible and ergodic with respect to an invariant probability measure μ on some state space X. We denote by P_μ the stationary Markov process with marginal distribution μ. If $v(x)$ is square integrable with respect to μ and has mean zero, by the ergodic theorem

$$\lim_{t\to\infty} E^{P_\mu} \Big[\frac{1}{t} \int_0^t v(x(s))\, ds \Big]^2 = 0$$

We are interested in calculating

$$\lim_{t\to\infty} E^{P_\mu} \Big\{ \frac{1}{\sqrt{t}} \int_0^t v(x(s)) ds \Big\}^2 = \Delta(v,v) \tag{6.1}$$

if it exists. There is the symmetric bilinear form

$$\lim_{t\to\infty} E^{P_\mu} \Big\{ \frac{1}{\sqrt{t}} \int_0^t v(x(s))\, ds \cdot \frac{1}{\sqrt{t}} \int_0^t u(x(s))\, ds \Big\} = \Delta(u,v) \tag{6.2}$$

It is know that in this situation the limit (6.1) always exists. It is finite if and only if v is in the range of $(-L)^{1/2}$ and in that case

$$\Delta(v,v) = 2\langle (-L)^{-1/2} v, (-L)^{-1/2} v \rangle \tag{6.3}$$

By standard arguments if u and v are in the range of $(-L)^{1/2}$ then

$$\Delta(u,v) = 2\langle(-L)^{-1/2}u,(-L)^{-1/2}v\rangle \tag{6.4}$$

Here $\langle\ ,\ \rangle$ is the standard inner product in $L_2(\mu)$. By duality $\Delta(v,v) = 2c^2$ where c is the smallest constant such that

$$|\langle u,v\rangle| \le c[\mathcal{D}(u)]^{1/2} \quad \text{for all} \quad u \in \mathcal{D}(L)\ . \tag{6.5}$$

$\mathcal{D}(u)$ here is the Dirichlet form

$$\mathcal{D}(u) = \langle -Lu, u\rangle\ . \tag{6.6}$$

If $u = Lf$ for some f in $\mathcal{D}(L)$ and $\Delta(v,v) < \infty$ then

$$\Delta(u,v) = -2\langle f,v\rangle \tag{6.7}$$

In particular $\Delta(u,u) < \infty$ and

$$\Delta(u,u) = -2\langle f, Lf\rangle = 2\mathcal{D}(f)\ . \tag{6.8}$$

We want to explore these ideas in a specific context. We fix n and consider R^{2n+1} indexed by x_j for $|j| \le n$. On R^{2n+1} we have the product measure $e^{-\Sigma\phi(x_j)}\,dx$ and the conditioned measures $\mu_{n,y}(dx)$ on the hyperplanes $\frac{1}{2n+1}\sum x_j = y$. These are well defined by continuity for every y. We look at the Dirichlet form

$$\mathcal{D}_{n,y}(u) = \frac{1}{2}\int \sum_{-n\le i\le i+1\le n} a(x_i,x_{i+1})\Big(\frac{\partial u}{\partial x_{i+1}} - \frac{\partial u}{\partial x_i}\Big)^2 \mu_{n,y}(dx) \tag{6.9}$$

and the corresponding operator

$$\begin{aligned} L_n u = \frac{1}{2} &\sum_{-n\le i\le i+1\le n} \Big(\frac{\partial}{\partial x_{i+1}} - \frac{\partial}{\partial x_i}\Big)\, a(x_i,x_{i+1})\Big(\frac{\partial}{\partial x_{i+1}} - \frac{\partial}{\partial x_i}\Big) \\ -\frac{1}{2} &\sum_{-n\le i\le i+1\le n} a(x_i,x_{i+1})\,[\phi'(x_{i+1}) - \phi'(x_i)]\,\Big[\frac{\partial}{\partial x_{i+1}} - \frac{\partial}{\partial x_i}\Big] \end{aligned} \tag{6.10}$$

Because L_n is elliptic on the hyperplanes each $\mu_{n,y}(dx)$ is ergodic for the process with generator L_n. We will be looking at three classes of functions

$$A_n(x_{-n},\ldots,x_n) = \phi'(x_n) - \phi'(x_{-n}) \tag{6.11}$$
$$= \sum_{-n\le i\le i+1\le n} [\phi'(x_{i+1}) - \phi'(x_i)]$$

$$B_n(x_{-n},\ldots,x_n) = \sum_{-n\le i\le i+1\le n} W(x_i,x_{i+1}) \tag{6.12}$$

where

$$W(x_1,x_2) = -a_1(x_1,x_2) + a_2(x_1,x_2) - a(x_1,x_2)\left(\phi'(x_2) - \phi'(x_1)\right) \tag{6.13}$$

We have some auxiliary functions

$$H_n(x_{-n},\ldots,x_n) = \sum_{-n\le j-(k+1)\le j+(k+1)\le n} u(x_{j-(k+1)},\ldots,x_{j+(k+1)}) \tag{6.14}$$

where k is a fixed integer and

$$u(x_{-(k+1)},\ldots,x_{(k+1)}) = LF(x_{-k},\ldots,x_k) \tag{6.15}$$

Here $F(x_{-k},\ldots,x_k)$ is a smooth function of $2k+1$ variables and L can be any L_n with $n \ge k+1$.

The three types of function A_n, B_n, H_n are similar in the sense that they are sums of translations of local functions. It is easily seen that all three functions have mean zero with respect to $\mu_{n,y}$ for every n and y. An elementary calculation yields

$$W(x_i,x_{i+1}) = 2L_n(x_{i+1} + \cdots + x_n)$$

and therefore

$$B_n(x_{-n},\ldots,x_n) = 2L_n\Big(\sum_{i=-n}^{n} i\,x_i\Big)$$

On the other hand

$$\int u\,[\phi'(x_n) - \phi'(x_{-n})]\; \mu_{n,y}(dx)$$
$$= \int (\frac{\partial u}{\partial x_n} - \frac{\partial u}{\partial x_{-n}})\; \mu_{n,y}(dx)$$
$$= \int \sum_{-n\le i\le i+1\le n} (\frac{\partial u}{\partial x_{i+1}} - \frac{\partial u}{\partial x_i})\; \mu_{n,y}(dx)$$
$$\le C_1(2n)^{1/2}(\mathcal{D}_{n,y}(u))^{1/2}\,. \tag{6.16}$$

It is clear then that A_n, B_n and H_n are all in the range of $(-L_n)^{1/2}$ so that the central limit theorem variances and covariances exist.

We denote these by $\Delta_{n,y}(\ ,\ )$. We want to study these six terms and calculate their asymptotic behavior as $n \to \infty$. These will be locally uniform in y. Let us remark that according to [2]

$$\lim_{n\to\infty} \int [\frac{1}{2n+1} \sum g(x_{i-k}, \dots, x_{i+k})]\; d\mu_{n,y} = \int g(x_{-k}, \dots, x_k)\, dQ_y$$

where Q_y is the product measure $\Pi e^{-\psi_\lambda(x_i)}\, dx_i$ on Ω with $\psi_\lambda(x) = \phi(x) - \lambda x + \log M(\lambda)$ and $\lambda = h'(y)$.

Lemma 6.1. *For some constant* C

$$\limsup_{n\to\infty} \frac{1}{2n} \Delta_{n,y}(A_n, A_n) \le C$$

locally uniformly in y.

Proof: This is contained in the estimate (6.16) and since (6.16) was obtained by Schwartz's inequality from the line preceding it we have a simple bound for C_1 and passing to the limit as $n \to \infty$ we have the lemma with $C = 4E^{Q_y}\{\frac{1}{a(x_1,x_2)}\}$.

Lemma 6.2.

$$\lim_{n\to\infty} \frac{1}{2n} \Delta_{n,y}(B_n, B_n) = 4E^{Q_y}\{a(x_0, x_1)\}$$

locally uniformly in y.

Proof: Because $B_n = 2L_n(\sum_{i=-n}^n i\,x_i)$,

$$\Delta_{n,y}(B_n, B_n) = 8\mathcal{D}_{n,y}(\sum_{i=-n}^{n} i\,x_i)$$
$$= 4 \int \sum a(x_i, x_{i+1})\, \mu_{n,y}(dx)\,.$$

and we are done.

Lemma 6.3.

$$\lim_{n\to\infty} \frac{1}{2n} \Delta_{n,y}(H_n, H_n) = E^{Q_y}\left[a(x_0, x_1)\left(\frac{\partial \psi}{\partial x_1} - \frac{\partial \psi}{\partial x_0}\right)^2\right]$$

locally uniformly in y, where

$$\psi(w) = \widetilde{F}(w) = \sum_{i=-\infty}^{\infty} F(x_{i-k}, \ldots, x_{i+k})$$

Proof: Again because $H_n = \mathcal{L}\psi_n$ where

$$\psi_n = \sum_{-n \le i-k \le i+k \le n} F(x_{i-k}, \ldots, x_{i+k})$$

$$\begin{aligned}\Delta_{n,y}(H_n, H_n) &= 2D_{n,y}(\psi_n) \\ &= \int \sum a(x_i, x_{i+1})\left(\frac{\partial \psi_n}{\partial x_{i+1}} - \frac{\partial \psi_n}{\partial x_i}\right)^2 \mu_{n,y}(dx) .\end{aligned}$$

It is not difficult to see that

$$\lim_{n\to\infty} \frac{1}{2n} \Delta_{n,y}(H_n, H_n) = E^{Q_y}\left[a(x_0, x_1)\left(\frac{\partial \psi}{\partial x_1} - \frac{\partial \psi}{\partial x_0}\right)^2\right] .$$

The next three lemmas are equally simple and involve the cross terms. We use relation (6.7).

Lemma 6.4.

$$\lim_{n\to\infty} \frac{1}{2n} \Delta_{n,y}(A_n, H_n) = 0$$

locally uniformly in y.

Proof:

$$\begin{aligned}\Delta_{n,y}(A_n, H_n) &= -2\int \left[\phi'(x_n) - \phi'(x_n)\right] \psi(x)\, \mu_{n,y}(dx) \\ &= -2\int \left(\frac{\partial \psi}{\partial x_n} - \frac{\partial \psi}{\partial x_{-n}}\right) \mu_{n,y}(dx)\end{aligned}$$

Clearly $\Delta_{n,y}(A_n, H_n)$ is bounded uniformly in n and y for fixed F.

Lemma 6.5.

$$\lim_{n\to\infty} \frac{1}{2n} \Delta_{n,y}(A_n, B_n) = -4 ,$$

locally uniformly in y.

Proof:

$$\Delta_{n,y}(A_n, B_n) = -4 \int [\phi'(x_n) - \phi'(x_{-n})] \, (\sum i\, x_i) \, \mu_{n,y}(dx)$$
$$= -4 \cdot 2n = -8n \ .$$

Lemma 6.6.

$$\lim_{n\to\infty} \frac{1}{2n} \Delta_{n,y}(B_n, H_n) = 2E^{Q_y} \left[a(x_0, x_1) \, (\frac{\partial \psi}{\partial x_1} - \frac{\partial \psi}{\partial x_0}) \right]$$

locally uniformly in y.

Proof:

$$\Delta_{n,y}(B_n, H_n) = 2 \int \sum a(x_i, x_{i+1}) \, (\frac{\partial \psi_n}{\partial x_{i+1}} - \frac{\partial \psi_n}{\partial x_i}) \, \mu_{n,y}(dx)$$

We can divide by $2n$ and let $n \to \infty$.

Finally we can obtain a lower bound for

$$\liminf_{n\to\infty} \frac{1}{2n} \Delta_{n,y}(A_n, A_n) \ .$$

We use the inequality

(6.17)
$$|\Delta_{n,y}(A_n, \xi_n)|^2 \le \Delta_{n,y}(A_n, A_n) \, \Delta_{n,y}(\xi_n, \xi_n)$$

and pick $\xi_n = B_n - H_n$. Using Lemmas 6.2–6.5 and (6.17) we can get a lower bound

(6.18)
$$\liminf_{n\to\infty} \frac{1}{2n} \Delta_{n,y}(A_n, A_n) \ge \frac{4}{\sigma^2(y, F)}$$

where

(6.19)
$$\sigma^2(y, F) = E^{Q_y} \{ a(x_0, x_1) [1 - (\frac{\partial \psi}{\partial x_1} - \frac{\partial \psi}{\partial x_0})]^2 \}$$
$$\psi(\omega) = \widetilde{F}(w) = \sum_i F(x_{i-k}, \dots, x_{k+k})$$

Since the left hand side of (6.18) is independent of F we have

$$\liminf_{n\to\infty} \frac{1}{2n} \Delta_{n,y}(A_n, A_n) \geq \frac{4}{\hat{a}(y)} \tag{6.20}$$

where

$$\hat{a}(y) = \inf_{F\in\mathcal{A}} E^{Q_y}\{a(x_0,x_1)\left[1-(\frac{\partial\psi}{\partial x_1}-\frac{\partial\psi}{\partial x_0})\right]^2\} \tag{6.21}$$

Moreover the limit in (6.20) is locally uniform in y.

The final theorem of this section is the following

Theorem 6.7. *The function $\hat{a}(y)$ defined by (6.21) is continuous in y and*

$$\lim_{n\to\infty} \frac{1}{2n} \Delta_{n,y}(A_n, A_n) = \frac{4}{\hat{a}(y)} \tag{6.22}$$

uniformly in bounded y intervals.

Proof: From the definiton (6.21) $\hat{a}(y)$ is upper semicontinuous: Let us define $\ell(y)$ by the relation

$$\ell(y) = \limsup_{\substack{n\to\infty \\ y'\to y}} \frac{1}{2n} \Delta_{n,y'}(A_n, A_n) \; .$$

If we prove that $\hat{a}(y) \leq 4/\ell(y)$ then it follows that $\hat{a}(y) = 4/\ell(y)$ and by definition $\ell(y)$ is upper semicontinuous. We are thus done.

The proof of Theorem 6.7 will be carried out in several steps. Assume that there exists a sequence y_n such that $y_n \to y$ and

$$\limsup_{\substack{n\to\infty \\ y_n\to y}} \frac{1}{2n} \Delta_{n,y_n}(A_n, A_n) \geq \theta$$

Because of (6.5) this means the existence of a function $u_n(x_{-n},\dots,x_n)$ such that

$$\mathcal{D}_{n,y_n}(u_n) \leq 2n \tag{6.23}$$

and

$$\liminf_{n\to\infty} \frac{1}{2n} \int u_n \cdot A_n \, \mu_{n,y_n}(dx) \geq \sqrt{\frac{\theta}{2}} \, . \tag{6.24}$$

We can renormalize things so that instead of (6.23) and (6.24) we have

$$\frac{1}{2n}\int \Big(\frac{\partial u_n}{\partial x_n} - \frac{\partial u_n}{\partial x_{-n}}\Big)\,\mu_{n,y_n}(dx) = 1 \tag{6.25}$$

and

$$\limsup_{n\to\infty} \frac{1}{2n}\int \sum a(x_i,x_{i+1})\Big(\frac{\partial u_n}{\partial x_{i+1}} - \frac{\partial u_n}{\partial x_i}\Big)^2\,\mu_{n,y_n}(dx) \le \frac{4}{\theta} \tag{6.26}$$

The aim now is to construct some sort of weak limit out of $u_n(x_{-n},\ldots,x_n)$ which will lead to a function $\xi(w)$ defined relative to Q_y such that

$$E^{Q_y}\big[\xi(w)\big] = 1$$
$$E^{Q_y}\big[a(x_0,x_1)\xi^2(w)\big] \le \frac{4}{\theta}$$

and if we define $\xi_{i,i+1}(w) = \xi(T^i w)$ then $\{\xi_{i,i+1}(w)\}$ saisfies (5.6). Then we can write $\xi = 1-\xi'$ with $E^{Q_y}\xi' = 0$. We can now conclude by the use of Theorem 5.2 that $\xi' \in \mathcal{H}_{0,1}$. Therefore

$$\hat{a}(y) \le \frac{4}{\theta}$$

by definition. This in turn implies

$$\limsup_{\substack{n\to\infty\\ y'\to y}} \frac{1}{2n}\Delta_{n,y'}(A_n,A_n) \le \frac{4}{\hat{a}(y)}$$

and our proof is finished.

The weak limit ξ, with the required properties is constructed in three steps. We will formualte them as lemmas.

Lemma 6.8. *Assume that for some $\theta > 0$, we have a sequence $y_n \to y$ satisfying* (6.25) *and* (6.26). *Then for each $k < \infty$ there exists a sequence $u_n^{(k)}$ of functions depending on the variables $x_{-k},\ldots,x_k$ such that*

$$E^{\mu_{n,y_n}}\Big\{\frac{\partial u_n^{(k)}}{\partial x_k} - \frac{\partial u_n^{(k)}}{\partial x_{-k}}\Big\} = 2k$$

$$\limsup_{n\to\infty} E^{\mu_{n,y_n}}\Big\{\sum_{-k\le i\le i+1\le k} a(x_i,x_{i+1})\Big(\frac{\partial u_n^{(k)}}{\partial x_i} - \frac{\partial u_n^{(k)}}{\partial x_{i+1}}\Big)^2\Big\} \le \frac{4}{\theta}\,.$$

Proof: Let us define

$$\alpha_{i,i+1}^{(n)} = E^{\mu_n,\nu_n}\{\frac{\partial u_n}{\partial x_{i+1}} - \frac{\partial u_n}{\partial x_i}\}$$
$$\beta_{i,i+1}^{(n)} = E^{\mu_n,\nu_n}\{a(x_i,x_{i+1})(\frac{\partial u_n}{\partial x_{i+1}} - \frac{\partial u_n}{\partial x_i})^2\}$$

Our information is that for any $\epsilon > 0$ and any n_0, we can find $n > n_0$ such that

(6.27)
$$\sum_{-n\le i\le i+1\le n} \alpha_{i,i+1}^{(n)} = 2n$$

(6.28)
$$\sum_{-n\le i\le i+1\le n} \beta_{i,i+1}^{(n)} \le 2n(\frac{4}{\theta} + \epsilon)$$

Let $\epsilon > 0$ and n_0 be given. Pick such an n.

We have the elementary bound

$$\alpha_{i,i+1}^{(n)} \le c\sqrt{\beta_{i,i+1}^{(n)}} \le c\sqrt{\Sigma\beta_{i,i+1}^{(n)}} \le c'\sqrt{n}\ .$$

let us divide the bonds in $[-n,n]$ into blocks of size $2k$, call them $b_1,\ldots,b_\ell$ and denote by E the partial block left out at the end. Let $\sum_{(i,i+1)\in b_r} \alpha_{i,i+1}^{(n)} = a_r$ and $\sum_{(i,i+1)\in b_r} \beta_{i,i+1}^{(n)} = A_r$ and define

(6.29)
$$R = \sum_{1\le r\le \ell} \frac{a_r}{\sqrt{A_r}} \cdot \frac{1}{\sqrt{2k}}$$

Then

$$a_r \le R\sqrt{A_n}\sqrt{sk} \quad \text{for } r = 1,2,\ldots \ell\ .$$
$$\sum a_r \le R\sqrt{2k}\sum\sqrt{A_r}$$
$$\le R\sqrt{2k}\sqrt{\ell}\sqrt{\Sigma A_r} \le R\sqrt{2n}(\frac{4}{\theta} + \epsilon)\sqrt{2n}$$

We can use the bound that a single $\alpha_{i,i+1}^{(n)}$ cannot exceed $c'\sqrt{n}$ to conclude that

$$2n \le 2k \cdot c'\sqrt{n} + R2n(\frac{4}{\theta} + \epsilon)^{1/2}\ ,$$

which in turn implies

$$R \geq \frac{2n - 2kc'\sqrt{n}}{(\frac{4}{\theta} + \epsilon)^{1/2} \cdot 2n}$$

For k fixed and n large this means that

$$R \geq \frac{1}{(\frac{4}{\theta} + \epsilon')^{1/2}}$$

where $\epsilon' \to 0$ as $\epsilon \to 0$. By normalization there is some block of $2k$ bonds for which $a_r = 2k$ and $A_r \leq 2k(\frac{4}{\theta} + \epsilon)$.

We now replace $u_n(x_{-n}, \ldots, x_n)$ by

$$E^{\mu_{n,y_n}}\{u_n \,|\, \mathcal{F}\}$$

where $\mathcal{F}$ is the σ-field corresponding to the block of $2k$ bonds constructed above. Since the measure $\mu_{n,y}$ is symmetric in $2n+1$ variables any block has the same distribution as any other block. Therefore the conditional expectation

$$E^{\mu_{n,y_n}}\{u_n \,|\, \mathcal{F}\}$$

can be viewed as a function $u_n^{(k)}(x_{-k}, \ldots, x_k)$ satisfying

$$E^{\mu_{n,y_n}}\{\frac{\partial u_n^{(k)}}{\partial x_k} - \frac{\partial u_n^{(k)}}{\partial x_{-k}}\} = 2k$$

and

$$E^{\mu_{n,y_n}}\{\sum_{-k \leq i \leq i+1 \leq k} a(x_i, x_{i+1})(\frac{\partial u_n^{(k)}}{\partial x_{i+1}} - \frac{\partial u_n^{(k)}}{\partial x_i})^2\} \leq 2k(\frac{4}{\theta} + \epsilon)$$

The next step is to take a weak limit of the functions $u_n^{(k)}$ and $n \to \infty$.

Lemma 6.9. *Let Q_y be the product measure. Then for each k, there exists a function $u_k(x_{-k}, \ldots, x_k)$ of $2k+1$ variables such that*

$$E^{Q_y}\{\frac{\partial u_k}{\partial x_k} - \frac{\partial u_k}{\partial x_{-k}}\} = 2k$$

and

$$E^{Q_y}\{\sum_{-k \leq i \leq i+1 \leq k} a(x_i, x_{i+1})(\frac{\partial u_k}{\partial x_{i+1}} - \frac{\partial u_k}{\partial x_i})^2\} \leq 2k\frac{4}{\theta} .$$

Proof: Let us look at the linear functionals

$$\Lambda^{(n)}_{i,i+1}(\omega) = E^{\mu_{n,y_n}}\{(\frac{\partial u^{(k)}_n}{\partial x_{i+1}} - \frac{\partial u^{(k)}_n}{\partial x_i})\omega\}$$

where ω is a function of $x_{-k}, \ldots, x_k$.

$$|\Lambda^{(n)}_{i,i+1}(\omega)| \leq C\Big(\int \omega^2 d\mu_{n,y_n}\Big)^{1/2} .$$

By taking a subsequence and passing to the limit we get linear functionals $\Lambda_{i,i+1}$ satisfying

$$|\Lambda_{i,i+1}(\omega)| \leq C\Big(\int \omega^2 \, dQ_y\Big)^{1/2}$$

It is not difficult to show that

$$\Lambda_{i,i+1}(\omega) = \int (\frac{\partial u}{\partial x_{i+1}} - \frac{\partial u}{\partial x_i})\omega \, dQ_y ,$$

for some u, satisfying

$$\int \sum a(x_i, x_{i+1})\,(\frac{\partial u}{\partial x_{i+1}} - \frac{\partial u}{\partial x_i})^2 \, dQ_y \leq 2k\frac{4}{\theta}$$

and

$$\int (\frac{\partial u}{\partial x_k} - \frac{\partial u}{\partial x_{-k}}) \, dQ_y = 2k .$$

u_k is this u.

Lemma 6.10. *One can arrange so that for a sequence of values of k going to ∞ the functions $u_k(x_{-k}, \ldots, x_k)$ of Lemma 6.9 have the additional property*

$$E^{Q_y}\{(\frac{\partial u_k}{\partial x_{j+1}} - \frac{\partial u_k}{\partial x_j})^2\} = o(k)$$

for $j = -k, -k+1$, and $j = (k-1), (k-2), \ldots$ i.e. the contribution to the Dirichlet form of u_k from any single bond near the boundary goes to zero as a proportion of the total Dirichlet form.

Proof: The proof is quite elementary. let us look at $k^{1/8}$ blocks of size $k^{1/8}$ each near the two edges. Since the total Dirichlet form of u_k is of order k there must be at least one block near each end such that the Dirichlet form from bonds in that block of size $k^{1/8}$ is at most of order $k^{7/8}$. If we reduce $[-k,k]$ to the interval $[-k',k']$ so that the ends now consist of "good" blocks. Replacing $u_k(x_{-k},\dots,x_k)$ by $\hat{u}_{k'}(x_{-k'},\dots,x_{k'})$ where

$$\hat{u}_{k'} = E^{Q_y}\left[u_k \,|\, \mathcal{F}_{-k'}^{k'}\right]$$

we get a function $\hat{u}_{k'}$ that satisfies our requirement. We have to renormalize so that

$$E^{Q_y}\left[\frac{\partial \hat{u}_{k'}}{\partial x_{k'}} - \frac{\partial \hat{u}_{k'}}{\partial x_{-k'}}\right] = 2k'$$

This is easy because for each omitted bond

$$\left|E^{Q_y}\left\{\frac{\partial u_k}{\partial x_{i+1}} - \frac{\partial u_k}{\partial x_i}\right\}\right| \le C\sqrt{k}\ .$$

Moreover

$$|k-k'| \le Ck^{1/8}\ .$$

Finally out of these functions $u_k(x_{-k},\dots,x_k)$ we produce a function ξ.

Lemma 6.11. *There exists a function ξ in $L_2(Q_y)$ satisfying*

$$E^{Q_y}\xi = 1$$

$$E^{Q_y}a(x_0,x_1)\xi^2 \le \frac{4}{\theta}$$

and if we define $\xi_{i,i+1} = \xi(T^i\omega)$, then

$$\left(\frac{\partial}{\partial x_{j+1}} - \frac{\partial}{\partial x_j}\right)\xi_{i,i+1} = \left(\frac{\partial}{\partial x_{i+1}} - \frac{\partial}{\partial x_i}\right)\xi_{j,j+1}$$

Proof: Consider

$$\xi_{(n)} = \frac{1}{2n} \sum_{-k_1 \le i \le i+1 \le k_n} \left(\frac{\partial u_n}{\partial x_{i+1}} - \frac{\partial u_n}{\partial x_i} \right) (T^{-i}\omega)$$

Then

$$E^{Q_\nu} \xi_{(n)} = 1$$

Moreover

$$\begin{aligned} E^{Q_\nu} a(x_0, x_1) \xi_{(n)}^2 &= E^{Q_\nu} a(x_0, x_1) \Big(\frac{1}{2n} \sum_{-n \le i \le i+1 \le n} (\frac{\partial u_n}{\partial x_{i+1}} - \frac{\partial u_n}{\partial x_i})(T^{-i}\omega) \Big)^2 \\ &\le E^{Q_\nu} a(x_0, x_1) \frac{1}{2n} \sum (\frac{\partial u_n}{\partial x_{i+1}} - \frac{\partial u_n}{\partial x_i})^2 (T^{-i}\omega) \\ &= \frac{1}{2n} E^{Q_\nu} \sum a(x_i\, x_{i+1}) (\frac{\partial u_n}{\partial x_{i+1}} - \frac{\partial u_n}{\partial x_i})^2 \\ &\le \frac{4}{\theta}\,. \end{aligned}$$

In addition, an elementary calculation shows that if we define

$$\xi_{(n)}^{i,i+1} = \xi_{(n)}(T^i\omega)$$

then any discrepancy between

$$\left(\frac{\partial}{\partial x_{r+1}} - \frac{\partial}{\partial x_r} \right) \xi_{(n)}^{s,s+1}$$

and

$$\left(\frac{\partial}{\partial x_{s+1}} - \frac{\partial}{\partial x_s} \right) \xi_{(n)}^{r,r+1}$$

arises only from a finite number of terms near the edge. Since according to Lemma 6.10 we can assume these to be negligible the final part of the lemma is satisfied as well.

7. Identification of the flux term

The aim of this section is to establish formula (2.9) which identifies the flux term. A uniqueness theorem is available for weak solutions of the Cauchy problem, for our nonlinear heat equation that comes up as the hydrodynamic equation, within the class of solutions satisfying the bounds of Theorem 3.9. We can therefore immediately establish the main result concerning the hydrodynamic behavior.

This identification of the flux term will be carried out in several steps. Let $J(\theta, t)$ be a smooth function on $S \times [0,T]$. Sometimes we will want to ignore the role of t and then take $J(\theta)$ to be just a smooth function on S. Given such a function J we consider an induced function on R^N depending on a parameter $\epsilon > 0$.

$$
\begin{aligned}
(7.1)\quad \xi_N(\epsilon,t,x) = {} & N\sum_{i=1}^{N} J(\frac{i}{N},t)\, W(x_i,x_{i+1}) \\
& + N\sum_{i=1}^{N} J(\frac{i}{N},t)\,\hat{a}(\frac{x_{i-N\epsilon}+\cdots+x_{i+N\epsilon}}{2N\epsilon+1})(\frac{\phi'(x_{i-N\epsilon})-\phi'(x_{i+N\epsilon})}{2N\epsilon}) \\
& - 2N\sum_{i=1}^{N} J(\frac{i}{N},t)(LF^i)(x) \\
& - 2\sum_{i=1}^{N} J^2(\frac{i}{N},t)\, A(F, \frac{x_{i-N\epsilon}+\cdots+x_{i+N\epsilon}}{2N\epsilon+1})
\end{aligned}
$$

Here $W(x,y)$ is the flux across a bond

$$W(x,y) = \frac{\partial}{\partial y}a(x,y) - \frac{\partial}{\partial x}a(x,y) + a(x,y)\left[\phi'(x)-\phi'(y)\right]$$

that was already defined in (1.18). The function $\hat{a}(y)$ is the diffusion coefficient defined in (6.21). The function F is an arbitrary function of a finite set $x_{-(\ell-1)},\dots,x_{(\ell-1)}$ of $2\ell-1$ variables, that is assumed to be smooth. L is the basic infinitesimal generator given by (1.1) without the scaling factor N^2. and if N is large the periodic arrangement of lattice sites incorporated in the definition of L will have no effect and LF will be a local function of the variables $x_{-\ell},\dots,x_\ell$. $(LF^i)(x)$ is obtained in a similar fashion where $F^i(x)$ is the shifted function

$$F^i(x) = F(x_{i-(\ell-1)},\dots,x_{i+(\ell-1)})$$

$(LF^i)(x)$ can be obtained by shifting $(LF^0)(x) = (LF)(x)$ by i sites.

Finally for such a function F we have the formal sum

$$\psi = \psi_F = \sum_i F(x_{i-(\ell-1)}, \ldots, x_{i+(\ell-1)}) \tag{7.2}$$

and

$$A(F, y) = E^{Q_y}\{a(x_1, x_2)[1 - (\frac{\partial \psi}{\partial x_1} - \frac{\partial \psi}{\partial x_2})]^2\} - \hat{a}(y) \tag{7.3}$$

From the definition of $\hat{a}(y)$ it is clear that

$$0 \le A(F, y) \le C_F \tag{7.4}$$

where C_F is a constant independent of y, and

$$\inf_F A(F, y) = 0 \quad \text{for every } y\ . \tag{7.5}$$

One of the main theorems of the section that will help in the identification of the flux term is the following:

Theorem 7.1. *For any* $0 < T < \infty$

$$\limsup_{\epsilon \to 0} \limsup_{N \to \infty} \frac{1}{N} \log E^{P_N}\{\exp \int_0^T \xi_N(\epsilon, t, x(t))\, dt\} \le 0\ .$$

Proof: Let us remind the reader that E^{P_N} refers to the process in the speeded up time initialized to start from the equilibrium, Φ_N on N sites. The proof will be broken up into several lemmas.

Lemma 7.2. *For any positive constant C let us denote $\mathcal{F}_C$ the set of probability densitites f, relative to the equilibrium Φ_N, that satisfy the bounds*

$$D(\sqrt{f}) = \frac{1}{8} \int \frac{1}{f} \sum a(x_i, x_{i+1}) (\frac{\partial f}{\partial x_i} - \frac{\partial f}{\partial x_{i+1}})^2 \Phi_N\, dx \le \frac{C}{N} \tag{7.6}$$

and

$$\int \sum \omega(x_i)\, f\, \Phi_N\, dx \le CN \tag{7.7}$$

where $w(x)$ is the function defined in (1.5) and (1.6). Then for any bounded continuous function $g(y)$ on R we have

$$\limsup_{\substack{\epsilon\to 0\\ k\to\infty}} \limsup_{N\to\infty} \sup_{f\in\mathcal{F}_C} \int \frac{1}{N}\sum_i |g(\frac{x_{i-N\epsilon}+\cdots+x_{i+N\epsilon}}{2N\epsilon+1})$$
$$- g(\frac{x_{i-k}+\cdots+x_{i+k}}{2k+1})| f\,\Phi_N\,dx = 0$$

Proof: This has been proved in [2]. See Theorem 4.7 in [2]. It is true that the Dirichlet for there was with $a(x,y)\equiv 1$. Since $a(x,y)$ is bounded above and below the Dirichlet form here is equivalent to the one $a(x,y)\equiv 1$ and those results do hold in our context.

The above lemma will be useful in replacing the term $A(F, \frac{x_{i-N\epsilon}+\cdots+x_{i+N\epsilon}}{2N\epsilon+1})$ in the definition (7.1) of ξ_N by $A(F, \frac{x_{i-k}+\cdots+x_{i+k}}{2k+1})$. We wish also to replace

$$N\sum J(\frac{i}{N},t)\,\hat{a}(\frac{x_{i-N\epsilon}+\cdots+x_{i+N\epsilon}}{2N\epsilon+1})(\frac{\phi'(x_{i-N\epsilon})-\phi'(x_{i+N\epsilon})}{2N\epsilon})$$

by

$$N\sum J(\frac{i}{N},t)\,\hat{a}(\frac{x_{i-k}+\cdots+x_{i+k}}{2k+1})(\frac{\phi'(x_{i-k})-\phi'(x_{i+k})}{2k})$$

The difference between these two functions can be written as

$$g_N(\epsilon,k,t,x) = g_N^{(1)}(\epsilon,k,t,x) + g_N^{(2)}(\epsilon,k,t,x)$$

where

$$g_N^{(1)}(\epsilon,k,t,x) = N\sum J(\frac{i}{N},t)\,\hat{a}(\frac{x_{i-N\epsilon}+\cdots+x_{i+N\epsilon}}{2N\epsilon+1})$$
$$[\frac{\phi'(x_{i-N\epsilon})-\phi'(x_{i+N\epsilon})}{2N\epsilon} - \frac{\phi'(x_{i-k})-\phi'(x_{i+k})}{2k}]$$

and

$$g_N^{(2)}(\epsilon,k,t,x) = N\sum J(\frac{i}{N},t)\,[\hat{a}(\frac{x_{i-N\epsilon}+\cdots+x_{i+N\epsilon}}{2N\epsilon+1})$$
$$- \hat{a}(\frac{x_{i-k}+\cdots+x_{i+k}}{2k+1})]\,[\frac{\phi'(x_{i-k})-\phi'(x_{i+k})}{2k}]$$

We will at this point drop t and just think of the above as functions of the rest of the variables. The function $g_N^{(1)}$ has the form

$$g_N^{(1)}(\epsilon,k,x) = N\sum_i b_i(x)\,[\phi'(x_i)-\phi'(x_{i+1})]$$

where

$$b_i(x) = \frac{1}{2N\epsilon} \sum_{|j-i|\le N\epsilon} \hat{a}\Big(\frac{x_{j-N\epsilon} + \cdots + x_{j+N\epsilon}}{2N\epsilon+1}\Big) J\Big(\frac{j}{N}\Big) - \frac{1}{2k} \sum_{|j-i|\le k} \hat{a}\Big(\frac{x_{j-N\epsilon} + \cdots + x_{j+N\epsilon}}{2N\epsilon+1}\Big) J\Big(\frac{j}{N}\Big)$$

The function $b_i(x)$ are uniformly bounded and the dependence of b_i on x_i and x_{i+1} is only through the sum x_i and x_{i+1}.

Lemma 7.3. *Let $\delta > 0$ and $\rho > 0$ be two arbitrary but fixed constants. Then*

$$\limsup_{\substack{\epsilon\to 0\\ k\to\infty}} \limsup_{N\to\infty} \frac{1}{N} \sup_f \left[E^f\{g_N^{(1)}(\epsilon,k,x) - \delta \sum \omega(x_i)\} - N^2 \rho\, D(\sqrt{f})\right] = 0$$

Proof: Since b_i are uniformly bounded and $(\frac{\partial}{\partial x_i} - \frac{\partial}{\partial x_{i+1}})b_i = 0$ we can write

$$\begin{aligned} E^f\left[g_N^{(1)}(\epsilon,k,x)\right] &= E^f\left[N \sum b_i(\phi'(x_i) - \phi'(x_{i+1}))\right] \\ &= N \int \sum_i b_i \Big(\frac{\partial f}{\partial x_i} - \frac{\partial f}{\partial x_{i+1}}\Big) \Phi_N\, dx \\ &= N \int \sum_i b_i \frac{1}{\sqrt{f}} \Big(\frac{\partial f}{\partial x_i} - \frac{\partial f}{\partial x_{i+1}}\Big) \Phi_N \sqrt{f}\, dx \\ &\le N \Big(\int \sum_i b_i^2 f\, dx\Big)^{1/2} (2D(\sqrt{f}))^{1/2} \\ &\le C N^{3/2} \left[D(\sqrt{f})\right]^{1/2} \end{aligned}$$

Therefore the supremum can be limited to densities f with $D(\sqrt{f}) \le C/N$ for some a priori C. That gives a bound of the form

$$E^f[g_N^{(1)}(\epsilon,k,x)] \le C'N$$

and that in turn implies a bound

$$E^f[\delta \sum \omega(x_i)] \le C''N$$

or

$$E^f[\sum \omega(x_i)] \le C_\delta N\ .$$

In order to complete the proof of Lemma 7.3 we have only to establish

$$\limsup_{\substack{\epsilon\to 0\\ k\to\infty}} \limsup_{N\to\infty} \frac{1}{N} \sup_{f\in\mathcal{F}_C} \int \frac{1}{N}\sum b_i^2\, f\, \Phi_N\, dx = 0\ .$$

This is again the same idea as in Lemma 7.2. According to the results referred to in [1] or [2] once $f\in\mathcal{F}_C$, the averages $\frac{x_{i-k}+\cdots+x_{i+k}}{2k+1}$ for k large and $\frac{x_{i-N\epsilon}+\cdots+x_{i+N\epsilon}}{2N\epsilon+1}$ for ϵ small are close to each other and show no fluctuations in scales below macroscopic scale. We note that $\hat{a}(y)$ is continuous and $J(\cdot)$ is smooth.

We can prove in exactly the same fashion using exactly the same ideas the following lemma.

Lemma 7.4. *Let $\delta>0$ and $\rho>0$ be two arbitrary but fixed constants. Then*

$$\limsup_{\substack{\epsilon\to 0\\ k\to\infty}} \limsup_{N\to\infty} \frac{1}{N} \sup_{f} \left[E^f\{g_N^{(2)}(\epsilon,k,x)-\delta\sum\omega(x_i)\} - \rho N^2 D(\sqrt{f})\right] = 0$$

Finally we want to replace

$$N\sum J(\frac{i}{N})W(x_i,x_{i+1})$$

and

$$N\sum J(\frac{i}{N})(LF)^i(x)$$

by

$$N\sum J(\frac{i}{N})\frac{1}{2k}\sum_{j=i-k}^{i+k-1} W(x_j,x_{j+1})$$

and

$$N\sum J(\frac{i}{N})\frac{1}{2(k-\ell)+1}\sum_{j=i-(k-\ell)}^{i+(k-\ell)} (LF)^j(x)$$

and if we denote by $g_N^{(3)}(k,x)$ the error in any of these replacements, one can prove

Lemma 7.5. *For any $\rho>0$ and fixed k,*

$$\limsup_{N\to\infty} \frac{1}{N} \sup_{f\in\mathcal{F}} \left[E^f\{g_N^{(3)}(k,x)\} - \rho N^2 D(\sqrt{f})\right] = 0\ .$$

Proof: By a summation by parts the error $g_N^{(3)}(k,x)$ can be written in the form

$$g_N^{(3)}(k,x) = N \sum J_{i,N}\, W(x_i, x_{i+1})$$

or

$$g_N^{(3)}(k,x) = N \sum J_{i,N}\, (LF)^i(x)$$

A simple estimate yields $|J_{i,N}| \le \frac{C}{N^2}$ for fixed k. It is elementary to do an integration by parts and bound $E^f[g_N^{(3)}(k,x)]$ by the Dirichlet form with powers of N to spare.

Our final lemmas in this sequence conerns the function

$$\begin{aligned} V_k(x) &= \frac{1}{2k} \sum_{i=-k}^{k-1} W(x_i, x_{i+1}) \\ &+ \hat{a}\Big(\frac{x_{-k} + \cdots + x_k}{2k+1}\Big)\Big(\frac{\phi'(x_{-k}) - \phi'(x_k)}{2k}\Big) \\ &- \frac{2}{2(k-\ell)+1} \sum_{i=-(k-\ell)}^{k-\ell} (LF)^i(x) \end{aligned}$$

Let λ be a real constant. Let θ and δ be positive constants. For each fixed y consider the hyperplane $\frac{x_{-k}+\cdots+x_k}{2k+1} = y$ and denote by $\mathcal{F}_y$ all densities supported on that hyperplane. The Dirichlet form $D^k(\sqrt{f})$ is naturally defined for such densities. Let $\sigma_k^2(V_k, y)$ be the central limit theorem variance of V_k for the ergodic diffusion with generator $\mathcal{L}^k = \sum_{i=-k}^{k-1} L_{i,i+1}$ corresponding to the interior bonds. The following lemma is an easy consequence of the results of Section 4.

Lemma 7.6. *For fixed θ, δ and uniformly for λ in a bounded interval we have*

$$\limsup_{N\to\infty} N^2 \sup_{f\in\mathcal{F}_y} \big[E^f\{\frac{\lambda}{N} V_k(x) - \frac{\delta}{N^2}\sum \omega(x_i)\} - \theta D^k(\sqrt{f})\big] \le \frac{\lambda^2}{2\theta}\sigma_k^2(V_k, y)\,.$$

Proof: This is essentially Theorem 4.1.

Now we return to the proof of Theorem 7.1.

We add and subtract a term $-\delta \sum \omega(x_i)$ to ξ_N so that

$$\begin{aligned}\exp\Big[\int_0^T \xi_N(\epsilon,t,x(t))\,dt\Big] &= \exp\Big[\int_0^T \xi_N(\epsilon,t,x(t)) \\ &\quad - \delta\int_0^T \sum \omega(x_j(t))\,dt\Big]\exp\Big[\delta\int_0^T \sum \omega(x_j(t))\,dt\Big] \\ &= A(T)\,B(T)\,.\end{aligned}$$

We estimate $E^{P_N}[A(T)\,B(T)]$ by Hölder's inequality with some p and q with $\frac{1}{p}+\frac{1}{q}=1$.

$$\frac{1}{N}\log E^{P_N}\big[A(T)\,B(t)\big] \le \frac{1}{Np}\log E^{P_N}\big[[A(T)]^p\big] + \frac{1}{Nq}\log E^{P_N}\big[[B(T)]^q\big]\,. \tag{7.8}$$

The second term is estimated as follows

$$\begin{aligned}E^{P_N}\Big\{\exp\int_0^T q\delta\sum\omega(x_j(t))\,dt\Big\} &= E^{P_N}\Big\{\exp\frac{1}{T}\int_0^T q\delta T\sum\omega(x_j(t))\,dt\Big\} \\ &\le E^{P_N}\Big\{\frac{1}{T}\int_0^T \exp\Big\{\sum q\delta T\,\omega(x_j(t))\Big\}\,dt\Big\} \\ &= \int e^{q\delta T\sum\omega(x_j)-\sum\phi(x_j)}\prod dx_j \\ &= \Big(\int e^{q\delta T\,\omega(x)-\phi(x)}\,dx\Big)^N\end{aligned}$$

Therefore

$$\frac{1}{Nq}\log E^{P_N}\{[B(T)]^q\} \le \frac{1}{q}\log\int e^{qT\delta\,\omega(x)-\phi(x)}\,dx \quad\text{provided}\quad qT\delta\le 1\,. \tag{7.9}$$

We now turn to the first term in (7.8). By a combination of Feynman-Kac formula and the spectral theorem

$$\log E^{P_N}\{[A(T)]^p\} \le \int_0^T \Lambda_N(t)\,dt$$

where $\Lambda_N(t)$ is the eigenvalue

$$\Lambda_N(t) = \sup_{f\in\mathcal{F}}\Big\{E^f\big[p\xi_N(\epsilon,t,x) - p\delta\sum\omega(x_j)\big] - N^2D(\sqrt{f})\Big\}$$

We can peel off terms $\frac{\delta}{2}\sum\omega(x_j)$ and $\rho N^2 D(\sqrt{f})$ and use it to control all error terms according to Lemmas 7.2, 7.3, 7.4 and 7.5. Moreover

$$\int_0^T \Lambda_N(t)\,dt \le T\sup_{0\le t\le T}\Lambda_N(t)\,.$$

Since the dependence on t is only through $J(\theta,t)$ we may drop it because our estimates will be uniform in t for $0 \le t \le T$. If we define

$$\tilde{\xi}_N(k,x) = N\sum_i J(\frac{i}{N})V_k(x_{i-k},\ldots,x_{i+k}) - 2\sum_i J^2(\frac{i}{N})A(F,\frac{x_{i-k}+\cdots+x_{i+k}}{2k+1}) - \frac{\delta}{2}\sum_i \frac{1}{2k+1}\sum_{|j-i|\le k}\omega(x_j) \tag{7.10}$$

then in order to prove the theorem it is sufficient to prove that for each $\delta > 0$

$$\limsup_{\rho\to 0}\limsup_{k\to\infty}\limsup_{N\to\infty}\frac{1}{N}\sup_f\left[E^f\{\tilde{\xi}_N(k,x)\} - (1-\rho)N^2 D(\sqrt{f})\right] \le 0 \tag{7.11}$$

If we denote by $f^{i,k}$ the marginal of f on a block of size $2k+1$ centered at i and by $D^k(\sqrt{f})$ the Dirichlet form on the interior bonds of the block we know

$$\sum_i D^k(\sqrt{f^{i,k}}) \le (2k)\,D(\sqrt{f}) \tag{7.12}$$

Combining (7.10) and (7.12) we can estimate

$$\begin{aligned}
&\frac{1}{N}\sup_f\left[E^f\{\tilde{\xi}_N(k,x)\} - (1-\rho)N^2 D(\sqrt{f})\right] \\
&\le \sup_{|\lambda|\le M}\sup_f\Big[E^f\{N\lambda V_k - 2\lambda^2 A(F,\frac{x_{-k}+\cdots+x_k}{2k+1}) \\
&\quad - \frac{\delta}{2}\cdot\frac{1}{2k+1}\sum_{j=-k}^{k}\omega(x_j)\} - \frac{N^2(1-\rho)}{2k}D^k(\sqrt{f})\Big]
\end{aligned}$$

we have repalced the sup of a sum with a factor of $1/N$ by the sup of the sups. The only variable is $\lambda = J(i/N)$ which varies over a bounded set. If we denote by $\mathcal{F}_y$ the space of densities on the hyperplane of level $\frac{1}{2k+1}\sum x_i = y$, we have to evaluate

$$\begin{aligned}
&\limsup_{N\to\infty}\sup_{|\lambda|\le M}\sup_y\Big[N^2\sup_{f\in\mathcal{F}_y}\Big[E^f\{\frac{\lambda}{N}V_k \\
&- \frac{\delta}{2(2k+1)N^2}\sum_{j=-k}^{k}\omega(x_j)\} - \frac{1-\rho}{2k}D^k(\sqrt{f})\Big] - 2\lambda^2 A(F,y)\Big]
\end{aligned}$$

Let us fix δ. Then it is easily seen that we need to look at only y in a bounded region $|y| \le C_\delta$ for the sup because

$$\frac{1}{2k+1}\sum \omega(x_j) \ge |y| \quad \text{on the hyperplane } y = \frac{1}{2k+1}\sum x_i$$

and

$$E^f\{\frac{\lambda}{N}V_k\} - \frac{1-\rho}{2\kappa}D^k(\sqrt{f}) \le C$$

uniformly if $|\lambda| \le M$. We can use Lemma 7.6 to conclude that

$$\begin{aligned}&\limsup_{N\to\infty}\frac{1}{N}\sup_{f\in\mathcal{F}}\left[E^f\{\tilde{\xi}_N(k,x)\} - (1-\rho)N^2\, D(\sqrt{f})\right]\\ &\le \sup_{|y|\le C_\delta}\ \{\frac{\lambda^2 2k}{2(1-\rho)}\sigma_k^2(V_k,y) - 2\lambda^2 A(F,y)\}\end{aligned}$$

According to the results of section 6

$$\lim_{k\to\infty} 2k\sigma_k^2(V_k,y) = 4A(F,y)$$

uniformly on bounded sets. Therefore

$$\begin{aligned}&\limsup_{k\to\infty}\limsup_{N\to\infty}\frac{1}{N}\sup_{f\in\mathcal{F}}\left[E^f\{\tilde{\xi}_N(k,x)\} - (1-\rho)N^2\, D(\sqrt{f})\right]\\ &\le \sup_{|y|\le C_\delta}\ \{2\lambda^2(\frac{1}{1-\rho}-1)\cdot A(F,y)\}\\ &\le 2C\lambda^2(\frac{1}{1-\rho}-1)\end{aligned}$$

We can now let $\rho \to 0$ and $\delta \to 0$ to complete the proof.

Let us now start from nonequilibrium and assume that the initial density f_N^0 has entropy relative to Φ_N that is bounded by CN. Then the process Q_N starting from f_N^0 has also the same entropy relative to the process P_N.

Lemma 7.7. *For any choice of F*

$$E^{Q_N}\left|\int_0^T \sum J(\frac{i}{N}, t)(LF)^i(x(t))\,dt\right| \to 0$$

as $N \to \infty$.

Proof: Let us look at

$$\xi(t) = \frac{1}{N^2}\sum_{i=1}^{N} J(\frac{i}{N},t)(F^i)(x(t))$$

Then by Itô's formula

$$d\xi(t) = \sum_{i=1}^{N} J(\frac{i}{N},t)(LF)^i(x(t))\,dt \tag{7.13}$$
$$+\frac{1}{N^2}\sum_{i=1}^{N} J_t(\frac{i}{N},t)\,F^i(x(t))\,dt$$
$$+\frac{1}{N}\sum J(\frac{i}{N},t)\sum_r (\frac{\partial F^i}{\partial x_r}-\frac{\partial F^i}{\partial x_{r-1}})\,\sigma(x_{r-1}(t),x_r(t))\,d\beta_r$$

Assuming F is bounded, clearly

$$|\xi(t)| \le \frac{C}{N}$$

so that $\xi(T)-\xi(0)\to 0$ as $N\to\infty$. Of the three terms on the right-hand side of (7.13) the second and the third clearly go to zero when integrated from 0 to T. Therefore the first term must go to zero as well. These results are true for any starting configuration and therefore for Q_N.

We can use entropy bounds to conclude that

$$E^{Q_N}[\psi] \le \log E^{P_N}[e^{\psi}] + H(Q_N\ ,\ P_N)$$

or replacing ψ by $N\psi$ and dividing by N

$$E^{Q_N}[\psi] \le \frac{1}{N}\log E^{P_N}[e^{N\psi}] + \frac{1}{N}H(Q_N\,|\,P_N)$$
$$E^{Q_N}[\psi] \le \frac{1}{N}\log E^{P_N}[e^{N\psi}] + C$$

If we take

$$\psi = \int_0^T \sum_{i=1}^{N} J(\frac{i}{N},t)\,W(x_i(t),x_{i+1}(t))\,dt$$
$$+\frac{1}{2N\epsilon}\int_0^T\sum_{i=1}^{N} J(\frac{i}{N},t)[\phi'(x_{i-N\epsilon}(t))-\phi'(x_{i+N\epsilon}(t))]\,\hat{a}(\frac{x_{i-N\epsilon}(t)+\cdots+x_{i+N\epsilon}(t)}{2N\epsilon+1})\,dt$$
$$-2\int_0^T\sum_{i=1}^{N} J(\frac{i}{N},t)(LF)^i(x(t))\,dt$$
$$-2\frac{1}{N}\int_0^T\sum_{i=1}^{N} J^2(\frac{i}{N},t)\,A(F,\frac{x_{i-N\epsilon}(t)+\cdots+x_{i+N\epsilon}(t)}{2N\epsilon+1})\,dt$$

then we can pass to the limit and we get

$$E^{\widehat{Q}}[\tilde{\psi}] \le C \tag{7.14}$$

where $\widehat{Q}$ is the limit obtained by compactness arguments of Section 3. Let us note that $\widehat{Q}$ is a measure on $\eta(t,\theta)$ and $m(t,\theta)$ and (7.14) reads

$$\begin{aligned}\limsup_{\epsilon\to 0} E^{\widehat{Q}}\Big[\int_0^T\int_S & J(\theta,t)\{\eta(\theta,t)+\frac{1}{2\epsilon}[h'(m(\theta-\epsilon,t))\\ &-h'(m(\theta+\epsilon,t))]\,\hat{a}(\frac{1}{2\epsilon}\int_{\theta-\epsilon}^{\theta+\epsilon} m(\theta',t)\,d\theta')\}\,d\theta\,dt\\ &-2\int_0^T\int_S J^2(\theta,t)\,A(F,\frac{1}{2\epsilon}\int_{\theta-\epsilon}^{\theta+\epsilon} m(\theta',t)\,d\theta')\,d\theta\,dt\Big] \le C\end{aligned}$$

The only part that needs an explanation is how

$$\frac{1}{2N\epsilon}\int_0^T\sum_{i=1}^N J(\frac{i}{N},t)[\phi'(x_{i-N\epsilon}(t))-\phi'(x_{i+N\epsilon}(t))]\,\hat{a}\left(\frac{x_{i-N\epsilon}(t)+\cdots+x_{i+N\epsilon}(t)}{2N\epsilon}\right)dt$$

can be replaced by

$$\frac{1}{2\epsilon}\int_0^T\int_S J(\theta,t)\Big[h'(m(\theta-\epsilon,t))-h'(m(\theta+\epsilon,t))\Big]\,\hat{a}\left(\frac{1}{2\epsilon}\int_{\theta-\epsilon}^{\theta+\epsilon} m(\theta',t)\,d\theta'\right)d\theta\,dt\,.$$

Clearly if we did not have $\hat{a}$ then

$$\frac{1}{2N\epsilon}\int_0^T\sum_{i=1}^N J(\frac{i}{N},t)[\phi'(x_{i-N\epsilon}(t))-\phi'(x_{i+N\epsilon}(t))]\,dt$$

can be replaced by

$$\frac{1}{2\epsilon}\int_0^T\int_S J(\theta,t)[h'(m(\theta-\epsilon,t))-h'(m(\theta+\epsilon,t))]\,d\theta\,dt$$

according to the results in [2]. But the terms $\hat{a}(\frac{x_{i-N\epsilon}(t)+\cdots+x_{i+N\epsilon}(t)}{2N\epsilon})$ are not oscillatory and converge strongly to $\hat{a}(\frac{1}{2\epsilon}\int_{\theta-\epsilon}^{\theta+\epsilon} m(\theta',t)\,d\theta')$. So the introduction of $\hat{a}$ can be accomplished. Because of estimates in Theorem 3.9 we can let $\epsilon\to 0$ and conclude

$$\begin{aligned}E^{\widehat{Q}}\Big[\int_0^T\int_S & J(\theta,t)\{\eta(\theta,t)-(h'(m(\theta,t)))_\theta\,\hat{a}(m(\theta,t))\}\,d\theta\,dt\\ &-2\int_0^T\int_S J^2(\theta,t)\,A(F,m(\theta,t))\,d\theta\,dt\Big] \le C\end{aligned} \tag{7.15}$$

It is necessary to widen the choice of $F = F(x_{-\ell+1}, \ldots x_{\ell-1})$ to allow it to depend on an extra variable $y \in R$, so that $F = F(y; x_{-\ell+1}, x_{\ell-1})$. For any $\epsilon > 0$, this function is viewed as a function on R^N by

$$F_\epsilon(x) = F\left(\frac{x_{-N\epsilon} + \cdots + x_{N\epsilon}}{2N\epsilon + 1}\ ;\ x_{-\ell+1}, \ldots, x_{\ell-1}\right)$$

In Lemma 7.7 we can consider

$$\xi(t) = \frac{1}{N^2} \sum_{i=1}^{N} J(\frac{i}{N}, t)\, F_\epsilon^i(x(t))$$

where

$$F_\epsilon^i(x) = F\left(\frac{x_{i-N\epsilon} + \cdots + x_{i+N\epsilon}}{2N\epsilon}\ ;\ x_{i-\ell+1}, \ldots, x_{i+\ell-1}\right)$$

Essentially all the steps can be repeated with this choice. We have to make sure that the extra terms arising in $d\xi(t)$ due to variations of $\dfrac{x_{i-N\epsilon}(t) + \cdots + x_{i+N\epsilon}(t)}{2N\epsilon + 1}$ are negligible as $N \to \infty$. There are three types of additional terms. The stochastic integral terms are of the form

$$\frac{1}{\epsilon N^2} \int_0^t \sum_{i=1}^{N} \psi_i(s)\, d\beta_i(s)$$

where $\psi_i(s)$ are all uniformly bounded. This is clearly negligible as $N \to \infty$ for any initial condition. The second type of terms looks like

$$\frac{1}{(N\epsilon)^2} \int_0^t \sum_{i=1}^{N} H_i(s)\, ds$$

with H_i uniformly bounded and these are uniformly small as $N \to \infty$ for any $\epsilon > 0$. The final remaining terms are of the form

$$\frac{1}{N\epsilon} \int_0^t \sum_{i=1}^{N} G_i\left(\frac{x_{i-N\epsilon} + \cdots + x_{i+N\epsilon}}{2N\epsilon + 1}\ ;\ x_{i-\ell+1}, \ldots, x_{i+\ell-1}\right) W(x_{i-N\epsilon-1}, , x_{i-N\epsilon})\, ds$$

Estimating these can be reduced by entropy bounds and the Feynman-Kac formula to demonstrating

$$\lim_{N\to\infty} \frac{1}{N} \sup_f \Big\{E^f\Big[\sum G_i\Big(\frac{x_{i-N\epsilon} + \cdots + x_{i+N\epsilon}}{2N\epsilon + 1}\ ;\ x_{i-\ell+1}, \ldots, x_{i+\ell-1}\Big) W(x_{i-N\epsilon-1}, x_{i+N\epsilon+1})\Big] - \frac{N^2}{2}\mathcal{D}(\sqrt{f}\Big\} = 0$$

This is easily done by integrating by parts.

We can therefore now have (7.15) for arbitrary $F(y; x_{-\ell+1}, \ldots, x_{\ell-1})$ and $A(F, y)$ is interpreted as $A(F(y,\cdot), y)$. By standard large deviation arguments one can replace the integrand by the supremum over any arbitrary finite set of J's and F's and one would still have

$$E^{\widehat{Q}}\Big[\sup_{J,F}\Big\{\int_0^T\int_S J(\theta,t)\{\eta(\theta,t)-(h'(m(\theta,t)))_\theta\,\hat{a}(m(\theta,t))\}\,d\theta\,dt$$
$$-2\int_0^T\int_S J^2(\theta,t)\,A(F,m(\theta,t))\,d\theta\,dt\Big\}\Big]\leq C\,.$$

Since $\inf_F A(F(\cdot,m),m)\equiv 0$ in m, we conclude from the above that

$$\eta(\theta,t)-(h'(m(\theta,t)))_\theta\,\hat{a}(m(\theta,t))=0 \text{ a.e. } \widehat{Q}$$

or

$$\eta=\hat{a}(m)\,(h'(m))_\theta\,.$$

This completes the identification of flux.

References

1. Donsker, M. D. and Varadhan, S. R. S., Large deviations from a hydrodynamic scaling limit, Comm. Pure Appl. Math. XLII, 243–270 (1989).
2. Guo, M. Z., Papanciolaou, G. C. and Varadhan, S. R. S., Nonlinear diffusion limit for a system with nearest neighbor interactions, Commun. Math. Phys. 118, 31–59 (1988).
3. Quastel, Jeremy, Diffusion of color in the simple exclusion process, thesis, New York Univ., 1990.
4. Stroock, D. W. and Varadhan, S. R. S., Multidimensional diffusion processes, Springer-Verlag, New York, 1979.
5. Varadhan, S. R. S. and Yau, H. T., in preparation.

This research was supported in part by the National Science Foundation, grant DM-S 89 01682.

Courant Institute of Mathematical Sciences

New York University

251 Mercer Street

New York, NY 10012

USA

The Dynkin Festschrift, Progr. Probab. no. 34, Birkhauser, 1994, 387-397

Regularity of Self-Diffusion Coefficient

S.R.S. Varadhan†

1. Introduction

We consider the infinite particle system in Z^d known commonly as symmetric random walk with simple exclusion. This consists of particles in Z^d, each trying to execute a more or less independent random walk with jump probabilities $p(x)$. We assume that $p(x)$ is symmetric, i.e., $p(x) = p(-x)$, and that it has finite range. If $d = 1$, we assume, in addition, that $p(1) + p(-1) < 1$, so that only the one-dimensional nearest neighbor random walk is excluded. Particles wait for an exponential time, pick a random jump size x with probability $p(x)$, and try to jump to a new site defined by a jump equal to x. If the site is free, the jump is completed and things start all over again. Otherwise the jump is disallowed and things start all over again with the particle remaining in the old site. All the particles are doing this simultaneously and independently of each other. See [3] for a discussion of the model.

We can take Ω to be the configuration space of particles in Z^d . A point $\omega \in \Omega$ is really $\{\eta(x) : x \in Z^d\}$. $\eta(x) = 1$ means that there is a particle at x and $\eta(x) = 0$ if there is no particle at x. The process that we have described is a Markov process on Ω and has an infinitesimal generator that can be written down. First some notation. Let $\tau_{\alpha,\beta}$ be defined as the following map of Ω into Ω for α, β in Z^d.

$$\begin{aligned}
\eta(\tau_{\alpha,\beta}\omega, x) &= \eta(\omega, x) \quad if \quad x \neq \alpha \ or \ \beta, \\
\eta(\tau_{\alpha,\beta}\omega, \alpha) &= \eta(\omega, \beta), \\
\eta(\tau_{\alpha,\beta}\omega, \beta) &= \eta(\omega, \alpha).
\end{aligned}$$

$\tau_{\alpha,\beta}$ exchanges the situation at sites α and β.

---

† Courant Institute, New York University. The author was partially supported by NSF grant DMS-9201222 and ARO grant DAAL03-92-G-0317.

The infinitesimal generator can be written as

$$(LF)(\omega) = \lambda \sum_{\alpha} \eta(\alpha) \sum_{\beta} (1 - \eta(\beta))[F(\tau_{\alpha,\beta}\omega) - F(\omega)]p(\beta - \alpha).$$

Here λ is the exponential holding rate and is essentially a normalization constant for the time scale. Because $\tau_{\alpha,\beta} = \tau_{\beta,\alpha}$, we can rewrite the summation over unordered pairs α, β,

$$\begin{aligned}(LF)(\omega) &= \lambda \sum_{\alpha,\beta} [\eta(\alpha)(1 - \eta(\beta) + \eta(\beta)(1 - \eta(\alpha)][F(\tau_{\alpha,\beta}\omega) - F(\omega)]p(\beta - \alpha) \\ &= \lambda \sum_{\alpha,\beta} [F(\tau_{\alpha,\beta}\omega) - F(\omega)]p(\beta - \alpha).\end{aligned}$$

We shall, for simplicity, assume $d \geq 2$, $p(\alpha) = (2d)^{-1}$ if $\|\alpha\| = 1$ and $\lambda = 2d$. This is only a matter of convenience. In such a case the diffusion matrix for a single isolated particle is $2I$ and we will denote this by S_0.

The infinite particle evolution on Ω has several invariant measures. They are characterized by the density ρ of particles. For each ρ, $0 < \rho < 1$, the Bernoulli product measure P_ρ such that $P_\rho[\eta(x) = 1] = \rho \ \forall x$ is an invariant measure. We shall be interested in the motion of a tagged particle starting from the origin. To begin with, we assume that we have a particle at 0 in Z^d and that in $Z^d - 0$ we have a random configuration from a Bernoulli measure P_ρ for some ρ, $0 < \rho < 1$. We trace the motion of the tagged particle in Z^d. It was proved in [1] that this motion, under usual scaling, converges to a Brownian motion with covariance specified by some symmetric matrix $S(\rho)$. This matrix, called the self-diffusion matrix, really depends on ρ and one expects $S(\rho)$ to converge to 0 as ρ tends to 1 and to S_0 as ρ tends to 0. In [2] a variational formula was given for $S(\rho)$. In order to describe the variational formula, we look at the evolution viewed from the tagged particle. There are two kinds of motion: when a particle other than the tagged particle jumps and when the tagged particle jumps. In the latter case it is interpreted as the entire universe shifting around the tagged particle. This introduces the map $\omega \to \tilde{\tau}_\alpha \omega$

$$\begin{aligned}\eta(\tilde{\tau}_\alpha\omega, x) &= \eta(\omega, x + \alpha) \ \ if \ \ x \neq -\alpha, \\ &= \eta(\omega, \alpha) \ \ if \ \ x = -\alpha.\end{aligned}$$

Of course $\eta(\omega, 0) = 1$ always, and is irrelevent. The infinitesimal generator of the motion of the particles seen from the tagged particle is an evolution on

$$\Omega_0 = \{\omega : \eta(\omega, 0) = 1\}$$

and has the generator

$$L_0 u = \sum_{|a|=1} [u(\tilde{\tau}_a \omega) - u(\omega)](1 - \eta(a)) + \sum_{\{\alpha,\beta\},|\alpha-\beta|=1} [u(\tau_{\alpha,\beta}\omega) - u(\omega)].$$

The variational formula of [2] can be expressed in the following manner. For each vector $\theta \in R^d$,

$$\langle S(\rho)\theta, \theta\rangle = \inf_u E^{P_\rho}[\sum_{|a|=1} [u(\tilde{\tau}_a \omega) - u(\omega) - \langle a, \theta\rangle]^2 (1 - \eta(a))$$
$$+ \sum_{\{\alpha,\beta\},|\alpha-\beta|=1} [u(\tau_{\alpha,\beta}\omega) - u(\omega)]^2].$$

The aim of this chapter is to establish certain qualitative properties of $S(\rho)$ as a function of ρ. Some of these results are proved for all d and some only for $d \geq 3$.

2. Properties valid for all d

Let us remember that we have ruled out the case of the nearest neighbor walk in $d = 1$. In this case it is known that $S(\rho) \equiv 0$ for all $0 < \rho < 1$ and there is nothing else to say.

2.1. Lemma. *For all ρ, $0 \leq S(\rho) \leq (1 - \rho)S_0$.*

Proof. In the variational formula we can take $u = 0$ to obtain the above inequality. ∎

2.2. Lemma. *If $0 < \rho_1 < \rho_2 < 1$, we have*

$$\frac{S(\rho_1)}{(1 - \rho_1)} \geq \frac{S(\rho_2)}{(1 - \rho_2)}.$$

Before we establish the lemma we need a construction which is a coupling of the two equilibria P_{ρ_1} and P_{ρ_2}. Let us denote by $\delta = \rho_2 - \rho_1$ and consider two types of particles, red and green. At each site $x \neq 0$ we may have either a red or green particle or no particle. We let $r(x) = 1$ *or* 0 depending on whether there is a red particle at x or not. $g(x)$ is similarly defined for the gren particle. $\eta(x) = r(x) + g(x)$ indicates the presence or otherwise of any particle at x. P_{ρ_1,ρ_2} is a Bernoulli product measure where at each site with probability ρ_1 we have a red particle. With probability $\delta = \rho_2 - \rho_1$, we have a green particle and with the remaining probabilty

$1-\rho_2$ we have no particle. On the configuration space $\tilde{\Omega}_0$ we have this measure and if we consider only $\{\eta(x)\}$ we have the colorblind configuration with Bernoulli measure P_{ρ_2}.

Proof of Lemma 2.2. We first note that

$$\begin{aligned}\langle S(\rho_2)\theta,\theta\rangle &= \inf_u E^{P_{\rho_1,\rho_2}}\Big[\sum_{|a|=1}[u(\tilde{\tau}_\alpha\omega)-u(\omega)-\langle a,\theta\rangle]^2(1-\eta(a))\\ &\quad +\sum[u(\tau_{\alpha,\beta}\omega)-u(\omega)]^2\Big]. \blacksquare\end{aligned}$$

Although the infimum is carried out over all u's that depend on r and g, i.e., color as well as configuration since η is colorblind, we can replace any u by its conditional expectation given η and so the infimum may be taken over u depending only on η and, in such a case, P_{ρ_1,ρ_2} can be replaced by P_{ρ_2}. By a similar reasoning,

$$\begin{aligned}\langle S(\rho_1)\theta,\theta\rangle &= \inf_u E^{P_{\rho_1,\rho_2}}\Big[\sum_{|a|=1}[u(\tilde{\tau}_\alpha\omega)-u(\omega)-\langle a,\theta\rangle]^2(1-r(a))\\ &\quad +\sum[u(\tau_{\alpha,\beta}\omega)-u(\omega)]^2\Big]\end{aligned}$$

with the infimum being taken essentially over functions u depending on the red configuration $r(\cdot)$. If we look at the two variational formulae for $\langle S(\rho_2)\theta,\theta\rangle$ and $\langle S(\rho_1)\theta,\theta\rangle$, we observe certain things. The variational formula for $S(\rho_2)$ is carried over u's that are functions of $\eta(.)$. So, we may write

$$\begin{aligned}\langle S(\rho_2)\theta,\theta\rangle &= \inf_{u=u(\eta)} E^{P_{\rho_1,\rho_2}}\Big[\sum_{|a|=1}[D_{a,\theta}(u)(1-\eta(a))]\\ &\quad +\sum_{\alpha,\beta} Q_{\alpha,\beta}(u)[1_{\{\eta(\alpha)\neq\eta(\beta)\}}\Big].\end{aligned}$$

This is because unless $\eta(\alpha)\neq\eta(\beta)$, $Q_{\alpha,\beta}=0$. If we limit ourselves to functions $u=u(r(\cdot))$ of red configurations alone we will do worse. We can compute

$$\begin{aligned}E^{P_{\rho_1,\rho_2}}[(1-\eta(a))|r(.)] &= \frac{(1-\rho_2)}{(1-\rho_1})(1-r(a))\\ E^{P_{\rho_1,\rho_2}}[1_{\{r(\alpha)\neq r(\beta)\}}|r(\cdot)] &= \frac{1-\rho_2}{1-\rho_1}[1_{\{r(\alpha)\neq\ r(\beta)\}}]\\ &\quad +\frac{2\delta}{1-\rho_1}(1-r(\alpha))(1-r(\beta))].\end{aligned}$$

If $u = u(r(\cdot))$, then $Q_{\alpha,\beta}(u) = 0$ on the set $r(\alpha) = r(\beta)$. Therefore,

$$\langle S(\rho_2)\theta, \theta\rangle \leq \inf_{u=u(r(.))} [E^{P_{\rho_1,\rho_2}} \frac{(1-\rho_2)}{(1-\rho_1)} (\sum_a D_{a,\theta}(u)(1-r(a)) + Q_{\alpha,\beta}(u))]$$

$$= \frac{(1-\rho_2)}{(1-\rho_1)} \langle S(\rho_1)\theta, \theta\rangle.$$

2.3. Lemma. *There exists a constant C such that*

$$S_0 - S(\rho) \leq C\rho.$$

Proof. By expanding $D_{a,\theta}(u)$, we obtain

$$D_{a,\theta}(u) = \langle a, \theta\rangle^2 - 2(u(\tilde{\tau}_a\omega) - u(\omega))\langle a, \theta\rangle + (u(\tilde{\tau}_a\omega) - u(\omega))^2,$$

$$E^{P_\rho}[(u(\tilde{\tau}_a\omega) - u(\omega))(1 - \eta(a))\langle a, \theta\rangle] = E^{P_\rho}[u(\omega)\langle a, \theta\rangle(\eta(a) - \eta(-a))].$$

If we now define $l(a) = \langle a, \theta\rangle\ (\eta(a) - \eta(-a))$ and $D_a(u) = (u(\tilde{\tau}_a\omega - u(\omega))^2$, we have

$$\langle S(\rho)\theta, \theta\rangle = (1-\rho)\langle S_0\theta, \theta\rangle - \sup_u [E^{P_\rho}\{2u(\omega) \sum_{|a|=1} l(a)\}$$
$$- E^{P_\rho}\{\sum D_a(u) + \sum_{\{\alpha,\beta\}\neq 0} Q_{\alpha,\beta}(u)\}]$$
$$\geq (1-\rho)\langle S_0\theta, \theta\rangle - \sup_u [E^{P_\rho}\{2u(\omega) \sum_{|a|=1} l(a)$$
$$- \sum Q_{\alpha,\beta}(u)\}]. \blacksquare$$

We can estimate

$$E^{P_\rho}\lambda[\eta(a) - \eta(-a)]u = \frac{1}{2} E^{P_\rho}\lambda[\eta(a) - \eta(-a)][u(\tau_{a,-a}\omega) - u(\omega)].$$

Because there is a path from "a" to "$-a$" that avoids 0, we have

$$|E^{P_\rho}\lambda(\eta(a) - \eta(-a))u| \leq C|\lambda|[E^{P_\rho}(\eta(a) - \eta(-a))^2]^{\frac{1}{2}} [E^{P_\rho} \sum Q_{\alpha,\beta}(u)]^{\frac{1}{2}},$$

or

$$S(\rho) \geq (1-\rho)S_0 - C\rho(1-\rho).$$

2.4. Lemma. *There exists a constant C such that*

$$S(\rho) \geq C(1-\rho).$$

Proof. The lower bound that we get from Lemma 2.3 is of the form

$$\langle S(\rho)\theta, \theta\rangle \geq (1-\rho)\langle S_0\theta, \theta\rangle - \sup_u [E^{P_\rho} l(a)u(\omega) - E^{P_\rho}\{T_1 + T_2\}].$$

Here T_1 and T_2 are two types of Dirichlet forms involving, respectively, the tagged particle and other particles. From the definition

$$\sup_u E^{P_\rho}[ul(a) - T_1] \leq (1-\rho)\langle S_0\theta, \theta\rangle$$

and from what we did in Lemma 2.3,

$$\sup_u E^{P_\rho}[ul(a) - T_2] \leq C\rho(1-\rho)\langle \theta, \theta\rangle.$$

It is now elementary to split $ul(a)$ optimally into two pieces, balance it against T_1 and T_2 and get

$$\sup_u E^{P_\rho}[ul(a) - T_1 - T_2] \leq \alpha(1-\rho)\langle S_0\theta, \theta\rangle,$$

where α is some constant $\alpha < 1$. ∎

3. Results for $d \geq 3$

We look at configurations on $Z^d - \{0\}$. We have the Dirichlet form

$$D_\rho(u) = E^{P_\rho}[\sum Q_{\alpha,\beta}(u)].$$

Let us denote by σ_x the flip operator at x,

$$\begin{aligned} \eta(\sigma_x\omega, y) &= \eta(y) \quad for \quad y \neq x \\ &= 1 - \eta(x) \quad for \quad y = x. \end{aligned}$$

The next lemma follows from the results of the appendix.

3.1. Lemma. *There exists a constant C independent of ρ such that for all u and x,*

$$\begin{aligned} E^{P_\rho}[|u(\sigma_x\omega) - u(\omega)|^2(1-\eta(x))] &\leq \frac{C}{\rho} D_\rho(u) \\ E^{P_\rho}[|u(\sigma_x\omega) - u(\omega)|^2\eta(x)] &\leq \frac{C}{1-\rho} D_\rho(u) \\ \text{and} \qquad E^{P_\rho}[|u(\sigma_x\omega) - u(\omega)|^2] &\leq \frac{C}{\rho(1-\rho)} D_\rho(u). \end{aligned}$$

Finally we turn to the main result on the Lipschitz continuity of S(ρ). We take $0 < \rho_1 < \rho_2 < 1$ and denote by $\delta = \rho_2 - \rho_1$,

$$\begin{aligned}\langle S(\rho_1)\theta, \theta\rangle &= \inf_u E^{P_{\rho_1}}[\sum |u(\tilde{\tau}_a\omega) - u(\omega) - \langle a, \theta\rangle|^2(1-\eta(a)) \\ &\quad + \sum |u(\tau_{\alpha,\beta}\omega) - u(\omega)|^2] \\ &= \inf_{u=u(r(.))} E^{P_{\rho_1,\rho_2}}[\sum D_{a,\theta}(u)(1-r(a)) + \sum Q_{\alpha,\beta}(u)] \\ &= \inf_{u=u(r(.),g(.))} E^{P_{\rho_1,\rho_2}}[\sum D_{a,\theta}(u)(1-r(a)) + \sum Q_{\alpha,\beta}(u)] \\ &\le \inf_{u=u(\eta(.))} E^{P_{\rho_1,\rho_2}}[\sum D_{a,\theta}(u)(1-r(a)) + \sum Q_{\alpha,\beta}(u)].\end{aligned}$$

We can compute

$$E^{P_{\rho_1,\rho_2}}[1 - r(a)|\eta(.)] = (1-\eta(a)) + \frac{\delta}{\rho_2}\eta(a).$$

Continuing on,

$$\begin{aligned}\langle S(\rho_1)\theta, \theta\rangle \le &\inf_{u=u(\eta(.))} E^{P_{\rho_1,\rho_2}}[\sum D_{a,\theta}(u)(1-\eta(a)) + \sum Q_{\alpha,\beta}(u) \\ &+ \frac{\delta}{\rho_2}\sum D_{a,\theta}(u)\eta(a)].\end{aligned}$$

We estimate each of the three terms on the right-hand side of the inequality

$$\begin{aligned}E^{P_{\rho_1,\rho_2}}[\sum D_{a,\theta}(u)\eta(a)] &= E^{P_{\rho_2}}[\sum D_{a,\theta}(u)\eta(a)] \\ &\le 3E^{P_{\rho_2}}[\sum[|u(\sigma_a\omega) - u(\omega)|^2 \\ &\quad + |u(\tilde{\tau}_a\sigma_a\omega) - u(\sigma_a\omega) - \langle a, \theta\rangle|^2 \\ &\quad + |u(\sigma_{-a}\tilde{\tau}_a\sigma_a\omega) - u(\tilde{\tau}_a\sigma_a\omega)|^2]\eta(a)].\end{aligned}$$

From Lemma 3.1, $E^{P_{\rho_2}}[|u(\sigma_a\omega) - u(\omega)|^2\eta(a)] \le \frac{C}{(1-\rho_2)}D_{\rho_2}(u)$ and

$$\begin{aligned}&E^{P_{\rho_2}}|u(\sigma_{-a}\tilde{\tau}_a\sigma_a\omega) - u(\tilde{\tau}_a\sigma_a\omega)|^2\eta(a) \\ &\qquad = \frac{\rho_2}{(1-\rho_2)}E^{P_{\rho_2}}|u(\sigma_{-a}\omega) - u(\omega)|^2(1-\eta(a)) \\ &\qquad \le \frac{\rho_2}{(1-\rho_2)}\frac{C}{\rho_2}D_{\rho_2}(u) \\ &\qquad = \frac{C}{(1-\rho_2)}D_{\rho_2}(u).\end{aligned}$$

And in addition,

$$E^{P_{\rho_2}}[|u(\tilde{\tau}_a\sigma_a\omega) - u(\sigma_a\omega) - \langle a, \theta\rangle|^2\eta(a)] = \frac{\rho_2}{(1-\rho_2)}E^{P\rho_2}[D_{a,\theta}(u)(1-\eta(a))].$$

Finally,

$$\langle S(\rho_1)\theta,\theta\rangle \le \inf_u\{E^{P_{\rho_2}}[\sum D_{a,\theta}(u)(1-\eta(a))+\sum Q_{\alpha,\beta}(u)] + \frac{\delta}{\rho_2}[\frac{C}{(1-\rho_2)}E^{P_{\rho_2}}\sum Q_{\alpha,\beta}(u) + \frac{\rho_2}{(1-\rho_2)}E^{P_{\rho_2}}[\sum D_{a,\theta}(u)(1-\eta(a))]\}.$$

We now conclude that

$$\langle S(\rho_1)\theta,\theta\rangle \le \langle S(\rho_2)\theta,\theta\rangle + \frac{C\delta}{\rho_2(1-\rho_2)}[(1-\rho_2)\langle(S_0-S(\rho_2))\theta,\theta\rangle] + \frac{\delta}{(1-\rho_2)}\langle S(\rho_2)\theta,\theta\rangle.$$

Finally we have,

$$\frac{S(\rho_1)-S(\rho_2)}{\delta} \le \frac{C}{\rho_2(1-\rho_2)}[(1-\rho_2)][S_0-S(\rho_2)] + \frac{S(\rho_2)}{(1-\rho_2)}$$

This immediately implies our main theorem.

3.4. Theorem. *If $d \ge 3$, the self-diffusion matrix $S(\rho)$ is a Lipschitz function of the density ρ uniformly in the interval $0 \le \rho \le 1$.*

4. Appendix

In this section we shall prove the basic inequalities that lead to the estimate in Lemma 3.1.

4.1. Lemma. *Let $f(x) : Z^d \to R$ be a real valued function on Z^d for some $d \ge 3$, such that*

$$\sum_{x\in Z^d} |f(x)|^2 = C < \infty.$$

Then for any $x_0 \in Z^d$, there exists a sequence $x_k \in Z^d$ such that x_k are distinct, $|x_k| \to \infty$ $|x_k - x_{k+1}| = 1$ for $k \ge 0$ and

$$\sum_{k=0}^{\infty} |f(x_k)| \le A\sqrt{C},$$

where A is a constant depending only on d.

Proof. Suppose we can constuct a possibly correlated random walk $\{S_n\}$ on Z^d such that $S_0 = x_0$, $|S_n - S_{n+1}| = 1$ and

$$G(x_0, x) = \sum_{n=0}^{\infty} P[S_n = x]$$

satisfies

$$\sum_{x \in Z^d} |G(x_0, x)|^2 \leq A^2.$$

Then

$$E[\sum_{k=0}^{\infty} |f(S_k)|] = \sum_{x} G(x_0, x)|f(x)|$$
$$\leq A\sqrt{C}.$$

There is clearly a random path for which

$$\sum_{k=0}^{\infty} |f(S_n)| \leq A\sqrt{C}.$$

We can erase the loops and we have the lemma for the resultant self-avoiding walk. We can assume, without loss of generality, that $x_0 = 0$. The usual symmetric random walk does not quite make it.

$$G(0, x) \sim ||x||^{d-2} \; in \; Z^d$$

and if $d = 3$,

$$\sum_{x} |G(0, x)|^2 = \infty.$$

We could take an asymmetric random walk with

$$P(X_k = e_i) = \frac{1}{2d} + \frac{\theta_i}{2}, \qquad P(X_k = -e_i) = \frac{1}{2d} - \frac{\theta_i}{2},$$

where θ_i, $1 \leq i \leq d$ are numbers between $-\frac{1}{2d}$ and $\frac{1}{2d}$. The Green's function has the Fourier transform $[1 - \psi(\theta, t)]^{-1}$ where

$$\psi(\theta, t) = \frac{1}{d}[\sum \cos t_j] + i \sum \theta_j t_j.$$

If we pick $\{\theta_i\}$ to be random and uniformly distributed over the sphere in R^d of radius $\frac{1}{4d}$, we can compute explicitly the Fourier transform of the Green's function and show that its singularity at 0 is square integrable. ∎

Remark. We can delete from Z^d the origin and insist that all our paths avoid the origin. Since there is a uniform positive lower bound that all our random walks avoid the origin, we have exactly the same type of bounds with a somewhat larger constant.

4.2. Lemma. *Let $D_\rho(u)$ be the Dirichlet form*

$$D_\rho(u) = E^{P_\rho}[\sum_{\{\alpha,\beta\}\neq 0; |\alpha-\beta|=1} |u(\tau_{\alpha,\beta}\omega) - u(\omega)|^2],$$

where P_ρ is the Bernoulli product measure on $Z^d - \{0\}$ with density ρ. Let σ_a be the flip at site "a". There is a constant A independent of ρ such that, for all functions u,

$$\int |u(\sigma_a\omega) - u(\omega)|^2 \eta(a) dP_\rho \leq \frac{A}{1-\rho} D_\rho(u)$$

and

$$\int |u(\sigma_a\omega) - u(\omega)|^2 (1-\eta(a)) dP_\rho \leq \frac{A}{\rho} D_\rho(u).$$

Proof. For each site x let us define

$$f(x) = \sum_{|\alpha-x|=1} ||u(\tau_{x,\alpha}(.) - u(.)||_2.$$

Then,

$$\sum_{x\neq 0} |f(x)|^2 \leq C_1 D_\rho(u).$$

According to Lemma 4.1 along some path $x_k \to \infty$,

$$\sum_k f(x_k) \leq C_2 \sqrt{D_\rho(u)}.$$

In particular,

$$\sum ||u(\tau_{x_j,x_{j+1}}(.) - u(.)||_2 \leq C_2 \sqrt{D_\rho(u)}.$$

Therefore,

$$\sup_n ||u(\tau_{x_0,x_n}.) - u(.)||_2 \leq C_2 \sqrt{D_\rho(u)}.$$

An elementary calculation yields

$$\lim_{n\to\infty} ||u(\tau_{x_0,x_n}.) - u(.)||_2 = \rho(1-\rho) \int [u(\sigma_a\omega) - u(\omega)]^2 dP_\rho.$$

Hence,

$$\int |u(\sigma_a\omega) - u(\omega)|^2 dP_\rho \leq \frac{C}{\rho(1-\rho)} D_\rho(u).$$

It follows in a similar fashion that,

$$\int |u(\sigma_a\omega) - u(\omega)|^2 \eta(a) dP_\rho \leq \frac{C}{1-\rho} D_\rho(u)$$

and

$$\int |u(\sigma_a \omega) - u(\omega)^2 (1 - \eta(a)) dP_\rho \leq \frac{C}{\rho} D_\rho(u) \blacksquare.$$

References

[1] C. Kipnis and S.R.S. Varadhan, Central limit theorem for additive functionals of reversible Markov processes and application to simple exclusions, *Comm. Math. Phys.* Vol. 104, (1986), 1–19.

[2] J. Quastel, Diffusion of color in simple exclusion process, *Comm. Pure Appl. Math.* Vol. XLV (1992), 623–680.

[3] H. Spohn, *Large Scale Dynamics of Interacting Particles*, Springer-Verlag, Berlin, 1991.

Proceedings of the International Congress of Mathematicians, Vol. 1 (Zürich, 1994), Birkhaüser, 1995, 196-208

Entropy Methods in Hydrodynamic Scaling

S. R. S. VARADHAN*

Courant Institute
251 Mercer Street
New York, NY 10012, USA

1 Introduction

Hydrodynamic scaling is a procedure that attempts rigorously to derive large scale behavior of complex interacting systems from laws governing its evolution that are specified at a smaller scale. The procedure involves statistical averaging over the small scales and can be viewed as part of nonequilibrium statistical mechanics.

The basic example is the classical problem of starting with a Hamiltonian system of interacting particles and deriving from it, after rescaling, the Euler equations of compressible gas dynamics.

Let us consider a Hamiltonian system of N particles moving in a rather large physical space, for instance, the 3-dimensional torus of size ℓ. These particles are governed by a pair interaction $V(x-y)$ between particles. $V(\cdot)$ is an even function that is nonnegative and has support in a fixed compact set independent of ℓ. N and ℓ will be large with $N = \ell^3$ so that the interparticle distance is of order 1 and each particle will typically see only a few particles at any given time. The phase space is $(T_\ell^3 \times R^3)^N$ and the equations of motion for the positions and velocities $(x_i(t), u_i(t))$ are

$$\begin{cases} \dfrac{dx_i^\alpha(t)}{dt} = \dfrac{\partial H}{\partial u_i^\alpha} \\ \dfrac{du_i^\alpha(t)}{dt} = -\dfrac{\partial H}{\partial x_i^\alpha} \end{cases} \tag{1.1}$$

Here i is the particle number, $1 \le i \le N$, and α is the coordinate index, $\alpha = 1, 2, 3$. The Hamiltonian $H(x_1, \ldots, x_N; u_1, \ldots, u_N)$ is, of course, given by

$$H(x;u) = \frac{1}{2}\sum_i \|u_i\|^2 + \frac{1}{2}\sum_{i \ne j} V(x_i - x_j). \tag{1.2}$$

The system has five conserved quantities: the number of particles N, the momenta $\sum u_i^\alpha$, and the total energy H. Suppose we rescale the torus to have size 1 and rescale time by a similar factor ℓ, then quantities of the form

$$\frac{1}{N}\sum J\left(\frac{x_i(\ell t)}{\ell}\right) \tag{1.3}$$

*) Supported by NSF grant DMS-9201222 and ARO grant DAAL03-92-G-0317.

$$\frac{1}{N}\sum J\left(\frac{x_i(\ell t)}{\ell}\right) u_i^\alpha(\ell t) \tag{1.4}$$

and

$$\frac{1}{N}\sum J\left(\frac{x_i(\ell t)}{\ell}\right) e_i(\ell t) \tag{1.5}$$

where

$$e_i = \frac{1}{2}\|u_i\|^2 + \frac{1}{2}\sum_{j:j\neq i} V(x_i - x_j)$$

change in a reasonable manner in t. As N and $\ell \to \infty$ in the manner specified, one should think of (1.3), (1.4), and (1.5) as converging to

$$\int J(y)\rho(y,\tau)\,dy \tag{1.6}$$

$$\int J(y)v^\alpha(y,\tau)\rho(y,\tau)\,dy \tag{1.7}$$

and

$$\int J(y)e(y,\tau)\rho(y,\tau)\,dy. \tag{1.8}$$

Here y and τ are rescaled space and time. $\rho(y,\tau)$ is the density at a given point of space time. $\{v^\alpha(y,\tau)\}$ are the local fluid velocities and $e(y,\tau)$ is the energy density, related to the temperature at a given point. The equations of gas dynamics in this context are a system of symmetric hyperbolic conservation laws that one can write down for the evolution of ρ, v^α, and e. These equations are somewhat different from the usual Euler equation one derives from the Boltzmann equation. In the Boltzmann limit the real density is small; thus, the Euler equation derived from it is linearization in ρ of our equations.

This classical model is deterministic and any randomness has to be in the initial configuration. A precise formulation of the problem has to be done carefully. The goal is to establish some rigorous connection between the Hamiltonian equations on one hand and the Euler equations on the other. Randomness is important because some averaging has to be done with respect to small scales and one needs some information as to how the particles will arrange themselves locally in phase space if we only know their local density, local average velocity, and local temperature. One expects the arrangement to be given by an appropriate Maxwell-Gibbs distribution and formally the equations are derived under that assumption. To justify it, at the least, one needs a reasonable ergodic theory and for that more noise is better.

We will first look at two other models where the evolution is stochastic, say something about these models, and return at the end to our classical model.

The next example is referred to as simple exclusion. The physical space will be the periodic d-dimensional integral lattice of period ℓ. After scaling by a factor of ℓ, this will be viewed as living inside the unit d-torus. We will have a certain number $N = \rho\ell^d$ of particles located at some of the lattice sites. Each site can have at most one particle. A particle at the lattice site x waits for a random exponential waiting time with mean 1 and then picks a random new site x' to which it tries to jump. If the new site already has a particle, then the jump cannot be completed and our particle waits again for a new exponential random time. If the site x' is free, it jumps and starts afresh. The choice of x' is made with probability $\pi(x'-x)$. The probability distribution $\pi(z)$ is of jump sizes, is independent of ℓ, and is assumed to have finite support. All the particles are doing this at the same time and independently of each other. Because of continuous time there will be no ties to resolve.

The only conserved quantity here is the number of particles and we look at the density. We want to study the behavior of

$$\frac{1}{\ell^d}\sum_{I=1}^{M} J\left(\frac{x_i(\tau)}{\ell}\right) \tag{1.9}$$

where $\tau=\ell^\alpha t$. The rescaling of time can be with either $\alpha=1$ or 2.

The mean $m=\sum z\pi(z)$ plays an important role. If $m\neq 0$, the motion is convective and one has to take $\alpha=1$. If we think of (1.9) as converging to

$$\int J(y)\rho(\tau,y)\,dy \tag{1.10}$$

then ρ is expected to satisfy the scalar conservation law

$$\frac{\partial\rho}{\partial\rho}+\nabla\cdot m\rho(1-\rho)=0. \tag{1.11}$$

If $m=0$, then we need to take $\alpha=2$. The scaling is diffusive and one then expects ρ to satisfy a nonlinear diffusion equation

$$\frac{\partial\rho}{\partial\tau}=\frac{1}{2}\nabla\cdot a(\rho(\tau,y)\nabla\rho). \tag{1.12}$$

If $\pi(z)=\pi(-z)$; i.e., π is symmetric, then the problem is rather easy and, in fact, one can verify that

$$a(\rho)\equiv a=\sum z\otimes z\;\pi(z) \tag{1.13}$$

is a constant matrix and is, in fact, the covariance matrix of $\pi(\cdot)$. The limiting equation in this case is the linear heat equation.

The third example we will consider is referred to as the Ginzburg-Landau model and is a lattice field model. We again start with the periodic lattice of size ℓ in d dimensions. We scale it down and think of it inside the unit torus. At each lattice site x, we have a variable $\xi(x,t)$ that is real valued and changes in time. The collection $\{\xi(x,t)\}$ is an ℓ^d-dimensional diffusion process and can be described either through a set of stochastic differential equations or though its infinitesimal generator. We will do the former. If $e_1,\ldots,e_d$ are the d positive coordinate directions, then $x\pm e_i$ are the neighbors of x.

$$d\xi(x,t)=\sum_{i=1}^{d}[d\eta(x-e_i,x,t)-d\eta(x,x+e_i,t)] \tag{1.14}$$

$$d\eta(x, x+e_i, t) = \frac{1}{2}\left[\phi'(\xi(x,t)) - \phi'(\xi(x+e_i,t))\right] dt + d\beta_{x,x+e_i}(t) \tag{1.15}$$

The equation (1.14) tells us that the way $\xi(x,t)$ changes is by "stuff" coming in or going out along the bonds. We orient the bonds using the positive coordinate directions and the net change is an algebraic sum. Equation (1.15) tells us that the flow along a bond is proportional to some nonlinear gradient modified by white noise. Here $\phi'(\xi)$ is a nonlinear function that satisfies some natural assumptions. We use ϕ', the derivative of $\phi(\xi)$, for convenience. If we again take $\tau = \ell^2 t$ and consider

$$\frac{1}{\ell^d}\sum J\left(\frac{x}{\ell}\right)\xi(x,\tau) \tag{1.16}$$

as an approximation of

$$\int J(y)m(y,\tau)\,dy \tag{1.17}$$

where $m(y,\tau)$ is the limiting "density" of "stuff", then $m(\cdot,\cdot)$ is supposed to satisfy an equation of the form

$$\frac{\partial m}{\partial t} = \frac{1}{2}\Delta\lambda(m)$$

where $\lambda(m)$ is to be determined in terms of ϕ.

We shall look at our three examples in some detail in our next three sections and end with some comments.

2 Ginzburg-Landau Model

It is convenient to write down the infinitesimal generator of our diffusion process on R^{ℓ^d}.

$$\begin{aligned}(\mathcal{L}_\ell F) =& \frac{\ell^2}{2}\sum_{i,x}\left(\frac{\partial}{\partial\xi(x+e_i)} - \frac{\partial}{\partial\xi(x)}\right)^2 F \\ &- \frac{\ell^2}{2}\sum_{i,x}\left[\phi'(\xi(x+e_i)) - \phi'(\xi(x))\right]\left[\frac{\partial}{\partial\xi(x+e_i)} - \frac{\partial}{\partial\xi(x)}\right] F.\end{aligned} \tag{2.1}$$

The factor ℓ^2 appears due to rescaling of time. The object we want to study is

$$G(t) = \frac{1}{\ell^d}\sum J\left(\frac{x}{\ell}\right)\xi(x,t). \tag{2.2}$$

Using the stochastic differential equations one can write

$$dG(t) = A(t)dt + dM(t) \tag{2.3}$$

where

$$\begin{aligned}A(t) &= \mathcal{L}_\ell\left[\frac{1}{\ell^d}\sum J\left(\frac{x}{\ell}\right)\xi(x)\right](t) \\ &\simeq \frac{1}{2\ell^d}\sum(\Delta J)\left(\frac{x}{\ell}\right)\phi'(\xi(x,t))\end{aligned} \tag{2.4}$$

and

$$dM(t) \simeq 0$$

because the Brownian noise terms cancel each other out by averaging. The difficulty now, which is typical of all of these problems, is that due to the nonlinearity of ϕ' the equations do not close and one has to represent

$$\frac{1}{\ell^d}\sum(\Delta J)\left(\frac{x}{\ell}\right)\phi'(\xi(x,t)) \tag{2.5}$$

in terms of $m(x,t)$ as $\ell \to \infty$. These are, of course, defined by (1.16) and (1.17). These require a knowledge of how $\xi(x,t)$ are "distributed" for a given value of "m". If we know this, then the volume average of ϕ' can be replaced by a mean value of ϕ' for a given m.

The evolution governed by $\mathcal{L}_\ell$ is reversible or symmetric with respect to the weight

$$e^{-\sum_x \phi(\xi(x))} = \Phi_\ell(\xi). \tag{2.6}$$

(We normalize ϕ so that $\int e^{-\phi(\xi)}\, d\xi = 1$.) However, the process is not ergodic because

$$m = \frac{1}{\ell^d}\sum_x \xi(x) \tag{2.7}$$

is conserved under our evolution. The conditional distribution $\nu_{m,\ell}(d\xi)$ of $\{\xi(x)\}$, given the average m, are the invariant ergodic pieces. By an "equivalence of ensembles" type theorem we can show

$$\nu_{m,\ell}(d\xi) \to \prod_x e^{-\phi_m(\xi(x))} d\xi(x)$$

and

$$\phi_m(\xi) = \phi(\xi) - \lambda(m)\xi + p(\lambda(m)).$$

Here

$$p(\lambda) = \log \int \exp\left[\lambda\xi - \phi(\xi)\right] d\xi$$

and $\lambda(m)$ solves

$$p'(\lambda(m)) = m$$

and equals

$$\lambda(m) = \frac{d}{dm}\sup_\lambda\left[\lambda m - p(\lambda)\right].$$

One finally verifies that

$$\int \phi'(\xi) e^{-\phi_m(\xi)} d\xi = \lambda(m).$$

This yields our equation

$$\frac{\partial m}{\partial t} = \frac{1}{2}\Delta\lambda(m). \tag{2.8}$$

In order to justify this one has to prove that averages of the form

$$\frac{1}{|B|}\sum_{x\in B}\phi'(\xi(x))$$

are nearly equal to

$$\lambda\left(\frac{1}{|B|}\sum_{x\in B}\xi(x)\right)$$

most of the time with probability nearly 1. The size of the block is important. It should be of size $\ell\epsilon$ with $\epsilon << 1$ but fixed. In [5] with Guo and Papanicolaou we developed a method for handling the problem by using entropy and entropy production as tools. We assume that initially the field $\{\xi(x,0)\}$ is random and is given by a density $f_\ell^0(\xi)\Phi_\ell(\xi)$ satisfying an entropy bound

$$\int f_\ell^0(\xi)\log f_\ell^0(\xi)\Phi_\ell(\xi)\,d\xi \leq C\ell^d. \tag{2.9}$$

Such a bound is natural and is satisfied in most cases because one can think of C as the bound for average entropy per site. Then the distribution of $\{\xi(x,t)\}$ will have a density $f_\ell^t(\xi)$ satisfying

$$H_\ell(t) = \int f_\ell^t(\xi)\log f_\ell^t(\xi)\Phi_\ell(\xi)\,d\xi \leq H_\ell(0)$$

and, in fact,

$$\frac{dH_\ell(t)}{dt} = -\frac{\ell^2}{2}\int\sum_{i,x}\left(\frac{\partial f_\ell^t}{\partial\xi(x+e_i)} - \frac{\partial f_\ell^t}{\partial\xi(x)}\right)^2\frac{1}{f_\ell^t}\Phi_\ell(\xi)\,d\xi\,dt. \tag{2.10}$$

Because $H_\ell(t) \geq 0$ one gets a trivial bound

$$\int_0^\infty\sum_{i,x}\int\left(\frac{\partial f_\ell^t}{\partial\xi(x+e_i)} - \frac{\partial f_\ell^t}{\partial\xi(x)}\right)^2\frac{1}{f_\ell^t}\Phi_\ell(\xi)\,d\xi\,dt \leq 2C\ell^{d-2}.$$

If we fix a finite time T and consider

$$\bar{f}_\ell = \frac{1}{T}\int_0^T f_\ell^t\,dt,$$

by convexity

$$\int\sum_{i,x}\left(\frac{\partial\bar{f}_\ell}{\partial\xi(x+e_i)} - \frac{\partial\bar{f}_\ell}{\partial\xi(x)}\right)^2\frac{1}{\bar{f}_\ell}\Phi_\ell(\xi)\,d\xi \leq C_T\ell^{d-2}.$$

We showed in [5] that the above estimate was sufficient to justify the averaging lemma and establish equation (2.8) rigorously. The following theorem was proved.

Assume $f_\ell^0(\xi)$ satisfies the bound (2.8). Assume further that for some density $m_0(y)$ and all smooth test functions $J(\cdot)$

$$\frac{1}{\ell^d}\sum J\left(\frac{x}{\ell}\right)\xi(x) \to \int J(y)m_0(y)\,dy$$

in probability under $f_\ell^0(\xi)\Phi_\ell(\xi)d\xi$. Then for any $t>0$

$$\frac{1}{\ell^d}\sum J\left(\frac{x}{\ell}\right)\xi(x,t) \to \int J(y)m(t,y)\,dy$$

in probability, where $m(t,y)$ solves (2.8) with initial condition $m_0(y)$.

In [14] a modified approach was developed by Yau that has much wider applicability. One could guess that the density at time t should look like

$$g_\ell^t(\xi)\Phi_\ell(\xi) = \frac{1}{Z}\exp\left[\sum\lambda\left(t,\frac{x}{\ell}\right)\xi(x)\right]\Phi_\ell(\xi)$$

where $Z=\exp[\sum p\left(t,\frac{x}{\ell}\right)]$ and $\lambda(t,x)=\lambda(m(t,x))$ and m solves (2.8). In general, the density is $\Phi_\ell(\xi)f_\ell^t(\xi)$ where $f_\ell^t(\xi)$ is obtained by solving the Fokker-Planck equation and is not g_ℓ^t, even if we start off with $f_\ell^0(\xi)=g_\ell^0(\xi)$.

The question is: How close are they?

If we define the specific entropy $s(f_\ell^t,g_\ell^t)$ by

$$s(f_\ell^t,g_\ell^t) = \frac{1}{\ell^d}\int f_\ell^t(\xi)\log\frac{f_\ell^t(\xi)}{g_\ell^t(\xi)}\Phi_\ell(\xi)\,d\xi, \tag{2.11}$$

the theorem of Yau in [14] is that uniformly in $0\le t\le T$,

$$\lim_{\ell\to\infty} s(f_\ell^t,g_\ell^t)=0.$$

This is enough to justify the hydrodynamic scaling and arrive at the same theorem as in [5]. One assumes more initially but one obtains a stronger conclusion.

The model that we have studied is very special because it is a "gradient" model. While computing

$$dG(t) = d\left(\frac{1}{\ell^d}\sum J\left(\frac{x}{\ell}\right)\xi(x,t)\right)$$

we obtained

$$\begin{aligned}
&\mathcal{L}_\ell\left(\frac{1}{\ell^d}\sum J\left(\frac{x}{\ell}\right)\xi(x)\right)\\
&= -\frac{1}{2}\frac{\ell^2}{\ell^d}\sum\left[J\left(\frac{x+e_i}{\ell}\right)-J\left(\frac{x}{\ell}\right)\right][\phi'(\xi(x+e_i))-\phi'(\xi(x))]\cdot\\
&\simeq \frac{1}{2\ell^d}\sum(\Delta J)\left(\frac{x}{\ell}\right)\phi'(\xi(x))
\end{aligned}$$

We were able to carry out summation by parts twice. The local "flux" turned out to be

$$\phi'(\xi(x+e)) - \phi'(\xi(x)),$$

which is a gradient and is amenable to another summation by parts. With that the global flux turned out to be

$$\frac{1}{2\ell^d}\sum(\Delta J)\left(\frac{x}{\ell}\right)\xi(x),$$

which is an order 1 quantity.

This is not true in general. To see this let us modify our evolution somewhat. Our operator corresponds to the following Dirichlet form relative to the weight $\Phi_\ell(\xi)$

$$\mathcal{D}_\ell(F) = \frac{\ell^2}{2}\int\sum_{x,i}\left(\frac{\partial F}{\partial\xi(x+e_i)} - \frac{\partial F}{\partial\xi(x)}\right)^2\Phi_\ell(\xi)d\xi.$$

We modify it by

$$\widetilde{\mathcal{D}}_\ell(F) = \frac{\ell^2}{2}\int\sum_{x,i}a(\xi(x+e_i),\xi(x))\left(\frac{\partial F}{\partial\xi(x+e_i)} - \frac{\partial F}{\partial\xi(x)}\right)^2\Phi_\ell(\xi)\,d\xi$$

where $a(\xi_1,\xi_2)$ is a smooth positive function bounded above and below. For simplicity, let us take $d=1$. The operator $\widetilde{\mathcal{L}}_\ell$ is given by

$$\begin{aligned}\widetilde{\mathcal{L}}_\ell F =& \frac{\ell^2}{2}\sum a\left(\xi(x+1),\xi(x)\right)\left(\frac{\partial}{\partial\xi(x+1)} - \frac{\partial}{\partial\xi(x)}\right)^2 F\\ &-\frac{\ell^2}{2}\sum W(\xi(a+1),\xi(x))\left(\frac{\partial}{\partial\xi(x+1)} - \frac{\partial}{\xi(x)}\right)F\end{aligned}$$

where

$$\begin{aligned}W(\xi_1,\xi_2) =& a(\xi_1,\xi_2)\left(\phi'(\xi_1)-\phi'(\xi_2)\right)\\ &+a_2(\xi_1,\xi_2) - a_1(\xi_1,\xi_2)\end{aligned}$$

and

$$a_i(\xi_1,\xi_2) = \frac{\partial}{\partial\xi_i}a(\xi_1,\xi_2)$$

for $i=1,2$.

Now, if we compute

$$\widetilde{\mathcal{L}}_\ell\left(\frac{1}{\ell}\sum J\left(\frac{x}{\ell}\right)\xi(x)\right)$$

we get

$$-\frac{1}{2}\sum J'\left(\frac{x}{\ell}\right)W\left(\xi(x+1),\xi(x)\right). \tag{2.12}$$

The term in (2.12) is a big term. The mean value of W is zero for any given value of m and so averaging produces a meaningless product of zero and infinity.

In [12] we developed a method for handling problems of this kind. We showed that there is a function $D(m)$ such that

$$W(\xi_1,\xi_2) - D(m)\,[\phi'(\xi_1) - \phi'(\xi_2)]$$

is negligible in a somewhat complicated but precise sense. With this modification the correct diffusion equation is

$$\begin{aligned}\frac{\partial m}{\partial t} &= \frac{1}{2}\frac{\partial}{\partial y}D(m(t,y))\frac{\partial}{\partial y}\lambda(m(t,y))\\ &= \frac{1}{2}\frac{\partial}{\partial y}\widetilde{D}(m(t,y))\frac{\partial m}{\partial y}\\ \widetilde{D}(m) &= D(m)\lambda'(m).\end{aligned}$$

For $D(m)$ we provided a variational formula that replaces the traditional Green-Kubo formula. In the Green-Kubo formula the diffusion coefficient $D(m)$ is a space-time integral (sum) of the autocorrelation function in equilibrium, and this is not convenient. Our variational formula is much more convenient and is given explicitly by

$$D(m) = \inf_f E_m\{a(\xi_1,\xi_2)(1-G_f)^2\}.$$

Here E_m refers to the expectation with respect to the infinite product measure

$$\exp[-\sum \phi_m(\xi(x))]\prod d\xi_x$$

and the infimum is taken over all tame test functions depending on a finite number of coordinates. For each such f, G_f is the gradient defined by

$$(G_f(\xi)) = \sum_x (\frac{\partial}{\partial\xi(2)} - \frac{\partial}{\partial\xi(1)})\tau_x f$$

and τ_x is the canonical shift operator on the product space.

In principle one could try to carry out the method of Yau in [14] and one would have to work with a trial function of the form

$$g_N^t(\xi) \simeq \exp\left[\sum\lambda\left(t,\frac{x}{\ell}\right)\xi(x) + \frac{1}{\ell}\sum_x \psi(t,x,\xi)\right]$$

for $g_N^t(\xi)$. One has to choose a suitable corrector ψ in order to carry out the analysis. This has been done very recently by Funaki, Uchiyama, and Yau.

3 Simple Exclusion

We will now return to simple exclusion models. The state of the system can be described by a configuration $\eta = \{\eta(x)\}$, $x \in Z_\ell$, where Z_ℓ is the lattice of integers modulo ℓ. For simplicity, we have taken $d=1$. $\eta(x)=1$ if there is a particle at x and 0 otherwise. The generator can be written as

$$(\mathcal{L}_\ell F)(\eta)=\ell^\alpha\sum_{x,x'}\eta(x)(1-\eta(x'))\pi(x'-x)\left[F\left(\eta^{x,x'}\right)-F(\eta)\right]. \tag{3.1}$$

Here $\eta^{x,x'}$ refers to the new configuration obtained when the particle at x jumps to x'. α will be either 1 or 2, depending on the scaling used for time.

The case $\alpha=1$.

$$\begin{aligned}\mathcal{L}_\ell\left[\frac{1}{\ell}\sum J\left(\frac{x}{\ell}\right)\eta(x)\right]&=\ell\sum\left[J\left(\frac{x'}{\ell}\right)-J\left(\frac{x}{\ell}\right)\right](x'-x)\eta(x)(1-\eta(x'))\\&\simeq m\int J'(y)\rho(y,t)(1-\rho(y,t))\,dy\end{aligned}$$

where m is the mean $\sum x\,\pi(x)$ of the jump distribution π. The last step is justified because one expects the probability that a site x is occupied to be the local density $\rho(\frac{x}{\ell},t)$, with occupancy of different sites being independent.
This leads to the equation

$$\frac{\partial\rho}{\partial t}+m(\rho(1-\rho))_y=0 \tag{3.2}$$

for the density ρ. The method of [5] does not work here. The method of relative entropy contained in [14] will work, but needs the solution $\rho(t,x)$ to be smooth. It is known that for most initial data, sooner or later, discontinuities will develop and so the method applies only up to the first shock. There are other coupling methods that establish convergence to the correct weak solution of (3.2). See, for instance, Rezakhanlou [10] for the best results in the case of attractive dynamics.

The case $\alpha=2$. In the symmetric case; i.e., $\pi(z)=\pi(-z)$, we always have $m=0$ and we take $\alpha=2$.

$$\begin{aligned}\mathcal{L}_\ell\frac{1}{\ell}\sum J\left(\frac{x}{\ell}\right)\eta(x)&=\frac{\ell^2}{\ell}\sum\left[J\left(\frac{x'}{\ell}\right)-J\left(\frac{x}{\ell}\right)\right]\eta(x)(1-\eta(x'))\pi(x'-x)\\&=\frac{\ell^2}{2\ell}\sum\left(J\left(\frac{x'}{\ell}\right)-J\left(\frac{x}{\ell}\right)\right)[\eta(x)-\eta(x')]\,\pi(x'-x)\\&\sim\frac{1}{2}\sum J''\left(\frac{x}{\ell}\right)\eta(x)\left(\sum z^2\pi(z)\right)\\&=\frac{D}{2}\sum J''\left(\frac{x}{\ell}\right)\eta(x).\end{aligned}$$

This yields

$$\frac{\partial\rho}{\partial t}=\frac{D}{2}\rho_{xx} \tag{3.3}$$

with

$$D = \sum z^2 \pi(z).$$

The nonlinearity miraculously cancels out and the equations close. No averaging is needed.

If we change the problem by coloring the particles so that some are green and some are red, we can ask how the colors spread. One should compute

$$\mathcal{L}_\ell \left[\frac{1}{\ell} \sum A\left(\frac{x}{\ell}\right) \eta_g(x) + \frac{1}{\ell} \sum B\left(\frac{x}{\ell}\right) \eta_r(x) \right]$$

and proceed from there. Here $\eta_g(x) = 1$ if there is a green particle at x and similarly for $\eta_r(x)$. $\eta_g(x) + \eta_r(x) \leq 1$ due to exclusion. The dynamics is color blind. The analysis is hard because the system turns out to be nongradient. The method of [12] was applied to this situation by Quastel in [9] and he obtained an elliptic system for the pair $\rho_g(t, y)$ and $\rho_r(t, y)$. One first solves for

$$\rho(t, y) = \rho_g(t, y) + \rho_r(t, y)$$

by the heat equation (3.3). Then $\rho_g(t, y)$ is solved by

$$\frac{\partial \rho_g}{\partial t} = \frac{1}{2} \frac{\partial}{\partial y} S(\rho) \frac{\partial}{\partial y} \rho_g - (b(t, y)\rho_g)_y$$

where

$$b(t, y) = \frac{1}{2} \frac{(S(\rho) - D)\rho_y}{\rho}$$

is the pressure due to density gradient. $S(\rho)$ is the self diffusion coefficient determined in [7] and really depends on ρ, with $S(\rho) \to D$ as $\rho \to 0$ as $S(\rho) \to 0$ as $\rho \to 1$. The case $\pi(1) = \pi(-1) = \frac{1}{2}$ is special and $S(\rho) \equiv 0$ in that case.

The case where $\pi(z) \neq \pi(-z)$ but still $m = 0$ is more complex. This is nongradient and nonreversible. The methods of [12] and [9] have to be modified. This was carried out by Xu in [13] who established a limiting equation of the form

$$\frac{\partial \rho}{\partial t} = \frac{1}{2} \frac{\partial}{\partial y} D(\rho) \frac{\partial \rho}{\partial y}$$

with $D(\rho) \geq D = \sum z^2 \pi(z)$. In general, $D(\rho) > D$ for $0 < \rho < 1$ with $D(\rho) \to D$ as $\rho \to 0$ or 1.

Navier-Stokes corrections. Let us return to the case $m \neq 0$, but start very close to equilibrium. We start with a density $\rho(y) = \frac{1}{2} + \frac{1}{\ell} q_0(y)$; i.e.,

$$P[\eta(x) = 1] = \frac{1}{2} + \frac{1}{\ell} q_0\left(\frac{x}{\ell}\right).$$

Now we can rescale with $\alpha = 2$. Recently Esposito, Marra, and Yau [4] have shown that when the dimension $d \geq 3$, the empirical density at the rescaled time t is of the form $\frac{1}{2}+\frac{1}{\ell}q(t,\frac{x}{\ell})$ to within an accuracy $o(\frac{1}{\ell})$, and $q(t,y)$ is given as the solution of a Navier-Stokes type equation; i.e., Burgers' equation with viscosity.

4 Hamiltonian System

Let us return finally to our classical Hamiltonian system. If we start with a random initial configuration given by a density g_N^0 at time $t = 0$, and let the configuration evolve by the Hamiltonian motion, one can obtain the density of the configuration at time t by solving the Liouville flow

$$\frac{\partial f_N}{\partial t} = \mathcal{L}_N f_N \qquad f_N = g_N^0 \text{ at } t = 0. \tag{4.1}$$

If we are given functions $\rho(y), v^\alpha(y)$, and $T(y)$ on T^3 representing local density, average velocity, and temperature, assuming that those values avoid regions of possible phase transitions, we can associate a density $g_N(y_1,\ldots,y_N;u_1,\ldots,u_N)$ in the phase space that is a family of slowly varying Maxwell-Gibbs distributions strung together. If we calculate averages like $\frac{1}{N}\sum J(y_i)$ we get $\int J(y)\rho(y)\,dy$ in probability as $N\to\infty$. If we pick ρ, v^α, and T to be, as functions of t and y, a solution of the Euler equation that is quite smooth in some interval, we can use $\rho(t,y), v^\alpha(t,y)$, and $T(t,y)$ to construct a time dependent family $g_N(t,y_1,\ldots,y_N;u_1,\ldots u_N)$ of such densities.

We would like to establish that the density obtained by the Euler equation; i.e., g_N is close to the density f_N obtained by solving the Hamiltonian or equivalently the Liouville flow. An ideal theorem will say that as $N\to\infty$ the specific (per particle) relative entropy

$$s(f_N,g_N) = \frac{1}{N}\int \log\frac{f_N}{g_N}\cdot f_N \quad dy\ du \to 0. \tag{4.2}$$

We note that if the specific entropy were to tend to zero, by the usual large deviation estimates, the local density, velocities, and temperature would be the same for f_N and g_N. This would establish the hydrodynamic limit. But we cannot quite prove such a theorem. We have problems at two levels. First there is difficulty with large velocities. This can be overcome by changing the kinetic energy in the Hamiltonian to a function $\phi(u)$ with a bounded gradient instead of $\frac{1}{2}\|u\|^2$. (One choice that will work is the relativistic kinetic energy.) This is a technical point. The more serious problem is the hunger for noise. It is needed to establish some ergodicity. But only very little is needed. We put it in as an additional noisy exchange of velocities between pairs of particles that conserves momenta and energy. The strength of this noisy term is much smaller than the exchange of velocities provided by the Hamiltonian equations. This is then a small second order perturbation of the Liouville operator that does not destroy the conservation laws. The Euler equations are still the same. With (4.1) now replaced by a modified Fokker-Planck equation, one can establish (4.2). The details can be found in our work [8] with Olla and Yau . A key step is that whereas the noise is responsible for keeping the velocitiy distributions locally Maxwellian, for a Hamiltonian dynamics the positions are then shown to satisfy the necessary ergodic behavior with the correct Gibbs distributions.

5 Comments

We have limited our discussion essentially to work that uses entropy-related methods to the study of problems of hydrodynamic scaling. There is considerable work that uses other methods to study similar problems and we have not described them. The monographs [2] and [11] are excellent sources for a much wider collection of material.

We have also not discussed issues of large deviation. In some sense entropy-related methods are intimately related to techniques of large deviation theory and the two often go hand in hand. See, for instance, [6] and [2] for connections to the methods of [5]. As for Hamiltonian systems, there is the earlier work of Boldrighini, Dobrusin, and Sukov [1], which deals with the case of hard rods in one dimension with elastic collision.

References

[1] C. Boldrighini, R. L. Dobrusin and Yu M. Suhov, *One-dimensional hard rod caricatures of hydrodynamics*, J. Stat. Phys. **31** (1983), 577–616.

[2] A. DeMasi and E. Presutti, Mathematical Methods for Hydrodynamical Limits, Lecture Notes in Math., **1501**, Springer Verlag, Berlin-Heidelberg-New York, 1991.

[3] M. D. Donsker and S. R. S. Varadhan, *Large deviations from a hydrodynamic scaling limit*, Comm. Pure. Appl. Math. **42** (1989), 243–270.

[4] R. Esposito, R. Marra and H. T. Yau, *Diffusive limit of asymmetric simple exclusion*, preprint.

[5] M. Z. Guo, G. C. Papanicolaou and S. R. S. Varadhan, *Nonlinear diffusion limit for a system with nearest neighbor interactions*, Comm. Math. Phys. **118** (1988), 31–59.

[6] C. Kipnis, S. Olla and S. R. S. Varadhan, *Hydrodynamics and large deviation for simple exclusion process*, Comm. Pure Appl. Math. **42** (1989), 115–137.

[7] C. Kipnis and S. R. S. Varadhan, *Central limit theorem for additive functionals of reversible Markov processes and application to simple exclusions*, Comm. Math. Phys. **104** (1986), 1–19.

[8] S. Olla, S. R. S. Varadhan and H. T. Yau, *Hydrodynamical limit for a Hamiltonian system with weak noise*, Comm. Math. Phys. **155** (1993), 523–560.

[9] J. Quastel, *Diffusion of color in simple exclusion process*, Comm. Pure Appl. Math. **45** (1992), 623–680.

[10] F. Rezakhanlou, *Hydrodynamical limit for attractive particle systems on* z^d, Comm. Math. Phys. **140** (1991), 417–448.

[11] H. Spohn, Large Scale Dynamics of Interacting Particles, Texts and Monographs in Physics, Springer Verlag, Berlin-Heidelberg-New York, 1991.

[12] S. R. S. Varadhan, *Nonlinear diffusion limit for a system with nearest neighbor interactions, II, Asymptotic Problems in Probability Theory*, in: Stochastic Models and Diffusions on Fractals (K. D. Elworthy and N. Ikeda, eds.), Pitman Res. Notes Math. Ser. **283**, 1991, 75–130.

[13] Lin Xu, Hydrodynamics for asymmetric mean zero simple exclusion, Ph.D. thesis, New York University, 1993.

[14] H. T. Yau, *Relative entropy and the hydrodynamics of Ginzburg-Landau models*, Lett. Math. Phys. **22** (1991), 63–80.

The Annals of Probability
1996, Vol. 24, No. 4, 1871–1902

SPECTRAL GAP FOR ZERO-RANGE DYNAMICS

BY C. LANDIM, S. SETHURAMAN AND S. VARADHAN[1]

IMPA and CNRS, Courant Institute and Courant Institute

We give a lower bound on the spectral gap for symmetric zero-range processes. Under some conditions on the rate function, we show that the gap shrinks as n^{-2}, independent of the density, for the dynamics localized on a cube of size n^d. We follow the method outlined by Lu and Yau, where a similar spectral gap is proved for Kawasaki dynamics.

1. Introduction and results. One of the basic estimates required for the hydrodynamical limit of nongradient systems is a sharp lower bound on the spectral gap of the finite coordinate process [13]. What is needed is that the gap, for the process confined to cubes of linear size n, shrinks at a rate n^{-2}. Up to constants, this is heuristically the best possible lower bound, in view of a comparison with random walk which prevents larger bounds.

Another application of such a bound is to establish an invariance principle for conservative particle dynamics [11]. In that article, the bound is used to estimate the rate of convergence of the canonical to the grand canonical ensemble. Among the models discussed there are the zero-range processes. Our intention in this paper is to prove the required lower bound for this type of dynamics.

The symmetric zero-range processes consist of infinitely many particles moving on the lattice Z^d according to a Markovian law. The evolution of the particles may be informally described as follows. Denote by N the set of nonnegative integers, fix a nonnegative function $c\colon N \to R_+$ such that $c(0) = 0 < c(i)$ for $i \geq 1$ and fix a symmetric transition measure $p(\cdot)$ on Z^d. If there are k particles at a site x of Z^d, one of them jumps to y at rate $c(k)p(y-x)$. This happens independently at each site. To fix ideas, we shall consider in this paper only nearest-neighbor interactions: $p(x) = 1/2$ if $|x| = 1$ and 0 otherwise.

At this point some notation is required. The sites of Z^d are denoted by x, y and z, the state space N^{Z^d} by the symbol Σ and the configurations by the Greek letters η and ξ. In this way η_x stands for the total number of particles at site x for the configuration η.

These so-called zero-range processes are Markov processes with infinitesimal generator L defined by its action on cylinder functions ϕ:

$$(L\phi)(\eta) = \frac{1}{2} \sum_{|y-x|=1} c(\eta_x)(\phi(\eta^{x,y}) - \phi(\eta)),$$

Received September 1995; revised March 1996.

[1]Research partially supported by National Science Foundation Grant DMS-92-01222 and ARO Grant DAAL 03-92-G-0317.

AMS 1991 *subject classifications*. Primary 60K35; secondary 60F05.

Key words and phrases. Particle systems, ergodic measure, zero-range process, Dirichlet form.

where

$$(\eta^{x,y})_z = \begin{cases} \eta_x - 1, & \text{if } z = x, \\ \eta_y + 1, & \text{if } z = y, \\ \eta_z, & \text{if } z \neq x, y \end{cases}$$

provided $\eta_x \geq 1$ and $x \neq y$; otherwise, $\eta^{x,y} \equiv \eta$.

To ensure that the process is well defined on the infinite lattice N^{Z^d} we shall assume throughout this article a Lipschitz condition on the rate:

(LG) $\sup_k |c(k+1) - c(k)| \leq a_1 < \infty$.

We refer to [1] for the details on the existence of this process.

As a conservative system where particles are neither created nor destroyed, it is expected that the process possesses a family of invariant measures supported on configurations of fixed density. In order to describe these measures, define the partition function $Z(\cdot)$ on R_+ by

$$Z(\alpha) = \sum_{k \geq 0} \frac{\alpha^k}{c(1) \cdots c(k)}.$$

It is clear that $Z(\cdot)$ is an increasing function. Let α^* denote the radius of convergence of Z:

$$\alpha^* = \sup\{\alpha;\ Z(\alpha) < \infty\}.$$

In order to avoid degeneracy we assume that the partition function Z diverges at the boundary of its domain of definition:

$$\lim_{\alpha \to \alpha^*} Z(\alpha) = \infty. \tag{1.1}$$

For $0 \leq \alpha < \alpha^*$, let $\bar{P}_\alpha$ be the translation invariant product measure on N^{Z^d} with marginals μ_α given by

$$\mu_\alpha\{\eta_x = k\} = \frac{1}{Z(\alpha)} \frac{\alpha^k}{c(1) \cdots c(k)} \quad \text{for } k \geq 0,\ x \in Z^d. \tag{1.2}$$

The family $\{\bar{P}_\alpha,\ 0 \leq \alpha < \alpha^*\}$ is a one parameter set of reversible measures for $\{\eta_t,\ t \geq 0\}$.

A more intuitive parameterization is through the particle density. Let $\rho(\alpha)$ be the density of particles for the measure $\bar{P}_\alpha$,

$$\rho(\alpha) = \bar{E}_\alpha[\eta_0],$$

where $\bar{E}_\alpha$ refers to expectation with respect to $\bar{P}_\alpha$.

From assumption (1.1) it follows that ρ: $[0, \alpha^*) \to R_+$ is a smooth strictly increasing bijection. Since $\rho(\alpha)$ has a physical meaning as the density of particles, instead of parameterizing the above family of measures by α, we parameterize it in terms of the density ρ and write for $\rho \geq 0$,

$$P_\rho = \bar{P}_{\alpha(\rho)}.$$

Under this convention it follows then that for $\rho \geq 0$,

$$\alpha(\rho) = E_\rho[c(\eta_0)],$$

where E_ρ refers to expectation with respect to P_ρ. Moreover, a simple computation shows that

$$\alpha'(\rho) = \frac{\alpha(\rho)}{\sigma(\alpha(\rho))^2}, \tag{1.3}$$

where $\sigma(\alpha(\rho))^2$ stands for the variance of η_0 under $\bar{P}_\alpha$: $\sigma(\alpha(\rho))^2 = \bar{E}_\alpha[(\eta_0 - \rho(\alpha))^2]$.

The associated Dirichlet form $D_\rho(\phi) = -E_\rho[\phi(\eta)(L\phi)(\eta)]$ is defined by its action on the test function ϕ:

$$D_\rho(\phi) = \tfrac{1}{4} \sum_{|x-y|=1} E_\rho\big[c(\eta_x)(\phi(\eta^{x,y}) - \phi(\eta))^2\big].$$

We now describe more precisely what we mean by the term "spectral gap." Consider the finite volume, finite particle zero-range process. This model governs the behavior of K particles jumping about in a finite cube, say to fix ideas, $\Lambda_n = \{1, 2, \dots, n\}^d$. The state space is then given as $\Sigma_{n,K} = \{\eta \in N^{\Lambda_n}: \sum_{x\in\Lambda_n} \eta_x = K\}$. For configurations $\eta \in \Sigma_{n,K}$ and test functions ϕ, the generator of this finite process takes the form

$$(L_n\phi)(\eta) = \tfrac{1}{2} \sum_{\substack{|x-y|=1 \\ x,y\in\Lambda_n}} c(\eta_x)[\phi(\eta^{x,y}) - \phi(\eta)].$$

The process defined by the generator L_n on the state space $\Sigma_{n,K}$ is an ergodic, reversible finite state Markov chain possessing discrete real-valued spectrum. The operator L_n in $L^2(P_{n,K})$ is negative definite and its largest eigenvalue is 0. The absolute value of the next largest eigenvalue is the so-called spectral gap of the process.

It is easily calculated that the ergodic measures $P_{n,K}$ are equal to the conditioned measure of the infinite volume invariant state on the restricted hyperplane $\Sigma_{n,K}$:

$$P_{n,K}(\cdot) = P_\rho\Big(\cdot \,\Big|\, \sum_{x\in\Lambda_n} \eta_x = K\Big).$$

Note that these definitions are independent of ρ.

The measures $P_{n,K}$, as alluded to earlier, are reversible and a simple calculation yields

$$c(r+1)P_{n,K}\{\eta;\ \eta_{e_1} = r+1,\ \eta_{2e_1} = l\} = c(l+1)P_{n,K}\{\eta;\ \eta_{e_1} = r,\ \eta_{2e_1} = l+1\}$$

or, what will be used later,

$$E_{n,K}[c(\eta_x)|\eta_y = r] = c(r+1)\frac{P_{n,K}\{\eta_{e_1} = r+1\}}{P_{n,K}\{\eta_{e_1} = r\}} \tag{1.4}$$

for every $x \neq y$ in Λ_n. Here $E_{n,K}$ denotes expectation with respect to the measure $P_{n,K}$ and $\{e_i,\ 1 \le i \le d\}$ stands for the canonical basis of R^d. The density of the measure $P_{n,K}$ is denoted by $\rho = K/n^d$.

The Dirichlet form for the finite process is defined as

$$\begin{aligned} D_{n,K}(\phi) &= -E_{n,K}[\phi(L_n\phi)] \\ &= \frac{1}{4} \sum_{\substack{|x-y|=1 \\ x,y\in\Lambda_n}} E_{n,K}\big[c(\eta_x)(\phi(\eta^{x,y}) - \phi(\eta))^2\big]. \end{aligned}$$

Consider now the quantity $W(n, K)$ appearing in the Poincaré inequality

$$E_{n,K}\big[(f - E_{n,K}[f])^2\big] \le W(n, K)D_{n,K}(f) \tag{1.5}$$

for every $f: \Sigma_{n,K} \to (-\infty, \infty)$ in $L^2(P_{n,K})$, where $W(n, K)$ is a constant depending only on n and K.

This inequality gives another way to evaluate the spectral gap of the finite volume process. In fact, as the inequality deals with mean-zero functions, the smallest possible value of $W(n, K)$ is the reciprocal of the gap.

We are interested in establishing the bound $W(n, K) < W_0 n^2$ for a constant W_0 independent of n and the total number of particles K. This would then imply that the spectral gap is of $O(n^{-2})$.

Similar uniform $O(n^2)$ spectral gap estimates have been found for the simple exclusion process by Quastel [10], and for more general exclusion processes by Lu and Yau [8]. The aim of this article is to determine for a class of zero-range processes the spectral gap bound $W(n, K) < W_0 n^2$, where W_0 is a uniform constant. Such a demand rules out many zero-range processes, as shown by the following example.

EXAMPLE 1.1. Let $c(r) = I(r \ge 1)$. By a clever transform in dimension $d = 1$ [4], the corresponding zero-range process is mapped onto the simple exclusion process where the gap is well known [10]. Inverting back, we determine that the zero-range process possesses a spectral gap of order $(n + K)^{-2}$ which is clearly dependent on the density ρ.

Another illuminating example is the zero-range process where particles evolve independently.

EXAMPLE 1.2. Consider $c(r) = r$, the identity function. With this rate, the corresponding zero-range dynamics are nothing more than a collection of mutually independent symmetric random walks on $\{1, \dots, n\}^d$. For this process, the n^{-2} spectral gap estimate, independent of ρ, is well known.

To establish a spectral gap bound with constant W_0 independent of n and K we will impose a second assumption that rules out Example 1.1.

(M) There exists $k_0 \in N$ and $a_2 > 0$ such that $c(k) - c(j) \geq a_2$ for all $k \geq j + k_0$.

Notice that, under assumptions (LG) and (M), α^* is actually infinite. We are now in a position to state the main theorem of this article.

THEOREM 1.1. *Given the conditions* (LG) *and* (M), *there exists a constant* W_0 *independent of n and K such that* (1.5) *holds with* $W(n, K) = W_0 n^2$ *for the corresponding nearest-neighbor zero-range processes. This implies a spectral gap of at least* $\{W_0 n^2\}^{-1}$ *on a cube of volume* n^d.

We conclude this section with a few comments on the method of proof. The approach we present here to prove a spectral gap independent of the total number of particles is easily adapted to generalized exclusion processes introduced in [6]. For attractive systems, a simpler proof can be given based on coupling and positivity (cf. [3]). On the other hand, for the rate described in Example 1.1 and for similar rates, we do not have yet methods to determine a sharp spectral gap estimate. The difficulty arises in that the spectral gap depends on the total number of particles and our approach is not adapted to this situation.

To prove Theorem 1.1, we adapt Lu and Yau's method [8] to the context of zero-range processes where, by a clever induction argument, the Poincaré inequality on f, ostensibly a function of n^d coordinates, is reduced to an estimate involving a function of only one coordinate. At this point we prove a spectral gap for functions that depend only on one site. This one site spectral gap reduces the problem to the estimation of two terms. The first will yield the desired Dirichlet form, while the second will give a smaller variance to be absorbed into the variance on the left-hand side of (1.5).

Section 2 develops Lu and Yau's basic ideas. Section 3 details the estimates of the first and second terms and specifies the iteration. In Section 4 we prove a spectral gap for functions depending only on one site. As a consequence Theorem 1.1 is established for the initial case with two sites only. In Section 5 we prove some estimates on the one site marginal of the grand canonical zero-range distribution important in the induction step of the proof and necessary to obtain local central limit theorems uniform in the density. This last question is the main theme of Section 6.

2. Proof of the main estimate. For simplicity we will prove Theorem 1.1 for dimension $d = 1$; the extension to higher dimensions is not difficult (cf. [5]). We will employ an iteration aided by an induction. We will assume that we have a spectral gap independent of the total number of particles for intervals of length $m \leq n - 1$: $W(m) = \sup_K W(m, K)$ is finite. Our efforts now will be to use this hypothesis to set up a recursive equation for $W(n)$. For n large, say $n > n_0$, we will give such a relation for $W(n)$. However, for $n \le n_0$ slightly different arguments are used to establish a recursion for $W(n)$. The results of these two relations are then combined, giving that $W(n) \le W_0 n^2$ is valid for all n. The initial induction case $n = 2$ is a consequence of the one site spectral gap and is further discussed in Section 4.

STEP 1. Let us write the identity

$$f - E_{n,K}[f] = (f - E_{n,K}[f|\eta_1]) + (E_{n,K}[f|\eta_1] - E_{n,K}[f]).$$

Through this decomposition we may express the variance given in (1.5) as

$$\begin{aligned} &E_{n,K}[(f - E_{n,K}[f])^2] \\ &\quad = E_{n,K}\left[(f - E_{n,K}[f|\eta_1])^2\right] + E_{n,K}\left[(E_{n,K}[f|\eta_1] - E_{n,K}[f])^2\right]. \end{aligned} \tag{2.1}$$

The first term on the right-hand side is easily analyzed through the induction assumption and a simple computation on the Dirichlet form. We write

$$\begin{aligned} &E_{n,K}[(f - E_{n,K}[f|\eta_1])^2] \\ &\quad = E_{n,K}[E_{n,K}[(f - E_{n,K}[f|\eta_1])^2 \mid \eta_1]] \\ &\quad = E_{n,K}[E_{n-1,K-\eta_1}[(f(\eta_1,\cdot) - E_{n-1,K-\eta_1}[f(\eta_1,\cdot)])^2]]. \end{aligned}$$

By the induction assumption this last expectation is bounded above by

$$\sum_{r=0}^{K} P_{n,K}\{\eta_1 = r\} W(n-1) D_{n-1,K-r}(f(r,\cdot)) \le W(n-1) D_{n,K}(f).$$

STEP 2. The second term in (2.1) is nothing more than the variance of $E_{n,K}[f|\eta_1]$, a function of one variable. For each fixed K and n, let $P^1_{n,K}$ be the one site marginal of the canonical measure $P_{n,K}$:

$$P^1_{n,K}(r) = P_{n,K}(\eta_1 = r) = P_\rho\left(\eta_1 = r \,\middle|\, \sum_{x=1}^{n} \eta_x = K\right).$$

Expectation with respect to $P^1_{n,K}$ is written $E^1_{n,K}$.

On $\{0,\dots,K\}$ consider the birth and death process which jumps from r to $r \pm 1$ with rates $p(r, r \pm 1)$ given by

$$p(r, r-1) = c(r) \quad \text{and} \quad p(r, r+1) = E^1_{n-1,K-r}[c(\eta_0)].$$

Denote the generator of this process by $\mathscr{L}_{n,K}$. It is elementary to check that $P^1_{n,K}$ is reversible for $\mathscr{L}_{n,K}$. Moreover, if $H : \Sigma_{n,K} \to R$ is a function of only one site [say $H(\eta) = H(\eta_1)$], then

$$D_{n,K}(H) = -(1/2) E^1_{n,K}[H \mathscr{L}_{n,K} H] =: D^1_{n,K}(H).$$

In order to estimate a spectral gap for zero-range processes, our method requires that the associated birth and death processes with generator $\mathscr{L}_{n,K}$ exhibit a spectral gap with magnitude independent of n and K. We shall prove in Section 4 the following one site spectral gap lemma.

LEMMA 2.1. *There exists a constant $B_0 = B_0(a_1, a_2, k_0)$ such that*

$$E^1_{n,K}\big[(H - E^1_{n,K}[H])^2\big] \le B_0 D^1_{n,K}(H)$$

for all $n \ge 1$, $K \ge 1$ and H in $L^2(P^1_{n,K})$.

This lemma applied to the function $E_{n,K}[f|\eta_1]$ shows that the second term of (2.1) is bounded above by $B_0 D^1_{n,K}(E_{n,K}[f|\eta_1])$. A few calculations simplify the one-coordinate Dirichlet form:

LEMMA 2.2. *For every $H = H(\eta_1)$ in $L^2(P_{n,K})$ we have*

$$D^1_{n,K}(H) = (1/2)\sum_{r=0}^{K-1} P_{n,K}\{\eta_1 = r+1\}c(r+1)(H(r+1) - H(r))^2.$$

PROOF. Recall $P^1_{n,K}(r) = P_{n,K}\{\eta_1 = r\}$. A simple computation shows that the Dirichlet form $4D^1_{n,K}(H(\eta_1))$ is equal to

$$E_{n,K}\big[c(\eta_1)(H(\eta_1-1)-H(\eta_1))^2\big]+E_{n,K}\big[E_{n,K}[c(\eta_2)|\eta_1](H(\eta_1+1)-H(\eta_1))^2\big].$$

In addition, taking advantage of the reversibility relation (1.4), we have that

$$E_{n,K}[c(\eta_2)|\eta_1 = r] = \frac{c(r+1)P^1_{n,K}(r+1)}{P^1_{n,K}(r)}. \tag{2.2}$$

Now, substituting this last identity into the previous formula, we obtain that

$$D^1_{n,K}(H(\eta_1)) = (1/2)\sum_{r=0}^{K-1} P^1_{n,K}(r+1)c(r+1)(H(r+1) - H(r))^2.$$

The lemma is thus proved. □

STEP 3. Recall that our intention is to apply Lemmas 2.1 and 2.2 to the one variable function $H(\eta_1) = E_{n,K}[f|\eta_1]$. In this respect, we derive a simpler expression for the difference

$$E_{n,K}[f|\eta_1 = r+1] - E_{n,K}[f|\eta_1 = r],$$

taking advantage of the reversibility of $P_{n,K}$.

LEMMA 2.3. *Let $M(\eta)$ be the function defined by*

$$M(\eta) = \frac{P^1_{n,K}(\eta_1)}{P^1_{n,K}(\eta_1+1)c(\eta_1+1)}\frac{1}{n-1}\sum_{x=2}^{n} c(\eta_x).$$

Then, for every $0 \le r \le K-1$, *the difference*

$$E_{n,K}[f|\eta_1 = r+1] - E_{n,K}[f|\eta_1 = r]$$

is equal to

$$\frac{1}{P^1_{n,K}(r+1)c(r+1)} \frac{1}{n-1} \sum_{x=2}^{n} E_{n,K}[c(\eta_x)(f(\eta^{x,1}) - f(\eta))I(\eta_1 = r)] \qquad (2.3)$$
$$+ E_{n,K}[M(\eta);\ f(\eta)|\eta_1 = r],$$

where $E_{n,K}[g;\ h|\eta_1 = r]$ *is the covariance of* g *and* h *with respect to* $E_{n,K}[\cdot|\eta_1 = r]$.

PROOF. Notice first that due to the reversibility criterion (2.2),

$$E_{n,K}[M(\eta)|\eta_1 = r] = 1 \qquad (2.4)$$

for every $0 \le r \le K-1$. On the other hand, exploiting again the reversibility (1.4), we obtain that

$$E_{n,K}[f|\eta_1 = r+1] = \frac{1}{P^1_{n,K}(r+1)c(r+1)} E_{n,K}[f(\eta^{x,1})c(\eta_x)I(\eta_1 = r)]$$

for every $2 \le x \le n$. Adding and subtracting an appropriate term and then averaging over x, this expression may be rewritten as

$$\frac{1}{P^1_{n,K}(r+1)c(r+1)} \frac{1}{n-1} \sum_{x=2}^{n} E_{n,K}[(f(\eta^{x,1}) - f(\eta))c(\eta_x)I(\eta_1 = r)]$$
$$+ E_{n,K}[f(\eta)M(\eta)|\eta_1 = r].$$

The observation (2.4) concludes the proof of the lemma. □

STEP 4. From Lemmas 2.1 and 2.2, the second term on the right-hand side of (2.1) is bounded above by

$$\tfrac{1}{2}B_0 \sum_{r=0}^{K-1} P^1_{n,K}(r+1)c(r+1)\{E_{n,K}[f|\eta_1 = r+1] - E_{n,K}[f|\eta_1 = r]\}^2.$$

From Lemma 2.3 and the Schwarz inequality, this last sum is bounded above by

$$B_0 \sum_{r=0}^{K-1} \frac{1}{P^1_{n,K}(r+1)c(r+1)}$$
$$\times \left\{E_{n,K}\left[\frac{1}{n-1} \sum_{x=2}^{n} c(\eta_x)(f(\eta^{x,1}) - f(\eta))I(\eta_1 = r)\right]\right\}^2$$
$$+ B_0 \sum_{r=0}^{K-1} P^1_{n,K}(r+1)c(r+1)(E_{n,K}[M;\ f|\eta_1 = r])^2.$$

We denote the first line by $B_0 A_1(n, K, f)$ and the second by $B_0 A_2(n, K, f)$.

In Lemma 3.1 of Section 3 we establish that

$$A_1(n, K, f) \le (n/2)D_{n,K}(f).$$

In Lemma 3.2 we prove under assumptions (LG) and (M) that

$$A_2(n, K, f) \le a_1^2 B_0 W(n-1) D_{n,K}(f).$$

This inequality shall be used to perform the iteration for small values of n. On the other hand, using extensively the uniform Edgeworth expansion presented in Section 6, in Proposition 3.1, under assumptions (LG) and (M), we prove that for all $\varepsilon > 0$, there exist finite $n_0(\varepsilon)$ and $C(\varepsilon)$ such that

$$A_2(n, K, f) \le C(\varepsilon) D_{n,K}(f) + \varepsilon n^{-1} E_{n,K}[(f - E_{n,K}(f))^2]$$

for $n \ge n_0(\varepsilon)$.

In conclusion, from Step 1 and these estimates, the variance of f satisfies the following relation:

$$E_{n,K}[(f - E_{n,K}(f))^2] \le \{[1 + a_1^2 B_0^2] W(n-1) + (nB_0/2)\} D_{n,K}(f)$$

for $n \ge 2$ and

$$E_{n,K}[(f - E_{n,K}(f))^2] \le \left\{W(n-1) + \left(\frac{nB_0}{2}\right) + B_0 C(\varepsilon)\right\} D_{n,K}(f) + \frac{\varepsilon B_0}{n} E_{n,K}[(f - E_{n,K}(f))^2]$$

for $n \ge n_0(\varepsilon)$. In other words, we have the following recurrence relations for the sequence $W(n)$:

$$W(n) \le [1 + a_1^2 B_0^2] W(n-1) + \frac{nB_0}{2} \quad \text{for } n \ge 2,$$

$$W(n) \le \left(1 - \frac{\varepsilon B_0}{n}\right)^{-1} \left\{W(n-1) + \left(\frac{nB_0}{2}\right) + B_0 C(\varepsilon)\right\} \quad \text{for } n \ge n_0(\varepsilon).$$

This yields immediately that $W(n) \le W_0 n^2$ for some universal constant W_0. This concludes the proof of Theorem 1.1. □

3. Technical bounds. The aim of this section is to estimate the two terms A_1 and A_2 and thereby establish the iteration stated in Section 2. The first term will be bounded by the full Dirichlet form multiplied by a factor of n. This is accomplished in Lemma 3.1. The second term will be estimated by the full variance multiplied by εn^{-1}, where ε is an arbitrarily small number, and the Dirichlet form. This takes place through several lemmas below. With this bound, this "small" variance may be absorbed into the original variance. We first estimate $A_1(n, K, f)$.

LEMMA 3.1. *For every $n \ge 2$ and positive integer K,*

$$A_1(n, K, f) \le (n/2)D_{n,K}(f).$$

PROOF. Recall the definition of $A_1(n, K, f)$. It is equal to

$$\sum_{r=0}^{K-1} \frac{1}{P^1_{n,K}(r+1)c(r+1)}\left\{E_{n,K}\left[\frac{1}{n-1}\sum_{x=2}^{n} c(\eta_x)(f(\eta^{x,1}) - f(\eta))I(\eta_1 = r)\right]\right\}^2.$$

By the Schwarz inequality this sum is bounded above by

$$\sum_{r=0}^{K-1} E_{n,K}\left[\frac{1}{n-1}\sum_{x=2}^{n} c(\eta_x)(f(\eta^{x,1}) - f(\eta))^2 I(\eta_1 = r)\right]$$
$$= E_{n,K}\left[\frac{1}{n-1}\sum_{x=2}^{n} c(\eta_x)(f(\eta^{x,1}) - f(\eta))^2\right]$$

since by reversibility we have that

$$E_{n,K}[c(\eta_x)I(\eta_1 = r)] = P^1_{n,K}(r+1)c(r+1)$$

for every $2 \le x \le n$, $0 \le r \le K-1$. Straightforward computations show that the last sum is bounded above by $n/2$ times the Dirichlet form of f. This completes the proof of the lemma. □

We now turn to the proof of an upper bound for $A_2(n, K, f)$. The naive way to approach this estimate is immediately to apply the Schwarz inequality to $A_2(n, K, f)$ and then try to bound the object [recall the definition of $M = M(\eta)$]

$$\frac{P^1_{n,K}(r+1)c(r+1)}{P^1_{n,K}(r)} E_{n,K}[M;\ M|\eta_1 = r]$$

uniformly on r. The Edgeworth expansion gives a bound on this quantity of type $C(a_1)n^{-1}$ and consequently $A_2(n, K, f)$ is bounded above by $C(a_1)n^{-1}E_{n,K}[f; f]$. This is not enough to ensure that the iteration in the previous section produces the estimate $W(n) < Cn^2$. In order for the iteration to succeed, an inequality such as $A_2(n, K, f) < C_2 n^{-1} E_{n,K}[f; f]$, where C_2 is a small universal constant, is required. The naive approach is not successful as there is no a priori smallness condition on the constant $C(a_1)$ arising from the Edgeworth expansion. A more subtle analysis is therefore needed. This is facilitated by the next two results. To carry this out we rewrite $A_2(n, K, f)$ more appropriately.

Notice that by reversibility (2.2), $M(\eta)$ may be rewritten as

$$\frac{1}{E_{n-1,K-\eta_1}[c(\eta_1)]}\frac{1}{n-1}\sum_{x=2}^{n} c(\eta_x)$$

and $A_2(n, K, f)$ as

$$E_{n,K}\left[\frac{1}{E_{n-1,K-\eta_1}[c(\eta_1)]}\left(E_{n-1,K-\eta_1}\left[f;\ \frac{1}{n-1}\sum_{x=2}^{n} c(\eta_x)\right]\right)^2\right]. \tag{3.1}$$

We shall first estimate $A_2(n, K, f)$ for n small using asumptions (LG) and (M).

LEMMA 3.2. *Under conditions* (LG) *and* (M) *we have, for* $n \geq 2$,

$$A_2(n, K, f) \leq a_1^2 B_0 W(n-1) D_{n, K}(f).$$

PROOF. By the Schwarz inequality

$$\left(E_{n-1, K-\eta_1}\left[f; \frac{1}{n-1}\sum_{x=2}^{n} c(\eta_x)\right]\right)^2 \\ \leq E_{n-1, K-\eta_1}[f; f] E_{n-1, K-\eta_1}\left[\frac{1}{n-1}\sum_{x=2}^{n} c(\eta_x); \frac{1}{n-1}\sum_{x=2}^{n} c(\eta_x)\right].$$

A simple computation exploiting Lemmas 2.1 and 2.2 shows that for every $m \geq 1$ and $L \geq 0$,

$$E_{m, L}\left[\frac{1}{m}\sum_{x=1}^{m} c(\eta_x); \frac{1}{m}\sum_{x=1}^{m} c(\eta_x)\right] \leq E_{m, L}[(c(\eta_1) - E_{m, L}[c])^2] \\ \leq \frac{1}{2} B_0 E_{m, L}[c(\eta_1)(c(\eta_1 - 1) - c(\eta_1))^2].$$

Now by assumption (LG), $|c(\eta_1) - c(\eta_1 - 1)| \leq a_1$. Therefore $A_2(n, K, f)$ is bounded by

$$\tfrac{1}{2} a_1^2 B_0 E_{n, K}[E_{n-1, K-\eta_1}[f; f]].$$

By the induction assumption, $E_{n-1, K-\eta_1}[f; f]$ is bounded by $W(n-1) \times D_{n-1, K-\eta_1}(f)$. Since $E_{n, K}[D_{n-1, K-\eta_1}(f)] \leq D_{n, K}(f)$, the lemma is proved. □

We now turn to the case where n is large enough that Edgeworth expansions can be used to estimate expectations involving the canonical measures. Recall the alternative formulation (3.1) for $A_2(n, K, f)$. To keep notation simple, we shall fix $m = n - 1$ and $L = K - \eta_1$ and bound the expression inside the expectation.

PROPOSITION 3.1. *For every* $\varepsilon > 0$, *there exists* n_0 *in* N *and a finite constant* $C(\varepsilon)$ *so that*

$$\frac{1}{E_{m, L}[c(\eta_1)]}\left(E_{m, L}\left[f; \frac{1}{m}\sum_{x=1}^{m} c(\eta_x)\right]\right)^2 \leq C(\varepsilon) D_{m, L}(f) + \frac{\varepsilon}{m} E_{m, L}[f; f]$$

for all L *and* $m \geq n_0$.

The proof of this result is divided into several lemmas for purposes of clarity. We first single out the case of small density.

LEMMA 3.3. *For each $\varepsilon > 0$, there exists a positive integer n_0 and a density ρ_0 such that*

$$\frac{1}{E_{m,L}[c(\eta_1)]} E_{m,L}\left[\frac{1}{m}\sum_{x=1}^{m} c(\eta_x);\ \frac{1}{m}\sum_{x=1}^{m} c(\eta_x)\right] \le \frac{\varepsilon}{m} \tag{3.2}$$

for all $m \ge n_0$ and $L/m \le \rho_0$.

PROOF. First we compute the left-hand side in (3.2) using the reversibility relation (1.4) and obtain that it is equal to

$$E_{m,L-1}[c(\eta_1)] - E_{m,L}[c(\eta_1)] + \frac{1}{m} E_{m,L-1}[c(\eta_1 + 1) - c(\eta_1)].$$

The lemma follows from Corollary 6.3. □

It remains to consider the case when the density is away from 0. Let n_0 be as specified in Corollary 6.4.

LEMMA 3.4. *For each $\rho_0 > 0$, there exists a finite constant $C(\rho_0)$ and $n_1(\rho_0) \in N$ such that the scaled square of the covariance*

$$\frac{1}{E_{m,L}[c(\eta_1)]}\left(E_{m,L}\left[f;\ \frac{1}{m}\sum_{x=1}^{m} c(\eta_x)\right]\right)^2$$
$$\le C(\rho_0)\frac{1}{m}\left\{\frac{1}{l}E_{m,L}[f;\ f] + (l+1)W(l+1)D_{m,L}(f)\right\}$$

for each $L/m \ge \rho_0$, $n_0 \le l < \sqrt{m}$ and $m \ge n_1$.

PROOF. Our strategy is to decompose the scaled covariance into two terms, the first of which will produce the Dirichlet form multiplied by a controlled factor; the second will yield the original variance with a small coefficient. In order to take these factors and coefficients as universal numbers, required by our iteration scheme in the previous section, we will need to use our assumptions (LG) and (M) on the rate $c(\cdot)$.

In what follows, we abbreviate L/m by ρ. For a fixed positive integer l, we divide the interval $\{1, \dots, m\}$ into $[m/l]$ adjacent intervals of length l or $l+1$. We denote by B_ι the ιth interval, by $|B_\iota|$ the total number of sites in B_ι and by y_ι the density of particles in B_ι: $y_\iota = |B_\iota|^{-1}\sum_{x\in B_\iota}\eta_x$. To keep notation simple, we denote by $E_{|B_\iota|, y_\iota}$ the expectation with respect to the canonical measure on $|B_\iota|$ sites and $y_\iota|B_\iota|$ particles.

By the Schwarz inequality, we have

$$\left(E_{m,L}\left[f;\ \frac{1}{m}\sum_{x=1}^{m} c(\eta_x)\right]\right)^2 \le 2\left(E_{m,L}\left[f;\ \frac{1}{m}\sum_{\iota=1}^{[m/l]}\sum_{x\in B_\iota}\{c(\eta_x) - E_{|B_\iota|, y_\iota}[c]\}\right]\right)^2$$
$$+ 2\left(E_{m,L}\left[f;\ \frac{1}{m}\sum_{\iota}|B_\iota|\, E_{|B_\iota|, y_\iota}[c]\right]\right)^2.$$

Let I_1 be the first term on the right-hand side and let I_2 be the second.

We may estimate the first term through the elementary inequality $2ab \le \varepsilon^{-1}a^2 + \varepsilon b^2$ and the induction assumption as follows. Taking conditional expectation with respect to y_ι, rewrite $\sqrt{(1/2)I_1}$ as

$$\begin{aligned} &\left|\frac{1}{m}\sum_\iota |B_\iota| E_{m,L}\left[E_{|B_\iota|,y_\iota}\left[f;\frac{1}{|B_\iota|}\sum_{x\in B_\iota} c(\eta_x)\right]\right]\right| \\ &\quad\le \frac{\varepsilon}{2m}\sum_\iota |B_\iota| W(|B_\iota|) E_{m,L}[D_{|B_\iota|,y_\iota}(f)] \qquad (3.3)\\ &\qquad + \frac{\varepsilon^{-1}}{2m}\sum_\iota |B_\iota| E_{m,L}\left[\left(\frac{1}{|B_\iota|}\sum_{x\in B_\iota} c(\eta_x)-\alpha(\rho)\right)^2\right]. \end{aligned}$$

Here, the inequality follows from the induction assumption.

On the one hand, a straightforward computation shows that

$$\sum_\iota E_{m,L}[D_{|B_\iota|,y_\iota}(f)]$$

is bounded above by $D_{m,L}(f)$. In particular the first term on the right-hand side is bounded by $\varepsilon(l+1)W(l+1)(2m)^{-1}D_{m,L}(f)$. On the other hand, by the Schwarz inequality, the second term is bounded above by

$$(2\varepsilon)^{-1}E_{m,L}[(c(\eta_1)-\alpha(\rho))^2].$$

Since the density L/m is assumed to be bounded below by $\rho_0 > 0$, Corollary 6.1(c) is invoked and the previous variance is bounded by

$$E_2(2\varepsilon)^{-1}E_\rho[(c(\eta_1)-\alpha(\rho))^2] \le E_2(2\varepsilon)^{-1}a_1\alpha(\rho)$$

for some universal constant E_2 because by (5.1) the variance of $c(\eta_1)$ is dominated by $a_1\alpha(\rho)$. Therefore the second term on the right-hand side of (3.3) is dominated by $(2\varepsilon)^{-1}E_2 a_1\alpha(\rho)$. Minimizing in ε, we have that I_1 is bounded above by $C(a_1,\rho_0)(l+1)W(l+1)m^{-1}\alpha(\rho)D_{m,L}(f)$.

Finally, by Corollary 6.4 and (5.2), $E_{m,L}[c(\eta_1)]$ is bounded below by $C(\rho_0)\alpha(\rho)$ for some positive constant $C(\rho_0)$ provided $\rho \ge \rho_0$. This shows that

$$I_1 \le C(a_1,\rho_0)(l+1)W(l+1)m^{-1}E_{m,L}[c(\eta_1)]D_{m,L}(f).$$

We turn now to I_2. We may rewrite this expression as

$$2\left(E_{m,L}\left[f;\frac{1}{m}\sum_\iota |B_\iota|\left\{E_{|B_\iota|,y_\iota}[c]-\alpha(\rho)-\alpha'(\rho)\left[y_\iota-\frac{L}{m}\right]\right\}\right]\right)^2.$$

For $1\le \iota\le [m/l]$, let $F(y_\iota) = E_{|B_\iota|,y_\iota}[c]-\alpha(\rho)-\alpha'(\rho)[y_\iota - L/m]$. By the Schwarz inequality this last expression is bounded by

$$2E_{m,L}[f;f]E_{m,L}\left[\left(\frac{1}{m}\sum_\iota |B_\iota| F(y_\iota)\right)^2\right]. \qquad (3.4)$$

To conclude the proof of the lemma it remains to show that the the second expectation is bounded by $\varepsilon m^{-1}E_{m,L}[c(\eta_1)]$ for sufficiently large l.

Assume, to keep notation simple, that $|B_\iota| = l$ for all ι. In this case the expectation is equal to

$$\frac{l}{m} E_{m,L}[F(y_1)^2] + \left(1 - \frac{l}{m}\right) E_{m,L}[F(y_1)F(y_2)].$$

Since, by assumption, the density $L/m \ge \rho_0$, we have that by Corollary 6.1(b), $E_{m,L}[F(y_1)^2]$ and $E_{m,L}[F(y_1)F(y_2)]$ are, respectively, dominated by

$$E_{L/m}[F(y_1)^2] + \frac{E(\rho_0)l}{m}\{E_{L/m}[F(y_1)^4]\}^{1/2} \tag{3.5}$$

and

$$(E_{L/m}[F(y_1)])^2 + \frac{E(\rho_0)l}{m} E_{L/m}[F(y_1)^2]. \tag{3.6}$$

Notice that $E_\rho[F(y_1)] = 0$. In particular the second variance in (3.4) is bounded by

$$\frac{E(\rho_0)l}{m} E_\rho[F^2] + \frac{E(\rho_0)l^2}{m^2}\{E_\rho[F^4]\}^{1/2}. \tag{3.7}$$

We shall consider these two terms separately.

Recall from (1.3) that $\alpha'(\rho) = \alpha(\rho)/\sigma(\alpha(\rho))^2$. We prove in Section 5 [cf. (5.2)] that $\alpha(\rho)/\sigma(\alpha(\rho))^2$ and therefore $\alpha'(\rho)$ is uniformly bounded on R_+. Moreover, we show in Corollary 6.4 that $|E_{l,y_1}[c] - \alpha(y_1)|$ is bounded by $C_1 l^{-1}\sqrt{1+\alpha(y_1)}$ for some universal constant C_1. Therefore we may bound F by $C(\|\alpha'\|_\infty, \rho_0)|l^{-1}\sum_{1\le x\le l}(\eta_x - \rho)| + C_1 l^{-1}\sqrt{1+\alpha(\rho)}$ and obtain that

$$E_\rho[F^4] \le C(\|\alpha'\|_\infty, \rho_0, a_1)\frac{1}{l^2}\sigma(\alpha(\rho))^4$$

because, by Lemma 5.2, $\sigma(\alpha)^{-4}\bar{E}_\alpha[(\eta_0 - \rho(\alpha))^4]$ is bounded for α away from the origin and, by (5.2), $\sigma(\alpha)^2/\alpha$ is bounded below by a strictly positive constant and above by a finite constant. In particular the second term in (3.7) is bounded by $C(\rho_0)lm^{-2}\sigma(\alpha(\rho))^2$.

We turn now to the first term of (3.7). By the Schwarz inequality the expectation is bounded above by

$$2E_\rho[(E_{l,y_1}[c] - \alpha(y_1))^2] + 2E_\rho[(\alpha(y_1) - \alpha(\rho) - \alpha'(\rho)[y_1 - \rho])^2]. \tag{3.8}$$

Again by Corollary 6.4, we may bound the first term in the last expression by

$$\frac{C_1}{l^2}\{1 + \alpha(\rho) + E_\rho[\alpha(y_1) - \alpha(\rho)]\}.$$

Much in the same way we estimated the second term in (3.7) we can prove that this last expression is less than or equal to

$$\frac{C(\rho_0)}{l^2}\{1 + \alpha(\rho) + \sigma(\alpha(\rho))l^{-1/2}\} \le \frac{C(\rho_0)}{l^2}\{1 + \sigma(\alpha(\rho))^2\}.$$

The second expectation in (3.8) is slightly more complicated.

Denote $l^{-1}\sum_{1\le x\le l}(\eta_x-\rho)/\sigma(\alpha(\rho))$ as Z_σ and $\alpha(y_1)-\alpha(\rho)-\alpha'(\rho)[y_1-\rho]$ as F_1. We shall decompose the second expectation in (3.8) according to the value of Z_σ. Fix some small parameter β. On the set where $|Z_\sigma|>\beta$, since $|F_1|$ is bounded by $2\|\alpha'\|_\infty\sigma|Z_\sigma|$, the expectation is less than or equal to

$$C_2\sigma(\alpha)^2E_\rho[Z_\sigma^2 1_{|Z_\sigma|>\beta}]\le C_2\frac{\sigma(\alpha)^2}{\beta^2}E_\rho[Z_\sigma^4]\le C(\rho_0)\frac{\sigma(\alpha)^2}{l^2\beta^2}$$

because $\sigma(\alpha)^{-4}\bar{E}_\alpha[(\eta_0-\rho(\alpha))^4]$ is bounded for α away from the origin by Lemma 5.2.

On the set $|Z_\sigma|\le\beta$, by Taylor's expansion, $F(y)$ is bounded by

$$(1/2)\sigma(\alpha(\rho))^2Z_\sigma^2\sup_{|y-\rho|\le\beta\sigma}|\alpha''(y)|.$$

We will prove below in Lemma 3.5 that for all $\rho_1>0$, $\sup_{y\ge\rho_1}\sigma(\alpha(y))|\alpha''(y)|\le C(\rho_1)$ and that

$$\sup_{\rho\ge\rho_1}\sup_{|y-\rho|\le\beta\sigma(\alpha(\rho))}\sigma(\alpha(\rho))\sigma(\alpha(y))^{-1}\le 2$$

for $\beta=\beta(\rho_1)$ small enough. In particular, on the set $|Z_\sigma|\le\beta$ the expectation of F^2 is bounded above by

$$C(\rho_0)\sigma(\alpha(\rho))^2E_\rho[Z_\sigma^4]\le C(\rho_0)\sigma(\alpha(\rho))^2l^{-2}$$

because, by Lemma 5.2, $\sigma(\alpha)^{-4}\bar{E}_\alpha[(\eta_0-\rho(\alpha))^4]$ is bounded for α away from the origin.

In conclusion, we have shown that the second expectation in (3.8) is bounded above by $C(\rho_0)\sigma(\alpha(\rho))^2l^{-2}$. Therefore the first expectation in (3.7) is bounded above by

$$\frac{C(\rho_0)}{ml}\{1+\sigma(\alpha(\rho))^2\}.$$

In particular, (3.7) is bounded by $C(\rho_0)(ml)^{-1}\{1+\sigma(\alpha(\rho))^2\}$ for $m\ge l^2$.

Since by Corollary 6.1(b) and (5.1), $E_{m,L}[c(\eta_1)]$ is bounded below by $C(\rho_0)\alpha(\rho)$ and since by (5.2), $\sigma(\alpha(\rho))^2/\alpha(\rho)\le C$, we have now demonstrated that I_2 is dominated by $C(\rho_0)(ml)^{-1}E_{m,L}[c(\eta_1)]\langle f;f\rangle_{m,L}$. This concludes the proof of the lemma. □

LEMMA 3.5. (a) *For all $\rho_1>0$, there exists $C(\rho_1)<\infty$ such that*

$$\sup_{\rho\ge\rho_1}\sigma(\alpha(\rho))|\alpha''(\rho)|\le C(\rho_1).$$

(b) *For all $\rho_1>0$, there exists a positive constant $\beta=\beta(\rho_1)$, so that*

$$\sup_{\rho\ge\rho_1}\sup_{|y-\rho|\le\beta\sigma(\alpha(\rho))}\frac{\sigma(\alpha(\rho))}{\sigma(\alpha(y))}\le 2.$$

PROOF. A simple computation shows that

$$\alpha''(\rho) = \frac{\{\sigma(\alpha(\rho))^2 - m_3(\alpha(\rho))\}\alpha(\rho)}{\sigma(\alpha(\rho))^6}$$

provide $m_3(\alpha(\rho))$ stands for $\bar{E}_\alpha[(\eta_0 - \rho(\alpha))^3]$. The first part of the lemma follows therefore from inequality (5.2) and Lemma 5.2.

To prove (b), notice that $\partial_\rho \sigma(\alpha(\rho)) = m_3(\alpha(\rho))/2\sigma(\alpha(\rho))^3$. By Lemma 5.2 this last quotient is bounded by $C_1(\rho_1)$ for $\rho \geq \rho_1$. In particular, on the set $|y - \rho| \leq \beta\sigma(\alpha(\rho))$, $\sigma(\alpha(y))$ is at least $(1 - C_1(\rho_1)\beta)\sigma(\alpha(\rho))$. It remains to set $\beta = (2C_1(\rho_1))^{-1}$. □

4. One site spectral gaps. In this section we will examine a one-coordinate process and show that assumptions (LG) and (M) imply the one site spectral gap for this model. The dynamics of the model are defined implicitly by the Dirichlet form $D_{n,K}(H)$ acting solely on functions of the type $H = H(\eta_1)$. After a few calculations we can recover the generator:

$$\begin{aligned} D_{n,K}(H(\eta_1)) &= -E_{n,K}[H(\eta_1)(L_n H)(\eta_1)] \\ &= -\tfrac{1}{2}E_{n,K}[c(\eta_1)(H(\eta_1 - 1) - H(\eta_1))H(\eta_1)] \\ &\quad - \tfrac{1}{2}E_{n,K}[c(\eta_2)(H(\eta_1 + 1) - H(\eta_1))H(\eta_1)]. \end{aligned}$$

The second term is then computed as

$$\begin{aligned} &E_{n,K}[c(\eta_2)(H(\eta_1 + 1) - H(\eta_1))H(\eta_1)] \\ &\quad = E_{n,K}[E_{n,K}[c(\eta_2)|\eta_1](H(\eta_1 + 1) - H(\eta_1))H(\eta_1)] \\ &\quad = E_{n,K}[E_{n-1,K-\eta_1}[c(\eta_2)](H(\eta_1 + 1) - H(\eta_1))H(\eta_1)]. \end{aligned}$$

We conclude from these calculations that the generator specifies a biased nearest-neighbor random walk, a birth–death process X_t, on the state space $\{0, 1, \ldots, K\}$ with rates

$$\begin{aligned} &(1/2)E_{n-1,K-r}[c(\cdot)] \quad &&\text{for jumps } r \to r+1, \\ &(1/2)c(r) &&\text{for jumps } r \to r-1, \end{aligned}$$

and associated Dirichlet form $D^1_{n,K}$ given by

$$D^1_{n,K}(f) = (1/2)E^1_{n,K}[c(X)[f(X-1) - f(X)]^2].$$

In the same way, if instead of the canonical measures $P_{n,K}$ we consider the grand canonical measures P_ρ, we obtain a birth and death process X_t on N with rates

$$\begin{aligned} &(1/2)\alpha(\rho) \quad &&\text{for jumps } r \to r+1, \\ &(1/2)c(r) &&\text{for jumps } r \to r-1, \end{aligned}$$

and associated Dirichlet form D^1_ρ given by

$$D^1_\rho(f) = (1/2)E^1_\rho[c(X)[f(X-1) - f(X)]^2].$$

Here E^1_ρ indicates expectation with respect to the one site marginal of the product measure P_ρ.

The purpose of this section is to prove a spectral gap independent of n, K and ρ for birth and death processes with the stated rates.

LEMMA 4.1. *Under assumptions* (LG) *and* (M), *there exists a universal constant* $B_0 = B_0(a_1, a_2, k_0)$ *such that*

$$E^1_{n,K}[(H - E^1_{n,K}[H])^2] \le B_0 D^1_{n,K}(H)$$

for all $n \ge 1$, $K \ge 1$ *and* H *in* $L^2(P^1_{n,K})$.

LEMMA 4.2. *Under assumptions* (LG) *and* (M), *there exists a universal constant* $B_0' = B_0'(a_1, a_2, k_0)$ *such that*

$$E_\rho[(f(\eta_1) - E_\rho[f])^2] \le B_0' D^1_\rho(f)$$

for all functions $f\colon N \to R$ *in* $L_2(P_\rho)$.

Our proof relies on the following two general propositions. The proof of the first one can be found in Lemma 3.11 and Lemma 3.12 of [12]. Consider a continuous time Markov process Y_t on some countable set $\mathscr{E}$. For a site r in $\mathscr{E}$, denote by τ^Y_r the hitting time of r for the process Y.

PROPOSITION 4.1. *Let there exist* r_0 *in* $\mathscr{E}$ *and* $\lambda > 0$ *such that* $E_r[\exp(\lambda\tau^Y_{r_0})] < \infty$ *for all* r. *Then the spectral gap is larger than* λ.

PROPOSITION 4.2. *Denote by* L *the generator of* Y_t. *Fix some* r_0 *in* $\mathscr{E}$ *and suppose* $u, v\colon \mathscr{E} \to R_+$ *are functions satisfying* $Lv + uv = 0$. *If* $u(r) \ge \lambda$ *for* $r \ne r_0$ *and* $v(r) \ge \gamma$ *for all* r *in* $\mathscr{E}$, *then* $E_x[\exp(\lambda\tau^Y_{r_0})] \le v(x)/\gamma$.

PROOF OF PROPOSITION 4.2. Define the martingale

$$m(t) = v(Y_t)\exp\biggl(\int_0^t u(Y_s)\,ds\biggr),$$

with respect to the usual σ-fields. By Doob's stopping theorem, $E_x[m(\tau^Y_{r_0}\wedge t)] \le v(x)$ for all $t \ge 0$. The result follows from the inequality $E_x[m(\tau^Y_{r_0} \wedge t)] \ge \gamma E_x[\exp\lambda(\tau^Y_{r_0} \wedge t)]$ and Fatou's lemma. □

We shall prove now a spectral gap for birth and death processes on N with death rate equal to $c(\cdot)$ and some birth rate $b(\cdot)$ that we shall assume to be Lipschitz.

From hypotheses (LG) and (M) we have that

$$c(k) - c(j) \ge a_2\biggl[\frac{k-j}{k_0}\biggr] - a_1 k_0 \ge \frac{a_2}{k_0}(k-j) - a_2 - a_1 k_0$$

for every $k \ge j$. Here $[r]$ denotes the integer part of r. Denote a_2/k_0 by J_1 and $a_2 + a_1 k_0$ by e_2.

LEMMA 4.3. *Let Y_t be a birth and death process on N with death rate $c(\cdot)$ and birth rate b: $N \to R_+$. Assume that $\sup_k |b(k+1) - b(k)| \le J_2 < J_1$. The spectral gap λ is bounded below by a strictly positive constant $C = C(J_2, a_1, a_2, k_0, c^*)$, where $c^* = \min_k c(k)$.*

Notice that Lemma 4.2 follows from this proposition since the birth rate is constant.

PROOF OF LEMMA 4.3. Denote by $D(r)$ the inward drift at r:

$$D(r) = c(r) - b(r).$$

Notice that $D(0) = -b(0) \le 0$. Let r_0 be the first integer with nonnegative inward drift: $r_0 = \min\{r \in N;\ D(r) \ge 0\}$. For $r \ge r_0$, we have that

$$D(r) = c(r) - c(r_0) + D(r_0) + b(r_0) - b(r) \ge (J_1 - J_2)(r - r_0) - e_2$$

because $D(r_0) \ge 0$ by definition of r_0. For $r \ge r_2 = r_0 + 2e_2/(J_1 - J_2)$, $D(r) \ge I_1(r - r_0)$, provided $I_1 = (J_1 - J_2)/2$. Moreover, for $r_0 \le r \le r_2$, $b(r) \le c(r_0) + I_2$ and $c(r) \ge c(r_0) - I_3$, where $I_2 = J_2(r_2 - r_0)$ and $I_3 = J_1(r_2 - r_0)$. Of course $c(r_0) - I_3$ could be negative. In this case we have the alternative lower bound $c(r) \ge c^*$.

In possession of the above estimates, it is not difficult to find functions u and v satisfying assumptions of Proposition 4.2 for some $\gamma > 0$ and some $\lambda = \lambda(J_2, a_1, a_2, k_0, c^*)$. We leave the details to the reader and just indicate that v can be chosen to be quadratic on $\{r_0, \dots, r_2\}$ and linear on $\{r_2, \dots, \infty\}$.

A similar argument applies to the set $\{0, \dots, r_0\}$. Proposition 4.1 concludes the argument. □

PROOF OF LEMMA 4.1. Set $J_2 = J_1/2$. By Proposition 4.3, there exists n_0 in N so that $|E_{n,K+1}[c(\eta_0)] - E_{n,K}[c(\eta_0)]|$ is bounded by J_2 for $n \ge n_0$. In particular, in this case Lemma 4.1 follows from Lemma 4.3. For $n \le n_0$ and $K \ge K_0$, the proof of Lemma 4.1 is similar to the proof of Lemma 4.3 and relies on the following inequality proved in Corollary 4.2:

$$E_{n,K}[c(\eta_0)] \le E_{n,K+B}[c(\eta_0)] + B_4(c)$$

for all $B \ge B_1 n$. We leave the details to the reader. Finally, since there are only a finite number of cases for $n \le n_0$ and $K \le K_0$ and since for each fixed n and K the birth and death process is a finite state ergodic Markov process, the lemma is proved. □

We conclude this section with some estimates on the birth rate $E_{n,K}[c(\eta_0)]$ that were used above.

PROPOSITION 4.3. *There exists a constant $B_3 = B_3(a_1, a_2, k_0)$ such that*

$$\big|E_{n,K+1}[c(\eta_0)] - E_{n,K}[c(\eta_0)]\big| \le \frac{B_3}{n}$$

for all $n \ge 1$ and $K \ge 0$.

We first show that the difference is bounded in absolute value by a constant independent of n and K. The proof of this statement relies on a coupling argument.

LEMMA 4.4. *There exists a constant $B_1 = B_1(a_1, a_2, k_0)$ such that*

$$P_{n,K} \leq P_{n,K+B} \quad \text{for all } B \geq B_1 n \text{ and } K \text{ in } N.$$

PROOF. To clarify the proof, we shall assume that $c(k)-c(j) \geq 0$ if $k-j \geq 2$. It is not difficult to adapt this proof to assumption (M).

Fix n, K and some positive integer B. Consider the zero range process on $\{1, \dots, n\}$ where particles jump to any site with rate $c(\cdot)$. Its generator $\tilde{L}_n$ acts on functions as

$$(\tilde{L}_n f)(\eta) = \sum_{1 \leq x, y \leq n} c(\eta_x)[f(\eta^{x,y}) - f(\eta)].$$

Notice that the canonical measures $P_{n,K}$ are ergodic and reversible for this Markov process.

Fix two configurations η and ξ, respectively, on $\Sigma_{n,K}$ and $\Sigma_{n,K+B}$ such that $\eta \leq \xi$. The proof consists of finding a coupled process $(\eta(t), \xi(t))$ with three properties. We require that $\eta(0) = \eta$, $\xi(0) = \xi$, both marginals to evolve as zero-range processes with generator $\tilde{L}_n$, and that $\eta(t) \leq \xi(t)$ for all $t \geq 0$. This can be easily done in the case where the jump rate c is a nondecreasing function (cf. [1]). We extend here the coupling for processes satisfying assumptions (LG) and (M).

It is enough to show that for any two configurations (η, ξ) in $\Sigma_{n,K} \times \Sigma_{n,K+B}$ such that $\eta \leq \xi$, we may couple the η-particles' jumps with the ξ-particles' jumps for the order to be maintained after any possible jump.

Fix therefore two such configurations. Denote by b_0 (b_1) the total number of sites where the number of η and ξ particles differ by 0 (1). Since $\eta \leq \xi$,

$$b_0 = \big|\{x;\ \eta(x) = \xi(x)\}\big| \quad \text{and} \quad b_1 = \big|\{x;\ \eta(x) = \xi(x) - 1\}\big|.$$

Since particles jump indifferently to any site and since $c(k) - c(j) \geq 0$ as long as $k \geq j + 2$, by assumption (LG), the rate at which an uncoupled η particle appears is bounded above by $b_0 b_1 a_1$.

To conclude the proof it remains to show that for $B \geq B_1 n$, the rate at which one uncoupled ξ particle in the remaining $b = n - b_0 - b_1$ sites jumps to one of the b_0 sites is bounded below by $b_0 b_1 a_1$. Denote by $u_1, \dots, u_b$ the number of uncoupled ξ particles in the remaining b sites. Notice that $\sum_{1 \leq i \leq b} u_i = B - b_1$. By assumption (M), the rate at which one uncoupled ξ particle jumps to one of the b_0 sites with equal numbers of η and ξ particles is bounded below by

$$b_0 \sum_{i=1}^{b} \left\{ a_2 \left[\frac{u_i}{k_0} \right] - a_1 k_0 \right\} \geq b_0 \sum_{i=1}^{b} \left\{ a_2 \left\{ \frac{u_i}{k_0} - 1 \right\} - a_1 k_0 \right\}$$
$$= \frac{a_2 b_0}{k_0}(B - b_1) - a_2 b b_0 - a_1 k_0 b b_0$$
$$\geq \frac{a_2 b_0}{k_0} B - b_0 n \left\{ a_2 + \frac{a_2}{k_0} + a_1 k_0 \right\}.$$

We just have to take $B_1 = a_1 a_2^{-1} k_0 (k_0 + 1) + k_0 + 1$. □

COROLLARY 4.1. *There exists a constant* $B_2 = B_2(k_0, a_1, a_2)$ *such that*

$$\left| E_{n,K+1}[c(\eta_0)] - E_{n,K}[c(\eta_0)] \right| \leq B_2$$

for all $n \geq 1$ *and* $K \geq 0$.

PROOF. Let B_1 be the constant given by the previous lemma. Set $K_0 = B_1 n + 1$. The absolute value in the statement of the lemma is bounded by

$$\left| E_{n,K}[c(\eta_0)] - E_{n,K+K_0}[c(\eta_0)] \right| + \left| E_{n,K+1}[c(\eta_0)] - E_{n,K+K_0}[c(\eta_0)] \right|.$$

We concentrate on the first expression. The second one is estimated in the same way.

Since $P_{n,K} \leq P_{n,K+K_0}$, there exists (cf. [7], Theorem II.2.4) a coupled measure P_{n,K,K_0} on the product space $N^n \times N^n$ with first marginal equal to $P_{n,K}$, second marginal equal to $P_{n,K+K_0}$ and concentrated on the configurations (η, ξ) above the diagonal: $P_{n,K,K_0}\{(\eta, \xi),\ \eta \leq \xi\} = 1$. With this notation, we may bound the first expression in the last formula by

$$E_{n,K,K_0}[|c(\eta_0) - c(\xi_0)|] \leq a_1 E_{n,K,K_0}[|\eta_0 - \xi_0|].$$

Since the probability measure P_{n,K,K_0} is ordered, the last expression is equal to

$$a_1\{E_{n,K+K_0}[\xi_0] - E_{n,K}[\eta_0]\} = a_1 K_0 n^{-1}.$$

The corollary is thus proved because $K_0 = B_1 n + 1$. □

COROLLARY 4.2. *There exists a finite constant* B_4 *such that*

$$E_{n,K}[c(\eta_0)] \leq E_{n,K+B}[c(\eta_0)] + B_4$$

for all $B \geq B_1 n$.

PROOF. By assumption (LG) and (M), the jump rate $c(\cdot)$ may be decomposed as $c = c_1 + c_2$, where c_1 is a nondecreasing function and c_2 is a bounded function. Take for instance $c_1(k) = c([k/k_0]k_0)$. Since by Lemma 4.4, $P_{n,K} \leq P_{n,K+B}$ for $B \geq B_1 n$, we have that

$$E_{n,K}[c(\eta_1)] \leq E_{n,K}[c_1(\eta_1)] + \|c_2\|_\infty \leq E_{n,K+B}[c(\eta_1)] + 2\|c_2\|_\infty. \qquad \square$$

We are now ready to prove Proposition 4.3.

PROOF OF PROPOSITION 4.3. The proof consists of three different parts. We first consider the case of a large number of sites and bounded density, then the case of a large number of sites and densities bounded away from 0 and finally the case of a small number of sites.

Fix some density $\rho_1 > 0$. By Corollary 6.4 there exists $n_0 > 0$ such that

$$|E_{n,K+1}[c(\eta_0)] - E_{n,K}[c(\eta_0)]| \le n^{-1}\Big\{\sup_{0\le\rho\le\rho_1}|\alpha'(\rho)| + 2E_3\sqrt{1+\alpha(\rho_1)}\Big\}$$

for all $n \ge n_0$ and $\rho = K/n \le \rho_1$ because $\alpha(\cdot)$ is an increasing function. Since by inequality (5.1), $\alpha'(\rho)$ [that is equal to $\alpha(\rho)/\sigma(\alpha(\rho))^2$] is bounded above by a_1, this estimate proves the proposition for $n \ge n_0$ and $\rho \le \rho_1$.

We turn now to the case $\rho \ge \rho_1$. From identity (1.4), we have that

$$E_{n,K}[c(\eta_0)] = \alpha(\rho)\frac{P_\rho[\sum_{x=1}^n \eta_x = K-1]}{P_\rho[\sum_{x=1}^n \eta_x = K]}.$$

We may therefore rewrite the difference $E_{n,K+1}[c(\eta_0)] - E_{n,K}[c(\eta_0)]$ as

$$\alpha(\rho)\frac{((\nabla P_\rho)[\sum_{x=1}^n \eta_x = K-1])^2 - P_\rho[\sum_{x=1}^n \eta_x = K-1](\Delta P_\rho)[\sum_{x=1}^n \eta_x = K]}{P_\rho[\sum_{x=1}^n \eta_x = K]P_\rho[\sum_{x=1}^n \eta_x = K+1]}.$$

In this formula ∇ and Δ stand, respectively, for the discrete derivative and Laplacian and are applied to K, the total number of particles (cf. notation introduced just before Theorem 6.2). It follows from Theorem 6.2, taking $k_1 = 3$, that there exists n_0 and a constant $E = E(n_0, \rho_1)$ such that this ratio is bounded by $En^{-1}\sigma(\alpha(\rho))^{-2}$ for $n \ge n_0$ and $\rho = K/n \ge \rho_1$. This proves that the difference $E_{n,K+1}[c(\eta_0)] - E_{n,K}[c(\eta_0)]$ is bounded above by $C(\rho_1)n^{-1}$ because $\alpha(\rho)\sigma(\alpha(\rho))^{-2} \le a_1$ by inequality (5.2).

It remains to consider the case where the total number of sites is bounded, but this follows from the estimate stated in Corollary 4.1 provided we take B_3 large enough. □

5. Properties of the zero-range marginal. We prove in this section some properties of zero-range distributions used in the previous three sections. For each $k \in N$, denote by $\gamma_k(\alpha)$ the kth cumulant of the probability $\bar{P}_\alpha$; the kth cumulant is a polynomial in the normalized moments $m_i(\alpha)$ for $0 \le i \le k$, where

$$m_i(\alpha) = \bar{E}_\alpha[(\eta_0 - \rho(\alpha))^i].$$

We shall abbreviate $\gamma_2(\alpha)$ by $\sigma(\alpha)^2$.

We start proving that $\sigma(\alpha)^2\alpha^{-1}$ is bounded above by a finite constant and below by a positive constant uniformly on R_+. By a change of variables and assumption (LG) we have that

$$\begin{aligned} &E_\rho[\{c(\eta_0) - \alpha(\rho)\}^2] = \alpha(\rho)E_\rho[c(\eta_0+1) - c(\eta_0)] \le a_1\alpha(\rho), \\ &E_\rho[\{c(\eta_0) - \alpha(\rho)\}\{\eta_0 - \rho\}] = \alpha(\rho). \end{aligned} \tag{5.1}$$

Therefore, by the Schwarz inequality,

$$\begin{aligned}\alpha(\rho) = E_\rho[\{c(\eta_0) - \alpha(\rho)\}\{\eta_0 - \rho\}] &\le a_1^{1/2}\sigma(\alpha(\rho))\alpha(\rho)^{1/2} \\ &\le a_1\sigma^2(\alpha(\rho)).\end{aligned}$$

The reverse inequality follows from the one site spectral gap for the birth and death process on N associated with the grand canonical measure.

LEMMA 5.1. *Let B_0' be the constant given by Lemma 4.2. Then*

$$\frac{\sigma^2(\alpha(\rho))}{\alpha(\rho)} \le \frac{1}{2}B_0'.$$

PROOF. Apply Lemma 4.2 to the function η_1 and notice that the variance of η_1 is $\sigma^2(\alpha(\rho))$ and that its Dirichlet form is $(1/2)\alpha(\rho)$. □

In conclusion, there exist constants C_1 and C_2 depending only on a_1, a_2 and k_0 such that

$$0 < C_1 \le \frac{\sigma^2(\alpha)}{\alpha} \le C_2 < \infty \tag{5.2}$$

for all α in R_+. Notice that the zero-range processes described in Example 1.1 do not satisfy the upper bound.

The second main goal of this section is to derive some estimates required in the proof of local central limit theorems that are uniform in density. Denote by $v_\alpha(t)$ the normalized characteristic function associated with the distribution $\bar{P}_\alpha$:

$$v_\alpha(t) = \bar{E}_\alpha[\exp\{it(\eta_0 - \rho(\alpha))/\sigma(\alpha)\}].$$

In the next section we shall prove uniform Edgeworth expansions under the following set of assumptions. For each $\bar{\alpha} > 0$ and $\bar{k}$ in N, let the following statements hold:

(CL1) There exists a finite constant K_0 such that

$$\sup_{\alpha\ge\bar{\alpha}} m_{2k}(\alpha)/\sigma(\alpha)^{2k} \le K_0 \quad \text{for } 1 \le k \le \bar{k}.$$

(CL2) For every $\delta > 0$, there exists $C(\delta) < 1$ such that

$$\sup_{\alpha\ge\bar{\alpha}} \sup_{\delta\le|t|\le\pi\sigma(\alpha)} |v_\alpha(t)| \le C(\delta).$$

(CL3) There exists $\kappa > 0$ so that

$$\sup_{\alpha\ge\bar{\alpha}} \int_{|t|\le\pi\sigma(\alpha)} |v_\alpha(t)|^\kappa\, dt \le C < \infty.$$

Notice that these hypotheses are satisfied for all $\bar{\alpha} > 0$ as soon as they are satisfied by any positive parameter $\bar{\alpha}_0 > 0$. It is also a simple task to verify that both zero-range processes defined in Examples 1.1 and 1.2 satisfy (CL1)–(CL3).

We now deduce assumptions (CL1)–(CL3) from hypotheses (LG) and (M). We start with (CL1). Recall that $\gamma_k(\alpha)$ denotes the kth cumulant of the occupation variable with respect to the probability $\bar{P}_\alpha$.

LEMMA 5.2. *Assume hypotheses* (LG) *and* (M). *For all* $k \geq 1$, *there exists a finite constant* $C(k)$ *such that*

$$m_{2k}(\alpha) \leq C(k)\sigma(\alpha)^{2k}$$

for all $\alpha \geq 1$.

PROOF. To keep notation simple, denote $\rho(\alpha)$ simply by ρ. We shall prove this lemma by induction on k. Notice that by the Schwarz inequality and Lemma 4.2,

$$\begin{aligned} m_{2k}(\alpha) &= \bar{E}_\alpha[((\eta_0-\rho)^k - m_k(\alpha) + m_k(\alpha))^2] \\ &\leq 2B_0' D_\rho^1((\eta_0-\rho)^k) + 2C(k)^2\sigma^{2k}(\alpha). \end{aligned}$$

The Dirichlet form in the first term is equal to

$$D_\rho^1((\eta_0-\rho)^k) = (\alpha/2)E_\rho[((\eta_0-\rho)^k - (\eta_0-\rho+1)^k)^2].$$

Computing the kth power inside the expectation and applying the Schwarz inequality and the induction assumption, we obtain that the right-hand side is bounded above by

$$C(k)\alpha\sum_{j=0}^{k-1}\sigma^{2j}(\alpha).$$

On the one hand, we have by (5.2) that $\alpha \leq C_1^{-1}\sigma(\alpha)^2$ and on the other $\sigma^{2j}(\alpha) \leq C\sigma^{2(k-1)}(\alpha)$ for $0 \leq j \leq k-1$ and some finite universal constant C because $\alpha \geq 1$ and $\sigma(\alpha)^2 \geq C_1\alpha$. □

We turn now to assumptions (CL2) and (CL3). For $\rho \geq 0$, define

$$p_\rho(k) = \frac{1}{Z(\alpha(\rho))}\frac{\alpha(\rho)^k}{c(1)\cdots c(k)} \quad \text{for } k \in N.$$

Most of the time we will omit the index ρ to keep notation simple. We begin with an estimate which provides a bound on the characteristic functions of the occupation variable.

LEMMA 5.3. *Under assumption* (LG), *for every* $\rho > 0$,

$$\sum_{k\geq 0}|p_\rho(k+1) - p_\rho(k)| \leq \frac{\sqrt{a_1}}{\sqrt{\alpha(\rho)}}.$$

PROOF. First note that $p(k+1)-p(k)$ is equal to $p(k)\{\alpha(\rho)c(k+1)^{-1}-1\}$. In particular the above sum is equal to

$$E_\rho\left[\left|\frac{\alpha(\rho)}{c(\eta_0+1)}-1\right|\right]=\frac{1}{\alpha(\rho)}E_\rho\left[c(\eta_0)\left|\frac{\alpha(\rho)}{c(\eta_0)}-1\right|\right].$$

By the Schwarz inequality this last expectation is bounded above by $\alpha(\rho)^{-1}E_\rho[(c(\eta_0)-\alpha)^2]^{1/2}$. By inequality (5.1), this expression is less than or equal to $a_1^{1/2}\alpha(\rho)^{-1/2}$. □

Recall that we denoted v_α as the normalized characteristic function of the occupation variable under $P_{\rho(\alpha)}$. The previous lemma implies a certain decay of $v_\alpha(t)$, uniform over α.

LEMMA 5.4. *Assume hypothesis* (LG). *There exists a constant C_0 depending only on a_1, $c(1)$ and C_2 such that*

$$|v_\alpha(t)|\le\frac{C_0}{|t|}$$

for each $\alpha>0$ and $0<|t|\le\pi\sigma(\alpha)$.

PROOF. Fix $\delta<1$ and $\alpha>0$. Define $\tilde v_\alpha(t)=\bar E_\alpha[\exp\{it(\eta_0-\rho)\}]$,

$$\tilde v_\alpha^\delta(t)=\sum_{k\ge0}\delta^k e^{itk}p(k)\quad\text{and}\quad S(k)=\sum_{l\ge k}\delta^l e^{itl}.$$

Notice that the absolute value of $S(k)$ is bounded by $|1-\delta\exp(it)|^{-1}$, uniformly on k. Summing by parts we write

$$\begin{aligned}\tilde v_\alpha^\delta(t)&=\sum_{k\ge0}\{S(k)-S(k+1)\}\,p(k)\\&=S(0)p(0)+\sum_{k\ge0}\{p(k+1)-p(k)\}\,S(k+1).\end{aligned}$$

Therefore,

$$|\tilde v_\alpha^\delta(t)|\le\frac{1}{|1-\delta e^{it}|}\left\{p(0)+\sum_{k\ge0}|p(k+1)-p(k)|\right\}.$$

Notice that $Z(\alpha)\ge1+\alpha c(1)^{-1}$. In particular, $Z(\alpha)\ge C\alpha^{1/2}$ for some constant C that depends on $c(1)$ only and

$$p(0)=\frac{1}{Z(\alpha)}\le\frac{C}{\sqrt{\alpha}}.$$

Hence, by the previous lemma, we have for each $\delta<1$ that

$$|\tilde v_\alpha^\delta(t)|\le\frac{1}{|1-\delta e^{it}|}\frac{C(a_1,c(1))}{\sqrt{\alpha}}.$$

Since $|1 - e^{it}| \geq (\pi/2)|t|$ for $0 \leq t \leq \pi$, allowing $\delta \uparrow 1$, we obtain that

$$|\tilde{v}_\alpha(t)| \leq \frac{C(a_1, c(1))}{|t|\sqrt{\alpha}}.$$

To conclude the proof, we need only comment that $v_\alpha(t) = \tilde{v}_\alpha(t\sigma^{-1})$ and that $\sigma^2\alpha^{-1}$ is bounded above by C_2 by (5.2). □

Assumption (CL3) now follows from the estimate obtained in the previous lemma and the fact that absolute value of the characteristic function is always bounded by 1. It remains to deduce hypothesis (CL2). We turn to a series of lemmas devised to bound the characteristic function $v_\alpha(t)$ near the origin. We start with a general result.

Consider a density function $f\colon R \to R_+$ on the real line and a cosine wave function $W\colon R \to [-1, 1]$ with period l and amplitude 1. Let $I = [a, b]$ be an interval and assume that the following statements hold:

ASSUMPTION (a). There is $\gamma > 0$ such that $\int_I f(x)\,dx \geq \gamma$.

ASSUMPTION (b). There is a finite constant C_3 such that $\sup_{x,\, y\in I} f(x)/f(y) \leq C_3$.

ASSUMPTION (c). The period l satisfies $100l \leq b - a$.

LEMMA 5.5. *Under Assumptions* (a), (b) *and* (c) *there is a strictly positive constant* $\delta = \delta(\gamma, C_3)$ *such that*

$$\int f(x)W(x)\,dx \leq 1 - \delta.$$

PROOF. Since f is a density, it is enough to show that

$$\int_I f(x)(1 - W(x))\,dx \geq \delta.$$

Let E be the closed set defined by $E = \{x\colon 1 - W(x) \geq 1/10\}$. Since there are at least 98 full periods in I, there exist universal constants $0 < A_1, A_2 < \infty$ such that

$$A_1 < \frac{|E \cap I|}{|E^c \cap I|} < A_2. \tag{5.3}$$

In this formula, for a measurable set B, $|B|$ stands for its Lebesgue measure. It is now easy to obtain a lower bound for the previous integral. Since W is bounded above by 1 and $1 - W \geq 1/10$ on E, we have that

$$\int_I f(1 - W)\,dx \geq \int_{I\cap E} f(1 - W)\,dx \geq \tfrac{1}{10}\int_{I\cap E} f\,dx.$$

To complete the proof we replace the last integral by $\int_I f\,dx$ and take advantage of Assumption (a). In order to perform this replacement, we rely on Assumption (b) and on inequalities (5.3). Indeed, by Assumption (b),

$$\int_{I\cap E} f\,dx \geq C_3^{-1} f(a)|I\cap E|.$$

By inequality (5.3) and Assumption (b), this last expression is at least

$$C_3^{-1}A_1 f(a)|I\cap E^c| \geq C_3^{-2}A_1\int_{I\cap E^c} f\,dx.$$

In conclusion,

$$\int_{I\cap E} f\,dx \geq \{1+C_3^2A_1^{-1}\}^{-1}\int_I f\,dx \geq \gamma\{1+C_3^2A_1^{-1}\}^{-1}.$$

This completes the proof with $\delta = 10^{-1}\gamma[1+C_3^2A_1^{-1}]^{-1}$. □

A version of this lemma for discrete probabilities on the integers is valid provided the amplitude l is not too small, say $l > 100$. Our intention is to apply this result to bound the characteristic function $v_\alpha(t)$ near the origin.

LEMMA 5.6. *Under assumptions* (LG) *and* (M), *for every* $R_0 > 0$, *there exists* $\alpha_0 \geq 1$ *with the following property. For every* $\varepsilon > 0$, *there exists a strictly positive constant* $\delta = \delta(\varepsilon)$ *such that*

$$\sup_{\varepsilon<|t|\leq 2\pi R_0} |v_\alpha(t)| \leq 1-\delta.$$

for all $\alpha > \alpha_0$.

Before we prove this lemma, we show how to derive (CL2) from (LG) and (M).

COROLLARY 5.1. *Assumption* (CL2) *follows from hypotheses* (LG) *and* (M).

PROOF. Fix $\varepsilon > 0$. We wish to bound $\sup_{\varepsilon<|t|\leq\pi\sigma}|v_\alpha(t)|$ by a constant strictly smaller than 1. In Lemma 5.4 we bounded $|v_\alpha(t)|$ by $C_0|t|^{-1}$ uniformly over the parameter α. In particular, for $|t|\geq 2C_0$, $|v_\alpha(t)|\leq 1/2$ for all $\alpha\geq 1$. It remains to consider the interval $[\varepsilon, 2C_0]$. Take $R_0 = \pi^{-1}C_0$ in Lemma 5.6. By this result, there exists α_0, independent of ε, and $\delta=\delta(\varepsilon)$ strictly positive so that

$$\sup_{\varepsilon<|t|\leq 2C_0} |v_\alpha(t)| \leq 1-\delta.$$

This concludes the proof. □

We conclude this section by proving Lemma 5.6.

PROOF OF LEMMA 5.6. Rewrite $v_\alpha(t)$ as $|v_\alpha(t)| \exp\{i\theta(t)\}$ for some real continuous function θ (that may depend on the parameter α). Then

$$|v_\alpha(t)| = \sum_k p(k) \cos\big(t\{k - \rho(\alpha)\}\sigma(\alpha)^{-1} - \theta(t)\big).$$

Fix t such that $\varepsilon \le |t| \le 2\pi R_0$ and let $W(x) = \cos(t\{x - \rho(\alpha)\}\sigma(\alpha)^{-1} - \theta(t))$. Denote by $l = 2\pi\sigma|t|^{-1}$ the period of $W(\cdot)$. Since $|t| \le 2\pi R_0$, the period l is bounded below by $\sigma(\alpha)R_0^{-1}$. In particular, l is bounded below by 100 provided $\alpha_0 = \alpha_0(R_0)$ is chosen large enough and $\alpha > \alpha_0$. Fix such parameter α.

Let $b_0 = \max\{50, 2R_0\}$ and define the interval I by

$$I = [\rho(\alpha) - b_0 l, \rho(\alpha) + b_0 l].$$

Assumption (c) is obviously satisfied. It is now a matter of verifying the first two assumptions. This is the subject of the following three lemmas. □

LEMMA 5.7. *There exists a strictly positive constant $\gamma = \gamma(R_0)$, independent of α, such that*

$$\sum_{k\in I} p(k) \ge \gamma$$

for all $|t| \le 2\pi R_0$.

PROOF. This follows from Chebyshev's inequality, the fact that $|t|$ is bounded above by $2\pi R_0$, and that $2R_0 \le b_0$. This argument gives $\gamma = 3/4$. □

For the next lemma we will linearly interpolate the jump rate $c(\cdot)$ to the positive real line.

LEMMA 5.8. *Assume hypotheses* (LG) *and* (M). *There exists a constant C_4 depending only on a_1, a_2 and k_0 such that*

$$|c(\rho) - \alpha(\rho)| \le C_4\sqrt{\alpha(\rho)}$$

for all $\rho > 0$.

PROOF. By the Schwarz inequality and assumption (LG),

$$|\alpha(\rho) - c(\rho)| \le E_\rho[|c(\eta_0) - c(\rho)|] \le a_1 E_\rho[|\eta_0 - \rho|] \le a_1\sigma(\alpha(\rho)).$$

Inequality (5.2) completes the proof. □

LEMMA 5.9. *Under assumptions* (LG) *and* (M), *there exists a finite constant C_5, depending only on ε, R_0, a_1, a_2 and k_0 such that*

$$\sup_{x,\, y\in I} \frac{p(x)}{p(y)} < C_5.$$

PROOF. Fix $x < y$. Since $\log(1+u) \le u$ for all $u > -1$,

$$\log \frac{p(x)}{p(y)} = \log \prod_{k=x}^{y-1} \frac{p(k)}{p(k+1)} = \sum_{k=x}^{y-1} \log\left(1 + \left\{\frac{p(k)}{p(k+1)} - 1\right\}\right)$$
$$\le \sum_{k=x}^{y-1} \left| \frac{p(k)}{p(k+1)} - 1 \right|.$$

From the definition of $p(k) = p_\alpha(k)$, $p(k)/p(k+1)$ is equal to $c(k+1)\alpha^{-1}$. Therefore, since x and y belong to the interval I and the linearly interpolated function $c\colon R \to R_+$ is a_1-Lipschitz continuous, this last sum is bounded above by

$$\sum_{i=-b_0 l}^{b_0 l} \left| \frac{c(\rho+i)}{\alpha} - 1 \right| + \frac{a_1}{\alpha} \le \frac{C_4}{\sqrt{\alpha}} + 2a_1 \sum_{i=1}^{b_0 l} \frac{i+1}{\alpha}.$$

In the last inequality, we have used the previous lemma to bound the difference $|c(\rho) - \alpha(\rho)|$ by $C_4\sqrt{\alpha}$. Since $\alpha \ge \alpha_0 \ge 1$ and $|t| \ge \varepsilon$, the last sum is bounded by $C\sigma^2(\alpha\varepsilon^2)^{-1} + C$ for some constant C which depends only on R_0 and a_1. By inequality (5.2), $\sigma^2\alpha^{-1}$ is bounded above by some constant C_2. Therefore the last sum is dominated by the expression $C_5(a_1, a_2, k_0, R_0)\varepsilon^{-2}$. A similar argument follows for $x > y$. □

6. Uniform local central limit theorem. This section is devoted to the study of local central limit theorems for zero range distributions uniformly in the parameter α. As a by-product we will also derive a few results important for the induction step in the proof of Theorem 1.1.

Diaconis and Freedman [2] considered uniform Edgeworth expansions up to the second order for exponential families under slightly different assumptions, not verifiable for zero-range marginals corresponding to small values of the parameter α. More precisely, recall that $m_k(\alpha)$ and $\gamma_k(\alpha)$ denote, respectively, the kth moment and the kth cumulant of the distribution $\overline{P}_\alpha$ and $\sigma(\alpha)^2 = \gamma_2(\alpha)$. For α close to 0, a simple computation shows that $m_k(\alpha) = c(1)^{-1}\alpha + O(\alpha^2)$ and, therefore, $\gamma_k(\alpha)/\sigma(\alpha)^k$ diverges as α approaches 0 for $k \ge 3$. In particular, the Edgeworth expansion with an error uniform in the parameter α cannot be correct close to the origin. In fact, as we shall see below, these errors are of order $(n\sigma(\alpha)^2)^{-k}$.

To state the main theorem of this section we recall several classical definitions. For $m \ge 0$, denote by $H_m(x)$ the Hermite polynomial of degree m:

$$H_m(x) = (-1)^m \exp\left(\frac{x^2}{2}\right) \frac{d^m}{dx^m} \exp\left(\frac{-x^2}{2}\right) = m! \sum_{k=0}^{[m/2]} (-1)^k \frac{x^{m-2k}}{k!(m-2k)!2^k}.$$

Let $q_0(x)$ denote the density of the normalized Gaussian distribution and, for $j \ge 1$, let

$$q_j(x) = \frac{1}{\sqrt{2\pi}} \exp\left(\frac{-x^2}{2}\right) \sum H_{j+2a}(x) \prod_{m=1}^{j} \frac{1}{k_m!} \left(\frac{\gamma_{m+2}}{(m+2)!\sigma^{m+2}}\right)^{k_m},$$

where the summation is carried out over all nonnegative integer solutions of $k_1 + 2k_2 + \cdots + jk_j = j$ and $k_1 + k_2 + \cdots + k_j = a$. We are now ready to state the main result of this section.

THEOREM 6.1. (a) *For all* $\alpha_0 > 0$ *and* $k_0 \in N$, *there exist finite constants* $E_0 = E_0(\alpha_0, k_0)$ *and* $A = A(\alpha_0, k_0)$ *such that*

$$\left|\sqrt{n\sigma^2}\bar{P}_\alpha\left[\sum_{i=1}^{n}\eta_i = K\right] - \sum_{j=0}^{k_0-2}\frac{1}{n^{j/2}}q_j(x)\right| \le \frac{E_0}{(\sigma^2(\alpha)n)^{(k_0-1)/2}}$$

uniformly over all parameters $\alpha \le \alpha_0$ *such that* $\sigma^2(\alpha)n \ge A$. *In this formula* x *stands for* $(K - n\rho(\alpha))/\sigma(\alpha)\sqrt{n}$.

(b) *Assume hypotheses* (CL1)–(CL3) *with* $\bar{\alpha} = \alpha_1$ *and* $\bar{k} = [k_1/2] + 1$. *There exist a constant* $E_0 = E_0(k_1, \alpha_1)$ *and* $n_0 = n_0(k_1, \alpha_1)$ *such that*

$$\sup_\alpha\left|\sqrt{n\sigma^2}\bar{P}_\alpha\left[\sum_{i=1}^{n}\eta_i = K\right] - \sum_{j=0}^{k_1-2}\frac{1}{n^{j/2}}q_j(x)\right| \le \frac{E_0}{n^{(k_1-1)/2}}$$

for all $n \ge n_0$ *where the supremum is taken over all* $\alpha \ge \alpha_1$.

The proof of this theorem is omitted since it follows closely the classical arguments given for instance in Petrov (Theorem VII.12 of [9]). There is only the slight problem of controlling the integral I_3 in this theorem. In part (a) this is not difficult since, if $\tilde{v}_\alpha(t) = \bar{E}_\alpha[\exp\{it(X - \rho(\alpha))\}]$ denotes the characteristic function of $\eta_0 - \rho(\alpha)$ under $\bar{P}_\alpha$, we may write

$$\begin{aligned}\left|\tilde{v}_\alpha(t)\right|^2 - 1 &\le \bar{P}_\alpha[\eta_0 = 0]\bar{P}_\alpha[\eta_0 = 1]\{\cos t - 1\}\\ &\le C(\alpha_0)\alpha\{\cos t - 1\}\end{aligned}$$

for some universal constant $C(\alpha_0)$ because

$$\bar{P}_\alpha[\eta_0 = 0] = 1 + O(\alpha) \quad \text{and} \quad \bar{P}_\alpha[\eta_0 = 1] = c(1)^{-1}\alpha + O(\alpha^2).$$

These estimates permit us to bound the integral I_3.

On the other hand, for part (b), hypotheses (CL2) and (CL3) are built to take care of the integral I_3.

Denote by ∇ and by Δ, respectively, the discrete derivative and Laplacian

$$(\nabla f)(i) = f(i+1) - f(i), \qquad (\Delta f)(i) = \{f(i+1) + f(i-1) - 2f(i)\}$$

for each $f\colon Z \to R$. In the next theorem we use the notation

$$(\nabla\bar{P}_\alpha)\left[\sum_{i=1}^{n}\eta_i = K\right] = \bar{P}_\alpha\left[\sum_{i=1}^{n}\eta_i = K+1\right] - \bar{P}_\alpha\left[\sum_{i=1}^{n}\eta_i = K\right],$$

$$(\nabla q_j)(x) = q_j\left(\frac{K + 1 - n\rho(\alpha)}{\sigma(\alpha)\sqrt{n}}\right) - q_j\left(\frac{K - n\rho(\alpha)}{\sigma(\alpha)\sqrt{n}}\right)$$

and similar notation with Δ replacing ∇.

THEOREM 6.2. *Under the assumptions of Theorem* 6.1(b), *there exist constants* $E_0' = E_0'(k_1, \alpha_1)$ *and* $n_0' = n_0'(k_1, \alpha_1)$ *such that*

$$\left|\sqrt{n\sigma^2}(\nabla \bar{P}_\alpha)\left[\sum_{i=1}^n \eta_i = K\right] - \sum_{j=0}^{k_1-2} \frac{1}{n^{j/2}}(\nabla q_j)(x)\right| \le \frac{E_0'}{\sigma(\alpha) n^{(k_1-1)/2}}$$

and

$$\left|\sqrt{n\sigma^2}(\Delta \bar{P}_\alpha)\left[\sum_{i=1}^n \eta_i = K\right] - \sum_{j=0}^{k_1-2} \frac{1}{n^{j/2}}(\Delta q_j)(x)\right| \le \frac{E_0'}{\sigma(\alpha)^2 n^{(k_1-1)/2}}$$

for all $n \ge n_0'$ *and all* $\alpha \ge \alpha_1$. *In these formulas the discrete derivatives and Laplacians are applied to* K.

The proof of Theorem 6.2 follows closely that of Theorem 6.1 and is therefore omitted. Notice that the estimate improves as the parameter α increases since $\sigma^2(\alpha)$ behaves as α by inequality (5.2).

We conclude this section by stating several corollaries important to the proof of the spectral gap result, Theorem 1.1.

COROLLARY 6.1. *Fix* $f: N^\ell \to R$ *for some fixed positive integer* ℓ.

(a) *There exists a constant* $E_1 = E_1(\alpha_0)$ *such that*

$$\begin{aligned}&|E_{n,K}[f] - E_\rho[f]| \\ &\quad\le \frac{E_1\ell}{n}\left\{\frac{1}{\sigma^2(\alpha(\rho))}E_\rho[|f - \langle f\rangle_\rho|] + \frac{1}{\sigma(\alpha(\rho))}\sqrt{E_\rho[(f - \langle f\rangle_\rho)^2]}\right\}\end{aligned}$$

uniformly over all n *and* K *such that* $n \ge 2\ell$, $K/n \le \rho(\alpha_0)$ *and* $\sigma^2(\alpha(\rho))n \ge A(\alpha_0, 3)$, *where* A *is given by Theorem* 6.1(a). *In this formula* ρ *stands for the density* K/n.

(b) *Assume hypotheses* (CL1)–(CL3) *with* $\bar{k} = 5$ *and* $\bar{\alpha} = \alpha_1$. *There exists a constant* $E_1 = E_1(\alpha_1)$ *such that*

$$|E_{n,K}[f] - E_\rho[f]| \le \frac{E_1\ell}{n}\sqrt{E_\rho[(f - \langle f\rangle_\rho)^2]}$$

uniformly over all $n \ge \max\{2\ell, n_0(3, \alpha_1)\}$ *and* K *so that* $K/n \ge \rho(\alpha_1)$.

(c) *Under the same assumptions of* (b), *there exist a universal constant* E_2 *and* $n_0 = n_0(\alpha_1)$ *so that*

$$E_{n,K}[f] \le E_2 E_{K/n}[|f|]$$

for all $n \ge n_0(\alpha_1)$ *and all* K *so that* $K/n \ge \rho(\alpha_1)$.

The proof is left to the reader. It is a simple consequence of the Edgeworth expansion up to the second order.

To keep notation simple, we will denote the variance of a function f with respect to the probability $\bar{P}_\alpha$ as $\sigma_f(\alpha)^2$. Completing the Edgeworth expansion up to the third order we obtain the following corollary.

COROLLARY 6.2. *Fix $f\colon N^\ell \to R$ for some fixed positive integer ℓ. There exists a constant $E_1 = E_1(\ell, \alpha_0)$ such that*

$$\left| E_{n,K}[f] - E_\rho[f] - \frac{1}{2n}\left\{ \frac{\gamma_3}{\sigma^4}\left\langle f;\, \sum_{i=1}^{\ell}(\eta_i - \rho)\right\rangle_\rho - \frac{1}{\sigma^2}\left\langle f;\, \left\{\sum_{i=1}^{\ell}(\eta_i - \rho)\right\}^2\right\rangle_\rho \right\}\right| \le \frac{E_1 \sigma \sigma_f}{(n\sigma^2)^{3/2}}$$

for all n and K such that $n \ge 2\ell$, $K/n \le \rho(\alpha_0)$ and $\sigma^2(\alpha(\rho))n \ge A(\alpha_0, 4)$, where A is given by Theorem 6.1(a). *In this formula ρ denotes the density K/n and $\langle f;\, g\rangle_\rho$ stands for the covariance of f and g with respect to P_ρ.*

Allowing $\ell = 1$ and $f(\eta) = c(\eta_0)$ in Corollary 6.2, we obtain the following result.

COROLLARY 6.3. *For every $\varepsilon > 0$, there exist $n_0 > 0$ and $\alpha_0 > 0$ such that*

$$\left| E_{n,K-1}[c(\eta_0)] - E_{n,K}[c(\eta_0)] + \frac{1}{n} E_{n,K-1}[c(\eta_0 + 1) - c(\eta_0)] \right| \le \frac{\varepsilon}{n}$$

for all $n \ge n_0$ and $K/n < \rho(\alpha_0)$.

PROOF. Fix some parameter $\alpha_0 > 0$ and $\varepsilon > 0$. For n large enough and K such that $\sigma^2(\alpha(\rho))n \ge C(\alpha_0, \varepsilon)$, the estimate follows from the previous corollary. On the other hand, $\sigma^2(\alpha(\rho))n \le C(\alpha_0, \varepsilon)$ implies that the total number of particles K is bounded by some constant $C_1 = C_1(\alpha_0, \varepsilon)$. It is therefore easy to complete the proof by inspection. □

With similar arguments we obtain from Corollary 6.1(a) and (b) the following result.

COROLLARY 6.4. *There exist a universal constant E_3 and $n_0 > 0$ such that*

$$|E_{n,K}[c(\eta_0)] - E_{K/n}[c(\eta_0)]| \le \frac{E_3}{n}\sqrt{1 + \alpha(\rho)}$$

for all $n \ge n_0$.

Acknowledgment. We thank Professor H. T. Yau for several enlightening discussions.

REFERENCES

[1] ANDJEL, E. D. (1982). Invariant measures for the zero range process. *Ann. Probab.* **10** 525–547.

[2] DIACONIS, P. and FREEDMAN, D. A. (1988). Conditional limit theorems for exponential families and finite versions of de Finetti's theorem. *J. Theoret. Probab.* **1** 381–410.

[3] KARLIN, S. (1968). *Total Positivity*. Stanford Univ. Press.

[4] KIPNIS, C. (1986). Central limit theorems for infinite series of queues and applications to simple exclusion. *Ann. Probab.* **14** 397–408.

[5] KIPNIS, C. and LANDIM, C. (1995). *Hydrodynamical Limit of Interacting Particle Systems*. Preprint.

[6] KIPNIS, C., LANDIM, C. and OLLA, S. (1994). Hydrodynamical limit for a non gradient system: the generalized symmetric exclusion process. *Comm. Pure Appl. Math.* **47** 1475–1545.

[7] LIGGETT, T. M. (1985). *Interacting Particle Systems*. Springer, New York.

[8] LU, S. L. and YAU, H. T. (1993). Spectral gap and logarithmic Sobolev inequality for Kawasaki and Glauber dynamics. *Comm. Math. Phys.* **156** 399–433.

[9] PETROV, V. V. (1975). *Sums of Independent Random Variables*. Springer, New York.

[10] QUASTEL, J. (1992). Diffusion of color in the simple exclusion process. *Comm. Pure Appl. Math.* **45** 623–679.

[11] SETHURAMAN, S. and XU, L. (1996). A central limit theorem for conservative particle dynamics. *Ann. Probab.* **24** 1622–1650.

[12] SOKAL, A. D. and THOMAS, L. E. (1989). Exponential convergence to equilibrium for a class of random-walk models. *J. Statist. Phys.* **54** 799–828.

[13] VARADHAN, S. R. S. (1994). Nonlinear diffusion limit for a system with nearest neighbor interactions II. In *Asymptotic Problems in Probability Theory: Stochastic Models and Diffusion on Fractals* (K. Elworthy and N. Ikeda, eds.) 75–128. Wiley, New York.

C. LANDIM
IMPA
ESTRADA DONA CASTORINA 110
CEP 22 460 JARDIM BOTANICO
RIO DE JANEIRO
BRAZIL
E-MAIL: landim@impa.br

S. SETHURAMAN AND S. VARADHAN
COURANT INSTITUTE
251 MERCER STREET
NEW YORK, NEW YORK 10012
E-MAIL: sethuraman@math.ethz.ch

Itô's Stochastic Calculus and Probability Theory, Springer, 1996, 385-400

The complex story of simple exclusion

S.R.S. Varadhan

Courant Institute of Mathematical Sciences, New York University,
251 Mercer Street, New York, NY 10012, USA

1. Introduction and description of the model

The simple exclusion model serves as perhaps the simplest model of an interacting system of particles that still exhibits some of the complex behavior one expects from an interacting system. The aim of this expository article is to describe, for various simple exclusion models, the different types of results that are known that fall within the framework of hydrodynamic scaling limits.

First we start with a description of the models. The physical space is Z^d, the lattice of integers in d dimensions. We will have particles located at some of the sites i.e. points of Z^d. Any given site can have atmost one particle. The quantity $\eta(x)$ is defined to be equal to 1 if there is a particle at site x and 0 otherwise. The function $\eta : Z^d \to \{0,1\}$ will then completely describe the state of our system and tell us exactly which sites are locations of particles. Of course in this picture the particles are indistinguishable and only the configuration of occupied sites is relevent. The state space then is the set

$$\Omega = \{\eta : Z^d \to \{0,1\}\}$$

of all configurations of particles on Z^d.

The particles move in time by random jumps. Each particle waits for a random exponential time and at the end of this time tries to jump to a new site. The jump size z is randomly chosen from Z^d according to a probability distribution $p(z)$. If the particle is located currently at x, and z is chosen as the jump size, an attempt is made to jump to $x+z$. The attempt is successful and the jump is executed if the site $x+z$ is unoccupied. If the site $x+z$ is occupied the jump is disallowed and the particle starts waiting for a new exponential time. If the jump to $x+z$ is completed the system starts afresh. Of course all the particles are doing this simultaneously and independently of each other. Since we are dealing with continuous time there will be no ties to resolve. However the possibility of having an infinite number of particles requires a careful construction of the process. See the book [5] for details in this regard. One can describe our motion as a Markovian evolution on the state space Ω. We define the operator $\nabla_{x,x'}$ acting on any function f defined on Ω by,

$$(\nabla_{x,x'} f)(\eta) = f(\eta^{x,x'}) - f(\eta) \tag{1.1}$$

The author was supported by NSF grant DMS- 9503419 and ARO grant DAAL03-92-G-0317

where $\eta^{x,x'}$ is the new configuration obtained from η by the rule

$$\eta^{x,x'}(a) = \begin{cases} \eta(a) & \text{if } a \neq x \text{ or } x' \\ \eta(x') & \text{if } a = x \\ \eta(x) & \text{if } a = x'. \end{cases} \tag{1.2}$$

In other words $\nabla_{x,x'}$ interchanges the situation at x and x' leaving things alone at all oter sites. If both the sites x and x' are occupied, or if both are empty, $\nabla_{x,x'}$ does nothing and if only one of the sites is occupied, it moves the particle from the occupied site to the free site. $\nabla_{x,x'}$ is symmetric in x and x' i.e. $\nabla_{x,x'} \equiv \nabla_{x',x}$.

Let $f(\eta)$ be a local function, i.e. a function that depends only on a finite number of coordiantes, that is to say on $\{\eta(x) : x \in F\}$ for some finite subset $F \subset Z^d$. The infinitesimal generator of our Markov Process acts on such local functions and expresses in mathematical terms the verbal description that was provided earlier.

$$(\mathcal{L}f)(\eta) = \sum_{x,y} p(y-x)\eta(x)(1-\eta(y))(\nabla_{x,y}f)(\eta) \tag{1.3}$$

We can assume without loss of generality that $p(0) = 0$.

Since Z^d is infinite we could have some technical problems in proving the existence of the process and in establishing some of the results. To simplify matters let us take a large integer N and consider the finite periodic lattice Z_N^d, of integers modulo N, in every coordinate. We shall make the hypothesis that $p(z)$ is local i.e. $p(z) = 0$ for $z \notin F$ for some fixed finite set F. The definition of $\mathcal{L}$ does not really depend on N in any essential way. Our process then will have a finite, but large, state space of 2^{N^d} points. The problems of hydrodynamic scaling have to do with the behavior of the system over large spatial and temporal scales. Space is scaled by imbedding the lattice Z_N^d in the d-dimensional torus T^d by sending $x \to \frac{x}{N}$ viewed as a point of the unit torus T^d. The lattice spacing then is $\frac{1}{N}$. The time scale is changed by speeding up time by a factor of N^α and this shows up as an extra factor multiplying the infinitesimal generator. The value of α is either 1 or 2 depending on the circumstance. There will be three main classes into which our models will fall depending on the nature of $p(\cdot)$.

Case1. $p(\cdot)$ is symmetric i.e. $p(z) = p(-z)$.

Case2. $p(\cdot)$ is asymmetric, but $\sum zp(z) = 0$.

Case3. $\sum zp(z) = b \neq 0$

No matter what $p(\cdot)$ is, the invariant measures of the process are easy to describe. For any fixed N the total number K of the particles is conserved in our dynamics and for each N and $K \leq N^d$, there will be an inavariant measure $\mu_{N,K}$ for our process. There are $\binom{N^d}{K}$ possible configurations of size

K among N^d sites and $\mu_{N.K}$ is the uniform distribution among them. These are precisely all extremal or ergodic inavariant distributions. As $K, N \to \infty$, with $\frac{K}{N^d} \to \rho$, we get as limits invariant measures μ_ρ for the evolution on Z^d, which are the Bernoulli product measures on Ω with

$$\mu_\rho[\eta(x) = 1] = \rho \quad \text{for each} \quad x \in Z^d \tag{1.4}$$

In the infinite volume case there may be other non translation invariant measures which are also invariant measures for the process. But we do not have to deal with them. The problem of hydrodynamic scaling is to study the behavior of the system in the speeded up time scale especially when it starts from an initial state that is far away from any equilibrium configuration. Our questions will deal with the collective behavior of our system and we will have more or less explicit answers for our questions.

2. The symmetric case: the behavior of the density

For each configuration $\eta(\cdot)$ we can associate the empirical density

$$\nu_N = \frac{1}{N^d} \sum \delta_{\frac{x}{N}} \eta(x) \tag{2.1}$$

and view it as a random measure on T^d. Its total mass is $\frac{K}{N^d}$ and is a constant in time. Suppose that our evolution starts at time 0 from a configuration that may or may not be random and for some nonrandom density $\rho_0(y)$

$$\lim_{N\to\infty} \frac{1}{N^d} \sum \delta_{\frac{x}{N}} \eta(x) = \lim_{N\to\infty} \nu_N(0) = \rho_0(y)dy \tag{2.2}$$

(in probabiliy in the weak topology). Let us also suppose that the time scale has been speeded up by a factor of N^2, which is the appropriate rescaling factor in this case. Then we may ask if for any $t > 0$ (in the new time scale)

$$\lim_{N\to\infty} \frac{1}{N^d} \sum \delta_{\frac{x}{N}} \eta_t(x) = \lim_{N\to\infty} \nu_N(t) = \rho(t, y)dy \tag{2.3}$$

also exists and if so how to calculate $\rho(t, y)$ in terms of $\rho_0(y)$. In particular since $\rho_0(\cdot)$ is deterministic, one expects $\rho(t, \cdot)$ to be deterministic as well. Since weak convergence is tested by integrating against smooth test functions, we need to compute

$$d\left(\frac{1}{N^d} \sum J(\frac{x}{N}) \eta_t(x)\right) = d\xi_J(t)$$

By standard Markov Process theory

$$d\xi_J(t) = A_J(t)dt + dM_J(t)$$

where $M_J(t)$ is a martingale. One can make explicit calculations and verify that for each $t > 0$,

$$E\{[M_J(t)]^2\} \to 0 \quad \text{as} \quad N \to \infty$$

The problem then reduces to calculating

$$A_J(t) = G_N(\eta_t(\cdot))$$

where
(2.4)

$$\begin{aligned} &G_N(\eta(\cdot)) \\ &= N^2 \mathcal{L}\left(\frac{1}{N^d}\sum J(\frac{x}{N})\eta(x)\right) \\ &= N^2 \frac{1}{N^d}\sum p(x'-x)\eta(x)(1-\eta(x'))[J(\frac{x'}{N}) - J(\frac{x}{N})] \\ &= \frac{N^2}{2N^d}\sum_{x,x'} p(x'-x)[J(\frac{x'}{N}) - J(\frac{x}{N})][\eta(x)(1-\eta(x')) - \eta(x')(1-\eta(x))] \\ &= \frac{N^2}{2N^d}\sum_{x,x'} p(x'-x)[J(\frac{x'}{N}) - J(\frac{x}{N})][\eta(x) - \eta(x')] \\ &\simeq \frac{1}{2N^d}\sum_{x}(\mathcal{A}J)(\frac{x}{N})\eta(x). \end{aligned}$$

The operator $\mathcal{A}$ is given by

$$(\mathcal{A}J)(y) = \sum C_{i,j}\frac{\partial^2 J}{\partial y_i \partial y_j}(y) = \nabla C \nabla J$$

with

$$C_{i,j} = \sum_z z_i z_j p(z) \quad ; \quad z = (z_1, \cdots, z_d)$$

being the covariance matrix. From this observation it is routine to show that (2.3) is in fact true if (2.2) is valid, and that $\rho(t,y)$ can be obtained by solving the heat equation

$$\frac{\partial \rho}{\partial t} = \frac{1}{2}\nabla C \nabla \rho \tag{2.5}$$

with the initial condition

$$\rho(0,y) = \rho_0(y) \tag{2.7}$$

This calculation was made possible by a lucky break. The nonlinearity in (2.4) canceled out, and that allowed us to sum by parts twice, getting rid of the two powers of N in front of the generator.

3. Weak Asymmetry

Let us change things a little bit. Suppose we destroy the symmetry by adding a term $\frac{1}{N}q(x'-x)$ to $p(x'-x)$. Think of $\frac{1}{N}q$ as a small anti-symmetric modification of p. Then the new $G_N(\eta(\cdot))$ has an additional term

$$\begin{aligned}(3.1)\qquad &\frac{N^2}{N^d}\sum_{x,x'}\eta(x)(1-\eta(x'))\frac{q(x'-x)}{N}[J(\frac{x'}{N})-J(\frac{x}{N})]\\ &\simeq \frac{1}{N^d}\sum_{x,x'}\eta(x)(1-\eta(x'))\langle(\nabla J)(\frac{x}{N})\ ,\ b\ \rangle\end{aligned}$$

where $b=\sum zq(z)$. We are now stuck with the non-linearity. We invoke a general principle that if the density at some point of space and time is $\rho(t,y)$ then the joint distribution of $\{\eta_t(x)\}$ around the macroscopic point y i.e when $\frac{x}{N}\sim y$ will be nearly the Bernoulli measure $\mu_{\rho(t,y)}$. In particular $\eta(x)(1-\eta(x+z))$ can be replaced by $\rho(t,y)(1-\rho(t,y))$, and we end up with the equation

$$(3.2)\qquad \frac{\partial\rho}{\partial t}=\frac{1}{2}\nabla C\nabla\rho-\nabla.(\rho(1-\rho)b)$$

This in fact has been rigorously established in [2].

4. Large Deviations

One of the problems we can raise is the estimation of the probability of large deviations from the hydrodynamical behavior, which is after all a form of law of large numbers. More precisely, if we start from some configuration $\eta(\cdot)$ on Z^d, (which may or may not be random) and follow the empirical measures $\nu_t(\cdot)$ in some time interval, say $[0,T]$, what is the probability that ν_t stays close to an arbitrary given path $\lambda(t)$ in that time interval $[0,T]$? In particular, if it is exponentially small, i.e. $\exp[-N^dI(\lambda(\cdot))+o(N^d)]$, then we should determine the functional $I(\lambda(\cdot))$ explicitly.

First let us suppose that $\lambda(t,dy)=\lambda(t,y)dy$. Otherwise the rate function would be clearly infinite. Typically, there would be an initial rate function $I_0(\lambda)$ that controls the large deviation rate for the initial profile. If the initial conditions were deterministic then $I_0(\lambda)$ would be 0 for the correct macroscopic profile ρ_0 and $+\infty$ for any other profile. We will now concentrate on the large deviation arising out of the dynamical part. One way to generate deviant behavior, is to assume that the model behaves like it had some small asymmetry q that could depend on t and x as well, i.e. $\frac{1}{N}q(t,y,x'-x)$. This is unlikely, but exactly how unlikely, can be measured by relative entropy calculated via a Girsanov type formula. A simple calculation yields the following estimate for the relative entropy H

$$H \simeq E\left\{\frac{1}{2}\int_0^1 \sum_{x,x'} \eta_t(x)(1-\eta_t(x'))\frac{q^2(t,\frac{x}{N},x'-x)}{p(x'-x)}dt\right\} \tag{4.1}$$

$$\simeq \frac{N^d}{2}\int_0^1\int_{T^d}\rho(t,y)(1-\rho(t,y))e(t,y)dtdy$$

where

$$e(t,y)=\sum_z \frac{q^2(t,y,z)}{p(z)}$$

This will produce a density ρ that will be the solution of

$$\frac{\partial\rho}{\partial t}=\frac{1}{2}\nabla C\nabla\rho-\nabla.\rho(1-\rho)b \tag{4.2}$$

with $b(t,y)=\sum zq(t,y,z)$. It definitely behooves us to minimize $\sum\frac{q^2(z)}{p(z)}$ subject to $\sum zq(z)=b$. An elementary calculation provides the answer as $\langle b\ ,C^{-1}\ b\ \rangle$. The best lower bound then is

$$\Psi(b)=\frac{1}{2}\int_0^1\int_{T^d}\rho(t,y)(1-\rho(t,y))\langle b(t,y)\ ,C^{-1}\ b(t,y)\rangle dtdy \tag{4.3}$$

Finally the best relative entropy that we can achieve is

$$\inf_{b\in\mathcal{B}_\rho}\Psi(b)$$

where

$$\mathcal{B}_\rho=\{b:\frac{\partial\rho}{\partial t}=\frac{1}{2}\nabla C\nabla\rho-\nabla.\rho(1-\rho)b\} \tag{4.4}$$

One can see that the infimum is attained when b is of the form $b=C\nabla\varphi$ for some φ and the rate function is computed to be

$$I_0(\lambda_0)+\frac{1}{2}\int_0^1\|\frac{\partial\rho}{\partial t}-\frac{1}{2}\nabla C\nabla\rho\|^2_{-1,\rho(t,\cdot)(1-\rho(t,\cdot))C}\ dt \tag{4.5}$$

where

$$\|u\|^2_{-1,A(\cdot)}=\sup_J\left[2\int u(y)J(y)dy-\int\langle(\nabla J)(y),A(y)(\nabla J)(y)\rangle dy\right] \tag{4.6}$$

In [2] it was shown that this best possible lower bound is also an upper bound so that (4.5) is the exact rate function. This describes the large deviations as well as the way they arise.

5. Tagged particles in equilibrium and nonequilibrium

In a system with a large number of interacting particles, like in our simple exclusion process, we can tag one particular particle as it interacts and moves amidst the rest of the particles. Let us suppose that we start our system in equlibrium on Z^d, at a density ρ. Let us condition our system to have initially a particle at the site 0, and let us tag that particle. If we watch our tagged particle, we might reasonably expect it to wander diffusively and converge to a Brownian motion after a suitable rescaling of space and time. The limiting Brownian motion will have a covariance matrix $S(\rho)$ depending on the density ρ. That this is infact the case was demonstrated in [3]. The diffusion matrix $S(\rho)$ was shown to be nondegenerate in all cases except the one dimensional nearest neighbor case. In this case, the particles block each other completely, and $S(\rho) \equiv 0$ for all $0 < \rho \leq 1$. In the nonexceptional cases one can see that $S(\rho) \to C$ as $\rho \to 0$ and $S(\rho) \to 0$ as $\rho \to 1$.

We could start far away from equilibrium and still ask how a tagged particle that starts somewhere in this mileu will evolve in time. Let us return to our model on Z_N^d, tag a particle that starts at time 0 from the origin 0. We will assume that the rest of the particles have an initial configuration that has asymptotically an initial density profile $\rho_0(y)$. If we look at the speeded up time scale, as we saw earlier, the density will evolve according to the heat equation (2.5). Let us suppose that, after rescaling time by a factor N^2 and space by N to get the torus T^d, our tagged particle is at a point y on the torus at time t. Then it will find itself surrounded by an environment somewhat similar to the equilibrium environment with density $\rho(t, y)$. The tagged particle might then diffuse like a tagged particle in equilibrium at density $\rho(t, y)$, i.e. with a diffusion coefficient that is equal to $S(\rho(t, y))$. However the description may not be accurate enough and we may have an as yet undetermined first order term or drift that is present. An educated guess for the motion of our tagged particle then is a time dependent diffusion with the backward Kolmogorov generator

$$\mathcal{L}_t = \frac{1}{2}\nabla S(\rho(t, y))\nabla + c(t, y).\nabla \tag{5.1}$$

for some suitable choice of c. The corresponding forward equation is given by

$$\frac{\partial \tilde{\rho}}{\partial t} = \frac{1}{2}\nabla S(\rho(t, y))\nabla \tilde{\rho} - \nabla.c(t, y)\tilde{\rho} \tag{5.2}$$

and the solution at time t will give the marginal distribution at that time of our tagged particle. If we pick at random one of the other particles and tag it, (while untagging our particle) the equation (5.2) does not change. The initial condition for $\tilde{\rho}$ at $t = 0$ bcomes ρ_0 normalized to have total integral 1 on T^d. In fact if the tagged particle is randomly chosen, its density at any time t, is nothing other than $\rho(t,y)$ normalized to have integral 1. In other words, since the equation (5.2) is linear, ρ must satisfy it.

$$\frac{\partial \rho}{\partial t} = \frac{1}{2}\nabla S(\rho(t,y))\nabla \rho - \nabla.c(t,y)\rho \tag{5.3}$$

Now comparing (2.5) and (5.3), we obtain

$$\frac{1}{2}\nabla S(\rho(t,y))\nabla \rho - \nabla.c(t,y)\rho = \frac{1}{2}\nabla C \nabla \rho$$

Continuing on,

$$\nabla.c\rho = \frac{1}{2}(\nabla S(\rho)\nabla\rho - \nabla C\nabla\rho).$$

Perhaps with some luck,

$$c\rho = \frac{1}{2}(S(\rho)\nabla\rho - C\nabla\rho)$$

or

$$c = \frac{1}{2}[S(\rho) - C]\frac{\nabla\rho}{\rho}.$$

One can then reasonably expect the tagged particle in nonequlibrium to behave, after rescaling, like a diffusion with a time dependent backward Kolmogorov generator

$$\mathcal{L}_t = \frac{1}{2}\nabla S(\rho(t,y))\nabla + \frac{1}{2\rho}(C - S(\rho(t,y))\nabla\rho(t,y).\nabla. \tag{5.4}$$

To prove this for every particle in every initial configuration appears to be hard and has not been done yet. If we denote by Q_0 the measure on the path space $C[[0,T];T^d]$ corresponding to our generator (5.4) with initial (sub probability) density $\rho_0(y)$ one may consider the empirical process

$$R_{N,\omega} = \frac{\delta_{y_1(\cdot)} + \delta_{y_2(\cdot)} + \cdots + \delta_{y_K(\cdot)}}{N^d}$$

where $y_i(\cdot)$ are the trajectories of the K particles that are individually tagged. $R_{N,\omega}$ is viewed as a random measure on $D[[0,T];T^d]$ and one can ask if as $N \to \infty$ the measures $R_{N,\omega} \Rightarrow Q_0$. This was answered in the affirmative in [9] and we will return to this later.

6. Several colors and nongradient systems

We may also decide that there are n types of particles distinguished by colors and may want to keep track of the colors while the evolution is colorblind. We define

$$\eta_j : Z_N^d \to \{0,1\}$$

where $\eta_j(x)$ is 1 if there is a particle of color j at site x. Otherwise it is 0. Of course

$$\sum_{j=1}^{n} \eta_j(x) = \eta(x).$$

If we denote the typical element of the state space by

$$\zeta = (\eta_1, \dots, \eta_n)$$

then the n-color generator is given by

$$(6.1) \quad (\mathcal{L}_N^n f)(\zeta) = N^2 \sum_x \sum_z p(z)\eta(x)(1-\eta(x+z)) \left(f(\zeta^{x,x+z}) - f(\zeta)\right)$$

or in a more symmetric form by,

(6.2)

$$(\mathcal{L}_N^n f)(\zeta) = \frac{N^2}{2} \sum_{x,x'} p(x'-x)[\eta(x)(1-\eta(x')) + (1-\eta(x))\eta(x')] \left(f(\zeta^{x,x'}) - f(\zeta)\right)$$

Here $\zeta^{x,x'}$ is the configuration which is defined in a manner similar to the way $\eta^{x,x'}$ was defined in (1.2) by interchanging the situation at x and x' without changing the color of particles, if any, at these sites. The trace of the ζ process on the torus will be a measure valued process with n components

$$\tilde{\nu}_N(t) = \left\{\nu_N^j(t)\right\}$$

where

$$\nu_N^j(t) = \frac{1}{N^d} \sum_x \delta_{\frac{x}{N}} \eta_{j,t}(x)$$

of total masses K_j/N^d, where K_j is the number of particles of color j. The marginal of the empirical process $R_{N,\omega}$ at time t is then $\nu_N(t)$.

Let us define C to be the convex set in R^n consisting of $\tilde{\rho} = (\rho_1, \rho_2, \dots, \rho_n)$ such that $\rho_j \geq 0$ for j=1, 2, ... , n and $\rho = \sum_{j=1}^n \rho_j \leq 1$

We define the following matrices A and χ which are really to be thought of as $n \times n$ matrices corresponing to color types whose entries are $d \times d$ matrices that correspond to coordinates in R^d. If j, j' are two color types then

$$(6.3) \quad \begin{aligned} A_{j,j'} &= A^{(1)}_{j,j'} + A^{(2)}_{j,j'} \\ &= S(\rho)\left(\rho_j \delta_{j,j'} - \frac{\rho_j \rho_{j'}}{\rho}\right) + C\left(\rho_j \rho_{j'} \frac{(1-\rho)}{\rho}\right) \end{aligned}$$

Each entry of $A_{j,j'}$, as a $d \times d$ matrix, is a linear combination of $S(\rho)$ and C with weights that have been given in (6.3). Similarly

$$(6.4) \quad \chi_{j,j'} = I\left(\frac{\delta_{j,j'}}{\rho_j} + \frac{1}{(1-\rho)}\right)$$

where I is the identity matrix of size $d \times d$. Let us note that although χ is undefined when $\rho = 1$, A and $A\chi$ are well defined on $\mathcal{C}$. In [6] it was shown that the empirical measures $\tilde{\nu}_N(\cdot)$ satisfy a law of large numbers. Assuming that the initial configurations, whether they are random are deterministic, satisfy

$$\tilde{\nu}_N(\cdot) \Rightarrow \tilde{\rho}^0(\theta)d\theta$$

in probability, in the weak topology on T^d, it was shown that for all $t > 0$,

$$\tilde{\nu}_N(\cdot) \Rightarrow \tilde{\rho}(t,\theta)d\theta$$

The limit $\tilde{\rho}(t,\theta)$ is characterized as the unique solution of

$$\frac{\partial \tilde{\rho}}{\partial t} = \frac{1}{2}\nabla \cdot A\chi(\tilde{\rho})\nabla \tilde{\rho} \tag{6.5}$$

with initial condition

$$\tilde{\rho}(0,\theta) = \tilde{\rho}^0(\theta) \tag{6.6}$$

The equations (6.5) and (6.6) have the following interpretation. The sum

$$\rho(t,\theta) = \sum_{j=1}^{n} \rho^j(t,\theta)$$

satisfies the heat equation

$$\frac{\partial \rho}{\partial t} = \frac{1}{2}\nabla \cdot C\nabla \rho \tag{6.7}$$

with initial condition

$$\tilde{\rho}(0,\theta) = \rho^0(\theta) = \sum_{1}^{n} \rho_j^0(\theta)$$

and, given the solution of (6.7), each component of (6.5) can be recast as

$$\frac{\partial \rho_j}{\partial t} = \mathcal{A}_\rho\, \rho_j$$

with $\rho_j(0,\theta) = \rho_j^0(\theta)$, where

$$\mathcal{A}_\rho f = \frac{1}{2}\nabla \cdot S(\rho(t,\theta))\nabla f + \frac{1}{2}\nabla \cdot [(C - S(\rho(t,\theta))]\,\frac{\nabla \rho(t,\theta)}{\rho(t,\theta)} f \tag{6.8}$$

This is compatible with the expectation for the diffusive behavior of the tagged particle keping in mind that coloring is a form of tagging.

The proof proceeds as follows. For the color j, if we compute as we did earlier

$$d\left(\frac{1}{N^d}\sum_x J(\frac{x}{N})\eta_{j,t}(x)\right) = A_J(t)dt + dM_J(t)$$

where $E\{[M_J(t)]^2\} \to 0$ as $N \to \infty$ as before, but now $A_J(t)$ is given by

$$\begin{aligned} A_J(t) &= \frac{N^2}{N^d}\sum_{x,x'} \eta_j(x)(1-\eta(x'))[J(\frac{x'}{N}) - J(\frac{x}{N})]p(x'-x) \\ &\simeq \frac{N}{N^d}\sum_x \eta_j(x)(\nabla J)(\frac{x}{N}).\sum_{x'}(1-\eta(x'))(x'-x)p(x'-x) \end{aligned}$$

Unlike the case of particles with single color we cannot perform summation by parts for a second time and we are saddled with a term of order N. The term does have mean 0. Some methods were developed in [10], for handling such situations referred to as nongradient models. There are several terms, also of order N that are completely negligible because of the rapid time scale. Our term is not one of them. Neither are the density gradients. Let us define the currents that constitute our terms,

$$W_{i,j} = \eta_j(0)\sum_{x'} p(x')(1-\eta(x'))(e_i, x')$$

and the density gradients

$$\xi_{i,j} = (\nabla_{e_i}\eta_j)(0) = \eta_j(e_i) - \eta_j(0)$$

where $\{e_1, \cdots, e_d\}$ is the standard basis for R^d. Relative to any equlibrium $\mu_{\tilde{\rho}}$ for the n color evolution, $W_{i,j}$ can be effectively replaced by a linear combinations of $\{\xi_{i,j}\}$, because the differences are negligible. However the coefficients depend on the density $\tilde{\rho}$. It is proved in [6] that the linear combinations are given by

$$W_{i,j} \sim \sum_{i',j'}[A(\tilde{\rho})\chi(\tilde{\rho})]^{i,j}_{i',j'}\xi_{i',j'}$$

If we do the formal substitution, we can then do another summation by parts and the equation (6.5) emerges. Actual details are rather complex and involve analysis in equlibrium at every $\mu_{\tilde{\rho}}$, and transition to nonequlibrium that involves methods from large deviation theory. This is typical of nongradient systems and was carried out in [6]. The result of Rezakhanlou that we quoted earlier on the behavior of $R_{N.\omega}$ was deduced from the n color behavior. We increase the number of colors indefinitely and color code past history. This way, modulo some difficult compactness estimates proved in [9], one can infer the behavior of $R_{N,\omega}$ from that of $\tilde{\nu}_N(\cdot)$.

7. Large deviations for several colors and empirical proceses

We will now describe the large deviation theory for the empirical measure valued process $\tilde{\nu}(\cdot)$. If P_N is the probability distribution of $\tilde{\nu}_N(\cdot)$, it has a standard large deviation property with the following rate function $I_n(\tilde{\rho}(\cdot,\cdot))$ defined for functions

$$\tilde{\rho}(\cdot) : [0,T] \times T^d \to C$$

such that $\int \rho_j(t,\theta)d\theta$ is independent of t for each j. Given a function $q(\theta)$ on T^d taking values in the space of symmetric positive semidefinite $d \times d$ matrices and a function $\tilde{r}(\cdot) : T^d \to R^n$ with mean zero we define

$$(7.1) \quad \|\tilde{r}\|_{-1,q(\cdot)} = \sup_{\Phi} \left\{ 2\int_{T^d} \langle \tilde{r}(\theta), \Phi(\theta)\rangle \, d\theta - \int_{T^d} \langle \nabla\Phi(\theta), q(\theta)\nabla\Phi(\theta)\rangle \, d\theta \right\}$$

where the supremum is over C^∞ maps from T^d to R^n. We then define

$$I_0^n(\tilde{\rho}(.,.)) = \sup_{\Phi} \left\{ 2\int_0^T \int_{T^d} \langle (A\chi)(\tilde{\rho})\nabla\tilde{\rho}, \nabla\Phi\rangle dt d\theta - \int_0^T \int_{T^d} \langle \nabla\Phi, (A)(\tilde{\rho})\nabla\Phi\rangle dt d\theta \right\}$$

where the supremum is now taken over C^∞ functions on $[0,T] \times T^d$ with values in R^{nd}. When it makes sense,

$$(7.2) \qquad \bar{I}^n(\tilde{\rho}) = \int_0^T \int_{T^d} \langle \nabla\tilde{\rho}, (\chi A\chi)(\tilde{\rho})\nabla\tilde{\rho}\rangle \, dt d\theta$$

One can explicitly calculate $\chi A\chi$ and the quantity in (7.2) is finite if and only if

$$\int_0^T \int_{T^d} (1-\rho) \sum_j \frac{|\nabla\rho_j|^2}{\rho_j} d\theta dt \; < \; \infty$$

and

$$\int_0^T \int_{T^d} \frac{|\nabla\rho|^2}{(1-\rho)} d\theta dt \; < \; \infty$$

The dynamical part $I_1^n(\tilde{\rho}(\cdot,\cdot))$ is defined to be $+\infty$ unless $\bar{I}^n(\tilde{\rho}(\cdot,\cdot)) < \infty$ and in such a case

$$I_1^n(\tilde{\rho}(\cdot,\cdot)) = \frac{1}{2}\int_0^T \left\| \frac{\partial\tilde{\rho}}{\partial t} - \frac{1}{2}\nabla \cdot A\chi\nabla\tilde{\rho} \right\|^2_{-1,A(\tilde{\rho}(t,\cdot))} dt$$

If the initial conditions are deterministic and the law of large numbers holds initially with profile $\tilde{\rho}^0(\cdot)$ then we define

$$I_0^n(\tilde{\rho}(0,\cdot)) = \begin{cases} 0 & \text{if } \tilde{\rho}(0,\cdot) = \tilde{\rho}^0(\cdot) \\ +\infty & \text{otherwise.} \end{cases}$$

For other suitable initial conditions we pick the corresponding rate function for the initial profile and this shall be our $I_0^n(\tilde{\rho}(0,\cdot))$. The full rate function is given by

$$I^n(\tilde{\rho}(\cdot,\cdot)) = I_0^n(\tilde{\rho}(0,\cdot)) + I_1^n(\tilde{\rho}\cdot,\cdot)).$$

Large deviation estimates with this rate function have been obtained for dimension ≥ 3 in [7]. The restriction on the dimension is dictated by technical demands, namely Lipschitz continuity of $S(\rho)$ as a function of ρ, a fact that is only availble if $d \geq 3$. The transition from the above result to the large deviation result for the empirical process is in principle straight forward and is a matter of careful calculation and identification. First a superexponential tightness estimate for the empirical process $R_{N,\omega}$ is proved that allows one to interchange limits freely and provides a cover for various technical points along the route. The rate function for the empirical process is then calculated by computing rate functions for finite dimensional projections. The torus T^d is divided into a finite number of cells and the large deviation principle is reduced to a large deviation principle for the proportions of particles that lie in a specified sequence of cells at time points $0 < t_1 < t_2 < \cdots < T$. This is a weak form of tagging and reduces the problem to one with a finite but large number of colors, by assigning different colors to particles with different histories. When the dust settles, it is a matter of identification of the rate function by chasing through the limiting procedure as the number of cells in the partition and the number of inspection times become infinite. This is carried out in [7]. The big surprise is the reemergence of entropy in the final formula and the physical interpretation that the rate function has because of it. This is not evident in the formula for the large deviation rate for n-colors. Finally let us describe the rate functional. Given Q, we can compute the marginal densities $\rho(t,\theta)$ for $0 \leq t \leq T$. There is the initial rate $I_0(\rho(0,\cdot))$ which depends on how the choice of the initial condition is made. The rate function is given by

$$I(Q) = I_0(\rho(0,.)) + \inf_{b\in\mathcal{B}_{\rho(\cdot,\cdot)}} \left\{ \frac{1}{2}\int_0^T \int_{T^d} \rho(1-\rho)\left\langle C^{-1}b,b\right\rangle dtdy + H(Q;P_\rho^b) \right\}$$

Here P_ρ^b is a time dependent diffusion process determined by b and will have one dimensional marginal densities $\rho(t,y)$. The class $\mathcal{B}_{\rho(\cdot,\cdot)}$ is given by (4.4). The diffusion process P_ρ^b is defined for each admissible ρ, and admissible b in $\mathcal{B}_{\rho(\cdot,\cdot)}$ and has for its backward Kolmogorov generator

$$\mathcal{L}_t^{\rho,b} = \frac{1}{2}\nabla S(\rho(t,y))\nabla + \frac{1}{2\rho}(S(\rho(t,y)) - C)\nabla\rho(t,y).\nabla + (1-\rho)b.\nabla$$

8. The asymmetric mean zero case

The case where $p(\cdot)$ is asymmetric but has the property $\sum_z zp(z) = 0$ is again a nongradient system and is therefore difficult. Even if one is only interested in the simplest of questions, namely the evolution of the particle density in hydrodynamic scales, the analysis is still hard. An added complication is the nonreversible nature of the dynamics due to the asymmetry of $p(.)$. In the application of the large deviation techniques to handle nonequilibrium, as well in establishing some of the properties of the system in equlibrium this causes difficulties. The fact that the mean is zero provides an important control in the form of a bound for the asymmetric part in terms of the symmetric part. At some level it is a bounded perturbation. This was exploited in [12], to show that the analog in this context of the heat equation (2.5) that appears in the symmetric case, is a nonlinear diffusion equation of the form

$$\frac{\partial \rho}{\partial t} = \frac{1}{2}\nabla \bar{C}(\rho)\nabla \rho \tag{8.1}$$

where the diffusion matrix $\bar{C}$ really depends on ρ and can be shown to be strictly larger than the matrix C that would result in the symmetric case provided $0 < \rho < 1$ and $p(.)$ is not symmetric.

9. The case of nonzero mean

The case where $\sum_x zp(z) = b \neq 0$ is qualitatively very different. The motion of the densities is hyperbolic rather than diffusive. The scaling reflects this and time is rescaled by a factor of N instead. A calculation shows

$$d\left(\frac{1}{N^d}\sum_x J(\frac{x}{N})\eta_t(x)\right) = A_J(t)dt + dM_J(t) \tag{9.1}$$

where

$$\begin{aligned} A_J(t) &= \frac{N}{N^d}\sum_{x,x'}\eta(x)(1-\eta(x'))p(x'-x)[J(\frac{x'}{N}) - J(\frac{x}{N})] \\ &\simeq \frac{1}{N^d}\sum_{x,x'}\eta(x)(1-\eta(x'))\left\langle(\nabla J)(\frac{x}{N}), b\right\rangle \\ &\simeq \int_{T^d}\rho(t,y)(1-\rho(t,y))\,\langle(\nabla J)(y), b\rangle\, dy \end{aligned}$$

Here $\rho(t,\cdot)$ is the macroscopic density of the system in the limit at time t. This leads to Burgers' equation for ρ

$$\frac{\partial \rho}{\partial t} + \langle b, \nabla\rho(1-\rho)\rangle = 0 \tag{9.2}$$

This equation in general develops shocks and the physical solution is characterized as the unique solution that satisfies certain entropy conditions. It has been shown that if the initial conditions are suitably chosen, for example, if the initial distribution of particles is a product measure with

$$P[\eta_0(x) = 1] = \rho_0(\frac{x}{N})$$

then for $t > 0$ and as $N \to \infty$ the empirical density converges in probability to the entropy solution of (9.2) with the given initial condition. For the best results in the general case see [8]. More detailed information is available in the 1-dimensional case. In the nearest neighbor totally asymmetric case, i.e.[p(1)=1] one can handle arbitrary initial data as was shown in [11].

Finally one can ask if viscosity corrections to Burgers' equation appear in some manner from the particle model. One can answer in the affirmative in two distinct ways, both for dimensions $d \geq 3$. One can ask for a small correction to Burgers' equation so that the empirical density is approximated by this modified equation to a higher order of accuracy. Another possibility is to start close to equlibrium i.e a small perturbation of constant or uniform initial density. Then this initial perturbation will move with the constant speed of $(1 - 2\rho)b$ in the time scale that was speeded up by N. It is possible that the perturbation dissipates in the time scale N^2 after recentering. And this dissipation should be governed by Burgers' equation with viscosity. Both questions have affirmative answers as was shown respectively in [4] and [1]. The viscosity coefficients are the same in both cases and identification of this coefficient is carried out by techniques developed for nongradient models. The situation is much more complex than in [12], because now the asymmetric part is far more dominant.

References

1. Esposito, R., Marra, R., and Yau, H. T., Diffusive limit of asymmetric simple exclusion, Reviews in Mathematical Physics, Vol 6. No 5A, (1994), 1233-1267.
2. Kipnis, C., Olla, S., and Varadhan, S. R. S., Hydrodynamics and large deviation for simple exclusion process, Comm. Pure Appl. Math. XLII (1989), 115–137.
3. Kipnis, C., and Varadhan, S. R. S., Central limit theorem for additive functionals of reversible Markov processes and application to simple exclusions, Comm. Math. Phys.
reversible Markov processes and application to simple exclusions, Comm. Math. Phys.
104 (1986), 1–19.
4. Landim, Claudio., Olla, S. and Yau, H.T., First order correction for the hydrodynamic limit of asymmetric simple exclusion processes in dimension $d \geq 3$, to appear in Communications in Pure and Applied Mathematics.
5. Ligget, Thomas M. Interacting Particle Systems, Springer-Verlag, New York, (1985).
6. Quastel, J., Diffusion of color in simple exclusion process, Comm. Pure Appl. Math. XLV (1992), 623–680.
7. Quastel, Jeremy.,Rezakhanlou, F., and Varadhan, S. R. S., Large Deviations for the symmetric simple exclusion process.(Preprint) 1995.
8. Rezakhanlou, F., Hydrodynamical limit for attractive particle systems on Z^d, Comm. Math. Phys. 140 (1991), 417–448.
9. Rezakhanlou, F., Propagation of Chaos for Symmetric Simple Exclusions, Communications on Pure and applied mathematics, Vol XLVII, (1994), 943-958.
10. Varadhan, S. R. S., Nonlinear diffusion limit for a system with nearest neighbor interactions, II, Asymptotic Problems in Probability Theory: Stochastic Models and Diffusions

on Fractals, K. D. Elworthy and N. Ikeda (Editors), Pitman Research Notes in Mathematics Series 283, 1991, 75–130.
11. Venkatasubramani, R., Ph.D. Thesis., New York University, 1995.
12. Xu, Lin, Ph.D. Thesis, New York University, 1993.

Analysis, Geometry and Probability, Text Read. Math. 10, Hindustan Book Agency, 1996, 397-416

Nongradient Models in Hydrodynamic Scaling

S.R.S. Varadhan[1]
Courant Institute of Mathematical Sciences
New York University

1. Introduction.

The aim of this expository article is to explain, by means of some well understood examples, the difference between gradient and nongradient systems in the context of diffusive hydrodynamic scaling limits.

In hydrodynamic scaling we study a rather large system of interacting components. The system could involve actual particles, in which case, they evolve collectively according to some dynamics with interaction between particles. Or alternatively the system could consist of variables that represent the values of a certain random field at locations that constitute a discrete lattice. The collective dynamics of all the field variables could be specified in some manner. The dynamics will be stochastic and will involve some noise. The starting point of the study therefore is a Markov Process with a state space that is of large , if not infinite, dimension.

Another important factor in these problems, that in fact characterizes the term "hydrodynamic scaling" is the existence of conserved quantities in the stochastic evolution. In the case of dynamics involving actual particles, assuming that particles are neither created nor destroyed, the total number of particles is clearly a conserved quantity. In the case of field variables the type of stochastic dynamics that we study will leave a certain quantity, usually the sum of all the field variables or equivalently the average field strength conserved.

The physical space in these problems is a finite dimensional space, and for convenience we will take it to be one dimensional and periodic, i.e. the circle **S** of unit circumference. There is a parameter in the problem that defines the scale. In the case of actual particles it would be the typical

[1]Supported by NSF grant DMS-9503419 and ARO grant DAAH 04-95-1-0666

interparticle distance and because we are in 1 dimension this is $\frac{1}{N}$ where N is the total number of particles. In the case of field variables we will look at an imbedded finite lattice in $\mathbf{S}$ and the field variable would correspond to the lattice sites. The scale parameter then is the lattice width, which again because of 1 dimension is of the same order as $\frac{1}{N}$, where N now is the number of lattice sites. We can also consider the hybrid of the two situations, where the particles actually live on the lattice sites and the Markov model will involve the jumping of these particles from one lattice site to another in some specified random fashion. If the number of particles per site remains order 1, then these models fit in both categories. As we mentioned earlier the lattice field evolutions will conserve the sum of the field variables over all the sites of the finite lattice.

Before we proceed any further let us describe some classes of examples that we will look at in order to explain the concepts. While the methods are general enough that they can be used widely, we will concentrate on a few examples to illustrate the methodology. The examples will fall into two groups.

In the first group of examples, known as random walks with simple exclusion we will have k particles that live in the lattice $\mathbf{Z}_N \subset \mathbf{S}$ of width $\frac{1}{N}$. The number of particles k will satisfy $k \leq N$ and it will be forbidden to have more than one particle per site. The state space will be the subset $\mathbf{Z}_N^{(k)}$ of the k-fold product $\mathbf{Z}_N^k$ of $\mathbf{Z}_N$ consisting of distinct k-tuples. Our Markov process on this state space will depend on an underlying random walk on the integers. The transitions or jumps of this random walk will be given by a probability distributuion $p(z)$ that is independent of the scaling parameter N. We will assume that $p(\cdot)$ has finite range, i.e. $p(z) = 0$ if $z \notin F$ for some finite set F of integers. A particle that is currently at site $x \in \mathbf{Z}_N$ can jump to the site $x + \frac{z}{N}$ (addition is modulo 1) at rate $p(z)$ provided the site $x + \frac{z}{N}$ is empty. Jumps are of course not possible to a site that is already occupied by a particle. The part of the generator that corresponds to this transition can be written in the following notation: a point in the state space is given by $(x_1, \cdots, x_k)$ the positions of the k particles. The functions

$\chi_i(x)$ are the incidence functions that are equal to 1 if the i-th particle is at site x and 0 otherwise. The quantity $\eta(x) = \sum_{i=1}^{k} \chi_i(x)$, which counts the number of particles at site x, is either 0 or 1 and tells us if the site is empty or not. If $f(x_1, \cdots, x_k)$ is a function on the state space, then the generator corresponding to the jump of size $\frac{z}{N}$ by the i-th particle is

$$\begin{aligned}(\mathcal{A}_{i,z}f)(x_1,\cdots,x_i,\cdots,x_k) &= (1-\eta(x_i+\frac{z}{N})[f(x_1,\cdots,x_i+\frac{z}{N},\cdots,x_k) \\ &\qquad - f(x_1,\cdots,x_i,\cdots,x_k)]\end{aligned}$$

and the full generator is

$$\mathcal{A}_N f = \sum_{i,z} p(z)\mathcal{A}_{i,z} f$$

that corresponds to all permissible jumps being active at their corresponding rates. If we do not want to distinguish the particles, we could reduce the state space to Ω_N^k the space of subsets of cardinality k of $\mathbf{Z}_N$. We will then know only which sites have particles and the labels of the particles will disappear. It is more convenient to deal with $\cup_k \Omega_N^k$ which can be identified with the space of maps $\eta : \mathbf{Z}_N \to \{0,1\}$, $\eta(x) = 1$ signifying the presence of a particle at x and $\eta(x) = 0$ implying that the site x is empty. The generator for the process on Ω_N is expressed using the following notation. For x, y in $\mathbf{Z}_N$ let us define the map $\sigma^{x,y}$ 0f $\Omega_N \to \Omega_N$

$$(\sigma^{x,y}\eta)(u) = \begin{cases} \eta(u) & \text{for } u \neq x \text{ or } y \\ \eta(x) & \text{for } u = y \\ \eta(y) & \text{for } u = x \end{cases} \tag{1.1}$$

In other words the map $\sigma^{x,y}$ interchanges the situation at x and y. The generator $\mathcal{L}_N$ can then be written as

$$(\mathcal{L}_N f)(\eta) = \sum_{x,z} p(z)\eta(x)(1-\eta(x+\frac{z}{N}))[f(\sigma^{x,x+\frac{z}{N}}\eta) - f(\eta)]$$

A slight but significant modification of the above version is when we assume that there are two types of particles, say particles of two different colors red and green. We can describe the state by two maps $g, r : \mathbf{Z}_N \to \{0,1\}$. If

$g(x) = 1$ then there is a green particle at x and similarly for the red particles. Of course $\eta(x) = g(x) + r(x)$ for all $x \in \mathbf{Z}_N$. Although the dynamics is color blind we do want to keep track of the colors. The generator $\mathcal{L}_N$ can be recast in the form

$$(\mathcal{L}_N f)(g,r) = \sum_{x,z} p(z)\eta(x)(1-\eta(x+\frac{z}{N}))[f(\sigma^{x,x+\frac{z}{N}}g, \sigma^{x,x+\frac{z}{N}}r) - f(g,r)]$$

where $\sigma^{x,x+\frac{z}{N}}g$ and $\sigma^{x,x+\frac{z}{N}}r$ are defined in a manner analogous to (1.1). These models have lots of invariant measures. In the color free version, if we fix the number of particles at k, there is an invariant measure $\mu_{N,k}$ which is seen to be the uniform distribution on the $\binom{N}{k}$ subsets of $\mathbf{Z}_N$ of cardinality k. Similarly in the two color version the invariant distributions μ_{N,k_1,k_2} are the uniform distributions on all partitions of $\mathbf{Z}_N$ into sets with cardinality k_1, k_2 and $N - k_1 - k_2$ corresponding to green, red and empty sites. Each one of these probability distributions is an ergodic or extremal invariant measure so that if we start in any configuration with given particle numbers, for fixed N, as $t \to \infty$, the system will converge to the equilibrium corresponding to the given particle numbers. However the convergence will be rather slow for large N. For instance if all the particles are loaded into one part of the circle, before we even think of reaching equilibrium, lots of particles have to jump through sites whose number is proportional to N. This will take a time of order N if the particles do not wander or may be even order N^2 if they wander diffusively. The goal of the theory is to say something about the approach to equilibrium. In the proper time scale which may be N^2 or N depending on the nature of $p(\cdot)$, (mean 0 or not) a local equilibrium is reached quickly, chararcterized by the local density in the colorless case or by two local densities, one for each color, in the two color case. This local density is to be thought of as a function of t and x, where t is the rescaled time and x is a point in the continuum $\mathbf{S}$. The theory derives a suitable evolution equation in the time variable t for the densities as functions of the spatial variable x.

The second class of examples deal with a field model. Since at the level N there are N lattice sites in $\mathbf{Z}_N \subset \mathbf{S}$, our state will be a vector $\{\xi_x : x \in \mathbf{Z}_N\}$

represented by a point in R^N. The Markov process in this model, referred to as Ginzburg-Landau model, is a diffusion in R^N. As we said earlier the diffusion will conserve the sum or equivalently the average $\frac{1}{N}\sum_{x\in \mathbf{Z}_N}\xi_x$. These models are best described by first specifying the inavariant measure and then the process by specifying a Dirichlet form. The process will then be symmetric or reversible with respect to the measure and the generator can be easily written down. The invariant measure μ_N on R^N is given by a density with respect to the Lebesgue measure

$$d\mu_N = \exp[-\sum_x \phi(\xi_x)]\ \Pi_x d\xi_x$$

The basic processes are "transfers" of "ξ" from a site x to a neighboring site y. They correspond to a term in the Dirichlet form of the type

$$\frac{1}{2}\int [\frac{\partial u}{\partial \xi_x} - \frac{\partial u}{\partial \xi_y}]^2 d\mu_N$$

where (x, y) are a pair of neighboring sites, which we will denote by $x \sim y$. We could put in a rate that depends on the state by changing the Dirichlet form to

$$\frac{1}{2}\int a(\xi_x, \xi_y)[\frac{\partial u}{\partial \xi_x} - \frac{\partial u}{\partial \xi_y}]^2 d\mu_N$$

where a is a nice positive function on $R \times R$. The full Dirichlet form is given by

$$\frac{1}{2}\sum_{x\sim y}\int a(\xi_x, \xi_y)[\frac{\partial u}{\partial \xi_x} - \frac{\partial u}{\partial \xi_y}]^2 d\mu_N.$$

The generators are easily computed as

$$\mathcal{L}_N f = \frac{1}{2}\sum_{x\sim y} a(\xi_x, \xi_y)\left[\frac{\partial}{\partial \xi_x} - \frac{\partial}{\partial \xi_y}\right]^2 f - \frac{1}{2}\sum_{x\sim y} W(\xi_x, \xi_y)\left[\frac{\partial}{\partial \xi_x} - \frac{\partial}{\partial \xi_y}\right] f$$

where

$$W(\xi_1, \xi_2) = a(\xi_1, \xi_2)(\phi'(\xi_1) - \phi'(\xi_2)) - a_1(\xi_1, \xi_2) + a_2(\xi_1, \xi_2)$$

a_1, a_2 representing the partial derivatives of a with respect to the variables ξ_1 and ξ_2 respectively. Just as in the earlier example we have a conserved quantity $m = \frac{1}{N}\sum_x \xi_x$. Therefore the ergodic invariant measures are the restrictions (or conditionals to be more precise) of μ_N to the hyperplanes $m = \frac{1}{N}\sum_x \xi_x$. We will denote these by $\mu_{N,m}$. Again the convergence to equilibrium is slowed down for large N and the time scale is of the order N^2. One expects, in this model as well, after time is speeded up by a factor of N^2, to describe the approach to equlibrium by the evolution in t of the local mean $m(t,x)$ as function of the point $x \in \mathbf{S}$.

2. Gradient Models.

The simplest example to analyze is the Random Walk with Simple Exclusion model with a symmetric distribution $p(\cdot)$ for the jump size, i.e. $p(z) = p(-z)$ for all z. For the Markov Process on the state space Ω_N, that is speeded up by a scale factor of N^2, the infinitesimal generator is

$$(2.1)\qquad (\mathcal{L}_N f)(\eta) = N^2 \sum_{x,z} p(z)\eta(x)(1-\eta(x+\frac{z}{N}))[f(\sigma^{x,x+\frac{z}{N}}\eta) - f(\eta)]$$

It is more convenient to introduce $p_N(u) = p(Nu)$ on the imbedded lattice and rewrite (2.1) as

$$(\mathcal{L}_N f)(\eta) = N^2 \sum_{x,y} p_N(y-x)\eta(x)(1-\eta(y))[f(\sigma^{x,y}\eta) - f(\eta)]$$

which simplifies, because of the symmetry of $p(\cdot)$ to,

(2.2)

$$(\mathcal{L}_N f)(\eta) = \frac{N^2}{2}\sum_{x,y} p_N(y-x)[\eta(x)(1-\eta(y))+\eta(y)(1-\eta(x))][f(\sigma^{x,y}\eta)-f(\eta)].$$

If $\eta(x) = \eta(y)$, the situation at x and y are the same and $\sigma^{x,y}$ does nothing and therefore $f(\sigma^{x,y}\eta) - f(\eta) = 0$ on that set. Then (2.2) can be simplified further to give

$$(\mathcal{L}_N f)(\eta) = \frac{N^2}{2}\sum_{x,y} p_N(y-x)[f(\sigma^{x,y}\eta) - f(\eta)].$$

If we follow the configuration $\eta(t,\cdot) \in \Omega_N$, through time, by tracking the empirical distribution

$$\nu_N(t) = \frac{1}{N}\sum_x \eta(t,x)\delta_x$$

we would get a process with values in the space $\mathcal{M}$ of measures on $\mathbf{S}$. Let us fix a time interval $[0,T]$ (in the new speeded up time scale) and consider the induced measure Q_N on the Skorohod space $D[[0,T];\mathcal{M}]$ of maps $[0,T] \to \mathcal{M}$ that are right continuous with left limits. The initial conditions are determined by the initial configuration η_N that may be far away from any equilibrium. If as $N \to \infty$, for any smooth function $J : \mathbf{S} \to R$,

$$\lim_{N\to\infty} \int_{\mathbf{S}} J(x)d\nu_N(0) = \lim_{N\to\infty} \frac{1}{N}\sum_x J(x)\eta(x) = \int_{\mathbf{S}} J(x)\rho_0(x)dx \tag{2.3}$$

exists then we say that the initial configurations correspond to a macroscopic profile with density ρ_0. The exclusion property forces any ρ_0 to satisfy $0 \le \rho_0(x) \le 1$ for all $x \in \mathbf{S}$. The configuration η_N could be random as well, in which case we would expect the limit in (2.3) to be in probability and we would always insist that the profiles ρ_0 be nonrandom. We would like to establish generally the following:

If the initial configurations correspond to a profile ρ_0, then for any time $t > 0$, and for any smooth function $J : \mathbf{S} \to R$,

$$\lim_{N\to\infty} \int_{\mathbf{S}} J(x)d\nu_N(t) = \lim_{N\to\infty} \frac{1}{N}\sum_x J(x)\eta(t,x) = \int_{\mathbf{S}} J(x)\rho(t,x)dx$$

exists, where $\rho(t,x)$ the profile at time t is obtained by solving some evolution equation with initial data ρ_0. Part of the problem is to determine the evolution equation that yields the correct profile.

The analysis is usually carried out following these steps. First show that the family Q_N of measures on $D[[0,T];\mathcal{M}]$ is compact and that any weak limit Q lives actually in the space $C[[0,T];\mathcal{R}]$ of continuous maps into nonnegative densities bounded by 1. Moreover show that Q is supported by the set of solutions (usually in a weak sense) of a suitable evolution equation, of course satisfying the correct initial condition ρ_0. If we then establish a uniqueness theorem for weak solutions of our evolution equation, Q can be supported by only one point and has to be a δ measure at that point. This is enough to give us the law of large numbers that we want for any $t > 0$.

In our context let us look at the functional

$$U_N(\eta) = \frac{1}{N}\sum_x J(x)\eta(x)$$

on Ω_N. $U_N(\cdot)$ is in the domain of $\mathcal{L}_N$ and if we denote by

$$V_N(\eta) = (\mathcal{L}_N U_N)(\eta)$$

then

$$U_N(\eta(t,\cdot)) - U_N(\eta(0,\cdot)) = \int_0^t V_N(\eta(s,\cdot))ds + M_N(t)$$

for some martingale $M_N(t)$. An elementary calculation yields

$$\begin{aligned} V_N(\eta) &= \frac{N^2}{2}\sum_{x,y} p_N(y-x)[U_N(\sigma^{x,y}\eta) - U_N(\eta)] \\ &= \frac{N^2}{2}\frac{1}{N}\sum_{x,y} p_N(y-x)[J(y)-J(x)][\eta(x)-\eta(y)] \\ &= -\frac{1}{2N}\sum_x J'(x)\sum_y N(y-x)p_N(y-x)N[\eta(x)-\eta(y)] \\ &\sim \frac{D}{2}\frac{1}{N}\sum_{x,y} J''(x)\eta(x) \end{aligned}$$

with $D = \sum_z z^2 p(z)$, and another easy calculation gives

$$\lim_{N\to\infty} E\{M_N(t)^2\} = 0.$$

Now one can see that Q lives on the set of weak solutions of the heat equation

$$\frac{\partial\rho}{\partial t} = \frac{D}{2}\rho_{xx}$$

with initial condition ρ_0.

This example was really too easy. The diffusion coefficient D turned out to be independent of the density ρ and hence the final equation was a linear heat equation. In general this rarely happens for interacting systems.

Let us now consider the Ginzburg-Landau model with $a \equiv 1$. Let us recall that the generator in this case is, after rescaling by a factor of N^2,

$$\mathcal{A}_N = \frac{N^2}{2}\sum_{x\sim y}[\frac{\partial}{\partial\xi_x} - \frac{\partial}{\partial\xi_y}]^2 - \frac{N^2}{2}\sum_{x\sim y}[\phi'(\xi_x) - \phi'(\xi_y)][\frac{\partial}{\partial\xi_x} - \frac{\partial}{\partial\xi_y}]$$

Just as in the Simple Exclusion model we look at the function

$$U_N(\xi) = \frac{1}{N}\sum_x J(x)\xi_x$$

and calculate

$$\begin{aligned} V_N(\xi) = (\mathcal{A}_N U_N)(\xi) &= -\frac{N^2}{2}\frac{1}{N}\sum_{x\sim y}[J(x) - J(y)][\phi'(\xi_x) - \phi'(\xi_y)] \\ &\sim \frac{1}{2N}\sum_x J''(x)\phi'(\xi_x). \end{aligned}$$

We are now stuck because ϕ is not necessarily quadratic and so ϕ' is not linear. It is not possible to express the averages of ϕ' in terms of the averages of ξ. We appeal to a general principle valid for problems of this kind to bail us out. While the system may not be in global equlibrium, we expect it to be locally in equilibrium, the precise equlibrium at any location being dictated by the average value of ξ around that location. By a theorem of Cramer if we condition the joint distribution of $(X_1, \cdots, X_\ell)$ of ℓ indepent identically distributed random variables with a common distribution α by specifying the average $\frac{1}{\ell}\sum_i X_i = m$ then, for large ℓ, the distribution of $(X_1, \cdots, X_\ell)$ is well approximated by the product measure with components β where β is related to α and m by Cramer's relation

$$\frac{d\beta}{d\alpha}(\xi) = \exp[\theta\xi - \psi(\theta)].$$

Here $\psi(\theta)$ is the normalization constant and the value of θ is picked so that

$$\int \xi d\beta(\xi) = \psi'(\theta) = m.$$

In our context this reduces to

$$\psi'(\theta) = m \quad \text{with} \quad \psi(\theta) = \log\int \exp[\theta\xi - \phi(\xi)]\, d\xi$$

By standard convex analysis, if we define the dual convex function

$$h(m) = \sup_\theta[\theta m - \psi(\theta)]$$

then

$$\psi'(\theta) = m \text{ if and only if } \theta = h'(m)$$

and the average value or the expected value of $\phi'(\xi)$ if we fix the value of the expected value of ξ at m is calculated by

$$\int \phi'(\xi) \exp[\theta\xi - \phi(\xi)] d\xi = \theta = h'(m).$$

This in turn suggests that if the averages

$$\frac{1}{N} \sum_x \xi_x \delta_x \sim m(x)dx$$

then

$$\frac{1}{N} \sum_x \phi'(\xi_x) \sim h'(m(x))dx.$$

With a certain amount of work one can establish in this context that the limiting measure lives on the set of weak solutions of

$$\frac{\partial m}{\partial t} = \frac{1}{2}[h'(m)]_{xx}.$$

The relevant evolution equation now is governed by the non-linear diffusion operator

$$\frac{1}{2} D_x h''(m) D_x$$

with the diffusion coefficient depending on the dependent variable m. The actual proof that the averages behave in the right manner is based on the control of the Dirichlet form, which in turn is controlled by the initial entropy. The deatils of the proof can be found in [2].

It is important to understand why these are examples of gradient systems. When we calculated the differential of a quantity $\Psi_{N,J}(t)$ like

$$\frac{1}{N} \sum_x J(x)\eta(t,x)$$

or

$$\frac{1}{N} \sum_x J(x)\xi_x(t)$$

the answer at the first step turned out to be of the form

$$\frac{1}{N}\sum_x J'(x)NW_x\ dt + dM_N(t)$$

with the martingale term becoming negligible due to a "law of large numbers" effect. In the case of simple exclusion the term W_x was given by

$$\begin{aligned} W_x &= -\frac{1}{2}\sum_y N(y-x)p_N(y-x)[\eta(y)-\eta(x)] \\ &= -\frac{1}{2}\sum_z zp(z)[\nabla_z\eta(x)] \end{aligned}$$

where ∇F is the gradient $F(\tau_z\cdot) - F(\cdot)$, with τ_z being the shift by z steps on the lattice. The "current" represented by W_x is a lattice gradient of some function and this allows another summation by parts to be performed. In this process we get rid of a burdensome power of N which we do not want, and W_x gets replaced by F_x of which it is the gradient. This leads to an expression of the form

$$d\Psi_{N,J}(t) = \frac{1}{N}\sum_x J''(x)F_x$$

and F is basically η agian and gives us the diffusion equation which is linear. In the Ginzburg-Landau model F_x is essentially $\phi'(\xi_x)$ and has to be replaced by its mean value in the equlibruim with mean field strength m producing thereby a nonlinear diffusion equation for our evolution and this averaging needed justification. The commomon feature of both models was the crucial fact that the current W_x was a linear combination of gradients and this allowed us to do a summation by parts.

3. Nongradient Systems.

Now that we have seen what gradient systems are, we are ready to look at the next level of complexity in problems of diffusive hydrodynamic scaling. The first example that we will examine is the Ginzburg-Landau model that we introduced earlier, but without the simplifying assumption that $a \equiv 1$.

In this case when we compute $d\Psi_{N,J}(t)$, we get

$$d\Psi_{N,J}(t) = \frac{1}{N}\sum_x J'(x)[NW_x(t)]dt + dM_N(t)$$

with

$$\begin{aligned} W_x &= \frac{1}{2}W(\xi_x, \xi_{x+e}) \\ &= \frac{1}{2}[a(\xi_x, \xi_{x+e})(\phi'(\xi_x) - \phi'(\xi_{x+e})) - a_1(\xi_x, \xi_{x+e}) + a_2(\xi_x, \xi_{x+e})]. \end{aligned}$$

Here $x + e$ denotes lattice site that is adjacent to x in the anticlockwise direction. Now we are stuck becuse we have no way to get rid of the unwelcome factor of N. Unlike the case of $a \equiv 1$, W_x is not a gradient of the form $F_{x+e} - F_x$, for any F and we cannot perform a summation parts. The averaging principle which is equally valid with any reasonable a is not of much use. W_x has mean zero in every local equilibrium and at some formal level the problem is to understand $E[NW_x]$ which requires a knowledge of $E[W_x]$ with an accuracy of $O(\frac{1}{N})$. A similar calculation for the simple exclusion process with two colors involves:

$$\Psi_{N.A,B}(t) = \frac{1}{N}\sum A(x)g(x,t) + \frac{1}{N}\sum_x B(x)r(x,t)$$

and

$$\begin{aligned} d\Psi_{N.A,B}(t) &= \frac{1}{N}N^2\sum_{x,y} p_N(y-x)[1-\eta(y,t)] \\ &\qquad [g(x,t)(A(y)-A(x)) + g(y,t)(B(y)-B(x))]dt \\ &\sim \frac{1}{N}\sum_x [A'(x)[NW_x^g(t)] + B'(x)[NW_x^r(t)]]dt \end{aligned}$$

with

$$W_x^g = g(x)\sum_z zp(z)[1-\eta(x+\frac{z}{N})]$$

and

$$W_x^r = r(x)\sum_z zp(z)[1-\eta(x+\frac{z}{N})].$$

It is important to note that while W_x^g and W_x^r are not gradients, their sum $W_x^\eta = W_x^g + W_x^r$ is a gradient as we saw earlier, for the total current is color blind. These difficulties of non gradient models can already be felt at the level of equilibrium fluctuations and an understanding at that level will enable us to find a solution. We open up the lattice Z_N and consider the problem on the full lattice Z. In the Ginzburg-Landau model we will have an infinite dimensional diffusion on R^Z with generator

$$\mathcal{L} = \frac{1}{2}\sum_{x\sim y} a(\xi_x,\xi_y)\left(\frac{\partial}{\partial \xi_x} - \frac{\partial}{\partial \xi_y}\right)^2 - \frac{1}{2}\sum_{x\sim y} W(\xi_x,\xi_y)\left(\frac{\partial}{\partial \xi_x} - \frac{\partial}{\partial \xi_y}\right).$$

These will have a family of translation invariant reversible invariant measures indexed by λ and for each λ this measure will be the product measure with marginal density $\exp[\lambda\xi - \phi(\xi) - \psi(\lambda)]$. Let us fix a value of λ, for instance $\lambda = 0$, and assume in addition that

$$\int_{-\infty}^{\infty} \xi \exp[-\phi(\xi)]d\xi = 0.$$

We denote by μ the invariant measaure and by P the stationary Markov process with μ as the invariant measure. If we look at the field averages

$$\langle J, H\rangle == \sqrt{N}\Psi_{N,J} = \frac{1}{\sqrt{N}}\sum_x J(\frac{x}{N})\xi_x,$$

as $N \to \infty$, under the measure μ the random field H approaches white noise on R. We can look at the time evolution $H(t)$ of H, speeded up by the factor N^2. We will first examine what we get in the gradient case. We take $a \equiv 1$, and write

$$d\langle J, H(N^2t)\rangle = \frac{1}{2\sqrt{N}}\sum_x J''(\frac{x}{N})\phi'(\xi_x(N^2t)) + dM_{N,J}(t).$$

If we use a stronger version of the averaging principle that replaces

$$\frac{1}{\sqrt{N}}\phi'(\xi_x)$$

by

$$\frac{1}{\sqrt{N}}h''(0)\sum_x \xi_x$$

(think of it as the linearization of the relation $E[\phi'(\xi)] = h'(E[\xi])$), then we get in the limit an Ornstein-Uhlenbeck process with white noise as invariant measure. The dissipative part is controlled by

$$\frac{D}{2}\frac{d^2}{dx^2}$$

where the diffusion coefficiet D is given by

$$D = h''(0) = \sigma^2 = \int_{-\infty}^{\infty} \xi^2 \exp[-\phi(\xi)]d\xi.$$

The martingales converge to a Brownian motion in time, with a covariance in space that balances the dissipative part and maintains the white noise as the inavriant measure for the Ornstein-Uhlenbeck process. Now we would like to do this in the nongradient situation as well. Calculating once again we get

$$d\langle J, H(N^2t)\rangle = \frac{1}{2\sqrt{N}}\sum_x J'(x)[NW_x(t)]dt + dM_{N,J}(t)$$

with

$$\begin{aligned} W_x &= \frac{1}{2}W(\xi_x, \xi_{x+e}) \\ &= \frac{1}{2}[a(\xi_x, \xi_{x+e})(\phi'(\xi_x) - \phi'(\xi_{x+e})) - a_1(\xi_x, \xi_{x+e}) + a_2(\xi_x, \xi_{x+e})] \end{aligned}$$

and we are in a similar predicament. The clue is to use cetain fluctuations that seem irrelevant. Suppose we consider a sum of the form

$$\mathbf{F}_N(t) = \frac{1}{N\sqrt{N}}\sum_x J(\frac{x}{N})F_x(N^2t)$$

and compute

$$d\mathbf{F}_N(t) = \frac{1}{\sqrt{N}}\sum_x J(\frac{x}{N})[NG_x](N^2t)dt + dM_{N,F}(t)$$

where F is a smooth local function and F_x represents its translate by x. For J with compact support, $\mathbf{F}_N$ is uniformly bounded by $\frac{1}{\sqrt{N}}$ and is negligible. Replacing J by J' we can add the two terms to get

$$d\langle J, H(N^2t)\rangle = \frac{1}{\sqrt{N}}\sum_x J'(x)[NW_x(t) + NG_x(t)]dt + dM_{N,J}(t) + dM_{N,F}(t).$$

If we denote by $\mathcal{G}$ the space of such G's, it will turn out that

$$W_x = c(\phi'(\xi_x) - \phi'(\xi_{x+e})) \qquad \text{modulo } \mathcal{G}$$

for some constant c. Once we identify the constant c, the dissipation term is then

$$\frac{1}{2}ch''(0)\frac{d^2}{dx^2}$$

and the (Green-Kubo) diffusion coefficient is $D = ch''(0)$. Of course the martingales $M_{N,F}(t)$ produced in the process will cancel part of $M_N(t)$ and in the end the remaining noise and the diffusion coefficient will have to match with the white noise invariant measure to get an Ornstein-Uhlenbeck process. What we are trying to do is to project out some rapidly decaying fluctuations from the current W_x to get a gradient term $c(\phi'(\xi_x) - \phi'(\xi_{x+e}))$ and we would like to set this up as a projection theorem in a Hilbert space. We will describe this in the next section and conclude with some remarks in the last section.

4. Analysis of Nongradient Systems.

Let us now describe the process by which the diffusion coefficient is calculated for nongradient systems. We consider the invariant measure μ and the stationary Markov process P, with generator $\mathcal{L}$ and invariant measure μ on R^Z. Suppose we have a local function G on R^Z depending only on a finite number of coordinates which has mean zero with respect to the invariant measure μ. We think of ξ as the point in R^Z that is evolving and consider the central limit theorem for

$$\frac{1}{\sqrt{t}}\int_0^t G(\xi(s))ds.$$

Let us suppose that the central limit theorem is valid and that the limiting variance is given by $[G;G]$. Since the process as well as the dynamics are invariant under translations of the lattice Z, we could consider the translates $\tau_x G = G_x$ as well. We will be interested in calculating

$$\langle G;G\rangle = \lim_{\ell\to\infty}\frac{1}{2\ell+1}[\sum_{|x|\le\ell} G_x \; ; \sum_{|x|\le\ell} G_x].$$

Roughly speaking our Hilbert space will consist of $\{G\}$ with the norm $\langle G; G\rangle$. Admissible G's include

$$\begin{aligned}\mathcal{G}_0 &= \{G : G = \mathcal{L}F \text{ for some } F\}\\ \mathcal{G}_1 &= \{kW_0 : k \in R\} \quad \text{one dimensional subspace}\\ \mathcal{G}_2 &= \{k(\phi'(\xi_0) - \phi'(\xi_1)) : k \in R\} \quad \text{another one dimensional subspace}\end{aligned}$$

We will find a convenient space $\mathcal{G}$ to imbed isometrically all of these spaces and we will show that in that space $\mathcal{G}$, $\mathcal{G}_0$ is of codimension 1 and that $\mathcal{G}$ is spanned by $\mathcal{G}_0$ and $\mathcal{G}_2$. It is then a simple matter to write $W_0 = c(\phi'(\xi_0) - \phi'(\xi_e))$ modulo $\mathcal{G}_0$. Such a space is provided by the following construction. For any smooth local function F let us form the formal sum

$$\tilde{F} = \sum_{x\in Z} \tau_x F = \sum_{x\in Z} F_x.$$

Although the sum does not quite make sense the derivatives $\frac{\partial \tilde{F}}{\partial \xi_x}$ make sense because for each x only a finite number of summands involve ξ_x. We would like to consider the difference

$$g = \frac{\partial \tilde{F}}{\partial \xi_1} - \frac{\partial \tilde{F}}{\partial \xi_0}$$

as an element of $L_2(\mu)$. An easy calculation yields the relationship, valid for arbitrary Markov processes, between the variance in the central limit theorem and the Dirichlet form

$$\lim_{t\to\infty} \frac{1}{t} E^P \left[\int_0^t [Lu(x(s)]ds\right]^2 = 2D(u)$$

where the Dirichlet form $D(u)$ is given by

$$D(u) = -(Lu, u) = -\int uLud\mu.$$

This allows us to make the identification

$$\langle G; G\rangle = \|g\|^2 = \int a(\xi_0, \xi_e)|g|^2 d\mu.$$

The set of such g's of course is far from closed in $L_2(\mu)$. For a Cauchy sequence g_n in $L_2(\mu)$ we have no control on the corresponding local F_n that produce g_n. They may start depending on more and more sites and may grow large in norm as well. On the other hand one can define abstractly g's that qualify as gradients for some $\tilde{F}$ that may or may not exist. These are g in $L_2(\mu)$ that satisfy

$$(\frac{\partial}{\partial \xi_y} - \frac{\partial}{\partial \xi_{y+e}})\tau_x g = (\frac{\partial}{\partial \xi_x} - \frac{\partial}{\partial \xi_{x+e}})\tau_y g$$

in the sense of distributions. Such abstract gradients define a Hilbert space $\mathcal{G}$ and the concrete g's that come from F for some F are in $\mathcal{G}$. We denote by $\mathcal{G}_0$ the closure in $L_2(\mu)$ of g's that come from local F's. The main theorem that allows us to complete the proof states that $\mathcal{G}_0$ is of codimension 1 in $\mathcal{G}$ and a complementary one dimensional space is spanned by the image g_D of $(\phi'(\xi_0) - \phi'(\xi_e))$. The quantity W_0 is similarly imbedded as some element g_C in $\mathcal{G}$ and all we need to do to calculate the diffusion coefficient c is to carry out the projection

$$g_C = c g_D \mod \mathcal{G}_0.$$

While this is sufficient for the analysis of equilibrium fluctuations there is still some work to be done do derive the nonlinear diffusion equation when we start from nonequlibrium. The diffusion constant c was derived by fluctuation considerations in a specific equilibrium determined by λ or equivalently m. Therefore $c = c(m)$. One then goes on to show, by use of techniques from the theory of large deviations, that

$$\frac{1}{N}\int_0^t \sum_x J'(x)[NW_x](s)ds$$

can be replaced by

$$\frac{1}{N}\int_0^t \sum_x J'(x)[Nc(\bar{\xi}_{k,x})(\phi'(\xi_x(s)) - (\phi'(\xi_{x+e}(s))]ds$$

with a large k or

$$\frac{1}{N}\int_0^t \sum_x J'(x)[Nc(\bar{\xi}_{N\epsilon,x})(\phi'(\xi_x(s)) - (\phi'(\xi_{x+e}(s))]ds$$

with a small ϵ. Here $\bar{\xi}_{x,\ell}$ denotes the average of ξ_y over the block $\{y : |y-x| \leq \ell\}$. Once this is done, it is onle one more step to show that the above quantity can be replaced in the limit with

$$-\frac{1}{2}\int_0^t \int_S J'(x)c(m(s,x))[h'(m(s,x))]'dx$$

leading to the nonlinear diffusion equation

$$\frac{\partial m}{\partial t} = \frac{1}{2}\frac{\partial}{\partial x}c(m(t,x))h''(m(t,x))\frac{\partial m}{\partial x}.$$

We outlined the method in the case of the Ginzburg-Landau model. But the problem of two colors in simple exclusion is treated in a similar fashion. Let us fix densities ρ_g and ρ_r with $\rho = \rho_g + \rho_r < 1$ and define μ to be the Bernoulli product measure on $[g, r, e]^Z$ where each site may have a green particle, a red particle or be empty with probabilities ρ_g, ρ_r and $\rho_e = 1 - \rho_g - \rho_r < 1$ respectively. Again for any local function $F = F(g,r)$ depending on a finite number of sites we form the formal sum

$$\tilde{F} = \sum_{x \in Z} \tau_x F$$

and define the gradients

$$g_{x,y} = \nabla_{x,y}\tilde{F} = \tilde{F}(\sigma^{x,y}g, \sigma^{x,y}r) - \tilde{F}(g,r)$$

that make sense for any pair x, y of lattice sites and any local function F. The functions $g_{x,y}$ have a lot of covariance properties with respect to translations and all of them can be completely determined by algebraic formulae starting from $g = g_{0,e}$. One has here the space of such abstract g's in $L_2(\mu)$ with the norm

$$\|g\|^2 = \int \sum_x p(x)[1 - \eta(x)]|g_{0,x}|^2 d\mu$$

The ones that come from F form a subspace $\mathcal{G}_0$ which is shown to have codimension two in this case, with a complementary subspace provided by the images of the two density gradients $g(e) - g(0)$ and $r(e) - r(0)$. The currents W_g and W_r are then projected modulo $\mathcal{G}_0$ to get a replacement of the form

$$[NW_g, NW_r] \Rightarrow \mathbf{B}[N(g(e) - g(0)), N(r(e) - r(0))]$$

where $\mathbf{B} = \mathbf{B}(\rho_g, \rho_r)$ is a 2×2 matrix that is determined in the process. We then get a parabolic system of two equations for the two densities $\rho_g(t, x)$ and $\rho_r(t, x)$.

5. Remarks.

The first results in the analysis of nongradient models were obtained for the Ginzburg-Landau model and the methods outlined here appeared in [6]. The work in [6] however is based on the earlier analysis of the gradient version carried out in [2]. The two color problem for symmetric exclusion was worked out in [5]. If we replace the symmetric jump distributuion $p(z)$ in the simple exclusion model by an asymmetric distribution $p(\cdot)$ but one that has mean 0, i.e. $\sum zp(z) = 0$, then one gets a diffusive non gradient model even for a single color. This was studied in [7]. Here the underlying particle system is not reversible and that causes some problems in the analysis. But the asymmetric part in this model is in some sense a controlled perturbation of the symmetric part and that helps. One gets a nonlinear diffusion in the limit as opposed to the symmetric case where we saw that the diffusion was linear. Another model that is a modification of symmetric simple exclusion , but one that is a nongradient model was studied in [3]. Finally, if we look at really asymmetric simple exclusion in Z^d with $\sum zp(z) \neq 0$, the scaling is convective or hyperbolic rather than diffusive. One obtains as limit Burgers equation with out viscosity. If $d \geq 3$, one can obtain some correction terms and they are governed by Burgers Equation with small viscosity. Also, in some special cases, near equilibrium where one can look directly at diffusive time scales, one gets as limits Burgers equation with a viscosity term. This analysis is done in [1] and [4] and uses the nongradient techniques that we outlined but is much harder because the situation is strongly asymmetric and the asymmetric part is not an easily controlled perturbation of the symmetric piece.

6. References.

1. Esposito, R., Marra, R., and Yau, H. T., Diffusive limit of asymmetric simple exclusion, Reviews in Mathematical Physics, Vol 6, No 5A, (1994), 1233-1267.

2. Guo, M. Z., Papanicolaou, G. C., and Varadhan, S. R. S., Nonlinear diffusion limit for a system with nearest neighbor interactions, Comm. Math. Phys. 118 (1988), 31–59.

3. Kipnis, C., Landim, C., and Olla, S., Hydrodynamic limit for a non-gradient system: the generalized simple exclusion process. Comm. Pure Appl. Math. Vol XLVII (1994), 1475-1545.

4. Landim, C., and Olla, S., and Yau H.T., First Order Correction for the Hydrodynamic Limit of Asymmetric Simple Exclusion Processes in Dimension $d \geq 3$, submitted to Communications on Pure and Applied Mathematics.

5. Quastel, J., Diffusion of color in simple exclusion process, Comm. Pure Appl. Math. 45 (1992), 623–680.

6. Varadhan, S. R. S., Nonlinear diffusion limit for a system with nearest neighbor interactions, II, Asymptotic Problems in Probability Theory: Stochastic Models and Diffusions on Fractals, K. D. Elworthy and N. Ikeda (Editors), Pitman Research Notes in Mathematics Series 283, 1991, 75–130.

7. Xu, Lin, Hydrodynamics for asymmetric mean zero simple exclusion, Ph.D. Thesis, New York University, 1993.

J. Math. Kyoto Univ. 36 (1996) 869-875

Relative entropy and mixing properties of interacting particle systems

By

A. F. RAMIREZ and S. R. S. VARADHAN

1. Introduction

Consider a Markov semigroup S_t on a compact metric space X that has the Feller property. If we start the process with some initial distribution μ, then it is not always true that for large t the distribution $\mu_t = \mu S_t$ is close to an invariant distribution for the process. While any limit point as $T \to \infty$ of the time average $\frac{1}{T}\int_0^t \mu_t dt$ is always an invariant measure, the same can not be claimed for limit points of μ_t itself. The simplest examples are provided by deterministic flows. However for any Markov chain on a finite state space, continuous time rules out periodic behavior and μS_t has a limit as $t \to \infty$ and this is always an invariant measure.

The natural question that arises is to determine if under some suitable conditions on the Markov semigroup S_t one can still claim that all possible limit points of μS_t are invariant measures. Such a result in conjunction with a uniqueness theorem for invariant measures will establish the convergence of μS_t to the unique invariant measure giving us a mixing result.

It has been conjectured that in the context of interacting particle systems the answer is in the affirmative under some very mild restrictions. Let $X = F^{\mathbf{Z}^d}$, where F is a finite set. The state η of the system is described by its values $\eta(x)$ for $x \in \mathbf{Z}^d$. The infinitesimal generator of the particle system is given by

$$\Omega f(\eta) = \sum_{T \subset \mathbf{Z}^d} \int_{F^T} c_T\ (d\xi, \eta)\ (f(\eta^\xi) - f(\eta)) \tag{1}$$

where the summation runs over all finite subsets of $\mathbf{Z}^d$. Here $c_T(d\xi, \eta)$ describes the rates for Poisson events that change the current configuration η to a new configuration η^ξ that has been altered on the finite index set $T \subset \mathbf{Z}^d$ from η to ξ. A whole family of such Poisson events are taking place simultaneously and the infinitesimal generator reflects that. Of course a whole lot of these $c_T(\cdot,\cdot)$ may be 0. We say that a particle system has bounded flip rates if there is a bound on the sum of all the Poisson rates that could affect a loca-

Communicated by Prof. S. Watanabe, March 22, 1996

Supported by NSF grant 9503419 and ARO grant DAAL03-92-G-0317

tion $x \in \mathbf{Z}^d$, i. e.,

$$\sup_{x \in Z^d} \left\{ \sum_{T \ni x} \sup_{\eta \in X} \int_{F^T} c_T(d\xi, \eta) \right\} < \infty \tag{2}$$

And we will say that a particle system has finite range R if the following two conditions are satisfied:
a) $c_T(\cdot,\cdot) \equiv 0$ whenever the diameter of T exceeds R, i. e. $\sup_{x,y \in T}|x-y| \geq R$.
b) For every $A \subset F^T$ the function $c_T(A, \eta)$ does not depend on $\eta(u)$ if dist $(T, u) \geq R$.
Under these conditions the operator Ω is well defined for $f \in \mathfrak{D}$, the space of 'tame functions' i.e functions that depend on $\{\eta(x)\}$ only for x in some finite set.

The operator Ω on $\mathfrak{D}$ has a natural extension that defines a Markov generator (see theorem 3.9 of Liggett [1]).

The following theorem was recently proved by Mountford [2] when the dimension $d=1$.

Theorem 1. *Let η_t be an infinite particle system on $X=F^{\mathbf{Z}}$ with bounded flip rates and of finite range. Let μ be some probability measure on X such that*

$$\lim_{n\to\infty} \mu S_{t_n} = \nu$$

exists, where $t_n \nearrow \infty$. Then ν is invariant.

We provide a different proof in this note. The proof of Mountford uses ABL coupling. We replace it by a relative entropy argument. Our proof is shorter and we believe that it can bemodified and applied to other situations.

idea of proof. In order to illustrate the idea behind the proof we will show that if S_t is a Markov semigroup with a bounded generator

$$\Omega f(x) = \int_X [f(y) - f(x)] \pi(x, dy) \tag{3}$$

then for any μ and any fixed $0 < \tau < \infty$, $\|\mu S_{\tau+t} - \mu S_t\| \to 0$ as $t \to \infty$, where $\|\alpha\|$ denotes the total variation norm of the signed measure α. If we take some $\lambda > 0$ and consider the generator $(1+\lambda)\Omega$ instead of Ω, by a Girsanov type formula it is seen that the new process is absolutely continuous with respect to the old one and one can get a bound on the relative entropy of the new process with respect to the original one during any finite time interval $[0, t]$. A simple estimate, for small λ, gives the bound $H \leq Ct\lambda^2$ for the relative entropy where C is a bound for the operator Ω. By making the choice of $\lambda = \frac{\tau}{t}$ we get $H = o(1)$ as $t \to \infty$ for fixed τ. Since the marginal at time t of the speededup process with generator $\left(1+\left(\frac{\tau}{t}\right)\right)\Omega$ is $\mu S\tau_{+t}$, and the relative entropy controls the variational distance the proof is complete.

Our proof of Mountford's theorem is basically a modification of the above idea. In order to have a bounded Ω we need to put things in a finite box and then try to control the influence of effects coming from outside the box. This is done through the finite range condition. There is a fight between the smallness of the entropy bound and the growth of the domain that influences what goes on in a finite box. When $d=1$ the entropy bound wing, giving us the result. The details are carried out in the next two sections.

it is enough to show that for any arbitrary finite M, local function f depending on sites $[-M, M]$, and any $\tau>0$,

$$\lim_{t\to\infty}\left|\int S_t f d\mu-\int S_{t+\tau} f d\mu\right|=0 \tag{4}$$

To do this we define a new process governed by the time dependent infinitesimal generator

$$\Omega^{t,\tau}(s)=\lambda^{t,\tau}(s)\,\Omega \tag{5}$$

where $\lambda^{t,\tau}(s)$: $[0,\infty)\rightarrow[1,\infty)$ is chosen so that

$$\int_0^t \lambda^{t,\tau}(s)\; ds=t+\tau \tag{6}$$

The new proces will then have transition operators $S_{0,t}^{\lambda(\cdot)}=S_{t+\tau}$. Because Ω is an unbounded operator, processes with generators $\Omega(s)=\lambda(s)\,\Omega$ are not absolutely continuous with respect to the original one. But this situation can be remedied by truncating the processes.

Suppose we have a function $h(s)\geq M$, defined for $0\leq s\leq t$, we truncate the processes to $[-h(s), h(s)]$ at time $s\in[0, t]$. We define the generators

$$\Omega^h(s)f(\eta)=\sum_{T\subset[-[h(s)],[h,(s)]]}\int_{F^T} c_T(d\xi,\zeta)\,(f(\zeta^\xi)-f(\zeta))$$

$$\Omega^{\lambda,h}(s)f(\zeta)=\lambda^{t,\tau}(s)\sum_{T\subset[-[h(s)],[h,(s)]]}\int_{F^T} c_T(d\xi,\zeta)\,(f(\zeta^\xi)-f(\zeta))$$

We define four processes P, P^λ, P^h and $P^{\lambda,h}$ corresponding to the generators Ω, $\Omega^\lambda(s)$, $\Omega^h(s)$ and $\Omega^{\lambda,h}(s)$ respectively all having the same initial distribution μ. The processes P and P^λ have $\mu_t=\mu S_t$ and $\mu_{t+\tau}=\mu S_{t+\tau}$ for marginal distributions at time t. P^h and $P^{\lambda,h}$ are governed by truncated generators and the relative entropy of $P^{\lambda,h}$ with respect to P^h can be estimated by means of a Girsanov formula and will be carried out in section 3. In section 2 we will estimate the effect of the truncation. If we denote by μ_t, μ_t^λ, μ_t^h and $\mu_t^{\lambda,h}$ the marginals at time t of the four processes P, P^λ, P^h and $P^{\lambda,h}$, the relative entropy estimate in section 3, will tell us how close $\mu_t^{\lambda,h}$ is to λ_t^h. In section 2 we will estimate the total variation on the σ-field corresponding to $[-M, M]$ of the differences

$\mu_t - \mu_t^h$ and $\mu_t^\lambda - \mu_t^{\lambda,h}$.

2. Construction of truncated processes

A theorem of Liggett (theorem 3.9 (c) [1]) provides the basis for estimating the difference between the truncated and untruncated processes. First some notation: for any bounded function f on $F^{\mathbf{Z}}$ let us define

$$\Delta_f(x) = \sup\{[f(\eta) - f(\eta']; \eta, \eta' : \eta(y) = \eta'(y) \quad \forall y \neq x\} \tag{7}$$

and

$$D(X) = \{f: \sum_{x \in \mathbf{Z}} \Delta_f(x) < \infty\} \tag{8}$$

Theorem 2. *Let S_t be the semigroup of a finite range, bounded flip rate particle system on $F^{\mathbf{Z}}$. Then for $f \in D(X)$*

$$\Delta_{S_t f}(x) \leq e^{t\Gamma} \Delta_f(x) \tag{9}$$

Here Γ is an operator defined on $l_1(\mathbf{Z})$, the Banach space of all functions β on $\mathbf{Z}$ such that $\|\beta\|_{l_1(\mathbf{Z})} := \sum_x |\beta(x)| < \infty$, by

$$\Gamma\beta(y) = \sum_x \beta(x)\gamma(x, y)$$

where $c_T(y) = \sup_{\eta_1, \eta_2: \eta_1(x) = \eta_2(x), \forall x \neq y} \{\|c_T(\eta_1, d\xi) - c_T(\eta_2, d\xi)\|_T\}$ *and* $\gamma(x, y) = \sum_{T \ni x} c_T(y)$

We assume with out loss of generality that $\sup_{x \in \mathbf{Z}} \{\sum_{T \ni x} \sup_{\eta \in \mathbf{Z}} c_T(F^T, \eta)\} \leq 1$ in what follows.

Corollary 1. *Let Ω be the generator of a particle system with bounded flip rates and dinite range R. Then ior each local function f depending on sites on $[-M, M]$ we have*

$$\Delta_{S_t f}(x) \leq 2\|f\|_\infty \exp\left[-y \log\frac{y}{ebt}\right] \tag{10}$$

whenver $|x| \geq M + 2Rbt$. *Here* $b = 2(4R+1)$, $y = \frac{|x| - M}{2R}$, *and* $\|f\|_\infty = \sup_{\eta \in F^{\mathbf{Z}}} |f(\eta)|$.

Proof. Note that

$$e^{t\Gamma}(x) = \sum_{n=0}^{\infty} \frac{(t\Gamma)^n}{n!} \Delta_f(x)$$

And since $\gamma(x, y) = 0$ whenever $|x - y| > 2R$, we have

$$\Gamma^n \Delta_f(x) \leq 2^n (4R+1)^n \sup_{y: |y - x| \leq 2Rn} \Delta_f(y).$$

But f is a function depending only on sites on $[-M, M]$, so that

$$\sup_{y:|y-x|\leq 2Rn} \Delta_f(y) \leq 2\|f\|_\infty \chi(M+2nR-x),$$

where $\chi(y)$ is the indicator function of the interval $[0, \infty)$. Therefore

$$\Delta_{S_tf}(x) \leq 2\|f\|_\infty \sum_{n\geq[(|x|-M)/2R]} \frac{(2t(4R+1))^n}{n!}$$

from which (10) follows.

Theorem 3. *Let ζ be a finite range, bounded flip rate particle system with state space $X=F_{\mathbf{Z}}$. Let ζ^h be the truncated process, with the same initial distribution, corresponding to the function $h(s)$ with the property that $h(s)\geq Rc(t-s)+R+Rk+M$ for some k that we may choose later to depend on t. The difference in the variational norm between μ_t and μ_t^h, the distributions at time t of the two processes ζ and ζ^h on the σ-field corresponding to the sites $[-M, M]$ is bounded by*

$$\|\mu_t^h-\mu_t\|\leq\frac{2}{Rb}e^{-kR}$$

provided $c=2ebe^R$.

Proof. Let $u(s, \zeta)=S_{t-s}f(\zeta)$, where f is any local function depending on sites on $[-M, M]$. Then since

$$u(s, \zeta_s^h)-u(0, \zeta_0^h)-\int_0^s\left(\frac{\partial u(\sigma, \zeta_\sigma^h)}{\partial\sigma}+(\Omega_\sigma^h u)(\sigma, \zeta_\sigma^h)\right)dv$$

is a martingale and $\frac{\partial u}{\partial s}+\Omega_s^h u=(\Omega_s^h-\Omega)u$, we have

$$(S_{0,t}^h f)(\zeta)-(S_tf)(\zeta)=\int_0^t(S_s^h(\Omega_s^h-\Omega)S_{t-s}f)(\zeta ds) \tag{11}$$

Now for any $g\in d(X)$ we have

$$|(\Omega_s^h-\Omega)g(\zeta)|\leq\sum_{|x|\geq h(s)-R}\Delta_g(x)$$

and since $S_{t-s}f\in D(X)$, we have

$$\sup_{\zeta\in X}|(S_{0,t}^h f)(\zeta)-(S_tf)(\zeta)|\leq\int_0^t\sum_{|x|\geq h(s)-R}\Delta_{S_{t-s}f}(x)ds. \tag{12}$$

But from corollary 1 and the fact that $c=2ebe^R=4e(4R+1)e^R$,

$$\int_0^t\sum_{|x|\geq h(s)-R}\Delta_{S_{t-s}f}(x)ds\leq\|f\|_\infty\int_0^t\sum_{n\geq Rcs+kR}e^{-n}ds$$

$$\leq \frac{2\|f\|_\infty}{bRe^R(e-1)}e^{-kR}$$

The proof is now complete.

Remarks. Exactly the same type of estimates are valid for comparing the speeded up process P^λ with its truncated version. One slight difference of no consequence is that the flip rates may not be bounded by 1. Since our speedups are going to be very minor we assume that $0 \leq \lambda(s) \leq 2$. If we had assumed intially that all flip rates are bounded by $\frac{1}{2}$ then Theorem 3 would apply for the speeded up version as well. So we have

$$\|\mu_t^{\lambda,h} - \mu^\lambda\| \leq \frac{2}{bRe^R(e-1)}e^{-kR} \tag{13}$$

3. Relative entropy of truncated processes

The processes P^h and $P^{\lambda,h}$ have the same initial distribution μ and time dependent gnerators $\Omega^h(s)$ and $\Omega^{\lambda,h}(s) = \lambda(s)\Omega^h(s)$ repectively. Since $\Omega^h(s)$ is a bounded operator on $[0, t]$, Girsanov formula can be used to calculate the relative entropy of $P^{\lambda,h}$ with respect to P^λ on the σ-field $\mathcal{F}_t^0$ generated by $\{\eta(s): 0 \leq s \leq t\}$. Since

$$H(\mu_t^{\lambda,h}|\mu_t^h) \leq H(P^{\lambda,h}|P^h)$$

if we can estimate $H(P^{\lambda,h}|P^h)$, and show that it tends to zero for some reasonable choices of k, λ, and h, we will be done.

Lemma 1.

$$H(P^{\lambda,h}|P^h) \leq 2\int_0^t h(s)(\lambda^{t,\tau}(s)-1)^2 ds \tag{14}$$

Proof. Denote by N (s) the total number of jumps ocurring up to time s. Then the Radon-Nikodym derivative of $P^{\lambda,h}$ respect to P^h on $\mathcal{F}_t^0$ is given by the following formula

$$\frac{dP^{\lambda,h}}{dP^h} = \exp\left\{\int_0^t \log \lambda^{t,\tau}(s)dN(s) - \int_0^t \sum_{T\subset[-h(s),h(s)]} c_T(F^T, \zeta_s)(\lambda^{t,\tau}(s)-1)ds\right\}$$

and a direct computation yields

$$H(P^{\lambda,h}|P^h) = \int_0^t E_{P^{\lambda,h}}\Big(\sum_{T\subset[-h(s),h(s)]} c_T(F^T,\zeta_s)\Big)(\lambda^{t,\tau}(s)\log\lambda^{t,\tau}(s) - (\lambda^{t,\tau}(s)-1))ds$$

Equation (14) now follows from the assumption of bounded flip rates and the inequality $\ln x \leq x-1$ for $x \leq 1$.

The following lemma is elementary.

Lemma 2. *Let* $\mathscr{A}_{t,\tau}=\{\lambda^{t,\tau}(\cdot)\in L^1([0,t]):\int_0^t(\lambda^{t,\tau}(s)-1)ds=\tau\}$ *and* $h(s)=Rc(t-s)+R+Rk+M$. *Then*

$$\inf_{\lambda\in\mathscr{A}_{t,\tau}}\int_0^t h(s)\lambda^{t,\tau}(s)-1)^2\,ds=Rc\gamma^2\left\{\log\frac{Rct+Rk+M+R}{Rk+M+R}\right\}^{-1} \tag{15}$$

and the infimum is attained at $\lambda^{t,\tau}(s)=1+\dfrac{\alpha}{h(s)}$ *with*

$$\alpha=Rc\tau\left\{\log\frac{Rct+Rk+M+R}{Rk+M+R}\right\}^{-1}.$$

COURANT INSTITUTE OF MATHEMATICAL SCIENCES,
251 MERCER STREET, NEW YORK,
NY 10012. , U.S.A

References

[1] T. M. Liggett, Interacting particle systems. Springer-Verlag, New York, 1985.

[2] T. S. Mountford, A coupling of infinite particle systems, J. Math. Kyoto Univ. **35**-1 (1995), 43-52.

ASIAN J. MATH.
Vol. 1, No. 4, pp. 623–678, December 1997

001

DIFFUSIVE LIMIT OF LATTICE GAS WITH MIXING CONDITIONS*

S.R.S. VARADHAN† AND HORNG-TZER YAU‡

Abstract. We prove, under certain mixing conditions, that the hydrodynamical limit of a stochastic lattice gas on the cubic lattice $\mathbb{Z}^d$ is governed by a nonlinear diffusion equation. Following [V1], we characterize the diffusion coefficient by a variational formula, which is equivalent to the Green-Kubo formula. The fluctuation-dissipation equation is established rigorously as an important step of the proof. Our mixing conditions are implied by the Dobrushin-Shlosman mixing conditions which are always valid at high temperatures.

1. Introduction. Classical or Newtonian mechanics is described by a system of ordinary differential equations involving usually a two-body interaction potential. When the number of particles involved is large, rather than describe the motion of every particle, one usually tries to describe the collective behavior of these particles. The typical formulation involves the consideration of local averages of conserved quantities like density, average velocity, and energy. Since these local averages of conserved quantities tend to vary slowly in time, under some suitable space-time rescaling, referred to as hydrodynamical scaling, we expect to obtain a closed system of equations for these conserved quantities, as functions of macroscopic space and time. While such a transition is reasonably well understood from a physical point of view, these problems appear to be far beyond the current available methods for a rigorous mathematical treatment.

One way to make the problem tractable is to introduce some noise or randomness into the dynamics and in this way we obtain stochastic interacting particle systems. There are many interesting models that have been studied successfully, and we wish to mention two of them. The first one is a Hamiltonian system perturbed by a weak noise considered in [OVY], that leads to the Euler equation under hydrodynamical scaling. The scaling there is the hyperbolic scaling $x \to \varepsilon x$ and $t \to \varepsilon^{-1}t$. The second model is one of interacting Brownian motions. Here the dynamics is controlled by the infinitesimal generator for the Markovian motion of these interacting Brownian particles. The infinitesimal generator involves a second order term which is the Laplace operator defining the noise, perturbed by a drift term or a first order term that comes from two particle interactions like in the Hamiltonian case. The similarity extends further in that the invariant measures are Gibbs measures very much like in the Hamiltonian case.

The natural scaling here is however the diffusive scaling $x \to \varepsilon x$, $t \to \varepsilon^{-2}t$ and the limiting equation is expected to be a nonlinear diffusion equation. This is rigorously

*Received September 26, 1997; accepted for publication November 19, 1997.

†Courant Institute of Mathematical Sciences, New York University, 251 Mercer Street, New York, NY 10012, USA (varadhan@cims.nyu.edu). Research partially supported by U. S. National Science Foundation grant NSF-DMS-9503419 and Army Research Office grant ARO-DAAH04-95-1-0666.

‡Courant Institute of Mathematical Sciences, New York University, 251 Mercer Street, New York, NY 10012, USA (yau@math.nyu.edu). Research partially supported by U. S. National Science Foundation grants DMS-9403462 and DMS-9703752, as well as by a David and Lucile Packard Foundation Fellowship.

established in [V2] for dimension $d = 1$. For dimension $d \geq 2$ and away from phase transition, one can prove it by applying the relative entropy argument of [Y1].

If we replace the Brownian motions by random walks, we obtain lattice gas models. The invariant measures are thus Gibbs states on $\mathbb{Z}^d$. One would expect that the hydrodynamical limit should be much easier to establish for lattice gases than for interacting Brownian motions. This turns out to be wrong. The main reason is that while the interacting Brownian motions model satisfies the 'gradient condition', lattice gas models in general do not. We will now explain briefly what the gradient condition means and why it is important in establishing hydrodynamical limits under diffusive scaling. To keep the notation simple let us take $d = 1$.

Denote by $\eta = (\eta_x)_{x\in\mathbb{Z}}$ with $\eta_x \in \{0,1\}$ a typical configuration of of particles in $\mathbb{Z}$ with η_x denoting the number of particles at x. From the dynamics of our lattice gas model, we can write down the microscopic conservation law

$$d\eta_x = [w_{x-1,x} - w_{x,x+1}]dt + dM_x(t)$$

where $w_{x,x+1}$ is the current along the bond $(x, x+1)$ and dM_x is a martingale term due to the noise. Because our goal is to derive a diffusive limit, it is most convenient if the current is itself a gradient. In other words, $w_{x,x+1}(\eta) = h(\tau_x\eta) - h(\tau_{x+1}\eta)$ for some local function h; here τ_x is the shift by x. This is the condition that is called the 'gradient condition'[KLS]. An analogous condition is satisfied in the interacting Brownian motions model but the condition is in general not satisfied for lattice gas models. Let us take a quick look at the role played by the 'gradient condition'. If we want to study the change of density as a function of space and time under diffusive scaling we need to study

$$d\frac{1}{N}\sum_x J(\frac{x}{N})\eta_x(N^2t) = \frac{N}{N}\sum_x J'(\frac{x}{N})\, w_{x,x+1}(\eta)dt + dM_N(t) \tag{1.1}$$

where J is a test function. If the 'gradient condition' holds we can simplify (1.1) to get

$$d\frac{1}{N}\sum_x J(\frac{x}{N})\eta_x(N^2t) \simeq \frac{1}{N}\sum_x J''(\frac{x}{N})\;\; h(\tau_x\eta)dt + dM_N(t) \tag{1.2}$$

If we believe in the validity of a local averaging principle, then the term $h(\tau_x\eta)$ can be replaced by $\hat{h}(\rho(t,x))$, where $\hat{h}(\rho)$ is the expected value of $h(\cdot)$ in the Gibbs state with density ρ. The martingale term becomes negligible for large N and we end up with the nonlinear diffusion equation

$$\frac{\partial\rho}{\partial t} = [h(\rho(t,x))]_{xx} = [h'(\rho(t,x))\rho_x(t,x)]_x \tag{1.3}$$

It is not clear how we would ever arrive at an equation like (1.3) without the gradient condition.

One way to bypass the gradient condition is to obtain a fluctuation-dissipation equation. A difficult step for us will be to establish the relation

$$w_{0,1}(\eta) = D(\rho)(\eta_0 - \eta_1) + \mathcal{L}g \tag{1.4}$$

where $\mathcal{L}$ is the generator of the dynamics and $\mathcal{L}g$ represents fluctuation in a certain sense. One then obtains a diffusive limit

$$\frac{\partial\rho}{\partial t} = [D(\rho(t,x))\rho_x(t,x)]_x \tag{1.5}$$

This step, while unnecessary for gradient models, is the key step for nongradient models.

A curious fact concerning lattice gases is the difficulty of constructing gradient models [S]. Except when the dimension d is 1 or when the Gibbs measures degenerate into product measures, we are not aware of any gradient lattice gas models. If, instead of the lattice gas models, we consider Ginzburg-Laudau models, there are plenty of examples satisfying the gradient condition. The hydrodynamical limit in this case is established in [R], following the approach of [GPV], and is valid even in the phase transition region. On the other hand, one can easily construct nongradient versions of Ginzburg-Laudau models or interacting Brownian motions. Indeed, the hydrodynamical limit for a nongradient Ginzburg-Landau model at infinite temperature is established by [V1] where a basic outline for the analysis of nongradient models is first given. It is then extended to lattice gases at infinite temperature in [Q1].

In this approach, one requires an accurate estimate on the spectral gap of the generator in a finite volume. This estimate is obtained in [LY] under certain mixing assumptions, summarized as assumptions A.1-A.3 in [LY]. These assumptions are imposed on the *canonical* Gibbs states and are hard to check. They are replaced in [Y2] by some mixing conditions on the grand canonical Gibbs states, to be stated as assumption (A) in Section 2. Indeed, in that paper a much stronger inequality, the logarithmic Sobolev inequality, is obtained. The assumption A is the usual Dobrushin-Shlosman mixing condition and holds, e.g., for ferromagnetic Ising models up to the critical point [MOS, N] in dimension $d = 2$ and at sufficiently high temperatures for general lattice gas models.

Even with an accurate estimate on the spectral gap, the hydrodynamical limit for lattice gases at a finite temperature is by no means a straightforward extension of the earlier work. This is because one has to solve the equation (1.4) and the generator $\mathcal{L}$ exhibits the effect of interaction between particles, an effect that is not present in any of the earlier work on lattice gas models that deal exclusively with the infinite temperature case where the Gibbs measures are product measures. Besides the approach of [V1] and [Q], we also find that the formulation of in [EMY, LOY1], designed mainly for nonreversible nongradient systems, provides a convenient framework for this paper.

Two problems remain outstanding. One is establishing hydrodynamical limits of lattice gases without assuming mixing conditions; the other is hydrodynamical limit for nongradient models in continuum, namely, nongradient interacting Brownian motions. For the second problem, one has to derive accurate estimates of spectral gap for interacting Brownian motions in a finite volume. It seems that the approach of [LY, Y2] might be useful here. A more ambitious goal is to establish hydrodynamical limit without using any estimates on the spectral gap. We are not aware of any such approach that works for nongradient models.

This paper is organized as follows: In Section 2 we state the main result. An outline of proof is given in Section 3. The tightness is proved in Section 4, and an energy estimate is obtained in Section 5. The two-block estimate is proved in Section 6, and an eigenvalue estimate is in Section 7. In Section 8 and 9 we solve the fluctuation-dissipation equation (1.4) in a precise sense. Finally, we prove some mixing conditions needed in this paper in Section 10.

2. Statement of the main result. Let $\Lambda = \Lambda_L$ be a cube of width $2L+1$ in Z^d and let $\eta = (\eta_x)_{x \in \Lambda_L}$ denote the configuration of a lattice gas where for each x, $\eta_x \in \{0, 1\}$. $\eta_x = 1$ if there is a particle at x and $\eta_x = 0$ means that the site x is empty. For any two points in Z^d we define two notions of distance:

$$|x-y|_1 := \sum_{i=1}^{d} |x_i - y_i|; \quad |x-y| := \max_{i=1,\cdots d} |x_i - y_i| \tag{2.1}$$

By the boundary $\partial\Lambda$ of Λ we shall mean the complement Λ^c of Λ although in practice we need only be concerned with those sites in Λ^c that are within a certain fixed distance (the range of the interaction) of Λ. A boundary condition ω is a configuration on $\partial\Lambda$ and $\eta \cup \omega$ is a configuration on $\Lambda \cup \partial\Lambda$. Let $F(\eta)$ be a local function depending on the configuration in some finite box of width $2R+1$. Formally the Hamiltonian H is given by

$$H(\eta) = \sum_x F_x(\eta) \tag{2.2}$$

Let us recall that $F_x(\eta) = F(\tau_x \eta)$ where τ_x is the translation $(\tau_x\eta)_y = \eta_{x+y}$. The energy H_ω of a configuration η on Λ with boundary condition ω is given by

$$H_\omega(\eta) = \sum_{x\in\Lambda} F_x(\eta \cup \omega). \tag{2.3}$$

We note that $H_\omega(.)$ depends on ω only through the configuration at sites in $\partial\Lambda$ that are with in a distance $2R$ of Λ. We denote by $\mu_{\Lambda,\omega,\lambda}$ the finite volume Gibbs measure on Λ with Hamiltonian H, boundary condition ω and chemical potential λ, namely

$$\mu_{\Lambda,\omega,\lambda}(\eta) = Z^{-1}_{\Lambda,\omega,\lambda} \exp[-H_\omega(\eta) + \lambda \sum_{x\in\Lambda} \eta_x\,], \tag{2.4}$$

Here $Z_{\Lambda,\omega,\lambda}$ is the normalization constant (the partition function)

$$Z_{\Lambda,\omega,\lambda} = \sum_\eta \exp[-H_\omega(\eta) + \lambda \sum_{x\in\Lambda} \eta_x\,]$$

so that $\mu_{\Lambda,\omega,\lambda}$ is a probability measure on the space of configurations on Λ. For simplicity, we assume that the inverse temperature β has been absorbed into F. When $\lambda = 0$ we will drop the suffix λ.

Recall that the pressure $p(\lambda)$ of the lattice gas is defined by

$$p(\lambda) = \lim_{\Lambda\to\infty} |\Lambda|^{-1} \log Z_{\Lambda,\omega,\lambda}\,. \tag{2.5}$$

Elementary theory of thermodynamics assures us that $p(\lambda)$ exists and is independent of the choice of boundary condition ω. We also have the usual definition of the free energy $h(m)$, as the Legendre transform of p :

$$h(m) = \sup_\lambda [\lambda m - p(\lambda)]. \tag{2.6}$$

It is well known that the infinite volume Gibbs measures that are limits of the finite volume ones given by (2.4) exist, are characterized by the DLR equations, and in general may not be unique when there are phase transitions. Our resells are restricted to the single-phase region. Indeed, we need certain mixing conditions.

Assumption A. *let $\mu_{\Lambda,\omega,\lambda}$ denote a Gibbs measure on Λ with boundary condition ω and chemical potential λ. Denote the corresponding density by $\rho = \rho(L, \lambda, \omega)$. Then there are constants γ_1, γ_2 and $\gamma_3 \geq R+1$ such that for any two functions f and g with supports S_f and S_g we have*

$$|E_{\Lambda,\omega,\lambda}[f;g]| \leq \gamma_1 \rho(1-\rho) \exp[-\gamma_2 \; dist(S_f, S_g)] \|f\|_\infty \|g\|_\infty \tag{2.7}$$

provided that the diameters of S_f and S_g are bounded by γ_3. Note that the constants are independent of the size of the cube Λ, the chemical potential and the boundary condition.

This exponential mixing condition is stronger than what we need in this paper and the exponential decay can be replaced by a weaker decay property. Since for most Gibbs measures, one usually obtains mixing conditions via the Dobrushin uniqueness condition or the cluster expansions that automatically yield exponential estimates, we will not attempt to find the optimal mixing condition.

We also need the concept of canonical Gibbs states. Let ρ be a fixed positive number. Then the canonical Gibbs state with total density ρ and boundary condition ω is the conditional probability of $\mu_{\Lambda_L,\omega}$ given that the total number of particles is equal to $L^d\rho$, namely, $\bar\eta := L^{-d}\sum_{x\in\Lambda}\eta_x = \rho$. Certainly, ρ has to be a rational number of some special form for the total number of particles to be an integer. One can avoid trouble by requiring that the total number of particles equal $[L^d\rho]$ rather than $L^d\rho$. Symbolically, we write the canonical measure as

$$d\mu_{\Lambda,\omega,\rho} = d\mu_{\Lambda,\omega}\big|_{\bar\eta=\rho} \tag{2.8}$$

Note that since $\bar\eta$ is fixed we can replace $\mu_{\Lambda,\omega}$ by $\mu_{\Lambda,\omega,\lambda}$ with any λ. Define the canonical partition function

$$Z^c_{\Lambda,\omega,\rho} = \sum_{\eta:\bar\eta=\rho} \exp[-H_{\Lambda,\omega}(\eta)] \,, \tag{2.9}$$

where $H_{\Lambda,\omega} = H_{\Lambda,\omega,0}$. It should be emphasized that the mixing condition **A** by itself does not imply any mixing property for the canonical Gibbs measures. It is proved in [Y2] that the mixing condition **A** does imply some mixing property, though not as strong, for the canonical Gibbs measure. More importantly, exact estimate on the spectral gap as well as a Logarithmic Sobolev inequality are obtained under assumption **A**.

Denote by b an unoriented bond $(x, y) \in \Lambda \times \Lambda$ with $|x-y|_1 = 1$. All bonds in this paper are unoriented unless otherwise specified. We have limited ourselves to nearest neighbor bonds mainly to simplify notation. As long as bonds with a fixed finite bound on length are used, the proofs will remain the same. Let S_b be defined by

$$\begin{aligned}(S_b\eta)_z := (\eta^b)_z = (\eta^{x,y})_z &= \eta_y \quad \text{if } z = x \\ &= \eta_x \quad \text{if } z = y \\ &= \eta_z \quad \text{otherwise}\end{aligned} \tag{2.10}$$

and define T_b by

$$T_b f(\eta) = f(\eta^b) - f(\eta) \,. \tag{2.11}$$

If b happens to be the bond $(0, e)$ for some $e > 0$ we will denote T_b by T_e. Define $\mathcal{L}_b$ to be the symmetric generator defined by

$$-\int f \, \mathcal{L}_b \, g d\mu = \frac{1}{2} \int c_b(\eta) \, [\, T_b f(\eta) \,] [\, T_b g(\eta) \,] \, d\mu \tag{2.12}$$

and $\mathcal{L} = \sum_{b \in \Lambda} \mathcal{L}_b$. The rate $c_b(\eta)$ is assumed to be a local function that is translation invariant i.e. $c_{\tau_x b}(\tau_x \eta) = c_b(\eta)$ and bounded away from zero and infinity, i.e., $0 < \delta < c_b(\eta) < \delta^{-1} < \infty$ for some constant δ. Explicitly, $\mathcal{L}_b$ is given by

$$\mathcal{L}_b f = A(b, \eta) T_b f(\eta), \tag{2.13}$$

where

$$A(b, \eta) = (1/2) \, [\, \exp \{ (T_b \log c_b)(\eta) - (T_b H)(\eta) \} + c_b(\eta) \,] \tag{2.14}$$

When $b = (x, y)$ we sometimes use the notation $A(x, y, \eta)$ instead of $A(b, \eta)$. The operator $\mathcal{L}$ can serve as the infinitesimal generator of a Markov process $\eta(t)$ of configurations changing in time. We could consider either the infinite lattice $\mathbb{Z}^d$, a finite box Λ with boundary conditions ω or a periodic lattice of size L. In all these cases the Markov Process will be reversible with respect to the corresponding Gibbs measure (in the infinite volume situation, with respect to any measure that satisfies the DLR equations). We shall for the present be concerned mainly with the periodic case with $L \simeq \varepsilon^{-1}$ and we denote by μ_ε the corresponding Gibbs measure with $\lambda = 0$. We will speed time up by a factor of ε^{-2} so that the generator is $\varepsilon^{-2}\mathcal{L}$. Let us start the system initially with a distribution having density f_0^ε with respect to μ_ε. We denote by $P_\varepsilon^{f_0}$ the corresponding Markov Process with generator $\varepsilon^{-2}\mathcal{L}$ and initial distribution $f_0^\varepsilon d\mu_\varepsilon$ on the space of trajectories $\eta(\cdot)$ in a fixed macroscopic time interval $[0, T]$.

If f_t^ε is the density of the system at time t relative to the measure μ_ε. Then f_t^ε satisfies the forward equation

$$\partial_t f_t^\varepsilon = \epsilon^{-2} \mathcal{L} f_t^\varepsilon. \tag{2.15}$$

We have, of course, already rescaled time with a diffusive rescaling $t \to \epsilon^{-2} t$ and by considering the evolution of empirical measure defined by

$$\nu_\epsilon(\theta, t) = \varepsilon^d \sum_{z \in \Lambda} \delta(\theta - \epsilon z) \eta_z(t), \tag{2.16}$$

where $\theta \in \mathbf{T}^d$, we are rescaling space by a factor of ε . Because $\eta \in \{0, 1\}$, the relation (2.16) induces (from P_ε) a distribution, $Q_\epsilon^{f_0}$, of $\nu_\epsilon(\cdot, \cdot)$ on the Skorohod space $X = D([0, T] \to M_1[\mathbf{T}^d])$, where $M_1(\mathbf{T}^d)$ is the space of nonnegative measures on $\mathbf{T}^d$ with total mass bounded by 1. The space $M_1(\mathbf{T}^d)$ is compact under the topology of weak convergence. We now describe the hydrodynamical limit. As $\epsilon \to 0$, one expects the distribution $Q_\varepsilon^{f_0}$ of empirical measures to converge to $\delta_{m(\theta, \cdot) d\theta}$, the Dirac measure concentrated on the trajectory $m(\theta, \cdot) \, d\theta$ where $m(t, \theta)$ satisfies the following nonlinear diffusion equation:

$$\frac{\partial}{\partial t} m(\theta, t) = \nabla_\theta \left(\, D(m(\theta \, , t)) \nabla_\theta m(\theta, t) \, \right) \; ; \quad m(\theta, 0) = m_0(\theta) \tag{2.17}$$

Here $m_0(\theta)$ is the initial density determined by the initial distribution $f_0^\varepsilon d\mu_\varepsilon$ and $D(m)$ is the diffusion (matrix) coefficient given by the Green-Kubo formula to be described later. Clearly, in order for this to make sense, we need at least that the initial distribution $f_0^\varepsilon d\mu_\varepsilon$ satisfies

$$\limsup_{\varepsilon\to 0} P^{f_0^\varepsilon}\left[\left|\int \nu_\varepsilon(\theta,0)J(\theta)d\theta - \int m_0(\theta)J(\theta)d\theta\right| > \delta\right] = 0 \tag{2.18}$$

for any smooth test function $J(\cdot)$ on $\mathbf{T}^d$ and $\delta > 0$. We shall make this assumption for the rest of this paper. Since it is the only condition we need on the initial distribution, we may as well have deterministic initial conditions and suppose that at 0 we start from a nonrandom configuration ξ^ε. We shall denote by P_ε^ξ the distribution of the process with this initial condition and the corresponding distribution on X by Q_ε^ξ. When no confusion arises, we shall drop the label ε. The condition (2.18) takes the form

$$\limsup_{\varepsilon\to 0}\left|\frac{1}{L^d}\sum_z J(\frac{z}{L})\,\xi_z - \int m_0(\theta)J(\theta)d\theta\right| = 0 \tag{2.19}$$

The Green-Kubo formula for the diffusion coefficient can be written in several ways [S]. Let τ_x denote the translation by x, namely,

$$(\tau_x\eta)(y) = \eta(y-x)\,, \qquad \text{and } \tau_x g(\eta) = g(\tau_x\eta). \tag{2.20}$$

Denote by μ_m the infinite volume Gibbs state with density m and $\chi(m)$ its compressibility defined by

$$\chi(m) = \sum_{x\in\mathbb{Z}^d} E^{\mu_m}[\,\eta_0\,;\,\eta_x\,]; \tag{2.21}$$

here the truncated correlation is defined by

$$E^{\mu_m}[\,f\,;\,g\,] = E^{\mu_m}[fg] - E^{\mu_m}[f]E^{\mu_m}[g]$$

The following characterization is due to [V1]. For all vector α we have

$$<\alpha, D(m)\alpha> = \frac{1}{2\chi(m)}\inf_g E^{\mu_m}\left[\sum_{|e|=1,e>0} c_e(\eta)\left(\alpha_e\nabla_e\eta_0 - T_e\sum_{x\in\mathbb{Z}^d}\tau_x g\right)^2\right] \tag{2.22}$$

where the infimum is over all local functions g, T is defined in (2.11) and ∇_e is defined by

$$(\nabla_e)h_x = h_{x+e} - h_x \tag{2.23}$$

for any function on $\mathbb{Z}^d$. Since g is local, $T_e\sum_{x\in\mathbb{Z}^d}\tau_x g$ makes sense even though the infinite sum $\sum_{x\in\mathbb{Z}^d}\tau_x g$ does not. Since the right hand side of (2.22) is a quadratic expression in the α and g, and the infimum is taken over g belonging to the linear subspace of all local functions, the left hand side is quadratic in α and defines a symmetric nonnegative definite matrix $D(m)$ for each m. Moreover because (2.22) defines the diffusion coefficient as an infimum we get upper bounds for $D(m)$. It is much harder to obtain a lower bound for the diffusion coefficient. Such a bound is obtained in [SY] which states that in matrix sense

$$D(m) \geq C[\chi(m)]^{-1}I \tag{2.24}$$

for some constant depending on the Hamiltonian. We remark that this bound is independent of phase transition and does not require any mixing conditions. However, we do need the mixing conditions to establish rigorously that the hydrodynamic limit can be taken to get a nonlinear diffusion equation with coefficients given by (2.22).

The uniqueness of (2.17) is quite subtle. If the diffusion coefficient is a scalar, then the uniqueness follows easily. Another situation where uniqueness holds is when the diffusion coefficient is Lipshitz continuous in m and satisfies uniform ellipticity bounds. Under this assumption, the general theory guarantees the existence of Lipshitz continuous solutions. Suppose v is such a solution and u is a weak solution satisfying an energy estimate. Then we have

$$\begin{aligned}\partial_t \int (u-v)^2(\theta)d\theta &= -\int \nabla(u-v)(\theta)\,[\,D(u)\nabla u - D(v)\nabla v\,]\,(\theta)d\theta \\ &= -\int \nabla(u-v)(\theta)D(u(\theta))\,[\,\nabla u - \nabla v\,]\,(\theta)d\theta \\ &\quad + \int \nabla(u-v)(\theta)\,[\,D(u)-D(v)\,]\,(\theta)\nabla v(\theta)d\theta\end{aligned}$$

From the Schwarz inequality the last term is bounded by

$$\begin{aligned}&\int \nabla(u-v)\,[\,D(u)-D(v)\,]\,\nabla v d\theta \\ &\leq \gamma\int [\,\nabla(u-v)\,]^2\,d\theta + \gamma^{-1}\int [\,D(u)-D(v)\,]^2\,(\nabla v)^2 d\theta\end{aligned}$$

Since v is Lipshitz continuous, the sup-norm of ∇v is bounded. Together with the Lipshitz continuity of D, the last term is bounded by

$$C\gamma^{-1}\int [\,D(u)-D(v)\,]^2\,d\theta \leq C\gamma^{-1}\int (u-v)^2 d\theta$$

By choosing γ small enough, we have thus proved that

$$\partial_t \int (u-v)^2 d\theta \leq C\gamma^{-1}\int (u-v)^2 d\theta$$

This proves the uniqueness. Our main result of this paper is the following theorem:

THEOREM 2.1 *Suppose that the Gibbs measure satisfies the mixing condition* **A**. *Suppose the initial data ξ^ε satisfies the condition (2.18) and that (2.17) with the initial condition specified by (2.18) has a unique solution $m(\theta,t)$ in the class of weak solutions satisfying the energy estimate. Then*

$$\limsup_{\varepsilon\to 0} P_\varepsilon^\xi\left[\;\left|\varepsilon^d\sum_{x\in\Lambda} J(\varepsilon x)\eta_x(t) - \int J(\theta)m(\theta,t)\,d\theta\right| > \delta\;\right] = 0$$

Here P_ε^ξ refers to the process with initial data ξ^ε.

Note that one can start from a deterministic initial data as long as (2.18) holds. The Lipshitz continuity of the diffusion coefficient has not been proved, though it is proved for the self-diffusion coefficient in [V3] and for asymmetric simple exclusion in

[LOY2]. In the special case $d = 1$ or when the Hamiltonian is isotropic (Lemma ♣diag), the diffusion coefficient is a scalar and Theorem 2.1 is then always valid.

As a technical remark, all error terms arising in the proof of Theorem 2.1 will be shown to be superexponentially small. Hence our estimates already give an upper bound to the large deviation probability in the hydrodynamical limit. One needs a corresponding lower bound to complete the large deviation theory. Since this has been carried out carefully in [DV, Q2] for Ginzburg-Landau models, in [QRV] for the symmetric simple exclusion processes with several colors and in [QY2] for lattice gas models for the incompressible Navier-Stokes equation, we will not pursue this direction here.

3. Outline of the proof. Our basic approach for proving the hydrodynamical limit consists of three steps: establishing the tightness of the measures $\{Q_\varepsilon^\xi : \varepsilon > 0\}$, deriving an energy estimate that provides some regularity for functions in the support of any weak limit Q of Q_ε^ξ as $\varepsilon \to 0$ and identification of the support of Q as weak solutions of (2.17). To prove tightness, from Prohorov's theorem, it suffices to prove the following estimate:

LEMMA 3.1 (TIGHTNESS) *For any initial data ξ^ε, any smooth test function J and any $\delta > 0$ we have*

$$\limsup_{\alpha\to 0}\limsup_{\varepsilon\to 0} P_\varepsilon^\xi\left[\sup_{|s-t|\le\alpha, 0\le s,t\le T} \Big|\, \varepsilon^d \sum_x J(\varepsilon x)\eta_x(t) - \varepsilon^d \sum_x J(\varepsilon x)\eta_x(s)\,\Big| \ge \delta\right] = 0 \tag{3.1}$$

Note that

$$\limsup_{\varepsilon\to 0} |\varepsilon^d \sum_x J(\varepsilon x)\eta_x| \le \| J \|_1$$

Hence any limit point Q of Q_ε^ξ is supported on measure valued functions on $[0, T]$, that are absolutely continuous w.r.t. the Lebegue measure, with densities $m(\theta, t)$ bounded between 0 and 1. At this point we only know that $m(\theta, t)$ is a measurable function. The next step is to prove that it satisfies a basic energy estimate.

LEMMA 3.2 (ENERGY ESTIMATE) *Suppose Q is any limit point for Q_ε^ξ. There is a constant C such that*

$$E^Q\left\{\int_0^T dt \int_{\mathbf{T}^d} d\theta(\nabla_\theta m(\theta,t))^2\right\} \le C.$$

Finally, we have to prove that Q is supported on densities m that satisfy the equation (2.17) in the weak sense. Denote by $\bar\eta_{x,\ell}$ the average density of η in a cube of width $2\ell + 1$, namely,

$$\bar\eta_{x,\ell} = (2\ell+1)^{-d} \sum_{y:|y-x|\le\ell} \eta_y := Av_{y:|y-x|\le\ell}\,\eta_y \tag{3.2}$$

This operation of averaging over a cube of side $2\ell + 1$ centered at x will be denoted by $Av_{y:|y-x|\le\ell}$ and can be performed on other random fields defined on the lattice.

THEOREM 3.3 (IDENTIFICATION OF THE EQUATION) *For any $\delta > 0$, let*

$$B^J_{a,b,\delta}r = \Big\{\eta(\cdot) : \sup_{0\le t\le T}\Big| \varepsilon^d \sum_x J(\varepsilon x)\eta_x(t) - \varepsilon^d \sum_x J(\varepsilon x)\eta_x(0)$$

$$+ \int_0^t (2b)^{-1}\varepsilon^d \sum_{x,e,e'} [\eta_{x+b\varepsilon^{-1}e}(s) - \eta_{x-b\varepsilon^{-1}e}(s)] D_{e,e'}(\bar\eta_{x,a\varepsilon^{-1}}(s))(\nabla_{e'}J)(\varepsilon x)ds\Big| > \delta\Big\}$$

denote the event that leads to the violation of the limiting equation for the smooth test function J; here D is the diffusion coefficient given by (2.22). Then

$$\limsup_{a\to 0}\limsup_{b\to 0}\limsup_{\epsilon\to 0} P^\xi_\varepsilon[B^J_{a,\delta}] = 0 \tag{3.3}$$

Clearly, Theorem 2.1 follows from these three results and the uniqueness of the weak solution for the equation (2.17) satisfying the energy estimate.

The tightness will be proved in next section. The energy estimate will be proved in section 5. We now outline the proof of Theorem 3.3.

Proof of Theorem 3.3.

Step 1: Recall that the current is defined by the equation

$$\mathcal{L}\eta_x = -\sum_{e>0} \nabla^-_e w_{x,x+e}(\eta)$$

where $\sum_{e>0}$ denotes the summation over the unit vectors in the positive coordinate directions and for any (possibly random) function $h(x,\cdot) = h_x(\cdot)$ on the lattice

$$\nabla^-_e h(x,\cdot) = h(x,\cdot) - h(x-e,\cdot).$$

We shall follow the convention of using e to denote a unit vector in the positive coordinate direction. Explicitly, the current is given by

$$w_{x,x+e}(\eta) = A(x, x+e, \eta)(\eta_x - \eta_{x+e}) \tag{3.4}$$

where A is the jump rate defined in (2.14). We have the summation by parts identity

$$\sum_x g(x)\nabla^-_e h(x) = -\sum_x (\nabla_e g(x))h(x)$$

where ∇ is defined in (2.23) and $\sum_x$ denotes the summation over the lattice sites on the periodic lattice with width ε^{-1}. This convention will be followed in the rest of this paper unless otherwise stated. From stochastic calculus,

$$\begin{aligned} \varepsilon^d \sum_x J(\varepsilon x)\eta_x(t) - \varepsilon^d \sum_x J(\varepsilon x)\eta_x(0) &= \int_0^t U(\eta(s))ds + M_N(t), \\ U(\eta) &= \varepsilon^{d-1}\sum_x\sum_{e>0}(\varepsilon^{-1}\nabla_e J(\varepsilon x))w_{x,x+e}(\eta) \\ M_N(t) &= \int_0^t \varepsilon^d \sum_x\sum_{e>0} \nabla_e J(\varepsilon x)\eta_x(s)dM_{x,x+e}(s) \end{aligned} \tag{3.5}$$

where $M_{x,x+e}(t)$ is a martingale with quadratic variation

$$d < M_{x,x+e}(t)\,, M_{x,x+e}(t) >= \varepsilon^{-2} A(x, x+e, \eta)\, dt.$$

Clearly,

$$d < M_N(t)\,, M_N(t) >= \varepsilon^{2d} \sum_x (\varepsilon^{-1} \nabla_e^- J(\varepsilon x))^2 \eta_x(t) A(x, x+e, \eta) dt \le C(J) \varepsilon^d\, dt$$

Hence we can neglect the martingale term.
Step 2 (Fluctuation-dissipation equation in equilibrium) It remains to identify the current as $-D(\eta_{x,a\varepsilon^{-1}})\nabla\eta_x(s)$. This step is very subtle and is the key step in nongradient system. In fact, we can not expect the substitution $w_{x,x+e} \to -\sum_{e'} D_{e,e'}(\eta_{x,a\varepsilon^{-1}})$ $\nabla_{e'}\eta_x$ to hold as it stands. The correct statement is that the relation

$$w_{x,x+e} = -\sum_{e'} D_{e,e'}(\eta_{x,a\varepsilon^{-1}}) \nabla_{e'} \eta_x + \mathcal{L} g_e \tag{3.6}$$

holds for some suitable choice of local functions $\{g_e\}$. For processes in equilibrium, a precise statement is given by the following Theorem ♣3.6 . We will first need some notation.

Following [LOY1] , we define the space

$$\mathcal{G} = \big\{ h : h \text{ is a local function and } h \text{ satisfies } E^{\mu}[h|\mathcal{F}_s] = 0 \text{ for some } s \ \big\}, \tag{3.7}$$

where $\mathcal{F}_s = \mathcal{F}_{0,s}$ and

$$\mathcal{F}_{x,s} = \ \text{the } \sigma\text{-algebra generated by } \{\bar{\eta}_s\} \cup \{\eta_y : |y-x| > s\}. \tag{3.8}$$

For latter usage, we also define the σ-algebra

$$\mathcal{K}_x^s = \ \sigma\text{-algebra generated by } \{\eta_y | |x-y| \le s\}\ . \tag{3.9}$$

For any local function $h \in \mathcal{G}$ we define the variance

$$V_\ell(h, y, \omega) = \ell^d \big\langle Av_{|x|\le \ell_1} h_x\,, (-\mathcal{L}_\ell)^{-1} Av_{|x|\le \ell_1} h_x \big\rangle_{\mu_{\ell,y,\omega}} \tag{3.10}$$

where $h_x = \tau_x h$, $\ell_1 = \ell - \sqrt{\ell}$ and $\mu_{\ell,y,\omega}$ is the canonical Gibbs measure on Λ_ℓ with the boundary condition ω and total density y. From the choice of ℓ_1 it follows that for any fixed $h \in \mathcal{G}$, for sufficiently large ℓ, $\tau_x h$ depends only on the configuration inside the cube Λ_ℓ provided $x \in \Lambda_{l_1}$. Since the boundary condition ω on Λ_ℓ^c and total density y in Λ are specified by the $\sigma-$algebra $\mathcal{F}_\ell$, the quantity $V_\ell(h, y, \omega)$, is an $\mathcal{F}_\ell$ measurable function $V_\ell(h, \eta)$ of the configuration η on $\mathbb{Z}^d$. We define

$$V(h, \beta) = \limsup_{\substack{\beta' \to \beta \\ \ell \to \infty}} E^{\mu_{\beta'}}[V_\ell(h, \eta)]. \tag{3.11}$$

Here μ_β is the (unique) infinite volume Gibbs state with density β. For a local function $h \notin \mathcal{G}$, i.e. not satisfying the mean 0 condition $E^{\mu}[h|\mathcal{F}_s] = 0$, we define $q_{x,k}(\omega) =$

$E^{\mu_{\ell,y,\omega}}[\, h_x|\mathcal{F}_{x,k}\,]$ where $\mathcal{F}_{x,k}$ defined in (3.8) is the σ-algebra generated by $\bar\eta_{x,k}$ and $\{\eta_y : |y-x|>k\}$. Then we define

$$V(h,\beta) = \limsup_{\substack{k\to\infty \\ \beta'\to\beta}} V(h-q_k,\beta'). \tag{3.12}$$

We now state the Theorem.

THEOREM 3.4 *Fix a density β. Suppose $\mathbf{g}=\{g_e(\eta)\}$ is a local function (with d components) of η. Let*

$$\phi_e(\mathbf{g}) = w_{0,e} + \mathcal{L}\, g_e(\eta) + \sum_{e'} D_{e,e'}(\beta)\nabla_{e'}\eta_0, \tag{3.13}$$

where the diffusion $D(\beta)$ is given by (2.22) . Then for any $\{\alpha_e\}$,

$$\inf_{\mathbf{g}} V\left(\sum_e \alpha_e\phi_e(\mathbf{g})\,,\beta\right) = 0. \tag{3.14}$$

Assuming this result, we continue to prove Theorem 3.3. The minimizing sequence for (3.14) depends on the density β. Since, from the definition, the variance V is upper semi-continuous, we have the following Corollary:

COROLLARY 3.5 *For any $\{\alpha_e\}$ and $\delta>0$ there exists $\mathbf{g}=\{g_e(\beta\,,\eta)\}$, which is smooth as a function of β and is local as a function of η, such that*

$$\sup_{\beta} V\left(\sum_e \alpha_e\phi_e(\beta,\mathbf{g})\,,\beta\right) \le \delta; \tag{3.15}$$

where

$$\phi_e(\beta,\mathbf{g}) = w_{0,e} + \mathcal{L}\, g_e(\beta,\eta) + \sum_{e'} D_{e,e'}(\beta)\nabla_{e'}\eta_0. \tag{3.16}$$

Step 3 (Insertion of local functions) To use theorem 3.5, we have to extend it to nonequilibrium process and insert the local function $\mathbf{g}$ into the stochastic equation (3.5). We first replace J by $J*\omega_{\ell_1}$ where ω_{ℓ_1} is the normalized characteristic function for a cube of size $2\ell_1+1$ with $\ell_1=\ell-\sqrt{\ell}$, namely,

$$\omega_{\ell_1}(y) = \begin{cases} (2\ell_1+1)^{-d} & \text{if } \ |y|\le \ell_1 \\ 0 & \text{otherwise} \end{cases}$$

Let $\mathbf{\Phi}^{\mathbf{g}}_x$ be defined as the vector with components

$$\Phi^{\mathbf{g}}_{x,e} = Av_{|y-x|\le\ell_1}\,[\, w_{y,y+e} + \mathcal{L}g_e(\bar\eta_{x,\ell},\tau_y\eta)\,] \tag{3.17}$$

Note that Φ depends additionally on ℓ which does not appear explicitly in the indices. Then

$$\varepsilon^d\sum_x (J(\varepsilon x)*\omega_{\ell_1})\eta_x(t) - \varepsilon^d\sum_x (J(\varepsilon x)*\omega_{\ell_1})\eta_x(0) + \Omega_{\mathbf{g}}(\eta(t)) - \Omega_{\mathbf{g}}(\eta(0)) = \int_0^t U(\eta(s),\mathbf{g})ds + M_{\mathbf{g}}(t) \tag{3.18}$$

where

$$\Omega_{\mathbf{g}}(\eta) = \varepsilon^{d+1} \sum_{x,e>0} [(\varepsilon^{-1}\nabla_e J(\varepsilon x)) * \omega_{\ell_1}] g_e(\bar{\eta}_{x,\ell}, \tau_y \eta)$$
$$U(\eta, \mathbf{g}) = \varepsilon^{d-1} \sum_{x,e>0} (\varepsilon^{-1}\nabla_e J(\varepsilon x)) \Phi^{\mathbf{g}}_{x,e} \tag{3.19}$$

The explicit form of $M_{\mathbf{g}}(t)$, the martingale term, is not important. One can check easily that

$$d < M_{\mathbf{g}}(t) \,, M_{\mathbf{g}}(t) > \leq C(J, \mathbf{g})\varepsilon^d \, dt$$

and thus it is negligible for proving Theorem 3.3. Also, for any local function g

$$\sup_{\eta} |\Omega_{\mathbf{g}}(\eta)| \leq C\varepsilon$$

and is negligible too.

The generator $\mathcal{L}$ in (3.17) acts on the variables η as well as $\bar{\eta}_{x,\ell}$ when applied to the function g_e. We first prove that the contribution of $\mathcal{L}$ acting on $\bar{\eta}_{x,\ell}$ is negligible. For this purpose, we need the next lemma relating the nonequilibrium process to equilibrium one.

LEMMA 3.6 *Recall that P^{ξ}_{ε} denotes the process starting from ξ and $P_{\varepsilon} := P^{\mu_{\varepsilon}}$ is the equilibrium process starting from μ_{ε} defined in section 2 after (2.16). Then there is a constant κ depending only on the Hamiltonian such that*

$$\log \| \frac{dP^{\xi}_{\varepsilon}}{dP_{\varepsilon}} \|_p \leq \kappa\varepsilon^{-d}, \quad 1 \leq p \leq \infty \,. \tag{3.20}$$

As a consequence,

$$P^{\xi}_{\varepsilon}(A) \leq \exp[\kappa\varepsilon^{-d}] P_{\varepsilon}(A) \tag{3.21}$$

In particular, for any events A_{ε} which are superexponentially small in equilibrium, i.e.,

$$\limsup_{\varepsilon \to 0} \varepsilon^d \log P_{\varepsilon}(A) = -\infty$$

we have

$$\limsup_{\varepsilon \to 0} \varepsilon^d \log P^{\xi}_{\varepsilon}(A) = -\infty$$

The proof of this Lemma is trivial. One simply notes that, since the dynamics of the two processes are the same, $dP^{\xi}_{\varepsilon}/dP_{\varepsilon} = d\mu_{\delta(\xi)}/d\mu_{\varepsilon}$; here $\mu_{\delta(\xi)}$ is the delta measure of the configuration ξ. From the explicit expression of $d\mu_{\varepsilon}$, we can get a lower bound on $\mu_{\varepsilon}(\xi)$ of the form $\log \mu_{\varepsilon}(\xi) \geq -\kappa\varepsilon^{-d}$ and (3.20) follows and we have proved this Lemma.

This lemma shows that the notion of superexponentially small events is the same under equilibrium and nonequilibrium processes. The following lemma is a simple consequence of the Feynman-Kac formula and allows us to transform equilibrium exponential estimates into eigenvalue problems.

LEMMA 3.7 *Let L be the generator of a Markov process $x(t)$ with a reversible invariant measure ν. Let*

$$u(x, t) = E_x \left[\exp \left\{ \int_0^T V(x(s)) ds \right\} \right], \tag{3.22}$$

where $E_x[\cdot]$ denotes the expectation with respect to the process starting at x. Then

$$\begin{aligned}\frac{1}{T}\log\int u(x,T)\,d\nu(x) &\le \sup spec\,\{V-(-\mathcal{L})\}\\ &= \sup_{\int f d\nu=1,\ f\ge 0}\left\{\int V f\,d\nu-\mathcal{D}(\sqrt{f})\right\}\end{aligned} \tag{3.23}$$

Here $\mathcal{D}(f)=-\int fLf\,d\nu$ is the Dirichlet form.

Proof. By the Feynman-Kac formula, u solves the equation $\frac{\partial u}{\partial t}=Lu+Vu$ with the initial condition $u(0,x)\equiv 1$. Multiplying the equation by u and integrating by parts we obtain

$$\frac{\partial}{\partial t}\frac{1}{2}\int u^2\,d\nu=\int Vu^2\,d\nu-\mathcal{D}(u)\le \sup spec[V-(-\mathcal{L})]\int u^2\,d\nu \tag{3.24}$$

Therefore

$$\frac{1}{T}\log\int u\,d\nu\le\frac{1}{2T}\log\int u^2\,d\nu\le\sup spec[V-(-\mathcal{L})]$$

Applying these two lemmas to our setting, we have

LEMMA 3.8 *The event $A=\{|\int_0^T V(\eta(s))ds|>\delta\}$ is super exponentially small if and only if for all $\gamma>0$*

$$\limsup_{\varepsilon\to 0}\sup spec\,\{V-\gamma^{-1}\varepsilon^{-2+d}(-\mathcal{L})\}\le 0 \tag{3.25}$$

Proof. Suppose the process is in equilibrium. Then the corollary follows from Lemma 3.7 and Chebyshev's inequality. For a general nonequilibrium process, we only have to use Lemma 3.6 to prove the Lemma. Note that we do not have to specify the initial distribution of the process.

We now return to the local function $\mathbf{g}$. Let

$$\mathcal{L}_{y,s}=\sum_{b\in\Lambda_{y,s}}\mathcal{L}_b$$

where $b\in\Lambda_{y,s}$ means that both ends of the bond b lie in the box $\Lambda_{y,s}$ of width $2s+1$ centered at y. We have

$$\mathcal{L}g_e(\bar\eta_{x,\ell},\tau_y\eta)=\mathcal{L}_{y,\sqrt{\ell}}\,g_e(\bar\eta_{x,\ell},\tau_y\eta)+\sum_{\substack{b\in\partial\Lambda_{x,\ell}\\ b\notin\Lambda_{y,\sqrt{\ell}}}}\mathcal{L}_b\,g_e(\bar\eta_{x,\ell},\tau_y\eta) \tag{3.26}$$

provided $\sqrt{\ell}$ is bigger than the range of the local function g_e, which we shall assume. By $b\in\partial\Lambda$, we mean that the bond straddles Λ and Λ^c. By the definition of $\mathcal{L}$ and Schwarz's inequality, we have for any function g,

$$\begin{aligned}\varepsilon^{-1}\int f\,\mathcal{L}_b\,g\,d\mu_\varepsilon &=\varepsilon^{-1}\int c_b(\eta)\,[\,f(\eta^b)-f(\eta)\,]\,T_b g\,d\mu_\varepsilon\\ &\le\varepsilon^{-2\ell-d}\int c_b(\eta)\left[\,\sqrt{f}(\eta^b)-\sqrt{f}(\eta)\,\right]^2 d\mu_\varepsilon\\ &\quad+\ell^d\int c_b(\eta)\left[\,\sqrt{f}(\eta^b)+\sqrt{f}(\eta)\,\right]^2(T_b g)^2\,d\mu_\varepsilon\end{aligned}$$

Since g_e is smooth as a function of its first variable, we have $T_b g_e(\bar\eta_{x,\ell}, \tau_y\eta) \le C\ell^{-d}$ for any boundary bond $b \in \partial\Lambda_{x,\ell}$. Note that T_b does not act on $\tau_y\eta$ since g_e is a local function and ℓ is large enough. Since the number of boundary terms is of order ℓ^{d-1}, we have for any smooth function J and any $\gamma > 0$,

$$\limsup_{\ell,\varepsilon} \sup spec \left[\epsilon^{-1} \sum_x \left(J(\varepsilon x) \sum_{\substack{b\in\Lambda_{x,\ell} \\ b\notin\Lambda_{y,\sqrt{\ell}}}} \mathcal{L}_b g(\bar\eta_{x,\ell}, \tau_y\eta) \right) - \gamma^{-1}\epsilon^{-2} Av_b(-\mathcal{L}_b) \right] \le 0$$

where

$$\limsup_{\ell,\varepsilon} = \limsup_{\ell\to\infty} \limsup_{\epsilon\to 0} \tag{3.27}$$

We will use this notation repeatedly when we want to take successive limsups in a specified order. Hence we can replace $\mathcal{L}$ in (3.17) by $\mathcal{L}_{y,\sqrt{\ell_1}}$, namely, we can assume that $\mathcal{L}$ in (3.17) does not act on $\bar\eta_{x,\ell}$. *Throughout the rest of this paper, while computing $\mathcal{L}g_e(\bar\eta_{x,\ell}, \tau_y\eta)$, we will therefore be justified in keeping only the first term on the right side of (3.26), and effectively ignoring the action of the generator on the first variable of g_e.*

Step 4 (Eigenvalue estimate) Our goal is to prove that (1.4) holds in nonequilibrium. The basic tool is Lemma 3.8. In order to apply this lemma, we will need to estimate some eigenvalues, to be proved in section 7.

THEOREM 3.9 (EIGENVALUE ESTIMATE) *Suppose $\mathbf{J} = \{J_e\}$ is a smooth vector valued function and $\mathbf{g} = \{g_e(\beta, \eta)\}$ is smooth as a function of β and is a local function of η. Recall the definitions of $\Phi^{\mathbf{g}}$ (3.17) and ϕ_e (3.13). Let us define the density gradients $\mathbf{\Psi}_{\varepsilon,c}(x)$ componentwise by $\Psi_{\varepsilon,c,e}(x) = \varepsilon(2c)^{-1}\left[\eta_{y+c\varepsilon^{-1}e} - \eta_{y-c\varepsilon^{-1}e}\right]$. Then for any $\gamma > 0$, there exists a constant $C(\gamma)$ such that*

$$\limsup_{\ell,a,c,\varepsilon} \sup spec \left[\epsilon^{-1}(\mathbf{J}(\varepsilon x), \Phi^{\mathbf{g}}_x + D(\bar\eta_{x,a\varepsilon^{-1}}) Av_{|y-x|\le\ell_1}\left[\mathbf{\Psi}_{\varepsilon,c}(y) \right] - \gamma\epsilon^{-2} Av_b(-\mathcal{L}_b) \right]$$
$$\le C(\gamma) \parallel J \parallel_\infty^2 \sup_{\sum_e \alpha_e^2=1} \sup_\beta V\left(\sum_e \alpha_e \phi_e(\beta, \mathbf{g}), \beta \right); \tag{3.28}$$

here $(\mathbf{J}, \tilde{\mathbf{J}}) = Av_x \sum_{e>0,|e|=1} J_e(\varepsilon x)\tilde J_e(\varepsilon x)$ and the meaning of the limit is given in (3.27).

We can now conclude Theorem 3.3. From (3.14) we can choose g so that the variance on the right side of (3.28) is arbitrarily small for any fixed γ . Hence the infimum over g of the left hand side of (3.28) is bounded above by zero. From Lemma 3.8 we can prove Theorem 3.3.

4. Tightness. In this section, we shall prove the tightness Lemma 3.1. Corresponding results are obtained in [GPV] and [V1] via the Garsia, Rodemich and Rumsey Lemma. We will however provide a more direct martingale argument that establishes tightness.

Recall the definition of current in (3.4). We have the exponential martingale: $e^{Z(\beta,t)}$ where

$$Z(\beta,t) = \beta\sum_x J(\varepsilon x)\eta_x(t) - \beta\sum_x J(\varepsilon x)\eta_x(0) - \int_0^t \left(\beta U(\eta(s)) + \Omega(\eta(s)) \right) ds, \tag{4.1}$$

$$U(\eta) = \varepsilon^{-1} \sum_x (\varepsilon^{-1} \nabla_e^- J(\varepsilon x)) w_{x,x+e}(\eta)$$

and

$$\Omega(\eta) = \exp(-\beta \sum_x J(\varepsilon x)\eta_x)\, \mathcal{L} \,\exp(\beta \sum_x J(\varepsilon x)\eta_x) - \beta U(\eta)$$

Since J is smooth,

$$\sup_\eta |\Omega(\eta)| \le C(J)\beta^2\varepsilon^{-d}, \quad \sup_\eta |U(\eta)| \le C(J)\beta^2\varepsilon^{-1-d}. \tag{4.2}$$

LEMMA 4.1 *For each smooth J, for sufficiently small ε we have*

$$P_\varepsilon^\xi \left[\sup_{0\le j\le T\varepsilon^{-2}} \sup_{j\varepsilon^2 \le t \le (j+1)\varepsilon^2} \left| \varepsilon^d \sum_x J(\varepsilon x)\eta_x(t) - \varepsilon^d \sum_x J(\varepsilon x)\eta_x(j\varepsilon^2) \right| \ge \delta \right] \tag{4.3}$$
$$\le \exp(-\delta\varepsilon^{-d}/8)$$

Proof. Let $\beta = 1$. From the martingale inequality we have

$$P_\varepsilon^\xi \left[\sup_{0\le s\le t} Z(1,s) \ge \varepsilon^{-d}\delta \right] \le \exp(-\delta\varepsilon^{-d}) E_\varepsilon^\xi\,[e^{Z(1,t)}\,] = \exp(-\delta\varepsilon^{-d})$$

From (4.2), we have, assuming without loss of generality that $\varepsilon \le 1$,

$$P_\varepsilon^\xi \left[\sup_{0\le s\le t} \left| \varepsilon^d \sum_x J(\varepsilon x)\eta_x(t) - \varepsilon^d \sum_x J(\varepsilon x)\eta_x(0) \right| \ge \delta \right]$$
$$\le 2\exp[2C(J)\varepsilon^{-1-d}t - \delta\varepsilon^{-d}]$$

Let $t = \varepsilon^2$. Then, for ε sufficiently small, the right hand side is bounded by the quantity $\exp(-\delta\varepsilon^{-d}/2)$. We divide the interval $[0,T]$ into ε^{-2} subintervals of size ε^2. We can apply previous argument in every subinterval. Taking the intersection of these events we obtain, for sufficiently small ε,

$$P_\varepsilon^\xi \left[\sup_{0\le j\le T\varepsilon^{-2}} \sup_{j\varepsilon^2 \le t \le (j+1)\varepsilon^2} \left| \varepsilon^d \sum_x J(\varepsilon x)\eta_x(t) - \varepsilon^d \sum_x J(\varepsilon x)\eta_x(j\varepsilon^2) \right| \ge \delta \right]$$
$$\le T\varepsilon^{-2}\exp(-\delta\varepsilon^{-d}/2)$$
$$\le \exp(-\delta\varepsilon^{-d}/8)$$

which establishes (4.3).

Equation (4.3) provides some continuity. Hence in order to prove Lemma 3.1, we can assume that the time t and s in (3.1) are of the form $\varepsilon^2 j$ for some integer j. Hence it suffices to prove

LEMMA 4.2 *For δ small enough,*

$$P_\varepsilon^\xi \left[\sup_{\substack{\varepsilon^2|i-j|\le\delta^4 \\ 0\le i,j\le T\varepsilon^{-2}}} \left| \varepsilon^d \sum_x J(\varepsilon x)\eta_x(\varepsilon^2 i) - \varepsilon^d \sum_x J(\varepsilon x)\eta_x(\varepsilon^2 j) \right| \ge \delta \right] \le \exp(-C\varepsilon^{-d}) \tag{4.4}$$

here i and j are integers and C is some positive constant.

Proof. From Lemma 3.6, we only have to prove (4.4) for the equilibrium process as long as the constant C can be chosen arbitrary large. From the definition of $Z(\beta, t)$ and Schwarz's inequality, we have for any t fixed and β positive ,

$$\begin{aligned} & E^{\mu_\varepsilon} \exp\left[\frac{\beta}{2} \left(\sum_x J(\varepsilon x)\eta_x(t) - \sum_x J(\varepsilon x)\eta_x(0) \right) \right] \\ & \leq \left\{ E^{\mu_\varepsilon}\left[e^{Z(\beta,t)} \right] \right\}^{1/2} \left\{ E^{\mu_\varepsilon}\left[\exp\left[\int_0^t (\ \beta U(\eta(s)) + \Omega(\eta(s))\ )\, ds \right] \right] \right\}^{1/2} \end{aligned} \tag{4.5}$$

From the exponential martingale $E^{\mu_\varepsilon}\left[e^{Z(\beta,t)}\right] = 1$. We can bound Ω by (4.2). Also, from Lemma 3.7 we have

$$\varepsilon^d \log E^{\mu_\varepsilon} \left[\exp\{ \int_0^t \beta\varepsilon^{-d} U(\eta(s)) ds \} \right] \leq t \sup spec\ [\ \beta U(\eta) - \epsilon^{-2} Av_b(-\mathcal{L}_b)\ ] \tag{4.6}$$

The current (3.4) satisfies an identity ($b = (x, x+e)$)

$$\int f\, w_{x,x+e}\, d\mu_\varepsilon = \int [\ f(\eta^b) - f(\eta)\ ]\, c_b(\eta)(\eta_{x+e} - \eta_x)\, d\mu_\varepsilon. \tag{4.7}$$

Hence by and Schwarz's inequality, we have for any function f,

$$\begin{aligned} \varepsilon^{-1}\beta \int f\, w_{x,x+e}\, d\mu_\varepsilon \leq (\varepsilon^{-2}/4) \int c_b(\eta) \left[\sqrt{f}(\eta^b) - \sqrt{f}(\eta) \right]^2 d\mu_\varepsilon \\ + 4\beta^2 \int \left[\sqrt{f}(\eta^b) + \sqrt{f}(\eta) \right]^2 c_b(\eta)(\eta_x - \eta_{x+e})^2\, d\mu_\varepsilon \end{aligned}$$

The last term is bounded by $C \int f d\mu$. Hence, for the eigenvalue problem, we get the bound $Ct\beta^2\varepsilon^{-d}$. Therefore, together with the bound of Ω in (4.2) the last expectation in (4.5) is bounded by $Ct\beta^2\varepsilon^{-d}$. From Chebyshev's inequality,

$$P^{\mu_\varepsilon} \left[\left| \varepsilon^d \sum_x J(\varepsilon x)\eta_x(t) - \varepsilon^d \sum_x J(\varepsilon x)\eta_x(0) \right| \geq \delta \right] \leq \exp(-\delta\beta\varepsilon^{-d}/4 + Ct\beta^2\varepsilon^{-d})$$

Let $\beta = \delta^{-2}$. For any $t \leq \delta^4$ we have

$$P^{\mu_\varepsilon} \left[\left| \varepsilon^d \sum_x J(\varepsilon x)\eta_x(t) - \varepsilon^d \sum_x J(\varepsilon x)\eta_x(0) \right| \geq \delta \right] \leq \exp(-\delta^{-1}\varepsilon^{-d}/8)$$

Clearly, (4.4) follows from this estimate and we conclude Lemma 3.1.

5. The energy estimate. Let us first recall a lemma concerning perturbation theory of eigenvalues applied to our setting. Because we are interested in lattice gases in a finite volume, it is only ordinary perturbation theory for eigenvalues of matrices and the following lemma can be proved by writing down the usual perturbation expansion for eigenvalues and using the assumption on the spectral gap. This lemma actually holds in considerable generality [RS] and an elementary proof is given in [JY].

LEMMA 5.1 *Let ν be a probability measure on a finite set B. Denote by $\langle\, ,\rangle$ the inner product in $L^2(\nu)$ defined by $\langle f, g\rangle = E^\nu[\ fg\ ]$. Let $\mathcal{L}$ be the generator of* *a Markov process on B, which is symmetric with respect to ν. Then $A = -\mathcal{L}$ is a nonnegative definite symmetric operator on $L^2(\nu)$. Suppose that the lowest eigenvalue 0 of A, with the corresponding eigenvector (the ground state) $g_0 \equiv 1$ has a spectral gap of $\delta > 0$. Let X be a bounded function on B considered as a multiplication operator on $L^2(\nu)$. Let $\| X \|_\infty$ denote the sup norm of X. Assume that $E^\nu\{X\} = \langle 1, X \rangle = 0$. Then there is a universal constant C such that for any constant $\gamma > 0$ satisfying*

$$\gamma \| X \|_\infty \le \frac{\delta}{4} \tag{5.1}$$

we have

$$\sup spec\, X - A \le C\gamma^2 \langle X, A^{-1}X \rangle + C\gamma^3 \frac{\| X \|_\infty^3}{\delta^2} .$$

In particular,

$$\sup spec\, X - A \le C\gamma^2 \frac{\| X \|_\infty^2}{\delta} .$$

We can now state the following corollary of Lemma 5.1 in our context. Let $\mu_{\ell,y,\omega}$ denote the (canonical) Gibbs measure with density y and a fixed boundary condition ω on a cube Λ_ℓ of width ℓ. Recall the definition of $\mathcal{L}_b$ given in (2.12) and (2.13) representing a jump across the bond b. The rates are determined in such a way that $\mathcal{L}_b$ is symmetric with respect to the measure $\mu_{\ell,y,\omega}$. If the bond b is close to the boundary, the explicit coefficients of $\mathcal{L}_b$ depend on the boundary condition. The boundary condition will not be specified because our results are all uniform w.r.t. the boundary conditions. In order to apply Lemma 5.1, we will need an estimate on the spectral gap. A key analytical input is provided by a result from [LY] stating that the gap of the operator $\sum_{b\in\Lambda_\ell}(-\mathcal{L}_b)$ is of order ℓ^{-2}.

LEMMA 5.2 *Suppose h is a local function in $\mathcal{G}$ (3.7). Let $\ell_1 = \ell - \sqrt{\ell}$ and $V_\ell(h,y)$ be given by (3.10). Recall also that the mixing condition Assumption (A) is satisfied. Then there is a universal constant C such that for any $\gamma > 0$ we have the following estimate on the eigenvalue*

$$\sup spec \Big[\epsilon^{-1} Av_{|x|\le \ell_1} h - \gamma \epsilon^{-2} Av_{b\in\Lambda_\ell}(-\mathcal{L}_b) - CV_\ell(h,y) \Big] \le 0 \tag{5.2}$$

provided

$$\varepsilon \ll 1, \quad \ell \gg 1 \quad \text{and} \quad \ell\, \epsilon^{\frac{1}{d+2}} \ll 0. \tag{5.3}$$

Furthermore,

$$V_\ell(h,y) \le C\ell^{d+2} E^{\mu_{\ell,y,\omega}} \left[\, (Av_{|x|\le\ell_1} \tau_x h)^2 \, \right] . \tag{5.4}$$

Proof. From [LY], the spectral gap of the operator $\epsilon^{-2} Av_{b\in\Lambda_\ell}(-\mathcal{L}_b)$ is of order $\epsilon^{-2}\ell^{-2-d}$. It is easy to verify that (5.3) implies the condition (5.1). Hence the sup-norm of the first term in (5.2) which is of order ϵ^{-1} is much smaller than the spectral gap of the operator appearing as the second term in (5.2) provided (5.3) is satisfied. Hence (5.2) is just a simple consequence of Lemma 5.1. The bound (5.4) just restates that the spectral gap is of order ℓ^{-2}.

The following theorem combines Lemma 5.2 with multi-scale analysis and provides a key estimate needed in estimating eigenvalues in connection with Theorem 3.9 and the energy estimate Lemma 3.2.

Let us recall that $\Lambda_{\varepsilon^{-1}}$ is the periodic lattice of side ε^{-1} and μ_ε is the Gibbs measure on it. For any k, $\mathcal{F}_k$ is the σ-field generated by the configuration in Λ_k^c i.e. $\{\eta_x : x \in \Lambda_k^c\}$ and the density $\bar{\eta}_k = Av_{\Lambda_k}\eta_x$ in Λ_k. The complement Λ_k^c is relative to the periodic lattice $\Lambda_{\varepsilon^{-1}}$ in this context. Otherwise it could be with respect to Z^d.

THEOREM 5.3 *Let h be a local function satisfying*

$$\begin{aligned} &\sup_\eta |h(\eta)| \le C_h \\ &\sup_\eta |E^{\mu_\varepsilon}[h|\Lambda_k^c, \bar{\eta}_k]| \le C_h k^{-d/2-1-2\alpha} \end{aligned} \tag{5.5}$$

with constants C_h and $\alpha > 0$ and for all k in the range $1 \le k \le \epsilon^{-\frac{2}{2+d}}$. Then there are positive constants C, κ and a function F such that for any h, $\gamma > 0$,

$$\sup spec \left[\epsilon^{-1}h - \gamma\,\epsilon^{-2} \sum_{b\in\Lambda_{\varepsilon^{-1}}} (-\mathcal{L}_b)|b-0|^{-\alpha-d} \right] \le C\gamma^{-1}C_h^2 + \varepsilon^\kappa F(\gamma, C_h) \tag{5.6}$$

on $L^2(\mu_\varepsilon)$ where μ_ε is the Gibbs measure defined in section 2. Furthermore, (5.5) is satisfied for all $h \in \mathcal{G}$, as well as for $h = w_{x,x+e}$, $h = \nabla_e\eta_x$ and $h = \mathcal{L}g$ for some local function g.

Let a nonlocal function h' be of the special form $h'(\eta) = G\,h$, where, for some $\delta > 0$, $G = G(\eta, \bar{\eta}_{\delta\varepsilon^{-1}})$ depends only on the configuration $\eta|\Lambda^c_{\delta\varepsilon^{-1}}$ and the density $\bar{\eta}_{0,\delta\varepsilon^{-1}}$ and h is a local function satisfying (5.5). Then (5.6) remains valid in a slightly modified form

$$\sup spec \left[\epsilon^{-1}h' - C_0\,\gamma^{-1}G^2 - \gamma\,\epsilon^{-2} \sum_{b\in\Lambda_{\varepsilon^{-1}}} (-\mathcal{L}_b)|b-0|^{-\alpha-d} \right] \le \varepsilon^\kappa F(\gamma, C_G) \tag{5.7}$$

where C_0 is a universal constant and C_H is the L_∞ bound for G.

If we replace h by $h_s = E^\mu\big[h|\mathcal{F}_{0,s}\big]$, then for any $\gamma > 0$,

$$\limsup_{s\to\infty} \limsup_{\varepsilon\to 0} \sup spec \left[\varepsilon^{-1}h_s - \gamma\varepsilon^{-2}\sum_b (-\mathcal{L}_b)\,|b-x|^{-\alpha-d} \right] \le 0 \tag{5.8}$$

REMARK. For the rest of this paper, the constant α will be fixed and can be chosen to be any small positive number.

Proof of Part I. Fix an integer ℓ. Let us denote by $\mathcal{F}_n$ the σ-field $\mathcal{F}_{\ell^n}$ and by Λ_n the cube Λ_{ℓ^n}. Define h_n by

$$h_n = E^{\mu_\varepsilon}[h|\mathcal{F}_n] \tag{5.9}$$

as the conditional expectation of h w.r.t. $\mathcal{F}_n$. By definition, $h_0 = h$. Rewrite the function h appearing on the left side of (5.7) as

$$h = \sum_{n=0}^{M-1} (h_n - h_{n+1}) + h_M \tag{5.10}$$

where M is the largest integer such that $\ell^{M[(d/2)+1]} \le (q\varepsilon)^{-1}$, so that

$$\ell^{M[(d/2)+1]} \le (q\varepsilon)^{-1} \le \ell^{(M+1)[(d/2)+1]} \tag{5.11}$$

with a constant q to be determined later. From the assumption (5.5) we have

$$\sup_{\eta} | h_M(\eta) | \leq C_h \ell^{-M(d/2+1+2\alpha)}.$$

With (5.11),

$$\varepsilon^{-1} \sup_{\eta} |h_M(\eta)| \leq C_h \ell^{(d/2+1+2\alpha)} \varepsilon^{-1} (q\varepsilon)^{\kappa} \tag{5.12}$$

where $\kappa = 1 + \frac{4\alpha}{2+d} > 1$.

We compare the two sides of ♣conv2 in operator sense. For some $C < \infty$,

$$\sum_n k^{-\alpha} Av_{b\in\Lambda_{n+1}}(-\mathcal{L}_b) \leq C \sum_b (-\mathcal{L}_b)|b-0|^{-\alpha-d} \tag{5.13}$$

where $k = \ell^{n+1}$. Because each $-\mathcal{L}_b$ is a symmetric positive semidefinite operator establishing (5.13) amounts to verifying

$$\sum_{n:\Lambda_{n+1}\ni b} k^{-\alpha-d} \leq C \sum_b |b-0|^{-\alpha-d} \tag{5.14}$$

for all b. This is easily seen to be true with a constant $C < \infty$, that depends only on ℓ and α. We can clearly estimate the term on the left hand side of (5.6) by decomposing both the function h and the operator according to (5.10) and (5.13) to obtain

$$\begin{aligned} &\sup spec \left[\epsilon^{-1} h - \gamma \epsilon^{-2} \sum_{b\in\Lambda_{\varepsilon^{-1}}} (-\mathcal{L}_b)|b-0|^{-\alpha-d} \right] \\ &\leq \sup spec \left[\epsilon^{-1} h - \frac{\gamma}{C} \epsilon^{-2} \sum_n k^{-\alpha} Av_{b\in\Lambda_{n+1}}(-\mathcal{L}_b) \right] \\ &\leq \sum_{n=0}^{M-1} \mathcal{E}_{n+1} + \sup_{\eta} |h_M(\eta)| \end{aligned} \tag{5.15}$$

where

$$\mathcal{E}_{n+1} = \sup spec \left[\varepsilon^{-1}(h_n - h_{n+1}) - \frac{\gamma}{C} \epsilon^{-2} k^{-\alpha} Av_{b\in\Lambda_{n+1}}(-\mathcal{L}_b) \right] \tag{5.16}$$

By limiting the operator to bonds in some Λ_{n+1} we have introduced considerable degeneracy. The extremal invariant measures are precisely the canonical Gibbs measures μ_{n+1} in the box Λ_{n+1} and are parameterized by ω, representing the boundary condition on Λ^c_{n+1} and the average density in Λ_{n+1}. Since the spectrum is calculated in each $L_2(\mu_{n+1})$ we obtain an upper bound for $\mathcal{E}_{n+1}$ by estimating it in each $L_2(\mu_{n+1})$ and then taking supremum over ω. Hence we only need to get bounds on (5.16) in $L_2(\mu_{n+1})$ that are uniform in ω. Note that once $\mathcal{F}_{n+1}$ is given, from [LY] the spectral gap of $\frac{\gamma}{C} \epsilon^{-2} k^{-\alpha} Av_{b\in\Lambda_{n+1}}(-\mathcal{L}_b)$ is bounded below by $\frac{c\gamma}{C} \epsilon^{-2} k^{-2-d-\alpha}$ for some $c > 0$. We note in addition that $\int (h_n - h_{n+1}) d\mu_{n+1} = 0$. For the perturbation theory to work, by Lemma 5.1, we need

$$\epsilon^{-1} \parallel h_n - h_{n+1} \parallel_\infty \leq \frac{c\gamma}{4C} \epsilon^{-2} \ell^{-(n+1)(2+d+\alpha)} \tag{5.17}$$

to hold for all $0 \le n \le M-1$. From our assumption (5.5) we see that

$$\varepsilon^{-1}\|h_n - h_{n+1}\|_\infty \le 2C_h \varepsilon^{-1}\ell^{-n(d/2+1+2\alpha)}$$

so that (5.17) holds provided

$$2C_h\varepsilon^{-1}\ell^{-n(d/2+1+2\alpha)} \le \frac{c\gamma}{4C}\epsilon^{-2}\ell^{-(n+1)(2+d+\alpha)}$$

which is implied by

$$\ell^{M(d/2+1)} \le \frac{c\gamma}{8CC_h}\varepsilon^{-1} \tag{5.18}$$

so that the choice of $q = \max(1, \frac{8CC_h}{c\gamma})$ will work in (5.11) as well as (5.18). By Lemma 5.2 one can now bound $\mathcal{E}_{n+1}$ by

$$\begin{aligned}\mathcal{E}_{n+1} &\le C\gamma^{-1}\epsilon^{-2}\epsilon^2 k^{d+2+\alpha} \parallel h_n - h_{n+1} \parallel_\infty^2 \\ &\le C\,\gamma^{-1}k^{-\alpha}C_h^2 = C\gamma^{-1}C_h^2\ell^{-\alpha(n+1)}.\end{aligned} \tag{5.19}$$

We have denoted by C, a constant that may change from line to line, but will always remain independent of the parameters $\varepsilon\,,\gamma$ and h.. Summing over n, we conclude that

$$\begin{aligned}\sum_{n=0}^{M-1}\mathcal{E}_{n+1} + \sup_\eta |h_M(\eta)| &\le C\gamma^{-1}C_h^2 + CC_h\varepsilon^{\frac{4\alpha}{2+d}}(1 + \frac{8CC_h}{c\gamma})^{(1+\frac{4\alpha}{2+d})} \\ &= C\gamma^{-1}C_h^2 + \varepsilon^\kappa F(\gamma\,,C_h)\end{aligned}$$

concluding the proof of part I of Theorem 5.3.

Proof of Part II. Clearly, (5.5) holds for $h \in \mathcal{G}$. From (4.7) $w_{x,x+e} \in \mathcal{G}$ and (5.5) holds. For any local function g, there is an s large enough such that g is measurable w.r.t. the σ-algebra generated by $\{\eta_x, |x| \le s\}$. Let $\mathcal{L}_s = \sum_{b\in\Lambda_s}\mathcal{L}_b$. Then we have $\mathcal{L}g = \mathcal{L}_s g$ and also $E^{\mu_\varepsilon}[\mathcal{L}_s g|\mathcal{F}_s] = 0$. In other words, $\mathcal{L}g \in \mathcal{G}$. This proves (5.5) when $h = \mathcal{L}g$. We now establish (5.5) for $h = \nabla_e\eta_x$, namely,

$$E^{\mu_\varepsilon}[\nabla_e\eta_0|\mathcal{F}_k] \le Ck^{-d/2-1-\alpha}. \tag{5.20}$$

This is clearly a type of mixing condition. Unfortunately, the mixing condition we have is for the grand canonical measure (2.5) rather than the canonical one that we are interested in here. We shall however prove in Section 10, that (5.20) holds .

We now prove (5.7) for $h' = Gh$. Note that in the argument in the part I, G is effectively a constant and replacing h by Gh only changes C_h into GC_h. Therefore (5.7) is not all that different from (5.6). Finally to see (5.8) we can assume for simplicity that $l^{n_0} = s$ and sum (5.17) for $n \ge n_0$. This concludes the proof of Theorem 5.3.

Lemma 5.3 states that, in an average sense, any h_x satisfying (5.5) $E^f\{\epsilon^{-1}h_x\}$ is of order 1 for any probability density f satisfying

$$\epsilon^{-2}Av_{b\in\Lambda}D_b(\sqrt{f}) \le C \tag{5.21}$$

We now prove the energy estimate.

Proof of Lemma 3.2. As we have already established tightness of the empirical measure valued process $\varepsilon^d\sum_x \delta_{\varepsilon\,x}\eta_x(\cdot)$, we know that any limit point of Q^ε is a measure Q on the space $C[0,T];\mathcal{M}]]$ of weakly continuos maps $t \to m(t\,,\theta)$ of densities on $\mathbf{T}^d$

that are bounded by 1. The energy estimate is established by proving, for some finite C,

$$E^Q\left\{\sup_J\left[\int_0^T\int_{\mathbf{T}^d}(D_eJ)(t,\theta)m(t,\theta)\,dtd\theta-\frac{C}{2}\int_0^T\int_{\mathbf{T}^d}|J(t,\theta)|^2\,dtd\theta\right]\right\}\le C \quad (5.22)$$

where the supremum is taken over all smooth functions of t and θ. If we ignore the supremum, for the approximating processes, this amounts to proving

$$\limsup_{\varepsilon\to 0}E^{Q_\varepsilon}\left\{\varepsilon^d\int_0^T\sum_x(\nabla_eJ)(t,\frac{x}{N})\eta_x(t)\,dt-\frac{C}{2}\varepsilon^d\int_0^T\sum_x|J(t,\frac{x}{N})|^2\,dt\right\}\le C \quad (5.23)$$

or

$$\limsup_{\varepsilon\to 0}E^{Q_\varepsilon}\left\{\varepsilon^{d-1}\int_0^T\sum_x J(t,\frac{x}{N})\nabla_e\eta_x(t)\,dt-\frac{C}{2}\varepsilon^d\int_0^T\sum_x|J(t,\frac{x}{N})|^2\,dt\right\}\le C \quad (5.24)$$

In view of Jensen's inequality and the relative entropy bound $H(Q^\varepsilon\,;P^\varepsilon)\le C\varepsilon^{-d}$, it is sufficient to establish

$$\begin{aligned}\limsup_{\varepsilon\to 0}\varepsilon^d\log E^{P_\varepsilon}\Bigg\{\exp\Bigg[\varepsilon^{-1}\int_0^T\sum_x J(t,\frac{x}{N})\nabla_e\eta_x(t)\,dt\\ -\frac{C}{2}\int_0^T\sum_x|J(t,\frac{x}{N})|^2\,dt\Bigg]\Bigg\}\le 0\end{aligned} \quad (5.25)$$

or equivalently

$$\limsup_{\varepsilon\to 0}\varepsilon^d\log E^{P_\varepsilon}\left\{\exp\left[\varepsilon^{-1}\int_0^T\sum_x J(t,\frac{x}{N})\nabla_e\eta_x(t)\,dt\right]\right\}\le\frac{C}{2}\int_0^T\int_{\mathbf{T}^d}|J(t,\theta)|^2\,dtd\theta \quad (5.26)$$

Since P^ε is the equilibrium and the functional inside the expectation is a Feynman-Kac functional with a time dependent potential we can estimate (5.26) by estimating the largest eigenvalue. If we denote by

$$\lambda_\varepsilon(t)=\sup spec\big[\varepsilon^{-1}\sum_x J(t,\frac{x}{N})\nabla_e\eta_x-\varepsilon^{-2}\sum_b(-\mathcal{L}_b)\big]$$

it is sufficient to prove

$$\limsup_{\varepsilon\to 0}\varepsilon^d\int_0^T\lambda_\varepsilon(t)\,dt\le\frac{C}{2}\int_0^T\int_{\mathbf{T}^d}|J(t,\theta)|^2\,dtd\theta \quad (5.27)$$

For estimating the eigenvalue λ_ε we decompose

$$\varepsilon^{-1}\sum_x J(t,\frac{x}{N})\nabla_e\eta_x-\varepsilon^{-2}\sum_b(-\mathcal{L}_b)$$

and, for some $\gamma>0$, estimate $\varepsilon^d\lambda_\varepsilon(t)$ from above by the sum

$$\varepsilon^d\lambda_\varepsilon(t)\le\varepsilon^d\sum_x\sup spec\left[\varepsilon^{-1}J(t,\frac{x}{N})\nabla_e\eta_x-\gamma\sum_b|b-x|^{-\alpha-d}(-\mathcal{L}_b)\right]$$

We have in addition the bound

$$\sup_{0\le t\le T} \varepsilon^d \lambda_\varepsilon(t) \le \sup_{0\le t\le T} \varepsilon^d \sum_x \sup spec\Big[\varepsilon^{-1} J(t\,,\frac{x}{N})\nabla_e \eta_x - \gamma \sum_b |b-x|^{-\alpha-d}(-\mathcal{L}_b)\Big]$$
$$\le \sup_{|\beta|\le C} \sup_x \sup spec\Big[\varepsilon^{-1}\beta\,\nabla_e \eta_x - \gamma \sum_b |b-x|^{-\alpha-d}(-\mathcal{L}_b)\Big]$$

where C is an upper bound on $|J(\cdot\,,\cdot)|$. Now (5.6) provides a uniform bound

$$\sup_{0\le t\le T} \varepsilon^d \lambda_\varepsilon(t) \le C$$

as well as a bound on the limit

$$\lim_{\varepsilon\to 0} \varepsilon^d \int_0^T \lambda_\varepsilon(t)\, dt \le \frac{C}{2}\int_0^T |J(t\,,\theta)|^2\, dt d\theta$$

concluding our estimate. In order to complete the proof of Lemma 3.2 we only have to remark that whenever we use large deviation bounds and Jensen's inequality to pass from

$$\limsup_{\varepsilon\to 0} \varepsilon^d \log E^{P^\varepsilon}\Big\{ \exp\big[\varepsilon^{-d} F_\alpha\big]\Big\} \le 0 \tag{5.28}$$

and

$$H(Q^\varepsilon; P^\varepsilon) \le C\varepsilon^{-d}$$

to

$$E^Q\big[F_\alpha\big] \le C \tag{5.29}$$

we actually can obtain

$$E^Q\big[\sup_\alpha F_\alpha\big] \le C \tag{5.30}$$

because, while one cannot sneak the supremum inside the expectation to go from (5.29) to (5.30), to do it in (5.28) is trivial and for any finite set of α's yields (5.30) with the same constant C, independently of the finite set. The rest is a routine application of the monotone convergence theorem.

6. Two blocks estimate. In this section, we prove the two block estimate Theorem ♣2b . It will be used in the next section to prove Theorem 3.9. The first step toward proving a two block estimate is to get a bound on the Dirichlet form involving "long jumps" in terms of the usual Dirichlet form with only the "nearest-neighbor jumps". This was one of the main difficulties in establishing the hydrodynamical limit of lattice gas models in dimension greater than 1. If μ is a Gibbs state on the space of configurations on a lattice Λ and z, y are two lattice sites that are a considerable distance away from each other, the problem is one of estimating $E^\mu\{|(T_{z,y}u)(\eta)\,|^2\} = E^\mu\{|u(\eta^{z,y}) - u(\eta)|^2\}$ in terms of the usual Dirichlet form involving $E^\mu\{|u(\eta^{x,x+e}) - u(\eta)|^2\}$. Let us introduce the notation $S_{x,y}\eta = \eta^{x,y}$. The natural way to make this estimation is to write

$$S_{z,y} = S_{z,x_1} S_{x_1,x_2} \cdots S_{x_{n-1},x_n} S_{x_n,y} S_{x_{n-1},x_n} \cdots S_{x_1,x_2} S_{z,x_1}$$
$$= S_{b_{2n-1}} \cdots S_{b_1}$$

as a product of exchanges over shorter bonds and estimate $T_{x,y}u$ by a telescoping sum

$$|u(S_{x,y}\eta) - u(\eta)| \leq \sum_{j=1}^{2n-1} |u(S_{b_j} \cdots S_{b_1}\eta) - u(S_{b_{j-1}} \cdots S_{b_1}\eta)|$$

and conclude that

$$\begin{aligned} E^{\mu}\{|u(S_{x,y}\eta) - u(\eta)|^2\} &\leq (2n-1) \sum_{j=1}^{2n-1} E^{\mu}\{|u(S_{b_j} \cdots S_{b_1}\eta) - u(S_{b_{j-1}} \cdots S_{b_1}\eta)|^2\} \\ &= (2n-1) \sum_{j=1}^{2n-1} E^{\mu}\{|u(S_{b_j}S^j\eta) - u(S^j\eta)|^2\} \end{aligned}$$

where $S^j = S_{b_j} \cdots S_{b_1}$ is some more complex permutation along the way. If μ were a Bernoulli measure or some other permutation invariant measure like a Bernoulli measure conditioned on the value of the sum $\sum_x \eta_x$ then

$$(2n-1) \sum_{j=1}^{2n-1} E^{\mu}\{|u(S_{b_j}S^j\eta) - u(S^j\eta)|^2\} = (2n-1) \sum_{j=1}^{2n-1} E^{\mu}\{|u(S_{b_j}\eta) - u(\eta)|^2\}$$

and, if we choose our intermediary chain of sites to be nearest neighbors, we are done. Difficulties arise because, in general, μ is not invariant under permutation. We will pickup a constant factor which is an upper bound on the Radon-Nikodym derivatives when μ is transformed by the permutation S^j. We note that if we are able to bound the Radon-Nikodym derivatives in both directions then conditioning to a set of the form $\sum_x \eta_x = m$ causes no problem. While the factor $2n-1$ which is comparable to the distance from x to y is acceptable, if not taken care of, the bound for the Radon Nikodym derivative will grow exponentially in n and is unacceptable for our purposes. The following theorem therefore requires a careful proof. Without loss of generality we will assume that the jump rate $c_b = 1$ in (2.12).

THEOREM 6.1 *Denote by μ a Gibbs state on a cube Λ and let D_b denote the Dirichlet form w.r.t μ for a nearest neighbor bond $b \in \Lambda$. For any two sites $x, y \in \Lambda$ let γ_{xy} denote the nearest neighbor path that goes from x to y, moving successively as far as it has to in each of the coordinate directions, following the natural order for the different coordinate directions. For every function u on $\{0,1\}^{|\Lambda|}$ define the Dirichlet form along the path by*

$$D^{\gamma_{zy}}(u) = \sum_{b \in \gamma_{xy}} D_b(u)$$

Then we have

$$E^{\mu}[(T_{zy}u)^2] \leq \quad c_0|\gamma_{zy}|D^{\gamma_{zy}}(u), \tag{6.1}$$

where c_0 is a constant depending on the dimension d and the Hamiltonian. Here $|\gamma_{zy}|$ is the length of the path. Note that no mixing condition is needed.

Proof. **Step 1.** With out loss of generality we can assume that $y = z + \ell e_1$. The general path is just a finite sequence of such straight lines $(z, x_1), (x_1, x_2) \cdots (x_{d-1}, y)$ along different coordinate directions going through at most $d-1$ intermediary corners $x_1, \cdots, x_{d-1}$. We can estimate the left side of (6.1) by a telescoping sum. Although the Radon-Nikodym derivatives could cause trouble here, because the only transformations that are involved in the telescoping sum are permutations among the $d+1$ sites $z, x_1, \cdots, x_{d-1}, y$ the extra factor is just a constant depending on the Hamiltonian and d.

Step 2. We can assume with out loss of generality that $y = z + \ell k e_1$ for some ℓ with k fixed at some positive integer so that two sites that are at a distance at least k, do not both interact with any third site in the Hamiltonian. This involves again the same idea. If $z - y = \ell e_1$ where ℓ is not a multiple of k, we can go through an intermediate point that is a multiple and the difference is at most a fixed finite distance away.

Step 3. The choice of k in the previous step guarantees that the conditional distribution of $\{\eta_{z+jke_1} : 0 \le j \le \ell\}$ under the canonical Gibbs measure given the configuration on the complement of the sites $\{z + jke_1\}$ is a product measure conditioned on the sum $\sum_{j=1}^{\ell} \eta_{z+jke_1}$. Of course the conditional probability $p_\omega(j) = \mu_\omega[\eta_{z+jke_1} = 1]$ is given by the DLR formula, and although different for each j takes only a fixed finite number of values depending on the total number of different configurations possible in the neighborhood of any one site. If we can establish (6.1) for the conditional distributions μ_ω with a uniform constant independent of ω, we can integrate with respect to μ at the end.

Step 4. The problem is now reduced to the following. We just have sites $0, 1 \cdots, \ell$. We have a measure μ which is the product measure with $\Pr[\eta_j = 1] = p_j$ conditioned on the set $\sum_{j=0}^{\ell} \eta_j = m$. If we assume that $\{p_j\}$ takes only a finite number r of distinct values and they all satisfy $\delta \le p_j \le 1 - \delta$, then with a constant c_0 depending only on r and δ, the analog of (6.1) is satisfied. This problem came up in the context of one dimensional lattice gases with random magnetization and the proof given in [QY1] is reproduced below. The proof is by induction on r where r is redefined as the number of distinct values among $\{p_j\}$ with $1 \le j \le \ell - 1$. If $r = 1$, except for the end points all probabilities are the same. In this case the factor produced by the Radon-Nikodym derivative is easily seen to be bounded by a constant independent of ℓ. Any permutation can only change the probabilities at most four sites. Now let us prove the induction step. Assume that if there are r distinct values we have a bound with a constant c_r independent of ℓ . Suppose now that there are $r+1$ distinct values. Let $1 \le j_1 < j_2 < \cdots < j_s \le \ell - 1$ be an enumeration of all the sites where p_j is equal to one of these distinct values. If we write

$$S_{0,\ell} = S_{0,j_1} S_{j_1,j_2} \cdots S_{j_{s-1},j_s} S_{j_s,\ell} S_{j_{s-1},j_s} \cdots S_{j_1,j_2} S_{0,j_1}$$

all but the two endsites have the same probability and therefore, for all the intermediate permutations, the Radon-Nikodym factor is controlled uniformly in s by a constant C. By Schwarz's inequality

$$\begin{aligned} &\frac{1}{\ell} E^\mu\{|T_{0,\ell}\, u|^2\} \\ &\le C\left[\frac{1}{j_1} E^\mu\{|T_{0,j_1}\, u|^2\} + \sum_{r=2}^{s} \frac{1}{j_r - j_{r-1}} E^\mu\{|T_{j_{r-1},j_r}\, u|^2\} + \frac{1}{\ell - j_s} E^\mu\{|T_{j_s,\ell}\, u|^2\}\right] \end{aligned}$$

By the induction hypothesis each summand satisfies the analog of (6.1). We obtain $c_{r+1} \le C c_r$ and the theorem is established.

Using this Theorem, it is not hard to prove the two block estimate. The following proof is similar to the one given in [GPV, R].

THEOREM 6.2 (TWO BLOCKS ESTIMATE) *For any bounded continuous function h and any $\gamma > 0$,*

$$\limsup_{k,a,\varepsilon} \sup spec \left\{ Av_x Av_{y:|y-x|\le a\varepsilon^{-1}} \left[\, h(\bar\eta_{x,k}) - h(\bar\eta_{y,k}) \, \right]^2 - \gamma\epsilon^{-2} Av_b(-\mathcal{L}_b) \right\} = 0 \tag{6.2}$$

where $\limsup_{k,a,\varepsilon} = \limsup_{k\to\infty} \limsup_{a\to 0} \limsup_{\epsilon\to 0}$. *In particular,*

$$\limsup_{k,a,\varepsilon} \sup spec \left\{ Av_x [h(\bar\eta_{x,a\epsilon^{-1}}) - h(\bar\eta_{x,k})]^2 - \gamma\epsilon^{-2} Av_b(-\mathcal{L}_b) \right\} = 0 \tag{6.3}$$

Proof. Let $L_{x,y}$ be the generator associated with the Dirichlet form $E^\mu[(T_{xy}u)^2]$, namely they are related via (2.12). For every pair x, y of sites in Λ let us choose a canonical path $\gamma_{x,y}$, consisting of d straightlines that lie along the d coordinate directions following their natural order. $|\gamma_{x,y}|$ is of order $|x-y|$. Recall that R is the range of the Hamiltonian defined in section 2. Hence we can rewrite (6.1) as an operator inequality

$$-L_{w,z} \le C|w-z| \sum_{b\in\gamma_{wz}} (-\mathcal{L}_b).$$

Taking the average of w with $|w-x| \le k$ and z with $|z-y| \le k$ and then averaging over $x \in \Lambda$ and y with $2R+k < |y-x| \le a\epsilon^{-1}$, we have

$$Av_x Av_{y:2R+k<|y-x|\le a\epsilon^{-1}} Av_{w:|w-x|\le k} Av_{z:|z-y|\le k} (-\mathcal{L}_{w,z}) \le C\epsilon^{-2}a^2 Av_b(-\mathcal{L}_b) \tag{6.4}$$

A similar inequality holds if we replace $|z-y| \le k$ by $|z-x| \le k$ or replace $|w-x| \le k$ by $|w-y| \le k$. Let Λ_x^k denote a cube of width k and centered at x. Hence we have

$$Av_x Av_{y:2R+k<|y-x|\le a\epsilon^{-1}} Av_{w,z\in\Lambda_x^k\cup\Lambda_y^k} (-\mathcal{L}_{w,z}) \le C\epsilon^{-2}a^2 Av_b(-\mathcal{L}_b) \tag{6.5}$$

We can replace the average of y in (6.2) by the average over y with $2R+k < |y-x| \le a\epsilon^{-1}$ since the error term is negligible. Together with (6.5), we can bound the eigenvalue of (6.2) by

$$\begin{aligned} Av_x \, Av_{y:2R+k<|y-x|\le a\epsilon^{-1}} \, \sup spec \Big\{ & \left[\, h(\bar\eta_{x,k}) - h(\bar\eta_{y,k}) \, \right]^2 \\ & - Av_{w,z\in\Lambda_x^k\cup\Lambda_y^k} Ca^{-2}(-\mathcal{L}_{w,z}) \Big\} \end{aligned} \tag{6.6}$$

Clearly we obtain an upper bound by replacing the average over x and y with sup over x and y and over the boundary conditions of $\Lambda_x^k \cup \Lambda_y^k$. Denote the configuration on $U = \Lambda^k \cup \Lambda^k$ by (ξ, ζ). Then (6.6) is bounded above by the supremum of

$$\sup spec \left\{ \left[\, h(\bar\xi_k) - h(\bar\zeta_k) \, \right]^2 - Ca^{-2} Av_{w,z\in U}(-\mathcal{L}_{w,z}) \right\} \tag{6.7}$$

as an eigenvalue problem on $L^2(\mu_{U,\omega,\rho})$ with boundary condition ω and density ρ.

We now fix a boundary condition ω and a density ρ. The operator

$$Ca^{-2} Av_{w,z\in U}(-\mathcal{L}_{w,z})$$

on $L^2(\mu_{U,\omega,\rho})$ clearly has a gap $C(k)a^{-2}$ and the ground state is the constant function. For any k fixed, and $a \to 0$, the gap diverges to ∞. Since h is bounded, for any k fixed from Lemma 5.1 we can bound the limsup of (6.7) as $a \to 0$ by

$$\sup_{\omega,\rho} E^{\mu_{U,\omega,\rho}} \left[\, h(\bar{\xi}_k) - h(\bar{\zeta}_k) \, \right]^2$$

Taking the limit $k \to \infty$, we have $\mu_{U,\omega,\rho} \to \mu_\rho \times \mu_\rho$, where μ_ρ denotes the infinite volume Gibbs measure on $\mathbb{Z}^d$ with density ρ. Clearly, (6.2) and (6.3) follow from the law of large number of Gibbs states. This proves Theorem 6.2.

7. Eigenvalue estimate. We are now ready to prove Theorem 3.9.

Proof of Theorem 3.9. Step 1. Let us recall that $\Phi^{\mathbf{g}}_{x,e}$, defined in (3.17), and $\Psi_{\varepsilon,c,e}$ defined in Theorem 3.9 are given by

$$\Phi^{\mathbf{g}}_{x,e} = Av_{|y-x|\le \ell_1} \left[\, w_{y,y+e} + \mathcal{L} g_e(\bar{\eta}_{x,\ell}, \tau_y \eta) \, \right]$$

$$\Psi_{\varepsilon,c,e}(x) = \varepsilon (2c)^{-1} Av_{|y-x|\le \ell_1} \left[\eta_{x+c\varepsilon^{-1}e} - \eta_{x-c\varepsilon^{-1}e} \right].$$

Here we continue to follow the convention set up after (3.26) in step 3 of the proof of Theorem 3.3 that the generator $\mathcal{L}$ does not apply to the variable $\bar{\eta}_{x,\ell}$ because the error terms were proved to be negligible. We can rewrite

$$\Phi^{\mathbf{g}}_{x,e} + \sum_{e'} D_{e,e'}(\bar{\eta}_{x,a\varepsilon^{-1}}) \Psi_{\varepsilon,\,c,\,e'} = \Phi^{\mathbf{g},s}_{x,e} + \Psi^{1,s}_{x,e} + \Psi^{2,s}_{x,e} + \Psi^{3}_{x,e},$$

where

$$\begin{aligned}
\Phi^{\mathbf{g},s}_{x,e} &= Av_{|y-x|\le \ell_1} \phi^{g,y,s}_{x,\varepsilon} \\
\phi^{g,y,s}_{x,\varepsilon} &= \left[w_{y,y+e} + \mathcal{L}\, g_e(\bar{\eta}_{x,\ell}, \tau_y \eta) + \sum_{e'} D_{e,e'}(\bar{\eta}_{x,\ell}) (\nabla_{e'} \eta_y - E^{\mu}[\nabla_{e'} \eta_y | \mathcal{F}_{y,s}]) \right] \\
\Psi^{1,s}_{x,e} &= \sum_{e'} D_{e,e'}(\bar{\eta}_{x,a\varepsilon^{-1}}) Av_{|y-x|\le \ell_1} E^{\mu}[\nabla_{e'} \eta_y | \mathcal{F}_{y,s}], \\
\Psi^{2,s}_{x,e} &= \sum_{e'} [D_{e,e'}(\bar{\eta}_{x,a\epsilon^{-1}}) - D_{e,e'}(\bar{\eta}_{x,\ell})] [Av_{|y-x|\le \ell_1} \left(\, \nabla_{e'} \eta_y - E^{\mu}[\nabla_{e'} \eta_y | \mathcal{F}_{y,s}] \, \right)] \\
\Psi^{3}_{x,e} &= \sum_{e'} D_{e,e'}(\bar{\eta}_{x,a\varepsilon^{-1}}) \left[Av_{|y-x|\le \ell_1} [\Psi_{\varepsilon,c,e'}(y) - \nabla_{e'} \eta_y] \right]
\end{aligned} \tag{7.1}$$

Hence we can establish the eigenvalue estimate of (3.28) by showing

$$\limsup_{s,\,a,\,c,\,\ell\,,\varepsilon} \Omega_0 \le C(\gamma) \parallel J \parallel_\infty^2 \sup_{\sum_e \alpha_e^2 = 1} \sup_{\beta} V \left(\sum_e \alpha_e \phi_e(\beta, \mathbf{g}), \beta \right) \tag{7.2}$$

and for $i = 1, 2, 3$,

$$\limsup_{s,\,a,\,c,\,\ell\,,\varepsilon} \Omega_i = 0 \tag{7.3}$$

where

$$\begin{aligned}
\Omega_0 &= \sup spec \, \left[\, \epsilon^{-1}(\mathbf{J}, \Phi^{\mathbf{g},s}) - (\gamma/4)\epsilon^{-2} Av_b(-\mathcal{L}_b) \, \right] \\
\Omega_i &= \sup spec \, \left[\, \epsilon^{-1}(\mathbf{J}, \Psi^{i,s}) - (\gamma/4)\epsilon^{-2} Av_b(-\mathcal{L}_b) \, \right], \; i = 1, 2 \\
\Omega_3 &= \sup spec \, \left[\, \epsilon^{-1}(\mathbf{J}, \Psi^{3} \,) - (\gamma/4)\epsilon^{-2} Av_b(-\mathcal{L}_b) \, \right]
\end{aligned} \tag{7.4}$$

and the inner product $(,)$ is as defined in Theorem 3.9.

We first consider Ω_1. From (5.8), we have (7.3) holds for $i=1$. For $i=2$, we apply (5.7) to reduce the problem to showing

$$\limsup_{s,\ell,a,\varepsilon} \sup spec \left[C(s,\gamma,J) Av_x (D(\bar{\eta}_{x,a\epsilon^{-1}}) - D(\bar{\eta}_{x,\ell}))^2 - (\gamma/8)\epsilon^{-2} Av_b(-\mathcal{L}_b) \right] = 0 \tag{7.5}$$

which is the familiar form of two block estimate. From (6.3) we conclude (7.5). Note that we take $\ell \to \infty$ before taking the limit $s \to \infty$.

To deal with Ω_3 we rewrite

$$(J, \Psi^3) = \sum_{e'} Av_x [G_{x,\varepsilon'}(J,\eta) \nabla_{e'} \eta_x]$$

and use (7.6) to control $\limsup_{a,c,\ell,\varepsilon} \Omega_3$. G is easily seen to satisfy the hypotheses needed to apply (7.7). The problem reduces to one of showing

$$\limsup_{a,c,\ell,\varepsilon} \sup spec [\, Av_x G^2_{x,\varepsilon} - (\gamma/8)\varepsilon^{-2} Av_b(-\mathcal{L}_b \,] \le 0$$

which is easily seen to follow from the two block estimate.

Finally we have to bound Ω_0. Note that $E^{\mu}[\phi^{\mathbf{g},y,s}_{x,e} | \mathcal{F}_{y,s}] = 0$ if $|y-x| \le \ell_1$ and hence from Lemma 5.2 we have

$$\Omega_0 \le C \sup spec \left[Av_x V_\ell \left(\sum_{e>0} J_e(\varepsilon x) \phi^{\mathbf{g},x,s}_{x,e} \,,\, \bar{\eta}_{x,\ell} \right) - (\gamma/8)\epsilon^{-2} Av_b(-\mathcal{L}_b) \right] + o(1)$$

with $\lim_{s,\ell,a,\varepsilon} o(1) = 0$. Note we have kept some Dirichlet form. For $k \ge \ell$ let

$$U_{x,k} = E^{\mu_\varepsilon} \left[V_\ell \left(\sum_{e>0} J_e(\varepsilon x) \phi^{\mathbf{g},x,s}_{x,e} \,,\, \bar{\eta}_{x,\ell} \right) \middle| \mathcal{F}_{x,k} \right],$$

and we can bound $\Omega_0 \le \Omega_4 + \Omega_5 + o(1)$ with

$$\Omega_4 = C \sup spec \left[Av_x \left\{ V_\ell \left(\sum_{e>0} J_e(\varepsilon x) \phi^{\mathbf{g},x,s}_{x,e} , \bar{\eta}_{x,\ell} \right) - U_{x,k} \right\} - (\gamma/8)\epsilon^{-2} Av_b(-\mathcal{L}_b) \right]$$

$$\Omega_5 = \sup_{x,\mathcal{F}_{x,k}} U_{x,k}$$

Using Lemma 5.2 and noting that the factor ϵ^{-1} in front of h in (5.2) is not present in this calculation, we have

$$\lim_{\varepsilon \to 0} \Omega_4 = 0\,.$$

It remains to bound Ω_4. By the translational invariance,

$$\Omega_5 \le \sup_{\omega,\rho} E^{\mu_{\Lambda_{2k+1},\omega,\rho}} \left[V_\ell (\sum_{e>0} J_e(\varepsilon x) \phi^{\mathbf{g},0,s}_{0,e} , \bar{\eta}_\ell) \middle| \mathcal{F}_k \right] \tag{7.8}$$

where $\mu_{\Lambda_{2k+1},\omega,\rho}$ denote the canonical Gibbs measure on Λ_{2k+1} with boundary condition ω and density ρ. For every ℓ fixed, we can take the limit $k \to \infty$ and any limiting point of the right hand side of (7.8) is given by

$$E^{\mu_\beta}\left[V_\ell(\sum_e J(\varepsilon x)\phi_{0,e}^{\mathbf{g},0,s}, \bar{\eta}_\ell) \right] \tag{7.9}$$

where μ_β is the unique Gibbs state with density β given by the limit of $\bar{\eta}_\ell$. Note that from our mixing assumption, there is only one Gibbs state for a given density. From the law of large number of Gibbs state, we can replace the density $\bar{\eta}_{0,\ell}$ in the definition of $\phi_{0,e}^{\mathbf{g},0,s}$ in (7.9) by the total density β. Recall $\phi_e^{\mathbf{g}}(\beta)$ is given by (3.13) and by definition of $\phi_{0,e}^{\mathbf{g},0,s}$ we have

$$\phi_{0,e}^{\mathbf{g},0,s} = \phi_e^{\mathbf{g}}(\beta) - E^{\mu_\varepsilon}\left[\phi_e^{\mathbf{g}}(\beta) \mid \mathcal{F}_s \right]$$

for s large enough; where the density $\bar{\eta}_{0,\ell}$ in $\phi_{0,e}^{\mathbf{g},0,s}$ is replaced by β. We can now take the limit $\ell \to \infty$ and then $s \to \infty$ to have

$$\limsup_{\ell\to\infty} E^{\mu_\beta}\left[V_\ell(\sum_{e>0} J_e(\varepsilon x)\phi_{0,e}^{\mathbf{g},0,s}, \bar{\eta}_\ell) \right] \le V\left(\sum_{e>0} J_e(\varepsilon x)\phi_e^{\mathbf{g}}(\beta), \beta \right) \tag{7.10}$$

where we have used the definition (3.12) of V This proves Theorem 3.9.

8. Computation of variance. Let us recall the definition of $\mathcal{G}$

$$\mathcal{G} = \{h : E^\mu[h|\mathcal{F}_s] = 0 \quad \text{for some } s\}$$

We would like to define the variance

$$V(h, h; \beta) = \lim_{\ell\to\infty} E^{\mu_\beta}\left[< h^{(\ell)}, \mathcal{L}_{(\ell)}^{-1} h^{(\ell)} >_{\ell,\omega,y} \right] \tag{8.1}$$

as a uniform limit in β. Since we cannot show the existence of this limit at this time we define the upper and lower limits

$$V^*(h, \beta) = \limsup_{\substack{\ell\to\infty \\ \beta'\to\beta}} E^{\mu_{\beta'}}\left[< h^{(\ell)}, \mathcal{L}_{(\ell)}^{-1} h^{(\ell)} >_{\ell,\omega,y} \right] \tag{8.2}$$

and

$$V_*(h, \beta) = \liminf_{\substack{\ell\to\infty \\ \beta'\to\beta}} E^{\mu_{\beta'}}\left[< h^{(\ell)}, \mathcal{L}_{(\ell)}^{-1} h^{(\ell)} >_{\ell,\omega,y} \right] \tag{8.3}$$

Here

$$h^{(\ell)} = \frac{1}{(2\ell_1+1)^d} \sum_{x\in\Lambda_{l_1}} \tau_x h$$

with $\ell_1 = \ell - \sqrt{\ell}$ and

$$\mathcal{L}_{(\ell)} = \frac{1}{(2\ell+1)^d} \sum_{b\in\Lambda_\ell} \mathcal{L}_b.$$

For any $h \in \mathcal{G}$, for sufficiently large ℓ, $E^{\mu_{\ell,\omega,y}}[h^{(\ell)}] = 0$. The following calculation is a typical way to estimate (8.2). From $E^{\mu_{s,\omega,y}}[h] = 0$, we conclude that

$$h = \mathcal{L}_{(s)} u_s$$

for some u_s. If we denote by C_s^{-1} a lower bound on the spectral gap for a box of size s, for any function v of the configurations in the box Λ_ℓ

$$\begin{aligned} |E[h_{(\ell)}\, v]| &\le \frac{1}{(2\ell_1+1)^d} \sum_{x\in\Lambda_{\ell_1}} |E[\tau_x h\, v]| \\ &\le \frac{\sqrt{C_s}}{(2\ell_1+1)^d} \sum_{x\in\Lambda_{\ell_1}} \sqrt{E[[\tau_x h]^2]} \sqrt{\sum_{b\in\Lambda_{x,s}} D_b(v)} \\ &\le \sqrt{C_s}(2s+1)^{\frac{d}{2}} \frac{(2\ell+1)^{\frac{d}{2}}}{(2\ell_1+1)^{\frac{d}{2}}} \|h\|_{2,\ell} < v\,, \mathcal{L}_{(\ell)}\, v >^{\frac{1}{2}} \end{aligned}$$

providing us the bound

$$< h^{(\ell)}, \mathcal{L}_{(\ell)}^{-1} h^{(\ell)} >_{\omega,y} \le C_s \left(\frac{\ell}{\ell_1}\right)^d (2s+1)^d \|h\|_{2,\ell}^2$$

where E denotes expectation with respect to any canonical Gibbs measure $\mu_{\ell,\omega,y}$ and

$$\|h\|_{2,\ell}^2 = \frac{1}{(2\ell_1+1)^2} E[\sum_{x\in\Lambda_{\ell_1}} |\tau_x h|^2]$$

Taking expectations with respect to $\mu^{\beta'}$, letting $\ell \to \infty$ and $\beta' \to \beta$ we have

$$V^*(h\,,\beta) \le C_s(2s+1)^d \|h\|_{2,\beta}^2 \tag{8.4}$$

Because we defined $V^*(h\,,\beta)$ as the limsup, and did not show the existence of the limit, it is not quite clear that $V^*(h\,,\beta)$ is a quadratic form in h, and consequently it is not possible to generate the inner product from it by the standard polarization formula:

$$V^*(h\,;g\,,\beta) = \frac{1}{4}[V^*(g+h\,,\beta) - V^*(g-h\,,\beta)]$$

Because $V_\ell(h\,;\omega\,,y)$ is actually a quadratic form, it is possible to polarize it and try to calculate directly the quantity

$$V(g\,,h\,;\beta) = \lim_{\substack{\ell\to\infty \\ \beta'\to\beta}} E^{\mu_{\beta'}}\{V_\ell(g\,,h\,;\omega\,,y)\} \tag{8.5}$$

We will show that the limit exists and calculate it explicitly for a large class of local functions g and h. Let us define the shifts τ_x for $x \in \mathbb{Z}^d$ by

$$(\tau_x\eta)_y = \eta_{y-x}, \quad \tau_x b = b - x, \quad (\tau_x g)(\eta) = g(\tau_x \eta)$$

on configurations η, bonds b and functions g. We have the basic exchange operator S_b defined by $S_b\eta = \eta^b = \eta^{x,y}$ for a bond $b = (x,y)$ in (2.10). The operator T_b was already defined as $(T_b g)(\eta) = g(S_b \eta) - g(\eta)$. The shift operators τ_x and the exchange operators S_b satisfy the commutation relations

$$\tau_x S_b \eta = S_{b+x} \tau_x \eta, \quad (S_b \tau_x g)(\eta) = (\tau_x S_{b+x} g)(\eta)$$

leading to the identities

$$(T_b \tau_x g)(\eta) = (\tau_x T_{b+x} g)(\eta) = (\tau_x T_{\tau_{-x} b} g)(\eta)$$

Suppose $g \in \mathcal{G}$, i.e it is a local functions such that $E^{\mu}\{g|\mathcal{F}_s\} = 0$ for some s, then for any other local function h, $E^{\mu}[g \tau_x h] = 0$ provided $\tau_x h$ does not depend on any of the variables $\{\eta_z : z \in \Lambda_s\}$. In particular for any h which is a local function and $g \in \mathcal{G}$ the series

$$\sum_x E^{\mu}[g \tau_x h] =< g , h >_0$$

is well defined as a finite sum. If it happens that $h \in \mathcal{G}$ as well, then

$$< g , h >_0 (\beta) = \sum_x E^{\mu_\beta}[g \tau_x h] = \sum_x E^{\mu_\beta}[g \tau_x h] =< h , g >_0 (\beta)$$

is a continuous function of β and

$$< g , g >_0 (\beta) = \lim_{\ell \to \infty} \frac{1}{(2\ell+1)^d} E^{\mu_\beta}\big[\big(\sum_{x \in \Lambda_\ell} \tau_x g\big)^2\big] \geq 0$$

so that for each β, $< , >_0$ is a positive semidefinite inner product. We will also need the quantities

$$t_e(g , \beta) = \sum_x (e , x) E^{\mu_\beta}[\eta_x \, g]$$

which are again only finite sums for $g \in \mathcal{G}$. Let u be a local function with $\mathcal{L}u = g$. Clearly $g \in \mathcal{G}$ and $\mathcal{L}\tau_x u = \tau_x g$ for all $x \in \mathbb{Z}^d$. In particular

$$\mathcal{L}_\ell\big(\sum_{x \in \Lambda_{\ell_1}} \tau_x u\big) = \mathcal{L}\big(\sum_{x \in \Lambda_{\ell_1}} \tau_x u\big) = \sum_{x \in \Lambda_{\ell_1}} \tau_x g$$

If v is now another local function with $\mathcal{L}v = h$, a direct calculation yields

$$V_\ell(g , h ; \omega , y) = -\frac{1}{(2\ell_1+1)^d} E^{\mu_{\ell,\omega,y}}\Big\{\sum_{x,y \in \Lambda_{\ell_1}} \tau_y g \, \tau_x v\Big\}$$

and one can easily show that the limit

$$V(g , h ; \beta) = \lim_{\substack{\ell \to \infty \\ \beta' \to \beta}} E^{\mu_{\beta'}}\big\{V_\ell(g , h ; \omega , y)\big\} = -\sum_{x \in \mathbb{Z}^d} E^{\mu_\beta}\big\{g \, \tau_x v\big\} = - < g , v >_0 (\beta) \tag{8.6}$$

exists.

The following calculation is easy to carry out. For each local function u let us define the formal sum

$$\mathbf{u} = \sum_x \tau_x u$$

Although $\mathbf{u}$ is not well defined,

$$\xi_b = T_b \mathbf{u}$$

is well defined for every b and

$$\tau_x \xi_b = \xi_{b-x}$$

for every x and b. If we denote $T_{0,e}$ by T_e, and $\xi_{0,e}$ by ξ_e, we have

$$< g\,, g >_0 (\beta) = - < \mathcal{L}u\,, u >_0 (\beta) = \frac{1}{2} E^{\mu_\beta}\big[\sum_e c_e(\eta)[\xi_e]^2\big]$$

and by polarization

$$< g\,, h >_0 (\beta) = - < \mathcal{L}u\,, v >_0 (\beta) = \frac{1}{2} E^{\mu_\beta}\big[\sum_e c_e(\eta)\,\xi_e\,\zeta_e\big]$$

where

$$\zeta_e = T_e \mathbf{v} = T_e \sum_x \tau_x v$$

We next take $g = w_e = w_{0,e}$. From the definition of the current w_b in (3.4) it is easy to establish the following identity, which is essentially an integration by parts formula, for any μ which is either μ_β for some β or any canonical Gibbs measure $\mu_\Lambda^{\omega,y}$ on a box Λ that contains the bond $(x, x+e)$.

$$E^\mu\big[w_{x,x+e}\, u\big] = -E^\mu\big[(\eta_{x+e} - \eta_x)c_{x,x+e}(\eta)(T_{x,x+e}u)(\eta)\big]$$

If we take any set $B \subset \Lambda$, it follows that

$$|E^{\mu_\Lambda^{\omega,y}}\big[(\sum_{b\in B} w_b)\, u\big]|^2 \le E^{\mu_\Lambda^{\omega,y}}\big[\sum_{b\in B} c_b(\eta)(\nabla_b\eta)^2\big] E^{\mu_\Lambda^{\omega,y}}\big[\sum_{b\in B} c_b[(T_b\, u)(\eta)]^2\big]$$

In particular

$$< \sum_{b\in B} w_b\,; \mathcal{L}_\ell^{-1} \sum_{b\in B} w_b >_{\omega\,,y} \le C|B|$$

Therefore in the definition of $V(w_e\,, w_{e'}\,; \beta)$ on can take

$$V_\ell(w_e\,, w_{e'}\,; \omega\,, y) =< Av_{\Lambda_\ell} w_e\,, [Av_{\Lambda_\ell}\mathcal{L}_b]^{-1} Av_{\Lambda_\ell} w_e >_{\omega,y}$$

and because

$$\mathcal{L}_{\Lambda_\ell} \sum_{x\in\Lambda_\ell} < x\,, e > \eta_x = \sum_{x\in\Lambda_\ell} w_{x,x+e}$$

it is easy to calculate that

$$V(w_e\,, w_{e'}\,; \beta) = \frac{\delta_{e,e'}}{2} E^{\mu_\beta}[c_e(\eta)(\nabla_e\eta)^2], \tag{8.7}$$

and for any local u

$$V(w_e\,, \mathcal{L}u\,; \beta) = \frac{1}{2} E^{\mu_\beta}\big[c_e(\eta)(T_e \sum_x \tau_x u)\nabla_e\eta\big], \tag{8.8}$$

and in fact for any $g \in \mathcal{G}$

$$V(w_e\,, g\,; \beta) = t_e(g\,, \beta). \tag{8.9}$$

We collect these results as a lemma.

LEMMA 8.1 *If we define the class $\mathcal{G}_0 \subset \mathcal{G}$ of functions as those which are of the form $g = \mathcal{L}u + \sum_e \alpha_e w_e$ for some local u, then $V(g\,, h\,; \beta)$, given by (8.6) -(8.9), is a well defined, possibly degenerate inner product on $\mathcal{G}_0$. In addition $V(g\,, h\,; \beta)$ exists as a limit in the sense of (8.5) so long as $g, h \in \mathcal{G}$ and either g or h is in $\mathcal{G}_0$.*

Our next goal is to show that $\mathcal{G}$ is not all that much larger than $\mathcal{G}_0$. We want to show that, for any β, $\mathcal{G}$ is contained in the completion $\overline{\mathcal{G}}_0$ of $\mathcal{G}_0$ relative to the inner product $V(\cdot\,, \cdot\,; \beta)$. More precisely with $V(h\,, h\,; \beta)$ defined as the limsup for $h \in \mathcal{G}$ we want to show that

$$\inf_{g \in \mathcal{G}_0} V^*(h - g\,, h - g\,; \beta) = 0 \tag{8.10}$$

Notice that although V^*, defined by(8.2), may not be a quadratic form it is still true that for $g \in \mathcal{G}_0$ and $h \in \mathcal{G}$

$$V^*(h - g\,, h - g\,; \beta) = V^*(h\,, h\,; \beta) - 2V(h\,, g\,; \beta) + V(g\,, g\,; \beta)$$

so that (8.10) reduces to proving the variational formula

$$V^*(h\,, h\,; \beta) = \sup_{g \in \mathcal{G}_0} \left\{ 2V(h\,, g\,; \beta) - V(g\,, g\,; \beta) \right\} \tag{8.11}$$

The variational formula (8.11) also proves that for V_* defined by (8.3)

$$\begin{aligned} V_*(h\,, h\,; \beta) &= V_*(h - g\,, h - g\,; \beta) + 2V(h - g\,, g\,; \beta) + V(g\,, g\,; \beta) \\ &\geq \sup_{g \in \mathcal{G}_0} \left[2V(h\,, g\,; \beta) - V(g\,, g\,; \beta) \right] \\ &= V^*(h\,, h\,; \beta) \end{aligned}$$

establishing the equality $V^* = V_*$ and therefore $V_* = V^* = V(h\,, h\,; \beta)$ exists as a limit and defines a quadratic form for all $h \in \mathcal{G}$. Let us remark that from the definition V^* is upper semicontinuous in β, while V_* is lower semicontinuous. In particular the equality implies the continuity in β. We now state these results as a theorem. Since β will be fixed from now on, we will drop all references to it.

THEOREM 8.2 *For $h \in \mathcal{G}$ the variance $V(h\,, h)$ satisfies the variational formula*

$$\begin{aligned} \frac{1}{2} V(h\,, h) &= \sup_{\alpha_e \in R, u \in \mathcal{G}} \left[V\Big(h, \sum_{e>0} \alpha_e w_e - \mathcal{L}u \Big) - \frac{1}{2} V\Big(\sum_{e>0} \alpha_e w_e - \mathcal{L}u \Big) \right] \\ &= \sup_{\alpha_e \in R, u \in \mathcal{G}} \left\{ \sum_{e>0} \alpha_e t_e(h) + \langle h, u \rangle_0 \right. \\ &\qquad \left. - 1/4 \sum_{e>0} E^\mu \left[c_e(\eta) \Big(\alpha_e \nabla_e \eta + T_e \sum_x \tau_x u \Big)^2 \right] \right\} \end{aligned} \tag{8.12}$$

The equivalence of the two formulae in (8.12) follows from Lemma 8.1. Hence we only have to prove the first identity in (8.12). This will be the main thrust of the rest of this paper.

We now assume Theorem 8.2 and conclude the proof of Theorem 3.4. In general $\nabla_e\eta \notin \mathcal{G}$. However for any e and s, $h_{e,s} = \nabla_e\eta - E^\mu[\nabla_e\eta\,|\mathcal{F}_s] \in \mathcal{G}$ and for $s_1 < s_2$, by (8.4) and the estimate on the spectral gap

$$V(h_{e,s_1} - h_{e,s_2}\,, h_{e,s_1} - h_{e,s_2}) \le C s_2^{d+2}\, s_1^{-(d+2+\alpha)}.$$

This is enough to prove that $h_{e,s}$ is a Cauchy sequence (uniformly with respect to β) in $\mathcal{G}$ and letting $s \to \infty$, $h = \lim_{s\to\infty} h_{e,s}$ gets represented in $\overline{\mathcal{G}}$. Using (8.6) and (8.9) we can explicitly calculate

$$V(w_e\,, \nabla_{e'}\eta) = t_e(\nabla_{e'}\eta) = \delta_{e\,,e'} \sum_x < \eta_0\,; \eta_x >_\mu = \delta_{e\,,e'}\chi \tag{8.13}$$

and

$$V(\mathcal{L}u\,, \nabla_e\eta) = 0 \tag{8.14}$$

Proof of Theorem 3.4. Let

$$\mathcal{G}^{(0)} = \{\sum_{e>0} \alpha_e \nabla_e \eta_0\} \tag{8.15}$$

We claim that $\mathcal{G}^{(0)} + \mathcal{L}\mathcal{G}$ is also dense in $\overline{\mathcal{G}}$ where $\mathcal{L}\mathcal{G}$ denotes the linear space $\{\mathcal{L}u\}$ as u varies over all local functions. From (8.14), $\mathcal{G}^{(0)} \perp \mathcal{L}\mathcal{G}$. From (8.13) we see that the projection of the space $\mathcal{G}^{(w)}$ onto $\mathcal{G}^{(0)}$ has rank equal to the dimension of $\mathcal{G}^{(w)}$. Therefore the dimensions of $\mathcal{G}^{(0)}$ and $\mathcal{G}^{(w)}$ are both d. Since $\mathcal{G}^{(w)} + \mathcal{L}\mathcal{G}$ is dense in $\overline{\mathcal{G}}$, it follows that $\overline{\mathcal{G}} = \mathcal{G}^{(0)} \oplus \overline{\mathcal{L}\mathcal{G}}$.

Let ξ_e denote the orthogonal projection of w_e onto the space $\overline{\mathcal{L}\mathcal{G}}$. Hence there is a diffusion matrix D such that

$$w_e = \frac{1}{2} \sum_{e*>0} D_{e,e*} \nabla_{e*}\eta_0 + \xi_e, \tag{8.16}$$

as an identity in $\overline{\mathcal{G}}$. Clearly, the projection of $\sum_e \gamma_e w_e$ is $\sum_e \gamma_e \xi_e$. By definition, ξ_e can be approximated by elements of the form $\mathcal{L}\, g_e$ and thus $\sum_e \gamma_e \xi_e$ can be approximated by $\mathcal{L} \sum_e \gamma_e g_e$. Therefore, (3.14) holds for some diffusion matrix D. Our final task is to prove the characterization (2.22).

Using the current notation, we can rewrite the diffusion coefficient (2.22) as

$$\frac{1}{2}\gamma \cdot D \cdot \gamma = \chi^{-1} \inf_{\zeta \in \mathcal{L}\mathcal{G}} V(\gamma \cdot w - \zeta) \tag{8.17}$$

The minimizer is simply the projection of $\gamma \cdot w$ onto the space $\overline{\mathcal{L}\mathcal{G}}$, namely, $\sum_e \gamma_e \xi_e$. Taking inner product of (8.16) with $w_{e'}$, we have

$$V(w_e, w_{e'}) = \frac{1}{2} \sum_{e*} D_{e,e*} V(\nabla_{e*}\eta_0, w_{e'}) + V(\xi_e, w_{e'})$$

We can compute $V(\nabla_{e*}\eta_0, w_{e'})$ by (8.13) and thus

$$V(w_e, w_{e'}) = \frac{1}{2} D_{e,e'}\chi + V(\xi_e, w_{e'})$$

Hence for any vector γ we have

$$\frac{1}{2}\gamma \cdot D \cdot \gamma = \chi^{-1} [\, V(\gamma \cdot w) - V(\gamma \cdot \xi, \gamma \cdot w) \,] \tag{8.18}$$

Since $\gamma \cdot \xi$ is the projection of $\gamma \cdot w$ onto the space $\overline{\mathcal{LG}}$, we have

$$V(\gamma \cdot \xi,\ \gamma \cdot \xi - \gamma \cdot w) = 0$$

Therefore we can rewrite (8.18) as

$$\frac{1}{2}\gamma \cdot D \cdot \gamma = \chi^{-1} [\, V(\gamma \cdot w) - 2V(\gamma \cdot \xi\ ,\ \gamma \cdot w) + V(\gamma \cdot \xi) \,] = V(\gamma \cdot w - \gamma \cdot \xi) \tag{8.19}$$

establishing the equivalence of (8.19) and (8.17). This proves Theorem 3.4 assuming Theorem 8.2.

We now prove that the diffusion matrix is diagonal if the model is isotropic, i.e., the Hamiltonian and the jump rate A in (2.14) are isotropic.

LEMMA 8.3 *Under the previous assumption, the diffusion coefficient is diagonal*

Proof. Fix two coordinate direction e and e'. Taking inner product of (8.16) with $\nabla_{e'}\eta_0$, we have

$$V(w_e, \nabla_{e'}\eta_0) = \frac{1}{2}\sum_{e*} D_{e,e*} V(\nabla_{e*}\eta_0, \nabla_{e'}\eta_0)$$

We can compute $V(\nabla_{e*}\eta_0, w_{e'})$ by (8.13). Hence we only have to prove that $V(\nabla_{e*}\eta_0, \nabla_{e'}\eta_0)$ is diagonal. Denote by $\theta \colon \mathbb{Z} \to \mathbb{Z}$ the reflection with respect to the origin along the e' direction. We may extend θ to the space of configurations in the natural way: $(\theta\eta)(x) = \eta(\theta(x))$ and to the space of continuous functions: $(\theta f)(\eta) = f(\theta\eta)$. Clearly, $V(f,g) = V(\theta f, \theta g)$ since our model is isotropic. Notice that

$$\theta(\nabla_{e'}\eta(0)) = -\tau_{-e'}\nabla_{e'}\eta(0)\ ; \quad \theta(\nabla_e\eta(0)) = \nabla_e\eta(0), \quad \text{if } e \neq e'$$

Hence $V(\nabla_e\eta_0, \nabla_{e'}\eta_0) = -V(\nabla_e\eta_0, \nabla_{e'}\eta_0) = 0$. This proves the Lemma.

Return to the proof of Theorem 8.2. We shall follow the approach of [V1] and [Q1]. We first need the notion of "closed forms". This is some what parallel to the usual notion of closed forms in differential calculus. Our goal is to characterize the 1-form arising from

$$\omega_b = T_b g \tag{8.20}$$

for some function g. Clearly, there are compatibility conditions for $\omega = \{\omega_b\}$. One can check easily that the following conditions hold:

$$T_b\omega_q = T_q\omega_b, \ \text{if } q \cap b = \phi\ . \tag{8.21}$$

Unlike in differential calculus, (8.21) fails if $q \cap b \neq \phi$ because the operators are not local. Hence (8.21) is not enough to characterize the forms arising in this manner. We therefore use instead the definition: $\omega = \{\omega_b\}$ is "closed" provided for any sequence of nearest neighbor bonds $\{b_1, b_2, ..., b_n\}$ such that $S_{b_n} S_{b_{n-1}} \dots S_{b_2} S_{b_1} = I$ one has

$$\sum_{i=1}^{n} \omega_{b_i}(S_{b_{i-1}} S_{b_{i-2}} \dots S_{b_1}\eta) = 0, \tag{8.22}$$

where $S_b\eta = \eta^b$. Let $P_i\eta$ denotes the new configuration obtained by applying the permutation $P_i = S_{b_i} S_{b_{i-1}} \dots S_{b_1}$ to η. One can check easily that if ω is given by (8.20) then (8.22) holds. To prove this, we substitute $\omega_b = T_b g$ into the equation and we only have to check

$$\sum_{i=1}^{n} T_{b_i} g(P_{i-1}\eta) = 0, \tag{8.23}$$

By definition,

$$T_{b_i} g(P_{i-1}\eta) = g(S_{b_i} P_{i-1}\eta) - g(P_{i-1}\eta) = g(P_i\eta) - g(P_{i-1}\eta)$$

We see that (8.23) is a telescoping sum vanishes because $P_n = P_0 = I$. Conversely, if (8.22) holds, we can reconstruct the function g as long as we are in a finite cube Λ with a fixed total number of particles. This can be achieved in a standard way.

For any two configuration η and ζ with the same total number of particles, i.e., $|\eta| = |\zeta|$, we can construct a sequence $S_{b_i}, i = 1, \cdots, n$ so that $\zeta = S_{b_n} \cdots S_{b_1}\eta$. Starting from a configuration η, we can define the value of the function g at any configuration ζ with $|\eta| = |\zeta|$ by

$$g(\zeta) = \sum_{i=1}^{n} \omega_{b_i}(S_{b_{i-1}} S_{b_{i-2}} \dots S_{b_1}\eta)$$

The condition (8.22) is exactly the compatibility condition that two different ways to reach ζ will yield the same value for $g(\zeta)$. Note that configurations with different total number of particles are completely unrelated. Hence in general we can require that

$$E^{\mu}[g|\mathcal{F}_L] = 0 \tag{8.24}$$

if the cube Λ is of width $2L + 1$; here $\mathcal{F}_L$ is the σ-algebra defined in (3.8). This procedure can be performed in a finite cube and thus the function g always exists locally, i.e., if we condition on the configuration outside a cube Λ, the condition (8.22) implies the existence of a function g such that $S_b g = \omega_b$ for all bonds b inside Λ. Therefore, (8.22) implies (8.21). In addition to the existence, we have the following bound

$$E^{\mu}[\, g^2 \,] \le CL^2 \sum_{b\in\Lambda_L} E^{\mu}[\, \omega_b^2 \,] \tag{8.25}$$

For the whole lattice $\mathbb{Z}^d$, it is more subtle and, indeed, the reconstruction is not necessarily valid. We shall discuss this in details in the next section. For now, we define the space of (translationally covariant) closed forms , $\mathcal{G}_C$, by

$$\mathcal{G}_C = \big\{\omega = \{\omega_b\} : \omega \text{ is translationally covariant and is closed}\big\} \tag{8.26}$$

where ω is said to be translationally covariant if it satisfies

$$\tau_x\omega_b(\eta) = \omega_b(\tau_x\eta) = \omega_{\tau_x b}(\eta) = \omega_{b-x}(\eta) \tag{8.27}$$

The existence of translationally covariant closed 1-forms is not entirely obvious. The idea is that, if we start with a local function u and define the formal sum

$$\mathbf{u} = \sum_{x\in\mathbb{Z}^d} \tau_x u\,,$$

as we saw before, although $\mathbf{u}$ does not really exist as a translation invariant function, its formal "gradients"

$$\omega_b = T_b \mathbf{u}$$

are well defined local functions and constitute a translationally invariant closed 1-form. These are to be thought of as exact forms. On the other hand for each e the form

$$\omega^e_{x,x+e'}(\eta) = \delta_{e,e'}[\eta_{x+e'} - \eta_x]$$

is a "closed" 1-form that is not of the exact type. The following Lemma is the first step in proving Theorem 8.2 and provides a variational formula for the variance in terms of "closed forms". The proof will be completed in the next section where we will prove that $\overline{\mathcal{G}}_0 = \mathcal{G}_C$ in a suitable sense. .

LEMMA 8.4 *For $h \in \mathcal{G}$ the variance $V^*(h\,,h)$ satisfies the variational principle*

$$\frac{1}{2}V^*(h\,,h) \le \sup_{\xi \in \mathcal{G}_C} E^\mu[\sum_{e>0} c_e \Phi^h_e(\eta)\xi_e] - \frac{1}{4}E^\mu[\sum_{e>0} c_e(\xi_e)^2], \tag{8.28}$$

where, for each e, $\Phi^h_e(\cdot)$ is a local function depending on h to be defined in ♣7.13 and ♣7.14 . In addition there is also the counterpart

$$\frac{1}{2}V_*(h\,,h) \ge \sup_{\xi \in \mathcal{G}_0} E^\mu[\sum_{e>0} c_e \Phi^h_e(\eta)\xi_e] - \frac{1}{4}E^\mu[\sum_{e>0} c_e(\xi_e)^2], \tag{8.29}$$

with the same functions $\Phi^h_e(\cdot)$.

Proof. First we prove (8.29), which is the easy part. Clearly

$$\frac{1}{2}V_\ell(h\,,h\,;\omega\,,y) = \sup_v \left\{ E^{\mu_{\ell,\omega\,y}}\left[[Av_{x\in\Lambda_{\ell_1}}\tau_x h]v\right] - \frac{1}{4}E^{\mu_{\ell,\omega\,y}}\left[Av_{b\in\Lambda_\ell}\left[c_b(\eta)(T_b\,v)^2\right]\right]\right\}$$

By choosing

$$v = \frac{1}{(2\ell+1)^d}\left[\sum_{x\in\Lambda_\ell} <\alpha\,,x> \eta_x - \sum_{x\in\Lambda_{\ell_1}} \tau_x u\right]$$

for some local u, taking expectations with respect $\mu_{\beta'}$, letting $\ell \to \infty$ while $\beta' \to \beta$, we get

$$-\frac{1}{4}E^\mu\left[\sum_e c_e(\eta)\xi_e^2\right]$$

for the second term, where

$$\xi_e = \alpha_e \nabla_e \eta_0 - T_e \sum_{x\in\mathbb{Z}^d} \tau_x u.$$

Since $h \in \mathcal{G}$ we can write

$$h = \mathcal{L}_s H = \sum_{b\in\Lambda_s} \mathcal{L}_b H \tag{8.30}$$

with H measurable with respect to some $\mathcal{K}_{s+R}$ (see (3.9)), and calculate

$$\begin{aligned} E^\mu E^{\mu_{\ell,\omega\,y}}\left[[Av_{x\in\Lambda_{\ell_1}}\tau_x h]v\right] &= E^\mu\left[\left(Av_{x\in\Lambda_{\ell_1}}\tau_x \sum_{b\in\Lambda_s}\mathcal{L}_b H\right)v\right] \\ &= -\frac{1}{2}\sum_{b\in\Lambda_s} Av_{x\in\Lambda_{\ell_1}} E^\mu\left[\tau_x[c_b(\eta)(T_b H)](T_{b-x}v)\right] \end{aligned}$$

It is now easily seen that the first term reduces to

$$-\frac{1}{2}E^{\mu}\Big[\sum_e c_e(\eta)\big(\sum_{y:\tau_y e\in\Lambda_s}\tau_y T_{\tau_y e}H\big)\xi_e = E^{\mu}\Big[\sum_e c_e(\eta)\Phi_e^h(\eta)\xi_e(\eta)\Big]$$

with

$$\Phi_e^h(\eta) = -\frac{1}{2}\sum_{y:\tau_y e\in\Lambda_s}\tau_y T_{\tau_y e}H \tag{8.31}$$

thereby proving (8.32) . We now turn to the proof of (8.28).
Step1. Let

$$V^* = \limsup_{\substack{\ell\to\infty\\ \beta'\to\beta}} E^{\mu_{\beta'}}[V_\ell(h\,,h\,;\omega\,,y)]$$

and suppose that $\ell \gg 1$ and $\beta' \sim \beta$ have been chosen so that, for some small $\delta > 0$, with $\mu' = \mu_{\beta'}$,

$$E^{\mu'}[V_\ell(h\,,h\,;\omega\,,y)] \geq V^* - \delta.$$

We have the usual variational principle

$$\frac{1}{2}V_\ell(h\,,h\,;\omega\,,y) = \sup_u\left\{E^{\mu_{\ell\,\omega\,y}}[u\,Av_{|x|\leq\ell_1}h] - \frac{1}{4}E^{\mu_{\ell\,\omega\,y}}\left[Av_{|b|\leq\ell}c_b(\eta)(T_b\,u)^2\right]\right\}$$

that is valid for each ω. Integration with respect to the infinite volume Gibbs measure μ' yields

$$\frac{1}{2}E^{\mu'}[V_\ell(h\,,h\,;\omega\,,y)] = \sup_u\left\{Av_{|x|\leq\ell_1}E^{\mu'}[(\tau_x h)u] - \frac{1}{4}Av_{|b|\leq\ell}E^{\mu'}[c_e(\eta)(T_b u)^2]\right\} \tag{8.33}$$

Hence, there exists a function $u = u^{(\ell)}$ such that

$$\frac{1}{2}E^{\mu'}[V_\ell(h\,,h\,;\omega\,,y)] \leq Av_{|x|\leq\ell_1}E^{\mu'}[(\tau_x h)u] - \frac{1}{4}Av_{|b|\leq\ell}E^{\mu'}[c_e(\eta)(T_b u)^2] + \delta, \tag{8.34}$$

where $\ell_1 = \ell - \sqrt{\ell}$. Moreover we can assume without loss of generality that u is a function of the configurations on $\Lambda_{\ell+R}$ and satisfies

$$Av_{|x|\leq\ell_1}E^{\mu'}[(\tau_x h)u] - \frac{1}{4}Av_{|b|\leq\ell}E^{\mu'}[c_e(\eta)(T_b u)^2] \geq 0 \tag{8.35}$$

Since h is a local function, it is measurable w.r.t. $\mathcal{K}_s$ for some s large enough. Because $h \in \mathcal{G}$ (3.7), as we saw earlier, $h = \mathcal{L}_s H$ for some H that is $\mathcal{K}_{s+R}$ measurable. Therefore, there is a constant $C(h)$ such that, for any constant $\gamma > 0$, and $x \in \Lambda_{\ell_1}$

$$E^{\mu'}[(\tau_{-x}h)u] \leq (4\gamma)^{-1}\sum_{|b-x|\leq s}E^{\mu'}\left[c_b(\eta)(T_b u)^2\right] + \frac{\gamma}{2}C(h) \tag{8.36}$$

Let $\gamma = 2(2s+1)^d$. Summing (8.36) over x with $|x| \leq \ell_1$ and using (8.35), we have the following energy bound that holds uniformly in ℓ:

$$Av_{|b|\leq\ell}E^{\mu}\left[c_b(\eta)(T_b u)^2\right] \leq C_s C(h) \tag{8.37}$$

Step2. For $h \in \mathcal{G}$ (3.7), $\{\Phi_e^h(\cdot)\}$ were defined by (8.30) and (8.31). From

$$E^{\mu'}[uh|\mathcal{F}_s] = E^{\mu'}[u\,\mathcal{L}_s H|\mathcal{F}_s]$$

we deduce that

$$E^{\mu'}[uh|\mathcal{F}_s] = -\frac{1}{2}\sum_{b\in\Lambda_s} E^{\mu'}[c_b(\eta)(T_bH)(\eta)(T_bu)(\eta)|\mathcal{F}_s], \tag{8.38}$$

and

$$\begin{aligned} &Av_{|x|\le\ell_1}E^{\mu'}\left[\ (\tau_x h)u\ \right] \\ =&-\frac{1}{2}(2\ell_1+1)^{-d}E^{\mu'}\left[\sum_{|x|\le\ell_1}\sum_{b+x\in\Lambda_s} c_b(\eta)(T_b\tau_x H)(\eta)(T_bu)(\eta)\right] \\ =&(2\ell_1+1)^{-d}E^{\mu'}\left[\sum_{b\in\Lambda_\ell} c_b(\eta)\tilde{\Phi}_b(\eta)(T_bu)(\eta)\right] \end{aligned}$$

where for $b = \tau_y e \in \Lambda_\ell$,

$$\tilde{\Phi}_b(\eta) = \sum_{\substack{x\in\Lambda_{\ell_1}\\ b+x\in\Lambda_s}} (T_b\tau_x H)(\eta) = \tau_y\Phi_e^h(\eta) + \Omega_b(\eta)$$

where Ω_b are boundary terms that satisfy $\Omega_b = 0$ for $b \in \Lambda_{\ell-2\sqrt{\ell}}$ and $|\Omega_b| \le C(h)$ otherwise. Clearly, from Schwarz's inequality, and (8.37)

$$\frac{1}{(2\ell_1+1)^d}\sum_{\substack{b\in\Lambda_\ell\\ b\notin\Lambda_{\ell-2\sqrt{\ell}}}} E^{\mu'}\left[\ c_b(\eta)\Omega_b(\eta)(\tau_x h)T_bu\ \right] \le C\ell^{-\frac{1}{2}}$$

and we can therefore replace

$$Av_{|x|\le\ell_1}E^{\mu'}\left[\ (\tau_x h)u\ \right]$$

in (8.33) with

$$Av_{|x|\le\ell_1}E^{\mu'}\left[\ \tau_x[c_e(\eta)\Phi_e^h(\eta)]T_{\tau_x e}u\ \right]$$

so that

$$Av_{|x|\le\ell_1}E^{\mu'}\left[\ \tau_x[c_e(\eta)\Phi_e^h(\eta)]T_{\tau_x e}u]\ \right] - \frac{1}{4}Av_{|b|\le\ell}E^{\mu'}\left[\ c_b(\eta)(T_bu)^2\ \right] \ge V^* - 3\delta$$

Step 3. Let us pick $k \gg 1$ and fix it. There must be a good box of size k, i.e with $\Lambda_{x,k+R} \subset \Lambda_{\ell_1}$ such that

$$Av_{y\in\Lambda_{x,k}}E^{\mu'}\left[\ \tau_y[c_e(\eta)\Phi_e^h(\eta)]T_{\tau_x e}u]\ \right] - \frac{1}{4}Av_{|b|\in\Lambda_{x,k}}E^{\mu'}\left[\ c_b(\eta)(T_bu)^2\ \right] \ge V^* - 4\delta$$

If we now replace $u = u^{(\ell)}$ by $\bar{u}^{(\ell)} = E^{\mu'}\left[\ u^{(\ell)}|\mathcal{K}_{x,s+R}\ \right]$ we still have

$$Av_{y\in\Lambda_{x,k}}E^{\mu'}\left[\ \tau_y[c_e(\eta)\Phi_e^h(\eta)]T_{\tau_x e}\bar{u}^{(\ell)}]\ \right] - \frac{1}{4}Av_{|b|\in\Lambda_{x,k}}E^{\mu'}\left[\ c_b(\eta)(T_b\bar{u}^{(\ell)})^2\ \right] \ge V^* - 4\delta$$

by taking conditional expectations first. It is easy to pass to the limit as $\ell \to \infty$ and $\mu' \to \mu$, taking a subsequence if needed. We produce in the limit a function u_k such that it is $\mathcal{K}_k$ measurable and satisfies

$$Av_{y\in\Lambda_k} E^{\mu}\left[\sum_e \tau_y[c_e(\eta)\Phi_e^h(\eta)]\, T_{\tau_x e} u_k\right] - \frac{1}{4} Av_{|b|\in\Lambda_k} E^{\mu}[\, c_b(\eta)(T_b u_k)^2\,] \geq V^* - 4\delta$$

or

$$E^{\mu}\left[\sum_e c_e(\eta)\Phi_e^h(\eta)\, T_e[Av_{y\in\Lambda_k}\tau_y u_k]\right] - \frac{1}{4} Av_{|b|\in\Lambda_k} E^{\mu}[\, c_b(\eta)(T_b u_k)^2\,] \geq V^* - 4\delta$$

If we define $\xi_e^k = T_e Av_{y\in\Lambda_k}[\tau_y\, u_k]$ then

$$\begin{aligned} &E^{\mu}\big[\sum_e c_e(\eta)(\xi_e^k)^2\big] \\ =&E^{\mu}\big[\sum_e c_e(\eta)\,[Av_{y\in\Lambda_k} T_{\tau_y e} u_k]^2\big] \leq E^{\mu}\big[\sum_e c_e(\eta)\,[Av_{y\in\Lambda_k}(T_{\tau_y e} u_k)^2]\big] \\ =&Av_{|b|\in\Lambda_k} E^{\mu}[\, c_b(\eta)(T_b u_k)^2\,] \end{aligned}$$

We conclude that

$$E^{\mu}\big[\sum_e c_e(\eta)\Phi_e^h(\eta)\xi_e^k\big] - \frac{1}{4}E^{\mu}\big[\sum_e c_e(\eta)(\xi_e^k)^2\big] \geq V^* - 4\delta$$

While ξ_e^k is not a covariant gradient it is routine to show that any limit point of ξ_e^k as $k \to \infty$ is one. This concludes the proof.

It remains to establish that $\mathcal{G}_0$ is dense in $\mathcal{G}$ in a suitable sense, and this will be taken up in the next section.

9. Structure of closed forms. Our goal is to prove that any translationally covariant closed form $\xi = \{\xi_e\}$ can be approximated in $L^2(\mu)$ by forms of the type $\{\alpha_e \nabla_e \eta + T_e \sum_x \tau_x g\}$ with suitable scalars $\{\alpha_e : e > 0\}$.

More precisely, let

$$\mathcal{G}_E = \{\xi : \xi_b = T_b \sum_{x\in\mathbb{Z}^d} \tau_x g\,,\ \text{g is a local function}\} \tag{9.1}$$

be the space of translationally covariant exact forms and $\mathcal{G}^{(1)}$ be the d-dimensional space of translationally covariant closed forms given by

$$\mathcal{G}^{(1)} = \{\xi^{\alpha} : \xi_b^{\alpha} = \sum_{e>0} \alpha_e \delta_{b,e} \nabla_b \eta \quad \text{for some} \quad \alpha_e \in \mathbb{R}\} \tag{9.2}$$

where $\delta_{b,e} = 1$ if the bond is in the same direction as e, and 0 otherwise. Formally one can think of

$$\xi_b^{\alpha} = -T_b \sum_{x\in\mathbb{Z}^d} < \alpha, x > \eta_x$$

Clearly, these two spaces are subspaces of $\mathcal{G}_C$. Define the L^2 norm of a closed form by

$$\| \xi \|_2^2 = E^\mu \left[\sum_{e>0} \xi_e^2 \right]$$

To simplify notation we use ξ_e to denote ξ_b with $b = (0, e)$. Then our goal is prove the main Theorem of this section.

THEOREM 9.1 *The following decomposition holds in $L_2(\mu)$:*

$$\mathcal{G}_C = \overline{\mathcal{G}_E + \mathcal{G}^{(1)}} \tag{9.3}$$

Together with Lemma 8.4 this proves Theorem 8.2. Theorem 9.1 has a parallel version if the conservative (Kawasaki) dynamics is replaced by a nonconservative (Glauber) one. For any configuration η and $x \in \mathbb{Z}^d$ we define

$$\eta_y^x = \begin{cases} \eta_y & \text{if} \quad x \neq y \\ 1 - \eta_x & \text{if} \quad x = y \end{cases}$$

i.e the spin is flipped at the site x. The Glauber gradient σ_x at site x of a function u is defined as

$$(\sigma_x u)(\eta) = u(\eta^x) - u(\eta)$$

For any local function g, although the formal sum

$$u = \sum_{z \in \mathbb{Z}^d} \tau_z g$$

does not exist, the Glauber gradients

$$\omega_x = \sigma_x u$$

exist and are in fact local functions. Moreover we have the translational covariance

$$\omega_e = \tau_x \omega_x \tag{9.4}$$

along with the obvious identities

$$\sigma_x \omega_y = \sigma_y \omega_x \qquad \text{for all} \quad x \neq y \tag{9.5}$$

and

$$\sigma_x \omega_x = -2\omega_x \qquad \text{for all} \quad x \in \mathbb{Z}^d \tag{9.6}$$

Let $\mathcal{G}_C^G$ be space of covariant closed forms (in the Glauber sense), i.e. $\{\omega_x\}$ that satisfy (9.4) , (9.5) and (9.6). Equip $\mathcal{G}_C^G$ with the L^2 norm defined by

$$\| \omega \|_2^2 = E^\mu [\, \omega_0^2 \,]$$

Let $\mathcal{G}_E^G$ be the space of translationally covariant exact forms, namely

$$\mathcal{G}_E^G = \{ \omega : \omega_x = \sigma_x \sum_{z \in \mathbb{Z}^d} \tau_y g, \ g \text{ is a local function } \}. \tag{9.7}$$

We have the following approximation Theorem.

THEOREM 9.2 *In $L^2(\mu)$:*

$$\mathcal{G}_C^G = \overline{\mathcal{G}_E^G} \tag{9.8}$$

Our interest in the Glauber dynamics arises from the observation that the spin flip and the spin exchange are related in the following manner. Given a pair of functions u and v let Q denote the function

$$Q_{x,x+e}(u,v) = [\eta_x(1-\eta_{x+e}) + \eta_{x+e}(1-\eta_x)][\sigma_{x+e}u + u + v] \tag{9.9}$$

Then for any function F,

$$T_{x,x+e}F = Q_{x,x+e}(\sigma_x F, \sigma_{x+e}F). \tag{9.10}$$

We now prove Theorems 9.1 and 9.2. The first few steps of the proofs are very similar and the steps leading to Theorem 9.1 are harder. We shall provide the details for this case and just sketch the proofs for the easier version.

Proof. In general, given functions ξ_b defined for bonds in a box Λ_ℓ satisfying the compatibility conditions for being a closed (Kawasaki) form, we can "integrate" it, and find a function g such that

$$\xi_b = T_b g \quad \text{for all} \quad b \in \Lambda_\ell$$

The function g is not unique, but can be determined uniquely by imposing a mean 0 condition

$$E^{\mu_{\ell,\omega,y}}[\, g\,] = 0$$

that makes the solution depend on the external boundary condition ω and the particle density y in Λ_ℓ. For such a choice we have the estimate

$$E^{\mu_{\ell,\omega,y}}[\,|g|^2\,] \leq C\ell^2 E^{\mu_{\ell,\omega,y}}[\sum_{b\in\Lambda_\ell} \xi_b^2] \tag{9.11}$$

as a consequence of the estimate on the spectral gap. The constant C is of course independent of ℓ, ω and y. We can perform this step even if g depends on the external boundary condition ω, so long as the compatibility is satisfied for each such ω. There is a similar result on integrating Glauber forms $\{\xi_x : x \in \Lambda_\ell\}$ with the normalization $E^{\mu_{\ell,\omega,\lambda}}[\, g\,] = 0$ along with the estimate

$$E^{\mu_{\ell,\omega,\lambda}}[\,|g|^2\,] \leq C E^{\mu_{\ell,\omega,\lambda}}[\sum_{x\in\Lambda_\ell} \xi_x^2] \tag{9.12}$$

again based on the spectral estimate, but now for the Glauber dynamics.

Let us recall that R is the range of interaction in the Gibbs measure. For each box Λ_M let $\partial\Lambda_M$ denote the "boundary" of Λ_M, consisting of sites in Λ_M that are within a distance R from some site in Λ_M^c (the complement of Λ_M) as well sites in Λ_M^c that are within a distance R of some site in Λ_M. Let $\Lambda_M^0 = \Lambda_M \backslash \partial\Lambda_M$ denote the "interior" of Λ_M. Given a (Kawasaki) closed form $\{\xi_b\}$ we can try to truncate it over a finite volume Λ_M by defining

$$\xi_b^M = E^\mu[\,\xi_b|\mathcal{K}_M\,] \quad \text{for each} \quad b \in \Lambda_M \tag{9.13}$$

While the conditioning interferes with the gradient near the boundary, if we restrict ξ_b^M to $b \in \Lambda_M^0$, it does in fact define a (Kawasaki) closed form and in particular if $\xi_b = T_b g$ for some g, then $\xi_b^M = T_b g^M$ where $g^M = E^\mu [\, g|\mathcal{K}_M \,]$ so long as $b \in \Lambda_M^0$. The identical statement is valid for Glauber forms as well.

Suppose we are given a translationally covariant (Kawasaki) form $\{\xi_b\}$. For some $L \gg 1$ we define the truncations ξ_b^{3L} according to (9.13) for a box of size $3L$. We can use these to construct a function g_{3L} measurable with respect to $\mathcal{K}_{3L}$ such that $T_b g_{3L} = \xi_b^{3L}$ for $b \in \Lambda_{3L}^0$. We then define

$$h_L = E^\mu [\, g_{3L}|\mathcal{K}_L \,]$$

and finally an approximation that is translationally covariant and exact by

$$\tilde{\xi}_b^L = \frac{1}{(2L+1)^d} T_b \sum_{z \in \mathbb{Z}^d} \tau_z h_L$$

In the Glauber context the approximation takes the form

$$\bar{\omega}_x^L = \frac{1}{(2L+1)^d} \sigma_x \sum_{z \in \mathbb{Z}^d} \tau_z h_L$$

with a different but similarly constructed h_L.

LEMMA 9.3 *The Kawasaki form $\tilde{\xi}_e^L$ can be decomposed as the sum $\tilde{\xi}_e^L = \tilde{\Omega}_{L,e}^{(1)} + \tilde{\Omega}_{L,e}^{(2)}$ coming respectively from the interior terms and the boundary terms. For each $e > 0$, they satisfy*

$$\lim_{L\to\infty} E^\mu \Big[\, |\xi_e - \tilde{\Omega}_{L,e}^{(1)}|^2 \,\Big] = 0 \tag{9.14}$$

and

$$\sup_L E^\mu \Big[\, |\tilde{\Omega}_{L,e}^{(2)}|^2 \,\Big] < \infty \tag{9.15}$$

The Glauber form $\bar{\omega}_0^L$ admits a similar decomposition as $\bar{\omega}_0^L = \overline{\Omega}_{L,0}^{(1)} + \overline{\Omega}_{L,0}^{(2)}$ coming from the interior terms and the boundary terms and they satisfy

$$\lim_{L\to\infty} E^\mu \Big[\, |\omega_0 - \overline{\Omega}_{L,0}^{(1)}|^2 \,\Big] = 0 \tag{9.16}$$

and the stronger estimate

$$\sup_L L E^\mu \Big[\, |\overline{\Omega}_{L,0}^{(2)}|^2 \,\Big] < \infty \tag{9.17}$$

REMARK. In the Glauber case, from (9.16) and (9.17) it follows that $\bar{\omega}_0^L \to \omega_0$ as $L \to \infty$, and this proves Theorem 9.2 . In the Kawasaki case we will take a weak limit of $\tilde{\Omega}_{L,0}^{(2)}$ which will again be a covariant Kawasaki closed form and analyze it further and show that it is made up of density gradients. This will complete the proof of Theorem 9.1.

Proof. Denote by $\Omega^{(1)}$ the "interior part" and $\Omega^{(2)}$ the "boundary part", namely,

$$\tilde{\Omega}_{L,e}^{(1)} = \frac{1}{(2L+1)^d} T_e \sum_{y:y,\, y+e \in \Lambda_L^0} \tau_y h_L$$

$$\tilde{\Omega}^{(2)}_{L,e} = \frac{1}{(2L+1)^d} T_e \sum_{y:\{y,\,y+e\}\cap\partial\Lambda_L\neq\phi} \tau_y h_L$$

It is important to note at this point that, if $y,\ y+e \in \Lambda_L^0$ then

$$T_e \tau_y h_L = \tau_y T_{y,\,y+e}\, h_L = \tau_y E^{\mu}[\ \xi_{y,\,y+e}|\mathcal{K}_L\ ]$$

Since $\xi_e \in L_2(\mu)$, for large k, $E^{\mu}[\ |\xi_e - E^{\mu}[\ \xi_e|\mathcal{K}_k\ ]|^2\ ]$ is small. By translational invariance $E^{\mu}\left[\ \{\xi_{\tau_y e} - E^{\mu}\left[\ \xi_{\tau_y e}|\mathcal{K}_{y,k}\ \right]\}^2\ \right]$ is independent of y and is therefore uniformly small for large k. For most $y \in \Lambda_L$, $\Lambda_{y,k} \subset \Lambda_L^0$ and consequently for most bonds b (with density close to 1 for large M), $E^{\mu}\left[\ (\ \xi_b - E^{\mu}[\ \xi_b|\mathcal{K}_L\ ]\ )^2\ \right]$ is small. This proves (9.14). The proof of (9.16) is identical. The following lemma provides a key estimate.

LEMMA 9.4 *Let h be a function in the cube Λ_{3L}. Suppose Λ_L is the centered subcube of width $2L+1$, $\overline{h} = E^{\mu}[h|\mathcal{K}_L]$ and z is a site in Λ_L. Denote the configuration by $\eta = (\xi, \eta_z, \zeta)$ where ζ denote the configuration on $\Lambda_L \setminus \{z\}$ and ξ denote the configuration on $\Lambda_{3L} \setminus \Lambda_L$. Denote by $Av_y = Av_{y\in\Lambda_{3L-R}\setminus\Lambda_{L+R}}$. Let μ be a Gibbs measure on Λ_{3L} or an infinite volume Gibbs state such that for any local function F the following mixing property holds:*

$$E^{\mu}[\ Av_y\, F_y; Av_y\, F_y\ ] \le C_F L^{-d} \tag{9.18}$$

Then

$$E^{\mu}\left[\ (\sigma_z \overline{h}(\eta_z, \zeta))^2\ \right] \le CL^{-d} E^{\mu}[\ h^2(\eta)\ ] + C Av_y E^{\mu}[\ (T_{z,y} h)^2(\eta)\ ] \tag{9.19}$$

where the constant C may depend on the density of the Gibbs measure.

We will defer the proof of the lemma to the end of the section and proceed with our main goal. We first consider the Glauber case.

By definition, $\overline{\Omega}^{(2)}_{L,0}$ has contribution from the boundary consisting of

$$\overline{\Omega}^{(2)}_{L,0} = \frac{1}{(2L+1)^d} \sigma_0 \sum_{y\in\partial\Lambda_L} \tau_y h_L = \frac{1}{(2L+1)^d} \sum_{y\in\partial\Lambda_L} \tau_y \sigma_y h_L$$

and (9.17) will follow provided we show that

$$\sup_L \sup_{y\in\Lambda_L} E^{\mu}[|\sigma_y h_L|^2] \le C < \infty \tag{9.20}$$

Clearly, (9.20) follows readily from (9.19) with the choice of $h = g_{3L}$. We only have to note that in the Glauber case we have from the spectral gap bounds (9.12)

$$E^{\mu}[\ g_{3L}^2(\eta)\ ] \le CL^d \tag{9.21}$$

and

$$E^{\mu}[\ (T_{z,\,y}\, g_{3L})^2(\eta)\ ] \le C E^{\mu}[\ (\sigma_z\, g_{3L})^2(\eta) + (\sigma_y\, g_{3L})^2(\eta)\ ] \le C \tag{9.22}$$

so long as z and y remain in Λ_{3L}^0. This concludes the proof of Theorem 9.2

We now turn to the Kawasaki case. We observe that instead of (9.21) and (9.22) we obtain from (9.11)

$$E^{\mu}[\ g_{3L}^2(\eta)\ ] \le CL^{d+2} \tag{9.23}$$

and

$$E^{\mu}\left[\,(T_{z,\,y}\,g_{3L})^2(\eta)\,\right] \le CL^2 \tag{9.24}$$

so long as z and y remain in Λ_{3L}^0. This provides the estimate

$$\sup_{y\in\Lambda_L} E^{\mu}\left[\,(\sigma_y\,h_L)^2(\eta)\,\right] \le CL^2 \tag{9.25}$$

with a constant C independent of L. Because the boundary terms are only $O(L^{d-1})$ in number, from the definition of $\tilde{\Omega}^{(2)}_{L,\varepsilon}$ the estimate (9.25) is sufficient to establish (9.15). This concludes the proof of Lemma 9.3.

Return to the proof of Theorem 9.1. The proof is broken up into several steps.

Step 1. The boundary in d dimensions comes in $2d$ faces with some overlap at the corners. The overlap involves only $O(L^{d-2})$ sites and hence negligible in view of the bound (9.25). We can therefore analyze the contribution from each face by itself. Since they are all similar we will just consider one of them. Depending on the range of the interaction, for some integer $q \ge 1$, the face consists of points $z = (z_1, \cdots, z_d) \in \mathbb{Z}^d$ satisfying $-L \le z_1 \le -L+q$ and $-L \le z_i \le L$ for $2 \le i \le d$. The face consists naturally of layers

$$B_L^j = \{z = (z_1, \cdots, z_d) : z_1 = -L+j\,; -L \le z_i \le L \ \text{ for } \ 2 \le i \le d\}$$

indexed by $0 \le j \le q$. We need to analyze the terms

$$\xi_e^{B,L} = \frac{1}{(2L+1)^d} \sum_{z\in B_L} \tau_z T_{z,\,z+e}\, h_L = \frac{1}{(2L+1)^d}\, T_e \sum_{z\in B_L} \tau_z\, h_L$$

where B_L is the slightly thickened boundary

$$B_L = \cup_{-1\le j\le q} B_L^j$$

We can rewrite

$$\xi_e^{B,L} = T_e \sum_{-1\le j\le q} \tau_j^1 \bar{h}_L$$

where

$$\bar{h}_L = \frac{1}{(2L+1)^d} \sum_{z\in B_L^0} \tau_z\, h_L$$

and $\tau_j^1 = \tau_{je_1}$ is the translation by j steps in the x_1 direction. For each site in $x = (x_1, \cdots, x_d) \in \mathbb{Z}^d$ let us consider

$$\omega_x^{B,L} = \sigma_x\, \bar{h}_L$$

Since $\omega_x^{B,L}$ is bounded in $L_2(\mu)$, we can take a weak limit as $L \to \infty$, with one subsequence working simultaneously for all x, and call the limt ω_x^B.

Step 2. By construction, as a weak limit of Glauber gradients, $\{\omega_x^B\}$ is again a Glauber gradient although it may not be covariant. Because averaging is done in all directions except in the first coordinate direction, it is covariant, in all directions except possibly in that direction. Moreover $\omega_x^B = 0$ if $x_1 < 0$, because none of the functions depend on the configuration at any site x with $x_1 < 0$. Let b be a bond that does not intersect the infinite strip $B_\infty = \{y = (y_1, \cdots, y_d) : 0 \le y_1 \le q\}$, as well as the site x. Then $T_b\bar{h}_L \to 0$, and σ_x commutes with T_b. Therefore $T_b\omega_x^B = 0$. This implies that ω_x^B can only depend on the configuration in $B_\infty \cup \{x\}$. In particular for $y \notin B_\infty \cup \{x\}$, $\sigma_y\omega_x^B = 0$. As a Glauber gradient $\sigma_y\omega_x^B = \sigma_x\omega_y^B$ for all x, y and hence we also have $\sigma_x\omega_y^B = 0$ for such pairs x and y. It is easy to arrive at the following conclusion: for each $x \in B_\infty$, ω_x^B depends only on the configuration in B_∞. On the other hand for x with $x_1 > q$, ω_x^B can only depend on η_x and as a gradient must equal $c(x)(2\eta_x - 1)$ for some constant $c(x)$. Covariance in all directions except the first coordinate means that $c(x)$ is a function of the first coordinate only i.e $c(x) = c(x_1)$

Step 3. If we denote by ξ_e^B the weak limit of $\xi_e^{B,L}$, then

$$\begin{aligned}
\xi_e^B &= \lim_{L\to\infty}[\eta_0(1-\eta_e)+\eta_e(1-\eta_0)]\Big[(\sigma_0\sigma_e+\sigma_0+\sigma_e)\big(\sum_{-1\le j\le q}\tau_j^1\bar{h}_L\big)\Big] \\
&= \lim_{L\to\infty}[\eta_0(1-\eta_e)+\eta_e(1-\eta_0)]\Big[\sigma_0\big(\sum_{-1\le j\le q}\tau_j^1\sigma_{e+je_1}\bar{h}_L\big)+\big(\sum_{-1\le j\le q}\tau_j^1\sigma_{je_1}\bar{h}_L\big) \\
&\qquad +\big(\sum_{-1\le j\le q}\tau_j^1\sigma_{e+je_1}\bar{h}_L\big)\Big] \\
&= [\eta_0(1-\eta_e)+\eta_e(1-\eta_0)]\Big[\sigma_0\big(\sum_{-1\le j\le q}\tau_j^1\omega_{e+je_1}^B\big)+\big(\sum_{-1\le j\le q}\tau_j^1\omega_{e+je_1}^B\big) \\
&\qquad +\big(\sum_{-1\le j\le q}\tau_j^1\omega_{je_1}^B\big)\Big]
\end{aligned}$$

If ω_x is a Glauber gradient then

$$\bar{\omega}_x = \sum_{z\in\mathbb{Z}^d}\tau_z\omega_{x+z}$$

is a covariant Glauber gradient. If ω_x is covariant in all directions except the first, we can redefine

$$\bar{\omega}_x = \sum_j \tau_j^1\omega_{x+je_1}$$

and get a covariant Glauber gradient. Let us define a new Glauber gradient

$$\tilde{\omega}_x^B = \omega_x^B - c(x_1)(2\eta_x - 1)$$

where $c(x_1)$ is defined in Step 2 for $x_1 > q$ and is taken to be 0 otherwise. Then

$$\overline{\omega}_x^B = \sum_j \tau_j^1\tilde{\omega}_{x+je_1}^B$$

is a covariant Glauber gradient and so can be approximated by exact Glauber gradients according to Theorem 9.2. Therefore the Kawasaki gradient

$$\overline{\xi}_e^B = [\eta_0(1-\eta_e)+\eta_e(1-\eta_0)]\big[\sigma_e\overline{\omega}_0^B+\overline{\omega}_0^B+\overline{\omega}_e^B\big]$$

is covariant and can also be approximated by exact forms. We now calculate the difference

$$\xi_e^B - \overline{\xi}_e^B = [\eta_0(1-\eta_e)+\eta_e(1-\eta_0)]\Big[\sigma_0(u^B-\overline{\omega}_0^B)+(u^B-\overline{\omega}_0^B)+(v^B-\overline{\omega}_e^B)\Big]$$

where

$$
\begin{aligned}
u^B &= \sum_{-1\le j\le q} \tau_j^1 \omega^B_{je_1} \\
\overline{\omega}^B_0 &= \sum_j \tau_j^1 [\omega^B_{je_1} - c_j(2\eta_{je_1} - 1)] \\
v^B &= \sum_{-1\le j\le q} \tau_j^1 \omega^B_{e+je_1} \\
\overline{\omega}^B_e &= \sum_j \tau_j^1 [\omega^B_{e+je_1} - c_{j+\delta_{e,e_1}}(2\eta_{e+je_1} - 1)]
\end{aligned}
$$

A straight forward calculation reveals that $u^B = \overline{\omega}^B_0$ and if $e \neq e_1$ we also have $v^B = \overline{\omega}^B_e$. However if $e = e_1$, $v^B - \overline{\omega}^B_{e_1} = c_{j+1}(2\eta_{e_1} - 1)$ and then $\xi^B_{e_1} - \overline{\xi}^B_{e_1}$ is seen to equal $c\nabla_{e_1}\eta_0$, a multiple of the density gradient. This completes the proof of Theorem 9.2.

Proof of Lemma 9.4. By definition,

$$
\overline{h}(\eta_z, \zeta) = E^\mu [\, h \,|\, \eta_z, \zeta \,] = Z(\eta_z, \zeta)^{-1} \sum_\xi h(\eta) \exp[-H(\eta)]
$$

where $Z(\eta_z, \zeta) = \sum_\xi \exp[-H(\xi, \eta_z, \zeta)]$ is the partition function. Let

$$
\bar{\xi} = Av_{y \in \Lambda_{3L-R} \setminus \Lambda_{L+R}} \xi_y
$$

and denote $E^\mu \left[\, \bar{\xi} \,|\, \eta_z, \zeta \,\right]$ by $\rho(\eta_z, \zeta)$. From the definition of truncated correlation function, we have $\overline{h}(\eta_z, \zeta) = B_2(\eta_z, \zeta) - B_1(\eta_z, \zeta)$ where

$$
\begin{aligned}
B_1(\eta_z, \zeta) &= \rho(\eta_z, \zeta)^{-1} E^\mu \left[\, h\, ;\, \bar{\xi} \,|\, \eta_z, \zeta \,\right] \\
B_2(\eta_z, \zeta) &= \rho(\eta_z, \zeta)^{-1} E^\mu \left[\, h\, \bar{\xi} \,|\, \eta_z, \zeta \,\right]
\end{aligned}
$$

Explicitly,

$$
B_2(0, \zeta) = \rho(0, \zeta)^{-1} Z(0, \zeta) \sum_\xi \bar{\xi}\, h(\xi, 0, \zeta) \exp[-H(\xi, 0, \zeta)]
$$

Replacing $h(\xi, 0, \zeta)$ by $Av_y\, \xi_y\, h(\sigma_y \xi, 1, \zeta)$, we have $B_2(0, \zeta) = B_3 + B_4$ where

$$
\begin{aligned}
B_3(\zeta) &= \rho(0, \zeta)^{-1} Z(0, \zeta)^{-1} \sum_\xi Av_y\, \xi_y\, [\, h(\xi, 0, \zeta) - h(\sigma_y \xi, 1, \zeta)\,] \exp[-H(\xi, 0, \zeta)] \\
B_4(\zeta) &= \rho(0, \zeta)^{-1} Z(0, \zeta)^{-1} \sum_\xi Av_y\, \xi_y\, h(\sigma_y \xi, 1, \zeta) \exp[-H(\xi, 0, \zeta)]
\end{aligned}
\tag{9.26}
$$

By definition,

$$
(\, \xi_y\, [\, h(\xi, 0, \zeta) - h(\sigma_y \xi, 1, \zeta)\,]\,)^2 \le (T_{z,y} h)(\xi, 0, \zeta)^2
$$

Hence we can bound B_3 by

$$
E^\mu \left[\, (\, B_3(\zeta)\,)^2 \,\right] \le C Av_y E^\mu [\, (T_{z,y} h)(\eta)^2 \,] \tag{9.27}
$$

By changing the variable $\sigma_y\xi \to \xi$, B_4 equals to

$$\begin{aligned} B_4(\zeta) &= \rho(0,\zeta)^{-1}Z(1,\zeta)^{-1}\sum_{\xi}[\,(Av_y\,F_y)\,h(\xi,1,\zeta)\,]\exp[-H(\xi,1,\zeta)] \\ &= \rho(0,\zeta)^{-1}E^{\mu}[\,(Av_y\;F_y)\,h\,|\,1,\zeta\,] \end{aligned} \tag{9.28}$$

where F_y is the local function

$$F_y = \exp[\,H(\xi,1,\zeta) - H(\sigma_y\xi,0,\zeta)\,].$$

and Because the sites y and z are sufficiently apart G does not depend on y. By definition of the corrected correlation function,

$$E^{\mu}[\,(Av_y\,F_y)\,h\,|\,1,\zeta\,] = E^{\mu}[\,(Av_y\,F_y)\,;\,h\,|\,1,\zeta\,] + E^{\mu}[\,(Av_y\,F_y)\,|\,1,\zeta\,]\,E^{\mu}[\,h\,|\,1,\zeta\,] \tag{9.29}$$

To compute $E^{\mu}[\,(Av_y\,F_y)\,|\,1,\zeta\,]$, consider h independent of ξ. From (9.26), $B_4(\zeta) = h(1,\zeta)$. From (9.28) and (9.29), $B_4(\zeta) = \rho(0,\zeta)^{-1}E^{\mu}[\,(Av_y\;F_y)\,|\,1,\zeta\,]\,h(1,\zeta)$. Hence $\rho(0,\zeta)^{-1}E^{\mu}[\,(Av_y\;F_y)\,|\,1,\zeta\,] = 1$ and thus

$$B_4(\zeta) = \rho(0,\zeta)^{-1}E^{\mu}[\,(Av_y\,F_y)\,;\,h\,|\,1,\zeta\,] + \bar{h}(1,\zeta)$$

We have thus proved that

$$\bar{h}(0,\zeta) - \bar{h}(1,\zeta) = -B_1(0,\zeta) + B_3(\zeta) + \rho(0,\zeta)^{-1}E^{\mu}[\,(Av_y\,F_y)\,;\,h\,|\,1,\zeta\,]$$

From the mixing assumption (9.18) the last term on the right hand side satisfies the bound

$$\begin{aligned} &E^{\mu}\left\{E^{\mu}[\,(Av_y\,F_y)\,;\,h\,|\,1,\zeta\,]^2\right\} \\ &\le CE^{\mu}\left\{E^{\mu}[(Av_yF_y);(Av_yF_y)|1,\zeta]^2\right\}E^{\mu}\{E^{\mu}[h^2|(1,\,\zeta)]\} \\ &\le CL^{-d}E^{\mu}[h^2] \end{aligned}$$

Also, B_1 satisfies a similar bound. Together with (9.27), this concludes Lemma 9.4.

10. Mixing properties of grand canonical and canonical gibbs measures.

Let us recall our basic mixing condition.

ASSUMPTION A. *Let $\mu_{\Lambda,\omega,\lambda}$ denote a Gibbs measure on Λ with boundary condition ω and chemical potential λ. Denote the corresponding density by $\rho = \rho(L,\,\lambda,\,\omega)$. Then there are constants γ_1, γ_2 and $\gamma_3 \ge R+1$ such that for any two functions f and g with supports S_f and S_g we have*

$$|E_{\Lambda,\omega,\lambda}[f;g]| \le \gamma_1\rho(1-\rho)\,\exp[-\gamma_2\;dist(S_f,S_g)]\|f\|_\infty\|g\|_\infty \tag{10.1}$$

provided that the diameters of S_f and S_g are bounded by γ_3. Note that the constants are independent of the size L of the cube Λ, the value of λ of the chemical potential and the boundary condition ω.

REMARK. The term $\rho(1-\rho)$ is effective only for densities ρ that are close to 0 or 1. The effect of L and ω is minimal. In fact for any local interaction, by low density cluster expansion, (10.1) is seen to be always satisfied provided $|\lambda|$ is sufficiently large. The factor $\rho(1-\rho)$ in (10.1) is therefore available for free.

From this assumption, one immediately obtains the following lemma.

LEMMA 10.1 *There is a positive function C on $\mathbb{Z}$ such that the following holds. Suppose ω is a boundary conditions on $\partial\Lambda$ and $z \in \partial\Lambda$. For any g in Λ,*

$$|E_{\Lambda,\omega,\lambda}[g] - E_{\Lambda,\sigma_z\omega,\lambda}[g]| \leq C(|S_g|)\exp[-\gamma_2\ dist(z, S_g)]\|g\|_\infty\ . \tag{10.2}$$

where $\sigma_z\omega$ denote the boundary condition obtained by changing the configuration at the site z and $|S_g|$ denotes the number of the sites in S_g.

Furthermore, there is a universal function $C(|S_f|, |S_g|)$ such that (10.1) holds with no restriction on the diameter of S_f or S_g provided that the constant γ_1 is replaced by $C(|S_f|, |S_g|)$.

We now sketch a proof of this lemma. By definition, $|E_{\Lambda,\omega,\lambda}[g] - E_{\Lambda,\sigma_z\omega,\lambda}[g]|$ can be estimated by correlation between g and a local function around the site z. Hence (10.2) follows from (10.1) provided that the diameter of the support g is bounded by γ_3. Next, we prove (10.1) if the diameter of S_g is bounded by γ_3. Let $\Omega \subset \Lambda \setminus S_f$ be a cube of size smaller than 2 $\text{dist}(S_f, S_g)$ and containing S_g such that $\text{dist}(\partial\Omega \setminus \partial\Lambda, S_f) \geq \text{dist}(S_f, S_g) - 2$. Then

$$E_{\Lambda,\omega,\lambda}[f; g] = E_{\Lambda,\omega,\lambda}[\ f\ ;\ E_{\Lambda,\omega,\lambda}[\ g|\eta_x, x \in \Lambda \setminus \Omega\ ]\ ]$$

Since the diameter of S_g is bounded by γ_3, we can apply the estimate (10.2). Hence $E_{\Lambda,\omega,\lambda}[\ g|\eta_x, x \in \Lambda \setminus \Omega\ ]$ is independent of the boundary condition $\eta_x, x \in \Lambda \setminus \Omega$, up to exponentially small error. This proves (10.1). Repeating the argument in the beginning of the sketch, we have (10.2) holds without restriction on the diameter of the support S_g. It follows that (10.1) holds without restriction on S_f or S_g by repeating the previous argument.

We now recall some of the notation that will be used in this section. We denote by Λ_L a cube of width $2L+1$. The (grand canonical) Gibbs measure with the chemical potential λ and boundary condition ω is denoted by $\mu_{L,\omega,\lambda}$. If, instead of fixing the chemical potential λ, we fix the number of particles N in Λ_L or equivalently the density $y = \bar{\eta} = NL^{-d}$ in Λ_L, the resulting canonical measure is denoted by $\mu_{L,\omega,y}$. We denote by $M = (2L+1)^d = |\Lambda_L|$. Using the elementary Fourier analysis the canonical measure $\mu_{L,\omega,y}$ can be represented as

$$d\mu_{L,\omega,y} = \frac{1}{p_y}\left\{\frac{1}{2\pi}\int_{-\pi}^{\pi} \exp[\ i\theta M(\bar{\eta} - y)\ ]\, d\mu_{L,\omega,\lambda}\, d\theta\right\} \tag{10.3}$$

where

$$p_y = \mu_{L,\omega,\lambda}[\ \bar{\eta} = y\ ]. \tag{10.4}$$

By definition, $M\bar{\eta}$ is an integer and thus My has to be an integer for $\mu_{L,\omega,y}$ to be meaningful. We shall always assume this to be the case. The chemical potential λ is usually chosen such that

$$E^{\mu_{L,\omega,\lambda}}[\ \bar{\eta}\ ] = y + o(1), \tag{10.5}$$

where $\limsup_{L\to\infty} o(1) = 0$. We shall make the choice of λ so that the right hand side is actually equal to y.

The boundary conditions will be fixed in this section and we will drop the subscript ω. The main result of this section is the following Lemma.

LEMMA 10.1 *Suppose Λ_L is a cube of width $2L+1$ centered at 0. Let the mixing condition (10.1) hold.. Then for any $s > 0$, there is a constant C_s such that the following estimate holds*

$$\sup_{\omega}\sup_{y}\left| E^{\mu_{L,\omega,y}}[\eta_0 - \eta_e]\right| < C_s L^{-s} \tag{10.6}$$

for any $e > 0$.

A version of Lemmas 10.1 was proved in [Y2] using expansion techniques in local limit theorems, for $s \le 3/2$. Our proof, although partly based on this approach, is somewhat simpler and yields a stronger result in the special case considered here. Since a local limit theorem is proved in [Y2], we have in particular

$$p_y \ge C(y) L^{-d/2} \tag{10.7}$$

under the mixing assumption (10.1) . Here the chemical potential λ is chosen such that (10.5) holds without the error term $o(1)$. We will need a lower bound that holds uniformly in the range $1 - L^{\varepsilon - d} \ge y \ge L^{\varepsilon - d}$. Since, in principle, the lower bound should get better at the edges we will provide a simple proof of a uniform version of this estimate while proving Lemma 10.1 below.

Proof of Lemma 10.1. Case 1: $1 - L^{\varepsilon - d} \ge y \ge L^{\varepsilon - d}$.
Step 1: Cutoff.

By definition, the expectation w.r.t. the canonical measure is given by

$$E^{\mu_{L,y}}[(\eta_0 - \eta_e)] = \frac{1}{p_y}\frac{1}{2\pi}\int_{-\pi}^{\pi} E^{\mu_{L,\lambda}}\{(\eta_0 - \eta_e)\exp[\, i\,\theta\, M(\bar\eta - y)\,]\}\, d\theta$$

We first prove that the contribution to the integral from the range $\{\theta : |\theta| \ge L^{-\delta}\}$ is exponentially small, provided $4\delta < \varepsilon$. Let Γ be such that it contains 0 as well as e, and for some positive constant c, $\Lambda \setminus \Gamma$ consists of cM widely separated sites that become independent when conditioned on $\Gamma \cup \Lambda_L^c$. Since

$$\begin{aligned}
|E^{\mu_{L,\lambda}}\{(\eta_0 - \eta_e)\exp\big[i\,\theta\, M(\bar\eta - y)\big]\}| &= |E^{\mu_{L,\lambda}}\{(\eta_0 - \eta_e)\exp\big[i\,\theta \sum_{x\in\Lambda_L}\eta_x\big]\}| \\
&= |E^{\mu_{L,\lambda}}\{(\eta_0 - \eta_e)\exp\big[i\,\theta \sum_{x\in\Gamma}\eta_x\big] E^{\mu_{\Gamma,\lambda}}\{\exp\big[i\,\theta \sum_{x\in\Lambda\setminus\Gamma}\eta_x\}\}| \\
&\le \sup_{\omega} |E^{\mu_{\Gamma,\lambda}}\{\exp\big[i\,\theta \sum_{x\in\Lambda\setminus\Gamma}\eta_x\}|
\end{aligned}$$

it is enough to prove that

$$|E^{\mu_{\Gamma,\lambda}}\{\exp\big[i\,\theta \sum_{x\in\Lambda\setminus\Gamma}\eta_x\}| \le \exp(-CL^{\alpha}) \tag{10.8}$$

for the expectation with respect $\mu_{\Gamma,\lambda}$, the distribution obtained by conditioning on the configuration in the domain Γ. Let $p_x^\Gamma = E^{\mu_{\Gamma,\lambda}}[\eta_x]$. Since the interaction between η_x and its neighbors is uniformly bounded for each fixed x, there is a constant C such that

$$C^{-1} \le \frac{p_x^\Gamma}{y} \le C \qquad \text{and} \qquad C^{-1} \le \frac{1 - p_x^\Gamma}{1 - y} \le C, \tag{10.9}$$

where y is the total density. Note that this bound is independent of the shape of Γ. Since under $\mu_{\Gamma,\lambda}$, $\{\eta_x\}$ are independent random variables one has

$$E^{\mu_{\Gamma,\lambda}}\big[\exp\big[\sum_{x\in\Lambda\setminus\Gamma} i\theta\eta_x\big]\big] = \prod_{x\Lambda\setminus\Gamma}\big[\, p_x^\Gamma e^{i\theta} + (1 - p_x^\Gamma\,)\,\big]\,.$$

For any $0 \leq a \leq 1$ and $-\pi \leq \theta \leq \pi$ we have the elementary bound

$$|ae^{i\theta} + 1 - a|^2 = 1 - 2a(1-a)(1 - cos\theta) \leq \exp(-a(1-a)\theta^2).$$

We have thus proved

$$|E^{\mu_{L,\lambda}}\{(\eta_0 - \eta_e)\exp\left[i\,\theta\, M(\bar\eta - y)\right]\}| \leq \exp\left[-C^{-1}L^{\varepsilon-d}cML^{-2\delta}\right] = \exp\left[-cC^{-1}\,L^{\varepsilon-2\delta}\right]$$

Step 2. We can now assume that

$$|\theta| \leq L^{-\delta} \tag{10.10}$$

Choose a cube I of width L^α with α to be chosen depending on ε and centered at 0. Hence

$$\begin{aligned} &E^{\mu_{L,\lambda}}\{(\eta_0 - \eta_e)\exp\left[i\,\theta \sum_{x\in\Lambda_L}(\eta_x - y)\right]\} \\ &= E^{\mu_{L,\lambda}}\{\exp\left[i\,\theta\sum_{x\in I^c}(\eta_x - y)\right]U(\eta)\} \end{aligned} \tag{10.11}$$

where

$$U(\eta) = E^{\mu_{L,\lambda}}\left[(\eta_0 - \eta_e)\exp\left(i\,\theta\sum_{x\in I}(\eta_x - y)\right)|\,\eta_x, x\in I^c\right]$$

From the Taylor expansion and the bound (10.10) on θ , for any integer ℓ, we have

$$U(\eta) = \sum_{j=0}^{\ell} U_j(\eta) + O(L^{-(\ell+1)(\delta - d\alpha)})$$

where

$$U_j(\eta) = \frac{(i\,\theta)^j}{j!}\sum_{x_i\in I, i=1,\cdots,j} E^{\mu_{L,\lambda}}\left[\,(\eta_{x_1} - y)\ldots(\eta_{x_j} - y)(\eta_0 - \eta_e)\,\middle|\, I^c\,\right] \tag{10.12}$$

We claim that

$$\sup_\eta |U_i(\eta)| \leq C\exp[-L^{-\beta}] \tag{10.13}$$

for some $\beta > 0$.

Step 3. We now prove this bound, e.g., for $j = \ell$. Given a box Λ_L, we pick an integer γ such that $L \geq (\ell+1)\gamma$ and consider boxes $\Lambda^j = \{x : |x| \leq j\gamma\}$ for $1 \leq j \leq \ell$, and $\Lambda^{\ell+1} = \Lambda_L$. We have $(\ell+1)$ annuli of width at least γ. Given ℓ sites $x_1, \cdots, x_\ell$ at least one annuli $A_j = \Lambda^j \setminus \Lambda^{j-1}$ is empty. Let us write the set of all possible configurations as a disjoint union of $B_1, \cdots, B_{\ell+1}$ where B_j is all configurations with A_j having no points and each A_i for $i > j$ having at least one site from among $x_1, \cdots, x_\ell$. We fix the boundary condition ω and denote by

$$f_L^\omega(x, x_1, \cdots, x_\ell) = E^{L,\omega}\left[(\eta_x - y)(\eta_{x_1} - y)\cdots(\eta_{x_\ell} - y)\right]$$

for $x = 0, e$ and $x_1, \cdots, x_\ell \in \Lambda_L$. Our goal is to estimate the sum

$$\Delta(\omega, L) = \sum_{x_1,\cdots,x_\ell\in\Lambda_L}\left[f_L^\omega(1, x_1, \cdots, x_\ell) - f_L^\omega(0, x_1, \cdots, x_\ell)\right]$$

and show that it decays uniformly in ω faster than a stretched exponential in L. It is clearly sufficient to show that every $j \le \ell + 1$

$$\Delta_j(\omega, L) = \sum_{x_1,\cdots,x_\ell \in B_j} \left[f_L^\omega(1, x_1, \cdots, x_\ell) - f_L^\omega(0, x_1, \cdots, x_\ell)\right]$$

has the same property. The set B_j has no sites in the jth annulus and can further be split up in to $\ell + 1$ sets depending on how many sites are inside and how many are outside. We denote the infinite volume expectations by

$$f(x, x_1, \cdots, x_\ell) = E\left[\eta_x \eta_{x_1} \cdots \eta_{x_\ell}\right]$$

If there are m inside sites $y_1 \cdots, y_m$ and n outside variables $z_1, \cdots, z_n$ with $\ell = m + n$ and $(x_1, \cdots, x_\ell) = (y_1, \cdots, y_m, z_1, \cdots, z_n)$, from the mixing conditions it follows that

$$f_L^\omega(x, x_1, \cdots, x_\ell) - f(x, y_1 \cdots, y_m) f_L^\omega(z_1, \cdots, z_n)$$

is exponentially small. Hence up to exponential small error we have

$$\Delta_j(\omega, L) \sim \sum_{y_1,\cdots,y_m \in \Lambda_{j-1}} \left[f(1, y_1 \cdots, y_m) - f(0, y_1 \cdots, y_m)\right] \sum_{z_1,\cdots,z_n \in \Lambda \setminus \Lambda_j} f_L^\omega(z_1, \cdots, z_n)$$

Since the number of terms is only polynomial in L, it follows that our task can be achieved if show that for every L the sum

$$\Delta(L) = \sum_{x_1,\cdots,x_\ell \in \Lambda_L} \left[f(1, x_1, \cdots, x_\ell) - f(0, x_1, \cdots, x_\ell)\right]$$

is exponentially small. We prove this by induction on ℓ and we use the completely corrected correlation functions.

Recall that the definition of the completely corrected correlation function can be defined inductively by

$$g_k(x_1, \cdots, x_k) = f_k(x_1, \cdots, x_k) - \sum_{A,B} g_{|A|}(A) g_{|B|}(B)$$

where A, B is a proper partition of the variables $x_1, \cdots, x_k$ into two disjoint subsets with A and B nonempty. It is well-known that under the mixing assumption A the function g decays in all directions, i.e.,

$$|g_k(x_1, \cdots, x_k)| \le Ce^{-c|x_i - x_j|}$$

for all i, j. Since f is expressed in terms of g, it suffices to prove that $\Delta(L)$ is exponentially small if f is replaced by g_k for $k \le \ell + 1$. Since $g_k(x, x_1, \cdots, x_k)$ is exponentially small unless $|x_j| \le L^\epsilon$ for some $\epsilon > 0$, we can drop the restriction that $x_i \in \Lambda_L$ because the added terms are exponentially small. But from the translational invariance of the Gibbs state, we have

$$\sum_{x_1,\cdots,x_k} \left[g_k(1, x_1, \cdots, x_k) - g_k(0, x_1, \cdots, x_k)\right] = 0$$

This concludes the proof.

Step 4. Lower bound.

Let y be given in the range $1 - L^{\varepsilon-d} \geq y \geq L^{\varepsilon-d}$ and $\lambda = \lambda(L, y, \omega)$ be picked to satisfy

$$E^{\mu_{\Lambda,\lambda,\omega}}\{M^{-1} \sum_{x\in\Lambda_L} \eta_x\} = y.$$

Let find a set Γ such that conditioning on $\Gamma \cup \Lambda^c$ makes $\{\eta_x : x \in \Lambda \setminus \Gamma\}$ mutually independent. The cardinality of $\Lambda \setminus \Gamma$ is about cM for some $c > 0$. Let us denote by n_Γ and n_{Γ^c} the number of particles in Γ and Γ^c respectively and for $x \in \Gamma^c$ by p_x^Γ the conditional probability that $\eta_x = 1$, given the configuration in $\Gamma \cup \Lambda^c$. By a conditioning argument

$$\mu_{\Lambda,\lambda,\omega}\{\bar{\eta} = y\} = E^{\mu_{\Lambda,\lambda,\omega}}\{\mu_{\Lambda,\lambda,\omega}\{n_{\Gamma^c} = My - n_\Gamma | \Gamma \cup \Lambda^c\}\}$$

The quantity n_{Γ^c} is a sum of independent Bernoulli random variables with expectations p_x^Γ that satisfy $cy \leq p_x^\Gamma \leq Cy$, one can easily derive the standard local limit theorem that provides the lower bound

$$\mu_{\Lambda,\lambda,\omega}\{n_{\Gamma^c} = My - n_\Gamma | \Gamma \cup \Lambda^c\} \geq \frac{c}{\sigma_L(\omega)} \exp[-\frac{(My - n_\Gamma - \mu_L(\omega))^2}{2\sigma_L^2(\omega)}]$$

provided

$$(My - n_\Gamma - \mu_L(\omega))^2 \leq C\sigma_L^2(\omega)$$

where

$$\mu_L(\omega) = \sum_{x\in\Lambda\setminus\Gamma} p_x^\Gamma$$

and

$$\sigma_L^2(\omega) = \sum_{x\in\Lambda\setminus\Gamma} p_x^\Gamma(1 - p_x^\Gamma)$$

Since $\sigma_L^2(\omega) \simeq C\, M\, y$ uniformly in ω, it is sufficient to obtain a uniform lower bound

$$\mu_{\Lambda,\lambda,\omega}\{(My - n_\Gamma - \mu_L(\omega))^2 \leq Cy\, M\} \geq \frac{1}{2}$$

for some fixed C. By our choice of λ

$$E^{\mu_{\Lambda,\lambda,\omega}}\{n_\Gamma + \mu_L(\omega)\} = My$$

and by orthogonality

$$E^{\mu_{\Lambda,\lambda,\omega}}\{(n_\Gamma + \mu_L(\omega) - My)^2\} \leq E^{\mu_{\Lambda,\lambda,\omega}}\{M^2(\bar{\eta} - y)^2\} \leq CMy$$

An application of Chbechev's inequality completes the proof of Lemma 10.1 for the case $1 - L^{\varepsilon-d} \geq y \geq L^{\varepsilon-d}$.

Case 2. $y \leq L^{\varepsilon-d}$. [*Case 3.* $y \geq 1 - L^{\varepsilon-d}$ is similar.]

We will actually prove the much stronger assertion

$$E^{\mu_{\Lambda,y,\omega}}[\eta_0 - \eta_e] \equiv 0 \tag{10.14}$$

which is of course a purely combinatorial fact. It is convenient to normalize the energy $\mathbf{H} = \sum_{x \in \Lambda_L} H(\tau_x \eta)$ by assuming that the local function $H(\eta)$ is 0 for the empty configuration. For any boundary condition ω and particle density $y = NL^{-d}$, we define the canonical partition function

$$Z_x^\omega = \sum_{\eta : \bar{\eta} = y} \eta(x) \exp[\mathbf{H}(\eta)]$$

and prove that for large enough L, $Z_0^\omega \equiv Z_e^\omega$.

We shall construct a graph with vertices in Λ_L, connecting every site with a particle to all the sites in a cube of width $2R+1$ centered at the site with the particle. Since the total number of particles is small, the cube will split up into several connected clusters. each cluster has a protective layer of empty sites preventing any site outside the union of all the clusters with interacting with any other site. Denoting the interior clusters by $C_1, \cdots, C_q$ and the clusters that interact with the boundary by $D_1 \cdots, D_p$ the Hamiltonian takes the form

$$\mathbf{H}(\eta) = \sum_{0 \le i \le q} H(C_i) + \sum_{0 \le i \le p} H_\omega(D_i)$$

The clusters are actually slightly more complicated animals, because one has to keep in mind that inside each cluster there are some occupied sites and some empty sites. We shall fix the boundary clusters, but translate the interior clusters by arbitrary translations. For fixed $\mathcal{C} = C_1, \cdots, C_q, D_1, \cdots, D_p$ we define

$$Z_x^{\mathcal{C}} = \sum_{x_1, \cdots, x_q : \Delta_x} \exp\big[\sum_{0 \le i \le q} H(\tau_{x_i} C_i) + \sum_{0 \le i \le p} H_\omega(D_i)\big]$$

where Δ_x is the set of $(x_1 \cdots, x_q)$ such that $\{\tau_{x_i} C_i, 1 \le i \le q\}$ and $\{D_j, 1 \le j \le p\}$ are all mutually disjoint and $x \in \cup_{1 \le i \le q} C_i$. It is clearly sufficient to prove that for large L, and for any $\mathcal{C}$ and boundary condition ω,

$$Z_0^{\mathcal{C}} = Z_e^{\mathcal{C}}$$

Since the energy depends only on the class $\mathcal{C}$ it is sufficient to prove that

$$|\Delta_0| = |\Delta_e|$$

This will be carried out by careful counting, using the inclusion exclusion formula.

First, we note that the number of particles in any cluster must be at least a fixed fraction of its diameter. Since the total number of particles is of the order L^ε with $\varepsilon < 1$, we can never connect any boundary cluster to the sites 0 or e by a chain of overlapping clusters that are translates of C_i. To simplify the notation let us think of a graph G with $p+q$ vertices corresponding to C_i and D_j. We will have bonds connecting two vertices that correspond to two different C's or between C-vertex and D-vertex but not between two D-vertices. A bond represents the intersection of the corresponding clusters after arbitrary translations of the C's. For each $z = (x_1, \cdots, x_q)$ we have the corresponding collection of bonds $\Gamma(z)$. If $z = (x_1, \cdots, x_q)$ is to be counted in Δ_x, $\Gamma(z)$ must be empty and $x \in \cup_{1 \le i \le q} \tau_{x_i} C_i$. Therefore

$$|\Delta_x| = \#\{z : \Gamma(z) = \emptyset, x \in \cup_{1 \le i \le q} \tau_{x_i} C_i\} = \sum_{1 \le i \le q} \#\{z : \Gamma(z) = \emptyset, x \in \tau_{x_i} C_i\}$$

because x cannot belong to different C's without creating an intersection. For each summand

$$|\Delta_x^i| = \#\{z : \Gamma(z) = \emptyset, x \in \tau_{x_i} C_i\}$$

by inclusion exclusion formula

$$\begin{aligned} |\Delta_x^i| &= \sum_r (-1)^r \sum_{\Gamma:|\Gamma|=r} \#\{z : \Gamma(z) \supset \Gamma, x \in \tau_{x_i} C_i\} \\ &= \sum_r (-1)^r \sum_{\Gamma:|\Gamma|=r} |\Delta_x^{i,\Gamma}| \end{aligned}$$

It is now a simple matter to complete the proof by establishing

$$|\Delta_0^{i,\Gamma}| = |\Delta_e^{i,\Gamma}|$$

for all i and Γ. The bonds Γ divide the graph G into connected components and the vertex i is in some component G_0. G_0 cannot contain any of the D vertices due to the absence of long chains. If we shift by e all the clusters that correspond to vertices in G_0, this establishes a one to one map between $\Delta_0^{i,\Gamma}$ and $\Delta_e^{i,\Gamma}$, and we are done.

REFERENCES

[DV] Donsker, M.D., and Varadhan, S. R. S., *Large deviations from a hydrodynamic scaling limit*, Comm. Pure Appl. Math. **42** , 243–270, (1989).

[EMY] Esposito, R., Marra, R., and Yau, H.T., *Diffusive limit of asymmetric simple exclusion*, Rev. in Math. Physics **6**, 1233-1267, (1994).

[GPV] Guo, M.G., Papanicolau, G.C., and Varadhan, S. R. S., *Nonlinear diffusion limit for a system with nearest neighbor interactions*, Comm. Math. Phys. **118**, 31–59, (1988).

[JY] Jensen, L., Yau, H.T., *Hydrodynamic scaling limits of simple exclusion models*, Park City/IAS Mathematics Series.

[KLS] Kats, S.,Lebowitz, J.L., and Spohn, H., *Non equilibrium steady state of stochastic lattice gas models of fast ionic conductors*. J. Stat Phys. **34**, 497–537, (1984).

[LOY1] Landim, C., Olla, S., and Yau, H.T., *First order correction for the hydrodynamic limit of asymmetric simple exclusion processes in dimension* $d \geq 3$, Commun. Pure. Appl. Math. Vol L, 0149-0203, (1997).

[LOY2] C. Landim , S. Olla and H. T. Yau, *Some properties of the diffusion coefficient for asymmetric simple exclusion processes*, Ann. Prob. **24**, 1779–1808, (1996).

[LY] Lu, S.L., and Yau, H.T., *Spectral gap and Logarithmic Sobolev Inequality for Kawasaki and Glauber dynamics*, Commun. Math. Phys. **156**, 399-433 (1993).

[MOS] Martinelli, F., Olivieri, E. and Schomann, R., *For 2-D lattice spin systems Weak Mixing Implies Strong Mixing*, Commun. Math. Phys. **165**, 33–47 (1994).

[N] Newman, C.: *Private communication.*

[OVY] Olla, S., Varadhan, S.R.S., and Yau, H.-T.; *Hydrodynamical limit for a Hamiltonian system with weak noise*, Comm. Math. Phys., **155**, 523-560, (1993).

[Q1] Quastel, J., *Diffusion of colour in the simple exclusion process*, Comm. Pure Appl. Math., **45**, 321–379, (1992).

[Q2] Quastel, J., *Large deviations from a hydrodynamic scaling limit for a nongradient system*, Ann. Prob. **23** No. 2, 724–742 (1995).

[QRV] Rezakhanlou, F., Quastel, J., and Varadhan, S.R.S., *Large Deviations for the symmetric simple exclusion process in dimension* $d \geq 3$. (submitted to Probability Theory and Related Fields)

[QY1] Quastel, J., and Yau, H.T., *Bulk diffusion in a system with site disorder*, in preperation.

[QY2] Quastel, J., and Yau, H.T., *Lattice gases, large deviations, and the incompressible Navier-Stokes equations*, Preprint.

[RS] Reed, M., Simon, B., *Methods of Modern Mathematical Physics, Vol. I-IV*, Academic Press (1972).

[R] Rezakhanlou, F., *Hydrodynamic limit for a system with finite range interactions*, Comm. Math. Phys. **129**, 445-480 (1990).

[S] Spohn, H., *Large Scale Dynamics of Interacting Particles*, Springer-Verlag New York (1991).

[SY] Spohn, H., and Yau, H.T., *Bulk diffusivity of lattice gases close to criticality*, Jour. Stat. Phys., **79**, 231-241, (1995).

[V1] Varadhan, S.R.S., *Nonlinear diffusion limit for a system with nearest neighbor interactions II*, in *Asymptotic Problems in Probability Theory : Stochastic Models and Diffusion on Fractals*, edited by K. Elworthy and N. Ikeda, Pitman Research Notes in Mathematics 283, Wiley, (1994). 75-128

[V2] Varadhan, S. R. S. *Scaling limits for interacting diffusions*, Comm. Math. Phys. **135** 313–353, (1991).

[V3] S. R. S. Varadhan, *Regularity of self diffusion coefficient*, "The Dynkin Festschrift, Markov processes and their applications", edited by M. Freidlin, Birkhäuser, (1994).

[Y1] Yau, H.T., *Relative Entropy and Hydrodynamics of a Ginzburg-Landau Model*, Lett. Math. Phys. **22**, 63-80 (1991).

[Y2] Yau, H.T., *Logarithmic Sobolev inequality for lattice gases with mixing conditions*, Commun. Math. Phys,**181**, 367-408, (1997).

Probab. Theory Relat. Fields **113**, 1–84 (1999)

Probability Theory and Related Fields

Large deviations for the symmetric simple exclusion process in dimensions $d \geq 3$

J. Quastel[1], **F. Rezakhanlou**[2], **S.R.S. Varadhan**[3]

[1]Department of Mathematics, University of California, Davis, CA 95616. Present address: Departments of Mathematics and Statistics, University of Toronto, 100 St. George Street, Toronto, Ontario, M5S 3G3, Canada. Partially supported by NSF grant DMS-9504791

[2]Department of Mathematics, University of California, Berkeley, CA 94720. Partially supported by NSF grant DMS-9424270

[3]Courant Institute, 251 Mercer St., New York, NY 10012. Partially supported by NSF grant DMS-9503419 and ARO grant DAAH04-95-1-0666

Received: 7 September 1997 / Revised version: 15 May 1998

Abstract. We consider symmetric simple exclusion processes with $L = \bar{\rho}N^d$ particles in a periodic d-dimensional lattice of width N. We perform the diffusive hydrodynamic scaling of space and time. The initial condition is arbitrary and is typically far away form equilibrium. It specifies in the scaling limit a density profile on the d-dimensional torus. We are interested in the large deviations of the empirical process, $N^{-d}[\sum_1^L \delta_{x_i(\cdot)}]$ as random variables taking values in the space of measures on $D[0.1]$. We prove a large deviation principle, with a rate function that is more or less universal, involving explicity besides the initial profile, only such canonical objects as bulk and self diffusion coefficients.

Mathematics Subject Classification (1991): 60K35 (60F10)

1. Introduction

A classical theorem of Sanov identifies the entropy function as the rate functional for the large deviations of the empirical measure. More precisely if $X_1, X_2, \ldots, X_n, \ldots$ is a sequence of independent and identically distributed random variables with values in some Polish space

D, having a common distribution P, then the strong law of large numbers asserts that, in the weak topology on the space of probability measures on D,

$$\lim_{L\to\infty}\frac{1}{L}\sum_{i=1}^{L}\delta_{X_i}=P \tag{1.1}$$

with probability one. Moreover the probabilities of deviations of the left hand side of (1.1) from P are exponentially small in the size L. Roughly speaking one has

$$\mathrm{Prob}\left[\frac{1}{L}\sum_{i=1}^{L}\delta_{X_i}\text{ is close to }Q\right]\simeq\exp\{-LH(Q,P)\} \tag{1.2}$$

where $H(\cdot,\cdot)$ is the entropy functional

$$H(Q,P)=\begin{cases}\int\log\frac{dQ}{dP}\,dQ & \text{if } Q\ll P\\ +\infty & \text{otherwise}\end{cases}. \tag{1.3}$$

The aim of this article is to prove a theorem similar in spirit in the context of an interacting particle system. If we have a system of L interacting particles and denote by $x_1(\cdot),\dots,x_L(\cdot)$ their trajectories in some finite interval $[0,T]$, we can form the empirical process

$$R_{L,\omega}=\frac{1}{L}\left[\delta_{x_1(\cdot)}+\cdots+\delta_{x_L(\cdot)}\right] \tag{1.4}$$

which will be a random measure on a suitable function space of trajectories on $[0,T]$. As $L\to\infty$, assuming that the model and the scaling have been properly chosen, one can expect a law of large numbers type result asserting the convergence of $R_{L,\omega}$ to a non-random limit P. We are interested in the rate function for large deviations from this convergence and our goal is the explicit calculation of the rate function in terms of physical quantities arising out of the model.

Questions of this sort are relevent for large classes of models. There are some examples of previous work related to particular aspects of the problem for certain models. For instance, in [BS], Benois and Saada have studied the large deviation behavior of a tagged particle in the case of one dimensional nearest neighbor symmetric simple exclusion. Because of the special nature of this example, this was done by an analysis of the large deviation of the empirical density from its hydrodynamic limit. In [GJL], Gabrielli et al. have studied the large deviation behavior of the two-color empirical density for the zero range process. We could have developed our large deviation principle for the empirical process in this model. Because the multi-color zero range process is still a gradient model, as was pointed out by the referee, it would no doubt have simplified our analysis considerably. But it would have also raised the issue of the validity of these results for other models without the simplifying features of the zero range model.

We will restrict our attention in this article to a widely studied special class of models known as random walks with symmetric simple exclusion. Our physical space will be the d-dimensional integer lattice modulo N, consisting of N^d lattice sites. The lattice sites will be denoted by letters x, y, etc. We will scale the periodic lattice by $\frac{1}{N}$, obtaining for each point on the lattice a corresponding point on the d-torus and these will be denoted by θ. We will have a certain number L of particles on the lattice and their locations will be denoted by $x_1, \ldots, x_L$. Since the particles move, their locations are functions of t and will be denoted by $x_1(t), \ldots, x_L(t)$. Each particle waits for an exponential time at its current site and then tries to jump to a new site. The new site is picked randomly, with the probability of attempting a jump from x to $x+z$ being given by $p(z)$. The probability distribution $p(\cdot)$ of the jump size is assumed to be symmetric, irreducible and of finite range. The jumps however are not always accomplished. Simple exclusion means that the particle is not allowed to jump to a site that is already occupied. In such a situation the particle does not jump and waits for a new exponential time. Of course, all particles are doing this simultaneously and independently of each other. Because of continuous time there will be no ties to resolve.

If we start our L particles with at most one particle per site, then it will remain so for all times. For any initial configuration $(x_1, \ldots, x_L)$ this defines a stochastic process on the space $D[[0,\infty),(Z_N^d)^L]$. We will rescale the process by defining

$$y_i(t) = \tfrac{1}{N} x_i(N^2 t) \tag{1.5}$$

and view $(y_1(t), \ldots, y_L(t))$ as taking values in $D\big[[0,T], \mathbf{T}^d\big]$ for some fixed time T. The corresponding empirical measure then is

$$\hat{R}_{L,\omega} = \tfrac{1}{L}\big[\delta_{y_1(\cdot)} + \cdots + \delta_{y_L(\cdot)}\big] \ . \tag{1.6}$$

We let $N \to \infty$ in such a way that

$$\lim_{N\to\infty} \frac{L}{N^d} = \bar{\rho} \tag{1.7}$$

exists and the initial configuration $\xi = (x_1, \ldots, x_L)$ is assumed to behave in such a way that

$$\lim_{N\to\infty} \frac{1}{N^d}\big[\delta_{y_1(0)} + \cdots + \delta_{y_L(0)}\big] = \rho_0(\theta)\, d\theta \tag{1.8}$$

exists weakly on $\mathbf{T}^d$. Note that because of the exclusion $0 \leq \rho_0(\theta) \leq 1$ and $\int_{\mathbf{T}^d} \rho_0(\theta)\, d\theta = \bar{\rho}$.

One can ask if under these conditions $\hat{R}_{L,\omega}$ satisfies a law of large numbers with a non-random weak limit P. It will be more convenient to replace $\hat{R}_{L,\omega}$ by $R_{N,\omega}$ where

$$R_{N,\omega} = \frac{1}{N^d}\left[\delta_{y_1(\cdot)} + \cdots + \delta_{y_L(\cdot)}\right] . \tag{1.9}$$

and in view of (1.7) they are essentially the same.

It is relatively straightforward to prove that the marginal distribution

$$R_{N,\omega,t} = \frac{1}{N^d}\left[\delta_{y_1(t)} + \cdots + \delta_{y_L(t)}\right] \tag{1.10}$$

of the empirical process at time t satisfies a law of large numbers and converges weakly to a non-random limit $\rho(t,\theta)d\theta$ on $\mathbf{T}^d$. The density $\rho(t,\theta)$ is determined as the solution of the linear heat equation

$$\frac{\partial \rho}{\partial t} = \frac{1}{2}\nabla \cdot D\nabla\rho \tag{1.11}$$

with $\rho(0,\theta) = \rho_0(\theta)$. The diffusion matrix D is nothing but the covariance matrix of the distribution $p(\cdot)$ of the jump sizes. The symmetry of $p(\cdot)$ makes the exclusion irrelevent in this context, and the result is basically the central limit theorem for the symmetric random walk. One should note however that this happens because the particles are not distinguishable at this level. On the other hand a law of large numbers also holds for the empirical process $R_{N,\omega}$ and now the particles are tagged and one has as the weak limit an inhomogeneous diffusion process on $\mathbf{T}^d$ with initial density $\rho_0(\theta)$. This diffusion process has a time dependent generator

$$\frac{1}{2}\nabla \cdot S(\rho(t,x))\nabla + \frac{1}{2}[S(\rho(t,x)) - D]\frac{(\nabla\rho)(t,x)}{\rho(t,x)} \cdot \nabla \tag{1.12}$$

where $\rho(t,\theta)$ is given by (1.11). Here $S = S(\rho)$ is the self diffusion coefficient that describes the diffusive behavior of a tagged particle in equilibrium at density ρ. The existence of this diffusion matrix was established in [KV] and the law of large numbers was proved in [R] based on the earlier work in [Q1].

Here a qualitative difference arises between the nearest neighbor case in $d = 1$, i.e. $p(1) = p(-1) = \frac{1}{2}$ in $d = 1$, and all other cases. In the one dimensional nearest neighbor case, particles block each other completely and therefore $S(\rho) \equiv 0$ for $\rho > 0$. In all other cases, $S(\rho) > 0$ for $0 \leq \rho < 1$ with $S(\rho) \to D$ as $\rho \to 0$, and to 0 as $\rho \to 1$.

Although the special case of $S(\rho) \equiv 0$ is relatively simple, we shall focus our attention exclusively on the alternate situation. Most of our results are valid in all dimensions. But at some crucial steps we need to know that $S(\rho)$ is Lipschitz continuous in ρ and this is known only when $d \geq 3$ [V2]. So the complete results are valid only when $d \geq 3$.

Now we turn to the large deviation behavior of the empirical process (1.9). As a first step we may look at the large deviation behavior of the empirical distribution (1.10) as measure-valued functions on $[0, T]$. This was carried out in [KOV] where it was shown that a large deviation principle is valid for $R_{N,t,\omega}$. If $q(t, \theta)$ is an arbitrary nonnegative function of t and θ with $\int q(t,\theta)\, d\theta \equiv \int q(0,\theta)\, d\theta$, the large deviation rate function for $R_{N,t,\omega}$ to be close to $q(\cdot,\theta)\, d\theta$ in $[0, T]$ is given by

$$I_{\mathrm{KOV}}(q(\cdot,\cdot)) = I_{\mathrm{init}}(q(\cdot,\cdot)) + I_{\mathrm{dyn}}(q(\cdot,\cdot)) \ . \tag{1.13}$$

Here $I_{\mathrm{init}}(q(\cdot,\cdot))$ is the large deviation rate for the initial profile $q(0,\cdot)$ which depends on the choice of the initial measure. If we were to start with a Bernoulli product measure with $\rho_0(x/N)$ as the probability of a site x being occupied then

$$I_{\mathrm{init}}(q(\cdot,\cdot)) = \int_{\mathbf{T}^d} h(q(0,\theta), \rho_0(\theta))\, d\theta \tag{1.14}$$

where

$$h(a,b) = a \log\frac{a}{b} + (1-a)\log\frac{1-a}{1-b} \ . \tag{1.15}$$

The second piece, $I_{\mathrm{dyn}}(q)$, is the dynamical part. It is infinite unless $\hat{I}(q(\cdot,\cdot)) < \infty$, where

$$\hat{I}(q(\cdot,\cdot)) = \int_0^T\!\!\int_{\mathbf{T}^d} \frac{|\nabla q|^2}{q(1-q)}\, d\theta\, dt \ . \tag{1.16}$$

For such q, one can consider the class $\mathscr{B}_q$ of $b(\cdot,\cdot)$ which satisfy

$$\frac{\partial q}{\partial t} = \frac{1}{2}\nabla\cdot D\nabla q - \nabla\cdot q(1-q)Db$$

and

$$\mathscr{E}[b] = \frac{1}{2}\int_0^T\!\!\int_{\mathbf{T}^d} q(1-q)b\cdot Db\, d\theta\, dt < \infty \ . \tag{1.17}$$

Then

$$I_{\mathrm{dyn}}(q) = \inf_{b\in\mathscr{B}_q} \mathscr{E}[b] \ . \tag{1.18}$$

It could happen that $\mathscr{B}_q$ is empty and then $I_{\text{dyn}}(q) = \infty$. It turns out that, if it is finite, the infimum in (1.18) is attained at some $b_0 = \nabla\phi$ for some $\phi : [0,T] \times \mathbf{T}^d \to \mathbf{R}$, and the choice of b is unique.

We are now ready to define the rate function $\mathscr{I}(Q)$ for the empirical process. Given an arbitrary subprobability measure Q on $D\big[[0,T];\mathbf{T}^d\big]$, we can compute the marginal densities $q(t,\theta)$ for $0 \le t \le T$. There is the initial rate $I_{\text{init}}(q(\cdot,\cdot))$ which depends on how the choice of the initial condition ξ is chosen. $I_{\text{init}}(q(\cdot,\cdot))$ would be given by (1.14) for a Bernoulli choice, or, if ξ is chosen deterministically, $I_{\text{init}}(q(\cdot,\cdot))$ would be infinite except when $q(0,\cdot) = \rho_0(\cdot)$, in which case it would be zero.

For each $b \in \mathscr{B}_q$ we have a well defined diffusion process P_q^b with one dimensional marginal densities $q(t,\theta)$ and with time dependent generator

$$\mathscr{A}_{q,b} = \frac{1}{2}\nabla \cdot S(q)\nabla + \frac{1}{2}(S(q) - D)\frac{\nabla q}{q} \cdot \nabla + (1-q)Db \cdot \nabla \ .$$

For each q with $I_{\text{dyn}}(q) < \infty$ the existence and uniqueness of such a process has been established in [QV]. One can also verify that the relative entropy $H(Q, P_q^b)$ is either identically infinite for all $b \in \mathscr{B}_q$, or is finite for every member of $\mathscr{B}_q$. The rate function is infinite in the former case. In the latter case a special $b = b_Q$ is singled out in $\mathscr{B}_q$ by the property that

$$E^Q\left[\int_0^T \phi(t,x(t)) \cdot dx(t)\right] = E^{P_q^{b_Q}}\left[\int_0^T \phi(t,x(t)) \cdot dx(t)\right]$$

for all smooth functions $\phi : [0,T] \times \mathbf{T}^d \to \mathbf{R}^d$. Note that in general $b_Q \ne b_0$. The full rate function for the large deviations of the empirical process is given by

$$\mathscr{I}(Q) = I_{\text{init}}(q(\cdot,\cdot)) + H(Q, P_q^{b_Q}) + \mathscr{E}[b_Q] \ .$$

Let $P_N \in \mathscr{M}_1\big(\mathscr{M}\big(D[[0,T];\mathbf{T}^d]\big)\big)$ be the distribution of the empirical process $R_{N,\omega}$. The topology is the usual topology of weak convergence for measures and the uniform topology on $D[[0,T];\mathbf{T}^d]$. We can now state the main result.

Theorem 1.1. *For $d \ge 3$, the distribution P_N of the empirical process satisfies the large deviation principle with rate function $\mathscr{I}(Q)$. In particular, $\mathscr{I}$ is lower semicontinuous and has compact level sets. We have for each closed set $\mathscr{C}$,*

$$\limsup_{N\to\infty} \frac{1}{N} \log P_N(\mathscr{C}) \le -\inf_{Q \in \mathscr{C}} \mathscr{I}(Q) \ ,$$

and for each open set $\mathcal{O}$,

$$\liminf_{N\to\infty}\frac{1}{N}\log P_N(\mathcal{O}) \geq -\inf_{Q\in\mathcal{O}}\mathcal{I}(Q) \ .$$

We will establish the large deviation principle by first studying the finite dimensional marginals of $R_{N,\omega}$ which correspond to the n-color version of the exclusion process. We then use certain superexponential estimates to pass from the finite dimensional distributions to the full empirical process.

Finally, a word about the organization of the material. In section 2, we provide precise statements of results and an outline of the methods of proof. Sections 3–6 deal with the n-color version of the exclusion process and the remaining sections deal with the transition from the finite dimensional marginals to the empirical process.

2. Notation and summary

Let Z_N denote the integers modulo N, i.e. $\{1,2,\ldots,N\}$ with $N+1$ identified periodically with 1. Z_N^d is the d-fold product of Z_N. We have L tagged particles and the positions of these L particles are described by a point $(x_1,x_2,\ldots,x_L)$ in $(Z_N^d)^L$. Since we have simple exclusion, which means that at most one particle can occupy any one site at a given time, our effective state space is the part of $(Z_N^d)^L$ where $x_1,x_2,\ldots,x_L$ are distinct points of Z_N^d.

Our Markov process can be specified by the following infinitesimal generator. If $f(x_1,x_2,\ldots,x_L)$ is a function on the state space,

$$\begin{aligned}(L_Nf)(x_1,x_2,\ldots,x_L) = N^2\sum_z\sum_{i=1}^{L}p(z)(1-\eta(x_i+z))\\ \big(f(x_1,x_2,\ldots,x_i+z,\ldots,x_L)\\ -f(x_1,x_2,\ldots,x_i,\ldots,x_L)\big)\end{aligned} \tag{2.1}$$

Here the factor N^2 represents a speeding up or rescaling of time by a factor N^2. The probability distribution $p(\cdot)$ of the jump size is independent of N, is assumed to be symmetric, i.e. $p(z)=p(-z)$, and of finite range which means that $p(z)\neq 0$ only for $z\in F$, for some finite set F. We will also assume that the random walk with transition probabilities $p(\cdot)$ is irreducible. Finally if $d=1$ we assume that $p(1)<1/2$ to exclude the nearest neighbor case. The function $\eta(x)$ is the particle count at x

$$\eta(x) = \sum_{i=1}^{L} \chi_x(x_i) \tag{2.2}$$

Because our occupied sites are distinct, $\eta(x)$ is either 0 or 1 and the factor $(1 - \eta(x+z))$ in (2.1) indicates that a jump from x_i to $x_i + z$ of the ith particle is allowed only when $\eta(x_i + z) = 0$, i.e, the site $x_i + z$ is free.

Since the generator is symmetric in the particles we can reduce the state space to just the set of occupied sites ignoring the labels of particles. The state space can then be the set of functions

$$\eta : Z_N^d \to \{0, 1\} \tag{2.3}$$

where $\eta(x)$ is given by (2.2) and tells us which sites are occupied. The reduced generator is given by

$$(L_N f)(\eta) = N^2 \sum_x \sum_z p(z)\eta(x)(1 - \eta(x+z))(f(\eta^{x,x+z}) - f(\eta)) \tag{2.4}$$

where $\eta^{x,y}$ is the new configuration obtained by interchanging the situation at the sites x and y.

We may also decide that there are n types of particles distinguished by colors and may want to keep track of just colors and not individual labels. We then define

$$\eta_j : Z_N^d \to \{0, 1\}, \qquad \eta_j(x) = \sum_{i=1}^{L} \chi_x^j(x_i) \ , \tag{2.5}$$

where χ_x^j counts the number of particles of color j at site x. Of course

$$\sum_{j=1}^{n} \eta_j(x) = \eta(x) \ . \tag{2.6}$$

If we denote the typical element of the state space by

$$\zeta = (\eta_1, \ldots, \eta_n)$$

then the n-color generator is given by

$$(L_N^n)(\zeta) = N^2 \sum_x \sum_z p(z)\eta(x)(1 - \eta(x+z))\big(f(\zeta^{x,x+z}) - f(\zeta)\big)$$

or in a more symmetric form by,

$$(L_N^n)(\zeta) = \frac{N^2}{2} \sum_x \sum_y p(y-x)[\eta(x)(1 - \eta(y)) + (1 - \eta(x))\eta(y)](f(\zeta^{x,y}) - f(\zeta)) \tag{2.7}$$

Here $\zeta^{x,y}$ is the configuration obtained by interchanging the situation at x and y without changing the color of particles, if any, at these sites. Of course if there are particles at both x and y the prefactor kills this transition. If both the sites x and y are empty, then nothing happens. So the only things that can happen is that a particle jumps from x to y or from y to x depending on where the particle is, and the colors of particles never change in our model. More formally one can define $\zeta^{x,y}$ by

$$\begin{aligned} \zeta^{x,y}(u) &= \zeta(u) \quad \text{if } u \neq x \text{ or } y; \\ \zeta^{x,y}(x) &= \zeta(y); \\ \zeta^{x,y}(y) &= \zeta(x) \ . \end{aligned} \tag{2.8}$$

We now imbed Z_N^d in $\mathbf{T}^d$, the d-dimensional torus by the map $y = x/N$. The L particles will evolve as a stochastic process

$$y(t) = (y_1(t), \dots, y_L(t)) \ . \tag{2.9}$$

The η process can be described by the measure-valued process

$$\nu_N(t) = \frac{1}{N^d} \sum_{i=1}^{L} \delta_{y_i(t)} \ , \tag{2.10}$$

on $\mathbf{T}^d$ of total mass $\frac{L}{N^d}$. If we denote the empirical process by

$$R_N = \frac{1}{N^d} \sum_{i=1}^{L} \delta_{y_i(\cdot)} \tag{2.11}$$

then $\nu_N(t)$ is the marginal at time t of R_N. The ζ process can, in a similar manner, be described by the measure-valued process with n components

$$\tilde{\nu}_N(t) = \left\{ \nu_N^j(t);\ 1 \leq j \leq n \right\} \tag{2.12}$$

of total masses L_j/N^d, where L_j is the number of particles of color j. The measures $\nu_N^j(t)$ are defined by restricting the sum in (2.10) to particles of color j. Let us choose an arbitrary $T < \infty$ and fix it. We will consider all processes in the time interval $[0, T]$. $\nu_N(\cdot)$ is then viewed as a stochastic process with values in $D[[0, T], \mathscr{M}]$ of measure-valued functions on $[0, T]$. The large deviation theory for $\nu_N(\cdot)$ was established in [KOV]. Since we outlined it in the introduction we will not repeat it here. Our first task is to extend the large deviation results to $\tilde{\nu}_N(\cdot)$ of n-tuples of measure-valued functions on $[0, T]$.

In [KV] the diffusive behavior of a tagged particle in equilibrium was established for the simple exclusion process with self diffusion coefficients given by a density dependent matrix $S(\rho)$, which is a positive semidefinite symmetric matrix of size $d \times d$. Of course $S(\rho)$ really depends on ρ with $S(\rho) \to D$ as $\rho \to 0$ and $S(\rho) \to 0$ as $\rho \to 1$.

Let us denote by $\mathscr{C}$ the set of possible color densities, which is the convex set in $\mathbf{R}^n$ consisting of $\tilde{\rho} = (\rho_1, \rho_2, \ldots, \rho_n)$ such that $\rho_i \geq 0$ for $i = 1, 2, \ldots, n$ and $\rho = \sum_{i=1}^n \rho_i \leq 1$.

We define the following matrices A and χ which are really to be thought of as $n \times n$ matrices corresponding to color types whose entries are $d \times d$ matrices that correspond to coordinates in $\mathbf{R}^d$. If j, j' are two color types then

$$\begin{aligned} A_{j,j'} &= A^{(1)}_{j,j'} + A^{(2)}_{j,j'} \\ &= S(\rho)\left(\rho_j \delta_{j,j'} - \frac{\rho_j \rho_{j'}}{\rho}\right) + D\left(\rho_j \rho_{j'} \frac{(1-\rho)}{\rho}\right) \end{aligned} \tag{2.13}$$

Each entry of $A_{j,j'}$, as a $d \times d$ matrix, is a linear combination of $S(\rho)$ and D with weights that have been given in (2.13). Similarly

$$\chi_{j,j'} = I\left(\frac{\delta_{j,j'}}{\rho_j} + \frac{1}{(1-\rho)}\right) \tag{2.14}$$

where I is the identity matrix of size $d \times d$. We will denote the inner products in $\mathbf{R}^n$, $\mathbf{R}^d$ and $\mathbf{R}^{nd}$ by $\langle , \rangle$, $\cdot$ and $\langle ; \rangle$ respectively. We remark that if we define the entropy function

$$h(\tilde{\rho}) = \sum_{i=1}^n \rho_i \log \rho_i + (1-\rho)\log(1-\rho)$$

then

$$\lambda(\tilde{\rho}) = (\nabla h)(\tilde{\rho})$$

and χ is the Hessian

$$\chi = (\nabla \lambda)(\tilde{\rho}) = (\nabla\nabla h)(\tilde{\rho})$$

Let us note that although χ is undefined when $\rho = 1$, A and $A\chi$ are well defined on $\mathscr{C}$. In [Q1] it was shown that the empirical measures $\tilde{v}_N(\cdot)$ satisfy a law of large numbers. Assuming that the initial configurations, whether they are random or deterministic, satisfy

$$\tilde{v}_N(\cdot) \Rightarrow \tilde{\rho}^0(\theta)\, d\theta \tag{2.15}$$

in probability, in the weak topology on $\mathbf{T}^d$, it was shown that for all $t > 0$,

$$\tilde{v}_N(\cdot) \Rightarrow \tilde{\rho}(t, \theta)\, d\theta \tag{2.16}$$

The limit $\tilde{\rho}(t, \theta)$ is characterized as the unique solution of

$$\frac{\partial \tilde{\rho}}{\partial t} = \frac{1}{2} \nabla \cdot A\chi \nabla \tilde{\rho} \tag{2.17}$$

with initial condition

$$\tilde{\rho}(0, \theta) = \tilde{\rho}^0(\theta) \tag{2.18}$$

The equations (2.17) and (2.18) have the following interpretation. The sum $\rho(t, \theta) = \sum_{j=1}^n \rho_j(t, \theta)$ satisfies the heat equation

$$\frac{\partial \rho}{\partial t} = \frac{1}{2} \nabla \cdot D \nabla \rho \tag{2.19}$$

with initial condition

$$\rho(0, \theta) = \rho^0(\theta) = \sum_1^n \rho_j^0(\theta) \tag{2.20}$$

and, given the solution of (2.19), each component of (2.17) can be recast as

$$\frac{\partial \rho_j}{\partial t} = \mathscr{A}_\rho^* \rho_j \tag{2.21}$$

with $\rho_j(0, \theta) = \rho_j^0(\theta)$, where

$$\mathscr{A}_\rho^* f = \frac{1}{2} \nabla \cdot S(\rho(t, \theta)) \nabla f + \frac{1}{2} \nabla \cdot [(D - S(\rho(t, \theta))] \frac{\nabla \rho(t, \theta)}{\rho(t, \theta)} f \ . \tag{2.22}$$

We will now describe the large deviation theory for the empirical measure-valued process $\tilde{v}_N(\cdot)$. If P_N is the probability distribution of $\tilde{v}_N(\cdot)$ it has a standard large deviation property with the rate function $I^n(\tilde{\rho}(\cdot, \cdot))$ defined in (2.33) for functions

$$\tilde{\rho}(\cdot) : [0, T] \times \mathbf{T}^d \to \mathscr{C} \tag{2.23}$$

such that $\int \rho_j(t, \theta)\, d\theta$ is independent of t for each j. Given a function $\mathbf{q}(\theta)$ on $\mathbf{T}^d$ taking values in symmetric positive semidefinite $n \times n$ matrices whose entries are $d \times d$ matrices and a function $\tilde{r}(\cdot) : \mathbf{T}^d \to \mathbf{R}^n$ with mean zero, we define

$$\|\tilde{r}\|_{-1,\mathbf{q}(\cdot)}^2 = \sup_\Phi \left\{ 2 \int_{\mathbf{T}^d} \langle \tilde{r}(\theta) \, , \Phi(\theta) \rangle \, d\theta - \int_{\mathbf{T}^d} \langle \nabla \Phi(\theta) \, ; \mathbf{q}(\theta) \nabla \Phi(\theta) \rangle \, d\theta \right\} \tag{2.24}$$

where the supremum is over C^∞ maps Φ from $\mathbf{T}^d$ to $\mathbf{R}^n$. Let us also define

$$\hat{I}^n(\tilde{\rho}(\cdot,\cdot)) = \sup_{\Phi}\left\{2\int_0^T\int_{\mathbf{T}^d}\langle (A\chi)(\tilde{\rho})\nabla\tilde{\rho}\,;\nabla\Phi\rangle\,dt\,d\theta - \int_0^T\int_{\mathbf{T}^d}\langle\nabla\Phi\,;A(\tilde{\rho})\nabla\Phi\rangle\,dt\,d\theta\right\} \tag{2.25}$$

where the supremum is now taken over C^∞ functions on $[0,T]\times\mathbf{T}^d$ with values in $\mathbf{R}^n$. When it makes sense, we can choose $\Phi = \lambda(\tilde{\rho})$, so that $\nabla\Phi = \chi\nabla\tilde{\rho}$, and obtain

$$\hat{I}^n(\tilde{\rho}(\cdot,\cdot)) = \int_0^T\int_{\mathbf{T}^d}\langle\nabla\tilde{\rho}\,;\chi A\chi(\tilde{\rho})\nabla\tilde{\rho}\rangle\,dt\,d\theta \tag{2.26}$$

One can explicitly calculate $\chi A\chi$ and the quantity in (2.26) is finite if and only if

$$\hat{I}(\tilde{\rho}(\cdot,\cdot)) + \sum_{j=1}^n \hat{I}_j(\tilde{\rho}(\cdot,\cdot)) < \infty\ ,$$

where $\hat{I}$ is given by

$$\hat{I}(\rho(\cdot\,,\cdot)) = \int_0^T\int_{\mathbf{T}^d}\frac{|\nabla\rho|^2}{\rho(1-\rho)}\,d\theta\,dt \tag{2.27}$$

as in (1.16) and for $1\le j\le n$,

$$\hat{I}_j(\tilde{\rho}(\cdot\,,\cdot)) = \int_0^T\int_{\mathbf{T}^d}(1-\rho)\frac{|\nabla\rho_j|^2}{\rho_j}\,d\theta\,dt\ . \tag{2.28}$$

The equivalence follows from the following properties which were established in [V2]:

$$D - S(\rho) \le c\rho \tag{2.29}$$

and

$$\delta(1-\rho)I \le S(\rho) \le (1-\rho)D \tag{2.30}$$

where δ and c are two positive constants. The dynamical rate function $I^n_{\mathrm{dyn}}(\tilde{\rho}(\cdot,\cdot))$ is defined to be $+\infty$ unless $\hat{I}^n(\tilde{\rho}(\cdot,\cdot)) < \infty$ and in such a case,

$$I^n_{\mathrm{dyn}}(\tilde{\rho}(\cdot,\cdot)) = \frac{1}{2}\int_0^T\left\|\frac{\partial\tilde{\rho}}{\partial t} - \frac{1}{2}\nabla\cdot A\chi\nabla\tilde{\rho}\right\|^2_{-1,A(\tilde{\rho}(t,\cdot))}dt\ . \tag{2.31}$$

If the initial conditions are deterministic and the law of large numbers holds initially with profile $\tilde{\rho}^0(\cdot)$ then

$$I^n_{\text{init}}(\tilde{\rho}(\cdot,\cdot)) = \begin{cases} 0 & \text{if } \tilde{\rho}(0,\cdot) = \tilde{\rho}^0(\cdot); \\ +\infty & \text{otherwise} \end{cases} . \tag{2.32}$$

For other suitable initial conditions we pick the corresponding rate function for the initial profile and this shall be our $I^n_{\text{init}}(\tilde{\rho}(\cdot,\cdot))$. The full rate function is given by

$$I^n(\tilde{\rho}(\cdot,\cdot)) = I^n_{\text{init}}(\tilde{\rho}(\cdot,\cdot)) + I^n_{\text{dyn}}(\tilde{\rho}(\cdot,\cdot)) . \tag{2.33}$$

A major step towards the full large deviation principle for the empirical process is the following theorem proved in sections 3–6.

Theorem 2.1. *Assuming that the initial distribution satisfies a large deviation principle with a rate funcion I^n_{init}, the measures P_N on the space $E_n = D\big[[0,T];\mathscr{M}(\mathbf{T}^d)\big]^n$ satisfy a large deviation principle with rate function I^n given in (2.33). Moreover, I^n is lower semicontinuous, has compact level sets and is supported on the continuous maps in E_n. More precisely,*

$$\limsup_{N\to\infty} \frac{1}{N^d} \log P_N(C) \le - \inf_{\tilde{\rho}(\cdot,\cdot)\in C} I^n(\tilde{\rho}(\cdot,\cdot)) \tag{2.34}$$

for closed $C \subset E_n$. For $G \subset E_n$ open, we have

$$\liminf_{N\to\infty} \frac{1}{N^d} \log P_N(G) \ge - \inf_{\tilde{\rho}(\cdot,\cdot)\in G} I^n(\tilde{\rho}(\cdot,\cdot)) \tag{2.35}$$

Although the transition from the above theorem to the large deviation result for the empirical process is in principle straightforward it requires careful calculation and identification. The first step is to prove a superexponential estimate for the empirical process $R_{N,\omega}$. As usual in large deviation theory, this estimate provides enough compactness and control to be able to interchange certain limits in the calculation of the rate function as well as in establishing the large deviation principle. The rate function for the empirical process is then calculated by computing rate functions for certain finite dimensional projections and passing to the limit. We divide the torus $\mathbf{T}^d$ into a finite number of cells and establish a large deviation principle for the proportions of particles that lie in a specified sequence of cells at time points $0 \le t_1 < t_2 < \cdots < t_k \le T$. If $E_1, \ldots, E_r$ is the partition of $\mathbf{T}^d$, we consider events of the form

$$\{\omega : \{\pi_{i_1,\ldots,i_k}\} \in A\}$$

where

$$\pi_{i_1,\ldots,i_k} = R_{N,\omega}[x(t_1) \in E_{i_1}, \ldots, x(t_k) \in E_{i_k}]$$

and $A \subset R^{r^k}$ is subset of possible relative frequencies $\pi = \{\pi_{i_1,\dots,i_k}\}$. If we have a rate functions $I_{k,r}(\pi)$ for the large deviations in this context, we could let r and k go to ∞ allowing the partitions to get finer and the time points to become dense. The rate function for the empirical process that we seek will be the limit of $I_{k,r}(\pi)$. While these rate functions are not explicit, they are contractions of rate functions for large deviation from hydrodynamic limit of multi-color systems. To see this we proceed by induction. Depending on the cells the particles fall into at times $t_1 < t_2 \cdots < t_{p-1}$, they are given one of r^{p-1} colors. The collection $\{\pi_{i_1,\dots,i_p}\}$ is nothing but the empirical measure for the multicolor process evaluated at the sets E_{i_p} of the various colors at time t_p are precisely. Therefore the rate functions $I_{k,r}(\pi)$ can in principle be obtained from the multi-color rate functions. This is a somewhat difficult task to carry out and we finesse it to some extent. We 'guess' the rate function and in section 7 connect it to the n-color rate function with a contraction formula. In section 9 we finally establish the large deviation principle for the empirical process using the contraction relationship and the induction procedure that we just outlined.

The big surprise is the reemergence of entropy in the final formula and the physical interpretation that the rate function has because of it. This is not evident in the formula for the large deviation rate for n-colors. Finally let us remark that it suffices to prove the large deviation principle for deterministic initial conditions. Since we can integrate at the end with respect to the initial distribution, any large deviation property of the initial distribution can be incorporated at the very end and contributes a term I^n_{init}.

3. Superexponential estimates

The proof of the large deviation principle for the n-color model follows essentially the methods of [Q1] and [Q2]. There are however some serious problems as $\rho \to 1$. In [Q1] while establishing the law of large numbers, these issues could be avoided because the total density ρ evolves according to the heat equation, and is bounded away from 1 for $t > 0$. In the context of large deviations, the density $\rho(t, \theta)$ can be 1 in a region of space and time, and still have a finite rate function. So the question of ρ being 1 in a region has to be faced.

In addition to the operator L^n_N for the evolution of our basic n-color process on Z^d_N we will consider the generator L^n on the entire lattice Z^d. We will also have occasion to consider the operator L^n_K which will operate in a finite box $|x| \leq K$ disallowing any jumps ever to the exterior of the box. The evolution in the box with a given number of particles of each color can have certain pathological ergodicity properties. For example there are irreducible symmetric random walks involving a single particle in the plane that have no way of getting out of a bad corner in a cube of any size. Also, if the box is full, nothing can move and every color configuration is frozen: The ergodic invariant measures are clearly all the degenerate distributions with mass 1 at an arbitrary color configuration. On the other hand, if all nearest neighbor jumps can occur with positive probability, and if there is at least one empty site, then for boxes with sides larger than a fixed number that depends only on the number n of colors, the ergodic invariant measures are precisely the uniform distributions on all compatible configurations. In fact, it is always true that, for some ℓ_0, any invariant distibution for a box of size ℓ with at least one empty site, when restricted to smaller box of size $\ell - \ell_0$, is a mixture of $\pi_{\ell-\ell_0,\tilde{\rho}}$, the uniform distributions on configurations in the smaller box with given particle numbers for each color.

Let $f(\zeta)$ be a function that is constant on each ergodic component of the configurations in a box of size l, and let us fix the densities of each color in the box. This may not determine f but there will be functions $f^+(\tilde{\rho})$ and $f^-(\tilde{\rho})$ such that $f^+ \geq f \geq f^-$. Throughout this section and the next the number of colors n will be fixed and so we will drop all references to it in our notation.

It is convenient to symmetrize the operator to some extent and for this we introduce the definition

$$a_{x,y}(\zeta) = [\eta(x)(1-\eta(y)) + \eta(y)(1-\eta(x)] \tag{3.1}$$

and rewrite our operators in the form

$$L = \frac{1}{2}\sum_{x,y} p(y-x)a_{x,y}(\zeta)[F(\zeta^{x,y}) - F(\zeta)]$$

$$L_K = \frac{1}{2}\sum_{x,y\in B_K} p(y-x)a_{x,y}(\zeta)[F(\zeta^{x,y}) - F(\zeta)]$$

where B_K denotes the box $|x| \leq K$.

There are two basic superexponential estimates, the one block and two block estimates that are important and have been proved in [Q1]. The one block estimate is:

Theorem 3.1. *Suppose $f(t,\theta,\zeta)$ is a smooth function of t and θ on $[0,t]\times \mathbf{T}^d$ which is a local function depending on a fixed box B and which vanishes identically in t and θ when the box is full. Let us denote by $\hat{f}(t,\theta,\tilde{\rho})$ the expected value of f under the Bernoulli measure $\pi_{\tilde{\rho}}$. Then*

$$\limsup_{K\to\infty}\limsup_{N\to\infty}\frac{1}{N^d}\log E^{P_N}\left[\exp\left\{\int_0^T\sum_x\Big|\frac{1}{(2K+1)^d}\right.\right.$$

$$\left.\left.\times\sum_{y:|y-x|\le K} f(t,\frac{y}{N},\tau_y\zeta(t))-\hat f(t,\frac{x}{N},\tau_x\bar\zeta_K)\Big|\,dt\right\}\right]\le 0\ .$$

Here $\bar\zeta_K$ refers to the average of $\zeta(x)$ over the box B_K and τ_x is the shift by x. The two block estimate is:

Theorem 3.2. *Suppose $f(t,\theta,\tilde\rho,\tilde\rho')$ is a continuous function on $[0,T]\times\mathbf{T}^d\times\mathscr{C}\times\mathscr{C}$ that vanishes identically on $[0,T]\times\mathbf{T}^d$ when $\rho=\rho'=1$ or $\tilde\rho=\tilde\rho'$. Then*

$$\limsup_{\epsilon\to0}\limsup_{K\to\infty}\limsup_{N\to\infty}\frac{1}{N^d}$$

$$\times\log E^{P_N}\left[\exp\left\{\int_0^T\sum_x f(t,\frac{x}{N},\tau_x\bar\zeta_{N\epsilon},\tau_x\bar\zeta_K)\,dt\right\}\right]\le 0\ .$$

We now turn to some of the details specific to nongradient systems. Suppose $v(\zeta)$ is a local function depending only on configurations in some finite box B_l, which has mean zero with respect to every invariant measure for the evolution restricted to that box. This will of course force the function to vanish for every configuration that fills the box. We denote the class of such functions by $\mathscr{N}_l$. A function defined on a box B_l may have this property not for B_l but only for a larger box $B_{l'}$ in which case it will be in $\mathscr{N}_{l'}$. We denote by $\mathscr{N}$, the increasing union of $\mathscr{N}_l$ over l.

For any $v\in\mathscr{N}_l$ we form the sum

$$\sum_{|x|\le K-l} v(\tau_x\zeta)=V_K(\zeta)$$

and it defines a function on B_K that belongs to $\mathscr{N}_K$. We define the quantity $f_K(\zeta)$, for configurations of the box B_K that belong to the support of any extremal invariant measure to equal $\langle(-L_K)^{-1}V_K,V_K\rangle$ relative to that invariant measure. On the set of configurations where every site is filled we may take f_K to be zero. The function f_K of course will be constant on each ergodic component. We should expect it to be purely a function of densities $\tilde\rho$ of various colors in the box. However ergodicity can sometimes breakdown and we can only bound f_K above and below by such functions $f_K^{\pm}$ that depend only on the densities. For $v\in\mathscr{N}$, it is easy to get an upper bound for f_K by writing the dual variational formula

$$f_K = \langle (-L_K)^{-1} V_K, V_K \rangle = \sup_u \{ 2E^{\zeta}[u(\cdot)V_K(\cdot)] - \langle -L_K u, u \rangle \} \qquad (3.2)$$

where E^{ζ} refers to expectation with respect to the extremal invariant measure supported on the irreducible ergodic set containing the configuration ζ in the box B_K. We note that if $v \in \mathcal{N}_l$, it can be written as $L_K w$ for some B_l measurable w. Then

$$V_K(\zeta) = \sum_{|x| \le K - l} L_{B+x} \tau_x w$$

and it is straightforward to perform a summation by parts and obtain, using Schwarz's inequality,

$$f_K \le C(w)(2K+1)^d$$

Here $C(w)$ can be bounded by

$$C(w) \le C|B| \sup_{x,\zeta} E^{\zeta}\{[\tau_x w]^2\}$$

where x ranges over allowable translates i.e. $|x| \le K - l$, and E^{ζ} ranges over all extremal invariant measures for the box of size K. In particular

$$\limsup_{K \to \infty} \frac{1}{(2K+1)^d} \sup_{\zeta} f_K(\zeta) \le C(w)$$

A subset $\mathcal{V}_l \subset \mathcal{N}_l$ of these functions v are of the form

$$\sum_r u_r(\eta) v_r(\zeta) \qquad (3.3)$$

for some functions u_r that are "color blind" and functions v_r that are linear functions of the color variables $\zeta(x)$. This class of functions is preserved under the action of the generator. In [Q1] it is shown that the spectral gap of the generator, acting on this smaller class of functions, on a box of side length l with either periodic or reflecting boundary conditions, decays as l^{-2}. In the future, for the most part, we will restrict v to the class $\mathcal{V}$ which is defined as the (increasing) union of $\mathcal{V}_l$ over l. It was shown in [Q1] that

$$\lim_{K \to \infty} \frac{1}{(2K+1)^d} f_K^{\pm}(\tilde{\rho}) = [v, v]_{\tilde{\rho}} \qquad (3.4)$$

exists locally uniformly on $\mathscr{C}^0$ defined as the subset of the color densities $\mathscr{C}$ where $\rho < 1$.

To carry out these calculations we need to introduce certain Hilbert spaces. For any $\tilde{\rho}$ in $\mathscr{C}^0$ let $\pi_{\tilde{\rho}}$ be the Bernoulli measure on the configurations in Z^d. Let $g(\zeta)$ be a local function and consider the formal sum

$$G(\zeta) = \sum_{x \in Z^d} g(\tau_x \zeta) \ . \tag{3.5}$$

While the sum G clearly does not exist, the quantities

$$(\nabla_{0,z} G)(\zeta) = G(\zeta^{0,z}) - G(\zeta) = \xi_z \tag{3.6}$$

exist and involve only a finite sum. We consider the collection $\{\xi_z\}$ as some sort of gradient and view them as elements in a Hilbert space $H^0_{\bar{\rho}}$ with respect to the norm

$$\langle\langle \xi, \xi \rangle\rangle_{\bar{\rho}} = \frac{1}{4} E^{\pi_{\bar{\rho}}} \left[\sum_z p(z) a_{0,z}(\zeta) |\xi_z|^2 \right] \ . \tag{3.7}$$

A gradient $\{\xi_z\}$ created in this way (3.6) will possess certain natural algebraic properties. For example, if for some k a finite sequence $(x_i, x_i + z_i)$, $i = 1, \ldots, k$ of bonds forms a closed loop in Z^d, then $\sum_{i=1}^k \tau_{x_i} \xi_{z_i} = 0$. One can then define abstract gradients $\{\xi_z\}$ that possess these same algebraic properties. Under the same norm, this will define a slightly larger Hilbert space $H_{\bar{\rho}}$ that is exactly *nd* dimensions larger than $H^0_{\bar{\rho}}$ and a relative (non-orthogonal) complement to $H^0_{\bar{\rho}}$ is spanned by $\Psi = \{\psi^{i,j}\}$,where

$$\psi^{i,j}_z = \langle z, e_i \rangle [\eta^j(z) - \eta^j(0)] \tag{3.8}$$

for $1 \le i \le d$ and $1 \le j \le n$.

Now we discuss how to compute $[v, v]_{\bar{\rho}}$. The general v is in fact a linear combination of three basic types and we discuss the computation type by type. If we have a function v of the form

$$v = Lg \tag{3.9}$$

where g is a local function which is linear in color, i.e. of the form (3.3), then one can compute $[v, v]_{\bar{\rho}}$ as the norm in $H_{\bar{\rho}}$ of the image ξ of v through the map $g \to G \to \xi$ defined in (3.5), (3.6) and one finds that

$$[v, v]_{\bar{\rho}} = \langle\langle \xi, \xi \rangle\rangle_{\bar{\rho}} \tag{3.10}$$

On the other hand we can have a v which under an appropriate shift takes the form,

$$v = \sum_z a_{0,z}(\zeta) p(z) [F(\zeta^{0,z}) - F(\zeta)] \tag{3.11}$$

where F is a local function linear in color. Let us briefly consider the space $\bar{H}_{\bar{\rho}}$ of functions $\{h_z\}$ that do not necessarily satisfy the algebraic relations, but equipped with the same norm

$$\langle\langle h, h \rangle\rangle_{\bar{\rho}} = \frac{1}{4} E^{\pi_{\bar{\rho}}} \left[\sum_z p(z) a_{0,z}(\zeta) |h_z|^2 \right] \ . \tag{3.12}$$

Such an $\{h_z\}$ can be projected orthogonally into $H_{\tilde{\rho}}$ to get a $\{\xi_z\}$. If v is of the form (3.11) then for any $\tilde{\rho} \in \mathscr{C}^0$ the way to compute the value of $[v, v]_{\tilde{\rho}}$ is as follows: First associate with v the following $\{h_z\}$ in $\bar{H}_{\tilde{\rho}}$:

$$h_z(\zeta) = -2[F(\zeta^{0,z}) - F(\zeta)] \ . \tag{3.13}$$

Then project $\{h_z\}$ into $H_{\tilde{\rho}}$ and obtain $\{\xi_z\}$. Finally $[v, v]_{\tilde{\rho}}$ is calculated as $\langle\langle \xi, \xi \rangle\rangle_{\tilde{\rho}}$. One can verify that this is consistent with the map $g \to G \to \xi$ defined earlier.

The "current" term $\mathbf{W}$ given by

$$W^{i,j} = \frac{1}{2} \sum_z a_{0,z}(\zeta) p(z) \langle z, e_i \rangle (\eta_j(z) - \eta_j(0)) \tag{3.14}$$

is easily seen to be imbedded in this manner in $H_{\tilde{\rho}}$ as $\Psi = \{\psi_z^{i,j}\}$ where the $\psi_z^{i,j}$ are defined in (3.8) for $1 \le i \le d$ and $1 \le j \le n$.

The following calculation was carried out in [Q1]: For $\tilde{\rho}$ in $\mathscr{C}^0$ and any matrix B

$$\inf_{\xi \in H^0_{\tilde{\rho}}} E^{\pi_{\tilde{\rho}}} \left[\frac{1}{4} \sum_z a_{0,z}(\zeta) p(z) (\langle B ; \Psi_{\mathbf{z}} \rangle - \xi_z)^2 \right] = \frac{1}{2} \langle B ; A(\tilde{\rho}) B \rangle \tag{3.15}$$

Although the result was proved for $\tilde{\rho}$ in the interior $\mathscr{C}^0$ where $\rho < 1$, we have an easy estimate that all the norms concerned for v in $\mathscr{V}$ are bounded by a multiple of $1 - \rho$. In fact, everything extends to the boundary easily except the following family of density gradients $\nabla\zeta$ which are *not* in $\mathscr{V}$:

$$\nabla_e \zeta = \{\eta_j(e_i) - \eta_j(0); \ 1 \le i \le d, 1 \le j \le n\} \tag{3.16}$$

It was shown in [Q1] that these can be imbedded in $H_{\tilde{\rho}}$ as well, for $\rho < 1$, as

$$\mathbf{f} = \{f_{i,j}\} \tag{3.17}$$

and in fact belong to the orthogonal complement of $H^0_{\tilde{\rho}}$ providing us a natural *nd* dimensional subspace orthogonal to $H^0_{\tilde{\rho}}$. A simple calculation using L_K^{-1} establishes

$$[\langle b ; \mathbf{W}\alpha \rangle, \langle b' ; (\nabla\zeta)\alpha' \rangle]_{\tilde{\rho}} = \langle b \otimes \alpha ; \chi^{-1}(\tilde{\rho}) \, b' \otimes \alpha' \rangle$$

Since χ is nonsingular this proves that modulo $H^0_{\tilde{\rho}}$

$$\Psi = \frac{1}{2} A \chi \mathbf{f} \tag{3.18}$$

which essentially captures the diffusion law.

Finally there are some large deviation results that are proved in [Q1] that are restated here for convenience. Suppose v is a function is class $\mathscr{N}$. Let $Q(\tilde{\rho})$ be a continuous function on $\mathscr{C}$ that satisfies

$$\limsup_{\substack{K\to\infty\\ \zeta\sim\tilde{\rho}}} \frac{1}{(2K+1)^d} f_K(\zeta) \le Q(\tilde{\rho})$$

In particular if $v \in \mathscr{V}$ we can take $Q(\tilde{\rho}) = |||v|||_{\tilde{\rho}}^2 = [v, v]_{\tilde{\rho}}$ which defines a continuous function on $\mathscr{C}^0$ and because it is dominated by $1-\rho$ it extends continuously to $\mathscr{C}$.

Theorem 3.3. *If we denote by $\hat{E}$ the expectation with respect to any equilibrium process then*

$$\limsup_{K\to\infty} \limsup_{N\to\infty} \frac{1}{N^d} \log \hat{E}\left[\exp\left\{N\int_0^T \sum_x \tau_x\, v\, ds - \int_0^T \sum_x \tau_x Q(\bar{\zeta}_K)\right\}\right] \le 0$$

where $\bar{\zeta}_K$ is the average of ζ over the box of side length $(2K+1)$.

Remark 1. Since the entropy of any deterministic initial condition is bounded by C^{N^d} for some constant C, for arbitrary initial conditions the above theorem is valid with a universal constant C replacing zero on the right as a bound for the limsup.

Remark 2. The theorem is valid in exactly the same form if v is replaced by $v(t, \frac{x}{N}, \tau_x\zeta)$ provided it depends continuously on t and θ as they range over $[0, T] \times \mathbf{T}^d$ and takes values in $\mathscr{V}_l$ for some value of l.

Remark 3. The right hand side of (3.15) is a continuous function of $\tilde{\rho} \in \mathscr{C}^0$, the subset of color densities with $\rho < 1$. If we limit the infimum to ξ that come from a g that is local and defined in a box of size l and then increase l, we get a nondecreasing sequence of continuous functions that converges on $\mathscr{C}^0$ to a continuous limit. By Dini's theorem the convergence is uniform. It now follows easily that the minimizers ξ_z that are defined on different spaces, have the property

$$E^{P_{\tilde{\rho}}}\left[\sum_z a_{0,z}(\zeta)p(z) < B_z(\zeta)\,;\,\xi_z >\right]$$

are continuous on $\mathscr{C}^0$ provided B_z are local functions.

4. Upper and lower bounds for the *n*-color system

Like many large deviation results, the upper bound depends on our ability to estimate $E[\exp\{N^d F_\alpha(\zeta(\cdot))\}]$ for a suitable family of functions F_α. It is important that the functions F_α have representations in terms of the measure-valued process $\tilde{\nu}(\cdot)$ that makes them bounded and continuous in the topology of weak (as measures on $\mathbf{T}^d$) uniform convergence on the interval $[0, T]$. Although we can, and will, use many of the exponential martingales that are available, we cannot do so directly because of the above restriction. The plan for the upper bound is then to exhibit a family of functionals $F_\alpha(\zeta(\cdot))$ that are representable as bounded continuous functionals $G_\alpha(\tilde{v}(\cdot))$ of the measure valued process $\tilde{v}(\cdot)$ and have the property

$$\limsup_{n\to\infty} \frac{1}{N^d} \log E[\exp\{N^d F_\alpha(\zeta(\cdot))\}] \leq 0$$

This will then give immediately an upper bound for compact sets with the rate function $I(\tilde{\rho}(\cdot,\cdot))$, if we can show that $I(\tilde{\rho}(\cdot,\cdot))$ is lower semicontinuous and satisfies

$$I(\tilde{\rho}(\cdot,\cdot)) \leq \bar{I}(\tilde{\rho}(\cdot,\cdot)) = \sup_\alpha G_\alpha(\tilde{\rho}(\cdot))$$

We will often misuse the notation and not distinguish between F and G. If we have superexponential tightness estimates, then the upper bound will extend from compact sets to closed sets. The lower semicontinuous functional I should be shown to have compact level sets.

The lower bounds depend on establishing laws of large numbers, in hydrodynamic scaling, for certain weak perturbations and controlling by the Girsanov formula the relative entropies of the processes that go with these perturbations. Since the methods for establishing laws of large numbers are fairly close to the ones for providing upper bounds for large deviations, we prefer to establish the upper bounds for a class of models that includes the weak perturbations that we will use for the lower bound.

Our basic infinitesimal generator L_N^n is modified by a time dependent lower order term A_N^n given by

$$A_N^n f(\zeta) = \frac{N}{2} \sum_{x,z} a_{x,x+z}(\zeta) p(z) b\Big(t, \frac{x}{N}, z, \tau_x \zeta\Big) (f(\zeta^{x,x+z}) - f(\zeta)) \tag{4.1}$$

The new term b provides a correction to the old jump rate $p(z)$, and the modified jump rate is of the form $p(z)(1 + \frac{1}{N} b(t, \frac{x}{N}, z, \tau_x \zeta))$. The functions $b(t, \theta, z, \cdot)$ are all assumed to be local functions of ζ that depend on the configuration in some fixed finite box. Moreover, for each jump size z, and each configuration ζ, they are assumed to be continuous functions of t and θ on $[0, T] \times \mathbf{T}^d$. We will also assume that they have a uniform bound.

We will denote by $P_{N,b}$ the measure corresponding to the n-color process on $[0, T]$ with generator $L_N^n + A_N^n$ and with arbitrary initial conditions. $E_{N,b}$ will denote expectation with respect to $P_{N,b}$. We will continue to denote by the same $P_{N,b}$ the induced measure on the space of measure–valued functions $\tilde{\nu}(\cdot)$ on $[0, T]$. If b is 0 it will be omitted. An elementary calculation with the Girsanov formula for the Radon-Nikodym derivative of $P_{N,b}$ with respect to P_N provides an upper bound for any L_p norm of this Radon-Nikodym derivative. In particular, since we have assumed a uniform bound for b, one can easily prove the following lemma.

Lemma 4.1. *Uniformly with respect to all initial conditions, we have*

$$\lim_{p\to 1}\limsup_{N\to\infty}\frac{1}{N^d}\log E_N\left[\left|\frac{dP_{N,b}}{dP_N}\right|^p\right]\leq 0\ .$$

Remark. We could also take the initial condition to be some reasonable equilibrium distribution. Changing over from such an equilibrium to any initial condition again involves a Radon-Nikodym derivative, for the initial condition, that clearly is bounded uniformly by $\exp[CN^d]$. In particular superexponential estimates obtained in equilibrium for the unperturbed process remain valid for all initial conditions and for all of our perturbations by a simple application of Hölder's inequality.

We will prove in section 8 a superexponential tightness estimate for the family Q_N of measures on the space of processes. This will automatically yield a similar estimate for $P_{N,b}$. We denote by $D[[0, T], \mathscr{M}]$ the space of measure-valued maps with discontinuities of the first kind only. E_n denotes the space of n-tuples of such maps on which the measures $P_{N,b}$ live and where the large deviation results are to be established. Moreover by the simple exclusion property we have an apriori bound of 1 for the total density. The rate function for the large deviation will then live on the space $\mathscr{D}_n$ of weakly continuous L_1 maps $\tilde{\rho}$ from $[0, T]\times\mathbf{T}^d$ into $\mathscr{C}=\{(\rho_1,\ldots,\rho_n):\rho_i\geq 0, \rho=\sum_{i=1}^n\rho_i\leq 1\}$. We will need to prove certain preliminary estimates in order to limit the support of the rate function. We define $\mathscr{D}_n^0$ as those maps for which $\hat{I}^n(\tilde{\rho}(\cdot\,,\,\cdot))$, given by (2.26), is finite or equivalently $\hat{I}(\tilde{\rho}(\cdot\,,\,\cdot))$ and $\{\hat{I}_j(\tilde{\rho}(\cdot\,,\,\cdot)); 1\leq j\leq n\}$, given by (2.27) and (2.28), are finite.

We define the rate function $I^{n,b}_{dyn}(\tilde{\rho}(\cdot\,,\,\cdot))$ by

$$I^{n,b}_{\text{dyn}}(\tilde{\rho}(\cdot,\cdot)) = \begin{cases}\frac{1}{2}\int_0^T\|\frac{\partial\tilde{\rho}}{\partial t}-\frac{1}{2}\nabla\cdot(A\chi)(\tilde{\rho})\nabla\tilde{\rho}-\nabla\cdot H\|^2_{-1,A(\tilde{\rho}(t,\cdot))}\,dt & \text{if } \tilde{\rho}(\cdot\,,\cdot)\in\mathscr{D}_n^0\\ +\infty & \text{if } \tilde{\rho}(\cdot\,,\cdot)\notin\mathscr{D}_n^0\ .\end{cases} \tag{4.2}$$

Here $H = H_{i,j}(t, x, \tilde{\rho})$ depends in a nontrivial way on the perturbation b. In fact $H = \hat{\mathbf{a}}[\mathbf{b}]$ is given by (4.51), but for now we do not need the specific form. All we need is that H is continuous in $\tilde{\rho}$ and satisfies the bound

$$\sup_{t,x,j} |H_{i,j}(t, x, \tilde{\rho})| \leq C\rho_i(1-\rho) \ . \tag{4.3}$$

For $\tilde{\rho}(\cdot,\cdot) \in \mathscr{D}_n^0$, by duality, $I_{dyn}^{n,b}(\tilde{\rho}(\cdot,\cdot))$ has a variational formula,

$$I_{\text{dyn}}^{n,b}(\tilde{\rho}(\cdot,\cdot)) = \sup_{\Phi}\left\{\int_0^T\int_{\mathbf{T}^d}\left[-\left\langle\frac{\partial\Phi}{\partial t},\tilde{\rho}\right\rangle+\frac{1}{2}\langle\nabla\Phi;(A\chi)(\tilde{\rho})\nabla\tilde{\rho}\rangle + \langle\nabla\Phi;H\rangle\right]dt\,dx - \frac{1}{2}\int_0^T\int_{\mathbf{T}^d}\langle\nabla\Phi;A(\tilde{\rho})\nabla\Phi\rangle\,dt\,dx\right\} \tag{4.4}$$

which makes sense for $\tilde{\rho} \in \mathscr{D}_n^0$, because χ^{-1} is bounded.

The space of functions $\tilde{\rho}$ with a uniform bound on their rate functions enjoy certain compactness properties that will be used in several occasions.

Lemma 4.2. *Let ρ_n and R_n be two sequences of functions such that for all n,*

$$\int_0^T\int\{|\nabla\rho_n|^2 + (1-\rho_n)|\nabla R_n|^2\}\,dx\,dt + \int_0^T\left\|\frac{\partial\rho_n}{\partial t}\right\|_{-1}^2 + \left\|\frac{\partial R_n}{\partial t}\right\|_{-1}^2 dt \leq C \ .$$

Suppose $0 \leq \rho, R \leq 1$ and that $\rho_n \rightharpoonup \rho$, $R_n \rightharpoonup R$ weakly in $L^2([0,T]\times\mathbf{T}^d)$. Then $\rho_n \to \rho$ and $(1-\rho_n)R_n \to (1-\rho)R$ strongly in $L^1([0,T]\times\mathbf{T}^d)$.

Proof. Let $\phi(x)$ be a smooth mollifier and put $\rho_n^\delta = \rho_n * \phi^\delta$, where $\phi^\delta(x) = \delta^{-d}\phi(\frac{x}{\delta})$ and the convolution is in the x variable. The gradient estimate on ρ_n implies

$$\lim_{\delta\to 0}\sup_n \int_0^T\int|\rho_n^\delta - \rho_n|\,dx\,dt = 0 \ . \tag{4.5}$$

On the other hand, the H_{-1} bound on the $\partial_t\rho_n$ implies that for each δ, the full gradient $(\partial_t\rho_n^\delta, \nabla\rho_n^\delta)$ is uniformly bounded in n. From this and the weak convergence of ρ_n to ρ we deduce that ρ_n^δ converges strongly to $\rho^\delta = \rho * \phi^\delta$. We can now let δ go to zero and use (4.5) to conclude the strong convergence of ρ_n to ρ.

For any integrable function g, let $u_n(g,t,x) = \int R_n(t,y)\, g(y,t)\phi^\delta(x-y)\, dy$. If g is smooth, the H_{-1} bound on $\partial_t R_n$ and the first assertion of the lemma imply the L^1 convergence of $u_n(g,t,x)$ to $u(g,t,x) = \int R(y,t)g(y,t)\phi^\delta$. Since

$$\int_0^T \int |u_n(g') - u_n(g)|\, dx\, dt \le \int_0^T \int |g' - g|\, dx\, dt \ ,$$

we also have that $u_n(g)$ converges strongly to $u(g)$ for every integrable g. Repeating the same reasoning we deduce that if g_n converges to g in L^1 sense, then $u_n(g_n)$ converges to $u(g)$ in L^1 sense too. We now choose $g_n = 1 - \rho_n$. Therefore we have the following L^1 convergence:

$$\lim_{n\to\infty} [(1-\rho_n)R_n] * \phi^\delta = [(1-\rho)R] * \phi^\delta \ .$$

On the other hand, we have a uniform bound on the L^2 norm of the spatial gradient of the sequence $(1-\rho_n)R_n$ which allows us to send δ to zero and drop the convolution with ϕ^δ.

Theorem 4.3. *There are constants C_1, C_2 and C_3 such that the rate function $I^{n,b}_{dyn}(\tilde\rho(\cdot,\cdot))$ defined by* (4.2) *satisfies the bounds*

$$\hat I^n\,(\tilde\rho(\cdot\,,\cdot)) \le C_1 + C_2 I^{n,b}_{\text{dyn}}(\tilde\rho(\cdot\,,\cdot)) + C_3 \int_0^T \int_{\mathbf{T}^d} \big\langle H\,; A^{-1}(\tilde\rho)H \big\rangle\, dt\, dx \tag{4.6}$$

and

$$\int_0^T \left\| \frac{\partial\tilde\rho}{\partial t} \right\|^2_{-1,A(\tilde\rho(t\,,\cdot))} dt \le C_1 + C_2 I^{n,b}_{\text{dyn}}(\tilde\rho(\cdot\,,\cdot)) + C_3 \int_0^T \int_{\mathbf{T}^d} \big\langle H\,; A^{-1}(\tilde\rho)H \big\rangle\, dt\, dx \tag{4.7}$$

Moreover the function $I^{n,b}_{\text{dyn}}$ is a lower semicontinuous function with compact level sets. Finally, for the measures $P_{N,b}$ the large deviation rate function is infinite outside $\mathscr{D}^0_n$.

Proof. The function $h(\tilde\rho) = \sum_i \rho_i \log \rho_i + (1-\rho)\log(1-\rho)$ is bounded on $\mathscr{C}$ and an easy calculation of

$$\frac{\partial}{\partial t} \int_{\mathbf{T}^d} h(\tilde\rho(t,x))\, dx$$

followed by an integration by parts yields the bound (4.6) and (4.7) follows from (4.6). The bounds (4.6) and (4.7) are sufficient to establish compact level sets. We next prove the lower semicontinuity of $I_{dyn}^{n,b}(\tilde{\rho}(\cdot\,,\cdot))$. Let $\tilde{\rho}^k(t\,,\cdot) \Rightarrow \tilde{\rho}(t\,,\cdot)$ uniformly in t. Under the assumption that $I_{dyn}^{n,b}(\tilde{\rho}^k(\cdot\,,\cdot))$ is uniformly bounded by some constant, we will establish that $A\chi(\tilde{\rho}^k(\cdot\,,\cdot))\nabla\tilde{\rho}^k$, $A(\tilde{\rho}^k(\cdot\,,\cdot))$ and $H(\tilde{\rho}^k(\cdot,\cdot))$ converge weakly in $L^2([0,T]\times\mathbf{T}^d)$ to $A\chi(\tilde{\rho}(\cdot\,,\cdot))$, $A(\tilde{\rho}(\cdot\,,\cdot))$ and $H(\tilde{\rho}(\cdot,\cdot))$ respectively. Then we can take the limit in (4.4) and the variational formula will establish the lower semicontinuity. The formula for $A(\tilde{\rho})$ is given in (2.13). An explicit calculation yields

$$[A\chi(\tilde{\rho})\nabla\tilde{\rho}]_i = S(\rho)\nabla\rho_i + \frac{\rho_i}{\rho}(D - S(\rho))\nabla\rho\ . \tag{4.8}$$

By observation of these various functions it is clear that we need: 1. Strong convergence of the total densities $\rho^k(\cdot\,,\cdot)$ to $\rho(\cdot\,,\cdot)$, 2. Weak convergence of $(1-\rho^k)\nabla\rho_i^k$ to $(1-\rho)\nabla\rho_i$, 3. Strong convergence of $\rho_i^k(1-\rho^k)$ to $\rho_i(1-\rho)$. By (4.3) we have a uniform bound on $\langle H\,; A^{-1}H\rangle$. Let us note that the bounds (4.6) and (4.7) imply uniform bounds on

$$\int_0^T\int_{\mathbf{T}^d}\frac{|\nabla\rho^k|^2}{\rho^k(1-\rho^k)}\,dt\,dx \text{ and } \int_0^T\left\|\frac{\partial\rho^k}{\partial t}\right\|_{-1}^2 dt$$

implying 1. While we do not have quite the same bounds on the densities of individual colors, we do have bounds on

$$\int_0^T\int_{\mathbf{T}^d}\frac{(1-\rho^k)}{\rho^k}|\nabla\rho_i^k|^2\,dt\,dx \text{ and } \int_0^T\left\|\frac{\partial\rho_i^k}{\partial t}\right\|_{-1}^2 dt$$

implying 2 and 3.

We now turn to proving that the large deviation rate for $P_{N,b}$ is infinite outside $\mathscr{D}_n^0$. Part of it, namely that $\hat{I}$ given by (2.27) has to be finite was established in [KOV]. We now concentrate on (2.28). Let e be a coordinate direction and let $f = \tau_e F - F$ be the difference between a local function F and its shift in one of the coordinate directions by one step. We shall assume that the local function F depends on a finite box, is linear in color, and vanishes identically for all configurations that fill that box. It is easily seen that f, which is then defined in a slightly larger box, is from the class $\mathscr{V}$ and can in fact be represented in the form $f = L_B\,g$ for some g and some box B. If f vanishes identically on a set A of configurations inside the box B which is invariant under permutations inside B, such a set will be an ergodic component under L_B and g can be taken to be zero on A. Moreover, if f is bounded by 1, there is a constant C that depends only on the size of the box B that will bound g.

In the next few steps, as we proceed with the reasoning, this constant may change but will always be replaced by another constant that only depends on the size of the box B. We shall consider the expectation $\hat{E}$ in equilibrium:

$$\hat{E}\left[\exp\left\{N\int_0^T\sum_x \Phi\left(t,\frac{x}{N}\right)\tau_x f(\zeta(t))\,dt - C\int_0^T\sum_x \tau_x\chi_A(\zeta(t))\left|\Phi\left(t,\frac{x}{N}\right)\right|^2 dt\right\}\right]. \tag{4.9}$$

Because f is of the form $f = L_B g$, the variational quantity

$$\sup_\phi\left\{E^\phi\left[N\sum_x \Phi\left(t,\frac{x}{N}\right)\tau_x f(\zeta) - C\sum_x \tau_x\chi_A(\zeta(t))\left|\Phi\left(t,\frac{x}{N}\right)\right|^2 - \frac{N^2}{4}\sum_{x,y} a_{x,y}(\zeta)p(y-x)\frac{1}{\phi}\left|\nabla_{x,y}\sqrt{\phi}\right|^2\right]\right\}$$

is seen to be nonpositive for a suitable choice of C. This yields a bound of 1, by an application of Feynman-Kac formula, to the quantity (4.9). After performing a summation by parts we can apply Theorems 3.1 and 3.2. The remarks about the Radon-Nikodym derivatives now allow us to construct functions of the form

$$F_\alpha(\zeta(\cdot)) = \int_0^T \frac{1}{N^d}\left[\sum_x (\nabla_e\Phi)\left(t,\frac{x}{N}\right)\hat{F}(\tau_x\bar{\zeta}_{N\epsilon}) - C\sum_x \hat{\chi}_A(\tau_x\bar{\zeta}_{N\epsilon})\left|\Phi\left(t,\frac{x}{N}\right)\right|^2\right] dt$$

that depend on $\alpha = \{\Phi, \epsilon\}$, that are weakly continuous and correspond to the functionals

$$G_\alpha(\tilde{\rho}(\cdot)) = \int_0^T\int_{\mathbf{T}^d}\left[(\nabla_e\Phi)(t,\theta)\hat{F}(\tilde{\rho}_\epsilon(t,\theta)) - C\,\hat{\chi}_A(\tilde{\rho}_\epsilon(t,\theta))|\Phi(t,\theta)|^2\right] dt\,d\theta$$

Here $\tilde{\rho}_\epsilon$ represents the convolution of $\tilde{\rho}$ with the normalized indicator function of a box of size ϵ on the torus $\mathbf{T}^d$. The functionals have the property that for some fixed $0 < \lambda < 1$ and $C < \infty$, that are independent of α

$$\limsup_{\epsilon\to 0}\limsup_{N\to\infty}\frac{1}{N^d}\log E\left[\exp\left\{N^d\lambda F_\alpha(\zeta(\cdot))\right\}\right] \le C$$

for every Φ. This provides a lower bound for the large deviation rate $\bar{I}$.

$$\bar{I}(\tilde{\rho}(\cdot,\cdot)) \ge \sup_\Phi \lim_{\epsilon\to 0}\{\lambda\, G_\alpha(\tilde{\rho}_\epsilon(\cdot,\cdot)) - C\} \tag{4.10}$$

Now we take F to be $F_j^l = \chi_l(\zeta)\eta_j(0)$ where χ_l is the indicator of those configurations with at least one empty site in the box of side $2l+1$ around the origin. While we will let $l \to \infty$ later on, for the present proof we only need some l so that for all i, j, $\nabla_{e_i} F_j^l$ are in $\mathscr{V}_{l'}$ for some l' and then hold l and l' fixed. F_j^l vanishes unless there is at least one empty site and at least one particle of color j in the bigger box of size l'. Taking A to be the set of these configurations we get the inequality $\hat{\chi}_A(\tilde{\rho}) \le C\rho_j(1-\rho)$. Moreover it is seen that $\hat{F}_j^l(\tilde{\rho}) = E_{\pi_{\tilde{\rho}}}[F_j^l] = \rho_j c_l(\rho)$ where $c_l(\rho)$ and $\frac{dc_l(\rho)}{d\rho}$ behave like $(1-\rho)$ and 1 as $\rho \to 1$. Since we know that the quantity defined by (2.27) has to be finite in order for the rate function to be finite , it now follows that the similar property holds for the quantity defined by (2.28).

We now turn towards proving upper bounds. We need to prove them only at those points that are in $\mathscr{D}_n^0$. If $U(t,\zeta)$ is any function of time and the n-color configuration then

$$\exp\left\{ U(t,\zeta(t)) - U(0,\zeta(0)) - \int_0^t U_s(s,\zeta(s))\,ds \right.$$
$$\left. - N^2 \int_0^t (\exp[-U]\;[L_N^n + A_N^n]\;\exp[U])(\zeta(s))\,ds \right\} \tag{4.11}$$

is a martingale with respect to the n-color process $P_{N,b}$. We can, without loss of generality, assume that the initial condition is deterministic and leads to a profile $\tilde{\rho}^0$ of densities at time 0. The basic U's that we want to look at are of the form

$$U(t,\zeta) = \sum_x \left\langle \Phi\left(t, \frac{x}{N}\right), \zeta(x) \right\rangle - \frac{1}{N}\sum_x g\left(t, \frac{x}{N}, \tau_x \zeta\right) \tag{4.12}$$

where Φ is a continuously differentiable function on $[0,T] \times \mathbf{T}^d$ with values in $\mathbf{R}^n$ and g is a local function depending continuously on $[0,T] \times \mathbf{T}^d$. We will later pick g to be of the form $g = \sum_{i,j} (\nabla\Phi)_{i,j} g_{i,j}$. In addition to Φ and g and the functions F_j^l that depend on the box size l, we will have certain auxiliary parameters: four small positive numbers $\epsilon_1, \epsilon_2, \epsilon, \delta$, and a continuous map J from $\mathscr{C}$ into $N \times d$ matrices. We will assume that J vanishes if $\rho = 1$.

Assume that F is a local function defined on some finite box, is linear in color, and is zero if the box is full. Let us define $\nabla_e F = \tau_e F - F$ where e is a coordinate direction and $\bar{F}_{x,k}$ as the average of $\tau_y F$ for y in a box of side $2k+1$ centered at x. For any function G we define the lattice gradient in the direction e as

$$\nabla_{N\epsilon,e} G = \frac{1}{2\epsilon}[\tau_{N\epsilon e} G - \tau_{-N\epsilon e} G].l$$

Remark. Although we use ∇ to denote different types of objects like ∇, ∇_e, $\nabla_{x,y}$ and $\nabla_{N\epsilon,e}$ it should be clear from the context what it denotes.

The next three lemmas are stated for the scalar case, with functions φ and h. They are typically one of the components of J, Φ or $\nabla\Phi$ and as a component of J, h will be zero on $\rho = 1$. The local function F will be linear in color and vanish if all sites are filled.

Lemma 4.4. *For any F, any coordinate direction e and for any $\epsilon_1 > 0$*

$$\limsup_{\epsilon\to 0} \limsup_{\epsilon_2\to 0} \limsup_{N\to\infty} \frac{1}{N^d} \log \hat{E} \times \left[\exp\left\{\int_0^T \sum_x \varphi\left(t, \frac{x}{N}\right) h(\bar{\zeta}_{x,N\epsilon_1}) \times [N\nabla_e(\tau_x F) - \nabla_{N\epsilon,e}\bar{F}_{x,N\epsilon_2}]\, dt\right\}\right] \le 0 \tag{4.13}$$

Proof. We rewrite

$$\begin{aligned}
&\sum_x \varphi\left(t, \frac{x}{N}\right) h(\bar{\zeta}_{x,N\epsilon_1})[N\nabla_e(\tau_x F) - \nabla_{N\epsilon,e}\bar{F}_{x,N\epsilon_2}] \\
&= \sum_x c_{N,\epsilon,\epsilon_1,\epsilon_2}\left(t, \frac{x}{N}, \zeta\right) N\tau_x f_e \\
&= \sum_x c_{N,\epsilon,\epsilon_1,\epsilon_2}\left(t, \frac{x}{N}, \zeta\right) N L_{B_x}[\tau_x g]
\end{aligned}$$

where $f_e = \tau_e F - F = L_B\, g$ for some box B. Since $c_{N,\epsilon,\epsilon_1,\epsilon_2}(t, \frac{x}{N}, \zeta)$ is unaltered by any rearrangement of particles inside B_x, it is easy to estimate the left hand side of (4.13) by

$$\limsup_{\epsilon\to 0} \limsup_{\epsilon_2\to 0} \limsup_{N\to\infty} \frac{1}{N^d} \sup_\zeta C \sum_x \left| c_{N,\epsilon,\epsilon_1,\epsilon_2}\left(t, \frac{x}{N}, \zeta\right)\right|^2$$

which is easily computed to be 0.

Lemma 4.5. *Let $\hat{F}(\tilde{\rho})$ be the expected value of F under the product measure $\pi_{\tilde{\rho}}$. Then for any $\epsilon < \frac{1}{2}\epsilon_1$ and coordinate direction e,*

$$\limsup_{\epsilon_2\to 0} \limsup_{N\to\infty} \frac{1}{N^d} \log \hat{E}\left[\exp\left\{\int_0^T \sum_x |\nabla_{N\epsilon,e}\bar{F}_{x,N\epsilon_2} - \frac{1}{2\epsilon}[\hat{F}(\bar{\zeta}_{x+N\epsilon e,N\epsilon_2}) - \hat{F}(\bar{\zeta}_{x-N\epsilon e,N\epsilon_2})]|\, dt\right\}\right] \le 0$$

Proof. Since $\epsilon > 0$ is fixed, it is sufficient to establish

$$\limsup_{\epsilon_2 \to 0} \limsup_{N \to \infty} \frac{1}{N^d} \log \hat{E}\left[\exp\left\{\int_0^T \sum_x |\bar{F}_{x,N\epsilon_2} - \hat{F}(\bar{\zeta}_{x,N\epsilon_2})|\, dt\right\}\right] \le 0$$

which is an easy consequence of Theorems 3.1 and 3.2.

Lemma 4.6. *For any F, φ and h,*

$$\limsup_{\substack{\epsilon \to 0 \\ k \to \infty}} \limsup_{N \to \infty} \frac{1}{N^d} \log \hat{E}\left[\exp\left\{\int_0^T \sum_x \varphi\left(t, \frac{x}{N}\right)[h(\bar{\zeta}_{x,N\epsilon}) .\right.\right.$$

$$\left.\left. - h(\bar{\zeta}_k)](N\tau_x \nabla_e F)\, dt\right\}\right] \le 0$$

Proof. We can apply Schwarz's inequality, and an application of Feynman-Kac formula and the variational formula yields

$$\limsup_{\substack{\epsilon \to 0 \\ k \to \infty}} \limsup_{N \to \infty} \frac{1}{N^d} \log \hat{E}\left[\exp\left\{\int_0^T \sum_x \varphi\left(t, \frac{x}{N}\right)[h(\bar{\zeta}_{x,N\epsilon}) - h(\bar{\zeta}_k)](N\tau_x \nabla_e F)\right.\right.$$

$$\left.\left. - C\sum_x [h(\bar{\zeta}_{x,N\epsilon}) - h(\bar{\zeta}_k)]^2 dt\right\}\right] \le 0$$

We are therefore left with proving

$$\limsup_{\substack{\epsilon \to 0 \\ k \to \infty}} \limsup_{N \to \infty} \frac{1}{N^d} \log \hat{E}\left[\exp\left\{C\int_0^T \sum_x [h(\bar{\zeta}_{x,N\epsilon}) - h(\bar{\zeta}_k)]^2\, dt\right\}\right] \le 0$$

which follows from Theorem 3.2.

Although the Lemmas 4.4, 4.5 and 4.6 are only stated for $\hat{E}$ they are valid for $E_{N,b}$ as well.

Recall that for v in $\mathscr{V}$ we defined

$$|\!|\!| v |\!|\!|^2(\tilde{\rho}) = [v, v]_{\tilde{\rho}} \quad . \tag{4.14}$$

If v depends on other parameters so will $|\!|\!| v |\!|\!|$.

Lemma 4.7. *Let $v(\tilde{\rho}, \cdot)$ be in some fixed $\mathscr{V}_l$ for every $\tilde{\rho}$ in $\mathscr{C}$ and be continuous on $\mathscr{C}$. Then for sufficiently large k, $v(\bar{\zeta}_k, \zeta)$ is in $\mathscr{N}$. Consequently, so is $v_k(\tilde{\rho}, \zeta) = v(\bar{\zeta}_k, \zeta) - v(\tilde{\rho}, \zeta)$ and if we define $f_K(k, \tilde{\rho}, \zeta)$ as $\langle (-L_K)^{-1}(\sum_{|x| \le K-k} v_k), (\sum_{|x| \le K-k} v_k \rangle$, there exist $\delta_k \to 0$ satisfying*

$$\limsup_{\substack{K \to \infty \\ \zeta \sim \tilde{\rho}}} \frac{1}{(2K+1)^d} f_K(\zeta) \le \delta_k \quad .$$

Proof. We can write $v(\tilde{\rho}, \cdot) = L_B w(\tilde{\rho}, \cdot)$ with w depending continuously on $\tilde{\rho}$ as well. If $k > l$, where l is the linear size of the box B, then $v_k(\bar{\zeta}_{x,k}, \tau_x\zeta) = L_{B+x}[w(\bar{\zeta}_{x,k}, \tau_x\zeta) - v(\tilde{\rho}, \tau_x\zeta)]$ and δ_k can be taken to be

$$\sup_{\tilde{\rho}} E^{\pi_{\tilde{\rho}}}\{\sup_{\zeta}[w(\bar{\zeta}_{0,k}, \zeta) - w(\tilde{\rho}, \zeta)]^2\}$$

which goes to 0.

We now calculate carefully the exponent in the exponential martingale (4.11), making the choice $g = \sum_{i,j}(\nabla\Phi)_{i,j} g_{i,j}$, and paying particular attention to various powers of N. We denote by Θ the exponent and write

$$\Theta = \Theta_1 + \Theta_2 + \Theta_3 + \Theta_4 - \Theta_5 + \Theta_6 \tag{4.15}$$

where

$$\Theta_1 = \sum_x \left\langle \Phi\left(T, \frac{x}{N}\right), \zeta(T,x)\right\rangle - \sum_x \left\langle \Phi\left(0, \frac{x}{N}\right), \zeta(0,x)\right\rangle - \int_0^T \sum_x \left\langle \Phi_t\left(t, \frac{x}{N}\right), \zeta(t,x)\right\rangle dt \tag{4.16}$$

$$\Theta_2 = N \int_0^T \sum_x \left\langle \nabla\Phi\left(t, \frac{x}{N}\right); \tau_x \mathbf{W}(t) + \tau_x L\mathbf{g}\left(t, \frac{x}{N}, \cdot\right)\right\rangle dt \tag{4.17}$$

Here $\mathbf{W}$ has components

$$W_{i,j} = \frac{1}{2}\sum_y a_{0,y}(\zeta)p(y)(y, e_i)[\eta_j(y) - \eta_j(0)] \tag{4.18}$$

and L is applied to each component $g_{i,j}$ of $\mathbf{g}$ keeping t and x fixed.

$$\Theta_3 = \frac{1}{2}\int_0^T \sum_x \left\langle \nabla\Phi\left(t, \frac{x}{N}\right); \tau_x\left\{\sum_z a_{0,z}(\zeta)p(z)b\left(t, \frac{x}{N}, z, \zeta\right) \times \left[z \otimes (\zeta(z) - \zeta(0)) + \nabla_{0,z}\mathbf{G}\left(t, \frac{x}{N}, \zeta\right)\right]\right\}\right\rangle dt \tag{4.19}$$

where $\mathbf{G}(t, \theta, \zeta) = \sum_x \tau_x \mathbf{g}(t, \theta, \zeta)$.

$$\Theta_4 = \frac{1}{4}\int_0^T \sum_x \sum_z a_{x,x+z}(\zeta)\langle (z \otimes z)\nabla^2\Phi, \zeta(x+z) - \zeta(x)\rangle \, dt \tag{4.20}$$

Let us note that every term of (4.20) above has mean zero under every $\pi_{\tilde{\rho}}$.

$$\Theta_5 = \int_0^T \sum_x q\left(t, \frac{x}{N}, \tau_x\zeta\right) dt \tag{4.21}$$

with

$$q(t,\theta,\zeta) = \frac{1}{4}\sum_y a_{0,y}(\zeta)p(y)\Big\langle \nabla\Phi\,,[y\otimes(\zeta(y)-\zeta(0)) + \nabla_{0,y}\mathbf{G}\Big(t,\frac{x}{N},\cdot\Big)]\Big\rangle^2 \,. \tag{4.22}$$

One can verify that

$$E_{\pi_{\tilde{\rho}}}[q(t,\theta,\cdot)] = \|\langle\nabla\Phi;\mathbf{W}+L\mathbf{G}\rangle\|^2(t,\theta,\tilde{\rho}) \tag{4.23}$$

where $\|\cdot\|$ is as defined in (4.14). We combine all the remaining terms as,

$$\Theta_6 = o(N^d)\ .$$

The last term is the error term. The sources are the exponential expansion, Taylor expansion for Φ and the slow variation in x when we pretend that $\mathbf{g}$ does not depend on θ in the definition of $\mathbf{G}$.

Let us now define a family of functions depending on $\delta > 0$, $0 < \lambda < \infty$, Φ , ϵ_1, ϵ, ϵ_2, J and l:

$$\mathbf{F} = \mathbf{F}_1 + \mathbf{F}_2 + \mathbf{F}_3 + \mathbf{F}_4 - (1+\delta)\mathbf{F}_5 - 2\lambda^2\mathbf{F}_6 \tag{4.24}$$

where

$$\begin{aligned}\mathbf{F}_1 = \Theta_1 = &\sum_x\Big\langle\Phi\Big(T,\frac{x}{N}\Big),\zeta(T,x)\Big\rangle - \sum_x\Big\langle\Phi\Big(0,\frac{x}{N}\Big),\zeta(0,x)\Big\rangle\\ &-\int_0^T\sum_x\Big\langle\Phi_t\Big(t,\frac{x}{N}\Big),\zeta(t,x)\Big\rangle\,dt\end{aligned} \tag{4.25}$$

$$\mathbf{F}_2 = \int_0^T\sum_x\Big\langle\nabla\Phi;J(\bar{\zeta}_{x,N\epsilon_1})\Big(\frac{1}{2\epsilon}\Big)[\hat{F}^l(\bar{\zeta}_{x+N\epsilon e,N\epsilon_2}) - \hat{F}^l(\bar{\zeta}_{x-N\epsilon e,N\epsilon_2})]\Big\rangle\,dt \tag{4.26}$$

$$\mathbf{F}_3 = \frac{1}{2}\int_0^T\sum_x\Big\langle\nabla\Phi\Big(t,\frac{x}{N}\Big);\mathbf{a}[\mathbf{b},\mathbf{g}]\Big(t,\frac{x}{N},\zeta_{x,N\epsilon_1}\Big)\Big\rangle\,dt\ . \tag{4.27}$$

Here $\mathbf{a}[\mathbf{b},\mathbf{g}]$ has components

$$\begin{aligned}(\mathbf{a}[\mathbf{b},\mathbf{g}])_{i,j}(t,\theta,\tilde{\rho}) = \frac{1}{2}E^{\pi_{\tilde{\rho}}}\Big[&\sum_z a_{0,z}(\zeta)p(z)b(t,\theta,z,\zeta)\{\langle z,e_i\rangle(\eta_j(z)\\ &-\eta_j(0)) + \nabla_{0,z}G_{i,j}(t,\theta,\zeta)\}\Big]\end{aligned} \tag{4.28}$$

where G and $\nabla_{0,z}G$ are defined in terms of g in (3.5), (3.6).

$$\mathbf{F}_4 = 0 \tag{4.29}$$

$$\mathbf{F}_5 = \int_0^T \sum_x \|\langle \nabla\Phi, \mathbf{W} + L\mathbf{g}\rangle\|^2 \left(t, \frac{x}{N}, \zeta_{x,N\epsilon_1}\right) dt \tag{4.30}$$

and

$$\mathbf{F}_6 = \int_0^T \sum_x \|\langle \nabla\Phi; \mathbf{W} + L\mathbf{g} - J(\tilde{\rho})\nabla_{\mathbf{e}} F^l\rangle\|^2 \left(t, \frac{x}{N}, \zeta_{x,N\epsilon_1}\right) dt \tag{4.31}$$

with $\nabla_{\mathbf{e}} F^l$ representing entries $\nabla_{e_i} F^l_j$. The notation here is that $\|.\|$ is a function of t, θ and $\tilde{\rho}$, and $\frac{x}{N}$ and $\zeta_{x,N\epsilon_1}$ are substituted for the last two.

Let us denote by

$$\mathbf{F}_1 - \Theta_1 = \mathbf{K}_1 = 0 \tag{4.32}$$

$$\mathbf{F}_2 - \Theta_2 - 2\lambda^2\mathbf{F}_6 = \mathbf{K}_2 \tag{4.33}$$

$$\mathbf{F}_3 - \Theta_3 = \mathbf{K}_3 \tag{4.34}$$

$$\mathbf{F}_4 - \Theta_4 = -\Theta_4 = \mathbf{K}_4 \tag{4.35}$$

$$(1+\delta)(\Theta_5 - \mathbf{F}_5) = \mathbf{K}_5 \tag{4.36}$$

We write

$$\mathbf{F} = \Theta_1 + \Theta_2 + \Theta_3 + \Theta_4 - (1+\delta)\Theta_5 + \mathbf{K}_1 + \mathbf{K}_2 + \mathbf{K}_3 + \mathbf{K}_4 + \mathbf{K}_5 \tag{4.37}$$

Theorem 4.8. *For any $\delta > 0$ there exist a $\lambda = \lambda(\delta) > 0$, depending only on δ, such that $\lambda(\delta) \to \infty$ as $\delta \to 0$ and*

$$\limsup_{N\to\infty} \frac{1}{N^d} \log E_{N,b}[\exp\{\mathbf{F}\}] \le C(\lambda, \epsilon, \epsilon_1, \epsilon_2) + \sigma(\lambda) \tag{4.38}$$

where

$$\lim_{\lambda\to\infty} \sigma(\lambda) = 0 \tag{4.39}$$

and for every positive λ,

$$\limsup_{\epsilon_1\to 0} \limsup_{\epsilon\to 0} \limsup_{\epsilon_2\to 0} C(\lambda, \epsilon, \epsilon_1, \epsilon_2) \le 0 \tag{4.40}$$

Proof. We estimate $E_{N,b}[\exp\{\mathbf{F}\}]$ by Hölder's inequality with a factor of $(1+\delta)$ multiplying the five Θ terms and a corresponding λ multiplying each onc of the $\mathbf{K}$ terms.

The first of these is basically the exponential martingale with two provisos. We have $(1+\delta)\Phi$ instead of Φ. The term Θ_6 which is uniformly negligible is missing. Therefore

$$\limsup_{N\to\infty}\frac{1}{N^d}\log E_{N,b}[\exp\{(1+\delta)(\Theta_1+\Theta_2+\Theta_3+\Theta_4) \\ -(1+\delta)^2\Theta_5\}]\le 0\ . \tag{4.41}$$

It remains to show only that

$$\limsup_{N\to\infty}\frac{1}{N^d}\log E_{N,b}[\exp\{\lambda\mathbf{K}_i\}]\le C(\lambda,\epsilon_1,\epsilon,\epsilon_2)+\sigma(\lambda) \tag{4.42}$$

for $i=1,2,3,4$ and 5, with C and σ satisfying (4.40) and (4.39). We go from $E_{N,b}$ to $\hat{E}$ with the help of the Hölder inequality using $p=\frac{\lambda}{\lambda-1}$ for the Radon-Nikodym derivative and $q=\lambda$ for the various exponentials. We therefore need only prove that

$$\limsup_{N\to\infty}\frac{1}{N^d}\log\hat{E}\ [\exp\{\lambda^2\mathbf{K}_i\}]\le C(\lambda,\epsilon_1,\epsilon,\epsilon_2) \tag{4.43}$$

for some C satisfying (4.40). Since $\mathbf{K}_1=0$ this is trivial for $i=1$. For $i=3,4$ and 5 this is an easy consequence of the one and two block estimates Theorems 3.1 and 3.2. We now concentrate on $\mathbf{K}_2$.

$$\begin{aligned}\mathbf{K}_2=&\int_0^T\sum_x\left\langle\nabla\Phi;J(\bar{\zeta}_{x,N\epsilon_1})\left(\frac{1}{2\epsilon}\right)[\hat{F}^l(\bar{\zeta}_{x+N\epsilon,N\epsilon_2})-\hat{F}^l(\bar{\zeta}_{x-N\epsilon,N\epsilon_2})]\right\rangle dt\\&-\int_0^T\sum_x\left\langle\nabla\Phi;[\tau_x\mathbf{W}+(L\mathbf{g})\left(t,\frac{x}{N},\tau_x\zeta\right)]\right\rangle dt\\&-2\lambda^2\int_0^T\sum_x\|\langle\nabla\Phi;\mathbf{W}+L\mathbf{g}-J(\tilde{\rho})\nabla_{\mathbf{e}}F^l\rangle\|^2\left(t,\frac{x}{N},\zeta_{x,N\epsilon_1}\right)dt\end{aligned} \tag{4.44}$$

is written as the sum of four terms $\mathbf{K}_{21}+\mathbf{K}_{22}+\mathbf{K}_{23}+\mathbf{K}_{24}$ given by

$$\begin{aligned}\mathbf{K}_{21}=&\int_0^T\sum_x\left\langle\nabla\Phi;J(\bar{\zeta}_{x,N\epsilon_1})\left(\frac{1}{2\epsilon}\right)[\hat{F}^l(\bar{\zeta}_{x+N\epsilon,N\epsilon_2})-\hat{F}^l(\bar{\zeta}_{x-N\epsilon,N\epsilon_2})]\right\rangle dt\\&-\int_0^T\sum_x\left\langle\nabla\Phi;J(\bar{\zeta}_{x,N\epsilon_1})\nabla_{N\epsilon,\mathbf{e}}[\bar{F}^l_{x,N\epsilon_2}]\right\rangle dt\end{aligned} \tag{4.45}$$

$$\begin{aligned}\mathbf{K}_{22}=&\int_0^T\sum_x\left\langle\nabla\Phi;J(\bar{\zeta}_{x,N\epsilon_1})\nabla_{N\epsilon,\mathbf{e}}[\bar{F}^l_{x,N\epsilon_2}]\right\rangle dt\\&-\int_0^T\sum_x\left\langle\nabla\Phi;J(\bar{\zeta}_{x,N\epsilon_1})(N\nabla_e)(\tau_xF^l)\right\rangle dt\end{aligned} \tag{4.46}$$

$$\mathbf{K}_{23} = \int_0^T \sum_x \Big\langle \nabla\Phi; J(\bar{\zeta}_{x,N\epsilon_1})(N\nabla_e)(\tau_x F^l)\Big\rangle\, dt - \int_0^T \sum_x \Big\langle \nabla\Phi; J(\bar{\zeta}_{x,k})(N\nabla_e)(\tau_x F^l)\Big\rangle\, dt \tag{4.47}$$

$$\mathbf{K}_{24} = \int_0^T \sum_x \Big\langle \nabla\Phi; [J(\bar{\zeta}_{x,k})(N\nabla_e)(\tau_x F^l) - \tau_x \mathbf{W} - (L\mathbf{g})\Big(t, \frac{x}{N}, \tau_x\zeta\Big)\Big\rangle\, dt - 2\lambda^2 \int_0^T \sum_x \Big\|\Big\langle \nabla\Phi; \mathbf{W} + L\mathbf{g} - J(\tilde{\rho})\nabla_{\mathbf{e}} F^l\Big\rangle\Big\|^2\Big(t, \frac{x}{N}, \zeta_{x,N\epsilon_1}\Big)\, dt \tag{4.48}$$

For $\mathbf{K}_{21}$, (4.43) is a consequence of Lemma 4.7. For $\mathbf{K}_{22}$ it follows from lemma 4.3 and Lemma 4.7 takes care of $\mathbf{K}_{23}$. Finally it follows from theorems 3.1, 3.2 and 3.3 that

$$\limsup_{k\to\infty,\epsilon_1\to 0}\ \limsup_{N\to\infty} \frac{1}{N^d} \log \hat{E}[\exp\{2\lambda^2 \mathbf{K}_{24}\}] \le 0 \tag{4.49}$$

We have to apply the Hölder inequality pull the exponents apart. We have allowed for an extra factor of 2 in $\mathbf{K}_{24}$ and the other terms that are linear in F can absorb any constant. We have now established a lower bound for the n-color "rate function" $\hat{I}$.

Theorem 4.9. *The rate function $\hat{I}(\tilde{\rho}(\cdot,\cdot))$ has the following lower bound.*

$$\begin{aligned}
\hat{I}(\tilde{\rho}(\cdot,\cdot)) \ge &\int_{\mathbf{T}^d} \langle\Phi(T,\theta), \tilde{\rho}(T,\theta))\rangle\, d\theta - \int_{\mathbf{T}^d} \langle\Phi(0,\theta), \tilde{\rho}(0,\theta))\rangle\, d\theta \\
&- \int_0^T\int_{\mathbf{T}^d} \langle\Phi_t(t,\theta); \tilde{\rho}(t,\theta))\rangle\, dt\, d\theta \\
&+ \int_0^T\int_{\mathbf{T}^d} \big\langle\nabla\Phi(t,\theta); (J\nabla\hat{F}^l)(\tilde{\rho}(t,\theta))\big\rangle\, dt\, d\theta \\
&+ \int_0^T\int_{\mathbf{T}^d} \langle\nabla\Phi(t,\theta); \mathbf{a}[\mathbf{b},\mathbf{g}](t,\theta,\tilde{\rho}(t,\theta))\rangle\, dt\, d\theta \\
&- (1+\delta)\int_0^T\int_{\mathbf{T}^d} \|\langle\nabla\Phi; \mathbf{W} + L\mathbf{g}\rangle\|^2(t,\theta,\tilde{\rho}(t,\theta))\, dt\, d\theta \\
&- \lambda^2\int_0^T\int_{\mathbf{T}^d} \big\|\big\langle\nabla\Phi; \mathbf{W} + L\mathbf{g} - J(\tilde{\rho})\nabla_e F^l\big\rangle\big\|^2(t,\theta,\tilde{\rho}(t,\theta))\, dt\, d\theta \\
&- \sigma(\lambda)
\end{aligned} \tag{4.50}$$

From this point we pick $J(\tilde{\rho})$ to be of the form $h(\rho)\ \frac{1}{2}A(\tilde{\rho})\ \chi(\tilde{\rho})$ where h is a cutoff function vanishing near $\rho = 1$. First we remark that because of Lemma 4.7 we can assume that $\mathbf{g}$ can depend on $\tilde{\rho}$ as well. Calculations in "gradient space" allows us to optimize with respect to $\mathbf{g}$ which essentially projects $\mathbf{W}$ into $\frac{1}{2}A\chi$ times the image of the density gradient $\mathbf{f}$. This projection is carried out explicitly in the Hilbert space $H_{\tilde{\rho}}$. If we denote by $\tilde{\Psi}$ and $\tilde{\mathbf{f}}$ the images in $H_{\tilde{\rho}}$ of $\mathbf{W}$ and $\mathbf{f}$ respectively then $\tilde{\Psi} = \frac{1}{2}A(\tilde{\rho})\chi(\tilde{\rho})\tilde{\mathbf{f}}$ modulo $H^0_{\tilde{\rho}}$ and while the minimizing g may not exist there will be a unique minimizing ξ in $H^0_{\tilde{\rho}}$. So the functions $\mathbf{a}[\mathbf{b},\mathbf{g}](t,\theta,\tilde{\rho})$ will have a limit along any minimizing sequence of g's and the limit can be calculated as

$$\hat{\mathbf{a}}[\mathbf{b}]_{i,j}(t,\theta,\tilde{\rho}) = \frac{1}{2}E^{\pi_{\tilde{\rho}}}\Bigg\{\sum_z a_{0,z}(\zeta)p(z)b(t,\theta,z,\zeta) \times [\langle z,e_i\rangle(\eta_j(z)-\eta_j(0)) + \xi_z]\Bigg\} \tag{4.51}$$

which, by Remark 3 at the end of Section 3, is a continuous function of t,θ and $\tilde{\rho}$ for our choice of b.

A simple calculation provides a new lower bound.

$$\begin{aligned}\hat{I}(\tilde{\rho}(\cdot,\cdot)) \geq &\int_{\mathbf{T}^d}\langle\Phi(T,\theta),\tilde{\rho}(T,\theta))\rangle\,d\theta - \int_{\mathbf{T}^d}\langle\Phi(0,\theta),\tilde{\rho}(0,\theta)\rangle\,d\theta\\ &- \int_0^T\!\!\int_{\mathbf{T}^d}\langle\Phi_t(t,\theta),\tilde{\rho}(t,\theta)\rangle\,dt\,d\theta\\ &+ \int_0^T\!\!\int_{\mathbf{T}^d}\left\langle\nabla\Phi(t,\theta);\left(\frac{1}{2}hA\chi(\nabla\mathbf{F}^l\right)(\tilde{\rho}(t,\theta))\right\rangle dt\,d\theta\\ &+ \int_0^T\!\!\int_{\mathbf{T}^d}\langle\nabla\Phi(t,\theta);\hat{\mathbf{a}}[\mathbf{b}](t,\theta,\tilde{\rho}(t,\theta))\rangle\,dt\,d\theta\\ &- \frac{(1+\delta)}{2}\int_0^T\!\!\int_{\mathbf{T}^d}\langle\nabla\Phi(t,\theta);A(\tilde{\rho}(t,\theta))\nabla\Phi(t,\theta)\rangle\,dt\,d\theta\\ &- \frac{\lambda^2}{4}\int_0^T\!\!\int_{\mathbf{T}^d}\|\langle\nabla\Phi;[A\chi\mathbf{f}-hA\chi\mathbf{f}^l]\rangle\|^2_{H_{\tilde{\rho}}}(t,\theta,\tilde{\rho}(t,\theta))\,dt\,d\theta\\ &- \sigma(\lambda)\end{aligned} \tag{4.52}$$

We next show that we can let $l\to\infty$ and $\mathbf{f}^l$ converges nicely to $\mathbf{f}$ away from $\rho = 1$. Because of the cutoff h this limit can be taken. Then the cutoff is removed allowing h to tend to 1. The estimates (4.6) and (4.7) provided in Theorem 4.3 allow us to show that the term that is multiplied by λ goes to zero for any λ. Finally we let $\delta\to 0$ and $\lambda\to\infty$ and obtain

$$
\begin{aligned}
\hat{I}(\tilde{\rho}(\cdot,\cdot)) \geq & \int_{\mathbf{T}^d} \langle \Phi(T,\theta), \tilde{\rho}(T,\theta)\rangle \, d\theta - \int_{\mathbf{T}^d} \langle \Phi(0,\theta), \tilde{\rho}(0,\theta)\rangle \, d\theta \\
& - \int_0^T \int_{\mathbf{T}^d} \langle \Phi_t(t,\theta), \tilde{\rho}(t,\theta)\rangle dt \, d\theta \\
& + \int_0^T \int_{\mathbf{T}^d} \left\langle \nabla\Phi(t,\theta); \left(\frac{1}{2}A\chi\right)(\tilde{\rho}(t,\theta))\nabla\tilde{\rho}(t,\theta) \right\rangle dt \, d\theta \\
& + \int_0^T \int_{\mathbf{T}^d} \langle \nabla\Phi(t,\theta); \hat{\mathbf{a}}[\mathbf{b}](t,\theta,\tilde{\rho}(t,\theta))\rangle \, dt \, d\theta \\
& - \frac{1}{2}\int_0^T \int_{\mathbf{T}^d} \langle \nabla\Phi(t,\theta); A(\tilde{\rho}(t,\theta))\nabla\Phi(t,\theta)\rangle \, dt \, d\theta \ . \qquad (4.53)
\end{aligned}
$$

Taking the supremum over Φ we obtain

$$
\hat{I}(\tilde{\rho}(\cdot,\cdot)) \geq I^{n,b}_{\text{dyn}}(\tilde{\rho}(\cdot,\cdot)) = \frac{1}{2}\int_0^T \left\| \frac{\partial\tilde{\rho}}{\partial t} - \frac{1}{2}\nabla\cdot A\chi(\tilde{\rho})\nabla\tilde{\rho} - \nabla\cdot\hat{\mathbf{a}}[\mathbf{b}] \right\|^2_{-1,A(\tilde{\rho})} dt \qquad (4.54)
$$

Theorem 4.10. *The measures $P_{N,b}$ have a large deviation upper bound with the rate function $I^{n,b}_{dyn}(\tilde{\rho}(\cdot,\cdot))$. More precisely on the space E_n the family $P_{N,b}$ satisfies the large deviation upper bound so that for every closed set C*

$$
\limsup_{N\to\infty} \frac{1}{N^d} \log P_{N,b}[C] \leq - \inf_{\tilde{\rho}(\cdot,\cdot)\in C_0} I^{n,b}_{\text{dyn}}(\tilde{\rho}(\cdot,\cdot)) \qquad (4.55)
$$

where $C_0 = C \cap \{\tilde{\rho}(\cdot,\cdot) : \tilde{\rho}(0,\cdot) = \tilde{\rho}_0(\cdot)\}$ are those profiles in C that are compatible with the initial profile at time 0.

In particular we can take $b \equiv 0$ and we obtain the upper bounds for P_N. A further consequence is the law of large numbers for the sequence $P_{N,b}$:

Theorem 4.11. *For any b the family $P_{N,b}$ is tight on the space E_n and any weak limit of $P_{N,b}$ will be concentrated in the subset of D^0_n that are weak solutions of*

$$
\frac{\partial\tilde{\rho}}{\partial t} - \frac{1}{2}\nabla\cdot A\chi(\tilde{\rho})\nabla\tilde{\rho} = \nabla\cdot\hat{\mathbf{a}}[b] \qquad (4.56)
$$

with initial condition $\tilde{\rho}_0$.

We will now attempt to prove the large deviation lower bounds for P_N using as our tool Theorem 4.9. Suppose $\mathbf{H}(t,\theta)$ is a continuous map from $[0,T]\times\mathbf{T}^d$ into $\mathbf{R}^n\otimes\mathbf{R}^d$ and suppose we are given a profile $\tilde{\rho}(t,\theta)$ that satisfies

$$\frac{\partial \tilde{\rho}}{\partial t} - \frac{1}{2} \nabla \cdot A\chi(\tilde{\rho}) \nabla \tilde{\rho} = \nabla \cdot A(\tilde{\rho}) \mathbf{H} \tag{4.57}$$

in the weak sense. Let us suppose moreover that the above equation has a unique solution $\tilde{\rho}$ in this class $\mathscr{D}_n^0$ for the given $\mathbf{H}$ and initial profile $\tilde{\rho}_0$. We look for functions $b(t, \theta, z, \zeta)$ such that $\hat{\mathbf{a}}[\mathbf{b}] = A(\tilde{\rho}(t, \theta))\mathbf{H}$. We may denote this class by $\mathscr{B}_{\tilde{\rho},\mathbf{H}}$. For each b in this class the perturbed process $P_{N,b}$ will have a certain relative entropy with respect to P_N and we call this $Ent(N, b)$. We now show

Theorem 4.12.

$$\inf_{b \in \mathscr{B}_{\tilde{\rho},\mathbf{H}}} \limsup_{N \to \infty} \frac{1}{N^d} Ent(N, b) \leq \frac{1}{2} \int_0^T \int_{\mathbf{T}^d} \langle \mathbf{H}(t, \theta), A(\tilde{\rho}(t, \theta)) \mathbf{H}(t, \theta) \rangle \, dt \, d\theta \tag{4.58}$$

Proof. If we make the choice

$$b(t, \theta, z, \zeta) = \langle \mathbf{H}, z \otimes (\zeta(z) - \zeta(0)) \rangle \tag{4.59}$$

then a straightforward calculation yields,

$$\hat{\mathbf{a}}[\mathbf{b}](t, \theta, \tilde{\rho}) = \frac{1}{2} E^{\pi_{\tilde{\rho}}} [\sum_z p(z) a_{0,z}(\zeta) \langle \mathbf{H}; z \otimes (\zeta(z) - \zeta(0)) \rangle (z \otimes (\zeta(z) - \zeta(0)) + \xi_z)]$$

$$= \frac{1}{2} E^{\pi_{\tilde{\rho}}} [\sum_z p(z) a_{0,z}(\zeta) \langle \mathbf{H}; \psi_z \rangle (\psi_z + \xi_z)] \tag{4.60}$$

$\hat{\mathbf{a}}[\mathbf{b}](t, \theta, \tilde{\rho})$

$$= \frac{1}{2} E^{\pi_{\tilde{\rho}}} \left[\sum_z p(z) a_{0,z}(\zeta) \langle \mathbf{H}; z \otimes (\zeta(z) - \zeta(0)) \rangle (z \otimes (\zeta(z) - \zeta(0)) + \xi_z) \right]$$

$$= \frac{1}{2} E^{\pi_{\tilde{\rho}}} \left[\sum_z p(z) a_{0,z}(\zeta) \langle \mathbf{H}; \psi_z \rangle (\psi_z + \xi_z) \right] \tag{4.60}$$

where $\{\xi_z\}$ is the projection of $\{\psi_z\} = z \otimes (\zeta(z) - \zeta(0))$ onto $H_{\tilde{\rho}}^0$. From (3.15) this is seen to equal $A(\tilde{\rho})\mathbf{H}$. Actually we could make the choice

$$b(t, \theta, z, \zeta) = \langle \mathbf{H}; z \otimes (\zeta(z) - \zeta(0)) \rangle + \nabla_{0,z} G \tag{4.61}$$

for some $G = \sum_x g(\tau_x \zeta)$ and due to the orthogonality of $\{\psi_z + \xi_z\}$ and $\nabla_{0,z} G$, $\hat{\mathbf{a}}[\mathbf{b}]$ will be unaffected by this modification. On the other hand, using the Girsanov formula the relative entropy $Ent(N, b)$ can be calculated explicitly and the answer depends on the modification $g = g(t, \theta, \zeta)$ used. An easy computation yields

$$\limsup_{N\to\infty}\frac{1}{N^d}Ent(N,b)$$
$$\leq \int_0^T\int_{\mathbf{T}^d}\frac{1}{4}E^{\pi_{\tilde\rho(t,\theta)}}\Big[\sum_z a_{0,z}(\zeta)p(z)\big(\langle\mathbf{H};\psi_z\rangle-\nabla_{0,z}G\big)^2\Big]\,dt\,d\theta\ . \tag{4.62}$$

By the variational formula (3.15) for A, minimization over g gives

$$\frac{1}{2}\int_0^T\int_{\mathbf{T}^d}\langle\mathbf{H}(t,\theta);A(\tilde\rho(t,\theta))\mathbf{H}(t,\theta)\rangle\,dt\,d\theta \tag{4.63}$$

which is the required result.

Let us describe a class $\mathscr{S}$ of profiles. For a given initial profile $\tilde\rho_0(\cdot)$, these will satisfy the following: $\tilde\rho(0,\theta)=\tilde\rho_0(\theta)$ on $\mathbf{T}^d$. $\tilde\rho(\cdot,\cdot)$ is an infinitely differentiable function of t and θ in the open interval $(0,T]\times\mathbf{T}^d$ and avoids all the boundaries of $\mathscr{C}$ i.e. ρ_j and $(1-\rho)$ are all bounded away from zero so long as t remains positive. Finally for some $t_0>0$ that is allowed to depend on the choice of $\tilde\rho(\cdot,\cdot)$ the profile $\tilde\rho(\cdot,\cdot)$ satisfies weakly the equation

$$\frac{\partial\tilde\rho}{\partial t}=\frac{1}{2}\nabla\cdot A\chi\nabla\tilde\rho \tag{4.64}$$

in the interval $[0,t_0]$.

In section 5 we will prove that each $\tilde\rho(\cdot,\cdot)$ in $\mathscr{S}$ is the unique weak solution in $\mathscr{D}_n^0$ of an equation

$$\frac{\partial\tilde\rho}{\partial t}=\frac{1}{2}\nabla\cdot A\chi\nabla\tilde\rho+\nabla\cdot A(\tilde\rho(t,\cdot))\nabla\mathscr{H} \tag{4.65}$$

for some continuously differentiable $\mathscr{H}$, that of course vanishes for sufficiently small t, i.e $t<t_0$.

In section 6, we will prove that $\mathscr{S}$ is dense in $\mathscr{D}_n^0$ in the proper sense. More precisely we will prove that for any $\tilde\rho(\cdot,\cdot)$ in $\mathscr{D}_n^0$ such that $I^n_{dyn}(\tilde\rho(\cdot,\cdot))<\infty$ there exists a sequence $\tilde\rho_k(\cdot,\cdot)$ in $\mathscr{S}$ such that, they all have the same initial profile $\tilde\rho_0(\cdot)$,

$$\lim_{k\to\infty}\tilde\rho_k(\cdot,\cdot)=\tilde\rho_k(\cdot,\cdot)\quad\text{in}\quad C[[0,T],\mathscr{M}]$$

and

$$\liminf_{k\to\infty}I^n_{\text{dyn}}(\tilde\rho_k(\cdot,\cdot))\leq I^n_{\text{dyn}}(\tilde\rho(\cdot,\cdot))\ .$$

Once these uniqueness and approximation results are established the lower bounds for the large deviation principle follows.

Theorem 4.13. *For any deterministic initial conditions compatible with any given profile $\tilde\rho_0(\cdot)$ the large deviation principle holds for P_N with rate function*

$$I(\tilde{\rho}(\cdot,\cdot)) = \frac{1}{2}\int_0^T \left\| \frac{\partial \tilde{\rho}}{\partial t} - \frac{1}{2}\nabla \cdot A\chi(\tilde{\rho}(t,\cdot))\nabla\tilde{\rho}(t,\cdot) \right\|^2_{-1,A(\tilde{\rho}(t,\cdot))} dt \tag{4.66}$$

on those $\tilde{\rho}(\cdot,\cdot)$ that are compatible with the initial profile $\tilde{\rho}_0(\cdot)$ at time 0.

5. Uniqueness for driven systems

The aim of this section is to prove a theorem regarding uniqueness of weak solutions for a system of equations of the form

$$\frac{\partial \tilde{\rho}}{\partial t} = \frac{1}{2}\nabla \cdot A\chi(\tilde{\rho}(t,\cdot))\nabla\tilde{\rho}(t,\cdot) + \nabla \cdot A(\tilde{\rho}(t,\cdot))\mathbf{H} \tag{5.1}$$

with initial conditions

$$\tilde{\rho}(0,\cdot) = \tilde{\rho}_0(\cdot) \ , \tag{5.2}$$

for a matrix $\mathbf{H}_{i,j}(t,x)$ each of whose components are uniformly bounded. Since we will be dealing exclusively with functions defined on $[0,T]\times\mathbf{T}^d$, in this section as well as the next one we will find it more convenient to denote by x and y points on the torus $\mathbf{T}^d$. Our uniqueness theorem will be for $\tilde{\rho}(\cdot,\cdot)$ in the class $\mathscr{D}_n^0$ of functions $\tilde{\rho}(\cdot,\cdot)$ mapping $[0,T]\times\mathbf{T}^d$ into $\mathscr{C}$, that satisfy the inital condition (5.2), the regularity conditions

$$\begin{aligned} &\int_0^T\int_{\mathbf{T}^d} \frac{|\nabla\rho|^2}{(1-\rho)}\,dx\,dt < \infty \\ &\int_0^T\int_{\mathbf{T}^d} (1-\rho)\frac{|\nabla\rho_j|^2}{\rho_j}\,dx\,dt < \infty \end{aligned} \tag{5.3}$$

or, equivalently,

$$\int_0^T\int_{\mathbf{T}^d} \langle\nabla\tilde{\rho};\chi A\chi\nabla\tilde{\rho}\rangle\,dx\,dt < \infty \tag{5.4}$$

and the equation (5.1) in weak form: For all smooth $\Phi : [0,T]\times\mathbf{T}^d \to \mathbf{R}^n$,

$$\begin{aligned} &\int_{\mathbf{T}^d}\langle\Phi(T,\cdot),\tilde{\rho}(T,\cdot)\rangle\,dx - \int_{\mathbf{T}^d}\langle\Phi(0,\cdot),\tilde{\rho}(0,\cdot)\rangle\,dx \\ &\quad - \int_0^T\int_{\mathbf{T}^d}\langle\Phi_t(t,\cdot),\tilde{\rho}(t,\cdot)\rangle\,dx\,dt \\ &\quad + \frac{1}{2}\int_0^T\int_{\mathbf{T}^d}\langle\nabla\Phi(t,\cdot);A\chi(\tilde{\rho}(t,\cdot))\nabla\tilde{\rho}(t,\cdot)\rangle\,dx\,dt \\ &\quad + \int_0^T\int_{\mathbf{T}^d}\langle\nabla\Phi(t,\cdot);A(\tilde{\rho}(t,\cdot))\mathbf{H}(t,\cdot)\rangle\,dx\,dt = 0 \end{aligned} \tag{5.5}$$

Our uniqueness theorem is a little bit complicated to state. We will have a much smaller class $\mathscr{D}_n^1 \subset \mathscr{D}_n^0$ of solutions $\tilde{R}$ of (5.1) in the interval $[0, T]$ that satisfy in addition

$$\sup_{x,t} |\nabla \tilde{R}(t,x)| < \infty \tag{5.6}$$

and for some $\delta > 0$,

$$\begin{aligned} &\sup_{j,x,t} R_j(t,x) \leq (1-\delta) \\ &\inf_{j,x,t} R_j(t,x) \geq \delta \end{aligned} \tag{5.7}$$

Before we state and prove our uniqueness theorem we will establish a preliminary estimate on $\rho = \sum_{i=1}^n \rho_i$ for any solution $\tilde{\rho} = (\rho_1, \ldots, \rho_n)$ of (5.1). Let $\tilde{\rho}$ be a weak solution of (5.1) and $u = 1 - \rho$. Then u is a weak solution of an equation of the form

$$\frac{\partial u}{\partial t} = \frac{1}{2} \nabla \cdot D \nabla + \nabla \cdot bu \tag{5.8}$$

where $b_i = -\sum_{j=1}^n \rho_j \sum_{i'=1}^d D_{i,i'} H_{i',j}$ is uniformly bounded in absolute value.

Lemma 5.1. *Let $u(t,x)$ be the density at time t of a diffusion process on the torus $\mathbf{T}^d$ with the backward generator*

$$\tfrac{1}{2} \nabla \cdot D \nabla + b \cdot \nabla \tag{5.9}$$

where $\|b\|_\infty < \infty$ and with initial density $u(0,x)$ bounded uniformly away from 0. Then for each $T > 0$ there exists $\delta > 0$ depending only on $(T, \|b\|_\infty)$ such that for all $t \in [0,T]$, $u(t,x) \geq \delta$.

Proof. It is more convenient, for computational reasons, to extend b periodically in space and look at the same problem on $\mathbf{R}^d$. The lower bound on $\mathbf{T}^d$ can be easily deduced from any reasonable lower bound on $\mathbf{R}^d$.

Let $q(0,x,t,y)$ be the transition density on $\mathbf{R}^d$ of the process with drift $b \equiv 0$. By the Cameron-Martin-Girsanov formula we have an explicit representation for p in terms of q and the Brownian bridge $Q_t^{x,y}$ (with covariance D) on paths $z(s)$ starting at x at time 0 and conditioned to end at y at time t,

$$p(0,x,t,y) = q(0,x,t,y) E^{Q_t^{x,y}} \left[\exp\left(\int_0^t \langle D^{-1} b,\, dz(s) \rangle - \frac{1}{2} \int_0^t \langle D^{-1} b,\, b \rangle\, ds \right) \right] \tag{5.10}$$

The second term in the exponent in uniformly bounded. By Jensen's inequality and a straightforward stochastic calculus computation,

$$E^{Q_t^{x,y}}\left[\exp\left(\int_0^t \langle D^{-1}b,\, dz(s)\rangle\right)\right] \geq \exp\left(E^{Q_t^{x,y}}\left[\int_0^t \langle D^{-1}b,\, dz(s)\rangle\right]\right)$$
$$= \exp\left(E^{Q_t^{x,y}}\left[\int_0^t \frac{\langle D^{-1}b\ ,y - z(s)\rangle}{(t-s)}\, ds\right]\right) \tag{5.11}$$

By Schwarz's inequality,

$$E^{Q_t^{x,y}}|y - z(s)| \leq \left[E^{Q_t^{x,y}}|y - z(s)|^2\right]^{1/2} = \left[|x-y|^2\left(\frac{t-s}{t}\right)^2 + \frac{s(t-s)}{t}\right]^{1/2} \tag{5.12}$$

and

$$\left|E^{Q_t^{x,y}}\int_0^t \frac{\langle D^{-1}b\ ,y - z(s)\rangle}{t-s}\, ds\right|$$
$$\leq C\|b\|_\infty \int_0^t \frac{1}{(t-s)}\left[|x-y|\left(\frac{t-s}{t}\right) + \left(\frac{s(t-s)}{t}\right)^{1/2}\right] ds$$
$$\leq \|b\|_\infty [C_1|x-y| + C_2] \tag{5.13}$$

for some constants that could possibly depend on T. This proves the lemma.

We are now ready to state and prove the main uniqueness theorem.

Theorem 5.2. *Suppose that the function* $\mathbf{H}$ *is bounded on* $[0,T]\times\mathbf{T}^d$ *and equation* (5.1) *admits a solution* $\tilde{R}$ *that is in* $\mathscr{D}_n^1$. *Then any weak solution of* (5.1) *from the class* $\mathscr{D}_n^0$ *is identical to* $\tilde{R}$.

Proof. Suppose we have two weak solutions: a solution $\tilde{R}$, in $\mathscr{D}_n^1$ and another $\tilde{\rho}$ in $\mathscr{D}_n^0$. We use the notation $A = A(\tilde{R})$, $a = A(\tilde{\rho})$, $X = \chi(\tilde{R})$, $\chi = \chi(\tilde{\rho})$; $\Lambda = \lambda(\tilde{R})$ is the chemical potential corresponding to $\tilde{R}$ given by $\Lambda_j = \log\frac{R_j}{(1-R)}$ for $1 \leq j \leq n$. Let

$$F(\lambda) = \log\left(1 + \sum_{i=1}^n e^{\lambda_i}\right) \tag{5.14}$$

be the free energy, and

$$h(\tilde{\rho}) = \sum_{i=1}^n \rho_i \log\rho_i + (1-\rho)\log(1-\rho) \tag{5.15}$$

be the corresponding entropy functional. Note that the relative entropy is given by

$$h(\tilde{\rho};\tilde{R}) = h(\tilde{\rho}) - \tilde{\rho}\cdot\Lambda + F(\Lambda) \tag{5.16}$$

Because $\tilde{R}$ is a solution in $\mathscr{D}_n^1$, $1-R$ and R_j are bounded away from 0 and since $\tilde{\rho}$ has the same initial value as $\tilde{R}$, $1-\rho_0$ is bounded away from 0 as well. By lemma 5.1 and (5.8), for our weak solution $\tilde{\rho}$ from the class $\mathscr{D}_n^0$, $1-\rho$ is bounded away from zero uniformly in the interval $[0,T]$. A computation gives

$$\begin{aligned}
&\int_{\mathbf{T}^d}[h(\tilde{\rho}) - \tilde{\rho}\cdot\Lambda + F(\Lambda)]\,dx\Big|_0^T \\
&= -\int_0^T\int_{\mathbf{T}^d}\Big[\big\langle(\chi\nabla\tilde{\rho} - X\nabla\tilde{R});(\tfrac{1}{2}a\chi\nabla\tilde{\rho} + a\mathbf{H})\big\rangle \\
&\quad + \big\langle\nabla(\tilde{R}-\tilde{\rho}); X\big(\tfrac{1}{2}AX\nabla\tilde{R} + A\mathbf{H}\big)\big\rangle \\
&\quad + \big\langle\nabla\tilde{R}X'\cdot(\tilde{R}-\tilde{\rho});\big(\tfrac{1}{2}AX\nabla\tilde{R} + A\mathbf{H}\big)\big\rangle\Big]\,dx\,dt
\end{aligned} \tag{5.17}$$

Completing the square and defining E by

$$\chi^{-1} = X^{-1} + X^{-1}(X'\cdot(\tilde{R}-\tilde{\rho}))X^{-1} + E \tag{5.18}$$

this becomes

$$\begin{aligned}
&\int_0^T\int_{\mathbf{T}^d}\Bigg[-\frac{1}{2}\Big|(\chi a\chi)^{1/2}\nabla\tilde{\rho} \\
&\quad - \frac{1}{2}(\chi a\chi)^{-1/2}(\chi aX + XAX)\nabla\tilde{R}\Big|^2 \\
&\quad - \frac{1}{8}\Big\langle\nabla\tilde{R};(XAX - Xa\chi)(\chi a\chi)^{-1}(XAX - \chi aX)\nabla\tilde{R}\Big\rangle \\
&\quad - \Big\langle(\chi a\chi)^{1/2}\nabla\tilde{\rho} - \frac{1}{2}(\chi a\chi)^{-1/2}(\chi aX + XAX)\nabla\tilde{R};(\chi a\chi)^{-1/2}(\chi a - XA)\mathbf{H}\Big\rangle \\
&\quad - \Big\langle\nabla\tilde{R};\frac{1}{2}(XAX - Xa\chi)(\chi a\chi)^{-1}(\chi a - XA)\mathbf{H}\Big\rangle \\
&\quad + \frac{1}{2}\langle\nabla\tilde{R};XEXAX\nabla\tilde{R}\rangle + \langle\nabla\tilde{R};(XEXA)\mathbf{H}\rangle\Bigg]\,dx\,dt
\end{aligned} \tag{5.19}$$

By Schwarz's inequality the sum of the first four terms in the integrand is less than or equal to a fixed constant times

$$\langle\mathbf{H};(a\chi - AX)(\chi a\chi)^{-1}(\chi a - XA)\mathbf{H}\rangle \tag{5.20}$$

This is quadratic in $\tilde{R} - \tilde{\rho}$ with a bounded constant because ρ and R are bounded away from 1. The same is true of the last two terms because E is quadratic in $\tilde{R} - \tilde{\rho}$. We can therefore find a constant C depending only on $\|\nabla \tilde{R}\|_\infty$, the lower bound δ for R_j, $1 - R$, and $1 - \rho$, and the Lipschitz norm of $S(\rho)$ such that the integrand in (5.19) is bounded by

$$C[h(\tilde{\rho}) - \tilde{\rho} \cdot \Lambda + F(\Lambda)] \ . \tag{5.21}$$

The uniqueness follows from Gronwall's inequality.

The main drawback of this uniqueness theorem is that it requires uniform bounds (5.6) and (5.7) on $\nabla \tilde{R}$, $\tilde{R}$ and $(1 - R)$. It forces the initial condition to satisfy these bounds uniformly in x. In particular if we consider the class of profiles $\tilde{\rho}(\cdot, \cdot)$ that have a given "bad" initial profile $\tilde{\rho}_0(\cdot)$, it may contain no profiles from the class $\mathscr{D}_n^1$. This can remedied by allowing solutions of (5.1) with $\mathbf{H} \equiv 0$. Suppose $\mathbf{H}$ is identically zero for $t \in [0, \delta]$ for some $\delta > 0$, and is uniformly bounded on $[\delta, T] \times \mathbf{R}^d$. Then, because a solution of (5.1) with $\mathbf{H} \equiv 0$ is unique in the class $\mathscr{D}_n^0$, we would have uniqueness in the time interval $[0, \delta]$. Moreover the solution $\tilde{\rho}(\delta, \cdot)$ at time δ will have a bounded gradient and $\rho_j(\delta, \cdot)$ as well as $(1 - \rho(\delta, \cdot))$ will be bounded away from zero. One could then expect the uniqueness to hold up to time T because a solution would exist on $[\delta, T]$ in the class $\mathscr{D}_n^1$.

We formally state this as a theorem.

Theorem 5.3. *Suppose $\tilde{R}$ is in the class $\mathscr{D}_n^0$ and solves* (5.1) *with a bounded* $\mathbf{H}$ *that satisfies* $\mathbf{H} \equiv 0$ *on* $[0, \delta]$ *for some* $\delta > 0$. *Suppose in addition, for $t \geq \delta$, $\tilde{R}$ is a solution satisfying the analogues of* (5.6) *and* (5.7), *i.e. a solution in the class $\mathscr{D}_n^1$ for the interval* $[\delta, T]$. *Then any weak solution $\tilde{\rho}$ of* (5.1), *belonging to the class $\mathscr{D}_n^0$ in* $[0, T]$, *and having the same initial value as $\tilde{R}$ in $\mathscr{D}_n^0$, coincides with $\tilde{R}$ in the interval* $[0, T]$.

6. Approximation theorems for the *n*-color rate function

The initial profile $\tilde{\rho}_0(\cdot)$ will be arbitrary and fixed throughout this section. We denote by $\mathscr{E}_n$ profiles that satisfy the initial condition

$$\tilde{\rho}(0, \cdot) = \tilde{\rho}_0(\cdot) \tag{6.1}$$

that are weakly continuous maps from $[0, T] \times \mathbf{T}^d \to \mathscr{C}$, and have a finite value for the rate function

$$I_{\text{dyn}}(\tilde{\rho}) = \frac{1}{2} \int_0^T \left\| \frac{\partial \tilde{\rho}}{\partial t} - \frac{1}{2} \nabla \cdot A\chi(\tilde{\rho}) \nabla \tilde{\rho} \right\|_{-1, A(\tilde{\rho})}^2 dt \tag{6.2}$$

In particular they satisfy (5.3), or equivalently (5.4). We denote by $\mathscr{E}_n^0$ the class of evolving profiles $\tilde{\rho}(\cdot,\cdot)$ that are weak solutions of,

$$\frac{\partial \tilde{\rho}}{\partial t} = \frac{1}{2}\nabla \cdot A\chi(\tilde{\rho})\nabla\tilde{\rho} + \nabla \cdot A(\tilde{\rho})\mathbf{H} \tag{6.3}$$

with the initial condition (6.1) for some bounded $\mathbf{H}$, and have the following additional properties: for some $\delta > 0$ depending on $\tilde{\rho}(\cdot,\cdot)$

$$\mathbf{H} \equiv 0 \quad \text{on} \quad [0,\delta] \tag{6.4}$$

$$\sup_{x\in\mathbf{T}^d, t\in[\delta,T]} |\nabla\tilde{\rho}(t,x)| < \infty \tag{6.5}$$

$$\inf_{x\in\mathbf{T}^d, t\in[\delta,T]} (1-\rho(t,x)) > 0 \tag{6.6}$$

$$\inf_{1\le j\le n,\, x\in\mathbf{T}^d, t\in[\delta,T]} \rho_j(t,x)) > 0\ . \tag{6.7}$$

$\tilde{\rho}(\cdot,\cdot)$ will then be a weak solution in the class $\mathscr{D}_n^0$ on $[0,T]$ and a weak solution in the class $\mathscr{D}_n^1$ on $[\delta,T]$. In order to complete the proof of the lower bound of section 4 we need to prove

Theorem 6.1. *For each fixed initial condition $\tilde{\rho}_0(\cdot)$, $\mathscr{E}_n^0$ is properly dense in $\bar{\mathscr{E}}_n$. That is, for any profile $\tilde{\rho}$ in $\bar{\mathscr{E}}_n$ with $I_{dyn}(\tilde{\rho}) < \infty$, there exists a sequence $\tilde{\rho}_k$ in $\mathscr{E}_n^0$ with*

$$\lim_{k\to\infty} \tilde{\rho}_k = \tilde{\rho} \tag{6.8}$$

uniformly on $[0,T]$ in the weak topology and

$$\lim_{k\to\infty} I_{\text{dyn}}(\tilde{\rho}_k) = I_{\text{dyn}}(\tilde{\rho}) \tag{6.9}$$

The proof of Theorem 6.1 is carried out in several steps each of which is formulated as a Theorem. Let us define $\mathscr{E}_n^2$ to be the subset of those $\tilde{\rho}(\cdot,\cdot) \in \bar{\mathscr{E}}_n$ that are weak solutions of

$$\frac{\partial \tilde{\rho}}{\partial t} = \frac{1}{2}\nabla \cdot A\chi(\tilde{\rho})\nabla\tilde{\rho} \tag{6.10}$$

in the interval $0 \le t \le \delta$ for some $\delta > 0$.

Theorem 6.2. *The set $\mathscr{E}_n^2$ is properly dense in $\bar{\mathscr{E}}_n$.*

Proof. Let $\rho \in \bar{\mathscr{E}}_n$ and consider the following construction. We have the solution $\tilde{R}$ of (6.10) that starts from our initial data $\tilde{\rho}_0(\cdot)$. For $\delta > 0$ let us define $\tilde{\rho}_\delta$ by the following formula

$$\begin{aligned}\tilde{\rho}_\delta(t,\cdot) &= \tilde{R}(t,\cdot) \quad &&\text{for} \quad 0 \le t \le \delta \\ &= \tilde{R}(2\delta - t,\cdot) \quad &&\text{for} \quad \delta \le t \le 2\delta \\ &= \tilde{\rho}(t - 2\delta,\cdot) \quad &&\text{for} \quad 2\delta \le t \le T\end{aligned} \tag{6.11}$$

where $\tilde{R}$ is the solution of (6.10) with initial condition $\tilde{\rho}_0$. It is clear that $\tilde{\rho}_\delta \to \tilde{\rho}$ as $\delta \to 0$ and we now have to estimate $I_{dyn}(\tilde{\rho}_\delta)$ as $\delta \to 0$. Because of lower semicontinuity only the upper bound has to be established. $I_{dyn}(\tilde{\rho}_\delta)$ is given as in (6.1) by an integral from 0 to T

$$\frac{1}{2}\int_0^T \left\| \frac{\partial \tilde{\rho}_\delta}{\partial t} - \frac{1}{2}\nabla \cdot A\chi(\tilde{\rho}_\delta)\nabla\tilde{\rho}_\delta \right\|^2_{-1,A(\tilde{\rho}_\delta)} dt \tag{6.12}$$

as the sum over the two intervals $\int_0^{2\delta} + \int_{2\delta}^T$. The second of these is clearly the same as the integral over $[0, T - 2\delta]$ in (6.2) and is dominated by $I_{dyn}(\tilde{\rho})$. We are therefore left with the task of estimating

$$\frac{1}{2}\int_0^{2\delta} \left\| \frac{\partial \tilde{\rho}_\delta}{\partial t} - \frac{1}{2}\nabla \cdot A\chi(\tilde{\rho}_\delta)\nabla\tilde{\rho}_\delta \right\|^2_{-1,A(\tilde{\rho}_\delta)} dt \tag{6.13}$$

and showing that it goes to 0 as $\delta \to 0$. Since $\tilde{\rho}_\delta$ is a solution (6.10) in the interval $[0, \delta]$, the integrand is zero on that interval. Since the profile $\tilde{\rho}_\delta$ in $[\delta, 2\delta]$ is the profile in $[0, \delta]$ run backwards in time, estimating (6.13) reduces to estimating

$$\frac{1}{2}\int_0^{\delta} \left\| \frac{\partial \tilde{R}}{\partial t} + \frac{1}{2}\nabla \cdot A\chi(\tilde{R})\nabla\tilde{R} \right\|^2_{-1,A(\tilde{R})} dt \tag{6.14}$$

or equivalently

$$\frac{1}{2}\int_0^{\delta} \|\nabla \cdot A\chi(\tilde{R})\nabla\tilde{R}\|^2_{-1,A(\tilde{R})} dt = \frac{1}{2}\int_0^{\delta}\int_{\mathbf{T}^d} \langle \nabla\tilde{R}, (\chi A\chi)(\tilde{R})\nabla\tilde{R}\rangle \, dx \, dt \ . \tag{6.15}$$

It is now clear that the quantity in (6.15) tends to 0 as δ tends to 0.

If $\mathbf{H} = 0$ in an interval $[0, \delta]$ then the solution $\tilde{\rho}$ of (6.10) will have the following properties. For each δ' satisfying $0 < \delta' < \delta$, $\rho = \sum_j \rho_j$ is a solution of the heat equation with constant coefficients D and is therefore smooth and bounded away from 0 and 1 in the interval $[\delta', \delta]$. Once ρ is determined ρ_j are solutions of the linear equation

$$\frac{\partial \rho_j}{\partial t} = \frac{1}{2}\nabla \cdot S(\rho(t,x))\nabla\rho_j + \frac{1}{2}\nabla \cdot (D - S(\rho(t,x)))\frac{\nabla\rho}{\rho}\rho_j \tag{6.16}$$

with $S(\rho(t,x))$ being Lipschitz and uniformly elliptic. From elliptic regularity theory this will provide us a uniform positive lower bound for ρ_j as well as a uniform upper bound on $\nabla\rho_j$ in the interval $[\delta_1,\delta]$.

Let us next define a class $\mathscr{E}_n^1$ of profiles $\tilde{\rho}(\cdot,\cdot)$ in $\mathscr{E}_n^2$ that possess the following properties. There are positive numbers $0<\delta_1<\delta_2<T$ such that in $[0,\delta_1]$, $\tilde{\rho}$ is a solution of (6.10) with initial data $\tilde{\rho}_0$, and in the class $\mathscr{D}_n^1$ in the interval $[\delta_1,\delta_2]$. Finally there is a positive constant α such that for every color j, $\rho_j(t,x)\geq\alpha\rho(t,x)$ uniformly on $[\delta_1,T]\times\mathbf{T}^d$. We then have,

Theorem 6.3. *The class $\mathscr{E}_n^1$ is properly dense in $\mathscr{E}_n^2$.*

Proof. Any given profile $\tilde{\rho}$ in $\mathscr{E}_n^2$ is an actual solution of (6.10) in some interval $[0,\delta]$. Let us take a smooth function $\alpha(t)$ that is equal to zero for $t\leq\frac{1}{4}\delta$, equal to one for $\frac{3}{4}\delta\leq t\leq T$ and goes from zero to one in between. We define

$$\tilde{\rho}_k(t,\cdot)=\left(1-\frac{\alpha(t)}{k}\right)\tilde{\rho}(t,\cdot)+\frac{\alpha(t)}{k}\frac{\tilde{c}}{c}\rho(t,\cdot) \tag{6.17}$$

Here $\tilde{c}$ is the vector of total densities

$$c_j=\int_{\mathbf{T}^d}\rho_j(t,x)\,dx \tag{6.18}$$

and

$$c=\sum_j c_j=\int_{\mathbf{T}^d}\rho(t,x)\,dx \tag{6.19}$$

Because of conservation of color and particles, these are independent of t. For every k, $\tilde{\rho}_k$ is in the class $\mathscr{E}_n^1$. We are now left with proving

$$\limsup_{k\to\infty} I_{\mathrm{dyn}}(\tilde{\rho}_k)\leq I_{\mathrm{dyn}}(\tilde{\rho}) \tag{6.20}$$

The quantity that we need to estimate is one of two types. When we compute the time derivative in

$$\tilde{e}_k=\frac{\partial\tilde{\rho}_k}{\partial t}-\frac{1}{2}\nabla\cdot A\chi(\tilde{\rho}_k)\nabla\tilde{\rho}_k \tag{6.21}$$

we may either differentiate $\alpha(\cdot)$ or $\tilde{\rho}_k(\cdot,\cdot)$. They lead to the following terms:

$$\tilde{e}_k=\tilde{e}_k^{(1)}+\tilde{e}_k^{(2)} \tag{6.22}$$

where

$$\tilde{e}_k^{(1)} = -\frac{1}{k}\ \alpha'(t)\ \tilde{\rho} + \frac{\tilde{c}}{c}\ \frac{1}{k}\ \alpha'(t)\ \rho$$
$$= \frac{1}{k}\alpha'(t)\left[\frac{\tilde{c}}{c}\ \rho - \tilde{\rho}\right] \tag{6.23}$$

Note that $\alpha'(t)$ is nonzero only in a time interval where ρ_j and $(1-\rho)$ are bounded away from zero, and $\|A^{-1}(\tilde{\rho}_k)\| \le C$ on that interval. On the other hand $[\frac{\tilde{c}}{c}\rho - \tilde{\rho}]$ is a vector of bounded mean zero functions. Therefore

$$\lim_{k\to\infty}\int_0^T \|\tilde{e}_k^{(1)}(t,\cdot)\|^2_{-1,A(\tilde{\rho}_k)}\,dt = 0 \tag{6.24}$$

We now turn our attention to $\tilde{e}_k^{(2)}$.

$$\tilde{e}_k^{(2)} = \left(1-\frac{m(t)}{k}\right)\left(\frac{\partial\tilde{\rho}}{\partial t} - \frac{1}{2}\nabla A\chi(\tilde{\rho})\nabla\tilde{\rho}\right) + \left(\frac{m(t)}{k}\right)\left(\frac{\partial\rho}{\partial t} - \frac{1}{2}\nabla D\nabla\rho\right)$$
$$= \left(1-\frac{m(t)}{k}\right)\nabla\cdot\mathbf{G} + \left(\frac{\tilde{c}}{c}\frac{m(t)}{k}\right)\nabla G \tag{6.25}$$

for some $\mathbf{G} = \{G^j\}$ and $G = \sum_j G^j$. Moreover

$$I_{\text{dyn}}(\tilde{\rho}) = \frac{1}{2}\int_0^T\int_{\mathbf{T}^d}\left(\sum_j \frac{\langle S^{-1}(\rho)G^j, G^j\rangle}{\rho_j} + \frac{\langle [D^{-1} - (1-\rho)S^{-1}(\rho)]G, G\rangle}{\rho(1-\rho)}\right)dx\,dt < \infty \tag{6.26}$$

Since

$$I_{\text{dyn}}(\tilde{\rho}_k) \le \frac{1}{2}\int_0^T\int_{\mathbf{T}^d}\left(\sum_j \frac{\langle S^{-1}(\rho)G_k^j\ , G_k^j\rangle}{\rho_{k,j}} + \frac{\langle [D^{-1} - (1-\rho)S^{-1}(\rho)]G_k\ , G_k\rangle}{\rho(1-\rho)}\right)dx\,dt \tag{6.27}$$

where

$$G_k^j = \left(1-\frac{m(t)}{k}\right)G^j + \frac{c_j}{c}\ \frac{m(t)}{k}\ G \quad \text{and} \quad G_k = \sum_j G_k^j = G\ , \tag{6.28}$$

and

$$\rho_k^j = \left(1-\frac{m(t)}{k}\right)\rho_j + \left(\frac{m(t)}{k}\right)\rho\ , \tag{6.29}$$

it is straightforward to see that (6.20) is valid.

The last step is

Theorem 6.4. *The class $\mathscr{E}_n^0$ is properly dense in $\mathscr{E}_n^1$.*

Proof. For a given $\tilde{\rho}(\cdot,\cdot)$ in $\mathscr{E}_n^1$ there are $0 < \delta_1 < \delta_2 < T$ with properties listed earlier. Let us pick a function $\epsilon(t)$ that is zero for $t \le \delta_1$ and increases smoothly to one over the interval $[\delta_1, \delta_2]$ remaining equal to one on $[\delta_2, T]$. We define $\epsilon_k(t) = \frac{1}{k}\epsilon(t)$ and take

$$\tilde{\rho}_k(t,\cdot) = \tilde{\rho}(t,\cdot) * \phi_{\epsilon_k(t)}(\cdot) \tag{6.30}$$

where ϕ_ϵ is the heat kernel on $\mathbf{T}^d$ at time ϵ^2 which we use as a mollifier. Clearly $\tilde{\rho}_k(\cdot,\cdot)$ is in $\mathscr{E}_n^0$ and in order to complete our program we need only show that

$$\limsup_{k\to\infty} I_{\text{dyn}}(\tilde{\rho}_k) \le I_{\text{dyn}}(\tilde{\rho}) \tag{6.31}$$

Before we take up the proof of Theorem 6.4 we will establish Lemmas 6.5–6.8. □

Lemma 6.5. *Let $f \in W^{1,2}(\mathbf{T}^d)$. The following families of functions are uniformly integrable:*

$$\text{(i)} \quad g_a(x) = \frac{|f(x+a) - f(x)|^2}{|a|^2}, \quad a \in \mathbf{R}^d;$$

$$\begin{aligned}\text{(ii)} \quad h_\epsilon(x) = \epsilon^{-2} \int_{\mathbf{T}^d}\int_{\mathbf{T}^d} &|f(x+y) \\ &- f(x+z)|^2 \phi_\epsilon(y)\phi_\epsilon(z)\,dy\,dz, \quad \epsilon > 0;\end{aligned} \tag{6.32}$$

$$\text{(iii)} \quad \tilde{h}_\epsilon(x) = \int_{\mathbf{T}^d}\int_{\mathbf{T}^d} |f(x+y) - f(x+z)|^2 \frac{|\nabla\phi_\epsilon|^2}{\phi_\epsilon}(y)\phi_\epsilon(z)\,dy\,dz, \quad \epsilon > 0\ .$$

Here $|\cdot|$ is used for distance on $\mathbf{R}$.

Proof. Recall that a family of functions $\{g_\gamma\}_{\gamma\in\Gamma}$ is uniformly integrable if and only if there exists a function $\Psi : \mathbf{R} \to \mathbf{R}_+$ such that

$$\begin{aligned} &(a) \quad \lim_{|x|\to\infty} \frac{\Psi(x)}{|x|} = \infty \\ &(b) \quad \sup_{\gamma\in\Gamma} \int \Psi(g_\gamma(x))\,dx < \infty\ . \end{aligned} \tag{6.33}$$

Furthermore if g is any function in L^1 then there exists an increasing, convex function Ψ satisfying (a) such that $\Psi(|g|) \in L^1$. Choose such a Ψ for $|\nabla f|^2$. By Jensen's inequality,

$$\begin{aligned}\int \Psi(g_a(x))\,dx &= \int \Psi\left(\frac{|f(x+a)-f(x)|^2}{|a|^2}\right) dx \\ &\le \int \Psi\left(\left[\int_0^1 |\nabla f(x+ta)|dt\right]^2\right) dx \\ &\le \int\int_0^1 \Psi\left(|\nabla f(x+ta)|^2\right) dt\,dx \\ &= \int \Psi\left(|\nabla f|^2\right) dx. \end{aligned} \tag{6.34}$$

The last term has a bound independent of a. Therefore the family $\{g_a\},\ a \in \mathbf{T}^d$ is uniformly integrable. In a similar fashion,

$$\begin{aligned}\int \Psi(h_\epsilon(x))dx &= \int \Psi\left(\epsilon^{-2}\int\int |f(x+y)-f(x+z)|^2\phi_\epsilon(y)\phi_\epsilon(z)\,dy\,dz\right)dx \\ &= \int \Psi\left(\int_{\mathbf{R}^d}\int_{\mathbf{R}^d} \frac{|f(x+y)-f(x+z)|^2}{|y-z|^2}\right. \\ &\quad \left.\times\left\{\frac{|y-z|^2}{\epsilon^2}\phi_\epsilon(y)\phi_\epsilon(z)\right\}dy\,dz\right)dx \end{aligned} \tag{6.35}$$

and

$$\begin{aligned}&\int \Psi\left(\tilde{h}_\epsilon(x)\right)\,dx \\ &= \int \Psi\left(\int_{\mathbf{T}^d}\int_{\mathbf{T}^d} |f(x+y)-f(x+z)|^2\frac{|\nabla\phi_\epsilon|^2}{\phi_\epsilon}(y)\phi_\epsilon(z)\,dy\,dz\right)\,dx \\ &\le \int \Psi\left(\int_{\mathbf{R}^d}\int_{\mathbf{R}^d} \frac{|f(x+y)-f(x+z)|^2}{|y-z|^2}\right. \\ &\quad \left.\times\left\{|y-z|^2\frac{|\nabla\phi_\epsilon|^2}{\phi_\epsilon}(y)\phi_\epsilon(z)\right\}\,dy\,dz\right)\,dx \end{aligned} \tag{6.36}$$

An easy computation shows that there is a constant C independent of ϵ such that

$$\int_{\mathbf{R}^d}\int_{\mathbf{R}^d} |y-z|^2\frac{|\nabla\phi_\epsilon|^2}{\phi_\epsilon}(y)\phi_\epsilon(z)\,dy\,dz \le C$$

The lemma follows by Jensen's inequality. The next lemma is the following.

Lemma 6.6. *Suppose that r and ρ are measurable functions mapping* $\mathbf{T}^d$ *into* $[0, 1]$ *such that*

$$\int_{\mathbf{T}^d} (1-\rho)\frac{|\nabla r|^2}{r}\,dx < \infty\ , \tag{6.37}$$

and

$$\int_{\mathbf{T}^d} \frac{|\nabla \rho|^2}{1-\rho}\,dx < \infty\ . \tag{6.38}$$

Then

$$\lim_{\epsilon\to 0}\int_{\mathbf{T}^d} (1-\rho_\epsilon)\frac{|\nabla r_\epsilon|^2}{r_\epsilon}\,dx = \int_{\mathbf{T}^d} (1-\rho)\frac{|\nabla r|^2}{r}\,dx\ . \tag{6.39}$$

Proof. The strategy is to bound the integrand in terms of objects which are uniformly integrable. First of all note that

$$\begin{aligned}
&(1-\rho_\epsilon)|\nabla r_\epsilon|^2\\
&= \left[\int_{\mathbf{T}^d} \sqrt{1-\rho_\epsilon}(x)\nabla r(x+y)\phi_\epsilon(y)\,dy\right]^2\\
&= \left[\int_{\mathbf{T}^d} \sqrt{1-\rho_\epsilon}(x)\ r(x+y)\nabla\phi_\epsilon(y)\,dy\right]^2\\
&\leq 2\left[\int_{\mathbf{T}^d} \sqrt{1-\rho}(x+y) r(x+y)\nabla\phi_\epsilon(y)\,dy\right]^2\\
&\quad + 2\left[\int_{\mathbf{T}^d} \left(\sqrt{1-\rho_\epsilon}(x) - \sqrt{1-\rho}(x+y)\right) r(x+y)\nabla\phi_\epsilon(y)\,dy\right]^2 .
\end{aligned} \tag{6.40}$$

The assumptions of the lemma imply in particular that

$$\int_{\mathbf{T}} \frac{|\nabla(r\sqrt{1-\rho})|^2}{r}\,dx < \infty. \tag{6.41}$$

Therefore

$$\frac{\left[\int_{\mathbf{T}^d}\sqrt{1-\rho}(x+y)r(x+y)\nabla\phi_\epsilon(y)\,dy\right]^2}{\int_{\mathbf{T}^d} r(x+y)\phi_\epsilon(y)\,dy} = \frac{|(\nabla(r\sqrt{1-\rho}))_\epsilon|^2}{r_\epsilon}(x) \tag{6.42}$$

is uniformly integrable. Now we consider the final term of (6.40). By Schwarz's inequality,

$$\frac{\left[\int_{\mathbf{T}^d}\left(\sqrt{1-\rho_\epsilon}(x)-\sqrt{1-\rho}(x+y)\right)r(x+y)\nabla\phi_\epsilon(y)\,dy\right]^2}{\int_{\mathbf{T}^d} r(x+y)\phi_\epsilon(y)\,dy}$$

$$\leq \int_{\mathbf{T}^d}\left|\sqrt{1-\rho_\epsilon}(x)-\sqrt{1-\rho}(x+y)\right|^2\frac{|\nabla\phi_\epsilon|^2}{\phi_\epsilon}(y)\,dy. \tag{6.43}$$

By Jensens's inequality,

$$\left|\sqrt{1-\rho_\epsilon}(x)-\sqrt{1-\rho}(x+y)\right|^2$$

$$\leq 2\left|\sqrt{1-\rho_\epsilon}(x)-\left(\sqrt{1-\rho}\right)_\epsilon(x)\right|^2+2\left|\left(\sqrt{1-\rho}\right)_\epsilon(x)-\sqrt{1-\rho}(x+y)\right|^2$$

$$\leq 2\int_{\mathbf{T}^d}\int_{\mathbf{T}^d}\left|\sqrt{1-\rho}(x+y)-\sqrt{1-\rho}(x+z)\right|^2\phi_\epsilon(y)\phi_\epsilon(z)\,dy\,dz$$

$$+2\int_{\mathbf{T}^d}\left|\sqrt{1-\rho}(x+y)-\sqrt{1-\rho}(x+z)\right|^2\phi_\epsilon(z)\,dy\,dz. \tag{6.44}$$

Therefore

$$\frac{\left[\int_{\mathbf{T}^d}\left(\sqrt{1-\rho_\epsilon}(x)-\sqrt{1-\rho}(x+y)\right)r(x+y)\nabla\phi_\epsilon(y)\,dy\right]^2}{\int_{\mathbf{T}^d} r(x+y)\phi_\epsilon(y)\,dy}$$

$$\leq 2\int_{\mathbf{T}^d}\int_{\mathbf{T}^d}\left|\sqrt{1-\rho}(x+y)-\sqrt{1-\rho}(x+z)\right|^2\phi_\epsilon(y)\phi_\epsilon(z)\,dy\,dz$$

$$\times\int_{\mathbf{T}^d}\frac{|\nabla\phi_\epsilon|^2}{\phi_\epsilon}(y)\,dy+2\int_{\mathbf{T}^d}\int_{\mathbf{T}^d}\left|\sqrt{1-\rho}(x+y)-\sqrt{1-\rho}(x+z)\right|^2$$

$$\times\frac{|\nabla\phi_\epsilon|^2}{\phi_\epsilon}(y)\phi_\epsilon(z)\,dy\,dz. \tag{6.45}$$

By assumption $\sqrt{1-\rho}\in W^{1,2}$ and therefore, by the previous lemma, these last two terms are uniformly integrable. □

Lemma 6.7. *Suppose that* $I_0(\tilde{\rho}(\cdot,\cdot))<\infty$ *and, for some* $\alpha>0$, $\rho_j(t,x)\geq\alpha\,\rho(t,x)$ *for all* $j=1,\ldots,m$, $t\in[0,T]$ *and* $x\in\mathbf{T}^d$. *Let*

$$\rho_{j,\epsilon}(t,x)=\int_{\mathbf{T}^d}\rho_j(t,x+y)\phi_\epsilon(y)\,dy. \tag{6.46}$$

Then, as $\epsilon\to 0$,

$$I_0(\tilde{\rho}_\epsilon(\cdot,\cdot))\to I_0(\tilde{\rho}(\cdot,\cdot)). \tag{6.47}$$

Proof. We write $I_0(\tilde{\rho}(\cdot,\cdot))$ explicitly as

$$\int_0^T \int_{\mathbf{T}^d} \sum_{j=1}^n \frac{\nabla\rho_j \cdot S(\rho)\nabla\rho_j}{\rho_j} + \frac{\nabla\rho \cdot D\nabla\rho}{\rho(1-\rho)} - \frac{\nabla\rho \cdot S(\rho)\nabla\rho}{\rho}\, dx\, dt\ . \tag{6.48}$$

Note that the third integrand is bounded pointwise by either of the previous two. Therefore the whole integral is finite only if each of the three integrals is finite separately. Recall

$$\delta(1-\rho)I \le S(\rho) \le (1-\rho)D\ , \tag{6.49}$$

in the sense of positive definite matrices. Also by assumption, $\alpha\rho \le \rho_j \le \rho$. Hence

$$\int_0^T \int_{\mathbf{T}^d} \frac{1-\rho}{\rho} |\nabla\rho_j|^2\, dx\, dt \le c^{-1} \int_0^T \int_{\mathbf{T}^d} \frac{\nabla\rho_j \cdot S(\rho)\nabla\rho_j}{\rho_j}\, dx\, dt < \infty\ . \tag{6.50}$$

By the previous lemma, $\frac{1-\rho_\epsilon}{\rho_\epsilon}|\nabla\rho_{j,\epsilon}|^2$ is uniformly integrable. Since

$$\frac{\nabla\rho_{j,\epsilon} \cdot S(\rho_\epsilon)\nabla\rho_{j,\epsilon}}{\rho_{j,\epsilon}} \le \alpha^{-1}\frac{1-\rho_\epsilon}{\rho_\epsilon}|\nabla\rho_{j,\epsilon}|^2 \tag{6.51}$$

it is also uniformly integrable. Therefore, as $\epsilon \to 0$,

$$\int_0^T \int_{\mathbf{T}^d} \frac{\nabla\rho_{j,\epsilon} \cdot S(\rho_\epsilon)\nabla\rho_{j,\epsilon}}{\rho_{j,\epsilon}}\, dx\, dt \to \int_0^T \int_{\mathbf{T}^d} \frac{\nabla\rho_j \cdot S(\rho)\nabla\rho_j}{\rho_j}\, dx\, dt\ . \tag{6.52}$$

For the second term, by convexity on the one hand, and lower semicontinuity on the other,

$$\int_0^T \int_{\mathbf{T}^d} \frac{\nabla\rho_\epsilon \cdot D\nabla\rho_\epsilon}{\rho_\epsilon(1-\rho_\epsilon)}\, dx\, dt \to \int_0^T \int_{\mathbf{T}^d} \frac{\nabla\rho \cdot D\nabla\rho}{\rho(1-\rho)}\, dx\, dt\ . \tag{6.53}$$

The third term is pointwise bounded by the first. The uniform integrability of

$$\frac{\nabla\rho_\epsilon \cdot S(\rho_\epsilon)\nabla\rho_\epsilon}{\rho_\epsilon} \tag{6.54}$$

follows. □

Lemma 6.8. *Suppose that* $I_{\mathrm{dyn}}(\tilde\rho(\cdot,\cdot)) < \infty$ *and, for some* $\alpha > 0$, $\rho_j(t,x) \ge \alpha\rho(t,x)$ *for all* $j = 1, \ldots, m$, $t \in [0,T]$ *and* $x \in \mathbf{T}^d$. *Let*

$$\rho_{j,\epsilon}(t,x) = \int_{\mathbf{T}^d} \rho_j(t,x+y)\phi_\epsilon(y)\, dy\ . \tag{6.55}$$

Then, as $\epsilon \to 0$,

$$I_{\rm dyn}(\tilde{\rho}_\epsilon(\cdot,\cdot)) \to I_{\rm dyn}(\tilde{\rho}(\cdot,\cdot)) \ . \tag{6.56}$$

Proof. The finiteness of $I_{\rm dyn}(\tilde{\rho}(\cdot,\cdot))$ implies in particular that $\frac{\partial\tilde{\rho}}{\partial t}$ can be written as $\nabla\cdot P$ for some P with $\int_0^T\int_{\mathbf{T}^d}\langle P;A^{-1}P\rangle\,dx\,dt<\infty$. Expanding the $H_{-1,A}$ norm,

$$I_{\rm dyn}(\rho(\cdot,\cdot)) = \frac{1}{2}\int_0^T\int_{\mathbf{T}^d}\langle P;A^{-1}P\rangle\,dx\,dt - \frac{1}{2}\int_0^T\int_{\mathbf{T}^d}\langle P;\chi\nabla\rho\rangle\,dx\,dt + \frac{1}{8}I_0(\tilde{\rho}) \ . \tag{6.57}$$

By assumption, $I_0(\tilde{\rho}(\cdot,\cdot))<\infty$. Therefore each of the three terms is finite on its own. Note also that

$$\langle P;\chi\nabla\tilde{\rho}\rangle \le \langle P;A^{-1}P\rangle + \frac{1}{4}\langle\chi\nabla\tilde{\rho};A\chi\nabla\tilde{\rho}\rangle \tag{6.58}$$

so it suffices to show that

$$\int_0^T\int_{\mathbf{T}^d}\langle P;A^{-1}P\rangle\,dx\,dt<\infty \tag{6.59}$$

and

$$\int_0^T\int_{\mathbf{T}^d}\langle\chi\nabla\tilde{\rho};A\chi\nabla\tilde{\rho}\rangle<\infty \tag{6.60}$$

imply the uniform integrability of

$$\langle P_\epsilon;A^{-1}(\tilde{\rho}_\epsilon)P_\epsilon\rangle \tag{6.61}$$

and

$$\langle\chi(\tilde{\rho}_\epsilon)\nabla\tilde{\rho}_\epsilon;A(\tilde{\rho}_\epsilon)\chi(\tilde{\rho}_\epsilon)\nabla\tilde{\rho}_\epsilon\rangle. \tag{6.62}$$

The second of these is the content of the previous lemma. To prove the first, we write it more explicitly as

$$\langle P;A^{-1}P\rangle = \sum_{j=1}^n \frac{P_j\cdot S^{-1}(\rho)P_j}{\rho_j} + \frac{\left(\sum_{j=1}^n P_j\right)\cdot\left(S^{-1}(\rho)-(D(1-\rho))^{-1}\right)\left(\sum_{j=1}^n P_j\right)}{\sum_{j=1}^n\rho_j} \ . \tag{6.63}$$

Now

$$\int_0^T\int_{\mathbf{T}^d}\frac{|P_j|^2}{\rho(1-\rho)}\,dx\,dt \le \int_0^T\int_{\mathbf{T}^d}\frac{P_j\cdot S^{-1}(\rho)P_j}{\rho_j}\,dx\,dt<\infty \tag{6.64}$$

so

$$\frac{P_{j,\epsilon} \cdot S{-}1(\rho_\epsilon)P_{j,\epsilon}}{\rho_{j,\epsilon}} \le (c\alpha)^{-1} \frac{|P_{j,\epsilon}|^2}{\rho_\epsilon(1-\rho_\epsilon)} \le (c\alpha)^{-1} \left(\frac{|P_j|^2}{\rho(1-\rho)} \right)_\epsilon \tag{6.65}$$

is uniformly integrable. Also

$$\frac{\left(\sum_{j=1}^n P_{j,\epsilon}\right) \cdot S^{-1}(\rho_\epsilon)\left(\sum_{j=1}^n P_{j,\epsilon}\right)}{\sum_{j=1}^n \rho_{j,\epsilon}} \le \sum_{j=1}^n \frac{P_{j,\epsilon} \cdot S^{-1}(\rho_\epsilon)P_{j,\epsilon}}{\rho_{j,\epsilon}} \tag{6.66}$$

and

$$\frac{\left(\sum_{j=1}^n P_{j,\epsilon}\right) \cdot (D(1-\rho_\epsilon))^{-1}\left(\sum_{j=1}^n P_j\right)}{\sum_{j=1}^n \rho_{j,\epsilon}} \le \text{const.} \sum_{j=1}^n \frac{|P_{j,\epsilon}|^2}{\rho_\epsilon(1-\rho_\epsilon)} \tag{6.67}$$

so these terms are also uniformly integrable.

We now return to complete the proof of theorem 6.4. When we start with $\tilde{\rho}$ and convolute it with a time dependent mollifier $\phi_{\frac{\epsilon(\cdot)}{k}}$ to obtain $\tilde{\rho}_k$ in the calculation of $I_{\text{dyn}}(\tilde{\rho}_k)$ there are two types of terms. One of them for each time t is obtained by differentiating with respect to t in the $\tilde{\rho}$ and this leads to exactly the type of term handled by lemma 6.8. The other term obtained by differentiating with respect to the t in $\epsilon(t)$ leads to a term that is of the form

$$\frac{1}{k^2} \int_0^T [\epsilon'(t)]^2 \|\Delta\tilde{\rho}\|_{-1,A(\tilde{\rho}_k)}\, dt \ . \tag{6.68}$$

The changeover in $\epsilon(\cdot)$ from 0 to 1 occurs inside the interval $[\delta_1, \delta_2]$ where $\nabla\tilde{\rho}$ is bounded with mean 0. Moreover ρ_j and $1-\rho$ have a uniform lower bound and this implies the same bounds for $\tilde{\rho}_k$ as well, leading to a uniform bound for $\|\Delta\tilde{\rho}\|_{-1,A(\tilde{\rho}_k)}$. This completes the proof of Theorem 6.4.

7. Contraction principle

The identification of the rate function $\mathscr{I}(Q)$ defined in (7.21) for the large deviation of the empirical process is carried out in this section. In order to formulate a crucial induction step in the proof we need to define a slightly more general rate function $\mathscr{I}_n(Q_1, \ldots, Q_n)$ defined in (7.22) which is the large deviation rate function for the empirical processes of particles that are classified as being one of n possible colors. If we look at the marginal densities $\rho_1(t,\cdot), \ldots, \rho_n(t,\cdot)$ of the n empirical processes $Q_1, \ldots, Q_n$, we have the empirical densities for the

n-color dynamics and our candidate $\mathscr{I}_n(Q_1, \ldots, Q_n)$ should contract to the n-color rate function defined for instance in (4.2). This is the content of Theorem 7.5 of this section. It turns out that given the form of the functionals $\mathscr{I}_n(Q_1, \ldots, Q_n)$ and $I_{\text{dyn}}(\tilde{\rho}(\cdot\,,\cdot))$ this contraction relationship is enough to completely identify $\mathscr{I}(Q)$ as the rate function for the large deviations of the empirical process given that $I_{\text{dyn}}(\tilde{\rho}(\cdot\,,\cdot))$ is already known to control the large deviations of the n-color densities. This final step is actually carried out in section 9 and constitutes the proof of the large deviation principle for the empirical process.

We begin this section with some preliminary analysis and in Theorem 7.1 establish a dual variational formula for $I_{\text{dyn}}(\tilde{\rho}(\cdot\,,\cdot))$. This is standard convex analysis. We then carefully define $\mathscr{I}_n(Q_1, \ldots, Q_n)$ and $\mathscr{I}(Q)$ and establish some of their properties. Finally in Theorem 7.5 the dual variational formula is used to establish the contraction relationship between $\mathscr{I}_n(Q_1, \ldots, Q_n)$ and $I_{\text{dyn}}(\tilde{\rho}(\cdot\,,\cdot))$.

Let us consider $\tilde{\rho}(\cdot\,,\cdot) : [0\,,T] \times \mathbf{T}^d \to \mathscr{C}$ that satisfies

$$\hat{I}_j(\tilde{\rho}(\cdot\,,\cdot)) = \int_0^T \int_{\mathbf{T}^d} (1-\rho) \frac{|\nabla \rho_j|^2}{\rho_j} \, dx \, dt < \infty \tag{7.1}$$

for $j = 1, \ldots, n$ and

$$\hat{I}(\rho(\cdot,\cdot)) = \int_0^T \int_{\mathbf{T}^d} \frac{|\nabla \rho|^2}{\rho(1-\rho)} \, dx \, dt < \infty \tag{7.2}$$

where $\rho = \sum_j \rho_j$.

We have the n color rate function $I_{\text{dyn}}(\tilde{\rho}(\cdot\,,\cdot))$ defined for such maps by

$$\begin{aligned} I_{\text{dyn}}(\tilde{\rho}(\cdot\,,\cdot)) &= \frac{1}{2} \int_0^T \left\| \frac{\partial \tilde{\rho}}{\partial t} - \frac{1}{2} \nabla A \chi(\tilde{\rho}) \nabla \tilde{\rho} \right\|^2_{-1, A(\tilde{\rho})} dt \\ &= \sup_{\Phi} \left[\int_0^T \int_{\mathbf{T}^d} \left\langle \frac{\partial \tilde{\rho}}{\partial t} - \mathscr{A}^*_\rho \tilde{\rho}, \Phi \right\rangle dx \, dt \right. \\ &\quad \left. - \frac{1}{2} \int_0^T \int_{\mathbf{T}^d} \langle \nabla \Phi, A(\tilde{\rho}) \nabla \Phi \rangle \, dx \, dt \right] \end{aligned} \tag{7.3}$$

where the supremum is taken over all Φ that map $[0, T] \times \mathbf{T}^d$ smoothly into R^n and $\mathscr{A}^*_\rho$ is the forward or adjoint operator

$$\mathscr{A}^*_\rho = \frac{1}{2} \nabla \cdot S(\rho) \nabla + \frac{1}{2} \nabla \cdot (D - S(\rho)) \frac{\nabla \rho}{\rho} \tag{7.4}$$

corresponding to the diffusion generator

$$\mathscr{A}_\rho f = \frac{1}{2}\nabla\cdot S(\rho)\nabla f + \frac{1}{2}(S(\rho)-D)\frac{\nabla\rho}{\rho}\cdot\nabla f \tag{7.5}$$

If $\tilde{\rho}$ does not satisfy either (7.1) or (7.2) we define

$$I_{\text{dyn}}(\tilde{\rho}(\cdot\,,\cdot)) = +\infty$$

The variational formula (7.3) has to be carefully interpreted. There is no apriori assumption of regularity on $\tilde{\rho}(\cdot\,,\cdot)$. Since Φ is smooth we can attempt an integration by parts and the best that we can do is

$$\int_0^T\int_{\mathbf{T}^d}\left(\frac{\partial\tilde{\rho}_i}{\partial t} - \mathscr{A}^*_\rho\tilde{\rho}_i\right)\Phi_i\,dx\,dt$$

$$= \int_{\mathbf{T}^d}\Phi_i(T,x)\rho_i(T,x)\,dx - \int_{\mathbf{T}^d}\Phi_i(0,x)\rho_i(0,x)\,dx$$

$$-\int_0^T\int_{\mathbf{T}^d}\frac{\partial\Phi_i}{\partial t}(t,x)\rho_i(t,x)\,dx\,dt + \frac{1}{2}\int_0^T\int_{\mathbf{T}^d}\nabla\rho_i\cdot S(\rho)\nabla\Phi_i\,dx\,dt$$

$$+\frac{1}{2}\int_0^T\int_{\mathbf{T}^d}(D-S(\rho))\frac{\nabla\rho}{\rho}\cdot\nabla\Phi_i(t,x)\rho_i(t,x)\,dx\,dt.$$

We therefore need to have some information on $\nabla\rho_i$. It turns out that (7.1) and (7.2) are sufficient. If we have an upper bound C on $I_{\text{dyn}}(\tilde{\rho})$ then we can get bounds on

$$\hat{I}_j(\tilde{\rho}(\cdot\,,\cdot)) = \int_0^T\int_{\mathbf{T}^d}(1-\rho)\frac{|\nabla\rho_j|^2}{\rho_j}\,dx\,dt \le C' \tag{7.6}$$

for $j = 1,\ldots,n$, and

$$\hat{I}(\rho(\cdot,\cdot)) = \int_0^T\int_{\mathbf{T}^d}\frac{|\nabla\rho|^2}{\rho(1-\rho)}\,dx\,dt \le C' \tag{7.7}$$

with C' depending only on C.

We start with a dual variational formula. Let $\mathscr{G}_{\tilde{\rho}} = \{(b,b_1,\ldots,b_n)\}$ that satisfy

$$\frac{\partial\rho_j}{\partial t} - \mathscr{A}^*_\rho\rho_j + \nabla\cdot\rho_j(1-\rho)Db + \nabla\cdot\rho_j b_j = 0;\ j=1,\ldots,n \tag{7.8}$$

$$\mathscr{E}[b] = \frac{1}{2}\int_0^T\int_{\mathbf{T}^d}\rho(1-\rho)\langle b\,,\,Db\rangle\,dx\,dt < \infty\ . \tag{7.9}$$

$$\mathscr{E}_n[b_1\ldots,b_n] = \sum_{j=1}^n\frac{1}{2}\int_0^T\int_{\mathbf{T}^d}\langle b_j\,,S^{-1}(\rho)b_j\rangle\,\rho_j\,dx\,dt < \infty \tag{7.10}$$

$$\sum_{j=1}^{n} \rho_j b_j \equiv 0 \tag{7.11}$$

We note that $(b, b_1, \ldots, b_n) \in \mathscr{G}_{\tilde{\rho}}$ implies in particular that

$$\frac{\partial \rho}{\partial t} - \frac{1}{2}\nabla D \nabla \rho = -\nabla \cdot \rho(1-\rho) D b \tag{7.12}$$

or $b \in \mathscr{B}_\rho$ where we define $\mathscr{B}_\rho$ as those b that satisfy (7.9) and (7.12).

Theorem 7.1. *For any* $\tilde{\rho} : [0, T] \times \mathbf{T}^d \to \mathscr{C}$

$$\begin{aligned} &I_{\text{dyn}}(\tilde{\rho}) \\ &= \sup_{\Phi}\left[\int_0^T\int_{\mathbf{T}^d}\left\langle \frac{\partial \tilde{\rho}}{\partial t} - \mathscr{A}^*_\rho \tilde{\rho}, \Phi \right\rangle dx\,dt - \frac{1}{2}\int_0^T\int_{\mathbf{T}^d}\langle \nabla\Phi; A(\tilde{\rho})\nabla\Phi\rangle\,dx\,dt\right] \\ &= \inf_{(b,b_1,\ldots,b_n)\in\mathscr{G}_{\tilde{\rho}}} \frac{1}{2}\int_0^T\int_{\mathbf{T}^d}\left\{\sum_{j=1}^{n}\rho_j\langle b_j, S^{-1}(\rho)\,b_j\rangle + \rho(1-\rho)\,\langle b, Db\rangle\right\} dx\,dt \end{aligned} \tag{7.13}$$

Proof. If $\tilde{\rho} \in \mathscr{E}_n^0$, we can solve

$$\frac{\partial \tilde{\rho}}{\partial t} - \mathscr{A}^*_\rho \tilde{\rho} = \nabla A(\tilde{\rho}) \nabla H$$

with a function $H \in L_2([0, T]; H_1)$. The two sides can then be shown to be equal by using the fact that the matrix $A(\rho)$ can be explicitly inverted and its inverse is seen to correspond to the quadratic form

$$\langle \mathbf{f}; A^{-1}(\rho)\mathbf{f}\rangle = \sum_{j=1}^{n}\frac{f_j \cdot S^{-1}(\rho) f_j}{\rho_j} - \frac{f \cdot S^{-1}(\rho) f}{\rho} + \frac{f \cdot D^{-1} f}{\rho(1-\rho)}$$

where $f = \sum_j f_j$. The clue to the explicit inversion is the fact that $A(\rho)$ is a rank one perturbation of $S(\rho) \otimes \Lambda$ where $\Lambda = \text{diag}\{\rho_1, \ldots, \rho_n\}$.

It is not hard to show that the left-hand side of (7.13) is less than its right-hand side. For the other direction, let us first note that the quantity defined by the formula on the left-hand of (7.13) is lower semicontinuous as a functional of $\tilde{\rho}$. We now show the right hand side is lower semicontinuous as well. Let us take a sequence $\tilde{\rho}^m$ converging weakly to $\tilde{\rho}$ with uniform bound ℓ for the right-hand side. As a result, we have a uniform bound ℓ for the left-hand side which in turn imply

$$\int_0^T\int_{\mathbf{T}^d}\frac{|\nabla \rho^m|^2}{\rho^m(1-\rho^m)}\,dx\,dt \le C,$$

$$\int_0^T \left\|\frac{\partial \rho}{\partial t}\right\|_{-1}^2 + \left\|\frac{\partial \rho_i^m}{\partial t}\right\|_{-1}^2 dt \le C,$$

$$\int_0^T \int_{\mathbf{T}^d} (1-\rho^m)\frac{|\nabla \rho_i^m|^2}{\rho_i^m}\, dx\, dt \le C\ .$$

for some constant $C < \infty$. We then use Lemma 4.2 to deduce the convergence in $L^1([0\,,T] \times \mathbf{T}^d)$ of ρ^m to ρ as well as that of $(1-\rho^m)\rho_i^m$ to $(1-\rho)\rho_i$

Since $S(\rho) \le C(1-\rho)$ we obtain a bound on

$$\int_0^T \int_{\mathbf{T}^d} \frac{\rho_i^m}{(1-\rho^m)}\, |b_i^m|^2\, dx\, dt$$

and therefore we can take a weak limit c_i in $L_2[[0\,,T] \times \mathbf{T}^d\,]$ of

$$c_i^m = \sqrt{\frac{\rho_i^m}{(1-\rho^m)}}\, b_i^m$$

and write

$$b_i = \sqrt{\frac{(1-\rho)}{\rho_i}}\, c_i$$

on the set where $\rho_i > 0$ and 0 elsewhere. Since we can write

$$\rho_i^m b_i^m = \sqrt{\rho_i^m(1-\rho^m)}\, c_i^m$$

we have the weak convergence in $L_1[[0\,,T] \times \mathbf{T}^d\,]$ of $\rho_i^m b_i^m$ to $\rho_i b_i$. It is now easy to produce a b such that $(b, b_1, \ldots, b_n) \in \mathcal{G}_{\tilde{\rho}}$ and

$$\frac{1}{2}\sum_j \int_0^T \int_{\mathbf{T}^d} \rho_j b_j \cdot S^{-1}(\rho)\, b_j\, dx\, dt + \frac{1}{2}\int_0^T \int_{\mathbf{T}^d} \rho(1-\rho)\, b \cdot Db\, dx\, dt \le \ell$$

thereby proving the lower semicontinuity of right-hand side.

By Theorem 6.1 we can approximate arbitrary $\tilde{\rho}$ nicely by $\tilde{\rho}^m \in \mathcal{E}_n^0$ so that the right hand side converges to the corresponding value of the right hand side. By lower semicontinuity of right hand side, we must have that the right hand side is less than or equal to the left hand side in the limit.

The following theorem summarizes results that can be found in [QV] on the question of existence and uniqueness of the diffusion processes that we need.

Theorem 7.2. *For each $b \in \mathscr{B}_\rho$ we have a unique diffusion process P^b that has $\rho(t, \cdot)$ as its marginal density at time t and has as its generator*

$$\mathscr{A}_{\rho,b} = \frac{1}{2}\nabla \cdot S(\rho(t,x))\nabla + \frac{1}{2}(S(\rho(t,x)) - D)\frac{(\nabla\rho)(t,x)}{\rho(t,x)} \cdot \nabla$$

$$+ (1-\rho)Db \cdot \nabla \ . \tag{7.14}$$

The family $\{P^b : b \in \mathscr{B}_\rho\}$ has the following additional properties:

By the orthogonality relations, the minimizer b_0 of $\mathscr{E}[b]$ in the class $\mathscr{B}_\rho$, is of gradient type. We have

$$\mathscr{E}[b] = \mathscr{E}[b_0] + \mathscr{E}[b - b_0]$$

and of course $\nabla \cdot \rho(1-\rho)D(b - b_0) = 0$ for any other $b \in \mathscr{B}_\rho$.
The relative entropy $H(P^b, P^{b_0})$ satisfies

$$\begin{aligned} &H(P^b; P^{b_0}) \\ &= \frac{1}{2}\int_0^T \int_{\mathbf{T}^d} (S^{-1}(\rho)(1-\rho)D(b-b_0) \cdot (1-\rho)D(b-b_0))\rho \, dx \, dt \\ &\leq C\mathscr{E}[b - b_0] \end{aligned}$$

for some constant C, because of the lower bound

$$S(\rho) \geq c(1-\rho)D$$

which is valid with some positive constant c.

Any weak solution of the forward equation

$$\frac{\partial R}{\partial t} = \mathscr{A}^*_{\rho,b} R$$

satisfying the bound

$$0 \leq R(t,x) \leq C\rho(t,x) \tag{7.15}$$

for all $(t,x) \in [0,T] \times \mathbf{T}^d$ with some constant C, also satisfies the energy estimate

$$\int_0^T \int_{\mathbf{T}^d} (1-\rho)\frac{|\nabla R|^2}{R} \, dx \, dt \leq C'$$

where $C' < \infty$ depends only on the constant C in equation (7.15), *the bound on the gradient estimate on ρ specified in equation* (7.2) *and a bound on the quantity $\mathscr{E}[b]$ defined in equation* (7.9).

In addition let $0 \leq R(t,x) \leq 1$ satisfy

$$\int_0^T \int_{\mathbf{T}^d} (1-\rho)\frac{|\nabla R|^2}{R} < \infty \tag{7.16}$$

and be a weak solution of

$$\frac{\partial R}{\partial t} = \mathscr{A}^*_{\rho,b,\tilde{b}} R.$$

Here $\mathscr{A}^*_{\rho,b,\tilde{b}}$ *is the adjoint of*

$$\mathscr{A}_{\rho,b,\tilde{b}} = \mathscr{A}_{\rho,b} + \tilde{b} \cdot \nabla$$

with a $\tilde{b}$ *satisfying the bound*

$$\frac{1}{2}\int_0^T \int_{\mathbf{T}^d} (\tilde{b} \cdot S^{-1}(\rho)\tilde{b})\, R\, dx\, dt < \infty$$

Then there is a unique diffusion corresponding to $\mathscr{A}_{\rho,b,\tilde{b}}$ *that has* $R(\cdot\,,\cdot)$ *for its marginals. If we denote this diffusion by* $P^{b,\tilde{b}}$, *it is absolutely continuous with respect to* $P^b_{R_0}$ *the diffusion corresponding to* $\mathscr{A}_{\rho,b}$, *but with initial density* $R(0\,,\cdot)$.

We will use the next theorem in the following two contexts.

1. We will have ρ satisfying (7.2) and $\mathscr{L}_t$ will be the operator given in (7.14), with $b \in \mathscr{B}_\rho$, written in nondivergence form. The process Q will have ρ as one dimensional marginals and will be absolutely continuous with respect to P^b with finite relative entropy $H(Q, P^b)$ for some $b \in \mathscr{B}_\rho$. We will find a unique $b_Q \in \mathscr{B}_\rho$ such that

$$H(Q, P^{b_Q}) \le H(Q, P^b) < \infty$$

and the process P^{b_Q} satisfies

$$E^Q\left[\int_0^T \phi(t, x(t)) \cdot dx(t)\right] = E^{P^{b_Q}}\left[\int_0^T \phi(t\,, x(t)) \cdot dx(t)\right] \tag{7.17}$$

for all smooth ϕ.

2. In addition to ρ satisfying (7.2) and $b \in \mathscr{B}_\rho$, we will have a function $c : [0, T] \times \mathbf{T}^d \to \mathbf{R}^d$ satisfying

$$\int_0^T \int_{\mathbf{T}^d} (c \cdot S^{-1}(\rho)c)\, dx\, dt < \infty\ .$$

The process $P^{b,c}$ will correspond to some initial distribution $R(0, \cdot)$ satisfying $R(0, \cdot) \le \rho(0, \cdot)$ and generator $\mathscr{L}_t$, the nondivergence form version of

$$\mathscr{A}_{\rho,b,c} = \frac{1}{2}\nabla \cdot S(\rho)\nabla + (S(\rho) - D)\frac{\nabla\rho}{\rho} \cdot \nabla + (1-\rho)Db \cdot \nabla + c \cdot \nabla$$

The process Q will have marginals $R(\cdot\,,\cdot)$ coinciding at time 0 with the initial density $R(0, \cdot)$ of $P^{b,c}$. Moreover $R(\cdot\,,\cdot)$ will satisfy (7.16). The process Q will also be absolutely continuous with respect to $P^{b,c}$

with finite relative entropy $H(Q, P^{b,c})$. We want to replace c by c_Q, so that

$$E^Q\left[\int_0^T \phi(t,x(t)) \cdot dx(t)\right] = E^{P^{b,c_Q}}\left[\int_0^T \phi(t,x(t)) \cdot dx(t)\right] \tag{7.18}$$

with $H(Q, P^{b,c_Q}) \le H(Q, P^{b,c}) < \infty$ and

$$\frac{1}{2}\int_0^T\int_{\mathbf{T}^d}((c-c_Q) \cdot S^{-1}(\rho)(c-c_Q))R(t,x)\,dx\,dt < \infty$$

Let P be a diffusion process on $[0,T]$ with marginals $p(t,\cdot)$ that corresponds to the generator

$$\mathscr{L}_t = \frac{1}{2}\sum_{i,j=1}^d a_{i,j}(t,x)\frac{\partial^2}{\partial x_i \partial x_j} + \sum_{j=1}^d b_j(t,x)\frac{\partial}{\partial x_j} \ .$$

We will assume that

$$\int_0^T\int_{\mathbf{T}^d}\left[\sum_{i,j=1}^d |a_{i,j}(t,x)| + \sum_{j=1}^d |b_j(t,x)|\right] p(t,x)\,dx\,dt < \infty$$

so that

$$y(t) = \int_0^t dx(s) - \int_0^t b(s,x(s))\,ds$$

is well defined as a square integrable martingale with quadratic variation

$$\int_0^t a_{i,j}(s\,,x(s))\,ds \ .$$

Let Q be the law of another process that has marginals $q(t,\cdot)$. Suppose that Q is absolutely continuous with respect to P and the relative entropy $H(Q,P)$ is finite. Assume further that

$$\int_0^T\int_{\mathbf{T}^d}\left[\sum_{i,j=1}^d |a_{i,j}(t,x)| + \sum_{j=1}^d |b_j(t,x)|\right] q(t\,,x)\,dx\,dt < \infty \ .$$

We then have,

Theorem 7.3. *There exist coefficients $\tilde{b}(t,x)$ and a diffusion process $\widehat{P}$ with generator*

$$\tilde{\mathscr{L}}_t = \frac{1}{2}\sum_{i,j=1}^d a_{i,j}(t,x)\frac{\partial^2}{\partial x_i \partial x_j} + \sum_{j=1}^d [b_j + \tilde{b}_j](t,x)\frac{\partial}{\partial x_j}$$

such that

$$E^P\left[\int_0^T f(t,x(t))\,dt + \phi(t,x(t)) \cdot dx(t)\right] = E^{\hat{P}}\left[\int_0^T f(t,x(t))\,dt + \phi(t,x(t)) \cdot dx(t)\right]$$

for all smooth functions f and ϕ. In particular $\hat{P}$ has the same marginals $p(t,\cdot)$ as P. We have the estimate

$$H(\hat{P},Q) = \frac{1}{2}\int_0^T\int_{\mathbf{T}^d} (\tilde{b} \cdot a^{-1}\tilde{b})\,p(t,x)\,dx\,dt \le H(P,Q) \tag{7.19}$$

Proof. The proof is fairly routine. We note that

$$E^P\left[\exp\left\{\int_0^T \phi(t,x(t)) \cdot dy(t) - \frac{1}{2}\int_0^T (\phi \cdot a\,\phi)(t,x(t))\,dt\right\}\right] = 1$$

and by Jensen's inequality

$$E^Q\left[\int_0^T \phi(t,x(t)) \cdot dy(t) - \frac{1}{2}\int_0^T (\phi \cdot a\phi)(t,x(t))\,dt\right] \le H(Q,P)$$

providing us with an estimate

$$\left|E^Q\left[\int_0^T \phi(t,x(t)) \cdot dy(t)\right]\right| \le H(Q,P) + \frac{1}{2}\int_0^T\int_{\mathbf{T}^d} (\phi \cdot a\phi)(t,x)\,q(t,x)\,dx\,dt$$

By the Riesz representation theorem there is a $\tilde{b}$ satisfying (7.19) for which

$$E^Q\left[\int_0^T \phi(t,x(t)) \cdot dy(t)\right] = \int_0^T\int_{\mathbf{T}^d} (\phi \cdot \tilde{b})(t,x)\,q(t,x)\,dx\,dt$$

In particular

$$E^Q\left[\int_0^T \phi(t,x(t)) \cdot dx(t)\right] = \int_0^T\int_{\mathbf{T}^d} (\phi \cdot (b+\tilde{b}))(t,x)\,q(t,x)\,dx\,dt \tag{7.20}$$

Ito's formula is valid almost surely with respect to P and therefore almost surely with respect to Q,

$$f(t,x(t)) - f(0,x(0)) = \int_0^t \left[\frac{\partial f}{\partial s} + \frac{1}{2}\sum_{i,j=1}^n a_{i,j}\frac{\partial^2 f}{\partial x_i \partial x_j}\right](s,x(s))\,ds + \int_0^t \nabla f(s,x(s)) \cdot dx(s)$$

It follows, by taking expectations with respect to Q and using (7.20), that $q(\cdot,\cdot)$ is a weak solution of

$$\frac{\partial q}{\partial t} = \tilde{\mathscr{L}}^{*} q$$

where $\tilde{\mathscr{L}}^{*}$ is the adjoint or forward operator corresponding to the generator

$$\tilde{\mathscr{L}}_t = \frac{1}{2}\sum_{i,j=1}^{n} a_{i,j}(t,x)\frac{\partial^2}{\partial x_i \partial x_j} + \sum_{j=1}^{n}[b_j + \tilde{b}_j](t,x)\frac{\partial}{\partial x_j} \ .$$

At this point we can assert the existence of a unique diffusion with marginals q and generator $\tilde{\mathscr{L}}_t$. We denote this by $\widehat{Q}$ and clearly $\widehat{Q}$ works.

Let a measure Q on $C[[0,T];\mathbf{T}^d]$ with total mass $\bar{\rho}$ be given. We define the following rate function $\mathscr{I}(Q)$. If the marginals of Q have densities $\rho(t,x)$ with $0 \le \rho(t,x) \le 1$ satisfying (7.2) and if Q is absolutely continuous with respect to P^b for some $b \in \mathscr{B}_\rho$ with finite relative entropy $H(Q, P^b)$, we define

$$\mathscr{I}(Q) = H(Q, P^{b_Q}) + \mathscr{E}[b_Q] \tag{7.21}$$

where b_Q and P^{b_Q} are as in (7.17) and $\mathscr{E}[b]$ is as defined in (1.17). Otherwise $\mathscr{I}(Q) = +\infty$.

It will be convenient to introduce a more general version of the same rate function. Given n measures $Q_1, \ldots, Q_n$ and their sum $Q = \sum_{j=1}^{n} Q_j$, assume that their marginal densities $\rho_1, \ldots, \rho_n$ and ρ satisfy (7.1) and (7.2). Let us construct $b_Q \in \mathscr{B}_\rho$ and P^{b_Q} as in (7.17). Let $P_j^{b_Q}$ be the diffusion corresponding to the generator $\mathscr{A}_{\rho, b_Q}$ with initial density that given by Q_j at time 0. We define

$$\mathscr{I}_n(Q_1, \ldots, Q_n) = \sum_{j=1}^{n} H(Q_j, P_j^{b_Q}) + \mathscr{E}[b_Q] \tag{7.22}$$

when it is well defined. Otherwise $\mathscr{I}_n(Q_1, \ldots, Q_n) = +\infty$. Note that for $n = 1$ it reduces to $\mathscr{I}(Q)$.

We now establish some properties of the functional $\mathscr{I}$.

Theorem 7.4. *The functional $\mathscr{I}(Q)$ is lower semicontinuous and has compact level sets.*

Proof. Let us suppose that we have a sequence Q_k such that $I(Q_k) \le \ell$. We denote by $\rho_k(t,x)$ the marginal densities of Q_k. Let us denote b_{Q_k} by b_k and P^{b_k} by P_k. From the definition of $I(Q)$ we have the following relations

$$\frac{\partial \rho_k}{\partial t} = \frac{1}{2} \nabla \cdot D \nabla \rho_k - \nabla \cdot \rho_k (1 - \rho_k) D b_k \tag{7.23}$$

and

$$\mathscr{E}[b_k] = \frac{1}{2} \int_0^T \int_{\mathbf{T}^d} \rho_k (1 - \rho_k)(b_k \cdot D b_k)\, dx\, dt$$

and

$$H(Q_k, P_k) + \mathscr{E}[b_k] \le \ell\ . \tag{7.24}$$

We write the generator of P_k in nondivergence form

$$\mathscr{A}_k = \frac{1}{2} S(\rho_k(t,x)) \nabla \cdot \nabla + c_k(t,x) \cdot \nabla \tag{7.25}$$

with

$$c_k = \frac{1}{2} S'(\rho_k) \nabla \rho_k + \frac{1}{2} [S(\rho_k) - D] \frac{\nabla \rho_k}{\rho_k} + (1 - \rho_k) D b_k$$

We shall complete the proof of the theorem in several steps.

Step1. The bound (7.24) yields the following estimate on $\nabla \rho_k(t,x)$:

$$\sup_k \int_0^T \int_{\mathbf{T}^d} \frac{|\nabla \rho_k(t,x)|^2}{\rho_k(t,x)(1 - \rho_k(t,x))}\, dx\, dt \le C_\ell \tag{7.26}$$

with a constant C_ℓ that depends only on ℓ. This in turn provides the estimate,

$$\sup_k \int_0^T \int_{\mathbf{T}^d} |c_k(t,x)|^2 \rho_k(t,x)\, dt\, dx < \infty \tag{7.27}$$

For each k, P_k is a diffusion process on $C[[0,T]; \mathbf{T}^d]$ with the generator given by (7.25). In nondivergence form, its second order diffusion coefficients are uniformly bounded by the bound on $S(\rho)$ and its first order term satisfies (7.27), $\rho_k(t,x)$ being the marginal density of P_k at time t. This establishes the compactness of the sequence P_k.

Since we have a uniform bound on the relative entropy $H(Q_k, P_k)$ the compactness of Q_k follows. Using the bound (7.24) and the weak compactness of Q_k and P_k, we can take a subsequence and assume that weak limits b of b_k and Q and P of Q_k and P_k exist. (The precise definition of b will be given after (7.29) in Step 2.) Since P_k and Q_k have the same marginal densities ρ_k, the same is true of their weak limits P and Q and we denote their common marginal by $\rho(t,x)$. From general arguments of semicontinuity it will follow that

$$H(Q,P) + \frac{1}{2} \int_0^T \int_{\mathbf{T}^d} (b(t,x) \cdot Db(t,x)) \rho(t,x)(1 - \rho(t,x))\, dt\, dx \le \ell$$

The proof of the theorem will be completed by showing that the convergence of $\rho_k(t,x)$ to $\rho(t,x)$ is strong, $b \in \mathscr{B}_\rho$, $P = P^b$ and $b_Q = b$ which will then establish that $I(Q) \leq \ell$.

Step 2. In order to simplify notation we will ignore the need for a subsequence and assume that all weak limits exist. From (7.26) we get the estimate

$$\sup_k \int_0^T \int_{\mathbf{T}^d} |\nabla \rho_k(t,x)|^2 \, dx \, dt \leq C_\ell \tag{7.28}$$

From (7.23)–(7.26), we get the estimate

$$\sup_k \int_0^T \left\| \frac{\partial \rho_k}{\partial t} \right\|_{-1}^2 dt < \infty \tag{7.29}$$

On account of Lemma 4.2, the estimates (7.28) and (7.29) are sufficient to obtain strong convergence of ρ_k to ρ in $L^1([0,T] \times \mathbf{T}^d)$. It is not hard to show that the sequence $\sqrt{\rho_k(1-\rho_k)}$ also converges to $\sqrt{\rho(1-\rho)}$ in L^1. We can use this to take some sort of weak limit of b_k. First, by the uniform L^1 bound on the $b_k^2 \rho_k(1-\rho_k)$, a subsequnce of $b_k\sqrt{\rho_k(1-\rho_k)}$ converges weakly in L^2 to a function h. We claim that indeed $h = b\sqrt{\rho(1-\rho)}$ for some function b. To see this, observe that by the strong convergence of $\sqrt{\rho_k(1-\rho_k)}$, we also obtain the weak convergence of $b_k \rho_k(1-\rho_k)$ to $u = h\sqrt{\rho(1-\rho)}$ and evidently u vanishes whenever $\rho = 0$ or 1. Hence b can be defined to be $u[\rho(1-\rho)]^{-1}$.

Since the convergence of ρ_k to ρ is strong in L^1 there is no problem in passing to the limit in

$$\frac{\partial \rho_k}{\partial t} - \frac{1}{2} \nabla D \nabla \rho_k = -\nabla \cdot \rho_k(1-\rho_k) D b_k$$

and showing that $b \in \mathscr{B}_\rho$. Moreover, if R_k is a solution of

$$\frac{\partial R_k}{\partial t} = \mathscr{A}_k^* R_k \,, \tag{7.30}$$

with $0 \leq R_k \leq C\rho_k$, similar estimates i.e. (7.6) and (7.16) are valid. These imply

$$\sup_k \int_0^T \int_{\mathbf{T}^d} \left| \nabla \left(R_k \sqrt{\frac{1-\rho_k}{\rho_k}} \right) \right|^2 dt \, dx < \infty \,. \tag{7.31}$$

We can use the equation (7.30) to get an estimate on $\frac{\partial R_k}{\partial t}$.

$$\sup_k \int_0^T \left\| \frac{\partial R_k}{\partial t} \right\|_{-1}^2 dt < \infty \,. \tag{7.32}$$

Estimates (7.6), (7.31) and (7.32) imply

$$\lim_{k\to\infty}\int_0^T\int_{\mathbf{T}^d}\left|\sqrt{\frac{1-\rho_k(t,x)}{\rho_k(t,x)}}R_k(t,x)-\sqrt{\frac{1-\rho(t,x)}{\rho(t,x)}}R(t,x)\right|dt\,dx=0\ .$$

We can now pass to the limit in equation (7.30) and see that

$$\frac{\partial R}{\partial t}=\mathscr{A}^*_{\rho,b}R \tag{7.33}$$

From the uniqueness of solutions to (7.33), we conclude that $P=P^b$.

Step 3. The last step is the identification $b=b_Q$. This requires establishing the relation

$$E^Q\left[\int_0^T\phi(t,x(t))\cdot dx(t)\right]=E^P\left[\int_0^T\phi(t,x(t))\cdot dx(t)\right]$$

by passing to the limit as $k\to\infty$ in both sides of

$$E^{Q_k}\left[\int_0^T\phi(t,x(t))\cdot dx(t)\right]=E^{P_k}\left[\int_0^T\phi(t,x(t))\cdot dx(t)\right]\ .$$

The right hand side is an easy calculation because P_k is a diffusion and we know the first order term of the generator $\mathscr{A}_k$ in nondivergence form. The left hand side is somewhat more complicated because, although we know that $Q_k\Rightarrow Q$, the functional

$$F(\omega)=\int_0^T\phi(t,x(t))\cdot dx(t)$$

is neither bounded nor continuous. We shall first establish the uniform integarability of $F(\cdot)$ with respect to $\{Q_k\}$. We write $F=G_k+(F-G_k)$, where

$$G_k(\omega)=\int_0^T\phi(t,x(t))\cdot c_k(t,x(t))\,dt$$

and

$$c_k=\frac{1}{2}S'(\rho_k)\nabla\rho_k+\frac{1}{2}[S(\rho_k)-D]\frac{(\nabla\rho_k)}{\rho_k}+(1-\rho_k)Db_k$$

It is easy to bound

$$E^{Q_k}\left[|G_k(\omega)|^2\right]\le TC(\phi)\int_0^T\int_{\mathbf{T}^d}|c_k|^2\rho_k\,dt\,dx$$

by a quantity independent of k. The functional $F-G_k$ is a stochastic integral with respect to a continuous martingale of uniformly bounded quadratic variation with respect to P_k having a uniformly bounded integrand. It has finite exponential moments with any constant relative to P_k. Since we have uniform bound on the relative entropy $H(Q_k, P_k)$, we obtain the uniform integrability of $(F - G_k)$ with respect to Q_k. Finally for $h > 0$ we will construct continuous functionals F_h such that

$$\lim_{h\to 0} \sup_k E^{Q_k}\left[|F_h(\omega) - F(\omega)|\right] = 0$$

This will complete the proof.

Let us do the usual approximation by simple functions by defining

$$F_h(\omega) = \int_0^T \phi(\tau_h(t), x(\tau_h(t))) \cdot dx(t)$$

where

$$\tau_h(t) = rh \text{ if } rh \le t < (r+1)h \text{ for integral } r \ge 0 \ .$$

If we use the earlier decomposition

$$dx(t) = c_k(t, x(t))\,dt + [dx(t) - c_k(t, x(t))\,dt] = dz_k(t) + dy_k(t)$$

and split our approximations as

$$F_h(\omega) = \int_0^T \phi\,(\tau_h(t),\, x(\tau_h(t))) \cdot dz_k(t) + \int_0^T \phi\,(\tau_h(t), x(\tau_h(t))) \cdot dy_k(t)$$

and correspondingly

$$F(\omega) = \int_0^T \phi(t, x(t)) \cdot dz_k(t) + \int_0^T \phi(t, x(t)) \cdot dy_k(t)$$

it is routine to estimate the difference between the two first terms and control it by the tightness of the family Q_k and the bound

$$\sup_k \int_0^T \int_{\mathbf{T}^d} |c_k(t,x)|^2 \rho_k(t,x)\,dt\,dx < \infty$$

Standard methods in the theory of stochastic integration will allow us to estimate the difference between the two second terms with respect to P_k to conclude that for every $\lambda > 0$,

$$\lim_{h\to 0} \sup_k E^{P_k}\left[\exp\left\{\lambda\left|\int_0^T [\phi\,(\tau_h(t), x(\tau_h(t))) - \phi(t, x(t))] \cdot dy_k(t)\right|\right\}\right] = 1.$$

By the entropy bounds this implies

$$\lim_{h\to 0} \sup_k E^{Q_k}\left[\left|\int_0^T [\phi(\tau_h(t), x(\tau_h(t))) - \phi(t, x(t))] \cdot dy_k(t)\right|\right] = 0$$

and we are done.

Now we turn to the relation of the rate function $\mathscr{I}_n(Q_1,\ldots,Q_n)$ to the n-color rate function $I_{\text{dyn}}(\tilde{\rho}(\cdot\,,\cdot))$ that was defined earlier. Let $\tilde{\rho}(\cdot\,,\cdot) : [0\,,T] \times \mathbf{T}^d \to \mathscr{C}$ be given, satisfying (7.1) and (7.2). For any given $\rho : [0\,,T] \times \mathbf{T}^d \to [0\,,1]$ we define $\mathscr{M}(\rho)$ to be the set of measures Q on $C[[0,T]\,;\mathbf{T}^d]$ that have $\rho(t,\cdot)$ as marginals at time t.

Theorem 7.5. *The n color rate function* $I_{\text{dyn}}(\tilde{\rho}(\cdot\,,\cdot))$ *is given by the contraction formula*

$$I_{\text{dyn}}(\tilde{\rho}(\cdot\,,\cdot)) = \inf_{\substack{Q_1 \in \mathscr{M}(\rho_1)\\ \cdots \\ Q_n \in \mathscr{M}(\rho_n)}} \mathscr{I}_n(Q_1,\ldots,Q_n)$$

Proof. We have the variational formula of Theorem 7.1 for $I_{\text{dyn}}(\tilde{\rho})$,

$$I_{\text{dyn}}(\tilde{\rho}) = \inf_{(b,b_1,\ldots,b_n)\in\mathscr{G}_{\tilde{\rho}}} \frac{1}{2}\int_0^T \int_{\mathbf{T}^d} \left\{ \sum_{j=1}^n \rho_j \langle b_j, S^{-1}(\rho)\, b_j\rangle + \rho(1-\rho)\,(b\cdot Db)\right\} dx\, dt$$

Because $Q_i \in \mathscr{M}(\rho_i)$ for $i = 1,\ldots,n$, we conclude that $\sum_{j=1}^n Q_j = Q \in \mathscr{M}(\rho)$. Recall that $\rho = \sum_{j=1}^n \rho_j$. Let b_Q be as in Theorem 7.3. For any diffusion process P we denote by $\mathscr{M}(P)$ the family of processes Q that satisfy $H(Q,P) < \infty$, have the same one dimensional marginals as P and satisfy in addition

$$E^Q\left[\int_0^T \phi(t,x(t))\cdot dx(t)\right] = E^P\left[\int_0^T \phi(t,x(t))\cdot dx(t)\right]$$

for all smooth test functions ϕ. We have the diffusions $P^{b_Q}, P_1^{b_Q},\ldots,P_n^{b_Q}$ with generator

$$\begin{aligned}\mathscr{A}_{\rho,b_Q} &= \frac{1}{2}\nabla\cdot S(\rho(t,x))\nabla + \frac{1}{2}(S(\rho(t,x)) - D)\left(\frac{\nabla\rho}{\rho}\right)(t,x)\cdot\nabla \\ &\quad + (1-\rho(t,x))Db_Q(t,x)\cdot\nabla\end{aligned}$$

and initial densities $\rho(0,\cdot), \rho_1(0,\cdot),\ldots,\rho_n(0,\cdot)$ respectively. By definition

$$\inf_{\substack{Q_1 \in \mathscr{M}(\rho_1)\\ \cdots \\ Q_n \in \mathscr{M}(\rho_n)}} \mathscr{I}_n(Q_1,\ldots,Q_n) = \inf_{\substack{Q_1 \in \mathscr{M}(\rho_1)\\ \cdots \\ Q_n \in \mathscr{M}(\rho_n)}} \left\{\sum_{j=1}^n H(Q_j, P_j^{b_Q}) + \mathscr{E}[b_Q]\right\}$$

Having already chosen b_Q, we can use Theorem 7.3 to find $(b_1,\ldots,b_n)$ so that the processes $P_j^{b_Q,b_j}$ with generators

$$\mathscr{A}_{\rho,b_Q,b_j} = \mathscr{A}_{\rho,b_Q} + b_j(x,t)\cdot\nabla$$

satisfy

$$\frac{\partial \rho_j}{\partial t} = \mathscr{A}^*_{\rho,b_Q,b_j}\rho_j$$

and $Q_j \in \mathscr{M}(P_j^{b_Q,b_j})$. The latter property and Girsanov Formula imply

$$E^{Q_j}\left[\log \frac{dP_j^{b_Q,b_j}}{dP_j^{b_Q}}\right] = H(P_j^{b_Q,b_j}, P_j^{b_Q}).$$

Hence

$$\begin{aligned} H(Q_j, P_j^{b_Q}) &= H(Q_j, P_j^{b_Q,b_j}) + H(P_j^{b_Q,b_j}, P_j^{b_Q}) \\ &\geq H(P_j^{b_Q,b_j}, P_j^{b_Q}) \\ &= \frac{1}{2}\int_0^T \int_{\mathbf{T}^d} (b_j \cdot S^{-1}(\rho) b_j)\rho_j(t,x)\,dx\,dt \end{aligned}$$

From $Q_j \in \mathscr{M}(P_j^{b_Q,b_j})$ for $1 \leq j \leq n$ and $Q \in \mathscr{M}(P^{b_Q})$ it follows that $\sum_{j=1}^n \rho_j b_j \equiv 0$. If we replace Q_j by $P_j^{b_Q,b_j}$ in the variational formula we get

$$\inf_{\substack{Q_1 \in \mathscr{M}(\rho_1) \\ \cdots \\ Q_n \in \mathscr{M}(\rho_n)}} \left\{ \sum_{j=1}^n H(Q_j, P_j^{b_Q}) + \mathscr{E}[b_Q] \right\} \geq \inf_{(b,b_1,\ldots,b_n)\in \mathscr{G}_{\bar{\rho}}}$$

$$\times \frac{1}{2}\int_0^T \int_{\mathbf{T}^d} \left\{ \sum_{j=1}^n \rho_j (b_j \cdot S^{-1}(\rho)\, b_j) + \rho(1-\rho)\,(b \cdot Db) \right\} dx\,dt$$

Since any choice of $(b, b_1, \ldots, b_n)$ leads to a choice of $\{P_j^b\}$ and $\{Q_j\}$, the reverse inequality is immediate. This proves the theorem.

A special situation is when $\rho_j(0,\cdot)$ are supported on disjoint subsets of $\mathbf{T}^d$. In this case $Q_1, \ldots, Q_n$ are mutually orthogonal. Then it is easily seen that $\mathscr{I}_n(Q_1, \ldots, Q_n) = \mathscr{I}(Q)$ where $Q = \sum_{j=1}^n Q_j$. The variational formula becomes

$$I_{\text{dyn}}(\tilde{\rho}) = \inf_{\substack{Q_1 \in \mathscr{M}(\rho_1) \\ \cdots \\ Q_n \in \mathscr{M}(\rho_n)}} \mathscr{I}(Q)$$

It is not difficult to understand the significance of the term $\mathscr{E}[b_Q]$. Let us leave our model and look instead at noninteracting Brownian motions on the circle starting from equilibrium. If we want the empirical process to resemble a Brownian Motion with a drift a, we are going to have a large deviation rate of $\bar{\rho} T \frac{a^2}{2}$ accounted for by a rate of $\frac{a^2}{2}$ per particle per unit time. The marginal densities are constants and the drift $b(t,x) = a$ will produce the right marginals as well as the right current.

8. Superexponential tightness

One of the key steps in any large deviation estimate is the superexponential estimate. If Q_n is a family of measures on a Polish space X, the natural definition of superexponential tightness is that for any $\ell < \infty$ there is a compact set C_ℓ in X such that

$$Q_n[X - C_\ell] \le e^{-\ell n} \text{ for all } n \tag{8.1}$$

The construction of such a C_ℓ is usually awkward, involving lots of intersections and unions. Ideally our measures Q_n will be supported on a fixed compact set C. The next best thing is that they will be supported on some neighborhood C^{ϵ_n} of a compact set with $\epsilon_n \to 0$ as $n \to \infty$. Failing that, the least one should demand is that for each $\epsilon > 0$ and $\ell < \infty$ there exists a compact set $C_{\ell,\epsilon}$ such that

$$\limsup_{n\to\infty} \frac{1}{n} \log Q_n\left[X - C^{\epsilon}_{\ell,\epsilon}\right] \le -\ell \tag{8.2}$$

It is a general fact that (8.2) implies what looks like the stronger definition (8.1). Here is a proof. Pick a sequence $\epsilon_k \downarrow 0$. For each ℓ and k we can clearly find a compact set C_{ℓ,ϵ_k} such that

$$Q_n\left[X - C^{\epsilon_k}_{\ell,\epsilon_k}\right] \le e^{-(\ell+1)n} \text{ for } n \ge n_0(\ell, k)$$

We can assume without loss of generality that $n_0(\ell, k) \ge k$ so that

$$Q_n\left[X - C^{\epsilon_k}_{\ell,\epsilon_k}\right] \le e^{-k} e^{-\ell n} \text{ for } n \ge n_0(\ell, k)$$

For each ℓ, k and $n = 1, 2, \ldots, n_0(\ell, k)$ by Prohorov's theorem we can find a compact set $D_{\ell,k}$ such that

$$Q_n\left[X - D_{\ell,k}\right] \le e^{-k} e^{-\ell n}$$

If we now define $B_{\ell,k} = D_{\ell,k} \cup C_{\ell,\epsilon_k}$, clearly $B_{\ell,k}$ is compact and

$$Q_n\left[X - B^{\epsilon_k}_{\ell,k}\right] \le e^{-k} e^{-\ell n} \text{ for all } k \text{ and } n \ .$$

If we now define K_ℓ to be the closure of $\cap_k B^{\epsilon_k}_{\ell,k}$, as the closure of a totally bounded set it is compact and

$$Q\left[X - K_\ell\right] \le \sum_{k\ge 1} e^{-k} e^{-\ell n} \le e^{-\ell n}$$

which is (8.1).

In our context our basic Polish space is the space of all probability measures on the Skorohod space $\Omega = D[[0, T]; \mathbf{T}^d]$ although the jumps once scaled are only of size $\frac{1}{N}$. The large deviations are going to happen only at measures supported on $C[[0, T]; \mathbf{T}^d]$. The set

$$A_{\delta,\epsilon} = \left\{ x(\cdot) : \sup_{\substack{|s-t|\leq\delta \\ 0\leq s,t\leq T}} |x(t) - x(s)| \leq \epsilon \right\}$$

is contained in $C_\ell^{2\epsilon}$ where C_ℓ is the space of functions that are Lipschitz with a Lipschitz constant $\ell = 2\frac{\epsilon}{\delta}$. Here by A^ϵ we mean the space of functions within a distance ϵ from A. The set $\{Q : Q[\Omega - A_\delta, \epsilon] \leq \epsilon\}$ is within an ϵ neighborhood of a compact set in the Prohorov metric. So the superexponential estimate reduces to the following theorem.

Theorem 8.1. *Let P_N be the measure corresponding to the L interacting trajectories starting from an arbitrary initial configuration. Then for every $\epsilon > 0$,*

$$\limsup_{\delta\to 0} \limsup_{N\to\infty} \frac{1}{N^d} \log P_N$$
$$\times \left[\text{card}\{j : \sup_{\substack{|s-t|\leq\delta \\ 0\leq s,t\leq T}} |x_j(s) - x_j(t)| \geq \epsilon\} \geq \epsilon N \right] = -\infty \ .$$

We will break the proof into several steps.

Although each $x_i(t)$ takes values on the torus $\mathbf{T}^d$ it is convenient to think of it as the projection onto the torus of an $\mathbf{R}^d$ - valued process. Since the jumps of $x_i(\cdot)$ are all of size $\frac{C}{N}$, for sufficiently large N, each path $x_i(\cdot)$ can be lifted to a unique path $y_i(\cdot)$ with values in $\mathbf{R}^d$, at least if we pick the starting point $y_i(0)$ in some canonical fashion. To be meticulously correct we need to consider the two sets of processes together because the $y_i(\cdot)$ are governed by the configuration $\eta(\cdot)$ of the $x_i(\cdot)$ on $\mathbf{T}^d$ and $x_i(\cdot)$ shadow $y_i(\cdot)$. The $y_i(\cdot)$ are semimartingales and have the decomposition

$$y_i(t) = A_i(t) + M_i(t)$$

as the sum of a continuous function of bounded variation and a martingale. The function $A_i(t)$ is seen to be

$$A_i(t) = N \int_0^t \sum_z z\, p(z)(1 - \eta(x_i(t) + z, s))\, ds \qquad (8.3)$$

Each term in the summation corresponding to jumps of size $\frac{z}{N}$, to the sites $x_i(t) + z$ on the torus that are currently empty, that could potentially take place with rates $N^2 p(z)$ in the speeded up time scale. The superexponential estimate that we are seeking for $\{x_i(\cdot)\}$ naturally breaks up into two parts, for $\{A_i(\cdot)\}$ and for $\{M_i(\cdot)\}$: par 1. For every $\epsilon > 0$,

$$\limsup_{\delta\to 0}\limsup_{N\to\infty}\frac{1}{N^d}\log P_N$$

$$\times\left[\operatorname{card}\{j : \sup_{\substack{|s-t|\le\delta\\0\le s,t\le T}}|A_j(s)-A_j(t)|\ge s\epsilon\}\ge\epsilon N\right]=-\infty\ , \qquad (8.4)$$

and 2. For every $\epsilon > 0$,

$$\limsup_{\delta\to 0}\limsup_{N\to\infty}\frac{1}{N^d}\log P_N$$

$$\times\left[\operatorname{card}\{j : \sup_{\substack{|s-t|\le\delta\\0\le s,t\le T}}|M_j(s)-M_j(t)|\ge\epsilon\}\ge\epsilon N\right]=-\infty\ . \qquad (8.5)$$

If we define for any trajectory $f(\cdot)$

$$S(f)=\sup_{0<\delta\le T}\ \sup_{\substack{|t-s|\le\delta\\0\le s,t\le T}}\frac{|f(t)-f(s)|}{\sqrt{\delta}\,\log\frac{1}{\delta}}$$

the following inequality was proved in [R: Lemma 2.4 inequality (2.11)]

$$\lim_{\ell\to\infty}\limsup_{N\to\infty}\frac{1}{N^d}\log P_N\left[\sum_j S(A_j(\cdot))\ge\ell N^d\right]=-\infty\ . \qquad (8.6)$$

Equation (8.4) is an easy consequence of (8.6), because

$$|f(s)-f(t)|\le S(f)\sqrt{\delta}\,\log\frac{1}{\delta}$$

implies that

$$\operatorname{card}\left\{j : \sup_{\substack{|s-t|\le\delta\\0\le s,t\le T}}|A_j(s)-A_j(t)|\ge\epsilon\right\}\le\operatorname{card}\{j : S(A_j(\cdot))\ge\theta(\delta,\epsilon)\}$$

where

$$\theta(\delta,\epsilon)=\frac{\epsilon}{\sqrt{\delta}\,\log\frac{1}{\delta}}\to\infty\quad\text{as }\delta\to 0$$

The rest of this section will be devoted to the proof of (8.5). As a jump Markov process our process of L interacting particles comes with certain basic counting processes

$$F_N(i,z;t)=\text{ number of jumps of size }\frac{z}{N}\text{ of }x_i(\cdot)\text{ during }0\le s\le t$$

that have a jump rate

$$\lambda_N(i, z; t) = N^2 p(z)(1 - \eta(x_i(t) + z, t))$$

and our martingales $M_i(t)$ have the representation

$$M_i(t) = \frac{1}{N}\sum_z z\left[F_N(i, z; t) - \int_0^t \lambda_N(i, z; s)\, ds\right]$$
$$= \sum_z z\, M_N(i, z; t)$$

as a finite sum of basic martingales that are mutually orthogonal for different z and i. It is important initially to estimate for each z,

$$P_N\left[H \geq \epsilon N\right]$$

where

$$H = \text{card}\left\{ j : \sup_{0 \leq s \leq \delta} |M_N(j, z; s) - M_N(j, z; 0)| \geq \epsilon \right\}$$

If it were not for exclusion, $M_N(j, z; t)$ will be independent centered Poisson processes and the corresponding probability will be given by a simple binomial probability

$$\bar{P}[H \geq \epsilon N] = \sum_{r \geq \epsilon N} \binom{L}{r} [\beta_N(\delta, \epsilon)]^r [1 - \beta_N(\delta, \epsilon)]^{L-r}$$

with

$$\beta_N(t, \epsilon) = \Pr\left[\sup_{0 \leq s \leq t} \left| \frac{1}{N} Z_N(s) - Np \right| \geq \epsilon \right]$$

Here $Z_N(\cdot)$ is a standard Poisson process with intensity $N^2 p$ and $p = p(z)$. Our key step is the following theorem.

Theorem 8.2. *For every N, every initial configuration, and every ℓ, t and $\delta > 0$ we have*

$$P_N[H \geq \ell] \leq \bar{P}[H \geq \ell].$$

Proof. The idea is that for the actual exclusion process the rates are sometimes less than $N^2 p(z)$ i.e. 0 when blocked by another particle. We use ideas from control theory or dynamic programming to show that the Poisson case is the worst case. Without loss of generality, we assume that $N^2 p(z) = 1$. The proof of the theorem will be broken up into several lemmas.

Let $Z(\cdot)$ be the standard Poisson process with intensity 1. Let $\alpha(s) = Z(s) - s$ be the centered version. It has the infinitesimal generator

$$(\Omega f)(\alpha) = f(\alpha + 1) - f(\alpha) - \frac{\partial f}{\partial \alpha}(\alpha)$$

Given $\alpha, t \in \mathbf{R}$ we can define a version that starts from α at time t by defining $\alpha(s) = \alpha + Z(s - t) - (s - t)$ for $s \geq t$. Given $m > 0$ we consider the stopping time τ_m which is the exit time of the process from the finite interval $(-m, m)$. For any $h(t)$ that is a smooth nonincreasing function on $\mathbf{R}$, with $0 \leq h(\cdot) \leq 1$ we define

$$q(\alpha, t) = \mathbf{E}_{\alpha,t}\left[h(\tau_m)\right] .$$

Lemma 8.3. *The function q defined on $(-m, m) \times \mathbf{R}$ is continuously differentiable on $[-m, m) \times \mathbf{R}$ and satisfies*

$$\begin{cases} 0 \leq q \leq 1 \\ \frac{\partial q}{\partial t} \leq 0 \\ \frac{\partial q}{\partial t}(\alpha, t) + (\Omega q)(\alpha, t) = 0 \end{cases} \tag{8.7}$$

with "boundary" conditions

$$\begin{cases} q(-m, t) = h(t) & \text{for} \quad t \in \mathbf{R} \\ q(\alpha, t) = h(t) & \text{for} \quad \alpha \in [m, m+1) \quad \text{and} \quad t \in \mathbf{R} \end{cases} .$$

Proof. By solving the Dirichlet problem

$$\frac{\partial q}{\partial t} + \Omega q = 0 \tag{8.8}$$

for the space time generator on $(-m, m) \times \mathbf{R}$ with boundary data $h(t)$ specified on the boundary $\{\{-m\} \cup [m, m+1)\} \times \mathbf{R}$, we see that q solves the equation (8.8). Since we can differentiate the equation with respect to t and obtain the same equation with boundary data $h'(t)$ it is clear by the maximum principle that we can get a bound on $\frac{\partial q}{\partial t}$ in terms of the bound on $h'(t)$. From the equation we have an apriori bound on $\frac{\partial q}{\partial t} - \frac{\partial q}{\partial \alpha}$. This yields bounds on both partial derivatives of q. Finally, because of the maximum principle, $h'(t) \leq 0$ implies that $\frac{\partial q}{\partial t} \leq 0$.

Given k numbers $x_1, \ldots, x_k$ with $0 \leq x_j \leq 1$, for each $1 \leq \ell \leq k$ we define

$$v_\ell^k(x_1, \ldots, x_k) = \sum_{\mathbf{A}} \prod_{j \in \mathbf{A}} x_j \prod_{j \notin \mathbf{A}} (1 - x_j)$$

where the summation extends over all subsets $\mathbf{A} \subset \{1, \ldots, k\}$ of cardinality at least ℓ. If we interpret v_ℓ^k as the probability of getting at least ℓ heads in k independent coin tosses, x_j being the probability of head for the j-th toss, it is clear that v_ℓ^k is a multilinear function of $x_1, \ldots, x_k$ which is nondecreasing in each x_j. Let us note that if $x_r = 1$, then the summation extends only over those **A** that contain r and the function coincides on $x_r = 1$ with the function $v_{\ell-1}^{k-1}$ of the remaining $\ell - 1$ variables.

Lemma 8.4. *Suppose* $\{\alpha_j(\cdot) : 1 \le j \le L\}$ *are* L *independent centered Poisson processes as in the previous lemma, having the generator*

$$\tilde{\Omega}_L u = \sum_j \Omega_j u$$

where for each j, Ω_j *is the generator* Ω *acting on the variable* α_j *that was defined in the previous lemma. If we define*

$$u_\ell^L(\alpha_1, \ldots, \alpha_L, t) = v_\ell^L(q(\alpha_1, t), \ldots, q(\alpha_L, t))$$

then u_ℓ^L *is sufficiently regular and satisfies:*

$$\frac{\partial u_\ell^L}{\partial t} + \tilde{\Omega}_L u_\ell^L = 0 \quad \text{in} \quad (-m, m)^L \times \mathbf{R}$$

and

$$\Omega_j u_\ell^L \ge 0$$

for every j.

Proof. Follows immediately from Lemma 8.3 and the fact that that v_ℓ^L is multilinear in its variables and nondecreasing with respect to each one.

At this point we can take limits and replace the smooth function $h(\cdot)$ by the function $h_T(t)$ which is 1 if $t \le T$ and 0 otherwise. If we do that, in the limit,

$$q(\alpha, t) = \bar{P}_{\alpha,t}[\tau_m \le T]$$

in $(-m, m)$ and equals 1 outside. Recall that $\bar{P}_{\alpha,t}$ denotes the probability associated with the centered Poisson process $\alpha(s), s \ge t$ with $\alpha(t) = \alpha$. If we write $\bar{P}$ for the law of the process $(\alpha_1(\cdot), \ldots, \alpha_L(\cdot))$ generated by $\tilde{\Omega}$ with $\alpha_i(0) = \alpha_i$, then

$$u_\ell^L(\alpha_1, \ldots, \alpha_L, 0) = \bar{P}(H \ge \ell) \ .$$

We see from (8.8) that the functions v_ℓ^L reduce to $v_{\ell-k}^{L-k}$ if k of the α_j's are outside $(-m, m)$ and the rest remain inside.

We now turn to the proof of Theorem 8.2. If we take L mutually orthogonal counting processes $\{c_j(t)\}$, i.e. no two of which will jump at the same time, having intensities $\{\lambda_j(t, \omega)\}$ satisfying $0 \le \lambda_j(t, \omega) \le 1$ for all j and if $\alpha_j(t) = c_j(t) - \int_0^t \lambda_j(s, \omega)\, ds$ are their centered versions, we can stop these processes one after another when they get out of

$(-m, m)$ until ℓ of them escape before time T. Of course this may not happen. From Lemma 8.4 and the comments immediately following it, we see that $U = u_\ell^L(\alpha_1(t), \dots, \alpha_L(t), t)$ remains a supermartingale until ℓ processes have escaped or until time T whichever is sooner. In particular, the probability that at least ℓ particles exit from $(-m, m)$ before time T for any L arbitrary centered pairwise orthogonal counting processes with intensities bounded by 1, is dominated by the corresponding probability for L independent centered canonical Poisson processes with intensity 1. This concludes the proof of Theorem 8.2.

We now take up the proof of our main result, Theorem 8.1. For each $\epsilon > 0$ and $\delta > 0$ we want to count the number of particles that "misbehave": i.e.

$$\sup_{\substack{|s-t|\le\delta \\ 0\le s,t\le T}} |M_i(s) - M_i(t)| \ge \epsilon$$

Let us divide the interval $[0, T]$ into k intervals $\{I_1, \dots, I_k\}$ of length $\delta = \frac{T}{k}$ with $I_r = [\frac{(r-1)T}{k}, \frac{rT}{k}]$. If for any trajectory $M(\cdot)$ and any $\epsilon > 0$

$$\sup_{t\in I_r} \left| M(t) - M\left(\frac{(r-1)T}{k}\right) \right| \le \frac{\epsilon}{3} \tag{8.9}$$

for every $1 \le r \le k$, then

$$\sup_{\substack{|s-t|\le\delta(k) \\ 0\le s,t\le T}} |M(s) - M(t)| \le \epsilon$$

where $\delta(k) = \frac{T}{k}$. Let K_r be the number of particles that violate (8.9). The total number of particles K that misbehave can clearly be estimated by

$$K \le \sum_r K_r$$

If we denote by $\pi(N, k, \frac{\epsilon}{3})$ the probability that a normalized centered Poisson process

$$\alpha_N(t) = \frac{1}{N}[Z(N^2 t) - N^2]$$

violates (8.9) in an interval I_r of length $\frac{T}{k}$, by Theorem 8.2, the processes M_i's are uniformly conditionally stochastically dominated by L independent binomials with probability $\pi(N, k, \frac{\epsilon}{3})$. Let γ be the cardinality of the set of possible jump sizes i.e the cardinality of the set $\{z : p(z) > 0\}$ and $\Gamma = \sum_{z:p(z)>0} |z|$. From the representation

$$M_i(t) = \sum_z z M_N(i, z; t) \ ,$$

we can bound the oscillation of $M_i(\cdot)$ by

$$|M_i(t) - M_i(s)| \le \Gamma \sup_z |M_N(i,z;t) - M_N(i,z;s)| \ .$$

With all of this we have the following inequality:

$$P_N[K \ge \ell] \le F\left(Lk\gamma, \pi\left(N, \frac{T}{k}, \frac{\epsilon}{3\Gamma}\right), \ell\right)$$

where $F(n, \pi, \ell)$ is the probability of at least ℓ successes in n binomial trials with probability π of success in a single trial. We conclude now that with $\ell = N^d \epsilon$

$$\limsup_{N\to\infty} \frac{1}{N^d} \log P_N\left[K \ge N^d \epsilon\right] \le -k\gamma\bar{\rho} \sup_{p \ge \theta(k)} \left[p \log\frac{p}{q} + (1-p)\log\frac{1-p}{1-q}\right] \tag{8.10}$$

Here

$$\bar{\rho} = \lim_{N\to\infty} \frac{L}{N^d}$$

is the overall density of particles,

$$q = q(k) = \lim_{N\to\infty} \pi\left(N, \frac{T}{k}, \frac{\epsilon}{3\Gamma}\right) = \Pr\left[\sup_{0\le t\le \frac{T}{k}} |\beta(t) - \beta(0)| \ge \frac{\epsilon}{3\Gamma}\right]$$

β being the standard Brownian motion, and

$$\theta(k) = \lim_{N\to\infty} \frac{\ell}{Lk\gamma} = \frac{\epsilon}{k\bar{\rho}\gamma}$$

In this estimate $\bar{\rho}$, γ and ϵ are fixed and letting $\delta \to 0$ amounts to letting $k \to \infty$. Then $\log q(k) \simeq -c_1(\epsilon)k$ and $\theta(k) \simeq \frac{c_2(\epsilon)}{k}$. The right hand side of (8.10) is then clearly behaving like

$$-\gamma\bar{\rho}k\theta(k)\log\frac{\theta(k)}{q(k)} \simeq -c_3(\epsilon)k$$

and can be made large with k.

9. Proof of the main theorem

We are now ready to prove our main theorem on the large deviations of the empirical process. Let us recall the notation. $D[[0,T];\mathbf{T}^d]$ is the space of trajectories on $[0,T]$ with values in $\mathbf{T}^d$ with discontinuities, if any, of only the first kind. The space $\mathscr{M}(D[[0,T];\mathbf{T}^d])$ is the space of nonnegative measures on $D[[0,T];\mathbf{T}^d]$. Given the L rescaled trajectories $\{y_1(\cdot), \dots, y_L(\cdot)\}$ of our interacting particles, we define the empirical process as

$$R_{N,\omega} = \frac{1}{N^d} \sum_{i=1}^{L} \delta_{y_i(\cdot)}$$

with values in the space $\mathcal{M}(D[[0,T];\mathbf{T}^d])$. We will denote by P_N^ξ the measure induced on this space from the underlying particle system with an arbitrary initial configuration $\xi = \{y_1, \dots, y_L\}$ of L particles in the lattice Z_N^d imbedded in $\mathbf{T}^d$. If the initial configuration is random with a distribution μ_N it is assumed to satisfy a large deviation principle on the space of macroscopic density profiles $q(\cdot)$ with a rate function $I_{init}(q(\cdot))$. We want to show that a large deviation principle holds for the distributions

$$P_N = \int P_N^\xi \mu_N(d\xi)$$

of $R_{N,\omega}$, and that the rate function for this large deviation principle is the one given in section 7. For any measure $Q \in \mathcal{M}(D[[0,T];\mathbf{T}^d])$, $\mathcal{I}(Q)$ is infinite unless Q is supported on the set of continuous trajectories $(C[[0,T];\mathbf{T}^d])$. Moreover if we denote by $q(t,\cdot)$ the marginal density of Q at time t, we must have

$$\int_0^T \int_{\mathbf{T}^d} \frac{(\nabla q \cdot D\nabla q)}{q(1-q)}\, dt\, d\theta < \infty \tag{9.1}$$

For any $q = q(\cdot,\cdot)$, satisfying (9.1), $\mathcal{B}_q$ is defined as $b(\cdot,\cdot)$ that satisfy

$$\mathcal{E}(b) = \frac{1}{2} \int_{\mathbf{T}^d} \int_0^T (Db \cdot b)\, q(1-q)\, d\theta\, dt < \infty \tag{9.2}$$

and solve in weak sense the equation

$$\frac{\partial q}{\partial t} - \frac{1}{2} \nabla \cdot D\nabla q + \nabla \cdot Dbq(1-q) = 0 \tag{9.3}$$

We know that for $b \in \mathcal{B}_q$, we have a unique time inhomogeneous diffusion process with marginals equal to q and generator

$$\frac{1}{2}\nabla \cdot S(q(t,\theta))\nabla + \frac{1}{2}(S(q(t,\theta)) - D)\frac{\nabla q(t,\theta)}{q(t,\theta)} \cdot \nabla + Db(t,\theta)(1 - q(t,\theta)) \cdot \nabla \tag{9.4}$$

We denote this diffusion processe by P^b. As a Markov Process, we can start this diffusion with any initial distribution ρ that is dominated by $q(0,\cdot)$ and we will denote these processes by P_ρ^b. Let us note that these are not necessarily probability measures and will instead have their total masses equal to that of the initial distribution.

The rate function $\mathscr{I}(Q) = \infty$ unless $Q \ll P^b$ with finite relative entropy $H(Q, P^b)$ for some $b \in \mathscr{B}_Q$. In that case, according to Theorem 7.2, there is a unique $b \in \mathscr{B}_Q$, which we denote by b_Q, that will satisfy

$$E^Q\left[\int_0^T \phi(t, x(t)) \cdot dx(t)\right] = E^{P^{b_Q}}\left[\int_0^T \phi(t, x(t)) \cdot dx(t)\right] \qquad (9.5)$$

for every smooth function $\phi : [0, T] \times \mathbf{T}^d \to \mathbf{R}^d$. Moreover $H(Q, P^{b_Q}) < \infty$ and

$$\mathscr{I}(Q) = I_{[0,T]}(Q) = I_{\text{init}}(q(0, \cdot)) + \mathscr{E}[b_Q] + H(Q, P^{b_Q}) \ . \qquad (9.6)$$

If the initial configuration of the particle system were nonrandom with a macroscopic profile of $\rho_0(\cdot)$, then the large deviation rate function of the empirical process would be given by

$$\mathscr{I}(Q) = \mathscr{I}_{[0,T]}(Q; \rho_0) = \mathscr{E}[b_Q] + H(Q, P^{b_Q}) \qquad (9.7)$$

provided Q is such that $q(0, \cdot) = \rho_0(\cdot)$, and $\mathscr{I}(Q) = +\infty$ otherwise. The transition from this deterministic version to the more general version with random initial data is routine. The proof of our main theorem will depend on proving large deviations for certain finite dimensional projections.

Let us take a partition $\mathscr{P}$ of the torus $\mathbf{T}^d$ into r sets $B_1, B_2, \ldots, B_r$ having boundaries with zero Lebesgue measure. Let $\mathscr{A}^k$ denote the set of indices $\{a_1, \ldots, a_k\}$ where each $a_i \in \{1, \ldots, r\}$. For a choice $\mathbf{a} \in \mathscr{A}^k$, and $k + 1$ time points $0 = t_1 < t_2 < \cdots < t_{k+1} \leq T$, we let

$$E_{\mathbf{a}} = \{y(\cdot) : y(t_i) \in B_{a_i} \quad \text{for} \quad i = 1, \ldots, k\}$$

and define $n = r^k$ empirical measures $v^k_{\mathbf{a}} \in \mathscr{M}(\mathbf{T}^d)$ on $\mathbf{T}^d$ that correspond to each of the r^k indices $\mathbf{a} \in \mathscr{A}^k$,

$$v^k_{\mathbf{a}} = \frac{1}{N^d} \sum_{i : y_i(\cdot) \in E_{\mathbf{a}}} \delta_{y_i(t_{k+1})}$$

We have simplified the past history by partitioning the space of trajectories on $[0, t_k]$ into r^k groups and we have painted the particles in each group a different color. We are then looking at the empirical distributions at time $t = t_{k+1}$ of particles with r^k different colors. As a first step, we fix r, the sets $B_1, \ldots, B_r$, k, and the time points $0 = t_1 < t_2 < \cdots < t_k < t_{k+1} \leq T$ and we are interested in establishing a large deviation theorem for the collection $\{v^k_{\mathbf{a}}, \, \mathbf{a} \in \mathscr{A}^k\}$ of r^k measures on $\mathbf{T}^d$.

Theorem 9.1. *The large deviation principle is valid for* $\{v^k_{\mathbf{a}}\}$ *on the space* $[\mathscr{M}(\mathbf{T}^d)]^{r^k}$ *with a rate function*

$$I^{(k)}(\{p_{\mathbf{a}}\}) = \inf_{Q \in \mathscr{M}_k(\{p_{\mathbf{a}}\})} \mathscr{I}(Q) \tag{9.8}$$

where $\mathscr{M}_k(\{p_{\mathbf{a}}\})$ is the set of measures Q on $C[[0,T];\mathbf{T}^d]$ with the property that for all $\mathbf{a} \in \mathscr{A}^k$, and for all Borel sets $A \subset \mathbf{T}^d$,

$$Q[\cap_{1\le j\le k}\{y(t_j) \in B_{a_j}\} \cap \{y(t_{k+1}) \in A\}]$$

$$= Q[E_{\mathbf{a}} \cap \{y(t_{k+1}) \in A\}] = \int_A p_{\mathbf{a}}(\theta)\, d\theta \ .$$

Remark. Given any Q, $\mathbf{a} \in \mathscr{A}^k$ and $F \in \mathscr{F}^k = \sigma\{y(s) : t_k \le s \le T\}$ we define

$$Q_{\mathbf{a}}[F] = Q[\cap_{1\le j\le k}\{y(t_j) \in B_{a_j}\} \cap F]$$

then $Q \in \mathscr{M}_k(\{p_{\mathbf{a}}\})$ if and only if for each $\mathbf{a} \in \mathscr{A}^k$, the marginal at time t_{k+1} of $Q_{\mathbf{a}}$ has density $p_{\mathbf{a}}$.

Remark. Although we are mainly interested in the large deviation rates for the quantitities $\{v^k_{\mathbf{a}}(B_a)\}$, in order to make the induction in k work it is important to carry along complete information about the empirical distributions $\{v^k_{\mathbf{a}}\}$ of particles of different colors at time t_{k+1}.

Proof. The proof of the theorem will proceed by induction on k. For $k = 1$, there is nothing to prove since it all happens at time 0 and I_{init} controls everything. For $k = 2$ we can color the particles with r colors depending on which B_a they belong to in the partition at time $t_1 = 0$. Since B_a are continuity sets for the Lebesgue measure, the large deviation principle can be deduced from Theorem 4.13, with a rate function that can be obtained by the contraction principle of Theorem 7.5.

$$I^{(1)}(\{p_a\}) = \inf_{\tilde\rho(\cdot\,,\cdot)\in\mathbf{C}} \left[I_{init}(\tilde\rho(0\,,\cdot)) + I_{\mathrm{dyn}}(\tilde\rho(\cdot\,,\cdot))\right]$$

$$= \inf_{Q\in\mathscr{M}_1(\{p_a\})} \left[I_{init}((q(0,\cdot)) + \mathscr{I}_{[t_1,t_2]}(Q)\right] \tag{9.9}$$

where $\mathbf{C}$ is the set of densities

$$\mathbf{C} = \{\tilde\rho(\cdot\,,\cdot) : \rho_i(t_1\,,\theta) = 0 \quad \text{if } \theta \notin B_i,$$
$$\text{and } \rho_i(t_2\,,\cdot) = p_i(\cdot) \quad \text{for } 1 \le i \le r\}$$

If the initial configuration is deterministic and corresponds to the densities $\{\rho_a\}$ for the different colors $a \in \mathscr{A}$, with the support of ρ_a contained B_a for each $a \in \mathscr{A}$, then the rate function becomes

$$I^{(1)}(\{p_a\}\,;\{\rho_a\}) = \inf_{\tilde{\rho}(\cdot\,,\cdot)\in \mathbf{C}} I_{\text{dyn}}(\tilde{\rho}(\cdot\,,\cdot))$$

$$= \inf_{Q\in\mathscr{M}_1(\{p_a\}\,;\{\rho_a\})} \mathscr{I}_{[t_1,t_2]}(Q) \tag{9.10}$$

with

$$\mathscr{M}_1(\{p_a\}\,;\{\rho_a\}) = \mathscr{M}_1(\{p_a\}) \cap \left\{Q : q(0\,,\cdot) = \sum_a \rho_a(\cdot)\right\}$$

Let us assume that the result is valid for $k+1$ time points $0 = t_0 < t_1 < \cdots < t_k < t_{k+1} \le T$ and try to establish the induction. We add an extra time point t_{k+2} at the end so that we now have $0 = t_1 < t_2 < \cdots < t_{k+1} < t_{k+2} \le T$.

The basic ingredient in the induction step is the following elementary property concerning relative entropy. Suppose we consider the three σ-fields $\mathscr{F}_t = \sigma\{x(s) : 0 \le s \le t\}$, $\mathscr{F}^t = \sigma\{x(s) : t \le s \le T\}$ and $\mathscr{F}_t^t = \sigma\{x(t)\}$, in $\Omega = C[[0\,,T];\mathbf{T}^d]$. Let Q be a measure on $(\Omega\,,\mathscr{F}_T)$. Let $\{P_{t,x}\}$ be some inhomogeneous diffusion in the time interval $[0\,,T]$. The measure Q is not necessarily a probability measure, but has a marginal $q(0\,,dx)$ at time 0. The diffusion initialized with the measure $q(0\,,\cdot)$ at $t=0$ will be denoted by $P_{0\,,q}$. Let us suppose that $H(Q\,,P_{0\,,q}) < \infty$. The regular conditional probabilities of Q given the σ-fields $\mathscr{F}_t$ and $\mathscr{F}_t^t$ are denoted respectively by $Q_\omega^{[0\,,t]}$ and Q_ω^t. These are of course probability measures. Let us replace the measure Q by $\bar{Q}$ defined by, $\bar{Q} = Q$ on $\mathscr{F}_t$ and the regular conditional probability $\bar{Q}_\omega^{[0,t]}$ of $\bar{Q}$ given $\mathscr{F}_t$ is taken to be Q_ω^t on $\mathscr{F}^t$. In other words we force the Markov property at time t. Then $\bar{Q} = Q$ on both the σ-fields $\mathscr{F}_t$ and $\mathscr{F}^t$.

$$H(\bar{Q}\,,P_{0\,,q}) \le H(Q\,,P_{0\,,q})$$

We now proceed with the induction argument. Suppose Q is given as a measure on $(\Omega\,,\mathscr{F}_T)$. Let $E_{\mathbf{a},a}$ be the partition of Ω indexed by $(\mathbf{a},a) \in \mathscr{A}^{k+1}$ according to the colors at times $0 = t_1 < \cdots < t_{k+1} \le T$ and $t_{k+2} \in (t_{k+1}\,,T]$ is the next inspection time. We denote by

$$Q_{\mathbf{a},a}(\cdot) = Q(E_{\mathbf{a},a} \cap \cdot)$$

the restriction of Q to the various subsets of the partition. If we replace each $Q_{\mathbf{a},a}$ by the corresponding $\bar{Q}_{\mathbf{a},a}$ at time t_{k+1} and resum to get

$$\bar{Q} = \sum_{\mathbf{a},a} \bar{Q}_{\mathbf{a},a}$$

we can make the following observations.

1) For each $(\mathbf{a}, a) \in \mathscr{A}^{k+1}$, $Q_{\mathbf{a},a}$ and $\bar{Q}_{\mathbf{a},a}$ agree on both $\mathscr{F}_{t_{k+1}}$ and $\mathscr{F}^{t_{k+1}}$. In particular $Q \in \mathscr{M}_{k+1}(\{p_{\mathbf{a},a}\})$ if and only if $\bar{Q} \in \mathscr{M}_{k+1}(\{p_{\mathbf{a},a}\})$.
2) The two measures Q and $\bar{Q}$ have the same marginals for every t.
3) The quantities b_Q and $b_{\bar{Q}}$ are the same. They are determined by the expected values of stochastic integrals and so do not depend on the fact that the conditionals across t_{k+1} have been changed. In particular $P^{b_Q} = P^{b_{\bar{Q}}}$.
4) Therefore in the definition of $\mathscr{I}(Q)$ the two terms $\mathscr{E}(b_Q)$ and $\mathscr{E}(b_{\bar{Q}})$ are equal.
5) The entropy terms satisfy

$$H(\bar{Q}, P^{b_Q}) \leq H(Q, P^{b_{\bar{Q}}})$$

6) In the infimum

$$\inf_{Q \in \mathscr{M}_{k+1}(\{p_{\mathbf{a},a}\})} \mathscr{I}(Q)$$

we can replace Q by $\bar{Q}$.

Let us now compute $\mathscr{I}(\bar{Q})$. The term $\mathscr{E}(b)$ is additive over disjoint time intervals. By the Markov property that was forced at time t_{k+1} on $\bar{Q}$,

$$\begin{aligned} H_{[0,t_{k+2}]}(\bar{Q}, P^{b_Q}) &= H_{[0,t_{k+1}]}(Q, P^{b_Q}) + \sum_{\mathbf{a},a} H_{[t_{k+1},t_{k+2}]}(\hat{Q}_{\mathbf{a},a}, P^{b_Q}_{\mathbf{a},a}) \\ &= H_{[0,t_{k+1}]}(Q, P^{b_Q}) + \sum_{\mathbf{a}} H_{[t_{k+1},t_{k+2}]}(\hat{Q}_{\mathbf{a}}, P^{b_Q}_{\mathbf{a}}) \end{aligned}$$

where

$$\hat{Q}_{\mathbf{a}} = \sum_a \hat{Q}_{\mathbf{a},a} = \int_{\mathbf{T}^d} Q^{t_{k+1}}_{\mathbf{a},x}\, p_{\mathbf{a}}(x)\, dx$$

are just the restrictions of $Q_{\mathbf{a}}$ to $\mathscr{F}^{t_{k+1}}$ represented as integrals of the conditionals given the initial position at time t_{k+1} with respect to the marginals at time t_{k+1} and

$$P^{b_Q}_{\mathbf{a}} = \int_{\mathbf{T}^d} P^{b_Q}_{t_{k+1},x} p_{\mathbf{a}}(x)\, dx$$

is the diffusion with initial density $p_{\mathbf{a}}(\cdot)$ at time t_{k+1}. We can now calculate

$$\begin{aligned} \inf_{Q \in \mathscr{M}_{k+1}(\{p_{\mathbf{a},a}\})} \mathscr{I}(Q) &= \inf_{Q \in \mathscr{M}_{k+1}(\{p_{\mathbf{a},a}\})} \mathscr{I}_{[0,t_{k+2}]}(Q) \\ &\geq \inf_{\substack{\{p_{\mathbf{a}}\} \in \mathscr{D} \\ Q \in \mathscr{M}_k(\{p_{\mathbf{a}}\}) \\ \{\bar{Q}_{\mathbf{a}}\} \in \mathscr{M}}} \left\{ \mathscr{I}_{[0,t_{k+1}]}(Q) + \sum_{\mathbf{a}} H_{[t_{k+1},t_{k+2}]}(\bar{Q}_{\mathbf{a}}, P^{b_{\bar{Q}}}_{\mathbf{a}}) + \mathscr{E}_{[t_{k+1},t_{k+2}]}(b_{\bar{Q}}) \right\} \end{aligned}$$

where

$$\mathscr{D} = \left\{ \{p_{\mathbf{a}}\} : \int_{B_a} p_{\mathbf{a}}(x)\, dx = \bar{Q}_{\mathbf{a}}(B_a) \quad \text{for all} \quad \mathbf{a} \in \mathscr{A}^k \right\}$$

and

$$\mathscr{M} = \left\{ \{\bar{Q}_{\mathbf{a}}\} : \bar{Q}_{\mathbf{a}}\left[\{y(t_{k+1}) \in B_a\} \cap \{y(t_{k+2}) \in A\}\right] = \int_A p_{\mathbf{a},a}(x)\, dx \right\}$$

for all $\mathbf{a}\,, a$ and measurable $A \subset \mathbf{T}^d$. The inequality is actually an equality because, starting from Q on $\mathscr{F}_{t_{k+1}}$ and $\bar{Q}_{\mathbf{a}}$ that are compatible, i.e. occur in the infimum, we can patch together a measure Q on $\mathscr{F}_{t_{k+2}}$ that figures in the original infimum.

Now it is easy to see that the induction works. If at time t_{k+1} we had r^k colors and they had density profiles $\{p_{\mathbf{a}}\}$ and we refined the colors at that time to have r^{k+1} colors, we will have the profiles $\{\rho_{\mathbf{a},a}\}$ where $\rho_{\mathbf{a},a}(x) = p_{\mathbf{a}}(x)\chi_{B_a}(x)$. The rate function for them to have profiles $p_{\mathbf{a},a}$ at time t_{k+2} is given by

$$\inf_{\{\bar{Q}_{\mathbf{a}}\}\in\mathscr{M}} \left\{ \sum_{\mathbf{a}} H_{[t_{k+1},t_{k+2}]}(\bar{Q}_{\mathbf{a}}\,, P_{\mathbf{a}}^{b_{\bar{Q}}}) + \mathscr{E}_{[t_{k+1},t_{k+2}]}(b_{\bar{Q}}) \right\}$$

according to the induction step with $k = 2$. To remove the condition of fixing the profiles at time t_{k+1}, we have to add the large deviation rate for $\{p_{\mathbf{a}}\}$, which is equal to

$$\inf_{Q\in\mathscr{M}_k(\{p_{\mathbf{a}}\})} \mathscr{I}_{[0\,,t_{k+1}]}(Q)$$

by the induction hypothesis and we are done.

We can now prove our main theorem that the large deviation principle holds for $R_{N,\omega}$ with the rate function $I(Q)$. In Theorem 9.1 keeping track of the distribution at the last time t_k was needed only for for the purpose of carrying out the induction.

By contraction we have the following large deviation principle.

Corollary 9.2. *The large deviation principle is valid for*

$$\{\pi_{\mathbf{a}}\} = \frac{1}{N^d}\{\text{number of particles} \quad i : y(t_i) \in B_{a_i} \quad \textit{for} \quad 1 \le i \le k\}$$

on the space $\mathbf{R}^{r^k}$ *with a rate function*

$$I^{(k)}[\pi] = \inf_{Q\in\mathscr{M}_k(\{\pi_{\mathbf{a}}\})} I(Q)$$

where $\mathscr{M}_k(\{\pi_{\mathbf{a}}\})$ *is the set of measures* Q *on* $C[[0,T];\mathbf{T}^d]$ *with the property*

$$Q[\cap_{1\le j\le k}\{y(t_j) \in B_{a_j}\}] = Q[E_{\mathbf{a}}] = \pi_{\mathbf{a}}$$

for all $\mathbf{a}$.

The proof of the main theorem is now routine by refining the partition and making the sampling times dense in $[0,T]$. We have superexponential tightness and compactness of the level sets of the rate function $I(Q)$. Because the rate function $I(Q)$ is finite only for measures supported on $C[[0,T];\mathbf{T}^d]$ and the sets in our partition are continuity sets for the marginals of Q with $I(Q)<\infty$ there are no technical difficulties in carrying out the limiting procedure and establishing our main theorem.

Acknowledgements. J.Q. and F.R. would like to thank S.R.S. Varadhan, H.-T. Yau and the Courant Institute for their warm hospitality while a part of this work was done. F.R. was partially supported by a Sloan Foundation Fellowship.

References

[BS] Benois, O., Saada, E.: Large deviations of a tagged particle in the one dimensional nearest neighbor symmetric simple exclusion model. (to appear)

[GJL] Gabrielli, D., Jona-Lasinio, G., Landim, C.: Onsager reciprocity relations without microscopic reversibility. Phys. Rev. Lett., **77**, no 7, 1202–1205 (1996)

[KOV] Kipnis, C., Olla, S., Varadhan, S.R.S.: Hydrodynamics and large deviations for simple exclusion processes. Comm. Pure Appl. Math. **42**, 115–137 (1989)

[KV] Kipnis, C., Varadhan, S.R.S.: Central limit theorem for additive functionals of reversible Markov processes and applications to simple exclusion. Comm. Math. Phys. 106, 1–19

[Q1] Quastel, J.: Diffusion of color in the simple exclusion process. Comm. Pure Appl. Math. **45**, 623–679 (1992)

[Q2] Quastel, J.: Large deviations from a hydrodynamic scaling limit for a nongradient system. Ann. Prob. **23** No. 2, 724–742 (1995)

[QV] Quastel, J., Varadhan, S.R.S.: Diffusion semigroups and diffusion processes corresponding to degenerate divergence form operators. Comm. Pure Appl. Math. **50**, 667–706 (1997)

[R] Rezakhanlou, F.: Propagation of chaos for symmetric simple exclusion. Comm. Pure Appl. Math. **117**, 943–957 (1994)

[V1] Varadhan, S.R.S.: Nonlinear diffusion limit for a system with nearest neighbor interaction II, in Asymptotic Problems in Probability Theory; Stochastic Models and Diffusions on Fractals. Proceedings of the Taniguchi Symposium, Sanda and Kyoto 1990, 75–130, Longman, Essex, England (1990)

[V2] Varadhan, S.R.S.: Regularity of self–diffusion coefficient. The Dynkin Festschrift, Prog. Prob. 34, 387–397, Birkhauser, Boston (1994)

Perplexing Problems in Probability, Progr. Probab. 44, Birkhaüser, 1999, 373-383

Large Deviations for Interacting Particle Systems

S.R.S. Varadhan

ABSTRACT We consider large systems of interacting particles. We obtain a large deviation principle for the empirical process viewed as a random measure on the path space. The precise rate function is obtained. While the formula is fairly general, the explicit dependence on the model is seen through quantities like bulk and self diffusion coefficients.

Keywords: Large deviations, interacting particle systems.

AMS Subject Classifications: 60K35, 60F10.

20.1 Introduction

We will consider large systems of interacting particles, i.e., a large number of particles interacting with each other and moving randomly in R^d or Z^d. The interactions will be local in the sense that the motion of each particle will be affected only by presence and positions of a few nearby particles that happen to be within a fixed finite range of it. We will be studying space-time rescalings of such systems, with a diffusive or parabolic relation between the space and time scales. For technical reasons, we will start with periodic boundary conditions on a finite but large cube, so that after rescaling, the space can be considered to be a fixed torus or some finely imbedded lattice inside one. We will give several classes of examples.

One important feature of all these examples is the validity of a hydrodynamical limit under diffusive scaling. We are rescaling systems with conserved quantities, like the total number of particles in our examples. The conserved quantities change slowly while the system rapidly equilibrates to a local equilibrium that is determined by the local average of the conserved quantity. Consequently in our examples, for large values of the scaling parameter, the system is essentially fully described by the density as a function of rescaled space and time. The density itself will evolve according to some (usually) nonlinear heat

NOTE: The present article contains a summary of results obtained jointly with J. Quastel and F. Rezakhnalou. The details can be found in an article that is to appear in a forthcoming issue of Probability Theory and Related Fields.

equation. One can use the hydrodynamical scaling principle to reduce many questions, including the macroscopic large deviation behavior and the diffusive motion of a tracer particle in nonequilibrium, to considerations of density alone.

20.2 Some Classes of Examples

The scaling parameter is a large number N that represents the "size" of the system. In physical terms, $\frac{1}{N}$ is the typical interparticle distance and this requires the number of particles to be of order N^d. The actual number of particles k_N will be assumed to be asymptotically equal to $\bar{\rho}N^d$ for some fixed constant $\bar{\rho}$.

Model 1 The scaled interacting diffusion model in $\mathbf{T}^d$ consists of a system of k_N stochastic differential equations for the positions $\{x_i(t)\}$ at time t of the k_N particles in $\mathbf{T}^d$.

$$dx_i(t) = d\beta_i(t) - N\sum_j V'(N(x_i(t) - x_j(t)))dt.$$

Here β_i are independent d-dimensional Brownian motions and the interaction consists of a drift term which is the gradient of the potential

$$V_N = \sum_{i<j} V(N(x_i - x_j))$$

where $V(x)$ is a repulsive pair potential with compact support.

Model 2 Zero Range Processes describe the motion of k_N particles that are located on the sites of a sublattice with width $h = \frac{1}{N}$. Each particle waits for a random exponential time and jumps to a new site. The jumps themselves are random and mutually independent, with probability $p(z)$ for a jump of size z in the unscaled lattice or of size hz in the rescaled one. It is natural to assume that $\sum_z zp(z) = 0$ or more strongly that $p(z) \equiv p(-z)$. The interaction is introduced through the jump rate that is allowed to depend on the number of particles that share the same site as the particle that is to jump. The model then is fully specified by $\{p(z)\}$ and a function $c(k)$. The jump rate of the i-th particle is $N^2c(k(x_i))$. Here x_i is the location of the i-th particle and $k(x_i)$, is the total number of particles at that site. The factor N^2 is the effect of rescaling time. The case where $c(k) \equiv 1$ is the non-interacting case of scaled independent random walks.

Model 3 The simple exclusion process is a slight modification of the second model. No more than one particle is allowed to be at any given site at any given time. The jump rates are all N^2, except that if there is already a particle present at the selected site where the particle is to jump, the jump is disallowed and the particle is forced to wait for a new exponential waiting time. The effective jump rate then is $N^2\sum' p(z)$ where the summation is restricted to jumps that lead to empty sites.

20.3 The Problem

In all of these problems, for each fixed time T, we have the collection of k_N trajectories $\{x_1(\cdot), x_2(\cdot), \ldots, x_{k_N}(\cdot)\}$ defined on the interval $[0, T]$. The empirical process, which is a measure on the path space $D[[0, T]; \mathbf{T}^d]$, is defined by

$$R_N = \frac{1}{N^d} \sum_{i=1}^{k_N} \delta_{x_i(\cdot)}.$$

It is a random measure and its distribution is determined by the model and the initial configuration $(x_1, \ldots, x_{k_N})$ of the k_N particles.

The basic problem is the asymptotic behavior of R_N. It is natural to assume that $(x_1, \ldots, x_{k_N})$ is such that the initial configuration has an asymptotic empirical distribution. For simplicity we could assume that the initial conditions are deterministic and that

$$\lim_{N\to\infty} \frac{1}{N^d} \sum_{i=1}^{k_N} \delta_{x_i} = \nu_0(dx) = \rho_0(x)dx \tag{20.3.1}$$

exists. One expects the asymptotic behavior of R_N to be governed for the most part by $\rho_0(\cdot)$. We will be concerned with the laws of large numbers asserting that

$$\lim_{N\to\infty} R_N = Q \quad \text{in probability}$$

for a Q that is to be determined, as well as calculating the rates for large deviations from it. One interesting aspect is a degree of universality in the form of the large deviation rate function, depending only on a few physically meaningful objects, like bulk and self diffusion coefficients, that are to be calculated specifically for each model.

We will describe the results in the context of simple exclusion processes, the third set of models that we described. At the end we will provide a brief summary of what is known for the other models.

20.4 Laws of Large Numbers

The first step in these problems is to look at the laws of large numbers for the family of empirical densities

$$\nu_t^N(dx) = \frac{1}{N^d} \sum_i \delta_{x_i(t)} \tag{20.4.1}$$

viewed as random maps from $[0, T]$ into the space $\mathcal{M}(T^d)$ of measures on T^d. In the context of the symmetric simple exclusion process it is quite elementary to establish the following law of large numbers. If at time $t = 0$, the initial conditions are such that (20.3.1) holds in probability, then

$$\lim_{n\to\infty} \nu_t^N(dx) = \nu_t(dx) = \rho(t,x)dx$$

exists in probability, where $\rho(t,x)$ is the solution of the linear heat equation

$$\frac{\partial \rho}{\partial t} = \frac{1}{2}\nabla D \nabla \rho \quad \text{for } t > 0 \tag{20.4.2}$$

with the initial condition

$$\rho(0,x) = \rho_0(x).$$

Here D is the diffusion matrix, which is just the covariance matrix of the jump distribution

$$\langle Du, u\rangle = \sum_z p(z)\langle u, z\rangle^2.$$

Although in other models the bulk diffusion matrix D is a function of the density ρ, for the symmetric simple exclusion model it turns out not to depend on ρ. To see this we make the following simple calculation that is almost a proof. The generator can be written in terms of the basic exchange operators

$$(\nabla_{x,y} f)(\eta) = f(\eta^{x,y}) - f(\eta)$$

where $\eta^{x,y}$ is the new configuration obtained by making an exchange between sites x and y.

$$\eta^{x,y}(z) = \begin{cases} \eta(z) & if\ \ z \neq x, y \\ \eta(y) & if\ \ z = x \\ \eta(x) & if\ \ z = y. \end{cases}$$

The generator is written as

$$\mathcal{L}_N f = N^2 \sum_x \eta(x) \sum_z (1 - \eta(x+z)) p(z) \nabla_{x,x+z} f$$

which can be rewritten using the symmetry of $p(\cdot)$ as

$$\mathcal{L}_N f = \frac{N^2}{2} \sum_{x,y} [\eta(x)(1-\eta(y)) + \eta(y)(1-\eta(x)] p(y-x) \nabla_{x,y} f.$$

Because $\nabla_{x,y} f(\eta) = 0$ unless $\eta(x)(1-\eta(y)) + \eta(y)(1-\eta(x)) = 1$, we can simplify

$$\mathcal{L}_N f = \frac{N^2}{2} \sum_{x,y} p(y-x) \nabla_{x,y} f$$

and it looks quite noninteracting. In fact, if we now compute

$$\mathcal{L}_N \frac{1}{N^d} \sum J(\frac{x}{N}) \eta(x),$$

it is not hard to see that

$$\mathcal{L}_N \frac{1}{N^d}\sum J(\frac{x}{N})\eta(x) \simeq \frac{1}{2}\sum_x (\nabla D \nabla J)(\frac{x}{N})\eta(x).$$

To complete the proof we only have to show that the martingale term in

$$d\frac{1}{N^d}\sum J(\frac{x}{N})\eta(x,t)$$

goes to zero with N.

The next step is to investigate the behavior of a tagged particle. In other words, at time zero one particular particle is tagged and its motion is followed. In equilibrium we expect the tagged particle to diffuse like a Brownian motion. In fact for a wide class of models this is now a well understood phenomenon [KV]. The covariance matrix for the limiting Brownian motion will in general depend on the equilibrium that is used. In our model on Z^d, the equilibria are the various Bernoulli or product measures μ_ρ with particle density

$$\rho = \mu_\rho\{\eta(x) = 1\}.$$

We will denote by $S(\rho)$ the limiting covariance usually referred to as the self-diffusion matrix. In the noninteracting case it will not depend on ρ. But for symmetric simple exclusion, if we exclude the very special case of nearest neighbor model in one dimension, it is not surprising that at low density, i.e., near $\rho = 0$, $S(\rho) \simeq D$ because the interaction is negligible and near high density, i.e., $\rho = 1$, $S(\rho) \simeq 0$ essentially due to the gridlock that develops when nearly every site is filled.

The behavior of a tagged particle in nonequilibrium is far from obvious. The density ρ is no longer a constant, but a function $\rho(t, x)$ of t and x, given by the solution of equation (20.4.2). One might guess that a tagged particle located at $x(t)$ at time t, being myopic, will only see its immediate neighborhood and behave as if it is in equlibrium at density $\rho = \rho(t, x)$. As a first approximation we could expect $x(t)$ to behave like a time dependent diffusion with generator

$$\frac{1}{2}\nabla S(\rho(t, x))\nabla,$$

but in reality it is more likely that the generator needs to be modified by a first order term and takes the form

$$\frac{1}{2}\nabla S(\rho(t, x))\nabla + b(t, x)\nabla$$

for some drift term $b(\cdot, \cdot)$. Because we have an extra as yet undetermined first order term, it does not matter if we write the higher order term in divergence form or not. We have chosen the divergence form for convenience.

If we picked initially a particle at random and tagged it, its distribution is given, subject to normalization, by a density evolving according to the forward (adjoint) equation corresponding to (20.4.2).

$$\frac{\partial \rho}{\partial t}(t,x) = \frac{1}{2}\nabla S(\rho(t,x))\nabla\rho(t,x) - \nabla\cdot(b(t,x)\rho(t,x) \quad \text{for } t>0 \tag{20.4.3}$$

with the initial condition

$$\rho(0,x) = \rho_0(x).$$

Tagging is only an exercise of the mind. The particles are totally oblivious to it. Since the empirical density is the same for the tagged or untagged system, the solutions of (20.4.2) and (20.4.3) must coincide. This yields

$$\frac{1}{2}\nabla S(\rho(t,x))\nabla\rho(t,x) - \nabla\cdot(b(t,x)\rho(t,x)) = \frac{1}{2}\nabla D\nabla\rho(t,x).$$

While it does not follow, it is quite likely that

$$\frac{1}{2}S(\rho(t,x))\nabla\rho(t,x) - b(t,x)\rho(t,x) = \frac{1}{2}D\nabla\rho(t,x)$$

which can be solved to yield

$$b(t,x) = \frac{1}{2\rho(t,x)}[S(\rho(t,x)) - D]\nabla\rho(t,x).$$

If we now denote by Q a Markov process (with total mass $\bar{\rho}$) having initial density $\rho_0(x)$ and time dependent (backward) generator

$$L_t = \frac{1}{2}\nabla S(\rho(t,x))\nabla + \frac{1}{2}\frac{[S(\rho(t,x)) - D]\nabla\rho(t,x)}{\rho(t,x)}\cdot\nabla, \tag{20.4.4}$$

then one expects

$$\lim_{N\to\infty} R_N = Q \tag{20.4.5}$$

in probability. For simple exclusion processes (20.4.5) was indeed proved by Rezakhanlou [R]. Once the distribution of the tagged particle is shown to converge to a diffusion with generator L_t given by (20.4.4), is not very hard to prove (20.4.5). One needs to show that the motion of two different tagged particles are asymptotically independent. However if we start with an initial configuration that is far away from equilibrium, like an arbitrary deterministic initial condition, to prove the limit theorem when an *arbitrary* particle in the initial configuration is tagged is very hard. On the other hand proving (20.4.5) requires only that the limit theorem be valid for *nearly all* particles in the initial configuration.

The proof of (20.4.5) in [R] uses earlier results in [Q] that study a situation that mediates between tagged and untagged versions. Let us divide the particles into a finite number of types that we will think of as colors. Colors do not affect the motion which is as before. We simply keep track of the colors. At any time t, we have the empirical distribution of each one of the k colors. In the scaled limit we will have k densities $\bar{\rho}(t,x) = \{\rho_j(t,x) : 1 \le j \le k\}$. They should evolve as a nonlinear system

$$\frac{\partial \tilde{\rho}(t,x)}{\partial t} = \nabla \mathbf{B}(\tilde{\rho}(t,x))\nabla\tilde{\rho}(t,x) \tag{20.4.6}$$

with suitable initial conditions. The system (20.4.6) is equivalent to solving first the linear equation (20.4.2) for the total density

$$\rho(t,x) = \sum_{j=1}^{k} \rho_j(t,x)$$

and then at the next step solving a set of (identical) equations that are again linear with the correct initial conditions for the densities of individual colors

$$\frac{\partial \rho_j(t,x)}{\partial t} = L_t^* \rho_j(t,x),$$

which is just the forward equation for the probability density of the position of the tagged particle.

In [Q] the multicolor system was viewed in its own right and the law of large numbers under hydrodynamical scaling was established for it. For the simple exclusion model, this turns out to be a nongradient system and the analysis becomes considerably harder. To see this let us consider k different colors. Denote the configuration by $\zeta(x) = \{\xi_j(x)\}$. The variable $\xi_j(x)$ is 1 if the site x contains a particle of color j. Exclusion still applies so that $\eta(x) = \sum_j \xi_j(x) \leq 1$. The generator $\mathcal{L}_N$ now takes the form

$$(\mathcal{L}_N f)(\zeta) = N^2 \sum_{x,y} \eta(x)(1-\eta(y))p(y-x)[f(\zeta^{x,y}) - f(\zeta)].$$

If we compute

$$\mathcal{L}_N \frac{1}{N^d}\sum_x J(\frac{x}{N})\xi_j(x)$$

we get

$$\frac{N^2}{N^d}\sum_{x,y} \xi_j(x)(1-\eta(y))p(y-x)[J(\frac{y}{N}) - J(\frac{x}{N})]$$

which is nearly equal to

$$\frac{N}{N^d}\sum_{x,y} \xi_j(x)(1-\zeta(y))p(y-x)\langle \nabla J(\frac{x}{N}), y-x\rangle.$$

Although $p(\cdot)$ is symmetric, the possibility of having the sites x, y filled by different colors makes the situation asymmetric, and we cannot simplify the expression above and do summation by parts once more, to neutralize the factor N. The microscopic currents for the different colors in the different coordinate directions

$$W_x^{j,i} = N\sum_y \xi_j(x)(1-\zeta(x+z))p(z)\langle e_i\ , z\rangle$$

can be replaced only asymptotically by linear combinations of density gradients of the form

$$W_x^{j,i} = -N\mathbf{B}_{r,s}^{j,i}(\mathbf{q})[\xi_r(x+e_s) - \xi_r(x)]$$

where $\mathbf{q} = \{q_1, \ldots, q_k\}$ is the empirical local density. This is the hard part in the analysis of nongradient models. Once this is done (20.4.6) is easy to establish.

The transition from the multicolor system to (20.4.5) is essentially a question of compactness. Let $\mathcal{P} = \{A_1, \ldots, A_p\}$ be an arbitrary but fixed partition of the $\mathbf{T}^d$ into nice sets. If E is a subset of the form

$$E = \{x : x(t_1) \in A_{i_1}, \ldots, x(t_n) \in A_{i_n}\}$$

and the convergence in probability of $R_N(E)$ to $Q(E)$ can be proved by induction on n. For $n = 1$, this just involves the empirical density and a single color. If we assume the result for n time points, we have p^n possible past histories and we code them by p^n distinct colors. The empirical density of the different colors at the next time t_{n+1} is sufficient to yield information on subsets E at stage $n+1$. Given the law of large numbers for the multicolor system, the induction proceeds smoothly. Since the partition is arbitrary, if we have compactness, we can pass from finite dimensional distributions to weak convergence of the processes R_N. This was carried out in [R].

20.5 Large Deviations

We now turn to the question of Large Deviation rates. We examine first the empirical densities ν_t^N.

The probability distribution of the empirical density satisfies a large deviation principle with the following rate functional $I(\rho(\cdot,\cdot))$ on the space of weakly continuous maps ρ from $[0, T]$ into $L_1(T^d)$ that satisfy $0 \le \rho(t, x) \le 1$.

$$I(\rho(\cdot,\cdot)) = I_0(\rho(0,\cdot) + \frac{1}{2}\int_0^T ||\frac{\partial \rho}{\partial t} - \frac{1}{2}\nabla D \nabla \rho||^2_{-1,\rho(t,\cdot)(1-\rho(t,\cdot))} dt. \quad (20.5.1)$$

The first term I_0 is the rate function for deviations of the initial profile from its appropriate limit. For deterministic initial conditions it is 0 for the true value and ∞ for all others. Otherwise it depends on the statistics of the initial configuration. The second term, which is the dynamical part, measures in some way by how much ρ fails to satisfy the heat equation (20.4.2). For any $\rho(\cdot)$ the norm $||\ ||_{-1,\rho(t,\cdot)(1-\rho(t,\cdot))}$ refers to the dual of the weighted Sobolev norm

$$||f||^2_{1,\rho(\cdot)} = \int_{T^d} \langle D\nabla f(x), \nabla f(x)\rangle \rho(x)(1-\rho(x)dx.$$

This was proved in [KOV] and it is necessary for us to understand how these large deviations arise. One can introduce a small bias in the system by perturbing the symmetric jump rates by a weak asymmetric term, i.e., the probabilities of transition from x to $x+z$ at time t are changed from $p(z)$, by a small perturbation, to $p(z) + \frac{1}{N}q(t, \frac{x}{N}, z)$. This has an entropy cost as well as a macroscopic effect. To calculate these quantities we need to know that the system is locally close to an equilibrium (some Bernoulli measure) and the empirical density dictates which equilibrium we have to pick. Quantities not explicitly dependent on the empirical densities are replaced by their expectations in equilibrium which means by suitable functions of the empirical density.

Macroscopically, with weak asymmetry, the equation (20.4.2) gets replaced by (20.5.2) below.

$$\frac{\partial \rho}{\partial t} = \frac{1}{2}\nabla D \nabla \rho - \nabla \cdot b\rho(1-\rho) \quad \text{for } t > 0 \tag{20.5.2}$$

with the initial condition

$$\rho(0, x) = \rho_0(x).$$

Here

$$b(t, x) = \sum_z zq(t, x, z)$$

and the entropy cost (after optimizing over $q(\cdot, \cdot, \cdot)$ for a given $b(\cdot, \cdot)$ and normalization by a factor of N^{-d}) is given by

$$\mathcal{E}(b(\cdot, \cdot)) = \frac{1}{2}\int_0^T \int_{T^d} \langle Db(t, x), b(t, x)\rangle \rho(t, x)(1 - \rho(t, x))dxdt.$$

For a given $\rho(\cdot, \cdot)$ we consider

$$\mathcal{B}_{\rho(\cdot,\cdot)} = \{b(\cdot, \cdot) : (20.5.2) \text{ holds}\}.$$

The dynamical term in (20.5.1) is the result of optimizing $\mathcal{E}(b(\cdot, \cdot))$ over $b(\cdot, \cdot) \in \mathcal{B}_{\rho(\cdot,\cdot)}$.

The weak asymmetry has an effect on the behavior of a tagged particle which can be described through $b(\cdot, \cdot)$. The generator L_t gets replaced by

$$L_{t,b(t,\cdot)} = L_t + (1 - \rho(t, x))b(t, x) \cdot \nabla$$

and the tagged particle process with the new generator will be denoted by Q_b. As long as $b(\cdot, \cdot) \in \mathcal{B}_{\rho(\cdot,\cdot)}$, Q_b will have $\rho(t, \cdot)$ for its marginal density at time t. Suppose Q is a general stochastic process on $C[[0, T]; T^d]$ with total mass $\bar{\rho}$. Its marginal densities can be denoted by $q(t, \cdot)$. The 'current' is the collection of expected values of the stochastic integrals

$$E^Q\Big[\int_0^T \langle(t, x(t)), dx(t)\rangle\Big]$$

and we can try to find $b_Q \in \mathcal{B}(q(\cdot, \cdot))$ such that the marginals as well as the 'current' of Q match those of the tagged particle distribution Q_{b_Q} that we will obtain under our perturbation. The measure Q will then determine b_Q and Q_{b_Q}. The large deviation rate function for R_N is the functional

$$\mathcal{I}(Q) = I_0(q(0, \cdot)) + \mathcal{E}(b_Q(\cdot, \cdot)) + H(Q; Q_{b_Q}). \tag{20.5.3}$$

The important point here is the interpretation of the terms. The first term is clearly the contribution of the statistics of the initial condition. The second term depends on the hydrodynamical scaling behavior of the model and depends on the objects that describe them. The last term is pure noise, and says that once the correct macroscopic behavior of density and current are assured the tagged particle motions are nearly independent.

The proof of this theorem again proceeds through the intermediate step of multicolor systems. A large deviation principle is established for them with a rate function that is similar to the one for single color.

$$I(\rho(\cdot,\cdot)) = I_0(\rho(0,\cdot) + \frac{1}{2}\int_0^T ||\frac{\partial\rho}{\partial t} - \nabla\mathbf{B}(\rho)\nabla\rho||^2_{-1,Q((\rho(t,\cdot))}dt. \qquad (20.5.4)$$

Here $\rho = \{\rho_j\}$ are the densities of the various colors and Q is a matrix of diffusion coeffecients closely related to $\mathbf{B}(\rho)$ and are determined in [Q]. We then partition T^d into a finite number p of cells and consider where the individual labeled particles fall at a finite number of specified times $0 \le t_1 < t_2 < \cdots < t_n \le T$. The large deviation rates for these p^n relative frequencies can be calculated by induction on n. The induction step depends on the large deviation property of a multicolor system. We have to let p and n go to ∞, and a careful calculation with the rate functions through the limiting procedure yields formula (20.5.3). The details can be found in the article [QRV].

20.6 Related Work

While we have described in some detail the work related to symmetric simple exclusion models, there are partial results concerning models 1 and 2.

In the case of interacting Brownian motions, there is the possibility of phase transitions in dimensions larger than one. In dimension one and at densities below phase transition in higher dimensions, there are results concerning laws of large numbers. There are general results in the one dimensional case [V] and the methods of [Y] are applicable in any dimension when the density is restricted to a nice region free of phase transition . For a special model in one dimension, with a $V(x)$ that is like a δ-function, the distribution of any tagged particle in nonequilibrium has been worked out in [G]. Little else is known regarding tagged particles in nonequilibrium.

A lot is known about Zero Range Processes, especially regarding laws of large numbers and tagged particles. See for instance [GJL] and a forthcoming book [KL]. In some sense this is a simpler class of models because the multicolor system is still a gradient system. The analysis is therefore simpler and the results that we have described for the empirical processes of simple exclusion models should have analogs here.

Acknowledgments: This research was supported by NSF grant DMS-9803140 and ARO grant DAAH04-95-1-0666.

References

[G] I. Grigorescu, New York University Thesis, 1997.

[GJL] D. Gabrielli, G. Jona-Lasinio, and C. Landim, Onsager reciprocity relations without microscopic reversibility, *Phys. Rev. Lett.* **77**, no. 7, 1202–1205 (1996).

[KL] C. Kipnis and C. Landim, *Scaling Limits of Interacting Particle Systems*, Springer-Verlag, Berlin, Heidelberg, 1999.

[KOV] C. Kipnis, S. Olla, and S.R.S. Varadhan, Hydrodynamics and large deviations for simple exclusion processes, *Comm. Pure Appl. Math.* **42**, 115–137 (1989).

[KV] C. Kipnis and S.R.S. Varadhan, Central limit theorem for additive functionals of reversible Markov processes and applications to simple exclusion, *Comm. Math. Phys.* **106**, 1–19.

[Q] J. Quastel, Diffusion of color in the simple exclusion process, *Comm. Pure Appl. Math.* **45**, 623–679 (1992).

[QRV] J. Quastel, F. Rezakhanlou, and S.R.S. Varadhan, Large deviations for the symmetric simple exclusion process in dimensions $d \geq 3$, *Probab. Theory Relat. Fields* **113**, 1–84 (1999).

[R] F. Rezakhanlou, Propagation of chaos for symmetric simple exclusion, *Comm. Pure Appl. Math.* **117**, 943–957 (1994).

[V] S.R.S. Varadhan, Scaling limit for interacting diffusions, *Comm. Math. Phys.* **135**, 313–353 (1991).

[Y] Yau, H.T., Relative entropy and the hydrodynamics of Ginzburg–Landau models, *Lett. Math. Phys.* **22**, 63–80 (1991).

Courant Institute of Mathematical Science
Department of Mathematics
New York University
251 Mercer Street
New York, NY 10012
varadhan@cims.nyu.edu

Mathematical Physics, Imp. Coll. Press, London, 2000, 306-315

INFINITE PARTICLE SYSTEMS AND THEIR SCALING LIMITS

S.R.S. VARADHAN

Courant Institute of Mathematical Sciences, New York University, 251 Mercer Street, New York, NY 10012, USA

E-mail: varadhan@cims.nyu.edu

The subject of nonequlibrium statistical mechanics deals with the question of how, a complex system with one or more conserved quantities, approaches its equilibrium. Although initially investigated in the context of classical mechanical systems that conserve mass, momenta and energy, its paradigm is relevent in the study of stochastic paticle systems with conserved quantities and mutiple equilibria. In fact the inherent noise in the stochastic systems provides a degree of stability that is absent in classical systems. During last twenty years several such models have been studied, with varying amounts of noise, and a body of rigorous results has been obtained.

A convenient class of examples go by the general name of lattice gas models. The physical space is the lattice Z^d of points with integer coordinates. The sites $x \in Z^d$ may be occupied by particles. The state of the system is described by specifying the number of particles $\eta(x)$ present at the site x. It is also possible that there are several types of particles, in which case $\eta(x)$ has multiple components specifying the number of particles of each type. The state space then is the collection Ω of all possible such functions $\omega = \eta(\cdot)$. The dynamics is a stochastic process, indeed a Markov process, with Ω as its state space. The infinitesimal generator of this process specifies the dynamics. Since the state space is essentially discrete, the generator describes all possible potential changes that can happen in the system and the rate at which these changes occur. The rate is the intensity parameter of the Poisson process whose events trigger the change. The rate of a certain change σ will in general depend on the current state $\omega = \eta(\cdot)$ and a full specification of the dynamics is achieved by the family of rate functions $\lambda(\sigma, \omega)$. If $f : \Omega \to R$ is a function on Ω the infinitesimal generator is given by,

$$(\mathcal{A}f)(\omega) = \sum_{\sigma} \lambda(\sigma, \omega)[f(\sigma\omega) - f(\omega)]$$

where $\sigma\omega$ is the new state after the change. While the list of possible potential changes is usually infinite, if f is a local function, all but a finite number will leave the value of $f(\sigma\omega)$ unchanged. But still, actually constructing the dynamics from the infinitesimal generator requires doing some analysis, not unlike integrating systems with infinite degrees of freedom. Liggett [1] is good source to see how this is done. One could replace Z^d by its periodic version Z_N^d where the coordinates are considered modulo N. Then the new state space Ω_N is a finite set and the analytical problem of constructing the dynamics goes away. The randomness in the dynamics comes from the random occurrences of the underlying Poisson times when the changes occur and the interaction inherent in making the rates depend on

ω adds a layer of complication making the order in which events occur important.

Let us suppose for simplicity that there are two types of particles and the changes involve one of the particles migrating to a neighboring site. Since the total number of particles of each type in the system is conserved, we expect that in Ω_N the equilibria will consist of a family μ_{n_1,n_2} of probability distributions on Ω_N corresponing to specified numbers of particles n_1, n_2 of each type. If the dynamics is translation invariant, then the equilibria will be too. We now let $N \to \infty$ in such a way that

$$\lim_{N\to\infty} \frac{n_i}{N^d} = \rho_i$$

exists, and expect (in the absence of any phase transitions) limits μ_{ρ_1,ρ_2} on Ω which should be a family of equilibria for the dynamics on Ω.

If we start from an arbitrary initial state in Ω_N with n_1 and n_2 particles of each type, it may take a long time before the statistical equlibrium μ_{n_1,n_2} is reached. Translation invariance requires the particles to be more or less uniformly distributed and this may require roughly $O(N^{d+1})$ migrations. The most optimistic estimate of the time needed for this is $O(N)$. If the particles wander aimlessly or diffuse we may have to wait for a much longer time of $O(N^2)$. Depending therefore on the situation, in timescale Nt or N^2t, some thing interesting may happen. The system may reach a local equlibrium, and if we rescale Z_N^d by $y = \frac{x}{N}$, near $y \in T^d$, and at time $N^\alpha t$ (with $\alpha = 1$ or 2) as the case may be, the local equlibrium may be specified by $\rho_1(t, y), \rho_2(t, y)$ and in the microscale of lattice length the joint statistics of $\{\eta(x, N^\alpha t)\}$ when $\frac{x}{N}$ is near y is presumably close to the ones specified by the stationary process $\mu_{\rho_1(t,y),\rho_2(t,y)}$ on Ω.

This raises the related issue of how to determine $\rho(t, y)$ as functions of t and y. Often one can establish a partial differential equation (a system in our case), that governs the evolution of ρ_1 and ρ_2. The equation is the analog of the Euler equation. If $\alpha = 1$ it usually a nonlinear hyperbolic system and if $\alpha = 2$ it is parabolic. The initial conditions of the microscopic evolution determine the initial conditions of the partial differential equation (macroscopic dynamics).

The primary goal is to determine the appropriate differential equation and then establish a rigorous connection between microscopic and macroscopic dynamics. Because of the stochastic nature of the problem, there are questions of fluctuation theory as well as large deviations that are interesting but often prove difficult.

We shall give a brief description of what is known and what is not for some specific models.

The simplest model is the asymmetric simple exclusion model on Z where the possible states of the system consist of having either one particle or no particle (of a single type) at each site. It can be described by either specifying the set A of occupied sites or the function $\eta(x)$ which equals 1 if there is a particle at x and 0 otherwise. The possible changes are the movement of a single particle from an occupied site x to an empty site x'. There is a limit r such that this can happen only if $|x - x'| \le r$. The rate at which this change occurs is $p(x' - x)$. We normalize so that $\sum_{|z|\le r} p(z) = 1$ and of course $p(z) = 0$ if $|z| > r$. It is convenient to introduce the notation

$$\eta^{x,x'}(z) = \begin{cases} \eta(x) & \text{if } z = x' \\ \eta(x') & \text{if } z = x \\ \eta(z) & \text{if } z \neq x, x' \end{cases}$$

so that the infinitesimal generator can be written as

$$(\mathcal{A}f)(\eta) = \sum_{x,x'} p(x'-x)\eta(x)(1-\eta(x'))[f(\eta^{x,x'}) - f(\eta)]$$

The quantity

$$m = \sum_z z p(z)$$

plays an important role. If we assume that $m \neq 0$, say $m > 0$ to be specific, the limiting macroscopic equation is

$$\frac{\partial \rho}{\partial t} + m\frac{\partial[\rho(1-\rho)]}{\partial y} = 0 \tag{1}$$

If we start with the physical space of Z and the configuration space of Ω, equation (1) is to be considered on $[0,\infty) \times R$. If we start from Z_N and Ω_N then we need to view the equation (1) with periodic bounadry conditions $\rho(t,0) = \rho(t,1)$ on $[0,\infty) \times [0,1]$ or on $[0,\infty) \times T^1$ where T^1 is the 1-torus. The initial condition $\rho(0,\cdot) = \rho_0(\cdot)$ for equation (1), is determined by the initial configuration of the microscopic evolution $\eta_0(\cdot)$, which may or may not be random. In any case, it is natural to assume that if the probability distribution of $\eta_0(\cdot)$ is specfied by $\mu_N(0)$ on Ω or Ω_N it is consistent with $\rho_0(\cdot)$ in the sense that for any test function $J(\cdot)$, with compact support on R or on T^1, as the case may be

$$\lim_{N\to\infty} \frac{1}{N}\sum J(\frac{x}{N})\eta(x) = \int J(y)\rho_0(y)dy \tag{2}$$

in probabilty with respect to $\mu_N(0)$. The stochastic dynamics gives us a distribution $\mu_N(Nt)$ at microscopic time Nt or macroscopic time t. Equation (1) gives us (hopefully!) a unique solution $\rho(t,\cdot)$ at the macroscopic time t. The first step in any theory is to establish a rigorous connection between the two. The least one should expect is that for suitable test functions

$$\lim_{N\to\infty} \frac{1}{N}\sum J(\frac{x}{N})\eta(x) = \int J(y)\rho(t,y)dy \tag{3}$$

in probabilty with respect to $\mu_N(Nt)$. One could formulate stronger versions of this relationship.

First a few words about how equation (1) is formally arrived at. If we take the functional

$$f_N(\eta) = \frac{1}{N}\sum J(\frac{x}{N})\eta(x)$$

and compute $N(\mathcal{A}f_N)(\eta)$, (the prefactor N is to account for the speedup to macro-time scale) one gets asymptotically

$$N(\mathcal{A}f_N)(\eta) \simeq \frac{1}{N}\sum_{x,x'} J'(\frac{x}{N})\eta(x)(1-\eta(x')(x'-x)p(x'-x)$$

We now remark that the one parameter family of equilibria on Ω relevent for this problem are the Bernoulli product measures $\{\mu_\rho\}$ with $\mu_\rho[\eta(x)=1]=\rho$. If we now assume that we are nearly in a local equlibrium and the relation (3) is valid, then it is easily seen that

$$N(\mathcal{A}f_N)(\eta) \simeq m\int J'(y)\rho(t,y)(1-\rho(t,y))dy.$$

It is not difficult to go from here to equation (1). By Markov process theory we have

$$f(x(t)) - f(x(0)) = \int_0^t (\mathcal{A}f)(x(s))ds + M(t)$$

where the noise term $M(t)$ (a martingale) is easily seen, in our context, to become negligible for large N, by an explicit estimation of its mean square.

The equation (1) however is not so nice. Even with smooth initial data $\rho_0(\cdot)$ regular solutions exist only upto some time. One is therefore forced into the consideration of weak solutions. While weak solutions exist, uniqueness is not valid in the class of all weak solutions. There is however a natural notion of entropy solutions. Within the class of entropy solutions there is existence and uniqueness for the initial value problem. Moreover smooth solutions are always entropy solutions. One expects equation (3) to be valid i.e. the entropy solution $\rho(t,y)$ to be the correct choice.

This has been proved in a wide variety of situations, but not in all cases. A natural initial distribution $\mu_N(0)$, is the product measure with

$$\mu_N(0)[\eta(x)=1] = N\int_{\frac{2x-1}{2N}}^{\frac{2x+1}{2N}} \rho_0(y)dy$$

This will automatically satisfy (2). With this initial condition it is known by Rezakhanlou [15] that (3) holds for $\mu_N(Nt)$, with $\rho(t,y)$ being the entropy solution with initial condition $\rho(\cdot)$. A stronger version stating that the specific relative entropy

$$H_N(t) = \frac{1}{N}h(\mu_N(Nt)|\lambda_N(t))$$

satisfies

$$\lim_{N\to\infty}\sup_{0\le t\le T} H_N(t) = 0$$

can be proved using the methods of Yau [23], provided the solution $\rho(t,y)$ is smooth on $[0,T]\times T^1$ and satisfies $a \le \rho(t,y) \le 1-a$ for some $a>0$. Here $\lambda_N(t)$ is constructed from $\rho(t,\cdot)$ the same way as $\mu_N(0)$ was constructed from $\rho_0(\cdot)$. That is to say, it is a product measure with

$$\lambda_N(t)[\eta(x)=1] = N\int_{\frac{2x-1}{2N}}^{\frac{2x+1}{2N}} \rho(t,y)dy$$

In the special case of totally asymmetric simple exclusion, that is when $p(1) = 1$ and $p(z) = 0$ for $z \neq 1$, a stronger result is valid. According to Seppalainen [17], for any initial condition satisfying (2), (3) always holds for the entropy solution $\rho(t,y)$.

The problem of large deviations are currently being worked out by Jensen [9] for the special case of totally asymmetric simple exclusion process. If we start from an arbitrary determinstic initial condition that corresponds to the macroscopic profile in the sense of (2), we can ask how small the probabilities of large deviation from the entropy solution will be. We would like them to be no smaller than e^{-CN} for some finite C. It turns out that at this level the only possible deviations are weak solutions $\rho(t,y)$ of equation (1) that do not necessarily satisfy the entropy condition. The precise exponential constant C depends on the deviation $\rho(t,y)$ and measures the amount by which the entropy condition is violated. More precisely, with the choice of the convex entropy $h(\rho) = \rho\log\rho + (1-\rho)\log(1-\rho)$ and the corresponding flux $f(\rho)$ determined by $f'(\rho) = h'(\rho)(1-2\rho)$, smooth solutions satisfy

$$\frac{d}{dt}h(\rho(t,y) + \frac{d}{dx}f(\rho(t,y) = 0 \tag{4}$$

The entropy condition is that as a distribution

$$\psi = \frac{d}{dt}h(\rho(t,y) + \frac{d}{dx}f(\rho(t,y) \leq 0 \tag{5}$$

The rate function for large deviation turns out to be finite, only when the distribution ψ in (5) is a signed measure of bounded variation, in which case it is equal to the total mass of the positive part in the Jordan decomposition of ψ.

The entropy condition, roughly speaking, says that for $t > 0$ if there is a simple discontinuity in the solution $\rho(t,y)$ at some point $y = y_0$, then $\rho(t, y_0 - 0) < \rho(t, y_0 + 0)$, i.e. a shock must necessarily have higher density in front and a lower density behind. An interesting result concerning the corresponding microscopic system is that while there is some noise, presumably of order $\sqrt{N}$, on the location of the shock in the microscopic scale, its extent is microscopically of order $O(1)$. The exact statistical description of the shock profile has been calculated in Derrida-Lebowitz -Speer [4] for the special case of nearest neighbour jumps. An important but easy consequence of the entropy condition is that the total specific entropy

$$H(t) = \int h(\rho(t,y))dy$$

which is conserved for smooth solutions and is always nonincreasing. More over its decrease is accounted for by the shocks with each shock making its own contribution to the reduction rate of the total entropy. (Note that the physical entropy is $-H(t)$). In addition the shock profile creates an entropy loss at the microscopic level that matches the macroscopic loss due to the shock. A nonphysical shock, i.e. one with $\rho(t, y_0 - 0) > \rho(t, y_0 + 0)$ pumps in entropy and in order to produce it, the corresponding amount of entropy has to be put into the microscopic particle system. This explains the rate function for large deviations.

We now move on to the case when $m = \sum_z zp(z) = 0$. This requires diffusive scaling and the macroscopic time scale is N^2t. The limiting equation will be a nonlinear parabolic equataion of the form

$$\frac{\partial \rho}{\partial t} = \frac{d}{dx}[D(\rho)\frac{d\rho}{dx}] \tag{6}$$

and part of the problem is the determination of the function $D(\rho)$ in terms of $p(\cdot)$. If $p(\cdot)$ is symmetric, thanks to some cancellation, it is a very easy computation to show that in this case (6) is valid, with the choice of

$$D(\rho) \equiv D = \frac{1}{2}\sum_z z^2 p(z)$$

The effect of the interaction due to exclusion is not felt at all. The reason is that because of symmetry the space of linear functions $\sum_x a(x)\eta(x)$ is left invariant by the generator and therefore the evolution. This makes possible a fair amount of explicit calculation. Equations for densities that normally should not close do close giving us the linear heat equation. By using the symmetry of $p(\cdot)$ we see that

$$\begin{aligned} N^2 \mathcal{A}\, \frac{1}{N}\sum J(\frac{x}{N})\eta(x) &\\ = \frac{N^2}{N}\sum_{x,x'} p(x'-x)\eta(x)(1-\eta(x'))[J(\frac{x'}{N}) - J(\frac{x}{N})] &\\ = \frac{N^2}{2N}\sum_{x,x'} p(x'-x)[\eta(x)-\eta(x')][J(\frac{x'}{N}) - J(\frac{x}{N})] &\\ \simeq \frac{1}{2N}\sum_{z,x} z^2 p(z)\eta(x)J''(\frac{x}{N}) & \qquad (7) \end{aligned}$$

In this model, because of diffusive scaling, although there is determinstic behaviour of the density profile, each individual particle exhibits random diffusive behavior. If we tagged a particle and followed its motion in equilibrium (the case of $p(1) = p(-1) = 0$ has to be excluded on Z because the particles are all locked in their relative positions.) it was shown in Kipnis-Varadhan [11] that there is diffusive behavior with a diffusive constant $S(\rho)$ that depends on the equlibrium density ρ. If we start from an arbitrary initial configuration that produces a density profile of $\rho(t,y)$ which is a solution of the heat equation (6) with $D(\rho) \equiv D$, one can expect that the tagged particle has now an inhomogeneous diffusive behavior with a Kolmogorov backward generator of the form

$$\mathcal{L}_t = \frac{1}{2}a(t,y)\frac{d^2}{dy^2} + b(t,y)\frac{d}{dy} \tag{8}$$

One can expect that $a(t,y) = S(\rho(t,y))$. Self consistency reuires that $\rho(\cdot,\cdot)$ should also be a solution of the Fokker-Planck equation corresponding to 8 and this can be used to determine $b(\cdot,\cdot)$. While there is no general proof of this for every particle, it is result of Rezakhanlou [16] that this is indeed true for most of the particles.

If we take the periodic case with a total number of $\bar{\rho}N$ particles and look at the empirical process

$$R_N = \frac{1}{N}\sum_i \delta_{y_i(N^2\cdot)}$$

as random measures on the space of trajectories on $D[[0,T],T]$, the law of large numbers asserts the convergence in probability of R_N to the measure Q corresponding to the process with generator $\mathcal{L}_t$ and initial distribution $\rho_0(\cdot)$.

Large deviations in this context have been worked out in Quastel- Rezakhanlou-Varadhan [13] (some of the results require technical assumptions on the regularity of the function $S(\rho)$ that have so far been established only when $d \geq 3$).

Generalizations to other lattice gas models that are reversible, relative to Gibbs measures with interactions that satisfy mixing conditions have been considered in Varadhan-Yau [21]. While they still lead, under diffusive scaling, to densities evolving according equations of the form (6), the dfficulties encountered in establishing such results are considerable. In the calculations carried out in (7) we were fortunate on two accounts. First, we were able to do summation by parts twice and get rid of two factors of N. Secondly the object we ended up with was again linear so the equations closed. There are models (called gradient types) that allow the first step. The paradigm about local equilibria allow us to replace the nonlinear terms by their expectations leading to the more general equation of the type given by (6) with a nonlinear $D(\cdot)$. This is not so bad. Entropy methods of Guo-Papanicolaou-Varadhan [7], as well as relative entropy methods of Yau [23] can be applied. However for most models the first reduction is not possible. The conservation property allows us to perform summation by parts only once, leading to

$$\begin{aligned} N^2 \mathcal{A}\, &\frac{1}{N}\sum J(\frac{x}{N})\eta(x) \\ &= \frac{N^2}{N}\sum_{x,x'} p(x'-x)W_{x,x',\eta}[J(\frac{x'}{N}) - J(\frac{x}{N})] \\ &\simeq \frac{N}{N}\sum_x J''(\frac{x}{N})W_x(\eta) \end{aligned}$$

$W_x(\cdot)$ is the current density and has mean 0 under every equilibrium. Earlier W was itself a gradient that made another summation by parts possible. Now one has to replace $W_x(\eta)$ by $D(\eta(x) - \eta(x+1))$ not just at the level of expectations but rather at the level of fluctuations. The constant $D(\rho)$ will in general depend on ρ. The calculation of $D(\rho)$ which is essentially the Green-Kubo diffusion invloves the projection (in a rather fancy Hilbert Space) of the current W into the one dimensional space of density fluctuations, very much in the spirit of Boltzmann-Gibbs principle and is carried out in every global equilibrium μ_ρ. One uses precise estimates Yau [24] on the spectral gap in finite volumes, that inevitably get worse as the volume increases due to the presence of mutiple equilibria. A fair amount of the machinery of large deviations has to be used in order to make the results proved in equlibrium to be applicable when the process is far away from any global equilibrium. See Varadhan [19], Quastel [12] or Kipnis-Landim-Olla [10] for examples of this approach. One has to develop so much machinery just to establish the basic evolution equation of the densities, that proving results on large devistions in these situations should not be much harder.

Fluctuations in equlibrium are also accessible at this point. An Ornstein-Uhlenbeck limit under appropriate scaling can be established. In the non gradient case, since this involves the fluctuation-dissipation relation, the Green-Kubo diffusion $D(\rho)$ will appear. For instance, see Quastel [7], or Spohn [18] for examples.

Fluctuation results in nonequlibrium appear to be very inaccessible at this time. There is only one example, in Chang-Yau [3] of a one-dimensional model. There are conceptual difficulties because the normalization $N^{\frac{d}{2}}$ used to scale the fluctuations in d- dimensions gets stronger with increasing dimension.

One has more or less the same situation when we consider the simple exclusion model that we started out with, but with $p(\cdot)$ being asymmetirc and with mean 0. This is a nongradient model and can be analysed in a similar manner as was done by Xu [22]. There is an additional complication due to the nonreversible nature of the dynamics, even in equlibrium, making the projection necessary to compute the Green-Kubo coefficient non-orthogonal.

If we go from lattice models to continuum models the situation is far less satisfactory. For the interacting Brownian motions model, the basic density behavior is known in $d = 1$ where there no phase-transitions (see Varadhan [20]) and the relative entropy methods of Yau [23] are applicable in higher dimensions away from phase tansitions. For a very special interaction, the tagged particle motion in nonequlibrium has been studied in Grigorescu [8]. This interaction interpolates between the case hard rods of zero size (reflecting) and the free noninteracting case. A fair amount of explicit calculation is done, including a closed form formula for $S(\rho)$. The equlibrium fluctuation theory exists below phase trnasitions Spohn [18].

Finally let us return to the totally asymmetric simple exclusion model of $p(1) = 1$ that we considered earlier. If we start with an initial density $\rho_0(y) = \frac{1}{2}+\frac{u_0(y)}{N}$ that is nearly constant and speed up time by an additonal factor of N then formally the solution takes the form $\rho(t,y) = \frac{1}{2} + \frac{u(t,y)}{N}$ with $u(t,y)$ satisfying the equation

$$\frac{\partial u}{\partial t} = u\frac{\partial u}{\partial y}$$

Clearly at this stage the dissipation term should reemerge and the correct equation could be

$$\frac{\partial u}{\partial t} = u\frac{\partial u}{\partial y} + \frac{D}{2}\frac{\partial^2 u}{\partial y^2}$$

While this is suspect in $d = 1$ the analogous results for $d \geq 3$ have been established in Esposito-Marra-Yau [5]. Models having different types of particles with preferences to go in different directions and interactions that change types by collisions while conserving momenta, have been studied in Esposito-Marra-Yau [6] and Quastel-Yau [14], leading to incompressible Navier-Stokes equations.

References

1. T.M. Liggett, *Interacting Particle Systems*, Springer, 1985.
2. C.C. Chang, *Equilibrium fluctuations of gradient reversible particle systems* Probab. Theory Related Fields,**100** (1994), 269–283.
3. C.C. Chang, and H.T. Yau, *Fluctuations of one-dimensional Ginzburg-Landau models in nonequilibrium*, Comm. Math. Phys. **145** (1992), 209–234.
4. B. Derrida, J.L Lebowitz and E.R. Speer, *Shock Profiles for the asymmetric simple exclusion process in one dimension*, J. Statis. Phys. **89** (1997), 135-167.
5. R. Esposito, R. Marra and H.T.Yau, *Diffusive limit of asymmetric simple exclusion*, Rev. Math. Phys.**6** (1994), 1233-1267.
6. R. Esposito, R. Marra and H.T. Yau, *Navier-Stokes equations for stochastic particle systems on the lattice*, Comm. Math. Phys, **182** (1996), 395–456.
7. M.Z. Guo, G.C. Papanicolaou and S.R.S Varadhan, *Nonlinear diffusion limit for a system with nearest neighbor interactions*, Comm. Math. Phys. **118** (1988), 31–59.
8. I. Grigorescu, *Self-Diffusion for Brownian Motions with Local Interaction* (to appear in Ann. Prob.)
9. L. Jensen, *Large Deviations for the asymmetric simple exclusion process.* (Ph.D thesis under preparation).
10. C.Kipnis, C.Landim and S.Olla, *Hydrodynamic limit for a non-gradient system : The generalized symmetric symmetric exclusion process*, Comm. Pure Appl. Math. **47** (1994), 1475-1545.
11. C. Kipnis, S.R.S. Varadhan, *Central limit theorem for additive functionals of reversible Markov processes and applications to simple exclusions*, Comm. Math. Phys. **104** (1986), 1-19.
12. J. Quastel, *Diffusion of color in simple exclusion process*, Comm. Pure Appl. Math. **45** (1992), 623-679
13. J. Quastel, F.Rezakhanlou, and S.R.S. Varadhan, *Large Deviations for the symmetric simple exclusion process in dimensions $d \geq 3$.*, Prob. Th. Rel. Fields **113** (1999), 1-84
14. J. Quastel, J and H.T. Yau, *Lattice gases, large deviations, and the incompressible Navier-Stokes equations*, Ann. of Math. **148**(1998), 51–108.
15. F. Rezakhanlou, *Hydrodynamic limit for attractive particle systems on Z^d*, Comm. Math. Phys. **140** (1991), 417-448.
16. F. Rezakhanlou, *Propagation of chaos for symmetric simple exclusion*, Comm. Pure Appl. Math. **47** (1994), 943-957
17. T. Seppäläinen, *Existence of hydrodynamics for the totally asymmetric K-exclusion process*, Ann. Prob. **27** (1999), 361-415
18. H. Spohn, *Equilibrium fluctuations for interacting Brownian particles*, Comm. Math. Phys. **103** (1986), 1–33.
19. S.R.S. Varadhan, *Nonlinear diffusion limit for a system with nearest neighbor interactions, II*, Asymptotic Problems in Probability Theory: Stochastic Models and Diffusions on Fractals, K. D. Elworthy and N. Ikeda (Editors), Pitman Research Notes in Mathematics Series 283, (1991), 75–130.
20. S.R.S. Varadhan, *Scaling limits for interacting diffusions*, Comm. Math. Phys.

135 (1991), 313–353.

21. S.R.S. Varadhan and H.T. Yau, *Diffusive limit of lattice gas with mixing conditions*, Asian J. Math. **1** (1997),
22. L. Xu, *Hydrodynamics for asymmetric mean zero simple exclusion*, (1993) Ph.D. Thesis, New York University.
23. H.T. Yau, *Relative entropy and the hydrodynamics of Ginzburg–Landau models*, Lett. Math. Phys. **22** (1991), 63–80.
24. H.T. Yau, *Logarithmic Sobolev inequality for lattice gases with mixing conditions*, Comm. Math. Phys. **181** (1996), 367–408.

Fields Institute Communications
Volume **27**, 2000

Lectures on Hydrodynamic Scaling

S.R.S. Varadhan
Courant Institute of Mathematical Sciences
New York University
251 Mercer Street
New York, NY 10012
varadhan@cims.nyu.edu

Abstract. Examples of hydrodynamic scaling, illustrating different methods including the relative entropy method are given. The use of the Dirichlet form as well as techniques from the theory of large deviations are demonstrated. Gradient and nongradient examples are considered.

1 From classical mechanics to Euler equations

The basic example of hydrodynamical scaling is naturally hydrodynamics itself. Let us start with a collection of $N \simeq \bar{\rho}\ell^3$ classical particles in a large periodic cube Λ_ℓ of side ℓ in R^3. The motion of the particles are governed by the equations of motion of a classical Hamiltonian dynamical system with energy given by

$$H(p\,,q) = \frac{1}{2}\sum_{i=1}^{N} \|p_i\|^2 + \frac{1}{2}\sum_{i\neq j} V(q_i - q_j)$$

Here, $q_i \in \Lambda_\ell$ is the position of the i-th particle and $p_i \in R^3$ is its velocity. The coordinates $k = 1, 2, 3$ refer to the three components of position or velocity. The repulsive potential $V \geq 0$ is an even function and has compact support in R^d. The interaction in particular is short range. The equations of motion are

$$\begin{aligned} \frac{dq_i^k}{dt} &= \frac{\partial H(p\,,q)}{\partial p_i^k} = p_i^k \\ \frac{dp_i^k}{dt} &= -\frac{\partial H(p\,,q)}{\partial q_i^k} = -\sum_{j=1}^{N} V_k(q_i - q_j) \end{aligned} \tag{1.1}$$

where

$$V_k(q) = \frac{\partial V(q)}{\partial q^k}\,; \quad k = 1, 2, 3$$

are the three components of the gradient of V. The dynamical system has five conserved quantities. The total number N of particles, the total momenta $\sum_{i=1}^{N} p_i^k$

2000 *Mathematics Subject Classification.* Primary 60K35; Secondary 82C22.

for $k = 1, 2, 3$ and the total energy $H(p, q)$. The hydrodynamic scaling in this context consists of rescaling space and time by a factor of ℓ. The rescaled space is the unit torus $\mathbf{T}^3$ in 3-dimensions. The macroscopic quantities to be studied correspond to the five conserved quantities. The first one of these is the density, and is measured by a function $\rho(t, x)$ of t and x. For each $\ell < \infty$ it is approximated by $\rho_\ell(t, x)$, defined by

$$\int_{\mathbf{T}^d} J(x)\rho_\ell(t,x)dx = \frac{1}{\ell^3}\sum_{i=1}^{N} J(\frac{q_i(\ell t)}{\ell}) \tag{1.2}$$

A straight forward differentiation with respect to t yields

$$\begin{aligned} \frac{d}{dt}\int_{\mathbf{T}^d} J(x)\rho_\ell(t,x)dx &= \frac{d}{dt}\frac{1}{\ell^3}\sum_{i=1}^{N} J(\frac{q_i(\ell t)}{\ell}) \\ &= \frac{1}{\ell^3}\sum_{i=1}^{N}(\nabla J)(\frac{q_i(\ell t)}{\ell})\cdot p_i(\ell t) \\ &\simeq \int_{\mathbf{T}^d}(\nabla J)(x)\cdot \rho_\ell(t,x)u_\ell(t,x)dx \end{aligned} \tag{1.3}$$

where $u_\ell(t,x) = u_\ell^k(t,x)$, $k = 1, 2, 3$ are the components of the 'average' velocity of the fluid at the rescaled space time point x, t. This introduces three other macroscopic variables, which represent the three coordinates of the momenta that are conserved.

We can now write down the first of our five equations

$$\frac{\partial \rho}{\partial t} + \nabla\cdot(\rho u) = 0 \tag{1.4}$$

To derive the next three equations, using a test functions J, we differentiate for $k = 1, 2, 3$

$$\begin{aligned} \frac{d}{dt}\frac{1}{\ell^3}\sum_{i=1}^{N} J(\frac{q_i(\ell t)}{\ell})p_i^k(\ell t) &= \frac{1}{\ell^3}\sum_{i=1}^{N} p_i^k(\ell t)(\nabla J)(\frac{q_i(\ell t)}{\ell})\cdot p_i(\ell t) \\ &\quad - \frac{1}{\ell^2}\sum_{i=1}^{N}\sum_{j=1}^{N} J(\frac{q_i(\ell t)}{\ell})V_k(q_i(\ell t) - q_j(\ell t)) \end{aligned} \tag{1.5}$$

If we now use the skew-symmetry of $V_k = \frac{\partial V}{\partial q_k}$, we can rewrite the second term of the right hand side of (1.5) as

$$\begin{aligned} &-\frac{1}{2\ell^2}\sum_{i=1}^{N}\sum_{j=1}^{N}(J(\frac{q_i(\ell t)}{\ell}) - J(\frac{q_j(\ell t)}{\ell}))V_k(q_i(\ell t) - q_j(\ell t)) \\ &\simeq -\frac{1}{2\ell^3}\sum_{i=1}^{N}\sum_{j=1}^{N} J_r(\frac{q_i(\ell t)}{\ell})(q_i^r(\ell t) - q_j^r(\ell t))V_k(q_i(\ell t) - q_j(\ell t)) \\ &= \frac{1}{\ell^3}\sum_{i=1}^{N}\sum_{j=1}^{N} J_r(\frac{q_i(\ell t)}{\ell})\psi_k^r(q_i(\ell t) - q_j(\ell t)) \end{aligned} \tag{1.6}$$

with

$$\psi_k^r(q) = -\frac{1}{2}q^r V_k(q)$$

The next step is rather mysterious and requires considerable explanation. The quantities

$$\sum_{i=1}^{N} p_i^k p_i^r, \quad \sum_{i,j=1}^{N} \psi_k^r(q_i(t) - q_j(t))$$

are not conserved. They depend on combinations of individual velocities that are not conserved and on spacings between particles both of which change in the microscopic time scale and therefore do so rapidly in the macroscopic time scale. They can therefore be replaced by their space-time averages. By appealing to an 'Ergodic Theorem' they can be replaced by their averages with repect to their equilibrium distributions. The equilibrium 'ensemble' consists of an infinite collection of points $\{p_\alpha, q_\alpha\}$, in the phase space $R^3 \times R^3$. There is a natural five parameter family of measures $\mu_{\rho,u,T}$ that are invariant under spatial translations as well as the Hamiltonian dynamics. The points $\{p_\alpha\}$ are distributed according to a Gibbs Distribution with density ρ and formal interaction energy

$$\frac{1}{2T}\sum_{\alpha,\beta} V(q_\alpha - q_\beta)$$

In other words $\{q_\alpha\}$ is a point process obtained by taking infinite volume limit of $N = \ell^3\rho$ particles distributed in the cube of side ℓ in R^3 according to the joint density

$$\frac{1}{Z}\exp\Big[-\frac{1}{2T}\sum_{1\le i\ne j\le N} V(q_i - q_j)\Big]$$

where Z is the normalization constant. The velocities $\{p_\alpha\}$ are distributed independently of each other as well as of $\{q_\alpha\}$, having a common 3-dimensional Gaussian distribution with mean u and covariance TI. Assuming that the infinite volume limit exists in a reasonable sense it will be a point process defined as an infinite volume Gibbs measure $\mu_{\rho,T}$. The velocities $\{p_\alpha\}$ will be an independent Gaussian ensemble $\nu_{u,T}$.

In the first term the quantities $p_i^k p_i^r$ are replaced by their expectations

$$u^k(t,x)u^r(t,x) + \delta_{k,r}T(t,x)$$

and in the second term $\psi_{k,r}$ are replaced by their expectations that involve the 'pressure' per unit volume in the Gibbs ensemble

$$\mathbf{P}_k^r(\rho,T) = \lim_{\ell\to\infty} E^{\mu_{\rho,T}}\Big\{\frac{1}{\ell^3}\sum_{\substack{|q_\alpha|\le\ell\\ |q_\beta|\le\ell}} \psi_k^r(q_\alpha - q_\beta)\Big\}$$

This leads to the equation

$$\begin{aligned}\frac{d}{dt}\int_{\mathbf{T}^d} J(x)u^k(t,x)dx &= \int_{\mathbf{T}^d}\sum_{r=1}^{3}\frac{\partial J}{\partial x_r}(x).(u^k(t,x)u^r(t,x) + \delta_{k,r}T(t,x))dx \\ &+ \int_{\mathbf{T}^d}\sum_{r=1}^{3}\frac{\partial J}{\partial x_r}(x)\mathbf{P}_k^r(\rho(t,x),T(t,x))dx\end{aligned} \tag{1.7}$$

We now integrate by parts, remove the test function J and obtain from (1.7)

$$\frac{d}{dt}(\rho u) + \nabla \cdot (\rho\, u \otimes u + \rho\, T I + \mathbf{P}(\rho\,, T)) = 0 \tag{1.8}$$

There is an equation of state that expresses the total energy per unit volume e as

$$e(\rho\,, u\,, T) = \frac{1}{2}\rho(|u|^2 + 3T) + f(\rho\,, T) \tag{1.9}$$

where $f(\rho\,, T)$, the potential energy per unit volume, is given by

$$f(\rho\,, T) = \lim_{\ell \to \infty} E^{\mu_{\rho,T}}\left\{\frac{1}{2\ell^3}\sum_{\substack{|q_\alpha| \le \ell \\ |q_\beta| \le \ell}} V(q_\alpha - q_\beta)\right\}$$

Although we will not derive it, there is a similar equation for $e(t, x)$ that is obtained by differentiating

$$\frac{d}{dt}\frac{1}{2\ell^3}\sum_{i=1}^{N} J(\frac{q_i(\ell t)}{\ell})\left[|p_i(\ell t)|^2 + \sum_{j=1}^{N} V(q_i(\ell t) - q_j(\ell t))\right]$$

and proceeding in a similar fashion. It looks like

$$\frac{de}{dt} + \nabla \cdot \left[(e + T)u + \mathbf{P}(\rho\,, T)u\right] = 0 \tag{1.10}$$

The five equations one for density given by (1.4), the three for velocities contained in (1.8) and finally the enrgy equation (1.10) constitute a first order system of non-linear hyperbolic conservation laws in the six variables $[\rho, u, T, e\,]$ with one relation between them given by (1.9). Given smooth initial data they have local solutions.

Rigorous derivation of these equations does not exist. The ergodic theory is definitely plausible, but hard to establish. The reason is that we have essentially an infinite system of ordinary differential equations representing a classical deterministic dynamical system and the ergodicity properties are nearly impossible to establish in any general context. However if we had some noise in the system, i.e. stochastic dynamics instead of deterministic dynamics, then we will be concerned with the ergodic theory of Markov Processes of some kind, which is far more accessible. This is essentially the focus of our lectures.

2 Some examples

The simplest example is a system of noninteracting particles undergoing independent motions. For example we could have on $\mathbf{T}^3$, $L \simeq \bar{\rho} N^3$ particles all behaving like independent Brownian Particles. If the initial configuration of the L particles is such that the empirical distribution

$$\nu_0(dx) = \frac{1}{N^3}\sum_i \delta_{x_i}$$

has a deterministic limit $\rho_0(x)dx$, then the empirical distribution

$$\nu_t(dx) = \frac{1}{N^3}\sum_i \delta_{x_i(t)}$$

of the configuration at time t, has a deterministic limit $\rho(t,x)dx$ as $N \to \infty$ and $\rho(t,x)$ can be obtained from $\rho_0(x)$ by solving the heat equation

$$\frac{\partial \rho}{\partial t} = \frac{1}{2}\Delta\rho$$

with the initial condition $\rho(0,x) = \rho_0(x)$. The proof is an elementary law of large numbers argument involving a calculation of two moments. Let $f(x)$ be a continuous function on $\mathbf{T}^3$ and let us calculate for

$$U = \frac{1}{N^3}\sum_i f(x_i(t)) \tag{2.1}$$

the first two moments given the initial configuration $(x_1, \cdots, x_L)$.

$$E(U) = \frac{1}{N^3}\sum \int_{\mathbf{T}^3} f(y)p(t,x_i,y)dy$$

and an elementary calculation reveals that the expectation converges to the constant:

$$\int_{\mathbf{T}^3}\int_{\mathbf{T}^3} f(y)p(t,x,y)\rho_0(x)dydx = \int_{\mathbf{T}^3} f(y)\rho(t,y)dy \tag{2.2}$$

Although the summamnds in (2.1) are not identically distributed, the independence clearly provides a uniform upper bound of the order CN^{-3} for the variance and goes to 0. Of course on $\mathbf{T}^3$ we could have had a process obtained by rescaling a random walk on a large torus of size N. Then the hydrodynamic scaling limit would be a consequence of the central limit theorem for a single particle under space time rescaling. Our result on the behavior of the empirical distribution of the particles will continue to be a consequence of the law of large numbers resulting from the averaging over L independently moving particles. The situation could be different if the particles interacted with each other.

The next class of examples are called simple exclusion processes. They make sense on any finite or countable set X and for us it will be either the integer lattice $\mathbf{Z}^d$ in d-dimensions or $\mathbf{Z}_N^d$ obtained from it as a quotient by considering each coordinate modulo N. At any given time a subset of these lattice sites will be occupied by particles, with atmost one particle at each site. In other words some sites are empty while others are occupied with one particle. The particles move randomly. Each particle waits for an exponential random time and then tries to jump from its current site x to a new site y. The new site y is picked randomly according to a probability distribution $\pi(x,y)$. In particular $\sum_y \pi(x,y) = 1$ for every x. Of course a jump to y is not always possible. If the site is empty the jump is possible and is carried out. If the site already has a particle, the jump cannot be carried out and the particle forgets about it and waits for another chance, i.e. waits for a new exponential waiting time.

If we normalize so that all waiting times have mean 1, the generator of the process can be written down as

$$(\mathcal{A}f)(\eta) = \sum_{x,y} \eta(x)(1-\eta(y))\pi(x,y)[f(\eta^{x,y}) - f(\eta)] \tag{2.3}$$

where η represents the configuration with $\eta(x) = 1$ if there is a particle at x and $\eta(x) = 0$ otherwise. For each configuration η and a pair of sites x, y the new configuration $\eta^{x,y}$ is defined by

$$\eta^{x,y}(z) = \begin{cases} \eta(y) & \text{if } z = x \\ \eta(x) & \text{if } z = y \\ \eta(z) & \text{if } z \neq x\,,y \end{cases} \tag{2.4}$$

We will be mainly concerned with the situation where the set X is $\mathbf{Z}^d$ or $\mathbf{Z}^d_N$, viewed naturally as an Abelian group with $\pi(x\,,y)$ being translation invariant and given by $\pi(x\,,y) = p(y-x)$ for some peobability distribution p. It is convenient to assume that p has finite support. There are various possibilities.

$$p \text{ is symmetric i.e. } p(x) = p(-x)$$

or more generally

$$p \text{ has mean zero i.e. } \sum_y yp(y) = 0$$

and finally

$$\sum_y y\,p(y) = m \neq 0$$

We shall first concentrate on the symmetric case. Let us look at the function

$$V_J(\eta) = \sum J(x)\eta(x)$$

and compute

$$\begin{aligned}(\mathcal{A}V_J)(\eta) &= \sum_{x,y} \eta(x)(1-\eta(y))p(y-x)(J(y)-J(x)) \\ &= \sum_{x,y} \eta(x)p(y-x)(J(y)-J(x)) \\ &= \sum_{x,y} \eta(x)[(\mathbf{P}-I)J](x) \\ &= V_{(\mathbf{P}-I)J}(\eta)\end{aligned} \tag{2.5}$$

The space of linear functionals is left invariant by the generator. It is not difficult to see that

$$E_\eta\big[V_J(\eta(t))\big] = V_{J(t)}(\eta)$$

where

$$J(t) = \exp[t(\mathbf{P}-I)]J$$

is the solution of

$$\frac{d}{dt}J(t\,,x) = (\mathbf{P}-I)J(t\,,x)$$

It is almost as if the interaction has no effect and in fact for the calculation of expectations of 'one particle' functions it clearly does not. Let us start with a configuration on $\mathbf{Z}^d_N$ and scale space by N and time by N^2. The generator becomes $N^2\mathcal{A}$ and the particles can be visualized as moving on a lattice imbedded in the unit torus $\mathbf{T}^d$, with a spacing of $\frac{1}{N}$, and becoming dense as $N \to \infty$.

Let us consider the functional

$$\xi(t) = \frac{1}{N^d}\sum_x J(\frac{x}{N})\eta_t(x)$$

and we can write

$$\xi(t) - \xi(0) = \int_0^t V_N(\eta(s))ds + M_N(t)$$

where

$$V_N(\eta) = (N^2 \mathcal{A} V_J)(\eta) = V_{J_N}(\eta)$$

with

$$(J_N)(x) = N^2 \sum \left[J(x + \frac{z}{N}) - J(x)\right] p(z)$$
$$\simeq \frac{1}{2}(\Delta_C J)(x)$$

Here Δ_C refers to the Laplacian

$$\sum_{i,j} C_{i,j} \frac{\partial^2}{\partial x_i \partial x_j}$$

with the covariance matrix C given by

$$C_{i,j} = \sum_x x_i x_j p(x)$$

$M_N(t)$ is a martingale and a very elementary calculation yields

$$E\left\{[M_N(t)]^2\right\} \leq C\, t\, N^{-d}$$

essentially completing the proof in this case. Technically the empirical distribution $\nu_N(t)$ is viewed as a measure on $\mathbf{T}^d$ and $\nu_N(\cdot)$ is viewed as a stochastic process with values in the space $\mathcal{M}(\mathbf{T}^d)$ of nonnegative measures on $\mathbf{T}^d$. In the limit it lives on the set of weak solutions of the heat equation

$$\frac{\partial \rho}{\partial t} = \frac{1}{2} \Delta_C \, \rho \tag{2.6}$$

and the uniqueness of such weak soultions for given initial density establishes the validity of the hydrodynamic limit. We could have have computed two moments as in the noninteracting case. The expectation would have been no diferent from the noninteracting case since it involves only one particle functions. The variance involves two particle functions and would have involved slighlty more work, because the independence is not there. The martingale argument however is more general.

Let us make the problem slightly more complicated by adding a small bias. Let $q(x)$ be an odd function with $q(-x) = -q(x)$ and we will modify the problem by making p depend on N in the form

$$p_N(x) = p(x) + \frac{1}{N} q(x) \tag{2.7}$$

Assuming that q is nonzero only when p is so, p_N will be an admissible transition probability for large enough N. A calculation yields that in the slightly modified model referred to as weakly asymmetric simple exclusion model V_N is given by

$$V_N(\eta) \simeq V_{J_N}(\eta) + \frac{1}{N^d} \sum \eta(x)(1 - \eta_x) < m \,, \nabla J(x) > \tag{2.8}$$

with

$$m = N \sum_x x \, p_N(x) = \sum_x x \, q(x)$$

If one thinks of $\rho(t,x)$ as the density of particles at the (macroscopic) time t and space x the first term on the right hand side of (2.8) clearly wants to have the limit

$$\int_{\mathbf{T}^d} \frac{1}{2}(\Delta_C J)(x)\rho(t,x)dx$$

It is not so clear what to do with the second term. The 'invariant' measures in this model are the Bernoulli measures with various densities ρ and upon 'averaging' the second term should converge to

$$\int_{\mathbf{T}^d} < m,(\nabla J)(x) > \rho(t,x)(1-\rho(t,x)dx$$

replacing the linear heat equation (2.6) by the nonlinear equation

$$\frac{\partial \rho}{\partial t} = \frac{1}{2}\Delta_C \rho - \nabla \cdot m\rho(1-\rho) \tag{2.9}$$

This requires justification and that will be the content of our next lecture.

Let us now turn to the case where p has mean zero but is not symmetric. In this case

$$V_N(\eta) = N^{2-d}\sum_{x,y}\eta(x)(1-\eta(y))p(y-x)[J(\frac{y}{N}) - J(\frac{x}{N})] \tag{2.10}$$

and we get stuck at this point. If p is symmetric, as we saw, we gain a factor of N^{-2}. Otherwise the gain is only a factor of N^{-1} which is not enough. We seem to end up with

$$\begin{aligned} N^{-d}\sum_x\sum_y \eta(x) &< \frac{1}{2}[(\nabla J)(\frac{x}{N}) + (\nabla J)(\frac{y}{N})], N(1-\eta(y))(y-x)p(y-x) > \\ &= \frac{1}{2N^d}\sum_x (\nabla J)(\frac{x}{N})N\Psi_x \end{aligned} \tag{2.11}$$

where

$$\begin{aligned} \Psi_0 &= \frac{1}{2}[\eta(0)\sum_z(1-\eta(z))z\,p(z) + (1-\eta(0))\sum_z \eta(-z)z\,p(z)] \\ &= \frac{1}{2}[-\eta(0)\sum_z \eta(z)z\,p(z) + (1-\eta(0))\sum_z \eta(-z)z\,p(z)] \\ &= \frac{1}{2}[\sum_z \eta(-z)z\,p(z) - \eta(0)\sum_z(\eta(z)+\eta(-z))z\,p(z)] \end{aligned} \tag{2.12}$$

The second sum is zero in the symmetric case and Ψ_0 can then be written as a 'gradient'

$$\Psi_0 = \sum_j \tau_{e_j}\xi_j - \xi_j$$

where τ_{e_j} are shifts in the coordinate directions. This allows us to do summation by parts and gain a factor of N^{-1}. When this is not the case, we have a 'nongradient' model and the hydrodynamic limit can no longer be established by simple averaging.

We will leave this somewhat difficult case till the end for lectures 5 and 6 but take up the third case

$$\sum_z zp(z) = m \neq 0$$

that needs hyperbolic scaling and leads to Burgers equation with zero viscosity in lecture 4.

3 Gradient models, Dirichlet forms and large deviations

The important ingredient in the analysis of gradient models is the ability to do averaging and replace quantities by their expected values calculated under various equilibrium distributions. Suppose $\mu_N$is a probability measure on the space Ω_N of configurations η on the periodic lattice $\mathbf{Z}_N^d$. We wish to think of μ_N as being a Bernoulli measure with some density ρ. The density ρ is not quite a constant, but a slowly varying function on $\mathbf{Z}_N^d$, and in fact a function of the macroscopic variable $\frac{x}{N}$. Let $g = g(\eta)$ be a local function depending on the configuration in some finite box around 0. By Av_ℓ we will denote the averaging process over the block(cube) $B_\ell = \{z : |z_j| \leq \ell;\ 1 \leq j \leq d\}$ of side $2\ell + 1$

$$(Av_\ell g)(\eta) = \frac{1}{(2\ell+1)^d} \sum_{z \in B_\ell} g(\tau_z \eta) \tag{3.1}$$

If $g(\eta) = \eta(0)$ this produces the empirical density $\bar{\eta}_{\ell,0}$. For any local g we can calculate its expected value with repect to the Bernoulli measure with density ρ and get

$$\hat{g}(\rho) = E^{P_\rho}\big[g(\eta)\big] \tag{3.2}$$

The validity of the averaging principle means that we can replace $(Av_\ell g)(\eta)$ by $\hat{g}(\bar{\eta}_{\ell,0})$. More generally if we denote by τ_x the translation by x on the lattice then we can replace $(Av_\ell g)(\tau_y \eta)$ by $\hat{g}(\bar{\eta}_{\ell,y})$ or more precisely showing that the error

$$\delta_{N,\ell} = E^{\mu_N}\left[\frac{1}{N^d} \sum_x |(Av_\ell g)(\tau_x \eta) - \hat{g}(\bar{\eta}_{\ell,x})|\right] \tag{3.3}$$

goes to zero as $N \to \infty$ so long as $\ell \to \infty$ and $\frac{\ell}{N} \to 0$. What property of the sequence μ_N will allow us to make such a conclusion?

Dirichlet Form. Given the density $\{f(\eta)\}$ with respect to the uniform distribution of a probability distribution μ on $\mathbf{Z}_N^d$, the Dirichlet form $\mathcal{D}_N(\mu)$ is the quantity

$$\mathcal{D}_N(\mu) = \frac{1}{2^{N^d}} \sum_\eta \sum_{\substack{x,y \in \mathbf{Z}_N^d \\ |x-y|=1}} [\sqrt{f(\eta^{x,y})} - \sqrt{f(\eta)}|^2 \tag{3.4}$$

Remark: One should think of the inner summation as being over all the nearest neighbor bonds. The outer summation with the factor $\frac{1}{2^{N^d}}$ is integration with respect to the uniform distribution on Ω_N. Since f is a nonnegative L_1 function, its square root will be naturally an L_2 function and we are really dealing with the Dirichlet form of $\sqrt{f}$.

Theorem 3.1 (Averaging principle) *If μ_N on Ω_N is a sequence of probability distributions such that*

$$\mathcal{D}_N(\mu_N) \le CN^{d-2} \tag{3.5}$$

for some constant C independent of N, then for every local g

$$\lim_{\epsilon\to 0} \limsup_{N\to\infty} \delta_{N,\epsilon N} = 0 \tag{3.6}$$

Proof The proof is carried out in two steps. First we show that

$$\lim_{\ell\to\infty} \limsup_{N\to\infty} \delta_{N,\ell} = 0 \tag{3.7}$$

and then

$$\lim_{\substack{\epsilon\to 0\\ \ell\to\infty}} E^{\mu_N}\left[\frac{1}{N^d}\sum_{x\in\mathbf{Z}_N^d} |\bar{\eta}_{\ell,x} - \bar{\eta}_{N\epsilon,x}|\right] = 0 \tag{3.8}$$

Being a polynomial in ρ, $\hat{g}(\rho)$ is a continuos function of ρ and (3.6) will follow from (3.7) and (3.8). □

Step 1: Since the details of the proof can be found in [KOV] we will only sketch it here. Suppose B_ℓ is a block of size $2\ell+1$ and $q_\ell(\eta)$ is an assignment of probabilities for the $2^{(2\ell+1)^d}$ possible configurations. We view the block non-periodically and we have $(2\ell)^d$ interior nearest neighbor bonds. The Dirichlet form is given by

$$\mathcal{D}_\ell(q_\ell) = \sum_{\substack{x,y\in B_\ell\\ |x-y|=1}} |\sqrt{q_\ell(\eta^{x,y})} - \sqrt{q_\ell(\eta)}|^2$$

Given a probability distribution μ_N on Ω_N, we denote its marginal on a block of size $2\ell+1$ centered at x by $\nu_{\ell,N,x}$ and by $\bar{\nu}_{\ell,N}$ we denote the average

$$\bar{\nu}_{\ell,N} = \frac{1}{N^d}\sum_{x\in\mathbf{Z}_N^d} \nu_{\ell,N,x}$$

over all the sites x. If we denote by $q_{\ell,N}(\eta)$ the individual probabilities the following inequality is a consequence of the convexity of $\mathcal{D}(\nu)$ in ν

$$\mathcal{D}_\ell(q_{\ell,N}) \le (\frac{2\ell}{N})^d \mathcal{D}_N(\mu_N) \le (\frac{2\ell}{N})^d CN^{d-2} = \frac{C(2\ell)^d}{N^2} \tag{3.9}$$

Let us denote the set of limit points of $q_{\ell,N}$ as $N\to\infty$ by $\mathcal{M}_\ell$. For any $\nu\in\mathcal{M}_\ell$ the associated probability function q satisfies $\mathcal{D}_\ell(q) = 0$. Therefore it is invariant under exchanges between neighboring sites, and consequently under all permutations. The conditional distribution under ν of occupied sites, given the total number of particles k, is therefore the uniform distribution over all subsets of cardinality k. If we denote this uniform distribution by $\lambda_{\ell,k}$, then every $\nu\in\mathcal{M}_\ell$ is of the form

$$\nu = \sum_k \lambda_{\ell,k}\pi_\nu(k) \tag{3.10}$$

where $\pi_\nu(k)$ is the probability that under ν the total number of particles in the block of size $(2\ell+1)^d$ is equal to k. Since $\lambda_{\ell,k}$ converges to the Bernoulli with density ρ, if $\frac{k}{(2\ell+1)^d} \to \rho$ it follows that

$$\begin{aligned}\lim_{\ell\to\infty} \sup_{\nu\in\mathcal{M}_\ell} & E^\nu[|Av_\ell\, g(\eta) - \hat{g}(Av_\ell\, \eta)|] \\ & \le \lim_{\ell\to\infty} \sup_{0\le k\le(2\ell+1)^d} E^{\lambda_{k,\ell}}[|Av_\ell\, g(\eta) - \hat{g}(\frac{k}{(2\ell+1)^d})|] \\ & = 0\end{aligned} \tag{3.11}$$

establishing (3.7).

Step 2: For step 2, we need only to establish that the local density $\bar{\eta}_{\ell,x}$ does not fluctuate over small macroscopic length scales. This will follow from the estimate

$$\limsup_{\substack{\ell\to\infty\\ \epsilon\to 0}} \limsup_{N\to\infty} \sup_{|y|\le N\epsilon} E^{\mu_N}\left[\frac{1}{N^d}\sum_{x\in\mathbf{Z}_N^d} |\bar{\eta}_{\ell,x} - \bar{\eta}_{\ell,x+y}|\right] = 0 \tag{3.12}$$

If we take two blocks of size $2\ell+1$ centered at x and $x+y$ there is no bond connecting x and $x+y$ which are far from each other. We can introduce a Dirichlet form for this bond.

$$\mathcal{D}_{x,x+y}(q) = \sum_\eta |\sqrt{q(\eta^{x,x+y})} - \sqrt{q(\eta)}|^2$$

and this can be estimated, by writing the difference as a telescopic sum over exchanges involving nearest neighbors and applying Schwartz's inequlaity, interms of Dirichlet forms for nearest neighbor bonds. If we denote by $\nu_{\ell,N,x,x+y}$ the joint marginal over two widely seperated blocks of size $2\ell+1$ centered at x and $x+y$, and by

$$\bar{\nu}_{\ell,N,y} = \frac{1}{N^d}\sum_{x\in\mathbf{Z}_N^d} \nu_{\ell,N,x,x+y}$$

the average over all rigid spatial translations of the pair of blocks

$$\mathcal{D}_{\ell,0,y}(\bar{\nu}_{\ell,N,y}) \le C|y|^2 N^{-2} \le C\epsilon^2$$

It follows, by pure thought, (once $\epsilon\to 0$, the two averages are essentially averages over two halves of a combined system of $2(2\ell+1)^d$ sites which is in equilibrium) that

$$\limsup_{\ell\to\infty} \limsup_{\epsilon\to 0} \limsup_{N\to\infty} \sup_{|y|\le N\epsilon} E^{\mu_N}\left[\frac{1}{N^d}\sum_{x\in\mathbf{Z}_N^d} |\bar{\eta}_{\ell,x} - \bar{\eta}_{\ell,x+y}|\right] = 0$$

Because $\bar{\eta}_{N\epsilon,x}$ is not very different from the average $Av_{|y|\le N\epsilon}\, \bar{\eta}_{\ell,x+y}$ (3.8) follows.

Superexponential Estimates: Let us denote by P_N the measure corresponding to the process on $D[[0,T];\Omega_N]$ starting from some initial distribution. We want to replace additive functionals of the form

$$\int_0^T \frac{1}{N^d}\sum_x J(\frac{x}{N}) g((\tau_x\eta)(s))ds$$

by

$$\int_0^T \frac{1}{N^d}\sum_x J(\frac{x}{N})\hat{g}(\bar{\eta}_{N\epsilon,x})(s))ds$$

and for the difference

$$F_N = \int_0^T \frac{1}{N^d}\sum_x J(\frac{x}{N})g((\tau_x\eta)(s))ds - \int_0^T \frac{1}{N^d}\sum_x J(\frac{x}{N})\hat{g}(\bar{\eta}_{N\epsilon,x})(s))ds$$

we obtain the following superexponential bound.

Theorem 3.2 *For any* $\delta > 0$

$$\limsup_{\epsilon\to 0}\limsup_{N\to\infty}\frac{1}{N^d}\log P_N[|F_N|\geq\delta] = -\infty$$

Proof It suffices to prove that for any $\lambda > 0$

$$\limsup_{\epsilon\to 0}\limsup_{N\to\infty}\frac{1}{N^d}\log E^{P_N}\Big[\exp\big[\lambda N^d F_N\big]\Big] = 0 \tag{3.13}$$

If we denote by $\bar{P}_N$ the process that is initialized to start from Bernoulli measure with density $\frac{1}{2}$ which is an invariant measure then the Radon-Nikodym derivative $\frac{dP_N}{d\bar{P}_N}$ is bounded by 2^{N^d}. Therefore

$$E^{P_N}\Big[\exp\big[\lambda N^d F_N\big]\Big] \leq 2^{N^d} E^{\bar{P}_N}\Big[\exp\big[\lambda N^d F_N\big]\Big] \tag{3.14}$$

If we can show that under equilibrium we have

$$\limsup_{\epsilon\to 0}\limsup_{N\to\infty}\frac{1}{N^d}\log E^{\bar{P}_N}\Big[\exp\big[\lambda N^d F_N\big]\Big] = 0 \tag{3.15}$$

then from (3.14)

$$\limsup_{\epsilon\to 0}\limsup_{N\to\infty}\frac{1}{N^d}\log E^{P_N}\Big[\exp\big[\lambda N^d F_N\big]\Big] \leq \log 2$$

By Hölder's inequality

$$\log E^{P_N}\Big[\exp\big[\lambda N^d F_N\big]\Big] \leq \frac{1}{a}\log E^{P_N}\Big[\exp\big[a\lambda N^d F_N\big]\Big] \leq \frac{\log 2}{a}$$

and since $a > 0$ can be made arbitrarily large we obtain (3.13)

By Feynamn-Kac (3.15) can be reduced to an estimation of an eigen value. More precisely we only need to show that for every $\lambda > 0$

$$\limsup_{\epsilon\to 0}\limsup_{N\to\infty}\frac{1}{N^d}\Big[\sup_\nu E^\nu\big[N^d\lambda F_N\big] - cN^2\mathcal{D}_N(\nu)\Big] = 0$$

or

$$\limsup_{\epsilon\to 0}\limsup_{N\to\infty}\Big[\sup_\nu E^\nu\big[\lambda F_N\big] - \frac{c}{N^{d-2}}\mathcal{D}_N(\nu)\Big] = 0 \tag{3.16}$$

The apperance of the constant c is because the actual Dirichlet form involves $p(\cdot)$ and if we assume irreducibility we can get a bound. The factor N^2 is due to the speed up of time. If F_N is bounded by A, we will only see ν with

$$\mathcal{D}_N(\nu) \leq \frac{\lambda A}{c} N^{d-2} = CN^{d-2}$$

(3.16) now follows from Theorem 3.1. This concludes the proof. □

Application 1: Weak asymmetry. If Q_N is the process with weak asymmetry then

$$R_N = \frac{dQ_N}{dP_N}$$

can be explicitly calculated by Girsanov type formula and

$$E^{P_N}\left[R_N^p\right] \leq \exp\left[CN^d\right]$$

Therefore Q_N inherits the super-exponential bounds. This establishes the hydrodynamic limit for weakly asymmetric perturbations.

In fact if we perturb $p(\cdot)$ by a skew symmetric $\frac{q}{N}$ with q of the form $q(t, \frac{x}{N}, \cdot)$ so that the generator looks like

$$N^2 \sum_{x,y} \eta(x)(1-\eta(y))p(y-x)[f(\eta^{x,y}) - f(\eta)]$$
$$+ N \sum_{x,y} \eta(x)(1-\eta(y))q(t, \frac{x}{N}, y-x)[f(\eta^{x,y}) - f(\eta)]$$

and the means are calculated as

$$b(t,\theta) = \sum_z z\, q(t,\theta,z),$$

then the hydrodynamic limit is

$$\frac{\partial \rho}{\partial t} = \frac{1}{2}\nabla \cdot C\nabla\rho - \nabla \cdot b\rho(1-\rho) \tag{3.17}$$

Application 2: Large deviations. The Entropy cost of the perturbation is calculated easily by the use of Girsanov's formula

$$H(Q_N, P_N) = E^{Q_N}\left[\int_0^T \sum_{x,y} \left[\eta_t(x)(1-\eta_t(y))c_N(t, \frac{x}{N}, y-x)\right] dt\right]$$

where c is the relative entropy of one Poisson distribution with respect to another

$$\begin{aligned} C_N(t,\theta,z) &= (N^2p(z) + Nq(t,\theta,z)) \log \frac{N^2p(z) + Nq(t,\theta,z)}{N^2p(z)} \\ &\quad - (N^2p(z) + Nq(t,\theta,z)) + N^2p(z) \\ &\simeq \frac{q^2(t,\theta,z)}{2p(z)} \end{aligned}$$

It is clearly in our interest to minimize

$$\frac{1}{2}\sum_z \frac{q^2(t,\theta,z)}{p(z)}$$

subject to

$$\sum_z z\, q(t,\theta,z) = b$$

and the minimum is seen to be

$$\frac{1}{2} < b\,, C^{-1}\, b >$$

so that the minimal entropy cost for the large deviation of the empirical density from the solution of the normal hydrodynamic limiting equation

$$\frac{\partial \rho}{\partial t} = \frac{1}{2}\nabla \cdot C\nabla \rho \tag{3.18}$$

to a ρ that is the solution of (3.17) is asymptotically

$$N^d \mathcal{E}(b) = \frac{1}{2} N^d \int_0^T \int_{\mathbf{T}^d} \rho(t\,,\theta)(1-\rho(t\,,\theta)) < b(t,\theta)\,, C^{-1} b(t,\,\theta) > dt d\theta$$

We have now a large deviation lower bound for a ρ not satisfying (3.18). Find the class $\mathcal{B}(\rho)$ of b's that satisfy (3.17) and optimize $\mathcal{E}(b)$ over $\mathcal{B}(\rho)$. In other words we have established the large deviation lower bound with the rate function

$$I(\rho) = \inf_{b \in \mathcal{B}(\rho)} \mathcal{E}(b) \tag{3.19}$$

We now work on the upper bound. We start with a class of exponential martingales. For any smooth test function $J(t,\theta)$

$$E^{P_N}\Bigg[\exp\Bigg[\sum_x J(T,\frac{x}{N})\,\eta_T(x) - \sum_x J(0,\frac{x}{N})\,\eta_0(x) - \int_0^T \sum_x J_t(t,\frac{x}{N})\eta_t(x)dt$$
$$- N^2 \int_0^T \sum_{x,y} p(y-x)\eta_t(x)(1-\eta_t(y))\big[\exp[J(t,\frac{y}{N}) - J(t,\frac{x}{N})] - 1\big]\Bigg]dt\Bigg] = 1$$

The quantity

$$N^2 \sum_{x,y} p(y-x)\eta_t(x)(1-\eta_t(y))\big[\exp[J(t,\frac{y}{N}) - J(t,\frac{x}{N})] - 1\big]$$

simplifies to

$$\frac{1}{2}\sum_x (\nabla\cdot C\nabla J)(t,\frac{x}{N})\eta_t(x) + \frac{1}{2}\sum_{x,y} p(y-x)\eta_t(x)(1-\eta_t(y))\,|< y-x\,,(\nabla J)(t,\frac{x}{N}) >|^2$$

Because of super exponential bounds we can effectively average the second term, even for large deviation purposes and interms of the density ρ, the expression looks like

$$N^d\Bigg[\frac{1}{2}\int_0^T \int_{\mathbf{T}^d} \nabla\cdot C\nabla J(t,\theta)\rho(t,\theta)dt d\theta$$
$$+ \frac{1}{2}\int_0^T \int_{\mathbf{T}^d} < \nabla J(t,\theta), C\nabla J(t,\theta) > \rho(t,\theta)(1-\rho(t,\theta))dt d\theta\Bigg]$$

Formally we are looking at

$$E^{P_N}\Big[\exp\big[N^d F_N\big]\Big] = 1$$

where

$$F_N = F_{N,J}(\rho(\cdot,\cdot))$$
$$\simeq \int_{\mathbf{Z}^d} J(T,\theta)\rho(T,\theta)d\theta - \int_{\mathbf{Z}^d} J(0,\theta)\rho(0,\theta)d\theta - \int_0^T \int_{\mathbf{Z}^d} J_t(t,\theta)\rho(t,\theta)dtd\theta$$
$$-\frac{1}{2}\int_0^T \int_{\mathbf{T}^d} \nabla \cdot C\nabla J(t,\theta)\rho(t,\theta)dtd\theta$$
$$-\frac{1}{2}\int_0^T \int_{\mathbf{T}^d} < \nabla J(t,\theta), C\nabla J(t,\theta) > \rho(t,\theta)(1-\rho(t,\theta))dtd\theta$$

By standard large deviation theory one gets an upper bound with a rate function

$$\sup_J F_{N,J}(\rho(\cdot,\cdot))$$

that is easily seen to equal (3.19).

4 Relative entropy method

We will look at the situation where $m = \sum_z z\,p(z) \neq 0$. In order to convey the idea it is sufficient to restrict ourselves to the case wher $d = 1$, and $p(1) = 1$ and $p(z) = 0$ for all $z \neq 0$. This is the totally asymmetric nearest neighbor simple exclusion model and of course $m = 1$. The method that we describe here can be found in [Y] where it is introduced in connection with a different model.

In this case the rescaling in time is done by a factor of N and the generator is

$$(\mathcal{A}_N f)(\eta) = N \sum_{x\in\mathbf{Z}_N} \eta(x)(1-\eta(x+1))[f(\eta^{x,x+1}) - f(\eta)] \tag{4.1}$$

We can easily calculate

$$d[\frac{1}{N}\sum_{x\in\mathbf{Z}_N} J(\frac{x}{N})\eta_t(x)]$$
$$= [\sum_{x\in\mathbf{Z}_N} [J(\frac{x+1}{N}) - J(\frac{x}{N})]\eta_t(x)(1-\eta_t(x+1))]dt + M_N(t) \tag{4.2}$$
$$\simeq \frac{1}{N}[\sum_{x\in\mathbf{Z}_N} J'(\frac{x}{N})\eta_t(x)(1-\eta_t(x+1))]dt + o(1)$$

The martingale term is negligible and if do simple 'averaging' we get the equation

$$\frac{d}{dt}\int_{\mathbf{T}} J(\theta)\rho(t,\theta)d\theta = \int_{\mathbf{T}} J'(\theta)\rho(t,\theta)(1-\rho(t,\theta))d\theta$$

or equivalently

$$\frac{\partial\rho}{\partial t} + \frac{\partial[\rho(1-\rho)]}{\partial\theta} = 0 \tag{4.3}$$

Let us consider a smooth solution of the above equation in $[0,T]\times\mathbf{T}$. Let us assume $c \le \rho(t,\theta) \le 1-c$ for some $c > 0$. On $\mathbf{Z}_N$ we start with an initial distribution α_N where $\eta(x)$ are independent under α_N with $\alpha_N[\eta(x) = 1] = \rho(0, \frac{x}{N})$. At any time $t \in [0,T]$ we have the actual distribution of the particle system obtained by solving the Kolmogorov forward equation

$$\frac{d\mu_N(t)}{dt} = \mathcal{A}_N^*\mu_N(t)$$

with $\mu_N(0) = \alpha_N$. We can define $\alpha_N(t)$ from $\rho(t,\cdot)$ the same way we defined α from $\rho(0,\cdot)$ at time 0. Although $\alpha_N(t)$ and $\mu_N(t)$ match by our choice at $t = 0$, in general they do not do so for $t > 0$. However we do have

Theorem 4.1 *The relative entropy $H(\mu_N(t);\alpha_N(t))$ of $\mu_N(t)$ with respect to $\alpha_N(t)$ satisfies*

$$\lim_{N\to\infty} \sup_{0\le t\le T} \frac{H(\mu_N(t);\alpha_N(t))}{N} = 0$$

By simple large deviation estimates this does imply the validity of the hydrodynamic scaling limit. In particular

$$\limsup_{N\to\infty} \sup_{o\le t\le T} E^{\mu_N(t)}\left|\frac{1}{N}\sum_x J(\frac{x}{N})\eta(x) - \int_{\mathbf{T}} J(\theta)\rho(t,\theta)d\theta\right| = 0$$

The proof involves careful reduction to Gronwall's lemma. The first step is to compute

$$\begin{aligned}
\frac{d}{dt}H(\mu_N(t);\alpha_N(t)) &= \frac{d}{dt}\sum_\eta \mu_N(t,\eta)\log\frac{\mu_N(t,\eta)}{\alpha_N(t,\eta)}\\
&= \frac{d}{dt}\sum_\eta \mu_N(t,\eta)[\log\mu_N(t,\eta) - \log\alpha_N(t,\eta)]\\
&= \sum_\eta (\mathcal{A}_N^*\mu_N(t,\cdot))(\eta)[\log\mu_N(t,\eta) - \log\alpha_N(t,\eta)]\\
&\quad + \sum_\eta \mu_N(t,\eta)[\frac{\frac{d}{dt}\mu_N(t,\eta)}{\mu_N(t,\eta)} - \frac{d}{dt}\log\alpha_N(t,\eta)]\\
&= \sum_\eta \mu_N(t,\eta)[\mathcal{A}_N(\log\frac{\mu_N(t,\cdot)}{\alpha_N(t,\cdot)})](\eta)\\
&\quad - \sum_\eta \mu_N(t,\eta)(\frac{d}{dt}\log\alpha_N(t,\eta))\\
&\le \sum_\eta \mu_N(t,\eta)\left[\frac{[\mathcal{A}_N\frac{\mu_N(t,\cdot)}{\alpha_N(t,\cdot)}](\eta)}{\frac{\mu_N(t,\eta)}{\alpha_N(t,\eta)}}\right]\\
&\quad - \sum_\eta \mu_N(t,\eta)(\frac{d}{dt}\log\alpha_N(t,\eta))\\
&= \sum_\eta \alpha_N(t,\eta)[\mathcal{A}_N\frac{\mu_N(t,\cdot)}{\alpha_N(t,\cdot)}](\eta)\\
&\quad - \sum_\eta \mu_N(t,\eta)(\frac{d}{dt}\log\alpha_N(t,\eta))\\
&= \sum_\eta \mu_N(t,\eta)\frac{[(\mathcal{A}_N^* - \frac{d}{dt})\alpha_N(t,\cdot)](\eta)}{\alpha_N(t,\eta)}\\
&= E^{\mu_N(t)}\left[\frac{[(\mathcal{A}_N^* - \frac{d}{dt})\alpha_N(t,\cdot)](\eta)}{\alpha_N(t,\eta)}\right]
\end{aligned}$$

Explicitly

$$\begin{aligned}\alpha_N(t,\eta) &= \prod_x \rho(t,\frac{x}{N})^{\eta(x)}[1-\rho(t,\frac{x}{N})]^{1-\eta(x)} \\ &= \exp\sum_x \eta(x)\log\rho(t,\frac{x}{N}) + (1-\eta(x))\log(1-\rho(t,\frac{x}{N})) \\ &= \exp\psi_N(t,\eta)\end{aligned}$$

We compute

$$\frac{[\mathcal{A}_N^*\alpha_N(t,\cdot)](\eta)}{\alpha_N(t,\eta)} = N\sum_x(1-\eta(x))\eta(x+1)\big[\exp[\psi(t,\eta^{x,x+1})-\psi(t,\eta)]-1\big]$$

and

$$\begin{aligned}&\exp[\log\rho(t,\frac{x}{N}) + \log(1-\rho(t,\frac{x+1}{N})) - \log(1-\rho(t,\frac{x}{N})) - \log\rho(t,\frac{x+1}{N})] - 1 \\ &= -\frac{1}{N}\frac{\rho'}{\rho}(t,\frac{x}{N}) - \frac{1}{N}\frac{\rho'}{1-\rho}(t,\frac{x}{N}) + o(\frac{1}{N})\end{aligned}$$

Therefore

$$\begin{aligned}\frac{[\mathcal{A}_N^*\alpha_N(t,\cdot)](\eta)}{\alpha_N(t,\eta)} &= -\sum_x(1-\eta(x))\eta(x+1)[\frac{\rho'}{\rho}(t,\frac{x}{N}) + \frac{\rho'}{1-\rho}(t,\frac{x}{N})] + o(N) \\ &= -\sum_x(1-\eta(x))\eta(x+1)\frac{\rho'}{\rho(1-\rho)}(t,\frac{x}{N})\end{aligned}$$

and

$$-\frac{d}{dt}\psi_n(t,x) = -\sum_x \eta(x)\frac{\rho_t}{\rho}(t,\frac{x}{N}) + \sum_x(1-\eta(x))\frac{\rho_t}{1-\rho}(t,\frac{x}{N})$$

There is local equilibrium over large microscopic blocks under $\mu_N(t)$. We can therefore replace $(1-\eta(x))\eta(x+1)$ by $\bar{\eta}_{k,x}(1-\bar{\eta}_{k,x})$. We get using $\rho_t = -[\rho(1-\rho)]' =$

$(2\rho-1)\rho'$

$$\begin{aligned}
&\frac{d}{dt}\frac{H(\mu_N(t)\,;\alpha_N(t))}{N}\\
&\quad\le E^{\mu_N(t)}\Bigg[\frac{1}{N}\sum_x \bar\eta_{k,x}(1-\bar\eta_{k,x})\frac{-\rho'}{\rho(1-\rho)}(t\,,\frac{x}{N})\\
&\qquad -\bar\eta_{k,x}\frac{(2\rho-1)\rho'}{\rho}(t\,,\frac{x}{N})+(1-\bar\eta_{k,x})\frac{(2\rho-1)\rho'}{1-\rho}(t\,,\frac{x}{N})\Bigg]+\epsilon_{k,N}\\
&\quad= E^{\mu_N(t)}\Bigg[\frac{1}{N}\sum_x \bar\eta_{k,x}(1-\bar\eta_{k,x})\frac{-\rho'}{\rho(1-\rho)}(t\,,\frac{x}{N})\\
&\qquad -\bar\eta_{k,x}\frac{(2\rho-1)\rho'}{\rho}(t\,,\frac{x}{N})-\bar\eta_{k,x}\frac{(2\rho-1)\rho'}{1-\rho}(t\,,\frac{x}{N})\Bigg]+\epsilon_{k,N}\\
&\quad= E^{\mu_N(t)}\Bigg[\frac{1}{N}\sum_x \bar\eta_{k,x}(1-\bar\eta_{k,x})\frac{-\rho'}{\rho(1-\rho)}(t\,,\frac{x}{N})\\
&\qquad -\bar\eta_{k,x}\frac{(2\rho-1)\rho'}{\rho(1-\rho)}(t\,,\frac{x}{N})\Bigg]+\epsilon_{k,N}\\
&\quad\le \frac{1}{N}\log E^{\alpha_N(t)}\Bigg[\exp\big[\delta V_{N,k}(\eta)\big]\Bigg]+\frac{1}{\delta}H(\mu_N(t)\,;\alpha_N(t))+\epsilon_{k,N}
\end{aligned}$$

where

$$\begin{aligned}
V_{N,k}(\eta)&=\sum_x \bar\eta_{k,x}(1-\bar\eta_{k,x})\frac{-\rho'}{\rho(1-\rho)}(t\,,\frac{x}{N})-\bar\eta_{k,x}\frac{(2\rho-1)\rho'}{\rho(1-\rho)}(t\,,\frac{x}{N})\\
&=\sum_x \rho'(t\,,\frac{x}{N})F(\bar\eta_{k,x}\,,\rho(t\,,\frac{x}{N}))
\end{aligned}$$

and

$$F(m\,,\rho)=-m\frac{(2\rho-1)}{\rho(1-\rho)}-\frac{m(1-m)}{\rho(1-\rho)}$$

The important fact is

$$F_m(m\,,\rho)=-\frac{(2\rho-1)}{\rho(1-\rho)}-\frac{1-2m}{\rho(1-\rho)}=0$$

if $\rho=m$. Therefore

$$|F(m\,,\rho)-F(\rho\,,\rho)|\le C|m-\rho|^2$$

Since we could easily subtract from $V_{N,k}(\eta)$ a sum of the type

$$\sum_x \rho'(t\,,\frac{x}{N})F(\rho(t\,,\frac{x}{N})\,,\rho(t\,,\frac{x}{N}))=o(N)$$

we conclude that

$$|V_{N,k}(\eta)|\le C\sum_x|\bar\eta_{k,x}-\rho(t\,,\frac{x}{N})|^2+o(N)$$

Now putting all the pieces together, if we can establish that for some $\delta>0$,

$$\limsup_{k\to\infty}\limsup_{N\to\infty}\frac{1}{N}\log E^{\alpha_N(t)}\Bigg[\exp\big[\delta\sum_x|\bar\eta_{k,x}-\rho(t\,,\frac{x}{N})|^2\big]\Bigg]=0$$

we would be done. This is done by large deviations. It is easy to estimate the above term by

$$\limsup_{k\to\infty} \frac{1}{k} \sup_{c\le\rho\le 1-c} \log E^{B_\rho}\big[\exp \delta\, k|\bar{\eta}_k-\rho|^2\big] = \sup_{c\le\rho\le 1-c} \sup_m [\delta|m-\rho|^2 - h(m\,,\rho)] = 0$$

for δ sufficiently small. Note that for some $a > 0$,

$$h(m\,,\rho) \ge a(m-\rho)^2$$

It finally remains to establish local equilibrium.

Lemma 4.2 *Let $\mu_i(t)$ be the probabilities at time t for the states i of a finite Markov Chain with transition rate matrix $A = \{a_{i,j}\}$. Assume that α_i are the invariant probabilities and*

$$p_i(t) = \frac{\mu_i(t)}{\alpha_i}$$

are the densities. Then

$$2\int_0^\infty D(\sqrt{p(t)}\,)dt \le H(0).$$

Proof

$$\begin{aligned}
\frac{dH(t)}{dt} &= \frac{d}{dt}\sum_i p_i(t)\log p_i(t)\alpha_i \\
&= \sum_i [\log p_i(t) - 1]p_i'(t)\alpha_i \\
&= \sum_{i,j} p_j(t)\, a_{j,i} \log p_i(t)\alpha_j \\
&= \sum_{i,j} \alpha_j p_j(t)\, a_{j,i} \log p_i(t) \\
&= 2\sum_{i,j} \alpha_j p_j(t)\, a_{j,i} \log \sqrt{p_i(t)} \\
&\le 2\sum_j \alpha_j \sqrt{p_j(t)}\, \sum_i a_{j,i}\sqrt{p_i(t)} \\
&= 2 < A\sqrt{p},\ \sqrt{p} > \\
&= -2\mathcal{D}(\sqrt{p(t)}\,)
\end{aligned}$$

Once we have control over the Dirichlet form, it does not matter what the dynamics is. The situation is the same as in the symmetric case with one big difference. There is a factor of N missing. The Dirichlet form in one dimension is bounded by C rather than CN^{-1}. This does not affect the local equilibrium principle as long we use the averages over long microscopic blocks. □

Remark Actually much stronger results are known in this special case. Instead of smooth solutions which exist only for short times, convergence to the physical (entropic) weak solution has been established for a wide class of models including the current one. However they all require some attractiveness assumptions due to the fact that coupling methods are used at some point. The relative entropy method works more or less universally, but needs both smooth solutions and special random initial conditions.

5 Nongradient systems

We now turn to the analysis of the asymmetric mean 0 case. Let us assume that we are in one dimension in order to simplify things a bit.

When we calculated

$$d\Big[\frac{1}{N}\sum_x J\Big(\frac{x}{N}\Big)\eta_t(x)\Big]$$

we ended up with

$$\begin{aligned}&\frac{1}{N}\sum_x\sum_y \eta(x) < \frac{1}{2}\Big[J'\Big(\frac{x}{N}\Big)+J'\Big(\frac{y}{N}\Big)\Big], N(1-\eta(y))(y-x)p(y-x) > dt + M_N(t)\\ &=\Big[\frac{1}{2N}\sum_x J'\Big(\frac{x}{N}\Big)N\Psi_x\Big]dt + M_N(t)\end{aligned} \tag{5.1}$$

with $\Psi_x = W(\tau_x\eta)$ and

$$\begin{aligned}W(\eta) &= [\eta(0)\sum_z(1-\eta(z))z\,p(z) + (1-\eta(0))\sum_z \eta(-z)z\,p(z)]\\ &= [\eta(0)\sum_z(1-\eta(z))z\,p(z) - (1-\eta(0))\sum_z \eta(z)z\,p(-z)]\end{aligned} \tag{5.2}$$

Moreover $M_N(t)$ was negligible. However at this point we cannot do summation by parts and so we have a 'nongradient' model and the hydrodynamic limit can no longer be established by simple averaging. Some of the details of nongradient analysis can be found in [Q], where symmetric simple exclusion models with multiple types of particles are considered. The asymmetric mean zero case was considered in [L].

The analysis for the asymmetric mean zero case depends on an important fact that is stated and proved as Lemma 5.1. A n-cycle $C=\{0=x_0,x_1,x_2,\cdots,x_n=0\}$ is a loop through $n-1$ distinct non zero integers that begins and ends at zero. With any such cycle we can associate a probability distribution π_C with mean 0:

$$\pi_C = \frac{1}{n}\sum_{i=1}^{n}\delta_{x_i - x_{i-1}}$$

While cycles of length 2 lead to symmetric distributions, in general, cycles of length 3 or more lead to asymmetric mean zero distributions. Since some of the increments $x_i - x_{i-1}$ could be repeated π_C is not necessarily the uniform distribution on some finite set.

Lemma 5.1 *Any probability distribution $p(\cdot)$ that is supported on a finite set of integers and has mean 0, can be written as a finite convex combination of distributions π_C arising from a collection of cycles.*

$$p = \sum_j \alpha_j \pi_{C_j}$$

Proof Let us denote the support of p by S. Suppose S consists of k points $a_1,\cdots,a_k$. We will show that there exist nonnegative integers $n_1,n_2,\cdots,n_k$ such that $\sum_j n_j a_j = 0$. In fact, because $d=1$, it is obvious that we must have at least two a_j's say a_1 and a_2 that are of opposite signs. The choice of $n_1=|a_2|, n_2=|a_1|$

does it and gives us a cycle C of size $n_1 + n_2$ (n_1 steps of size a_1 and n_2 steps of size a_2 in any order). The support of π_C is contained in S. We can pick a weight α such that α is the largest possible positive number such that

$$\alpha\pi_C \leq p$$

Such a choice will allow us to write

$$p = \alpha\pi_C + (1-\alpha)p'$$

with p' being another mean zero probability distribution but supported on a set S' of cardinality at least one less than that of S. We can repeat the process till we are done. □

Given a cycle $C = \{0 = x_0, x_1, \cdots, x_n = 0\}$ of length n there is a special simple exclusion generator that corresponds to it

$$\mathcal{A}_C = \sum_x \mathcal{L}_{C+x}$$

that is the sum over translates of

$$(\mathcal{L}_C\, g)(\eta) = \frac{1}{n}\sum_{i=1}^{n} \eta(x_i)(1-\eta(x_{i+1}))[g(\eta^{x_i,x_{i+1}}) - g(\eta)] \tag{5.3}$$

For each C, $\mathcal{A}_C$ is the generator of a special asymmetric mean zero simple exclusion. For instance the cycle $\{0,1,2,0\}$ produces $p(1) = \frac{2}{3}$ and $p(-2) = \frac{1}{3}$. Finally the lemma allows us to write for any p a representation

$$\mathcal{A} = \sum_j \alpha_j \mathcal{A}_{C_j} = \sum_{j,x} \alpha_j \mathcal{L}_{C_j+x} \tag{5.4}$$

The important point here is each $\mathcal{L}_{C_j+x}$, while slightly more complicated than simple exchanges $[g(\eta^{x,y}) - g(\eta)]$ across a bond, are still finite dimensional objects and our invariant measures are invariant under these.

For each k the uniform distribution ν_k on the $\binom{n}{k}$ subsets of cardinality k is an invariant measure for both $\mathcal{L}_C$ and $\mathcal{L}_{C,sym}$. The Dirichlet form is given by

$$D_{C,k}(f) = \int \frac{1}{4n}\sum_i [f(\eta^{x_{i-1},x_i}) - f(\eta)]^2 d\nu_k$$

or more generally for any invariant measure μ

$$D_{C,\mu}(f) = \int \frac{1}{4n}\sum_i [f(\eta^{x_{i-1},x_i}) - f(\eta)]^2 d\mu \tag{5.5}$$

Since the set of invariant measures for $\mathcal{L}_{C,sym}$ and $\mathcal{L}_C$ are the same, by linear algebra, the equation

$$\mathcal{L}_{C,sym} f_C = \mathcal{L}_C f \tag{5.6}$$

can be solved and we can have estimates of the form

$$D_{C,k}(f_C) \leq C D_{C,k}(f) \tag{5.7}$$

$$\|f_C\|_\infty \leq C\|\mathcal{L}_C f\|_\infty \tag{5.8}$$

with a constant that depends only on n.

We can use the representation (5.4) and the existence of solutions to (5.6) with estimates (5.7) and (5.8) to estimate various objects. For instance we have,

Theorem 5.2 *There exists a constant C depending only on $p(\cdot)$ such that*

$$|\int(\mathcal{A}f)(\eta)g(\eta)d\mu(\eta)| \le C\sqrt{D_\mu(f)}\sqrt{D_\mu(g)}.$$

Proof We write $\mathcal{A}$ as the sum

$$\mathcal{A} = \sum_{j,x} \alpha_j \mathcal{L}_{C_j+x}$$

and try to estimate

$$\int(\mathcal{L}_{C_j,x}f)(\eta)g(\eta)d\mu(\eta)$$

by first solving

$$\mathcal{L}_{C_j+x,sym}\ f_{j,x} = \mathcal{L}_{C_j+x}f$$

and getting good estimates on $f_{C_j,x}$ of the form (5.7). Then

$$\begin{aligned}
|\int(\mathcal{A}f)(\eta)g(\eta)d\mu(\eta)| &= \sum_{x,j}\alpha_j|\int(\mathcal{L}_{C_j+x}f)(\eta)g(\eta)d\mu(\eta)| \\
&= \sum_{x,j}\alpha_j|\int(\mathcal{L}_{C_j+x,sym}f_{j,x})(\eta)g(\eta)d\mu(\eta)| \\
&\le \sum_{j,x}\alpha_j\sqrt{D_{j,x}(f_{j,x})}\sqrt{D_{j,x}(g)} \\
&\le \sqrt{\sum_{j,x}\alpha_j D_{j,x}(f_{j,x})}\sqrt{\sum_{j,x}\alpha_j D_{j,x}(g)} \\
&\le C\sqrt{\sum_{j,x}\alpha_j D_{j,x}(f)}\sqrt{\sum_{j,x}\alpha_j D_{j,x}(g)} \\
&\le C\sqrt{D(f)}\sqrt{D(g)}
\end{aligned}$$

□

Remark This estimate is valid in the infinite volume for any Bernoulli measure μ_ρ with a constant C independent of ρ or in the periodic case $\mathbf{Z}_N$ for the uniform distribution of k particles, with a constant independent of N and k.

Remark The Dirichlet form $D(f)$ depends on $p(\cdot)$. But as long as the Random Walk induced by the symmetrized version of p is irreducible, they are all equivalent.

Another use of the representation is an estimate on the Feynaman-Kac eigenvalue. But first we need an elementary Lemma.

Lemma 5.3 *Let P_μ be a Markov Process with generator $\mathcal{A}$ which is in equlibrium having for its marginals an inavriant measure μ.*

$$E^{P_\mu}\Big[\exp\big[\int_0^t f(\eta(s))ds\big]\Big] \le \exp[t\,\sigma(f)]$$

where

$$\sigma(f) = \sup_{\|g\|_{2,\mu}=1}\Big[E^\mu[f(\eta)g^2(\eta)] - D(g)\Big]$$

$$= \sup_{\lambda \in \mathcal{M}} \left[\sum f(\eta)\lambda(\eta) - I(\lambda) \right]$$

where the supremum is taken over all probability distributions λ and $I(\lambda)$ is defined as

$$I(\lambda) = D\left(\sqrt{\frac{d\lambda}{d\mu}}\right) = -\langle \frac{\mathcal{A}+\mathcal{A}^*}{2} \sqrt{\frac{d\lambda}{d\mu}}, \sqrt{\frac{d\lambda}{d\mu}} \rangle_\mu$$

Remark Note that although the expectation involves a process with generator $\mathcal{A}$ which may not be symmetric with respect to μ, the variational formula involves only the symmetric part of $\mathcal{A}$.

Proof If we define

$$u(t,x) = E_x\left[\exp\left[\int_o^t f(x(s))ds \right] \right]$$

then $u(t,x)$ solves

$$\frac{\partial u}{\partial t} = \mathcal{A}u + fu; \quad u(0,x) = 1$$

and therfore

$$\begin{aligned} \frac{d}{dt}\int u^2 d\mu &= \int 2u\frac{du}{dt}d\mu = \int 2u[\mathcal{A}u + fu]d\mu \\ &= 2[\int fu^2 d\mu - D(u)] \\ &\le 2\|u\|_{2,\mu}^2 \sigma(f) \end{aligned}$$

providing us the estimate

$$\|u(t,\cdot)\|_{2,\mu} \le \exp[t\,\sigma(f)]$$

Since

$$E^\mu\left[\exp\left[\int_o^t f(x(s))ds \right] \right] = \int u(t,x)d\mu \le \|u(t,\cdot)\|_{2,\mu}$$

we are done. □

Theorem 5.4 *Let f be arbitrary and $g = \mathcal{A}f$. Then*

$$\sigma(g) = \sigma(\mathcal{A}f) = \sup_{\|\phi\|_{2,\mu}=1} \left[\int g\phi^2 d\mu - D(\phi) \right]$$

satisfies the bound

$$\sigma(\mathcal{A}f) \le C \sum_{x,j} \|\mathcal{L}_{x,j} f\|_\infty^2$$

Proof For arbitrary ϕ with $\|\phi\|_{2,\mu} = 1$ we need to estimate

$$\begin{aligned}
&\int (\mathcal{A}f)\phi^2 d\mu - D(\phi) \\
&= \sum_{x,j} \alpha_j \Big[\int (\mathcal{L}_{x,j} f)\phi^2 d\mu - D_{x,j}(\phi) \Big] \\
&= \sum_{x,j} \alpha_j \Big[\int (\mathcal{L}_{x,j,sym} f_{x,j})\phi^2 d\mu - D_{x,j}(\phi) \Big] \\
&= \sum_{x,j} \alpha_j \Big[\int [\sum_{b \in C_{x,j}} \frac{1}{2n_j} [f_{x,j}(\eta^b) - f_{x,j}(\eta)]] \phi^2 d\mu - D_{x,j}(\phi) \Big] \\
&= -\frac{1}{2} \sum_{x,j} \alpha_j \Big[\int [\sum_{b \in C_{x,j}} \frac{1}{2n_j} [f_{x,j}(\eta^b) - f_{x,j}(\eta)]] [\phi^2(\eta^b) - \phi^2(\eta)] d\mu - D_{x,j}(\phi) \Big] \\
&= -\frac{1}{2} \sum_{x,j} \alpha_j \Big[\int [\sum_{b \in C_{x,j}} \frac{1}{2n_j} [f_{x,j}(\eta^b) - f_{x,j}(\eta)]] [\phi(\eta^b) - \phi(\eta)][\phi(\eta^b) + \phi(\eta)] d\mu \\
&\qquad\qquad - D_{x,j}(\phi) \Big] \\
&\le C \sum_{x,j} \|\mathcal{L}_{x,j} f\|_\infty^2
\end{aligned}$$

We have represented $\mathcal{L}_{x,j} f$ as $\mathcal{L}_{x,j,sym} f_{x,j}$ and estimated $\|f_{x,j}\|_\infty$ interms of $\|\mathcal{L}_{x,j} f\|_\infty$ by using (5.8). □

Now we return to the analysis of (5.1). It would be nice if W were a gradient, perhaps even of the form $a(\eta(0) - \eta(1))$ for some constant $a > 0$. A summation by parts will give us the heat equation. At least we can hope that $W - a(\eta(0) - \eta(1))$ is negligible in some sense and so can be thrown away. It is indeed true that there is such a constant a, but it depends on the local density ρ and is really a function $a(\rho)$. The final equation then is

$$\frac{\partial \rho(t,\theta)}{\partial t} = \frac{1}{2} \frac{\partial}{\partial \theta} a(\rho(t,\theta)) \frac{\partial \rho(t,\theta)}{\partial \theta} \tag{5.9}$$

The first question is the sense in which $W - a(\eta(0) - \eta(1))$ is negligible. This will be formulated by means of a property that will hold relative to the processes $\{P_\rho\}$ in equilibria with marginals $\{\mu_\rho\}$ i.e. the Bernoulli meaasure with density ρ on $[0,1]^{\mathbf{Z}}$. Let us consider the class $\mathcal{F}$ of local functions $f(\eta)$ that have mean zero under every μ_ρ. If we denote by $\mu_{N,n}$, the uniform distribution on all configurations on N sites that have n occupied sites, for any local function f measurable relative to these sites, we have

$$E^\rho[f] = \sum \binom{N}{n} \rho^n (1-\rho)^{N-n} E^{\mu_{N,n}}[f]$$

Therefore $E^\rho[f] \equiv 0$ in ρ if and only if $E^{\mu_{N,n}}[f] \equiv 0$ for all $0 \le n \le N$ and N large enough for f to be measurable. Examples of such functions are W, $\eta(0) - \eta(1)$ and a whole family $f = \mathcal{A}g$ where $\mathcal{A}$ is the generator and g is any local function. Suppose $W = \mathcal{A}g + a(\eta(0) - \eta(1))$ for some local g. The only problem in replacing

W by $\alpha(\eta(0) - \eta(1))$ is the disposition of the difference

$$\int_0^t \mathcal{A}[\sum_x J'(\frac{x}{N})g(\tau_x\eta_s)]ds \tag{5.10}$$

We apply Ito's formula to the function

$$\frac{1}{N^2}\sum_x J(\frac{x}{N})g(\tau_x\eta)$$

and because the generator has an extra factor of N^2 due to the speeded up time scale we get

$$\begin{aligned}\frac{1}{N^2}\sum_x J'(\frac{x}{N})g(\tau_x\eta_t) - \frac{1}{N^2}\sum_x J'(\frac{x}{N})g(\tau_x\eta_0)\\ = \int_0^t \mathcal{A}[\sum_x J'(\frac{x}{N})g(\tau_x\eta_t)]dt + M_N(t)\end{aligned} \tag{5.11}$$

An elementary computation shows that the martingale term is negligible for large N. So we see that the term (5.10) can be safely ignored. What we really need is a theorem to the effect that there is a function $a(\rho)$ and local functions $g_\rho(\eta)$ such that it is almost true that $W = \mathcal{A}g_\rho + a(\rho)(\eta(0) - \eta(1))$ relative to P_ρ (what ever that means!)

Let $f \in \mathcal{F}$ be measurable with respect to sites on $[-\ell, \ell]$. We can consider the central limit theorem for

$$\frac{1}{\sqrt{t}}\int_0^t f(\eta_s)ds$$

under P_ρ or some $P_{\mu_{N,n}}$. We shall assume that the dynamics is some reversible version of our processs, perhaps even just the nearest neighbor symmetric simple exclusion process. We will have an asymptotic variance that we denote by $\sigma^2(N, n, f)$. We can replace $f(\eta)$ by $f_N(\eta) = \sum_{|i|\le N-\ell}^n f(\tau_i\,\eta)$ where ℓ is chosen so that f is measurable on $[-\ell, \ell]$. The corresponding variance will be denoted by $\sigma_N^2(n\,, f)$. The main step in the analysis of nongradient systems is the following Theorem 5.5 which will be proved in the next lecture.

Theorem 5.5 *Let $\mathcal{G}$ denote the set of functions $g = g(\rho\,, \eta) = g_\rho(\eta)$ that are local and depend smoothly on ρ. There exists a continuous function $a(\rho)$ such that*

$$\inf_{g\in\mathcal{G}} \limsup_{N\to\infty} \sup_{0\le n\le 2N+1} \frac{\sigma_N^2(n\,, W - a(\frac{n}{2N+1})[\eta(0) - \eta(1)] - \mathcal{A}g_{\frac{n}{2N+1}})}{2N+1} = 0$$

Remark Note that because $N \to \infty$, this is a theorem concerning the behavior under various stationary equilibrium processes $\{P_\rho\}$ although we need to consider nonequilibrium situations while trying to establish hydrodynamic limits.

We will proceed with our analysis and see why Theorem 5.5 is sufficient.

Let P_N be the process on the time interval $[0,T]$ starting from some configuration η_0. We define

$$\xi_{N,\epsilon,\delta} = \int_0^T \sum_x J'(\frac{x}{N})W(\tau_x\eta_t)dt - \int_0^T \mathcal{A}\big[\sum_x J'(\frac{x}{N})g(\bar{\eta}_{N\epsilon,x,t}\,,\tau_x\eta_t)\big]\,dt$$
$$- \int_0^T \frac{1}{2N\delta}\sum_x J'(\frac{x}{N})a(\bar{\eta}_{N\epsilon,x,t})[\eta_t(x-N\delta)-\eta_t(x+N\delta)]dt$$

It is clearly sufficient to show that

$$\inf_{g\in\mathcal{G}}\limsup_{\epsilon\to 0}\limsup_{\delta\to 0}\limsup_{N\to\infty} E^{P_N}[|\xi_{N,\epsilon,\delta}|]=0 \tag{5.12}$$

It is not difficult to see that the term

$$\int_0^T \mathcal{A}\big[\sum_x J(\frac{x}{N})g(\bar{\eta}_{N\epsilon,x,t}\,,\tau_x\eta_t)\big]\,dt$$

continues to be negligible and therefore

$$\frac{1}{2}\int_0^T \sum_x J'(\frac{x}{N})W(\tau_x\eta_t)dt$$

can be replaced by

$$\lim_{\epsilon\to 0}\lim_{\delta\to 0}\frac{1}{2}\int_0^T \frac{1}{2N\delta}\sum_x J'(\frac{x}{N})a(\bar{\eta}_{N\epsilon,x,t})[\eta_t(x-N\delta)-\eta_t(x+N\delta)]dt$$
$$=\lim_{\epsilon\to 0}\lim_{\delta\to 0}\frac{1}{2}\int_0^T\int_{\mathbf{T}}\frac{1}{2\delta}J'(\theta)a\Big(\frac{1}{2\epsilon}\int_{\theta-\epsilon}^{\theta+\epsilon}\rho(t\,,w)dw\Big)[\rho(t\,,\theta-\delta)-\rho(t\,,\theta+\delta)]d\theta dt$$
$$=\lim_{\epsilon\to 0}-\frac{1}{2}\int_0^T\int_{\mathbf{T}}J'(\theta)a\Big(\frac{1}{2\epsilon}\int_{\theta-\epsilon}^{\theta+\epsilon}\rho(t\,,w)dw\Big)\rho'(t\,,\theta)d\theta dt$$
$$=-\frac{1}{2}\int_0^T\int_{\mathbf{T}}J'(\theta)a(\rho(t\,,\theta))\rho'(t\,,\theta)d\theta dt$$

giving us a weak formulation of (5.9).

An actual proof will proceed in the following manner. We consider the family of processes P_N on $\Omega_N = D[[0,T];[0,1]^{\mathbf{Z}_N}]$ with generator $N^2\mathcal{A}$ and an arbitrary initial configuration $\eta_o(\cdot)$. We view the process of empirical distributions

$$\nu_N(t)=\frac{1}{N}\sum_x \delta_{\frac{x}{N}}\eta_t(x) \qquad 0\le t\le T$$

as a stochastic process with values in $\mathcal{M}(\mathbf{T})$, the space of measures on $\mathbf{T}$ and call the induced distribution $\widehat{P}_N$. We prove that the family $\widehat{P}_N$ is tight and if $\widehat{P}$ is any limit point it is supported on measure valued maps $\nu(t)$ that are weakly continuous in t and have densities $\rho(t,\theta)$ satisfying $0\le\rho(\cdot,\cdot)\le 1$. Moreover

$$E^P\Big[\int_0^T\int_{\mathbf{T}}|\rho_\theta(t,\theta)|^2dtd\theta\Big]<\infty \tag{5.13}$$

Once we have (5.12) and the estimate (5.13) we see that any limit point $\widehat{P}$ is concentrated on the set of weak solutions of (5.9) satisfying the energy estimate

$$\int_0^T \int_{\mathbf{T}} |\rho_\theta(t,\theta)|^2 dt d\theta < \infty \tag{5.14}$$

Moreover if the initial condition $\eta_0 = \eta_0^N(\cdot)$ is such that

$$\lim_{N\to\infty} \frac{1}{N} \sum_x J(\frac{x}{N}) \eta_0^N(x) = \int_{\mathbf{T}} J(\theta)\rho_0(\theta) d\theta$$

then $\widehat{P}$ is supported by weak solutions with initial condition $\rho_0(\cdot)$. Such solutions are unique according to Lemma 5.6. and the hydrodynamic limit is established because we see that $\widehat{P}$ is supported on the single unique solution.

Therefore it all comes down to proving tightness, (5.12) and (5.13). First we will prove the tightness of the family $\widehat{P}_N$. Since the martingale part is small in the mean square sense, by Doob's inequlaity it is small uniformly in t for every starting configuration. Tightness will follow from superexponential estimates on the modulus of continuity of

$$\zeta_J(t) = \int_0^t \sum_x J(\frac{x}{N}) \Psi_x(\eta(s)) ds$$

in equlibrium. Clearly an estimate of the form

$$E^N \Big[\exp \big[N \int_0^t \sum_x J(\frac{x}{N}) \Psi_x(\eta(s)) ds \big] \Big] \leq \exp[N C_J t]$$

on the equilibrium process will do the trick. This in turn can be derived from the variational estimate

$$\sup_{\|\phi\|_{\mu_N}=1} \Big[E^{\mu_N} \big[\phi^2 \sum_x N J(\frac{x}{N}) \Psi_x(\eta) \big] - N^2 D_N(\phi) \Big] \leq N C_J t$$

using theorem 5.4. Basically $f = \sum_y y\eta(y)$. The same method provides the estimate

$$\sup_{\|\phi\|_{\mu_N}=1} \Big[E^{\mu_N} \big[\phi^2 \sum_x N J(\frac{x}{N}) [\eta(x+1) - \eta(x)] \big] - N^2 D_N(\phi) \Big] \leq N C_J t$$

The constant C_J is essentially $C \int J^2(\theta) d\theta$. Since, for any initial distribution, the entropy relative to the equilibrium μ_N is atmost N, the standard entropy inequality implies that

$$E^{\widehat{P}_N} \Big[\int_0^T \frac{1}{N} \sum_x J'(\frac{x}{N}) \eta(x,s) ds \Big] \leq C + \frac{T}{2} \int J^2(\theta) d\theta$$

In fact one can do the time dependent potentials as well and get

$$E^{\widehat{P}_N} \Big[\int_0^T \frac{1}{N} \sum_x J'(t,\frac{x}{N}) \eta(x,s) ds \Big] \leq C + \frac{1}{2} \int \int_0^T J^2(t,\theta) d\theta$$

and (5.13) folows not so routinely from this estimate. If we are going to use entropy inequality, then

$$\begin{aligned}E^Q[\max(X,Y)] &\le \log E^P[\max(e^X,e^Y)] + h(Q,P)\\ &\le \log E^P[e^X + e^Y] + h(Q,P)\\ &\le \log 2 + \max\{\log E^P[e^X], \log E^P[e^Y]\} + h(Q,P)\end{aligned}$$

and if we are going to divide by N and pass to the limit $\log 2$ disappears from sight. It is almost as though the supremum can be moved inside the integral freely. So for the limit instead of just

$$\sup_J \left[E^{\widehat{P}}\left[\int_0^T \int J'(t,\theta)\rho(\theta,s)dsd\theta - \frac{1}{2}\int\int_0^T J^2(t,\theta)d\theta\right]\right] \le C$$

we actually can get

$$E^{\widehat{P}}\left[\sup_J \left[\int_0^T \int J'(t,\theta)\rho(\theta,s)dsd\theta - \frac{1}{2}\int\int_0^T J^2(t,\theta)d\theta\right]\right] \le C$$

which is just (5.13)

We will prove (5.12) by showing

$$\inf_{g\in\mathcal{G}} \limsup_{\epsilon\to 0} \limsup_{\delta\to 0} \limsup_{N\to\infty} \frac{1}{N} \log E^{Q_N}\left[\exp\left[\lambda\, N\, \xi_{N,\epsilon,\delta}\right]\right] = 0 \qquad (5.15)$$

for every real λ. Since we can replace J by λJ there is no loss of generality in taking $\lambda = 1$. We can as well replace J' by an arbitrary J and consider

$$\xi_{N,\epsilon,\delta} = \int_0^T \sum_x J(\frac{x}{N}) G_{N,\epsilon,\delta}(\tau_x \eta_t)dt$$

with

$$G_{N,\epsilon,\delta}(\eta) = W(\eta) - \mathcal{A}g(\bar{\eta}_{N\epsilon,0}, \eta) - \frac{1}{2N\delta} a(\bar{\eta}_{N\epsilon,0})[\eta(-N\delta) - \eta(N\delta)]$$

Since Q_N is starting from equlibrium we have the following estimates for the Feynman-Kac integrals by the variational formula. See Lemma 5.3.

$$\begin{aligned}\frac{1}{N}\log E^{Q_N}&\left[\exp\left[N\,\xi_{N,\epsilon,\delta}\right]\right]\\ &\le \frac{T}{N}\sup_{\|\phi\|_{2,\mu}=1}\left[E^\mu[\phi^2(\eta)\sum_x N\, J(\frac{x}{N})G_{N,\epsilon,\delta}(\tau_x\eta)] - N^2\,\mathcal{D}_N(\phi)\right]\\ &= T\sup_{\|\phi\|_{2,\mu}=1}\left[E^\mu[\phi^2(\eta)\sum_x J(\frac{x}{N})G_{N,\epsilon,\delta}(\tau_x\eta)] - N\mathcal{D}_N(\phi)\right]\end{aligned} \qquad (5.16)$$

Basically we need to show that

$$\inf_{g\in\mathcal{G}} \limsup_{\epsilon\to 0} \limsup_{\delta\to 0} \limsup_{N\to\infty} \sup_{\|\phi\|_{2,\mu}=1}\left[E^\mu[\phi^2(\eta)\sum_x J(\frac{x}{N})G_{N,\epsilon,\delta}(\tau_x\eta)] - N\mathcal{D}_N(\phi)\right] = 0 \qquad (5.17)$$

This is a matter of controlling the first term by the second term. The first term and the Dirichlet form may be split into pieces, with control over individual pieces. We also note that the Dirichlet form $\mathcal{D}_N(\phi)$ naturally splits as the sum $\sum_x \mathcal{D}_{N,x,x+1}(\phi)$ of contributions over the individual bonds. We can pretend that our Dirichlet form is the nearest neighbor one because they are in any case equivalent. Since J is arbitrary and can be replaced by a constant multiple of itself, for obtaining estimates, it does not matter if we replace the Dirichlet form by an equivalent one. We replace $G_{N,\epsilon,\delta}$ by $H_{N,k}$ where

$$\begin{aligned} H_{N,k}(\eta) = \frac{1}{2k+1}\sum_{|x|\le k-\ell} W(\tau_x\eta) - \frac{1}{2k+1}\sum_{|x|\le k-\ell} \mathcal{A}g(\bar\eta_{k,0},\tau_x\eta) \\ - \frac{1}{2k+1} a(\bar\eta_{k,0})[\eta(-k)-\eta(k)] \end{aligned}$$

Here ℓ is fixed integer depending on the local function W and g and is picked so that the function $H_{N,k}(\eta)$ depends only the $\{\eta(x)\}$ with $|x|\le k$ For the difference $F_{N,k,\epsilon,\delta} = G_{N,\epsilon,\delta} - H_{N,k}$ we show that for every $\beta > 0$,

$$\begin{aligned} \limsup_{\epsilon\to 0}\limsup_{\delta\to 0}\limsup_{k\to\infty}\limsup_{N\to\infty}\sup_{\|\phi\|_{2,\mu}=1} \Big[E^{\mu}\big[\phi^2(\eta)\sum_x J(\frac{x}{N})F_{N,\epsilon,\delta,k}(\tau_x\eta)\big] \\ - \beta N\mathcal{D}_N(\phi)\Big] = 0 \end{aligned} \tag{5.18}$$

To see this we rewrite $F_{N,k.\epsilon,\delta}$ as the sum of several terms and control each one of them in terms of the Dirichlet form. The first one takes the form

$$\sum_x \Big[\frac{1}{2k+1}\sum_{|y-x|\le k}[J(\frac{y}{N}) - J(\frac{x}{N})]\Big] W(\tau_x\eta)$$

that is easily controlled. The second term is

$$\sum_x J(\frac{x}{N})\big[\frac{1}{2N\delta}a(\bar\eta_{N\epsilon,x})[\eta(x-N\delta)-\eta(x+N\delta)] - \frac{1}{2k}a(\bar\eta_{k,x})[\eta(x-k)-\eta(x+k)]\big]$$

that can be rewritten as

$$\sum_x A_N(\epsilon,\delta,k,x)[\eta(x-1)-\eta(x)]$$

where

$$\begin{aligned} A_N(\epsilon,\delta,k,x) = \frac{1}{2N\delta}\sum_{x-N\delta+1\le y\le x+N\delta} J(\frac{y}{N})a(\bar\eta_{N\epsilon,y}) \\ - \frac{1}{2k+1}\sum_{x-k+1\le y\le x+k} J(\frac{y}{N})a(\bar\eta_{k,y}). \end{aligned}$$

For any $\beta > 0$ there exists a constant C_β such that

$$\begin{aligned} \sup_{\|\phi\|_{2,\mu_N}=1} E^{\mu_N}\Big[A_N(\epsilon,\delta,k,x)\phi^2(\eta)[\eta(x+1)-\eta(x)] - \beta N\mathcal{D}_{N,x,x+1}(\phi) \\ - \frac{C_\beta}{N}[A_N(\epsilon,\delta,k,x)]^2\Big] \le 0 \end{aligned} \tag{5.19}$$

and therefore with the help of (5.19), we can reduce (5.18) to estimating

$$\sup_{\|\phi\|_{2,\mu_N}=1}\left[E^{\mu_N}\left[\phi^2(\eta)\frac{1}{N}\sum_x [A_N(\epsilon,\delta,k,x)]^2\right]-\beta N\mathcal{D}_N(\phi)\right]$$

This can be done using the continuity of $a(\cdot)$ and $J(\cdot)$ and the two block estimate. The last term is the difference

$$S_{\epsilon,k}(\eta)=\sum_x J(\frac{x}{N})\Big[\frac{1}{2k+1}\sum_{y:|y-x|\le k-\ell}[\mathcal{A}g(\bar\eta_{k,y},\tau_y\eta)-\mathcal{A}g(\bar\eta_{N\epsilon,x},\tau_x\eta)]\Big]$$

An elmentary calculation shows that

$$\|\mathcal{L}_{x,j}S_{\epsilon,k}\|_\infty\le C\left[\frac{1}{N\epsilon}+\frac{1}{k}\right]$$

and we can use Theorem 5.4 to prove

$$\limsup_{\epsilon\to 0}\limsup_{k\to\infty}\limsup_{N\to\infty}\sup_{\|\phi\|=1}\left[E^{\mu}\left[\phi^2(\eta)S_{\epsilon,k}(\eta)\right]-\beta N\mathcal{D}_N(\phi)\right]=0 \tag{5.20}$$

In order to complete the argument we will have to see why Theorem 5.5 implies that

$$\inf_{g\in\mathcal{G}}\limsup_{k\to\infty}\limsup_{N\to\infty}\sup_{\|\phi\|_{2,\mu_N}=1}\left[E^{\mu}_N\left[\phi^2(\eta)\sum_x J(\frac{x}{N})H_{N,k}(\tau_x\eta)\right]-N\mathcal{D}_N(\phi)\right]=0 \tag{5.21}$$

We can write

$$D_N(\phi)=\frac{1}{2k}\sum_x D_{x,k}(\phi)=\frac{1}{2k}\sum_x\sum_{b\in(x-k,x+k)}D_b(\phi)$$

and

$$\begin{aligned}
&\sup_{\|\phi\|_{2,\mu_N}=1}\left[E^{\mu_N}\left[\phi^2(\eta)\sum_x J(\frac{x}{N})H_{N,k}(\tau_x\eta)\right]-N\mathcal{D}_N(\phi)\right]\\
&=\frac{1}{2k+1}\sup_{\|\phi\|_{2,\mu_N}=1}\Bigg[E^{\mu_N}\left[\phi^2(\eta)\sum_x J(\frac{x}{N})(2k+1)H_{N,k}(\tau_x\eta)\right]\\
&\qquad\qquad-N(2k+1)\mathcal{D}_N(\phi)\Bigg]\\
&\le\frac{1}{2k+1}\sum_x\sup_{\|\phi\|_{2,\mu_N}=1}\left[E^{\mu_N}\left[\phi^2(\eta)J(\frac{x}{N})(2k+1)H_{N,k}(\tau_x\eta)\right]-N\mathcal{D}_{x,k}(\phi)\right]\\
&\le\frac{N}{2k+1}\sum_x\sup_{\|\phi\|_{2,\mu_N}=1}\left[E^{\mu_N}\left[\phi^2(\eta)\frac{1}{N}J(\frac{x}{N})(2k+1)H_{N,k}(\tau_x\eta)\right]-\mathcal{D}_{x,\ell}(\phi)\right]
\end{aligned}$$

Taking limits as $N\to\infty$ and using the perturbation expansion of the eigenvalue in the small parameter $\frac{1}{N}$, we estimate the limsup of left hand side by

$$C_J\sup_n\frac{1}{2k+1}\sigma_k^2(n\,,(2k+1)H_{N,k}(\eta))$$

where

$$C_J=\int J^2(\theta)d\theta$$

and $\sigma_k^2\big(n\,,(2k+1)H_{N,k}(\eta)\big)$ is the variance of

$$\sum_{|x|\le k-\ell} W(\tau_x\eta) - \sum_{|x|\le k-\ell} \mathcal{A}g(\bar{\eta}_{k,0},\tau_x\eta) - a(\bar{\eta}_{k,0})[\eta(-k)-\eta(k)]$$

calculated for the uniform distribution on the configurations in a block of size $2k+1$ with n particles. If $k\to\infty$ and $\frac{n}{2k+1}\to\rho$ we are back in the situation of Theorem 5.5

Remark We have the genrator $\mathcal{A}$ and its symmetric version $\mathcal{A}_{sym}$ that share the same Dirichlet form. The variational formula, actually gives the eigenvalues for $\mathcal{A}_{sym}$ and perturbatively for large N they are controlled by the variance in the central limit theorem for the symmetric process. It is a fact for the mean zero case, that if we write $\mathcal{A}_N = \mathcal{A}_{sym,N} + \mathcal{A}_{skew,N}$, then from theorem 5.2

$$\|\mathcal{A}_{skew,N}\|_{H^1\to H^{-1}} \le C$$

uniformly in N, and this in turn means a uniform bound of the form

$$\sigma^2(f) \le \sigma^2_{sym}(f) \le C\sigma^2(f)$$

Therefore while we have to use $\mathcal{A}$ for the exact eavluation of $a(\rho)$, for the purpose of estimation we can use $\mathcal{A}_{symm}$ and we will not lose anything in the process.

6 Nongradient case (continuation)

It only remains to prove theorem 5.5.

Theorem 5.5 *Let $\mathcal{G}$ denote the set of functions $g = g(\rho\,,\eta) = g_\rho(\eta)$ that are local and depend smoothly on ρ. There exists a function $a(\rho)$ such that*

$$\inf_{g\in\mathcal{G}} \limsup_{N\to\infty} \sup_{0\le n\le 2N+1} \frac{\sigma_N^2(n\,, W - a(\frac{n}{2N+1})[\eta(0)-\eta(1)] - \mathcal{A}g_{\frac{n}{2N+1}})}{2N+1} = 0.$$

σ_N^2 can be taken relative to any reasonable symmetric simple exclusion process and we will for simplicity pick the nearest neighbor one. The big reduction now is that we have only to deal asymptotically, as $N\to\infty$, with different equilibria having n particles among $2N+1$ sites.

We will prove a stronger version, namely

Theorem 6.1 *For each $f\in\mathcal{F}$ the limit*

$$\lim_{\substack{N\to\infty\\ \frac{n}{2N+1}\to\rho}} \frac{\sigma_N^2(n,f)}{2N+1} = \langle\langle f,f\rangle\rangle_\rho$$

exists and for each f there exists $a(f,\rho)$ such that

$$\inf_{g\in\mathcal{G}} \langle\langle f - a(f,\rho)(\eta(0)-\eta(1)) - \mathcal{A}g, f - a(f,\rho)(\eta(0)-\eta(1)) - \mathcal{A}g\rangle\rangle_\rho = 0$$

In particular $a(\rho) = a(W,\rho)$. Moreover if we take $g = g(\rho,\eta)$ from $\mathcal{G}_s$, the class of smooth local functions, we have

$$\inf_{g\in\mathcal{G}_s} \sup_\rho \langle\langle f - a(f,\rho)(\eta(0)-\eta(1)) - \mathcal{A}g, f - a(f,\rho)(\eta(0)-\eta(1)) - \mathcal{A}g\rangle\rangle_\rho = 0.$$

There is a basic formalism which is important. Let us fix ρ and the Bernoulli measure μ_ρ which is an invariant measure for the simple exclusion dynamics. Given a local function $f(\eta)$ we consider the formal sum

$$F(\eta) = \sum_x f(\tau_x \eta).$$

F does not really exist. But, for any two sites a, b

$$\xi_f^{a,b}(\eta) = F(\eta^{a,b}) - F(\eta)$$

makes perfectly good sense. Moreover, they are covariant, in the sense that

$$\xi_f^{x+a,x+b}(\eta) = \xi_f^{a,b}(\tau_x \eta)$$

and in fact the entire collection $\xi_f^{a,b}$ can be determined algebraically from $\xi_f^{0,1}(\eta)$ which itself must satisfy some simple algebraic conditions. We can look at the space of such abstract objects $\{\xi^{0,1}\}$ as a closed subspace $\mathcal{H} \subset L_2(\mu_\rho)$.

In addition to the ξ_f generated by local functions f, there is an additional object in $\mathcal{H}$ that is important.

$$\xi^{a,b} = (b-a)[\eta(a) - \eta(b)]$$

this comes formally from the nonlocal nonexistent f

$$f(\eta) = \sum_{x \le 0} \eta(x)$$

or equally formally, with

$$F(\eta) = \sum_x f(\tau_x \eta) = \sum_x x\, \eta(x)$$

Although $F(\tau\eta) \neq F(\eta)$ we do have $F(\tau\eta) - F(\eta) = -\sum_x \eta(x)$ which is formally conserved. The important fact is that $F(\eta^{0,1}) - F(\eta) = \xi_f^{0,1} = \eta(0) - \eta(1) \in \mathcal{H}$.

Proof of Theorem 6.1 Given $f \in \mathcal{F}$ which is a local function on $[-k, k]$ we solve in an interval $[-N, N]$ the equation

$$(\mathcal{A}_N^0 v_N)(\eta) = \frac{1}{2} \sum_{(x,x+1) \in [-N,N]} [v_N(\eta^{x,x+1}) - v_N(\eta] = \sum_{|x| \le N-k} f(\tau_x \eta)$$

where $\mathcal{A}_N^0$ is the restriction of the nearest neighbor symmetric simple exclusion process to the interval $[-N, N]$. The set of configurations with a given number of particles are the invariant subsets, with uniform distributions over all such classes of configurations being the invariant measures.

$$\begin{aligned}
\sigma_N^2(n, f) &= -2 < (\mathcal{A}_N^0)^{-1} \sum_x f(\tau_x \eta)\, , \sum_x f(\tau_x \eta) >_{N,n} \\
&= -2 < v_N\, , \mathcal{A}_N^0 v_N >_{N,n} \\
&= 2 D_{N,n}(v_N) \\
&= E^{\nu_{N,n}} \left[\sum_{(x,x+1) \in [-N,N]} [v_N(\eta^{x,x+1}) - v_N(\eta)]^2 \right]
\end{aligned}$$

Let us try some examples. Suppose $f = \mathcal{A}^0 g$ for some local $g \in \mathcal{G}$. Then

$$\mathcal{A}_N^0 \sum_x \tau_x g = \sum_x \tau_x f$$

and

$$\lim_{\substack{N\to\infty \\ \frac{n}{2N+1}\to\rho}} \frac{1}{2N}\sigma_N^2(n,f) = E^{\mu_\rho}[\xi_g^2]$$

or in polarized form if $f_i = \mathcal{A}^0 g_i$ for $i = 1,2$

$$\lim_{\substack{N\to\infty \\ \frac{n}{2N+1}\to\rho}} \frac{1}{2N} Cov_N(n, f_1, f_2) = E^{\mu_\rho}[\xi_{g_1}\xi_{g_2}]$$

Also if we consider

$$g_N = 2\sum_{|x|\le N} x\,\eta(x)$$

then

$$\mathcal{A}_N^0 g_N = \sum_{(x,x+1)\in[-N,N]} [\eta(x) - \eta(x+1)] = [\eta(-N) - \eta(N)]$$

and for $f = \eta(0) - \eta(1)$

$$\lim_{\substack{N\to\infty \\ \frac{n}{2N+1}\to\rho}} \frac{1}{2N}\sigma_N^2(n,f) = \lim_{\substack{N\to\infty \\ \frac{n}{2N+1}\to\rho}} \frac{1}{2N} D_{N,n}(g_N) = E^{\mu_\rho}[\xi_d^2]$$

with $\xi_d = 2(\eta(0) - \eta(1))$. In fact by poalrization we also have

$$\lim_{\substack{N\to\infty \\ \frac{n}{2N+1}\to\rho}} \frac{1}{2N} Cov_N(n, \mathcal{A}^0 g, f) = E^{\mu_\rho}[\xi_g \xi_d]$$

We can write any $f \in \mathcal{F}$ as a finite sum

$$\sum [u_a(\eta^{a,a+1}) - u_a(\eta)]$$

and let us assume that we have $f_0 = u_0(\eta^{a,a+1}) - u_0(\eta)$ for some a and $u_0 \in \mathcal{G}$. We shall compute

$$\lim_{\substack{N\to\infty \\ \frac{n}{2N+1}\to\rho}} \frac{1}{2N} Cov_N(n, f_0, f) = E^{\mu_\rho}[\xi_0 \xi]$$

Where $\xi = \xi_g$ if $f = \mathcal{A}^0 g$ and $\xi = \xi_d = 2(\eta(0) - \eta(1)$ if $f = \eta(0) - \eta(1)$. This is not a hard calculation. Denoting by $G_N = \sum_{|x|\le N-k} \tau_x g$

$$\begin{aligned}
&\lim_{\substack{N\to\infty \\ \frac{n}{2N+1}\to\rho}} \frac{1}{2N} Cov_N(n, f_0, \mathcal{A}^0 g) \\
&= \lim_{\substack{N\to\infty \\ \frac{n}{2N+1}\to\rho}} \frac{2}{2N} < -(\mathcal{A}_N^0)^{-1}\mathcal{A}_N^0 G_N\,, \sum_x [(\tau_x u_0)(\eta^{x+a,x+a+1}) \\
&\qquad\qquad - (\tau_x u_0)(\eta)] >_{N,n} \\
&= \lim_{\substack{N\to\infty \\ \frac{n}{2N+1}\to\rho}} \frac{2}{2N} < -\sum_x [G_N(\eta^{a+x,a+x+1}) - G_N(\eta)]\,, u_0(\eta)] >_{N,n} \\
&= E^{\mu_\rho}[(-2\tau_{-a} u_0)\xi_g]
\end{aligned}$$

In exactly the same way the choice of $G_N = \sum_{|x|\le N} x\,\eta(x)$ produces

$$\lim_{\substack{N\to\infty \\ \frac{n}{2N+1}\to\rho}} \frac{1}{2N} Cov_N(n, f_0, \eta(0)-\eta(1)) = E^{\mu_\rho}[(-2\tau_{-a}u_0)\xi_d]$$

If we replace $-2\tau_{-a}u_0$ with its projection $\xi_0 = \mathbf{P}[-2\tau_{-a}u_0]$ onto $\mathcal{H}$, we get

$$\lim_{\substack{N\to\infty \\ \frac{n}{2N+1}\to\rho}} \frac{1}{2N} Cov_N(n, f_0, f) = E^{\mu_\rho}[\xi_o \xi]$$

where $f \to \xi$ is a linear map from $\mathcal{F}$ into $\mathcal{H}$. Once we prove Theorem 6.2, the range of this linear map will be dense in $\mathcal{H}$ and we will have

$$\liminf_{\substack{N\to\infty \\ \frac{n}{2N+1}\to\rho}} \frac{1}{2N} \sigma_N^2(n, f_0) \ge E^{\mu_\rho}[\xi_0^2]$$

We now try to control

$$\limsup_{\substack{N\to\infty \\ \frac{n}{2N+1}\to\rho}} \frac{1}{2N} \sigma_N^2(n, f)$$

If v_N solves $\mathcal{A}_N^0 v_N = \sum_{|x|\le N-k} \tau_x f$

$$\begin{aligned} D_{N,n}(v_N) &= - < v_N(\eta), \sum_x [(\tau_x u_0)(\eta^{x+a,x+a+1}) - (\tau_x u_0)(\eta)] >_{N,n} \\ &= -\sum_x < [v_N(\eta^{x+a,x+a+1}) - v_N(\eta)], (\tau_x u_0)(\eta) >_{N,n} \\ &\le \sqrt{2D_{N,n}(v_N)} \sqrt{\sum_x \|\tau_x u_0\|_{N,n}^2} \end{aligned}$$

providing the estimate

$$D_{N,n}(v_N) \le 2(2N+1)\|u_0\|_{N,n}^2$$

and therefore

$$\limsup_{\substack{N\to\infty \\ \frac{n}{2N+1}\to\rho}} \frac{1}{2N} \sigma_N^2(n, f) \le 4E^{\mu_\rho}[u_0^2]$$

To improve this bound we use a duality argument. We shall only sketch the proof.

$$\begin{aligned} &\frac{1}{\sqrt{2N}} \sigma_N(n, f) \\ &= [\frac{1}{2N} \sigma_N^2(n, f)]^{\frac{1}{2}} \\ &= \frac{2}{2N} \sup_{w: D_{N,n}(w) \le 2N} E^{\nu_{N,n}} \left[\sum_x [(\tau_x u_0)(\eta^{x+a,x+a+1}) - (\tau_x u_0)(\eta)] w(\eta) \right] \\ &= \frac{2}{2N} \sup_{w: D_{N,n}(w) \le 2N} E^{\nu_{N,n}} \left[\sum_x (\tau_x u_0)(\eta) [w(\eta^{x+a,x+a+1}) - w(\eta)] \right] \\ &= \frac{2}{2N} E^{\nu_{N,n}} \left[\sum_x (\tau_x u_0)(\eta) [w_N(\eta^{x+a,x+a+1}) - w_N(\eta)] \right] \end{aligned}$$

for some W_N with $D_{N,n}(w_N) \leq 2N$. On the average $w_N(\eta^{y,y+1}) - w_N(\eta)$ has L_2 norm 1, and a weak limit of averages of suitably chosen long blocks of

$$\tau_{-y}[w_N(\eta^{y,y+1}) - w_N(\eta)]$$

will produce a $\xi \in \mathcal{H}$ such that $E^{\mu_\rho}[\xi^2] \leq 1$ and

$$\limsup_{\substack{N\to\infty \\ \frac{n}{2N+1}\to\rho}} \frac{1}{2N}\sigma_N^2(n,f) = |E^{\mu_\rho}[2(\tau_{-a}u_0)\xi]|^2 \leq E^{\mu_\rho}[|\mathbf{P}(2(\tau_{-a}u_0))|^2]$$

so that f can be represented by the projection of $-2\tau_{-a}u_0$ □

Theorem 6.2 *The span $\mathcal{H}_0$ of $\xi_f^{0,1}$ in $L_2(\mu)$ as f varies over all local functions is of codimension one in $\mathcal{H}$ and a complementary (orthogonal) one dimensional subspace is provided by the span of $\eta(0) - \eta(1)$.*

Proof Let us start with a ξ in $\mathcal{H}$. We denote the translations $\tau_x\xi$ by $\xi^{x,x+1}$ and truncate them by taking conditional expectation given the σ-field generated by $\{\eta(x)\}$ in a finite interval $[-N, N]$ and get a $\xi_N^{x,x+1}$ that approximates $\xi^{x,x+1}$ quite well if $N - |x| >> 0$. We then solve simultaneously the equations

$$[v_N(\eta^{x,x+1}) - v_N(\eta)] = \xi_N^{x,x+1}(\eta) \quad \text{for all} \quad x$$

for all $(x, x+1) \in [-N, N]$, which can be done. If we fix a value n for the total number of particles in $[-N, N]$, then the equations can be 'integrated' upto a constant on each of the spaces $\Omega_{N,n}$ to a get solution v_N normalized to have mean 0 with repect to every uniform distribution $\nu_{N,n}$ on $\Omega_{N,n}$. We can get some estimates by thinking of v_N as the solution of the scalar (Poisson) equation.

$$\sum_{(x,x+1)\in[-N,N]} [v_N(\eta^{x,x+1}) - v_N(\eta)] = \sum_{(x,x+1)\in[-N,N]} \xi_N^{x,x+1}(\eta) \tag{6.1}$$

We get

$$\mathcal{D}_N(v_N) \leq CN$$

If we take $f_N = N^{-1}v_N$ and try with

$$F_N(\eta) = \sum_x \tau_x f_N$$

Then $F_N(\eta^{0,1}) - F_N(\eta)$ almost works except for some boundary terms. The two boundary terms are

$$f_N(\eta(0), \eta(2), \ldots, \eta(2N+1)) - f_N(\eta(1), \eta(2), \ldots, \eta(2N+1))$$

and a simlar one coming from the other end. Denote them by g_N. If we had an estimate of the form

$$\|g_N\|_2 \leq C$$

then the boundary terms will be bounded in $L_2(\mu)$ and yield a weak limit. From the estimates $D_{x,x+1}(v_N) \leq 1$ it is clear that $D_{x,x+1}(f_N) \to 0$ and this inturn can be shown to imply that any weak limit of g_N has to be of the form $c(\eta(0) - \eta(1))$ establishing that $\xi = c(\eta(0) - \eta(1))$ modulo $\mathcal{H}_0$.

Such an estimate does not exist, because the spectral gap on $[-N, N]$ for the operator appearing in (6.1) which is the generator of the symmetric simple exclusion process is cN^{-2} and the best we can do is

$$\|v_N\|_2^2 \le CN^3$$

There is no possibility of cancellation because the difference involves comparing the values of f_N on $\Omega_{N,n}$ for diffrent n and the Dirichlet form cannot help. We overcome the difficulty by solving (6.1) on $[-2N, 2N]$ to get v_{2N} and then integrating all the $\eta(x)$ outside $[-N, N]$ to obtain its conditional expectaion w_N defined on $\cup_n \Omega_{N,n}$. For w_N obtained in this manner Lemma 6.3 provides the estimate

$$\|w_N(\eta(0), \eta(2), \dots, \eta(2N+1)) - w_N(\eta(1), \eta(2), \dots, \eta(2N+1))\|_2 \le C$$

and this would complete the proof.

The orthogonality of ξ_f to $\eta(0) - \eta(1)$ is easily seen.

$$\begin{aligned}
&E^\mu\big[(F(\eta^{0,1}) - F(\eta))(\eta(0) - \eta(1))\big]\\
&= E^\mu\big[F(\eta^{0,1})\eta(0) - F(\eta^{0,1})\eta(1) - F(\eta)\eta(0) + F(\eta)\eta(1)\big]\\
&= E^\mu\big[F(\eta)\eta(1) - F(\eta)\eta(0) - F(\eta)\eta(0) + F(\eta)\eta(1)\big]\\
&= 2E^\mu\big[F(\eta)(\eta(1) - \eta(0))\big]\\
&= 2E^\mu\big[(F(\eta) - F(\tau\eta))\eta(1)\big]\\
&= 0
\end{aligned}$$

by the formal translation invariance of F. A rigorous proof involves truncating the sum that is F and passing to the limit. □

Lemma 6.3 *Let v_{2N} be the solution of*

$$\sum_{(x,x+1)\in[-2N,2N]} [v_{2N}(\eta^{x,x+1}) - v_{2N}(\eta)] = \sum_{(x,x+1)\in[-2N,2N]} \xi_{2N}^{x,x+1}$$

Define w_N by

$$w_N(\eta) = E^\mu\Big[w_{2N}(\eta) | \{\eta(x)\} : |x| \le N\Big]$$

Then

$$\|w_N(\eta(0), \eta(2), \dots, \eta(2N+1)) - w_N(\eta(1), \eta(2), \dots, \eta(2N+1))\|_2 \le C$$

Proof The main step consists of the following idea. We have two functions $f(0, \xi_1, \dots, \xi_n)$ and $f(1, \xi_1, \dots, \xi_n)$ and need to estimate

$$\sum_\xi [f(1, \xi_1, \dots, \xi_n) - f(0, \xi_1, \dots, \xi_n)]\rho^k(1-\rho)^{n-k}$$

where $k = k(\xi)$ is the number of 1's among $\xi_1, \dots, \xi_n$. We rewrite this as

$$\sum_k \binom{n}{k} \rho^k (1-\rho)^{n-k}[a_k - b_k]$$

where

$$a_k = \frac{1}{\binom{n}{k}} \sum_{k(\xi)=k} f(1, \xi_1, \dots, \xi_n)$$

and

$$b_k = \frac{1}{\binom{n}{k}} \sum_{k(\xi)=k} f(0, \xi_1, \dots, \xi_n)$$

While one cannot estimate $|a_k - b_k|$ it is relatively easy to estimate $|a_k - b_{k+1}|$ because this involves interchanging a 0 and 1 and is controlled by the Dirichlet form. In fact we can write

$$|a_k - b_{k+1}| \le \frac{1}{n\binom{n-1}{k-1}} \sum_{\substack{k(\xi)=k-1 \\ 1\le i\le n}} |f(0, \xi_1, \dots, \xi_{i-1}, 1, \xi_{i+1}, \dots, \xi_n) - f(1, \xi_1, \dots, \xi_{i-1}, 0, \xi_{i+1}, \dots, \xi_n)|$$

which leads quickly to the kind of estimate that we need.

We now return to complete the proof of Theorem 6.1. The second statement is really the key. The final statement is a routine improvement of the second that we will not prove.

There are three different generators that we will consider. The generator $\mathcal{A}$ that is our primary concern, its symmetrized version $\mathbf{A}$ and the generator of the symmetric nearest neighbor exclusion process $\mathcal{A}^0$

For any local function $f \in \mathcal{F}$ we associated a vector $\xi = Tf \in \mathcal{H}$ such that

$$\lim_{\substack{N\to\infty \\ \frac{n}{2N}\to\rho}} \frac{1}{2N}\sigma^2_{\mathcal{A}^0}(N, n, f) = E^\rho[\|\xi\|^2]$$

The same method would have yielded a similar result for $\mathbf{A}$, with $\xi = T_{\mathbf{A}} f$

$$\lim_{\substack{N\to\infty \\ \frac{n}{2N}\to\rho}} \frac{1}{2N}\sigma^2_{\mathbf{A}}(N, n, f) = E^\rho[\|\xi\|^2_{\mathbf{A}}]$$

where

$$\|\xi\|^2_{\mathbf{A}} = \sum p(z)\|\xi_{0,z}\|^2$$

and some elementary algebra allows one to express $\xi_{0,z}$ in terms of $\xi = \xi_{0,1}$. Formally $f = \mathbf{A}g$, $F = \sum_x \tau_x g$ and

$$\xi_{0,x} = F(\eta^{0,x}) - F(\eta)$$

Because of the equivalence of the Dirichlet Forms corresponding to $\mathcal{A}^0$ and $\mathbf{A}$, $E^\rho[\|\xi\|^2]$ and $E^\rho[\|\xi\|^2_{\mathbf{A}}]$ are equivalent. We need to prove that for some c

$$\inf_g E^\rho[|T_{\mathbf{A}}(W - \mathcal{A}g - c(\eta(1) - \eta(0)))|^2] = 0$$

which would follow if we show that the closure in $\mathcal{H}$ of $\{T_{\mathbf{A}}\mathcal{A}g\}$ as g varies over local functions is of codimension 1 and does not contain $\eta(1) - \eta(0)$. The similar statement with $\{T_{\mathbf{A}}\mathbf{A}g\}$ replacing $\{T_{\mathbf{A}}\mathcal{A}g\}$, which we proved for $\mathcal{A}^0$ can be established in the same manner. The orthogonality of $T_{\mathbf{A}}\mathbf{A}g$ and $\eta(1) - \eta(0)$ will now hold for the inner product in the Hilbert space with norm $E^\rho[\|\xi\|^2_{\mathbf{A}}]$.

Let us consider the family $\mathcal{A}_t = t\mathcal{A} + (1-t)\mathbf{A}$ that corresponds to simpe exclusion with $p_t(z) = \frac{1+t}{2}p(z) + \frac{1-t}{2}p(-z)$. Denoting by $\mathcal{H}_t$ the range of $\{T_{\mathbf{A}}(\mathcal{A}_t g + c(\eta(1) - \eta(0))\}$ as g and c vary, we can map $\mathcal{H} \to \mathcal{H}_t$ by the linear map S_t sending $T_{\mathbf{A}}(\mathcal{A}_0 g + c(\eta(1) - \eta(0)) \to T_{\mathbf{A}}(\mathcal{A}_t g + c(\eta(1) - \eta(0))$. Note that $\mathcal{A}_0 = \mathbf{A}$. The boundedness of $\mathcal{A}$ from $H_1 \to H_{-1}$ makes this a bounded family of perturbations in the uniform norm. Since $\mathcal{H}_0$ is dense in $\mathcal{H}$, S_t extends as continuous perturbation of $S_0 = I$. If we show that $\|S_t\xi\| \geq C\|\xi\|$ uniformly in $0 \leq t \leq 1$ for some $C > 0$, we will succeed in our goal.

Let

$$u_t = c(\eta(1) - \eta(0)) + \mathcal{A}_t g$$

We need a uniform lower bound on $<< u_t, u_t >>_\rho$. From the variational formula we see that

$$\begin{aligned}
<< u_t, u_t >>_\rho &\geq \lim_{N\to\infty} \frac{1}{2N} E\Big[\sum_{|x|\leq N-k} \tau_x u_t \cdot \sum_{|x|\leq N-k} \tau_x g - \frac{1}{2} D(\sum_{|x|\leq N-k} \tau_x g)\Big] \\
&= \lim_{N\to\infty} \frac{1}{2N} E\Big[\mathcal{A}_t \sum_{|x|\leq N-k} \tau_x g \cdot \sum_{|x|\leq N-k} \tau_x g - \frac{1}{2} D(\sum_{|x|\leq N-k} \tau_x g)\Big] \\
&= \frac{1}{2} \lim_{N\to\infty} \frac{1}{2N} D(\sum_{|x|\leq N-k} \tau_x g) \\
&\geq \delta << \mathbf{A}g, \mathbf{A}g >>_\rho \\
&= \delta << u_0 - c(\eta(1) - \eta(0)), u_0 - c(\eta(1) - \eta(0)) >>_\rho
\end{aligned}$$

for some $\delta > 0$. From $\|S_t x + y\| \geq \delta \|x\|$ one easily obtains $\|S_t x + y\| \geq C_\delta \|x + y\|$. Take $x = u_0$ and $y = c(\eta(1) - \eta(0))$.

References

[1] Kipnis, C., and Landim, C. [1999] *Scaing Limits of Interacting Particle Systems*, Springer-Verlag, Heidelberg.

[2] Kipnis, C., Olla, S., and Varadhan, S. R. S. [1989] *Hydrodynamics and large deviation for simple exclusion process*, Comm. Pure Appl. Math. **42**, 115–137.

[3] Lin, Xu. [1993] *Hydrodynamics for asymmetric mean zero simple exclusion, Ph.D. Thesis, New York University.*

[4] Quastel, J. [1992] *Diffusion of color in simple exclusion process*, Comm. Pure Appl. Math. **45**, 623–680.

[5] Yau, H. T. [1991] *Relative entropy and the hydrodynamics of Ginzburg–Landau models*, Lett. Math. Phys. **22**, 63–80.

The Mathematical Sciences After the Year 2000, World Sci. Publ., 2000, 144-158

SCALING LIMITS OF LARGE INTERACTING SYSTEMS

S.R.S. VARADHAN
Courant Institute, New York University, 251 Mercer Street, New York, NY, 10012, USA
E-mail: varadhan@cims.nyu.edu

1 From Classical Mechanics to Euler Equations

The basic example of a scaling limit is derivation of the 'Euler Equations 'from the equations of classical mechanics. Let us start with a collection of $N \simeq \bar{\rho}\ell^3$ particles in a large periodic cube Λ_ℓ of side ℓ in R^3. The motion of the particles is governed by the equations of motion of a classical Hamiltonian dynamical system with energy given by

$$H(p,q) = \frac{1}{2}\sum_{i=1}^{N} \|p_i\|^2 + \frac{1}{2}\sum_{i\neq j} V(q_i - q_j) \tag{1}$$

Here, $q_i \in \Lambda_\ell$ is the position of the i-th particle and $p_i \in R^3$ is its velocity. We will use indices $k = 1, 2, 3$ to refer to the three components of position or velocity. The repulsive potential $V \geq 0$ is an even function and has compact support in R^d. The interaction in particular is short range. The equations of motion are

$$\begin{aligned} \frac{dq_i^k}{dt} &= \frac{\partial H(p,q)}{\partial p_i^k} = p_i^k \\ \frac{dp_i^k}{dt} &= -\frac{\partial H(p,q)}{\partial q_i^k} = -\textstyle\sum_{j=1}^{N} V_k(q_i - q_j) \end{aligned} \tag{2}$$

where $V_k(q) = \frac{\partial V(q)}{\partial q^k}$ for $k = 1, 2, 3$ are the three components of the gradient of V. The dynamical system has five conserved quantities. The total number N of particles, the total momenta $\sum_{i=1}^{N} p_i^k$ for $k = 1, 2, 3$ and the total energy $H(p,q)$. The hydrodynamic scaling in this context consists of rescaling space and time by a factor of ℓ. The rescaled space is the unit torus $\mathbf{T}^3$ in 3-dimensions. The macroscopic quantities to be studied correspond to the five conserved quantities. The first one of these is the density, and is measured by a function $\rho(t,x)$ of t and x. For each $\ell < \infty$ it is approximated by $\rho_\ell(t,x)$, defined by

$$\int_{\mathbf{T}^d} J(x)\rho_\ell(t,x)dx = \frac{1}{\ell^3}\sum_{i=1}^{N} J(\frac{q_i(\ell t)}{\ell}) \tag{3}$$

Differentiation with respect to t yields

$$\begin{aligned} \frac{d}{dt}\int_{\mathbf{T}^d} J(x)\rho_\ell(t,x)dx &= \frac{d}{dt}\frac{1}{\ell^3}\sum_{i=1}^{N} J(\frac{q_i(\ell t)}{\ell}) \\ &= \frac{1}{\ell^3}\sum_{i=1}^{N}(\nabla J)(\frac{q_i(\ell t)}{\ell})\cdot p_i(\ell t) \\ &\simeq \int_{\mathbf{T}^d}(\nabla J)(x)\cdot \rho_\ell(t,x)u_\ell(t,x)dx \end{aligned} \tag{4}$$

where $u_\ell(t,x) = u_\ell^k(t,x)$, $k = 1,2,3$ are the components of the 'average' velocity of the fluid at the rescaled space time point x,t. This introduces three other macroscopic variables corresponding to the three components of the momenta that are conserved. We can now write down the first of our five equations

$$\frac{\partial \rho}{\partial t} + \nabla\cdot(\rho u) = 0 \tag{5}$$

To derive the next three equations, using a test functions J, we differentiate for $k = 1,2,3$

$$\begin{aligned} \frac{d}{dt}\frac{1}{\ell^3}\sum_{i=1}^{N} J(\frac{q_i(\ell t)}{\ell})p_i^k(\ell t) &= \frac{1}{\ell^3}\sum_{i=1}^{N} p_i^k(\ell t)(\nabla J)(\frac{q_i(\ell t)}{\ell})\cdot p_i(\ell t) \\ &-\frac{1}{\ell^2}\sum_{i=1}^{N}\sum_{j=1}^{N} J(\frac{q_i(\ell t)}{\ell})V_k(q_i(\ell t)-q_j(\ell t)) \end{aligned} \tag{6}$$

If we now use the skew-symmetry of $V_k = \frac{\partial V}{\partial q_k}$, we can rewrite the second term of the right hand side of equation 6 as

$$\begin{aligned} &-\frac{1}{2\ell^2}\sum_{i=1}^{N}\sum_{j=1}^{N}(J(\frac{q_i(\ell t)}{\ell}) - J(\frac{q_j(\ell t)}{\ell})V_k(q_i(\ell t)-q_j(\ell t)) \\ &\simeq -\frac{1}{2\ell^3}\sum_{i=1}^{N}\sum_{j=1}^{N} J_r(\frac{q_i(\ell t)}{\ell})(q_i^r(\ell t)-q_j^r(\ell t))V_k(q_i(\ell t)-q_j(\ell t)) \\ &= \frac{1}{\ell^3}\sum_{i=1}^{N}\sum_{j=1}^{N} J_r(\frac{q_i(\ell t)}{\ell})\psi_k^r(q_i(\ell t)-q_j(\ell t)) \end{aligned} \tag{7}$$

with

$$\psi_k^r(q) = -\frac{1}{2}q^r V_k(q)$$

The next step is rather mysterious and requires considerable explanation. The quantities

$$\sum_{i=1}^{N} p_i^k p_i^r, \quad \sum_{i,j=1}^{N} \psi_k^r(q_i(t)-q_j(t))$$

are not conserved. They depend on combinations of individual velocities and on spacings between particles both of which change in the microscopic time scale and therefore do so rapidly in the macroscopic scale. They should therefore be replaced by their space-time averages. By appealing to an 'Ergodic Theorem' they can be replaced instead by their averages with repect to their equilibrium distributions. The equilibrium 'ensemble' consists of an infinite collection of points $\{p_\alpha, q_\alpha\}$, in the phase space $R^3 \times R^3$. There is a natural five parameter family of measures $\mu_{\rho,u,T}$ that are invariant under spatial translations as well as the Hamiltonian dynamics. The points $\{p_\alpha\}$ are distributed according to a Gibbs Distribution with density ρ and formal interaction energy

$$\frac{1}{2T}\sum_{\alpha,\beta} V(q_\alpha - q_\beta)$$

In other words $\{q_\alpha\}$ is a point process obtained by taking infinite volume limit of $N = \ell^3\rho$ particles distributed in the cube of side ℓ in R^3 according to the joint density

$$\frac{1}{Z}\exp\Big[-\frac{1}{2T}\sum_{1\le i\ne j\le N} V(q_i - q_j)\Big]$$

where Z is the normalization constant. The velocities $\{p_\alpha\}$ are distributed independently of each other as well as of $\{q_\alpha\}$, having a common three dimensional Gaussian distribution with mean u and covariance TI. Assuming that the infinite volume limit exists in a reasonable sense, it will be a point process defined as an infinite volume Gibbs measure $\mu_{\rho,T}$. The velocities $\{p_\alpha\}$ will be an independent Gaussian ensemble $\nu_{u,T}$. In the first term the quantities $p_i^k p_i^r$ are replaced by their expectations

$$u^k(t,x)u^r(t,x) + \delta_{k,r}T(t,x)$$

and in the second term $\psi_{k,r}$ are replaced by their expectations that involve the 'pressure' per unit volume in the Gibbs ensemble

$$\mathbf{P}_k^r(\rho,T) = \lim_{\ell\to\infty} E^{\mu_{\rho,T}}\Big\{\frac{1}{\ell^3}\sum_{\substack{|q_\alpha|\le\ell\\|q_\beta|\le\ell}} \psi_k^r(q_\alpha - q_\beta)\Big\}$$

This leads to the equation

$$\begin{aligned}&\tfrac{d}{dt}\textstyle\int_{\mathbf{T}^d} J(x)u^k(t,x)dx\\ &\quad= \textstyle\int_{\mathbf{T}^d}\sum_{r=1}^3 \frac{\partial J}{\partial x_r}(x).(u^k(t,x)u^r(t,x) + \delta_{k,r}T(t,x))dx \\ &\qquad+ \textstyle\int_{\mathbf{T}^d}\sum_{r=1}^3 \frac{\partial J}{\partial x_r}(x)\mathbf{P}_k^r(\rho(t,x),T(t,x))dx\end{aligned} \tag{8}$$

We now integrate by parts, remove the test function J and obtain from equation 8

$$\frac{d}{dt}(\rho u) + \nabla \cdot (\rho\, u \otimes u + \rho\, T I + \mathbf{P}(\rho\,, T)) = 0 \tag{9}$$

There is an equation of state that expresses the total energy per unit volume e as

$$e(\rho\,, u\,, T) = \frac{1}{2}\rho(|u|^2 + 3T) + f(\rho\,, T) \tag{10}$$

where $f(\rho\,, T)$, the potential energy per unit volume, is given by

$$f(\rho\,, T) = \lim_{\ell \to \infty} E^{\mu_{\rho,T}}\left\{\frac{1}{2\ell^3} \sum_{\substack{|q_\alpha| \leq \ell \\ |q_\beta| \leq \ell}} V(q_\alpha - q_\beta)\right\}$$

Although we will not derive it, there is a similar equation for $e(t, x)$ that is obtained by differentiating

$$\frac{d}{dt}\frac{1}{2\ell^3}\sum_{i=1}^{N} J(\frac{q_i(\ell t)}{\ell})\left[|p_i(\ell t)|^2 + \sum_{j=1}^{N} V(q_i(\ell t) - q_j(\ell t))\right]$$

and proceeding in a similar fashion. It looks like

$$\frac{de}{dt} + \nabla \cdot \left[(e + T)u + \mathbf{P}(\rho\,, T)u\right] = 0 \tag{11}$$

The five equations, one for density given by equation 5, the three for velocities contained in equation 9 and finally the enrgy equation 11 constitute a first order symmetric system of non-linear hyperbolic conservation laws in the six variables $[\rho, u, T, e\,]$ with one relation between them given by equation 10. Given smooth initial data they have local solutions. Rigorous derivation of these equations does not exist. The ergodic theory is definitely plausible, but hard to establish. The reason is that we have essentially an infinite system of ordinary differential equations representing a classical deterministic dynamical system and the ergodicity properties are nearly impossible to establish in any general context. See [S] for a discussion of these and other related ideas. However if we had some noise in the system, i.e. stochastic dynamics instead of deterministic dynamics, then we will be concerned with the ergodic theory of Markov Processes of some kind, which is far more accessible. This is essentially the focus of our lectures.

2 Some Stochastic Models.

We will consider a class of examples that are called simple exclusion processes. They make sense on any finite or countable set X and for us X will be either the integer lattice $\mathbf{Z}^d$ in d-dimensions or $\mathbf{Z}^d_N$ obtained from it as a quotient by considering each coordinate modulo N. At any given time a subset of these lattice sites will be occupied by particles, with atmost one particle at each site. In other words some sites are empty while others are occupied with one particle. The particles move randomly. Each particle waits for an exponential random time and then tries to jump from its current site x to a new site y. The new site y is picked randomly according to a probability distribution $\pi(x\,,y)$. In particular $\sum_y \pi(x\,,y) = 1$ for every x. Of course a jump to y is not always possible. If the site is empty the jump is possible and is carried out. If the site already has a particle, the jump cannot be carried out and the particle forgets about it and waits for another chance, i.e. waits for a new random exponential time.

If we normalize so that all waiting times have mean 1, the generator of the process can be written down as

$$(\mathcal{A}f)(\eta) = \sum_{x,y} \eta(x)(1-\eta(y))\pi(x\,,y)[f(\eta^{x,y}) - f(\eta)] \tag{12}$$

where η represents the configuration with $\eta(x) = 1$ if there is a particle at x and $\eta(x) = 0$ otherwise. For each configuration η and a pair of sites x, y the new configuration $\eta^{x,y}$ is defined by

$$\eta^{x,y}(z) = \begin{cases} \eta(y) & \text{if } z = x \\ \eta(x) & \text{if } z = y \\ \eta(z) & \text{if } z \neq x\,, y \end{cases} \tag{13}$$

We will mainly be concerned with the situation where the set X is $\mathbf{Z}^d_N$, viewed naturally as an Abelian group with $\pi(x\,,y)$ being translation invariant and given by $\pi(x\,,y) = p(y-x)$ for some probability distribution p that does not sepend on N. It is natural to assume that p has finite support. There are various possibilities. p is symmetric i.e. $p(x) = p(-x)$ or more generally p has mean zero i.e.$\sum_y yp(y) = 0$ and finally $\sum_y y\,p(y) = m \; not = 0$. We shall first concentrate on the symmetric case. Let us look at the function

$$V_J(\eta) = \sum J(x)\eta(x)$$

and compute

$$\begin{aligned}(\mathcal{A}V_J)(\eta) &= \textstyle\sum_{x,y} \eta(x)(1-\eta(y))p(y-x)(J(y)-J(x)) \\ &= \textstyle\sum_{x,y} \eta(x)p(y-x)(J(y)-J(x)) \\ &= \textstyle\sum_{x} \eta(x)[(\mathbf{P}-I)J](x) \\ &= V_{(\mathbf{P}-I)J}(\eta)\end{aligned} \tag{14}$$

The space of linear functionals is left invariant by the generator. It is not difficult to see that

$$E_\eta\left[V_J(\eta(t))\right] = V_{J(t)}(\eta)$$

where

$$J(t) = \exp[t(\mathbf{P}-I)]J$$

is the solution of

$$\frac{d}{dt}J(t,x) = (\mathbf{P}-I)J(t,x)$$

It is almost as if the interaction had no effect and in fact for the calculation of expectations of 'one particle' functions it clearly does not. Let us start with a configuration on $\mathbf{Z}_N^d$ and scale space by N and time by N^2. The generator becomes $N^2\mathcal{A}$ and the particles can be visualized as moving on a lattice imbedded in the unit torus $\mathbf{T}^d$, with a spacing of $\frac{1}{N}$, and becoming dense as $N \to \infty$.

We will be concerned with the asymptotic behavior of the empirical distribution

$$\nu_N(t) = \frac{1}{N^d}\sum_x \delta_{\frac{x}{N}}\eta(x,t) \tag{15}$$

that does not keep track of the identity of the particle as well as that of the empirical process

$$R_n = \frac{1}{N^d}\sum_i \delta_{x_i(\cdot)} \tag{16}$$

Let us consider the functional

$$\xi(t) = \frac{1}{N^d}\sum_x J(\frac{x}{N})\eta_t(x)$$

and we can write

$$\xi(t) - \xi(0) = \int_0^t V_N(\eta(s))ds + M_N(t)$$

where

$$V_N(\eta) = (N^2 \mathcal{A} V_J)(\eta) = V_{J_N}(\eta)$$

with

$$\begin{aligned}(J_N)(x) &= N^2 \sum \left[J(x + \tfrac{z}{N}) - J(x)\right]p(z) \\ &\simeq \tfrac{1}{2}(\Delta_C J)(x)\end{aligned} \tag{17}$$

Here Δ_C refers to the Laplacian

$$\sum_{i,j} C_{i,j} \frac{\partial^2}{\partial x_i \partial x_j}$$

with the covariance matrix C given by

$$C_{i,j} = \sum_x x_i x_j p(x)$$

$M_N(t)$ is a martingale and a very elementary calculation yields

$$E\left\{[M_N(t)]^2\right\} \leq C\,t\,N^{-d}$$

essentially completing the proof in this case. Technically the empirical distribution $\nu_N(t)$ defined in equation 15 is viewed as a measure on $\mathbf{T}^d$ and $\nu_N(\cdot)$ is viewed as a stochastic process with values in the space $\mathcal{M}(\mathbf{T}^d)$ of nonnegative measures on $\mathbf{T}^d$. In the limit it lives on the set of weak solutions of the heat equation

$$\frac{\partial \rho}{\partial t} = \frac{1}{2}\Delta_C\, \rho \tag{18}$$

and the uniqueness of such weak soultions for given initial density establishes the validity of the scaling limit. We could have have computed two moments in order to prove the convergence which is essentially a large numbers. The first moment will involve nothing more than the calculation done in 14 The variance involves two particle functions and need a little bit of work because of the interaction. The martingale argument however is more general.

Let us make the problem slightly more complicated by adding a small bias. Let $q(x)$ be an odd function with $q(-x) = -q(x)$ and we will modify the problem by making p depend on N in the form

$$p_N(x) = p(x) + \frac{1}{N}q(x) \tag{19}$$

Assuming that q is nonzero only when p is so, p_N will be an admissible transition probability for large enough N. A calculation in this slightly different model referred to as weakly asymmetric simple exclusion model shows that V_N is now modified by

$$V_N(\eta) \simeq V_{J_N}(\eta) + \frac{1}{N^d}\sum \eta(x)(1-\eta_x) < m\,,\nabla J(x) > \tag{20}$$

with

$$m = N\sum_x x\,p_N(x) = \sum_x x\,q(x)$$

If one thinks of $\rho(t\,,x)$ as the density of particles at the (macroscopic) time t and space x the first term on the right hand side of equation 20 clearly wants to have the limit

$$\int_{\mathbf{T}^d} \frac{1}{2}(\Delta_C J)(x)\rho(t\,,x)dx$$

It is not obvious what to do with the second term. The 'invariant' measures in this model are the Bernoulli measures with various densities ρ and upon 'averaging' the second term should converge to

$$\int_{\mathbf{T}^d} < m\,,(\nabla J)(x) > \rho(t\,,x)(1-\rho(t\,,x)dx$$

replacing the linear heat equation 18 by the nonlinear equation

$$\frac{\partial\rho}{\partial t} = \frac{1}{2}\Delta_C\rho - \nabla\cdot m\rho(1-\rho) \tag{21}$$

This requires justification that can be provided, because the 'Ergodic Theory 'can be carried out.

Let us now turn to the case where p has mean zero but is not symmetric. In this case

$$V_N(\eta) = N^{2-d}\sum_{x,y}\eta(x)(1-\eta(y))p(y-x)[J(\frac{y}{N}) - J(\frac{x}{N})] \tag{22}$$

and we get stuck at this point. If p is symmetric, as we saw, we gain a factor of N^{-2}. Otherwise the gain is only a factor of N^{-1} which is not enough. We seem to end up with

$$\begin{aligned} N^{-d}\textstyle\sum_x\sum_y \eta(x) &< \tfrac{1}{2}[(\nabla J)(\tfrac{x}{N}) \\ &\quad +(\nabla J)(\tfrac{y}{N})]\,,N(1-\eta(y))(y-x)p(y-x) > \\ = \tfrac{1}{2N^d}\textstyle\sum_x(\nabla J)(\tfrac{x}{N})N\Psi_x & \end{aligned} \tag{23}$$

where

$$\begin{aligned}\Psi_0 &= \tfrac{1}{2}[\eta(0)\textstyle\sum_z(1-\eta(z))z\,p(z) + (1-\eta(0))\sum_z \eta(-z)z\,p(z)]\\ &= \tfrac{1}{2}[-\eta(0)\textstyle\sum_z \eta(z)z\,p(z) + (1-\eta(0))\sum_z \eta(-z)z\,p(z)]\\ &= \tfrac{1}{2}[\textstyle\sum_z \eta(-z)z\,p(z) - \eta(0)\sum_z(\eta(z)+\eta(-z))z\,p(z)]\end{aligned} \tag{24}$$

The second sum is zero in the symmetric case and Ψ_0 can then be written as a 'gradient' $\Psi_0 = \sum_j \tau_{e_j}\xi_j - \xi_j$ where τ_{e_j} are shifts in the coordinate directions. This allows us to do summation by parts and gain a factor of N^{-1}. When this is not the case, we have a 'nongradient' model and the scaling limit can no longer be established by simple averaging.

We will now consider the third case $\sum_z zp(z) = m \neq 0$ that needs hyperbolic scaling and leads to Burgers equation with zero viscosity. In order to convey the idea it is sufficient to restrict ourselves to the case wher $d = 1$, and $p(1) = 1$ and $p(z) = 0$ for all $z \neq 1$. This is the totally asymmetric nearest neighbor simple exclusion model and of course $m = 1$.

In this case the rescaling in time is done by a factor of N and the generator is

$$(\mathcal{A}_N f)(\eta) = N \sum_{x\in \mathbf{Z}_N} \eta(x)(1-\eta(x+1))[f(\eta^{x,x+1}) - f(\eta)] \tag{25}$$

We can easily calculate

$$\begin{aligned}&d[\tfrac{1}{N}\textstyle\sum_{x\in\mathbf{Z}_N} J(\tfrac{x}{N})\eta_t(x)]\\ &\quad= [\textstyle\sum_{x\in\mathbf{Z}_N}[J(\tfrac{x+1}{N}) - J(\tfrac{x}{N})]\eta_t(x)(1-\eta_t(x+1))]dt + M_N(t)\\ &\quad\simeq \tfrac{1}{N}[\textstyle\sum_{x\in\mathbf{Z}_N} J'(\tfrac{x}{N})\eta_t(x)(1-\eta_t(x+1))]dt + o(1)\end{aligned} \tag{26}$$

The martingale term is negligible and if do simple 'averaging' we get the equation

$$\frac{d}{dt}\int_{\mathbf{T}} J(\theta)\rho(t,\theta)d\theta = \int_{\mathbf{T}} J'(\theta)\rho(t,\theta)(1-\rho(t,\theta))d\theta$$

or equivalently

$$\frac{\partial\rho}{\partial t} + \frac{\partial[\rho(1-\rho)]}{\partial\theta} = 0 \tag{27}$$

Let us consider a smooth solution of the above equation in $[0,T]\times\mathbf{T}$. Let us assume $c \le \rho(t,\theta) \le 1-c$ for some $c > 0$. On $\mathbf{Z}_N$ we start with an initial distribution α_N where $\eta(x)$ are independent under α_N with $\alpha_N[\eta(x) = 1] =$

$\rho(0, \frac{x}{N})$. At any time $t \in [0,T]$ we have the actual distribution of the particle system obtained by solving the forward equation of Kolmogorov

$$\frac{d\mu_N(t)}{dt} = \mathcal{A}_N^* \mu_N(t)$$

with $\mu_N(0) = \alpha_N$. We can define $\alpha_N(t)$ from $\rho(t, \cdot)$ the same way we defined α from $\rho(0, \cdot)$ at time 0. Although $\alpha_N(t)$ and $\mu_N(t)$ match by our choice at $t = 0$, in general they do not do so for $t > 0$. However we can prove that the relative entropy $H(\mu_N(t); \alpha_N(t))$ of $\mu_N(t)$ with respect to $\alpha_N(t)$ satisfies

$$\lim_{N\to\infty} \sup_{0\le t\le T} \frac{H(\mu_N(t); \alpha_N(t))}{N} = 0$$

By simple large deviation estimates this does imply the validity of the scaling limit. In particular

$$\limsup_{N\to\infty} \sup_{o\le t\le T} E^{\mu_N(t)} \left| \frac{1}{N} \sum_x J(\frac{x}{N})\eta(x) - \int_{\mathbf{T}} J(\theta)\rho(t, \theta)d\theta \right| = 0$$

The proof involves what is known as the relative entropy method and involves a careful reduction to an application Gronwall's lemma. See [JY] or [KL] for details ragarding material presented here.

3 Tagged Particles.

The next step is to investigate the behavior of tagged particles. First we need to consider the motion of a tagged particle in equilibrium. In other words at time zero one particular particle is tagged and its motion is followed. In equilibrium we expect the tagged particle to diffuse like a Brownian motion. In fact for a wide class of models this is now a well understood phenomenon [KV]. The covariance matrix for the limiting Brownian motion will in general depend on the equlibrium that is used. In our model on Z^d the equilibria are the various Bernoulli or product mesures μ_ρ with particle density

$$\rho = \mu_\rho\{\eta(x) = 1\}.$$

The limiting covariance will be denoted $S(\rho)$ which is called the self-diffusion matrix. In the noninteracting case it will not depend on ρ. But for symmetric simple exclusion, if we ignore the very special case of nearest neighbor model in one dimension, it is not surprising that at low density, i.e near $\rho = 0$, $S(\rho) \simeq D$ because the interaction is negligible and near high density, i.e. $\rho = 1$, $S(\rho) \simeq 0$ essentially due to the gridlock that develops when nearly every site is filled.

The behavior of a tagged particle in nonequilibrium is far from obvious. The density ρ is no longer a constant, but a function $\rho(t,x)$ of t and x, given by the solution of equation 18. One might guess that a tagged particle located at $x(t)$ at time t, being myopic, will only see its immediate neighborhood and behave as if it is in equilibrium at density $\rho = \rho(t,x)$. As a first approximation we could expect $x(t)$ to behave like a time dependent diffusion with generator

$$\frac{1}{2}\nabla S(\rho(t,x))\nabla$$

but in reality it is more likely that the generator needs to be modified by a first order term and takes the form

$$\frac{1}{2}\nabla S(\rho(t,x))\nabla + b(t,x)\nabla \tag{28}$$

for some drift term $b(\cdot,\cdot)$. Because we have an extra as yet undetermined first order term it does not matter if we write the higher order term in divergence form or not. We have chosen the divergence form for convenience.

If we picked initially a particle at random and tagged it, its distribution is given, subject to normalization, by a density evolving according to the forward (adjoint) equation corresponding to equation 28.

$$\frac{\partial \rho}{\partial t}(t,x) = \frac{1}{2}\nabla S(\rho(t,x))\nabla\rho(t,x) - \nabla\cdot(b(t,x)\rho(t,x) \quad \text{for } t>0 \tag{29}$$

with the initial condition

$$\rho(0,x) = \rho_0(x).$$

Tagging is only an exercise of the mind. The particles are totally oblivious to it. Since the empirical density is the same for the tagged or untagged system the solutions of equations 18 and 29 must coincide. This yields

$$\frac{1}{2}\nabla S(\rho(t,x))\nabla\rho(t,x) - \nabla\cdot(b(t,x)\rho(t,x)) = \frac{1}{2}\nabla D\nabla\rho(t,x).$$

While it does not follow, it is quite likely that

$$\frac{1}{2}S(\rho(t,x))\nabla\rho(t,x) - b(t,x)\rho(t,x) = \frac{1}{2}D\nabla\rho(t,x)$$

which can be solved to yield

$$b(t,x) = \frac{1}{2\rho(t,x)}[S(\rho(t,x)) - D]\nabla\rho(t,x)$$

If we now denote by Q a Markov process (total mass $\bar{\rho}$) having initial density $\rho_0(x)$ and time dependent (backward) generator

$$L_t = \frac{1}{2}\nabla S(\rho(t,x))\nabla + \frac{1}{2}\frac{[S(\rho(t,x)) - D]\nabla\rho(t,x)}{\rho(t,x)} \cdot \nabla \tag{30}$$

then one expects the empirical process defined in 16 to have the limit

$$\lim_{N\to\infty} R_N = Q \tag{31}$$

in probability. For simple exclusion processes 31 was indeed proved by Rezakhanlou [R]. Once the distribution of the tagged particle is shown to converge to a diffusion with generator L_t given by 30 is not very hard to prove 31. One needs to show that the motion of two different tagged particles are asymptotically independent. However if we start with an initial configuration that is far away from equilibrium, like an arbitrary deterministic initial condition, to prove the limit theorm when an **arbitrary** particle in the initial configuration is tagged is a very hard result to prove. On the other hand proving 31 requires only that the limit theorem be valid for **nearly all** particles in the initial configuration.

The proof of 31 in [R] uses earlier results in [Q] that study a situation that mediates between tagged and untagged versions. Let us devide the particles into a finite number of types that we will think of as colors. Colors do not affect the motion which is as before. We simply keep track of the colors. At any time t, we have the empirical distribution of each one of the k colors. In the scaled limit we will have k densities $\tilde{\rho}(t,x) = \{\rho_j(t,x) : 1 \le j \le k\}$. They should evolve as a nonlinear system

$$\frac{\partial \tilde{\rho}(t,x)}{\partial t} = \nabla \mathbf{B}(\tilde{\rho}(t,x))\nabla\tilde{\rho}(t,x) \tag{32}$$

with suitable initial conditions. The system 32 is equivalent to solving first the *linear* equation 18 for the total density

$$\rho(t,x) = \sum_{j=1}^{k} \rho_j(t,x)$$

and then at the next step solving a set of (identical) equations that are again linear with the correct initial conditions for the densities of individual colors

$$\frac{\partial \rho_j(t,x)}{\partial t} = L_t^* \rho_j(t,x)$$

which is just the forward equation for the probability density of the position of the tagged particle.

In [Q] the multicolor system was viewed in its own right and the law of large numbers under diffusive scaling was established for it. For the simple exclusion model, this turns out to be a nongradient system and the analysis becomes considerably harder. The transition from the multicolor system to 31 is essentially a question of compactness. Let $\mathcal{P} = \{A_1, \cdots, A_q\}$ be an arbitrary but fixed partition of the $\mathbf{T}^d$ into nice sets. If E is a subset of the form

$$E = \{x : x(t_1) \in A_{i_1}, \cdots, x(t_n) \in A_{i_n}\}$$

and the convergence in probability of $R_N(E)$ to $Q(E)$ can be proved by induction on n. For $n = 1$, this just involves the empirical density and a single color. If assume the result for n time points we have q^n possible past histories and we code them by q^n distict colors. The emprical density of the diffrent colors at the next time t_{n+1} is sufficient to yield information on subsets E at stage $n+1$. Given the law of large numbers for the multicolor system the induction proceeds smoothly. Since the partition is arbitrary, if we have compactness, we can pass from finite dimensional distributions to weak convergence of the processes R_N. This was carried out in [R].

4 Large Deviations.

We now turn to the question of Large Deviation rates. We examine first the empirical densities ν_t^N.

The probability distribution of the empirical density satisfies a large deviation principle with the folowing rate functional $I(\rho(\cdot,\cdot))$ on the space of weakly continuous maps ρ from $[0,T]$ into $L_1(T^d)$ that satisfy $0 \le \rho(t,x) \le 1$.

$$I(\rho(\cdot,\cdot)) = I_0(\rho(0,\cdot) + \frac{1}{2}\int_0^T ||\frac{\partial \rho}{\partial t} - \frac{1}{2}\nabla D \nabla \rho||^2_{-1,\rho(t,\cdot)(1-\rho(t,\cdot))} dt \tag{33}$$

The first term I_0 is the rate function for deviations of the initial profile from its appropriate limit. For deterministic initial conditions it is 0 for the true value and ∞ for all others. Otherwise it depends on the statistics of the initial configuration. The second term which is the dynamical part measures in some way by how much ρ fails to satisfy the heat equation 18. For any $\rho(\cdot)$ the norm $||\ ||_{-1,\rho(t,\cdot)(1-\rho(t,\cdot))}$ refers to the dual of the weighted Sobolev norm

$$||f||^2_{1,\rho(\cdot)} = \int_{T^d} < D\nabla f(x), \nabla f(x) > \rho(x)(1-\rho(x))dx$$

This was proved in [KOV] and it is necessary for us to understand how these large deviations arise. One can introduce a small bias in the system by perturbing the symmetric jump rates by a weak asymmeric term i.e. the probabilities of transition from x to $x+z$ at time t, are changed from $p(z)$, by a small perturbation, to $p(z)+\frac{1}{N}q(t,\frac{x}{N},z)$. This has an enrtopy cost as well as a macroscopic effect. To calculate these quantities we need to know that the system is locally close to an equlibrium (some Bernoulli measure) and the empirical density dictates which equlibrium we have to pick. Quntities not explicitly dependent on the empirical densities are replaced by their expectations in equlibrium which means by suitable functions of the empirical density.

Macroscopically, with weak asymmetry, we saw that the equation 18 gets replaced by 21 and the entropy cost (after optimizing over $q(\cdot,\cdot,\cdot)$ for a given $b(\cdot,\cdot)$ and normalization by a factor of N^{-d}) is given by

$$\mathcal{E}(b(\cdot,\cdot)) = \frac{1}{2}\int_0^T\int_{T^d} < Db(t,x), b(t,x) > \rho(t,x)(1-\rho(t,x))dxdt$$

For a given $\rho(\cdot,\cdot)$ we consider

$$\mathcal{B}_{\rho(\cdot,\cdot)} = \{b(\cdot,\cdot) : (5.2) \text{ holds } \}$$

The dynamical term in 33 is the result of optimizing $\mathcal{E}(b(\cdot,\cdot))$ over $b(\cdot,\cdot)\in\mathcal{B}_{\rho(\cdot,\cdot)}$.

The weak asymmetry has an efect on the behavior of a tagged particle which can be described through $b(\cdot,\cdot)$. The generator L_t gets replaced by

$$L_{t,b(t,\cdot)} = L_t + (1-\rho(t,x))b(t,x)\cdot\nabla$$

and the tagged particle process with the new generator will be denoted by Q_b. As long as $b(\cdot,\cdot)\in\mathcal{B}_{\rho(\cdot,\cdot)}$, Q_b will have $\rho(t,\cdot)$ for its marginal density at time t.

Suppose Q is a general stochastic process on $C[[0,T];T^d]$ with total mass $\bar{\rho}$. Its marginal densities can be denoted by $q(t,\cdot)$. The 'current' is the collection of expected values of the stochastic integrals

$$E^Q\left[\int_0^T < f(t,x(t)), dx(t) > \right]$$

and we can try to find $b_Q\in\mathcal{B}(q(\cdot,\cdot))$ such that the marginals as well as the 'current' of Q match those of the tagged particle distribution Q_{b_Q} that we will obtain under our perturbation. The measure Q will then determine b_Q and Q_{b_Q}. The large deviation rate function for R_N is the functional

$$\mathcal{I}(Q) = I_0(q(0,\cdot)) + \mathcal{E}(b_Q(\cdot,\cdot)) + H(Q; Q_{b_Q}) \tag{34}$$

The important point here is the interpretation of the terms. The first term is clearly the contribution of the statistics of the initial condition. The second term depends on the hydrodynamical scaling behavior of the model and depends on the objects that describe them. The last term is pure noise, and says that once the correct macroscopic behavior of density and current are assured the tagged particle motions are nearly independent. The details can be found in [QRV]

References

1. L. Jensen and H.T. Yau, Hydrodynamical Scaling Limits of Simple Exclusion Models *IAS/Park City Mathematics Series* 167-221 1998
2. C. Kipnis and S.R.S Varadhan, Central Limit Theorem for Additive Functionals of Reversible Markov Processes and Applications to Simple Exclusions. *Commun. Math. Phys.* 104 1-19. 1986
3. C. Kipnis, S. Olla and S.R.S. Varadhan, Hydrodynamics and large deviations for simple exclusion processes *Comm. Pure Appl. Math.* 42 115–137 1989
4. J. Quastel, Diffusion of color in the simple exclusion process *Comm. Pure Appl. Math.* 45 623–679 1992
5. J. Quastel, F. Rezakhanlou and S.R.S. Varadhan, Large Deviations for the Symmetric Simple Exclusion Process *Prob. Th. and Rel. Fields* 113 1–84 1999
6. F. Rezakhanlou, Propagation of chaos for symmetric simple exclusion *Comm. Pure Appl. Math.* 117 943–957 1994
7. H. Spohn, Large Scale Dynamics of Interacting Particles *Text and Monographs in Physics* Springer-Verlag 1991

Nova Série
BOLETIM
DA SOCIEDADE BRASILEIRA DE MATEMÁTICA

Bol. Soc. Bras. Mat., Vol.31, No. 3, 241-275

Asymptotic behavior of a tagged particle in simple exclusion processes

C. Landim, S. Olla and S. R. S. Varadhan

Abstract. We review in this article central limit theorems for a tagged particle in the simple exclusion process. In the first two sections we present a general method to prove central limit theorems for additive functional of Markov processes. These results are then applied to the case of a tagged particle in the exclusion process. Related questions, such as smoothness of the diffusion coefficient and finite dimensional approximations, are considered in the last section.

Keywords: central limit theorem, interacting particles systems, tagged particle.

1 Introduction

In the early 80's Kipnis and Varadhan [6] proved an invariance principle for the position of a tagged particle in a symmetric simple exclusion process in equilibrium. Their proof relies on a general central limit theorem for additive functionals of reversible Markov processes. Time reversibility and translation invariance of the system are the basic ingredients of this method, which in principle can be applied to any system with these two symmetries. Later it has been extended to non-reversible processes that satisfy a sector condition, [21], or a graded sector condition, [19].

The effective diffusion matrix of the limiting Brownian Motion is a function $D(\alpha)$ of the density α of the particles. These diffusion coefficients are usually expressed in terms of integrals of time correlation functions (Green-Kubo formulas), or as infinite dimensional variational formulas. They also appear in the diffusive equations that govern the non-equilibrium evolution of the conserved quantities of the system. In order to have regular strong solution to these diffusive

Received 1 October 2000.

equations it is important to establish the regularity of these diffusion coefficients as functions of the conserved quantities.

In this article we review some recent results on the central limit theorem for a tagged particle in the simple exclusion process and on the smoothness of the diffusion coefficient.

In the next section, we present a general method to prove a central limit theorem for an additive functional of Markov ergodic processes and show that the proof is reduced to the verification of a bound in $\mathcal{H}_{-1}$ for the solution of the resolvent equation. In the third section we show that this estimate can be deduced if the generator of the Markov process has some properties, called the sector and the graded sector condition. In the fourth section, we apply these results to prove a central limit theorem for the tagged particle for mean zero exclusion processes and for asymmetric exclusion processes in dimension $d \geq 3$. In the last section we show that the covariance matrix depends smoothly on the density of particles and we present some extensions.

2 Central limit theorem for Markov processes

The purpose of this section is to find conditions which guarantee a central limit theorem for an additive functional of a Markov process. The idea is to represent the additive functional as the sum of a martingale with a small term, that vanishes in the limit, and to use the well known central limit theorem for martingales that we now recall.

On a probability space $(\Omega, P, \mathcal{F})$, consider a square-integrable martingale $\{M_t : t \geq 0\}$ which vanishes at time 0 and denote by $< M, M >_t$ its quadratic variation.

Lemma 2.1. *Assume that the increments of the martingale M_t are stationary and that its quadratic variation converges in $L^1(P)$ to some positive constant σ^2: for every $t \geq 0$, $n \geq 1$ and $0 \leq s_0 < \cdots < s_n$,*

$$(M_{s_1} - M_{s_0}, \ldots, M_{s_n} - M_{s_{n-1}}) \;=\; (M_{t+s_1} - M_{t+s_0}, \ldots, M_{t+s_n} - M_{t+s_{n-1}})$$

in distribution and

$$\lim_{t\to\infty} E\Big[\Big|\frac{< M, M >_t}{t} - \sigma^2\Big|\Big] = 0\,.$$

Then, $M_t/\sqrt{t}$ converges in distribution to a mean zero Gaussian law with variance σ^2.

In this statement E stands for the expectation with respect to P. The proof of this result is a simple consequence of the well known central limit theorem for sums of stationary and ergodic square integrable martingale differences.

It follows from the stationarity assumption that $\sigma^2 = E[M_1^2]$ because

$$\begin{aligned} E[< M, M >_n] &= \sum_{0 \le j < n} E[< M, M >_{j+1} - < M, M >_j] \\ &= \sum_{0 \le j < n} E[(M_{j+1} - M_j)^2] = nE[M_1^2] . \end{aligned}$$

Consider now a Markov process X_t taking values in a complete separable metric space E endowed with its Borel σ-algebra $\mathcal{E}$. Assume that there exists a stationary ergodic state π. Denote by L the generator of the Markov process in $L^2(\pi)$ and by $\mathcal{D}(L)$ its domain. Let L^* be the adjoint of L in $L^2(\pi)$. Since π is stationary, L^* is itself the generator of a Markov process. Assume that there exists a core $C \subset \mathcal{D}(L) \cap \mathcal{D}(L^*)$ for both generators L and L^*. We denote by $\mathbb{P}_\pi$ the measure on the path space $D(\mathbb{R}_+, E)$ induced by the Markov process X_t starting from π and by $\mathbb{E}_\pi$ expectation with respect to $\mathbb{P}_\pi$.

Fix a function $V: E \to \mathbb{R}$ in $L^2(\pi)$. The object of this section is to find conditions on V which guarantee a central limit theorem for

$$\frac{1}{\sqrt{t}} \int_0^t V(X_s)\, ds .$$

Assume first that there exists a solution f in $\mathcal{D}(L)$ of the Poisson equation

$$-Lf = V . \tag{2.1}$$

In this case a central limit theorem follows from the central limit theorem for martingales stated in Lemma 2.1. Indeed, since f belongs to the domain of the generator,

$$M_t = f(X_t) - f(X_0) - \int_0^t (Lf)(X_s)\, ds$$

is a martingale with quadratic variation given by

$$< M, M >_t = \int_0^t ds \left\{ (Lf^2(X_s) - 2f(X_s)(Lf)(X_s) \right\} .$$

Since f is the solution of the Poisson equation (2.1), we may write the additive functional in terms of the martingale M_t:

$$\frac{1}{\sqrt{t}} \int_0^t V(X_s)\, ds = \frac{M_t}{\sqrt{t}} + \frac{f(X_0) - f(X_t)}{\sqrt{t}} .$$

Since f belongs to $L^2(\pi)$ and the measure is stationary, $[f(X_0) - f(X_t)]/\sqrt{t}$ vanishes in $L^2(\mathbb{P}_\pi)$ as $t \uparrow \infty$. It remains to check that the martingale M_t satisfies the assumptions of Lemma 2.1.

The increments of the martingale M_t are stationary because X_t under $\mathbb{P}_\pi$ is itself a stationary Markov process. On the other hand, since π is ergodic, in view of the formula for the quadratic variation of the martingale in terms of f, $t^{-1} < M, M >_t$ converges in $L^1(\mathbb{P}_\pi)$, as $t \uparrow \infty$, to $E_\pi[Lf^2 - 2f(Lf)] = 2 < f, (-L)f >_\pi$, where $< \cdot, \cdot >_\pi$ stands for the inner product in $L^2(\pi)$. Since the martingale M_t vanishes at time 0, by Lemma 2.1, $t^{-1/2} M_t$, and therefore $t^{-1/2} \int_0^t V(X_s)\, ds$, converges in distribution to a Gaussian law with mean zero and variance $\sigma^2 = 2 < f, (-L)f >_\pi$.

Of course the existence of a solution f in $L^2(\pi)$ of the Poisson equation (2.1) is too strong and should be weakened. To state the main result of this section, we need to introduce the Sobolev spaces $\mathcal{H}_1$, $\mathcal{H}_{-1}$ associated to the generator L.

2.1 The Sobolev spaces $\mathcal{H}_1$, $\mathcal{H}_{-1}$

Consider the semi-norm $\| \cdot \|_1$ defined in the domain of the generator $\mathcal{D}(L)$ by

$$\|f\|_1^2 = < f, (-L)f >_\pi \ . \tag{2.2}$$

Let $\sim_1$ be the equivalence relation in $\mathcal{D}(L)$ defined by $f \sim_1 g$ if $\|f - g\|_1 = 0$ and denote by $\mathcal{G}_1$ the normed space $(\mathcal{D}(L)|_{\sim_1}, \| \cdot \|_1)$. It is easy to see from definition (2.2) that the norm $\| \cdot \|_1$ satisfies the parallelogram law so that $\mathcal{H}_1$, the completion of $\mathcal{G}_1$ with respect to the norm $\| \cdot \|_1$, is a Hilbert space with inner product $< \cdot, \cdot >_1$ given by polarization:

$$< f, g >_1 = \frac{1}{4}\Big\{ \|f + g\|_1^2 - \|f - g\|_1^2 \Big\} \ .$$

Notice that in this definition only the symmetric part of the generator, $S = (1/2)(L + L^*)$, plays a role because

$$\|f\|_1^2 = < f, (-L)f >_\pi = < f, (-S)f >_\pi \ .$$

It is also easy to check that for any f, g in the domain of L,

$$< f, g >_1 = < f, (-S)g >_\pi$$

and that $\|c\|_1 = 0$ for any constant c.

Associated to the Hilbert space $\mathcal{H}_1$, is the dual space $\mathcal{H}_{-1}$ defined as follows. For f in $L^2(\pi)$, let

$$\|f\|_{-1}^2 = \sup_{g \in \mathcal{D}(L)} \left\{ 2 < f, g >_\pi - \|g\|_1^2 \right\} .$$

Denote by $\mathcal{G}_{-1}^0$ the subspace of $L^2(\pi)$ of all functions with finite $\|\cdot\|_{-1}$ norm. Introduce in $\mathcal{G}_{-1}^0$ the equivalence relation $\sim_{-1}$ by stating that $f \sim_{-1} g$ if $\|f - g\|_{-1} = 0$ and denote by $\mathcal{G}_{-1}$ the normed space $(\mathcal{G}_{-1}^0|_{\sim_{-1}}, \|\cdot\|_{-1})$. The completion of $\mathcal{G}_{-1}$ with respect to the norm $\|\cdot\|_{-1}$, denoted by $\mathcal{H}_{-1}$, is again a Hilbert space with inner product defined through polarization.

Before we state the main result of this section, we summarize some properties of the spaces $\mathcal{H}_1$, $\mathcal{H}_{-1}$ that will be repeatedly used. It is easy to check from the variational formula for the $\mathcal{H}_{-1}$ norm that for every function f in $\mathcal{D}(L)$ and every functions g in $L^2(\pi) \cap \mathcal{H}_{-1}$

$$\Big| < f, g >_\pi \Big| \le \|f\|_1 \, \|g\|_{-1} . \tag{2.3}$$

The same variational formula permits to show that a function in $\mathcal{D}(L)$ belongs to $\mathcal{H}_{-1}$ if and only if there exists a finite constant C_0 such that

$$< f, g >_\pi \le C_0 \|g\|_1 \tag{2.4}$$

for every g in $\mathcal{D}(L)$. In this case, $\|f\|_{-1} \le C_0$. Finally, it is not difficult to show (cf. [15], Lemma 2.5) that a function f belongs to $\mathcal{H}_{-1}$ if and only if there exists h in the domain of $\sqrt{-S}$ such that $\sqrt{-S}h = f$. In this case $\|f\|_{-1} = \|h\|_0$, which means that

$$\|f\|_{-1}^2 = < h, h >_\pi = < (-S)^{-1/2} f, (-S)^{-1/2} f >_\pi = < (-S)^{-1} f, f >_\pi$$

because S is symmetric.

We may now state the main result of this section. Fix a function V in $L^2(\pi) \cap \mathcal{H}_{-1}$, $\lambda > 0$ and consider the resolvent equation

$$\lambda f_\lambda - L f_\lambda = V . \tag{2.5}$$

Theorem 2.2. *Suppose that*

$$\sup_{0 < \lambda \le 1} \|L f_\lambda\|_{-1} < \infty . \tag{2.6}$$

Then, $t^{-1/2}\int_0^t V(X_s)\,ds$ *converges in law to a mean zero Gaussian distribution with variance*

$$\sigma^2 = \lim_{\lambda\to 0} \| f_\lambda \|_1^2 .$$

Notice that

$$\sup_{0<\lambda\le 1} \|Lf_\lambda\|_{-1} < \infty \quad \text{if and only if} \quad \sup_{0<\lambda\le 1} \|\lambda f_\lambda\|_{-1} < \infty \tag{2.7}$$

because V belongs to $\mathcal{H}_{-1}$.

The proof of this theorem is divided into two parts. We first compute in the next subsection the limiting variance of $t^{-1/2}\int_0^t V(X_s)\,ds$ and show that it is bounded by a multiple of the $\mathcal{H}_{-1}$ norm of V. If the generator is symmetric, i.e., if π is a reversible measure for the Markov process, the limiting variance is equal to $2\|V\|_{-1}^2$. In subsection 2.3, we prove that a central limit theorem holds for $t^{-1/2}\int_0^t V(X_s)\,ds$, V satisfying the assumptions of Theorem 2.2, provided the following two conditions are satisfied:

$$\lim_{\lambda\to 0} \lambda \| f_\lambda \|_0^2 = 0 \quad \text{and} \quad \lim_{\lambda\to 0} \| f_\lambda - f \|_1 = 0$$

for some f in $\mathcal{H}_1$. Finally, in subsection 2.4, we show that the bound (2.6) implies the previous two conditions.

2.2 The limiting variance

We estimate in this subsection the limiting variance of the integral $\int_0^t W(X_s)ds$ for a mean zero function W in $L^2(\pi)$. Let

$$\sigma(W)^2 = \limsup_{t\to\infty} \mathbb{E}_\pi\Big[\Big(\frac{1}{\sqrt{t}}\int_0^t W(X_s)\,ds\Big)^2\Big] .$$

Denote by P_t the semi-group of the Markov process X_t. Since π is invariant, a change of variables shows that for each fixed t the expectation on the right hand side of the previous formula is equal to

$$\begin{aligned}\frac{1}{t}\int_0^t ds \int_0^t dr\, \mathbb{E}_\pi[W(X_{|s-r|})W(X_0)] &= \frac{1}{t}\int_0^t ds \int_0^t dr \, < P_{|s-r|}W, W >_\pi \\ &= 2\int_0^t ds\, [1-(s/t)] < P_s W, W >_\pi .\end{aligned}$$

Denote by a^+ the positive part of a: $a^+ = \max\{0, a\}$, so that

$$\sigma(W)^2 = 2\limsup_{t\to\infty}\int_0^\infty ds\,[1-(s/t)]^+ < P_s W, W >_\pi . \tag{2.8}$$

In the general case, it is not clear whether this limsup is in fact a limit or wheter it is finite without some restrictions on W. However, in the reversible case, the sequence is increasing in t because $< P_s W, W >_\pi = < P_{s/2} W, P_{s/2} W >_\pi$ is a positive function. Hence, in the reversible case, by the monotone convergence theorem,

$$\begin{aligned} \sigma(W)^2 &= 2 \lim_{t\to\infty} \int_0^\infty ds\, [1-(s/t)]^+ < P_s W, W >_\pi \\ &= 2 \int_0^\infty ds\ < P_s W, W >_\pi \ . \end{aligned}$$

In the general case, one can show that $\sigma(W)^2$ defined in (2.8) is finite provided the function W belongs to the Sobolev space $\mathcal{H}_{-1}$: there exists a universal constant C_0 such that

$$\sigma(W)^2 \ \le\ C_0 \, \|W\|_{-1}^2 \tag{2.9}$$

for all functions W in $\mathcal{H}_{-1} \cap L^2(\pi)$. The main difference between $\sigma(W)^2$ and $\|W\|_{-1}^2$ is that while in the second term only the symmetric part of the generator is involved, in the first term the full generator appears. Formally,

$$\begin{aligned} \sigma(W)^2 &= 2 \int_0^\infty < P_s W, W >_\pi \ = \ 2 < W, (-L)^{-1} W >_\pi \\ &= 2 < W, [(-L)^{-1}]^s W >_\pi \ , \\ \text{and} \quad \|W\|_{-1}^2 &= < W, (-S)^{-1} W >_\pi \ . \end{aligned}$$

In this formula and below, M^s represents the symmetric part of the operator M and $A = (1/2)(L - L^*)$ is the asymmetric part of the generator L. Since,

$$\Big\{[(-L)^{-1}]^s\Big\}^{-1} \ = \ -S \ + \ A^*(-S)^{-1} A \ \ge\ -S \ ,$$

we have that $[(-L)^{-1}]^s \le (-S)^{-1}$, from what it follows that $\sigma(W)^2 \le 2\|W\|_{-1}^2$. We now present a rigorous proof of this informal argument.

Lemma 2.3. *Fix $T > 0$ and a function $f\colon [0,T] \times E \to \mathbb{R}$. Assume that $f(t,\cdot)$ belongs to $L^2(\pi)$ for each $0 \le t \le T$ and that $f(\cdot, x)$ is smooth for each x in E. There exists a finite universal constant C_0 such that,*

$$\mathbb{E}_\pi\Big[\sup_{0\le t\le T} \Big(\int_0^t f(s, X_s)\, ds \Big)^2 \Big] \ \le\ C_0 \int_0^T ds\, \|f(s,\cdot)\|_{-1}^2 \ .$$

The proof of this lemma relies on a representation formula for an additive functional of a Markov process by forward and backward martingales. The proof can be found in [3], Lemma 4.3 or in [19], Theorem 2.2. This lemma applied to the function $f(s, \cdot) = W(\cdot)$ proves (2.9).

Remark 2.4. *We just proved the existence of a universal finite constant C_0 such that $\sigma(W)^2 \le C_0 \|W\|_{-1}^2$. It has been proved ([18], Theorem 1.1) that for the asymmetric simple exclusion process in dimensions 1 and 2, there are local functions W* **not** *in $\mathcal{H}_{-1}$ and for which $\sigma(W)^2 < \infty$ We suspect that a central limit theorem holds for these functions with the usual scaling $t^{-1/2}$.*

2.3 The resolvent equation

We assume from now on that V belongs to $\mathcal{H}_{-1} \cap L^2(\pi)$. Taking inner product with respect to f_λ on both sides of the resolvent equation (2.5), we get that

$$\lambda < f_\lambda, f_\lambda >_\pi + \|f_\lambda\|_1^2 = < V, f_\lambda > . \tag{2.10}$$

Since f_λ belongs to $\mathcal{D}(L)$ and V belongs to $L^2(\pi) \cap \mathcal{H}_{-1}$, by Schwarz inequality (2.3), the right hand side is bounded above by $\|V\|_{-1}\|f_\lambda\|_1$. In particular, $\|f_\lambda\|_1 \le \|V\|_{-1}$ so that $\|V\|_{-1}\|f_\lambda\|_1 \le \|V\|_{-1}^2$ and

$$\lambda < f_\lambda, f_\lambda >_\pi + \|f_\lambda\|_1^2 \le \|V\|_{-1}^2 . \tag{2.11}$$

Therefore,

$$\lim_{\lambda \to 0} \lambda f_\lambda = 0 \text{ in } L^2(\pi) \quad \text{and} \quad \sup_{0<\lambda\le 1} \|f_\lambda\|_1 < \infty . \tag{2.12}$$

The purpose of this subsection is to show that a central limit theorem for $t^{-1/2}\int_0^t V(X_s)ds$ holds provided we can prove the following stronger statements:

$$\lim_{\lambda \to 0} \lambda \|f_\lambda\|_0^2 = 0 \quad \text{and} \quad \lim_{\lambda \to 0} \|f_\lambda - f\|_1 = 0 \tag{2.13}$$

for some f in $\mathcal{H}_1$.

Proposition 2.5. *Fix a function V in $L^2(\pi) \cap \mathcal{H}_{-1}$ and assume (2.13). Then, $t^{-1/2}\int_0^t V(X_s)\, ds$ converges in law to a mean zero Gaussian distribution with variance*

$$\sigma(V)^2 = 2 \lim_{\lambda \to 0} \|f_\lambda\|_1^2 .$$

It follows from (2.13) and (2.10) that

$$\sigma(V)^2 = 2\lim_{\lambda\to 0} \| f_\lambda \|_1^2 = 2\lim_{\lambda\to 0} < V, f_\lambda >_\pi . \tag{2.14}$$

The idea of the proof of Proposition 2.5 is again to express $\int_0^t V(X_s)\,ds$ as the sum of a martingale and a term which vanishes in the limit. This is proved in (2.16) and Lemma 2.7 below. We start taking advantage of the resolvent equation (2.5) to build up a martingale closely related to $\int_0^t V(X_s)\,ds$.

For each fixed $\lambda > 0$, let M_t^λ be the martingale defined by

$$M_t^\lambda = f_\lambda(X_t) - f_\lambda(X_0) - \int_0^t (Lf_\lambda)(X_s)\,ds$$

so that

$$\int_0^t V(X_s)\,ds = M_t^\lambda + f_\lambda(X_0) - f_\lambda(X_t) + \lambda \int_0^t f_\lambda(X_s)\,ds . \tag{2.15}$$

Lemma 2.6. *The martingale M_t^λ converges in $L^2(\pi)$, as $\lambda \downarrow 0$, to a martingale M_t and $\lambda \int_0^t f_\lambda(X_s)\,ds$ vanishes.*

Proof. We prove first that M_t^λ is a Cauchy sequence in $L^2(\pi)$. Indeed, for λ, $\lambda' > 0$, since π is an invariant state, the expectation of the quadratic variation of the martingale $M_t^\lambda - M_t^{\lambda'}$ is

$$\mathbb{E}_\pi\Big[\int_0^t ds\, \Big\{Lf_{\lambda,\lambda'}(X_s)^2 - 2f_{\lambda,\lambda'}(X_s)Lf_{\lambda,\lambda'}(X_s)\Big\}\Big]$$
$$= 2t < f_\lambda - f_{\lambda'}, (-L)f_\lambda(X_s) - f_{\lambda'} >_\pi = 2t \, \| f_\lambda - f_{\lambda'} \|_1^2 .$$

In this formula, $f_{\lambda,\lambda'} = f_\lambda - f_{\lambda'}$. By assumption (2.13), f_λ converges in $\mathcal{H}_1$. In particular, M_t^λ is a Cauchy sequence in $L^2(\pi)$ and converges to a martingale M_t. This proves the first statement. The second assertion of the lemma follows from (2.12) and Schwarz inequality. □

It follows from this result and from identity (2.15) that $f_\lambda(X_t) - f_\lambda(X_0)$ also converges in $L^2(\pi)$ as $\lambda \downarrow 0$. Denote this limit by R_t so that

$$\int_0^t V(X_s)\,ds = M_t + R_t . \tag{2.16}$$

Lemma 2.7. *$t^{-1/2}R_t$ vanishes in $L^2(\pi)$ as $t \uparrow \infty$.*

Proof. Putting together equation (2.15) with (2.16), we get that

$$\frac{R_t}{\sqrt{t}} = \frac{1}{\sqrt{t}}\Big\{M_t^\lambda - M_t + f_\lambda(X_0) - f_\lambda(X_t) + \lambda\int_0^t f_\lambda(X_s)\,ds\Big\}\,. \tag{2.17}$$

We consider separately each term on the right hand side of this expression.

Since M_t^λ converges in $L^2(\pi)$ to M_t,

$$\frac{1}{t}\,\mathbb{E}_\pi\Big[\Big(M_t^\lambda - M_t\Big)^2\Big] = \frac{1}{t}\lim_{\lambda'\to 0}\mathbb{E}_\pi\Big[\Big(M_t^\lambda - M_t^{\lambda'}\Big)^2\Big]\,.$$

In the previous Lemma, we computed the expectation of the quadratic variation of the martingale $M_t^\lambda - M_t^{\lambda'}$. This calculation shows that the previous expression is equal to

$$\lim_{\lambda'\to 0}\|f_\lambda - f_{\lambda'}\|_1^2 = \|f_\lambda - f\|_1^2\,.$$

In the last step we used assumption (2.13) which states that f_λ converges in $\mathcal{H}_1$ to some f.

We now turn to the second term in (2.17). Since π is invariant, the expectation of its square is bounded by

$$2t^{-1}\mathbb{E}_\pi\Big[f_\lambda(X_t)^2\Big] + 2t^{-1}\mathbb{E}_\pi\Big[f_\lambda(X_0)^2\Big] = 4t^{-1}\|f_\lambda\|_0^2\,.$$

On the other hand, by Schwarz inequality, the expectation of the square of the third term in (2.17) is bounded by $t\lambda^2\|f_\lambda\|_0^2$.

Putting together all previous estimates, we obtain that

$$\frac{1}{t}\,\mathbb{E}_\pi[R_t^2] \le 3\,\|f_\lambda - f\|_1^2 + 3(4t^{-1} + \lambda^2)\,\|f_\lambda\|_0^2$$

for all $\lambda > 0$. Set $\lambda = t^{-1}$ to conclude the proof of the lemma in view of hypotheses (2.13). □

We may now prove Proposition 2.5. Recall equation (2.16). By the previous lemma the second term on the right hand side divided by $\sqrt{t}$ vanishes in $L^2(\pi)$ (and therefore in probability) as $t \uparrow \infty$. On the other hand, by the martingale convergence theorem, $M_t/\sqrt{t}$ converges in law to a mean zero Gaussian distribution with variance

$$\begin{aligned}\sigma^2 = \mathbb{E}_\pi[M_1^2] &= \lim_{\lambda\to 0}\mathbb{E}_\pi[(M_1^\lambda)^2]\\ &= \lim_{\lambda\to 0}\mathbb{E}_\pi[< M^\lambda, M^\lambda >_1] = 2\lim_{\lambda\to 0}\|f_\lambda\|_1^2\,.\end{aligned}$$

The last identity follows from the computation of the expectation of the quadratic variation of the martingale M_t^λ performed in the proof of Lemma 2.6.

2.4 An $\mathcal{H}_{-1}$ estimate

In the previous subsection we showed that the central limit theorem for the additive functional $t^{-1/2}\int_0^t V(X_s)ds$ follows from conditions (2.13) if V belongs to $L^2(\pi)\cap\mathcal{H}_{-1}$. In the present section we prove that (2.13) follows from the bound (2.6) on the solution of the resolvent equation (2.5).

Lemma 2.8. *Fix a function V in $\mathcal{H}_{-1}\cap L^2(\pi)$ and denote by $\{f_\lambda,\ \lambda>0\}$ the solution of the resolvent equation (2.5). Assume that* $\sup_{\lambda>0}\|Lf_\lambda\|_{-1}\le C_0$ *for some finite constant C_0. Then, there exists f in $\mathcal{H}_1$ such that*

$$\lim_{\lambda\to 0}\lambda < f_\lambda, f_\lambda > = 0 \quad \textit{and} \quad \lim_{\lambda\to 0} f_\lambda = f$$

strongly in $\mathcal{H}_1$.

Proof. We already proved in (2.11) that

$$\sup_{0<\lambda\le 1}\|f_\lambda\|_1 \le \|V\|_{-1} \quad \text{and} \quad \sup_{0<\lambda\le 1}\lambda < f_\lambda, f_\lambda > \le \|V\|_{-1}^2 .$$

In particular, λf_λ converges to 0 in $L^2(\pi)$, as $\lambda\downarrow 0$.

Since $\sup_{\lambda>0}\|Lf_\lambda\|_{-1}$ is bounded, for any sequence $\lambda_n\downarrow 0$, there exists a subsequence, still denoted by λ_n, such that Lf_{λ_n} converges weakly in $\mathcal{H}_{-1}$ to some function U. We claim that the weak limit is unique and equal to $-V$. Indeed, fix a function g in $L^2(\pi)\cap\mathcal{H}_1$. Since g belongs to $\mathcal{H}_1$ and Lf_{λ_n} converges weakly to U, $< U, g >_\pi = \lim_n < Lf_{\lambda_n}, g >_\pi$. On the other hand, since f_λ is the solution of the resolvent equation, $\lim_n < Lf_{\lambda_n}, g > = - < V, g > + \lim_n < \lambda_n f_{\lambda_n}, g >$. This latter expression is equal to $- < V, g >$ because g belongs to $L^2(\pi)$ and λf_λ converges strongly to 0 in $L^2(\pi)$, as $\lambda\downarrow 0$. Thus, $< U, g > = - < V, g >$ for all functions g in $L^2(\pi)\cap\mathcal{H}_1$. Since this set is dense in $\mathcal{H}_1$, $U=-V$, proving the claim.

In the same way, since $\sup_{\lambda>0}\|f_\lambda\|_1$ is bounded, each sequence $\lambda_n\downarrow 0$ has a subsequence still denoted by λ_n, for which f_{λ_n} converges weakly in $\mathcal{H}_1$ to some function, denoted by W. We claim that any such limit W satisfies the relation $\|W\|_1^2 = < W, V >$. To check this identity, through convex combinations of the sequences f_{λ_n}, Lf_{λ_n}, we obtain sequences v_n, Lv_n which converge strongly to W, $-V$, respectively. On the one hand, since v_n (resp. Lv_n) converges strongly in $\mathcal{H}_1$ (resp. $\mathcal{H}_{-1}$) to W (resp. $-V$), $< v_n, Lv_n >$ converges to $- < W, V >$. On the other hand, since $- < v_n, Lv_n > = \|v_n\|_1^2$, it converges to $\|W\|_1^2$. Therefore, $\|W\|_1^2 = < W, V >$.

We have now all elements to prove the first part of the lemma. Suppose by contradiction that $\lambda < f_\lambda, f_\lambda >$ does not converge to 0 as $\lambda \downarrow 0$. In this case there exists $\varepsilon > 0$ and a subsequence $\lambda_n \downarrow 0$ such that $\lambda_n < f_{\lambda_n}, f_{\lambda_n} > \geq \varepsilon$ for all n. We have just shown the existence of a sub-subsequence $\lambda_{n'}$ for which $f_{\lambda_{n'}}$ converges weakly in $\mathcal{H}_1$ to some W satisfying the relation $< W, V >= \|W\|_1^2$. Since f_λ is solution of the resolvent equation,

$$\limsup_{n'\to\infty} \|f_{\lambda_{n'}}\|_1^2 \leq \limsup_{n'\to\infty} \left\{\lambda_{n'}\|f_{\lambda_{n'}}\|^2 + \|f_{\lambda_{n'}}\|_1^2\right\}$$

$$= \limsup_{n'\to\infty} < f_{\lambda_{n'}}, V > = < W, V > = \|W\|_1 \leq \limsup_{n'\to\infty} \|f_{\lambda_{n'}}\|_1^2 \cdot$$

This contradicts the fact that $\lambda_n < f_{\lambda_n}, f_{\lambda_n} > \geq \varepsilon$ for all n, so that $\lim_{\lambda\to 0}\lambda < f_\lambda, f_\lambda >= 0$.

It follows also from the previous argument that $f_{\lambda_{n'}}$ converges to W strongly in $\mathcal{H}_1$. In particular, all sequences λ_n have subsequences $\lambda_{n'}$ for which $f_{\lambda_{n'}}$ converges strongly in $\mathcal{H}_1$. To show that f_λ converges strongly, it remains to check uniqueness of the limit.

Consider two decreasing sequences λ_n, μ_n, vanishing as $n \uparrow \infty$. Denote by W_1, W_2 the strong limit in $\mathcal{H}_1$ of f_{λ_n}, f_{μ_n}, respectively. Since f_λ is the solution of the resolvent equation,

$$< \lambda_n f_{\lambda_n} - \mu_n f_{\mu_n}, f_{\lambda_n} - f_{\mu_n} >_\pi + \|f_{\lambda_n} - f_{\mu_n}\|_1^2 = 0$$

for all n. Since f_{λ_n}, f_{μ_n} converges strongly to W_1, W_2 in $\mathcal{H}_1$,

$$\lim_{n\to\infty} \|f_{\lambda_n} - f_{\mu_n}\|_1^2 = \|W_1 - W_2\|_1^2 \,.$$

On the other hand, since $\lambda\|f_\lambda\|^2$ vanishes as $\lambda \downarrow 0$,

$$\lim_{n\to\infty} < \lambda_n f_{\lambda_n} - \mu_n f_{\mu_n}, f_{\lambda_n} - f_{\mu_n} >_\pi$$

$$= - \lim_{n\to\infty} \left\{ < \lambda_n f_{\lambda_n}, f_{\mu_n} >_\pi + < \mu_n f_{\mu_n}, f_{\lambda_n} >_\pi \right\}.$$

Each of these terms vanish as $n \uparrow \infty$. Indeed,

$$\lambda_n < f_{\lambda_n}, f_{\mu_n} >_\pi = \lambda_n < f_{\lambda_n}, f_{\mu_n} - W_2 >_\pi + \lambda_n < f_{\lambda_n}, W_2 >_\pi$$

By Schwarz inequality (2.3), the first term on the right hand side is bounded above by $\|\lambda_n f_{\lambda_n}\|_{-1}\|f_{\mu_n} - W_2\|_1$, which vanishes because λf_λ is bounded in $\mathcal{H}_{-1}$ and f_{μ_n} converges to W_2 in $\mathcal{H}_1$. The second term of the previous formula also vanishes in the limit because W_2 belongs to $\mathcal{H}_1$ and λf_λ converges weakly to 0 in $\mathcal{H}_{-1}$. This concludes the proof of the lemma. □

Theorem 2.2 follows from this lemma and Proposition 2.5.

3 Some examples

Fix a function V in $L^2(\pi) \cap \mathcal{H}_{-1}$. In this section we present three conditions which guarantee that the solution f_λ of the resolvent equation (2.5) satisfies the bound (2.6).

3.1 Reversibility

Assume that the generator L is self-adjoint in $L^2(\pi)$. In this case, by Schwarz inequality,

$$\Big| < Lf, g >_\pi \Big| \;\le\; \|f\|_1 \, \|g\|_1 \, .$$

Therefore, in view of the variational formula (2.4) for the $\mathcal{H}_{-1}$ norm, for any f in $L^2(\pi) \cap \mathcal{H}_1$, Lf belongs to $\mathcal{H}_{-1}$ and

$$\|Lf\|_{-1} \;\le\; \|f\|_1 \, .$$

In particular, in the reversible case (2.6) follows from the elementary estimate (2.12).

3.2 Sector condition

Assume now that the generator L satisfies the sector condition

$$< f, Lg >_\pi^2 \le C_0 < f, (-L)f >_\pi < g, (-L)g >_\pi \tag{3.1}$$

for some finite constant C_0 and every functions f, g in the domain of the generator. In view of (2.4), for any function g in $\mathcal{D}(L)$,

$$\|Lg\|_{-1} \;\le\; C_0 \|g\|_1$$

and condition (2.6) follows from estimate (2.12).

The previous inequality states that the generator L is a bounded operator from $\mathcal{H}_1$ to $\mathcal{H}_{-1}$. Since S, the symmetric part of the generator, has certainly this property, L is bounded if and only if A, the asymmetric part of the generator, is a bounded operator from $\mathcal{H}_1$ to $\mathcal{H}_{-1}$, i.e., if

$$< g, A^*(-S)^{-1}Ag >_\pi \;=\; \|Ag\|_{-1}^2 \;\le\; C_0 \|g\|_1^2 \;=\; C_0 < g, (-S)g >_\pi$$

for all functions g in $\mathcal{D}(L)$. Hence, the sector condition requires that

$$A^*(-S)^{-1}A \;\le\; C_0(-S)$$

for some finite constant C_0. This inequality states that the asymmetric part of the generator can be estimated by the symmetric part. Furthermore, in this case, in view of the computations performed just after (2.9),

$$(-S) \le (-S) + A^*(-S)^{-1}A \le (1+C_0)(-S)$$

so that

$$C_1^{-1}\sigma(V)^2 \le \|V\|_{-1}^2 \le C_1\sigma(V)^2$$

for some finite constant C_1. This means that under the sector condition, the limiting variance is finite if and only if the function belongs to $\mathcal{H}_{-1}$.

3.3 Graded sector condition

Now, instead of assuming that the generator satisfies a sector condition on the all space, we decompose $L^2(\pi)$ as a direct sum of orthogonal spaces $\mathcal{A}_n$ and assume that on each subspace $\mathcal{A}_n$, the generator satisfies a sector condition with a constant which may be different on each $\mathcal{A}_n$.

Assume that $L^2(\pi)$ can be decomposed as a direct sum $\oplus_{n\ge 0}\mathcal{A}_n$ of orthogonal spaces. Functions in $\mathcal{A}_n$ are said to have degree n. For $n \ge 0$, denote by π_n the orthogonal projection on $\mathcal{A}_n$ so that

$$f = \sum_{n\ge 0} \pi_n f \quad \text{and} \quad \pi_n f \in \mathcal{A}_n$$

for all $n \ge 0$, f in $L^2(\pi)$.

Suppose that that the generator L keeps the degree of a function or changes it by one: $L\colon \mathcal{D}(L)\cap\mathcal{A}_n \to \mathcal{A}_{n-1}\cup\mathcal{A}_n\cup\mathcal{A}_{n+1}$. Denote by L_- (resp. L_+ and L_0) the piece of the generator that decreases (resp. increases and keeps) the degree of a function. Assume that L_0 can be decomposed as $R_0 + B_0$, where $-R_0$ is a positive operator bounded by $-C_0L$ for some positive constant C_0:

$$0 \le \langle f, (-R_0)f\rangle_\pi \le C_0 \langle f, (-L)f\rangle_\pi \tag{3.2}$$

for all functions f in $\mathcal{D}(L)$.

Since $-R_0$ is a positive operator, repeating the steps of Subsection 2.1 with R_0 in place of L, we define the Sobolev spaces $\mathcal{H}_{0,1}$, $\mathcal{H}_{0,-1}$ and the norms $\|\cdot\|_{0,1}$, $\|\cdot\|_{0,-1}$ associated to R_0. Since R_0 keeps the degree of a function,

$$\begin{aligned}\|f\|_{0,1}^2 &= \langle f, (-R_0)f\rangle_\pi = \Big\langle \sum_{n\ge 0}\pi_n f, (-R_0)\sum_{n\ge 0}\pi_n f\Big\rangle_\pi \\ &= \sum_{n\ge 0}\langle \pi_n f, (-R_0)\pi_n f\rangle_\pi = \sum_{n\ge 0}\|\pi_n f\|_{0,1}^2 .\end{aligned}$$

for all functions f in the domain of the generator. By the same reasons, for a function f in $L^2(\pi)$,

$$\|f\|_{0,-1}^2 = \sup_{g\in\mathcal{D}(L)} \{2 < f, g >_\pi - \|g\|_{0,1}^2\} = \sum_{n\geq 0} \|\pi_n f\|_{0,-1}^2 .$$

In terms of the new norm $\|\cdot\|_{0,1}$, (3.2) translates to

$$\|f\|_{0,1} \leq \sqrt{C_0}\, \|f\|_1$$

for all function f in the domain of the generator and some finite constant C_0. It follows from this inequality and from the variational formula for the $\mathcal{H}_{-1}$, $\mathcal{H}_{0,-1}$ norms that

$$\|f\|_{-1} \leq \sqrt{C_0}\, \|f\|_{0,-1} \tag{3.3}$$

for all function f in $L^2(\pi)$ and the same finite constant C_0.

Suppose now that a sector condition holds on each subspace $\mathcal{A}_n$ with a constant which depends on n: there exists $\beta < 1$ and a finite constant C_0 such that

$$\begin{aligned} &< f, (-L_+)g >_\pi^2 \leq C_0 n^{2\beta} < f, (-R_0)f >_\pi < g, (-R_0)g >_\pi , \\ &< g, (-L_-)f >_\pi^2 \leq C_0 n^{2\beta} < f, (-R_0)f >_\pi < g, (-R_0)g >_\pi \end{aligned} \tag{3.4}$$

for all g in $\mathcal{D}(L) \cap \mathcal{A}_n$ and f in $\mathcal{D}(L) \cap \mathcal{A}_{n+1}$. It follows from the previous assumptions and from the variational formula for the $\|\cdot\|_{-1,0}$ norm that

$$\|L_+ g\|_{0,-1} \leq \sqrt{C_0}\, n^\beta\, \|g\|_{0,1} , \quad \|L_- f\|_{0,-1} \leq \sqrt{C_0}\, n^\beta\, \|f\|_{0,1} \tag{3.5}$$

for all g in $\mathcal{D}(L) \cap \mathcal{A}_n$ and f in $\mathcal{D}(L) \cap \mathcal{A}_{n+1}$. The proof of Lemma 3.1 below, due to [14], [19], shows that the restriction $\beta < 1$ is crucial.

Lemma 3.1 *Let V be a function in $L^2(\pi)$ such that*

$$\sum_{n\geq 0} n^{2k} \|\pi_n V\|_{0,-1}^2 < \infty .$$

Denote by f_λ the solution of the resolvent equation (2.5). There exists a finite constant C_1 depending only on β, k and C_0 such that

$$\sum_{n\geq 0} n^{2k} \|\pi_n f_\lambda\|_{0,1}^2 \leq C_1 \sum_{n\geq 0} n^{2k} \|\pi_n V\|_{0,-1}^2 .$$

Proof. Consider an increasing sequence $\{t_n : n \geq 0\}$, to be fixed later, and denote by $T: L^2(\pi) \to L^2(\pi)$ the operator which is a multiple of the identity on each subspace $\mathcal{A}_n$:

$$Tf = \sum_{n\geq 0} t_n \pi_n f \, .$$

Apply T to both sides of the resolvent equation and take the inner product with respect to Tf_λ on both sides of the identity to obtain that

$$\begin{aligned}\lambda < Tf_\lambda, Tf_\lambda >_\pi \; &- \; < Tf_\lambda, LTf_\lambda >_\pi \\ =< Tf_\lambda, TV > \; &- \; < Tf_\lambda, [L, T]f_\lambda >_\pi \; .\end{aligned}$$

In this formula, $[L, T]$ stands for the comutator of L and T and is given by $LT - TL$. By assumption (3.2), the left hand side is bounded below by

$$C_0^{-1} < Tf_\lambda, (-R_0)Tf_\lambda >_\pi = C_0^{-1}\|Tf_\lambda\|_{0,1}^2 = C_0^{-1}\sum_{n\geq 0} t_n^2 \, \|\pi_n f_\lambda\|_{0,1}^2 \, .$$

Let $\delta > 0$. We now estimate the scalar product $< Tf_\lambda, [L, T]f_\lambda >_\pi$ in terms of $\|Tf_\lambda\|_{0,1}^2$. Since T commutes with any operator that keeps the degree, $[L, T] = [L_+ + L_-, T]$. To fix ideas, consider the operator $[L_+, T]$, the other expression being estimated in a similar way. Since L_+ increases the degree by one, by definition of the comutator,

$$\pi_n[L_+, T]f = L_+T\pi_{n-1}f - TL_+\pi_{n-1}f = (t_{n-1} - t_n)L_+\pi_{n-1}f$$

for all functions f in $\mathcal{D}(L)$. Therefore,

$$\begin{aligned}< Tf_\lambda, [L_+, T]f_\lambda >_\pi \; &= \; \sum_{n\geq 0} < \pi_n Tf_\lambda, \pi_n[L_+, T]f_\lambda >_\pi \\ &= \; \sum_{n\geq 0}(t_{n-1} - t_n)t_n < \pi_n f_\lambda, L_+\pi_{n-1}f_\lambda >_\pi \; .\end{aligned}$$

By (3.4) and since the sequence t_n is increasing, the previous expression is bounded below by

$$\begin{aligned}&\sum_{n\geq 0}(t_n - t_{n-1})t_n C_0 n^\beta \|\pi_n f_\lambda\|_{0,1}\|\pi_{n-1}f_\lambda\|_{0,1} \\ &\leq \frac{1}{2}\sum_{n\geq 0}(t_n - t_{n-1})t_n C_0 n^\beta \|\pi_n f_\lambda\|_{0,1}^2 \\ &+ \frac{1}{2}\sum_{n\geq 0}(t_n - t_{n-1})t_n C_0 n^\beta \|\pi_{n-1} f_\lambda\|_{0,1}^2 \, .\end{aligned}$$

Since $\beta < 1$, there exists $n_1 = n_1(C_0, \beta, \delta, k)$ such that

$$C_0 n^\beta \Big\{ 1 - \frac{(n-1)^{2k}}{n^{2k}} \Big\} \le \delta\,, \quad C_0 n^\beta \Big\{ \frac{n^{2k}}{(n-1)^{2k}} - 1 \Big\} \frac{n^{2k}}{(n-1)^{2k}} \le \delta$$

for all $n \ge n_1$. Fix $n_2 > n_1$ and set $t_n = n_1^{2k} \mathbf{1}\{n < n_1\} + n^{2k} \mathbf{1}\{n_1 \le n \le n_2\} + n^{2k} \mathbf{1}\{n > n_2\}$. With this definition, we obtain that the previous expression is bounded by

$$\delta \sum_{n \ge 0} t_n^2 \|\pi_n f_\lambda\|_{0,1}^2 = \delta\, \|T f_\lambda\|_{0,1}^2 \,.$$

It remains to estimate $< T f_\lambda, TV >_\pi$. By (2.3), and since $2ab \le A^{-1} a^2 + A b^2$ for every $A > 0$,

$$\begin{aligned} < T f_\lambda, TV >_\pi &= \sum_{n \ge 0} t_n^2 < \pi_n f_\lambda, \pi_n V >_\pi \le \sum_{n \ge 0} t_n^2 \, \|\pi_n f_\lambda\|_{0,1} \, \|\pi_n V\|_{0,-1} \\ &\le \delta \sum_{n \ge 0} t_n^2 \, \|\pi_n f_\lambda\|_{0,1}^2 + \delta^{-1} \sum_{n \ge 0} t_n^2 \, \|\pi_n V\|_{0,-1}^2 \\ &= \delta \|T f_\lambda\|_{0,1}^2 + \delta^{-1} \|TV\|_{0,-1}^2 \,. \end{aligned}$$

Putting together the previous three estimates, we obtain that

$$C_0^{-1} \|T f_\lambda\|_{0,1}^2 \le 3\delta \, \|T f_\lambda\|_{0,1}^2 + \delta^{-1} \|TV\|_{0,-1}^2$$

so that

$$\|T f_\lambda\|_{0,1}^2 \le 16 C_0^2 \, \|TV\|_{0,-1}^2$$

if we choose $\delta = 1/4C_0$. Recall the definition of the sequence t_n. This estimate holds uniformly in n_2. Let $n_2 \uparrow \infty$ and definite T' as the operator associated to the sequence t'_n, where $t'_n = n^{2k} \mathbf{1}\{n \ge n_1\} + n_1^{2k} \mathbf{1}\{n < n_1\}$, to deduce that

$$\begin{aligned} \sum_{n \ge 0} n^{2k} \|\pi_n f_\lambda\|_{0,1}^2 &\le \sum_{n \ge 0} (t'_n)^2 \|\pi_n f_\lambda\|_{0,1}^2 \\ &\le 16 C_0^2 \sum_{n \ge 0} (t'_n)^2 \|\pi_n V\|_{0,-1}^2 \\ &\le 16 C_0^2 n_1^{2k} \sum_{n \ge 0} n^{2k} \|\pi_n V\|_{0,-1}^2 \,. \end{aligned}$$

To conclude the proof of the lemma, it remains to recall that we fixed $\delta = 1/4C_0$ and that $n_1 = n_1(C_0, k, \beta, \delta)$. □

Assume now that $L_0 = R_0 + B_0$ satisfies a sector condition on each subset $\mathcal{A}_n$:

$$< g, (-L_0)f >_\pi^2 \le C_0 n^{2\gamma} < f, (-R_0)f >_\pi < g, (-R_0)g >_\pi \tag{3.6}$$

for some $\gamma > 0$ and all functions f, g in $\mathcal{D}(L) \cap \mathcal{A}_n$. Notice that we do not impose any condition on γ. By the variational formula for the norm $\| \cdot \|_{0,-1}$,

$$\|L_0 f\|_{0,-1} \le \sqrt{C_0}\, n^\gamma \|f\|_{0,1} \tag{3.7}$$

for all functions f in $\mathcal{D}(L) \cap \mathcal{A}_n$.

Lemma 3.2. *Suppose that the generator L satisfies (3.2), (3.4) and (3.6). Fix a function V such that*

$$\sum_{n\ge 0} n^{2k} \|\pi_n V\|_{0,-1}^2 < \infty$$

for some $k \ge (\beta \vee \gamma)$. Let f_λ be the solution of the resolvent equation (2.5). Then,

$$\sup_{0<\lambda\le 1} \|Lf_\lambda\|_{0,-1} < \infty .$$

Proof. It follows from (3.3) that

$$\|Lf_\lambda\|_{-1}^2 \le \|Lf_\lambda\|_{0,-1}^2 = \sum_{n\ge 0} \|\pi_n L f_\lambda\|_{0,-1}^2 . \tag{3.8}$$

Fix $n \ge 0$. Since $\pi_n L f_\lambda = L_- \pi_{n+1} f_\lambda + L_0 \pi_n f_\lambda + L_+ \pi_{n-1} f_\lambda$, by (3.5), (3.7),

$$\begin{aligned} \|\pi_n L f_\lambda\|_{0,-1} &\le \|L_- \pi_{n+1} f_\lambda\|_{0,-1} + \|L_0 \pi_n f_\lambda\|_{0,-1} + \|L_+ \pi_{n-1} f_\lambda\|_{0,-1} \\ &\le C_0 n^\beta \|\pi_{n+1} f_\lambda\|_{0,1} + C_0 n^\gamma \|\pi_n f_\lambda\|_{0,1} + C_0 n^\beta \|\pi_{n-1} f_\lambda\|_{0,1} . \end{aligned}$$

In particular, by Schwarz inequality, by Lemma 3.1 and since $k \ge (\beta \vee \gamma)$, the right hand side of (3.8) is bounded above by

$$C_1 \sum_{n\ge 0} n^{2k} \|\pi_n f_\lambda\|_{0,1}^2 \le C_1 \sum_{n\ge 0} n^{2k} \|\pi_n V\|_{0,-1}^2$$

for some finite constant C_2 depending only on C_0, β and γ. This proves the lemma. □

Therefore, to prove a central limit theorem for an additive functional of a Markov process, it is enough to check whether its generator satisfies the graded sector conditions (3.2), (3.4) and (3.6).

4 Tagged particle in simple exclusion process

We prove in this section a central limit theorem for the position of a tagged particle in the simple exclusion process with the method presented in the previous section.

Among the simplest and most widely studied interacting particle systems is the simple exclusion process. It represents the evolution of random walks on the lattice $\mathbb{Z}^d$ with a hard-core interaction that prevents more than one particle per site and may be described as follows. Fix a probability measure $p(\cdot)$ on $\mathbb{Z}^d$ and distribute particles on the lattice in such a way that each site is occupied by at most one particle. Particles evolve on $\mathbb{Z}^d$ as random walks with translation–invariant transition probability $p(x, y) = p(y - x)$. Each time a particle tries to jump over a site already occupied, the jump is suppressed to respect the exclusion rule.

This informal description corresponds to a Markov process on $\mathcal{X}_d = \{0, 1\}^{\mathbb{Z}^d}$ whose generator L is given by

$$(Lf)(\eta) = \sum_{x,z\in\mathbb{Z}^d} \eta(x)[1 - \eta(x + z)]\, p(z)\, [f(\sigma^{x,x+z}\eta) - f(\eta)] \,. \tag{4.1}$$

Here, η stands for a configuration of $\mathcal{X}_d$ so that $\eta(x)$ is equal to 1 (resp. 0) if the site x is occupied (resp. unoccupied) for the configuration η. f is a cylinder function, which means that it depends on η only through a finite number of coordinates, and $\sigma^{x,y}\eta$ is the configuration obtained from η by interchanging the occupation variables $\eta(x)$, $\eta(y)$:

$$(\sigma^{x,y}\eta)(z) = \begin{cases} \eta(z) & \text{if } z \neq x, y \,, \\ \eta(y) & \text{if } z = x \,, \\ \eta(x) & \text{if } z = y \,. \end{cases}$$

The simple exclusion process is said to be symmetric if the transition probability is symmetric ($p(z) = p(-z)$) and to be mean zero if the transition probability is not symmetric but has zero average: $\sum_z zp(z) = 0$. All other cases are said to be asymmetric

To avoid degeneracies, we assume that the transition probability $p(\cdot)$ is irreducible in the sense that the set $\{x : p(\pm x) > 0\}$ generates $\mathbb{Z}^d$, i.e., that for any pair of sites x, y in $\mathbb{Z}^d$, there exists $M \geq 1$ and a sequence $x = x_0, \dots, x_M = y$ such that $p(x_{i+1} - x_i) + p(x_i - x_{i+1}) > 0$ for $0 \leq i \leq M - 1$. We also suppose that the transition probability is of finite range: there exists A_0 in $\mathbb{N}$ such that $p(z) = 0$ for all sites z outside the cube $[-A_0, A_0]^d$.

Notice that the total number of particles is conserved by the dynamics. This conservation is reflected in the existence of a one parameter family of invariant measures. For $0 \le \alpha \le 1$, denote by ν_α the Bernoulli product measure of parameter α. This means that under ν_α the variables $\{\eta(x),\ x \in \mathbb{Z}^d\}$ are independent with marginals given by

$$\nu_\alpha\{\eta(x) = 1\} \ = \ \alpha \ = \ 1 - \nu_\alpha\{\eta(x) = 0\} \ .$$

An elementary computation shows that the Bernoulli measures $\{\nu_\alpha \ , 0 \le \alpha \le 1\}$ are invariant for simple exclusion processes. Denote by L^* the generator defined by (4.1) associated to the transition probability $p^*(x) = p(-x)$. L^* is the adjoint of L in $L^2(\nu_\alpha)$. In particular, symmetric simple exclusion processes are self-adjoint with respect to each ν_α.

For $t \ge 0$, denote by η_t the state at time t of the Markov process. Among all particles, tag one of them and denote by X_t its position at time t. X_t by itself is not a Markov process because its evolution depends on the position of the other particles. However, (η_t, X_t) is a Markov process on $\mathcal{X}_d \times \mathbb{Z}^d$.

Denote by $\{\tau_x,\ x \in \mathbb{Z}^d\}$ the group of translations on $\mathcal{X}_d$. For x, y in $\mathbb{Z}^d$ and a configuration η in $\mathcal{X}_d$, $(\tau_x \eta)(y) = \eta(x + y)$. The action of the translation group is naturally extended to functions and measures.

Denote by ξ_t the state of the process at time t as seen form the tagged particle: $\xi_t = \tau_{X_t}\eta_t$. Notice that the origin is always occupied because $\xi_t(0) = (\tau_{X_t}\eta_t)(0) = \eta_t(X_t) = 1$. In particular, we can consider either ξ as a configuration of $\mathcal{X}_d$ with a particle at the origin or ξ as a configuration of $\mathcal{X}_d^* = \{0, 1\}^{\mathbb{Z}_*^d}$, where $\mathbb{Z}_*^d = \mathbb{Z}^d - \{0\}$. We adopt here the latter convention. It is also not difficult to show that ξ_t is a Markov process on $\mathcal{X}_d^*$ with generator $\mathcal{L}$ given by

$$\begin{aligned}(\mathcal{L}f)(\xi) = & \sum_{x,y \in \mathbb{Z}_*^d} p(y - x)\xi(x)[1 - \xi(y)][f(\sigma^{x,y}\xi) - f(\xi)] \\ & + \sum_{z \in \mathbb{Z}_*^d} p(z)[1 - \xi(z)][f(\theta_z \xi) - f(\xi)] \ .\end{aligned} \tag{4.2}$$

The first part of the generator takes into account the jumps of the environment, while the second one corresponds to jumps of the tagged particle. In the above formula, $\theta_y \xi$ stands for the configuration where the tagged particle, sitting at the origin, is first transferred to site z and then all the configuration is translated by $-z$: for all y in $\mathbb{Z}_*^d$

$$(\theta_z \xi)(y) = \begin{cases} \xi(z) & \text{if } y = -z, \\ \xi(y + z) & \text{for } y \ne -z. \end{cases}$$

For $0 \le \alpha \le 1$, denote by ν_α^* the Bernoulli product measure on $\mathcal{X}_d^*$, by $\mathcal{L}^*$ the generator defined by (4.2) with the transition probability $p^*(y) = p(-y)$ in place of p and by $< \cdot, \cdot >_\mu$ the inner product in $L^2(\mu)$, for a probability measure μ. An elementary computation shows that ν_α^* is an invariant state for the Markov process ξ_t and that its adjoint in $L^2(\nu_\alpha^*)$ is $\mathcal{L}^*$. In particular, $\mathcal{L}$ is self-adjoint in the symmetric case.

We have seen that the $\mathcal{H}_1$ plays an important role in the investigation of the central limit theorem. For the simple exclusion process as seen from a tagged particle, a simple computation shows that for any function f in the domain of the generator,

$$\begin{aligned}\langle f, (-\mathcal{L}) f\rangle_{\nu_\alpha^*} &\\ = (1/2) \sum_{x,y\in\mathbb{Z}_d^*} p(y-x) \int \xi(x)[1-\xi(y)]\,[f(\sigma^{x,y}\xi) - f(\xi)]^2\, d\nu_\alpha^* & \qquad (4.3)\\ + \ (1/2) \sum_{z\in\mathbb{Z}_d^*} p(z) \int [1-\xi(z)]\,[f(\theta_z\xi) - f(\xi)]^2\, d\nu_\alpha^* \, . &\end{aligned}$$

The first question on the asymptotic behavior of the tagged particle concerns the law of large numbers. For $0 < \alpha < 1$, denote by $\mathbb{P}_{\nu_\alpha^*}$ the measure on the path space $D(\mathbb{R}_+, X_d^*)$ induced by the Markov process with generator $\mathcal{L}$ starting from ν_α^*. Saada proved in [17] the following result.

Theorem 4.1. *For every* $0 \le \alpha \le 1$

$$\lim_{t\to\infty} \frac{X_t}{t} \;=\; [1-\alpha]\,\gamma$$

in $\mathbb{P}_{\nu_\alpha^*}$ *probability, where* $\gamma = \sum_{z\in\mathbb{Z}_*^d} z p(z)$.

To investigate the central limit theorem, denote by Z_t the re-scaled position of the tagged particle:

$$Z_t \;=\; \frac{X_t - t\,\gamma\,(1-\alpha)}{\sqrt{t}} \, .$$

For each z such that $p(z) > 0$ and for $s < t$, denote by $N_{[s,t]}^z$ the total number of jumps of the tagged particle from the origin to z in the time interval $[s,t]$. Let $N_t^z = N_{[0,t]}^z$. It is not difficult to check that that

$$M_t^z \;=\; N_t^z \;-\; \int_0^t p(z)[1-\xi_s(z)]\,ds \quad \text{and} \quad (M_t^z)^2 \;-\; \int_0^t p(z)[1-\xi_s(z)]\,ds$$

are martingales vanishing at $t = 0$. In the same way, for y, z in $\mathbb{Z}_*^d$ such that $p(z-y) > 0$, $s < t$, denote by $N_{[s,t]}^{y,z}$ the number of jumps of a particle from y

to z in the interval $[s, t]$. Let $N_t^{y,z} = N_{[0,t]}^{y,z}$. As before

$$M_t^{y,z} = N_t^{y,z} - \int_0^t p(z-y)\xi_s(y)[1-\xi_s(z)]\,ds$$

$$\text{and} \quad (M_t^{y,z})^2 - \int_0^t p(z-y)\xi_s(y)[1-\xi_s(z)]\,ds$$

are martingales.

Since $\{M_t^z,\ p(z) > 0\}$, $\{M_t^{y,y+z},\ y \in \mathbb{Z}_*^d,\ p(z) > 0\}$ are pure jumps martingales and do not have common jumps, they are orthogonal in the sense that the product of two such martingales is still a martingale.

To obtain the position at time t of the tagged particle, we just need to sum the number of jumps multiplied by the size of the jumps: $X_t = \sum_z zN_t^z$ so that

$$X_t = \sum_{z\in\mathbb{Z}^d} zN_t^z = \sum_{z\in\mathbb{Z}^d} zM_t^z + \sum_{z\in\mathbb{Z}^d}\int_0^t ds\, z\, p(z)[1-\xi_s(z)]\,.$$

In particular, for any vector a in $\mathbb{R}^d$,

$$(a\cdot Z_t) = \frac{M_t^a}{\sqrt{t}} + \frac{1}{\sqrt{t}}\int_0^t V_a(\xi_s)\,ds\,,$$

where M_t^a is the one-dimensional martingale defined by

$$M_t^a = \sum_{z\in\mathbb{Z}_*^d} (a\cdot z)M_t^z$$

and V_a is the mean-zero cylinder function

$$V_a(\xi) = \sum_{z\in\mathbb{Z}_*^d} (a\cdot z)p(z)[\alpha - \xi(z)]\,. \tag{4.4}$$

In these formulas and below $(a\cdot b)$ stands for the inner product in $\mathbb{R}^d$.

To prove a central limit theorem for the tagged particle, we need to represent $\int_0^t V_a(\xi_s)\,ds$ as a martingale $\tilde{M}_t$ plus a small term and to compute the limiting variance of $M_t^a + \tilde{M}_t$. We have seen in Theorem 2.2 that such a representation is possible provide the solution f_λ of the resolvent equation

$$\lambda f_\lambda - \mathcal{L} f_\lambda = V_a \tag{4.5}$$

satisfies (2.6). In this case $\tilde{M}_t$ is the limit, as $\lambda \downarrow 0$ of the martingale M_t^λ given by

$$M_t^\lambda = f_\lambda(\xi_t) - f_\lambda(\xi_0) - \int_0^t (\mathcal{L} f_\lambda)(\xi_s)\, ds \,.$$

This martingale can be expressed in terms of the elementary martingales M_t^z, $M_t^{x,y}$ introduced above. Since

$$\begin{aligned} f_\lambda(\xi_t) - f_\lambda(\xi_0) &= \sum_{x,y\in\mathbb{Z}_*^d} \int_0^t [f_\lambda(\sigma^{x,y}\xi_{s-}) - f_\lambda(\xi_{s-})]\, dN_s^{x,y} \\ &+ \sum_{z\in\mathbb{Z}_*^d} \int_0^t [f_\lambda(\theta_z\xi_{s-}) - f_\lambda(\xi_{s-})]\, dN_s^z \,, \end{aligned}$$

an elementary computation shows that

$$\begin{aligned} M_t^\lambda &= \sum_{x,y\in\mathbb{Z}_*^d} \int_0^t [f_\lambda(\sigma^{x,y}\xi_{s-}) - f_\lambda(\xi_{s-})]\, dM_s^{x,y} \\ &+ \sum_{z\in\mathbb{Z}_*^d} \int_0^t [f_\lambda(\theta_z\xi_{s-}) - f_\lambda(\xi_{s-})]\, dM_s^z \,. \end{aligned}$$

By Theorem 2.2, if (2.6) holds for the solution of the resolvent equation,

$$(a \cdot Z_t) = \frac{1}{\sqrt{t}} M_t^a + \frac{1}{\sqrt{t}} M_t^\lambda + R_t^\lambda \,,$$

where

$$\lim_{t\to\infty} \lim_{\lambda\to 0} R_t^\lambda = 0 \quad \text{in} \quad L^2(\nu_\alpha^*)\,.$$

Since the martingale $M_t^a + \tilde{M}_t$ satisfies the assumptions of Lemma 2.1 for every a in $\mathbb{R}^d$, under assumption (2.6), Z_t converges in law to a mean zero Gaussian distribution with co-variance $D(\alpha)$ given by

$$\begin{aligned} a \cdot D(\alpha) a &= E_{\nu_\alpha^*}[(M_1^a + \tilde{M}_1)^2] = \lim_{\lambda\to 0} E_{\nu_\alpha^*}[(M_1^a + M_1^\lambda)^2] \\ &= \lim_{\lambda\to 0} \Big\{ E_{\nu_\alpha^*}\Big[\Big(\sum_{x,y\in\mathbb{Z}_*^d} \int_0^1 [f_\lambda(\sigma^{x,y}\xi_{s-}) - f_\lambda(\xi_{s-})]\, dM_s^{x,y} \Big)^2\Big] \\ &+ E_{\nu_\alpha^*}\Big[\Big(\sum_{z\in\mathbb{Z}_*^d} \int_0^1 \Big\{ (a\cdot z) + [f_\lambda(\theta_z\xi_{s-}') - f_\lambda(\xi_{s-})] \Big\}\, dM_s^z \Big)^2\Big] \Big\} \end{aligned}$$

$$= \lim_{\lambda\to 0}\Big\{ \sum_{x,y\in\mathbb{Z}_*^d} p(y-x)E_{\nu_\alpha^*}\Big[\xi(x)[1-\xi(y)][f_\lambda(\sigma^{x,y}\xi)-f_\lambda(\xi)]^2\Big]$$

$$+\sum_{z\in\mathbb{Z}_*^d} p(z)E_{\nu_\alpha^*}\Big[[1-\xi(z)]\Big\{(a\cdot z)+[f_\lambda(\theta_z\xi)-f_\lambda(\xi)]\Big\}^2\Big]\Big\}\,.$$

Here we used extensively the fact that the martingales M_t^z, $M_t^{x,y}$ are orthogonal and the explicit form of their quadratic variation. Developing the square, we get that for each fixed λ the previous expectation is equal to

$$(1-\alpha)\sum_{z\in\mathbb{Z}_*^d} p(z)\,(a\cdot z)^2$$

$$+\,2\sum_{z\in\mathbb{Z}_*^d}(a\cdot z)\,p(z)E_{\nu_\alpha^*}\Big[[1-\xi(z)]\,[f_\lambda(\theta_z\xi)-f_\lambda(\xi)]\Big]$$

$$+\sum_{x,y\in\mathbb{Z}_*^d} p(y-x)E_{\nu_\alpha^*}\Big[\xi(x)[1-\xi(y)][f_\lambda(\sigma^{x,y}\xi)-f_\lambda(\xi)]^2\Big]$$

$$+\sum_{z\in\mathbb{Z}_*^d} p(z)E_{\nu_\alpha^*}\Big[[1-\xi(z)][f_\lambda(\theta_z\xi)-f_\lambda(\xi)]^2\Big]\,.$$

In view of (4.3), the last two terms are equal to $2\|f_\lambda\|_1^2$. A change of variables $\zeta=\theta_z\xi$ in the the expectation $E_{\nu_\alpha^*}[[1-\xi(z)]f_\lambda(\theta_z\xi)]$ permits to write the second term as $2<W_a, f_\lambda>_{\nu_\alpha^*}$, where

$$W_a \;=\; \sum_{z\in\mathbb{Z}_*^d}(a\cdot z)p(z)\{\xi(z)-\xi(-z)\}\,. \tag{4.6}$$

In conclusion,

$$\begin{aligned} a\cdot D(\alpha)a \;&=\; (1-\alpha)\sum_{z\in\mathbb{Z}_*^d} p(z)\,(a\cdot z)^2 \\ &\quad+\;\lim_{\lambda\to 0}\Big\{2<W_a, f_\lambda>_{\nu_\alpha^*}+2\,\|f_\lambda\|_1^2\Big\} \end{aligned} \tag{4.7}$$

for every a in $\mathbb{R}^d$. Recall from (2.14) that $\lim_{\lambda\to 0}\|f_\lambda\|_1^2=\lim_{\lambda\to 0}<V_a, f_\lambda>$. Since, on the other hand, $V_a+W_a=\sum_z(a\cdot z)p(z)[\alpha-\xi(-z)]$,

$$a\cdot D(\alpha)a \;=\; (1-\alpha)\sum_{z\in\mathbb{Z}_*^d} p(z)\,(a\cdot z)^2 \;+\; 2\lim_{\lambda\to 0}<\tilde{W}_a, f_\lambda>_{\nu_\alpha^*}\,, \tag{4.8}$$

where

$$\tilde{W}_a = \sum_z (a \cdot z) p(z)[\alpha - \xi(-z)] .$$

Up to this point we have shown that a central limit theorem for the tagged particle in the simple exclusion process holds provided that (2.6) is in force for the solution of the resolvent equation (4.5). In this case the limiting variance is given by (4.8). In the next three subsections, we prove condition (2.6) in different contexts.

4.1 Symmetric case

Assume that p is symmetric. In this case the generator $\mathcal{L}$ is self-adjoint. To apply the method presented in sections 2 and 3 and the results proved in subsection 3.1, we first need to examine whether V_a belongs to the Sobolev space $\mathcal{H}_{-1}$ associated to the generator $\mathcal{L}$.

Fix a function f in $L^2(\nu_\alpha^*)$. Since V_a has mean zero, if $< f >_{\nu_\alpha^*}$ stands for the expectation of f with respect to ν_α^*,

$$\begin{aligned} E_{\nu_\alpha^*}[V_a f] &= \sum_{z \in \mathbb{Z}_*^d} (a \cdot z)\, p(z) \int [\alpha - \xi(z)]\, [f - < f >_{\nu_\alpha^*}]\, d\nu_\alpha^* \\ &= \sum_{z \in \mathbb{Z}_*^d} (a \cdot z)\, p(z) \int [1 - \xi(z)]\, [f - < f >_{\nu_\alpha^*}]\, d\nu_\alpha^* . \end{aligned}$$

Write this last expression as the sum of two halfs. In one of the sums, perform the change of variables $\zeta = \theta_z \xi$, which is possible because the indicator $[1 - \xi(z)] = \mathbf{1}\{\xi(z) = 0\}$ guarantees that there are no particles at z. After these operations the last term becomes

$$\begin{aligned} & (1/2) \sum_{z \in \mathbb{Z}_*^d} (a \cdot z)\, p(z) \int [1 - \xi(z)]\, [f - < f >_{\nu_\alpha^*}]\, d\nu_\alpha^* \\ & \quad + (1/2) \sum_{z \in \mathbb{Z}_*^d} (a \cdot z)\, p(z) \int [1 - \xi(-z)]\, [f(\theta_{-z}\xi) - < f >_{\nu_\alpha^*}]\, d\nu_\alpha^* . \end{aligned}$$

Change variables $z' = -z$ in the second sum, recall that p is symmetric and add the two terms to obtain that the previous sum is equal to

$$(1/2) \sum_{z \in \mathbb{Z}_*^d} (a \cdot z)\, p(z) \int [1 - \xi(z)]\, [f(\xi) - f(\theta_z \xi)]\, d\nu_\alpha^* .$$

It remains to apply Schwarz inequality to bound the square of this expression by

$$(1/4)(1-\alpha)\Big(\sum_{z\in\mathbb{Z}^d_*}(a\cdot z)^2 p(z)\Big)\sum_{z} p(z)\int[1-\xi(z)]\,[f(\xi)-f(\theta_z\xi)]^2\,d\nu^*_\alpha\,.$$

In view of formula (4.3) for the Dirichlet form of f, we have just proved that

$$\Big\{ < V_a, f >_{\nu^*_\alpha} \Big\}^2 \;\le\; (1/2)(1-\alpha)\sum_{z\in\mathbb{Z}^d_*}(a\cdot z)^2 p(z) < f, (-\mathcal{L})f >_{\nu^*_\alpha}\,.$$

This proves not only that V_a belongs to $\mathcal{H}_{-1}$ but gives also the bound ·

$$\|V_a\|^2_{-1} \;\le\; (1/2)(1-\alpha)\sum_{z\in\mathbb{Z}^d_*}(a\cdot z)^2 p(z) \tag{4.9}$$

for the $\mathcal{H}_{-1}$ norm of V_a.

We have just proved that V_a belongs to $\mathcal{H}_{-1}$. Since, on the other hand, the generator is self-adjoint, in view of subsection 3.1, the assumptions of Theorem 2.2 are in force. This proves a central limit theorem for the tagged particle in the reversible context, originally proved by Kipnis and Varadhan [6]:

Theorem 4.2. *Assume that the transition probability $p(\cdot)$ is symmetric. Then, Z_t converges in distribution, as $t \uparrow \infty$, to a mean zero Gaussian law with matrix co-variance $D(\alpha)$ characterized by (4.8).*

4.2 Mean zero case

Consider now the mean zero case. Varadhan in [21], Theorem 5.1, proved a sector condition for this model. He showed the existence of a finite constant C_0 such that

$$\Big\{ < f, (-\mathcal{L})g >_{\nu^*_\alpha} \Big\}^2 \;\le\; C_0 < f, (-\mathcal{L})f >_{\nu^*_\alpha} < g, (-\mathcal{L})g >_{\nu^*_\alpha}$$

for all functions in the domain of the generator. He proved furthermore that the local function V_a given by (4.4) belongs to $\mathcal{H}_{-1}$. In view of the results presented in subsection 3.2, (2.6) holds for the solution of the resolvent equation (4.5). We have therefore the following theorem due to Varadhan [21]:

Theorem 4.3. *Assume that the transition probability $p(\cdot)$ has mean zero. Then, Z_t converges in distribution, as $t \uparrow \infty$, to a mean zero Gaussian law with co-variance matrix $D(\alpha)$ given by (4.8).*

4.3 Asymmetric exclusion process

Assume now that $p(\cdot)$ is asymmetric. For each $n \geq 0$, denote by $\mathcal{E}_{*,n}$ the subsets of $\mathbb{Z}^d_*$ with n points and let $\mathcal{E}_* = \cup_{n\geq 0}\mathcal{E}_{*,n}$ be the class of finite subsets of $\mathbb{Z}^d_*$. For each A in $\mathcal{E}_*$, let Ψ_A be the local function

$$\Psi_A = \prod_{x\in A} \frac{\xi(x)-\alpha}{\sqrt{\chi(\alpha)}} ,$$

where $\chi(\alpha) = \alpha(1-\alpha)$. By convention, $\Psi_\phi = 1$. It is easy to check that $\{\Psi_A,\ A \in \mathcal{E}_*\}$ is an orthonormal basis of $L^2(\nu^*_\alpha)$. For each $n \geq 1$, denote by $\mathcal{G}_n$ the subspace of $L^2(\nu^*_\alpha)$ generated by $\{\Psi_A,\ A \in \mathcal{E}_{*,n}\}$, so that $L^2(\nu^*_\alpha) = \oplus_{n\geq 0}\mathcal{G}_n$. Functions of $\mathcal{G}_n$ are said to have degree n.

Consider a local function f. Since $\{\Psi_A : A \in \mathcal{E}_*\}$ is a basis of $L^2(\nu^*_\alpha)$, we may write

$$f = \sum_{n\geq 0} \sum_{A\in\mathcal{E}_{*,n}} \mathfrak{f}(A)\Psi_A = \sum_{n\geq 0} \pi_n f.$$

Here we have denoted by π_n the orthogonal projection on $\mathcal{G}_n$. Notice that the coefficients $\mathfrak{f}(A)$ depend not only on f but also on the density α: $\mathfrak{f}(A) = \mathfrak{f}(\alpha, A)$. Since f is a local function, $\mathfrak{f}\colon \mathcal{E}_* \to \mathbb{R}$ is a function of finite support.

For a subset A of $\mathbb{Z}^d$ and x, y in $\mathbb{Z}^d$, denote by $A_{x,y}$, $S_y A$ the sets defined by

$$A_{x,y} = \begin{cases} (A\backslash\{x\}) \cup \{y\} & \text{if } x \in A,\ y \notin A, \\ (A\backslash\{y\}) \cup \{x\} & \text{if } y \in A,\ x \notin A, \\ A & \text{otherwise ;} \end{cases} \tag{4.10}$$

$$S_y A = \begin{cases} A - y & \text{if } y \notin A, \\ (A-y)_{0,-y} & \text{if } y \in A\ . \end{cases}$$

In this formula, $B + z$ is the set $\{x + z;\ x \in B\}$. Therefore, to obtain $S_y A$ from A in the case where y belongs to A, we first translate A by $-y$ (getting a new set that contains the origin) and we then remove the origin and add site $-y$.

Denote by $s(\cdot)$ (resp. $a(\cdot)$) the symmetric (resp. asymmetric) part of the transition probability p:

$$s(x) = (1/2)\{p(x) + p(-x)\} , \quad a(x) = (1/2)\{p(x) - p(-x)\} .$$

A simple computation shows that

$$(\mathcal{L}f) = \sum_{A\in\mathcal{E}_*} \Big\{(\mathfrak{L}_{0,\alpha}\mathfrak{f})(A) + (\mathfrak{L}_{\tau,\alpha}\mathfrak{f})(A)\Big\}\Psi_A ,$$

where $\mathfrak{L}_{0,\alpha}$ is an operator that can be decomposed as $\mathfrak{L}_{0,\alpha} = \mathfrak{L}_0^1 + (1-2\alpha)\mathfrak{L}_0^2 + 2\sqrt{\chi(\alpha)}(\mathfrak{L}_0^+ - \mathfrak{L}_0^-)$, with

$$
\begin{aligned}
(\mathfrak{L}_0^1\mathfrak{f})(A) &= (1/2) \sum_{x,y\in\mathbb{Z}_*^d} s(y-x)[\mathfrak{f}(A_{x,y}) - \mathfrak{f}(A)] , \\
(\mathfrak{L}_0^2\mathfrak{f})(A) &= \sum_{\substack{x\in A, y\notin A \\ x\neq 0, y\neq 0}} a(y-x)[\mathfrak{f}(A_{x,y}) - \mathfrak{f}(A)] , \\
(\mathfrak{L}_0^-\mathfrak{f})(A) &= \sum_{\substack{x\notin A, y\notin A \\ x\neq 0, y\neq 0}} a(y-x)\mathfrak{f}(A\cup\{x\}) , \\
(\mathfrak{L}_0^+\mathfrak{f})(A) &= \sum_{\substack{x\in A, y\in A \\ x\neq 0, y\neq 0}} a(y-x)\mathfrak{f}(A-\{y\}) ;
\end{aligned}
\tag{4.11}
$$

and $\mathfrak{L}_{\tau,\alpha}$ is an operator which can be decomposed as $\alpha\mathfrak{L}_\tau^1 + (1-\alpha)\mathfrak{L}_\tau^2 + \sqrt{\chi(\alpha)}(\mathfrak{L}_\tau^+ + \mathfrak{L}_\tau^-)$, where

$$
\begin{aligned}
(\mathfrak{L}_\tau^1\mathfrak{f})(A) &= \sum_{y\in A} p(y)[\mathfrak{f}(S_yA) - \mathfrak{f}(A)] , \\
(\mathfrak{L}_\tau^2\mathfrak{f})(A) &= \sum_{y\notin A} p(y)[\mathfrak{f}(S_yA) - \mathfrak{f}(A)] , \\
(\mathfrak{L}_\tau^+\mathfrak{f})(A) &= \sum_{y\in A} p(y)[\mathfrak{f}(A-\{y\}) - \mathfrak{f}(S_yA-\{-y\})] , \\
(\mathfrak{L}_\tau^-\mathfrak{f})(A) &= \sum_{y\notin A} p(y)[\mathfrak{f}(A\cup\{y\}) - \mathfrak{f}(S_yA\cup\{-y\})] .
\end{aligned}
\tag{4.12}
$$

We may therefore decompose the generator $\mathcal{L}$ as

$$\mathcal{L} = \mathcal{L}_- + \mathcal{R}_0 + \mathcal{B}_0 + \mathcal{L}_+ ,$$

where

$$
\begin{aligned}
\mathcal{L}_- f &= \sqrt{\chi(\alpha)} \sum_{A\in\mathcal{E}_*} \{-2\mathfrak{L}_0^- + \mathfrak{L}_\tau^-\}\mathfrak{f}(A)\Psi_A , \\
\mathcal{L}_+ f &= \sqrt{\chi(\alpha)} \sum_{A\in\mathcal{E}_*} \{2\mathfrak{L}_0^+ + \mathfrak{L}_\tau^+\}\mathfrak{f}(A)\Psi_A , \\
\mathcal{B}_0 f &= \sum_{A\in\mathcal{E}_*} \{(1-2\alpha)\mathfrak{L}_0^2 + \alpha\mathfrak{L}_\tau^1 + (1-\alpha)\mathfrak{L}_\tau^2\}\mathfrak{f}(A)\Psi_A , \\
\mathcal{R}_0 f &= \sum_{A\in\mathcal{E}_*} \mathfrak{L}_0^1\mathfrak{f}(A)\Psi_A .
\end{aligned}
$$

The space $L^2(\nu_\alpha^*)$ and the generator $\mathcal{L}$ have therefore exactly the structure presented in subsection 3.3. Denote by $\mathcal{H}_{0,1}$, $\mathcal{H}_{0,-1}$ the Sobolev spaces induced by the local functions and the symmetric positive operator $\mathcal{R}_0$.

To conclude the proof of the central limit theorem for the tagged particle, we need only to check that the local function V_a given by (4.4) satisfies the assumption of Lemma 3.2 and that the generator satisfies hypotheses (3.2), (3.4), (3.6).

Sethuraman, Varadhan and Yau [19] proved that in dimension $d \geq 3$ all mean zero local functions belong to $\mathcal{H}_{-1}$. In particular, since V_a has mean zero,

$$\left| < V_a, f >_{\nu_\alpha^*} \right| \leq C_0 \| f \|_1$$

for some finite constant C_0 and all functions f in the domain of the generator. Notice that we only need to consider functions f having degree one because V_a has degree one. In this case, by Lemma 4.4 in [8], $\|f\|_1 \leq C_0 \|f\|_{0,1}$ for some finite constant C_0 so that V_a belongs to $\mathcal{H}_{0,-1}$ and the first assumption is fulfilled.

We now turn to the second set of hypotheses of Lemma 3.2. We need to check assumptions (3.2), (3.4) and (3.6). In view of formula (3.5) in [8], (3.2) holds for the asymmetric simple exclusion process. On the other hand, putting together Lemma 4.1 and formula (3.1) in [19] with Lemma 5.1 of [7], we show that (3.4) holds with $\beta = 1/2$ for the asymmetric exclusion process in dimension $d \geq 3$. Finally, in respect to (3.6), observe that $\mathcal{B}_0$ has two pieces. The first one, corresponding to $\alpha \mathfrak{L}_\tau^1 + (1-\alpha)\mathfrak{L}_\tau^2$, satisfies (3.6) with $\gamma = 1$ in view of [8], Lemma 4.2. However, the piece which corresponds to $(1-2\alpha)\mathfrak{L}_0^2$ does not satisfy (3.6). What we can prove instead, [9], is that

Lemma 4.4. *Let Φ be a function such that $\|\pi_n \Phi\|_{0,-1} < \infty$ for each $n \geq 1$. Let h_λ be the solution of the resolvent equation*

$$\lambda h_\lambda - \mathcal{L} h_\lambda = \Phi \, .$$

There exists a finite constant C_0, independent of λ and α, such that

$$\begin{aligned} \|\mathcal{L}_0^2 h_{\lambda,n}\|_{0,-1} \leq & \frac{C_0\sqrt{n}}{\alpha} \|\pi_n \Phi\|_{0,-1} \\ & + \frac{C_0 n^{3/2}}{\alpha} \sum_{j=n-1}^{n+1} \|\pi_j h_\lambda\|_{0,1} \end{aligned}$$

for all $n \geq 1$, $0 < \alpha < 1$ and $\lambda > 0$.

It very easy to check that this estimate may replace condition (3.6) in the proof of Lemma 3.2. In view of the previous estimates, we have the following result due to Sethuraman, Varadhan and Yau [19]:

Theorem 4.5. *Assume that the transition probability $p(\cdot)$ is asymmetric. Then, in dimension $d \geq 3$, Z_t converges in distribution, as $t \uparrow \infty$, to a mean zero Gaussian law with co-variance matrix $D(\alpha)$ characterized by (4.8).*

5 Comments and extensions

We list in this section some results related to the problem of the central limit theorem for the tagged particle.

5.1 Invariance principle

With a little more effort, one can prove in fact that the tagged particle converges to a d-dimensional Brownian motion with diffusion coefficient characterized by (4.8).

Remark 5.1. *In the conditions of Theorem 4.2, Theorem 4.3 or Theorem 4.5, $Z_t^N = Z_{Nt}$ converges, as $N \uparrow \infty$, to a Brownian motion with diffusion coefficient given by (4.8).*

We never excluded the possibility that the variance vanishes. In fact one can prove that $D(\alpha)$ is strictly positive in all cases but one. In the nearest-neighbor one-dimensional symmetric simple exclusion process $D(\alpha) = 0$. In fact, in this case, $\sqrt{t}$ is not the correct renormalization and this is easy to understand. Since particles cannot jump over the others, if we want the tagged particle to move from the origin up to N, we need also to move all particles which were originally between the origin and N to the right of N. The displacement of the tagged particle is thus much more rigid. Relating the exclusion process to the symmetric nearest-neighbor one-dimensional zero-range process, Arratia [1] and Rost and Vares [16] proved that $X_{tN}/N^{1/4}$ converges in distribution to a fractionary Brownian motion.

In a similar spirit, Landim, Olla and Volchan [10] and Landim and Volchan [13] considered the evolution of an asymmetric tagged particle in $\mathbb{Z}$, jumping to the right with intensity $p > 1/2$ and to the left with intensity $1 - p$, evolving as a random walk with exclusion in a medium of symmetric particles. They proved a law of large numbers and an equilibrium central limit theorem for the position of a tagged particle.

5.2 Smoothness of the diffusion coefficient

Recall from (4.8) the characterization of the self-diffusion coefficient $D(\cdot)$ and notice that it depends on α, the density of particles in the environment. Based on the duality introduced in Subsection 4.3, Landim, Olla and Varadhan [8] proved that this dependence is smooth:

Theorem 5.2. *In the symmetric case the self-diffusion coefficient $D(\cdot)$ is of class C^∞ on $[0, 1]$ and in the asymmetric case, in dimension $d \geq 3$, it is of class C^∞ on $(0, 1]$.*

It is not yet clear whether the lack of smoothness at the origin comes from the method (essentially the factor α^{-1} appearing in the statement of Lemma 4.4) or whether it is intrinsic to the problem.

Here is the idea of the proof. Recall equation (4.8) for the self-diffusion matrix. Let $R_a = \chi(\alpha)^{-1/2} V_a$ be the cylinder function given by

$$R_a(\xi) \;=\; \frac{1}{\sqrt{\alpha(1-\alpha)}} \sum_{y \in \mathbb{Z}_*^d} p(y)(y \cdot a)[\alpha - \xi(y)]\,.$$

With the notation introduced in the previous section, we may write R_a as

$$R_a(\xi) \;=\; -\sum_{y \in \mathbb{Z}_*^d} (y \cdot a) p(y) \Psi_y\,,$$

where $\Psi_z = \Psi_{\{z\}}$ for z in $\mathbb{Z}_*^d$. For $\lambda > 0$, denote by g_λ the solution of the resolvent equation:

$$\lambda g_\lambda - \mathcal{L} g_\lambda \;=\; R_a\,.$$

Of course, $g_\lambda = \chi(\alpha)^{-1/2} f_\lambda$. In view of (4.8),

$$a \cdot D(\alpha) a \;=\; (1-\alpha) \sum_{z \in \mathbb{Z}_*^d} p(z)\,(a \cdot z)^2 \;+\; 2\chi(\alpha) \lim_{\lambda \to 0} < S_a, g_\lambda >_{\nu_\alpha^*}\,, \tag{5.1}$$

where $S_a = -\sum_z (a \cdot z) p(z) \Psi_{-z}$.

Denote by $\mathfrak{g}_\lambda(\alpha, A)$ the coefficients of g_λ on the basis $\{\Psi_A,\ A \in \mathcal{E}_*\}$. Writing both g_λ and R_a on the basis $\{\Psi_A,\ A \in \mathcal{E}_*\}$, we obtain an equation for the coefficients $\mathfrak{g}_\lambda(\alpha, A)$:

$$\lambda \mathfrak{g}_\lambda - \mathfrak{L}(\alpha) \mathfrak{g}_\lambda \;=\; \mathfrak{R}_a\,. \tag{5.2}$$

Here, $\mathfrak{L}(\alpha)$ is the operator $\mathfrak{L}_{0,\alpha} + \mathfrak{L}_{\tau,\alpha}$ defined in (4.11), (4.12) and, for each a in $\mathbb{R}^d$, $\mathfrak{R}_a = \mathfrak{R}_a^{\cdot}$ is the real function defined on $\mathcal{E}_*$ by $\mathfrak{R}_a(\{y\}) = -(y \cdot a)p(y)$, $\mathfrak{R}_a(A) = 0$ for $|A| > 1$ and $A = \phi$.

In view of (5.1), to prove that D is smooth, we just need to show that

$$\lim_{\lambda \to 0} < S_a, g_\lambda >_{\nu_\alpha^*} = - \lim_{\lambda \to 0} \sum_{z \in \mathbb{Z}_*^d} p(z)(a \cdot z)\mathfrak{g}_\lambda(\alpha, \{z\})$$

is a smooth function in the density α. We need thus to show that there exists a subsequence $\lambda_k \downarrow 0$ such that $\{\mathfrak{g}_{\lambda_k}(\alpha, \{y\}) \, : \, k \geq 1\}$ converges uniformly to a smooth function for all y with $p(y) > 0$.

To prove that $\mathfrak{g}_\lambda(\alpha, \{y\})$ is a sequence of smooth functions, observe from equation (5.2) and from the explicit form of the operator $\mathfrak{L}(\alpha)$ given in (4.11), (4.12) that $\mathfrak{g}_\lambda(\alpha, \cdot)$ is the solution of a elliptic equation for each fixed α. The density α is now a parameter of the equation and we want to prove that the solutions depend smoothly on this parameter.

In the case where the operator $\mathfrak{L}(\alpha)$ doesn't change the degree of a function, we would have a one-parameter family of finite dimensional elliptic equation. To show that the solutions depend smoothly on the parameter α, we would first deduce the equations satisfied by the derivatives of $\mathfrak{g}_\lambda$ and then obtain estimates, uniform in λ, on the L^∞ norm of these derivatives to conclude the existence of a subsequence λ_k for which $\mathfrak{g}_{\lambda_k}$ converges to a smooth function.

In our case, the operator $\mathfrak{L}(\alpha)$ changes the degree of a function by at most one. To apply the previous ideas, one need first to show that the solution is such that the high degrees are small in some sense. This is exactly the content of Lemma 3.1. Details of the proof can be found in [8], [9].

Remark 5.3. *The approach just presented to prove smoothness of the self-diffusion coefficient provides Taylor expansions at any order of the co-variance matrix through the inversion of finite-dimensional parabolic operators (cf. [8]).*

5.3 Bulk diffusion

The method presented above is quite general and can be used to prove that the Bulk diffusion coefficient of nongradient [20] interacting particle systems are smooth (cf. [2]).

These results have an important application. There are essentially two general methods to prove the hydrodynamic behavior of an interacting particle system. The first one, introduced by Guo, Papanicolau and Varadhan, [5], requires uniqueness of weak solutions of the partial differential equation which describes the macroscopic behavior of the system. The second one, called the relative entropy method and due to Yau, [22], requires the existence of smooth solutions.

For some nongradient systems, the differential equation is of parabolic type and the diffusion matrix is given by a variational formula, similar to the ones derived in this article. In order to apply the relative entropy method, one needs to check that this diffusion matrix is regular in order to guarantee the existence of smooth solutions. The approach presented here may therefore validate the relative entropy method for nongradient systems. For instance, in the case of the Navier-Stokes correction [4], [11], [12] for the asymmetric simple exclusion process, the same method permits to prove that the bulk diffusion coefficient in dimension $d \geq 3$ is smooth in the interval $[0, 1/2) \cup (1/2, 1]$ (cf. [9]).

5.4 Finite dimensional approximations

Fix $N \geq 1$ and consider a finite dimensional version of the symmetric exclusion process on the torus $\{-N, \dots, N\}^d$ (i.e. with periodic boundary conditions, preserving in this manner the translation symmetry). Since we want to work with an ergodic process, we also fix the total number K of particles. Consider now a tagged particle in this finite system. If N is much larger than the size of a single jump, the motion of the tagged particle has a unique canonical lifting to $\mathbb{Z}^d$. We get in this manner a process $X_N(t)$ with values in $\mathbb{Z}^d$. Let us denote by $D_{[N,K]}$ the variance of the Brownian motion which is the limit of the scaled process $\varepsilon X_N(\varepsilon^{-2}t)$ as $\varepsilon \to 0$. It is proved in [7] that

$$\lim_{\substack{N\to\infty \\ K/(2N)^d \to \alpha}} D_{[N,K]} = D(\alpha) .$$

References

[1] Arratia R. The motion of a tagged particle in the simple symmetric exclusion system in $\mathbb{Z}$. *Ann. Prob.*, **11:** (1983), 362-373.

[2] Bernardin C. Regularity of the diffusion coefficient for the lattice gas reversible under Bernoulli measure. (2001), preprint.

[3] Chang C. C., Landim C. and Olla S. Equilibrium fluctuations of asymmetric exclusion processes in dimension $d \geq 3$. To appear in Probability Theory and Related Fields.

[4] Esposito, R., Marra, R. and Yau, H.T. Diffusive limit of asymmetric simple exclusion. *Rev. Math. Phys.*, **6:** (1994), 1233-1267.

[5] Guo M., Papanicolaou G. C. and Varadhan S. R. S. Non linear diffusion limit for a system with nearest neighbor interactions. *Comm. Math. Phys.*, **118:** (1988), 31-59.

[6] Kipnis C. and Varadhan S. R. S. Central limit theorem for additive functionals of reversible Markov processes and applications to simple exclusion. *Commun. Math. Phys.*, **106:** (1986), 1-19.

[7] Landim C., Olla S. and Varadhan S. R. S. Finite-dimensional approximation of the self-diffusion coefficient for the exclusion process. (2000), preprint.

[8] Landim C., Olla S. and Varadhan S. R. S. Symmetric simple exclusion process: regularity of the self-diffusion coefficient. (2000), preprint.

[9] Landim C., Olla S. and Varadhan S. R. S. On the viscosity of the asymmetric simple exclusion process in dimension $d \geq 3$. In preparation.

[10] Landim C., Olla S. and Volchan S. Driven tracer particle in one dimensional symmetric simple exclusion process nearest neighbor symmetric simple exclusion process. *Commun. Math. Phys.* **192:** (1998), 287-307.

[11] Landim C., Olla S. and Yau H. T. First order correction for the hydrodynamic limit of asymmetric simple exclusion processes in dimension $d \geq 3$. *Communications on Pure and Applied Mathematics*, **50:** (1997), 149-203.

[12] Landim C., Olla S. and Yau H. T. Some properties of the diffusion coefficient for asymmetric simple exclusion processes. *The Annals of Probability*, **24:** (1996), 1779-1807.

[13] Landim C. and Volchan S. Equilibrium fluctuation of a driven tracer particle dynamics. *Stoch. Proc. Appl.*, **85:** (2000), 139-158.

[14] Landim C. and Yau H. T. Fluctuation-dissipation equation of asymmetric simple exclusion processes. *Probab. Th. Rel. Fields*, **108:** (1997), 321-356.

[15] Olla S. *Homogenization of Diffusion Processes in Random Fields*. Ecole Polytechnique, France, (1994).

[16] Rost H. and Vares M. E. Hydrodynamics of a one dimensional nearest neighbor model. *Contemp. Math.*, **41:** (1985), 329-342.

[17] Saada E. A limit theorem for the position of a tagged particle in a simple exclusion process. *Ann. Probab.*, **15:** (1987), 375-381.

[18] Sethuraman S. Central Limit Theorem for additive functionals of the simple exclusion process. (1999), preprint.

[19] Sethuraman S., Varadhan S. R. S. and Yau H. T. Diffusive limit of a tagged particle in asymmetric exclusion process. preprint.

[20] Varadhan S.R.S. Non-linear diffusion limit for a system with nearest neighbor interactions II. In *Asymptotic Problems in Probability Theory: Stochastic Models and diffusion on Fractals*, Elworthy K.D. and Ikeda N. editors. Pitman Research Notes in Mathematics vol. 283, 75-128, John Wiley & Sons, New York, (1994).

[21] Varadhan S.R.S. Self diffusion of a tagged particle in equilibrium for asymmetric mean zero random walks with simple exclusion. *Ann. Inst. H. Poincaré (Probabilités)*, **31:** 273-285.

[22] Yau H. T. Relative entropy and hydrodynamics of Ginsburg-Landau models Letters. *Math. Phys.*, **22:** (1991), 63-80.

C. Landim
IMPA, Estrada Dona Castorina 110
22460 Rio de Janeiro
Brasil and
CNRS UPRES-A 6085, Université de Rouen
76128 Mont Saint Aignan
France
E-mail: landim@impa.br

S. Olla
Université de Cergy Pontoise
Département de Mathématiques
2 Av.Adolphe Chauvin, B.P.222, Pontoise
Cergy-Pontoise-Cedex
E-mail: olla@math.u-cergy.fr

S.R.S Varadhan
Courant Institute of Mathematical Sciences
New York University, 251 Mercer street
New York, NY, 10012
U.S.A.
E-mail: varadhan@cims.nyu.edu

Advanced Studies in Pure Mathematics 31, 2001
pp. 265-286

Large Deviation and Hydrodynamic Scaling

Srinivasa R. S. Varadhan

§1. What are Large Deviations?

The theory of large deviations is, roughly speaking, a method of describing the rapidity with which probability distributions depending on a parameter approach the degenerate distribution at some point as the parameter becomes large.

Let us suppose that for each n there is a probabilty measure P_n on some space Ω_n defined on some σ-field Σ_n. There is a complete separable metric space X with its Borel sets $\mathcal{B}$, such that for each n there is a measurable map Φ_n of Ω_n into X. We denote the induced measure $P_n\Phi_n^{-1}$ on $(X, \mathcal{B})$ by Q_n. Actually it is the situation $(X, \mathcal{B}, Q_n)$ that will be of interest to us. As $n \to \infty$ the measures Q_n will converge weakly to a limit which will be degenrate at some point x_0 of X. This is usually a 'law of large numbers', statement. In particular for any closed set $A \subset X$, with $x_0 \notin A$,

$$\lim_{n\to\infty} Q_n(A) = 0. \tag{1.1}$$

If the parametrization has been chosen properly, the convergence in the limit (1.1) will often be exponentially fast and

$$\lim_{n\to\infty} \frac{1}{n} \log Q_n(A) = -\Psi(A) \tag{1.2}$$

will exist atleast for a large class nice sets. Since the exponential behavior of a sum is the same as that of the larger of the summands

$$\Psi(A \cup B) = \min\{\Psi(A), \Psi(B)\}$$

and one can expect $\Psi(A)$ to be given by a formula of the type

$$\Psi(A) = \inf_{x\in A} I(x)$$

Received December 2, 1999.

for some function $I(\cdot): X \to [0, \infty]$. The theory of large deviations is a large collection of interesting examples that fit this model. What makes the class of models interesting is the ability to identify the rate function $I(\cdot)$ in specific cases.

We say that the family Q_n on X satisfies the 'large deviation principle' with rate function $I(\cdot)$ if

$$\limsup_{n\to\infty} \frac{1}{n} \log Q_n(A) \le - \inf_{x\in A} I(x) \quad \text{for closed sets } A \in X, \tag{1.3}$$

$$\liminf_{n\to\infty} \frac{1}{n} \log Q_n(G) \ge - \inf_{x\in G} I(x) \quad \text{for open sets } G \in X. \tag{1.4}$$

Of course if $E \subset X$ is nice enough that

$$\inf_{x\in E^0} I(x) = \inf_{x\in E} I(x) = \inf_{x\in \overline{E}} I(x)$$

we get

$$\lim_{n\to\infty} \frac{1}{n} \log Q_n(E) = - \inf_{x\in E} I(x). \tag{1.5}$$

It is important that the function $I(\cdot)$ that can take the value $+\infty$ be lower semi-continuous and have compact level sets, i.e., for each $\ell < \infty$, the set

$$K_\ell = \{x : I(x) \le \ell\} \tag{1.6}$$

be a compact (closed and totally bounded) subset of X.

We will begin with some simple examples.

Example. Let α be a probability measure on R. Let P_n on $\Omega = R^n$ be the product measure $\alpha \times \alpha \times \cdots \times \alpha$. Let Φ_n be the map

$$\Phi_n(x_1, \ldots, x_n) = \frac{x_1 + \cdots + x_n}{n}.$$

The law of large numbers asserts that $Q_n \to \delta_a$ with $a = \int x\, d\alpha$. According to a theorem of Cramér [1] Q_n satisfies a large deviation principle with rate function

$$I(y) = \sup_\sigma [\sigma y - \log M(\sigma)] \tag{1.7}$$

where

$$M(\sigma) = \int e^{\sigma x} \alpha(dx). \tag{1.8}$$

Another example is the following.

Example. Let F be a finite alphabet $\mathcal{A}$ consisting of letters $\{a_1, \ldots, a_k\}$. Let Ω_n consist of words $W = \{x_1, \ldots, x_n\}$ of length n in $\mathcal{A}$. The probabilities $P_n(W)$ are all equal and since there are k^n words of length n

$$P_n(W) = \frac{1}{k^n}$$

for every word $W \in \Omega_n$. X is the space of probability distributions on $\{1, \ldots, k\}$, i.e., $\{p_1, \ldots, p_k : p_i \geq 0 \text{ and } \sum_i p_i = 1\}$. The map Φ_n is the empirical distibution

$$\Phi_n(x_1, \ldots, x_n) = \left\{\frac{\sum_j \delta_{a_i, x_j}}{n}\right\} = \left\{\frac{n_i}{n}\right\} \tag{1.9}$$

where n_i is the number of times the letter a_i occurs in the word $W = \{x_1, \ldots, x_n\}$. Again, by the law of large numbers, Q_n converges to $\delta_{\{1/k, \ldots, 1/k\}}$.

$$\begin{aligned} Q_n[p_1, \ldots, p_n] &\simeq \frac{n!}{(np_1)! \cdots (np_k)!} \frac{1}{k^n}, \\ I(p) &= -\lim_{n\to\infty} \frac{1}{n} \log Q_n[p_1, \ldots, p_n] \\ &= \log k + \sum p_i \log p_i \\ &= \sum_i p_i \log \frac{p_i}{\frac{1}{k}}. \end{aligned}$$

A slightly more general form of the example is Sanov's theorem.

Example. If we define

$$P_n(W) = \frac{n!}{n_1! \cdots n_k!} \pi_1^{n_1} \cdots \pi_k^{n_k}$$

we get

$$I_\pi(p) = -\sum_i p_i \log \pi_i + \sum_i p_i \log p_i = \sum_i p_i \log \frac{p_i}{\pi_i}$$

or even more generally

Example. We take $\Omega_n = X \times X \times \cdots \times X$, $P_n = \alpha \times \alpha \times \cdots \times \alpha$ and Φ_n the map of Ω into the space $\mathcal{M}$ of all probability measures on X defined by

$$\Phi_n(x_1, \ldots, x_n) = \frac{\delta_{x_1} + \cdots + \delta_{x_n}}{n}.$$

In this example $I_\alpha(\mu) < \infty$ only if $\mu \ll \alpha$ and

$$\frac{d\mu}{d\alpha} \log \frac{d\mu}{d\alpha} \in L_1(\alpha).$$

Then

$$I_\alpha(\mu) = \int_X \log \frac{d\mu}{d\alpha}\, d\mu = \int_X \frac{d\mu}{d\alpha} \log \frac{d\mu}{d\alpha}\, d\alpha. \tag{1.10}$$

Otherwise $I_\alpha(\mu) = +\infty$.

There are some general principles in the theory which are relatively easy to establish. Here is one. A general property known as 'Contarction Principle' is the following:

Theorem 1.1. *Let P_n satisfy the large deviation property with rate function $I(\cdot)$ on X. Let $f : X \to Y$ be a continuous map into Y. Then $Q_n = P_n f^{-1}$ satifies a large deviation principle on Y with rate function $J(y) = \inf[I(x)\,;\, x : f(x) = y]$.*

We will illustrate the Contarction Principle by showing that Cramér's theorem can be obtained from Sanov's theorem. Consider the map $f : \mathcal{M} \to R$ defined by

$$f(\mu) = \int_R x\, d\mu. \tag{1.11}$$

Then the sample mean $(x_1 + \cdots + x_n)/n$ can be thought of as

$$\frac{x_1 + \cdots + x_n}{n} = f(\Phi_n(x_1, \ldots, x_n))$$

where Φ_n is the empirical distribution. A calculation shows that

$$I(y) = \inf_{\mu : \int x\, d\mu = y} I_\alpha(\mu)$$

which is the contraction principle. Actually Sanov's theorem can be sort of seen as a version of Cramér's theorem as well. We can repalce R by the locally convex topological vector space $\mathcal{M}(R)$ and replace α on R by the distribution β induced on $\mathcal{M}(R)$ by the map $x \to \delta_x$. The empirical distribution is just the sum of n independent $\mathcal{M}(R)$ valued random vectors with the common distribution β. The moment generating function is replaced by

$$M(V) = \int_{\mathcal{M}(R)} e^{\langle V, \mu\rangle}\, d\beta = \int_R e^{V(x)}\, d\alpha \tag{1.12}$$

and

$$\text{(1.13)} \qquad H(\mu;\alpha) = I_\alpha(\mu) = \sup_{V(\,\cdot\,)} \left[\int_R V(x)\, d\mu - \log M(V) \right] = \int_R \frac{d\mu}{d\alpha} \log \frac{d\mu}{d\alpha}\, d\alpha.$$

In particular

$$\text{(1.14)} \qquad \int_R V(x)\, d\mu \le \log \int_R e^{V(x)}\, d\alpha + H(\mu;\alpha)$$

or for any $\sigma > 0$, replacing V by σV, we get

$$\text{(1.15)} \qquad \int_R V(x)\, d\mu \le \frac{1}{\sigma} \log \int_R e^{\sigma V(x)}\, d\alpha + \frac{1}{\sigma} H(\mu;\alpha).$$

Another general principle is the following theorem on the exponential growth rate of integrals. It is basically a fancy version of the simple fact that for $a, b > 0$ we have

$$\lim_{n\to\infty} [a^n + b^n]^{1/n} = \max(a,b).$$

Theorem 1.2. *If the large deviation principle holds for some Q_n on X, with a rate function $I(\,\cdot\,)$, then for any real valued bounded continuous function $F(\,\cdot\,)$ on X*

$$\lim_{n\to\infty} \frac{1}{n} \log \int_X \exp[n\,F(x)]\, dQ_n = \sup_{x\in X} [F(x) - I(x)].$$

The book [2] is a good source for a discussion of these topics as well as for additional references.

§2. Hydrodynamic Scaling

The basic example of Hydrodynamic scaling is the derivation of Euler equations from the equations of classical mechanics. Let us start with a collection of $N \simeq \bar{\rho}\ell^3$ classical particles in a large periodic cube Λ_ℓ of side ℓ in R^3. The motion of the particles are governed by the equations of motion of a classical Hamiltonian dynamical system with energy given by

$$\text{(2.1)} \qquad H(p,q) = \frac{1}{2} \sum_{i=1}^{N} \|p_i\|^2 + \frac{1}{2} \sum_{i\ne j} V(q_i - q_j).$$

Here, $q_i \in \Lambda_\ell$ is the position of the i-th particle and $p_i \in R^3$ is its velocity. The coordinates $k = 1, 2, 3$ refer to the three components of position or velocity. The repulsive potential $V \geq 0$ is an even function that is not identically zero and has compact support in R^3. The interaction in particular is short range. The classical equations of motion are

$$\frac{dq_i^k}{dt} = \frac{\partial H(p,q)}{\partial p_i^k} = p_i^k, \tag{2.2}$$

$$\frac{dp_i^k}{dt} = -\frac{\partial H(p,q)}{\partial q_i^k} = -\sum_{j=1}^{N} V_k(q_i - q_j), \tag{2.3}$$

where $V_k(q) = \partial V(q)/\partial q^k$ for $k = 1, 2, 3$ are the three components of the gradient of V. The dynamical system has five conserved quantities. The total number N of particles, the total momenta $\sum_{i=1}^N p_i^k$ for $k = 1, 2, 3$ and the total energy $H(p, q)$. The hydrodynamic scaling in this context consists of rescaling space and time by a factor of ℓ. The rescaled space is the unit torus $\mathbf{T}^3$ in 3-dimensions. The macroscopic quantities to be studied correspond to the five conserved quantities. The first one of these is the density, and is measured by a function $\rho(t, x)$ of t and x. For each $\ell < \infty$ it is approximated by $\rho_\ell(t, x)$, defined by

$$\int_{\mathbf{T}^3} J(x)\rho_\ell(t,x)\,dx = \frac{1}{\ell^3}\sum_{i=1}^{N} J\Big(\frac{q_i(\ell t)}{\ell}\Big). \tag{2.4}$$

A straight forward differentiation with respect to t yields

$$\begin{aligned} \frac{d}{dt}\int_{\mathbf{T}^3} J(x)\rho_\ell(t,x)\,dx &= \frac{d}{dt}\frac{1}{\ell^3}\sum_{i=1}^{N} J\Big(\frac{q_i(\ell t)}{\ell}\Big) \\ &= \frac{1}{\ell^3}\sum_{i=1}^{N}(\nabla J)\Big(\frac{q_i(\ell t)}{\ell}\Big)\cdot p_i(\ell t) \\ &\simeq \int_{\mathbf{T}^3}(\nabla J)(x)\cdot \rho_\ell(t,x)u_\ell(t,x)\,dx \end{aligned} \tag{2.5}$$

where $u_\ell(t, x) = u_\ell^k(t, x)$, $k = 1, 2, 3$ are the components of the 'average' velocity of the fluid at the rescaled space time point x, t. This introduces three other macroscopic variables, which represent three coordinates of the momenta that are conserved. We can now write down the first of our five equations

$$\frac{\partial \rho}{\partial t} + \nabla\cdot(\rho u) = 0. \tag{2.6}$$

To derive the next three equations, using a test functions J, we differentiate for $k = 1, 2, 3$

$$\text{(2.7)} \quad \frac{d}{dt}\frac{1}{\ell^3}\sum_{i=1}^{N} J\Big(\frac{q_i(\ell t)}{\ell}\Big)p_i^k(\ell t) = \frac{1}{\ell^3}\sum_{i=1}^{N} p_i^k(\ell t)(\nabla J)\Big(\frac{q_i(\ell t)}{\ell}\Big)\cdot p_i(\ell t)$$
$$- \frac{1}{\ell^2}\sum_{i=1}^{N}\sum_{j=1}^{N} J\Big(\frac{q_i(\ell t)}{\ell}\Big)V_k(q_i(\ell t) - q_j(\ell t)).$$

If we now use the skew-symmetry of $V_k = \partial V/\partial q_k$, we can rewrite the second term of the right hand side of equation (2.7) as

(2.8)

$$- \frac{1}{2\ell^2}\sum_{i=1}^{N}\sum_{j=1}^{N}\Big(J\Big(\frac{q_i(\ell t)}{\ell}\Big) - J\Big(\frac{q_j(\ell t)}{\ell}\Big)\Big)V_k(q_i(\ell t) - q_j(\ell t))$$
$$\simeq - \frac{1}{2\ell^3}\sum_{i=1}^{N}\sum_{j=1}^{N} J_r\Big(\frac{q_i(\ell t)}{\ell}\Big)(q_i^r(\ell t) - q_j^r(\ell t))V_k(q_i(\ell t) - q_j(\ell t))$$
$$= \frac{1}{\ell^3}\sum_{i=1}^{N}\sum_{j=1}^{N} J_r\Big(\frac{q_i(\ell t)}{\ell}\Big)\psi_k^r(q_i(\ell t) - q_j(\ell t))$$

with

$$\psi_k^r(q) = -\frac{1}{2}q^r V_k(q).$$

The next step is rather mysterious and requires considerable explanation. The quantities

$$\sum_{i=1}^{N} p_i^k p_i^r, \quad \sum_{i,j=1}^{N} \psi_k^r(q_i(t) - q_j(t))$$

are not conserved. They depend on combinations of individual velocities that are not conserved and on spacings between particles both of which change in the microscopic time scale and therefore do so rapidly in the macroscopic time scale. They should therefore be replaced by their space-time averages. By appealing to an 'Ergodic Theorem' they can be replaced by their averages with repect to their equilibrium distributions. The equilibrium 'ensemble' consists of an infinite collection of points $\{p_\alpha, q_\alpha\}$, in the phase space $R^3 \times R^3$. There is a natural five parameter family of measures $\mu_{\rho,u,T}$ that are invariant under spatial translations as well as the Hamiltonian dynamics. The points $\{p_\alpha\}$ are distributed according to a Gibbs Distribution with density ρ and formal interaction energy

$$\frac{1}{2T}\sum_{\alpha,\beta} V(q_\alpha - q_\beta).$$

In other words $\{q_\alpha\}$ is a point process obtained by taking infinite volume limit of $N = \ell^3\rho$ particles distributed in the cube of side ℓ in R^3 according to the joint density

$$\frac{1}{Z}\exp\left[-\frac{1}{2T}\sum_{1\le i\ne j\le N} V(q_i - q_j)\right]$$

where Z is the normalization constant. The velocities $\{p_\alpha\}$ are distributed independently of each other as well as of $\{q_\alpha\}$, having a common three dimensional Gaussian distribution with mean u and covariance TI. Assuming that the infinite volume limit exists in a reasonable sense it will be a point process defined as an infinite volume Gibbs measure $\mu_{\rho,T}$. The velocities $\{p_\alpha\}$ will be an independent Gaussian ensemble $\nu_{u,T}$. In the first term the quantities $p_i^k p_i^r$ are replaced by their expectations

$$u^k(t,x)u^r(t,x) + \delta_{k,r}T(t,x)$$

and in the second term $\psi_{k,r}$ are replaced by their expectations that involve the 'pressure' per unit volume in the Gibbs ensemble

$$\mathbf{P}_k^r(\rho,T) = \lim_{\ell\to\infty} E^{\mu_{\rho,T}}\left\{\frac{1}{\ell^3}\sum_{|q_\alpha|,|q_\beta|\le\ell} \psi_k^r(q_\alpha - q_\beta)\right\}.$$

This leads to the equation

$$\begin{aligned}\frac{d}{dt}&\int_{\mathbf{T}^3} J(x)u^k(t,x)\,dx \\ &= \int_{\mathbf{T}^3}\sum_{r=1}^3 \frac{\partial J}{\partial x_r}(x)(u^k(t,x)u^r(t,x) + \delta_{k,r}T(t,x))\,dx \\ &\quad + \int_{\mathbf{T}^3}\sum_{r=1}^3 \frac{\partial J}{\partial x_r}(x)\mathbf{P}_k^r(\rho(t,x),T(t,x))\,dx.\end{aligned} \tag{2.9}$$

We now integrate by parts, remove the test function J and obtain from equation (2.9)

$$\frac{d}{dt}(\rho u) + \nabla\cdot(\rho u\otimes u + \rho TI + \mathbf{P}(\rho,T)) = 0. \tag{2.10}$$

There is an equation of state that expresses the total energy per unit volume e as

$$e(\rho, u, T) = \frac{1}{2}\rho(|u|^2 + 3T) + f(\rho, T) \tag{2.11}$$

where $f(\rho, T)$, the potential energy per unit volume, is given by

$$f(\rho, T) = \lim_{\ell \to \infty} E^{\mu_{\rho,T}}\left\{\frac{1}{2\ell^3} \sum_{|q_\alpha|,|q_\beta| \le \ell} V(q_\alpha - q_\beta)\right\}.$$

Although we will not derive it, there is a similar equation for $e(t, x)$ that is obtained by differentiating

$$\frac{d}{dt}\frac{1}{2\ell^3}\sum_{i=1}^{N} J\Big(\frac{q_i(\ell t)}{\ell}\Big)\Big[|p_i(\ell t)|^2 + \sum_{j=1}^{N} V(q_i(\ell t) - q_j(\ell t))\Big]$$

and proceeding in a similar fashion. It looks like

$$\frac{de}{dt} + \nabla \cdot \big[(e + T)u + \mathbf{P}(\rho, T)u\big] = 0. \tag{2.12}$$

The five equations one for density given by equation (2.6), the three for velocities contained in equation (2.10) and finally the enrgy equation (2.12) constitute a first order system of non-linear hyperbolic conservation laws in the six variables $[\rho, u, T, e]$ with one relation between them given by equation (2.11). Given smooth initial data they have local solutions. Rigorous derivation of these equations does not exist.

We have made a basic assumption in the above derivation. If we take a small volume in space around the point (x, t) in macroscopic space-time and blow up the space by a factor of ℓ we will see a bunch of particles with velocities. The positions of these particles will form a point process in a big domain in R^3. The statistics of these points is assumed to be a Gibbs distribution $\mu_{\rho,T}$ corresponding to the density $\rho = \rho(t, x)$ and 'Temperature', $T = T(t, x)$. Given the positions, the velocities are assumed to be mutually independent and have a common Gaussian distribution with mean $u = u(t, x)$ and covariance $TI = T(t, x)I$. The five parameters (ρ, u, T) locally detemine a Gibbs-Gaussian equilibrium. The equations are derived under the assumption that this picture holds asymptotically for large ℓ. There is no known proof of this. While it is possible to prepare the initial state so that this property of loacl equlibrium holds at time $t = 0$, there is no guarantee that this property persists at positive macroscopic times. If ρ, u, T are constants independent of

x at time 0, then we have a global equilibrium and that persists. But hydrodynamically this is the uninteresting case.

While the validity of the principle of local equlibrium is very hard to establish for the Hamiltonian system, it is not nearly so hard for stochastic systems of comparable type. Noise helps to establish local equilibria. This in many cases can be rigorously established and thence the corresponding hydrodynamical equations can be derived with full mathematical rigor.

We will consider a class of stochastic models that are called simple exclusion processes. They make sense on any finite or countable set X and for us it will be either the integer lattice $\mathbf{Z}^d$ in d-dimensions or $\mathbf{Z}_N^d$ obtained from it as a quotient by considering each coordinate modulo N. At any given time a subset of these lattice sites will be occupied by particles, with atmost one particle at each site. In other words some sites are empty while others are occupied with one particle. The particles move randomly. Each particle waits for an exponential random time and then tries to jump from its current site x to a new site y. The new site y is picked randomly according to a probability distribution $\pi(x,y)$. In particular $\sum_y \pi(x,y) = 1$ for every x. Of course a jump to y is not always possible. If the site is empty the jump is possible and is carried out. If the site already has a particle, the jump cannot be carried out and the particle forgets about it and waits for another chance, i.e., waits for a new exponential waiting time. If we normalize so that all waiting times have mean 1, the generator of the process can be written down as

$$(\mathcal{A}f)(\eta) = \sum_{x,y} \eta(x)(1-\eta(y))\pi(x,y)[f(\eta^{x,y}) - f(\eta)] \tag{2.13}$$

where η represents the configuration with $\eta(x) = 1$ if there is a particle at x and $\eta(x) = 0$ otherwise. For each configuration η and a pair of sites x, y the new configuration $\eta^{x,y}$ is defined by

$$\eta^{x,y}(z) = \begin{cases} \eta(y), & \text{if } z = x, \\ \eta(x), & \text{if } z = y, \\ \eta(z), & \text{if } z \neq x, y. \end{cases} \tag{2.14}$$

We will mainly be concerned with the situation where the set X is $\mathbf{Z}^d$ or $\mathbf{Z}_N^d$, viewed naturally as an Abelian group with $\pi(x,y)$ being translation invariant and given by $\pi(x,y) = p(y-x)$ for some probability distribution p. It is convenient to assume that p has finite support. There are various possibilities. We will first consider the case $\sum_z zp(z) = m \neq 0$ that needs hyperbolic scaling and leads to Burgers equation with zero viscosity. In order to convey the idea it is sufficient to restrict ourselves to the case where $d = 1$, $p(1) = 1$ and $p(z) = 0$ for all $z \neq 1$. This is the totally asymmetric nearest neighbor simple exclusion model and of course $m = 1$. In this case the rescaling in time is done by a factor of N and the generator is

$$(\mathcal{A}_N f)(\eta) = N \sum_{x \in \mathbf{Z}_N} \eta(x)(1-\eta(x+1))[f(\eta^{x,x+1}) - f(\eta)].$$

We can easily calculate

(2.16)

$$\begin{aligned} d&\left[\frac{1}{N}\sum_{x\in\mathbf{Z}_N} J\Big(\frac{x}{N}\Big)\eta_t(x)\right] \\ &= \left[\sum_{x\in\mathbf{Z}_N}\left[J\Big(\frac{x+1}{N}\Big) - J\Big(\frac{x}{N}\Big)\right]\eta_t(x)(1-\eta_t(x+1))\right]dt + M_N(t) \\ &\simeq \frac{1}{N}\left[\sum_{x\in\mathbf{Z}_N} J'\Big(\frac{x}{N}\Big)\eta_t(x)(1-\eta_t(x+1))\right]dt + o(1). \end{aligned}$$

The martingale term is negligible and we need to do simple 'averaging'. The equlibria are the Bernoulli measures μ_ρ indexed by density. Since there is only a single invariant quantity, i.e., the number of particles, we can replace $\eta_t(x)(1-\eta_t(x+1))$ by its expected value $\rho(t,x)(1-\rho(t,x))$.

$$\frac{d}{dt}\int_{\mathbf{T}^1} J(\theta)\rho(t,\theta)\,d\theta = \int_{\mathbf{T}^1} J'(\theta)\rho(t,\theta)(1-\rho(t,\theta))\,d\theta$$

or equivalently

$$\frac{\partial \rho}{\partial t} + \frac{\partial[\rho(1-\rho)]}{\partial \theta} = 0. \tag{2.17}$$

A different situation occurs when p is symmetric, i.e., $p(x) = p(-x)$. Let us look at the function

$$V_J(\eta) = \sum J(x)\eta(x)$$

and compute

$$(2.18)\qquad \begin{aligned}(\mathcal{A}V_J)(\eta) &= \sum_{x,y} \eta(x)(1-\eta(y))p(y-x)(J(y)-J(x))\\ &= \sum_{x,y} \eta(x)p(y-x)(J(y)-J(x))\\ &= \sum_{x,y} \eta(x)[(\mathbf{P}-I)J](x)\\ &= V_{(\mathbf{P}-I)J}(\eta).\end{aligned}$$

The space of linear functionals is left invariant by the generator. It is not difficult to see that

$$E_\eta\big[V_J(\eta(t))\big] = V_{J(t)}(\eta)$$

where

$$J(t) = \exp[t(\mathbf{P}-I)]J$$

is the solution of

$$\frac{d}{dt}J(t,x) = (\mathbf{P}-I)J(t,x).$$

It is almost as if the interaction had no effect and in fact for the calculation of expectations of 'one particle' functions it clearly does not. Let us start with a configuration on $\mathbf{Z}_N^d$ and scale space by N and time by N^2. The generator becomes $N^2\mathcal{A}$ and the particles can be visualized as moving on a lattice imbedded in the unit torus $\mathbf{T}^d$, with a spacing of $1/N$, that becomes dense as $N \to \infty$.

Let us consider the functional

$$\xi(t) = \frac{1}{N^d}\sum_x J\Big(\frac{x}{N}\Big)\eta_t(x).$$

We can write

$$\xi(t)-\xi(0) = \int_0^t V_N(\eta(s))\,ds + M_N(t)$$

where

$$V_N(\eta) = (N^2\mathcal{A}V_J)(\eta) = V_{J_N}(\eta)$$

with

$$\begin{aligned}(J_N)(x) &= N^2\sum\Big[J\Big(x+\frac{z}{N}\Big)-J(x)\Big]p(z)\\ &\simeq \frac{1}{2}(\Delta_C J)(x).\end{aligned}$$

Here Δ_C refers to the Laplacian

$$\sum_{i,j} C_{i,j} \frac{\partial^2}{\partial x_i \partial x_j}$$

with the covariance matrix C given by

$$C_{i,j} = \sum_x x_i x_j p(x).$$

$M_N(t)$ is a martingale and a very elementary calculation yields

$$E\{[M_N(t)]^2\} \leq CtN^{-d}$$

essentially completing the proof in this case. Technically the empirical distribution $\nu_N(t)$ is viewed as a measure on $\mathbf{T}^d$ and $\nu_N(\cdot)$ is viewed as a stochastic process with values in the space $\mathcal{M}(\mathbf{T}^d)$ of nonnegative measures on $\mathbf{T}^d$. In the limit it lives on the set of weak solutions of the heat equation

$$\frac{\partial \rho}{\partial t} = \frac{1}{2}\Delta_C \rho \tag{2.19}$$

with the initial condition $\rho(0,x) = \rho_0(x)$ determined by

$$\int_{\mathbf{T}^d} J(x)\rho_0(x)\,dx = \lim_{n\to\infty} \frac{1}{N^d} \sum_{x\in\mathbf{Z}_N^d} J\Big(\frac{x}{N}\Big)\eta_0(x) \tag{2.20}$$

and the uniqueness of such weak soultions for given initial density establishes the validity of the scaling limit. We could have have computed two moments as in the noninteracting case. The expectation would have been no different from the noninteracting case since it involves only one particle functions. The variance involves two particle functions and would have involved slighlty more work, because the independence is not there. The martingale argument however is more general.

Let us now turn to the case where p has mean zero but is not symmetric. In this case

$$V_N(\eta) = N^{2-d} \sum_{x,y} \eta(x)(1-\eta(y))p(y-x)\Big[J\Big(\frac{y}{N}\Big) - J\Big(\frac{x}{N}\Big)\Big] \tag{2.21}$$

and we get stuck at this point. If p is symmetric, as we saw, we gain a factor of N^{-2}. Otherwise the gain is only a factor of N^{-1} which is not enough. We seem to end up with

$$
\begin{aligned}
&N^{-d}\sum_x\sum_y \eta(x)\\
&\qquad\times\Big\langle \frac{1}{2}\Big[(\nabla J)\Big(\frac{x}{N}\Big)+(\nabla J)\Big(\frac{y}{N}\Big)\Big], N(1-\eta(y))(y-x)p(y-x)\Big\rangle\\
&=\frac{1}{2N^d}\sum_x(\nabla J)\Big(\frac{x}{N}\Big)N\Psi_x
\end{aligned}
$$

where

$$
\begin{aligned}
\Psi_0 &= \frac{1}{2}\Big[\eta(0)\sum_z(1-\eta(z))zp(z)+(1-\eta(0))\sum_z\eta(-z)zp(z)\Big]\\
&=\frac{1}{2}\Big[-\eta(0)\sum_z\eta(z)zp(z)+(1-\eta(0))\sum_z\eta(-z)zp(z)\Big]\\
&=\frac{1}{2}\Big[\sum_z\eta(-z)zp(z)-\eta(0)\sum_z(\eta(z)+\eta(-z))zp(z)\Big].
\end{aligned}
$$

The second sum is zero in the symmetric case and Ψ_0 can then be written as a 'gradient' $\Psi_0 = \sum_j \tau_{e_j}\xi_j - \xi_j$ where τ_{e_j} are shifts in the coordinate directions. This allows us to do summation by parts and gain a factor of N^{-1}. When this is not the case, we have a 'nongradient' model and the scaling limit can no longer be established by simple averaging.

Exactly the same situation arises in the symmetric case if we make the probabilities of jumps $p(x) = 1/2d$ for the $2d$ nearest neighbors and 0 otherwise, but change the rates so that the generator reads

$$
(\mathcal{A}f)(\eta) = \sum_{|x-y|=1} a_{x,y}(\eta)[f(\eta^{x,y}) - f(\eta)] \tag{2.22}
$$

where $a_{x,y}(\eta)$ are translation invariant and satisfy the 'detailed balance', conditions relative to the Bernoulli measures.

There are several good sources for this and related material. In particular the book [4], the monograph [10] and the notes [3] contain all of this material as well as more references.

§3. Large Deviation Methods in Hydrodynamic Scaling

A rigorous proof of the validity of the hydrodynamic scaling limit depends on establishing some sort of a local ergodic theorem. There are several ways of carrying this out depending on the circumstances.

But the methods that are fairly general use ideas from large deiviations in some form. We will consider the example of the totally asymmetric simple exclusion process in one dimension. Suppose we are give a smooth function $\rho(t,x)$ on $[0,T]\times \mathbf{T}$ that satisfies $0 < c \le \rho(t,x) \le 1-c < 1$ and solves Burgers equation (2.17)

$$\frac{\partial \rho}{\partial t} + \frac{\partial}{\partial x}[\rho(t,x)(1-\rho(t,x)] = 0.$$

For any smooth $\rho(x)$ from $\mathbf{T} \to (0,1)$ we can associate a local or slowly varying equlibrium state

$$f^N_{\rho(\,\cdot\,)}(\eta) = \prod_{x=1}^{N} \rho\Big(\frac{x}{N}\Big)^{\eta_x}\Big(1-\rho\Big(\frac{x}{N}\Big)\Big)^{1-\eta_x}. \tag{3.1}$$

We could guess that the state at time t is more or less

$$f_N(t,\eta) = \prod_{x=1}^{N} \rho\Big(t,\frac{x}{N}\Big)^{\eta_x}\Big(1-\rho\Big(t,\frac{x}{N}\Big)\Big)^{1-\eta_x}. \tag{3.2}$$

Even if we start at time 0 with initial distribution $f_N(0,\eta)$ the true state at time t is the solution $g_N(t,\eta)$ of the Kolmogorov forward equation

$$\frac{\partial g_N}{\partial t} = \mathcal{L}^*_N g_N \tag{3.3}$$

with the initial condition

$$g_N(0,\eta) = f_N(0,\eta). \tag{3.4}$$

We wish to compare the true solution g_N to our guess f_N. They match at $t=0$. What about $t>0$?

The comparison is done by

$$H_N(t) = H(g_N(t,\,\cdot\,); f_N(t,\,\cdot\,)). \tag{3.5}$$

It is controlled by establishing a Gronwall type of inequality

$$\frac{dH_N(t)}{dt} \le CH_N(t) + \text{“}error\text{”} \tag{3.6}$$

that leads to

Theorem 3.1. *We have*

$$\lim_{N\to\infty} \sup_{0\le t\le T} \frac{1}{N} H_N(t) = 0 \tag{3.7}$$

which in turn implies the validity of the hydrodynamic limit

$$\lim_{N\to\infty} \frac{1}{N} \sum_{x\in \mathbf{Z}_N} J\Big(\frac{x}{N}\Big)\eta_t(x) = \int_{\mathbf{T}} J(x)\rho(t,x)\,dx.$$

One key ingredient is the validity of the local equilibrium principle which needs to be established. This takes the following form. Let ℓ be an intermediate scale, i.e., $1 \ll \ell \ll N$. For any local function $g(\eta)$ let us define

$$\hat{g}(\rho) = E^{\mu_\rho}[g(\eta)] \tag{3.8}$$

where μ_ρ is the Bernoulli measure with density ρ. We look at the difference

(3.9)

$$D_{\ell,N,g}(\eta) = \frac{1}{N}\sum_x \left| \frac{1}{2\ell+1} \sum_{y:|y-x|\le \ell} g(\tau_y \eta) - \hat{g}\Big(\frac{1}{2\ell+1}\sum_{y:|y-x|\le\ell} \eta(y)\Big)\right|.$$

Establishing hydrodynamic limit requires

$$\lim_{\epsilon\to 0} \limsup_{N\to\infty} E^{P_N}\left[\int_0^T D_{\epsilon N,N,g}(\eta_t)\,dt\right] = 0 \tag{3.10}$$

where P_N is the process starting from an arbitrary initial configuration. The relative entropy considerations reduce this to proving estimates of the form

$$\lim_{\ell\to\infty} \limsup_{N\to\infty} E^{P_N}[D_{\ell,N,g}(g)] = 0. \tag{3.11}$$

This can be reduced to proving a much stronger estimate for the process Q_N in equilibrium.

$$\lim_{\ell\to\infty} \limsup_{N\to\infty} \frac{1}{N} \log E^{Q_N}\left[\exp N\int_0^T D_{\epsilon N,N,g}(\eta_t)\,dt\right] = 0 \tag{3.12}$$

for every g.

This in turn can be estimated in terms of Feynman-Kac representation and variational formulae involving Dirichlet forms.

In the case of diffusive scaling this approach is often powerful enough to yield equation (3.10).

In the last example alluded to in the previous section, i.e., nongradient models, the analysis involves writing the 'current', in the form

$$W_{x,x+1} = c(\eta(x) - \eta(x+1)) + \text{'}\textit{negligible terms}\text{'} \tag{3.13}$$

corresponding to a projection in $L_2(P_\rho)$. The left handside can be thought of as a closed one form while the negligible terms are the exact ones. One has to prove that the codimension of the negligible terms is one and can be represented by the density gradient term. Because the analysis is carried out seperately in each equilibrium this determines $c = c(\rho)$ and one ends up with an equation of the form

$$\frac{\partial \rho}{\partial t} = \frac{1}{2}\frac{\partial}{\partial x}\left[c(\rho)\frac{\partial \rho}{\partial x}\right]. \tag{3.14}$$

The proofs agian involve establishing superexponetial estimates in equlibrium and use Jensen type inequality (1.15) to go from equilibrium to nonequilibrium.

§4. Large Deviations in Hydrodynamic Scaling

Let us consider $k_N \simeq N^d$ independent random walks on the lattice $\mathbf{Z}_N^d$ of $\mathbf{Z}^d$ modulo N. If we denote their trajectories by $\{x_1(\cdot), \ldots, x_{k_N}(\cdot)\}$ and rescale them as $\frac{1}{N}x_i(N^2 t) = y_i(t)$ we have k_N noninteracting rescaled random walks and the empirical process

$$R_{N,\omega} = \frac{1}{N^d}\sum_{i=1}^{k_N} \delta_{y_i(\cdot)} \tag{4.1}$$

will converge on the Skorohod space of trajectories $D[[0,T];\mathbf{T}^d]$ to a Brownian motion with covariance

$$\langle Ca, b\rangle = \sum_z p(z)\langle z, a\rangle\langle z, b\rangle \tag{4.2}$$

where $p(\cdot)$ is the probability distribution of a single step and it is assumed to be symmetric. Of course we need to assume that the initial distribution

$$\nu_{N,\omega} = \frac{1}{N^d}\sum_{i=1}^{k_N} \delta_{y_i(0)} \tag{4.3}$$

has a limit μ and if we take Q to be the Brownian motion with Covariance C and initial distribution μ then

$$\lim_{N\to\infty} R_{N,\omega} = Q \tag{4.4}$$

in probability. Because of the way we have normalized, the total mass of Q, which is the same as total mass of μ, is given by

$$\bar{\rho} = \lim_{N\to\infty} \frac{k_N}{N^d}.$$

We can ask about the probabilities of large deviations in this context. It is a minor variation of Sanov's theorem and the rate function with normalization by N^d is given by the relative entropy

$$\mathcal{I}(R) = H(R;Q). \tag{4.5}$$

We would like to see how this changes if we go from the context of independent random walks to an interacting model like simple exclusion. We will keep the jump distribution as the same $p(\cdot)$.

We saw before that the hydrodynamic limt in this case was still given by equation (2.19). However the behavior of $R_{N,\omega}$ is more complex. For that we have to understand how a tagged particle will behave in equlibrium as well as nonequlibrium. It is known, (see [6]) that a tagged particle in equlibrium will diffuse like a Brownian Motion with some covariane $S(\rho)$ that depends on the density ρ. This is to be expected, because in low density there is very little interaction and one expects $S(\rho) \to C$ as $\rho \to 0$. On the other hand at high density, i.e., when $\rho \to 1$ there is gridlock and one should expect $S(\rho) \to 0$. (The one dimensional nearest neighbor case is different due to blocking and $S(\rho) \equiv 0$ in that one case.)

One expects therefore that if the initial condition is a random configuration chosen from equilibrium with density ρ then, in probability,

$$\lim_{N\to\infty} R_{N,\omega} = Q_\rho \tag{4.6}$$

where Q_ρ is the Browninan motion (with total mass ρ) having covariance $S(\rho)$ and initial density ρ.

In nonequlibrium the situation is a lot more complicated. First of all the density itself is given by the solution $\rho(t,x)$ of the heat equation (2.19) with initial condition $\rho(x)$ which is determined as the limit of the empirical distribution of the initial configuration in the sense of equation (3.4). The tagged particle will only see its immediate neighborhood and will behave as if it is in equilibrium at density $\rho(t,x)$ if it finds itself at time t at the point x. It is reasonable then to expect it to behave like a diffusion with the second order or diffusion coefficients equal to $S(\rho(t,x))$. It could have an additional first order or drift term. It is more convenient to write the expected backward generator in divergence form as

$$\mathcal{L}_t = \frac{1}{2}\nabla S(\rho(t,x))\nabla + c(t,x)\nabla. \tag{4.7}$$

Of course we can tag any are all of the particles and the empirical process is the same whether they or tagged or not. Therefore the solution $\rho(t,x)$ of the heat equation (2.19) must also be a solution of the forward equation corresponding to (4.7), i.e.,

$$\frac{\partial \rho(t,x)}{\partial t} = \frac{1}{2}\nabla S(\rho(t,x))\nabla\rho(t,x) - \nabla\cdot c(t,x)\rho(t,x). \tag{4.8}$$

This means

$$\frac{1}{2}\nabla C\nabla\rho(t,x) = \frac{1}{2}\nabla S(\rho(t,x))\nabla\rho(t,x) - \nabla\cdot c(t,x)\rho(t,x).$$

One can guess (with fingers crossed) that

$$c(t,x) = [S(\rho(t,x) - C]\frac{\nabla\rho(t,x)}{2\rho(t,x)}. \tag{4.9}$$

That this is indeed true is a result in [9] which is based on results of [7].

We now turn to large deviations. To simplify the presentation we assume that the initial configuration is deterministic. Otherwise we have to factor in the large deviation behavior of the initial profile and this adds an extra term to all the rate functions.

Large deviations are invariably obtained by perturbing the dynamics in such a way that the modification produces the needed deviation. The modified process will have, after suitable normalization, some entropy relative to the original process. This can be thought of as 'cost' of the modification. It is conceivable that there are lots of modifications with different costs that produce the same desired deviation. The rate function is always the minimum of such costs. If one can run through a large class of modifications one gets a large deviation lower bound which is the minimum of the costs over that class of modifications. One tries then to match it with an upper bound by some other method.

In our example the possible perturbations are of the jump rates $N^2p(z)$ of the speeded up dynamics. If the magnitude of the perturbation is $\lambda_N \ll N^2$ then the magnitude of the entropy 'cost' is $\lambda_N^2 N^{-2}$

per particle. This suggests a perurbation of order N to obtain a total entropy 'cost' of order N^d. We therefore consider a perturbed generator of the form

$$(\mathcal{A}_{t,N,q(\cdot,\cdot,\cdot)}f)(\eta) = \sum_{x,y} \eta(x)(1-\eta(y)) \times \Big[N^2 p(y-x) + Nq\Big(t,\frac{x}{N},\cdot\Big)\Big][f(\eta^{x,y}) - f(\eta)]$$

where $q(t,x,z)$ is a nice function of t, x and z. If we denote by

$$b(t,x) = \sum zq(t,x,z) \tag{4.11}$$

the effect of the perturbation is to produce a solution of the following modified equation as the hydrodynamic limit.

$$\frac{\partial\rho(t,x)}{\partial t} = \frac{1}{2}\nabla C \nabla \rho(t,x) - \nabla\cdot[b(t,x)\rho(t,x)(1-\rho(t,x))] \tag{4.12}$$

with the same initial condition given by (2.20). Given $\rho(\cdot,\cdot)$ we view (4.12) as an equation for $b(\cdot,\cdot)$ and denote the set of solutions by $\mathcal{B}_{\rho(\cdot,\cdot)}$. For a given $b(\cdot,\cdot)$ the set of $q(\cdot,\cdot,\cdot)$ that satisfy (4.11) is denoted by $\mathcal{Q}_{b(\cdot,\cdot)}$. The entropy cost when divided by N^d converges to

$$\frac{1}{2}\int_0^T \int_{T^d} \Big[\sum_z \frac{q(t,x,z)^2}{p(z)}\Big]\rho(t,x)(1-\rho(t,x))\,dxdt. \tag{4.13}$$

Minimizing (4.13) over $\mathcal{Q}_{b(\cdot,\cdot)}$ yields
(4.14)

$$\mathcal{E}(b(\cdot,\cdot)) = \frac{1}{2}\int_0^T \int_{T^d} \langle b(t,x), C^{-1}b(t,x)\rangle \rho(t,x)(1-\rho(t,x))\,dxdt$$

Minimizing $\mathcal{E}(b(\cdot,\cdot))$ over $\mathcal{B}_{\rho(\cdot,\cdot)}$ gives us the rate function for the large deviation of the empirical density which is the family of one dimensionaal marginals of $R_{N,\omega}$. This was done in [5]. Next, we need to consider the effect of the perturbation on the motion of the tagged particle. This will produce for $R_{N,\omega}$, a weak limit Q_b in probability, with the same initial distribution but with the new backward generator

$$\mathcal{L}_b = \frac{1}{2}\nabla S(\rho(t,x)\nabla + c(t,x)\cdot\nabla + b(t,x)(1-\rho(t,x))\cdot\nabla. \tag{4.15}$$

Finally we can write down the rate function $I(R)$ for the large deviations of $R_{N,\omega}$. From R in addition to its one dimensional marginals

$\rho(\cdot,\cdot)$ we can consider the 'currents'

$$\Lambda_R(f) = E^R\left[\int_0^T \langle f(s,x(s)), dx(s)\rangle\right]. \tag{4.16}$$

From $\mathcal{B}_{\rho(\cdot,\cdot)}$ we look for a $\bar{b}(\cdot,\cdot)$ such that

$$\Lambda_R(f) \equiv \Lambda_{Q_{\bar{b}}}(f)$$

and call it $\overline{R}$. The rate function turns out to be

$$I(R) = \mathcal{E}(\bar{b}(\cdot,\cdot)) + H(R;\overline{R}). \tag{4.17}$$

If the marginal of R does not match the initial density or if we have trouble defining (4.17) at any stage then $I(R)$ is $+\infty$. It turns out that $\bar{b}(\cdot,\cdot)$ if it exists is unique. Details of these results can be found in [8].

References

[1] H. Cramér, Sur un nouveau théeorème-limite de la théorie des probabilités, Actualités Scientifiques et Industrielles, number **736**, Hermann, Paris (1938), 5–23.

[2] J.-D. Dueschel and W. D. Stroock, "Large Deviations", Academic Press, Boston, 1989.

[3] L. Jensen and H.-T. Yau, "Hydrodynamic Scaling Limits of Simple Exclusion Models, Probability Theory and Applications", (P. H. Elton and S. R. S Varadhan, eds.), IAS-Park City Mathematics Series, Vol. **6**, American Mathematical Society, Providence, Rhode Island, 1999, pp. 167–225.

[4] C. Kipnis and C. Landim, "Scaling limits of interacting particle systems", Springer-Varlag, Berlin, 1999.

[5] C. Kipnis, S. Olla and S. R. S. Varadhan, Hydrodynamics and large deviations for simple exclusion processes, Comm. Pure Appl. Math., **42** (1989), 115–137.

[6] C. Kipnis and S. R. S. Varadhan, Central Limit Theorem for additive functionals of reversible Markov processes and applications to simple exclusion, Comm. Math. Phys., **104** (1986), 1–19.

[7] J. Quastel, Diffusion of color in the simple exclusion process, Comm. Pure Appl. Math., **45** (1992), 623–679.

[8] J. Quastel, F. Rezakhanlou and S. R. S. Varadhan, Large deviations for the symmetric simple exclusion process in dimensions $d \geq 3$, Probab. Theory Related Fields, **113,** no. 1 (1999), 1–84.

[9] F. Rezakhanlou, Propagation of chaos for symmetric simple exclusions, Comm. Pure Appl. Math., **47,** no. 7 (1994), 943–957.

[10] H. Spohn, "Large Scale Dynamics of Interacting Particles", Springer-Verlag, New York, 1991.

Courant Institute of Mathematical Sciences
251 Mercer Street
New York, NY 10012
U.S.A.
varadhan@cims.nyu.edu

Commun. Math. Phys. 224, 307 – 321 (2001)

Communications in
Mathematical Physics

Symmetric Simple Exclusion Process: Regularity of the Self-Diffusion Coefficient

C. Landim[1,2,⋆], **S. Olla**[3], **S. R. S. Varadhan**[4,⋆⋆]

[1] IMPA, Estrada Dona Castorina 110, CEP 22460 Rio de Janeiro, Brasil. E-mail: landim@impa.br
[2] CNRS UPRES-A 6085, Université de Rouen, 76128 Mont Saint Aignan, France
[3] Université de Cergy Pontoise, Département de Mathématiques, CNRS UPRES-A 8088, 2 Av. Adolphe Chauvin, B.P.222, Pontoise, 95302 Cergy-Pontoise-Cedex, France. E-mail: olla@math.u-cergy.fr
[4] Courant Institute of Mathematical Sciences, New York University, 251 Mercer Street, New York, NY 10012, USA. E-mail: varadhan@cims.nyu.edu

Received: 6 December 2000 / Accepted: 6 April 2001

Dedicated to Joel Lebowitz on his seventieth birthday

Abstract: We prove that the self-diffusion coefficient of a tagged particle in the symmetric exclusion process in Z^d, which is in equilibrium at density α, is of class C^∞ as a function of α in the closed interval $[0, 1]$. The proof provides also a recursive method to compute the Taylor expansion at the boundaries.

1. Introduction

In the course of the study of macroscopic behavior of large particle systems, effective diffusion coefficients which are functions of the parameters (associated to the conserved quantities) that define the equilibrium measures of the system often appear. These diffusion coefficients are usually expressed in terms of integrals of time correlations functions (Green–Kubo formulas), or through (infinite dimensional) variational formulas. They also appear as coefficients in the diffusive equations that govern the non-equilibrium evolution of the conserved quantities of the system. In order to study the existence and regularity of solutions to these equations it is important to establish first the regularity of these diffusion coefficients as functions of the parameters.

In this article we develop a method for proving smooth dependence, on the density, of the self diffusion coefficient of a tagged or tracer particle in symmetric simple exclusion particle systems that are in equilibrium. It is based on the duality properties of the symmetric simple exclusion process. This method, with modifications, can also be applied to study the smooth dependence on the density of other diffusion coefficients that arise in the study of more general simple exclusion processes. But this will be taken up elsewhere.

The paper is organized along the following lines. In Sect. 2 we introduce the notation and state the main theorem. In Sect. 3 we describe the generalized duality and discuss

⋆ Partially supported by CNPq grant 300358/93-8, FAPERJ grant E26/150.940/99 and PRONEX grant 41.96.0923.00.

⋆⋆ Supported by the National Science Foundation grant DMS-9803140.

several operators and norms that appear in the dual representation. Section 4 is devoted to some key estimates that are used in Sect. 5 to prove the main result on the smoothness of the self diffusion coefficient of the tagged particle in the case of the symmetric simple exclusion process. At the end, in Remark 5.3, we expose a recursive method to compute the Taylor expansion at the boundaries.

Very few results exist at the present time about the regularity of diffusion coefficients. Continuous dependence on the density has been established in different contexts (cf. [2]). Generally proving continuity does not seem to be considerably harder than establishing the existence of a diffusion coefficient. In [6], Lipschitz continuity of the selfdiffusion coefficient for the tagged particle in the symmetric simple exclusion is proved in dimensions $d \geq 3$.

2. Notation and Results

Let us fix a symmetric finite-range probability distribution $p(\cdot)$ on $\mathbb{Z}^d$. Consider the symmetric simple exclusion process associated with p. We assume, without loss of generality, that the subgroup generated by the support of p is all of $\mathbb{Z}^d$. In addition we assume that we are not dealing with the trivial situation of $d = 1$ and $p(\pm 1) = \frac{1}{2}$, i.e. the one dimensional nearest neighbor case where the self diffusion coefficient is identically zero.

The simple exclusion process is the Markov process on $\mathcal{X} = \{0, 1\}^{\mathbb{Z}^d}$ whose generator L acts on cylinder functions f as

$$
\begin{aligned}
(Lf)(\eta) &= \sum_{x,y\in\mathbb{Z}^d} p(y-x)\eta(x)[1-\eta(y)][f(\sigma^{x,y}\eta) - f(\eta)] \\
&= \frac{1}{2}\sum_{x,y\in\mathbb{Z}^d} p(y-x)[f(\sigma^{x,y}\eta) - f(\eta)].
\end{aligned}
\tag{2.1}
$$

Here and below the configurations of $\mathcal{X}$ are denoted by Greek letters. In particular, for x in $\mathbb{Z}^d$, $\eta(x)$ is equal to 1 if the site x is occupied in the configuration η and is equal to 0 if it is not. Moreover, for a configuration η and x, y in $\mathbb{Z}^d$, $\sigma^{x,y}\eta$ is the configuration obtained from η by exchanging the occupation variables $\eta(x)$, $\eta(y)$:

$$
(\sigma^{x,y}\eta)(z) = \begin{cases} \eta(y) & \text{if } z = x, \\ \eta(x) & \text{if } z = y, \\ \eta(z) & \text{otherwise.} \end{cases}
\tag{2.2}
$$

Fix $0 \leq \alpha \leq 1$ and denote by μ_α the Bernoulli product measure on $\mathcal{X}$. This is the probability measure on $\mathcal{X}$ obtained by placing a particle with probability α at each site x, independently from the other sites. It easy to check that the one-parameter family of probability measures $\{\mu_\alpha,\ 0 \leq \alpha \leq 1\}$ are stationary, reversible and ergodic for the Markov process with generator L.

We examine in this article the evolution of a single tagged particle in the symmetric simple exclusion process. Let η be an initial configuration with a particle at the origin, i.e. with $\eta(0) = 1$. Tag this particle and denote by η_t (resp. X_t) the state of the process (resp. the position of the tagged particle) at time t. We shall refer to η_t as the environment. Let ξ_t be the state of the environment as seen from the tagged particle: $\xi_t = \theta_{X_t}\eta_t$. Here, for x in $\mathbb{Z}^d$ and a configuration η, θ_x stands for the translation of η by x, i.e.

$(\theta_x \eta)(y) = \eta(x+y)$. Notice that the origin is always occupied (by the tagged particle) for the environment as seen from the tagged particle. For this reason, we shall consider the process ξ_t as taking values in $\{0,1\}^{\mathbb{Z}^d_*}$, where $\mathbb{Z}^d_* = \mathbb{Z}^d \backslash \{0\}$.

Whereas X_t is not a Markov process due to the presence of the environment, (X_t, ξ_t) and ξ_t are. A simple computation shows that the generator $\mathcal{L}$ of the Markov process ξ_t is given by $\mathcal{L} = \mathcal{L}_0 + \mathcal{L}_\tau$, where

$$
\begin{aligned}
(\mathcal{L}_0 f)(\xi) &= \sum_{x,y \in \mathbb{Z}^d_*} p(y-x)\xi(x)[1-\xi(y)][f(\sigma^{x,y}\xi) - f(\xi)], \\
&= \frac{1}{2} \sum_{x,y \in \mathbb{Z}^d_*} p(y-x)[f(\sigma^{x,y}\xi) - f(\xi)], \\
(\mathcal{L}_\tau f)(\xi) &= \sum_{z \in \mathbb{Z}^d_*} p(z)[1-\xi(z)][f(\tau_z \xi) - f(\xi)].
\end{aligned}
\tag{2.3}
$$

The first part of the generator takes into account the jumps in the environment, while the second one corresponds to jumps of the tagged particle. In the above formula, $\tau_z \xi$ stands for the configuration where the tagged particle, sitting at the origin, is first transferred to the (empty) site z and then the entire environment is translated by $-z$: for all y in $\mathbb{Z}^d_*$,

$$
(\tau_z \xi)(y) = \begin{cases} \xi(z) & \text{if } y = -z, \\ \xi(y+z) & \text{for } y \neq -z. \end{cases}
$$

For $0 \le \alpha \le 1$, denote by μ_α the Bernoulli product measure on $\mathcal{X}_* = \{0,1\}^{\mathbb{Z}^d_*}$. A simple computation shows that μ_α is a reversible and ergodic stationary measure for the Markov process ξ_t.

In this context Kipnis and Varadhan ([1]) proved a central limit theorem for the position of the tagged particle starting with an initial environment chosen randomly from the equilibrium μ_α. They showed that $\varepsilon X_{t\varepsilon^{-2}}$ converges, as $\varepsilon \downarrow 0$, to a Brownian motion with diffusion coefficient $D(\alpha)$ which we will describe in more detail in the next section.

This result has been generalized by Varadhan ([6]) to the asymmetric case with 0-mean ($\sum_y y p(y) = 0$). More recently, for the general asymmetric case in dimension $d \ge 3$, if $\sum_y y p(y) = m \neq 0$, in Sethuraman-Varadhan-Yau ([5]) it is proved that $\varepsilon[X_{t\varepsilon^{-2}} - mt(1-\alpha)\varepsilon^{-2}]$ converges, as $\varepsilon \downarrow 0$, to a Brownian motion with another diffusion coefficient.

In this article we limit ourselves to the symmetric case and study the regularity properties of $D(\alpha)$ as a function of α. The main result is

Theorem 2.1. *The self-diffusion coefficient $D(\alpha)$, as a function of α, is of class C^∞ in the closed interval* $[0,1]$.

3. Generalized Duality

The proof of Theorem 2.1 relies on the duality properties of the symmetric exclusion process that we will now describe.

We have the Hilbert space $L_2(\mu_\alpha)$ with its natural inner product $\langle\cdot,\cdot\rangle_\alpha$. The operator $\mathcal{L}$ is self adjoint and the natural Dirichlet inner products will be denoted by

$$\langle f,g\rangle_{1,\alpha} = \langle -\mathcal{L}f, g\rangle_\alpha ,$$
$$\langle f,g\rangle_{1,env,\alpha} = \langle -\mathcal{L}_0 f, g\rangle_\alpha .$$

The dual norms $\|f\|_{-1,\alpha}$ and $\|f\|_{-1,env,\alpha}$ are defined by

$$\|f\|^2_{-1,\alpha} = \sup_g \Big\{ 2\,\langle f,g\rangle - \langle g,g\rangle_{1,\alpha} \Big\},$$
$$\|f\|^2_{-1,env,\alpha} = \sup_g \Big\{ 2\,\langle f,g\rangle - \langle g,g\rangle_{1,env,\alpha} \Big\}.$$

For each $n \ge 0$, denote by $\mathcal{E}_{*,n}$ the subsets of $\mathbb{Z}^d_*$ with n points and let $\mathcal{E}_* = \cup_{n\ge 0}\mathcal{E}_{*,n}$ be the class of all finite subsets of $\mathbb{Z}^d_*$. Let us consider an abstract Hilbert space $\mathfrak{H}$ with a complete orthonormal basis consisting of $\{\mathfrak{e}_A : A \in \mathcal{E}_*\}$. The space $\mathfrak{H}$ can be viewed as the space of square summable maps $\mathfrak{f}$ of $\mathcal{E}_* \to \mathbb{R}$. In a natural way $\mathfrak{H} = \oplus_{n\ge 0}\mathfrak{G}_n$, where $\mathfrak{G}_n$ is spanned by $\{\mathfrak{e}_A : A \in \mathcal{E}_{*,n}\}$. For each A in $\mathcal{E}_*$, let the local function in $L_2(\mu_\alpha)$ be defined by

$$\Psi_A = \Psi_A(\alpha,\xi) = \prod_{x\in A} \frac{\xi(x)-\alpha}{\sqrt{\chi(\alpha)}},$$

where $\chi(\alpha) = \alpha(1-\alpha)$. By convention, $\Psi_\phi = 1$. It is easy to check that $\{\Psi_A,\ A \in \mathcal{E}_*\}$ is an orthonormal basis of $L^2(\mu_\alpha)$. For each $n \ge 0$, denote by $\mathcal{G}_n$ the subspace of $L^2(\mu_\alpha)$ generated by $\{\Psi_A,\ A \in \mathcal{E}_{*,n}\}$, so that $L^2(\mu_\alpha) = \oplus_{n\ge 0}\mathcal{G}_n$. Functions of $\mathcal{G}_n$ are said to have degree n. The main property of the symmetric simple exclusion process that will be used here is that part of the generator, i.e. $\mathcal{L}_0$, preserves the degree of the functions.

Consider a local function f. Since $\{\Psi_A : A \in \mathcal{E}_*\}$ is a basis of $L^2(\mu_\alpha)$, we may write

$$f = \sum_{n\ge 0}\sum_{A\in\mathcal{E}_{*,n}} \mathfrak{f}(A)\Psi_A = \sum_{n\ge 0}\pi_n f.$$

Here we have denoted by π_n the orthogonal projection onto $\mathcal{G}_n$. Notice that the coefficients $\mathfrak{f}(A)$ depend not only on f but also on the density α: $\mathfrak{f}(A) = \mathfrak{f}(A,\alpha)$. Since f is a local function, $\mathfrak{f}\colon \mathcal{E}_* \to \mathbb{R}$ has finite support. In other words we have a unitary isomorphism, $f \sim \sum \mathfrak{f}(A)\mathfrak{e}_A$ between $L_2(\mu_\alpha)$ and $\mathfrak{H}$ that takes local functions in $L_2(\mu_\alpha)$ onto finite linear combinations of the basis elements. Of course this establishe also an isomorphism between $\mathcal{G}_n$ and $\mathfrak{G}_n$. We now conclude this section by expressing the operators $\mathcal{L}$ and $\mathcal{L}_0$ as well as their Dirichlet forms, through this isomorphism, in the basis $\{\mathfrak{e}_A\}$ of $\mathfrak{H}$. To begin with, because the isomorphism is unitary, we have

$$\langle f,g\rangle_\alpha = \langle \mathfrak{f},\mathfrak{g}\rangle = \sum_{A\in\mathcal{E}_*} \mathfrak{f}(A)\mathfrak{g}(A),$$

where

$$f \sim \sum \mathfrak{f}(A)\mathfrak{e}_A \quad\text{and}\quad g \sim \sum \mathfrak{g}(A)\mathfrak{e}_A .$$

The norm in $\mathfrak{H}$ will be denoted by $\|\mathfrak{f}\|_0$.

For a subset A of $\mathbb{Z}_*^d$ and x, y in $\mathbb{Z}_*^d$, denote by $A_{x,y}$, $S_y A$ the sets defined by

$$A_{x,y} = \begin{cases} (A\backslash\{x\}) \cup \{y\} & \text{if } x \in A,\ y \notin A, \\ (A\backslash\{y\}) \cup \{x\} & \text{if } y \in A,\ x \notin A, \\ A & \text{otherwise;} \end{cases}$$
$$S_z A = \begin{cases} A - z & \text{if } z \notin A, \\ ((A\backslash\{z\}) - z) \cup \{-z\} & \text{if } z \in A. \end{cases} \tag{3.1}$$

In this formula, $B + z$ is the set $\{x + z;\ x \in B\}$. Therefore, to obtain $S_z A$ from A in the case where z belongs to A, we first remove z to get a set not containing z, then translate $A\backslash\{z\}$ by $-z$ and finally add the site $-z$.

Recall the definition of the generators $\mathcal{L}_0$, $\mathcal{L}_\tau$ given in (2.3). A simple computation shows that

$$(\mathcal{L}_0 f) \sim \sum_{A \in \mathcal{E}_*} (\mathfrak{L}_0 \mathfrak{f})(A) \mathfrak{e}_A, \quad (\mathcal{L}_\tau f) \sim \sum_{A \in \mathcal{E}_*} (\mathfrak{L}_{\tau,\alpha} \mathfrak{f})(A) \mathfrak{e}_A,$$

where

$$(\mathfrak{L}_0 \mathfrak{f})(A) = \frac{1}{2} \sum_{x, y \in \mathbb{Z}_*^d} p(y - x)[\mathfrak{f}(A_{x,y}) - \mathfrak{f}(A)] \tag{3.2}$$

and $\mathfrak{L}_{\tau,\alpha}$ is an operator which can be decomposed as

$$\mathfrak{L}_{\tau,\alpha} = \alpha \mathfrak{L}_\tau^1 + (1 - \alpha)\mathfrak{L}_\tau^2 + \sqrt{\chi(\alpha)}(\mathfrak{L}_\tau^+ + \mathfrak{L}_\tau^-),$$

where

$$\begin{aligned} (\mathfrak{L}_\tau^1 \mathfrak{f})(A) &= \sum_{y \in A} p(y)[\mathfrak{f}(S_y A) - \mathfrak{f}(A)], \\ (\mathfrak{L}_\tau^2 \mathfrak{f})(A) &= \sum_{y \notin A} p(y)[\mathfrak{f}(S_y A) - \mathfrak{f}(A)], \\ (\mathfrak{L}_\tau^+ \mathfrak{f})(A) &= \sum_{y \in A} p(y)[\mathfrak{f}(A\backslash\{y\}) - \mathfrak{f}(S_y A\backslash\{-y\})], \\ (\mathfrak{L}_\tau^- \mathfrak{f})(A) &= \sum_{y \notin A} p(y)[\mathfrak{f}(A \cup \{y\}) - \mathfrak{f}(S_y A \cup \{-y\})]. \end{aligned} \tag{3.3}$$

Notice that $\mathcal{L}$ on $L_2(\mu_\alpha)$ is represented on $\mathfrak{H}$ by $\mathfrak{L}_\alpha = \mathfrak{L}_0 + \mathfrak{L}_{\tau,\alpha}$:

$$\mathfrak{L}_\alpha = \mathfrak{L}_0 + \alpha \mathfrak{L}_\tau^1 + (1 - \alpha)\mathfrak{L}_\tau^2 + \sqrt{\chi(\alpha)}[\mathfrak{L}_\tau^+ + \mathfrak{L}_\tau^-]. \tag{3.4}$$

We mentioned earlier that the main property to be exploited here is that the generator of the symmetric exclusion process preserves the degree of local functions. It is easy to check that the operators $\mathfrak{L}_0$, $\mathfrak{L}_\tau^1$, $\mathfrak{L}_\tau^2$ preserve the degree of a function, i.e. they map $\mathfrak{G}_n$ into itself. Moreover, $\mathfrak{L}_\tau^+$ increases the degree of a function by one while $\mathfrak{L}_\tau^-$ decreases it by one.

For a function $\mathfrak{f}\colon \mathcal{E}_* \to \mathbb{R}$ and $n \geq 0$, denote by $\pi_n \mathfrak{f}$ or by $\mathfrak{f}_n$ its restriction to $\mathcal{E}_{n,*}$: $(\pi_n \mathfrak{f})(A) = \mathfrak{f}(A)\mathbf{1}\{A \in \mathcal{E}_n\}$.

For local functions $f, g \colon \mathcal{X}_* \to \mathbb{R}$, a long but elementary computation shows that, if we define

$$
\begin{aligned}
2\,\langle \mathfrak{f}, \mathfrak{g}\rangle_{1,\alpha} &= \sum_{x,y\in\mathbb{Z}_*^d}\sum_{A\in\mathcal{E}} p(y-x)[\mathfrak{f}(A_{x,y})-\mathfrak{f}(A)][\mathfrak{g}(A_{x,y})-\mathfrak{g}(A)] \\
&+ \sum_{y\in\mathbb{Z}_*^d}\sum_{A\in\mathcal{E}} p(y)r_y(A)[\mathfrak{f}(S_yA)-\mathfrak{f}(A)][\mathfrak{g}(S_yA)-\mathfrak{g}(A)] \\
&- \sqrt{\chi(\alpha)}\sum_{y\in\mathbb{Z}_*^d}\sum_{\substack{A\in\mathcal{E}\\ y\notin A}} p(y)[\mathfrak{f}(S_yA)-\mathfrak{f}(A)][\mathfrak{g}(S_y[A\cup\{y\}])-\mathfrak{g}([A\cup\{y\}])] \\
&- \sqrt{\chi(\alpha)}\sum_{y\in\mathbb{Z}_*^d}\sum_{\substack{A\in\mathcal{E}\\ y\notin A}} p(y)[\mathfrak{f}(S_y[A\cup\{y\}])-\mathfrak{f}([A\cup\{y\}])][\mathfrak{g}(S_yA)-\mathfrak{g}(A)].
\end{aligned}
\tag{3.5}
$$

with $r_y(A)$ is equal to α if y belongs to A and is equal to $1-\alpha$ if y does not belong to A, then

$$\langle f, g\rangle_{1,\alpha} = \langle \mathfrak{f}, \mathfrak{g}\rangle_{1,\alpha}$$

Notice that the last three terms can be recombined to give a positive expression when $\mathfrak{f} = \mathfrak{g}$. The corresponding norm will be denoted by $\|\mathfrak{f}\|_{1,\alpha}$ which of course is equal to $\|f\|_{1,\alpha}$. By completing the space of finitely supported functions with this norm we obtain the Dirichlet space $\mathfrak{H}_1$.

Let $\mathfrak{H}_{-1}$ be the dual of $\mathfrak{H}_1$ with respect to the standard inner product on $\mathfrak{H}$. This is the Hilbert space generated by finitely supported functions and the norm $\|\cdot\|_{-1,\alpha}$ defined by

$$\|\mathfrak{f}\|^2_{-1,\alpha} = \sup_{\mathfrak{g}} \Big\{ 2\,\langle \mathfrak{f}, \mathfrak{g}\rangle - \langle \mathfrak{g}, \mathfrak{g}\rangle_{1,\alpha} \Big\},$$

where the supremum is carried over all finitely supported functions $\mathfrak{g}$. It follows from the isomorphism that $\|f\|_{-1,\alpha} = \|\mathfrak{f}\|_{-1,\alpha}$.

The Dirichlet form corresponding to $\mathcal{L}_0$ is much simpler to calculate in the $\mathfrak{H}$ representation. Denote by $\|\cdot\|_{1,env}$ and $\|\cdot\|_{-1,env}$ respectively the Dirichlet norm and its dual associated to the generator $\mathfrak{L}_0$:

$$
\begin{aligned}
\|\mathfrak{g}\|^2_{1,env} &= \langle \mathfrak{g}, (-\mathfrak{L}_0)\mathfrak{g}\rangle \\
&= \frac{1}{2}\sum_{x,y\in\mathbb{Z}_*^d}\sum_{A\in\mathcal{E}} p(y-x)[\mathfrak{g}(A_{x,y})-\mathfrak{g}(A)]^2 \\
&= \sum_{n\ge 0} \|\pi_n\mathfrak{g}\|^2_{1,env}
\end{aligned}
$$

and

$$
\begin{aligned}
\|\mathfrak{g}\|^2_{-1,env} &= \sup_{\mathfrak{f}} \Big\{ 2\,\langle \mathfrak{f}, \mathfrak{g}\rangle - \langle \mathfrak{f}, (-\mathfrak{L}_0)\mathfrak{f}\rangle \Big\} \\
&= \sum_{n\ge 0} \|\pi_n\mathfrak{g}\|^2_{-1,env},
\end{aligned}
\tag{3.6}
$$

where the supremum is carried over all finitely supported functions. In contrast to the norms $\|\cdot\|_{1,\alpha}$, $\|\cdot\|_{-1,\alpha}$, the norms $\|\cdot\|_{1,env}$, $\|\cdot\|_{-1,env}$ do not depend explicitly on the parameter α. Moreover, since $\langle \mathfrak{f}, (-\mathfrak{L}_0)\mathfrak{f}\rangle \le \langle \mathfrak{f}, (-\mathfrak{L}_\alpha)\mathfrak{f}\rangle$, it follows that $\|\mathfrak{g}\|_{1,env} \le \|\mathfrak{g}\|_{1,\alpha}$ and $\|\mathfrak{g}\|_{-1,\alpha} \le \|\mathfrak{g}\|_{-1,env}$. In Lemma 4.4, we estimate $\|\mathfrak{g}\|_{1,\alpha}$ and $\|\mathfrak{g}\|_{-1,env}$ in terms of $\|\mathfrak{g}\|_{1,env}$ and $\|\mathfrak{g}\|_{-1,\alpha}$, respectively.

Finally, for any $k \ge 0$, let us define

$$\begin{aligned} &|||\mathfrak{f}|||^2_{0,k} = \sum_{n\ge 0} n^{2k}\|\pi_n\mathfrak{f}\|_0^2, \quad |||\mathfrak{f}|||^2_{1,k} = \sum_{n\ge 0} n^{2k}\|\pi_n\mathfrak{f}\|^2_{1,env}, \\ &|||\mathfrak{f}|||^2_{-1,k} = \sum_{n\ge 0} n^{2k}\|\pi_n\mathfrak{f}\|^2_{-1,env}. \end{aligned} \tag{3.7}$$

If T is the operator that acts as scalar multiplication by n on the space $\mathfrak{G}_n$ of degree n, these are the quadratic forms $\|T^k\mathfrak{f}\|^2$, $< T^k\mathfrak{f}, (-\mathfrak{L}_0)T^k\mathfrak{f} >$ and $< T^k\mathfrak{f}, (-\mathfrak{L}_0)^{-1}T^k\mathfrak{f} >$ respectively. Note that $\mathfrak{L}_0$ commutes with T. The completion under these norms will be denoted by $\mathfrak{H}_{0,k}$, $\mathfrak{H}_{1,k}$ and $\mathfrak{H}_{-1,k}$ respectively.

4. Some Estimates

Since $\mathfrak{L}_\alpha$ is self adjoint, for the solution $\mathfrak{u}_\lambda$ of the resolvent equation

$$\lambda\mathfrak{u}_\lambda - \mathfrak{L}_\alpha\mathfrak{u}_\lambda = \mathfrak{f}, \tag{4.1}$$

we have the basic estimate

$$\|\mathfrak{u}_\lambda\|_{1,\alpha} \le \|\mathfrak{f}\|_{-1,\alpha}$$

that implies

$$\|\mathfrak{u}_\lambda\|_{1,env} \le \|\mathfrak{f}\|_{-1,env}$$

or

$$|||\mathfrak{u}_\lambda|||_{1,0} \le |||\mathfrak{f}|||_{-1,0}.$$

The following regularity result follows from Eq. (5.5) of [4].

Lemma 4.1. *Let $k \ge 1$ be given. Let $\mathfrak{f}$ be a function such that $|||\mathfrak{f}|||_{-1,k} < \infty$. For $\lambda > 0$, let $\mathfrak{u}_\lambda$ be the solution of the resolvent equation* (4.1). *Then,*

$$|||\mathfrak{u}_\lambda|||_{1,k} \le C(k)|||\mathfrak{f}|||_{-1,k} \tag{4.2}$$

for a finite constant $C(k)$ independent of α and λ.

In fact the proof of (4.2) given in [4] extends immediately to non-local $\mathfrak{f}$.

We now state some bounds on the restrictions of $\mathfrak{L}^1_\tau$, $\mathfrak{L}^2_\tau$, $\mathfrak{L}^+_\tau$ and $\mathfrak{L}^2_\tau$ on $\mathfrak{G}_n$. These bounds will grow linearly with n. Notice that $\mathfrak{L}^j_\tau$, $j = 1, 2$ are symmetric operators, while $\mathfrak{L}^+_\tau$ is the adjoint of $\mathfrak{L}^-_\tau$:

$$\langle \mathfrak{L}^+_\tau\mathfrak{f}, \mathfrak{g}\rangle = \langle \mathfrak{f}, \mathfrak{L}^-_\tau\mathfrak{g}\rangle, \quad \left\langle \mathfrak{L}^j_\tau\mathfrak{f}, \mathfrak{g}\right\rangle = \left\langle \mathfrak{f}, \mathfrak{L}^j_\tau\mathfrak{g}\right\rangle$$

for $j = 1, 2$ and $\mathfrak{f}$, $\mathfrak{g}$ in $L^2(\mathcal{E}_*)$. Moreover,

$$\begin{aligned} \left\langle \mathfrak{L}^1_\tau\mathfrak{f}, \mathfrak{f}\right\rangle &= (1/2)\sum_{A\in\mathcal{E}_*}\sum_{y\in A} p(y)[\mathfrak{f}(S_yA) - \mathfrak{f}(A)]^2, \\ \left\langle \mathfrak{L}^2_\tau\mathfrak{f}, \mathfrak{f}\right\rangle &= (1/2)\sum_{A\in\mathcal{E}_*}\sum_{y\notin A} p(y)[\mathfrak{f}(S_yA) - \mathfrak{f}(A)]^2. \end{aligned}$$

Lemma 4.2. *There exists a finite constant C_0 depending only on the transition probability p such that*

$$\left\langle (-\mathfrak{L}_\tau^j)\mathfrak{f}, \mathfrak{f} \right\rangle \le; C_0 n \left\langle (-\mathfrak{L}_0)\mathfrak{f}, \mathfrak{f} \right\rangle \tag{4.3}$$

for $j = 1, 2$, all $n \ge 1$ and all $\mathfrak{f}$ in $\mathfrak{H}_n$. Moreover

$$\left\langle (-\mathfrak{L}_\tau^{\pm})\mathfrak{f}, \mathfrak{g} \right\rangle^2 \le C_0^2 n^2 \left\langle (-\mathfrak{L}_0)\mathfrak{f}, \mathfrak{f} \right\rangle \left\langle (-\mathfrak{L}_0)\mathfrak{g}, \mathfrak{g} \right\rangle \tag{4.4}$$

for all $n \ge 1$ and all $\mathfrak{f}$ in $\mathfrak{G}_n$, $\mathfrak{g}$ in $\mathfrak{G}_{n\pm 1}$. On the other hand for $j = 1, 2$,

$$\|\mathfrak{L}_\tau^j \mathfrak{f}\|_0^2 \le 4\|\mathfrak{f}\|_0^2 \quad \textit{and} \quad \|\mathfrak{L}_\tau^{\pm} \mathfrak{f}\|_0^2 \le 4\|\mathfrak{f}\|_0^2 \tag{4.5}$$

for all $\mathfrak{f}$ in $\mathfrak{H}$.

Proof. The first estimate (4.3) follows immediately from Lemma 5.1 in [4].

We first prove that for all $\mathfrak{f}$, $\mathfrak{g}$ in $L^2(\mathcal{E}_*)$,

$$\left\langle \mathfrak{L}_\tau^{\pm}\mathfrak{f}, \mathfrak{g} \right\rangle^2 \le \left\langle (-\mathfrak{L}_\tau^1)\mathfrak{f}, \mathfrak{f} \right\rangle \left\langle (-\mathfrak{L}_\tau^2)\mathfrak{g}, \mathfrak{g} \right\rangle. \tag{4.6}$$

Fix $\mathfrak{f}$, $\mathfrak{g}$ in $L^2(\mathcal{E}_*)$. By the explicit formula for $\mathfrak{L}_\tau^+$, we have that

$$\left\langle (-\mathfrak{L}_\tau^+)\mathfrak{f}, \mathfrak{g} \right\rangle = \sum_y p(y) \sum_{A \ni y} \mathfrak{g}(A) \big\{ \mathfrak{f}(S_y A \backslash \{-y\}) - \mathfrak{f}(A \backslash \{y\}) \big\}.$$

Rewrite this expression as twice one half of it. In one of the pieces, we perform the change of variables $B = S_y A$, $z = -y$ to obtain that it is equal to

$$-(1/2) \sum_y p(y) \sum_{A \ni y} \mathfrak{g}(S_y A) \big\{ \mathfrak{f}(S_y A \backslash \{-y\}) - \mathfrak{f}(A \backslash \{y\}) \big\}.$$

Here we used the fact that $p(\cdot)$ is symmetric. Adding the two expressions we get that $\left\langle (-\mathfrak{L}_\tau^+)\mathfrak{f}, \mathfrak{g} \right\rangle$ is equal to

$$-(1/2) \sum_y p(y) \sum_{A \ni y} \big\{ \mathfrak{g}(S_y A) - \mathfrak{g}(A) \big\} \big\{ \mathfrak{f}(S_y (A \backslash \{y\})) - \mathfrak{f}(A \backslash \{y\}) \big\}.$$

By Schwarz's inequality, this expression is bounded above by

$$\frac{1}{4\beta} \sum_y p(y) \sum_{A \ni y} \big\{ \mathfrak{g}(S_y A) - \mathfrak{g}(A) \big\}^2 + \frac{\beta}{4} \sum_y p(y) \sum_{A \ni y} \big\{ \mathfrak{f}(S_y A \backslash \{-y\}) - \mathfrak{f}(A \backslash \{y\}) \big\}^2$$

for all $\beta > 0$. By the identities presented just before the statement of the lemma, the first term is $(1/2\beta) \left\langle (-\mathfrak{L}_\tau^1)\mathfrak{g}, \mathfrak{g} \right\rangle$. A change of variables $B = A - \{y\}$ shows that the second is bounded by $(\beta/2) \left\langle (-\mathfrak{L}_\tau^2)\mathfrak{f}, \mathfrak{f} \right\rangle$. Minimizing over β, we conclude the proof of (4.6).

We may now prove the second estimate of the lemma. Fix $n \geq 1$, and functions $\mathfrak{f}$ and $\mathfrak{g}$ of degree n and $n+1$, respectively. By (4.6), $\langle \mathfrak{L}_\tau^+ \mathfrak{f}, \mathfrak{g}\rangle^2$ is bounded above by $\langle (-\mathfrak{L}_\tau^1)\mathfrak{f}, \mathfrak{f}\rangle \langle (-\mathfrak{L}_\tau^2)\mathfrak{g}, \mathfrak{g}\rangle$. By the first part of the lemma, this product is bounded by

$$C_0^2 n^2 \langle (-\mathfrak{L}_0)\mathfrak{f}, \mathfrak{f}\rangle \langle (-\mathfrak{L}_0)\mathfrak{g}, \mathfrak{g}\rangle$$

This proves (4.4) for $\mathfrak{L}_\tau^+$. The proof for $\mathfrak{L}_\tau^-$ is similar.

The last estimate (4.5) is elementary and follows from Schwarz's inequality and the explicit formulas for the operators $\mathfrak{L}_\tau^1$, $\mathfrak{L}_\tau^2$, $\mathfrak{L}_\tau^+$, and $\mathfrak{L}_\tau^-$. □

Lemma 4.3. *For every $k \geq 0$, there exists a finite constant C_k such that for $j = 1, 2, +, -$,*

$$|||\mathfrak{L}_\tau^j \mathfrak{f}|||_{-1,k} \leq C_k |||\mathfrak{f}|||_{1,k+1},$$

so that $\mathfrak{L}_\tau^j$ maps $\mathfrak{H}_{1,k+1}$ boundedly into $\mathfrak{H}_{-1,k}$

Proof. Follows immediately from the preceding lemma. □

Lemma 4.4. *There exists a finite constant C_0 such that for all $n \geq 1$,*

$$\|\mathfrak{f}\|_{1,\alpha} \leq C_0 n \|\mathfrak{f}\|_{1,\text{env}}, \quad \|\mathfrak{f}\|_{-1,\text{env}} \leq C_0 n \|\mathfrak{f}\|_{-1,\alpha}$$

for all α in $[0, 1]$, and all $\mathfrak{f}$ in $\mathfrak{G}_n$.

Proof. Fix $n \geq 1$ and $\mathfrak{f}$ in $\mathfrak{G}_n$. By (3.5) and Schwarz's inequality, $< \mathfrak{f}, \mathfrak{f} >_{1,\alpha}$ is bounded above by

$$\|\mathfrak{f}\|_{1,\text{env}}^2 + 2 \sum_{A \in \mathcal{E}_*} \sum_{y \in \mathbb{Z}_*^d} p(y)[\mathfrak{f}(S_y A) - \mathfrak{f}(A)]^2$$
$$+ \sum_{A \in \mathcal{E}_*} \sum_{y \notin A} p(y)[\mathfrak{f}(S_y[A \cup \{y\}]) - \mathfrak{f}([A \cup \{y\}])]^2$$

because $|r_y(A)| \leq 1$ and $\chi(\alpha) \leq 1$. Since $\mathfrak{f}$ belongs to $\mathfrak{G}_n$, we may restrict the second sum to sets A in $\mathcal{E}_{n,*}$. A change of variables permits us to estimate the third sum by the second one. In conclusion,

$$\langle \mathfrak{f}, \mathfrak{f}\rangle_{1,\alpha} \leq \|\mathfrak{f}\|_{1,\text{env}}^2 + 3 \sum_{A \in \mathcal{E}_{*,n}} \sum_{y \in \mathbb{Z}_*^d} p(y)[\mathfrak{f}(S_y A) - \mathfrak{f}(A)]^2.$$

By Lemma 4.2, the second term on the right-hand side is less than or equal to $C_0 n \|\mathfrak{f}\|_{1,\text{env}}^2$ because $\mathfrak{f}$ belongs to $\mathfrak{G}_n$. The second estimate of the lemma is obtained by duality. □

5. The Self-Diffusion Coefficient

By [1], the self–diffusion coefficient $D(\alpha)$ in the direction v is given by the variational formula :

$$v \cdot D(\alpha)v = \inf_f \Big\{ \sum_{z\in\mathbb{Z}^d_*} p(z) E_{\mu_\alpha}\Big[[1-\xi(z)]\{v\cdot z - [f(\tau_z\xi) - f(\xi)]\}^2\Big] + \sum_{x,y\in\mathbb{Z}^d_*} p(x-y) E_{\mu_\alpha}\Big[\xi(x)[1-\xi(y)]\{f(\sigma^{x,y}\xi) - f(\xi)\}^2\Big]\Big\},$$

where the infimum is carried over all cylinder functions f. A simple computation shows that

$$v \cdot D(\alpha)v = (1-\alpha)\sum_{z\in\mathbb{Z}^d_*}(z\cdot v)^2 p(z) - \alpha(1-\alpha)\|f_v\|^2_{-1,\alpha} \tag{5.1}$$

for each v in $\mathbb{R}^d$. Here f_v is the cylinder function given by

$$f_v(\xi) = \frac{1}{\sqrt{\alpha(1-\alpha)}}\sum_{y\in\mathbb{Z}^d_*} p(y)(y\cdot v)[1-\xi(y)] = \frac{1}{\sqrt{\alpha(1-\alpha)}}\sum_{y\in\mathbb{Z}^d_*} p(y)(y\cdot v)[\alpha-\xi(y)]$$

because p has mean zero. With the notation introduced in the previous section, we may write f_v as

$$f_v(\xi) = -\sum_{y\in\mathbb{Z}^d_*}(y\cdot v)p(y)\Psi_y,$$

where $\Psi_z = \Psi_{\{z\}}$ for z in $\mathbb{Z}^d_*$.

We are now in a position to state the main result of this section. Theorem 2.1 follows from this result in view of formula (5.1).

Theorem 5.1. *As a function of α, $\|f_v\|^2_{-1,\alpha}$ is of class C^∞ on* $[0, 1]$.

The proof is based on the lemmas at the end of the previous section. To explain the strategy of the proof we introduce the resolvent equation associated to f_v: for $\lambda > 0$, denote by u_λ the solution of the resolvent equation:

$$\lambda u_\lambda - \mathcal{L}u_\lambda = f_v.$$

We will use the dual representation and carry out the estimates in $\mathfrak{H}$. Let $u_\lambda \sim \mathfrak{u}_\lambda$ through the unitary isomorphism. Of course $\mathfrak{u}_\lambda = \mathfrak{u}_\lambda(\alpha)$ depends on α,

$$f_v \sim \mathfrak{f}_v = -\sum_{z\in\mathcal{E}_*}(z\cdot v)p(z)\mathfrak{e}_{\{z\}}$$

is independent of α and is actually in $\mathfrak{H}_{-1}$. We have

$$\lambda\mathfrak{u}_\lambda(\alpha) - \mathfrak{L}_\alpha\mathfrak{u}_\lambda = \mathfrak{f}_v. \tag{5.2}$$

It follows from [1] that

$$\begin{aligned}\|f_v\|_{-1,\alpha}^2 &= \lim_{\lambda\to 0}\langle f_v, u_\lambda\rangle_\alpha = -\lim_{\lambda\to 0}\sum_{z\in\mathbb{Z}_*^d}(z\cdot v)p(z)\mathfrak{u}_\lambda(\{z\},\alpha) \\ &= \lim_{\lambda\to 0}\frac{1}{2}\sum_{z\in\mathbb{Z}_*^d}(z\cdot v)p(z)\left[\mathfrak{u}_\lambda(\{-z\},\alpha)-\mathfrak{u}_\lambda(\{z\},\alpha)\right]\end{aligned} \tag{5.3}$$

because $p(\cdot)$ is symmetric. In view of this identity, to prove Theorem 5.1 we just need to show that there exists a subsequence $\lambda_k \downarrow 0$ such that, for each z with $p(z) > 0$, $\{\mathfrak{u}_{\lambda_k}(\alpha,\{z\}) - \mathfrak{u}_{\lambda_k}(\alpha,\{-z\}),\ k \geq 1\}$ converges uniformly in α to a smooth function. To prove the existence of such a subsequence, it is enough to show that the functions $\{\mathfrak{u}_\lambda(\alpha,\{z\})\}$ are smooth for each $\lambda > 0$ and, for each z and $j \geq 0$, to obtain the uniform bounds

$$\sup_{0<\lambda\leq 1}\ \sup_{0\leq\alpha\leq 1}|\mathfrak{u}_\lambda^{(j)}(\alpha,\{-z\}) - \mathfrak{u}_\lambda^{(j)}(\alpha,\{z\})| < \infty. \tag{5.4}$$

Here $\mathfrak{u}_\lambda^{(j)}$ stands for the jth derivative of $\mathfrak{u}_\lambda$ with respect to the density α.

By Schwarz's inequality,

$$\left(\sum_z p(z)|\mathfrak{u}_\lambda^{(j)}(\alpha,\{-z\}) - \mathfrak{u}_\lambda^{(j)}(\alpha,\{z\})|\right)^2 \leq \sum_z p(z)[\mathfrak{u}_\lambda^{(j)}(\alpha,\{-z\}) - \mathfrak{u}_\lambda^{(j)}(\alpha,\{z\})]^2.$$

Since the support of p generates $\mathbb{Z}^d$, and we can exclude the one dimensional nearest neighbor case, there exists a path $z_0 = -z, z_1, \dots, z_n = z$, avoiding 0, such that $p(z_{i+1} - z_i) > 0$ for $0 \leq i < n$. Rewriting the difference $\mathfrak{u}_\lambda^{(j)}(\alpha,\{-z\}) - \mathfrak{u}_\lambda^{(j)}(\alpha,\{z\})$ as $\sum_{0\leq i<n}\mathfrak{u}_\lambda^{(j)}(\alpha,\{z_{i+1}\}) - \mathfrak{u}_\lambda^{(j)}(\alpha,\{z_i\})$ and applying Schwarz's inequality, we prove that the previous expression is bounded above by

$$C_0\|\pi_1\mathfrak{u}_\lambda^{(j)}\|_{1,env}^2 \leq C_0|||\mathfrak{u}_\lambda^{(j)}|||_{1,0}^2. \tag{5.5}$$

By (5.5), in order to prove (5.4) it is enough to obtain for each $j \geq 0$, the bound

$$\sup_{0<\lambda\leq 1}\ \sup_{0\leq\alpha\leq 1}|||\mathfrak{u}_\lambda^{(j)}|||_{1,0} < \infty.$$

Notice that the coefficients of $\mathfrak{L}_\alpha$ are not smooth at the boundary of [0, 1]. For this reason, we reparametrize the family of equations by $\alpha = \sin^2 t$, $t \in [0,\pi/2]$ to get

$$\mathfrak{L}(t) = \mathfrak{L}_0 + (\sin^2 t)\ \mathfrak{L}_\tau^1 + (\cos^2 t)\ \mathfrak{L}_\tau^2 + (\sin t\cos t)\ \left[\mathfrak{L}_\tau^+ + \mathfrak{L}_\tau^-\right]$$

and consider the resolvent equation

$$\lambda\mathfrak{v}_\lambda(t) - \mathfrak{L}(t)\mathfrak{v}_\lambda(t) = \mathfrak{f}_v.$$

Since $\mathfrak{f}_v$ does not depend on α we have $\mathfrak{u}_\lambda(\alpha(t)) = \mathfrak{v}_\lambda(t)$. To prove that the sequences $\{\mathfrak{u}_\lambda^{(j)}(A,\alpha),\ \lambda > 0\}$, $j \geq 0$, are uniformly bounded in the $|||\cdot|||_{1,0}$ norm, we first prove such a statement for the sequences $\{\mathfrak{v}_\lambda^{(j)}(A,t),\ \lambda > 0\}$, $j \ge 0$. From this result and the relation between $\mathfrak{u}_\lambda$ and $\mathfrak{v}_\lambda$, t and α, we deduce boundedness in $|\!|\!| \cdot |\!|\!|_{1,0}$ norm of $\{\mathfrak{u}_\lambda^{(j)}(A,\alpha),\ \lambda > 0\}$ in the interior of the domain. An extra argument, presented at the end of the proof, extends the smoothness up to the boundary.

We start by observing that the function $\mathfrak{f}$ has finite $\mathfrak{H}_{-1,k}$ norm for all $k \ge 0$, i.e. there exists a finite constant C_0 such that

$$|\!|\!| \mathfrak{f}_v |\!|\!|_{-1,k} \le C_0 \tag{5.6}$$

for all $k \ge 0$. The proof of this claim is elementary. Since $\mathfrak{f}_v$ has degree 1, $|\!|\!| \mathfrak{f}_v |\!|\!|_{-1,k} = |\!|\!| \mathfrak{f}_v |\!|\!|_{-1,0} = \|\mathfrak{f}_v\|_{-1,env}$ is finite as soon as $\|\mathfrak{f}\|_{-1,env}$ is finite. To prove that $\|\mathfrak{f}_v\|_{-1,env}$ is finite, recall the variational formula (3.6) for the $\|\cdot\|_{-1,env}$ norm and fix a finite supported function $\mathfrak{g}$. Since $\mathfrak{L}_0$ does not change the degree of a function and since $\mathfrak{f}_v$ has degree one, we may assume that $\mathfrak{g}$ has degree one. Since p is symmetric,

$$\langle \mathfrak{f}, \mathfrak{g}\rangle = \frac{1}{2}\sum_z p(z)(z\cdot v)[\mathfrak{g}(\{-z\}) - \mathfrak{g}(\{z\})].$$

By Schwarz's inequality, the square of this expression is bounded by

$$\frac{1}{4}\sum_z p(z)|(z\cdot v)|^2 \sum_z p(z)[\mathfrak{g}(\{-z\}) - \mathfrak{g}(\{z\})]^2.$$

Now we proceed as for the bound (5.5): there exists a path $z_0 = -z, z_1, \dots, z_n = z$, avoiding 0, such that $p(z_{i+1} - z_i) > 0$ for $0 \le i < n$. Rewriting the difference $\mathfrak{g}(\{-z\}) - \mathfrak{g}(\{z\})$ as $\sum_{0\le i<n} \mathfrak{g}(\{z_{i+1}\}) - \mathfrak{g}(\{z_i\})$ and applying Schwarz's inequality, we prove that the previous expression is bounded above by $C_0 < \mathfrak{g}, (-\mathfrak{L}_0\mathfrak{g}) >$, which proves the claim (5.6) in view of the variational formula (3.6) for the $\|\cdot\|_{-1,env}$ norm.

We now start our way through the proof that $\mathfrak{v}_\lambda$ is a sequence of smooth functions with bounded derivatives. Lemma 4.1 applied to $\mathfrak{f}$ shows that

$$\sup_{0<\lambda\le 1}\ \sup_{0\le t\le \pi/2} |\!|\!| \mathfrak{v}_\lambda(t) |\!|\!|_{1,k}$$

is finite for all $k \ge 1$.

We now turn to the proof of the differentiability of $\mathfrak{v}_\lambda(\cdot)$. We say that a function $\mathfrak{g}(t)$ with values in $\mathfrak{H}$ is differentiable at t if $\gamma^{-1}[\mathfrak{g}(t+\gamma) - \mathfrak{g}(t)]$ converges, as $\gamma \downarrow 0$, strongly in $\mathfrak{H}$ to some function that we denote by $\mathfrak{g}'$. Notice that differentiating formally $\mathfrak{L}(t)$ in t we get the operator

$$\mathfrak{L}'(t) = (2\sin t\cos t)\ (\mathfrak{L}_\tau^1 - \mathfrak{L}_\tau^2) + (\cos^2 t - \sin^2 t)\ \left[\mathfrak{L}_\tau^- + \mathfrak{L}_\tau^+\right].$$

Lemma 5.2. *Suppose that $\mathfrak{f}(t)$ is a differentiable function of t. Let $\mathfrak{u}_\lambda$ be the solution of the resolvent equation*

$$\lambda\mathfrak{u}_\lambda(t) - \mathfrak{L}(t)\mathfrak{u}_\lambda(t) = \mathfrak{f}(t).$$

Then, $\mathfrak{u}_\lambda(t)$ is differentiable and its derivative is the solution $\mathfrak{u}_\lambda'(t)$ of

$$\lambda\mathfrak{u}_\lambda' - \mathfrak{L}(t)\mathfrak{u}_\lambda' = \mathfrak{f}'(t) + \mathfrak{L}'(t)\,\mathfrak{u}_\lambda. \tag{5.7}$$

Proof. The proof of the differentiability of $\mathfrak{u}_\lambda(t)$ is standard; all we need to control is that $\mathfrak{L}'(t)\mathfrak{u}_\lambda(t)$ is in $\mathfrak{H}$, which follows from (4.5) of Lemma 4.2 and the boundedness of the coefficients of $\mathfrak{L}'(t)$. □

The previous lemma applied to $\mathfrak{f} = \mathfrak{f}_v$ shows that the family of functions $\mathfrak{u}_\lambda$ is differentiable for each fixed λ and that the derivative $\mathfrak{u}'_\lambda$ satisfies some resolvent-type equation.

Proof of Theorem 5.1. We first show that $\{\mathfrak{u}_\lambda(t),\ \lambda > 0\}$ is a family of smooth functions whose derivatives satisfy for each $k \geq 0$,

$$\sup_{0<\lambda\leq 1}\ \sup_{0\leq t\leq \frac{\pi}{2}} \|\!|\mathfrak{u}'_\lambda(t)|\!\|_{1,k} < \infty. \tag{5.8}$$

By (5.6) $\|\!|\mathfrak{f}|\!\|_{-1,k}$ is bounded uniformly in t. Hence, by Lemma 4.1, $\|\!|\mathfrak{u}_\lambda|\!\|_{1,k}$ is bounded, uniformly in λ and t. Since $\mathfrak{f}$ does not depend on t, by Lemma 5.2, $\mathfrak{u}_\lambda$ is differentiable and its derivative $\mathfrak{u}'_\lambda$ satisfies

$$\lambda \mathfrak{u}'_\lambda \ -\ \mathfrak{L}(t)\mathfrak{u}'_\lambda = \mathfrak{L}'(t)\,\mathfrak{u}_\lambda.$$

By Lemma 4.3 and the explicit form of the operator $\mathfrak{L}(t)$,

$$\|\!|\mathfrak{L}'(t)\,\mathfrak{u}_\lambda|\!\|_{-1,k} \leq 2 \sum_{j=1,2,+,-} \|\!|\mathfrak{L}^j(t)\,\mathfrak{u}_\lambda|\!\|_{-1,k} \leq 8C_k \|\!|u_\lambda|\!\|_{1,k+1}$$

then by Lemma 4.1 $\|\!|\mathfrak{L}'(t)\,\mathfrak{u}_\lambda|\!\|_{-1,k}$ is bounded for each $k \geq 1$, uniformly in λ and t. We may therefore apply again Lemma 4.1 to show that $\|\!|\mathfrak{u}'_\lambda(t)|\!\|_{1,k}$ is uniformly bounded in (t, λ) for all $k \geq 1$.

To iterate the argument, we just need to prove by induction the existence of constants $\{a_{n,i},\ n \geq 1, 0 \leq i < n\}$ such that

$$\lambda \mathfrak{u}_\lambda^{(j)} \ -\ \mathfrak{L}(t)\mathfrak{u}_\lambda^{(j)} = \sum_{i=0}^{j-1} a_{j,i}\,\mathfrak{L}^{(j-i)}(t)\,\mathfrak{u}_\lambda^{(i)}, \tag{5.9}$$

where $\mathfrak{u}_\lambda^{(i)}$, $\mathfrak{L}^{(i)}(t)$ stands for the ith derivative of $\mathfrak{u}_\lambda(t)$, $\mathfrak{L}(t)$. This is elementary and left to the reader.

The previous argument shows that $\mathfrak{u}_\lambda(t)$ is a sequence of smooth functions on $[0, 1]$ with their derivatives having the uniform bounds

$$\sup_{0<\lambda\leq 1}\ \sup_{0\leq t\leq \pi/2} \|\!|\mathfrak{u}_\lambda^{(j)}(t)|\!\|_{1,k} < \infty$$

for each $j \geq 0$. We have seen just after (5.3) that these uniform estimates guarantee the smoothness of $\|\mathfrak{f}\|_{-1,\alpha(t)}$ as a function of t defined in $[0, \pi/2]$. Since $\alpha = \sin^2 t$, this translates immediately into smoothness in α for $\alpha \in (0, 1)$. Regularity at the boundary requires the following extra argument.

We claim that the odd derivatives of $< \mathfrak{f}, \mathfrak{u}_\lambda(t) >$ vanish at $t = 0$ and $\pi/2$. We consider the case $t = 0$, the other being similar. To keep notation simple, let $U_\lambda(t) =< \mathfrak{f}, \mathfrak{u}_\lambda(t) >$. Since $\mathfrak{f}$ does not depend on α, for $j \geq 0$, $U_\lambda^{(j)}(0) = < \mathfrak{f}, \mathfrak{u}_\lambda^{(j)}(0) >$. Since $\mathfrak{f}$ is a function of degree 1, to prove that the odd derivatives of $U_\lambda(t)$ vanish at 0, it is enough to prove that $\mathfrak{u}_\lambda^{(2j+1)}(0)$ is a function of even degree. We prove this statement by induction on j.

Observe that $\mathfrak{L}(0) = \mathfrak{L}_0 + \mathfrak{L}_\tau^2$, which are operators that preserve the degree of a function. On the other hand, since $\sin^2 t$, $\cos^2 t$ are even functions and since $\sin t \cos t$ is an odd function, there exist constants a_j, b_j, c_j such that

$$\mathfrak{L}^{(2j)}(0) = a_j \mathfrak{L}_\tau^1 + b_j \mathfrak{L}_\tau^2, \quad \mathfrak{L}^{(2j+1)}(0) = c_j [\mathfrak{L}_\tau^+ + \mathfrak{L}_\tau^-]$$

for $j \geq 0$. In particular, while $\mathfrak{L}^{(2j)}(0)$ preserves the degree of a function, $\mathfrak{L}^{(2j+1)}(0)$ changes it by one.

To prove that $\mathfrak{u}_\lambda^{(2j+1)}(0)$ (resp. $\mathfrak{u}_\lambda^{(2j)}(0)$), $j \geq 0$, are functions of even (resp. odd) degree, notice first that $\mathfrak{u}_\lambda(0)$ is the solution of

$$[\lambda - (\mathfrak{L}_0 + \mathfrak{L}_\tau^2)]\mathfrak{u}_\lambda(0) = \mathfrak{f}.$$

Since $\mathfrak{f}$ is a function of degree 1, $\mathfrak{u}_\lambda(0)$ is also of degree 1. This proves the claim for $j = 0$. It is easy to conclude the proof by induction using formula (5.9) and the fact that $\mathfrak{L}^{(2j)}(0)$ preserves the degree, while $\mathfrak{L}^{(2j+1)}(0)$ changes it by one.

Since $\mathfrak{u}_\lambda^{(2j+1)}(0)$ are functions of even degree, $U_\lambda^{(2j+1)}(0) =< \mathfrak{f}, \mathfrak{u}_\lambda^{(2j+1)}(0) >$ vanishes because $\mathfrak{f}$ has degree one. Since we proved uniform convergence of a subsequence $\mathfrak{u}_{\lambda_k}(t)$ and its derivatives, the limit $U(t) = \|\mathfrak{f}\|_{-1,\alpha(t)}^2$ of $U_\lambda(\cdot)$ inherits these properties. In particular, $U^{2j+1}(0) = 0$. Elementary analytic considerations show that $U(t)$ is in fact a smooth function of t^2 and hence of $\sin^2 t = \alpha$. □

Remark 5.3. The proof of the smoothness at the boundary provides a recursive method to compute the Taylor expansion at the origin of the diffusion coefficient. Recall that $U(t) = \|\mathfrak{f}\|_{-1,\alpha(t)}$. By Theorem 5.1, $U(0) = \lim_{\lambda \to 0} < \mathfrak{f}, \mathfrak{u}_\lambda(0) > = < \mathfrak{f}, \mathfrak{u}(0) >$, where $\mathfrak{u}(0)$ is the solution of

$$-[\mathfrak{L}_0 + \mathfrak{L}_\tau^2]\,\mathfrak{u}(0) = \mathfrak{f}. \tag{5.10}$$

Since $\mathfrak{f}$ has degree one and since $\mathfrak{L}_0$, $\mathfrak{L}_\tau^2$ preserve the degree, this equation can be solved in $\mathfrak{H}_1$. In this space both $\mathfrak{L}_0$, $\mathfrak{L}_\tau^2$ are essentially Laplace operators and this equation may be solved. Knowing $\mathfrak{u}(0)$, we may examine the equation

$$-[\mathfrak{L}_0 + \mathfrak{L}_\tau^2]\,\mathfrak{u}^{(1)}(0) = \mathfrak{L}^{(1)}(0)\mathfrak{u}(0).$$

As noticed earlier, the right hand side is a function of degree 0 and 2 so that $\mathfrak{u}^{(1)}(0)$ has this property. By induction we may obtain $\mathfrak{u}^{(j)}(0)$ for all $j \geq 1$ by inverting an operator which is essentially a Laplacian. This permits us to compute the Taylor expansion of U around the origin because $U^{(j)}(0) =< \mathfrak{f}, \mathfrak{u}^{(j)}(0) >$. In particular, from (5.1),

$$v \cdot D(\alpha)v = (1-\alpha) \sum_{z \in \mathbb{Z}_*^d} (z \cdot v)^2 p(z) - \alpha(1-\alpha) < \mathfrak{u}(0), \mathfrak{f}_v > + O(\alpha^2)$$

includes the first order correction, where $\mathfrak{u}(0)$ is the solution of (5.10).

References

1. Kipnis, C., Varadhan, S.R.S.: Central limit theorem for additive functionals of reversible Markov processes and applications to simple exclusion. Commun. Math. Phys. **106**, 1–19 (1986)
2. Landim, C., Olla, S., Yau, H.T.: Some properties of the diffusion coefficient for asymmetric simple exclusion processes. Ann. of Probab. **24**, 1779–1807 (1996)
3. Landim C., Yau, H.T.: Fluctuation–dissipation equation of asymmetric simple exclusion processes. Probab. Th. Rel. Fields **108**, 321–356 (1997)
4. Landim C., Olla S., Varadhan S.R.S.: Finite-dimensional approximation of the self-diffusion coefficient for the exclusion process. Preprint
5. Sethuraman, S., Varadhan, S.R.S., Yau, H. T.: Diffusive limit of a tagged particle in asymmetric exclusion process. Comm Pure Appl. Math. **53**, 972–1006 (2000)
6. Varadhan, S.R.S.: Regularity of the self-diffusion coefficient. In: *The Dynkin Festschrift*, Progr. Probab. **34**, Boston, MA: Birkhäuser Boston 1994, pp. 387–397

Communicated by H.-T. Yau

The Annals of Probability
2002, Vol. 30, No. 2, 483–508

FINITE-DIMENSIONAL APPROXIMATION OF THE SELF-DIFFUSION COEFFICIENT FOR THE EXCLUSION PROCESS

BY C. LANDIM,[1] S. OLLA AND S. R. S. VARADHAN[2]

IMPA and CNRS UPRES, Université de Cergy CNRS UPRES and Courant Institute, New York University

We show that for the symmetric simple exclusion process on $\mathbb{Z}^d$ the self-diffusion coefficient of a tagged particle is stable when approximated by simple exclusion processes on large periodic lattices. The proof depends on a similar stability property of the asymptotic variance of additive functionals of mean 0. This requires establishing a property for the Dirichlet space known as the Liouville-D property.

1. Introduction. In the early 1980s, Kipnis and Varadhan proved an invariance principle for the position of a tagged particle in a symmetric simple exclusion process in equilibrium (cf. [6]). Their proof relied on a general central limit theorem for additive functionals of a reversible Markov process that they established. Time reversibility and translation invariance of the system are the basic ingredients of this method, and, in principle, it can be applied to any system with these two symmetries. Later it was extended to nonreversible processes that satisfy a strong sector condition (cf. [10] and [13]).

The (effective) diffusion matrix of the limiting Brownian motion is a function $D(\alpha)$ of the density α of the particles, and in the reversible case it is given by a variational formula.

This method works directly in infinite systems, and it raises the natural question of the stability of the effective diffusion matrix under finite-dimensional approximations. More precisely, consider a finite-dimensional version of the same process on the torus $\{-N, \dots, 0, \dots, N\}^d$ (i.e., with periodic boundary conditions, preserving in this manner the translation symmetry). Since we want to work with an ergodic process, we also fix the total number K of particles. Consider now a tagged particle in this finite system. If N is much larger than the size of a single jump, the motion of the tagged particle has a unique canonical lifting to $\mathbb{Z}^d$. We get in this manner a process $X_N(t)$ with values in $\mathbb{Z}^d$. Let us denote by $D_{[N,K]}$

Received October 2000; revised June 2001.

[1]Supported in part by CNPq Grant 300358/93-8, FAPERJ Grant E26/150.940/99 and PRONEX Grant 41.96.0923.00.

[2]Supported by NSF Grant DMS-98-03140.

AMS 2000 subject classification. 60K35.

Key words and phrases. Self-diffusion, tagged particle, exclusion process, Liouville property.

the variance of the Brownian motion which is the limit of the scaled process $\varepsilon X_N(\varepsilon^{-2}t)$ as $\varepsilon \to 0$. We expect that

$$\lim_{\substack{N\to\infty \\ K/(2N)^d\to\alpha}} D_{[N,K]} = D(\alpha). \tag{1.1}$$

Related questions arise in other contexts, for example, the effective conductivity in percolation regime (cf. [1] and [4]) and fluctuations in massless interface models (cf. [2]).

In general, it is easy to prove that

$$\limsup_{\substack{N\to\infty \\ K/(2N)^d\to\alpha}} D_{[N,K]} \le D(\alpha).$$

In this paper we prove equality (1.1) for symmetric simple exclusion processes on $\mathbb{Z}^d$.

A related problem that is somewhat easier to handle is the scaling behavior of an additive functional $A(t) = \int_0^t V(\eta(s))\,ds$, where V is a suitable local function that depends on the configuration at a few sites. There is again a limiting Brownian motion for $\varepsilon A(\varepsilon^{-2}t)$ with variance $\sigma_{N,K}^2$. The problem is to show that

$$\lim_{\substack{N\to\infty \\ K/(2N)^d\to\alpha}} \sigma_{N,K}^2 = \sigma^2(\alpha), \tag{1.2}$$

where $\sigma^2(\alpha)$ is the corresponding variance in the infinite-volume situation where the central limit theorem can be established directly. In this context the nontrivial part is the inequality

$$\limsup_{\substack{N\to\infty \\ K/(2N)^d\to\alpha}} \sigma_{N,K}^2 \le \sigma^2(\alpha).$$

The proofs depend on *approximating* the *solutions* of the infinite-dimensional variational problem for the limiting variance by solutions of the corresponding finite-dimensional problem.

A basic step in this approximation is to prove a certain regularity property of the Dirichlet form associated with the simple exclusion process. Let us describe this regularity property in a more general context.

Suppose we have a collection of invertible measure-preserving transformations $\{T_z\}$ on some probability space $(\Omega, \mathcal{F}, \mu)$. Then we can define a Dirichlet form

$$D(u) = \frac{1}{2}\int_\Omega \sum_z [u(T_z\omega) - u(\omega)]^2\,d\mu$$

corresponding to the generator

$$Lu = \sum_z [u(T_z\omega) - u(\omega)]$$

of a process consisting of random stirrings. If we define $v_z(\omega) = (\nabla_z u)(\omega) = u(T_z\omega) - u(\omega)$, there is natural map

$$u \to \delta u = \{v_z(\omega)\} \in \bigoplus_z L_2(\mu).$$

The problem is to characterize the closure of the range of this map by relations that remain stable under suitable approximations done with finitely supported functions.

In Section 3 we prove this regularity property for the symmetric simple exclusion processes, and in Section 4 for the version with a tagged particle. The main tool in the proof is duality, which allows us to reduce the problem to a finite-dimensional situation (cf. [8] and [11]).

A similar question arises in the context of a hydrodynamic limit for nongradient models (cf. [5], Appendix 3, and [14]). But there the approximation is needed in a different norm and the proof relies usually on a suitable estimate of the spectral gap for the corresponding dynamics in a finite volume.

2. Notation and results. Let us fix a symmetric, finite-range probability measure $p(\cdot)$ on $\mathbb{Z}^d$: $p(0) = 0$, $p(z) = p(-z)$ and $p(z) = 0$ for $|z|$ large. Let us assume that the random walk with transition probability $p(\cdot)$ is irreducible; that is, the finite set $\{z : p(z) > 0\}$ generates the group $\mathbb{Z}^d$. Consider the symmetric simple exclusion process associated with $p(\cdot)$. This is the Markov process on $\mathcal{X} = \{0, 1\}^{\mathbb{Z}^d}$ whose generator L acting on cylinder functions f is given by

$$\begin{aligned}(Lf)(\eta) &= \sum_{x,y\in\mathbb{Z}^d} p(y-x)\eta(x)[1-\eta(y)][f(\sigma^{x,y}\eta) - f(\eta)]\\ &= \tfrac{1}{2}\sum_{x,y\in\mathbb{Z}^d} p(y-x)[f(\sigma^{x,y}\eta) - f(\eta)].\end{aligned} \tag{2.1}$$

Here and below, a configuration of particles in $\mathbb{Z}^d$, that is, a point in $\mathcal{X}$, will be denoted by η. In particular, for x in $\mathbb{Z}^d$, $\eta(x)$ is equal to 1 if the site x is occupied and 0 otherwise. Moreover, for a configuration η and x, y in $\mathbb{Z}^d$, $\sigma^{x,y}\eta$ is the configuration obtained from η by exchanging the occupation variables $\eta(x)$ and $\eta(y)$:

$$(\sigma^{x,y}\eta)(z) = \begin{cases}\eta(y), & \text{if } z = x,\\ \eta(x), & \text{if } z = y,\\ \eta(z), & \text{otherwise.}\end{cases} \tag{2.2}$$

Note that $\sigma^{x,y} = \sigma^{y,x}$ and we can think of either one as being σ^e, where e is an edge or a bond connecting x and y. Fix $0 \le \alpha \le 1$ and denote by ν_α the Bernoulli product measure on $\mathcal{X}$. This is the probability measure on $\mathcal{X}$ obtained by placing a particle with probability α at each site x, and independently for different sites. It is easy to check that the one-parameter family of probability measures

$\{\nu_\alpha, 0 \le \alpha \le 1\}$ are reversible invariant measures for the process with generator L on $\mathcal{X}$. The Dirichlet form is given by

$$D(f) = \frac{1}{4} \int_{\mathcal{X}} \sum_{x,y \in \mathbb{Z}^d} p(y-x)[f(\sigma^{x,y}\eta) - f(\eta)]^2 \, d\nu_\alpha.$$

We will think of $\mathbb{Z}^d$ as a connected graph with an edge e connecting x, y being present whenever $p(y-x) > 0$. Since $p(\cdot)$ is symmetric, one can view $p(y-x)$ as defined on edges with $p(e) = p(y-x)$. Denote by $\mathbb{E}$ the set of such edges. For each local function f, denote by δf the gradient obtained from f. This is the collection $\{(\nabla_e f)(\eta)\}$ of functions on $\mathcal{X}$ defined for $e = (x, y) \in \mathbb{E}$ by

$$(\nabla_e f)(\eta) = f(\sigma^{x,y}\eta) - f(\eta).$$

Since $\sigma^{x,y} = \sigma^{y,x}$ we can define σ^e for $e \in \mathbb{E}$. If $e_1, \dots, e_n$ are n edges in $\mathbb{E}$ such that $\sigma^{e_1} \cdots \sigma^{e_n}$ is the identity permutation, then

$$\begin{aligned} 0 &= f(\sigma^{e_1} \cdots \sigma^{e_n} \eta) - f(\eta) \\ &= \sum_{j=1}^{n} [f(\sigma^{e_j} \cdots \sigma^{e_n} \eta) - f(\sigma^{e_{j+1}} \cdots \sigma^{e_n} \eta)] \\ &= \sum_{j=1}^{n} (\nabla_{e_j} f)(\sigma^{e_{j+1}} \cdots \sigma^{e_n} \eta) \\ &= \sum_{j=1}^{n} \sigma^{e_{j+1}} \cdots \sigma^{e_n} (\nabla_{e_j} f)(\eta). \end{aligned}$$

It is understood here that for $j = n$ the expression is just $(\nabla_{e_n} f)(\eta)$. The above conditions are called the compatibility conditions for a gradient and they permit us to recognize when an arbitrary collection $\Phi = \{\Phi_e, e \in \mathbb{E}\}$ of $L^2(\nu_\alpha)$ functions could be a gradient. We say that such a collection is a formal gradient if each Φ_e is square integrable and whenever $e_1, \dots, e_n$ is a finite set of edges in $\mathbb{E}$ such that $\sigma^{e_1} \cdots \sigma^{e_n}$ is equal to the identity permutation, we have

$$\sum_{j=1}^{n} \sigma^{e_{j+1}} \cdots \sigma^{e_n} \Phi_{e_j} = 0. \tag{2.3}$$

As before, for $j = n$ this expression is Φ_{e_n}. We associate a Hilbert space with norm defined by

$$\|\Phi\|_\alpha^2 = \sum_{e \in \mathbb{E}} p(e) \langle \Phi_e^2 \rangle_{\nu_\alpha}. \tag{2.4}$$

Any Φ for which the norm is finite will be called a square-integrable formal gradient. Although we have defined $\nabla_e f$ only for edges $e = (x, y)$ with $p(y-x) > 0$, if the random walk with transition probability $p(\cdot)$ is irreducible then knowing ∇_e for edges in $\mathbb{E}$ determines it uniquely for all edges (x, y). The Dirichlet forms corresponding to different irreducible finitely supported probabilities $p(\cdot)$ are equivalent. For qualitative statements we could take as a Dirichlet form simply

$$\frac{1}{4}\int \sum_{\substack{x,y \\ y-x\in F}} |f(\sigma^{x,y}\eta) - f(\eta)|^2 \, d\nu_\alpha,$$

where F is any symmetric set that generates $\mathbb{Z}^d$ as a group. The natural question in this context is whether any square-integrable formal gradient can be approximated by actual gradients in the L^2 sense. That is, for each $\Phi = \{\Phi_e\}$, does there exist a sequence of local functions f_n such that

$$\lim_{n\to\infty} \|\Phi - \delta f_n\|_\alpha = 0? \tag{2.5}$$

The next result states that this is indeed possible.

THEOREM 2.1. *For the symmetric exclusion process on $\mathbb{Z}^d$, any square-integrable formal gradient can be approximated by gradients of local functions in the sense of* (2.5).

COROLLARY 2.2. *For a symmetric simple exclusion process on $\mathbb{Z}^d$, the limit* (1.2) *is valid for any local function V which has mean* 0 *with respect to every ν_α.*

To study the evolution of a tagged particle, we must consider simple exclusion on the lattice $\mathbb{Z}^d_* = \mathbb{Z}^d - \{0\}$. We have analogous results in that case. Of course, in dealing with simple exclusion on $\mathbb{Z}^d_*$, $\mathbb{E}$ contains only edges (x, y) such that $y - x \in F$ and $x, y \in \mathbb{Z}^d_*$ and the Dirichlet form is now given by

$$D(f) = \frac{1}{4}\int_{\mathcal{X}} \sum_{\substack{x,y\in\mathbb{Z}^d_* \\ y-x\in F}} p(y-x)[f(\sigma^{x,y}\eta) - f(\eta)]^2 \, d\nu_\alpha.$$

THEOREM 2.3. *For the symmetric exclusion process* (*other than the one-dimensional nearest neighbor walk*) *on $\mathbb{Z}^d_*$, any square-integrable formal gradient can be approximated by gradients of local functions in the sense of* (2.5).

COROLLARY 2.4. *For a symmetric simple exclusion process* (*other than the one-dimensional nearest neighbor walk*) *on $\mathbb{Z}^d_*$, the limit* (1.2) *is valid for any local function V which has mean* 0 *with respect to every ν_α.*

We will apply Theorem 2.3 in order to study the evolution of a tagged particle in the symmetric simple exclusion process. Let η be an initial configuration with a particle at the origin: $\eta(0) = 1$. Tag this particle and denote by η_t and X_t the state of the process and the position of the tagged particle, respectively, at time t. We shall refer to η_t as the environment. Let ξ_t be the state of the environment as seen from the tagged particle: $\xi_t = \tau_{X_t}\eta_t$. Here, for x in $\mathbb{Z}^d$ and a configuration η, τ_x stands for the translation of η by x units: $(\tau_x\eta)(y) = \eta(x+y)$. Notice that the origin is always occupied (by the tagged particle) for the environment as seen from the tagged particle. For this reason, we shall consider the process ξ_t as taking values in $\{0,1\}^{\mathbb{Z}^d_*}$.

While X_t is not a Markov process due to the presence of the environment, (X_t, η_t) and ξ_t are Markov processes. A simple computation shows that the generator $\mathcal{L}$ of the Markov process ξ_t is given by $\mathcal{L} = \mathcal{L}_0 + \mathcal{L}_\tau$, where

$$\begin{aligned}(\mathcal{L}_0 f)(\xi) &= \sum_{x,y\in\mathbb{Z}^d_*} p(y-x)\xi(x)[1-\xi(y)][f(\sigma^{x,y}\xi) - f(\xi)]\\ &= \frac{1}{2}\sum_{x,y\in\mathbb{Z}^d_*} p(y-x)[f(\sigma^{x,y}\xi) - f(\xi)],\\ (\mathcal{L}_\tau f)(\xi) &= \sum_{z\in\mathbb{Z}^d_*} p(z)[1-\xi(z)][f(\tau_z\xi) - f(\xi)].\end{aligned}$$

The first part of the generator takes into account the jumps of the environment, while the second one corresponds to jumps of the tagged particle. In the above formula, $\tau_z\xi$ stands for the configuration where the tagged particle, sitting at the origin, is first transferred to site z and then the entire configuration is translated by $-z$; that is, for $x \neq 0$,

$$(\tau_z\xi)(x) = \begin{cases} 0, & \text{if } x = -z,\\ \xi(x+z), & \text{for } x \neq -z.\end{cases}$$

For $0 \le \alpha \le 1$, denote by μ_α the Bernoulli product measure on $\mathcal{X}_* = \{0,1\}^{\mathbb{Z}^d_*}$. A simple computation shows that μ_α is reversible for the Markov process ξ_t.

Kipnis and Varadhan [6] proved a central limit theorem for the position of the tagged particle when the environment process is in equilibrium with marginal distribution μ_α. They showed that $\varepsilon X_{t\varepsilon^{-2}}$ converges, as $\varepsilon \downarrow 0$, to a Brownian motion with self-diffusion coefficient $D(\alpha)$ that can be described as follows. Fix λ in $\mathbb{R}^d$. Then the quadratic form $\langle\lambda, D(\alpha)\lambda\rangle$ is given by the variational formula

$$\begin{aligned}\langle\lambda, D(\alpha)\lambda\rangle = \inf_f \bigg\{ &\sum_{z\in\mathbb{Z}^d_*} p(z)\langle[1-\xi(z)]\{\lambda z - T_z f\}^2\rangle_\alpha\\ &+ \sum_{x,y\in\mathbb{Z}^d_*} p(y-x)\langle\{T_{x,y}f\}^2\rangle_\alpha\bigg\}.\end{aligned} \tag{2.6}$$

In this formula the infimum is taken over all cylinder functions f, $\langle\cdot\rangle_\alpha$ stands for the expectation with respect to the measure μ_α, T_z, $T_{x,y}$, x, y, z in $\mathbb{Z}^d_*$, are the operators defined by

$$(T_z f)(\xi) = [1 - \xi(z)]\{f(\tau_z\xi) - f(\xi)\},$$
$$(T_{x,y} f)(\xi) = \{f(\sigma^{x,y}\xi) - f(\xi)\}$$

and the second summation is performed over all bonds of $\mathbb{Z}_*^d$.

The diffusion coefficient $D(\alpha)$ can also be written as

$$\langle \lambda \cdot D(\alpha)\lambda \rangle = (1-\alpha) \sum_{z \in \mathbb{Z}_*^d} (z \cdot \lambda)^2 p(z) - 2\|\psi_\lambda\|_{-1,\alpha}^2, \tag{2.7}$$

where ψ_λ is the cylinder function given by

$$\psi_\lambda(\xi) = \sum_{z \in \mathbb{Z}_*^d} p(z)(z \cdot \lambda)[1 - \xi(z)] \tag{2.8}$$

and $\|\psi_\lambda\|_{-1,\alpha}^2$ is the H_{-1} norm of ψ_λ, which is given by

$$\|\psi\|_{-1,\alpha}^2 = \sup_f \{2\langle f, \psi_\lambda\rangle_\alpha - \langle f, (-\mathcal{L})f\rangle_\alpha\}. \tag{2.9}$$

In this formula the supremum is carried over all cylinder functions f in $\mathcal{X}_*$ and $\langle \cdot, \cdot \rangle_\alpha$ stands for the inner product in $L^2(\mu_\alpha)$.

Fix a positive integer N and denote by $\mathbb{T}_N^d = \{-N, \dots, N\}^d$ the discrete d-dimensional torus with N and $-N$ identified, consisting of $(2N)^d$ points. For the same transition probability $p(\cdot)$, one can consider the symmetric simple exclusion process evolving on $\mathbb{T}_N^d$. The state space is now $\mathcal{X}_N = \{0,1\}^{\mathbb{T}_N^d}$ and the generator L_N acting on any function f is given by

$$(L_N f)(\eta) = \sum_{x,y \in \mathbb{T}_N^d} p(y-x)\eta(x)[1-\eta(y)][f(\sigma^{x,y}\eta) - f(\eta)].$$

In a similar way, we can examine the evolution of a tagged particle on the torus $\mathbb{T}_N^d$. The environment as seen from the tagged particle is a Markov process with generator $\mathcal{L}_N = \mathcal{L}_{0,N} + \mathcal{L}_{\tau,N}$, where

$$\begin{aligned}(\mathcal{L}_{0,N} f)(\xi) &= \sum_{x,y \in \mathbb{T}_{N,*}^d} p(y-x)\xi(x)[1-\xi(y)][f(\sigma^{x,y}\xi) - f(\xi)],\\ (\mathcal{L}_{\tau,N} f)(\xi) &= \sum_{z \in \mathbb{T}_{N,*}^d} p(z)[1-\xi(z)][f(\tau_z\xi) - f(\xi)].\end{aligned} \tag{2.10}$$

Let the total number K of particles be fixed and, of course, $1 \le K \le (2N)^d$. A simple computation shows that the measure $\mu_{N,K}$, which is uniform over the configurations of $\mathcal{X}_{N,*} = \{0,1\}^{\mathbb{T}_{N,*}^d}$ with $K-1$ particles, is reversible with respect to $\mathcal{L}_N$. Since we are in a finite state space, it is not hard to prove an invariance principle for the motion X_t^N of the tagged particle lifted to $\mathbb{Z}^d$ in this context:

$\varepsilon X^N_{t\varepsilon^{-2}}$ converges as $\varepsilon \downarrow 0$ to a Brownian motion with diffusion coefficient $D_{N,K}$ given by

$$\text{(2.11)}\quad \langle \lambda, D_{N,K}\,\lambda\rangle = \left(1 - \frac{K-1}{(2N)^d - 1}\right) \sum_{z\in\mathbb{Z}^d_*} (z\cdot\lambda)^2 p(z) - 2\|\psi_\lambda\|^2_{-1,N,K},$$

where $\|\psi_\lambda\|^2_{-1,N,K}$ stands for the H_{-1} norm of ψ_λ in $L^2(\mu_{N,K})$, given by the variational formula

$$\text{(2.12)}\qquad \|\psi_\lambda\|^2_{-1,N,K} = \sup_f\{2\langle\psi_\lambda, f\rangle_{N,K} - \langle f, (-\mathcal{L}_N)f\rangle_{N,K}\}.$$

Here the supremum is carried over all functions in $L^2(\mu_{N,K})$.

The main result of this article concerns the behavior of the finite-volume diffusion coefficient $D_{N,K}$ as $N \uparrow \infty$:

THEOREM 2.5. *Fix α in $(0,1)$ and a sequence K_N such that $K_N/(2N)^d$ converges to α as $N\uparrow\infty$. Then*

$$\lim_{N\to\infty} D_{N,K_N} = D(\alpha).$$

We conclude this section with a sketch of the proof of Theorem 2.5. In view of formulas (2.7) and (2.11), Theorem 2.5 follows from the convergence of the H_{-1} norms of ψ_λ. It is enough to show that

$$\text{(2.13)}\qquad \lim_{\substack{N\uparrow\infty\\ K_N/(2N)^d\to\alpha}} \langle\psi_\lambda, (-\mathcal{L}_N)^{-1}\psi_\lambda\rangle_{N,K_N} = \langle\psi_\lambda, (-\mathcal{L})^{-1}\psi_\lambda\rangle_\alpha.$$

Here, to emphasize the dependence of the H_{-1} norm on the generators, we have denoted the H_{-1} norms $\|\cdot\|^2_{-1,N,K}$ and $\|\cdot\|^2_{-1,\alpha}$, respectively, by $\langle\cdot, (-\mathcal{L}_N)^{-1}\cdot\rangle_{N,K}$ and $\langle\cdot, (-\mathcal{L})^{-1}\cdot\rangle_\alpha$.

The generator $\mathcal{L}$ has two pieces. The first one, $\mathcal{L}_0$, corresponds to jumps of the environment and the second one, $\mathcal{L}_\tau$, corresponds to jumps of the tagged particle. From Corollary 2.4 we conclude that

$$\text{(2.14)}\qquad \lim_{\substack{N\uparrow\infty\\ K_N/(2N)^d\to\alpha}} \langle\psi_\lambda, (-\mathcal{L}_{0,N})^{-1}\psi_\lambda\rangle_{N,K_N} = \langle\psi_\lambda, (-\mathcal{L}_0)^{-1}\psi_\lambda\rangle_\alpha.$$

In Section 4 we prove, in addition, that for each $\varepsilon > 0$, we may express the local function ψ_λ as the sum of $\mathcal{L}g_\varepsilon + u_\varepsilon$, where g_ε, u_ε are local functions, u_ε has mean zero with respect to all invariant measures and $\|u_\varepsilon\|^2_{0,-1} = \langle u_\varepsilon, (-\mathcal{L}_0)^{-1}u_\varepsilon\rangle_\alpha \le \varepsilon^2$. Since $\|\cdot\|_{-1} \le \|\cdot\|_{0,-1}$ and (2.13) is obvious if we replace ψ_λ by a local function of the form $\mathcal{L}f$, (2.13) follows from the previous decomposition and (2.14). Details are given in Section 4.

3. Formal gradients of the symmetric exclusion process. We prove in this section Theorems 2.1 and 2.3 and Corollaries 2.2 and 2.4. The proofs rely on the duality properties of the symmetric exclusion process that we now explain. Let $\mathbb{X}$ be a countable set. For us it will be either $\mathbb{Z}^d$ or $\mathbb{Z}^d - \{0\}$. We denote by $\mathcal{X} = \{0, 1\}^{\mathbb{X}}$ the set of configurations η on $\mathbb{X}$. For $0 \le \alpha \le 1$, ν_α is the Bernoulli product measure on $\mathcal{X}$ with $P[\eta(x) = 1] = \alpha$ for each $x \in \mathbb{X}$. Averages or norms with respect to ν_α will be denoted by the subscript α. Let $\mathbb{E}$ be a collection of edges $e = (x, y)$ connecting some pairs of points in $\mathbb{X}$. Assume that the resulting graph is connected and that every vertex in it has a finite degree. There will also be given transition rates $\{p(x, y)\colon \mathbb{X} \times \mathbb{X} \to [0, \infty)\}$ that are symmetric and satisfy $\sum_{y\in\mathbb{X}} p(x , y) \le 1$ for every $x \in \mathbb{X}$. Moreover, $\mathbb{E} = \{e = (x, y) : p(x, y) > 0\}$. In our context $p(x , y) = p(y - x)$ for $x, y \in \mathbb{Z}^d$ or $x, y \in \mathbb{Z}^d_*$ as the case may be. We have a simple exclusion process on $\mathcal{X}$ with generator

$$(\mathcal{L} f)(\eta) = \sum_{x,y\in\mathbb{X}} p(x, y)\eta(x)\bigl(1 - \eta(y)\bigr)[f(\sigma^{x,y}\eta) - f(\eta)]$$

and Dirichlet form

$$D(f) = \tfrac{1}{4} \sum_{x,y\in\mathbb{X}} p(x , y)\| f(\sigma^{x,y}\eta) - f(\eta)\|_\alpha^2.$$

For each $n \ge 0$, denote by $\mathcal{E}_n$ the subsets of $\mathbb{X}$ with n points. For each A in $\mathcal{E}_n$, let Ψ_A be the local function

$$\Psi_A = \prod_{x\in A} \frac{\eta(x) - \alpha}{\sqrt{\alpha(1 - \alpha)}}. \tag{3.1}$$

By convention, $\Psi_\phi = 1$. It is easy to check that $\{\Psi_A, A \subset \mathbb{X}\}$ is an orthonormal basis of $L^2(\nu_\alpha)$. For each $n \ge 0$, denote by $\mathcal{H}_n$ the subspace of $L^2(\nu_\alpha)$ generated by $\{\Psi_A, A \in \mathcal{E}_n\}$, so that $L^2(\nu_\alpha) = \bigoplus_{n\ge0} \mathcal{H}_n$. Functions of $\mathcal{H}_n$ are said to have degree n.

For $n \ge 0$, denote by π_n the projection of $L^2(\nu_\alpha)$ on $\mathcal{H}_n$. It follows from definition (2.3) of formal gradients that $\{\pi_n\Phi_e, e \in \mathbb{E}\}$ is a formal gradient if $\{\Phi_e, e \in \mathbb{E}\}$ is one:

LEMMA 3.1. *Let $\{\Phi_e, e \in \mathbb{E}\}$ be a formal gradient. Then, for every $n \ge 0$, $\{\pi_n\Phi_e, e \in \mathbb{E}\}$ is also a formal gradient.*

PROOF. The proof is elementary. We verify that the projection π_n commutes with the exchanges σ^e, which follows from a direct computation. □

This lemma allows us to decompose a formal gradient into components of different degrees. We shall say that a formal gradient $\{\Phi_e, e \in \mathbb{E}\}$ has degree n if each function Φ_e has degree n. The proofs of Theorems 2.1 and 2.3 are thus reduced to proving that a square-integrable formal gradient of degree n can be approximated by gradients of local functions of degree n.

Let us compute $\sigma^e \Psi_A$. If $e = (x, y)$, $\sigma^e \Psi_A = 0$ unless e connects A and A^c. If $x \in A$ and $y \notin A$, then $\sigma^e \Psi_A = \Psi_{(A\setminus\{x\})\cup\{y\}}$, and if $y \in A$ and $x \notin A$, then $\sigma^e \Psi_A = \Psi_{(A\setminus\{y\})\cup\{x\}}$. In other words, $\sigma^e \Psi_A = \Psi_{\sigma^e A}$, with the obvious definition of $\sigma^e A = (A\setminus\{x\}) \cup \{y\}$ or $(A\setminus\{y\}) \cup \{x\}$. This suggests that we should turn the space $\mathbb{X}_n$ of subsets A of cardinality n in $\mathbb{X}$ into a graph with edges $\mathbb{E}_n$ by connecting two vertices A and B (i.e., subsets of $\mathbb{X}$ of cardinality n) in $\mathbb{X}_n$ with an edge if $B = (A\setminus\{x\}) \cup \{y\}$ for some $x \in A$ and $y \notin A$ that are connected by an edge in $\mathbb{X}$. The transition probabilities $p_n(A, B)$ on $\mathbb{X}_n$ are given by $p(x, y)$ if $B = (A\setminus\{x\}) \cup \{y\}$ for some $x \in A$ and $y \notin A$ and 0 otherwise.

Any local function f of degree n has a representation as

$$f(\eta) = \sum_A f(A)\Psi_A(\eta)$$

and

$$(\sigma^e f)(\eta) = f(\sigma^e \eta) = \sum_A f(A)\sigma^e \Psi_A = \sum_A f(\sigma^e A)\Psi_A.$$

The Dirichlet form $D(f)$ for a function of degree n reduces to

$$\frac{1}{2} \sum_{\substack{e\in\mathbb{E} \\ A\in\mathbb{X}_n}} p(e)|f(\sigma^e A) - f(A)|^2.$$

The factor $1/2$ is explained by the sum over the unoriented edges. The total Dirichlet form of f can be written as the sum

$$\frac{1}{2}\sum_{n\geq 1} \sum_{\substack{e\in\mathbb{E} \\ A\in\mathbb{X}_n}} p(e)|\pi_n f(\sigma^e A) - \pi_n f(A)|^2,$$

where

$$\pi_n f = \sum_{A\in\mathbb{X}_n} f(A)\Psi_A.$$

The approximation of any square-integrable formal gradient in $\mathbb{X}$ by gradients of local functions reduces to a similar property for each of the graphs $\mathbb{X}_n$. Note that for the purpose of the approximation result we can replace the Dirichlet form by any equivalent one.

Let $\mathcal{G}$ be a connected graph with vertices $V = \{x\}$ and edges $\mathbb{E} = \{e\}$. We assume that the graph is countable and that there is an upper bound D on the degree of all of its vertices. Let us pick for each edge an arbitrary orientation as the positive direction but keep the edges in both directions so we have the set $\widehat{\mathbb{E}}$ of oriented edges. Given a function $f\colon V \to \mathbb{R}$, we can define $g = \delta f\colon \widehat{\mathbb{E}} \to \mathbb{R}$

by $(\delta f)(e) = f(y) - f(x)$, where $(x, y) = e$ is an oriented edge. It is clear that $g\colon \widehat{\mathbb{E}} \to \mathbb{R}$ satisfies

$$\sum_{e \in \ell} g(e) = 0 \tag{3.2}$$

for any loop ℓ of oriented edges. We can turn the set of g's satisfying (3.2) into a Hilbert space $\mathcal{H}_1$ by defining

$$|||g|||^2 = \sum_{e \in \mathbb{E}} |g(e)|^2 = \frac{1}{2} \sum_{e \in \widehat{\mathbb{E}}} |g(e)|^2. \tag{3.3}$$

Given any g satisfying (3.2), we can construct an f, unique up to an additive constant, such that $g = \delta f$. However, in general, we have no control on the growth of f at ∞. We say that the graph has property **LD**, if, for any given $g \in \mathcal{H}_1$, we can find a sequence of functions f_n, each of which vanishes outside a finite set of vertices, such that

$$\lim_{n \to \infty} |||g - \delta f_n||| = 0$$

(cf. [3]). Not every graph has this property: a counterexample is given by the homogeneous tree of degree 3 (cf. [12], example (3.19), page 37).

The following lemma will allow us to deduce property **LD** for a graph from a closely related one.

LEMMA 3.2. *Let $\mathcal{G}$ be a graph with the degree of its vertices bounded by D. Let $\mathcal{G}_1$ be obtained from $\mathcal{G}$ by deleting a set of vertices and retaining all the edges of $\mathcal{G}$ that are contained in $\mathcal{G}_1$. Assume that there are finite constants $c(k)$, depending only on $(k, \mathcal{G}_1)$, such that any two vertices in $\mathcal{G}_1$ that are connected by a path of k edges in $\mathcal{G}$ are connected by a path of at most $c(k)$ edges contained in $\mathcal{G}_1$. Assume also that there is a bound ℓ such that any deleted vertex of $\mathcal{G}$ is connected to an undeleted vertex in $\mathcal{G}_1$ by a path of edges in $\mathcal{G}$ of length at most ℓ. Then if $\mathcal{G}$ has property* **LD** *so does $\mathcal{G}_1$.*

PROOF. The proof proceeds along the following lines. Given a g defined on the oriented edges of $\mathcal{G}_1$ that satisfies (3.2), we can find f on the vertices of $\mathcal{G}_1$ such that $g(e) = (\delta f)(e)$. Assume that we can extend f as a function h defined on the vertices of $\mathcal{G}$ in such a way that $|||\delta h|||_{\mathcal{G}}^2 \le C |||g|||_{\mathcal{G}_1}^2$. Because $\mathcal{G}$ has property **LD**, we can find a sequence of functions h_n defined on the vertices of $\mathcal{G}$, each h_n having finite support such that $|||\delta h_n - \delta h|||_{\mathcal{G}}^2 \to 0$ as $n \to \infty$. It follows that the restriction f_n of h_n to $\mathcal{G}_1$ satisfies $|||\delta f_n - g|||_{\mathcal{G}_1}^2 \to 0$, proving that $\mathcal{G}_1$ has the property **LD** as well.

This shows that all we need to do is to construct such an extension h of f. In order to do this, let us proceed as follows. For any x that is a deleted vertex, there is at least one vertex x' of $\mathcal{G}_1$ within a distance ℓ. Pick any one. Define $h(x) = f(x')$.

If $e = (x, y)$ is a deleted edge, then $(\delta h)(e) = f(x') - f(y')$, where x', y', are vertices in $\mathcal{G}_1$. The graph distance between x' and y' can be at most $2\ell + 1$ in $\mathcal{G}$ and therefore at most $c(2\ell + 1)$ in $\mathcal{G}_1$. It is clear that there is a constant $K = K(\ell, c(\cdot))$ such that all the edges involved in the path from x' to y' through the vertices of $\mathcal{G}_1$ are within a distance K from x and y. For any edge e let us define

$$S_K(e) = \sum_{\substack{e':\ d(e,e') \le K \\ e' \in \mathcal{G}_1}} |(\delta f)(e')|^2 ,$$

where the sum extends only over the edges of $\mathcal{G}_1$ within a distance K of e. We can define the distance between two edges as the largest of the four possible distances between the two pairs of vertices. They differ from one another by at most 2. By an application of Schwarz's inequality and the crudest upper bound,

$$|(\delta h)(e)|^2 \le c(2\ell + 1) S_K(e)$$

and

$$|||\delta h|||_{\mathcal{G}}^2 \le c(2\ell + 1) \sum_e S_K(e) = Cc(2\ell + 1) |||\delta f|||_{\mathcal{G}_1}^2 ,$$

where

$$C = \sup_e \{ |e' : d(e, e') \le K| \}.$$

Clearly, $C \le C(K, D) = (D + D^2 + \cdots + D^K)$, where D is a bound on the degree of any vertex. □

LEMMA 3.3. *Any symmetric random walk on $\mathbb{Z}^d$ has property* **LD**.

PROOF. We can assume without loss of generality that the walk is the nearest neighbor one. On any graph, if δf with f having finite support is not dense in the Hilbert space $\mathcal{H}_1$, there will be a $g \in \mathcal{H}_1$ that is orthogonal to δf for all finitely supported f. In other words, if we represent $g = \delta h$, after a summation by parts

$$\sum_e (\delta h)(e)(\delta f)(e) = \sum_e (\delta^* \delta h)(e) f(e) = 0 \quad \text{for all } f.$$

For the operator

$$L = \delta^* \delta,$$

we have a solution of $Lh \equiv 0$ with $\delta h \in \mathcal{H}_1$. We will show that this is impossible for the simple random walk on any $\mathbb{Z}^d$ (except for the trivial solution $h = 0$). Define $h_i(x) = h(x + z_i) - h(x)$ where $z_i : 1 \le i \le d$ are the unit vectors in the coordinate directions. Then (3.2) implies $(e^{i\theta_j} - 1)\hat{h}_k(\theta) = (e^{i\theta_k} - 1)\hat{h}_j(\theta)$ a.e. for $1 \le j < k \le d$, where $\hat{h}_j$ is the Fourier transform of the ℓ_2 function $h_j(x)$.

Therefore there exists a function $u(\theta)$ on the d-torus such that, for $1 \le j \le d$, $\hat{h}_j(\theta) = (e^{i\theta_j} - 1)u(\theta)$. It is easy to see that $Lh = 0$ now implies $k(\theta)u(\theta) = 0$, where $k(\theta) = \sum_j (1 - \cos\theta_j) > 0$ for $\theta \neq 0$. Thus $\hat{h}_j \equiv 0$. □

Clearly, the proof only depends on the property

$$\sum_x p(x)e^{i\,\theta\cdot x} = \sum_x p(x)\cos(\theta\cdot x) < 1$$

for $\theta \neq 0$ which is the same as irreducibility.

LEMMA 3.4. *Any symmetric exclusion process involving a finite number k of particles on* $\mathbb{Z}_*^d$ *has property* **LD**.

PROOF. Without exclusion and allowing 0, we have essentially a symmetric random walk on $\mathbb{Z}^{kd}$ which has property **LD**. (We only have to deal with symmetric functions of the k positions $x_1, \dots, x_k$.) We now delete the set of vertices $A = \{(x_1, \dots, x_k)\}$, where either some $x_i = 0$ or $x_i = x_j$ for some pair i, j. It is easy to verify that A satisfies the conditions of Lemma 3.2 except when $d = 1$ and the walk is the nearest neighbor one. □

REMARK 3.5. Tom Liggett pointed out to us that this result can also be obtained by using coupling techniques due to Spitzer and Liggett, as presented in [9], Chapter 8, Section 1.

We have thus proved Theorems 2.1 and 2.3. We shall now prove Corollaries 2.2 and 2.4. Since the proofs are identical we will only deal with Corollary 2.2.

PROOF OF COROLLARY 2.2. Consider a local function h such that $\langle h\rangle_\alpha \equiv 0$ for any $\alpha \in [0, 1]$. This is equivalent to $\langle h\rangle_{N,K} \equiv 0$ for any K and N sufficiently large. We define for such a function

$$\|h\|_{-1,\alpha}^2 = \sup_f \{2\langle h, f\rangle_\alpha - \langle f, (-\mathcal{L})f\rangle_\alpha\},$$

$$\|h\|_{-1,N,K}^2 = \sup_f \{2\langle h, f\rangle_{N,K} - \langle f, (-\mathcal{L}_N)f\rangle_{N,K}\}.$$

In the first formula one uses only local functions f. The second formula is in a finite-dimensional space and we are free to use arbitrary functions f. We need to prove

$$\lim_{\substack{N\to\infty \\ K_N/(2N)^d \to \alpha}} \|h\|_{-1,N,K_N}^2 = \|h\|_{-1,\alpha}^2. \tag{3.4}$$

It is easy to see that

$$\|h\|_{-1,\alpha}^2 \le \liminf_{\substack{N\to\infty \\ K_N/(2N)^d\to\alpha}} \|h\|_{-1,N,K_N}^2. \tag{3.5}$$

Indeed, by definition, for every $\varepsilon > 0$, there exists a local function f_ε such that

$$\|h\|_{-1,\alpha}^2 \le 2\langle h, f_\varepsilon\rangle_\alpha - \langle f_\varepsilon, (-\mathcal{L}) f_\varepsilon\rangle_\alpha + \varepsilon.$$

Since h and f_ε are local functions, the right-hand side is equal to

$$\begin{aligned}
&\lim_{\substack{N\to\infty \\ K_N/(2N)^d\to\alpha}} \{2\langle h, f_\varepsilon\rangle_{N,K} - \langle f_\varepsilon, (-\mathcal{L}) f_\varepsilon\rangle_{N,K}\} + \varepsilon \\
&\quad = \lim_{\substack{N\to\infty \\ K_N/(2N)^d\to\alpha}} \{2\langle h, f_\varepsilon\rangle_{N,K} - \langle f_\varepsilon, (-\mathcal{L}_N) f_\varepsilon\rangle_{N,K}\} + \varepsilon \\
&\quad \le \liminf_{\substack{N\to\infty \\ K_N/(2N)^d\to\alpha}} \|h\|_{-1,N,K}^2 + \varepsilon.
\end{aligned}$$

The reverse inequality is more demanding. Our goal is to prove that

$$\limsup_{\substack{N\to\infty \\ K_N/(2N)^d\to\alpha}} \|h\|_{-1,N,K_N} \le \|h\|_{-1,\alpha}. \tag{3.6}$$

It is more convenient to express the norms $\|h\|_{-1,N,K}$ and $\|h\|_{-1,\alpha}$ as

$$\|h\|_{-1,\alpha} = \sup\{\langle h, f\rangle_\alpha : \langle f, -\mathcal{L} f\rangle_\alpha \le 1\}, \tag{3.7}$$

$$\|h\|_{-1,N,K} = \sup\{\langle h, f\rangle_\alpha : \langle f, -\mathcal{L}_N f\rangle_{N,K} \le 1\}, \tag{3.8}$$

where the suprema are taken over f from the unit balls in the corresponding Dirichlet spaces.

It is easy to see that the following lemma is equivalent to (3.6). In fact, we can choose f_N as the solution of the variational problem (3.8) and applying the Lemma 3.6 we obtain (3.6). □

LEMMA 3.6. *Let $K_N/(2N)^d \to \alpha$. Assume that for some $A \ge 0$ we have a sequence of functions f_N in $L_2(\mu_{N,K_N})$ that satisfies*

$$\begin{aligned}
&\sup_N \langle f_N, -\mathcal{L}_N f_N\rangle_{N,K_N} \le 1, \\
&\lim_{N\to\infty} \langle h, f_N\rangle_{N,K_N} = A.
\end{aligned} \tag{3.9}$$

Then, for any $\varepsilon > 0$, there is a local function f_ε such that

$$\langle f_\varepsilon, h\rangle_\alpha \ge A - \varepsilon \quad \text{and} \quad \langle f_\varepsilon, -\mathcal{L} f_\varepsilon\rangle_\alpha \le 1. \tag{3.10}$$

PROOF. Fix a positive integer ℓ large enough so that $\Lambda_\ell = \{-\ell, \dots, \ell\}^d$ contains the support of h. Since h has mean 0 with respect to all the canonical measures $\mu_{\ell,K}$, from linear algebra, $H = \mathcal{L}_{\Lambda_\ell}^{-1} h$ is a well-defined cylinder function. Here $\mathcal{L}_{\Lambda_\ell}$ is the generator of the exclusion process restricted to Λ_ℓ. Notice the difference between $\mathcal{L}_{\Lambda_\ell}$ and $\mathcal{L}_\ell$. While the first generator stands for the exclusion process on the cube Λ_ℓ, the second generator stands for the exclusion process on Λ_ℓ with periodic boundary conditions, that is, on $\mathbb{T}_\ell^d$. Since $h = \mathcal{L}_{\Lambda_\ell} H$, for any $N > \ell$,

$$\langle h, f_N \rangle_{N,K_N} = -\tfrac{1}{2} \sum_{x,y \in \Lambda_\ell} p(y-x) \langle T_{x,y} f_N, T_{x,y} H \rangle_{N,K_N}. \tag{3.11}$$

The summation in the above formula is carried over all sites in the cube Λ_ℓ (without periodic boundary condition). Fix a cube Λ_L with $L \geq \ell$. Denote by $f_{N,L}$ the conditional expectation of f_N given $\mathcal{F}_{\Lambda_L} = \sigma(\eta(x), x \in \Lambda_L)$. A simple computation shows that $E[T_{x,y} f_N | \mathcal{F}_{\Lambda_L}] = T_{x,y} E[f_N | \mathcal{F}_{\Lambda_L}] = T_{x,y} f_{N,L}$ for all bonds $\{x, y\}$ in Λ_ℓ. Therefore, since H is $\mathcal{F}_{\Lambda_L}$-measurable, taking conditional expectations with respect to $\mathcal{F}_{\Lambda_L}$, we obtain that the right-hand side of (3.11) is equal to

$$-\tfrac{1}{2} \sum_{x,y \in \Lambda_\ell} p(y-x) \langle T_{x,y} f_{N,L}, T_{x,y} H \rangle_{N,K_N}. \tag{3.12}$$

On the other hand, by Jensen's inequality and (3.9),

$$\begin{aligned} \tfrac{1}{2} \sum_{x,y \in \Lambda_L} p(x-y) \langle (T_{x,y} f_{N,L})^2 \rangle_{N,K_N} &\leq \tfrac{1}{2} \sum_{x,y \in \Lambda_L} p(x-y) \langle (T_{x,y} f_N)^2 \rangle_{N,K_N} \\ &= \langle f_N, -\mathcal{L}_N f_N \rangle_{N,K_N} \\ &\leq 1. \end{aligned} \tag{3.13}$$

Since $f_{N,L}$ is a cylinder function, the state space $\{0, 1\}^{\Lambda_L}$ is finite and we have a uniform bound on the $L^2(\mu_{N,K_N})$ norms of the gradients of $f_{N,L}$, there exists a subsequence N' and a cylinder function g_L such that the gradients of $f_{N',L}$ converge to the gradients of g_L pointwise. On the other hand, by the equivalence of ensembles, as $N \uparrow \infty$ and $K_N/(2N)^d \to \alpha$, the finite-dimensional marginals of the canonical measure μ_{N,K_N} converge to the corresponding marginals of μ_α. Therefore, as $N' \uparrow \infty$ and as $K_N/(2N')^d \to \alpha$, the expression in (3.12) converges to

$$-\tfrac{1}{2} \sum_{x,y \in \Lambda_\ell} p(y-x) \langle T_{x,y} H, T_{x,y} g_L \rangle_\alpha. \tag{3.14}$$

Moreover, since $T_{x,y} f_{N,L}$ converges to $T_{x,y} g_L$, by (3.13),

$$\frac{1}{2}\sum_{x,y\in\Lambda_L} p(x-y)\langle\{T_{x,y}g_L\}^2\rangle_\alpha = \frac{1}{4}\sum_{x,y\in\Lambda_L} p(x-y)\langle\{\nabla_{x,y}g_L\}^2\rangle_\alpha \le 1 \tag{3.15}$$

uniformly in L.

For each x, y in $\mathbb{Z}^d$ such that $y-x\in F$ where $F=\{z:p(z)>0\}$, $T_{x,y}g_L$ is a bounded sequence in $L^2(\mu_\alpha)$. There exists, therefore, a subsequence L' for which $T_{x,y}g_{L'}$ converges weakly for each x, y in $\mathbb{Z}^d$ such that $y-x\in F$. Denote by $\Phi_{x,y}$ the weak $L^2(\mu_\alpha)$ limit. By construction, as $L'\uparrow\infty$, (3.14) converges to

$$-\frac{1}{2}\sum_{x,y\in\Lambda_\ell} p(x-y)\langle T_{x,y}H\,,\Phi_{x,y}\rangle_\alpha.$$

Moreover, the collection of $L^2(\mu_\alpha)$ functions $\Phi_{x,y}$ satisfy conditions (2.3) of a formal gradient.

At this point we have proved that

$$-\frac{1}{2}\sum_{x,y\in\Lambda_\ell} p(x-y)\langle T_{x,y}H,\Phi_{x,y}\rangle_\alpha \ge A \tag{3.16}$$

and from (3.15) it follows that

$$\frac{1}{4}\sum_{x,y\in\mathbb{Z}^d} p(x-y)\langle\Phi^2_{x,y}\rangle_\alpha \le 1. \tag{3.17}$$

We can now apply Theorem 2.1 and approximate $\{\Phi_{x,y}\}$ by δf_ε, thereby establishing (3.10). □

4. Approximations for the tagged particle. Given $1\le K\le(2N)^d$, denote by $\mathcal{X}_{N,K}$ the space of configurations of $\mathcal{X}_N$ with $K-1$ particles: $\mathcal{X}_{N,K}=\{\xi\in\mathcal{X}_N,\sum_{x\in\mathbb{T}^d_{N,*}}\xi(x)=K-1\}$. Recall from Section 2 that $\mu_{N,K}$ stands for the uniform measure on $\mathcal{X}_{N,K}$ and that $\langle\cdot\rangle_{N,K}$ represents the expectation with respect to $\mu_{N,K}$.

Denote by $\mathcal{C}$ (resp. $\mathcal{C}_0$) the space of local functions (resp. local functions that have mean 0 with respect to all canonical measures). Thus a local function h belongs to $\mathcal{C}_0$ provided $\langle h\rangle_{N,K}=0$ for all large enough N and all $1\le K\le(2N)^d$. Examples of subspaces of $\mathcal{C}_0$ are $\mathcal{L}_0\mathcal{C}=\{\mathcal{L}_0h,\,h\in\mathcal{C}\}$, $\mathcal{L}_\tau\mathcal{C}=\{\mathcal{L}_\tau h,\,h\in\mathcal{C}\}$. Also $\{\psi_\lambda,\,\lambda\in\mathbb{R}^d\}$ defined by (2.8) are functions in $\mathcal{C}_0$.

On the space of local functions $\mathcal{C}$, consider the inner product $\langle\cdot,\cdot\rangle_{0,\alpha,1}$ defined by

$$\langle f,g\rangle_{0,1,\alpha}=\langle-\mathcal{L}_0f,g\rangle_\alpha.$$

Here the index 0 signifies that only the jump part of the generator enters in the definition. Denote by $\mathcal{H}^0_{1,\alpha}$ the Hilbert space obtained by completing the space of local functions endowed with the above inner product and by $D_{0,\alpha}$ the Dirichlet form associated with the generator $\mathcal{L}_0$ and the probability measure ν_α, that is, $D_{0,\alpha}(f) = \langle -\mathcal{L}_0 f, f\rangle_\alpha$. For a local function f, let

$$\|f\|_{0,-1,\alpha}^2 = \sup_{g\in\mathcal{C}}\{2\langle f, g\rangle_\alpha - D_{0,\alpha}(g)\}$$

be the dual H_{-1} norm.

REMARK 4.1. We could have defined the H_{-1} norm as

$$\|f\|_{0,-1,\alpha} = \sup\{\langle f, g\rangle_\alpha : D_{0,\alpha}(g) \le 1\},$$

where the supremum is taken over either local functions or functions in $L_2(\nu_\alpha)$ with Dirichlet norm bounded by 1.

We denote by $\|f\|_{0,-1,\alpha}$ the $\mathcal{H}^0_{-1,\alpha}$ norm of f. It is known that every local function in $\mathcal{C}_0$ has finite $\mathcal{H}^0_{-1,\alpha}$ norm (cf. [5], Section 7.2). On the other hand, constants have infinite $\mathcal{H}^0_{-1,\alpha}$ norm. For two local functions f, g with finite $\mathcal{H}^0_{-1,\alpha}$ norm, define by polarization the $\mathcal{H}^0_{-1,\alpha}$ inner product of f, g, denoted by $\langle f, g\rangle_{0,-1,\alpha}$. Let $\mathcal{H}^0_{-1,\alpha}$ be the Hilbert space induced by the local functions endowed with the inner product $\langle f, g\rangle_{0,-1,\alpha}$.

In a similar way, for each $N \ge 1$ and $1 \le K \le (2N)^d$, we define the spaces $\mathcal{H}^0_{1,N,K}$ and $\mathcal{H}^0_{-1,N,K}$ and the norms $\|\cdot\|_{0,1,N,K}$ and $\|\cdot\|_{0,-1,N,K}$. According to Corollary 2.4, for any local function in $h \in \mathcal{C}_0$,

$$\lim_{\substack{N\to\infty\\ K_N/(2N)^d\to\alpha}} \|h\|_{0,-1,N,K_N}^2 = \|h\|_{0,-1,\alpha}^2. \tag{4.1}$$

Since the density α is fixed throughout this section, to keep the notation simple, for $n = -1, 1$, we denote the norms $\|\cdot\|_{0,n,\alpha}$ and $\|\cdot\|_{n,\alpha}$ by $\|\cdot\|_{0,n}$ and $\|\cdot\|_n$ and we denote the inner product $\langle\cdot,\cdot\rangle_\alpha$ simply by $\langle\cdot,\cdot\rangle$. The main result of this section states that any $h \in \mathcal{C}_0$ can be approximated in $\|\cdot\|_{0,-1}$ norm by $\mathcal{L}f$ with a local function f. Notice that while the H_{-1} norm corresponds to the generator $\mathcal{L}_0$, the generator that appears in front of f is the full generator $\mathcal{L} = \mathcal{L}_0 + \mathcal{L}_\tau$.

THEOREM 4.2. *Let h be a local function in $\mathcal{C}_0$. For every $\varepsilon > 0$, there exist local functions u_ε, f_ε such that $h = \mathcal{L}f_\varepsilon + u_\varepsilon$ and $\|u_\varepsilon\|_{0,-1} \le \varepsilon$.*

Notice that u_ε belongs to $\mathcal{C}_0$ because $\mathcal{L}f_\varepsilon$ belongs automatically to $\mathcal{C}_0$. Theorem 4.2 is based on the following two results. Fix a local function h in $\mathcal{C}_0$. For each $\lambda > 0$, let u_λ be the solution of the resolvent equation

$$\lambda u_\lambda - \mathcal{L}u_\lambda = -h. \tag{4.2}$$

PROPOSITION 4.3. *For every $\varepsilon > 0$, there exists $\lambda_0 > 0$ such that $\|\mathcal{L}u_\lambda - h\|_{0,-1} \le \varepsilon$ for all $\lambda < \lambda_0$. That is,*

$$\lim_{\lambda\to 0}\|\mathcal{L}u_\lambda - h\|_{0,-1} = 0. \tag{4.3}$$

PROPOSITION 4.4. *Fix a local function h in $\mathcal{C}_0$ and let u_λ be the solution of the resolvent equation* (4.2). *For every $\varepsilon > 0$ and every $\lambda > 0$, there exists a local function $g_{\lambda,\varepsilon}$ such that $\|\mathcal{L}(u_\lambda - g_{\lambda,\varepsilon})\|_{0,-1} \le \varepsilon$.*

PROOF OF THEOREM 4.2. Since

$$h = \mathcal{L}g_{\lambda,\varepsilon} + \mathcal{L}(u_\lambda - g_{\lambda,\varepsilon}) + (h - \mathcal{L}u_\lambda)$$

and we can make

$$\|\mathcal{L}(u_\lambda - g_{\lambda,\varepsilon})\|_{0,-1} + \|(h - \mathcal{L}u_\lambda)\|_{0,-1} < \varepsilon$$

by first picking a small λ according to Proposition 4.3 to make the second term small and then by picking ε small according to Proposition 4.4 so that the first term is small, Theorem 4.2 follows from Propositions 4.3 and 4.4. □

We now turn to the proofs of Propositions 4.3 and 4.4, which are based on the duality properties of the symmetric exclusion process explained in Section 3. Denote by $D_0(\cdot)$ the Dirichlet form associated with the jump part of the generator, by $D_\tau(\cdot)$ the Dirichlet form associated with the translation part of the generator and by $D(\cdot)$ the full Dirichlet form:

$$D_0(f) = \langle f, (-\mathcal{L}_0)f\rangle,$$
$$D_\tau(f) = \langle f, (-\mathcal{L}_\tau)f\rangle,$$
$$D(f) = D_0(f) + D_\tau(f).$$

Recall the definition of the set $A_{x,y}$ given in Section 3 and notice that $\sigma^{x,y}\Psi_A = \Psi_{A_{x,y}}$. In particular, if $f = \sum_{n\ge1}\sum_{A\in\mathcal{E}_n} f(A)\Psi_A$,

$$\begin{aligned} D_0(f) &= \frac{1}{4}\sum_{x,y\in\mathbb{Z}^d_*} p(y-x)\int [f(\sigma^{x,y}\xi) - f(\xi)]^2\, d\nu_\alpha \\ &= \frac{1}{4}\sum_{n\ge1}\sum_{x,y\in\mathbb{Z}^d_*} p(y-x)\sum_{A\in\mathcal{E}_n}\{f(A_{x,y}) - f(A)\}^2. \end{aligned}$$

Actually, if F is any symmetric finite set that generates $\mathbb{Z}^d$, $D_0(f)$ is seen to be equivalent to the form

$$\begin{aligned} \widehat{D}_0(f) &= \frac{1}{2}\sum_{\substack{x,y\in\mathbb{Z}^d_* \\ y-x\in F}} \int [f(\sigma^{x,y}\xi) - f(\xi)]^2\, d\nu_\alpha \\ &= \frac{1}{2}\sum_{n\ge1}\sum_{\substack{x,z\in\mathbb{Z}^d_* \\ y-x\in F}}\sum_{A\in\mathcal{E}_n}\{f(A_{x,y}) - f(A)\}^2 \qquad (4.4) \\ &= \sum_{n\ge1}\sum_{e\in\mathbb{E}_n}\sum_{A\in\mathcal{E}_n}\{f(\sigma^e A) - f(A)\}^2, \end{aligned}$$

where $\mathbb{E}_n$ is the set of edges in $\mathcal{E}_n$ with an edge e connecting A and B present if $A = \sigma^{x,y} B$ for some $x, y \in \mathbb{Z}_*^d$ such that $y - x \in F$. A consequence of the equivalence of the two Dirichlet forms is their equivalence on all of the spaces $\mathcal{H}_n$ with constants that are independent of n.

On the other hand, for a set A of $\mathcal{E}_n$, denote by $\tau_z A$ the set defined by

$$(\tau_z A) = \begin{cases} A + z, & \text{if } -z \notin A, \\ (A + z) \setminus \{0\} \cup \{z\}, & \text{if } -z \in A, \end{cases}$$

where $B + z$ is the set $\{x + z;\ x \in B\}$. Therefore, in the case where $-z$ belongs to A, we first translate A by z (obtaining a new set that contains the origin) and then we remove the origin and we add the site z. Of course, $\tau_z\colon \mathcal{E}_n \to \mathcal{E}_n$ is a one-to-one function. We denote its inverse by τ_z^{-1}, which is seen to be τ_{-z}. A straightforward computation shows that $\tau_z \Psi_A = \Psi_{\tau_z A}$.

We now use the results in the Appendix to prove Proposition 4.3 along the lines of the proofs of Lemma 2.5 of [8] or Theorem 5.1 of [11].

PROOF OF PROPOSITION 4.3. From the results in [6], one can show that

$$\lim_{\lambda \to 0} \|\mathcal{L}u_\lambda - h\|_{-1} = 0. \tag{4.5}$$

Our goal is to prove that

$$\lim_{\lambda \to 0} \|\mathcal{L}u_\lambda - h\|_{0,-1} = 0.$$

To this end, we write $u_\lambda = \sum_n u_{n,\lambda}$, where $u_{n,\lambda} = \pi_n u_\lambda$ is the projection of u_λ into $\mathcal{H}_n$, the space functions of degree n and a similar decomposition for $\mathcal{L}u_\lambda = v_\lambda = \sum_n v_{n,\lambda}$ as well as $h = \sum h_n$. Since h is local, the decomposition for h is only a finite sum, and since the decomposition is orthogonal in $\mathcal{H}^0_{-1,\alpha}$,

$$\|\mathcal{L}u_\lambda - h\|_{0,-1}^2 = \sum_{n \le n_0} \|v_{n,\lambda} - h_n\|_{0,-1}^2 + \sum_{n > n_0} \|v_{n,\lambda}\|_{0,-1}^2.$$

From (4.5) and Lemma A.1,

$$\|f\|_1^2 \le Cn \|f\|_{0,1}^2,$$

valid for $f \in \mathcal{H}_n$, it follows that, for each fixed n and any function $u \in L^2$,

$$\begin{aligned} \|\pi_n u\|_{0,-1} &= \sup_{\substack{f \in \mathcal{H}_n \\ \|f\|_{0,1} \le 1}} |\langle u, f\rangle| \le \sup_{\|f\|_1 \le \sqrt{Cn}} |\langle u, f\rangle| \\ &\le \sqrt{Cn} \|u\|_{-1}. \end{aligned} \tag{4.6}$$

Then we have

$$\begin{aligned} \|v_{n,\lambda} - h_n\|_{0,-1} &\le \sqrt{Cn} \|\mathcal{L}u_\lambda - h\|_{-1} \\ &\to 0 \quad \text{as } \lambda \to 0. \end{aligned}$$

It is therefore sufficient to control

$$\sum_n \sup_\lambda \|v_{n,\lambda}\|_{0,-1} < \infty.$$

On the other hand, by the tridiagonal property of $\mathcal{L}$ and using Lemma A.1 and (4.6), we have

$$\begin{aligned}\|v_{n,\lambda}\|_{0,-1} &= \|\pi_n \mathcal{L}(u_{n-1,\lambda} + u_{n,\lambda} + u_{n+1,\lambda})\|_{0,-1}\\ &\le \sqrt{Cn}\|\mathcal{L}(u_{n-1,\lambda} + u_{n,\lambda} + u_{n+1,\lambda})\|_{-1}\\ &= \sqrt{Cn}\|u_{n-1,\lambda} + u_{n,\lambda} + u_{n+1,\lambda}\|_1\\ &\le C'n\{\|u_{n-1,\lambda}\|_{0,1} + \|u_{n,\lambda}\|_{0,1} + \|u_{n+1,\lambda}\|_{0,1}\}.\end{aligned}$$

In view of estimate (A.5), the proof of the proposition is complete. □

PROOF OF PROPOSITION 4.4. Let u_λ be the solution of the resolvent equation. In the proof of Proposition 4.3, we saw that

$$\mathcal{L}u_\lambda = \mathcal{L}\left[\sum_{n\ge 1} u_{\lambda,n}\right] = \sum_{n\ge 1} \mathcal{L}u_{\lambda,n}$$

can be evaluated term by term and the second sum converges rapidly in $\mathcal{H}_{0,-1}$. It is therefore enough to show that, if $g = \sum_{j=1}^n u_{\lambda,j}$, we can find for any given ε a local function g_ε such that $\|\mathcal{L}(g - g_\varepsilon)\|_{0,-1} \le \varepsilon$. If $u \in \bigoplus_{j=1}^n \mathcal{H}_j$, with components $\{u_j\}$,

$$\|\mathcal{L}u\|_{0,-1} \le \sum_{j=1}^n \|\mathcal{L}u_j\|_{0,-1} \le C\sum_{j=1}^n j\|u_j\|_1 \le C_n\|u\|_{0,1}.$$

All we need to do now is to find a local function $g_\varepsilon \in \bigoplus_{j=1}^n \mathcal{H}_j$, such that

$$\|g - g_\varepsilon\|_{0,1} < \varepsilon.$$

Note that g is in $L_2(\nu_\alpha)$ and the approximation can be done in each $\mathcal{H}_j$. □

We now prove the main result of the article.

PROOF OF THEOREM 2.5. In view of (3.5), which holds with $\|\cdot\|_{-1,\alpha}$ in place of $\|\cdot\|_{0,-1,\alpha}$, to prove Theorem 2.5, it is enough to show that

$$\limsup_{\substack{N\to\infty\\ K_N/(2N)^d\to\alpha}} \|h\|_{-1,N,K_N}^2 \le \|h\|_{-1,\alpha}^2 \tag{4.7}$$

for every local function h in $\mathfrak{C}_0$. Notice first that this statement is elementary if $h = \mathcal{L}h_0$ for some local function h_0.

Fix $\varepsilon > 0$. By Theorem 4.2, there exist local functions f_ε and u_ε such that $h = \mathcal{L} f_\varepsilon + u_\varepsilon$ and $\|u_\varepsilon\|_{0,-1,\alpha} \le \varepsilon$. Let N be larger than the supports of $\mathcal{L} f_\varepsilon$ and u_ε. In this case,

$$\|h\|_{-1,N,K} = \|\mathcal{L} f_\varepsilon + u_\varepsilon\|_{-1,N,K} \le \|\mathcal{L} f_\varepsilon\|_{-1,N,K} + \|u_\varepsilon\|_{-1,N,K}.$$

Since (4.7) is elementary for functions of type $\mathcal{L}g$, as $N \uparrow \infty$ and $K_N/(2N)^d \to \alpha$, the first term on the right-hand side converges to $\|\mathcal{L} f_\varepsilon\|^2_{-1,\alpha}$. By Corollary 2.4, since u_ε belongs to $\mathcal{C}_0$, as $N \uparrow \infty$ and $K_N/(2N)^d \to \alpha$, the second term converges to $\|u_\varepsilon\|_{0,-1,\alpha}$, which is at most ε. Therefore,

$$\limsup_{\substack{N\to\infty \\ K_N/(2N)^d \to \alpha}} \|h\|_{-1,N,K_N} \le \|\mathcal{L} f_\varepsilon\|_{-1,\alpha} + \varepsilon \le \|h\|_{-1,\alpha} + 2\varepsilon.$$

Since $\varepsilon > 0$ is arbitrary we are done. □

APPENDIX

We begin with a lemma.

LEMMA A.1. *There exists a finite constant depending only on the transition probability p such that*

$$D_\tau(f) \le C n D_0(f)$$

for all functions f in $\mathcal{H}_n$.

PROOF. We note that, on $\mathcal{H}_n$, $D_0(f)$ can be replaced by the expression

$$\widehat{\mathcal{D}}_0(f) = \frac{1}{2} \sum_{e\in\mathbb{E}_n} \sum_{A\in\mathcal{E}_n} \{f(\sigma^e A) - f(A)\}^2$$

because of its equivalence to $D_0(f)$ which is, in fact, uniform in n. Here $\mathbb{E}_n$ is the set of edges in $\mathcal{E}_n$ connecting A and B if $A = \sigma^{x,y} B$ for some $x, y \in \mathbb{Z}^d_*$ such that $y - x \in F$. Recall that F can be taken to be the set of points in $\mathbb{Z}^d$ where every coordinate is bounded by 2 in absolute value.

An easy calculation of

$$\int (1 - \xi(z))[f(\tau_z \xi) - f(\xi)]^2 \, d\nu_\alpha$$

for $f \in \mathcal{H}_n$ of the form $f = \sum_{|A|=n} f(A)\Psi_A$ leads to

$$D_\tau(f) = \tfrac{1}{2} \sum_z \sum_{|A|=n} p(z) r_z(A)[f(\tau_{-z} A) - f(A)]^2, \tag{A.1}$$

with $r_z(A) = \alpha$ if $z \in A$ and $1 - \alpha$ otherwise. Since $r_z(A) \le 1$ and (A.1) is really a finite sum over $\{z : p(z) > 0\}$, it is basically a question of estimating

$$\sum_{A \in \mathcal{E}_n} \big(f(\tau_{-z}A) - f(A)\big)^2$$

for each z. We can assume without loss of generality that $z = (1, 0, \ldots)$, the unit vector in the direction of the first coordinate axis.

The problem is now reduced to the following: we are given a function $g : \mathcal{E}_n \to \mathbb{R}$. We think of $\mathcal{E}_n$ as a graph with edges $\mathbb{E}_n$. The Dirichlet form is

$$\widehat{\mathcal{D}}_0(g) = \tfrac{1}{2} \sum_{e \in \mathbb{E}_n} |(\delta g)(e)|^2$$

and we are claiming an estimate of the form

$$\tfrac{1}{2} \sum_{A \in \mathcal{E}_n} |g(\tau_z A) - g(A)|^2 \le Cn \widehat{\mathcal{D}}_0(g),$$

with a constant independent of n. It is clear that for any A and $B = \tau_z A$ one can move from A to B along the edges of the graph $\mathcal{E}_n$, using only the edges in $\mathbb{E}_n$. We will verify that we can assign a set of edges $\mathbb{E}_A \subset \mathbb{E}_n$ such that (i) for every A one can use the edges of $\mathbb{E}_A$ to go from A to $\tau_z A$, (ii) for any A there are at most n edges in $\mathbb{E}_A$ and (iii) the subsets $\{\mathbb{E}_A\}$ are mutually disjoint as A varies over $\mathcal{E}_n$. Then it is easy to see that

$$|g(\tau_z A) - g(A)|^2 = \left| \sum_{e \in \mathbb{E}_A} (\delta g)(e) \right|^2 \le n \sum_{e \in \mathbb{E}_A} |(\delta g)(e)|^2,$$

and, summing over $A \in \mathcal{E}_n$, because the $\mathbb{E}_A$ are disjoint, we can establish the lemma.

To construct the paths from A to $\tau_z A$, we totally order the points of $\mathbb{Z}_*^d$ by lexicographic ordering. We say that $z = (z_1, \ldots, z_d) \in \mathbb{Z}_*^d$ is *positive* if either $z_1 > 0$ or $z_1 = 0, z_2 = 0, \ldots, z_{j-1} = 0$ and $z_j > 0$ for some j, $2 \le j \le d$. The total ordering declares $y > x$ if $y - x$ is *positive*. Let the set $A \in \mathcal{E}_n$ consist of the n points $(x_1, x_2, \ldots, x_n)$ of $\mathbb{Z}_*^d$. We can assume that they are ordered so that $x_1 > x_2 > \cdots > x_n$. Then $\tau_z A = (x_1^*, \ldots, x_n^*)$, where $x_j^* = x_j + z$ unless $x_j = -z$ in which case $x_j^* = x_j + 2z = z$. We use the edges in $\mathbb{E}_n$ to shift successively each x_i to x_i^* starting from x_1 and proceeding in order and ending with shifting x_n. Any edge that is used goes from some A_1 to $A_2 = \sigma^{x,x^*} A_1$. Since the shifts were made in lexicographic order we can determine without ambiguity which points of A_1 have already been shifted and which have not been. In other words, the paths from any two different A to the corresponding $\tau_y A$ do not share a common edge. It is also clear that exactly n edges are used. Any edge corresponds to $x^* - x = z$ or $2z$, both of which are in F. □

We want to look at the resolvent equation

$$\lambda u_\lambda - \mathcal{L} u_\lambda = -h \tag{A.2}$$

and obtain some decay estimates on the contributions to the Dirichlet form from terms of large degree. We follow essentially the method carried out in [8] or [11]. The lattice is $\mathbb{Z}_*^d$ and we have the Bernoulli measure ν_α on $\mathcal{X} = \{0,1\}^{\mathbb{Z}_*^d}$. We consider the orthonormal basis $\{\Psi_A\}$ introduced in (3.1) and the decomposition of $L_2(\nu_\alpha)$ as $\bigoplus_j \mathcal{H}_j$. We will consider operators $\mathbb{T}$ acting on $L_2(\nu_\alpha)$ as scalar multiplication by $t(j)$ on $\mathcal{H}_j$. Assume that $t(j)$ is constant for $j \le n_1$. These are the operators that commute with the projection operators π_j onto the subspaces $\mathcal{H}_j$. The operator $\mathcal{L}$ can be written as $\mathcal{L}_0 + \mathcal{L}_\tau$, the sum of the parts corresponding to the motions of untagged particles and the tagged particle. The operator $\mathcal{L}_0$ commutes with $\mathbb{T}$. Corresponding to the decomposition $L_2(\nu_\alpha) = \oplus \mathcal{H}_j$, we can write a block decomposition for $\mathcal{L}_\tau$ involving a tridiagonal block matrix $\mathfrak{L}_{i,j}$. If we write $u = \sum_j u_j$ where $u_j = \pi_j u$, then we can compute the commutator

$$\begin{aligned}
[\mathbb{T}, \mathcal{L}] &= \mathbb{T}\mathcal{L} - \mathcal{L}\mathbb{T} \\
&= \sum_j t(j)[\mathfrak{L}_{j-1,j} u_{j-1} + \mathfrak{L}_{j,j} u_j + \mathfrak{L}_{j+1,j} u_{j+1}] \\
&\quad - \sum_j \bigl[t(j-1)\mathfrak{L}_{j-1,j} u_{j-1} + t(j)\mathfrak{L}_{j,j} u_j + t(j+1)\mathfrak{L}_{j+1,j} u_{j+1}\bigr] \\
&= \sum_j \bigl[[t(j) - t(j-1)]\mathfrak{L}_{j-1,j} u_{j-1} + [t(j) - t(j+1)]\mathfrak{L}_{j+1,j} u_{j+1}\bigr].
\end{aligned}$$

We let $t(j) u_j = v_j$ and $t(j) - t(j-1) = s(j)$. From the reversibility of $\mathcal{L}_\tau$, we can conclude that

$$\mathfrak{L}_{j-1,j}^* = \mathfrak{L}_{j,j-1}.$$

We then use it to estimate the following quadratic form

$$\begin{aligned}
&\langle [\mathbb{T}, \mathcal{L}]\mathbb{T}^{-1}\mathbb{T}u, \mathbb{T}u \rangle \\
&\quad = \sum_j \bigl\langle [t(j) - t(j-1)]\mathfrak{L}_{j-1,j} u_{j-1} \\
&\qquad\qquad + [t(j) - t(j+1)]\mathfrak{L}_{j+1,j} u_{j+1}, t(j) u_j \bigr\rangle \\
&\quad = \sum_j \frac{s(j)}{t(j-1)} \langle \mathfrak{L}_{j-1,j}\, v_{j-1}, v_j \rangle
\end{aligned} \tag{A.3}$$

$$
\begin{aligned}
&- \sum_j \frac{s(j+1)}{t(j+1)} \langle \mathfrak{L}_{j+1,j} v_{j+1}, v_j \rangle \\
&= \sum_j \frac{s(j)}{t(j-1)} \langle \mathfrak{L}_{j-1,j} v_{j-1}, v_j \rangle \\
&\quad - \sum_j \frac{s(j+1)}{t(j+1)} \langle \mathfrak{L}_{j,j+1} v_j, v_{j+1} \rangle \\
&= \sum_j \frac{s(j)^2}{t(j-1)t(j)} \langle \mathfrak{L}_{j-1,j} v_{j-1}, v_j \rangle \\
&= \sum_j \frac{s(j)^2}{t(j-1)t(j)} \langle \mathcal{L}_\tau v_{j-1}, v_j \rangle \\
&\le \sum_j \frac{s(j)^2}{t(j-1)t(j)} \sqrt{\langle -\mathcal{L}_\tau v_{j-1}, v_{j-1} \rangle} \sqrt{\langle -\mathcal{L}_\tau v_j, v_j \rangle} \\
&\le \frac{1}{2} \sum_j \frac{s(j)^2}{t(j-1)t(j)} [\langle -\mathcal{L}_\tau v_{j-1}, v_{j-1} \rangle + \langle -\mathcal{L}_\tau v_j, v_j \rangle].
\end{aligned}
$$

Since $-\mathcal{L}_\tau$ is symmetric and positive definite, the penultimate inequality is valid. We can now combine the last inequality of our calculation in (A.3) with Lemma A.1 to arrive at

$$
\langle [\mathbb{T}, \mathcal{L}] \mathbb{T}^{-1} \mathbb{T} u, \mathbb{T} u \rangle \le \sum_j \frac{C}{2} \frac{j s(j)^2}{t(j-1)t(j)} [\langle -\mathcal{L}_0 v_{j-1}, v_{j-1} \rangle + \langle -\mathcal{L}_0 v_j, v_j \rangle].
$$

Let us suppose that

$$
\sup_j \frac{C j s(j)^2}{t(j-1)t(j)} \le \delta.
$$

Then

$$
\langle [\mathbb{T}, \mathcal{L}] \mathbb{T}^{-1} \mathbb{T} u, \mathbb{T} u \rangle \le \delta \langle -\mathcal{L}_0 v, v \rangle \le \delta \langle -\mathcal{L} v, v \rangle = \delta \langle -\mathcal{L} \mathbb{T} u, \mathbb{T} u \rangle. \tag{A.4}
$$

We start with the resolvent equation

$$
\lambda u_\lambda - \mathcal{L} u_\lambda = -h,
$$

operate by $\mathbb{T}$ and take inner product with $v = \mathbb{T} u_\lambda$. We get

$$
\lambda \langle v, v \rangle - \langle \mathbb{T} \mathcal{L} \mathbb{T}^{-1} v, v \rangle = -\langle h, v \rangle,
$$

which can be rewritten using (A.4) as

$$\begin{aligned}\lambda\langle v, v\rangle - \langle \mathcal{L}v, v\rangle &= \langle [\mathbb{T}\mathcal{L} - \mathcal{L}\mathbb{T}]\mathbb{T}^{-1}v, v\rangle - \langle h, v\rangle \\ &\le \delta\langle -\mathcal{L}v, v\rangle - \langle h, v\rangle.\end{aligned}$$

This implies the estimate

$$\|v\|_1^2 = \langle -\mathcal{L}v, v\rangle \le (1-\delta)^{-1}|\langle h, v\rangle| \le \|\mathbb{T}h\|_{-1}\|v\|_1 \le \|\mathbb{T}h\|_{0,-1}\|v\|_1.$$

In particular,

$$\sup_{\lambda>0} \|\mathbb{T}u_\lambda\|_{0,1}^2 \le (1-\delta)^{-1}\|\mathbb{T}h\|_{0,-1}.$$

It is not hard to construct sequences $t(j)$ that satisfy

$$\sup_j \frac{Cjs(j)^2}{t(j-1)t(j)} \le \delta$$

and increase rapidly to ∞. We can get up to $e^{c\sqrt{n}}$ for some small $c > 0$ that depends on C and δ. This a priori estimate can easily be formally established to show that for any local $h \in \mathcal{H}_{0,-1}$ there is a constant $C(k, h)$ such that

$$D_0(\pi_n u_\lambda) \le C(k, h)n^{-k} \tag{A.5}$$

for all $\lambda > 0$ and $n \ge 1$. Note that the mean zero property implies that $\pi_0(u_\lambda) = 0$.

REFERENCES

[1] DE MASI, A., FERRARI, P. A., GOLDSTEIN, S. and WICK, W. D. (1989). An invariance principle for reversible Markov processes: applications to random motions in random environments. *J. Statist. Phys.* **55** 787–855.

[2] GIACOMIN, G., OLLA, S. and SPOHN, H. (2001). Equilibrium fluctuations for $\nabla\phi$ interface models. *Ann. Probab.* **29** 1138–1172.

[3] GRIGOR'YAN, A. A. (1988). On Liouville theorems for harmonic functions with finite Dirichlet integral. *Math. USSR-Sb.* **60** 485–504.

[4] KESTEN, H. (1982). *Percolation Theory for Mathematicians*. Birkhäuser, Boston.

[5] KIPNIS, C. and LANDIM, C. (1999). *Scaling Limit of Interacting Particle Systems*. Springer, Berlin.

[6] KIPNIS, C. and VARADHAN, S. R. S. (1986). Central limit theorem for additive functionals of reversible Markov processes and applications to simple exclusion. *Comm. Math. Phys.* **106** 1–19.

[7] LANDIM, C. (1998). Decay to equilibrium in L^∞ of asymmetric simple exclusion processes in infinite volume. *Markov Process. Related Fields* **4** 517–534.

[8] LANDIM, C. and YAU, H. T. (1997). Fluctuation–dissipation equation of asymmetric simple exclusion processes. *Probab. Theory Related Fields* **108** 321–356.

[9] LIGGETT, T. (1985). *Interacting Particles Systems*. Springer, Berlin.

[10] OSADA, H. and SAITOH, T. (1995). An invariance principle for non-symmetric Markov processes and reflecting diffusions in random domains. *Probab. Theory Related Fields* **101** 45–63.

[11] SETHURAMAN, S., VARADHAN, S. R. S. and YAU, H. T. (2000). Diffusive limit of a tagged particle in asymmetric exclusion process. *Comm. Pure Appl. Math.* **53** 972–1006.

[12] SOARDI, P. M. (1994). *Potential Theory on Infinite Networks. Lecture Notes in Math.* **1590**. Springer, New York.

[13] VARADHAN, S. R. S. (1995). Self diffusion of a tagged particle in equilibrium for asymmetric mean zero random walks with simple exclusion. *Ann. Inst. H. Poincaré Probab. Statist.* **31** 273–285.

[14] VARADHAN, S. R. S. (1994). Non-linear diffusion limit for a system with nearest neighbor interactions II. In *Asymptotic Problems in Probability Theory: Stochastic Models and Diffusion on Fractals* (K. D. Elworthy and N. Ikeda, eds.). 75–128. Wiley, New York.

C. LANDIM
IMPA
ESTRADA DONA CASTORINA 110
CEP 22460 RIO DE JANEIRO
BRAZIL
AND
CNRS UPRES-A 6085
UNIVERSITÉ DE ROUEN
76128 MONT SAINT AIGNAN
FRANCE
E-MAIL: landim@impa.br

S. OLLA
DÉPARTEMENT DE MATHÉMATIQUES
UNIVERSITÉ DE CERGY-PONTOISE
2 AV. ADOLPHE CHAUVIN
B.P. 222
PONTOISE 95.302 CERGY-PONTOISE CEDEX
FRANCE
AND
CENTRE DE MATHÉMATIQUES APPLIQUÉES
ECOLE POLYTECHNIQUE
91128 PALAISEAU CEDEX
FRANCE
E-MAIL: olla@u-cergy.fr

S. R. S. VARADHAN
COURANT INSTITUTE OF MATHEMATICAL SCIENCES
NEW YORK UNIVERSITY
251 MERCER STREET
NEW YORK, NEW YORK 10012
E-MAIL: varadhan@cims.nyu.edu

Advanced Studies in Pure Mathematics 39, 2004
pp. 1-27

Large Deviations for the Asymmetric Simple Exclusion Process

Srinivasa R.S. Varadhan

Abstract.

We explain the large deviation behavior of the totally asymmetric simple exclusion process in one dimension.

§1. Introduction

So far, large deviations from hydrodynamic scaling have been worked out only for systems under diffusive scaling. Large deviation results are presented here for the Totally Asymmetric Simple Exclusion Process or TASEP in one dimension. This work was carried out by Leif Jensen [2] in his PhD dissertation submitted to New York University in the year 2000 and is available at the website

http://www.math.columbia.edu/˜jensen/thesis.html

We will present here a detailed sketch of the derivation of the upper bound and a rough outline of how the lower bound is established.

§2. Hydrodynamic limit of TASEP

The Model.

We have a particle system on the integers $\mathbf{Z}$ or (in the periodic case) on $\mathbf{Z}_N$, the integers moduo N. The configuration is $\eta = \{\eta_x : x \in \mathbf{Z}\}$ or $\{\eta_x : x \in \mathbf{Z}_N\}$. The evolution of $\eta(t) = \{\eta_x(t)\}$ is governed by the generator

$$(\mathcal{L}f)(\eta) = \sum_x \eta_x(1 - \eta_{x+1})[f(\eta^{x,x+1}) - f(\eta)]$$

Received December 31, 2002.
Revised May 19, 2003.
Partially Supported by NSF grant 0104343.

where

$$\eta_z^{x,y} = \begin{cases} \eta_z & if \quad z \neq x, y \\ \eta_y & if \quad z = x \\ \eta_x & if \quad z = y \end{cases}$$

This corresponds to the process where the particles independently wait for an exponential time and then jump one step to the right if the site is free. Otherwise they wait for another exponential time. All the particles are doing this simultaneously and independently.

The Scaling.

For each N we consider an initial configuration $\eta_{x,N}$, that may or may not be random. We consider these models for $N \to \infty$. Assume that for some deterministic density function $\rho_0(\xi)$, $0 \leq \rho_0(\cdot) \leq 1$, and every test function $J(\cdot)$,

$$\lim_{N\to\infty} \frac{1}{N} \sum J(\frac{x}{N})\eta_{x,N} = \int J(\xi)\rho_0(\xi)d\xi$$

The limit is taken in probability in the random case. The class of test functions are continuous functions with compact support in $\mathbf{R}$, if we started with $\mathbf{Z}$ and the periodic unit interval $\mathbf{S}$, if we started with $\mathbf{Z}_N$.

Time is speeded up by a factor of N, i.e. the process is viewed at time Nt or equivalently the generator is multiplied by a factor of N. This introduces in a natural way a probability measure P_N on the space of trajectories $\{\eta_x(t) : x \in Z_N \text{ or } Z, t \geq 0\}$.

Theorem 2.1. *(The law of large numbers.) For any $t > 0$, there exists a deterministic density function $\rho(t, \cdot)$, on $\mathbf{R}$ or $\mathbf{S}$ as the case may be, such that*

$$\lim_{N\to\infty} \frac{1}{N} \sum J(\frac{x}{N})\ \eta_x(t) = \int J(\xi)\ \rho(t,\xi)d\xi$$

in probability for every suitable test function. The density $\rho(t, x)$ is determined as the unique weak solution of

$$\rho_t(t,\xi) + [\rho(t,\xi)(1 - \rho(t,\xi))]_\xi = 0 \tag{1}$$

with initial condition $\rho(0, \cdot) = \rho_0(\cdot)$, that satisfies the 'entropy condition'.

Remark 2.2. *The entropy condition can be stated in many equivalent forms. For example if $\rho(t, \cdot)$ is a smooth solution, then for any smooth function $h(r)$*

$$[h(\rho(t,\xi))]_t = h'(\rho(t,\xi))\rho_t(t,\xi) = -h'(\rho(t,\xi))(1 - 2\rho(t,\xi))\rho_\xi(t,\xi)$$

or

$$[h(\rho(t,\xi))]_t + [g(\rho(t,\xi))]_\xi = 0 \tag{2}$$

where g and h are related by

$$g'(r) = h'(r)(1-2r) \tag{3}$$

If $\rho(t,\cdot)$ is only a weak solution, then equation (2) may not hold even weakly. A weak solution of equation (1) is said to satisfy the entropy condition if for every convex function h and the corresponding g defined by equation (3),

$$[h(\rho(t,\xi))]_t + [g(\rho(t,\xi))]_\xi \le 0 \tag{4}$$

holds as a distribution on $[0,T]\times\mathbf{R}$ or $[0,T]\times\mathbf{S}$ as the case may be. Then for any initial value, the weak solution satisfying the entropy condition exists and is unique. The density profile of the TASEP converges to this unique solution.

We will not prove it here. For the special case when the sites are $\mathbf{Z}$ and $\eta_{x,N}(0) = 1$ for $x \le 0$ and 0 otherwise was carried out by Rost [4], who proved that in this case the solution $\rho(t,\xi)$ is the rarefaction wave,

$$\rho(t,\xi) = \begin{cases} 1 & if \quad \xi \le -t \\ \frac{t-\xi}{2t} & if \quad -t \le \xi \le t \\ 0 & if \quad \xi \ge t \end{cases}$$

and the density of the TASEP converges to it. Seppäläinen in [5] obtained a representation of the TASEP with arbitrary initial conditions in terms of a family of coupled processes with initial conditions of Rost type and was able to reduce the general case to the Rost case.

If we look at special solutions of the form

$$\rho(t,\xi) = \begin{cases} \rho & if \quad \xi \le 0 \\ 1-\rho & if \quad \xi \ge 0 \end{cases}$$

then this will be an entropic solution only when $\rho \le \frac{1}{2}$. In particular if $\rho = 1$, although the initial profile in the Rost case is a stationary weak solution it is not entropic. On the other hand if we hold the lead particle from jumping, then nothing can move. So with probability e^{-Nt}, the Rost initial profile can remain intact up to time t. This illustrates that non-entropic solutions are relevant for large deviations.

§3. Large Deviations. Some super exponential estimates

The validity of hydrodynamical scaling depends on some basic facts. We will state them in the periodic case. The needed modifications when we have the entire **Z** are obvious. The 'one block estimate' allows one to replace the microscopic flux by its expectations, given the densities over blocks of size $2k+1$. If

$$\mathcal{E}(N,k,t) = \frac{1}{N}\int_0^t \sum_x |e_{N,k,x}(s)|ds$$

where

$$e_{N,k,x}(s) = |\frac{1}{2k+1}\sum_{y:|y-x|\leq k} \eta_y(s)(1-\eta_{y+1}(s)) - \bar{\eta}_x^k(s)(1-\bar{\eta}_x^k(s))|$$

and $\bar{\eta}_x^k = \frac{1}{2k+1}\sum_{y:|y-x|\leq k}\eta_y$, then

$$\lim_{k\to\infty}\lim_{N\to\infty} E^{P_N}\Big[\mathcal{E}(N,k,t)\Big] = 0$$

The expectation is taken here with respect to the measure P_N that corresponds to some initial profile on the periodic lattice Z_N and evolves according to TASEP dynamics in the speeded up time scale. Then the two block estimate allows one to replace $\bar{\eta}_x^k$ with large k by $\bar{\eta}_x^{N\epsilon}$ with a small ϵ. One can exhibit this in many ways. For instance, if we define,

$$\mathcal{D}(N,\epsilon,k,t) = \int_0^t \big[\frac{1}{N}\sum_x [\bar{\eta}_{x,N}^k(s)]^2 - \frac{1}{N}\sum_x [\bar{\eta}_{x,N}^{N\epsilon}(s)]^2\big]ds$$

then, by proving

$$\lim_{\epsilon\to 0}\lim_{k\to\infty}\lim_{N\to\infty} E^{P_N}\Big[\mathcal{D}(N,\epsilon,k,t)\Big] = 0$$

one can establish that any limit of the empirical density is a weak solution of equation (1).

Remark 3.1. *Because of finite propagation speed, basically the effect of any change in a region is only felt over a finite macroscopic domain. This allows us to go back and forth between the periodic and the nonperiodic cases without much effort. If we take the domain large enough then the probability of any effect outside is superexponentially small. So even for large deviations, one can go back and forth.*

Theorem 3.2. *One has the super exponential 'one block' and 'two block estimates'. For any $\delta > 0$,*

$$\limsup_{k\to\infty}\limsup_{N\to\infty}\frac{1}{N}\log P\Big[\mathcal{E}(N,k,t)\geq\delta\Big]=-\infty \tag{5}$$

$$\limsup_{\epsilon\to 0}\limsup_{k\to\infty}\limsup_{N\to\infty}\frac{1}{N}\log P\Big[\mathcal{D}(N,\epsilon,k,t)\geq\delta\Big]=-\infty \tag{6}$$

Sketch of proof: We look at the periodic case. The Dirichlet form

$$D(p)=\sum_{x,\eta}[\sqrt{p(\eta^{x,x+1})}-\sqrt{p(\eta)}]^2$$

can be used in conjunction with the Feynman-Kac formula to provide the first estimate. This is not any different from the symmetric case. The fact that the scaling factor is N and not N^2 does not affect the estimate. It only matters that it is large.

The second estimate on the other hand is a bit tricky. In the symmetric case the proof uses the full strength of the factor N^2, and does not work here. Instead the proof is carried out in several steps. First one proves that there is an exponential error bound, for large deviations from the hydrodynamical limit in the Rost case, by explicit computation. This is not hard and can be done by just following Rost's proof carefully. Then this is extended to arbitrary initial conditions by following through Seppäläinen's proof. One then notices that, by convexity, if $\mathcal{D}(N,\epsilon,k,t)$ does not go to zero, and the one block estimate holds, then the hydrodynamic limit cannot hold. Therefore the two block estimate holds with exponential error probability. Finally a bootstrap argument is used to improve the exponential error probability to a superexponential estimate. The space time region of size $N\times N$ is divided into ℓ^2 grids of size $\frac{N}{\ell}\times\frac{N}{\ell}$. The probability of a significant violation in the two block estimate is $e^{-c\frac{N}{\ell}}$ for one grid. The grids do not influence each other that much. Now the usual Bernoulli large deviation estimate yields a multiplication of the exponent by a factor ℓ^2, that equals the number of grids. If we pick ℓ large we are done.

Corollary 3.3. *Outside the set of weak solutions the probability measure P_N decays superexponentially fast.*

It is then natural to expect that the rate function for large deviations will be a measure of how 'nonentropic' the weak solution is.

§4. Macroscopic and Microscopic Entropies.

A microstate on the configurations on $\mathbf{Z}_N$ is a probability distribution $p_N(\eta)$ on the configurations $\eta \in \{0,1\}^{\mathbf{Z}_N}$. Its entropy (relative to the uniform distribution) is defined as

$$H_N(p_N) = N \log 2 + \sum_{\eta} p_N(\eta) \log[p_N(\eta)]$$

For a macroscopic density profile $\rho(\xi)$, the corresponding entropy function is defined by

$$\mathcal{H}(\rho(\cdot)) = \log 2 + \int_{\mathbf{S}} [\rho(\xi) \log \rho(\xi) + (1 - \rho(\xi)) \log(1 - \rho(\xi))] d\xi$$

If p_N has asymptotic profile ρ, in the sense that

$$\lim_{N \to \infty} \frac{1}{N} \sum J(\frac{x}{N}) \eta_x = \int J(\xi) \rho(\xi) d\xi$$

in probability with respect to p_N, then by Jensen's inequality

$$\liminf_{N \to \infty} \frac{1}{N} H_N(p_N) \geq \mathcal{H}(\rho(\cdot))$$

We need a result of Kosygina [3] that asserts that under certain additional conditions the equality holds, i.e.

$$\lim_{N \to \infty} \frac{1}{N} H_N(p_N) = \mathcal{H}(\rho(\cdot))$$

Two conditions are needed.

- The Dirichlet form is "small"

$$D_N(p_N) = \sum_{x,\eta} [\sqrt{p(\eta^{x,x+1})} - \sqrt{p(\eta)}]^2 = o(N)$$

- The two block estimate holds.

$$\lim_{\epsilon \to 0} \lim_{k \to \infty} \lim_{N \to \infty} E^{p_N} \Big[\mathcal{D}(N, \epsilon, k) \Big] = 0$$

where

$$\mathcal{D}(N, \epsilon, k) = \frac{1}{N} \sum_{x} [\bar{\eta}^{k}_{x,N}]^2 - \frac{1}{N} \sum_{x} [\bar{\eta}^{N\epsilon}_{x,N}]^2$$

The proof uses the fact that the control of Dirichlet form allows us to estimate $\frac{1}{N}H_N(p_N)$ by

$$\log 2 + E^{P_N}\left[\frac{1}{N}\sum_x [\bar{\eta}_x^k \log \bar{\eta}_x^k + (1-\bar{\eta}_x^k)\log(1-\bar{\eta}_x^k)]\right]$$

and the two block estimate allows k to be replaced by $N\epsilon$ and if the law of large numbers holds then we easily pass to $\mathcal{H}(\rho(\cdot))$, providing the upper bound. The lower bound as we mentioned is essentially Jensen's inequality.

With some additional work the following theorem due to Kosygina can be proved.

Theorem 4.1. *Consider the evolution according to TASEP in the periodic case with any initial conditions. Suppose the hydrodynamic limit holds with some profile $\rho(t,\xi)$. Then for any $\delta > 0$*

$$\limsup_{N\to\infty} \sup_{\delta\le s\le t} |\frac{1}{N}H_N(p_N(s)) - \mathcal{H}(\rho(s,\cdot))| = 0$$

Idea of proof: The discussion above will allow us to control it for most times s. But the entropy is monotone and cannot fluctuate wildly.

Remark 4.2. *Actually the theorem Kosygina will continue to hold even if we modify the dynamics by changing the rates, replacing in the speeded up scale N by $N\lambda_{x,x+1}(s,\eta)$, provided the relative entropy of the modified process with respect to the unperturbed process remains bounded by CN. This is because the estimates on the Dirichlet form, usually obtained by differentiating the entropy at time t, with respect to t can still be derived. Because the two block estimates has superexponential error estimates for the unperturbed process, they will continue to hold for the perturbed process which has relative entropy bounded by CN. Since the proof of Theorem 4.1 depends only on estimates on the Dirichlet form and two block estimates, the Theorem will continue to hold even when we perturb.*

Remark 4.3. *If for some p_N with profile ρ the entropy relation*

$$\limsup_{N\to\infty} |\frac{1}{N}H_N(p_N) - \mathcal{H}(\rho(\cdot))| = 0$$

holds, then from the super additivity of the entropy function over disjoint blocks, one has for the marginal $p_{N,B}$ of p_N on any block of size $N(b-a)$

say from $[Na, Nb]$,

$$\limsup_{N\to\infty} |\frac{1}{N} H_N(p_{N,B}) - \int_a^b h(\rho(\xi))d\xi| = 0$$

§5. Large Deviation. The rate function

The basic space on which we will carry out the large deviation is the space $\Omega = C[[0,T],\mathcal{M}]$ of continuous maps $\rho(t,d\xi)$ of $[0,T]$ into the space $\mathcal{M}$ of nonnegative measures on S. Although under P_N, $\rho(t,d\xi)$ consists of atoms with mass $\frac{1}{N}$, because of exclusion any conceivable limit will be supported on $\rho(t,d\xi)$ that have densities $\rho(t,\xi)d\xi$ that satisfy $0 \le \rho(t,\xi) \le 1$ for all $(t,\xi) \in [0,T] \times \mathbf{S}$ and are weakly continuous as mappings of $[0,T] \to \mathcal{M}$.

The rate function $\mathcal{I}(\rho(\cdot,\cdot))$ is defined as $+\infty$ if $\rho(\cdot,\cdot)$ is not a weak solution of

$$\rho_t + [\rho(1-\rho)]_\xi = 0$$

If it is a weak solution, then

$$\begin{aligned}
\mathcal{I}(\rho(\cdot,\cdot)) &= \int_{0+0}^{T-0}\int_{\mathbf{S}} \big[[h(\rho(\cdot,\cdot))]_t + [g(\rho(\cdot,\cdot))]_\xi\big]^+ dtd\xi \\
&= \sup_{J\in\mathcal{J}} \int_0^T\int_{\mathbf{S}} J(t,\xi)\big[[h(\rho(\cdot,\cdot))]_t + [g(\rho(\cdot,\cdot))]_\xi\big] dtd\xi \\
&= -\inf_{J\in\mathcal{J}} \int_0^T\int_{\mathbf{S}} \big[J_t(t,\xi)h(\rho(\cdot,\cdot)) + J_\xi(t,\xi)g(\rho(\cdot,\cdot))\big] dtd\xi
\end{aligned}$$

Here $h(r) = r\log r + (1-r)\log(1-r)$ and $g(r)$ as defined by equation (3) is

$$g(r) = r(1-r)\log\frac{r}{(1-r)} - r$$

and

$$\mathcal{J} = \{J(\cdot,\cdot) : 0 \le J(\cdot,\cdot) \le 1, J(0,\cdot) \equiv J(T,\cdot) \equiv 0\}$$

It is interesting to note that the set of weak solutions of nonlinear equations is in general not weakly closed. However a result on compensated compactness, that can be found in Tartar [6], tells us that the set C_ℓ of weak solutions for which $\mathcal{I}(\rho(\cdot,\cdot)) \le \ell$ is in fact compact in the strong topology, guaranteeing that the rate function is indeed lower semi continuous. It is easy to check uniform modulus of continuity in time in the weak topology. So the rate function in fact does have compact level sets.

§6. Upper Bounds

For upper bounds we will use the formulation of Ellis and Dupuis [1]. Suppose $\eta_{x,N}$ is a deterministic initial condition with a profile $\rho_0(\xi)$.

Theorem 6.1. *Suppose P_N is the measure on the configuration space $\{\eta_x(t)\}$ induced by the TASEP and Q_N is such that $Q_N << P_N$ and the measure $\widehat{Q}_N$ induced by Q_N on Ω converges to the degenerate distribution at $\rho(\cdot,\cdot) \in \Omega$. Then*

$$\liminf_{N\to\infty} \frac{1}{N} H(Q_N|P_N) \geq \mathcal{I}(\rho(\cdot,\cdot))$$

Remark 6.2. *This is easily seen to be equivalent to the standard upper bound LDP estimate.*

The proof is broken up into several lemmas.

Lemma 6.3. *Without loss of generality we can assume that Q_N is Markov with rates $N\widehat{\lambda}(t,x,\eta)$.*

Proof. Consider the probability distribution $q_N(t,\eta)$ of $\eta(t)$ at time t under Q_N. We have

$$\frac{1}{N}\sum_x J(\frac{x}{N})\eta_x \to \int J(\xi)\rho(t,\xi)d\xi$$

in probability with respect to $q_N(t,\eta)$. The process Q_N has some rates $N\lambda_N(t,x,\omega)$ of particles jumping from x to $x+1$, that may depend on the past history upto time t. This comes from general martingale theory. One can write the formal generator

$$(\mathcal{L}_{t,\omega}f)(\eta) = N\sum_x \lambda(t,x,\omega)\eta_x(1-\eta_{x+1})[f(\eta^{x,x+1}) - f(\eta)]$$

and with respect to Q_N,

$$f(\eta(t)) - f(\eta(0)) - \int_0^t (\mathcal{L}_{s,\omega}f)(\eta(s))ds$$

will be martingales. By Girsanov formula one can calculate on $[0,T]$,

$$\frac{1}{N}H(Q_N|P_N) = E^{Q_N}\left[\int_0^T \big[\sum_x \eta_x(t)(1-\eta_{x+1}(t))\theta(\lambda(t,x,\omega))\big]dt\right]$$

where $\theta(\lambda) = \lambda\log\lambda - \lambda + 1$. If we replace $\lambda(t,x,\omega)$ by its conditional expectation

$$\widehat{\lambda}(t,x,\eta) = E^{Q_N}\left[\lambda(t,x,\omega)|\eta(t)\right]$$

we see that

$$E^{Q_N}[f(\eta(t)) - f(\eta(0))] = E^{Q_N}\left[\int_0^T (\widehat{\mathcal{L}}_t f)(\eta(t))dt\right]$$

with

$$(\widehat{\mathcal{L}}_t f)(\eta) = N\sum_x \widehat{\lambda}(t,x,\eta)\eta_x(1-\eta_{x+1})[f(\eta^{x,x+1}) - f(\eta)]$$

In other words $q_N(t,\eta)$ is the solution of the forward equation corresponding to $\widehat{\mathcal{L}}$. On the other hand, since $\theta(\lambda)$ is a convex function of λ, by Jensen's inequality,

$$\begin{aligned}E^{Q_N}&\left[\eta_x(t)(1-\eta_{x+1}(t))\theta(\lambda(t,x,\omega))\right]\\ &\geq E^{Q_N}\left[\eta_x(t)(1-\eta_{x+1}(t))\theta(\widehat{\lambda}(t,x,\omega))\right]\end{aligned}$$

The Markov process with $\widehat{\mathcal{L}}_t$ as generator has the same marginals at time t as Q_N and will work as well. In other words for our theorem we can assume with out loss of generality that Q_N is Markov with rates $N\widehat{\lambda}(x,t,\eta)$. Q.E.D.

Consider the joint probability distribution $q_{N,x,k}(t,\eta)$ at the $2k+1$ sites $[x-k,\ldots,x+k]$ of $\{\eta_y\}$ under $q_N(t,\eta)$. We think of it as function of η that depends on the variables $\{\eta_y : |y-x| \leq k\}$.

We let

$$H(N,x,k,t) = \frac{1}{N}\sum_{\eta\in[0,1]^{2k+1}} q_{N,x,k}(t,\eta)\log q_{N,x,k}(t,\eta)$$

and compute

$$\begin{aligned}H_t(N,x,k,t) &= \frac{1}{N}\sum_{\eta\in[0,1]^{2k+1}} \dot{q}_{N,x,k}(t,\eta)[1+\log q_{N,x,k}(t,\eta)]\\ &= \frac{1}{N}\sum_{\eta\in[0,1]^{2k+1}} \dot{q}_{N,x,k}(t,\eta)\log q_{N,x,k}(t,\eta)\\ &= \frac{1}{N}\sum_{\eta\in[0,1]^{N}} \dot{q}_{N}(t,\eta)\log q_{N,x,k}(t,\eta)\end{aligned}$$

Using the forward equation $\dot{q}_N(t,\eta) = N(\mathcal{L}_t^* q_N)(t,\eta)$, we get

$$\begin{aligned} & H_t(N,x,k,t) \\ &= \sum_{\eta\in[0,1]^N} q_N(t,\eta)\mathcal{L}_t[\log q_{N,x,k}(t,\eta)] \\ &= \sum_{\substack{\eta\in[0,1]^N \\ x-k-1\le y\le x+k}} q_N(t,\eta)\widehat{\lambda}(t,y,\eta)e_{y,y+1}(\eta)\log\frac{q_{N,x,k}(t,\eta^{y,y+1})}{q_{N,x,k}(t,\eta)} \end{aligned}$$

where $e_{y,y+1}(\eta) = \eta_y(1-\eta_{y+1})$. We use the inequality

$$\lambda\log y \le \lambda\log\lambda - \lambda + 1 + e^a y - 1 - a\lambda$$

with the choice of $a = a_{N,x,y,k}(\eta)$ to be made later.

$$\begin{aligned} & H_t(N,x,k,t) \le \\ & \sum_{\substack{\eta\in[0,1]^N \\ x-k-1\le y\le x+k}} q_N(t,\eta)e_{y,y+1}(\eta)[\widehat{\lambda}(t,y,\eta)\log\widehat{\lambda}(t,y,\eta) - \widehat{\lambda}(t,y,\eta)+1] \\ & + \sum_{\substack{\eta\in[0,1]^N \\ x-k-1\le y\le x+k}} q_N(t,\eta)e_{y,y+1}(\eta)\frac{e^{a_{N,x,y,k}}q_{N,x,k}(t,\eta^{y,y+1}) - q_{N,x,k}(t,\eta)}{q_{N,x,k}(t,\eta)} \\ & - \sum_{\substack{\eta\in[0,1]^N \\ x-k-1\le y\le x+k}} q_N(t,\eta)e_{y,y+1}(\eta)\widehat{\lambda}(t,y,\eta)a_{N,x,y,k} \end{aligned}$$

We rewrite this as

$$\begin{aligned} & \sum_{\substack{\eta\in[0,1]^N \\ x-k-1\le y\le x+k}} q_N(t,\eta)e_{y,y+1}(\eta)[\widehat{\lambda}(t,y,\eta)\log\widehat{\lambda}(t,y,\eta) - \widehat{\lambda}(t,y,\eta)+1] \\ & \qquad \ge H_t(N,x,k,t) - A_1(N,x,k,t) + A_2(N,x,k,t) \end{aligned}$$

where

$$\begin{aligned} & A_1(N,x,k,t) = \\ & \sum_{\substack{\eta\in[0,1]^N \\ x-k-1\le y\le x+k}} q_N(t,\eta)e_{y,y+1}(\eta)\frac{e^{a_{N,x,y,k}}q_{N,x,k}(t,\eta^{y,y+1}) - q_{N,x,k}(t,\eta)}{q_{N,x,k}(t,\eta)} \end{aligned}$$

and

$$A_2(N,x,k,t) = \sum_{\substack{\eta\in[0,1]^N \\ x-k-1\le y\le x+k}} q_N(t,\eta)e_{y,y+1}(\eta)\widehat{\lambda}(t,y,\eta)a_{N,x,y,k}$$

We now multiply by $J(t,\frac{x}{N})$, where $J\in\mathcal{J}$, sum over x, integrate with respect to t from 0 to T and finally multiply by $\frac{1}{(2k+2)N}$,

$$\begin{aligned}
&\frac{1}{N}H(Q_N|P_N)\\
&\ge \int_0^T \frac{1}{(2k+2)N}\sum_{x\in\mathbf{Z}_N} J(t,\frac{x}{N})d\Big[\sum_{\eta\in[0,1]^{2k+1}} q_{N,x,k}(t,\eta)\log q_{N,x,k}(t,\eta)\Big]\\
&\quad - E^{Q_N}\Bigg[\int_0^T\Bigg[\frac{1}{2k+2}\sum_{\substack{x,y\\ x-k-1\le y\le x+k}} J(t,\frac{x}{N})e_{y,y+1}(\eta)\times\\
&\qquad\qquad \frac{e^{a_{N,x,y,k}}q_{N,x,k}(t,\eta^{y,y+1})-q_{N,x,k}(t,\eta)}{q_{N,x,k}(t,\eta)}\Bigg]dt\Bigg]\\
&\quad + E^{Q_N}\Bigg[\int_0^T\Bigg[\frac{1}{2k+2}\sum_{\substack{x,y\\ x-k-1\le y\le x+k}} J(t,\frac{x}{N})e_{y,y+1}(\eta)\times\\
&\qquad\qquad \widehat{\lambda}(t,y,\eta)a_{N,x,y,k}\Bigg]dt\Bigg]\\
&= T_1(N,J(\cdot,\cdot),k) - T_2(N,J(\cdot,\cdot),k) + T_3(N,J(\cdot,\cdot),k)
\end{aligned}$$

Now we have to analyse the terms on the right. Let us look at

$$T_1(N,J(\cdot,\cdot),k) = \int_0^T \frac{1}{(2k+2)N}\sum_{x\in\mathbf{Z}_N} J(t,\frac{x}{N})d\Big[\sum_{\eta\in[0,1]^{2k+1}} q_{N,x,k}(t,\eta)\log q_{N,x,k}(t,\eta)\Big]$$

Integrating by parts,

$$T_1(N,J(\cdot,\cdot),k) = -\int_0^T \sum_{x\in\mathbf{Z}_N} J_t(t,\frac{x}{N})\Big[\frac{1}{(2k+2)N}\sum_{\eta\in[0,1]^{2k+1}} q_{N,x,k}(t,\eta)\log q_{N,x,k}(t,\eta)\Big]dt$$

We pick $k=N\epsilon$ and let $\epsilon\to 0$ at the end.

Lemma 6.4. *If for $t \in [0,T]$, $q_N(t,\eta)$ leads to the profile $\rho(t,\cdot)$, then*

$$\lim_{\epsilon\to 0}\lim_{N\to\infty}$$
$$\int_0^T \sum_{x\in \mathbf{Z}_N} J_t(t,\frac{x}{N})\left[\frac{1}{2N^2\epsilon}\sum_{\eta\in[0,1]^{2N\epsilon}} q_{N,x,N\epsilon}(t,\eta)\log q_{N,x,N\epsilon}(t,\eta)\right]dt$$
$$= \int_0^T\int_{\mathbf{S}} J_t(t,\xi)h(\rho(t,\xi))dtd\xi$$

Proof. Let us consider the quantity

$$H_N(t,x,\epsilon) = \log 2 + \frac{1}{2N\epsilon}\sum_{\eta\in[0,1]^{2N\epsilon+1}} q_{N,x,N\epsilon}(t,\eta)\log q_{N,x,N\epsilon}(t,\eta)$$

and the measure

$$\mu_N(t,\epsilon) = \frac{1}{N}\sum_x H_N(t,x,\epsilon)\delta_{\frac{x}{N}}$$

We need to prove the weak convergence of

$$\lim_{\epsilon\to 0}\lim_{N\to\infty}\mu_N(t,\epsilon)dt = h(\rho(t,\xi))dtd\xi$$

Since we are looking at relative entropy with respect to a product measure, i.e. uniform distribution on $[0,1]^{\mathbf{Z}_N}$, it is easy to see that

$$\liminf_{\epsilon\to 0}\liminf_{N\to\infty}\mu_N(t,\epsilon)dt \ge h(\rho(t,\xi))dtd\xi$$

in view of the remark at the end of the last section. On the other hand the total mass of $\mu_N(t,\epsilon)$ is dominated by the total entropy

$$\log 2 + \frac{1}{N}\sum_{\eta\in[0,1]^{\mathbf{Z}_N}} q_N(t,\eta)\log q_N(t,\eta)$$

and we are done. Q.E.D.

Now we try to control $T_2(N,J(\cdot,\cdot),N\epsilon) - T_2(N,J(\cdot,\cdot),N\epsilon)$ which is more difficult. The interior terms with $x-k-1<y<x+k$ are easy. We

choose $a_{N,x,y,k} = 0$.

$$E^{Q_N}\left[\frac{1}{2N\epsilon}\sum_{x-k-1<y<x+k} e_{y,y+1}(\eta)\frac{q_{N,x,N\epsilon}(t,\eta^{y,y+1}) - q_{N,x,N\epsilon}(t,\eta)}{q_{N,x,N\epsilon}(t,\eta)}\right]$$

$$= \sum_{\eta\in[0,1]^{\mathbf{Z}_{2N\epsilon}}} \frac{1}{2N\epsilon}\sum_{x-k-1<y<x+k} e_{y,y+1}(\eta)\times$$

$$[q_{N,x,N\epsilon}(t,\eta^{y,y+1}) - q_{N,x,N\epsilon}(t,\eta)]$$

$$= \sum_{\eta\in[0,1]^{\mathbf{Z}_{2N\epsilon}}} \frac{1}{2N\epsilon}\sum_{x-k-1<y<x+k} [\eta_{y+1}-\eta_y]q_{N,x,N\epsilon}(t,\eta)$$

If we carry out a summation by parts in x and integration over t, this leads in the limit to

$$-\int_0^T\int_{\mathbf{S}} J_\xi(t,\xi)\rho(t,\xi)dtd\xi$$

We look next at the boundary terms. Note that $k=[N\epsilon]$. The boundary terms equal $B = B_1 + B_2 + B_3 + B_4$

$$B_1 = -\,E^{Q_N}\left[\int_0^T\left[\frac{1}{2N\epsilon}\sum_x J(t,\frac{x}{N})\eta_{x-k-1}(1-\eta_{x-k})\times\right.\right.$$

$$\left.\left.\frac{e^{a_{N,x,-,k}}q_{N,x,k}(t,\eta^{x-k-1,x-k}) - q_{N,x,k}(t,\eta)}{q_{N,x,k}(t,\eta)}\right]dt\right]$$

$$B_2 = -\,E^{Q_N}\left[\int_0^T\left[\frac{1}{2N\epsilon}\sum_x J(t,\frac{x}{N})\eta_{x+k}(1-\eta_{x+k+1})\times\right.\right.$$

$$\left.\left.\frac{e^{a_{N,x,+,k}}q_{N,x,k}(t,\eta^{x+k,x+k+1}) - q_{N,x,k}(t,\eta)}{q_{N,x,k}(t,\eta)}\right]dt\right]$$

$$B_3 = +\,E^{Q_N}\left[\int_0^T\left[\frac{1}{2N\epsilon}\sum_x J(t,\frac{x}{N})\eta_{x-k-1}(1-\eta_{x-k})\times\right.\right.$$

$$\left.\left.\widehat{\lambda}(t,x-k-1,\eta)a_{N,x,-,k}\right]dt\right]$$

$$B_4 = + E^{Q_N}\left[\int_0^T \left[\frac{1}{2N\epsilon}\sum_x J(t, \frac{x}{N})\eta_{x+k}(1-\eta_{x+k+1}) \times \right.\right.$$
$$\left.\left.\widehat{\lambda}(t, x+k, \eta)a_{N,x,+,k}\right]dt\right]$$

We would like to make the choice of $a_{N,x,-,k} = -u(t, \frac{x-k-1}{N})$ and $a_{N,x,+,k} = u(t, \frac{x+k}{N})$ for some smooth u. We can combine B_3 and B_4 and write

$$\begin{aligned}
&B_3 + B_4 \\
&= E^{Q_N}\left[\int_0^T \left[\frac{1}{2N\epsilon}\sum_x [J(t, \frac{x+k+1}{N}) - J(t, \frac{x-k}{N})] \times \right.\right. \\
&\qquad \left.\left.\eta_x(1-\eta_{x+1})\widehat{\lambda}(t,x,\eta)u(t,\frac{x}{N})\right]dt\right] \\
&= E^{Q_N}\left[\int_0^T \left[\frac{1}{2N\epsilon}\sum_x [J(t, \frac{x+k+1}{N}) - J(t, \frac{x-k}{N})] \times \right.\right. \\
&\qquad \left.\left.\eta_x(1-\eta_{x+1})u(t,\frac{x}{N})\right]dt\right] \\
&\quad + E^{Q_N}\left[\int_0^T \left[\frac{1}{2N\epsilon}\sum_x [J(t, \frac{x+k+1}{N}) - J(t, \frac{x-k}{N})] \times \right.\right. \\
&\qquad \left.\left.[\widehat{\lambda}(t,x,\eta)-1]u(t,\frac{x}{N})\right]dt\right] \\
&\simeq \frac{1}{2\epsilon}\int_0^T\int_{\mathbf{S}} [J(t,\xi+\epsilon) - J(t,x-\epsilon)]\rho(t,\xi)(1-\rho(t,\xi))u(t,\xi)dtd\xi \\
&\quad + Error
\end{aligned}$$

The error term is dominated by

$$CE^{Q_N}\left[\int_0^T \left[\frac{1}{N}\sum_x |\widehat{\lambda}(t,x,\eta(t)) - 1|\right]dt\right]$$

For any $\theta > 0$, there is a constant C_θ such that

$$|\lambda - 1| \le \theta + C_\theta[\lambda\log\lambda - \lambda + 1]$$

Therefore

$$Error \le C\theta + \frac{C_\theta}{N^2}H(Q_N|P_N)$$

We will get an estimate on B_1. The term B_2 is similar. B_1 is estimated by

$$\frac{1}{2N\epsilon}E^{Q_N}\left[\int_0^T \sum_x \eta_{x-1}(1-\eta_x)\left|e^{-u(t,\frac{x}{N})}\frac{q_{N,x,x+2k}(t,\eta^{x-1,x})}{q_{N,x,x+2k}(t,\eta)}-1\right|dt\right]$$

The quantity

$$R_N = \eta_{x-1}(1-\eta_x)\frac{q_{N,x,x+2k}(t,\eta^{x-1,x})}{q_{N,x,x+2k}(t,\eta)}$$

has to be looked at carefully. Take $x=0$. If we denote $q_{N,x,x+2k}(t,\eta)$ by $f_N(\eta_0,\eta_1,\cdots,\eta_{2k})$ then

$$\begin{aligned} R_N &= \eta_1(1-\eta_0)\frac{f_N(1,\eta_1,\cdots,\eta_{2k})}{f_N(0,\eta_1,\cdots,\eta_{2k})} = \eta_1(1-\eta_0)\frac{p_N(1|\eta_1,\cdots,\eta_{2k})}{p_N(0|\eta_1,\cdots,\eta_{2k})} \\ &\simeq \eta_1(1-\eta_0)\frac{\rho(0)}{1-\rho(0)} \end{aligned}$$

Therefore it follows that

$$\limsup_{N\to\infty} B_1 \le \frac{1}{2\epsilon}\int_0^T\int_{\mathbf{S}} \rho(t,\xi)(1-\rho(t,\xi))\left|\frac{e^{-u(t,\xi)}\rho(t,\xi)}{1-\rho(t,\xi)}-1\right|dtd\xi$$

and similarly

$$\limsup_{N\to\infty} B_2 \le \frac{1}{2\epsilon}\int_0^T\int_{\mathbf{S}} \rho(t,\xi)(1-\rho(t,\xi))\left|\frac{e^{u(t,\xi)}(1-\rho(t,\xi))}{\rho(t,\xi)}-1\right|dtd\xi$$

If we let u approach $\log\frac{\rho(t,\xi)}{1-\rho(t,\xi)}$ both B_1 and B_2 tend to 0 for any positive ϵ. Finally we let $\epsilon\to 0$.

$$\lim_{\epsilon\to 0}\lim_{N\to\infty}[B_3+B_4] = \int_0^T\int_{\mathbf{S}} J_\xi(t,\xi)\rho(t,\xi)(1-\rho(t,\xi))\log\frac{\rho(t,\xi)}{1-\rho(t,\xi)}dtd\xi$$

This proves

$$\begin{aligned} &\liminf_{N\to\infty}\frac{1}{N}H(Q_N|P_N) \ge \\ &\quad -\int_0^T\int_{\mathbf{S}} J_t(t,\xi)h(\rho(t,\xi))dtd\xi + \int_0^T\int_{\mathbf{S}} J_\xi(t,\xi)\rho(t,\xi)dtd\xi \\ &\quad -\int_0^T\int_{\mathbf{S}} J_\xi(t,\xi)\rho(t,\xi)(1-\rho(t,\xi))\log\frac{\rho(t,\xi)}{1-\rho(t,\xi)}dtd\xi \\ &\quad = -\int_0^T\int_{\mathbf{S}}[J_t(t,\xi)h(\rho(t,\xi)) + J_\xi(t,\xi)g(\rho(t,\xi))]dtd\xi \end{aligned}$$

Since J is arbitrary, we are done.

§7. Lower Bounds

The situation with the lower bounds is not totally satisfactory. Ideally one should construct an explicit perturbation of the rates that produces a particular profile, the entropy cost of such a perturbation being approximately equal to the rate function for such a large deviation. This one is not able to do at this time. The best we can do is to prove the existence of such perturbations and construct them implicitly. Even this we can do only to produce a single non-entropic shock traveling at a constant speed. By patching together, one can possibly handle a finite number of shocks of varying speeds, even crossing each other forming caustics. However one does not see at the moment how to produce a 'general' non-entropic weak solution, partly because one does not know what it is. Ideally there would be an approximation theorem allowing us to pass from a solution with a finite number of shocks to a general weak solution with a finite large deviation rate.

We will sketch the proof for the simple case of a stationary non-entropic shock at 0 starting from a special initial configuration. Suppose we are given on $\mathbf{Z}$ an initial configuration of particles where every site $x \leq 0$ is filled and every site $x > 0$ is empty. We wish to perturb the standard speeded up TASEP dynamics with new rates $N\lambda_N(t, x, \eta)$, such that for the modified process Q_N, for every test function J with compact support and every $t \in [0, T]$, we have in probability,

$$\lim_{N\to\infty} \sum_x J(\frac{x}{N})\eta_x(t) = \int_{-\infty}^{\infty} J(\xi)\rho(t,\xi)d\xi$$

where $\rho(t, \xi)$ is the following special weak solution.

$$\rho(t,\xi) = \begin{cases} 1 & if \quad x \leq -t \\ \frac{t-x}{2t} & if \quad -t \leq x \leq -t(2\rho-1) \\ \rho & if \quad -t(2\rho-1) \leq x \leq 0 \\ 1-\rho & if \quad 0 \leq x \leq t(2\rho-1) \\ \frac{t-x}{2t} & if \quad t(2\rho-1) \leq x \leq t \\ 0 & if \quad x \geq t \end{cases} \tag{7}$$

Here $\rho > \frac{1}{2}$ and there is a non-entropic shock at 0 where the density jumps from a higher value of $\rho > \frac{1}{2}$ to the lower value of $1-\rho < \frac{1}{2}$. The rate function for this profile in the interval $[0, T]$ is proportional to T, i.e. equals $c(\rho)T$, where

$$c(\rho) = 2\rho - 1 - 2\rho(1-\rho)\log\frac{\rho}{1-\rho} \tag{8}$$

The problem is to find rates $N\lambda_N(x,\eta)$ such that the process with these new rates has a law of large numbers with the profile $\rho(t,\xi)$ given by (7) and achieve this with an entropy cost that is roughly $c(\rho)N$. We know from the upper bound that we cannot do better. Since we want to slow down particles at or near 0, ideally cutting down the rate at 0 should do it. If we slow down the rate at 0 to $N\lambda$ with some fixed $\lambda < 1$, holding all other rates at N, we will produce a profile of the type we want with some $\rho = \rho(\lambda)$ that is hard to determine, except in the trivial case of $\lambda = 0, \rho = 1, c(0) = 1$. The cost is surely not going to be optimal. We can however lower the rate at several points around 0, depending on the current configuration of particles. The new rates $N\lambda_N(x,\eta)$ will do the trick. We will implicitly construct them. We will then have to see how this will work for any initial condition. After that we need to modify the construction for shocks moving with constant velocity. Then patch things together for one or more shocks with non constant jumps and non constant velocities that may cross each other.

The idea for a single shock is simple enough. A non-entropic shock is entropic if time is reversed. Let us begin with a generator of a TASEP with jumps to the left rather than to the right. The generator is given by

$$(\widehat{\mathcal{L}}f)(\eta) = N\sum_x \eta_{x+1}(1-\eta_x)[f(\eta^{x,x+1}) - f(\eta)]$$

We start with some initial configuration at $t = 0$, that produces the density profile of $\rho(T,\xi)$ specified by equation (7). The hydrodynamical scaling limit will be an entropic solution of

$$\widehat{\rho}_t - [\widehat{\rho}(1-\widehat{\rho})]_x = 0$$

with $\widehat{\rho}_T(0,\xi) = \rho(T,\xi)$. This is seen to be the time reversal of the profile given in (7).

$$\widehat{\rho}_T(t,\xi) = \rho(T-t,\xi)$$

for $0 \le t \le T$ and $\xi \in R$. If we now take the process Q_N corresponding to $\widehat{\mathcal{L}}$ and reverse time to get trajectories $\eta(T-t)$, the new process will have some generator $\mathcal{L}_{N,T,t}$ that is time inhomogeneous and nearly impossible to compute. However it does have the advantage that it has a hydrodynamical limit with a profile that is the time reversed version of $\widehat{\rho}_T(t,\xi)$ which is of course $\rho(t,\xi)$. The entropy will match, because while the forward motion is losing entropy at the shock, the time reversed motion will put it back and this is done by the new rates for the reversed process. If we do not waste entropy at microscopic scale, then the book keeping at micro and macro levels match and will give us the lower bound for large deviations. However the rates for $\widehat{\mathcal{L}}_{N,T,t}$ are too messy and one has to make it independent of T, N and t, and localize it, so that it is transportable and can be used as a module that we can use at any place and time to slow the flow, which is all that any non-entropic solution is expected to do. We start with a fairly general simple calculation.

Let P be a time homogeneous Markov process with trajectories $x(t)$ in a finite time interval $[0,T]$, on a finite state space with generator

$$(Af)(x) = \sum_y c(x,y)f(y).$$

Let $\pi(t,x)$ be the marginal distributions in the time interval $[0,T]$. We denote by $C(x) = -c(x,x) = \sum_{y\neq x} c(x,y)$. Let $\widehat{P}_T$ be the process that corresponds to the time reversed trajcetories $y(t) = x(T-t)$. Although $\widehat{P}_T$ is a Markov process, it is in general time inhomogeneous and will have a generator that depends on the marginals $\pi(\cdot,\cdot)$. We denote its time dependent generator by

$$(\widehat{\mathcal{A}}_{T,t}f)(x) = \sum_{y\neq x} \widehat{c}_T(t,x,y)f(y)$$

and

$$\widehat{C}_T(x) = -\widehat{c}_T(t,x,x) = \sum_y \widehat{c}_T(t,x,y)$$

We can also reverse the generator and define $\widehat{\mathcal{A}}$ as

$$(\widehat{\mathcal{A}}f)(x) = \sum_y \widehat{c}(x,y)f(y)$$

with $\widehat{c}(x,y) = c(y,x)$ for $x \neq y$ and

$$\widehat{C}(x) = -\widehat{c}(x,x) = \sum_{y\neq x} \widehat{c}(x,y) = \sum_{y\neq x} c(y,x)$$

We denote by $\widehat{Q}_T$, the process with generator $\widehat{\mathcal{A}}$ with initial distribution $\pi(T,\cdot)$.

Theorem 7.1. *We have the following simple formula connecting the function*

$$H(t) = \sum_x \pi(t,x) \log \pi(t,x)$$

and the relative entropy $H(\widehat{P}_T, \widehat{Q}_T)$.

$$(9) \quad H(\widehat{P}_T|\widehat{Q}_T) = H(0) - H(T) + E^{\widehat{Q}_T}\left[\int_0^T [\widehat{C}(x(t)) - C(x(t))]dt\right]$$

Proof. The probabilities $\pi(t,x)$ satisfy the forward equation

$$\frac{d\pi(t,y)}{dt} = \sum_{x \neq y} c(x,y)\pi(t,x) - C(y)\pi(t,y)$$

The time reversed process $\widehat{P}_T$ defined by $y(t) = x(T-t)$ will have marginals $\pi(T-t,y)$ and some generator

$$(\widehat{\mathcal{A}}_{T,t} f)(x) = \sum_y \widehat{c}_T(t,x,y) f(y)$$

Of course

$$\begin{aligned}\frac{d\pi(T-t,y)}{dt} &= -\sum_{x \neq y} c(x,y)\pi(T-t,x) + C(y)\pi(t,y) \\ &= \sum_{x \neq y} \widehat{c}_T(t,x,y)\pi(T-t,x) - \widehat{C}_T(t,y)\pi(T-t,y)\end{aligned}$$

Actually it is not hard to see that for $x \neq y$

$$\pi(T-t,x)\widehat{c}_T(t,x,y) = \widehat{c}(x,y)\pi(T-t,y)$$

and

$$\widehat{C}_T(t,x) = \frac{1}{\pi(T-t,x)} \sum_{y \neq x} \widehat{c}(x,y)\pi(T-t,y)$$

Our goal is to compute the relative entropy

$$\begin{aligned}H(\widehat{P}_T|\widehat{Q}_T) = \int_0^T \sum_x \pi(T-t,x) \sum_{y: y \neq x} [c_T(t,x,y) \log \frac{c_T(t,x,y)}{\widehat{c}(x,y)} \\ - c_T(t,x,y) + \widehat{c}(x,y)]dt\end{aligned}$$

$$
\begin{aligned}
&H(\widehat{P}_T|\widehat{Q}_T)\\
&= \int_0^T \Big[\sum_x \pi(T-t,x)\Big[\Big[\sum_{y:y\neq x} \frac{c(y,x)\pi(T-t,y)}{\pi(T-t,x)}\log\frac{\pi(T-t,y)}{\pi(t-t,x)}\Big]\\
&\qquad\qquad - C_T(t,x) + \widehat{C}(x)\Big]\Big]dt\\
&= \int_0^T \sum_{x\neq y} c(y,x)\Big[\pi(T-t,y)\log\frac{\pi(T-t,y)}{\pi(T-t,x)}\\
&\qquad\qquad - \pi(T-t,y) + \pi(T-t,x)\Big]dt\\
&= \int_0^T \sum_{x\neq y} c(y,x)\Big[\pi(t,y)\log\frac{\pi(t,y)}{\pi(t,x)} - \pi(t,y) + \pi(t,x)\Big]dt
\end{aligned}
$$

We begin by differentiating $H(t) = \sum_y \pi(t,y)\log\pi(t,y)$.

$$
\begin{aligned}
H'(t) &= \frac{d}{dt}\sum_y \pi(t,y)\log\pi(t,y) = \sum_y \pi(t,y)(\mathcal{A}\log\pi(t,\cdot))(y)\\
&= \sum_y \pi(t,y)\Big[[\sum_{x\neq y} c(y,x)\log\pi(t,x)] - C(y)\log\pi(t,y)\Big]\\
&= \sum_{x\neq y} c(y,x)[\pi(t,y)\log\pi(t,x) - \pi(t,y)\log\pi(t,y)]\\
&= -\sum_{x\neq y} c(y,x)\pi(t,y)\log\frac{\pi(t,y)}{\pi(t,x)}
\end{aligned}
$$

This proves (9). Q.E.D.

Let us start the backward TASEP $\widehat{\mathcal{L}}$, with an initial distribution μ_N concentrated on the finite set

$$
\Omega_{N,L} = \Big\{\eta : \eta_x = 1 \text{ for } x < -NL \quad\text{and}\quad \eta_x = 0 \text{ for } x \geq NL\Big\}
$$

for some $L \geq T$. Our initial distribution μ_N will be a Bernoulli with $\mu_N[\eta_x = 1] = \rho(T, \frac{x}{N})$ given in equation (7). Assume $L \geq T$. Then the

TASEP will have profile $\rho(T-t,\xi)$ and at time $t=T$ end up at $\rho(0,\xi)$. The time reversed process will now be a perturbation of the TASEP going in the right direction with the profile we need. (7.1). We note that

$$\begin{aligned}\widehat{C}(\eta)-C(\eta)&=N\sum_x\eta_x(1-\eta_{x+1})-N\sum_x\eta_{x+1}(1-\eta_x)\\&=N\sum_x[\eta_x-\eta_{x+1}]\equiv N.\end{aligned}$$

Moreover $H(T)=0$ and

$$\begin{aligned}\frac{1}{NT}H(0)&\simeq\frac{1}{T}\int_{-T}^{T}[\rho(T,\xi)\log\rho(T,\xi)+(1-\rho(T,\xi))\log(1-\rho(T,\xi))]d\xi\\&=2(2\rho-1)[\rho\log\rho+(1-\rho)\log(1-\rho)]\\&\quad+2\int_{2\rho-1}^{1}[\frac{1-\xi}{2}\log\frac{1-\xi}{2}+\frac{1+\xi}{2}\log\frac{1+\xi}{2}]d\xi\\&=2\rho-2-2\rho(1-\rho)\log\frac{\rho}{1-\rho}\end{aligned}$$

The relative entropy can now be computed using formula (9) and is seen to be asymptotic to CTN with

$$\begin{aligned}C=c(\rho)&=2\rho-2-2\rho(1-\rho)\log\frac{\rho}{1-\rho}+1\\&=2\rho-1-2\rho(1-\rho)\log\frac{\rho}{1-\rho}\end{aligned}$$

agreeing with (8).

This perturbation is neither stationary in time nor local in nature. We need to modify it.

The special initial configuration of particles at every site $x\le 0$ and no particles at any site $x>0$ will be denoted by $\bar{\eta}$. Let $N(T)$ be the number of transitions from 0 to 1 during $[0,T]$ for the TASEP. Let P be the unpertubed TASEP from this special configurartion. Our initial goal is to construct, for each given ρ a perturbed measure Q_ρ such that $Q_\rho << P$, with

$$E^{Q_\rho}[N(T)]\simeq\rho(1-\rho)T$$

and

$$H(Q_\rho|P)\simeq Tc(\rho)$$

with $c(\rho)$ given by (8). We wish to do this with a local, time independent perturbation at least approximately. We can work out the algebra and restate it as trying to make for $a < \frac{1}{4}$,

$$E^{Q_\rho}[N(T)] \simeq aT$$

with an entropy cost not exceeding

$$I(a) = \sqrt{1-4a} - 2a \log \frac{1+\sqrt{1-4a}}{1-\sqrt{1-4a}} \tag{10}$$

We consider for $\sigma > 0$,

$$U(\sigma, t, \eta) = E^{\eta}[e^{-\sigma N(t)}]$$

where η is the initial configurartion. First we note that by a simple coupling argument

$$U(\sigma, t, \bar{\eta}) \le U(\sigma, t, \eta) \le U(\sigma, t, \bar{\eta}) e^{\sigma g(\eta)}$$

with

$$g(\eta) = \sum_{x \le 0} (1 - \eta_x) + \sum_{x > 0} \eta_x$$

for all configurations η with only a finite number of ocupied sites x with $x > 0$ and finite number of empty sites x with $x \le 0$. By Markov property, if

$$A(\sigma, t) = \inf_{\eta} U(\sigma, t, \eta) = U(\sigma, t, \bar{\eta})$$

then $A(\sigma, t+s) \ge A(\sigma, t) A(\sigma, s)$ and $-\log A(t)$ is subadditive and

$$\lim_{t \to \infty} \frac{\log A(\sigma, t)}{t} = \sup_{t} \frac{\log A(\sigma, t)}{t} = -\lambda(\sigma) \tag{11}$$

exists. Moreover

$$e^{-t\lambda(\sigma) - \theta(t)} \le U(\sigma, t, \bar{\eta}) \le U(\sigma, t, \eta) \le e^{-t\lambda(\sigma) + \sigma g(\eta)}$$

where $\theta(t) = o(t)$ as $t \to \infty$. We can write down a differential equation satisfied by $U(\sigma, t, \eta)$

$$\frac{\partial U(\sigma, t, \eta)}{\partial t} = (\mathcal{L}_\sigma U)(\sigma, t, \eta)$$

with

$$U(\sigma, 0, \eta) = 1$$

The genertaor $\mathcal{L}_\sigma$ is obtained by a combination of Girsanov formula and Feynman-Kac formula. It takes the form

$$(\mathcal{L}_\sigma U)(\eta) = \sum_x c_{x,x+1}\eta_x(1-\eta_{x+1})[U(\eta^{x,x+1}) - U(\eta)] - (1-e^{-\sigma})U(\eta)$$

where $c_{x,x+1} = 1$ for $x \neq 0$ and $c_{0,1} = e^{-\sigma}$.

Theorem 7.2. *Let $\sigma > 0$ be given. For each $\epsilon > 0$, there exists a positive local function $V = V_{\sigma,\epsilon}(\eta)$ that satisfies*

$$(\mathcal{L}_\sigma V)(\eta) + (\lambda(\sigma)+\epsilon)V(\eta) \geq 0$$

for all η.

Proof. As a first step we produce a function that is continuous, i.e. depends weakly on far away sites and then approximate it to get a local function. Our first choice is

$$W(\eta) = \frac{1}{t_0}\int_0^{t_0} \exp[(\lambda(\sigma)+\frac{\epsilon}{2})t]U(\sigma,t,\eta)dt$$

Because of the lower bound on U we can assume that t_0 is large enough so that for all η,

$$e^{[\lambda(\sigma)+\frac{\epsilon}{2}]t_0}U(\sigma,t_0,\eta) \geq 1$$

Let us compute $\mathcal{L}_\sigma W$.

$$\begin{aligned}(\mathcal{L}_\sigma W)(\eta) &= \frac{1}{t_0}\int_0^{t_0} \exp[(\lambda(\sigma)+\frac{\epsilon}{2})t](\mathcal{L}_\sigma U)(\sigma,t,\eta)dt \\ &= \frac{1}{t_0}\int_0^{t_0} \exp[(\lambda(\sigma)+\frac{\epsilon}{2})t]U_t(\sigma,t,\eta)dt \\ &= \frac{1}{t_0}[e^{[\lambda(\sigma)+\frac{\epsilon}{2}]t_0}U(\sigma,t_0,\eta)-1] - (\lambda(\sigma)+\frac{\epsilon}{2})W(\eta) \\ &\geq -(\lambda(\sigma)+\frac{\epsilon}{2})W(\eta)\end{aligned}$$

Since t_0 is finite, W depends weakly on far away sites, and can be nicely approximated by a V that is local. Q.E.D.

The next step is to use $V = V_{\sigma,\epsilon}$ to construct our perturbations. These perturbations cost entropy but will limit the flow between 0 and 1. There is a trade off and σ is the parameter that will control this trade off. Opitmality in the trade off is reached as $\epsilon \to 0$. We begin by defining the rates

$$c_{x,x+1}(\sigma,\epsilon,\eta) = c_{x,x+1}(\sigma)\frac{V(\eta^{x,x+1})}{V(\eta)}$$

Note that $c_{x,x+1}(\sigma) = 1$ for $x \neq 0$ and $c_{0,1} = e^{-\sigma}$. The corresponding perturbed evolution

$$(\mathcal{L}_{\sigma,\epsilon} f)(\eta) = \sum_x c_{x,x+1}(\sigma, \epsilon, \eta)\eta_x(1 - \eta_{x+1})[f(\eta^{x,x+1}) - f(\eta)]$$

is local. Let us speed up by N, and do the hydrodynamic scaling for this perturbation. Let Q_N be the perturbed process and P_N be the original process both rescaled and with the special initial configuration $\bar{\eta}$.

Theorem 7.3. *For every N, we have on the interval $[0, T]$,*

$$H(Q_N|P_N) \leq E^{Q_N}\left[\log V(\eta(T)) - \log V(\eta(0)) - \sigma N(T)\right] + NT[\lambda(\sigma) + \epsilon]$$

Proof.

$$\frac{1}{N} H(Q_N|P_N) = E^{Q_N}\left[\int_0^T \phi_{\sigma,\epsilon}(\eta(t))dt\right]$$

where

$$\phi_{\sigma,\epsilon}(\eta) = e_{x,x+1}(\eta)[c_{x,x+1}(\sigma, \epsilon, \eta) \log c_{x,x+1}(\sigma, \epsilon, \eta) \quad c_{x,x+1}(\sigma, \epsilon, \eta) + 1]$$

We use our definition of $c_{x,x+1}(\sigma, \epsilon, \eta)$ and an easy calculation to get

$$\phi_{\sigma,\epsilon}(\eta) = (\mathcal{L}_{\sigma,\epsilon} \log V)(\eta) - \sigma\, e_{0,1}(\eta) c_{0,1}(\sigma, \epsilon, \eta) - \frac{(\mathcal{L}_\sigma V)(\eta)}{V(\eta)}$$

The proof is completed by integrating with respect to t and taking expectations, noting

$$N E^{Q_N}\left[\int_0^T (\mathcal{L}_{\sigma,\epsilon} \log V)(\eta(t))dt\right]$$
$$= E^{Q_N}\left[\int_0^T [\log V(\eta(T)) - \log V(\eta(0))]\right]$$

Q.E.D.

Now the rest of the argument is relatively straight forward. First, we need a lemma.

Lemma 7.4. *The limit $\lambda(\sigma)$ defined in (11) satisfies*

$$\lambda(\sigma) = \inf_{\rho \geq \frac{1}{2}} [\sigma\rho(1 - \rho) + c(\rho)] = \inf_{a \leq \frac{1}{4}} [a\sigma + I(a)]$$

where $I(a)$ is as in (10).

Proof. *Upper bound*: If $F(\rho(\cdot,\cdot))$ is the flow through the origin during $[0,T]$ for a weak solution $\rho(\cdot,\cdot)$, then from the upper bound already established

$$T\,\lambda(\sigma) \geq -\inf_{\rho(\cdot,\cdot)}\Big[\sigma\,F(\rho(\cdot,\cdot)) + I(\rho(\cdot,\cdot))\Big]$$

If one fixes $F(\rho(\cdot,\cdot)) = aT = \rho(1-\rho)T$, the infimum of $I(\rho(\cdot,\cdot))$ is shown by a variational argument to equal $Tc(\rho) = TI(a)$.

Lower bound: By a simple calculation using Jensen's inequality

$$\begin{aligned}
\log E^{P_N}\Big[e^{-\sigma N(T)}\Big] &= \log E^{Q_N}\Big[e^{-\sigma N(T)}\frac{dP_N}{dQ_N}\Big]\\
&= \log E^{Q_N}\Big[e^{-\sigma N(T)+\log[\frac{dP_N}{dQ_N}]}\Big]\\
&= \log E^{Q_N}\Big[e^{-\sigma N(T)-\log[\frac{dQ_N}{dP_N}]}\Big]\\
&\geq -E^{Q_N}\Big[\sigma N(T) + \log[\frac{dQ_N}{dP_N}]\Big]\\
&= -E^{Q_N}\big[\sigma N(T)\big] - H(Q_N|P_N)
\end{aligned}$$

We can take any Q_N and we pick it as the time reversal of the backward TASEP. Our earlier calculations establish the lower bound of $Tc(\rho)$ for the relative entropy and $T\rho(1-\rho)$ for the flow. One checks that $I(a)$ is a strictly convex function of a. Q.E.D.

This proves that if we perturb by $\mathcal{L}_{\sigma,\epsilon}$ and take the limit as $\epsilon \to 0$, the profiles we get will satisfy the entropy condition, will have flow at 0 limited by Ta and the realtive entropy will be bounded by $TI(a)$, where a is dual to σ.

The final step in the proof is to prove that the only profile, that satisfies the entropy condition away from 0, has the rate function bounded by $TI(a)$ and the flow through the origin bounded by Ta, is given by (7) with $\rho > \frac{1}{2}$ chosen so that $a = \rho(1-\rho)$.

References

[1] Dupuis, Paul.; Ellis, Richard S. A weak convergence approach to the theory of large deviations. Wiley Series in Probability and Statistics: A Wiley-Interscience Publication. John Wiley & Sons, Inc., New York,

[2] Jensen, Leif. Large deviations of TASEP. PhD thesis, NYU, 2000.

[3] Kosygina, Elena. The behavior of the specific entropy in the hydrodynamic scaling limit. Ann. Probab. 29 (2001), no. 3, 1086–1110.

[4] Rost, H. Nonequilibrium behaviour of a many particle process: density profile and local equilibria. Z. Wahrsch. Verw. Gebiete 58 (1981), no. 1, 41–53.

[5] Seppäläinen, T. Coupling the totally asymmetric simple exclusion process with a moving interface. I Brazilian School in Probability (Rio de Janeiro, 1997). Markov Process. Related Fields 4 (1998), no. 4, 593–628.

[6] Tartar, L. Compacité par compensation: résultats et perspectives. (French) [Compensated compactness: results and perspectives] Nonlinear partial differential equations and their applications. Collège de France Seminar, Vol. IV (Paris, 1981/1982), 350–369, Res. Notes in Math., 84, Pitman, Boston, MA, 1983.

Courant Institute of mathematical Sciences
New York University
New York
USA

Advanced Studies in Pure Mathematics 39, 2004
pp. 307-324

Diffusive Behaviour of the Equilibrium Fluctuations in the Asymmetric Exclusion Processes

Claudio Landim, Stefano Olla and Srinivasa R. S. Varadhan

Abstract.

We consider the asymmetric simple exclusion process in dimension $d \geq 3$. We review some results concerning the equilibrium bulk fluctuations and the asymptotic behaviour of a second class particle.

§1. Introduction

The asymmetric simple exclusion process is the simplest model of a driven lattice gas. This model is given by the dynamics of infinitely many particles moving on $\mathbb{Z}^d$ as asymmetric random walks with an exclusion rule: when a particle attempts to jump on a site occupied by another particle the jump is suppressed. Of course we consider initial configurations where there is at most one particle per site. We denote the configurations by $\eta \in \{0,1\}^{\mathbb{Z}^d}$ so that $\eta(x) = 1$ if site x is occupied for the configuration η and $\eta(x) = 0$ if site x is empty. The number of particles is conserved and the Bernoulli product measures $\{\nu_\alpha,\ \alpha \in [0,1]\}$ are the ergodic invariant measures.

Rezakhanlou, in [11], proved that the empirical field of particles, after a hyperbolic rescaling of space and time by a parameter ϵ

$$\pi_t^\epsilon = \epsilon^d \sum_x \eta_{t\epsilon^{-1}}(x)\delta_{\epsilon x} \tag{1.1}$$

converges weakly, as $\epsilon \to 0$, to the (entropic) solution of the Burgers equation

$$\begin{cases} \partial_t \rho \; + \; \gamma \cdot \nabla[(1-\rho)\rho] \; = \; 0\,, \\ \rho(0,\cdot) = \rho_0(\cdot)\,. \end{cases}$$

Received August 16, 2003.
Revised May 5, 2003.

Here ρ_0 is the initial asymptotic profile, i.e.

$$\pi_0^\epsilon \underset{\epsilon\to 0}{\rightharpoonup} \rho_0 \qquad \text{in probability,} \tag{1.2}$$

while γ is the average velocity of a single particle. In fact in [11] the initial distribution of the particles is assumed to be an inhomogeneous product measure with a slowly varing profile $\nu_{\rho_0(\epsilon\cdot)}$, such that $\nu_{\rho_0(\epsilon\cdot)}(\eta(x)) = \rho_0(\epsilon x)$.

Suppose now that we start the system with a stationary measure ν_α, for a certain density $\alpha \in]0,1[$. The invariant measure ν_α has macroscopic Gaussian uncorrelated fluctuations, i.e. the fluctuation field

$$\xi_0^\epsilon = \epsilon^{d/2} \sum_{x\in\mathbb{Z}^d} \delta_{\epsilon x}(\eta_0(x) - \alpha) \tag{1.3}$$

converges in law to a white noise field ξ with covariance

$$\mathbb{E}(\xi(q)\xi(q')) = \alpha(1-\alpha)\delta(q-q') \tag{1.4}$$

It is not hard to prove in this context that the macroscopic evolution of these fluctuations, at time $t\epsilon^{-1}$, will converge to the solution of the linearized equation

$$\partial_t \xi + (1-2\alpha)\gamma \cdot \nabla \xi = 0 \tag{1.5}$$

Of course this equation (and the following ones) is to be intended in the weak sense, since ξ is only a distribution-valued field on $\mathbb{R}^d$. This means that an initial fluctuation will evolve macroscopically by a deterministic translation with velocity $(1-2\alpha)\gamma$. In simple exclusion processes there is a simple way to keep track of density fluctuations. Let us condition the stationary measure ν_α to have the site 0 empty and put in this site a *second class particle*, i.e. a particle that has the same jump rates as the other particles but when a normal (*first class*) particle attempts to jump in the site occupied by a second class particle, the particles exchange sites. Then this second class particles evolves like a density fluctuation (cf. section 6) and equation (1.5) corresponds to a law of large numbers for the position X_t of the second class particle

$$\frac{X_t}{t} \underset{t\to\infty}{\longrightarrow} (1-2\alpha)\gamma\ . \tag{1.6}$$

The natural question is about the fluctuation around this law of large numbers. The answer is that, in dimension $d \geq 3$, the recentered position of the second class particle $X_t - (1-2\alpha)\gamma t$ behaves diffusively and

$\epsilon(X_{t\epsilon^{-2}} - (1-2\alpha)\gamma t\epsilon^{-2})$ converges in law to a Brownian motion with a diffusion matrix $D(\alpha)$ (cf. Theorem 6.2).

In dimension $d = 1$ and 2 the asymptotic behavior of the second class particle is superdiffusive (cf. [9, 14]), and one of the most interesting and challenging problem is to determine the corresponding limit process in these cases.

We review in this article results concerning the dimension $d \geq 3$. The diffusive behaviour of the density fluctuations is then stated in the following way. Consider the recentered density fluctuation field at diffusive scaling

$$Y_t^\epsilon = \epsilon^{d/2} \sum_{x \in \mathbb{Z}^d} \delta_{\epsilon x - vt\epsilon^{-1}} (\eta_{t\epsilon^{-2}}(x) - \alpha) \tag{1.7}$$

where $v = (1-2\alpha)\gamma$. We show in section 4 that, as a process with values on the distributions on $\mathbb{R}^d$, it converges (in law) to the solution of the linear stochastic partial differential equation

$$\partial_t Y = \nabla \cdot D(\alpha) \nabla Y + \sqrt{2\alpha(1-\alpha)}\sqrt{D(\alpha)}\nabla \cdot W \tag{1.8}$$

where $W(x,t) = (W_1, \ldots, W_d)$ are independent standard white noises on $\mathbb{R}^{d+1}$.

This result was proved in [1] for the asymmetric exclusion process in $d \geq 3$ and in finite macroscopic volume. The purpose of this article is to give a simpler proof in infinite volume based on the fluctuation-dissipation theorem as stated in [8]. The fluctuation-dissipation theorem, which was first proved in this context in [7], is in fact the the core of the proof for the macroscopic density fluctuation, as we explain in section 4. This theorem states that the *space-time* fluctuations of the current associated to the density are equivalent to to the fluctuations of a gradient of the density times the matrix $D(\alpha)$, in the sense that the variance of the difference is asymptotically small (cf. Theorem 3.2 in section 3).

This fluctuation-dissipation theorem was also applied in order to study the diffusive incompressible limit (cf. [3]) and the first order corrections to the hydrodynamic limit (cf. [5]). It has also been used to study corresponding results for a lattice system of particles with exclusion rule with conservation of number of particles, velocity and energy (cf. [2] and references therein).

The fluctuation-dissipation theorem for the exclusion processes can be proved by using the duality properties of these models (as explained in [8]), that permits to control the size of these fluctuation by estimates on the Green functions of simple random walks (see also [12] where this method is applied in the study of the diffusive behaviour of a tagged particle). At the moment there are not any result of this type for systems that do not have such nice duality.

§2. Density fluctuations

Fix a probability distribution $p(\cdot)$ supported on a finite subset of $\mathbb{Z}^d_* = \mathbb{Z}^d \backslash \{0\}$ and denote by L the generator of the simple exclusion process on $\mathbb{Z}^d$ associated to $p(\cdot)$. L acts on local functions f on $\mathcal{X}_d = \{0,1\}^{\mathbb{Z}^d}$ as

$$(2.1) \qquad (Lf)(\eta) = \sum_{x,y\in\mathbb{Z}^d} p(y-x)\eta(x)\{1-\eta(y)\}[f(\sigma^{x,y}\eta) - f(\eta)] \, ,$$

where $\sigma^{x,y}\eta$ stands for the configuration obtained from η by exchanging the occupation variables $\eta(x)$, $\eta(y)$:

$$(\sigma^{x,y}\eta)(z) = \begin{cases} \eta(z) & \text{if } z \neq x, y \;, \\ \eta(x) & \text{if } z = y \;, \\ \eta(y) & \text{if } z = x \;. \end{cases}$$

Denote by $s(\cdot)$ and $a(\cdot)$ the symmetric and the anti-symmetric parts of the probability $p(\cdot)$:

$$s(x) = (1/2)[p(x) + p(-x)] \, , \quad a(x) = (1/2)[p(x) - p(-x)]$$

and denote by L^s, L^a the symmetric part and the anti-symmetric part of the generator L. L^s, L^a are obtained by replacing p by s, a in the definition of L.

For α in $[0,1]$, denote by ν_α the Bernoulli product measure on $\mathcal{X}$ with $\nu_\alpha[\eta(x)=1] = \alpha$. Measures in this one-parameter family are stationary and ergodic for the simple exclusion dynamics and in the symmetric case, i.e. $p(x) = p(-x)$, these measures are reversible. Expectation with respect to ν_α is represented by $< \cdot >_\alpha$ and the scalar product in $L^2(\nu_\alpha)$ by $< \cdot, \cdot >_\alpha$.

Fix $\varepsilon > 0$. For a configuration η, denote by $\pi^\varepsilon = \pi^\varepsilon(\eta)$ the empirical measure associated to η. This is the measure on $\mathbb{R}^d$ obtained assigning mass ε^d to each particle of η :

$$\pi^\varepsilon = \varepsilon^d \sum_{x\in\mathbb{Z}^d} \eta(x)\delta_{x\varepsilon} \, ,$$

where δ_u stands for the Dirac measure concentrated on u. For a measure π on $\mathbb{R}^d$ and a continuous function $G\colon \mathbb{R}^d \to \mathbb{R}$, denote by $< \pi, G >$ the integral of G with respect to π. In particular, $< \pi^\varepsilon, G >$ is equal to $\varepsilon^d \sum_{x\in\mathbb{Z}^d} G(x\varepsilon)\eta(x)$.

Consider a sequence of probability measures μ^ε on the configuration space $\mathcal{X}_d$ and assume that, under μ^ε, π^ε converges in probability to an absolutely continuous measure $\rho_0(u)du$. This means that for every continuous function $G\colon \mathbb{R}^d \to \mathbb{R}$ and every $\delta > 0$,

$$\lim_{\varepsilon\to 0} \mu^\varepsilon \Big\{ \Big| < \pi^\varepsilon, G > - \int G(u)\rho_0(u)\, du \Big| > \delta \Big\} = 0 . \tag{2.2}$$

Consider the hyperbolic equation

$$\begin{cases} \partial_t \rho + \gamma \cdot \nabla [\rho(1-\rho)] = 0 , \\ \rho(0,\cdot) = \rho_0(\cdot) . \end{cases} \tag{2.3}$$

In this formula γ stands for the drift : $\gamma = \sum_y yp(y)$. For $t > 0$, denote by π_t^ε the empirical measure associated to the state of the process at time t : $\pi_t^\varepsilon = \pi^\varepsilon(\eta_t)$. Rezakhanlou [11] proved that, starting from a inhomogeneous product measure $\nu_{\rho(\varepsilon\cdot)}$ satisfying (2.2), then, for every $t \ge 0$, $\pi^\varepsilon_{t\varepsilon^{-1}}$ converges in probability to the measure $\rho(t,u)du$, where the density ρ is the entropy solution of the hyperbolic equation (2.3). More precisely, for a measure μ on $\mathcal{X}_d$, denote by $\mathbb{P}_\mu$ the measure on the path space $D(\mathbb{R}_+, \mathcal{X}_d)$ induced by the Markov process η_t and the measure μ. Then, for every $t \ge 0$, every continuous function G and every $\delta > 0$,

$$\lim_{\varepsilon\to 0} \mathbb{P}_{\nu_{\rho(\varepsilon\cdot)}} \Big\{ \Big| < \pi^\varepsilon_{t\varepsilon^{-1}}, G > - \int G(u)\rho(t,u)\, du \Big| > \delta \Big\} = 0 ,$$

where the density ρ is the entropy solution of the hyperbolic equation (2.3). Notice the Euler rescaling of time in the previous formula.

Let us now consider, for a fixed density $\alpha \in (0,1)$, the system in equilibrium with distribution $\mathbb{P}_{\nu_\alpha}$ on the path space. Let $Y^\varepsilon_\cdot$ be the density fluctuation field that acts on smooth functions H as

$$Y_t^\varepsilon(H) = \varepsilon^{d/2} \sum_{x\in\mathbb{Z}^d} H(x\varepsilon - vt\varepsilon^{-1})(\eta_{t\varepsilon^{-2}}(x) - \alpha) , \tag{2.4}$$

where $v = \gamma(1-2\alpha)$. Notice the diffusive rescaling of time on the right hand side of this identity.

For any $k \ge 0$ and $f, g \in C^\infty(\mathbb{R}^d)$ consider the scalar product

$$(g,f)_k = \int_{\mathbb{R}^d} g(q) \left(|q|^2 - \Delta\right)^k f(q)\, dq \tag{2.5}$$

and denote by $\mathcal{H}_k$ the corresponding closure. For any positive k we denote by $\mathcal{H}_{-k}$ its dual space with respect to the $L^2(\mathbb{R}^d) \equiv \mathcal{H}_0$ scalar product.

We will show in section 5 that the probability distribution Q^ϵ of $Y^\varepsilon_\cdot$ under $\mathbb{P}_{\nu\alpha}$, is supported and is tight in $D([0,T],\mathcal{H}_{-k})$ for any $k > d+1$. Observe that in [1] the tightness is proved for any $k > d/2+1$. This difference is due to the fact that here we are dealing with distribution on the infinite volume $\mathbb{R}^d$, while in [1] we were working in the finite d-dimensional torus.

We state now the main theorem. Assume that $d \geq 3$. Let $D_{i,j}(\alpha)$ a strictly positive symmetric matrix. Denote by $\mathcal{A}$, $\mathcal{B}$ the differential operators defined by $\mathcal{A} = \sum_{1\leq i,j\leq d} D_{i,j}(\alpha)\partial^2_{u_i,u_j}$, $\mathcal{B} = \sqrt{2\alpha(1-\alpha)}\,\sigma\,\nabla$ (where the matrix σ is the positive square root of D). Denote by $\{T_t,\, t \geq 0\}$ the semigroup in $L^2(\mathbb{R}^d)$ associated to the operator $\mathcal{A}$. Fix a positive integer $k_0 > d+1$. Let Q be the probability measure concentrated on $C([0,T],\mathcal{H}_{-k_0})$ corresponding to the stationary generalized Ornstein–Uhlenbeck process with mean 0 and covariance

$$E_Q\Big[Y_t(H)Y_s(G)\Big] \;=\; \chi(\alpha)\int_{\mathbb{R}^d} du\,(T_{|t-s|}H)(u)\,G(u)$$

for every $0 \leq s \leq t$ and H, G in $\mathcal{H}_{k_0}$. Here $\chi(\alpha)$ stands for the static compressibility given by $\chi(\alpha) = \mathbf{Var}_{\nu_\alpha}[\eta(0)] = \alpha(1-\alpha)$.

Theorem 2.1. *There exists a strictly positive symmetric matrix $D_{i,j}(\alpha)$ such that the sequence Q^ε converges weakly to the probability measure Q of the corresponding Ornstein–Uhlenbeck process.*

Formally, Y_t is the solution of the stochastic differential equation

$$dY_t \;=\; \mathcal{A}Y_t dt \;+\; dB_t^\nabla\;,$$

where B_t^∇ is a mean zero Gaussian field with covariances given by

$$E_Q\Big[B_t^\nabla(H)B_s^\nabla(G)\Big] \;=\; 2\chi(\alpha)(s\wedge t)\int_{\mathbb{R}^d}(\nabla H)\cdot D\,(\nabla G)\,.$$

In section 6 we show that the viscosity matrix $D(\alpha)$ is identified as

(2.6)
$$D_{i,j}(\alpha) = \frac{1}{\alpha(1-\alpha)}\lim_{t\to\infty}\frac{1}{t}\sum_{x\in\mathbb{Z}^d} x_i x_j \mathbb{E}_{\nu_\alpha}\left((\eta_t(x-vt)-\alpha)(\eta_0(0)-\alpha)\right),$$

which is also the asymptotic variance of the second class particle. Other expressions for this matrix can be given (cf. [6]), while in [8] we prove that $D_{i,j}(\alpha)$ are smooth functions of the density $\alpha\in[0,1]$.

§3. The Fluctuation-Dissipation Theorem

We recall in this section some results proved in [8] (cf. also [7]). For local functions u, v, define the scalar product $\ll \cdot\, , \cdot \gg$ by

$$\ll u, v \gg = \sum_{x \in \mathbb{Z}^d} \{ < \tau_x u, v > - < u >< v > \} , \tag{3.1}$$

where $\{\tau_x,\ x \in \mathbb{Z}^d\}$ is the group of translations and $< \cdot >$ stands for the expectation with respect to the measure ν_α. That this is in fact an inner product can be seen by the relation

$$\ll u, v \gg = \lim_{V \uparrow \mathbb{Z}^d} \frac{1}{|V|} < \sum_{x \in V} \tau_x (u - < u >), \sum_{x \in V} \tau_x (v - < v >) > .$$

Since $\ll u - \tau_x u\, ,\, v \gg = 0$ for all x in $\mathbb{Z}^d$, this scalar product is only positive semidefinite. Denote by $L^2_{\ll \cdot, \cdot \gg}(\nu_\alpha)$ the Hilbert space generated by the local functions and the inner product $\ll \cdot, \cdot \gg$.

Denote by L^s the symmetric part of the generator. For two local functions u, v, let

$$\ll u, v \gg_1 = \ll u, (-L^s) v \gg$$

and let H_1 be the Hilbert space generated by local functions and the inner product $\ll \cdot, \cdot \gg_1$. To introduce the dual Hilbert spaces of H_1, for a local function u, consider the semi-norm $\| \cdot \|_{-1}$ given by

$$\|u\|_{-1} = \sup_v \Big\{ 2 \ll u, v \gg - \ll v, v \gg_1 \Big\} ,$$

where the supremum is carried over all local functions v. Denote by H_{-1} the Hilbert space generated by the local functions and the semi-norm $\| \cdot \|_{-1}$.

Notice that the function $\eta(0) - \alpha$ does not belong to H_{-1}. Indeed, due to the translations, $\|\eta(0) - \alpha\|_1 = 0$ so that $\|\eta(0) - \alpha\|_{-1} = \infty$. One can show, however, that these linear functions are essentially the only zero-mean local functions that do not belong to H_{-1}.

To make the previous statement precise, we need to introduce some notation. For a local function f, denote by $\tilde{f} : [0,1] \to \mathbb{R}$ the polynomial defined by

$$\tilde{f}(\alpha) = E_{\nu_\alpha}[f] .$$

Denote by $\mathcal{C}_0 = \mathcal{C}_0(\alpha)$ the collection of local functions such that

$$\tilde{f}(\alpha) = E_{\nu_\alpha}[f] = 0 , \quad \tilde{f}'(\alpha) = \frac{d}{d\beta} E_{\nu_\beta}[f] \Big|_{\beta=\alpha} = 0 .$$

It is proved in [8] that all functions in $\mathcal{C}_0$ have finite H_{-1} norm:

$$\|f\|_{-1} < \infty \quad \text{if } f \in \mathcal{C}_0 \ . \tag{3.2}$$

Denote by $\mathcal{C}_1$ the space of local functions f in $\mathcal{C}_0$ which are orthogonal to all linear functions:

$$< f, \eta(x) - \alpha > = 0 \quad \text{for all } x \text{ in } \mathbb{Z}^d.$$

The next result states that a local function f in $\mathcal{C}_1$ has a finite space-time variance in the diffusive scaling.

Theorem 3.1. *Fix $T > 0$, a vector v_0 in $\mathbb{R}^d$, a smooth function $G : \mathbb{R}^d \to \mathbb{R}$ with compact support and a local function f in $\mathcal{C}_1$. There exists a finite constant C_0 such that*

$$\limsup_{\varepsilon \to 0} \mathbb{E}_{\nu_\alpha}\Big[\sup_{0 \le t \le T} \Big(\varepsilon^{d/2+1} \int_0^{t\varepsilon^{-2}} \sum_{x \in \mathbb{Z}^d} G(\varepsilon[x - rv_0]) f(\tau_x \eta_r)\, dr \Big)^2 \Big] \\ \le C_0 T \|G\|_{L^2}^2 \|f\|_{-1}^2 \ .$$

The theorem is not correct if we replace $\mathcal{C}_1$ by $\mathcal{C}_0$ because $\eta(e_1) - \eta(0)$ has H_{-1} norm equal to 0 and a finite, strictly positive space-time variance.

It is proved in [8] that any local function in $\mathcal{C}_0$ may be approximated in H_{-1} by a local function in the range of the generator. More precisely, for every local function f in $\mathcal{C}_0$ and every $\varepsilon > 0$, there exists a local function u_ε, which may be taken in $\mathcal{C}_1$, such that

$$\|Lu_\varepsilon - f\|_{-1}^2 \le \varepsilon \ . \tag{3.3}$$

The fluctuation-dissipation theorem stated below follows from this result, the estimate stated in Theorem 3.1 and some elementary computations.

Theorem 3.2. *Fix $T > 0$, a vector v_0 in $\mathbb{R}^d$, a local function w in $\mathcal{C}_1$ and a smooth function $G : \mathbb{R}^d \to \mathbb{R}$ with compact support. There exist a sequence of local functions u_m and $D_z(\alpha)$ such that*

$$\limsup_{m \to \infty} \limsup_{\varepsilon \to 0} \\ \mathbb{E}_{\nu_\alpha}\Big[\sup_{0 \le t \le T} \Big(\varepsilon^{d/2+1} \int_0^{t\varepsilon^{-2}} \sum_{x \in \mathbb{Z}^d} G(\varepsilon[x - rv_0]) \tau_x W_m(\eta_s)\, ds \Big)^2 \Big] = 0 \ ,$$

where

$$W_m(\eta) = w - Lu_m + \sum_{z \in \mathbb{Z}^d} a(z) D_z(\alpha) \{\eta(z) - \eta(0)\} \ .$$

The idea of the proof is quite simple. By (3.3), we may approximate w by some local function Lu_m in the range of the generator. However, Lu_m might have linear terms and therefore might not be in $\mathcal{C}_1$. Subtracting expressions of type $\eta(z)-\eta(0)$, we convert Lu_m in a $\mathcal{C}_1$-function and apply Theorem 3.1.

§4. Time evolution of density fluctuations

We show in this section how to deduce from the fluctuation-dissipation theorem the equilibrium fluctuations for the density field defined in (2.4).

Fix a smooth function $G : \mathbb{R}^d \to \mathbb{R}$ with compact support and define the martingale $M_t^{1,\varepsilon}$ by the time evolution equation

$$Y_t^\varepsilon(G) - Y_0^\varepsilon(G) = \int_0^t \left(\partial_s + \varepsilon^{-2}L\right) Y_s^\varepsilon(G)\, ds + M_t^{1,\varepsilon} .$$

An elementary computation shows that

(4.1)

$$\begin{aligned}(\partial_t + \varepsilon^{-2}L)Y_t^\varepsilon(G) = &-\varepsilon^{d/2-1}\sum_x (v\cdot\nabla G)[\varepsilon(x - vt\varepsilon^{-2})](\eta(x)-\alpha)\\ &- \varepsilon^{d/2-2}\sum_x G[\varepsilon(x-vt\varepsilon^{-2})]\sum_y j_{x,y} ,\end{aligned}$$

where $j_{x,y}$, the instantaneous current between sites x and y, is given by

$$\begin{aligned} j_{x,y} &= -s(x-y)[\eta(y)-\eta(x)]\\ &\quad - a(x-y)\Big\{\eta(y)(1-\eta(x)) + \eta(x)(1-\eta(y))\Big\} .\end{aligned}$$

Let $G_t^\varepsilon(x) = G[\varepsilon(x - vt\varepsilon^{-2})]$. Since the current is anti-symmetric ($j_{x,y} = -j_{y,x}$) and since its expectation with respect to ν_α is equal to $2a(y-x)\alpha(1-\alpha)$, an elementary computation gives that

$$\begin{aligned} -\sum_x G_t^\varepsilon(x)\sum_y j_{x,y} &= \sum_{x,y} s(x-y)[G_t^\varepsilon(y) - G_t^\varepsilon(x)][\eta(x)-\alpha]\\ &\quad - \sum_{x,y} a(x-y)[G_t^\varepsilon(x)-G_t^\varepsilon(y)][\eta(y)-\alpha][\eta(x)-\alpha]\\ &\quad + (1-2\alpha)\sum_{x,y} a(x-y)[G_t^\varepsilon(x)-G_t^\varepsilon(y)][\eta(x)-\alpha]\end{aligned}$$

because

$$\begin{aligned}&\eta(y)(1-\eta(x)) \;+\; \eta(x)(1-\eta(y)) \;-\; 2\alpha(1-\alpha)\\ &\quad = -2(\eta(y)-\alpha)(\eta(x)-\alpha) \;+\; (1-2\alpha)\Big\{(\eta(y)-\alpha) + (\eta(x)-\alpha)\Big\} .\end{aligned}$$

Finally, since $a(\cdot)$ is anti-symmetric, a Taylor expansion and a change of variables in the summation permit to conclude that

$$
\begin{aligned}
(4.2)\quad & (\partial_t + \varepsilon^{-2} L) Y_t^\varepsilon(G) \\
&= \varepsilon^{d/2} \sum_x \sum_{i,j} \sigma_{i,j} (\partial_i \partial_j G)[\varepsilon(x - vt\varepsilon^{-2})](\eta(x) - \alpha) \\
&\quad - \varepsilon^{d/2-1} \sum_{x,z} a(z)(z \cdot \nabla G)[\varepsilon(x - vt\varepsilon^{-2})] \Phi_{x,x+z} + R_\varepsilon(\eta) .
\end{aligned}
$$

In this formula, $\sigma_{i,j}$ is the symmetric matrix defined by

$$
\sigma_{i,j}^s = \sum_z s(z) z_i z_j ,
$$

$\Phi_{x,y}$ is the zero-mean local function defined by

$$
\Phi_{x,y} = (\eta(x) - \alpha)(\eta(y) - \alpha)
$$

and $R_\varepsilon(\eta)$ is a remainder which vanishes in $L^2(\nu_\alpha)$ as $\varepsilon \downarrow 0$. In fact $< R_\varepsilon(\eta)^2 >= O(\varepsilon^2)$.

Since $\Phi_{x,y}$ belongs to $\mathcal{C}_1$, by Theorem 3.2, there exist a sequence of local functions $\{v_m,\ m \geq 1\}$ in $\mathcal{C}_1$ and constants $D_{z,z'}$ such that

$$
\begin{aligned}
\lim_{m\to\infty} \lim_{\varepsilon\to 0} \mathbb{E}_{\nu_\alpha} \Big[\sup_{0\leq t\leq T} \Big| \varepsilon^{d/2-1} \int_0^t \sum_{x,z} a(z)(z\cdot\nabla G)[\varepsilon(x - vt\varepsilon^{-2})] \\
\Big\{ \Phi_{0,z} - Lv_m - \sum_{z'} a(z') D_{z,z'} [\eta(z') - \eta(0)] \Big\} (\tau_x \eta_{\varepsilon^{-2}s})\, ds \Big|^2 \Big] = 0 .
\end{aligned}
$$

This result shows that we may replace $\Phi_{x,x+z}$ in the second term on the right hand side of (2.4) by $Lv_m - \sum_{z'} a(z') D_{z,z'}[\eta(z') - \eta(0)]$. The difference $\eta(z') - \eta(0)$ enables a second summation by parts which cancels a factor ε^{-1}, while the term Lv_m produces an extra martingale.

Let $F(x) = \sum_z a(z)(z \cdot \nabla G)(x)$. For each $m \geq 1$, consider the martingale $M_t^{2,m,\varepsilon}$ defined by

$$
\begin{aligned}
M_t^{2,m,\varepsilon} =& \varepsilon^{d/2+1} \sum_x \int_0^t F(\varepsilon(x - vs\varepsilon^{-2})) \varepsilon^{-2} L v_m (\tau_x \eta_{\varepsilon^{-2}s})\, ds \\
&+ \varepsilon^{d/2+1} \sum_x \int_0^t \partial_s F(\varepsilon(x - vs\varepsilon^{-2})) v_m(\tau_x \eta_{\varepsilon^{-2}s})\, ds \\
&- \varepsilon^{d/2+1} \sum_x \Big\{ F(\varepsilon(x - vt\varepsilon^{-2})) v_m(\tau_x \eta_{\varepsilon^{-2}t}) - F(\varepsilon x) v_m(\tau_x \eta_0) \Big\} .
\end{aligned}
$$

Since v_m are local functions, it is easy to see that the L^2 norm of the third term of the right hand side vanishes as $\varepsilon \to 0$. The second term is equal to

$$\varepsilon^{d/2} \sum_x \int_0^T (v \cdot \nabla F)(\varepsilon(x - vt\varepsilon^{-2})) v_m(\tau_x \eta_{\varepsilon^{-2}t}) \, dt \, .$$

Since v_m belongs to $\mathcal{C}_1$ for each $m \geq 1$, it has finite H_{-1} norm in view of (3.2). In particular, by Theorem 3.1, the expectation of the square goes to 0 as $\varepsilon \to 0$.

In conclusion, we have shown so far that

$$\int_0^t \left(\partial_s + \varepsilon^{-2} L\right) Y_s^\varepsilon(G) \, ds = \int_0^t Y_s^\varepsilon(\mathcal{A}G) \, ds + M_t^{2,m,\varepsilon} + R_{m,\varepsilon}(t) \, ,$$

where

$$\lim_{m \to \infty} \lim_{\varepsilon \to 0} E_{\nu_\alpha} \Big(\sup_{0 \leq t \leq T} \Big| R_{m,\varepsilon}(t) \Big|^2 \Big) = 0$$

and the second order differential operator $\mathcal{A}$ is given by

$$\mathcal{A} = \sum_{i,j}^{d} D_{i,j} \partial_i \partial_j \, ,$$

with the matrix $D_{i,j}$ given by

$$D_{i,j} = \sigma_{i,j} \; - \; \sum_{z,z'} a(z) a(z') z_i z'_j D_{z,z'} \, .$$

We turn now to the calculation of the quadratic variation of the martingale $M_t^{1,\varepsilon} + M_t^{2,n,\varepsilon}$. This is given by

(4.3)

$$\int_0^t \sum_y \sum_z p(z) \eta(y) [1 - \eta(y+z)] \varepsilon^{-2} \left[H_s^\varepsilon(\eta_{s\varepsilon^{-2}}^{y,y+z}) - H_s^\varepsilon(\eta_{s\varepsilon^{-2}}) \right]^2 \, ds \, ,$$

where

$$H_s^\varepsilon(\eta) = Y_s^\varepsilon(G) - \varepsilon^{d/2+1} \sum_x \int_0^T F(\varepsilon(x - vt\varepsilon^{-2})) v_m(\tau_x \eta_{\varepsilon^{-2}t}) \, .$$

Elementary computations show that (4.3) is equal to

(4.4)

$$\int_0^t \varepsilon^d \sum_y \sum_z p(z)\eta(y)[1-\eta(y+z)] \times \Big(\varepsilon^{-1}[G_s^\varepsilon(y+z)-G_s^\varepsilon(y)] - \sum_x F_s^\varepsilon(x)\left(v_m(\tau_x\sigma^{y,y+z}\eta_{s\varepsilon^{-2}}) - v_m(\tau_x\eta_{s\varepsilon^{-2}})\right)\Big)^2 ds\,,$$

where $F_s^\varepsilon(x) = F(\varepsilon(x - vt\varepsilon^{-2}))$.

Since v_m is a local function, the sum inside the square in the above expression extends over a finite number of x depending only on the support of v_m. We can therefore substitute $F_s^\varepsilon(x)$ by $F_s^\varepsilon(y)$, with an error that is small in view of Theorem 3.1. In the same way, we replace the discrete derivative of G by the actual derivative, obtaining that (4.3) is equal to

$$\int_0^t \varepsilon^d \sum_y \sum_z p(z)\eta(y)[1-\eta(y+z)] \times \Big[(z\cdot\nabla G)(\varepsilon(y-vt\varepsilon^{-2})) - F_s^\varepsilon(y)\sum_x \left(v_m(\tau_x\sigma^{y,y+z}\eta_{s\varepsilon^{-2}}) - v_m(\tau_x\eta_{s\varepsilon^{-2}})\right)\Big]^2 ds$$

plus a remainder $R_\varepsilon(t)$ which vanishes in L^2 as $\varepsilon \downarrow 0$. Recall the definition of F and take the limit as $\varepsilon \to 0$. By the law of large numbers, we obtain that the previous expression converges to

$$t\int dy \sum_z p(z)(z\cdot\nabla G)(y)^2 \times \left\langle \eta(0)(1-\eta(z))\Big(1-a(z)\Big[\Gamma_{v_m}(\eta^{0,z})-\Gamma_{v_m}(\eta)\Big]\Big)^2\right\rangle\,,$$

where $\Gamma_{v_m}(\eta)$ denotes the formal sum $\sum_x v_m(\tau_x\eta)$.

Since we performed this calculations in equilibrium and since for the invariant product measure the static fluctuations converges to the Gaussian field with covariance operator $\alpha(1-\alpha)(-\Delta)^{-1}$, if

$$b_{i,j} = \lim_{m\to\infty}\sum_z p(z) \times \left\langle \eta(0)(1-\eta(z))\Big(1-a(z)\Big[\Gamma_{v_m}(\eta^{0,z})-\Gamma_{v_m}(\eta)\Big]\Big)^2\right\rangle z_i z_j\,,$$

the fluctuation-dissipation relation for the limit Ornstein-Uhlenbeck process gives

$$b_{i,j} = \alpha(1-\alpha)D_{i,j} \, .$$

§5. Tightness

Recall that we have defined, for any $k \in \mathbb{R}$, $\mathcal{H}_k$ as the closure of $C^\infty(\mathbb{R}^d)$ with respect to the scalar product

$$(g,f)_k = \int_{\mathbb{R}^d} g(q) \left(|q|^2 - \Delta\right)^k f(q)\, dq$$

It is convenient to represent the scalar product $(\cdot,\cdot)_k$ in the orthonormal basis of the Hermite polynomials, which are the eigenfunctions of $|q|^2 - \Delta$. Let $\vec{n}$ be a multi-index of $(\mathbb{Z}^+)^d$ and $|\vec{n}| = \sum_{i=1}^d n(i)$. We denote by $\lambda_{n(i)} = 2n(i)+1$ for $n(i) \in \mathbb{Z}^+$ and $\lambda_{\vec{n}} = \sum_{i=1}^d \lambda_{n(i)}$. Define $h_{\vec{n}}(q) = \prod_{i=1}^d h_{n(i)}(q_i)$ where h_m is the mth normalized Hermite polynomial of order m in $\mathbb{R}$. We have then for every $k \geq 0$ and $f \in L^2$

$$\|f\|_k^2 = \int_{\mathbb{R}^d} f(q)(|q|^2 - \Delta)^k f(q)\, dq = \sum_{\vec{n}\in(\mathbb{Z}^+)^d} \lambda_{\vec{n}}^k \left(\int_{\mathbb{R}^d} f(q) h_{\vec{n}}(q)\, dq\right)^2$$

This is valid also for negative k. So the $\mathcal{H}_{-k}$-norm of a distribution ξ on $\mathbb{R}^d$ can be written as

$$\|\xi\|_{-k}^2 = \sum_{\vec{n}\in(\mathbb{Z}^+)^d} \lambda_{\vec{n}}^{-k}\, \xi(h_{\vec{n}})^2 \tag{5.1}$$

Observe that, for $k' > k$, the injection J of $\mathcal{H}_{-k}$ in $\mathcal{H}_{-k'}$ is compact. In fact it can be approximated by the finite range operators $J_m\xi = \sum_{|\vec{n}|\leq m} \xi(h_{\vec{n}}) h_{\vec{n}}$, and it is easy to see that the operator norm of the difference is bounded by

$$\|J - J_m\| \leq (2m+d)^{-(k'-k)}$$

By the compactness of the injections $\mathcal{H}_{-k} \hookrightarrow \mathcal{H}_{-k'}$ for $k < k'$, and standard compactness arguments, the tightness of the distribution of Y_t^ϵ is a consequence of the following proposition.

Proposition 5.1. *For any $k > d+1$ and every $T > 0$, we have that*

1.

$$\sup_{\varepsilon\in(0,1)} \mathbb{E}_\mu \left(\sup_{t\in[0,T]} ||Y_t^\varepsilon||_{-k}^2 \right) < +\infty$$

2. For any $R > 0$,

$$\lim_{\delta\to 0} \limsup_{\varepsilon\to 0} \mathbb{P}_\mu \left(\sup_{\substack{t,s\in[0,T]\\ |t-s|\le\delta}} \|Y_t^\varepsilon - Y_s^\varepsilon\|_{-k} > R \right) = 0$$

It is easy to see, by using (5.1) and that $k > d+1$, that Proposition 5.1 is a consequence of the following

Proposition 5.2. *For any smooth function* G *on* $\mathbb{R}^d$ *with compact support*

$$\sup_\varepsilon \mathbb{E}_\mu \left(\sup_{t\in[0,T]} Y_t^\varepsilon(G)^2 \right) \le TC_G \tag{5.2}$$

$$\lim_{\delta\to 0} \limsup_{\varepsilon\to 0} \mathbb{P}_\mu \left(\sup_{\substack{t,s\in[0,T]\\ |t-s|\le\delta}} |Y_t^\varepsilon(G) - Y_s^\varepsilon(G)| > R \right) = 0 \tag{5.3}$$

Let just sketch the proof of (5.3), the proof of (5.2) will follow a similar argument (cf. [1] for details).

By the same calculation made in the previous section we have

$$\begin{aligned} Y_t^\varepsilon(G) - Y_s^\varepsilon(G) &= \int_s^t \left(\partial_\tau + \varepsilon^{-2}L\right) Y_\tau^\varepsilon(G)\, d\tau + M_{s,t}^{1,\varepsilon} \\ &= \int_s^t Y_\tau^\varepsilon(\mathcal{A}G)\, d\tau + M_{s,t}^{1,\varepsilon} + M_{s,t}^{2,m,\varepsilon} + R_{m,\varepsilon}(s,t)\,, \end{aligned} \tag{5.4}$$

where $M_{s,t}^{1,\varepsilon}$ and $M_{s,t}^{2,m,\varepsilon}$ are the differences of the corresponding martingales defined in the previous section, and

$$\lim_{m\to\infty} \lim_{\varepsilon\to 0} E_{\nu_\alpha} \Big(\sup_{0\le s\le t\le T} \Big| R_{m,\varepsilon}(s,t) \Big|^2 \Big) = 0$$

The first term is easy to deal since

$$\mathbb{E}_\mu \left(\sup_{\substack{t,s\in[0,T]\\ |t-s|\le\delta}} \Big| \int_s^t Y_\tau^\varepsilon(\mathcal{A}G)\, d\tau \Big|^2 \right) \le \delta T \left\langle Y^\varepsilon(\mathcal{A}G)^2 \right\rangle \le \delta TC \|\mathcal{A}G\|_2^2. \tag{5.5}$$

About the difference martingale $\widetilde{M}^m_{s,t} = M^{1,\varepsilon}_{s,t} + M^{2,m,\varepsilon}_{s,t}$, for any finite m has a bounded quadratic variation given by (4.3) or (4.4), and it is not difficult to show exponential bounds for it. So it follows that for any m

$$\lim_{\delta\to 0}\limsup_{\varepsilon\to 0}\mathbb{P}_\mu\left(\sup_{\substack{t,s\in[0,T]\\|t-s|\le\delta}}|\widetilde{M}^m_{s,t}| > R\right) = 0.$$

§6. The second class particle

The second class particle is a special particle that has the same jump rates of the other particles (with the exclusion rule), furthermore when a normal (first class) particle attempts to jump in the site occupied by the second class particle, the particles exchange site. Therefore the evolution of the first class particles is unaffected by the presence of the second class particles. We will show here that the asymptotic evolution of the second class particle is closely related to the equilibrium fluctuations of the density.

Let ν^0_α the Bernoulli measure on $\mathbb{Z}^d$ with parameter α conditioned to have the site 0 empty. Let X_t be the position at time t of the second class particle, one starting at time 0 from 0, when the other particles are distributed initially with ν^0_α.

Proposition 6.1.

$$E^{\nu^0_\alpha}(1_{[X_t=x]}) = \frac{1}{\alpha(1-\alpha)}\left[E^{\nu_\alpha}(\eta_t(x)\eta_0(0)) - \alpha^2\right] \tag{6.1}$$

Proof. Let $\sigma_0\eta(x) = \eta(x)$ if $x \neq 0$ and $\sigma_0\eta(0) = 0$.

$$\begin{aligned}
E^{\nu_\alpha}(\eta_t(x)\eta_0(0)) &= \int \nu_\alpha(d\eta)\eta(0)E^\eta(\eta_t(x))\\
&= \int \nu_\alpha(d\eta)\eta(0)\left[E^\eta(\eta_t(x)) - E^{\sigma_0\eta}(\eta_t(x))\right]\\
&\quad + \int \nu_\alpha(d\eta)\eta(0)E^{\sigma_0\eta}(\eta_t(x))\\
&= \int \nu_\alpha(d\eta)\eta(0)E^\eta(1_{[X_t=x]}) + \int \nu_\alpha(d\xi)\frac{\alpha}{1-\alpha}(1-\xi(0))E^\xi(\eta_t(x))
\end{aligned}$$

(where in the last term we have performed the change of variable $\xi = \sigma_0\eta$)

$$= \alpha E^{\nu^0_\alpha}(1_{[X_t=x]}) + \frac{\alpha}{1-\alpha}\alpha - \frac{\alpha}{1-\alpha}E^{\nu_\alpha}(\eta_t(x)\eta_0(0))$$

Reordering the term one obtains (6.1). □

Theorem 6.2. *$\varepsilon(X_{t\varepsilon^{-2}} - vt\varepsilon^{-2})$ converges in law to a zero mean Gaussian r.w. with covariance matrix $tD^s_{i,j}$, where D^s is the symmetric part of the viscosity matrix D introduced in the previous section.*

Proof. Let $H(y)$ a bounded continuous function on $\mathbb{R}^d$ with compact support. Let $G(y)$ a probability density on $\mathbb{R}^d$ and let ν^y_α the Bernoulli measure conditioned to have the site y empty. Then

$$\begin{aligned}\varepsilon^d \sum_y G(\varepsilon y) E^{\nu^y_\alpha}\left(H(\varepsilon(X^y_{t\varepsilon^{-2}} - vt\varepsilon^{-2}))\right)\\ = \varepsilon^d \sum_y G(\varepsilon y) \sum_x H(\varepsilon(x - vt\varepsilon^{-2})) E^{\nu^y_\alpha}(1_{[X^y_{t\varepsilon^{-2}}=x]}).\end{aligned}$$

In this formula, X^y_t stands for the position at time t of a second class particle initially at y. By (6.1) this is equal to

$$\begin{aligned}&\frac{1}{\alpha(1-\alpha)}\varepsilon^d \sum_y \sum_x H(\varepsilon(x - vt\varepsilon^{-2}))G(\varepsilon y)\left[E^{\nu_\alpha}(\eta_{t\varepsilon^{-2}}(x)\eta_0(y)) - \alpha^2\right]\\ &= \frac{1}{\alpha(1-\alpha)} E^{\nu_\alpha}\left[\varepsilon^d \sum_x H(\varepsilon(x - vt\varepsilon^{-2}))\right.\\ &\qquad \left.\times(\eta_{t\varepsilon^{-2}}(x) - \alpha)\sum_y G(\varepsilon y)(\eta_0(y) - \alpha)\right]\\ &= \frac{1}{\alpha(1-\alpha)} E^{\nu_\alpha}\left[Y^\varepsilon_t(H) Y^\varepsilon_0(G)\right]\end{aligned}$$

and by the convergence result for Y^ε_t proved in the previous section, this converge as $\varepsilon \to 0$ to

$$\int_{\mathbb{R}^d} G(u)\, e^{t\mathcal{A}} H(u)\, du.$$

□

An immediate consequence of the above result is the following formula for the symmetric part of the viscosity matrix:

$$D^s_{i,j} = \lim_{t\to\infty} \frac{1}{t}\frac{1}{\alpha(1-\alpha)} \sum_x x_i x_j \left[E^{\nu_\alpha}(\eta_t(x + tv)\eta_0(0)) - \alpha^2\right].$$

References

[1] C. C. Chang, C. Landim, S. Olla: Equilibrium fluctuations of asymmetric exclusion processes in dimension $d \geq 3$. Probability Theory and Related Field, **119**, 381-409, (2001).

[2] O. Benois, R. Esposito, R. Marra: Equilibrium fluctuations for lattice gases. To appear in Ann. Inst. H. Poincaré, Probabilités et Statistiques (2003).

[3] R. Esposito, R. Marra, H. T. Yau: Diffusive limit of asymmetric simple exclusion. Rev. Math. Phys. **6**, 1233-1267 (1994).

[4] C. Kipnis, C. Landim: *Scaling Limit of Interacting Particle Systems*, Springer Verlag, Berlin (1999).

[5] C. Landim, S. Olla, H. T. Yau; First order correction for the hydrodynamic limit of asymmetric simple exclusion processes in dimension $d \geq 3$. Communications on Pure and Applied Mathematics **50**, 149–203, (1997).

[6] C. Landim, S. Olla, H. T. Yau; Some properties of the diffusion coefficient for asymmetric simple exclusion processes. The Annals of Probability **24**, 1779–1807, (1996).

[7] C. Landim, H. T. Yau; Fluctuation–dissipation equation of asymmetric simple exclusion processes, Probab. Th. Rel. Fields **108**, 321–356, (1997)

[8] C. Landim, S. Olla, S. R. S. Varadhan; On viscosity and fluctuation-dissipation in exclusion processes. To appear in J. Stat. Phys. (2003).

[9] C. Landim, J. Quastel, M. Salmhofer, H.T. Yau; Superdiffusivity of asymmetric exclusion process in dimension 1 and 2, to appear in Commun. Math. Phys., arXiv:math.PR/0201317, 2002.

[10] T. M. Liggett; *Interacting Particle Systems*, Springer Verlag, New York, 1985.

[11] F. Rezakhanlou: Hydrodynamic limit for attractive particle systems on Z^d. Commun. Math. Phys. **140**, 417-448, (1990).

[12] S. Sethuraman, S. R. S. Varadhan, H. T. Yau: Diffusive limit of a tagged particle in asymmetric exclusion process, Comm.Pure Appl.Math. **53**, 972-1006, (2000).

[13] H. Spohn: *Large Scale Dynamics of Interacting Particles*, Springer 1991.

[14] H. T. Yau: $(\log t)^{2/3}$ law of the two dimensional asymmetric simple exclusion process, preprint, arXiv:math-ph/0201057 v1, 2002.

Claudio Landim
IMPA
Estrada Dona Castorina 110
CEP 22460 Rio de Janeiro, Brasil
and
CNRS UMR 6085, Université de Rouen
76128 Mont Saint Aignan, France.
E-mail address: `landim@impa.br`

Stefano Olla
Ceremade
UMR CNRS 7534
Université de Paris IX - Dauphine
Place du Maréchal De Lattre De Tassigny
75775 Paris Cedex 16 - France.
E-mail address: olla@ceremade.dauphine.fr

Srinivasa R. S. Varadhan
Courant Institute of Mathematical Sciences
New York University
251 Mercer street
New York, NY, 10012
U.S.A.
E-mail address: varadhan@cims.nyu.edu

Journal of Statistical Physics, Vol. 115, Nos. 1/2, April 2004 (© 2004)

On Viscosity and Fluctuation-Dissipation in Exclusion Processes*

C. Landim,[1] S. Olla,[2] and S. R. S. Varadhan[3]

Received February 12, 2003; accepted April 1, 2003

We consider the asymmetric simple exclusion process. We review some results in dimension $d \geqslant 3$ concerning the fluctuation-dissipation theorem and we prove regularity of viscosity coefficients.

KEY WORDS: Exclusion process; fluctuation-dissipation theorem; regularity of viscosity coefficients.

1. INTRODUCTION

In this paper, we will review some of the results concerning the space-time fluctuations in equilibrium of the asymmetric simple exclusion process in dimension $d \geqslant 3$. These are examples of nonreversible models of interacting particle systems on $\mathbb{Z}^d$ with conservation of particles and a family of ergodic equilibrium distributions indexed by a single parameter, i.e., the density. The model is simple enough to admit a certain amount of explicit calculations.

Let $\hat{\xi}(x)$ be stationary random scalar field on $\mathbb{R}^d$ with a distribution ν that is invariant under translations by $x \in \mathbb{R}^d$. Under some reasonable conditions the distribution μ_ϵ of the field rescaled by

$$\xi_\epsilon(x) = \epsilon^{-\frac{d}{2}}[\hat{\xi}(\epsilon^{-1}x) - \alpha]$$

* Dedicated to Gianni Jona Lasinio on his 70th birthday.

[1] IMPA, Estrada Dona Castorina 110, CEP 22460 Rio de Janeiro, Brasil and CNRS UMR 6085, Université de Rouen, 76128 Mont Saint Aignan, France; e-mail: landim@impa.br

[2] Ceremade, UMR CNRS 7534, Université de Paris IX, Dauphine, Place du Maréchal De Lattre De Tassigny, 75775 Paris Cedex 16, France; e-mail: Stefano.Olla@ceremade.dauphine.fr

[3] Courant Institute of Mathematical Sciences, New York University, 251 Mercer Street, New York, New York 10012; e-mail: varadhan@cims.nyu.edu

where α is the mean $E^{\nu}[\hat{\xi}(x)]$, will converge, as $\epsilon \to 0$, to white noise μ with variance

$$E^{\mu}[\xi(x)\,\xi(y)] = \sigma^2\delta(y-x)$$

If there is a Markovian evolution of the field $\hat{\xi}(\cdot)$ with ν as invariant measure then we have space time process $\hat{\xi}(x, t)$ with distribution $\hat{P}$ with marginals ν for any fixed time. If we now do a space-time rescaling of the form

$$\xi_\epsilon(x, t) = \epsilon^{-\frac{d}{2}}[\xi(\epsilon^{-1}x, \epsilon^{-2}t) - \alpha]$$

the scaled process $\xi_\epsilon(x, t)$ with distribution P_ϵ is expected to converge to an Ornstein–Uhlenbeck type fluctuation process $\xi(t, x)$ with distribution P, satisfying

$$d\xi(t, x) = (\mathscr{A}\xi)(t, x)\,dt + d\beta(x, t)$$

expressed in the weak form as a linear stochastic differential equation

$$d\langle G, \xi(t)\rangle = \langle \mathscr{A}G, \xi(t)\rangle\,dt + d\beta_G(t).$$

Usually $\mathscr{A}$ takes the form of an elliptic second order differential operator with constant coefficients

$$\mathscr{A}G = \sum_{i,j} a_{i,j} D_i D_j G$$

where $\{a_{i,j}\}$ is s symmetric positive definite matrix. The fluctuation-dissipation relation asserts that the family of Brownian motions $\beta_G(\cdot)$ that depend linearly on the test function G satisfy

$$E[[\beta_G(t)]^2] = \sigma^2 t \int_{\mathbb{R}^d} \langle a\,\nabla G, \nabla G\rangle\,dx$$

The proof usually follows the martingale methods. We can write a martingale decomposition (Itô's formula) of the form

$$d\langle G, \xi_\epsilon(t)\rangle = \langle G, \Psi_\epsilon(\xi_\epsilon(t))\rangle\,dt + dM_{\epsilon, G}(t)$$

We only have to check that

$$\lim_{\epsilon \to 0} \langle G, \Psi_\epsilon(\xi(\cdot))\rangle = \langle \mathscr{A}G, \xi\rangle$$

$$\lim_{\epsilon \to 0} E^{P_\epsilon}[[M_{\epsilon, G}(t)]^2] = \sigma^2 t \int_{\mathbb{R}^d} \langle a\,\nabla G, \nabla G\rangle$$

A central limit theorem for Martingales ensures that any limit of $M_{\epsilon, G}$ is a Brownian motion. Except for some technical issues this is the basic proof.

However there are models where something quite different happens. The term $\Psi_\epsilon(\xi)$ becomes big but its integral

$$\int_0^t \langle G, \Psi_\epsilon(\xi(s))\rangle \, ds \tag{1.1}$$

stays finite. Ψ_ϵ is represented as the sum $\phi_\epsilon^{(1)} + \phi_\epsilon^{(2)}$. The big piece $\phi_\epsilon^{(1)}$ is transformed, by a central limit theorem of sorts, into Brownian noise and combines with the noise $M_{\epsilon, G}(t)$ to provide the new $\beta_G(t)$. The remaining part $\phi_\epsilon^{(2)}$ stays finite and becomes

$$\int_0^t \langle \mathscr{A} G, \xi(s)\rangle \, ds$$

where $\mathscr{A}$ again is of the same form, i.e., a second order elliptic operator. This splitting has to done by a carefully constructed decomposition and a new formula provided for the coefficients $\{a_{i,j}\}$ that define $\mathscr{A}$.

The simple exclusion models studied here provide examples of both situations. The symmetric case is an easy text book case, while the asymmetric versions exhibit the more complex behavior alluded to earlier.

2. NOTATION AND RESULTS

Fix a probability distribution $p(\cdot)$ supported on a finite subset of $\mathbb{Z}_*^d = \mathbb{Z}^d \setminus \{0\}$ and denote by L the generator of the simple exclusion process on $\mathbb{Z}^d$ associated to $p(\cdot)$. L acts on local functions f on $\mathscr{X} = \{0, 1\}^{\mathbb{Z}^d}$ as

$$(Lf)(\eta) = \sum_{x, y \in \mathbb{Z}^d} p(y-x)\, \eta(x)\{1-\eta(y)\}[f(\sigma^{x, y}\eta) - f(\eta)] \tag{2.1}$$

where $\sigma^{x, y}\eta$ stands for the configuration obtained from η by exchanging the occupation variables $\eta(x)$, $\eta(y)$:

$$(\sigma^{x, y}\eta)(z) = \begin{cases} \eta(z) & \text{if } z \neq x, y \\ \eta(x) & \text{if } z = y \\ \eta(y) & \text{if } z = x \end{cases}$$

The model describes a possibly infinite system of particles, with at most one particle per site on the lattice $\mathbb{Z}^d$ that evolves according to the following simple random motion. Each particle waits for a standard exponential time, i.e., for a Poisson event of rate 1 to occur, and then tries to jump to a new site that is picked at random. If the original site is x the probability of picking y is $p(y-x)$. If the new site is empty the jump is completed. If the site y is occupied the jump cannot be carried out and the particle is forced to wait for another exponential time. All the particles are performing this motion independently and concurrently. This is described formally by the generator (2.1).

For this model one can check that Bernoulli product measures with $\nu_\alpha[\eta(x)=1]=\alpha$ for $0\leqslant\alpha\leqslant 1$ are equilibrium distributions. This amounts to verifying that

$$\int_X (Lf)(\eta)\, \nu_\alpha(d\eta)=0$$

for every local function $f\colon X\to R$. It is not hard to check that these are actually extremal or ergodic for the evolution. If we fix $0<\alpha<1$, there is a stationary process η_t with values in X, which has ν_α as the distribution at any time t. If $p(\cdot)$ is symmetric about the origin, the generator (2.1) is self-adjoint with respect to ν_α and therefore the process η_t is reversible. In any case if $f(\eta)$ is a local function on X with mean 0 with respect to ν_α then the ergodic theorem asserts that

$$\lim_{T\to\infty}\frac{1}{T}\int_0^T f(\eta_t)\,dt=0$$

almost surely with respect to the process P_α, the stationary Markov process with evolution determined by (2.1) and invariant distribution ν_α. The existence of the conserved quantity results in the absence of a uniform rate in the ergodic theorem. In the reversible case this manifests itself in the absence of a spectral gap for the operaror L given by (2.1) in $L_2(\nu_\alpha)$. Let us start with the density field

$$\sum_{x\in\mathbb{Z}^d}\eta(x)\,\delta_x$$

Define its normalized and rescaled version

$$\xi_\epsilon(\cdot)=\epsilon^{\frac{d}{2}}\sum_x\,[\eta(x)-\alpha]\,\delta_{\epsilon x}$$

which is to be thought of as density fluctuation over a spatial scale of size ϵ^{-1} and express it in weak form as

$$g_\epsilon(\eta)=\langle G,\xi_\epsilon\rangle=\epsilon^{\frac{d}{2}}\sum_x G(\epsilon x)[\eta(x)-\alpha]$$

A computation reveals

$$(Lg_\epsilon)(\eta) = \epsilon^{\frac{d}{2}} \sum_x \sum_y p(y-x)\, \eta(x)(1-\eta(y))[G(\epsilon y) - G(\epsilon x)]$$

If we now assume that $p(\cdot)$ is symmetric,

$$\begin{aligned}(Lg_\epsilon)(\eta) &= \frac{\epsilon^{\frac{d}{2}}}{2} \sum_x \sum_y p(y-x)\, [\eta(x)(1-\eta(y)) - \eta(y)(1-\eta(x))][G(\epsilon y) - G(\epsilon x)] \\ &= \frac{\epsilon^{\frac{d}{2}}}{2} \sum_x \sum_y p(y-x)\, [\eta(x) - \eta(y)][G(\epsilon y) - G(\epsilon x)] \\ &= \epsilon^{\frac{d}{2}} \sum_x \sum_y p(y-x)\, \eta(x)\, [G(\epsilon y) - G(\epsilon x)] \\ &\simeq \epsilon^{\frac{d}{2}+2} \sum_x [\eta(x) - \alpha]\, \mathscr{A} G(\epsilon x)\end{aligned}$$

where $\mathscr{A}$ is the elliptic operator

$$\tfrac{1}{2} \sum_{i,j} a_{i,j} D_{i,j}$$

with

$$a_{i,j} = \sum_x p(x)\, x_i x_j$$

the covariance matrix of $p(\cdot)$. Therefore the density fluctuation

$$\xi_\epsilon = \epsilon^{\frac{d}{2}} \sum_x [\eta(x) - \alpha]\, \delta_{\epsilon x} \tag{2.2}$$

has an approximate dissipation that equals

$$\epsilon^2 \mathscr{A} \xi_\epsilon$$

Speeding up time by the factor ϵ^{-2} takes us to diffusive scaling, replacing L by $\epsilon^{-2} L$, and now the dissipation will be approximately $\mathscr{A}\xi_\epsilon$. There is noise that is built in, in the Poisson process generating the jumps, which scales in the diffusive scale to a Brownian noise, leading to an Ornstein–Uhlenbeck process for the density fluctuations in the diffusive scale. The equilibrium fluctuations are that of white noise

$$E[\langle \xi(t), G \rangle^2] = \alpha(1-\alpha) \int_{R^d} |G|^2\, dx$$

and the evolution of the fluctuations in the diffusive scale takes the form

$$d\langle \xi(t) , G\rangle = \langle G, \mathscr{A}\xi(t)\rangle \, dt + d\beta_G(t)$$

with

$$E[d\beta_G^2] = \alpha(1-\alpha)\left[\left[\int_{R^d} \sum_{i,j} a_{i,j}(D_iG)(D_jG)\, dx\right]\right] dt$$

If we drop the assumption of symmetry on $p(\cdot)$, the situation is much more complex. Let us still suppose that

$$\sum_z zp(z) = 0$$

We cannot proceed beyond

$$(Lg_\epsilon)(\eta) = \epsilon^{\frac{d}{2}} \sum_x \sum_y p(y-x)\, \eta(x)(1-\eta(y))[G(\epsilon y) - G(\epsilon x)]$$

which we can rewrite as

$$(Lg_\epsilon)(\eta) \simeq \epsilon^{\frac{d}{2}+1} \sum_x \sum_y p(y-x)[\eta(x)(1-\eta(y)) - \alpha(1-\alpha)]\langle y-x, \nabla G(\epsilon x)\rangle$$

The rescaling, which should still be by a factor of ϵ^{-2} leads us to a term

$$\begin{aligned}\Psi_\epsilon(\xi) &= \epsilon^{\frac{d}{2}-2} \sum_x \sum_y p(y-x)\, \eta(x)(1-\eta(y))[G(\epsilon y) - G(\epsilon x)] \\ &\simeq \epsilon^{\frac{d}{2}-1} \sum_x \langle \nabla G(\epsilon x), \tau_x W(\eta)\rangle\end{aligned}$$

with $W(\eta) = \{W_i(\eta)\}$ given by

$$W_i(\eta) = \eta(0) \sum_z p(z)\, z_i(1-\eta(z))$$

which is of magnitude ϵ^{-1}. In the reversible case $W_i(\eta)$ is of a special form

$$W_i(\eta) = \sum_z c_i(z)[\eta(z) - \eta(0)]$$

allowing for another summation by parts, getting rid of the troublesome ϵ^{-1} factor and yielding, as earlier,

$$\Psi_\epsilon(\xi) \simeq \langle \mathscr{A}G, \xi_\epsilon\rangle$$

for some second order elliptic differential operator $\mathscr{A}$ that can be calculated. More generally if we have a *gradient system*, i.e., if W_i takes the form

$$W_i(\eta) = \sum_z c_i(z)[\tau_z h(\eta)) - h(\eta)]$$

for some local h with zero mean, then

$$\Psi_\epsilon(\xi) = \epsilon^{\frac{d}{2}} \sum_x (\mathscr{A}G)(\epsilon x)\, \tau_x h(\eta)$$

In such a case one can replace Ψ_ϵ by

$$\hat{\Psi}_\epsilon = c\epsilon^{\frac{d}{2}} \sum_x (\mathscr{A}G)(\epsilon x)[\eta(x) - \alpha]$$

where the constant c is calculated as

$$c = \frac{d}{d\alpha} E^{\nu_\alpha}[h(\eta)]$$

In general if $W(\eta)$ is not of gradient type then we show that $W(\eta)$ can be replaced by $\sum_z c(W, z)[\eta(z) - \eta(0)]$ for a suitable choice of $c(W, z)$.

Finally when $\sum_z zp(z) = m \neq 0$, it turns out that if we speed up time by ϵ^{-1}, then there is a limit for the fluctuations and it is

$$d\xi(x, t) + (1 - 2\alpha)\langle m, \nabla\xi(x, t)\rangle = 0$$

In that scaling the Poisson noise become ineffective and disappears leaving just a translation. If we denote $\gamma = m(1 - 2\alpha)$, then we can center the translations and consider

$$\xi_\epsilon(x - \gamma\epsilon^{-1}t, \epsilon^{-2}t)$$

This will now have a scaling limit and the analysis is very similar to the one for the mean 0 asymmetric case.

In order to carry out all the steps we need to understand the behavior of

$$\int_0^T \sum_{z \in V} \tau_z f(\eta_t)\, dt$$

over large volumes and large times. We can take V to be the box $\{x: |x_i| \leqslant \epsilon^{-1}; 1 \leqslant i \leqslant d\}$. We can take T to be either ϵ^{-1} or ϵ^{-2}. With the choice of $T = \epsilon^{-1}$, we can show that

$$\epsilon^{\frac{d}{2}} \int_0^T \left[\sum_{z \in V} \tau_z(f(\eta_t) - c(f, \alpha)\, \eta(0)) \right] dt$$

is negligible if we choose $c(f, \alpha)$ to be

$$c(f, \alpha) = \frac{d}{d\alpha} E^{\nu_\alpha}[f(\eta)]$$

Now we turn to $T = \epsilon^{-2}$. The main result is Theorem 5.3, known as the fluctuation-dissipation theorem. There are two distinct types of fluctuations. If u is a local function, then fluctuations of the form

$$\int_0^t f_\epsilon(\eta_{s\epsilon^{-2}})\, ds$$

where

$$f_\epsilon(\eta) = \sum_x G(\epsilon x)(\tau_x Lu)(\eta)$$

satisfy a central limit theorem and converge to a Brownian Motion when scaled by $\epsilon^{1+\frac{d}{2}}$.

On the other hand if $v = \sum_z b(z)[\eta(z) - \eta(0)]$ for some b supported on a finite subset of $\mathbb{Z}^d_*$, then

$$\begin{aligned} g_\epsilon &= \sum_x G(\epsilon x)(\tau_x v)(\eta) = \sum_{x,z} G(\epsilon x)\, b(z)[\eta(x+z) - \eta(x)] \\ &= \sum_x G(\epsilon x)\, \tau_x h(\eta) \end{aligned}$$

with $h(\eta) = \sum_z b(z)[\eta(z) - \eta(0)]$ and as we saw before, a summation by parts helps. Consequently fluctuations of the form

$$\int_0^t g_\epsilon(\eta_s \epsilon^{-2})\, ds$$

when normalized again by $\epsilon^{1+\frac{d}{2}}$ correspond to a time average of density fluctuations.

The analysis of *non-gradient systems* depends on our ability to write any W that satisfies some conditions as

$$W = Lu + h$$

While u and h may not be local functions, what we do show is that it is possible to approximate W well enough by $Lu+h$ with some local u and h.

The fluctuation dissipation theorem asserts that if $w \in \bigoplus_{n \geqslant 2} \mathscr{G}_n$ (see the end of the section for a definition) is a local function then there are coefficients $b(z, w)$ such that the fluctuation of $w - \sum_z b(z, w)[\eta(z) - \eta(0)]$ is well approximated by fluctuations of the first type. The constants $b(z, w)$ are defined for each α. We prove in Section 7 that these are C^∞ functions of α.

The study of fluctuations therefore depends on controlling space-time correlations of the form

$$E^{\nu_\alpha}\left[\left|\int_0^T \sum_{z \in V} \tau_z f(\eta_t)\, dt\right|^2\right]$$

for large times T and large volumes V.

The natural way to control such objects is to invert the generator of the process L, i.e., to solve the equation

$$Lu = \sum_{z \in V} \tau_z f \tag{2.3}$$

perhaps with $u = \sum_{z \in V} \tau_z v$ for some v.

Our goal is to show that we can approximate, in a proper weak sense, the solution of equation (2.3) by such functions. Our interest is to perform such approximation in a diffusive space-time scaling limit (i.e., $T \sim |V|^{2/d}$).

We will characterize the functions f such that the corresponding asymptotic space-time variance is finite and then we will prove that this variance is a smooth function of the density α.

There is a natural orthonormal basis for $L_2(\nu_\alpha)$ indexed by finite subsets A of Z^d that commutes with τ_z. Moreover there is a corresponding decomposition of $L_2(\nu_\alpha)$ as $\bigoplus_{n \geqslant 0} \mathscr{G}_n$ as the direct sum of orthogonal subspaces according to the cardinality n of A. Since, in this context, one cannot distinguish between f and $\tau_z f$, A and $\tau_z A$ can be identified. A reasonable cross section, one with a multiplicity of n for sets of cardinality n can be found for the equivalence classes. Another important point is that that in the symmetric case L leaves each $\mathscr{G}_n$ invariant so that the inversion needs to be done only on each $\mathscr{G}_n$, which is no harder than the analysis of random walk of $n-1$ particles on Z^d. If $d \geqslant 3$, the asymmetric case can be treated as a perturbation of the symmetric one. Although this is a somewhat singular perturbation the transience of the random walk in $d \geqslant 3$ and the fact that the perturbation only causes $\mathscr{G}_n$ to be mapped into $\mathscr{G}_{n-1} \oplus \mathscr{G}_n \oplus \mathscr{G}_{n+1}$ help the analysis. The perturbation is controlled with the use of weighted norms with weights growing polynomially in the degree n of $\mathscr{G}_n$.

Duality for the symmetric simple exclusion was first observed by Spitzer and was broadly used in this context (cf. ref. 12). The use of a dual base in order to study fluctuations in the asymmetric simple exclusion in dimension $d \geqslant 3$ was introduced in ref. 8 (where the fluctuation-dissipation theorem was first proved) and in ref. 13. The fluctuation dissipation theorem was then applied in order to study the diffusive incompressible limit (cf. ref. 3), the first order corrections to the hydrodynamic limit (cf. ref. 6) and the equilibrium fluctuations for the density field.[2]

Always with the use of this dual base, in ref. 11 we prove the regularity of the selfdiffusion coefficient (as function of the density) for the *symmetric* simple exclusion (reversible). Using this approach in ref. 1 is proved the regularity of the bulk diffusion coefficient for a non-gradient speed change exclusion that has ν_α as equilibrium (reversible) measures.

3. DUALITY

We examine in this section the action of the symmetric part L^s of the generator on the space of local functions endowed with a particular scalar product $\langle\langle \cdot, \cdot \rangle\rangle$. Fix once for all a density α in $(0, 1)$. All expectations in this section are taken with respect to ν_α and we omit all sub-indices.

3.1. The Dual Space

For each $n \geqslant 0$, denote by $\mathscr{E}_n$ the subsets of $\mathbb{Z}$ with n points and let $\mathscr{E} = \bigcup_{n \geqslant 0} \mathscr{E}_n$ be the class of finite subsets of $\mathbb{Z}^d$. For each A in $\mathscr{E}$, let Ψ_A be the local function

$$\Psi_A = \prod_{x \in A} \frac{\eta(x) - \alpha}{\sqrt{\chi(\alpha)}}$$

where $\chi(\alpha) = \alpha(1-\alpha)$. By convention, $\Psi_\phi = 1$. It is easy to check that $\{\Psi_A, A \in \mathscr{E}\}$ is an orthonormal basis of $L^2(\nu_\alpha)$. For each $n \geqslant 0$, denote by $\mathscr{G}_n$ the subspace of $L^2(\nu_\alpha)$ generated by $\{\Psi_A, A \in \mathscr{E}_n\}$, so that $L^2(\nu_\alpha) = \bigoplus_{n \geqslant 0} \mathscr{G}_n$. Functions in $\mathscr{G}_n$ are said to have degree n.

Consider a local function f. Since $\{\Psi_A\colon A \in \mathscr{E}\}$ is a basis of $L^2(\nu_\alpha)$, we may write

$$f = \sum_{n \geq 0} \sum_{A \in \mathscr{E}_n} \mathfrak{f}(A)\, \Psi_A$$

Note that the coefficients $\mathfrak{f}(A)$ depend not only on f but also on the density α: $\mathfrak{f}(A) = \mathfrak{f}(A, \alpha)$. Since f is a local function, $\mathfrak{f}\colon \mathscr{E} \to \mathbb{R}$ is a function of finite support.

For local functions u, v, define the scalar product $\langle\!\langle \cdot\,, \cdot \rangle\!\rangle$ by

$$\langle\!\langle u, v \rangle\!\rangle = \sum_{x \in \mathbb{Z}^d} \{\langle \tau_x u, v \rangle - \langle u \rangle \langle v \rangle\} \tag{3.1}$$

where $\{\tau_x, x \in \mathbb{Z}^d\}$ is the group of translations. That this is in fact an inner product can be seen by the relation

$$\langle\!\langle u, v \rangle\!\rangle = \lim_{V \uparrow \mathbb{Z}^d} \frac{1}{|V|} \left\langle \sum_{x \in V} \tau_x(u - \langle u \rangle), \sum_{x \in V} \tau_x(v - \langle v \rangle) \right\rangle.$$

Since $\langle\!\langle u - \tau_x u\,, v \rangle\!\rangle = 0$ for all x in $\mathbb{Z}^d$, this scalar product is only semidefinite. Denote by $L^2_{\langle\!\langle \cdot, \cdot \rangle\!\rangle}(\nu_\alpha)$ the Hilbert space generated by the local functions and the inner product $\langle\!\langle \cdot, \cdot \rangle\!\rangle$. The scalar product of two local functions u, v can be written in terms of the Fourier coefficients of u, v through a simple formula. To this end, fix two local functions u, v and write them in the basis $\{\Psi_A, A \in \mathscr{E}\}$:

$$u = \sum_{A \in \mathscr{E}} \mathfrak{u}(A)\, \Psi_A, \qquad v = \sum_{A \in \mathscr{E}} \mathfrak{v}(A)\, \Psi_A$$

An elementary computation shows that

$$\langle\!\langle u, v \rangle\!\rangle = \sum_{x \in \mathbb{Z}^d} \sum_{n \geq 1} \sum_{A \in \mathscr{E}_n} \mathfrak{u}(A)\, \mathfrak{v}(A + x)$$

In this formula, $B + z$ is the set $\{x + z; x \in B\}$. The summation starts from $n = 1$ due to the centering by the mean in the definition of the inner product $\langle\!\langle \cdot, \cdot \rangle\!\rangle$.

We say that two finite subsets A, B of $\mathbb{Z}^d$ are equivalent if one is the translation of the other. This equivalence relation is denoted by $\sim$ so that $A \sim B$ if $A = B + x$ for some x in $\mathbb{Z}^d$. Let $\tilde{\mathscr{E}}_n$ be the quotient of $\mathscr{E}_n$ with respect to this equivalence relation: $\tilde{\mathscr{E}}_n = \mathscr{E}_n/_\sim$, $\tilde{\mathscr{E}} = \mathscr{E}/_\sim$. If, for some n, $\mathfrak{f}\colon \mathscr{E}_n \to \mathbb{R}$ is a summable function,

$$\sum_{A \in \mathscr{E}_n} \mathfrak{f}(A) = \sum_{\tilde{A} \in \tilde{\mathscr{E}}_n} \tilde{\mathfrak{f}}(\tilde{A})$$

where, for any equivalence class $\tilde{A}$ and a summable function $\mathfrak{f}: \mathscr{E} \to \mathbb{R}$,

$$\tilde{\mathfrak{f}}(\tilde{A}) = \sum_{z \in \mathbb{Z}^d} \mathfrak{f}(A+z) \tag{3.2}$$

A being any representative from $\tilde{A}$.

In particular, for two local functions u, v,

$$\begin{aligned} \langle\!\langle u, v \rangle\!\rangle &= \sum_{x, z \in \mathbb{Z}^d} \sum_{n \geq 1} \sum_{\tilde{A} \in \tilde{\mathscr{E}}_n} \mathfrak{u}(A+z)\, \mathfrak{v}(A+x+z) \\ &= \sum_{n \geq 1} \sum_{\tilde{A} \in \tilde{\mathscr{E}}_n} \tilde{\mathfrak{u}}(\tilde{A})\, \tilde{\mathfrak{v}}(\tilde{A}) \end{aligned}$$

We say that a function $\mathfrak{f}: \mathscr{E} \to \mathbb{R}$ is translation invariant if $\mathfrak{f}(A+x) = \mathfrak{f}(A)$ for all sets A in $\mathscr{E}$ and all sites x of $\mathbb{Z}^d$. Of course, functions $\tilde{\mathfrak{f}}$ on $\tilde{\mathscr{E}}$ are the same as translation invariant functions on $\mathscr{E}$. Fix a subset A of $\mathbb{Z}^d$ with n points. There are n sets in the class of equivalence of A that contain the origin.

Therefore, summing a translation invariant function $\mathfrak{f}$ over all equivalence classes $\tilde{A}$ in $\tilde{\mathscr{E}}_n$ is the same as summing $\mathfrak{f}$ over all sets B in $\mathscr{E}_n$ which contain the origin and dividing by n:

$$\sum_{\tilde{A} \in \tilde{\mathscr{E}}_n} \mathfrak{f}(\tilde{A}) = \frac{1}{n} \sum_{\substack{A \in \mathscr{E}_n \\ A \ni 0}} \mathfrak{f}(A)$$

provided $\mathfrak{f}(A) = \mathfrak{f}(A+x)$ for all A, x. Let $\mathscr{E}_*$ be the class of all finite subsets of $\mathbb{Z}^d_* = \mathbb{Z}^d \setminus \{0\}$ and let $\mathscr{E}_{*,n}$ be the class of all subsets of $\mathbb{Z}^d_*$ with n points. Then, we may write

$$\begin{aligned} \langle\!\langle u, v \rangle\!\rangle &= \sum_{n \geq 1} \frac{1}{n} \sum_{\substack{A \in \mathscr{E}_n \\ A \ni 0}} \tilde{\mathfrak{u}}(A)\, \tilde{\mathfrak{v}}(A) \\ &= \sum_{n \geq 0} \frac{1}{n+1} \sum_{A \in \mathscr{E}_{*,n}} \tilde{\mathfrak{u}}(A \cup \{0\})\, \tilde{\mathfrak{v}}(A \cup \{0\}) \end{aligned}$$

In conclusion, if for a finitely supported function $\mathfrak{f}: \mathscr{E} \to \mathbb{R}$, we define $\mathfrak{T}\mathfrak{f}: \mathscr{E}_* \to \mathbb{R}$ by

$$(\mathfrak{T}\mathfrak{f})(A) = \tilde{\mathfrak{f}}(A \cup \{0\}) = \sum_{z \in \mathbb{Z}^d} \mathfrak{f}([A \cup \{0\}] + z)$$

we have that

$$\langle\langle u, v \rangle\rangle = \sum_{n \geqslant 0} \frac{1}{n+1} \sum_{A \in \mathscr{E}_{*,n}} \mathfrak{T}\mathfrak{u}(A)\, \mathfrak{T}\mathfrak{v}(A) \tag{3.3}$$

For $n \geqslant 0$, denote by π_n the projection that corresponds to $\mathscr{E}_{*,n}$ i.e., $(\pi_n \mathfrak{f})(A) = \mathfrak{f}(A)\, \mathbf{1}\{A \in \mathscr{E}_{*,n}\}$ and denote by $\langle \cdot, \cdot \rangle$ the usual scalar product on each set $\mathscr{E}_{*,n}$: for $\mathfrak{f}$, $\mathfrak{g}$: $\mathscr{E}_{*,n} \to \mathbb{R}$,

$$\langle \mathfrak{f}, \mathfrak{g} \rangle = \sum_{A \in \mathscr{E}_{*,n}} \mathfrak{f}(A)\, \mathfrak{g}(A)$$

In view of formula (3.3), it is natural to introduce, for an integer $k \geqslant -1$, the Hilbert spaces $L^2(\mathscr{E}_*, k)$ generated by finite supported functions $\mathfrak{f}$: $\mathscr{E}_* \to \mathbb{R}$ and the scalar product $\langle\langle \cdot, \cdot \rangle\rangle_{0,k}$ defined by

$$\langle\langle \mathfrak{f}, \mathfrak{g} \rangle\rangle_{0,k} = \sum_{n \geqslant 0} (n+1)^k \langle \pi_n \mathfrak{f}, \pi_n \mathfrak{g} \rangle$$

With this notation, for local functions f, g in $\bigoplus_{n \geqslant 1} \mathscr{G}_n$,

$$\langle\langle f, g \rangle\rangle = \langle\langle \mathfrak{T}\mathfrak{f}, \mathfrak{T}\mathfrak{g} \rangle\rangle_{0,-1}$$

To summarize some observations on the transformation $\mathfrak{T}$, we need some notation. For a subset A of $\mathbb{Z}_*^d$ and x, y, z in $\mathbb{Z}_*^d$, denote by $A_{x,y}$ the set defined by

$$A_{x,y} = \begin{cases} (A \backslash \{x\}) \cup \{y\} & \text{if } \; x \in A,\ y \notin A \\ (A \backslash \{y\}) \cup \{x\} & \text{if } \; y \in A,\ x \notin A \\ A & \text{otherwise} \end{cases} \tag{3.4}$$

and denote by $S_z A$ the set defined by

$$S_z A = \begin{cases} A - z & \text{if } \; z \notin A \\ (A - z)_{0,-z} & \text{if } \; z \in A \end{cases} \tag{3.5}$$

Therefore, to obtain $S_z A$ from A in the case where z belongs to A, we first translate A by $-z$, getting a new set which contains the origin, and we then remove the origin and add site $-z$.

Remark 3.1.

(a) The restriction of $\mathfrak{f}$ to $\mathscr{E}_1$ is irrelevant for the definition of $\mathfrak{T}\mathfrak{f}(A)$ if A is nonempty.

(b) Not every function $\mathfrak{f}_*: \mathscr{E}_* \to \mathbb{R}$ is the image by $\mathfrak{T}$ of some function $\mathfrak{f}: \mathscr{E} \to \mathbb{R}$ since

$$(\mathfrak{T}\mathfrak{f})(A) = (\mathfrak{T}\mathfrak{f})(S_z A) \tag{3.6}$$

for all z in A.

(c) Let $\mathfrak{f}_*: \mathscr{E}_* \to \mathbb{R}$ be a finitely supported function satisfying (3.6): $\mathfrak{f}_*(A) = \mathfrak{f}_*(S_z A)$ for all z in A. Define $\mathfrak{f}: \mathscr{E} \to \mathbb{R}$ by

$$\mathfrak{f}(B) = \begin{cases} |B|^{-1}\, \mathfrak{f}_*(B \setminus \{0\}) & \text{if} \quad B \ni 0 \\ 0 & \text{otherwise} \end{cases} \tag{3.7}$$

An elementary computations shows that $\mathfrak{T}\mathfrak{f} = \mathfrak{f}_*$. With this choice, which is natural but not unique, $\mathfrak{f}(0) = \mathfrak{f}_*(\phi)$.

(d) $\mathfrak{T}$ maps $\mathscr{E}_n$ into $\mathscr{E}_{*,n-1}$ lowering the degree of a function by one. Thus the translations in the inner product $\langle\langle \cdot, \cdot \rangle\rangle$ effectively reduce the degree by one while replacing the space $\mathbb{Z}^d$ by $\mathbb{Z}^d_*$.

(e) Formula (3.3) shows also that a local function f is in the kernel of the inner product $\langle\langle \cdot, \cdot \rangle\rangle$ if and only if $\mathfrak{T}\mathfrak{f}$ vanishes, i.e., if and only if

$$\sum_{x \in \mathbb{Z}^d} \mathfrak{f}(A + x) = 0$$

for all finite subsets A such that $|A| \geqslant 1$. Examples of such functions are the constants or the difference of the translations of a local function: $\tau_y f - \tau_x f$.

To keep notation simple, most of the time, for a function $\mathfrak{f}: \mathscr{E} \to \mathbb{R}$, we denote $\mathfrak{T}\mathfrak{f}$ by $\bar{\mathfrak{f}}$. Real functions on $\mathscr{E}$ or on $\mathscr{E}_*$ are indistinctively denoted by the symbols $\mathfrak{f}$, $\mathfrak{g}$.

3.2. Some Hilbert Spaces

We investigate in this subsection the action of the symmetric part of the generator L on the basis $\{\Psi_A, A \in \mathscr{E}_*\}$.

Fix a function u of degree $n \geqslant 1$ and denote by $\mathfrak{u}$ its Fourier coefficients. A straightforward computation shows that

$$L^s u = \sum_{A \in \mathscr{E}_n} (\mathscr{L}_s \mathfrak{u})(A)\, \Psi_A \tag{3.8}$$

where $\mathscr{L}_s$ is the generator of finite symmetric random walks evolving with exclusion on $\mathbb{Z}^d$:

$$(\mathscr{L}_s \mathfrak{u})(A) = (1/2) \sum_{x,\, y \in \mathbb{Z}^d} s(y-x)[\mathfrak{u}(A_{x,y}) - \mathfrak{u}(A)] \tag{3.9}$$

and $A_{x,y}$ is the set defined by (3.4). Furthermore, an elementary computation, based on the fact that

$$\sum_{z \in \mathbb{Z}^d} \mathfrak{f}([B \cup \{y\}] + z) = \mathfrak{T}\mathfrak{f}(S_y B)$$

for all subsets B of $\mathbb{Z}^d_*$, sites y not in B and finitely supported functions $\mathfrak{f}\colon \mathscr{E} \to \mathbb{R}$, shows that for every set B in $\mathscr{E}_*$

$$\mathfrak{T}\mathscr{L}_s \mathfrak{u}\,(B) = \mathfrak{L}_s \mathfrak{T}\mathfrak{u}\,(B) \tag{3.10}$$

where

$$(\mathfrak{L}_s \bar{\mathfrak{u}})\,(B) = (1/2) \sum_{x,\, y \in \mathbb{Z}^d_*} s(y-x)[\bar{\mathfrak{u}}(B_{x,y}) - \bar{\mathfrak{u}}(B)]$$
$$+ \sum_{y \notin B} s(y)[\bar{\mathfrak{u}}(S_y B) - \bar{\mathfrak{u}}(B)]$$

This computation should be understood as follows. We introduced an equivalence relation in $\mathscr{E}$ when we decided not to distinguish between a set and its translations. This is the same as assuming that all sets contain the origin. If n particles evolve as exclusion random walks on $\mathbb{Z}^d$, one of them fixed to be at the origin, two things may happen. Either one of the particles which is not at the origin jumps or the particle we assumed to be at the origin jumps. In the first case, this is just a jump on $\mathbb{Z}^d_*$ and is taken care by the first piece of the generator $\mathfrak{L}_s$. In the second case, however, since we are imposing the origin to be always occupied, we need to translate back the configuration to the origin. This part corresponds to the second piece of the generator $\mathfrak{L}_s$.

We are now in a position to define the Hilbert space induced by the local functions $\mathscr{C}$, the symmetric part of the generator L and the scalar product $\langle\!\langle \cdot, \cdot \rangle\!\rangle$. For two local functions u, v, let

$$\langle\!\langle u, v \rangle\!\rangle_1 = \langle\!\langle u, (-L^s)\, v \rangle\!\rangle$$

and let $H_1 = H_1(\mathscr{C}, L^s, \langle\!\langle \cdot, \cdot \rangle\!\rangle)$ be the Hilbert space generated by local functions f and the inner product $\langle\!\langle \cdot, \cdot \rangle\!\rangle_1$. By (3.8), (3.3), and (3.10) the previous scalar product is equal to

$$-\sum_{n\geqslant 0} \frac{1}{n+1} \sum_{A\in\mathscr{E}_{*,n}} \bar{\mathfrak{u}}(A)\, \overline{\mathscr{L}_s \mathfrak{v}}(A) = -\sum_{n\geqslant 0} \frac{1}{n+1} \sum_{A\in\mathscr{E}_{*,n}} \bar{\mathfrak{u}}(A)\, (\mathfrak{L}_s \bar{\mathfrak{v}})(A)$$

$$= \sum_{n\geqslant 0} \frac{1}{n+1} \langle \pi_n \bar{\mathfrak{u}}\,,\, (-\mathfrak{L}_s)\, \pi_n \bar{\mathfrak{v}} \rangle$$

because $\mathfrak{L}_s$ keeps the degree of the functions mapping $\mathscr{E}_{*,n}$ in $\mathscr{E}_{*,n}$. This formula leads to the following definitions. For each $n\geqslant 0$, denote by $\langle \cdot, \cdot \rangle_1$ the scalar product on $\mathscr{E}_{*,n}$ defined by

$$\langle \mathfrak{f}, \mathfrak{g} \rangle_1 = \langle \mathfrak{f}, (-\mathfrak{L}_s)\, \mathfrak{g} \rangle$$

and denote by $\mathfrak{H}_1(\mathscr{E}_{*,n})$ the Hilbert space on $\mathscr{E}_{*,n}$ induced by the finitely supported functions and the scalar product $\langle \cdot, \cdot \rangle_1$. The associated norm is denoted by $\|\mathfrak{f}\|_1^2 = \langle \mathfrak{f}, (-\mathfrak{L}_s)\, \mathfrak{f} \rangle$. Furthermore, for an integer $k \geqslant -1$, denote by $\mathfrak{H}_{1,k} = \mathfrak{H}_1(\mathscr{E}_*, \mathfrak{L}_s, k)$ the Hilbert space induced by the finite supported functions $\mathfrak{f}, \mathfrak{g}\colon \mathscr{E}_* \to \mathbb{R}$ and scalar product

$$\langle\!\langle \mathfrak{f}, \mathfrak{g} \rangle\!\rangle_{1,k} = \langle\!\langle \mathfrak{f}, (-\mathfrak{L}_s)\, \mathfrak{g} \rangle\!\rangle_{0,k} = \sum_{n\geqslant 0} (n+1)^k \langle \pi_n \mathfrak{f}, (-\mathfrak{L}_s)\, \pi_n \mathfrak{g} \rangle$$

The associated norm is denoted by $\|\cdot\|_{1,k}$ so that $\|\mathfrak{f}\|_{1,k}^2 = \langle\!\langle \mathfrak{f}, \mathfrak{f} \rangle\!\rangle_{1,k}$

Three observations are in order. First of all, in the definition of the scalar product $\langle\!\langle \cdot, \cdot \rangle\!\rangle_{1,k}$, because $\mathfrak{L}_s \mathfrak{f}(\phi) = 0$ for all $\mathfrak{f}$, it is irrelevant if the summation starts from $n = 0$ or from $n = 1$. On the other hand, it follows from the previous relation that

$$\|\mathfrak{f}\|_{1,k}^2 = \sum_{n\geqslant 0} (n+1)^k\, \|\pi_n \mathfrak{f}\|_1^2$$

Finally, for every local function u, v,

$$\langle\!\langle u, v \rangle\!\rangle_1 = \langle\!\langle \mathfrak{T} u, \mathfrak{T} v \rangle\!\rangle_{1,-1}$$

vanishes for functions u and v of degree 0 or 1. Moreover, for every $n\geqslant 1$ and every finitely supported functions $\mathfrak{f}, \mathfrak{g}\colon \mathscr{E}_{*,n} \to \mathbb{R}$,

$$\langle \mathfrak{f}, \mathfrak{g} \rangle_1 = \tfrac{1}{4} \sum_{x,\, y\neq 0} s(y-x) \sum_{A\in\mathscr{E}_{*,n}} [\mathfrak{g}(A_{x,y}) - \mathfrak{g}(A)]\, [\mathfrak{f}(A_{x,y}) - \mathfrak{f}(A)]$$

$$+ \tfrac{1}{2} \sum_{y\in\mathbb{Z}^d} s(y) \sum_{\substack{A\in\mathscr{E}_{*,n}\\ A\not\ni y}} [\mathfrak{g}(S_y A) - \mathfrak{g}(A)]\, [\mathfrak{f}(S_y A) - \mathfrak{f}(A)]$$

To introduce the dual Hilbert spaces of H_1, $\mathfrak{H}_1$, for a local function u, consider the semi-norm $\| \cdot \|_{-1}$ given by

$$\|u\|_{-1} = \sup_{v} \{2\langle\!\langle u, v \rangle\!\rangle - \langle\!\langle v, v \rangle\!\rangle_1\}$$

where the supremum is taken over all local functions v. Denote by $H_{-1} = H_{-1}(\mathscr{C}, L^s, \langle\!\langle \cdot, \cdot \rangle\!\rangle)$ the Hilbert space generated by the local functions and the semi-norm $\| \cdot \|_{-1}$.

Recall the definition of the spaces $\mathscr{G}_n$ introduced at the beginning of Section 3.1. Since L^s keeps the degree of a function and since the spaces $\mathscr{G}_n$ are orthogonal, for local functions of fixed degree, we may restrict the supremum to local functions of the same degree so that

$$\|f\|_{-1}^2 = \sum_{n \geqslant 1} \|\pi_n f\|_{-1}^2$$

where $\pi_n f$ stands for the projection of f on $\mathscr{G}_n$. Moreover, we will see later in (4.1) that functions of degree 1 which do not vanish in $L^2(\langle\!\langle \cdot, \cdot \rangle\!\rangle)$ have infinite H_{-1} norm.

In the same way, for an integer $n \geqslant 1$ and a finitely supported function $\mathfrak{u}: \mathscr{E}_{*,n} \to \mathbb{R}$, let

$$\|\mathfrak{u}\|_{-1}^2 = \sup_{\mathfrak{v}} \{2\langle \mathfrak{u}, \mathfrak{v} \rangle - \langle \mathfrak{v}, \mathfrak{v} \rangle_1\}$$

where the supremum is carried over all finitely supported functions $\mathfrak{v}: \mathscr{E}_{*,n} \to \mathbb{R}$. Denote by $\mathfrak{H}_{-1} = \mathfrak{H}_{-1}(\mathscr{E}_{*,n})$ the Hilbert space induced by the finitely supported functions $\mathfrak{u}: \mathscr{E}_{*,n} \to \mathbb{R}$ and the semi-norm $\| \cdot \|_{-1}$.

For a integer $k \geqslant -1$, define the $\mathfrak{H}_{-1,k} = \mathfrak{H}_{-1}(\mathscr{E}_*, \mathfrak{L}_s, k)$ norm of a finite supported function $\mathfrak{u}: \mathscr{E}_+ \to \mathbb{R}$ by

$$\|\mathfrak{u}\|_{-1,k} = \sup_{\mathfrak{v}} \{2\langle\!\langle \mathfrak{u}, \mathfrak{v} \rangle\!\rangle_{0,k} - \langle\!\langle \mathfrak{v}, (-\mathfrak{L}_s)\, \mathfrak{v} \rangle\!\rangle_{0,k}\}$$

where the supremum is carried over all finitely supported functions $\mathfrak{v}: \mathscr{E}_* \to \mathbb{R}$. Denote by $\mathfrak{H}_{-1,k} = \mathfrak{H}_{-1}(\mathscr{E}_*, \mathfrak{L}_s, k)$ the Hilbert space induced by this semi-norm and the space of finite supported functions. Here again, since $\mathfrak{L}_s$ does not change the degrees of a function, for every finitely supported $\mathfrak{u}: \mathscr{E}_* \to \mathbb{R}$,

$$\|\mathfrak{u}\|_{-1,k}^2 = \sum_{n \geqslant 1} (n+1)^k \, \|\pi_n \mathfrak{u}\|_{-1}^2$$

and for any local function u,

$$\|u\|_{-1} = \|\mathfrak{T}\mathfrak{u}\|_{-1,-1}$$

3.3. The Fourier Coefficients of the Generator *L*

We conclude this section deriving explicit expressions for the generator L on the basis $\{\Psi_A, A \subset \mathbb{Z}^d\}$. A long and elementary computation gives the following dual representation: For every local function $u = \sum_{A \in \mathscr{E}} \mathfrak{u}(A)\,\Psi_A$,

$$Lu = \sum_{A \in \mathscr{E}} (\mathscr{L}_\alpha \mathfrak{u})(A)\,\Psi_A,$$

where $\mathscr{L}_\alpha = \mathscr{L}_s + (1-2\alpha)\,\mathscr{L}_d + \sqrt{\chi(\alpha)}\,(\mathscr{L}_+ + \mathscr{L}_-)$,

$$(\mathscr{L}_d \mathfrak{u})(A) = \sum_{x \in A,\, y \notin A} a(y-x)\{\mathfrak{u}(A_{x,y}) - \mathfrak{u}(A)\}$$

$$(\mathscr{L}_+ \mathfrak{u})(A) = 2 \sum_{x \in A,\, y \in A} a(y-x)\,\mathfrak{u}(A \setminus \{y\})$$

$$(\mathscr{L}_- \mathfrak{u})(A) = 2 \sum_{x \notin A,\, y \notin A} a(y-x)\,\mathfrak{u}(A \cup \{y\})$$

and $\mathscr{L}_s$ is defined by (3.9). Furthermore, for any function $\mathfrak{u}\colon \mathscr{E} \to \mathbb{R}$, $\mathfrak{T}\mathscr{L}_\alpha \mathfrak{u} = \mathfrak{L}_\alpha \mathfrak{T}\mathfrak{u}$, provided

$$\mathfrak{L}_\alpha = \mathfrak{L}_s + (1-2\alpha)\,\mathfrak{L}_d + \sqrt{\chi(\alpha)}\,\{\mathfrak{L}_+ + \mathfrak{L}_-\}$$

and, for $A \in \mathscr{E}_*$, $\mathfrak{v}\colon \mathscr{E}_* \to \mathbb{R}$ a finitely supported function,

$$(\mathfrak{L}_d \mathfrak{v})(A) = \sum_{\substack{x \in A,\, y \notin A \\ x,\, y \neq 0}} a(y-x)\{\mathfrak{v}(A_{x,y}) - \mathfrak{v}(A)\} + \sum_{\substack{y \notin A \\ y \neq 0}} a(y)\{\mathfrak{v}(S_y A) - \mathfrak{v}(A)\}$$

$$(\mathfrak{L}_+ \mathfrak{v})(A) = 2 \sum_{x \in A,\, y \in A} a(y-x)\,\mathfrak{v}(A \setminus \{y\})$$

$$+ 2 \sum_{x \in A} a(x)\{\mathfrak{v}(A \setminus \{x\}) - \mathfrak{v}(S_x[A \setminus \{x\}])\}$$

$$(\mathfrak{L}_- \mathfrak{v})(A) = 2 \sum_{\substack{x \notin A,\, y \notin A \\ x,\, y \neq 0}} a(y-x)\,\mathfrak{v}(A \cup \{y\})$$

4. APPROXIMATIONS IN H_{-1}

From this section on, we work in dimension $d \geqslant 3$. All finite constants denoted by C_0 are allowed to depend on the transition probability $p(\cdot)$ and only on $p(\cdot)$.

The main goal of this section is to show that finitely supported functions can be approximated in H_{-1} by finitely supported functions in the image of the operator $\mathfrak{L}_\alpha$. We start proving that all finitely supported functions $\mathfrak{w}: \mathscr{E}_* \to \mathbb{R}$ such that $\mathfrak{w}(\phi) = 0$ are in H_{-1}.

Several estimates on the operators $\mathfrak{L}_d$, $\mathfrak{L}_-$, $\mathfrak{L}_+$ are assumed here and proved in Section 7.

Theorem 4.1. A finitely supported function $\mathfrak{w}: \mathscr{E}_* \to \mathbb{R}$ belongs to $\mathfrak{H}_{-1,k}(\mathscr{E}_*, \mathfrak{L}_s)$ for every $k \geqslant 1$ provided $\mathfrak{w}(\phi) = 0$.

This result states that any local function w in $\bigoplus_{n\geqslant 2} \mathscr{G}_n$ belongs to $H_{-1}(\mathscr{C}, L^s, \langle\!\langle \cdot, \cdot \rangle\!\rangle)$. Notice that we are requiring the degree to be greater or equal to 2. The reason is simple. Any local function f of degree one has H_1 norm equal to 0: if $f = \sum_{x\in\mathbb{Z}^d} c_x \Psi_x$,

$$\begin{aligned}\langle\!\langle f, Lf \rangle\!\rangle &= \langle\!\langle f, L^s f \rangle\!\rangle \\ &= \sum_{x,y\in\mathbb{Z}^d} c_x c_y \sum_{z,w\in\mathbb{Z}^d} s(w) \langle \Psi_{x+z}, \Psi_{y+w} - \Psi_y \rangle = 0 \end{aligned} \tag{4.1}$$

There are thus two possibilities. Either $\sum_x c_x = 0$, in which case $f = 0$ both in $L^2(\langle\!\langle \cdot, \cdot \rangle\!\rangle)$ and in H_1 or $\sum_x c_x \neq 0$ in which case $f = 0$ in H_1 but not in L^2. In this later circumstance, f does not belong to H_{-1}. Therefore, a function of degree one belongs to H_{-1} only if it vanishes in L^2, i.e., only if $\sum_x c_x = 0$. This explains the restriction on the degree in the previous theorem.

Theorem 4.1 is proved in the same way as Lemma 2.1 in[13]. One has to show that the Green function associated to the generator $\mathfrak{L}_s$ restricted to $\mathscr{E}_{*,n}$, for some $n \geqslant 1$, is finite. This is done by comparing the Green function with the one associated to independent random walks, which is finite because $d \geqslant 3$.

Recall from Remark 3.1 in the previous section that for a finitely supported function $\mathfrak{f}: \mathscr{E} \to \mathbb{R}$, $(\mathfrak{T}\mathfrak{f})(A) = (\mathfrak{T}\mathfrak{f})(S_z A)$ for all z in A. Thus, for $n \geqslant 0$, denote by $\mathscr{I}_n$ the closed subspace of $L^2(\mathscr{E}_{*,n})$ of all functions $\mathfrak{f}$ for which (3.6) holds and let $\mathscr{I} = \bigoplus_{n\geqslant 0} \mathscr{I}_n$.

We are now in a position to state the main result of this section. We have just seen that all finitely supported functions $\mathfrak{w}: \mathscr{E}_* \to \mathbb{R}$ such that

$\mathfrak{w}(\phi) = 0$ belong to H_{-1}. We now prove that these functions can be approximated in H_{-1} by finitely supported functions in the image of the operator $\mathfrak{L}_\alpha$.

Theorem 4.2. Fix a finitely supported function $\mathfrak{w}: \mathscr{E}_* \to \mathbb{R}$ in $\mathscr{I}$ such that $\mathfrak{w}(\phi) = 0$. For each $\varepsilon > 0$ and $k \geqslant -1$, there exists a finitely supported function $\mathfrak{f}: \mathscr{E}_* \to \mathbb{R}$ such that

$$\|\mathfrak{L}_\alpha \mathfrak{f} + \mathfrak{w}\|_{-1,k} \leqslant \varepsilon$$

Moreover, we may take $\mathfrak{f}$ in $\mathscr{I}$ and $\mathfrak{f}(\phi) = 0$.

The proof of this result requires several estimates on the resolvent equation (4.2). The existence of a solution in L^2 of the resolvent equation stated in the next lemma requires a proof because $\mathfrak{L}_\alpha$ is not the generator of a Markov process.

Lemma 4.3. For each $\lambda > 0$, there exists a function $\mathfrak{u}_\lambda$ in $\mathscr{I}$ which solves the resolvent equation

$$\lambda \mathfrak{u}_\lambda - \mathfrak{L}_\alpha \mathfrak{u}_\lambda = \mathfrak{w} \tag{4.2}$$

Proof. For $n \geqslant 1$, let Π_n be the projection on $\bigcup_{j=1}^n \mathscr{E}_{*,j}$: $\Pi_n = \sum_{1 \leqslant j \leqslant n} \pi_j$ and let $\mathfrak{M}_n = \Pi_n(\mathfrak{L}_+ + \mathfrak{L}_-)\, \Pi_n$. We first prove the existence of a solution in L^2 of the truncated resolvent equation

$$\lambda \mathfrak{u}_{\lambda,n} - \{\mathfrak{L}_s + (1-2\alpha)\, \mathfrak{L}_d + \sqrt{\chi(\alpha)}\, \mathfrak{M}_n\}\, \mathfrak{u}_{\lambda,n} = \mathfrak{w} \tag{4.3}$$

By Lemma 7.1, the operators $\mathfrak{L}_s$, $\mathfrak{L}_d$, and $\mathfrak{L}_\pm$ are bounded on $L^2(\bigcup_{j=1}^n \mathscr{E}_{*,j})$ for each fixed $n \geqslant 1$. There exists, in particular, a solution for λ large enough. $\mathfrak{L}_s$ is a symmetric negative semi-definite operator while, by Lemma 7.2, $\mathfrak{L}_d$ and $\mathfrak{L}_+ + \mathfrak{L}_-$ are anti-symmetric. Since Π_n is a projection, for any $\mathfrak{f}$ in $L^2(\mathscr{E}_*)$, $\langle \mathfrak{f}, \mathfrak{M}_n \mathfrak{f}\rangle = \langle \mathfrak{f}, \Pi_n(\mathfrak{L}_+ + \mathfrak{L}_-)\, \Pi_n \mathfrak{f}\rangle = \langle \Pi_n \mathfrak{f}, (\mathfrak{L}_+ + \mathfrak{L}_-)\, \Pi_n \mathfrak{f}\rangle = 0$, so that $\mathfrak{M}_n$ is also anti-symmetric. Therefore, taking the scalar product on both sides of the previous identity with respect to $\mathfrak{u}_{\lambda,n}$ we obtain by Schwarz inequality that

$$\lambda\, \|\mathfrak{u}_{\lambda,n}\|_0 \leqslant \|\mathfrak{w}\|_0.$$

Here $\|\cdot\|_0$ stand for the $L^2(\mathscr{E}_*)$ norm. By the proof of Proposition I.2.8 (b) in ref. 12, there exists a solution of (4.3) for every $\lambda > 0$. Moreover, $\mathfrak{u}_{\lambda,n}$ belongs to $\mathscr{I}$ because $\mathfrak{w}$ does.

Up to this point we proved the existence of a solution of the resolvent equation (4.3). The previous estimate shows that the sequence $\{\mathfrak{u}_{\lambda,n}, n \geqslant 1\}$

is uniformly bounded in L^2 for each $\lambda > 0$. Moreover, by Lemma 7.7 and the proof of Lemma 2.5 in ref. 8 or the proof of Theorem 5.1 in ref. 13, since $\mathfrak{w}$ is finitely supported, for every $k \geqslant 1$,

$$\lambda \sum_{j \geqslant 1} j^k \|\pi_j \mathfrak{u}_{\lambda, n}\|_0 \leqslant C(k, \mathfrak{w})$$

uniformly over n. Let $\mathfrak{f}$ be a limit point of the sequence $\{\mathfrak{u}_{\lambda, n}, n \geqslant 1\}$. $\mathfrak{f}$ inherits the previous bound and belongs therefore to the domain of the operators $\mathfrak{L}_s$, $\mathfrak{L}_d$, $\mathfrak{L}_{\pm}$. Furthermore, taking scalar products with finitely supported functions, it is easy to show that any limit point of this sequence is a solution of the resolvent equation (4.2). Finally, $\mathfrak{f}$ belongs to $\mathscr{I}$ because each function $\mathfrak{u}_{\lambda, n}$ belongs and $\mathscr{I}$ is closed. This proves the lemma. ∎

By the proof of Lemma 2.5 in ref. 8 or the proof of Theorem 5.1 in ref. 13 we may deduce from Lemma 7.7 the following bound on the solution $\mathfrak{u}_\lambda$ of the resolvent equation (4.2).

Theorem 4.4. Let $\mathfrak{u}_\lambda$ be the solution of the resolvent equation (4.2). For any positive integer k, there exists a finite constant C_k depending only on k such that

$$\sup_{\lambda \geqslant 0} \{\lambda \|\mathfrak{u}_\lambda\|_{0,k}^2 + \|\mathfrak{u}_\lambda\|_{1,k}^2\} \leqslant C_k \|\mathfrak{w}\|_{-1,k}^2 \tag{4.4}$$

The following estimate on the asymmetric part of the generator that preserves the degree is needed in the proof of Theorem 4.2.

Lemma 4.5. Let $\mathfrak{u}_\lambda$ be the solution of the resolvent equation (4.2). For any $k \geqslant 1$, there exists a finite constant C_k, depending only on k, such that

$$\|\mathfrak{L}_d \mathfrak{u}_\lambda\|_{-1,k}^2 \leqslant C_k \|\mathfrak{w}\|_{-1,k+3}^2.$$

Proof. Let $\mathfrak{w}_1 = \mathfrak{w} + \sqrt{\chi(\alpha)}\,[\mathfrak{L}_+ + \mathfrak{L}_-]\,\mathfrak{u}_\lambda$ so that

$$\lambda \mathfrak{u}_\lambda - \{\mathfrak{L}_s + (1-2\alpha)\,\mathfrak{L}_d\}\,\mathfrak{u}_\lambda = \mathfrak{w}_1.$$

By Lemma 7.7 there exists a finite constant C_0 such that

$$\|\pi_n \mathfrak{w}_1\|_{-1}^2 \leqslant 2\,\|\pi_n \mathfrak{w}\|_{-1}^2 + C_0 n \sum_{j=n-1}^{n+1} \|\pi_j \mathfrak{u}_\lambda\|_1^2 \tag{4.5}$$

for all $n \geqslant 1$.

Notice that the operator $\mathfrak{L}_s + (1-2\alpha)\,\mathfrak{L}_d$ does not change the degree of a function. We may therefore examine equation (4.6) on each set $\mathscr{E}_{*,n}$:

$$\lambda \pi_n \mathfrak{u}_\lambda - \{\mathfrak{L}_s + (1-2\alpha)\,\mathfrak{L}_d\}\,\pi_n \mathfrak{u}_\lambda = \pi_n \mathfrak{w}_1 \tag{4.6}$$

Since n is fixed until estimate (4.10), we omit the operator π_n in the next formulas.

Following Section 6 of ref. 13, we approximate this operator by a convolution operator that can be analyzed through Fourier transforms. Fix $n \geqslant 1$ and let $\mathscr{X}_n = (\mathbb{Z}^d)^n$. We consider a set A in $\mathscr{E}_{*,n}$ as an equivalent class on $n!$ sets of distinct points of $\mathbb{Z}^d_*$. A function $\mathfrak{f}$: $\mathscr{E}_{*,n} \to \mathbb{R}$ can be lifted into a symmetric function $\mathfrak{B}\mathfrak{f}$ on $\mathscr{X}_n$, that vanishes on $\mathscr{X}_n \setminus \mathscr{E}_{n,*}$: $\mathfrak{B}\mathfrak{f}(x_1,..., x_n) = 0$ if $x_i = x_j$ for some $i \neq j$ or if $x_i = 0$ for some $1 \leqslant i \leqslant n$. The operators $\mathfrak{L}_s$, $\mathfrak{L}_d$ can also be extended in a natural way to $\mathscr{X}_n$. Denote by $\{\mathbf{e}_j, 1 \leqslant j \leqslant n\}$ the canonical basis of $\mathbb{R}^n$, by $\mathbf{1}$ the vector $\sum_{1 \leqslant j \leqslant n} \mathbf{e}_j$ and consider on $\mathscr{X}_n$ the operators $\tilde{\mathfrak{L}}_s$, $\tilde{\mathfrak{L}}_d$ defined by

$$\begin{aligned}(\tilde{\mathfrak{L}}_s \mathfrak{f})(\mathbf{x}) &= \sum_{\substack{1 \leqslant j \leqslant n \\ z \in \mathbb{Z}^d}} s(z)\{\mathfrak{f}(\mathbf{x} + z\mathbf{e}_j) - \mathfrak{f}(\mathbf{x})\} + \sum_{z \in \mathbb{Z}^d} s(z)\{\mathfrak{f}(\mathbf{x} - z\mathbf{1}) - \mathfrak{f}(\mathbf{x})\} \\ (\tilde{\mathfrak{L}}_d \mathfrak{f})(\mathbf{x}) &= \sum_{\substack{1 \leqslant j \leqslant n \\ z \in \mathbb{Z}^d}} a(z)\{\mathfrak{f}(\mathbf{x} + z\mathbf{e}_j) - \mathfrak{f}(\mathbf{x})\} + \sum_{z \in \mathbb{Z}^d} a(z)\{\mathfrak{f}(\mathbf{x} - z\mathbf{1}) - \mathfrak{f}(\mathbf{x})\}\end{aligned} \tag{4.7}$$

In this formula and below, $\mathbf{x} = (x_1,..., x_n)$ is an element of $\mathscr{X}_n$, so that each x_j belongs to $\mathbb{Z}^d$ and $\mathbf{x} + z\mathbf{e}_j = (x_1,..., x_{j-1}, x_j + z, x_{j+1},..., x_n)$, $\mathbf{x} + z\mathbf{1} = (x_1 + z,..., x_n + z)$.

Denote by $\|\cdot\|_{\mathscr{X}_n, 1}$ the H_1 norm associated to the generator $\tilde{\mathfrak{L}}_s$: for each function $\mathfrak{f}$: $\mathscr{X}_n \to \mathbb{R}$,

$$\|\mathfrak{f}\|^2_{\mathscr{X}_n, 1} = \frac{1}{n!} \sum_{\mathbf{x} \in \mathscr{X}_n} \mathfrak{f}(\mathbf{x})(-\tilde{\mathfrak{L}}_s)\,\mathfrak{f}(\mathbf{x})$$

and denote by $\|\cdot\|_{\mathscr{X}_n, -1}$ its dual norm defined by

$$\|\mathfrak{f}\|^2_{\mathscr{X}_n, -1} = \frac{1}{n!} \sum_{\mathbf{x} \in \mathscr{X}_n} \mathfrak{f}(\mathbf{x})(-\tilde{\mathfrak{L}}_s)^{-1}\,\mathfrak{f}(\mathbf{x})$$

Lifting the resolvent equation (4.6) to $\mathscr{X}_n$ and adding and subtracting $\tilde{\mathfrak{L}}_s \mathfrak{B}\mathfrak{u}_\lambda + (1-2\alpha)\,\tilde{\mathfrak{L}}_d \mathfrak{B}\mathfrak{u}_\lambda$, we obtain that

$$\lambda \mathfrak{B}\mathfrak{u}_\lambda - \{\tilde{\mathfrak{L}}_s + (1-2\alpha)\,\tilde{\mathfrak{L}}_d\}\,\mathfrak{B}\mathfrak{u}_\lambda = \mathfrak{w}_2 \tag{4.8}$$

where

$$\mathfrak{w}_2 = \mathfrak{B}\mathfrak{w}_1 + \{\mathfrak{B}\mathfrak{L}_s - \tilde{\mathfrak{L}}_s\mathfrak{B}\}\ \mathfrak{u}_\lambda + (1-2\alpha)\{\mathfrak{B}\mathfrak{L}_d - \tilde{\mathfrak{L}}_d\mathfrak{B}\}\ \mathfrak{u}_\lambda$$

We claim that $\mathfrak{w}_2$ has finite $H_{-1}(\mathscr{X}_n)$ norm. Indeed, for each $n \geqslant 1$, by (7.5) and Lemma 7.6 below, there exists a finite constant C_0 such that

$$\begin{aligned}
\|\pi_n \mathfrak{B}\mathfrak{w}_1\|^2_{\mathscr{X}_n, -1} &\leqslant \|\pi_n \mathfrak{w}_1\|^2_{-1} \\
\|\mathfrak{B}\mathfrak{L}_s\pi_n\mathfrak{u}_\lambda - \tilde{\mathfrak{L}}_s\mathfrak{B}\pi_n\mathfrak{u}_\lambda\|^2_{\mathscr{X}_n, -1} &\leqslant C_0 n^2\ \|\pi_n\mathfrak{u}_\lambda\|^2_1 \\
\|\mathfrak{B}\mathfrak{L}_d\pi_n\mathfrak{u}_\lambda - \tilde{\mathfrak{L}}_d\mathfrak{B}\pi_n\mathfrak{u}_\lambda\|^2_{\mathscr{X}_n, -1} &\leqslant C_0 n^2\ \|\pi_n\mathfrak{u}_\lambda\|^2_1
\end{aligned}$$

so that

$$\|\pi_n \mathfrak{w}_2\|^2_{\mathscr{X}_n, -1} \leqslant 2\ \|\pi_n\mathfrak{w}_1\|^2_{-1} + C_0 n^2\ \|\pi_n \mathfrak{u}_\lambda\|^2_1 \tag{4.9}$$

for some finite constant C_0.

It remains to examine the resolvent equation (4.8) through Fourier analysis. Let $\mathbb{T}_{n,d} = [-\pi, \pi]^{nd}$ and denote by $\hat{\mathfrak{u}}_\lambda \colon (\mathbb{T}^d)^n \to \mathbb{C}$ the Fourier transform of $\mathfrak{B}\mathfrak{u}_\lambda$:

$$\hat{\mathfrak{u}}_\lambda(\mathbf{k}) = \sum_{\mathbf{x}\in\mathscr{X}_n} e^{i\mathbf{x}\cdot\mathbf{k}}\ (\mathfrak{B}\mathfrak{u}_\lambda)(\mathbf{x}).$$

In this formula, $\mathbf{x}\cdot\mathbf{k} = \sum_{1\leqslant j\leqslant n} x_j \cdot k_j$. It follows from the resolvent equation (4.8) that $\hat{\mathfrak{u}}_\lambda$ is the solution of

$$\lambda\hat{\mathfrak{u}}_\lambda(\mathbf{k}) - \{\hat{\mathfrak{L}}_s(\mathbf{k}) + (1-2\alpha)\ \hat{\mathfrak{L}}_d(\mathbf{k})\}\ \hat{\mathfrak{u}}_\lambda(\mathbf{k}) = \hat{\mathfrak{w}}_2(\mathbf{k})$$

where $\hat{\mathfrak{L}}_s$, $\hat{\mathfrak{L}}_d$ are the functions associated to the operators $\tilde{\mathfrak{L}}_s$, $\tilde{\mathfrak{L}}_d$:

$$-\hat{\mathfrak{L}}_s(\mathbf{k}) = 2\sum_{\substack{1\leqslant j\leqslant n\\ z\in\mathbb{Z}^d}} s(z)\{1-\cos(k_j\cdot z)\} + 2\sum_{z\in\mathbb{Z}^d} s(z)\left\{1-\cos\left(\sum_{j=1}^n k_j\cdot z\right)\right\}$$

$$-\hat{\mathfrak{L}}_d(\mathbf{k}) = 2i\sum_{\substack{1\leqslant j\leqslant n\\ z\in\mathbb{Z}^d}} a(z)\sin(k_j\cdot z) - 2i\sum_{z\in\mathbb{Z}^d} a(z)\sin\left(\sum_{j=1}^n k_j\cdot z\right)$$

The $H_{-1}(\mathscr{X}_n, \hat{\mathfrak{L}}_s)$ norm of a function $\mathfrak{v}\colon \mathscr{X}_n \to \mathbb{R}$ has a simple and explicit expression in terms of the Fourier transform:

$$\|\mathfrak{v}\|^2_{\mathscr{X}_n, -1} = \frac{-1}{n!(2\pi)^{nd}} \int_{\mathbb{T}_{n,d}} d\mathbf{k}\ |\hat{\mathfrak{v}}(\mathbf{k})|^2 \frac{1}{\hat{\mathfrak{L}}_s(\mathbf{k})}$$

Since $\mathfrak{B}\mathfrak{u}_\lambda$ is the solution of the resolvent equation (4.8), for every $\lambda > 0$,

$$\|\tilde{\mathfrak{L}}_d \mathfrak{B}\mathfrak{u}_\lambda\|^2_{\mathscr{X}_n,-1} = \frac{-1}{n!(2\pi)^{nd}} \int_{\mathbb{T}_{n,d}} \left| \frac{\hat{\mathfrak{L}}_d(\mathbf{k})}{\lambda - \hat{\mathfrak{L}}_s(\mathbf{k}) - (1-2\alpha)\, \hat{\mathfrak{L}}_d(\mathbf{k})} \right|^2 \frac{|\hat{\mathfrak{w}}_2(\mathbf{k})|^2}{\hat{\mathfrak{L}}_s(\mathbf{k})}\, d\mathbf{k}$$

It follows from the explicit formulas for the functions $\hat{\mathfrak{L}}_s$, $\hat{\mathfrak{L}}_d$ and a Taylor expansion for $|\mathbf{k}|$ small that the previous expression is bounded by

$$\frac{-C_0}{n!(2\pi)^{nd}} \int_{\mathbb{T}_{n,d}} \frac{|\hat{\mathfrak{w}}_2(\mathbf{k})|^2}{\hat{\mathfrak{L}}_s(\mathbf{k})}\, d\mathbf{k} = C_0\, \|\mathfrak{w}_2\|^2_{\mathscr{X}_n,-1}$$

for some finite constant C_0. We have thus proved that

$$\|\tilde{\mathfrak{L}}_d \mathfrak{B}\mathfrak{u}_\lambda\|^2_{\mathscr{X}_n,-1} \leqslant C_0\, \|\mathfrak{w}_2\|^2_{\mathscr{X}_n,-1} \tag{4.10}$$

We may now conclude the proof of the Lemma 4.5. Fix $n \geqslant 1$. By (7.5), by the estimate presented just before (4.9) and by the inequality just derived, there exists a finite constant C_0, which may change from line to line, such that

$$\begin{aligned} \|\pi_n \mathfrak{L}_d \mathfrak{u}_\lambda\|^2_{-1} &= \|\mathfrak{L}_d \pi_n \mathfrak{u}_\lambda\|^2_{-1} \leqslant C_0 n\, \|\mathfrak{B}\mathfrak{L}_d \pi_n \mathfrak{u}_\lambda\|^2_{\mathscr{X}_n,-1} \\ &\leqslant C_0 n \{ n^2\, \|\pi_n \mathfrak{u}_\lambda\|_1^2 + \|\tilde{\mathfrak{L}}_d \mathfrak{B} \pi_n \mathfrak{u}_\lambda\|^2_{\mathscr{X}_n,-1} \} \\ &\leqslant C_0 \{ n^3\, \|\pi_n \mathfrak{u}_\lambda\|_1^2 + n\, \|\pi_n \mathfrak{w}_2\|^2_{\mathscr{X}_n,-1} \} \end{aligned}$$

In particular, by (4.9) and (4.5),

$$\begin{aligned} \|\pi_n \mathfrak{L}_d \mathfrak{u}_\lambda\|^2_{-1} &\leqslant C_0 \{ n^3\, \|\pi_n \mathfrak{u}_\lambda\|_1^2 + n\, \|\pi_n \mathfrak{w}_1\|^2_{\mathscr{X}_n,-1} \} \\ &\leqslant C_0 \left\{ n^3 \sum_{j=n-1}^{n+1} \|\pi_j \mathfrak{u}_\lambda\|_1^2 + n\, \|\pi_n \mathfrak{w}\|^2_{\mathscr{X}_n,-1} \right\} \end{aligned} \tag{4.11}$$

It remains to recall the definition of the norm $\|\cdot\|_{-1,k}$ and the statement of Theorem 4.4 to conclude the proof. ∎

Notice that the constant C_k, which appears in the statement of Lemma 4.5, depends on k only because the one which appears in Theorem 4.4 depends on k. We have now all elements to prove Theorem 4.2.

Proof of Theorem 4.2. The proof is done in two steps. We first use the resolvent equation (4.2) to obtain a sequence $\{\mathfrak{v}_j;\, j \geqslant 1\}$ of functions in $L^2(\mathscr{E}_*, k)$ such that

$$\lim_{j \to \infty} \|\mathfrak{L}_\alpha \mathfrak{v}_j + \mathfrak{w}\|_{-1,k} = 0$$

The sequence $\mathfrak{v}_j$ is obtained through convex combinations (in λ) of the solutions $\mathfrak{u}_\lambda$ of the resolvent equation (4.2). Details can be found at the end of the proof of Lemma 2.1 in ref. 8 or at the beginning of the proof of Lemma 2.8 in ref. 10.

At this point, it remains to show the existence, for each fixed j and $\varepsilon > 0$, of a finitely supported function $\mathfrak{f}\colon \mathscr{E}_* \to \mathbb{R}$ such that $\|\mathfrak{L}_\alpha f - \mathfrak{L}_\alpha \mathfrak{v}_j\|_{-1,k} \leqslant \varepsilon$. To prove the existence of such function $\mathfrak{f}$, assume that $\mathfrak{f}$ vanishes on $\bigcup_{j\geqslant n} \mathscr{E}_{*,j}$ and recall the decomposition of the operator $\mathfrak{L}_\alpha$ to deduce that

$$\begin{aligned} &\|\mathfrak{L}_\alpha \mathfrak{f} - \mathfrak{L}_\alpha \mathfrak{v}_j\|_{-1,k} \\ &\quad \leqslant \|\mathfrak{L}_+(\mathfrak{f} - \Pi_n \mathfrak{v}_j)\|_{-1,k} + \|\mathfrak{L}_-(\mathfrak{f} - \Pi_n \mathfrak{v}_j)\|_{-1,k} \\ &\qquad + \|\mathfrak{L}_d(\mathfrak{f} - \Pi_n \mathfrak{v}_j)\|_{-1,k} + \|\mathfrak{L}_s(\mathfrak{f} - \Pi_n \mathfrak{v}_j)\|_{-1,k} + \|\mathfrak{L}_\alpha(I - \Pi_n)\, \mathfrak{v}_j)\|_{-1,k} \end{aligned} \tag{4.12}$$

where $\Pi_n = \sum_{j\leqslant n} \pi_j$ and I is the identity. We estimate each term on the right hand side separately. By Lemma 7.7, the first term on the right hand side is such that

$$\begin{aligned} \|\mathfrak{L}_+(\mathfrak{f} - \Pi_n \mathfrak{v}_j)\|_{-1,k}^2 &= \sum_{\ell=0}^{n} (\ell+1)^k \|\mathfrak{L}_+ \pi_\ell \{\mathfrak{f} - \mathfrak{v}_j\}\|_{-1}^2 \\ &\leqslant C_0 \sum_{\ell=1}^{n+1} \ell^{k+1} \|\pi_\ell \mathfrak{f} - \pi_\ell \mathfrak{v}_j\|_1^2 \leqslant C_0 \sum_{\ell=1}^{n+1} \ell^{k+2} \|\pi_\ell \mathfrak{f} - \pi_\ell \mathfrak{v}_j\|_0^2 \end{aligned}$$

The second term on the right hand side of (4.12) is estimated in the same way. By Lemma 7.4, the third one is such that

$$\|\mathfrak{L}_d(\mathfrak{f} - \Pi_n \mathfrak{v}_j)\|_{-1,k}^2 = \sum_{\ell=1}^{n} \ell^k \|\mathfrak{L}_d \pi_\ell(\mathfrak{f} - \mathfrak{v}_j)\|_{-1}^2 \leqslant C_0 \sum_{\ell=1}^{n} \ell^{k+1} \|\pi_\ell(\mathfrak{f} - \mathfrak{v}_j)\|_0^2$$

for some finite constant C_0. The fourth member on the right hand side of (4.12) is easily estimated by exactly the same arguments and by using again Lemma 7.4.

Finally, since $\mathfrak{v}_j$ is a convex combination of the solutions of the resolvent equation (4.2), by (4.11),

$$\begin{aligned} \|\mathfrak{L}_\alpha(I - \Pi_n)\, \mathfrak{v}_j)\|_{-1,k}^2 &= \sum_{\ell > n} \ell^k \|\mathfrak{L}_\alpha \pi_\ell \mathfrak{v}_j\|_{-1}^2 \\ &\leqslant \sum_{\ell \geqslant n} \ell^{k+3} \|\pi_\ell \mathfrak{v}_j\|_1^2 + \sum_{\ell \geqslant n} \ell^{k+1} \|\pi_\ell \mathfrak{w}\|_{-1}^2 \end{aligned}$$

Now, for $\varepsilon > 0$ fixed, since $\mathfrak{w}$ is finitely supported, by Theorem 4.4 and Theorem 4.1 there exists $n_0 > 0$ large enough for the last quantity to be bounded by ε. For this fixed n_0, find a finitely supported function $\mathfrak{f}: \bigcup_{n \leqslant n_0} \mathscr{E}_{*,n} \to \mathbb{R}$ for which all previous expressions are bounded by ε, which is possible because $\mathfrak{v}_j$ belongs to $L^2(\mathscr{E}_*, k)$.

It remains to check that we may take $\mathfrak{f}$ in $\mathscr{I}$ with $\mathfrak{f}(\phi) = 0$. The first property follows from the fact, easy to verify, that the operators $\mathfrak{L}_s$, $\mathfrak{L}_d$, $\mathfrak{L}_+$, $\mathfrak{L}_-$ map the closed subspace $\mathscr{I}$ of $L^2(\mathscr{E}_*)$ into $\mathscr{I}$. In particular, the solutions of the resolvent equations, as well as their convex combinations, belong to $\mathscr{I}$ so that $\mathfrak{f}$ can be taken in $\mathscr{I}$.

The second requirement follows from the fact that $(\mathfrak{L}_* \mathfrak{f})(\phi) = 0$ for any finitely supported function $\mathfrak{f}$ in $\mathscr{I}$, where $\mathfrak{L}_*$ stands for any of the four operators $\mathfrak{L}_s$, $\mathfrak{L}_d$, $\mathfrak{L}_+$, $\mathfrak{L}_-$ and from the fact that $(\mathfrak{L}_+ \mathfrak{f})(\{x\}) = 0$ for all x in $\mathbb{Z}^d_*$. These two properties show that we may set the value of $\mathfrak{f}$ at ϕ to be 0 without changing $\mathfrak{L}_\alpha \mathfrak{f}$.

The special features of the operators $\mathfrak{L}_*$ just used are proved in the beginning of Section 7. This concludes the proof. ▮

5. THE FLUCTUATION-DISSIPATION THEOREM

We consider in this section the general asymmetric simple exclusion in dimension $d \geqslant 3$. Here again, all finite constants C_0 which appear in the statement of the theorems may depend only on the transition probability $p(\cdot)$.

We prove in this section a fluctuation–dissipation theorem, Theorem 5.3 below, also called the Boltzmann–Gibbs principle. It allows the replacement of a local function w in $\bigoplus_{n \geqslant 2} \mathscr{G}_n$ by the sum of gradients $\eta(x+e_j) - \eta(x)$ and local functions in the range of the generator. This result is the main step in the proof of Gaussian fluctuations of the empirical measure around the hydrodynamic limit (cf. ref. 4).

Denote by $\mathbb{E}_{\nu_\alpha}$ the expectation on the path space $D(\mathbb{R}_+, \mathscr{X})$ corresponding to the Markov process with generator given by (2.1) and starting from the stationary measure ν_α. The first result states that a local function f in $H_{-1}(L^s, \langle\!\langle \cdot, \cdot \rangle\!\rangle)$ has a finite space-time variance in the diffusive scaling:

Theorem 5.1. Fix a smooth function $G: \mathbb{R}^d \to \mathbb{R}$ with compact support, a local function f of degree $n \geqslant 2$, $T > 0$ and a vector v in $\mathbb{R}^d$. There exists a finite constant C_0 such that

$$\limsup_{\epsilon \to 0} \mathbb{E}_{\nu_\alpha} \left[\sup_{0 \leqslant t \leqslant T} \left(\epsilon^{d/2+1} \int_0^{t\epsilon^{-2}} \sum_{x \in \mathbb{Z}^d} G(\epsilon(x - rv))\, f(\tau_x \eta_r)\, dr \right)^2 \right]$$
$$\leqslant C_0 T \, \|G\|^2_{L^2} \, \|\mathfrak{T}\mathfrak{f}\|^2_{-1}$$

Proof. Set $G_r^\epsilon(x) = G(\epsilon(x - rv))$ and recall Lemma 4.3 in ref. 2 to obtain that

$$\mathbb{E}_{\nu_\alpha}\left[\sup_{0\leqslant t\leqslant T}\left(\epsilon^{d/2+1}\int_0^{t\epsilon^{-2}} dr \sum_{x\in\mathbb{Z}^d} G_r^\epsilon(x)\, f(\tau_x\eta_r)\, dr\right)^2\right]$$
$$\leqslant 14\int_0^T dr\, \epsilon^d \left\langle \sum_{x\in\mathbb{Z}^d} G_r^\epsilon(x)\, \tau_x f, (-L^s)^{-1} \sum_{x\in\mathbb{Z}^d} G_r^\epsilon(x)\, \tau_x f\right\rangle \qquad (5.1)$$

Fix a local function g of degree $n\geqslant 2$. Let $\mathscr{L}_s$ be the generator defined in (3.9) and $\mathfrak{g}$ be the Fourier coefficient of g. Then,

$$\langle g, (-L^s)^{-1} g\rangle = \langle \mathfrak{g}, (-\mathscr{L}_s)^{-1}\mathfrak{g}\rangle$$

where the last scalar product is on $\mathscr{E}_n$. Recall from Section 4 that $\mathscr{X}_n = (\mathbb{Z}^d)^n$. The proof of Lemma 7.5 shows that the previous expression is bounded by

$$\frac{C_0}{(n-1)!}\langle \mathfrak{B}\mathfrak{g}, (-\widetilde{\mathscr{L}}_0^s)^{-1}\mathfrak{B}\mathfrak{g}\rangle_{\mathscr{X}_n}$$

for some finite constant C_0, where $\widetilde{\mathscr{L}}_0^s$ stands for the generator of n independent random walks on $\mathbb{Z}^d$ with transition probability $s(\cdot)$ and $\mathfrak{B}\mathfrak{g}$ for the extension of $\mathfrak{g}$ to $\mathscr{X}_n$. Denote by $\mathfrak{G}_n(\cdot,\cdot)$ the Green function associated to the generator $\widetilde{\mathscr{L}}_0^s$ restricted to $\mathscr{X}_n$. The previous expression can be written as

$$\frac{C_0}{(n-1)!}\sum_{\mathbf{x},\mathbf{y}\in\mathscr{X}_n} (\mathfrak{B}\mathfrak{g})(\mathbf{x})\, \mathfrak{G}_n(\mathbf{x},\mathbf{y})(\mathfrak{B}\mathfrak{g})(\mathbf{y})$$

In particular, for $g = \sum_{x\in\mathbb{Z}^d} G_r^\epsilon(x)\,\tau_x f$, the right hand side of (5.1) is bounded by

$$\frac{C_0}{(n-1)!}\int_0^T dr\,\epsilon^d \sum_{z,w\in\mathbb{Z}^d} G_r^\epsilon(z)\, G_r^\epsilon(w) \sum_{\mathbf{x},\mathbf{y}\in\mathscr{X}_n} \mathfrak{B}\mathfrak{f}(\mathbf{x}+z\mathbf{1})\, \mathfrak{G}_n(\mathbf{x},\mathbf{y})\, \mathfrak{B}\mathfrak{f}(\mathbf{y}+w\mathbf{1})$$

A change of variables permits to rewrite the expression inside the integral as

$$\epsilon^d \sum_{z,w\in\mathbb{Z}^d} G_r^\epsilon(z)\, G_r^\epsilon(z-w) \sum_{\mathbf{x},\mathbf{y}\in\mathscr{X}_n} \mathfrak{B}\mathfrak{f}(\mathbf{x})\, \mathfrak{G}_n(\mathbf{x},\mathbf{y}+w\mathbf{1})\, \mathfrak{B}\mathfrak{f}(\mathbf{y})$$

Since f is a local function, $\mathfrak{f}$ has finite support. In particular, the sums over $\mathbf{x}$, $\mathbf{y}$ are carried over finite sets. On the other hand, replacing $G_r^\epsilon(z-w)$ by $G_r^\epsilon(z)$, we write the previous expression as

$$\epsilon^d \sum_{z \in \mathbb{Z}^d} G_r^\epsilon(z)^2 \sum_{\mathbf{x}, \mathbf{y} \in \mathscr{X}_n} \mathfrak{B}\mathfrak{f}(\mathbf{x}) \sum_{w \in \mathbb{Z}^d} \mathfrak{G}_n(\mathbf{x}, \mathbf{y}+w\mathbf{1}) \, \mathfrak{B}\mathfrak{f}(\mathbf{y})$$

$$+\epsilon^d \sum_{z, w \in \mathbb{Z}^d} G_r^\epsilon(z)\{G_r^\epsilon(z-w)-G_r^\epsilon(z)\} \sum_{\mathbf{x}, \mathbf{y} \in \mathscr{X}_n} \mathfrak{B}\mathfrak{f}(\mathbf{x}) \, \mathfrak{G}_n(\mathbf{x}, \mathbf{y}+w\mathbf{1}) \, \mathfrak{B}\mathfrak{f}(\mathbf{y})$$

We claim that the second term is bounded above by $C(f, G)\,\epsilon^{1/2}$ for some finite constant depending on f and G. Indeed, since G has a bounded derivative and $\mathfrak{B}\mathfrak{f}$ a finite support, the second line is less than or equal to

$$C(\|\nabla G\|_\infty, f)\,\epsilon^{1/2} \sup_{\mathbf{x} \in \mathscr{X}_{\mathfrak{f}}^-} \sum_{w \in \mathbb{Z}^d} |w|^{1/2} \, \mathfrak{G}_n(0, \mathbf{x}+w\mathbf{1}). \tag{5.2}$$

In this formula, $\mathscr{X}_{\mathfrak{f}}^-$ is the set of all sites which can be written as difference of two points in the support of $\mathfrak{B}\mathfrak{f}$: $\mathscr{X}_{\mathfrak{f}}^- = \{\mathbf{y}-\mathbf{x}, \mathfrak{B}\mathfrak{f}(\mathbf{x})\,\mathfrak{B}\mathfrak{f}(\mathbf{y}) \neq 0\}$. Since $\mathfrak{G}_n(0, \mathbf{z})$ is the Green function of n independent random walks on $\mathbb{Z}^d$, it decays as $|\mathbf{z}|^{2-nd}$. The sum over w is thus finite, uniformly over $\mathbf{x}$, as soon as $(n-1)\,d > 5/2$, inequality which is fulfilled because we are assuming $d \geqslant 3$ and $n \geqslant 2$. This proves (5.2).

Collecting all previous estimates we obtain that the expectation which appears in the statement of the lemma is bounded above by

$$\frac{C_0}{(n-1)!} \int_0^T dr\, \epsilon^d \sum_{z \in \mathbb{Z}^d} G_r^\epsilon(z)^2 \sum_{\mathbf{x}, \mathbf{y} \in \mathscr{X}_n} \mathfrak{B}\mathfrak{f}(\mathbf{x}) \sum_{w \in \mathbb{Z}^d} \mathfrak{G}_n(\mathbf{x}, \mathbf{y}+w\mathbf{1}) \, \mathfrak{B}\mathfrak{f}(\mathbf{y})$$

plus a remainder which converges to 0 as $\epsilon \downarrow 0$. Denote by $\tilde{\mathfrak{G}}_n(\cdot,\cdot)$ the Green function associated to the generator $\tilde{\mathfrak{L}}_s$ defined in (4.7) and restricted to $\mathscr{X}_n$. An elementary computation shows that

$$\sum_{\mathbf{x}, \mathbf{y} \in \mathscr{X}_n} \mathfrak{B}\mathfrak{f}(\mathbf{x}) \sum_{w \in \mathbb{Z}^d} \mathfrak{G}_n(\mathbf{x}, \mathbf{y}+w\mathbf{1}) \, \mathfrak{B}\mathfrak{f}(\mathbf{y}) = \sum_{\mathbf{x}, \mathbf{y} \in \mathscr{X}_{n-1}} \mathfrak{B}\bar{\mathfrak{f}}(\mathbf{x}) \, \tilde{\mathfrak{G}}_{n-1}(\mathbf{x}, \mathbf{y}) \, \mathfrak{B}\bar{\mathfrak{f}}(\mathbf{y})$$
$$= (n-1)! \, \|\pi_{n-1} \mathfrak{B}\bar{\mathfrak{f}}\,\|_{-1, \mathscr{X}_{n-1}}^2$$

On the right hand side, $\bar{\mathfrak{f}}$ stands for $\mathfrak{T}\mathfrak{f}$. Notice that n is replaced by $n-1$ in the right hand side because we are fixing a particle at the origin. The last time integral is thus equal to

$$C_0 \int_0^T dr\, \epsilon^d \sum_{z \in \mathbb{Z}^d} G_r^\epsilon(z)^2 \, \|\mathfrak{B}\bar{\mathfrak{f}}\|_{\mathscr{X}_n, -1}^2 \leqslant C_0 \int_0^T dr\, \epsilon^d \sum_{z \in \mathbb{Z}^d} G_r^\epsilon(z)^2 \, \|\bar{\mathfrak{f}}\|_{-1}^2$$

where in the last step we used Lemma 7.5. As $\epsilon \downarrow 0$ this integral converges to

$$C_0 T \, \|G\|_{L^2}^2 \, \|\bar{\mathfrak{f}}\|_{-1}^2$$

which concludes the proof of the lemma. ▮

Corollary 5.2. Fix a smooth function $G: \mathbb{R}^d \to \mathbb{R}$ with compact support, a local function f in $\bigoplus_{n \geq 2} \mathscr{G}_n$, $T > 0$ and a vector v in $\mathbb{R}^d$. There exists a finite constant C_0 such that

$$\limsup_{\epsilon \to 0} \mathbb{E}_{\nu_\alpha} \left[\sup_{0 \leq t \leq T} \left| \epsilon^{d/2+1} \int_0^{t\epsilon^{-2}} \sum_{x \in \mathbb{Z}^d} G(\epsilon(x - rv)) \, f(\tau_x \eta_r) \, dr \right|^2 \right] \leq C_0 T \, \|G\|_{L^2}^2 \, \|\mathfrak{T}\mathfrak{f}\|_{-1,0}^2$$

Proof. Since the symmetric part of the generator does not change the degree of local functions, for a local function g in $\bigoplus_{n \geq 2} \mathscr{G}_n$,

$$\langle g, (-L^s)^{-1} g \rangle = \sum_{n \geq 2} \langle \pi_n g, (-L^s)^{-1} \pi_n g \rangle = \sum_{n \geq 2} \langle \pi_n \mathfrak{g}, (-\mathscr{L}_s)^{-1} \pi_n \mathfrak{g} \rangle$$

and we may proceed as in the proof of Theorem 5.1. ▮

Fix a local function w in $\bigoplus_{n \geq 2} \mathscr{G}_n$ and denote by $\mathfrak{w}$ its Fourier coefficients and by $\bar{\mathfrak{w}}$ the finitely supported function $\mathfrak{T}\mathfrak{w}: \mathscr{E}_* \to \mathbb{R}$. For each $\lambda > 0$, let $\mathfrak{u}_\lambda$ be the solution of the resolvent equation

$$\lambda \mathfrak{u}_\lambda - \mathfrak{L}_\alpha \mathfrak{u}_\lambda = \bar{\mathfrak{w}}$$

We proved in Section 4 the existence of the solution $\mathfrak{u}_\lambda$ of the resolvent equation and some of its properties. In the next section, among several other properties, we show the existence of a subsequence λ_k for which the sequence $\mathfrak{u}_{\lambda_k}(\alpha, z)$ converges, as $k \uparrow \infty$, uniformly in α in $[0, 1]$, to some limit, denoted by $D_z(\alpha)$:

$$D_z(\alpha) = \lim_{k \to \infty} \mathfrak{u}_{\lambda_k}(\alpha, \{z\}) \tag{5.3}$$

for each z in $\mathbb{Z}_*^d$.

Theorem 5.3. Fix a local function w in $\bigoplus_{n \geq 2} \mathscr{G}_n$ and a smooth function $G: \mathbb{R}^d \to \mathbb{R}$ with compact support. There exists a sequence of local functions u_m such that

$$\limsup_{m \to \infty} \limsup_{\epsilon \to 0} \mathbb{E}_{\nu_\alpha} \left[\sup_{0 \leq t \leq T} \left(\epsilon^{d/2+1} \int_0^{t\epsilon^{-2}} \sum_{x \in \mathbb{Z}^d} G(\epsilon x) \, \tau_x W_m(\eta_s) \, ds \right)^2 \right] = 0$$

where

$$W_m(\eta) = W_{\alpha,m}(\eta) = w - Lu_m + \sqrt{\chi(\alpha)} \sum_{z \in \mathbb{Z}^d} a(z)\, D_z(\alpha)\{\Psi_z - \Psi_0\}$$

We will refer to Theorem 5.3 as the *fluctuation-dissipation theorem*. It will be the basic ingredient to study the equilibrium fluctuations for the density field.

Proof. Recall that $\bar{\mathfrak{w}} = \mathfrak{T}\mathfrak{w}$. Since w belongs to $\bigoplus_{n\geqslant 2}\mathcal{G}_n$, by Theorem 4.2 with $k=0$, there exists a sequence of finitely supported functions $\mathfrak{v}_m\colon \mathcal{E}_* \to \mathbb{R}$ such that $\mathfrak{v}_m(\phi) = 0$, $\mathfrak{v}_m$ belongs to $\mathcal{I}$ and

$$\lim_{m\to\infty} \|\bar{\mathfrak{w}} - \mathfrak{L}_\alpha \mathfrak{v}_m\|_{-1} = 0$$

Since $\mathfrak{v}_m$ satisfies (3.6), in view of (3.7), there exists a finitely supported function $\mathfrak{u}_m\colon \mathcal{E} \to \mathbb{R}$ such that $\mathfrak{T}\mathfrak{u}_m = \mathfrak{v}_m$. Moreover, $\mathfrak{u}_m(A) \neq 0$ only if A contains the origin and $\mathfrak{u}_m(\{0\}) = 0$ because $\mathfrak{v}_m(\phi) = 0$. Let u_m be the local function defined by $u_m = \sum_{A\in\mathcal{E}} \mathfrak{u}_m(A)\,\Psi_A$ and let $\hat{W}_m$ be the local function in $\bigoplus_{n\geqslant 2}\mathcal{G}_n$ defined by

$$\hat{W}_m = w - \{Lu_m - \pi_1 Lu_m\}$$

By Corollary 5.2,

$$\limsup_{\epsilon\to 0} \mathbb{E}_{\nu_\alpha}\left[\sup_{0\leqslant t\leqslant T}\left(\epsilon^{d/2+1}\int_0^{t\epsilon^{-2}} \sum_{x\in\mathbb{Z}^d} G(\epsilon x)\,\tau_x \hat{W}_m(\eta_s)\,ds\right)^2\right]$$
$$\leqslant C_0 T\,\|G\|_{L^2}^2\,\|\mathfrak{T}\hat{W}_m\|_{-1}^2 \tag{5.4}$$

for some finite constant C_0. An elementary computation gives that

$$\begin{aligned}\pi_1(Lu) = &-\sum_{x,y\in\mathbb{Z}^d} s(y-x)\{\mathfrak{u}(y)-\mathfrak{u}(x)\}\{\Psi_y-\Psi_x\}\\ &-\frac{1-2\alpha}{2}\sum_{x,y\in\mathbb{Z}^d} a(y-x)\{\mathfrak{u}(y)+\mathfrak{u}(x)\}\{\Psi_y-\Psi_x\}\\ &+\sqrt{\chi(\alpha)}\sum_{x,y\in\mathbb{Z}^d} a(y-x)\,\mathfrak{u}(x,y)\{\Psi_y-\Psi_x\}\end{aligned}$$

Since $\mathfrak{u}_m(\{x\}) = 0$ for every x in $\mathbb{Z}^d$, $\mathfrak{u}_m(A) \neq 0$ only if A contains the origin, $\mathfrak{u}_m(A) = |A|^{-1}\,\mathfrak{v}_m(A\setminus\{0\})$, we have that

$$\pi_1(Lu) = \sqrt{\chi(\alpha)}\sum_{x\in\mathbb{Z}^d} a(x)\,\mathfrak{v}_m(x)\{\Psi_x - \Psi_0\}$$

which vanishes in $L^2(\chi, \langle\!\langle \cdot,\cdot \rangle\!\rangle)$ so that $\mathfrak{T}\pi_1(Lu)=0$. In particular, $\mathfrak{T}\hat{W}_m = \bar{\mathfrak{w}} - \mathfrak{L}_\alpha \mathfrak{v}_m$ and the right hand side of (5.4) is bounded above by

$$C_0 T \, \|G\|_{L^2}^2 \, \|\bar{\mathfrak{w}} - \mathfrak{L}_\alpha \mathfrak{v}_m\|_{-1}^2$$

Since this expression vanishes as $m \uparrow \infty$, all we need to prove is that

$$\limsup_{m\to\infty} \limsup_{\epsilon\to 0} \mathbb{E}_{\nu_\alpha}\left[\sup_{0\leqslant t\leqslant T} \left(\epsilon^{d/2+1} \int_0^{t\epsilon^{-2}} \sum_{x\in\mathbb{Z}^d} G(\epsilon x)\, \tau_x \tilde{W}_m(\eta_s)\, ds \right)^2 \right] = 0$$

where

$$\tilde{W}_m = \pi_1 L u_m - \sqrt{\chi(\alpha)} \sum_{z\in\mathbb{Z}^d} a(z)\, D_z(\alpha) \{\Psi_z - \Psi_0\}$$

By the explicit formula for $\pi_1 L u_m$ obtained above,

$$\tilde{W}_m = \sqrt{\chi(\alpha)} \sum_{x\in\mathbb{Z}^d} a(x) \{\mathfrak{v}_m(x) - D_x(\alpha)\} \{\Psi_x - \Psi_0\}$$

Since ν_α is a stationary state, by Schwarz inequality, the previous expectation is bounded above by

$$\epsilon^{d-2} T^2 E_{\nu_\alpha}\left[\left(\sum_{x\in\mathbb{Z}^d} G(\epsilon x)\, \tau_x \tilde{W}_m(\eta) \right)^2 \right]$$

A change of variables gives that this expression is equal to

$$\epsilon^{d-2} \chi(\alpha)\, T^2 \sum_{x\in\mathbb{Z}^d} \left(\sum_{y\in\mathbb{Z}^d} a(y) \{G(\epsilon[x-y]) - G(\epsilon x)\} \{\mathfrak{v}_m(y) - D_y(\alpha)\} \right)^2$$

A Taylor expansion shows that this expression converges, as $\epsilon \downarrow 0$, to

$$\chi(\alpha)\, T^2 \int_{\mathbb{R}^d} da\, \{(\nabla G)(a) \cdot R_m\}^2$$

where $R_m = (R_m^1, \ldots, R_m^d)$, $R_m^j = \sum_{y\in\mathbb{Z}^d} a(y)\, y_j \{\mathfrak{v}_m(y) - D_y(\alpha)\}$. By the definition of $D_y(\alpha)$, $R_m^j \to 0$ as $m \uparrow \infty$ because $\mathfrak{v}_m$ is constructed as convex combinations of $\mathfrak{u}_{\lambda_k}$, the solution of the resolvent equation (4.2). The integral therefore tends to 0. ∎

6. REGULARITY OF VISCOSITY COEFFICIENTS

Fix a local function w in $\bigoplus_{n\geqslant 2}\mathscr{G}_n$ and denote by $\mathfrak{w}$ its Fourier coefficients and by $\bar{\mathfrak{w}}$ the finitely supported function $\mathfrak{T}\mathfrak{w}$: $\mathscr{E}_* \to \mathbb{R}$. For each $\lambda > 0$, let $\mathfrak{u}_\lambda$ be the solution of the resolvent equation

$$\lambda \mathfrak{u}_\lambda - \mathfrak{L}_\alpha \mathfrak{u}_\lambda = \bar{\mathfrak{w}}$$

We proved in Section 4 the existence of the solution $\mathfrak{u}_\lambda$ of the resolvent equation. We prove in this section the existence of a subsequence λ_k for which $\mathfrak{u}_{\lambda_k}(\cdot, z)$ converges uniformly in $[0, 1]$, as well as all its derivatives, to a smooth function $D_z(\cdot)$.

The proof is very close to the one presented in ref. 11 for the self diffusion in the symmetric case, so we only show here the main points.

We want to show that there exists a subsequence $\lambda_k \downarrow 0$, such that, for each z such that $|a(z)| > 0$, $\mathfrak{u}_{\lambda_k}(\alpha, \{z\})$ converges uniformly in α to a smooth function. To prove the existence of such subsequence it is enough to show that $\mathfrak{u}_\lambda(\alpha, \{z\})$ are smooth function of α for each $\lambda > 0$ and each z, and

$$\sup_{\lambda>0}\ \sup_{0\leqslant\alpha\leqslant 1} |\mathfrak{u}_\lambda^{(j)}(\alpha, \{z\})| < \infty$$

where $\mathfrak{u}_\lambda^{(j)}(\alpha, \{z\})$ stands for the jth derivative of $\mathfrak{u}_\lambda$.

By Lemma 3.1 of ref. 13, we have the bound

$$|\mathfrak{u}_\lambda^{(j)}(\alpha, \{z\})| \leqslant C_0\ \|\mathfrak{u}_\lambda^{(j)}(\alpha, \cdot)\|_1$$

for some finite constant C_0 depending only on $p(\cdot)$. With the iterated use of Theorem 4.4, we will prove that, for any k and j,

$$\sup_{\lambda>0}\ \sup_{0\leqslant\alpha\leqslant 1} \|\mathfrak{u}_\lambda^{(j)}(\alpha, \cdot)\|_{1,k} < \infty$$

Since the coefficients of $\mathfrak{L}_\alpha$ are not smooth at the boundary of $[0, 1]$, we reparameterize by $\alpha = \sin^2 t$, $t \in [0, 2\pi]$, as done in ref. 11. We obtain

$$\mathfrak{L}(t) = \mathfrak{L}_s + (\cos^2 t - \sin^2 t)\ \mathfrak{L}_d + (\sin t \cos t)\{\mathfrak{L}_+ + \mathfrak{L}_-\}$$

and then we consider the equation

$$\lambda \mathfrak{v}_\lambda(t) - \mathfrak{L}(t)\ \mathfrak{v}_\lambda(t) = \mathfrak{w}$$

Since $\mathfrak{w}$ does not depends on α, we have $\mathfrak{u}_\lambda(\alpha(t)) = v_\lambda(t)$. So if we prove that $v_\lambda^{(j)}(t)$ are uniformly bounded in the $\|\cdot\|_{1,k}$ norm, we obtain the boundedness in the same norm for $\mathfrak{u}_\lambda(\alpha)$ for α in the interior of $[0, 1]$.

The extra argument to extend this smoothness up to the boundary is identical to the one used in ref. 11 (see the end of the proof of Theorem 5.1 in there).

Differentiating formally $\mathfrak{L}(t)$ in t, we obtain

$$\mathfrak{L}'(t) = -4(\sin t \cos t)\, \mathfrak{L}_d + (\cos^2 t - \sin^2 t)\{\mathfrak{L}_+ + \mathfrak{L}_-\}$$

By Lemma 5.2 of ref. 11, $\mathfrak{v}_\lambda(t)$ is differentiable in t, and its derivative $\mathfrak{v}'_\lambda(t)$ satisfies

$$\lambda \mathfrak{v}'_\lambda(t) - \mathfrak{L}(t)\, \mathfrak{v}'_\lambda(t) = \mathfrak{L}'(t)\, \mathfrak{v}_\lambda(t)$$

As a consequence of Lemma 7.7, Lemma 4.5 and the explicit form of $\mathfrak{L}(t)$, there exists a constant C_k depending only on k

$$\|\mathfrak{L}'(t)\, \mathfrak{v}_\lambda(t)\|_{-1,k} \leqslant C_k\, \|\bar{\mathfrak{w}}\|_{-1,k+3}$$

Then if we apply Theorem 4.4 to Eq. (6.1), we obtain the bound $\|\mathfrak{v}'_\lambda(t)\|_{-1,k}$ uniform in t and λ. The argument can be iterated exactly as done in ref. 11, obtaining similar bounds for all the derivatives $v_\lambda^{(j)}(t)$.

7. ESTIMATES ON THE OPERATORS $\mathfrak{L}_d$, $\mathfrak{L}_+$, $\mathfrak{L}_-$

We prove in this section some elementary identities or estimates involving the operators $\mathfrak{L}_d$, $\mathfrak{L}_+$, $\mathfrak{L}_-$. Recall that all functions $\mathfrak{f}\colon \mathscr{E}_* \to \mathbb{R}$ which come from a local function f through the transformation $\mathfrak{T}$ of the Fourier coefficients of f are such that $\mathfrak{f}(S_z A) = \mathfrak{f}(A)$ for all z in A. Recall also the definition of the spaces $\mathscr{I}_n$ given just before the statement of Theorem 4.2.

A simple computation shows that the space $\mathscr{I}$ is left invariant by the operators $\mathfrak{L}_s$, $\mathfrak{L}_d$, $\mathfrak{L}_+$, $\mathfrak{L}_-$: For every $n \geqslant 1$ and every $\mathfrak{f}\colon \mathscr{I}_n \to \mathbb{R}$,

$$\mathfrak{L}_s \mathfrak{f} \in \mathscr{I}_n, \qquad \mathfrak{L}_d \mathfrak{f} \in \mathscr{I}_n, \qquad \mathfrak{L}_- \mathfrak{f} \in \mathscr{I}_{n-1}, \qquad \mathfrak{L}_+ \mathfrak{f} \in \mathscr{I}_{n+1} \tag{7.1}$$

This claim can be proved in two different ways. Either by a direct computation or by reconstructing a local function f from $\mathfrak{f}$. More precisely, to prove that $\mathfrak{L}_s \mathfrak{f}_*$ belongs to $\mathscr{I}_n$ if $\mathfrak{f}_*$ belongs to $\mathscr{I}_n$, let $\mathfrak{f}$ be given by (3.7) so that $\mathfrak{T}\mathfrak{f} = \mathfrak{f}_*$. By (3.10), $\mathfrak{T}\mathscr{L}_s \mathfrak{f} = \mathfrak{L}_s \mathfrak{f}_*$, which proves that $\mathfrak{L}_s \mathfrak{f}_*$ belongs to $\mathscr{I}$.

We turn now to an elementary identity to illustrate the fact that the space $\mathscr{I}$ enjoys some special properties. For every $\mathfrak{f}\colon \mathscr{E}_{*,1} \to \mathbb{R}$,

$$(\mathfrak{L}_- \mathfrak{f})(\phi) = -2 \sum_{x \neq 0} a(x)\, \mathfrak{f}(\{x\})$$

In particular, $(\mathfrak{L}_-\mathfrak{f})(\phi)=0$ for all $\mathfrak{f}$ in $\mathscr{I}_1$ because in this space $\mathfrak{f}(\{x\})=\mathfrak{f}(\{-x\})$ and $a(\cdot)$ is anti-symmetric. In contrast, $(\mathfrak{L}_+\mathfrak{g})(\{x\})=0$ for all functions $\mathfrak{g}\colon \mathscr{E}_{*,0}\to\mathbb{R}$ so that, for all $\mathfrak{f}$ in $\mathscr{I}_1$ and all $\mathfrak{g}\colon \mathscr{E}_{*,0}\to\mathbb{R}$,

$$\mathfrak{L}_-\mathfrak{f}=0, \qquad \mathfrak{L}_+\mathfrak{g}=0 \tag{7.2}$$

We turn now to the proof of some estimates involving the operators $\mathfrak{L}_s$, $\mathfrak{L}_d$, $\mathfrak{L}_-$, and $\mathfrak{L}_+$. The first lemma states that the operators $\mathfrak{L}_s$, $\mathfrak{L}_d$, $\mathfrak{L}_-$, and $\mathfrak{L}_+$ are bounded in $L^2(\mathscr{E}_{*,n})$ for each fixed $n\geqslant 1$.

Lemma 7.1. There exists a finite constant C_0 such that

$$\|\mathfrak{A}\mathfrak{f}\|_0^2 \leqslant C_0 n^2 \,\|\mathfrak{f}\|_0^2$$

for each $\mathfrak{f}$ in $L^2(\mathscr{E}_{*,n})$, were the operator $\mathfrak{A}$ stands for $\mathfrak{L}_s$, $\mathfrak{L}_d$, $\mathfrak{L}_+$, or $\mathfrak{L}_-$.

Proof. We only prove the estimate concerning $\mathfrak{L}_-$, the other ones being elementary. Fix a function $\mathfrak{f}\colon \mathscr{E}_{*,n}\to\mathbb{R}$ in L^2 and keep in mind that $\mathfrak{L}_-\mathfrak{f}$ maps $\mathscr{E}_{*,n-1}$ in $\mathbb{R}$. By the explicit expression for $\mathfrak{L}_-$ and a change of variables,

$$\|\mathfrak{L}_-\mathfrak{f}\|_0^2 \leqslant 4 \sum_{A\in\mathscr{E}_{*,n-1}} \Big\{ \sum_{\substack{x,\,y\notin A\\ x,\,y\neq 0}} a(y-x)\,\mathfrak{f}(A\cup\{y\}) \Big\}^2$$

Since $\sum_{x\in\mathbb{Z}^d} a(x)=0$, the previous expression is less than or equal to

$$8 \sum_{A\in\mathscr{E}_{*,n-1}} \Big\{ \sum_{\substack{y\notin A\\ y\neq 0}} a(y)\,\mathfrak{f}(A\cup\{y\}) \Big\}^2 + 8 \sum_{A\in\mathscr{E}_{*,n-1}} \Big\{ \sum_{\substack{y\notin A,\, y\neq 0\\ x\in A}} a(y-x)\,\mathfrak{f}(A\cup\{y\}) \Big\}^2$$

By Schwarz inequality, since $a(\cdot)$ is absolutely bounded and $\sum_{x\in\mathbb{Z}^d}|a(x)|$ is finite, this expression is less than or equal to

$$C_0 n \sum_{A\in\mathscr{E}_{*,n-1}} \sum_{\substack{y\notin A\\ y\neq 0}} \mathfrak{f}(A\cup\{y\})^2$$

for some finite constant C_0. To sum over A in $\mathscr{E}_{*,n-1}$ and over $y\neq 0$, $y\notin A$ is the same as to sum over B in $\mathscr{E}_{*,n}$ with a multiplicity factor n because all sets are counted n times. The previous expression is thus equal to

$$C_0 n^2 \sum_{A\in\mathscr{E}_{*,n}} \mathfrak{f}(A)^2$$

which concludes the proof of the lemma. ▮

It follows from the next statement that the operators $\mathfrak{L}_d$ and $\mathfrak{L}_+ + \mathfrak{L}_-$ are anti-symmetric on $\mathscr{I}\cap L^2_{0,-1}(\mathscr{E}_*)$.

Lemma 7.2. For every $n \geqslant 1$ and every finitely supported functions $\mathfrak{u}, \mathfrak{v}: \mathscr{E}_{*,n} \to \mathbb{R}$

$$\langle \mathfrak{L}_d \mathfrak{u}, \mathfrak{v} \rangle = -\langle \mathfrak{u}, \mathfrak{L}_d \mathfrak{v} \rangle$$

For every finitely supported functions $\mathfrak{f}$, $\mathfrak{g}$ in $\mathscr{I}_{n-1}$, $\mathscr{I}_n$ respectively,

$$\frac{1}{n+1} \langle \mathfrak{L}_+ \mathfrak{f}, \mathfrak{g} \rangle = -\frac{1}{n} \langle \mathfrak{f}, \mathfrak{L}_- \mathfrak{g} \rangle.$$

Proof. The first identity is elementary and relies on the fact that $\sum_{x, y \in A} a(y-x) = 0$. Note, however, that both pieces of the operator are needed.

The proof of the second statement is more demanding. Fix finitely supported functions $\mathfrak{f}$, $\mathfrak{g}$ in $\mathscr{I}_{n-1}$, $\mathscr{I}_n$, respectively. By the explicit form of $\mathfrak{L}_+$,

$$\begin{aligned} \langle \mathfrak{g}, \mathfrak{L}_+ \mathfrak{f} \rangle = 2 & \sum_{A \in \mathscr{E}_{*,n}} \sum_{x, y \in A} a(y-x)\, \mathfrak{g}(A)\, \mathfrak{f}(A \backslash \{y\}) \\ & + 2 \sum_{A \in \mathscr{E}_{*,n}} \sum_{x \in A} a(x)\, \mathfrak{g}(A) \{ \mathfrak{f}(A \backslash \{x\}) - \mathfrak{f}(S_x[A \backslash \{x\}]) \} \end{aligned}$$

Since $S_x[A \backslash \{x\}] = S_x A \backslash \{-x\}$ and since $\mathfrak{g}(S_x A) = \mathfrak{g}(A)$ for x in A because $\mathfrak{g}$ belongs to $\mathscr{I}_n$, a change of variables $B = S_x A$, $x' = -x$ in the second piece of the second term permits to rewrite the second term on the right hand side as

$$4 \sum_{x \neq 0} a(x) \sum_{\substack{A \in \mathscr{E}_{*,n} \\ A \ni x}} \mathfrak{g}(A)\, \mathfrak{f}(A \backslash \{x\})$$

because $a(-x) = -a(x)$. We claim that

$$\sum_{x \neq 0} a(x) \sum_{\substack{A \in \mathscr{E}_{*,n} \\ A \ni x}} \mathfrak{g}(A)\, \mathfrak{f}(A \backslash \{x\}) = \frac{1}{n-1} \sum_{x, y \neq 0} a(y-x) \sum_{\substack{A \in \mathscr{E}_{*,n} \\ A \ni x, y}} \mathfrak{g}(A)\, \mathfrak{f}(A \backslash \{y\}) \tag{7.3}$$

We conclude the proof of the lemma assuming (7.3), whose proof is presented at the end. It follows from identity (7.3) and the previous expression for $\langle \mathfrak{g}, \mathfrak{L}_+ \mathfrak{f} \rangle$ that

$$\begin{aligned} \langle \mathfrak{g}, \mathfrak{L}_+ \mathfrak{f} \rangle = 2 \left(1 + \frac{1}{n} \right) & \sum_{A \in \mathscr{E}_{*,n}} \sum_{x, y \in A} a(y-x)\, \mathfrak{g}(A)\, \mathfrak{f}(A \backslash \{y\}) \\ & + 2 \left(1 + \frac{1}{n} \right) \sum_{A \in \mathscr{E}_{*,n}} \sum_{x \in A} a(x)\, \mathfrak{g}(A)\, \mathfrak{f}(A \backslash \{x\}) \end{aligned}$$

The first term of the right hand side, which can be written as

$$2\left(1+\frac{1}{n}\right)\sum_{y\neq 0}\sum_{\substack{A\in\mathscr{E}_{*,n}\\A\ni y}}\mathfrak{g}(A)\,\mathfrak{f}(A\setminus\{y\})\sum_{x\in A}a(y-x)$$

is equal to

$$-2\left(1+\frac{1}{n}\right)\sum_{y\neq 0}a(y)\sum_{\substack{A\in\mathscr{E}_{*,n}\\A\ni y}}\mathfrak{g}(A)\,\mathfrak{f}(A\setminus\{y\})$$
$$-2\left(1+\frac{1}{n}\right)\sum_{x,\,y\neq 0}a(y-x)\sum_{\substack{A\in\mathscr{E}_{*,n}\\A\ni y,\,A\not\ni x}}\mathfrak{g}(A)\,\mathfrak{f}(A\setminus\{y\})$$

because $\sum_{x\in A}a(y-x)=-a(y)-\sum_{x\neq 0,\,x\notin A}a(y-x)$. The first term of this formula cancels with the second one in the last expression for $\langle\mathfrak{g},\mathfrak{L}_+\mathfrak{f}\rangle$. Therefore,

$$\langle\mathfrak{g},\mathfrak{L}_+\mathfrak{f}\rangle=-2\left(1+\frac{1}{n}\right)\sum_{x,\,y\neq 0}a(y-x)\sum_{\substack{A\in\mathscr{E}_{*,n}\\A\ni y,\,A\not\ni x}}\mathfrak{g}(A)\,\mathfrak{f}(A\setminus\{y\})$$

To conclude the proof of the lemma, it remains to change variables $B=A\setminus\{y\}$ and to recall the definition of the operator $\mathfrak{L}_-$.

We turn now to the proof of Claim (7.3). Since for y in A, $\mathfrak{g}(A)=\mathfrak{g}(S_yA)$ and since $|A|=n$, the left hand side of (7.3) is equal to

$$\frac{1}{n}\sum_{x\neq 0}a(x)\sum_{\substack{A\in\mathscr{E}_{*,n}\\A\ni x}}\sum_{\substack{y\in A\cup\{0\}\\y\neq x}}\mathfrak{g}(S_yA)\,\mathfrak{f}(A\setminus\{x\})$$
$$=\frac{1}{n}\sum_{\substack{x,\,y\neq 0\\y\neq x}}a(x)\sum_{\substack{A\in\mathscr{E}_{*,n}\\A\ni x,\,y}}\mathfrak{g}(S_yA)\,\mathfrak{f}(A\setminus\{x\})+\frac{1}{n}\sum_{x\neq 0}a(x)\sum_{\substack{A\in\mathscr{E}_{*,n}\\A\ni x}}\mathfrak{g}(A)\,\mathfrak{f}(A\setminus\{x\})$$

Notice that the second term on the right hand side is precisely the original one. Consider the first term. Perform a change of variables $B=S_yA$, rewrite $(S_{-y}A)\setminus\{x\}$ as $S_{-y}(A\setminus\{x-y\})$ and recall that $\mathfrak{f}(S_{-y}(A\setminus\{x-y\})=\mathfrak{f}(A\setminus\{x-y\})$ if $-y$ belongs to A because $\mathfrak{f}$ is in $\mathscr{I}_{n-1}$, to rewrite this expression as

$$\frac{1}{n}\sum_{\substack{x,\,y\neq 0\\y\neq x}}a(x)\sum_{\substack{A\in\mathscr{E}_{*,n}\\A\ni x-y,\,-y}}\mathfrak{g}(A)\,\mathfrak{f}(A\setminus\{x-y\})$$

A change of variables $x' = x - y$, $y' = -y$, shows that this expression is equal to

$$\frac{1}{n} \sum_{x, y \neq 0} a(x-y) \sum_{\substack{A \in \mathscr{E}_{*,n} \\ A \ni x, y}} \mathfrak{g}(A)\, \mathfrak{f}(A \setminus \{x\})$$

To prove (7.3), it remains to recollect all previous identities. ∎

It follows from this result that the operator $\mathfrak{L}_+ + \mathfrak{L}_-$ is anti-symmetric in $L^2_{0,-1}(\mathscr{E}_*) \cap \mathscr{I}$:

Corollary 7.3. The operator $\mathfrak{L}_+ + \mathfrak{L}_-$ is anti-symmetric in $L^2_{0,-1}(\mathscr{E}_*) \cap \mathscr{I}$:

$$\langle\!\langle \mathfrak{f}, (\mathfrak{L}_+ + \mathfrak{L}_-)\, \mathfrak{g} \rangle\!\rangle_{0,-1} = -\langle\!\langle (\mathfrak{L}_+ + \mathfrak{L}_-)\, \mathfrak{f}, \mathfrak{g} \rangle\!\rangle_{0,-1}$$

for all finitely supported functions $\mathfrak{f}$, $\mathfrak{g}$ in $\mathscr{I}$. The same statement remains in force if $\mathfrak{L}_+ + \mathfrak{L}_-$ is replaced by $\Pi_n(\mathfrak{L}_+ + \mathfrak{L}_-)\, \Pi_n$ for every $n \geqslant 1$, where $\Pi_n = \sum_{1 \leqslant j \leqslant n} \pi_j$.

The proof of Corollary 7.3 is elementary and left to the reader. One needs only to recall identities (7.2). The next result states that $\mathfrak{L}_d$ is a bounded operator from $L^2(\mathscr{E}_{*,n})$ to $\mathfrak{H}_{-1}(\mathscr{E}_{*,n})$.

Lemma 7.4. There exists a finite constant C_0, independent of n, such that for each $n \geqslant 1$ and for any finitely supported functions $\mathfrak{f}$, $\mathfrak{g}$: $\mathscr{E}_{*,n} \to \mathbb{R}$,

$$\langle \mathfrak{L}_d \mathfrak{f}, \mathfrak{g} \rangle \leqslant C_0 \sqrt{n}\, \|\mathfrak{g}\|_1\, \|\mathfrak{f}\|_0$$

In particular,

$$\|\mathfrak{L}_d \mathfrak{f}\|_{-1} \leqslant C_0 \sqrt{n}\, \|\mathfrak{f}\|_0$$

The very same estimates remain in force if $\mathfrak{L}_d$ is replaced by $\mathfrak{L}_s$.

The proof of Lemma 7.4 is elementary and left to the reader. Next estimate is Lemma 2.3 of ref. 8. Recall from Section 4 the definition of the operator $\mathfrak{B}$.

Lemma 7.5. There exists a finite constant C_0 such that for any function $\mathfrak{f}$: $\mathscr{E}_{n,*} \to \mathbb{R}$ in $\mathfrak{H}_1$,

$$\|\mathfrak{f}\|_1^2 \leqslant \|\mathfrak{B}\mathfrak{f}\|_{\mathscr{X}_n, 1}^2 \leqslant C_0 n\, \|\mathfrak{f}\|_1^2$$

Proof. The first inequality is elementary and follows from the explicit formulas for the respective H_1 norms. To prove the second inequality, for $\mathbf{x} = (x_1,..., x_n) \in \mathscr{E}_{n,*}$, let

$$W_1(\mathbf{x}) = \sum_{j=1}^{n} s(x_j), \qquad W_2(\mathbf{x}) = \sum_{i,j=1}^{n} s(x_i - x_j)$$

We also denote these quantities by $W_1(A)$ and $W_2(A)$, for $A = \{x_1,..., x_n\}$. A simple computation shows that there exists a finite constant C_0 such that

$$|\mathfrak{B}\mathfrak{L}_s\mathfrak{f}(\mathbf{x}) - \tilde{\mathfrak{L}}_s\mathfrak{B}\mathfrak{f}(\mathbf{x})| \leqslant C_0\{W_1(\mathbf{x}) + W_2(\mathbf{x})\}\ |\mathfrak{B}\mathfrak{f}(\mathbf{x})| \tag{7.4}$$

for every $\mathbf{x}$ in $\mathscr{E}_{n,*}$ and $\mathfrak{f}\colon \mathscr{E}_{n,*} \to \mathbb{R}$.

We are now in a position to prove the second bound. By definition,

$$\|\mathfrak{B}\mathfrak{f}\|^2_{\mathfrak{X}_n,1} = -\frac{1}{n!} \sum_{\mathbf{x} \in \mathfrak{X}_n} (\mathfrak{B}\mathfrak{f})(\mathbf{x})(\tilde{\mathfrak{L}}_s\mathfrak{B}\mathfrak{f})(\mathbf{x})$$

Since $\mathfrak{B}\mathfrak{f}$ vanishes outside $\mathscr{E}_{n,*}$, we may restrict the sum to $\mathscr{E}_{n,*}$. Now, adding and subtracting $(\mathfrak{B}\mathfrak{L}_s\mathfrak{f})(\mathbf{x})$ in this expression and recalling (7.4), we obtain that

$$\begin{aligned}\|\mathfrak{B}\mathfrak{f}\|^2_{\mathfrak{X}_n,1} &\leqslant \|\mathfrak{f}\|_1^2 + \frac{C_0}{n!} \sum_{\mathbf{x} \in \mathscr{E}_{n,*}} \{W_1(x) + W_2(x)\}\{\mathfrak{B}\mathfrak{f}(\mathbf{x})\}^2 \\ &= \|\mathfrak{f}\|_1^2 + C_0 \sum_{A \in \mathscr{E}_{n,*}} \{W_1(A) + W_2(A)\}\ \mathfrak{f}(A)^2\end{aligned}$$

By Theorem 4.7 in ref. 8 or Lemma 3.7 in ref. 13 the second term of the previous formula is bounded by $C_0 n\ \|\mathfrak{f}\|_1^2$, which concludes the proof of the lemma. ▮

It follows from this result that

$$\frac{1}{C_0 n} \|\mathfrak{f}\|_{-1}^2 \leqslant \|\mathfrak{B}\mathfrak{f}\|^2_{\mathfrak{X}_n,-1} \leqslant \|\mathfrak{f}\|_{-1}^2 \tag{7.5}$$

Lemma 7.6. There exists a finite constant C_0 such that

$$\|\mathfrak{B}\mathfrak{L}_s\mathfrak{f} - \tilde{\mathfrak{L}}_s\mathfrak{B}\mathfrak{f}\|^2_{\mathfrak{X}_n,-1} \leqslant C_0 n^2\ \|\mathfrak{f}\|_1^2, \qquad \|\mathfrak{B}\mathfrak{L}_d\mathfrak{f} - \tilde{\mathfrak{L}}_d\mathfrak{B}\mathfrak{f}\|^2_{\mathfrak{X}_n,-1} \leqslant C_0 n^2\ \|\mathfrak{f}\|_1^2$$

for all $n \geqslant 1$ and all functions $\mathfrak{f}\colon \mathscr{E}_{n,*} \to \mathbb{R}$.

Proof. We prove the first estimate and leave to the reader the details of the second. Fix $n \geqslant 1$ and a function $\mathfrak{h}$: $\mathscr{X}_n \to \mathbb{R}$. We need to estimate the scalar product

$$\frac{1}{n!} \sum_{\mathbf{x} \in \mathscr{X}_n} \mathfrak{h}(\mathbf{x}) \{ \mathfrak{B} \mathfrak{L}_s \mathfrak{f}(\mathbf{x}) - \tilde{\mathfrak{L}}_s \mathfrak{B} \mathfrak{f}(\mathbf{x}) \} \tag{7.6}$$

in terms of the $H_1(\mathscr{X}_n)$ norm of $\mathfrak{h}$ and the $H_1(\mathscr{E}_{n,*})$ norm of $\mathfrak{f}$. There are two possible cases. Either $\mathbf{x}$ belongs to $\mathscr{E}_{n,*}$ or $\mathbf{x}$ does not belong to $\mathscr{E}_{n,*}$.

In the first case, by (7.4), the expression inside braces in previous formula is absolutely bounded by $C_0 W_*(\mathbf{x}) \, |\mathfrak{B}\mathfrak{f}(\mathbf{x})|$ for some finite constant C_0, where $W_*(\mathbf{x}) = W_1(\mathbf{x}) + W_2(\mathbf{x})$. Therefore, the corresponding piece in the previous formula is bounded above by

$$\frac{1}{n!} \sum_{\mathbf{x} \in \mathscr{E}_{n,*}} W_*(\mathbf{x}) \, |\mathfrak{h}(\mathbf{x})| \, |\mathfrak{B}\mathfrak{f}(\mathbf{x})|$$

$$\leqslant \frac{1}{2\ell} \frac{1}{n!} \sum_{\mathbf{x} \in \mathscr{E}_{n,*}} W_*(\mathbf{x}) \, \mathfrak{h}(\mathbf{x})^2 + \frac{\ell}{2} \sum_{A \in \mathscr{E}_{n,*}} W_*(A) \, \mathfrak{f}(A)^2$$

for every $\ell > 0$.

If $\mathbf{x}$ does not belong to $\mathscr{E}_{n,*}$, the corresponding piece of the scalar product writes

$$-\frac{2}{n!} \sum_{\substack{\mathbf{x} \notin \mathscr{X}_n \\ z \in \mathbb{Z}^d \\ 1 \leqslant j \leqslant n}} s(z) \, \mathfrak{h}(\mathbf{x}) \, \mathfrak{B}\mathfrak{f}(\mathbf{x} + z\mathbf{e}_j) - \frac{2}{n!} \sum_{\substack{\mathbf{x} \notin \mathscr{X}_n \\ z \in \mathbb{Z}^d}} s(z) \, \mathfrak{h}(\mathbf{x}) \, \mathfrak{B}\mathfrak{f}(\mathbf{x} + z\mathbf{1})$$

because in this case $\mathfrak{B}\mathfrak{L}_s\mathfrak{f}(\mathbf{x}) = 0$.

We estimate the first term and claim that the second can be handled in the same way. Since $\mathfrak{B}\mathfrak{f}$ vanishes outside $\mathscr{E}_{n,*}$, it is implicit in the previous formula that the first sum is restricted to all $\mathbf{x}$ such that $\mathbf{x} + z\mathbf{e}_j$ belongs to $\mathscr{E}_{n,*}$. Since $\mathbf{x} + z\mathbf{e}_j \in \mathscr{E}_{n,*}$ and $\mathbf{x} \notin \mathscr{E}_{n,*}$, either $x_j = 0$ or $x_j = x_k$ for some k. In particular, since $2ab \leqslant \ell a^2 + \ell^{-1} b^2$ for every $\ell > 0$, a change of variables gives that the first term of the previous formula is bounded above by

$$\frac{1}{n!\ell} \sum_{\mathbf{x} \in \mathscr{X}_n} \mathfrak{h}(\mathbf{x})^2 \, \tilde{W}_*(\mathbf{x}) + \frac{\ell}{n!} \sum_{\mathbf{x} \in \mathscr{E}_{n,*}} \mathfrak{B}\mathfrak{f}(\mathbf{x})^2 \, W_*(\mathbf{x})$$

where $\tilde{W}_*(\mathbf{x}) = \sum_{1 \leqslant j \leqslant n} \mathbf{1}\{x_j = 0\} + \sum_{j \neq k} \mathbf{1}\{x_j = x_k\}$. We may of course replace the sum over $\mathscr{X}_n$ by a sum over $\mathscr{E}_{n,*}$ in the second term, loosing the factor $n!$.

Adding together all previous estimates, we obtain that the scalar product (7.6) is bounded above by

$$\frac{C_0}{n!\ell}\sum_{\mathbf{x}\in\mathscr{X}_n}\mathfrak{h}(\mathbf{x})^2\{\tilde{W}_*(\mathbf{x})+W_*(\mathbf{x})\}+C_0\ell\sum_{A\in\mathscr{E}_{n,*}}\mathfrak{f}(A)^2 W_*(A).$$

The same proof of Theorem 4.7 in ref. 8 or Lemma 3.7 in ref. 13 shows that in dimension $d\geqslant 3$, the first term is bounded above by $C_1 nA^{-1}\|\mathfrak{h}\|^2_{\mathscr{X}_n,1}$ for some finite constant C_1, while the second, by the quoted results, is less than or equal to $C_1 nA\|\mathfrak{f}\|_1^2$. To conclude the proof of the lemma, it remains to minimize over A and to recall the variational formula for the H_{-1} norm. ∎

We conclude this section with a central estimate involving the operators $\mathfrak{L}_+$, $\mathfrak{L}_-$. This result is Theorem 4.4 in ref. 8. The reader can find in Lemma 4.1 of ref. 13 a clearer proof.

Lemma 7.7. There exists a finite constant C_0 depending only on the transition probability p such that

$$\left\{\sum_{A\in\mathscr{E}_{*,n+1}}(\mathfrak{L}_+\mathfrak{f})(A)\,\mathfrak{g}(A)\right\}^2\leqslant C_0 n\,\|\mathfrak{f}\|_1^2\,\|\mathfrak{g}\|_1^2$$

$$\left\{\sum_{A\in\mathscr{E}_{*,n}}\mathfrak{f}(A)(\mathfrak{L}_-\mathfrak{g})(A)\right\}^2\leqslant C_0 n\,\|\mathfrak{f}\|_1^2\,\|\mathfrak{g}\|_1^2$$

for all $n\geqslant 1$ and all finite supported functions $\mathfrak{f}\colon\mathscr{E}_{*,n}\to\mathbb{R}$, $\mathfrak{g}\colon\mathscr{E}_{*,n+1}\to\mathbb{R}$. In particular,

$$\|\mathfrak{L}_+\mathfrak{f}\|_{-1}^2\leqslant C_0 n\,\|\mathfrak{f}\|_1^2,\qquad \|\mathfrak{L}_-\mathfrak{g}\|_{-1}^2\leqslant C_0 n\,\|\mathfrak{g}\|_1^2$$

REFERENCES

1. C. Bernardin, Regularity of the diffusion coefficient for lattice gas reversible under Bernoulli measures, *Stochastic Process. Appl.* **101**:43–68 (2002).
2. C. C. Chang, C. Landim, and S. Olla, Equilibrium fluctuations of asymmetric exclusion processes in dimension $d\geqslant 3$, *Probab. Theory Related Fields* **119**:381–409 (2001).
3. R. Esposito, R. Marra, and H. T. Yau, Diffusive limit of asymmetric simple exclusion, *Rev. Math. Phys.* **6**:1233–1267 (1994).
4. C. Kipnis and C. Landim, *Scaling Limit of Interacting Particle Systems* (Springer Verlag, Berlin, 1999).
5. C. Kipnis and S. R. S. Varadhan, Central limit theorem for additive functionals of reversible Markov processes and applications to simple exclusion, *Comm. Math. Phys.* **106**:1–19 (1986).

6. C. Landim, S. Olla, and H. T. Yau, First order correction for the hydrodynamic limit of asymmetric simple exclusion processes in dimension $d \geqslant 3$, *Communications on Pure and Applied Mathematics* **50**:149–203 (1997).
7. C. Landim, S. Olla, and H. T. Yau, Some properties of the diffusion coefficient for asymmetric simple exclusion processes, *Ann. Probab.* **24**:1779–1807 (1996).
8. C. Landim and H. T. Yau, Fluctuation–dissipation equation of asymmetric simple exclusion processes, *Probab. Theory Related Fields* **108**:321–356 (1997).
9. C. Landim, S. Olla, and S. R. S. Varadhan, Finite-dimensional approximation of the self-diffusion coefficient for the exclusion process, *Ann. Probab.* **30**:1–26 (2002).
10. C. Landim, S. Olla, and S. R. S. Varadhan, Asymptotic behavior of a tagged particle in simple exclusion processes, *Bol. Soc. Bras. Mat.* **31**:241–275 (2000).
11. C. Landim, S. Olla, and S. R. S. Varadhan, Symmetric simple exclusion process: regularity of the self-diffusion coefficient, *Comm. Math. Phys.* **224**:302–321 (2001).
12. T. M. Liggett, *Interacting Particle Systems* (Springer Verlag, New York, 1985).
13. S. Sethuraman, S. R. S. Varadhan, and H. T. Yau, Diffusive limit of a tagged particle in asymmetric exclusion process, *Comm. Pure Appl. Math.* **53**:972–1006 (2000).
14. H. Spohn, *Large Scale Dynamics of Interacting Particles* (Springer, 1991).
15. S. R. S. Varadhan, Non-linear diffusion limit for a system with nearest neighbor interactions. II, in *Asymptotic Problems in Probability Theory: Stochastic Models and Diffusion on Fractals*, Pitman Research Notes in Mathematics, Vol. 283, K. D. Elworthy and N. Ikeda, eds. (Wiley, New York, 1994), pp. 75–128.
16. S. R. S. Varadhan, Self diffusion of a tagged particle in equilibrium for asymmetric mean zero random walks with simple exclusion, *Ann. Inst. H. Poincaré (Probabilités)* **31**:273–285 (1995).

The Annals of Probability
2012, to appear

LARGE DEVIATIONS FOR THE CURRENT AND TAGGED PARTICLE IN 1D NEAREST-NEIGHBOR SYMMETRIC SIMPLE EXCLUSION

SUNDER SETHURAMAN AND S.R.S. VARADHAN

ABSTRACT. Laws of large numbers, starting from certain nonequilibrium measures, have been shown for the integrated current across a bond, and a tagged particle in one dimensional symmetric nearest-neighbor simple exclusion (Jara-Landim (2006)). In this article, we prove corresponding large deviation principles, and evaluate the rate functions, showing different growth behaviors near and far from their zeroes which connect with results in Derrida-Gerschenfeld (2009).

1. INTRODUCTION AND RESULTS

The one dimensional nearest-neighbor symmetric simple exclusion process follows a collection of nearest-neighbor random walks on the lattice $\mathbb{Z}$ which move equally likely left or right except in that jumps to already occupied sites are suppressed. More precisely, the model is a Markov process $\eta_t = \{\eta_t(x) : x \in \mathbb{Z}\}$ evolving on the configuration space $\Sigma = \{0,1\}^{\mathbb{Z}}$ with generator,

$$(L\phi)(\eta) = (1/2)\sum_x \left[\eta(x)(1-\eta(x+1)) + \eta(x+1)(1-\eta(x))\right]\left(\phi(\eta^{x,x+1}) - \phi(\eta)\right)$$

where $\eta^{x,y}$, for $x \neq y$, is the configuration obtained from η by exchanging the values at x and y,

$$\eta^{x,y}(z) = \begin{cases} \eta(z) & \text{when } z \neq x, y \\ \eta(x) & \text{when } z = y \\ \eta(y) & \text{when } z = x. \end{cases}$$

A detailed treatment can be found in Liggett [23].

As the process is 'mass conservative,' that is no birth or death, one expects a family of invariant measures corresponding to particle density. In fact, for each $\rho \in [0,1]$, the product over $\mathbb{Z}$ of Bernoulli measures ν_ρ which independently puts a particle at locations $x \in \mathbb{Z}$ with probability ρ, that is $\nu_\rho(\eta_x = 1) = 1 - \nu_\rho(\eta_x = 0) = \rho$, are invariant. We will denote E_ρ as expectation under ν_ρ.

Consider now the integrated current across the bond $(-1,0)$, and a distinguished, or tagged particle, say initially at the origin. Let $J_{-1,0}(t)$ and X_t be the current and position of the tagged particle at time t, respectively. The problem of characterizing the asymptotic behavior of the current and tagged particle in interacting systems

---

2000 *Mathematics Subject Classification.* primary 60K35.

Key words and phrases. symmetric, exclusion, current, tagged particle, large deviations, one dimensional.

Supported in part by NSF-0906713 and NSA Q H982301010180, and NSF-0904701 and NSF-0730136.

has a long history (cf. Spohn [38, Chapters 8.I, 6.II]), and was mentioned in Spitzer's seminal paper [37].

The goal of this paper is to investigate the large deviations of $J_{-1,0}(t)$ and X_t when the initial distribution of particles is part of a large class of nonequilibrium measures. Our initial motivation was to understand better laws of large numbers (LLN) and central limit theorems (CLT) in Jara-Landim [16] for the current and tagged particle when the process starts from a class of 'local equilibrium' initial measures. It turns out that recent formal expansions of the large deviation 'pressure' for the current in Derrida-Gerschenfeld [10], [11] might also be recovered in such a study.

The article [16] is a nonequilibrium generalization of CLTs in Arratia [1], Rost-Vares [32], and De Masi-Ferrari [8] which established 'subdiffusive' behaviors in the 1D nearest-neighbor symmetric simple exclusion model. Namely, starting under an equilibrium ν_ρ, $t^{-1/4}J_{-1,0}(t) \Rightarrow \mathrm{N}(0,\sigma_J^2)$ and $t^{-1/4}X_t \Rightarrow \mathrm{N}(0,\sigma_X^2)$, where $\sigma_J^2 = \sqrt{2/\pi}(1-\rho)\rho$ and $\sigma_X^2 = \sqrt{2/\pi}(1-\rho)/\rho$. Physically, the 'subdiffusive' scale in the CLT is explained as due to 'trapping' induced from the nearest-neighbor dynamics which enforces a rigid ordering of particles. Recently, the CLTs were extended to an invariance principle with respect to a fractional Brownian motion, $\lambda^{-1/4}J_{-1,0}(\lambda t) \Rightarrow \sigma_J fBM_{1/4}(t)$ and $\lambda^{-1/4}X_{\lambda t} \Rightarrow \sigma_X fBM_{1/4}(t)$, in Peligrad-Sethuraman [26].

We now specify the class of initial measures considered, that is 'deterministic initial configurations' and 'local equilibrium product measures.' Let M_1 be the space of functions $\gamma : \mathbb{R} \to [0,1]$, and let $M_1(\rho_*,\rho^*)$ be those functions in M_1 which equal ρ_* for all $x \leq x_*$, and which equal ρ^* for all $x \geq x^*$, for some $x_* \leq x^*$.

We will consider on M_1 the topology induced by $C_K(\mathbb{R})$, the set of continuous compactly supported functions on $\mathbb{R}$, with the duality $\langle\cdot;\cdot\rangle$ where $\langle\gamma;G\rangle = \int G(x)\gamma(x)dx$ for $\gamma \in M_1$ and $G \in C_K(\mathbb{R})$. This topology, if M_1 is thought of as a measure space, is the vague topology which is metrizable.

Local equilibrium measure (LEM). For $0 < \rho_*, \rho^* < 1$, let $\gamma \in M_1(\rho_*,\rho^*)$ be a piecewise continuous function such that $0 < \gamma(x) < 1$ for all $x \in \mathbb{R}$. With respect to γ and a scaling parameter $N \geq 1$, we define a sequence of local equilibrium product measures $\nu^{(N)}_{\gamma(\cdot)}$ as those formed from the marginals $\nu^{(N)}_{\gamma(\cdot)}(\eta(x)=1) = \gamma(x/N)$ for $x \neq 0$, and $\nu^{(N)}_{\gamma(\cdot)}(\eta(0)=1) = 1$.

Deterministic initial configuration (DIC). For $0 < \rho_*, \rho^* < 1$, let γ be a piecewise continuous function in $M_1(\rho_*,\rho^*)$. Then, a sequence of deterministic initial configurations $\xi^{\gamma,N}$ is one such that $\xi^{\gamma,N}(0) = 1$ and, for all continuous, compactly supported G, $\lim_{N\to\infty} \frac{1}{N}\sum_x \xi^{\gamma,N}(x)G(x/N) = \int G(x)\gamma(x)dx$.

We remark particular examples of local equilibrium measures $\nu^{(N)}_{\gamma(\cdot)}$ are the equilibrium measures $\nu_\rho(\cdot|\eta(0)=1)$ conditioned to have a particle at the origin for $0 < \rho < 1$. Suitable deterministic configurations $\xi^{\gamma,N}$ for instance include the 'alternating' configuration where every other vertex is occupied corresponding to $\gamma(x) \equiv 1/2$. Nonequilibrium initial measures corresponding to step profiles $\gamma(x) = \rho_* 1_{(-\infty,0]}(x) + \rho^* 1_{(0,\infty)}(x)$ can also be constructed. The condition that the origin is occupied in these configurations allows to distinguish the corresponding particle as the 'tagged' particle.

In a sense, the profiles γ associated to the local equilibria and deterministic profiles above are 'non-degenerate' in that γ is asymptotically bounded strictly between 0 and 1. Also, the property that $\gamma(x)$ is constant for large $|x|$, and with respect to (LEM) specifications that $0 < \gamma < 1$, is useful to establish later Proposition 1.3, although some modifications, for instance in terms of profiles sufficiently close to being constant for large $|x|$, should be possible with more work. However, under 'degenerate' profiles, different current and tagged particle large deviation behaviors might occur. See comments after Theorem 1.7 and Theorem 1.8 for an 'example'.

We now describe the LLNs proved in Jara-Landim [16] (stated under a class of local equilibrium measures, but the same proof also works starting from the initial measures above):

$$\lim_{N\to\infty} \frac{1}{N} J_{-1,0}(N^2 t) \;=\; v_t \quad\text{and}\quad \lim_{N\to\infty} \frac{1}{N} X_{N^2 t} \;=\; u_t, \tag{1.1}$$

in probability, where v_t and u_t satisfy

$$\frac{dv_t}{dt} = -\frac{1}{2}\partial_x \rho(t,0) \quad\text{and}\quad \frac{du_t}{dt} = -\frac{1}{2}\frac{\partial_x \rho(t,u_t)}{\rho(t,u_t)}$$

and $\partial_t \rho = (1/2)\partial_{xx}\rho$ and $\rho(0,x) = \gamma(x)$, that is $\rho(t,x) = \sigma_t * \gamma(x)$ where $\sigma_t(y) = (2\pi t)^{-1/2}\exp\{-y^2/2t\}$. Note that

$$v_t \;=\; -\frac{1}{2}\int_0^t \partial_x\rho(s,0)ds \;=\; \int_0^\infty \big[\rho(t,x) - \rho(0,x)\big]dx$$

and u_t is also the unique number α where

$$\int_0^\alpha \rho(t,x)dx \;=\; -\frac{1}{2}\int_0^t \partial_x\rho(s,0)ds \;=\; \int_0^\infty \big[\rho(t,x) - \rho(0,x)\big]dx.$$

To explain the last equation, the right-side, as already indicated, is the integrated macroscopic current across the origin up to time t. As the microscopic dynamics is nearest-neighbor with enforced ordering of particles, the tagged particle, initially at the origin, will be at the head of the flow through the origin. So, to compute its macroscopic position u_t at time t, we find α so that the mass at time t between positions $x = 0$ and $x = \alpha$, the left-side of the equation, equals the integrated current, and conclude $u_t = \alpha$.

We remark, starting from a class of local equilibrium measures, corresponding invariance principles in subdiffusive $t^{1/4}$ scale, in the sense of finite-dimensional distributions, with respect to fractional Brownian motion-type Gaussian processes was also proved in [16]. Also, for the current, starting from a large class of product measures, self-normalized CLTs have been shown in Liggett [24] and Vandenberg-Rodes [39].

In this context, we derive large deviation principles (LDPs) (Theorem 1.5), in diffusive scale, corresponding to the laws of large numbers (1.1) when starting from (LEM) or (DIC) measures. We give also lower and upper bounds on the associated rate functions starting from various non-degenerate initial conditions (Theorems 1.6). A consequence of these rate function bounds, say when starting from deterministic initial configurations, is that the following growth structure can be deduced: Namely, the rate functions are quadratic near their zeroes, but are third order far away from the zeroes.

In particular, the third order asymptotics we derive confirm the formal third order expansions in Derrida-Gerschenfeld [10], for the probability distribution of the current across the origin at large times (cf. discussion after Theorem 1.6). On the other hand, starting from a 'degenerate' deterministic initial configuration with $\gamma(x) = 1_{[-1,1]}(x)$, we show that the large deviations behavior is at most quadratic (Theorem 1.8).

Moreover, in Theorem 1.7, starting under deterministic configurations when $\gamma(x) \equiv \rho$, we find the exact asymptotic behavior of the rate functions near their zeroes.

The main idea for the LDPs is to relate, through several 'entropy' and 'energy' estimates, the current and tagged particle deviations to those established in Kipnis-Olla-Varadhan [19], Landim [20], and Landim-Yau [22] with respect to the hydrodynamic limit of the process empirical density (cf. Propositions 1.1, 1.4). The growth order asymptotics are proved in part by estimations of currents and calculus of variations arguments.

At this point, we remark that the behavior of the tagged particle, in contrast to the subdiffusive $d = 1$ nearest-neighbor result, scales differently in symmetric exclusion models in $d \geq 2$, and also in $d = 1$ when the underlying jump probability is not nearest-neighbor, that is when particles are free to pass by other particles. Namely, in Kipnis-Varadhan [18], starting under an equilibrium $\nu_\rho\{\cdot|\eta(0) = 1)$, in diffusive scale, invariance principles for the tagged particle to Brownian motion were proved. Later, in Rezakhanlou [31], starting from local equilibrium measures, in diffusive scale, an invariance principle with respect to a diffusion with a drift given in terms of the profile γ is proved for the 'averaged' tagged particle position, averaging over all the positions of $O(N)$ particles in a sequence of tori with N vertices. In Quastel-Rezakhanlou-Varadhan [30], in $d \geq 3$, a corresponding large deviations principle is proved for the 'averaged' tagged particle position with rate function which is finite on processes with finite relative entropy with respect to diffusions which in some sense add an additional drift to the limit diffusion in [31]. This LDP for the 'averaged' tagged particle would seem also to hold in $d \leq 2$ (non-nearest-neighbor in $d = 1$) given regularity results on the self-diffusion coefficient in Landim-Olla-Varadhan [21] not available when [30] was written.

We also mention, other large deviation works with respect to empirical densities and currents in related interacting systems are Benois-Landim-Kipnis [2], Bertini-DeSole-Gabrielli-Jona Lasinio-Landim [3], [4], Bertini-Landim-Mourragui [5], Farfan-Landim-Mourragui [12], Quastel [29], and Grigorescu [15]; see also Kipnis-Landim [17][Chapter 10] and references therein. Also, we note, with respect to totally asymmetric nearest-neighbor exclusion in $d = 1$, large deviation 'lower tail' bounds for tagged particles are found in Seppäläinen [35].

We now give the hydrodynamic limit and rate function for the process empirical density $\mu^N(s, x; \eta) \in D([0, T]; M_1)$,

$$\mu^N(s, x; \eta) = \sum_{k \in \mathbb{Z}} \eta_{N^2 s}(k) 1_{[k/N,(k+1)/N)}(x)$$

where $x \in \mathbb{R}$, $s \in [0, T]$, and $0 < T < \infty$ is a fixed time.

Proposition 1.1. *Starting from local equilibrium measures, or deterministic configurations, we have, for $t \in [0, T]$, $\epsilon > 0$ and smooth, compactly supported ϕ, that*

$$\lim_{N \uparrow \infty} P\Big\{\Big| \int \phi(x)\mu^N(t, x)dx - \int \phi(x)m(t, x)dx\Big| > \epsilon\Big\} = 0$$

where m satisfies $\partial_t m = (1/2)\partial_{xx} m$ with initial data $m(0,x) = \gamma(x)$.

A reference for the proof of Proposition 1.1, among other places, is Theorem 8.1 Seppäläinen [36].

The rate functions for the process empirical density differ depending on the type of initial distribution. First, following [19], [20], suppose the process starts from a local equilibrium measure $\nu^{(N)}_{\gamma(\cdot)}$. For $\mu \in D([0,T]; M_1)$, define the linear functional on $C_K^{1,2}([0,T]\times\mathbb{R})$:

$$\begin{aligned} l(\mu; G) &= \int G(T,x)\mu_T(x)dx - \int G(0,x)\mu_0(x)dx \\ &\quad - \int_0^T \int \mu_t(x)\Big(\frac{\partial}{\partial t} + \frac{1}{2}\frac{\partial^2}{\partial x^2}\Big)G(t,x)dxdt. \end{aligned}$$

Let

$$\begin{aligned} I_0(\mu) &= \sup_{G\in C_K^{1,2}([0,T]\times\mathbb{R})} \Big\{ l(\mu;G) - \frac{1}{2}\int_0^T \int \mu_t(1-\mu_t)(x)G_x^2(t,x)dxdt \Big\}, \\ h(\mu_0;\gamma) &= \sup_{\phi_0,\phi_1\in C_K(\mathbb{R})} \Big\{ \int \mu_0(x)\phi_0(x)dx + \int (1-\mu_0(x))\phi_1(x)dx \\ &\quad - \int \log[\gamma(x)e^{\phi_0(x)} + (1-\gamma(x))e^{\phi_1(x)}]dx \Big\}, \end{aligned}$$

and form the rate function

$$I_\gamma^{\mathrm{LE}}(\mu) = I_0(\mu) + h(\mu_0;\gamma).$$

Here, $C_K^{\alpha,\beta}$ is the space of compactly supported functions, α and β-times continuously differentiable in t and x respectively. In addition, we will use the notation $\mu_t(x) = \mu(t,x)$.

Next, starting from deterministic configurations $\xi^{\gamma,N}$, the rate function in [22] (written for zero-range systems, but the methods straightforwardly apply to our exclusion context) is given by

$$I_\gamma^{\mathrm{DC}}(\mu) = \left\{ \begin{array}{cl} I_0(\mu) & \text{when } \mu_0 = \gamma \\ \infty & \text{otherwise.} \end{array} \right\}$$

To simplify notation, we call both I_γ^{LE} and I_γ^{DC} as I_γ, omitting the super scripts 'LE' and 'DC,' when statements apply to both and the context clear. For $0 \le \alpha, \beta \le 1$, let $h_d(\alpha;\beta) = \alpha\log[\alpha/\beta] + (1-\alpha)\log[(1-\alpha)/(1-\beta)]$ with usual conventions $0\log 0 = 0/0 = 0$ and $\log 0 = -\infty$.

From the definition, I_γ is a convex function. Also, a main point in [19] was to note that when $I_\gamma(\mu) < \infty$ is finite that first

$$h(\mu_0;\gamma) = \int h_d(\mu_0(x);\gamma(x))dx < \infty.$$

[Of course, starting from deterministic configurations, $\mu_0 = \gamma$.] Also second, μ corresponds to a function $H_x \in L^2([0,T]\times\mathbb{R}, \mu(1-\mu)dxdt)$ and satisfies a 'weakly asymmetric hydrodynamic equation,'

$$\partial_t \mu = \frac{1}{2}\partial_{xx}\mu - \partial_x\big[H_x\mu(1-\mu)\big] \tag{1.2}$$

in the weak sense. That is, for $G \in C_K^{1,2}([0,T] \times \mathbb{R})$, we have

$$l(\mu; G) = \int_0^T \int G_x H_x \mu(1-\mu)(t,x)dxdt \tag{1.3}$$

and

$$I_0(\mu) = \frac{1}{2}\int_0^T \int H_x^2 \mu(1-\mu)dxdt. \tag{1.4}$$

Reciprocally, if for a density $\mu \in D([0,T]; M_1)$ there exists $H_x \in L^2([0,T]\times\mathbb{R}, \mu(1-\mu)dxdt)$ such that μ satisfies (1.2) weakly, then $I_0(\mu)$ is given by (1.4).

Recall, a function $\mathcal{I} : \mathcal{X} \to [0,\infty]$ on a complete, separable metric space $\mathcal{X}$ is a rate function if it has closed level sets $\{x : \mathcal{I}(x) \leq a\}$. It is a *good* rate function if the level sets are also compact. Also, a sequence $\{X_n\}$ of random variables with values in $\mathcal{X}$ satisfies a large deviation principle (LDP) with speed n and rate function $\mathcal{I}$ if for every Borel set $U \in \mathcal{B}_{\mathcal{X}}$,

$$\begin{aligned} -\inf_{x\in\bar{U}} \mathcal{I}(x) &\geq \limsup_{n\to\infty} \frac{1}{n}\log \Pr(X_n \in U) \\ &\geq \liminf_{n\to\infty} \frac{1}{n}\log \Pr(X_n \in U) \geq -\inf_{x\in U^\circ} \mathcal{I}(x) \end{aligned}$$

where U° is the interior of U and $\bar{U}$ is the closure of U.

Let $\mathcal{A} = \mathcal{A}(\gamma)$ be the space of all densities μ such that $I_\gamma(\mu) < \infty$ which can be approximated in $D([0,T]; M_1)$ by a sequence of densities $\{\mu^n\}$ satisfying (1.2) corresponding to $\{H_x^n\} \subset C_K^{1,2}([0,T] \times \mathbb{R})$ such that $I_\gamma(\mu^n) \to I_\gamma(\mu)$.

For general local equilibrium measures (LEM) and deterministic initial configurations (DIC), only a weak large deviation principle is available. The next proposition follows straightforwardly from the methods of [19] (see also [17][Chapter 10]), and replacement estimates in [22], namely Theorem 6.1 and Claims 1,2 [22][Section 6].

Proposition 1.2. *With respect to initial local equilibrium measures (LEM) or deterministic configurations (DIC), corresponding to profile γ, I_γ is a good convex rate function, and for $U \subset D([0,T]; M_1)$,*

$$\begin{aligned} -\inf_{\mu\in\bar{U}} I_\gamma(\mu) &\geq \limsup_{N\uparrow\infty} \frac{1}{N}\log P\big[\mu^N \in U\big] \\ &\geq \liminf_{N\uparrow\infty} \frac{1}{N}\log P\big[\mu^N \in U\big] \geq -\inf_{\mu\in U^\circ\cap\mathcal{A}} I_\gamma(\mu). \end{aligned}$$

The last proposition raises the question when $\mathcal{A}(\gamma)$ is large enough so that the lowerbound matches the upperbound. However, with respect to the profiles considered, the following containment is true, so that as a corollary the full LDP holds.

Proposition 1.3. *With respect to profiles γ associated to local equilibrium measures (LEM) and deterministic configurations (DIC),*

$$\mathcal{A}(\gamma) \supset \{\mu : I_\gamma(\mu) < \infty\}.$$

Corollary 1.4. *With respect to initial local equilibrium mesures (LEM) and deterministic configurations (DIC), the LDP with speed N holds for $\{\mu^N\}$ with good convex rate function I_γ.*

We note Proposition 1.3, for continuous profiles $\gamma \in M_1(\rho,\rho)$ with $0 < \rho < 1$ and $0 < \gamma(\cdot) < 1$ corresponding to local equilibrium measures, was stated in [20], and the associated LDP in Corollary 1.4 with respect to these initial measures is

Theorems 3.2, 3.3 [20]. In section 5, we prove Proposition 1.3, generalizing the initial states allowed.

It will be convenient to rewrite (1.2) in terms of a macroscopic 'current' or 'flux' J: That is, when $I_\gamma(\mu) < \infty$, define J so that weakly,

$$\partial_x J + \partial_t \mu \;=\; 0; \quad J \;=\; -\frac{1}{2}\partial_x \mu + H_x \mu(1-\mu).$$

It turns out such currents have nice properties and relations (cf. Propositions 2.4 and 2.6). Namely, the time integrated current $x \mapsto \int_0^T J(x,t)dt$ is well defined function on $\mathbb{R}$. Also, the limit

$$\int_0^\infty \big[\mu_T(x) - \mu_0(x)\big]dx \;:=\; \lim_{L\to\infty}\int_0^L \big[\mu_T(x) - \mu_0(x)\big]dx \quad \text{converges} \tag{1.5}$$

and

$$\int_0^T J(0,t)dt \;=\; \int_0^\infty \mu_T(x) - \mu_0(x)dx. \tag{1.6}$$

In addition, for $\alpha, \beta \in \mathbb{R}$, $\int_0^T \big[J(\beta,t) - J(\alpha,t)\big]dt = \int_\beta^\alpha \big[\mu_T(x) - \mu_0(x)\big]dx$.

We now write the current and tagged particle rate function in terms of I_γ. Define the functions $\mathbb{J} = \mathbb{J}_\gamma$ and $\mathbb{I} = \mathbb{I}_\gamma$, for $a \in \mathbb{R}$, by

$$\begin{aligned}\mathbb{J}(a) \;&=\; \inf\Big\{I_\gamma(\mu) : \int_0^T J(0,t)dt = a\Big\}\\ &=\; \inf\Big\{I_\gamma(\mu) : \int_0^\infty \mu_T(x) - \mu_0(x)dx = a\Big\}\end{aligned}$$

and

$$\begin{aligned}\mathbb{I}(a) \;&=\; \inf\Big\{I_\gamma(\mu) : \int_0^T J(0,t)dt = \int_0^a \mu_T(x)dx\Big\}\\ &=\; \inf\Big\{I_\gamma(\mu) : \int_0^\infty \mu_T(x) - \mu_0(x)dx = \int_0^a \mu_T(x)dx\Big\}.\end{aligned}$$

When starting from (LEM) or (DIC) initial conditions, we sometimes distinguish the corresponding rate functions by adding a superscript.

It follows from the definitions that

$$\mathbb{I}_\gamma^{LE}(a) \;\le\; \mathbb{I}_\gamma^{DC}(a) \;\text{ and }\; \mathbb{J}_\gamma^{LE}(a) \;\le\; \mathbb{J}_\gamma^{DC}(a). \tag{1.7}$$

We also observe that the restriction in the infimum in the definition of $\mathbb{I}$ may take different form. For instance, when $\int_0^T J(0,t)dt = \int_0^a \mu_T(x)dx$, by the relation $\int_0^T J(0,t) - J(a,t)dt = \int_0^a \mu_T(x) - \mu_0(x)dx$, one obtains the following restriction which could be used instead: $\int_0^T J(a,t)dt = \int_0^a \mu_0(x)dx$.

In addition, by translation-invariance, considering $\mu'(t,x) = \mu(t,x+a)$, $J'(x,t) = J(x+a,t)$ and $\gamma'(x) = \gamma(x+a)$, we see, starting from a (DIC) initial state, that

$$\begin{aligned}\mathbb{I}_\gamma^{DC}(a) \;&=\; \inf\Big\{I_\gamma^{DC}(\mu) : \int_0^T J(a,t)dt \;=\; \int_0^a \gamma(x)dx\Big\}\\ &=\; \mathbb{J}_{\gamma'}^{DC}\left(\int_0^a \gamma(x)dx\right).\end{aligned} \tag{1.8}$$

Although one can readily see $\mathbb{J}$ is convex, given I_γ is convex and the constraint in the defiition of $\mathbb{J}$ is linear in μ and a, it is not so easily seen whether $\mathbb{I}$ is convex from this sort of argument. However, as seen later in Theorems 1.6 and 1.7, near their zeroes, both $\mathbb{J}$ and $\mathbb{I}$ behave quadratically.

Also, it is perhaps curious to note that $\mathbb{J}$ and $\mathbb{I}$ can be written completely in terms of densities μ, a consequence of the enforced ordering of particles in the nearest-neighbor $d = 1$ setting. In contrast, the large deviation rate function for the 'averaged' tagged particle position in [30] involves an auxiliary current in its description.

We now give some properties of $\mathbb{J}$ and $\mathbb{I}$ and state the large deviation principles.

Theorem 1.5. *With respect to (DIC) or (LEM) initial measures,*

(i) $\mathbb{J}$ and $\mathbb{I}$ are finite on $\mathbb{R}$, $\lim_{|a|\uparrow\infty} \mathbb{J}(a) = \lim_{|a|\uparrow\infty} \mathbb{I}(a) = \infty$, and $\mathbb{J}$ and $\mathbb{I}$ are a good rate functions. Further, $\mathbb{J}$ and $\mathbb{I}$ have unique zeroes at the LLN constants v_T and u_T respectively.

(ii) The scaled quantities $\{J_{-1,0}(N^2T)/N\}$ and $\{X_{N^2T}/N\}$ satisfy LDPs in scale N with respective rate functions $\mathbb{J}$ and $\mathbb{I}$.

A natural question at this point is to calculate the rate functions $\mathbb{J}$ and $\mathbb{I}$. Although this appears difficult, some bounds (with non-optimal constants) are possible under various conditions.

Theorem 1.6. *Starting under (DIC) or (LEM) initial conditions, there is a constant $c_1 = c_1(\gamma)$ such that*

$$\limsup_{a\to v_T} \frac{\sqrt{T}}{(a-v_T)^2}\mathbb{J}(a),\ \limsup_{|a|\uparrow\infty} \frac{T}{|a|^3}\mathbb{J}(a) \ \leq\ c_1$$

$$\limsup_{a\to u_T} \frac{\sqrt{T}}{(a-u_T)^2}\mathbb{I}(a),\ \limsup_{|a|\uparrow\infty} \frac{T}{|a|^3}\mathbb{I}(a) \ \leq\ c_1.$$

Also, starting under (DIC) initial conditions, there is a constant $c_2 = c_2(\gamma) > 0$ such that

$$\liminf_{a\to v_T} \frac{\sqrt{T}}{(a-v_T)^2}\mathbb{J}(a),\ \liminf_{|a|\uparrow\infty} \frac{T}{|a|^3}\mathbb{J}(a) \ \geq\ c_2$$

$$\liminf_{a\to u_T} \frac{\sqrt{T}}{(a-u_T)^2}\mathbb{I}(a),\ \liminf_{|a|\uparrow\infty} \frac{T}{|a|^3}\mathbb{I}(a) \ \geq\ c_2.$$

We remark the quadratic asymptotics for $\mathbb{J}(a)$ and $\mathbb{I}(a)$ near their zeroes recalls Gaussian expansions, and the CLTs in [16], [24] and [39]. On the other hand, the cubic bounds for large $|a|$ in Theorem 1.6 seem intriguing, perhaps connected with totally asymmetric nearest-neighbor exclusion (TASEP) effects. That is, for the current or tagged particle to deviate to a far level aN, order $O(|a|N)$ particles must be driven far away from their initial positions, so that perhaps the process behaves like a driven system like TASEP.

We remark on these last points that in Derrida-Gerschenfeld [10], [11], starting from a local equilibrium measure with step profile $\gamma^{\rho_1,\rho_2}(x) = \rho_l 1_{(-\infty,0]} + \rho_r 1_{(0,\infty)}$, the large deviation 'pressure' of the current $J_{0,1}(t)$ across the bond $(0,1)$, $\lim_{t\uparrow\infty} t^{-1/2}\log E[\exp\{\lambda J_{0,1}(t)\}] = F(\rho_l,\rho_r,\lambda)$, is found. Also, formal asymptotics with F give $P(J_{0,1}(t) = a) \sim \exp[\sqrt{t}\{-\frac{\pi^2}{12}a^3 + \cdots\}]$, for large t and large $a > 0$ (cf. p. 980 [11]).

In this context, the large deviation principle in Theorem 1.5 and bounds in Theorem 1.6 prove the form of this expression with respect to the dominant third order term when starting from (DIC) initial conditions: Namely, for large a,

$$\begin{aligned} -c_0|a|^3 &\geq -\inf_{|x|\geq a}\mathbb{J}(x) \geq \limsup_{t\to\infty}\frac{1}{\sqrt{t}}P(|J_{0,1}(t)|\geq a) \\ &\geq \liminf_{t\to\infty}\frac{1}{\sqrt{t}}\log P(|J_{0,1}(t)|\geq a) \geq -\inf_{|x|>a}\mathbb{J}(x) \geq -c_1|a|^3. \end{aligned}$$

This addresses in part a question in [10] whether the large $|a|$ asymptotics would extend to non-step profiles. See also Hurtado-Garrido [14].

Also, with respect to the current and tagged particle, fluctuations in the 'KPZ' class are discussed in Praehofer-Spohn [28], Ferrari-Spohn [13] and Sasamoto [34] with respect to TASEP starting initial conditions with step or constant profiles. In particular, the scaling limits of the current and tagged particle are of 'Tracy-Widom' or 'Airy' process types whose marginal distribution have upper tail on order $e^{-c_0|x|^3}$ as $x\uparrow\infty$, and lower tail on order $e^{-c_1|x|^{3/2}}$ as $x\downarrow\infty$, for some constants c_0, c_1. In our context, starting from (DIC) initial conditions, we have from Theorem 1.6 that $\mathbb{J}(a), \mathbb{I}(a)$ are on cubic order $|a|^3$ for large $|a|$. Formally, one is tempted to link this cubic order in terms of the TASEP scaling limit process exponents. It would be interesting to investigate such analogies.

We now refine the behavior of $\mathbb{J}(a)$ and $\mathbb{I}(a)$ near their zeroes $v_T = u_T = 0$ when the deterministic initial condition has constant profile $\gamma\equiv\rho$. Arratia's CLT variances σ_J^2 and σ_X^2, mentioned earlier, can be computed by adding static and dynamic contributions, due to initial configuration and later motion fluctuations respectively. However, starting from deterministic initial configurations, only the dynamical contributions would be present, and we show later in Proposition 4.5 that these parts of the variances are $\sigma^2_{J,dyn} = \sqrt{\pi}\rho(1-\rho)$ and $\sigma^2_{X,dyn} = \sqrt{\pi}(1-\rho)/\rho$.

Theorem 1.7. *For $\rho\in(0,1)$, starting from (DIC) initial configurations with profile $\gamma\equiv\rho$, we have*

$$\begin{aligned} \lim_{|a|\downarrow 0}\frac{1}{a^2}\mathbb{J}(a) &= \frac{1}{2\sigma^2_{J,dyn}\sqrt{T}} = \frac{\sqrt{\pi}}{2\sqrt{T}}\rho(1-\rho) \quad \text{and} \\ \lim_{|a|\downarrow 0}\frac{1}{a^2}\mathbb{I}(a) &= \frac{1}{2\sigma^2_{X,dyn}\sqrt{T}} = \frac{\sqrt{\pi}}{2\sqrt{T}}\frac{\rho}{1-\rho}. \end{aligned}$$

At this point, one might ask about the large deviation behavior starting from initial conditions with 'degenerate' profiles. In this case, diffusive scaling may not always capture for the tagged particle nontrivial LLN's as in (1.1) or large deviations as in Theorem 1.5. For instance, starting under $\xi^{\gamma,N}$ where $\gamma(x) = 1_{(-\infty,0]}(x)$ is the step profile, in Arratia [1] it is shown that $t^{-1/2}x(t) - \sqrt{\log(t)}\to 0$ a.s. which shows that the tagged particle diverges at rate $\sqrt{t\log(t)}$. With respect to large deviations, it is clear the tagged particle, initially at the origin, cannot travel to negative locations. Also, for $a\geq 0$, the condition in $\mathbb{I}(a)$ reduces to $\int_a^\infty \mu_T(x)dx = 0$ which, given that $\mu(t,x)$ satisfies (1.2), is impossible since the density formally becomes positive on $\mathbb{R}$ as soon as $t>0$. Hence, starting from this step profile configuration, formally $\mathbb{I}=\infty$. However, for the current, starting from this initial condition, in diffusive scaling, $v_T<\infty$, and a corresponding CLT is proved in [24].

On the other hand, when the degenerate initial profile has a density of particles around the tagged particle, diffusive scaling would still seem appropriate to establish an LDP for the tagged particle and current. Here, as a contrast to the results in

Theorem 1.6 and to argue this last sentiment, we show quadratic upper bounds for the current and tagged particle large deviations starting from the degenerate configuration $\xi^{\gamma_1,N}$ where $\xi^{\gamma_1,N}(x) = 1$ for $|x| \leq N$ and $\xi^{\gamma_1,N}(x) = 0$ otherwise. Here, $\gamma_1(x) = 1_{[-1,1]}(x)$. Note the associated LLN speeds $v_T = u_T = 0$.

Theorem 1.8. *Starting under $\xi^{\gamma_1,N}$, there exists $c_1(T) > 0$ such that, for $a \geq 0$,*

$$\limsup_{N\uparrow\infty} \frac{1}{N} \log P\left(|J_{-1,0}(N^2T)|/N \geq a\right) \leq \begin{cases} -c_1a^2 & \text{for } 0 \leq a \leq 1 \\ -\infty & \text{for } a > 1, \end{cases}$$

$$\limsup_{N\uparrow\infty} \frac{1}{N} \log P\left(|X(N^2T)|/N \geq a\right) \leq -c_1a^2.$$

The interpretation, for instance with respect to the tagged particle, is that in configurations $\xi^{\gamma_1,N}$ although it is trapped in the middle of a large segment of particles, to displace large distances, as there are only $O(N)$ number of particles in the system, the cost is not as great as under $\xi^{\rho,N}$ where there are an infinite number of particles. At the same time, there is a positive density of particles to the left and right of the origin, unlike for the profile $\gamma(x) = 1_{(-\infty,0]}(x)$, which slows down the tagged particle so that deviations to $a \in \mathbb{R}$ have finite cost in diffusive scale. With respect to the current, a similar explanation applies; we note however current levels larger than N cannot happen, and so they are given infinite cost.

Finally, we remark on some natural questions. 1. As indicated by Theorem 1.8, different large deviation behaviors might arise when starting from degenerate initial conditions. It would be of interest to investigate these phenomena and providing estimates for the corresponding rate functions. When starting from a degenerate initial profile, with a density of mass around the initial tagged particle position, although the basic argument of Theorem 1.5 (ii) in Section 3 holds, main obstacles are to extend approximation Propositions 1.3 and 2.1, energy estimate Proposition 2.4, first bounds and development of the rate functions in subsection 2.4, and exponential tightness Lemma 3.2.

2. Also, a joint large deviations principle for the current and tagged particle, with rate

$$\mathbb{K}(a,b) := \inf\left\{I_\gamma(\mu) : \int_0^\infty \big(\mu_T(x) - \mu_0(x)\big)dx = \int_0^b \mu_T(x)dx = a\right\},$$

should hold by the methods of the article. In this case, asymptotics of the rate function $\mathbb{K}(a,b)$ for (a,b) near (v_T, u_T) might be studied.

The plan of the paper is now to develop preliminary estimates in section 2. In section 3, we prove Theorem 1.5. Then, in section 4, we prove Theorems 1.6, 1.7. and 1.8. These last two sections can be read independently of each other. Finally, in section 5, as remarked earlier, we prove Proposition 1.3, and other approximations.

2. Preliminary estimates

We develop, in several subsections, 'energy' and current estimates with respect to finite rate densities, and also prove that $\mathbb{J}$ and $\mathbb{I}$ are a finite-valued rate functions.

2.1. Approximation and limit estimates.

We state an approximation result derived in the course of the proof of Proposition 1.3, and also certain useful limits at infinity. Proofs of these results are given in Section 5.

Proposition 2.1. *Let μ be a density such that $I_0(\mu) < \infty$. Then, for all $\epsilon > 0$ there is $\mu^+ \in D([0,T]; M_1)$ such that*

(i) $\exists\, 0 < \delta < 1$ *such that* $\delta \le \mu^+(t,x) \le 1-\delta$ *for* $(t,x) \in [0,T] \times \mathbb{R}$,
(ii) $\mu^+ \in C^\infty([0,T] \times \mathbb{R})$,
(iii) $H_x^+ \in C_K^\infty([0,T] \times \mathbb{R})$, *and*
(iv) $\|\partial_x^{(k)}\partial_t^{(l)}\mu^+\|_{L^\infty([0,T]\times\mathbb{R})} < \infty$ *for* $k, l \ge 1$.
(v) *If* $\mu_0 \equiv \gamma \in M_1(\rho_*, \rho^*)$, *then* $\mu_0^+ = \sigma_\alpha * \gamma$ *for an* $\alpha > 0$. *In particular, if* $\mu_0(x) \equiv \rho$, *then* $\mu_0^+(x) \equiv \rho$.
(vi) *Also, Skorohod distance* $d(\mu^+, \mu) < \epsilon$ *in* $D([0,T]; M_1)$,
(vii) $|I_0(\mu^+) - I_0(\mu)| < \epsilon$.
(viii) *Also, suppose* $\hat\gamma \in M_1(\rho_*, \rho^*)$ *is piecewise continuous, and* $0 < \hat\gamma(x) < 1$ *for* $x \in \mathbb{R}$. *Then, if* $h(\mu_0; \hat\gamma) < \infty$, *we have* $|h(\mu_0^+; \hat\gamma) - h(\mu_0; \hat\gamma)| \le \epsilon$.

We remark, of course, Proposition 2.1 implies if $I_0(\mu) < \infty$ there is a sequence of densities μ^n satisfying properties (i) - (viii) which converges to μ in $D([0,T]; M_1)$.

Lemma 2.2. *Let $\hat\gamma \in M_1(\rho_*, \rho^*)$, and μ be a smooth density such that $h(\mu_0; \hat\gamma) < \infty$, $I_0(\mu) < \infty$, and which also satisfies (i)-(iv) in Proposition 2.1 Then, we have*

$$\lim_{|y|\uparrow\infty} \sup_{t\in[0,T]} |\mu(t,y) - \hat\gamma(y)| \;=\; 0.$$

The next lemma will be used in the proof of Theorem 1.7.

Lemma 2.3. *Let $\{\mu\}$ be a smooth density such that $\mu_0(x) \equiv \rho$, $I_0(\mu) < \infty$, and which satisfies (i)-(iv) in Proposition 2.1. Then,*

$$\sup_{0\le t\le T} \int (\mu_t - \rho)^2 dx \;\le\; 8 I_0(\mu).$$

2.2. **'Energy' and current estimates.** We give a formula for the rate $I_0(\mu)$, bounds on the 'energy' $\|\partial_x \mu\|_{L^2}$, and relations with the current.

Proposition 2.4. *Let μ be a smooth density, with finite rate $I_0(\mu)$, satisfying (i)-(iv) in Proposition 2.1. Suppose also there is a smooth $\hat\gamma \in M_1(\rho_*, \rho^*)$, strictly bounded between 0 and 1, such that $h(\mu_0; \hat\gamma) < \infty$. Then,*

$$\begin{aligned} I_0(\mu) \;&=\; \frac{1}{8}\int_0^T \int \frac{(\partial_x\mu)^2}{\mu(1-\mu)} dxdt + \frac{1}{2}\left[h(\mu_T;\hat\gamma) - h(\mu_0;\hat\gamma)\right] \\ &\quad + \frac{1}{2}\int \frac{\partial_x\hat\gamma}{\hat\gamma(1-\hat\gamma)} \int_0^T J dt dx + \frac{1}{2}\int_0^T \int \frac{J^2}{\mu(1-\mu)} dxdt, \end{aligned}$$

$$\frac{1}{4}\|\partial_x\mu\|_{L^2}^2 \;\le\; h(\mu_0;\hat\gamma) + I_0(\mu) + T\|\partial_x\hat\gamma/(\hat\gamma(1-\hat\gamma))\|_{L^2}^2 \quad \text{and} \tag{2.1}$$

$$\int_0^T J(a,t)dt - \int_0^T J(b,t)dt \;=\; \int_a^b \mu_T(x) - \mu_T(0)dx, \quad \text{for } a,b \in \mathbb{R}. \tag{2.2}$$

Proof. First, as $J = -(1/2)\partial_x\mu + H_x\mu(1-\mu)$, we have

$$\begin{aligned} I_0(\mu) \;&=\; \frac{1}{2}\int_0^T \int H_x^2 \mu(1-\mu) dxdt \\ &=\; \frac{1}{8}\int_0^T \int \frac{(\partial_x\mu)^2}{\mu(1-\mu)} dxdt + \frac{1}{2}\int_0^T \int \frac{J\partial_x\mu}{\mu(1-\mu)} dxdt + \frac{1}{2}\int_0^T \int \frac{J^2}{\mu(1-\mu)} dxdt. \end{aligned}$$

We now find a suitable expression for the middle term. Let G_L be a smooth, nonnegative, compactly supported function in $[-L, L]$, bounded by 1, which equals 1 on $[-L+1, L-1]$, and $\sup_L \int_{A_L} (G'_L)^2/G_L dx < \infty$ where $A_L = [L-1, L] \cup [-L, -L+1]$. Then,

$$\begin{aligned} &\partial_t \int G_L(x) h_d(\mu_t(x); \hat{\gamma}) dx \qquad (2.3) \\ &= -\frac{1}{2} \int G_L(x) \frac{(\partial_x \mu_t)^2}{\mu_t(1-\mu_t)} dx + \int G_L(x) H_x \partial_x \mu_t dx \\ &+\frac{1}{2} \int G_L(x) \frac{\partial_x \hat{\gamma} \partial_x \mu}{\hat{\gamma}(1-\hat{\gamma})} dx - \int G_L(x) H_x \mu(1-\mu) \frac{\partial_x \hat{\gamma}}{\hat{\gamma}(1-\hat{\gamma})} dx \\ &+ \int_{A_L} G'_L(x) \Big[-(1/2)\partial_x \mu_t + H_x(\mu_t(1-\mu_t)) \Big] \log \frac{\mu_t}{1-\mu_t} \frac{1-\hat{\gamma}}{\hat{\gamma}} dx. \end{aligned}$$

Hence, by Schwarz inequality and $0 \le \mu \le 1$, we can bound, with respect to a universal constant C,

$$\begin{aligned} &\int G_L(x) h_d(\mu_T(x); \hat{\gamma}(x)) dx + \frac{1}{4} \int_0^T \int G_L(x) \frac{(\partial_x \mu_s)^2}{\mu_s(1-\mu_s)} dx ds \\ &\le \int G_L(x) h_d(\mu_0(x); \hat{\gamma}(x)) dx + C \int_0^T \int (H_x)^2 \mu_s(1-\mu_s) dx ds \\ &+CT \int G_L(x) \frac{(\partial_x \hat{\gamma})^2}{\hat{\gamma}^2(1-\hat{\gamma})^2} dx + C \int_{A_L} [(G'_L)^2/G_L][\log \frac{\mu_t}{1-\mu_t} \frac{1-\hat{\gamma}}{\hat{\gamma}}]^2 dx. \end{aligned}$$

We can take $L \uparrow \infty$, so that the last term vanishes by Lemma 2.2. Then, by monotone convergence, with respect to a universal constant C,

$$\begin{aligned} &h(\mu_T; \hat{\gamma}) + \frac{1}{4} \int_0^T \int \frac{(\partial_x \mu_s)^2}{\mu_s(1-\mu_s)} dx ds \\ &\le h(\mu_0; \hat{\gamma}) + C \int_0^T \int (H_x)^2 \mu_s(1-\mu_s) dx ds + CT \|\partial_x \hat{\gamma}\|_{L^2}, \end{aligned}$$

and as $0 < \mu, \hat{\gamma} < 1$, we have $\|\partial_x \mu\|_{L^2}, \|J\|_{L^2} < \infty$.

Hence, integrating (2.3) and taking limit on L, the middle term equals

$$\int_0^T \int \frac{J \partial_x \mu}{\mu(1-\mu)} dx dt = h(\mu_T; \hat{\gamma}) - h(\mu_0; \hat{\gamma}) + \int_0^T \int \frac{J \partial_x \hat{\gamma}}{\hat{\gamma}(1-\hat{\gamma})} dx dt.$$

The desired bound on $\|\partial_x \mu\|_{L^2}$ now follows. Since $\mu(1-\mu) \le 1/4$ and $\|H_x \mu(1-\mu)\|_{L^2}^2 \le I_0(\mu)$ (cf. (1.4)), by Schwarz inequality, we may write

$$\begin{aligned} \|\partial_x \mu\|_{L^2}^2 &\le h(\mu_0; \hat{\gamma}) + \frac{1}{2}\|J\|_{L^2}^2 + \frac{T}{2}\|\partial_x \hat{\gamma}/(\hat{\gamma}(1-\hat{\gamma}))\|_{L^2}^2 + 2I_0(\mu) \\ &\le h(\mu_0; \hat{\gamma}) + \frac{2}{8}\|\partial_x \mu\|_{L^2}^2 + \frac{2}{4} I_0(\mu) + 2I_0(\mu) + T\|\partial_x \hat{\gamma}/(\hat{\gamma}(1-\hat{\gamma}))\|_{L^2}^2. \end{aligned}$$

Finally, (2.2) expresses that the difference of the currents across a and b up to time T is equal to the difference in the masses in the interval $[a, b]$ from times T to 0. This is obtained by integrating $\partial_x J = -\partial_t \mu$. □

Corollary 2.5. *Let μ be a density with finite rate $I_\gamma(\mu) < \infty$. Let also $\{\mu^n\}$ be a sequence converging to μ with properties (i)-(viii) in Proposition 2.1. Then, $\partial_x \mu^n$*

and J^n are uniformly bounded in $L^2([0,T]\times\mathbb{R})$ and $\partial_x\mu^n \to \partial_x\mu$, $J^n \to J$ weakly in $L^2([0,T]\times\mathbb{R})$; consequently, $\partial_x\mu, J \in L^2([0,T]\times\mathbb{R})$.

Proof. Let $\hat\gamma$ be a smooth function in $M(\rho_*,\rho^*)$ such that $0<\gamma_*<\hat\gamma<\gamma^*<1$ for some constants γ_*,γ^*, and $h(\gamma;\hat\gamma)<\infty$. Then, by property (viii) Proposition 2.1, as $h(\mu_0;\hat\gamma)<\infty$, we have $h(\mu_0^n;\hat\gamma)\to h(\mu_0;\hat\gamma)$ and in particular $\{h(\mu_0^n;\hat\gamma)\}$ is uniformly bounded.

Also, as $I_0(\mu)<\infty$, by property (vii) Proposition 2.1, we have $I_0(\mu^n)\to I_0(\mu)$ and $\{I_0(\mu^n)\}$ is uniformly bounded. In particular, we conclude $\{\|H_x^n\mu^n(1-\mu^n)\|_{L^2}\}$ is uniformly bounded.

Hence, as $\partial_x\hat\gamma/(\hat\gamma(1-\hat\gamma)) \in L^2$, and by (2.1) in Proposition 2.4, we have $\{\|\partial_x\mu^n\|_{L^2}\}$ is uniformly bounded. Also, since $J^n = (1/2)\partial_x\mu^n + H_x^n\mu^n(1-\mu^n)$, we also conclude $\{\|J^n\|_{L^2}\}$ is uniformly bounded.

We can then extract subsequences $\partial_x\mu^{n_k}$ and J^{n_k} converging weakly to ζ and ϕ respectively. Given $\mu^{n_k}\to\mu$ in $D([0,T]\times M_1)$, for smooth, compactly supported G, we have $\int G\partial_x\mu^{n_k}dxds = \int -G_x\mu^{n_k}dxds$ converges to both $\int G\zeta dxds$ and $\int -G_x\mu dxds$. Then, $\partial_x\mu$ exists weakly in L^2 and $\partial_x\mu=\zeta$. Hence, the whole sequence $\partial_x\mu^n\to\partial_x\mu$ weakly in L^2.

Similarly, noting Skorohod convergence $\mu^n\to\mu$ implies at the endpoints that μ_0^n,μ_T^n converge to μ_0,μ_T respectively, and $\partial_t\mu^n+\partial_xJ^n=0$, we have $\phi_x=-\partial_t\mu$ weakly in L^2. Then, $\phi_x = (-1/2)\partial_{xx}\mu + \partial_x[H_x\mu(1-\mu)]$ weakly in L^2, and so $\phi = (-1/2)\partial_x\mu + H_x\mu(1-\mu)+C(t)$ with respect to a function $C(t)$ not dependent on x. But, given $\phi,\partial_x\mu, H_x\mu(1-\mu)\in L^2([0,T]\times\mathbb{R})$, we conclude $C(t)\equiv 0$. In particular, $\phi = J = -(1/2)\partial_x\mu + H_x\mu(1-\mu)\in L^2$, and the sequence $J^n\to J$ weakly in L^2. □

2.3. **Current-mass relation.** We give some properties of the integrated current $\int_0^T J(x,t)dt$ and prove the current-mass relation indicated in the Introduction.

Proposition 2.6. *Let μ be a density such that $I_\gamma(\mu)<\infty$. Let $\{\mu^n\}$ be a sequence converging to μ with properties (i)-(viii) in Proposition 2.1. Then, $x\mapsto\int_0^T J(x,t)dt$ is a Lipschitz function, $\lim_{|x|\uparrow\infty}\int_0^T J(x,t)dt = 0$, and pointwise for $x\in\mathbb{R}$,*

$$\lim_{n\to\infty}\int_0^T J^n(x,t)dt = \int_0^T J(x,t)dt.$$

In addition, the convergence (1.5), and the 'current-mass relation (1.6) hold.

Proof. First, from (2.2) in Proposition 2.4, we have

$$\int_0^T J^n(a,t)dt - \int_0^T J^n(b,t)dt = \int_a^b \mu_T^n(x) - \mu_T^n(0)dx.$$

Hence $|\int_0^T J^n(a,t)dt - \int_0^T J^n(b,t)dt| \le |b-a|$ as $0\le\mu^n\le 1$. In particular, $\int_0^T J^n(a,t)dt$ is Lipschitz in a. Moreover, a subsequence, $\int_0^T J^{n_k}(\cdot,t)dt \to \psi(\cdot)$ converges uniformly on compact subsets to a Lipschitz function ψ. Given $J^n\to J$ weakly in $L^2([0,T]\times\mathbb{R})$ by Corollary 2.5, we conclude by a limit argument with respect to $G\in L^2(\mathbb{R})$ that $\int G(a)\int_0^T J(a,t)dtda = \int G(a)\psi(a)da$ and so $\psi(a) = \int_0^T J(a,t)dt$. In particular, the whole sequence $\int_0^T J^n(\cdot,t)dt \to \int_0^T J(\cdot,t)dt$, and the limit $\int_0^T J(\cdot,t)dt$ is Lipschitz.

Therefore, since

$$\int \Big[\int_0^T J(x,t)dt\Big]^2 dx \ \le \ T\int\int_0^T J^2(x,t)dtdx \ < \ \infty,$$

we obtain the pointwise limit $\int_0^T J(x,t)dt \to 0$ as $|x| \uparrow \infty$.

Finally, given Skorohod convergence $\mu^n \to \mu$ we have μ_0^n and μ_T^n converge respectively to μ_0 and μ_T. Then, by taking limits, we can write

$$\int_0^T J(0,t)dt - \int_0^T J(L,t)dt \ = \ \int_0^L \mu_T(x) - \mu_0(x)dx.$$

Now, since $\lim_{L\to\infty}\int_0^T J(L,t)dt = 0$, we obtain (1.5) and (1.6). □

2.4. **First estimates on $\mathbb{J}$ and $\mathbb{I}$.** We develop some first bounds on $\mathbb{J}$ and $\mathbb{I}$, and at the end show they are rate functions.

Recall $\sigma_t(x) = (2\pi t)^{-1/2}\exp\{-x^2/2t\}$, and consider a C^∞ smooth function, supported on $[-1,1]$, say

$$\psi_0(x) \ = \ \exp\big\{ -1/(1-x^2)\big\}.$$

Define the smooth, anti-symmetric function,

$$\psi(x) \ = \ \begin{cases} -\psi_0(2(x+1/2)) & \text{for } x\le 0 \\ \ \ \psi_0(2(x-1/2)) & \text{for } x \ge 0 \end{cases}$$

and also the anti-derivative $\Psi(x) = \int_{-1}^x \psi(y)dy$, both supported on $[-1,1]$.

Let $\gamma \in M_1(\rho_*,\rho^*)$ be a profile associated to an initial (LEM) local equilibrium measure or a (DIC) deterministic configuration. Recall, when $I_\gamma(\mu) < \infty$, it has explicit representation (cf. 1.4). Recall, also that v_T and u_T are the LLN speeds associated to γ (cf. (1.1)).

Since $\mathbb{J}$ and $\mathbb{I}$ are given through infima, it is natural to look for explicit densities where computations can be made. Consider the density

$$\mu(s,x) \ = \ \sigma_s * \gamma(x) + (\lambda\epsilon(s/T))\psi(x/L)$$

where $\epsilon(t)$ is a smooth, increasing function which vanishes for $0 \le t \le 1/10$, and $\epsilon(1) = 1$, and $L \neq 0$. At time $s = T/10$, $0 < \gamma_* < \sigma_s * \gamma < \gamma^* < 1$ for some constants γ_*, γ^*. We will take $0 \le \lambda < \min\{\gamma_*, 1-\gamma^*\}/2$, small enough so that $\gamma_*/2 \le \mu \le (1-\gamma^*)/2$ for $T/10 \le t \le T$.

Then, as μ follows the heat equation for $[0,T/10]$, μ satisfies (1.2) with respect to H_x, supported on $[T/10,T]\times[-|L|,|L|]$ given by

$$H_x \ = \begin{cases} \frac{1}{\mu(1-\mu)}\Big[\frac{\lambda\epsilon(s/T)}{2L}\psi'\big(\frac{x}{L}\big) - \frac{\lambda L\epsilon'(s/T)}{T}\Psi\big(\frac{x}{L}\big)\Big] & \text{for } \frac{T}{10}\le s\le T, |x|\le |L| \\ 0 & \text{otherwise .}\end{cases}$$

Also, as $\mu_0 = \gamma$, we have $h(\mu_0;\gamma) = 0$, and

$$\begin{aligned} I_0(\mu) \ &= \ \frac{1}{2}\int_{T/10}^T\int \frac{1}{\mu(1-\mu)}\Big[\frac{\lambda\epsilon(s/T)}{2L}\psi'\Big(\frac{x}{L}\Big) - \frac{\lambda L\epsilon'(s/T)}{T}\Psi\Big(\frac{x}{L}\Big)\Big]^2 dxds \\ &\le \ \frac{4\epsilon^*}{\gamma_*(1-\gamma^*)}\Big[\frac{\lambda^2 T}{4|L|}\int_{-1}^1 \psi'(x)^2dx + \frac{\lambda^2|L|^3}{T}\int_{-1}^1 \Psi(x)^2dx\Big] \end{aligned} \tag{2.4}$$

where $\epsilon^* = 1 + \|\epsilon'\|^2_{L^\infty}$. Compute now

$$\begin{aligned}\int_0^\infty \big[\mu_T(x) - \mu_0(x)\big]dx &= \lambda L \int_0^1 \psi(x)dx + v_T \\ &= \lambda L \int_0^1 \psi(x)dx + \int_0^{u_T} \sigma_T * \gamma(x)dx,\end{aligned}$$

and, for $\mathfrak{c} \in \mathbb{R}$,

$$\int_0^{\mathfrak{c}} \mu_T(x)dx = \int_0^{\mathfrak{c}} \sigma_T * \gamma(x)dx + \lambda L \int_0^{|\mathfrak{c}|/|L|} \psi(x)dx.$$

Then, the restriction specified in the definition of $\mathbb{J}(\mathfrak{c})$, $\int_0^T J(0,t)dt = \mathfrak{c}$, is the same as

$$\lambda L \int_0^1 \psi(x)dx = \mathfrak{c} - v_T \tag{2.5}$$

and the restriction listed in $\mathbb{I}(\mathfrak{c})$, $\int_0^T J(0,t)dt = \int_0^\infty \mu_T(x) - \mu_0(x)dx = \int_0^{\mathfrak{c}} \mu_T(x)dx$, is equivalent to

$$\lambda L \int_{|\mathfrak{c}|/|L|}^1 \psi(x)dx = \int_{u_T}^{\mathfrak{c}} \sigma_T * \gamma(x)dx. \tag{2.6}$$

Lemma 2.7. *For $\mathfrak{c} \in \mathbb{R}$, $\mathbb{J}(\mathfrak{c}), \mathbb{I}(\mathfrak{c}) < \infty$, and in particular $\mathbb{J}(v_T) = \mathbb{I}(u_T) = 0$. Moreover, on any interval $[a,b] \subset \mathbb{R}$, $\sup_{\mathfrak{c}\in[a,b]} \mathbb{J}(c), \sup_{\mathfrak{c}\in[a,b]} \mathbb{I}(c) < \infty$.*

Proof. For $\mathfrak{c} \in \mathbb{R}$, given the bound (2.4), we need only demonstrate that restrictions (2.5) and (2.6), with respect to $\mathbb{J}$ and $\mathbb{I}$, hold with respective choices of λ and L. If $\mathfrak{c} = v_T$ or u_T, we may take $\lambda = 0$, and so clearly $\mathbb{J}(v_T) = \mathbb{I}(u_T) = 0$.

For $\mathfrak{c} \neq v_T$, let $\lambda > 0$, and note the left-side of (2.5) can be made equal to the right-side $v_T - \mathfrak{c}$ with a proper choice of L. Similarly, when $\mathfrak{c} \neq u_T$, let $\lambda > 0$, and note that the left-side of (2.6) vanishes for $|L| \leq |\mathfrak{c}|$ and diverges to $\pm\infty$ as $L \to \pm\infty$. Hence, a proper choice of L allows to verify (2.6) also.

In particular, we can see, by varying L, with respect to $\mathfrak{c} \in [a,b]$ in any finite interval, we obtain $\sup_{\mathfrak{c}\in[a,b]} \mathbb{J}(\mathfrak{c}), \sup_{\mathfrak{c}\in[a,b]} \mathbb{I}(\mathfrak{c}) < \infty$. □

Lemma 2.8. *With respect to local equilibrium measures or deterministic initial configurations, $\mathbb{J}$ and $\mathbb{I}$ are lower semi-continuous.*

Proof. We give the proof for $\mathbb{I}$; the argument for $\mathbb{J}$ is analogous. We first consider when starting from a local equilibrium measure and $I_\gamma = I_\gamma^{LE}$. Let $\{a_n\}$ be a convergent sequence $a_n \to a$. From Proposition 2.7, we have $\sup_n \mathbb{I}(a_n) < \infty$. Then, by Propositions 2.1 and 2.6, we can find densities $\{\mu^n\}$ so that $|I_\gamma^{LE}(\mu^n) - \mathbb{I}(a^n)| < n^{-1}$ and $|\int_0^T J^n(0,t)dt - \int_0^{a_n} \mu_T^n(x)dx| \leq n^{-1}$.

As I_γ^{LE} is a good rate function and $\{I_\gamma^{LE}(\mu^n)\}$ is uniformly bounded, a subsequence can be found where μ^{n_k} converges to a density $\hat{\mu}$ in $D([0,T]; M_1)$ and $\liminf \mathbb{I}(a^n) = \lim \mathbb{I}(a^{n_k}) = \lim I_\gamma^{LE}(\mu^{n_k})$.

By Proposition 2.6, we have $\int_0^T J^{n_k}(0,t)dt \to \int_0^T \hat{J}(0,t)dt$. Also, as $\mu_T^{n_k} \to \hat{\mu}_T$, and $a_n \to a$, we have $\int_0^{a_n} \mu_T^{n_k}(x)dx \to \int_0^a \hat{\mu}_T(x)dx$. Then, $\int_0^T \hat{J}(0,t)dt = \int_0^a \hat{\mu}_T(x)dx$, and hence $\hat{\mu}$ satisfies the infimum restriction in the definition of $\mathbb{I}(a)$.

By lower semi-continuity of I_γ^{LE}, the desired lower semi-continuity of $\mathbb{I}$ follows as $\liminf \mathbb{I}(a_n) = \lim I_\gamma^{LE}(\mu^{n_k}) \geq I_\gamma^{LE}(\hat{\mu}) \geq \mathbb{I}(a)$.

Starting from a deterministic configuration, we can repeat the steps with I^{LE}_γ replaced by I_0. The densities $\{\mu^{n_k}\}$, by Proposition 2.1, also are such that $\mu^{n_k}_0$ converges to γ. Hence, the limit $\hat{\mu}$ satisfies $\hat{\mu}_0 = \gamma$ and so $I_0(\hat{\mu}) = I^{DC}_\gamma(\hat{\mu})$. Therefore, $\mathbb{I}$ is also lower semi-continuous in this case. □

Corollary 2.9. *With respect to local equilibrium measures or deterministic initial conditions, $\mathbb{J}$ and $\mathbb{I}$ are finite-valued rate functions. In addition, $\mathbb{J}(a') = 0$ and $\mathbb{I}(a) = 0$ exactly when $a' = v_T$ and $a = u_T$.*

Proof. We concentrate on the proof with respect to $\mathbb{I}$, as a similar argument holds for $\mathbb{J}$. First, that $\mathbb{I} : \mathbb{R} \to \mathbb{R}$, $\mathbb{I}(u_T) = 0$, and $\mathbb{I}$ is a rate function follows from Lemmas 2.7 and 2.8. We need only show that u_T is the only zero of $\mathbb{I}$.

When $a \neq u_T$, if $\mathbb{I}(a)$ vanishes, out of a minimizing sequence of densities, through Propositions 2.1 and 2.6, one can find a subsequence converging to a minimizing μ satisfying the restriction $\int_0^T J(0,t)dt = \int_0^a \mu_T(x)dx$.

With respect to local equilibrium measures, by lower semi-continuity of $h(\cdot;\gamma)$ and $I_0(\cdot)$, we have $h(\mu_0;\gamma) = I_0(\mu) = 0$. Under deterministic initial conditions, since the subsequence at time 0 converges to γ, we have $\mu_0 = \gamma$, and by lower semi-continuity, $I_0(\mu) = 0$.

Then, in either case, $\mu_0 = \gamma$ a.s. and, noting (1.4), $H^2_x\mu(1-\mu) = 0$ a.s. In particular, $\mu_t = \sigma_t * \gamma$ is the unique bounded solution of the weak heat equation with initial data γ. However, then $\int_0^T J(0,t)dt = \int_0^{u_T} \mu_T(x)dx$ which does not equal $\int_0^a \mu_T(x)dx$ since μ_T is positive and $a \neq u_T$. This is a contradiction. □

3. Proof of Theorem 1.5

The proofs follow in several steps which are divided into subsections. The first step is to describe key relations between a tagged particle and the current across the bond $(-1,0)$, which will allow us later to invoke large deviations of the empirical density. Next, a super-exponential inequality is given. Then, exponential tightness is established, and weak upper and lower large deviation bounds are proved. Finally, Theorem 1.5 is shown.

3.1. **Tagged particle and current relations.** For $x \in \mathbb{Z}$ and $t \geq 0$, define $J_{x,x+1}(t)$ as the integrated current up to time t across the bond $(x, x+1)$, that is the number of particles which crossed from x to $x+1$ up to time t minus the number of particles which moved from $x+1$ to x in time t. It is well known (cf. Liggett [23], DeMasi-Ferrari [8]) that, for integers $r > 0$,

$$\{X_t \geq r\} = \{J_{-1,0}(t) \geq \sum_{x=0}^{r-1} \eta_t(x)\}. \tag{3.1}$$

Similarly, for $r < 0$,

$$\{X_t \leq r\} = \{J_{-1,0}(t) \leq -\sum_{x=r}^{-1} \eta_t(x)\} \tag{3.2}$$

and

$$\{X_t \leq 0\} = \{J_{-1,0}(t) \leq 0\}.$$

Also, from a moment's thought, we have

$$J_{x-1,x}(N^2t) - J_{x,x+1}(N^2t) = \eta_{N^2t}(x) - \eta_0(x).$$

We would like to make a summation-by-parts,

$$J_{-1,0}(N^2t) = \sum_{x\geq 0} J_{x-1,x}(N^2t) - J_{x,x+1}(N^2t) = \sum_{x\geq 0} \eta_{N^2t}(x) - \eta_0(x)$$

to write the current across the bond $(-1,0)$ in terms of the empirical process. However, the above display is only formal as the sum on the right may not converge. To treat it carefully, we introduce a 'cutoff' function as in Rost-Vares [32]. For $n \geq 1$, let

$$G_n(u) = 1_{[0,n]}(u)(1-u/n).$$

Also, denote for a function $G \in C_K^\infty(\mathbb{R})$,

$$Y_t^N(G) = \frac{1}{N}\sum_x G(x/N)\eta_{N^2t}(x).$$

Then,

$$\begin{aligned} Y_t^N(G_n) - Y_0^N(G_n) &= \frac{1}{N}\sum_x G_n(x/N)\Big(J_{x-1,x}(N^2t) - J_{x,x+1}(N^2t)\Big) \\ &= \frac{1}{N}\sum_x \Big(G_n(x/N) - G_n(x-1/N)\Big)J_{x-1,x}(N^2t) \\ &= \frac{1}{N}J_{-1,0}(N^2t) - \frac{1}{N}\sum_{x=1}^{nN}\frac{1}{nN}J_{x-1,x}(N^2t). \end{aligned}$$

This implies

$$\frac{1}{N}J_{-1,0}(N^2t) = Y_t^N(G_n) - Y_0^N(G_n) + \frac{1}{N}\sum_{x=1}^{nN}\frac{1}{nN}J_{x-1,x}(N^2t).$$

Hence, for $a > 0$,

$$\begin{aligned} \Big\{X_{N^2t}/N \geq a\Big\} &= \Big\{\frac{1}{N}J_{-1,0}(N^2t) \geq \frac{1}{N}\sum_{x=0}^{\lfloor aN\rfloor}\eta_{N^2t}(x)\Big\} \qquad (3.3) \\ &= \Big\{Y_t^N(G_n) - Y_0^N(G_n) + \frac{1}{nN^2}\sum_{x=1}^{nN}J_{x-1,x}(N^2t) \geq \frac{1}{N}\sum_{x=0}^{\lfloor aN\rfloor}\eta_{N^2t}(x)\Big\}. \end{aligned}$$

A similar statement holds for $a \leq 0$, namely,

$$\begin{aligned} \Big\{X_{N^2t}/N \leq a\Big\} &= \Big\{Y_t^N(G_n) - Y_0^N(G_n) \\ &\quad + \frac{1}{nN^2}\sum_{x=1}^{nN}J_{x-1,x}(N^2t) \leq -\frac{1}{N}\sum_{x=\lfloor aN\rfloor}^{-1}\eta_{N^2t}(x)\Big\} \end{aligned}$$

where for $a = 0$ we take $\sum_{x=0}^{-1}\eta_{N^2t}(x) = 0$.

Therefore, heuristically, the tagged particle large deviations should be given in terms of the rate for the empirical density I_γ under a certain restriction, as long as the contribution from the term $(1/nN^2)\sum_{x=1}^{nN} J_{x-1,x}(N^2t)$ is superexponentially small as $n, N \uparrow \infty$.

3.2. **Superexponential estimate.** In relation to (3.3), the superexponential estimate needed is implied by the following estimate.

Proposition 3.1. *For each $\lambda > 0$, starting from (LEM) or (DIC) initial states,*

$$\lim_{n\uparrow\infty}\lim_{N\uparrow\infty}\frac{1}{N}\log E\exp\left|\frac{\lambda N}{nN^2}\sum_{x=1}^{nN}J_{x-1,x}(N^2t)\right| \; = \; 0.$$

Proof. By the inequality $e^{|x|} \leq e^x + e^{-x}$, we can remove the absolute value in the last display. Now, note that

$$\exp\left\{\frac{\lambda N}{nN^2}\sum_{x=1}^{nN}J_{x-1,x}(N^2t) - \sum_{x=1}^{nN}(e^{\lambda/nN}-1)\int_0^{N^2t}\eta_{x-1}(1-\eta_x)(s)ds\right.$$
$$\left. - \sum_{x=1}^{nN}(e^{-\lambda/nN}-1)\int_0^{N^2t}\eta_x(1-\eta_{x-1})(s)ds\right\}$$

is a martingale with mean 1. Then, together the second and third terms in the exponent equal

$$\sum_{x=1}^{nN}\left[(e^{\lambda/nN}-\lambda/nN-1)\int_0^{N^2t}\eta_{x-1}(1-\eta_x)(s)ds\right.$$
$$\left.+(e^{-\lambda/nN}+\lambda/nN-1)\int_0^{N^2t}\eta_x(1-\eta_{x-1})(s)ds\right]$$
$$+\frac{\lambda}{nN}\int_0^{N^2t}(\eta_0-\eta_{nN})(s)ds$$
$$\leq \frac{2e^{\lambda/nN}\lambda^2}{n^2N^2}(nN)(N^2t) + \frac{\lambda}{nN}(N^2t) \quad \leq \quad \frac{C(t,\lambda)N}{n}$$

which gives the result with standard manipulations. □

3.3. **Exponential tightness estimate.** We now show that the scaled tagged particle positions are exponential tight.

Lemma 3.2. *Starting from (LEM) or (DIC) initial states, we have*

$$\lim_{a\uparrow\infty}\lim_{N\uparrow\infty}\frac{1}{N}\log P\Big\{|J_{-1,0}(N^2T)|/N \;\geq a\Big\}$$
$$= \lim_{a\uparrow\infty}\lim_{N\uparrow\infty}\frac{1}{N}\log P\Big\{|X_{N^2T}|/N \;\geq\; a\Big\} \;=\; -\infty.$$

Proof. We give the argument for the tagged particle, as the proof for the current is similar, and somewhat easier. From (3.3), we need only super-exponentially estimate, for a positive (as a similar argument works for $a<0$) and n fixed,

$$P\Big\{Y_T^N(G_n) - Y_0^N(G_n) + \frac{1}{nN^2}\sum_{x=1}^{nN}J_{x-1,x}(N^2T) \geq Y_T^N(1_{[0,a]})\Big\}.$$

We need only estimate

$$E\Big[\exp\Big\{N\Big[Y_T^N(G_n) - Y_0^N(G_n) + \frac{1}{nN^2}\sum_{x=1}^{nN}J_{x-1,x}(N^2T) - Y_T^N(1_{[0,a]})\Big]\Big\}\Big]$$
$$= E[e^{Q_1}e^{Q_2}e^{Q_3}e^{Q_4}]$$

where $Q_1 = NY_T^N(G_n)$, $Q_2 = -NY_0^N(G_n)$, $Q_3 = (nN)^{-1}\sum_{x=1}^{nN} J_{x-1,x}(N^2t)$, and $Q_4 = -\sum_{x=0}^{\lfloor aN\rfloor} \eta_{N^2T}(x)$. By Chebychev, we can estimate the exponential terms separately. For fixed n, $\lim N^{-1}\log E[e^{4Q_3}]$ is bounded from Proposition 3.1, and as $Q_1 \leq nN$ by properties of G_n, $\lim N^{-1}\log E[e^{4Q_1}]$ is also bounded. In addition, as $\exp\{4Q_2\} \leq 1$, this term can be neglected.

Finally, by Borcea-Branden-Liggett [6][Theorem 5.2], as the initial measure of type (LEM) or (DIC) is a product measure (of degenerate Bernoulli's under (DIC) initial configurations), the coordinates $\{\eta_{N^2T}(x)\}$ are negatively associated. Hence, $E[e^{4Q_4}] \leq \prod_{x=1}^{\lfloor aN\rfloor} E[e^{-4\eta_{N^2T}(x)}]$, and using $\log(1-x) \leq -x$ for $0 \leq x \leq 1$, we write

$$\begin{aligned}
\frac{1}{N}\log E[e^{4Q_4}] &\leq \frac{1}{N}\sum_{x=1}^{\lfloor aN\rfloor} \log E[e^{-4\eta_{N^2T}(x)}] \\
&\leq \frac{1}{N}\sum_{x=1}^{\lfloor aN\rfloor} \log\left[(e^{-4}-1)P(\eta_{N^2T}(x)=1)+1\right] \\
&\leq \frac{e^{-4}-1}{N} E\left[\sum_{x=1}^{\lfloor aN\rfloor} \eta_{N^2T}(x)\right] \to (e^{-4}-1)\int_0^a m(T,x)dx
\end{aligned}$$

where $m(T,x) = \sigma_T * \gamma(x)$ is the solution of the hydrodynamic equation (Proposition 1.1). Since $\sigma_{T/10} * \gamma(x) \geq \gamma_* > 0$ as $\gamma \in M_1(\rho_*, \rho^*)$ for $\rho_*, \rho^* > 0$, the right-side is bounded above by $(e^{-4}-1)\gamma_* a \downarrow -\infty$ as $a \uparrow \infty$. □

3.4. **Weak LDP upperbounds.** The weak upperbound for the tagged particle deviations, starting from local equilibrium measures or deterministic initial configuration, follows in several steps and is stated in Step 6. As the same argument works for the current, we also state its associated weak upperbound in Step 6 below. For the convenience of the reader, we indicate the modifications needed in Step 1; the other steps involve similar changes.

Step 1. Consider an interval $[a,b]$ for $0 < a < b$; subsequent arguments carry over straightforwardly to all intervals $[a,b] \subset \mathbb{R}$ using (3.2) by splitting at the origin if necessary. Now, divide $[a,b]$ into m equal intervals $A_k = [c_k, c_{k+1}]$. Then, by the union of events estimate,

$$\limsup_{N\to\infty} \frac{1}{N}\log P\Big(X_{N^2T}/N \in [a,b]\Big) \leq \max_k \limsup_{N\to\infty} \frac{1}{N}\log P\Big(X_{N^2T}/N \in A_k\Big).$$

Then, from (3.3) and Proposition 3.1, we have that

$$\begin{aligned}
&\limsup_{N\uparrow\infty} \frac{1}{N}\log P\Big(X_{N^2T}/N \in [a,b]\Big) \\
&\leq \limsup_{m\uparrow\infty}\limsup_{\delta\downarrow 0}\limsup_{n\uparrow\infty}\max_{1\leq k\leq m}\limsup_{N\uparrow\infty} \\
&\quad \frac{1}{N}\log P\Big(Y_T^N(G_n) - Y_0^N(G_n) \in \Big[Y_T^N(1_{[0,c_k]}) - \delta, Y_T^N(1_{[0,c_{k+1}]}) + \delta\Big]\Big).
\end{aligned}$$

Since the maps $\mu \mapsto \int G(x)\mu_T dx$, $\int G(x)\mu_0 dx$, $\int_0^c \mu_T dx$, for compactly supported G and constants c, are continuous in the Skorohod topology on $D([0,T]; M_1)$, from

Corollary 1.4, we conclude, for fixed k, n and δ that

$$\limsup_{N\to\infty}\frac{1}{N}\log P\Big(Y_T^N(G_n)-Y_0^N(G_n)ds\in\Big[Y_T^N(1_{[0,c_k]})-\delta, Y_T^N(1_{[0,c_{k+1}]})+\delta\Big]\Big)$$
$$\leq\ -\inf\Big\{I_\gamma(\mu);\int G_n(x)\big[\mu_T(x)-\mu_0(x)\big]dx \qquad (3.4)$$
$$\in\Big[\int_0^{c_k}\mu_T(x)dx-\delta,\int_0^{c_{k+1}}\mu_T(x)dx+\delta\Big]\Big\}.$$

We now indicate the modifications needed for the current in this step. For $0<a<b$, from (3.3) and Proposition 3.1, we have

$$\limsup_{N\uparrow\infty}\frac{1}{N}\log P\Big(J_{-1,0}(N^2T)/N\in[a,b]\Big)$$
$$\leq\ \limsup_{m\uparrow\infty}\limsup_{\delta\downarrow 0}\limsup_{n\uparrow\infty}\max_{1\leq k\leq m}\limsup_{N\uparrow\infty}$$
$$\frac{1}{N}\log P\Big(Y_T^N(G_n)-Y_0^N(G_n)\in[c_k-\delta,c_{k+1}+\delta]\Big).$$

From continuity of the maps $\mu\mapsto\int G\mu_T dx$ and $\mu\mapsto\int G\mu_0 dx$, and Corollary 1.4, we further bound

$$\limsup_{N\to\infty}\frac{1}{N}\log P\Big(Y_T^N(G_n)-Y_0^N(G_n)ds\in[c_k-\delta,c_{k+1}+\delta]\Big)$$
$$\leq\ -\inf\Big\{I_\gamma(\mu);\int G_n(x)\big[\mu_T(x)-\mu_0(x)\big]dx\in[c_k-\delta,c_{k+1}+\delta]\Big\}.$$

Step 2. Next, we give a uniform upperbound of the infimum in (3.4). We exhibit a density $\mu^{\mathfrak{c}}$ satisfying, for each $\delta>0$ and all large n,

$$\int G_n(x)[\mu_T^{\mathfrak{c}}(x)-\mu_0^{\mathfrak{c}}(x)]dx\in\Big[\int_0^{\mathfrak{c}}\mu_T^{\mathfrak{c}}(x)dx-\delta,\int_0^{\mathfrak{c}}\mu_T^{\mathfrak{c}}(x)dx+\delta\Big]$$

and $\sup_{\mathfrak{c}\in[a,b]}I_\gamma(\mu^{\mathfrak{c}})<B_0<\infty$ where B_0 is independent of n and δ.

This is accomplished by the constructions in subsection 2.4, namely one can take $\mu^{\mathfrak{c}}=\sigma_t*\gamma+\lambda\epsilon(t/T)\psi(x/L)$ with λ and L chosen so that $\lambda L\int_{|\mathfrak{c}/L|}^1\psi(x)dx=\int_{u_T}^{\mathfrak{c}}\sigma_T*\gamma(x)dx$. Let $J^{\mathfrak{c}}$ be its current, and $H_x^{\mathfrak{c}}$ be the associated function with respect to (1.2).

Proposition 2.7 gives $I_\gamma(\mu^{\mathfrak{c}})$ is uniformly bounded for $\mathfrak{c}\in[a,b]$. Now compute

$$\int G_n(x)\big[\mu_T^{\mathfrak{c}}(x)-\mu_0^{\mathfrak{c}}(x)\big]dx \qquad (3.5)$$
$$=\ \int_0^T\int G_n(x)\,[(1/2)\partial_{xx}\mu^{\mathfrak{c}}-\partial_x H_x^{\mathfrak{c}}\mu^{\mathfrak{c}}(1-\mu^{\mathfrak{c}})]\,dxdt$$
$$=\ \int_0^T -(1/2)\partial_x\mu^{\mathfrak{c}}(t,0)+H_x^{\mathfrak{c}}\mu^{\mathfrak{c}}(1-\mu^{\mathfrak{c}})(t,0)dt$$
$$+\frac{1}{n}\int_0^T\int_0^n\big[(1/2)\partial_x\mu^{c}-H_x^{\mathfrak{c}}\mu^{\mathfrak{c}}(1-\mu^{\mathfrak{c}})\big]dxdt.$$

Since $\int_0^{\mathfrak{c}} \mu_T^{\mathfrak{c}}(x)dx = \int_0^T J^{\mathfrak{c}}(0,t)dt$ and $J^{\mathfrak{c}}(0,t) = -(1/2)\partial_x \mu^{\mathfrak{c}}(t,0) + H_x^{\mathfrak{c}}\mu^{\mathfrak{c}}(1-\mu^{\mathfrak{c}})(t,0)$, we have

$$\begin{aligned}\sup_{\mathfrak{c}\in[a,b]} & \Big|\int G_n(x)\big[\mu_T^{\mathfrak{c}}(x)-\mu_0^{\mathfrak{c}}(x)\big]dx - \int_0^{\mathfrak{c}} \mu_T^{\mathfrak{c}}(x)dx\Big| \\ &\le \frac{1}{n}\Big|\int_0^T\int_0^n (1/2)\partial_x\mu^{\mathfrak{c}} - H_x^{\mathfrak{c}}\mu^{\mathfrak{c}}(1-\mu^{\mathfrak{c}})dxdt\Big| \\ &\le \sup_{\mathfrak{c}\in[a,b]}\Big|\frac{1}{2n}\int_0^T(\mu_t^{\mathfrak{c}}(n)-\mu_t^{\mathfrak{c}}(0))dt\Big| + \frac{1}{\sqrt{n}}\|H_x^{\mathfrak{c}}\mu^{\mathfrak{c}}(1-\mu^{\mathfrak{c}})\|_{L^2([0,T]\times\mathbb{R})}.\end{aligned}$$

Since $\|H_x^{\mathfrak{c}}\mu^{\mathfrak{c}}(1-\mu^{\mathfrak{c}})\|_{L^2}^2 \le 2I_0(\mu^{\mathfrak{c}})$, the right-side is $O(n^{-1/2})$ by Lemma 2.7.

Step 3. As I_γ is a good rate function, by the uniform bounds in step 2, out of minimizers $\nu^{k,n,\delta}$ over $k=k(m)$, and n,δ in the infimum in (3.4), by the uniform bound on $I_\gamma(\nu^{k,n,\delta})$, we can extract a subsequence, on which the limsup of (3.4) is attained as $\delta\downarrow 0$ and $n,m\uparrow\infty$, and which converges in $D([0,T];M_1)$ to a $\bar\mu$.

By Proposition 2.1, the subsequence, labeled $\nu^{k,n,\delta}$ itself for simplicity, may be approximated by $\{\mu^{k,n,\delta}\}$ so that $\mu^{k,n,\delta}$ is smooth, strictly bounded between 0 and 1, $H_x^{k,n,\delta}\in C^\infty([0,T]\times\mathbb{R})$, Skorohod distance $d(\mu^{k,n,\delta},\nu^{k,n,\delta})\downarrow 0$, $|I_0(\nu^{k,n,\delta}) - I_0(\mu^{k,n,\delta})|\downarrow 0$, and when $\hat\gamma\in M_1(\rho_*,\rho^*)$ is piecewise continuous and $0<\hat\gamma(x)<1$ for $x\in\mathbb{R}$, $|h(\nu_0^{k,n,\delta};\hat\gamma)-h(\mu_0^{k,n,\delta};\hat\gamma)|\downarrow 0$. Also, as $[a,b]$ is compact, the subsequence can be chosen so that c_{k+1} converges to a $\mathfrak{c}\in[a,b]$.

Given $\nu^{k,n,\delta}$ satisfies the restriction in (3.4), we may also arrange

$$\begin{aligned}\int_0^{c_k}\mu_T^{k,n,\delta}(x)dx - 2\delta &\le \int G_n(x)\big[\mu_T^{k,n,\delta}(x)-\mu_0^{k,n,\delta}(x)\big]dx \\ &\le \int_0^{c_{k+1}}\mu_T^{k,n,\delta}(x)dx + 2\delta.\end{aligned} \tag{3.6}$$

With these specifications, by lower semi-continuity, we have (3.4) is less than, in the case of starting from a local equilibrium measure,

$$\lim_{m\uparrow\infty}\lim_{\delta\downarrow 0}\lim_{n\uparrow\infty}\max_k -I_\gamma^{LE}(\mu^{k,n,\delta}) \le -I_\gamma^{LE}(\bar\mu).$$

When starting from a deterministic configuration, noting $\nu_0^{k,n,\delta}=\bar\mu_0=\gamma$, (3.4) is less than

$$\lim_{m\uparrow\infty}\lim_{\delta\downarrow 0}\lim_{n\uparrow\infty}\max_k -I_0(\mu^{k,n,\delta}) \le -I_0(\bar\mu) = -I_\gamma^{DC}(\bar\mu).$$

Step 4. We now show that $\bar\mu$ satisfies

$$\int_0^T \bar J(0,t)dt = \int_0^{\mathfrak{c}}\bar\mu_T(x)dx. \tag{3.7}$$

As convergence in $D([0,T];M_1)$ implies $\mu_T^{k,n,\delta}\to\bar\mu_T$, $c_{k+1}-c_k=m^{-1}$, and $0\le\mu_T^{k,n,\delta}(x)\le 1$, we have both

$$\int_0^{c_k}\mu_T^{k,n,\delta}(x)dx,\ \int_0^{c_{k+1}}\mu_T^{k,n,\delta}(x)dx \to \int_0^{\mathfrak{c}}\bar\mu_T(x)dx.$$

Also, following the sequence (3.5),

$$\int G_n(x)\big[\mu_T^{k,n,\delta}(x) - \mu_0^{k,n,\delta}(x)\big]dx = \int_0^T J^{k,n,\delta}(0,t)dt$$
$$+ \frac{1}{n}\int_0^T \int_0^n \big[(1/2)\partial_x \mu_t^{k,n,\delta} - H_x^{k,n,\delta}\mu^{k,n,\delta}(1-\mu^{k,n,\delta})(t,x)\big]dxdt.$$

Since $\|H_x^{k,n,\delta}\mu^{k,n,\delta}(1-\mu^{k,n,\delta})\|_{L^2}^2 \le 2I_0(\mu^{k,n,\delta})$ is uniformly bounded, the last term is bounded uniformly by $n^{-1}T + (nT)^{-1/2}\sqrt{2I_0(\mu^{k,n,\delta})}$. On the other hand, $\int_0^T J^{k,n,\delta}(0,t)dt \to \int_0^T \bar{J}(0,t)dt$ by Proposition 2.6.

Hence, noting (3.6), we obtain (3.7) immediately.

Step 5. Therefore,

$$\limsup_{m\uparrow\infty}\limsup_{\delta\downarrow 0}\limsup_{n\uparrow\infty}\max_{1\le k\le m}$$
$$-\inf\Big\{I_\gamma(\mu); \int G_n(x)\big[\mu_T(x)-\mu_0(x)\big]dx$$
$$\in \Big[\int_0^{c_k}\mu_T(x)dx - \delta, \int_0^{c_{k+1}}\mu_T(x)dx + \delta\Big]\Big\}$$
$$\le -I_\gamma(\bar\mu) \le -\min_{\mathfrak{c}\in[a,b]}\mathbb{I}(\mathfrak{c}).$$

Step 6. The weak LDP upperbound with respect to the tagged particle, for compact $K \subset \mathbb{R}$,

$$\limsup_{N\uparrow\infty}\frac{1}{N}P\Big(X_{N^2t}/N \in K\Big) \le -\inf_{a\in K}\mathbb{I}(a), \tag{3.8}$$

is now standard given $\mathbb{I}$ is lower semi-continuous (Lemma 2.8).

Similarly, we have the weak upperbound for the current:

$$\limsup_{N\uparrow\infty}\frac{1}{N}P\Big(J_{-1,0}(N^2t)/N \in K\Big) \le -\inf_{a\in K}\mathbb{J}(a). \tag{3.9}$$

3.5. **LDP Lowerbound.** As before, we concentrate on the tagged particle deviations, as the proof for the current is analogous. For the first step, the scheme for the weak upper bound is used. Let $O \subset \mathbb{R}$ be a nonempty open set, and suppose $a \in O$. We also assume $a > 0$ as a similar argument works for $a \le 0$ by focusing on a subinterval to the left of the origin. Let $\epsilon > 0$ be such that $a - \epsilon > 0$ and $(a-\epsilon, a+\epsilon) \subset O$.

Then, for $\theta > 0$,

$$\lim_{N\uparrow\infty}\frac{1}{N}\log P\Big(X_{N^2T}/N \in O\Big) \ge \lim_{N\uparrow\infty}\frac{1}{N}P\Big(X_{N^2T}/N \in (a-\epsilon, a+\epsilon)\Big)$$
$$\ge \lim_{n\uparrow\infty}\lim_{N\to\infty}\frac{1}{N}\log P\Big(Y_T^N(1_{[0,a-\epsilon]}) < Y_T^N(G_n) - Y_0^N(G_n) \tag{3.10}$$
$$+\frac{1}{nN^2}\sum_{x=1}^{nN} J_{x-1,x}(N^2T) < Y_T^N(1_{[0,a+\epsilon]}), \text{ and } Y_T^N(1_{[a-\epsilon,a+\epsilon]}) > \theta\Big).$$

From Proposition 3.1 and Corollary 1.4, the left-side of (3.10) is greater than

$$\begin{aligned}
&\lim_{\theta\downarrow 0}\lim_{\delta\downarrow 0}\lim_{n\uparrow\infty}\lim_{N\to\infty}\frac{1}{N}\log P\Big(Y_T^N(1_{[0,a-\epsilon]})+\delta \\
&< Y_T^N(G_n)-Y_0^N(G_n) < Y_T^N(1_{[0,a+\epsilon]})-\delta,\ \text{and } Y_T^N(1_{[a-\epsilon,a+\epsilon]})>\theta\Big) \\
&\geq \lim_{\theta\downarrow 0}\lim_{\delta\downarrow 0}\lim_{n\uparrow\infty} \\
&\quad -\inf\Big\{I_\gamma(\mu):\int_0^{a-\epsilon}\mu_T(x)dx+\delta<\int G_n(x)\big[\mu_T(x)-\mu_0(x)\big]dx \qquad (3.11)\\
&\qquad\qquad <\int_0^{a+\epsilon}\mu_T(x)dx-\delta,\ \text{and } \int_{a-\epsilon}^{a+\epsilon}\mu_T(x)dx>\theta\Big\}.
\end{aligned}$$

Now, for $\alpha>0$, let $\bar\mu$ be a density such that $|I_\gamma(\bar\mu)-\mathbb{I}(a)|<\alpha$, and

$$\int_0^T \bar J(0,t)dt \;=\; \int_0^a \bar\mu_T(x)dx.$$

By the method used for (3.5) and (3.8) in the last section, through approximations of $\bar\mu$ with smooth μ^n by Proposition 2.1, we can show that

$$\lim_n\int_0^\infty G_n(x)\big[\bar\mu_T(x)-\bar\mu_0(x)\big]dx \;=\; \int_0^T \bar J(0,t)dt. \qquad (3.12)$$

We will need now to approximate $\bar\mu$ as follows to ensure a certain positivity. Let $\chi=\sigma_s*\gamma+\lambda\epsilon(t/T)\psi(x/L)$ from subsection 2.4 where λ, L are chosen so that $\int_0^T J^\chi(0,t)dt=\int_0^a\chi_T(x)dx$. Recall $I_\gamma(\chi)<\infty$, and note (3.12), with χ and J^χ replacing $\bar\mu$ and $\bar J$, also holds by the explicit construction. For $0<b<1$, define $\mu^b=(1-b)\chi+b\bar\mu$. Clearly, $\lim_{b\uparrow 1}\mu^b=\bar\mu$ uniformly, and so in $D([0,T];M_1)$. In fact, $\lim_{b\uparrow 1}I_\gamma(\mu^b)=I_\gamma(\bar\mu)$: By lower semi-continuity, $\liminf I_\gamma(\mu^b)\geq I_\gamma(\bar\mu)$ and, by convexity, $\limsup I_\gamma(\mu^b)\leq I_\gamma(\bar\mu)$. Now, for given $\beta>0$, let b be such that $|I_\gamma(\mu^b)-I_\gamma(\bar\mu)|<\beta$.

With $\theta>0$, noting

$$\lim_n\int_0^\infty G_n(x)\big[\mu_T^b(x)-\mu_0^b(x)\big]dx \;=\; \int_0^a\mu_T^b(x)dx,$$

we have for $n\geq N(\theta,\bar\mu,\chi)$ that

$$\begin{aligned}
&\int_0^{a-\epsilon}\mu_T^b(x)dx+b\int_{a-\epsilon}^a\bar\mu_T(x)dx+(1-b)\int_{a-\epsilon}^a\chi_T(x)dx-\theta\\
&\quad\leq \int G_n(x)\big[\mu_T^b(x)-\mu_0^b(x)\big]dx\\
&\quad\leq \int_0^{a+\epsilon}\mu_T^b(x)dx-b\int_a^{a+\epsilon}\bar\mu_T(x)dx-(1-b)\int_a^{a+\epsilon}\chi_T(x)dx+\theta.
\end{aligned}$$

By the construction of χ, $\int_{a-\epsilon}^a\chi_T(x)dx, \int_a^{a+\epsilon}\chi_T(x)dx\geq c\epsilon$ for a constant $c>0$. Hence, we can choose $\theta=\theta(\epsilon,b,\chi)$ so that for all small δ,

$$(1-b)\int_a^{a+\epsilon}\chi_T(x)dx-\theta,\ (1-b)\int_{a-\epsilon}^a\chi_T(x)dx-\theta \;>\; \delta.$$

Therefore, as $\bar{\mu}$ is nonnegative, μ^b satisfies the restriction in the infimum in (3.11). In particular, we have

$$\lim_{N\uparrow\infty}\frac{1}{N}P\Big(X_{N^2t}/N\in O\Big)\ \ge\ -I_\gamma(\mu^b)\ \ge\ -I_\gamma(\bar{\mu})-\beta\ \ge\ -\mathbb{I}(a)-\alpha-\beta.$$

Hence,

$$\lim_{N\uparrow\infty}\frac{1}{N}P\Big(X_{N^2t}/N\in O\Big)\ \ge\ -\inf_{a\in O}\mathbb{I}(a). \tag{3.13}$$

Analogously, we have weak lowerbound large deviations for the current:

$$\lim_{N\uparrow\infty}\frac{1}{N}P\Big(J_{-1,0}(N^2t)/N\in O\Big)\ \ge\ -\inf_{a\in O}\mathbb{J}(a). \tag{3.14}$$

Proof of Theorem 1.5. First, the functions $\mathbb{J}$ and $\mathbb{I}$ are finite-valued rate functions which vanish exactly at v_T and u_T respectively by Corollary 2.9.

Next, a 'weak' LDP is found from (3.9) and (3.14) with respect to rate function $\mathbb{J}$, and (3.8) and (3.13) with respect to $\mathbb{I}$. Standard arguments, given exponential tightness (Lemma 3.2), extend the 'weak' LDP to the full large deviation principle.

Finally, given the LDP and exponential tightness, it follows that (1) $\mathbb{J}$ and $\mathbb{I}$ are good rate functions by Lemma 1.2.18 [9], and also that (2) $\lim_{|a|\uparrow\infty}\mathbb{J}(a)=\lim_{|a|\uparrow\infty}\mathbb{I}(a)=\infty$. □

4. Asymptotic evaluations

We prove Theorems 1.6, 1.7 and 1.8 in succeeding subsections.

4.1. Proof of Theorem 1.6.

We first prove the upperbounds which are implied by the following lemma, and then the lowerbounds.

Lemma 4.1. *Starting from (DIC) or (LEM) initial conditions, there are constants c_0, c_1, c_2, c_3 depending only on γ such that when respectively $|a-v_T|/\sqrt{T}\ge c_0$ and $|a-u_T|/\sqrt{T}\ge c_0$, we have in turn*

$$\mathbb{J}(a)\ \le\ \frac{c_1|a-v_T|^3}{T}\quad\text{and}\quad\mathbb{I}(a)\ \le\ \frac{c_1|a-u_T|^3}{T}.$$

Also, when respectively $|a-v_T|/\sqrt{T}\le c_2$ and $|a-u_T|/\sqrt{T}\le c_2$, we have correspondingly

$$\mathbb{J}(a)\ \le\ \frac{c_3(a-v_T)^2}{\sqrt{T}}\quad\text{and}\quad\mathbb{I}(a)\ \le\ \frac{c_3(a-u_T)^2}{\sqrt{T}}.$$

Proof. We prove the estimates for the current rate function, and deduce corresponding bounds for the tagged particle rate function. Let also $a>v_T$ as the argument for $a<v_T$ is analogous. For the reader's convenience, we recall the estimate (2.4) and write,

$$\mathbb{J}(a)\ \le\ \frac{4\epsilon^*}{\gamma_*(1-\gamma^*)}\Big[\frac{\lambda^2T}{4|L|}\int_{-1}^{1}\psi'(x)^2dx+\frac{\lambda^2|L|^3}{T}\int_{-1}^{1}\Psi(x)^2dx\Big],$$

and the restriction equation (2.6) when $\mathfrak{c}=a$,

$$\lambda\ =\ \frac{a-v_T}{L\int_0^1\psi dx},$$

subject to $0 < \lambda \leq \min\{\gamma_*, 1-\gamma^*\}/2$. The requirement on λ holds when

$$L \geq |a - v_T|/[\frac{1}{2}\min\{\gamma_*, (1-\gamma^*)\}\int_0^1 \psi dx] := \kappa_0|a - v_T|.$$

Now take L in the form $L = \kappa\sqrt{T}$. Substituting into the bound for $\mathbb{J}(a)$, we obtain

$$\mathbb{J}(a) \leq \frac{(a-v_T)^2}{\sqrt{T}}\frac{4\epsilon^*\kappa}{\gamma_*(1-\gamma^*)}\Big[\frac{1}{4\kappa^4}\int_{-1}^1 \psi'(x)^2 dx + \int_{-1}^1 \Psi(x)^2 dx\Big].$$

Hence, when a is large, say $\kappa = |a - v_T|\kappa_0/\sqrt{T} \geq 1$, we have $\mathbb{J}(a) \leq c(\gamma)|a - v_T|^3/T$. Correspondingly, when a is such that $|a - v_T|\kappa_0/\sqrt{T} \leq 1$, we choose $\kappa = 1$ to get $\mathbb{J}(a) \leq c(\gamma)|a - v_T|^2/\sqrt{T}$.

The bounds on the tagged particle rate function $\mathbb{I}$ follow from the current rate function bounds. First, by (1.7), $\mathbb{I}^{LE}_\gamma(a) \leq \mathbb{I}^{DC}_\gamma(a)$. Also, by (1.8), with $\gamma'(x) = \gamma(x+a)$, $\mathbb{I}^{DC}_\gamma(a) = \mathbb{J}^{DC}_{\gamma'}(\int_0^a \gamma dx)$. For fixed a, let now $v_T(\gamma')$ be the LLN integrated current through the origin starting from γ'. Then,

$$\begin{aligned}\int_0^a \gamma dx - v_T(\gamma') &= \int_0^a \gamma dx - \int_0^\infty \sigma_T * \gamma' - \gamma' dx \\ &= \int_0^a \sigma_T * \gamma dx - \int_0^\infty \sigma_T * \gamma - \gamma dx = \int_{u_T}^a \sigma_T * \gamma dx.\end{aligned}$$

Hence, $\gamma_*|a - u_T| \leq \left|\int_0^a \gamma dx - v_T(\gamma')\right| \leq \gamma^*|a - u_T|$. Since γ_*, γ^* are uniform lower and upper bounds on $\sigma_{T/10} * \gamma$ (and hence on $\sigma_{T/10} * \gamma'$), the desired estimates on $\mathbb{I}(a)$ are derived from the bounds on $\mathbb{J}_{\gamma'}(\int_0^a \gamma dx)$. □

The lowerbounds in Theorem 1.6 are implied by the following two estimates.

Lemma 4.2. *Starting from a (DIC) condition with profile γ, there are constants $c_0 = c_0(\gamma, T), c_1 = c_1(\gamma)$ such that for $|a| \geq c_0$ we have*

$$\mathbb{J}(a),\ \mathbb{I}(a) \geq \frac{c_1|a|^3}{T}.$$

Lemma 4.3. *Starting from a (DIC) condition with profile γ, there is a constant $c_1(\gamma)$ such that for $a \in \mathbb{R}$ we have*

$$\mathbb{J}(a) \geq \frac{c_1(a-v_T)^2}{\sqrt{T}} \quad \text{and} \quad \mathbb{I}(a) \geq \frac{c_1(a-u_T)^2}{\sqrt{T}}.$$

Proof of Lemma 4.2. We concentrate first on the current calculation. Suppose $a > 0$, as the argument for $a < 0$ is similar. Let $\hat{\gamma} \in M_1(\rho_*, \rho^*)$ be a smooth density, strictly bounded away from 0 and 1, such that $h(\gamma;\hat{\gamma}) < \infty$. For $\epsilon > 0$, by Proposition 2.1, let μ be a smooth density such that $\mu_0 = \sigma_\alpha * \gamma$, $|h(\mu_0;\hat{\gamma}) - h(\gamma;\hat{\gamma})| < \epsilon$, and $|I_0(\mu) - \mathbb{J}(a)| \leq \epsilon$. Noting Proposition 2.6, we can in addition impose on the approximating density that $\left|\int_0^T J(0,t)dt - a\right| \leq \epsilon$.

Now, noting (2.2) in Proposition 2.4, we have the Lipschitz bound,

$$\Big|\int_0^T J(x,t)dt - \int_0^T J(0,t)dt\Big| = \Big|\int_0^x \mu_T(z) - \mu_0(z)dz\Big| \leq |x|.$$

Then, for $0 \le x \le a-\epsilon$, we have $\int_0^T J(x,t)dt \ge \int_0^T J(0,t)dt - x \ge a-\epsilon-x$ so that

$$\begin{aligned}(a-\epsilon)^3/3 &= \int_0^{a-\epsilon} [a-\epsilon-x]^2 dx \\ &\le \int_0^{a-\epsilon} \Big[\int_0^T J(x,t)dt\Big]^2 dx \;\le\; T\int\int_0^T J^2 dt dx.\end{aligned}$$

Hence, as $\mu(1-\mu), \hat{\gamma}(1-\hat{\gamma}) \le 1/4$, from the formula for $I_0(\mu)$ in Proposition 2.4 and simple computations,

$$\begin{aligned}\mathbb{J}(a) \;\ge\; I_0(\mu)-\epsilon &\ge \int\int_0^T J^2 dtdx - \frac{1}{2}h(\mu_0;\hat{\gamma}) - T\int \frac{(\partial_x\hat{\gamma})^2}{\hat{\gamma}^2(1-\hat{\gamma})^2}dx - \epsilon \\ &\ge \frac{(a-\epsilon)^3}{3T} - \frac{1}{2}h(\gamma;\hat{\gamma}) - T\int \frac{(\partial_x\hat{\gamma})^2}{\hat{\gamma}^2(1-\hat{\gamma})^2}dx - \frac{3}{2}\epsilon. \qquad (4.1)\end{aligned}$$

For the tagged particle rate function, from (1.8), we have $\mathbb{I}_\gamma^{DC}(a) = \mathbb{J}_{\gamma'}^{DC}(\int_0^a \gamma dx)$ where $\gamma'(x) = \gamma(x+a)$. Since $\gamma(x) \ge \min\{\rho_*, \rho^*\}$ for all large $|x|$, $|\int_0^a \gamma dx| \ge c(\gamma)|a|$ for all large $|a|$ where $c(\gamma) > 0$. Also, as $\gamma', \hat{\gamma} \in M_1(\rho_*, \rho^*)$, by calculation $h(\gamma';\hat{\gamma}) = O(|a|)$. Hence, plugging into (4.1), we obtain the desired estimate on $\mathbb{I}(a)$. □

Proof of Lemma 4.3. We focus first on the current rate function computation. By Proposition 2.1 and 2.6, let μ be a smooth density with properties (i)-(viii) such that $\mu_0 = \sigma_\alpha * \gamma$, $|\mathbb{J}(a) - I_0(\mu)| < \epsilon$ and $|\int_0^T J(t,0)dt - a| < \epsilon$. Let $v_T(\alpha)$ be the LLN speed starting from profile $\sigma_\alpha * \gamma$, and note $\lim_{\alpha\downarrow 0}|v_T - v_T(\alpha)| = 0$.

Consider now solutions of the equations $\partial_t\rho = (1/2)\rho_{xx}$ and $\partial_t = (1/2)\partial_{xx}\mu - \partial_x(H_x\mu(1-\mu))$ both with initial value $\sigma_\alpha * \gamma$. The difference $U = \rho - \mu$ satisfies equation $\partial_t U = (1/2)\partial_{xx}U - \partial_x(H_x\mu(1-\mu))$ with $U(0,x) \equiv 0$. Integrating once in the space variable, noting properties of μ, $S(t,x) = \int_{-\infty}^x U(t,y)dy$ satisfies $\partial_t S = (1/2)\partial_{xx}S - H_x\mu(1-\mu)$. Hence, we have

$$\begin{aligned}S(t,x) &= \sigma_t * S(0,x) + \int_0^t\int \frac{1}{\sqrt{2\pi(t-s)}}e^{\frac{-(x-y)^2}{2(t-s)}}[-H_x\mu(1-\mu)](s,y)dyds \\ &= -\int_0^t\int \frac{1}{\sqrt{2\pi t}}e^{\frac{-(x-y)^2}{2t}}H_x\mu(1-\mu)(t-s,y)dyds.\end{aligned}$$

Now, the difference in integrated macroscopic currents across x up to time t with respect to ρ and μ is $-S(t,x)$ (cf. above (1.5)). Therefore, by Schwarz inequality and $0 \le \mu \le 1$, when $x = 0$, we have for small α that

$$(v_T - a + O(\epsilon))^2 \;\le\; \left[\int_0^T \Big(\int \sigma_t^2(y)dy\Big)^{1/2}\Big(\int H_x^2\mu(1-\mu)(y,t)dy\Big)^{1/2} dt\right]^2.$$

Since $\|\sigma_t^2\|_{L^2(\mathbb{R})}^2 \le Ct^{1/2}$, a further upper bound of the right-side is $2C\sqrt{T}I_0(\mu) \le 2C\sqrt{T}(\mathbb{J}(a) + O(\epsilon))$ for some universal constant C.

We now use relations (1.8) to analyze the tagged particle rate function. Indeed, let $u_T(\alpha)$ be the corresponding tagged particle LLN speed starting from profile $\rho_0 = \sigma_\alpha * \gamma$, and note $\lim_{\alpha\downarrow 0} u_T(\alpha) = u_T$. As before, by Proposition 2.1, let μ be a smooth density such that $\mu_0 = \rho_0$, $|\mathbb{I}(a) - I_0(\mu)| < \epsilon$ and by Proposition 2.6, $|\int_0^T J(a,t)dt - \int_0^a \rho_0(x)dx| < \epsilon$.

Note that with respect to density ρ, the current across a equals $\int_a^\infty \sigma_T * \rho_0 - \rho_0 dx$, and the current across the origin equals $\int_0^\infty \sigma_T * \rho_0 - \rho_0 dx = \int_0^{u_T(\alpha)} \sigma_T * \rho_0 dx$. Then, for small α, the square of the difference in integrated currents with respect to ρ and μ across a equals $\left(\int_{u_T}^a \sigma_T * \gamma(x)dx + O(\epsilon)\right)^2 \geq \gamma_*^2(a - u_T + O(\epsilon))^2$ where $\sigma_T * \gamma \geq \gamma_* > 0$. But, on the other hand, as before $\|\sigma_t(\cdot - a)\|^2_{L^2(\mathbb{R})} \leq Ct^{1/2}$, and the square current difference is still bounded by $2C\sqrt{T}I_0(\mu) \leq 2C\sqrt{T}(\mathbb{I}(a) + O(\epsilon))$. This finishes the proof. □

4.2. **Proof of Theorem 1.7.** Starting from a (DIC) state, since $\gamma(x) \equiv \rho$, noting (1.8), we observe that $\gamma'(x) \equiv \rho$, $\int_0^a \gamma dx = a\rho$, and $\mathbb{I}(a) = \mathbb{J}(a\rho)$. Hence, we need only give the argument for the current rate function $\mathbb{J}$, as the estimate for the tagged particle rate function $\mathbb{I}$ follows directly.

We now make some useful reductions. Recall, when starting under a deterministic configuration with profile $\gamma(x) \equiv \rho$, in order for $I_\gamma^{\mathrm{DC}}(\nu) < \infty$, ν must satisfy $\nu_0(x) \equiv \rho$ and $I_\gamma^{\mathrm{DC}}(\nu) = I_0(\nu) < \infty$. By Proposition 2.1 and Proposition 2.6, for each $\epsilon > 0$, we can find a smooth density μ such that $\mu_0(x) \equiv \rho$ and

$$\mathbb{J}(a) \ \geq \ I_0(\mu) - \epsilon\frac{a^2}{\sqrt{T}}.$$

In addition, we may also impose that

$$\Big|\int_0^T J(0,t)dt - a\Big| < a\epsilon.$$

For such a density μ, by Proposition 2.4 (applied with $\gamma(x) \equiv \rho$),

$$I_0(\mu) \ = \ \frac{1}{8}\int\int_0^T \frac{(\partial_x\mu)^2}{\mu(1-\mu)}dtdx + \frac{1}{2}h(\mu_T;\rho) + \frac{1}{2}\int\int_0^T \frac{J^2}{\mu(1-\mu)}dtdx. \tag{4.2}$$

Consider now a sequence $\{\mu^a\}$ of such $\epsilon a^2/\sqrt{T}$-minimizers of $\mathbb{J}(a)$ as $|a| \downarrow 0$. The upper and lower bounds in Lemmas 4.1 and 4.3, as $v_T = 0$, gives $I_0(\mu^a) = O(a^2/\sqrt{T})$. Then, by Lemma 2.3, we have $\mu^a \to \rho$ in $L^2([0,T] \times \mathbb{R})$, and in fact

$$\sup_{0 \leq t \leq T} \int (\mu^a(t,x) - \rho)^2 dx \ = \ O(a^2).$$

We now deduce that there are functions $r(t,x)$ and $j(t,x)$ on $[0,T] \times \mathbb{R}$ such that $r(0,x) \equiv 0$, $\partial_t r + \partial_x j = 0$ weakly in $L^2([0,T] \times \mathbb{R})$, $\int_0^T j(t,0)dt = 1$, and

$$\begin{aligned} &\rho(1-\rho) \times \liminf_{a \downarrow 0} \mathbb{J}(a)/a^2 \\ &\quad \geq \ \frac{1}{8}\int_0^T \int (\partial_x r)^2 dxdt + \frac{1}{4}\int |r(T,x)|^2 dx + \frac{1}{2}\int_0^T \int j^2(t,x)dxdt. \end{aligned} \tag{4.3}$$

Consider a function $\lambda^a(t,x) = \psi(\mu^a(t,x))$ where $\psi'(x) = \min\{(x(1-x))^{-1/2}, M\}$ for some $M \geq 2(\rho(1-\rho))^{-1/2}$. Then, $\partial_x\lambda^a = \psi'(\mu^a)\partial_x\mu^a \leq (\mu^a(1-\mu^a))^{-1/2}\partial_x\mu^a$, and so

$$\int_0^T \int (\partial_x\lambda^a)^2 dxdt \ \leq \ \int_0^T \int \frac{(\partial_x\mu^a)^2}{\mu^a(1-\mu^a)}dxdt.$$

At this point, let us take weak $L^2([0,T] \times \mathbb{R})$ limits of $a^{-1}\partial_x\lambda^a$, $a^{-1}(\mu^a - \rho)$ and $a^{-1}J^a$, and label them as u, r and j respectively. Also, take a weak $L^2(\mathbb{R})$ limit of $a^{-1}(\mu^a(T,x)-\rho)$ and call it q. Using suitable truncations, and Fatou's Lemma, given $\mu^a \to \rho$ strongly, we have

$$\begin{aligned}\int_0^T\int u^2dxdt &\leq \liminf \frac{1}{a^2}\int_0^T\int(\partial_x\lambda^a)^2dxdt\\ \frac{1}{2\rho(1-\rho)}\int|q(x)|^2dx &\leq \liminf\frac{1}{a^2}\int h_d(\mu_T^a(x);\rho)dx\\ \frac{1}{\rho(1-\rho)}\int_0^T\int j^2dxdt &\leq \liminf\frac{1}{a^2}\int_0^T\int\frac{(J^a)^2}{\mu^a(1-\mu^a)}dxdt.\end{aligned}$$

We may identify (a) $\partial_x r = \sqrt{\rho(1-\rho)}u$, (b) $r(T,x)=q(x)$, and (c) $\partial_t r+\partial_x j=0$ weakly in $L^2([0,T]\times\mathbb{R})$. The last two (b-c) follow from weak limits and properties of μ^a. However, (a) also holds given the weak limits since $\partial_x\mu^a = \psi'(\mu^a)^{-1}\partial_x\lambda^a$ and $\psi'(\mu^a)^{-1}\to\sqrt{\rho(1-\rho)}$ strongly in L^2.

Now, define

$$K(t,x) = \int_0^t j(s,x)ds.$$

Then, the right-side of (4.3) becomes

$$\mathcal{K} = \frac14\int|\partial_x K(T,x)|^2dx+\frac12\int_0^T\int|\partial_t K(t,x)|^2dxdt+\frac18\int_0^T\int|\partial_{xx}K(t,x)|^2dxdt.$$

By scaling, $M(t,x)=K(tT,x\sqrt{T})$, we obtain

$$\liminf_{|a|\downarrow 0}\frac{\sqrt{T}}{a^2}\mathbb{J}(a) \geq [\rho(1-\rho)]^{-1}\inf\mathcal{M}$$

where the infimum is over $M\in C^{1,2}([0,1]\times\mathbb{R})$ such that $M(0,x)\equiv 0$ and $M(1,0)=1$, and

$$\mathcal{M} = \frac14\int|M_x(1,x)|^2dx+\frac12\int_0^1\int|M_t(t,x)|^2dxdt+\frac18\int_0^1\int|M_{xx}(t,x)|^2dxdt.$$

On the other hand, the upperbound

$$\limsup_{|a|\downarrow 0}\frac{\sqrt{T}}{a^2}\mathbb{I}(a) \leq [\rho(1-\rho)]^{-1}\inf\mathcal{M} \tag{4.4}$$

also follows by a similar strategy: In Proposition 4.4 below, we evaluate $\inf\mathcal{M}$ and find a minimizer. One can find a smooth ϵ approximating M with bounded derivatives and trace back to obtain the corresponding density μ^a satisfying $a^{-1}(\rho-\mu^a)=\partial_x K$, $a^{-1}J^a=\partial_t K$, $a^{-1}\partial_x\mu^a=\partial_{xx}K$ with $\int_0^T J^a(0,t)dt = a$ and $\mu_0^a(x)\equiv\rho$. Given $\|\partial_x M\|_{L^\infty([0,T]\times\mathbb{R})}<\infty$, we have $\|\mu^a-\rho\|_{L^\infty([0,T]\times\mathbb{R})}\leq |a|\|\partial_x K\|_{L^\infty} = (|a|/\sqrt{T})\|\partial_x M\|_{L^\infty}=O(|a|)$. The argument to derive (4.4) now follows from standard approximations with respect to (4.2).

Hence, the proof of Theorem 1.7 will follow from the evaluations $\inf_M\mathcal{M}=\sqrt{\pi}/2$, $\sigma^2_{X,dyn}=(1-\rho)/(\rho\sqrt{\pi})$ and $\sigma^2_{J,dyn}=\rho(1-\rho)/\sqrt{\pi}$ in Propositions 4.4 and 4.5 below. □

Proposition 4.4. *We have*

$$\inf_M\mathcal{M} = \frac{\sqrt{\pi}}{2}$$

where the infimum is over $M \in C^{1,2}([0,1]\times\mathbb{R})$ such that $M(0,x) \equiv 0$ and $M(1,0) =$ 1.

Proof. The argument is in three steps. (A) We first minimize

$$\int_0^1 \int_{-\infty}^{\infty} \frac{1}{2}|M_t(t,x)|^2 + \frac{1}{8}|M_{xx}(t,x)|^2 \, dxdt \tag{4.5}$$

when $M(0,x) \equiv 0$ and $M(1,x)$ is a given compactly supported $C^4(\mathbb{R})$ function. The Euler equation is

$$M_{tt} = \frac{1}{4}M_{xxxx} \tag{4.6}$$

with the boundary conditions at $t = 0, 1$.

One can verify the solution of (4.6), which is smooth and classical, in terms of Fourier transform with respect to the x variable but not transforming the t variable, is given by

$$\hat{M}(t,y) = \hat{M}(1,y)\frac{e^{\frac{ty^2}{2}} - e^{\frac{-ty^2}{2}}}{e^{\frac{y^2}{2}} - e^{\frac{-y^2}{2}}} \tag{4.7}$$

where

$$\hat{M}(1,y) = \frac{1}{\sqrt{2\pi}} \int e^{iyx} M(1,x) dx.$$

The corresponding value of (4.5), through Plancherel's formula, is expressed as

$$\int_{-\infty}^{\infty} |\hat{M}(1,y)|^2 k(y) dy$$

where

$$\begin{aligned} k(y) &= \int_0^1 \frac{y^4}{8} \frac{\left[e^{\frac{ty^2}{2}} + e^{\frac{-ty^2}{2}}\right]^2 + \left[e^{\frac{ty^2}{2}} - e^{\frac{-ty^2}{2}}\right]^2}{\left[e^{\frac{y^2}{2}} - e^{\frac{-y^2}{2}}\right]^2} dt \\ &= \int_0^1 \frac{y^4}{4} \frac{e^{ty^2} + e^{-ty^2}}{\left[e^{y^2/2} - e^{-y^2/2}\right]^2} dt \\ &= \frac{y^2}{4} \frac{e^{y^2} - e^{-y^2}}{\left[e^{y^2/2} - e^{-y^2/2}\right]^2} = \frac{y^2}{4} \frac{e^{y^2/2} + e^{-y^2/2}}{e^{y^2/2} - e^{-y^2/2}}. \end{aligned}$$

Given the integrand in (4.5) is a strict convex function of M_t and M_{xx}, the solution (4.7) is the unique minimizer of (4.5) (by say straightforward modifications of the proof of [7, Theorem 2.1]).

(B) Now, we consider the term

$$\frac{1}{4}\int_{-\infty}^{\infty} |M_x(1,x)|^2 dx = \frac{1}{4}\int_{-\infty}^{\infty} y^2 |\hat{M}(1,y)|^2 dy$$

and minimize, over $M \in L^2([0,1]\times\mathbb{R})$,

$$\begin{aligned} &\frac{1}{4}\int_{-\infty}^{\infty} |\hat{M}(1,y)|^2 y^2 \Big[1 + \frac{e^{y^2/2} + e^{-y^2/2}}{e^{y^2/2} - e^{-y^2/2}}\Big] dy \\ &= \frac{1}{2}\int_{-\infty}^{\infty} |\hat{M}(1,y)|^2 y^2 \frac{e^{y^2/2}}{e^{y^2/2} - e^{-y^2/2}} dy \end{aligned}$$

subject to

$$\frac{1}{\sqrt{2\pi}}\int_{-\infty}^{\infty}\hat{M}(1,y)dy = 1.$$

Recall that the minimizer of

$$\int |g(y)|^2 K(y)dy$$

subject to $\int g(y)dy = a$ is given by $g(y) = cK(y)^{-1}$ and $c = a[\int K(y)^{-1}dy]^{-1}$, with minimum value $a^2[\int K(y)^{-1}dy]^{-1}$. Hence, with $a = \sqrt{2\pi}$ and

$$K(y) = \frac{y^2}{2}\frac{e^{y^2/2}}{e^{y^2/2} - e^{-y^2/2}},$$

we identify $M(1,y)$ through its transform $\hat{M}(1,y) = cK(y)^{-1}$. Denote $\widetilde{M}$ as the function in (4.7) with this choice of $M(1,y)$.

(C) Let now M^* be a compactly supported $C^{2,4}([0,1]\times\mathbb{R})$ function such that $M^*(0,x)\equiv 0$ and $M^*(1,0) = 1$ which approximates $\inf_M \mathcal{M}$. From steps (A) and (B), we obtain a lower bound of the infimum value which is actually achieved by the smooth $C^{2,4}([0,1]\times\mathbb{R})$ function $\widetilde{M}$. Therefore, $\widetilde{M}$ is a minimizer.

Finally, given

$$\int K(y)^{-1}dy = 2\int\frac{1-e^{-y^2}}{y^2}dy = 2\int_0^1\int e^{-ty^2}dydt = 2\int_0^1\sqrt{\frac{\pi}{t}}dt = 4\sqrt{\pi},$$

we obtain the infimum, $\inf_M \mathcal{M} = 2\pi/(4\sqrt{\pi}) = \sqrt{\pi}/2$ as desired. □

Proposition 4.5. *The dynamical parts of the limiting variances of $T^{-1/4}J_{-1,0}(T)$ and $T^{-1/4}x(T)$ under ν_ρ are*

$$\sigma^2_{J,dyn} := \lim_{T\to\infty}\frac{1}{\sqrt{T}}E_{\nu_\rho}\Big[\Big(J_{-1,0}(T) - E_\eta[J_{-1,0}(T)]\Big)^2\Big] = \frac{\rho(1-\rho)}{\sqrt{\pi}} \quad \text{and}$$

$$\sigma^2_{X,dyn} := \lim_{T\to\infty}\frac{1}{\sqrt{T}}E_{\nu_\rho}\Big[\Big(x(T) - E_\eta[x(T)]\Big)^2\Big] = \frac{1-\rho}{\rho}\frac{1}{\sqrt{\pi}}.$$

Proof. First, we recall the limit distribution and variance of both $T^{-1/4}x(T)$ and $\rho^{-1}T^{-1/4}J_{-1,0}(T)$ are the same, namely $N(0,\sigma^2)$ with $\sigma^2 = \sqrt{2/\pi}(1-\rho)/\rho$ (cf. [1]). Moreover, one may show $\lim_{T\uparrow\infty}E_{\nu_\rho}[(T^{-1/4}X(T) - \rho^{-1}T^{-1/4}J_{-1,0}(T))^2] = 0$ since $(T^{-1/4}X(T) - \rho^{-1}T^{-1/4}J_{-1,0}(T))^2$ vanishes in probability, and also is uniformly integrable (cf. [8][equation (28)], or [1][p. 368] and [26][Proposition 4.2 and proof of Lemma 3.2]).

Then, we need only show

$$\lim_{T\to\infty}\frac{1}{\sqrt{T}}E_{\nu_\rho}\Big[\Big(E_\eta[J_{-1,0}(T)]\Big)^2\Big] = \rho(1-\rho)\frac{\sqrt{2}-1}{\sqrt{\pi}}$$

which given the form of the limiting variance of the scaled current, and

$$E_{\nu_\rho}[(J_{-1,0}(T))^2] = E_{\nu_\rho}[(J_{-1,0}(T) - E_\eta[J_{-1,0}(T)])^2] + E_{\nu_\rho}[(E_\eta[J_{-1,0}(T)])^2]$$

implies the desired resuls.

Now, the current $J_{-1,0}$ has martingale decomposition (cf. Section 2 [26]),

$$J_{-1,0}(t) = M(t) + \frac{1}{2}\int_0^t \eta_s(-1) - \eta_s(0)ds.$$

Also, for $x \in \mathbb{Z}$, from 'duality' (cf. Liggett [23][Section VIII.1 p. 363]),

$$E_\eta[\eta_t(x)] \;=\; \sum_i p(t, i-x)\eta(i)$$

where $p(t,j) = P(S_t = j)$ is the probability a continuous time random walk, starting from the origin, travels to j in time t. Then,

$$\begin{aligned} E_\eta[J_{-1,0}(T)] &= \frac{1}{2}\int_0^T E_\eta[\eta_t(-1)] - E_\eta[\eta_t(0)]dt \\ &= \frac{1}{2}\sum_i \eta(i)\int_0^T p(t,i+1) - p(t,i)dt \\ &= \frac{1}{2}\sum_i (\eta(i)-\rho)\int_0^T p(t,i+1) - p(t,i)dt. \end{aligned}$$

Therefore, from independence of coordinates $\{\eta(i)\}$,

$$Q_0(T) \;:=\; E_{\nu_\rho}\Big[\big(E_\eta[J_{-1,0}(T)]\big)^2\Big] \;=\; \rho(1-\rho)\sum_i \Big|\frac{1}{2}\int_0^T p(t,i+1)-p(t,i)dt\Big|^2.$$

Now, as a priori the variance $Q_0(u) \le E_{\nu_\rho}[J^2_{-1,0}(u)] = O(\sqrt{u})$, we need only find the limit of

$$\begin{aligned} Q_1(T) &= \frac{\rho(1-\rho)}{\sqrt{T}}\sum_i \Big|\frac{1}{2}\int_{\epsilon T}^T p(t,i+1)-p(t,i)dt\Big|^2 && (4.8) \\ &= \frac{\rho(1-\rho)}{4\sqrt{T}}\sum_i \int_{[\epsilon T,T]^2} \big[p(t,i+1)-p(t,i)\big]\big[p(s,i+1)-p(s,i)\big]dsdt \end{aligned}$$

To estimate the integrand, from Doob's inequality, note

$$p(t,x) \;=\; \mathbb{E}[P(S_{N_t}=x)] \;=\; \mathbb{E}[P(S_{N_t}=x), \sup_{t\in[\epsilon T,T]}|N_t/t-1|\le\epsilon] + O(T^{-10}) \quad (4.9)$$

where N_t is a Poisson process with rate 1 independent of the discrete time random walk $\{S_k\}$, $N_t/t-1$ is a martingale, and $\mathbb{E}$ refers to expectation with respect to N_t. Further, (since we could not find an appropriate continuous time version) from the local limit theorem (Petrov [27][Theorem VII.13; p. 205]), uniformly over x, with respect to the discrete time walk, we have for $N_t \ge 1$ that

$$\begin{aligned} P(S_{N_t}=x) &= \frac{1}{\sqrt{2\pi N_t}}e^{-\frac{x^2}{2N_t}} + \frac{1}{\sqrt{2\pi}}e^{-\frac{x^2}{2N_t}}\frac{q_2(x/\sqrt{N_t})}{N_t^{3/2}} + o(N_t^{-3/2}) \\ &= \frac{1}{\sqrt{2\pi N_t}}e^{-\frac{x^2}{2N_t}} + O(N_t^{-3/2}) && (4.10) \end{aligned}$$

where $q_2(y) = (\gamma_4/24\theta^4)(y^4-6y^2+3)$, γ_k is the kth order cumulant and θ^2 is the variance of the symmetric Bernoulli variable. [In our case, in Petrov's formula, $q_1(y) = (\gamma_3/6\theta^3)(y^3-3y) \equiv 0$ as $\gamma_3 = 0$.]

Let $p^N(t,x) = P(S_{N_t}=x)$ and $p^R(s,x) = P(S_{R_s}=x)$ where R_s is an independent Poisson process also with rate 1. We now argue that only the leading terms in (4.9) and (4.10) are significant.

Since $\sum_x p(u,x), \sum_x p^N(u,x) \le 1$, the error term on order $O(T^{-10})$ in (4.9) can be neglected in estimating (4.8). Indeed,

$$\frac{O(T^{-10})}{\sqrt{T}}\int_{[\epsilon T,T]^2}\sum_i p(s,i)dsdt = \frac{O(T^{-10})}{\sqrt{T}}\int_{[\epsilon T,T]^2}\sum_i p^N(t,i)dsdt = o(1).$$

Also, note the error term of order $O(N_t^{-3/2})$ in (4.10) is not significant with respect to (4.8). Indeed,

$$\begin{aligned}\sum_x \frac{1}{\sqrt{2\pi N_t}}\left|e^{-\frac{(x+1)^2}{2N_t}} - e^{-\frac{x^2}{2N_t}}\right| &= \sum_x \frac{1}{\sqrt{2\pi N_t}}\left|e^{-(2x+1)/2N_t} - 1\right| e^{-x^2/2N_t}\\ &\le Ce^{-\sqrt{N_t}/4} + C\sum_{|x|\le N_t^{3/4}} \frac{1}{\sqrt{2\pi N_t}}\frac{|x|}{N_t}e^{-x^2/2N_t} \le \frac{C}{\sqrt{N_t}}\end{aligned}$$

for some constants C. Then, given $|N_t/t-1|, |R_s/s-1| \le \epsilon$ for $s,t\in[\epsilon T,T]$, a product of $\sum_i (2\pi N_t)^{-1/2}|e^{-(i+1)^2/2N_t} - e^{-i^2/2N_t}|$ and the error term with respect to the s-integration, for instance, leads to bounding

$$\begin{aligned}&\frac{1}{\sqrt{T}}\int_{[\epsilon T,T]^2}\sum_i \frac{1}{R_s^{3/2}}\frac{1}{\sqrt{2\pi N_t}}\left|e^{-\frac{(i+1)^2}{2N_t}} - e^{-\frac{i^2}{2N_t}}\right| dsdt\\ &\qquad\le \frac{1}{\sqrt{T}}\int_{[\epsilon T,T]^2}\frac{C}{R_s^{3/2}\sqrt{N_t}}dsdt \le O(T^{-1/2}).\end{aligned}$$

Therefore, focusing on the leading order terms,

$$\begin{aligned}&\frac{1}{4\sqrt{T}}\sum_i\int_{[\epsilon T,T]^2}\left[p^N(t,i+1)-p^N(t,i)\right]\left[p^R(s,i+1)-p^R(s,i)\right]dsdt\\ &= o(1) + \frac{1}{8\pi\sqrt{T}}\sum_i\int_{[\epsilon T,T]^2}\frac{1}{\sqrt{R_sN_t}}\left[e^{-(i+1)^2/2N_t} - e^{-i^2/2N_t}\right]\\ &\qquad\times\left[e^{-(i+1)^2/2R_s} - e^{-i^2/2R_s}\right]dsdt.\end{aligned}$$

Now, using again $|N_t/t-1|, |R_s/s-1| \le \epsilon$ for $s,t\in[\epsilon T,T]$, we further evaluate the integral on the right-side as

$$\begin{aligned}&o(1) + \frac{1}{8\pi\sqrt{T}}\sum_{|i|\le T^{3/4}}\int_{[\epsilon T,T]^2}\frac{i^2}{R_sN_t\sqrt{R_sN_t}}e^{\frac{-i^2}{2}\left[\frac{1}{N_t}+\frac{1}{R_s}\right]}dsdt\\ &= o(1) + \frac{1}{8\pi\sqrt{T}}\int_{[\epsilon T,T]^2}\int_{-\infty}^{\infty}\frac{x^2}{R_sN_t\sqrt{R_sN_t}}e^{\frac{-x^2}{2}\left[\frac{1}{N_t}+\frac{1}{R_s}\right]}dxdsdt\\ &= o(1) + \frac{\sqrt{2}}{8\sqrt{\pi T}}\int_{[\epsilon T,T]^2}(N_t+R_s)^{-3/2}dsdt =: Q_2(T,\epsilon).\end{aligned}$$

Finally, we have $Q_2(T,\epsilon)$ satisfies

$$\lim_{T\uparrow\infty}\left|Q_2(T,\epsilon) - \frac{\sqrt{2}-1}{\sqrt{\pi}}\right| \le c(\epsilon)$$

where $c(\epsilon)$ vanishes as $\epsilon\downarrow 0$. □

4.3. **Proof of Theorem 1.8.** We concentrate on the argument for the tagged particle, as a similar proof holds for the current. By symmetry,

$$P(|X(N^2T)|/N \geq a) \;=\; 2P(X(N^2T)/N \geq a).$$

From (3.1) and noting $J_{-1,0}(t) - J_{\lfloor aN\rfloor,\lfloor aN\rfloor+1}(t) = \sum_{x=0}^{\lfloor aN\rfloor} \eta_t(x) - \eta_0(x)$ by the development of subsection 3.1, we have

$$\{X(N^2t) \geq aN\} \;=\; \left\{ J_{\lfloor aN\rfloor,\lfloor aN\rfloor+1}(N^2t) \geq \sum_{x=0}^{\lfloor aN\rfloor} \eta_0(x) \right\}.$$

We now rewrite currents in terms of the standard Harris stirring process $\{\xi_t^x\}$. Namely, at time $t=0$, a particle is put at each $x \in \mathbb{Z}$. Then, to bonds $(x, x+1)$ in $\mathbb{Z}$, associate independent Poisson clocks with parameter $1/2$. When the clock rings at a bond, interchange the positions of the particles at the bond's vertices. Let ξ_t^x be the position at time t of the particle initially at x. Then, the exclusion process, starting from initial configuration η, satisfies $\eta_t(x) = 1\big\{x \in \{\xi_t^i : \eta(i) = 1\}\big\}$. More details and constructions can be found in Chapter VIII [23].

Then, for $0 \leq a \leq 1$,

$$J_{\lfloor aN\rfloor,\lfloor aN\rfloor+1}(N^2t) \;=\; \sum_{x\leq\lfloor aN\rfloor} \eta_0(x)1_{[\xi^x_{N^2t}>\lfloor aN\rfloor]} - \sum_{x>\lfloor aN\rfloor} \eta_0(x)1_{[\xi^x_{N^2t}\leq\lfloor aN\rfloor]}.$$

Write, given the initial profile η_0 is deterministic, by Chebychev, that

$$\begin{aligned}\frac{1}{N}\log P\left(X(N^2t) \geq aN\right) \;&\leq\; \frac{1}{N}\log E\exp\left\{-\lambda\sum_{x=0}^{\lfloor aN\rfloor}\eta_0(x)\right\} \\ &\quad+\frac{1}{N}\log E\exp\left\{\lambda J_{\lfloor aN\rfloor,\lfloor aN\rfloor+1}(N^2t)\right\}.\end{aligned} \tag{4.11}$$

The first term on the right-side tends to $-\lambda a$ as $N\uparrow\infty$. The second term is bounded, by Chebychev, Liggett [23][Proposition VIII.1.7] noting $e^{\alpha\sum_{i=k}^{l}1_{[x_i\in A]}}$ is positive definite for any $\alpha\in\mathbb{R}$, and $\log(1+x)\leq x$ for $x\geq 1$, by

$$\begin{aligned}&\frac{1}{2N}\log E\exp\left\{2\lambda\sum_{x\leq\lfloor aN\rfloor}\eta_0(x)1_{[\xi^x_{N^2t}>\lfloor aN\rfloor]}\right\} \\ &\qquad+\frac{1}{2N}\log E\exp\left\{-2\lambda\sum_{x>\lfloor aN\rfloor}\eta_0(x)1_{[\xi^x_{N^2t}\leq\lfloor aN\rfloor]}\right\} \\ &\leq\;\frac{1}{2N}\sum_{x\leq\lfloor aN\rfloor}(e^{2\lambda\eta_0(x)}-1)P\left(\xi^x_{N^2t}>\lfloor aN\rfloor\right) \\ &\qquad+\frac{1}{2N}\sum_{x>\lfloor aN\rfloor}(e^{-2\lambda\eta_0(x)}-1)P\left(\xi^x_{N^2t}\leq\lfloor aN\rfloor\right).\end{aligned}$$

Given $\eta_0(x) = 1_{[|x|\leq N]}$ and $\xi^x_{N^2t}$ marginally is the position of a simple random walk started at x at time N^2t, as $N\uparrow\infty$, we have

$$\frac{1}{2N}\sum_{x\leq\lfloor aN\rfloor}(e^{2\lambda\eta_0(x)}-1)P\left(\xi^x_{N^2t}>\lfloor aN\rfloor\right) \;\to\; \frac{e^{2\lambda}-1}{2}\int_{-1}^{a}P(N(0,t)>a-x)dx$$

and

$$\frac{1}{2N}\sum_{x>\lfloor aN\rfloor}(e^{-2\lambda\eta_0(x)}-1)P\left(\xi^x_{N^2t}\le\lfloor aN\rfloor\right)\ \to\ \frac{e^{-2\lambda}-1}{2}\int_a^1 P(N(0,t)\le a-x)dx$$

where $N(0,t)$ is a Normal distribution with mean 0 and variance t.

Hence, combining the estimates, we have (4.11) is less than

$$-\lambda a+\frac{e^{2\lambda}-1}{2}\int_{-1}^{a}P(N(0,t)>a-x)dx\ +\ \frac{e^{-2\lambda}-1}{2}\int_a^1 P(N(0,t)\le a-x)dx.$$

Choosing $\lambda=\epsilon a$ for small $\epsilon>0$, we obtain further that (4.11) is bounded by

$$-\epsilon a^2\left[1-\frac{1}{a}\int_{1-a}^{1+a}P(N(0,t)>y)dy\right]+O(\epsilon^2a^2)\ \le\ -Ca^2$$

for a constant C noting $1>a^{-1}\int_{1-a}^{1+a}P(N(0,t)>y)dy$ for $0<a\le 1$.

For $a\ge 1$, we write

$$J_{\lfloor aN\rfloor,\lfloor aN\rfloor+1}(t)\ =\ \sum_{|x|\le N}\eta_0(x)1_{[\xi^x_{N^2t}>\lfloor aN\rfloor]}.$$

Then, as above,

$$\begin{aligned}P\left(X(N^2t)\ge aN\right)\ &\le\ e^{-\lambda\sum_{x=0}^{\lfloor aN\rfloor}\eta_0(x)}E\exp\left\{\lambda\sum_{|x|\le N}\eta_0(x)1_{[\xi^x_{N^2t}>\lfloor aN\rfloor]}\right\}\\ &\le\ e^{-\lambda N}\prod_{|x|\le N}E\exp\left\{\lambda 1_{[\xi^x_{N^2t}>\lfloor aN\rfloor]}\right\}.\end{aligned}$$

Taking logarithm, dividing by N, and taking the limit, we obtain

$$\begin{aligned}&\limsup_{N\uparrow\infty}\frac{1}{N}\log P\left(X(N^2t)\ge aN\right)\\ &\quad\le\ -\lambda+\limsup_{N\uparrow\infty}\frac{1}{N}\sum_{|x|\le N}(e^\lambda-1)P(\xi^x_{N^2t}>\lfloor aN\rfloor)\\ &\quad\le\ -\lambda+(e^\lambda-1)\int_{-1}^{1}P(N(0,t)\ge a-x)dx.\end{aligned}$$

Optimizing on λ, the right-side of the above display is bounded by

$$\log\int_{-1}^{1}P(N(0,t)>a-x)dx+1-\int_{-1}^{1}P(N(0,t)>a-x)dx\ <\ 0.$$

However, for a large, this expression is bounded by $-Ca^2$.

Working with the $0\le a\le 1$ and $a>1$ bounds, we obtain the desired quadratic order estimate. □

5. Proofs of Approximations

We give the proofs of Propositions 1.3 and 2.1, and Lemmas 2.2 and 2.3.

5.1. **Proofs of Propositions 1.3 and 2.1.** The proofs are through a series of lemmas inspired by the scheme in [20] (see also Oelschläger [25], and Bertini-Landim-Mourragui [5]). As several of the steps are different, we give some details.

To this end, let μ be a density such that $I_0(\mu) < \infty$. The first lemma states that finite rate densities μ, when integrated against smooth test functions, are uniformly continuous in time (cf. Lemma 4.4 [5]).

Lemma 5.1. *Let $\eta \in D([0,T]; M_1)$ be a density such that $I_0(\eta) < \infty$, and let $\mathfrak{J} \in C_K^2(\mathbb{R})$. Then, $s \mapsto \langle \eta_s, \mathfrak{J}\rangle = \int \mathfrak{J}(x)\eta_s(x)dx$ is a uniformly continuous function.*

Proof. Let $G \in C_K^{1,2}([0,T] \times \mathbb{R})$. As $I_0(\eta) < \infty$, from (1.3), we infer

$$l^2(\eta; G) \ \le\ 2I_0(\eta) \int_0^T \int G_x^2(t,x)\eta_t(x)(1-\eta_t(x))dxdt.$$

Let F^δ be a smooth approximations of the indicator $1_{[s,t]}(u)$. Then, by applying the previous inequality with $G^\delta = F^\delta \mathfrak{J}$, we obtain

$$\begin{aligned}\left|\int \eta_t \mathfrak{J} dx - \int \eta_s \mathfrak{J} dx\right| \ &=\ \lim_{\delta\downarrow 0}\left\{ l(\eta; G^\delta) + \frac{1}{2}\int_0^T \int G^\delta_{xx}\eta_u(x)dxdu\right\} \\ &\le\ |t-s|\|\mathfrak{J}''\|_{L^1} + \sqrt{2I_0(\eta)}|t-s|^{1/2}\|\mathfrak{J}'\|_{L^2},\end{aligned}$$

finishing the proof. □

For the remainder of the subsection, let $\mu \in D([0,T]; M_1)$ be a density with finite rate, $I_0(\mu) < \infty$. We now build a succession of approximating densities in the next lemmas with special properties.

Lemma 5.2. *For each $\epsilon > 0$, there exists a density $\hat{\mu}$, smooth in the space variable, such that (1) the Skorohod distance $d(\hat{\mu}; \mu) < \epsilon$, (2) there is $0 < \delta_\epsilon < 1$ such that $\delta_\epsilon < \hat{\mu}(t,x) < 1 - \delta_\epsilon$ for $(t,x) \in [0,T] \times \mathbb{R}$, and (3) $|I_0(\hat{\mu}) - I_0(\mu)| < \epsilon$.*

In addition, (4) if $\hat{\gamma} \in M_1(\rho_*, \rho^*)$ is piecewise continuous, $0 < \hat{\gamma}(x) < 1$ for $x \in \mathbb{R}$, and $h(\mu_0; \hat{\gamma}) < \infty$, then also $|h(\hat{\mu}_0; \hat{\gamma}) - h(\mu_0; \hat{\gamma})| < \epsilon$.

Proof. For $0 < \rho_*, \rho^* < 1$, let $\gamma \in M_1(\rho_*, \rho^*)$ be a function. Consider

$$\mu^{b,\alpha} \ =\ \sigma_{t+\alpha} * \gamma + b(\sigma_\alpha * \mu - \sigma_{t+\alpha} * \gamma) \tag{5.1}$$

for $0 \le b \le 1$ and $\alpha \ge 0$. Clearly, $\mu^{b,\alpha}$ is smooth in the space variable when $\alpha > 0$.

Next, for fixed $\alpha > 0$ and $0 < b < 1$, there is $0 < \delta < 1$ such that $\delta < \mu^{b,\alpha} < 1-\delta$ as $\sigma_{t+\alpha} * \gamma$ is strictly bounded between 0 and 1 for $t \in [0,T]$.

Now, $\mu^{b,\alpha} \to \mu^{1,\alpha}$ as $b \uparrow 1$ in $D([0,T]\times\mathbb{R})$. Also, noting $\lim_{\alpha\downarrow 0} \|\sigma_\alpha * G - G\|_{L^1(\mathbb{R})} = 0$ for $G \in L^1(\mathbb{R})$, we have also have the Skorohod convergence $\mu^{1,\alpha} \to \mu$ as $\alpha \downarrow 0$.

By lower semi-continuity of I_0,

$$\liminf I_0(\mu^{b,\alpha}) \ \ge\ I_0(\mu).$$

On the other hand, by convexity of $I_0(\nu)$, we have

$$I_0(\mu^{b,\alpha}) \ \le\ (1-b)I_0(\sigma_{t+\alpha} * \gamma) + bI_0(\sigma_\alpha * \mu).$$

Note that $I_0(\sigma_{t+\alpha} * \gamma) = 0$, and by translation-invariance and convexity, the right-side in the display is less than

$$b\int \sigma_\alpha(y) I_0(\mu(t, x-y))dy \ =\ bI_0(\mu) \ \uparrow\ 1 \ \text{ as } b \uparrow 1.$$

Similarly, if $\hat{\gamma} \in M_1(\rho_*, \rho^*)$ is piecewise continuous, $0 < \hat{\gamma} < 1$, and $h(\mu_0; \hat{\gamma}) < \infty$, then by lower semi-continuity and convexity of $h(\cdot; \hat{\gamma})$ we have

$$h(\mu_0; \hat{\gamma}) \leq \liminf_{b\uparrow 1, \alpha\downarrow 0} h(\mu_0^{b,\alpha}; \hat{\gamma}) \quad \text{and} \quad h(\mu_0^{b,\alpha}; \hat{\gamma}) \leq (1-b)h(\sigma_\alpha * \gamma; \hat{\gamma}) + bh(\sigma_\alpha * \mu_0; \hat{\gamma}).$$

Also, once more by convexity,

$$h(\sigma_\alpha * \mu_0; \hat{\gamma}) \ \leq \ \int dy \sigma_\alpha(y) \int dx h_d(\mu_0(x); \hat{\gamma}(x-y)).$$

The right-side, since $|h(\mu_0(\cdot); \hat{\gamma}(\cdot - y)) - h(\mu_0; \hat{\gamma})| \leq C|y|$ by properties of $\hat{\gamma}$, converges to $h(\mu_0; \hat{\gamma})$ as $\alpha \downarrow 0$. By the same argument, $\lim_{\alpha\downarrow 0} h(\sigma_\alpha * \gamma; \hat{\gamma}) = h(\gamma; \hat{\gamma})$. Hence, $\lim_{b\uparrow 1, \alpha\downarrow 0} h(\mu_0^{b,\alpha}; \hat{\gamma}) = h(\mu_0; \hat{\gamma})$.

Therefore, statements (1)-(4) hold for $\hat{\mu} = \mu^{b,\alpha}$ when $b \sim 1$, $\alpha \sim 0$. □

Lemma 5.3. *Let $\hat{\mu}$ be the density constructed in Lemma 5.2. Then, (1) for each $\epsilon > 0$, there exists a smooth density $\tilde{\mu}$ such that $\tilde{\mu}_0 = \hat{\mu}_0$, (2) the Skorohod distance $d(\tilde{\mu}; \hat{\mu}) < \epsilon$, and (3) $|I_0(\tilde{\mu}) - I_0(\hat{\mu})| < \epsilon$. Also, (4) all partial derivatives of $\tilde{\mu}$ are uniformly bounded in $[0,T] \times \mathbb{R}$.*

Proof. To obtain a smooth density, we need only approximate $\hat{\mu}$ by smoothing in the time variable. Define for $\beta > 0$ a density which is constant in time on a short time interval:

$$\nu^\beta(t,x) \ = \ \begin{cases} \hat{\mu}_0(x) & \text{for } 0 \leq t < \beta \\ \hat{\mu}(t-\beta, x) & \text{for } \beta \leq t \leq T + \beta. \end{cases}$$

Let $\kappa_\varepsilon \in C_K^\infty(\mathbb{R})$ be smooth approximations of the identity in $L^1(\mathbb{R})$ such that $\kappa_\varepsilon \geq 0$, $\int \kappa_\varepsilon(x)dx = 1$, $\text{Supp}(\kappa_\varepsilon) \subset (0, \varepsilon)$ and for $f \in L^1(\mathbb{R})$, $f * \kappa_\varepsilon \to f$ as $\varepsilon \downarrow 0$ in L^1. Form the convolution, for $0 < \varepsilon \leq \beta$,

$$\nu^{\beta,\varepsilon}(t,x) \ = \ \int_0^T \nu^\beta(t+s, x)\kappa_\varepsilon(s)ds.$$

It is clear, by continuity of $\hat{\mu}$ in time (Lemma 5.1), that $\lim_{\beta\downarrow 0}\lim_{\varepsilon\downarrow 0} \nu^{\beta,\varepsilon} = \hat{\mu}$ in $D([0,T]; M_1)$. By construction, $\nu^{\beta,\varepsilon}$ is smooth, and also $\nu_0^{\beta,\varepsilon} = \hat{\mu}_0$.

From lower semi-continuity and convexity

$$\liminf_{\beta,\varepsilon} I_0(\nu^{\beta,\varepsilon}) \geq I_0(\hat{\mu}), \quad \text{and} \quad I_0(\nu^{\beta,\varepsilon}) \leq \int_0^T \kappa_\varepsilon(s) I_0(\nu^\beta(t+s,x))ds.$$

Using the variational definition of I_0, noting $\hat{\mu}_0 = \sigma_\alpha * (\gamma + b(\mu_0 - \gamma))$, the rate of ν^β on the interval $[0,\beta]$ is bounded by

$$\sup_{G\in C_K^{1,2}} \frac{1}{2}\int_0^\beta \int G_x\partial_x\hat{\mu}_0 - G_x^2\hat{\mu}_0(1-\hat{\mu}_0)dxdt \ \leq \ \frac{\beta}{8}\int \frac{(\partial_x\hat{\mu}_0)^2}{\hat{\mu}_0(1-\hat{\mu}_0)}dx$$

which vanishes as $\beta \downarrow 0$. On the other hand, by the formula (1.4), the rate of ν^β on the interval $[\beta, T]$ converges to $I_0(\hat{\mu})$ as $\beta \downarrow 0$. We can conclude then that $\lim_{\beta,\varepsilon\downarrow 0} I_0(\nu^{\beta,\varepsilon}) \to I_0(\hat{\mu})$.

Moreover, by differentiating the convolutions, since $\|\nu^{\beta,\varepsilon}\|_{L^\infty} \leq 1$, we have $\|\partial_x^{(k)}\partial_t^{(l)}\tilde{\mu}\|_{L^\infty} \leq \|\partial_x^{(k)}\sigma_\alpha\|_{L^1}\|\partial_t^{(l)}\kappa_\varepsilon\|_{L^1} < \infty$.

Hence, to find the desired density, we can take $\tilde{\mu} = \nu^{\beta,\varepsilon}$ for β, ε small. □

We now continue to adjust the approximation so that the associated function 'H_x' of the approximating density has desired properties.

Lemma 5.4. *Let $\tilde{\mu}$ be the density constructed in Lemma 5.3, and let $\tilde{H}_x$ be associated to it via (1.2). Then, (1) $\tilde{H}_x \in C^\infty([0,T] \times \mathbb{R})$, (2) $\|\tilde{H}_x\|_{L^2([0,T]\times\mathbb{R})} < \infty$, and (3) $\|\tilde{H}_x\|_{L^\infty([0,T]\times\mathbb{R})} < \infty$.*

Proof. By construction, we recall, for a $\delta > 0$, that $\delta < \tilde{\mu} < 1 - \delta$, $\tilde{\mu}$ is smooth with uniformly bounded derivatives on $[0,T] \times \mathbb{R}$ of all orders, and

$$\partial_t \tilde{\mu} \;=\; \frac{1}{2}\partial_{xx}\tilde{\mu} - \partial_x\big[\tilde{H}_x\tilde{\mu}(1-\tilde{\mu})\big]. \tag{5.2}$$

Then, as $2I_0(\tilde{\mu}) = \int_0^T \int (\tilde{H}_x)^2\tilde{\mu}(1-\tilde{\mu})dxdt < \infty$, we obtain the L^2 bound on $\tilde{H}_x$ and, by solving for $\tilde{H}_x$ in (5.2), we have that $\tilde{H}_x$ is smooth.

We now deduce that $\tilde{H}_x$ is bounded in L^∞. This bound will follow from the L^2 bound on $\tilde{H}_x\tilde{\mu}(1-\tilde{\mu})$ and $\delta < \tilde{\mu} < 1-\delta$, if we show that $\tilde{H}_x\tilde{\mu}(1-\tilde{\mu})$ is Lipschitz in both space and time variables with uniform constant over $[0,T] \times \mathbb{R}$. However, from (5.2), $\tilde{H}_x\hat{\mu}(1-\hat{\mu})$ is Lipschtiz in the space variable with uniform constant as $\partial_t\tilde{\mu}$ and $\partial_{xx}\tilde{\mu}$ are bounded on $[0,T] \times \mathbb{R}$.

To show $\tilde{H}_x\tilde{\mu}(1-\tilde{\mu})$ is also Lipschitz in the time variable t uniformly over $[0,T] \times \mathbb{R}$, write

$$\tilde{H}_x\tilde{\mu}(1-\tilde{\mu})(x,t) = \tilde{H}_x\tilde{\mu}(1-\tilde{\mu})(0,t) + (1/2)\partial_x\tilde{\mu}(x,t) - (1/2)\partial_x\tilde{\mu}(0,t) - \int_0^x \partial_t\tilde{\mu}dy.$$

The first three terms on the right-side are clearly uniformly Lipschitz in t as their partial derivatives in time are bounded on $[0,T]$.

To treat the last term, consider a smooth G compactly supported in $[-\epsilon, x+\epsilon]$ which equals 1 on $[0,x]$. Since $\partial_{tt}\tilde{\mu}$ is bounded, we have

$$\Big|\int_0^x \partial_{tt}\tilde{\mu}(u,y)dy\Big| \;\le\; \Big|\int G(y)\partial_{tt}\tilde{\mu}(u,y)dy\Big| + 2C\epsilon.$$

Now, by construction in the proof of Lemma 5.3, $\tilde{\mu} = \kappa_\varepsilon * \nu^\beta$, and so

$$\int G(y)\partial_{tt}\tilde{\mu}(u,y)dy \;=\; \int_0^T \int G(y)\kappa_\varepsilon''(s)\nu^\beta(u+s,y)dyds.$$

As $I_0(\nu^\beta) < \infty$, we can associate via (1.2) an H_x^β to the density ν^β. From the weak formulation (1.3), and $\kappa_\varepsilon'(0) = \kappa_\varepsilon'(T) = 0$, the right-side equals

$$-\frac{1}{2}\int_0^T \int G''(y)\kappa_\varepsilon'(s)\nu^\beta(u+s,y)dyds$$
$$- \int_0^T \int G'(y)\kappa_\varepsilon'(s)H_x^\beta\nu^\beta(1-\nu^\beta)(u+s,y)dyds.$$

The first integral, because ν^β is bounded and $G' \neq 0$ on a set of width at most 2ϵ is uniformly bounded in time u and space x. Similarly, the second integral, as $\|H_x^\beta\nu^\beta(1-\nu^\beta)\|_{L^2} \le 2I_0(\nu^\beta) < \infty$, is also both uniformly bounded in u and x. □

The function $\tilde{H}_x$ associated to $\tilde{\mu}$ in Lemma 5.4, although smooth, does not necessarily have compact support. Let $H_x^m \in C_K^\infty((0,T] \times \mathbb{R})$ be smooth approximations of $\tilde{H}_x$ with the following properties:

$$\|H_x^m - \tilde{H}_x\|_{L^2([0,T]\times\mathbb{R})} \le m^{-1}, \quad \text{and} \quad \sup_{\substack{t\in[0,T]\\ x\in[-m,m]}} |H_x^m - \tilde{H}_x| \le m^{-1}.$$

Denote $w^m \in D([0,T];M_1)$ as the smooth density with initial condition $w_0^m = \tilde{\mu}_0$ which satisfies the equation

$$\partial_t v \;=\; (1/2)\partial_{xx} v - \partial_x(H_x^m v(1-v)).$$

Existence, for instance, follows from the hydrodynamic limit for weakly asymmetric exclusion processes in [19] using the replacement estimates Theorem 6.1 and Claims 1,2 [22][Section 6]; see also Theorem 3.1 [20]. Uniqueness in the class of bounded solutions follows by the method of Proposition 3.5 [25].

We now show that w^m, whose associated function $H_x^m \in C_K^\infty$, is close to $\tilde{\mu}$. Hence, w^m will turn out to be a suitable candidate with respect to Proposition 2.1. In addition, we will be able to deduce that $\mu \in \mathcal{A}$ under (LEM) initial distributions.

Lemma 5.5. *The sequence w^m converges uniformly to $\tilde{\mu}$ on compact subsets of $[0,T]\times\mathbb{R}$, and hence in $D([0,T];M_1)$. Also, $I_0(w^m)\to I_0(\tilde{\mu})$.*

Proof. Suppose that we have proven $w^m \to \tilde{\mu}$ uniformly on compact subsets. As $\|H_x^m - \hat{H}_x\|_{L^2}\to 0$, we would then conclude $I_0(w^m)\to I_0(\tilde{\mu})$. In the following, the constant C may change line to line.

Now, given $\partial_t\sigma_t(x) = (1/2)\partial_{xx}\sigma_t(x)$, we have, for $t,h>0$, that

$$\begin{aligned}\sigma_h * w_t^m(y) &- \sigma_{t+h} * w_0^m(y)\\ &= \int_0^t\int H_x^m w^m(1-w^m)(s,z)\frac{-(z-y)}{t+h-s}\sigma_{t-s+h}(z-y)dzds.\end{aligned}$$

By properties of w^m, H_x^m and $(|z-y|/\sqrt{u})\exp(-(z-y)^2/4u)\le 1$,

$$\begin{aligned}|H_x^m w^m(1-w^m)|(s,z)&\frac{|z-y|}{t+h-s}\sigma_{t+h-s}(z-y)\\ &\le\; C|t-s|^{-1/2}\sigma_{2T}(z-y)\;\in\; L^1([0,t]\times\mathbb{R}).\end{aligned}$$

Hence, taking $h\downarrow 0$, we obtain

$$w_t^m(y) \;=\; \sigma_t * w_0^m(y) + \int_0^t\int H_x^m w^m(1-w^m)(s,z)\frac{-(z-y)}{t-s}\sigma_{t-s}(z-y)dzds. \quad (5.3)$$

The equation (5.3) also holds with respect to $\tilde{\mu}$.

Let now $|y|\le m/2$. We have then, using again $(|z-y|/\sqrt{u})\exp(-(z-y)^2/4u)\le 1$, and $w_0^m = \tilde{\mu}_0$, that

$$\begin{aligned}|w_t^m(y) - \tilde{\mu}_t(y)| \;&\le\; \sigma_t * |w_0^m - \tilde{\mu}_0|(y) \qquad (5.4)\\ &+\int_0^t\int \big|H_x^m w^m(1-w^m) - \hat{H}_x\tilde{\mu}(1-\tilde{\mu})\big|(s,z)\frac{|z-y|}{t-s}\sigma_{t-s}(z-y)dzds\\ &\le C\int_0^t\int \big|H_x^m w^m(1-w^m) - \tilde{H}_x\tilde{\mu}(1-\tilde{\mu})\big|(s,z)(t-s)^{-1/2}\sigma_{2(t-s)}(z-y)dzds.\end{aligned}$$

We now estimate the last line in two parts, noting

$$\begin{aligned}\big|H_x^m w^m(1-w^m) &- \tilde{H}_x\tilde{\mu}(1-\tilde{\mu})\big|(s,z)\\ &\le\; |H_x^m - \tilde{H}_x|(s,z) + |\tilde{H}_x|\big|\tilde{\mu}(1-\hat{\mu}) - w^m(1-w^m)\big|(s,z).\end{aligned}$$

The first part, noting $|y| \leq m/2$, by properties of H_x^m, $\|\sigma_t(x)\|_{L^2([0,T]\times\mathbb{R})} \leq CT^{1/4}$, and $\sup_{t\in(0,T]} t^{-1/2}\sigma_{4t}(1) \leq C$, is bounded for large m, as

$$\begin{aligned}
&\int_0^t \int |H_x^m - \tilde{H}_x|(s,z)(t-s)^{-1/2}\sigma_{2(t-s)}(z-y)dzds \\
&\leq \int_0^t \int_{|z|\geq m} |H_x^m - \tilde{H}_x|(s,z)(t-s)^{-1/2}\sigma_{2(t-s)}(z-y)dzds + m^{-1}\sqrt{t} \\
&\leq C\int_0^t \int_{|z|\geq m} |H_x^m - \tilde{H}_x|(s,z)\sigma_{4(t-s)}(z-y)dzds + m^{-1}\sqrt{t} \\
&\leq Cm^{-1}T^{1/4} + m^{-1}\sqrt{T}.
\end{aligned}$$

The second part is decomposed as the sum of three terms,

$$\begin{aligned}
&\int_0^t \int |\tilde{H}_x||\tilde{\mu}(1-\tilde{\mu}) - w^m(1-w^m)|(s,z)(t-s)^{-1/2}\sigma_{2(t-s)}(z-y)dzds \\
&\qquad = D_1 + D_2 + D_3
\end{aligned}$$

where D_1, D_2, D_3 is the integral over $[0,t]\times\{|z|\geq m/2+\epsilon\}$, $[0,t]\times\{m/2\leq |z|\leq m/2+\epsilon\}$ and $[0,t]\times\{|z|\leq m/2\}$ respectively, for $\epsilon > 0$.

The term D_1, noting $\sup_{t\in(0,T]} t^{-1/2}\sigma_{4t}(\epsilon) \leq C_\epsilon$, is bounded by

$$\begin{aligned}
&2\int_0^t \int_{|z|\geq m/2+\epsilon} |\tilde{H}_x|(s,z)(t-s)^{-1/2}\sigma_{2(t-s)}(z-y)dzds \\
&\qquad \leq C(\epsilon,T)\|\tilde{H}_x\|_{L^2([0,T]\times\{z:|z|\geq m/2\})}.
\end{aligned}$$

The second term D_2 is bounded by

$$\begin{aligned}
&2\int_0^t \int_{m/2\leq|z|\leq m/2+\epsilon} |\tilde{H}_x||t-s|^{-1/2}\sigma_{2(t-s)}(z-y)dzds \\
&\qquad \leq C\|\tilde{H}_x\|_{L^\infty}\int_0^t s^{-3/4}\int_0^\epsilon s^{-1/4}e^{-z^2/4s}dzds \\
&\qquad \leq C\|\tilde{H}_x\|_{L^\infty}T^{1/4}\sqrt{\epsilon}.
\end{aligned}$$

The third term D_3 is bounded, with respect to a $\tau \geq t$, by

$$\begin{aligned}
&2\int_0^t \int_{|z|\leq m/2} |\tilde{H}_x||\tilde{\mu} - w^m|(s,z)|t-s|^{-1/2}\sigma_{2(t-s)}(z-y)dzds \\
&\qquad \leq 2\sqrt{t}\|\tilde{H}_x\|_{L^\infty}\sup_{\substack{|z|\leq m/2\\ s\leq\tau}}|\tilde{\mu}_s(z) - w_s^m(z)|. \qquad (5.5)
\end{aligned}$$

Hence, for $\tau > 0$ small enough but fixed, which satisfies $2\sqrt{\tau}\|\tilde{H}_x\|_{L^\infty} = 1/2$, or $\tau = (16\|\tilde{H}_x\|_{L^\infty})^{-1}$, and $L < m/2$, we have

$$\begin{aligned}
&\sup_{\substack{|z|\leq L\\ t\leq\tau}}|\tilde{\mu}(z,t) - w^m(z,t)| \leq \sup_{\substack{|z|\leq m/2\\ t\leq\tau}}|\tilde{\mu}(z,t) - w^m(z,t)| \\
&\quad \leq C(T)m^{-1} + 2C(\epsilon,T)\|\tilde{H}_x\|_{L^2([0,T]\times\{z:|z|\geq m/2\})} + 2C\|\tilde{H}_x\|_{L^\infty}T^{1/4}\sqrt{\epsilon}.
\end{aligned}$$

Here, we absorbed the right-side of (5.5) into the left-side above.

We may repeat the same scheme, starting from time τ, where now the initial difference (5.4) is taken into account:

$$\begin{aligned}\sup_{|y|\leq m/3} \sigma_t * |w^m_\tau - \tilde{\mu}_\tau|(y) &\leq \sup_{|z|\leq m/2} |w^m_\tau - \tilde{\mu}_\tau|(z) + \sup_{|y|\leq m/3}\int_{|z|>m/2} \sigma_t(y-z)dz \\ &\leq \sup_{|z|\leq m/2} |w^m_\tau - \tilde{\mu}_\tau|(z) + e^{-Cm^2/T}.\end{aligned}$$

With a finite number of iterations of such type, say $N_\tau = [T/\tau]+1$ iterations, when $L < m/N_\tau$, we obtain uniform convergence, as $m \uparrow \infty$, for $|z| \leq L$ and $0 \leq s \leq T$. □

Proof of Proposition 2.1. The proof follows by applying Lemmas 5.2, 5.3, and 5.5 to build a density $\mu^+ = w^m$ which satisfies specifications (i)-(viii). We remark property (v) is shown as follows: When $\mu_0 = \gamma$, by construction in (5.1), we have $w^m_0 = \tilde{\mu}_0 = \hat{\mu}_0 = \mu_0^{b,\alpha} = \sigma_\alpha * \gamma$. When $\gamma(x) \equiv \rho$, this reduces to $\tilde{\mu}_0(x) \equiv \rho$. □

Starting under (DIC) initial conditions, however, to prove Proposition 1.3, we will need to specify that w^m can be approximated by a suitable density with initial value equal to $\mu_0 = \gamma \in M_1(\rho_*, \rho^*)$.

Lemma 5.6. *Recall w^m from Lemma 5.5. Suppose $\mu_0 = \gamma \in M_1(\rho_*, \rho^*)$. Then, for $\epsilon > 0$, $\exists M$ such that $\forall m \geq M$, there is a density $\bar{\chi} \in C^\infty((0,T]\times\mathbb{R})$ such that (1) equation (1.2) is satisfied with respect to $\bar{H}_x \in C_K^\infty([0,T]\times\mathbb{R})$, (2) initial value $\bar{\chi}_0 = \gamma$, (3) the Skorohod distance $d(\bar{\chi}, w^m) < \epsilon$, and (4) $|I_0(\bar{\chi}) - I_0(w^m)| < \epsilon$.*

Proof. Consider w^m_0 from Lemma 5.5. From the assumption $\mu_0 = \gamma$, we have $w^m_0 = \tilde{\mu}_0 = \sigma_\alpha * \gamma$ from (5.1). Form the density $\bar{\chi}$ as follows.

$$\bar{\chi} = \begin{cases} \sigma_t * \gamma & \text{for } 0 \leq t \leq \alpha \\ w^m_{t-\alpha} & \text{for } \alpha \leq t \leq T. \end{cases}$$

Since H^m_x is supported on a compact subset of $(0,T]\times\mathbb{R}$, $\bar{\chi} \in C^\infty((0,T]\times\mathbb{R})$, and satisfies (1.2) with respect to $\bar{H}_x \in C_K^\infty([0,T]\times\mathbb{R})$ given by

$$\bar{H}_x = \begin{cases} 0 & \text{for } (t,x) \in [0,\alpha]\times\mathbb{R} \\ H^m_x(t-\alpha, x) & \text{for } (t,x) \in [\alpha, T]\times\mathbb{R}. \end{cases}$$

Now, $2I_0(\bar{\chi}) = \int_0^T \int \bar{H}_x^2 \bar{\chi}(1-\bar{\chi})dxdt = \int_0^{T-\alpha}\int (H^m_x)^2 w^m(1-w^m)dxdt$. Then, the difference

$$2I_0(\bar{\chi}) - 2I_0(w^m) = \int_{T-\alpha}^T \int (H^m_x)^2 w^m (1-w^m)dxdt.$$

To estimate the right-side, recall from Lemma 5.5 that $\|H^m_x - \tilde{H}_x\|_{L^2} \leq m^{-1}$, and $w^m \to \tilde{\mu}$ uniformly on compact subsets. Then,

$$\begin{aligned}&\int_{T-\alpha}^T \int (H^m_x)^2 w^m(1-w^m)dxdt \\ &\leq 2\|H^m_x - \tilde{H}_x\|^2_{L^2([0,T]\times\mathbb{R})} + 2\int_0^T \int_{|x|\geq L} \tilde{H}_x^2 dxdt \\ &\quad +4\int_0^T\int_{|x|\leq L} \tilde{H}_x^2|w^m - \tilde{\mu}|dxdt + 2\int_{T-\alpha}^T \int \tilde{H}_x^2 \tilde{\mu}(1-\tilde{\mu})dxdt \\ &= B_1 + B_2 + B_3 + B_4.\end{aligned}$$

Choose $L = L(\tilde{H}_x)$ large so that $B_2 \le \epsilon/4$, and take $m = m(\tilde{H}_x, L)$ large enough so that both $B_1, B_3 \le \epsilon/4$.

The term $B_4/4$ is the rate of $\tilde{\mu}$ on the time interval $[T-\alpha, T]$. Since $\tilde{\mu}$ and $\tilde{H}_x$ depend on α, we bound B_4 in terms of H_x (which does not depend on α) to show that it is small when α is small. By the construction of $\tilde{\mu}$ in Lemma 5.3, convexity of the the rate, translation-invariance, and that the rate of $\sigma_{t+\alpha} * \gamma$ vanishes, we estimate

$$
\begin{aligned}
B_4 &\le 2b \int \sigma_\alpha(z) \int_0^T \kappa_\varepsilon(s) \int_{T-\alpha}^T \int H_x^2 \mu(1-\mu)(t+s-\beta, x-z) dx dt ds dz \\
&\le 2 \int_{T-2\alpha}^T \int H_x^2 \mu(1-\mu) dx dt,
\end{aligned}
$$

when $\beta \le \alpha \le T-\alpha$. Then, as $I_0(\mu) < \infty$, $B_4 \downarrow 0$ as $\alpha \downarrow 0$.

Hence, with α small enough, there is M so that for $m \ge M$, we have $|I_0(\bar{\chi}) - I_0(w^m)| < \epsilon$. Also, by Lemma 5.5, $I_0(w^m) \le I_0(\mu)+1$, and so by uniform continuity (Lemma 5.1), the Skorohod distance $d(\bar{\chi}; w^m) < \epsilon$. □

Proof of Proposition 1.3. Let γ be a profile associated to an (LEM) or (DIC) measure, and let μ be such that $I_\gamma(\mu) < \infty$. By successively applying Lemmas 5.2, 5.3, 5.5 and 5.6, we can approximate μ by an appropriate density μ^+ to verify $\mu \in \mathcal{A}$. Specifically, under an (LEM) initial measure, when $I_\gamma(\mu) = I_\gamma^{LE}(\mu)$, $\mu^+ = w^m$ in Lemma 5.5 with appropriate choice of parameters $b, \alpha, \beta, \varepsilon$ and m. Under a (DIC) initial configuration, when $I_\gamma(\mu) = I_\gamma^{DC}(\mu)$, $\mu^+ = \bar{\chi}$ in Lemma 5.6 again with suitable parameters. □

5.2. **Proof of Lemmas 2.2, 2.3.** We prove the lemmas in succession.

Proof of Lemma 2.2. Note that

$$
\begin{aligned}
|\mu_t(x) - \hat{\gamma}(x)| &= \lim_{h\downarrow 0} |\sigma_h * (\mu_t - \hat{\gamma})(x)| \\
&\le \lim_{h\downarrow 0} |\sigma_{t+h} * (\mu_0 - \hat{\gamma})(x)| + \int_0^t \int |H_x| \mu(1-\mu)(s,z) |\partial_z \sigma_{t-s+h}(z-x)| dz ds.
\end{aligned}
$$

Since H_x has compact support in $[0,T] \times \mathbb{R}$, the second term on the right-side is bounded by

$$
C_H \int_0^t \int_{|z|\le M_H} \frac{|x-z|}{t-s} \sigma_{t-s}(z-x) dz ds
$$

for some constants C_H, M_H. Since $(|y|/\sqrt{s}) e^{-y^2/4s} \le 1$, when $|x| \ge M_H$, we can bound it further by $C_H \sqrt{T} e^{-(x-M_H)^2/8T}$ which vanishes as $|x| \uparrow \infty$.

The first term however is bounded as follows:

$$
\begin{aligned}
&|\sigma_{t+h} * (\mu_0 - \hat{\gamma})(x)| \\
&\le \sup_{|z|\le l} |\mu_0 - \hat{\gamma}|(z-x) \cdot \int_{|z|\le l} \sigma_t(z) dz + \sqrt{2} e^{-l^2/4T} \int_{|z|\ge l} \sigma_{2t}(z) dz \\
&\le \sup_{|z|\le l} |\mu_0 - \hat{\gamma}|(z-x) + \sqrt{2} e^{-l^2/4T}.
\end{aligned}
$$

Now, since $h(\mu_0; \hat{\gamma}) < \infty$, $\hat{\gamma} \in M_1(\rho_*, \rho^*)$, $\|\partial_x \mu_0\|_{L^\infty} < \infty$, we conclude for fixed l that $\lim_{|x|\uparrow\infty} \sup_{|z|\le l} |\mu_0 - \hat{\gamma}|(z-x) = 0$. This finishes the proof. □

Proof of Lemma 2.3. Consider Hellinger's inequality $(\sqrt{\alpha} - \sqrt{\beta})^2 \leq h_d(\alpha;\beta)$. [Let $H(\alpha;\beta) = (\sqrt{\alpha} - \sqrt{\beta})^2 + (\sqrt{1-\alpha} - \sqrt{1-\beta})^2$. By Jensen inequality and $\log(1-x) \leq -x$ for $0 \leq x < 1$, $h_d(\alpha;\beta) \geq -2\log[1-(1/2)H(\alpha;\beta)] \geq H(\alpha;\beta)$.] We write then $(\mu_t(x) - \rho)^2 \leq 2(\sqrt{\mu} - \sqrt{\rho})^2 \leq 2h_d(\mu_t(x);\rho)$. Hence, by Proposition 2.4 (with respect to the density on $[0,t] \times \mathbb{R}$ with $\hat{\gamma} \equiv \rho$),

$$\begin{aligned} \int (\mu_t(x) - \rho)^2 dx &\leq 2\int h_d(\mu_t(x);\rho)dx \\ &\leq 4\int_0^t \int (H_x^n)^2 \mu(1-\mu) dx ds \leq 8I_0(\mu) \end{aligned}$$

uniformly in $0 \leq t \leq T$. □

Acknowledgements. We would like to thank Claudio Landim, K. Ravishankar, and Ellen Saada for helpful conversations. Thanks also to the referee for constructive comments with respect to extensions at end of Section 1, and in other parts.

References

[1] Arratia, R. (1983) The motion of a tagged particle in the simple symmetric exclusion system on $\mathbb{Z}^1$. *Ann. Probab.* **11** 362-373.

[2] Benois, O.; Landim, C.; Kipnis, C. (1995) Large deviations from the hydrodynamical limit of mean zero asymmetric zero range processes. *Stoch. Proc. Appl.* **55** 65-89.

[3] Bertini, L.; De Sole, A.; Gabrielli, D.; Jona-Lasinio, G.; Landim, C. (2006) Large deviations of the empirical current in interacting particle systems. *Teor. Veroyatn. Primen.* **51** 144–170; translation in *Theory Probab. Appl.* **51** 2–27

[4] Bertini, L.; De Sole, A.; Gabrielli, D.; Jona-Lasinio, G.; Landim, C. (2006) Large deviation approach to non equilibrium processes in stochastic lattice gases. *Bull. Braz. Math. Soc. (N.S.)* **37** 611-643.

[5] Bertini, L.; Landim, C.; Mourragui, M. (2009) Dynamical large deviations for the boundary driven weakly asymetric exclusion process. *Ann. Probab.* **37** 2357-2403.

[6] Borcea, J.; Branden, P.; Liggett, T.M. (2009) Negative dependence and the geometry of polynomials. *J. American Mathematical Society* **22** 521-567.

[7] Dacorogna, B. (2009) *Introduction to the calculus of variations.* Second Ed. Imperial College Press, London.

[8] De Masi, A.; Ferrari, P.A. (2002) Flux fluctuations in the one dimensional nearest neighbors symmetric simple exclusion process. *J. Stat. Phys.* **107** 677-683.

[9] Dembo, A.; Zeitouni, O. (1998) *Large Deviations Techniques and Applications.* Springer-Verlag, New York.

[10] Derrida, B.; Gerschenfeld, A. (2009) Current fluctuations of the one dimensional symmetric simple exclusion process with step initial condition. *J. Stat. Phys.* **136** 1-15.

[11] Derrida, B.; Gerschenfeld, A. (2009) Current fluctuations in one dimensional diffusive systems with a step initial density profile *J. Stat. Phys.* **137** 978-1000.

[12] Farfan, J.; Landim, C.; Mourragui, M. (2009) Hydrostatics and dynamical large deviations of boundary driven gradient symmetric exclusion. Available at arXiv.org 0903.5526v1.pdf

[13] Ferrari, P.L.; Spohn, H. (2006) Scaling limit for the space-time covariance of the stationary totally asymmetric simple exclusion process. *Commun. Math. Phys.* **265** 1-44.

[14] Hurtado, P.I.; Garrido, P.L. (2010) Large fluctuations of the macroscopic current in diffusive systems: A numerical test of the additivity principle *Phys. Rev. E* **81** 041102

[15] Grigorescu, I. (2007) Large deviations for a catalytic Fleming-Viot branching system. *Commun. Pure Appl. Math.* **60** 1056-1080.

[16] Jara, M.; Landim, C. (2006) Nonequilibrium central limit theorem for a tagged particle in symmetric simple exclusion. *Ann. Inst. Henri Poincare Prob. et Statistiques* **42** 567-577.

[17] Kipnis, C.; Landim, C. (1999) *Scaling limits of interacting particle systems.* Grundlehren der Mathematischen Wissenschaften [Fundamental Principles of Mathematical Sciences], 320. Springer-Verlag, Berlin.

[18] Kipnis, C., Varadhan, S. R. S. (1986) Central limit theorem for additive functionals of reversible Markov processes. *Commun. Math. Phys.* **104** 1-19.

[19] Kipnis, C.; Olla, S.; Varadhan, S.R.S. (1989) Hydrodynamics and large deviation for the simple exclusion process. *Comm. Pure. Appl. Math.* **42** 115-137.

[20] Landim, C. (1992) Occupation time large deviations for the symmetric simple exclusion process. *Ann. Probab.* **20** 206-231.

[21] Landim, C.; Olla, S.; Varadhan, S. R. S. (2004) Symmetric simple exclusion process: regularity of the self-diffusion coefficient. *Comm. Math. Phys.* **224** 307-321.

[22] Landim, C; Yau, H.T. (1995) Large deviations of interacting particle systems in infinite volume. *Comm. Pure Appl. Math.* **48** 339-379.

[23] Liggett, T. M. (1985) *Interacting Particle Systems* Grundlehren der Mathematischen Wissenschaften **276** Springer-Verlag, New York.

[24] Liggett, T. M. (2009) Distributional limits for the symmetric exclusion process. Stochastic Process. Appl. **119** 115.

[25] Oelschläger, K (1985) A law of large numbers for moderately interacting diffusion processes. *Z. Wahrsch. Verw. Gebiete* **69** 279-322.

[26] Peligrad, M., Sethuraman, S. (2008) On fractional Brownian motion limits in one dimensional nearest-neighbor symmetric simple exclusion. *ALEA Lat. Am. J. Prob. Stat.* **4** 245-255.

[27] Petrov, V.V. (1975) *Sums of Independent Random Variables.* Ergebnisse der Mathematik und ihrer Grenzgebiete, **82** Springer-Verlag, New York.

[28] Praehofer, M.; Spohn, H. (2002) Current fluctuations for the totally asymmetric simple exclusion process. In and out of equilibrium (Mambucaba, 2000), *Progr. Probab.*, **51**, 185-204, Birkhuser Boston, Boston, MA,

[29] Quastel, J. (1995) Large deviations from a hydrodynamic scaling limit for a nongradient system. *Ann. Probab.* **23** 724-742.

[30] Quastel, J., Rezakhanlou, F., Varadhan, S.R.S. (1999) Large deviations for the symmetric simple exclusion process in dimensions $d \geq 3$. *Probab. Theory Related Fields* **113** 1-84.

[31] Rezakhanlou, F. (1994) Propagation of chaos for symmetric simple exclusions. *Comm. Pure Appl. Math.* **47** 943-957.

[32] Rost, H.; Vares, M.E. (1985) Hydrodynamics of a one dimensional nearest neighbor model. *Contemp. Math.* **41** 329-342.

[33] Saada, E. (1987) A limit theorem for the position of a tagged particle in a simple exclusion process. *Ann. Probab.* **15, #1** 375-381.

[34] Sasamoto, T. (2008) Exact results for the 1D asymmetric exclusion process and KPZ fluctuations. *Eur. Phys. J. B* **64** 373-377.

[35] Seppäläinen, T. (1998) Coupling the totally asymmetric simple exclusion process with a moving interface. *Markov Proc. Rel. Fields* **4** 593-628.

[36] Seppäläinen, T. (2008) *Translation Invariant Exclusion Processes.* Manuscript available at http://www.math.wisc.edu/ seppalai/excl-book/etusivu.html

[37] Spitzer, F. (1970) Interaction of Markov processes. *Adv. Math.* **5** 246-290.

[38] Spohn, H. (1991) *Large Scale Dynamics of Interacting Particles.* Springer-Verlag, Berlin.

[39] Vandenberg-Rodes, A. (2010) A limit theorem for particle current in the symmetric exclusion process. *Elec. Comm. Probab.* **15** 240-252.

Department of Mathematics, 396 Carver Hall, Iowa State University, Ames, IA 50011
E-mail: sethuram@iastate.edu

Courant Institute, 212 Mercer St., New York, NY 10012
E-mail: texttvaradhan@cims.nyu.edu

List Of Publications of S.R.S. Varadhan

Books

1. *Stochastic Processes.* S. R. S. Varadhan, Courant Institute of Mathematical Sciences, 1968.
2. *Mathematical Statistics.* S. R. S. Varadhan (notes by M. Levandowsky and N. Rubin), Courant Institute of Mathematical Sciences, 1974.
3. *Multidimensional Diffusion Processes.* D. W. Stroock and S. R. S. Varadhan, Grundlehren der Mathematischen Wissenschaften, 233.Springer, 1979.
4. *Lectures On Diffusion Problems And Partial Differential Equations.* S. R. S. Varadhan (notes by P. Muthuramalingam and T. R. Nanda), Tata Institute of Fundamental Research, Bombay, 1980.
5. *Large Deviations And Applications.* S. R. S. Varadhan, CBMS-NSF Regional Conference Series in Applied Mathematics, 46, SIAM, 1984.
6. *Probability Theory.* S. R. S. Varadhan, Courant Lecture Notes in Mathematics, 7, American Mathematical Society, 2001.
7. *Stochastic Processes.* S. R. S. Varadhan, Courant Lecture Notes in Mathematics, 16, American Mathematical Society, 2007.

Books Edited

1. *Topics In Probability Theory.* Eds. D. W. Stroock and S. R. S. Varadhan, Courant Institute of Mathematical Sciences, 1973.
2. *Ergodic Theory.* Eds. J. Moser, E. Phillips and S. R. S. Varadhan, Courant Institute of Mathematical Sciences, 1975.
3. *Itô, Kiyosi: Selected Papers.* Eds. S. R. S. Varadhan and D. W. Stroock, Springer, 1987.
4. *Colloque Paul Lévy Sur Les Processus Stochastiques.* Eds. M. Métivier, J. Neveu and S. R. S. Varadhan, *Astérisque No. 157-158 (1988).* Société Mathèmatique de France, 1988.
5. *Probability Theory And Applications.* Eds. E. P. Hsu and S. R. S. Varadhan, *IAS/Park City Mathematics Series, 6.* American Mathematical Society, 1999.

6. *Differential Equations: La Pietra 1996.* Eds. M. Giaquinta, J. Shatah and S. R. S. Varadhan, *Proceedings of Symposia in Pure Mathematics, 65.* American Mathematical Society, 1999.
7. *Hydrodynamic Limits And Related Topics.* Eds. S. Feng, A. T. Lawniczak and S. R. S. Varadhan, *Fields Institute Communications, 27.* American Mathematical Society, 2000.

Articles

1. K. R. Parthasarathy, R. Ranga Rao and S. R. S. Varadhan. *On the category of indecomposable distributions on topological groups.* Trans. Amer. Math. Soc. 102 (1962) 200 – 217.
2. S. R. S. Varadhan. *Limit theorems for sums of independent random variables with values in a Hilbert space.* Sankhya Ser. A 24 (1962) 213 – 238.
3. K. R. Parthasarathy, R. Ranga Rao and S. R. S. Varadhan. *Probability distributions on locally compact abelian groups.* Illinois J. Math. 7 (1963) 337 – 369.
4. K. R. Parthasarathy and S. R. S. Varadhan. *Extension of stationary stochastic processes.* Teor. Verojatnost. i Primenen. 9 (1964) 72 – 78.
5. S. R. S. Varadhan. *Limit theorems in probability.* Math. Student 32 (1964) 17 – 21.
6. S. R. S. Varadhan. *Asymptotic probabilities and differential equations.* Comm. Pure Appl. Math. 19 (1966) 261 – 286.
7. S. R. S. Varadhan. *On the behavior of the fundamental solution of the heat equation with variable coefficients.* Comm. Pure Appl. Math. 20 (1967) 431 – 455.
8. S. R. S. Varadhan. *Diffusion processes in a small time interval.* Comm. Pure Appl. Math. 20 (1967) 659 – 685.
9. D. W. Stroock and S. R. S. Varadhan. *Diffusion processes with continuous coefficients - I.* Comm. Pure Appl. Math. 22 (1969) 345 – 400.
10. D. W. Stroock and S. R. S. Varadhan. *Diffusion processes with continuous coefficients - II.* Comm. Pure Appl. Math. 22 (1969) 479 – 530.
11. D. W. Stroock and S. R. S. Varadhan. *Diffusion processes with boundary conditions.* Comm. Pure Appl. Math. 24 (1971) 147 – 225.
12. D. W. Stroock and S. R. S. Varadhan. *Diffusion processes and martingales - I.* Lecture Notes in Math. 190, Springer, 1971, 60 – 66.
13. D. W. Stroock and S. R. S. Varadhan. *Diffusion processes and martingales - II.* Lecture Notes in Math. 190, Springer, 1971, 67 – 75.
14. D. W. Stroock and S. R. S. Varadhan. *On degenerate elliptic-parabolic operators of second order and their associated diffusions.* Comm. Pure Appl. Math. 25 (1972) 651 – 713.

15. D. W. Stroock and S. R. S. Varadhan. *On the support of diffusion processes with applications to the strong maximum principle.* Proceedings of the Sixth Berkeley Symposium on Mathematical Statistics and Probability, Vol. III: Probability theory, Univ. California Press, 1972, 333 – 359.
16. D. W. Stroock and S. R. S. Varadhan. *Diffusion processes.* Proceedings of the Sixth Berkeley Symposium on Mathematical Statistics and Probability, Vol. III: Probability theory, Univ. California Press, 1972, 361 – 368.
17. G. Papanicolaou and S. R. S. Varadhan. *A limit theorem with strong mixing in Banach space and two applications to stochastic differential equations.* Comm. Pure Appl. Math. 26 (1973) 497 – 524.
18. D. W. Stroock and S. R. S. Varadhan. *Limit theorems for random walks on Lie groups.* Sankhya Ser. A 35 (1973) 277 – 294.
19. S. R. S. Varadhan. *Strassen's version of the law of the iterated logarithm.* Topics in Probability Theory, Courant Institute of Mathematical Sciences, 1973, 15 – 31.
20. D. W. Stroock and S. R. S. Varadhan. *Martingales - I, II, III.* Topics in Probability Theory, Courant Institute of Mathematical Sciences, 1973, 113 – 161.
21. D. W. Stroock and S. R. S. Varadhan. *Probability theory and the strong maximum principle.* Partial differential equations, Proc. Sympos.Pure Math., Vol. XXIII, Amer. Math. Soc., 1973, 215 – 220.
22. D. W. Stroock and S. R. S. Varadhan. *A probabilistic approach to $H^p(R^d)$.* Trans. Amer. Math. Soc. 192 (1974) 245 – 260.
23. M. D. Donsker and S. R. S. Varadhan. *On a variational formula for the principal eigenvalue for operators with maximum principle.* Proc. Nat. Acad. Sci. U.S.A. 72 (1975) 780 – 783.
24. M. D. Donsker and S. R. S. Varadhan. *Asymptotic evaluation of certain Markov process expectations for large time - I.* Comm. Pure Appl. Math. 28 (1975) 1 – 47.
25. M. D. Donsker and S. R. S. Varadhan. *Asymptotic evaluation of certain Markov process expectations for large time - II.* Comm. Pure Appl. Math. 28 (1975) 279 – 301.
26. M. D. Donsker and S. R. S. Varadhan. *Asymptotic evaluation of certain Wiener integrals for large time.* Functional Integration and its Applications, Clarendon Press, 1975, 15 – 33.
27. M. D. Donsker and S. R. S. Varadhan. *Asymptotics for the Wiener sausage.* Comm. Pure Appl. Math. 28 (1975) 525 – 565.

28. M. D. Donsker and S. R. S. Varadhan. *Erratum: "Asymptotics for the Wiener sausage"*. Comm. Pure Appl. Math. 28 (1975) 677 – 678.

29. M. D. Donsker and S. R. S. Varadhan. *Large deviations for Markov processes and the asymptotic evaluation of certain Markov process expectations for large times.* Lecture Notes in Math. 451, Springer, 1975, 82 – 88.

30. M. D. Donsker and S. R. S. Varadhan. *On some problems of large deviations for Markov processes.* Bull. Inst. Internat. Statist. 46 (1975) 409 – 416.

31. M. D. Donsker and S. R. S. Varadhan. *Asymptotic evaluation of certain Markov process expectations for large time - III.* Comm. Pure Appl. Math. 29 (1976) 389 – 461.

32. M. D. Donsker and S. R. S. Varadhan. *On the principal eigenvalue of second-order elliptic differential operators.* Comm. Pure Appl. Math. 29 (1976) 595 – 621.

33. S. R. S. Varadhan. *Book Review: "Stochastic differential equations and applications" by A. Friedman.* Bull. Amer. Math. Soc. 82 (1976), 842 – 846.

34. M. D. Donsker and S. R. S. Varadhan. *On laws of the iterated logarithm for local times.* Comm. Pure Appl. Math. 30 (1977) 707 – 753.

35. G. C. Papanicolaou, D. W. Stroock and S. R. S. Varadhan. *Martingale approach to some limit theorems.* Papers from the Duke Turbulence Conference, Duke Univ., 1977.

36. M. D. Donsker and S. R. S. Varadhan. *Some problems of large deviations.* Symposia Mathematica, Vol. XXI, Academic Press, 1977, 313 – 318.

37. M. D. Donsker and S. R. S. Varadhan. *On the number of distinct sites visited by a random walk.* Proc. Internat. Conf., Northwestern Univ., Evanston, III., Academic Press, 1978, 57 – 62.

38. M. D. Donsker and S. R. S. Varadhan. *On the principal eigenvalue of elliptic second order differential operators.* Proceedings of the International Symposium on Stochastic Differential Equations, Wiley, 1978, 41 – 47.

39. M. D. Donsker and S. R. S. Varadhan. *On the number of distinct sites visited by a random walk.* Comm. Pure Appl. Math. 32 (1979) 721 – 747.

40. K. L. Chung and S. R. S. Varadhan. *Kac functional and Schrödinger equation.* Studia Math. 68 (1980) 249 – 260.

41. M. D. Donsker and S. R. S. Varadhan. *A law of the iterated logarithm for total occupation times of transient Brownian motion.* Comm. Pure Appl. Math. 33 (1980) 365 – 393.

42. G. Papanicolaou and S. R. S. Varadhan. *Diffusion in regions with many small holes.* Lecture Notes in Control and Inform. Sci. no. 25, 1980, 190 – 206.
43. S. R. S. Varadhan. *Some problems of large deviations.* Proceedings of the International Congress of Mathematicians (Helsinki, 1978), Acad. Sci. Fennica, 1980, 755 – 762.
44. M. D. Donsker and S. R. S. Varadhan. *The polaron problem and large deviations.* Phys. Rep. 77 (1981) 235 – 237.
45. M. D. Donsker and S. R. S. Varadhan. *Some problems of large deviations.*. Lecture Notes in Control and Inform. Sci. no. 36, Springer, 1981, 41 –46.
46. G. Papanicolaou and S. R. S. Varadhan. *Boundary value problems with rapidly oscillating random coefficients.* Random Fields, Vol. II, Math. Soc. János Bolyai 27, 1981, 835 – 873.
47. G. Papanicolaou and S. R. S. Varadhan. *Diffusions with random coefficients.* Statistics and Probability: Essays in honor of C. R. Rao, North-Holland, 1982, 547 – 552.
48. S. R. S. Varadhan. *Large deviations.* Advances in filtering and optimal stochastic control, Lecture Notes in Control and Inform. Sci. 42, Springer, 1982, 382 – 392.
49. M. D. Donsker and S. R. S. Varadhan. *Asymptotic evaluation of certain Markov process expectations for large time - IV.* Comm. Pure Appl. Math. 36 (1983) 183 – 212.
50. M. D. Donsker and S. R. S. Varadhan. *Asymptotics for the polaron.* Comm. Pure Appl. Math. 36 (1983) 505 – 528.
51. S. R. S. Varadhan. *Random walks among random scatterers.* Lecture Notes in Control and Inform. Sci. 49, Springer, 1983, 282 – 283.
52. M. D. Donsker and S. R. S. Varadhan. *Large deviations for stationary Gaussian processes.* Comm. Math. Phys. 97 (1985) 187 – 210.
53. M. D. Donsker and S. R. S. Varadhan. *Large deviations for stationary Gaussian processes.* Lecture Notes in Control and Inform. Sci. no. 69, Springer, 1985, 108 – 112.
54. C. Kipnis and S. R. S. Varadhan. *Central limit theorems for additive functionals of reversible Markov chains and applications.* Astérisque 132 (1985) 65 – 70.
55. G. Papanicolaou and S. R. S. Varadhan. *Ohrnstein-Uhlenbeck process in a random potential.* Comm. Pure Appl. Math. 38 (1985) 819 – 834.
56. S. R. S. Varadhan. *Large deviations and applications.* Expo. Math. 3 (1985) 251 – 272.
57. S. R. S. Varadhan and R. J. Williams. *Brownian motion in a wedge with oblique reflection.* Comm. Pure Appl. Math. 38 (1985) 405 – 443.

58. C. Kipnis and S. R. S. Varadhan. *Central limit theorem for additive functionals of reversible Markov processes and applications to simple exclusions.* Comm. Math. Phys. 104 (1986) 1 – 19.

59. A. S. Sznitman and S. R. S. Varadhan. *A multidimensional process involving local time.* Probab. Theory Related Fields 71 (1986) 553 – 579.

60. S. R. S. Varadhan. *Stochastic differential equations - large deviations.* Phénomènes critiques, systèmes aléatoires, théories de jauge, Part I, II, North-Holland, 1986, 625 – 678.

61. M. D. Donsker and S. R. S. Varadhan. *Large deviations for noninteracting infinite-particle systems.* J. Statist. Phys. 46 (1987) 1195 – 1232.

62. M. Z. Guo, G. C. Papanicolaou and S. R. S. Varadhan. *Nonlinear diffusion limit for a system with nearest neighbor interaction.* Comm. Math. Phys. 118 (1988) 31 – 59.

63. S. R. S. Varadhan. *Large deviations and applications* (1988), Lecture notes in Math. 1362, Springer, 1988, 1 – 49.

64. C. Kipnis, S. Olla and S. R. S. Varadhan. *Hydrodynamics and large deviation for simple exclusion processes.* Comm. Pure Appl. Math. 42 (1989) 115 – 137.

65. J. A. Gubner, B. Gopinath and S. R. S. Varadhan. *Bounding functions of Markov processes and the shortest queue problem.* Adv. In Appl. Probab. 21 (1989) 842 – 860.

66. M. D. Donsker and S. R. S. Varadhan. *Large deviations from a hydrodynamic scaling limit.* Comm. Pure Appl. Math. 42 (1989) 243 – 270.

67. S. R. S. Varadhan. *On the derivation of conservation laws for stochastic dynamics.* Analysis, et cetera, Academic Press, 1990, 677 – 694.

68. J. R. Baxter, N. C. Jain and S. R. S. Varadhan. *Some familiar examples for which the large deviation principle does not hold.* Comm. Pure Appl. Math. 44 (1991) 911 – 923.

69. S. R. S. Varadhan. *Scaling limits for interacting diffusions.* Comm. Math. Phys. 135 (1991) 313 – 353.

70. S. Olla and S. R. S. Varadhan. *Scaling limit for interacting Ornstein-Uhlenbeck processes.* Comm. Math. Phys. 135 (1991) 355 – 378.

71. S. R. S. Varadhan. *Book Review: "Large deviations" by J. D. Deuschel and D. W. Stroock.* Bull. Amer. Math. Soc. (N.S.) 24 (1991), 448 – 451.

72. S. R. S. Varadhan. *Entropy methods in hydrodynamical scaling.* Mathematical Physics, X, Springer, 1992, 103 – 112.

73. H. Berestycki, L. Nirenberg and S. R. S. Varadhan. *État fondamental et principe du maximum pour les opérateurs elliptiques du second ordre dans des domaines généraux.* [*The ground state and maximum principle*

for second-order elliptic operators in general domains], C. R. Acad. Sci. Paris Sér, I Math 317, 1993, 51 –56.

74. S. Olla, S. R. S. Varadhan and H. -T. Yau. *Hydrodynamical limit for a Hamiltonian system with weak noise.* Comm. Math. Phys. 155 (1993) 523 – 560.
75. S. R. S. Varadhan. *Entropy methods in hydrodynamic scaling.* Lecture Notes in Math. no. 1551, Springer, 1993,112 – 145.
76. S. R. S. Varadhan. *Nonlinear diffusion limit for a system with nearest neighbor interactions - II.* Pitman Res. Notes Math. Ser., 283, Longman Sci. Tech., 1993, 75 – 128.
77. S. R. S. Varadhan. *Relative entropy and hydrodynamic limits.* Stochastic processes, A Festschrift in honor of Gopinath Kallianpur, Springer, 1993, 329 – 336.
78. H. Berestycki, L. Nirenberg and S. R. S. Varadhan. *The principal eigenvalue and maximum principle for second-order elliptic operators in general domains.* Comm. Pure Appl. Math. 47 (1994) 47 – 92.
79. T. Digernes, V. S. Varadarajan and S. R. S. Varadhan. *Finite approximations to quantum systems.* Rev. Math. Phys. 6 (1994) 621 – 648.
80. S. R. S. Varadhan. *Regularity of self-diffusion coefficient.* The Dynkin Festschrift, Progr. Probab. no.34, Birkhäuser, 1994, 387 – 397.
81. S. R. S. Varadhan. *Self-diffusion of a tagged particle in equilibrium for asymmetric mean zero random walk with simple exclusion.* Ann. Inst. H. Poincaré Probab. Statist. 31 (1995) 273 – 285.
82. S. R. S. Varadhan. *The work of Pierre-Louis Lions.* Proceedings of the International Congress of Mathematicians, Vol. 1 (Zürich, 1994), Birkhäuser, 1995, 6 – 10.
83. S. R. S. Varadhan. *Entropy methods in hydrodynamic scaling.* Proceedings of the International Congress of Mathematicians, Vol. 1 (Zürich, 1994), Birkhäuser, 1995, 196 – 208.
84. C. Landim, S. Sethuraman and S. R. S. Varadhan. *Spectral gap for zero-range dynamics.* Ann. Probab. 24 (1996) 1871 – 1902.
85. A. F. Ramirez and S. R. S. Varadhan. *Relative entropy and mixing properties of interacting particle systems.* J. Math. Kyoto Univ. 36 (1996) 869 – 875.
86. S. R. S. Varadhan. *The complex story of simple exclusion.* Itô's Stochastic Calculus and Probability Theory, Springer, 1996, 385 – 400.
87. S. R. S. Varadhan. *Nongradient models in hydrodynamic scaling.* Analysis, Geometry and Probability, Text Read. Math. 10, Hindustan Book Agency, 1996,397 – 416.

88. J. Quastel and S. R. S. Varadhan. *Diffusion semigroups and diffusion processes corresponding to degenerate divergence form operators.* Comm. Pure Appl. Math. 50 (1997) 667 – 706.

89. S. R. S. Varadhan. *The work of Pierre-Louis Lions.* Fields Medallists' Lectures, World Sci. Publ., 1997, 555 – 559.

90. S. R. S. Varadhan and H.-T. Yau. *Diffusive limit of lattice gas with mixing conditions.* Asian J. Math. 1 (1997) 623 – 678.

91. S. R. S. Varadhan and H.-T. Yau. *Scaling limits for lattice gas models.* Lecture Notes in Pure and Appl. Math. 186, Dekker, 1997, 39 – 43.

92. J. Quastel, F. Rezakhanlou and S. R. S. Varadhan. *Large deviations for the symmetric simple exclusion process in dimensions* $d \geq 3$. Probab. Theory Related Fields 113 (1999) 1 – 84.

93. S. R. S. Varadhan. *Large deviations for interacting particle systems.* Perplexing problems in probability, Progr.Probab. no. 44, Birkhäuser, 1999, 373 – 383.

94. C. Landim, S. Olla and S. R. S. Varadhan. *Asymptotic behavior of a tagged particle in simple exclusion processes.* Bol. Soc. Brasil. Mat. (N.S.) 31 (2000) 241 – 275.

95. S. Sethuraman, S. R. S. Varadhan and H.-T. Yau. *Diffusive limit of a tagged particle in asymmetric simple exclusion processes.* Comm. Pure Appl. Math. 53 (2000) 972 – 1006.

96. S. R. S. Varadhan. *Infinite particle systems and their scaling limits.* Mathematical Physics, Imp. Coll. Press, London, 2000, 306 – 315.

97. S. R. S. Varadhan. *Lectures on hydrodynamic scaling.* Fields Inst. Commun. no. 27, American Mathematical Society, 2000, 3 – 40.

98. S. R. S. Varadhan. *Scaling limits of large interacting systems.* The Mathematical Sciences After the Year 2000, World Sci. Publ., 2000, 144 – 158.

99. S. R. S. Varadhan. *Diffusion processes.* Stochastic processes: theory and methods, Handbook of Statist. 19, North-Holland, 2001, 853 – 872.

100. C. Landim, S. Olla and S. R. S. Varadhan. *Symmetric simple exclusion process: regularity of the self-diffusion coefficient.* Comm. Math. Phys. 224 (2001) 307 – 321.

101. S. R. S. Varadhan. *Large deviation and hydrodynamic scaling.* Adv. Stud. Pure Math. 31, Math. Soc. Japan, 2001, 265 – 286.

102. C. Landim, S. Olla and S. R. S. Varadhan. *Finite-dimensional approximation of the self-diffusion coefficient for the exclusion process.* Ann. Probab. 30 (2002) 483 – 508.

103. S. R. S. Varadhan. *Rare events, large deviations.* Mathematical finance, Bachelier Congress, 2000, Springer, 2002, 85 – 92.

104. S. R. S. Varadhan. *Large deviations and entropy.* Entropy, Princeton Univ. Press, 2003, 199 – 214.

105. S. R. S. Varadhan. *Large deviations for random walks in a random environment.* Comm. Pure Appl. Math. 56 (2003) 1222 – 1245.

106. S. R. S. Varadhan. *Particle systems and partial differential equations.* New Stud. Adv. Math. no. 2, Int. Press, 2003, 217 – 222.

107. S. R. S. Varadhan. *Stochastic analysis and applications.* Bull. Amer. Math. Soc. (N.S.) 40 (2003) 89– 97.

108. S. R. S. Varadhan. *Large deviations for the asymmetric simple exclusion process.* Adv. Stud. Pure Math. 39, Math. Soc. Japan, 2004, 1 – 27.

109. C. Landim, S. Olla and S. R. S. Varadhan. *Diffusive behaviour of the equilibrium fluctuations in the asymmetric exclusion processes.* Adv. Stud. Pure Math. 39, Math. Soc. Japan, 2004, 307 –324.

110. C. Landim, S. Olla and S. R. S. Varadhan. *On viscosity and fluctuation-dissipation in exclusion processes.* J. Statist. Phys. 115 (2004) 323 – 363.

111. S. R. S. Varadhan. *Random walks in a random environment.* Proc. Indian Acad. Sci. Math. Sci. 114 (2004) 309 – 318.

112. S. Sethuraman and S. R. S. Varadhan. *A martingale proof of Dobrushin's theorem for non-homogeneous Markov chains.* Electron. J. Probab. 10 (2005) 1221 – 1235.

113. E. Kosygina, F. Rezakhanlou and S. R. S. Varadhan. *Stochastic homogenization of Hamilton-Jacobi-Bellman equations.* Comm. Pure Appl. Math. 59 (2006) 1489 – 1521.

114. S. R. S. Varadhan. *Homogenization.* Math. Student 76 (2007) 129 – 136.

115. S. R. S. Varadhan. *Large deviations.* Ann. Probab. 36 (2008) 397 – 419.

116. E. Kosygina and S. R. S. Varadhan. *Homogenization of Hamilton-Jacobi-Bellman equations with respect to time-space shifts in a stationary ergodic medium.* Comm. Pure Appl. Math. 61 (2008) 816 – 847.

117. S. R. S. Varadhan and N. Zygouras. *Behavior of the solution of a random semilinear heat equation.* Comm. Pure Appl. Math. 61 (2008) 1298 – 1329.

118. S. R. S. Varadhan. *Homogenization of random Hamilton-Jacobi-Bellman equations.* Probability, geometry and integrable systems, Math. Sci. Res. Inst. Publ., Cambridge Univ. Press, 2008, 397 – 403.

119. S. R. S. Varadhan. *Large deviations and scaling limit.* Lett. Math. Phy. 88 (2009) 175 – 185.

120. S. R. S. Varadhan. *The role of weak convergence in probability theory.* Symmetry in Mathematics and Physics, Contemp. Math.490, Amer. Math. Soc.,2009, 3 – 10.

121. S. R. S. Varadhan. *Scaling limits.* Perspectives in mathematical sciences - I, Stat. Sci. Interdiscip. Res., no. 7, World Sci. Publ.,2009, 247 – 262.

122. S. R. S. Varadhan. *Workshop on large deviations.* Bull. Kerala Math Assoc. (2009) 1 – 14.

123. S. Chatterjee and S. R. S. Varadhan. *The large deviation principle for the Erdös-Rènyi random graph.* European J. Comb. 32 (2011) 1000 – 1017.

124. Y. Kifer and S. R. S. Varadhan. *Nonconventional limit theorems in discrete and continuous time via martingales.* Submitted to *Ann. Probab. in Dec. 2010.* arXiv:1012.2223v1 [math.PR].

125. S. Sethuraman and S. R. S. Varadhan. *Large deviations for the current and tagged particle in 1-d nearest neighbor symmetric simple exclusion.* To appear in *Ann. Probab.*

126. S. Chatterjee and S. R. S. Varadhan. *Large deviations for random matrices.* Communications on Stochastic Analysis. 6 (2012) 1 – 13.

Acknowledgements

The editors thank the publishers and copyright holders for granting permission to reprint the following papers in these volumes.

American Mathematical Society

1. K. R. Parthasarathy, R. Ranga Rao and S. R. S. Varadhan. *On the category of indecomposable distributions on topological groups.* Trans. Amer. Math. Soc. 102 (1962) 200 – 217.
2. D. W. Stroock and S. R. S. Varadhan. *A probabilistic approach to* $H^p(R^d)$. Trans. Amer. Math. Soc. 192 (1974) 245 – 260.
3. S. R. S. Varadhan. *Lectures on hydrodynamic scaling.* Fields Inst. Commun. no. 27, American Mathematical Society, 2000, 3 – 40.
4. S. R. S. Varadhan. *Stochastic analysis and applications.* Bull. Amer. Math. Soc. (N.S.) 40 (2003) 89 – 97.
5. S. R. S. Varadhan. *The role of weak convergence in probability theory.* Symmetry in Mathematics and Physics, Contemp. Math. 490, Amer. Math. Soc., 2009, 3 – 10.

Applied Probability Trust

6. J. A. Gubner, B. Gopinath and S. R. S. Varadhan. *Bounding functions of Markov processes and the shortest queue problem.* Adv. In Appl. Probab. 21 (1989) 842 – 860.

Clarendon Press/Oxford University Press

7. M. D. Donsker and S. R. S. Varadhan. *Asymptotic evaluation of certain Wiener integrals for large time.* Functional Integration and its Applications, Clarendon Press, 1975, 15 – 33.

Duke University Press

8. A. F. Ramirez and S. R. S. Varadhan. *Relative entropy and mixing properties of interacting particle systems.* J. Math. Kyoto Univ. 36 (1996) 869 – 875.

Elsevier

9. M. D. Donsker and S. R. S. Varadhan. *Some problems of large deviations.* Symposia Mathematica, Vol. XXI, Academic Press, 1977, 313 – 318.
10. M. D. Donsker and S. R. S. Varadhan. *The polaron problem and large deviations.* Phys. Rep. 77 (1981) 235 – 237.

11. G. Papanicolaou and S. R. S. Varadhan. *Boundary value problems with rapidly oscillating random coefficients.* Random Fields, Vol. II, Math. Soc. János Bolyai 27, 1981, 835 – 873.
12. G. Papanicolaou and S. R. S. Varadhan. *Diffusions with random coefficients.* Statistics and Probability: Essays in honor of C. R. Rao, North-Holland, 1982, 547 – 552.
13. S. R. S. Varadhan. *On the derivation of conservation laws for stochastic dynamics.* Analysis, et cetera, Academic Press, 1990, 677 – 694.
14. S. R. S. Varadhan. *Self-diffusion of a tagged particle in equilibrium for asymmetric mean zero random walk with simple exclusion.* Ann. Inst. H. Poincaré Probab. Statist. 31 (1995) 273 – 285.
15. S. R. S. Varadhan. *Diffusion processes.* Stochastic processes: theory and methods, Handbook of Statist. 19, North-Holland, 2001, 853 – 872.
16. S. Chatterjee and S. R. S. Varadhan. *The large deviation principle for the Erdös-Rènyi random graph.* European J. Comb. 32 (2011) 1000 – 1017.

Finnish Academy of Science and Letters

17. S. R. S. Varadhan. *Some problems of large deviations.* Proceedings of the International Congress of Mathematicians (Helsinki, 1978), Acad. Sci. Fennica, 1980, 755 – 762.

Indian Academy of Sciences

18. S. R. S. Varadhan. *Random walks in a random environment.* Proc. Indian Acad. Sci. Math. Sci. 114 (2004) 309 – 318.

Indian Mathematical Society

19. S. R. S. Varadhan. *Limit theorems in probability.* Math. Student 32 (1964) 17 – 21.

Indian Statistical Institute

20. S. R. S. Varadhan. *Limit theorems for sums of independent random variables with values in a Hilbert space.* Sankhya Ser. A 24 (1962) 213 – 238.
21. D. W. Stroock and S. R. S. Varadhan. *Limit theorems for random walks on Lie groups.* Sankhya Ser. A 35 (1973) 277 – 294.

Institute of Mathematics, Polish Academy of Sciences

22. K. L. Chung and S. R. S. Varadhan. *Kac functional and Schrödinger equation.* Studia Math. 68 (1980) 249 – 260.

Institute of Mathematical Statistics

23. C. Landim, S. Sethuraman and S. R. S. Varadhan. *Spectral gap for zero-range dynamics.* Ann. Probab. 24 (1996) 1871 – 1902.

24. C. Landim, S. Olla and S. R. S. Varadhan. *Finite-dimensional approximation of the self-diffusion coefficient for the exclusion process.* Ann. Probab. 30 (2002) 483 – 508.

25. S. R. S. Varadhan. *Large deviations.* Ann. Probab. 36 (2008) 397 – 419.

26. S. Sethuraman and S. R. S. Varadhan. *A martingale proof of Dobrushin's theorem for non-homogeneous Markov chains.* Electron. J. Probab. 10 (2005) 1221 – 1235.

27. S. Sethuraman and S. R. S. Varadhan. *Large deviations for the current and tagged particle in 1-d nearest neighbor symmetric simple exclusion.* To appear in *Ann. Probab.*

International Press

28. S. R. S. Varadhan and H.-T. Yau. *Diffusive limit of lattice gas with mixing conditions.* Asian J. Math. 1 (1997) 623 – 678.

John Wiley & Sons Inc.

29. S. R. S. Varadhan. *Asymptotic probabilities and differential equations.* Comm. Pure Appl. Math. 19 (1966) 261 – 286.

30. S. R. S. Varadhan. *On the behavior of the fundamental solution of the heat equation with variable coefficients.* Comm. Pure Appl. Math. 20 (1967) 431 – 455.

31. S. R. S. Varadhan. *Diffusion processes in a small time interval.* Comm. Pure Appl. Math. 20 (1967) 659 – 685.

32. D. W. Stroock and S. R. S. Varadhan. *Diffusion processes with continuous coefficients - I.* Comm. Pure Appl. Math. 22 (1969) 345 – 400.

33. D. W. Stroock and S. R. S. Varadhan. *Diffusion processes with continuous coefficients - II.* Comm. Pure Appl. Math. 22 (1969) 479 – 530.

34. D. W. Stroock and S. R. S. Varadhan. *Diffusion processes with boundary conditions.* Comm. Pure Appl. Math. 24 (1971) 147 – 225.

35. D. W. Stroock and S. R. S. Varadhan. *On degenerate elliptic-parabolic operators of second order and their associated diffusions.* Comm. Pure Appl. Math. 25 (1972) 651 – 713.

36. G. Papanicolaou and S. R. S. Varadhan. *A limit theorem with strong mixing in Banach space and two applications to stochastic differential equations.* Comm. Pure Appl. Math. 26 (1973) 497 – 524.

37. M. D. Donsker and S. R. S. Varadhan. *Asymptotic evaluation of certain Markov process expectations for large time - I.* Comm. Pure Appl. Math. 28 (1975) 1 – 47.

38. M. D. Donsker and S. R. S. Varadhan. *Asymptotic evaluation of certain Markov process expectations for large time - II.* Comm. Pure Appl. Math. 28 (1975) 279 – 301.

39. M. D. Donsker and S. R. S. Varadhan. *Asymptotics for the Wiener sausage.* Comm. Pure Appl. Math. 28 (1975) 525 – 565.
40. M. D. Donsker and S. R. S. Varadhan. *Erratum: "Asymptotics for the Wiener sausage"* Comm. Pure Appl. Math. 28 (1975) 677 – 678.
41. M. D. Donsker and S. R. S. Varadhan. *On the principal eigenvalue of second-order elliptic differential operators.* Comm. Pure Appl. Math. 29 (1976) 595 – 621.
42. M. D. Donsker and S. R. S. Varadhan. *On laws of the iterated logarithm for local times.* Comm. Pure Appl. Math. 30 (1977) 707 – 753.
43. M. D. Donsker and S. R. S. Varadhan. *On the principal eigenvalue of elliptic second order differential operators.* Proceedings of the International Symposium on Stochastic Differential Equations, Wiley, 1978, 41 – 47.
44. M. D. Donsker and S. R. S. Varadhan. *On the number of distinct sites visited by a random walk.* Comm. Pure Appl. Math. 32 (1979) 721 – 747.
45. M. D. Donsker and S. R. S. Varadhan. *A law of the iterated logarithm for total occupation times of transient Brownian motion.* Comm. Pure Appl. Math. 33 (1980) 365 – 393.
46. M. D. Donsker and S. R. S. Varadhan. *Asymptotic evaluation of certain Markov process expectations for large time - IV.* Comm. Pure Appl. Math. 36 (1983) 183 – 212.
47. M. D. Donsker and S. R. S. Varadhan. *Asymptotics for the polaron.* Comm. Pure Appl. Math. 36 (1983) 505 – 528.
48. G. Papanicolaou and S. R. S. Varadhan. *Ornstein-Uhlenbeck process in a random potential.* Comm. Pure Appl. Math. 38 (1985) 819 – 834.
49. S. R. S. Varadhan and R. J. Williams. *Brownian motion in a wedge with oblique reflection.* Comm. Pure Appl. Math. 38 (1985) 405 – 443.
50. M. D. Donsker and S. R. S. Varadhan. *Large deviations from a hydrodynamic scaling limit.* Comm. Pure Appl. Math. 42 (1989) 243 – 270.
51. C. Kipnis, S. Olla and S. R. S. Varadhan. *Hydrodynamics and large deviation for simple exclusion processes.* Comm. Pure Appl. Math. 42 (1989) 115 – 137.
52. J. R. Baxter, N. C. Jain and S. R. S. Varadhan. *Some familiar examples for which the large deviation principle does not hold.* Comm. Pure Appl. Math. 44 (1991) 911 – 923.
53. H. Berestycki, L. Nirenberg and S. R. S. Varadhan. *The principal eigenvalue and maximum principle for second-order elliptic operators in general domains.* Comm. Pure Appl. Math. 47 (1994) 47 – 92.
54. J. Quastel and S. R. S. Varadhan. *Diffusion semigroups and diffusion processes corresponding to degenerate divergence form operators.* Comm. Pure Appl. Math. 50 (1997) 667 – 706.

55. S. Sethuraman, S. R. S. Varadhan and H.-T. Yau. *Diffusive limit of a tagged particle in asymmetric simple exclusion processes.* Comm. Pure Appl. Math. 53 (2000) 972 – 1006.
56. S. R. S. Varadhan. *Large deviations for random walks in a random environment.* Comm. Pure Appl. Math. 56 (2003) 1222 – 1245.
57. E. Kosygina, F. Rezakhanlou and S. R. S. Varadhan. *Stochastic homogenization of Hamilton-Jacobi-Bellman equations.* Comm. Pure Appl. Math. 59 (2006) 1489 – 1521.
58. E. Kosygina and S. R. S. Varadhan. *Homogenization of Hamilton-Jacobi-Bellman equations with respect to time-space shifts in a stationary ergodic medium.* Comm. Pure Appl. Math. 61 (2008) 816 – 847.
59. S. R. S. Varadhan and N. Zygouras. *Behavior of the solution of a random semilinear heat equation.* Comm. Pure Appl. Math. 61 (2008) 1298 – 1329.

Longman Scientific Technical /Taylor & Francis

60. S. R. S. Varadhan. *Nonlinear diffusion limit for a system with nearest neighbor interactions - II.* Pitman Res. Notes Math. Ser., 283, Longman Sci. Tech., 1993, 75 – 128.

Mathematical Society of Japan

61. S. R. S. Varadhan. *Large deviation and hydrodynamic scaling.* Adv. Stud. Pure Math., Math. Soc. Japan, 2001, 265 – 286.
62. C. Landim, S. Olla and S. R. S. Varadhan. *Diffusive behaviour of the equilibrium fluctuations in the asymmetric exclusion processes.* Adv. Stud. Pure Math. no. 39, 2004, 307–324.
63. S. R. S. Varadhan. *Large deviations for the asymmetric simple exclusion process.* Adv. Stud. Pure Math. no. 39, 2004, 1 – 27.

Serial Publications

64. S. Chatterjee and S. R. S. Varadhan. *Large deviations for random matrices.* Communications on Stochastic Analysis. 6 (2012) 1 – 13.

Society for Industrial and Applied Mathematics

65. K. R. Parthasarathy and S. R. S. Varadhan. *Extension of stationary stochastic processes.* Teor. Verojatnost. i Primenen. 9 (1964) 72 – 78.

Springer Science + Business Media

66. G. Papanicolaou and S. R. S. Varadhan. *Diffusion in regions with many small holes.* Lecture Notes in Control and Inform. Sci. no. 25, Springer, 1980, 190 – 206.
67. M. D. Donsker and S. R. S. Varadhan. *Large deviations for stationary Gaussian processes.* Comm. Math. Phys. 97 (1985) 187 – 210.
68. C. Kipnis and S. R. S. Varadhan. *Central limit theorem for additive functionals of reversible Markov processes and applications to simple exclusions.* Comm. Math. Phys. 104 (1986) 1 – 19.

69. A. S. Sznitman and S. R. S. Varadhan. *A multidimensional process involving local time.* Probab. Theory Related Fields 71 (1986) 553 – 579.

70. M. D. Donsker and S. R. S. Varadhan. *Large deviations for noninteracting infinite-particle systems.* J. Statist. Phys. 46 (1987) 1195 – 1232.

71. M. Z. Guo, G. C. Papanicolaou and S. R. S. Varadhan. *Nonlinear diffusion limit for a system with nearest neighbor interaction.* Comm. Math. Phys. 118 (1988) 31 – 59.

72. S. R. S. Varadhan. *Scaling limits for interacting diffusions.* Comm. Math. Phys. 135 (1991) 313 – 353.

73. S. Olla and S. R. S. Varadhan. *Scaling limit for interacting Ornstein-Uhlenbeck processes.* Comm. Math. Phys. 135 (1991) 355 – 378.

74. S. R. S. Varadhan. *Entropy methods in hydrodynamical scaling.* Mathematical Physics, X, Springer, 1992, 103 – 112.

75. S. Olla, S. R. S. Varadhan and H. -T. Yau. *Hydrodynamical limit for a Hamiltonian system with weak noise.* Comm. Math. Phys. 155 (1993) 523 – 560.

76. S. R. S. Varadhan. *Regularity of self-diffusion coefficient.* The Dynkin Festschrift, Progr. Probab. no.34, Birkhäuser, 1994, 387 – 397.

77. S. R. S. Varadhan. *Entropy methods in hydrodynamic scaling.* Proceedings of the International Congress of Mathematicians, Vol. 1 (Zürich, 1994), Birkhäuser, 1995, 196 – 208.

78. S. R. S. Varadhan. *The complex story of simple exclusion.* Itô's Stochastic Calculus and Probability Theory, Springer, 1996, 385 – 400.

79. J. Quastel, F. Rezakhanlou and S. R. S. Varadhan. *Large deviations for the symmetric simple exclusion process in dimensions* $d \geq 3$. Probab. Theory Related Fields 113 (1999) 1 – 84.

80. S. R. S. Varadhan. *Large deviations for interacting particle systems.* Perplexing problems in probability, Progr.Probab. no. 44, Birkhäuser, 1999, 373 – 383.

81. C. Landim, S. Olla and S. R. S. Varadhan. *Asymptotic behavior of a tagged particle in simple exclusion processes.* Bol. Soc. Brasil. Mat. (N.S.) 31 (2000) 241 – 275.

82. C. Landim, S. Olla and S. R. S. Varadhan. *Symmetric simple exclusion process: regularity of the self-diffusion coefficient.* Comm. Math. Phys. 224 (2001) 307 – 321.

83. C. Landim, S. Olla and S. R. S. Varadhan. *On viscosity and fluctuation-dissipation in exclusion processes.* J. Statist. Phys. 115 (2004) 323 – 363.

University of Illinois at Urbana-Champaign

84. K. R. Parthasarathy, R. Ranga Rao and S. R. S. Varadhan. *Probability distributions on locally compact abelian groups.* Illinois J. Math. 7 (1963) 337 – 369.

World Scientific Publishers

85. T. Digernes, V. S. Varadarajan and S. R. S. Varadhan. *Finite approximations to quantum systems.* Rev. Math. Phys. 6 (1994) 621 – 648.
86. S. R. S. Varadhan. *Scaling limits of large interacting systems.* The Mathematical Sciences After the Year 2000, World Sci. Publ., 2000, 144 – 158.
87. S. R. S. Varadhan. *Infinite particle systems and their scaling limits.* Mathematical Physics, Imp. Coll. Press, London, 2000, 306 – 315.

Collected Papers of
S. R. S. Varadhan

Volume 3
Large Deviations

Collected Papers of S. R. S. Varadhan

Volume 1: Limit Theorems,
Review Articles

Volume 2: PDE, SDE, Diffusions,
Random Media

Volume 3: Large Deviations

Volume 4: Particle Systems and their Large Deviations

Collected Papers of
S. R. S. Varadhan

Volume 3
Large Deviations

Editors

Rajendra Bhatia
Abhay G. Bhatt
K. R. Parthasarathy

HINDUSTAN
BOOK AGENCY

Published by:

Hindustan Book Agency (India)
P 19 Green Park Extension
New Delhi 110 016
India

email: info@hindbook.com
www.hindbook.com

ISBN 978-93-80250-39-7 (4 volume set)

Printed and bound by Gopsons Papers Ltd., Noida

Contents

Volume 1

Limit Theorems

Review Articles

Volume 2

PDE, SDE, Diffusions

Random Media

Volume 3

Large Deviations

Volume 4

Particle Systems and their Large Deviations

Large Deviations

COMMUNICATIONS ON PURE AND APPLIED MATHEMATICS, VOL. XIX, 261–286 (1966)

Asymptotic Probabilities and Differential Equations*

S. R. S. VARADHAN

Indian Statistical Institute and Courant Institute of Mathematical Sciences

1. Recently Donsker [1] considered the behavior of the solution u_ε for the initial value problem

$$u_t = uu_x + \frac{\varepsilon^2}{2} u_{xx} + p(x, t) , \tag{1.1}$$

$u(0, x) = u(x)$; as $\varepsilon \to 0$. It was shown that, under mild conditions, $\lim_{\varepsilon \to 0} u_\varepsilon = u$ exists and that u is the natural solution for (1.1) with $\varepsilon = 0$. The method used was to transform (1.1) by a transformation of Hopf [3] to a linear equation of the form

$$V_t = \frac{\varepsilon^2}{2} V_{xx} + \frac{P(x, t)}{\varepsilon^2} V . \tag{1.2}$$

Since (1.2) can be solved as a function space integral, the problem reduces to the analysis of the asymptotic behavior of some function space integrals. This was carried out by Schilder [8] and the development resembles the "saddle point" method of one dimensional integrals.

One can consider variations of (1.1) of the form

$$u_t = [f(u)]_x + p(t, x) , \tag{1.3}$$

where $f(u)$ is a convex function of u. The question that can be asked is whether one can somehow alter (1.3) slightly so that the resulting equation can be made linear, by a simple transformation, in such a way that it can be solved easily and naturally. It turns out as will be shown here that it is possible to do this in such a way that the solution to (1.3) is approximated by the solution to the altered equations. The approach is similar to that of Donsker and Schilder. But one encounters more general families of measures on function space whose asymptotic behavior is to be investigated.

---

* This paper represents results obtained at the Courant Institute of Mathematical Sciences, New York University, under the sponsorship of the Sloan Foundation Grant for Probability and Statistics.

A similar approach in some special cases was carried out by Lax [5], [6]. Sections 2 and 3 deal with the asymptotic properties of measures and integrals based on these measures in function space. Sections 4 and 5 give the application to the problem described earlier.

2. Let $f(\theta)$ be a convex function of θ. We shall suppose that $|f(\theta)| < \infty$ for all θ. We further assume that $f(\theta)$ is not a linear function of θ and that $f(0) = 0$. Define the conjugate convex function $h(y)$ by the relation

$$h(y) = \max_{\theta} [\theta y - f(y)] \; ; \tag{2.1}$$

$h(y)$ has the following properties:

(i) $h(y)$ is a convex function of y,

(ii) $h(y) \geqq 0$,

(iii) $h(y) < \infty$ for y in some nondegenerate interval,

(iv) $f(\theta) = \max_{y} [\theta y - h(y)]$,

(v) $\lim_{|y| \to \infty} \frac{|h(y)|}{|y|} = +\infty$,

(vi) if f^+ and f^- are the left and right derivatives of $f(\theta)$ at $\theta = 0$, then $h(y)$ is nonincreasing for $y \leqq f^-$, equals zero on $f^- \leqq y \leqq f^+$ and is nondecreasing for $y \geqq f^+$.

Let $\{a_n\}$ be any sequence of non-negative numbers chosen and fixed in advance such that

$$\lim_{n \to \infty} \frac{a_n}{n} = \infty .$$

Define a distribution F_n by the formula

$$dF_n = c_n \exp\left\{-\frac{a_n}{n} h(y)\right\} dy , \qquad -\infty < y < \infty \; ; \tag{2.2}$$

c_n is chosen such that

$$c_n^{-1} = \int_{-\infty}^{\infty} \exp\left\{-\frac{a_n}{n} h(y)\right\} dy .$$

Lemma 2.1. *c_n is well defined and* $\lim_{n \to \infty} \frac{n}{a_n} \log c_n = 0$.

Proof: Using properties (ii)–(vi) listed above this can be derived by standard estimation.

Lemma 2.2.

$$\lim_{n \to \infty} \frac{n}{a_n} \log\left[\int c_n \exp\left\{\frac{a_n}{n} [\theta y - h(y)]\right\} dy\right] = f(\theta) .$$

Proof: This can be proved by familiar saddle point considerations.

For each n, $\{X_{n,j}\}$ is a sequence of independent and identically distributed random variables all having the common distribution F_n. For each n, the partial sums $S_{n,k}$ are defined as

$$S_{n,0} = 0,$$

$$S_{n,k} = \sum_{j=1}^{k} X_{n,j}.$$

Let $[0, T]$ be some finite time interval. Ω is the space of functions on $[0, T]$ which have left and right limits at each point, vanish at 0 and are right continuous at every point $t \in [0, T]$. Ω is endowed with the topology of uniform convergence on the interval $[0, T]$. For each n a measure P_n is constructed on Ω as follows. Let us define a stochastic process $z_n(t)$ by

$$z_n(t) = \frac{1}{n} S_{n,[nt]}, \qquad 0 \leqq t \leqq T, \tag{2.3}$$

$[nt]$ denoting the integral part of nt. P_n is the measure on Ω corresponding to $z_n(t)$.

Our interest is in the asymptotic behavior of P_n. We shall show that in some suitable sense

$$dP_n \simeq \exp\left\{-a_n \int_0^T h(z'(t))\, dt\right\} \prod dz(t).$$

The precise formulation of the above statement is as follows. Let $\Omega_0 \subseteq \Omega$ be the set of those functions z in Ω, which are absolutely continuous with respect to t and for which

$$\int_0^T h(z'(t))\, dt < \infty.$$

Define a function $I(z)$ for $z \in \Omega$ by

$$I(z) = \begin{cases} \int_0^T h(z'(t))\, dt & \text{for} \quad z \in \Omega_0, \\ \infty & \text{for} \quad z \in \Omega - \Omega_0, \end{cases}$$

with the uniform topology on Ω; $I(z)$ is lower semicontinuous. We shall prove in this section the following theorem.

THEOREM 2.1. *If C is a closed set in Ω and G is an open set in Ω, then*

$$\limsup_{n\to\infty} \frac{1}{a_n} \log P_n(C) \leqq -\inf_{z\in C} I(z),$$

$$\liminf_{n\to\infty} \frac{1}{a_n} \log P_n(G) \geqq -\inf_{z\in G} I(z).$$

We shall prove the theorem by establishing several lemmas.

LEMMA 2.3. *Let C be a subset of Ω of the form*

$$C = [z\colon z(t_2) - z(t_1) \geqq x]\,, \qquad t_2 > t_1, \delta = t_2 - t_1\,.$$

Then

$$\limsup_{n\to\infty} \frac{1}{a_n} \log P_n(C) \leqq -\inf_{x\geqq y} \delta h\left(\frac{y}{\delta}\right).$$

Proof:

$$z_n(t_2) - z_n(t_1) = \xi_n = \frac{S_n[nt_2] - S_n[nt_1]}{n}\,.$$

If $\lambda_n = [nt_2] - [nt_1]$, ξ_n has a distribution which is identical with that of $(S_n, \lambda_n)/n$. Hence

$$Ee^{\theta\xi_n} = \left[c_n \int \exp\left\{\frac{\theta y}{n} - \frac{a_n}{n} h(y)\right\}\right]^{\lambda_n},$$

$$\begin{aligned} P_n(C) &= P\{\xi_n \geqq x\} \leqq e^{-\theta x} Ee^{\theta\xi_n} \qquad \text{for} \qquad \theta \geqq 0 \\ &= e^{-\theta x}\left[c_n \int \exp\left\{\frac{\theta y}{n} - \frac{a_n}{n} h(y)\right\} dy\right]^{\lambda_n}. \end{aligned}$$

Replacing θ by $a_n\theta$, we have for every $\theta > 0$

$$P_n(C) \leqq e^{-a_n\theta x}\left[c_n \int \exp\left\{\frac{a_n}{n}[\theta y - h(y)]\right\} dy\right]^{\lambda_n},$$

or setting

$$\phi_n(\theta) = \frac{n}{a_n} \log\left[\int c_n \exp\left\{\frac{a_n}{n}[\theta y - h(y)]\right\} dy\right],$$

$$\log P_n(C) \leqq a_n\theta x + \frac{\lambda_n}{a} \cdot \frac{a_n}{n} \phi_n(\theta)\,,$$

$$\frac{1}{a_n} \log P_n(C) \leqq \frac{\lambda_n}{n} \phi_n(\theta) - \theta x\,. \tag{2.4}$$

Using Lemma 2.2, we see that, since $\lambda_n/n \to \delta$ as $n \to \infty$,

$$\limsup_{n\to\infty} \frac{1}{a_n} \log P_n(C) \leqq \delta f(\theta) - \theta x\,. \tag{2.5}$$

Since (2.5) is true for all $\theta \geqq 0$,

$$\limsup_{n\to\infty} \frac{1}{a_n} \log P_n(C) \leqq \inf_{\theta\geqq 0}[\delta f(\theta) - \theta x] = -\delta \sup_{\theta\geqq 0}\left[\theta \frac{x}{\delta} - f(\theta)\right]. \tag{2.6}$$

CASE i: $x > \delta f^+(0)$.

If $\theta < 0$,

$$f(\theta) \geqq \theta f^- \geqq \theta f^+ \geqq \frac{\theta x}{\delta},$$

or

$$\frac{\theta x}{\delta} - f(\theta) \leqq 0 .$$

Hence

$$\sup_{\theta \geqq 0} \left[\theta \frac{x}{\delta} - f(\theta)\right] = \sup_{\theta} \left[\theta \frac{x}{\theta} - f(\theta)\right] = h\left(\frac{x}{\delta}\right) .$$

Therefore (2.6) reduces to

$$\limsup_{n \to \infty} \frac{1}{a_n} \log P_n(C) \leqq - \delta h\left(\frac{x}{\delta}\right) .$$

Since $h(y)$ is nondecreasing for $y \geqq f^+$,

$$\inf_{y \geqq x} h\left(\frac{y}{\delta}\right) = h\left(\frac{x}{\delta}\right)$$

if $x > \delta f^+$, and the lemma is proved.

CASE ii: $x \leqq \delta f^+(0)$.

Since $h(y) = 0$ for $f^- \leqq y \leqq f^+$,

$$\inf_{y \geqq x} h\left(\frac{y}{\delta}\right) = 0$$

if $a \leqq \delta f^+$, and the lemma is valid trivially.

LEMMA 2.4. *Let C be a closed set of the form*

$$C = [z(t_2) - z(t_1) \leqq x] , \qquad t_2 > t_1 , \delta = t_2 - t_1 .$$

Then

$$\limsup_{n \to \infty} \frac{1}{a_n} \log P_n(C) \leqq -\inf_{y \leqq x} \delta h\left(\frac{y}{\delta}\right) .$$

The proof is similar to that of Lemma 2.3. We now prove a lemma to be used later.

LEMMA 2.5. *Let $\varepsilon > 0$ be given. Then there exists a $\delta_0 > 0$ such that for all t_1, t_2, with $0 \leqq t_2 - t_1 \leqq \delta_0$,*

$$P_n[z: | z(t_2) - z(t_1)| \geqq \varepsilon] \leqq \tfrac{1}{2}$$

for all $n \geqq n_0$.

Proof: From equation (2.4), for $\theta \geqq 0$,

$$P_n[z\colon z(t_2) - z(t_1) \geqq \varepsilon] \leqq \exp\left\{-a_n\theta\varepsilon + \lambda_n \frac{a_n}{n}\,\theta_n(\theta)\right\}. \tag{2.7}$$

Take $\theta = 1$. Since $\theta_n(1) \to f(1)$, $\theta_n(1) \leqq M$ for all n, $\lambda_n \leqq n\delta_0 + 1$. Using these in (2.7), we obtain

$$P_n[z\colon z(t_2) - z(t_1) \geqq \varepsilon] \leqq \exp\left\{-\varepsilon a_n + a_n\left(\delta_0 + \frac{1}{n}\right)M\right\}.$$

If n is such that $1/n \leqq \delta_0$, we have

$$P_n[z\colon z(t_2) - z(t_1) \geqq \varepsilon] \leqq \exp\{-\varepsilon a_n + 2M\delta_0 a_n\}.$$

For any $\varepsilon > 0$, choosing δ_0 small enough and n large enough, we can achieve that

$$P_n[z\colon z(t_2) - z(t_1) > \varepsilon] \leqq \tfrac{1}{4}.$$

Similarly we can make

$$P_n[z\colon z(t_2) - z(t_1) \leqq -\varepsilon] \leqq \tfrac{1}{4}.$$

Combining the two inequalities we see that the lemma is true.

LEMMA 2.6. *Let C be a closed set of the form*

$$C = [z\colon x_1 \leqq z(t_2) - z(t_1) \leqq x_2], \quad t_2 > t_1, t_2 - t_1 = \delta.$$

Then

$$\limsup_{n\to\infty} \frac{1}{a_n} \log P_n(C) \leqq -\inf_{x_1 \leqq y \leqq x_2} \delta h\left(\frac{y}{\delta}\right).$$

Proof: CASE i: $x_1 \geqq \delta f^+$.

Since

$$P_n(C) \leqq P_n[z\colon z(t_2) - z(t_1) \geqq x_1],$$

$$\begin{aligned}\limsup_{n\to\infty} \frac{1}{a_n} \log P_n(C) &\leqq \limsup_{n\to\infty} \frac{1}{a_n} \log P_n[z\colon z(t_2) - z(t_1) \geqq x_1] \\ &\leqq -\inf_{y \geqq x_1} \delta h\left(\frac{y}{\delta}\right) \\ &= -\inf_{x_2 \geqq y \geqq x_1} \delta h\left(\frac{y}{\delta}\right).\end{aligned}$$

We use Lemma 2.3 and the fact that $h(y)$ is nondecreasing for $y \geqq f^+$.

CASE ii: $x_2 \leqq \delta f^-$.

This is treated similar to case i using Lemma 2.4.

CASE iii: $[\delta f^-, \delta f^+]$ and $[x_1, x_2]$ have a point in common.

In this case if y is a common point, $y/\delta \in [f^-, f^+]$. Consequently

$$\inf_{x_1 \leqq y \leqq x_2} \delta h\left(\frac{y}{\delta}\right) = 0,$$

and the lemma is trivially true.

We now introduce some notation. Let k points $0 < t_1 < t_2 < \cdots < t_k \leqq T$ be chosen. Let Π denote the map that sends $z \in \Omega$ to the point $\{z(t_1), z(t_2) - z(t_1), \cdots, z(t_k) - z(t_{k-1})\}$ of R_k. We denote by y a point $(y_1, \cdots, y_k)$ of R_k and by $\psi(y)$ the function

$$\psi(y) = \sum_{j=1}^{k} (t_j - t_{j-1}) h\left(\frac{y_j}{t_j - t_{j-1}}\right).$$

For any set $A \subseteq R_k$, $\Pi^{-1} A$ is the subset of Ω defined by

$$\Pi^{-1} A = [z: z(t_1), z(t_2) - z(t_1), \cdots, z(t_k) - z(t_{k-1}) \in A].$$

LEMMA 2.7. *Let A be a closed subset of R_k. Then*

$$\limsup_{n\to\infty} \frac{1}{a_n} \log P_n(\Pi^{-1} A) \leqq -\inf_{y \in A} \psi(y).$$

Proof: Let first A be a rectangle of the form

$$\begin{aligned} A &= [a_1 \leqq y_1 \leqq b_1 \leqq \cdots \leqq a_k \leqq y \leqq b_k] \\ &= A_1 \cap A_2 \cap \cdots \cap A_k. \end{aligned}$$

Then

$$\begin{aligned} \limsup_{n\to\infty} \frac{1}{a_n} \log P_n(\Pi^{-1} \bigcap_{j=1}^{k} A_j) &= \limsup_{n\to\infty} \frac{1}{a_n} \log P_n(\bigcap_{j=1}^{k} \delta\, \Pi^{-1} A_j) \\ &= \limsup_{n\to\infty} \frac{1}{a_n} \sum \log P_n(\Pi^{-1} A_j) \\ &\leqq \sum \limsup_{n\to\infty} \frac{1}{a_n} \log P_n(\Pi^{-1} A_j) \\ &\leqq -\sum \inf_{a_j \leqq y \leqq b_j} \delta_j h\left(\frac{y_j}{\delta_j}\right) \\ &= -\inf_{y \in A} \psi(y). \end{aligned}$$

Next let A be a closed bounded set. Let A^ε be the sphere of radius ε around A. We can find closed rectangles $A_1, \cdots, A_N$ such that

$$A \subseteq A_1 \cup \cdots \cup A_N \subseteq A^\varepsilon.$$

Then

$$\limsup_{n\to\infty} \frac{1}{a_n} \log P_n(\Pi^{-1} A) \leqq \limsup_{n\to\infty} \frac{1}{a_n} \log P_n(\cup \Pi^{-1} A_j)$$

$$\leqq \limsup_{n\to\infty} \frac{1}{a_n} \log \sum P_n(\Pi^{-1} A_j)$$

$$\leqq \limsup_{n\to\infty} \frac{1}{a_n} \log [N \max P_n(\Pi^{-1} A_j)]$$

$$= \limsup_{n\to\infty} \frac{1}{a_n} \max \log [P_n(\Pi^{-1} A_j)] \tag{2.8}$$

$$= \max \limsup_{n\to\infty} \frac{1}{a_n} \log [P_n(\Pi^{-1} A_j)]$$

$$\leqq \max [-\inf_{y\in A_j} \psi(y)]$$

$$= -\inf_{y\in A_j} \psi(y)$$

$$\leqq -\inf_{y\in A^\varepsilon} \psi(y) .$$

Since $\psi(y)$ is lower semicontinuous and since A is a closed bounded set,

$$\liminf_{\varepsilon\downarrow 0}\ \inf_{y\in A^\varepsilon} \psi(y) = \inf_{y\in A} \psi(y) .$$

Since ε is arbitrary in (2.8) we have, letting $\varepsilon \downarrow 0$,

$$\limsup_{n\to\infty} \frac{1}{a_n} \log P_n(\Pi^{-1} A) \leqq -\inf_{y\in A} \psi(y) .$$

Now let A be an arbitrary closed set. Define

$$E_N = \{y\colon |y_j| \leqq N \text{ for all } j\} ,$$

$$F_N = \{y\colon |y_j| \geqq N \text{ for some } j\} ,$$

$$A_N = A \cap E_N ,$$

$$F^{(1)}_{N,j} = \{y\colon y_j \geqq N\} ,$$

$$F^{(2)}_{N,j} = \{y\colon y_j \leqq -N\} ,$$

$$\delta_j = t_j - t_{j-1} , \qquad t_0 = 0 ,$$

$$\limsup_{n\to\infty} \frac{1}{a_n} \log P_n(\Pi^{-1} F^{(1)}_{n,j}) \leqq -\inf_{y\geqq N} \delta_j h\left(\frac{y}{\delta_j}\right) ,$$

$$\limsup_{n\to\infty} \frac{1}{a_n} \log P_n(\Pi^{-1} F^{(2)}_{n,j}) \leqq -\inf_{y\leqq -N} \delta_j h\left(\frac{y}{\delta_j}\right) .$$

Since $F_N = \bigcup_j F_{n,j}^{(1)} \bigcup_j F_{n,j}^{(2)}$, we can get as before

$$\limsup_{n\to\infty} \frac{1}{a_n} \log P_n[\Pi^{-1} F_N] \leqq -\inf_j \inf_{|y|\geqq N} \delta_j \delta h\left(\frac{y}{\delta_j}\right) = -\rho_N . \tag{2.9}$$

Since $h(y) \to \infty$ as $|y| \to \infty$, we have

$$\lim_{N\to\infty} \rho_N = +\infty . \tag{2.10}$$

Now if A is an arbitrary closed set in R_k, it follows that for any N

$$\begin{aligned} \limsup_{n\to\infty} \frac{1}{a_n} \log P_n[\Pi^{-1} A] &\leqq \limsup_{n\to\infty} \frac{1}{a_n} \log P_N[\Pi^{-1} A_N \cup F_N] \\ &\leqq \max\,[-\inf_{y\in A_N} \psi(y),\ -\rho_N] \\ &\leqq \max\,[-\inf_{y\in A} \psi(y),\ -\rho_N] . \end{aligned} \tag{2.11}$$

Since $\rho_N \to \infty$ as $N \to \infty$, we have, by letting $N \to \infty$ in (2.11),

$$\limsup_{n\to\infty} \frac{1}{a_n} \log P_n[\Pi^{-1} A] \leqq -\inf_{y\in A} \psi(y)$$

which proves the lemma.

Let τ be the map which sends a point $z \in \Omega$ to a point $[z(t_1), \cdots, z(t_k)]$ in R_k. Let $\Phi(y)$ be the function

$$\Phi(y) = \sum (t_j - t_{j-1}) h\left(\frac{y_j - y_{j-1}}{t_j - t_{j-1}}\right),$$

$y_0 = t_0 = 0$. Then the following lemma holds.

LEMMA 2.8. *For any closed set A in R_k*

$$\limsup_{n\to\infty} \frac{1}{a_n} \log P_n[\tau^{-1} A] \leqq \inf_{y\in A} \Phi(y) .$$

Proof: Let f be the map of R_k onto R_k defined by

$$f(y_1, \cdots, y_k) = y_1, y_1 + y_2, \cdots, y_1 + y_2 \cdots + y_k .$$

Then

$$\tau = f\pi \qquad \text{and} \qquad \tau^{-1} A = \pi^{-1} f^{-1} A .$$

Moreover, f is continuous and $f^{-1}A$ is closed if A is closed. Hence

$$\begin{aligned}\limsup_{n\to\infty} \frac{1}{a_n} \log P_n[\tau^{-1}A] &= \limsup_{n\to\infty} \frac{1}{a_n} \log P_n[\pi^{-1}f^{-1}A] \\ &\leqq -\inf_{y\in f^{-1}A} \psi(y) \\ &= -\inf_{y\in A} \psi(f^{-1}y) \\ &= -\inf_{y\in A} \Phi(y) .\end{aligned}$$

We now prove an elementary lemma about sums of independent random variables. Let $x_1, x_2, \cdots$ be independent random variables and $S_j = x_1 + \cdots + x_j$.

LEMMA 2.9. *Let* $P\{|S_n - S_j| \geqq T\} \leqq \frac{1}{2}$ *for* $j = 1, 2, \cdots, n$. *Then*

$$P\{\sup_{1\leqq j\leqq n} |S_j| \geqq 2T\} \leqq 2P\{|S_n| \geqq T\} .$$

Proof: Define

$$E = \{\sup_{1\leqq j\leqq n} |S_j| \geqq 2T\} ,$$

$$E_j = \{\sup_{1\leqq i\leqq j-1} |S_i| < 2T, |S_j| \geqq 2T\} .$$

Then E is the disjoint union of E_j, and

$$\begin{aligned}P\{E, |S_n| \leqq T\} &= \sum P\{E_j, |S_n| \leqq T\} \\ &\leqq \sum P\{E_j, |S_n - S_j| \geqq T\} \\ &= \sum P(E_j)P\{|S_n - S_j| \geqq T\} \\ &\leqq \tfrac{1}{2} \sum P(E_j) \\ &= \tfrac{1}{2}P(E) .\end{aligned} \tag{2.12}$$

On the other hand,

$$P\{E, |S_n| > T\} \leqq P\{|S_n| \geqq T\} . \tag{2.13}$$

Adding (2.12) and (2.13), we have

$$P(E) \leqq \tfrac{1}{2}P(E) + P\{|S_n| \geqq T\} ,$$

or

$$P(E) = P\{\sup_{1\leqq j\leqq n} |S_j| \geqq 2T\} \leqq 2P\{|S_n| \geqq T\} .$$

We are now ready to estimate $P_n(C)$ for an arbitrary closed set C in Ω. For $z \in \Omega$ let us define a map θ_n of $\Omega \to \Omega$ as follows.

The interval $[0, T]$ is divided into N equal parts; the points of division are $T/N, 2T/N, \cdots, dT/N, \cdots, NT/N$. I_j is the interval $(j-1)T/N \leqq t \leqq jT/N$. For z in Ω, $\theta_N z$ is that element ω of Ω which equals z at the points of division and is linear on each I_j. $\|z\|$ denotes the maximum on $[0, T]$ of the function $|z(t)|$.

LEMMA 2.10. *For any* $\varepsilon > 0$,

$$\lim_{N\to\infty} \limsup_{n\to\infty} \frac{1}{a_n} \log P_n[z\colon \|\theta_N z - z\| \geqq \varepsilon] = -\infty .$$

Proof:

$$\|\theta_N z - z\| \leqq \max_j \sup_{t\in I_j} |z(t_j) - z(t)| , \tag{2.14}$$

where t_j is the left end point of I_j. Let, in view of Lemma 2.5, N be chosen such that

$$\sup_{t\in I_j} P\{|z(t_{j+1}) - z(t)| \geqq \tfrac{1}{4}\varepsilon\} \leqq \tfrac{1}{2} .$$

Then using Lemma 2.9,

$$P_n\{z\colon \sup_{t\in I_j} |z(t_j) - z(t)| \geqq \tfrac{1}{2}\varepsilon\} \leqq 2P_n\{z\colon |z(t_{j+1}) - z(t_j)| \geqq \tfrac{1}{4}\varepsilon\} .$$

Using (2.14) for N large enough,

$$\begin{aligned}
&\limsup_{n\to\infty} \frac{1}{a_n} \log P_n[z\colon \|\theta_N z - z\| \geqq \varepsilon] \\
&\leqq \limsup_{n\to\infty} \frac{1}{a_n} \log P_n\left[z\colon \max_j \sup_{t\in I_j} |z(t_j) - z(t)| \geqq \tfrac{1}{2}\varepsilon\right] \\
&\leqq \limsup_{n\to\infty} \frac{1}{a_n} \log \left[\sum_j 2P_n\{z\colon |z(t_{j+1}) - z(t_j)| \geqq \tfrac{1}{4}\varepsilon\}\right] \\
&= \limsup_{n\to\infty} \frac{1}{a_n} \log \left[\max_j P_n[z\colon |z(t_{j+1}) - z(t_j)| \geqq \tfrac{1}{4}\varepsilon\right] \\
&\leqq -\inf_{|y|\geqq\frac{1}{4}\varepsilon} \frac{T}{N} h\left(\frac{\varepsilon N}{4T}\right) .
\end{aligned}$$

Property (v) of $h(y)$ now ensures that

$$\lim_{N\to\infty} \limsup_{n\to\infty} \frac{1}{a_n} \log P_n[z\colon \|\theta_N z - z\| \geqq \varepsilon] = -\infty .$$

LEMMA 2.11. *Let C be any closed subset of Ω. Let C^ε be the sphere around C of radius ε. Then*

$$\limsup_{n\to\infty} \frac{1}{a_n} \log P_n(C) \leqq -\lim_{\varepsilon\downarrow 0} \inf_{z\in C^\varepsilon} I(z) .$$

Proof:

$$P_n(C) \leqq P_n[z\colon \|\theta_N z - z\| \geqq \varepsilon] + P_n[z\colon \theta_N z \in C^\varepsilon] . \tag{2.15}$$

We denote by L_N the range of θ_N and by τ_N the map of $\Omega \to R_N$ sending ω

to the vector $\omega(jT/N), j = 1, 2, \cdots, N$; τ_N maps L_N onto R_N in a bicontinuous manner. Further $\tau_N \theta_N = \tau_N$. Hence

$$\begin{aligned} P_n[z\colon \theta_N z \in C^\varepsilon] &\leqq P_n[z\colon \theta_N z \in \overline{C^\varepsilon}] \\ &= P_n[z\colon \theta_N z \in \overline{C^\varepsilon} \cap L_N] \\ &= P_n[z\colon \tau_N \theta_N z \in \tau_N(\overline{C^\varepsilon} \cap L_N)] \\ &= P_n[z\colon \tau_N z \in \tau_N(\overline{C^\varepsilon} \cap L_N)]\,, \end{aligned}$$

and therefore,

$$\begin{aligned} &\limsup_{n\to\infty} \frac{1}{a_n} \log P_n[z\colon \theta_N z \in C^\varepsilon] \\ &\qquad \leqq \limsup_{n\to\infty} \frac{1}{a_n} \log P_n[\tau_N^{-1}\tau_N(\overline{C^\varepsilon} \cap L_N)] \\ &\qquad \leqq -\inf_{y \in \tau_N(\overline{C^\varepsilon} \cap L_N)} \Phi(y) \\ &\qquad = -\inf_{z \in \overline{C^\varepsilon} \cap L_N} \Phi(\tau_N z)\,. \end{aligned}$$

But for z in L_N, $\Phi(\tau_N z) = I(z)$. Hence

$$\begin{aligned} \limsup_{n\to\infty} \frac{1}{a_n} \log P_n[z\colon \theta_N z \in C^\varepsilon] &\leqq -\inf_{z \in C^\varepsilon \cap L_N} I(z) \\ &\leqq -\inf_{z \in C^\varepsilon} I(z)\,. \end{aligned}$$

In view of Lemma 2.10 and (2.15),

$$\limsup_{n\to\infty} \frac{1}{a_n} \log P_n(C) \leqq -\inf_{z\in\overline{C^\varepsilon}} I(z)\,.$$

Since this is true for every $\varepsilon > 0$ and since $\overline{C^\varepsilon} \subseteq C^{2\varepsilon}$,

$$\limsup_{n\to\infty} \frac{1}{a_n} \log P_n(C) \leqq -\lim_{\varepsilon\downarrow 0} \inf_{z\in C^\varepsilon} I(z)\,.$$

LEMMA 2.12. *The set $[z\colon I(z) \leqq M]$ is a compact subset of continuous functions in Ω.*

Proof: Since $I(z) < \infty$ implies that z is continuous, it suffices to show that the set in question is an equicontinuous family of functions. Let us denote the set by A. For $z \in A$ write

$$\begin{aligned} z_1(t) &= \int_0^t \alpha(z'(s))\, ds\,, \\ z_2(t) &= \int_0^t \beta(z'(s))\, ds\,, \end{aligned}$$

where

$$\alpha(x) = \begin{cases} 0 & \text{if} \quad |x| > \lambda, \\ x & \text{if} \quad |x| > \lambda. \end{cases}$$

Clearly $\alpha + \beta \equiv x$ and $z_1 + z_2 \equiv z$. As z varies over A the first derivative of z_1 is bounded by λ and hence is equicontinuous. We shall complete the proof by showing that , by choosing λ large, $\|z_2\|$ can be made uniformly small over A.

Using property (v) of $h(y)$, let λ be chosen such that

$$\frac{h(y)}{|y|} \geqq \frac{1}{\varepsilon} \qquad \text{for} \qquad |y| \geqq \lambda .$$

Then

$$\begin{aligned} \|z_2\| &\leqq \int_0^T |\beta(z'(s))|\, ds \\ &\leqq \varepsilon \int_0^T h(z'(s))\, ds \\ &\leqq \varepsilon M . \end{aligned}$$

$I(z)$ is lower semicontinuous; therefore if the $[z\colon I(z) \leqq M]$ are compact sets,

(2.16) $$\lim_{\varepsilon \downarrow 0} \inf_{z \in C^\varepsilon} I(z) = \inf_{z \in C} I(z)$$

for any closed C. Lemmas 2.11 and 2.12 complete the proof of one part of Theorem 2.1.

LEMMA 2.13. *Let G be the subset of Ω defined by*

$$G = [z\colon |z(t) - a| < \varepsilon],$$

where ε is some positive number. Then

$$\liminf_{n \to \infty} \frac{1}{a_n} \log P_n(G) \geqq -h\left(\frac{a}{t}\right).$$

Proof: Let $\lambda_n = [nt]$. Then

$$\begin{aligned} P_n[G] &= P\left[\left|\frac{x_{n1} + \cdots + x_{n\lambda_n}}{n} - a\right| < \varepsilon\right] \\ &\geqq P\left[\left|x_{nj} - \frac{an}{\lambda_n}\right| < \varepsilon \text{ for all } j\right] \\ &\geqq P\left[\left|x_{nj} - \frac{a}{t}\right| < \tfrac{1}{2}\varepsilon \text{ for all } j\right] \quad \text{if } n \text{ is large enough} \\ &= \left\{c_n \int_{|y - a/t| < \frac{1}{2}\varepsilon} \exp\left[-\frac{a_n}{n}\, n(y)\right] dy\right\}^{\lambda_n}. \end{aligned}$$

The function $h(y)$ is continuous except for jumps to $+\infty$; therefore, if $h(a/t) < \infty$, $h(y)$ is continuous at that point either from the left or from the right so that, for small enough Δ, there is a one-sided interval around a/t, where $h(y) \geqq h(a/t) - \Delta$. Hence

$$P_n[G] \geqq c_n^{\lambda_n}\delta^{\lambda_n} \exp\left\{-\frac{a_n}{n}\left[h\left(\frac{a}{t}\right) - \Delta\right]\lambda_n\right\} .$$

Since

$$\lim_{n\to\infty} \frac{\lambda_n}{n} = t ,$$

$$\lim_{n\to\infty} a\frac{n}{a_n} \log c_n = 0 ,$$

$$\liminf_{n\to\infty} \frac{1}{a_n} \log P_n[G] \geqq -t\left[h\left(\frac{a}{t}\right) - \Delta\right] . \tag{2.17}$$

Now letting $\Delta \to 0$, we have

$$\liminf_{n\to\infty} \frac{1}{a_n} \log P_n[G] \geqq -th\left(\frac{a}{t}\right) .$$

The following lemmas are derived easily from Lemma 2.13 in a manner which is similar to that used in the derivation of Lemmas 2.7 and 2.8.

LEMMA 2.14. *Let A be a sphere around a point y in R, τ the map in Lemma* 2.8, *Φ the function defined there. Let $G = \tau^{-1}A$, then*

$$\liminf_{n\to\infty} \frac{1}{a_n} \log P_n[G] \geqq -\Phi(y) .$$

LEMMA 2.15. *Let G be the sphere around a point f in Ω of radius ε. Then*

$$\liminf_{n\to\infty} \frac{1}{a_n} \log P_n[G] \geqq -I(f) .$$

Proof: If f is not continuous, $I(f) = +\infty$ and there is nothing to prove. Let f be continuous, then

$$\lim_{N\to\infty} \|\theta_N f - f\| = 0 .$$

Let N be chosen large enough so that

$$\|\theta_N f - f\| < \tfrac{1}{3}\varepsilon .$$

Then

$$\begin{aligned} P_n[G] &= P_n[z: \|z - f\| < \varepsilon] \\ &\geqq P_n[z: \|\theta_N z - \theta_N f\| < \tfrac{1}{3}\varepsilon, \|\theta_N z - z\| < \tfrac{1}{3}\varepsilon] \qquad (2.18) \\ &\geqq P_n[z: \|\theta_N z - \theta_N f\| < \tfrac{1}{3}\varepsilon] - P_n[z: \|\theta_N z - z\| \geqq \tfrac{1}{3}\varepsilon] , \end{aligned}$$

and hence

$$\begin{aligned}\liminf_{n\to\infty}\frac{1}{a_n}\log P_n[z\colon \|\theta_N z-\theta_N f\|<\tfrac{1}{3}\varepsilon]\\ &=\liminf_{n\to\infty}\frac{1}{a_n}\log P_n[z\colon \|\tau_N z-\tau_N f\|<\tfrac{1}{3}\varepsilon]\\ &=\liminf_{n\to\infty}\frac{1}{a_n}\log P_n[\tau_N^{-1}(y\colon \|y-\tau_N f\|<\tfrac{1}{3}\varepsilon]\\ &\geqq -\Phi(\tau_N f)\\ &= -I(\theta_N f)\\ &\geqq -I(f)\,.\end{aligned}$$

On the other hand, Lemma 2.10 implies that if N is large enough, the second term of (2.18) with a negative sign is negligible compared to the first term. Therefore,

$$\liminf_{n\to\infty}\frac{1}{a_n}\log P_n[G]\geqq -I(f)\,.$$

Now let G be any open set. If $f\in G$, there exists a sphere around f contained in G. If S is the sphere, then

$$\begin{aligned}\liminf_{n\to\infty}\frac{1}{a_n}\log P_n[G] &\geqq \liminf_{n\to\infty}\frac{1}{a_n}\log P_n[S]\\ &\geqq -I(f)\,;\end{aligned}$$

since this is true for every f in G,

$$\liminf_{n\to\infty}\frac{1}{a_n}\log P_n[G]\geqq -\inf_{z\in G} I(z)\,.$$

This completes the proof of Theorem 2.1.

COROLLARY 2.1. *If $\bar{A}$, A^0 are the closure and interior of a set A in Ω and if*

$$\inf_{z\in A^0} I(z)=\inf_{z\in\bar{A}} I(z)=m\,,$$

then

$$\lim_{n\to\infty}\{P_n[A]\}^{1/a_n}=e^{-m}\,.$$

3. Let Ω be a regular topological space, and P_n a sequence of probability measures on Ω. The σ-field of subsets of Ω on which P_n is defined is assumed to include all open and closed subsets of Ω. Let a_n be a sequence of non-negative numbers such that

$$a_n \to \infty ,$$

$$\text{(3.1)} \qquad \limsup_{n\to\infty} \frac{1}{a_n} \log P_n(C) \leqq -\inf_{z\in C} I(z) ,$$

$$\liminf_{n\to\infty} \frac{1}{a_n} \log P_n(G) \geqq -\inf_{z\in G} I(z) ,$$

where G is open in Ω and C is closed in Ω. $I(z)$ is assumed to have the following properties:

(i) $0 \leqq I(z) \leqq \infty$,

(ii) $I(z)$ is lower semicontinuous everywhere in Ω,

(iii) $[z : I(z) \leqq M]$ is a compact subset of Ω for every finite M.

Let $F_n(z)$ be a suitable function on Ω. By $J_n(A)$ we denote the integral

$$J_n(A) = \int_A \exp\{a_n F_n(z)\}\, dP_n ,$$

$$\text{(3.2)}$$

$$J_n = J_n(\Omega) = \int_\Omega \exp\{a_n F_n(z)\}\, dP_n .$$

We shall investigate the asymptotic behavior of J_n. We make the following assumptions on the sequence $F_n(z)$.

(i) Given $\varepsilon > 0$, there exists, for any z_0 such that $I(z_0) < \infty$, a neighborhood N_{z_0} of z_0 in Ω such that, for all sufficiently large n and z in N_{z_0},

$$F_n(z) \leqq F(z_0) + \varepsilon .$$

(ii) $F_n(z) \leqq L$ for all n and z in Ω.

THEOREM 3.1. *Under conditions* (i) *and* (ii) *on* $F_n(z)$,

$$\limsup_{n\to\infty} \frac{1}{a_n} \log J_n \leqq \sup_{z\in\Omega} [F(z) - I(z)] .$$

Proof: Let

$$\Lambda = \sup_{z\in\Omega} [F(z) - I(z)] .$$

Let $f \in \Omega$ be an arbitrary point in Ω such that $I(f) < \infty$. Then for any $\varepsilon > 0$ there exists (from property (ii) of $I(z)$ and property (i) of $F_n(z)$) a neighborhood N_f of f such that

$$F_n(z) \leqq F(f) + \varepsilon \qquad \text{for} \qquad z \in N_f ,$$

$$I(z) \geqq I(f)) - \varepsilon \qquad \text{for} \qquad z \in N_f .$$

Let G_f be another neighborhood of f such that

$$\bar{G}_f = C_f \subset N_f ,$$

then

$$\begin{aligned} J_n(G_f) &\leqq \exp \{\sup_{z\in G_f} a_n F_n(z)\} P_n(G_f) \\ &\leqq \exp \{a_n \sup_{z\in N_f} a_n F_n(z)\} P_n(C_f) \\ &\leqq \exp \{a_n[F(f) + \varepsilon]\} P_n(C_f) . \end{aligned}$$

Therefore,

$$\begin{aligned} \limsup_{n\to\infty} \frac{1}{a_n} \log J_n(G_f) &\leqq F(f) + \varepsilon - \inf_{z\in C_f} I(z) \\ &\leqq F(f) - I(f) + 2\varepsilon \qquad (3.3) \\ &\leqq \Lambda + 2\varepsilon . \end{aligned}$$

Let K be the set $\{z\colon I(z) \leqq M\}$. For a given $\varepsilon > 0$, and any $f \in K$, there exists a neighborhood G_f of f such that

$$\limsup_{n\to\infty} \frac{1}{a_n} \log J_n(G_f) \leqq \Lambda + 2\varepsilon .$$

Since by property (iii) of $I(z)$, K is compact in Ω, there exists a finite union of $\{G_f\}, f \in K$, which covers K. Let us denote by G this finite union. Then

$$\begin{aligned} \limsup_{n\to\infty} \frac{1}{a_n} \log J_n(G) &\leqq \max_{f=f_1,\cdots,f_N} \limsup_{n\to\infty} \frac{1}{a_n} \log J_n(G_f) \\ &\leqq \Lambda + 2\varepsilon . \end{aligned}$$

Let $C = \Omega - G$. Then

$$J_n(C) \leqq \exp \{a_n L\} P_n(C) .$$

$$\limsup_{n\to\infty} \frac{1}{a_n} \log J_n(C) \leqq L - \inf_{z\in C} I(z) \leqq L - M .$$

If we choose M large enough so that $L - M \leqq \Lambda$,

$$\limsup_{n\to\infty} \frac{1}{a_n} \log J_n \leqq \max [\Lambda + 2\varepsilon, L - M] = \Lambda + 2\varepsilon .$$

Since $\varepsilon > 0$ is arbitrary,

$$\limsup_{n\to\infty} \frac{1}{a_n} \log J_n \leqq \Lambda .$$

We now replace the assumption that $\{F_n(z)\}$ is uniformly bounded from above by the less restrictive condition

$$\lim_{L\to\infty} \limsup_{n\to\infty} \frac{1}{a_n} \log \left[\int_{F_n(z)\geqq L} \exp \{a_n F_n(z)\}\, dP_n \right] = -\infty . \qquad (3.4)$$

THEOREM 3.2. *If $\{F_n(z)\}$ satisfies condition* (i) *and* (3.4), *then*

$$\limsup_{n\to\infty} \frac{1}{a_n} \log J_n \leqq \sup_z [F(z) - I(z)] .$$

Proof: Let us denote by $\bar{F}_n(z)$ the function defined to be

$$\bar{F}_n(z) = \min [F_n(z), L] .$$

Then $\bar{F}_n(z)$ satisfies the conditions of Theorem 3.1 and, hence, if

$$\bar{J}_n = \int_\Omega \exp \{a_n \bar{F}_n(z)\} \, dP_n ,$$

$$\limsup_{n\to\infty} \frac{1}{a_n} \log \bar{J}_n \leqq \sup [\bar{F}(z) - I(z)] \leqq \sup [F(z) - I(z)] .$$

Condition (3.4) insures that for large L

$$\limsup_{n\to\infty} \frac{1}{a_n} \log J_n = \limsup_{n\to\infty} \frac{1}{a_n} \log \bar{J}_n .$$

This proves the theorem.

Remark. Condition (i) is equivalent to the following:
Whenever $z_n \to z$ and $I(z) < \infty$,

$$\limsup_{n\to\infty} F_n(z_n) \leqq F(z) .$$

This is always true if $F_n = F$ for all n, and F is upper semicontinuous.

Let $f \in \Omega$ be such that for any $\varepsilon > 0$ there exists a neighborhood G_f of f in Ω such that, for n suitably large and z in G_f,

$$F_n(z) \geqq F(f) - \varepsilon . \tag{3.5}$$

Then

$$J_n \geqq J_n(G_f) \geqq \exp \{a_n[F(f) - \varepsilon]\} P_n(G_f) ,$$

$$\liminf_{n\to\infty} \frac{1}{a_n} \log J_n \geqq F(f) - \varepsilon - \inf_{z \in G_f} I(z) \geqq F(f) - I(f) - \varepsilon .$$

Since $\varepsilon > 0$ is arbitrary,

$$\liminf_{n\to\infty} \frac{1}{a_n} \log J_n \geqq F(f) - I(f) . \tag{3.6}$$

If Ω_0 is defined as the set of points f in Ω for which, given any $\varepsilon > 0$, there exists a neighborhood G_f such that (3.5) holds for suitably large n, we have from (3.6) the following theorem.

THEOREM 3.3.

$$\liminf_{n\to\infty} \frac{1}{a_n} \log J_n \geqq \sup_{z\in\Omega_0} [F(z) - I(z)] .$$

We can now summarize Theorems 3.2 and 3.3 in the following manner. Let $\{F_n(z)\}$ be a sequence of functions satisfying condition (3.4). Assume further:

(i) For every z in Ω such that $I(z) < \infty$ and $z_n \to z$,

$$\limsup_{n\to\infty} F_n(z_n) \leqq F(z) .$$

(ii) For every z in Ω_0 and $z_n \to z$,

$$\liminf_{n\to\infty} F_n(z_n) \geqq F(z) .$$

(iii) $\Lambda = \sup_{z\in\Omega} [F(z) - I(z)] = \sup_{z\in\Omega_0} [F(z) - I(z)]$.

Then the following theorem holds.

THEOREM 3.4.

$$\lim_{n\to\infty} \frac{1}{a_n} \log J_n = \Lambda .$$

Let A be any closed subset of Ω. Let $\{F_n(z)\}$ satisfy the conditions of Theorem 3.2. Define $\bar{F}_n(z) = F_n(z)$ for $z \in A$, $\bar{F}_n(z) = -\infty$ for $z \in \Omega - A$, $\bar{F}(z) = F(z)$ for $z \in A$ and $\bar{F}(z) = -\infty$ for $z \in \Omega - A$. Then $\{\bar{F}_n(z)\}$ satisfies the conditions and one obtains

THEOREM 3.5.

$$\limsup_{n\to\infty} \frac{1}{a_n} \log J_n(A) \leqq \sup_{z\in A} [F(z) - I(z)] .$$

Let us now suppose that $\{F_n(z)\}$ is a sequence of functions satisfying the condition (3.4) and such that:

(i) Whenever $I(z) < \infty$ and $z_n \to z$,

$$\limsup_{n\to\infty} F_n(z_n) \leqq F(z) .$$

(ii) The supremum $\Lambda = \sup_{z\in\Omega} [F(z) - I(z)]$ is attained at a point f in Ω.

(iii) $\sup_{\|z-f\|\geqq\varepsilon} [F(z) - I(z)] < \Lambda$ for every $\varepsilon > 0$.

(iv) $\liminf_{n\to\infty} F_n(z_n) \geqq F(f)$ for every sequence $z_n \to f$.

We define a measure Q_n on Ω by the formula

$$Q_n(A) = \frac{1}{J_n} J_n(A) \qquad \text{for every} \qquad A. \tag{3.7}$$

THEOREM 3.6. *If $g(z)$ is a bounded function on Ω which is continuous at $z = f$, then*

$$\lim_{n\to\infty} \int g(z)\, dQ_n = g(f).$$

Proof: Since $g(z)$ is continuous at $z = f$, it suffices to show that

$$\lim_{n\to\infty} Q_n(A) = 0,$$

for every set A of the form $\|z - f\| \geqq \varepsilon$. Since A is a closed set,

$$\limsup_{n\to\infty} \frac{1}{a_n} \log J_n(A) < \Lambda$$

and

$$\liminf_{n\to\infty} \frac{1}{a_n} \log J_n = \Lambda.$$

This proves the theorem.

4. We now go back to the case considered in Section 2. Let P_n be the measures defined there. Let $\{F_n(z)\}$ be any sequence of functions such that

$$F_n(z) \leqq A + B\,\|z\|$$

for some constants A and B, independent of n. Then we can show that condition (3.4) is satisfied.

LEMMA 4.1. *If $f_n(z) \leqq A + B\,\|z\|$ for all n and z,*

$$\lim_{L\to\infty} \limsup_{n\to\infty} \frac{1}{a_n} \log\left[\int_{F_n(z)\geqq L} \exp\{a_n F_n(z)\}\, dP_n\right] = -\infty.$$

Proof:

$$\int_{F_n(z)\geqq L} \exp\{a_n F_n(z)\}\, dP_n \leqq \int_{F_n(z)\geqq L} \exp\{a_n[A + B\,\|z\|]\}\, dP_n$$

$$\leqq \int_{\|z\|\geqq (L-A)/B} \exp\{a_n[A + B\,\|z\|]\}\, dP_n.$$

It remains to prove that

$$\lim_{L\to\infty} \limsup_{n\to\infty} \frac{1}{a_n} \log\left|\int_{\|z\|\geqq L} \exp\{a_n B\,\|z\|\}\, dP_n\right| = -\infty.$$

For this purpose we need to estimate $P_n[\|z\| \geqq L]$. It is easy to see that, for L suitably large, the conditions of Lemma 2.9 are satisfied and, hence,

$$P_n[\|z\| \geqq L] \leqq 2P_n[|z(T)| \geqq \tfrac{1}{2}L] .$$

Therefore it suffices to prove that, for every B,

$$\lim_{L\to\infty} \limsup_{n\to\infty} \frac{1}{a_n} \log \left[\int_{|z(T)| \geqq L} \exp\{a_n B\,|z(T)|\}\,dP_n \right] = -\infty . \tag{4.1}$$

From equation (2.4) we have

$$P_n[z(T) \geqq M] \leqq \exp\{-a_n \max_{\theta \geqq 0} [\theta M - TQ_n(\theta)]\} \leqq \exp\{-a_n[\theta_0 M - K]\} ,$$

where K is a constant depending on θ_0. Since there exists a similar estimate for $P_n[z(T) \leqq -M]$, we can ensure that for a given value of θ_0 there exists a constant K such that

$$P_n[|z(T)| \geqq M] \leqq K \exp\{-a_n\theta_0\} .$$

By choosing θ_0 bigger than B, one can establish (4.1).

Let $P(\tau, x)$ be a continuous function of x for $0 \leqq t \leqq T$ and $-\infty < x < \infty$. Let $U(x)$ be a continuous function of x in $-\infty < x < \infty$. Moreover, let

(i) $U(x) \leqq A + B\,|x|$ for all x ,

(ii) $P(\tau, x) \leqq A + B\,|x|$ for all τ and x .

Consider the functions $F_n(z)$ on Ω defined by

$$F_n(z) = \int_0^{t_n} P(\tau, z(t_n - \tau))\,d\tau + U(z(t_n)) ,$$

and let

$$F(z) = \int_0^{t} P(\tau, z(t - \tau))\,d\tau + U(z(t)) .$$

Since z is continuous in $0 \leqq t \leqq T$, it is not difficult to see that, if $z_n \to z$ and $t_n \to t$ as $n \to \infty$,

$$\lim_{n\to\infty} F_n(z_n) = F(z)$$

and

$$F_n(z) \leqq A_0 + B_0\,\|z\| ,$$

for all z with $I(z) < \infty$. We therefore have

THEOREM 4.1.

$$\lim_{n\to\infty} \frac{1}{a_n} \log \left[\int_\Omega \exp\{a_n F_n(z)\}\,dP_n \right] = \sup_{z\in\Omega} [F(z) - I(z)] ,$$

where

$$I(z) = \int_0^T h(z'(\tau))\,d\tau .$$

COROLLARY 4.1. *Since $F(z)$ depends only on the path up to time t, we can write*

$$\sup_{z\in\Omega} [F(z) - I(z)] = \sup_{z\in\Omega} \left[F(z) - \int_0^t h(z'(\tau))\, d\tau\right].$$

LEMMA 4.2. *Let $F(z) \leqq A + B\,\|z\|$ and let $F(z)$ be upper semicontinuous. Then for any closed set C the supremum*

$$\Lambda = \sup_{z\in C} [F(z) - I(z)]$$

is attained.

Proof: Since $\Lambda > -\infty$ we can assume, without loss of generality, that $z_n \in C$ is such that

$$F(z_n) - I(z_n) \to \Lambda .$$

If we show that z_n is compact, then, in view of the proper semicontinuity assumptions, the supremum will be attained at any limit point (which are necessarily within the closed set C). Since

$$F(z_n) - I(z_n) \to \Lambda > -\infty ,$$

there are two possibilities. Either $I(z_n) \to +\infty$, or there exists a subsequence with $I(z_n)$ bounded, which would mean that there exists at least one limit point. So let us suppose that $I(z_n) \to +\infty$. Let $L > B$ and let M be chosen such that if $|y| \geqq M$, $|n(y)| \geqq L\,|y|$,

$$\begin{aligned} I(z_n) &= \int_0^T h(z_n'(\tau))\, d\tau \\ &\geqq L \int_{[t:|z_n'(t)|\geqq M]} |z_n'(\tau)|\, d\tau \\ &\geqq L \int_0^T |z_n'(\tau)|\, d\tau - TML, \\ &\geqq L\,\|z_n\| - TML , \end{aligned}$$

then

$$F(z_n) - I(z_n) \leqq (B - L)\,\|z_n\| + TML + A .$$

Since $F(z_n) - I(z_n) \to \Lambda > -\infty$, and $L > B$, we must have $\|z_n\|$ bounded or $F(z_n)$ bounded from above. This in turn would imply that $I(z_n)$ is bounded and hence the lemma.

Let $\{F_n(z)\}$, $F(z)$ be as described in Theorem 4.1. Moreover, let the supremum

$$\Lambda = \sup_{z\in\Omega} F(z) - \int_0^t h(z'(\tau))\, d\tau$$

be attained at a unique function f in the space of functions on the interval $[0, t]$. Let $g(z)$ be a function of z for z in Ω depending only on $z(\tau)$ for $0 \leqq \tau \leqq t$. Let g be continuous at $z = f$. Then

THEOREM 4.2.

$$\lim_{n\to\infty} \frac{\int g(z) \exp\{a_n F_n(z)\}\, dP_n}{\int \exp\{a_n F_n(z)\}\, dP_n} = g(f)\,.$$

Proof: The proof is not quite the same as that of Theorem 3.6, because the supremum over Ω may not be unique. However the conditions ensure that any function $\bar{f}$ in Ω where the sup is attained must be an extension of f defined on $0 \leqq \tau \leqq t$. The function $g(z)$ depends only on $z(\tau)$ for $0 \leqq \tau \leqq t$. Hence taking C_ε to be the closed set

$$C_\varepsilon = [z\colon \sup_{0\leqq\tau\leqq t} |z(\tau) - f(\tau)| \geqq \varepsilon]\,,$$

it suffices to show that for every $\varepsilon > 0$

$$\lim_{n\to\infty} \frac{J_n[C_\varepsilon]}{J_n} = 0\,.$$

But

$$\limsup_{n\to\infty} \frac{1}{a_n} \log J_n[C_\varepsilon] \leqq \sup_{z\in C_\varepsilon} [F(z) - I(z)] < \Lambda\,.$$

From now on the theorem is proved in the same way as Theorem 3.6.

5. Let us start with the functions (continuous in x)

$$U(x) \leqq A + B|x|\,,$$

$$P(t, x) \leqq A + B|x|\,,$$

and consider the integral

$$V_n(t, x) = \int_\Omega H_n(t, z, x)\, dP_n\,, \tag{5.1}$$

where

$$H_n(t, z, x) = \exp\left\{a_n\left[\int_0^t P(t-\tau, z(\tau) + x)\, d\tau + U(z(t) + x)\right]\right\}.$$

It is easy to check, by taking suitable conditional expectations in (5.1), that

$$V_n\left(\frac{j+1}{n}, x\right) = \exp\left\{a_n \int_{j/n}^{(j+1)/n} P(\tau, x)\, d\tau\right\} \int V_n\left(\frac{j}{n}, x + \frac{y}{n}\right) dF_n(y)\,, \tag{5.2}$$

$$V_n(0, x) = \exp\{a_n U(x)\}$$

Here P_n and F_n are as in Section 2. If we take log on both sides of (5.2) and put

$$U_n(t, x) = \frac{1}{a_n} \log V_n(t, x)\,,$$

then

$$U_n\left(\frac{j+1}{n}, x\right) - U_n\left(\frac{j}{n}, x\right) = \int_{j/n}^{(j+1)/n} P(\tau, x)\, d\tau$$

$$+ \frac{1}{a_n} \log \int \exp\left\{a_n \left[U_n\left(\frac{j}{n}, x + \frac{U}{n}\right) - U_n\left(\frac{j}{n}, x\right)\right]\right\} dF_n(y) .$$

Multiplying by n and taking the limit as $n \to \infty$,

$$U_t(t, x) = P(t, x) + f(U_x(t, x)) .$$

Hence, formally denoting $u = U_x$,

$$u_t = p(t, x) + [f(u)]_x .$$

We shall first examine the behavior of $U_n(t, x)$. Since the recurrence relation (5.2) yields the value of $U_n(t, x)$ for t of the form j/n, we can choose j_n such that $j_n/n \to t$.

THEOREM 5.1.

$$\lim_{n\to\infty} U_n(t_n, x) = U(t, x)$$

exists, whenever $t_n \to t$ as $n \to \infty$, and the limit is given by

$$U(t, x) = \sup_{z\in\Omega} \left[\int_0^t P(t - \tau, z(\tau) + x)\, d\tau + U(z(t) + x) - \int_0^t h(z'(t))\, d\tau\right] .$$

Proof:

$$V_n(t_n, x) = \int_\Omega \exp\{a_n F_n(z)\}\, dP_n ,$$

where

$$F_n(z) = \int_0^{t_n} P(t_n - \tau, z(\tau) + x)\, d\tau + U(z(t_n) + x)$$

$$= \int_0^{t_n} P(\tau, z(t_n - \tau) + x)\, d\tau + U(z(t_n) + x) .$$

Since Theorem 4.1 is applicable now (the translation by x does not affect the validity) the theorem is proved.

THEOREM 5.2. *Let*

$$U(x) = \int_0^x u(y)\, dy ,$$

$$P(t, x) = \int_0^x p(t, y)\, dy ,$$

$$u_n(t, x) = \frac{d}{dx} U_n(t, x) ,$$

where u and p are continuous functions. Then, if for a given t and x the supremum in Theorem 5.1 is attained at a unique function $\omega(\tau)$, then

$$\lim_{n\to\infty} u_n(t_n, x) = \int_0^t p(t-\tau, \omega(\tau)+x)\, d\tau + u(\omega(t)+x)\,.$$

Proof: This follows immediately from the definition of $u_n(t, x)$ and Theorem 4.2.

We can summarize the results as follows. Suppose we are interested in the equation

$$u_t = [f(u)]_x + p(t, x) \tag{5.3}$$

with initial condition $u_0(x) = u(x)$. We form the functions $U(x)$, $P(t, x)$ which are indefinite integrals of $u(x)$ and $p(t, x)$, respectively. We then construct V_n successively by the linear model (5.2). We define

$$U_n = \frac{1}{a_n} \log V_n\,.$$

Then under suitable conditions, $U_n \to U$, where U is given in Theorem 5.1. U_x is known to solve the equation (5.3) in the weak sense, [4], [7].

Under additional conditions we even have the convergence of dU_n/dx for specific values of t and x.

An important special case is when $f(\theta)$ is of the form

$$f(\theta) = c\left[\log \int_{-\infty}^{\infty} e^{\theta y}\, dF(y)\right].$$

In this case F_n can be taken to be equal to F for all n and a_n can be chosen to be n.

The estimates of Section 2 can be obtained in a similar way starting instead from the known results for large deviations of arithmetic means. This includes the case studied by Lax [6]. In that case F is a discrete measure on a discrete set of equally spaced points.

If $f(\theta)$ has the further special form

$$f(\theta) = c\left[\log \int_{-\infty}^{\infty} e^{\theta y}\, dF(y)\right],$$

where F is an infinitely divisible distribution, the discrete manner in which P_n is constructed could be replaced by a continuous process and instead of (5.2) we would obtain

$$\frac{dV_n}{dt} = a_n V_n P + L_n[V_n]\,,$$

where L_n is the generator of a suitable process with independent increments. The case where $f = \frac{1}{2}\theta^2$ leads through the Gaussian distribution to Brownian motion and the second derivative for L_n.

For the detailed exposition of processes with independent increments and connections with semigroups and their generators see Hille and Phillips [2].

Bibliography

[1] Donsker, M. D., *Function space integrals and non-linear differential equations,* to appear.

[2] Hille, E., and Phillips, R. S., *Functional Analysis and Semigroups,* Rev. Ed., Amer. Math. Soc., Providence, R. I., 1957.

[3] Hopf, E., *On the partial differential equation* $u_t + uu_x = \mu u_{xx}$, Comm. Pure Appl. Math., Vol. 3, 1950, pp. 201–230.

[4] Conway, E., and Hopf, E., *Hamilton's theory and the generalized solutions of the Hamilton Jacobi equation,* J. Math. Mech., Vol. 13, 1964, pp. 939–986.

[5] Lax, P. D., *Hyperbolic systems of conservation laws, II,* Comm. Pure Appl. Math., Vol. 10, 1957, pp. 537–566.

[6] Lax, P. D., *On difference schemes for solving initial value problems for conservation laws,* Proc. Rome Symposium, Provisional International Computation Centre, Libreria Eredi Virgilio Veschi, Roma, 1958, pp. 69–78.

[7] Oleinik, O. A., *On discontinuous solutions of non-linear differential equations,* Uspekhi Mat. Nauk, Vol. 12, 1957, pp. 3–73. (In Russian.)

[8] Schilder, M., *On a Laplace asymptotic formula for Wiener integrals,* to appear in the Trans. Amer. Math. Soc.

Received November, 1965.

COMMUNICATIONS ON PURE AND APPLIED MATHEMATICS, VOL. XX, 431–455 (1967)

On the Behavior of the Fundamental Solution of the Heat Equation with Variable Coefficients*

S. R. S. VARADHAN

1. Introduction

The solution $q(t, x, y)$ of the equation

$$\frac{\partial q}{\partial t} = \frac{1}{2} \sum_{i=1}^{k} \frac{\partial^2 q}{\partial x_i^2} \tag{1.1}$$

with the boundary condition $q(t, x, y) \to \delta_x(y)$ as $t \to 0$ can of course be written explicitly as

$$q(t, x, y) = (2\pi t)^{-k/2} \exp\left\{-\frac{1}{2t} \|x - y\|^2\right\}, \tag{1.2}$$

where $\|x - y\|$ denotes the Euclidean distance. Looking at (1.2) one sees immediately that

$$\lim_{t \to 0} [-2t \log q(t, x, y)] = \|x - y\|^2 . \tag{1.3}$$

We shall consider the analogue of (1.1). Let $p(t, x, y)$ be the solution of the equation

$$\frac{\partial p}{\partial t} = \frac{1}{2} \sum_{i,j=1}^{k} a_{ij}(x) \frac{\partial^2 p}{\partial x_i \partial x_j} \tag{1.4}$$

with the boundary condition $p(t, x, y) \to \delta_x(y)$ as $t \to 0$. We prove a formula similar to (1.3):

$$\lim_{t \to 0} [-2t \log p(t, x, y)] = d^2(x, y) , \tag{1.5}$$

where $d(x, y)$ is the distance induced by a Riemannian metric derived from the coefficients $\{a_{ij}(x)\}$.

* This paper represents results obtained at the Courant Institute of Mathematical Sciences, New York University, under the sponsorship of the Office of Naval Research, Contract Nonr-285(46), and a Sloan Foundation Grant for Probability and Statistics.

If G is a region with a boundary B, we consider the solution $\phi(x, \lambda)$ of the equation

$$\tfrac{1}{2}\sum a_{ij}(x)\frac{\partial^2\phi}{\partial x_i\,\partial x_j} = \lambda\phi \qquad \text{for} \qquad x \in G$$

with the boundary value $\phi = 1$ on B. We prove that

$$\lim_{\lambda\to\infty}[-(2\lambda)^{-1/2}\log\phi(x, \lambda)] = d(x, B)\,, \tag{1.6}$$

where x is any point of G and $d(x, B)$ is the shortest distance to the boundary B from x.

The behavior (1.5) of $p(t, x, y)$ is derived from the behavior (1.6) of $\phi(x, \lambda)$ which is proved first. Although these results are closely related to certain properties of Markov processes, the probabilistic connections will be explored separately, [3].

Section 2 covers the preliminaries. The assumptions and the main theorems are stated there. The later sections cover the actual proof.

2. Preliminaries

L stands for the following differential operator acting on smooth functions on R_k:

$$Lf = \tfrac{1}{2}\sum a_{ij}(x)\frac{\partial^2 f}{\partial x_i\,\partial x_j}\,. \tag{2.1}$$

The matrix $\{a(x)\}$ of coefficients is of course symmetric and positive definite and the following assumptions are made concerning $a(x)$:

(i) Uniform Hölder condition, i.e., for all i and j we have

$$|a_{ij}(x) - a_{ij}(y)| \leqq M\,\|x - y\|^{\alpha} \text{ for all } x, y, \qquad \alpha > 0.$$

(2.2) (ii) Uniform ellipticity condition, i.e., there exist constants α and A such that for any vector $\xi_1, \cdots, \xi_k$, and x

$$A\sum \xi_i^2 \geqq \sum a_{ij}(x)\xi_i\xi_j \geqq \alpha\sum \xi_i^2\,.$$

We then consider the equation

$$\frac{\partial p}{\partial t} = Lp \tag{2.3}$$

with the boundary condition $p(t, x, y) \to \delta_x(y)$ as $t \to 0$.

The following theorem concerning the existence of a solution $p(t, x, y)$ for (2.3) can be found in [1], [2].

THEOREM 2.1. *Under the hypothesis* (2.2) *on the coefficients there exists a solution* $p(t, x, y)$ *for equation* (2.3) *and it has the following properties:*

(i) $p(t, x, y) \geqq 0$,

(ii) $\int p(t, x, y)p(s, y, z)\, dy = p(t + s, x, z)$,

(iii) $p(t, x, y)$ *is continuous in* t, x, *and* y,

(iv) *there are constants* M *and* α *such that*

$$p(t, x, y) \leqq Mt^{-k/2} \exp\left\{-\frac{\alpha}{2t}\|x - y\|^2\right\},$$

(v) *there are positive constants* α_1, α_2, M_1, M_2, λ *such that*

$$p(t, x, y) \geqq M_1 t^{-k/2} \exp\left\{-\frac{\alpha_1}{2t}\|x - y\|^2\right\}$$

$$-M_2 t^{-k/2+\lambda} \exp\left\{-\frac{\alpha_2}{2t}\|x - y\|^2\right\}.$$

Further $p(t, x, y)$ *is unique.*

We now introduce the metric d. Let $p(\tau)$, $0 \leqq \tau \leqq 1$, be a smooth path in R_k. Then the length of such a path is defined as

$$l(p) = \int_0^1 [\dot{p}(\tau)a^{-1}(p(\tau))\dot{p}(\tau)]^{1/2}\, d\tau,$$

where $\dot{p}(\tau)$ stands for $dp(\tau)/d\tau$, $a^{-1}(p(\tau))$ for the matrix inverse to $a(p(\tau))$ and $(\theta a \theta)$ for the quadratic form $\Sigma a_{ij}\theta_i\theta_j$; $l(p)$ is the natural length in a metric defined locally as

$$ds^2 = \sum a^{ij}(x)\, dx_i\, dx_j.$$

The operator L modified by a first order term is Laplacian in this metric. The global distance $d(x, y)$ induced by this metric is defined as

$$d(x, y) = \inf_{\substack{p:\, p(0)=x \\ p(1)=y}} l(p). \tag{2.4}$$

We can now state the first main theorem concerning the behavior of $p(t, x, y)$ introduced in Theorem 2.1 for small t.

THEOREM 2.2.

$$\lim_{t \to 0} [-2t \log p(t, x, y)] = d^2(x, y)$$

uniformly over x *and* y *such that* $d(x, y)$ *is bounded.*

We consider an open set G in R_k. Let B be the boundary of G. It is assumed that every point b on B is a limit point of the exterior of G.

Let $\phi(x, \lambda)$ be the solution of the equation

$$(2.5)\qquad \begin{aligned} L\phi &= \lambda\phi \qquad \text{for} \qquad x \in G\,, \\ \phi &= 1 \qquad \text{for} \qquad b \in B\,. \end{aligned}$$

The solution exists for $\lambda \geqq 0$, if the boundary value is properly interpreted. The next theorem is concerned with the behavior of the solution $\phi(x, \lambda)$ for large λ.

THEOREM 2.3. *With the assumptions stated above,*

$$\lim_{\lambda\to\infty}\left[-\frac{1}{\sqrt{2\lambda}}\log\phi(x, \lambda)\right] = d(x, B)$$

uniformly over compact subsets of $G \cup B$.

We state some notation and a few elementary consequences concerning the distance $d(x, y)$. The distance $[(x-y)\Lambda(x-y)]^{1/2}$ is denoted by $\|x-y\|_\Lambda$, where Λ is a positive definite matrix. If Λ is chosen as $a^{-1}(z)$, then $\|x-y\|$ is denoted by $\|x-y\|_z$; $\|x-y\|$ is of course the Euclidean distance. It is clear from the assumptions that d, $\|\ \ \|$, $\|\ \ \|_z$ are all equivalent in the sense that the ratio of any two is bounded by a universal constant. We shall denote by $\Delta(x, y)$ any one of these metrics. If a and b are positive definite matrices, we shall mean by $a \geqq b$ that $a - b$ is positive semidefinite. The following is a consequence of the Hölder condition in (2.2).

LEMMA 2.4. *There exists a function* $\delta(\varepsilon) \downarrow 0$ *as* $\varepsilon \downarrow 0$ *such that if* $\Delta(x, y) \leqq \varepsilon$, *then*

$$[1-\delta(\varepsilon)]a(x) \leqq a(y) \leqq [1+\delta(\varepsilon)]a(x)\,,$$

$$[1-\delta(\varepsilon)]a^{-1}(x) \leqq a^{-1}(y) \leqq [1+\delta(\varepsilon)]a^{-1}(x)\,.$$

From Lemma 2.4 and the definition of $d(x, y)$ we deduce

LEMMA 2.5. *There exists a function* $\delta(\varepsilon) \downarrow 0$ *as* $\varepsilon \downarrow 0$ *such that if* $\Delta(x, y) \leqq \varepsilon$ *and* $\Delta(x, z) \leqq \varepsilon$, *then*

$$[1-\delta(\varepsilon)]\,\|x-y\|_z \leqq d(x, y) \leqq [1+\delta(\varepsilon)]\,\|x-y\|_z\,.$$

3. Elliptic Equation

In this section we prove Theorem 2.3. The method consists of comparing our solution with the solution of an equation having constant coefficients in a small domain and then piecing together the estimate with the help of the maximum principle.

Let y be an arbitrary point in R_k. Let the set S_ε and its boundary B_ε be defined as follows:

$$S_\varepsilon = [x: \|x - y\|_y < \varepsilon],$$

$$B_\varepsilon = [x: \|x - y\|_y = \varepsilon].$$

We denote by L_y the operator

$$L_y f = \tfrac{1}{2} \sum a_{ij}(y) \frac{\partial^2 f}{\partial x_i \, \partial x_j}.$$

It is an operator with constant coefficients. Let $\psi_\varepsilon(x, \lambda)$ be the solution to the equation

$$\begin{aligned} L_y \psi_\varepsilon &= \lambda \psi_\varepsilon \qquad \text{for} \qquad x \in S_\varepsilon, \\ \psi_\varepsilon &= 1 \qquad \text{for} \qquad x \in B_\varepsilon. \end{aligned} \tag{3.1}$$

THEOREM 3.1. *For every $\lambda > 0$, $\psi(x, \lambda)$ is explicitly given by the formula below for $x \in S_\varepsilon$:*

$$\psi_\varepsilon(x, \lambda) = \frac{\cosh(\sqrt{2\lambda}\, r)}{\cosh(\sqrt{2\lambda}\, \varepsilon)} \qquad \text{for} \qquad k = 1,$$

$$\psi_\varepsilon(x, \lambda) = \frac{\int_0^\pi \cosh(\sqrt{2\lambda}\, r \cos\theta) \sin^{k-2}\theta \, d\theta}{\int_0^\pi \cosh(\sqrt{2\lambda}\, \varepsilon \cos\theta) \sin^{k-2}\theta \, d\theta} \qquad \text{for} \qquad k \geqq 2,$$

where $r = \|x - y\|_y$. For all k, $\psi_\varepsilon(x, \lambda)$ is a convex function of x.

Proof: The proof is elementary and consists of direct verification.

THEOREM 3.2. *For every $\rho > 0$ there exists a constant $M_\rho < \infty$ depending only on the dimension k such that, for all y, ε and λ,*

$$\psi_\varepsilon(y, \lambda) \leqq M_\rho \exp\{-\varepsilon(1 - \rho)\sqrt{2\lambda}\}.$$

Proof: When $x = y$ or $r = 0$ the numerator in Theorem 3.1 reduces to a dimension constant. As for the denominator, it can be estimated from below by limiting the range of integration to the region $0 \leqq \theta \leqq \cos^{-1}(1 - \rho)$.

We now consider the sets

$$T_\varepsilon = [x: d(x, y) < \varepsilon],$$

$$D_\varepsilon = [x: d(x, y) = \varepsilon].$$

Then in view of Lemma 2.5, there exists a function $\delta(\varepsilon) \downarrow 0$ as $\varepsilon \downarrow 0$ such that the set

$$S = [x: \|x - y\|_y < \varepsilon(1 - \delta(\varepsilon))]$$

is contained in T_ε. Let B be the boundary of S:

$$B = [x: \|x - y\|_y = \varepsilon(1 - \delta(\varepsilon))] .$$

Obviously B is contained in $T_\varepsilon \cup D_\varepsilon$.

Let $f_\varepsilon(x, \lambda)$, $g_\varepsilon(x, \lambda)$, and $h_\varepsilon(x, \lambda)$ denote the solutions of the following equations:

$$\begin{aligned} Lf &= \lambda f && \text{for} \quad x \in T_\varepsilon , \\ f &= 1 && \text{for} \quad x \in D_\varepsilon , \end{aligned} \tag{3.2}$$

$$\begin{aligned} Lg &= \lambda g && \text{for} \quad x \in S , \\ g &= 1 && \text{for} \quad x \in B , \end{aligned} \tag{3.3}$$

$$\begin{aligned} L_y h &= \lambda h && \text{for} \quad x \in S , \\ h &= 1 && \text{for} \quad x \in B . \end{aligned} \tag{3.4}$$

Lemma 3.3.

$$f(y, \lambda) \leqq g(y, \lambda) .$$

Proof: From equation (3.2) it is obvious that $f \leqq 1$ in $T_\varepsilon \cup D_\varepsilon$. Since $B \subset T_\varepsilon \cup D_\varepsilon$, it follows that $f \leqq 1$ on B. But f and g satisfy the same equation in S and on the boundary B of S, $g = 1$ and $f \leqq 1$. Hence $f \leqq g$ everywhere in S. Therefore in particular, $f(y, \lambda) \leqq g(y, \lambda)$.

Lemma 3.4. *There exists a function $\delta(\varepsilon) \downarrow 0$ as $\varepsilon \downarrow 0$ such that*

$$g(y, \lambda) \leqq h(y, \lambda(1 - \delta(\varepsilon))) .$$

Proof: By Theorem 3.1, h is convex and by Lemma 2.4

$$a(x) \leqq \frac{1}{1 - \delta(\varepsilon)} a(y) .$$

Therefore,

$$\begin{aligned} \tfrac{1}{2} \sum a_{ij}(x) \frac{\partial^2 h}{\partial x_i \, \partial x_j} &\leqq \frac{1}{2(1 - \delta(\varepsilon))} \sum a_{ij}(y) \frac{\partial^2 h}{\partial x_i \, \partial x_j} \\ &= \frac{1}{(1 - \delta(\varepsilon))} L_y h \\ &= \frac{\lambda}{(1 - \delta(\varepsilon))} h , \end{aligned}$$

or

$$Lh(x, \lambda(1 - \delta(\varepsilon))) \leqq \lambda h \quad \text{in} \quad S .$$

But

$$Lg(x, \lambda) = \lambda g \quad \text{in} \quad S ,$$

and h and g have the same boundary value on B. Hence the solution $g(x, \lambda)$ is smaller than $h(x, \lambda(1 - \delta(\varepsilon)))$ in S. In particular,

$$g(y, \lambda) \leqq h(y, \lambda(1 - \delta(\varepsilon))) .$$

THEOREM 3.5. *The solution $f_\varepsilon(x, \lambda)$ of equation (3.2) has the following property. There exist a function $\delta(\varepsilon) \downarrow 0$ as $\varepsilon \downarrow 0$ and a constant $M_\rho < \infty$ for every $\rho > 0$ such that, for all ε, y and λ,*

$$f_\varepsilon(y, \lambda) \leqq M_\rho \exp\{-(1 - \delta(\varepsilon))(1 - \rho)\varepsilon\sqrt{2\lambda}\} .$$

The proof is immediate from Theorem 3.2 and Lemmas 3.3 and 3.4.

Let G be any arbitrary region with a boundary B. Let $\phi(x, \lambda)$ be the solution to the equation

$$\begin{aligned} L\phi &= \lambda\phi \quad \text{for} \quad x \in G , \\ \phi &= 1 \quad \text{for} \quad x \in B . \end{aligned} \tag{3.5}$$

If G is unbounded, we determine ϕ by the additional restriction that $0 \leqq \phi \leqq 1$. If G is bounded, it is automatically true that $0 \leqq \phi \leqq 1$. Let

$$d(x, B) = \inf_{b \in B} d(x, b) . \tag{3.6}$$

We then have the following estimate concerning $\phi(x, \lambda)$.

THEOREM 3.6. *For every $\rho > 0$, there exists a function $C_\rho(x)$ such that*

$$\phi(x, \lambda) \leqq C_\rho(x) \exp\{-(1 - \rho)d(x, B)\sqrt{2\lambda}\} .$$

$C_\rho(x)$ depends only on $d(x, B)$ and is uniformly bounded whenever x varies in such a way that $d(x, B)$ remains bounded.

Proof: Let $x \in G$ be arbitrary. Then the set

$$S_R = [y : d(x, y) < d(x, B)]$$

is obviously contained in G and its boundary is

$$B_R = [y : d(x, y) = d(x, B)] .$$

Hence if $\psi(x, \lambda)$ is the solution of

$$\text{(3.7)}\qquad \begin{aligned} L\psi &= \lambda\psi \qquad \text{for} \quad y \in S_R , \\ \psi &= 1 \qquad \text{for} \quad y \in B_R , \end{aligned}$$

then

$$\phi(x, \lambda) \leqq \psi(x, \lambda) .$$

To estimate $\psi(x, \lambda)$ we choose ε small so that $n\varepsilon = d(x, B)$ and define the sets $S_1, S_2, \cdots S_n$ as follows:

$$S_j = [y: d(x,y) < j\varepsilon] ,$$

$$B_j = [y: d(x,y) = j\varepsilon] .$$

Let ψ_j be the solution to the equation

$$\text{(3.8)}\qquad \begin{aligned} L\psi_j &= \lambda\psi_j \quad \text{for } y \text{ in } S_j , \\ \psi_j &= 1 \quad \text{for } y \text{ on } B_j . \end{aligned}$$

We estimate ψ_{j+1} on B_j. It is obvious that if $b_j \in B_j$, the set $S = [y: d(b_j, y) < \varepsilon]$ is contained in S_{j+1} and hence

$$\psi_{j+1}(b_j, \lambda) \leqq f_\varepsilon(b_j, \lambda) ,$$

where f_ε refers to the function defined in equation (3.2). Therefore,

$$\text{(3.9)}\qquad \psi_{j+1}(b_j, \lambda) \leqq M_\rho \exp\{-[1 - \delta(\varepsilon)](1 - \rho)\sqrt{2\lambda}\,\varepsilon\} .$$

Now we can use the inequality

$$\phi(b_j, \lambda) \leqq \psi_{j+1}(b_j, \lambda) \sup_{b_{j+1} \in B_{j+1}} \phi(b_{j+1}, \lambda)$$

which with (3.9) leads to

$$\text{(3.10)}\qquad \phi(x, \lambda) \leqq [M_\rho]^n \exp\{-[1 - \delta(\varepsilon)](1 - \rho)n\varepsilon\sqrt{2\lambda}\} .$$

Since $\delta(\varepsilon) \downarrow 0$ as $\varepsilon \downarrow 0$ and since $n\varepsilon = d(x, B)$, this leads immediately to the theorem.

For the rest of the section we shall be concerned with obtaining a lower bound. Let y be an arbitrary point in R_k, let S be the set

$$S = [x: \|x - y\|_y < \varepsilon]$$

and B its boundary

$$B = [x: \|x - y\|_y = \varepsilon] .$$

Γ is that part of the boundary B defined by

$$\Gamma = [x \colon \|x - y\|_y = \varepsilon\,,\ \|x - z\|_y < \varepsilon\sqrt{2\eta}\,]\,,$$

where z is a point of B and η is a positive number. So Γ is a small portion of the boundary around z.

Let $\phi(x, \lambda)$ be the solution to the equation

$$\begin{aligned} L\phi &= \lambda\phi \quad \text{for } x \text{ in } S\,, \\ \phi &= 1 \quad \text{for } x \text{ on } \Gamma\,, \\ \phi &= 0 \quad \text{for } x \text{ on } B - \Gamma\,. \end{aligned} \tag{3.11}$$

We have the following theorem giving us a lower bound on $\phi(y, \lambda)$.

THEOREM 3.7. *There exists a function $\delta(\varepsilon) \downarrow 0$ as $\varepsilon \downarrow 0$ and a constant M_ρ for every $\rho > 0$ such that for all y and λ*

$$\phi(y, \lambda) \geqq \exp\{-\varepsilon(1 + \delta(\varepsilon))\sqrt{2\lambda}\}[1 - M_\rho \exp\{-\varepsilon(\eta - \rho - \delta(\varepsilon))\sqrt{2\lambda}\}]\,.$$

Proof: Let θ be a vector in R_k. For any z in R_k, θz is the scalar product $\Sigma\, \theta_i z_i$. For x in S we have

$$\begin{aligned} Le^{\theta(x-y)} &= \tfrac{1}{2}[\theta a(x)\theta]e^{\theta(x-y)} \\ &\geqq \tfrac{1}{2}(1 - \delta(\varepsilon))(\theta\Lambda\theta)e^{\theta(x-y)}\,, \end{aligned}$$

where $\Lambda = a(y)$. Putting $\lambda = \frac{1}{2}(1 - \delta(\varepsilon))[\theta\Lambda\theta]$, we have

$$Le^{\theta(x-y)} \geqq \lambda e^{\theta(x-y)} \quad \text{for } x \text{ in } S\,.$$

Hence if $g(x, \lambda)$ is the solution to

$$\begin{aligned} Lg &= \lambda g \quad \text{in } S\,, \\ g &= e^{\theta(x-y)} \quad \text{for } x \text{ on } B\,, \end{aligned} \tag{3.12}$$

then

$$e^{\theta(x-y)} \leqq g(x, \lambda)\,. \tag{3.13}$$

Let $\psi(x, \lambda)$ be the solution to

$$\begin{aligned} L\psi &= \lambda\psi \quad \text{for } x \text{ in } S\,, \\ \psi &= 1 \quad \text{for } x \text{ on } B\,. \end{aligned} \tag{3.14}$$

Then from Theorem 3.2, there is, for every $\rho > 0$, a constant M_ρ such that

$$\psi(y, \lambda) \leqq M_\rho \exp\{-\varepsilon(1 - \rho)\sqrt{2\lambda}\}\,. \tag{3.15}$$

Let us now take the vector $\theta = k\Lambda^{-1}(z - y)$, where the constant k is to be chosen later, such that

$$\lambda = \tfrac{1}{2}(1 - \delta(\varepsilon))[\theta\Lambda\theta] . \tag{3.16}$$

The following estimates are easily obtained:

$$\sup_{x\in B} \theta(x - y) = k\varepsilon^2 ,$$

$$\sup_{x\in B-\Gamma} \theta(x - y) = k\varepsilon^2(1 - \eta) .$$

Hence on B,

$$e^{\theta(x-y)} \leqq e^{k\varepsilon^2(1-\eta)+k\varepsilon^2}\chi_\Gamma(x) , \tag{3.17}$$

where $\chi_\Gamma(x)$ is the indicator function of Γ. Let us recall that $\phi(x, \lambda)$ satisfies the equation

$$\begin{aligned} L\phi &= \lambda\phi \quad \text{for } x \text{ in } S , \\ \phi &= \chi_\Gamma(x) \quad \text{for } x \text{ on } B , \end{aligned} \tag{3.18}$$

and compare the equations (3.12), (3.14) and (3.18). The functions g, ψ, and ϕ satisfy the same equation in S but with different boundary values $e^{\theta(x-y)}$, 1, and $\chi_\Gamma(x)$, respectively, on B. Since (3.17) is valid on the boundary, the following inequality is valid throughout S:

$$g(x, \lambda) \leqq e^{k\varepsilon^2(1-\eta)}\psi(x, \lambda) + e^{k\varepsilon^2}\phi(x, \lambda) . \tag{3.19}$$

Setting $x = y$ in (3.19) and using (3.13), we obtain

$$e^{k\varepsilon^2(1-\eta)}\psi(y, \lambda) + e^{k\varepsilon^2}\phi(y, \lambda) \geqq 1 .$$

Using now the estimate (3.15) for $\psi(y, \lambda)$, we get

$$\phi(y, \lambda) \geqq e^{-k\varepsilon^2} - e^{-k\varepsilon^2}\eta M_\rho e^{-\varepsilon(1-\rho)\sqrt{2\lambda}} . \tag{3.20}$$

Now (3.16) leads to the value $\varepsilon k = [1 + \delta(\varepsilon)]\sqrt{2\lambda}$. Substitution of this value of k in (3.20), a little simplification and replacement of $\delta(\varepsilon)$ by a larger function which again tends to zero as $\varepsilon \to 0$ yields

$$\phi(y, \lambda) \geqq e^{-\varepsilon(1+\delta(\varepsilon))\sqrt{2\lambda}}(1 - M_\rho e^{-\varepsilon(\eta-\rho-\delta(\varepsilon))\sqrt{2\lambda}}) .$$

This completes the proof.

Let y and z be two arbitrary points in R_k with $d(x,y) = \varepsilon$. Let S be the set

$$S = [x: d(x, z) \leqq 2\varepsilon\sqrt{\eta}] .$$

We consider the solution $\phi(x, \lambda)$ of the equation

$$\begin{aligned} L\phi &= \lambda\phi \quad \text{for } x \text{ in the exterior of } R_k - S\,, \\ \phi &= 1 \quad \text{for } x \text{ on the boundary } B \text{ of } S\,. \end{aligned} \tag{3.21}$$

We further assume that η is small enough so that y is in the exterior of S. Then we have

THEOREM 3.8. *There exists a function $\delta(\varepsilon) \downarrow 0$ as $\varepsilon \downarrow 0$, and a constant M_ρ for every $\rho > 0$ such that, for all y and z,*

$$\phi(y, \lambda) \geqq \exp\{-\varepsilon(1 + \delta(\varepsilon))\sqrt{2\lambda}\}[1 - M_\rho \exp\{-(\eta - \rho - \delta(\varepsilon))\sqrt{2\lambda}\}]\,.$$

Proof: Let $\varepsilon' = \|y - z\|_v$. Choose η' such that the set

$$E = [x\colon \|x - z\|_v \leqq \varepsilon'\sqrt{2\eta'}]$$

is contained in S. Let F denote the boundary

$$F = [x\colon \|x - z\|_v = \varepsilon'\sqrt{2\eta'}]$$

of S. Let $\psi(x, \lambda)$ be the solution to the equation

$$\begin{aligned} L\psi &= \lambda\psi \quad \text{for } x \text{ in the exterior of } E\,, \\ \psi &= 1 \quad \text{for } x \text{ on } F\,. \end{aligned} \tag{3.22}$$

Let S' be the set

$$S' = [x\colon \|x - y\|_v < \varepsilon']$$

with boundary

$$B' = [x\colon \|x - y\|_v = \varepsilon']$$

and Γ' the set $\Gamma' = B' \cap E$. Let $g(x, \lambda)$ be the solution of

$$\begin{aligned} Lg &= \lambda g \quad \text{for } x \text{ in } S'\,, \\ g &= \chi_{\Gamma'}(x) \quad \text{for } x \text{ on } B'\,. \end{aligned} \tag{3.23}$$

Since $E \subset S$ and ψ is the solution to (3.22) in the exterior of E, ψ is defined on the boundary B of S and $0 \leqq \psi \leqq 1$ on B. But ϕ is the solution of (3.21) in the exterior of S and therefore for x in the exterior of S

$$\phi(x, \lambda) \geqq \psi(x, \lambda)\,. \tag{3.24}$$

Let A be the set $S' \cap$ (exterior of E). Its boundary is given by $[B' \cap$ (exterior of E)$] \cup (F \cap S')$. Then g and ψ are both solutions of the same equation $Lf = \lambda f$ in A. As for their boundary values,

$$\begin{aligned} &\text{on } B' \cap (\text{exterior of } E), && g = 0\,, \quad \psi \geqq 0\,, \\ &\text{on } F \cap S', && g \leqq 1\,, \quad \psi = 1\,. \end{aligned}$$

Hence throughout A, $\psi \geqq g$. Since y is in A, we have with (3.24)

$$\phi(y, \lambda) \geqq g(y, \lambda) .$$

Using Theorem 3.7 to estimate $g(y, \lambda)$, we have

$$g(y, \lambda) \geqq \exp\{-\varepsilon'(1 + \delta(\varepsilon'))\sqrt{2\lambda}\}[1 - M_\rho \exp\{-(\eta' - \rho - \delta(\varepsilon'))\sqrt{2\lambda}\}] .$$

It is easy to estimate using Lemma 2.5:

$$\varepsilon' \leqq \varepsilon(1 + \delta(\varepsilon))$$

and

$$\eta' \geqq 2\eta\left(\frac{1 - \delta(\varepsilon)}{1 + \delta(\varepsilon)}\right)^2 .$$

Therefore for ε small enough, $\eta' \geqq \eta$ and this proves the theorem.

Let y, z be two points with $d(y, z) = d$. Let n be an integer such that $n\varepsilon = d$. Let G be the sphere around z of radius $4\varepsilon\sqrt{\eta}$:

$$G = [x\colon d(x, z) \leqq 4\varepsilon\sqrt{\eta}] .$$

Let $\phi(x, \lambda)$ be the solution of the equation

$$\begin{aligned} L\phi &= \lambda\phi \quad \text{for } x \text{ in the exterior of } G , \\ \phi &= 1 \quad \text{on the boundary of } G . \end{aligned} \tag{3.25}$$

THEOREM 3.9. *For the solution ϕ of equation* (3.25) *there exist a function $\delta(\varepsilon) \downarrow 0$ as $\varepsilon \downarrow 0$ and a constant M_ρ for every $\rho > 0$ such that*

$$\phi(y, \lambda) \geqq \exp\{-d(1 + 4\sqrt{\eta})(1 + \delta(\varepsilon))\sqrt{2\lambda}\} \times [1 - M_\rho \exp\{-\varepsilon(1 - 4\sqrt{\eta})(\eta - \rho - \delta(\varepsilon))\sqrt{2\lambda}\}]^n .$$

Proof: Let the points $y = y_0, y_1, y_2, \cdots, y_n = z$ be chosen such that $d(y_i, y_{i+1}) = \varepsilon$; $n\varepsilon$ being equal to d. Let $\varepsilon > 0$. Construct the sets C_i successively as follows:

$$C_1 = \overline{S(y_1, 2\varepsilon\sqrt{\eta})} ,$$

$$C_2 = \overline{\bigcup_{z_1 \in C_1} S(y_2, 2\sqrt{\eta}\, d(z_1, y_2))} ,$$

$$C_{j+1} = \overline{\bigcup_{z_j \in C_j} S(y_{j+1}, 2\sqrt{\eta}\, d(z_j, y_{j+1}))} ,$$

and let

$$\Delta_j = \sup_{z_j \in C_j} d(z_j, y_j),$$

$$\rho_{j-1} = \sup_{z_{j-1} \in C_{j-1}} d(z_{j-1}, y_j),$$

$$\mu_{j-1} = \inf_{z_{j-1} \in C_{j-1}} d(z_{j-1}, y_j).$$

Then

$$\rho_{j-1} \leqq \varepsilon + \Delta_{j-1},$$

$$\mu_{j-1} \geqq \varepsilon - \Delta_{j-1},$$

$$\Delta_j \leqq 2\sqrt{\eta}\, \rho_{j-1},$$

$$\rho_0 = \varepsilon, \qquad \Delta_0 = 0.$$

It follows from these equations that

$$\begin{aligned}\Delta_j &\leqq \varepsilon[2\sqrt{\eta} + (2\sqrt{\eta})^2 + \cdots (2\sqrt{\eta})^j] \\ &\leqq \varepsilon 2\sqrt{\eta} \frac{1}{1 - 2\sqrt{\eta}} \\ &\leqq 4\varepsilon\sqrt{\eta} \quad \text{for} \quad 2\sqrt{\eta} \leqq \tfrac{1}{2}, \qquad j = 1, 2, \cdots n.\end{aligned}$$

If ε_j is a typical distance from C_j to y_{j+1}, then

$$\varepsilon - \Delta_j \leqq \varepsilon_j \leqq \varepsilon + \Delta_j,$$

so that

$$\varepsilon(1 - 4\sqrt{\eta}) \leqq \varepsilon_j \leqq \varepsilon(1 + 4\sqrt{\eta}).$$

Now we can apply Theorem 3.8 and obtain a uniform estimate for x in C_{n-1}. Repeated application of Theorem 3.8 leads to

$$\begin{aligned}\phi(y, \lambda) \geqq \exp\{-d(1 + 4\sqrt{\eta})(1 + \delta(\varepsilon))\sqrt{2\lambda}\} \\ \times [1 - M_\rho \exp\{-\varepsilon(1 - 4\sqrt{\eta})(\eta - \rho - \delta(\varepsilon))\sqrt{2\lambda}\}]^n.\end{aligned}$$

THEOREM 3.10. *Let z be a point in R_k, S the sphere $S = [x\colon d(x, z) < \delta]$. Let $\phi(x, \lambda)$ be the solution to the equation $L\phi = \lambda\phi$ in the exterior of S with $\phi = 1$ on the boundary of S. Then*

$$\liminf_{\lambda \to \infty} \frac{1}{\sqrt{2\lambda}} \log \phi(x, \lambda) \geqq -d(x, z)$$

uniformly over all x, z and δ, provided δ remains bounded away from zero and $d(x, z)$ remains bounded.

The proof follows at once from Theorem 3.9. One chooses η first and then ρ and ε suitably; n remains bounded for any $\varepsilon > 0$ if d is bounded.

Let G be a region with exterior E. B is the boundary of G. It is assumed that B is also the boundary of E.

LEMMA 3.11. *Let $x \in G$ and let $d = d(x, B)$. Then for any $\varepsilon > 0$ there exists a $\delta > 0$ and a point $z \in E$ such that*

(i) $d(x, z) \leqq d + \varepsilon$,

(ii) $[y: d(y, z) < \delta] \subset E$.

Moreover δ can be chosen uniformly when x varies over a bounded subset of G.

Proof: Let $x \in G \cup B$. Then there exists a y on B such that $d(x, B) = d(x, y)$. The sphere around y of radius $\frac{1}{2}\varepsilon$ has a non-empty intersection with E and there exists a sphere around some points z of some radius δ completely contained inside the intersection. Obviously $d(x, z) \leqq d(x, y) + \frac{1}{2}\varepsilon$. To show uniformity we point out that if x' is in $G \cup B$ with $d(x, x') < \frac{1}{2}\varepsilon$, then

$$\begin{aligned} d' = d(x', B) &\geqq d(x, B) - d(x, x') \\ &\geqq d - \tfrac{1}{2}\varepsilon \\ &\geqq d(x, z) - \tfrac{1}{2}\varepsilon - \tfrac{1}{2}\varepsilon \\ &= d(x, z) - \varepsilon , \end{aligned}$$

or

$$d(x, z) \leqq d' + \varepsilon .$$

In other words, the sphere of radius δ around z serves not only for x but for a neighborhood around x as well. A routine compactness argument using finite subcovers proves the lemma for compact subsets of $G \cup B$ or for bounded subsets of G.

Now we can prove the main theorem:

THEOREM 3.12. *Let G be any region with boundary B. B is assumed to be the boundary of the exterior E of G as well. If $\phi(x, \lambda)$ is the solution of the equation*

$$L\phi = \lambda\phi \quad \text{for} \quad x \quad \text{in} \quad G ,$$

$$\phi = 1 \quad \text{on} \quad B ,$$

then

$$\lim_{\lambda \to \infty} \left[-\frac{1}{\sqrt{2\lambda}} \log \phi(x, \lambda) \right] = d(x, B)$$

uniformly for x varying over bounded subsets of G.

The proof follows from Theorems 3.6, 3.10, and Lemma 3.11.

4. The Fundamental Solution

Let $F(t)$ be a nondecreasing function of t for $t \geqq 0$ with $0 \leqq F(t) \leqq 1$ and $F(0) = 0$. Let $\phi(\lambda)$ be the Laplace transform

$$\phi(\lambda) = \int_0^\infty e^{-\lambda\tau}\, dF(\tau)\,. \tag{4.1}$$

We assume that the $\{\phi_\alpha(\lambda)\}$ are Laplace transforms of $\{F_\alpha(t)\}$ and that

$$\lim_{\lambda\to\infty}\left[-\frac{1}{\sqrt{2\lambda}}\log\phi_\alpha(\lambda)\right] = d_\alpha\,, \tag{4.2}$$

where the d_α are such that $0 \leqq d_\alpha \leqq D < \infty$. We want to conclude that if the limit in (4.2) is uniform with respect to α, then

$$\lim_{t\to 0}\left[-2t\log F_\alpha(t)\right] = d_\alpha^2 \tag{4.3}$$

uniformly in α.

Initially we assume that (4.2) holds uniformly and that in addition $0 < d \leqq d_\alpha \leqq D < \infty$. The following lemmas are conclusions regarding $F_\alpha(t)$.

Lemma 4.1. *For every $\varepsilon > 0$ there exists a constant $A_\varepsilon < \infty$ such that, for all $t \geqq 0$,*

$$F_\alpha(t) \leqq A_\varepsilon \exp\left\{-\frac{(1-\varepsilon)}{2t}\,d_\alpha^2\right\}.$$

Proof: Since $0 < d \leqq d_\alpha \leqq D < \infty$ we can conclude, from the fact that (4.2) holds uniformly, that for every $\varepsilon > 0$ there exists a constant A_ε such that for all $\lambda \geqq 0$

$$\phi_\alpha(\lambda) \leqq A_\varepsilon \exp\{-\sqrt{1-\varepsilon}\,\sqrt{2\lambda}\,d\alpha\}\,. \tag{4.4}$$

It follows from (4.1) that

$$\phi_\alpha(\lambda) \geqq e^{-\lambda t}F_\alpha(t)\,,$$

or

$$F_\alpha(t) \leqq e^{\lambda t}\phi_\alpha(\lambda)\,. \tag{4.5}$$

Choosing $\lambda = (1-\varepsilon)d_\alpha^2/2t$ and combining (4.5) and (4.4) we get

$$F_\alpha(t) \leqq A_\varepsilon \exp\left\{-\frac{(1-\varepsilon)}{2t}\,d_\alpha^2\right\}.$$

COROLLARY 4.2.

$$\limsup_{t\to 0} 2t \log F_\alpha(t) \leqq -d_\alpha^2$$

uniformly in α.

LEMMA 4.3. *For every* $k > 1$,

$$\int_{kt}^{\infty} \exp\left\{-\frac{d_\alpha^2 \tau}{2t^2}\right\} F_\alpha(\tau)\, d\tau \leqq \frac{2A_\varepsilon t^2}{\varepsilon d^2} \exp\left\{-\frac{(1-\varepsilon)d_\alpha^2}{2t} - \frac{\rho_k}{t}\right\},$$

where $\rho_k = \frac{1}{2}d^2(k + 1/k - 2)$.

Proof: Using Lemma 4.1, we have

$$\begin{aligned}\int_{kt}^{\infty} \exp\left\{-\frac{d_\alpha^2 \tau}{2t}\right\} F_\alpha(\tau)\, d\tau &\leqq A_\varepsilon \int_{kt}^{\infty} \exp\left\{-\frac{d_\alpha^2 \tau}{2t^2}\right\} \exp\left\{-\frac{(1-\varepsilon)d_\alpha^2}{2t}\right\} d\tau \\ &= A_\varepsilon t \int_k^{\infty} \exp\left\{-\frac{1}{2t} d_\alpha^2\left[u + \frac{(1-\varepsilon)}{u}\right]\right\} du \\ &= A_\varepsilon t \int_k^{\infty} \exp\left\{-\frac{\varepsilon d_\alpha^2}{2t}\right\} \exp\left\{-\frac{(1-\varepsilon)d_\alpha^2}{2t}\left[u + \frac{1}{u}\right]\right\} du \\ &\leqq \frac{2A_\varepsilon t^2}{\varepsilon d^2} \exp\left\{-\frac{(1-\varepsilon)d_\alpha^2}{t} - \frac{\rho_k}{t}\right\}.\end{aligned}$$

LEMMA 4.4. *For every* $k' < 1$,

$$\int_{0}^{k't} \exp\left\{-\frac{d_\alpha^2 \tau}{2t^2}\right\} F_\alpha(\tau)\, d\tau \leqq \frac{2A_\varepsilon t^2}{\varepsilon d^2} \exp\left\{-\frac{(1-\varepsilon)d_\alpha^2}{t} - \frac{\rho_{k'}}{t}\right\},$$

where $\rho_{k'} = \frac{1}{2}d^2[k' + 1/k' - 2]$.

The proof is similar to that of Lemma 4.3.

LEMMA 4.5.

$$\lim_{t\to 0}\left[-2t \log F_\alpha(t)\right] = d_\alpha^2$$

uniformly in α.

Proof: Because of (4.2), there exists, for every $\varepsilon > 0$, a constant B_ε such that for all $\lambda \geqq 0$

$$\phi_\alpha(\lambda) \geqq B_\varepsilon \exp\{-(1+\varepsilon)\sqrt{2\lambda}\, d_\alpha\}. \tag{4.6}$$

However,

$$\phi_\alpha(\lambda) = \lambda \int_0^\infty e^{-\lambda\tau} F_\alpha(\tau)\, d\tau$$

$$(4.7) \qquad = \lambda\left[\int_0^{k't} + \int_{k't}^{kt} + \int_{kt}^{\infty}\right] e^{-\lambda\tau} F_\alpha(\tau)\, d\tau\,.$$

Therefore, substituting $\lambda = d_\alpha^2/2t^2$ and combining (4.6) and (4.7) with Lemmas 4.3 and 4.4, we have

$$(4.8) \quad \int_{k't}^{kt} \exp\left\{-\frac{d_\alpha^2 \tau}{2t^2}\right\} F_\alpha(\tau)\, d\tau \geqq C_\varepsilon t^2 \exp\left\{\frac{(1+\varepsilon)\, d_\alpha^2}{t}\right\}\left(1 - k_\varepsilon \exp\left\{\frac{2\varepsilon D^2 - \rho}{t}\right\}\right),$$

where $\rho = \min(\rho_k, \rho_{k'})$.

For a given k and k' with $k' < 1 < k$ one can find ε small enough so that $2\varepsilon D^2 - \rho < 0$. Therefore, we conclude from (4.8) that, for every k, k' with $k' < 1 < k$,

$$(4.9) \qquad \liminf_{t\to 0}\, t \log\left[\int_{k't}^{kt} \exp\left\{-\frac{d_\alpha^2}{2t^2}\right\} F_\alpha(\tau)\, d\tau\right] \geqq -d_\alpha^2$$

uniformly in α.

On the other hand,

$$\int_{k't}^{kt} \exp\left\{-\frac{d_\alpha^2 \tau}{2t^2}\right\} F_\alpha(\tau)\, d\tau \leqq F_\alpha(kt) \exp\left\{-\frac{k' d_\alpha^2}{2t}\right\}$$

which with (4.9) yields

$$\liminf_{t\to 0}\, t \log F_\alpha(kt) \geqq -d_\alpha^2 + \tfrac{1}{2} k' d_\alpha^2\,,$$

or

$$\liminf_{t\to 0}\, 2t \log F_\alpha(t) \geqq -2kd_\alpha^2 + kk' d_\alpha^2$$

uniformly in α. Since k and k' can be chosen close to 1, we have

$$\liminf_{t\to 0}\, 2t \log F_\alpha(t) \geqq -d_\alpha^2$$

uniformly in α. This is one part of the lemma and the other part is Corollary 4.2.

Now we drop the assumption that $d > 0$. We assume only that $0 \leqq d_\alpha \leqq D$.

LEMMA 4.6. *Lemma 4.5 still holds if we assume only that* $0 \leqq d_\alpha \leqq D$.

Proof: It is of course enough to show that if, for $\lambda \geqq \lambda_0$,

$$-\varepsilon \leqq -\frac{1}{\sqrt{2\lambda}} \log \phi_\alpha(\lambda) \leqq \varepsilon,$$

then there exists a t_0 and a function $\eta(\varepsilon) \to 0$ as $\varepsilon \to 0$ such that

$$-\eta \leqq -2t \log F_\alpha(t) \leqq \eta.$$

Since $0 \leqq \phi_\alpha(\lambda) \leqq 1$ and $0 \leqq F_\alpha(t) \leqq 1$, one side is trivial. If

$$\phi_\alpha(\lambda) \geqq e^{-\varepsilon\sqrt{2\lambda}} \qquad \text{for} \qquad \lambda \geqq \lambda_0,$$

we shall show that for ε small enough there exist η and t_0 such that

$$F_\alpha(t) \geqq e^{-\eta/t} \qquad \text{for} \qquad t \leqq t_0.$$

To prove this we notice that

$$\phi_\alpha(\lambda) \leqq F_\alpha(t) + e^{-\lambda t},$$

or

$$\begin{aligned} F_\alpha(t) &\geqq \phi_\alpha(\lambda) - e^{-\lambda t} \\ &\geqq e^{-\varepsilon\sqrt{2\lambda}} - e^{-\lambda t} \qquad \text{for} \qquad \lambda \geqq \lambda_0. \end{aligned}$$

Choose $\lambda = 2/t^2$. If $t \leqq t_0 = \sqrt{2/\lambda_0}$ and $\varepsilon < \frac{1}{2}$, then

$$f_\alpha(t) \geqq e^{-2\varepsilon/t}(1 - e^{-1/t})$$

which concludes the proof.

We have, therefore, proved the following theorem:

THEOREM 4.7. *Let* $\phi_\alpha(\lambda) = \int_0^\infty e^{-\lambda t}\, dF_\alpha(\tau)$. *If*

$$\lim_{\lambda\to\infty} \left[-\frac{1}{\sqrt{2\lambda}} \log \phi_\alpha(\lambda) \right] = d_\alpha$$

uniformly in α *and if* $0 \leqq d_\alpha \leqq D < \infty$, *then*

$$\lim_{t\to 0} \left[-2t \log F_\alpha(t) \right] = d_\alpha^2$$

uniformly in α.

Let $G \subset R_k$ be a region with a boundary B. It is assumed that B is the boundary of the exterior of G as well. Let S be the cylinder in $R_k \times (0, \infty)$ with base G.

That is,

$$S = G \times (0, \infty) .$$

The boundary of S is $\{B \times (0, \infty)\} \cup \{G \times \{0\}\}$. We consider the solution $u(x, t)$ to the equation

$$\begin{aligned} u_t - Lu &= 0 \quad &&\text{for} \quad (x, t) \in G \times (0, \infty) , \\ u(x, 0) &= 0 \quad &&\text{for} \quad x \in G, \\ u(x, t) &= 1 \quad &&\text{for} \quad x \in B, \quad t > 0 . \end{aligned} \tag{4.10}$$

The solution to the equation exists and has the properties:

$u(x, t)$ is nondecreasing in t for $t \geqq 0$,

$0 \leqq u(x, t) \leqq 1$.

THEOREM 4.8.

$$\limsup_{t \to 0} 2t \log u(x, t) \leqq -d^2(x, B)$$

uniformly over all regions G and points x in G so long as $d(x, B)$ remains bounded.

Proof: If $\phi(x, \lambda) = \int_0^\infty e^{-\lambda\tau} u(x, d\tau)$, then $\phi(x, \lambda)$ satisfies the equation

$$\begin{aligned} L\phi &= \lambda\phi \quad &&\text{for} \quad x \in G , \\ \phi &= 1 \quad &&\text{for} \quad x \in B . \end{aligned} \tag{4.11}$$

From Theorem 3.6, there exists for every $\rho > 0$, a $C_\rho < \infty$ such that

$$\phi(x, \lambda) \leqq C_\rho \exp \{-(1 - \rho)\sqrt{2\lambda}\, d(x, B)\}$$

so long as $d(x, B)$ remains bounded. From equation (4.5), we have

$$u(x, t) \leqq e^{\lambda t}\phi(x, \lambda)$$

for every $\lambda \geqq 0$. Choosing $\lambda = (1 - \rho)^2 d^2(x, B)/2t^2$,

$$u(x, t) \leqq C_\rho \exp \left\{\frac{(1 - \rho)^2 d^2(x, B)}{2t}\right\}$$

which proves the theorem.

THEOREM 4.9. *The fundamental solution $p(t, x, y)$ of Theorem 2.1 has the property*

$$\limsup_{t \to 0} 2t \log p(t, x, y) \leqq -d^2(x, y)$$

uniformally over all x, y such that $d(x, y)$ is bounded.

Proof: Let $d(x,y) = d$. Define, for any $\varepsilon > 0$,

$$G = [z: d(x, z) < d - \varepsilon] ,$$
$$B = [z: d(x, z) = d - \varepsilon] .$$

Then $p(t, x, y)$ satisfies the equation

$$\begin{aligned} p_t - Lp &= 0 \quad \text{for} \quad (x, t) \text{ in } G \times (0, \infty) , \\ p(0, x, y) &= 0 \quad \text{for} \quad x \in G , \\ p(t, x, y) &= p(t, b, y) \quad \text{for} \quad b \in B . \end{aligned} \tag{4.12}$$

If $u(x, t)$ is the solution to

$$\begin{aligned} u_t - Lu &= 0 \quad \text{for} \quad (x, t) \text{ in } G \times (0, \infty) , \\ u &= 0 \quad \text{for} \quad x \in G \text{ and } t = 0 , \\ u &= 1 \quad \text{for} \quad x \in B \text{ and } t > 0 , \end{aligned} \tag{4.13}$$

then clearly

$$p(t, x, y) \leqq \Big[\sup_{\substack{0 \leqq \tau < \infty \\ b \in B}} p(\tau, b, y) \Big] u(x, t) . \tag{4.14}$$

Since $d(b, y) \geqq \varepsilon$, we have from property (iv) of Theorem 2.1

$$\sup_{0 < \tau < \infty} \sup_{b \in B} p(\tau, b, y) \leqq M_\varepsilon < \infty . \tag{4.15}$$

Hence, combining (4.14) and (4.15) with Theorem 4.8, we obtain

$$\lim_{t \to 0} 2t \log p(t, x, y) \leqq -(d - \varepsilon)^2 .$$

Since ε is arbitrary, this proves the theorem. However, apparently the uniformity holds only in regions of the form $0 < \varepsilon \leqq d(x,y) \leqq D < \infty$. But the direct estimate in Theorem 2.1,

$$p(t, x, y) \leqq Ct^{-k/2} \exp \left\{ \frac{-\alpha \|x - y\|^2}{2t} \right\} ,$$

establishes the uniformity over all x, y such that $d(x, y)$ remains bounded.

We now proceed to obtain a lower bound on $p(t, x, y)$ for small t.

THEOREM 4.10. *There exists a function $\psi(\varepsilon)$ such that $\psi(\varepsilon) \to 0$ as $\varepsilon \to 0$ and*

$$\liminf_{t \to 0} \inf_{x, y : d(x, y) \leqq \varepsilon} 2t \log p(t, x, y) \geqq -\psi(\varepsilon) .$$

Proof: Let $s > 0$. Define G_s and its boundary B_s by

$$G_s = [z: d(y, z) > s],$$
$$B_s = [z: d(y, z) = s].$$

Let $u_s(x, t)$ be the solution to the equation

$$(4.16)\qquad \begin{aligned} u_t - Lu &= 0 \quad \text{in} \quad G_s \times (0, \infty), \\ u &= 1 \quad \text{on} \quad B_s \times (0, \infty), \\ u &= 0 \quad \text{on} \quad G_s \times \{0\}. \end{aligned}$$

Let $\phi_s(x, \lambda) = \int_0^\infty e^{\lambda\tau} u_s(x, d\tau)$. Then $\phi_s(x, \lambda)$ satisfies the equation

$$(4.17)\qquad \begin{aligned} L\phi &= \lambda\phi \quad \text{in} \quad G_s, \\ \phi &= 1 \quad \text{on} \quad B_s. \end{aligned}$$

From Theorem 3.9,

$$(4.18)\qquad \phi_s(x, \lambda) \geqq \exp\{-d(1 + 4\sqrt{\eta})(1 + \delta(\varepsilon))\sqrt{2\lambda}\} \times [1 - M_\rho \exp\{-\varepsilon(1 - 4\sqrt{\eta})(\eta - \rho - \delta(\varepsilon))\sqrt{2\lambda}\}]^n,$$

where $d = d(x, y)$, ε, η, ρ are arbitrary, $n\varepsilon = d$, with the restriction that $s \geqq 4\varepsilon\sqrt{\eta}$. Since

$$u_s(x, t) \geqq \phi_s(x, \lambda) - e^{-\lambda t}$$

for every λ, let us choose $\lambda = 8/t^2$, $\varepsilon = s$ and η small enough. We further put $s = \sqrt{t}$. Then using (4.18) we get

$$(4.19)\qquad u_{\sqrt{t}}(x, t) \geqq \exp\{-\frac{4d}{t}(1 + \theta)\}(1 - M_\rho e^{-C/\sqrt{t}})^{d/\sqrt{t}} - e^{-8/t},$$

where θ is a small number which can be chosen arbitrarily small. This function is defined for x such that $d(x, y) \geqq s = \sqrt{t}$. It is clear from (4.19) that if $\varepsilon \leqq 1$

$$(4.20)\qquad \liminf_{t\to 0} \inf_{\substack{x,y: d(x,y) \geqq \sqrt{t} \\ d(x,y) \leqq \varepsilon}} t \log u_{\sqrt{t}}(x, t) \geqq -4\varepsilon.$$

We now consider the solution $u_s(x, t_0, t)$ to the equation

$$(4.21)\qquad \begin{aligned} u_t - Lu &= 0 \quad \text{in} \quad G_s \times (0, \infty), \\ u &= 0 \quad \text{on} \quad B_s \times (0, t_0), \\ u &= 1 \quad \text{on} \quad B_s \times (t_0, \infty), \\ u &= 0 \quad \text{on} \quad G_s \times \{0\}. \end{aligned}$$

It is clear that

$$u_s(x, t_0, t) = 0 \qquad \text{for} \qquad t \leqq t_0,$$

$$u_s(x, t_0, t) = u_s(x, t - t_0) \qquad \text{for} \qquad t > t_0.$$

We are interested in $u_{\sqrt{t}}(x, \frac{1}{2}t, t) = u_{\sqrt{t}}(x, \frac{1}{2}t)$. From (4.20) we deduce

(4.22)
$$\liminf_{t\to 0} \inf_{\substack{(x,y):d(x,y)\geqq\sqrt{t}\\ d(x,y)\leqq\varepsilon}} t \log u_{\sqrt{t}}(x, \tfrac{1}{2}t, t) \geqq -8\varepsilon .$$

Going back to $p(t, x, y)$ we see that it satisfies an equation similar to (4.21) in $G_s \times (0, \infty)$ with different boundary values. However, since the solution $u_s(x, t_0, t)$ depends only on the boundary values on $B_s \times (0, \infty)$ for $0 < \tau < t$ we conclude by direct comparison that

$$p(t, x, y) \geqq u_s(x, t_0, t) \inf_{\substack{b\in B_s\\ t_0\leqq\tau\leqq t}} p(\tau, b, y) .$$

Since t_0 and t are arbitrary, we choose $t_0 = \frac{1}{2}t$ and $s = \sqrt{t}$. Therefore,

(4.23)
$$p(t, x, y) \geqq u_{\sqrt{t}}(x, \tfrac{1}{2}t, t) \inf_{\substack{b\in B_{\sqrt{t}}\\ \frac{1}{2}t\leqq\tau\leqq t}} p(\tau, b, y) .$$

For t sufficiently small we have from property (v) of Theorem 2.1

$$\inf_{\substack{b\in B_{\sqrt{t}}\\ \frac{1}{2}t\leqq\tau\leqq t}} p(\tau, b, y) \geqq \gamma > 0 .$$

This with (4.22) and (4.23) yields

$$\liminf_{t\to 0} \inf_{\substack{x,y:d(x,y)\leqq\varepsilon\\ d(x,y)\geqq\sqrt{t}}} 2t \log p(t, x, y) \geqq -8\varepsilon .$$

On the other hand, if $d(x, y) \leqq \sqrt{t}$, again from property (v) of Theorem 2.1,

$$\liminf_{t\to 0} \inf_{x,y:d(x,y)\leqq\sqrt{t}} 2t \log p(t, x, y) \geqq 0.$$

This completes the proof.

THEOREM 4.11. *Let $\varepsilon > 0$ and let G_y be the set $[z: d(y, z) < \varepsilon]$. If $k(x, t)$ is defined as*

$$k(x, t, y) = \int_{G_y} p(t, x, z)\, dz ,$$

then

$$\liminf_{t\to 0} 2t \log k(x, t, y) \geqq -d^2(x, y)$$

uniformly over all x, y such that $d(x, y)$ is bounded.

Proof: Let

$$H_y = [z : d(y, z) > \tfrac{1}{2}\varepsilon] ,$$
$$B_y = [z : d(y, z) = \tfrac{1}{2}\varepsilon] .$$

From property (iv) of Theorem 2.1 it follows that

$$k(x, t, y) \to 1 \quad \text{as} \quad t \to 0 \tag{4.24}$$

uniformly for x on B_y.

Assume that $u(x, t)$ solves the equation

$$\begin{aligned} u_t - Lu &= 0 \quad \text{in} \quad H_y \times (0, \infty) , \\ u &= 1 \quad \text{on} \quad B_y \times (0, \infty) , \\ u &= 0 \quad \text{on} \quad H_y \times \{0\} . \end{aligned} \tag{4.25}$$

Then the Laplace transform $\phi(x, \lambda)$ solves the equation

$$\begin{aligned} L\phi &= \lambda\phi \quad \text{for } x \text{ in } H_y , \\ \phi &= 1 \quad \text{on} \quad B_y . \end{aligned}$$

From Theorem 3.12 it follows that

$$\lim_{t\to\infty} \left[-\frac{1}{\sqrt{2\lambda}} \log \phi(x, \lambda) \right] = d(x, B_y) .$$

Moreover, from Theorems 3.6 and 3.10 it is clear that the above limit is uniform whenever $d(x,y)$ remains bounded. Of course $\phi(x, \lambda)$ is defined only when $d(x,y) \geqq \frac{1}{2}\varepsilon$. Now from Theorem 4.7 we obtain

$$\lim_{t\to\infty} 2t \log u(x, t) \geqq -d^2(x, B_y) \geqq -d^2(x, y) \tag{4.26}$$

uniformly whenever $d(x,y)$ remains bounded. Standard comparison of $k(x, t, y)$ which satisfies an equation similar to (4.25) with $u(x, t)$ shows that

$$k(x, t, y) \geqq u(x, t) \inf_{\substack{0 \leqq \tau \leqq t \\ b \in B_y}} k(b, \tau, y) . \tag{4.27}$$

Inequalities (4.26) and (4.27) complete the proof insofar as the region $d(x,y) \geqq \frac{1}{2}\varepsilon$ is concerned. However, property (iv) of Theorem 2.1 shows that

$$k(x, t, y) \to 1 \quad \text{as} \quad t \to 0$$

uniformly over the set $d(x,y) \leqq \frac{1}{2}\varepsilon$.

THEOREM 4.12.

$$\liminf_{t\to 0} 2t \log p(t, x, y) \geqq -d^2(x, y)$$

uniformly over all x, y such that $d(x, y)$ is bounded.

Proof: If G_y is the set $[z: d(y, z) \leqq \varepsilon]$, then

$$p(t, x, y) = \int p([1 - \delta]t, x, z)p(\delta t, z, y)\, dz$$

$$\geqq \int_{G_y} p([1 - \delta]t, x, z)p(\delta t, z, y)\, dz$$

$$\geqq k(x, [1 - \delta]t, y) \inf_{z\in G_y} p(\delta t, z, y)\,.$$

Therefore, for every $\varepsilon > 0$ and $\delta > 0$, applying Theorems 4.10 and 4.11 we get

$$\liminf_{t\to 0} 2t \log p(t, x, y) \geqq -\left[\frac{d^2(x, y)}{1 - \delta} + \frac{\psi(\varepsilon)}{\delta}\right].$$

Since δ is arbitrary, let us choose

$$\delta = \frac{[\psi(\varepsilon)]^{1/2}}{d(x, y) + [\psi(\varepsilon)]^{1/2}}\,.$$

Then

$$\liminf_{t\to 0} 2t \log p(t, x, y) \geqq -[d(x, y) + [\psi(\varepsilon)]^{1/2}]\,.$$

Since ε is arbitrary and $\psi(\varepsilon) \to 0$ as $\varepsilon \to 0$,

$$\liminf_{t\to 0} 2t \log p(t, x, y) \geqq -d^2(x, y)\,.$$

Since $d(x, y)$ is bounded, the only possible trouble regarding uniformity can occur when $\delta \to 1$ or $d(x, y)$ is very small. But then Theorem 4.10 can be used directly to show that it is uniform even around $d(x, y) = 0$.

We have finally proved

THEOREM 4.13.

$$\lim_{t\to 0}\left[-2t \log p(t, x, y)\right] = d^2(x, y)$$

uniformly in x, y such that $d(x, y)$ is bounded.

Remarks. The assumption in Theorem 3.12 that B is the common boundary of G and of its exterior can be weakened further. It suffices to assume that, for a dense set $B_0 \subset B$, every sufficiently small sphere S around $b_0 \in B_0$ is disconnected by B.

We can consider equation (2.3) in a domain $G \subset R_k$ with the extra condition that the solution is zero on the boundary. Then there is a formula similar to Theorem 2.2, where $d(x, y)$ is replaced by $d_G(x, y)$ which is the length of the shortest path from x to y not leaving G.

These results are best proved using probabilistic methods and are therefore left to [3].

Acknowledgement. The methods used by the author in an earlier version of this paper were heavily probabilistic. The author expresses his thanks to Prof. L. Nirenberg whose suggestions were useful in replacing the probabilistic methods by analytical ones.

Bibliography

[1] Friedman, A., *Partial Differential Equations of Parabolic Type*, Prentice-Hall, Englewood Cliffs, N.J., 1964.
[2] Miranda, C., *Equazioni alle Derivate Parziali di Tipo Ellittico*, Springer, Berlin, 1955.
[3] Varadhan, S. R. S., *Diffusion processes in a small time interval*, to appear in Comm. Pure. Appl. Math., Vol. 20, 1967.

Received August, 1966.

COMMUNICATIONS ON PURE AND APPLIED MATHEMATICS, VOL. XX, 659–685 (1967)

Diffusion Processes in a Small Time Interval*

S. R. S. VARADHAN

1. Introduction

Let $X(t)$ be a diffusion process in k dimensions with diffusion coefficients $\{a_{ij}(x)\}$. The transition probability densities of the homogeneous Markov process $X(t)$ are given by the fundamental solution $p(t, x, y)$ of the equation

$$\frac{\partial p}{\partial t} = \tfrac{1}{2} \sum a_{ij}(x) \frac{\partial^2 p}{\partial x_i \, \partial x_j} . \tag{1.1}$$

Our interest is in the behavior of the process $X(t)$ for small t. For that purpose we introduce a small parameter $\varepsilon > 0$ and define a process $X_\varepsilon(t) = X(\varepsilon t)$. We consider in detail the process $X_\varepsilon(t)$ in a fixed time interval $[0, T]$. This of course corresponds to studying the behavior, as $\varepsilon \to 0$, of the diffusion processes corresponding to the equations

$$\frac{\partial p}{\partial t} = \tfrac{1}{2}\varepsilon \sum a_{ij}(x) \frac{\partial^2 p}{\partial x_i \, \partial x_j} .$$

A similar analysis for Brownian motion was carried out by M. Schilder [4] and such an analysis has been used by Donsker in [2]. The same approach in connection with processes with independent increments was considered by the author, [6].

Section 2 outlines the contents of the paper and describes some preliminary material. The later sections contain the actual proofs.

2. Preliminaries and Summary

We denote by L_ε the operator

$$L_\varepsilon f = \tfrac{1}{2}\varepsilon \sum a_{ij}(x) \frac{\partial^2 f}{\partial x_i \, \partial x_j} \tag{2.1}$$

acting on smooth functions on R_k. The following assumptions are made regarding the coefficients $\{a_{ij}(x)\}$:

(A) $\{a_{ij}(x)\}$ is a symmetric positive definite matrix for each x.

* This paper represents results obtained under a Sloan Foundation grant for Probability and Statistics.

(B) The coefficients satisfy a uniform Hölder condition:

$$|a_{ij}(x) - a_{ij}(y)| \leqq M\,|x - y|^{\alpha}\,.$$

(C) The matrices $\{a_{ij}(x)\}$ satisfy a uniform ellipticity condition:

$$\alpha_1 \sum \xi_j^2 \leqq \sum a_{ij}(x)\xi_i\xi_j \leqq \alpha_2 \sum \xi_j^2$$

for some $0 < \alpha_1 \leqq \alpha_2 < \infty$.

Under these assumptions there exists a fundamental solution $p_\varepsilon(t, x, y)$ to the equation

$$\frac{\partial p}{\partial t} = \tfrac{1}{2}\varepsilon \sum a_{ij}(x)\,\frac{\partial^2 p}{\partial x_i\,\partial x_j}\,. \tag{2.2}$$

Moreover, if we denote L_ε for $\varepsilon = 1$ by L, and the corresponding $p_\varepsilon(t, x, y)$ by $p(t, x, y)$, it is obvious that

$$p_\varepsilon(t, x, y) = p(\varepsilon t, x, y)\,. \tag{2.3}$$

Let us choose and fix a $T < \infty$. We denote by Ω the space of all continuous functions $\omega(t)$ on $[0, T]$ with values in R_k. The topology in Ω is that of uniform convergence. The measurable sets are the Borel sets. We denote by Ω_x all functions $\omega \in \Omega$ with $\omega(0) = x$, where x is a point in R_k. For any set $A \subset \Omega$, A_x will be the intersection of A and Ω_x, $\|\omega\|$ will denote the sup norm.

For every $\varepsilon > 0$, $p_\varepsilon(t, x, y)$ satisfies the Chapman-Kolmogorov equations (semigroup property)

$$p_\varepsilon(t + s, x, y) = \int p_\varepsilon(t, x, z)p_\varepsilon(s, z, y)\,dz\,. \tag{2.4}$$

We can therefore use these functions as probability densities and construct a Markov process. For each $\varepsilon > 0$ they lead to a measure P_x^ε on Ω corresponding to the process starting at time $t = 0$ at the point x. In other words,

$$P_x^\varepsilon[\Omega_x] = 1\,. \tag{2.5}$$

If $A \subset \Omega$ is a cylinder set of the form

$$A = \{\omega : \omega(t_1), \cdots \omega(t_n) \in B\}\,,$$

where $0 < t_1 < t_2 \cdots < t_n \leqq T$ and B is a measurable subset of the n-fold product of R_k, then $P_x^\varepsilon[A]$ is defined as

$$P_x^\varepsilon[A] = \int_B p_\varepsilon(t_1, x, y_1)p_\varepsilon(t_2 - t_1, y_1, y_2)p_\varepsilon(t_n - t_{n-1}, y_{n-1}, y_n)\,dy_1 \cdots dy_n\,. \tag{2.6}$$

The relation (2.6) completely determines P_x^ε on Ω. The family $\{P_x^\varepsilon\}$ is referred to as the Markov process corresponding to the operator L_ε. For more details and for further properties of $\{P_x^\varepsilon\}$, such as the strong Markov property, the reader is referred to Dynkin [3].

We are interested in the behavior of P_x^ε as $\varepsilon \to 0$. For this purpose we introduce a functional $I(\omega)$ on Ω. Let us denote by $a^{-1}(x)$ the matrix inverse of the coefficient matrix $\{a_{ij}(x)\}$ at the point x. If θ is a vector, then $\theta a^{-1}(x)\theta$ will denote the quadratic form using the matrix $a^{-1}(x)$. For any $\omega \in \Omega$ which is differentiable, $\dot{\omega}(\tau)$ will be the vector $d\omega(\tau)/d\tau$. We now define $I(\omega)$ as follows:

If ω is absolutely continuous with a square integrable derivative on $[0, T]$, then

$$I(\omega) = \tfrac{1}{2}\int_0^T [\dot{\omega}(\tau)a^{-1}(\omega(\tau))\dot{\omega}(\tau)]\, d\tau\,, \tag{2.7a}$$

otherwise,

$$I(\omega) = +\infty\,. \tag{2.7b}$$

We have the following theorem concerning $I(\omega)$.

THEOREM 2.1. *The function $I(\omega)$ is lower semicontinuous on all of Ω. Moreover, for any finite constant N, the set $\{\omega : I(\omega) \leq N\}$ is an equicontinuous family of continuous functions on $[O, T]$.*

The proof is straightforward and is omitted.

We next introduce the global Riemannian distance $d(x, y)$ on R_k. For smooth functions $\omega(\tau)$ we first define the length

$$l(\omega) = \int_0^T \sqrt{\dot{\omega}(\tau)a^{-1}(\omega(\tau))\dot{\omega}(\tau)}\, d\tau\,, \tag{2.8}$$

and then we define

$$d(x, y) = \inf_{\omega:\,\substack{\omega(0)=x\\ \omega(T)=y}} l(\omega)\,.$$

It is clear that d is independent of T and is the natural geodesic distance; $l(\omega)$ is of course the geodesic length.

There is a relation between $I(\omega)$ and $l(\omega)$. The geodesic length $l(\omega)$ is independent of the parametrization of the curve, whereas $I(\omega)$ depends on the parametrization. For, a class of ω all representing the same curve $I(\omega)$ is minimum when the curve is parametrized so as to cover equal lengths in equal time. This corresponds to the fact that for the minimizing parametrization the integrand in (2.8) is a constant. We therefore obtain

LEMMA 2.2. *For* $0 \leqq \alpha < \beta \leqq T$,

$$\inf_{\omega:\substack{\omega(\alpha)=x\\ \omega(\beta)=y}} I(\omega) = \frac{d^2(x, y)}{2(\beta - \alpha)} .$$

An obvious extension is

LEMMA 2.3. *For* $0 \leqq t_1 < t_2 < t_3 \cdots < t_n \leqq T$,

$$\inf_{\omega:\substack{\omega(t_j)=x_j\\ j=1,2,\cdots n}} I(\omega) = \frac{1}{2}\sum_{j=1}^{n-1} \frac{d^2(x_{j+1}, x_j)}{t_{j+1} - t_j} .$$

We are now ready to state the main theorem of Section 3.

THEOREM 2.4. *Let* $C \subset \Omega$ *be closed and* $G \subset \Omega$ *be open. Then*

$$\limsup_{\substack{\varepsilon\to 0\\ y\to x}} \varepsilon \log P_y^\varepsilon[C] \leqq -\inf_{\omega\in C_x} I(\omega) ,$$

$$\liminf_{\substack{\varepsilon\to 0\\ y\to x}} \varepsilon \log P_y^\varepsilon[G] \geqq -\inf_{\omega\in G_x} I(\omega) .$$

In proving Theorem 2.4 the following theorem proved in [7] plays an important role.

THEOREM 2.5.

$$\lim_{t\to 0} - 2t \log p(t, x, y) = d^2(x, y)$$

uniformly over all x, y *such that* $d(x, y)$ *is bounded.*

The rest of this paper, namely Sections 4, 5, 6 are applications of Theorem 2.4. In Section 4 the following problem is considered: Let $D \subset R_k$ be an open subset with a boundary B. Let us consider the solution $p_D(t, x, y)$ of equation (1.1) but with the boundary condition

$$p_D(t, x, y) \to 0 \qquad \text{as} \qquad x \to B .$$

In other words, the process is killed on reaching the boundary B of D. We want to consider the behavior of

$$\lim_{t\to 0} \frac{p_D(t, x, y)}{p(t, x, y)} \tag{2.9}$$

for points x, y in D. We introduce the distance

$$d_D(x,y) = \inf_{\substack{\omega:\omega(0)=x \\ \omega(T)=y \\ \omega(\tau)\in D \text{ for } 0\leq\tau\leq T}} l(\omega) .$$

For suitably smooth regions D it is shown that

$$\lim_{t\to 0} - 2t \log p_D(t, x, y) = d_D^2(x, y) .$$

Moreover, the ratio (2.9) behaves as $t \to 0$ as follows: If the geodesics from x to y do not leave D, the ratio goes to 1. Otherwise it goes to 0. This is a rough form of the result. The precise version will be found in Section 4.[1] Section 5 is concerned with some applications to certain differential equations. Section 6 deals with the case when the operator L has an additional first order term. Essentially the same results can be obtained in this case. In fact, they are deduced from the present case.

3. Behavior of $\{P_x^\varepsilon\}$

Let $0 = t_0 < t_1 < t_2 \cdots < t_n = T$ be some time points. We shall denote by π this partition of the interval $[O, T]$. T_π will denote the map from Ω to the $(n + 1)$-fold product of R_k defined by

$$T_\pi\omega = \{\omega(t_0), \omega(t_1), \cdots, \omega(t_n)\} . \tag{3.1}$$

For any set A in the range of T_π the inverse image $T_\pi^{-1}A$ is defined in the usual manner. We then have from (2.6)

$$P_y^\varepsilon[T_\pi^{-1}A] = \int_{A_y} \prod_{j=1}^{n} p_\varepsilon(t_j - t_{j-1}, y_{j-1}, y_j)\, dy_j , \tag{3.2}$$

where A_y is the section of A with $y_0 = y$.

LEMMA 3.1. *Let $C \subset \Omega$ be a closed set of the form $C = T_\pi^{-1}A$, where A is closed. Then*

$$\limsup_{\substack{\varepsilon\to 0\\ y\to x}} \varepsilon \log P_y^\varepsilon[C] \leq -\inf_{\omega\in C_x} I(\omega) .$$

Proof: First let us suppose that A is closed and bounded. Then

$$P_y^\varepsilon[C] = \int_{A_y} \prod_{j=1}^{n} p_\varepsilon(t_j - t_{j-1}, y_{j-1}, y_j)\, dy_j . \tag{3.3}$$

[1] A more detailed investigation of this behavior for the case of Brownian motion was carried out by Ciesielski [1].

From Theorem 2.5 and Lemma 2.3 we see that

$$\limsup_{\substack{y\to x\\ \varepsilon\to 0}} \varepsilon \log P_y^\varepsilon[C] \leqq \limsup_{y\to x}\left[-\inf_{y_1,y_n\in A_y} \frac{1}{2}\sum \frac{d^2(y_{j-1}, y_j)}{t_j - t_{j-1}}\right]$$

$$\leqq -\inf_{y_1,y_n\in A_x} \frac{1}{2}\sum \frac{d^2(y_{j-1}, y_j)}{t_j - t_{j-1}}$$

$$= -\inf_{\omega\in C_x} I(\omega) .$$

The first step follows from straight-forward estimation of the integral (3.3) using the uniformity assured by Theorem 2.5. Since A is closed and bounded, points chosen in A_y have limit points in A_x as y approaches x. This justifies the second step. The last step is the identification furnished by Lemma 2.3.

If A is unbounded, we choose a Λ large enough so that

$$\Lambda > \inf_{\omega\in C_x} I(\omega) . \tag{3.4}$$

There is a standard estimate (Theorem 0.5, page 229, Volume II of [3]) of the form

$$p_\varepsilon(t, x, y) \leqq \frac{M}{(t\varepsilon)^{k/2}} \exp\left\{-\frac{\alpha}{2t\varepsilon}|x-y|^2\right\} . \tag{3.5}$$

We now define

$$S_\rho = \{|y_0| \leqq \rho, \cdots, |y_n| \leqq \rho\}$$

and S_ρ^* as the complement of S_ρ. From (3.5) we see that we can choose ρ large enough so that for all y in a compact set K

$$P_y^\varepsilon[T_\pi^{-1}S_\rho^*] \leqq Me^{-\Lambda/\varepsilon} . \tag{3.6}$$

Let us set

$$C_\rho = T_\pi^{-1}[A \cap S_\rho] ,$$

$$C_\rho^* = T_\pi^{-1}[S_\rho^*] .$$

Since $A \cap S_\rho$ is bounded and closed,

$$\limsup_{\substack{\varepsilon\to 0\\ y\to x}} \varepsilon \log P_y^\varepsilon[C_\rho] \leqq -\inf_{\substack{\omega\in C_\rho\\ \omega\in C_x}} I(\omega)$$

$$\leqq -\inf_{\omega\in C_x} I(\omega) .$$

From (3.6) we have

$$\limsup_{\substack{\varepsilon\to 0\\ y\to x}} \varepsilon \log P^\varepsilon_y[C^*_\rho] \leqq -\Lambda .$$

Since $P^\varepsilon_y[C] \leqq P^\varepsilon_y[C_\rho] + P^\varepsilon_y[C^*_\rho]$,

$$\limsup_{\substack{\varepsilon\to 0\\ y\to x}} \varepsilon \log P^\varepsilon_y[C] \leqq -\inf_{\omega\in C_x} I(\omega) .$$

We now choose for any integer n equally spaced time points $0 = t_0 < t_1 < \cdots t_n = T$. The partition has a length $h_n = T/n$. We denote by ω_π the trajectory obtained from ω by taking the points $\{\omega(t_j)\}$ and joining the successive ones by geodesics with natural parametrization. If there are several geodesics between two such points it is immaterial which one is chosen. In this way we obtain for each ω a ω_π which is piecewise geodesic and

$$\omega(t_j) = \omega_\pi(t_j) , \qquad j = 0, 1, 2, \cdots, n .$$

LEMMA 3.2. *For every $\delta > 0$,*

$$\limsup_{n\to\infty} \limsup_{\varepsilon\to 0} \sup_x \varepsilon \log P^\varepsilon_x[\|\omega - \omega_\pi\| \geqq \delta] = -\infty .$$

Proof: In [7] it was proved (Theorem 3.5) that if $\phi(\lambda, z)$ is the solution to the equation

$$L\phi = \lambda\phi \quad \text{for} \quad z \in U ,$$

$$\phi = 1 \quad \text{for} \quad z \in \delta U ,$$

where U is the set

$$U = \{z : d(x, z) < \eta\} ,$$

with a boundary δU, then for every $\mu > 0$ there is a constant M_μ such that, for all $\eta \leqq \eta_0$ and for all $x \in R_k$,

$$\phi(\lambda, x) \leqq M_\mu \exp\{-\sqrt{2\lambda}\,\eta(1 - \mu)\} . \tag{3.7}$$

Noting that if τ is the first exit time for the process starting at x for the set U, we then have

$$\phi(\lambda, x) = E_x[e^{-\lambda\tau}] .$$

We conclude that the expression

$$\phi(\varepsilon, \lambda, x) = \int e^{-\lambda\tau}\, dP^\varepsilon_x \tag{3.8}$$

satisfies the equation

$$\varepsilon L\phi = \lambda\phi \quad \text{in} \quad U,$$

$$\phi = 1 \quad \text{on} \quad \delta U.$$

Consequently, from (3.7), we have

$$\phi(\varepsilon, \lambda, x) \leqq M_\mu \exp\left\{-\sqrt{\frac{2\lambda}{\varepsilon}}\,\eta(1-\mu)\right\}.$$

Choosing $\mu = 1 - 1/\sqrt{2}$, we obtain

$$\phi(\varepsilon, \lambda, x) \leqq M \exp\left\{-\sqrt{\frac{\lambda}{\varepsilon}}\,\eta\right\}.$$

Using (3.8), a Tchebechev type of estimate yields

$$P_x^\varepsilon[\tau \leqq h] \leqq M \exp\left\{-\frac{\eta^2}{4\varepsilon h}\right\}.$$

We notice that

$$P_x^\varepsilon[\tau \leqq h] = P_x^\varepsilon\left[\sup_{0\leqq t\leqq h} d(\omega(t), \omega(0)) \geqq \eta\right]$$

and we therefore obtain, for all x and $\eta \leqq \eta_0$,

$$P_x^\varepsilon\left[\sup_{0\leqq t\leqq h} d(\omega(t), \omega(0)) \geqq \eta\right] \leqq M \exp\left\{-\frac{\eta^2}{4\varepsilon h}\right\}. \tag{3.9}$$

We want to estimate

$$P_x^\varepsilon\left[\sup_{0\leqq t\leqq T} d(\omega(t), \omega_\pi(t)) \geqq 2\eta\right].$$

Since $\omega(t_j) = \omega_\pi(t_j)$ for $j = 0, 1, \cdots, n$, and ω_π is piecewise geodesic for t in (t_j, t_{j+1}),

$$\begin{aligned} d(\omega_\pi(t), \omega_\pi(t_j)) &\leqq d(\omega_\pi(t_{j+1}), \omega_\pi(t_j)) = d(\omega(t_{j+1}), \omega(t_j)) \\ &\leqq \sup_{t_j\leqq t\leqq t_{j+1}} d(\omega(t_j), \omega(t)). \end{aligned}$$

We now have

$$\begin{aligned} \sup_{0\leqq t\leqq T} d(\omega(t), \omega_\pi(t)) &= \sup_{0\leqq j\leqq n-1}\ \sup_{t_j\leqq t\leqq t_{j+1}} d(\omega(t), \omega_\pi(t)) \\ &\leqq \sup_{0\leqq j\leqq n-1}\ \sup_{t_j\leqq t\leqq t_{j+1}} [d(\omega(t), \omega(t_j)) + d(\omega_\pi(t), \omega_\pi(t_j))] \\ &\leqq 2 \sup_{0\leqq j\leqq n-1}\ \sup_{t_j\leqq t\leqq t_{j+1}} d(\omega(t), \omega(t_j)). \end{aligned} \tag{3.10}$$

From the Markov property and the inequality (3.10), we obtain

$$P_x^\varepsilon\Big[\sup_{0\leqq t\leqq T} d(\omega(t), \omega_\pi(t)) \geqq 2\eta\Big] \leqq P_x^\varepsilon\Big[\sup_{0\leqq j\leqq n-1}\ \sup_{t_j\leqq t\leqq t_{j+1}} d(\omega(t), \omega(t_j)) \geqq \eta\Big]$$

$$\leqq \sum_{j=0}^{n-1} P_x^\varepsilon\Big[\sup_{t_j\leqq t\leqq t_{j+1}} d(\omega(t), \omega(t_j)) \geqq \eta\Big]$$

$$\leqq n \sup_x P_x^\varepsilon\Big[\sup_{0\leqq t\leqq h} d(\omega(t), \omega(0)) \geqq \eta\Big]$$

$$\leqq nM \exp\left\{-\frac{\eta^2}{4\varepsilon h}\right\}.$$

Since $|x - y| \leqq \beta\, d(x,y)$ for some β, we have for $\delta \leqq \delta_0$

$$P_x^\varepsilon[\|\omega - \omega_\pi\| \geqq \delta] \leqq nM \exp\left\{-\frac{\delta^2}{16\beta^2\varepsilon h}\right\}.$$

Consequently,

$$\limsup_{n\to\infty}\ \limsup_{\varepsilon\to 0}\ \sup_x\ \varepsilon \log P_x^\varepsilon[\|\omega - \omega_\pi\| \geqq \delta] = -\infty .$$

THEOREM 3.3. *For any closed set* $C \subset \Omega$,

$$\limsup_{\substack{\varepsilon\to 0\\ y\to x}} \varepsilon \log P_y^\varepsilon[C] \leqq -\inf_{\omega\in C_x} I(\omega) .$$

Proof: Let us define for $\delta > 0$ the sets C_δ and C_δ^δ as

$$C_\delta = \{\omega : \omega \in C\ |\omega(0) - x| < \delta\} ,$$

$$C_\delta^\delta = \{z : \|z - \omega\| < \delta \text{ for some } \omega \in C_\delta\} .$$

We introduce the function

$$I^\delta(\omega) = \inf_{z:\|z-\omega\|<\delta} I(z) .$$

Then $I^\delta(\omega)$ is finite everywhere. Moreover if

$$T_\delta = \inf_{\omega\in C_\delta^\delta} I(\omega) ,$$

then $\omega \in C_\delta \Rightarrow I^\delta(\omega) \geqq T_\delta$. Since $y \to x$ we can assume $|y - x| < \delta$. We would therefore obtain

$$P_y^\varepsilon[C] = P_y^\varepsilon[C_\delta] \leqq P_y^\varepsilon[I^\delta(\omega) \geqq T_\delta] .$$

Further,

$$P_y^\varepsilon[I^\delta(\omega) \geqq T_\delta] \leqq P_y^\varepsilon[\|\omega - \omega_\pi\| \geqq \delta] + P_y^\varepsilon[I(\omega_\pi) \geqq T_\delta] \tag{3.11}$$

and

$$\lim_{n\to\infty} \limsup_{\varepsilon\to 0} \sup_x \varepsilon \log P_y^\varepsilon[\|\omega - \omega_\pi\| \geqq \delta] = -\infty . \tag{3.12}$$

The set $[I(\omega_\pi) \geqq T_\delta]$ is equal to

$$\left\{\omega : \frac{1}{2}\sum_{j=0}^{n-1} \frac{d^2(\omega(t_j), \omega(t_{j+1}))}{t_{j+1} - t_j} \geqq T_\delta\right\} .$$

Hence for every n,

$$\limsup_{\substack{\varepsilon\to 0\\ y\to x}} \varepsilon \log P_y^\varepsilon[I(\omega_\pi) \geqq T_\delta] \leqq -T_\delta . \tag{3.13}$$

It follows from (3.11), (3.12) and (3.13) that

$$\limsup_{\substack{\varepsilon\to 0\\ y\to x}} \varepsilon \log P_y^\varepsilon[C] \leqq -T_\delta .$$

This is true for every $\delta > 0$ and $\lim_{\delta\to 0} T_\delta = T$, where

$$T = \inf_{\omega \in C_x} I(\omega) .$$

The proof of the theorem is now complete.

LEMMA 3.4. *Let $f \in \Omega$ and $f(0) = x$. Let V be the set $V = \{\omega : \|\omega - f\| < \delta\}$, where $\delta > 0$. Then*

$$\liminf_{\substack{\varepsilon\to 0\\ y\to x}} \varepsilon \log P_y^\varepsilon[U] \geqq -I(f) .$$

Proof: We can assume without loss of generality that δ is small and that $I(f) = a < \infty$. Let V_π denote the set

$$V_\pi = \{\omega : |\omega(t_j) - f(t_j)| \leqq \tfrac{1}{2}\delta \text{ for } j = 0, 1, 2, \cdots, n\} ,$$

then

$$P_y^\varepsilon[V_\pi] = \int_{A_\pi} \prod p_\varepsilon(t_j - t_{j-1}, y_{j-1}, y_j)\, dy_j ,$$

where A_π is the set

$$A_\pi = \{y : |y_j - f(t_j)| \leqq \tfrac{1}{2}\delta \text{ for } j = 0, 1, \cdots, n\} .$$

Using Theorem 2.3, one can obtain

$$\begin{aligned}\liminf_{\substack{\varepsilon\to 0\\ y\to x}} \varepsilon \log P_y^\varepsilon[V_\pi] &\geqq -I(f_\pi)\\ &\geqq -I(f)\,.\end{aligned}$$

It is therefore sufficient to prove that

$$\lim_{n\to\infty} \limsup_{\substack{\varepsilon\to 0\\ y\to x}} \varepsilon \log P_y^\varepsilon[U_\pi - U] = -\infty\,. \tag{3.14}$$

We see that $\omega \in V_\pi - V \Rightarrow |\omega(t_j) - f(t_j)| \leqq \frac{1}{2}\delta$ for all j but $\|\omega(t) - f(t)\| \geqq \delta$. This implies that, for some interval $t_j \leqq t \leqq t_{j+1}$,

$$|\omega(t_j) - f(t_j)| < \tfrac{1}{2}\delta\,, \qquad |\omega(t_{j+1}) - f(t_{j+1})| < \tfrac{1}{2}\delta\,,$$

but $|\omega(\tau) - f(\tau)| \geqq \delta$ for some τ in $t_j \leqq \tau \leqq t_{j+1}$. Further, we can choose n large enough so that, for the resulting π, $|f(\tau) - f(t_j)| < \frac{1}{4}\delta$ for all τ in $t_j \leqq \tau \leqq t_{j+1}$. We conclude from all this that, if $\omega \in V_\pi - V$, then $|\omega(\tau) - \omega(t_j)| > \frac{1}{4}\delta$ for τ in (t_j, t_{j+1}) for some j. Hence $d(\omega(\tau), \omega(t_j)) \geqq \beta\frac{1}{4}\delta$ for some $\beta > 0$. This in turn implies that $I(\omega) \geqq \beta^2\delta^2/32h$. Moreover, $V_\pi - V$ is closed and therefore, by Theorem 3.3,

$$\lim_{n\to\infty} \limsup_{\substack{\varepsilon\to 0\\ y\to x}} \varepsilon \log P_y^\varepsilon[V_\pi - V] = -\infty\,.$$

This establishes (3.14) and hence the lemma is proved.

THEOREM 3.5. *Let $G \subset \Omega$ be open. Then*

$$\liminf_{\substack{\varepsilon\to 0\\ y\to x}} \varepsilon \log P_y^\varepsilon[G] \geqq -\inf_{\omega\in G_x} I(\omega)\,.$$

Proof: Since $G \subset \Omega$ is open for $f \in \Omega$, there is a sphere V around f contained in Ω. By Lemma 3.4,

$$\liminf_{\substack{\varepsilon\to 0\\ y\to x}} \varepsilon \log P_y^\varepsilon[G] \geqq -I(f)\,.$$

This is true for all $f \in \Omega$ with $f(0) = x$. Hence the theorem is true. Theorems 3.3 and 3.5 contain Theorem 2.4.

We conclude this section with a corollary to Theorem 3.5.

COROLLARY 3.6. *Let $A \subset \Omega$ be measurable, A° and $\bar{A}$ the interior and closure of A. Assume that for $x \in K$ (K compact)*

$$\phi(x) = \inf_{\omega\in\bar{A}\cap\Omega_x} I(\omega) = \inf_{\omega\in A^\circ\cap\Omega_x} I(\omega)\,.$$

Then $\phi(x)$ *is continuous on* K *and*

$$\lim_{\varepsilon \to 0} \varepsilon \log P_x^\varepsilon[A] = -\phi(x)$$

uniformly over K.

4. Fundamental Solution in a Subdomain

$D \subset R_k$ is an open set and we kill the process at the boundary B of D. We consider the fundamental solution $q_\varepsilon(t, x, y)$ of the equation

$$\frac{\partial q}{\partial t} = \tfrac{1}{2}\varepsilon \sum a_{ij}(x) \frac{\partial^2 q}{\partial x_i \, \partial x_j} \qquad \text{for} \qquad x \in D \,, \tag{4.1}$$

$$q_\varepsilon(t, x, y) \to 0 \qquad \text{as} \qquad x \to B \,.$$

It is clear that

$$q_\varepsilon(t, x, y) \leqq p_\varepsilon(t, x, y) \,.$$

Moreover, if we define the set function

$$q_\varepsilon(t, x, A) = \int_A q_\varepsilon(t, x, y) \, dy \tag{4.2}$$

for sets $A \subset D$, then the connection between (4.2) and the measures P_x^ε is

$$q_\varepsilon(t, x, A) = P_x^\varepsilon\{\omega : \omega(t) \in A, \, \omega(\tau) \in D \text{ for } 0 \leqq \tau \leqq t\} \,. \tag{4.3}$$

Let us introduce two subsets of Ω,

$$G = \{\omega : \omega(\tau) \in D \text{ for } 0 \leqq \tau \leqq T\} \,,$$

and

$$\bar{G} = \text{closure of } G \,.$$

Then $\bar{G} \subset \{\omega : \omega(\tau) \in D \cup B \text{ for } 0 \leqq \tau \leqq T\}$. In some cases the inclusion is proper and that is relevant. We introduce the distances

$$\underline{d}_D(x, y) = \inf_{\substack{\omega(0)=x \\ \omega(T)=y \\ \omega \in \bar{G}}} l(\omega) \,,$$

and

$$\bar{d}_D(x, y) = \inf_{\substack{\omega(0)=x \\ \omega(T)=y \\ \omega \in G}} l(\omega) \,.$$

The set D is called "smooth" if $\underline{d}_D(x, y) = \bar{d}_D(x, y)$. This means that if a trajectory of length l can be approximated by trajectories in D, then it can be also approximated by trajectories within D whose lengths approach l. In such a case we shall denote by $d_D(x, y)$ the common value of $\underline{d}_D$ and $\bar{d}_D$. We can check that the following relations hold:

$$\inf_{\substack{\omega \in G \\ \omega(0)=x \\ \omega(T)=y}} I(\omega) = \frac{1}{2T} \bar{d}_D^2(x, y) ,$$

(4.4)

$$\inf_{\substack{\omega \in \bar{G} \\ \omega(0)=x \\ \omega(T)=y}} I(\omega) = \frac{1}{2T} \underline{d}_D^2(x, y) .$$

LEMMA 4.1. *For $A \subset D$ and closed.*

$$\limsup_{\substack{\varepsilon \to 0 \\ y \to x}} \varepsilon \log q_\varepsilon(T, y, A) \leqq -\inf_{z \in A} \frac{\underline{d}_D^2(x, z)}{2T} .$$

Proof: Let us take $C \subset \Omega$ as

$$C = \{\omega : \omega(T) \in A, \omega \in \bar{G}\} .$$

Then from relation (4.3), we have

$$\limsup_{\substack{\varepsilon \to 0 \\ y \to x}} \varepsilon \log q_\varepsilon(T, y, A) \leqq \limsup_{\substack{\varepsilon \to 0 \\ y \to x}} \varepsilon \log P_y^\varepsilon[C]$$

$$\leqq -\inf_{\omega \in C_x} I(\omega) = -\frac{1}{2T} \inf_{z \in A} \underline{d}_D^2(x, z) .$$

Lemma 4.1 is only a weak version of what we want to prove. To study the density $q_\varepsilon(t, x, y)$ we use a slightly different approach.

LEMMA 4.2. *Let x, y be points in D. Then*

$$\limsup_{\substack{\varepsilon \to 0 \\ x' \to x \\ y' \to y}} \varepsilon \log q_\varepsilon(T, y, y') \leqq -\frac{1}{2T} \underline{d}_D^2(x, y) .$$

Proof: Let us take a neighborhood S_δ around y of radius δ and a neighborhood $S_{2\delta}$ around S_δ of radius δ. Let τ be the first time the set $S_{2\delta}$ is reached starting from x'. Then for $y' \in S_\delta$,

(4.5)
$$q_\varepsilon(T, x', y') = \int_{\substack{\tau < T \\ \omega \in G}} q_\varepsilon(t - \tau, \omega(\tau), y') P_{x'}^\varepsilon(d\omega) .$$

Since $\omega(\tau)$ is in the boundary of $S_{2\delta}$ and y' is in S_δ, we have $|\omega(\tau) - y'| \geqq \delta$. Moreover, $q_\varepsilon(t, x, y) \leqq p_\varepsilon(t, x, y)$ and, if x and y are kept away, the functions are bounded. Hence for all $\omega(\tau)$ and y',

$$q_\varepsilon(T - \tau, \omega(\tau), y') \leqq M_\delta < \infty .$$

Continuing with (4.5) this yields

$$q_\varepsilon(T, x', y') \leqq M_\delta P^\varepsilon_{x'}[\omega : \omega \in G, \tau \leqq T] .$$

Therefore,

$$\limsup_{\substack{\varepsilon \to 0 \\ x' \to x \\ y' \to y}} \varepsilon \log q_\varepsilon(T, x', y') \leqq - \frac{1}{2T} \inf_{z \in S_{2\delta}} \underline{d}^2_D(x, z) .$$

Since this is true for every $\delta > 0$, we prove the lemma by letting $\delta \to 0$.

LEMMA 4.3. *For $A \subset D$ and open,*

$$\liminf_{\substack{\varepsilon \to 0 \\ y \to x}} \varepsilon \log q_\varepsilon(T, y, A) \geqq -\inf_{z \in A} \frac{\bar{d}^2_D(x, z)}{2T} .$$

Proof: We have

$$q_\varepsilon(T, y, A) = P^\varepsilon_y[\omega : \omega \in G, \omega(T) \in A] .$$

The set on the right is open. A direct application of Theorem 3.5 yields

$$\begin{aligned} \liminf_{\substack{\varepsilon \to 0 \\ y \to x}} \varepsilon \log q_\varepsilon(T, y, A) &\geqq -\inf_{\substack{\omega \in G_x \\ \omega(T) \in A}} I(\omega) \\ &= -\frac{1}{2T} \inf_{z \in A} \bar{d}^2_D(x, z) . \end{aligned}$$

LEMMA 4.4. *Let $K \subset D$ be compact. Then for $\Delta > 0$,*

$$\liminf_{\substack{\varepsilon \to 0 \\ x, y \in K \\ |x - y| \to 0}} \varepsilon \log q_\varepsilon(\Delta, x, y) \geqq 0 .$$

Proof: We use the formula

$$q_\varepsilon(\Delta, x, y) = p_\varepsilon(\Delta, x, y) - \int_{\tau \leqq \Delta} p_\varepsilon(\Delta - \tau, \omega(\tau), y) P^\varepsilon_x(d\omega) ,$$

where τ is the first exit time from D. Since $y \in K \subset D$ and K is compact, $|y - \omega(\tau)| \geqq \eta > 0$ for all $y \in K$ and $\omega(\tau) \in B$. Therefore,

$$q_\varepsilon(\Delta, x, y) \geqq p_\varepsilon(\Delta, x, y) - \sup_{\substack{b \in K \\ 0 \leqq u \leqq \Delta \\ y \in B}} p_\varepsilon(u, b, y) \tag{4.6}$$

$$\geqq p_\varepsilon(\Delta, x, y) - Me^{-c\eta^2/\varepsilon} .$$

Since by Theorem 2.5

$$\liminf_{\substack{\varepsilon \to 0 \\ |x-y| \to 0}} \varepsilon \log p_\varepsilon(\Delta, x, y) \geqq 0 ,$$

it follows from (4.6) that

$$\liminf_{\substack{\varepsilon \to 0 \\ x, y \in K \\ |x-y| \to 0}} \varepsilon \log q_\varepsilon(\Delta, x, y) \geqq 0 .$$

LEMMA 4.5. *Let x, y be points in D. Then*

$$\liminf_{\substack{\varepsilon \to 0 \\ x' \to x \\ y' \to y}} \varepsilon \log q_\varepsilon(T, x', y') \geqq -\frac{1}{2T} d_D^2(x, y) .$$

Proof: Since $q_\varepsilon(t, x, y)$ satisfies the semigroup property,

$$q_\varepsilon(T, x', y') = \int_D q_\varepsilon(T - \Delta, x', z) q_\varepsilon(\Delta, z, y')\, dz$$

$$\geqq \int_{z \in S} q_\varepsilon(T - \Delta, x', z) q_\varepsilon(\Delta, z, y')\, dz$$

$$\geqq q_\varepsilon(T - \Delta, x', S) \inf_{z \in S} q_\varepsilon(\Delta, z, y') .$$

We take S to be a small sphere around y of radius δ. Then,

$$\liminf_{\substack{\varepsilon \to 0 \\ x' \to x \\ y' \to y}} \varepsilon \log q_\varepsilon(T : x', y') \geqq \liminf_{\substack{\varepsilon \to 0 \\ x' \to x}} \varepsilon \log q_\varepsilon(T - \Delta, x', S)$$
$$+ \liminf_{\substack{\varepsilon \to 0 \\ y' \to y}} \inf_{z \in S} \varepsilon \log q_\varepsilon(\Delta, z, y') .$$

Combining Lemmas 4.3 and 4.4, we have if we let $\delta \to 0$

$$\liminf_{\substack{\varepsilon \to 0 \\ x' \to x \\ y' \to y}} \varepsilon \log q_\varepsilon(T, x', y') \geqq -\frac{1}{2(T - \Delta)} d_D^2(x, y) .$$

Since $\Delta > 0$ is arbitrary, we complete the proof of Lemma 4.5 by letting $\Delta \to 0$.
Lemmas 4.2 and 4.5 yield

THEOREM 4.6. *Let D be smooth. Then*

$$\lim_{\varepsilon \to 0} \varepsilon \log q_\varepsilon(T, x, y) = -\frac{1}{2T} d_D^2(x, y)$$

uniformly for $x, y \in K$, where $K \subset D$ is any compact set.

COROLLARY 4.7. *For D smooth,*

$$\lim_{t \to 0} t \log q(t, x, y) = -\tfrac{1}{2} d_D^2(x, y) .$$

Let us set

$$R_\varepsilon(t, x, y) = p_\varepsilon(t, x, y) - q_\varepsilon(t, x, y) , \tag{4.7}$$

$$d_D^*(x, y) = \inf_{\substack{\omega(0)=x \\ \omega(T)=y \\ \omega \notin G}} l(\omega) . \tag{4.8}$$

Then

$$\inf_{\substack{\omega(0)=x \\ \omega(T)=y \\ \omega \notin G}} I(\omega) = \frac{1}{2T} d_D^{*2}(x, y) .$$

Moreover, it is easily verified that

$$d_D^*(x, y) = \inf_{z \in B} [d(x, z) + d(z, y)] .$$

LEMMA 4.8.

$$\limsup_{\substack{\varepsilon \to 0 \\ x' \to x \\ y' \to y}} \varepsilon \log R_\varepsilon(T, x', y') \leqq -\frac{1}{2T} d_D^{*2}(x, y) .$$

Proof: This is proved in a manner similar to Lemma 4.2. In order to get to y' from x' by a trajectory not in G, we have to reach the boundary first and then come to a sphere around y. From this point on the proof is exactly like that of Lemma 4.2. Since τ is defined differently, we end up with

$$R_\varepsilon(T, x', y') = \int_{\tau \leqq T} p_\varepsilon(t - \tau, \omega(\tau), y') P_{x'}^\varepsilon(d\omega) ,$$

where τ is the first time the trajectory enters a set $S_{2\delta}$ around y after reaching the boundary B of D. Hence, by suitably letting $\delta \to 0$, we have

$$\limsup_{\substack{\varepsilon\to 0\\ x'\to x\\ y'\to y}} \varepsilon \log R_\varepsilon(T, x', y') \leqq -\frac{1}{2T} d_D^*(x, y) .$$

For any region D with boundary B, Lemmas 4.8, 4.2 and Theorem 2.5 imply

$$\begin{aligned} &\lim_{t\to 0} 2t \log p(t, x, y) = -d^2(x, y) , \\ (4.9)\qquad &\limsup_{t\to 0} 2t \log q(t, x, y) \leqq \underline{d}_D^2(x, y) , \\ &\limsup_{t\to 0} 2t \log R(t, x, y) \leqq -d_D^{*2}(x, y) . \end{aligned}$$

An immediate consequence of (4.9) is

THEOREM 4.9. *Let x, y in D be such that all the geodesics from x, y lie completely within D. Then the same is true for all a, b near x, y, respectively, and*

$$\lim_{t\to 0} \frac{q(t, a, b)}{p(t, a, b)} = 1$$

uniformly for a, b near x, y. On the other hand, let x, y in D be such that no geodesic from x to y can be approximated uniformly by paths completely in D. Then the same is true for all a, b near x, y, respectively, and

$$\lim_{t\to 0} \frac{q(t, a, b)}{p(t, a, b)} = 0$$

uniformly for a, b near x, y.

5. Some Associated Differential Equations

Let $D \subset R_k$ be a domain with boundary B. We assume throughout this section that the set of points $b \in B$, with the property that every sufficiently small neighborhood N_b around b is disconnected by B, is dense in B.

Let $\phi(x, \lambda)$ be the solution of the equation

$$(5.1)\qquad \begin{aligned} &\tfrac{1}{2} \sum a_{ij}(x) \frac{\partial^2 \phi}{\partial x_i\, \partial x_j} = \lambda\phi \qquad \text{for} \quad x \in D , \\ &\phi = 1 \qquad \text{on} \quad B . \end{aligned}$$

We then have

THEOREM 5.1.

$$\lim_{\lambda\to\infty} -\frac{1}{\sqrt{2\lambda}} \log \phi(x, \lambda) = d(x, B)$$

uniformly in x over compact subsets of $D \cup B$.

Proof: Let τ be the exit time from the set D. Then the solution $\phi(x, \lambda)$ of equation (5.1) has a representation

$$\phi(x, \lambda) = \int e^{-\lambda\tau} P_x(d\omega)$$

$$= \int e^{-\lambda\tau} F(x, dt) ,$$

where

$$F(x, t) = P_x[\tau \leqq t] .$$

In order to estimate $\phi(x, \lambda)$ for large λ we need estimate $F(x, t)$ for small t:

$$F(x, \varepsilon) = P_x[\tau \leqq \varepsilon] = P_x^\varepsilon[\tau \leqq 1] .$$

Let C be the set

$$C = \{\omega : \omega(s) \in B \text{ for some } s, \quad 0 \leqq s \leqq 1\}.$$

Then C is closed and $C = \{\omega : \tau \leqq 1\}$. Therefore,

$$\limsup_{\substack{\varepsilon\to 0\\ y\to x}} \varepsilon \log F(y, \varepsilon) = \limsup_{\substack{\varepsilon\to 0\\ y\to x}} \varepsilon \log P_y^\varepsilon[C] \tag{5.2}$$
$$\leqq -\inf_{C_x} I(\omega) = -\tfrac{1}{2}d^2(x, B) .$$

Let G be defined as the set of trajectories ω for which there are two points $0 \leqq u_1 < u_2 \leqq 1$ in the time interval $0 \leqq s \leqq 1$ such that, in the interval $u_1 \leqq \tau \leqq u_2$, $\omega(\tau)$ lies completely in one of the disconnected neighborhoods N_b and $\omega(u_1)$ and $\omega(u_2)$ belong to two different components of the disconnected set N_b. It is clear that G is open and $G \subset C$. Therefore,

$$\liminf_{\substack{\varepsilon\to 0\\ y\to x}} \varepsilon \log P_y^\varepsilon[C] \geqq \liminf_{\substack{\varepsilon\to 0\\ y\to x}} \varepsilon \log P_y^\varepsilon[G] \geqq -\inf_{\omega\in G_x} I(\omega) .$$

Let $b \in B$ and N_b be a small neighborhood around b which is disconnected. Consider the trajectory $\omega(\tau)$, $0 \leqq \tau \leqq 1$, which is a geodesic from x to b, to be naturally parametrized. Then $I(\omega) = \frac{1}{2}d^2(x, b)$. The trajectory $\omega(\tau)$ can be extended slightly so that it penetrates into two different components of N_b and lies completely in N_b in between. This can always be achieved by choosing N_b small with only a slight increase in $I(\omega)$. The parametrization is to be adjusted by rescaling time so that the time interval is still $0 \leqq s \leqq 1$. The final conclusion is

$$\inf_{\omega \in G_x} I(\omega) \leqq \inf_{b \in B_0} \tfrac{1}{2} d^2(x, b) ,$$

where B_0 is the set of points in B for which the construction is possible. By assumption, B_0 is dense in B. Consequently,

$$\liminf_{\substack{\varepsilon \to 0 \\ y \to x}} \varepsilon \log F(y, \varepsilon) \geqq -\tfrac{1}{2} d^2(x, B) .$$

Along with (5.2) this implies

$$\lim_{\substack{\varepsilon \to 0 \\ y \to x}} \varepsilon \log F(y, \varepsilon) = -\tfrac{1}{2} d^2(x, B) ,$$

since

$$\phi(x, \lambda) = \int e^{-\lambda t} F(x, dt) .$$

Standard estimation yields

$$\lim_{\substack{\lambda \to \infty \\ y \to x}} \frac{1}{\sqrt{2\lambda}} \log \phi(y, \lambda) = -d(x, B) .$$

We still assume that we have a domain D with a boundary B satisfying the condition described early in the section. Let us define for any set $\Gamma \subset B$ (Γ measurable) the function $\psi(x, \lambda, \Gamma)$ as the solution of the equation

$$\begin{aligned} L\psi &= \lambda\psi && \text{for} \quad x \in D , \\ \psi &= 1 && \text{for} \quad x \in \Gamma , \\ \psi &= 0 && \text{for} \quad x \in B - \Gamma . \end{aligned} \tag{5.3}$$

Then $\psi(x, \lambda, \Gamma)$ has a representation

$$\psi(x, \lambda, \Gamma) = \int e^{-\lambda t} F(x, \Gamma, dt) . \tag{5.4}$$

If τ is the first exit time from D, then

$$F(x, \Gamma, t) = P_x[\tau \leqq t, \omega(\tau) \in \Gamma] . \tag{5.5}$$

We assume that D is smooth in the sense of Section 4. Moreover, we assume that the property extends to the boundary B. The actual assumptions are as follows: If

$$G = \{\omega : \omega(\tau) \in D \quad \text{for} \quad 0 \leqq \tau < T\},$$

$$\bar{G} = \text{closure of } G,$$

$$\bar{d}_D(x,y) = \inf_{\substack{\omega \in G \\ \omega(0)=x \\ \omega(T)=y}} l(\omega),$$

$$\underline{d}_D(x,y) = \inf_{\substack{\omega \in \bar{G} \\ \omega(0)=x \\ \omega(T)=y}} l(\omega),$$

then $\bar{d}_D(x,y) = \underline{d}_D(x,y)$ for all $x \in D$ and $y \in D \cup B$. We denote the common value by $d_D(x,y)$ and by $d_D(x,b)$ if b is on the boundary.

LEMMA 5.2. *For $\Gamma \subset B$ open and $x \in D$,*

$$\liminf_{\substack{\lambda\to\infty \\ y\to x}} \frac{1}{\sqrt{2\lambda}} \log \psi(y, \lambda, \Gamma) \geqq -\inf_{b\in\Gamma} d_D(x, b).$$

For $\Gamma \subset B$ closed and $x \in D$,

$$\limsup_{\substack{\lambda\to\infty \\ y\to x}} \frac{1}{\sqrt{2\lambda}} \psi(y, \lambda, \Gamma) \leqq -\inf_{b\in\Gamma} d_D(x, b).$$

Proof: As in Theorem 5.1 we study

$$F(x, \Gamma, \varepsilon) = P_x[\tau \leqq \varepsilon, \omega(\tau) \in \Gamma] = P_x^\varepsilon[\tau \leqq 1, \omega(\tau) \in \Gamma].$$

Let U be the set

$$U = \{\omega : \tau \leqq 1, \omega(\tau) \in \Gamma\}.$$

In order to show that

$$\liminf_{\substack{\varepsilon\to 0 \\ y\to x}} \varepsilon \log F(y, \Gamma, \varepsilon) \geqq -\tfrac{1}{2} \inf_{b\in\Gamma} d_D^2(x, b), \tag{5.6}$$

it suffices to show that for any given $b \in \Gamma$ and $\delta > 0$ there exists an interior point ω_0 of U (U is not open in general) such that

$$I(\omega_0) \leqq \tfrac{1}{2} d_D^2(x, b) + \delta. \tag{5.7}$$

Let $b \in \Gamma$ be given. Then there is a b' near b for which the disconnectedness condition holds. Let us take a ω' in G with $\omega'(1) = b$ and $I(\omega')$ nearly equal to

$\frac{1}{2}d_D^2(x, b)$. We take a point near b on ω' inside D and connect the point with b' by a geodesic not necessarily inside D. Then we repeat the construction of Theorem 5.1 to get a ω_0 which penetrates into two different components of a small disconnected neighborhood $N_{b'}$ around b'. Any trajectory close to ω_0 will have to exit at a point near b or b'. Since Γ is open, things can be managed so that all these trajectories exit in Γ. $I(\omega_0)$ of course is not much larger than $\frac{1}{2}d_D^2(x, b)$. This proves (5.7) and hence (5.6). Now let Γ be closed. Then the set

$$C = \{\omega : \omega(s) \in \Gamma \text{ for some } 0 \leqq s \leqq 1, \omega \in \bar{G}\}$$

is closed and obviously

$$\limsup_{\substack{\varepsilon \to 0 \\ y \to x}} \varepsilon \log F(y, \Gamma, \varepsilon) \leqq \limsup_{\substack{\varepsilon \to 0 \\ y \to x}} \varepsilon \log P_y^\varepsilon[C] \tag{5.8}$$

$$\leqq -\inf_{\omega \in G_x} I(\omega) = -\tfrac{1}{2} \inf_{b \in \Gamma} d_D^2(x, b) .$$

The lemma is then proved using the relation (5.4) and the properties (5.6) and (5.8).

We now consider the solution $\phi(x, \lambda)$ to the equation

$$\begin{aligned} L\phi &= \lambda\phi \qquad \text{for} \quad x \in D , \\ \phi &= h(b, \lambda) \qquad \text{for} \quad b \in B , \end{aligned} \tag{5.9}$$

where

$$h(b, \lambda) = \exp \{\sqrt{2\lambda}\, h(b)\} \tag{5.10}$$

and $h(b)$ is a bounded continuous function on B.

THEOREM 5.3.

$$\lim_{\lambda \to \infty} \frac{1}{\sqrt{2\lambda}} \log \phi(x, \lambda) = \sup_{b \in B} [h(b) - d_D(x, b)]$$

uniformly over compact subsets $K \subset D$.

Proof: We use the formula

$$\phi(x, \lambda) = \int h(b, \lambda)\psi(x, \lambda, db) .$$

Lemma 5.2 is all we need to be able to use Theorem 3.4 of [6] and the theorem follows. It is elementary to see that the uniformity in Lemma 5.2 carries over to the uniformity asserted in the theorem.

Finally we turn to the following initial value problem: $u_\varepsilon(t, x)$ is the solution of the equation

$$
\begin{aligned}
\frac{\partial u_\varepsilon}{\partial t} &= \tfrac{1}{2}\varepsilon L u_\varepsilon + \tfrac{1}{2}\sum a_{ij}(x)\,\frac{\partial u_\varepsilon}{\partial x_i}\,\frac{\partial u_\varepsilon}{\partial x_j} + P(t, x)\,, \\
u_\varepsilon(0, x) &= V(x)\,.
\end{aligned}
\tag{5.11}
$$

Let us suppose for simplicity that $V(x)$ and $P(t, x)$ are continuous and bounded. We are interested in the behavior of $u_\varepsilon(t, x)$ as $\varepsilon \to 0$.

THEOREM 5.4.

$$\lim_{\varepsilon \to 0} u_\varepsilon(t, x) = u(t, x)$$

exists and

$$u(t, x) = \sup_{\omega \in \Omega_x} \left[V(\omega(t)) + \int_0^t P(t - \tau, \omega - \tau)\, d\tau - I(\omega) \right].$$

Proof: Let us set

$$v_\varepsilon(t, x) = \exp\left\{\frac{1}{\varepsilon}\, u_\varepsilon(t, x)\right\}.$$

Then the equation satisfied by v_ε is

$$
\begin{aligned}
\frac{\partial v_\varepsilon}{\partial t} &= \tfrac{1}{2}\varepsilon\, L v_\varepsilon + \frac{1}{\varepsilon}\, v_\varepsilon P\,, \\
v_\varepsilon(0, x) &= \exp\left\{\frac{1}{\varepsilon}\, V(x)\right\};
\end{aligned}
\tag{5.12}
$$

hence the solution is given by the function space integral

$$v_\varepsilon(t, x) = \int \exp\left\{\frac{1}{\varepsilon}\, V(\omega(t)) + \frac{1}{\varepsilon}\int_0^t P(t - \tau, \omega(\tau))\, d\tau\right\} P_x^\varepsilon(d\omega)\,.$$

We now appeal to Theorem 3.4 in [5] and conclude that

$$
\begin{aligned}
\lim_{\varepsilon \to 0} u_\varepsilon(t, x) &= \lim_{\varepsilon \to 0} \varepsilon \log v_\varepsilon(t, x) \\
&= \sup_{\omega \in \Omega_x} \left[V(\omega(t)) + \int_0^t P(t - \tau, \omega(\tau))\, d\tau - I(\omega) \right].
\end{aligned}
$$

Remark. With a little more effort one can show that the limit in Theorem 5.4 is actually uniform over compact subsets of t and x. One can also replace the boundedness of P and V by reasonable growth conditions. A similar approach for the case when L is the Laplacian was carried out in [2] using the results in [4].

6. The Case With a First Order Term

We assume that the operator (2.1) has an additional first order term:

$$L_0 f = \tfrac{1}{2}\sum a_{ij}(x) \frac{\partial^2 f}{\partial x_i \, \partial x_j} + \sum b_j(x) \frac{\partial f}{\partial x_j} .$$

The coefficients $\{a_{ij}(x)\}$ satisfy the assumptions postulated earlier. In addition we assume that

$$|b_j(x)| \leqq M < \infty ,$$

$$|b_j(x) - b_j(y)| \leqq M |x - y|^\alpha .$$

Let $q(t, x, y)$ be the fundamental solution for the equation

$$\frac{\partial q}{\partial t} = L_0 q .$$

If we replace L_0 by εL_0, we get a family

$$q_\varepsilon(t, x, y) = q(\varepsilon t, x, y) .$$

Using the densities $q_\varepsilon(t, x, y)$ we obtain a family of measures Q_x^ε. The following Theorem is proved in [5].

THEOREM 6.1. *For every x and ε the measure Q_x^ε on Ω is absolutely continuous with respect to P_x^ε and*

$$\frac{dQ_x^\varepsilon}{dP_x^\varepsilon} = \exp\left\{\int_0^T b(\omega(\tau))a^{-1}(\omega(\tau))\, d\omega(\tau) - \tfrac{1}{2}\varepsilon\int_0^T [b(\omega(\tau))a^{-1}(\omega(\tau))b(\omega(\tau))]\, d\tau\right\} .$$

Remark. Although the assumptions in [5] are stronger (the Hölder exponent α is assumed there to be 1) the theorem can be proved under less restrictive conditions. For instance in [3], Volume II, Chapter XIV, Section 6, it is carried out by a different method when $a_{ij}(x) = \delta_{ij}$. The same method will work in the present case. It is elementary to obtain the following estimate.

LEMMA 6.2. *If we denote by $F(\varepsilon, \omega)$ the exponent in the Radon-Nikodym derivative of Theorem 6.1, then*

$$\int \exp\{\lambda F(\varepsilon, \omega)\} P_x^\varepsilon(d\omega) \leqq \exp\{C\varepsilon\tfrac{1}{2}\lambda^2\} .$$

Proof: One uses the fact that

$$\exp\left\{\lambda\int_0^t b(\omega(\tau))a^{-1}(\omega(\tau))\,d\omega(\tau) - \tfrac{1}{2}\lambda^2\varepsilon\int_0^t [b(\omega(\tau))a^{-1}(\omega(\tau))b(\omega(\tau))]\,d\tau\right\}$$

is a martingale relative to P_x^ε.

THEOREM 6.3. *For $G \subset \Omega$ open and $C \subset \Omega$ closed,*

$$\liminf_{\substack{\varepsilon\to 0\\ y\to x}} Q_y^\varepsilon[G] \geqq -\inf_{\omega\in G_x} I(\omega)\,,$$

$$\limsup_{\substack{\varepsilon\to 0\\ y\to x}} Q_y^\varepsilon[C] \leqq -\inf_{\omega\in C_x} I(\omega)\,.$$

Proof: We have

$$\begin{aligned} Q_y^\varepsilon[G] &= \int_G \exp\{F(\varepsilon,\omega)\}P_y^\varepsilon[d\omega] \\ &\geqq e^{-A}P_y^\varepsilon[G \cap \{F(\varepsilon,\omega) \geqq -A\}] \\ &\geqq e^{-A}P_y^\varepsilon[G] - e^{-A}P_y^\varepsilon[|F(\varepsilon,\omega)| \geqq A]\,. \end{aligned}$$

It therefore suffices to show that

$$\lim_{A\to\infty}\limsup_{\varepsilon\to 0} \varepsilon \log P_y^\varepsilon[|F(\varepsilon,\omega)| \geqq A] = -\infty \tag{6.1}$$

uniformly in y. Similarly,

$$\begin{aligned} Q_y^\varepsilon[C] &= \int_C \exp\{F(\varepsilon,\omega)\}P_y^\varepsilon(d\omega) \\ &\leqq e^{A}P_y^\varepsilon[C] + \int_{[F(\varepsilon,\omega)\geqq A]} \exp\{F(\varepsilon,\omega)\}P_y^\varepsilon(d\omega)\,. \end{aligned}$$

Hence we need only show that

$$\lim_{A\to\infty}\limsup_{\varepsilon\to 0} \varepsilon \log \int_{[F\geqq A]} \exp\{F(\varepsilon,\omega)\}P_y^\varepsilon(d\omega) = -\infty \tag{6.2}$$

uniformly in y. Relations (6.1) and (6.2) follow easily from Lemma 6.2.

THEOREM 6.4.

$$\lim_{t\to 0} -2t \log q(t,x,y) = d^2(x,y)$$

uniformly in x and y over compact sets.

Proof: We have the following estimates from Theorem 0.5, page 229, Volume II of [3]:

$$q(t, x, y) \leqq Mt^{-k/2} \exp\left\{-\frac{\alpha}{2t}|x-y|^2\right\}, \tag{6.3}$$

$$\begin{aligned} q(t, x, y) &\geqq M_1 t^{-k/2} \exp\left\{-\frac{\alpha_1}{2t}|x-y|^2\right\} \\ &\quad - M_2 t^{-k/2+\lambda} \exp\left\{-\frac{\alpha_2}{2t}|x-y|^2\right\}. \end{aligned} \tag{6.4}$$

UPPER BOUND. Let x, y be two points in R_k. We choose a sphere S_δ around y of radius δ and a sphere $S_{2\delta}$ of radius δ around $S_{2\delta}$. Let B denote the boundary of $S_{2\delta}$. We assume that $y' \to y$ and hence $y' \in S_\delta$. Let τ be the first entrance time into the set B for a process starting at x'. Then,

$$\begin{aligned} q(t, x', y') &= \int_{\tau \leqq t} q(t-\tau, \omega(\tau), y') P_{x'}(d\omega) \\ &\leqq \sup_{\substack{0 \leqq u \leqq t \\ |z'-y'| \geqq \delta}} q(u, z', y') \cdot P_{x'}[\tau \leqq t]. \end{aligned}$$

From (6.3) we have

$$\sup_{\substack{0 \leqq u \leqq t \\ |z'-y'| \geqq \delta}} q(u, z', y') \leqq M < \infty.$$

Hence Theorem 6.3 yields

$$\begin{aligned} \limsup_{\substack{t \to 0 \\ x' \to x \\ y' \to y}} 2t \log q(t, x', y') &\leqq \limsup_{\substack{t \to 0 \\ x' \to x}} t \log P_{x'}[\tau \leqq t] \\ &\leqq -\inf_{z \in B} \tfrac{1}{2} d^2(x, z). \end{aligned}$$

Letting $\delta \to 0$, we find

$$\limsup_{\substack{t \to 0 \\ x' \to x \\ y' \to y}} 2t \log q(t, x', y') \leqq -\tfrac{1}{2} d^2(x, y).$$

LOWER BOUND. Equation (6.4) gives, for every $\Delta > 0$,

$$\operatorname*{Inf}_{0 \leqq t \leqq t_0} \inf_{|x-y| \leqq \Delta t} q(t, x, y) = \theta > 0 \quad \text{for} \quad t_0 \quad \text{small}. \tag{6.5}$$

Let x, y be points in R_k. Let $x' \to x$ and $y' \to y$. We define the sets

$$U_t = \{z : |z - y'| \leqq t\},$$

and

$$G_t = \{\omega : \omega(t) \in U_t\} .$$

Then,

$$\int_{G_t} p(t, x', v)\, dv = \int_{|z-y'| \leqq t} p(t, x', v)\, dv$$

$$\geqq ct^k \inf_{v \in G_t} p(t, x', v) .$$

Consequently, from Theorem 2.5, we obtain

$$\liminf_{\substack{t \to 0 \\ x' \to x \\ y' \to y}} t \log P_{x'}[G_t] \geqq -\tfrac{1}{2} d^2(x, y) .$$

Rewriting this inequality by introducing the measures P_x^ε, we have

$$\liminf_{\substack{\varepsilon \to 0 \\ x' \to x \\ y' \to y}} \varepsilon \log P_{x'}^\varepsilon[G_\varepsilon] \geqq -\tfrac{1}{2} d^2(x, y) ,$$

where we redefine

$$G_\varepsilon = \{\omega : \omega(1) \in U_\varepsilon\} .$$

From relation (6.1), it follows that

$$\liminf_{\substack{\varepsilon \to 0 \\ x' \to x \\ y' \to y}} \varepsilon \log Q_{x'}^\varepsilon[G_\varepsilon] \geqq -\tfrac{1}{2} d^2(x, y) ,$$

or equivalently that

$$\liminf_{\substack{t \to 0 \\ x' \to x}} t \log \int_{G_t} q(t, x', v)\, dv \geqq -\tfrac{1}{2} d^2(x, y) . \tag{6.6}$$

Now,

$$q(t, x', y') = \int q(t(1 - \Delta), x', v) q(\Delta t, v, y')\, dv$$

$$\geqq \int_{G_t} q(t(1 - \Delta), x', v) q(\Delta t, v, y')\, dv$$

$$\geqq \inf_{v \in G_t} q(\Delta t, v, y') \int_{G_t} q(t(1 - \Delta), x', v)\, dv .$$

Using (6.5) and (6.6) we arrive at

$$\liminf_{\substack{t \to 0 \\ x' \to x \\ y' \to y}} t \log q(t, x', y') \geqq -\frac{1}{2(1 - \Delta)} d^2(x, y) .$$

Allowing $\Delta \to 0$ we prove the theorem.

Bibliography

[1] Ciesielski, Z., *Heat conduction and the principle of not feeling the boundary*, Bull. Acad. Polon. Sci. Ser. Sci. Math. Astronom. Phys., Vol. 15, 1966, pp. 435–440.

[2] Donsker, M. D., *An application of function space integrals to a non-linear equation*, to appear.

[3] Dynkin, E. B., *Markov Processes*, Vols. I, II, Academic Press, New York, 1965.

[4] Schilder, M., *On a Laplace asymptotic formula for Wiener integrals*, Trans. Amer. Math. Soc., to appear.

[5] Skorokhod, A. V., *On the differentiability of measures corresponding to random processes, II, Markov processes*, Theor. Probability Appl., Vol. 5, 1960, pp. 40–49.

[6] Varadhan, S. R. S., *Asymptotic probabilities and differential equations*, Comm. Pure Appl. Math., Vol. 19, 1966, pp. 261–286.

[7] Varadhan, S. R. S., *On the behavior of the fundamental solution of the heat equation with variable coefficients*, Comm. Pure Appl. Math., Vol. 20, 1967, pp. 431–455.

Received January, 1967.

Proc. Nat. Acad. Sci. USA
Vol. 72, No. 3, pp. 780–783, March 1975

On a Variational Formula for the Principal Eigenvalue for Operators with Maximum Principle

(semigroups/Rayleigh–Ritz formula/Feynman–Kac formula)

MONROE D. DONSKER AND S. R. S. VARADHAN

Courant Institute of Mathematical Sciences, New York University, New York, N.Y. 10012

Communicated by Louis Nirenberg, December 11, 1974

ABSTRACT In this paper a variational formula is obtained for the principal eigenvalue for operators with maximum principle. This variational formula does not require the operators to be self-adjoint. But if they are self-adjoint this formula reduces to the classical Rayleigh-Ritz formula.

Let X be a compact metric space, and for $t \geq 0$ let T_t be a strongly continuous semigroup mapping $C(X) \to C(X)$ having the properties that if $f \geq 0$, then $T_t f \geq 0$ and $T_t 1 = 1$. Under these hypotheses the infinitesimal generator L of T_t will have domain $\mathfrak{D}$ dense in $C(X)$, and L will satisfy the maximum principle. Examples of operators L arising from such semigroups are

(*a*) X is a compact manifold and L is a second order elliptic operator with reasonable coefficients.

(*b*) $(Lf)(x) = \int_X [f(y) - f(x)]\pi(x,dy)$ where $\pi(x,dy)$ is a nonnegative measure on X for each $x \in X$ and is weakly continuous in x.

In fact, the most general L satisfying our assumptions is a limit of examples of type (*b*).

Let $V(x) \in C(X)$ and let λ_V be the principal eigenvalue of the operator $L+V$. If the operator L is self-adjoint with respect to a measure ν on X, then there is a classical variational formula (Rayleigh-Ritz) for λ_V, namely,

$$\lambda_V = \sup_{\substack{f \in L^2(\nu) \\ \|f\|_2 = 1}} \left[\int_X V(x)f^2(x)\nu(dx) + \langle Lf,f\rangle\right]. \quad [1]$$

In this note we obtain a variational formula for λ_V for all L considered above whether L is self-adjoint or not. In the self-adjoint case the new variational formula reduces to [1].

Let $\mathfrak{D}^+$ denote the functions $u \in \mathfrak{D}$ that are positive and let $\mathfrak{M}$ be the space of all probability measures on X. For each $\mu \in \mathfrak{M}$ we define

$$I(\mu) = -\inf_{u \in \mathfrak{D}^+} \int_X \left(\frac{Lu}{u}\right)(x)\mu(dx). \quad [2]$$

It is easy to see that $I(\mu)$ is nonnegative, lower semicontinuous, and convex.

The operator $L+V$ is the infinitesimal generator of a semigroup T_t^V given by a family of measures $p_V(t,x,dy)$, i.e.,

$$T_t^V f(x) = \int_X f(y) p_V(t,x,dy)$$

and we note

$$\|T_t^V\| = \sup_{x \in X} \int_X p_V(t,x,dy).$$

Moreover, if we let $\phi(t,V) = \log \|T_t^V\|$, we see that $\phi(t,V)$ is subadditive in t and therefore

$$\phi(V) = \lim_{t\to\infty} \frac{\phi(t,V)}{t} = \lim_{t\to\infty} \frac{1}{t} \log \sup_{x\in X} \int_X p_V(t,x,dy) = \lambda_V$$

exists. Because T_t is a positive semigroup, λ_V is in the spectrum of $L+V$ and is in fact the principal eigenvalue. We will prove

THEOREM. *The principal eigenvalue* λ_V of $L+V$ is given by*

$$\lambda_V = \sup_{\mu \in \mathfrak{M}} \left[\int_X V(x)\mu(dx) - I(\mu)\right]$$

where $I(\mu)$ is defined by [2]. *The theorem follows from Lemmas 1 and 2 below. Let us define*

$$\phi(V) = \lim_{t\to\infty} \frac{1}{t} \log \sup_{x\in X} \int_X p_V(t,x,dy),$$

$$\psi_1(V) = \sup_{\mu\in\mathfrak{M}} \left[\int_X V(x)\mu(dx) - I(\mu)\right],$$

$$\psi_2(V) = \inf_{u\in\mathfrak{D}^+} \sup_{x\in X} \left[V(x) + \left(\frac{Lu}{u}\right)(x)\right].$$

The theorem states that $\phi(V) = \psi_1(V)$. We will first prove in Lemma 1 that $\psi_1(V) \leq \phi(V) \leq \psi_2(V)$ and then in Lemma 2 that $\psi_2(V) = \psi_1(V)$.

* In general $\phi(V)$ is not an eigenvalue. However, it belongs to the spectrum of $L+V$ which is contained in $[z{:}\mathrm{Re}\ z \leq \phi(V)]$. We therefore call it the principal eigenvalue λ_V.

LEMMA 1.

$$\psi_1(V) \le \phi(V) \le \psi_2(V).$$

Proof: Let $\psi_2(V) = l$ and $\epsilon > 0$ be given. Then there exists $u \in \mathfrak{D}^+$ such that for all $x \in X$

$$V(x) + \left(\frac{Lu}{u}\right)(x) \le l+\epsilon.$$

For this u, let $v(t,x) = u(x)e^{(l+\epsilon)t}$ and we have

$$(L+V)v(t,x) = \left[\left(\frac{Lu}{u}\right)(x) + V(x)\right]v(t,x) \le (l+\epsilon)u(x)e^{(l+\epsilon)t}.$$

Thus, for all x and t,

$$\frac{\partial v(t,x)}{\partial t} \ge (L+V)v(t,x)$$

so that from the maximum principle we conclude

$$v(t,x) \ge \int_X v(0,y)p_V(t,x,dy) = \int_X u(y)p_V(t,x,dy).$$

Since X is compact and $u \in \mathfrak{D}^+$ we have $\inf_{x\in X} u(x) > 0$, and therefore for all x and t

$$\int_X p_V(t,x,dy) \le \frac{v(t,x)}{\inf_{x\in X} u(x)} \le \frac{\sup_{x\in X} u(x)}{\inf_{x\in X} u(x)} e^{(l+\epsilon)t}.$$

Thus,

$$\phi(V) = \lim_{t\to\infty} \frac{1}{t} \log \sup_{x\in X} \int_X p_V(t,x,dy) \le l+\epsilon,$$

and hence $\phi(V) \le \psi_2(V)$.

Now suppose $\phi(V) = h$. Let $u(t,x)$ be the solution for $t \ge 0$ of

$$\frac{\partial u}{\partial t} = Lu + Vu,$$

$$u(0,x) = 1,$$

i.e.,

$$u(t,x) = \int_X p_V(t,x,dy).$$

Since $\phi(V) = h$, corresponding to any $\epsilon > 0$ there is a t_0 such that $t \ge t_0$ implies

$$\sup_{x\in X} u(t,x) \le e^{(h+\epsilon)t}. \quad [3]$$

Let $V^\epsilon(x) = V(x) - h - \epsilon$ and $u_\epsilon(t,x)$ be the solution of

$$\frac{\partial u_\epsilon}{\partial t} = Lu_\epsilon + V^\epsilon u_\epsilon,$$

$$u_\epsilon(0,x) = 1.$$

We see that $u_\epsilon(t,x) = u(t,x)e^{-(h+\epsilon)t}$ and from [3] we have for all $t \ge t_0$

$$\sup_{x\in X} u_\epsilon(t,x) \le 1.$$

In particular, $u_\epsilon(t_0,x) \le 1$ for all $x \in X$. If we consider then $u_\epsilon(t,x)$ for $x \in X$ and $0 \le t \le t_0$, we have

$$\frac{\partial u_\epsilon}{\partial t} = Lu_\epsilon + V^\epsilon u_\epsilon,$$

$$u_\epsilon(0,x) = 1,$$

$$u_\epsilon(t_0,x) \le 1,$$

and $u_\epsilon(t,x)$ is bounded below. Thus,

$$\frac{\partial \log u_\epsilon}{\partial t} = \frac{Lu_\epsilon}{u_\epsilon} + V^\epsilon,$$

so that for any $\mu \in \mathfrak{M}$ and $0 \le t \le t_0$

$$\int_X \frac{\partial \log u_\epsilon(t,x)}{\partial t}\mu(dx) = \int_X \left(\frac{Lu_\epsilon}{u_\epsilon}\right)(t,x)\mu(dx) + \int_X V^\epsilon(x)\mu(dx).$$

Therefore,

$$\int_0^{t_0} dt\left[\int_X \left(\frac{Lu_\epsilon}{u_\epsilon}\right)(t,x)\mu(dx) + \int_X V^\epsilon(x)\mu(dx)\right] = \int_X \log u_\epsilon\,(t_0,x)\mu(dx) - \int_X \log u_\epsilon(0,x)\mu(dx) = \int_X \log u_\epsilon(t_0,x)\mu(dx) \le 0.$$

This implies that there is some point $0 \le t' \le t_0$ such that

$$\int_X \left(\frac{Lu_\epsilon}{u_\epsilon}\right)(t',x)\mu(dx) + \int_X V^\epsilon(x)\mu(dx) \le 0,$$

and hence, letting $u(x) = u_\epsilon(t',x)$, we see that for each $\mu \in \mathfrak{M}$ there is a $u \in \mathfrak{D}^+$ such that

$$\int_X \left(\frac{Lu}{u}\right)(x)\mu(dx) + \int_X V^\epsilon(x)\mu(dx) \le 0.$$

This means

$$\sup_{\mu\in\mathfrak{M}}\left[\inf_{u\in\mathfrak{D}^+}\int_X \left(\frac{Lu}{u}\right)(x)\mu(dx) + \int_X V^\epsilon(x)\mu(dx)\right] \le 0,$$

and thus, from the definition of $I(\mu)$, we have

$$\sup_{\mu\in\mathfrak{M}}\left[-I(\mu) + \int_X V^\epsilon(x)\mu(dx)\right] \le 0,$$

i.e., $\psi_1(V^\epsilon) \le 0$. But obviously $\psi_1(V^\epsilon) = \psi_1(V) - h - \epsilon$, so that $\psi_1(V) \le h+\epsilon$ which implies $\psi_1(V) \le \phi(V)$.

LEMMA 2.

$$\psi_1(V) = \psi_2(V).$$

Proof: We must show $\psi_2(V) \le \psi_1(V)$. Let $\psi_1(V) = l$ and $\epsilon > 0$ be given. We then want to find $u_\epsilon \in \mathfrak{D}^+$ such that for all $x \in X$

$$\left(\frac{Lu_\epsilon}{u_\epsilon}\right)(x) + V(x) \leq l+\epsilon.$$

Since $\psi_1(V) = l$, we have from the definition of $I(\mu)$

$$\sup_{\mu\in\mathfrak{M}} \inf_{u\in\mathfrak{D}^+} \left[\int_X \left(\frac{Lu}{u}\right)(x)\mu(dx) + \int_X V(x)\mu(dx)\right] = l.$$

Thus, for any $\mu \in \mathfrak{M}$ and our given $\epsilon > 0$, there exists $u_{\epsilon,\mu} \in \mathfrak{D}^+$ such that

$$\int_X \left(\frac{Lu_{\epsilon,\mu}}{u_{\epsilon,\mu}}\right)(x)\mu(dx) + \int_X V(x)\mu(dx) \leq l + \frac{\epsilon}{8}.$$

Since the left side of this last inequality is continuous in the weak topology on $\mathfrak{M}$, there exists a neighborhood N_μ of μ such that if $\lambda \in N_\mu$,

$$\int_X \left(\frac{Lu_{\epsilon,\mu}}{u_{\epsilon,\mu}}\right)(x)\lambda(dx) + \int_X V(x)\lambda(dx) \leq l + \frac{\epsilon}{4}.$$

Thus $\bigcup_{\mu\in\mathfrak{M}} N_\mu$ is an open covering of the compact space $\mathfrak{M}$. Hence, there exists a finite covering $N_{\mu_1}, N_{\mu_2}, \ldots, N_{\mu_k}$. Let $u_i = u_{\epsilon,\mu_i}$ for $i = 1,2,\ldots,k$ so that we now have

$$\sup_{\mu\in\mathfrak{M}} \inf_{1\leq i\leq k} \left[\int_X \frac{Lu_i}{u_i}(x)\mu(dx) + \int_X V(x)\mu(dx)\right] \leq l + \frac{\epsilon}{4}. \quad [4]$$

From the definition of the infinitesimal generator, $\dfrac{T_h u_i - u_i}{h} \to Lu_i$ uniformly in x as $h \to 0$. Since u_i has a positive lower bound, and

$$\int_X \left(\frac{T_h u_i - u_i}{hu_i}\right)(x)\mu(dx) \to \int_X \left(\frac{Lu_i}{u_i}\right)(x)\mu(dx)$$

uniformly in μ, we conclude from [4] that there exists an h_0 such that if $h \leq h_0$

$$\sup_{\mu\in\mathfrak{M}} \inf_{1\leq i\leq k} \left[\int_X \left(\frac{T_h u_i - u_i}{hu_i}\right)(x)\mu(dx) + \int_X V(x)\mu(dx)\right] \leq l + \frac{\epsilon}{3}. \quad [5]$$

Since u_i is continuous and is bounded above and bounded below by a positive constant, we can write $u_i = e^{g_i}$ where $g_i \in C(X)$. Let G be the convex hull of $\{g_1, g_2, \ldots, g_k\}$. From [5] we obtain for $h \leq h_0$

$$\sup_{\mu\in\mathfrak{M}} \inf_{g\in G} \left[\int_X \left(\frac{T_h e^g - e^g}{he^g}\right)(x)\mu(dx) + \int_X V(x)\mu(dx)\right] \leq l + \frac{\epsilon}{3}. \quad [6]$$

Since T_h is bounded and nonnegative it is easy to see that $\dfrac{T_h e^g - e^g}{he^g}$ is a convex functional of g for each $x \in X$. Hence, we have from one of the mini-max theorems (see e.g., ref. 1, Thm. 3.4) that

$$\inf_{g\in G} \sup_{\mu\in\mathfrak{M}} \left[\int_X \left(\frac{T_h e^g - e^g}{he^g}\right)(x)\mu(dx) + \int_X V(x)\mu(dx)\right] \leq l + \frac{\epsilon}{3}, \quad [7]$$

which implies

$$\inf_{g\in G} \sup_{x\in X} \left[\left(\frac{T_h e^g - e^g}{he^g}\right)(x) + V(x)\right] \leq l + \frac{\epsilon}{3}.$$

Since G is compact the infimum is attained at some point of G, call it g_h, i.e., for all $x \in X$

$$\left[\left(\frac{T_h e^{g_h} - e^{g_h}}{he^{g_h}}\right)(x) + V(x)\right] \leq l + \frac{\epsilon}{3}.$$

Letting $u_h = e^{g_h}$ we have for all $x \in X$ and $h \leq h_0$

$$\left[\left(\frac{T_h u_h - u_h}{hu_h}\right)(x) + V(x)\right] \leq l + \frac{\epsilon}{3}. \quad [8]$$

Since the family $\{u_h\}$ for $h \leq h_0$ belongs to a compact set, we can assume without loss of generality that $u_h \to u_0$ as $h \to 0$ where $u_0 \in C^+(X)$ (continuous functions on X bounded below by a positive number) and $u_0 = e^g$ where $g \in G$. Without loss of generality we can assume $u_h \in \mathfrak{D}^+$ because $\mathfrak{D}^+$ is dense in $C^+(X)$. Hence, we get from [8]

$$\left[\left(\frac{T_h u_h - u_h}{hu_h}\right)(x) + V(x)\right] \leq l + \frac{\epsilon}{2}$$

for all $x \in X$, $h \leq h_0$, and where $u_h \to u_0$ as $h \to 0$ with $u_h \in \mathfrak{D}^+$. Since $u_h \in \mathfrak{D}^+$ we can rewrite this last as

$$\left[\frac{L\left(\frac{1}{h}\int_0^h T_s u_h ds\right)}{u_h}\right](x) + V(x) \leq l + \frac{\epsilon}{2} \quad [9]$$

for all $x \in X$ and $h \leq h_0$. Let

$$v_h = \frac{1}{h}\int_0^h T_s u_h ds \in \mathfrak{D}^+.$$

Since $u_h \to u_0$ and T_s is a strongly continuous semigroup, we have $v_h \to u_0$ as $h \to 0$. Now from [9] we get for all $x \in X$ and $h \leq h_0$

$$\left(\frac{Lv_h}{v_h u_h} \cdot v_h\right)(x) + V(x) \leq l + \frac{\epsilon}{2}$$

or

$$\left(\frac{Lv_h}{v_h}\right)(x) + V(x) \leq \left(l + \frac{\epsilon}{2}\right) + \left(\frac{u_h(x)}{v_h(x)} - 1\right) \times \left(l + \frac{\epsilon}{2}\right) + V(x)\left(1 - \frac{u_h(x)}{v_h(x)}\right)$$

$$\leq \left(l + \frac{\epsilon}{2}\right) + \left[\left(l + \frac{\epsilon}{2}\right) + \|V\|\right]\left(\left\|\frac{u_h}{v_h} - 1\right\|\right).$$

But $u_h \to u_0$ uniformly and, as noted earlier, $v_h \to u_0$ so that $\frac{u_h}{v_h} \to 1$ uniformly, i.e., for h sufficiently small

$$\left\|\frac{u_h}{v_h} - 1\right\|\left[\left(l + \frac{\epsilon}{2}\right) + \|V\|\right] \leq \frac{\epsilon}{2}.$$

Thus, there is an h_1 such that for all $x \in X$

$$\left(\frac{Lv_{h_1}}{v_{h_1}}\right)(x) + V(x) \leq l + \epsilon$$

which gives $\psi_2(V) \leq \psi_1(V)$. This completes the proof of *Lemma* 2 and the theorem.

Now in ref. 2 the authors proved (Section 4 of ref. 2) that if L is self-adjoint with respect to a reference measure ν on X and if $I(\mu) < \infty$ for some $\mu \in \mathfrak{M}$, then under mild assumptions $\mu \ll \nu$ and letting $g = \frac{d\mu}{d\nu}$ and $f = g^{1/2}$ we have

$$I(\mu) = -\langle Lf, f\rangle.$$

This shows that in the case where L is self-adjoint the variational expression in the theorem reduces to the classical formula [1].

The relation between $\phi(V)$ and $I(\mu)$ is clearly that they are conjugate convex functionals. The relation $\phi(V) = \psi_2(V)$ has been noticed before. See, for instance, ref. 3.

The work reported in this paper was supported by the Office of Naval Research, Contract N00014-67-A-L467-0031 and the National Science Foundation, Grant NSF-GP-41985.

1. Sion, M. (1958) "On general minimax theorem," *Pac. J. Math.* **8,** 171–176.
2. Donsker, M. D. & Varadhan, S. R. S. (1975) "Asymptotic evaluation of certain Markov process expectations for large time—I," *Comm. Pure Appl. Math.*, in press.
3. Protter, M. H. & Weinberger, H. F. (1966) "On the spectrum of general second order operators," *Bull. Amer. Math. Soc.* **72,** 251–255.

COMMUNICATIONS ON PURE AND APPLIED MATHEMATICS, VOL. XXVIII, 1–47 (1975)

Asymptotic Evaluation of Certain Markov Process Expectations for Large Time, I*

M. D. DONSKER AND S. R. S. VARADHAN

1. Introduction

Let Ω_x be the space of real-valued continuous functions $z(s)$, $0 \leqq s < \infty$, with $z(0)=x$. Let P_x denote Brownian motion measure (Wiener measure) on Ω_x and $E_x\{\ \}$ integration with respect to P_x (Wiener integral). For $t>0$ and $\omega \in \Omega_x$, let

$$L(t, \omega, y)=\frac{1}{t}\lambda\{s : 0 \leqq s \leqq t, z(s) \leqq y\},$$

where $\lambda\{\ \}$ denotes Lebesgue measure, i.e., $L(t, \omega, y)$ is the proportion of time on $[0, t]$ for which a particular path $\omega = z(\cdot)$ is less than or equal to y. For each $t>0$ and each $\omega \in \Omega_x$, $L(t, \omega, y)$ is a probability distribution function on $-\infty < y < \infty$ possessing a density function which is often called the "local time". Let $\mathcal{F}$ denote the space of all probability distribution functions F on R_1. In [1] the authors proved that, for a wide class of functionals Φ defined on $\mathcal{F}$,

$$(1.1) \quad \lim_{t\to\infty} \frac{1}{t} \log E_x\{\exp\{-t\,\Phi(L(t, \omega, \cdot))\}\} = -\inf_{f \in \mathcal{F}_0} \left[\frac{1}{8}\int_{-\infty}^{\infty} \frac{[f'(y)]^2}{f(y)}\, dy + \Phi(f)\right],$$

where $\mathcal{F}_0$ denotes the class of probability density functions f which are continuously differentiable on $-\infty<y<\infty$ and either $f(y)>0$ for all $-\infty<y<\infty$ or f has compact support and is strictly positive on the interior of that support. In (1.1), $\Phi(f)$ means $\Phi(F)$, where F is the distribution function corresponding to the density f.

If, for example, in (1.1) we let $\Phi(F)=\int_{-\infty}^{\infty} V(y)\, dF(y)$, where $V(y) \geqq 0$ is continuous and $V(y) \to +\infty$ as $y \to \pm\infty$, then Φ satisfies the hypotheses

* The research for this paper was performed at the Courant Institute. The first author was supported by the Office of Naval Research under Contract No. N00014-67A-0467-0031 and the second by the National Science Foundation under Grant No. NSF-GP-41985. Reproduction in whole or in part is permitted for any purpose of the United States Government.

imposed in [1] and (1.1) becomes

$$\lim_{t\to\infty}\frac{1}{t}\log E_x\left\{\exp\left\{-\int_0^t V(z(s))\,ds\right\}\right\} \tag{1.2}$$
$$=-\inf_{f\in\mathcal{F}_0}\left[\frac{1}{8}\int_{-\infty}^{\infty}\frac{[f'(y)]^2}{f(y)}\,dy+\int_{-\infty}^{\infty}V(y)f(y)\,dy\right].$$

Now, on the right of (1.2) let $\psi(y)=\sqrt{f(y)}$ so that $\psi(y)\in L_2(R_1)$ and, indeed, $\int_{-\infty}^{\infty}\psi^2(y)\,dy=1$. Moreover,

$$\int_{-\infty}^{\infty}V(y)f(y)\,dy=\int_{-\infty}^{\infty}V(y)\psi^2(y)\,dy\,,$$

and

$$\frac{1}{8}\int_{-\infty}^{\infty}\frac{[f'(y)]^2}{f(y)}\,dy=\frac{1}{2}\int_{-\infty}^{\infty}[\psi'(y)]^2\,dy\,.$$

Thus, the right side of (1.2) becomes

$$-\inf_{\substack{\psi\in L_2(R_1)\\ \int_{-\infty}^{\infty}\psi^2(y)dy=1}}\left[\frac{1}{2}\int_{-\infty}^{\infty}[\psi'(y)]^2\,dy+\int_{-\infty}^{\infty}V(y)\psi^2(y)\,dy\right]$$

which is $-\lambda_1$, λ_1 being the lowest eigenvalue of

$$\tfrac{1}{2}\psi''(y)-V(y)\psi(y)=-\lambda\psi(y)\,.$$

Hence the special case of (1.1) when $\Phi(F)=\int_{-\infty}^{\infty}V(y)\,dF(y)$ is the old result of Kac [2] that, for $V(y)$ as described above,

$$\lim_{t\to\infty}\frac{1}{t}\log E_x\left\{\exp\left\{-\int_0^t V(z(s))\,ds\right\}\right\}=-\lambda_1\,.$$

The point of (1.1) is that it holds for functionals Φ of much greater generality—in particular for functionals Φ where the corresponding expectations on the left of (1.1) are not solutions of partial differential equations as obtains in the special case (1.2).

In the present paper the authors consider the analogous asymptotic behavior for expectations on Markov processes—both discrete parameter and continuous parameter. In that generality we shall, however, confine ourselves in this paper to Markov processes having a state space which is a compact metric space (whereas in [1] the Brownian motion state space was all of R_1).

It is clear that in the Brownian motion case (1.1) a special role is played by the functional

$$\frac{1}{8}\int_{-\infty}^{\infty}\frac{[f'(y)]^2}{f(y)}\,dy\,.$$

We shall make additional comments about this functional now and more so later in this introduction. Let $\mathcal{U}$ be the space of functions $u(y)$, $-\infty<y<\infty$, having two bounded continuous derivatives, each function having two positive constants α and β such that, for all y, $0<\alpha\leqq u(y)\leqq\beta<\infty$. For $F\in\mathcal{F}$ (defined above), let

$$I(F)=-\inf_{u\in\mathcal{U}}\int_{-\infty}^{\infty}\frac{u''(y)}{2u(y)}\,dF(y)\,. \tag{1.3}$$

It is easy to see that, as a functional on $\mathcal{F}$, $I(F)$ is non-negative, lower semicontinuous in the weak topology on $\mathcal{F}$, and convex. In [1] (Lemma 2) we proved that if f is a density function which is positive and continuously differentiable on $-\infty<y<\infty$, then for its corresponding distribution function F

$$I(F)=\frac{1}{8}\int_{-\infty}^{\infty}\frac{[f'(y)]^2}{f(y)}\,dy\,, \tag{1.4}$$

in the sense that if one side is finite so is the other, and they are equal.

We shall now define the corresponding I-functions for discrete and continuous parameter Markov processes.

Let X be a compact metric space and let $\lambda(dx)$ be a probability measure on X. Let $X_0, X_1, X_2, \cdots$ be a stationary Markov process whose state space is X, with $X_0=x$, having transition probability function $\pi(x, dy)$ about which we assume:

1. $\pi(x, dy)=\pi(x, y)\lambda(dy)$,

2. there exist constants a and A such that $0<a\leqq\pi(x, y)\leqq A<\infty$ for all $x\in X$ and almost all (λ-measure) $y\in X$,

3. for any function $u(y)\in L_1(\lambda)$,

$$\int_X \pi(x, y)u(y)\lambda(dy)$$

is a continuous function of x.

Let $\mathcal{U}_1$ be the set of functions $u \in C(X)$ such that $u > 0$ on X. With

$$\pi u(x) = \int_X u(y)\pi(x, dy),$$

we define, for any probability measure μ on X,

$$I(\mu) = -\inf_{u \in \mathcal{U}_1} \int_X \log\left(\frac{\pi u}{u}\right)(x)\mu(dx). \tag{1.5}$$

Let $\mathcal{M}$ be the space of all probability measures on X, and impose the weak topology on $\mathcal{M}$. Denote by Ω_x the space of all sequences $X_0, X_1, X_2, \cdots$ with $X_0 = x$ and $X_j \in X$. We have a probability measure on Ω_x induced by $\pi(x, dy)$ which we shall call P_x. For each $\omega \in \Omega_x$, each positive integer n, and A a Borel set in X, let

$$L_n(\omega, A) = \frac{1}{n}\sum_{j=0}^{n-1} \chi_A(X_j). \tag{1.6}$$

We note that L_n maps Ω_x into $\mathcal{M}$. Let $Q_{n,x}$ be the measure induced on $\mathcal{M}$ by L_n from P_x, i.e., $Q_{n,x} = P_x L_n^{-1}$. Thus, if B is a set of measures in $\mathcal{M}$, we define for each positive integer n and each $x \in X$

$$Q_{n,x}(B) = P_x\{\omega : L_n(\omega, \cdot) \in B\}.$$

In Section 2 of this paper, with the I-function defined in (1.5), we shall prove

THEOREM 1. *For any closed set* $C \subset \mathcal{M}$,

$$\overline{\lim_{n\to\infty}} \frac{1}{n} \log Q_{n,x}(C) \leqq -\inf_{\mu \in C} I(\mu). \tag{1.7}$$

For any open set $G \subset \mathcal{M}$,

$$\underline{\lim_{n\to\infty}} \frac{1}{n} \log Q_{n,x}(G) \geqq -\inf_{\mu \in G} I(\mu). \tag{1.8}$$

A corollary of this theorem, which follows from it by the arguments in Section 3 of [3], is that if $\Phi(\mu)$ is a weakly continuous functional on $\mathcal{M}$, then

$$\lim_{n\to\infty} \frac{1}{n} \log E^{Q_{n,x}}\{e^{-n\Phi(\mu)}\} = -\inf_{\mu \in \mathcal{M}} [\Phi(\mu) + I(\mu)]. \tag{1.9}$$

Or, going back to the P_x-measure on Ω_x, we have equivalently

THEOREM 2. *If* Φ *is a real-valued weakly continuous functional on* $\mathcal{M}$, *then*

$$\lim_{n\to\infty}\frac{1}{n}\log E^{P_x}\{\exp\{-n\Phi(L_n(\omega,\cdot))\}\}=-\inf_{\mu\in\mathcal{M}}[\Phi(\mu)+I(\mu)]\,, \tag{1.10}$$

where $I(\mu)$ *is defined in* (1.5).

We see that Theorem 2 is the analogue of (1.1) for discrete parameter Markov processes. Let us now examine the situation for continuous parameter Markov processes.

Let X be a compact metric space with a probability measure $\lambda(dx)$. Let $C(X)$ be continuous functions on X and T_t a strongly continuous Markovian semigroup such that T_t maps $C(X)\to C(X)$ (Feller semigroup). We assume the associated transition probabilities $p(t, x, dy)$ have densities relative to $\lambda(dx)$, i.e.,

$$p(t, x, dy)=p(t, x, y)\lambda(dy)\,,$$

where for each $t>0$

$$0<a(t)\leqq p(t, x, y)\leqq A(t)<\infty\,.$$

Let L be the infinitesimal generator of the semigroup T_t having domain $\mathcal{D}\subset C(X)$. Our assumptions above imply that $\mathcal{D}$ is dense in $C(X)$.

Let Ω_x be the space of functions $x(t)$, $0\leqq t<\infty$, which have values in X, $x(0)=x$, are right continuous and have left limits at every point t. The process described above induces a measure P_x on Ω_x.

For each $t>0$, $\omega\in\Omega_x$, and Borel set $A\subset X$, let

$$L_t(\omega, A)=\frac{1}{t}\int_0^t \chi_A(x(s))\,ds\,. \tag{1.11}$$

In other words, $L_t(\omega, A)$ is the proportion of time up to t that a particular sample path $\omega=x(\cdot)$ spends in the set A. We note that, for each $t>0$ and each ω, $L_t(\omega,\cdot)$ is a probability measure on X.

Let $\mathcal{M}$ be the space of all probability measures on X. For each $x\in X$ and each $t>0$, let $Q_{x,t}$ be the measure induced on $\mathcal{M}$ by L_t from P_x, i.e., $Q_{x,t}=P_xL_t^{-1}$ or, in other words, if $B\subset\mathcal{M}$,

$$Q_{x,t}(B)=P_x\{\omega\in\Omega_x : L_t(\omega,\cdot)\in B\}\,.$$

We endow $\mathcal{M}$ with the weak topology.

Now, for any $\mu \in \mathcal{M}$, we define

$$I(\mu) = -\inf_{\substack{u>0 \\ u\in\mathcal{D}}} \int_{X} \left(\frac{Lu}{u}\right)(x)\mu(dx) . \tag{1.12}$$

In Section 3 of this paper we shall prove

THEOREM 3. *For any closed set* $C \subset \mathcal{M}$,

$$\overline{\lim_{t\to\infty}} \frac{1}{t} \log Q_{x,t}(C) \leqq -\inf_{\mu\in C} I(\mu) . \tag{1.13}$$

For any open set $G \subset \mathcal{M}$,

$$\underline{\lim_{t\to\infty}} \frac{1}{t} \log Q_{x,t}(G) \geqq -\inf_{\mu\in G} I(\mu) . \tag{1.14}$$

As before, a corollary of Theorem 3 is that if Φ is a real-valued weakly continuous functional on $\mathcal{M}$, then

$$\lim_{t\to\infty} \frac{1}{t} \log E^{Q_{x,t}}\{e^{-t\Phi(\mu)}\} = -\inf_{\mu\in\mathcal{M}} [\Phi(\mu) + I(\mu)] . \tag{1.15}$$

Going back to the P_x-measure on Ω_x we have equivalently

THEOREM 4. *If* Φ *is a real-valued weakly continuous functional on* $\mathcal{M}$, *then*

$$\lim_{t\to\infty} \frac{1}{t} \log E^{P_x}\{\exp\{-t\Phi(L_t(\omega, \cdot))\}\} = -\inf_{\mu\in\mathcal{M}} [\Phi(\mu) + I(\mu)] , \tag{1.16}$$

where $I(\mu)$ *is defined by* (1.12) *and* $L_t(\omega, A)$ *is defined by* (1.11).

Since, for Brownian motion in R_1, the infinitesimal generator L is $\frac{1}{2}\partial^2/\partial x^2$, we see the correspondence between (1.3) and (1.12). Now, we want to take a closer look at the I-function and in particular determine conditions for a general Markov process under which one has a simple expression for the I-function as happens for Brownian motion in (1.4). Section 4 of this paper will be devoted to this topic, and we now describe the assumptions and results.

Let X be an arbitrary space and $\mathcal{B}$ a σ-field of subsets of X. Let $p(t, x, dy)$ be the transition probabilities of a Markov process with $(X, \mathcal{B})$ as state space. Let T_t be the corresponding semigroup on $B(X)$, the space of bounded measurable functions on X. Let B_0 be the strongly continuous center of T_t. We assume the following:

1. There exists a σ-finite measure $\lambda(dx)$ such that $p(t, x, dy)$ has a density $p(t, x, y)$ with respect to λ for each $t>0$; the density $p(t, x, y)$ is jointly measurable in x and y and satisfies

$$p(t, x, y) = p(t, y, x) \quad \text{almost everywhere} \quad (\lambda \times \lambda).$$

2. The center B_0 is large enough in the sense that $B_0 \cap L_2(\lambda)$ is dense in $L_2(\lambda)$.

3. For any probability measure μ on X and any $f \in B(X)$ there is a sequence $\{f_n\} \in B_0$ such that $\|f_n\| \leqq \|f\|$ and $f_n \to f$ almost everywhere with respect to μ.

We note that if X is a locally compact separable metric space, if λ is a Borel measure which is finite on compact sets, and if B_0 contains continuous functions which have compact support, then assumptions 2 and 3 are automatically satisfied.

Under the assumptions above, T_t generates a strongly continuous contraction semigroup in $L_2(\lambda)$ which in addition is *selfadjoint* for each t. Thus we have the infinitesimal generators L on a domain $\mathscr{D}$ dense in $B_0 \subset B(X)$ and $\tilde{L}$ on a domain $\tilde{\mathscr{D}}$ dense in $L_2(\lambda)$. Since the T_t are contractions, $\tilde{L}$ is selfadjoint, negative semidefinite, and we denote by $\sqrt{-\tilde{L}}$ the canonical positive semidefinite square root of the selfadjoint positive semidefinite operator $-\tilde{L}$. Let $\tilde{\mathscr{D}}_{1/2}$ be the natural domain of $\sqrt{-\tilde{L}}$.

Denote by $\mathscr{D}^+$ those functions (bounded) in $\mathscr{D} \subset B(X)$ which have a positive lower bound. Since constants belong to $\mathscr{D}$, $\mathscr{D}^+$ is quite rich. We define, as before, for any probability measure μ on X,

$$I(\mu) = -\inf_{u \in \mathscr{D}^+} \int_X \left(\frac{Lu}{u}\right)(x) \mu(dx).$$

In Section 4 of this paper we shall prove

THEOREM 5. *$I(\mu) < \infty$ if and only if $\mu \ll \lambda$ and $(d\mu/d\lambda)^{1/2} \in \tilde{\mathscr{D}}_{1/2}$. Moreover, letting $f = d\mu/d\lambda$ and $g = f^{1/2}$, we have*

$$I(\mu) = \|\sqrt{-\tilde{L}}\, g\|^2. \tag{1.17}$$

2. Discrete Parameter Markov Process

In this section we shall prove Theorem 1 of the introduction. We first prove (1.7).

Let $u\in\mathcal{U}_1$, $V=\pi u$, and $e^{-W}=u/V$. From our hypotheses on $\pi(x, dy)$ we see that W is a bounded continuous function on $\mathfrak{X}$. Denoting integration on Ω_x with respect to P_x measure by $E^{P_x}\{\ \}$, we show first that

$$E^{P_x}\{\exp\{-[W(X_0)+W(X_1)+\cdots+W(X_{n-1})]\}V(X_{n-1})\}=u(x)\,, \tag{2.1}$$

for all $x\in\mathfrak{X}$ and any positive integer n. Now,

$$E^{P_x}\{\exp\{-[W(X_0)+\cdots+W(X_{n-1})]\}V(X_{n-1})\}=E^{P_x}\left\{\frac{u(X_0)\cdots u(X_{n-2})}{V(X_0)\cdots V(X_{n-2})}u(X_{n-1})\right\}.$$

Since the process is Markovian,

$$\begin{aligned}E^{P_x}\{u(X_{n-1})\mid X_0, X_1,\cdots, X_{n-2}\}&=E^{P_x}\{u(X_{n-1})\mid X_{n-2}\}\\&=\int_{\mathfrak{X}} u(y)\pi(X_{n-2}, dy)=V(X_{n-2})\,,\end{aligned}$$

and hence

$$E^{P_x}\left\{\frac{u(X_0)\cdots u(X_{n-2})}{V(X_0)\cdots V(X_{n-2})}u(X_{n-1})\right\}=E^{P_x}\left\{\frac{u(X_0)\cdots u(X_{n-3})}{V(X_0)\cdots V(X_{n-3})}u(X_{n-2})\right\}.$$

Thus,

$$E^{P_x}\{\exp\{-[W(X_0)+\cdots+W(X_{n-1})]\}V(X_{n-1})\}=E^{P_x}\{u(X_0)\}=u(x)$$

which is (2.1).

Since u is bounded above and V is bounded below, it follows from (2.1) that there exists a constant $M>0$ such that

$$E^{P_x}\{\exp\{-[W(X_0)+W(X_1)+\cdots+W(X_{n-1})]\}\}\leqq\frac{u(x)}{\inf_{x\in\mathfrak{X}} V(x)}\leqq M\,.$$

But,

$$\begin{aligned}\exp\{-[W(X_0)+W(X_1)+\cdots+W(X_{n-1})]\}&=\exp\{-n[[W(X_0)+\cdots+W(X_{n-1})]/n]\}\\&=\exp\left\{-n\int_{\mathfrak{X}} W(y)L_n(\omega, dy)\right\},\end{aligned}$$

and therefore, from the definition of $Q_{n,x}$ measure on $\mathcal{M}$,

$$E^{Q_{n,x}}\left\{\exp\left\{-n\int_{\mathfrak{X}} W(y)l(dy)\right\}\right\}=E^{P_x}\{\exp\{-[W(X_0)+\cdots+W(X_{n-1})]\}\}\leqq M\,. \tag{2.2}$$

On the left of (2.2), $l(dy)$ is a measure in $\mathscr{M}$ and is, of course, the variable of integration.

For the moment, let C be any measurable set in $\mathscr{M}$. Then

$$E^{Q_{n,x}}\left\{\exp\left\{-n\int_{\mathscr{X}} W(y)l(dy)\right\}\right\} \geqq Q_{n,x}(C)\exp\left\{-n\sup_{l\in C}\int_{\mathscr{X}} W(y)l(dy)\right\},$$

and from (2.2) we have

$$Q_{n,x}(C) \leqq M\exp\left\{n\sup_{l\in C}\int_{\mathscr{X}} W(y)l(dy)\right\}.$$

Thus,

$$\begin{aligned}\overline{\lim_{n\to\infty}}\frac{1}{n}\log Q_{n,x}(C) &\leqq \sup_{l\in C}\int_{\mathscr{X}} W(y)l(dy)\\ &= \sup_{l\in C}\int_{\mathscr{X}} \log\left(\frac{\pi u}{u}\right)(y)l(dy)\,.\end{aligned} \tag{2.3}$$

Since, in (2.3), u was an arbitrary element of $\mathscr{U}_1$, we have

$$\overline{\lim_{n\to\infty}}\frac{1}{n}\log Q_{n,x}(C) \leqq \inf_{u\in\mathscr{U}_1}\sup_{l\in C}\int_{\mathscr{X}} \log\left(\frac{\pi u}{u}\right)(y)l(dy)\,. \tag{2.4}$$

Now, if $C \subset \bigcup_{j=1}^{k} C_j$, then

$$\overline{\lim_{n\to\infty}}\frac{1}{n}\log Q_{n,x}(C) \leqq \sup_{1\leqq j\leqq k}\inf_{u\in\mathscr{U}_1}\sup_{l\in C_j}\int_{\mathscr{X}} \log\left(\frac{\pi u}{u}\right)(y)l(dy)\,,$$

and hence, considering all possible finite unions covering C,

$$\overline{\lim_{n\to\infty}}\frac{1}{n}\log Q_{n,x}(C) \leqq \inf_{\substack{C_1,C_2,\cdots,C_k\\ C\subset\bigcup_{j=1}^{k}C_j}}\sup_{1\leqq j\leqq k}\inf_{u}\sup_{l\in C_j}\int_{\mathscr{X}} \log\left(\frac{\pi u}{u}\right)(y)l(dy)\,. \tag{2.5}$$

In (2.5), C is an arbitrary measurable set in $\mathscr{M}$. If C is compact, then it is easy to show[1] that the right side of inequality (2.5) is, in fact, just

$$\sup_{l\in C}\inf_{u\in\mathscr{U}_1}\int_{\mathscr{X}} \log\left(\frac{\pi u}{u}\right)(y)l(dy) = -\inf_{l\in C}\left[-\inf_{u\in\mathscr{U}_1}\int_{\mathscr{X}} \log\left(\frac{\pi u}{u}\right)(y)l(dy)\right] = -\inf_{l\in C} I(l)\,.$$

[1] The argument can be found in [1] (just after (2.6) of that paper) where a closely related situation was encountered.

Since $\mathfrak{X}$ is compact, $\mathcal{M}$ is compact in the weak topology, and hence the hypothesis for (1.7) that C be closed implies it is compact. From (2.5) and the last remark we have (1.7).

The lower estimate (1.8) is much more delicate and we prove first a succession of nine lemmas.

LEMMA 2.1. *Let $(X, \mathcal{S})$ be a metric space and $\mathcal{S}$ its Borel sets. Let α and β be probability measures on X, and denote by $\mathcal{U}^*$ the set of functions $u(x)$ which are continuous on X and for each of which there exist constants c_1 and c_2 such that $0<c_1\leqq u(x)\leqq c_2<\infty$ for all $x\in X$. Define*

$$I_\beta(\alpha)=-\inf_{u\in\mathcal{U}^*}\left[\log\int_X u(x)\beta(dx)-\int_X \log u(x)\alpha(dx)\right]. \tag{2.6}$$

Then, $I_\beta(\alpha)<\infty$ if and only if $\alpha\ll\beta$ with a density $a(x)=d\alpha/d\beta$ such that

$$\int_X a(x)\,|\log a(x)|\,\beta(dx)<\infty.$$

Moreover, in this case,

$$I_\beta(\alpha)=\int_X a(x)\log a(x)\beta(dx)=\int_X \log a(x)\alpha(dx).$$

Proof: Assume first that $\alpha\ll\beta$ and the Radon-Nikodym derivative $a(x)=d\alpha/d\beta$ is such that

$$\int_X a(x)\,|\log a(x)|\,\beta(dx)<\infty.$$

Let $\int_X a(x)\log a(x)\beta(dx)=l$. If we show that, for all $u\in\mathcal{U}^*$,

$$\log\int_X u(x)\beta(dx)-\int_X \log u(x)\alpha(dx)\geqq -l, \tag{2.7}$$

then we can conclude that

$$I_\beta(\alpha)\leqq l=\int_X a(x)\log a(x)\beta(dx). \tag{2.8}$$

So, let $u \in \mathcal{U}^*$, and we have

$$\begin{aligned}\int_X \log u(x)\alpha(dx) &= \int_X (\log u(x))a(x)\beta(dx)\\ &= \int_X \left(\log\left[\frac{u(x)}{a(x)}\,a(x)\right]\right)a(x)\beta(dx)\\ &= \int_{a(x)>0} \left(\log\frac{u(x)}{a(x)}\right)a(x)\beta(dx)+l,\end{aligned}$$

because $a(x)>0$ almost everywhere (α-measure).

From Jensen's inequality we then have

$$\begin{aligned}\int_X \log u(x)\alpha(dx) &\leqq \log\int_{a(x)>0} \frac{u(x)}{a(x)}\cdot a(x)\beta(dx)+l\\ &\leqq \log\int_X u(x)\beta(dx)+l,\end{aligned}$$

and therefore (2.7), and hence (2.8), holds for any $u \in \mathcal{U}^*$.

Now let us assume $I_\beta(\alpha)<\infty$. We want to show that $d\alpha/d\beta = a(x)$ exists and satisfies

$$\int_X a(x)\,|\log a(x)|\,\beta(dx)<\infty\,;$$

moreover,

$$\int_X a(x)\log a(x)\beta(dx)\leqq I_\beta(\alpha)\,.$$

Let $I_\beta(\alpha)=l<\infty$ so that

$$\log\int_X u(x)\beta(dx)-\int_X \log u(x)\alpha(dx)\geqq -l, \tag{2.9}$$

for all $u \in \mathcal{U}^*$. Let $B^+(X)$ be the space of all bounded measurable functions u on X which are bounded away from zero and infinity. From Lusin's theorem we know that (2.9) holds not just for all $u \in \mathcal{U}^*$ but for all $u \in B^+(X)$ as well.

Now assume $A \in \mathcal{S}$, and let, for any $k>0$, $u(x)=k$ on A and $u(x)=1$ on $\tilde{A}$ so that $u \in B^+(X)$ and hence, from (2.9),

$$(\log k)\alpha(A)\leqq \log[k\beta(A)+\beta(\tilde{A})]+l\,. \tag{2.10}$$

Consequently, if $\beta(A)=0$, (2.10) implies that

$$\alpha(A)\leqq\frac{l}{\log k},$$

and hence, letting $k\to\infty$, we obtain $\alpha(A)=0$, i.e., $\alpha\ll\beta$. Let $a(x)=d\alpha/d\beta$ and, for any positive integer n, define

$$u_n(x)=\left(a(x)\vee\frac{1}{n}\right)\wedge n.$$

We see that $u_n\in B^+(X)$ and, moreover,

$$\lim_{n\to\infty}\int_X|u_n(x)-a(x)|\,\beta(dx)=0. \tag{2.11}$$

From (2.9)

$$\int_X\log u_n(x)\alpha(dx)\leqq\log\int_X u_n(x)\beta(dx)+l,$$

and (2.11) implies that the right side of this last inequality approaches l as $n\to\infty$. Hence,

$$\overline{\lim_{n\to\infty}}\int_X\log u_n(x)\alpha(dx)\leqq l. \tag{2.12}$$

Writing $\log u_n=(\log u_n)^+-(\log u_n)^-$, we note that $(\log u_n)^+\uparrow(\log a)^+$, $(\log u_n)^-\uparrow(\log a)^-$, and (2.12) becomes

$$\overline{\lim_{n\to\infty}}\left[\int_X(\log u_n)^+\alpha(dx)-\int_X(\log u_n)^-\alpha(dx)\right]\leqq l. \tag{2.13}$$

Since $\int_X a(x)(\log a(x))^-\beta(dx)$ is clearly finite, we have by monotone convergence

$$\int_X(\log u_n(x))^-\alpha(dx)=\int_X a(x)(\log u_n(x))^-\beta(dx)\uparrow\int_X a(x)(\log a(x))^-\beta(dx).$$

Thus, from (2.13)

$$\overline{\lim_{n\to\infty}}\int_X(\log u_n(x))^+\alpha(dx)\leqq l+\int_X a(x)(\log a(x))^-\beta(dx).$$

Again, using the monotone convergence theorem, we have

$$\int_X (\log a(x))^+ \alpha(dx) \leqq l + \int_X a(x)(\log a(x))^- \beta(dx).$$

We therefore conclude that

$$\int_X a(x)\,|\log a(x)|\,\beta(dx) < \infty,$$

and moreover,

$$\int_X a(x) \log a(x) \beta(dx) \leqq l,$$

which completes Lemma 2.1.

COROLLARY. *The function*

$$I_\beta(\alpha) = \int_X a(x) \log a(x) \beta(dx) = \int_X \log a(x) \alpha(dx)$$

is a lower semicontinuous function of α *for fixed* β *in the weak topology of measures on* X.

Proof: For fixed $u \in \mathcal{U}^*$,

$$\log \int_X u(x)\beta(dx) - \int_X \log u(x)\alpha(dx)$$

is a continuous functional of α for fixed β in the weak topology on X. Therefore,

$$\inf_{u \in \mathcal{U}^*} \left[\log \int_X u(x)\beta(dx) - \int_X \log u(x)\alpha(dx) \right]$$

is upper semicontinuous and hence $I_\beta(\alpha)$, from its definition (2.6), is lower semicontinuous.

Using the definitions and notation of the introduction just before the statement of Theorem 1, we now state

LEMMA 2.2. *Assume* $\mu \in \mathcal{M}$ *is such that* $d\mu/d\lambda = m(x)$ *exists and satisfies*

$$0 < \eta \leqq m(x) \leqq \xi < \infty.$$

Let $\mathcal{M}_\mu$ be the set of all probability measures $\alpha(dx_1, dx_2)$ on $\mathbf{X}\times\mathbf{X}$ such that both marginal measures are μ. Let $\beta(dx_1, dx_2)$ be the probability measure on $\mathbf{X}\times\mathbf{X}$ defined by

$$\begin{aligned}\beta(dx_1, dx_2) &= \pi(x_1, x_2)\mu(dx_1)\lambda(dx_2)\\ &= \pi(x_1, x_2)m(x_1)\lambda(dx_1)\lambda(dx_2)\,.\end{aligned}$$

Then, $\inf_{\alpha\in\mathcal{M}_\mu} I_\beta(\alpha)$ is attained at some point of $\mathcal{M}_\mu$—call it α_0. Moreover, $\alpha_0 \ll \beta$ and $d\alpha_0/d\beta = a_0(x_1, x_2)$ is strictly positive almost everywhere in $\mathbf{X}\times\mathbf{X}$ ($\lambda\times\lambda$-measure).

Proof: In the corollary to Lemma 2.1 we showed that $I_\beta(\alpha)$ is lower semicontinuous as a function of α for fixed β. For $\beta(dx_1, dx_2)$ as defined above and $\alpha\in\mathcal{M}_\mu$ it then follows, from the fact that $\mathcal{M}_\mu$ is compact in the weak topology of measures on $\mathbf{X}\times\mathbf{X}$, that $\inf_{\alpha\in\mathcal{M}_\mu} I_\beta(\alpha)$ is attained at some point of $\mathcal{M}_\mu$—call it α_0.

Now the product measure $\mu\times\mu$ is, of course, in $\mathcal{M}_\mu$ and

$$\begin{aligned}a(x_1, x_2) = a(x) = \frac{d(\mu\times\mu)}{d\beta} &= \frac{m(x_1)m(x_2)\lambda(dx_1)\lambda(dx_2)}{\pi(x_1, x_2)m(x_1)\lambda(dx_1)\lambda(dx_2)}\\ &= \frac{m(x_2)}{\pi(x_1, x_2)}\end{aligned}$$

is bounded above and below due to our assumptions on $m(x)$ and $\pi(x, y)$. Therefore,

$$\int_{\mathbf{X}\times\mathbf{X}} a(x)\,|\log a(x)|\,\beta(dx_1, dx_2) < \infty\,,$$

and from Lemma 2.1 this implies $I_\beta(\mu\times\mu)<\infty$. Thus, $I_\beta(\alpha_0)<\infty$. Again from Lemma 2.1 we now know that $\alpha_0\ll\beta$. All that remains to prove is that $a_0(x_1, x_2) = d\alpha_0/d\beta$ is positive almost everywhere ($\lambda\times\lambda$-measure).

Let $a(x_1, x_2)$ be any density such that

$$a(x_1, x_2)\lambda(dx_1)\lambda(dx_2) = \alpha(dx_1, dx_2)\in\mathcal{M}_\mu\,.$$

The marginal measures of any convex linear combination of probability measures in $\mathcal{M}_\mu$ are still μ. Thus, for any $\varepsilon>0$, $(1-\varepsilon)\alpha_0+\varepsilon\alpha$ is a measure in $\mathcal{M}_\mu$ and by the definition of α_0 we have

$$I_\beta(\alpha_0) \leqq I_\beta((1-\varepsilon)\alpha_0+\varepsilon\alpha)\,.$$

Since $I_\beta(\alpha)$ is a convex functional of α, this implies that $I_\beta((1-\varepsilon)\alpha_0+\varepsilon\alpha)$ is a nondecreasing function of ε in $0\leqq\varepsilon\leqq 1$ so that

$$\frac{d}{d\varepsilon}[I_\beta((1-\varepsilon)\alpha_0+\varepsilon\alpha)]\geqq 0\,, \qquad 0\leqq\varepsilon\leqq 1\,, \tag{2.14}$$

provided it exists. We have

$$\frac{d[(1-\varepsilon)\alpha_0+\varepsilon\alpha]}{d\beta}=(1-\varepsilon)a_0+\frac{\varepsilon a(x_1, x_2)}{m(x_1)\pi(x_1, x_2)}\,;$$

we shall call it $a_\varepsilon(x_1, x_2)$. Therefore, from Lemma 2.1,

$$I_\beta[(1-\varepsilon)\alpha_0+\varepsilon\alpha]=\int_{\mathcal{X}\times\mathcal{X}} a_\varepsilon(x_1, x_2)\log a_\varepsilon(x_1, x_2)\beta(dx_1, dx_2)\,.$$

Hence, for $0<\varepsilon<1$, $(d/d\varepsilon)I_\beta[(1-\varepsilon)\alpha_0+\varepsilon\alpha]$ exists and is seen to equal

$$\begin{aligned}&\int_{\mathcal{X}\times\mathcal{X}}\left[\frac{a(x_1, x_2)}{m(x_1)\pi(x_1, x_2)}-a_0(x_1, x_2)\right][\log a_\varepsilon(x_1, x_2)]\\ &\qquad\times m(x_1)\pi(x_1, x_2)\lambda(dx_1)\lambda(dx_2)\\ &\quad+\int_{\mathcal{X}\times\mathcal{X}} a_\varepsilon(x_1, x_2)\frac{[a(x_1, x_2)/m(x_1)\pi(x_1, x_2)-a_0(x_1, x_2)]}{a(x_1, x_2)}\\ &\qquad\times m(x_1)\pi(x_1, x_2)\lambda(dx_1)\lambda(dx_2)\,.\end{aligned} \tag{2.15}$$

The second integral on the right of (2.15) is just

$$\int_{\mathcal{X}\times\mathcal{X}} a(x_1, x_2)\lambda(dx_1)\lambda(dx_2)-\int_{\mathcal{X}\times\mathcal{X}} a_0(x_1, x_2)\beta(dx_1, dx_2)=1-1=0\,.$$

Thus from (2.14) and (2.15) we have, for $1>\varepsilon>0$,

$$\int_{\mathcal{X}\times\mathcal{X}} a(x_1, x_2)\log a_\varepsilon(x_1, x_2)\lambda(dx_1)\lambda(dx_2)\geqq\int_{\mathcal{X}\times\mathcal{X}} a_0(x_1, x_2)\log a_\varepsilon(x_1, x_2)\beta(dx_1, dx_2)\,.$$

As $\varepsilon\to 0$, $a_\varepsilon(x_1, x_2)\to a_0(x_1, x_2)$ and, by the dominated convergence theorem,

$$\begin{aligned}&\lim_{\varepsilon\to 0}\int_{\mathcal{X}\times\mathcal{X}} a_0(x_1, x_2)\log a_\varepsilon(x_1, x_2)\beta(dx_1, dx_2)\\ &\qquad=\int_{\mathcal{X}\times\mathcal{X}} a_0(x_1, x_2)\log a_0(x_1, x_2)\beta(dx_1, dx_2)=I_\beta(\alpha_0)\geqq 0\,.\end{aligned}$$

On the other hand, if $I_\beta(\alpha)<\infty$, then $I_\beta[\frac{1}{2}\alpha_0+\frac{1}{2}\alpha]<\infty$ by convexity and this implies that $|\log a_{1/2}(x_1, x_2)|$ is integrable with respect to $\frac{1}{2}\alpha_0+\frac{1}{2}\alpha$, and hence with respect to both α and α_0. In particular,

$$\begin{aligned}\log^+ a(x_1, x_2) &= \log^+\left[(1-\varepsilon)a_0(x_1, x_2)+\varepsilon\frac{a(x_1, x_2)}{m(x_1)\pi(x_1, x_2)}\right]\\ &\leqq \log^+\left[a_0(x_1, x_2)+\frac{a(x_1, x_2)}{m(x_1)\pi(x_1, x_2)}\right]\\ &\leqq \log 2+\log^+ a_{1/2}(x_1, x_2)\end{aligned}$$

is uniformly integrable with respect to α. It now follows from Fatou's lemma that

$$\int_{X\times X} a(x_1, x_2)\log^- a_0(x_1, x_2)\lambda(dx_1)\lambda(dx_2)>-\infty .$$

Therefore, $a_0(x_1, x_2)>0$ whenever $a(x_1, x_2)>0$. If we pick $\alpha(dx_1, dx_2)=\mu\times\mu$ so that $a(x_1, x_2)=m(x_1)m(x_2)$, then $a(x_1, x_2)$ never vanishes and moreover $I_\beta(\alpha)<\infty$. Hence $a_0(x_1, x_2)$ cannot vanish on a set of positive $\lambda\times\lambda$-measure which completes Lemma 2.2.

The following lemma is needed in the proof of Lemma 2.4.

LEMMA 2.3. *Let $(X, \mathscr{B}, \lambda)$ be a probability measure space. Assume there exist two sequences of functions $\{f_n(x_1)\}$, $\{g_n(x_2)\}$ such that, in the product space $X\times X$,*

$$\phi(x_1, x_2)=\lim_{n\to\infty} \text{in meas.}\,[f_n(x_1)+g_n(x_2)],$$

where the measure indicated on the right is the product measure $\lambda\times\lambda$. Then, there exists a subsequence n_j and a sequence of constants $\{a_{n_j}\}$ such that

$$\lim_{j\to\infty}[f_{n_j}(x_1)-a_{n_j}]=f(x_1)$$

exists for almost all x_1,

$$\lim_{j\to\infty}[g_{n_j}(x_2)+a_{n_j}]=g(x_2)$$

exists for almost all x_2, and

$$\phi(x_1, x_2)=f(x_1)+g(x_2)$$

for almost all (x_1, x_2) ($\lambda\times\lambda$-measure).

Proof: Let $\phi_n(x_1, x_2) = f_n(x_1) + g_n(x_2)$ so that our assumption is that $\phi(x_1, x_2) = \lim \text{in meas}_{n\to\infty} \phi_n(x_1, x_2)$. There exists a subsequence, which we again denote by $\{\phi_n(x_1, x_2)\}$, which converges to $\phi(x_1, x_2)$ almost everywhere ($\lambda \times \lambda$-measure). Let

$$E = \{(x_1, x_2) : f_n(x_1) + g_n(x_2) \to \phi(x_1, x_2)\}$$

so that $(\lambda \times \lambda)(E) = 1$. Let

$$E_{x_1} = \{x_2 : f_n(x_1) + g_n(x_2) \to \phi(x_1, x_2)\}.$$

By the Fubini theorem, $\lambda(E_{x_1}) = 1$ for almost all x_1 (λ-measure). Let $E^* = \{x_1 : \lambda(E_{x_1}) = 1$ so that $\lambda(E^*) = 1$ and hence $E^* \neq \varnothing$. Choose $\bar{x}_1 \in E^*$; then $f_n(\bar{x}_1) + g_n(x_2) \to \phi(\bar{x}_1, x_2)$ for all $x_2 \in E_{\bar{x}_1}$. Thus, if we let $a_n = f_n(\bar{x}_1)$, we see that

$$\lim_{n\to\infty} [g_n(x_2) + a_n]$$

exists for all $x_2 \in E_{\bar{x}_1}$ and $\lambda(E_{\bar{x}_1}) = 1$. But a_n is independent of x_2 and therefore there exists a function $g(x_2)$ such that

$$\lim_{n\to\infty} [g_n(x_2) + a_n] = g(x_2)$$

for almost all x_2 (λ-measure). Now

$$f_n(x_1) + g_n(x_2) = f_n(x_1) - a_n + g_n(x_2) + a_n$$

and so it follows from the fact that $g_n(x_2) + a_n \to g(x_2)$ for almost all (x_1, x_2) ($\lambda \times \lambda$-measure) that $f_n(x_1) - a_n$ converges for almost all (x_1, x_2) ($\lambda \times \lambda$-measure). But $f_n(x_1) - a_n$ is independent of x_2 and therefore $f_n(x_1) - a_n$ converges for almost all x_1 to a function of x_1—call it $f(x_1)$. Thus $f_n(x_1) + g_n(x_2) \to f(x_1) + g(x_2)$ for almost all (x_1, x_2) ($\lambda \times \lambda$-measure) which implies that $\phi(x_1, x_2) = f(x_1) + g(x_2)$ almost everywhere in $X \times X$ ($\lambda \times \lambda$-measure).

LEMMA 2.4. *Assume $\mu \in \mathcal{M}$ is such that $d\mu/d\lambda = m(x)$ exists and satisfies*

$$0 < \eta \leqq m(x) \leqq \xi < \infty .$$

Then there exists a function $V(y)$ on $\mathfrak{X}$ which is non-negative, not identically zero, an element of $L_1(\lambda)$, such that the discrete parameter Markov process with transition function

$$\pi'(x, dy) = \frac{\pi(x, y) V(y) \lambda(dy)}{\int_{\mathfrak{X}} \pi(x, y) V(y) \lambda(dy)}$$

has μ *for its invariant probability measure, i.e., for every Borel set* $A \subset \mathscr{X}$

$$\int_{\mathscr{X}} \pi'(x, A)\mu(dx) = \mu(A) .$$

Proof: Let $\mathscr{M}_\mu$ and $\beta(dx_1, dx_2)$ be defined as in Lemma 2.2. In that lemma we proved that $\inf_{\alpha \in \mathscr{M}_\mu} I_\beta(\alpha)$ was attained in $\mathscr{M}_\mu$ at some point α_0 and that $\alpha_0 \ll \beta$. Moreover, we established in Lemma 2.2 that $d\alpha_0/d\beta = a_0(x_1, x_2)$ was strictly positive almost everywhere on $\mathscr{X}\times\mathscr{X}$ ($\lambda\times\lambda$-measure). We now show that $a_0(x_1, x_2)$ factors, i.e., we find functions $U(x_1)$ and $V(x_2)$ such that $a_0(x_1, x_2) = U(x_1)V(x_2)$ for almost all (x_1, x_2) ($\lambda\times\lambda$-measure). We shall then prove that this function V fulfills the requirements of the present lemma. To show first of all that $a_0(x_1, x_2)$ factors in this way we make use of the Euler equation characterizing the fact that α_0 minimizes $I_\beta(\alpha)$ for α varying in $\mathscr{M}_\mu$.

Let $k(x_1, x_2)$ be a bounded function on $\mathscr{X}\times\mathscr{X}$ having the property

$$\int_{\mathscr{X}\times\mathscr{X}} k(x_1, x_2)a_0(x_1, x_2)[f(x_1)+g(x_2)]\beta(dx_1, dx_2) = 0 \tag{2.16}$$

for all continuous (hence bounded) functions f and g. Choose ε such that $|\varepsilon| \leqq 1/C$, where $|k| \leqq C$. Consider the density $a_\varepsilon(x_1, x_2) = a_0(x_1, x_2)(1+\varepsilon k(x_1, x_2))$ which is non-negative almost everywhere on $\mathscr{X}\times\mathscr{X}$ ($\lambda\times\lambda$-measure). We define a measure on $\mathscr{X}\times\mathscr{X}$ by $\alpha_\varepsilon(dx_1, dx_2) = a_\varepsilon(x_1, x_2)\beta(dx_1, dx_2)$. Now $\alpha_\varepsilon \in \mathscr{M}_\mu$ because $\alpha_0 \in \mathscr{M}_\mu$ and the defining property of $k(x_1, x_2)$ implies that for all continuous f and g

$$\int_{\mathscr{X}\times\mathscr{X}} a_\varepsilon(x_1, x_2)f(x_1)\beta(dx_1, dx_2) = \int_{\mathscr{X}\times\mathscr{X}} a_0(x_1, x_2)f(x_1)\beta(dx_1, dx_2) ,$$

$$\int_{\mathscr{X}\times\mathscr{X}} a_\varepsilon(x_1, x_2)g(x_2)\beta(dx_1, dx_2) = \int_{\mathscr{X}\times\mathscr{X}} a_0(x_1, x_2)g(x_2)\beta(dx_1, dx_2) .$$

To obtain the Euler equation for $I_\beta(\alpha)$ at $\alpha = \alpha_0$, we consider

$$\frac{\partial}{\partial\varepsilon} I_\beta(\alpha_\varepsilon)|_{\varepsilon=0} = 0 . \tag{2.17}$$

From Lemma 2.1 and the definition of $a_\varepsilon(x_1, x_2)$ we have

$$\begin{aligned} I_\beta(\alpha_\varepsilon) &= \int_{\mathscr{X}\times\mathscr{X}} a_\varepsilon(x_1, x_2) \log a_\varepsilon(x_1, x_2)\beta(dx_1, dx_2) \\ &= \int_{\mathscr{X}\times\mathscr{X}} [a_0(x_1, x_2)(1+\varepsilon k(x_1, x_2))] \log [a_0(x_1, x_2) \\ &\qquad \times (1+\varepsilon k(x_1, x_2)]\beta(dx_1, dx_2) . \end{aligned}$$

Hence,

$$\left.\frac{\partial I_\beta(\alpha_\varepsilon)}{\partial \varepsilon}\right|_{\varepsilon=0} = \int_{\mathcal{X}\times\mathcal{X}} a_0(x_1, x_2)k(x_1, x_2) \log a_0(x_1, x_2)\beta(dx_1, dx_2)$$
$$+ \int_{\mathcal{X}\times\mathcal{X}} a_0(x_1, x_2)k(x_1, x_2)\beta(dx_1, dx_2)\,.$$

From (2.16) with $f\equiv 1$ and $g\equiv 0$ we see that the second integral on the right in this last expression is zero. Thus equation (2.17) is just

(2.18) $$\int_{\mathcal{X}\times\mathcal{X}} k(x_1, x_2) \log a_0(x_1, x_2)\alpha_0(dx_1, dx_2)=0\,.$$

Our hypothesis on $k(x_1, x_2)$, (2.16), can be rewritten as

(2.19) $$\int_{\mathcal{X}\times\mathcal{X}} k(x_1, x_2)[f(x_1)+g(x_2)]\alpha_0(dx_1, dx_2)=0$$

for all continuous f and g. Therefore, what we have shown is that if $k(x_1, x_2)$ is a bounded function satisfying (2.19), then (2.18) holds. It now follows from the Hahn-Banach theorem that there exist sequences of continuous functions $\{f_n(x_1)\}$, $\{g_n(x_2)\}$, such that

(2.20) $$\log a_0(x_1, x_2)=\lim_{n\to\infty}[f_n(x_1)+g_n(x_2)]\,,$$

where the limit on the right is in the sense of $L_1(\alpha_0)$. Hence, there exist sequences of continuous functions $\{f_n(x_1)\}$, $\{g_n(x_2)\}$ such that

(2.21) $$\log a_0(x_1, x_2)=\lim_{n\to\infty} \text{ in prob } [f_n(x_1)+g_n(x_2)]\,,$$

where the probability measure indicated on the right of (2.21) is, of course, $\alpha_0(dx_1, dx_2)$. However, from Lemma 2.2 and our hypotheses on $\pi(x_1, x_2)$ and $m(x)$ we see that $\alpha_0 \sim \beta \sim \lambda\times\lambda$ so that in fact

(2.22) $$\log a_0(x_1, x_2)=\lim_{n\to\infty} \text{ in prob } [f_n(x_1)+g_n(x_2)]\,,$$

where now the probability measure on the right of (2.22) is $\lambda\times\lambda$. Therefore, from Lemma 2.3 we conclude that there exist functions $f(x_1)$ and $g(x_2)$ such that

(2.23) $$\log a_0(x_1, x_2)=f(x_1)+g(x_2)$$

for almost all (x_1, x_2) ($\lambda\times\lambda$-measure).

Thus, letting $U(x_1)=\exp\{f(x_1)\}$ and $V(x_2)=\exp\{g(x_2)\}$, we have

$$a_0(x_1, x_2)=U(x_1)V(x_2) \tag{2.24}$$

for almost all (x_1, x_2) ($\lambda\times\lambda$-measure).

Now since $\beta(dx_1, dx_2)=\pi(x_1, x_2)m(x_1)\lambda(dx_1)\lambda(dx_2)$ and $d\alpha_0/d\beta=a_0(x_1, x_2)$, (2.24) implies that

$$\alpha_0(dx_1, dx_2)=U(x_1)V(x_2)\pi(x_1, x_2)m(x_1)\lambda(dx_1)\lambda(dx_2)\,. \tag{2.25}$$

Since $\alpha_0\in\mathcal{M}_\mu$ and $d\mu/d\lambda=m(x)$, we get from (2.25)

$$U(x_1)m(x_1)\int_{\mathcal{X}} V(x_2)\pi(x_1, x_2)\lambda(dx_2)=m(x_1) \tag{2.26}$$

for almost all x_1 (λ-measure).

In (2.26) alter U on a set of measure zero so that the equation is valid for all x_1 instead of almost all x_1. This revised U could just as well have been used instead of the original U in (2.24) since the set of measure zero on which we have changed U does not affect the Radon-Nikodym derivative. With this new U we then have

$$\int_{\mathcal{X}} V(x_2)\pi(x_1, x_2)\lambda(dx_2)=\frac{1}{U(x_1)} \tag{2.27}$$

for all x_1. By its definition we know that $a_0\in L_1(\beta)$, and, since $d\beta/d(\lambda\times\lambda)=\pi(x_1, x_2)m(x_1)$ is bounded above and below, we have $a_0\in L_1(\lambda\times\lambda)$. Therefore, from the Fubini theorem and (2.24) we see that $V\in L_1(\lambda)$. It follows then from (2.27) and our hypothesis on $\pi(x_1, x_2)$ that U is continuous. We now define

$$\pi'(x, dy)=U(x)\pi(x, y)V(y)\lambda(dy) \tag{2.28}$$

and note from (2.27) that it is a transition probability.

Finally, we want to show that μ is an invariant probability measure for $\pi'(x, dy)$. Consider, for any Borel set $A\subset\mathcal{X}$,

$$\begin{aligned}\int_{\mathcal{X}} \pi'(x, A)\mu(dx) &= \int_{\mathcal{X}}\int_A U(x)\pi(x, y)V(y)\lambda(dy)\mu(dx)\\ &= \int_{\mathcal{X}}\int_A U(x)\pi(x, y)V(y)m(x)\lambda(dy)\lambda(dx)\\ &= \int_{\mathcal{X}}\int_A \alpha_0(dx, dy)=\mu(A)\,,\end{aligned}$$

since $\alpha_0\in\mathcal{M}_\mu$. This completes Lemma 2.4.

COROLLARY. *The V of Lemma 2.4 is bounded above and below. In fact,*

$$\frac{\eta}{\left(\sup_{x\in\mathfrak{X}} U(x)\right)\xi A} \leqq V(y) \leqq \frac{\xi}{\left(\inf_{x\in\mathfrak{X}} U(x)\right)a\eta}.$$

Proof: From (2.27) and the fact that $V\in L_1(\lambda)$ we know that, for any $x\in\mathfrak{X}$ and $x'\in\mathfrak{X}$,

$$\frac{U(x')}{U(x)} = \frac{\int_{\mathfrak{X}} V(y)\pi(x, y)\lambda(dy)}{\int_{\mathfrak{X}} V(y)\pi(x', y)\lambda(dy)} \leqq \frac{A}{a}.$$

Thus $\sup_{x\in\mathfrak{X}} U(x)/\inf_{x\in\mathfrak{X}} U(x) \leqq A/a$. Since $\alpha_0\in\mathcal{M}_\mu$, we have for almost all y (λ-measure)

$$m(y) = \int_{\mathfrak{X}} U(x)V(y)\pi(x, y)m(x)\lambda(dx).$$

Moreover, since, by Lemma 2.2, $a_0(x_1, x_2)>0$ almost everywhere ($\lambda\times\lambda$-measure), $V(y)>0$ almost everywhere (λ-measure). Thus

$$\frac{m(y)}{V(y)} = \int_{\mathfrak{X}} U(x)\pi(x, y)m(x)\lambda(dx)$$

$$\leqq \left(\sup_{x\in\mathfrak{X}} U(x)\right)A\xi,$$

so that

$$V(y) \geqq \frac{\eta}{\left(\sup_{x\in\mathfrak{X}} U(x)\right)A\xi}.$$

Similarly,

$$\frac{m(y)}{V(y)} \geqq \left(\inf_{x\in\mathfrak{X}} U(x)\right)a\eta,$$

so that

$$V(y) \leqq \frac{\xi}{\left(\inf_{x\in\mathfrak{X}} U(x)\right)a\eta}.$$

The remaining lemmas are concerned with the function $I(\mu)$ defined in (1.5).

LEMMA 2.5. *$I(\mu)=0$ if and only if μ is the invariant measure of $\pi(x, dy)$.*

Proof: Since $u\equiv 1$ is in $\mathcal{U}_1$ and, for this u, $\log(\pi u/u)(x)=0$, it clearly suffices to show that

$$\int_{\mathfrak{X}} \log\left(\frac{\pi u}{u}\right)(x)\mu(dx)\geqq 0 \tag{2.29}$$

for all $u\in\mathcal{U}_1$ if and only if for every Borel set $A\subset\mathfrak{X}$

$$\int_{\mathfrak{X}} \pi(x, A)\mu(dx)=\mu(A)\,. \tag{2.30}$$

First, assume (2.29); then as just noted the left side of (2.29) is zero if $u\equiv 1$. Therefore, if we consider

$$\int_{\mathfrak{X}} \log\left(\frac{\pi(1+\varepsilon u)}{1+\varepsilon u}\right)(x)\mu(dx)=\int_{\mathfrak{X}} \log\left(\frac{1+\varepsilon\pi u}{1+\varepsilon u}\right)(x)\mu(dx)$$

as a function of ε for continuous u such that $|u|\leqq c$ and $|\varepsilon|<1/c$, we see that it has a minimum at $\varepsilon=0$. Hence, setting the derivative with respect to ε equal to zero at $\varepsilon=0$, we obtain for all u as considered above

$$\int_{\mathfrak{X}} (\pi u)(x)\mu(dx)=\int_{\mathfrak{X}} u(x)\mu(dx)$$

which implies (2.30).

Now assume (2.30). Let $u\in\mathcal{U}_1$ and define h by $u=e^h$. From Jensen's inequality, $\pi e^h\geqq e^{\pi h}$ and therefore $\log\pi e^h\geqq\pi h$. This last and (2.30) imply that

$$\begin{aligned}\int_{\mathfrak{X}} \log(\pi e^h)(x)\mu(dx) &\geqq \int_{\mathfrak{X}} (\pi h)(x)\mu(dx)\\ &= \int_{\mathfrak{X}} h(x)\mu(dx)\end{aligned}$$

which is (2.29).

LEMMA 2.6. *Let* $\mathcal{U}_2$ *be the set of measurable functions* u *on* $\mathfrak{X}$ *for each of which there exist constants* c_1 *and* c_2 *such that*

$$0<c_1\leqq u(x)\leqq c_2<\infty$$

for all $x\in\mathfrak{X}$. *Then, for any* $\mu\in\mathcal{M}$,

$$I(\mu)=-\inf_{u\in\mathcal{U}_2}\int_{\mathfrak{X}}\log\left(\frac{\pi u}{u}\right)(x)\mu(dx)\,.$$

Proof: By definition,

$$I(\mu)=-\inf_{u\in\mathcal{U}_1}\int_{\mathfrak{X}}\log\left(\frac{\pi u}{u}\right)(x)\mu(dx)\,.$$

Let $I(\mu)=l$ and let

$$l'=-\inf_{u\in\mathcal{U}_2}\int_{\mathfrak{X}}\log\left(\frac{\pi u}{u}\right)(x)\mu(dx)\,.$$

Since $\mathcal{U}_1\subset\mathcal{U}_2$, $l\leqq l'$ and we must show that $l'\leqq l$. Now, for all $u\in\mathcal{U}_1$,

$$\int_{\mathfrak{X}}\log(\pi u)(x)\mu(dx)-\int_{\mathfrak{X}}\log u(x)\mu(dx)\geqq -l\,.$$

Let

$$\mathcal{U}_3=\left\{u\in\mathcal{U}_2:\int_{\mathfrak{X}}\log(\pi u)(x)\mu(dx)-\int_{\mathfrak{X}}\log u(x)\mu(dx)\geqq -l\right\},$$

so that $\mathcal{U}_3\supset\mathcal{U}_1$. Let $\{u_n(x)\}$ be a sequence in $\mathcal{U}_3$ converging for each $x\in\mathfrak{X}$ to a function $u(x)$. It is clear from the bounded convergence theorem that $u(x)\in\mathcal{U}_3$. Hence, $\mathcal{U}_2=\mathcal{U}_3$ and $l'\leqq l$.

LEMMA 2.7. *Assume* $\mu\in\mathcal{M}$ *is such that* $\mu\ll\lambda$ *and* $d\mu/d\lambda=m(x)$ *satisfies* $0<\eta\leqq m(x)\leqq\xi<\infty$ *for all* $x\in\mathfrak{X}$. *Then,*

$$\begin{aligned}I(\mu)&=-\int_{\mathfrak{X}}\log\left(\frac{\pi V}{V}\right)(x)\mu(dx)\\&=\int_{\mathfrak{X}}\log[U(x)V(x)]\mu(dx)\end{aligned}$$

for the $U(x)$ *and* $V(x)$ *determined in Lemma* 2.4.

Proof: By definition,

$$I(\mu)=-\inf_{u\in\mathcal{U}_1}\int_{\mathcal{X}}\log\left(\frac{\pi u}{u}\right)(x)\mu(dx)\,.$$

In Lemma 2.6 we saw that

$$I(\mu)=-\inf_{u\in\mathcal{U}_2}\int_{\mathcal{X}}\log\left(\frac{\pi u}{u}\right)(x)\mu(dx)\,, \tag{2.31}$$

so that the present lemma states that in (2.31) the infimum is in fact attained at V which, by Lemma 2.4 and its corollary, we know to be an element of $\mathcal{U}_2$.

In Lemma 2.4 we defined $\pi'(x, y)=U(x)V(y)\pi(x, y)$, and therefore

$$\begin{aligned}(\pi u)(x)&=\int_{\mathcal{X}}u(y)\pi(x, y)\lambda(dy)\\&=\int_{\mathcal{X}}\frac{\pi(x, y)}{\pi'(x, y)}\pi'(x, y)u(y)\lambda(dy)\\&=\int_{\mathcal{X}}\frac{1}{U(x)V(y)}\pi'(x, y)u(y)\lambda(dy)\\&=\frac{1}{U(x)}\left(\pi'\left(\frac{u}{V}\right)\right)(x)\\&=\frac{1}{U(x)}\int_{\mathcal{X}}\frac{u(y)}{V(y)}\pi'(x, dy)\,.\end{aligned}$$

From this and Lemma 2.6 we have

$$\begin{aligned}I(\mu)&=-\inf_{u\in\mathcal{U}_2}\int_{\mathcal{X}}\log\left(\frac{\pi u}{u}\right)(x)\mu(dx)\\&=-\inf_{u\in\mathcal{U}_2}\int_{\mathcal{X}}\log\left(\frac{\pi'(u/V)}{uU}\right)(x)\mu(dx)\\&=-\inf_{u\in\mathcal{U}_2}\int_{\mathcal{X}}\log\left(\frac{\pi'(u/V)}{(u/V)\cdot VU}\right)(x)\mu(dx)\\&=-\inf_{u\in\mathcal{U}_2}\left[\int_{\mathcal{X}}\log\left(\frac{\pi'(u/V)}{u/V}\right)(x)\mu(dx)-\int_{\mathcal{X}}\log[U(x)V(x)]\mu(dx)\right]\\&=\int_{\mathcal{X}}\log[U(x)V(x)]\mu(dx)-\inf_{u\in\mathcal{U}_2}\int_{\mathcal{X}}\log\left(\frac{\pi'(u/V)}{u/V}\right)(x)\mu(dx)\,.\end{aligned}$$

Now if we can show that

$$\inf_{u\in\mathcal{U}_2}\int_{\mathcal{X}}\log\left(\frac{\pi'(u/V)}{u/V}\right)(x)\mu(dx)=0, \tag{2.32}$$

then indeed

$$I(\mu)=\int_{\mathcal{X}}\log\left[U(x)V(x)\right]\mu(dx)$$

and the lemma is completed by recalling from (2.27) that

$$\left(\frac{\pi V}{V}\right)(x)=\frac{1}{U(x)V(x)}$$

so that

$$-\int_{\mathcal{X}}\log\left(\frac{\pi V}{V}\right)(x)\mu(dx)=\int_{\mathcal{X}}\log\left[U(x)V(x)\right]\mu(dx).$$

To see that (2.32) holds, consider first

$$I'(\mu)=-\inf_{u\in\mathcal{U}_1}\int_{\mathcal{X}}\log\left(\frac{\pi' u}{u}\right)(x)\mu(dx),$$

i.e., the function $I(\mu)$ based on the transition function $\pi'(x, dy)$ instead of $\pi(x, dy)$. From Lemma 2.4 we know that μ is the invariant measure for $\pi'(x, dy)$ so that Lemma 2.5 implies $I'(\mu)=0$. From the argument of Lemma 2.6 we have therefore

$$\inf_{u\in\mathcal{U}_2}\int_{\mathcal{X}}\log\left(\frac{\pi' u}{u}\right)(x)\mu(dx)=0.$$

Since $V\in\mathcal{U}_2$, the infimum is not affected if we replace u by u/V, i.e.,

$$\inf_{u\in\mathcal{U}_2}\int_{\mathcal{X}}\log\left(\frac{\pi'(u/V)}{u/V}\right)(x)\mu(dx)=0$$

which is (2.32).

LEMMA 2.8. *Let* $\mu\in\mathcal{M}$. *Then,*

$$I_\lambda(\mu)-\log A\leqq I(\mu)\leqq I_\lambda(\mu)-\log a, \tag{2.33}$$

so that a necessary and sufficient condition for $I(\mu)<\infty$ *is that* $I_\lambda(\mu)<\infty$.

Proof: For $u\in\mathcal{U}_1$,

$$\int_{\mathcal{X}} \log\left(\frac{\pi u}{u}\right)(x)\mu(dx)=\int_{\mathcal{X}} \log \pi u(x)\mu(dx)-\int_{\mathcal{X}} \log u(x)\mu(dx),$$

where, since $\pi u(x)=\int_X \pi(x, y)u(y)\lambda(dy)$, we have for all $x\in\mathcal{X}$

$$a\int_{\mathcal{X}} u(y)\lambda(dy)\leqq \pi u(x)\leqq A\int_{\mathcal{X}} u(y)\lambda(dy).$$

Thus,

$$\log a+\log\int_{\mathcal{X}} u(y)\lambda(dy)-\int_{\mathcal{X}} \log u(x)\mu(dx)$$

$$\leqq\int_{\mathcal{X}} \log\left(\frac{\pi u}{u}\right)(x)\mu(dx)$$

$$\leqq\log A+\log\int_{\mathcal{X}} u(y)\lambda(dy)-\int_{\mathcal{X}} \log u(x)\mu(dx).$$

Taking infimum over all $u\in\mathcal{U}_1$ we get (2.33).

COROLLARY. *Let* $\mu\in\mathcal{M}$. *Then* $I(\mu)<\infty$ *if and only if* $\mu\ll\lambda$ *and* $d\mu/d\lambda=m(x)$ *is such that*

$$\int_{\mathcal{X}} m(x)\,|\log m(x)|\,\lambda(dx)<\infty.$$

Proof: The corollary follows immediately from Lemma 2.1 and Lemma 2.8.

LEMMA 2.9. *Let* G *be an open set in* $\mathcal{M}$ *and let* $\mathcal{M}_0\subset\mathcal{M}$ *be the set of probability measures* μ *on* $\mathcal{X}$ *such that* $d\mu/d\lambda=m(x)$ *exists and satisfies* $0<\eta\leqq m(x)\leqq\xi<\infty$ (ξ *and* η *depend on* μ). *Then,*

$$\inf_{\mu\in G} I(\mu)=\inf_{\mu\in G\cap\mathcal{M}_0} I(\mu).$$

Proof: It will suffice to show that for any $\mu\in\mathcal{M}$ such that $I(\mu)<\infty$ there exists a sequence $\{\mu_n\}\in\mathcal{M}_0$ with

$$\mu_n\Rightarrow\mu,$$

$$\lim_{n\to\infty} I(\mu_n)=I(\mu).$$

So let $\mu \in \mathcal{M}$ be such that $I(\mu)<\infty$, and let $\mathcal{M}_1=\{\mu \in \mathcal{M} : \mu \ll \lambda$ and $d\mu/d\lambda=m(x)$ satisfies $0<\eta \leqq m(x) \leqq \infty\}$. We shall show first that there exists a sequence $\{\mu_n\} \in \mathcal{M}_1$ such that $\mu_n \Rightarrow \mu$ and $\lim_{n\to\infty} I(\mu_n)=I(\mu)$.

For any positive integer n, let $\mu_n=(1-1/n)\mu+(1/n)\lambda$. Since $I(\mu)<\infty$, we know from the corollary to Lemma 2.8 that $\mu \ll \lambda$ and therefore $\mu_n \ll \lambda$. Moreover, $d\mu_n/d\lambda \geqq 1/n$ so that the sequence $\{\mu_n\} \in \mathcal{M}_1$. Clearly, $\mu_n \Rightarrow \mu$ and since I is lower semicontinuous we have

$$\varliminf_{n\to\infty} I(\mu_n) \geqq I(\mu) .$$

But λ itself is in $\mathcal{M}_0$ so again by the corollary to Lemma 2.8 we have $I(\lambda)<\infty$. Therefore, since due to the convexity of I

$$I(\mu_n) \leqq \left(1-\frac{1}{n}\right) I(\mu)+\frac{1}{n} I(\lambda) ,$$

we have

$$\varlimsup_{n\to\infty} I(\mu_n) \leqq I(\mu)$$

so that indeed $\lim_{n\to\infty} I(\mu_n)=I(\mu)$.

Next we take a $\mu \in \mathcal{M}_1$ for which $I(\mu)<\infty$ and show there exists a sequence $\{\mu_n\} \in \mathcal{M}_0$ such that $\mu_n \Rightarrow \mu$ and $\lim_{n\to\infty} I(\mu_n)=I(\mu)$. Since $\mu \in \mathcal{M}_1$ and $I(\mu)<\infty$, we know from the corollary to Lemma 2.8 that $d\mu/d\lambda=m(x)$ satisfies

$$\int_{\mathfrak{X}} m(x)\,|\log m(x)|\,\lambda(dx)<\infty . \tag{2.34}$$

Since $\mu \in \mathcal{M}_1$, there exists an $\eta>0$ such that $\eta \leqq m(x)$ for all $x \in \mathfrak{X}$. Define

$$m_n(x)=\frac{m(x) \wedge n}{\int_{\mathfrak{X}} [m(x) \wedge n] \lambda(dx)} ,$$

where n is chosen greater than or equal to η and so large that

$$\int_{\mathfrak{X}} [m(x) \wedge n] \lambda(dx) \geqq \tfrac{1}{2} .$$

Thus $\eta \leqq m_n(x) \leqq 2n$; hence defining $\mu_n(dx)=m_n(x)\lambda(dx)$ we see that $\mu_n \Rightarrow \mu$ and each $\mu_n \in \mathcal{M}_0$. By lower semicontinuity,

$$\varliminf_{n\to\infty} I(\mu_n) \geqq I(\mu) ;$$

consequently it remains to show that

$$\overline{\lim_{n\to\infty}} I(\mu_n) \leqq I(\mu) . \tag{2.35}$$

Since $\mu_n \in \mathcal{M}_0$, it follows from Lemma 2.7 that there exists a $V_n \in \mathcal{U}_2$ (the existence of the V_n is guaranteed by Lemma 2.4) such that

$$\begin{aligned} I(\mu_n) &= -\int_{\mathcal{X}} \log\left(\frac{\pi V_n}{V_n}\right)(x)\mu_n(dx) \\ &= -\int_{\mathcal{X}} \log\left(\frac{\pi V_n}{V_n}\right)(x) m_n(x)\lambda(dx) . \end{aligned}$$

On the other hand, from the definition of $I(\mu)$ and from Lemma 2.6 we know that

$$\begin{aligned} I(\mu) &\geqq -\int_{\mathcal{X}} \log\left(\frac{\pi V_n}{V_n}\right)(x)\mu(dx) \\ &= -\int_{\mathcal{X}} \log\left(\frac{\pi V_n}{V_n}\right)(x) m(x)\lambda(dx) . \end{aligned}$$

Thus

$$\begin{aligned} I(\mu) &\geqq I(\mu_n) + \int_{\mathcal{X}} \left[\log\left(\frac{\pi V_n}{V_n}\right)(x)\right][m_n(x) - m(x)]\lambda(dx) \\ &\geqq I(\mu_n) - \int_{\mathcal{X}} \left|\log\left(\frac{\pi V_n}{V_n}\right)(x)\right| |m_n(x) - m(x)|\, \lambda(dx) , \end{aligned}$$

and (2.35) will follow if we show that

$$\lim_{n\to\infty} \int_{\mathcal{X}} \left|\log\left(\frac{\pi V_n}{V_n}\right)(x)\right| |m_n(x) - m(x)|\, \lambda(dx) = 0 . \tag{2.36}$$

The corollary to Lemma 2.4 and its proof yield

$$\begin{aligned} \frac{\sup_{x\in\mathcal{X}} V(x)}{\inf_{x\in\mathcal{X}} V(x)} &\leqq \frac{\xi}{\left(\inf_{x\in\mathcal{X}} U(x)\right) a\eta} \times \frac{\left(\sup_{x\in\mathcal{X}} U(x)\right)\xi A}{\eta} \\ &\leqq \frac{\xi^2 A^2}{a^2 \eta^2} . \end{aligned}$$

Therefore, applying this last to V_n (where the corresponding $\xi = 2n$), we have

$$\left|\log\left(\frac{\pi V_n}{V_n}\right)(x)\right| \leqq \log\frac{4n^2A^2}{a^2\eta^2} = 2\log n + c\,.$$

Hence

$$\int_{\mathscr{X}} \left|\log\left(\frac{\pi V_n}{V_n}\right)(x)\right| \, |m_n(x)-m(x)|\,\lambda(dx) \tag{2.37}$$
$$\leqq [2\log n + c]\int_{\mathscr{X}} |m_n(x)-m(x)|\,\lambda(dx)\,.$$

But, remembering that $\int_{\mathscr{X}} [m(x)\wedge n]\lambda(dx) \geqq \frac{1}{2}$, we obtain

$$\int_{\mathscr{X}} |m_n(x)-m(x)|\,\lambda(dx) = \int_{\mathscr{X}} \left|\frac{[m(x)\wedge n]}{\int_{\mathscr{X}}[m(x)\wedge n]\lambda(dx)} - m(x)\right| \lambda(dx)$$

$$\leqq \int_{\mathscr{X}} \left|\frac{[m(x)\wedge n]}{\int_{\mathscr{X}}[m(x)\wedge n]\lambda(dx)} - [m(x)\wedge n]\right| \lambda(dx) + \int_{\mathscr{X}} |[m(x)\wedge n]-m(x)|\,\lambda(dx)$$

$$\leqq \int_{\mathscr{X}} [m(x)\wedge n]\left|\frac{1}{\int_{\mathscr{X}}[m(x)\wedge n]\lambda(dx)} - 1\right| \lambda(dx) + \int_{\mathscr{X}} |[m(x)\wedge n]-m(x)|\,\lambda(dx)$$

$$\leqq \frac{1-\int_{\mathscr{X}}[m(x)\wedge n]\lambda(dx)}{\int_{\mathscr{X}}[m(x)\wedge n]\lambda(dx)} + \int_{\mathscr{X}} |[m(x)\wedge n]-m(x)|\,\lambda(dx)$$

$$\leqq 2\int_{\mathscr{X}} (m(x)-[m(x)\wedge n])\lambda(dx) + \int_{\mathscr{X}} |[m(x)\wedge n]-m(x)|\,\lambda(dx)$$

$$\leqq 3\int_{\mathscr{X}} |m(x)-[m(x)\wedge n]|\,\lambda(dx)$$

$$\leqq 3\int_{m(x)\geqq n} m(x)\lambda(dx) \leqq \frac{3}{\log n}\int_{m(x)\geqq n} m(x)\,|\log m(x)|\,\lambda(dx)$$

$$= \frac{3}{\log n}\,o(1)\,,$$

because of (2.34). Thus from (2.37)

$$\int_{\mathfrak{X}} \left|\log\left(\frac{\pi V_n}{V_n}\right)(x)\right| |m_n(x)-m(x)|\,\lambda(dx) \leqq [2\log n + c]\left(\frac{3}{\log n}\,o(1)\right) \to 0$$

as $n\to\infty$ which gives (2.36) and hence (2.35) and the lemma.

Now we are ready to return to the proof of (1.8). Let $\mu \in \mathcal{M}_0$, i.e., $\mu \ll \lambda$ and $d\mu/d\lambda = m(x)$ satisfies the condition $0<\eta\leqq m(x)\leqq\xi<\infty$ for all $x\in\mathfrak{X}$. For any $\varepsilon>0$, let $S(\mu,\varepsilon)$ be the sphere with center at μ of radius ε in the weak topology on $\mathcal{M}$. Consider

$$Q_{n,x}[S(\mu,\varepsilon)] = P_x\{L_n(\omega,\cdot)\in S(\mu,\varepsilon)\}.$$

Let

$$E_{n,\mu,\varepsilon} = \{\omega : L_n(\omega,\cdot)\in S(\mu,\varepsilon)\},$$

and, for any $\varepsilon'>0$, let

$$F_{n,\mu,\varepsilon'} = \left\{\omega : \left|\frac{W(X_0)+W(X_1)+\cdots+W(X_{n-1})}{n} - \int_{\mathfrak{X}} W(y)\mu(dy)\right| < \varepsilon'\right\},$$

where $W(y)=\log V(y)+\log U(y)$ with the functions U and V as determined in Lemma 2.4. We note from Lemma 2.7 that

$$\int_{\mathfrak{X}} W(y)\mu(dy) = I(\mu).$$

Now

$$\begin{aligned} Q_{n,x}[S(\mu,\varepsilon)] &= \int_{E_{n,\mu,\varepsilon}} dP_x \\ &= \int_{E_{n,\mu,\varepsilon}} \pi(x,x_1)\pi(x_1,x_2)\cdots\pi(x_{n-2},x_{n-1})\lambda(dx_1)\cdots\lambda(dx_{n-1}) \\ &= \int_{E_{n,\mu,\varepsilon}} \pi'(x,x_1)\pi'(x_1,x_2)\cdots\pi'(x_{n-2},x_{n-1}) \\ &\quad\times\left[\frac{\pi(x,x_1)\cdots\pi(x_{n-2},x_{n-1})}{\pi'(x,x_1)\cdots\pi'(x_{n-2},x_{n-1})}\right]\lambda(dx_1)\cdots\lambda(dx_{n-1}). \end{aligned}$$

From (2.28)

$$\frac{\pi(x,y)}{\pi'(x,y)} = \frac{1}{U(x)V(y)},$$

and, therefore,

$$Q_{n,x}[S(\mu,\varepsilon)]=\int_{E_{n,\mu,\varepsilon}} \frac{1}{U(x)V(x_1)}\frac{1}{U(x_1)V(x_2)}\cdots\frac{1}{U(x_{n-2})V(x_{n-1})}$$
$$\times\pi'(x,x_1)\cdots\pi'(x_{n-2},x_{n-1})\lambda(dx_1)\cdots\lambda(dx_{n-1})$$
$$=\int_{E_{n,\mu,\varepsilon}} \frac{1}{U(x)V(x_1)}\frac{1}{U(x_1)V(x_2)}\cdots\frac{1}{U(x_{n-2})V(x_{n-1})}\,dP'_x\,,$$

where P'_x is the probability measure on Ω_X induced by the transition probability $\pi'(x, dy)$. Hence

$$Q_{n,x}[S(\mu,\varepsilon)]=\int_{E_{n,\mu,\varepsilon}} V(x)U(x_{n-1})\exp\{-[W(x_0)+W(x_1)+\cdots+W(x_{n-1})]\}\,dP'_x$$
$$=\int_{E_{n,\mu,\varepsilon}} V(x)U(x_{n-1})\exp\{-n[(W(x_0)+W(x_1)+\cdots$$
$$+W(x_{n-1}))/n]\}\,dP'_x$$
$$\geqq\int_{E_{n,\mu,\varepsilon}\cap F_{n,\mu,\varepsilon'}} V(x)U(x_{n-1})\exp\{-n[(W(x_0)+\cdots+W(x_{n-1}))/n]\}\,dP'_x\,.$$

Since, on $F_{n,\mu,\varepsilon'}$,

$$\left|\frac{W(x_0)+W(x_1)+\cdots+W(x_{n-1})}{n}-I(\mu)\right|<\varepsilon'\,,$$

we have

$$Q_{n,x}[S(\mu,\varepsilon)]\geqq\exp\{-n[I(\mu)+\varepsilon']\}\int_{E_{n,\mu,\varepsilon}\cap F_{n,\mu,\varepsilon'}} V(x)U(x_{n-1})\,dP'_x\,.$$

But both U and V are bounded below so that we get, for some constant C,

$$Q_{n,x}[S(\mu,\varepsilon)]\geqq C\exp\{-n[I(\mu)+\varepsilon']\}P'_x\{E_{n,\mu,\varepsilon}\cap F_{n,\mu,\varepsilon'}\}\,. \tag{2.38}$$

In Lemma 2.4 we showed that μ is the invariant measure for π', and therefore by the ergodic theorem

$$\lim_{n\to\infty} P'_x\{E_{n,\mu,\varepsilon}\}=1 \quad\text{for all}\quad \varepsilon>0\,,$$

$$\lim_{n\to\infty} P'_x\{F_{n,\mu,\varepsilon'}\}=1 \quad\text{for all}\quad \varepsilon'>0\,.$$

Thus from (2.38) we have

$$\underline{\lim}_{n\to\infty} \frac{1}{n} \log Q_{n,x}[S(\mu, \varepsilon)] \geqq -I(\mu) - \varepsilon' ,$$

and, since $\varepsilon'>0$ was arbitrary, we obtain on letting $\varepsilon'\to 0$

$$\underline{\lim}_{n\to\infty} \frac{1}{n} \log Q_{n,x}[S(\mu, \varepsilon)] \geqq -I(\mu) . \tag{2.39}$$

Now let $\mu\in G\cap\mathcal{M}_0$, where G is an open set in $\mathcal{M}$. Since G is open, there exists an $\varepsilon>0$ such that the sphere $S(\mu,\varepsilon)\subset G$. Thus, from (2.39),

$$\underline{\lim}_{n\to\infty} \frac{1}{n} \log Q_{n,x}(G) \geqq \underline{\lim}_{n\to\infty} \frac{1}{n} \log Q_{n,x}[S(\mu, \varepsilon)] \geqq -I(\mu) .$$

This last is true for every $\mu\in G\cap\mathcal{M}_0$ and therefore

$$\underline{\lim}_{n\to\infty} \frac{1}{n} \log Q_{n,x}(G) \geqq -\inf_{\mu\in G\cap\mathcal{M}_0} I(\mu) .$$

But in Lemma 2.9 we showed that

$$\inf_{\mu\in G\cap\mathcal{M}_0} I(\mu) = \inf_{\mu\in G} I(\mu) ,$$

and thus we obtain

$$\underline{\lim}_{n\to\infty} \frac{1}{n} \log Q_{n,x}(G) \geqq -\inf_{\mu\in G} I(\mu)$$

which is (1.8). This completes the proof of Theorem 1.

3. Continuous Parameter Markov Process

In this section we shall prove Theorem 3. The notation and definitions will be those of the introduction just prior to the statement of Theorem 3. We recall that $\mathcal{M}$, the space of probability measures on $\mathcal{X}$, was given the weak topology. This topology on $\mathcal{M}$ is induced by a metric, the Prokhorov metric, which we now recall:

Let $\rho(x, y)$ be the metric on $\mathcal{X}$. For any closed set $C\subset\mathcal{X}$ and any $\delta>0$, let

$$C^{\delta} = \{y\in\mathcal{X}, \rho(x, y)<\delta \quad \text{for} \quad \text{some} \quad x\in C\} .$$

Given two measures μ_1 and μ_2 in $\mathcal{M}$ the Prokhorov metric $d(\mu_1, \mu_2)$ is defined by

$$d(\mu_1, \mu_2)=\inf_{\delta>0}\{\mu_1(c)\leqq\mu_2(c^\delta)+\delta,\ \mu_2(c)\leqq\mu_1(c^\delta)+\delta \quad \text{for all closed sets } C\subset \mathfrak{X}\}.$$

$\mathcal{M}$ then becomes a compact metric space.

In addition to the definition of $I(\mu)$ given in (1.12) we find it useful to define, for any $h>0$ and any $\mu\in\mathcal{M}$,

$$\begin{aligned} I_h(\mu)&=-\inf_{\substack{u>0\\ u\in C(\mathfrak{X})}}\int_{\mathfrak{X}}\log\left(\frac{T_h u}{u}\right)(x)\mu(dx) \\ &=-\inf_{\substack{u>0\\ u\in\mathcal{D}}}\int_{\mathfrak{X}}\log\left(\frac{T_h u}{u}\right)(x)\mu(dx). \end{aligned} \tag{3.1}$$

We shall first prove (1.14) which will require four lemmas.

LEMMA 3.1. *For any $h>0$ and $\mu\in\mathcal{M}$,*

$$I_h(\mu)\leqq hI(\mu). \tag{3.2}$$

Moreover, for every $\mu\in\mathcal{M}$,

$$\lim_{h\to 0}\frac{1}{h}I_h(\mu)=I(\mu). \tag{3.3}$$

Proof: Let $u>0$ and $u\in\mathcal{D}$. Consider, for any $\mu\in\mathcal{M}$ and $0\leqq t<\infty$,

$$\phi(t)=\int_{\mathfrak{X}}\log\left(\frac{T_t u}{u}\right)(x)\mu(dx).$$

Since T_t maps positive elements of $\mathcal{D}$ into positive elements of $\mathcal{D}$, and since $dT_tu/dt=LT_tu$, we have

$$\frac{d\phi(t)}{dt}=\int_{\mathfrak{X}}\left(\frac{LT_tu}{T_tu}\right)(x)\mu(dx)\geqq -I(\mu).$$

Now $T_0u=u$, $\phi(0)=0$, and hence for any $h>0$

$$\phi(h)=\int_0^h\frac{d\phi(s)}{ds}\,ds\geqq-\int_0^h I(\mu)\,ds=-hI(\mu).$$

Thus

$$\int_{\mathscr{X}} \log\left(\frac{T_h u}{u}\right)(x)\mu(dx) \geqq -hI(\mu)$$

holds for any $h>0$ and any positive element u of $\mathscr{D}$ which implies (3.2).

Let $\mu\in\mathscr{M}$ and consider, for any positive element u of $\mathscr{D}$,

$$\frac{1}{h}I_h(\mu)\geqq -\frac{1}{h}\int_{\mathscr{X}} \log\left(\frac{T_h u}{u}\right)(x)\mu(dx)\,. \tag{3.4}$$

Since $u\in\mathscr{D}$, $T_h u=u+hLu+o(h)$, where $o(h)$ is uniform in x. Since $\mathscr{X}$ is compact, u is actually bounded away from zero, and so

$$\frac{T_h u}{u}=1+h\frac{Lu}{u}+o(h)\,,$$

$$\log\left(\frac{T_h u}{u}\right)=\log\left(1+h\frac{Lu}{u}+o(h)\right)$$

$$=h\frac{Lu}{u}+o(h)\,,$$

from which we obtain

$$\frac{1}{h}\log\left(\frac{T_h u}{u}\right)=\frac{Lu}{u}+o(1)\,,$$

where $o(1)$ is uniform in x. Thus, from (3.4),

$$\varliminf_{h\to 0}\frac{1}{h}I_h(\mu)\geqq -\int_{\mathscr{X}}\left(\frac{Lu}{u}\right)(x)\mu(dx)\,, \tag{3.5}$$

and since (3.5) is true for every positive element of $\mathscr{D}$, we have

$$\varliminf_{h\to 0}\frac{1}{h}I_h(\mu)\geqq -\inf_{\substack{u>0\\ u\in\mathscr{D}}}\int_{\mathscr{X}}\left(\frac{Lu}{u}\right)(x)\mu(dx)=I(\mu)\,.$$

But (3.2) implies

$$\varlimsup_{h\to 0}\frac{1}{h}I_h(\mu)\leqq I(\mu)\,,$$

and hence we have (3.3).

LEMMA 3.2. *Let (Ω, Σ, P) be any probability space and let ξ and η be two maps of Ω into $\mathfrak{X}$. If we define $\alpha = P\xi^{-1}$, $\beta = P\eta^{-1}$ (so that α and β are in $\mathcal{M}$), then for any $\delta > 0$*

$$P\{\omega : \rho(\xi, \eta) \geqq \delta\} \leqq \delta \quad \textit{implies} \quad d(\alpha, \beta) \leqq \delta .$$

Proof: Let C be any closed set in $\mathfrak{X}$. By the definition of C^{δ} and the assumption that $P\{\omega : \rho(\xi, \eta) \geqq \delta\} \leqq \delta$, we have

$$\begin{aligned} \alpha(C) &= P\{\omega : \xi(\omega) \in C\} \\ &\leqq P\{\omega : \eta(\omega) \in C^{\delta}\} + P\{\omega : \rho(\xi, \eta) \geqq \delta\} \\ &\leqq \beta(C^{\delta}) + \delta . \end{aligned}$$

By a symmetric argument, $\beta(C) \leqq \alpha(C^{\delta}) + \delta$. Since these two inequalities hold for any closed set C, the definition of the Prokhorov metric implies that $d(\alpha, \beta) \leqq \delta$, which is the lemma.

For any $h > 0$, let $X_j^{(h)} = x(jh)$, $j = 0, 1, 2, \cdots$, so that $X_0^{(h)}, X_1^{(h)}, \cdots$ is a discrete parameter Markov process whose transition probability function is $\pi^{(h)}(x, dy) = p(h, x, dy)$. Note that, for a positive element u of $C(\mathfrak{X})$,

$$T_h u = \pi^{(h)} u = \int_{\mathfrak{X}} u(y) p(h, x, y) \lambda(dy) ,$$

and therefore the function $I_h(\mu)$ defined in (3.1) for any $\mu \in \mathcal{M}$ is just the $I(\mu)$ of Section 2 (as defined by (1.5)) for the discrete parameter Markov process $X_0^{(h)}, X_1^{(h)}, \cdots$.

For any positive integer n and any Borel set $A \subset \mathfrak{X}$, let

$$L_n^{(h)}(\omega, A) = \frac{1}{n} \sum_{j=0}^{n-1} \chi_A(X_j^{(h)}) .$$

Now, let $t > 0$ be given and write $t = nh$, where n is a positive integer and $h > 0$. We define the process $\{x^h(s), 0 \leqq s \leqq t\}$ as follows:

$$x^h(s) = \begin{cases} X_j^{(h)} & \text{if } \ jh \leqq s < (j+1)h , \\ X_{n-1}^{(h)} & \text{if } \ s = t = nh . \end{cases} \qquad j = 0, 1, 2, \cdots, n-1 ,$$

LEMMA 3.3. *Let $t > 0$ be given and write $t = nh$, where n is a positive integer and $h > 0$. For any $\delta > 0$ and any $\omega \in \Omega_x$, let*

$$N_t^{\delta}(\omega) = \left\{ \#j\text{'s such that} \sup_{jh \leqq s \leqq (j+1)h} \rho(X_j^{(h)}, x(s)) > \delta, \ 0 \leqq j \leqq n-1 \right\} .$$

Then

$$P_x\{d(L_t(\omega,\cdot), L_n^{(h)}(\omega,\cdot))>\delta\}\leqq P_x\left\{\frac{N_t^\delta(\omega)}{n}>\delta\right\}. \tag{3.6}$$

Proof: Both $\{x(s), 0\leqq s\leqq t\}$ and $\{x^h(s), 0\leqq s\leqq t\}$ map $[0,t]$ into X. On $[0,t]$ impose the measure $\nu=ds/t$ so that

$$L_t(\omega,\cdot)=\nu[X(\cdot)]^{-1},$$
$$L_n^{(h)}(\omega,\cdot)=\nu[X^h(\cdot)]^{-1}.$$

Then, by Lemma 3.2, for any $\omega\in\Omega_x$, $\nu\{s:\rho(x(s),x^h(s))\geqq\delta\}\leqq\delta$ implies $d(L_t(\omega,\cdot), L_n^{(h)}(\omega,\cdot))\leqq\delta$.

But, for every $\omega\in\Omega_x$,

$$\nu\{s:\rho(x(s),x^h(s))\geqq\delta\}\leqq\frac{h}{t}N_t^\delta(\omega)=\frac{1}{n}N_t^\delta(\omega),$$

and hence

$$P_x\{d(L_t(\omega,\cdot), L_n^{(h)}(\omega,\cdot))>\delta\}\leqq P_x\{\nu\{s:\rho(x(s),x^h(s))\geqq\delta\}>\delta\}\leqq P_x\left\{\frac{N_t^\delta(\omega)}{n}>\delta\right\}.$$

LEMMA 3.4. *For every* $\varepsilon>0$,

$$\overline{\lim_{h\to0}}\,\overline{\lim_{n\to\infty}}\frac{1}{nh}\log P_x\{d(L_{nh}(\omega,\cdot), L_n^{(h)}(\omega,\cdot))>\varepsilon\}=-\infty. \tag{3.7}$$

Proof: By Lemma 3.3 it suffices to show that

$$\overline{\lim_{h\to0}}\,\overline{\lim_{n\to\infty}}\frac{1}{nh}\log P_x\left\{\frac{N_{nh}^\varepsilon(\omega)}{n}>\varepsilon\right\}=-\infty \tag{3.8}$$

for every $\varepsilon>0$. Let, for $h>0$ and $\varepsilon>0$,

$$\phi^\varepsilon(h)=\sup_{x\in X}P_x\left\{\sup_{0\leqq s\leqq h}\rho(x(s),x)\geqq\varepsilon\right\}.$$

Our hypotheses on the process $\{X(s), 0\leqq s<\infty\}$ imply that, for every $\varepsilon>0$, $\phi^\varepsilon(h)\to0$ as $h\to0$.

For $j=0,1,2,\cdots,n-1$, let

$$\chi_j^\varepsilon(\omega)=\begin{cases}1 & \text{if } \sup_{jh\leqq s<(j+1)h}\rho(x(s),x_j^{(h)})\geqq\varepsilon,\\ 0 & \text{otherwise}.\end{cases}$$

For $\theta>0$, consider

$$E^{P_x}\{\exp\{\theta N^{\varepsilon}_{nh}(\omega)\}\}=E^{P_x}\left\{\exp\left\{\theta\sum_{j=0}^{n-1}\chi^{\varepsilon}_j(\omega)\right\}\right\}.$$

Since the process is Markovian,

$$\begin{aligned}E^{P_x}\left\{\exp\left\{\theta\sum_{j=0}^{n-1}\chi^{\varepsilon}_j(\omega)\right\}\right\}&=E^{P_x}\left\{\exp\left\{\theta\sum_{j=0}^{n-2}\chi^{\varepsilon}_j(\omega)\right\}[E^{P_{x((n-2)h)}}\{\exp\{\theta\chi^{\varepsilon}_{n-1}(\omega)\}\}]\right\}\\&\leqq E^{P_x}\left\{\exp\left\{\theta\sum_{j=0}^{n-2}\chi^{\varepsilon}_j(\omega)\right\}\sup_{y\in\mathcal{X}}E^{P_y}\{\exp\{\theta\chi^{\varepsilon}_{n-1}(\omega)\}\}\right\}\\&\leqq[e^{\theta}\phi^{\varepsilon}(h)+(1-\phi^{\varepsilon}(h))]E^{P_x}\left\{\exp\left\{\theta\sum_{j=0}^{n-2}\chi^{\varepsilon}_j(\omega)\right\}\right\}\\&\leqq[e^{\theta}\phi^{\varepsilon}(h)+(1-\phi^{\varepsilon}(h))]^n.\end{aligned}$$

Thus,

$$E^{P_x}\{\exp\{\theta N^{\varepsilon}_{nh}(\omega)\}\}\leqq[e^{\theta}\phi^{\varepsilon}(h)+(1-\phi^{\varepsilon}(h))]^n.$$

On the other hand,

$$\begin{aligned}E^{P_x}\{e^{\theta N^{\varepsilon}_{nh}(\omega)}\}&\geqq E^{P_x}_{N^{\varepsilon}_{nh}(\omega)>\varepsilon n}\{e^{\theta N^{\varepsilon}_{nh}(\omega)}\}\\&\geqq e^{n\theta\varepsilon}P_x\left\{\frac{N^{\varepsilon}_{nh}(\omega)}{n}>\varepsilon\right\},\end{aligned}$$

so that

$$P_x\left\{\frac{N^{\varepsilon}_{nh}(\omega)}{n}>\varepsilon\right\}\leqq e^{-n\varepsilon\theta}[e^{\theta}\phi^{\varepsilon}(h)+(1-\phi^{\varepsilon}(h))]^n.$$

Therefore,

$$\overline{\lim_{n\to\infty}}\frac{1}{nh}\log P_x\left\{\frac{N^{\varepsilon}_{nh}(\omega)}{n}>\varepsilon\right\}\leqq-\frac{\varepsilon\theta}{h}+\frac{1}{h}\log[e^{\theta}\phi^{\varepsilon}(h)+(1-\phi^{\varepsilon}(h))]. \tag{3.9}$$

As noted earlier, $\phi^{\varepsilon}(h)\to 0$ as $h\to 0$, and hence the right side of (3.9) behaves as follows as $h\to 0$:

$$\frac{1}{h}\{\log[e^{\theta}\phi^{\varepsilon}(h)+(1-\phi^{\varepsilon}(h))]-\varepsilon\theta\}\sim-\frac{\varepsilon\theta}{h}.$$

This gives us (3.8) and the lemma. We shall find it convenient to use the obvious

COROLLARY. *Assume* $\mu\in\mathcal{M}$ *is such that* $I(\mu)<\infty$. *There exists an* h_0 *such that for given* $\varepsilon>0$

$$\overline{\lim_{n\to\infty}}\frac{1}{nh_0}\log P_x\{d(L_{nh_0}(\omega,\cdot),L_n^{(h_0)}(\omega,\cdot))>\varepsilon\}<-I(\mu)-1\,. \tag{3.10}$$

Now we are ready to prove (1.14). Let G be an open set in $\mathcal{M}$. Let $\mu\in G$ be such that $I(\mu)<\infty$, and let $S(\mu,\varepsilon)$ be a sphere with center at μ and radius ε such that $S(\mu,\varepsilon)\subset G$. It will suffice to show

$$\underline{\lim_{t\to\infty}}\frac{1}{t}\log Q_{x,t}[S(\mu,\varepsilon)]=\underline{\lim_{t\to\infty}}\frac{1}{t}\log P_x\{L_t(\omega,\cdot)\in S(\mu,\varepsilon)\}\geqq -I(\mu)\,. \tag{3.11}$$

Let h_0 be as in the corollary to Lemma 3.4 so that (3.10) holds. For any $\varepsilon>0$, it is clear that

$$\begin{aligned}&P_x\{L_{nh_0}(\omega,\cdot)\in S(\mu,\tfrac{1}{2}\varepsilon)\}\\ &\qquad\geqq P_x\{L_n^{(h_0)}(\omega,\cdot)\in S(\mu,\tfrac{1}{4}\varepsilon)\}-P_x\{d(L_{nh_0}(\omega,\cdot),L_n^{(h_0)}(\omega,\cdot))\geqq\tfrac{1}{4}\varepsilon\}\end{aligned} \tag{3.12}$$

for all positive integers n. By (2.39) and Lemma 3.1,

$$\underline{\lim_{n\to\infty}}\frac{1}{nh_0}\log P_x\{L_n^{(h_0)}(\omega,\cdot)\in S(\mu,\tfrac{1}{4}\varepsilon)\}\geqq-\frac{1}{h_0}I_{h_0}(\mu)\geqq-I(\mu)\,. \tag{3.13}$$

From (3.13) and (3.10) we see that (3.12) implies that, for any $\varepsilon>0$,

$$\underline{\lim_{n\to\infty}}\frac{1}{nh_0}\log P_x\{L_{nh_0}(\omega,\cdot)\in S(\mu,\tfrac{1}{2}\varepsilon)\}\geqq-I(\mu)\,. \tag{3.14}$$

Now, let $n=[t/h_0]$. For any Borel set $A\subset X$ and any $\omega\in\Omega_x$,

$$\begin{aligned}|L_t(\omega,A)-L_{nh_0}(\omega,A)|&=\left|\frac{1}{t}\int_0^t\chi_A(x(s))\,ds-\frac{1}{nh_0}\int_0^{nh_0}\chi_A(x(s))\,ds\right|\\ &=\left|\frac{1}{nh_0}\int_0^t\chi_A(x(s))\,ds+\left(\frac{1}{t}-\frac{1}{nh_0}\right)\int_0^t\chi_A(x(s))\,ds\right.\\ &\qquad\left.-\frac{1}{nh_0}\int_0^{nh_0}\chi_A(x(s))\,ds\right|\\ &=\left|\frac{1}{nh_0}\int_{nh_0}^t\chi_A(x(s))\,ds+\frac{nh_0-t}{nh_0t}\int_0^t\chi_A(x(s))\,ds\right|\\ &\leqq\frac{2(t-nh_0)}{nh_0}\leqq\frac{2}{n}=\frac{2}{[t/h_0]}\,.\end{aligned}$$

Thus, there exists a t_0 such that if $t>t_0$ we have, for all $\omega\in\Omega_x$,

$$d(L_t(\omega,\cdot),L_{nh_0}(\omega,\cdot))<\tfrac{1}{2}\varepsilon\,.$$

Hence since, for any $\varepsilon>0$,

$$P_x\{L_t(\omega,\cdot)\in S(\mu,\varepsilon)\}\geqq P_x\{L_{nh_0}(\omega,\cdot)\in S(\mu,\tfrac{1}{2}\varepsilon)\}-P_x\{d(L_t(\omega,\cdot),L_{nh_0}(\omega,\cdot))\geqq\tfrac{1}{2}\varepsilon\}\,,$$

we see that if $t>t_0$ and $n=[t/h_0]$

$$P_x\{L_t(\omega,\cdot)\in S(\mu,\varepsilon)\}\geqq P_x\{L_{nh_0}(\omega,\cdot)\in S(\mu,\tfrac{1}{2}\varepsilon)\}$$

or

$$\frac{1}{t}\log P_x\{L_t(\omega,\cdot)\in S(\mu,\varepsilon)\}\geqq\frac{nh_0}{t}\cdot\frac{1}{nh_0}\log P_x\{L_{nh_0}(\omega,\cdot)\in S(\mu,\tfrac{1}{2}\varepsilon)\}\,.$$

Therefore, from (3.14) and the fact that $nh_0/t\to 1$ for $n=[t/h_0]$ as $t\to\infty$, we have for all $\varepsilon>0$

$$\varliminf_{t\to\infty}\frac{1}{t}\log P_x\{L_t(\omega,\cdot)\in S(\mu,\varepsilon)\}\geqq -I(\mu)$$

which is (3.11) and hence (1.14).

Now we prove (1.13). For $u>0$ and $u\in\mathcal{D}$, let

$$\psi(x,t)=E^{P_x}\left\{u(x(t))\exp\left\{-\int_0^t(Lu/u)(x(s))\,ds\right\}\right\}$$

for $x\in\mathcal{X}$ and $t>0$. From the Feynmann-Kac formula we know that $\psi(x,t)$ satisfies

$$\frac{\partial\psi}{\partial t}=L\psi-\left(\frac{Lu}{u}\right)(x)\psi\,,$$

$$\psi(x,0)=u(x)\,,$$

but since the solution of this last is clearly $\psi(x,t)=u(x)$, we have

$$E^{P_x}\left\{u(x(t))\exp\left\{-\int_0^t(Lu/u)(x(s))\,ds\right\}\right\}=u(x)\,. \tag{3.15}$$

Since $u>0$ and $\mathcal{X}$ is compact, there is a constant α such that $u(y)\geqq\alpha>0$ for all $y\in\mathcal{X}$. Thus

$$E^{P_x}\left\{\exp\left\{-\int_0^t(Lu/u)(x(s))\,ds\right\}\right\}\leqq\frac{u(x)}{\alpha}\,. \tag{3.16}$$

From the definition of $L_t(\omega, \cdot)$ and of $Q_{x,t}$-measure on $\mathcal{M}$,

$$E^{P_x}\left\{\exp\left\{-\int_0^t (Lu/u)(x(s))\,ds\right\}\right\}=E^{P_x}\left\{\exp\left\{-t\int_{\mathcal{X}} (Lu/u)(y)L_t(\omega, dy)\right\}\right\}$$
$$=E^{Q_{x,t}}\left\{\exp\left\{-t\int_{\mathcal{X}} (Lu/u)(y)\mu(dy)\right\}\right\}.$$

Thus, (3.16) implies that, for any measurable set $C\subset\mathcal{M}$,

$$Q_{x,t}(C)\leqq\frac{u(x)}{\alpha}\exp\left\{t\sup_{\mu\in C}\int_{\mathcal{X}} (Lu/u)(y)\mu(dy)\right\},$$

and hence

$$\overline{\lim_{t\to\infty}}\frac{1}{t}\log Q_{x,t}(C)\leqq\sup_{\mu\in C}\int_{\mathcal{X}}\left(\frac{Lu}{u}\right)(y)\mu(dy)\,. \tag{3.17}$$

Since (3.17) holds for any positive $u\in\mathcal{D}$,

$$\overline{\lim_{t\to\infty}}\frac{1}{t}\log Q_{x,t}(C)\leqq\inf_{\substack{u>0\\u\in\mathcal{D}}}\sup_{\mu\in C}\int_{\mathcal{X}}\left(\frac{Lu}{u}\right)(y)\mu(dy)\,. \tag{3.18}$$

If $C\subset\bigcup_{j=1}^{k} C_j$, then this last implies that

$$\overline{\lim_{t\to\infty}}\frac{1}{t}\log Q_{x,t}(C)\leqq\sup_{1\leqq j\leqq k}\inf_{\substack{u>0\\u\in\mathcal{D}}}\sup_{\mu\in C_j}\int_{\mathcal{X}}\left(\frac{Lu}{u}\right)(y)\mu(dy)\,.$$

Taking the infimum over all such finite unions which cover C, we have

$$\overline{\lim_{t\to\infty}}\frac{1}{t}\log Q_{x,t}(C)\leqq\inf_{\substack{C_1,C_2\cdots C_k\\ C\subset\bigcup_{j=1}^{k} C_j}}\sup_{1\leqq j\leqq k}\inf_{\substack{u>0\\u\in\mathcal{D}}}\sup_{\mu\in C_j}\int_{\mathcal{X}}\left(\frac{Lu}{u}\right)(y)\mu(dy)\,. \tag{3.19}$$

Now by an elementary argument (see footnote 1) one can show that if C is compact, then (3.19) implies that

$$\overline{\lim_{t\to\infty}}\frac{1}{t}\log Q_{x,t}(C)\leqq\sup_{\mu\in C}\inf_{\substack{u>0\\u\in\mathcal{D}}}\int_{\mathcal{X}}\left(\frac{Lu}{u}\right)(y)\mu(dy)$$
$$=-\inf_{\mu\in C} I(\mu)$$

which is (1.13). The hypothesis of (1.13) is that C is a closed set in $\mathcal{M}$ but $\mathcal{M}$ is compact and therefore C is compact. This completes the proof of Theorem 3.

4. The *I*-Function for Selfadjoint Generators

Let $\pi(x, dy)$ be the transition probability of a discrete parameter Markov process on a state space X. For any probability measure μ on X we define, as in (1.5),

$$I(\mu) = -\inf_{u\in\mathcal{U}} \int_X \log\left(\frac{\pi u}{u}\right)(x)\mu(dx)\,, \tag{4.1}$$

where $\mathcal{U}$ consists of all positive measurable functions which have an upper bound and a positive lower bound. In (4.1), $\pi u(x)$ stands, as before, for

$$(\pi u)(x) = \int_X u(y)\pi(x, dy)\,.$$

Let $\mathcal{M}$ be the space of all probability measures on X and consider the norm topology on $\mathcal{M}$. Thus, if μ_1 and μ_2 are two measures in $\mathcal{M}$, $\|\mu_1 - \mu_2\| = 2\sup_A |\mu_1(A) - \mu_2(A)|$. By $\mu\pi(A)$ we mean the measure

$$(\mu\pi)(A) = \int_X \pi(x, A)\mu(dx)\,.$$

Now in Lemma 2.5 we proved that a necessary and sufficient condition for $I(\mu) = 0$ is that $\mu\pi(A) = \mu(A)$ for every Borel set A, i.e., that μ is the invariant measure for $\pi(x, dy)$. The following lemma sharpens the result that if $I(\mu) = 0$, then $\mu\pi = \mu$, by giving an estimate on $\|\mu\pi - \mu\|$ in terms of a function $\phi(I(\mu))$ (which indeed vanishes when its argument does). Note that the function ϕ in the lemma is universal—same for all π and μ.

LEMMA 4.1. *There exists a function $\phi(l)$, such that $0 \leqq \phi(l) \leqq 2$ for $l \geqq 0$ and $\phi(l) \to 0$ as $l \to 0$, having the property that if $\pi(x, dy)$ is any transition probability and μ is any probability measure on X with the corresponding $I(\mu) = l$, then*

$$\|\mu\pi - \mu\| \leqq \phi(l)\,.$$

Proof: Since $I(\mu) = l$, we have

$$\int_X \log\left(\frac{\pi u}{u}\right)(x)\mu(dx) \geqq -l \tag{4.2}$$

for every bounded measurable u which is bounded below by a positive number. Let A be a Borel set contained in X and consider the function $u=a+1$ on A and $u=1$ on $\tilde{A}$, where $a>0$. From (4.2) we have, for all $a>0$,

$$\int_X \log\left[\frac{a\pi(x, A)+1}{a\chi_A(x)+1}\right]\mu(dx) \geqq -l$$

or

$$\int_X \log(1+a\pi(x, A))\mu(dx) \geqq -l+\mu(A)\log(1+a)\,. \tag{4.3}$$

Since $\log x$ is a concave function of x,

$$\log \int_X (1+a\pi(x, A))\mu(dx) \geqq \int_X \log(1+a\pi(x, A))\mu(dx)\,. \tag{4.4}$$

Combining (4.3) and (4.4), we get

$$\log[1+a\mu\pi(A)] \geqq -l+\mu(A)\log(1+a)\,.$$

Since $\log(1+\xi)\leqq\xi$ for all ξ, we obtain

$$a\mu\pi(A) \geqq -l+\mu(A)\log(1+a)\,; \tag{4.5}$$

this inequality holds for all Borel sets A and all $a>0$. Subtracting $\mu(A)$ on both sides of (4.5) after dividing by a, we get

$$\mu\pi(A)-\mu(A) \geqq \frac{-l+\mu(A)[\log(1+a)-a]}{a}\,.$$

Since $\mu(A)\leqq 1$ for all A, again using the fact that $\log(1+a)\leqq a$ for all $a>0$, we have

$$\mu\pi(A)-\mu(A) \geqq -\frac{l+a-\log(1+a)}{a}\,. \tag{4.6}$$

Since (4.6) is true for every Borel set A, and since $\mu\pi(\tilde{A})-\mu(\tilde{A})= -[\mu\pi(A)-\mu(A)]$, we obtain

$$\sup_A |\mu\pi(A)-\mu(A)| \leqq \frac{l+a-\log(1+a)}{a}\,.$$

Finally, since $\|\mu\pi-\mu\|=2\sup_A|\mu\pi(A)-\mu(A)|$, we get

$$\|\mu\pi-\mu\|\leqq 2\left[\frac{l+a-\log(1+a)}{a}\right],$$

which implies, since $a>0$ is arbitrary, that

$$\|\pi\mu-\mu\|\leqq 2\inf_{a>0}\left[\frac{l+a-\log(1+a)}{a}\right].$$

Thus, if we define

$$\phi(l)=2\inf_{a>0}\left[\frac{l+a-\log(1+a)}{a}\right],$$

then

$$\|\mu\pi-\mu\|\leqq\phi(l)$$

and $\phi(l)$ can easily be seen to have the properties stated in the lemma.

Now suppose $p(t,x,dy)$ is the transition probability of a continuous parameter Markov process whose state space is X. Then we can define, for each $h>0$, $\pi^{(h)}(x,dy)$ to be $p(h,x,dy)$ and use it to define $I_h(\mu)$ with $\pi^{(h)}$ playing the role of π in (4.1).

We now prove

LEMMA 4.2. *Let B_0 be the strongly continuous center of the semigroup T_t associated with the given Markov process acting on the space $B(X)$ of bounded measurable functions on X. We assume that for any measure μ on X and any $u\in B(X)$ there exists a sequence $\{u_n\}\in B_0$ which converges almost everywhere (μ) to u (assumption 3 in the introduction just preceding Theorem 5). Let L be the infinitesimal generator of the semigroup T_t having domain $\mathfrak{D}$ (which is dense in B_0), and let $\mathfrak{D}^+$ be the subset of $\mathfrak{D}$ consisting of those functions which have a positive lower bound. Let*

$$I(\mu)=-\inf_{u\in\mathfrak{D}^+}\int_X\left(\frac{Lu}{u}\right)(x)\mu(dx).$$

Then, for any $h>0$,

$$I_h(\mu)\leqq hI(\mu).\tag{4.7}$$

Proof: As in Lemma 3.1 we can obtain

$$\int_X\log\pi^{(h)}u\mu(dx)-\int_X\log u\mu(dx)\geqq -hI(\mu)$$

for every u in B_0^+, where $B_0^+ \subset B_0$ consists of those functions in B_0 which have a positive lower bound.

Consider the measure $\mu+\pi^{(h)}\mu$, where

$$(\pi^{(h)}\mu)(A)=\int_X \pi^{(h)}(x, A)\mu(dx) .$$

Let $u\in B^+$, where $B^+\subset B$ consists of those bounded measurable functions which have a positive lower bound. By our hypothesis there exists a sequence $\{u_n\}\in B_0$ which converges almost everywhere $(\mu+\pi^{(h)}\mu)$ to u. Without loss of generality we can assume that $\{u_n\}\in B_0^+$ and that $\{u_n\}$ is uniformly bounded above and below by the bounds on u. Let E be the exceptional set where $u_n \not\to u$ so that $(\mu+\pi^{(h)}\mu)(E)=0$, i.e., $\mu(E)=0$ and $(\pi^{(h)}\mu)(E)=0$. Consider

$$(\pi^{(h)}u_n)(x)=\int_X u_n(y)\pi^{(h)}(x, dy) ,$$

and let $F=\{x : \pi^{(h)}(x, E)=0\}$. Then, for $x\in F$, $\pi^{(h)}u_n\to\pi^{(h)}u$. On the other hand, $0=(\pi^{(h)}\mu)(E)=\int_X \pi^{(h)}(x, E)\mu(dx)$ and therefore $\mu(F)=1$, i.e., $\pi^{(h)}u_n\to\pi^{(h)}u$ almost everywhere (μ). Thus we can let $n\to\infty$ in

$$\int_X \log \pi^{(h)}u_n\mu(dx)-\int_X \log u_n\mu(dx)\geqq -hI(\mu) ,$$

and get for every $u\in B^+$

$$\int_X \log \pi^{(h)}u\mu(dx)-\int_X \log u\mu(dx)\geqq -hI(\mu) ,$$

which implies the lemma.

Lemmas 4.1 and 4.2 give us the

COROLLARY. *Let T_t be any Markov semigroup on any state space X satisfying the assumption of Lemma 4.2. If $I(\mu)=l<\infty$, then*

$$\|\mu T_t-\mu\|\leqq\phi(tl)$$

for the ϕ defined in Lemma 4.1. Hence

$$\|\mu T_t-\mu\|\to 0 \qquad as \qquad t\to 0 .$$

We now proceed to the proof of Theorem 5 in the introduction. The notation, definitions, and hypotheses are stated in the introduction just before the statement of that theorem.

Suppose, first of all, that $I(\mu)=l<\infty$. Then, from the corollary to Lemma 4.2, we know that $\|\mu T_t-\mu\|\to 0$ as $t\to 0$. Since $\mu T_t \ll \lambda$ for $t>0$, by assumption 1 on the process, we conclude that $\mu \ll \lambda$.

Let B^+ denote those elements u in $B(x)$ which have positive lower bounds. For $h>0$, let

$$I_h(\mu)=-\inf_{u\in B^+}\int_X \log\left(\frac{\pi^{(h)}u}{u}\right)(x)\mu(dx)\,.$$

Then, from Lemma 4.2, $I_h(\mu)\leqq hI(\mu)$. Hence, for any $u\in B^+$,

$$\int_X \log\left(\frac{\pi^{(h)}u}{u}\right)(x)\mu(dx)\geqq -hI(\mu)\,. \tag{4.8}$$

But

$$\log\left(\frac{\pi^{(h)}u}{u}\right)=\log\left(1-\frac{u-\pi^{(h)}u}{u}\right)$$
$$=\log\left(1-\frac{(I-\pi^{(h)})u}{u}\right)$$

and, since $u>0$ implies $(I-\pi^{(h)})u/u\leqq 1$, and $\log(1-x)\leqq -x$ for $-\infty<x\leqq 1$, we have from (4.8) that, for any $u\in B^+$,

$$-\int_X \left[\frac{(I-\pi^{(h)})u}{u}\right](x)\mu(dx)\geqq -hI(\mu)\,. \tag{4.9}$$

With $f=d\mu/d\lambda$ we get from (4.9), for any $u\in B^+$,

$$\int_X \left[\frac{(I-\pi^{(h)})u}{u}\right](x)f(x)\lambda(dx)\leqq hI(\mu)\,. \tag{4.10}$$

Now, let $\{u_n\}$ be a sequence such that $0\leqq u_n\leqq\sqrt{f}$ and $u_n\to\sqrt{f}$ with u_n bounded (e.g. we could take $u_n=\sqrt{f}\wedge n$). Let $\varepsilon>0$. Since $u_n+\varepsilon\in B^+$, we have from (4.10) that for all n

$$\int_X \left[\frac{(I-\pi^{(h)})u_n}{\varepsilon+u_n}\right](x)f(x)\lambda(dx)\leqq hI(\mu)\,.$$

Letting $n\to\infty$, this last yields

$$\int_X (I-\pi^{(h)})\sqrt{f(x)}\left(\frac{f(x)}{\varepsilon+\sqrt{f(x)}}\right)\lambda(dx)\leqq hI(\mu),$$

and then, letting $\varepsilon\to 0$, we get

$$\int_X \sqrt{f(x)}(I-\pi^{(h)})\sqrt{f(x)}\lambda(dx)\leqq hI(\mu), \tag{4.11}$$

i.e., if $E(\Lambda)$ is the spectral measure corresponding to $\sqrt{f}$ relative to the operator $-\tilde{L}$, then

$$\int_0^\infty (1-e^{-h\Lambda})E(d\Lambda)\leqq hI(\mu). \tag{4.12}$$

If in (4.12) we divide by h and then let $h\to 0$, we get, using Fatou's lemma,

$$\int_0^\infty \Lambda E(d\Lambda)\leqq I(\mu).$$

This proves that $g=\sqrt{f}\in\tilde{\mathscr{D}}_{1/2}$ and

$$\|\sqrt{-\tilde{L}}\,g\|^2\leqq I(\mu).$$

To prove the converse assume $g=\sqrt{f}\in\tilde{\mathscr{D}}_{1/2}$ where $f=d\mu/d\lambda$, and that $\|\sqrt{-\tilde{L}}\sqrt{f}\|^2=l$. We wish to show $I(\mu)\leqq l$.

Consider the semigroup

$$S_t^\phi f=E_x\left\{f(x(t))\exp\left\{-\int_0^t (L\phi/\phi)(x(s))\,ds\right\}\right\},$$

where $\phi\in\mathscr{D}^+$, and $E_x\{\ \}$ denotes expectation on the process $\{x(s), 0\leqq s<\infty\}$ (with $x(0)=x$) corresponding to our given transition probability $p(t,x,dy)$. Now $V=L\phi/\phi$ is bounded and S_t^ϕ is a selfadjoint semigroup in $L_2(\lambda)$ with a generator $\tilde{L}_\phi$ given by

$$\tilde{L}_\phi u=\tilde{L}u-Vu.$$

Since

$$\sup_t\sup_x E_x\left\{\exp\left\{-\int_0^t (L\phi/\phi)(x(s))\,ds\right\}\right\}\leqq\frac{\sup\phi}{\inf\phi}=C<\infty,$$

and S_t^ϕ is selfadjoint, it follows from Schwartz' inequality that $\|S_t^\phi f\|_2 \leqq C\|f\|_2$, and hence $\|S_t^\phi\| \leqq C$. Since S_t^ϕ is selfadjoint, $\|S_t^\phi\| \leqq 1$ so that it is a contraction. Thus, $\tilde{L}_\phi$ is negative semidefinite, i.e., $\langle \tilde{L}u - Vu, u\rangle \leqq 0$. In other words, $(\tilde{L}u, u) \leqq \langle Vu, u\rangle$ or $(Vu, u) \geqq (\tilde{L}u, u)$ for $u \in \tilde{\mathfrak{D}}$. Thus $(Vu, u) \geqq -(-\tilde{L}u, u) = -\|\sqrt{-\tilde{L}}\, u\|^2$ for $u \in \tilde{\mathfrak{D}}_{1/2}$. In particular, let $u = \sqrt{f}$ so that $(V\sqrt{f}, \sqrt{f}) \geqq -l$, i.e.,

$$\int_X \left(\frac{L\phi}{\phi}\right)(x) f(x) \lambda(dx) = \int_X \left(\frac{L\phi}{\phi}\right)(x) \mu(dx) \geqq -l$$

for all $\phi \in \mathfrak{D}^+$, which implies $I(\mu) \leqq l$; hence we have completed the proof of Theorem 5.

Bibliography

[1] Donsker, M. D., and Varadhan, S. R. S., *Asymptotic evaluation of certain Wiener integrals for large time*, to appear in Proceedings of International Conference on Function Space Integration, Oxford, 1974.

[2] Kac, M., *On some connections between probability theory and differential and integral equations*, Proceedings of the Second Berkeley Symposium, 1950, pp. 189–215.

[3] Varadhan, S. R. S., *Asymptotic probabilities and differential equations*, Comm. Pure Appl. Math., Vol. XIX, No. 3, 1966, pp. 261–286.

Received September, 1974.

COMMUNICATIONS ON PURE AND APPLIED MATHEMATICS, VOL. XXVIII, 279–301 (1975)

Asymptotic Evaluation of Certain Markov Process Expectations for Large Time, II*

M. D. DONSKER AND S. R. S. VARADHAN

1. Introduction

Let $\mathcal{X}$ be a compact metric space. Let $p(t, x, dy)$ be the transition probability function of a homogeneous Markov Process having $\mathcal{X}$ as its state space. Denote by T_t the corresponding semigroup which we shall assume to be a strongly continuous Feller semigroup on $C(\mathcal{X})$. We denote by L the infinitesimal generator of the semigroup and by $\mathcal{D}$ the domain of L. Without loss of generality we can assume that the trajectories of the process are right-continuous and have left limits at every point of time $0 \leqq t < \infty$. For each starting point $x \in \mathcal{X}$ we get a measure P_x on the space Ω_x consisting of those trajectories $\omega = x(\cdot)$ satisfying $x(0) = x$.

Let $\mathcal{M}$ be the space of all probability measures on $\mathcal{X}$. We shall regard $\mathcal{M}$ as a compact metric space endowed with the weak topology. For each $t > 0$ there is a canonical map $\Omega_x \to \mathcal{M}$ provided by the occupation distribution. For each $\omega \in \Omega_x$ and $t > 0$, we define the occupation distribution $L_t(\omega, A)$ by

$$L_t(\omega, A) = \frac{1}{t} \text{ Lebesgue measure } \{\sigma : x(\sigma) \in A, 0 \leqq \sigma \leqq t\},$$

where $A \subset \mathcal{X}$ is a Borel subset of $\mathcal{X}$. Clearly $L_t(\omega, A)$ is the proportion of time up to t that a particular trajectory $\omega = x(\cdot)$ spends in the set A, and $L_t(\omega, \cdot) \in \mathcal{M}$. It is easy to see that, for each $\omega \in \Omega_x$ and $t > 0$,

$$\int_{\mathcal{X}} V(y) L_t(\omega, dy) = \frac{1}{t} \int_0^t V(x(\sigma))\, d\sigma,$$

where V is any bounded continuous function on $\mathcal{X}$.

If $\mathcal{X}$ is one-dimensional and $p(t, x, dy)$ is a suitable diffusion, then $L_t(\omega, dy)$ is almost surely absolutely continuous with respect to the Lebesgue measure on $\mathcal{X}$, and the density is a multiple of the local time.

* Results obtained at the Courant Institute of Mathematical Sciences, New York University, with the Office of Naval Research, Contract N00014-67-A-0467-0031 and the National Science Foundation, Grant NSF-GP-41985. Reproduction in whole or in part is permitted for any purpose of the United States Government.

For each $x \in X$ and $t>0$, let $Q_{x,t}$ be the measure on $\mathcal{M}$ induced by the map L_t from $\Omega_x \to \mathcal{M}$, i.e., $Q_{x,t} = P_x L_t^{-1}$. Thus, if B is a set of measures in $\mathcal{M}$,

$$Q_{x,t}(B) = P_x\{\omega : L_t(\omega, \cdot) \in B\}.$$

Let $\mathcal{U}$ be those elements of the domain $\mathcal{D}$ which are strictly positive. Since every element $u \in \mathcal{U}$ is continuous and X is compact, u satisfies bounds of the form $0<a \leqq u(x) \leqq b < \infty$.

For each $\mu \in \mathcal{M}$, we define

$$I(\mu) = -\inf_{u \in \mathcal{U}} \int_X \left(\frac{Lu}{u}\right)(x)\mu(dx). \tag{1.1}$$

It is easy to see that $I(\mu)$ is a convex, lower semicontinuous, non-negative functional on $\mathcal{M}$.

We shall now assume that the transition probability $p(t, x, dy)$ satisfies the following:

Assumption A. There is a finite reference measure λ on X such that $p(t, x, dy)$ is absolutely continuous with respect to λ for each $x \in X$ and $t>0$. Moreover, the density $p(t, x, y)$ satisfies bounds of the form

$$0 < a(t) \leqq p(t, x, y) \leqq A(t) < \infty$$

for each $t>0$.

Under Assumption A, the following theorem relating the asymptotic behavior of $Q_{x,t}$ for large t to the functional $I(\mu)$ was proved in [1].

Theorem 1.1. *If $C \subset \mathcal{M}$ is closed and $G \subset \mathcal{M}$ is open (in the weak topology on $\mathcal{M}$), then for each $x \in X$*

$$\overline{\lim_{t\to\infty}} \frac{1}{t} \log Q_{x,t}(C) \leqq -\inf_{\mu \in C} I(\mu),$$

$$\underline{\lim_{t\to\infty}} \frac{1}{t} \log Q_{x,t}(G) \geqq -\inf_{\mu \in G} I(\mu).$$

Such a theorem has a corollary as was proved in [2] (Section 3).

Corollary. *If $\Phi(\mu)$ is a weakly continuous functional on $\mathcal{M}$, then*

$$\lim_{t\to\infty} \frac{1}{t} \log E^{Q_{x,t}}\{e^{t\Phi(\mu)}\} = \sup_{\mu} [\Phi(\mu) - I(\mu)],$$

or equivalently

$$\lim_{t\to\infty} \frac{1}{t} \log E^{P_x}\{e^{t\Phi(L_t(\omega,\cdot))}\} = \sup_{\mu} [\Phi(\mu) - I(\mu)] .$$

In some interesting applications (see e.g. [3] for the sausage problem arising in statistical mechanics) the functional Φ of interest is not continuous in the weak topology on $\mathcal{M}$ but perhaps has continuity properties in the norm topology on $\mathcal{M}$. This means that one needs a theorem like Theorem 1.1 but where C and G are, respectively, closed and open in the norm topology rather than in the weak topology. There are difficulties in this situation because the occupation distribution $L_t(\omega, dy)$ can have a density with respect to a reference measure only in the one-dimensional case. In higher dimensions, although $Q_{x,t}$ will be concentrated mostly on singular measures, in the limit as $t\to\infty$ the variational formula $\sup_\mu [\Phi(\mu)-I(\mu)]$ involves only the values of $\Phi(\mu)$ for measures μ such that $I(\mu)$ is finite and these in general are absolutely continuous with respect to a reference measure. Such a transition in the limit from singular to absolutely continuous measures can easily occur in the weak topology but cannot occur in the norm topology. To derive results that are valid in the norm topology a certain amount of mollification is necessary. We shall state shortly Theorem 1.2 which is the analogue of Theorem 1.1 but for the norm topology. First we must consider the mollification of the occupation distribution.

We start with a family $k_\varepsilon(x, y)$ of functions on $\mathfrak{X}\times\mathfrak{X}$ satisfying Assumption B which follows. The λ measure which appears in Assumption B is the same finite reference measure as in Assumption A.

Assumption B.

(i) The family $k_\varepsilon(x, y)$ is non-negative and jointly measurable in x and y for each $\varepsilon>0$.

(ii) $\int_{\mathfrak{X}} k_\varepsilon(x, y)\lambda(dy)=1$ for each $\varepsilon>0$ and $x\in\mathfrak{X}$.

(iii) For each $\varepsilon>0$, the map $x\to k_\varepsilon(x,\cdot)$ is a continuous map of $\mathfrak{X}$ into $L_1(\lambda)$.

(iv) For each $\phi\in C(\mathfrak{X})$,

$$\lim_{\varepsilon\to 0} \int_{\mathfrak{X}} k_\varepsilon(x, y)\phi(y)\lambda(dy) = \phi(x)$$

uniformly for $x\in\mathfrak{X}$.

(v) For each $f \in L_1(\lambda)$,

$$\lim_{\varepsilon \to 0} \int_{\mathcal{X}} k_\varepsilon(x, y) f(x) \lambda(dx) = f(y)$$

in $L_1(\lambda)$.

We then use $k_\varepsilon(x, y)$ as a mollifier and define

$$l_t^\varepsilon(\omega, y) = \int_{\mathcal{X}} k_\varepsilon(x, y) L_t(\omega, dx) .$$

We see that $l_t^\varepsilon(\omega, y)$ is a probability density on $\mathcal{X}$ relative to λ.

Let γ be the subset of $\mathcal{M}$ consisting of those probability measures on $\mathcal{X}$ which are absolutely continuous with respect to λ. We can think of each element of γ in terms of its density f relative to λ and hence identify γ with the subset of $L_1(\lambda)$ consisting of non-negative elements having total integral 1. Thus for each ω, $\varepsilon > 0$, and $t > 0$, $l_t^\varepsilon(\omega, y) \in \gamma$.

Now, we make a choice of ε for each t and denote the choice by $\varepsilon(t)$. Assume that $\varepsilon(t) \to 0$ as $t \to \infty$ in an appropriate manner. Assume also that the measure $R_{x,t}$ on γ is defined by $R_{x,t} = Q_{x,t} k_{\varepsilon(t)}^{-1}$, that is, if A is a Borel subset of γ,

$$R_{x,t}(A) = Q_{x,t}\left\{\mu : \int_{\mathcal{X}} k_{\varepsilon(t)}(x, \cdot) \mu(dx) \in A\right\}$$

$$= P_x\{\omega : l_t^{\varepsilon(t)}(\omega, \cdot) \in A\} .$$

We would like to prove that $R_{x,t}$ obeys the following

THEOREM 1.2. *If $C \subset \gamma$ is closed and $G \subset \gamma$ is open (in the norm topology on γ), then*

$$\overline{\lim_{t \to \infty}} \frac{1}{t} \log R_{x,t}(C) \leqq -\inf_{f \in C} I(f) ,$$

$$\underline{\lim_{t \to \infty}} \frac{1}{t} \log R_{x,t}(G) \geqq -\inf_{f \in G} I(f) .$$

In Theorem 1.2,

$$I(f) = -\inf_{u \in \mathcal{U}} \int_{\mathcal{X}} \left(\frac{Lu}{u}\right)(x) f(x) \lambda(dx)$$

which is just $I(\mu)$ as previously defined but for $\mu \in \gamma$.

Just as we had a corollary to Theorem 1.1 we have a corollary to Theorem 1.2.

COROLLARY. *If $\Phi(f)$ is a continuous functional on γ which is bounded above, then*

$$\lim_{t\to\infty}\frac{1}{t}\log E^{R_{x,t}}\{e^{t\Phi(f)}\}=\sup_{f\in\gamma}[\Phi(f)-I(f)],$$

or equivalently

$$\lim_{t\to\infty}\frac{1}{t}\log E^{P_x}\{\exp\{t\Phi(l_t^{\varepsilon(t)}(\omega,\cdot))\}\}=\sup_{f\in\gamma}[\Phi(f)-I(f)].$$

Remark. If μ_0 is the unique invariant measure for the process, then the ergodic theorem asserts that $L_t(\omega, dy)$ converges weakly to μ_0 as $t\to\infty$ for almost all ω. Theorem 1.2 will imply that $l_t^{\varepsilon(t)}(\omega, y)$ converges to $f_0=d\mu_0/d\lambda$ in $L_1(\lambda)$ for almost all ω. Therefore, Theorem 1.2 is a non-trivial strengthening of Theorem 1.1 and should require among its assumptions that $\varepsilon(t)$ does not go to zero too rapidly as $t\to\infty$. We shall state below Assumption C which is of this nature.

Assumption B on $k_\varepsilon(x, y)$ implies that the operator K_ε defined by

$$(K_\varepsilon\phi)(x)=\int_{\mathfrak{X}} k_\varepsilon(x, y)\phi(y)\lambda(dy)$$

is a compact operator mapping the unit ball of $C(\mathfrak{X})$ into an equicontinuous and bounded family of functions on $\mathfrak{X}$. Let S_ε be the image of the unit ball under K_ε. Since S_ε is totally bounded, it can be covered by a finite number of spheres in $C(\mathfrak{X})$ of any given positive radius δ. Let $N_\delta(\varepsilon)$ be the smallest number of such spheres required to cover S_ε. Our assumption on the rate of convergence of $\varepsilon(t)$ to 0 is

ASSUMPTION C. $\varepsilon(t)\to 0$ as $t\to\infty$ in such a manner that

$$\overline{\lim_{t\to\infty}}\frac{1}{t}\log N_\delta(\varepsilon(t))<\infty$$

for each $\delta>0$.

Finally, we need a regularity assumption on the semigroup T_t which we state as

ASSUMPTION D. For each $t>0$, the transition probability density $p(t, x, y)$ gives a continuous map $x\to p(t, x, \cdot)$ of $\mathfrak{X}$ into $\gamma\subset L_1(\lambda)$.

Sections 2, 3, and 4 of this paper contain the proof of Theorem 1.2 under Assumptions A, B, C, and D. In Section 5 we treat a class of examples which are used in [3] for the sausage problem.

2. Some Preliminary Estimates

Let $\phi(x)\in C(\mathfrak{X})$ and consider, for $\varepsilon>0$ and $\delta>0$, the function

$$\phi_{\varepsilon,\delta}(x)=(K_\varepsilon(T_\delta-I)\phi)(x).$$

Define for each ϕ

$$\lambda_{\varepsilon,\delta}(\phi)=\sup_{\mu\in\mathcal{M}}\left[\int_{\mathfrak{X}}\phi_{\varepsilon,\delta}(x)\mu(dx)-I(\mu)\right].$$

We have

LEMMA 2.1. *For each* $0<\rho<\infty$,

$$\varlimsup_{\substack{\varepsilon\to0\\ \delta\to0}}\sup_{\|\phi\|\leqq\rho}\lambda_{\varepsilon,\delta}(\phi)\leqq0.$$

Proof: For any $\phi(x)$ which is bounded and continuous,

$$\lambda(\phi)=\sup_{\mu\in\mathcal{M}}\left[\int_{\mathfrak{X}}\phi(x)\mu(dx)-I(\mu)\right]\leqq\sup_x\phi(x)\leqq\|\phi\|.$$

For each $\phi\in C(\mathfrak{X})$, $\|K_\varepsilon(T_\delta-I)\phi\|\to0$ as $\varepsilon,\delta\to0$ and therefore

$$\varlimsup_{\substack{\varepsilon\to0\\ \delta\to0}}\lambda_{\varepsilon,\delta}(\phi)\leqq0.$$

However, the convergence of $K_\varepsilon(T_\delta-I)\phi$ is not uniform over bounded sets because $\|K_\varepsilon(T_\delta-I)\|$ does not tend to zero. So we must estimate more carefully.

Since $|\phi_{\varepsilon,\delta}(x)|\leqq2\rho$ for all x if $\|\phi\|\leqq\rho$, we have

$$\int_{\mathfrak{X}}\phi_{\varepsilon,\delta}(x)\mu(dx)-I(\mu)\leqq0$$

if $I(\mu)\geqq2\rho$. To prove the lemma it is therefore sufficient to show that

$$\lim_{\substack{\varepsilon\to0\\ \delta\to0}}\sup_{\mu:I(\mu)\leqq2\rho}\left|\int_{\mathfrak{X}}\phi_{\varepsilon,\delta}(x)\mu(dx)\right|=0. \tag{2.1}$$

Since

$$\int_{\mathfrak{X}} \phi_{\varepsilon,\delta}(x)\mu(dx)=\int_{\mathfrak{X}} (K_\varepsilon(T_\delta-I)\phi)(x)\mu(dx)$$
$$=\int_{\mathfrak{X}} \phi(x)[\mu K_\varepsilon(T_\delta-I)](dx)\,,$$

it is enough to show that

$$\sup_{\mu:I(\mu)\leqq 2\rho} \|\mu K_\varepsilon(T_\delta-I)\|\to 0 \tag{2.2}$$

as ε and $\delta\to 0$. From Lemma 4.1 of [1] and its corollary we have

$$\lim_{\delta\to 0}\sup_{\mu:I(\mu)\leqq 2\rho} \|\mu T_\delta-\mu\|=0\,. \tag{2.3}$$

By Assumption D, for each $\delta>0$ the range of μT_δ as μ varies over $\mathcal{M}$ is a totally bounded subset of γ in the norm topology. Combining this with (2.3) the set of probability measures μ satisfying $I(\mu)\leqq 2\rho$ is a totally bounded subset of γ. Since K_ε is a contraction and $\|\mu K_\varepsilon-\mu\|\to 0$ as $\varepsilon\to 0$ for each $\mu\in\gamma$, we conclude from the fact that $\{\mu : I(\mu)\leqq 2\rho\}$ is totally bounded in γ that

$$\sup_{\mu:I(\mu)\leqq 2\rho} \|\mu K_\varepsilon-\mu\|\to 0 \quad \text{as} \quad \varepsilon\to 0\,. \tag{2.4}$$

Now, we can write

$$\|\mu K_\varepsilon(T_\delta-I)\|=\|\mu K_\varepsilon T_\delta-\mu K_\varepsilon\|$$
$$\leqq\|(\mu K_\varepsilon-\mu)T_\delta\|+\|\mu T_\delta-\mu\|+\|\mu-\mu K_\varepsilon\|$$
$$\leqq 2\,\|\mu K_\varepsilon-\mu\|+\|\mu T_\delta-\mu\|\,,$$

and therefore (2.3) and (2.4) imply (2.2) and the lemma.

LEMMA 2.2. *For each $0<\rho<\infty$, there is a constant C_ρ such that if $V(x)$ is any continuous function on $\mathfrak{X}$ bounded in absolute value by ρ and if*

$$\lambda_V=\sup_{\mu\in\mathcal{M}}\left[\int_{\mathfrak{X}} V(x)\mu(dx)-I(\mu)\right],$$

then, for each $x \in \mathbf{X}$ *and* $t \geqq 0$,

$$E_x\left\{\exp\left\{\int_0^t V(x(s))\,ds\right\}\right\} = E^{P_x}\left\{\exp\left\{\int_0^t V(x(s))\,ds\right\}\right\} \leqq C_p e^{\lambda_V t}.$$

Proof: It was proved in [1] (see the corollary to Theorem 1.1 in the introduction of the present paper) that for any $V(x)$ which is bounded and continuous

$$\begin{aligned}\lambda_V &= \lim_{t\to\infty} \frac{1}{t} \log E_x\left\{\exp\left\{\int_0^t V(x(s))\,ds\right\}\right\} \\ &= \sup_{\mu\in\mathcal{M}}\left[\int V(x)\mu(dx) - I(\mu)\right].\end{aligned} \tag{2.5}$$

Let $v(t, x) = E_x\left\{\exp\left\{\int_0^t V(x(s))\,ds\right\}\right\}$, and $v^+(t) = \sup_x v(t, x)$, $v^-(t) = \inf_x v(t, x)$, so that

$$v^-(t) \leqq v(t, x) \leqq v^+(t). \tag{2.6}$$

By the Markov property,

$$v^+(t+s) \leqq v^+(t)v^+(s)$$

and

$$v^-(t+s) \geqq v^-(t)v^-(s).$$

Therefore, from the property of subadditive functions,

$$\lim_{t\to\infty} \frac{\log v^+(t)}{t} = \inf_{t>0} \frac{\log v^+(t)}{t}$$

and

$$\lim_{t\to\infty} \frac{\log v^-(t)}{t} = \sup_{t>0} \frac{\log v^-(t)}{t}.$$

From (2.5) and (2.6) it follows that, for every $t \geqq 0$,

$$v^-(t) \leqq e^{\lambda_V t} \leqq v^+(t). \tag{2.7}$$

Now, let $t \geqq 1$. Then for any pair x, y of points in $\mathfrak{X}$, we have using Assumption A

$$
\begin{aligned}
E_x\left\{\exp\left\{\int_0^t V(x(s))\,ds\right\}\right\} &= E_x\left\{\exp\left\{\int_0^1 V(x(s))\,ds\right\}\exp\left\{\int_1^t V(x(s))\,ds\right\}\right\} \\
&= E_x\left\{\exp\left\{\int_0^1 V(x(s))\,ds\right\}E_{x(1)}\left\{\exp\left\{\int_0^{t-1} V(x(s))\,ds\right\}\right\}\right\} \\
&= E_x\left\{\exp\left\{\int_0^1 V(x(s))\,ds\right\}v(t-1, x(1))\right\} \\
&\leqq e^{\rho}E_x\{v(t-1, x(1))\} \\
&= e^{\rho}\int_{\mathfrak{X}} v(t-1, z)p(1, x, dz) \\
&= e^{\rho}\int_{\mathfrak{X}} v(t-1, z)p(1, x, z)\lambda(dz) \\
&= e^{\rho}\int_{\mathfrak{X}} v(t-1, z)\frac{p(1, x, z)}{p(1, y, z)}p(1, y, z)\lambda(dz) \\
&\leqq e^{\rho}\frac{A(1)}{a(1)}\int_{\mathfrak{X}} v(t-1, z)p(1, y, z)\lambda(dz) \\
&\leqq e^{2\rho}\frac{A(1)}{a(1)}E_y\left\{\exp\left\{\int_0^t V(x(s))\right\}\right\}.
\end{aligned}
$$

Since x and y are arbitrary points in $\mathfrak{X}$, we obtain for $t \geqq 1$

$$v^+(t) \leqq e^{2\rho}\frac{A(1)}{a(1)}v^-(t)\,.$$

On the other hand, for $0 \leqq t \leqq 1$,

$$v^+(t) \leqq e^{t\rho} \leqq e^{\rho}\,,$$

$$v^-(t) \geqq e^{-t\rho} \geqq e^{-\rho}\,.$$

Thus, for $0 \leqq t \leqq 1$, $v^+(t) \leqq e^{2\rho} v^-(t)$ and, since $A(1)/a(1) \geqq 1$, we can take $C_\rho = e^{2\rho} A(1)/a(1)$ and have

$$v^+(t) \leqq C_\rho v^-(t) \quad \text{for all} \quad t \geqq 0 .$$

Using (2.7), we obtain from this last inequality

$$v(t, x) \leqq v^+(t) \leqq C_\rho v^-(t) \leqq C_\rho e^{\lambda_V t} ,$$

which is the lemma.

3. The Main Estimate

Let us suppose that a choice of $\varepsilon(t)$ has been made for each t in such a manner that Assumption C is satisfied. We recall the definitions:

$$L_t(\omega, A) = \frac{1}{t} \text{ Lebesgue measure } \{\sigma : x(\sigma) \in A,\ 0 \leqq \sigma \leqq t\} ,$$

$$l_t^\varepsilon(\omega, y) = \int_{\mathbf{X}} k_\varepsilon(x, y) L_t(\omega, dx) .$$

For convenience we denote $l_t^{\varepsilon(t)}(\omega, y)$ by $g_t(\omega, y)$ and define, for any $\delta > 0$,

$$g_t^\delta(\omega, y) = \int_{\mathbf{X}} p(\delta, x, y) g_t(\omega, x) \lambda(dx)$$

and

$$\Delta_t^\delta(\omega) = \int_{\mathbf{X}} |g_t^\delta(\omega, y) - g_t(\omega, y)|\, \lambda(dy) .$$

We now prove

THEOREM 3.1. *For any $\theta > 0$ and $x \in \mathbf{X}$,*

$$\overline{\lim_{\delta \to 0}}\ \overline{\lim_{t \to \infty}} \frac{1}{t} \log P_x\{\Delta_t^\delta(\omega) \geqq \theta\} = -\infty .$$

Proof: We note that

$$\begin{aligned}
\Delta_t^{\delta}(\omega) &= \sup_{\|\phi\|\leq 1} \int_{\mathscr{X}} \phi(y)[g_t^{\delta}(\omega, y) - g_t(\omega, y)]\lambda(dy) \\
&= 2 \sup_{\|\phi\|\leq 1/2} \int_{\mathscr{X}} \phi(y)[g_t^{\delta}(\omega, y) - g_t(\omega, y)]\lambda(dy) \\
&= 2 \sup_{\|\phi\|\leq 1/2} \int_{\mathscr{X}} \phi(y)[L_t(\omega, \cdot)K_{\varepsilon(t)}(T_\delta - I)](dy) \\
&= 2 \sup_{\|\phi\|\leq 1/2} \int_{\mathscr{X}} (K_{\varepsilon(t)}(T_\delta - I)\phi)(y)L_t(\omega, dy) \\
&= 2 \sup_{\|\phi\|\leq 1/2} \frac{1}{t}\int_0^t (K_{\varepsilon(t)}(T_\delta - I)\phi)(x(s))\, ds \\
&= 2 \sup_{\psi\in A_{\varepsilon(t),\delta}} \frac{1}{t}\int_0^t \psi(x(s))\, ds\,,
\end{aligned}$$

where $A_{\varepsilon(t),\delta}$ is the set of functions of the form

$$\psi = \phi_{\varepsilon,\delta} = K_\varepsilon(T_\delta - I)\phi\,,$$

where ϕ is continuous and $\|\phi\|\leq\frac{1}{2}$.

If A is any set in a metric space and if $A\subset B$, where B can be covered by a finite number N of spheres of radius θ, then A can be covered by the same number N of spheres of radius 2θ with centers in A. To see this, let $s(\xi_1, \theta), \cdots, s(\xi_N, \theta)$ be the covering of B. For each $i=1, 2, \cdots, N$, we pick a_i from $s(\xi_i, \theta)\cap A$ if it is non-empty. Otherwise, we pick any a_i from A. Then $s(a_i, 2\theta)$, $1\leq i\leq N$, is a covering of A.

The set $A_{\varepsilon,\delta}$ defined above is clearly contained in the set B_ε which is the image of $\{\phi : \|\phi\|\leq 1\}$ under K_ε. Since B_ε can be covered by a finite number $N_{\theta/8}(\varepsilon)$ of spheres of radius $\frac{1}{8}\theta$, $A_{\varepsilon,\delta}$ can be covered by the same number of spheres of radius $\frac{1}{4}\theta$ with centers $\psi_1, \psi_2, \cdots, \psi_N$ in $A_{\varepsilon,\delta}$. Here N stands for $N_{\theta/8}(\varepsilon(t))$. We can now write

$$\sup_{\psi\in A_{\varepsilon(t),\delta}} \frac{1}{t}\int_0^t \psi(x(s))\, ds \leq \tfrac{1}{4}\theta + \sup_{1\leq i\leq N} \frac{1}{t}\int_0^t \psi_i(x(s))\, ds\,.$$

Therefore,

$$
\begin{aligned}
P_x\{\Delta_t^\delta(\omega)\geqq\theta\}&=P_x\left\{\sup_{\psi\in A_{\varepsilon,\delta}}\frac{1}{t}\int_0^t\psi(x(s))\,ds\geqq\tfrac{1}{2}\theta\right\}\\
&\leqq P_x\left\{\sup_{1\leqq i\leqq N}\frac{1}{t}\int_0^t\psi_i(x(s))\,ds\geqq\tfrac{1}{4}\theta\right\}\\
&\leqq\sum_{i=1}^{N}P_x\left\{\frac{1}{t}\int_0^t\psi_i(x(s))\,ds\geqq\tfrac{1}{4}\theta\right\}\\
&\leqq N\sup_{1\leqq i\leqq N}P_x\left\{\frac{1}{t}\int_0^t\psi_i(x(s))\,ds\geqq\tfrac{1}{4}\theta\right\}.
\end{aligned}
\tag{3.1}
$$

But

$$
P_x\left\{\frac{1}{t}\int_0^t\psi_i(x(s))\,ds\geqq\tfrac{1}{4}\theta\right\}\leqq e^{-\rho t\theta/4}E_x\left\{\exp\left\{\int_0^t\rho\psi_i(x(s))\,ds\right\}\right\},
\tag{3.2}
$$

where $\rho>0$ is an arbitrary constant. Moreover, since $\|\psi_i\|\leqq 1$, we have $\|\rho\psi_i\|\leqq\rho$ and therefore from Lemma 2.2

$$
E_x\left\{\exp\left\{\int_0^t\rho\psi_i(x(s))\,ds\right\}\right\}\leqq C_\rho e^{t\lambda_{\rho\psi_i}},
$$

where

$$
\lambda_{\rho\psi_i}=\sup_{\mu\in\mathcal{M}}\left[\int_X\rho\psi_i(x)\mu(dx)-I(\mu)\right].
$$

Let

$$
\lambda_{\varepsilon,\delta,\rho}=\sup_{\psi\in A_{\varepsilon,\delta}}\lambda_{\rho\psi}
$$

so that

$$
\sup_{1\leqq i\leqq N}E_x\left\{\exp\left\{\int_0^t\rho\psi_i(x(s))\,ds\right\}\right\}\leqq C_\rho e^{t\lambda_{\varepsilon(t),\delta,\rho}}.
$$

Combining this with (3.1) and (3.2) we have

$$
\overline{\lim_{t\to\rho}}\frac{1}{t}\log P_x\{\Delta_t^\delta(\omega)\geqq\theta\}\leqq\overline{\lim_{t\to\infty}}\frac{1}{t}\log N_{\theta/8}(\varepsilon(t))+\overline{\lim_{t\to\rho}}\lambda_{\varepsilon(t),\delta,\rho}-\tfrac{1}{4}\theta\rho.
$$

Recalling that $\psi \in A_{\varepsilon,\delta}$ if it is of the form $\phi_{\varepsilon,\delta}=K_\varepsilon(T_\delta-I)\phi$, where ϕ is continuous and $\|\phi\|\leqq\frac{1}{2}$, we see from Lemma 2.1 that

$$\overline{\lim_{\delta\to0}}\,\overline{\lim_{t\to\infty}}\,\lambda_{\varepsilon(t),\delta,\rho}\leqq 0$$

for each fixed $0<\rho<\infty$. Thus

$$\overline{\lim_{\delta\to0}}\,\overline{\lim_{t\to\infty}}\,\frac{1}{t}\log P_x\{\Delta_t^\delta(\omega)\geqq\theta\}\leqq\overline{\lim_{t\to\infty}}\,\frac{1}{t}\log N_{\theta/8}(\varepsilon(t))-\tfrac{1}{4}\theta\rho$$

for each $0<\rho<\infty$. The first term on the right of this inequality is independent of ρ and by Assumption C is finite for each $\theta>0$. Hence keeping $\theta>0$ fixed and letting $\rho\to\infty$ we conclude that

$$\overline{\lim_{\delta\to0}}\,\overline{\lim_{t\to\infty}}\,\frac{1}{t}\log P_x\{\Delta_t^\delta(\omega)\geqq\theta\}=-\infty$$

which is the theorem.

4. Proof of Theorem 1.2

We now turn to the proof of Theorem 1.2. First we recall our notation. For each $x\in X$, P_x is the basic measure on Ω_x corresponding to the Markov process starting at time 0 from the point x. We denoted by $L_t(\omega,\cdot)$ the map from $\Omega_x\to\mathcal{M}$ defined by the occupation distribution, where $\mathcal{M}$ is the space of all probability measures on X. We defined a mollifying kernel $k_\varepsilon(x, y)$ which led to a map K_ε of $C(X)\to C(X)$ and the dual map also denoted by K_ε from $\mathcal{M}\to\gamma$ given by

$$(\mu K_\varepsilon)(y)=\int_X k_\varepsilon(x, y)\mu(dx)\,,$$

where γ is the subset of $\mathcal{M}$ consisting of those measures which are absolutely continuous with respect to the reference measure λ. We identified γ with the set of probability densities relative to λ. We note that, for any $\mu\in\mathcal{M}$, if $I(\mu)<\infty$, then $\mu\in\gamma$ (cf. Lemma 2.8 and its corollary in [1]). Moreover, $\{\mu\in\mathcal{M} : I(\mu)\leqq l\}$ is a compact subset of γ in the norm or L_1 topology for each finite l. We also consider the maps T_δ induced by the Markov semigroup

$$(\mu T_\delta)(y)=\int_X p(\delta, x, y)\mu(dx)$$

which also map $\mathcal{M}\to\gamma$.

Now $\mathcal{M}$ comes with two topologies. Weak convergence makes $\mathcal{M}$ a compact metric space, and $\mathcal{M}$ can also be considered as a metric space under the norm topology. In the norm topology, $\gamma\subset\mathcal{M}$ is closed.

We denote by $Q_{x,t}$ the measure on $\mathcal{M}$ induced by P_x via the occupation distribution map $L_t(\omega, \cdot)$ from $\Omega_x\to\mathcal{M}$. The map $K_{\varepsilon(t)}$ from $\mathcal{M}\to\gamma$ induces a measure $R_{x,t}$ on γ—namely $R_{x,t}=Q_{x,t}K^{-1}_{\varepsilon(t)}$. Similarly we consider the measure $R^{\delta}_{x,t}$ on γ given by $R^{\delta}_{x,t}=R_{x,t}T^{-1}_{\delta}$.

We want to prove Theorem 1.2 which is the analogue of Theorem 1.1 in the norm topology for the measures $R_{x,t}$. First we prove the theorem in the weak topology.

THEOREM 4.1. *Let $C\subset\mathcal{M}$ be closed and $G\subset\mathcal{M}$ be open (in the weak topology on $\mathcal{M}$). Then*

$$\varlimsup_{t\to\infty}\frac{1}{t}\log R_{x,t}(C)\leqq-\inf_{\mu\in C}I(\mu)\,,$$

$$\varliminf_{t\to\infty}\frac{1}{t}\log R_{x,t}(G)\geqq-\inf_{\mu\in G}I(\mu)\,.$$

Proof: It is easy to see that $\mu K_\varepsilon\Rightarrow\mu$ (weakly) as $\varepsilon\to 0$ uniformly over all $\mu\in\mathcal{M}$. Let C_1 and G_1 be, respectively, closed and open sets in $\mathcal{M}$ in the weak topology such that $C\subset C_1^0$ and $\bar{G}_1\subset G$, where C_1^0 is the interior of C_1 and $\bar{G}_1$ is the closure of G_1. Since $\mu K_\varepsilon\Rightarrow\mu$ uniformly over $\mathcal{M}$ as $\varepsilon\to 0$, there exists $\varepsilon_0>0$ such that, for all $\varepsilon<\varepsilon_0$, $\mu K_\varepsilon\in C$ implies $\mu\in C_1$ and $\mu\in G_1$ implies $\mu K_\varepsilon\in G$. Thus, since $\varepsilon(t)\to 0$ as $t\to\infty$, for $t\geqq t_0(\varepsilon_0)$, $R_{x,t}(C)\leqq Q_{x,t}(C_1)$ and $R_{x,t}(G)\geqq Q_{x,t}(G_1)$. From Theorem 1.1 we have then

$$\varlimsup_{t\to\infty}\frac{1}{t}\log R_{x,t}(C)\leqq-\inf_{\mu\in C_1}I(\mu)$$

and

$$\varliminf_{t\to\infty}\frac{1}{t}\log R_{x,t}(G)\geqq-\inf_{\mu\in G_1}I(\mu)\,.$$

Since C_1 is arbitrary except that $C_1^0\supset C$ and G_1 is arbitrary except that $\bar{G}_1\subset G$, we conclude that

$$\varlimsup_{t\to\infty}\frac{1}{t}\log R_{x,t}(C)\leqq-\sup_{C_1:C_1^0\supset C}\inf_{\mu\in C_1}I(\mu)$$

and

$$\varliminf_{t\to\infty}\frac{1}{t}\log R_{x,t}(G)\geqq-\inf_{G_1:\bar{G}_1\subset G}\inf_{\mu\in G_1}I(\mu)\,.$$

It is easy to see, using the lower semicontinuity of $I(\mu)$ and the compactness of $\mathcal{M}$, that

$$\sup_{C_1:C_1^0\supset C}\inf_{\mu\in C_1} I(\mu)=\inf_{\mu\in C} I(\mu)\,.$$

On the other hand, it is obvious that

$$\inf_{G_1:\bar{G}_1\subset G}\inf_{\mu\in G_1} I(\mu)=\inf_{\mu\in G} I(\mu)\,.$$

This proves the theorem.

We now prove a version of Theorem 1.2 for $R^{\delta}_{x,t}$ in the norm topology.

THEOREM 4.2. *Let C be closed in γ and G be open in γ in the norm topology. Then, for each $\delta>0$,*

$$\overline{\lim_{t\to\infty}}\frac{1}{t}\log R^{\delta}_{x,t}(C)\leqq -\inf_{\mu:\mu T_\delta\in C} I(\mu)\,,$$

$$\underline{\lim_{t\to\infty}}\frac{1}{t}\log R^{\delta}_{x,t}(G)\geqq -\inf_{\mu:\mu T_\delta\in G} I(\mu)\,.$$

Proof: We note that, for any set A, $R^{\delta}_{x,t}(A)=R_{x,t}(\tilde{A}_\delta)$, where $\tilde{A}_\delta=\{\mu:\mu T_\delta\in A\}$. Moreover, for each $\delta>0$, the map $\mu\to\mu T_\delta$ maps $\mathcal{M}$ with the weak topology continuously into γ with the norm topology. Therefore, $\{\mu:\mu T_\delta\in C\}$ and $\{\mu:\mu T_\delta\in G\}$ are, respectively, weakly closed and weakly open in $\mathcal{M}$ so long as C and G are norm closed and norm open in γ. We can now deduce Theorem 4.2 from Theorem 4.1.

We turn to the proof of Theorem 1.2. The transition from $R^{\delta}_{x,t}$ to $R_{x,t}$ is carried out by letting $\delta\to 0$ and using the estimates provided by Theorem 3.1.

Proof of Theorem 1.2: Let C be norm closed in γ. Let $C_\theta=\{\beta\in\gamma:\|\beta-\alpha\|<\theta \text{ for some } \alpha\in C\}$ and denote its closure by $\bar{C}_\theta$. We have, for any $t>0$, $\delta>0$, $\theta>0$,

$$\{\omega: g_t(\omega,\cdot)\in C\}\subset\{\omega: g^{\delta}_t(\omega,\cdot)\in\bar{C}_\theta\}\cup\{\omega:\|g^{\delta}_t(\omega,y)-g_t(\omega,y)\|\geqq\theta\}$$

so that

$$P_x\{g_t(\omega,\cdot)\in C\}\leqq P_x\{g^{\delta}_t(\omega,\cdot)\in\bar{C}_\theta\}+P_x\{\Delta^{\delta}_t(\omega)\geqq\theta\}\,,$$

or

$$R_{x,t}(C)\leqq R^{\delta}_{x,t}(\bar{C}_\theta)+P_x\{\Delta^{\delta}_t(\omega)\geqq\theta\}\,. \tag{4.1}$$

Hence,

$$\overline{\lim_{t\to\infty}} \frac{1}{t} \log R_{x,t}(C) \leqq \max \left[\overline{\lim_{t\to\infty}} \frac{1}{t} \log R^{\delta}_{x,t}(\bar{C}_\theta), \overline{\lim_{t\to\infty}} \frac{1}{t} \log P_x\{\Delta^{\delta}_t(\omega) \geqq \theta\} \right].$$

From Theorems 3.1 and 4.2, we obtain on letting $\delta \to 0$

$$\overline{\lim_{t\to\infty}} \frac{1}{t} \log R_{x,t}(C) \leqq -\overline{\lim_{\delta\to 0}} \inf_{\mu:\mu T_\delta \in \bar{C}_\theta} I(\mu) . \tag{4.2}$$

Since the left side of (4.2) is independent of $\theta > 0$, we can let $\theta \to 0$ and conclude that

$$\overline{\lim_{t\to\infty}} \frac{1}{t} \log R_{x,t}(C) \leqq -\overline{\lim_{\theta\to 0}}\, \overline{\lim_{\delta\to 0}} \inf_{\mu:\mu T_\delta \in \bar{C}_\theta} I(\mu) . \tag{4.3}$$

We have to show that

$$\overline{\lim_{\theta\to 0}}\, \overline{\lim_{\delta\to 0}} \inf_{\mu:\mu T_\delta \in \bar{C}_\theta} I(\mu) \geqq \inf_{\mu \in C} I(\mu) . \tag{4.4}$$

In fact, we shall show that

$$\underline{\lim}_{\substack{\theta\to 0\\ \delta\to 0}} \inf_{\mu:\mu T_\delta \in \bar{C}_\theta} I(\mu) \geqq \inf_{\mu \in C} I(\mu) . \tag{4.5}$$

Suppose there is a sequence $\{\mu_n\}$ such that $I(\mu_n) \to l$ and $\mu_n T_{\delta_n} \in \bar{C}_{\theta_n}$ for some sequence δ_n, θ_n going to zero. Since $I(\mu_n)$ is bounded, we know that $\{\mu_n\} \in \gamma$ is a totally bounded sequence in γ in the norm topology. We also know from the corollary to Lemma 4.1 of [1] that

$$\sup_n \|\mu_n T_\delta - \mu_n\| \to 0$$

as $\delta \to 0$. Therefore, $\{\mu_n\}$ has a subsequence converging in norm to a limit μ and, since $\mu_n T_{\delta_n} \to \mu$ in norm, we conclude that $\mu \in C$. By lower semicontinuity, $I(\mu) \leqq l$. This proves (4.5) and from (4.3) we have

$$\overline{\lim_{t\to\infty}} \frac{1}{t} \log R_{x,t}(C) \leqq -\inf_{\mu \in C} I(\mu) .$$

To derive the lower bound of Theorem 1.2, we let $G \subset \gamma$ be open in the norm topology, and we must prove

$$\underline{\lim}_{t\to\infty} \frac{1}{t} \log R_{x,t}(G) \geqq -\inf_{\mu \in G} I(\mu) . \tag{4.6}$$

It is clearly sufficient to prove that, for any $\mu \in G$,

$$\varliminf_{t\to\infty} \frac{1}{t} \log R_{x,t}(G) \geqq -I(\mu) . \tag{4.7}$$

To prove (4.7) we note that, since G is open and $\mu \in G$, there exists a sphere $S(\mu, \theta)$ of some positive radius θ around μ such that $S(\mu, \theta) \subset G$. Clearly,

$$R_{x,t}(G) \geqq R_{x,t}(S(\mu, \theta)) \geqq R^{\delta}_{x,t}(S(\mu, \tfrac{1}{2}\theta)) - P_x(\Delta^{\delta}_t(\omega) \geqq \tfrac{1}{2}\theta) .$$

We again use Theorems 3.1 and 4.2 and obtain

$$\varliminf_{t\to\infty} \frac{1}{t} \log R_{x,t}(G) \geqq -\varliminf_{\delta\to 0} \inf_{\alpha : \alpha T_\delta \in S(\mu, \theta/2)} I(\alpha) . \tag{4.8}$$

Since $I(\mu)$ can be assumed to be finite, $\|\mu T_\delta - \mu\| \to 0$ as $\delta \to 0$ by the corollary to Lemma 4.1 of [1]. Therefore, for sufficiently small δ,

$$\inf_{\alpha : \alpha T_\delta \in S(\mu, \theta/2)} I(\alpha) \leqq I(\mu) .$$

This combined with (4.8) gives us (4.7) and hence (4.6) which completes the proof of Theorem 1.2.

5. A Class of Examples

We shall now consider in detail a class of examples for Theorem 1.2 which we use later in [3] to study the sausage problem.

Let R^d be d-dimensional Euclidean space. Let $M>0$ and let G^M_d be the discrete subgroup of R^d consisting of vectors having for each coordinate an integral multiple of M. We denote by T^M_d the quotient of R^d modulo G^M_d. With reference to Theorem 1.2, we shall take as our state space X the space T^M_d which is a compact torus.

For the Markov process with transition probability $p(t, x, dy)$ and state space X, we take a process with independent increments, i.e., we have a one-parameter convolution semigroup π_t of probability measures on X and

$$p(t, x, A) = \pi_t(A - x) .$$

If, for $t>0$, $\pi_t(A)$ has a density $\nu_t(y)$ with respect to the Haar measure (Lebesgue measure) on X and if the density $\nu_t(y)$ is continuous and strictly positive on X, then Assumptions A and D are clearly satisfied with the Haar measure as the reference measure λ.

Let us take a family $\tilde{k}_\varepsilon(x)$ of probability densities on R^d such that $\tilde{k}_\varepsilon(x)$ converges weakly as $\varepsilon \to 0$ to the degenerate distribution at the origin. We can form the induced density on $\mathfrak{X}$ by

$$k_\varepsilon(x) = \sum_{g \in G_d^M} \tilde{k}_\varepsilon(x+g)$$

and now $k_\varepsilon(x)$ can be used as a mollifier on $\mathfrak{X}$, i.e., $k_\varepsilon(x-y)$ satisfies Assumption B. We shall further assume that $\tilde{k}_\varepsilon(x)$ depends on ε by a scale change. That is,

$$\tilde{k}_\varepsilon(x) = \varepsilon^{-d} \tilde{k}\left(\frac{x}{\varepsilon}\right),$$

where $\tilde{k}(x)$ is a fixed probability density on R^d.

Let us find the impact of Assumption C so that we can use Theorem 1.2.

LEMMA 5.1. *Assumption C holds if $\varepsilon(t) \to 0$ as $t \to \infty$ in such a manner that $t^{1/d}\varepsilon(t)$ is bounded away from zero.*

Proof: We must compute, for fixed $\delta > 0$ and $\varepsilon > 0$, the number of δ spheres required to cover the image of the unit ball of $C(\mathfrak{X})$ by the map

$$(K_\varepsilon \phi)(x) = \int_{\mathfrak{X}} k_\varepsilon(x-y)\phi(y)\, dy .$$

Since $\|\phi\| \leqq 1$, denoting by ψ the function $\psi = K_\varepsilon \phi$, we have

$$|\psi(x) - \psi(x')| \leqq \int_{\mathfrak{X}} |k_\varepsilon(x-y) - k_\varepsilon(x'-y)|\, dy$$

$$\leqq \int_{R^d} |\tilde{k}_\varepsilon(x-y) - \tilde{k}_\varepsilon(x'-y)|\, dy .$$

If we introduce the L_1-modulus of continuity

$$W_\varepsilon(h) = \sup_{\|z\| \leqq h} \int_{R^d} |\tilde{k}_\varepsilon(x+z) - \tilde{k}_\varepsilon(x)|\, dx ,$$

then the image S_ε of the unit ball by K_ε is contained in the set of functions on $\mathfrak{X}$ bounded by 1 and having a modulus of continuity dominated by $W_\varepsilon(h)$. Given a $\delta > 0$ and $\varepsilon > 0$ we pick $h_0 = h_0(\varepsilon, \delta)$ such that

$$W_\varepsilon(h_0) < \tfrac{1}{2}\delta .$$

Then the torus X can be covered by a finite number $N(h_0)$ of sets each of which has diameters less than h_0. We can assume these sets to be disjoint and call them $T_1, T_2, \cdots, T_N$. By the modulus of continuity estimate, any $\psi \in S_\varepsilon$ can be approximated by a simple function which is constant on each T_i uniformly to within $\frac{1}{2}\delta$. Each such simple function can be approximated to within $\frac{1}{2}\delta$ by a simple function which takes its values only from a finite grid on $[-1, 1]$ with gap size at most $\frac{1}{2}\delta$. The number of points on the grid can be taken to be $[\frac{1}{4}\delta]+1$. We have, therefore, an estimate on the number of spheres of radius δ needed to cover S_ε. It is an estimate of the form

$$N \leqq ([\tfrac{1}{4}\delta]+1)^{N(h_0)}\,.$$

It is easy to see that Assumption C means that $\varepsilon(t) \to 0$ as $t \to \infty$ in such a way that $N(h_0)$ grows at most linearly with t. We know that, as $h \to 0$,

$$N(h) \leqq Ch^{-d}\,,$$

where C is a constant depending only on the size M of the torus. Therefore, we have to be sure that

$$\overline{\lim_{t\to\infty}} \frac{1}{[h(\varepsilon(t), \delta)]^d} \cdot \frac{1}{t} < \infty\,,$$

or equivalently that

$$\underline{\lim_{t\to\infty}}\, t^{1/d} h(\varepsilon(t), \delta) > 0\,. \tag{5.1}$$

If $\tilde{k}_\varepsilon$ is obtained from $\tilde{k}$ by a scale change, then

$$\begin{aligned} W_\varepsilon(h) &= \sup_{\|z\| \leqq h} \int_{R^d} \varepsilon^{-d} \left| \tilde{k}\left(\frac{x+z}{\varepsilon}\right) - \tilde{k}\left(\frac{x}{\varepsilon}\right) \right| dx \\ &= \sup_{\|z\| \leqq h} \int_{R^d} \left| \tilde{k}\left(x + \frac{z}{\varepsilon}\right) - \tilde{k}(x) \right| dx \\ &= \sup_{\|z\| \leqq h/\varepsilon} \int_{R^d} |\tilde{k}(x+z) - \tilde{k}(x)|\, dx \\ &= W_1\left(\frac{h}{\varepsilon}\right). \end{aligned}$$

So, for fixed $\delta > 0$, since $W_1(h) \to 0$ as $h \to 0$, we can pick $h(\delta)$ such that

$$W_1(h(\delta)) < \tfrac{1}{2}\delta\,,$$

so that

$$h_0(\varepsilon, \delta) = \varepsilon h(\delta) .$$

Therefore, condition (5.1) reduces to

$$\varliminf_{t\to\infty} t^{1/d}\varepsilon(t) > 0$$

which proves the lemma.

One way of generating processes with independent increments on $\boldsymbol{X}$ is to start from a process with independent increments on R^d and project it onto $\boldsymbol{X}$. We shall now look at a class of such processes and see how our theorem applies to them.

Let $0<\alpha\leqq 2$. By a symmetric stable process of order α, we mean a process with homogeneous independent increments $\tilde{x}(t)$ such that

$$E\{e^{i\langle\xi, \tilde{x}(t)\rangle}\} = e^{-tQ(\xi)} ,$$

where

$$Q(\xi) = \int_0^\infty \int_{S^{d-1}} (1-\cos\langle\xi, rs\rangle)\frac{\Omega(ds)}{r^{1+\alpha}}\,dr \qquad \text{if} \qquad 0<\alpha<2 ,$$

$$= \tfrac{1}{2}\sum a_{ij}\xi_i\xi_j \qquad \text{if} \qquad \alpha = 2 .$$

If $0<\alpha<2$, then $\Omega(ds)$ is a symmetric finite mass distribution on the unit sphere S^{d-1}, $r>0$ is a positive scalar, and rs is interpreted as a vector in R^d. If $\alpha=2$, then $\{a_{ij}\}$ is a symmetric positive semidefinite matrix. An elementary computation yields

$$Q(\xi) = Q(-\xi)$$

and

$$Q(\lambda\xi) = |\lambda|^\alpha Q(\xi)$$

for real λ.

We make the non-degeneracy assumption: for $0<\alpha<2$, the support S_0 of $\Omega(ds)$ spans R^d as a vector space; for $\alpha=2$, we assume $\{a_{ij}\}$ is positive definite. Under these assumptions, it follows that

$$\inf_{\|\xi\|=1} Q(\xi) > 0 .$$

Therefore,

$$e^{-tQ(\xi)} \leqq e^{-tc\|\xi\|^{\alpha}}$$

and is rapidly decreasing. It follows from this that

$$\tilde{\nu}_t(y) = \left(\frac{1}{2\pi}\right)^d \int_{R^d} e^{-tQ(\xi)} e^{-i\langle\xi,y\rangle} \, d\xi$$

is bounded on all of R^d and is smooth. The same is true of

$$\begin{aligned} \nu_t(y) &= \sum_{g\in G_d^M} \tilde{\nu}_t(y+g) \\ &= c \sum_{\xi\in \tilde{G}_d^M} e^{-tQ(\xi)} e^{-i\langle\xi,y\rangle} . \end{aligned}$$

Here $\tilde{G}_d^M$ is the lattice which is dual to G_d^M. One can also check that, under the non-degeneracy assumption, $\tilde{\nu}_t(x)>0$ for each $t>0$ and x. Therefore the same is true of $\nu_t(x)$. The space $\boldsymbol{X}$ being compact, $\nu_t(x)$ has a positive lower bound for each $t>0$.

If we take a probability density $\tilde{k}(x)$ on R^d and project it onto $\boldsymbol{X}$ by

$$k(x) = \sum_{g\in G_d^M} \tilde{k}(x+g) ,$$

then the family $k_\varepsilon(x)$ obtained from

$$\tilde{k}_\varepsilon(x) = \varepsilon^{-d} \tilde{k}\left(\frac{x}{\varepsilon}\right)$$

can be used as a mollifier. If we take $\varepsilon(t)=\varepsilon/t^{-1/d}$ for some $\varepsilon>0$, then Theorem 1.2 is applicable. In order to have a convenient reference for future use in the sausage problem (cf. [3]) we shall restate Theorem 1.2 as it applies to this case.

Let $\tilde{x}(t)$ be a symmetric stable process on R^d of order α, $0<\alpha\leqq 2$. It is characterized either by an Ω if $0<\alpha<2$, or by $\{a_{ij}\}$ if $\alpha=2$ (Brownian motion in R^d). We shall assume that it is non-degenerate. Let $x(t)$ be the process induced on the torus $\boldsymbol{X}$ by projection. Then $x(t)$ is again a process with independent increments on $\boldsymbol{X}$. Let $L_t(\omega, A)$ be the occupation distribution of the $x(\cdot)$ process. We take a mollifier $k_\varepsilon(x)$ on $\boldsymbol{X}$ by projecting down a mollifier $\tilde{k}_\varepsilon(x)$ from R^d. The dependence of $\tilde{k}_\varepsilon(x)$ on ε is by a scale change

$$\tilde{k}_\varepsilon(x) = \varepsilon^{-d} \tilde{k}\left(\frac{x}{\varepsilon}\right) .$$

We define

$$l_t^\varepsilon(\omega, y) = \int k_\varepsilon(y-x) L_t(\omega, dx)$$

and denote

$$g_t(\omega, y) = l_t^{\varepsilon(t)}(\omega, y),$$

where $\varepsilon(t) = \varepsilon t^{-1/d}$ with $\varepsilon > 0$ fixed. We denote by $R_{x,t}$ the distribution of $g_t(\omega, y)$ on γ and by $I(\mu)$ the I-functional corresponding to the given process. We then have the following theorem and its corollary.

THEOREM 5.1. *Let $C \subset \gamma$ be closed and $G \subset \gamma$ be open in the norm topology. Then*

$$\overline{\lim_{t\to\infty}} \frac{1}{t} \log R_{x,t}(C) \leqq -\inf_{\mu\in C} I(\mu),$$

$$\underline{\lim}_{t\to\infty} \frac{1}{t} \log R_{x,t}(G) \geqq -\inf_{\mu\in G} I(\mu).$$

COROLLARY. *Let $\Phi_t(f)$ be a family of functionals defined on γ such that, for each $f \in \gamma$ with $I(f) < \infty$ and each family $f_t \in \gamma$ converging to f in norm,*

$$\underline{\lim}_{t\to\infty} \Phi_t(f) \geqq \Phi(f),$$

where $\Phi(f)$ is a lower semicontinuous functional on γ. We assume further that $\Phi_t(f) \geqq 0$ for all t. Then,

$$\overline{\lim_{t\to\infty}} \frac{1}{t} \log E^{P_x}\{e^{-t\Phi_t(g_t(\omega,\cdot))}\} = \overline{\lim_{t\to\infty}} \frac{1}{t} \log E^{R_{x,t}}\{e^{-t\Phi_t(f)}\}$$

$$\leqq -\inf_{f\in\gamma} [\Phi(f) + I(f)].$$

Remark. If $f \in \gamma$, then f can be thought of as a measure μ on $\boldsymbol{X}$ and we use $I(\mu)$ and $I(f)$ interchangeably. The corollary follows from Theorem 5.1 by using Section 3 of [2].

Bibliography

[1] Donsker, M. D., and Varadhan, S. R. S., *Asymptotic evaluation of certain Markov process expectations for large time, I,* Comm. Pure App. Math. Vol. 28, 1975, pp. 1–47.

[2] Varadhan, S. R. S., *Asymptotic probabilities and differential equations,* Comm. Pure App. Math. Vol. 19, 1966, pp. 261–286.

[3] Donsker, M. D., and Varadhan, S. R. S., *Asymptotics for the Wiener sausage,* to appear in Comm. Pure App. Math., Vol. 28, issue 4, 1975.

Received October, 1974.

Functional Integration and its Applications, Clarendon Press, 1975, 15-33

Asymptotic evaluation of certain Wiener integrals for large time

M. D. DONSKER AND S. R. S. VARADHAN

2.1. Introduction

LET $V(y) \geqslant 0$ and continuous on $-\infty < y < \infty$. If $V(y) \to +\infty$ as $y \to \pm\infty$, the eigenvalue problem

$$\tfrac{1}{2}\psi''(y) - V(y)\psi(y) = -\lambda\psi(y) \tag{2.1}$$

has a discrete spectrum, and an old result of Kac [1] is that, for the least eigenvalue λ_1,

$$\lim_{t\to\infty} \frac{1}{t} \ln E_x\left\{\exp\left[-\int_0^t V(z(\tau))\,d\tau\right]\right\} = -\lambda_1 \tag{2.2}$$

where $E_x\{\ \}$ denotes expectation on the Brownian-motion process $\{z_\tau, 0 \leqslant \tau < \infty\}$ with $z(0) = x$.

The formal idea of his proof was that if, for any real y, we let $u(t, x, y)$ be the solution of

$$\frac{\partial u}{\partial t} = \frac{1}{2}\frac{\partial^2 u}{\partial x^2} - V(x)u,$$

$$u(x, 0) = \delta(x - y),$$

then, on the one hand, by the Feynman–Kac formula,

$$u(t, x, y) = E_x\left\{\exp\left[-\int_0^t V(z(\tau))\,d\tau\right]\delta(z(t) - y)\right\}$$

and, on the other hand,

$$u(t, x, y) = \sum_{j=1}^{\infty} \exp(-\lambda_j t)\psi_j(x)\psi_j(y)$$

where $\{\lambda_j\}$ and $\{\psi_j\}$ are respectively the eigenvalues and normalized eigenfunctions of (2.1). Thus, integrating on y

$$E_x\left\{\exp\left[-\int_0^t V(z(\tau))\,d\tau\right]\right\} = \sum_{j=1}^{\infty} \exp(-\lambda_j t)\psi_j(x)\int_{-\infty}^{\infty} \psi_j(y)\,dy$$

which gives (2.2).

A long-standing question has been the relation between (2.2) and the fact that we also know

$$\lambda_1 = \inf\left\{\int_{-\infty}^{\infty} V(y)\psi^2(y)\,dy + \tfrac{1}{2}\int_{-\infty}^{\infty} [\psi'(y)]^2\,dy\right\} \tag{2.3}$$

where the infimum is taken over all $\psi \in L^2$ such that $\int_{-\infty}^{\infty} \psi^2(y)\,dy = 1$. It is reasonable to expect that an expression like the right side of (2.3) should come from a direct asymptotic evaluation of $E_x\{\exp[-\int_0^t V(z(\tau))\,d\tau]\}$. Moreover, if such a direct connection can be made, one should be able to find the asymptotic behaviour of Brownian-motion expectations of more general functionals—including those for which there is no associated differential equation at all. Such a theorem is proved in this chapter.

Let $\mathscr{F}$ be the space of probability distribution functions $F(y)$, $-\infty < y < \infty$, with the Lévy metric:

$$\rho(F_1, F_2) = \inf_{h>0} [F_1(y-h) - h \leq F_2(y) \leq F_1(y+h) + h \text{ for all } y].$$

Let $\Phi(F)$ be a function on $\mathscr{F}$ satisfying:

(1) $0 \leq \Phi(F) \leq \infty$.

(2) $\Phi(F)$ is lower semicontinuous on $\mathscr{F}$.

(3) $\{F \in \mathscr{F} : \Phi(F) \leq K\}$ is a compact subset of $\mathscr{F}$ for every finite $K > 0$.

(4) If $F_n \Rightarrow F$ and the support of F_n is contained in an interval $[a, b]$ for all n, then $\Phi(F_n) \to \Phi(F)$.

(5) Let $F \in \mathscr{F}$ be such that $\Phi(F) < \infty$ and let $\{g_n(y)\}$ be a sequence of continuous functions such that $0 \leq g_n(y) \leq 1$, $-\infty < y < \infty$, with $g_n(y) = 1$ for $|y| \leq n$. If we define $F_n(y) = \int_{-\infty}^{y} g_n(\xi)\,dF(\xi) \Big/ \int_{-\infty}^{\infty} g_n(\xi)\,dF(\xi)$, then $\Phi(F_n) \to \Phi(F)$.

If $F \in \mathscr{F}$ has a density function f, we will sometimes for convenience write $\Phi(f)$ for $\Phi(F)$.

Let Ω_x be the space of continuous functions $z(\tau)$, $0 \leq \tau < \infty$, with $z(0) = x$. For $t > 0$ and $\omega \in \Omega_x$, we define

$$L(t, \omega, y) = \frac{1}{t} \lambda\{s : 0 \leq s \leq t,\ z(s) \leq y\},$$

where $\lambda\{\ \}$ denotes Lebesgue measure, i.e. $L(t, \omega, y)$ is the proportion of time on $[0, t]$ that a particular path $\omega = z(\cdot)$ is less than or equal to y. We note that for each ω and each $t > 0$, $L(t, \omega, .) \in \mathscr{F}$. Since each ω is bounded on $[0, t]$, $L(t, \omega, y)$ is a distribution function whose support is $[\inf_{0 \leq \tau \leq t} z(\tau), \sup_{0 \leq \tau \leq t} z(\tau)]$. $L(t, \omega, y)$ has a density function, $l(t, \omega, y)$, often called the 'local time', and conveniently expressed as

$$\frac{1}{t} \int_0^t \delta(z(\tau) - y)\,d\tau.$$

Let $\mathscr{f}$ denote the space of probability density functions f which are continuously differentiable on $-\infty < y < \infty$ and either $f(y) > 0$ for all $-\infty < y < \infty$ or f has compact support and is strictly positive on the interior of that support. The main result of this chapter is the following.

THEOREM 2.1. *Let $\Phi(F)$ satisfy hypotheses* (1)–(5) *above. Then*

$$\lim_{t\to\infty}\frac{1}{t}\ln E_x\{e^{-t\Phi[L(t,\omega,.)]}\} = -\inf_{t\in \mathscr{f}}\left\{\frac{1}{8}\int_{-\infty}^{\infty}\frac{[f'(y)]^2}{f(y)}\,dy+\Phi(f)\right\}, \tag{2.4}$$

where $E_x\{\ \}$ denotes integration on Ω_x with respect to Brownian-motion measure.

Let us look at some examples of this theorem. In the particular case, $\Phi(F)=\int_{-\infty}^{\infty}V(y)\,dF(y)$, where $V(y)\geq 0$ and is continuous, and $V(y)\to+\infty$ as $y\to\pm\infty$, it is easy to see that $\Phi(F)$ satisfies hypotheses (1)–(5).† In this case,

$$\Phi(\dot{L}(t,\omega,.))=\int_{-\infty}^{\infty}V(y)\,dL(t,\omega,y)=\frac{1}{t}\int_0^t V(z(\tau))\,d\tau,$$

and if, for any $f\in\mathscr{f}$, we let $\psi(y)=\sqrt{(f(y))}$, then $\psi\in L^2$ and, indeed, $\int_{-\infty}^{\infty}\psi^2(y)\,dy=1$. Moreover,

$$\int_{-\infty}^{\infty}V(y)\,dF(y)=\int_{-\infty}^{\infty}V(y)f(y)\,dy=\int_{-\infty}^{\infty}V(y)\psi^2(y)\,dy$$

and

$$\frac{1}{8}\int_{-\infty}^{\infty}\frac{[f'(y)]^2}{f(y)}\,dy=\frac{1}{2}\int_{-\infty}^{\infty}[\psi'(y)]^2\,dy.$$

Thus, in the special case $\Phi(F)=\int_{-\infty}^{\infty}V(y)dF(y)$, (2.4) becomes

$$\lim_{t\to\infty}\frac{1}{t}\ln E_x\left\{\exp\left[-\int_0^t V(z(\tau))\,d\tau\right]\right\}$$

$$=-\inf\left\{\int_{-\infty}^{\infty}V(y)\psi^2(y)\,dy+\frac{1}{2}\int_{-\infty}^{\infty}[\psi'(y)]^2\,dy\right\} \tag{2.5}$$

where the infimum is taken over all $\psi\in L^2$ such that $\int_{-\infty}^{\infty}\psi^2(y)\,dy=1$. This solves the problem mentioned earlier.

If we let

$$\Phi(F)=\int_{-\infty}^{\infty}\int_{-\infty}^{\infty}\rho(y_1,y_2,)\,dF(y_1)\,dF(y_2),$$

where $\rho(y_1,y_2)$ is continuous and $\rho(y_1,y_2)\to+\infty$ if $|y_1|+|y_2|\to\infty$, then $\Phi(F)$ satisfies hypotheses (1)–(5) and (2.4) becomes

$$\lim_{t\to\infty}\frac{1}{t}\ln E_x\left\{\exp\left[-\frac{1}{t}\int_0^t\int_0^t\rho(z(s),z(\tau))\,ds\,d\tau\right]\right\}$$

$$=-\inf_{f\in\mathscr{f}}\left\{\frac{1}{8}\int_{-\infty}^{\infty}\frac{[f'(y)]^2}{f(y)}\,dy+\int_{-\infty}^{\infty}\int_{-\infty}^{\infty}\rho(y_1,y_2)f(y_1)f(y_2)\,dy_1\,dy_2\right\}. \tag{2.6}$$

† The condition $V(y)\to+\infty$ as $y\to\pm\infty$ is used only for hypothesis (3).

If $\Phi(F)=\exp[\int_{-\infty}^{\infty} y^2\,\mathrm{d}F(y)]$, one obtains readily from (2.4) that

$$\lim_{t\to\infty}\frac{1}{t}\ln E_x\left(\exp\left\{-t\exp\left[\frac{1}{t}\int_0^t z^2(\tau)\,\mathrm{d}\tau\right]\right\}\right)=-\frac{\alpha+1}{8\alpha^2},$$

where $\alpha>0$ is the root of $\mathrm{e}^u=1/8u^2$.

Before we proceed with the proof we shall try to explain the idea behind the proof. By the ergodic theorem, since Brownian motion has no finite invariant measure on the line, as time increases, most of the Brownian paths will spend only a negligible proportion of time in a finite region. However, there will be a few Brownian paths, with very low probability, which hang around too long. Let us fix a nice probability density $f(y)$ and ask the following question. What is the probability that the occupation distribution $L(t,\omega,y)$ of a Brownian path is close to the distribution $F(y)$ with density $f(y)$? The probability will of course go to zero with t: the question is, at what rate? It turns out that, roughly,

$$\mathrm{Prob}\{L(t,\omega,.)\sim F(.)\}\sim\exp\left\{-\frac{t}{8}\int_{-\infty}^{\infty}\frac{[f'(y)]^2}{f(y)}\,\mathrm{d}y\right\}.$$

For practical purposes one can view the measure Q_x^t on the space of distribution functions as behaving like

$$Q_x^t\sim\exp\left\{-\frac{t}{8}\int_{-\infty}^{\infty}\frac{[f'(y)]^2}{f(y)}\,\mathrm{d}y\right\}\mathrm{d}f.$$

The theorem is then a Laplace asymptotic formula.

This type of reasoning generalizes to a wide class of Markov processes both with discrete and continuous time and will be dealt with in several forthcoming articles.

2.2. The upper estimate

Let P_x denote Brownian-motion measure on Ω_x. For each x and $t>0$ we define a probability measure $Q_{x,t}$ on $\mathscr{F}$ by

$$Q_{x,t}(C)=P_x\{\omega:L(t,\omega,.)\in C\}$$

for measurable sets $C\subset\mathscr{F}$, i.e. for which the right side is defined.

Denoting expectation with respect to the measure $Q_{x,t}$ by $E^{Q_{x,t}}\{\ \}$, we have

$$E_x\{\mathrm{e}^{-t\Phi[L(t,\omega,.)]}\}=E^{Q_{x,t}}(\mathrm{e}^{-t\Phi(F)}).\qquad(2.7)$$

In order to get an upper estimate on the left side of (2.7) we first get an upper estimate on $Q_{x,t}(C)$ for appropriate sets C. Let $\mathscr{U}$ be the space of functions $u(y)$, $-\infty<y<\infty$, having two bounded continuous derivatives and for each of which there are two positive numbers α and β such that for all y, $0<\alpha\leqslant u(y)\leqslant\beta<\infty$.

For $u \in \mathscr{U}$, define

$$\psi(x, t) = E_x\left\{u(z(t)) \exp\left[-\int_0^t \left(\frac{u''}{2u}\right)(z(s))\, ds\right]\right\}.$$

From the Feynman–Kac formula we see that $\psi(x, t)$ satisfies

$$\frac{\partial \psi}{\partial t} = \frac{1}{2}\frac{\partial^2 \psi}{\partial x^2} - \frac{u''(x)}{2u(x)}\quad \psi,$$

but, since the solution of this last is clearly $\psi(x, t) = u(x)$, we have

$$E_x\left\{u(z(t)) \exp\left[-\int_0^t \left(\frac{u''}{2u}\right)(z(s))\, ds\right]\right\} = u(x). \tag{2.8}$$

Since $u(y) \geqslant \alpha > 0$ for all y,

$$E_x\left\{\exp\left[-\int_0^t \left(\frac{u''}{2u}\right)(z(s))\, ds\right]\right\} \leqslant \frac{u(x)}{\alpha}.$$

But

$$E_x\left\{\exp\left[-\int_0^t \left(\frac{u''}{2u}\right)(z(s))\, ds\right]\right\} = E_x\left\{\exp\left[-t\int_{-\infty}^{\infty} \left(\frac{u''}{2u}\right)(y)\, dL(t, \omega, y)\right]\right\}$$
$$= E^{Q_{x,t}}\left\{\exp\left[-t\int_{-\infty}^{\infty} \left(\frac{u''}{2u}\right)(y)\, dF(y)\right]\right\},$$

and therefore for any measurable $C \subset \mathscr{F}$,

$$Q_{x,t}(C) \leqslant \frac{u(x)}{\alpha} \exp\left\{t \sup_{F \in C} \int_{-\infty}^{\infty} \left(\frac{u''}{2u}\right)(y)\, dF(y)\right\}. \tag{2.9}$$

Since (2.9) holds for any $u \in \mathscr{U}$, we obtain for any measurable $C \subset \mathscr{F}$,

$$\overline{\lim_{t\to\infty}} \frac{1}{t} \ln Q_{x,t}(C) \leqslant \inf_{u \in \mathscr{U}} \sup_{F \in C} \int_{-\infty}^{\infty} \left(\frac{u''}{2u}\right)(y)\, dF(y). \tag{2.10}$$

Let

$$I(F) = -\inf_{u \in \mathscr{U}} \int_{-\infty}^{\infty} \frac{u''(y)}{2u(y)}\, dF(y).$$

We see that $I(F)$ is lower semicontinuous on $\mathscr{F}$ and $0 \leqslant I(F) \leqslant \infty$. We will show now that for compact sets $C \subset \mathscr{F}$ it follows from (2.10) that

$$\overline{\lim_{t\to\infty}} \frac{1}{t} \ln Q_{x,t}(C) \leqslant \sup_{F \in C} \inf_{u \in \mathscr{U}} \int_{-\infty}^{\infty} \frac{u''(y)}{2u(y)}\, dF(y)$$
$$= -\inf_{f \in C} I(F)$$

First of all, for any measurable set $C \subset \mathscr{F}$ which is contained in a finite union $\bigcup_{j=1}^{k} C_j$ of measurable sets, we have from (2.10)

$$\overline{\lim_{t\to\infty}} \frac{1}{t} \ln Q_{x,t}(C) \leq \max_{1\leq j\leq k} \inf_{u\in\mathscr{U}} \sup_{F\in C_j} \int_{-\infty}^{\infty} \frac{u''(y)}{2u(y)} \, \mathrm{d}F(y),$$

and for any measurable set $C \subset \mathscr{F}$

$$\overline{\lim_{t\to\infty}} \frac{1}{t} \ln Q_{x,t}(C) \leq \inf_{\substack{C_1, C_2, \ldots, C_k \\ C \subset \bigcup_{j=1}^{k} C_j}} \max_{1\leq j\leq k} \inf_{u\in\mathscr{U}} \sup_{F\in C_j} \int_{-\infty}^{\infty} \frac{u''(y)}{2u(y)} \, \mathrm{d}F(y).$$

Let C be a compact subset of $\mathscr{F}$ and let

$$l = \sup_{F\in C} \inf_{u\in\mathscr{U}} \int_{-\infty}^{\infty} \frac{u''(y)}{2u(y)} \, \mathrm{d}F(y),$$

so that for every $F \in C$,

$$\inf_{u\in\mathscr{U}} \int_{-\infty}^{\infty} \frac{u''(y)}{2u(y)} \, \mathrm{d}F(y) \leq l.$$

Thus, given $\varepsilon > 0$, for every $F \in C$ there exists $u_F \in \mathscr{U}$ such that

$$\int_{-\infty}^{\infty} \frac{u_F''(y)}{2u_F(y)} \, \mathrm{d}F(y) \leq l + \varepsilon.$$

Since $u''(y)/2u(y)$ is bounded and continuous on $-\infty < y < \infty$ for every $u \in \mathscr{U}$, $\int_{-\infty}^{\infty} u_F''(y)/2u_F(y) \, \mathrm{d}F(y)$ is a continuous function on $\mathscr{F}$. Thus for every $F \in C$ there is a Levy neighbourhood N_F such that for every $G \in N_F$,

$$\int_{-\infty}^{\infty} \frac{u_F''(y)}{2u_F(y)} \, \mathrm{d}G(y) \leq l + \varepsilon.$$

Now $\bigcup_{F\in C} N_F$ is an open covering of C and therefore there exists $F_1, F_2, \ldots, F_k$ in C such that $C \subset \bigcup_{j=1}^{k} N_{F_j}$. Let $C_j = N_{F_j}$ and we have

$$\sup_{G\in C_j} \int_{-\infty}^{\infty} \frac{u_{F_j}''(y)}{2u_{F_j}(y)} \, \mathrm{d}G(y) \leq l + 2\varepsilon,$$

$$\inf_{u\in\mathscr{U}} \sup_{G\in C_j} \int_{-\infty}^{\infty} \frac{u''(y)}{2u(y)} \, \mathrm{d}G(y) \leq l + 2\varepsilon,$$

$$\max_{1\leq j\leq k} \inf_{u\in\mathscr{U}} \sup_{G\in C_j} \int_{-\infty}^{\infty} \frac{u''(y)}{2u(y)} \, \mathrm{d}G(y) \leq l + 2\varepsilon,$$

and finally

$$\inf_{\substack{C_1,C_2,\ldots,C_k\\ \bigcup_{j=1}^k C_j \supset C}} \max_{l\leqslant j\leqslant k} \inf_{u\in\mathscr{U}} \sup_{G\in C_j} \int_{-\infty}^{\infty} \frac{u''(y)}{2u(y)} \mathrm{d}G(y) \leqslant l + 2\varepsilon.$$

Since ε is arbitrary, we conclude that for compact sets C†

$$\inf_{\substack{C_1,C_2,\ldots,C_k\\ \bigcup_{j=1}^k C_j \supset C}} \max_{l\leqslant j\leqslant k} \inf_{u\in\mathscr{U}} \sup_{F\in C_j} \int_{-\infty}^{\infty} \frac{u''(y)}{2u(y)} \mathrm{d}F(y) \leqslant \sup_{F\in C} \inf_{u\in\mathscr{U}} \int_{-\infty}^{\infty} \frac{u''(y)}{2u(y)} \mathrm{d}F(y)$$

$$= -\inf_{F\in C} I(F). \tag{2.13}$$

Thus (2.13) and (2.12) imply (2.11).

For $0<K<\infty$, let $A_K=\{F\in\mathscr{F}:\Phi(F)\leqslant K\}$ which is compact from hypothesis (3) on Φ. From (2.1) we have

$$\begin{aligned} E_x\{\mathrm{e}^{-t\Phi[L(t,\omega,.)]}\} &= E^{Q_{x,t}}_{A_K}\{\mathrm{e}^{-t\Phi(F)}\} + E^{Q_{x,t}}_{A_K'}\{\mathrm{e}^{-t\Phi(F)}\} \\ &\leqslant E^{Q_{x,t}}_{A_K}\{\mathrm{e}^{-t\Phi(F)}\} + \mathrm{e}^{-tK}. \end{aligned} \tag{2.14}$$

Let $F^*\in A_K$ and let $\varepsilon>0$ be given. Assume first of all that $I(F^*)>\infty$. Then, since both $\Phi(F)$ and $I(F)$ are lower semicontinuous at F^*, there exists a neighbourhood N_{F^*} such that for all $F\in N_{F^*}$,

$$\Phi(F^*)\leqslant\Phi(F)+\varepsilon,$$

$$I(F^*)\leqslant I(F)+\varepsilon.$$

Let G_{F^*} be another neighbourhood of F^* such that $\bar{G}_{F^*}\subset N_{F^*}$. We have

$$E^{Q_{x,t}}_{\bar{G}_{F^*}\cap A_K}(\mathrm{e}^{-t\Phi(F)})\leqslant \mathrm{e}^{-t[\Phi(F^*)-\varepsilon]}Q_{x,t}(\bar{G}_{F^*}\cap A_K).$$

Since $\bar{G}_{F^*}\cap A_K$ is a closed subset of a compact set it is itself compact, and therefore from (2.11)

$$\begin{aligned} \overline{\lim_{t\to\infty}}\frac{1}{t}\ln E^{Q_{x,t}}_{\bar{G}_{F^*}\cap A_K}(\mathrm{e}^{-t\Phi(F)}) &\leqslant -\Phi(F^*)+\varepsilon-\inf_{F\in \bar{G}_{F^*}\cap A_K} I(F) \\ &\leqslant -\Phi(F^*)+\varepsilon-[I(F^*)-\varepsilon] \\ &\leqslant -\inf_{F\in\mathscr{F}}[\Phi(F)+I(F)]+2\varepsilon. \end{aligned} \tag{2.15}$$

On the other hand, if $I(F^*)=\infty$ we have from the lower semicontinuity of $I(F)$ that there exists a neighbourhood M_{F^*} such that for all $F\in M_{F^*}$, $I(F)\geqslant K$. Let G_{F^*} be another neighbourhood of F^* such that $\bar{G}_{F^*}\subset M_{F^*}$.

† In fact it is easy to show that for compact sets C equality holds in (2.13).

Then

$$E^{Q_{x,t}}_{G_{F^*}\cap A_K}(e^{-t\Phi(F)}) \leq Q_{x,t}(\bar{G}_{F^*}\cap A_K),$$

and again from (2.11) we have

$$\overline{\lim_{t\to\infty}} \frac{1}{t} \ln E^{Q_{x,t}}_{G_{F^*}\cap A_K}(e^{-t\Phi(F)}) \leq - \inf_{F\in\bar{G}_{F^*}\cap A_K} I(F) \leq -K. \tag{2.16}$$

Thus each point F^* of A_K is covered by a neighbourhood G_{F^*} and since A_K is compact there exists $F_1^*, F_2^*, \ldots, F_n^*$ such that $A_K \subset \bigcup_{j=1}^n G_{F^*_j}$. From (2.15) and (2.16) we have, therefore,

$$\overline{\lim_{t\to\infty}} \frac{1}{t} \ln E^{Q_{x,t}}_{A_K}(e^{-t\Phi(F)}) \leq -\min\{K, \inf_{F\in\mathcal{F}} [\Phi(F)+I(F)]-2\varepsilon\}. \tag{2.17}$$

From (2.14) and (2.17) we obtain

$$\overline{\lim_{t\to\infty}} \frac{1}{t} \ln E_x(e^{-t\Phi[L(t,\omega,.)]}) \leq -\min\{K, \inf_{F\in\mathcal{F}} [\Phi(F)+I(F)]-2\varepsilon\},$$

and letting $\varepsilon \to 0$ we get

$$\overline{\lim_{t\to\infty}} \frac{1}{t} \ln E_x(e^{-t\Phi[L(t,\omega,.)]}) \leq -\min\{K, \inf_{F\in\mathcal{F}} [\Phi(F)+I(F)]\}.$$

Finally, letting $K \to \infty$ we find

$$\overline{\lim_{t\to\infty}} \frac{1}{t} \ln E_x(e^{-t\Phi[L(t,\omega,.)]}) \leq - \inf_{F\in\mathcal{F}} [\Phi(F)+I(F)]. \tag{2.18}$$

In the next section we obtain a lower estimate on

$$\lim_{t\to\infty} \frac{1}{t} \ln E_x(e^{-t\Phi[L(t,\omega,.)]}),$$

and in § 2.4 we show that this lower estimate and the upper estimate on the right of (2.18) are equal and in fact are the same as the right side of (2.4). This will prove the theorem.

2.3. The lower estimate

Let $[a, b]$ be a finite closed interval such that $x \in (a, b)$. Let $f(y)$ be a density function having two continuous derivatives on $-\infty < y < \infty$, $f(y) > 0$ for $y \in (a, b)$, $f(y) = 0$ for y outside of $[a, b]$, and

$$\int_a^b \frac{[f'(y)]^2}{f(y)} \, dy < \infty.$$

Let $F(y)$ be the distribution function corresponding to $f(y)$.

For each $\omega = z(\,.\,)\in\Omega_x$ such that $a\leqslant z(s)\leqslant b$ for all $0\leqslant s\leqslant t$, $L(t,\omega,y)$ is a distribution function whose support is contained in $[a,b]$. Thus, by hypothesis (4) (p. 16) on Φ, for any $\varepsilon>0$ there exists a Lévy neighbourhood of F, say N_F, so that if $L(t,\omega,.\,)\in N_F$ for any ω such that $a\leqslant z(s)\leqslant b$ for all $0\leqslant s\leqslant t$, then

$$|\Phi(L(t,\omega,.\,))-\Phi(F)|\leqslant\varepsilon.$$

Thus, if we let $C'_{a,b}=\{z\in\Omega_x : a\leqslant z(s)\leqslant b \text{ for all } 0\leqslant s\leqslant t\}$,

$$\begin{aligned}E_x(\mathrm{e}^{-t\Phi[L(t,\omega,.)]}) &\geqslant E_x\{\mathrm{e}^{-t\Phi[L(t,\omega,.)]};\ L(t,\omega,.\,)\in N_F,\ z\in C'_{a,b}\}\\ &\geqslant \mathrm{e}^{-t[\Phi(F)+\varepsilon]}P_x\{L(t,\omega,.\,)\in N_F,\ z\in C'_{a,b}\}.\end{aligned}\tag{2.19}$$

In Lemma 2.1, which follows, we will show

$$\varliminf_{t\to\infty}\frac{1}{t}\ln P_x\{L(t,\omega,.\,)\in N_F,\ z\in C'_{a,b}\}\geqslant -\frac{1}{8}\int_a^b\frac{[f'(y)]^2}{f(y)}\,\mathrm{d}y.\tag{2.20}$$

From (2.19) we obtain that

$$\varliminf_{t\to\infty}\frac{1}{t}\ln E_x\{\mathrm{e}^{-t\Phi[L(t,\omega,.)]}\}\geqslant -\Phi(F)-\varepsilon-\frac{1}{8}\int_a^b\frac{[f'(y)]^2}{f(y)}\,\mathrm{d}y,$$

and since ε is arbitrary we obtain

$$\begin{aligned}\varliminf_{t\to\infty}\frac{1}{t}\ln E_x(\mathrm{e}^{-t\Phi[L(t,\omega,.)]}) &\geqslant -\left\{\Phi(F)+\frac{1}{8}\int_a^b\frac{[f'(y)]^2}{f(y)}\,\mathrm{d}y\right\}\\ &\geqslant -\inf_{f\in\mathcal{f}_x}\left\{\Phi(f)+\frac{1}{8}\int_a^b\frac{[f'(y)]^2}{f(y)}\,\mathrm{d}y\right\},\end{aligned}\tag{2.21}$$

where $\mathcal{f}_x$ is the set of all density functions f with two continuous derivatives for which there is a closed interval $[a,b]$ containing x in its interior, outside of which $f=0$, $f>0$ on (a,b), and

$$\int_a^b\frac{[f'(y)]^2}{f(y)}\,\mathrm{d}y<\infty.$$

Since (2.20) is of independent interest we prove it as a lemma.

LEMMA 2.1. *Let $f\in\mathcal{f}_x$ and let N_F be any Lévy neighbourhood of its corresponding distribution function. Then*

$$\varliminf_{t\to\infty}\frac{1}{t}\ln P_x\{L(t,\omega,.\,)\in N_F,\ a\leqslant z(s)\leqslant b\quad\text{for all}\quad 0\leqslant s\leqslant t\}$$
$$\geqslant -\frac{1}{8}\int_a^b\frac{[f'(y)]^2}{f(y)}\,\mathrm{d}y.$$

Proof. Let $b(y)=\frac{1}{2}f'(y)/f(y)$ and consider Brownian motion with drift b, i.e. the process whose generator is

$$\frac{1}{2}\frac{\partial^2}{\partial x^2}+b(x)\frac{\partial}{\partial x}.$$

Now, the solution of

$$\frac{1}{2}\frac{\partial^2 u}{\partial x^2}+b(x)\frac{\partial u}{\partial x}=0$$

being, in the case $b(x)=\frac{1}{2}f'(x)/f(x)$, simply $u'=1/f$, we have

$$u(y)=\int_\eta^y \frac{d\xi}{f(\xi)}$$

for $y\in(a,b)$, $\eta\in(a,b)$. Hence $u(y)\to-\infty$ as $y\to a$ and $u(y)\to+\infty$ as $y\to b$, which means according to Feller's test [2] that with probability 1 Brownian-motion paths with this drift never leave (a, b).

If we let $\tilde{P}_x$ denote the measure on Ω_x induced by this drift process, we have the Cameron–Martin formula [3]

$$P_x\{A\cap[z(s)\in[a,b]\quad\text{for all}\quad 0\leq s\leq t]\}$$
$$=E_A^{\tilde{P}_x}\left\{\exp\left[-\int_0^t b(z(s))\,dz(s)+\frac{1}{2}\int_0^t b^2(z(s))\,ds\right]\right\}.$$

In our case, this becomes

$$P_x\{L(t,\omega,.)\in N_F,\ z(s)\in[a,b]\quad\text{for all}\quad 0\leq s\leq t\}$$
$$=E^{\tilde{P}_x}\left\{\exp\left[-\frac{1}{2}\int_0^t\left(\frac{f'}{f}\right)(z(s))\,dz(s)+\frac{1}{8}\int_0^t\left(\frac{f'}{f}\right)^2(z(s))\,ds\right];\quad L(t,\omega,.)\in N_F\right\}. \tag{2.22}$$

Since f is twice continuously differentiable, we have by Ito's formula

$$d\ln f(z(s))=(\ln f)'(z(s))\,dz(s)+\tfrac{1}{2}(\ln f)''(z(s))\,ds,$$

and, integrating this from 0 to t using $z(0)=x$, we obtain

$$\ln f(z(t))-\ln f(x)$$
$$=\int_0^t\left(\frac{f'}{f}\right)(z(s))\,dz(s)+\frac{1}{2}\int_0^t\left(\frac{f''}{f}\right)(z(s))\,ds-\frac{1}{2}\int_0^t\left(\frac{f'}{f}\right)^2(z(s))\,ds. \tag{2.23}$$

Using (2.23) in (2.22) we obtain

$$P_x\{L(t,\omega,.)\in N_F,\ z(s)\in[a,b]\quad\text{for all}\quad 0\leqslant s\leqslant t\}$$
$$=E^{\tilde P_x}\left(\sqrt{\left[\frac{f(x)}{f(z(t))}\right]}\exp\left\{-\left[\frac{1}{8}\int_0^t\left(\frac{f'}{f}\right)^2(z(s))\,\mathrm{d}s-\frac{1}{4}\int_0^t\left(\frac{f''}{f}\right)(z(s))\,\mathrm{d}s\right]\right\};\right.$$
$$\left.L(t,\omega,.)\in N_F\right). \tag{2.24}$$

Let

$$J(y)=\frac{1}{8}\left(\frac{f'}{f}\right)^2(y)-\frac{1}{4}\left(\frac{f''}{f}\right)(y),$$

and let $\varepsilon>0$ be given. Define

$$S(t,\varepsilon)=\left\{z\in\Omega_x:\left|\int_a^b J(y)\,\mathrm{d}L(t,\omega,y)-\int_a^b J(y)f(y)\,\mathrm{d}y\right|<\varepsilon,\right.$$

and

$$S'(t,\varepsilon)=\{z\in\Omega_x:L(t,\omega,.)\in N_F\}\cap S(t,\varepsilon).$$

Since, for almost all $z\in\Omega_x$ ($\tilde P_x$-measure),

$$\int_a^b J(y)\,\mathrm{d}L(t,\omega,y)=\int_{-\infty}^{\infty}J(y)L(t,\omega,y)=\frac{1}{t}\int_0^t J(z(s))\,\mathrm{d}s,$$

we have from (2.24)

$$P_x\{L(t,\omega,.)\in N_F,\ z(s)\in[a,b]\quad\text{for all}\quad 0\leqslant s\leqslant t\}$$
$$=E^{\tilde P_x}\left(\sqrt{\left[\frac{f(x)}{f(z(t))}\right]}\exp\left\{-t\int_a^b J(y)L(t,\omega,y)\right];\quad L(t,\omega,.)\in N_F\right)$$
$$\geqslant\exp\left\{-t\left[\varepsilon+\int_a^b J(y)f(y)\,\mathrm{d}y\right]\right\}E^{\tilde P_x}\left\{\sqrt{\left[\frac{f(x)}{f(z(t))}\right]};\quad S'(t,\varepsilon)\right\}$$
$$=\exp\left\{-\varepsilon t-\frac{t}{8}\int_a^b\frac{[f'(y)]^2}{f(y)}\,\mathrm{d}y\right\}E^{\tilde P_x}\left\{\sqrt{\left[\frac{f(x)}{f(z(t))}\right]};\quad S'(t,\varepsilon)\right\}. \tag{2.25}$$

Because of the hypotheses on f there exists a constant γ such that

$$\sqrt{\left[\frac{f(x)}{f(z(t))}\right]}\geqslant\gamma>0$$

for almost all $z \in \Omega_x$ ($\tilde{P}_x$-measure). Therefore

$$E^{\tilde{P}_x}\left\{\sqrt{\left[\frac{f(x)}{f(z(t))}\right]};\quad S'(t,\varepsilon)\right\} \geqslant \gamma \tilde{P}_x(S'(t,\varepsilon))$$

$$= \gamma - \gamma \tilde{P}_x(\Omega_x - S'(t,\varepsilon)). \quad (2.26)$$

Now the reason the drift $b(y) = \frac{1}{2}f'(y)/f(y)$ was selected is because then f itself is the Invariant measure for the drift process. This is easy to see since if the drift is $\frac{1}{2}f'/f$ then $u = f$ is the solution of

$$\tfrac{1}{2}u'' - [b(x)u(x)]' = 0,$$

the adjoint equation. From the ergodic properties of Markov processes (see e.g. [2])

$$\lim_{t\to\infty} \tilde{P}_x(\Omega_x - S(t,\varepsilon))$$

$$= \lim_{t\to\infty} \tilde{P}_x\left(z \in \Omega_x : \left| \int_a^b J(y)\, \mathrm{d}L(t,\omega,y) - \int_a^b J(y) f(y)\, \mathrm{d}y \right| \geqslant \varepsilon\right)$$

$$= 0$$

and

$$\lim_{t\to\infty} \tilde{P}_x(z \in \Omega_x : L(t,\omega,.) \notin N_F) = 0.$$

These last two statements imply

$$\lim_{t\to\infty} \tilde{P}_x(\Omega_x - S'(t,\varepsilon)) = 0,$$

and, therefore, from (2.25) and (2.26) we obtain

$$\varliminf_{t\to\infty} \frac{1}{t} \ln P_x\{L(t,\omega,.) \in N_F,\ z(s) \in [a,b] \quad \text{for all} \quad 0 \leqslant s \leqslant t\}$$

$$\geqslant -\varepsilon - \frac{1}{8}\int_a^b \frac{[f'(y)]^2}{f(y)}\, \mathrm{d}y.$$

Finally, letting $\varepsilon \to 0$, we obtain (2.20).

2.4. The equality of the estimates

In this section we want to show that the expressions on the right of (2.18) and (2.21) are equal and the same as the expression on the right of (2.4). We shall first show Lemma 2.2.

LEMMA 2.2. *Let f be a density function such that $f>0$ on (a, b) where $-\infty \leqslant a<b \leqslant +\infty$, $f=0$ outside of $[a, b]$ and f is continuously differentiable on $-\infty<y<\infty$. Then, letting F be the corresponding distribution function,*

$$I(F)=\frac{1}{8}\int_a^b \frac{[f'(y)]^2}{f(y)}\,dy, \tag{2.27}$$

in the sense that if one side is finite so is the other and they are equal.

Proof. Assume first that

$$\int_a^b \frac{[f'(y)]^2}{f(y)}\,dy<\infty.$$

We show now that for every $u \in \mathcal{U}$

$$\int_{-\infty}^{\infty} \frac{u''(y)}{2u(y)}\,dF(y) \geqslant -\frac{1}{8}\int_a^b \frac{[f'(y)]^2}{f(y)}\,dy, \tag{2.28}$$

which will imply

$$I(F) \leqslant \frac{1}{8}\int_a^b \frac{[f'(y)]^2}{f(y)}\,dy<\infty. \tag{2.29}$$

So, for any $u \in \mathcal{U}$, let $h=\ln u$ and we have

$$\begin{aligned}\int_{-\infty}^{\infty} \frac{u''(y)}{2u(y)}\,dF(y) &= \frac{1}{2}\int_{-\infty}^{\infty} \{h''(y)+[h'(y)]^2\} f(y)\,dy \\ &= \frac{1}{2}\int_a^b \{h''(y)+[h'(y)]^2\} f(y)\,dy.\end{aligned} \tag{2.30}$$

Using elementary inequalities

$$\begin{aligned}\left|\int_a^b h'(y)f'(y)\,dy\right| &= \left|\int_a^b \frac{h'(y)f'(y)}{\sqrt{[f(y)]}}\sqrt{[f(y)]}\,dy\right| \\ &\leqslant \sqrt{\left[\int_a^b [h'(y)]^2 f(y)\,dy\right]}\sqrt{\left[\int_a^b \frac{[f'(y)]^2}{f(y)}\,dy\right]} \\ &= 2\sqrt{\left[\int_a^b [h'(y)]^2 f(y)\,dy\right]}\sqrt{\left[\frac{1}{4}\int_a^b \frac{[f'(y)]^2}{f(y)}\,dy\right]} \\ &\leqslant \int_a^b [h'(y)]^2 f(y)\,dy+\frac{1}{4}\int_a^b \frac{[f'(y)]^2}{f(y)}\,dy,\end{aligned}$$

and hence

$$\frac{1}{2}\left\{\int_a^b [h'(y)]^2 f(y)\,dy - \int_a^b h'(y)f'(y)\,dy\right\} \geq -\frac{1}{8}\int_a^b \frac{[f'(y)]^2}{f(y)}\,dy.$$

Integrating by parts we obtain (2.28) and hence (2.29).

To complete the proof it suffices to show that if F is such that $I(F)<\infty$, then

$$\frac{1}{8}\int_a^b \frac{[f'(y)]^2}{f(y)}\,dy \leq I(F). \tag{2.31}$$

From the definition of $I(F)$, we have for every $u \in \mathscr{U}$ and every $h = \ln u$

$$\int_a^b \frac{u''(y)}{2u(y)}\,dF(y) = \frac{1}{2}\int_a^b [h'(y)]^2 f(y)\,dy + \frac{1}{2}\int_a^b h''(y)f(y)\,dy \geq -I(F). \tag{2.32}$$

Now (2.32) holds in particular for these $u \in \mathscr{U}$ such that $h = \ln u$ are constant outside of a closed bounded interval contained in (a, b). If we let $\mathscr{H}$ denote this latter class of functions h, we have from (2.32) for any $h \in \mathscr{H}$

$$\frac{1}{2}\int_a^b [h'(y)]^2 f(y)\,dy - \frac{1}{2}\int_a^b h'(y)f'(y)\,dy \geq -I(F). \tag{2.33}$$

If $h \in \mathscr{H}$ then $\lambda h \in \mathscr{H}$ for any real λ, so from (2.33) we have for all $h \in \mathscr{H}$

$$\frac{\lambda^2}{2}\int_a^b [h'(y)]^2 f(y)\,dy - \frac{\lambda}{2}\int_a^b h'(y)f'(y)\,dy + I(F) \geq 0.$$

Thus, the discriminant of this quadratic in λ must be less than or equal to zero, i.e. for all $h \in \mathscr{H}$,

$$\left|\int_a^b h'(y)f'(y)\,dy\right| \leq \sqrt{(8I(F))}\sqrt{\left\{\int_a^b [h'(y)]^2 f(y)\,dy\right\}}. \tag{2.34}$$

If we let $g(y) = h'(y)$, then we have

$$\left|\int_a^b g(y)f'(y)\,dy\right| \leq \sqrt{[8I(F)]}\sqrt{\left[\int_a^b g^2(y)f(y)\,dy\right]} \tag{2.35}$$

is valid for all functions $g(y)$ which have compact support contained in (a, b) and which have a continuous derivative. By approximation we then have (2.35) valid for all functions $g(y)$ which are continuous and have compact support contained in (a, b). Let $\phi(y) \in C^\infty$ with compact support contained in (a, b) and $0 \leq \phi(y) \leq 1$ for all $-\infty < y < \infty$. We now apply (2.35) to

$g(y)=[f'(y)/f(y)]\phi(y)$, obtaining

$$\left|\int_a^b \frac{[f'(y)]^2}{f(y)}\phi(y)\,\mathrm{d}y\right| \leqslant \sqrt{(8I(F))}\sqrt{\left\{\int_a^b \frac{[f'(y)]^2}{f(y)}\phi^2(y)\,\mathrm{d}y\right\}}$$

$$\leqslant \sqrt{(8I(F))}\sqrt{\left\{\int_a^b \frac{[f'(y)]^2}{f(y)}\phi(y)\,\mathrm{d}y\right\}}.$$

Dividing and squaring we obtain

$$\frac{1}{8}\int_a^b \frac{[f'(y)]^2}{f(y)}\phi(y)\,\mathrm{d}y \leqslant I(F). \tag{2.36}$$

Since (2.36) holds for all $\phi(y)$ such that $0 \leqslant \phi(y) \leqslant 1$, $\phi(y) \in C^\infty$, $\phi(y)$ has compact support contained in (a, b) we conclude that in fact

$$\frac{1}{8}\int_a^b \frac{[f'(y)]^2}{f(y)}\,\mathrm{d}y \leqslant I(F).$$

We now prove Lemma 2.3.

LEMMA 2.3.

$$\inf_{f\in \mathscr{f}_x}\left\{\frac{1}{8}\int_a^b \frac{[f'(y)]^2}{f(y)}\mathrm{d}y + \Phi(f)\right\} = \inf_{F\in\mathscr{F}}[I(F)+\Phi(F)]. \tag{2.37}$$

Proof. Let $f \in \mathscr{f}_x$, then from Lemma 2.2 we have

$$\frac{1}{8}\int_a^b \frac{[f'(y)]^2}{f(y)}\mathrm{d}y = I(F)$$

where F is the distribution function with f. Thus, the left side of (2.37) is clearly greater than or equal to the right side. We prove the reverse inequality in successive stages.

Let $\mathscr{F}_c$ be the class of distribution functions having compact support. We show now

$$\inf_{F\in\mathscr{F}_c}[I(F)+\Phi(F)] = \inf_{F\in\mathscr{F}}[I(F)+\Phi(F)]. \tag{2.38}$$

It suffices to show the left side of (2.38) is less than or equal to the right side, and for this let $F \in \mathscr{F}$ such that $\Phi(F)=a_1<\infty$ and $I(F)=a_2<\infty$. We then show that for any $\varepsilon>0$ there exists a distribution function $G\in\mathscr{F}_c$ such that $\Phi(G)<a_1+\varepsilon$ and $I(G)<a_2+\varepsilon$.

Define a sequence of functions $\{h_n(y)\}$ such that $0 \leqslant h_n(y) \leqslant 1$ for $-\infty < y < \infty$, $h_n(h) = 1$ if $|y| \leqslant n$, $h_n(y) = 0$ if $|y| > 2n$, and $|h_n'(y)| < 1/n$ for $n < |y| \leqslant 3n$. For the given $F \in \mathscr{F}$ and $g_n(y) = h_n^2(y)$, define

$$F_n(y) = \frac{\int_{-\infty}^{y} g_n(\xi)\, dF(\xi)}{\int_{-\infty}^{\infty} g_n(\xi)\, dF(\xi)}.$$

For each n, $F_n \in \mathscr{F}_c$ and from hypothesis (5) (p. 16) on $\Phi(F)$, there exists an n_0 so that $n \geqslant n_0$ implies $\Phi(F_n) \leqslant \Phi(F) + \varepsilon = a_1 + \varepsilon$.

We show now that $\lim_{n\to\infty} I(F_n) = I(F)$. First, since I is lower semicontinuous on $\mathscr{F}$, $\underline{\lim}_{n\to\infty} I(F_n) \geqslant I(F)$, and so we want to show $\overline{\lim}_{n\to\infty} I(F_n) \leqslant I(F)$. For any $\eta > 0$, let ψ_η be the Gaussian distribution with mean 0 and variance η. For our given $F \in \mathscr{F}$, define $F_\eta = F * \psi_\eta$. By convexity†, $I(F_\eta) \leqslant I(F)$, and moreover, since F_η has a strictly positive density f_η with continuous derivatives, we have from Lemma 2.2,

$$I(F_\eta) = \frac{1}{8}\int_{-\infty}^{\infty} \frac{[f_\eta'(y)]^2}{f_\eta(y)}\, dy. \tag{2.39}$$

Let

$$F_{n,\eta}(y) = \frac{\int_{-\infty}^{y} g_n(\xi)\, dF_\eta(\xi)}{\int_{-\infty}^{\infty} g_n(\xi)\, dF_\eta(\xi)},$$

and note $F_{n,\eta}(y)$ has a strictly positive density with continuous derivative, namely,

$$\frac{g_n f_\eta}{\int_{-\infty}^{\infty} g_n(\xi) f_\eta(\xi)\, d\xi}.$$

Thus again from Lemma 2.2, and the definition of g_n,

$$\begin{aligned}
I(F_{n,\eta}) &= \frac{1}{8\int_{-\infty}^{\infty} g_n(\xi) f_\eta(\xi)\, d\xi} \int_{-\infty}^{\infty} \frac{[f_\eta(y) g_n(y)']^2}{f_\eta(y) g_n(y)}\, dy \\
&= \frac{1}{8\int_{-\infty}^{\infty} g_n(\xi) f_\eta(\xi)\, d\xi} \left\{ \int_{-\infty}^{\infty} \frac{[f_\eta'(y)]^2}{f_\eta(y)} g_n(y)\, dy \; + \right. \\
&\qquad \left. + \int_{-\infty}^{\infty} f_\eta(y) \frac{[g_n'(y)]^2}{g_n(y)}\, dy + 2\int_{-\infty}^{\infty} f_\eta'(y) g_n'(y)\, dy \right\} \\
&\leqslant \frac{1}{8\int_{-\infty}^{\infty} g_n(\xi) f_\eta(\xi) d\xi} \left\{ \int_{-\infty}^{\infty} \frac{[f_\eta'(y)]^2}{f_\eta(y)}\, dy + \frac{4}{n^2}\int_{-\infty}^{\infty} f_\eta(y)\, dy \right. \\
&\qquad \left. + \frac{2}{n}\int_{-\infty}^{\infty} |f_\eta'(y)| dy \right\}.
\end{aligned}$$

† From the definition of $I(F)$ it follows that it is translation invariant, is lower semicontinuous, and is convex. Therefore it decreases under convolution.

From (2.39) and the fact, noted earlier, that $I(F_\eta) \leqslant I(F)$,

$$\int_{-\infty}^{\infty} |f'_\eta(y)|\,\mathrm{d}y = \int_{-\infty}^{\infty} \frac{|f'_\eta(y)|}{\sqrt{[f_\eta(y)]}} \sqrt{[f_\eta(y)]}\,\mathrm{d}y$$

$$\leqslant \sqrt{\left\{\int_{-\infty}^{\infty} \frac{[f'_\eta(y)]^2}{f_\eta(y)}\right\}}\mathrm{d}y \sqrt{\left\{\int_{-\infty}^{\infty} f_\eta(y)\,\mathrm{d}y\right\}}$$

$$= \sqrt{\left\{\int_{-\infty}^{\infty} \frac{[f'_\eta(y)]^2}{f_\eta(y)}\right\}}\,\mathrm{d}y = \sqrt{[8I(F_\eta)]} \leqslant \sqrt{[8I(F)]}.$$

Therefore, again using (2.39), and $I(F_\eta) \leqslant I(F)$,

$$I(F_{n,\eta}) \leqslant \frac{1}{8\int_{-\infty}^{\infty} g_n(\xi) f_\eta(\xi)\,\mathrm{d}\xi} \left\{8I(F) + \frac{4}{n^2} + \frac{2}{n}\sqrt{[8I(F)]}\right\}. \tag{2.40}$$

Holding n fixed and letting $\eta \to 0$ we see that

$$\lim_{\eta \to 0} I(F_{n,\eta}) = I(F_\eta),$$

$$\lim_{\eta \to 0} \int_{-\infty}^{\infty} g_n(\xi) f_\eta(\xi)\,\mathrm{d}\xi = \int_{-\infty}^{\infty} g_n(\xi)\,\mathrm{d}\xi,$$

and therefore from (2.40), letting $\eta \to 0$,

$$I(F_n) \leqslant \frac{1}{8\int_{-\infty}^{\infty} g_n(\xi)\,\mathrm{d}\xi} \left\{8I(F) + \frac{4}{n^2} + \frac{2}{n}\sqrt{[8I(F)]}\right\}.$$

Thus $\overline{\lim}_{n\to\infty} I(F_n) \leqslant I(F)$, and so there exists an n_1 such that $n > n_1$ implies $I(F_n) \leqslant I(F) + \varepsilon = a_2 + \varepsilon$. For any $n > \max(n_0, n_1)$ choose $G(y) = F_n(y) \in \mathscr{F}_c$ and $\Phi(G) < a_1 + \varepsilon$ and $I(G) < a_2 + \varepsilon$ so that we have (2.38).

Now let $\mathscr{F}_{c'}$ be the space of distribution functions having compact support and possessing twice continuously differentiable density functions. We show now

$$\inf_{F \in \mathscr{F}_{c'}} [I(F) + \Phi(F)] = \inf_{F \in \mathscr{F}_c} [I(F) + \Phi(F)]. \tag{2.41}$$

Again, of course, it suffices to show the left side of (2.41) is less than or equal to the right side. Let $F \in \mathscr{F}_c$. For any $\varepsilon > 0$ we want to find $G \in \mathscr{F}_{c'}$ such that $I(G) < I(F) + \varepsilon$ and $\Phi(G) < \Phi(F) + \varepsilon$. For any $\eta > 0$, let $\eta(y)$ be a density function which is twice continuously differentiable with compact support $\subset (-\eta, \eta)$. For our $F \in \mathscr{F}_c$, define

$$g_\eta(x) = \int \eta(x - y)\,\mathrm{d}F(y).$$

Thus $g_\eta(x)$ is a density function which is twice continuously differentiable and has compact support, since both $\eta(y)$ and $F(y)$ have compact support. If we let G_η be the distribution function corresponding to the density g_η we have that for all $\eta>0$, $G_\eta \in \mathscr{F}_{c'}$. Moreover, by convexity, for all $\eta>0$,

$$I(G_\eta)\leqslant I(F)\leqslant I(F)+\varepsilon.$$

On the other hand, as $\eta\to 0$, $G_\eta \Rightarrow F$ and the support of G_η is for every $\eta>0$ contained in a fixed interval independent of η. Thus by hypothesis (4) (p. 16) on Φ we can choose η so small that $\Phi(G_\eta)\leqslant\Phi(F)+\varepsilon$. Hence, for any η so small, let $G=G_\eta$ and we have $I(G)\leqslant I(F)+\varepsilon$ and $\Phi(G)\leqslant\Phi(F)+\varepsilon$, establishing (2.41).

Finally, we show

$$\inf_{f\in\not{f}_x}\left\{\frac{1}{8}\int_a^b\frac{[f'(y)]^2}{f(y)}\mathrm{d}y+\Phi(f)\right\}=\inf_{F\in\mathscr{F}_{c'}}[I(F)+\Phi(F)]. \tag{2.42}$$

Since $\not{f}_x\subset\mathscr{F}_{c'}$ it again suffices to show that for $F^*\in\mathscr{F}_{c'}$ and given $\varepsilon>0$ there exists a density $f\in\not{f}_x$ such that

$$\frac{1}{8}\int_a^b\frac{[f'(y)]^2}{f(y)}\mathrm{d}y\leqslant I(F^*)+\varepsilon$$

and

$$\Phi(f)\leqslant\Phi(F^*)+\varepsilon.$$

Let $h\in\not{f}_x$ such that the interval $[a,b]\ni x$ and contains the support of F^* on the interior of which $h>0$. Let for any $\eta>0$,

$$g_\eta=(1-\eta)f^*+\eta h$$

where f^* is the density of F^*. Thus g_η has two continuous derivatives and $g_\eta>0$ on (a,b). Again by convexity, for any $\eta>0$

$$I(g_\eta)\leqslant(1-\eta)I(F^*)+\eta I(h)$$

so by letting $\eta\to 0$ and using hypothesis (4) (p. 16) on Φ we have for small enough η,

$$I(g_\eta)\leqslant I(F^*)+\varepsilon$$

and

$$\Phi(g_\eta)\leqslant\Phi(F^*)+\varepsilon.$$

Also from Lemma 2.2, since $I(g_\eta)<\infty$, we have

$$\int_a^b\frac{[g_\eta'(y)]^2}{g_\eta(y)}\mathrm{d}y<\infty.$$

This gives (2.42).

Now, since $\mathscr{f}_x \subset \mathscr{f} \subset \mathscr{F}$, we have finally that

$$\inf_{f \in \mathscr{f}_x} \left\{ \frac{1}{8} \int_a^b \frac{[f'(y)]^2}{f(y)} \mathrm{d}y + \Phi(f) \right\} = \inf_{f \in \mathscr{f}} \left\{ \frac{1}{8} \int_{-\infty}^{\infty} \frac{[f'(y)]^2}{f(y)} \mathrm{d}y + \Phi(f) \right\}$$

$$= \inf_{F \in \mathscr{F}} [I(F) + \Phi(F)].$$

This completes the proof of the theorem.

Acknowledgements

This research was supported by the Office of Naval Research, Contract N00014-67-A-0467-0031 and the National Science Foundation, Grant NSF-GP-41985.

References

1. KAC, M. On some connections between probability theory and differential and integral equations, *Proceedings of the 2nd Berkeley Symposium*, pp. 189–215 (1950).
2. ITO, K. and MCKEAN JR., H. P. *Diffusion processes and their sample paths.* Springer–Verlag, New York (1965).
3. CAMERON, R. H. and MARTIN, W. T. The transformation of Wiener integrals by non-linear transformations. *Trans. Am. math. Soc.* **75**, 552–75 (1953).

COMMUNICATIONS ON PURE AND APPLIED MATHEMATICS, VOL. XXVIII, 525–565 (1975)

Asymptotics for the Wiener Sausage*

M. D. DONSKER AND S. R. S. VARADHAN

(Dedicated to Mark Kac on the occasion of his sixtieth birthday.)

1. Introduction

Let $\{\beta(s),\ 0 \leqq s < \infty\}$ be the Brownian motion in d dimensions with $\beta(0) = x \in R^d$. For any $t > 0$, let $C_t = C_t(\beta(\cdot))$ be the image set of the particular path $\beta(s)$ for $0 \leqq s \leqq t$ so that $C_t \subset R^d$. For any set $A \subset R^d$ and any $\epsilon > 0$, let $A^\epsilon = \cup_{y \in A} S(y,\epsilon)$, where $S(y,\epsilon)$ is the sphere in R^d of radius ϵ with center at y. We shall denote the d-dimensional Lebesgue measure of a set A by $|A|$. In particular then we shall be interested in the set $C_t^\epsilon = C_t^\epsilon(\beta(\cdot))$ which has been called the Wiener sausage (see e.g. [1]). Its d-dimensional volume $|C_t^\epsilon(\beta(\cdot))|$ is clearly a functional of the Brownian motion path whose increase with t depends on how much the sausage intersects itself for that particular path as t grows.

We shall be principally concerned in this paper with the asymptotic behavior as $t \to \infty$ of

$$E_x\{\exp\{-\nu |C_t^\epsilon(\beta(\cdot))|\}\}, \tag{1.1}$$

where $\nu > 0$, $x \in R^d$, and $E_x\{\ \}$ denotes expectation on the Brownian motion paths in R^d starting at x. We shall prove

THEOREM 1. *For any $x \in R^d$ and any $\epsilon > 0$,*

$$\lim_{t\to\infty} \frac{1}{t^{d/(d+2)}} \log E_x\{\exp\{-\nu|C_t^\epsilon(\beta(\cdot))|\}\} = -k(\nu,d), \tag{1.2}$$

where

$$k(\nu,d) = \nu^{2/(d+2)}\left(\frac{d+2}{2}\right)\left(\frac{2\gamma_d}{d}\right)^{d/(d+2)},$$

and γ_d is the lowest eigenvalue of $-\frac{1}{2}\Delta$ for the sphere of unit volume in d dimensions with zero boundary values.

---

*This research was performed at the Courant Institute and supported by the U. S. Office of Naval Research, Contract N00014-67A-0467-0031 and by the National Science Foundation, Grant NSF-GP-41985. Reproduction in whole or in part is permitted for any purpose of the United States Government.

It is quite easy to show that

$$\text{(1.3)} \qquad \varliminf_{t\to\infty} \frac{1}{t^{d/(d+2)}} \log E_x \{\exp\{-\nu|C_t^\epsilon(\beta(\cdot))|\}\} \geqq -k(\nu,d)$$

and we do this in Section 2. This lower estimate has already been obtained by others (c.f. [1]), and, in fact, the result (1.2) has been conjectured on physical grounds (cf. [2]). However, to prove

$$\text{(1.4)} \qquad \varlimsup_{t\to\infty} \frac{1}{t^{d/(d+2)}} \log E_x \{\exp\{-\nu|C_t^\epsilon(\beta(\cdot))|\}\} \leqq -k(\nu,d)$$

is a much more delicate matter. We prove (1.4) in Section 3 of this paper making use of results recently obtained by the authors for the asymptotic evaluation of Markov process expectations as the time becomes infinite (cf. [3], [4], [5]). To see how these latter results apply to (1.1), it will be convenient to make some changes of variable. Since the expectation in (1.4) is independent of the starting point, we can take x equal to zero.

For any $\theta > 0$, let, as usual, θA denote the set of points $\{\theta y\}$ where $y \in A$. Then, for any $\lambda > 0$, we have, using the scaling properties of Brownian motion,

$$\begin{aligned} E_0\{\exp\{-\nu|C_t^\epsilon(\beta(\cdot))|\}\} &= E_0\Big\{\exp\Big\{-\nu|C_{t/\lambda}^\epsilon(\sqrt{\lambda}\ \beta(\cdot))|\Big\}\Big\} \\ &= E_0\Big\{\exp\Big\{-\nu|\sqrt{\lambda}\, C_{t/\lambda}^{\epsilon/\sqrt{\lambda}}(\beta(\cdot))|\Big\}\Big\} \\ &= E_0\Big\{\exp\Big\{-\nu\lambda^{d/2}|C_{t/\lambda}^{\epsilon 0\sqrt{\lambda}}(\beta(\cdot))|\Big\}\Big\}. \end{aligned}$$

Thus, if we let $\lambda = t^{2/(d+2)}$, we have

$$\text{(1.5)} \quad E_0\{\exp\{-\nu|C_t^\epsilon(\beta(\cdot))|\}\} = E_0\Big\{\exp\Big\{-\nu t^{d/(d+2)}|C_{t^{d/(d+2)}}^{\epsilon/t^{1/(d+2)}}(\beta(\cdot))|\Big\}\Big\},$$

which means that to examine the asymptotic behavior of (1.1) as $t\to\infty$ we can just as well examine the asymptotic behavior of

$$\text{(1.6)} \qquad E_0\Big\{\exp\Big\{-\nu t|C_t^{\epsilon/t^{1/d}}(\beta(\cdot))|\Big\}\Big\}$$

as $t\to\infty$, i.e., replace $t^{d/(d+2)}$ by t. Therefore, what we actually prove is

$$\text{(1.7)} \qquad \varliminf_{t\to\infty} \frac{1}{t} \log E_0\Big\{\exp\Big\{-\nu t|C_t^{\epsilon/t^{1/d}}(\beta(\cdot))|\Big\}\Big\} \geqq -k(\nu,d),$$

and

$$\overline{\lim_{t\to\infty}} \frac{1}{t} \log E_0\left\{\exp\left\{-\nu t|C_t^{\epsilon/t^{1/d}}(\beta(\cdot))|\right\}\right\} \leqq -k(\nu,d), \tag{1.8}$$

which then imply Theorem 1.

We now describe how we handle the left side of (1.8). For any $t>0$, Borel set $A\subset R^d$, and Brownian path $\omega=\beta(\cdot)$, let

$$L_t(\omega,A)=\frac{1}{t}\lambda\left\{s:0\leqq s\leqq t,\beta(s)\in A\right\},$$

where $\lambda\{\ \}$ denotes one-dimensional Lebesgue measure. $L_t(\omega,A)$ is then the proportion of time up to t that a particular Brownian path spends in the set A. For each $t>0$, and each Brownian path ω, $L_t(\omega,\cdot)$ is a probability measure on R^d the support of which is $C_t(\beta(\cdot))$.

For any $\epsilon>0$, let $\phi^{(\epsilon)}(y)$ be non-negative, an element of C^∞ on R^d, and such that $\{y:\phi^{(\epsilon)}(y)>0\}=\{y:|y|<\epsilon\}$ and $\int_{R^d}\phi^{(\epsilon)}(y)\,dy=1$. Let $\phi_t^{(\epsilon)}(y)=t\phi^{(\epsilon)}(t^{1/d}y)$ so that

$$\int_{R^d}\phi_t^{(\epsilon)}(y)dy=1 \text{ and } \left\{y:\phi_t^{(\epsilon)}(y)>0\right\}=\left\{y:|y|<\epsilon/t^{1/d}\right\}.$$

For any $\epsilon>0$, $t>0$, and Brownian path $\omega=\beta(\cdot)$, let

$$f_t^{(\epsilon)}(\omega,y)=\int_{R^d}\phi_t^{(\epsilon)}(y-x)L_t(\omega,dx). \tag{1.9}$$

We see that $f_t^{(\epsilon)}(\omega,\cdot)$ is a probability density on R^d, and, what is important, since the support of the convolution of $\phi_t^{(\epsilon)}$ and L_t is the closure of the sum of their supports,

$$\left|\left\{y:f_t^{(\epsilon)}(\omega,y)>0\right\}\right|\leqq|C_t^{\epsilon/t^{1/d}}(\beta(\cdot))|.$$

Thus, to show (1.8) it suffices to show that

$$\overline{\lim_{t\to\infty}} \frac{1}{t} \log E_0\left\{\exp\left\{-\nu t|\left\{y:f_t^{(\epsilon)}(\omega,y)>0\right\}|\right\}\right\} \leqq -k(\nu,d). \tag{1.10}$$

The earlier results we wish to use on (1.8) apply to Markov processes where the state space is a compact metric space while here the d-dimensional Brownian motion has all of R^d as its state space. We find it convenient, therefore, to consider Brownian motion on a d-dimensional torus and, as we shall see shortly, this suffices to obtain (1.8). For $M>0$, let T_d^M be the torus in R^d of size M, i.e., the set of equivalence classes defined by $x \sim y$ if $x \in R^d, y \in R^d$, and $x-y$ has coordinates which are integral multiples of M. We can, of course, think of T_d^M as $0 \leqq x_i \leqq M$, $i=1,2,3,\cdots,d$, with the sides identified. Let g_M be the canonical mapping of $R^d \to T_d^M$, i.e., any element of R^d maps into the equivalence class to which it belongs. If we have Brownian motion $\{\beta(s), 0 \leqq s<\infty\}$ on R^d, then by Brownian motion on the torus T_d^M we mean $\{g_M(\beta(s)), 0 \leqq s<\infty\}$. Now, as will be shown in Lemma 3.3, for any $M>0$,

$$(1.11)\quad E_0\{\exp\{-\nu t|C_t^{\epsilon/t^{1/d}}(\beta(\cdot))|\}\} \leqq E_0\{\exp\{-\nu t|C_t^{\epsilon/t^{1/d}}(g_M(\beta(\cdot)))|\}\},$$

so that, for any $M>0$,

$$(1.12)\quad \begin{aligned} &\overline{\lim_{t\to\infty}} \frac{1}{t}\log E_0\{\exp\{-\nu t|C_t^{\epsilon/t^{1/d}}(\beta(\cdot))|\}\} \\ &\leqq \overline{\lim_{t\to\infty}} \frac{1}{t}\log E_0\{\exp\{-\nu t|C_t^{\epsilon/t^{1/d}}(g_M(\beta(\cdot)))|\}\}. \end{aligned}$$

If we let

$$\tilde{L}_t(\omega,A)=\frac{1}{t}\lambda\left\{s:0 \leqq s \leqq t, g_M(\beta(s)) \in A\right\}$$

and

$$\psi_t^{(\epsilon)}(y)=\sum_{g \in G_d^M} \phi_t^{(\epsilon)}(y+g),$$

where G_d^M is the set of vectors in R^d each of the coordinates of which is an integral multiple of M, then

$$\tilde{f}_t^{(\epsilon)}(\omega,y)=\int_{T_d^M} \psi_t^{(\epsilon)}(y-x)\tilde{L}_t(\omega,dx)$$

is the projection of $f_t^{(\epsilon)}(\omega,y)$ (defined by (1.9)) on the torus T_d^M. Moreover, as before,

$$|\{y: \tilde{f}_t^{(\epsilon)}(\omega,y)>0\}| \leqq |C_t^{\epsilon/t^{1/d}}(g_M(\beta(\cdot)))|,$$

and therefore from (1.12) we have

$$\begin{aligned} &\overline{\lim_{t\to\infty}} \frac{1}{t}\log E_0\{\exp\{-\nu t|C_t^{\epsilon/t^{1/d}}(\beta(\cdot))|\}\} \\ &\qquad \leqq \overline{\lim_{t\to\infty}} \frac{1}{t}\log E_0^M\{\exp\{-\nu t|\{y:\tilde{f}_t^{(\epsilon)}(\omega,y)>0\}|\}\}, \end{aligned} \tag{1.13}$$

where $E_0^M\{\ \}$ denotes integration over Brownian motion paths on T_d^M starting at the origin.

We note that $\tilde{f}_t^{(\epsilon)}(\omega,\cdot)$ is a probability density on T_d^M. If we let $\mathfrak{F}_M$ be the space of all probability densities on T_d^M, then the functional $|\{y: f(y)>0\}|$ for $f\in\mathfrak{F}_M$ is lower semicontinuous in the L_1 topology on $\mathfrak{F}_M$.

But in [5] (corollary of Theorem 5.1) we proved that, for any non-negative functional $\Phi(f)$ which is lower semicontinuous in the L_1 topology on $\mathfrak{F}_M$,

$$\overline{\lim_{t\to\infty}} \frac{1}{t}\log E_0^M\{\exp\{-t\Phi(\tilde{f}_t^{(\epsilon)}(\omega,\cdot))\}\} \leqq -\inf_{f\in\mathfrak{F}_M}[\Phi(f)+I_M(f)], \tag{1.14}$$

where

$$I_M(f)=\frac{1}{8}\int_{T_d^M}\frac{|\nabla f|^2}{f}dy.$$

Thus, in particular, for $\Phi(f)=\nu|\{y: f(y)>0\}|$ we get

$$\begin{aligned} &\overline{\lim_{t\to\infty}} \frac{1}{t}\log E_0^M\{\exp\{-\nu t|\{y:\tilde{f}_t^{(\epsilon)}(\omega,y)>0\}|\}\} \\ &\qquad \leqq -\inf_{f\in\mathfrak{F}_M}[\nu|\{y: f(y)>0\}|+I_M(f)]. \end{aligned} \tag{1.15}$$

Hence, from (1.13) and (1.15), we have, for any $M>0$,

$$\begin{aligned} &\overline{\lim_{t\to\infty}} \frac{1}{t}\log E_0\{\exp\{-\nu t|C_t^{\epsilon/t^{1/d}}(\beta(\cdot))|\}\} \\ &\qquad \leqq -\inf_{f\in\mathfrak{F}_M}[\nu|\{y: f(y)>0\}|+I_M(f)]. \end{aligned} \tag{1.16}$$

Since (1.16) holds for all $M>0$, we get

$$\overline{\lim_{t\to\infty}}\frac{1}{t}\log E_0\left\{-\exp\left\{-\nu t\left|C_t^{\epsilon/t^{1/d}}(\beta(\cdot))\right|\right\}\right\}$$

(1.17)

$$\leqq -\sup_M \inf_{f\in\mathfrak{F}_M}\left[\nu|\{y: f(y)>0\}|+I_M(f)\right].$$

Let $\mathcal{G}$ be the space of all probability density functions on R^d, i.e., $g\in\mathcal{G}$ if $g\geqq 0$ on R^d and $\int_{R^d} g(y)dy=1$. Let

$$I(g)=\frac{1}{8}\int_{R^d}\frac{|\nabla g|^2}{g}dy.$$

In Lemma 3.5 we show that

$$-\sup_M \inf_{f\in\mathfrak{F}_M}\left[\nu|\{y: f(y)>0\}|+I_M(f)\right]$$

(1.18)

$$\leqq -\inf_{g\in\mathcal{G}}\left[\nu|\{y: g(y)>0\}|+I(g)\right],$$

and in Lemma 3.9 we show that

$$\inf_{g\in\mathcal{G}}\left[\nu|\{y: g(y)>0\}|+I(g)\right]=k(\nu,d), \tag{1.19}$$

where $k(\nu,d)$ is defined in Theorem 1. Therefore, from (1.17), (1.18), and (1.19) we obtain (1.8).

The function space integral (1.1), the asymptotic behavior of which as $t\to\infty$ is determined by Theorem 1, is actually a special case of what arises in statistical mechanics. Let $V(y)$ be non-negative on R^d and consider for $\nu>0$

$$E_x\left\{\exp\left\{-\nu\int_{R_d}\left[1-\exp\left\{-\int_0^t V(\beta(s)-y)\,ds\right\}\right]dy\right\}\right\}, \tag{1.20}$$

where $E_x\{\ \}$ denotes expectation on the Brownian motion paths $\beta(\cdot)$ in R^d starting at x. If, for $\epsilon>0$, we let $V(y)=\infty$ for $|y|<\epsilon$ and $V(y)=0$ for $|y|\geqq\epsilon$ (i.e., the potential for "hard spheres"), then

$$1-\exp\left\{-\int_0^t V(\beta(s)-y)\,ds\right\}$$

is just the indicator function of the Wiener sausage of size ϵ. Hence, for this particular V, (1.20) reduces to (1.1).

In Section 4 of this paper we shall show that, under very broad assumptions on $V(y)$, the asymptotic behavior of (1.20) as $t\to\infty$ is exactly the same as it is in the special case of the sausage. We shall prove

THEOREM 2. *Let V be non-negative but not identically zero. Then,*

$$\overline{\lim_{t\to\infty}}\frac{1}{t^{d/(d+2)}}\log E_x\left\{\exp\left\{-\nu\int_{R_d}\left[1-\exp\left\{-\int_0^t V(\beta(s)-y)\,ds\right\}\right]dy\right\}\right\} \leqq -k(\nu,d). \tag{1.21}$$

Moreover, if $V(y)=o(1/|y|^{d+2})$ as $|y|\to\infty$, then

$$\underline{\lim_{t\to\infty}}\frac{1}{t^{d/(d+2)}}\log E_x\left\{\exp\left\{-\nu\int_{R_d}\left[1-\exp\left\{-\int_0^t V(\beta(s)-y)\,ds\right\}\right]dy\right\}\right\} \geqq -k(\nu,d), \tag{1.22}$$

where $k(\nu,d)$ is the same as in Theorem 1.

The proof of Theorem 2 follows the same lines as Theorem 1, i.e., for (1.21) we go to the torus in R^d and use the general result in [5]. The proof of (1.22) is easier and analogous to the proof of (1.3) (or equivalently (1.7)), but the asymptotic condition on V seems to be necessary for (1.22).

Since Brownian motion is a particular symmetric stable process, it is interesting to ask how the asymptotic rate in Theorem 1 would be affected if instead of the Wiener sausage treated there we consider the general symmetric stable sausage. Only in the special case of the Wiener sausage will the paths be almost surely continuous which means that all other symmetric stable sausages come in disconnected links. However our methods still apply because of the scaling properties of symmetric stable processes in R^d and the generality of the results in [5].

Thus, if $\{x(s), 0 \leqq s < \infty\}$ is a symmetric stable process[1] in R^d with exponent $0 < \alpha \leqq 2$, and now $E_0\{\ \}$ denotes expectation on the paths of this process starting at 0, we have, for any $\lambda > 0$,

$$
\begin{aligned}
E_0\{\exp\{-\nu|C_t^\epsilon(x(\cdot))|\}\} &= E_0\Big\{\exp\Big\{-\nu|C_{t/\lambda}^\epsilon(\lambda^{1/\alpha}x(\cdot))|\Big\}\Big\} \\
&= E_0\Big\{\exp\Big\{-\nu|\lambda^{1/\alpha}C_{t/\lambda}^{\epsilon/(\lambda^{1/\alpha})}(x(\cdot))|\Big\}\Big\} \\
&= E_0\Big\{\exp\Big\{-\nu\lambda^{1/\alpha}|C_{t/\lambda}^{\epsilon/\lambda^{1/\alpha}}(x(\cdot))|\Big\}\Big\}.
\end{aligned}
$$

Hence, if we let $\lambda = t^{\alpha/(d+\alpha)}$, we have

$$
(1.23)\quad E_0\{\exp\{-\nu|C_t^\epsilon(x(\cdot))|\}\} = E_0\Big\{\exp\Big\{-\nu t^{d/(d+\alpha)}|C_{t^{d/(d+\alpha)}}^{\epsilon/t^{1/(d+\alpha)}}(x(\cdot))|\Big\}\Big\},
$$

which means that to examine the asymptotic behavior of the expectation on the left of (1.23) we can just as well examine the asymptotic behavior of

$$
(1.24)\qquad E_0\Big\{\exp\Big\{-\nu t|C_t^{\epsilon/t^{1/d}}(x(\cdot))|\Big\}\Big\}
$$

as $t \to \infty$, i.e., replace $t^{d/(d+\alpha)}$ by t. But (1.24) is exactly of the form we discussed earlier in the special case of Brownian motion and is susceptible to the same treatment since the asymptotic results in [5] apply to Markov processes in general.

In order to state precisely the asymptotic behavior of (1.24) (and hence of the left side of (1.23)) as $t \to \infty$ we must introduce the I-function for symmetric stable processes when $0 < \alpha < 2$, i.e., the function which plays the same role in the asymptotics as

$$
(1.25)\qquad I(g) = \frac{1}{8}\int_{R^d}\frac{|\nabla g|^2}{g}\,dy
$$

did in the special case of Brownian motion.

Let $Q(\xi)$, defined for $\xi \in R^d$, be such that $Q(-\xi) = Q(\xi)$, $Q(\lambda\xi) = \lambda^\alpha Q(\xi)$ for some α with $0 < \alpha \leqq 2$ and all $\lambda > 0$, and $\phi(t,\xi) = e^{-tQ(\xi)}$ is positive definite for $t > 0$. If we then let

$$
p(t,x) = \frac{1}{(2\pi)^d}\int_{R^d} e^{-i\langle\xi,x\rangle}\phi(t,\xi)\,d\xi,
$$

[1]For a discussion of these processes in the present context see [5], Section 5.

then $p(t,x)$ is the transition probability for a symmetric stable process in R^d of exponent α, $0<\alpha \leqq 2$. Now, if $\alpha=2$, then $Q(\xi)$ is of the form $\frac{1}{2}\sum a_{ij}\xi_i\xi_j$, where (a_{ij}) is a symmetric positive definite matrix. Brownian motion is the special case when $\alpha=2$ and the matrix (a_{ij}) is the identity, but, of course, this last can be accomplished by a suitable change of coordinates.

When $0<\alpha<2$, the infinitesimal generator L of the semigroup corresponding to $p(t,x)$ can be represented by the Lévy-Khintchine representation as

$$(Lf)(x)=\int_{R^d}\left[f(x+y)-f(x)-\frac{(y\cdot\nabla f(x))}{1+|y|^2}\right]\Omega(dy), \tag{1.26}$$

for suitably smooth functions f, where Ω is a σ-finite measure which is finite outside of any neighborhood of the origin such that

$$\int_{R^d}\frac{|y|^2}{1+|y|^2}\Omega(dy)<\infty.$$

In particular, $\Omega\{y:|y|\geqq M\}\to 0$ as $M\to\infty$. Thus a symmetric stable process in R^d of order α, where $0<\alpha<2$, is characterized by the measure Ω.

In our earlier work on the asymptotic evaluation of Markov process expectations, the essential role is played by the I-function (defined for general Markov processes in [4]). For Brownian motion in R^d the I-function is given by (1.25) for any $g\in\mathcal{G}$, the space of all probability densities on R^d. For any Markov process with a symmetric transition probability density it was proved in [4] (Theorem 5) that the I-function takes the form

$$I(\mu)=\|(-L)^{1/2}\sqrt{g}\,\|^2,$$

where L is the infinitesimal generator of the process viewed as a selfadjoint negative definite operator on square integrable functions. The formula above says that, in order that $I(\mu)$ be finite, μ must have a density g with respect to the reference measure (Lebesgue measure in the present case) and $\sqrt{g}$ must be in the domain of $(-L)^{1/2}$. In such a case $I(\mu)$, which will be denoted by $I(g)$, is given by $\|(-L)^{1/2}\sqrt{g}\,\|^2$. In the case of symmetric stable processes it is easy to see that

$$I(g)=\frac{1}{8}\int_{R^d}\frac{|\nabla g|^2}{g}\,dx$$

for Brownian motion, and

$$I(g)=\frac{1}{4}\int_{R^d}\int_{R^d}\left|\sqrt{g(x+y)}-\sqrt{g(x)}\right|^2\Omega(dy)\,dx \tag{1.27}$$

when $0<\alpha<2$. Here, $g\in\mathcal{G}$ and Ω is defined by (1.26).

Following along the same discussion as in the Brownian motion case we arrive at

THEOREM 3. *Let $\{x(s),\ 0\leqq s<\infty\}$ be a symmetric stable process in R^d of order $0<\alpha\leqq 2$, with infinitesimal generator L. For any $x\in R^d$, $\nu>0$, and $\epsilon>0$,*

$$\lim_{t\to\infty}\frac{1}{t^{d/(d+\alpha)}}\log E_x\left\{\exp\{-\nu|C_t^\epsilon(x(\cdot))|\}\right\}=-k(\nu,L), \tag{1.28}$$

where

$$k(\nu,L)=\nu^{d/(d+\alpha)}\left(\frac{d+\alpha}{\alpha}\right)\left(\frac{\alpha\lambda_L}{d}\right)^{d/(d+\alpha)}$$

and

$$\lambda_L=\inf_{G:|G|=1}\lambda(G),$$

the infimum being over all open sets G in R^d of unit volume with $\lambda(G)$ the smallest eigenvalue of $-L$ with zero boundary conditions on G.

In Lemma 3.10 we show that $\lambda_L>0$ for all L under consideration.

Theorem 1 is a special case of Theorem 3 because, in the case of Brownian motion, $L=-\frac{1}{2}\Delta$ and $\inf_{G:|G|=1}\lambda(G)$ is attained for the sphere of unit volume (see Lemma 3.13). Hence $\lambda_L=\gamma_d$, as defined in Theorem 1, and $k(\nu,L)$ is just the $k(\nu,d)$ of that theorem.

It will also be proved in Lemma 3.13 that if $0<\alpha<2$ and Ω is spherically symmetric, then again $\inf_{G:|G|=1}\lambda(G)$ is attained for the sphere of unit volume.

It is interesting to observe that the 2 that appears in $t^{d/(d+2)}$ in Theorem 1 is the 2 of the Gaussian to be replaced by α for a general symmetric stable process as in (1.28).

To prove Theorem 3 we use exactly the same arguments as outlined above for Brownian motion. In Lemma 3.9 we shall show that

$$\inf_{g\in\mathcal{G}}\left[\nu|\{y:\ g(y)>0\}|+I(g)\right]=k(\nu,L), \tag{1.29}$$

where

$$I(g)=\frac{1}{8}\int_{R^d}\frac{|\nabla g|^2}{g}\,dy$$

in the Brownian motion case and

$$I(g)=\frac{1}{4}\int_{R^d}\int_{R^d}\left|\sqrt{g(x+y)}-\sqrt{g(x)}\right|^2\Omega(dy)\,dx,$$

for the symmetric stable process with $0<\alpha<2$.

In what follows, all proofs are framed so as to include symmetric stable processes of order $0<\alpha\leqq 2$ unless the Brownian motion case is sufficiently simpler to justify a separate proof.

2. Lower Bounds

Let E be a bounded closed set in R^d and let $\mathcal{E}$ be the collection of such sets. Let $\{x(s), 0\leqq s<\infty\}$ be a symmetric stable process in R^d of order $0<\alpha\leqq 2$, and let $C_t^\epsilon(x(\cdot))$ be the symmetric stable sausage up to time t. Let $I(\mu)$ be the I-function for this symmetric stable process (given by (1.25) in the case of Brownian motion and by (1.27) if $0<\alpha<2$). In this section we prove

THEOREM 2.1. *For any* $\epsilon>0$, $x\in R^d$, *and* $\nu>0$,

$$\text{(2.1)}\qquad \begin{aligned}&\varliminf_{t\to\infty}\frac{1}{t}\log E_x\left\{\exp\left\{-\nu t\left|C_t^{\epsilon/t^{1/d}}(x(\cdot))\right|\right\}\right\}\\ &\qquad\geqq -\inf_{E\in\mathcal{E}}\left[\nu|E|+\inf_{\mu:\,\mathrm{supp}\,\mu\subset E}I(\mu)\right].\end{aligned}$$

Before proving this theorem we make a few comments. Let, as before, $\mathcal{G}$ be the set of all probability densities on R^d. In Section 3 we shall show that

$$\text{(2.2)}\qquad \inf_{E\in\mathcal{E}}\left[\nu|E|+\inf_{\mu:\,\mathrm{supp}\,\mu\subset E}I(\mu)\right]=\inf_{g\in\mathcal{G}}\left[\nu|\{y: g(y)>0\}|+I(g)\right],$$

and also in Section 3 we shall show that

$$\text{(2.3)}\qquad \inf_{g\in\mathcal{G}}\left[\nu|\{y: g(y)>0\}|+I(g)\right]=k(\nu,L),$$

where $k(\nu,L)$ is defined just after (1.28). Hence Theorem 2.1, together with (2.2) and (2.3), implies

$$\varliminf_{t\to\infty} \frac{1}{t}\log E_x\left\{\exp\left\{-\nu t\left|C_t^{\epsilon/t^{1/d}}(x(\cdot))\right|\right\}\right\} \geqq -k(\nu,L). \tag{2.4}$$

Proof of Theorem 2.1: Let $p(t,x,y)$ be the transition density of the symmetric stable process $\{x(s),\ 0\leqq s<\infty\}$. Let G be a bounded open set in R^d and A, a Borel set such that $A\subset G$. We define

$$q_G(t,x,A)=P_x\left\{x(t)\in A,\ x(s)\in G \quad \text{for} \quad \text{all} \quad 0\leqq s\leqq t\right\}, \tag{2.5}$$

where, of course, $P_x\{\ \}$ denotes the probability measure on the paths of the process starting at x. Now $q_G(t,x,\cdot)$ has a density $q_G(t,x,y)\leqq p(t,x,y)$ which satisfies the Chapman-Kolmogorov equation and is symmetric and bounded for $t>0$. Thus, $q_G(t,x,y)$ is a compact symmetric operator for each $t>0$ on $L^2(G)$ so that

$$q_G(t,x,y)=\sum_{j=1}^{\infty} e^{-\lambda_j^\delta t}\ \psi_j(x)\psi_j(y) \tag{2.6}$$

on $L^2(G\times G)$, where $\{\lambda_j\}$ and $\{\psi_j\}$ are, respectively, the eigenvalues and normalized eigenfunctions of L_G, the infinitesimal generator of the semigroup associated with $q_G(t,x,y)$. Since $p(t,x,y)$ is a contraction, so is $q_G(t,x,y)$, and therefore all the λ_j are positive.

Now let $E\in\mathfrak{E}$ and let $\delta>0$ be given. Since E^δ is a bounded open set, we have from the preceding discussion that

$$q_{E^\delta}(t,x,y)=\sum_{j=1}^{\infty} e^{-\lambda_j^\delta t}\,\psi_j^\delta(x)\psi_j^\delta(y) \tag{2.7}$$

on $L^2(E^\delta\times E^\delta)$, where now $\{\lambda_j^\delta\}$ and $\{\psi_j^\delta\}$ are the eigenvalues and eigenfunctions of L_{E^δ}, the infinitesimal generator of the semigroup associated with $q_{E^\delta}(t,x,y)$. Since the series on the right of (2.7) converges in $L^2(E^\delta\times E^\delta)$,

$$\int_{R^d} q_{E^\delta}(t,x,y)\psi_1^\delta(y)\,dy=e^{-\lambda_1^\delta t}\,\psi_1^\delta(x) \tag{2.8}$$

for almost all $x\in E^\delta$. Therefore we can pick an x, call it x_δ, such that

$$\int_{R^d} q_{E^\delta}(t,x_\delta,y)\psi_1^\delta(y)\,dy=e^{-\lambda_1^\delta t}\,\psi_1^\delta(x_\delta)$$

and $|\psi_1^\delta(x_\delta)| = c \neq 0$. Since $q_{E^\delta}(t,x,y)$ has its support on E^δ and is bounded there,

$$\left\| \int_{R^d} q_{E^\delta}(t,x,y)\psi_1^\delta(y)dy \right\| \leqq \text{const.}\|\psi_1^\delta\| \leqq \text{const.},$$

which means from (2.8) that $\psi_1^\delta(x)e^{-\lambda_1^\delta t}$ is bounded, i.e., $\sup_{x\in E^\delta}|\psi_1^\delta(x)| < \infty$. Now, since $q_{E^\delta}(t,x,y) = 0$ for $y \not\in E^\delta$,

$$\begin{aligned} ce^{-\lambda_1^\delta t} &= \left| \int_{R^d} q_{E^\delta}(t,x_\delta,y)\psi_1^\delta(y)dy \right| \\ &\leqq \sup_{y\in E^\delta} |\psi_1^\delta(y)| \int_{E^\delta} q_{E^\delta}(t,x_\delta,y)dy. \end{aligned}$$

Therefore,

$$P_{x_\delta}\left\{x(s)\in E^\delta \quad \text{for} \quad \text{all} \quad 0\leqq s\leqq t\right\} \geqq \frac{ce^{-\lambda_1^\delta t}}{\sup_{y\in E^\delta}|\psi_1^\delta(y)|}. \tag{2.9}$$

Since,

$$\begin{aligned} &E_{x_\delta}\left\{\exp\{-\nu t|C_t^{\epsilon/t^{1/d}}(x(\cdot))|\}\right\} \\ &\geqq E_{x_\delta}\left\{\exp\{-\nu t|C_t^{\epsilon/t^{1/d}}(x(\cdot))|\};x(s)\in E^\delta \quad \text{for} \quad \text{all} \quad 0\leqq s\leqq t\right\} \\ &\geqq \exp\{-\nu t|E^{\delta+\epsilon/t^{1/d}}|\}P_{x_\delta}\left\{x(s)\in E^\delta \quad \text{for} \quad \text{all} \quad 0\leqq s\leqq t\right\}, \end{aligned}$$

we have from (2.9) and the fact that the asymptotic behavior of

$$E_x\left\{\exp\{-\nu t|C_t^{\epsilon/t^{1/d}}(x(\cdot))|\}\right\}$$

is independent of the starting point x,

$$\varlimsup_{t\to\infty} \frac{1}{t}\log E_x\left\{\exp\{-\nu t|C_t^{\epsilon/t^{1/d}}(x(\cdot))|\}\right\} \geqq -\nu|E^{2\delta}| - \lambda_1^\delta. \tag{2.10}$$

Since $E \in \mathcal{E}$, $\lim_{\delta\to 0}|E^{2\delta}| = |E|$. Moreover, we prove below in Lemma 2.3, that

$$\varlimsup_{\delta\to 0} \lambda_1^\delta \leqq \inf_{\mu:\,\mathrm{supp}\,\mu\subset E} I(\mu). \tag{2.11}$$

Thus, from (2.10) and (2.11), we obtain on letting $\delta \to 0$

$$\overline{\lim_{t\to\infty}} \frac{1}{t} \log E_x \left\{ \exp\left\{ -\nu t \left| C_t^{\epsilon/t^{1/d}}(x(\cdot)) \right| \right\} \right\} \tag{2.12}$$
$$\geqq -\left[\nu|E| + \inf_{\mu:\, \mathrm{supp}\,\mu \subset E} I(\mu) \right].$$

But (2.12) holds for any set $E \in \mathcal{E}$ and hence we have (2.1).

Let G be a bounded open set in R^d and define $q_G(t,x,y)$ by (2.5). Let L_G be the infinitesimal generator associated with $q_G(t,x,y)$ and L be the infinitesimal generator associated with $p(t,x,y)$.

LEMMA 2.2. *Let $h \in C^\infty$ on R^d and $h = 0$ off of G. Then $h \in \mathfrak{D}(L_G)$, the domain of L_G, and moreover*

$$L_G h = Lh|_G .$$

Proof: We have to show that

$$\int_G h(y) q_G(t,x,y)\,dy - h(x) = \int_0^t ds \int_G (Lh)(y) q_G(s,x,y)\,dy, \tag{2.13}$$

i.e., $T_t^G h - h = \int_0^t T_s^G W\, ds$, where $W = Lh$. Let $\{x(s),\ 0 \leqq s < \infty\}$ be the symmetric stable process and consider

$$h(x(t)) - \int_0^t W(x(s))\,ds .$$

This is a martingale and the proof of the lemma is just an application of Doob's stopping theorem. By definition,

$$E_x\left\{ h(x(t)) - \int_0^t W(x(s))\,ds \right\}\Bigg|_{t=0} = h(x),$$

and by Doob's theorem,

$$E_x\left\{ h(x(\tau\wedge t)) - \int_0^{\tau\wedge t} W(x(s))\,ds \right\} = h(x) \tag{2.14}$$

for any stopping time τ. Thus, let, in particular, $\tau = \tau_G$ be the first time the process exits from G. From (2.14), with this τ,

$$E_x\{h(x(\tau\wedge t))\}-h(x)=E_x\left\{\int_0^{\tau\wedge t}W(x(s))ds\right\}. \tag{2.15}$$

First of all,

$$E_x\{h(x(\tau\wedge t))\}=E_x\{h(x(t)),\tau>t\}+E_x\left\{h(x(\tau)),\tau\leqq t\right\}.$$

But, from the definition of τ, $x(\tau)\in\tilde{G}$ (the complement of G) and since $h=0$ on $\tilde{G}$ we have $h(x(\tau))=0$. Thus,

$$\begin{aligned}E_x\{h(x(\tau\wedge t))\}&=E_x\{h(x(t)),\tau>t\}\\&=\int_G h(y)q_G(t,x,y)dy\,.\end{aligned} \tag{2.16}$$

Secondly, if we let $\overline{W}=W=Lh$ on G and $\overline{W}=0$ on $\tilde{G}$, then

$$\begin{aligned}E_x\left\{\int_0^{\tau\wedge t}W(x(s))ds\right\}&=E_x\left\{\int_0^t\overline{W}(x(\tau\wedge s))ds\right\}\\&=\int_0^t E_x\{\overline{W}(x(\tau\wedge s))\}ds\,.\end{aligned}$$

But, as before,

$$\begin{aligned}E_x\left\{\overline{W}(x(\tau\wedge s))\right\}&=E_x\left\{\overline{W}(x(s)),\tau>s\right\}+E_x\left\{\overline{W}(x(\tau)),\tau\leqq s\right\}\\&=E_x\left\{\overline{W}(x(s)),\tau>s\right\}\\&=\int_G\overline{W}(y)q_G(s,x,y)\,dy\\&=\int_G W(y)q_G(s,x,y)dy,\end{aligned}$$

so that

$$E_x\left\{\int_0^{\tau\wedge t}W(x(s))ds\right\}=\int_0^t ds\int_G W(y)q_G(s,x,y)dy\,. \tag{2.17}$$

Now from (2.15), (2.16), and (2.17) we have

$$\int_G h(y)q_G(t,x,y)dy - h(x) = \int_0^t ds \int_G W(y)q_G(s,x,y)dy$$

which is (2.13).

LEMMA 2.3. *Let $E \in \mathfrak{S}$. Then,*

$$\overline{\lim_{\delta \to 0}}\, \lambda_1^\delta \leq \inf_{\mu : \operatorname{supp} \mu \subset E} I(\mu).$$

Proof: Since λ_1^δ is the principal eigenvalue of L_{E^δ}, the infinitesimal generator of the semigroup associated with $q_{E^\delta}(t,x,y)$, we have, for any $\delta > 0$,

$$\lambda_1^\delta = \inf_{\substack{\|\psi\|=1 \\ \operatorname{supp} \psi \subset E^\delta}} \|(-L_{E^\delta})^{1/2}\psi\|^2. \tag{2.18}$$

Now, let μ be a probability measure on R^d such that $\mu(E)=1$ and such that $I(\mu)=l<\infty$. As noted in the introduction, $I(\mu)<\infty$ implies that the measure μ has a density f which now, moreover, is zero almost everywhere off of E. If we let $g=\sqrt{f}$ and again let L be the infinitesimal generator of the semigroup associated with $p(t,x,y)$, then, as was also noted in the introduction,

$$I(\mu) = \|(-L)^{1/2}g\|^2 = l. \tag{2.19}$$

For any $\delta > 0$, let $\phi_\delta(x) \in C^\infty$ on R^d, $\phi_\delta(x)>0$ for $|x|<\delta$, $\phi_\delta(x)=0$ if $|x|>\delta$ and $\int_{R^d} \phi_\delta(x)\, dx = 1$. We define the function $g_\delta(x)$ to be the convolution of g with this mollifier, i.e., $g_\delta(x) = (\phi_\delta * g)(x)$. We note that $g_\delta \in C^\infty$ on R^d and supp $g_\delta \subset E^\delta$. It is easy to see that $\|(-L)^{1/2}g_\delta\|^2 \leq l$ and $\|(-L)^{1/2}g_\delta\|^2 \to l$ as $\delta \to 0$. Thus, if we let $h_\delta(x) = g_\delta(x)/\|g_\delta\|$, we observe that supp $h_\delta \subset E^\delta$, $\|h_\delta\|=1$, $h_\delta \in C^\infty$ on R^d, and

$$\|(-L)^{1/2}h_\delta\|^2 = -\langle Lh_\delta, h_\delta\rangle \to l \quad \text{as} \quad \delta \to 0. \tag{2.20}$$

From Lemma 2.2 we have, therefore,

$$-\langle L_{E^\delta} h_\delta, h_\delta\rangle = -\langle Lh_\delta, h_\delta\rangle \to l \quad \text{as} \quad \delta \to 0. \tag{2.21}$$

But from (2.18), since $\|h_\delta\| = 1$ and supp $h_\delta \subset E^\delta$, we have

$$\lambda_1^\delta \leqq \|(-L_{E^\delta})^{1/2} h_\delta\|^2 = -\langle L_{E^\delta} h_\delta, h_\delta \rangle,$$

so that (2.21) implies

$$\overline{\lim_{\delta \to 0}}\, \lambda_1^\delta \leqq l = I(\mu). \tag{2.22}$$

Since (2.22) holds for any probability measure μ on R^d such that $\mu(E) = 1$, we have finally

$$\overline{\lim_{\delta \to 0}}\, \lambda_1^\delta \leqq \inf_{\mu : \operatorname{supp} \mu \subset E} I(\mu),$$

which is the lemma.

3. Proof of Theorems 3 and 1

As pointed out in the introduction, Theorem 1 is a special case of Theorem 3. So again let $\{x(s), 0 \leqq s < \infty\}$ be a symmetric stable process in R^d of order $0 < \alpha \leqq 2$, and let $C_t^\epsilon(x(\cdot))$ be the symmetric stable sausage up to time t. Let $I(\mu)$ be the I-function for this symmetric stable process (given by (1.25) in the case of Brownian motion and by (1.27) if $0 < \alpha < 2$). In this section we prove

THEOREM 3.1. *For any $\epsilon > 0$, $x \in R^d$, and $\nu > 0$,*

$$\begin{aligned}\overline{\lim_{t \to \infty}}\, \frac{1}{t} \log E_x \left\{ \exp\left\{ -\nu t \left| C_t^{\epsilon / t^{1/d}}(x(\cdot)) \right| \right\} \right\} \\ \leqq -\inf_{g \in \mathcal{G}} \left[\nu |\{ y : g(y) > 0 \}| + I(g) \right].\end{aligned} \tag{3.1}$$

We shall then prove, with $\mathcal{E}$ defined just before Theorem 2.1,

THEOREM 3.2.

$$\begin{aligned}\inf_{E \in \mathcal{E}} \left[\nu |E| + \inf_{\mu : \operatorname{supp} \mu \subset E} I(\mu) \right] \\ = \inf_{g \in \mathcal{G}} \left[\nu |\{ y : g(y) > 0 \}| + I(g) \right].\end{aligned} \tag{3.2}$$

From Theorems 2.1, 3.1, and 3.2 we deduce

$$\lim_{t\to\infty}\frac{1}{t}\log E_x\left\{\exp\left\{-\nu t\left|C_t^{\epsilon/t^{1/d}}(x(\cdot))\right|\right\}\right\}$$

(3.3)

$$=-\inf_{g\in\mathcal{G}}\left[\nu\left|\{y:g(y)>0\}\right|+I(g)\right].$$

Finally in Lemma 3.9 we show that

$$\inf_{g\in\mathcal{G}}\left[\nu\left|\{y:g(y)>0\}\right|+I(g)\right]=k(\nu,L),$$

as defined in Theorem 3, which gives us Theorems 3 and 1 as stated in the introduction.

We first prove

LEMMA 3.3. *Let $\{x(s),\ 0\leqq s<\infty\}$ be the symmetric stable process in R^d of order $0<\alpha\leqq 2$. Let $M>0$ and g_M be the canonical mapping of $R^d\to T_d^M$, the torus in R^d of size M. Then*

$$E_0\left\{\exp\left\{-\nu t\left|C_t^{\epsilon/t^{1/d}}(x(\cdot))\right|\right\}\right\}\leqq E_0\left\{\exp\left\{-\nu t\left|C_t^{\epsilon/t^{1/d}}(g_M(x(\cdot)))\right|\right\}\right\}.$$

Proof: This is really quite obvious, since if B is any set in R^d, we have clearly $|g_M(B)|\leqq|B|$ and hence, in particular, $|C_t^{\epsilon/t^{1/d}}(x(\cdot))|\geqq|C_t^{\epsilon/t^{1/d}}(g_M(x(\cdot)))|$ which gives the lemma.

LEMMA 3.4. *Let $f\in\mathcal{F}_M$ and $0<\lambda<M$. Define*

$$E_M^\lambda=\cup_{i=1}^d\left\{\left(0\leqq x_i\leqq\lambda\right)\cup\left(M-\lambda\leqq x_i\leqq M\right)\right\}.$$

Then there exists an $a\in R^d$ such that

$$\int_{E_M^\lambda}f(x-a)\,dx\leqq\frac{2\lambda d}{M}.$$

Proof: Let $M_d=\cap_{i=1}^d\ \{0\leqq x_i\leqq M\}$. We note that $E_M^\lambda\subset M_d$ and $|E_M^\lambda|\leqq 2\lambda dM^{d-1}$. Let $E_M^\lambda+a$ be the translation modulo M of E_M^λ by any $a\in R^d$, and consider

$$\int_{E_M^\lambda+a}f(x)\,dx$$

as a function of a. We obtain, for its mean value over M_d,

$$\frac{1}{|M_d|}\int_{M_d} da \int_{E_M^\lambda + a} f(x)\,dx = \frac{1}{|M_d|}\int_{M_d} da \int_{M_d} \chi_{E_M^\lambda}(x-a) f(x)\,dx$$

$$= \frac{1}{|M_d|}|E_M^\lambda| \int_{M_d} f(x)\,dx$$

$$\leq \frac{2\lambda d}{M},$$

and hence there exists an a such that

$$\int_{E_M^\lambda + a} f(x)\,dx \leqq \frac{2\lambda d}{M} \quad \text{or} \quad \int_{E_M^\lambda} f(x-a)\,dx \leqq \frac{2\lambda d}{M}.$$

LEMMA 3.5. *Let $M>0$, $\nu>0$, and let $\mathfrak{F}_M$ be the space of probability density functions on the torus T_d^M. Let $\mathcal{G}$ be the space of probability density functions on R^d. Let*

$$I_M(f) = \frac{1}{8}\int_{T_d^M} \frac{|\nabla f|^2}{f}\,dx,$$

and

$$I(g) = \frac{1}{8}\int_{R^d} \frac{|\nabla g|^2}{g}\,dx.$$

Then

$$\sup_M \inf_{f\in\mathfrak{F}_M} \left[\nu|\{y : f(y)>0\}| + I_M(f)\right]$$

(3.4)

$$\geqq \inf_{g\in\mathcal{G}} \left[\nu|\{y : g(y)>0\}| + I(g)\right].$$

Proof: Let $J = \sup_M \inf_{f\in\mathfrak{F}_M} [\nu|\{y:\ f(y)>0\}| + I_M(f)]$, and let $\epsilon>0$ be given. We must find $g\in\mathcal{G}$ such that $\nu|\{y : g(y)>0\}| + I(g) \leqq J+\epsilon$. Let

$l = J + \frac{1}{2}\epsilon$ and choose M so large that

$$\frac{1}{8(1-2d/\sqrt{M})}\left[\frac{16ld}{\sqrt{M}} + \frac{4d}{M} + \frac{4\sqrt{8dl}}{\sqrt{M}}\right] < \frac{1}{2}\epsilon. \tag{3.5}$$

For this fixed M, we know there exists an $f \in \mathcal{F}_M$ such that

$$\nu|\{y: f(y)>0\}| + I_M(f) \leqq l. \tag{3.6}$$

For this f which we just picked and any $a \in R^d$, let $f^*(x) = f(x-a)$. Then, $\nu|\{y: f^*(y)>0\}| + I_M(f^*) \leqq l$, and thus, taking $\lambda = \sqrt{M}$ in Lemma 3.4, we can without loss of generality assume that the $f \in \mathcal{F}_M$ which satisfies (3.6) also satisfies

$$\int_{E_M^{\sqrt{M}}} f(x)\,dx \leqq \frac{2d}{\sqrt{M}}, \tag{3.7}$$

where $E_M^{\sqrt{M}}$ is defined in Lemma 3.4. Now, let

$$\phi_M(x) = \begin{cases} 0 & \text{for } x \leqq 0, \\ \dfrac{x}{\sqrt{M}} & \text{for } 0 \leqq x \leqq \sqrt{M}, \\ 1 & \text{for } \sqrt{M} \leqq x \leqq M - \sqrt{M}, \\ -\dfrac{x}{\sqrt{M}} + \sqrt{M} & \text{for } M - \sqrt{M} \leqq x \leqq M, \\ 0 & \text{for } x \geqq M, \end{cases}$$

and $\hat{\phi}_M(x_1, x_2, \cdots, x_d) = \phi_M(x_1)\phi_M(x_2)\cdots\phi_M(x_d)$, so that

$$\left|\frac{\partial \hat{\phi}_M}{\partial x_i}\right| \leqq \frac{1}{\sqrt{M}} \quad \text{and hence} \quad |\nabla \hat{\phi}_M| \leqq \frac{\sqrt{d}}{\sqrt{M}}.$$

Let $\psi(x_1, x_2, \cdots, x_d) = [\hat{\phi}_M]^2 = \phi_M^2(x_1)\cdots\phi_M^2(x_d)$, and define $g \in \mathcal{G}$ by

$$g(x) = \frac{f(x)\psi(x)}{\int_{R^d} f(x)\psi(x)\,dx}$$

for the f which satisfies (3.6) and (3.7). We note from the definition of g that $\{y: g(y)>0\} \subset \{y: f(y)>0\}$ and hence

$$|\{y: g(y)>0\}| \leqq |\{y: f(y)>0\}|. \tag{3.8}$$

Let $\alpha = \int_{R^d} f(x)\psi(x)dx$. From the definition of g we have

$$\frac{1}{8}\int_{T_d^M} \frac{|\nabla g|^2}{g} dx = \frac{1}{8\alpha}\left[\int_{T_d^M} \frac{|\nabla f|^2}{f}\psi\, dx + \int_{T_d^M} \frac{|\nabla \psi|^2}{\psi} f dx + 2\int_{T_d^M} \langle \nabla\psi, f\rangle dx\right].$$

Since $0 \leqq \psi(x) \leqq 1$, $f \in \mathfrak{F}_M$, $|\nabla\psi| \leqq 2\sqrt{d}\,/\sqrt{M}$, and

$$\frac{|\nabla\psi|^2}{\psi} = \frac{4\hat{\phi}_M^2|\nabla\hat{\phi}_M|^2}{\hat{\phi}_M^2} = 4|\nabla\hat{\phi}_M|^2 \leqq \frac{4d}{M},$$

we have

$$\frac{1}{8}\int_{T_d^M} \frac{|\nabla g|^2}{g} dx \leqq \frac{1}{8\alpha}\left[\int_{T_d^M} \frac{|\nabla f|^2}{f} dx + \frac{4d}{M} + \frac{4\sqrt{d}}{\sqrt{M}}\int_{T_d^M} |\nabla f| dx\right]. \tag{3.9}$$

Now, from (3.6),

$$\int_{T_d^M} |\nabla f| dx = \int_{T_d^M} \frac{|\nabla f|}{\sqrt{f}}\sqrt{f}\, dx \leqq \sqrt{\int_{T_d^M} \frac{|\nabla f|^2}{f} dx} = \sqrt{8I_M(f)} \leqq \sqrt{8l},$$

and hence

$$\begin{aligned} I(g) &= \frac{1}{8}\int_{R^d} \frac{|\nabla g|^2}{g} dx = \frac{1}{8}\int_{T_d^M} \frac{|\nabla g|^2}{g} dx \\ &\leqq \frac{1}{8\alpha}\left[8I_M(f) + \frac{4d}{M} + \frac{4\sqrt{8dl}}{\sqrt{M}}\right]. \end{aligned} \tag{3.10}$$

We note from the definition of ψ and (3.7) that

$$\alpha=\int_{R^d} f(x)\psi(x)\,dx=\int_{T_d^M} f(x)\psi(x)\,dx \geqq \int_{\substack{\sqrt{M}\leqq x_i\leqq M-\sqrt{M},\\ i=1,2,\cdots,d}} f(x)\psi(x)\,dx$$

$$=\int_{\substack{\sqrt{M}\leqq x_i\leqq M-\sqrt{M}\\ i=1,2,,\cdots,d}} f(x)\,dx=1-\int_{E_M^{\sqrt{M}}} f(x)\,dx \geqq 1-\frac{2d}{\sqrt{M}}.$$

Thus from (3.8), (3.10), and this last inequality, we have

$$\nu|\{y:g(y)>0\}|+I(g)$$

$$\leqq \nu|\{y:f(y)>0\}|+\frac{1}{8(1-2d/\sqrt{M})}\left[8I_M(f)+\frac{4d}{M}+\frac{4\sqrt{8dl}}{\sqrt{M}}\right]$$

$$=\nu|\{y:f(y)>0\}|+I_M(f)+\left[\frac{I_M(f)}{(1-2d/\sqrt{M})}-I_M(f)\right]$$

$$+\frac{1}{8(1-2d/\sqrt{M})}\left[\frac{4d}{M}+\frac{4\sqrt{8dl}}{\sqrt{M}}\right].$$

Using (3.6) and (3.5), we obtain finally

$$\nu|\{y:g(y)>0\}|+I(g)\leqq l+\frac{1}{8(1-2d/\sqrt{M})}\left[\frac{16dl}{\sqrt{M}}+\frac{4d}{M}+\frac{4\sqrt{8dl}}{\sqrt{M}}\right]$$

$$\leqq l+\tfrac{1}{2}\epsilon=J+\epsilon,$$

which completes the proof of this lemma.

We now prove the analogue of Lemma 3.5 for the symmetric stable process in R^d with $0<\alpha<2$. Let $I(f)$ be the I-function for the symmetric stable process in R^d and let $I_M(f)$ be the corresponding I-function for the symmetric stable process in T_d^M. We want to show (3.4) with these I-functions instead of the Brownian motion I-functions which appear there. It is just as easy to prove

$$\text{(3.11)}\qquad \varlimsup_{M\to\infty}\inf_{f\in\mathcal{F}_M}\left[\nu|\{y:f(y)>0\}|+I_M(f)\right] \geqq \inf_{g\in\mathcal{G}}\left[\nu|\{y:g(y)>0\}|+I(g)\right],$$

and, as a matter of fact, (3.11) is implied by

LEMMA 3.6. *For each $l<\infty$ there exists $\epsilon_M \to 0$ as $M\to\infty$ with the property that if $f\in \mathcal{F}_M$ with $I_M(f) \leqq l$, then there is a $g\in \mathcal{G}$ such that*

$$|\{y:g(y)>0\}| \leqq |\{y:f(y)>0\}| \quad \textit{and} \quad I(g) \leqq I_M(f)+\epsilon_M.$$

Proof: Let $M>0$ and let $\hat{\phi}_M(x)$ be as defined in Lemma 3.5. With f as in the statement of the present lemma and satisfying (3.7), define

$$g(x)=\frac{f(x)[\hat{\phi}_M(x)]^2}{\int_{R^d} f(x)[\hat{\phi}_M(x)]^2 dx},$$

so that $g\in \mathcal{G}$, and $|\{y: g(y)>0\}| \leqq |\{y: f(y)>0\}|$. Now,

$$I(g)=\frac{1}{4}\int_{R^d}\int_{R^d}\left|\sqrt{g(x+y)}-\sqrt{g(x)}\right|^2 \Omega(dy)\,dx$$

and

$$I_M(f)=\frac{1}{4}\int_{T_d^M}\int_{T_d^M}\left|\sqrt{f(x\mathring{+}y)}-\sqrt{f(x)}\right|^2 \Omega_M(dy)\,dx,$$

where in this last expression $x\mathring{+}y$ means addition modulo M and

$$\Omega_M(A)=\sum_{\xi\in G_d^M}\Omega(A+\xi),$$

where, as before, G_d^M is the set of vectors in R^d each of the coordinates of which is an integral multiple of M.

To estimate $I(g)$ we decompose $R^d\times R^d$ into the four sets:

$$S_1=\{(x,y):x+y\in T_d^M,\ x\in T_d^M\},$$

$$S_2=\{(x,y):x+y\in T_d^M,\ x\not\in T_d^M\},$$

$$S_3=\{(x,y):x+y\not\in T_d^M,\ x\in T_d^M\},$$

$$S_4=\{(x,y):x+y\not\in T_d^M,\ x\not\in T_d^M\}.$$

Since the support of g is contained in T_d^M,

$$
\begin{aligned}
I(g) &= \frac{1}{4}\int_{S_1}\int \left|\sqrt{g(x+y)} - \sqrt{g(x)}\right|^2 \Omega(dy)\,dx \\
&\quad + \frac{1}{4}\int_{S_2}\int g(x+y)\Omega(dy)\,dx + \frac{1}{4}\int_{S_3}\int g(x)\Omega(dy)\,dx.
\end{aligned}
\tag{3.12}
$$

Letting $\alpha_M = \int_{R^d} f(x)[\hat{\phi}_M(x)]^2\,dx$, we estimate the first term on the right of (3.12),

$$
\begin{aligned}
&\frac{1}{4}\int_{S_1}\int \left|\sqrt{g(x+y)} - \sqrt{g(x)}\right|^2 \Omega(dy)\,dx \\
&\leqq \frac{1}{4\alpha_M}\int_{S_1}\int \left|\sqrt{f(x+y)}\,\hat{\phi}_M(x+y) - \sqrt{f(x)}\,\hat{\phi}_M(x)\right|^2 \Omega(dy)\,dx \\
&\leqq \frac{1}{4\alpha_M}\left[\sqrt{\int_{S_1}\int [\hat{\phi}_M(x+y)]^2\left|\sqrt{f(x+y)} - \sqrt{f(x)}\right|^2 \Omega(dy)\,dx} \right. \\
&\quad \left. + \sqrt{\int_{S_1}\int f(x)|\hat{\phi}_M(x+y) - \hat{\phi}_M(x)|^2\,\Omega(dy)\,dx}\,\right]^2.
\end{aligned}
\tag{3.13}
$$

We have

$$
\begin{aligned}
&\int_{S_1}\int |\hat{\phi}_M(x+y)|^2\left|\sqrt{f(x+y)} - \sqrt{f(x)}\right|^2 \Omega(dy)dx \\
&\qquad \leqq \int_{x+y\in T_d^M}\int_{x\in T_d^M}\left|\sqrt{f(x+y)} - \sqrt{f(x)}\right|^2 \Omega(dy)\,dx,
\end{aligned}
$$

and, if we extend f periodically off of T_d^M, the last integral is at most

$$
\begin{aligned}
&\int_{x\in T_d^M}\int_{x+y\in R^d}\left|\sqrt{f(x+y)} - \sqrt{f(x)}\right|^2 \Omega(dy)\,dx \\
&\quad = \int_{x\in T_d^M}\int_{y\in T_d^M}\left|\sqrt{f(x+y)} - \sqrt{f(x)}\right|^2 \Omega(dy)dx \\
&\quad = \int_{x\in T_d^M}\int_{y\in T_d^M}\left|\sqrt{f(x\dotplus y)} - \sqrt{f(x)}\right|^2 \Omega_M(dy)\,dx = 4I_M(f).
\end{aligned}
$$

Also

$$\int_{S_1}\int f(x)|\hat{\phi}_M(x+y)-\hat{\phi}_M(x)|^2\,\Omega(dy)\,dx$$

$$\leqq \int_{x\in T_d^M} f(x)\int_{y\in R^d}|\hat{\phi}_M(x+y)-\hat{\phi}_M(x)|^2\,\Omega(dy)\,dx$$

$$\leqq \rho\int_{T_d^M} f(x)\,dx=\rho,$$

where

$$\rho=\sup_{x\in R^d}\int_{R^d}|\hat{\phi}_M(x+y)-\hat{\phi}_M(x)|^2\,\Omega(dy).$$

Thus, from (3.13)

$$\frac{1}{4}\int_{S_1}\int\left|\sqrt{g(x+y)}-\sqrt{g(x)}\,\right|^2\Omega(dy)dx\leqq\frac{1}{4\alpha_M}\left[2\sqrt{I_M(f)}+\sqrt{\rho}\,\right]^2. \tag{3.14}$$

Taking the third integral on the right of (3.12),

$$\frac{1}{4}\int_{S_3}\int g(x)\,\Omega(dy)\,dx$$

$$\begin{aligned}
&=\frac{1}{4\alpha_M}\iint_{\substack{x+y\notin T_d^M\\ x\in T_d^M}} f(x)[\hat{\phi}_M(x)]^2\,\Omega(dy)\,dx\\
&=\frac{1}{4\alpha_M}\iint_{\substack{x+y\notin T_d^M\\ x\in T_d^M}} f(x)|\hat{\phi}_M(x+y)-\hat{\phi}_M(x)|^2\,\Omega(dy)\,dx\\
&\leqq\frac{1}{4\alpha_M}\int_{x\in T_d^M}\int_{R^d} f(x)|\hat{\phi}_M(x+y)-\hat{\phi}_M(x)|^2\,\Omega(dy)dx\leqq\frac{\rho}{4\alpha_M}.
\end{aligned} \tag{3.15}$$

Finally the second integral on the right of (3.12) satisfies,

$$\frac{1}{4}\int_{S_2}\int g(x+y)\,\Omega(dy)\,dx=\frac{1}{4\alpha_M}\iint_{\substack{x+y\in T_d^M\\ x\notin T_d^M}} f(x+y)[\hat{\phi}_M(x+y)]^2\,\Omega(dy)\,dx$$

$$=\frac{1}{4\alpha_M}\iint_{\substack{x+y\in T_d^M\\ x\notin T_d^M}} f(x+y)|\hat{\phi}_M(x+y)-\hat{\phi}_M(x)|^2\,\Omega(dy)\,dx.$$

In the last integral we keep y fixed and let $x' = x+y$, $y' = y$. Thus,

$$\frac{1}{4}\int_{S_2}\int g(x+y)\,\Omega(dy)\,dx = \frac{1}{4\alpha_M}\iint_{\substack{x'-y'\notin T_d^M\\ x'\in T_d^M}} f(x')|\hat{\phi}_M(x')-\hat{\phi}_M(x'-y')|^2\,\Omega(dy')\,dx.$$

Change y' to $-y'$ and recall that $\Omega(A)=\Omega(-A)$, so that

$$\frac{1}{4}\int_{S_2}\int g(x+y)\Omega(dy)dx = \frac{1}{4\alpha_M}\iint_{\substack{x+y\notin T_d^M\\ x\in T_d^M}} f(x)|\hat{\phi}_M(x+y)-\hat{\phi}_M(x)|^2\,\Omega(dy)dx,$$

which is an integral we estimated in handling the third term above. Hence,

$$\frac{1}{4}\int_{S_2}\int g(x+y)\Omega(dy)dx \leqq \frac{\rho}{4\alpha_M}, \tag{3.16}$$

and finally we have from (3.12), (3.13), (3.14) and (3.15)

$$I(g) \leqq \frac{1}{4\alpha_M}\left[2\sqrt{I_M(f)}+\sqrt{\rho}\right]^2 + \frac{2\rho}{4\alpha_M}. \tag{3.17}$$

We have

$$\begin{aligned}\rho=\rho_M &= \sup_{x\in R^d}\int_{R^d}|\hat{\phi}_M(x+y)-\hat{\phi}_M(x)|^2\,\Omega(dy)\\ &\leqq \sup_{x\in R^d}\left[\int_{|y|\leqq N}|\hat{\phi}_M(x+y)-\hat{\phi}_M(x)|^2\,\Omega(dy)+2\Omega\{|y|>N\}\right].\end{aligned}$$

We recall that $|\nabla\hat{\phi}_M| \leqq \sqrt{d}/\sqrt{M}$, and hence we have

$$\rho_M \leqq \frac{d}{M}\int_{|y|\leqq N}|y|^2\Omega(dy)+2\Omega\{|y|>N\}.$$

But, as noted earlier, the measure Ω is such that $\int_{|y|\leqq N}|y|^2\,\Omega(dy)<\infty$ for all $N>0$ and $\Omega\{|y|>N\}\to 0$ as $N\to\infty$. Thus, first choosing N large and then letting

$M \to \infty$, we see that $\rho_M \to 0$. Hence, from (3.17)

$$I(g) \leqq I_M(f) + \left(\frac{I_M(f)}{\alpha_M} - I_M(f) \right) + \frac{\sqrt{I_M(f)}\,\sqrt{\rho}}{\alpha_M} + \frac{3\rho}{4\alpha_M} \tag{3.18}$$

implies there is a sequence $\epsilon_M \to 0$ as $M \to \infty$ such that $I(g) \leqq I_M(f) + \epsilon_M$. In this last observation we make use of the fact that $I_M(f) \leqq l$, $\rho_M \to 0$, and, as in the preceding lemma, $\alpha_M \geqq 1 - 2d/\sqrt{M}$ because without loss of generality we picked $f \in \mathcal{F}_M$ so that (3.7) would be satisfied. This completes the lemma.

As noted in the introduction, Theorem 3.1 follows from Lemmas 3.3, 3.5, 3.6 and the corollary to Theorem 5.1 in [5].

We now wish to prove Theorem 3.2. First of all, it is clear that, if $\mathcal{B}$ is the class of Borel sets in R^d,

$$\begin{aligned} &\inf_{g \in \mathcal{G}} [\nu|\{y : g(y) > 0\}| + I(g)] \\ &\qquad = \inf_{A \in \mathcal{B}} \left[\nu|A| + \inf_{\mu : \mu(A) = 1} I(\mu) \right]. \end{aligned} \tag{3.19}$$

In Lemma 3.7 we show for $\mathcal{C}$, the class of closed sets in R^d,

$$\inf_{E \in \mathcal{C}} \left[\nu|E| + \inf_{\mu : \mu(E) = 1} I(\mu) \right] \leqq \inf_{A \in \mathcal{B}} \left[\nu|A| + \inf_{\mu : \mu(A) = 1} I(\mu) \right], \tag{3.20}$$

and then in Lemma 3.8 that

$$\inf_{E \in \mathcal{S}} \left[\nu|E| + \inf_{\mu : \operatorname{supp}\mu \subset E} I(\mu) \right] \leqq \inf_{E \in \mathcal{C}} \left[\nu|E| + \inf_{\mu : \mu(E) = 1} I(\mu) \right]. \tag{3.21}$$

Thus (3.19), (3.20), and (3.21) will give Theorem 3.2. To verify (3.20) it suffices to prove

LEMMA 3.7. *Let A be a Borel set, $|A| = a > 0$, and let f be a probability density in R^d which equals 0 off of A and is such that $I(f) = l < \infty$. Let $\epsilon > 0$ be given. There exists a closed set C_ϵ and a density f_ϵ supported on C_ϵ such that $|C_\epsilon| \leqq a + \epsilon$, and $I(f_\epsilon) \leqq l + \epsilon$.*

Proof: Let $\theta > 0$ and $\delta > 0$ be given and let $g(x) = \sqrt{f(x)}$. Let $g_\theta = g * \phi_\theta$ where ϕ_θ is some mollifier. Let $h_\theta^\delta = (g_\theta - \delta)^+$ and $E_\theta^\delta(x) = g(x) - h_\theta^\delta(x)$. We shall show first that

$$\lim_{\substack{\delta\to 0\\ \theta\to 0}} \int_{R^d} |E_\theta^\delta(x)|^2\, dx = 0, \tag{3.22}$$

$$\lim_{\substack{\delta\to 0\\ \theta\to 0}} \int_{R^d}\int_{R^d} |E_\theta^\delta(x+y) - E_\theta^\delta(x)|^2\, \Omega(dy)\, dx = 0. \tag{3.23}$$

If we write $E_\theta^\delta(x) = g(x) - g_\theta(x) + g_\theta(x) - h_\theta^\delta(x)$, we note that as $\theta \to 0$

$$\begin{aligned} &\int_{R^d} |g_\theta(x) - g(x)|^2\, dx \to 0, \\ &\int_{R^d} |(g_\theta - g)(x+y) - (g_\theta - g)(x)|^2\, \Omega(dy)\, dx \to 0; \end{aligned} \tag{3.24}$$

thus to verify (3.22) and (3.23) we must estimate $g_\theta(x) - h_\theta^\delta(x)$. For $\delta > 0$ and $x \in R^1$, let $\psi_\delta(x) = x \wedge \delta$ so that

$$\begin{aligned} g_\theta(x) - h_\theta^\delta(x) &= g_\theta(x) - (g_\theta(x) - \delta)^+ \\ &= g_\theta(x) \wedge \delta = \psi_\delta(g_\theta(x)). \end{aligned}$$

Let Φ be the class of functions ϕ on R^d such that

$$\int_{R^d} |\phi(x)|^2\, dx < \infty \quad \text{and} \quad \int_{R^d}\int_{R^d} |\phi(x+y) - \phi(x)|^2\, \Omega(dy)\, dx < \infty. \tag{3.25}$$

We show now that

$$\lim_{\delta\to 0} \int_{R^d} |\psi_\delta(\phi(x))|^2\, dx = 0, \tag{3.26}$$

$$\lim_{\delta\to 0} \int_{R^d}\int_{R^d} |\psi_\delta(\phi(x+y)) - \psi_\delta(\phi(x))|^2\, \Omega(dy)\, dx = 0, \tag{3.27}$$

both limits being uniform for ϕ in a compact set $K \subset \Phi$, where compactness is with respect to both norms in (3.25).

Take (3.26) first; if K is a compact set of ϕ's, then there exists an R independent of ϕ such that $\int_{|x|>R} \phi^2(x)\, dx < \epsilon$. Thus

$$\begin{aligned} \int_{R^d} [\phi(x) \wedge \delta]^2\, dx &= \int_{|x| \leqq R} [\phi(x) \wedge \delta]^2\, dx + \int_{|x|>R} [\phi(x) \wedge \delta]^2\, dx \\ &< M_R \cdot \delta + \epsilon, \end{aligned}$$

and hence we have (3.26).

Note that $|\psi_\delta(x)-\psi_\delta(y)| \leqq |x-y|$ and, of course, $|\psi_\delta(x)| \leqq \delta$ so that $\psi_\delta(x)\to 0$ as $\delta\to 0$ for each x. Since K is a compact family of ϕ's and the measure Ω is σ-finite, we can find ρ and R independent of ϕ such that

$$\int_{\tilde{E}}\int |\phi(x+y)-\phi(x)|^2\,\Omega(dy)dx<\epsilon,$$

E being the set in $R^d\times R^d$ where $|y| \geqq \rho$ and $|x| \leqq R$. Thus

$$\int_{R^d}\int_{R^d}|\psi_\delta(\phi(x+y))-\psi_\delta(\phi(x))|^2\,\Omega(dy)\,dx$$

$$\leqq 4\delta^2\int_E\int \Omega(dy)dx+\int_{\tilde{E}}\int |\phi(x+y)-\phi(x)|^2\,\Omega(dy)\,dx$$

$$\leqq 4\delta^2\int_E\int \Omega(dy)\,dx+\epsilon,$$

which establishes (3.27).

Now if we let $\phi(x)=g_\theta(x)$, we see that (3.24), (3.26), and (3.27) imply the double limits (3.22) and (3.23).

Let $k_\theta^\delta=h_\theta^\delta/\|h_\theta^\delta\|_2$ and let $f_\theta^\delta=(k_\theta^\delta)^2$ which is then a probability density. Moreover, (3.22) and (3.23) imply that $\lim_{\substack{\delta\to 0\\ \theta\to 0}} I(f_\theta^\delta)=l$. So let $\epsilon>0$ be given and choose $\theta_0>0$, $\delta_0>0$ such that

$$I\left(f_\theta^{\delta_0}\right)\leqq l+\epsilon \text{ for all } \theta\leqq\theta_0.$$

The support of $f_\theta^{\delta_0}$ is the same as the support of $h_\theta^{\delta_0}$ which is $\{x: g_\theta(x)\geqq\delta_0\}$. But $g_\theta\in C^\infty$ and therefore $\{x: g_\theta(x)\geqq\delta_0\}$ is closed—call it $C_\theta^{\delta_0}$. To complete the proof of this lemma we want to show that, for sufficiently small θ,

$$|C_\theta^{\delta_0}|\leqq A+\epsilon. \tag{3.28}$$

But, since $g=\sqrt{f}$, we have from Tchebycheff's inequality,

$$\left|\left\{x: g_\theta(x)\geqq\delta_0\right\}\right|\leqq\left|\left\{x: g(x)\geqq\frac{1}{2}\delta_0\right\}\right|+\left|\left\{x: |g(x)-g_\theta(x)|\geqq\frac{1}{2}\delta_0\right\}\right|$$

$$\leqq|A|+\frac{4\|g-g_\theta\|^2}{\delta_0^2}.$$

Since $\| g-g_\theta\|^2 \to 0$ as $\theta\to 0$, for θ small enough, $|C_\theta^{\delta_0}| \leqq |A|+\epsilon$ which completes the proof of the lemma.

Now to see that (3.21) holds it suffices to prove

LEMMA 3.8. *Let C be a closed set in R^d and $g\in \mathcal{G}$ such that* supp $g\subset C$ *and* $I(g)=l<\infty$. *For any $\epsilon>0$ there exists a probability density f on R^d with $I(f)\leqq l+\epsilon$ and* supp $f\subset K$ *(compact)* $\subset C$.

Proof: Let $\phi(x)=1$ for $|x|\leqq 1$, $\phi(x)=0$ for $|x|>2$, $\phi(x)\in C^\infty$ on R^d and $0\leqq \phi(x)\leqq 1$ for all $x\in R^d$. For $R>0$, let $\phi_R(x)=\phi(x/R)$ and define

$$C_R=\frac{1}{\int_{R^d}\phi_R^2(x)\,g(x)\,dx}.$$

We note that $C_R\to 1$ as $R\to\infty$. Let $f_R(x)=C_R\phi_R^2(x)\,g(x)$ so that $f_R(x)$ is a probability density on R^d with supp $f_R\subset C\cap B_{2R}$, where B_{2R} is the closed ball in R^d of radius $2R$. Note that $C\cap B_{2R}$ is a compact set contained in C. Thus, if we can show that

$$\overline{\lim_{R\to\infty}}\, I(f_R)\leqq l, \tag{3.29}$$

the proof is completed by merely choosing $f=f_R$ for large enough R. To establish (3.29) we use the explicit form of the I-function and therefore we shall consider the Brownian motion case and the symmetric stable case $(0<\alpha<2)$ separately.

CASE 1. Symmetric stable process $(0<\alpha<2)$.

We have

$$\begin{aligned}
I(f_R)&=\frac{1}{4}\int_{R^d}\int_{R^d}\left|\sqrt{f_R(x+y)}-\sqrt{f_R(x)}\right|^2\Omega(dy)\,dx\\
&=\frac{1}{4}C_R\int_{R^d}\int_{R^d}\left|(\phi_R\sqrt{g}\,)(x+y)-(\phi_R\sqrt{g}\,)(x)\right|^2\Omega(dy)\,dx\\
&\leqq\frac{1}{4}C_R\Bigg[\sqrt{\int_{R^d}\int_{R^d}\phi_R^2(x+y)\left|\sqrt{g(x+y)}-\sqrt{g(x)}\right|^2\Omega(dy)\,dx}\\
&\qquad+\sqrt{\int_{R^d}\int_{R^d}g(x)\left|\phi_R(x+y)-\phi_R(x)\right|^2\Omega(dy)\,dx}\,\Bigg]^2\\
&\leqq\frac{1}{4}C_R\left[\sqrt{\int_{R^d}\int_{R^d}\left|\sqrt{g(x+y)}-\sqrt{g(x)}\right|^2\Omega(dy)\,dx}+\sqrt{\rho}\,\right]^2,
\end{aligned}$$

where

$$\rho = \sup_{x \in R^d} \int_{R^d} |\phi_R(x+y) - \phi_R(x)|^2 \, \Omega(dy).$$

Now, for any $M > 0$,

$$\rho \leqq \frac{\text{const.}}{R^2} \int_{|y| \leqq M} |y|^2 \, \Omega(dy) + 2\Omega\{y : |y| > M\}.$$

But, as pointed out in the introduction, the measure Ω is such that, for every $M > 0$, $\int_{|y| \leqq M} |y|^2 \Omega(dy) < \infty$ and also $\Omega\{y : |y| > M\} \to 0$ as $M \to \infty$. Therefore, if we first choose M large enough and then R large, we can make ρ as small as we please. This together with the fact that $C_R \to 1$ as $R \to \infty$ gives us

$$\begin{aligned}\overline{\lim_{R\to\infty}} I(f_R) &\leqq \frac{1}{4} \int_{R^d} \int_{R^d} \left| \sqrt{g(x+y)} - \sqrt{g(x)} \right|^2 \Omega(dy)\, dx \\ &= I(g) = l\end{aligned}$$

which is (3.29).

Case 2. Brownian motion.
Again we wish to establish (3.29), where now

$$\begin{aligned}I(f_R) &= \frac{1}{8} \int_{R^d} \frac{|\nabla f_R|^2}{f_R}\, dx \\ &= \frac{1}{8} C_R \int_{R^d} \frac{\left[\phi_R^2(x) \nabla g(x) + 2g(x)\phi_R(x)\nabla\phi_R(x)\right]^2}{\phi_R^2(x)\, g(x)}\, dx \\ &= \frac{1}{8} C_R \int_{R^d} \phi_R^2(x) \frac{|\nabla g|^2}{g}\, dx + \frac{1}{4} C_R \int_{R^d} \phi_R(x) \langle \nabla g, \nabla \phi_R \rangle\, dx \\ &\quad + \frac{1}{2} C_R \int_{R^d} g(x) |\nabla \phi_R(x)|^2\, dx.\end{aligned}$$

Since $|\nabla\phi_R| \leqq \text{const.}/R$, the third term goes to zero as $R \to \infty$. The second term is at most $\frac{1}{4}C_R$ (const./R)$\int_{R^d}|\nabla g|dx$, and since

$$\int_{R^d}|\nabla g|dx = \int_{R^d}(|\nabla g|/\sqrt{g}\,)\cdot\sqrt{g}\,dx$$

$$\leq \sqrt{\int_{R^d}(|\nabla g|^2/g)dx}\;\sqrt{\int_{R^d}g(x)dx}\;<\infty,$$

we see that the second term goes to zero as $R\to\infty$. Since $C_R\to 1$ as $R\to\infty$ and $\phi_R(x)\to 1$ as $R\to\infty$, the first term converges to $I(g)$. This completes Lemma 3.8.

Finally we end this section by carrying out the computation of

$$\inf_{g\in\mathcal{G}}\big[\nu|\{y:g(y)>0\}|+I(g)\big]$$

as explicitly as possible. Let L be the generator of a symmetric stable process characterized either by the positive definite covariance matrix (a_{ij}) when $\alpha=2$ or by a suitable measure Ω when $0<\alpha<2$. Let G be an open set in R^d and let $\lambda(G)$ be the smallest eigenvalue of $-L$ with zero boundary conditions on G. We denote

$$\lambda_L = \inf_{G:|G|=1}\lambda(G).$$

LEMMA 3.9. *We have*

$$\inf_{g\in\mathcal{G}}\big[\nu|\{y:g(y)>0\}|+I(g)\big]=k(\nu,L),$$

where

$$k(\nu,L)=\nu^{\alpha/(d+\alpha)}\left(\frac{d+\alpha}{\alpha}\right)\left(\frac{\alpha\lambda_L}{d}\right)^{d/(d+\alpha)}$$

Proof: As we saw earlier, if we let $\mathcal{B}$ be the Borel sets in R^d, then

$$\inf_{g\in\mathcal{G}}\big[\nu|\{y:g(y)>0\}|+I(g)\big]$$

$$=\inf_{B\in\mathcal{B}}\left[\nu|B|+\inf_{\operatorname{supp} g\subset B}I(g)\right].$$

It follows from Lemma 2.3 that the above infimum is equal to

$$\inf_{G \text{ open}} \left[\nu|G|+\lambda(G)\right] = \inf_{a>0} \inf_{\substack{G \text{ open} \\ |G|=a}} \left[\nu|G|+\lambda(G)\right]$$

$$= \inf_{a>0} \left[\nu a + \inf_{\substack{G \text{ open} \\ |G|=a}} \lambda(G)\right].$$

From the scaling properties we have

$$\inf_{\substack{G \text{ open} \\ |G|=a}} \lambda(G) = \frac{1}{a^{\alpha/d}} \inf_{\substack{G \text{ open} \\ |G|=1}} \lambda(G) = \frac{1}{a^{\alpha/d}} \lambda_L,$$

so that

$$\inf_{g \in \mathcal{G}} \left[\nu|\{y: g(y)>0\}| + I(g)\right]$$

$$= \inf_{a>0} \left[\nu a + \frac{\lambda_L}{a^{\alpha/d}}\right],$$

and it is elementary to show that this equals

$$\nu^{\alpha/(d+\alpha)} \left(\frac{d+\alpha}{\alpha}\right) \left(\frac{\alpha\lambda_L}{d}\right)^{d/(d+\alpha)} = k(\nu, L).$$

LEMMA 3.10. *For any generator L under consideration*

$$\lambda_L > 0.$$

Proof: As noted in the introduction, the transition probability density is given by

$$p(t,x) = \frac{1}{(2\pi)^d} \int_{R^d} e^{-i\langle \xi, x\rangle} e^{-tQ(\xi)}\, d\xi,$$

and, since $\sup_x p(t,x) \to 0$ as $t \to \infty$, we can choose $t_0 < \infty$ such that

$$\sup_x p(t_0, x) \leq \tfrac{1}{2}.$$

Let G be an open set with $|G|=1$. Let $q_G(t,x,y)$ be as defined in (2.5), so that

$$\int_G q_G(t_0,x,y)\,dy \leqq \int_G p(t_0,y-x)\,dy \leqq \frac{1}{2}.$$

Thus, if τ_G is the exit time from G,

$$P_x\left\{\tau_G \geqq nt_0\right\} \leqq \frac{1}{2^n} \quad \text{for every} \quad n.$$

In view of (2.9), we have

$$\lambda(G) \geqq \frac{\log 2}{t_0}.$$

Since this is true for every G with $|G|=1$,

$$\lambda_L \geqq \frac{\log 2}{t_0},$$

which completes the lemma.

We conclude Section 3 by proving that if L is spherically symmetric, i.e., if L is either $C\Delta$ when $\alpha=2$, or if Ω is spherically symmetric for $0<\alpha<2$, then the infimum in the definition of λ_L is attained when G is the sphere of unit volume.

Let $f(x)$ be a non-negative function on R^d such that the d-dimensional Lebesgue measure of $\{x: f(x)>\lambda\}$ is finite for every $\lambda>0$. By the spherical symmetrization of f, denoted by f^*, we mean the unique function f^* such that for each $\lambda>0$ the set $\{x: f^*(x)>\lambda\}$ is the sphere around the origin of the same measure as $\{x: f(x)>\lambda\}$.

LEMMA 3.11. *Let f, g, h be three non-negative functions on R^d and f^*, g^*, h^* their symmetrizations; then*

$$\int_{R^d}\int_{R^d} f(x)\,g(y)h(x-y)\,dx\,dy \leqq \int_{R^d}\int_{R^d} f^*(x)\,g^*(x)h^*(x-y)\,dx\,dy.$$

Proof: This lemma, a generalization of a result of F. Riesz in one dimension, is due to Sobolev and can be found in [6].

LEMMA 3.12. *Let f be any density such that, for some spherically symmetric stable process, $I(f)=l<\infty$. If f^* is the spherical symmetrization of f, then $I(f^*) \leqq l$.*

Proof: This is classical for $\alpha=2$ and therefore we shall prove it for $0<\alpha<2$.

In this case,

$$I(f)=\frac{1}{4}\int_{R^d}\int_{R^d}\left|\sqrt{f(x+y)}-\sqrt{f(x)}\right|^2 \Omega(dy)\,dx.$$

Since $(f^*)^{1/2}=(f^{1/2})^*$, it is enough to show that

$$\int_{R^d}\int_{R^d}|f(x+y)-f(x)|^2\,\Omega(dy)\,dx \geqq \int_{R^d}\int_{R^d}|f^*(x+y)-f^*(x)|^2\,\Omega(dy)\,dx.$$

By hypothesis, $\Omega(dy)$ is of the form

$$\Omega(dy)=\Phi(r)dy,$$

where $\Phi(r)$ is a nonincreasing function of r. We can then consider

$$\Phi_\lambda(r)=\min[\Phi(r),\lambda]$$

which is an integrable function on R^d having the same nonincreasing properties. We shall show that, for every $\lambda>0$,

$$\begin{aligned}&\int_{R^d}\int_{R^d}|f(x+y)-f(x)|^2\,\Phi_\lambda(|y|)\,dy\,dx\\ &\qquad \geqq \int_{R^d}\int_{R^d}|f^*(x+y)-f^*(x)|^2\,\Phi_\lambda(|y|)\,dy\,dx\end{aligned} \tag{3.30}$$

and then obtain the lemma by letting $\lambda\to\infty$. But, to prove (3.30) means showing that

$$\begin{aligned}&\int_{R^d}\int_{R^d}f(x+y)f(x)\Phi_\lambda(|y|)\,dx\,dy\\ &\qquad \leqq \int_{R^d}\int_{R^d}f^*(x+y)f^*(x)\Phi_\lambda(|y|)\,dx\,dy,\end{aligned}$$

or equivalently that

$$\begin{aligned}&\int_{R^d}\int_{R^d}f(x)f(y)\Phi_\lambda(|x-y|)\,dx\,dy\\ &\qquad \leqq \int_{R^d}\int_{R^d}f^*(x)f^*(y)\Phi_\lambda(|x-y|)\,dx\,dy,\end{aligned}$$

and this last inequality follows from Lemma 3.11 noting that $\Phi^*_\lambda=\Phi_\lambda$.

LEMMA 3.13. *If L is spherically symmetric, then*

$$\lambda_L = \inf_{G:|G|=1} \lambda(G) = \lambda(S),$$

where S is the sphere of unit volume.

Proof: Let G be any set with $|G|=1$ and $\lambda(G)=\lambda$. Then there is an f supported in G such that $I(f) \leqq \lambda+\epsilon$, where $\epsilon>0$ is given. Let f^* be the symmetrization of f. Then, from Lemma 3.12, $I(f^*) \leqq \lambda+\epsilon$ and supp f^* is contained in S, the sphere of unit volume. Therefore, $\lambda(S) \leqq \lambda+\epsilon$ and by letting $\epsilon\to 0$ we get $\lambda(S) \leqq \lambda$, which gives the lemma.

4. Proof of Theorem 2

We shall first prove the upper bound (1.21). If $V_1 \leqq V_2$ on R^d, then

$$\begin{aligned} & E_x\left\{\exp\left\{-\nu\int_{R^d}\left[1-\exp\left\{-\int_0^t V_1(\beta(s)-y)\,ds\right\}\right]dy\right\}\right\} \\ (4.1)\qquad & \geqq E_x\left\{\exp\left\{-\nu\int_{R^d}\left[1-\exp\left\{-\int_0^t V_2(\beta(s)-y)\,ds\right\}\right]dy\right\}\right\}. \end{aligned}$$

Now, if V is non-negative but not identically zero, there exists a V_1 such that $0 \leqq V_1 \leqq V$ and $\int_{R^d} V_1(x)\,dx = a \neq 0$. Thus in proving (1.21) we can, without loss of generality, assume because of (4.1) that V is of the form aV, where V is a probability density on R^d and $a>0$. We find it convenient to make the same sort of scale changes that we made on (1.1). Also, since the asymptotic behavior of (1.20) as $t\to\infty$ is independent of the starting point, we set $x=0$. Thus, for any $\lambda>0$,

$$\begin{aligned} & E_0\left\{\exp\left\{-\nu\int_{R^d}\left[1-\exp\left\{-a\int_0^t V(\beta(s)-y)\,ds\right\}\right]dy\right\}\right\} \\ & = E_0\left\{\exp\left\{-\nu\int_{R^d}\left[1-\exp\left\{\lambda a\int_0^{t/\lambda} V(\beta(\lambda\sigma)-y)\,d\sigma\right\}\right]dy\right\}\right\} \\ & = E_0\left\{\exp\left\{-\nu\lambda^{d/2}\int_{R^d}\left[1-\exp\left\{-\lambda a\int_0^{t/\lambda} V\big(\sqrt{\lambda}\,(\beta(\sigma)-x)\big)\,d\sigma\right\}\right]dx\right\}\right\}. \end{aligned}$$

In particular, letting $\lambda = t^{2/(d+2)}$ we get

$$E_0\left\{\exp\left\{-\nu\int_{R^d}\left[1-\exp\left\{-a\int_0^t V(\beta(s)-y)\,ds\right\}\right]dy\right\}\right\}$$

$$=E_0\left\{\exp\left\{-\nu t^{d/(d+2)}\int_{R^d}\left[1-\exp\left\{-a t^{2/(d+2)}Q(x)\right\}\right]dx\right\}\right\}.$$

where

$$Q(x)=\int_0^{t^{d/(d+2)}} V\left(t^{1/(d+2)}\,(\beta(\sigma)-x)\right)d\sigma .$$

Thus, to examine the asymptotic behavior of (1.20) as $t\to\infty$, it suffices to examine the asymptotic behavior of

$$E_0\left\{\exp\left\{-\nu t\int_{R^d}\left[1-\exp\left\{-a t^{2/d}\int_0^t V\left(t^{1/d}(\beta(s)-y)\right)ds\right\}\right]dy\right\}\right\}$$

as $t\to\infty$, i.e., replace $t^{d/(d+2)}$ by t. To prove (1.21) we shall, therefore, prove

$$\overline{\lim_{t\to\infty}}\ \frac{1}{t}\log E_0\left\{\exp\left\{-\nu t\int_{R^d}\left[1-\exp\left\{-a t^{2/d}\int_0^t V\left(t^{1/d}(\beta(s)-y)\right)ds\right\}\right]dy\right\}\right\}$$

(4.2)

$$\leqq -k(\nu,d).$$

With $L_t(\omega,A)$ defined as before, i.e., the proportion of time up to t that a particular Brownian motion path $\omega=\beta(\cdot)$ spends in the Borel set A, we have

$$\int_0^t V\left(t^{1/d}(\beta(s)-y)\right)ds = t\int_{R^d} V\left(t^{1/d}(\xi-y)\right)L_t(\omega,d\xi). \tag{4.3}$$

Let $V_t(y)=tV(t^{1/d}y)$ and

$$f_t(\omega,y)=\int_{R^d} V_t(\xi-y)L_t(\omega,d\xi),$$

so that the left side of (4.2) becomes

$$\overline{\lim_{t\to\infty}}\ \frac{1}{t}\log E_0\left\{\exp\left\{-\nu t\int_{R^d}\left[1-\exp\left\{-a t^{2/d}f_t(\omega,y)\right\}\right]dy\right\}\right\}. \tag{4.4}$$

As in the proof of Theorem 1, let T_d^M be the torus in R^d of size M and let $\tilde{f}_t(\omega,y)$ be the projection of $f_t(\omega,y)$ on T_d^M. We show now that, for any $M>0$,

$$
\begin{aligned}
&E_0\left\{\exp\left\{-\nu t\int_{R^d}\left[1-\exp\{-at^{2/d}f_t(\omega,y)\}\right]dy\right\}\right\}\\
(4.5)\qquad &\leqq E_0^M\left\{\exp\left\{-\nu t\int_{T_d^M}\left[1-\exp\{-at^{2/d}\tilde{f}_t(\omega,y)\}\right]dy\right\}\right\},
\end{aligned}
$$

where $E_0^M\{\ \}$ denotes integration with respect to Brownian motion on the torus T_d^M. To see (4.5), first note that if $\xi_i \geqq 0$, then it is elementary that

$$1-\exp\left\{-\sum_i \xi_i\right\} \leqq \sum_i (1-\exp\{-\xi_i\}).$$

Thus, if $h(y)$ is an L_1 function on R^d and if $\tilde{h}(y)$ is its projection on T_d^M, i.e.,

$$\tilde{h}(y)=\sum_{g\in G_d^M} h(y+g),$$

we have, for any constant c,

$$
\begin{aligned}
\int_{T_d^M}[1-\exp\{-c\tilde{h}(y)\}]dy &= \int_{T_d^M}\left[1-\exp\left\{-c\sum_{g\in G_d^M} h(y+g)\right\}dy\right.\\
(4.6)\qquad &\leqq \int_{T_d^M}\sum_{g\in G_d^M}[1-\exp\{-ch(y+g)\}]dy\\
&= \int_{R^d}[1-\exp\{-ch(y)\}]dy.
\end{aligned}
$$

If in (4.6) we let $c=at^{2/d}$ and $h(y)=f_t(\omega,y)$, we obtain (4.5). This means that

$$
\begin{aligned}
&\overline{\lim_{t\to\infty}}\frac{1}{t}\log E_0\left\{\exp\left\{-\nu t\int_{R^d}\left[1-\exp\{-at^{2/d}f_t(\omega,y)\}\right]dy\right\}\right\}\\
(4.7)\qquad &\leqq \overline{\lim_{t\to\infty}}\frac{1}{t}\log E_0^M\left\{\exp\left\{-\nu t\int_{T_d^M}\left[1-\exp\{-at^{2/d}\tilde{f}_t(\omega,y)\}\right]dy\right\}\right\}.
\end{aligned}
$$

Now, for any $f \in \mathcal{F}_M$, let

$$\Phi_t(f) = \int_{T_d^M} \left[1 - \exp\{-at^{2/d} f(y)\}\right] dy.$$

We note that this family of functionals, $\Phi_t(f)$, on $\mathcal{F}_M$ has the property that if $f_t \in \mathcal{F}_M$ and f_t converges to f in L_1, then

$$\varliminf_{t \to \infty} \Phi_t(f_t) \geqq |\{y : f(y) > 0\}|.$$

Therefore, from the corollary to Theorem 5.1 in [5] we have

$$\text{(4.8)} \quad \begin{aligned} &\varlimsup_{t \to \infty} \frac{1}{t} \log E_0^M \left\{ \exp\left\{ -\nu t \int_{T_d^M} \left[1 - \exp\{-at^{2/d} \tilde{f}_t(\omega, y)\}\right] dy \right\} \right\} \\ &\qquad \leqq - \inf_{f \in \mathcal{F}_M} \left[\nu |\{y : f(y) > 0\}| + I_M(f)\right], \end{aligned}$$

where $I_M(f) = \frac{1}{8} \int_{T_d^M} (|\nabla f|^2 / f) dx$. The estimate in (4.8) is now exactly the same as we obtained in the special case of the sausage and we can proceed in the same way to obtain from (4.7) and (4.8)

$$\varlimsup_{t \to \infty} \frac{1}{t} \log E_0 \left\{ \exp\left\{ -\nu t \int_{R^d} \left[1 - \exp\{-at^{2/d} f_t(\omega, y)\}\right] dy \right\} \right\} \leqq -k(\nu, d)$$

which is (1.21).

Now we want to obtain (1.22). From the previous discussion it suffices to show that

$$\text{(4.9)} \quad \begin{aligned} &\varliminf_{t \to \infty} \frac{1}{t} \log E_0 \left\{ \exp\left\{ -\nu t \int_{R^d} \left[1 - \exp\left\{ -t^{2/d} \int_0^t V\left(t^{1/d}(\beta(s) - y)\right) ds \right\}\right] dy \right\} \right\} \\ &\qquad \geqq -k(\nu, d), \end{aligned}$$

where we are assuming that V is non-negative on R^d and satisfies

$$V(y) = o\left(\frac{1}{|y|^{d+2}}\right) \quad \text{as} \quad |y| \to \infty.$$

Let G be a bounded open set in R^d containing the origin. Then

$$E_0\left\{\exp\left\{-\nu t\int_{R^d}\left[1-\exp\left\{-t^{2/d}\int_0^t V\left(t^{1/d}(\beta(s)-y)\right)ds\right\}\right]dy\right\}\right\}$$

$$\geqq E_0\left\{\exp\left\{-\nu t\int_{R^d}\left[1-\exp\left\{-t^{2/d}\int_0^t V\left(t^{1/d}(\beta(s)-y)\right)ds\right\}\right]dy\right\};\right. \tag{4.10}$$

$$\left.\beta(s)\in G \text{ for all } s, 0\leqq s\leqq t\right\}.$$

If $\beta(s)\in G$ for $0\leqq s\leqq t$, then

$$\int_{R^d}\left[1-\exp\left\{-t^{2/d}\int_0^t V\left(t^{1/d}(\beta(s)-y)\right)ds\right\}\right]dy$$

$$\leqq \int_{R^d}\left[1-\exp\left\{-t^{2/d}\sup_{x\in G} tV\left(t^{1/d}(x-y)\right)\right\}\right]dy$$

$$=\int_{R^d}\left[1-\exp\left\{-t^{(d+2)/d}\sup_{x\in G} V\left(t^{1/d}(x-y)\right)\right\}\right]dy.$$

Therefore, from (4.10) we get

$$E_0\left\{\exp\left\{-\nu t\int_{R^d}\left[1-\exp\left\{-t^{2/d}\int_0^t V\left(t^{1/d}(\beta(s)-y)\right)ds\right\}\right]dy\right\}\right\}$$

$$\geqq \exp\left\{-\nu t\int_{R^d}\left[1-\exp\left\{-t^{(d+2)/d}\sup_{x\in G} V\left(t^{1/d}(x-y)\right)\right\}\right]dy\right\} \tag{4.11}$$

$$\times P_0\left\{\beta(s)\in G, 0\leqq s\leqq t\right\},$$

where $P_0\{\ \}$ indicates Brownian motion measure for paths starting at the origin in R^d.

Our hypothesis on V implies that, for any $\epsilon>0$, there exists T such that $t\geqq T$ implies

$$V\left(t^{1/d}(x-y)\right)\leqq\frac{\epsilon}{|t^{1/d}(x-y)|^{d+2}},$$

and hence it is easy to see that

$$\int_{R^d}\left[1-\exp\left\{-t^{(d+2)/d}\sup_{x\in G} V\left(t^{1/d}(x-y)\right)\right\}\right]dy \leqq |G|+o(1).$$

Using this in (4.11) we deduce

$$\lim_{t\to\infty}\frac{1}{t}\log E_0\left\{\exp\left\{-\nu t\int_{R^d}\left[1-\exp\left\{-t^{2/d}\int_0^t V\left(t^{1/d}(\beta(s)-y)\right)ds\right\}\right]dy\right\}\right\}$$

$$\geqq -\nu|G|-\lambda_1(g), \tag{4.12}$$

where $\lambda_1(G)$ is the least eigenvalue for the Laplacian, $-\frac{1}{2}\Delta$, on G. Since (4.12) holds for all bounded open sets G ,we can replace the right side of (4.12) by $-\inf_G\ [\nu|G|+\lambda_1(G)]$ and by our discussion in Section 3 we know that this is $k(\nu,d)$. This completes the proof of Theorem 2.

Bibliography

[1] Kac, M., and Luttinger, J. M., *Bose-Einstein condensation in the presence of impurities*, J. Math. Phys., Vol. 15, No. 2, February 1974, pp. 183–186.

[2] Lifschitz, I. M., *Theory of fluctuation levels in disordered systems*, Soviet Physics JETP, Vol. 26, pp. 462–469, (English translation).

[3] Donsker, M. D., and Varadhan, S. R. S., *Asymptotic evaluation of certain Wiener integrals for large time*, to appear in *Integration in Function Space*, Oxford Univ. Press.

[4] Donsker, M. D., and Varadhan, S. R. S., *Asymptotic evaluation of certain Markov process expectations for large time, I*, Comm. Pure Appl. Math., Vol. 28, 1975, pp. 1–47.

[5] Donsker, M. D., and Varadhan, S. R. S., *Asymptotic evaluation of certain Markov process expectations for large time*, II, Comm. Pure Appl. Math., Vol. 28, 1975, pp. 279–301.

[6] Sobolev, S., *On a theorem of functional analysis*, Math. Sbornik, Vol. 4 (46), 1938, pp. 471–497.

Received February, 1975.

ERRATA

Communications on Pure and Applied Mathematics. Issue 4, 1975

The following misprints which distort the meaning have crept in after the last proofs.

Reflection of Singularities at the Boundary
by Andrew Majda and Stanley Osher

p. 487–Equation (1.15), second line should read:

$$\operatorname{supp} \rho_k \cap \Gamma_j = \varnothing$$

p. 493–Line 5 from above should read:

$(\xi, \tau), Q_0(x_0, \xi, \tau) = Q_0^+(x_0(x_0, \xi, \tau) Q_0^-(x_0, \xi, \tau)$, where $Q_0^+(x_0, \xi, \tau)(Q_0^-(x_0, \xi, \tau))$ is a...

Decay of Solutions to Nondissipative Hyperbolic Systems on Compact Manifolds
by Jeffrey Rauch and Michael Taylor

p. 514, line 5–**R** should be $\mathbb{R}$.

Asymptotics for the Wiener Sausage
by M. D. Donsker and S. R. S. Varadhan:

p. 526–Last line of displayed equation above equation (1.5) should read:

$$= E_0\left\{\exp\left\{-\nu\lambda^{d/2}|C_{t/\lambda}^{\epsilon/\sqrt{\lambda}}(\beta(\cdot))|\right\}\right\}.$$

p. 531–In inequality (1.22)

$\overline{\lim}_{t\to\infty}$ should be replaced by $\underline{\lim}_{t\to\infty}$

p. 538–In inequality (2.12)

$\overline{\lim}_{t\to\infty}$ should be replaced by $\underline{\lim}_{t\to\infty}$

p. 560–Middle line of last equation should read:

$$= E_0\left\{\exp\left\{-\nu\int_{R^d}\left[1-\exp\left\{-\lambda\alpha\int_0^{t/\lambda}V(\beta(\lambda\sigma)-y)d\sigma\right\}\right]dy\right\}\right\}$$

p. 562–The right side of the first line of relation (4.6) should read:

$$= \int_{T_d^M}\left[1-\exp\left\{-c\sum_{g\in G_d^M}h(y+g)\right\}\right]dy$$

p. 562–In relation (4.7)

$\underline{\lim}_{t\to\infty}$ should be replaced by $\overline{\lim}_{t\to\infty}$

COMMUNICATIONS ON PURE AND APPLIED MATHEMATICS, VOL. XXIX, 389–461 (1976)

Asymptotic Evaluation of Certain Markov Process Expectations for Large Time—III*

M. D. DONSKER AND S. R. S. VARADHAN

1. Introduction

Before stating hypotheses and theorems precisely we give a brief description of the kind of results we are considering.

Let $X_0, X_1, X_2, \cdots$ be a discrete parameter Markov process with transition probabilities $\pi(x, dy)$ on some state space X. Each realization ω of the process is a sequence $x_0, x_1, x_2, \cdots$ where $x_i \in X$. If we denote by Ω the space of all such sequences, then, for each starting point $x \in X$, $\pi(x, dy)$ induces a probability measure P_x on Ω. For each $\omega \in \Omega$, positive integer n, and set $A \subset X$, let

$$L_n(\omega, A) = \frac{1}{n} \sum_{j=0}^{n-1} \chi_A(x_j) \tag{1.1}$$

be the proportion of time the process spends in the set A during the first n steps. We note that $L_n(\omega, \cdot)$ maps Ω into $\mathcal{M}$, the space of probability measures on X. For each $x \in X$ and each positive integer n we define a probability measure $Q_{n,x}$ on $\mathcal{M}$ by $Q_{n,x} = P_x L_n^{-1}$, i.e., if B is a set of measures in $\mathcal{M}$, then

$$Q_{n,x}(B) = P_x\{\omega \in \Omega : L_n(\omega, \cdot) \in B\}. \tag{1.2}$$

The basic problem is to analyze the asymptotic behavior of $Q_{n,x}$ for large n.

In [1] it was assumed that X is a compact metric space and $\pi(x, dy)$ is a Feller transition probability. It was further assumed that $\pi(x, dy)$ has a density $\pi(x, y)$ relative to a reference probability measure $\beta(dy)$ and that $\pi(x, y)$ is uniformly bounded above and below. Under these assumptions it was proved that the asymptotic behavior of $Q_{n,x}$ for large n is determined by the functional $I(\mu)$ defined for $\mu \in \mathcal{M}$ by

$$I(\mu) = -\inf_{u>0} \int_X \log \frac{(\pi u)(x)}{u(x)} \mu(dx), \tag{1.3}$$

---

* This research was performed at the Courant Institute and was partially supported by the National Science Foundation, Grant NSF-MPS-74-01921-A01. Reproduction in whole or in part is permitted for any purpose of the United States Government.

where $(\pi u)(x) = \int_X u(y)\pi(x, dy)$ and the infimum is taken over all positive u in $C(X)$. More precisely we proved in [1] that for any closed set $C \subset \mathcal{M}$ (weak topology on $\mathcal{M}$) and any open set $G \subset \mathcal{M}$,

$$\overline{\lim_{n\to\infty}} \frac{1}{n} \log Q_{n,x}(C) \leqq -\inf_{\mu\in C} I(\mu), \tag{1.4}$$

$$\underline{\lim_{n\to\infty}} \frac{1}{n} \log Q_{n,x}(G) \geqq -\inf_{\mu\in G} I(\mu). \tag{1.5}$$

It follows from (1.4) and (1.5) that if $\Phi(\mu)$ is a real-valued weakly continuous functional on $\mathcal{M}$ (cf. Theorem 2 in [1]), then

$$\lim_{n\to\infty} \frac{1}{n} \log E_x\{\exp\{-n\Phi(L_n(\omega, \cdot))\}\} = -\inf_{\mu\in\mathcal{M}} [\Phi(\mu) + I(\mu)]. \tag{1.6}$$

Corresponding results were obtained in [1] in the case of continuous time Markov processes where analogous assumptions are made on the transition probability $p(t, x, dy)$ and the state space X is again a compact metric space. Now (1.1) is replaced by

$$L_t(\omega, A) = \frac{1}{t} \int_0^t \chi_A(x(s))\, ds,$$

and (1.3) is replaced by

$$I(\mu) = -\inf_{u\in\mathcal{D}^+} \int_X \left(\frac{Lu}{u}\right)(x)\mu(dx),$$

where L is the infinitesimal generator of the semigroup T_t associated with $p(t, x, dy)$ and $\mathcal{D}^+$ is the set of positive functions u in the domain of L. With $Q_{t,x} = P_x L_t^{-1}$, theorems corresponding to (1.4), (1.5), and (1.6), where now $t \to \infty$, were obtained in [1] (Theorems 3 and 4).

The fact that in both the discrete and continuous parameter versions (1.6) holds for such a large class of functionals Φ is due to the assumption that the state space X is compact. If the state space X is *not* compact, then the class of functionals Φ for which a result like (1.6) holds depends on the properties imposed on $\pi(x, dy)$ or $p(t, x, dy)$. For example, in the special case of Brownian motion, which we treated in [2], where the state space is R_1, we obtained a result like (1.6) only under fairly stringent hypotheses on the functional Φ. In Section 9 of this paper we shall exhibit some examples to show that such conditions are necessary.

The present paper is concerned with the general case where the state space X is a complete separable metric space. In this generality, to obtain asymptotic results like (1.6) one must choose to prove the result either for a large class of functionals Φ by imposing special hypotheses on the transition probabilities or for a restricted class of functionals Φ but assuming much less about the transition probabilities. To examine this point further we consider separately the two basic results (1.4) and (1.5).

Under the assumption of this paper that X is a complete separable metric space, to get lower bounds, in both the discrete and continuous parameter cases, like (1.5), holding for each starting point x, requires some strong transitivity assumptions. In the discrete case, the transitivity assumption we make is that, for any two points x_1 and x_2, the resolvents

$$R(x_1, dy) = \sum_{n=1}^{\infty} \frac{1}{2^n} \Pi^{(n)}(x_1, dy) \quad \text{and} \quad R(x_2, dy) = \sum_{n=1}^{\infty} \frac{1}{2^n} \Pi^{(n)}(x_2, dy)$$

are mutually absolutely continuous. Under this assumption, called Hypothesis H in Section 3, the lower bounds are obtained. In Section 8 the lower bounds are proved in the case of continuous parameter Markov processes where we have to assume the continuous version of Hypothesis H above. In addition some mild assumptions involving the strong continuity of the semigroup of transition operators as $t \to 0$ are made.

In the case of upper bounds like (1.4), in both the discrete and continuous parameter cases, the problem is to obtain the upper estimate for *closed* sets, that is, if one adopts the point of view which we do here—to have (1.6) holding for as wide a class of functionals as possible and pay the price of imposing stronger conditions on the transition probabilities. The point is that (1.4) holds for *compact* sets just under the hypothesis that the transition probability satisfies the Feller property. Thus we see that if X is a compact space, (1.4) holds for closed sets also, just under that single hypothesis. But if X is not compact, to have (1.4) hold for closed sets in $\mathcal{M}$ requires a strong positive recurrence condition on the transition probability. This is carried out in Section 4 under Hypothesis H* of that section. In Section 7 we obtain upper bounds for continuous time Markov processes under the continuous analogue of Hypothesis H* which again allows us to obtain the upper estimate for closed, not just compact, sets.

In the special case of independent identically distributed random variables, the Hypothesis H* is always fulfilled. However since our state space is so general, this leads to interesting applications to identically distributed random variables with values in a Banach space. This is carried out in Sections 5 and 6 and yields new derivations of some known estimates on the tail probabilities of Gaussian processes.

To summarize some of the results in this paper, let X be a complete separable metric space and $\pi(x, dy)$ a transition probability satisfying the Feller property. If $\pi(x, dy)$ has a density $\pi(x, y)$ with respect to a reference measure $\beta(dy)$, and if, for each $x \in X$ and each $E \subset X$ such that $\beta(E) > 0$, there exists a $k = k(x, E)$ for which the k-step transition $\pi^{(k)}(x, E) > 0$, then, for any open set $G \subset \mathcal{M}$,

$$\lim_{n\to\infty} \frac{1}{n} \log Q_{n,x}(G) \geqq -\inf_{\mu\in G} I(\mu). \tag{1.7}$$

Inequality (1.7) follows from Corollary 3.4 in Section 3.

To get the proper upper bound we assume in addition to the fact that $\pi(x, dy)$ is a Feller transition probability that there exists a function $u_0(x)$ on X such that $u_0(x) \geqq 1$ for all x, that $v(x) = (\pi u_0)(x) = \int_X u_0(dy)\pi(x, dy)$ is bounded on every compact set and that, for every l, $\{x \in X : u_0(x)/v(x) \leqq l\}$ is a compact set. Then for C, a closed set in $\mathcal{M}$,

$$\overline{\lim_{n\to\infty}} \frac{1}{n} \log Q_{n,x}(C) \leqq -\inf_{\mu\in C} I(\mu). \tag{1.8}$$

Inequality (1.8) is a consequence of Theorem 4.4 in Section 4.

Using (1.7) and (1.8) it follows from Section 3 of [9] that, for real-valued weakly continuous functionals $\Phi(\mu)$ on $\mathcal{M}$,

$$\lim_{n\to\infty} \frac{1}{n} \log E_x\{\exp\{-n\Phi(L_n(\omega, \cdot))\}\} = -\inf_{\mu\in\mathcal{M}} [\Phi(\mu) + I(\mu)]. \tag{1.9}$$

For the continuous parameter case assume $p(t, x, dy)$ is a Feller transition probability on a complete separable metric space X. Let L be the infinitesimal generator of the semigroup associated with $p(t, x, dy)$ having domain $\mathcal{D}$. We assume there exists a $v(x)$ on X such that $\{x \in X : v(x) \geqq l\}$ is a compact set for each $l > -\infty$ and that there exists a sequence $\{u_n\}$ in $\mathcal{D}$ satisfying:

(i) $u_n(x) \geqq 1$ for all n and all $x \in X$;

(ii) for each compact set $W \subset X$,

$$\sup_{x\in W} \sup_n u_n(x) < \infty;$$

(iii) for each $x \in X$,

$$\lim_{n\to\infty} \left(\frac{Lu_n}{u_n}\right)(x) = v(x);$$

(iv) for some $A<\infty$,

$$\sup_{n,x}\left(\frac{Lu_n}{u_n}\right)(x)\leqq A.$$

Then, for any closed set $C\subset\mathcal{M}$,

$$\overline{\lim_{t\to\infty}}\frac{1}{t}\log Q_{x,t}(C)\leqq-\inf_{\mu\in C}I(\mu). \tag{1.10}$$

Inequality (1.10) follows from Theorem 7.2 in Section 7.

To get the proper lower bound in the continuous parameter case, we assume $p(t, x, dy)$ is a Feller transition probability having a density $p(t, x, y)$ with respect to a reference measure $\beta(dy)$. Let B_0 be the center of the semigroup T_t associated with $p(t, x, dy)$, i.e., $B_0=\{f\in C(X):\lim_{t\to 0}\|T_tf-f\|=0\}$, where the norm is the usual norm for $C(X)$. Let

$$B_{00}=\left\{f\in C(X):\lim_{t\to 0}\sup_x\int_X|f(y)-f(x)|\,p(t,x,dy)=0\right\},$$

so that $B_{00}\subset B_0\subset C(X)$. Let $\alpha\in\mathcal{M}$ be such that $I(\alpha)=l<\infty$ and let N_α be a neighborhood of α in $\mathcal{M}$ of the form

$$N_\alpha=\left\{\mu\in\mathcal{M}:\left|\int_X f_j(x)[\mu(dx)-\alpha(dx)]\right|<\varepsilon,\quad 1\leqq j\leqq k,\ \text{where}\quad f_j\in B_{00},\quad 1\leqq j\leqq k\right\}. \tag{1.11}$$

We assume that $p(t, x, dy)$ is such that every neighborhood of $\alpha\in\mathcal{M}$ contains a neighborhood of the form (1.11). We also assume that, for any $\mu\in\mathcal{M}$ and any bounded measurable function $f(x)$, there exists a sequence $\{f_n\}\in B_0$ such that $\|f_n\|\leqq\|f\|$ and $f_n\to f$ almost everywhere with respect to μ. Finally we impose on $p(t, x, dy)$ the condition that, for any $x\in X$, $\sigma>0$, and $E\subset X$ such that $\beta(E)>0$, we have $\int_0^\infty e^{-\sigma t}p(t, x, E)\,dt>0$.

Under these hypotheses we prove that, for any open set $G\subset\mathcal{M}$,

$$\lim_{t\to\infty}\frac{1}{t}\log Q_{x,t}(G)\geqq-\inf_{\mu\in G}I(\mu). \tag{1.12}$$

Inequality (1.12) follows from Theorem 8.1 in Section 8. From (1.10) and

(1.12) we obtain the result that, for weakly continuous functionals $\Phi(\mu)$ on $\mathcal{M}$,

$$\lim_{t\to\infty}\frac{1}{t}\log E_x\{\exp\{-t\Phi(L_t(\omega,\cdot))\}\} = -\inf_{\mu\in\mathcal{M}}[\Phi(\mu)+I(\mu)]. \tag{1.13}$$

In some of the theorems in this paper we actually prove somewhat more than is necessary for the results quoted above. This is because we need such estimates in a forthcoming paper on the laws of the iterated logarithm for the local time.

2. Some Preliminary Results

Let X be a complete separable metric space and let $\pi(x, dy)$ be the transition probability function of a discrete parameter Markov process with X as state space. We denote by $\mathcal{M}$ the space of probability measures on X.

Let $\mathcal{U}$ be the space of Borel measurable functions u on X for each of which there exist constants a and b such that $0<a\leqq u(x)\leqq b<\infty$ for all $x\in X$. With

$$(\pi u)(x) = \int_X u(y)\pi(x, dy),$$

we define, for each $\mu\in\mathcal{M}$,

$$I(\mu) = -\inf_{u\in\mathcal{U}}\int_X \log\left(\frac{\pi u}{u}\right)(x)\mu(dx). \tag{2.1}$$

If we let $\mathcal{U}_1$ be the subset of $\mathcal{U}$ consisting of functions u which are also continuous on X, then it is easy to see that

$$-\inf_{u\in\mathcal{U}_1}\int_X \log\left(\frac{\pi u}{u}\right)(x)\mu(dx) = I(\mu). \tag{2.2}$$

Let $\alpha\in\mathcal{M}$, $\beta\in\mathcal{M}$ and define

$$I(\alpha,\beta) = -\inf_{u\in\mathcal{U}_1}\left[\int_X \log(\pi u)(x)\alpha(dx) - \int_X \log u(x)\beta(dx)\right] \tag{2.3}$$

so that, in particular, $I(\alpha,\alpha)=I(\alpha)$.

Let $\mathcal{M}_{X\times X}$ be the space of all probability measures on $X\times X$, and let $\mathcal{M}_{\alpha,\beta}$ be the set of all $\lambda\in\mathcal{M}_{X\times X}$ with first marginal α and second marginal β. We denote by $\mathcal{M}_{\alpha,\cdot}$ the union of $\mathcal{M}_{\alpha,\beta}$ for all β. For any $\lambda \in \mathcal{M}_{X\times X}$, we define

$$(2.4)\ \bar{I}(\lambda) = -\inf_{u\in\mathcal{U}_2}\left[\log \iint_{X\times X} u(x, y)\lambda_0(dx, dy) - \iint_{X\times X} \log u(x, y)\lambda(dx, dy)\right],$$

where $\lambda_0(dx, dy) = \gamma(dx)\pi(x, dy)$, γ being the first marginal of λ. In (2.4), $\mathcal{U}_2$ is the set of functions $u(x, y)$ on $X \times X$ which are continuous and for each of which there are constants a and b such that $0 < a \leqq u(x, y) \leqq b < \infty$ for all $(x, y) \in X \times X$. Finally, we define, for any $\alpha \in \mathcal{M}$, $\beta \in \mathcal{M}$,

$$(2.5)\qquad J(\alpha, \beta) = \inf_{\lambda\in\mathcal{M}_{\alpha,\beta}} \bar{I}(\lambda)\,.$$

We now prove

THEOREM 2.1. *For all* $\alpha \in \mathcal{M}$, $\beta \in \mathcal{M}$,

$$I(\alpha, \beta) = J(\alpha, \beta)\,.$$

Proof: Since *this* theorem involves X only as a Borel structure and since by our assumption on X (complete separable metric) it is a standard space, by a theorem of Kuratowski [4] one can assume without loss of generality that X is a compact metric space.

We show first that $I(\alpha, \beta) \leqq J(\alpha, \beta)$. Assume $J(\alpha, \beta) = J_0 < \infty$, and let $\varepsilon > 0$ be given. There exists $\lambda \in \mathcal{M}_{\alpha,\beta}$ such that $\bar{I}(\lambda) < J_0 + \varepsilon$, i.e., such that, for every $u \in \mathcal{U}_2$,

$$\iint_{X\times X} \log u(x, y)\lambda(dx, dy) - \log \iint_{X\times X} u(x, y)\lambda_0(dx, dy) \leqq J_0 + \varepsilon\,.$$

In particular, if we set $u(x, y) = g(y)/(\pi g)(x)$ where $g \in \mathcal{U}_1$, we have

$$\iint_{X\times X} [\log g(y) - \log (\pi g)(x)]\lambda(dx, dy)$$
$$\leqq \log \iint_{X\times X} \frac{g(y)}{(\pi g)(x)}\lambda_0(dx, dy) + J_0 + \varepsilon\,.$$

Since $\lambda \in \mathcal{M}_{\alpha,\beta}$ and hence $\lambda_0(dx, dy) = \alpha(dx)\pi(x, dy)$, we obtain

$$\int_X \log g(y)\beta(dy) - \int_X \log (\pi g)(x)\alpha(dx) \leqq \log \int_X \frac{(\pi g)(x)}{(\pi g)(x)} \alpha(dx) + J_0 + \varepsilon .$$

Thus,

$$-\left[\int_X \log (\pi g)(x)\alpha(dx) - \int_X \log g(y)\beta(dy)\right] \leqq J_0 + \varepsilon ,$$

and this inequality holds for all $g \in \mathcal{U}_1$ which implies

$$-\inf_{g \in \mathcal{U}_1} \left[\int_X \log (\pi g)(x)\alpha(dx) - \int_X \log g(x)\beta(dx)\right] \leqq J_0 + \varepsilon .$$

Consequently, $I(\alpha, \beta) \leqq J(\alpha, \beta)$.

To show that $J(\alpha, \beta) \leqq I(\alpha, \beta)$ for all $\alpha \in \mathcal{M}$, $\beta \in \mathcal{M}$, assume that, for some fixed $\alpha \in \mathcal{M}$ and $\beta \in \mathcal{M}$, $J(\alpha, \beta) > J_0$; then it suffices to show $I(\alpha, \beta) \geqq J_0$ for that α and β. But $J(\alpha, \beta) > J_0$ means $\mathcal{M}_{\alpha,\beta} \cap \{\lambda \in \mathcal{M}_{\alpha,\cdot} : \bar{I}(\lambda) \leqq J_0\} = \varnothing$. Thus, $\mathcal{M}_{\alpha,\beta}$ and $\{\lambda \in \mathcal{M}_{\alpha,\cdot} : \bar{I}(\lambda) \leqq J_0\}$ are disjoint, non-empty, convex, compact subsets of $\mathcal{M}_{X \times X}$. By the separation theorem there exists a $v(x, y) \in C(X \times X)$ such that

$$\iint_{X \times X} v(x, y)\lambda(dx, dy) \leqq 0 \quad \text{for all} \quad \lambda \in \mathcal{M}_{\alpha,\beta} , \tag{2.6}$$

and

$$\iint_{X \times X} v(x, y)\lambda(dx, dy) \geqq 1 \quad \text{for all} \quad \lambda \in \{\lambda \in \mathcal{M}_{\alpha,\cdot} : \bar{I}(\lambda) \leqq J_0\} . \tag{2.7}$$

From (2.6) and Lemma 2.2 below we can choose, for any $\varepsilon > 0$, $f_\varepsilon(x) \in C(X)$ and $g_\varepsilon(x) \in C(X)$ such that $v(x, y) \leqq f_\varepsilon(x) - g_\varepsilon(y) + \varepsilon$ for all $(x, y) \in X \times X$ which moreover satisfy

$$\int_X f_\varepsilon(x)\alpha(dx) = \int_X g_\varepsilon(y)\beta(dy) .$$

Therefore, from (2.7) we see that for any $\varepsilon > 0$ there exist $f_\varepsilon(x) \in C(X)$, $g_\varepsilon(x) \in C(X)$ such that

$$\iint_{X \times X} [f_\varepsilon(x) - g_\varepsilon(y) + \varepsilon]\lambda(dx, dy) \geqq 1 \quad \text{for all} \quad \lambda \in \{\lambda \in \mathcal{M}_{\alpha,\cdot} : \bar{I}(\lambda) \leqq J_0\} \tag{2.8}$$

and also

$$\int_X f_\varepsilon(x)\alpha(dx) = \int_X g_\varepsilon(y)\beta(dy).$$

Let $0<\varepsilon<1$. Relation (2.8) shows that if $\lambda \in \mathcal{M}_{\alpha,\cdot}$ is such that

$$\iint_{X\times X} [f_\varepsilon(x) - g_\varepsilon(y)]\lambda(dx, dy) = 0, \tag{2.9}$$

then certainly, for that λ, $\bar{I}(\lambda) > J_0$. Hence, for our given $0<\varepsilon<1$,

$$\inf_{\substack{\lambda\in\mathcal{M}_{\alpha,\cdot} \\ \iint_{X\times X}[f_\varepsilon(x)-g_\varepsilon(y)]\lambda(dx,dy)=0}} \bar{I}(\lambda) \geqq J_0. \tag{2.10}$$

Using Lemma 2.3 below (with $\Omega = X\times X$, $W(x, y) = f_\varepsilon(x) - g_\varepsilon(y)$, $P = \alpha(dx)\pi(x, dy)$, and $a = 0$) we get from (2.10)

$$-\inf_\theta \log M(\theta) \geqq J_0, \tag{2.11}$$

where

$$M(\theta) = \iint_{X\times X} \exp\{\theta[f_\varepsilon(x) - g_\varepsilon(x)]\}\alpha(dx)\pi(x, dy).$$

Thus, for any $\eta>0$, there exists θ_η such that $\log M(\theta_\eta) \leqq -J_0 + \eta$, i.e.,

$$\iint_{X\times X} \exp\{\theta_\eta[f_\varepsilon(x) - g_\varepsilon(y)]\}\alpha(dx)\pi(x, dy) \leqq e^{-J_0+\eta}. \tag{2.12}$$

In (2.12) let $V(y) = \exp\{-\theta_\eta g_\varepsilon(y)\}$, $W(x) = \exp\{-\theta_\eta f_\varepsilon(x)\}$, then we have

$$\iint_{X\times X} \frac{V(y)}{W(x)}\alpha(dx)\pi(x, dy) = \int_X \left(\frac{\pi V}{W}\right)(x)\alpha(dx) \leqq e^{-J_0+\eta}.$$

Hence, from Jensen's inequality,

$$\int_X \log\left(\frac{\pi V}{W}\right)(x)\alpha(dx) \leqq \log\int_X \left(\frac{\pi V}{W}\right)(x)\alpha(dx) \leqq -J_0 + \eta. \tag{2.13}$$

Since we were able to choose $f_\varepsilon(x)$ and $g_\varepsilon(y)$ such that

$$\int_X f_\varepsilon(x)\alpha(dx) = \int_X g_\varepsilon(y)\beta(dy),$$

we see that

$$\int_X \log W(x)\alpha(dx) = \int_X \log V(y)\beta(dy),$$

and thus from (2.13) we get

$$\int_X \log(\pi V)(x)\alpha(dx) - \int_X \log V(x)\beta(dx) \leqq -J_0 + \eta. \tag{2.14}$$

But (2.14) implies that, for any $\eta > 0$,

$$\inf_{u\in\mathcal{U}_1}\left[\int_X \log(\pi u)(x)\alpha(dx) - \int_X \log u(x)\beta(dx)\right] \leqq -J_0 + \eta,$$

so that

$$-\inf_{u\in\mathcal{U}_1}\left[\int_X \log(\pi u)(x)\alpha(dx) - \int_X \log u(x)\beta(dx)\right] \geqq J_0,$$

and hence $I(\alpha, \beta) \geqq J_0$ which completes Theorem 2.1.

LEMMA 2.2. *Suppose X is compact and $v(x, y) \in C(X \times X)$ is such that*

$$\iint_{X\times X} v(x, y)\lambda(dx, dy) \leqq 0 \quad \textit{for all} \quad \lambda \in \mathcal{M}_{\alpha,\beta}. \tag{2.15}$$

Then, for any $\varepsilon > 0$, there exist $f_\varepsilon(x) \in C(X)$, $g_\varepsilon(x) \in C(X)$ such that $v(x, y) \leqq f_\varepsilon(x) - g_\varepsilon(y) + \varepsilon$ for all $(x, y) \in X \times X$ and

$$\int_X f_\varepsilon(x)\alpha(dx) = \int_X g_\varepsilon(y)\beta(dy).$$

Proof: Before proving this lemma we note that if for any $\varepsilon > 0$ there exist functions $f_\varepsilon(x)$ and $g_\varepsilon(y)$ such that $v(x, y) \leqq f_\varepsilon(x) - g_\varepsilon(y) + \varepsilon$ for all $(x, y) \in X \times X$ and also $\int_X f_\varepsilon(x)\alpha(dx) = \int_X g_\varepsilon(y)\beta(dy)$, then surely, for every

$\lambda \in \mathcal{M}_{\alpha,\beta}$, $\iint_{X\times X} v(x, y)\lambda(dx, dy) = 0$. Therefore, this lemma characterizes functions $v(x, y)$ on $X \times X$ satisfying (2.15).

Let $\phi(x) \in C(X)$, $\psi(x) \in C(X)$ and consider, for $\lambda \in \mathcal{M}_{X\times X}$ and $v(x, y)$ satisfying (2.15),

$$H[\lambda, \phi, \psi] = \iint\limits_{X\times X} v(x, y)\lambda(dx, dy) - \iint\limits_{X\times X} [\phi(x) - \psi(y)]\lambda(dx, dy)$$

$$+ \int\limits_X \phi(x)\alpha(dx) - \int\limits_X \psi(y)\beta(dy).$$

If $\lambda \in \mathcal{M}_{\alpha,\beta}$ then, clearly, $H[\lambda, \phi, \psi] \leqq 0$ for all $\phi \in C(X)$ and $\psi \in C(X)$. Moreover, if $\lambda \in \mathcal{M}_{X\times X}$ but $\lambda \notin \mathcal{M}_{\alpha,\beta}$, then

$$\inf_{\substack{\phi \in C(X)\\ \psi \in C(X)}} H[\lambda, \phi, \psi] = -\infty.$$

Thus, we have

$$\sup_{\lambda \in \mathcal{M}_{X\times X}} \inf_{\substack{\phi \in C(X)\\ \psi \in C(X)}} H[\lambda, \phi, \psi] \leqq 0. \tag{2.16}$$

Since X is compact, $\mathcal{M}_{X\times X}$ is compact, and hence by one of the mini-max theorems (cf. [7]) we have

$$\sup_{\lambda \in \mathcal{M}_{X\times X}} \inf_{\substack{\phi \in C(X)\\ \psi \in C(X)}} H[\lambda, \phi, \psi] = \inf_{\substack{\phi \in C(X)\\ \psi \in C(X)}} \sup_{\lambda \in \mathcal{M}_{X\times X}} H[\lambda, \phi, \psi],$$

so that

$$\inf_{\substack{\phi \in C(X)\\ \psi \in C(X)}} \sup_{\lambda \in \mathcal{M}_{X\times X}} H[\lambda, \phi, \psi] \leqq 0. \tag{2.17}$$

But

$$\sup_{\lambda \in \mathcal{M}_{X\times X}} H[\lambda, \phi, \psi]$$

$$= \sup_{\lambda \in \mathcal{M}_{X\times X}} \left[\iint\limits_{X\times X} \{v(x, y) - \phi(x) + \psi(y)\}\lambda(dx, dy) + \int\limits_X \phi(x)\alpha(dx) - \int\limits_X \psi(y)\beta(dy) \right]$$

$$= \sup_{(x,y) \in X\times X} [v(x, y) - \phi(x) + \psi(y)] + \int\limits_X \phi(x)\alpha(dx) - \int\limits_X \psi(y)\beta(dy),$$

and (2.17) becomes

$$\text{(2.18)}\quad \inf_{\substack{\phi\in C(X)\\ \psi\in C(X)}}\left\{\sup_{(x,y)\in X\times X}[v(x,y)-\phi(x)+\psi(y)] + \int_X \phi(x)\alpha(dx)-\int_X \psi(y)\beta(dy)\right\}\leqq 0 .$$

Consequently, for any $\varepsilon>0$, there exist $\phi_\varepsilon\in C(X)$ and $\psi_\varepsilon\in C(X)$ such that

$$\text{(2.19)}\quad \sup_{(x,y)\in X\times X}[v(x,y)-\phi_\varepsilon(x)+\psi_\varepsilon(y)]+\int_X \phi_\varepsilon(x)\alpha(dx)-\int_X \psi_\varepsilon(y)\beta(dy)\leqq \varepsilon .$$

If we let

$$l=\int_X \phi_\varepsilon(x)\alpha(dx)-\int_X \psi_\varepsilon(y)\beta(dy),$$

$f_\varepsilon(x)=\phi_\varepsilon(x)-l$, and $g_\varepsilon(y)=\psi_\varepsilon(y)$, then (2.19) becomes

$$\text{(2.20)}\quad \sup_{(x,y)\in X\times X}[v(x,y)-f_\varepsilon(x)+g_\varepsilon(y)]\leqq \varepsilon$$

and, moreover,

$$\int_X f_\varepsilon(x)\alpha(dx)=\int_X g_\varepsilon(y)\beta(dy) .$$

This completes the proof of the lemma.

The next lemma is a special case of Theorem 5.2 (of Section 5) and we refer to that theorem for the proof.

LEMMA 2.3. *Let (Ω,Σ,P) be a probability space and let W be a bounded random variable. Let $\mathcal{M}$ be the space of probability measures γ on Ω and $\Gamma_a\subset\mathcal{M}$ the set of probability measures γ on Ω such that $\int_\Omega W(x)\gamma(dx)=a$. For any $\gamma\in\mathcal{M}$, let*

$$I_P(\gamma)=-\inf_{u\in\mathcal{U}_1}\left[\log\int_\Omega u(x)P(dx)-\int_\Omega \log u(x)\gamma(dx)\right],$$

and

$$h(a)=\inf_{\gamma\in\Gamma_a} I_P(\gamma) .$$

Then,

$$h(a)=\sup_\theta[\theta a-\log M(\theta)],$$

where

$$M(\theta)=\int_\Omega e^{\theta W(x)}P(dx),$$

i.e., $h(a)$ is the conjugate convex function of $\log M(\theta)$.

In proving Theorem 2.1 we made minimal assumptions on the Markov process transition probability $\pi(x, dy)$. We now assume that $\pi(x, dy)$ is a Feller transition probability and that there exists a reference measure β on X such that $\pi(x, dy)=\pi(x, y)\beta(dy)$. Under this condition, if $\mu\in\mathcal{M}$ is such that $I(\mu)<\infty$, then μ is absolutely continuous with respect to the reference measure β.

Let $\mu\in\mathcal{M}$ and for convenience denote $\mathcal{M}_{\mu,\mu}$ by $\mathcal{M}_\mu$. From Theorem 2.1 we know that

$$I(\mu)=I(\mu,\mu)=J(\mu,\mu)=\inf_{\lambda\in\mathcal{M}_\mu}\bar{I}(\lambda). \tag{2.21}$$

Under the hypotheses made on $\pi(x, dy)$ above, $\bar{I}(\lambda)$ is lower semicontinuous and, since $\mathcal{M}_\mu$ is compact, the infimum in (2.21) is actually attained. Denote the element of $\mathcal{M}_\mu$ where this infimum is attained by $\bar{\lambda}$, i.e., $\bar{I}(\bar{\lambda})=I(\mu)$. We now write $\bar{\lambda}(dx, dy)=\mu(dx)\bar{\pi}(x, dy)$, where $\bar{\pi}(x, dy)$ is defined for almost all x relative to μ. However we can extend $\bar{\pi}$ arbitrarily so that we have a transition probability defined for all x. We shall continue to call this extension $\bar{\pi}(x, dy)$, and consider it as a transition probability for a new discrete parameter Markov process on X. We note that, for $\bar{\pi}$, μ is an invariant measure since μ is not only a first marginal for $\bar{\lambda}$ but also a second marginal. Denote by $\bar{P}$ the probabilities for the stationary discrete parameter Markov process having $\bar{\pi}(x, dy)$ as its transition probability and the invariant measure μ as initial distribution. We denote by $\bar{P}_x$ the corresponding probabilities for this process starting at x. In Section 3 we shall need the fact that the stationary process $\bar{P}$ is ergodic. The next two lemmas lead up to Theorem 2.6 where the ergodicity of $\bar{P}$ is established under the assumption that $\pi(x, y)>0$ almost everywhere with respect to the product measure $\mu\times\mu$.

LEMMA 2.4. *Let $\mu\in\mathcal{M}$ be such that $I(\mu)<\infty$, and let $\{u_n\}$ be a sequence in $\mathcal{U}_1$ such that*

$$\lim_{n\to\infty}\int_X \log\left(\frac{\pi u_n}{u_n}\right)(x)\mu(dx)=-I(\mu).$$

Let $\lambda_n(dx, dy)=\mu(dx)(u_n(y)/(\pi u_n)(x))\pi(x, dy)$, and let $\bar{\lambda}$ be that element of $\mathcal{M}_\mu$ for which $\bar{I}(\bar{\lambda})=I(\mu)$. Then, $\lim_{n\to\infty}\|\lambda_n-\bar{\lambda}\|=0$, where the norm is the variation norm.

Proof: It follows from [1] (Lemma 2.1) that $\bar\lambda \ll \mu(dx)\pi(x, dy)$, and since u_n and πu_n are bounded above and below, we have $\bar\lambda \ll \lambda_n$. Let

$$I_{\lambda_n}(\bar\lambda) = -\inf_{u\in\mathcal{U}_2}\left[\log \iint_{X\times X} u(x, y)\lambda_n(dx, dy) - \iint_{X\times X} \log u(x, y)\bar\lambda(dx, dy)\right].$$

Again, from [1] (Lemma 2.1) and recalling that $\lambda_0(dx, dy) = \mu(dx)\pi(x, dy)$, we have

$$\begin{aligned}
I_{\lambda_n}(\bar\lambda) &= \iint_{X\times X} \log\frac{d\bar\lambda}{d\lambda_n}\,\bar\lambda(dx, dy)\\
&= \iint_{X\times X} \log\left[\left(\frac{d\bar\lambda}{d\lambda_0}\right)\left(\frac{d\lambda_0}{d\lambda_n}\right)\right]\bar\lambda(dx, dy)\\
&= \iint_{X\times X} \log\frac{d\bar\lambda}{d\lambda_0}\,\bar\lambda(dx, dy) + \iint_{X\times X} \log\frac{d\lambda_0}{d\lambda_n}\,\bar\lambda(dx, dy)\\
&= \bar I(\bar\lambda) + \iint_{X\times X} \log\frac{(\pi u_n)(x)}{u_n(y)}\,\bar\lambda(dx, dy)\\
&= I(\mu) + \iint_{X\times X} \log(\pi u_n)(x)\bar\lambda(dx, dy) - \iint_{X\times X} \log u_n(y)\bar\lambda(dx, dy)\\
&= I(\mu) + \int_X \log(\pi u_n)(x)\mu(dx) - \int_X \log u_n(y)\mu(dy)\\
&= I(\mu) + \int_X \log\left(\frac{\pi u_n}{u_n}\right)(x)\mu(dx).
\end{aligned}$$

By hypothesis the second term on the right in this last expression converges to $-I(\mu)$; hence we conclude that

$$\lim_{n\to\infty} I_{\lambda_n}(\bar\lambda) = 0. \tag{2.22}$$

In [1] (Lemma 4.1) we showed the existence of a universal function $\phi(l)$ such that $\phi(l) \geqq 0$ for $l \geqq 0$ and $\phi(l) \to 0$ as $l \to 0$, having the property that if

$\pi(x, dy)$ is any transition probability and μ is any probability measure on X with the corresponding $I(\mu)=l$, then, in the variation norm, $\|\mu\pi-\mu\|\leqq\phi(l)$. Here $\mu\pi$ means the measure $(\mu\pi)(A)=\int_X \pi(x, A)\mu(dx)$. Consider a very special case of this, namely, the case where $\pi(x, dy)=\alpha(dy)$. Note first that, in this case, $(\mu\pi)(A)=\alpha(A)$ and second that the I-function, $I(\mu)$, becomes

$$I_\alpha(\mu)=-\inf_{u\in\mathcal{U}_1}\left[\log\int_X u(x)\alpha(dx)-\int_X \log u(x)\mu(dx)\right].$$

Thus Lemma 4.1 of [1] says that, whenever $I_\alpha(\mu)=l$ for any two measures α and μ, we have $\|\alpha-\mu\|\leqq\phi(l)$. Hence (2.22) and the fact that $\phi(l)\to 0$ as $l\to 0$ implies that $\lim_{n\to\infty}\|\lambda_n-\bar{\lambda}\|=0$.

LEMMA 2.5. *Let (X, Σ, α) and (Y, S, β) be two probability measure spaces, and let (Z, Γ, λ) be the product space with product measure. Let $\{h_n(x, y)\}$ be a sequence of non-negative functions in Z of the form $h_n(x, y)=f_n(x)/g_n(y)$ where, for each n, $f_n(x)$ and $g_n(y)$ are bounded strictly positive functions on X and Y, respectively. If* $\lim_{n\to\infty} h_n(x, y)=h(x, y)$ *exists almost everywhere in Z and $0\leqq h(x, y)<\infty$ almost everywhere in Z, then $h(x, y)=f(x)/g(y)$ where $0\leqq f(x)<\infty$ almost everywhere in X and $0<g(y)\leqq\infty$ almost everywhere in Y.*

Proof: Let $E^*=\{(x, y): f_n(x)/g_n(y)\to h(x, y)\}$ so that $\lambda(E^*)=1$. Let $E_x=\{y : (x, y)\in E^*\}$ so that $\alpha\{x : \beta(E_x)=1\}=1$. Without loss of generality we can assume $h(x, y)\not\equiv 0$. Therefore, there exists an x such that $h(x, y)\not\equiv 0$ in y. In fact, $\alpha\{x : \beta\{y : h(x, y)=0\}<1\}>0$. Now pick $\bar{x}$ such that $\beta(E_{\bar{x}})=1$ and $\beta\{y : h(\bar{x}, y)=0\}<1$.

For any $(x, y)\in E^*$,

$$\lim_{n\to\infty} h_n(x, y)=\lim_{n\to\infty}\frac{f_n(x)f_n(\bar{x})}{f_n(\bar{x})g_n(y)}=h(x, y),$$

and

$$\lim_{n\to\infty}\frac{f_n(\bar{x})}{g_n(y)}=h(\bar{x}, y) \quad \text{for all} \quad y\in E_{\bar{x}}.$$

Let $E_1=\{x : \beta(E_x)=1\}$. For any $x\in E_1$ we have $\lim_{n\to\infty} h_n(x, y)=h(x, y)$ almost everywhere in y (β-measure). On the other hand, for $x\in E_1$ and any y, $h_n(x, y)=(f_n(x)/f_n(\bar{x}))(f_n(\bar{x})/g_n(y))$; thus letting $l_n(y)=f_n(\bar{x})/g_n(y)$ and $a_n(x)=f_n(x)/f_n(\bar{x})$ we get that $\lim_{n\to\infty} l_n(y)=h(\bar{x}, y)$ almost everywhere in y (β-measure) and $\lim_{n\to\infty} a_n(x)l_n(y)=h(x, y)$ almost everywhere in y (β-measure). Moreover, $h(\bar{x}, y)\not\equiv 0$ in y. Therefore, $\lim_{n\to\infty} a_n(x)$ exists—call it

$f(x)$, and $h(x, y)=f(x)h(\bar{x}, y)$. Letting $g(y)=1/h(\bar{x}, y)$ we have $h(x, y)=f(x)/g(y)$. Clearly $0 \leqq f(x)<\infty$ almost everywhere in X and $0<g(y) \leqq \infty$ almost everywhere in Y.

THEOREM 2.6. *Let μ be such that $I(\mu)<\infty$. Let $\bar{P}$ be the stationary process that was constructed earlier. If $\pi(x, y)>0$ a.e. with respect to the product measure $\mu \times \mu$, then $\bar{P}$ is ergodic.*

Proof: As in Lemma 2.4, let $\{u_n\}$ be a sequence in $\mathcal{U}_1$ such that

$$\lim_{n\to\infty} \int_X \log\left(\frac{\pi u_n}{u_n}\right)(x)\mu(dx) = -I(\mu)$$

and let

$$\lambda_n(dx, dy)=\mu(dx)\frac{u_n(y)}{(\pi u_n)(x)}\pi(x, dy)=\frac{\mu(dx)u_n(y)}{(\pi u_n)(x)}\pi(x, y)\beta(dy).$$

From that lemma, $\lim_{n\to\infty} \lambda_n=\bar{\lambda}$ in variation norm, where $\bar{\lambda}$ is the element of $\mathcal{M}_\mu$ for which $\bar{I}(\bar{\lambda})=I(\mu)$. Hence

$$\lim_{n\to\infty} \frac{u_n(y)}{(\pi u_n)(x)}\pi(x, y)=\theta(x, y)$$

exists in $L_1(\mu\times\beta)$ and $\bar{\lambda}(dx, dy)=\theta(x, y)\mu(dx)\beta(dy)$. We can choose a subsequence if necessary and conclude, from the assumption $\pi(x, y)>0$ a.e. $\mu\times\mu$, that

$$\lim_{n\to\infty} \frac{u_n(y)}{(\pi u_n)(x)}=\frac{\theta(x, y)}{\pi(x, y)}=h(x, y) \quad \text{a.e.} \quad \mu\times\mu.$$

Lemma 2.5 implies that $h(x, y)=a(x)/b(y)$ a.e. $\mu\times\mu$, where $0\leqq a(x)<\infty$ a.e. μ and $0<b(y)\leqq\infty$ a.e. μ. So $\theta(x, y)=(a(x)/b(y))\pi(x, y)$ a.e. $\mu\times\mu$. Since $\mu \ll \beta$, let B be the set in X where $d\mu/d\beta>0$. Then $\mu(\tilde{B})=0$ and since $\bar{\lambda}$ has both marginals equal to μ, $\theta(x, y)$ must be equal to zero a.e. $\mu\times\beta$ on the set $X\times\tilde{B}$. Therefore we can extend $b(y)$ defined on B to be equal to $+\infty$ on $\tilde{B}$ and, with this extension,

$$\theta(x, y)=\frac{a(x)}{b(y)}\pi(x, y) \quad \text{a.e.} \quad \mu\times\beta.$$

Let $E=\{x : a(x)=0\}$ and $F=\{y : b(y)=\infty\}$. Since

$$\int_{E\times X} \bar{\lambda}(dx, dy)=0=\int_{X\times F} \bar{\lambda}(dx, dy),$$

we have $\bar{\lambda}(\tilde{E}\times\tilde{F})=1$, where $\tilde{E}$ denotes the complement of E. We shall now show that if $A\subset X$ with the property that $\bar{\lambda}(A\times\tilde{A})=\bar{\lambda}(\tilde{A}\times A)=0$, then either $\mu(A)=0$ or $\mu(\tilde{A})=0$. To see this, we first note that since

$$\iint_{A\times\tilde{A}} \frac{a(x)}{b(y)}\pi(x, y)\mu(dx)\beta(dy)=0\,,$$

and $\pi(x, y)>0$ almost everywhere (β) for $x\in X$, we must have $a(x)/b(y)=0$ almost everywhere ($\mu\times\beta$) on $A\times\tilde{A}$. This implies

$$(2.23)\quad (\mu\times\beta)\{(A\times\tilde{A})\cap(\tilde{E}\times\tilde{F})\}=0 \quad\text{and}\quad (\mu\times\beta)\{(\tilde{A}\times A)\cap(\tilde{E}\times\tilde{F})\}=0\,.$$

The first equation in (2.23) is the same as $\mu(A\cap\tilde{E})\beta(\tilde{A}\cap\tilde{F})=0$, and the second is the same as $\mu(\tilde{A}\cap\tilde{E})\beta(A\cap\tilde{F})=0$. Hence, we have four possibilities; either

$$\mu(A\cap\tilde{E})=0 \qquad\text{and}\qquad \mu(\tilde{A}\cap\tilde{E})=0\,,$$

or

$$\mu(A\cap\tilde{E})=0 \qquad\text{and}\qquad \beta(A\cap\tilde{F})=0\,,$$

or

$$\beta(\tilde{A}\cap\tilde{F})=0 \qquad\text{and}\qquad \mu(\tilde{A}\cap\tilde{E})=0\,,$$

or

$$\beta(\tilde{A}\cap\tilde{F})=0 \qquad\text{and}\qquad \beta(A\cap\tilde{F})=0\,.$$

The first and fourth of these possibilities are impossible since then either $\beta(\tilde{F})=0$ or $\mu(\tilde{E})=0$, and, since $\bar{\lambda}\ll(\mu\times\beta)$, this contradicts $\bar{\lambda}(\tilde{E}\times\tilde{F})=1$. Thus either the second or third of these possibilities holds. Since $\bar{\lambda}\in\mathcal{M}_\mu$ and $\bar{\lambda}(\tilde{E}\times\tilde{F})=1$, we see that $\mu(\tilde{E})=1$ and $\mu(\tilde{F})=1$. Therefore, if the second possibility holds, we conclude that $\mu(A)=0$, and if the third possibility holds we conclude that $\mu(\tilde{A})=0$. That this implies the ergodicity of the $\bar{P}$ process is now easy to see in the following way.

Since the $\bar{P}$ process is a Markov process with stationary distribution μ, if it were not ergodic, there would be a subset $A\subset X$ such that $0<\mu(A)<1$ and the process never leaves A and never leaves $\tilde{A}$, i.e., $\bar{\lambda}(A\times\tilde{A})=0=\bar{\lambda}(\tilde{A}\times A)$. But we just proved that this last equation implies either $\mu(A)=0$ or $\mu(\tilde{A})=0$ which contradicts $0<\mu(A)<1$. Hence the $\bar{P}$ process is ergodic.

3. Lower Bounds—Discrete Case

Let X be a complete separable metric space and $\pi(x, dy)$ the transition probability of a discrete parameter Markov process with X as state space. We assume $\pi(x, dy)$ is a Feller transition probability and that there exists a reference measure β on X such that $\pi(x, dy) = \pi(x, y)\beta(dy)$. Denote by Ω the space of all sequences $x_0, x_1, x_2, \cdots$ with $x_j \in X$. For each starting point $x \in X$, the transition probability $\pi(x, dy)$ induces a probability measure on Ω which we denote by P_x. For each $\omega \in \Omega$, positive integer n, and A a Borel set in X, let

$$L_n(\omega, A) = \frac{1}{n} \sum_{j=0}^{n-1} \chi_A(x_j).$$

We note that L_n maps Ω into $\mathcal{M}$, the space of probability measures on X. Let $Q_{n,x}$ be the measure induced on $\mathcal{M}$ by L_n from P_x, i.e., if B is a set of measures in $\mathcal{M}$, we define, for each $x \in X$ and each positive integer n,

$$Q_{n,x}(B) = P_x\{\omega \in \Omega : L_n(\omega, \cdot) \in B\}.$$

In this section we obtain asymptotic lower bounds on $Q_{n,x}$ as $n \to \infty$. In doing so we use the ergodicity of the $\bar{P}$ process proved in the preceding section. We begin with

THEOREM 3.1. *Let $\mu \in \mathcal{M}$ be such that $I(\mu) < \infty$ and let $A \subset X$ be any subset of the state space with $\mu(A) = 1$. Let N be a weak neighborhood of μ in $\mathcal{M}$, and denote by $\mathcal{M}(A)$ the set of all probability measures ν on X with $\nu(A) = 1$. If*

$$\pi(x, y) > 0 \quad \textit{almost everywhere} \quad (\mu \times \mu), \tag{3.1}$$

and if $x \in A$ is such that

$$\pi(x, dy) \quad \textit{is not orthogonal to} \quad \mu(dy), \tag{3.2}$$

then

$$\lim_{n \to \infty} \frac{1}{n} \log Q_{n,x}(N \cap \mathcal{M}(A)) \geqq -I(\mu). \tag{3.3}$$

If, in addition, $\pi(x, E)$ is a continuous function of x for each Borel set E, and K is a compact set of points in A for each of which (3.2) holds, then (3.3) holds uniformly over K.

Proof: Let $\mu \in \mathcal{M}$ be given such that $I(\mu) < \infty$ and $A \subset X$ be such that $\mu(A) = 1$. We note that

$$Q_{n,x}(N \cap \mathcal{M}(A)) = P_x\{\omega \in \Omega : L_n(\omega, \cdot) \in N; x_0, x_1, \cdots, x_{n-1} \in A\}.$$

As in the preceding section, we denote by $\bar{\lambda}=\bar{\lambda}(dx, dy)$ that element of $\mathcal{M}_\mu$ for which $\bar{I}(\bar{\lambda})=I(\mu)$ and define $\bar{\pi}(x, dy)$ by $\bar{\lambda}(dx, dy)=\mu(dx)\bar{\pi}(x, dy)$. Although $\bar{\pi}(x, dy)$ is defined only for almost all x with respect to μ, we can extend it arbitrarily and treat $\bar{\pi}(x, dy)$ as a transition probability. As noted in the proof of Lemma 2.4, $\bar{\lambda} \ll \mu(dx)\pi(x, dy)$, and therefore $\bar{\pi}(x, dy)$ is absolutely continuous with respect to $\pi(x, dy)$ for almost all x with respect to μ. If we write $\bar{\pi}(x, dy)=h(x, y)\pi(x, dy)$, then we have from [1] (Lemma 2.1) that

(3.4)
$$I(\mu)=\iint_{X\times X} \log h(x, y)\bar{\lambda}(dx, dy).$$

Let N_1 be a weak neighborhood of μ contained in N and let $\delta>0$ be so chosen that if $\nu\in N_1$ and $\|\nu'-\nu\|<\delta$, then $\nu'\in N$ where $\|\cdot\|$ denotes the variation norm. Set $E_n=\{\omega : L_n(\omega, \cdot)\in N_1\}$ and $D_n=\{\omega : x_j\in A, 0\leqq j\leqq n-1\}$. For each $\varepsilon>0$, let

$$F_{n,\varepsilon}=\left\{\omega : \frac{1}{n-1}\sum_{j=1}^{n-1} \log h(x_{j-1}, x_j)\leqq I(\mu)+\varepsilon\right\}.$$

We denote by $\bar{P}_x$ probabilities for the Markov process with $\bar{\pi}(x, dy)$ as transition function and starting from the point x. We denote by $\bar{P}$ probabilities for the Markov process with transition function $\bar{\pi}(x, dy)$ and initial distribution μ. We note that the $\bar{P}$ process is a stationary process and, by (3.1) and Theorem 2.6, is an ergodic process.

Let $L'_n(\omega, A)=\frac{1}{n}\sum_{i=1}^{n} \chi_A(x_i)$. If n is so large that $2/n<\delta$, then

$$\|L_n(\omega, \cdot)-L'_n(\omega, \cdot)\|<\delta,$$

and therefore, for large n,

$$\{\omega : L_n(\omega, \cdot)\in N\}\supset\{\omega : L'_n(\omega, \cdot)\in N_1\}.$$

Thus

$$\begin{aligned} &P_x\{\omega : L_n(\omega, \cdot)\in N;\ x_i\in A, 0\leqq i\leqq n-1\} \\ &\quad\geqq P_x\{\omega : L'_n(\omega, \cdot)\in N_1;\ x_i\in A, 0\leqq i\leqq n\} \\ &\quad=\int_A P_y\{E_n\cap D_n\}\pi(x, dy). \end{aligned}$$

But

$$P_y\{E_n \cap D_n\} = \int_{E_n \cap D_n} \pi(y, dx_1) \cdots \pi(x_{n-2}, dx_{n-1})$$

$$\geqq \int_{E_n \cap D_n} \frac{\pi(y, dx_1)}{\bar{\pi}(y, dx_1)} \cdots \frac{\pi(x_{n-2}, dx_{n-1})}{\bar{\pi}(x_{n-2}, dx_{n-1})} \bar{\pi}(y, dx_1) \cdots \bar{\pi}(x_{n-2}, dx_{n-1})$$

$$= \int_{E_n \cap D_n} \exp\left\{-\sum_{j=1}^{n-1} \log h(x_{j-1}, x_j)\right\} d\bar{P}_y$$

$$\geqq \int_{E_n \cap D_n \cap F_{n,\varepsilon}} \exp\{-(n-1)[I(\mu)+\varepsilon]\}\, d\bar{P}_y$$

$$= \exp\{-(n-1)[I(\mu)+\varepsilon]\} \bar{P}_y\{E_n \cap D_n \cap F_{n,\varepsilon}\}\,.$$

Since μ is an invariant measure for $\bar{\pi}(x, dy)$ and $\mu(A)=1$, we have $\bar{P}_y(D_n)=1$ for almost all y with respect to μ. Therefore, if we denote by $\alpha(x, dy)$ the part of $\pi(x, dy)$ that is absolutely continuous with respect to μ, then

$$(3.5)\quad Q_{n,x}(N \cap \mathcal{M}(A)) \geqq \exp\{-(n-1)[I(\mu)+\varepsilon]\} \int_A \bar{P}_y\{E_n \cap F_{n,\varepsilon}\} \alpha(x, dy)\,.$$

By the ergodic theorem, $\bar{P}_y\{E_n \cap F_{n,\varepsilon}\} \to 1$ as $n \to \infty$ for almost all y with respect to μ. Since α is absolutely continuous with respect to μ and $\alpha(x, A) > 0$ by (3.2), we conclude that

$$\varliminf_{n\to\infty} \frac{1}{n} \log Q_{n,x}(N \cap \mathcal{M}(A)) \geqq -[I(\mu)+\varepsilon]\,.$$

Since $\varepsilon > 0$ is arbitrary, we have (3.3). To prove uniformity over K we see that, by the properties of Lebesgue decomposition, $\alpha(x, E)$ is also continuous as a function of x for each E. Moreover, for $x \in K$, $\alpha(x, dy)$ is uniformly absolutely continuous with respect to μ. It is now routine to see that

$$\lim_{n\to\infty} \int_A \bar{P}_y\{E_n \cap F_{n,\varepsilon}\} \alpha(x, dy) = \alpha(x, A)$$

uniformly over K, and therefore (3.3) holds uniformly over K.

It will be useful for us to have the analogue of Theorem 3.1 for substochastic transition probabilities. This becomes important if one wants to estimate in the continuous case the probabilities relative to a process that is killed outside of a set. It is also advantageous to replace the strong transitivity assumptions (3.1) and (3.2) by slightly weaker versions.

Let $p(x, dy)$ be a substochastic transition probability on a state space X which we assume to be a complete separable metric space. We can think of $1-p(x, X)$ as the probability of extinction for a chain starting from x in a single step. We can adjoin a point ∞ to our state space (isolated point) and call $X'=X\cup\{\infty\}$. We can then redefine a stochastic transition probability on X' by

$$\pi(x, A)=p(x, A) \quad \text{for} \quad x\in X \quad \text{and} \quad A\subset X,$$

$$\pi(x, \{\infty\})=1-p(x, X),$$

$$\pi(\infty, \{\infty\})=1.$$

Even though $p(x, dy)$ is substochastic, we can define an I-function for it in the usual way:

$$I(\mu)=-\inf_{u\in\mathcal{U}_1}\int_X \log\frac{\int u(y)p(x, dy)}{u(x)}\mu(dx),$$

for any probability measure μ on X. For the stochastic transition function $\pi(x, dy)$ on X' we have the usual I-function defined for probability measures on X', i.e.,

$$I'(\mu)=-\inf_{u'\in\mathcal{U}_1'}\int\log\left(\frac{\pi u'}{u'}\right)(x)\mu(dx).$$

We can state

LEMMA 3.2. *For any measure μ on X' which is concentrated on X,*

$$I'(\mu)=I(\mu).$$

Proof: A function u' on X' can be thought of as a function u on X and a constant k which is the value of u' at ∞. Clearly,

$$I'(\mu)=-\inf_{u\in\mathcal{U}_1, k>0}\int_{X'}\log\left(\frac{\pi u'}{u'}\right)(x)\mu(dx).$$

Since $\mu(\{\infty\})=0$, the integration above is carried out over X only and for $x\in X$

$$(\pi u')(x)=\int_X u(y)p(x,dy)+k(1-p(x,X)).$$

Therefore,

$$I'(\mu)=-\inf_{u\in\mathcal{U}_1,k>0}\int_X \log\left(\frac{\pi u'}{u}\right)(x)\mu(dx),$$

but clearly $k=0$ produces the infimum for any $u\in\mathcal{U}_1$ so that

$$I'(\mu)=-\inf_{u\in\mathcal{U}_1}\int_X \log\frac{\int p(x,dy)u(y)}{u(x)}\mu(dx)=I(\mu).$$

We make the following assumptions on $p(x, dy)$.

Hypothesis H. There is a reference measure $\beta(dy)$ such that $p(x,dy)=p(x,y)\beta(dy)$ for each $x\in X$. Moreover, β can be so chosen that, for each $x\in X$ and $E\subset X$ with $\beta(E)>0$, there is an integer $k=k(x,E)$ such that $p^{(k)}(x,E)>0$. Here, $p^{(k)}(x,E)$ denotes the k-step transition probability from x into E.

Just as before, for each integer n we have the map $L_n(\omega,\cdot)$ from Ω into $\mathcal{M}$ and the measure $\rho_{n,x}$ induced on $\mathcal{M}$ from the $p(x,dy)$ chain. The only difference now is that the total mass of $\rho_{n,x}$ is less than or equal to 1 and may vary with n and x. Our aim is to prove

Theorem 3.3. *Under Hypothesis* H, *if μ is any measure with $I(\mu)<\infty$ and N is any weak neighborhood of μ in $\mathcal{M}$, then*

$$\lim_{n\to\infty}\frac{1}{n}\log\rho_{n,x}(N)\geqq -I(\mu).$$

Moreover, if $p(x,E)$ is continuous as a function of x for each $E\subset X$, then the conclusion above is uniform over every compact subset of X.

We shall reduce the proof of the theorem to an application of Theorem 3.1. First of all we carry out the construction just preceding Lemma 3.2 to obtain a stochastic transition probability $\pi(x,dy)$ on the augmented space $X'=X\cup\{\infty\}$. We can take for reference measure β' which is β on X and δ-measure at ∞. We denote by $\pi(x,y)$ the density of $\pi(x,dy)$ relative to β'.

For each ξ in $0<\xi<1$ we define a transition probability density (relative to β') on X' by

$$\begin{aligned}\pi_\xi(x, y) &= (\pi + \xi\pi^2 + \xi^2\pi^3 + \cdots)(1-\xi)\\ &= (1-\xi)\pi(1-\xi\pi)^{-1} = \pi R_\xi ,\end{aligned} \tag{3.6}$$

where $R_\xi = (1-\xi)(1-\xi\pi)^{-1}$. If $p(x, dy)$ satisfies Hypothesis H, then we note that $\pi_\xi(x, y)$ satisfies

$$\pi_\xi(x, y) > 0 \text{ almost everywhere } (\beta') \text{ on } X, \text{ for each } x \in X. \tag{3.7}$$

If $p(x, E)$ is a continuous function of x for each $E \subset X$, then

$$\pi_\xi(x, E) = \int_E \pi_\xi(x, y)\beta'(dy)$$

is again a continuous function of x for each $E \subset X'$. Let $I^{(\xi)}(\mu)$ be the I-function associated with π_ξ. We now have

LEMMA 3.4. *For each μ on X' with $I'(\mu) < \infty$,*

$$\lim_{\xi \to 0} I^{(\xi)}(\mu) = I'(\mu) .$$

In particular, if μ is concentrated on X, we have

$$\lim_{\xi \to 0} I^{(\xi)}(\mu) = I'(\mu) = I(\mu) .$$

Proof: Consider

$$\begin{aligned} I^{(\xi)}(\mu) &= -\inf_{u \in \mathcal{U}_1'} \int_{X'} \log\left(\frac{\pi_\xi u}{u}\right)(x)\mu(dx)\\ &= -\inf_{u \in \mathcal{U}_1'} \left[\int_{X'} \log\left(\frac{\pi R_\xi u}{R_\xi u}\right)(x)\mu(dx) + \int_{X'} \log\left(\frac{R_\xi u}{u}\right)(x)\mu(dx)\right]\\ &\leqq -\inf_{u \in \mathcal{U}_1'} \int_{X'} \log\left(\frac{\pi R_\xi u}{R_\xi u}\right)(x)\mu(dx) - \inf_{u \in \mathcal{U}_1'} \int_{X'} \log\left(\frac{R_\xi u}{u}\right)(x)\mu(dx) .\end{aligned}$$

Since R_ξ is itself a transition probability and $R_\xi u \geqq (1-\xi)u$, we conclude from this last inequality that, for each $0<\xi<1$, $I^{(\xi)}(\mu) \leqq I'(\mu) - \log(1-\xi)$. Thus, if $I'(\mu) < \infty$, we see that, for each $0<\xi<1$, $I^{(\xi)}(\mu) < \infty$. Moreover,

$\overline{\lim}_{\xi\to 0} I^{(\xi)}(\mu) \leqq I'(\mu)$. On the other hand,

$$\begin{aligned}\underline{\lim}_{\xi\to 0} I^{(\xi)}(\mu) &= \underline{\lim}_{\xi\to 0}\left[-\inf_{u\in\mathcal{U}_1'}\int_{X'}\log\left(\frac{\pi_\xi u}{u}\right)(x)\mu(dx)\right]\\ &= -\overline{\lim}_{\xi\to 0}\inf_{u\in\mathcal{U}_1'}\int_{X'}\log\left(\frac{\pi_\xi u}{u}\right)(x)\mu(dx)\\ &\geqq -\inf_{u\in\mathcal{U}_1'}\int_{X'}\log\left(\frac{\pi u}{u}\right)(x)\mu(dx)\\ &= I'(\mu),\end{aligned}$$

so that we have $\lim_{\xi\to 0} I^{(\xi)}(\mu) = I'(\mu)$. If $\mu(X)=1$, then the second conclusion of the lemma follows from Lemma 3.2.

For $0<\xi<1$ consider an integer-valued random variable τ haveing distribution

$$P\{\tau = i\} = (1-\xi)\xi^{i-1}, \qquad i = 1, 2, 3, \cdots. \tag{3.8}$$

Let $\tau_1, \tau_2, \cdots, \tau_n, \cdots$ be a sequence of independent random variables each having the same distribution (3.8) and, moreover, independent of our Markov process with transition probability $\pi(x, dy)$. Let $s_n = \tau_1+\tau_2+\cdots+\tau_n$ and $s_0=0$. The chain with transition probabilities $\pi_\xi(x, dy)$ can be thought of as the chain with transition probabilities $\pi(x, dy)$ sampled at times $s_0, s_1, \cdots, s_n, \cdots$. Since $\tau_i \geqq 1$ with probability 1, $s_0, s_1, \cdots$ are all distinct. If $s_{n-1} = l$, the sets $[s_0, s_1, \cdots, s_{n-1}]$ and $[0, 1, 2, \cdots, n-1]$ as subsets of the integers have at most $2(l-n+1)$ points which are in one set and not in the other. We denote by P_x and P_x^ξ the processes corresponding to $\pi(x, dy)$ and $\pi_\xi(x, dy)$, respectively. We denote by $Q_{n,x}$ and $Q_{n,x}^\xi$ the induced measures on $\mathcal{M}'$ by the function $L_n(\omega, \cdot)$. Now, if D is any subset of $\mathcal{M}'$ and $D^\varepsilon = \{\gamma\in\mathcal{M}' : \|\gamma-\alpha\|<\varepsilon$ for some $\alpha\in D\}$, then the above considerations yield the inequality

$$\begin{aligned}Q_{n,x}(D^\varepsilon) &\geqq Q_{n,x}^\xi(D) - P\left\{\frac{\tau_1+\tau_2+\cdots+\tau_{n-1}-n+1}{n}\geqq\tfrac{1}{2}\varepsilon\right\}\\ &\geqq Q_{n,x}^\xi(D) - P\left\{\frac{\tau_1+\cdots+\tau_n-n}{n}\geqq\tfrac{1}{2}\varepsilon\right\}.\end{aligned} \tag{3.9}$$

We first estimate the second term on the right which is independent of the starting point x of the chain.

LEMMA 3.5. *For any* $\varepsilon > 0$,

$$\lim_{\xi\to 0} \overline{\lim_{n\to\infty}} \frac{1}{n} \log P\left\{\frac{\tau_1+\tau_2+\cdots+\tau_n}{n} \geqq 1+\varepsilon\right\} = -\infty .$$

Proof: Let $\sigma > 0$ be arbitrary and choose $0 < \xi < 1$ such that $\xi e^{\sigma} < 1$. Clearly,

$$P\{\tau_1+\tau_2+\cdots+\tau_n \geqq n(1+\varepsilon)\} \leqq \exp\{-n(1+\varepsilon)\sigma\} \times E\{\exp\{\sigma(\tau_1+\tau_2+\cdots+\tau_n)\}\},$$

and since

$$E\{e^{\sigma\tau}\} = (1-\xi) \sum_{i=1}^{\infty} \xi^{i-1} e^{i\sigma} = \frac{(1-\xi)e^{\sigma}}{1-\xi e^{\sigma}},$$

we have

$$E\{\exp\{\sigma(\tau_1+\tau_2+\cdots+\tau_n)\}\} = \frac{(1-\xi)^n e^{n\sigma}}{(1-\xi e^{\sigma})^n}.$$

Therefore,

$$\overline{\lim_{n\to\infty}} \frac{1}{n} \log P\left\{\frac{\tau_1+\tau_2+\cdots+\tau_n}{n} \geqq 1+\varepsilon\right\} \leqq -\varepsilon\sigma + \log(1-\xi) - \log(1-\xi e^{\sigma}),$$

and hence

$$\lim_{\xi\to 0} \overline{\lim_{n\to\infty}} \frac{1}{n} \log P\left\{\frac{\tau_1+\tau_2+\cdots+\tau_n}{n} \geqq 1+\varepsilon\right\} \leqq -\varepsilon\sigma .$$

But $\varepsilon > 0$ and $\sigma > 0$ is arbitrary so that the lemma follows.

We now are ready for the

Proof of Theorem 3.3: We need first a slightly modified version of inequality (3.9). Let $\mathcal{M}'(X)$ be the set of measures in $\mathcal{M}'$ concentrated on X. If $D \subset \mathcal{M}'(X)$, then we actually have

$$(3.10) \qquad Q_{n,x}(D^{\varepsilon} \cap \mathcal{M}'(X)) \geqq Q^{\xi}_{n,x}(D) - P\left\{\frac{\tau_1+\tau_2+\cdots+\tau_n-n}{n} \geqq \tfrac{1}{2}\varepsilon\right\}.$$

This follows from the fact that ∞ is an absorbing point of the π chain, and if the chain remains in X at times $s_0, s_1, s_2, \cdots, s_{n-1}$, then, s_{n-1} being at least $n-1$, the chain remains in X for times up to $n-1$. If in addition $(\tau_1+\tau_2+\cdots+\tau_n-1)/n < \frac{1}{2}\varepsilon$, then of course $L_n(\omega,\cdot)$ is in D^{ε} as well as in $\mathcal{M}'(X)$.

So let $\mu \in \mathcal{M}$ be such that $I(\mu) < \infty$ and let N be a neighborhood of μ in $\mathcal{M}$. We seek a lower estimate for $\rho_{n,x}(N)$. Let $\varepsilon > 0$ be chosen so that we can find a smaller neighborhood N_1 of μ such that, if $\gamma \in N_1$ and $\|\alpha - \gamma\| < \varepsilon$, then $\alpha \in N$. We can think of N_1 as coming from a neighborhood U of μ in $\mathcal{M}'$, i.e., $N_1 = U \cap \mathcal{M}'(X)$. Then

$$\rho_{n,x}(N) = Q_{n,x}(N \cap \mathcal{M}'(X)), \tag{3.11}$$

and

$$N \supset N_1^\varepsilon, \tag{3.12}$$

where $N_1^\varepsilon = \{\gamma : \|\gamma - \alpha\| < \varepsilon \text{ for some } \alpha \in N_1\}$. Taking $D = U \cap \mathcal{M}'(X)$ in (3.10) we obtain

$$\rho_{n,x}(N) \geqq Q_{n,x}^\xi(U \cap \mathcal{M}'(X)) - P\left\{\frac{\tau_1 + \tau_2 + \cdots + \tau_n - n}{n} \geqq \tfrac{1}{2}\varepsilon\right\}. \tag{3.13}$$

We now pick $0 < \xi < 1$ small enough so that

$$I^{(\xi)}(\mu) \leqq I(\mu) + \varepsilon \tag{3.14}$$

and

$$\varlimsup_{n \to \infty} \frac{1}{n} \log P\left\{\frac{\tau_1 + \cdots + \tau_n - n}{n} \geqq \tfrac{1}{2}\varepsilon\right\} \leqq -2(I(\mu) + \varepsilon), \tag{3.15}$$

which we can do because of Lemmas 3.4 and 3.5. Since μ is supported in X, we see that, for each $\xi > 0$, $\pi_\xi(x, y)$ satisfies assumptions (3.1) and (3.2) of Theorem 3.1 (see (3.6) and (3.7)). Therefore, Theorem 3.1 can be applied and, if in that theorem we choose X for the set A, then we obtain from Theorem 3.1 and (3.14)

$$\varliminf_{n \to \infty} \frac{1}{n} \log Q_{n,x}^\xi(U \cap \mathcal{M}'(X)) \geqq -[I(\mu) + \varepsilon]. \tag{3.16}$$

In (3.13) we now use (3.15) and (3.16) to see that, since ε is arbitrary,

$$\varliminf_{n \to \infty} \frac{1}{n} \log \rho_{n,x}(N) \geqq -I(\mu),$$

which is Theorem 3.3.

If we assume that $p(x, E)$ is continuous in x for each E, then as noted earlier $\pi_\xi(x, E)$ is also continuous in x for each E. This gives the required uniformity in (3.16) which carries over to give the uniformity asserted in Theorem 3.3.

COROLLARY 3.4. *Let G be an open set in $\mathcal{M}$. If the substochastic transition function $p(x, dy)$ satisfies Hypothesis* H, *then*

$$\varliminf_{n\to\infty} \frac{1}{n} \log \rho_{n,x}(G) \geqq -\inf_{\mu\in G} I(\mu)\,. \tag{3.17}$$

Moreover, if $p(x, E)$ is continuous in x for each E, then (3.17) *holds uniformly over compact subsets of X.*

Proof: G can be thought of as a neighborhood of each point μ in G and Theorem 3.3 can be applied.

4. Upper Bounds—Discrete Case

The state space X will in general not be compact, and in order to obtain asymptotic upper bounds on the $Q_{n,x}$ measure appropriate to the asymptotic evaluation of expectations with respect to the Markov process with transition probability $\pi(x, dy)$, we would have to impose further hypotheses on $\pi(x, dy)$. However, we note that if K is a compact subset of $\mathcal{M}$, then we can prove without any further hypotheses, in exactly the same way as we did in [2], that

$$\varlimsup_{n\to\infty} \frac{1}{n} \log \sup_x Q_{n,x}(K) \leqq -\inf_{\mu\in K} I(\mu)\,. \tag{4.1}$$

In order to obtain the asymptotic theorems described in the introduction we need a result like (4.1) but for closed sets—not just compact sets. To obtain such a result in general we impose a new Hypothesis H* stated below, and subsequently we observe that in the special applications described in the introduction this hypothesis is automatically satisfied.

We now make the following hypothesis:

HYPOTHESIS H*. There exists a function $u_0(x)$ on X such that $u_0(x) \geqq 1$ for all $x \in X$, $(\pi u_0)(x) = V(x)$ is bounded on every compact set, and, for every l, $\{x \in X : u_0(x)/V(x) \leqq l\}$ is a compact set.

Hypothesis H* implies that π has an invariant measure—it forces the discrete parameter Markov process with transition probability $\pi(x, dy)$ to be positive recurrent. Before proving the analogue of (4.1) for closed sets we need three lemmas. First of all we note that H* implies that $\inf_{x\in X} u_0(x)/V(x)$ is positive. Let $\inf_{x\in X} u_0(x)/V(x)\geqq\rho>0$ for some $\rho<1$.

LEMMA 4.1. *For any compact set* $W\subset X$ *let* $K_l=W\cup\{x\in X:\ u_0(x)/V(x)\leqq l\}$. *For any* $\varepsilon>0$,

$$\overline{\lim_{n\to\infty}}\frac{1}{n}\log\sup_{x\in W}Q_{n,x}\{\mu\in\mathcal{M}:\ \mu(K_l)\leqq 1-\varepsilon\}\leqq-\log\rho-\varepsilon\log\frac{l}{\rho}.$$

Proof: Since the process $X_0, X_1, X_2,\cdots$ is Markovian, it is easy to see that (cf. (2.1) of [1])

$$E_x\left\{\frac{u_0(X_1)u_0(X_2)\cdots u_0(X_n)}{V(X_0)V(X_1)\cdots V(X_{n-1})}\right\}=1,$$

and because $X_0=x$, this is the same as

$$E_x\left\{\exp\left\{\sum_{i=1}^{n-1}\log(u_0/V)(X_i)\right\}u_0(X_n)\right\}=V(x).$$

But $u_0(x)\geqq 1$ for all $x\in X$, so

$$E_x\left\{\exp\left\{\sum_{i=1}^{n-1}\log(u_0/V)(X_i)\right\}\right\}\leqq V(x).\tag{4.2}$$

By the definition of $L_n(\omega,\cdot)$, for $X_0=x\in W\subset K_l$,

$$\begin{aligned}\sum_{i=1}^{n-1}\log\left(\frac{u_0}{V}\right)(X_i)&=\sum_{i=1}^{n-1}\log\left(\frac{u_0}{V}\right)(X_i)[\chi_{\tilde K_l}(X_i)+(1-\chi_{\tilde K_l}(X_i))]\\&\geqq(\log\rho)\sum_{i=1}^{n-1}(1-\chi_{\tilde K_l}(X_i))+\log l\sum_{i=1}^{n-1}\chi_{\tilde K_l}(X_i)\\&=(n-1)\log\rho+\log\frac{l}{\rho}\sum_{i=1}^{n-1}\chi_{\tilde K_l}(X_i).\end{aligned}$$

Since $X_0\in K_l$ and $\rho<1$, we have

$$\begin{aligned}\sum_{i=1}^{n-1}\log\left(\frac{u_0}{V}\right)(X_i)&\geqq n\log\rho+\log\frac{l}{\rho}\sum_{i=0}^{n-1}\chi_{\tilde K_l}(X_i)\\&=n\log\rho+n\log\frac{l}{\rho}L_n(\omega,\tilde K_l).\end{aligned}\tag{4.3}$$

From (4.2), (4.3) and the definition of $Q_{n,x}$,

$$E^{Q_{n,x}}[\exp\{n \log(l/\rho)\mu(\tilde{K}_l)\}] \leqq V(x)\exp\{-n \log \rho\}. \tag{4.4}$$

From (4.4) we get for $x \in W$

$$Q_{n,x}\{\mu \in \mathcal{M} : \mu(\tilde{K}_l) > \varepsilon\} \leqq V(x)\exp\{-n \log \rho - n\varepsilon \log(l/\rho)\}, \tag{4.5}$$

and this inequality gives us, since $V(x)$ is bounded on W,

$$\overline{\lim_{n\to\infty}} \frac{1}{n} \log \sup_{x\in W} Q_{n,x}\{\mu \in \mathcal{M} : \mu(K_l) \leqq 1-\varepsilon\} \leqq -\log \rho - \varepsilon \log\left(\frac{l}{\rho}\right).$$

LEMMA 4.2. *For each $l<\infty$, the set of measures $A_l = \{\mu \in \mathcal{M} : I(\mu) \leqq l\}$ is a tight family, i.e., for any $\varepsilon > 0$, there exists a compact set $K \subset X$ such that $\mu(\tilde{K}) < \varepsilon$ for all $\mu \in A_l$.*

Proof: If $\mu \in A_l$, then by the definition of $I(\mu)$

$$\int_X \log\left(\frac{\pi u}{u}\right)(x)\mu(dx) \geqq -l \tag{4.6}$$

for all $u \in \mathcal{U}$. By hypothesis H* there exists a function u_0 such that $u_0(x) \geqq 1$ for all $x \in X$, and, for any σ, $K_\sigma = \{x \in X : u_0(x)/(\pi u_0)(x) \leqq \sigma\}$ is compact. Let $u_n(x) = \min[u_0(x), n]$ for $n = 1, 2, 3, \cdots$, and consider $u_n(x)/(\pi u_n)(x)$. Clearly,

$$\lim_{n\to\infty} \frac{u_n(x)}{(\pi u_n)(x)} = \frac{u_0(x)}{(\pi u_0)(x)} \quad \text{for all} \quad x \in X. \tag{4.7}$$

Moreover, from (4.6) and the definition of u_n we get, for any $\mu \in A_l$,

$$\int_X \log\left(\frac{u_n}{\pi u_n}\right)(x)\mu(dx) \leqq l \quad \text{for all} \quad n. \tag{4.8}$$

We easily observe that $u_n(x)/(\pi u_n)(x)$ is uniformly bounded below because for each $x \in X$ either $1 \leqq u_n(x) < n$ or $u_n(x) = n$. In the first case, $u_n(x)/(\pi u_n)(x) = u_0(x)/(\pi u_n)(x) \geqq u_0(x)/(\pi u_0)(x) \geqq \rho$, and in the second case, $u_n(x)/(\pi u_n)(x) \geqq n/n = 1$. Therefore, from (4.7), (4.8) and Fatou's lemma, we get

$$\int_X \log\left(\frac{u_0}{\pi u_0}\right)(x)\mu(dx) \leqq l, \tag{4.9}$$

for all $\mu \in A_l$. Writing this last inequality as

$$\int_{K_\sigma} \log\left(\frac{u_0}{\pi u_0}\right)(x)\mu(dx) + \int_{\tilde{K}_\sigma} \log\left(\frac{u_0}{\pi u_0}\right)(x)\mu(dx) \leqq l,$$

we obtain, for all $\mu \in A_l$ and all σ,

$$(\log \rho)\mu(K_\sigma) + (\log \sigma)\mu(\tilde{K}_l) \leqq l. \tag{4.10}$$

If we now choose $\sigma > \rho$ we have

$$\mu(\tilde{K}_\sigma) \leqq \frac{l - \log \rho}{\log \sigma - \log \rho},$$

for all $\mu \in A_l$. Indeed, for any $\varepsilon > 0$ we can choose σ so large that $\mu(\tilde{K}_\sigma) < \varepsilon$ for all $\mu \in A_l$ which means, since K_σ is compact, that A_l is a tight family.

LEMMA 4.3. *For any $A < \infty$ and for any compact set $W \subset X$, there exists a compact set (a tight family of measures) $\bar{K}_A \subset \mathcal{M}$ such that*

$$\overline{\lim_{n\to\infty}} \frac{1}{n} \log \sup_{x \in W} Q_{n,x}(\tilde{\bar{K}}_A) \leqq -A.$$

Proof: Let $\{\sigma_i\}$, $i = 1, 2, 3, \cdots$, be an increasing sequence with $\lim_{i\to\infty} \sigma_i = \infty$ and let $\{\varepsilon_i\}$, $i = 1, 2, 3, \cdots$, be a decreasing sequence with $\lim_{i\to\infty} \varepsilon_i = 0$. With $K_{\sigma_i} = \{x \in X : u_0(x)/(\pi u_0)(x) \leqq \sigma_i\} \cup W$ consider

$$\bar{K}_{\sigma,\varepsilon} = \{\mu \in \mathcal{M} : \mu(K_{\sigma_i}) \geqq 1 - \varepsilon_i \text{ for all } i\}.$$

Now $\bar{K}_{\sigma,\varepsilon}$ is obviously a tight family of measures in $\mathcal{M}$ for every $\sigma = (\sigma_1, \sigma_2, \cdots)$ and $\varepsilon = (\varepsilon_1, \varepsilon_2, \cdots)$ with the described properties.

From (4.5) we have for $x \in W$

$$\begin{aligned} Q_{n,x}(\tilde{\bar{K}}_{\sigma,\varepsilon}) &\leqq \sum_{i=1}^{\infty} Q_{n,x}\{\mu \in \mathcal{M} : \mu(K_{\sigma_i}) < 1 - \varepsilon_i\} \\ &\leqq \sum_{i=1}^{\infty} V(x) \exp\{-n \log \rho - n\varepsilon_i \log(\sigma_i/\rho)\} \\ &= V(x) \exp\{-n \log \rho\} \sum_{i=1}^{\infty} \exp\{-n\varepsilon_i \log(\sigma_i/\rho)\} \\ &= \frac{V(x)}{\rho^n} \sum_{i=1}^{\infty} \left(\frac{\rho}{\sigma_i}\right)^{n\varepsilon_i}. \end{aligned} \tag{4.11}$$

We choose $\varepsilon_i = 1/i$ and without loss of generality we can assume that $\rho < 1$. From (4.11) we then have

$$Q_{n,x}(\bar{\bar{K}}_{\sigma,\varepsilon}) \leqq \frac{V(x)}{\rho^n} \sum_{i=1}^{\infty} \frac{1}{(\sigma_i^{1/i})^n} \tag{4.12}$$

in which we now take $\sigma_i = (A/\rho)^{i^2}$ noting that again without loss of generality we can choose $A > 1$. Thus

$$Q_{n,x}(\bar{\bar{K}}_{\sigma,\varepsilon}) \leqq \frac{V(x)}{\rho^n} \sum_{i=1}^{\infty} \left(\frac{\rho}{A}\right)^{ni} = \frac{V(x)}{A^n - \rho^n}. \tag{4.13}$$

For the above choices of $\sigma = (\sigma_1, \sigma_2, \cdots)$ and $\varepsilon = (\varepsilon_1, \varepsilon_2, \cdots)$, we let $\bar{K}_{\sigma,\varepsilon} = \bar{K}_A$ and we see from (4.13) that, since V is bounded on W,

$$\overline{\lim_{n\to\infty}} \frac{1}{n} \log \sup_{x\in W} Q_{n,x}(\bar{\bar{K}}_A) \leqq -A.$$

THEOREM 4.4. *Let C be a closed set in $\mathcal{M}$, and assume Hypothesis* H*. *Then, for any compact set $W \subset X$,*

$$\overline{\lim_{n\to\infty}} \frac{1}{n} \log \sup_{x\in W} Q_{n,x}(C) \leqq -\inf_{\mu\in C} I(\mu).$$

Proof: Let $A < \infty$ and let $\bar{K}_A$ be selected as in Lemma 4.3. Since

$$\begin{aligned} Q_{n,x}(C) &= Q_{n,x}(C \cap \bar{K}_A) + Q_{n,x}(C \cap \bar{\bar{K}}_A) \\ &\leqq Q_{n,x}(C \cap \bar{K}_A) + Q_{n,x}(\bar{\bar{K}}_A), \end{aligned}$$

we have

$$\overline{\lim_{n\to\infty}} \frac{1}{n} \log Q_{n,x}(C) \leqq \max\left[\overline{\lim_{n\to\infty}} \frac{1}{n} \log Q_{n,x}(C \cap \bar{K}_A), \overline{\lim_{n\to\infty}} \frac{1}{n} \log Q_{n,x}(\bar{\bar{K}}_A)\right].$$

Since $C \cap \bar{K}_A$ is compact in $\mathcal{M}$, we obtain from (4.1)

$$\overline{\lim_{n\to\infty}} \frac{1}{n} \log \sup_{x\in W} Q_{n,x}(C \cap \bar{K}_A) \leqq -\inf_{\mu\in C\cap\bar{K}_A} I(\mu),$$

and from Lemma 4.3,

$$\overline{\lim_{n\to\infty}} \frac{1}{n} \log \sup_{x\in W} Q_{n,x}(\tilde{\tilde{K}}_A) \leqq -A,$$

so that

$$\overline{\lim_{n\to\infty}} \frac{1}{n} \log \sup_{x\in W} Q_{n,x}(C) \leqq \max\left[-\inf_{\mu\in C\cap \tilde{\tilde{K}}_A} I(\mu), -A\right]$$

$$\leqq \max\left[-\inf_{\mu\in C} I(\mu), -A\right].$$

But this last inequality is true for every A, and therefore letting $A\to\infty$ we obtain the theorem.

Now we want to observe, and use in the sequel, the fact that in the special case of independent random variables with values in X, i.e., the transition probability $\pi(x, dy) = \beta(dy)$, the Hypothesis H*, imposed at the beginning of this section in order to obtain the upper bound, is automatically satisfied. To see this, we choose first of all a sequence of compact sets $K_j \subset X$ with $K_1 \subset K_2 \subset \cdots$ and $\beta(\tilde{K}_j) \leqq 1/2^j$. Then, for any $x \in X$ define $u_0(x) \equiv \inf\{j : x \in K_j\}$. Observe that $u_0(x) \geqq 1$ for all $x \in X$ and that

$$V(x) = (\pi u_0)(x) = \int_X u_0(x)\beta(dx) = m < \infty$$

so that $V(x)$ is trivially bounded on every compact set. Moreover, for every l, $\{x \in X : u_0(x)/V(x) \leqq l\} = \{x \in X : u_0(x) \leqq ml\}$ is a compact set and hence H* is satisfied.

We also note that, in the case of independent random variables with values in X, the Hypothesis H of Section 3 (on the lower bound) is also automatically satisfied. Thus for independent random variables with values in X both Theorems 3.4 and 4.4 hold without imposing Hypothesis H* for the upper bound or H for the lower bound.

In the case of independent random variables it is more natural to consider

$$L'_n(\omega, A) = \sum_{j=1}^{n} \chi_A(X_j). \tag{4.14}$$

In such a case, the distribution of L'_n is independent of the starting point $X_0 = x$ and we shall denote it by Q_n. Since $\|L_n - L'_n\| \leqq 2/n$ asymptotically, there is no difference between L_n and L'_n. Theorems 3.4 and 4.4 hold just as well with L'_n instead of L_n.

Moreover, in this case

$$\begin{aligned} I(\mu) &= -\inf_{u\in\mathcal{U}_1}\left[\int_X \log(\pi u)(x)\mu(dx) - \int_X \log u(x)\mu(dx)\right] \\ &= -\inf_{u\in\mathcal{U}_1}\left[\log\int_X u(x)\beta(dx) - \int_X \log u(x)\mu(dx)\right] \\ &= I_\beta(\mu). \end{aligned}$$

We then have the following corollary to Theorems 3.4 and 4.4.

THEOREM 4.5. *Let X be a complete separable metric space and let $X_1, X_2, \cdots$ be independent random variables with values in X and having common distribution $\beta(dy)$. For any closed set $C \subset \mathcal{M}$,*

$$\overline{\lim_{n\to\infty}} \frac{1}{n} \log Q_n(C) \leqq -\inf_{\mu\in C} I_\beta(\mu),$$

and, for any open set $G \subset \mathcal{M}$,

$$\underline{\lim_{n\to\infty}} \frac{1}{n} \log Q_n(G) \geqq -\inf_{\mu\in G} I_\beta(\mu).$$

5. X a Banach Space

For the purposes of the applications in the next section, we want to consider independent random variables with values in a Banach space.

Let X be a Banach space and λ a probability measure on X such that

$$\int_X e^{\sigma\|x\|}\lambda(dx) < \infty \quad \text{for all} \quad \sigma > 0. \tag{5.1}$$

As usual, let $\mathcal{M}$ be the space of all probability measures on X. Now, however, instead of considering general discrete parameter Markov processes with X as state space, we want to focus on the special case of independent random variables with values in X having common distribution $\lambda(dy)$. In other words, the transition probability function $\pi(x, dy)$ is $\lambda(dy)$. In this special case, the

I-function becomes, for any $\mu \in \mathcal{M}$,

$$
\begin{aligned}
I(\mu) &= -\inf_{u \in \mathcal{U}} \left[\int_X \log (\pi u)(x) \mu(dx) - \int_X \log u(x) \mu(dx) \right] \\
&= -\inf_{u \in \mathcal{U}} \left[\log \int_X u(x) \lambda(dx) - \int_X \log u(x) \mu(dx) \right] \qquad (5.2) \\
&= I_\lambda(\mu) .
\end{aligned}
$$

In the sequel we shall use several times results about $I_\lambda(\mu)$ obtained in [1] (Lemma 2.1), namely, $I_\lambda(\mu) < \infty$ if and only if $\mu \ll \lambda$ with a density $f(x) = d\mu/d\lambda$ such that

$$\int_X f(x) \, |\log f(x)| \, \lambda(dx) < \infty .$$

Moreover, in this case,

$$I_\lambda(\mu) = \int_X f(x) \log f(x) \lambda(dx) = \int_X \log f(x) \mu(dx) . \qquad (5.3)$$

We first prove

LEMMA 5.1. *Let $\{\lambda_n\}$ and $\{\mu_n\}$ be sequences in $\mathcal{M}$. If $\{\lambda_n\}$ is a tight sequence and $I_{\lambda_n}(\mu_n) \leqq l < \infty$ for all n, then $\{\mu_n\}$ is a tight sequence. If, moreover,*

$$\sup_n \int_X e^{\sigma \|x\|} \lambda_n(dx) < \infty \quad \textit{for all} \quad \sigma > 0 , \qquad (5.4)$$

then $\int_X \|x\| \, \mu_n(dx)$ is uniformly bounded in n, and

$$\lim_{A \to \infty} \sup_n \int_{\|x\| \geqq A} \|x\| \, \mu_n(dx) = 0 . \qquad (5.5)$$

Proof: The main content of this lemma is that if $\int_X e^{\sigma\|x\|}\lambda_n(dx)$ is uniformly bounded in n for every $\sigma>0$ and $I_{\lambda_n}(\mu_n)$ is uniformly bounded, then $\int_X \|x\|\,\mu_n(dx)$ is uniformly integrable.

First of all, if $\{\lambda_n\}$ is a tight sequence, by definition, for any $\varepsilon>0$ there exists a compact set $K\subset X$ such that $\lambda_n(\tilde{K})<\varepsilon$ for all n. By the remark just preceding this lemma and the hypothesis that $I_{\lambda_n}(\mu)\leqq l<\infty$, we see that $\mu_n \ll \lambda_n$ for every n. We show now that $\{\mu_n\}$ is a tight sequence. We have

$$l\geqq I_{\lambda_n}(\mu_n)=-\inf_{u\in\mathcal{U}}\left[\log\int_X u(x)\lambda_n(dx)-\int_X \log u(x)\mu_n(dx)\right],$$

and, therefore, for all $u\in\mathcal{U}$,

$$\log\int_X u(x)\lambda_n(dx)-\int_X \log u(x)\mu_n(dx)\geqq -l.$$

Assume $A\subset X$; if, in particular, we let $u(x)=1$ on A and $M>0$ on $\tilde{A}$, then this last inequality becomes

$$\mu_n(\tilde{A})\leqq\frac{1}{\log M}[l+\log\{1+\lambda_n(\tilde{A})(M-1)\}].$$

By hypothesis, $\{\lambda_n\}$ is a tight sequence and therefore, for any $\varepsilon>0$, there exists a compact set K such that $\lambda_n(\tilde{K})<\varepsilon$ for all n. If we set $A=K$ and $M=1+1/\varepsilon$ in the last displayed inequality, we get

$$\mu_n(\tilde{K})\leqq\frac{1}{\log(1+1/\varepsilon)}[l+\log 2]$$

which shows that $\{\mu_n\}$ is also a tight sequence.

For $\sigma>0,\ x\geqq 0,\ y\geqq 0$,

$$\inf_x[e^{\sigma x}-xy]=\frac{y}{\sigma}\left(1-\log\frac{y}{\sigma}\right),$$

and therefore

$$xy\leqq e^{\sigma x}+\frac{y}{\sigma}\log\frac{y}{\sigma}. \tag{5.6}$$

Using (5.6) and letting $d\mu_n/d\lambda_n = f_n(x)$, we have

$$
\begin{aligned}
\int_{\|x\|\geqq A} \|x\|\,\mu_n(dx) &= \int_{\|x\|\geqq A} \|x\|\, f_n(x)\lambda_n(dx) \\
&\leqq \int_{\|x\|\geqq A} \left(e^{\sigma\|x\|} + \frac{f_n(x)}{\sigma}\log\frac{f_n(x)}{\sigma}\right)\lambda_n(dx) \\
&\leqq \int_{\|x\|\geqq A} e^{\sigma\|x\|}\lambda_n(dx) + \frac{1}{\sigma}\int_{\|x\|\geqq A} f_n(x)\,|\log f_n(x)|\,\lambda_n(dx) \\
&\quad - \frac{\log\sigma}{\sigma}\int_{\|x\|\geqq A}\lambda_n(dx)\,.
\end{aligned}
\tag{5.7}
$$

If we take $\sigma \geqq 1$ in (5.7) so that $(\log\sigma)/\sigma \geqq 0$, we obtain

$$
\begin{aligned}
\sup_n \int_{\|x\|\geqq A} \|x\|\,\mu_n(dx) \leqq \sup_n \int_{\|x\|\geqq A} & e^{\sigma\|x\|}\lambda_n(dx) \\
&+ \frac{1}{\sigma}\sup_n \int_X f_n(x)\,|\log f_n(x)|\,\lambda_n(dx)\,.
\end{aligned}
\tag{5.8}
$$

The assumption (5.4) implies

$$\lim_{A\to\infty}\sup_n \int_{\|x\|\geqq A} e^{\sigma\|x\|}\lambda_n(dx) = 0 \quad \text{for all} \quad \sigma > 0\,,$$

and therefore to show (5.5) it suffices to prove

$$\sup_n \int_X f_n(x)\,|\log f_n(x)|\,\lambda_n(dx) < \infty\,, \tag{5.9}$$

since we can take σ as large as we please. Before doing this we note that inequalities (5.7) and (5.8) also hold if the region of integration is all of X instead of $\|x\| \geqq A$, and therefore the uniform boundedness of $\int_X \|x\|\,\mu_n(dx)$ will also follow from (5.9).

But

$$
\begin{aligned}
\int_X & f_n(x)\,|\log f_n(x)|\,\lambda_n(dx) \\
&= \int_{f_n\geqq 1} f_n(x)\log f_n(x)\lambda_n(dx) - \int_{f_n\leqq 1} f_n(x)\log f_n(x)\lambda_n(dx) \\
&= \int_X f_n(x)\log f_n(x)\lambda_n(dx) - 2\int_{f_n\leqq 1} f_n(x)\log f_n(x)\lambda_n(dx)\,,
\end{aligned}
\tag{5.10}
$$

and by (5.3) the first term on the right is $I_{\lambda_n}(\mu_n)\leqq l$. Thus, since $|\log x|\leqq c/\sqrt{x}$ for $|x|\leqq 1$,

$$\begin{aligned}\int_X f_n(x)\,|\log f_n(x)|\,\lambda_n(dx) &\leqq l+2\int_{f_n\leqq 1} f_n(x)\,|\log f_n(x)|\,\lambda_n(dx)\\ &\leqq l+2c\int_{f_n\leqq 1}\sqrt{f_n(x)}\,\lambda_n(dx)\\ &\leqq l+2c\sqrt{\int_X f_n(x)\lambda_n(dx)}\\ &= l+2c,\end{aligned}$$

which gives (5.9) and the lemma.

For $a\in X$ we define

$$h(a)=\inf_{\mu:\int x\mu(dx)=a} I(\mu),$$

and let X^* be the dual of X.

THEOREM 5.2.

(i) *$h(a)$ is non-negative and lower semicontinuous on X;*

(ii) *$h(a)=0$ if and only if $a=\int x\lambda(dx)$;*

(iii) *for any l, $\{a\in X : h(a)\leqq l\}$ is a compact subset of X;*

(iv) *$h(a)=\sup_{\theta\in X^*}[\langle\theta, a\rangle-\log M(\theta)]$, where*

$$M(\theta)=\int_X e^{\langle\theta,x\rangle}\lambda(dx).$$

Proof: (i) $I(\mu)$ itself is non-negative so that $h(a)$ is non-negative. Let $\{a_n\}$ be a sequence in X converging to a. We want to show that $h(a)\leqq \underline{\lim}_{n\to\infty} h(a_n)$. Let $l=\underline{\lim}_{n\to\infty} h(a_n)$. By definition there exists a sequence of measures $\{\mu_n\}$ such that $\int x\mu_n(dx)=a_n$ and $I(\mu_n)\to l$. Since $I(\mu_n)=I_\lambda(\mu_n)$ is bounded, we have from the preceding lemma that $\{\mu_n\}$ is a tight sequence. Thus without loss of generality we can assume μ_n converges weakly to a measure μ. Moreover, the uniform integrability proved in Lemma 5.1 implies that $\int x\mu_n(dx)\to\int x\mu(dx)=a$. But $I(\mu)\leqq l$ by the lower semicontinuity of $I(\mu)$ so that indeed $h(a)\leqq l$.

(ii) If $a=\int x\lambda(dx)$, then $h(a)=0$ since $I(\lambda)=I_\lambda(\lambda)=0$. Conversely, if $h(a)=0$ there exists a sequence $\{\mu_n\}$ in $\mathcal{M}$ such that $\int x\mu_n(dx)=a$ and $I(\mu_n)\to 0$. By the preceding lemma, there then exists a subsequence $\{\mu_{n_j}\}$ such that $\{\mu_{n_j}\}$ converges weakly to some measure $\mu\in\mathcal{M}$ and $I(\mu)=0$. By the uniform integrability part of the preceding lemma, this implies $\int x\mu(dx)=a$. But $I(\mu)=0$ if and only if $\mu=\lambda$ so that $a=\int x\lambda(dx)$.

(iii) Let $\{a_n\}$ be a sequence in X such that, for each n, $h(a_n)\leqq l$. Corresponding to the $\{a_n\}$ sequence there exists a sequence $\{\mu_n\}$ in $\mathcal{M}$ such that $\int x\mu_n(dx)=a_n$ and $I(\mu_n)\leqq l+1$. Since $I(\mu_n)=I_\lambda(\mu_n)$ is bounded we know from the preceding lemma that $\{\mu_n\}$ is tight, i.e., compact. Thus there exists a subsequence $\{\mu_{n_j}\}$ converging weakly to some measure $\mu\in\mathcal{M}$. Again because of the uniform integrability shown in the preceding lemma, we have $\int x\mu_{n_j}(dx)\to\int x\mu(dx)$ which implies $\{a_{n_j}\}$ is a convergent subsequence.

(iv) Let

$$\phi(a)=\sup_{\theta\in X^*}\left[\langle\theta,a\rangle-\log M(\theta)\right].$$

We prove first that $h(a)\geqq\phi(a)$. Let μ be such that $\int x\mu(dx)=a$ and $I(\mu)=l<\infty$. We must show that, for all $\theta\in X^*$, $I(\mu)\geqq\langle\theta,a\rangle-\log M(\theta)$. By definition,

$$\begin{aligned}I(\mu)&=-\inf_{u\in\mathcal{U}}\left[\int_X\log\frac{\int_X u(x)\lambda(dx)}{u(x)}\mu(dx)\right]\\&=\sup_{u\in\mathcal{U}}\left[\int_X\log u(x)\mu(dx)-\log\int_X u(x)\lambda(dx)\right].\end{aligned}$$

Thus, for all $u\in\mathcal{U}$,

$$I(\mu)\geqq\int_X\log u(x)\mu(dx)-\log\int_X u(x)\lambda(dx).$$

In particular, let $u(x)=e^{\langle\theta,x\rangle}$ ($e^{\langle\theta,x\rangle}$ is not in $\mathcal{U}$ but we can truncate and pass to the limit inside the integrals), and we get

$$I(\mu) \geqq \int_X \langle \theta, x \rangle \mu(dx) - \log M(\theta).$$

Since $\int x\mu(dx) = a$, this last inequality becomes $I(\mu) \geqq \langle \theta, a \rangle - \log M(\theta)$.

To show that $h(a) \leqq \phi(a)$ is more delicate, and we prove it first for $X = R_1$ and then use this result to show that $h(a) \leqq \phi(a)$ for X a Banach space. So consider first $X = R_1$ where now

$$\phi(a) = \sup_\theta [\theta a - \log M(\theta)] \quad \text{and} \quad M(\theta) = \int_X e^{\theta x} \lambda(dx).$$

Suppose, as a first case, that the supremum in $\phi(a)$ is actually attained. Then it is easy to see that $h(a) \leqq \phi(a)$. This is so because if θ_0 is the point where the supremum is attained, i.e., $\phi(a) = \theta_0 a - \log M(\theta_0)$, then $a = M'(\theta_0)/M(\theta_0)$ or

$$\frac{1}{M(\theta_0)} \int_X x e^{\theta_0 x} \lambda(dx) = a.$$

Hence, if we let μ be a probability measure on R_1 such that $d\mu/d\lambda = 1/M(\theta_0) e^{\theta_0 x}$, we have $\int_X x\mu(dx) = a$ and, what is important, from (5.3),

$$\begin{aligned} I(\mu) &= \int_X \log \left(\frac{d\mu}{d\lambda}\right)(x) \mu(dx) = \int_X \log \frac{e^{\theta_0 x}}{M(\theta_0)} \mu(dx) \\ &= \theta_0 a - \log M(\theta_0) = \phi(a) \end{aligned}$$

so that clearly $h(a) \leqq \phi(a)$.

Now, as a second case, still with $X = R_1$, suppose the supremum in $\phi(a)$ is not attained. For any $0 < \varepsilon < 1$ let $\lambda_\varepsilon = (1 - \varepsilon)\lambda + \varepsilon \nu$, where ν is Gaussian measure on R_1 with mean 0 and variance 1. Since

$$\begin{aligned} \log M_\varepsilon(\theta) &= \log \int_X e^{\theta x} \lambda_\varepsilon(dx) \\ &= \log [(1 - \varepsilon) M(\theta) + \varepsilon e^{\theta^2/2}] \\ &\geqq \log \varepsilon + \tfrac{1}{2}\theta^2, \end{aligned}$$

we see that $\sup_\theta[a\theta-\log M_\varepsilon(\theta)]$ is actually attained and therefore, by the preceding case,

$$\inf_{\mu:\int x\mu(dx)=a} I_{\lambda_\varepsilon}(\mu)=\sup_\theta[a\theta-\log M_\varepsilon(\theta)]. \tag{5.11}$$

But

$$\begin{aligned}\sup_\theta[a\theta-\log M_\varepsilon(\theta)]&=\sup_\theta[a\theta-\log\{(1-\varepsilon)M(\theta)+\varepsilon e^{\theta^2/2}\}]\\&\leqq\sup_\theta[a\theta-\log(1-\varepsilon)-\log M(\theta)]\\&=\sup_\theta[a\theta-\log M(\theta)]-\log(1-\varepsilon),\end{aligned}$$

and so the argument that $h(a)\leqq\phi(a)$ in R_1 will be completed if we show that

$$\inf_{\mu:\int x\mu(dx)=a} I_\lambda(\mu)\leqq\overline{\lim_{\varepsilon\to0}}\inf_{\mu:\int x\mu(dx)=a} I_{\lambda_\varepsilon}(\mu). \tag{5.12}$$

To do this, let μ_ε be such that

$$\inf_{\mu:\int x\mu(dx)=a} I_{\lambda_\varepsilon}(\mu)=l_\varepsilon=I_{\lambda_\varepsilon}(\mu_\varepsilon),$$

and let $l=\overline{\lim}_{\varepsilon\to0} l_\varepsilon$. If $l=\infty$, there is nothing to prove; so let $l<\infty$. Thus, for all ε, $I_{\lambda_\varepsilon}(\mu_\varepsilon)\leqq l$ and, since $\{\lambda_\varepsilon\}$ is a tight family, we have by Lemma 5.1 that $\{\mu_\varepsilon\}$ is a tight family. Moreover, the uniform integrability part of that same lemma implies that there exists a subsequence μ_{ε_j} with $\varepsilon_j\to0$ such that $\lim_{j\to\infty}\mu_{\varepsilon_j}=\bar\mu$ exists and

$$\int x\mu_{\varepsilon_j}(dx)\to\int x\bar\mu(dx).$$

Since, for each $\varepsilon>0$, $\int x\mu_\varepsilon(dx)=a$, we have $\int x\bar\mu(dx)=a$. $I_\lambda(\mu)$ is jointly lower semicontinuous in λ and μ and therefore

$$I_\lambda(\bar\mu)\leqq\underline{\lim}_{\varepsilon\to0} I_{\lambda_\varepsilon}(\mu_\varepsilon)\leqq\overline{\lim_{\varepsilon\to0}} I_{\lambda_\varepsilon}(\mu_\varepsilon)=l,$$

which implies that

$$\inf_{\mu:\int x\mu(dx)=a} I_\lambda(\mu)\leqq l$$

and (5.12).

Having shown that $h(a)\leqq\phi(a)$ when $X=R_1$, we are prepared to show the same inequality for X a Banach space. To recapitulate, we have a probability measure λ on a Banach space X such that

$$\int_X e^{\sigma\|x\|}\lambda(dx)<\infty$$

for all $\sigma>0$. For any $a\in X$, we have

$$h(a)=\inf_{\mu:\int x\mu(dx)=a} I(\mu)$$

and

$$\phi(a)=\sup_{\theta\in X^*}[\langle a,\theta\rangle-\log M(\theta)],$$

where

$$M(\theta)=\int_X e^{\langle\theta,x\rangle}\lambda(dx).$$

Thus we assume $h(a)>l>\phi(a)$ and we show that this leads to a contradiction.

Let

$$k_l=\left\{y\in X: y=\int x\mu(dx) \text{ for some } \mu \text{ with } I_\lambda(\mu)\leqq l\right\}.$$

We note that k_l is a compact, convex set and $a\notin k_l$. Therefore by the separation theorem there exists $\theta_0\in X^*$ such that $\langle\theta, a\rangle=a_0$ and $\langle\theta_0, y\rangle\geqq a_1$, for all $y\in k_l$ with $a_1>a_0$. For this θ_0 let λ_0, a probability measure on R_1, be the distribution of the linear functional $\langle\theta_0, x\rangle$, i.e., if $A\subset R_1$, then $\lambda_0(A)=\lambda\{x\in X:\langle\theta_0,x\rangle\in A\}$. Moreover, let μ_0 be a probability measure on R_1 such that $I_{\lambda_0}(\mu_0)\leqq l$. Then we know that $\mu_0\ll\lambda_0$; so let $(d\mu_0/d\lambda_0)(\xi)=f_0(\xi)$. Define μ on X by $d\mu/d\lambda=f_0(\langle\theta_0,x\rangle)$. Thus by (5.3),

$$\begin{aligned}I_\lambda(\mu)&=\int_X[\log f_0(\langle\theta_0,x\rangle)]f_0(\langle\theta_0,x\rangle)\lambda(dx)\\&=\int_{R_1}[\log f_0(\xi)]f_0(\xi)\lambda_0(d\xi)=I_{\lambda_0}(\mu_0)\leqq l.\end{aligned}\tag{5.13}$$

Therefore for this particular μ we have $\int x\mu(dx)\in k_l$ and so

$$\left\langle\theta_0,\int x\mu(dx)\right\rangle\geqq a_1,$$

i.e.,

$$\int_X\langle\theta_0,x\rangle\mu(dx)\geqq a_1.$$

But this means that

$$\int_X \langle \theta_0, x\rangle f_0(\langle \theta_0, x\rangle)\lambda(dx) = \int_{R_1} \xi f_0(\xi)\lambda_0(d\xi) = \int_{R_1} \xi\mu_0(d\xi) \geqq a_1 .$$

To summarize, we have shown that if $I_{\lambda_0}(\mu_0) \leqq l$, then $\int_{R_1} \xi\mu_0(d\xi) \geqq a_1 > a_0$. We conclude that

$$\inf_{\mu_0 : \int_{R_1} \xi\mu_0(d\xi) = a_0} I_{\lambda_0}(\mu_0) \geqq l. \tag{5.14}$$

But we have shown that, in one dimension, $h(a) = \phi(a)$ so that (5.14) implies

$$\sup_\sigma [a_0\sigma - \log M_0(\sigma)] \geqq l, \tag{5.15}$$

where

$$M_0(\sigma) = \int_{R_1} e^{\sigma\xi}\lambda_0(d\xi) .$$

Therefore, since $\langle \theta_0, a\rangle = a_0$ and

$$M(\sigma\theta_0) = \int_X e^{\sigma\langle \theta_0, x\rangle}\lambda(dx)$$

$$= \int_{R_1} e^{\sigma\xi}\lambda_0(d\xi) = M_0(\sigma) ,$$

we get from (5.15)

$$\sup_\sigma [\sigma\langle \theta_0, a\rangle - \log M(\sigma\theta_0)] \geqq l ,$$

or

$$\sup_\sigma [\langle \sigma\theta_0, a\rangle - \log M(\sigma\theta_0)] \geqq l , \tag{5.16}$$

which implies

$$\sup_{\theta\in X^*} [\langle \theta, a\rangle - \log M(\theta)] \geqq l. \tag{5.17}$$

This last inequality states that $\phi(a) \geqq l$ which contradicts our assumption that $h(a) > l > \phi(a)$; we have completed part (iv) of the theorem.

Let X be a Banach space and λ a probability measure on X such that $\int_X e^{\sigma\|x\|}\lambda(dx)<\infty$ for all $\sigma>0$. Let $\tilde{X}$ be a random variable with values in X having distribution λ, i.e., if $A\subset X$, $P\{\omega:\tilde{X}(\omega)\in A\}=\lambda(A)$. Let $\tilde{X}_1,\tilde{X}_2,\cdots,\tilde{X}_n$ be independent copies of $\tilde{X}$ and let ν_n be the distribution of $(\tilde{X}_1+\tilde{X}_2+\cdots+\tilde{X}_n)/n$. For any $a\in X$ let, as before,

$$\phi(a)=\sup_{\theta\in X^*}[\langle\theta,a\rangle-\log M(\theta)].$$

We now prove

THEOREM 5.3. *For any closed set $C\subset X$,*

$$\overline{\lim_{n\to\infty}}\frac{1}{n}\log\nu_n(C)\leqq-\inf_{a\in C}\phi(a), \tag{5.18}$$

and, for any open set $G\subset X$,

$$\underline{\lim_{n\to\infty}}\frac{1}{n}\log\nu_n(G)\geqq-\inf_{a\in G}\phi(a). \tag{5.19}$$

Proof: We prove this theorem in two parts—first we assume there exists an l such that $P\{\omega:\|\tilde{X}(\omega)\|\leqq l\}=\lambda\{x\in X:\|x\|\leqq l\}=1$, i.e., that the random variables are bounded, and then in the second part we remove this hypothesis by truncation.

Assuming therefore that $\lambda\{\|x\|\leqq l\}=1$ we let $\mathcal{M}_l=\{\mu\in\mathcal{M}:\mu(\|x\|\leqq l\}=1\}$ and note that $L_n'\in\mathcal{M}_l$ for each n. We denote by Q_n, which is a measure on $\mathcal{M}_l$, the distribution of L_n'. Consider the map $\Phi:\mathcal{M}_l\to X$ given by $\Phi(\mu)=\int x\mu(dx)$. It is clear that Φ is continuous with $\mathcal{M}_l$ given the weak topology and X its usual topology, and moreover $Q_n\Phi^{-1}=\nu_n$. If we let C be a closed set in X, we have

$$\overline{\lim_{n\to\infty}}\frac{1}{n}\log\nu_n(C)=\overline{\lim_{n\to\infty}}\frac{1}{n}\log Q_n[\Phi^{-1}C]. \tag{5.20}$$

But by Theorem 4.5 and Theorem 5.2 we then get

$$\begin{aligned}\overline{\lim_{n\to\infty}}\frac{1}{n}\log Q_n[\Phi^{-1}C]&\leqq-\inf_{\mu:\int x\mu(dx)\in C}I_\lambda(\mu)\\&=-\inf_{a\in C}\inf_{\mu:\int x\mu(dx)=a}I_\lambda(\mu)\\&=-\inf_{a\in C}h(a)=-\inf_{a\in C}\phi(a).\end{aligned} \tag{5.21}$$

Thus (5.20) and (5.21) give us (5.18) under the hypothesis that the random variables are bounded.

Similarly, to prove (5.19) under the hypothesis of boundedness, we again use Theorem 4.5 and Theorem 5.2 to see that

$$\tag{5.22}\begin{aligned}\underline{\lim}_{n\to\infty}\frac{1}{n}\log \nu_n(G) &= \underline{\lim}_{n\to\infty}\frac{1}{n}\log Q_n[\Phi^{-1}G]\\ &\geqq -\inf_{\mu:\int x\mu(x)\in G} I_\lambda(\mu)\\ &= -\inf_{a\in G}\inf_{\mu:\int x\mu(dx)=a} I_\lambda(\mu)\\ &= -\inf_{a\in G} h(a) = -\inf_{a\in G}\phi(a).\end{aligned}$$

Now we wish to remove the hypothesis that the random variables are bounded. For any $l>0$, define

$$\tilde{Y}_j^l=\begin{cases}\tilde{X}_j & \text{if } \|\tilde{X}_j\|\leqq l,\\ 0 & \text{if } \|\tilde{X}_j\|>l,\end{cases}$$

and

$$\tilde{Z}_j^l=\begin{cases}\tilde{X}_j & \text{if } \|\tilde{X}_j\|>l,\\ 0 & \text{if } \|\tilde{X}_j\|\leqq l,\end{cases}$$

so that $\tilde{X}_j=\tilde{Y}_j^l+\tilde{Z}_j^l$. Then, for C a closed set in X and any $\varepsilon>0$, $l>0$,

$$\tag{5.23}\begin{aligned}\nu_n(C) &= P\left\{\frac{\tilde{X}_1+\tilde{X}_2+\cdots+\tilde{X}_n}{n}\in C\right\}\\ &\leqq P\left\{\frac{\tilde{Y}_1^l+\tilde{Y}_2^l+\cdots+\tilde{Y}_n^l}{n}\in \bar{C}^\varepsilon\right\}+P\left\{\frac{\|\tilde{Z}_1^l+\cdots+\tilde{Z}_n^l\|}{n}>\varepsilon\right\},\end{aligned}$$

where C^ε is the set of all points in X the distance of which from C is less than ε and $\bar{C}^\varepsilon$ is its closure. From (5.23) and what we just proved for bounded random variables, we have

$$\tag{5.24}\begin{aligned}&\overline{\lim}_{n\to\infty}\frac{1}{n}\log \nu_n(C)\\ &\leqq \max\left[\overline{\lim}_{n\to\infty}\frac{1}{n}\log P\left\{\frac{\tilde{Y}_1^l+\tilde{Y}_2^l+\cdots+\tilde{Y}_n^l}{n}\in\bar{C}^\varepsilon\right\},\right.\\ &\qquad\qquad \left.\overline{\lim}_{n\to\infty}\frac{1}{n}\log P\left\{\frac{\|\tilde{Z}_1^l+\cdots+\tilde{Z}_n^l\|}{n}>\varepsilon\right\}\right]\\ &\leqq \max\left[-\inf_{a\in\bar{C}^\varepsilon}\phi^l(a),\ \overline{\lim}_{n\to\infty}\frac{1}{n}\log P\left\{\frac{\|\tilde{Z}_1^l+\cdots+\tilde{Z}_n^l\|}{n}>\varepsilon\right\}\right],\end{aligned}$$

where, if we let λ^l be the distribution of $\tilde{Y}^l$ and $M^l(\theta)=\int_X e^{\langle\theta,x\rangle}\lambda^l(dx)$,

$$\phi^l(a)=\sup_{\theta\in X^*}[\langle\theta,a\rangle-\log M^l(\theta)].$$

In Lemma 5.4 below we show that, for every $\varepsilon>0$,

$$\overline{\lim_{l\to\infty}}\ \overline{\lim_{n\to\infty}}\frac{1}{n}\log P\left\{\frac{\|\tilde{Z}_1^l+\tilde{Z}_2^l+\cdots+\tilde{Z}_n^l\|}{n}>\varepsilon\right\}=-\infty. \tag{5.25}$$

If, in (5.24), we first let $l\to\infty$ and then $\varepsilon\to 0$, we get

$$\overline{\lim_{n\to\infty}}\frac{1}{n}\log\nu_n(C)\leqq-\underline{\lim_{\varepsilon\to 0}}\ \underline{\lim_{l\to\infty}}\ \inf_{a\in\overline{C^\varepsilon}}\phi^l(a). \tag{5.26}$$

It follows from Lemma 5.5 below that, in fact,

$$\underline{\lim_{\varepsilon\to 0}}\ \underline{\lim_{l\to\infty}}\ \inf_{a\in\overline{C^\varepsilon}}\phi^l(a)\geqq\inf_{a\in C}\phi(a), \tag{5.27}$$

and hence (5.26) and (5.27) give us (5.18) in the general case.

For the lower bound (5.19) in the general case, we truncate exactly as for the upper bound and proceed as follows. Let G be an open set in X. Pick $a\in G$ such that $\phi(a)<\infty$ and $\delta>0$ such that $S(a,\delta)\subset G$. From Lemma 5.4 below we can pick l_0 so large that, for $l>l_0$,

$$\overline{\lim_{n\to\infty}}\frac{1}{n}\log P\left\{\frac{\|\tilde{Z}_1^l+\tilde{Z}_2^l+\cdots+\tilde{Z}_n^l\|}{n}\geqq\tfrac{1}{4}\delta\right\}\leqq-(\phi(a)+1). \tag{5.28}$$

In Lemma 5.6 below we shall show that if $\phi(a)<\infty$, there exist a_l such that $a_l\to a$ as $l\to\infty$ and $\phi^l(a_l)\leqq\phi(a)$ for all l. Therefore, we can assume that the l_0 we just picked for (5.28) is also so large that $\|a_l-a\|<\frac{1}{4}\delta$ for all $l>l_0$. Hence, for a fixed $l>l_0$,

$$\begin{aligned}\nu_n(G)&\geqq\nu_n(S(a,\delta))\geqq\nu_n(S(a_l,\tfrac{1}{2}\delta))\\ &=P\left\{\frac{\tilde{X}_1+\tilde{X}_2+\cdots+\tilde{X}_n}{n}\in S(a_l,\tfrac{1}{2}\delta)\right\} \qquad (5.29)\\ &\geqq P\left\{\frac{\tilde{Y}_1^l+\tilde{Y}_2^l+\cdots+\tilde{Y}_n^l}{n}\in S(a_l,\tfrac{1}{4}\delta)\right\}-P\left\{\frac{\|\tilde{Z}_1^l+\cdots+\tilde{Z}_n^l\|}{n}\geqq\tfrac{1}{4}\delta\right\}.\end{aligned}$$

Let $0<\eta<\frac{1}{2}$. By (5.28) we can take n so large that

$$P\left\{\frac{\|\tilde{Z}_1^l+\tilde{Z}_2^l+\cdots+\tilde{Z}_n^l\|}{n}\geqq\tfrac{1}{4}\delta\right\}\leqq\exp\{-n[\phi(a)+1-\eta]\}, \tag{5.30}$$

and since the $\tilde{Y}^l$ variables are bounded, we have from the first part of the proof that, for n large enough,

$$P\left\{\frac{\tilde{Y}_1^l+\tilde{Y}_2^l+\cdots+\tilde{Y}_n^l}{n}\in S(a_l,\tfrac{1}{4}\delta)\right\}\geqq \exp\{-n[\phi^l(a_l)+\eta]\}. \tag{5.31}$$

But, for all l, $\phi^l(a_l)\leqq\phi(a)$ (Lemma 5.6) and therefore by (5.31), for n large enough,

$$P\left\{\frac{\tilde{Y}_1^l+\tilde{Y}_2^l+\cdots+\tilde{Y}_n^l}{n}\in S(a_l,\tfrac{1}{4}\delta)\right\}\geqq \exp\{-n[\phi(a)+\eta]\}. \tag{5.32}$$

Using (5.30) and (5.32) in (5.29) we have, for n large enough,

$$\nu_n(G)\geqq\exp\{-n[\phi(a)+\eta]\}-\exp\{-n[\phi(a)+1-\eta]\},$$

and therefore, since $0<\eta<\frac{1}{2}$,

$$\underline{\lim}_{n\to\infty}\frac{1}{n}\log\nu_n(G)\geqq-(\phi(a)+\eta). \tag{5.33}$$

But η is arbitrary and a is any point in G so that (5.33) implies (5.19). This completes Theorem 5.3 except for three lemmas.

LEMMA 5.4. *For every* $\varepsilon>0$,

$$\overline{\lim_{l\to\infty}}\;\overline{\lim_{n\to\infty}}\frac{1}{n}\log P\left\{\frac{\|\tilde{Z}_1^l+\tilde{Z}_2^l+\cdots+\tilde{Z}_n^l\|}{n}>\varepsilon\right\}=-\infty.$$

Proof: The random variables $\tilde{Z}_1^l,\tilde{Z}_2^l,\cdots,\tilde{Z}_n^l$ are independent, identically distributed, and using obvious inequalities we have, for any $\varepsilon>0$ and any $\sigma>0$,

$$\begin{aligned}P\left\{\frac{\|\tilde{Z}_1^l+\tilde{Z}_2^l+\cdots+\tilde{Z}_n^l\|}{n}>\varepsilon\right\}&\leqq P\left\{\frac{\|\tilde{Z}_1^l\|+\|\tilde{Z}_2^l\|+\cdots+\|\tilde{Z}_n^l\|}{n}>\varepsilon\right\}\\&\leqq e^{-\sigma\varepsilon}E\left\{\exp\left\{\frac{\sigma}{n}\sum_{j=1}^n\|\tilde{Z}_j^l\|\right\}\right\}\\&=e^{-\sigma\varepsilon}\left[E\left\{\exp\left\{\frac{\sigma}{n}\|\tilde{Z}_1^l\|\right\}\right\}\right]^n\\&=e^{-n\sigma\varepsilon}[E\{\exp\{\sigma\|\tilde{Z}_1^l\|\}\}]^n.\end{aligned}\tag{5.34}$$

Thus,

$$\overline{\lim_{n\to\infty}} \frac{1}{n} \log P\left\{\frac{\|\tilde{Z}_1^l + \tilde{Z}_2^l + \cdots + \tilde{Z}_n^l\|}{n} > \varepsilon\right\} \leqq -\sigma\varepsilon + \log E\{\exp\{\sigma\|\tilde{Z}_1^l\|\}\},$$

and from the definition of $\tilde{Z}_1^l$ we conclude that

$$\overline{\lim_{l\to\infty}}\ \overline{\lim_{n\to\infty}} \frac{1}{n} \log P\left\{\frac{\|\tilde{Z}_1^l + \cdots + \tilde{Z}_n^l\|}{n} > \varepsilon\right\} \leqq -\sigma\varepsilon .$$

Since the last inequality is true for all $\sigma > 0$ and since $\varepsilon > 0$, we have the lemma.

LEMMA 5.5. *Let C be a closed subset of X and C^ε the set of points in X all of which are a distance less than ε from C. Then,*

$$\underline{\lim_{\substack{\varepsilon\to 0\\ l\to\infty}}} \inf_{a\in \overline{C^\varepsilon}} \phi^l(a) \geqq \inf_{a\in C} \phi(a) . \tag{5.35}$$

Proof: Denote the left side of (5.35) by q. Then there exist sequences $\{l_n\}\to\infty$, $\{\varepsilon_n\}\to 0$, $\{a_n\}\in \overline{C^{\varepsilon_n}}$ such that $\phi^{l_n}(a_n)\to q$. By Theorem 5.2 and the fact that the infima are attained, we have a sequence $\{\mu^{l_n}\}$ such that $I_{\lambda^{l_n}}(\mu^{l_n})\to q$ and $\int x\mu^{l_n}(dx) = a_n$. The family of measures $\{\lambda^{l_n}\}$ is tight and since $I_{\lambda^{l_n}}(\mu^{l_n})$ is uniformly bounded, we know by Lemma 5.1 that $\{\mu^{l_n}\}$ is tight. Thus there exists a subsequence $\{\mu^{l_{n_j}}\}$ which converges to some $\mu\in\mathcal{M}$ and $\int x\mu(dx)$ exists. Moreover, the uniform integrability part of Lemma 5.1 implies that $\int x\mu^{l_{n_j}}(dx)\to\int x\mu(dx)$. Thus $\{a_n\}$ is compact. Let a be any limit point of $\{a_n\}$. By Theorem 5.2, $\phi(a)\leqq I_\lambda(\mu)$ and since $I_\lambda(\mu)$ is lower semicontinuous, we have

$$\phi(a)\leqq I_\lambda(\mu)\leqq \underline{\lim_{j\to\infty}}\, I_{\lambda^{l_{n_j}}}(\mu^{l_{n_j}}) = q ,$$

which completes the lemma.

LEMMA 5.6. *Let $\phi(a)<\infty$. There exist a_l such that $a_l\to a$ as $l\to\infty$ and $\phi^l(a_l)\leqq\phi(a)$ for all l.*

Proof: Let $p=\phi(a)<\infty$. From Theorem 5.2,

$$p=\inf_{\mu:\int x\mu(dx)=a} I(\mu)$$

and this infimum is attained. Thus, there exists μ such that $I(\mu)=I_\lambda(\mu)=p$ and $\int x\mu(dx)=a$. Let μ^l be the distribution of $\tilde{y}^l$ with respect to μ measure and λ^l be the distribution of $\tilde{y}^l$ with respect to λ measure. Since $I_{\lambda^l}(\mu^l)\leqq I_\lambda(\mu)$, if we define $a_l=\int x\mu^l(dx)$, we have $a_l\to a$ as $l\to\infty$ and $\phi^l(a_l)\leqq I_{\lambda^l}(\mu^l)\leqq I_\lambda(\mu)=\phi(a)$ for all l.

6. Some Applications of Section 5

Let $C(0,1)$ be the space of continuous functions $z(\tau)$, $0\leqq\tau\leqq 1$, with $z(0)=0$. Let $F[z]$ be a real-valued function defined on $C(0,1)$. For a large class of F, Schilder [6] showed[1]

$$\lim_{n\to\infty}\frac{1}{n}\log E\{\exp\{-nF[z/\sqrt{n}]\}\}=-\inf_{z\in C(0,1)}\left[F[z]+\frac{1}{2}\int_0^1[z'(\tau)]^2\,d\tau\right], \tag{6.1}$$

where $E\{\ \}$ denotes expectation (Wiener integral) with respect to Brownian motion measure on $C(0,1)$, i.e., Gaussian measure on $C(0,1)$ with mean function 0 and covariance function $\rho(s,\tau)=\min(s,\tau)$. Later Pincus [5] (in an application of the asymptotics to Hammerstein nonlinear equations) showed that if in (6.1) instead of Brownian motion one has a suitable Gaussian process $\{y_\tau, 0\leqq\tau\leqq 1\}$ with mean function 0, covariance function $\rho(s,\tau)$, having continuous trajectories, then

$$\lim_{n\to\infty}\frac{1}{n}\log E\{\exp\{-nF[y/\sqrt{n}]\}\}=-\inf_{y\in C(0,1)}[F[y]+\tfrac{1}{2}\langle y,\rho^{-1}y\rangle]. \tag{6.2}$$

Obviously,

$$\langle y,\rho^{-1}y\rangle=\int_0^1[y'(\tau)]^2\,d\tau$$

when $\rho(s,\tau)=\min(s,\tau)$.

[1] Actually he obtained the whole asymptotic expansion with which we are concerned here only with the leading term.

For any Gaussian process $\{y_\tau, 0 \leqq \tau \leqq 1\}$ with continuous trajectories, Marcus and Shepp [3] proved

$$\lim_{\alpha\to\infty} \frac{1}{\alpha^2} \log P\left\{\sup_{0\leqq s\leqq 1} |y(s)| \geqq \alpha\right\} = -\frac{1}{2a}, \tag{6.3}$$

where $a = \sup_{0\leqq s\leqq 1} \rho(s, s)$.

We shall show how the results (6.1)–(6.3) can be obtained from the considerations of the preceding section. Our starting point is a lemma due to Skorohod [8].

LEMMA 6.1. *Let λ be a Gaussian measure on a separable Banach space X. Then*

$$\int_X e^{\sigma\|x\|} \lambda(dx) < \infty$$

for some $\sigma > 0$, and therefore, a fortiori for all $\sigma > 0$.

Proof: If ξ is a symmetric Gaussian random variable with values in a Banach space X and if $\xi_1, \xi_2, \cdots, \xi_n$ are independent copies of ξ, then the distribution of $(\xi_1 + \cdots + \xi_n)/\sqrt{n}$ is the same as that of ξ. Hence

$$\begin{aligned} E\{e^{\sigma\|\xi\|}\} &= E\{\exp\{(\sigma/\sqrt{n})\, \|\xi_1 + \cdots + \xi_n\|\}\} \\ &\leqq E\{\exp\{(\sigma/\sqrt{n})[\|\xi_1\| + \cdots + \|\xi_n\|]\}\} \\ &= (E\{\exp\{(\sigma/\sqrt{n})\, \|\xi\|\}\})^n . \end{aligned}$$

Therefore if $E\{e^{\sigma\|\xi\|}\} < \infty$ for some $\sigma > 0$, then the same is true for all $\sigma < \infty$.

For a symmetric Gaussian measure λ on a separable Banach space X we define λ_n as the distribution of $x/\sqrt{n}$. Estimating $\lambda_n(A)$ for various sets A is the same as estimating the tail of λ. We can also interpret λ_n as the distribution of $(\xi_1 + \cdots + \xi_n)/n$, where $\xi_1, \xi_2, \cdots, \xi_n$ are independent and have the common distribution λ. The results of Section 5 are now applicable. In particular we can deduce from Theorem 5.3 the following

THEOREM 6.2. *Let λ be a symmetric Gaussian distribution on a separable Banach space X with dual X^*. Let $M(\theta)$ be the moment generating function defined for $\theta \in X^*$ by*

$$M(\theta) = \int \exp\{\langle\theta, x\rangle\}\lambda(dx) = \exp\{\tfrac{1}{2}Q(\theta)\},$$

where

$$Q(\theta)=\int \langle \theta, x\rangle^2 \lambda(dx).$$

Let $f(x)$ be defined by

$$f(x)=\sup_{\theta}\,[\langle \theta, x\rangle-\tfrac{1}{2}Q(\theta)].$$

Then, for any closed set $C\subset X$,

$$\limsup_{n\to\infty}\frac{1}{n}\log \lambda\left[x : \frac{x}{\sqrt{n}}\in C\right]\leqq -\inf_{x\in C} f(x) \tag{6.4}$$

and, for any open set $G\subset X$,

$$\liminf_{n\to\infty}\frac{1}{n}\log \lambda\left[x : \frac{x}{\sqrt{n}}\in G\right]\geqq -\inf_{x\in G} f(x). \tag{6.5}$$

Combining this with Section 3 of [9] we have

COROLLARY 6.3. *Let $F(x)$ be a bounded continuous function on X. Then*

$$\lim_{n\to\infty}\frac{1}{n}\log\int \exp\left\{nF\left(\frac{x}{\sqrt{n}}\right)\right\}\lambda(dx)=\sup_{x}\,[F(x)-f(x)].$$

This is clearly the result of Schilder [6] and Pincus [5]. In order to connect Theorem 6.2 with the result of Marcus and Shepp [3] we have to perform the following calculation.

LEMMA 6.4. *We have the relation*

$$\inf_{\|x\|\geqq 1} f(x)=\frac{1}{2a},$$

where

$$a=\sup_{\|\theta\|\leqq 1} Q(\theta).$$

Proof: Let us look at $\lambda[\|x\|\geqq\sqrt{n}]$. Clearly, by (6.4),

$$\limsup_{n\to\infty}\frac{1}{n}\log \lambda[\|x\|\geqq\sqrt{n}]\leqq -\inf_{\|x\|\geqq 1} f(x). \tag{6.6}$$

On the other hand, for any θ with $\|\theta\| \leqq 1$,

$$\lambda[\|x\| \geqq \sqrt{n}] \geqq \lambda[|\langle x, \theta\rangle| \geqq \sqrt{n}],$$

and, by the tail behavior of the Gaussian random variables on the real line,

$$\limsup_{n\to\infty} \frac{1}{n} \log \lambda[|\langle \theta, x\rangle| \geqq \sqrt{n}] \geqq -\frac{1}{2Q(\theta)}. \tag{6.7}$$

Therefore, combining (6.6) and (6.7), we see that

$$\inf_{\|x\|\geqq 1} f(x) \leqq \frac{1}{2Q(\theta)},$$

for any θ with $\|\theta\| \leqq 1$. Since θ with $\|\theta\| \leqq 1$ is arbitrary,

$$\inf_{\|x\|\geqq 1} f(x) \leqq \inf_{\|\theta\|\leqq 1} \frac{1}{2Q(\theta)} = \frac{1}{2a},$$

where

$$a = \sup_{\|\theta\|\leqq 1} Q(\theta).$$

Now with a defined as above, $Q(\theta) \leqq a\|\theta\|^2$ for all θ. Therefore,

$$\begin{aligned} f(x) &= \sup_\theta [\langle \theta, x\rangle - \tfrac{1}{2}Q(\theta)] \\ &\geqq \sup_\theta [\langle \theta, x\rangle - \tfrac{1}{2}a\|\theta\|^2] \\ &= \sup_{k\geqq 0} \sup_{\|\theta\|=k} [\langle \theta, x\rangle - \tfrac{1}{2}ak^2] \\ &= \sup_{k\geqq 0} [k\|x\| - \tfrac{1}{2}ak^2] \\ &= \frac{1}{2a}\|x\|^2. \end{aligned}$$

Hence

$$\inf_{\|x\|\geqq 1} f(x) \geqq \frac{1}{2a}.$$

This completes the proof of the lemma.

7. Upper Bounds—Continuous Case

We return to the setting of Sections 3 and 4 except that now we want to consider a continuous parameter Markov process. Let X be a complete separable metric space, and let $p(t, x, dy)$ be the transition probability of a Markov process with X as state space. We shall assume $p(t, x, dy)$ is a Feller transition probability. Let T_t be the semigroup associated with $p(t, x, dy)$, i.e., for $f \in C(X)$, $(T_tf)(x) = \int_X f(y)p(t, x, dy)$. Let L be the infinitesimal generator of this semigroup and $\mathcal{D} = \mathcal{D}(L)$ its domain. Again let $\mathcal{M}$ be the space of probability measures on X.

We define the I-function in the continuous case for any $\mu \in \mathcal{M}$ by

$$I(\mu) = -\inf_{\substack{u>0 \\ u\in\mathcal{D}}} \int_X \left(\frac{Lu}{u}\right)(x)\mu(dx).$$

Let Ω be the space of right continuous functions $x(t)$, $0 \leqq t < \infty$, which have values in X. We assume that for each $x \in X$, $p(t, x, dy)$ induces a measure P_x on Ω. For each $t > 0$, $\omega \in \Omega$, and Borel set $A \subset X$, let

$$L_t(\omega, A) = \frac{1}{t}\int_0^t \chi_A(x(s))\, ds.$$

In other words, $L_t(\omega, A)$ is the proportion of time up to t that a particular sample path $\omega = x(\cdot)$ spends in the set A. For each $t > 0$ and each ω, $L_t(\omega, \cdot) \in \mathcal{M}$. For each $x \in X$ and each $t > 0$, let $Q_{x,t}$ be the measure induced on $\mathcal{M}$ by L_t from P_x, i.e., $Q_{x,t} = P_x L_t^{-1}$ or, in other words, if $B \subset \mathcal{M}$,

$$Q_{x,t}(B) = P_x\{\omega \in \Omega : L_t(\omega, \cdot) \in B\}.$$

In this section we shall prove, under hypotheses stated below, that for any closed set $C \subset \mathcal{M}$ and any compact set $W \subset X$

$$\varlimsup_{t\to\infty} \frac{1}{t} \log \sup_{x\in W} Q_{x,t}(C) \leqq -\inf_{\mu\in C} I(\mu), \tag{7.1}$$

i.e., the continuous analogue of Theorem 4.4. If we are content with having (7.1) for just compact sets K instead of closed sets C, then (7.1) can be proved without further hypotheses using exactly the method of [2]. The reason that we want (7.1) for closed sets is that then, in conjunction with the corresponding lower bound for open sets, one obtains a useful asymptotic theorem for expectations on the Markov process.

To prove (7.1), we now make the hypothesis on the semigroup T_t that there exists a function $V(x)$ such that $\{x \in X : V(x) \geqq l\}$ is a compact set for each $l > -\infty$ and that there exists a sequence $\{u_n\} \in \mathscr{D}$ such that

(i) $u_n \geqq 1$ for all n and all $x \in X$,
(ii) for each compact set $W \subset X$, $\sup_{x \in W} \sup_n u_n(x) < \infty$,
(iii) for each $x \in X$,

$$\lim_{n\to\infty} \left(\frac{Lu_n}{u_n}\right)(x) = V(x),$$

(iv) for some $A < \infty$,

$$\sup_{n,x} \left(\frac{Lu_n}{u_n}\right)(x) \leqq A.$$

We prove

LEMMA 7.1. *Under the hypotheses above, $\{\mu \in \mathscr{M} : I(\mu) \leqq l\}$ is compact for each l, and for any $N < \infty$ there exists a compact set $K \subset \mathscr{M}$ such that, for any compact set $W \subset X$,*

$$\overline{\lim_{t\to\infty}} \frac{1}{t} \log \sup_{x \in W} Q_{x,t}(\tilde{K}) \leqq -N. \tag{7.2}$$

Proof: Let $\mu \in \mathscr{M}$ be such that $I(\mu) \leqq l$. Then $\int_X (Lu/u)(x)\mu(dx) \geqq -l$ for all $u \in \mathscr{D}$ with $u > 0$. In particular, using the sequence $\{u_n\}$ of the hypothesis, we have $\int_X (Lu_n/u_n)(x)\mu(dx) \geqq -l$ for all n, and, by Fatou's lemma, this implies that $\int_X V(x)\mu(dx) \geqq -l$. Since $\lim_{n\to\infty} (Lu_n/u_n)(x) = V(x)$ for each $x \in X$ and $(Lu_n/u_n)(x) \leqq A$ uniformly in n and x, we have $V(x) \leqq A$ for all x. Thus, for any $h < 0$,

$$\begin{aligned}\int_X V(x)\mu(dx) &= \int_{V<h} V(x)\mu(dx) + \int_{V \geqq h} V(x)\mu(dx)\\ &\leqq h\mu\{x : V(x) < h\} + A.\end{aligned}$$

Hence we have $h\mu\{x : V(x) < h\} \geqq -(l+A)$ or $\mu\{x : V(x) < h\} \leqq (l+A)/|h|$. Let $K_h = \{x \in X : V(x) \geqq h\}$ which by hypothesis is compact. We have just seen that for any l

$$\{\mu \in \mathscr{M} : I(\mu) \leqq l\} \subset \{\mu \in \mathscr{M} : \mu(\tilde{K}_h) \leqq (l+A)/|h|\}. \tag{7.3}$$

For any $\varepsilon > 0$, we can choose $|h|$ so large that $\mu(\tilde{K}_h) \leqq \varepsilon$ which means that the family of measures on the right of (7.3) is tight and therefore a compact subset of $\mathcal{M}$. But, since $I(\mu)$ is lower semicontinuous, the family of measures on the left of (7.3) is closed and being a subset of a compact set is itself compact.

Let, for $x \in X$ and $t > 0$,

$$\psi_n(x, t) = E^{P_x}\left\{u_n(x(t)) \exp\left\{-\int_0^t (Lu_n/u_n)(x(s))\, ds\right\}\right\}. \tag{7.4}$$

From the Feynmann–Kac formula, $\psi_n(x, t)$ satisfies

$$\begin{aligned} \frac{\partial \psi_n}{\partial t} &= L\psi_n - \left(\frac{Lu_n}{u_n}\right)(x)\psi_n\,, \\ \psi_n(x, 0) &= u_n(x)\,, \end{aligned} \tag{7.5}$$

but, since the solution of (7.5) is clearly $\psi_n(x, t) \equiv u_n(x)$, we have

$$u_n(x) = E^{P_x}\left\{u_n(x(t)) \exp\left\{-\int_0^t (Lu_n/u_n)(x(s))\, ds\right\}\right\}. \tag{7.6}$$

By hypothesis, $u_n \geqq 1$ so that, for all $x \in X$,

$$E^{P_x}\left\{\exp\left\{-\int_0^t (Lu_n/u_n)(x(s))\, ds\right\}\right\} \leqq u_n(x)\,. \tag{7.7}$$

From the hypothesis and Fatou's lemma we see that

$$\varliminf_{n\to\infty} \left(-\int_0^t \left(\frac{Lu_n}{u_n}\right)(x(s))\, ds\right) \geqq -\int_0^t V(x(s))\, ds\,,$$

or

$$\varliminf_{n\to\infty} \exp\left\{-\int_0^t (Lu_n/u_n)(x(s))\, ds\right\} \geqq \exp\left\{-\int_0^t V(x(s))\, ds\right\},$$

which implies, using Fatou's lemma again, that

$$\varliminf_{n\to\infty} E^{P_x}\left\{\exp\left\{-\int_0^t (Lu_n/u_n)(x(s))\, ds\right\}\right\} \geqq E^{P_x}\left\{\exp\left\{-\int_0^t V(x(s))\, ds\right\}\right\}. \tag{7.8}$$

By the hypothesis, for each compact set $W \subset X$, $\{u_n(x)\}$ is uniformly bounded on W so that from (7.7) and (7.8) we get

$$E^{P_x}\left\{\exp\left\{-\int_0^t V(x(s))\,ds\right\}\right\} \leqq \varliminf_{n\to\infty} u_n(x) \leqq \sup_{x\in W}\sup_n u_n(x) \leqq c, \tag{7.9}$$

for each $x \in W$, where c is a constant independent of t. Recalling the definition of $Q_{x,t}$ measure and noting that

$$\int_0^t V(x(s))\,ds = t\int_X V(y)L_t(\omega, dy),$$

we see that (7.9) is the same as

$$E^{Q_{x,t}}\left\{\exp\left\{-t\int_X V(x)\mu(dx)\right\}\right\} \leqq c \quad \text{for} \quad x \in W. \tag{7.10}$$

For $\sigma > 0$ and any positive integer n, let $K_n^\sigma = \{x \in X : V(x) \geqq -n\sigma\}$. By hypothesis this is a compact set. For any $\varepsilon > 0$, let $M_{n,\sigma}^\varepsilon = \{\mu \in \mathcal{M} : \mu(\tilde{K}_n^\sigma) > \varepsilon\}$. We noted earlier that $V(x) \leqq A$ for all $x \in X$, and therefore, if $\mu \in M_{n,\sigma}^\varepsilon$, we have

$$\begin{aligned} \int_X V(x)\mu(dx) &= \int_{K_n^\sigma} V(x)\mu(dx) + \int_{\tilde{K}_n^\sigma} V(x)\mu(dx) \\ &\leqq A(1-\mu(\tilde{K}_n^\sigma)) - n\sigma\mu(\tilde{K}_n^\sigma) \leqq A - \varepsilon n\sigma. \end{aligned} \tag{7.11}$$

From (7.10) and (7.11) we see that, for $x \in W$,

$$Q_{x,t}(M_{n,\sigma}^\varepsilon) \leqq ce^{tA}e^{-t\varepsilon n\sigma},$$

and, in particular, $Q_{x,t}(M_{n^2,\sigma}^{1/n}) \leqq ce^{tA}e^{-n\sigma t}$, so that

$$Q_{x,t}\left(\bigcup_{n=1}^\infty M_{n^2,\sigma}^{1/n}\right) \leqq ce^{tA}\frac{e^{-\sigma t}}{1-e^{-\sigma t}}. \tag{7.12}$$

Let $J_\sigma = \widetilde{\bigcup_{n=1}^\infty M_{n^2,\sigma}^{1/n}}$; then (7.12) yields

$$\varlimsup_{t\to\infty}\frac{1}{t}\log\sup_{x\in W} Q_{x,t}(\tilde{J}_\sigma) \leqq A - \sigma. \tag{7.13}$$

But

$$J_\sigma = \bigcap_{n=1}^\infty \widetilde{M_{n^2,\sigma}^{1/n}} = \{\mu \in \mathcal{M} : \mu(\tilde{K}_{n^2}^\sigma) \leqq 1/n \text{ for all } n\},$$

and this family of measures is obviously tight and hence compact. Since we can choose σ as large as we please, the proof of the lemma is complete.

With this lemma we can now prove the upper bound in the continuous case completely, using a techsnique which is analogous to that for the discrete case.

THEOREM 7.2. *Let C be a closed set in $\mathcal{M}$, and assume the hypothesis stated above. Then, for any compact set $W \subset X$,*

$$\overline{\lim_{t\to\infty}} \frac{1}{t} \log \sup_{x\in W} Q_{x,t}(C) \leqq -\inf_{\mu\in C} I(\mu) .$$

Proof: Let $N<\infty$ and choose a compact set $K\subset \mathcal{M}$, as in Lemma 7.1, so that

$$\overline{\lim_{t\to\infty}} \frac{1}{t} \log \sup_{x\in W} Q_{x,t}(\tilde{K}) \leqq -N . \tag{7.14}$$

Since

$$\begin{aligned} Q_{x,t}(C) &= Q_{x,t}(C\cap K)+Q_{x,t}(C\cap \tilde{K}) \\ &\leqq Q_{x,t}(C\cap K)+Q_{x,t}(\tilde{K}) , \end{aligned}$$

we have

(7.15)

$$\overline{\lim_{t\to\infty}} \frac{1}{t} \log \sup_{x\in W} Q_{x,t}(C) \leqq \max \left[\overline{\lim_{t\to\infty}} \frac{1}{t} \log \sup_{x\in W} Q_{x,t}(C\cap K),\ \overline{\lim_{t\to\infty}} \frac{1}{t} \log \sup_{x\in W} Q_{x,t}(\tilde{K}) \right] .$$

But $C\cap K$ is compact and therefore by our earlier remark (just after (7.1)) we know that

$$\overline{\lim_{t\to\infty}} \frac{1}{t} \log \sup_{x\in W} Q_{x,t}(C\cap K) \leqq -\inf_{\mu\in C\cap K} I(\mu) . \tag{7.16}$$

Thus, from (7.14) and (7.15),

$$\begin{aligned} \overline{\lim_{t\to\infty}} \frac{1}{t} \log \sup_{x\in W} Q_{x,t}(C) &\leqq \max \left[-\inf_{\mu\in C\cap K} U(\mu), -N \right] \\ &\leqq \max \left[-\inf_{\mu\in C} I(\mu), -N \right] . \end{aligned}$$

The last inequality holding for all N, we obtain the theorem by letting $N\to\infty$.

8. Lower Bounds—Continuous Case

In order to obtain lower bounds in the continuous case, the state space X being a complete separable metric space, we consider several hypotheses on the transition probability $p(t, x, dy)$.

HYPOTHESIS H_1. We assume there exists a reference measure β on X such that, for all $x \in X$, $p(t, x, dy) = p(t, x, y)\beta(dy)$.

Let B_0 be the center of the semigroup T_t associated with $p(t, x, dy)$, i.e., $B_0 = \{f \in C(X) : \lim_{t\to 0} \|T_t f - f\| = 0\}$, where the norm is the usual norm for $C(X)$ and, of course,

$$(T_t f)(x) = \int_X f(y) p(t, x, dy) .$$

Further, let

$$B_{00} = \left\{ f \in C(X) : \limsup_{t\to 0} \sup_x \int_X |f(y) - f(x)| \, p(t, x, dy) = 0 \right\} ,$$

so that $B_{00} \subset B_0 \subset C(X)$. We note that if both f and f^2 are in B_0, then $f \in B_{00}$. Indeed, if B_0 contains any algebra of functions, then the same algebra is contained in B_{00}. Let $\alpha \in \mathcal{M}$ be such that $I(\alpha) = l < \infty$ and let N_α be a neighborhood of α in $\mathcal{M}$ of the form

$$N_\alpha = \left\{ \mu \in \mathcal{M} : \left| \int_X f_j(x)[\mu(dx) - \alpha(dx)] \right| < \varepsilon, \quad 1 \leqq j \leqq k, \right. \tag{8.1}$$

$$\left. \text{where} \quad f_j \in B_{00}, \quad i \leqq j \leqq k \right\} .$$

We make the second hypothesis on $p(t, x, dy)$:

HYPOTHESIS H_2. B_{00} is such that every neighborhood of $\alpha \in \mathcal{M}$ contains a neighborhood of the form (8.1).

We assume (cf. (3) just after the statement of Theorem 4 in the introduction of [1]):

HYPOTHESIS H_3. For any $\mu \in \mathcal{M}$ and any bounded measurable function $f(x)$ there is a sequence $\{f_n\} \in B_0$ such that $\|f_n\| \leqq \|f\|$ and $f_n \to f$ almost everywhere with respect to μ.

Let $\mu \in \mathcal{M}$ be such that $I(\mu) < \infty$ and let C be a closed set in X such that $\mu(C) = 1$. Let $G \supset C$ be an open set and define for $A \subset G$

$$p_G(t, x, A) = P_x\{x(s) \in G, 0 \leqq s \leqq t \text{ and } x(t) \in A\}.$$

For $\sigma > 0$, let

$$R_G^\sigma(x, A) = \int_0^\infty e^{-\sigma t} p_G(t, x, A)\, dt.$$

From Hypothesis H_1 we see that $R_G^\sigma(x, dy) \ll \beta(dy)$. We now make

Hypothesis H_4. If $x \in G$ and $E \subset G$ with $\beta(E) > 0$, then $R_G^\sigma(x, E) > 0$.

It is clear that H_4 means that, if the process starts from $x \in G$ and E is a subset of G having positive β measure, then with positive probability the process will spend positive time in E before exiting from G.

We shall also consider

Hypothesis H_5. For each $E \subset G$ and $\sigma > 0$, $R_G^\sigma(x, E)$ is a continuous function of x for $x \in G$.

We now state

Theorem 8.1. *Let $\mu \in \mathcal{M}$ be such that $I(\mu) < \infty$ and let N be a neighborhood of μ in $\mathcal{M}$. Let* $\operatorname{supp} \mu \subset G$ *for some open $G \subset X$. $\mathcal{M}(G)$ is the set of probability measures in $\mathcal{M}$ whose support is contained in G. Under Hypotheses* H_1–H_4,

$$\underline{\lim}_{t \to \infty} \frac{1}{t} \log Q_{x,t}(N \cap \mathcal{M}(G)) \geqq -I(\mu). \tag{8.2}$$

Moreover, if in addition H_5 holds, then (8.2) holds uniformly for x in compact subsets of G.

Proof: Because of Hypothesis H_2 we can assume N is of the form

$$N_\varepsilon = \left\{\alpha \in \mathcal{M} : \left|\int_X f_j(x)[\mu(dx) - \alpha(dx)]\right| < \varepsilon,\ 1 \leqq j \leqq k, \quad \text{where} \quad f_j \in B_{00},\ 1 \leqq j \leqq k\right\}.$$

Consider, then,

$$Q_{x,t}(N_\varepsilon \cap \mathcal{M}(G)) = P_x\{L_t(\omega, \cdot) \in N_\varepsilon\,;\, x(s) \in G, 0 \leqq s \leqq t\}. \tag{8.3}$$

If we choose $\Delta>0$ so that $2\Delta \max_{1\leq j\leq k}|f_j|<\frac{1}{2}\varepsilon$, then

$$P_x\{L_t(\omega,\cdot)\in N_\varepsilon\,;\, x(s)\in G, 0\leqq s\leqq t\} \geqq P_x\{L_{t(1+\Delta)}(\omega,\cdot)\in N_{\varepsilon/2}\,;\, x(s)\in G, 0\leqq s\leqq t\}. \tag{8.4}$$

For each $\sigma>0$, let $\tau_1, \tau_2, \cdots$ be a sequence of independent random variables which are independent of the process and have as common distribution $P\{\tau\in dt\}=\sigma e^{-\sigma t}\,dt$. Let $s_0=0$ and $s_n=\tau_1+\tau_2+\cdots+\tau_n$. For any Borel set $A\subset X$, we have our usual

$$L_t(\omega, A)=\frac{1}{t}\int_0^t \chi_A(x(s))\,ds,$$

but now we consider also

$$L_{n,\sigma}(\omega, A)=\frac{1}{n}\sum_{j=0}^{n-1}\chi_A(x(s_j)).$$

It is important to note that if we let $x_0=x(s_0)$, $x_1=x(s_1),\cdots, x_j=x(s_j),\cdots$, then $x_0, x_1, x_2, \cdots$ form a discrete parameter Markov process with transition probability

$$\pi_\sigma(x, dy)=\sigma\int_0^\infty e^{-\sigma t}p(t, x, dy)\,dt.$$

We see that, for all n and $\sigma>0$,

$$\begin{aligned} &P_x\{L_{t(1+\Delta)}(\omega,\cdot)\in N_{\varepsilon/2}\,;\, x(s)\in G, 0\leqq s\leqq t\}\\ &\quad\geqq P_x\{L_{n,\sigma}(\omega,\cdot)\in N_{\varepsilon/4}\,;\, x(s)\in G, 0\leqq s\leqq t\}\\ &\quad\quad -P_x\left\{\sup_{1\leqq j\leqq k}\left|\int_X f_j(x)L_{n,\sigma}(\omega, dx)-\int_X f_j(x)L_{t(1+\Delta)}(\omega, dx)\right|>\tfrac{1}{4}\varepsilon\right\}\\ &\quad\geqq P_x\{L_{n,\sigma}(\omega,\cdot)\in N_{\varepsilon/4}\,;\, x(s)\in G, 0\leqq s\leqq \tau_1+\tau_2+\cdots+\tau_n\}\\ &\quad\quad -P_x\left\{\sup_{1\leqq j\leqq k}\left|\int_X f_j(x)L_{n,\sigma}(\omega, dx)-\int_X f_j(x)L_{t(1+\Delta)}(\omega, dx)\right|>\tfrac{1}{4}\varepsilon\right\}\\ &\quad\quad -P\{\tau_1+\tau_2+\cdots+\tau_n\leqq t\}. \end{aligned} \tag{8.5}$$

Thus from (8.3), (8.4), and (8.5) we have, for any n and $\sigma>0$,

$$Q_{x,t}(N_\varepsilon \cap \mathcal{M}(G))$$

$$\geqq P_x\{L_{n,\sigma}(\omega,\cdot)\in N_{\varepsilon/4};\ x(s)\in G,\ 0\leqq s\leqq \tau_1+\tau_2+\cdots+\tau_n\}$$

(8.6)
$$-P_x\left\{\sup_{1\leqq j\leqq k}\left|\int_X f_j(x)L_{n,\sigma}(\omega,dx)-\int_X f_j(x)L_{t(1+\Delta)}(\omega,dx)\right|>\tfrac{1}{4}\varepsilon\right\}$$

$$-P\{\tau_1+\tau_2+\cdots+\tau_n\leqq t\}.$$

In particular, choose $n=\sigma t(1+\Delta)$. In Lemma 8.3 below we show that, for $f\in B_{00}$, $n=\sigma t$, and any $\varepsilon>0$,

$$\lim_{\sigma\to\infty}\overline{\lim_{t\to\infty}}\frac{1}{t}\log\sup_x P_x\left\{\left|\int_X f(y)L_t(\omega,dy)-\int_X f(y)L_{n,\sigma}(\omega,dy)\right|\geqq\varepsilon\right\}=-\infty.$$

This implies that, for $n=\sigma t(1+\Delta)$,

(8.7)
$$P_x\left\{\sup_{1\leqq j\leqq k}\left|\int_X f_j(x)L_{n,\sigma}(\omega,dx)-\int_X f_j(x)L_{t(1+\Delta)}(\omega,dx)\right|>\tfrac{1}{4}\varepsilon\right\}$$
$$\leqq\exp\{-t[h_{\varepsilon/4}(\sigma)+o(1)]\},$$

where $h_{\varepsilon/4}(\sigma)\to\infty$ as $\sigma\to\infty$ and $o(1)$ refers to $t\to\infty$.

With

$$R_G^\sigma(x,dy)=\int_0^\infty e^{-\sigma t}p_G(t,x,dy)\,dt,$$

we let $q_\sigma(x,dy)=\sigma R_G^\sigma(x,dy)$ and note that $q_\sigma(x,dy)$ is a substochastic transition probability on X. Let I_G^σ be the I-function coming from $q_\sigma(x,dy)$. If we let $\bar{Q}_{n,x}$ be the Q-measure coming from $q_\sigma(x,dy)$, then the important observation is that

$$P_x\{L_{n,\sigma}(\omega,\cdot)\in N_{\varepsilon/4};\ x(s)\in G,\ 0\leqq s\leqq\tau_1+\tau_2+\cdots+\tau_n\}-\bar{Q}_{n,x}(N_{\varepsilon/4}).$$

Since, by H_1 and H_4, $q_\sigma(x,dy)$ satisfies the hypotheses imposed in Theorem 3.3, that theorem implies that

(8.8)
$$\underline{\lim}_{n\to\infty}\frac{1}{n}\log P_x\{L_{n,\sigma}(\omega,\cdot)\in N_{\varepsilon/4};\ x(s)\in G,\ 0\leqq s\leqq\tau_1+\cdots+\tau_n\}\geqq -I_G^\sigma(\mu),$$

and indeed that (8.8) is uniform in x if H_5 obtains. Now (8.8) means that

$$P_x\{L_{n,\sigma}(\omega,\cdot)\in N_{\varepsilon/4}\,;\, x(s)\in G, 0\leqq s\leqq \tau_1+\cdots+\tau_n\}\geqq \exp\{-n[I_G^\sigma(\mu)+o(1)]\},$$

where here the $o(1)$ refers to $n\to\infty$. In Lemma 8.2 below we prove that $\sigma I_G^\sigma(\mu)\leqq I(\mu)$. In Lemma 8.4 we prove that, for $\Delta>0$ and $n=\sigma t(1+\Delta)$,

$$\lim_{\sigma\to\infty}\overline{\lim_{t\to\infty}}\frac{1}{t}\log P\{\tau_1+\cdots+\tau_n\leqq t\}=-\infty$$

which means, for $n=\sigma t(1+\Delta)$, that

$$P\{\tau_1+\cdots+\tau_n\leqq t\}\leqq \exp\{-t[r(\sigma,\Delta)+o(1)]\},$$

where $r(\sigma,\Delta)\to\infty$ for each $\Delta>0$ as $\sigma\to\infty$ and $o(1)$ refers to $t\to\infty$.

Putting all these estimates together into (8.6), we obtain, for $n=\sigma t(1+\Delta)$,

$$\begin{aligned}Q_{x,t}(N_\varepsilon\cap\mathcal{M}(G))\geqq{}&\exp\{-\sigma t(1+\Delta)[I_G^\sigma+o(1)]\}\\&-\exp\{-t[h_{\varepsilon/4}(\sigma)+o(1)]\}-\exp\{-t[r(\sigma,\Delta)+o(1)]\}.\end{aligned}$$

For any $\eta>0$, pick Δ_0 so small (still satisfying the condition imposed on Δ so that (8.4) holds) that $\Delta_0 I(\mu)<\eta$. Now choose σ_0 so large that $h_{\varepsilon/4}(\sigma_0)>2(I(\mu)+\eta+1)$ and $r(\sigma_0,\Delta_0)>2(I(\mu)+\eta+1)$. Then since $\sigma_0 I_G^{\sigma_0}(\mu)\leqq I(\mu)$, for large t,

$$\begin{aligned}Q_{x,t}(N_\varepsilon\cap\mathcal{M}(G))&\geqq\exp\{-t(1+\Delta_0)[I(\mu)+o(1)]\}\\&\quad-\exp\{-t[2I(\mu)+2\eta+2+o(1)]\}\\&\quad-\exp\{-t[2I(\mu)+2\eta+2+o(1)]\}\\&\geqq\exp\{-t[(I(\mu)+\eta+o(1)]\}\\&\quad-2\exp\{-t[2I[\mu)+2\eta+2+o(1)]\},\end{aligned}\tag{8.9}$$

where now all $o(1)$ refer to $t\to\infty$. Thus,

$$\underline{\lim}_{t\to\infty}\frac{1}{t}\log Q_{x,t}(N_\varepsilon\cap\mathcal{M}(G))\geqq -I(\mu)-\eta,$$

and since $\eta>0$ is arbitrary we have the inequality (8.2). Also, since under H_5

the estimates for all the terms on the right of (8.9) are uniform for x in a compact subset of G, we have (8.2) uniformly for x in such sets.

LEMMA 8.2. *Let $\mu \in \mathcal{M}$ be such that $I(\mu)<\infty$ and let C be a closed set in X such that $\mu(C)=1$. Let $G \supset C$ be an open set and define*

$$p_G(t, x, A) = P_x\{x(s) \in G, 0 \leqq s \leqq t, x(t) \in A\}.$$

Let

$$R_G^\sigma(x, A) = \int_0^\infty e^{-\sigma t} p_G(t, x, A)\, dt$$

and $q_\sigma(x, dy) = \sigma R_G^\sigma(x, dy)$. The I-function, I_G^σ, corresponding to $q_\sigma(x, dy)$ satisfies

$$\sigma I_G^\sigma(\mu) \leqq I(\mu).$$

Proof: The transition probability $p_G(t, x, dy)$ is substochastic as is $q_\sigma(x, dy)$. Denote the I-function corresponding to $p_G(t, x, dy)$ by $I_{G,t}$. We show first that

$$I_{G,t}(\mu) \leqq tI(\mu). \tag{8.10}$$

Let $I(\mu)=l$. From Hypothesis H_3 and [1] (Lemma 3.1) we have, for every u and $h>0$,

$$\int \log\left(\frac{T_h u}{u}\right)(x)\mu(dx) \geqq -lh, \tag{8.11}$$

where $(T_h u)(x) = \int_X u(y)p(h, x, dy)$. Let W and v be bounded measurable functions on X with strict positive lower bound and $W=1$ on C. Then if we let

$$u(x) = WT_h WT_h W \cdots T_h Wv,$$

we have from (8.11)

$$\begin{aligned} &\int_X \log(T_h WT_h W \cdots T_h Wv)(x)\mu(dx) \\ &\qquad - \int_X \log(WT_h WT_h W \cdots T_h Wv)(x)\mu(dx) \geqq -lh. \end{aligned} \tag{8.12}$$

Set $S_h = T_h W$; noting that $W=1$ on C and $\mu(C)=1$, we rewrite (8.12) as

$$\int_X \log (S_h^k v)(x)\mu(dx) - \int_X \log (S_h^{k-1} v)(x)\mu(dx) \geqq -lh.$$

Summing over k from 1 to n, we get

$$\int_X \log (S_h^n v)(x)\mu(dx) - \int_X \log v(x)\mu(dx) \geqq -nlh. \tag{8.13}$$

But

$$(S_h^n v)(x) = E_x\{v(x(nh))W(x(nh))W(x(n-1)h)) \cdots W(x(h))\},$$

and if we let W decrease to the indicator function of C, we have from (8.13) and monotone convergence that

$$\int_X \log E_x\{v(x(nh))\chi_C(x(h))\chi_C(x(2h)) \cdots \chi_C(x(nh))\}\mu(dx)$$
$$- \int_X \log v(x)\mu(dx) \geqq -nlh.$$

If in this last inequality we let $n \to \infty$ and $h \to 0$ so that $nh = t$ is fixed we get, by Fatou's lemma,

$$\int_X \log E_x\{v(x(t)); x(s) \in C, 0 \leqq s \leqq t\}\mu(dx) - \int_X \log v(x)\mu(dx) \geqq -lt,$$

i.e.,

$$\int_X \log \left(\int_X v(y) p_C(t, x, dy) \right) \mu(dx) - \int_X \log v(x)\mu(dx) \geqq -lt.$$

Since $G \supset C$ and v is positive, we conclude that

$$\int_X \log \left(\int_X v(y) p_G(t, x, dy) \right) \mu(dx) - \int_X \log v(x)\mu(dx) \geqq -lt$$

which, holding for all v in the class, implies $I_{G,t}(\mu) \leqq tI(\mu)$.

Consider

$$
\begin{aligned}
\int_X \log\left(\frac{q_\sigma u}{u}\right)(x)\mu(dx) &= \int_X \log (q_\sigma u)(x)\mu(dx) - \int_X \log u(x)\mu(dx) \\
&= \int_X \log\left[\int_X u(y)\int_0^\infty \sigma e^{-\sigma t} p_G(t, x, dy)\, dt\right]\mu(dx) \\
&\quad - \int_X \log u(x)\mu(dx) \qquad (8.14)\\
&= \int_X \log\left[\int_0^\infty (T_{t,G}u)(x)\sigma e^{-\sigma t}\, dt\right]\mu(dx) \\
&\quad - \int_X \log u(x)\mu(dx),
\end{aligned}
$$

where $(T_{t,G}u)(x) = \int_X u(y)p_G(t, x, dy)$. By Jensen's inequality,

$$
\log \int_0^\infty (T_{t,G}u)(x)\sigma e^{-\sigma t}\, dt \geqq \int_0^\infty [\log (T_{t,G}u)(x)]\sigma e^{-\sigma t}\, dt,
$$

for each $x \in X$, and therefore

$$
\int_X \log (q_\sigma u)(x)\mu(dx) \geqq \int_X \int_0^\infty [\log (T_{t,G}u)(x)]\sigma e^{-\sigma t}\, dt\, \mu(dx). \qquad (8.15)
$$

Using (8.15) in (8.14) we get

$$
\begin{aligned}
\int_X \log\left(\frac{q_\sigma u}{u}\right)(x)\mu(dx) &\geqq \int_X \int_0^\infty \left[\log\left(\frac{T_{t,G}u}{u}\right)(x)\right]\sigma e^{-\sigma t}\, dt\, \mu(dx) \\
&= \int_0^\infty \sigma e^{-\sigma t}\int_X \log\left(\frac{T_{t,G}u}{u}\right)(x)\mu(dx)\, dt \\
&\geqq -\int_0^\infty \sigma e^{-\sigma t} I_{G,t}(\mu)\, dt.
\end{aligned}
$$

Applying (8.10) to this last inequality, we obtain

$$\int_X \log\left(\frac{q_\sigma u}{u}\right)(x)\mu(dx) \geqq -\int_0^\infty \sigma t e^{-\sigma t}\, dtI(\mu)$$
$$= -\frac{1}{\sigma} I(\mu),$$

and since this holds for all eligible u, we have $\sigma I_G^\sigma(\mu) \leqq I(\mu)$.

LEMMA 8.3. *Let $f \in B_{00}$. For any $\varepsilon > 0$,*

$$\lim_{\sigma\to\infty} \overline{\lim_{t\to\infty}} \frac{1}{t} \log \sup_x P_x\left\{\left|\int_X f(y)L_t(\omega, dy) - \int_X f(y)L_{n,\sigma}(\omega, dy)\right| \geqq \varepsilon\right\} = -\infty,$$

where $n = \sigma t$.

Proof: Let $\{x_t, 0 \leqq t < \infty\}$ again denote the Markov process with transition probability $p(t, x, dy)$. Let f be a given element of B_{00}, and define a stochastic process $\{z_t, 0 \leqq t < \infty\}$ by $z(t) = f(x(t))$. Then

$$\int_X f(y)L_t(\omega, dy) = \frac{1}{t}\int_0^t f(x(s))\, ds = \frac{1}{t}\int_0^t z(s)\, ds,$$

(8.16)

$$\int_X f(y)L_{n,\sigma}(\omega, dy) = \frac{1}{n}\sum_{j=0}^{n-1} f(x(s_j)) = \frac{z(s_0) + z(s_1) + \cdots + z(s_{n-1})}{n}.$$

Since $t = n/\sigma$, (8.16) yields

$$\int_X f(y)L_t(\omega, dy) - \int_X f(y)L_{n,\sigma}(\omega, dy)$$

(8.17)

$$= \frac{\sigma}{n}\int_0^{n/\sigma} z(s)\, ds - \frac{1}{n}[z(s_0) + z(s_1) + \cdots + z(s_{n-1})]$$

$$= \frac{\sigma}{n}\int_0^{s_n} z(s)\, ds - \frac{1}{n}[z(s_0) + z(s_1) + \cdots + z(s_{n-1})] + E,$$

where $|E| \leqq (\sigma/n)\,|s_n - n/\sigma|\, A = (1/n)\,|\sigma s_n - n|\, A$ and $|f| \leqq A$. With $n = rl$, we can rewrite (8.17) as

$$\int_X f(y)L_t(\omega, dy) - \int_X f(y)L_{n,\sigma}(\omega, dy)$$

$$= \frac{1}{n}\sum_{j=1}^{r}\left(\sigma\int_{s_{l(j-1)}}^{s_{lj}} z(s)\,ds - [z(s_{l(j-1)}) + \cdots + z(s_{lj-1})]\right) + E$$

$$= \frac{1}{n}\sum_{j=1}^{r}\left(\sigma\int_{s_{l(j-1)}}^{s_{lj}} [z(s) - z(s_{l(j-1)})]\,ds + \sigma(s_{lj} - s_{l(j-1)})z(s_{l(j-1)}) - lz(s_{l(j-1)})\right. \tag{8.18}$$

$$\left. + [z(s_{l(j-1)}) - z(s_{l(j-1)})] + \cdots + [z(s_{l(j-1)}) - z(s_{lj-1})]\right) + E,$$

so that

$$\left|\int_X f(y)L_t(\omega, dy) - \int_X f(y)L_{n,\sigma}(\omega, dy)\right|$$

$$\leqq \frac{1}{n}\sum_{j=1}^{r}\left([\sigma\,|s_{lj} - s_{l(j-1)}| + l] \sup_{s_{l(j-1)} \leqq s \leqq s_{lj}} |z(s) - z(s_{l(j-1)})|\right. \tag{8.19}$$

$$\left. + A\,|\sigma(s_{lj} - s_{l(j-1)}) - l|\right) + \frac{A}{n}|\sigma s_n - n|.$$

But

$$|\sigma s_n - n| = \left|\sum_{j=1}^{r} [\sigma(s_{lj} - s_{l(j-1)}) - l]\right| \leqq \sum_{j=1}^{r} |\sigma(s_{lj} - s_{l(j-1)}) - l|,$$

or

$$\frac{A}{n}|\sigma s_n - n| \leqq \frac{1}{n}\sum_{j=1}^{r} A\,|\sigma(s_{lj} - s_{l(j-1)}) - l|.$$

Thus, from (8.19) we get

$$\left|\int_X f(y)L_t(\omega, dy) - \int_X f(y)L_{n,\sigma}(\omega, dy)\right|$$

$$\leqq \frac{1}{n}\sum_{j=1}^{r}\left([\sigma\,|s_{lj} - s_{l(j-1)}| + l] \sup_{s_{l(j-1)} \leqq s \leqq s_{lj}} |z(s) - z(s_{l(j-1)})|\right. \tag{8.20}$$

$$\left. + 2A\,|\sigma(s_{lj} - s_{l(j-1)}) - l|\right).$$

Moreover,

$$[\sigma\,|s_{lj}-s_{l(j-1)}|+l]\sup_{s_{l(j-1)}\leqq s\leqq s_{lj}}|z(s)-z(s_{l(j-1)})|$$

$$\leqq 2l\sup_{s_{l(j-1)}\leqq s\leqq s_{lj}}|z(s)-z(s_{l(j-1)})|+2A[\sigma\,|s_{lj}-s_{l(j-1)}|-l]$$

$$\leqq 2l\sup_{s_{l(j-1)}\leqq s\leqq s_{lj}}|z(s)-z(s_{l(j-1)})|+2A\,|\sigma(s_{lj}-s_{l(j-1)})-l|,$$

and hence, from (8.20), using $n=lr$, we have

$$\left|\int_X f(y)L_t(\omega,dy)-\int_X f(y)L_{n,\sigma}(\omega,dy)\right|$$

(8.21)
$$\leqq\frac{2l}{n}\sum_{j=1}^{r}\sup_{s_{l(j-1)}\leqq s\leqq s_{lj}}|z(s)-z(s_{l(j-1)})|+\frac{4A\sigma}{n}\sum_{j=1}^{r}\left|(s_{lj}-s_{l(j-1)})-\frac{l}{\sigma}\right|$$

$$=\frac{2}{r}\sum_{j=1}^{r}\sup_{s_{l(j-1)}\leqq s\leqq s_{lj}}|z(s)-z(s_{l(j-1)})|+\frac{4A}{r}\sum_{j=1}^{r}\left|\left(\frac{\sigma}{l}\right)(s_{lj}-s_{l(j-1)})1\right|.$$

From (8.21) we get, for each $x\in X$,

$$P_x\left\{\left|\int_X f(y)L_t(\omega,dy)-\int_X f(y)L_{n,\sigma}(\omega,dy)\right|>\varepsilon\right\}$$

$$=P_x\left\{\left|\frac{1}{t}\int_0^t z(s)\,ds-\frac{1}{n}[z(s_0)+z(s_1)+\cdots+z(s_{n-1})]\right|>\varepsilon\right\}$$

(8.22)
$$\leqq P_x\left\{\frac{2}{r}\sum_{j=1}^{r}\sup_{s_{l(j-1)}\leqq s\leqq s_{lj}}|z(s)-z(s_{l(j-1)})|\geqq\tfrac{1}{2}\varepsilon\right\}$$

$$+P\left\{\frac{4A}{r}\sum_{j=1}^{r}\left|\left(\frac{\sigma}{l}\right)(s_{lj}-s_{l(j-1)})-1\right|\geqq\tfrac{1}{2}\varepsilon\right\}$$

$$=F_1+F_2.$$

We now estimate F_1 and F_2. For any $\theta>0$, by the obvious estimate,

$$F_2 \leqq e^{-\theta\varepsilon/2} E\left\{\exp\left\{\frac{\theta 4A}{r}\sum_{j=1}^{r}\left|\frac{\sigma}{l}(s_{l_j}-s_{l(j-1)})-1\right|\right\}\right\}.$$

Replacing θ by $r\theta$ and then using the independence of $\tau_1, \tau_2, \cdots$, we obtain

$$F_2 \leqq e^{-r\theta\varepsilon/2}\left[E\left\{\exp\left\{\theta 4A\left|\frac{\sigma}{l}s_l-1\right|\right\}\right\}\right]^r. \tag{8.23}$$

As for F_1, let E_j, for $j=1,2,\cdots,r$, be the event

$$\left\{\sup_{s_{l(j-1)}\leqq s\leqq s_{lj}}|z(s)-z(s_{l(j-1)})|>\tfrac{1}{8}\varepsilon\right\},$$

and consider $(1/r)\sum_{j=1}^{r}\chi_{E_j}$. Clearly, the event $\left\{(1/r)\sum_{j=1}^{r}\chi_{E_j}>\varepsilon/16A\right\}$ contains the event

$$\left\{\frac{2}{r}\sum_{j=1}^{r}\sup_{s_{l(j-1)}\leqq s\leqq s_{lj}}|z(s)-z(s_{l(j-1)})|\geqq\tfrac{1}{2}\varepsilon\right\}$$

so that

$$F_1 \leqq P_x\left\{\frac{1}{r}\sum_{j=1}^{r}\chi_{E_j}\geqq\frac{\varepsilon}{16A}\right\}. \tag{8.24}$$

For any $\nu>0$, by the usual estimate,

$$P_x\left\{\frac{1}{r}\sum_{j=1}^{r}\chi_{E_j}\geqq\frac{\varepsilon}{16A}\right\}\leqq e^{-\nu\varepsilon/16A}E_x\left\{\exp\left\{\frac{\nu}{r}\sum_{j=1}^{r}\chi_{E_j}\right\}\right\},$$

and therefore replacing ν by $r\nu$ we get

$$F_1 \leqq e^{-r\varepsilon\nu/16A}E_x\left\{\exp\left\{\nu\sum_{j=1}^{r}\chi_{E_j}\right\}\right\}. \tag{8.25}$$

In handling F_2 at the corresponding stage we were able to exploit the independence of the τ_j's so that the expectation of the exponential factored (cf. (8.23)). Now, however, the events $E_1, E_2, \cdots$ are not independent and so we must be more delicate. Let $\sigma(E_1, E_2, \cdots, E_{j-1})$ be the σ-field generated by $E_1, E_2, \cdots, E_{j-1}$. Let

$$p=\sup_{1\leqq j\leqq r}\sup_{\omega}P\{E_j\,|\sigma(E_1,E_2,\cdots,E_{j-1})\},$$

where $\sup_\omega$ means the supremum over all possible conditionings. Then (8.25) yields

$$F_1 \leqq e^{-r\varepsilon\nu/16A}(pe^{\nu}+q)^r, \tag{8.26}$$

where $q=1-p$. To get an estimate on p, we note that

$$p=\sup_{1\leqq j\leqq r}\sup_{\omega} P\left\{\sup_{s_{l(j-1)}\leqq s\leqq s_{lj}} |z(s)-z(s_{l(j-1)})|>\tfrac{1}{8}\varepsilon \mid \sigma(E_1, E_2, \cdots, E_{j-1})\right\}$$

$$\leqq \sup_x P_x\left\{\sup_{0\leqq s\leqq s_l} |z(s)-z(0)|>\tfrac{1}{8}\varepsilon\right\}=g(l,\sigma,\varepsilon).$$

We shall show below that, for any $\varepsilon>0$ and fixed l,

$$\lim_{\sigma\to\infty} g(l,\sigma,\varepsilon)=0. \tag{8.27}$$

Now we fix l and σ and recall that $n=rl=\sigma t$, so that, by (8.26), $r\to\infty$, $t\to\infty$ but r/t is fixed at σ/l and

$$\overline{\lim_{t\to\infty}}\frac{1}{t}\log F_1 \leqq -\frac{\sigma}{l}\frac{\varepsilon\nu}{16A}+\frac{\sigma}{l}\log(pe^{\nu}+q). \tag{8.28}$$

Similarly from (8.23) we get

$$\overline{\lim_{t\to\infty}}\frac{1}{t}\log F_2 \leqq -\frac{\sigma}{l}\frac{\varepsilon\theta}{2}+\frac{\sigma}{l}\log E\left\{\exp\left\{\theta 4A\left|\frac{\sigma}{l}S_l-1\right|\right\}\right\}. \tag{8.29}$$

To complete the proof of the lemma, except for (8.27) which we show below, we must, in view of (8.22), (8.28) and (8.29), pick l as a function of σ and ε, θ as a function of σ and ε, and ν as a function of σ and ε so that the expressions on the right of (8.28) and (8.29) both go to $-\infty$ as $\sigma\to\infty$.

Let J_1 and J_2 be, respectively, the expressions on the right of inequalities (8.28) and (8.29). Recall that $s_l=\tau_1+\tau_2+\cdots+\tau_l$, where each τ_i has as its density $\sigma e^{-\sigma t}$. Let $\bar{s}_l=\mu_1+\mu_2+\cdots+\mu_l$, where the μ_i are independent random variables each having density e^{-t}. We see that

$$E\left\{\exp\left\{4\theta A\left|\frac{\sigma}{l}s_l-1\right|\right\}\right\}=E\{\exp\{4\theta A\,|\bar{s}_l/l-1|\}\},$$

and if we choose $\theta=\nu=1$ we have

$$J_2=-\frac{\sigma}{l}\left[\tfrac{1}{2}\varepsilon-\log E\{\exp\{4A\,|\bar{s}_l/l-1|\}\}\right],$$

(8.30)

$$J_1=-\frac{\sigma}{l}\left[\frac{\varepsilon}{16A}-\log(pe+q)\right].$$

For our given $\varepsilon>0$, we can choose l_0 so large that

$$\left[\tfrac{1}{2}\varepsilon-\log E\{\exp\{4A\,|\bar{s}_{l_0}/l_0-1|\}\}\right]>0\,.$$

By (8.27) we have, for this l_0, $\lim_{\sigma\to\infty} g(l_0,\sigma,\varepsilon)=0$. Since $p\leqq g(l_0,\sigma,\varepsilon)$, we see that $q\to 1$ as $\sigma\to\infty$, and therefore $[\varepsilon/16A-\log(pe+q)]$ becomes positive as $\sigma\to\infty$. Thus, with $l=l_0$, both J_1 and J_2 approach $-\infty$ as $\sigma\to\infty$. This completes the proof of the lemma except for (8.27).

To estimate $g(l,\sigma,\varepsilon)$ we first observe that, for fixed l and Δ,

$$g(l,\sigma,\varepsilon)=\sup_x P_x\left\{\sup_{0\leqq s\leqq s_l}|z(s)-z(0)|>\tfrac{1}{8}\varepsilon\right\}$$

$$\leqq\sup_x P_x\left\{\sup_{0\leqq s\leqq\Delta}|z(s)-z(0)|>\tfrac{1}{8}\varepsilon\right\}+P\{s_l>\Delta\}\,.$$

By Tchebycheff's inequality, $P\{s_l>\Delta\}\to 0$ as $\sigma\to\infty$ and hence, for fixed l and Δ,

$$\overline{\lim_{\sigma\to\infty}}\ g(l,\sigma,\varepsilon)\leqq\sup_x P_x\left\{\sup_{0\leqq s\leqq\Delta}|z(s)-z(0)|>\tfrac{1}{8}\varepsilon\right\}. \tag{8.31}$$

If (Ω,Σ,P) is a probability space and Σ_t an increasing family of σ-fields with $z(t)$ a stochastic process with right continuous paths such that $z(t)$ is Σ_t-measurable for every t, then

$$P\left\{\sup_{0\leqq t\leqq\Delta}|z(t)-z(0)|\geqq\varepsilon\right\}\leqq\frac{\rho_{\varepsilon/2}(\Delta)}{1-\rho_{\varepsilon/2}(\Delta)},$$

where

$$\rho_\varepsilon(\delta)=\sup_t\ \sup_{0\leqq h\leqq\delta}\ \operatorname*{ess\,sup}_\omega P\{|z(t+h)-z(t)|>\varepsilon\mid\Sigma_t\}\,.$$

If $z(t)$ were a process with independent increments this would be just Lévy's inequality. Since we have uniform conditional estimates, the proof of the above inequality is no different from that of Lévy's inequality. Thus, by (8.31), it will suffice to show that for our process $z(t)$ and with $f \in B_{00}$ we have $\rho_\varepsilon(\Delta) \to 0$ as $\Delta \to 0$.

First of all

$$E\{|f(x(t+h)) - f(x(t))| \,|\, \Sigma_t\} = \int |f(y) - f(x)| \, p(h, x, dy)\big|_{x = x(t)},$$

and therefore,

$$P\{|z(t+h) - z(t)| > \varepsilon \,|\, \Sigma_t\} \leqq \frac{1}{\varepsilon} \sup_x \int |f(y) - f(x)| \, p(h, x, dy). \tag{8.32}$$

The right side of $(8.32) \to 0$ as $h \to 0$ since $f \in B_{00}$. Thus $\rho_\varepsilon(\Delta) \to 0$ as $\Delta \to 0$ and we have (8.27) and the lemma.

LEMMA 8.4. *For $\Delta > 0$ and $n = \sigma t(1+\Delta)$,*

$$\lim_{\sigma \to \infty} \overline{\lim_{t \to \infty}} \frac{1}{t} \log P\{\tau_1 + \tau_2 + \cdots + \tau_n \leqq t\} = -\infty.$$

Proof: Since the τ_j are independent and $P\{\tau \in dt\} = \sigma e^{-\sigma t}\, dt$ we have

$$E\{\exp\{\theta(\tau_1 + \tau_2 + \cdots + \tau_n)\}\} = \left(1 - \frac{\theta}{\sigma}\right)^{-n},$$

and therefore, for $\theta > 0$,

$$P\{\tau_1 + \tau_2 + \cdots + \tau_n \leqq t\} \leqq e^{\theta t}\left(1 + \frac{\theta}{\sigma}\right)^{-n} = e^{\theta t}\left(1 + \frac{\theta}{\sigma}\right)^{-\sigma t(1 \geqq \Delta)}.$$

Consequently,

$$\overline{\lim_{t \to \infty}} \frac{1}{t} \log P\{\tau_1 + \tau_2 + \cdots + \tau_n \leqq t\} \leqq \theta - \sigma(1+\Delta) \log\left(1 + \frac{\theta}{\sigma}\right),$$

and since this is true for all $\theta > 0$,

$$\overline{\lim_{t \to \infty}} \frac{1}{t} \log P\{\tau_1 + \tau_2 + \cdots + \tau_n \leqq t\} \leqq \inf_{\theta > 0}\left[\theta - \sigma(1+\Delta) \log\left(1 + \frac{\theta}{\sigma}\right)\right].$$

But

$$\lim_{\sigma\to\infty}\inf_{\theta>0}\left[\theta-(1+\Delta)\sigma\log\left(1+\frac{\theta}{\sigma}\right)\right]\leqq\inf_{\theta>0}\lim_{\sigma\to\infty}\left[\theta-(1+\Delta)\sigma\log\left(1+\frac{\theta}{\sigma}\right)\right]$$
$$=\inf_{\theta>0}[-\Delta\theta]=-\infty,$$

and hence the lemma.

9. Examples and Counterexamples

In this section we exhibit some examples and counterexamples to illustrate the scope of our hypotheses. First of all without some sort of hypothesis like H* we cannot have an upper bound for all closed sets. To see this consider the Markov chain on the integers $\{1, 2, \cdots\}$ with transition probability $\pi_{ij}=\delta_{i+1,j}$. The chain is purely deterministic and moves one step to the right each time. One can verify that $I(\mu)=\infty$ for all μ. If we take for the closed set C the whole space, i.e., all probability measures on the positive integers, then

$$\overline{\lim_{n\to\infty}}\frac{1}{n}\log Q_{n,x}(C)=0.$$

On the other hand, $\inf_{\mu\in C}I(\mu)=\infty$, so that (1.4) cannot be valid in this case for all closed sets.

In the case of continuous time, if we take the process with generator

$$\frac{1}{2}\frac{\partial^2}{\partial x^2}+a\frac{\partial}{\partial x},$$

i.e., Brownian motion with a constant drift a, then

$$I(\mu)=\frac{1}{8}\int_{-\infty}^{\infty}\frac{[f'(y)]^2}{f(y)}\,dy+\tfrac{1}{2}a^2,$$

where $\mu(dx)=f(x)\,dx$. Clearly, $\inf_{\mu\in\mathcal{M}}I(\mu)=\frac{1}{2}a^2>0$. We see the continuous analogue of (1.4) cannot hold when C is the whole space.

These two examples are also counterexamples to (1.6) and its continuous version with $\Phi(\mu)\equiv 0$.

To see that some sort of transitivity condition is needed, we consider the case of a finite state space Markov chain with transition probability $\pi_{ij}=\delta_{i,j}$. Again the chain is deterministic and never moves. In this case, $I(\mu)=0$ for all μ and so if we take for the open set G the set of all nondegenerate probability distributions on the state space, then $Q_{n,x}(G)=0$ for all n and x. On the other hand, if the state space is at least two points, $\inf_{\mu\in G}I(\mu)=0$ so that (1.5) cannot hold.

Finally, we give an example to illustrate when Condition H* is valid. If we take the one-dimensional Ornstein–Uhlenbeck process, i.e., the process whose generator is

$$\frac{1}{2}\frac{\partial^2}{\partial x^2}-x\frac{\partial}{\partial x},$$

we see that Condition H* is satisfied with

$$u_n(x)=\cosh n\theta\left(\frac{x}{n}\right),$$

$$V(x)=\tfrac{1}{2}-x\sinh x,$$

where

$$\theta(x)=\begin{cases}x & \text{for } 0\leqq x\leqq 1,\\ 2 & \text{for } x\geqq 2,\\ \text{smooth and increasing} & \text{for } 1\leqq x\leqq 2,\end{cases}$$

and $\theta(-x)=-\theta(x)$.

Bibliography

[1] Donsker, M. D., and Varadhan, S. R. S., *Asymptotic evaluation of certain Markov process expectations for large time—I,* Comm. Pure Appl. Math., Vol. 27, 1975, pp. 1–47.

[2] Donsker, M. D., and Varadhan, S. R. S., *Asymptotic evaluation of certain Wiener integrals for large time,* Functional Integration and Its Applications, Proceedings of the International Conference held at the Cumberland Lodge, Windsor Great Park, London in April 1974. A. M. Arthurs (editor), Clarendon Press, Oxford, 1975.

[3] Marcus, M. B., and Shepp, L. A., *Sample behavior of Gaussian processes,* Sicth Berkeley Symposium on Math. Stat. and Probability, Vol. II, 1972, pp. 423–439.

[4] Kuratowski, C., *Topologie I,* Warszawa-Lwow, 1933.

[5] Pincus, M., *Gaussian processes and Hammerstein integral equations,* Trans. Amer. Math. Soc., Vol. 134, 1968, pp. 193–216.

[6] Schilder, M., *Some asymptotic formulae for Wiener integrals,* Trans. Amer. Math. Soc., Vol. 125, 1966, pp. 63–85.

[7] Sion, M., *On general minimax theorem,* Pacific Journal of Math., Vol. VIII, 1958, pp. 171–176.

[8] Skorohod, A. V., *A note on Gaussian measures in a Bãnach space,* Theory of Prob. and Its Appl., Vol. XV, 1970, p. 508.

[9] Varadhan, S. R. S., *Asymptotic probabilities and differential equations,* Comm. Pure Appl. Math., Vol. 19, 1966, pp. 261–286.

Received April, 1976.

COMMUNICATIONS ON PURE AND APPLIED MATHEMATICS, VOL. XXIX, 595-621 (1976)

On the Principal Eigenvalue of Second-Order Elliptic Differential Operators*

M. D. DONSKER AND S. R. S. VARADHAN

1. Introduction

Let X be a compact metric space and T_t a strongly continuous semigroup of operators on $C(X)$ such that $T_t f$ is non-negative for each non-negative f and $T_t 1 = 1$ for all t. Let L be the infinitesimal generator with domain $\mathcal{D}$ and let $\mathcal{D}_+$ denote the strictly positive functions in $\mathcal{D}$. We define

$$I(\mu) = -\inf_{u \in \mathcal{D}_+} \int_X \left(\frac{Lu}{u}\right)(x)\mu(dx), \tag{1.1}$$

where μ is a probability measure on X. The function $I(\mu)$ is lower semicontinuous on the space $\mathcal{M}$ of all probability measures on X.

We proved in [1] that if $V \in C(X)$, then the principal eigenvalue λ_V of the operator $L + V$ is given by

$$\lambda_V = \sup_{\mu \in \mathcal{M}} \left[\int_X V(x)\mu(dx) - I(\mu)\right]. \tag{1.2}$$

If the operator L is selfadjoint with respect to some reference measure on X, then $I(\mu)$ can be readily computed and formula (1.2) reduces to the classical variational formula for the principal eigenvalue. If L is not selfadjoint, then in general $I(\mu)$ is difficult to evaluate explicitly.

In this paper we investigate the situation when X is R^d (it can be replaced by any connected manifold without boundary) and L is an elliptic second-order differential operator. $I(\mu)$ can be defined as the exact analogue of (1.1) but we need some compactness for (1.2) and therefore we consider the operator $L + V$ only on bounded regions with Dirichlet boundary conditions. After establishing the analogue of (1.2) in such cases, we look at the problem of when two operators L_1 and L_2 of the type we are dealing with can have identical I-functions. Equivalently, we seek the relation between L_1 and L_2 if

* This research was performed at the Courant Institute and partially supported by the National Science Foundation, Grant NSF-MPS-74-01921-A01. Reproduction in whole or in part is permitted for any purpose of the United States Government.

L_1+V and L_2+V have the same principal eigenvalue for all regions G and all potentials V. The answer to this question is the content of Theorem 3.1.

In order to prove that theorem we need to evaluate the I-function as explicitly as we can for operators of the type described. It turns out that, for elliptic second-order differential operators, $I(\mu)$ can be calculated quite explicitly (cf. Lemma 3.3). For example, if X is the circle parametrized by real numbers x (mod 1) and L is the operator

$$L=\frac{1}{2}\frac{\partial^2}{\partial x^2}+b(x)\frac{\partial}{\partial x}, \tag{1.3}$$

then L is selfadjoint with respect to some measure if and only if $\int_0^1 b(x)\,dx=0$. In any case, the I-function for (1.3) on the circle is given by

$$I(f)=\frac{1}{8}\int_0^1\frac{[f'(x)]^2}{f(x)}\,dx+\frac{1}{2}\int_0^1[b^2(x)+b_x(x)]f(x)\,dx-\frac{1}{2}\frac{\left(\int_0^1 b(x)\,dx\right)^2}{\int_0^1\frac{dx}{f(x)}}, \tag{1.4}$$

where $f(x)$ is a probability density on the circle.

2. The Principal Eigenvalue

Consider the operator

$$L=\frac{1}{2}\sum_{i,j=1}^{d}a_{ij}(x)\frac{\partial^2}{\partial x_i\,\partial x_j}+\sum_{j=1}^{d}b_j(x)\frac{\partial}{\partial x_j}, \tag{2.1}$$

where $a_{ij}(x)$ and $b_j(x)$ are in $C^\infty(R^d)$ for $1\leqq i,\ j\leqq d$, and the matrix $(a_{ij}(x))$ is positive definite for each $x\in R^d$. We shall assume, for Lemma 2.1 and Theorems 2.2 and 2.3 below, that, for large x, $a_{ij}(x)=\delta_{i,j}$ and $b_j(x)=0$, $1\leqq i$, $j\leqq d$. It will then be shown that for our purposes this hypothesis on the behavior of the coefficients for large x can be discarded.

Under the hypotheses above, L is the infinitesimal generator of the semigroup T_t associated with the fundamental solution of $\partial u/\partial t=Lu$. Let us denote the domain of L by $\mathfrak{D}$ and by $\mathfrak{D}_+$ those elements u in $\mathfrak{D}$ for each of which there exist constants c_1 and c_2 such that $0<c_1\leqq u(x)\leqq c_2<\infty$ for all $x\in R^d$. Let $\mathcal{M}$ be the space of probability measures on R^d. For any $\mu\in\mathcal{M}$ we define

$$I(\mu)=-\inf_{u\in\mathfrak{D}_+}\int_{R^d}\left(\frac{Lu}{u}\right)(x)\mu(dx). \tag{2.2}$$

For any $\mu \in \mathcal{M}$ we also define

$$\bar{I}(\mu) = -\inf_{u \in C_+^\infty(R^d)} \int_{R^d} \left(\frac{Lu}{u}\right)(x)\mu(dx), \tag{2.3}$$

where $C_+^\infty(R^d)$, analogous to $\mathcal{D}_+$, is the set of positive functions $u \in C^\infty(R^d)$ which are bounded above and below. We now prove

LEMMA 2.1. *For all* $\mu \in \mathcal{M}$,

$$I(\mu) = \bar{I}(\mu).$$

Proof: Clearly $\bar{I}(\mu) \leqq I(\mu)$; thus we must show the reverse inequality. Let $f \in C_+^\infty(R^d)$. Since $T_t f \in C_+^\infty(R^d)$, we have

$$\frac{\partial}{\partial t} \int_{R^d} \log\left(\frac{T_t f}{f}\right)(x)\mu(dx) = \int_{R^d} \left(\frac{LT_t f}{T_t f}\right)(x)\mu(dx) \geqq -\bar{I}(\mu). \tag{2.4}$$

When $t = 0$, $\log(T_t f/f) = 0$ so that integrating both sides of (2.4) from 0 to t we get

$$\int_{R^d} \log\left(\frac{T_t f}{f}\right)(x)\mu(dx) \geqq -t\bar{I}(\mu) \tag{2.5}$$

which holds for all $t \geqq 0$ and all $f \in C_+^\infty(R^d)$. Actually, if $f \in C_+(R^d)$, then (2.5) holds for this f also because we can find a sequence $\{f_n\} \in C_+^\infty(R^d)$ (even satisfying the same bounds) which converges to f uniformly on compact sets. Thus, if $u \in \mathcal{D}_+$, then $u \in C_+(R^d)$, and from (2.5) we have, for all $t \geqq 0$,

$$\int_{R^d} \log\left(\frac{T_t u}{u}\right)(x)\mu(dx) \geqq -t\bar{I}(\mu).$$

Both sides of this inequality vanish at $t = 0$, and hence the same inequality holds for the derivatives at $t = 0$, i.e.,

$$\int_{R^d} \left(\frac{Lu}{u}\right)(x)\mu(dx) \geqq -\bar{I}(\mu).$$

Since this last inequality holds for any $u \in \mathcal{D}_+$, we conclude that $I(\mu) \leqq \bar{I}(\mu)$ which proves the lemma.

Let G be an open set in R^d with compact closure $\bar{G}$ and having a smooth boundary ∂G. Let $V(x)$ be continuous on R^d and denote by $T_t^{V,G}$ the semigroup defined by

$$\text{(2.6)}\qquad \begin{aligned} &\frac{\partial u}{\partial t} = Lu + Vu && \text{on} \quad [0,\infty)\times G, \\ &u(t,x) = 0 && \text{on} \quad [0,\infty)\times \partial G, \\ &u(0,x) = f(x) && \text{on} \quad G. \end{aligned}$$

Then, $T_t^{V,G}$ defines a semigroup on $C(G)$ and it leaves $C_0(G)$, the space of continuous functions on $\bar{G}$ which vanish on ∂G, invariant; $T_t^{V,G}$ is strongly continuous on $C_0(G)$.

Since $\|T_t^{V,G}\| = \sup_{x\in\bar{G}} (T_t^{V,G}\cdot 1)(x)$ is submultiplicative in t, the following limit exists:

$$\text{(2.7)}\qquad \lambda_{V,G} = \lim_{t\to\infty} \frac{1}{t} \log \|T_t^{V,G}\|.$$

Let $L_{V,G}$ be the infinitesimal generator of $T_t^{V,G}$ on $C_0(G)$, and denote the spectrum of $L_{V,G}$ by $\sigma_{V,G}$.

THEOREM 2.2. *We always have* $\lambda_{V,G} \in \sigma_{V,G}$ *and if* $\mathcal{Re}\, z > \lambda_{V,G}$, *then* $z \notin \sigma_{V,G}$. *Moreover,*

$$\text{(2.8)}\qquad \lambda_{V,G} = \sup_{\mu:\mu(\bar{G})=1} \left[\int_{R^d} V(x)\mu(dx) - I(\mu)\right].$$

Proof: If $\mathcal{Re}\, z > \lambda_{V,G}$, then

$$\int_0^\infty e^{-zt} T_t^{V,G}\, dt$$

is a convergent integral and defines explicitly the resolvent so that $z \notin \sigma_{V,G}$.

Let

$$\text{(2.9)}\qquad l_{V,G} = \sup_{\mu:\mu(\bar{G})=1} \left[\int_{R^d} V(x)\mu(dx) - I(\mu)\right].$$

We show first that $\lambda_{V,G} \leqq l_{V,G}$. From Lemma 2.1,

$$
\begin{aligned}
l_{V,G} &= \sup_{\mu:\mu(\bar{G})=1} \left[\int_{R^d} V(x)\mu(dx) + \inf_{u\in C_+^\infty(R^d)} \int_{R^d} \left(\frac{Lu}{u}\right)(x)\mu(dx) \right] \\
&= \sup_{\mu:\mu(\bar{G})=1} \inf_{u\in C_+^\infty(R^d)} \left[\int_{R^d} V(x)\mu(dx) + \int_{R^d} \left(\frac{Lu}{u}\right)(x)\mu(dx) \right].
\end{aligned} \tag{2.10}
$$

For $u \in C_+^\infty(R^d)$ we can write $u = e^h$, where $h \in C^\infty(R^d)$. Moreover, from (2.1) we see that $(Le^h/e^h)(x) = [Lh + \frac{1}{2}\langle a\, \nabla h, \nabla h\rangle](x)$, where $a = a(x) = (a_{ij}(x))$. Thus, from (2.10) we get

$$
l_{V,G} = \sup_{\mu:\mu(\bar{G})=1} \inf_{h\in C^\infty(R^d)} \left[\int_{R^d} V(x)\mu(dx) + \int_{R^d} [Lh + \tfrac{1}{2}\langle a\, \nabla h, \nabla h\rangle](x)\mu(dx) \right].
$$

If we impose the C^2-topology on $C^\infty(R^d)$, then

$$
\int_{R^d} V(x)\mu(dx) + \int_{R^d} [Lh + \tfrac{1}{2}\langle a\, \nabla h, \nabla h\rangle](x)\mu(dx)
$$

is a continuous convex function of h for each fixed μ, and for each fixed h, it is a continuous linear function of μ. Since μ varies over a compact set, the mini-max theorem of Sion ([2], Corollary 3.3) applies, and we have

$$
\begin{aligned}
l_{V,G} &= \inf_{h\in C^\infty(R^d)} \sup_{\mu:\mu(\bar{G})=1} \left[\int_{R^d} V(x)\mu(dx) + \int_{R^d} [Lh + \tfrac{1}{2}\langle a\, \nabla h, \nabla h\rangle](x)\mu(dx) \right] \\
&= \inf_{u\in C_+^\infty(R^d)} \sup_{\mu:\mu(\bar{G})=1} \left[\int_{R^d} V(x)\mu(dx) + \int_{R^d} \left(\frac{Lu}{u}\right)(x)\mu(dx) \right] \\
&= \inf_{u\in C_+^\infty(R^d)} \sup_{x\in\bar{G}} \left[V(x) + \left(\frac{Lu}{u}\right)(x) \right].
\end{aligned}
$$

Therefore, given $\varepsilon > 0$, there exists $u \in C_+^\infty(R^d)$ such that

$$
V(x) + \left(\frac{Lu}{u}\right)(x) \leqq l_{V,G} + \varepsilon, \tag{2.11}
$$

for all $x \in \bar{G}$. For this ε and this u, let $W(t, x) = u(x)\exp\{(\varepsilon + l_{V,G})t\}$. Then, (2.11) implies that, for all $x \in \bar{G}$,

$$
\begin{aligned}
\frac{\partial W}{\partial t} - LW - VW &= (\varepsilon + l_{V,G})W - \exp\{(\varepsilon + l_{V,G})t\}Lu - \exp\{(\varepsilon + l_{V,G})t\}Vu \\
&= W\left[\varepsilon + l_{V,G} - \frac{Lu}{u} - V\right] \geqq 0 .
\end{aligned} \tag{2.12}
$$

For the function u in (2.12), let $c_1 = \inf_{x \in \bar{G}} u(x)$ and $c_2 = \sup_{x \in \bar{G}} u(x)$. For a given $f(x)$ defined on $\bar{G}$, let $U(t, x)$ be the solution of (2.6), and let $\|f\| = \sup_{x \in \bar{G}} |f(x)|$. From (2.12) we see that the function $(\|f\|/c_1)W$ satisfies

$$
\frac{\partial}{\partial t}\left(\frac{\|f\|}{c_1} W\right) - L\left(\frac{\|f\|}{c_1} W\right) - V\left(\frac{\|f\|}{c_1} W\right) \geqq 0 . \tag{2.13}
$$

Also, at $t = 0$, $\|f\|\, W/c_1 \geqq f$ on $\bar{G}$. Moreover, on $[0, \infty) \times \partial G$ we have $\|f\|\, W/c_1 \geqq 0$. Thus, from the maximum principle,

$$
\frac{\|f\|\, W}{c_1} \geqq U(t, x) . \tag{2.14}
$$

On the other hand, $\|f\|\, W/c_1 \leqq (\|f\|/c_1)c_2 \exp\{(\varepsilon + l_{V,G})t\}$, so that

$$
U(t, x) \leqq \frac{c_2}{c_1} \|f\| \exp\{(\varepsilon + l_{V,G})t\} . \tag{2.15}
$$

Since $U(t, x) = (T_t^{V,G} f)(x)$, we get from (2.15), by taking the supremum over all $x \in \bar{G}$,

$$
\|T_t^{V,G} f\| \leqq \frac{c_2}{c_1} \|f\| \exp\{(\varepsilon + l_{V,G})t\} ,
$$

and therefore

$$
\lim_{t \to \infty} \frac{1}{t} \log \|T_t^{V,G}\| \leqq \varepsilon + l_{V,G} . \tag{2.16}
$$

In (2.16), ε is arbitrary and we conclude that $\lambda_{V,G} \leqq l_{V,G}$.

To show that $l_{V,G} \leqq \lambda_{V,G}$ we first note that the supremum in (2.9) is actually attained because $\int_{R^d} V(x)\mu(dx)$ is continuous in μ, $I(\mu)$ is lower semicontinuous in μ, and $\{\mu : \mu(\bar{G})=1\}$ is a compact set. So let μ_0 be the μ, where the supremum is attained, i.e.,

$$l_{V,G}=\int_{R^d} V(x)\mu_0(dx)-I(\mu_0)\,. \tag{2.17}$$

For $h>0$ consider the operator $S_h=\exp\{hV(x)\}T_h$. With $u\in C_+^\infty(R^d)$ and k a positive integer, let $u_k=S_h^k u$. Then,

$$\begin{aligned}\int_{R^d}\log\left(\frac{S_h u_k}{u_k}\right)(x)\mu_0(dx)&=\int_{R^d}\log\left[\exp\{hV(x)\}\left(\frac{T_h u_k}{u_k}\right)(x)\right]\mu_0(dx)\\&=h\int_{R^d}V(x)\mu_0(dx)+\int_{R^d}\log\left(\frac{T_h u_k}{u_k}\right)(x)\mu_0(dx)\,.\end{aligned} \tag{2.18}$$

From Lemma 2.1 and its proof, we know that

$$\int_{R^d}\log\left(\frac{T_h u_k}{u_k}\right)(x)\mu_0(dx)\geqq -hI(\mu_0)\,,$$

and thus from (2.18) and (2.17) we get

$$\begin{aligned}\int_{R^d}\log\left(\frac{S_h u_k}{u_k}\right)(x)\mu_0(dx)&\geqq h\left[\int_{R^d}V(x)\mu_0(dx)-I(\mu_0)\right]\\&=hl_{V,G}\,.\end{aligned} \tag{2.19}$$

Since $S_h u_k=u_{k+1}$, we rewrite (2.19) as

$$\int_{R^d}\log\left(\frac{u_{k+1}}{u_k}\right)(x)\mu_0(dx)\geqq hl_{V,G}\,,$$

and summing on k from 0 to $n-1$, we obtain

$$\int_{R^d}\log\left(\frac{u_n}{u}\right)(x)\mu_0(dx)\geqq nhl_{V,G}\,. \tag{2.20}$$

In (2.20) we let $n \to \infty$ and $h \to 0$ so that $nh \to t$. First of all, we have by Trotter's formula

$$\lim_{\substack{n\to\infty \\ h\to 0 \\ nh\to t}} u_n = \lim_{\substack{n\to\infty \\ h\to 0 \\ nh\to t}} S_h^n u = \lim_{\substack{n\to\infty \\ h\to 0 \\ nh\to t}} [e^{hV(x)} T_h]^n u = T_t^V u,$$

where $T_t^V = T_t^{V,G}$, as defined by (2.6), but with $G = R^d$. Thus from (2.20) we obtain

$$\int_{R^d} \log\left(\frac{T_t^V u}{u}\right)(x)\mu_0(dx) \geqq t l_{V,G},$$

or

$$\int_{R^d} \log(T_t^V u)(x)\mu_0(dx) \geqq t l_{V,G} + \int_{R^d} \log u(x)\mu_0(dx). \tag{2.21}$$

Inequality (2.21) holds for any $u \in C_+^\infty(R^d)$ and we choose $u(x) \equiv 1$ on $\bar{G}$. Since $\mu_0(\bar{G}) = 1$, we get from (2.21), for such u,

$$\int_{R^d} \log(T_t^V u)(x)\mu_0(dx) \geqq t l_{V,G}. \tag{2.22}$$

From Jensen's inequality this gives us

$$\log \int_{R^d} (T_t^V u)(x)\mu_0(dx) \geqq t l_{V,G},$$

or

$$\int_{R^d} (T_t^V u)(x)\mu_0(dx) \geqq \exp\{t l_{V,G}\}. \tag{2.23}$$

In (2.23), $\mu_0(\bar{G}) = 1$ so that $l_{V,G} = \int_{R^d} V(x)\mu_0(dx) - I(\mu_0)$ is insensitive to the values of $V(x)$ for $x \notin \bar{G}$; we also recall that $u(x) \equiv 1$ on $\bar{G}$. Thus, since (2.23) holds for all $V(x)$ continuous on R^d, we can choose a sequence $\{V_n(x)\}$ of such functions converging to a V which is continuous on $\bar{G}$ and equal to $-\infty$ off $\bar{G}$. In this way we get

$$\int_{R^d} (T_t^{V,G} \cdot 1)(x)\mu_0(dx) \geqq \exp\{t l_{V,G}\}. \tag{2.24}$$

But (2.24) implies that

$$\|T_t^{V,G}\| = \sup_{x\in\bar{G}} (T_t^{V,G}\cdot 1)(x) \geqq \exp\{tl_{V,G}\},$$

and hence that $\lambda_{V,G} \geqq l_{V,G}$ which completes the proof that $\lambda_{V,G} = l_{V,G}$.

Finally, we want to show that $\lambda_{V,G} \in \sigma_{V,G}$. Assume $\lambda_{V,G} \notin \sigma_{V,G}$. Then, from the first part of the proof of this theorem we know that, for $\lambda > \lambda_{V,G}$,

$$(R_\lambda f)(x) = \int_0^\infty e^{-\lambda t}(T_t^{V,G}f)(x)\,dt$$

exists, and therefore assuming $\lambda_{V,G} \notin \sigma_{V,G}$ implies that

$$(R_{\lambda_{V,G}}f)(x) = \lim_{\lambda\downarrow\lambda_{V,G}} \int_0^\infty e^{-\lambda t}(T_t^{V,G}f)(x)\,dt$$

exists. If we take $f \geqq 0$, then by monotone convergence

$$(R_{\lambda_{V,G}}f)(x) = \int_0^\infty \exp\{-\lambda_{V,G}t\}(T_t^{V,G}f)(x)\,dt,$$

so that, in particular,

$$\|R_{\lambda_{V,G}}\| = \sup_{x\in\bar{G}} \int_0^\infty \exp\{-\lambda_{V,G}t\}(T_t^{V,G}\cdot 1)(x)\,dt < c < \infty.$$

But this implies

$$\int_{R^d} \mu_0(dx) \int_0^\infty \exp\{-\lambda_{V,G}t\}(T_t^{V,G}\cdot 1)(x)\,dt < c,$$

and, by Fubini's theorem,

$$\int_0^\infty \exp\{-\lambda_{V,G}t\}\,dt \int_{R^d} (T_t^{V,G}\cdot 1)(x)\mu_0(dx) < c. \tag{2.25}$$

Now (2.24) and (2.25) yield

$$\int_0^\infty \exp\{-\lambda_{V,G}t\}\exp\{tl_{V,G}\}\,dt < c < \infty,$$

and, having just proved that $\lambda_{V,G} = l_{V,G}$, we obtain a contradiction so that $l_{V,G} \in \sigma_{V,G}$. This completes the proof of Theorem 2.2.

With $T_t^{V,G}$ defined by (2.6) and $\lambda_{V,G}$ defined by (2.7) we now prove

THEOREM 2.3. *For any* μ *with* $\operatorname{supp} \mu \subset \bar{G}$,

$$I(\mu) = -\inf_{V \in C^\infty(R^d)} \left[\lambda_{V,G} - \int_{R^d} V(x)\mu(dx)\right]. \tag{2.26}$$

Proof: From (2.8) of Theorem 2.2 we know that the right side of (2.26) is

$$-\inf_{V \in C^\infty(R^d)} \left[\sup_{\alpha:\alpha(\bar{G})=1} \left\{\int_{R^d} V(x)\alpha(dx) - I(\alpha)\right\} - \int_{R^d} V(x)\mu(dx)\right]$$

$$= -\inf_{V \in C^\infty(R^d)} \sup_{\alpha:\alpha(\bar{G})=1} \left[-I(\alpha) + \int_{R^d} V(x)\alpha(dx) - \int_{R^d} V(x)\mu(dx)\right].$$

We use again the mini-max theorem of Sion ([2], Corollary 3.3) to see that the right side of (2.26) is

$$-\sup_{\alpha:\alpha(\bar{G})=1} \inf_{V \in C^\infty(R^d)} \left[-I(\alpha) + \int_{R^d} V(x)\alpha(dx) - \int_{R^d} V(x)\mu(dx)\right]. \tag{2.27}$$

But since $\operatorname{supp} \mu \subset \bar{G}$, we note that

$$\inf_{V \in C^\infty(R^d)} \left[-I(\alpha) + \int_{R^d} V(x)\alpha(dx) - \int_{R^d} V(x)\mu(dx)\right]$$

is clearly equal to $-\infty$ if $\alpha \neq \mu$ and is equal to $-I(\mu)$ if $\alpha = \mu$. Thus (2.27) is just $I(\mu)$ as claimed.

As noted at the beginning of this section, Theorems 2.2 and 2.3 have been proved under the hypothesis that the coefficients in the operator L behave suitably at infinity; namely, $a_{ij}(x) = \delta_{ij}$ and $b_j(x) = 0$ if $|x|$ is large, for $1 \leq i, j \leq d$. We needed this assumption in order to define the semigroup T_t which we used in the proofs.

Consider

$$L = \frac{1}{2} \sum_{i,j=1}^{d} a_{ij}(x) \frac{\partial^2}{\partial x_i\, \partial x_j} + \sum_{j=1}^{d} b_j(x) \frac{\partial}{\partial x_j}, \tag{2.28}$$

where $a_{ij}(x)$ and $b_j(x)$ are in $C^\infty(R^d)$ for $1 \leq i, j \leq d$ and the matrix $(a_{ij}(x))$ is positive definite for each $x \in R^d$ but where now we make no hypotheses on the behavior of the coefficients at infinity. If μ is a probability measure on R^d with compact support, we can still define

$$I(\mu) = -\inf_{\substack{u>0 \\ u \in C^\infty(R^d)}} \int_{R^d} \left(\frac{Lu}{u}\right)(x)\mu(dx). \tag{2.29}$$

The following lemma is then obvious.

LEMMA 2.4. *Suppose L and L' are both operators satisfying the hypotheses imposed on (2.28) and that, for some $R, L = L'$ on $|x| \leqq R$. If $\mu \in \mathcal{M}$ and $\mu(|x| \leqq R) = 1$, then we have*

$$I'(\mu) = I(\mu).$$

Thus if L and L' are two such operators and they agree on $\bar{G}$, it follows that $\lambda_{V,G} = \lambda'_{V,G}$ and $I'(\mu) = I(\mu)$. This means Theorems 2.2 and 2.3 are true for operators (2.28) without imposing any conditions on the behavior of the coefficients as $|x| \to \infty$.

3. The Inverse Problem

As in the last section, let

$$L = \frac{1}{2} \sum_{i,j=1}^{d} a_{ij}(x) \frac{\partial^2}{\partial x_i \, \partial x_j} + \sum_{j=1}^{d} b_j(x) \frac{\partial}{\partial x_j}, \tag{3.1}$$

where $a_{ij}(x)$ and $b_j(x)$ are in $C^\infty(R^d)$ for $1 \leqq i, j \leqq d$, and the matrix $(a_{ij}(x))$ is positive definite for each $x \in R^d$. If G is an open set in R^d with compact closure $\bar{G}$ and smooth boundary ∂G, and if $V(x)$ is a continuous function in R^d, then we saw in the last section that the principal eigenvalue $\lambda_{V,G}$ of the region G and potential V is given by a variational formula

$$\lambda_{V,G} = \sup_{\mu : \mu(\bar{G}) = 1} \left[\int_{R^d} V(x)\mu(dx) - I(\mu) \right]. \tag{3.2}$$

Here, $I(\mu)$ is defined for μ with compact support by

$$I(\mu) = -\inf_{\substack{u>0 \\ u \in C^\infty(R^d)}} \int_{R^d} \left(\frac{Lu}{u}\right)(x)\mu(dx). \tag{3.3}$$

We also saw in the last section that for any μ with $\operatorname{supp}\mu \subset \bar{G}$

$$(3.4) \qquad I(\mu) = -\inf_{V \in C^\infty(R^d)} \left[\lambda_{V,G} - \int_{R^d} V(x)\mu(dx)\right],$$

so that $\lambda_{V,G}$ and $I(\mu)$ are conjugate convex functions.

Suppose now we have two operators L_1 and L_2 of the type above. For each continuous potential $V(x)$ and each admissible open set G we have then the principal eigenvalue $\lambda^{(1)}_{V,G}$ corresponding to L_1 and the principal eigenvalue $\lambda^{(2)}_{V,G}$ corresponding to L_2. In this section we consider the question: If $\lambda^{(1)}_{V,G} = \lambda^{(2)}_{V,G}$ for all such regions G and all such potentials V, then what can be said about the operators L_1 and L_2? The main result is the following

THEOREM 3.1. *Let L_1 and L_2 be two diffusion operators satisfying the hypotheses imposed on* (3.1). *Suppose that for every open set $G \subset R^d$ with compact closure $\bar{G}$ and smooth boundary ∂G and for every function $V(x) \in C^\infty(R^d)$ we have*

$$(3.5) \qquad \lambda^{(1)}_{V,G} = \lambda^{(2)}_{V,G}.$$

Then, either there exists a positive harmonic function u for L_2 (i.e., $L_2 u = 0$ and $u > 0$) such that

$$L_1 f = \frac{L_2(uf)}{u},$$

or there exists a positive invariant density for L_2 (i.e., $L_2^* \rho = 0$ and $\rho > 0$) such that

$$L_1 f = \frac{L_2^*(\rho f)}{\rho}.$$

To prove this theorem we make use of the relation between $\lambda_{V,G}$ and $I(\mu)$ as exemplified by (3.4) and the special nature of $I(\mu)$ corresponding to an operator of the form (3.1). To the two operators L_1 and L_2 correspond $I_1(\mu)$ and $I_2(\mu)$ defined by (3.3) for μ with compact support. It is clear from (3.4) that the hypothesis $\lambda^{(1)}_{V,G} = \lambda^{(2)}_{V,G}$ of Theorem 3.1 implies that $I_1(\mu) = I_2(\mu)$ for all μ with compact support. Indeed, to conclude this, all we really need to assume is that $\lambda^{(1)}_{V,G} = \lambda^{(2)}_{V,G}$ for all $V \in C^\infty(R^d)$ and for an increasing sequence of bounded open sets G whose union is R^d. In any event, since (3.5) implies $I_1(\mu) = I_2(\mu)$ for all μ with compact support, we have in particular $I_1(\mu) = I_2(\mu)$ for all μ with compact support and having a density $\phi \in C^\infty(R^d)$. Hence, Theorem 3.1 will follow from

THEOREM 3.2. *If L_1 and L_2 are two operators satisfying the hypotheses imposed on* (3.1) *and if $I_1(\mu) = I_2(\mu)$ for all μ with compact support having a* *density $\phi \in C^\infty(R^d)$, then either there exists a positive harmonic function u for L_2 such that $L_1 f = L_2(uf)/u$ or there exists a positive invariant density ρ for L_2 such that $L_1 f = L_2^*(\rho f)/\rho$.*

To prove Theorem 3.2 we first prove a succession of lemmas. We shall find it convenient to express the operator (3.1) in divergence form, i.e.,

$$L = \tfrac{1}{2}\nabla \cdot a\nabla + \bar{b} \cdot \nabla . \tag{3.6}$$

We first prove

LEMMA 3.3. *Let L be given by (3.6) and let $I(\mu)$ be its I-function defined by (3.3). If μ is a probability measure on R^d with compact support such that $\mu(dx) = \phi(x)\,dx$, where $\phi(x) \in C^\infty(R^d)$, then*

$$\begin{aligned} I(\mu) &= \frac{1}{8}\int_{R^d} \frac{\langle a\,\nabla\phi, \nabla\phi\rangle}{\phi}\,dx + \frac{1}{2}\int_{R^d} [\nabla \cdot \bar{b} + \langle \bar{b}, a^{-1}\bar{b}\rangle]\phi\,dx \\ &\quad - \frac{1}{2}\inf_{W\in C^\infty(R^d)} \int_{R^d} (a^{-1}\bar{b} - \nabla W)^T a(a^{-1}\bar{b} - \nabla W)\phi\,dx . \end{aligned} \tag{3.7}$$

Remark. The last term on the right-hand side of (3.7) is clearly zero if $d = 1$.

Proof: First of all,

$$\begin{aligned} I(\mu) &= -\inf_{\substack{u>0\\ u\in C^\infty(R^d)}} \int_{R^d} \left(\frac{Lu}{u}\right)(x)\phi(x)\,dx \\ &= -\inf_{h\in C^\infty(R^d)} \int_{R^d} \left(\frac{Le^h}{e^h}\right)(x)\phi(x)\,dx \\ &= -\inf_{h\in C^\infty(R^d)} \int_{R^d} [Lh + \tfrac{1}{2}\langle a\,\nabla h, \nabla h\rangle]\phi\,dx . \end{aligned}$$

For $\varepsilon > 0$ and $h \in C^\infty(R^d)$, define $W(x) = \frac{1}{2}\log[\phi(x) + \varepsilon] - h(x)$ so that $W(x)$ is also in $C^\infty(R^d)$. An elementary calculation shows that

$$\begin{aligned} Lh + \tfrac{1}{2}\langle a\,\nabla h, \nabla h\rangle &= \frac{1}{4}\frac{\nabla \cdot a\,\nabla\phi}{(\phi+\varepsilon)} - \frac{1}{8}\frac{\langle a\,\nabla\phi, \nabla\phi\rangle}{(\phi+\varepsilon)^2} + \frac{\bar{b}\cdot\nabla\phi}{2(\phi+\varepsilon)} - \bar{b}\cdot\nabla W \\ &\quad - \tfrac{1}{2}\nabla \cdot a\,\nabla W + \tfrac{1}{2}\langle a\,\nabla W, \nabla W\rangle - \frac{\langle a\,\nabla\phi, \nabla W\rangle}{2(\phi+\varepsilon)} . \end{aligned}$$

Therefore,

$$I(\mu) = -\inf_{W\in C^\infty(R^d)}\left[\int_{R^d}\left\{\frac{\nabla\cdot a\nabla\phi}{4(\phi+\varepsilon)} - \frac{\langle a\nabla\phi,\nabla\phi\rangle}{8(\phi+\varepsilon)^2} + \frac{\tilde b\cdot\nabla\phi}{2(\phi+\varepsilon)} - \tilde b\cdot\nabla W\right.\right.$$
$$\left.\left. -\tfrac12\nabla\cdot a\nabla W + \tfrac12\langle a\nabla W,\nabla W\rangle - \frac{\langle a\nabla\phi,\nabla W\rangle}{2(\phi+\varepsilon)}\right\}\phi\,dx\right]$$

$$(3.8)\qquad = -\frac14\int_{R^d}\frac{\nabla\cdot a\nabla\phi}{(\phi+\varepsilon)}\phi\,dx + \frac18\int_{R^d}\frac{\langle a\nabla\phi,\nabla\phi\rangle}{(\phi+\varepsilon)^2}\phi\,dx - \frac12\int_{R^d}\frac{\tilde b\cdot\nabla\phi}{(\phi+\varepsilon)}\phi\,dx$$
$$-\inf_{W\in C^\infty(R^d)}\left[\int_{R^d}\left\{\tfrac12\langle a\nabla W,\nabla W\rangle - \tilde b\cdot\nabla W - \tfrac12\nabla\cdot a\nabla W\right.\right.$$
$$\left.\left. - \frac{\langle a\nabla\phi,\nabla W\rangle}{2(\phi+\varepsilon)}\right\}\phi\,dx\right].$$

Integrating by parts, we see that

$$\frac12\int_{R^d}(\nabla\cdot a\nabla W)\phi\,dx = -\frac12\int_{R^d}\langle a\nabla\phi,\nabla W\rangle\,dx,$$

and hence

$$-\frac12\int_{R^d}(\nabla\cdot a\nabla W)\phi\,dx - \frac12\int_{R^d}\frac{\langle a\nabla\phi,\nabla W\rangle}{\phi+\varepsilon}\phi\,dx = \tfrac12\varepsilon\int_{R^d}\frac{\langle a\nabla\phi,\nabla W\rangle}{\phi+\varepsilon}\,dx.$$

Thus

$$I(\mu) = -\frac14\int_{R^d}\frac{\nabla\cdot a\nabla\phi}{\phi+\varepsilon}\phi\,dx + \frac18\int_{R^d}\frac{\langle a\nabla\phi,\nabla\phi\rangle}{\phi+\varepsilon^{\;2}}\phi\,dx$$
$$(3.9)\qquad -\frac12\int_{R^d}\frac{\tilde b\cdot\nabla\phi}{\phi+\varepsilon}\phi\,dx - \inf_{W\in C^\infty(R^d)}\left[\int_{R^d}\{\tfrac12\langle a\nabla W,\nabla W\rangle - \tilde b\cdot\nabla W\}\phi\,dx\right.$$
$$\left. + \tfrac12\varepsilon\int_{R^d}\frac{\langle a\nabla\phi,\nabla W\rangle}{\phi+\varepsilon}\,dx\right].$$

Since $\phi\in C^\infty(R^d)$ with compact support, $\int_{R^d}\nabla\cdot a\nabla\phi\,dx = 0$, and therefore

$$\lim_{\varepsilon\to0}\left[-\frac14\int_{R^d}\frac{\nabla\cdot a\nabla\phi}{\phi+\varepsilon}\phi\,dx + \frac18\int_{R^d}\frac{\langle a\nabla\phi,\nabla\phi\rangle}{(\phi+\varepsilon)^2}\phi\,dx - \frac12\int_{R^d}\frac{\tilde b\cdot\nabla\phi}{\phi+\varepsilon}\phi\,dx\right]$$
$$(3.10)\qquad = \frac18\int_{R^d}\frac{\langle a\nabla\phi,\nabla\phi\rangle}{\phi}\,dx + \frac12\int_{R^d}(\nabla\cdot\tilde b)\phi\,dx.$$

We shall show below that

$$
\begin{aligned}
(3.11)\quad & \lim_{\varepsilon\to 0} \inf_{W\in C^\infty(R^d)} \left[\int_{R^d} \{\tfrac{1}{2}\langle a\,\nabla W, \nabla W\rangle - \tilde{b}\cdot\nabla W\}\phi\,dx + \tfrac{1}{2}\varepsilon\int_{R^d} \frac{\langle a\,\nabla\phi, \nabla W\rangle}{\phi+\varepsilon}\,dx\right] \\
&= \inf_{W\in C^\infty(R^d)} \lim_{\varepsilon\to 0} \left[\int_{R^d} \{\tfrac{1}{2}\langle a\,\nabla W, \nabla W\rangle - \tilde{b}\cdot\nabla W\}\phi\,dx + \tfrac{1}{2}\varepsilon\int_{R^d} \frac{\langle a\,\nabla\phi, \nabla W\rangle}{\phi+\varepsilon}\,dx\right] \\
&= \inf_{W\in C^\infty(R^d)} \int_{R^d} \{\tfrac{1}{2}\langle a\,\nabla W, \nabla W\rangle - \tilde{b}\cdot\nabla W\}\phi\,dx .
\end{aligned}
$$

Thus, if we use (3.10) and (3.11) in (3.9), we get

$$
I(\mu) = \frac{1}{8}\int_{R^d} \frac{\langle a\,\nabla\phi, \nabla\phi\rangle}{\phi}\,dx + \frac{1}{2}\int_{R^d} (\nabla\cdot\tilde{b})\phi\,dx - \inf_{W\in C^\infty(R^d)} \int_{R^d} \{\tfrac{1}{2}\langle a\,\nabla W, \nabla W\rangle - \tilde{b}\cdot\nabla W\}\phi\,dx ,
$$

and by adding and subtracting the term $\frac{1}{2}\int_{R^d} \langle\tilde{b}, a^{-1}\tilde{b}\rangle\phi\,dx$, we obtain

$$
I(\mu) = \frac{1}{8}\int_{R^d} \frac{\langle a\,\nabla\phi, \nabla\phi\rangle}{\phi}\,dx + \frac{1}{2}\int_{R^d} [\nabla\cdot\tilde{b} + \langle\tilde{b}, a^{-1}\tilde{b}\rangle]\phi\,dx - \frac{1}{2}\inf_{W\in C^\infty(R^d)} \int_{R^d} (a^{-1}\tilde{b} - \nabla W)^T a (a^{-1}\tilde{b} - \nabla W)\phi\,dx
$$

which is what the lemma states. It remains to prove (3.11). Let

$$
H[W,\varepsilon] = \int_{R^d} \{\tfrac{1}{2}\langle a\,\nabla W, \nabla W\rangle - \tilde{b}\cdot\nabla W\}\phi\,dx + \tfrac{1}{2}\varepsilon\int_{R^d} \frac{\langle a\,\nabla\phi, \nabla W\rangle}{\phi+\varepsilon}\,dx ,
$$

and since always $\overline{\lim}_{\varepsilon\to 0} \inf_{W\in C^\infty(R^d)} H[W,\varepsilon] \leqq \inf_{W\in C^\infty(R^d)} \lim_{\varepsilon\to 0} H[W,\varepsilon]$, it suffices to show that

$$
(3.12)\qquad \underline{\lim}_{\varepsilon\to 0} \inf_{W\in C^\infty(R^d)} H[W,\varepsilon] \geqq \inf_{W\in C^\infty(R^d)} \lim_{\varepsilon\to 0} H[W,\varepsilon] .
$$

We note that, for $W\equiv 0$, $H[W,\varepsilon]=0$ so that $\inf_{W\in C^\infty(R^d)} H[W,\varepsilon]\leqq 0$. Let $\mathscr{W}$ be the set of $W\in C^\infty(R^d)$ such that $H[W,\varepsilon]\leqq 0$. We show now that

$$
(3.13)\qquad \lim_{\varepsilon\to 0} \tfrac{1}{2}\varepsilon \int_{R^d} \frac{\langle a\,\nabla\phi, \nabla W\rangle}{\phi+\varepsilon}\,dx = 0
$$

uniformly for $W \in \mathscr{W}$ which clearly implies (3.12). If we denote the vectors ∇W and $\tilde{b} - \varepsilon a \nabla\phi/(2\phi(\phi+\varepsilon))$ by ξ and η, respectively, then

$$\begin{aligned} H[W,\varepsilon] &= \int_{R^d} \{\tfrac{1}{2}\langle a\xi,\xi\rangle - \langle \eta,\xi\rangle\}\phi\,dx \\ &= \int_{R^d} \{\tfrac{1}{4}\langle a\xi,\xi\rangle + \tfrac{1}{4}\langle a\xi,\xi\rangle - \langle\eta,\xi\rangle + \langle\eta,a^{-1}\eta\rangle - \langle\eta,a^{-1}\eta\rangle\}\phi\,dx \\ &= \int_{R^d} \{\tfrac{1}{4}\langle a\xi,\xi\rangle + \langle a(\tfrac{1}{2}\xi - a^{-1}\eta),(\tfrac{1}{2}\xi - a^{-1}\eta)\rangle - \langle\eta,a^{-1}\eta\rangle\}\phi\,dx \\ &\geqq \int_{R^d} \{\tfrac{1}{4}\langle a\xi,\xi\rangle - \langle\eta,a^{-1}\eta\rangle\}\phi\,dx. \end{aligned}$$

Thus, if $W \in \mathscr{W}$,

$$\int_{R^d} \tfrac{1}{4}\langle a\xi,\xi\rangle\phi\,dx \leqq \int_{R^d} \langle\eta,a^{-1}\eta\rangle\phi\,dx,$$

i.e.,

$$\tag{3.14} \begin{aligned} &\int_{R^d} \langle a\nabla W,\nabla W\rangle\phi\,dx \\ &\leqq 4\left[\int_{R^d}\langle\tilde{b},a^{-1}\tilde{b}\rangle\phi\,dx - \varepsilon\int_{R^d}\frac{\langle\tilde{b},\nabla\phi\rangle}{\phi+\varepsilon}\,dx + \tfrac{1}{4}\varepsilon^2\int_{R^d}\frac{\langle a\nabla\phi,\nabla\phi\rangle}{\phi(\phi+\varepsilon)^2}\,dx\right]. \end{aligned}$$

The right side of (3.14) is bounded in ε, and hence there exists a constant C such that, for all $W \in \mathscr{W}$,

$$\tag{3.15} \int_{R^d}\langle a\nabla W,\nabla W\rangle\phi\,dx \leqq C.$$

From (3.15) and obvious inequalities we get

$$\begin{aligned} \tfrac{1}{2}\varepsilon\int_{R^d}\frac{\langle a\nabla\phi,\nabla W\rangle}{\phi+\varepsilon}\,dx &\leqq \tfrac{1}{2}\varepsilon\int_{R^d}\frac{\langle a\nabla\phi,\nabla\phi\rangle^{1/2}\langle a\nabla W,\nabla W\rangle^{1/2}}{\sqrt{\phi}\,(\phi+\varepsilon)}\sqrt{\phi}\,dx \\ &\leqq \tfrac{1}{2}\varepsilon\sqrt{\int_{R^d}\frac{\langle a\nabla\phi,\nabla\phi\rangle}{\phi(\phi+\varepsilon)^2}\,dx}\sqrt{\int_{R^d}\langle a\nabla W,\nabla W\rangle\phi\,dx} \\ &\leqq \tfrac{1}{2}\varepsilon\sqrt{C}\sqrt{\int_{R^d}\frac{\langle a\nabla\phi,\nabla\phi\rangle}{\phi(\phi+\varepsilon)^2}\,dx} \\ &= \tfrac{1}{2}\sqrt{C}\sqrt{\int_{R^d}\frac{\langle a\nabla\phi,\nabla\phi\rangle\varepsilon^2}{\phi(\phi+\varepsilon)^2}\,dx}. \end{aligned}$$

But, as $\varepsilon \to 0$, $\varepsilon/(\phi+\varepsilon) \to 0$ and $\langle a\nabla\phi, \nabla\phi\rangle \varepsilon^2/\phi(\phi+\varepsilon)^2 \leqq \langle a\nabla\phi, \nabla\phi\rangle/\phi$ so that, by dominated convergence,

$$\lim_{\varepsilon\to 0} \tfrac{1}{2}\varepsilon \int_{R^d} \frac{\langle a\nabla\phi, \nabla W\rangle}{\phi+\varepsilon}\,dx = 0$$

uniformly for $W \in \mathscr{W}$ which is (3.13). This completes the proof of Lemma 3.3.

Now let L_1 and L_2 be two operators satisfying the hypotheses imposed on (3.1). We express L_1 and L_2 in divergence form:

$$L_1 = \tfrac{1}{2}\nabla \cdot a\nabla + \tilde{b} \cdot \nabla\,,$$

$$L_2 = \tfrac{1}{2}\nabla \cdot \alpha\nabla + \tilde{\beta} \cdot \nabla\,,$$

and denote their I-functions as defined by (3.3) with $I_1(\mu)$ and $I_2(\mu)$.

LEMMA 3.4. *If $I_1(\mu) = I_2(\mu)$ for all μ with compact support having a density $\phi \in C^\infty(R^d)$, then the matrices $a = (a_{ij}(x))$ and $\alpha = (\alpha_{ij}(x))$ are identical.*

Proof: From Lemma 3.3 we have for any such μ

$$I_1(\mu) = \frac{1}{8}\int_{R^d} \frac{\langle a\nabla\phi, \nabla\phi\rangle}{\phi}\,dx + \frac{1}{2}\int_{R^d} [\nabla\cdot\tilde{b} + \langle\tilde{b}, a^{-1}\tilde{b}\rangle]\phi\,dx$$

(3.16)

$$-\frac{1}{2}\inf_{W\in C^\infty(R^d)} \int_{R^d} (a^{-1}\tilde{b} - \nabla W)^T a (a^{-1}\tilde{b} - \nabla W)\phi\,dx\,,$$

and

$$I_2(\mu) = \frac{1}{8}\int_{R^d} \frac{\langle \alpha\nabla\phi, \nabla\phi\rangle}{\phi}\,dx + \frac{1}{2}\int_{R^d} [\nabla\cdot\tilde{\beta} + \langle\tilde{\beta}, \alpha^{-1}\tilde{\beta}\rangle]\phi\,dx$$

(3.17)

$$-\frac{1}{2}\inf_{W\in C^\infty(R^d)} \int_{R^d} (\alpha^{-1}\tilde{\beta} - \nabla W)^T \alpha (\alpha^{-1}\tilde{\beta} - \nabla W)\phi\,dx\,.$$

Let $\phi(x)$ be a probability density in $C^\infty(R^d)$ with compact support. For any $\lambda > 0$, $\phi_\lambda(x) = (1/\lambda^d)\phi(x/\lambda)$ is also a probability density in $C^\infty(R^d)$ with compact support. By hypothesis, $I_1(\phi_\lambda) = I_2(\phi_\lambda)$. Since

$$\frac{1}{8}\int_{R^d} \frac{\langle a(x)\nabla\phi_\lambda, \nabla\phi_\lambda\rangle}{\phi_\lambda}\,dx = \frac{1}{8\lambda^2}\int_{R^d} \frac{\langle a(\lambda x)\nabla\phi, \nabla\phi\rangle}{\phi}\,dx\,,$$

it is clear from (3.16) that

$$\lim_{\lambda\to 0} \lambda^2 I_1(\phi_\lambda) = \frac{1}{8}\int_{R^d} \frac{\langle a(0)\nabla\phi, \nabla\phi\rangle}{\phi}\,dx\,.$$

Similarly, from (3.17) we see that

$$\lim_{\lambda\to 0} \lambda^2 I_2(\phi_\lambda) = \frac{1}{8}\int_{R^d} \frac{\langle \alpha(0)\,\nabla\phi, \nabla\phi\rangle}{\phi}\,dx\,,$$

and hence that, for all ϕ,

$$\frac{1}{8}\int_{R^d} \frac{\langle a(0)\,\nabla\phi, \nabla\phi\rangle}{\phi}\,dx = \frac{1}{8}\int_{R^d} \frac{\langle \alpha(0)\,\nabla\phi, \nabla\phi\rangle}{\phi}\,dx\,.$$

This implies $a(0)=\alpha(0)$, i.e., $(a_{ij}(0))=(\alpha_{ij}(0))$. If we take $\phi_\lambda(x)=(1/\lambda^d)\phi((x-x_0)/\lambda)$, where x_0 is any point in R^d, the same argument leads to $(a_{ij}(x_0))=(\alpha_{ij}(x_0))$.

LEMMA 3.5. *If $I_1(\mu)=I_2(\mu)$ for all μ with compact support having a density $\phi\in C^\infty(R^d)$, then*

$$[\nabla\cdot\tilde{b}+\langle\tilde{b}, a^{-1}\tilde{b}\rangle](x)\equiv[\nabla\cdot\tilde{\beta}+\langle\tilde{\beta}, \alpha^{-1}\tilde{\beta}\rangle](x)\,. \tag{3.18}$$

Proof: From Lemmas 3.3 and 3.4, the hypothesis $I_1(\mu)=I_2(\mu)$ implies

$$\begin{aligned}\frac{1}{2}\int_{R^d}[\nabla\cdot\tilde{b}+\langle\tilde{b}, a^{-1}\tilde{b}\rangle]\phi\,dx-\frac{1}{2}\inf_{W\in C^\infty(R^d)}\int_{R^d}(a^{-1}\tilde{b}-\nabla W)^T a(a^{-1}\tilde{b}-\nabla W)\phi\,dx\\ =\frac{1}{2}\int_{R^d}[\nabla\cdot\tilde{\beta}+\langle\tilde{\beta}, \alpha^{-1}\tilde{\beta}\rangle]\phi\,dx\\ -\frac{1}{2}\inf_{W\in C^\infty(R^d)}\int_{R^d}(\alpha^{-1}\tilde{\beta}-\nabla W)^T\alpha(\alpha^{-1}\tilde{\beta}-\nabla W)\phi\,dx\,.\end{aligned} \tag{3.19}$$

If we take a sequence of probability densities $\phi_k\in C^\infty(R^d)$ having compact support and approaching δ_{x_0}, then (3.18) follows from (3.19) providing we can show

$$\lim_{k\to\infty}\frac{1}{2}\inf_{W\in C^\infty(R^d)}\int_{R^d}(a^{-1}\tilde{b}-\nabla W)^T a(a^{-1}\tilde{b}-\nabla W)\phi_k\,dx=0\,. \tag{3.20}$$

Now $(a^{-1}\tilde{b})(x_0)$ is a vector θ, and if we take $W_0=\langle\theta, x\rangle$ for x near x_0 and extend it arbitrarily but so that $W_0\in C^\infty(R^d)$, then clearly $(a^{-1}\tilde{b}-\nabla W_0)(x_0)=0$. Thus

$$\lim_{k\to\infty}\int_{R^d}(a^{-1}\tilde{b}-\nabla W_0)a(a^{-1}\tilde{b}-\nabla W_0)\phi_k(x)\,dx=0$$

which implies (3.20) and the lemma.

LEMMA 3.6. *If $I_1(\mu)=I_2(\mu)$ for all μ with compact support having a density $\phi\in C^\infty(R^d)$, then*

$$\left|\int_C \langle a^{-1}\tilde{b}, dx\rangle\right|^2=\left|\int_C \langle \alpha^{-1}\tilde{\beta}, dx\rangle\right|^2 \tag{3.21}$$

for any smooth simple closed contour C in R^d.

Remark. The lemma is vacuous if $d=1$. We assume that $d\geqq 2$.

Proof: Let

$$K_1(\phi)=\frac{1}{2}\inf_{W\in C^\infty(R^d)}\int_{R^d}(a^{-1}\tilde{b}-\nabla W)^T a(a^{-1}\tilde{b}-\nabla W)\phi\,dx$$

and

$$K_2(\phi)=\frac{1}{2}\inf_{W\in C^\infty(R^d)}\int_{R^d}(\alpha^{-1}\tilde{\beta}-\nabla W)^T \alpha(\alpha^{-1}\tilde{\beta}-\nabla W)\phi\,dx.$$

From Lemmas 3.3, 3.4, and 3.5 we see that the hypothesis $I_1(\mu)=I_2(\mu)$ implies $K_1(\phi)=K_2(\phi)$ for all probability densities $\phi\in C^\infty(R^d)$ with compact support.

Let C be a smooth simple closed contour in R^d. Consider a tubular neighborhood of C and foliate it as $D\times S^1$, where D is an open neighborhood of 0 in R^{d-1} and S^1 is the circle. For each $y\in D$ we have a simple closed contour C_y in the tubular neighborhood of C with C itself being $\{0\}\times S^1$. For any $y\in D$, consider the non-negative measure $\bar{\sigma}_y(dx)$ equal to the arc length on C_y in the metric $a^{ij}\,dx_i\,dx_j$, where $(a^{ij}(x))$ is the matrix inverse to $(a_{ij}(x))$. Let $l(y)$ be the total length of C_y in this metric, i.e., the total mass of $\bar{\sigma}_y(dx)$. In particular, $l(0)$ is the length of C in this metric. Since $I_1(\mu)=I_2(\mu)$ implies $(a_{ij}(x))=(\alpha_{ij}(x))$ for all x (Lemma 3.4), we note that $l(y)$ is the same whether based on the metric coming from a or α. We define $\sigma_y(dx)=\bar{\sigma}_y(dx)/l(y)$. Let $\psi_\varepsilon(y)$ be a family of smooth probability densities supported in D and such that $\psi_\varepsilon(y)\to\delta_0$ as $\varepsilon\to 0$. We define a probability measure μ_ε on R^d by setting, for $A\subset R^d$,

$$\mu_\varepsilon(A)=\int_D \frac{\bar{\sigma}_y(A)}{l(y)}\psi_\varepsilon(y)\,dy. \tag{3.22}$$

We note that μ_ε has compact support and has a density $\phi_\varepsilon\in C^\infty(R^d)$. We shall show that

$$\lim_{\varepsilon\to 0}K_1(\phi_\varepsilon)=\frac{1}{2}\frac{\left|\int_C\langle a^{-1}\tilde{b}, dx\rangle\right|^2}{[l(0)]^2}; \tag{3.23}$$

the same proof shows that

$$\lim_{\varepsilon\to 0} K_2(\phi_\varepsilon) = \frac{1}{2}\frac{\left|\int_C \langle \alpha^{-1}\tilde{\beta}, dx\rangle\right|^2}{[l(0)]^2}.$$

Since $K_1(\phi_\varepsilon) = K_2(\phi_\varepsilon)$ for all $\varepsilon > 0$ and, as noted above, $l(0)$ is the same whether based on a or α, we see that (3.21) follows. So we must show that (3.23) is true.

Fix an orientation on the contours C_y and let $T(x)$ be the oriented tangent vector at x to the contour through x normalized so that $\langle a^{-1}(x)T(x), T(x)\rangle = 1$. Consider this vector field $T(x)$. Choose a constant λ so that

$$\int_C \langle a^{-1}\tilde{b} - \lambda a^{-1}T(x), dx\rangle = 0, \tag{3.24}$$

i.e.,

$$\lambda = \frac{\int_C \langle a^{-1}\tilde{b}, dx\rangle}{\int_C \langle a^{-1}T(x), dx\rangle} = \frac{\int_C \langle a^{-1}\tilde{b}, dx\rangle}{l(0)}. \tag{3.25}$$

The last equality in (3.25) follows from the normalization of $T(x)$ because $dx = T(x)\,d\sigma$ and hence

$$\int_C \langle a^{-1}T(x), dx\rangle = \int_C \langle a^{-1}T(x), T(x)\rangle\,d\sigma = \int_C d\sigma = l(0).$$

Now

$$\left|\sqrt{\frac{1}{2}\inf_{W\in C^\infty(R^d)}\int_{R^d}(a^{-1}\tilde{b} - \nabla W)^T a(a^{-1}\tilde{b} - \nabla W)\phi\,dx} - \sqrt{\frac{1}{2}\inf_{W\in C^\infty(R^d)}\int_{R^d}(\lambda a^{-1}T(x) - \nabla W)^T a(\lambda a^{-1}T(x) - \nabla W)\phi\,dx}\right| \tag{3.26}$$
$$\leq \sqrt{\frac{1}{2}\inf_{W\in C^\infty(R^d)}\int_{R^d}[(a^{-1}\tilde{b} - \lambda a^{-1}T(x)) - \nabla W]^T a[(a^{-1}\tilde{b} - \lambda a^{-1}T(x)) - \nabla W]\phi\,dx}.$$

To verify inequality (3.26) notice that $f(y) = \sqrt{\int_{R^d} y^T a y \phi\,dx}$ is a norm; if we denote by M the linear manifold of functions on R^d which are gradients, then $g(y)=\inf_{z\in M} f(y-z)$ is again a norm and (3.26) is just the triangle inequality for this norm.

In (3.26) replace ϕ by the ϕ_ε defined just after (3.22). In Lemma 3.7 below we shall show that if $f=(f_1, f_2, \cdots, f_d)$ is a vector of smooth functions defined on C, then a necessary and sufficient condition for the existence of a function W on R^d such that $\nabla W=f$ on C is that $\int_C \langle f, dx\rangle=0$. Now we chose λ so that $\int_C \langle a^{-1}\tilde{b}-\lambda a^{-1}T(x), dx\rangle=0$ and hence, by Lemma 3.7, there exists W_0 on R^d such that, on C, $\nabla W_0=a^{-1}\tilde{b}-\lambda a^{-1}T(x)$. Moreover, with ϕ replaced by ϕ_ε, the integral over R^d on the right of (3.26) becomes the integral over C as $\varepsilon\to 0$. We conclude that

$$\overline{\lim_{\varepsilon\to 0}}\left|\sqrt{\frac{1}{2}\inf_{W\in C^\infty(R^d)}\int_{R^d}(a^{-1}\tilde{b}-\nabla W)^T a(a^{-1}\tilde{b}-\nabla W)\phi_\varepsilon\,dx}\right. \tag{3.27}$$
$$\left.-\sqrt{\frac{1}{2}\inf_{W\in C^\infty(R^d)}\int_{R^d}(\lambda a^{-1}T(x)-\nabla W)^T a(\lambda a^{-1}T(x)-\nabla W)\phi_\varepsilon\,dx}\right|=0\,.$$

Each radical is bounded because the integral under each radical is bounded for $W=0$. Thus we can remove the square roots and obtain

$$\overline{\lim_{\varepsilon\to 0}}\left|K_1(\phi_\varepsilon)-\frac{1}{2}\inf_{W\in C^\infty(R^d)}\int_{R^d}(\lambda a^{-1}T(x)-\nabla W)^T a(\lambda a^{-1}T(x)\right. \tag{3.28}$$
$$\left.-\nabla W)\phi_\varepsilon\,dx\right|=0\,.$$

We show now that, for each $\varepsilon>0$,

$$\frac{1}{2}\inf_{W\in C^\infty(R^d)}\int_{R^d}(\lambda a^{-1}T(x)-\nabla W)^T a(\lambda a^{-1}T(x)-\nabla W)\phi_\varepsilon\,dx=\tfrac{1}{2}\lambda^2 \tag{3.29}$$
$$=\frac{\left|\int_C\langle a^{-1}\tilde{b}, dx\rangle\right|^2}{2l^2(0)},$$

which, because of (3.28), proves (3.23).

To show (3.29) consider first of all

$$\int_{R^d}\langle T(x), \nabla W\rangle\phi_\varepsilon\,dx\,.$$

We note that

$$\int_{R^d} \langle T(x), \nabla W\rangle \phi_\varepsilon \, dx = \int_D \int_{R^d} \langle T(x), \nabla W\rangle \frac{\bar{\sigma}_y(dx)}{l(y)} \psi_\varepsilon(y)\, dy$$

$$= \int_D \frac{1}{l(y)}\left[\int_{R^d} \langle T(x), \nabla W\rangle \bar{\sigma}_y(dx)\right] \psi_\varepsilon(y)\, dy$$

$$= \int_D \frac{1}{l(y)}\left[\int_{C_y} \nabla W \, dx\right] \psi_\varepsilon(y)\, dy,$$

because $T(x)\bar{\sigma}_y(dx)$ is the element dx on C_y. But, for any W, $\int_{C_y} \nabla W\, dx = 0$ and hence

$$\int_{R^d} \langle T(x), \nabla W\rangle \phi_\varepsilon \, dx = 0\,. \tag{3.30}$$

Returning to the left side of (3.29) we have then

$$\frac{1}{2} \inf_{W\in C^\infty(R^d)} \int_{R^d} (\lambda a^{-1} T(x) - \nabla W) a (\lambda a^{-1} T(x) - \nabla W) \phi_\varepsilon \, dx$$

$$= \tfrac{1}{2}\lambda^2 \inf_{W\in C^\infty(R^d)} \int_{R^d} (a^{-1} T(x) - \nabla W) a (a^{-1} T(x) - \nabla W) \phi_\varepsilon \, dx$$

$$= \tfrac{1}{2}\lambda^2 \inf_{W\in C^\infty(R^d)} \left[\int_{R^d} \langle a^{-1} T(x), T(x)\rangle \phi_\varepsilon \, dx \right.$$

$$\left. + \int_{R^d} \langle a \nabla W, \nabla W\rangle \phi_\varepsilon \, dx - 2\int_{R^d} \langle T(x), \nabla W\rangle \phi_\varepsilon \, dx\right]$$

$$= \tfrac{1}{2}\lambda^2 \inf_{W\in C^\infty(R^d)} \left[\int_{R^d} \langle a^{-1} T(x), T(x)\rangle \phi_\varepsilon \, dx + \int_{R^d} \langle a \nabla W, \nabla W\rangle \phi_\varepsilon \, dx\right].$$

But both terms inside the brackets are non-negative for each $\varepsilon > 0$ so the infimum is taken on when $W = 0$. Thus, from the normalization of $T(x)$,

$$\frac{1}{2} \inf_{W\in C^\infty(R^d)} \int_{R^d} (\lambda a^{-1} T(x) - \nabla W) a (\lambda a^{-1} T(x) - \nabla W) \phi_\varepsilon \, dx$$

$$= \tfrac{1}{2}\lambda^2 \int_{R^d} \langle a^{-1} T(x), T(x)\rangle \phi_\varepsilon \, dx = \tfrac{1}{2}\lambda^2,$$

for each $\varepsilon > 0$. This proves (3.29) and the lemma except for

LEMMA 3.7. *Let $f = (f_1, f_2, \cdots, f_d)$ be a vector of smooth functions defined on C. In order that there exists a function W on R^d such that $\nabla W = f$ on C, it is necessary and sufficient that*

$$\int_C \langle f, dx \rangle = 0 .$$

Proof: The necessity is obvious and for the sufficiency it is clearly enough to construct W in a tubular neighborhood of C. We can change the coordinates in such a way that the tubular neighborhood is $G \times S^1$, where G is an open subset of R^{d-1} containing the origin and S^1 is the circle. The contour C itself is $\{0\} \times S^1$. Then, our hypothesis becomes

$$\int_{S^1} f_d(0, 0, 0, \cdots, \theta)\, d\theta = 0 .$$

There exists a function $F(\theta)$ such that $dF(\theta)/d\theta = f_d(0, 0, 0, \cdots, \theta)$ and so if we define $W(y, \theta)$ for $y \in G$ and $\theta \in S^1$ by

$$W(y, \theta) = F(\theta) + \sum_{j=1}^{d-1} y_j f_j(0, 0, 0, \cdots, \theta) ,$$

then $W(y, \theta)$ satisfies $\nabla W = f$ on C.

LEMMA 3.8. *Let f and g be two smooth R^d-valued functions on R^d such that*

$$\left| \int_C \langle f(x), dx \rangle \right|^2 = \left| \int_C \langle g(x), dx \rangle \right|^2$$

for all simple smooth closed contours C in R^d. Then either

$$\int_C \langle f(x), dx \rangle = \int_C \langle g(x), dx \rangle$$

for all closed rectifiable contours C in R^d or

$$\int_C \langle f(x), dx \rangle = -\int_C \langle g(x), dx \rangle$$

for all closed rectifiable contours C in R^d.

Remark. Again we can assume that $d \geqq 2$. The case $d = 2$ is the difficult one. The following proof covers all cases $d \geqq 2$.

Proof: First we show that, for any two nonintersecting simple smooth closed contours C_1 and C_2,

$$\int_{C_1} \langle f(x), dx\rangle \int_{C_2} \langle f(x), dx\rangle = \int_{C_1} \langle g(x), dx\rangle \int_{C_2} \langle g(x), dx\rangle . \tag{3.31}$$

To see this, let C_1 and C_2 be two such contours and connect them by an arc γ which has no additional intersections with C_1 and C_2. Then, with a suitable orientation of C_1 and C_2, the closed contour $C = C_1\gamma C_2\gamma^{-1}$ can be approximated by simple smooth closed contours. Therefore, from our hypothesis,

$$\left|\int_C \langle f(x), dx\rangle\right|^2 = \left|\int_C \langle g(x), dx\rangle\right|^2 ,$$

i.e.,

$$\left|\int_{C_1} \langle f(x), dx\rangle + \int_{C_2} \langle f(x), dx\rangle\right|^2 = \left|\int_{C_1} \langle g(x), dx\rangle + \int_{C_2} \langle g(x), dx\rangle\right|^2$$

which again using the hypothesis implies (3.31).

Now let C be any closed rectifiable contour in R^d. We can approximate C by contours C_n each of which is a finite union of mutually disjoint smooth simple closed contours. Therefore, from (3.31) and the hypothesis, it follows that

$$\left|\int_C \langle f(x), dx\rangle\right|^2 = \left|\int_C \langle g(x), dx\rangle\right|^2 \tag{3.32}$$

for all closed rectifiable contours C in R^d. If we now let C_1 and C_2 be any two closed rectifiable contours and let γ be an arc connecting them, then $C = C_1\gamma C_2\gamma^{-1}$ is again a closed rectifiable contour and from (3.32) we have

$$\int_{C_1} \langle f(x), dx\rangle \int_{C_2} \langle f(x), dx\rangle = \int_{C_1} \langle g(x), dx\rangle \int_{C_2} \langle g(x), dx\rangle .$$

This last equation clearly implies the lemma.

We are now ready for the

Proof of Theorem 3.2: We have two operators L_1 and L_2 satisfying the hypotheses imposed on (3.1) and for which $I_1(\mu) = I_2(\mu)$ for all μ with compact support having a density $\phi \in C^\infty(R^d)$. If we write L_1 and L_2 in divergence form:

$$L_1 = \tfrac{1}{2}\nabla \cdot a\, \nabla + \tilde{b} \cdot \nabla ,$$

$$L_2 = \tfrac{1}{2}\nabla \cdot \alpha\, \nabla + \tilde{\beta} \cdot \nabla ,$$

then the preceding lemmas showed that the hypothesis $I_1(\mu)=I_2(\mu)$ implies

$$a(x)=(a_{ij}(x))=(\alpha_{ij}(x))=\alpha(x) \quad \text{for all} \quad x\in R^d, \tag{3.33}$$

$$\nabla\cdot\tilde{b}+\tilde{b}^T a^{-1}\tilde{b}=\nabla\cdot\tilde{\beta}+\tilde{\beta}^T\alpha^{-1}\tilde{\beta} \quad \text{for all} \quad x\in R^d, \tag{3.34}$$

and either

$$\int_C \langle a^{-1}\tilde{b}, dx\rangle=\int_C \langle a^{-1}\tilde{\beta}, dx\rangle \text{ for all closed rectifiable contours } C \text{ in } R^d, \tag{3.35a}$$

or

$$\int_C \langle a^{-1}\tilde{b}, dx\rangle=-\int_C \langle a^{-1}\tilde{\beta}, dx\rangle \text{ for all closed rectifiable contours } C \text{ in } R^d. \tag{3.35b}$$

If $d=1$, (3.35) is a vacuous statement and we work only with (3.33) and (3.34). Equations (3.35) are used only to show that $a^{-1}(\tilde{b}-\tilde{\beta})$ is a gradient of a function. In case $d=1$ this of course is always true. Another feature of the one-dimensional situation is that the two alternatives $L_1f=L_2(uf)/u$ and $L_1f=L_2^*(\rho f)/\rho$ lead to the same class of operators L_1. This is because any L is selfadjoint with respect to some measure on the line.

Suppose the first alternative in (3.35) holds so that

$$\int_C \langle a^{-1}(\tilde{b}-\tilde{\beta}), dx\rangle=0$$

for all closed rectifiable contours C in R^d. This implies that there exists a function V on R^d such that $a^{-1}(\tilde{b}-\tilde{\beta})=\nabla V$ or $\tilde{b}-\tilde{\beta}=a\,\nabla V$. Thus from (3.33) and (3.34) we get

$$\begin{aligned}\nabla\cdot\tilde{\beta}+\tilde{\beta}^T\alpha^{-1}\tilde{\beta}&=\nabla\cdot\tilde{\beta}+\tilde{\beta}^T a^{-1}\tilde{\beta}\\ &=\nabla\cdot\tilde{b}+\tilde{b}^T a^{-1}\tilde{b}\\ &=\nabla\cdot(\tilde{\beta}+a\,\nabla V)+(\tilde{\beta}+a\,\nabla V)^T a^{-1}(\tilde{\beta}+a\,\nabla V)\\ &=\nabla\cdot\tilde{\beta}+\nabla\cdot a\,\nabla V+\tilde{\beta}^T a^{-1}\tilde{\beta}+(a\,\nabla V,\nabla V)+2(\tilde{\beta},\nabla V),\end{aligned}$$

from which we conclude that

$$\nabla\cdot a\,\nabla V+(a\,\nabla V,\nabla V)+2(\tilde{\beta},\nabla V)=0. \tag{3.36}$$

Let $V=\log H$, where $H>0$; then we get from (3.36)

$$\nabla \cdot a\frac{\nabla H}{H}+\left(a\frac{\nabla H}{H},\frac{\nabla H}{H}\right)+2\left(\tilde{\beta},\frac{\nabla H}{H}\right)=0$$

or

$$\frac{\nabla \cdot a\,\nabla H}{H}-\frac{(a\,\nabla H,\nabla H)}{H^2}+\frac{(a\,\nabla H,\nabla H)}{H^2}+\frac{2(\tilde{\beta},\nabla H)}{H}=0\,.$$

Since $H>0$, this last equation gives us

$$\tfrac{1}{2}\nabla \cdot a\,\nabla H+\tilde{\beta}\cdot \nabla H=\tfrac{1}{2}\nabla \cdot \alpha\,\nabla H+\tilde{\beta}\cdot \nabla H=0\,,$$

which says that $L_2 H=0$. Thus H is a positive harmonic function for L_2. We noted earlier that choosing the first alternative in (3.35) implies $\tilde{b}-\tilde{\beta}=a\,\nabla H/H$. If we use this fact together with (3.33), (3.34), and $L_2H=0$, an easy calculation shows that $L_1 f=L_2(Hf)/H$.

If the second alternative in (3.35) holds, then

$$\int_C \langle a^{-1}(\tilde{b}+\tilde{\beta}),dx\rangle=0$$

for all closed rectifiable contours C in R^d. Thus there exists a function V on R^d such that $\tilde{b}+\tilde{\beta}=a\,\nabla V$, and again using (3.33) and (3.34) we have

$$\begin{aligned}\nabla \cdot \tilde{\beta}+\tilde{\beta}^T\alpha^{-1}\tilde{\beta}&=\nabla \cdot \tilde{b}+\tilde{b}^Ta^{-1}\tilde{b}\\ &=\nabla \cdot(-\tilde{\beta}+a\,\nabla V)+(-\tilde{\beta}+a\,\nabla V)^Ta^{-1}(-\tilde{\beta}+a\,\nabla V)\\ &=-\nabla\tilde{\beta}+\nabla \cdot a\,\nabla V+\tilde{\beta}^Ta^{-1}\tilde{\beta}+(a\,\nabla V,\nabla V)-2(\tilde{\beta},\nabla V)\,.\end{aligned}$$

Hence

$$\nabla \cdot a\,\nabla V+(a\,\nabla V,\nabla V)-2(\tilde{\beta},\nabla V)-2\nabla \cdot \tilde{\beta}=0\,. \tag{3.37}$$

Let $V=\log \rho$, where $\rho>0$; then from (3.37) we get

$$\nabla \cdot a\frac{\nabla \rho}{\rho}+\left(a\frac{\nabla \rho}{\rho},\frac{\nabla \rho}{\rho}\right)-2\left(\tilde{\beta},\frac{\nabla \rho}{\rho}\right)-2\nabla \cdot \tilde{\beta}=0$$

or

$$\frac{\nabla \cdot a\,\nabla \rho}{\rho}-\frac{(a\,\nabla \rho,\nabla \rho)}{\rho^2}+\frac{(a\,\nabla \rho,\nabla \rho)}{\rho^2}-2\frac{(\tilde{\beta},\nabla \rho)}{\rho}-2\nabla \cdot \tilde{\beta}=0\,.$$

Since $\rho>0$, this last equation implies

$$\tfrac{1}{2}\nabla \cdot a\,\nabla \rho-(\tilde{\beta},\nabla \rho)-(\nabla \cdot \tilde{\beta})\rho=0\,,$$

i.e., $L_2^* \rho = 0$. Thus there exists a positive invariant density ρ for L_2. We saw above that the second alternative in (3.35) implies $\tilde{b} + \tilde{\beta} = a \nabla \rho / \rho$. If we use this fact together with (3.33), (3.34), and $L_2^* \rho = 0$, an elementary calculation shows that $L_1 f = L_2^*(\rho f)/\rho$. This completes the proof of Theorem 3.2.

Bibliography

[1] Donsker, M. D., and Varadhan, S. R. S., *On a variational formula for the principal eigenvalue for operators with maximum principle*, Proc. Nat. Acad. Sci. U.S.A., Vol. 72, No. 3, March 1975, pp. 780–783.

[2] Sion, M., *On general minimax theorem*, Pac. J. Math., Vol. 8, 1958, pp. 171–176.

Received May, 1976.

COMMUNICATIONS ON PURE AND APPLIED MATHEMATICS, VOL. XXX, 707–753 (1977)

On Laws of the Iterated Logarithm for Local Times*

M. D. DONSKER AND S. R. S. VARADHAN

1. Introduction

Let $\{\beta(s),\ s \geqq 0\}$ be the standard Brownian motion in one dimension. There are various types of laws of the iterated logarithm that are known in this case. For example, (see Chung [1])

$$\overline{\lim_{t\to\infty}} \frac{1}{(2t \log\log t)^{1/2}} \sup_{0 \leqq s \leqq t} |\beta(s)| = 1 \quad \text{a.e.}, \tag{1.1}$$

and (see Chung [1], Jain and Pruitt [2])

$$\underline{\lim_{t\to\infty}} \left(\frac{\log\log t}{t}\right)^{1/2} \sup_{0 \leqq s \leqq t} |\beta(s)| = \frac{\pi}{\sqrt{8}} \quad \text{a.e.} \tag{1.2}$$

These are really two different types of laws of the iterated logarithm. The typical Brownian path wanders a distance $\sqrt{t}$ in the time interval $[0, t]$. The first law, (1.1), concerns the possibility of the path wandering farther than $\sqrt{t}$ whereas the second law, (1.2), is concerned with the possibility of wandering much less.

Laws of the first type, of which (1.1) is only a special case, are all essentially contained in Strassen's version of the law of the iterated logarithm for Brownian motion. For each $t > 0$, we define

$$\beta_t(s) = \frac{1}{(t \log\log t)^{1/2}} \beta(st) \quad \text{for} \quad 0 \leqq s \leqq 1,$$

so that $\beta_t(s)$ is again a Brownian motion but with a variance at time s equal to $s/\log\log t$). As $t \to \infty$, although $\beta_t(\cdot)$ is nearly zero for most ω, the family $\beta_t(\cdot)$ is compact and has as limit points the set C of continuous trajectories

* The research for this paper was done at the Courant Institute of Mathematical Sciences and supported by the National Science Foundation under NSF Grant No. MPS 74-01921. Reproduction in whole or in part is permitted for any purpose of the United States Government.

$f(\cdot)$ such that $\frac{1}{2}\int_0^1 [f'(s)]^2\,ds \leqq 1$. In particular, this implies that, for any functional $\Phi(f)$ continuous on $C[0, 1]$,

$$\overline{\lim_{t\to\infty}}\,\Phi(\beta_t(\cdot)) = \sup_{f\in C} \Phi(f) \quad \text{a.e.} \tag{1.3}$$

This well-known result of Strassen [3] contains, of course, (1.1). In this paper we obtain a theorem like (1.3) but which covers the other type of law of the iterated logarithm of which (1.2) is a special case. Before stating this theorem we give another special case of it which is perhaps more revealing of the second type of law of the iterated logarithm. Let $\xi(t, x)$ be the local time at x of $\beta(\cdot)$ during $[0, t]$, i.e., $\xi(t, x) = \int_0^t \delta(\beta(\sigma) - x)\,d\sigma$. Then, Kesten [4] proved

$$\overline{\lim_{t\to\infty}}\,\frac{\xi(t)}{(2t \log\log t)^{1/2}} = 1 \quad \text{a.e.}, \tag{1.4}$$

where $\xi(t) = \sup_x \xi(t, x)$. The reason (1.4) should be considered as a law of the second type is that in order for the local time to build up to more than the normal $\sqrt{t}$, the path itself must remain around zero much more than usual.

In order to obtain a Strassen type law encompassing (1.2), (1.4) and other examples of the second type, we look at

$$\beta_t(s) = \left(\frac{\log\log t}{t}\right)^{1/2} \beta(st) \quad \text{for} \quad 0 \leqq s \leqq 1\,.$$

Now $\beta_t(s)$ is a Brownian motion on $[0, 1]$ with variance $s \log\log t$ and should dissipate as $t \to \infty$. To make it precise, we look at the occupation distribution

$$L_t(\omega, A) = \int_0^1 \chi_A(\beta_t(s))\,ds\,,$$

or its density $\hat{L}_t(\omega, y)$ determined by

$$L_t(\omega, A) = \int_A \hat{L}_t(\omega, y)\,dy\,.$$

One can view $\hat{L}_t$ as an element of all continuous subprobability densities on R_1 with the topology of uniform convergence on bounded intervals. One can show that, for almost all ω, $\{\hat{L}_t\}$ is compact and the set of limit points of $\hat{L}_t$

can be characterized as the set C of subprobability densities f such that

$$I(f)=\tfrac{1}{8}\int_{R_1}\frac{[f'(y)]^2}{f(y)}\,dy\leqq 1\,.$$

It then follows that, for suitable functionals $\Phi(\cdot)$,

$$\overline{\lim_{t\to\infty}}\,\Phi(\hat{L}_t(\omega,\cdot))=\sup_{f\in C}\Phi(f)\quad\text{a.e.}\tag{1.5}$$

By taking $\Phi(f)=f(0)$ we obtain (1.4) (cf. Corollary 3.12 below), and by taking

$$\Phi(f)=\inf\left\{l:\int_{-l}^{l}f(y)\,dy=1\right\}$$

we obtain (1.2). The I-function entering into (1.5), i.e.,

$$I(f)=\tfrac{1}{8}\int_{R_1}\frac{[f'(y)]^2}{f(y)}\,dy$$

is the special case for Brownian motion of the I-function introduced by the authors in a recent series of papers, [5]–[8], on asymptotics for Markov processes. In proving (1.5) essential use is made of these asymptotics.

Another feature of the second type of law of the iterated logarithm is that there is an analogous result for symmetric stable processes as well. If the stable law index α is larger than one, then only the normalization and the I-function have to be changed appropriately in (1.5). If $\alpha\leqq 1$, then, of course, local times do not exist and we have to work with L_t instead of $\hat{L}_t$. In this latter case we view L_t as an element of the space of subprobability measures on R_1, and the topology will be the vague topology. One then obtains the exact analogue of (1.5).

In Section 2, we treat this latter case, i.e., symmetric stable processes of index α, $0<\alpha\leqq 2$. In Section 3, we prove (1.5) in its more general form, i.e., not just for Brownian motion ($\alpha=2$), but for any symmetric stable process with index $\alpha>1$. In Section 4, we present a few examples to illustrate the general theorems.

2. The Weak Topology

Let $\{x(s),\ 0\leqq s<\infty\}$ with $x(0)=x$ be a symmetric stable process of index α, i.e., the process has stationary independent increments and

$$E_x\{\exp\{i\xi[x(t+h)-x(h)]\}\}=\exp\{-t\,|\xi|^\alpha\}\,,$$

where $0<\alpha\leqq 2$. We first prove

LEMMA 2.1. *For a fixed interval* $[a,b]$, *let*

$$f_\alpha^T(x)=P_x\{x(t)\in[a,b] \text{ for some } t, 0\leqq t\leqq T\}. \tag{2.1}$$

Then,

$$\int_{-\infty}^{\infty} f_\alpha^T(x)\,dx\leqq \gamma_1(T+1), \tag{2.2}$$

where γ_1 *depends only on a, b, and* α. *Furthermore, for any* $d>0$ *and* $0\leqq c\leqq d$,

$$\sum_{n=-\infty}^{\infty} f_\alpha^T(c+nd)\leqq \gamma_2(T+1), \tag{2.3}$$

where γ_2 *depends only on a, b, d, and* α.

Proof: Let

$$g_\alpha^T(x)=E_x\left\{\int_0^{T+1}\chi_{[a-1,b+1]}(x(s))\,ds\right\}$$

and let $\tau=\inf[t:x(t)\in[a,b]]$. Then, clearly,

$$g_\alpha^T(x)\geqq E_x\left\{\int_\tau^{\tau+1}\chi_{[x(\tau)-1,x(\tau)+1]}(x(s))\,ds;\quad \tau\leqq T\right\}. \tag{2.4}$$

By the strong Markov property,

$$\begin{aligned}E_x&\left\{\int_\tau^{\tau+1}\chi_{[x(\tau)-1,x(\tau)+1]}(x(s))\,ds;\tau\leqq T\right\}\\&=E_x\left\{E_{x(\tau)}\left\{\int_0^1\chi_{[x(\tau)-1,x(\tau)+1]}(x(s))\,ds\right\};\tau\leqq T\right\},\end{aligned} \tag{2.5}$$

but, since the process has independent increments,

$$\begin{aligned}E_{x(\tau)}\left\{\int_0^1\chi_{(x(\tau)-1,x(\tau)+1]}(x(s))\,ds\right\}&=E_0\left\{\int_0^1\chi_{[-1,1]}(x(s))\,ds\right\}\\&=p>0,\end{aligned} \tag{2.6}$$

where p depends only on α. Thus, from (2.4)–(2.6) we see that

$$g_\alpha^T(x) \geqq pP_x\{\tau \leqq T\} = pf_\alpha^T(x)\,. \tag{2.7}$$

Now

$$\begin{aligned} g_\alpha^T(x) &= E_x\left\{\int_0^{T+1} \chi_{[a-1,b+1]}(x(s))\,ds\right\} \\ &= E_0\left\{\int_0^{T+1} \chi_{[a-1-x,b+1-x]}(x(s))\,ds\right\}, \end{aligned}$$

and

$$\int_{-\infty}^{\infty} \chi_{[a-1-x,b+1-x]}(y)\,dx = b-a+2\,,$$

so that

$$\int_{-\infty}^{\infty} g_\alpha^T(x)\,dx = (b-a+2)(T+1)\,. \tag{2.8}$$

Thus, from (2.7) and (2.8) with $\gamma_1 = (b-a+2)/p$, we see that

$$\int_{-\infty}^{\infty} f_\alpha^T(x)\,dx \leqq \gamma_1(T+1)\,.$$

Similarly,

$$\sum_{n=-\infty}^{\infty} \chi_{[a-1-nd-c,b+1-nd-c]} \leqq \frac{b-a+2}{d}+1\,,$$

and therefore

$$\sum_{n=-\infty}^{\infty} g_\alpha^T(c+nd) \leqq \left(\frac{b-a+2}{d}\right)(T+1)\,. \tag{2.9}$$

Hence, (2.7) and (2.9) with $\gamma_2 = (1/p)((b-a+2)/d+1)$ imply

$$\sum_{n=-\infty}^{\infty} f_\alpha^T(c+nd) \leqq \gamma_2(T+1)\,.$$

Corresponding to the symmetric stable process under consideration, there is an infinitesimal generator $L=L_\alpha$. If we let U denote the space of functions $u\in C^\infty(R_1)$ which are strictly positive and are equal to some positive constant for large $|x|$, then U is contained in the domain of L.

Let M be the space of measures λ on R_1 such that $\lambda(R_1)\leqq 1$. The space M will be provided with the vague topology, i.e., the topology induced by defining $\lambda_n\to\lambda$ if $\int f\lambda_n(dx)\to\int f\lambda(dx)$ for all f continuous on R_1 with $f\to 0$ as $|x|\to\infty$. In the vague topology M is compact.

If L is the infinitesimal generator of the stable process, then it is easy to show that

$$(2.10)\qquad I(\lambda)=-\inf_{u\in U}\int_{R_1}\left(\frac{Lu}{u}\right)(x)\lambda(dx)=-\inf_{u\in\mathscr{D}^+}\int_{R_1}\left(\frac{Lu}{u}\right)(x)\lambda(dx),$$

where $\mathscr{D}^+$ consists of those functions in the domain of L which have a positive lower bound. Moreover, $I(\lambda)$ is a lower semicontinuous, convex functional on M which is homogeneous of degree 1.

For $t>0$, E a Borel set in R_1, and any sample path $\omega=x(\cdot)$, we define

$$L_t(\omega,E)=\frac{1}{t}\int_0^t\chi_E(x(s))\,ds,$$

i.e., the proportion of time up to t that the particular path $\omega=x(\cdot)$ spends in the set E. For each fixed t and ω, $L_t(\omega,\cdot)$ is a probability measure on R_1 so that, for fixed t and x, $L_t(\omega,\cdot)$ maps the sample paths starting at x into M. For each $t>0$, and $x\in R_1$, define for sets $A\subset M$

$$Q_{t,x}(A)=P_x\{L_t(\omega,\cdot)\in A\}.$$

If we compactify R_1 by adding a point at infinity, then M can be identified naturally as the space of probability measures on the compactified space by placing any missing mass at infinity. The vague topology we imposed on M is nothing but the weak topology in the compactified space. It is therefore easy to deduce from our earlier work

LEMMA 2.2. *For any closed set $A\subset M$,*

$$(2.11)\qquad \overline{\lim_{t\to\infty}}\frac{1}{t}\log\sup_x Q_{t,x}(A)\leqq-\inf_{\lambda\subset A}I(\lambda).$$

Let $\beta\in M$, $x\in R_1$, and let E be a Borel set in R_1. We define $\beta_x(E)=\beta(E-x)$ so that $\beta_x\in M$. If $A\subset M$ and $x\in R_1$, we define $A_x=\{\beta_x\}$ for $\beta\in A$.

We shall use the notation $D_A = \bigcup_x A_x$. We now prove

LEMMA 2.3. *For any set $A \subset M$, let $C_0(A) = \bigcup_{|x| \leqq 1/2} A_x$. Let $0 < a \leqq 1$ be given, and define*

$$C_0^a(A) = \left\{ \beta' \in M : \beta'(E) = \theta\beta\left(\frac{1}{\theta^{1/\alpha}} E\right) \text{ for all Borel sets } E \right.$$

$$\left. \text{and all } 1 \leqq \theta \leqq 2/a, \text{ where } \beta \in C_0(A) \right\}.$$

Then, for any $\lambda \in M$ and any $\varepsilon > 0$, there exists a neighborhood N of λ such that

$$\inf_{\beta \in C_0^a(N)} I(\beta) \geqq I(\lambda) - \varepsilon. \tag{2.12}$$

Proof: We note from the definition of $C_0^a(A)$ that if $A \subset B$, then $A \subset C_0^a(A) \subset C_0^a(B)$. From the lower semicontinuity of I, there exists a neighborhood N_1 of λ such that $\inf_{\beta \in N_1} I(\beta) \geqq I(\lambda) - \varepsilon$. Let N be a neighborhood of λ such that $N \subseteq \bar{N} \subset N_1$, so that $C_0^a(N) \subset C_0^a(\bar{N})$. Since M is compact, it is clear that if A is closed, then $C_0(A)$ and $C_0^a(A)$ are closed. Thus $C_0^a(\bar{N})$ is closed and $\overline{C_0^a(N)} \subseteq C_0^a(\bar{N})$. This implies that

$$\inf_{\beta \in \overline{C_0^a(N)}} I(\beta) \geqq \inf_{\beta \in C_0^a(\bar{N})} I(\beta).$$

Thus, since

$$\inf_{\beta \in \bar{N}} I(\beta) \geqq \inf_{\beta \in N_1} I(\beta) \geqq I(\lambda) - \varepsilon,$$

it suffices to show that

$$\inf_{\beta \in C_0^a(\bar{N})} I(\beta) = \inf_{\beta \in \bar{N}} I(\beta).$$

But this follows from the fact that $I(\beta)$ is translation invariant and that for symmetric stable processes the I-function has the property that

$$I\left(\theta\beta\left(\frac{1}{\theta^{1/\alpha}} \cdot\right)\right) = \theta I\left(\beta\left(\frac{1}{\theta^{1/\alpha}} \cdot\right)\right) = I(\beta).$$

This last identity for the I-function of a symmetric stable process of index α is the reflection in the I-function of the fundamental scaling property of such a process—namely, that $x(cs)$ has the same distribution as $c^{1/\alpha}x(s)$ for any constant c. This scaling property is shared by the semigroup, the infinitesimal generator, and hence also the I-function.

We state the following elementary property of the vague topology as

LEMMA 2.4. *Let $\lambda \in M$. For any neighborhood N of λ there exists a neighborhood V of λ and a $K<\infty$ such that if $\beta \in V$ and $\beta(A)=\beta'(A)$ for all Borel sets $A \subset [-K, K]$, then $\beta' \in N$.*

We recall the notation that for any set $A \subset M$, $D_A = \bigcup_x A_x$ and prove next

LEMMA 2.5. *Let $\lambda \in M$ and $\varepsilon>0$ be given. There exists a neighborhood $V=V_\varepsilon$ of λ such that*

$$\overline{\lim_{t\to\infty}} \frac{1}{t} \log Q_{t,0}(D_V) \leqq -I(\lambda)+\varepsilon . \tag{2.13}$$

Proof: First of all, if $\lambda(R_1)=0$ there is nothing to prove since in that case $I(\lambda)=0$. So assume $\lambda(R_1)=a>0$. For the given $\lambda \in M$, the given $\varepsilon>0$, and this $a=\lambda(R_1)$, we have from Lemma 2.3 that there exists a neighborhood N of λ such that, for the corresponding $C_0^a(N)$,

$$\inf_{\beta \in \overline{C_0^a(N)}} I(\beta) \geqq I(\lambda)-\varepsilon . \tag{2.14}$$

Corresponding to this neighborhood N of λ we have from Lemma 2.4 that there exists a neighborhood V of λ and a $K<\infty$ such that if $\beta \in V$ and $\beta(E)=\beta'(E)$ for all $E \subset [-K, K]$, then $\beta' \in N$. Without loss of generality we can assume that V and K have been selected so that if $\beta \in V$, then $\beta([-K, K]) \geqq \frac{1}{2}a$. It is this neighborhood V of λ which we shall show satisfies (2.13).

Let $B_0 = \bigcup_{|x|\leqq 1/2} V_x$ and, for any integer n, let $B_n = \bigcup_{|x-n|\leqq 1/2} V_x$. Thus, $D_V = \bigcup_x V_x = \bigcup_{n=-\infty}^{\infty} B_n$, and

$$Q_{t,0}(D_V) \leqq \sum_{n=-\infty}^{\infty} Q_{t,0}(B_n) = \sum_{n=-\infty}^{\infty} P_0\{L_t(\omega,\cdot) \in B_n\} . \tag{2.15}$$

If $\beta \in B_n$, then for some x with $|x-n| \leqq \frac{1}{2}$, $\beta \in V_x$ and hence $\beta_{-x} \in V$. From the way we selected V and K this means that for some x with $|x-n| \leqq \frac{1}{2}$ we have $\beta_{-x}([-K, K]) \geqq \frac{1}{2}a$, i.e., $\beta([-K+x, K+x]) \geqq \frac{1}{2}a$. If we let $K' = K + \frac{1}{2}$, this last inequality implies $\beta([-K'+n, K'+n]) \geqq \frac{1}{2}a$. Thus we see that if $\beta \in B_n$, then $\beta([-K'+n, K'+n]) \geqq \frac{1}{2}a$.

Let $\tau_n = \inf[t : x(t) \in [-K'+n, K'+n]]$, and the conclusion of the last paragraph implies

$$\{\omega : L_t(\omega, \cdot) \in B_n\} \subset \{\omega : \tau_n \leqq (1-\tfrac{1}{2}a)t\}. \tag{2.16}$$

Let

$$\beta(E) = L_t(\omega, E) = \frac{1}{t}\int_0^t \chi_E(x(s))\, ds,$$

and let

$$\beta'(E) = \frac{1}{t}\int_{\tau_n}^t \chi_E(x(s))\, ds.$$

Obviously, $\beta'(E) = \beta(E)$ for all $E \subset [-K'+n, K'+n]$. This means, with our choice of V and K and since $K' = K + \frac{1}{2}$, that

$$\{\omega : L_t(\omega, \cdot) \in B_n\} \subset \left\{\omega : \frac{1}{t}\int_{\tau_n}^t \chi_\cdot(x(s))\, ds \in C_n\right\}, \tag{2.17}$$

where $C_n = \bigcup_{|x-n| \leqq 1/2} N_x$. Hence, from (2.15)–(2.17), we see that

$$Q_{t,0}(D_V) \leqq \sum_{n=-\infty}^{\infty} P_0\left\{\tau_n \leqq (1-\tfrac{1}{2}a)t, \frac{1}{t}\int_{\tau_n}^t \chi_\cdot(x(s))\, ds \in C_n\right\}. \tag{2.18}$$

From the strong Markov property and the fact that the process has stationary independent increments,

$$\begin{aligned} &P_0\left\{\tau_n \leqq (1-\tfrac{1}{2}a)t, \frac{1}{t}\int_{\tau_n}^t \chi_\cdot(x(s))\, ds \in C_n\right\} \\ &= E_0\left\{\chi_{\tau_n \leq (1-a/2)t}(x(\cdot)) P_{x(\tau_n)}\left[\frac{1}{t}\int_0^{t-\tau_n} \chi_\cdot(x(s))\, ds \in C_n\right]\right\} \\ &\leqq \left(\sup_{\substack{x \\ 0 \leq \tau \leq t(1-a/2)}} P_x\left\{\frac{1}{t}\int_0^{t-\tau} \chi_\cdot(x(s))\, ds \in C_n\right\}\right) P_0\{\tau_n \leq t(1-\tfrac{1}{2}a)\} \\ &= \left(\sup_{\substack{x \\ 0 \leq \tau \leq t(1-a/2)}} P_x\left\{\frac{1}{t}\int_0^{t-\tau} \chi_\cdot(x(s))\, ds \in C_0\right\}\right) P_{-n}\{\tau_0 \leq t(1-\tfrac{1}{2}a)\}. \end{aligned} \tag{2.19}$$

Using the notation of Lemma 2.1, we have from (2.3),

$$\sum_{n=-\infty}^{\infty} P_{-n}\{\tau_0 \leqq t(1-\tfrac{1}{2}a)\} = \sum_{n=-\infty}^{\infty} P_n\{\tau_0 \leqq t(1-\tfrac{1}{2}a)\}$$

$$= \sum_{n=-\infty}^{\infty} P_n\{x(s)\in[-K', K'] \text{ for some } s, 0 \leqq s \leqq t(1-\tfrac{1}{2}a)\} \tag{2.20}$$

$$= \sum_{n=-\infty}^{\infty} f_\alpha^{t(1-a/2)}(n) \leqq \gamma_2(t-\tfrac{1}{2}at+1) \leqq \gamma_2(t+1),$$

where γ_2 depends on K' and α but not on t. Thus, from (2.18)–(2.20), we obtain

$$\overline{\lim_{t\to\infty}} \frac{1}{t}\log Q_{t,0}(D_V) \leqq \overline{\lim_{t\to\infty}} \frac{1}{t}\log \sup_{\substack{x \\ 0\leqq\tau\leqq t(1-a/2)}} P_x\left\{\frac{1}{t}\int_0^{t-\tau} \chi_\cdot(x(s))\,ds \in C_0\right\}. \tag{2.21}$$

The scaling property of a symmetric stable process of index α, i.e., that the distribution of $x(ks)$ is the same as the distribution of $k^{1/\alpha}x(s)$ for any positive constant k, implies that $L_{kt}(\omega,\cdot)$ has the same distribution as $L_t(\omega,(1/k^{1/\alpha})\cdot)$. This, together with the fact that

$$\left\{\omega : \frac{1}{t}\int_0^{t-\tau} \chi_\cdot(x(s))\,ds \in C_0\right\}$$

$$= \left\{\omega : \frac{1}{t-\tau}\int_0^{t-\tau} \chi_\cdot(x(s))\,ds \in \left(\frac{t}{t-\tau} C_0 \cap M\right)\right\}$$

$$= \left\{\omega : L_{t-\tau}(\omega,\cdot) \in \left(\frac{t}{t-\tau} C_0 \cap M\right)\right\},$$

gives us

$$\sup_{\substack{x \\ 0\leqq\tau\leqq t(1-a/2)}} P_x\left\{\frac{1}{t}\int_0^{t-\tau} \chi_\cdot(x(s))\,ds \in C_0\right\}$$

$$= \sup_{\substack{x \\ 0\leqq\tau\leqq t(1-a/2)}} P_x\left\{L_{t-\tau}(\omega,\cdot) \in \left(\frac{t}{t-\tau} C_0 \cap M\right)\right\}$$

$$= \sup_{\substack{x \\ 0\leqq\tau\leqq t(1-a/2)}} P_x\left\{L_t\left(\omega, \left(\frac{t}{t-\tau}\right)^{1/\alpha} \cdot\right) \in \left(\frac{t}{t-\tau} C_0 \cap M\right)\right\} \tag{2.22}$$

$$= \sup_{\substack{x \\ 0\leqq\tau\leqq t(1-a/2)}} P_x\{L_t(\omega,\cdot) \in F_{t,\tau}\},$$

where $F_{t,\tau}$ consists of all measures $\beta' \in M$ of the form $\beta'(A) = (t/(t-\tau))\beta(((t-\tau)/t)^{1/\alpha}A)$ with $\beta \in C_0$.

If now we let $F_{t,a} = \bigcup_{0 \leqq \tau \leqq t(1-a/2)} F_{t,\tau}$, then it follows from (2.22) that

$$\begin{aligned} \sup_{\substack{x \\ 0 \leqq \tau \leqq (1-a/2)t}} & P_x\left\{\frac{1}{t}\int_0^{t-\tau} \chi_\cdot(x(s))\, ds \in C_0\right\} \\ &\leqq \sup_x P_x\{L_t(\omega, \cdot) \in (F_{t,a})\} \\ &\leqq \sup_c P_x\{L_t(\omega, \cdot) \in \overline{F_{t,a}}\} \\ &= \sup_x Q_{t,x}(\overline{F_{t,a}}) . \end{aligned} \tag{2.23}$$

From Lemma 2.2 we see that

$$\overline{\lim_{t\to\infty}} \frac{1}{t} \log \sup_x Q_{t,x}(\overline{F_{t,a}}) \leqq -\inf_{\beta \in F_{t,a}} I(\beta) . \tag{2.24}$$

Putting together (2.21), (2.23), and (2.24) we obtain

$$\overline{\lim_{t\to\infty}} \frac{1}{t} \log Q_{t,0}(D_V) \leqq -\inf_{\beta \in \overline{F_{t,a}}} I(\beta) . \tag{2.25}$$

But $F_{t,a} = C_0^a(N)$ and therefore by Lemma 2.3 and the choice of N,

$$\inf_{\beta \in \overline{F_{t,a}}} I(\beta) = \inf_{\beta \in \overline{C_0^a(N)}} I(\beta) \geqq I(\lambda) - \varepsilon .$$

This completes the proof of Lemma 2.5.

We can now prove

THEOREM 2.6. *Let $A \subset M$ be closed and let $D_A = \bigcup_x A_x$. Then,*

$$\overline{\lim_{t\to\infty}} \frac{1}{t} \log Q_{t,0}(D_A) \leqq -\inf_{\beta \in A} I(\beta) . \tag{2.26}$$

Proof: Let $\inf_{\beta\in\lambda} I(\beta)=l$. Let $\lambda\in A$ and $\varepsilon>0$ be given. From Lemma 2.5 there exists a neighborhood V_λ of λ such that

$$\overline{\lim_{t\to\infty}}\frac{1}{t}\log Q_{t,0}(D_{V_\lambda})\leqq -I(\lambda)+\varepsilon$$
$$\leqq -l+\varepsilon.$$

The neighborhoods V_λ cover A, and since A is compact, there exists $\lambda_1, \lambda_2,\ldots,\lambda_n$ such that $V_{\lambda_1}, V_{\lambda_2},\ldots,V_{\lambda_n}$ cover A. Hence, $Q_{t,0}(D_A)\leqq \sum_{j=1}^n Q_{t,0}(D_{V\lambda_j})$, so that

$$\overline{\lim_{t\to\infty}}\frac{1}{t}\log Q_{t,0}(D_A)\leqq \max_{1\leqq j\leqq n}\left(\overline{\lim_{t\to\infty}}\frac{1}{t}\log Q_{t,0}(D_{V_{\lambda_j}})\right)$$
$$\leqq -l+\varepsilon,$$

and since $\varepsilon>0$ is arbitrary, the theorem follows.

For the next theorem we need the following lemma. We denote the complement of a set A by $\tilde{A}$.

LEMMA 2.7. *Let G be a neighborhood in M and N another neighborhood such that $N\subset\bar{N}\subset G$. Suppose $\{\alpha_n\}$ is a sequence of measures in M such that, for sufficiently large n, $\alpha_n\notin D_{\tilde{N}}$. Let $\{\gamma_n\}$ be a sequence of measure in M such that*

$$\lim_{n\to\infty}\|\alpha_n-\gamma_n\|=0,$$

where the norm is the variation norm. If $\{\lambda_n\}$ is a sequence of constants such that $\lim_{n\to\infty}\lambda_n=1$ and we define, for any Borel set A, $\beta_n(A)=\gamma_n(\lambda_n A)$, then, for sufficiently large n, $\beta_n\notin D_{\tilde{G}}$.

Proof: Assume the contrary, i.e., $\beta_{n_j}\in D_{\tilde{G}}$ for some sequence $\{n_j\}$ with $\lim_{j\to\infty} n_j=\infty$. Relabeling, our assumption means that $\beta_n\in D_{\tilde{G}}$ for all n. This implies that for each n there is an x_n such that $\beta'_n=\beta_n(\cdot+x_n)\in\tilde{G}$. For any Borel set A, define

$$\gamma'_n(A)=\beta'_n\left(\frac{1}{\lambda_n}A\right)=\beta_n\left(\frac{1}{\lambda_n}A+x_n\right)=\gamma_n(a+\lambda_n x_n),$$

and also define

$$\alpha'_n(A) = \alpha_n(A + \lambda_n x_n).$$

Obviously, $\|\gamma'_n - \alpha'_n\| = \|\gamma_n - \alpha_n\|$ and hence, $\lim_{n\to\infty} \|\gamma'_n - \alpha'_n\| = 0$. Since $\beta'_n \in \tilde{G}$ for all n, and $\gamma'_n(A) = \beta'_n((1/\lambda_n)A)$ with $\lambda_n \to 1$ as $n \to \infty$, we see that any limit point of γ'_n is also in $\bar{\tilde{G}}$. Thus, any limit point of α'_n is also in $\bar{\tilde{G}}$. Now, $\tilde{N}$ is a neighborhood of $\bar{\tilde{G}}$ so that $\alpha'_n \in \tilde{N}$ for sufficiently large n, i.e., $\alpha'_n \in \tilde{N}$ for sufficiently large n. Since α'_n is a translate of α_n, this last statement implies that $\alpha_n \in D_{\tilde{N}}$ for sufficiently large n which contradicts the hypothesis that $\alpha_n \notin D_{\tilde{N}}$ for sufficiently large n.

We defined earlier $L_t(\omega, A) = (1/t)\int_0^t \chi_A(x(s))\, ds$, and now we define

$$\hat{L}_t(\omega, A) = L_t(\omega, \rho_t A), \tag{2.27}$$

where $\rho_t = (t/\log\log t))^{1/\alpha}$. Thus,

$$\hat{L}_t(\omega, A) = \frac{1}{t}\int_0^t \chi_A\left(\left(\frac{\log\log t}{t}\right)^{1/\alpha} x(s)\right) ds. \tag{2.28}$$

We now prove

THEOREM 2.8. *Let $C = \{\beta \in M : I(\beta) \leqq 1\}$. Then, for almost all ω, the set of limit points as $t \to \infty$ of any translate of $\hat{L}_t(\omega, \cdot)$ is in C, i.e., if we let*

$$C_t^\omega = \{\hat{L}_t(\omega, \cdot + x),\, x \in R_1\},$$

then, for almost all ω,

$$\bigcap_T \overline{\bigcup_{t \geqq T} C_t^\omega} \subset C. \tag{2.29}$$

Proof: Let G be a neighborhood of C in M, and let N be a neighborhood of C such that $C \subset N \subset \bar{N} \subset G$. Using the same notation introduced earlier, let $D_{\tilde{G}} = \bigcup_x \tilde{G}_x$ and $D_{\tilde{N}} = \bigcup_x \tilde{N}_x$. If we set $\theta = \inf_{\beta \in \tilde{N}} I(\beta)$, then since I is lower semicontinuous and $\tilde{N}$ is compact in M, we have $\theta > 1$. Choose $0 < k < 1$ but such that $k\theta' = k(\frac{1}{2}(1+\theta)) > 1$. Define, for each integer n, $t_n = e^{n^k}$.

As noted earlier, for any positive constant λ, $L_{\lambda t}(\omega, A)$ has the same distribution as $L_t(\omega, (1/\lambda^{1/\alpha})^A)$. Thus

$$P_0\{\hat{L}_t(\omega,\cdot)\in D_{\tilde{N}}\} = P_0\{L_{\log\log t}(\omega,\cdot)\in D_{\tilde{N}}\} \tag{2.30}$$
$$= Q_{\log\log t, 0}(D_{\tilde{N}}),$$

and since $\tilde{N}$ is closed in M, we have from Theorem 2.6, with $\theta' = \frac{1}{2}(1+\theta)$ and for sufficiently large t,

$$Q_{\log\log t, 0}(D_{\tilde{N}}) \leqq \exp\{-(\log\log t)\theta'\}. \tag{2.31}$$

In particular, for large enough n, we see from (2.30) and (2.31) that

$$P_0\{\hat{L}_{t_n}(\omega,\cdot)\in D_{\tilde{N}}\} \leqq \exp\{-(\log\log t_n)\theta'\} = \frac{1}{n^{k\theta'}}. \tag{2.32}$$

Since k was chosen so that $k\theta' > 1$, we conclude that $\sum_{n=1}^{\infty} P_0\{\hat{L}_{t_n}(\omega,\cdot)\in D_{\tilde{N}}\} < \infty$, and hence from the Borel–Cantelli lemma we have:

$$\text{for almost all } \omega, \quad \hat{L}_{t_n}(\omega,\cdot)\in D_{\tilde{N}} \tag{2.33}$$
$$\text{for only finitely many } n.$$

Consider then a fixed ω for which $\hat{L}_{t_n}(\omega,\cdot)\in D_{\tilde{N}}$ for only finitely many n. For this ω we shall show below that, for sufficiently large t,

$$\hat{L}_t(\omega,\cdot)\notin D_{\tilde{G}}. \tag{2.34}$$

But (2.34) implies that no translate of $\hat{L}_t(\omega,\cdot)\in\tilde{G}$ for sufficiently large t. Thus all translates of $\hat{L}_t(\omega,\cdot)$ are in G for sufficiently large t, and since G is any neighborhood of C and M is compact metric, we conclude that all translates of $\hat{L}_t(\omega,\cdot)$ have limit points in C as $t\to\infty$. Hence we have (2.29) for almost all ω completing the proof of the theorem except for (2.34) which we show now.

Assume (2.34) *is false, then for a sequence* $\{\tau_j\}$ with $\lim_{j\to\infty}\tau_j = \infty$ we have $\hat{L}_{\tau_j}(\omega,\cdot)\in D_{\tilde{G}}$ for all $j = 1, 2, 3, \ldots$. We defined above the sequence $t_n = e^{n^k}$ with $k<1$. Let t_{n_j} be such that $t_{n_j-1}\leqq\tau_j\leqq t_{n_j}$. Define

$$\alpha_j = L_{t_{n_j}}(\omega,\cdot), \quad \beta_j = \hat{L}_{\tau_j}(\omega,\cdot),$$

$$\gamma_j = L_{\tau_j}\left(\omega, \left(\frac{t_{n_j}}{\log\log t_{n_j}}\right)^{1/\alpha}\cdot\right),$$

and

$$\lambda_j = \left(\frac{t_{n_j}}{\log\log t_{n_j}} \frac{\log\log \tau_j}{\tau_j}\right)^{-1/\alpha}.$$

We observe first of all, that by our choice of ω, $\alpha_j \notin D_{\tilde{N}}$ for sufficiently large j. Since $k<1$, $t_{n-1}/t_n = \exp\{(n-1)^k - n^k\} \to 1$ as $n \to \infty$, and therefore $t_{n_j-1}/t_{n_j} \to 1$ as $j \to \infty$. This implies, since $t_{n_j-1} \leqq \tau_j \leqq t_{n_j}$, that $\|L_{\tau_j}(\omega, \cdot) - L_{t_{n_j}}(\omega, \cdot)\| \to 0$ as $j \to \infty$, where the norm is the variation norm.

Since

$$\alpha_j(A) = \hat{L}_{t_{n_j}}(\omega, A) = L_{t_{n_j}}\left(\omega, \left(\frac{t_{n_j}}{\log\log t_{n_j}}\right)^{1/\alpha} A\right),$$

we see that $\|\alpha_j - \gamma_j\| \to 0$ as $j \to \infty$. Next, we note that, for any Borel set A,

$$\beta_j(A) = \hat{L}_{\tau_j}(\omega, A) = L_{\tau_j}\left(\omega, \left(\frac{\tau_j}{\log\log \tau_j}\right)^{1/\alpha} A\right) = \gamma_j(\lambda_j A).$$

Moreover, since $t_{n_j-1}/t_{n_j} \to 1$ as $j \to \infty$ and $t_{n_j-1} \leqq \tau_j \leqq t_{n_j}$,

$$\lim_{j\to\infty} \lambda_j = \lim_{j\to\infty} \left(\frac{t_{n_j}}{\log\log t_{n_j}} \frac{\log\log \tau_j}{\tau_j}\right)^{-1/\alpha} = 1.$$

Hence, our assumption that (2.34) is false, i.e., $\beta_j \in D_{\tilde{G}}$ for all $j = 1, 2, 3, \ldots$ contradicts Lemma 2.7. Thus (2.34) is true and the proof of Theorem 2.8 is complete.

In Theorem 2.8 we showed that for almost all ω, the set of limit points as $t \to \infty$ of any translate of $\hat{L}_t(\omega, \cdot)$ is in $C = \{\beta \in M : I(\beta) \leqq 1\}$, or in the notation of that theorem, $\bigcap_T \overline{\bigcup_{t \geqq T} C_t^\omega} \subset C$. Now we wish to prove

THEOREM 2.9. *Let $C = \{\beta \in M : I(\beta) \leqq 1\}$. Then, for almost all ω,*

$$\bigcap_T \overline{\bigcup_{t \geqq T} \{\hat{L}_t(\omega, \cdot)\}} \supset C. \tag{2.35}$$

Remark. Clearly, $\bigcap_T \overline{\bigcup_{t \leqq T} \{\hat{L}_t(\omega, \cdot)\}} \subset \bigcap_T \overline{\bigcup_{t \geqq T} C_t^\omega}$, and therefore from Theorems 2.8 and 2.9 we can conclude that, for almost all ω,

$$C \subset \bigcap_T \bigcup_{t \geqq T} \{\hat{L}_t(\omega, \cdot) \subset \bigcap_T \overline{\bigcup_{t \leqq T} C_t^\omega} \subset C. \tag{2.36}$$

To prove Theorem 2.9 we shall first prove several lemmas and another theorem. Let $a>0$ and denote by $\tau_{(-a,a)}$ the first exit time for the process from $(-a, a)$, i.e.,

$$\tau_{(-a,a)} = \tau_{(-a,a)}(\omega) = \inf\{t : x(t) \notin (-a, a)\}.$$

We define, for any Borel set $E \subset (-a, a)$,

$$\hat{p}(t, x, E) = P_x\{x(t) \in E, \tau_{(-a,a)} > t\}. \tag{2.37}$$

We write, as usual, $p(t, x, E) = P_x\{x(t) \in E\}$ and begin with

LEMMA 2.10. *Let $f(y)$ be a bounded measurable function on $(-a, a)$ and define*

$$u(t, x) = E_x\{f(x(t)); \tau_{(-a,a)} > t\}.$$

Then, $u(t, x)$ is jointly continuous in t and x for $t>0$ and $|x|<a$. In particular, $\hat{p}(t, x, E)$ is jointly continuous in t and x for $t>0$ and $|x|<a$ for any Borel set $E \subset (-a, a)$.

Proof: The last assertion about $\hat{p}(t, x, E)$ follows, of course from the first statement of the lemma by letting $f(y) = \chi_E(y)$.

We can assume without loss of generality that $|f| \leqq 1$, and we write

$$u(t, x) = \int f(y)\hat{p}(t, x, dy).$$

Fix a $t>0$ and choose $h>0$ such that $t-h>0$. Defining $u(t, y)=0$ if $|y|>a$, we have from the semigroup property,

$$\begin{aligned} u(t, x) &= \int u(t-h, y)\hat{p}(h, x, dy) \\ &= \int u(t-h, y)p(h, x, dy) - \int u(t-h, y)[p(h, x, dy) - \hat{p}(h, x, dy)]. \end{aligned}$$

Thus, since $|u(t-h, y)| \leqq 1$,

$$\begin{aligned} \left| u(t, x) - \int u(t-h, y)p(h, x, dy) \right| &\leqq \int [p(h, x, dy) - \hat{p}(h, x, dy)] \\ &\leqq P_x\{\tau_{(-a,a)} \leqq h\}. \end{aligned} \tag{2.38}$$

Let $t \geqq t_0 > 0$ and $0 < h \leqq t_0$. Define $V(t, h, x) = \int u(t-h, y)p(h, x, dy)$, and let K

be a closed set with $K \subset (-a, a)$. From (2.38) we have

$$\sup_{\substack{x_1, x_2 \in K \\ t \geqq t_0 \\ |x_1 - x_2| \leqq \delta}} |u(t, x_1) - u(t, x_2)| \tag{2.39}$$

$$\leqq \sup_{\substack{t \geqq t_0 \\ |x_1 - x_2| \leqq \delta}} |V(t, h, x_1) - V(t, h, x_2)| + 2 \sup_{x \in K} P_x\{\tau_{(-a,a)} \leqq h\}.$$

Since the process has stationary independent increments and $p(t, x, dy)$ has a density, we can write $V(t, h, x) = \int u(t-h, y) p(h, y-x)\, dy$ so that

$$|V(t, h, x_1) - V(t, h, x_2)| \leqq \int |p(h, y - x_1) - p(h, y - x_2)|\, dy.$$

Hence, from (2.39),

$$\sup_{\substack{x_1, x_2 \in K \\ t \geqq t_0 \\ |x_1 - x_2| \leqq \delta}} |u(t, x_1) - u(t, x_2)| \tag{2.40}$$

$$\leqq \sup_{|z| \leqq \delta} \int |p(h, y) - p(h, y - z)|\, dy + 2 \sup_{x \in K} P_x\{\tau_{(-a,a)} \leqq h\}.$$

Because $p(h, y)$ is a density, the first term on the right of inequality (2.40) approaches zero as $\delta \to 0$. Thus

$$\overline{\lim_{\delta \to 0}} \sup_{\substack{x_1, x_2 \in K \\ t \geqq t_0 \\ |x_1 - x_2| \leqq \delta}} |u(t, x_1) - u(t, x_2)| \leqq 2 \sup_{x \in K} P_x\{\tau_{(-a,a)} \leqq h\}. \tag{2.41}$$

The left side of inequality (2.41) does not involve h, and since $0 < h \leqq t_0$ is arbitrary and the right side of (2.41) approaches zero as $h \to 0$, we conclude that

$$\overline{\lim_{\delta \to 0}} \sup_{\substack{x_1, x_2 \in K \\ t \geqq t_0 \\ |x_1 - x_2| \leqq \delta}} |u(t, x_1) - u(t, x_2)| = 0. \tag{2.42}$$

Therefore, $u(t, x)$ is continuous in x and uniformly so for $t \geqq t_0 > 0$ and x in a compact subset of $(-a, a)$.

To show the continuity in t, let $t_2 > t_1 > t_0 > 0$ and $t_2 - t_1 = h$. Consider

$$\begin{aligned} u(t_2, x) - u(t_1, x) &= \int u(t_1, y)\hat{p}(h, x, dy) - u(t_1, x) \\ &= \int u(t_1, y)p(h, x, dy) - u(t_1, x) \\ &\quad - \int u(t_1, y)[p(h, x, dy) - \hat{p}(h, x, dy)] \\ &= \int [u(t_1, y) - u(t_1, x)]p(h, x, dy) \\ &\quad - \int u(t_1, y)[p(h, x, dy) - \hat{p}(h, x, dy)]. \end{aligned}$$

Since $|u(t_1, y)| \leqq 1$,

$$\begin{aligned} |u(t_2, x) - u(t_1, x)| \leqq \left| \int [u(t_1, y) - u(t_1, x)]p(h, x, dy) \right| \\ + P_x\{\tau_{(-a,a)} \leqq h\}. \end{aligned} \tag{2.43}$$

From the first part of this lemma, the first term on the right of (2.43) approaches zero as $h \to 0$, and the second term also approaches 0 as $h \to 0$. Hence $u(t, x)$ is continuous in t and uniformly so for x in compact subsets of $(-a, a)$.

LEMMA 2.11. *If $E \subset (-a, a)$ and has positive Lebesgue measure, then*

$$\int_0^\infty \hat{p}(t, x, E)e^{-t}\, dt > 0 \tag{2.44}$$

for every $x \in (-a, a)$.

Proof: If $\alpha = 2$, i.e., the symmetric stable process is Brownian motion, then this lemma is obvious because in that case $\hat{p}(t, x, y)$ is strictly positive. If, however, $0 < \alpha < 2$, then (2.44) requires proof.

Let $Q(x, E) = \int_0^\infty \hat{p}(t, x, E)e^{-t}\, dt$. Without loss of generality we can assume $E \subset (-a, a)$ is a compact set. We show first that $Q(x, E)$ is not identically zero in x. Suppose it were, then $\hat{p}(t, x, E)$ must be 0 for almost all t. In the preceding lemma we showed that $\hat{p}(t, x, E)$ is jointly continuous in t and x so that we infer $\hat{p}(t, x, E)=0$ for all $t>0$ and all $x\in(-a, a)$. But, since $|p(t, x, E)-\hat{p}(t, x, E)|\leqq P_x\{\tau_{(-a,a)}\leqq t\}$, the fact that $\hat{p}(t, x, E)=0$ for all $t>0$ and all $x\in(-a, a)$ implies

$$\sup_{x\in E} p(t, x, E)\leqq \sup_{x\in E} P_x\{\tau_{(-a,a)}\leqq t\}. \tag{2.45}$$

Since E is compact, the right side of inequality (2.45) approaches zero as $t\to 0$ and therefore $\lim_{t\to 0}\sup_{x\in E} p(t, x, E)=0$. However, it is a basic property of $p(t, x, E)$ that $\lim_{t\to 0} p(t, x, E)=\chi_E(x)$ in L_1. Thus, we have a contradiction and conclude that $Q(x, E)$ is not identically zero in x.

We show now that if $Q(x, E)=0$ for some x, then $Q(x, E)\equiv 0$ in x which will finish the lemma. To show this we prove that if $Q(x, E)>0$ for some x, then $Q(x, E)>0$ for all x. So let x' be such that $Q(x', E)>0$. Since $\hat{p}(t, x, E)$ is continuous in x, $Q(x, E)$ is also continuous in x, and therefore there exists a neighborhood U of x' such that $Q(\xi, E)\geqq\delta>0$ for all $\xi\in U$. It will suffice to show that $Q(x, U)>0$ for all $x\in(-a, a)$ since, by the choice of U and the semigroup property, we have

$$\begin{aligned} 0<\int_U Q(y, E)Q(x, dy) &\leqq \int Q(y, E)Q(x, dy) \\ &= \int_0^\infty t\hat{p}(t, x, E)e^{-t}\,dt, \end{aligned} \tag{2.46}$$

and (2.46) implies $\hat{p}(t, x, E)>0$ on some set of t having positive Lebesgue measure which in turn implies $Q(x, E)>0$. So we must show $Q(x, U)>0$ for all $x\in(-a, a)$. This means proving that

$$P_x\{x(t) \text{ spends positive time in } U \text{ before exiting } (-a, a)\}>0.$$

Since U is open and almost all paths of the process are right continuous, if the process enters U before exiting $(-a, a)$ it will spend positive time in U. Thus it suffices to show that the probability of entering U before exiting from $(-a, a)$, starting from x, is positive. Consider two open sets N_1 and N_2 such that $x\in N_1\subset(-a, a)$ and $N_1+N_2\subset U$. Now the probability of entering U starting from x is at least the probability that the stable process remains in N_1 up to a certain time and then has a jump whose size is in N_2. That this last probability is positive is obvious since the probability that the process remains in N_1 for some length of time is positive because N_1 is open and, on the other hand, the Lévy measure for stable processes gives support for all possible jumps including those whose size is in N_2. This completes the proof of Lemma 2.11.

We now prove

LEMMA 2.12. *If β is a probability measure such that $\beta(x : |x| \leqq a) = 1$ and G is a weak neighborhood of β, then, for $a' > a$ and $\tau_{(-a',a')}$ denoting the first exit time for the process from $(-a', a')$, we have*

$$\varliminf_{t\to\infty} \frac{1}{t} \log \inf_{|x|\leqq a} P_x\{L_t(\omega, \cdot) \in G, \tau_{(-a',a')} > t\} \geqq -I(\beta). \tag{2.47}$$

Proof: This lemma is a special case of Theorem 8.1 in [8], and Lemmas 2.10 and 2.11 are proved here to verify the hypotheses necessary for that theorem. In particular, Lemma 2.11 verifies hypothesis H_4 of Theorem 8.1. Hypotheses H_1—H_3 for Theorem 8.1 are obvious properties of any process with stationary independent increments and therefore, in particular, are satisfied by the symmetric stable processes considered here.

LEMMA 2.13. *Let (Ω, Σ, μ) be a probability space and $\mathcal{F}_n$ an increasing sequence of σ-fields. Let $\{f_n(\omega)\}$ be a sequence of functions which forms a martingale relative to $(\Omega, \mathcal{F}_n, \mu)$ such that $|f_n(\omega) - f_{n-1}(\omega)| \leqq 1$ for all n and ω. Then*

$$\mu\{\omega : \lim_{n\to\infty} f_n(\omega) \text{ exists and equals } +\infty\} = 0.$$

Proof: Normalize the sequence so that $f_0(\omega) = 0$ and consider an interval $(-a, b)$. Let $\tau = \tau(\omega) = \inf[n : f_n(\omega) \notin (-a, b)]$ and note that τ might be $+\infty$. However, $\tau \wedge n$ is a finite stopping time and so, by Doob's theorem, $\int f_{\tau\wedge n}(\omega)\, d\mu = 0$, i.e.,

$$\int_{\tau\leqq n} f_\tau(\omega)\, d\mu + \int_{\tau > n} f_n(\omega)\, d\mu = 0. \tag{2.48}$$

Let

$$J_1 = \int_{\substack{\tau\leqq n \\ f_\tau(\omega)\leqq -a}} f_\tau(\omega)\, d\mu, \qquad J_2 = \int_{\substack{\tau\leqq n \\ f_\tau(\omega)\geqq b}} f_\tau(\omega)\, d\mu,$$

and

$$J_3 = \int_{\tau > n} f_n(\omega)\, d\mu\,,$$

so that (2.48) implies $J_1 + J_2 + J_3 = 0$. Since $|f_n(\omega) - f_{n-1}(\omega)| \leqq 1$ for all n and ω, we have

$$J_1 \geqq -(a+1)\mu\{\tau \leqq n\,,\ f_\tau(\omega) \leqq -a\}\,,$$

$$J_2 \geqq b\mu\{\tau \leqq n\,,\ f_\tau(\omega) \geqq b\}\,,$$

and

$$J_3 \geqq -a\mu\{\tau > n\}\,.$$

Since $J_1 + J_2 + J_3 = 0$ and

$$\mu\{\tau \leqq n, f_\tau(\omega) \leqq -a\} + \mu\{\tau \leqq n, f_\tau(\omega) \geqq b\} + \mu\{\tau > n\} = 1\,,$$

we get

$$\begin{aligned}(a+1)\mu\{\tau \leqq n, f_\tau(\omega) \leqq -a\} &\geqq b\mu\{\tau \leqq n, f_\tau(\omega) \geqq b\} - a\mu\{\tau > n\} \\ &= b\mu\{\tau \leqq n, f_\tau(\omega) \geqq b\} \\ &\quad - a[1 - \mu\{\tau \leqq n, f_\tau(\omega) \leqq -a\} - \mu\{\tau \leqq n, f_\tau(\omega) \geqq b\}]\,.\end{aligned}$$

Hence,

$$a + \mu\{\tau \leqq n, f_\tau(\omega) \leqq -a\} \geqq (a+b)\mu\{\tau \leqq n, f_\tau(\omega) \geqq b\}\,,$$

and therefore, for every n,

$$\mu\{\tau \leqq n, f_\tau(\omega) \geqq b\} \leqq (a+1)/(a+b)\,,$$

which means that

$$\lim_{a\to\infty} \lim_{b\to\infty} \lim_{n\to\infty} \mu\{\tau \leqq n, f_\tau(\omega) \geqq b\} = 0\,.$$

This last equation implies

$$\mu\{\lim_{n\to\infty} f_n(\omega) = +\infty\} = 0$$

which is the lemma.

LEMMA 2.14. *Let (Ω, Σ, μ) be a probability space and let $\mathcal{F}_n$ be an increasing sequence of σ-fields. Let $E_n \in \mathcal{F}_n$, and define $p_{n-1}(\omega) = \mu\{E_n \mid \mathcal{F}_{n-1}\}$. Then, if for almost all ω,*

$$\sum p_{n-1}(\omega) = \infty ,$$

we have

$$\sum \chi_{E_n}(\omega) = \infty ,$$

for almost all ω.

Proof: Let $g_n(\omega) = -\chi_{E_n}(\omega) + p_{n-1}(\omega)$, and note that $E\{g_n \mid \mathcal{F}_{n-1}\} = 0$. Defining $f_n(\omega) = g_1(\omega) + \cdots + g_n(\omega)$, we see that $\{f_n\}$ is a martingale because

$$\begin{aligned} E\{f_n \mid \mathcal{F}_{n-1}\} &= g_1(\omega) + \cdots + g_{n-1}(\omega) + E\{g_n \mid \mathcal{F}_{n-1}\} \\ &= g_1(\omega) + \cdots + g_{n-1}(\omega) = f_{n-1}(\omega) . \end{aligned}$$

Now, $f_n(\omega) - f_{n-1}(\omega) = g_n(\omega)$ and $|g_n(\omega)| \leqq 1$ for all n and all ω. Therefore from Lemma 2.13 we have

$$\mu\{\lim_{n\to\infty} f_n(\omega) = \infty\} = 0 ,$$

or

$$\mu\left\{\sum g_n(\omega) = \infty\right\} = 0 .$$

Since $g_n(\omega) = -\chi_{E_n}(\omega) + p_{n-1}(\omega)$, and by hypothesis $\sum p_{n-1}(\omega) = \infty$ for almost all ω, we conclude that $\sum \chi_{E_n}(\omega) = \infty$ for almost all ω.

We are now ready to prove

THEOREM 2.15. *Let β be a probability measure supported on $(-a, a)$ such* *that* $I(\beta)<1$. *Let* $a'>a$ *and* G *be a weak neighborhood of* β. *With* $\rho_t = (t/\log\log t))^{1/\alpha}$, *define*

$$E_t = \{\omega : \hat{L}_t(\omega, \cdot) \in G, \tau_{(-a'\rho_t, a'\rho_t)} > t\}.$$

Then, for almost all ω, $\omega \in E_t$ *for a sequence of times increasing to* ∞.

Proof: Let G_1 be a weak neighborhood of β such that $G_1 \subset \bar{G}_1 \subset G$. By hypothesis, $I(\beta) = \theta < 1$. Choose $k>1$ but such that $(2k-1)(\frac{1}{2}(1+\theta))<1$. For this k, we let $t_n = e^{n^k}$ and for convenience we write $\rho_n = \rho_{t_n} = (t_n/(\log\log t_n))^{1/\alpha}$. By definition,

$$\hat{L}_{t_n}(\omega, A) = \frac{1}{t_n}\int_0^{t_n} \chi_{\rho_n A}(x(s))\, ds.$$

We introduce the notation

$$\tilde{L}_{t_{n-1},t_n}(\omega, A) = \frac{1}{t_n - t_{n-1}}\int_{t_{n-1}}^{t_n} \chi_{\rho_n A}(x(s))\, ds.$$

Since $k>1$, $t_{n-1}/t_n \to 0$ as $n\to\infty$, and since, in variation norm,

$$\|\tilde{L}_{t_{n-1},t_n}(\omega, \cdot) - \hat{L}_{t_n}(\omega, \cdot)\| \leqq \frac{2t_{n-1}}{t_n},$$

we see that, for sufficiently large n,

$$\tilde{L}_{t_{n-1},t_n}(\omega, \cdot) \in G_1 \Rightarrow \hat{L}_{t_n}(\omega, \cdot) \in G.$$

Thus, defining

$$F_n = \{\omega : \tilde{L}_{t_{n-1},t_n}(\omega, \cdot) \in G_1, \tau_{(-a\rho_n, a\rho_n)} > t_{n-1}, \tau_{(-a'\rho_n, a'\rho_n)} > t_n\},$$

it is clear that if F_n occurs infinitely often so does E_{t_n} occur infinitely often. This means that to prove the theorem it will suffice to show that, for almost all ω, F_n occurs infinitely often.

Let $\mathcal{F}_n = \sigma(x(s), 0 \leqq s \leqq t_n)$, i.e., let $\mathcal{F}_n$ be the σ-field generated by the process up to time t_n. We see that

$$\begin{aligned}&P_0\{F_n \mid \mathcal{F}_{n-1}\}\\ &= \chi_{[\tau_{(-a\rho_n, a\rho_n)} > t_{n-1}]}(\omega)\, P_{x(t_{n-1})}\{\tau_{(-a'\rho_n, a'\rho_n)} > (t_n - t_{n-1}), L'_{t_n - t_{n-1}, t_n}(\omega, \cdot) \in G_1\},\end{aligned} \tag{2.49}$$

where

$$L'_{t,s}(\omega, A) = \frac{1}{t}\int_0^t \chi_{\rho_s A}(x(\sigma))\,d\sigma = \frac{1}{t}\int_0^t \chi_A\left(\frac{x(\sigma)}{\rho_s}\right)d\sigma .$$

Let $B_{n-1} = \{\omega : \tau_{(-a\rho_n, a\rho_n)} > t_{n-1}\}$. Then, from (2.49), we have

$$\text{(2.50)}\quad P_0\{F_n | \mathcal{F}_{n-1} \geqq \chi_{B_{n-1}}(\omega) \inf_{|x| \leqq a\rho_n} P_x\{\tau_{(-a'\rho_n, a'\rho_n)} > (t_n - t_{n-1}),\ L'_{t_n - t_{n-1}, t_n}(\omega, \cdot) \geqslant G_1\}.$$

The process being symmetric stable of index α, the distribution of $x(s/\rho_n^\alpha)$ is the same as $(1/\rho_n)x(s)$. Therefore, letting $s_n = (t_n - t_{n-1})/\rho_n^\alpha$, we see that

$$\inf_{|x| \leqq a\rho_n} P_x\{\tau_{(-a'\rho_n, a'\rho_n)} > (t_n - t_{n-1}),\ L'_{t_n - t_{n-1}, t_n}(\omega, \cdot) \in G_1\}$$

$$\text{(2.51)}\quad = \inf_{|x| \leqq a\rho_n} P_x\left\{\sup_{0 \leqq s \leqq t_n - t_{n-1}} |x(s)| \leqq a'\rho_n, \frac{1}{t_n - t_{n-1}}\int_0^{t_n - t_{n-1}} \chi.\left(\frac{x(s)}{\rho_n}\right) ds \in G_1\right\}$$

$$= \inf_{|x| \leqq a} P_x\left\{\sup_{0 \leqq s \leqq s_n} |x(s)| \leqq a', L_{s_n}(\omega, \cdot) \in G_1\right\}.$$

Now

$$s_n = \frac{t_n - t_{n-1}}{\rho_n^\alpha} = \left(\frac{t_n - t_{n-1}}{t_n}\right)(\log\log t_n)$$

$$= \left(1 - \frac{t_{n-1}}{t_n}\right)(\log\log t_n),$$

and, as noted earlier, $t_{n-1}/t_n \to 0$ as $n \to \infty$. Thus, $s_n \to \infty$ as $n \to \infty$ and indeed $s_n \sim \log\log t_n$, i.e., $s_n \sim k \log n$ as $n \to \infty$. Hence, from Lemma 2.12 we have, for sufficiently large n,

$$\inf_{|x| \leqq a} P_x\left\{\sup_{0 \leqq s \leqq s_n} |x(s)| \leqq a', L_{s_n}(\omega, \cdot) \in G_1\right\}$$

$$\geqq \exp\{-s_n(\tfrac{1}{2}(1+\theta))\}$$

$$\text{(2.52)}\quad \geqq \exp\{-(2k-1)\tfrac{1}{2}(1+\theta)\log n\}$$

$$= \exp\{-\theta' \log n\} = 1/n^{\theta'},$$

where we recall that $k>1$ was picked so that $\theta'=(2k-1)(\frac{1}{2}(1+\theta))<1$.

From (2.50)–(2.52) we conclude that there exists an N such that, for all ω,

$$\sum_{n=N}^{\infty} P_0\{F_n \mid \mathfrak{F}_{n-1}\} \geqq \sum_{n=N}^{\infty} \frac{\chi_{B_{n-1}}(\omega)}{n^{\theta'}}. \tag{2.53}$$

If we show that, for almost all ω,

$$\sum \frac{\chi_{B_{n-1}}(\omega)}{n^{\theta'}} = \infty, \tag{2.54}$$

then (2.53) and Lemma 2.14 imply that $\sum \chi_{F_n}(\omega)=\infty$ for almost all ω which as noted above gives us the theorem because F_n occurring infinitely often for almost all ω implies that E_{t_n} occurs infinitely often for almost all ω. To prove (2.54) it will suffice, since $\theta'<1$, to demonstrate that, for almost all ω, $\Sigma\,[1-\chi_{B_{n-1}}(\omega)]<\infty$. From the Borel–Cantelli lemma, this last statement will follow if we show that

$$\sum P_0\{\tilde{B}_n\}<\infty. \tag{2.55}$$

It is a result of Lévy that, for symmetric stable processes of index α, $P_0\{\sup_{0\leqq s\leqq t} |x(s)| \geqq a\} = O(t)$ as $t\to 0$. But,

$$\begin{aligned} P_0\{\tilde{B}_n\} &= P_0\{\tau_{(-a\rho_{n+1},a\rho_{n+1})} \leqq t_n\} \\ &= P_0\Big\{\sup_{0\leqq s\leqq t_n} |x(s)| \geqq a\rho_{n+1}\Big\} = P_0\Big\{\sup_{0\leqq s\leqq t_n/\rho^{\alpha}_{n+1}} |x(s)|\geqq a\Big\} \end{aligned} \tag{2.56}$$

and, since for all n, $(n+1)^k - n^k \geqq n^{k-1}$, we see that

$$\begin{aligned} \frac{t_n}{\rho^{\alpha}_{n+1}} &= \left(\frac{t_n}{t_{n+1}}\right)(\log\log t_{n+1}) \\ &= \exp\{[n^k-(n+1)^k]\} \log\log t_{n+1} \\ &\leqq \exp\{-n^{k-1}\} k \log (n+1), \end{aligned}$$

which goes to zero as $n\to\infty$ since $k>1$. Hence,

$$P_0\{\tilde{B}_n\} = O\left(\frac{t_n}{\rho^{\alpha}_{n+1}}\right) \leqq O(\exp\{-n^{k-1}\}k\log(n+1)),$$

so that we have (2.55) and the proof of the theorem is complete.

Using the following lemma we will be able to prove Theorem 2.9.

LEMMA 2.16. *Let $\mathscr{D}$ be the set of probability measures β on R_1 for which $I(\beta)<1$ and for each of which there exists some $K<\infty$ such that $\beta\{|x|\leqq K\}=1$. Let $C=\{\beta\in M : I(\beta)\leqq 1\}$. Then,*

$$C\subset\bar{\mathscr{D}}\,. \tag{2.57}$$

Proof: Let $\beta\in C$, and for $\lambda>1$, define $\beta_\lambda(A)=\beta((1/\lambda)A)$. As noted earlier, the I-function for a symmetric stable process of index α has the property that $I(\beta_\lambda)=(1/\lambda^\alpha)I(\beta)$. Thus, since $I(\beta)\leqq 1$, we see that $I(\beta_\lambda)<1$. Moreover, as $\lambda\to 1$, $\beta_\lambda\to\beta$ so to prove (2.57) it suffices to show that $\{\beta\in M : I(\beta)<1\}\subset\bar{\mathscr{D}}$. Let ϕ be a smooth probability density with $I(\phi)<\infty$. Let $\beta\in M$ with $\beta(R_1)=a<1$ and $I(\beta)<1$. For $0<\lambda\leqq 1$, define γ_λ by $d\gamma_\lambda=d\beta+(1-a)\lambda\phi(\lambda x)\,dx$. Now γ_λ is a probability measure and, as $\lambda\to 0$, $\gamma_\lambda\to\beta$. Also, by convexity, $I(\gamma_\lambda)\leqq I(\beta)+(1-a)\lambda^\alpha I(\phi)$, so that $I(\gamma_\lambda)<1$ for small enough λ. Thus to prove (2.57) it suffices to show that

$$\{\beta\in\mathscr{M} : I(\beta)<1\}\subset\bar{\mathscr{D}}\,, \tag{2.58}$$

where $\mathscr{M}\subset M$ is the set of probability measures on R_1. To prove (2.58) we take a $\beta\in\mathscr{M}$ with $I(\beta)<1$ and we must produce a sequence $\{\beta_n\}\in\mathscr{M}$, each β_n having compact support, such that $I(\beta_n)<1$ and $\beta_n\to\beta$. That such a sequence exists follows from the proof of Lemma 3.8 in [5].

From the definition of $\mathscr{D}$ in Lemma 2.16 we see that $\mathscr{D}\subset C$ and since C is closed in M (because I is lower semicontinuous) we have $\bar{\mathscr{D}}\subset C$. Thus the preceding lemma implies $C=\bar{\mathscr{D}}$. We can now give the

Proof of Theorem 2.9: First of all, it follows from Theorem 2.15 that if β is a probability measure supported on $(-a,a)$ such that $I(\beta)<1$ and G is a weak neighborhood of β, then, for almost all ω,

$$\omega\in\{\omega : \hat{L}_t(\omega,\cdot)\in G\}$$

for a sequence of times increasing to infinity. This means that for almost all ω the set of limit points of $\hat{L}_t(\omega,\cdot)$ as $t\to\infty$ contains $\mathscr{D}$, i.e., for almost all ω,

$$\bigcap_T \overline{\bigcup_{t\geqq T}\{\hat{L}_t(\omega,\cdot)\}}\supset\mathscr{D}\,. \tag{2.59}$$

Since the set on the left of (2.59) is closed, and since we just showed that

$\bar{\mathscr{D}} = C$, we conclude that, for almost all ω,

$$\bigcap_T \overline{\bigcup_{t \geqq T} \{\hat{L}_t(\omega, \cdot)\}} \supset C \tag{2.60}$$

which is Theorem 2.9.

As noted in (2.36), we now know that, for almost all ω,

$$C \subset \bigcap_T \overline{\bigcup_{t \geqq T} \{\hat{L}_t(\omega, \cdot)\}} \subset \bigcap_T \overline{\bigcup_{t \geqq T} C_t^{\omega}} \subset C,$$

and in particular,

$$\bigcap_T \overline{\bigcup_{t \geqq T} \hat{L}_t(\omega, \cdot)} = C. \tag{2.61}$$

If Φ is a functional on M which is continuous in the vague topology, then (2.61) implies that the set of limit points of $\Phi(\hat{L}_t(\omega, \cdot))$ as $t \to \infty$ is exactly the set $\{\Phi(\beta)\}$ with $\beta \in C$. Thus we have proved

THEOREM 2.17. *Let $\{x(s),\ 0 \leqq s < \infty\}$ be a symmetric stable process of index α, $0 < \alpha \leqq 2$, and let*

$$\hat{L}_t(\omega, A) = \frac{1}{t} \int_0^t \chi_A\left(\left(\frac{\log \log t}{t}\right)^{1/\alpha} x(s)\right) ds.$$

Let Φ be a functional on M, the space of measures β on R_1 such that $\beta(R_1) \leqq 1$, which is continuous in the vague topology. Then, for almost all ω,

$$\overline{\lim_{t \to \infty}}\, \Phi(\hat{L}_t(\omega, \cdot)) = \sup_{\beta \in C} \Phi(\beta), \tag{2.62}$$

where $C = \{\beta \in M : I(\beta) \leqq 1\}$ and $I(\beta)$ is defined by (2.10).

From the preceding remarks and Theorem 2.17, the following corollaries are self-evident.

COROLLARY 2.18. *If Φ is a functional on M which is lower semicontinuous in the vague topology, then, for almost all ω,*

$$\overline{\lim_{t \to \infty}}\, \Phi(\hat{L}_t(\omega, \cdot)) \geqq \sup_{\beta \in C} \Phi(\beta), \tag{2.63}$$

and if Φ *is a functional on* M *which is upper semicontinuous in the vague topology, then, for almost all* ω,

$$\overline{\lim_{t\to\infty}}\, \Phi(\hat{L}_t(\omega,\cdot)) \leqq \sup_{\beta\in C} \Phi(\beta)\,. \tag{2.64}$$

COROLLARY 2.19. *For* $\beta \in M$, *let* $\beta_x(A)=\beta(A-x)$, *and* $\Phi_x(\beta)=\Phi(\beta_x)$. *Then, if* Φ *is a functional on* M *which is continuous in the vague topology, we have for almost all* ω

$$\overline{\lim_{t\to\infty}}\, \sup_x \Phi_x(\hat{L}_t(\omega,\cdot)) = \sup_{\beta\in C} \Phi(\beta)\,. \tag{2.65}$$

To see (2.65) we note from (2.36) that

$$\bigcap_T \overline{\bigcup_{t\geqq T} C_t^\omega} = C\,, \tag{2.66}$$

and (2.66) and the definition of C_t^ω gives us (2.65).

3. The Strong Topology

Again we consider a symmetric stable process $\{x(s),\ 0\leqq s<\infty\}$ of index α, but now we assume $1<\alpha\leqq 2$. *The reason for restricting* $\alpha>1$ is that in this case $L_t(\omega,\cdot)$ has a continuous density with respect to Lebesgue measure (cf. Boylan [10], Kesten and Getoor [11]) which we denote by $l_t(\omega,y)$. In this section we shall prove the laws of the iterated logarithm for $\hat{l}_t(\omega,y)$ which is a suitably scaled version of $l_t(\omega,y)$. For $l_t(\omega,y)$ the topology will be uniform convergence on compact subsets of the real line rather than the vague convergence which we employed for $L_t(\omega,\cdot)$. In the functional form of the theorem, analogous to Theorem 2.17 of the last section, this allows for a richer class of functionals.

We start with

LEMMA 3.1. *Let* $\phi(x)$ *vanish outside of* $|x|\leqq h$ *and satisfy* $\int_{-\infty}^{\infty}\phi(x)\,dx=0$, $\int_{-\infty}^{\infty}|\phi(x)|\,dx\leqq c<\infty$. *For a given* $1<\alpha\leqq 2$, *pick* $0<\theta<\alpha-1$ *and define* $\gamma=2-(\theta+2)/\alpha$. *Then, for any positive integer* n *and* $0<u\leqq 1$,

$$\sup_x E_x\left\{\left(\int_0^u \phi(x(s))\,ds\right)^{2n}\right\} \leqq \frac{(2n)!\,B^n h^{n\theta}}{\Gamma(1+n\gamma)}\,, \tag{3.1}$$

where B *is a constant independent of* n.

Proof: The transition probability density for a symmetric stable process of index α is given by $(1/2\pi)\int_{-\infty}^{\infty} e^{-i\xi(x-y)}e^{-t|\xi|^\alpha}\,d\xi$ and so is of the form $t^{-1/\alpha}\Phi_\alpha(t^{-1/\alpha}(x-y))$, where

$$\Phi(x)=\Phi_\alpha(x)=\frac{1}{2\pi}\int_{-\infty}^{\infty} e^{-i\xi x}e^{-|\xi|^\alpha}\,d\xi.$$

Thus, we can write

$$E_x\left\{\left(\int_0^u \phi(x(s))\,ds\right)^{2n}\right\}$$

$$=(2n)!\int\cdots\int_{0\leqq s_1\leqq\cdots\leqq s_{2^n}\leqq u}\int_{-\infty}^{\infty}\cdots\int_{-\infty}^{\infty}\prod_{j=0}^{2n-1}\frac{\phi(x_{j+1})}{(s_{j+1}-s_j)^{1/\alpha}} \tag{3.2}$$

$$\times\Phi\left(\frac{x_{j+1}-x_j}{(s_{j+1}-s_j)^{1/\alpha}}\right)dx_1\cdots dx_{2n}\,ds_1\cdots ds_{2n},$$

where $s_0=0$ and $x_0=x$. Since $\int_{-\infty}^{\infty}\phi(x_j)\,dx_j=0$ for all $1\leqq j\leqq 2n$, we can express (3.2) in the form

$$E_x\left\{\left(\int_0^u \phi(x(s))\,ds\right)^{2n}\right\}$$

$$=(2n)!\int\cdots\int_{0\leqq s_1\leqq s_2\leqq\cdots\leqq s_{2n}\leqq u}\int_{-\infty}^{\infty}\cdots\int_{-\infty}^{\infty}\prod_{j=0}^{2n-1}\frac{\phi(x_{j+1})}{(s_{j+1}-s_j)^{1/\alpha}}$$

$$\times\prod_{k=0}^{n-1}\left\{\Phi\left(\frac{x_{2k+1}-x_{2k}}{(s_{2k+1}-s_{2k})^{1/\alpha}}\right)\Phi\left(\frac{x_{2k+2}-x_{2k+1}}{(s_{2k+2}-s_{2k+1})^{1/\alpha}}\right)\right. \tag{3.3}$$

$$\left.-\Phi^2(0)\right\}dx_1\cdots dx_{2n}\,ds_1\cdots ds_{2n}.$$

Now $\Phi(x)$ is a bounded function on R_1, and for every $0\leqq\theta\leqq 2$ there exists a constant C_θ such that $|\Phi(x)-\Phi(0)|/|x|^\theta\leqq C_\theta$ for $0\leqq x<\infty$. Thus, if we denote the bound on $\Phi(x)$ by K, we have, for the θ picked in this lemma,

$$\left|\Phi\left(\frac{x_{2k+1}-x_{2k}}{(s_{2k+1}-s_{2k})^{1/\alpha}}\right)\Phi\left(\frac{x_{2k+2}-x_{2k+1}}{(s_{2k+2}-s_{2k+1})^{1/\alpha}}\right)-\Phi^2(0)\right|$$

$$\tag{3.4}\begin{aligned}&\leqq K\left\{\left|\Phi\left(\frac{x_{2k+1}-x_{2k}}{(s_{2k-1}-s_{2k})^{1/\alpha}}\right)-\Phi(0)\right|\right.\\&\qquad\left.+\left|\Phi\left(\frac{x_{2k+2}-x_{2k+1}}{(s_{2k+2}-s_{2k+1})^{1/\alpha}}\right)-\Phi(0)\right|\right\}\\&\leqq KC_\theta\left[\frac{|x_{2k+1}-x_{2k}|^\theta}{|s_{2k+1}-s_{2k}|^{\theta/\alpha}}+\frac{|x_{2k+2}-x_{2k+1}|^\theta}{|s_{2k+2}-s_{2k+1}|^{\theta/\alpha}}\right]\\&\leqq KC_\theta(2h)^\theta\left[\frac{1}{|s_{2k+1}-s_{2k}|^{\theta/\alpha}}+\frac{1}{|s_{2k+2}-s_{2k+1}|^{\theta/\alpha}}\right].\end{aligned}$$

By hypothesis, $\int_{-\infty}^{\infty}|\phi(x)|\,dx\leqq x<\infty$. Using this and (3.4) in (3.3) we obtain

$$\tag{3.5}\begin{aligned}&\sup_x E_x\left\{\left(\left(\int_0^u\phi(x(s))\,ds\right)^{2n}\right\}\right.\\&\leqq(2n)!A_\theta^n h^{n\theta}\int_{0\leqq s_1\leqq s_2\leqq\cdots\leqq s_{2n}\leqq 1}\cdots\int\prod_{j=0}^{2n-1}\frac{1}{(s_{j+1}-s_j)^{1/\alpha}}\\&\times\prod_{k=0}^{n-1}\left(\frac{1}{|s_{2k+1}-s_{2k}|^{\theta/\alpha}}+\frac{1}{|s_{2k+2}-s_{2k+1}|^{\theta/\alpha}}\right)ds_1\,ds_2\cdots ds_{2n},\end{aligned}$$

where A_θ is a constant depending on c, θ, and K but independent of n and where we used the fact that $0<u\leqq 1$. The right side of inequality (3.5) can be rewritten as

$$\tag{3.6}\begin{aligned}&(2n)!A_\theta^n h^{n\theta}\overset{(2n)}{\int_{0\leqq s_1\leqq s_2\leqq\cdots\leqq s_{2n}\leqq 1}\cdots\int}\prod_{j=0}^{n-1}\left[\frac{1}{(s_{2j+1}-s_{2j})^{\theta/\alpha}}+\frac{1}{(s_{2j+2}-s_{2j+1})^{\theta/\alpha}}\right]\\&\times\frac{1}{(s_{2j+1}-s_{2j})^{1/\alpha}}\frac{1}{(s_{2j+2}-s_{2j+1})^{1/\alpha}}\,ds_1\,ds_2\cdots ds_{2n}.\end{aligned}$$

If, in the integral

$$\int_x^y \left[\frac{1}{(z-x)^{\theta/\alpha}}+\frac{1}{(y-z)^{\theta/\alpha}}\right]\frac{1}{(z-x)^{1/\alpha}}\cdot\frac{1}{(y-z)^{1/\alpha}}\,dz,$$

we let $z=x+\xi(y-x)$ so that $0\leqq\xi\leqq1$, we obtain

$$\frac{1}{(y-x)^{(\theta/\alpha)+(2/\alpha)-1}}\int_0^1\left[\frac{1}{\xi^{\theta/\alpha}}+\frac{1}{(1-\xi)^{\theta/\alpha}}\right]\frac{1}{\xi^{1/\alpha}}\cdot\frac{1}{(1-\xi)^{1/\alpha}}\,d\xi. \tag{3.7}$$

Since $\theta+1<\alpha$,

$$B_{\theta,\alpha}=\int_0^1\left[\frac{1}{\xi^{\theta/\alpha}}+\frac{1}{(1-\xi)^{\theta/\alpha}}\right]\frac{1}{\xi^{1/\alpha}}\cdot\frac{1}{(1-\xi)^{1/\alpha}}\,d\xi$$

is a finite constant and using this and (3.7) in (3.6) we obtain

$$\sup_x E_x\left\{\left(\int_0^u\phi(x(s))\,ds\right)^{2n}\right\}$$

$$\leqq(2n)!\,A_\theta^n h^{n\theta}B_{\theta,\alpha}^n\underset{0\leqq s_2\leqq s_4\leqq\cdots\leqq s_{2n}\leqq1}{\int\overset{(n)}{\cdots}\int}\prod_{j=0}^{n-1}\frac{1}{(s_{2j+2}-s_{2j})^{\theta/\alpha+2/\alpha-1}}\,ds_2\,ds_4\cdots ds_{2n}$$

$$=(2n)!\,A_\theta^n h^{n\theta}B_{\theta,\alpha}^n\frac{[\Gamma(\gamma)]^n}{\Gamma(1+n\gamma)},$$

where $\gamma=1-(\theta/\alpha+2\alpha-1)=2-(\theta+2)/\alpha$. This completes the proof of the lemma.

LEMMA 3.2. *With the function ϕ and the constants θ and γ determined as in Lemma 3.1, we can find a constant $k_0>0$ such that, for all x and $0<u\leqq1$,*

$$E_x\left\{\exp\left\{k_0\left|\frac{1}{h^{\theta/2}}\int_0^u\phi(x(s))\,ds\right|^\rho\right\}\right\}\leqq M<\infty,$$

where $\rho=2/(2-\gamma)$.

Proof: From the definition of γ we see that $0<\gamma<1$ and therefore

$1<\rho<2$. Consider, for $k>0$,

$$E_x\left\{\exp\left\{k\left|\int_0^u \phi(x(s)\,ds\right|^\rho\right\}\right\}=E_x\left\{\sum_{n=0}^\infty \frac{k^n\left|\int_0^u \phi(x(s))\,ds\right|^{n\rho}}{n!}\right\}$$

$$\leqq \sum_{n=0}^\infty \frac{k^n}{n!}\left[E_x\left\{\left(\int_0^u \phi(x(s))\,ds\right)^{2n}\right\}\right]^{\rho/2}$$

$$\leqq \sum_{n=0}^\infty \frac{k^n}{n!}\left[\frac{(2n)!\,B^n h^{n\theta}}{\Gamma(1+n\gamma)}\right]^{\rho/2}.$$

Let $k_0>0$ and set $k=k_0h^{-\theta\rho/2}$. Then this last inequality becomes

$$E_x\left\{\exp\left\{k_0\left|\frac{1}{h^{\theta/2}}\int_0^u \phi(x(s))\,ds\right|^\rho\right\}\right\}\leqq \sum_{n=0}^\infty \frac{k_0^n B^{n\rho/2}[2n!]^{\rho/2}}{n![\Gamma(1+n\gamma)]^{\rho/2}}. \tag{3.8}$$

But $[2n!]^{\rho/2}/n!\,[\Gamma(1+n\gamma)]^{\rho/2}\sim n^{n\rho}/(n^{n+1/2}n^{n\gamma\rho/2})$, and since $n+n\gamma\rho/2=n\rho$ we see that by choosing $k_0>0$ small enough we can make the series on the right of (3.8) converge, proving the lemma.

LEMMA 3.3. *Let θ and ρ be determined as in the preceding lemma. There exists a constant k_1 such that for almost all ω*

$$\iint_{|y_1-y_2|\leqq 1}\left(\exp\left\{k_1\left(\frac{|l_1(\omega,y_1)-l_1(\omega,y_2)|}{|y_1-y_2|^{\theta/2}}\right)^\rho\right\}-1\right)dy_1\,dy_2=M(\omega)<\infty, \tag{3.9}$$

and moreover, $E\{M(\omega)\}\leqq K<\infty$.

Proof: By definition, $l_1(\omega,y)=\int_0^1 \delta(x(s)-y)\,ds$; thus if in Lemma 3.2 we choose $\phi(x)=\delta_{y_1}(x)-\delta_{y_2}(x)$, $h=\frac{1}{2}|y_1-y_2|$, and $u=1$, we have, for all x,

$$E_x\left\{\exp\left\{k_1\left(\frac{|l_1(\omega,y_1)-l_1(\omega,y_2)|}{|y_1-y_2|^{\theta/2}}\right)^\rho\right\}\right\}\leqq K_1<\infty, \tag{3.10}$$

where $k_1 = k_0 2^{\theta\rho/2}$. We shall show now that

$$\iint\limits_{|y_1-y_2|\leqq 1} E_0\left\{\left(\exp\left\{k_1\left(\frac{|l_1(\omega, y_1)-l_1(\omega, y_2)|}{|y_1-y_2|^{\theta/2}}\right)^\rho\right\}-1\right)\right\} dy_1\, dy_2 < \infty. \tag{3.11}$$

Since

$$\{y_1, y_2 : |y_1-y_2|\leqq 1\} \subset \bigcup_{n=-\infty}^{\infty} \{y_1, y_2 : |y_1-n|\leqq 1, |y_2-n|\leqq 1\},$$

to prove (3.11) it will suffice to show that

$$\sum_{n=-\infty}^{\infty} \iint\limits_{\substack{|y_1-n|\leqq 1\\ |y_2-n|\leqq 1}} E_0\left\{\exp\left\{k_1\left(\frac{|l_1(\omega, y_1)-l_1(\omega, y_2)|}{|y_1-y_2|^{\theta/2}}\right)^\rho\right\}-1\right\} dy_1\, dy_2 < \infty. \tag{3.12}$$

Fix n, y_1 and y_2 in such a way that $|y_1-n|\leqq 1$ and $|y_2-n|\leqq 1$. Let $\tau_n = \inf\{t : x(t)\in[n-1, n+1]\}$. Then, using the strong Markov property once again, we have

$$E_0\left\{\exp\left\{k_1\left(\frac{|l_1(\omega, y_1)-l_1(\omega, y_2)|}{|y_1-y_2|^{\theta/2}}\right)^\rho\right\}-1\right\}$$

$$= E_0\left\{\chi_{\tau_n<1}(\omega)\left(\exp\left\{k_1\left(\frac{|l_1(\omega, y_1)-l_1(\omega, y_2)|}{|y_1-y_2|^{\theta/2}}\right)^\rho\right\}-1\right)\right\} \tag{3.13}$$

$$= E_Q\left\{\chi_{\tau_n<1}(\omega) E_{x(\tau_n)}\left\{\exp\left\{k_1\left(\frac{|l'_{1-\tau_n}(\omega, y_1)-l'_{1-\tau_n}(\omega, y_2)|}{|y_1-y_2|^{\theta/2}}\right)^\rho\right\}-1\right\}\right\},$$

where $l'_t(\omega, y) = \int_0^t \delta(x(s)-y)\, ds$. We use Lemma 3.2 again with $\phi(x) = \delta_{y_1}(x)-\delta_{y_2}(x)$, $h=\frac{1}{2}|y_1-y_2|$, $u = 1-\tau_n$ to conclude that

$$E_{x(\tau_n)}\left\{\exp\left\{k_1\left(\frac{l'_{1-\tau_n}(\omega, y_1)-l'_{1-\tau_n}(\omega, y_2)|}{|y_1-y_2|^{\theta/2}}\right)^\rho\right\}-1\right\} \leqq K_2 < \infty.$$

Hence, from (3.13) we see that

$$E_0\left\{\exp\left\{k_1\left(\frac{|l_1(\omega, y_1)-l_1(\omega, y_2)|}{|y_1-y_2|^{\theta/2}}\right)^{\rho}\right\}-1\right\} \tag{3.14}$$

$$\leqq K_2 E_0\{\chi_{\tau_n<1}(\omega)\} = K_2 P_0\{\tau_n<1\}.$$

Thus,

$$\iint\limits_{\substack{|y_1-n|\leqq 1\\ |y_2-n|\leqq 1}} E_0\left\{\exp\left\{k_1\left(\frac{|l_1(\omega, y_1)-l_1(\omega, y_2)|}{|y_1-y_2|^{\theta/2}}\right)^{\rho}\right\}-1\right\} dy_1\, dy_2 \leqq 4K_2 P_0\{\tau_n<1\},$$

and to show that (3.12) (and hence (3.11)) holds it suffices to prove that $\sum_{n=-\infty}^{\infty} P_0\{\tau_n<1\}<\infty$. But this last follows from (2.3) of Lemma 2.1 so that we have established (3.11). Applying Fubini's theorem to (3.11), we conclude that, for almost all ω,

$$\iint\limits_{|y_1-y_2|\leqq 1} \left(\exp\left\{k_1\left(\frac{|l_1(\omega, y_1)-l_1(\omega, y_2)|}{|y_1-y_2|^{\theta/2}}\right)^{\rho}\right\}-1\right) dy_1\, dy_2 \leqq M(\omega)<\infty$$

and $E\{M(\omega)\}\leqq K<\infty$, which is the lemma.

In the following Theorem 3.5 we shall make essential use of a result of Garsia, Rodemich, and Rumsey [8] which we state as

LEMMA 3.4. *Let $f(x)$ be defined and continuous on $[a, b]$. Let $\psi(u)$ be a non-negative even function on $(-\infty, \infty)$ such that $\psi(u)$ is nondecreasing for $u\geqq 0$ and $\psi(\infty)=\infty$. Let $p(u)$ be a non-negative even continuous function on $[-1, 1]$ such that $p(0)=0$ and $p(u)$ is nondecreasing for $u\geqq 0$. If*

$$\int_a^b\int_a^b \psi\left(\frac{f(x)-f(y)}{p(x-y)}\right) dx\, dy \leqq B<\infty, \tag{3.15}$$

then, for all $s, t\in[a, b]$,

$$|f(s)-f(t)|\leqq 8\int_0^{|s-t|} \psi^{-1}\left(\frac{4B}{u^2}\right) dp(u). \tag{3.16}$$

We are now ready to prove

THEOREM 3.5. *For every* $a>0$,

$$\lim_{\delta\to 0}\overline{\lim_{t\to\infty}}\frac{1}{t}\log P\left\{\sup_{|y_1-y_2|\leq\delta}|l_t(\omega,y_1)-l_t(\omega,y_2)|\geqq a\right\}=-\infty. \tag{3.17}$$

Proof: First of all, using the scaling properties of symmetric stable processes of index α, we observe that

$$l_t(\omega,y)=\frac{1}{t}\int_0^t\delta(x(s)-y)\,ds=\int_0^1\delta(x(ts)-y)\,ds$$

is distributed in the same way as

$$\begin{aligned}\int_0^1\delta(t^{1/\alpha}x(s)-y)\,ds&=\int_0^1\delta\left(t^{1/\alpha}\left(x(s)-\frac{y}{t^{1/\alpha}}\right)\right)ds\\&=\frac{1}{t^{1/\alpha}}\int_0^1\delta\left(x(s)-\frac{y}{t^{1/\alpha}}\right)ds\\&=\frac{1}{t^{1/\alpha}}l_1\left(\omega,\frac{y}{t^{1/\alpha}}\right).\end{aligned}$$

Therefore, to show (3.17) is equivalent to showing

$$\lim_{\delta\to 0}\overline{\lim_{t\to\infty}}\frac{1}{t}\log P\left\{\sup_{|y_1-y_2|\leq\delta/t^{1/\alpha}}|l_1(\omega,y_1)-l_1(\omega,y_2)|\geqq at^{1/\alpha}\right\}=-\infty. \tag{3.18}$$

To prove (3.18) we use Lemma 3.4. Let $\psi(x)=\exp\{k_1|x|^\rho\}-1$, $p(y)=|y|^{\theta/2}$, and $f(y)=l_1(\omega,y)$, where k_1, ρ, and θ are as in Lemma 3.3. We note that $\psi(x)$ and $p(y)$ satisfy the hypotheses of Lemma 3.4. Moreover if we pick y_1 and y_2 such that $|y_1-y_2|\leqq\delta$, where δ is small enough, then hypothesis (3.15) of Lemma 3.4 for this ψ, p, and f follows from Lemma 3.3. We conclude from (3.16) that, for all y_1 and y_2 with $|y_1-y_2|\leqq\delta$,

$$|l_1(\omega,y_1)-l_1(\omega,y_2)|\leqq 8\int_0^\delta\psi^{-1}\left(\frac{4M(\omega)}{u^2}\right)dp(u), \tag{3.19}$$

where

$$M(\omega)=\iint_{|y_1-y_2|\leq 1}\left(\exp\left\{k_1\left(\frac{|l_1(\omega,y_1)-l_1(\omega,y_2)|}{|y_1-y_2|^{\theta/2}}\right)^\rho\right\}-1\right)dy_1\,dy_2.$$

Thus

$$\sup_{|y_1-y_2|\leqq\delta/t^{1/\alpha}} |l_1(\omega, y_1)-l_1(\omega, y_2)|\leqq 8\int_0^{\delta/t^{1/\alpha}} \psi^{-1}\left(\frac{4M(\omega)}{u^2}\right) dp(u)\,. \tag{3.20}$$

From the definition of the function $\psi(x)$ we see that $\psi^{-1}(4M(\omega)/u^2)=[1/k_1 \log(1+4M(\omega)/u^2)]^{1/\rho}$, and from the definition of $p(u)$, we see that (3.20) becomes

$$\begin{aligned}&\sup_{|y_1-y_2|\leqq\delta/t^{1/\alpha}} |l_1(\omega, y_1)-l_1(\omega, y_2)|\\ &\qquad\leqq\frac{4\theta}{k_1^{1/\rho}}\int_0^{\delta/t^{1/\alpha}}\left[\log\left(1+\frac{4M(\omega)}{u^2}\right)\right]^{1/\rho} u^{\theta/2-1}\,du\,.\end{aligned} \tag{3.21}$$

Therefore,

$$\begin{aligned}&P\left\{\sup_{|y_1-y_2|\leqq\delta/t^{1/\alpha}} |l_1(\omega, y_1)-l_1(\omega, y_2)|\geqq at^{1/\alpha}\right\}\\ &\qquad\leqq P\left\{\frac{4\theta}{k_1^{1/\rho}}\int_0^{\delta/t^{1/\alpha}}\left[\log\left(1+\frac{4M(\omega)}{u^2}\right)\right]^{1/\rho} u^{\theta/2-1}\,du\geqq at^{1/\alpha}\right\}.\end{aligned} \tag{3.22}$$

Now, in (3.18), a is arbitrary and we are letting $t\to\infty$ and $\delta\to 0$ so that in proving (3.18) it will suffice, in view of (3.22), to prove

$$\lim_{\delta\to 0}\overline{\lim_{t\to\infty}}\frac{1}{t}\log P\left\{\int_0^{\delta/t^{1/\alpha}}[\log 4M(\omega)]^{1/\rho}u^{\theta/2-1}\,du\geqq at^{1/\alpha}\right\}=-\infty\,. \tag{3.23}$$

Since $\theta>0$, we have, absorbing inessential constants into a, and noting that $\rho(1/\alpha+\theta/2\alpha)=\rho\cdot(2-\gamma)/2=1$,

$$\begin{aligned}&P\left\{\int_0^{\delta/t^{1/\alpha}}[\log 4M(\omega)]^{1/\rho}u^{\theta/2-1}\,du\geqq at^{1/\alpha}\right\}\\ &\qquad=P\left\{[\log 4M(\omega)]^{1/\rho}\left(\frac{\delta}{t^{1/\alpha}}\right)^{\theta/2}\geqq at^{1/\alpha}\right\}\\ &\qquad=P\left\{M(\omega)\geqq\tfrac{1}{4}\exp\left\{\frac{a^\rho t^{\rho(1/\alpha+\theta/2\alpha)}}{\delta^{\theta\rho/2}}\right\}\right\}\\ &\qquad=P\{M(\omega)\geqq\tfrac{1}{4}\exp\{a^\rho t/\delta^{\theta\rho/2}\}\,.\end{aligned} \tag{3.24}$$

But in Lemma 3.3 we showed that $E\{M(\omega)\} \leqq K < \infty$, and hence

$$P\{M(\omega) \geqq \tfrac{1}{4} \exp\{a^{\rho}t/\delta^{\theta\rho/2}\} \leqq \tfrac{1}{4} K \exp\{-a^{\rho}t/\delta^{\theta\rho/2}\}. \tag{3.25}$$

Since $\theta\rho > 0$, (3.25) and (3.24) imply (3.23) which we noted implies (3.18) and the theorem.

Again letting $\rho_t = (t/\log\log t))^{1/\alpha}$, we now define $\hat{l}_t(\omega, y) = \rho_t l_t(\omega, \rho_t y)$. We next prove

LEMMA 3.6. *Let $t_1 < t_2 < t_3 < \cdots$ be any sequence of times approaching infinity. Then, for all t satisfying $t_n \leqq t \leqq t_{n+1}$ and any $\delta > 0$,*

$$\begin{aligned}\sup_{|y_1 - y_2| \leqq \delta} |\hat{l}_t(\omega, y_1) - \hat{l}_t(\omega, y_2)| \leqq &\sup_{|y_1 - y_2| \leqq \rho_{t_{n+1}}\delta/\rho_{t_n}} |\hat{l}_{t_n}(\omega, y_1) - \hat{l}_{t_n}(\omega, y_2)| \\ &+ \frac{\rho t_n}{t_n}(t_{n+1} - t_n) \sup_y l_{t_n, t_{n+1}}(\omega, y),\end{aligned} \tag{3.26}$$

where $l_{s,t}(\omega, y) = (1/(t-s)) \int_s^t \delta(x(\sigma) - y)\, d\sigma$.

Proof: Clearly,

$$\begin{aligned}\hat{l}_t(\omega, y) &= \frac{\rho_t}{t} \cdot t l_t(\omega, \rho_t y) \\ &= \frac{\rho_t}{t}[t_n l_{t_n}(\omega, \rho_t y) + (t - t_n) l_{t_n, t}(\omega, \rho_t y)],\end{aligned}$$

so that

$$\begin{aligned}|\hat{l}_t(\omega, y_1) - \hat{l}_t(\omega, y_2)| &\leqq \frac{\rho_t}{t} t_n |l_{t_n}(\omega, \rho_t y_1) - l_{t_n}(\omega, \rho_t y_2)| + \frac{\rho_t(t - t_n)}{t} \sup_y l_{t_n, t}(\omega, y) \\ &= \frac{\rho_t t_n}{t\rho_{t_n}} \left| \rho_{t_n} l_{t_n}\left(\omega, \frac{\rho_t}{\rho_{t_n}} \rho_{t_n} y_1\right) - \rho_{t_n} l_{t_n}\left(\omega, \frac{\rho_t}{\rho_{t_n}} \rho_{t_n} y_2\right) \right| \\ &\quad + \frac{\rho_t(t - t_n)}{t} \sup_y l_{t_n, t}(\omega, y) \\ &= \frac{\rho_t t_n}{t\rho_{t_n}} \left| \hat{l}_{t_n}\left(\omega, \frac{\rho_t}{\rho_{t_n}} y_1\right) - \hat{l}_{t_n}\left(\omega, \frac{\rho_t}{\rho_{t_n}} y_2\right) \right| + \frac{\rho_t(t - t_n)}{t} \sup_y l_{t_n, t}(\omega, y).\end{aligned} \tag{3.27}$$

Thus,

$$\sup_{|y_1-y_2|\leqq\delta}|\hat{l}_t(\omega,y_1)-\hat{l}_t(\omega,y_2)|\leqq\frac{\rho_t t_n}{t\rho_{t_n}}\sup_{|y_1-y_2|<(\rho_t\delta/\rho_{t_n})}|\hat{l}_{t_n}(\omega,y_1)-\hat{l}_{t_n}(\omega,y_2)|$$

$$+\frac{\rho_t(t-t_n)}{t}\sup_y l_{t_n,t}(\omega,y).$$

For sufficiently large t, ρ_t/t is monotone decreasing in t and ρ_t is monotone increasing in t. Since $t\geqq t_n$, this implies $\rho_t t_n/t\rho_{t_n}<1$, and since also $t\leqq t_{n+1}$, we obtain from (3.27)

$$\sup_{t_n\leqq t\leqq t_{n+1}}\sup_{|y_1-y_2|\leqq\delta}|\hat{l}_t(\omega,y_1)-\hat{l}_t(\omega,y_2)|\leqq\sup_{|y_1-y_2|\leqq(\rho_{t_{n+1}}\delta/\rho_{t_n})}|\hat{l}_{t_n}(\omega,y_1)-\hat{l}_{t_n}(\omega,y_2)|$$

$$+\frac{\rho_{t_n}(t_{n+1}-t_n)}{t_n}\sup_y l_{t_n,t_{n+1}}(\omega,y)$$

which is the lemma.

LEMMA 3.7. *For every* $a>0$,

$$\overline{\lim_{t\to\infty}}\frac{1}{t}\log P_0\left\{\sup_y l_t(\omega,y)\geqq a\right\}<0.$$

Proof: Let $f(y)$ be any probability density, i.e., $f\geqq 0$ and $\int_{-\infty}^{\infty}f(y)\,dy=1$, and assume that, for some $\delta>0$ and some $l<\infty$,

$$\sup_{|y_1-y_2|\leqq\delta}|f(y_1)-f(y_2)|\leqq l. \tag{3.28}$$

Let $M=\sup_y f(y)$, then $(M-l)2\delta\leqq 1$. This is because, on some interval $(x-\delta,x+\delta)$, (3.28) implies that f must be at least $M-l$ and hence $\int_{x-\delta}^{x+\delta}f(y)\,dy\geqq(M-l)2\delta$ so that indeed $(M-l)2\delta\leqq 1$ or $M\leqq l+1/2\delta$. Taking $l=\frac{1}{2}a$ and $\delta=1/a$, this remark clearly yields

$$P_0\left\{\sup_y l_t(\omega,y)\geqq a\right\}\leqq P_0\left\{\sup_{|y_1-y_2|\leqq 1/a}|l_t(\omega,y_1)-l_t(\omega,y_2)|\geqq\tfrac{1}{2}a\right\}. \tag{3.29}$$

Now $l_t(\omega, y)$ has the same distribution as $(1/t^{1/\alpha})l_1(\omega, y/t^{1/\alpha})$ and therefore

$$P_0\left\{\sup_{|y_1-y_2|\leqq 1/a} |l_t(\omega, y_1)-l_t(\omega, y_2)|\geqq \tfrac{1}{2}a\right\} \tag{3.30}$$
$$= P_0\left\{\sup_{|y_1-y_2|\leqq 1/at^{1/\alpha}} |l_1(\omega, y_1)-l_1(\omega, y_2)|\geqq \tfrac{1}{2}at^{1/\alpha}\right\}.$$

From (3.22)–(3.25) we see that

$$\overline{\lim_{t\to\infty}} \frac{1}{t}\log P_0\left\{\sup_{|y_1-y_2|\leqq 1/at^{1/\alpha}} |l_1(\omega, y_1)-l_1(\omega, y_2)|\geqq \tfrac{1}{2}at^{1/\alpha}\right\} \leqq \frac{-(\frac{1}{2}a)^{\rho}}{(1/a)^{\theta\rho}}<0\,. \tag{3.31}$$

Thus the lemma follows from (3.29)–(3.31),

THEOREM 3.8. *For each $a>0$, there exists $\delta>0$ such that*

$$P_0\left\{\overline{\lim_{t\to\infty}}\ \sup_{|y_1-y_2|\leqq\delta} |\hat{l}_t(\omega, y_1)-\hat{l}_t(\omega, y_2)|\geqq a\right\}=0\,. \tag{3.32}$$

Proof: Let $t_n = e^{\sqrt{n}}$ and note that $\rho_{t_{n+1}}/\rho_{t_n}\to 1$ as $n\to\infty$. Due to Lemma 3.6 and the Borel–Cantelli lemma it will clearly suffice to prove the following two statements in order to prove (3.32):

For any $a>0$, there exists $\delta>0$ such that

$$\sum_n P_0\left\{\sup_{|y_1-y_2|\leqq\delta} |\hat{l}_{t_n}(\omega, y_1)-\hat{l}_{t_n}(\omega, y_2)|\geqq a\right\}<\infty\,, \tag{3.33}$$

and, for any $a>0$,

$$\sum_n P_0\left\{\sup_y l_{t_n,t_{n+1}}(\omega, y)\geqq \frac{at_n}{\rho_{t_n}(t_{n+1}-t_n)}\right\}<\infty\,. \tag{3.34}$$

To show (3.33), we first observe that $\hat{l}_t(\omega,\cdot)$ has the same distribution as $l_{\log\log t}(\omega,\cdot)$ and, therefore, we must estimate

$$P_0\left\{\sup_{|y_1-y_2|\leqq\delta} |l_{\log\log t_n}(\omega, y_1)-l_{\log\log t_n}(\omega, y_2)|\geqq a\right\}.$$

For the given $a>0$, we can find, by Theorem 3.5, a $\delta>0$ such that

$$P_0\left\{\sup_{|y_1-y_2|\leq\delta}|l_{\log\log t_n}(\omega, y_1)-l_{\log\log t_n}(\omega, y_2)|\geqq a\right\} \tag{3.35}$$
$$\leqq \exp\{-3\log\log t_n\}=\frac{1}{n^{3/2}},$$

and hence (3.33) follows.

To show (3.34), we first observe that, since the process has independent increments,

$$P_0\left\{\sup_y l_{t_n,t_{n+1}}(\omega, y)\geqq\frac{at_n}{\rho_{t_n}(t_{n+1}-t_n)}\right\} \tag{3.36}$$
$$=P_0\left\{\sup_y l_{t_{n+1}-t_n}(\omega, y)\geqq\frac{at_n}{\rho_{t_n}(t_{n+1}-t_n)}\right\}.$$

Also, since for any constant $\lambda>0$, $l_{\lambda t}(\omega, y)$ has the same distribution as $(1/\lambda^{1/\alpha})l_t(\omega,(1/\lambda^{1/\alpha})y)$ we see that $\sup_y l_{\lambda t}(\omega, y)$ has the same distribution as $(1/\lambda^{1/\alpha})\sup_y l_t(\omega, y)$ so that

$$P_0\left\{\sup_y l_{\lambda t}(\omega, y)\geqq a\right\}=P_0\left\{\sup_y l_t(\omega, y)\geqq a\lambda^{1/\alpha}\right\},$$

or equivalently

$$P_0\left\{\sup_y l_t(\omega, y)\geqq a\right\}=P_0\left\{\sup_y l_{t/\lambda}(\omega, y)\geqq a\lambda^{1/\alpha}\right\}. \tag{3.37}$$

In particular, let $\lambda=[\rho_{t_n}(t_{n+1}-t_n)/t_n]^\alpha$, we know from (3.37) that

$$P_0\left\{\sup_y l_{t_{n+1}-t_n}(\omega, y)\geqq\frac{at_n}{\rho_{t_n}(t_{n+1}-t_n)}\right\} \tag{3.38}$$
$$=P_0\left\{\sup_y l_{k_n}(\omega, y)\geqq a\right\},$$

where

$$k_n=\frac{(t_{n+1}-t_n)t_n^\alpha}{[\rho_{t_n}(t_{n+1}-t_n)]^\alpha}=\frac{t_n^\alpha}{\rho_{t_n}^\alpha}\frac{1}{(t_{n+1}-t_n)^{\alpha-1}}$$

$$=\frac{e^{\alpha\sqrt{n}}}{e^{\sqrt{n}}}\frac{\frac{1}{2}\log n}{(e^{\sqrt{n+1}}-e^{\sqrt{n}})^{\alpha-1}}$$

$$=\frac{e^{\alpha\sqrt{n}}}{e^{\sqrt{n}}(e^{\sqrt{n}})^{\alpha-1}}\frac{\frac{1}{2}\log n}{(e^{\sqrt{n+1}-\sqrt{n}}-1)^{\alpha-1}}$$

$$\geqq(\tfrac{1}{2}\log n)n^{(\alpha-1)/2}\cdot\text{const.}\geqq n^{(\alpha-1)/2}.$$

But now Lemma 3.7 together with (3.36)–(3.38) yield (3.34). This completes Theorem 3.8.

Let $\mathscr{A}$ be the set of uniformly continuous subprobability densities on R_1, i.e.,

$$\mathscr{A}=\left\{f: f\geqq 0, \int_{-\infty}^{\infty} f(y)\,dy\leqq 1 \text{ and } f \text{ is uniformly continuous on } (-\infty,\infty)\right\}.$$

We impose on $\mathscr{A}$ the topology of uniform convergence on compact subsets of R_1 and we note that $\hat{l}_t(\omega,\cdot)\in\mathscr{A}$.

We now define $B_{t,\omega}$ to be the set of all translates of $\hat{l}_t(\omega,\cdot)$, i.e.,

$$B_{t,\omega}=\{\hat{l}_t(\omega,\cdot+x), x\in R_1\},$$

and state

THEOREM 3.9. *For almost all ω, $B_{t,\omega}$ is conditionally compact in $\mathscr{A}$ as $t\to\infty$ and*

$$\bigcap_T\overline{\bigcup_{t\geqq T}B_{t,\omega}}\subset\{f\in\mathscr{A}: I(f)\leqq 1\}.$$

Moreover, for almost all ω,

$$\bigcap_T\overline{\bigcup_{t\geqq T}\{\hat{l}_t(\omega,\cdot)\}}\supset\{f\in\mathscr{A}: I(f)\leqq 1\}.$$

Proof: Pick a sequence $\{a_n\}$ such that $\lim_{n\to\infty}a_n=0$. For each a_n there exists a δ_n and a set Ω_n with $P_0\{\Omega_n\}=1$ such that, for $\omega\in\Omega_n$,

$$\overline{\lim_{t\to\infty}}\sup_{|y_1-y_2|\leqq\delta_n}|\hat{l}_t(\omega,y_1)-\hat{l}_t(\omega,y_2)|\leqq a_n.$$

This, of course, is precisely what we proved in Theorem 3.8. Let $\Omega_\infty = \bigcap_n \Omega_n$ and pick $\omega \in \Omega_\infty$. Let $\{t_k\}$ be a sequence such that $\lim_{k\to\infty} t_k = \infty$ and, for each k, pick $f_k \in B_{t_k,\omega}$. We note that

$$\lim_{\delta\to 0} \overline{\lim_{k\to\infty}} \sup_{|y_1-y_2|\leqq\delta} |f_k(y_1)-f_k(y_2)| = 0 .$$

Therefore, by Ascoli's theorem, the f_k are totally bounded in $\mathscr{A}$ and hence $B_{t,\omega}$ is conditionally compact in $\mathscr{A}$ as $t\to\infty$. Thus, the set of limit points in $\mathscr{A}$ (in the topology imposed above on $\mathscr{A}$) is the same as the set of limit points in the vague topology which we imposed on the corresponding measures in the preceding section. Hence, using Theorems 2.8 and 2.9 we complete the proof of this theorem.

We can now state the analogue of Theorem 2.17 and its corollaries for the strong topology in $\mathscr{A}$.

THEOREM 3.10. *Let $\{x(s), 0\leqq s<\infty\}$ be a symmetric stable process of index α, $1<\alpha\leqq 2$, and let*

$$\hat{l}_t(\omega, y) = \left(\frac{t}{\log\log t}\right)^{1/\alpha} \frac{1}{t}\int_0^t \delta\left(x(s) - \left(\frac{t}{\log\log t}\right)^{1/\alpha} y\right) ds .$$

Let Φ be a functional on $\mathscr{A}$, the space of substochastic uniformly continuous densities on R_1, which is continuous in the topology of uniform convergence on compact subsets of R_1. Then, for almost all ω,

$$\overline{\lim_{t\to\infty}} \Phi(\hat{l}_t(\omega, \cdot)) = \sup_{f\in\{f\in\mathscr{A}: I(f)\leqq 1\}} \Phi(f) \tag{3.39}$$

and $I(f)$ is defined by (2.10).

We have, as in the preceding section, the following

COROLLARY 3.11. *If Φ is a functional on $\mathscr{A}$ which is lower semicontinuous, then, for almost all ω,*

$$\overline{\lim_{t\to\infty}} \Phi(\hat{l}_t(\omega, \cdot)) \geqq \sup_{f\in\{f\in\mathscr{A}: I(f)\leqq 1\}} \Phi(f) , \tag{3.40}$$

and if Φ is a functional on $\mathscr{A}$ which is upper semicontinuous, then for almost all ω,

$$\overline{\lim_{t\to\infty}}\, \Phi(\hat{l}_t(\omega,\cdot)) \leqq \sup_{f\in\{f\in\mathscr{A}:I(f)\leqq 1\}} \Phi(f)\,. \tag{3.41}$$

We also have, as before,

Corollary 3.12. *For $f\in\mathscr{A}$, let $f_x(y)=f(y-x)$ and $\Phi_x(f)=\Phi(f_x)$. Then, if Φ is a functional on $\mathscr{A}$ which is continuous in the topology of $\mathscr{A}$, we have, for almost all ω,*

$$\overline{\lim_{t\to\infty}}\, \sup_x \Phi_x(\hat{l}_t(\omega,\cdot)) = \sup_{f\in\{f\in\mathscr{A}:I(f)\leqq 1\}} \Phi(f)\,. \tag{3.42}$$

4. Examples

In this section we present a few examples to illustrate the theorems of the last two sections. Since we treated symmetric stable processes of index $0<\alpha\leqq 2$ all at once, the special case of Brownian motion, when $\alpha=2$, has been normalized to have variance 2 at time 1 instead of the customary variance 1. This should be noted in comparing the results with known results in the Brownian motion case. As in the previous sections, M denotes the class of measures λ on R such that $\lambda(R)\leqq 1$ and $\mathscr{M}\subset M$ denotes the class of probability measures on R.

Example 1. Let $V(x)$ be a continuous function on R such that $V(x)\to 0$ as $|x|\to\infty$. For each $\beta\in M$ consider the functional

$$\Phi(\beta)=\int V(x)\beta(dx)\,.$$

Let $\{x(s),\ 0\leqq s<\infty\}$ be a symmetric stable process of index α, $0<\alpha\leqq 2$, and let $I_\alpha(\beta)$ be the I-function corresponding to this process. From Theorem 2.17 we obtain in this special case

$$\overline{\lim_{t\to\infty}}\, \frac{1}{t}\int_0^t V\!\left(x(s)\left(\frac{\log\log t}{t}\right)^{1/\alpha}\right) ds = \sup_{\beta\in C_\alpha}\int V(x)\beta(dx) \text{ a.e.}, \tag{4.1}$$

where $C_\alpha=\{\beta : I_\alpha(\beta)\leqq 1\}$.

EXAMPLE 2. Let $V(x)$ be a continuous function on R with $V(x)\to\infty$ as $|x|\to\infty$. Consider the functional

$$\Phi(\beta)=\int V(x)\beta(dx)$$

for $\beta\in\mathcal{M}$ provided $\int V(x)\beta(dx)$ exists and is finite; otherwise, define $\Phi(\beta)=+\infty$. If $\beta\in M-\mathcal{M}$, we define $\Phi(\beta)=+\infty$ as well. It then follows from Corollary 2.18 that

$$\lim_{t\to\infty}\frac{1}{t}\int_0^t V\left(x(s)\left(\frac{\log\log t}{t}\right)^{1/\alpha}\right)ds=\inf_{\beta\in C_\alpha\cap\mathcal{M}}\int V(x)\beta(dx)\text{ a.e.}\tag{4.2}$$

To see this, we first note that Corollary 2.18 implies that, for almost all ω,

$$\underline{\lim}_{t\to\infty}\frac{1}{t}\int_0^t V\left(x(s)\left(\frac{\log\log t}{t}\right)^{1/\alpha}\right)ds\geqq\inf_{\beta\in C_\alpha\cap\mathcal{M}}\int V(x)\beta(dx).\tag{4.3}$$

To show the reverse inequality, let $\beta_n\to\beta$, where all the β_n and β have a fixed compact support. Then $\Phi(\beta_n)\to\Phi(\beta)$ as $n\to\infty$, and from Theorem 2.15 it follows that for each β having compact support and such that $I_\alpha(\beta)<1$ we have, for almost all ω,

$$\underline{\lim}_{t\to\infty}\frac{1}{t}\int_0^t V\left(x(s)\left(\frac{\log\log t}{t}\right)^{1/\alpha}\right)ds\leqq\int V(x)\beta(dx),$$

so that, for almost all ω,

$$\underline{\lim}_{t\to\infty}\frac{1}{t}\int_0^t V\left(x(s)\left(\frac{\log\log t}{t}\right)^{1/\alpha}\right)ds\leqq\inf_{\substack{\beta\text{ such that }I_\alpha(\beta)<1,\\ \beta\text{ has compact support}}}\int V(x)\beta(dx).\tag{4.4}$$

It is routine to show that the right-hand side of inequality (4.4) is the same as the right-hand side of (4.2).

A special case of Example 2 is when $V(x)=|x|^\theta$ for some $\theta>0$. Then (4.2) becomes

$$\lim_{t\to\infty}\frac{(\log\log t)^{\theta/\alpha}}{t^{1+\theta/\alpha}}\int_0^t|x(s)|^\theta\,ds=A_{\alpha,\theta}\text{ a.e.},\tag{4.5}$$

where

$$A_{\alpha,\theta}=\inf_{\beta\in C_\alpha\cap\mathcal{M}}\int|x|^\theta\beta(dx).$$

In particular when $\theta=\alpha=2$, the constant $A_{2,2}$ can be computed explicitly ($A_{2,2}=\frac{1}{2}$), and so, for Brownian motion paths $x(\cdot)$,

$$\varliminf_{t\to\infty}\frac{\log\log t}{t^2}\int_0^t|x(s)|^2\,ds=\tfrac{1}{2}\text{ a.e.}\tag{4.6}$$

One should compare (4.6) with Strassen's limit version

$$\varlimsup_{t\to\infty}\frac{1}{t^2\log\log t}\int_0^t|x(s)|^2\,ds=\frac{16}{\pi^2}\text{ a.e.}\tag{4.7}$$

There is no analogue of (4.7) for the stable case.

EXAMPLE 3. Consider, for $l>0$, the functional

$$\Phi_l(\beta)=\beta\{x:|x|\leqq l\}.$$

Although $\Phi_l(\beta)$ is not continuous, it is upper semicontinuous and

$$\Phi_l'(\beta)=\beta\{x:|x|<l\}$$

is lower semicontinuous. Now, if $I_\alpha(\beta)<\infty$ for any $0<\alpha\leqq 2$, then β is necessarily absolutely continuous with respect to Lebesgue measure. Therefore,

$$\sup_{\beta\in C_\alpha}\Phi_l(\beta)=\sup_{\beta\in C_\alpha}\Phi_l'(\beta)=k_{\alpha,l}.$$

We can conclude from Corollary 2.18 that for a stable process of order α

$$\varlimsup_{t\to\infty}\frac{1}{t}\operatorname{meas}\left\{s:0\leqq s\leqq t,|x(s)|\leqq l\left(\frac{t}{\log\log t}\right)^{1/\alpha}\right\}=k_{\alpha,l}\text{ a.e.}\tag{4.8}$$

A computation of $k_{2,l}$ yields the following:

If $l<\frac{1}{2}\pi$, then

$$k_{2,l}=\frac{1+l^2(z_l^2-1)^2}{1+l^2z_l^2(z_l^2-1)^2},$$

where z_l is the root of the equation in z,

$$\frac{1}{z\tan lz}+l=lz^2.$$

If $l\geqq\frac{1}{2}\pi$, then $k_{2,l}=1$.

For $0<\alpha<2$, $k_{\alpha,l}$ is hard to compute but one can still conclude the existence of a smallest l_α such that, for $l \geqq l_\alpha$, $k_{\alpha,l}=1$.

EXAMPLE 4. If $l>l_\alpha$, we conclude from Example 3 that for the symmetric stable process of order α

$$\overline{\lim_{t\to\infty}} \frac{1}{t} \operatorname{meas}\left\{s : 0\leqq s\leqq t, |x(s)|\leqq l\left(\frac{t}{\log\log t}\right)^{1/\alpha}\right\} = 1 \text{ a.e.,} \tag{4.9}$$

and that l_α is the smallest value of l for which statement (4.9) is true. Using Theorem 2.15 we can see that

$$\underline{\lim_{t\to\infty}} \left(\frac{\log\log t}{t}\right)^{1/\alpha} \sup_{0\leqq s\leqq t} |x(s)| = l_\alpha \text{ a.e.} \tag{4.10}$$

As we saw in Example 3, $l_2=\frac{1}{2}\pi$ and (4.10) becomes (1.2), the so-called "other law of the iterated logarithm" (see [2] and the remark at the beginning of this section on Brownian motion normalization).

EXAMPLE 5. We now take $1<\alpha\leqq 2$ and consider the local time, $l_t(\omega, y)$, and the theorems of Section 3. In particular if we consider the functional $\Phi(f)=f(0)$, we have from Theorem 3.10

$$\overline{\lim_{t\to\infty}} \left(\frac{t}{\log\log t}\right)^{1/\alpha} \frac{1}{t} l_t(\omega, 0) = d_\alpha \text{ a.e.,} \tag{4.11}$$

where $d_\alpha = \sup_{\substack{f\in\mathscr{A} \\ I_\alpha(f)\leqq 1}} f(0)$. A computation of d_α can be carried out and yields

$$d_\alpha = \frac{\Gamma(1/\alpha)\Gamma(1-1/\alpha)}{\pi(\alpha-1)^{1-1/\alpha}},$$

for $1<\alpha\leqq 2$. We see that when $\alpha=2$, $d_\alpha=1$ and (4.11) becomes (1.4). Using Corollary 3.12 one can prove that

$$\overline{\lim_{t\to\infty}} \left(\frac{t}{\log\log t}\right)^{1/\alpha} \frac{1}{t} \sup_y l_t(\omega, y) = d_\alpha \text{ a.e.} \tag{4.12}$$

Bibliography

[1] Chung, K. L., *On the maximum partial sums of sequences of independent random variables*, T.A.M.S. 64, 1948, pp. 205–233.

[2] Jain, N. C., and Pruitt, W. E., *The other law of the iterated logarithm*, Annals of Prob. 3, 1975, pp. 1046–1049.

[3] Strassen, V., *An invariance principle for the law of the iterated logarithm*, Z. Wahrscheinlich. 3, 1964, pp. 227–246.
[4] Kesten, H., *An iterated logarithm law for local time*, Duke Math. J. 32 (3), 1965, pp. 447–456.
[5] Donsker, M. D., and Varadhan, S. R. S., *Asymptotic evaluation of certain Wiener integrals for large time*, in *Functional Integration and Its Applications*, Proceedings of the International Conference, London; Clarendon Press, Oxford, 1975, pp. 15–33.
[6] Donsker, M. D., and Varadhan, S. R. S., *Asymptotic evaluation of certain Markov process expectations for large time, I*, Comm. Pure Appl. Math. 28, 1975, pp. 1–47.
[7] Donsker, M. D., and Varadhan, S. R. S., *Asymptotic evaluation of certain Markov process expectations for large time, II*, Comm. Pure Appl. Math. 28, 1975, pp. 279–301.
[8] Donsker, M. D., and Varadhan, S. R. S., *Asymptotic evaluation of certain Markov process expectations for large time, III*, Comm. Pure Appl. Math. 29, 1976, pp. 389–461.
[9] Garsia, A. M., Rodemich, E., and Rumsey, H., *A real variable lemma and the continuity of paths of some Gaussian processes*, Indiana Univ. Math. J. 20 (6), 1970, pp. 565–578.
[10] Boylan, E., *Local times for a class of Markov processes*, Illinois J. Math. 8 (1) 1964, pp. 19–39.
[11] Getoor, R. K., and Kesten, H., *Continuity of local times for Markov processes*, Comp. Math. 24 (3), 1972, pp. 277–303.

Received March, 1977.

Symposia Mathematica, Vol. XXI, Academic Press, 1977, 313-318

SOME PROBLEMS OF LARGE DEVIATIONS (*) (**)

M. D. DONSKER and S. R. S. VARADHAN

Let $\xi_1, \xi_2, \ldots, \xi_n, \ldots$ be a sequence of independent identically distributed random variables with a common distribution function $F(x)$ on the line. Let $M(\theta)$ be the moment generating function

$$M(\theta) = \int \exp [\theta x]\, dF(x)$$

which we assume to be finite for all θ. The problem of estimating asymptotically the probability of large deviations for the sample mean

$$\eta_n = \frac{\xi_1 + \ldots + \xi_n}{n}$$

has been studied in great detail. Among other results Cramér proved in [1] that for each $a > m = \int x\, dF(x)$

$$\lim_{n\to\infty} \frac{1}{n} \log P[\eta_n \geqslant a] = -h^+(a)$$

exists and

$$h^+(a) = a\theta_1 - \log M(\theta_1)$$

where θ_1 is the root of $M'(\theta_1)/M(\theta_1) = a$. A similar formula holds for

$$\lim_{n\to\infty} \frac{1}{n} \log P[\eta_n \leqslant a] = -h^- \qquad \text{for } a < m\,.$$

(*) I risultati conseguiti in questo lavoro sono stati esposti da S. R. S. Varadhan nella conferenza tenuta il 23 giugno 1975.

(**) This work was supported by the National Science Foundation, Grant No. NFS-MPS74-01921-A01.

A convenient way of restating these results in a form that generalizes is to start with the function

$$h(a) = \sup_{\theta} [\theta a - \log M(\theta)] \tag{1}$$

and then Cramér's result becomes

THEOREM 1: For every closed set $C \subset R$

$$\limsup_{n\to\infty} \frac{1}{n} \log P[\eta_n \in C] \leqslant -\inf_{a\in C} h(a)$$

and for every open set $G \subset R$

$$\liminf_{n\to\infty} \frac{1}{n} \log P[\eta_n \in G] \geqslant -\inf_{a\in G} h(a) .$$

If $A \subset R$ is such that

$$\inf_{a\in A^0} h(a) = \inf_{a\in A} h(a) = \inf_{a\in \bar{A}} h(a)$$

where A^0 and $\bar{A}$ are respectively the interior and the closure of A then

$$\lim_{n\to\infty} \frac{1}{n} \log P[\eta_n \in A] = -\inf_{a\in A} h(a) .$$

Stated in the above manner the theorem can be generalized. Let V be a vector space and $\xi_1, \ldots, \xi_n, \ldots$ a sequence of independent identically distributed random variables with values in V, having a common distribution $\mu(dx)$. For each θ in V^*, the dual of V we can of course consider

$$M(\theta) = \int \exp [\langle \theta, x\rangle] \mu(dx)$$

and then define

$$h(a) = \sup_{\theta\in V^*} [\langle \theta, a\rangle - \log M(\theta)] . \tag{2}$$

The analog of Theorem 1 if V is finite dimensional is easy to prove and is carried out in [7]. If V is infinite dimensional additional assumptions are needed. We shall now assume that V is a separable Banach space and that $\mu(dx)$ on V is such that

$$\int \exp[\sigma\|x\|]\mu(dx) < \infty \quad \text{for all } \sigma > 0 .$$

Then the exact analog of Theorem 1 is valid.

THEOREM 2: Let $C \subset V$ be closed and $G \subset V$ be open. Then

$$\limsup_{n\to\infty} \frac{1}{n} \log P[\eta_n \in C] \leqslant - \inf_{a\in C} h(a) ,$$

$$\liminf_{n\to\infty} \frac{1}{n} \log P[\eta_n \in G] \geqslant - \inf_{a\in G} h(a) .$$

We shall give an interesting application of Theorem 2 and then sketch a proof of Theorem 2 based on some recent results of the authors [2].

Suppose the common distribution $\mu(dx)$ is a Gaussian distribution on V with mean zero and a covariance $S(\theta, \varphi) = E[\langle\theta, x\rangle\langle\varphi, x\rangle]$ then the distribution of η_n is also Gaussian. In fact its distribution is the same as that of $\xi_1/\sqrt{n}$, with mean zero and covariance $(1/n)S(\theta, \varphi)$. Therefore

$$P[\eta_n \in A] = \mu[\sqrt{n} A] .$$

Furthermore, from a Theorem of Skorohod [6] for any Gaussian μ, $\int \exp[\sigma\|x\|]\mu(dx) < \infty$ for some $\sigma > 0$. From the fact that the distribution of ξ_1 is the same as that of $\xi_1 + \ldots + \xi_n/\sqrt{n}$ one can read off that $\int \exp[\sigma\|x\|]\mu(dx) < \infty$ for all $\sigma > 0$. This in turn makes Theorem 2 applicable and leads to

THEOREM 3: If μ is a Gaussian measure on V with mean zero, then for any $A \subset V$ with

$$\inf_{a\in A^0} h(a) = \inf_{a\in A} h(a) = \inf_{a\in \bar{A}} h(a) ,$$

$$\lim_{n\to\infty} \frac{1}{n^2} \log \mu(nA) = - \inf_{a\in A} h(a) .$$

Since the moment generating function $M(\theta)$ is just $\exp[\frac{1}{2}S(\theta, \theta)]$ for a Gaussian measure one obtains

$$h(a) = \sup_{\theta} [\langle a, \theta\rangle - \tfrac{1}{2} S(\theta, \theta)] .$$

This makes $h(a)$ the dual quadratic form to S. If V is a Hilbert space $h(a)$ will be the quadratic form $\frac{1}{2}Q(a, a)$ where the operators representing Q and S are inverses of each other. Such a result for the case of the Wiener measure is contained in Schilder [5]. For a Gaussian process on $C[0, 1]$ with Hölder continuous covariance function it was proved by Pincus [4]. Here we see that it is valid for any Gaussian process on V.

If we specialize further and set $A = \{x: \|x\| \leqslant 1\}$ then A is closed and $A^0 = \{x: \|x\| > 1\}$. It is easy to check that

$$\inf_{a\in A} h(a) = \inf_{a\in A^0} h(a) .$$

Hence we have

THEOREM 4.

$$\lim_{n\to\infty} \frac{1}{n^2} \log \mu[\|x\| \geqslant n] = - \inf_{a\in A} h(a) .$$

The computation of $\inf_{a\in A} h(a)$ is interesting. If we let $C = \sup_{\|\theta\|\leqslant 1} S(\theta, \theta)$ then one can verify that

$$\inf_{a\in A} h(a) = \frac{1}{2C} .$$

This result was proved by Shepp and Marcus [3].

We now turn to the proof of Theorem 2. Suppose V for the moment is only looked at as a complete separable metric space. Then given a finite set $\xi_1, \ldots, \xi_n$ of random variables with values in V we can form the sample distribution

$$\lambda_n(\omega, A) = \frac{1}{n}[\chi_A(\xi_1) + \ldots + \chi_A(\xi_n)] .$$

In our case where $\xi_1, \xi_2, \ldots, \xi_n, \ldots$ are independent and have a common distribution μ the ergodic theorem asserts that $\lambda_n(\omega, \cdot)$ is close to $\mu(\cdot)$ weakly with probability close to 1. We therefore introduce the measure Q_n on the space $\mathcal{M}$ of all probability distributions on V. The space $\mathcal{M}$, with weak convergence of probability measures as the convergence notion is a complete separable metric space. We note that as $n \to \infty$, $Q_n \Rightarrow \delta_\mu$ the measure degenerate at the measure μ. We are now interested in its rate.

The role of the $h(\cdot)$ function is now played by the following function $I(\lambda)$. For any λ such that $\lambda \ll \mu$ and $f(x) = d\lambda/d\mu$ with

$$\int |f(x) \log f(x)| \mu(dx) < \infty$$

we define

$$I(\lambda) = \int f(x) \log f(x) \, \mu(dx) .$$

In all other cases we define

$$I(\lambda) = \infty .$$

With this definition $I(\lambda)$ is a nonnegative lower semicontinuous functional on $\mathcal{M}$. It has the additional property that for each $l < \infty$

$$\{\lambda : I(\lambda) \leqslant l\}$$

is a compact subset of $\mathcal{M}$. The following theorem is proved in [2].

THEOREM 5: For $C \subset \mathcal{M}$ closed and $G \subset \mathcal{M}$ open we have

$$\limsup_{n\to\infty} \frac{1}{n} \log Q_n(C) \leqslant - \inf_{\lambda \in C} I(\lambda) ,$$

$$\liminf_{n\to\infty} \frac{1}{n} \log Q_n(G) \geqslant - \inf_{\lambda \in G} I(\lambda) .$$

We shall now indicate how Theorem 2 follows from Theorem 5. Assume for the moment that μ is concentrated in some sphere of finite radius B. Then all the sample distributions will be concentrated on B and therefore we need consider only those measures λ on $\mathcal{M}$ with $\lambda(B) = 1$. Let us consider the map T from such measures λ on $\mathcal{M}$ into V defined by

$$T\lambda = \int x \, d\lambda .$$

Clearly if λ is the sample distribution then $T\lambda$ is the sample mean Since every distribution involved has support in B, T is clearly a continuous map from $\mathcal{M}$ into V. We also have

$$\text{Prob}\,\{\eta_n \in A\} = Q_n\{T^{-1}A\} .$$

Since T is continuous we have that $T^{-1}A$ is closed or open according to A is closed or open in V. By Theorem 5 we conclude that for C closed in V

$$\limsup_{n\to\infty} \frac{1}{n} \log Q_n(T^{-1} C) \leqslant - \inf_{\lambda \in T^{-1} C} I(\lambda) .$$

If G is open in V

$$\liminf_{n\to\infty}\frac{1}{n}\log Q_n(T^{-1}G)\geqslant -\inf_{\lambda\in T^{-1}G} I(\lambda)\,.$$

We can calculate

$$\inf_{\lambda\in T^{-1}A} I(\lambda)=\inf_{a\in A}\ \inf_{\lambda:\,T\lambda=a} I(\lambda)$$

and verify that

$$h(a)=\inf_{\lambda:\,T\lambda=a} I(\lambda)$$

so that

$$\inf_{\lambda\in T^{-1}A} I(\lambda)=\inf_{a\in A} h(a)\,.$$

The case when μ does not have bounded support is technically more involved. But standard truncation of μ to a large sphere and passing to the limit works. The details are carried out in [2].

We have shown that starting from Theorem 5, Theorems 2, 3 and 4 can be obtained after some elementary computation.

Testo pervenuto il 19 giugno 1975.
Bozze licenziate il 1° dicembre 1976.

REFERENCES

[1] H. Cramér, *On a new limit theorem in the theory of probability*; Colloquium on the Theory of Probability, 1937, Hermann and Co., Paris.

[2] M. D. Donsker and S. R. S. Varadhan, *Asymptotic evaluation of certain Markov process expectations for large time - III*, Comm. Pure App. Math. Vol. XXIX, 4 (1976), pp. 389-461.

[3] M. B. Marcus and L. A. Shepp, *Sample behaviour of certain Gaussian processes*, Proceedings of the Sixth Berkeley Symposium, 1970, Vol. II, pp. 423-441.

[4] M. Pincus, *Gaussian processes and Hammerstein integral equations*, Trans. Amer. Math. Soc., 134 (1968), pp. 193-216.

[5] M. Schilder, *Some asymptotic formulae for Wiener integrals*, Trans. Amer. Math. Soc., 125 (1966), pp. 63-85.

[6] A. V. Skorohod, *A note on Gaussian measures in Banach space*, Theory of Probability and its Applications (English translation), Vol. 15 (1970), p. 508.

[7] S. R. S. Varadhan, *Asymptotic probabilities and differential equations*, Comm. Pure Appl. Math., Vol. 19 (1966), pp. 261-286.

Comm. Pure Appl. Math. 32 (1979) 721-747

On the Number of Distinct Sites Visited by a Random Walk*

M. D. DONSKER AND S. R. S. VARADHAN

1. Introduction

Consider a spatially homogeneous random walk on Z^d with one-step transition probability distribution π, i.e., to each point $j \in Z^d$ is assigned a non-negative probability π_j, $\sum_{j\in Z^d} \pi_j = 1$. If we let $X_1, X_2, X_3, \ldots$ be independent random variables taking values in Z^d and having common distribution π, then $Y_n = X_1 + X_2 + \cdots + X_n$ is the position of the random walk after n steps. Let $D_n = \{j \in Z^d : j = Y_r \text{ for some } r,\ 1 \leqq r \leqq n\}$, and let $D_n^{\#}$ be the cardinality of D_n, i.e., $D_n^{\#}$ is the number of distinct sites in Z^d visited by the random walk in the first n steps.

For $\nu > 0$, consider $E[\exp\{-\nu D_n^{\#}\}]$, where $E\{\ \}$ denotes expectation over the random walks starting from the origin. In this paper we study the asymptotic behavior of $E[\exp\{-\nu D_n^{\#}\}]$ as $n \to \infty$. This behavior depends, of course, on what we assume about π; therefore we state now precisely what these assumptions are. It is convenient to make these hypotheses on the characteristic function of the distribution π. For $t \in R_d$, let $\hat{\pi}(t) = \sum_{j\in Z^d} \pi_j e^{i\langle j,t\rangle}$ which is then periodic of period 2π in each of its d variables, $t = (t_1, t_2, \cdots, t_d)$. About $\hat{\pi}(t)$ we assume the following:

(i) $\hat{\pi}(t) = 1$ if and only if $t = 2\pi(n_1, n_2, \cdots, n_d)$, where $n_1, n_2, \cdots, n_d$ are integers.

(ii) $\hat{\pi}(t) = 1 - \Omega(t) + o(|t|^\alpha)$ as $|t| \to 0$, where $0 < \alpha \leqq 2$ and $e^{-\Omega(t)}$ is the characteristic function of a symmetric stable law of index α in R_d which is non-degenerate, i.e.,

$$\Omega(t) = \begin{cases} \sum a_{ij} t_i t_j \text{ for some} & \\ \text{positive definite matrix } \{a_{ij}\}, & \alpha = 2, \\ \displaystyle\int_{S^{d-1}} \int_0^\infty [1 - \cos\langle t, ry\rangle] \frac{M(dy)\, dr}{r^{\alpha+1}}, & 0 < \alpha < 2. \end{cases}$$

* The research for this paper was done at the Courant Institute of Mathematical Sciences and supported by the National Science Foundation under NSF Grant No. MCS-77-02687. Reproduction in whole or in part is permitted for any purpose of the United States Government.

In the expression for $\Omega(t)$, M is a symmetric measure on the unit ball S^{d-1} in R_d $(M(dy) = M(d(-y)))$, and the assumption of non-degeneracy means that the support of M spans R_d.

Let $\{\pi_j^n\}$ be the n-step transition probabilities. It follows from hypotheses (i) and (ii) above that for any $j \in Z^d$ there exists an n_0 such that $\pi_j^{n_0} > 0$. Moreover, from hypothesis (ii) it follows that as $n \to \infty$

$$\left[\hat{\pi}\left(\frac{t}{c_n}\right)\right]^n \to e^{-k\Omega(t)},$$

provided $n/c_n^\alpha \to k$ as $n \to \infty$. In other words, hypotheses (i) and (ii) imply that the random walk is irreducible and that the distribution π belongs to the domain of normal attraction of a non-degenerate symmetric stable law of index α, $0 < \alpha \leqq 2$.

Let $\{x(s),\ 0 \leqq s < \infty\}$ be a symmetric stable process in R_d of index $\alpha, 0 < \alpha \leqq 2$, corresponding to $\Omega(t)$ and let L_Ω be its infinitesimal generator. Our main result is that under the hypotheses above we have

THEOREM 1. *For $\nu > 0$,*

$$\lim_{n\to\infty} \frac{1}{n^{d/(d+\alpha)}} \log E[\exp\{-\nu D_n^\#\}] = -k(\nu, L), \tag{1.1}$$

where

$$k(\nu, L) = k(\nu, L, d, \alpha) = \nu^{\alpha/(d+\alpha)} \left(\frac{d+\alpha}{\alpha}\right)\left(\frac{\alpha\lambda_L}{d}\right)^{d/(d+\alpha)},$$

and

$$\lambda_L = \inf_{G:|G|=1} \lambda(G),$$

the infimum being over all open sets G in R_d of unit volume and $\lambda(G)$ the smallest eigenvalue of $-L_\Omega$ with Dirichlet boundary conditions for G.

In [1] (Lemma 3.10), we showed that the constant λ_L appearing in Theorem 1 is always strictly positive. We note also that $k(\nu, L)$ in Theorem 1 is the same constant that appears in the asymptotics for the symmetric stable sausage (Theorem 3 in [1]) which is obviously the continuous analogue of the problem treated here. Consider the special case of Theorem 1 where $\alpha = 2$ and the infinitesimal generator $L_\Omega = \frac{1}{2}\Delta$. Then $\inf_{G:|G|=1} \lambda(G)$ is attained for the sphere of unit volume (Lemma 3.13) in [1]) so that in this special case

$$k(\nu, L) = \nu^{2/(d+2)} \left(\frac{d+2}{2}\right)\left(\frac{2\gamma_d}{d}\right)^{d/(d+2)}, \tag{1.2}$$

where γ_d is the lowest eigenvalue of $-\frac{1}{2}\Delta$ for the sphere of unit volume in d dimensions with zero boundary values. The prototype random walk in Z^d, where transition is possible only to adjacent lattice points with equal probability, satisfies hypotheses (i) and (ii) for $\alpha = 2$ and Theorem 1 applies with $k(\nu, L)$ given in this special case by (1.2).

It is clear that in the asymptotic evaluation of $E[\exp\{-D_n^\#\}]$ as $n \to \infty$ the random walks which contribute most to the integral are those for which $D_n^\#$ are small so that what is required here are estimates on the probabilities of large deviations for $D_n^\#$.

2. Notation, Local Limit Theorem, and Upper Estimates

Certain estimates required for proving Theorem 1 are best made when the random walk is on a compact space; thus we find it convenient to map the walk on Z^d into an appropriately selected lattice in the d-dimensional torus T_d. In this mapping the periodicity of the walk must be taken into account.

Let $p = \text{g.c.d.}\ \{n\colon \pi_0^n \neq 0\}$ be the period, and let $\mathcal{T}_0 = \{t \in R_d : |\hat{\pi}(t)| = 1\}$ so that $\mathcal{T}_0 \supset \{2\pi(n_1, n_2, \cdots, n_d)\}$ where $n_1, n_2, \cdots, n_d$ are all integers. If $H \subset Z^d$ is the annihilator of $\mathcal{T}_0$, i.e.,

$$H = \{j \in Z^d : e^{i\langle j,t\rangle} = 1 \text{ for all } t \in \mathcal{T}_0\},$$

then H is a subgroup of Z^d and Z^d/H is cyclic with p elements. Moreover, if j is such that $\pi_j > 0$, then $H, H+j, \cdots, H+(p-1)j$ are the cyclically moving subclasses of the random walk.

Let $\lambda > 0$ be given, and let $\{a_n\}$ be a sequence of integers such that $p \mid a_n$ for every n, and

$$\frac{a_n}{n^{1/(d+\alpha)}} \to \lambda \quad \text{as} \quad n \to \infty. \tag{2.1}$$

Consider $R_d/Z^d = T_d$, the d-dimensional torus. We can think of T_d as the d-dimensional cube, $|x_i| \leq \frac{1}{2}$, $i = 1, 2, \cdots, d$, with opposite sides identified. For each n, let

$$\mathscr{L}_d^{(n)} = \left\{x \in T_d : x = \frac{j}{a_n} \text{ for some } j \in z^d\right\},$$

and, for $x \in \mathscr{L}_d^{(n)}$, let

$$\tilde{\pi}_n(x) = \sum_{j : j/a_n = x (\text{mod } 1)} \pi_j .$$

Thus, $\tilde{\pi}_n(x)$ serves as the one-step transition probabilities for a random walk on the subgroup $\mathscr{L}_d^{(n)} \subset T_d$. Moreover, since $p \mid a_n$, the random walk on $\mathscr{L}_d^{(n)}$ is also periodic of period p, the cyclically moving subclasses being the images of $H, H+j, \cdots, H+(p-1)j$ under the canonical mapping of Z^d into $\mathscr{L}_d^{(n)}$. We shall denote these cyclically moving subclasses by $\tilde{H}_{d,r}^{(n)}$ for $r \geqq 1$ where $\tilde{H}_{d,r+p}^{(n)} = \tilde{H}_{d,r}^{(n)}$.

For any $\rho > 0$, let

$$\Phi(\rho, x) = \frac{1}{(2\pi)^d} \int_{R_d} e^{-\rho \Omega(t)} \cos \langle t, x\rangle \, dt ,$$

and let $\tilde{\Phi}(\rho, x) = \sum_{j \in Z^d} \Phi(\rho, x+j)$. For $x \in \mathscr{L}_d^{(n)}$, denote by $\tilde{\pi}_n^l(x)$ the l-step transition probability for the random walk on $\mathscr{L}_d^{(n)}$, i.e.,

$$\tilde{\pi}_n^l(x) = \sum_{\substack{j \in Z^d \\ j/a_n = x (\text{mod } 1)}} \pi_j^l .$$

From our assumptions, the following local limit theorem follows. The proof is omitted since it conforms to the usual pattern for local limit theorems.

THEOREM 2.

$$\lim_{\substack{l \to \infty \\ n \to \infty \\ ln^{\alpha/(\alpha+d)} \to \rho}} \sup_{x \in \tilde{H}_{d,l}^{(n)}} \left| a_n^d \tilde{\pi}_n^l(x) - p \tilde{\Phi}\left(\frac{\rho}{\lambda^\alpha}, x\right)\right| = 0 . \tag{2.2}$$

Let $\tilde{X}_1, \tilde{X}_2, \cdots$ be independent random variables taking values in $\mathscr{L}_d^{(n)}$ with common distribution $\tilde{\pi}_n(x)$, and let $\tilde{Y}_l = \tilde{X}_1 + \tilde{X}_2 + \cdots + \tilde{X}_l$ (addition on T_d). Denote by $\tilde{D}_n^{\#}(l)$ the number of distinct sites in $\mathscr{L}_d^{(n)}$ visited by

$\tilde{Y}_1, \tilde{Y}_2, \cdots, \tilde{Y}_l$. We recall that

$$D_n = \{j \in Z^d : j = Y_r \text{ for some } 1 \leqq r \leqq n\},$$

so that if we let

$$\bar{D}_n = \{x \in T_d : j/a_n = x \ (\text{mod } 1) \text{ for some } j \in D_n\},$$

we observe that $\bar{D}_n \subset \mathscr{L}_d^{(n)}$ and $\bar{D}_n^{\#} \leqq D_n^{\#}$. What is important is to note that the distribution of $\bar{D}_n^{\#}$ is the same as the distribution of $\tilde{D}_n^{\#}(n)$. We explain now why this is important for getting upper estimates on the expectation appearing in Theorem 1.

For A, a Borel set in T_d, we consider

$$\tilde{L}_l(\omega, A) = \frac{1}{l} \sum_{j=1}^{l} \chi_A(\tilde{Y}_l), \tag{2.3}$$

i.e., the proportion of visits to A made by the random walk in $\mathscr{L}_d^{(n)}$ in the first l steps. We note that, for fixed ω, l, and n, $\tilde{L}_l(\omega, \cdot)$ is a probability measure on T_d. Let $\mathscr{M}(T_d)$ be the space of all probability measures on T_d endowed with the weak topology. We shall subsequently consider the distribution of $\tilde{L}_l(\omega, \cdot)$ in $\mathscr{M}(T_d)$, i.e., let $P_l^{(n)}$ be a probability measure on $\mathscr{M}(T_d)$ defined by

$$P_l^{(n)}(C) = P\{\tilde{L}_l(\omega, \cdot) \in C\},$$

where C is a set of measures in $\mathscr{M}(T_d)$ and P is the probability measure on the random walks in $\mathscr{L}_d^{(n)}$ starting from the origin.

For $x \in T_d$, define

$$\tilde{L}_l^{(n)}(\omega, x) = \int_{T_d} \chi_n(x-y) \tilde{L}_l(\omega, dy), \tag{2.4}$$

where

$$\chi_n(x) = \begin{cases} a_n^d & \text{if } |x_i| \leqq \dfrac{1}{2a_n} \text{ for all } 1 \leqq i \leqq d, \\ 0 & \text{otherwise}. \end{cases} \tag{2.5}$$

Let $|x : \tilde{L}_l^{(n)}(\omega, x) > 0|$ denote the Lebesgue measure of $\{x : \tilde{L}_l^{(n)}(\omega, x) > 0\}$.

We observe that

$$|x : \tilde{L}_l^{(n)}(\omega, x) > 0| = \frac{1}{a_n^d} \tilde{D}_n^{\#}(l). \tag{2.6}$$

For fixed ω, l, and n, $\tilde{L}_l^{(n)}(\omega, \cdot)$ is a probability density on T_d. Let $L_1(T_d)$ be the space of all probability densities on T_d endowed with the L_1 topology. Let $Q_l^{(n)}$ be the probability measure on $L_1(T_d)$ defined as the distribution of $\tilde{L}_l^{(n)}(w, \cdot)$ on $L_1(T_d)$. In other words, if B is a set in $L_1(T_d)$ we define

$$Q_l^{(n)}(B) = P[\tilde{L}_l^{(n)}(\omega, \cdot) \in B], \tag{2.7}$$

where P is again the probability measure on the random walks in $L_d^{(n)}$ starting from the origin. Clearly,

$$Q_l^{(n)}[f \in L_1(T_d) : f \in B] = P_l^{(n)}[\mu \in \mathcal{M}(T_d) : \chi_n * \mu \in B].$$

From the comments and definitions above, we now see that

$$E\{\exp\{-\nu D_n^{\#}\}\} \leqq E\{\exp\{-\nu \tilde{D}_n^{\#}\}\} = E^{Q_n^{(n)}}\{\exp\{-\nu a_n^d\, |x : f(x) > 0|\}\}. \tag{2.8}$$

Corresponding to the $\Omega(t)$ which appears in the hypotheses on π, there is a symmetric stable process $\{x(s), 0 \leqq s \leqq \infty\}$ in R_d the infinitesimal generator of which acting on smooth functions on R_d is given by

$$L_\Omega u = \begin{cases} \sum a_{ij} \dfrac{\partial^2 u}{\partial x_i \partial x_j}, & \alpha = 2, \\ \displaystyle\int_{S^{d-1}} \int_0^\infty \left[\frac{u(x+ry) + u(x-ry)}{2} - u(x) \right] \frac{M(dy)\, dr}{r^{1+\alpha}}, & 0 < \alpha < 2. \end{cases} \tag{2.9}$$

The process $\{x(s), 0 \leqq s < \infty\}$ can be projected onto the torus T_d. Denote it by $\{\tilde{x}(s), 0 \leqq s < \infty\}$; then this is again a process with independent increments the generator $\tilde{L}_\Omega$ of which acts on smooth functions on T_d. The relation between L_Ω and $\tilde{L}_\Omega$ is very simple: Take a function u on T_d, view it as a periodic function on R_d, apply L_Ω to this periodic extension thereby getting a periodic function which is now $\tilde{L}_\Omega$ applied to u.

Let $\mathcal{U}$ be the space of positive functions $u \in C^\infty(T_d)$ each of which is strictly bounded away from zero. For $\mu \in \mathcal{M}(T_d)$, we define

$$I_{\tilde{L}}(\mu) = -\inf_{u \in \mathcal{U}} \int_{T_d} \left(\frac{\tilde{L}_\Omega u}{u} \right)(x) \mu(dx), \tag{2.10}$$

and if μ is given by a density f it will be convenient to denote $I_{\tilde{L}}(\mu)$ by $I_{\tilde{L}}(f)$.

For the λ picked just before (2.1) we define

$$I_\lambda(f) = \frac{1}{\lambda^{\alpha+d}} I_{\tilde{L}}(f).$$

We shall prove below (Theorem 6) that if $C \subset L_1(T_d)$ is closed, then

$$\overline{\lim_{n\to\infty}} \frac{1}{a_n^d} \log Q_n^{(n)}(C) \leqq -\inf_{f\in C} I_\lambda(f). \tag{2.11}$$

Since $|x : f(x)>0|$ is a lower semicontinuous function on $L_1(T_d)$, it follows from (2.11) (see corollary to Theorem 6 below) that for $\nu>0$

$$\begin{aligned}&\overline{\lim_{n\to\infty}} \frac{1}{a_n^d} \log E^{Q_n^{(n)}}\{\exp\{-\nu a_n^d |x : f(x)>0|\}\}\\ &\qquad \leqq -\inf_{g\in L_1(T_d)} \{\nu|x : g(x)>0| + I_\lambda(g)\}.\end{aligned} \tag{2.12}$$

Recalling (2.1) and using (2.8) we then conclude that

$$\overline{\lim_{n\to\infty}} \frac{1}{n^{d/(d+\alpha)}} \log E[\exp\{-\nu D_n^\#\}] \leqq -\lambda^d \inf_{g\in L_1(T_d)} \{\nu|x : g(x)>0| + I_\lambda(g)\}. \tag{2.13}$$

But (2.13) holds for all $\lambda>0$ so that

$$\begin{aligned}&\overline{\lim_{n\to\infty}} \frac{1}{n^{d/(d+\alpha)}} \log E[\exp\{-\nu D_n^\#\}]\\ &\qquad \leqq -\sup_{\lambda>0} \lambda^d \Big[\inf_{g\in L_1(T_d)} \{\nu|x : g(x)>0| + I_\lambda(g)\}\Big],\end{aligned} \tag{2.14}$$

and at the end of Section 4 we shall show that

$$\sup_{\lambda>0} \Big[\lambda^d \inf_{g\in L_1(T_d)} \{\nu|x : g(x)>0) + I_\lambda(g)\}\Big] - k(\nu, L), \tag{2.15}$$

where $k(\nu, L)$ is defined in the statement of Theorem 1. From (2.14) and (2.15) we have

$$\overline{\lim_{n\to\infty}} \frac{1}{n^{d/(d+\alpha)}} \log E[\exp\{-\nu D_n^\#\}] \leqq -k(\nu, L). \tag{2.16}$$

In Section 5 we shall show

$$\varliminf_{n\to\infty}\frac{1}{n^{d/(d+\alpha)}}\log E[\exp\{-\nu D_n^{\#}\}]\geqq -k(\nu, L) \tag{2.17}$$

and thus will have proved Theorem 1.

3. Preliminary Results

We first prove

THEOREM 3. *If C is closed in $\mathcal{M}(T_d)$, then*

$$\varlimsup_{n\to\infty}\frac{1}{a_n^d}\log P_n^{(n)}(C)\leqq -\inf_{\mu\in C} I_\lambda(\mu)\,. \tag{3.1}$$

Proof: Let $u\in\mathcal{U}$ and for $x\in T_d$ define the operator

$$(\tilde{\pi}_n u)(x)=\sum_{y\in\mathcal{L}_d^{(n)}}\tilde{\pi}_n(y)\,u(x+y)\,.$$

Let

$$V_n(x)=\log\left(\frac{\tilde{\pi}_n u}{u}\right)(x)=\log\left(1+\frac{\tilde{\pi}_n u-u}{u}\right). \tag{3.2}$$

From elementary properties of Markov processes we deduce (c.f. (2.2) in [2]) that, for each $u\in\mathcal{U}$ and all n,

$$E^{P_n^{(n)}}\left\{\exp\left\{-n\int_{T_d}V_n(x)\mu(dx)\right\}\right\}\leqq\frac{\sup_{x\in T_d}u(x)}{\inf_{x\in T_d}u(x)}. \tag{3.3}$$

With $L_\Omega u$ as defined in (2.9), we now assert that, for all $u\in\mathcal{U}$,

$$\tilde{\pi}_n u-u=\frac{1}{a_n^\alpha}\tilde{L}_\Omega u+o\left(\frac{1}{a_n^\alpha}\right) \tag{3.4}$$

uniformly for $x\in T_d$ as $n\to\infty$. To see (3.4), it suffices to verify it for functions u of the form $e^{i\langle t,x\rangle}$ on T_d, where t is a multiple of 2π since linear combinations of such functions are dense in $\mathcal{U}$. It is easy to check that, whether $\alpha=2$ or $0<\alpha<2$, for these special functions u, $\tilde{L}_\Omega u=-e^{i\langle t,x\rangle}\Omega(t)$

and $\tilde{\pi}_n u = e^{i\langle t,x\rangle}\hat{\pi}(t/a_n)$. Hence

$$\tilde{\pi}_n u - u = e^{i\langle t,x\rangle}\left(\hat{\pi}\left(\frac{t}{a_n}\right) - 1\right),$$

and verifying (3.4) for these special functions u is equivalent to showing

$$\hat{\pi}\left(\frac{t}{a_n}\right) - 1 = -\frac{1}{a_n^\alpha}\Omega(t) + o\left(\frac{1}{a_n^\alpha}\right). \tag{3.5}$$

But $\Omega(t)$ has the property that $\Omega(t/a_n) = (1/a_n^\alpha)\Omega(t)$, and therefore (3.5) follows from hypothesis (ii) on $\hat{\pi}(t)$ in the introduction.

From (3.4), for any $u \in \mathcal{U}$,

$$\frac{\tilde{\pi}_n u - u}{u} = \frac{1}{a_n^\alpha}\frac{\tilde{L}_\Omega u}{u} + o\left(\frac{1}{a_n^\alpha}\right)$$

and hence, as $n \to \infty$,

$$\begin{aligned} V_n(x) = \log\left(1 + \frac{\tilde{\pi}_n u - u}{u}\right) &= \frac{\tilde{\pi}_n u - u}{u} + O\left(\frac{1}{a_n^{2\alpha}}\right) \\ &= \frac{1}{a_n^\alpha}\frac{\tilde{L}_\Omega u}{u} + o\left(\frac{1}{a_n^\alpha}\right) \end{aligned}$$

uniformly for $x \in T_d$. Thus, for each $u \in \mathcal{U}$, there is a sequence of constants $\delta_n \to 0$ such that

$$V_n(x) \leqq \frac{1}{a_n^\alpha}\left[\frac{\tilde{L}_\Omega u}{u} + \delta_n\right]. \tag{3.6}$$

From (3.3) and (3.6), for each $u \in \mathcal{U}$,

$$E^{P_n^{(n)}}\left\{\exp\left\{-\frac{n}{a_n^\alpha}\int_{T_d}\left(\frac{\tilde{L}_\Omega u}{u}\right)(x)\mu(dx)\right\}\right\} \leqq \frac{\sup_{x\in T_d} u(x)}{\inf_{x\in T_d} u(x)} \exp\left\{\frac{n\delta_n}{a_n^\alpha}\right\}, \tag{3.7}$$

This implies that, for each $u \in \mathcal{U}$,

$$P_n^{(n)}(C) \leqq \frac{\sup_{x\in T_d} u(x)}{\inf_{x\in T_d} u(x)} \exp\left\{\frac{n\delta_n}{a_n^\alpha} + \frac{n}{a_n^\alpha}\sup_{\mu\in C}\int_{T_d}\left(\frac{\tilde{L}_\Omega u}{u}\right)(x)\mu(dx)\right\}$$

and, therefore, letting $b_n = n/a_n^\alpha$, we have

$$\overline{\lim_{n\to\infty}} \frac{1}{b_n} \log P_n^{(n)}(C) \leqq \sup_{\mu\in C} \int_{T_d} \left(\frac{\tilde{L}_\Omega u}{u}\right)(x)\mu(dx), \tag{3.8}$$

where we note from (2.1) that

$$b_n = \frac{n}{a_n^\alpha} \sim \frac{n}{\lambda^\alpha n^{\alpha/(d+\alpha)}} = \frac{n^{d/(d+\alpha)}}{\lambda^\alpha}. \tag{3.9}$$

Now (3.8) holds for all $u \in \mathcal{U}$ and hence

$$\overline{\lim_{n\to\infty}} \frac{1}{b_n} \log P_n^{(n)}(C) \leqq \inf_{u\in\mathcal{U}} \sup_{\mu\in C} \int_{T_d} \left(\frac{\tilde{L}_\Omega u}{u}\right)(x)\mu(dx). \tag{3.10}$$

Since C is closed in $\mathcal{M}(T_d)$ and $\mathcal{M}(T_d)$ is compact in the weak topology, we see that C is compact, and it follows (c.f. [4] just before (2.13)) that

$$\overline{\lim_{n\to\infty}} \frac{1}{b_n} \log P_n^{(n)}(C) \leqq \sup_{\mu\in C} \inf_{u\in\mathcal{U}} \int_{T_d} \left(\frac{\tilde{L}_\Omega u}{u}\right)(x)\mu(dx). \tag{3.11}$$

Thus, from (3.11) and the definition of $I_{\tilde{L}}(\mu)$ we get

$$\begin{aligned} \overline{\lim_{n\to\infty}} \frac{1}{b_n} \log P_n^{(n)}(C) &\leqq \sup_{\mu\in C} (-I_{\tilde{L}}(\mu)) \\ &= -\inf_{\mu\in C} I_{\tilde{L}}(\mu). \end{aligned} \tag{3.12}$$

From (3.9) and (2.1) we see that $b_n \sim a_n^d/\lambda^{\alpha+d}$ which allows us to rewrite (3.12) as

$$\begin{aligned} \overline{\lim_{n\to\infty}} \frac{1}{a_n^d} \log P_n^{(n)}(C) &\leqq -\inf_{\mu\in C} \frac{1}{\lambda^{\alpha+d}} I_{\tilde{L}}(\mu) \\ &= -\inf_{\mu\in C} I_\lambda(\mu), \end{aligned}$$

which completes the proof.

THEOREM 4. *Let C be closed in $\mathcal{M}(T_d)$. Then*

$$\overline{\lim_{n\to\infty}} \frac{1}{a_n^d} \log Q_n^{(n)}(C) \leqq -\inf_{\mu\in C} I_\lambda(\mu).$$

Proof: Let C be weakly closed in $\mathcal{M}(T_d)$. With χ_n defined by (2.5), we observe that $\chi_n * \mu$ converges weakly to μ as $n \to \infty$ uniformly for $\mu \in \mathcal{M}(T_d)$. Thus, for any $\varepsilon > 0$ there exists n_0 such that, for $n \geqq n_0$, $\chi_n * \mu \in C$ implies $\mu \in C^\varepsilon$, where C^ε is the set of measures in $\mathcal{M}(T_d)$ within distance ε (Prokhorov distance) of C. Therefore, for $n \geqq n_0$, $Q_n^{(n)}(C) \leqq P_n^{(n)}(C^\varepsilon)$, and letting $\overline{C^\varepsilon}$ denote the weak closure of C^ε we have from Theorem 3,

$$\begin{aligned}\overline{\lim_{n\to\infty}} \frac{1}{a_n^d} \log Q_n^{(n)}(C) &\leqq \overline{\lim_{n\to\infty}} \frac{1}{a_n^d} \log P_n^{(n)}(\bar{C}^\varepsilon) \\ &\leqq - \inf_{\mu \in \bar{C}^\varepsilon} I_\lambda(\mu)\,.\end{aligned} \tag{3.13}$$

This last holds for every $\varepsilon > 0$, and since $I_\lambda(\mu)$ is lower semicontinuous and $\mathcal{M}(T_d)$ is compact in the weak topology, we obtain the theorem by letting $\varepsilon \to 0$ in (3.13).

Let $\varepsilon > 0$ be given and let $\psi_\varepsilon(x)$ be a C^∞ function on T_d with $\psi_\varepsilon(x) = \psi_\varepsilon(-x)$, $\psi_\varepsilon(x) \geqq 0$, $\int_{T_d} \psi_\varepsilon(x)\,dx = 1$, and $\psi_\varepsilon(x) \to \delta(x)$ as $\varepsilon \to 0$. Let $K_\varepsilon : L_1(T_d) \to L_1(T_d)$ be defined by

$$(K_\varepsilon f)(x) = \int_{T_d} \psi_\varepsilon(x-y) f(y)\,dy\,. \tag{3.14}$$

We now prove

THEOREM 5. *For every $\delta > 0$,*

$$\overline{\lim_{\varepsilon\to 0}}\ \overline{\lim_{n\to\infty}} \frac{1}{a_n^d} \log Q_n^{(n)}\left[f : \int_{T_d} |(K_\varepsilon f)(x) - f(x)|\,dx \geqq \delta\right] = -\infty\,. \tag{3.15}$$

Proof: From the definition of $Q_n^{(n)}$ measure on $L_1(T_d)$ and $P_n^{(n)}$ measure on $\mathcal{M}(T_d)$, we have

$$\begin{aligned}&Q_n^{(n)}\left[f : \int_{T_d} |(K_\varepsilon f)(x) - f(x)|\,dx \geqq \delta\right] \\ &\quad = P_n^{(n)}\left[\mu : \int_{T_d} |(K_\varepsilon \chi_n * \mu)(x) - (\chi_n * \mu)(x)|\,dx \geqq \delta\right] \\ &\quad = P_n^{(n)}\left[\mu : \sup_{\substack{g \in C(T_d) \\ \|g\| \leqq 1}} \int_{T_d} [(K_\varepsilon \chi_n * \mu)(x) - (\chi_n * \mu)(x)] g(x)\,dx \geqq \delta\right].\end{aligned} \tag{3.16}$$

Writing $\chi_n * g = \chi_n g$ and using the symmetry of χ_n and ψ_ε, we see that the last expression in (3.16) is the same as

$$P_n^{(n)}\left[\mu : \sup_{\substack{g\in C(T_d)\\ \|g\|\leq 1}} \int_{T_d} [(K_\varepsilon - I)\chi_n g]\mu(dx) \geqq \delta\right]. \tag{3.17}$$

Moreover, if we let $M_{n,\varepsilon} \subset C(T_d)$ be the set of all functions $V(x)$ of the form $V=(K_\varepsilon - I)\chi_n g$, where $g\in C(T_d)$ with $\|g\|\leqq 1$, then (3.17) is the same as (see definition of $P_l^{(n)}$ measure just after (2.3))

$$P\left[\sup_{V\in M_{n,\varepsilon}} [V(\tilde{Y}_1)+V(\tilde{Y}_2)+\cdots+V(\tilde{Y}_n)]\geqq n\delta\right], \tag{3.18}$$

where, as before, $\tilde{Y}_k = \tilde{X}_1+\tilde{X}_2+\cdots+\tilde{X}_k$ and $\tilde{X}_1, \tilde{X}_2, \cdots$ are independent random variables with values in $\mathscr{L}_d^{(n)}$ having common distribution $\tilde{\pi}_n(x)$. Thus

$$\begin{aligned} Q_n^{(n)}&\left[f : \int_{T_d} |(K_\varepsilon f)(x) - f(x)|\, dx \geqq \delta\right] \\ &= P\left[\sup_{V\in M_{n,\varepsilon}} [V(\tilde{Y}_1)+V(\tilde{Y}_2)+\cdots+V(\tilde{Y}_n)]\geqq n\delta\right]. \end{aligned} \tag{3.19}$$

In Lemma 4.1 below we show that, given n, ε, δ, there exist functions V_1, $V_2, \cdots, V_J$ all in $M_{n,\varepsilon}$ such that, for any function $V\in M_{n,\varepsilon}$,

$$\inf_{1\leq i\leq J}\ \sup_{x\in \mathscr{L}_d^{(n)}} |V(x) - V_i(x)| \leqq \tfrac{1}{2}\delta,$$

and, moreover, $J=J(n,\varepsilon,\delta)\leqq([8/\delta]+1)^{a_n^d}$. In other words, in Lemma 4.1 we show that M_n is totally bounded, and we estimate the number of spheres of radius $\frac{1}{2}\delta$ it takes to cover $M_{n,\varepsilon}$. Thus

$$\begin{aligned} P&\left[\sup_{V\in M_{n,\varepsilon}} [V(\tilde{Y}_1)+V(\tilde{Y}_2)+\cdots+V(\tilde{Y}_n)]\geqq n\delta\right] \\ &\leqq P\left[\sup_{1\leq i\leq J} [V_i(\tilde{Y}_1)+V_i(\tilde{Y}_2)+\cdots+V_i(\tilde{Y}_n)]\geqq \tfrac{1}{2}n\delta\right] \\ &\leqq \left(\left[\frac{8}{\delta}\right]+1\right)^{a_n^d} \sup_{1\leq i\leq J} P[[V_i(\tilde{Y}_1)+V_i(\tilde{Y}_2)+\cdots+V_i(\tilde{Y}_n)]\geqq \tfrac{1}{2}n\delta]. \end{aligned} \tag{3.20}$$

In Lemma 4.4 below we shall show that there exists an n_0 such that $n \geqq n_0$ implies, for any $\theta > 0$, and any V of the type being considered here,

$$E[\exp\{\theta[V(\tilde{Y}_1)+V(\tilde{Y}_2)+\cdots+V(\tilde{Y}_n)]\}] < c\exp\{4\theta(n^{\alpha/(\alpha+d)}+p)\}\exp\{n\lambda_n(\theta V)\}, \tag{3.21}$$

where c is a constant and

$$\lambda_n(\theta V) = \sup_{\mu\in\mathcal{M}(T_d)}\left[\theta\int_{T_d} V(x)\mu(dx) - \tilde{I}_n(\mu)\right]$$

with

$$\tilde{I}_n(\mu) = -\inf_{u\in\mathcal{U}_1}\int_{T_d}\log\left(\frac{\tilde{\pi}_n u}{u}\right)(x)\mu(dx),$$

$\mathcal{U}_1$ being the set of functions $u\in C(T_d)$ such that $u>0$ on T_d. Thus, from (3.19), (3.20), and (3.21) we obtain, for $n\geqq n_0$ and any $\theta>0$,

$$Q_n^{(n)}\left[f:\int_{T_d}|(K_\varepsilon f)(x)-f(x)|\,dx \geqq \delta\right] \leqq \left(\left[\frac{8}{\delta}\right]+1\right)^{a_n^d} c\exp\{4\theta(n^{\alpha/(\alpha+d)}+p\}\exp\{-\tfrac{1}{2}n\delta\theta\}\sup_{1\leqq i\leqq J}\exp\{n\lambda_n(\theta V_i)\}. \tag{3.22}$$

In (3.22) choose $\theta = \tau/n^{\alpha/(d+\alpha)}$ where $\tau>0$; then (3.22) becomes

$$Q_n^{(n)}\left[f:\int_{T_d}|(K_\varepsilon f)(x)-f(x)|\,dx \geqq \delta\right] \leqq \left(\left[\frac{8}{\delta}\right]+1\right)^{a_n^d} c\exp\left\{4\tau+\frac{4p\tau}{n^{\alpha/(\alpha+d)}}\right\}\exp\{-\tfrac{1}{2}n^{d/(\alpha+d)}\tau\delta\} \times \sup_{1\leqq i\leqq J}\exp\left\{n\lambda_n\left(\frac{\tau}{n^{\alpha/(\alpha+d)}}V_i\right)\right\}. \tag{3.23}$$

Recalling that $|V|\leqq 2$ for all V under consideration, $\tilde{I}_n(\mu)>0$, and

$$\lambda_n\left(\frac{\tau}{n^{\alpha/(\alpha+d)}}V\right) = \sup_{\mu\in\mathcal{M}(T_d)}\left[\frac{\tau}{n^{\alpha/(\alpha+d)}}\int_{T_d}V(x)\mu(dx)-\tilde{I}_n(\mu)\right], \tag{3.24}$$

we see that

$$\frac{-2\tau}{n^{\alpha/(\alpha+d)}} \leqq \lambda_n\left(\frac{\tau}{n^{\alpha/(\alpha+d)}} V\right) \leqq \frac{2\tau}{n^{\alpha/(\alpha+d)}}.$$

Hence, in (3.24) nothing is changed if, in taking the supremum over all μ, we take it instead over all μ such that $\tilde{I}_n(\mu) \leqq \frac{\tau}{n^{\alpha/(\alpha+d)}}$, i.e.,

$$(3.25)\quad \lambda_n\left(\frac{\tau}{n^{\alpha/(\alpha+d)}} V\right) \leqq \sup_{0<\sigma\leqq\tau} \sup_{\mu:\tilde{I}_n(\mu)=\sigma/n^{\alpha/(\alpha+d)}} \left[\frac{\tau}{n^{\alpha/(\alpha+d)}} \int_{T_d} V(x)\mu(dx) - \frac{\sigma}{n^{\alpha/(\alpha+d)}}\right].$$

But, it follows from Lemma 4.5 below that, for any of the V under consideration, i.e., $V=(K-I)\chi_n g$ where $g \in C(T_d)$ with $\|g\| \leqq 1$, and any $t>0$,

$$(3.26)\quad \sup_{\mu:\tilde{I}_n(\mu)=\sigma/n^{\alpha/(\alpha+d)}} \frac{\tau}{n^{\alpha/(\alpha+d)}} \int_{T_d} V(x)\mu(dx) \leqq \frac{\tau}{n^{\alpha/(\alpha+d)}}[2h(t\sigma)+2\Delta_t(n)+k_t(\varepsilon)],$$

where $h(\sigma)\to 0$ as $\sigma\to 0$, $\Delta_t(n)\to 0$ as $n\to\infty$ for all $t>0$, and $k_t(\varepsilon)\to 0$ as $\varepsilon\to 0$ for all $t>0$. The functions $h(\sigma)$, $\Delta_t(n)$ and $k_t(\varepsilon)$ are all independent of V.

Thus, from (3.25) and (3.26), we get

$$(3.27)\quad \begin{aligned} \lambda_n\left(\frac{\tau}{n^{\alpha/(\alpha+d)}} V\right) &\leqq \sup_{0<\sigma\leqq\tau}\left[\frac{\tau}{n^{\alpha/(\alpha+d)}}\{2h(t\sigma)+2\Delta_t(n)+k_t(\theta)\} - \frac{\sigma}{n^{\alpha/(\alpha+d)}}\right] \\ &= \frac{1}{n^{\alpha/(\alpha+d)}} \sup_{0<\sigma\leqq t}[\tau\{2h(t\sigma)+2\Delta_t(n)+k_t(\varepsilon)\}-\sigma]. \end{aligned}$$

From (2.1), $n^{d/(d+\alpha)}/a_n^d \to 1/\lambda^d$ as $n\to\infty$, and hence from (3.23) and (3.27) we obtain, for all $t>0$ and all $\tau>0$,

$$(3.28)\quad \begin{aligned} &\overline{\lim_{n\to\infty}} \frac{1}{a_n^d} \log Q_n^{(n)}\left[f: \int_{T_d} |(K_\varepsilon f)(x)-f(x)|\,dx \geqq \sigma\right] \\ &\leqq \log\left(\left[\frac{8}{\delta}\right]+1\right) - \frac{\delta\tau}{2\lambda^d} + \frac{1}{\lambda^d} \overline{\lim_{n\to\infty}} \sup_{0<\sigma\leqq\tau}[\tau\{2h(t\sigma)+2\Delta_t(n)+k_t(\varepsilon)\}-\sigma] \\ &= \log\left(\left[\frac{8}{\delta}\right]+1\right) - \frac{\delta\tau}{2\lambda^d} + \frac{1}{\lambda^d} \sup_{0<\sigma\leqq\tau}[2\tau h(t\sigma)+\tau k_t(\varepsilon)-\sigma]. \end{aligned}$$

From (3.28) we get, for all $t>0$, $\tau>0$,

$$\varlimsup_{\varepsilon\to 0}\varlimsup_{n\to\infty}\frac{1}{a_n^d}\log Q_n^{(n)}\left[f:\int_{T_d}|(K_\varepsilon f)(x)-f(x)|\,dx\geqq\delta\right] \tag{3.29}$$
$$\leqq\frac{1}{\lambda^d}\left[\sup_{0<\sigma\leqq\tau}\{2\tau h(t\sigma)-\sigma\}-\tfrac{1}{2}\delta\tau\right]+\log\left(\left[\frac{8}{\delta}\right]+1\right).$$

In (3.29) let $\sigma'=\sigma/\tau$ and the right side of inequality (3.29) becomes

$$\frac{\tau}{\lambda^d}\left[\sup_{0<\sigma'\leqq 1}\{2h(t\tau\sigma')-\sigma'\}-\tfrac{1}{2}\delta\right]+\log\left(\left[\frac{8}{\delta}\right]+1\right)$$
$$\leqq\frac{\tau}{\lambda^d}\left[\sup_{0<\sigma'\leqq 1}2h(t\tau\sigma')-\tfrac{1}{2}\delta\right]+\log\left(\left[\frac{8}{\delta}\right]+1\right) \tag{3.30}$$
$$\leqq\frac{\tau}{\lambda^d}[2h(t\tau)-\tfrac{1}{2}\delta]+\log\left(\left[\frac{8}{\delta}\right]+1\right),$$

where the last inequality follows from the fact that without loss of generality we could pick the function h to go to 0 monotonically.

From (3.29) and (3.30), letting $t\to 0$, we get

$$\varlimsup_{\varepsilon\to 0}\varlimsup_{n\to\infty}\frac{1}{a_n^d}\log Q_n^{(n)}\left[f:\int_{T_d}|(K_\varepsilon f)(x)-f(x)|\,dx\geqq\sigma\right] \tag{3.31}$$
$$\leqq-\frac{\delta\tau}{2\lambda^d}+\log\left(\left[\frac{8}{\delta}\right]+1\right),$$

and finally, if we let $\tau\to\infty$ in (3.31), we obtain (3.15).

4. Lemmas for Theorem 5 and Proof of Theorem 6

We now prove five lemmas used in the proof of Theorem 5 in the preceding section.

LEMMA 4.1. *Let $g\in C(T_d)$ with $\|g\|\leqq 1$ and let $M_{n,\varepsilon}\subset C(T_d)$ be the set of all functions $V(x)$ of the form $V=(K_\varepsilon-I)\chi_n g$. For any $\delta>0$ there exist functions $V_1, V_2,\cdots, V_J$ in $M_{n,\varepsilon}$ such that, for any $V\in M_{n,\varepsilon}$,*

$$\inf_{1\leqq i\leqq J}\sup_{x\in\mathscr{L}_d^{(n)}}|V(x)-V_i(x)|\leqq\tfrac{1}{2}\delta, \tag{4.1}$$

and, moreover, $J=J(n,\varepsilon,\delta)\leqq([8/\delta]+1)^{a_n^d}$.

Proof: We note that $|V|\leqq 2$ for any $V\in M_{n,\varepsilon}$ and also that $\mathscr{L}_d^{(n)}$ has a_n^d points. Subdivide the interval $[-2,2]$ into subintervals of length $\frac{1}{2}\delta$ so that there are at most $[8/\delta]+1$ subintervals. Now $[-2,2]^{a_n^d}$ consists of cubes of sides $\frac{1}{2}\delta$ and there are at most $([8/\delta]+1)^{a_n^d}$ such cubes. But, any $V\in M_{n,\varepsilon}$, as a function on $\mathscr{L}_d^{(n)}$, can be thought of as an element of $[-2,2]^{a_n^d}$. Thus, $M_{n,\varepsilon}\subset[-2,2]^{a_n^d}$. Think of the subcubes constructed above as closed and, for a particular cube C, pick one of its elements if $M_{n,\varepsilon}\cap C\neq\emptyset$. In this way we pick $V_1, V_2,\cdots, V_J$, where J is the number of cubes having non-empty intersection with $M_{n,\varepsilon}$. Thus $J=J(n,\varepsilon,\sigma)\leqq([8/\delta]+1)^{a_n^d}$. Moreover, if $V\in M_{n,\varepsilon}$, then $V\in[-2,2]^{a_n^d}$ and must belong to some cube, i.e., the intersection of $M_{n,\varepsilon}$ with that cube is non-empty and hence we must have picked some element V_i from that cube which means (4.1) holds.

LEMMA 4.2. *Let $l=[n^{\alpha/(d+\alpha)}]$. There exists an integer n_0 and there exists a constant $c<\infty$ such that, for $n\geqq n_0$,*

$$\tilde{\pi}_n^l(x)\leqq c\tilde{\pi}_n^{l+q}(y) \tag{4.2}$$

for all $x\in\mathscr{L}_d^{(n)}$, $y\in\mathscr{L}_d^{(n)}$, $y-x\in\tilde{H}_{d,q}^{(n)}$, where $0\leqq q\leqq p-1$.

Proof: Suppose first that $x\notin\tilde{H}_{d,l}^{(n)}$; then $y\notin\tilde{H}_{d,l+q}^{(n)}$ so that $\tilde{\pi}_n^l(x)=0$ and $\tilde{\pi}_n^{l+q}(y)=0$ which means (4.2) holds. Next, suppose $x\in\tilde{H}_{d,l}^{(n)}$ which implies $y\in\tilde{H}_{d,l+q}^{(n)}$. From Theorem 2 (with $\rho=1$ and noting that $\tilde{\Phi}$ is strictly positive for $x\in T_d$) we can choose n_0 such that $n\geqq n_0$ implies

$$\sup_{x\in\tilde{H}_{d,l}^{(n)}}\left|a_n^d\tilde{\pi}_n^l(x)-p\tilde{\Phi}\left(\frac{1}{x^\alpha},x\right)\right|\leqq\tfrac{1}{8}\inf_{x\in T_d}p\tilde{\Phi}\left(\frac{1}{\lambda^\alpha},x\right),$$

and

$$\sup_{y\in\tilde{H}_{d,l+q}^{(n)}}\left|a_n^d\tilde{\pi}_n^{l+q}(y)-p\tilde{\Phi}\left(\frac{1}{\lambda^\alpha},y\right)\right|\leqq\tfrac{1}{8}\inf_{x\in T_d}p\tilde{\Phi}\left(\frac{1}{\lambda^\alpha},x\right).$$

Hence, for all $x\in\tilde{H}_{d,l}^{(n)}$, we have

$$a_n^d\tilde{\pi}_n^l(x)\leqq\tfrac{9}{8}\sup_{x\in T_d}p\tilde{\Phi}\left(\frac{1}{\lambda^\alpha},x\right),$$

and, for all $y\in\tilde{H}_{d,l+q}^{(n)}$, we have

$$a_n^d\tilde{\pi}_n^{l+q}(y)\geqq\tfrac{7}{8}\inf_{x\in T_d}p\tilde{\Phi}\left(\frac{1}{\lambda^\alpha},x\right).$$

Letting

$$c = \tfrac{9}{7}\frac{\sup_{x\in T_d}\bar{\Phi}(1/\lambda^\alpha, x)}{\inf_{x\in T_d}\bar{\Phi}(1/\lambda^\alpha, x)},$$

we have the lemma.

LEMMA 4.3. *Let x and y be any two points in $\mathscr{L}_d^{(n)}$. There exists an integer n_0 such that $n \geqq n_0$ implies, for any $\theta > 0$,*

$$\begin{aligned}(4.3)\qquad & E_y[\exp\{\theta[V(\tilde{Y}_1)+\cdots+V(\tilde{Y}_n)]\}] \\ &\qquad\leqq C_{n,\theta}E_x[\exp\{\theta[V(\tilde{Y}_1)+\cdots+V(\tilde{Y}_n)]\}],\end{aligned}$$

where $C_{n,\theta} = c\exp\{4\theta n^{\alpha/(\alpha+d)}+p\}$. The constant c is the constant of Lemma 4.2 and, indeed, n_0 is the same as picked in Lemma 4.2. The notation $E_x\{\ \}$ denotes expectation where the starting point is x.

Proof: Choose l and n_0 as in Lemma 4.2. Since x and y are in $\mathscr{L}_d^{(n)}$, $y-x\in\tilde{H}_{d,q}^{(n)}$ for some q, $0\leqq q\leqq p-1$. Using the fact that $|V|\leqq 2$, we have

$$\begin{aligned}(4.4)\qquad & E_y[\exp\{\theta[V(\tilde{Y}_1)+\cdots+V(\tilde{Y}_n)]\}]\leqq e^{2\theta l}E_y[\exp\{\theta[V(\tilde{Y}_{l+1})+\cdots+V(\tilde{Y}_n)]\}] \\ &\qquad = e^{2\theta l}\sum_{z\in\mathscr{L}_d^{(n)}}\tilde{\pi}_n^l(z-y)E_z[\exp\{\theta[V(\tilde{Y}_1)+\cdots+V(\tilde{Y}_{n-l})]\}].\end{aligned}$$

Since $(z-x)-(z-y)=y-x\in\tilde{H}_{d,q}^{(n)}$, we have from Lemma 4.2 and (4.4),

$$\begin{aligned} & E_y[\exp\{\theta[V(\tilde{Y}_1)+\cdots+V(\tilde{Y}_n)]\}] \\ &\qquad\leqq ce^{2\theta l}\sum_{z\in\mathscr{L}_d^{(n)}}\tilde{\pi}_n^{l+q}(z-x)E_z[\exp\{\theta[V(\tilde{Y}_1)+\cdots+V(\tilde{Y}_{n-l})]\}] \\ &\qquad = ce^{2\theta l}E_x[\exp\{\theta[V(\tilde{Y}_{l+q+1})+\cdots+V(\tilde{Y}_{n+q})]\}] \\ &\qquad\leqq ce^{4\theta(l+q)}E_x[\exp\{\theta[V(\tilde{Y}_1)+\cdots+V(\tilde{Y}_n)]\}] \\ &\qquad\leqq ce^{4\theta(l+p)}E_x[\exp\{\theta[V(\tilde{Y}_1)+\cdots+V(\tilde{Y}_n)]\}],\end{aligned}$$

which is the lemma.

LEMMA 4.4. *There exists an integer n_0 such that $n\geqq n_0$ implies, for any $\theta>0$,*

$$\begin{aligned}(4.5)\qquad E[\exp\{\theta[V(\tilde{Y}_1)+\cdots+V(\tilde{Y}_n)]\}] &= E_0[\exp\{\theta[V(\tilde{Y}_1)+\cdots+V(\tilde{Y}_n)]\}] \\ &\leqq C_{n,\theta}\exp\{n\lambda_n(\theta V)\},\end{aligned}$$

Where $C_{n,\theta}$ *is the constant in Lemma* 4.3, *and*

$$\lambda_n(\theta V) = \sup_{\mu \in \mathcal{M}(T_d)} \left[\theta \int_{T_d} V(x)\mu(dx) - \tilde{I}_n(\mu)\right] \tag{4.6}$$

with (*c.f.* [2])

$$\tilde{I}_n(\mu) = -\inf_{u \in \mathcal{U}_1} \int_{T_d} \log\left(\frac{\tilde{\pi}_n u}{u}\right)(x)\mu(dx), \tag{4.7}$$

and $\mathcal{U}_1$ *is the set of functions* $u \in C(T_d)$ *such that* $u>0$ *on* T_d.

Proof: From Lemma 4.3, there exists an integer n_0 such that $n \geqq n_0$ implies, for any $\theta>0$,

$$E[\exp\{\theta[V(\tilde{Y}_1)+\cdots+V(\tilde{Y}_n)]\}]$$
$$\leqq C_{n,\theta} \inf_{x \in T_d} E_x[\exp\{\theta[V(\tilde{Y}_1)+\cdots+V(\tilde{Y}_n)]\}]. \tag{4.8}$$

It follows from the proof of Theorem 1 in [2] that

$$\inf_{x \in T_d} E_x[\exp\{\theta[V(\tilde{Y}_1)+\cdots+V(\tilde{Y}_n)]\}] \leqq \exp\{n\lambda_n(\theta V)\},$$

and hence we have (4.5).

LEMMA 4.5. *Let* μ *be a probability measure on* $\mathcal{L}_d^{(n)}$ *such that* $\tilde{I}_n(\mu) \leqq \sigma/n^{\alpha/(\alpha+d)}$, where $\sigma>0$. *Let* $V=(K_\varepsilon - I)\chi_n g$, *where* $\|g\| \leqq B$. *Then, for any* $t>0$,

$$\left|\int_{T_d} V(x)\mu(dx)\right| \leqq B[2h(t\sigma)+2\Delta_t(n)+k_t(\varepsilon)], \tag{4.9}$$

where $h(\sigma) \to 0$ *as* $\sigma \to 0$, $\Delta_t(n) \to 0$ *as* $n \to \infty$ *for all* $t>0$, *and* $k_t(\varepsilon) \to 0$ *as* $\varepsilon \to 0$ *for all* $t>0$.

Proof: For any $t>0$, let $n_t = [tn^{\alpha/(d+\alpha)}]$ and define

$$\tilde{\pi}_{n,t}(x) = \frac{\tilde{\pi}_n^{n_t}(x) + \tilde{\pi}_n^{n_t-1}(x) + \cdots + \tilde{\pi}_n^{n_t-(p-1)}(x)}{p}.$$

Let

$$\tilde{I}_{n,t}(\mu) = -\inf_{u\in\mathcal{U}_1}\int_{T_d}\log\left(\frac{\tilde{\pi}_{n,t}u}{u}\right)(x)\mu(dx), \tag{4.10}$$

and let

$$\tilde{\phi}_{n,t}(x) = \sum_{y\in\mathcal{L}_d^{(n)}} \chi_n(x-y)\tilde{\pi}_{n,t}(y).$$

Since $\|g\|\leqq B$, we note that $|V(x)|\leqq 2B$ on T_d. Let $\mu\tilde{\pi}_{n,t}$ be defined by: for $x\in\mathcal{L}_d^{(n)}$

$$(\mu\tilde{\pi}_{n,t})(x) = \int_{T_d}\tilde{\pi}_{n,t}(x-y)\mu(dy), \tag{4.11}$$

and write

$$\int_{T_d} V(x)\mu(dx) = \int_{T_d} V(x)(\mu-\mu\tilde{\pi}_{n,t})(dx) + \int_{T_d} V(x)\mu\tilde{\pi}_{n,t}(dx). \tag{4.12}$$

The I-function has the property that if

$$I_1(\mu) = -\inf_{u\in\mathcal{U}_1}\int_{T_d}\log\left(\frac{\pi_1 u}{u}\right)\mu(dx),$$

$$I_2(\mu) = -\inf_{u\in\mathcal{U}_1}\int_{T_d}\log\left(\frac{\pi_2 u}{u}\right)\mu(dx),$$

and if $\pi_3 = a\pi_1 + (1-a)\pi_2$, where $0<a<1$, then

$$I_3(\mu) = -\inf_{u\in\mathcal{U}_1}\int_{T_d}\log\left(\frac{\pi_3 u}{u}\right)\mu(dx) \leqq aI_1(\mu) + (1-a)I_2(\mu). \tag{4.13}$$

Also, if $\pi_1^{(n)}$ indicates the n-step transition probabilities, then

$$I_1^{(n)}(\mu) = -\inf_{u\in\mathcal{U}_1}\int_{T_d}\log\left(\frac{\pi_1^{(n)} u}{u}\right)\mu(dx) \leqq nI_1(\mu). \tag{4.14}$$

Using (4.13) and (4.14) we see from (4.10) that

$$\tilde{I}_{n,t}(\mu) \leqq \sigma t. \tag{4.15}$$

By Lemma 4.1 in [2] it follows from (4.15) and (4.11) that (in the variation norm)

$$\|\mu\tilde{\pi}_{n,t}-\mu\|\leqq h(\sigma t), \tag{4.16}$$

where $h(\sigma)\to 0$ as $\sigma\to 0$. Thus, from (4.16) and (4.12), we get

$$\begin{aligned}\left|\int_{T_d} V(x)\mu(dx)\right| &\leqq 2Bh(\sigma t)+B\int_{T_d}\left|\int_{T_d}[\tilde{\phi}_{n,t}(x-y)-\tilde{\phi}_{n,t}(x)]\psi_\varepsilon(y)\,dy\right|dx\\ &\leqq 2Bh(\sigma t)+B\sup_{|y|\leqq\varepsilon}\int_{T_d}|\tilde{\phi}_{n,t}(x-y)-\tilde{\phi}_{n,t}(x)|\,dx\\ &\leqq 2Bh(\sigma t)+B\sup_{|y|\leqq\varepsilon}\left[\int_{T_d}\left|\tilde{\phi}_{n,t}(x-y)-\tilde{\Phi}\left(\frac{t}{\lambda^\alpha},x-y\right)\right|dx\right.\\ &\qquad+\int_{T_d}\left|\tilde{\phi}_{n,t}(x)-\tilde{\Phi}\left(\frac{t}{\lambda^\alpha},x\right)\right|dx\\ &\qquad\left.+\int_{T_d}\left|\tilde{\Phi}\left(\frac{t}{\lambda^\alpha},x-y\right)-\tilde{\Phi}\left(\frac{t}{\lambda^\alpha},x\right)\right|dx\right]\\ &\leqq 2Bh(\sigma t)+2B\Delta_t(n)+Bk_t(\varepsilon),\end{aligned} \tag{4.17}$$

where

$$\Delta_t(n)=\int_{T_d}\left|\tilde{\phi}_{n,t}(x)-\tilde{\Phi}\left(\frac{t}{\lambda^\alpha},x\right)\right|dx,$$

and

$$k_t(\varepsilon)=\sup_{|y|\leqq\varepsilon}\int_{T_d}\left|\tilde{\Phi}\left(\frac{t}{\lambda^\alpha},x-y\right)-\tilde{\Phi}\left(\frac{t}{\lambda^\alpha},x\right)\right|dx.$$

Since $\tilde{\Phi}$ is a density function on T_d, we see that, for each $t>0$, $k_t(\varepsilon)\to 0$ as $\varepsilon\to 0$.

To show that, for each $t>0$, $\Delta_t(n)\to 0$ as $n\to\infty$, it suffices to show, for each $t>0$,

$$\operatorname*{ess\,sup}_{x\in T_d}\left|\tilde{\phi}_{n,t}(x)-\tilde{\Phi}\left(\frac{t}{\lambda^\alpha},x\right)\right|\to 0 \quad\text{as}\quad n\to\infty, \tag{4.18}$$

i.e.,

$$\operatorname*{ess\,sup}_{x\in T_d}\left|\sum_{y\in\mathscr{L}_d^{(n)}}\chi_n(x-y)\tilde{\pi}_{n,t}(y)-\tilde{\Phi}\left(\frac{t}{\lambda^\alpha},x\right)\right|\to 0 \quad\text{as}\quad n\to\infty.$$

Let $A_n = \{x \in T_d : |x_i| < 1/2a_n,\ 1 \leqq i \leqq d\}$. On A_n, the function χ_n is equal to a_n^d; thus, if $x \in A_n + z$ for some $z \in \mathscr{L}_d^{(n)}$, then $\bar{\phi}_{n,t}(x) = a_n^d \bar{\pi}_{n,t}(z)$. Observe also that meas $(T_d - \bigcup_{z \in \mathscr{L}_d^{(n)}} \{A_n + z\}) = 0$. Since $z \in \mathscr{L}_d^{(n)}$, we have $z \in \tilde{H}_{d,q}^{(n)}$ for some q, $0 \leqq q \leqq p-1$. Consequently,

$$\bar{\phi}_{n,t}(x) = \frac{a_n^d}{p} \bar{\pi}_n^{n_t - r}(z)$$

for some r, $0 \leqq r \leqq p-1$, and $z \in \tilde{H}_{d,n_t-r}^{(n)}$. Now, from Theorem 2,

$$\begin{aligned} \bar{\phi}_{n,t}(x) &= \bar{\Phi}\left(\frac{t}{\lambda^{\alpha}}, z\right) + o(1) \quad \text{as} \quad n \to \infty \\ &= \left(\frac{t}{\lambda^{\alpha}}, x\right) + o(1) \quad \text{as} \quad n \to \infty, \end{aligned}$$

which gives us (4.18) and the lemma.

THEOREM 6. *Let $C \subset L_1(T_d)$ be closed. Then*

$$\overline{\lim_{n\to\infty}} \frac{1}{a_n^d} \log Q_n^{(n)}(C) \leqq -\inf_{f \in C} I_\lambda(f). \tag{4.19}$$

Proof: Consider the mapping from $\mathcal{M}(T_d)$ with its weak topology to $L_1(T_d)$ with its strong topology given by $\mu \to K_\varepsilon \mu$, where

$$(K_\varepsilon \mu)(x) = \int_{T_d} \psi_\varepsilon(x-y)\mu(dy)$$

with ψ_ε defined just before (3.14). This map is continuous and hence $\{\mu \in \mathcal{M}(T_d) : K_\varepsilon \mu \in C\}$ is closed in $\mathcal{M}(T_d)$. Thus, from Theorem 4 we get

$$\overline{\lim_{n\to\infty}} \frac{1}{a_n^d} \log Q_n^{(n)}[f : K_\varepsilon f \in C] \leqq -\inf_{f : K_\varepsilon f \in C} I_\lambda(f). \tag{4.20}$$

For any $\delta > 0$, let $C_\delta = \{g \in L_1(T_d) : \|g - f\| < \delta$ for some $f \in C\}$ and let $\bar{C}_\delta$ be the L_1 closure of C_δ. From (4.20), applied to this closed set $\bar{C}_\delta$ instead of C, we have

$$\overline{\lim_{n\to\infty}} \frac{1}{a_n^d} \log Q_n^{(n)}[f : K_\varepsilon f \in \bar{C}_\delta] \leqq -\inf_{f : K_\varepsilon f \in \bar{C}_\delta} I_\lambda(f). \tag{4.21}$$

Consequently, for any set C closed in $L_1(T_d)$, we have, for all $\delta>0$ and all $\varepsilon>0$,

$$\begin{aligned}
&\overline{\lim_{n\to\infty}}\frac{1}{a_n^d}\log Q_n^{(n)}(C)\\
&\leqq \overline{\lim_{n\to\infty}}\frac{1}{a_n^d}\log\{Q_n^{(n)}[f: K_\varepsilon f\in \bar{C}_\delta]+Q_n^{(n)}[f: \|K_\varepsilon f-f\|\geqq\delta]\}\\
&\leqq \max\left\{-\inf_{f:K_\varepsilon f\in\bar{C}_\delta} I_\lambda(f),\ \overline{\lim_{n\to\infty}}\frac{1}{a_n^d}\log Q_n^{(n)}[f: \|K_\varepsilon f-f\|\geqq\delta]\right\}.
\end{aligned}\tag{4.22}$$

To prove the theorem it suffices then to show that

$$\underline{\lim}_{\delta\to 0}\,\underline{\lim}_{\varepsilon\to 0}\max\left\{-\inf_{f:K_\varepsilon f\in\bar{C}_\delta} I_\lambda(f),\ \overline{\lim_{n\to\infty}}\frac{1}{a_n^d}\log Q_n^{(n)}[f: \|K_\varepsilon f-f\|\geqq\delta]\right\}\leqq -\inf_{f\in C} I_\lambda(f).\tag{4.23}$$

In view of Theorem 5, this means showing

$$\overline{\lim_{\delta\to 0}}\,\overline{\lim_{\varepsilon\to 0}}\inf_{f:K_\varepsilon f\in\bar{C}_\delta} I_\lambda(f)\geqq\inf_{f\in C} I_\lambda(f).\tag{4.24}$$

In fact, we prove even more, namely that

$$\underline{\lim}_{\substack{\varepsilon\to 0\\ \delta\to 0}}\inf_{f:K_\varepsilon f\in\bar{C}_\delta} I_\lambda(f)\geqq\inf_{f\in C} I_\lambda(f).\tag{4.25}$$

To see that (4.25) is true, let $\{f_n\}$ be a sequence in $L_1(T_d)$ such that $I_\lambda(f_n)\to l$ and $K_{\varepsilon}f_n\in\bar{C}_{\delta_n}$, where $\varepsilon_n\to 0$ and $\delta_n\to 0$; we must show that there exists $f\in C$ such that $I_\lambda(f)\leqq l$. If $l=\infty$, there is noting to prove; thus assume $l<\infty$. But, the set of f in $L_1(T_d)$ such that $I_\lambda(f)\leqq l$ is compact in $L_1(T_d)$, and the result follows from this compactness and the fact that $I_\lambda(f)$ is lower semicontinuous. This completes the proof of Theorem 6.

From Theorem 6, using the arguments of Section 3 in [3], (c.f. remarks just after (2.11) above) we obtain

COROLLARY TO THEOREM 6. *For any* $\nu>0$,

$$\overline{\lim_{n\to\infty}}\frac{1}{a_n^d}\log E^{Q_n^{(n)}}[\exp\{-\nu a_n^d\,|x: f(x)>0|\}]\leqq -\inf_{g\in L_1(T_d)}\{\nu\,|x: g(x)>0|+I_\lambda(g)\}.\tag{4.26}$$

As pointed out in (2.14), this implies

$$\overline{\lim_{n\to\infty}} \frac{1}{n^{d/(d+\alpha)}} \log E[\exp\{-\nu D_n^{\#}\}] \leqq -\sup_{\lambda>0}\left\{\lambda^d \inf_{g\in L_1(T_d)} [\nu|x : g(x)>0| + I_\lambda(g)]\right\}. \tag{4.27}$$

We shall now show that

$$\sup_{\lambda>0}\left\{\lambda^d \inf_{g\in L_1(T_d)} [\nu|x : g(x)>0| + I_\lambda(g)]\right\} \geqq k(\nu, L), \tag{4.28}$$

as defined in the statement of Theorem 1. Thus, we will have shown that

$$\overline{\lim_{n\to\infty}} \frac{1}{n^{d/(d+\alpha)}} \log E[\exp\{-\nu D_n^{\#}\}] \leqq -k(\nu, L). \tag{4.29}$$

To see that (4.28), holds we note first that, from the definition of $I_\lambda(g)$,

$$\sup_{\lambda>0}\left\{\lambda^d \inf_{g\in L_1(T_d)} [\nu|x : g(x)>0| + I_\lambda(g)]\right\} = \sup_{\lambda>0} \inf_{g\in L_1(T_d)} \left\{\nu\lambda^d|x : g(x)>0)| + \frac{1}{\lambda^\alpha} I_{\tilde{L}}(g)\right\}. \tag{4.30}$$

Let T_d^λ be the torus in $R_d : |x_1| \leqq \frac{1}{2}\lambda$, $i = 1, 2, \cdots, d$, with opposite sides identified. Let

$$h(x) = \frac{1}{\lambda^d} g\left(\frac{x}{\lambda}\right).$$

We can then rewrite the expression on the right of (4.30) as

$$\sup_{\lambda>0} \inf_{h\in L_1(T_d)} \{\nu|x : h(x)>0| + I_{\tilde{L}}^\lambda(g)\}, \tag{4.31}$$

where $I_{\tilde{L}}^\lambda(h)$ is the I-function for the symmetric stable process on the torus T_d^λ with generator $\tilde{L}$. It now follows from Lemmas 3.5, 3.6, and 3.9 of [1] that

$$\sup_{\lambda>0} \inf_{h\in I_1(T_d^\lambda)} \{\nu|h(x)>0| + I_{\tilde{L}}^\lambda\} \geqq k(\nu, L). \tag{4.32}$$

Hence from (4.30), (4.31), and (4.32) we obtain (4.28).

To prove Theorem 1 it remains to show that

$$\varliminf_{n\to\infty} \frac{1}{n^{d/(d+\alpha)}} \log E[\exp\{-\nu D_n^{\#}\}] \geqq -k(\nu, L) \tag{4.33}$$

which is the object of the next section.

5. The Lower Bound

Let G be a bounded open set in R^d containing the origin such that the Lebesgue measure of its boundary is zero. Consider the original random walk on Z^d and recall the definition

$$D_n = \{j \in Z^d : j = Y_r \text{ for some } r,\ 1 \leqq r \leqq n\}.$$

Let us pick an open set U with compact closure $\bar{U}$ such that

$$G \subset \bar{G} \subset U \subset \bar{U}$$

and $|\bar{U}| \leqq |G| + \eta$, where $\eta > 0$ is a given arbitrarily small positive number. Let $b_n = n^{1/(\alpha+d)}$ and define the set

$$b_n U = \{y \in R^d : y = b_n x \text{ for some } x \in U\};$$

we have, clearly, by letting $(b_n U)^{\#}$ denote the number of lattice points in the set $b_n U$,

$$\begin{aligned} E[\exp\{-\nu D_n^{\#}\}] &\geqq E[\exp\{-\nu D_n^{\#}\};\ D_n \subset b_n U] \\ &\geqq \exp\{-\nu (b_n U)^{\#}\} P[D_n \subset b_n U]. \end{aligned} \tag{5.1}$$

In Lemma 5.1 below we shall show that

$$\varliminf_{n\to\infty} \frac{1}{n^{d/(\alpha+d)}} \log P[D_n \subset b_n U] \geqq -\lambda(G). \tag{5.2}$$

where $\lambda(G)$ is the smallest eigenvalue of $-L_\Omega$ with Dirichlet boundary conditions for G. Using (5.2) we obtain from (5.1)

$$\begin{aligned} \varliminf_{n\to\infty} \frac{1}{n^{d/(d+\alpha)}} \log E[\exp\{-\nu D_n^{\#}\}] &\geqq -\nu \varlimsup_{n\to\infty} \frac{(b_n U)^{\#}}{b_n^d} - \lambda(G) \\ &\geqq -\nu |U| - \lambda(G) \\ &\geqq -\nu |G| - \lambda(G) - \nu\eta\ ; \end{aligned}$$

since $\eta > 0$ was arbitrary we conclude that

$$\varliminf_{n\to\infty} \frac{1}{n^{d/(\alpha+d)}} \log E[\exp\{-\nu D_n^{\#}\}] \geqq -\nu|G| - \lambda(G). \tag{5.3}$$

Denoting by $\mathscr{G}$ the class of all bounded open sets that are continuity sets for Lebesgue measures we conclude from (5.3) that

$$\varliminf_{n\to\infty} \frac{1}{n^{d/(\alpha+d)}} \log E[\exp\{-\nu D_n^{\#}\}] \geqq -\inf_{G\in\mathscr{G}} [\nu|G| + \lambda(G)]. \tag{5.4}$$

It follows from the proof of Lemma 3.9 in [1] that

$$\inf_{G\in\mathscr{G}} [\nu|G| + \lambda(G)] = k(\nu, L_\Omega), \tag{5.5}$$

as defined in the statement of Theorem 1. Thus (5.4) and (5.5) yield (4.33) and the proof of Theorem 1 is complete except for (5.2) which we state and prove as a lemma.

LEMMA 5.1.

$$\varliminf_{n\to\infty} \frac{1}{n^{d/(\alpha+d)}} \log P[D_n \subset b_n U] \geqq -\lambda(G).$$

Proof: For $j \in Z^d$ consider

$$P_j\left[\frac{Y_k}{b_n} \in U, \text{ for } 1 \leqq k \leqq l_n, \frac{Y_{l_n}}{b_n} \in G\right],$$

where $P_j[\ ]$ denotes probabilities over the random walk starting from j. Let

$$q(U, G, l_n, n) = \inf_{j\in b_n G} P_j\left[\frac{Y_k}{b_n} \in U, 1 \leqq k \leqq l_n, \frac{Y_{l_n}}{b_n} \in G\right]. \tag{5.6}$$

Here l_n is a sequence of integers tending to ∞ such that $l_n/n^{\alpha/(\alpha+d)} \to \lambda > 0$, N_n is a sequence of integers such that $N_n/n^{d/(\alpha+d)} \to 1/\lambda$ with $l_n N_n \geqq n$ for every n, and $\lambda > 0$ is an arbitrary positive number. It follows from the proof of Markov property that

$$\begin{aligned} P\{D_n \subset b_n U\} &= P\left[\frac{Y_k}{b_n} \in U \text{ for } 1 \leqq k \leqq n\right] \\ &\geqq [q(U, G, l_n, n)]^{N_n}. \end{aligned} \tag{5.7}$$

From (5.7) we get

$$\varliminf_{n\to\infty} \frac{1}{n^{d/(\alpha+d)}} \log P[D_n \subset b_n U] \geqq \frac{1}{\lambda} \varliminf_{n\to\infty} \log q(U, G, l_n, n). \tag{5.8}$$

If $\{x(s),\ 0 \leqq s < \infty\}$ is the symmetric stable process in R_d of index α corresponding to $\Omega(t)$, then from (5.6) and the invariance principle we see that

$$\varliminf_{n\to\infty} q(U, G, l_n, n) \geqq \inf_{x\in G} P_x[x(s)\in U,\ 0\leqq s\leqq\lambda,\ x(\lambda)\in G]. \tag{5.9}$$

Thus from (5.8) and (5.9), we have

$$\begin{aligned} &\varliminf_{n\to\infty} \frac{1}{n^{d/(\alpha+d)}} \log P[D_n \subset b_n U] \\ &\qquad \geqq \frac{1}{\lambda} \log \left[\inf_{X\in G} P_x[x(s)\in U,\ 0\leqq s\leqq\lambda,\ x(\lambda)\in G] \right]. \end{aligned} \tag{5.10}$$

The left-hand side being independent of λ,

$$\begin{aligned} &\varliminf_{n\to\infty} \frac{1}{n^{d/(\alpha+d)}} \log P\{D_n \subset b_n U\} \\ &\qquad \geqq \varlimsup_{\lambda\to\infty} \frac{1}{\lambda} \log \left[\inf_{x\in \bar G} P_x[x(s)\in U,\ 0\leqq s<\lambda,\ x(\lambda)\in G] \right]. \end{aligned} \tag{5.11}$$

Now

$$\begin{aligned} &\inf_{x\in\bar G} P_x[x(s)\in U, 0\leqq s\leqq\lambda,\ x(\lambda)\in G] \\ &\qquad \geqq \inf_{x\in\bar G} P_x[x(s)\in V, 0\leqq s\leqq\lambda-1] \inf_{x\in\bar V} P_x[x(s)\in U, 0\leqq s\leqq 1, x(1)\in G], \end{aligned}$$

where V is any open set such that

$$\bar G \subset V \subset \bar V \subset U.$$

Since the second term is positive for any stable process and does not depend on λ, we conclude that

$$\varliminf_{\lambda\to\infty} \frac{1}{\lambda} \log\left[\inf_{x\in\bar{G}} P_x[x(s)\in U, 0\leqq s\leqq\lambda, x(\lambda)\in G]\right]$$

$$\geqq \varliminf_{\lambda\to\infty} \frac{1}{\lambda} \log\left[\inf_{x\in\bar{G}} P_x[x(s)\in V, 0\leqq s\leqq\lambda-1]\right]$$

$$= \varliminf_{\lambda\to\infty} \frac{1}{\lambda} \log\left[\inf_{x\in\bar{G}} P_x[x(s)\in V, 0\leqq s\leqq\lambda]\right].$$

It now follows from Theorem (8.1) of [5] that

$$\varliminf_{n\to\infty} \frac{1}{n^{d/(d+\alpha)}} \log P\{D_n \subset b_n U\} \geqq -\lambda(G)$$

and the proof of Lemma 5.1 is complete.

Bibliography

[1] Donsker, M. D., and Varadhan, S. R. S., *Asymptotics for the Wiener sausage,* Comm. Pure Appl. Math. 28, 1975 pp. 525–565.

[2] Donsker, M. D., and Varadhan, S. R. S., *Asymptotic evaluation of certain Markov process expectations for large time, I,* Comm. Pure Appl. Math. 28, 1975, pp. 1–47.

[3] Varadhan, S. R. S., *Asymptotic probabilities and differential equations,* Comm. Pure Appl. Math. 19, 1966, pp. 261–286.

[4] Donsker, M. D., and Varadhan, S. R. S., *Asymptotic evaluation of certain Weiener integrals for large time,* Functional Integration and Its Applications, Proceedings of the International Conference, London; Clarendon Press, Oxford, 1975, pp. 15–33.

[5] Donsker, M. D., and Varadhan, S. R. S., *Asymptotic evaluation of certain Markov process expectations for large time,* III, Comm. Pure Appl. Math. 29. 1976, pp. 389–461.

Received November, 1978.

Comm. Pure Appl. Math. 33 (1980) 365-393

A Law of the Iterated Logarithm for Total Occupation Times of Transient Brownian Motion*

M. D. DONSKER AND S. R. S. VARADHAN

Dedicated to Fritz John on the Occasion of his 70th Birthday, June 14, 1980

1. Introduction

Let $\{\beta(s), 0 \leqq s < \infty\}$ be Brownian motion in R_d starting from the origin with $d \geqq 3$, and let $T_d(\lambda, \omega)$ be the total time that a particular path $\omega = \beta(\cdot)$ occupies the sphere with center at the origin of radius λ. In [1] Ciesielski and Taylor showed that, for almost all Brownian paths,

$$\overline{\lim_{\lambda \downarrow 0}} \frac{T_d(\lambda, \omega)}{\lambda^2 \log\log(1/\lambda)} = \frac{2}{p_d^2}, \tag{1.1}$$

where p_d is the first positive zero of $J_\nu(x)$ with $\nu = \frac{1}{2}d - 2$.

In this paper, motivated by (1.1), the authors prove a Strassen type law of the iterated logarithm (see Theorems 1.1 and 1.2 below) for total Brownian occupation times in three or more dimensions. These theorems involve the I-function introduced by the authors in their recent work and which we now describe in the present context.

Let $\{\beta(s), 0 \leqq s < \infty\}$ be Brownian motion in R_d starting from $x \in R_d$ with $d \geqq 3$. For E, a Borel set in R_d, define

$$\sigma(E, \omega) = \int_0^\infty \chi_E(\beta(s))\, ds,$$

i.e., the total time the set E is occupied by the particular path $\omega = \beta(\cdot)$. For each ω, $\sigma(\cdot, \omega)$ is a σ-finite measure on R_d. If we let M denote the space of all σ-finite measures on R_d, we can, for each positive integer n and each $x \in R_d$, define a probability measures Q_n^x on M as follows: if $A \subset M$, then

$$Q_n^x(A) = P_x\{\omega : (1/n)\sigma(\cdot, \omega) \in A\}, \tag{1.2}$$

where $P_x\{\ \}$ is the Brownian measure on the paths $\omega = \beta(\cdot)$ in R_d starting from x.

* The research for this paper was done at the Courant Institute of Mathematical Sciences and supported by the National Science Foundation under NSF Grant No. MCS-77-02687.

Let U be the space of functions $u(x) \in C^2$ on R_d such that

$$0 < \inf_{x \in R_d} u(x) \leqq \sup_{x \in R_d} u(x) < \infty$$

and Δu has compact support. For each σ-finite measure $\sigma \in M$, we define

$$I(\sigma) = -\inf_{u \in U} \int_{R_d} \left(\frac{\frac{1}{2}\Delta u}{u}\right)(x)\sigma(dx). \tag{1.3}$$

In Section 3 of this paper we show that the asymptotic behavior of the measure Q_n^x as $n \to \infty$ is described by the I-function of (1.3). Roughly speaking, as $n \to \infty$,

$$Q_n^x(A) \sim \exp\{-n \inf_{\sigma \in A} I(\sigma)\}, \tag{1.4}$$

so that the importance of the I-function is that it gives us the constant in the exponential decay of the Q_n^x measure as $n \to \infty$. The precise description of this asymptotic behavior is given in Section 3.

In Section 2 of the paper we show that the σ-finite measures $\sigma \in M$ for which $I(\sigma) < \infty$ have the property that $\sigma\{x : |x| \leqq R\} = o(R^2)$ as $R \to \infty$. Moreover (cf. Theorem 2.1 below), in this case, σ is absolutely continuous with respect to Lebesgue measure in R_d, and the I-function defined by (1.3) takes on a simple form; namely, if, in this case, we set $\sigma(dx) = \phi^2(x)\,dx$, we have

$$I(\sigma) = \frac{1}{2} \int_{R_d} |\nabla \phi|^2 \, dx. \tag{1.5}$$

On the space M we impose the vague topology, i.e., if $\sigma_0 \in M$, r is a positive integer, and $\phi_1, \phi_2, \cdots, \phi_r$ are functions in $C_0^\infty(R_d)$, then for $\varepsilon > 0$ a neighborhood of σ_0 is defined by

$$N(\sigma_0) = N(\sigma_0, r, \phi_1, \cdots, \phi_r, \varepsilon)$$
$$= \left\{\sigma \in M : \left|\int_{R_d} \phi_i(x)\sigma(dx) - \int_{R_d} \phi_i(x)\sigma_0(dx)\right| < \varepsilon, 1 \leqq i \leqq r\right\}.$$

For $\lambda \geqq 3$, let

$$Z_\lambda(A, \omega) = \frac{\sigma(\lambda A, \omega)}{\lambda^2 \log\log \lambda},$$

so that, for each $\omega = \beta(\cdot)$, $\{Z_\lambda(\cdot, \omega)\}$ is a family (in the parameter λ) of σ-finite measures on R_d. We show in Section 4 that, for almost all ω, $\{Z_\lambda(\cdot, \omega)\}$ is conditionally compact, and if we let C_ω be the cluster set for the family $\{Z_\lambda(\cdot, \omega)\}$ (as $\lambda \to \infty$), then the main result of Section 4 is to identify C_ω. Indeed, if we let

$$\mathcal{B} = \{\sigma \in M : I(\sigma) \leqq 1\},$$

then Lemma 4.3 and Theorem 4.4 show that, for almost all ω, $C_\omega = \mathcal{B}$.

This immediately gives us

THEOREM 1.1. *Let $\{\beta(s), 0 \leqq s < \infty\}$ be Brownian motion in R_d, $d \geqq 3$, starting from the origin, and for any Borel set $A \subset R_d$, let*

$$\sigma(A, \omega) = \int_0^\infty \chi_A(\beta(s))\, ds.$$

Define, for $\lambda \geqq 3$,

$$Z_\lambda(A, \omega) = \frac{\sigma(\lambda A, \omega)}{\lambda^2 \log\log\lambda}.$$

Let $\Phi(\sigma)$ be continuous (in the vague topology) at the points of $\mathcal{B} = \{\sigma \in M : I(\sigma) \leqq 1\}$. Then, for almost all Brownian paths ω,

$$\overline{\lim_{\lambda\to\infty}}\, \Phi(Z_\lambda(\cdot, \omega)) = \sup_{\sigma : I(\sigma) \leqq 1} \Phi(\sigma). \tag{1.6}$$

If we let

$$Z^*_\lambda(A, \omega) = \frac{\sigma(\lambda A, \omega)}{\lambda^2 \log\log(1/\lambda)},$$

then by a proof directly analogous to the proof of Theorem 4.4 we can show that as $\lambda \to 0$ the cluster set of the family $\{Z^*_\lambda(\cdot, \omega)\}$ is also the same set $\mathcal{B}$ as above. This yields the following theorem.

THEOREM 1.2. *Let $\{\beta(s), 0 \leqq s < \infty\}$ be Brownian motion in R_d, $d \geqq 3$, starting from the origin, and for any Borel set $A \subset R_d$, let*

$$\sigma(A, \omega) = \int_0^\infty \chi_A(\beta(s))\, ds.$$

Define

$$Z_\lambda^*(A, \omega) = \frac{\sigma(\lambda A, \omega)}{\lambda^2 \log\log(1/\lambda)}.$$

Let $\Phi(\sigma)$ be continuous (in the vague topology) at the points of $\mathcal{B} = \{\sigma \in M : I(\sigma) \leqq 1\}$. Then, for almost all Brownian paths ω,

$$\overline{\lim_{\lambda \to 0^+}} \Phi(Z_\lambda^*(\cdot, \omega)) = \sup_{\sigma : I(\sigma) \leqq 1} \Phi(\sigma). \tag{1.7}$$

If, in (1.7), we let $\Phi(\sigma) = \sigma(s_d(0, 1))$, where $s_d(0, 1)$ is the unit sphere in R_d with center at 0, then the left side of (1.7) becomes the left side of (1.1). As pointed out above, if $I(\sigma) < \infty$, then σ is absolutely continuous with respect to Lebesgue measure and so this $\Phi(\sigma)$ is continuous at the points of $\mathcal{B}$. It is an easy calculation, since $I(\sigma)$ on $\mathcal{B}$ is given by (1.5), to see that the variational problem on the right of (1.7) yields the right side of (1.1).

2. σ-Finite Measures and Corresponding *I*-Functions

In (1.3) we defined, for each σ-finite measure σ on R_d, the I-function, $I(\sigma)$. We observe easily from this definition that Lebesgue measure on R_d, $d \geqq 3$, does not have a finite I-function. In this section we determine the properties of a σ-finite measure on R_d, $d \geqq 3$, for which the corresponding I-function is finite, and moreover in this case we find a simple expression for the I-function. These results are needed in later sections of this paper. The following theorem gives exact conditions for the finiteness of the I-function.

THEOREM 2.1. *Let $\sigma \in M$ be such that $I(\sigma) < \infty$. Then,*

(i) $\sigma[x : |x| \leqq R] = o(R^2)$ *as* $R \to \infty$,

(ii) *there exists $\phi(x)$ on R_d such that $\sigma(dx) = \phi^2(x)\, dx$,*

and

(iii) $\frac{1}{2}\int_{R_d} |\nabla\phi|^2\, dx = I(\sigma)$.

Conversely, if $\phi \in H^1(\text{loc})$ on R_d, $\int_{R_d} |\nabla\phi|^2\, dx < \infty$, and $\int_{|x| \leqq R} \phi^2(x)\, dx = o(R^2)$ as $R \to \infty$, then defining $\sigma(dx) = \phi^2(x)\, dx$ we have $I(\sigma) = \frac{1}{2}\int_{R_d} |\nabla\phi|^2\, dx$.

Theorem 2.1 follows from a succession of lemmas. Statement (i) of the theorem is the Corollary to Lemma 2.3. Statement (ii) of the theorem follows from Lemma 2.9, and statement (iii) follows from (i) and Lemmas 2.7 and 2.9. The converse part of the theorem follows from Lemmas 2.7 and 2.9.

LEMMA 2.2. *Let $\sigma \in M$ and let $G(x, y)$ be Green's function for $-\frac{1}{2}\Delta$ on R_d. Choose a smooth function $V \geqq 0$ such that*

$$u(x) = \int_{R_d} G(x, y)\, V(y)\, dy < \infty .$$

Then,

$$\int_{R_d} \frac{V(x)}{u(x)}\, \sigma(dx) \leqq I(\sigma) . \tag{2.1}$$

Proof: Let $\psi(y)$ be a smooth function on R_d satisfying $0 \leqq \psi \leqq 1$, and

$$\psi(y) = \begin{cases} 1 & \text{for } |y| \leqq 1 , \\ 0 & \text{for } |y| \geqq 2 . \end{cases}$$

Let $R > 0$, $\varepsilon > 0$, and define $\psi_R(y) = \psi(y/R)$ and

$$u_{R,\varepsilon}(x) = \varepsilon + \int_{R_d} G(x, y)\, V(y)\, \psi_R(y)\, dy .$$

We note that $u_{R,\varepsilon} \in U$, and from the definition of $I(\sigma)$, we have

$$\int_{R_d} \frac{V(x)\, \psi_R(x)}{u_{R,\varepsilon}(x)}\, \sigma(dx) \leqq I(\sigma) . \tag{2.2}$$

Letting $R \to \infty$ and $\varepsilon \to 0$, we obtain (2.1) from (2.2) and Fatou's lemma.

LEMMA 2.3. *If $\sigma \in M$ is such that $I(\sigma) = l < \infty$, then*

$$\int_{R_d} \frac{\sigma(dx)}{r^2} \leqq \frac{8l}{(d-2)^2} , \tag{2.3}$$

where r is the radius vector in R_d.

Proof: Let $\varepsilon > 0$, $\alpha = \frac{1}{4}(2-d)$, and define $u_\varepsilon(x) = (\varepsilon^2 + r^2)^\alpha$. Since

$$\Delta f(r) = f''(r) + \frac{d-1}{r} f'(r) ,$$

we have

$$
\begin{aligned}
-\tfrac{1}{2}\Delta u_\varepsilon(x) &= -\tfrac{1}{2}\Big[2\alpha(\varepsilon^2+r^2)^{\alpha-1}+4\alpha r^2(\alpha-1)(\varepsilon^2+r^2)^{\alpha-2} \\
&\qquad +2\Big(\frac{d-1}{r}\Big)\alpha r(\varepsilon^2+r^2)^{\alpha-1}\Big] \\
&= (\varepsilon^2+r^2)^{\alpha-2}\Big[-\alpha(\varepsilon^2+r^2)-2\alpha r^2(\alpha-1)-\Big(\frac{d-1}{r}\Big)\alpha r(\varepsilon^2+r^2)\Big] \\
&= (\varepsilon^2+r^2)^{\alpha-2}[\tfrac{1}{4}d(d-2)\varepsilon^2+\tfrac{1}{8}(d-2)^2r^2].
\end{aligned}
$$

If we set $V_\varepsilon(x)=-\tfrac{1}{2}\Delta u_\varepsilon(x)$, then $V_\varepsilon(x)\geqq 0$ is a smooth function. Since $\alpha<0$, $u_\varepsilon(x)\to 0$ as $|x|\to\infty$, and therefore

$$u_\varepsilon(x)=\int_{R_d} G(x,y)\,V_\varepsilon(y)\,dy.$$

Hence, from Lemma 2.2,

$$\int_{R_d}\frac{[\tfrac{1}{4}d(d-2)\varepsilon^2+\tfrac{1}{8}(d-2)^2r^2]}{(\varepsilon^2+r^2)^2}\,\sigma(dx)\leqq I(\sigma). \tag{2.4}$$

Letting $\varepsilon\to 0$ we obtain (2.3) from (2.4) and Fatou's lemma.

COROLLARY. *If $\sigma\in M$ is such that $I(\sigma)=l<\infty$, then*

$$\sigma[x:|x|\leqq R]=o(R^2)\quad as\quad R\to\infty. \tag{2.5}$$

Proof: Let $\psi(R)=\sigma[x:|x|\leqq R]$. From Lemma 2.3 we see that

$$\int_{R_d}\frac{\sigma(dx)}{r^2}=\int_0^\infty\frac{d\psi(R)}{R^2}\leqq\frac{8l}{(d-2)^2}.$$

Hence, for any $R\geqq 1$,

$$\int_1^R\frac{d\psi(y)}{y^2}\leqq\frac{8l}{(d-2)^2},$$

and integrating by parts, we obtain

$$\frac{\psi(R)}{R^2}-\psi(1)+2\int_1^R\frac{\psi(y)}{y^3}\,dy\leqq\frac{8l}{(d-2)^2}.$$

Since $\psi(R)/R^2>0$, we conclude from this last inequality that $\int_1^\infty (\psi(y)/y^3)\,dy<\infty$. Let $\varepsilon>0$ be given and choose R so large that $\int_R^\infty (\psi(y)/y^3)\,dy<\varepsilon$. Since $\psi(y)$ is monotone increasing, we have

$$\frac{\psi(R)}{2R^2}=\psi(R)\int_R^\infty \frac{dy}{y^3}\leqq \int_R^\infty \frac{\psi(y)}{y^3}\,dy<\varepsilon\,,$$

which implies, since $\varepsilon>0$ is arbitrary, that $\psi(R)=o(R^2)$ as $R\to\infty$.

LEMMA 2.4. *Let $\sigma\in M$ be such that $I(\sigma)=l<\infty$, and assume $\sigma(dx)=f(x)\,dx$, where $f>0$, $f\in C^\infty$, and $f\to 0$ as $|x|\to\infty$. Then, if $\phi(x)=(f(x))^{1/2}$, we have*

(2.6) $$\tfrac{1}{2}\int_{R_d} |\nabla\phi|^2\,dx\leqq l\,.$$

Proof: For all $u\in U$,

(2.7) $$\tfrac{1}{2}\int_{R_d} \left(\frac{\Delta u}{u}\right)(x)f(x)\,dx\geqq -l\,.$$

If $h\in C_0^\infty$ and λ is real, then $u=e^{\lambda h}\in U$ so that from (2.7) we have

(2.8) $$\tfrac{1}{2}\left[\lambda\int_{R_d} (\Delta h)f\,dx+\lambda^2\int_{R_d} |\nabla h|^2 f\,dx\right]\geqq -l\,.$$

Since (2.8) holds for all real λ, we obtain from the obvious discriminant inequality and integration by parts:

$$\left|\int_{R_d} \langle\nabla h,\nabla f\rangle\,dx\right|^2\leqq 8l\int_{R_d} |\nabla h|^2 f\,dx\,,$$

or

(2.9) $$\left|\int_{R_d} \langle\nabla h,\nabla\phi\rangle\phi\,dx\right|^2\leqq 2l\int_{R_d} |\nabla h|^2\phi^2\,dx\,.$$

Now, for $n=1,2,3,\cdots$, let

$$\psi_n(y)=\begin{cases}0 & \text{for}\quad 0<y\leqq 1/2n\,,\\ 1/y & \text{for}\quad y\geqq 1/n\,,\end{cases}$$

where $0 \leqq \psi_n(y) \leqq 1/y$ for all y and ψ_n is smooth. Let $\Psi_n(y) = \int_0^y \psi_n(s)\,ds$ and define $h_n(x) = \Psi_n(\phi(x))$; then we observe that $h_n \in C_0^\infty$ and that it must satisfy (2.9), i.e.,

$$\left|\int_{R_d} \psi_n(\phi(x))\,|\nabla\phi|^2\phi\,dx\right|^2 \leqq 2l \int_{R_d} \psi_n^2(\phi(x))\,|\nabla\phi|^2\phi^2\,dx$$
$$\leqq 2l \int_{R_d} \psi_n(\phi(x))\,|\nabla\phi|^2\phi\,dx\,,$$

where the last inequality follows from the fact that $\psi_n(y) \leqq 1/y$. Hence,

$$\int_{R_d} \psi_n(\phi(x))\,|\nabla\phi|^2\phi\,dx \leqq 2l\,,$$

and letting $n \to \infty$ we obtain (2.6) from this last inequality and Fatou's lemma.

LEMMA 2.5. *If $\sigma \in M$ is such that $I(\sigma) < \infty$, then*

$$\lim_{|x|\to\infty} \sigma[B(x, R)] = 0 \tag{2.10}$$

for all $0 < R < \infty$, where $B(x, R)$ is the ball in R_d of radius R with center at x.

Proof: Assume the contrary, and let $a_n = 4^n + 2(R+1)$ for $n = 1, 2, 3, \cdots$. Then, there exists a sequence $\{x_n\}$ such that $|x_m - x_n| \geqq a_{n\vee m}$ and, for all n, $\sigma[B(x_n, R)] \geqq \delta > 0$. Set

$$f(x) = \begin{cases} 1 & \text{for} \quad x \in B(0, R)\,, \\ 0 & \text{for} \quad x \notin B(0, R+1)\,, \end{cases}$$

where $f \geqq 0$ for all x and is in C^∞ on R_d. Let $G(x, y)$ be Green's function for $-\frac{1}{2}\Delta$ on R_d, and define

$$u(x) = \int_{R_d} G(x, y) f(y)\,dy\,.$$

Since f has compact support, there exist constants C_1 and C_2 such that

$$\sup_{x \in R_d} u(x) \leqq C_1\,, \tag{2.11}$$
$$u(x) \leqq \frac{C_2}{|x|^{d-2}} \quad \text{for all} \quad x \in R_d\,.$$

Consider $\sum_{i=1}^{n} u(x-x_i)$. This sum is bounded for al x and it approaches 0 as $|x|\to\infty$. It is harmonic outside of $\bigcup_{j=1}^{n} B(x_j, R+1)$ so that, by the maximum principle,

$$\sup_{x\in R_d} \sum_{j=1}^{n} u(x-x_j) = \sup_{x\in \bigcup_{j=1}^{n} B(x_j, R+1)} \sum_{j=1}^{n} u(x-x_j)\,. \tag{2.12}$$

Since $a_n = 4^n + 2(R+1)$ and $|x_m - x_n| \geqq a_{n\vee m}$ we have from (2.11) that

$$\begin{aligned}\sup_{x\in \bigcup_{j=1}^{n} B(x_j, R+1)} \sum_{j=1}^{n} u(x-x_j) &\leqq C_1 + C_2 \sup_i \sum_{j\neq i} \frac{1}{(|x_i - x_j| - R - 1)^{d-2}} \\ &\leqq C_1 + C_2 \sup_i \sum_{j\neq 1} \frac{1}{(4^{j\vee i})^{d-2}} \\ &\leqq C_1 + C_2 \sum_{j=1}^{n} \frac{1}{4^{j(d-2)}} \leqq C_1 + C_2 \sum_{j=1}^{\infty} \frac{1}{4^{j(d-2)}}\,.\end{aligned} \tag{2.13}$$

Therefore, in view of (2.12) and (2.13), there exists a constant K independent of n and x such that $\sum_{j=1}^{n} u(x-x_j) \leqq K$.

Using Lemma 2.2 we conclude that, for all n,

$$\begin{aligned} I(\sigma) &\geqq \int_{R_d} \frac{\sum_{j=1}^{n} f(x-x_j)}{\sum_{j=1}^{n} u(x-x_j)}\,\sigma(dx) \\ &\geqq \frac{1}{K} \sum_{j=1}^{n} \sigma[B(x_j, R)] \geqq \frac{n\delta}{K}, \end{aligned}$$

which implies $I(\sigma) = \infty$ contradicting our hypothesis.

LEMMA 2.6. *If $f = \phi^2$ and $\frac{1}{2}\int_{R_d} |\nabla\phi|^2\,dx = l$, where ϕ has compact support, and if we define $\sigma(dx) = f(x)\,dx$, then*

$$I(\sigma) \leqq l\,. \tag{2.14}$$

Proof: For any $u\in U$ consider $\frac{1}{2}\int_{R_d} (\Delta u/u)(x) f(x)\,dx$ which becomes, on writing $u = e^h$, $\frac{1}{2}\int_{R_d} (\Delta h + |\nabla h|^2)\phi^2\,dx$. Since ϕ has compact support, we see that

$$\begin{aligned}\tfrac{1}{2}\int_{R_d}(\Delta h+|\nabla h|^2)\phi^2\,dx&=\int_{R_d}\langle\nabla h,\nabla\phi\rangle\phi\,dx+\tfrac{1}{2}\int_{R_d}|\nabla h|^2\,\phi^2\,dx\\&\geqq-\int_{R_d}|\nabla h|\,|\nabla\phi|\,\phi\,dx+\tfrac{1}{2}\int_{R_d}|\nabla h|^2\,\phi^2\,dx\\&\geqq-\Big(\int_{R_d}|\nabla h|^2\,\phi^2\,dx\Big)^{1/2}\Big(\int_{R_d}|\nabla\phi|^2\,dx\Big)^{1/2}\\&\quad+\tfrac{1}{2}\int_{R_d}|\nabla h|^2\,\phi^2\,dx.\end{aligned}$$

Thus, if we let $Q=\int_{R_d}|\nabla h|^2\,\phi^2\,dx$, we have for any $u\in U$, with h defined by $u=e^h$,

$$\begin{aligned}\tfrac{1}{2}\int_{R_d}\Big(\frac{\Delta u}{u}\Big)(x)\,f(x)\,dx&\geqq-(2lQ)^{1/2}+\tfrac{1}{2}Q\\&=-l+[(\tfrac{1}{2}Q)^{1/2}-l^{1/2}]^2\geqq-l.\end{aligned}\tag{2.15}$$

Since (2.15) holds for all $u\in U$, we conclude that $I(\sigma)\leqq l$.

LEMMA 2.7. *If* $f=\phi^2$,

$$\tfrac{1}{2}\int_{R_d}|\nabla\phi|^2\,dx=l\quad\text{and}\quad\int_{|x|\leqq R}\phi^2\,dx=o(R^2)$$

as $R\to\infty$, *letting* $\sigma(dx)=f(x)dx$, *we have* $I(\sigma)\leqq l$. *Moreover, there exists a sequence* $\{\phi_n\}$ *with compact support such that*

$$\tfrac{1}{2}\int_{R_d}|\nabla\phi_n|^2\,dx\to l\quad\text{as}\quad n\to\infty,$$

and

$$\int_B|\phi_n-\phi|^2\,dx\to 0\quad\text{as}\quad n\to\infty,$$

for all balls in R_d.

Proof: Let

$$\psi(x)=\begin{cases}1 & \text{for } |x|\leqq 1,\\ 0 & \text{for } |x|\geqq 2,\end{cases}$$

with $0 \leqq \psi(x) \leqq 1$ for all x and smooth. Define $\phi_n(x) = \phi(x)\,\psi(x/n)$. Then, $\nabla\phi_n = \psi(x/n)\,\nabla\phi + (1/n)\phi(\nabla\psi)(x/n)$ and, for a constant C,

$$
\begin{aligned}
\left(\int_{R_d} |\nabla\phi_n - \nabla\phi|^2\,dx\right)^{1/2} &\leqq \left(\int_{R_d} |\nabla\phi|^2 \left|1-\psi\!\left(\frac{x}{n}\right)\right|^2 dx\right)^{1/2} \\
&\quad + \left(\frac{1}{n^2}\int_{R_d} \phi^2 \left|\nabla\psi\!\left(\frac{x}{n}\right)\right|^2 dx\right)^{1/2} \\
&\leqq \left(\int_{|x|\geqq n} |\nabla\phi|^2\,dx\right)^{1/2} + \left(\frac{C^2}{n^2}\int_{|x|\leqq 2n} \phi^2(x)\,dx\right)^{1/2}.
\end{aligned}
\tag{2.16}
$$

The hypotheses on ϕ imply that the right side of (2.16) approaches 0 as $n \to \infty$, and we conclude that $\frac{1}{2}\int_{R_d} |\nabla\phi_n|^2\,dx \to l$ as $n \to \infty$. From the definition of ϕ_n we see that, for all balls B,

$$
\lim_{n\to\infty} \int_B |\phi_n - \phi|^2\,dx = 0\,.
$$

Finally, if we define $\sigma_n(dx) = \phi_n^2(x)\,dx$, then Lemma 2.6 implies that

$$
I(\sigma_n) \leqq \tfrac{1}{2}\int_{R_d} |\nabla\phi_n|^2\,dx\,.
$$

Thus, if we let $n \to \infty$ and use the lower semicontinuity of the I-function we see that $I(\sigma) \leqq l$.

Using Lemma 2.7 and Lemma 2.4 we can prove a sharper form of Lemma 2.4, i.e., the function f appearing in Lemma 2.4 is now assumed only to be non-negative whereas in Lemma 2.4 it must be strictly positive.

LEMMA 2.8. *Let $\sigma \in M$ be such that $I(\sigma) = l < \infty$. Assume $\sigma(dx) = f(x)\,dx$, where $f(x) \geqq 0$, $f \in C^\infty$, and $f \to 0$ as $|x| \to \infty$. Then, letting $\phi = f^{1/2}$, we have*

$$
\tfrac{1}{2}\int_{R_d} |\nabla\phi|^2\,dx \leqq l\,.
\tag{2.17}
$$

Proof: Let $f_0(x) = \exp\{-\frac{1}{2}|x|^2\}$ and let $\sigma_0(dx) = f_0(x)\,dx$. From Lemma 2.7 we conclude that $I(\sigma_0) = k < \infty$. For $\varepsilon > 0$ we obtain from the convexity property of the I-function that

$$
I(\sigma + \varepsilon\sigma_0) \leqq I(\sigma) + \varepsilon I(\sigma_0) = l + \varepsilon k\,.
$$

Hence, if we define $f_\varepsilon = f + \varepsilon f_0$, then f_ε satisfies the hypotheses of Lemma 2.4 so that $\phi_\varepsilon(x) = (f(x) + \varepsilon f_0(x))^{1/2}$ satisfies

$$\frac{1}{2}\int_{R_d} |\nabla\phi_\varepsilon|^2\, dx \leqq l + \varepsilon k\,.$$

If we let $\varepsilon \to 0$ in this last inequality, we obtain (2.17) from Fatou's lemma.

LEMMA 2.9. *Let $\sigma \in M$ be such that $I(\sigma) = l < \infty$. Then σ is absolutely continuous with respect to Lebesgue measure on R_d, $\sigma(dx) = f(x)\,dx$, and if we set $\phi(x) = (f(x))^{1/2}$,*

$$\frac{1}{2}\int_{R_d} |\nabla\phi|^2\, dx \leqq l\,.$$

Proof: Let $g(x) \geqq 0$, $g \in C^\infty$, $\operatorname{supp} g \subset B(0, 1)$, and $\int_{R_d} g(x)\,dx = 1$. For $\varepsilon > 0$, let $g_\varepsilon(x) = \varepsilon^{-d} g(x/\varepsilon)$ and define $\sigma_\varepsilon = \sigma * g_\varepsilon$. Then $\sigma_\varepsilon(dx) = f_\varepsilon(x)\,dx$, where

$$f_\varepsilon(x) = \int_{R_d} g_\varepsilon(x-y)\,\sigma(dy)\,.$$

From the fact that the I-function is translation invariant and convex we have $I(\sigma_\varepsilon) \leqq I(\sigma) = l$. We note that $f_\varepsilon \in C^\infty$, $f_\varepsilon \geqq 0$, and $f_\varepsilon \to 0$ as $|x| \to \infty$. Therefore, if $\phi_\varepsilon = (f_\varepsilon)^{1/2}$ we have, by Lemma 2.8,

$$\frac{1}{2}\int_{R_d} |\nabla\phi_\varepsilon|^2\, dx \leqq l$$

for all $\varepsilon > 0$. Hence, letting $\varepsilon \to 0$ we conclude that σ is absolutely continuous with respect to Lebesgue measure, i.e., for some f, $\sigma(dx) = f(x)\,dx$ and, moreover, if $\phi = f^{1/2}$, then

$$\frac{1}{2}\int_{R_d} |\nabla\phi|^2\, dx \leqq l\,.$$

3. Asymptotics

Let $\{\beta(s), 0 \leqq s < \infty\}$ be Brownian motion in R_d starting from $x \in R_d$ with $d \geqq 3$. For E, a Borel set in R_d, define

$$\sigma(E, \omega) = \int_0^\infty \chi_E(\beta(s))\,ds\,, \tag{3.1}$$

i.e., the total time the set E is occupied by the particular path $\omega = \beta(\cdot)$. Clearly, for each ω, $\sigma(\cdot, \omega)$ is a σ-finite measure on R_d, i.e., an element of M. For each $n = 1, 2, 3, \cdots$ and each $x \in R_d$ we define a probability measure Q_n^x on M as follows: If A is a set of σ-finite measures on R_d, i.e., $A \subset M$, then

$$Q_n^x(A) = P_x\{\omega : (1/n)\sigma(\cdot, \omega) \in A\}, \tag{3.2}$$

where $P_x\{\ \}$ denotes Brownian motion measure on the space of paths $\omega = \beta(\cdot)$ starting from x.

In this section we shall examine the asymptotic behavior of the measure Q_n^x as $n \to \infty$. This will involve the I-function defined by (1.3) for all σ-finite measures in M. Here all topological notions in M refer to the vague topology on M described in the introduction.

THEOREM 3.1. *For each $K \subset M$ that is compact,*

$$\overline{\lim_{n\to\infty}} \frac{1}{n} \log Q_n^x(K) \leqq - \inf_{\sigma \in K} I(\sigma), \tag{3.3}$$

uniformly in x.

Proof: Assume $u \in U$ and consider, for any $t > 0$,

$$u(\beta(t)) \exp\left\{-\int_0^t \tfrac{1}{2}\left(\frac{\Delta u}{u}\right)(\beta(s))\, ds\right\}. \tag{3.4}$$

From the Feynman–Kac formula we see that

$$E_x\left\{\exp\left\{-\int_0^t \tfrac{1}{2}\left(\frac{\Delta u}{u}\right)(\beta(s))\, ds\right\} u(\beta(t))\right\} = u(x), \tag{3.5}$$

where, as usual, $E_x\{\ \}$ denotes integration over the Brownian motion paths $\beta(\cdot)$ starting at x. Equation (3.5) is just the statement that (3.4) is a martingale. We conclude that for any $u \in U$,

$$E_x\left\{\exp\left\{-\int_0^t \tfrac{1}{2}\left(\frac{\Delta u}{u}\right)(\beta(s))\, ds\right\}\right\} \leqq \frac{\sup_{x\in R_d} u(x)}{\inf_{x\in R_d} u(x)}. \tag{3.6}$$

Since $d \geqq 3$, Brownian motion is transient and, since $u \in U$, Δu has compact support. Therefore, if we let $t \to \infty$ in (3.6) we obtain from Fatou's lemma

$$E_x\left\{\exp\left\{-\int_0^\infty \tfrac{1}{2}\left(\frac{\Delta u}{u}\right)(\beta(s))\,ds\right\}\right\} \leqq \frac{\sup_{x\in R_d} u(x)}{\inf_{x\in R_d} u(x)}. \tag{3.7}$$

From the definition of Q_n^x measure on M we can rewrite (3.7) as

$$E^{Q_n^x}\left\{\exp\left\{-n\int_{R_d} \tfrac{1}{2}\left(\frac{\Delta u}{u}\right)(x)\sigma(dx)\right\}\right\} \leqq \frac{\sup_{x\in R_d} u(x)}{\inf_{x\in R_d} u(x)}, \tag{3.8}$$

where $E^{Q_n^x}\{\ \}$ denotes integration over the space M with respect to Q_n^x measure so that in (3.8) the variable of integration is σ.

From (3.8) we obtain by the usual estimate that, for the set K,

$$Q_n^x(K) \leqq \frac{\sup_{x\in R_d} u(x)}{\inf_{x\in R_d} u(x)} \exp\left\{n \sup_{\sigma\in K}\int_{R_d} \tfrac{1}{2}\left(\frac{\Delta u}{u}\right)(x)\sigma(dx)\right\}. \tag{3.9}$$

Since (3.9) holds for all $u \in U$ and all $x \in R_d$, we see that

$$\begin{aligned}\overline{\lim_{n\to\infty}} \sup_{x\in R_d} \frac{1}{n}\log Q_n^x(K) &\leqq \sup_{\sigma\in K}\int_{R_d} \tfrac{1}{2}\left(\frac{\Delta u}{u}\right)(x)\,\sigma(dx)\\ &\leqq \inf_{u\in U}\sup_{\sigma\in K}\int_{R_d} \tfrac{1}{2}\left(\frac{\Delta u}{u}\right)(x)\,\sigma(dx).\end{aligned} \tag{3.10}$$

Since K is compact, we can use the argument of [2] (after (2.10)) to conclude that (3.10) implies

$$\begin{aligned}\overline{\lim_{n\to\infty}} \sup_{x\in R_d} \frac{1}{n}\log Q_n^x(K) &\leqq -\inf_{\sigma\in K}\left[-\inf_{u\in U}\int_{R_d} \tfrac{1}{2}\left(\frac{\Delta u}{u}\right)(x)\,\sigma(dx)\right]\\ &= -\inf_{\sigma\in K} I(\sigma).\end{aligned} \tag{3.11}$$

In the following we denote the complement of a set A by A^c.

THEOREM 3.2. *For each $l<\infty$, there exists a compact set $K\subset M$ such that*

$$\overline{\lim_{n\to\infty}} \frac{1}{n}\log \sup_{x\in R_d} Q_n^x(K^c) \leqq -l. \tag{3.12}$$

Proof: Let r denote the radius vector in R_d and define $u(x) = (1+r^2)^{(2-d)/4}+1$. As in the proof of Lemma 2.3,

$$-\tfrac{1}{2}\Delta u(x) = (1+r^2)^{(2-d)/4-2}\left[\tfrac{1}{4}d(d-2)+\tfrac{1}{8}(d-2)^2 r^2\right],$$

so that if we let $V(x)=\frac{1}{2}(\Delta u/u)(x)$, then

$$\frac{\frac{1}{4}d(d-2)+\frac{1}{8}(d-2)^2r^2}{(1+r^2)^2+(1+r^2)^{2+(d-2)/4}}>0\,.$$

As in Theorem 3.1, for this particular u and with $V(x)$ as defined, we see that

$$u(\beta(t))\exp\left\{\int_0^t V(\beta(s))\,ds\right\}$$

is a martingale, and we conclude, as in the previous theorem, that

$$E^{Q_n^x}\left\{\exp\left\{n\int_{R_d} V(x)\sigma(dx)\right\}\right\}\leqq\frac{\sup_{x\in R_d}u(x)}{\inf_{x\in R_d}u(x)}\,. \tag{3.11$'$}$$

From (3.11)$'$ we see that

$$Q_n^x\left\{\sigma:\int_{R_d} V(x)\,\sigma(dx)>l\right\}\leqq\frac{\sup_{x\in R_d}u(x)}{\inf_{x\in R_d}u(x)}\,e^{-nl}\,. \tag{3.12$'$}$$

If we define

$$K_l=\left\{\sigma:\int_{R_d} V(x)\sigma(dx)\leqq l\right\},$$

then we note that K_l is compact and from (3.12)$'$ that

$$\overline{\lim_{n\to\infty}}\,\frac{1}{n}\log\sup_{x\in R_d}Q_n^x(K_l^c)\leqq -l\,.$$

THEOREM 3.3. *For each $C\subset M$ that is closed,*

$$\overline{\lim_{n\to\infty}}\,\frac{1}{n}\log\sup_{x\in R_d}Q_n^x(C)\leqq-\inf_{\sigma\in C}I(\sigma)\,. \tag{3.13}$$

Proof: Let $l>0$ be given and choose K_l as in Theorem 3.2. Since $C\subset(C\cap K_l)\cup K_l^c$, $Q_n^x(C)\leqq Q_n^x(C\cap K_l)+Q_n^x(K_l^c)$. Thus, from Theorems 3.1 and 3.2 we see that

$$\begin{aligned}\overline{\lim_{n\to\infty}}\,\frac{1}{n}\log\sup_{x\in R_d}Q_n^x(C)&\leqq\max\left[-\inf_{\sigma\in C\cap K_l}I(\sigma),-l\right]\\&\leqq\max\left[-\inf_{\sigma\in C}I(\sigma),-l\right].\end{aligned} \tag{3.14}$$

Since (3.14) holds for all $l>0$, we obtain (3.13) if we let $l\to\infty$.

Next, we seek asymptotic lower estimates for the Q_n^x measure of open sets in M. We begin with the following theorem.

THEOREM 3.4. *Let N be a neighborhood of $0\in M$ (the zero measure in M). Then,*

$$\lim_{n\to\infty} Q_n^x(N)=1\,, \tag{3.15}$$

uniformly in x.

Proof: Set $C=N^c$; from Theorem 3.3 we have

$$\overline{\lim_{n\to\infty}}\frac{1}{n}\log\sup_{x\in R_d} Q_n^x(N^c)\leqq-\inf_{\sigma\in N^c} I(\sigma)\,. \tag{3.16}$$

From the properties of the I-function and Lemma 2.3 we see that $\inf_{\sigma\in N^c} I(\sigma)>0$. Therefore, (3.16) implies that $Q_n^x(N)\to 1$ uniformly in x as $n\to\infty$.

THEOREM 3.5. *Let N be a neighborhood of some $\sigma_0\in M$ with $I(\sigma_0)=l<\infty$. Then,*

$$\underline{\lim}_{n\to\infty}\frac{1}{n}\log Q_n^x(N)\geqq-l\,, \tag{3.17}$$

uniformly for x in compact sets of R_d.

Proof: Let $\varepsilon>0$ be given. In the neighborhood N of σ_0 there exists a measure σ_1 such that $I(\sigma_1)\leqq l+\varepsilon$ and $\sigma_1(R_d)=T<\infty$. This follows from Theorem 2.1 and Lemma 2.7 wherein we constructed an approximating sequence with compact support. Thus in proving Theorem 3.5 we can without loss of generality assume that σ_0 in the statement of the theorem is actually a finite measure, i.e., $\sigma_0(R_d)=T<\infty$. We can then write $\sigma_0=T\mu_0$, where μ_0 is a probability measure on R_d, and note from the definition of the I-function that if $I(\sigma_0)=l$, then $I(\mu_0)=l/T$.

Since N is a neighborhood of $\sigma_0=T\mu_0$ we can find a neighborhood G_1 of 0 in M and a neighborhood G_2 of μ_0 such that if $\sigma\in G_1$ and $\mu\in G_2$, then $\sigma+T\mu\in N$. Also, for any integer n we can write

$$\begin{aligned}\sigma(A,\omega)&=\int_0^\infty \chi_A(\beta(s))\,ds=\int_0^{nT}\chi_A(\beta(s))\,ds+\int_{nT}^\infty \chi_A(\beta(s))\,ds\\ &=\sigma_{[0,nT]}(A,\omega)+\sigma_{[nT,\infty]}(A,\omega)\,.\end{aligned} \tag{3.18}$$

Hence, with G_1 and G_2 as defined above, we observe that

$$(3.19)\quad P_x\left\{\frac{1}{n}\sigma(\cdot,\omega)\in N\right\}\geqq P_x\left\{\frac{1}{nT}\sigma_{[0,nT]}(\cdot,\omega)\in G_2,\frac{1}{n}\sigma_{[nT,\infty]}(\cdot,\omega)\in G_1\right\}.$$

But, from the Markov property, we have

$$(3.20)\quad \begin{aligned} &P_x\left\{\frac{1}{nT}\sigma_{[0,nT]}(\cdot,\omega)\in G_2,\frac{1}{n}\sigma_{[nT,\infty]}(\cdot,\omega)\in G_1\right\}\\ &\quad\geqq P_x\left\{\frac{1}{nT}\sigma_{[0,nT]}(\cdot,\omega)\in G_2\right\}\inf_{x\in R_d}P_x\left\{\frac{1}{n}\sigma(\cdot,\omega)\in G_1\right\}.\end{aligned}$$

Thus,

$$(3.21)\quad \begin{aligned} Q_n^x(N)&=P_x\left\{\frac{1}{n}\sigma(\cdot,\omega)\in N\right\}\\ &\geqq P_x\left\{\frac{1}{nT}\sigma_{[0,nT]}(\cdot,\omega)\in G_2\right\}\inf_{x\in R_d}P_x\left\{\frac{1}{n}\sigma(\cdot,\omega)\in G_1\right\}.\end{aligned}$$

But,

$$P_x\left\{\frac{1}{n}\sigma(\cdot,\omega)\in G_1\right\}=Q_n^x(G_1)$$

and G_1 is a neighborhood of 0 in M so, by Theorem 3.4, $\lim_{n\to\infty}Q_n^x(G_1)=1$ uniformly in x. Using this in (3.21) we obtain

$$(3.22)\quad \varliminf_{n\to\infty}\frac{1}{n}\log Q_n^x(N)\geqq\varliminf_{n\to\infty}\frac{1}{n}\log P_x\left\{\frac{1}{nT}\sigma_{[0,nT]}(\cdot,\omega)\in G_2\right\}.$$

Now in [2] (Lemma 1) we showed that

$$(3.23)\quad \varliminf_{n\to\infty}\frac{1}{n}\log P_x\left\{\frac{1}{nT}\sigma_{[0,nT]}(\cdot,\omega)\in G_2\right\}\geqq -TI(\mu_0)=-l,$$

uniformly for x in compact sets. Thus, (3.22) and (3.23) yield the theorem.

We have as an immediate corollary

COROLLARY. *If G is an open set in M, then*

$$\varliminf_{n\to\infty} \frac{1}{n} \log Q_n^x(G) \geqq -\inf_{\sigma\in G} I(\sigma)\,, \tag{3.24}$$

uniformly for x in compact sets.

4. Strong Limit Theorems

It is an elementary scaling property of Brownian motion that the probability distribution of $(1/\lambda)\beta(s)$ is the same as the probability distribution of $\beta(s/\lambda^2)$. Using the notation $\stackrel{d}{=}$ to mean "has the same distribution as" we observe that

$$\begin{aligned}\sigma(\lambda A, \omega) = \int_0^\infty \chi_{\lambda A}(\beta(s))\,ds &= \int_0^\infty \chi_A\Big(\frac{1}{\lambda}\beta(s)\Big)\,ds \\ &\stackrel{d}{=} \int_0^\infty \chi_A\Big(\beta\Big(\frac{s}{\lambda^2}\Big)\Big)\,ds \\ &\stackrel{d}{=} \lambda^2\sigma(A,\omega)\,.\end{aligned} \tag{4.1}$$

For $\lambda \geqq 3$, let

$$Z_\lambda(A,\omega) = \frac{\sigma(\lambda A,\omega)}{\lambda^2 \log\log\lambda}\,,$$

so that for each ω, i.e., each Brownian motion path $\beta(\cdot)$, $Z_\lambda(\cdot,\omega)$ is a family, in the parameter λ, of σ-finite measures on R_d. In this section we show, first of all, that, for almost all ω, $\{Z_\lambda(\cdot,\omega)\}$ is conditionally compact. Let C_ω, for each ω, denote the cluster set for the family $\{Z_\lambda(\cdot,\omega)\}$ of σ-finite measures.

As we saw in Section 2, if a σ-finite measure σ has a finite I-function, $I(\sigma)$, then σ has a density with respect to Lebesgue measure on R_d. Denoting this density by f and writing $I(f)$ instead of $I(\sigma)$ we consider the set $\mathcal{B} = \{f : I(f) \leqq 1\}$, and the main result of this section will be that, for almost all ω, $C_\omega = \mathcal{B}$. As indicated in the introduction, the strong limit theorem we seek then follows.

We start with the following lemma:

LEMMA 4.1. *For almost all* ω,

$$\sup_{\lambda \geqq 3} Z_\lambda(K, \omega) < \infty \tag{4.2}$$

for each compact set $K \subset R_d$.

Proof: It suffices to show the result for balls in R_d; so, as before, let $B = B(0, R)$ denote the ball in R_d with center at 0 of radius R. Let $\rho > 1$ and, for positive integers n, define

$$X_n = X_n(\omega) = \sup_{\rho^n \leqq \lambda \leqq \rho^{n+1}} Z_\lambda(B, \omega),$$

so that $\{X_n\}$ is a sequence of random variables (on the Brownian motion paths in R_d), and we want to show that this sequence is bounded almost surely.

We observe that

$$\begin{aligned} X_n = \sup_{\rho^n \leqq \lambda \leqq \rho^{n+1}} \frac{\sigma(\lambda B(0, R), \omega)}{\lambda^2 \log\log \lambda} &\leqq \frac{\sigma(B(0, \rho^{n+1}R), \omega)}{\rho^{2n} \log\log \rho^n} \\ &= \frac{\sigma(B(0, \rho^{n+1}R), \omega)}{\rho^{2n}(\log n + \log\log \rho)}. \end{aligned} \tag{4.3}$$

Let

$$I_0 = \inf_{f : \int_{B(0,R)} f dx \geqq 1} I(f).$$

From Lemma 2.3 we see that $I_0 > 0$. Pick α so that $\alpha I_0/\rho^2 > 1$, and from (4.3) and (4.1) we have

$$\begin{aligned} P\{X_n \geqq \alpha\} &\leqq P\left\{\frac{\sigma(B(0, \rho^{n+1}R), \omega)}{\rho^{2n+2}} \geqq \frac{\alpha}{\rho^2}(\log n + \log\log \rho)\right\} \\ &= P\left\{\sigma(B(0, R), \omega) \geqq \frac{\alpha}{\rho^2}(\log n + \log\log \rho)\right\}. \end{aligned} \tag{4.4}$$

Using the same kind of estimate as used in the proof of Theorem 3.1 we see that

$$P\left\{\sigma(B(0,R),\omega) \geqq \frac{\alpha}{\rho^2}(\log n + \log\log\rho)\right\} \leqq A \exp\left\{-\frac{\alpha}{\rho^2}(\log n + \log\log\rho)\right\} I_0, \tag{4.5}$$

where A is a constant.

Thus, (4.4) and (4.5) imply, since $\alpha I_0/\rho^2 > 1$, that $\sum P\{X_n \geqq \alpha\} < \infty$. It follows now from the Borel–Cantelli lemma that the sequence $\{X_n\}$ is bounded almost surely.

COROLLARY. *For almost all ω, the family of σ-finite measures $\{Z_\lambda(\cdot,\omega)\}$, $\lambda \geqq 3$, is conditionally compact.*

With C_ω and $\mathcal{B}$ as defined at the beginning of this section, we next show the inclusion $C_\omega \subset \mathcal{B}$ for almost all ω. To do this we use a lemma the proof of which is immediate from the definition of the vague topology.

LEMMA 4.2. *Given $\sigma \in M$ and a neighborhood N_1 of σ, there exists another neighborhood N of σ and an $\varepsilon > 0$ such that if $\sigma_1 \in N$, $|\alpha - 1| < \varepsilon$, and $|\beta - 1| < \varepsilon$, then*

$$\sigma_2(\cdot) \equiv \beta\sigma_1(\alpha\,\cdot) \in N_1\,. \tag{4.6}$$

LEMMA 4.3. *For almost all ω,*

$$C_\omega \subset \mathcal{B}. \tag{4.7}$$

Proof: Let $\sigma^* \notin \mathcal{B}$. We must show that there exists a neighborhood N of σ^* such that

$$\lim_{\lambda_0 \to \infty} P\{Z_\lambda(\cdot,\omega) \in N \quad \text{for some} \quad \lambda \geqq \lambda_0\} = 0\,. \tag{4.8}$$

Since $\sigma^* \notin \mathcal{B}$, $I(\sigma^*) > 1$, i.e., $I(\sigma^*) = 1 + \Delta$, where $\Delta > 0$. Since the I-function is lower-semicontinuous, there exists a neighborhood N_1 of σ^* such that $\inf_{\sigma' \in \bar{N}_1} I(\sigma') \geqq 1 + \frac{1}{2}\Delta$, where $\bar{N}_1$ denotes the vague closure of N_1. Let $\rho = 1 + \varepsilon_0$, where $\varepsilon_0 > 0$, and let n be a positive integer. Pick $\lambda \in [\rho^n, \rho^{n+1}]$ and

define

$$\sigma_1(\cdot, \omega) = Z_\lambda(\cdot, \omega) = \frac{\sigma(\lambda \cdot, \omega)}{\lambda^2 \log\log \lambda},$$

$$\sigma_2(\cdot, \omega) = Z_{\rho^{n+1}}(\cdot, \omega) = \frac{\sigma(\rho^{n+1} \cdot, \omega)}{\rho^{2(n+1)} \log\log \rho^{n+1}}.$$

Using (4.1) we see that

$$\sigma_2(\cdot, \omega) \stackrel{d}{=} \beta \sigma_1(\alpha \cdot, \omega), \tag{4.9}$$

where

$$\beta = \frac{\lambda^2 \log\log \lambda}{\rho^{2(n+1)} \log\log \rho^{n+1}} \quad \text{and} \quad \alpha = \frac{\rho^{n+1}}{\lambda}.$$

Since $\lambda \in [\rho^n, \rho^{n+1}]$ we see that $1 \leqq \alpha \leqq \rho$, and

$$\frac{1}{\rho^2} \frac{\log n + \log\log \rho}{\log(n+1) + \log\log \rho} \leqq \beta \leqq 1.$$

Hence, by choosing ε_0 close to 0 and by taking n large enough we can, according to (4.9) and Lemma 4.2, find a neighborhood N of σ^* such that if $\sigma_1 = Z_\lambda(\cdot, \omega) \in N$, then $\sigma_2 = Z_{\rho^{n+1}}(\cdot, \omega) \in N_1$. Thus, to show (4.8) it suffices to show that

$$\lim_{l \to \infty} P\{Z_{\rho^n}(\cdot, \omega) \in N_1 \quad \text{for some} \quad n \geqq l\} = 0. \tag{4.10}$$

We recall that N_1 is a neighborhood of σ^* such that $\inf_{\sigma' \in \bar{N}_1} I(\sigma') \geqq 1 + \frac{1}{2}\Delta$. Thus, using (4.1) and again the exponential estimates of Theorem 3.1 (or Theorem 3.4), we have, for sufficiently large n,

$$\begin{aligned} P\{Z_{\rho^n}(\cdot, \omega) \in N_1\} &= P\left\{\frac{\sigma(\rho^n \cdot, \omega)}{\rho^{2n} \log\log \rho^n} \in N_1\right\} \\ &= P\left\{\frac{\sigma(\cdot, \omega)}{\log\log \rho^n} \in N_1\right\} \\ &\leqq P\left\{\frac{\sigma(\cdot, \omega)}{\log\log \rho^n} \in \bar{N}_1\right\} \\ &\leqq A \exp\{-(\log\log \rho^n)(1 + \tfrac{1}{4}\Delta)\} \\ &\leqq \frac{A}{n^{(1+\Delta/4)}}, \end{aligned} \tag{4.11}$$

where A is a constant.

This last inequality implies that $\sum P\{Z_{\rho^n}(\cdot, \omega) \in N_1\}$ converges so that by the Borel–Cantelli lemma we have (4.10).

Now we wish to show that $\mathscr{B} \subset C_\omega$.

THEOREM 4.4. *If f is such that $I(f) \leqq 1$, then, for almost all ω, $f \in C_\omega$.*

Theorem 4.4 follows from a succession of lemmas all of which involve certain sequences and random times which we now define once and for all.

First of all we note that, since the cluster set C_ω is closed, the theorem follows if we show that for every f such that $I(f) < 1$, for almost all ω, $f \in C_\omega$. Since $I(f) < 1$, we can pick $\alpha > 1$ and $\delta > 0$ so that

$$\alpha(I(f) + \delta) < 1\,. \tag{4.12}$$

For this α and each positive integer n, let

$$\lambda_n = \exp\{n^\alpha\}, \qquad a_n = n^2 \exp\{n^\alpha\}, \qquad c_n = 2n^2 \exp\{n^\alpha\}\,, \tag{4.13}$$

so that

$$c_{n-1} < \lambda_n < a_n < c_n < \lambda_{n+1}\,. \tag{4.14}$$

Consider Brownian motion in R_d starting from the origin and let τ_{a_n} be the first exit time from the ball $B(0, a_n)$, and τ_{c_n} the first exit time from the ball $B(0, c_n)$. For any set $A \subset R_d$ and any positive integer n, define

$$\tag{4.15} \begin{aligned} \alpha_n'(A, \omega) &= \int_0^{\tau_{a_n}} \chi_A(\beta(s))\, ds\,, \\ \beta_n'(A, \omega) &= \int_{\tau_{c_{n-1}}}^{\tau_{a_n}} \chi_A(\beta(s))\, ds\,. \end{aligned}$$

Thus, $\alpha_n'(A, \omega)$ is the amount of time a particular Brownian motion path $\omega = \beta(\cdot)$ starting from 0 spends in the set A up to its first exit time from the ball $B(0, a_n)$, whereas $\beta_n'(A, \omega)$ is the amount of time the path spends in A between the first exit time from $B(0, c_{n-1})$ and the first exit time from the larger ball $B(0, a_n)$. Let

$$\tag{4.16} \begin{aligned} \alpha_n(A, \omega) &= \frac{\alpha_n'(\lambda_n A, \omega)}{\lambda_n^2 \log\log \lambda_n}\,, \\ \beta_n(A, \omega) &= \frac{\beta_n'(\lambda_n A, \omega)}{\lambda_n^2 \log\log \lambda_n}\,, \end{aligned}$$

and recall that

$$Z_{\lambda_n}(A,\omega)=\frac{\sigma(\lambda_n A,\omega)}{\lambda_n^2\log\log\lambda_n}.$$

Let N be a neighborhood of f. To prove Theorem 4.4 it clearly suffices to show that

(4.17) $$P\{Z_{\lambda_n}(\cdot,\omega)\in N \text{ for infinitely many } n\}=1,$$

and the following lemmas serve that purpose.

LEMMA 4.5. *Let N be a neighborhood of f, then*

(4.18) $$P\{Z_{\lambda_n}(\cdot,\omega)\notin N,\ \alpha_n(\cdot,\omega)\in N \textit{ for infinitely many } n\}=0.$$

Proof: From the Borel–Cantelli lemma it suffices to show that

$$\sum P\{Z_{\lambda_n}(\cdot,\omega)\notin N\},\ \alpha_n(\cdot,\omega)\in N\}<\infty.$$

Let us specify the vague neighborhood N of f, i.e., for some positive integer r, $\varepsilon>0$, and $\phi_i\in C_0^\infty(R_d)$, $i=1,2,\cdots,r$,

$$\begin{aligned}N&=N(f)=N(f;r,\varepsilon,\phi_1,\cdots,\phi_r)\\&=\left\{\sigma\in M:\left|\int_{R_d}\phi_i(x)\,\sigma(dx)-\int_{R_d}\phi_i(x)\,f(x)\,dx\right|<\varepsilon,\ i=1,2,\cdots,r\right\}.\end{aligned}$$

We note, first of all, that

(4.19) $$\begin{aligned}&P\{Z_{\lambda_n}(\cdot,\omega)\notin N,\ \alpha_n(\cdot,\omega)\in N\}\\&\le P\left\{\int_{R_d}\phi_i(x)\,Z_{\lambda_n}(dx,\omega)\neq\int_{R_d}\phi_i(x)\,\alpha_n(dx,\omega)\text{ for some }1\le i\le r\right\}.\end{aligned}$$

Also, the event

(4.20) $$\begin{aligned}&\left\{\int_{R_d}\phi_i(x)\,Z_{\lambda_n}(dx,\omega)\neq\int_{R_d}\phi_i(x)\alpha_n(dx,\omega)\text{ for some }1\le i\le r\right\}\\&\left\{\int_{R_d}\phi_i\left(\frac{x}{\lambda_n}\right)\sigma(dx,\omega)\neq\int_{R_d}\phi_i\left(\frac{x}{\lambda_n}\right)\alpha_n'(dx,\omega)\quad\text{for some}\quad 1\le i\le r\right\}\\&=\left\{\int_0^\infty\phi_i\left(\frac{\beta(s)}{\lambda_n}\right)ds\neq\int_0^{T_{a_n}}\phi_i\left(\frac{\beta(s)}{\lambda_n}\right)ds\quad\text{for some}\quad 1\le i\le r\right\}\\&=\left\{\int_{T_{a_n}}^\infty\phi_i\left(\frac{\beta(s)}{\lambda_n}\right)ds\neq 0\quad\text{for some}\quad 1\le i\le r\right\}.\end{aligned}$$

Let

$$R=\inf\{l:\phi_j(x)=0 \text{ for } |x|>l,\ j=1,2,\cdots,r\},$$

and let $\theta_{\lambda_n R}$ be the last exit time for the sphere of radius $\lambda_n R$. Since, for n large enough, $\lambda_n R<a_n$ and since the support of $\phi_i(x)\subset B(0,R)$ for all $1\leqq i\leqq r$, we see that the last event in (4.20) can occur only if $\theta_{\lambda_n R}>\tau_{a_n}$. Thus, (4.19), (4.20), and this last remark imply that, for n large enough,

$$P\{Z_{\lambda_n}(\cdot,\omega)\notin N,\ \alpha_n(\cdot,\omega)\in N\}\leqq P\{\theta_{\lambda_n R}>\tau_{a_n}\}. \tag{4.21}$$

Let $\tilde{\tau}_{\lambda_n R}$ denote the first hitting time of the ball $B(0,\lambda_n R)$. From the strong Markov property we obtain

$$P\{\theta_{\lambda_n R}>\tau_{a_n}\}=E\{P_{\beta(\tau_{a_n})}[\tilde{\tau}_{\lambda_n R}<\infty]\}. \tag{4.22}$$

It is an elementary property of d-dimensional Brownian motion that if we are given two spheres in R_d centered at the origin of radii R_1 and R_2 with $R_1<R_2$, then the probability of ever hitting the inner sphere of radius R_1 if the path starts on the outer sphere of radius R_2 is $(R_1/R_2)^{d-2}$. Therefore,

$$E\{P_{\beta(\tau_{a_n})}[\tilde{\tau}_{\lambda_n R}<\infty]\}=\left(\frac{\lambda_n R}{a_n}\right)^{d-2}, \tag{4.23}$$

and from (4.21), (4.22) and (4.23) we see that for n large enough

$$P\{Z_{\lambda_n}(\cdot,\omega)\notin N,\ \alpha_n(\cdot,\omega)\in N\}\leqq\left(\frac{\lambda_n R}{a_n}\right)^{d-2}. \tag{4.24}$$

Since $d\geqq 3$ and $\lambda_n/a_n=1/n^2$, we obtain (4.18) and the lemma.

In the preceding lemma we specified the neighborhood $N=N(f)$. Now define $N_1\subset N$ to be the neighborhood

$$N_1=\left\{\sigma\in M:\left|\int_{R_d}\phi_i(x)\sigma(dx)-\int_{R_d}\phi_i(x)f(x)\,dx\right|<\tfrac{2}{3}\varepsilon,\ i=1,2,\cdots,r\right\}.$$

LEMMA 4.6. *We have*

$$P\{\beta_n(\cdot,\omega)\in N_1,\ \alpha_n(\cdot,\omega)\notin N \text{ for infinitely many } n\}=0. \tag{4.25}$$

Proof: If, for some n and some ω, $\beta_n(\cdot,\omega)\in N_1$, but $\alpha_n(\cdot,\omega)\notin N$, then, for at least one $1\leqq i\leqq r$,

$$\left|\int_{R_d}\phi_i(x)\,\beta_n(dx,\omega)-\int_{R_d}\phi_i(x)\,\alpha_n(dx,\omega)\right|\geqq\tfrac{1}{3}\varepsilon\,,$$

which means

$$\left|\int_{\tau_{c_{n-1}}}^{\tau_{a_n}}\phi_i\left(\frac{\beta(s)}{\lambda_n}\right)ds-\int_0^{\tau_{a_n}}\phi_i\left(\frac{\beta(s)}{\lambda_n}\right)ds\right|\geqq\tfrac{1}{3}\varepsilon\lambda_n^2\log\log\lambda_n\,,$$

i.e.,

$$\left|\int_0^{\tau_{c_{n-1}}}\phi_i\left(\frac{\beta(s)}{n}\right)ds\right|\geqq\tfrac{1}{3}\varepsilon\lambda_n^2\log\log\lambda_n\,. \tag{4.26}$$

Let

$$\Lambda=\sup_{\substack{1\leqq i\leqq r\\ x\in R_d}}|\phi_i(x)|\,;$$

then, (4.26) implies that

$$\Lambda\tau_{c_{n-1}}\geqq\tfrac{1}{3}\varepsilon\lambda_n^2\log\log\lambda_n\,.$$

Hence, for each positive integer n,

$$P\{\beta_n(\cdot,\omega)\in N_1,\,\alpha_n(\cdot,\omega)\notin N\}\leqq P\left\{\tau_{c_{n-1}}\geqq\frac{\varepsilon}{3\Lambda}\lambda_n^2\log\log\lambda_n\right\}. \tag{4.27}$$

For Brownian motion starting at the origin it is well known that if τ_R is the first exit time from $B(0,R)$, then there exist constants $k_1>0$ and $k_2>0$ such that $P\{\tau_R>l\}\leqq k_2\exp\{-k_1 l/R^2\}$. Hence, for n large enough,

$$\begin{aligned}P\left\{\tau_{c_{n-1}}\geqq\frac{\varepsilon}{3\Lambda}\lambda_n^2\log\log\lambda_n\right\}&\leqq k_2\exp\left\{\frac{-\varepsilon k_1\lambda_n^2\log\log\lambda_n}{3\Lambda c_{n-1}^2}\right\}\\ &\leqq k_2\exp\left\{\frac{-\varepsilon k_1\lambda_n^2}{3\Lambda c_{n-1}^2}\right\}\\ &\leqq A\frac{c_{n-1}^2}{\lambda_n^2}\,,\end{aligned} \tag{4.28}$$

where A is a constant. But from (4.13) we see that $\sum(c_{n-1}^2/\lambda_n^2)<\infty$, and therefore from the Borel–Cantelli lemma, (4.27), and (4.28) we obtain (4.25).

Let $N_2 \subset N_1 \subset N$ be the neighborhood of f specified by

$$N_2=\left\{\sigma \in M : \left|\int_{R_d} \phi_i(x)\,\sigma(dx)-\int_{R_d} \phi_i(x)\,f(x)\,dx\right|<\tfrac{1}{3}\varepsilon, i=1,2,\cdots,r\right\}.$$

LEMMA 4.7.

$$\sum P\{Z_{\lambda_n}(\cdot,\omega)\in N_2\}=\infty. \tag{4.29}$$

Proof: First of all, from (4.1),

$$\begin{aligned} P\{Z_{\lambda_n}(\cdot,\omega)\in N_2\} &= P\left\{\frac{\sigma(\lambda_n\cdot,\omega)}{\lambda_n^2\log\log\lambda_n}\in N_2\right\} \\ &= P\left\{\frac{\sigma(\cdot,\omega)}{\log\log\lambda_n}\in N_2\right\}. \end{aligned}$$

From Theorem 3.5 we see that, for n large enough and for the $\delta>0$ picked in (4.12),

$$P\left\{\frac{\sigma(\cdot,\omega)}{\log\log\lambda_n}\in N_2\right\}\geqq \exp\{-(\log\log\lambda_n)(I(f)+\delta)\}. \tag{4.30}$$

Since $\lambda_n=\exp\{n^\alpha\}$, we conclude that

$$P\{Z_{\lambda_n}(\cdot,\omega)\in N_2\}\geqq \exp\{-(\alpha\log n)(I(f)+\delta)\}=\frac{1}{n^{\alpha[I(f)+\delta]}}. \tag{4.31}$$

Since α and δ were picked in (4.12) so that $\alpha[I(f)+\delta]<1$, we see that (4.29) holds.

LEMMA 4.8.

$$\sum P\{\alpha_n(\cdot,\omega)\in N_2\}=\infty. \tag{4.32}$$

Proof: Exactly as in Lemma 4.5 we see that

$$\sum P\{Z_{\lambda_n}(\cdot,\omega)\notin N_2,\ \alpha_n(\cdot,\omega)\in N_2\}<\infty,$$

and since

$$P\{\alpha_n(\cdot,\omega)\in N_2\}\geqq P\{Z_{\lambda_n}(\cdot,\omega)\in N_2\}-P\{Z_{\lambda_n}(\cdot,\omega)\notin N_2,\ \alpha_n(\cdot,\omega)\in N_2\}, \tag{4.33}$$

(4.32) follows from (4.33) and Lemma 4.7.

LEMMA 4.9.

$$\sum P\{\beta_n(\cdot,\omega)\in N_1\}=\infty. \tag{4.34}$$

Proof: The same argument that was used in Lemma 4.6 shows that

$$\sum P\{\beta_n(\cdot,\omega)\notin N_1,\, \alpha_n(\cdot,\omega)\in N_2\}<\infty.$$

Since,

$$P\{\beta_n(\cdot,\omega)\in N_1\}\geqq P\{\alpha_n(\cdot,\omega)\in N_2\}-P\{\beta_n(\cdot,\omega)\notin N_1,\, \alpha_n(\cdot,\omega)\in N_2\}, \tag{4.35}$$

we see that Lemma 4.8 and (4.35) imply (4.34).

LEMMA 4.10. *For $n=1,2,\cdots$, let $E_n=\{\omega:\beta_n(\cdot,\omega)\in N_1\}$. Then, for $n>m$ there exists a constant c, independent of m and n, such that*

$$P\{E_n\cap E_m\}\leqq cP\{E_m\}P\{E_n\}. \tag{4.36}$$

Proof: The event E_n occurs only between the times $\tau_{c_{n-1}}$ and τ_{a_n} whereas the event E_m occurs only between the times $\tau_{c_{m-1}}$ and τ_{a_m}. Since $n>m$, if we denote by $\gamma_{\tau_{a_m}}$ the σ-field corresponding to the stopping time τ_{a_m}, then we have, from the strong Markov property,

$$\begin{aligned} P\{E_n\cap E_m\} &= E\{P[E_n\cap E_m \mid \gamma_{\tau_{a_m}}]\} \\ &= E\{\chi_{E_m}(\omega)\cdot P_{\beta(\tau_{a_m})}[E_n]\} \\ &= E\{\chi_{E_m}(\omega)E_{\beta(\tau_{a_m})}\{P_{\beta(\tau_{c_{n-1}})}[E_n]\}. \end{aligned} \tag{4.37}$$

But,

$$E_{\beta(\tau_{a_m})}\{P_{\beta(\tau_{c_{n-1}})}[E_n]\}=\int_{S(0,c_{n-1})}\Pi^{c_{n-1}}(\beta(\tau_{a_m}),y)P_y[E_n]\,dy, \tag{4.38}$$

where $\Pi^{c_{n-1}}(x,y)$ is the Poisson kernel for the sphere of radius c_{n-1}, $S(0,c_{n-1})$ is the surface of that sphere, and dy is normalized so that the total surface measure is 1. If now we let

$$\Gamma_{m,n}=\sup_{\substack{|x|=a_m\\ |y|=c_{n-1}}}\Pi^{c_{n-1}}(x,y),$$

then (4.37) and (4.38) imply

$$P\{E_n \cap E_m\} \leqq \Gamma_{m,n} E\left\{\chi_{E_m}(\omega) \int_{S(0,c_{n-1})} P_y[E_n]\, dy\right\} \tag{4.39}$$
$$= \Gamma_{m,n} P\{E_m\}\, P\{E_n\}\,.$$

Since $a_{n-1}/c_{n-1}=\frac{1}{2}$, we see that $a_m/c_{n-1}<\frac{1}{2}$ for all $n>m$, and therefore from the Poisson kernel, $\Pi^{c_{n-1}}(x, y)$, we observe that there exists an absolute constant c such that $\Gamma_{m,n} \leqq c$ which yields the lemma.

LEMMA 4.11.

$$P\{\beta_n(\cdot, \omega) \in N_1 \textit{ for infinitely many } n\} = 1\,. \tag{4.40}$$

Proof: Lemmas 4.9 and 4.10 establish the nonobvious hypotheses of a refined form of the Borel–Cantelli lemma stated below and (4.40) is the conclusion of that lemma.

The following lemma is contained in [3] (p. 425) as well as in [1] (Theorem A).

LEMMA 4.12. *Let E_n, $n=1, 2, \cdots$, be a sequence of events such that $P\{E_n\} \to 0$ as $n \to \infty$, while $\sum_{n=1}^{\infty} P\{E_n\}$ diverges and* $P\{\limsup E_n\}=0$ *on* 1. *If there exists an absolute constant c such that, for $n \neq r$,*

$$P\{E_n \cap E_r\} \leqq cP\{E_n\}\, P\{E_r\}\,,$$

then infinitely many of the events E_n occur with probability 1.

Proof of Theorem 4.4: Combining Lemmas 4.11 and 4.6 we conclude that

$$P\{\alpha_n(\cdot, \omega) \in N \text{ for infinitely many } n\} = 1\,, \tag{4.41}$$

and (4.41) together with Lemma 4.5 imply

$$P\{Z_{\lambda_n}(\cdot, \omega) \in N \text{ for infinitely many } n\} = 1$$

which, as noted in (4.10), establishes the theorem.

Bibliography

[1] Ciesielski, Z., and Taylor, S. J., *First passage times and sojourn times for Brownian motion in space and the exact Hausdorff measure of the sample path*, Trans. Am. Math. Soc. 103, 1962, pp. 434–450.

[2] Donsker, M. D., and Varadhan, S. R. S., *Asymptotic evaluation of certain Wiener integrals for large time, in Proceedings of the International Conference on Functional Integration and its Applications,* London: Clarendon Press, Oxford, 1975, pp. 15–33.

[3] Erdös, P., *On the law of the iterated logarithm*, Ann. Math. 43, 1942, pp. 419–486.

Received November, 1979.

Proceedings of the International Congress of Mathematicians (Helsinki, 1978), Acad. Sci. Fennica, 1980, 755-762

Some Problems of Large Deviations

S. R. S. Varadhan

1. Introduction. The problem of large deviations arises naturally in probability theory in different contexts. We shall first look at some typical examples.

If $X_1, X_2, \ldots, X_n, \ldots$ is a sequence of independent and identically distributed random variables with mean zero, then for any $a>0$

$$\lim_{n\to\infty} P\left[\frac{X_1+\ldots+X_n}{n} \geqslant a\right] = 0 \tag{1.1}$$

by the law of large numbers. Under the assumption that for every real θ

$$M(\theta) = E[e^{\theta X_i}] < \infty$$

Cramér [3] showed that the probability in (1.1) has the specific behavior:

$$\lim_{n\to\infty} \left\{P\left[\frac{X_1+\ldots+X_n}{n} \geqslant a\right]\right\}^{1/n} = \varrho_a \tag{1.2}$$

exists and is given by

$$\varrho_a = \inf_{\theta \geqq 0} M(\theta) e^{-\theta a}.$$

For the second example we consider the process $x_\varepsilon(t)=\varepsilon\beta(t)$ defined in the time interval $0\leqslant t\leqslant 1$, where $\beta(\cdot)$ is the standard one-dimensional Brownian motion. We denote by P_ε the measure corresponding to $x_\varepsilon(t)$, which lives on the space $C_0[0, 1]$ of continuous functions vanishing at $t=0$. As $\varepsilon\to 0$, $x_\varepsilon(t)$ tends to the zero function uniformly and therefore

$$\lim_{\varepsilon\to 0} P_\varepsilon(A) = 0 \tag{1.3}$$

for sets A that are disjoint from some neighborhood of the zero function. One can again show that for a large class of sets $A \subset C_0[0, 1]$

(1.4) $$\lim_{\varepsilon \to 0} 2\varepsilon^2 \log P_\varepsilon(A) = -h(A)$$

exists where

(1.5) $$h(A) = \inf_{f \in A} \int_0^1 [f'(t)]^2 \, dt$$

and the infimum is taken over absolutely continuous functions in f in A with a square integrable derivative. Such a result in a slightly different form can be found in **[14]**.

For the third example we will consider a Markov chain on a finite state space X and transition probabilities π_{ij}. For simplicity we shall assume that $\pi_{ij} > 0$ for all i and j. Let $\vec{q} = \{q_i\}$ be the unique invariant probability vector. If $X_0, X_1, \ldots, X_n, \ldots$ is a realization of the Markov chain, we denote by $f_i^{(n)}$ the frequency of occurrence of the state i during the first n steps of the Markov chain and by $p_i^{(n)}$ the proportion of visits to the state i during the same time. The vector $\vec{p}^{(n)}$ with components $\{p_i^{(n)}\}$ is a random probability distribution on X and is the empirical distribution based on the first n steps. By the ergodic theorem, for large values of n, the vector $\vec{p}^{(n)}$ is close to the invariant probability vector $\vec{q}$ with a very high probability. The theorem on large deviations states in this context that for suitable sets A in space of probability distributions on X

(1.6) $$\lim_{n \to \infty} \frac{1}{n} \log P\{\vec{p}^{(n)} \in A\} = -h(A)$$

exists, where

(1.7) $$h(A) = \inf_{p \in A} I(\vec{p})$$

and the function $I(\vec{p})$ is defined for probability distributions $\vec{p}$ on X by

(1.8) $$I(\vec{p}) = -\inf_{u > 0} \sum_{i \in X} p_i \log\left(\frac{\sum_{j \in X} \pi_{ij} u_j}{u_i}\right)$$

where the infimum is taken over vectors $u = \{u_i\}$ which have strictly positive components. The function $I(\cdot)$, while it is not explicit, is a nonnegative convex function of $\vec{p}$ which vanishes only when $\vec{p}$ equals the invariant probability $\vec{q}$.

The first example can be generalized by allowing the random variables to take values in more general linear spaces. The second can be generalized to cover processes $x_\varepsilon(t)$ governed by stochastic differential equations of the form

$$dx_\varepsilon(t) = \varepsilon\sigma(x_\varepsilon(t))\, d\beta(t) + b(x_\varepsilon(t))\, dt,$$

$$x_\varepsilon(0) = x.$$

Actually one can even combine the two classes in a single large class of problems of large deviations for Markov processes. See for instance **[15]**–**[20]**, **[1]**, **[2]**, and **[13]**.

As I am hoping that Professor Ventcel will cover this ground in some detail I shall proceed to a discussion of the ideas connected with the third example.

2. Large deviation for occupation times. Let the state space X of the Markov chain be a complete separable metric space instead of a finite set. Let the transition probabilities be given by $\pi(x, dy)$. We shall denote by π, the corresponding operator on the space $B(X)$ of bounded measurable functions on X, defined by

$$(\pi f)(x) = \int f(y)\pi(x, dy). \tag{2.1}$$

We denote by $\mathcal{M}$ the space of all probability distributions on X and we will view $\mathcal{M}$ as a complete separable metric space with weak convergence as the underlying convergence notion. We make the following assumptions on $\pi(x, dy)$.

(H1) *(Feller Property).* If f is bounded and continuous then so is πf.

(H2) *(Strong transitivity).* There is a reference measure $\alpha(dy)$ on X such that $\alpha(dy)$ and $\pi(x, dy)$ are mutually absolutely continuous for every $x \in X$.

(The above condition can be relaxed somewhat.)

For $\mu \in \mathcal{M}$ we define $I(\mu)$ by

$$I(\mu) = -\inf_{u \in U} \int_X \log\left(\frac{\pi u}{u}\right)(x)\mu(dx) \tag{2.2}$$

where U consists of functions in the space $C(X)$ of bounded continuous functions which have a positive lower bound. One can verify that $I(\cdot)$ is a convex, nonnegative, lower semicontinuous functional on $\mathcal{M}$ which vanishes only at invariant probability distributions. The chain of course may not possess an invariant probability distribution in which case $I(\cdot)$ is never zero.

We shall denote by ω, a realization $X_0, X_1, \ldots, X_n, \ldots$ of the chain and by $L_n(\omega, \cdot)$ the empirical distribution based on the first n steps

$$L_n(\omega, E) = \frac{1}{n}\sum_{j=1}^{n} \chi_E(X_j), \quad E \subset X. \tag{2.3}$$

Fixing an arbitrary starting point $x \in X$ for the chain we have the measure P_x corresponding to the process on the space Ω of all possible realizations. One can view $L_n(\cdot,\cdot)$ as a map of Ω into $\mathcal{M}$ and this induces a measure $Q_{n,x}$ on $\mathcal{M}$ which is the distribution of the empirical distribution. If A is a subset of $\mathcal{M}$ then we are interested in the asymptotic behavior of $Q_{n,x}(A)$ as $n \to \infty$. We have the following results:

If $C \subset \mathcal{M}$ is compact, then

$$\limsup_{n\to\infty} \frac{1}{n} \log Q_{n,x}[C] \leqslant -\inf_{\mu \in C} I(\mu). \tag{2.4}$$

If $G \subset \mathcal{M}$ is open, then

$$\limsup_{n\to\infty} \frac{1}{n} \log Q_{n,x}[G] \geqslant -\inf_{\mu \in G} I(\mu). \tag{2.5}$$

The reason that (2.4) can be established only for compact sets is that we have no assumption of positive recurrence. Perhaps the measures $Q_{n,x}$ are dissipative and (2.4) really measures the rate of dissipation. It is even possible that $\inf_\mu I(\mu)>0$ and in that case (2.4) cannot hold when C is taken to be the whole space $\mathcal{M}$. However when X is not compact, we can impose a strong positive recurrence condition that will enable us to prove (2.4) for closed sets.

There are continuous time analogs to these results for Markov processes $x(t)$ with transition probabilities $p(t, x, dy)$ on a state space X. We have the corresponding operators

$$(T_t f)(x) = \int f(y)\, p(t, x, dy) \tag{2.6}$$

we make the following assumptions regarding the semigroup $\{T_t\}$.

(H1*) T_t maps the space of bounded continuous functions into itself.

(H2*) There is a reference measure $\alpha(dy)$ such that $\alpha(dy)$ and $p(t, x, dy)$ are mutually absolutely continuous for every $t>0$ and $x\in X$.

(H3*) The strongly continuous center of the semigroup T_t is sufficiently large.

We denote by L the infinitesimal generator acting on the domain $\mathcal{D}\subset C(X)$. We denote by $\mathcal{D}^+$ those functions in $\mathcal{D}$ with a positive lower bound. We define the I-function as the analog of (2.2) by

$$I(\mu) = -\inf_{u\in\mathcal{D}^+} \int \left(\frac{Lu}{u}\right)(x)\, \mu(dx). \tag{2.7}$$

We look at the occupation distribution

$$L_t(\omega, E) = \frac{1}{t}\int_0^t \chi_E(x(s))\, ds \tag{2.8}$$

as the analog of (2.3) and the distribution $Q_{t,x}$ of $L_t(\cdot,\cdot)$ as the analog of $Q_{n,x}$. We then have the exact analogs of (2.4) and (2.5).

For C compact in $\mathcal{M}$,

$$\limsup_{t\to\infty} \frac{1}{t}\log Q_{t,x}(C) \leqslant -\inf_{\mu\in C} I(\mu). \tag{2.9}$$

For G open in $\mathcal{M}$,

$$\liminf_{t\to\infty} \frac{1}{t}\log Q_{t,x}(G) \geqslant -\inf_{\mu\in G} I(\mu). \tag{2.10}$$

Again if X is not compact one can impose a strong recurrence condition and obtain (2.9) for closed sets instead of just compact sets.

If the transition probabilities $p(t, x, dy)$ are such that they have a symmetric density with respect to the reference measure $\alpha(dy)$ then the semigroup T_t is a family of self adjoint contractions in the Hilbert space of functions that are square integrable with respect to the measure $\alpha(dy)$. Then the infinitesimal generator L can be thought of as a nonpositive self adjoint operator. Therefore the operator $\sqrt{-L}$ is well defined. In such a context one can show that $I(\mu)$ is finite if and only if μ is absolutely continuous with respect to α and the square integrable function $(d\mu/d\alpha)^{1/2}$ is in the domain of $\sqrt{-L}$. In such a case

$$I(\mu) = \left\| \sqrt{-L}\left(\frac{d\mu}{d\alpha}\right)^{1/2} \right\|^2 \tag{2.11}$$

where $\| \ \|$ is the mean square norm with respect to α. For the Brownian motion in R^d this becomes

$$I(f) = I(\mu) = \frac{1}{8}\int \frac{|\nabla f|^2}{f}\,dx = \frac{1}{2}\int |\nabla \sqrt{f}|^2\,dx \tag{2.12}$$

where $\mu(dx)=f(x)\,dx$.

The details of these results can be found in [4], [5], [21].

3. Connections with the principal eigenvalue. Let us consider the case when X is compact. If $x(t)$ is the Markov process corresponding to the generator L and $I(\cdot)$ is the corresponding I-functional defined by (2.7) then one can show [see [15]] that

$$\begin{aligned} \lambda(V) &= \lim_{t\to\infty} \frac{1}{t} \log E_x\left[\exp\left\{\int_0^t V(x(s))\,dx\right\}\right] \\ &= \sup_{\mu\in\mathcal{M}} \left[\int V(x)\,\mu(dx) - I(\mu)\right] \end{aligned} \tag{3.1}$$

for all $V\in C(X)$. The quantity $\lambda(V)$ can also be identified as the principal eigenvalue (i.e. the point in the spectrum with the largest real part) of the operator $L+V$. From the maximum principle for L one can conclude that $\lambda(V)$ is a convex functional of V and $I(\cdot)$ is its conjugate convex functional. This is the explanation of why $I(\cdot)$ is readily computable in the self-adjoint case. The variational formula (3.1) in that case reduces to the classical Rayleigh–Ritz variational formula for the principal eigenvalue of self adjoint operators. See in this context [6] and [12].

4. Applications. If $\beta(s)$ is the d-dimensional Brownian motion then the Wiener sausage $C_\varepsilon^t(\omega)$ for the trajectory $\omega=\beta(\cdot)$ up to time t is the set

$$C_\varepsilon^t(\omega) = \left\{y\colon \inf_{0\leqq s\leqq t} |y-\beta(s)| \leqslant \varepsilon\right\}.$$

A problem that comes from statistical mechanics [11] is the behavior of

$$E[\exp\{-\nu|C_\varepsilon^t(\omega)|\}]$$

where $|A|$ stands for the Lebesgue measure of the set A. Using the techniques developed in the preceding section (see [7], [8]) one can show that

$$\lim_{t\to\infty} \frac{1}{t^{d/(d+2)}} \log E[\exp\{-\nu|C_\varepsilon^t(\omega)|\}] = -k(\nu, d) \tag{4.1}$$

exists and compute $k(\nu, d)$ as

$$k(\nu, d) = \inf_{\substack{f \in L_1(R^d) \\ f \geq 0}} [\nu|x: f(x) > 0| + I(f)] \tag{4.2}$$

where $I(f)$ is given by (2.12).

One can consider a random walk on the lattice Z^d in R^d of points with integral coordinates. If the distribution of the single step has mean zero and covariance identity then under mild irreducibility assumptions the number $D(n)$ of distinct sites visited by the random walk in the first n steps can be shown to satisfy

$$\lim_{n\to\infty} \frac{1}{n^{d/(d+2)}} \log E[\exp\{-\nu D(n)\}] = -k(\nu, d) \tag{4.3}$$

where $k(\nu, d)$ is again given by (4.2). There are analogs when the Brownian motion is replaced by a symmetric stable process (see [9] for details).

Another application is to derive laws of the iterated logarithm for local times. Let $\omega = \beta(\cdot)$ be one dimensional Brownian motion. Let

$$l(t, y) = \int_0^t \delta(\beta(s) - y)\, ds$$

be the local time of the Brownian path as a function of t and y. We define

$$L(t, y) = \frac{1}{\sqrt{t \log\log t}}\, l\left(t, y\sqrt{\frac{t}{\log\log t}}\right).$$

We can view $L(t, y)$ as a random probability density on the line. One can show that the set of limit points of $L(t, \cdot)$ as $t\to\infty$ coincides as functions of y (in the topology of uniform convergence on compact sets) almost surely with the set of subprobability densities $f(\cdot)$ satisfying

$$\int f(x)\, dx \leq 1 \quad \text{and} \quad \frac{1}{8}\int \frac{[f'(x)]^2}{f(x)}\, dx \leq 1.$$

These results and similar ones for certain stable processes can be found in [10].

5. General remarks. In the case of continuous time processes the I-functional plays the role of the Dirichlet integral to which it reduces in the self adjoint case. For instance $G \subset X$ is an open set with compact closure and one is interested in the exponential decay rate of $P_x[x(s) \in G, 0 \leq s \leq t]$ as $t\to\infty$. One can under mild conditions compute it as

$$\lambda(G) = \inf_{\substack{\mu \in \mathcal{M} \\ \mu(G)=1}} I(\mu).$$

There is a close connection, at least in spirit, to the theory developed by Lanford in [13] in the context of statistical mechanics.

If we start with a Markov process and reverse it in time using an invariant probability distribution then the new reversed Markov process has the same I-functional as the old one. It is an interesting question to examine how much information about the process can be recovered by a knowledge of its I-functional.

In case the process is transient one can consider the total occupation time i.e.

$$L(\omega, E) = \int_0^\infty \chi_E(x(s))\, ds$$

which will be a σ-finite measure on the state space. The tail behavior of its distribution on the space of all σ-finite measures on the state space should again be related to the I-functional. The details of the connection are being currently worked out.

References

1. R. Azencott and G. Ruget, *Mélanges d'équation différentielles et grand écart à la loi des grands nombres*, Z. Wahrscheinlichkeitstheorie und Verw. Gebiete **38** (1977), 1—54.

2. R. R. Bahadur and S. L. Zabell, *Large deviations of the sample mean in general vector spaces* (preprint).

3. H. Carmér, *On a new limit theorem in the theory of probability:* Colloquium on the Theory of Probability, Hermann, Paris, 1937.

4. M. D. Donsker and S. R. S. Varadhan, *Asymptotic evaluation of certain Markov process expectations for large time*, I, Comm. Pure Appl. Math. **28** (1975), 1—47.

5. ——, *Asymptotic evaluation of certain Markov process expectations for large time*, III, Comm. Pure Appl. Math. **29** (1977), 389—461.

6. ——, *On a variational formula for the principal eigenvalue for operators with maximum principle*, Proc. Nat. Acad. Sci. U.S.A. **72** (1975), 780—783.

7. ——, *Asymptotic evaluation of certain Markov process expectations for large time*, II, Comm. Pure Appl. Math. **28** (1975), 279—301.

8. ——, *Asymptotics for the Wiener sausage*, Comm. Pure Appl. Math. **28** (1975), 525—565.

9. ——, *On the number of distinct sites visited by a random walk* (preprint).

10. ——, *On laws of the iterated logarithms for local times*, Comm. Pure Appl. Math. **30** (1977), 707—753.

11. M. Kac, *Probabilistic methods in some problems of scattering theory*, Rocky Mountain J. Math. **4** (1974), 511—538.

12. T. G. Kurtz, *A variational formula for the growth rate of a positive operator semigroup* (preprint).

13. O. E. Lanford, *Entropy and equilibrium states in classical statistical mechanics*, Statistical Mechanics and Mathematical Problems, Lecture notes in Physics, vol. 20, Springer-Verlag, Berlin and New York, 1971, pp. 1—113.

14. M. Schilder, *Some asymptotic formulae for Wiener integrals*, Trans. Amer. Math. Soc. **125** (1966), 63—85.

15. S. R. S. Varadhan, *Asymptotic probabilities and differential equations*, Comm. Pure Appl. Math. **19** (1966), 261—286.

16. ——, *On the behavior of the fundamental solution of the heat equation with variable coefficients*, Comm. Pure Appl. Math. **20** (1967), 431—455.

17. ———, *Diffusion processes in a small time interval*, Comm. Pure Appl. Math. **20** (1967), 659—685.

18. A. D. Ventcel, *Rough limit theorems on large deviations for Markov stochastic processes*, I, Theor. Probability Appl. **21** (1976), 227—242.

19. ———, *Rough limit theorems on large deviations for Markov stochastic processes*, II, Theor. Probability Appl. **21** (1976), 499—512.

20. A. D. Ventcel and M. I. Freidlin, *On small random perturbations of dynamical systems*, Russian Math. Surveys **25** (1970), 1—55.

21. J. Gärtner, *On large deviations from the Invariant Measure*, Theor. Probability Appl. **22** (1977), 24—39.

Courant Institute of Mathematical Sciences
New York University
New York, New York 10012, U.S.A.

PHYSICS REPORTS (Review Section of Physics Letters) 77, No. 3 (1981) 235–237. North-Holland Publishing Company

The Polaron Problem and Large Deviations*

M.D. DONSKER and S.R.S. VARADHAN

Courant Institute of Mathematical Sciences, 251 Mercer Street, New York, N.Y. 10012, U.S.A.

Let $\{x_t(s), 0 \le s \le t\}$ be three dimensional Brownian motion tied down at the ends of the time interval, i.e., $x_t(0) = x_t(t) = 0$. Let $\alpha > 0$ and consider the following function space integral:

$$A(t, \alpha) = E\left\{\exp\left[\alpha \int_0^t \int_0^t \frac{e^{-|\sigma - s|}}{\|x_t(\sigma) - x_t(s)\|} \, ds \, d\sigma\right]\right\}. \tag{1}$$

The "Polaron Problem" which arises in statistical mechanics [1] is to evaluate $G(\alpha) = \lim_{t\to\infty}(1/t) \log A(t, \alpha)$. Now, $G(\alpha)$ is complicated to evaluate and even to estimate, but a conjecture by Pekar is that

$$\lim_{\alpha\to\infty} \frac{G(\alpha)}{\alpha^2} = c$$

where

$$c = \sup_{\substack{\varphi \in L^2(R^3) \\ \|\varphi\|_2 = 1}} \left[2 \int \int \frac{\varphi^2(x)\, \varphi^2(y)}{\|x - y\|} \, dx \, dy - \frac{1}{2} \int |\nabla \varphi|^2 \, dx\right].$$

Using methods developed by us in [2, 3], we succeeded in finding a "fairly explicit" expression for $G(\alpha)$ and then used this expression to prove rigorously the conjecture of Pekar. The details of that argument will be found in [4]. In this note we briefly describe some of the ideas lying behind our work on asymptotics.

In determining the asymptotic behavior of (1) as $t \to \infty$, it is clear that the major contribution is coming from those Brownian motion paths which tend to stay near themselves as time goes on. Since "typical" Brownian motion paths do not behave that way, determination of the asymptotics of (1) requires a large deviation result. In the Wiener sausage [5], and other applications [6–10], a common feature is that the asymptotic behavior results from the effect of atypical paths so that what is required is precise estimates of large deviations. Let us examine in a simplified context what we mean by a large deviation result.

Consider, for example, a complete separable metric space X and a sequence $\{Q_n\}$ of probability measures on it. As $n \to \infty$, Q_n behaves in one of two different ways:

*The research for this paper was done at the Courant Institute of Mathematical Sciences and supported by the National Science Foundation under NSF Grant No. MCS-77-02687.

Case 1. $Q_n \Rightarrow \delta_{x_0}$ for some $x_0 \in X$.
Case 2. Q_n is dissipative so that $Q_n(K) \to 0$ as $n \to \infty$ for every compact set $K \subset X$.
In most cases of interest in case 1, the space X is itself compact, whereas in case 2, X must necessarily be non-compact. Now, by a large deviation property for Q_n, we mean in case 1:

There exists a function $I(x)$: $X \to [0, \infty]$ such that

(i) $I(x)$ is lower semicontinuous.

(ii) $[x : I(x) \leq l]$ is compact in X for each $l < \infty$, and

$$\limsup_{n\to\infty} \frac{1}{n} \log Q_n(C) \leq -\inf_{x\in C} I(x) \qquad \text{for C closed},$$

$$\liminf_{n\to\infty} \frac{1}{n} \log Q_n(G) \geq -\inf_{x\in G} I(x) \qquad \text{for G open}.$$

In case 2 we demand only that there exists a function $I(x)$: $X \to [0, \infty]$ such that $I(x)$ is lower semicontinuous and

$$\limsup_{n\to\infty} \frac{1}{n} \log Q_n(K) \leq -\inf_{x\in K} I(x) \qquad \text{for K compact}$$

$$\liminf_{n\to\infty} \frac{1}{n} \log Q_n(G) \geq -\inf_{x\in G} I(x) \qquad \text{for G open}.$$

In case 1, if we have an $I(x)$ with the stated properties, then one can prove [11] that for any F: $X \to R$ that is bounded and continuous,

$$\lim_{n\to\infty} \frac{1}{n} \log \int \exp[n\,F(x)]\,dQ_n = \sup_x [F(x) - I(x)]. \tag{2}$$

In certain situations in case 2, (2) also holds but only for a more restricted class of functions F (see e.g. [12]).

Thus, in a particular context the problem is to determine the I-function. Let us illustrate in the simplest possible example. The reader is referred to our papers cited above for other contexts. Let $\xi_1, \xi_2, \dots$ be a sequence of independent random variables with values in $[0, 1]$ having a common distribution λ on $[0, 1]$. Let $a = E(\xi) = \int_0^1 x\,\lambda(dx)$, and let $\eta_n = (\xi_1 + \xi_2 + \cdots + \xi_n)/n$, so that the strong law of large numbers states that $\eta_n \to a$ with probability one. If we let Q_n be the probability distribution of η_n, then $Q_n \Rightarrow \delta_a$ on $[0, 1]$. Now, a case 1 large deviation result was found in this classical case by Cramér [13] with $I_1(x) = \sup_\theta [\theta x - \log \int e^{\theta x} \lambda(dx)]$. In this situation (2) becomes

$$\lim_{n\to\infty} \frac{1}{n} \log E\{\exp[nF(\eta_n)]\} = \sup_x [F(x) - I_1(x)]. \tag{3}$$

We can look at this classical problem at a more complex level. Given $\xi_1, \xi_2, \dots$, consider for each n the sample distribution, i.e., the random measure $\mu_{n,\omega} = (\delta_{\xi_1} + \delta_{\xi_2} + \cdots + \delta_{\xi_n})/n$ as a random variable in the space $\mathcal{M}$ of probability distributions on $[0, 1]$. Then, $\mu_{n,\omega}$ will have its probability distribution which we will denote by Q_n. Q_n is now a probability measure on $\mathcal{M}$. By the ergodic theorem (Glivenko–Cantelli theorem in this context), we know $\mu_{n,\omega} \to \lambda$ almost surely, and therefore $Q_n \Rightarrow \delta_\lambda$ on $\mathcal{M}$. We have again a case 1 situation, and now the I-function must be found as a function on $\mathcal{M}$, the space of probability measures on [0, 1]. In [2-III] the authors determined that $I_2(\mu) = \int \log(\mathrm{d}\mu/\mathrm{d}\lambda)\,\mathrm{d}\mu$ is the appropriate I-function. Hence (2) becomes

$$\lim_{n\to\infty} \frac{1}{n} \log E\{\exp[nF(\mu_{n,\omega})]\} = \sup_{\mu \in \mathcal{M}} [F(\mu) - I_2(\mu)]\,. \tag{4}$$

Similar large deviation results can be found for Markov chains instead of independent random variables as in the illustrations above. Whether we are in case 1 or in case 2 in the case of Markov chains depends on the recurrence properties of the chain. Instead of discrete time we can consider Markov processes with continuous time and then instead of n getting large it is t which becomes infinite. In all these cases the I-function can be found and asymptotic evaluation of the corresponding expectations obtained.

References

[1] R.P. Feynman and A.R. Hibbs, Quantum Mechanics and Path Integrals (1965).

[2] M.D. Donsker and S.R.S. Varadhan, Asymptotic evaluation of certain Markov process expectations for large time, I, Comm. Pure Appl. Math. 28 (1975) 1–47; II, Comm. Pure Appl. Math. 28 (1975) 279–301; III, Comm. Pure Appl. Math. 29 (1976) 389–461.

[3] M.D. Donsker and S.R.S. Varadhan, Asymptotic evaluation of certain Markov process expectations for large time, IV, to appear in Comm. Pure Appl. Math.

[4] M.D. Donsker and S.R.S. Varadhan, Asymptotics for the Polaron problem, to appear.

[5] M.D. Donsker and S.R.S. Varadhan, Asymptotics for the Wiener sausage, Comm. Pure Appl. Math. 28 (1975) 525–565.

[6] M.D. Donsker and S.R.S. Varadhan, On the number of distinct sites visited by a random walk, Comm. Pure Appl. Math. 32 (1979) 721–747.

[7] M.D. Donsker and S.R.S. Varadhan, On a variational formula for the principal eigenvalue for operators with maximum principle, Proc. Nat. Acad. Sci. U.S.A., Vol. 72, No. 3 (March 1975) pp. 780–783.

[8] M.D. Donsker and S.R.S. Varadhan, On the principal eigenvalue of second-order elliptic differential operators, Comm. Pure Appl. Math. 29 (1976) 595–621.

[9] M.D. Donsker and S.R.S. Varadhan, On laws of the iterated logarithm for local times, Comm. Pure Appl. Math. 30 (1977) 707–753.

[10] M.D. Donsker and S.R.S. Varadhan, A law of the iterated logarithm for total occupation times of transient Brownian motion, Comm. Pure Appl. Math. 33 (1980) 365–393.

[11] S.R.S. Varadhan, Asymptotic probabilities and differential equations, Comm. Pure Appl. Math. 19 (1966) 261, 286.

[12] M.D. Donsker and S.R.S. Varadhan, Asymptotic evaluation of certain Wiener integrals for large time, Functional Integration and its Applications, in: Proc. Intern. Conf. 1974, ed. A.M. Arthurs (Clarendon Press, Oxford, 1975).

[13] H. Cramer, On a new limit theorem in the theory of probability, Colloquium on the Theory of Probability, 1937 (Hermann and Co., Paris).

Comm. Pure Appl. Math. 36 (1983) 183-212

Asymptotic Evaluation of Certain Markov Process Expectations for Large Time. IV

M. D. DONSKER AND S. R. S. VARADHAN

1. Introduction

In this paper we extend our asymptotic results [1], [2] to a more general setting. Let Ω be a space of functions $\omega(\cdot)$ on $-\infty<t<\infty$ with values in a Polish space X. We assume Ω consists of functions with discontinuities only of the first kind, normalized to be right continuous, and with convergence induced by the Skorohod topology on bounded intervals. In this case, Ω is itself a Polish space. Denote by Ω_t^+ the corresponding space of functions on $[t, \infty)$ with values in X. We denote by $\mathscr{F}_t^s$ the σ-field in Ω generated by $\omega(\sigma)$ for $s \leqq \sigma \leqq t$. We denote the translation map on Ω by θ_t, i.e., $(\theta_t\omega)(s)=\omega(s+t)$.

Let $P_{0,x}$ be a Markov process on Ω_0^+ starting from $x \in X$ satisfying the hypothesis that the mapping $x \to P_{0,x}$ is weakly continuous (which implies the Feller property for the process $P_{0,x}$).

Let $\omega \in \Omega$ and, for each $t>0$, define ω_t by

$$\omega_t(s)=\omega(s), \qquad 0 \leqq s<t,$$

$$\omega_t(s+t)=\omega_t(s) \quad \text{for all} \quad s \in(-\infty, \infty).$$

Now, $(\theta_s\omega_t)(\tau)=\omega_t(s+\tau)$, $0 \leqq s \leqq t$, and for any set $A \subset \Omega$ we define

$$R_{t,\omega}(A)=\frac{1}{t} \int_0^t \chi_A(\theta_s\omega_t)\, ds. \tag{1.1}$$

If we let $\mathscr{M}_S(\Omega)$ denote the space of stationary processes on Ω, we see that, for each $\omega \in \Omega$ and each $t>0$, $R_{t,\omega}(\cdot) \in \mathscr{M}_S(\Omega)$. This is because, for any $\sigma>0$,

$$\begin{aligned} R_{t,\omega}(\theta_\sigma A) &=\frac{1}{t} \int_0^t \chi_{\theta_\sigma A}(\theta_s\omega_t)\, ds=\frac{1}{t} \int_0^t \chi_A(\theta_{s-\sigma}\omega_t)\, ds \\ &=\frac{1}{t} \int_{-\sigma}^{t-\sigma} \chi_A(\theta_s\omega_t)\, ds=\frac{1}{t} \int_0^t \chi_A(\theta_s\omega_t)\, ds=R_{t,\omega}(A). \end{aligned}$$

Thus, for fixed $t>0$, $R_{t,\omega}$ is a mapping of $\Omega \to \mathscr{M}_S(W)$ and as such is $\mathscr{F}_t^0$ measurable. For each $t>0$ and each $x \in X$, we use this mapping to induce a probability measure $\Gamma_{t,x}$ on $\mathscr{M}_S(\Omega)$ by defining $\Gamma_{t,x}=P_{0,x}R_{t,\omega}^{-1}$, i.e., if $B \subset \mathscr{M}_S(\Omega)$, then

$$\Gamma_{t,x}(B)=P_{0,x}\{\omega \in \Omega\colon R_{t,\omega} \in B\}. \tag{1.2}$$

If $P_{0,x}$ is ergodic with invariant measure $\nu(dx)$ on X, and if $\bar{Q} \in \mathcal{M}_S(\Omega)$ is the stationary Markov process with ν as its marginal distribution, then, by the ergodic theorem, as $t \to \infty$,

$$\Gamma_{t,x} \Rightarrow \delta_{\bar{Q}} \quad \text{for all} \quad x. \tag{1.3}$$

Here, $\delta_{\bar{Q}}$ is the Dirac measure on $\mathcal{M}_S(\Omega)$ concentrated at $\bar{Q}$. What this paper is concerned with are the probabilities of large deviations for the measure $\Gamma_{t,x}$. If $G \subset \mathcal{M}_S(\Omega)$ is open and contains $\bar{Q}$, then, as just noted, $\lim_{t\to\infty} \Gamma_{t,x}(G) = 1$. On the other hand, if A is a set whose closure does not contain $\bar{Q}$, $\Gamma_{t,x}(A) \to 0$ as $t \to \infty$ and the question we are concerned with here is the rate at which it goes to 0.

In Sections 2 and 3 of this paper we define, for any $Q \in \mathcal{M}_S(\Omega)$, the entropy, $H(Q)$, of the stationary process Q with respect to the Markov process $P_{0,x}$. We then show that this entropy function governs the rate at which $\Gamma_{t,x}(G) \to 0$ as described above. In particular, we show, under suitable hypotheses to be detailed below, that for open sets $G \subset \mathcal{M}_S(\Omega)$,

$$\varliminf_{t\to\infty} \frac{1}{t} \log \Gamma_{t,x}(G) \geqq -\inf_{Q\in G} H(Q). \tag{1.4}$$

We also show under suitable hypotheses to be detailed below that, for closed sets $C \subset \mathcal{M}_S(\Omega)$,

$$\varlimsup_{t\to\infty} \frac{1}{t} \log \Gamma_{t,x}(C) \leqq -\inf_{Q\in C} H(Q). \tag{1.5}$$

Now, $H(Q) \geqq 0$ and it turns out that $H(\bar{Q}) = 0$ so that (1.4) and (1.5) are consistent with (1.3).

If $F(Q)$ is a continuous real-valued function on $\mathcal{M}_S(\Omega)$, then (1.4) and (1.5) imply (see [3])

$$\lim_{t\to\infty} \frac{1}{t} \log E^{\Gamma_{t,x}}\{e^{tF(Q)}\} = \sup_{Q\in\mathcal{M}_S(\Omega)} [F(Q) - H(Q)]. \tag{1.6}$$

An equivalent form of (1.6), which is responsible for the title of this paper, is

$$\lim_{t\to\infty} \frac{1}{t} \log E^{P_{0,x}}\{\exp\{tF(R_{t,\omega})\}\} = \sup_{Q\in\mathcal{M}_S(\Omega)} [F(Q) - H(Q)]. \tag{1.7}$$

Before giving examples of (1.7) we want to make two remarks which will relate (1.7) to our earlier results.

First of all, our earlier asymptotic results involved functions of the occupation distribution: for $\omega \in \Omega$, $t > 0$, and $A \subset X$,

$$L_{t,\omega}(A) = \frac{1}{t} \int_0^t \chi_A(\omega(s))\, ds,$$

i.e., $L_{t,\omega}(A)$ is the proportion of time up to t that the particular path $\omega(\cdot)$ occupies the set $A \subset X$. For fixed $t > 0$, and $\omega \in \Omega$, $L_{t,\omega}(\cdot)$ is then a probability measure on X. We denote the space of probability measures on X by $\mathcal{M}(X)$. Now, the relation between $L_{t,\omega} \in \mathcal{M}(X)$ and $R_{t,\omega} \in \mathcal{M}_S(\Omega)$ is made clear by observing that the marginal distribution of $R_{t,\omega}$ is precisely $L_{t,\omega}$.

Furthermore, in our earlier papers the asymptotic rate involving functionals of $L_{t,\omega}$ was governed by the I-function (for the Markov process $P_{0,x}$) which was defined on $\mathcal{M}(X)$ (see, e.g., [2]). To see the relation between that I-function and the entropy function $H(Q)$ of this paper, we prove in Section 6 that

$$\inf_{\substack{Q \in \mathcal{M}_S(\Omega) \\ q(Q)=\mu}} H(Q) = I(\mu), \tag{1.8}$$

where, in (1.8), the notation $q(Q)=\mu$ means the marginal of the stationary measure Q is μ. We refer to relation (1.8) as the contraction principle.

We now mention a few examples to illustrate (1.7) and the remarks just made.

EXAMPLE 1. Let $V(\cdot)$ be a continuous real-valued function on X and define

$$F(Q) = E^Q\{V(\omega(0))\}.$$

Then,

$$F(R_{t,\omega}) = \int_\Omega V(\omega'(0)) R_{t,\omega}(d\omega') = \frac{1}{t}\int_0^t V(\omega(s))\, ds,$$

and (1.7) becomes, in this case,

$$\lim_{t\to\infty} \frac{1}{t} \log E^{P_{0,x}}\left\{\exp\left\{\int_0^t V(\omega(s))\, ds\right\}\right\} = \sup_{Q\in\mathcal{M}_S(\Omega)} [E^Q\{V(\omega(0))\} - H(Q)]. \tag{1.9}$$

Using (1.8), this becomes

$$\lim_{t\to\infty} \frac{1}{t} \log E^{P_{0,x}}\left\{\exp\left\{\int_0^t V(\omega)(s))\, ds\right\}\right\} = \sup_{\mu\in\mathcal{M}(X)} \left[\int V(x)\mu(dx) - I(\mu)\right], \tag{1.10}$$

where, of course, $I(\mu)$ in (1.10) is the I-function associated with the Markov process $P_{0,x}$. Formula (1.10) agrees with earlier results of the authors.

EXAMPLE 2. Let $F(Q) = E^Q\{V(\omega(0), \omega(1))\}$, where $V: X\times X \to R$ is bounded and continuous. Now,

$$F(R_{t,\omega}) = \frac{1}{t}\int_0^{t-1} V(\omega(s), \omega(s+1))\, ds + O\left(\frac{1}{t}\right) \quad \text{as} \quad t\to\infty,$$

and from this and (1.7) it follows that

$$\begin{aligned} &\log_{t\to\infty} \frac{1}{t} \log E^{P_{0,x}}\left\{\exp\left(\int_0^t V(\omega(s), \omega(s+1))\, ds\right)\right\} \\ &\quad = \sup_{Q\in\mathcal{M}_S(\Omega)} [E^Q\{V(\omega(0), \omega(1))\} - H(Q)]. \end{aligned} \tag{1.11}$$

EXAMPLE 3. Let P_t and $E^{P_t}\{\ \}$ denote, respectively, the probability measure and expectation with respect to three-dimensional Brownian motion $\omega(\cdot)$ tied down at both ends, i.e., $\omega(0)=\omega(t)=0$. For $\alpha>0$, let

$$G(\alpha,t)=E^{P_t}\left\{\exp\left\{\alpha\int_0^t\int_0^t \frac{e^{-|\sigma-s|}}{|\omega(\sigma)-\omega(s)|}\,d\sigma\,ds\right\}\right\}.$$

A long standing problem in statistical mechanics (cf. [4]), the "polaron problem" has been to show

$$\lim_{t\to\infty}\frac{1}{t}\log G(\alpha,t)=g(\alpha)\text{ exists,}\tag{1.12}$$

and, moreover, according to a conjecture of Pekar [5],

$$\lim_{\alpha\to\infty}\frac{g(\alpha)}{\alpha^2}=g_0\text{ exists,}\tag{1.13}$$

with

$$g_0=\sup_{\substack{\phi\in L_2(R_3)\\ \|\phi\|=1}}\left[2\int\int\frac{\phi^2(x)\phi^2(y)}{|x-y|}\,dx\,dy-\tfrac{1}{2}\int|\nabla\phi|^2\,dx\right].\tag{1.14}$$

Using the results of the present paper, the authors succeeded in proving (1.12) obtaining as an expression for $g(\alpha)$ a variational formula over all stationary processes as in (1.7). This infinite-dimensional variational formula turns out to be just explicit enough to allow us to also prove Pekar's conjecture (1.13) obtaining indeed the expression (1.14) for g_0. The details of this example will be found in a forthcoming paper [6].

Now, we wish to make precise the hypotheses imposed in order to prove (1.4) and (1.5). For the upper estimate (Theorem 4.6 below) we first comment that if the space X happens to be compact, then no further hypotheses on the Markov process $P_{0,x}$ need to be imposed in order to obtain (1.5), but if X is not compact we need to impose the following hypothesis on $P_{0,x}$.

There exists a sequence $\{u_n(x)\}$ of functions in $\mathscr{D}(L)$ (the domain of the infinitesimal generator L of the Markov process $P_{0,x}$) with the five properties:

(i) $u_n(x)\geqq c>0$ for all x and n.

(ii) There exists, for every compact set $K\subset X$, a constant C_K such that $\sup_{x\in K}\sup_n u_n(x)\leqq C_K$.

(iii) $-(Lu_n/u_n)(x)\equiv V_n(x)\geqq -C$ for all n and all x.

(iv) There exists a function $V(x)$ such that, for all $x\in X$, $\lim_{n\to\infty}V_n(x)=V(x)$.

(v) $\{x\colon V(x)\leqq l\}$ is compact in X for every $l<\infty$.

In Section 7 of this paper we show that the Ornstein–Uhlenbeck process in d-dimensions, for example, satisfies these 5 properties. In the polaron problem we use the upper estimate (1.5) for precisely this case of the Ornstein–Uhlenbeck process.

For the lower estimate (1.4) (see Theorem 5.5 below), we impose two restrictions on $P_{0,x}$. Let $p(t, x, dy)$ be the transition probability associated with $P_{0,x}$. We shall assume that there exists a density function for $p(1, x, dy)$ with respect to a reference measure α on X, i.e., for all $x \in X$ there exists a function $p(1, x, y) > 0$ for almost all y (α measure) such that for all $x \in X$:

I: $p(1, x, dy) = p(1, x, y)\alpha(dy)$,

and

II: $p(1, x, \cdot)$ as a mapping from $X \to L_1(\alpha)$ is continuous.

In the polaron problem, for example, we need the lower estimate (1.4) for the case of ordinary three-dimensional Brownian motion and this satisfies I and II.

As for the asymptotic formula (1.6) or equivalently (1.7) we see that this holds for any Markov process $P_{0,x}$ satisfying both (i)–(v) and I–II above.

2. Some Properties of Entropy

Let (X, Σ) be a measurable space and let λ and μ be the probability measures on (X, Σ). Let $\mathscr{B}(\Sigma)$ be the space of bounded measurable functions on (X, Σ). We define the entropy of μ with respect to λ by

$$h(\lambda : \mu) = \inf \left\{ c : \int \Phi(x)\mu(dx) \leq c + \log \int e^{\Phi(x)}\lambda(dx) \quad \text{for all} \quad \Phi \in \mathscr{B}(\Sigma) \right\}. \tag{2.1}$$

If X is a Polish space and Σ is the Borel σ-field, then replacing $\mathscr{B}(\Sigma)$ by $C(X)$ in (2.1) gives the same infimum (see [2-III]).

From the definition (2.1) we see that, for fixed λ, $h(\lambda ; \mu)$ is a non-negative, convex function of μ and $0 \leqq h(\lambda ; \mu) \leqq \infty$. Moreover, if X is a Polish space, then $h(\lambda ; \mu)$ is lower semicontinuous in μ in the weak topology.

The following basic theorem was proved in our earlier work (see [2-III]).

THEOREM 2.1. *For any (X, Σ) and any two probability measures λ, μ on (X, Σ), $h(\lambda ; \mu)$ is finite if and only if*

(a) $\mu \ll \lambda$,

and

(b) *$d\mu/d\lambda = f(x)$ is such that $f(x) \log f(x) \in L_1(\lambda)$.*

Under these circumstances,

$$h(\lambda ; \mu) = \int \log f(x)\mu(dx) = \int f(x) \log f(x)\lambda(dx).$$

Remark. Suppose $\mu \ll \lambda$ and $d\mu/d\lambda = f(x)$. Since $f(x)\log f(x)$ is bounded for small $f(x)$, there is no difficulty when $\log f(x)$ is negative, i.e., $f(x)(\log(f(x))^-$ is always integrable with respect to λ and thus $f(x)\log f(x) \in L_1(\lambda)$ if and only if $f(x)(\log f(x))^+ \in L_1(\lambda)$.

We shall also use the following lemma (proved in [2-III]) which we state here for convenience.

LEMMA 2.2. *Let $h(\lambda:\mu) \leqq c < \infty$ and let $A \subset X$ be any set with $\lambda(A) = \varepsilon > 0$. Then, $\mu(A) \leqq \phi(c, \varepsilon)$, where $\phi(c, \varepsilon)$ is a universal function tending to 0 as $\varepsilon \to 0$ for every $c < \infty$.*

If $\mathscr{F} \subset \Sigma$ is a sub σ-field in the measurable space (X, Σ), and if we take $\Phi \in \mathscr{B}(\mathscr{F})$ instead of $\Phi \in \mathscr{B}(\Sigma)$ in definition (2.1), then we shall denote the corresponding entropy by $h_{\mathscr{F}}(\lambda;\mu)$. This is, of course, the same as restricting λ and μ to $(X, \mathscr{F})$.

LEMMA 2.3. *Let (X, Σ) be a Polish space and let $\mathscr{F}_1 \subset \mathscr{F}_2 \subset \Sigma$ be sub σ-fields. Denote by λ_ω, μ_ω the regular conditional probability distributions (r.c.p.d.) of λ and μ given $\mathscr{F}_1$. Then,*

$$h_{\mathscr{F}_2}(\lambda;\mu) = h_{\mathscr{F}_1}(\lambda;\mu) + E^{\mu}\{h_{\mathscr{F}_2}(\lambda_\omega;\mu_\omega)\}. \tag{2.2}$$

Proof: If $h_{\mathscr{F}_1}(\lambda;\mu) = \infty$, then since $h_{\mathscr{F}_2}(\lambda;\mu) \geqq h_{\mathscr{F}_1}(\lambda;\mu)$ we have $h_{\mathscr{F}_2}(\lambda;\mu) = \infty$ also and there is nothing to prove. Thus, we assume $h_{\mathscr{F}_1}(\lambda;\mu) < \infty$ and thus, from Theorem 2.1, $\mu \ll \lambda$ on $\mathscr{F}_1$. This means λ_ω is defined almost everywhere with respect to μ measure and therefore the expectation on the right of (2.2) is well defined. Let $\Phi \in \mathscr{B}(\mathscr{F}_2)$ so that from the definition of $h_{\mathscr{F}_2}(\lambda_\omega;\mu_\omega)$ we have

$$\begin{aligned} E^{\mu}\{\Phi\} &= E^{\mu}\{E^{\mu_\omega}\{\Phi\}\}, \\ &\leqq E^{\mu}\{h_{\mathscr{F}_2}(\lambda_\omega;\mu_\omega) + \log E^{\lambda_\omega}\{e^{\Phi}\}\}. \end{aligned} \tag{2.3}$$

If we let $\psi(\omega) = \log E^{\lambda_\omega}\{e^{\Phi}\}$, then $\psi(\omega) \in \mathscr{B}(\mathscr{F}_1)$ and so from the definition of $h_{\mathscr{F}_1}(\lambda;\mu)$ and (2.3) we get

$$\begin{aligned} E^{\mu}\{\Phi\} &\leqq E^{\mu}\{h_{\mathscr{F}_2}(\lambda_\omega;\mu_\omega)\} + E^{\mu}\{\psi(\omega)\} \\ &\leqq E^{\mu}\{h_{\mathscr{F}_2}(\lambda_\omega;\mu_\omega)\} + h_{\mathscr{F}_1}(\lambda;\mu) + \log E^{\lambda}\{e^{\psi(\omega)}\} \\ &= E^{\mu}\{h_{\mathscr{F}_2}(\lambda_\omega;\mu_\omega)\} + h_{\mathscr{F}_1}(\lambda;\mu) + \log E^{\lambda}\{e^{\Phi}\}. \end{aligned} \tag{2.4}$$

Using again the definition of $h_{\mathscr{F}_2}(\lambda;\mu)$, we conclude from (2.4) that

$$h_{\mathscr{F}_2}(\lambda;\mu) \leqq E^{\mu}\{h_{\mathscr{F}_2}(\lambda_\omega;\mu_\omega)\} + h_{\mathscr{F}_1}(\lambda;\mu). \tag{2.5}$$

Now we must show this inequality in the reverse direction. The inequality desired is trivial if $h_{\mathscr{F}_2}(\lambda;\mu) = \infty$; thus assume $h_{\mathscr{F}_2}(\lambda;\mu) < \infty$ which means by Theorem

2.1 that $\mu \ll \lambda$ on $\mathcal{F}_2$. Let $g(\omega) = d\mu/d\lambda|_{\mathcal{F}_2}$. As for the regular conditional probability distributions, we see that $\mu_\omega \ll \lambda_\omega$ on $\mathcal{F}_2$ for almost all ω (λ measure). Moreover,

$$g(\omega) = \frac{d\mu}{d\lambda}\bigg|_{\mathcal{F}_1}(\omega)\,\frac{d\mu_\omega}{d\lambda_\omega}\bigg|_{\mathcal{F}_2}(\omega),$$

and hence, from Theorem 2.1,

$$\begin{aligned} E^\mu\{\log g(\omega)\} &= E^\mu\left\{\log \frac{d\mu}{d\lambda}\bigg|_{\mathcal{F}_1}\right\} + E^\mu\left\{\log \frac{d\mu_\omega}{d\lambda_\omega}\bigg|_{\mathcal{F}_2}\right\}, \\ &= E^\mu\left\{\log \frac{d\mu}{d\lambda}\bigg|_{\mathcal{F}_1}\right\} + E^\mu\left\{E^{\mu_\omega}\left\{\log \frac{d\mu_\omega}{d\lambda_\omega}\bigg|_{\mathcal{F}_2}\right\}\right\} \\ &= E^\mu\left\{\log \frac{d\mu}{d\lambda}\bigg|_{\mathcal{F}_1}\right\} + E^\mu\{h_{\mathcal{F}_2}(\lambda_\omega; \mu_\omega)\} \\ &= h_{\mathcal{F}_1}(\lambda; \mu) + E^\mu\{h_{\mathcal{F}_2}(\lambda_\omega; \mu_\omega)\}. \end{aligned} \tag{2.6}$$

Again using Theorem 2.1 we get from (2.6)

$$h_{\mathcal{F}_2}(\lambda; \mu) = h_{\mathcal{F}_1}(\lambda, \mu) + E^\mu\{h_{\mathcal{F}_2}(\lambda_\omega; \mu_\omega)\}.$$

3. Entropy of Processes

Let Ω be a space of functions $\omega(\cdot)$ on $-\infty < t < \infty$ with values in a Polish space X. We shall assume that either Ω consists of continuous functions with uniform convergence on bounded intervals or it consists of functions with discontinuities only of the first kind and with convergence induced by the Skorohod topology on bounded intervals. In either case, Ω itself is a Polish space. For each $\omega \in \Omega$, we denote by $\omega(t)$ the value of the function $\omega(\cdot)$ at time t. We denote by Ω_t^+ the corresponding space of functions on $[t, \infty)$ with values in X and we denote by $\mathcal{F}_t^s$ the σ-fields in Ω generated by $\omega(\sigma)$ for $s \leqq \sigma \leqq t$. The space Ω_t^+ can, of course, be naturally identified with the space Ω and the σ-field $\mathcal{F}_\infty^t$. We shall use the notation θ_t for the translation map on Ω, i.e., $(\theta_t\omega)(s) = \omega(t+s)$.

Let Q be a stationary process on Ω and let $\{P_{t,x}\}$ be a homogeneous Markov family of measures on Ω_t^+. We define $P_{t,\omega(t)}$ to be $P_{t,x}$ with the starting point $x = \omega(t)$. We denote by $\{Q_{t,\omega}\}$ the regular conditional probability distributions of Q given $\mathcal{F}_t^{-\infty}$. We now define the entropy of the stationary process Q with respect to the Markov process $P_{0,x}$ at time $t > 0$ by

$$H(t, Q) = E^Q\{h_{\mathcal{F}_t^0}(P_{0,\omega(0)}; Q_{0,\omega})\}. \tag{3.1}$$

Here, $h_{\mathcal{F}_t^0}(P_{0,\omega(0)}; Q_{0,\omega})$ is the entropy introduced in the previous section. It is easy to check that $H(t, Q)$ is well defined for all $t > 0$ with $0 \leqq H(t, Q) \leqq \infty$.

THEOREM 3.1. *Either $H(t, Q)\equiv\infty$ for all $t>0$ or there exists a constant $H(Q)<\infty$ such that*

$$H(t, Q)=tH(Q) \quad \text{for all} \quad t>0. \tag{3.2}$$

Proof: Pick $0<t_1<t_2$ and consider $h_{\mathscr{F}^0_{t_2}}(P_{0,\omega(0)}; Q_{0,\omega})$. From Lemma 2.3,

$$\begin{aligned} h_{\mathscr{F}^0_{t_2}}(P_{0,\omega(0)}; Q_{0,\omega}) &= h_{\mathscr{F}^0_{t_1}}(P_{0,\omega(0)}; Q_{0,\omega}) \\ &\quad + E^{Q_{0,\omega}}\{h_{\mathscr{F}^0_{t_2}}((P_{0,\omega(0)})_{t,\bar\omega}; (Q_{0,\omega})_{t,\bar\omega})\}, \end{aligned} \tag{3.3}$$

where $(P_{0,\omega(0)})_{t_1,\bar\omega}$ means the r.c.p.d. of $P_{0,\omega(0)}$ with respect to the σ-field $\mathscr{F}^0_{t_1}$ and where $\bar\omega$ is the variable, i.e., $E^{Q_{0,\omega}}\{\ \}$ indicates integration with respect to $\bar\omega$ under $Q_{0,\omega}$ measure. Of course, $(Q_{0,\omega})_{t_1,\bar\omega}$ is the analogous r.c.p.d.

It will be convenient to express (3.3) in an alternative notation. Let $\bar\omega$ be a trajectory in Ω, $t\in(-\infty,\infty)$, and P be a measure on $\mathscr{F}^s_\infty$ with $s\leqq t$. Suppose $P\{\omega: \omega(t)=\bar\omega(t)\}=1$. Then, we define $\delta_{\bar\omega}\otimes_t P$ to be a new measure on Ω such that

$$\begin{aligned} &(\delta_{\bar\omega}\otimes_t P)\{\omega(t_1)\in A_1, \omega(t_2)\in A_2, \cdots, \omega(t_n)\in A_n\} \\ &\quad = \chi_{A_1}(\bar\omega(t_1))\cdots\chi_{A_k}(\bar\omega(t_k))P\{\omega(t_{k+1})\in A_{k+1}, \cdots, \omega(t_n)\in A_n\}, \end{aligned}$$

where $t_1<t_2<\cdots<t_k\leqq t\leqq t_{k+1}<t_{k+2}<\cdots<t_n$. Now, $Q_{0,\omega}\{\bar\omega: \bar\omega(0)=\omega(0)\}=1$, and $(P_{0,\omega(0)})_{t_1,\bar\omega}=\delta_{\bar\omega}\otimes_{t_1}P_{t_1,\bar\omega}(t_1)$ on $\mathscr{F}^0_\infty$. Moreover, $(Q_{0,\omega})_{t_1,\bar\omega}=Q_{t_1,\bar\omega}=\delta_{\bar\omega}\otimes_{t_1}Q_{t_1,\bar\omega}$ on $\mathscr{F}^0_\infty$. Thus, we can rewrite (3.3) as

$$\begin{aligned} h_{\mathscr{F}^0_{t_2}}(Q_{0,\omega(0)}; Q_{0,\omega}) &= h_{\mathscr{F}^0_{t_1}}(P_{0,\omega(0)}; Q_{0,\omega}) \\ &\quad + E^{Q_{0,\omega}}\{h_{\mathscr{F}^0_{t_2}}(\delta_{\bar\omega}\otimes_{t_1}P_{t_1,\bar\omega(t_1)}; \delta_{\bar\omega}\otimes_{t_1}Q_{t_1,\bar\omega})\}. \end{aligned} \tag{3.4}$$

We now take expectations on both sides of (3.4) with respect to Q measure (ω is the variable of integration), and noting that

$$E^Q\{E^{Q_{0,\omega}}\{h_{\mathscr{F}^{t_1}_{t_2}}(P_{t_1,\bar\omega(t_1)}; Q_{t_1,\bar\omega})\}=E^Q\{h_{\mathscr{F}^{t_1}_{t_2}}(P_{t_1,\omega(t_1)}; Q_{t_1,\omega})\},$$

we obtain

$$\begin{aligned} E^Q\{h_{\mathscr{F}^0_{t_2}}(P_{0,\omega(0)}; Q_{0,\omega})\} &= E^Q\{h_{\mathscr{F}^0_{t_1}}(P_{0,\omega(0)}; Q_{0,\omega})\} \\ &\quad + E^Q\{h_{\mathscr{F}^{t_1}_{t_2}}(P_{t_1,\omega(t_1)}; Q_{t_1,\omega})\}. \end{aligned} \tag{3.5}$$

By definition (3.1), the left side of (3.5) is $H(t_2, Q)$ and the first term on the right of (3.5) is $H(t_1, Q)$. By the stationarity of the Q process and the homogeneity of the Markov process we see that the second term on the right of (3.5) is $H(t_2-t_1, Q)$. Thus, $H(t, Q)$ is additive in t and from its definition it is clearly monotone in t. Hence, $H(t, Q)$ is linear in t unless it is identically infinite.

We shall call $H(Q)$ the entropy of the stationary process Q with respect to the Markov process $P_{0,x}$.

THEOREM 3.2. *Let $P_\omega=\delta_\omega\otimes_0 P_{0,\omega(0)}$. For all $t>0$,*

$$H(t, Q)=\sup_{\Phi\in\mathscr{B}(\mathscr{F}_t^{-\infty})}[E^Q\{\Phi\}-E^Q\{\log E^{P_\omega}\{e^\Phi\}\}]. \tag{3.6}$$

Proof: Let $\Phi \in \mathscr{B}(\mathscr{F}_t^{-\infty})$. Using (2.1), (3.1), and the fact that $Q_{0,\omega} = \delta_\omega \otimes_0 Q_{0,\omega}$, we obtain

$$\begin{aligned} E^Q\{\Phi\} &= E^Q\{E^{Q_{0,\omega}}\{\Phi\}\} \\ &\leqq E^Q\{h_{\mathscr{F}_t^{-\infty}}(P_\omega; Q_{0,\omega})\} + E^Q\{\log E^{P_\omega}\{e^\Phi\}\} \\ &= E^Q\{h_{\mathscr{F}_t^0}(P_{0,\omega(0)}; Q_{0,\omega})\} + E^Q\{\log E^{P_\omega}\{e^\Phi\}\} \\ &= H(t, Q) + E^Q\{\log E^{P_\omega}\{e^\Phi\}\}. \end{aligned}$$

Thus,

$$\sup_{\Phi \in \mathscr{B}(\mathscr{F}_t^{-\infty})} [E^Q\{\Phi\} - E^Q\{\log E^{P_\omega}\{e^\Phi\}\}] \leqq H(t, Q). \tag{3.7}$$

To show the inequality the other way, let

$$\sup_{\Phi \in \mathscr{B}(\mathscr{F}_t^{-\infty})} [E^Q\{\Phi\} - E^Q\{\log E^{P_\omega}\{e^\Phi\}\}] = l.$$

Then, by Jensen's inequality,

$$\sup_{\Phi \in \mathscr{B}(\mathscr{F}_t^{-\infty})} [E^Q\{\Phi\} - \log E^Q\{E^{P_\omega}\{e^\Phi\}\}] \leqq l. \tag{3.8}$$

Let $\bar{Q}$ measure be defined by $\int P_\omega Q(d\omega)$. Then, from (3.8) we have, for every $\Phi \in \mathscr{B}(\mathscr{F}_t^{-\infty})$,

$$E^Q\{\Phi\} - \log E^{\bar{Q}}\{e^\Phi\} \leqq l,$$

which implies from (2.1) that

$$h_{\mathscr{F}_t^{-\infty}}(\bar{Q}; Q) \leqq l.$$

Using Lemma 2.3, we then get

$$h_{\mathscr{F}_t^{-\infty}}(\bar{Q}; Q) = h_{\mathscr{F}_0^{-\infty}}(\bar{Q}; Q) + E^Q\{h_{\mathscr{F}_t^{-\infty}}(\bar{Q}_\omega; Q_{0,\omega})\} \leqq l. \tag{3.9}$$

But, by the definition of P_ω, $\bar{Q} = Q$ on $\mathscr{F}_0^{-\infty}$, and therefore $h_{\mathscr{F}_0^{-\infty}}(\bar{Q}; Q) = 0$. Moreover, $\bar{Q}_\omega = P_\omega = \delta_\omega \otimes_0 P_{0,\omega(0)}$, and $Q_{0,\omega} = \delta_\omega \otimes_0 Q_{0,\omega}$ so that

$$E^Q\{h_{\mathscr{F}_t^0}(P_{0,\omega(0)}; Q_{0,\omega})\} = E^Q\{h_{\mathscr{F}_t^{-\infty}}(\bar{Q}_\omega; Q_{0,\omega})\}.$$

Hence, from (3.9) we obtain

$$H(t, Q) = E^Q\{h_{\mathscr{F}_t^0}(P_{0,\omega(0)}; Q_{0,\omega})\} \leqq l$$

which completes the proof.

THEOREM 3.3. *Under the hypothesis that the mapping $x \to P_{t,x}$ is weakly continuous, $H(Q)$ is lower semicontinuous and convex in Q.*

Proof: From Theorems 3.1 and 3.2,

$$H(Q) = H(1, Q) = \sup_{\Phi \in \mathscr{B}(\mathscr{F}_1^{-\infty})} [E^Q\{\Phi\} - E^Q\{\log E^{P_\omega}\{e^\Phi\}\}]. \tag{3.10}$$

The supremum on the right of (3.10) is not changed if we restrict $\Phi \in \mathcal{B}(\mathcal{F}_1^{-\infty}) \cap C(\Omega)$. Thus, to show the lower semicontinuity of $H(Q)$, it suffices to show that, for each $\Phi \in \mathcal{B}(\mathcal{F}_1^{-\infty}) \cap C(\Omega)$, $E^Q\{\Phi\} - E^Q\{\log E^{P_\omega}\{e^\Phi\}\}$ is continuous as a functional of Q. Clearly, $E^Q\{\Phi\}$ is a continuous functional of Q. Thus let $\psi(\omega) = E^{P_\omega}\{e^\Phi\}$ and it suffices to show that $\psi(\omega)$ is continuous almost everywhere with respect to Q measure for an arbitrary stationary Q. To show this it suffices to show that if $\omega_n \to \omega$ in Ω, then $P_{\omega_n} \Rightarrow P_\omega$. Assume, then that $\omega_n \to \omega$ and that ω has no jump at the origin, i.e., $\omega(0^+) = \omega(0^-)$. Then, $\omega_n(0) \to \omega(0)$ and hence $P_{\omega_n} \Rightarrow P_\omega$. It remains to observe that, for all stationary processes Q, $Q\{\omega : \omega$ has a jump at $0\} = 0$.

The convexity of $H(Q)$ follows from (3.10) when we note that $H(Q)$ is the supremum of a collection of linear functionals of Q.

Next (Theorem 3.5 below) we prove the surprising fact that $H(Q)$ is linear in Q, but first we need a lemma.

LEMMA 3.4. *Let $\mathcal{M}_\Omega$ be the space of probability measures on Ω. There exists a conditional probability $R_\omega : \Omega \to \mathcal{M}_\Omega$ such that $R'_\omega = \delta_\omega \otimes_0 R_\omega$ and, for all stationary measures Q on Ω, $Q_{0,\omega} = R_\omega$ almost everywhere with respect to Q measure.*

Proof: As before, let $\mathcal{M}_S(\Omega)$ be the space of stationary measures on Ω and let $\mathcal{M}_E(\Omega) \subset \mathcal{M}_S(\Omega)$ be the subset of $\mathcal{M}_S(\Omega)$ consisting of the ergodic measures. $\mathcal{M}_E(\Omega)$ is the set of extremals of $\mathcal{M}_S(\Omega)$. It follows from an argument of Oxtoby [7] that there exists a subset $\Omega_0 \subset \Omega$ which is $\mathcal{F}_0^{-\infty}$ measurable and a $\mathcal{F}_0^{-\infty}$ measurable map $\pi_\omega : \Omega_0 \to \mathcal{M}_E(\Omega)$ with the following properties: $Q(\Omega_0) = 1$ for all $Q \in \mathcal{M}_S(\Omega)$ and $Q\{\omega : \pi_\omega = Q\} = 1$ for all $Q \in \mathcal{M}_E(\Omega)$.

Now, denote by $R(Q, \omega)$ the r.c.p.d. of Q given $\mathcal{F}_0^{-\infty}$. It is easily seen that this r.c.p.d. can always be selected so that it is jointly measurable in Q and ω. We define the desired R_ω by $R_\omega = R(\pi_\omega, \omega)$. Then, for all $Q \in \mathcal{M}_E(\Omega)$,

$$\int R_\omega Q(d\omega) = \int R(Q, \omega) Q(d\omega) = Q.$$

But, $R_\omega = R(\pi_\omega, \omega)$ is independent of Q and so we have a linear relation true for extremals and hence true for all $Q \in \mathcal{M}_S(\Omega)$.

THEOREM 3.5. *$H(Q)$ is linear in Q.*

Proof: From Theorem 3.1 and the preceding lemma,

$$\begin{aligned} H(Q) = H(1, Q) &= E^Q\{h_{\mathcal{F}_1^0}(P_{0,\omega(0)}; Q_{0,\omega})\} \\ &= E^Q\{h_{\mathcal{F}_1^0}(P_{0,\omega(0)}; R_\omega)\}, \end{aligned} \tag{3.11}$$

where R_ω, as defined in the preceding lemma, is independent of Q. Thus, the integrand on the right of (3.11) is independent of Q and so $H(Q)$ is linear in Q.

In Theorem 3.2 we proved that, for $P_\omega=\delta_\omega\otimes_0 P_{0,\omega(0)}$ and for all $t>0$,

$$H(t,Q)=\sup_{\Phi\in\mathscr{B}(\mathscr{F}_t^{-\infty})}[E^Q\{\Phi\}-E^Q\{\log E^{P_\omega}\{e^\Phi\}\}].$$

Now define

$$\begin{aligned}\bar H(t,Q)&=\sup_{\Phi\in\mathscr{B}(\mathscr{F}_t^0)}[E^Q\{\Phi\}-E^Q\{\log E^{P_{0,\omega(0)}}\{e^\Phi\}\}]\\ &=\sup_{\Phi\in\mathscr{B}(\mathscr{F}_t^0)\cap C(\Omega)}[E^Q\{\Phi\}-E^Q\{\log E^{P_{0,\omega(0)}}\{e^\Phi\}\}].\end{aligned}\tag{3.12}$$

Clearly, $\bar H(t,Q)\leqq H(t,Q)=tH(Q)$.

THEOREM 3.6.

$$\lim_{T\to\infty}\frac{\bar H(t,Q)}{t}=H(Q).\tag{3.13}$$

Proof: Define $\bar P$ measure on $\mathscr{F}_\infty^0$ by $\bar P=\int P_{0,\omega(0)}Q(d\omega)$. By Jensen's inequality and (2.1),

$$\begin{aligned}\bar H(t,Q)&\geqq\sup_{\Phi\in\mathscr{B}(\mathscr{F}_t^0)}[E^Q\{\Phi\}-\log E^Q\{E^{P_{0,\omega(0)}}\{e^\Phi\}\}]\\ &=\sup_{\Phi\in\mathscr{B}(\mathscr{F}_t^0)}[E^Q\{\Phi\}-\log E^{\bar P}\{e^\Phi\}]\\ &=h_{\mathscr{F}_t^0}(\bar P;Q).\end{aligned}$$

Hence, to show (3.13), it suffices to show

$$\varliminf_{t\to\infty}\frac{1}{t}h_{\mathscr{F}_t^0}(\bar P;Q)\geqq H(Q).\tag{3.14}$$

From Lemma 2.3,

$$h_{\mathscr{F}_{t+1}^0}(\bar P;Q)-h_{\mathscr{F}_t^0}(\bar P;Q)=E^Q\{h_{\mathscr{F}_{t+1}^0}(\bar P_{t,\omega}^0;Q_{t,\omega}^0)\},\tag{3.15}$$

where $\bar P_{t,\omega}^0$ and $Q_{t,\omega}^0$ are the r.c.p.d. of $\bar P$ and Q, respectively, given $\mathscr{F}_t^0$. But,

$$\begin{aligned}E^Q\{h_{\mathscr{F}_{t+1}^0}(\bar P_{t,\omega}^0;Q_{t,\omega}^0)\}&=E^Q\{h_{\mathscr{F}_{t+1}^0}(\delta_\omega\otimes_t\bar P_{t,\omega}^0;\delta_\omega\otimes_t Q_{t,\omega}^0)\}\\ &=E^Q\{h_{\mathscr{F}_{t+1}^t}(\bar P_{t,\omega}^0;Q_{t,\omega}^0)\}\\ &=E^Q\{h_{\mathscr{F}_{t+1}^t}(P_{t,\omega(t)};Q_{t,\omega}^0)\}\\ &=E^Q\{h_{\mathscr{F}_1^0}(P_{0,\omega(0)};Q_{0,\omega}^{-t})\},\end{aligned}\tag{3.16}$$

using the homogeneity of the Markov family and the stationarity of the Q process. By the martingale convergence theorem, $Q_{0,\omega}^{-t}\Rightarrow Q_{0,\omega}$ as $t\to\infty$ almost everywhere with respect to Q measure. As noted earlier, $h(\lambda;\mu)$ is lower-semicontinuous in μ, and hence

$$\varliminf_{t\to\infty}h_{\mathscr{F}_1^0}(P_{0,\omega(0)};Q_{0,\omega}^{-t})\geqq h_{\mathscr{F}_1^0}(P_{0,\omega(0)};Q_{0,\omega}).\tag{3.17}$$

Thus, by Fatou's lemma, (3.15), (3.16), and (3.17),

$$\text{(3.18)} \qquad \varliminf_{t\to\infty} [h_{\mathscr{F}^0_{t+1}}(\bar{P}; Q) - h_{\mathscr{F}^0_t}(\bar{P}; Q)] \geqq E^Q\{h_{\mathscr{F}^0_1}(P_{0,\omega(0)}; Q_{0,\omega})\}$$

$$= H(1, Q) = H(Q).$$

Using the usual Caesaro argument, we obtain (3.14) and (3.18). This completes the proof of Theorem 3.6.

Let $C_S(\Omega)$ be the space of functions Φ on Ω which are bounded and measurable on Ω and such that, for all $Q \in \mathcal{M}_S(\Omega)$,

$$Q\{\text{discontinuities of } \Phi\} = 0.$$

For $\Phi \in C_S(\Omega)$, the linear functional of Q, $\int \Phi Q(d\omega)$, is then continuous in Q.

LEMMA 3.7. *Let* $Y_1 = \{\Phi: \Phi \in \mathscr{B}(\mathscr{F}^0_t),\ E^{P_{0,x}}\{e^{\Phi}\} \leqq 1$, *for all* $x\}$ *and let* $Y_2 = \{\Phi: \Phi \in \mathscr{B}(\mathscr{F}^0_t) \cap C_S(\Omega),\ E^{P_{0,x}}\{e^{\Phi}\} \leqq 1$, *for all* $x\}$. *Then, with* $\bar{H}(t, Q)$ *defined by* (3.12),

$$\text{(3.19)} \qquad \bar{H}(t, Q) = \sup_{\Phi \in Y_1} E^Q\{\Phi\} = \sup_{\Phi \in Y_2} E^Q\{\Phi\}.$$

Proof: From (3.12),

$$\bar{H}(t, Q) = \sup_{\Phi \in \mathscr{B}(\mathscr{F}^0_t)} [E^Q\{\Phi\} - E^Q\{\log E^{P_{0,\omega(0)}}\{e^{\Phi}\}\}]$$

$$\geqq \sup_{\Phi \in Y_1} E^Q\{\Phi\} \geqq \sup_{\Phi \in Y_2} E^Q\{\Phi\}.$$

Conversely, suppose $\sup_{\Phi \in Y_2} E^Q\{\Phi\} = l$. Let $\psi \in \mathscr{B}(\mathscr{F}^0_t) \cap C(\Omega)$ and define $\bar{\psi} = \log E^{P_{0,x}}\{e^{\psi}\}$. Clearly, $\bar{\psi} \in C(R_1)$ and therefore $\bar{\psi}(\omega(0))$ is continuous as a function of ω on the set where ω is continuous at 0. This is a set of ω of Q measure 1 for every $Q \in \mathcal{M}_S(\Omega)$, i.e., $\bar{\psi}(\omega(0)) \in C_S(\Omega)$. Moreover,

$$E^{P_{0,x}}\{e^{\psi - \bar{\psi}(\omega(0))}\} = e^{-\bar{\psi}(x)} E^{P_{0,x}}\{e^{\psi}\}$$

$$= e^{-\bar{\psi}(x)} e^{\bar{\psi}(x)} = 1.$$

Thus, if we let $\Phi = \psi(\omega) - \bar{\psi}(\omega(0))$, then $\Phi \in \mathscr{B}(\mathscr{F}^0_t) \cap C_S(\Omega)$ and, as we just noted, $E^{P_{0,x}}\{e^{\Phi}\} = 1$. Therefore, Φ, as defined, is in Y_2 and so $E^Q\{\Phi\} \leqq l$. Hence,

$$\text{(3.20)} \qquad \begin{aligned} E^Q\{\psi\} &= E^Q\{\Phi\} + E^Q\{\bar{\psi}(\omega(0))\} \\ &\leqq l + E^Q\{\bar{\psi}(\omega(0))\}. \end{aligned}$$

But,

$$\bar{H}(t, Q) = \sup_{\psi \in \mathscr{B}(\mathscr{F}^0_t) \cap C(\Omega)} [E^Q\{\psi\} - E^Q\{\log E^{P_{0,\omega(0)}}\{e^{\psi}\}\}]$$

$$= \sup_{\psi \in \mathscr{B}(\mathscr{F}^0_t) \cap C(\Omega)} [E^Q\{\psi\} - E^Q\{\bar{\psi}(\omega(0))\}].$$

Consequently, (3.20) implies $\bar{H}(t, Q) \leqq l$ and the proof is complete.

4. Upper Bounds

In this section we obtain upper bounds on $\Gamma_{t,x}(A)$ as $t\to\infty$ for appropriate sets A. The main results are Lemma 4.4 and Theorem 4.6. We need some preparatory lemmas.

LEMMA 4.1. *Let $\phi:\Omega\to R_1$ be F_T^0 measurable and such that $E^{P_{0,x}}\{e^{\phi(\omega)}\}\leqq 1$ for all x. Then,*

$$E^{P_{0,x}}\left\{\exp\left\{\frac{1}{T}\int_0^t \phi(\theta_s\omega)\,ds\right\}\right\}\leqq 1 \quad \text{for all} \quad t. \tag{4.1}$$

Proof: Define

$$\psi_s(\omega)=\sum_{\substack{k:k\geqq 0\\ s+kT\leqq t}}\phi(\theta_{s+kT}\omega)$$

so that we can rewrite the left side of (4.1) as $E^{P_{0,x}}\{\exp\{(1/T)\int_0^T\psi_s(\omega)\,ds\}\}$. From Jensen's inequality,

$$E^{P_{0,x}}\left\{\exp\left\{\frac{1}{T}\int_0^T \psi_s(\omega)\,ds\right\}\right\}\leqq E^{P_{0,x}}\left\{\frac{1}{T}\int_0^T \exp\{\psi_s(\omega)\}\,ds\right\}. \tag{4.2}$$

Since $\phi(\theta_{s+kT}\omega)$ is $\mathcal{F}_{s+(k+1)T}^{s+kT}$ measurable, we have from the hypothesis on ϕ and successive conditioning that

$$E^{P_{0,x}}\{\exp\{\psi_s(\omega)\}\}=E^{P_{0,x}}\left\{\exp\left\{\sum_k\phi(\theta_{s+kT}\omega)\right\}\right\}\leqq 1. \tag{4.3}$$

Inequalities (4.3) and (4.2) give us (4.1).

COROLLARY. *Let $\Phi\in\mathcal{B}(\mathcal{F}_T^0)$ and be such that $E^{P_{0,x}}\{e^{\Phi(\omega)}\}\leqq 1$ for all x. Then, for all t,*

$$E^{\Gamma_{t,x}}\left\{\exp\left\{\frac{t}{T}\int_\Omega \Phi(\omega)Q(d\omega)\right\}\right\}\leqq \exp\left\{2\sup_{\omega\in\Omega}|\phi(\omega)|\right\}. \tag{4.4}$$

Proof: By definition of $\Gamma_{t,x}$ measure,

$$E^{\Gamma_{t,x}}\left\{\exp\left\{\frac{t}{T}\int_\Omega \Phi(\omega)Q(d\omega)\right\}\right\}=E^{P_{0,x}}\left\{\exp\left\{\frac{t}{T}\int_\Omega \Phi(\omega')R_{t,\omega}(d\omega')\right\}\right\},$$

and since

$$\left|\int_0^t \Phi(\theta_s,\omega)\,ds-t\int_\Omega \Phi(\omega')R_{t,\omega}(d\omega')\right|\leqq 2T\sup_{\omega\in\Omega}|\Phi(\omega)|, \tag{4.5}$$

we have from Lemma 4.1, for all t,

$$E^{\Gamma_{t,x}}\left\{\exp\left\{\frac{t}{T}\int_\Omega \Phi(\omega)Q(d\omega)\right\}\right\}\leqq \exp\left\{2\sup_{\omega\in\Omega}|\Phi(\omega)|\right\}$$

In what follows we shall use the notation

$$J(A)=\overline{\lim_{t\to\infty}}\frac{1}{t}\log\sup_{x\in X}\Gamma_{t,x}(A)$$

for $A\subset\mathcal{M}_S(\Omega)$.

LEMMA 4.2. *Let E_T be the set of $\Phi\in\mathcal{B}(\mathcal{F}_T^0)\cap C_S(\Omega)$ such that, for all x, $E^{P_{0,x}}\{e^{\Phi(\omega)}\}\leqq 1$. Then, for any set $A\subset\mathcal{M}_S(\Omega)$,*

$$J(A)\leqq-\sup_{\substack{l:A_1,A_2,\cdots,A_l\\ A\subset\bigcup_{j=1}^l A_j}}\inf_{1\leqq j\leqq l}\sup_{T>0}\sup_{\Phi\in E_T}\inf_{Q\in A_j}\frac{1}{T}\int_\Omega\Phi(\omega)Q(d\omega). \tag{4.6}$$

Proof: From (4.4) we see that, for any set $A\subset\mathcal{M}_S(\Omega)$ and any $\Phi\in E_T$,

$$\Gamma_{t,x}(A)\leqq\exp\left\{2\sup_{\omega\in\Omega}|\Phi(\omega)|\right\}\cdot\exp\left\{-\frac{t}{T}\inf_{Q\in A}\int_\Omega\Phi(\omega)Q(d\omega)\right\}$$

Thus,

$$J(A)\leqq-\inf_{Q\in A}\frac{1}{T}\int_\Omega\Phi(\omega)Q(d\omega). \tag{4.7}$$

Since (4.7) holds for all $\Phi\in E_T$ and all $T>0$,

$$J(A)\leqq-\sup_{T>0}\sup_{\Phi\in E_T}\inf_{Q\in A}\frac{1}{T}\int_\Omega\Phi(\omega)Q(d\omega).$$

From the definition of $J(A)$ we see that $J(A\cup B)\leqq\max(J(A),J(B))$. Thus, if $A_1,A_2,\cdots,A_l$ are sets in $\mathcal{M}_S(\Omega)$ with $A\subset\bigcup_{j=1}^l A_j$, then $J(A)\leqq\max_{1\leqq j\leqq l}\{J(A_j)\}$. Finally, taking the supremum over all such finite coverings of A we obtain (4.6).

LEMMA 4.3. *Let A be compact in $\mathcal{M}_S(\Omega)$. For any $\varepsilon>0$, there exists an l and open sets $G_1,G_2,\cdots,G_l$ in $\mathcal{M}_S(\Omega)$ such that $A\subset\bigcup_{j=1}^l G_j$ and*

$$\inf_{1\leqq j\leqq l}\sup_{T>0}\sup_{\Phi\in E_T}\inf_{Q\in G_j}\frac{1}{T}\int_\Omega\Phi(\omega)Q(d\omega)\geqq\inf_{Q\in A}H(Q)-\varepsilon. \tag{4.8}$$

In particular, for any compact set A in $\mathcal{M}_S(\Omega)$,

$$\sup_{\substack{l:A_1,A_2,\cdots,A_l\\ A\subset\bigcup_{j=1}^l A_j}}\inf_{1\leqq j\leqq l}\sup_{T>0}\sup_{\Phi\in E_T}\inf_{\phi\in A_j}\frac{1}{T}\int_\Omega\Phi(\omega)Q(d\omega)\geqq\inf_{Q\in A}H(Q). \tag{4.9}$$

Proof: Let $\inf_{Q\in A}H(Q)=\eta$. From Theorem 3.6 and Lemma 3.7 it follows that, given $Q\in A$ and $\varepsilon>0$, there exists T_Q and $\Phi_Q\in E_{T_Q}$ such that

$$\frac{1}{T_Q}\int_\Omega\Phi_Q(\omega)Q(d\omega)\geqq\eta-\tfrac{1}{2}\varepsilon. \tag{4.10}$$

Since $\Phi_Q \in C_S(\Omega)$, the integral on the left of (4.10) is a continuous functional of Q which implies that there exists a neighborhood G_Q of Q in $\mathcal{M}_S(\Omega)$ such that

$$\frac{1}{T_Q}\int_\Omega \Phi_Q(\omega)Q(d\omega) \geqq \eta - \varepsilon \tag{4.11}$$

for all Q in G_Q. The neighborhoods $\{G_Q\}$ form an open covering of the compact set A and therefore there exists $G_{Q_1}, G_{Q_2}, \cdots, G_{Q_l}$ such that $A \subset \bigcup_{j=1}^{l} G_{Q_j}$. This and (4.11) yield (4.8) and hence (4.9).

LEMMA 4.4. *Let A be closed in $\mathcal{M}_S(\Omega)$ and such that the family of one-dimensional marginals of Q as Q varies over A forms a tight family of measures on X. Then*

$$J(A) \leqq -\inf_{Q\in A} H(Q). \tag{4.12}$$

Proof: Let $\mathcal{M}(X)$ be the space of probability measures on X. Let $A_M \subset \mathcal{M}(X)$ be the family of one-dimensional marginals of Q as Q varies over A. Since, by hypothesis, A_M is tight, given $\varepsilon_n \to 0$ there exist compact sets $K_n \subset X$ such that $\mu(K_n) \geqq 1-\varepsilon_n$ for all n and all $\mu \in A_M$. We assumed that $x \to P_{0,x}$ is weakly continuous, and therefore, given $\eta_n \to 0$ there exists a sequence of compact sets $C_n \subset D_X[0, 1]$ such that $P_{0,x}(C_n) \geqq 1-\eta_n$ for all $x \in K_n$. Thus, if we let $\nu > 0$, denote the complement of C_n by $\tilde{C}_n$, and take $x \in K_n$,

$$\begin{aligned} E^{P_{0,x}}\{\exp\{\nu\chi_{K_n}(\omega(0))\chi_{\tilde{C}_n}(\omega)\}\} &= E^{P_{0,x}}\{\exp\{\nu\chi_{\tilde{C}_n}(\omega)\}\} \\ &= e^\nu P_{0,x}\{\tilde{C}_n\} + P_{0,x}\{C_n\} \\ &= 1 + (e^\nu - 1)P_{0,x}\{\tilde{C}_n\} \\ &\leqq 1 + (e^\nu - 1)\eta_n. \end{aligned}$$

On the other hand, if $x \notin K_n$, $E^{P_{0,x}}\{\exp\{\nu\chi_{K_n}(\omega(0))\chi_{\tilde{C}_n}(\omega)\}\} = 1$. Hence, for all x and all $\nu > 0$,

$$E^{P_{0,x}}\{\exp\{\nu\chi_{K_n}(\omega(0))\chi_{\tilde{C}_n}(\omega)\}\} \leqq 1 + \eta_n(e^\nu - 1). \tag{4.13}$$

Let $\phi(\omega) = \nu\chi_{K_n}(\omega(0))\chi_{\tilde{C}_n}(\omega) - \log[1+\eta_n(e^\nu - 1)]$ so that from (4.13) we observe that $E^{P_{0,x}}\{e^{\phi(\omega)}\} \leqq 1$, and thus, using Lemma 4.1 (with $T = 1$), we obtain

$$E^{P_{0,x}}\left\{\exp\left\{\nu\int_0^t \chi_{K_n}(\omega(0))\chi_{\tilde{C}_n}(\theta_s\omega)\, ds\right\}\right\} \leqq \exp\{t\log[1+\eta_n(e^\nu - 1)]\}. \tag{4.14}$$

From (4.5),

$$\begin{aligned} \int_0^t \chi_{K_n}(\omega(s))\chi_{\tilde{C}_n}(\theta_s\omega)\, ds &= \int_0^t \chi_{\tilde{C}_n}(\theta_s\omega)\, ds - \int_0^t \chi_{\tilde{K}_n}(\omega(s))\chi_{\tilde{C}_n}(\theta_s\omega)\, ds \\ &\geqq \int_0^t \chi_{\tilde{C}_n}(\theta_s\omega)\, ds - \int_0^t \chi_{\tilde{K}_n}(\omega(s))\, ds \\ &\geqq tR_{t,\omega}(\tilde{C}_n) - 1 - tR_{t,\omega}\{\omega' : \omega'(0) \in \tilde{K}_n\}, \end{aligned}$$

which implies, in view of (4.14), and the definition of $\Gamma_{t,x}$ measure,

$$E^{\Gamma_{t,x}}\left\{\exp\left\{\nu t\left[Q(\tilde{C}_n)-\frac{1}{t}-Q\{\omega:\omega(0)\in\tilde{K}_n\}\right]\right\}\right\} \tag{4.15}$$
$$\leqq \exp\{t\log[1+\eta_n(e^{\nu}-1)]\}.$$

Since $Q\in A$, $Q\{\omega:\omega(0)\in\tilde{K}_n\}\leqq\varepsilon_n$ and, therefore, from (4.15), for each n,

$$\Gamma_{t,x}\left\{A\cap\left\{Q:Q(\tilde{C}_n)\geqq\frac{1}{t}+2\varepsilon_n\right\}\right\} \tag{4.16}$$
$$\leqq \exp\{t\log[1+\eta_n(e^{\nu}-1)]-\varepsilon_n\nu t\}.$$

Inequality (4.16) holds for any $\nu>0$, sequence $\varepsilon_n\to0$, and sequence $\eta_n\to0$. In particular, let us choose, for $\lambda>0$, $\nu=\lambda n^2$, $\varepsilon_n=1/n$, and $\eta_n=\exp\{-\lambda n^2\}$. Then,

$$t\log[1+\eta_n(e^{\nu}-1)]=t\log[1+\exp\{-\lambda n^2\}(\exp\{\lambda n^2\}-1)],$$

and $\varepsilon_n\nu t=(1/n)\cdot\lambda n^2 t=\lambda nt$. Thus, from (4.16), for each n,

$$\Gamma_{t,x}\left\{A\cap\left\{Q:Q(\tilde{C}_n)\geqq\frac{1}{t}+\frac{2}{n}\right\}\right\}$$
$$\leqq e^{t\log 2-\lambda nt}.$$

This implies

$$\Gamma_{t,x}\left\{A\cap\bigcup_{n=1}^{\infty}\left\{Q:Q(\tilde{C}_n)\geqq\frac{1}{t}+\frac{2}{n}\right\}\right\}\leqq e^{t\log 2}\sum_{n=1}^{\infty}e^{-\lambda nt}$$
$$=e^{t\log 2}\left(\frac{e^{-\lambda t}}{1-e^{-\lambda t}}\right).$$

If we let $A_t=\{Q:Q(\tilde{C}_n)\leqq 1/t+2/n \text{ for all } n\}$, then we have just shown

$$\Gamma_{t,x}\{A\cap\tilde{A}_t\}\leqq e^{t\log 2}\left(\frac{e^{-\lambda t}}{1-e^{-\lambda t}}\right). \tag{4.17}$$

We should note that the set A_t depends on λ since the sets C_n depend on the choice of η_n, and η_n was selected in terms of λ. From (4.17) we get

$$\overline{\lim_{t\to\infty}}\frac{1}{t}\log\sup_{x\in X}\Gamma_{t,x}(A\cap\tilde{A}_t)\leqq\log 2-\lambda. \tag{4.18}$$

Let $A_\infty=\bigcap_{t>0}A_t=\{Q\in\mathcal{M}_S(\Omega): Q(\tilde{C}_n)\leqq 2/n, \text{ for all } n\}$. The restriction of Q to $\mathcal{F}_1^0$ constitutes a tight family of measures on $D_X[0,1]$. But, for stationary processes, tightness when restricted to some interval implies overall tightness and hence A_∞ is a compact set in $\mathcal{M}_S(\Omega)$. Since A is closed, $A\cap A_\infty$ is compact in $\mathcal{M}_S(\Omega)$. Thus, from Lemmas 4.2 and 4.3, $J(A\cap A_\infty)\leqq-\inf_{Q\in A\cap A_\infty}H(Q)\leqq$

$-\inf_{Q\in A} H(Q)$ which implies that for any $\varepsilon>0$ there exists an open set $G_\varepsilon \supset A\cap A_\infty$ such that

$$J(G_\varepsilon)\leqq -\inf_{Q\in A} H(Q)+\varepsilon. \tag{4.19}$$

In Lemma 4.5 to follow, we show that, given any open set $G\subset \mathcal{M}_S(\Omega)$ such that $G\supset A\cap A_\infty$, there exists a t_0 such that $t\geqq t_0$ implies $A\cap A_t\subset G$. Using this, (4.18), and (4.19), we have, for any $\varepsilon>0$, $\lambda>0$, and $t\geqq t_0$,

$$\begin{aligned}\Gamma_{t,x}(A)&\leqq \Gamma_{t,x}(A\cap A_t)+\Gamma_{t,x}(A\cap \tilde{A}_t)\\ &\leqq \Gamma_{t,x}(G_\varepsilon)+\Gamma_{t,x}(A\cap\tilde{A}_t)\end{aligned}$$

which means

$$\begin{aligned}J(A)&\leqq \max\,(J(G_\varepsilon), \log 2-\lambda)\\ &\leqq \max\left(-\inf_{Q\in A} H(Q)+\varepsilon, \log 2-\lambda\right).\end{aligned} \tag{4.20}$$

Since the left side of (4.20) is independent of ε and λ we now let $\varepsilon\to 0$ and $\lambda\to\infty$ obtaining the desired (4.12).

LEMMA 4.5. *Let $A\subset\mathcal{M}_S(\Omega)$ be closed, $A_t=\{Q: Q(\tilde{C}_n)<2/n+1/t$ for all $n\}$, and $A_\infty=\bigcap_{t>0}A_t$. Let G be a neighborhood in $\mathcal{M}_S(\Omega)$ such that $G\supset A\cap A_\infty$. Then, there exists a t_0 such that $t\geqq t_0$ implies $A\cap A_t\subset G$.*

Proof: Let $\{t_k\}\to\infty$ and let $Q_k\in A\cap A_{t_k}$. Assume $Q_k\in\tilde{G}$ for every k. We want to show that this leads to a contradiction. Since $Q_k\in A_{t_k}$, we have $Q_k(\tilde{C}_n)< 2/n+1/t_k$ for all n. Thus $\overline{\lim}_{k\to\infty} Q_k(\tilde{C}_n)\leqq 2/n$ for each n. This last implies that the family of stationary measures $\{Q_k\}$ is tight on $[0, 1]$. As noted earlier, this means the family $\{Q_k\}$ is tight in $\mathcal{M}_S(\Omega)$. Thus, if $\bar{Q}$ is any limit point, we must have $\bar{Q}\in A\cap A_\infty$ since $A\cap A_\infty$ is closed, and also $\bar{Q}\in\tilde{G}$ since $\tilde{G}$ is closed. Hence, we have contradicted the assumption, $A\cap A_\infty\subset G$.

We should note that Lemma 4.4 gives us the desired upper estimate for closed sets A in $\mathcal{M}_S(\Omega)$, i.e., $J(A)\leqq -\inf_{Q\in A} H(Q)$ but only under the additional hypothesis that the family A_M of one-dimensional marginals of Q as Q varies over A forms a tight family of measures in X. Now, of course, if X is itself compact, then this last hypothesis on A_M is automatically satisfied. However, if X is not compact, then we can replace the hypothesis of tightness on A_M by imposing additional hypotheses on the Markov process $P_{0,x}$. We still obtain the upper estimate cited above for closed sets but we must give up a bit on the uniformity in x. The estimates obtained above involving $J(A)$ were (by definition of $J(A)$) uniform for all $x\in X$. Now, under the additional hypotheses to be imposed on $P_{0,x}$ (when X is not compact) we will get the same upper bounds but they will hold uniformly only for x in compact subsets of X.

THEOREM 4.6. *Assume the Markov process $P_{0,x}$ satisfies hypotheses* (i)–(v). *Let A be a closed set in $\mathcal{M}_S(\Omega)$. Then, for any compact set K in X,*

$$\overline{\lim_{t\to\infty}} \frac{1}{t} \log \sup_{x\in K} \Gamma_{t,x}(A) \leqq -\inf_{Q\in A} H(Q). \tag{4.21}$$

Proof: By hypothesis there exists a sequence $\{u_n(x)\}$ of functions in the domain of L, the infinitesimal generator of the Markov process $P_{0,x}$, with properties (i)–(v) (see the introduction). Consider, then, the equation for $u(x, t)$:

$$u_t = Lu + V_n(x)u,$$
$$u(x, 0) = u_n(x).$$

By the Feynman-Kac formula the solution is given by

$$u(x, t) = E^{P_{0,x}}\left\{\exp\left\{\int_0^t V_n(\omega(s))\, ds\right\} u_n(\omega(t))\right\},$$

but, from hypothesis (iii), $-(Lu_n/u_n)(x) = V_n(x)$ so that the solution $u(x, t) \equiv u_n(x)$, and thus

$$E^{P_{s,x}}\left\{\exp\left\{\int_s^t V_n(\omega(s))\, ds\right\} u_n(\omega(t))\right\} = u_n(x). \tag{4.22}$$

From hypothesis (ii), for any compact set K there exists a constant C_K such that $\sup_{x\in K} \sup_n u_n(x) \leqq C_K$. Also, from hypothesis (i), there exists a constant c such that $u_n(x) \geqq c > 0$ for all x and n. Thus, from (4.22) we conclude that, for any compact set K,

$$\sup_{x\in K} \sup_n E^{P_{0,x}}\left\{\exp\left\{\int_0^t V_n(\omega(s))\, ds\right\}\right\} \leqq \frac{\sup_{x\in K} \sup_n u_n(x)}{\inf_{x\in X} u_n(x)} \leqq \frac{C_K}{c}.$$

Using hypothesis (iv) we get from Fatou's lemma that, for all t,

$$\sup_{x\in K} E^{P_{0,x}}\left\{\exp\left\{\int_0^t V(\omega(s))\, ds\right\}\right\} \leqq \frac{C_K}{c} = M. \tag{4.23}$$

By the definition of $\Gamma_{t,x}$ measure we have

$$E^{\Gamma_{t,x}}\{e^{tF(Q)}\} = E^{P_{0,x}}\{\exp\{tF(R_{t,\omega})\}\},$$

hence, in particular, if

$$F(R_{t,\omega}) = \int_\Omega V(\omega'(0)) R_{t,\omega}(d\omega') = \frac{1}{t}\int_0^t V(\omega(s))\, ds,$$

then $F(Q) = \int_X V(y)\mu(dy)$, where μ is the one-dimensional marginal of Q. Thus,

(4.23) becomes

$$\sup_{x\in K} E^{\Gamma_{t,x}}\left\{\exp\left\{t\int_X V(y)\mu(dy)\right\}\right\}\leqq M. \tag{4.24}$$

Let $K_l=\{x\in X\colon V(x)\leqq l\}$. Using hypothesis (iii) and the constant C which appears there, we get from (4.24)

$$\sup_{x\in K} E^{\Gamma_{t,x}}\{\exp\{t[l\mu(\tilde{K}_l)-C]\}\}\leqq M.$$

Thus, for all $x\in K$,

$$\Gamma_{t,x}\{Q\colon \mu(\tilde{K}_l)\geqq\delta_l\}\leqq M\exp\{t(C-l\delta_l)\}. \tag{4.25}$$

In particular, take $\lambda>0$ and choose $l_n=Cn+\lambda n^2$, $\delta_{l_n}=1/n$. From (4.25) we get, for all $x\in K$,

$$\Gamma_{t,x}\left\{Q\colon \mu(\tilde{K}_{l_n})\geqq\frac{1}{n}\right\}\leqq Me^{-\lambda nt},$$

which implies that, for all $x\in K$,

$$\Gamma_{t,x}\left\{\bigcup_{n=1}^{\infty}\left\{Q\colon \mu(\tilde{K}_{l_n})\geqq\frac{1}{n}\right\}\right\}\leqq M\frac{e^{-\lambda t}}{1-e^{-\lambda t}}. \tag{4.26}$$

If we let

$$A_\lambda=\bigcap_{n=1}^{\infty}\left\{Q\colon \mu(\tilde{K}_{l_n})<\frac{1}{n}\right\},$$

then (4.26) says that

$$\Gamma_{t,x}(\tilde{A}_\lambda)\leqq M\frac{e^{-\lambda t}}{1-e^{-\lambda t}}. \tag{4.27}$$

Now, let A be a closed set in $\mathcal{M}_S(\Omega)$. From the definition of A_λ we see that the marginals of Q as Q varies over A_λ form a tight family. Since $A\cap A_\lambda$ is closed, we have from Lemma 4.4 that, for each $\lambda>0$,

$$J(A\cap A_\lambda)\leqq-\inf_{Q\in A\cap A_\lambda}H(Q)\leqq-\inf_{Q\in A}H(Q). \tag{4.28}$$

Since $A\subset(A\cap A_\lambda)\cup\tilde{A}_\lambda$, we conclude from (4.27) and (4.28) that, for every $\lambda>0$,

$$\overline{\lim_{t\to\infty}}\,\frac{1}{t}\log\sup_{x\in K}\Gamma_{t,x}(A)\leqq\max\left(-\inf_{Q\in A}H(Q),\,-\lambda\right).$$

If we let $\lambda\to\infty$ in this last inequality, we get (4.21).

5. Lower Bounds

We need additional hypotheses on the Markov process $P_{0,x}$ in order to obtain the desired lower bounds. Let $p(t, x, dy)$ be the transition probability associated with $P_{0,x}$. We shall assume that there exists a density function for $p(1, x, dy)$ with respect to a reference measure α on X, i.e., there exists, for all $x \in X$, a function $p(1, x, y) > 0$ for almost all y (α measure) such that for all $x \in X$:

I: $p(1, x, dy) = p(1, x, y)\alpha(dy)$,

and

II: $p(1, x, \cdot)$ as a mapping from $X \to L_1(\alpha)$ is continuous.

We begin with three lemmas. The main result of this section is Theorem 5.5.

LEMMA 5.1. *Let $Q \in \mathcal{M}_S(\Omega)$ be such that $H(Q) < \infty$ and let μ be the marginal distribution of Q. Then, $\mu \ll \alpha$.*

Proof: Let $A \subset X$ be such that $\alpha(A) = 0$. We want to show that if $Q \in \mathcal{M}_S(\Omega)$ such that $H(Q) < \infty$, then $\mu(A) = 0$, where μ is the marginal of Q. From hypothesis I, for all $x \in X$,

$$P_{0,x}(\omega : \omega(1) \in A) = \int_A p(1, x, y)\alpha(dy) = 0. \tag{5.1}$$

Since $H(Q) < \infty$ and $H(1, Q) = H(Q)$, we know from (3.1) and Theorem 2.1 that $Q_{0,\omega}$ is absolutely continuous with respect to $P_{0,\omega(0)}$ for almost all ω (Q measure). Thus, from (5.1) we conclude that for almost all ω (Q measure) $Q_{0,\omega}\{\omega' : \omega'(1) \in A\} = 0$. Hence, $E^Q\{Q_{0,\omega}\{\omega' : \omega'(1) \in A\}\} = 0$, i.e., $\mu(A) = 0$.

LEMMA 5.2. *The reference measure α on X is absolutely continuous with respect to $p(1, x, \cdot)$ uniformly for x in compact subsets of X. In particular, if $H(Q) < \infty$, and μ is the marginal of Q, then $\mu \ll p(1, x, \cdot)$ uniformly for x in compact sets.*

Proof: The second statement of the lemma follows from the first and Lemma 5.1; thus we want to prove $\alpha \ll p(1, x, \cdot)$ uniformly for x in compact sets. Let $A \subset X$ such that $\alpha(A) = \delta > 0$. By hypothesis *I*, $p(1, x, A) \geqq \varepsilon(x, \delta)$. By hypothesis II, $p(1, x, \cdot)$ as a mapping from $X \to L_1(\alpha)$ is continuous and therefore $p(1, x', A) \geqq \varepsilon(x, \delta)/2$ for all x' in a neighborhood of x. The result now follows by the usual compactness argument.

LEMMA 5.3. *Let $Q \in \mathcal{M}_S(\Omega)$ be such that $H(Q) < \infty$ and let μ be the marginal of Q. Let $0 \leqq \phi_n(y) \leqq 1$ for $n = 1, 2, \cdots$, where $y \in X$. Assume*

$$\int_X \phi_n(y)\mu(dy) \geqq c_1 > 0 \quad \text{for} \quad n = 1, 2, \cdots.$$

Then, for any compact set $K \subset X$, there exists $c_2 > 0$ such that, for all $x \in K$, and all $n = 1, 2, 3, \cdots$,

$$\int_X \phi_n(y) p(1, x, dy) \geqq c_2.$$

Proof: Assume the contrary, i.e., for some sequence $\{x_n\} \in K$,

$$\lim_{n\to\infty} \int_X \phi_n(y) p(1, x_n, dy) = 0.$$

This implies $p(1, x_n, \{y \in X: \phi_n(y) \geqq \varepsilon\}) \to 0$ as $n \to \infty$ for every $\varepsilon > 0$. From Lemma 5.2 we conclude that $\mu\{y \in X: \phi_n(y) \geqq \varepsilon\} \to 0$ as $n \to \infty$ for every $\varepsilon > 0$. Since $\phi_n(y) \leqq 1$, the bounded convergence theorem then implies $\int_X \phi_n(y)\mu(dy) \to 0$ as $n \to \infty$ which is a contradiction.

THEOREM 5.4. *Let $Q \in \mathcal{M}_S(\Omega)$ be ergodic and such that $H(Q) < \infty$. Let K_1 be a compact set in X such that $\alpha(K_1) > 0$, and let K_2 be any compact set in X. For any neighborhood N of Q in $\mathcal{M}_S(\Omega)$, we have uniformly, for $x \in K_2$,*

$$\varliminf_{t\to\infty} \frac{1}{t} \log P_{0,x}\{\omega: R_{t,\omega} \in N, \omega(t) \in K_1\} \geqq -H(Q). \tag{5.2}$$

Proof: As before, let $Q_{0,\omega}$ be the r.c.p.d. of Q given $\mathcal{F}_0^{-\infty}$. Let

$$\psi(\omega, t) = \log \left| \frac{dQ_{0,\omega}}{dP_{0,\omega(0)}} \right|_{\mathcal{F}_t^0},$$

so that we have

$$\psi(\omega, t+s) = \psi(\omega, t) + \psi(\theta_t\omega, s)$$

and also, from (3.1) and Theorem 2.1,

$$E^Q\{\psi(\omega, t)\} = tH(Q).$$

Hence, by the ergodic theorem, for almost all ω (Q measure)

$$\lim_{t\to\infty} \frac{\psi(\omega, t)}{t} = H(Q). \tag{5.3}$$

Stated in another way, (5.3) says that

$$\lim_{t\to\infty} \frac{\psi(\omega', t)}{t} = H(Q) \tag{5.4}$$

almost everywhere ω'($Q_{0,\omega}$ measure) for almost all ω (Q measure). Again, from the ergodic theorem, for almost all ω (Q measure), and any neighborhood N of Q,

$$\lim_{t\to\infty} Q_{0,\omega}\{\omega': R_{t,\omega'} \in N\} = 1. \tag{5.5}$$

With μ denoting the marginal of Q, let K_3 be a compact set in X such that $\mu(K_3) \geqq \frac{1}{2}$. Let $D_t = \{\omega: R_{t,\omega} \in N, \omega(t) \in K_3\}$ and consider $P_{0,\omega(0)}\{D_t\}$. Although $Q_{0,\omega}$ is absolutely continuous with respect to $P_{0,\omega(0)}$ the reverse is not necessarily true; however, for $\varepsilon > 0$,

$$P_{0,\omega(0)}\{D_t\} \geqq \int_{D_t} \frac{dP_{0,\omega(0)}}{dQ_{0,\omega}}\bigg|_{\mathcal{F}_t^0} dQ_{0,\omega}$$

$$= \int_{D_t} e^{\psi(\omega',t)} Q_{0,\omega}(d\omega')$$

$$\geqq e^{-t(H(Q)+\varepsilon)} Q_{0,\omega}\left\{D_t\left\{\omega': \frac{1}{t}\psi(\omega', t) \leqq H(Q)+\varepsilon\right\}\right\}.$$

If we let $F_t = D_t \cap \{\omega': (1/t)\psi(\omega', t) \leqq H(Q)+\varepsilon\}$, and $\phi(t, x) = [P_{0,x}\{D_t\} \exp\{t(H(Q)+\varepsilon)\}] \wedge 1$, then we have just shown that

$$\phi(t, \omega(0)) \geqq Q_{0,\omega}\{F_t\}. \tag{5.6}$$

Taking expectations on both sides of (5.6) with respect to Q measure we get

$$\int_X \phi(t, x)\mu(dx) \geqq E^Q\{Q_{0,\omega}\{F_t\}\} = Q\{F_t\}. \tag{5.7}$$

Moreover, it follows from (5.4) and (5.5) that

$$\lim_{t\to\infty} Q\{F_t\} = \mu(K_3)$$

and so, from (5.7) and the choice of K_3, we obtain

$$\varliminf_{t\to\infty} \int_X \phi(t, x)\mu(dx) \geqq \tfrac{1}{2}. \tag{5.8}$$

Given the neighborhood N of Q in $\mathcal{M}_S(\Omega)$, we can pick another neighborhood N_1 such that $\bar{N}_1 \subset N$ and we pick $T < \infty$ and $\varepsilon > 0$ such that, for any $Q' \in N_1$ and any Q'' satisfying $\|Q' - Q''\|_{\mathcal{F}_T^0} < \varepsilon$, it follows that $Q'' \in N$. Here the norm is the variation norm. Now $\|R_{t,\omega} - R_{t-2,\theta_1\omega}\|_{\mathcal{F}_T^0} \leqq (T+2)/t$ so that by taking t large enough this norm can be made less than ε. Therefore, for the neighborhood N_1 just picked and for t large enough,

$$\begin{aligned} &P_{0,x}\{R_{t,\omega} \in N, \omega(t) \in K_1\} \\ &\quad\geqq P_{0,x}\{R_{t-2,\theta_1\omega} \in N_1, \omega(t) \in K_1\} \\ &\quad= \int_X \int_X p(1, x, dy) P_{0,y}\{R_{t-2,\omega} \in N_1, \omega(t-2) \in dz\} p(1, z, K_1) \\ &\quad\geqq \left[\int_X p(1, x, dy) P_{0,y}\{R_{t-2,\omega} \in N_1, \omega(t-2) \in K_3\right] \cdot W, \end{aligned} \tag{5.9}$$

where $W=\inf_{z\in K_3} p(1,z,K_1)$. Using the notation introduced earlier in this proof we can rewrite (5.9) as

$$e^{t(H(Q)+\varepsilon)}P_{0,x}\{R_{t,\omega}\in N,\ \omega(t)\in K_1\}\geqq W\int_X \phi(t-2,y)p(1,x,dy). \tag{5.10}$$

From (5.8), (5.10), and Lemma 5.3 we have uniformly, for $x\in K_2$,

$$\varliminf_{t\to\infty}\frac{1}{t}\log P_{0,x}\{R_{t,\omega}\in N,\ \omega(t)\in K_1\}\geqq -H(Q). \tag{5.11}$$

In obtaining (5.11) we also used the fact that since $\alpha(K_1)>0$ and K_3 is compact we have $W>0$.

THEOREM 5.5. *Let $Q\in\mathcal{M}_S(\Omega)$ be such that $H(Q)<\infty$ and let N be any neighborhood of Q; then, uniformly for x in compact subsets of X,*

$$\varliminf_{t\to\infty}\frac{1}{t}\log\Gamma_{t,x}(N)\geqq -H(Q). \tag{5.12}$$

Thus, if G is open in $\mathcal{M}_S(\Omega)$, we have, uniformly for x in compact subsets of X,

$$\varliminf_{t\to\infty}\frac{1}{t}\log\Gamma_{t,x}(G)\geqq -\inf_{Q\in G}H(Q). \tag{5.13}$$

Proof: We can assume without loss of generality that $Q=\sum_{j=1}^{l}\pi_jQ_j$ where each Q_j is an ergodic member of $\mathcal{M}_S(\Omega)$ and the π_j are non-negative constants whose sum is 1. Since $H(Q)$ is linear by Theorem 3.5 we have $H(Q)=\sum_{j=1}^{l}\pi_jH(Q_j)$. Divide the interval $[0,t]$ into subintervals of length π_jt and let t_j, $j=1,2,\cdots,l$, be the right-hand endpoints of these subintervals. Let $K\subset X$ be compact. For each of the ergodic stationary processes Q_j, $j=1,2,\cdots l$, let N_j be a neighborhood of Q_j such that if $Q_j'\in N_j$ for $j=1,2,\cdots,l$ and $\|Q''-\sum_{j=1}^{l}\pi_jQ_j'\|_{\mathcal{F}_T^0}<\varepsilon$ for some choice of T and ε, then $Q''\in N$. Then,

$$\begin{aligned}\Gamma_{t,x}(N)&=P_{0,x}\{\omega: R_{t,\omega}\in N\}\\&\geqq P_{0,x}\{\omega: R_{t_j-t_{j-1}},\ \theta_{t_{j-1}}\omega\in N_j, j=1,2,\cdots,l\}\\&\geqq P_{0,x}\{\omega: R_{t_j-t_{j-1}},\ \theta_{t_{j-1}}\omega\in N_j,\ \omega(t_j)\in K, j=1,2,\cdots,l\}.\end{aligned}$$

Hence, if K_0 is the compact set in X for which we want to show uniformity in the estimate, we get from the preceding inequality

$$\begin{aligned}\inf_{x\in K_0}\Gamma_{t,x}(N)\geqq&\inf_{x\in K_0}P_{0,x}\{\omega: R_{t_1,\omega}\in N_1,\ \omega(t_1)\in K\}\\&\cdot\prod_{j=1}^{l-1}\inf_{x\in K}P_{0,x}\{\omega: R_{t_j-t_{j-1}},\ \omega\in N_j,\ \omega(t_j-t_{j-1})\in K\}.\end{aligned} \tag{5.14}$$

Now, if we apply Theorem 5.4 to (5.14) using the fact that the Q_i are ergodic and the previously noted linearity of $H(Q)$ we obtain (5.12).

6. The Contraction Principle

In this section we show the connection between the entropy function $H(Q)$ and the I-function introduced by the authors in their earlier papers. The relationship is clear from the following theorem. For the definition of the I-function see [2-III].

THEOREM 6.1. *For $Q \in \mathcal{M}_S(\Omega)$ let $q(Q)$ denote its one-dimensional marginal. Then,*

$$\inf_{Q:q(Q)=\mu} H(Q) = I(\mu). \tag{6.1}$$

Proof: Let $Q \in \mathcal{M}_S(\Omega)$ such that $H(Q) = l$ and $q(Q) = \mu$. Let $\Phi \in \mathcal{B}(\mathcal{F}_t^0)$ and be such that

$$E^{P_{0,x}}\{e^{\Phi}\} \leqq 1 \quad \text{for all} \quad x. \tag{6.2}$$

We earlier defined (in (3.12))

$$\bar{H}(t, Q) = \sup_{\Phi \in \mathcal{B}(\mathcal{F}_t^0)} [E^Q\{\Phi\} - E^Q\{\log E^{P_{0,\omega(0)}}\{e^{\Phi}\}\}],$$

and noted that $\bar{H}(t, Q) \leqq tH(Q)$. Thus, for the Φ selected,

$$E^Q\{\Phi\} \leqq lt. \tag{6.3}$$

In particular, let $\mathcal{U}$ be the space of functions $u \in \mathcal{B}(X)$ for each of which there exist constants c and C such that, for all x, $0 < c \leqq u \leqq C < \infty$ and, for $h > 0$ and $u \in \mathcal{U}$, take

$$\Phi(\omega) = \log \frac{u(\omega(h))}{(T_h u)(\omega(0))},$$

where T_h is the semigroup associated with the Markov process $P_{0,x}$. Now, this particular Φ is an element of $\mathcal{B}(\mathcal{F}_h^0)$ and direct calculation shows that Φ satisfies condition (6.2). Thus, from (6.3) we conclude

$$E^Q\left\{\log \frac{u(\omega(h))}{(T_h u)(\omega(0))}\right\} \leqq lh,$$

i.e.,

$$\begin{aligned} &E^Q\{\log u(\omega(h))\} - E^Q\{\log (T_h u)(\omega(0))\} \\ &= \int_X \log u(x)\mu(dx) - \int_X \log (T_h u)(x)\mu(dx) \qquad (6.4) \\ &= \int_X \log \frac{u(x)}{(T_h u)(x)} \mu(dx) \leqq lh. \end{aligned}$$

Since (6.4) holds for every $u \in \mathcal{U}$ we get

$$\sup_{u \in \mathcal{U}} \int_X \log \frac{u(x)}{(T_h u)(x)} \mu(dx) \leqq lh. \tag{6.5}$$

In the notation of [2-III], (6.5) states that $(1/h)I_h(\mu) \leqq l$. But in [2-III] we proved $\lim_{h\to 0}(1/h)I_h(\mu) = I(\mu)$, and therefore from (6.5) we conclude that $I(\mu) \leqq l$. Since this was true for any Q such that $q(Q) = \mu$ we have

$$I(\mu) \leqq \inf_{Q:q(Q)=\mu} H(Q). \tag{6.6}$$

We want now to show the inequality in the other direction. Let μ be a probability measure on X and suppose $I(\mu) = l < \infty$. With $I_h(\mu)$ defined as above in (6.5), we know from [2-III] that, for all $h > 0$, $I_h(\mu) \leqq hI(\mu) = hl$. In [2-III] we proved that there exists a bivariate distribution $\rho(dx, dy)$ on $X \times X$ such that both marginals of ρ are equal to μ, and if $\rho_x(dy)$ denotes the r.c.p.d. of the second component given the first, then for all $h > 0$

$$\int_X h(p(h, x, \cdot); \rho_x(\cdot))\mu(dx) = I_h(\mu) \leqq hl. \tag{6.7}$$

In (6.7), $h(\alpha; \beta)$ is, of course, the entropy of β with respect to α, as introduced in Section 2, and $p(t, x, \cdot)$ is the transition function for the Markov process $P_{0,x}$.

Let n be a positive integer, let $h = 1/n$ and consider the grid $\{jh\}$, $j = 0, \pm 1, \pm 2, \cdots$. Define a Markov chain on this grid with stationary marginal μ and transition probability $\rho_x(dy)$. Given a sample of this Markov chain, let us interpolate between the grid points by random trajectories whose distribution is $P^h_{x,y}$, where $P^h_{x,\omega(h)}$ is the r.c.p.d. of $P_{0,x}$ given $\mathcal{F}^h_h$ and where, in the interpolation, x and y are the end points, i.e., values of the Markov chain at adjacent grid points. Let $Q^{(h)}$ be the measure induced by this interpolation procedure from the Markov chain on the grid. We note that $P^h_{x,y}$ is defined for each x for almost all y with respect to $p(h, x, \cdot)$ measure, but $\rho_x(\cdot)$ is absolutely continuous with respect to $p(h, x, \cdot)$ for almost all x (μ measure) and hence $P^h_{x,y}$ is defined almost everywhere with respect to $\rho(dx, dy)$ measure. Thus the interpolation procedure is well defined.

Let P_μ be the Markov process on $\mathcal{F}^0_\infty$ with initial distribution μ, i.e., $P_\mu = \int P_{0,x}\mu(dx)$. Consider $h_{\mathcal{F}^0_1}(P_\mu; Q^{(h)})$. Let $\mathcal{F}^{(h)}$ be the σ-field generated by $\omega(jh)$, $0 \leqq j \leqq 1/h = n$. Then, by Lemma 2.3,

$$h_{\mathcal{F}^0_1}(P_\mu; Q^{(h)}) = h_{\mathcal{F}^{(h)}}(P_\mu; Q^{(h)}) + E^{Q^{(h)}}\{h_{\mathcal{F}^0_1}(P_{\mu,\omega}; Q^{(h)}_\omega)\}, \tag{6.8}$$

where $P_{\mu,\omega}$ and $Q^{(h)}_\omega$ are, respectively, the r.c.p.d. of P_μ and $Q^{(h)}$ given $\mathcal{F}^{(h)}$. Notice that, by construction, $P_{\mu,\omega} = Q^{(h)}_\omega$ on $\mathcal{F}^0_1$ for almost all ω with respect to $Q^{(h)}$ measure, and therefore $E^{Q^{(h)}}\{h_{\mathcal{F}^0_1}(P_{\mu,\omega}; Q^{(h)}_\omega)\} = 0$. Hence, from (6.8) we get

$$h_{\mathcal{F}^0_1}(P_\mu; Q^{(h)}) = h_{\mathcal{F}^{(h)}}(P_\mu; Q^{(h)}). \tag{6.9}$$

In Lemma 6.2 to follow we show, using (6.7), that $h_{\mathcal{F}^{(h)}}(P_\mu; Q^{(h)}) \leqq nhl = l$. Hence, from (6.9), we conclude that $h_{\mathcal{F}_1^0}(P_\mu; Q^{(h)}) \leqq l$. This last inequality and Lemma 2.2 imply that the family of measures $\{Q^{(h)}\}$ is tight. Thus, let Q be any weak limit of $Q^{(h)}$ as $h \to 0$ ($n \to \infty$) and we see that Q is stationary with marginal μ and, moreover, $H(Q) \leqq l$. Hence, $\inf_{Q: q(Q)=\mu} H(Q) \leqq l = I(\mu)$ and the proof is complete except for the lemma.

LEMMA 6.2. *Let $\pi(x, dy)$ be any Markov chain transition probability with state space X and let μ be a probability measure on X. Let $\rho(dx, dy)$ be a probability measure on $X \times X$ with both marginals equal to μ, conditional measure $\rho_x(dy)$, and let*

$$\tau = \int_X h(\pi(x, \cdot); \rho_x(\cdot))\mu(dx).$$

Let P_μ be the π-chain with initial distribution μ and let R_μ be the stationary ρ-chain with initial distribution μ. Let $\mathcal{F}_n$ be the σ-field generated by $x_0, x_1, \cdots, x_n$, the first n steps of the chain. Then

$$h_{\mathcal{F}_n}(P_\mu; R_\mu) = n\tau. \tag{6.10}$$

Proof: If $\tau = \infty$ the result is obvious, so assume $\tau < \infty$. This means $\rho_x(dy) \ll \pi(x, dy)$ for almost all x (μ measure). Let

$$\theta(x, y) = \frac{d\rho_x(\cdot)}{d\pi(x, \cdot)},$$

and we see that

$$\left.\frac{dR_\mu}{dP_\mu}\right|_{\mathcal{F}_n} = \theta(x_0, x_1)\theta(x_1, x_2) \cdots \theta(x_{n-1}, x_n).$$

Now,

$$h_{\mathcal{F}_n}(P_\mu; R_\mu) = E^{R_\mu}\left\{\log \left.\frac{dR_\mu}{dP_\mu}\right|_{\mathcal{F}_n}\right\}$$

$$= \sum_{i=1}^n E^{R_\mu}\{\log \theta(x_{i-1}, x_i)\}.$$

But R_μ is stationary, and hence

$$\sum_{i=1}^n E^{R_\mu}\{\log \theta(x_{i-1}, x_i)\} = nE^{R_\mu}\{\log \theta(x_0, x_1)\}$$

$$= n\tau.$$

We finish this section with two theorems which, although not used directly in this paper, are used in applying this theory to the polaron problem in [6]. We include them here since the contraction principle is used in the proof.

THEOREM 6.3. *Let $\mathcal{A}$ be a family of stationary processes satisfying*
(i) *the family $\mathcal{A}_M$ of one-dimensional marginals is tight,*
(ii) *$H(Q) \leqq l < \infty$ for all $Q \in \mathcal{A}$.*
Then, $\mathcal{A}$ is tight in $\mathcal{M}_S(\Omega)$.

Proof: Let $Q \in \mathcal{A} \subset \mathcal{M}_S(\Omega)$ with marginal μ and let $\Phi \in \mathcal{B}(\mathcal{F}_T^0)$. Then,

$$E^Q\{\Phi\} \leqq TH(Q) + \int \log E^{P_{0,x}}\{e^{\Phi}\}\mu(dx). \tag{6.11}$$

Let $K \subset X$ be compact and let $\varepsilon > 0$ be given. Since we have assumed everywhere that $x \to P_{0,x}$ is weakly continuous, there exists a set $C \subset \Omega$ which is compact in $D_X[0, 1]$ such that $P_{0,x}(\tilde{C}) < \varepsilon$ for all $x \in K$. As a special case of (6.11) choose, for $\lambda > 0$,

$$\Phi(\omega) = \lambda \int_0^{T-1} \chi_K(\omega(s))\chi_{\tilde{C}}(\theta_s\omega)\, ds.$$

Now, this Φ is an element of $\mathcal{B}(\mathcal{F}_T^0)$, and from (6.11)

$$E^Q\{\Phi\} \leqq Tl + \int \log E^{P_{0,x}}\left\{\exp\left\{\lambda \int_0^{T-1} \chi_K(\omega(s))\chi_{\tilde{C}}(\theta_s\omega)\, ds\right\}\right\}\mu(dx) \tag{6.12}$$

since $H(Q) \leqq l$. On the other hand, for this Φ,

$$\begin{aligned} E^Q\{\Phi\} &= \lambda \int_0^{T-1} E^Q\{\chi_K(\omega(s))\chi_{\tilde{C}}(\theta_s\omega)\}\, ds \\ &\geqq \lambda(T-1)[Q(\tilde{C}) - \mu(\tilde{K})]. \end{aligned} \tag{6.13}$$

From (6.12) and (6.13) we get

$$\begin{aligned} Q(\tilde{C}) &\leqq \mu(\tilde{K}) + \frac{Tl}{\lambda(T-1)} \\ &+ \frac{1}{\lambda(T-1)} \int \log E^{P_{0,x}}\left\{\exp\left\{\lambda \int_0^{T-1} \chi_K(\omega(s))\chi_{\tilde{C}}(\theta_s\omega)\, ds\right\}\right\}\mu(dx). \end{aligned} \tag{6.14}$$

In particular, take $T = 2$ and using the same estimates as in (4.13) we get

$$Q(\tilde{C}) \leqq \mu(\tilde{K}) + \frac{2l}{\lambda} + \frac{1}{\lambda}\log[1 + \varepsilon(e^{\lambda} - 1)]. \tag{6.15}$$

Since the one-dimensional marginals $\mathcal{A}_M$ form a tight family, one can clearly choose K, ε, λ to make the right side of (6.15) as small as one pleases for all $Q \in \mathcal{A}$ simultaneously, showing $\mathcal{A}$ to be tight in $\mathcal{M}_S(\Omega)$.

THEOREM 6.4. *Assume the Markov process $P_{0,x}$ satisfies hypotheses* (i)–(v) *(in the introduction). Let $\mathcal{A}$ be a family in $\mathcal{M}_S(\Omega)$ such that $H(Q) \leqq l < \infty$ for all $Q \in \mathcal{A}$. Then, $\mathcal{A}$ is tight in $\mathcal{M}_S(\Omega)$.*

Proof: If $\mu \in \mathscr{A}_M$, the family of one-dimensional marginals in $\mathscr{A}$, then, by the contraction principle,

$$I(\mu)=\inf_{Q:q(Q)=\mu} H(Q).$$

But since $\mu \in \mathscr{A}_M$, there is a Q with $H(Q)\leqq l$ and marginal μ so that $I(\mu)\leqq l$. This is true for all $\mu \in \mathscr{A}_M$. We proved in [2-III] that if, under hypotheses (i)–(v), for all $\mu \in \mathscr{A}_M$, $I(\mu)\leqq l$, then $\mathscr{A}_M$ is tight. The result now follows from Theorem 6.3.

7. Special Properties of the Ornstein-Uhlenbeck Process

In the case when X is not compact, to get appropriate upper bounds required imposing hypotheses (i)–(v) on the Markov process $P_{0,x}$, as we saw in Section 4. We show now that the O–U process in d-dimensions satisfies (i)–(v). Also in this section we prove (Theorem 7.1) an entropy identity for the O–U process which we use in the polaron problem in [6].

For the O–U process in d-dimensions, the infinitesimal generator is

$$L=\tfrac{1}{2}\Delta-kx\cdot\nabla, \tag{7.1}$$

where $k>0$ is an arbitrary positive constant.

To show (i)–(v) are satisfied, define $\theta(x)=x$ for $|x|\leqq 1$, bounded on R_1, θ' and θ'' uniformly bounded on R_1, $\theta'\geqq 0$, and θ chosen to be odd. Define

$$u_n(x)=\prod_{i=1}^{d}\cosh\,(n\theta(x_i/n)).$$

Clearly, $u_n(x)\geqq 1$ for all x and all n so that hypothesis (i) is satisfied. As $n\to\infty$, $u_n(x)\to\prod_{i=1}^{d}\cosh x_i$ uniformly on compact sets which means hypothesis (ii) is satisfied.

Using (7.1), we find, in $d=1$,

$$\begin{aligned}-\frac{Lu_n(x)}{u_n}&=V_n(x)\\ &=-\left\{\frac{1}{2}\left[\theta'\left(\frac{x}{n}\right)\right]^2+\frac{1}{2n}\theta''\left(\frac{x}{n}\right)\tanh\left(n\theta\left(\frac{x}{n}\right)\right)\right.\\ &\qquad\left.-kx\theta'\left(\frac{x}{n}\right)\tanh\left(n\theta\left(\frac{x}{n}\right)\right)\right\}\end{aligned}$$

and hence, showing that there is a constant c such that $V_n(x)\geqq -c$ for all n and all x is equivalent to showing that the expression inside the curly braces is bounded. But this follows from the fact that θ' and θ'' are bounded and $x\tanh(n\theta(x/n))\geqq 0$ for all x. A similar argument holds in d-dimensions. So hypothesis (iii) holds.

If we let $f(x)=\cosh(n\theta(x/n))$ and $\psi_n(x)=(\frac{1}{2}f''-kxf')/f$, then in d-dimensions, $V_n(x)=-\sum_{i=1}^d \psi_n(x_i)$. As $n\to\infty$,

$$V_n(x)\to -\sum_{i=1}^{d}(\tfrac{1}{2}-kx_i\tanh x_i).$$

Thus, if we define

$$V(x)=\sum_{i=1}^{d}(kx_i\tanh x_i-\tfrac{1}{2}),$$

we see that $V_n(x)\to V(x)$ as $n\to\infty$ for all x. Hence, hypothesis (iv) is satisfied. Hypothesis (v) is satisfied because there is a constant c_1 such that, for all x,

$$V(x)\geqq\sum_{i=1}^{d}k|x_i|-c_1.$$

THEOREM 7.1. *Let $H_k(Q)$ be the entropy of a d-dimensional stationary process Q with respect to the d-dimensional* O–U *process with constant $k>0$. Let $H_0(Q)$ be the entropy of Q with respect to d-dimensional Brownian motion. Then,*

$$H_k(Q)=H_0(Q)+\tfrac{1}{2}k^2\int\|x\|^2\mu(dx)-\tfrac{1}{2}dk, \tag{7.2}$$

where μ is the marginal of Q.

Proof: By definition,

$$H_k(Q)=E^Q\{h_{\mathscr{F}_1^0}(P^k_{0,\omega(0)};Q_{0,\omega})\},$$

where $P^k_{0,x}$ is the O–U process with parameter k. Brownian motion is $P^0_{0,x}$ and we now use the fact that $P^k_{0,x}$ and $P^0_{0,x}$ are mutually absolutely continuous. Indeed, for any pair $\alpha\approx\beta$,

$$\begin{aligned} h_\Sigma(\alpha;\lambda)&=E^\lambda\left\{\log\frac{d\lambda}{d\alpha}\bigg|_\Sigma\right\}\\ &=E^\lambda\left\{\log\frac{d\lambda}{d\beta}\bigg|_\Sigma+\log\frac{d\beta}{d\alpha}\bigg|_\Sigma\right\}\\ &=h_\Sigma(\beta,\lambda)+E^\lambda\left\{\log\frac{d\beta}{d\alpha}\bigg|_\Sigma\right\}.\end{aligned}$$

Apply this with $\lambda=Q_{0,\omega}$, $\alpha=P^k_{0,\omega(0)}$, $\beta=P^0_{0,\omega(0)}$ and we get $(\Sigma=\mathscr{F}_1^0)$

$$H_k(Q)=H_0(Q)+E^Q\left\{E^{Q_{0,\omega}}\left\{\log\frac{dP^0_{0,\omega(0)}}{dP^k_{0,\omega(0)}}\bigg|_{\mathscr{F}_1^0}\right\}\right\}. \tag{7.3}$$

From the Cameron-Martin formula,

$$\log\frac{dP^0_{0,\omega(0)}}{dP^k_{0,\omega(0)}}\bigg|_{\mathscr{F}_1^0}=\tfrac{1}{2}k\|\omega(1)\|^2+\tfrac{1}{2}k^2\int_0^1\|\omega(s)\|^2\,ds-\tfrac{1}{2}kd-\tfrac{1}{2}k\|\omega(0)\|^2$$

and since Q is stationary we have

$$E^Q\left\{E^{Q_{0,\omega}}\left\{\log \frac{dP^0_{0,\omega(0)}}{dP^k_{0,\omega(0)}}\bigg|_{\mathcal{F}^0_1}\right\}\right\} = E^Q\left\{\log \frac{dP^0_{0,\omega(0)}}{dP^k_{0,\omega(0)}}\bigg|_{\mathcal{F}^0_1}\right\}$$

$$= E^Q\left\{\tfrac{1}{2}k^2\int_0^1 \|\omega(s)\|^2\, ds\right\} - \tfrac{1}{2}kd$$

$$= \tfrac{1}{2}k^2 \int \|x\|^2 \mu(dx) - \tfrac{1}{2}kd$$

which gives us (7.2).

Acknowledgment. This work was supported by the National Science Foundation, Grant No. MCS-8002568.

Bibliography

[1] Donsker, M. D., and Varadhan, S. R. S., *Asymptotic Evaluation of Certain Wiener Integrals for Large Time*, Functional Integration and its Applications, Proceedings of the International Conference held at the Cumberland Lodge, Windsor Great Park, London, in April 1974. A. M. Arthurs, Ed., Clarendon, Oxford, 1975.

[2] Donsker, M. D., and Varadhan, S. R. S., *Asymptotic evaluation of certain Markov process expectations for large time, I*, Comm. Pure Appl. Math. 28, 1975, pp. 279–301; III, Comm. Pure Appl. Math. 29, 1976, pp. 389–461.

[3] Varadhan, S. R. S., *Asymptotic probabilities and differential equations*, Comm. Pure Appl. Math. 19, 1966, pp. 261–286.

[4] Feynman, R. P., *Statistical Mechanics, a Set of Lectures*, Benjamin, New York, 1972.

[5] Pekar, G. S., *Research on Electron Theory of Crystals*, Translation Series No. AEC-tr. 5575, Physics AEC, Division of Technical Information, Washington, DC, 1963.

[6] Donsker, M. D., and Varadhan, S. R. S., *Asymptotics for the polaron*, to appear.

[7] Oxtoby, J. C., *Ergodic sets*, Bull. Amer. Math. Soc. 58, 1952, pp. 116–136.

Received August, 1982.

Comm. Pure Appl. Math. 36 (1983) 505-528

Asymptotics for the Polaron

M. D. DONSKER

AND

S. R. S. VARADHAN

1. Introduction

Let P_t and $E^{P_t}\{\ \}$ denote, respectively, the probability measure and expectation with respect to three-dimensional Brownian motion $x(\cdot)$ tied down at both ends, i.e., with $x(0)=x(t)=0$. For $\alpha>0$, let

$$G(\alpha, t)=E^{P_t}\left\{\exp\left\{\alpha\int_0^t\int_0^t \frac{e^{-|\sigma-s|}}{|x(\sigma)-x(s)|}\,d\sigma\,ds\right\}\right\}.$$

A long standing problem in statistical mechanics (cf. [1]), the "polaron problem," has been to show that

$$\lim_{t\to\infty}\frac{1}{t}\log G(\alpha, t)=g(\alpha)\quad\text{exists,}\tag{1.1}$$

and, moreover, according to a conjecture of Pekar [2], that

$$\lim_{\alpha\to\infty}\frac{g(\alpha)}{\alpha^2}=g_0\quad\text{exists,}\tag{1.2}$$

with

$$g_0=\sup_{\substack{\phi\in L_2(R^3)\\ \|\phi\|=1}}\left[2\iint\frac{\phi^2(x)\phi^2(y)}{|x-y|}\,dx\,dy-\tfrac{1}{2}\int|\nabla\phi|^2\,dx\right].\tag{1.3}$$

In this paper we prove (1.1) obtaining an expression for $g(\alpha)$ which is explicit enough to allow us to prove also the conjecture (1.2) of Pekar, g_0 being indeed given by (1.3). We make use of large deviation results obtained in our earlier papers [3] and, in particular, [4]. That the polaron problem depends on sharp large deviation theorems is natural, since in determining the asymptotic behavior of $G(\alpha, t)$ for large t it is clear that the three-dimensional Brownian motion paths which contribute the most are those which make $|x(\sigma)-x(s)|$ small. However σ and s must not be so far apart that the contribution is killed by $e^{-|\sigma-s|}$. Thus, the influential paths are those which tend to stay awhile near where they have just been. Since this is not the way "typical" Brownian motion paths behave, we are dealing with probabilities of large deviations.

Our expression for $g(\alpha)$ is an infinite-dimensional variational formula. The fact that, as the coupling constant α becomes infinite, this variational formula behaves as indicated in (1.2) and (1.3) finally involving only a finite-dimensional variational formula is perhaps of significance in other problems of statistical mechanics, e.g., justification of mean field approximations.

The authors are grateful to Elliot Lieb who suspected that the polaron problem might yield to our methods and brought the problem to our attention. Recently in [5] Adamouski, Gerlach and Lesche provide a convincing argument of Pekar's conjecture using some of the earlier results of the present authors. For discussion of the existence and uniqueness of a function ϕ yielding the supremum in (1.3) see Lieb [6].

In Section 2 we give an outline of the proof of (1.1) and indicate the nature of the variational expression for $g(\alpha)$. Also in Section 2 we show formally how this last expression can be handled as $\alpha \to \infty$ yielding (1.2) and (1.3). In Sections 3 and 4 we present the proofs of the heuristic arguments described in Section 2.

2. Outline of the Results

To prove (1.1) we perform successive reductions of the problem. First of all we show that the asymptotic behavior of $G(\alpha, t)$ as $t \to \infty$ is the same if we replace tied-down Brownian motion P_t by standard Brownian motion P. This is justified because for large t the terminal condition makes no essential difference. The proof of these remarks is contained in Theorem 3.3 below.

Next we approximate the integrand in $G(\alpha, t)$ as follows:

$$\begin{aligned}\alpha \int_0^t \int_0^t \frac{e^{-|\sigma - s|}}{|x(\sigma) - x(s)|} d\sigma\, ds &= 2\alpha \int_0^t \int_s^t \frac{e^{-(\sigma - s)}}{|x(\sigma) - x(s)|} d\sigma\, ds \\ &\cong 2\alpha \int_0^t ds \int_s^\infty \frac{e^{-(\sigma - s)}}{|x(\sigma) - x(s)|} d\sigma \\ &= 2\alpha \int_0^t ds \int_0^\infty \frac{e^{-\tau}}{|x(s+\tau) - x(s)|} d\tau \\ &= \int_0^t F(\theta_s \omega)\, ds,\end{aligned}$$

where

$$F(\omega) = 2\alpha \int_0^\infty \frac{e^{-\tau}}{|x(\tau) - x(0)|} d\tau, \qquad \omega = x(\cdot),$$

and $\theta_s \omega = \omega(\cdot + s)$ is the translation operator. To prove (1.1) then becomes the problem of finding

$$\lim_{t \to \infty} \frac{1}{t} \log E^P\left\{\exp\left\{\int_0^t F(\theta_s \omega)\, ds\right\}\right\}, \tag{2.1}$$

where $E^P\{\ \}$ denotes expectation with respect to standard Brownian motion in R^3.

Let Ω be the space of trajectories on $-\infty<t<\infty$. Let Q be any stationary stochastic process with values in R^3 and let $\mathscr{F}_0^{-\infty}$ be the σ-field generated by the process $x(t)$ for $t\leqq 0$. Let $\mathscr{F}_T^0$ be the σ-field generated by the process $x(t)$ for $0\leqq t\leqq T$. We can think of $\mathscr{F}_0^{-\infty}$ and $\mathscr{F}_T^0$ as sub σ-fields of $\mathscr{F}$, the entire σ-field on which Q is a translation invariant measure.

Let Q_ω^T be the regular conditional probability of Q on $\mathscr{F}_T^0$ given $\mathscr{F}_0^{-\infty}$. Denote by $P_{\omega(0)}^T$ the measure on $\mathscr{F}_T^0$ corresponding to a Markov process P starting at time 0 from $\omega(0)$. Let

$$h_T(\omega)=\int \log R_\omega^T(\omega')Q_\omega^T(d\omega')$$

be the entropy where $R_\omega^T(\cdot)$ is the Radon-Nikodym derivative of Q_ω^T with respect to $P_{\omega(0)}^T$ on $\mathscr{F}_T^0$. We then define

$$h(T)=\int h_T(\omega)Q(d\omega).$$

It turns out that either $h(T)=\infty$ for all $T>0$ or $h(T)=HT$ for some constant H which depends on Q. We therefore write $h(T)=T\cdot H(Q)$ for some $0\leqq H(Q)\leqq\infty$ and call $H(Q)$ the entropy of the stationary process Q with respect to the Markov process P. All the details about this entropy are contained in [4] and what is important about $H(Q)$ is that it governs the probabilities of large deviations which is the main topic of [4]. Indeed, we show in [4] that, under suitable hypotheses,

$$\lim_{t\to\infty}\frac{1}{t}\log E^P\left\{\exp\left\{\int_0^t F(\theta_s\omega)\,ds\right\}\right\}=\sup_Q\,[E^Q\{F(\omega)\}-H(Q)], \tag{2.2}$$

where the supremum is taken over all stationary processes Q, $E^P\{\ \}$ is the expectation with respect to the Markov process P, and F is a bounded continuous function on Ω. What we mean by "suitable hypotheses" above is that the Markov process P must satisfy certain restrictions in order to obtain

$$\overline{\lim_{t\to\infty}}\frac{1}{t}\log E^P\left\{\exp\left\{\int_0^t F(\theta_s\omega)\,ds\right\}\right\}\leqq\sup_Q\,[E^Q\{F(\omega)\}-H(Q)] \tag{2.3}$$

and must satisfy certain other restrictions in order to obtain

$$\underline{\lim_{t\to\infty}}\frac{1}{t}\log E^P\left\{\exp\left\{\int_0^t F(\theta_s\omega)\,ds\right\}\right\}\geqq\sup_Q\,[E^Q\{F(\omega)\}-H(Q)]. \tag{2.4}$$

In [4] we show that (2.3) holds if the Markov process P satisfies some strong positive recurrence conditions which in that paper we list as hypotheses (i)–(v).

We show there that conditions (i)–(v) are satisfied by the Ornstein–Uhlenbeck process in particular and so in the present paper we use (2.3) for the Ornstein–Uhlenbeck process and then use that result to get the corresponding upper estimate on (2.1).

For (2.4) we found in [4] two hypotheses needed for the Markov process P which we called I–II. Ordinary Brownian motion satisfies these and therefore we use (2.4) directly to get a lower estimate on (2.1).

Thus, with

$$F(\omega)=2\alpha\int_0^\infty \frac{e^{-\tau}}{|x(\tau)-x(0)|}\,d\tau,$$

and P denoting Brownian motion, we establish in Section 3 below that (1.1) holds by showing

$$\begin{aligned}g(\alpha)&=\lim_{t\to\infty}\frac{1}{t}\log E^P\left\{\exp\left\{\int_0^t F(\theta_s\omega)\,ds\right\}\right\}\\ &=\sup_Q\{E^Q\{F(\omega)\}-H(Q)\},\end{aligned}\tag{2.5}$$

where the supremum is taken over all stationary processes and $H(Q)$ is the entropy of the stationary process Q with respect to ordinary Brownian motion.

In Section 4 of this paper we show that the variational formula for $g(\alpha)$ on the right of (2.5) can be used to verify Pekar's conjecture (1.2) and (1.3). First of all, using Brownian scaling we show in Section 4 that

$$g_0=\lim_{\alpha\to\infty}\frac{g(\alpha)}{\alpha^2}=\lim_{\alpha\to\infty}\sup_Q\left[E^Q\left\{2\int_0^\infty\frac{e^{-t}}{|x(\alpha^2 t)-x(0)|}\,dt\right\}-H(Q)\right].\tag{2.6}$$

In [4] we prove that $H(Q)$ is linear in Q and therefore the supremum involved in (2.6) is over linear functionals of Q and is therefore attained at an extremal which is an ergodic process. Using then the ergodic theorem, we show that roughly speaking

$$\begin{aligned}\lim_{\alpha\to\infty}E^Q\left\{2\int_0^\infty\frac{e^{-t}}{|x(\alpha^2t)-x(0)|}\,dt\right\}&=\lim_{\alpha\to\infty}2\left[\int_0^\infty e^{-t}\,dt\iint\frac{1}{|x-y|}\lambda_Q^{\alpha^2 t}(dx,dy)\right]\\ &=2\int_0^\infty e^{-t}\,dt\iint\frac{1}{|x-y|}\mu_Q(dx)\mu_Q(dy).\end{aligned}$$

Here $\lambda_Q^t(dx,dy)$ is the joint distribution of $x(0)$, $x(t)$ under Q and μ_Q is the marginal distribution of $x(\cdot)$ under Q. We prove in Section 4 that

$$g_0=\lim_{\alpha\to\infty}\frac{g(\alpha)}{\alpha^2}=\sup_Q\left[2\iint\frac{\mu_Q(dx)\mu_Q(dy)}{|x-y|}-H(Q)\right].\tag{2.7}$$

In [4] we proved the contraction principle for the entropy, namely

$$\inf_{Q:\mu_Q=\mu} H(Q)=I(\mu),$$

where $I(\mu)$ is the I function introduced by the authors in [3]. Here, of course, the I function is for Brownian motion and we have shown in the earlier papers that in that case $I(\mu)<\infty$ if and only if $\mu(dx)=\phi^2(x)\,dx$ for some $\phi\in L^2(R^3)$ with $\|\phi\|_2=1$ and $\int|\nabla\phi|^2\,dx<\infty$. In such a case, $I(\mu)=\frac{1}{2}\int|\nabla\phi|^2\,dx$. Thus, from (2.7),

$$\begin{aligned} g_0&=\sup_{\mu}\left[2\iint\frac{\mu(dx)\mu(dy)}{|x-y|}-I(\mu)\right]\\ &=\sup_{\substack{\phi\in L_2(R^3)\\ \|\phi\|=1}}\left[2\iint\frac{\phi^2(x)\phi^2(y)}{|x-y|}\,dx\,dy-\tfrac{1}{2}\int|\nabla\phi|^2\,dx\right].\end{aligned}$$

3. Proof of (1.1)

As before, we let P_t be the measure corresponding to tied-down $(x(0)=x(t)=0)$ Brownian motion in R^3 and P be the measure corresponding to free $(x(0)=0)$ Brownian motion in R^3. Let

$$z(\alpha,\lambda,t_1,t_2)=\alpha\int_{t_1}^{t_2}\int_{t_1}^{t_2}\frac{\lambda e^{-\lambda|\sigma-s|}}{|x(\sigma)-x(s)|}\,d\sigma\,ds,$$

where $\alpha\geqq 0$, $\lambda>0$, and $0\leqq t_1<t_2$. The object of this section is to prove

THEOREM 3.1.

$$\lim_{t\to\infty}\frac{1}{t}\log E^{P_t}\{\exp\{z(\alpha,\lambda,0,t)\}\}=\Phi(\alpha,\lambda),\tag{3.1}$$

where

$$\Phi(\alpha,\lambda)=\sup_{Q}\left[E^{Q}\left\{2\alpha\int_0^\infty\frac{\lambda e^{-\lambda\sigma}}{|x(\sigma)-x(0)|}\,d\sigma\right\}-H(Q)\right],\tag{3.2}$$

the supremum being taken over all stationary processes Q in R^3, and where $H(Q)$ is the entropy of the stationary process Q relative to the Brownian motion P (see Section 2 above for definitions).

It should be noted that statement (1.1) in the introduction is just (3.1) with $\lambda=1$, i.e., $g(\alpha)=\Phi(\alpha,1)$. To prove Theorem 3.1 it will suffice, as will be shown in Theorem 3.3, to prove

THEOREM 3.2

$$\lim_{t\to\infty}\frac{1}{t}\log E^{P}\{\exp\{z(\alpha,\lambda,0,t)\}\}=\Phi(\alpha,\lambda),\tag{3.3}$$

where $\Phi(\alpha,\lambda)$ is given by (3.2).

We first show

THEOREM 3.3. *Theorem* 3.2 *implies Theorem* 3.1.

The proof of this last theorem requires a few preliminaries.

LEMMA 3.4. *If ρ_1 and ρ_2 are covariance functions for two Gaussian processes in R^3 with mean* 0, *and if $\rho_1 \leqq \rho_2$ in the sense of quadratic forms, then*

$$E^{\rho_1}\{\exp\{z(\alpha,\lambda,t_1,t_2)\}\} \geqq E^{\rho_2}\{\exp\{z(\alpha,\lambda,t_1,t_2)\}\}. \tag{3.4}$$

Proof: Since $\alpha \geqq 0$, to show (3.4) it suffices to check that for each $n \geqq 1$,

$$E^{\rho_1}\left\{\left(\int_0^t\int_0^t \frac{e^{-\lambda|\sigma-s|}}{|x(\sigma)-x(s)|}\,d\sigma\,ds\right)^n\right\} \geqq E^{\rho_2}\left\{\left(\int_0^t\int_0^t \frac{e^{-\lambda|\sigma-s|}}{|x(\sigma)-x(s)|}\,d\sigma\,ds\right)^n\right\}.$$

But for this it suffices to show that

$$\begin{aligned}&E^{\rho_1}\left\{\frac{1}{|x(\sigma_1)-x(s_1)|}\cdot\frac{1}{|x(\sigma_2)-x(s_2)|}\cdots\frac{1}{|x(\sigma_n)-x(s_n)|}\right\}\\ &\geqq E^{\rho_2}\left\{\frac{1}{|x(\sigma_1)-x(s_1)|}\cdot\frac{1}{|x(\sigma_2)-x(s_2)|}\cdots\frac{1}{|x(\sigma_n)-x(s_n)|}\right\},\end{aligned} \tag{3.5}$$

for arbitrary choices of $\sigma_1, \sigma_2, \cdots, \sigma_n, s_1, s_2, \cdots, s_n$, in $[0, t]$.

For each $\delta > 0$, $1/(\delta^2+|x|^2)^{1/2}$ is positive definite and since the product of positive definite functions is also positive definite there exists a positive measure μ on R^{6n} such that

$$\begin{aligned}\phi(\xi) &= \frac{1}{(\delta^2+|x(\sigma_1)-x(s_1)|^2)^{1/2}}\cdots\frac{1}{(\delta^2+|x(\sigma_n)-x(s_n)|^2)^{1/2}}\\ &= \int e^{i\langle\xi,\eta\rangle}\mu(d\eta),\end{aligned}$$

where $\xi = (x(\sigma_1), x(s_1), \cdots, x(\sigma_n), x(s_n)) \in R^{6n}$. Thus

$$\begin{aligned}E^{\rho_1}\{\phi(\xi)\} &= \int e^{-\langle\rho_1\eta,\eta\rangle/2}\mu(d\eta)\\ &\geqq \int e^{-\langle\rho_2\eta,\eta\rangle/2}\mu(d\eta)\\ &= E^{\rho_2}\{\phi(\xi)\},\end{aligned}$$

and (3.5) follows by letting $\delta \to 0$.

We shall later use the following obvious

Corollary 3.5. *Let*

$$z_\delta(\alpha,\lambda,t_1,t_2) = \alpha\int_{t_1}^{t_2}\int_{t_1}^{t_2}\frac{\lambda\,e^{-\lambda|\sigma-s|}}{(\delta^2+|x(\sigma)-x(s)|^2)^{1/2}}\,d\sigma\,ds.$$

If ρ_1 and ρ_2 are as in the lemma, then

$$E^{\rho_1}\{\exp\{z_\delta(\alpha, \lambda, t_1, t_2)\}\} \geqq E^{\rho_2}\{\exp\{z_\delta(\alpha, \lambda, t_1, t_2)\}\}.$$

COROLLARY 3.6. *With P_t and P as defined at the beginning of this section,*

$$E^{P_t}\{\exp\{z(\alpha, \lambda, t_1, t_2)\}\} \geqq E^{P}\{\exp\{z(\alpha, \lambda, t_1, t_2)\}\}. \tag{3.6}$$

Proof: In view of Lemma 3.4 it suffices to show that the covariance function ρ_t of the P_t process is less than or equal to the covariance function ρ of the P process in the sense of quadratic forms. But, $\rho(s, u)_{ij} = \delta_{ij} \min(s, u)$ and $\rho_t(s, u)_{ij} = \delta_{ij}[\min(s, u) - su/t]$. Thus, for any $f^i(s)$, $i = 1, 2, 3$,

$$\sum_{i,j=1}^{3} \int_0^\infty \int_0^\infty [\rho(s, u)_{ij} - \rho_t(s, u)_{ij}] f^i(s) f^j(u)\, ds\, du$$

$$= \sum_{i=1}^{3} \frac{1}{t} \left(\int_0^\infty s f^i(s)\, ds \right)^2 \geqq 0.$$

We are now ready for

Proof of Theorem 3.3: From Corollary 3.6 and (3.3) we have

$$\varliminf_{t\to\infty} \frac{1}{t} \log E^{P_t}\{\exp\{z(\alpha, \lambda, 0, t)\}\} \geqq \Phi(\alpha, \lambda);$$

thus we want to show that (3.3) implies the corresponding upper estimate,

$$\varlimsup_{t\to\infty} \frac{1}{t} \log E^{P_t}\{\exp\{z(\alpha, \lambda, 0, t)\}\} \leqq \Phi(\alpha, \lambda). \tag{3.7}$$

To do this, we note first of all that

$$\begin{aligned} z(\alpha, \lambda, 0, t) &= z(\alpha, \lambda, 0, t-1) + z(\alpha, \lambda, t-1, t) \\ &\quad + 2\alpha \int_0^{t-1} \int_{t-1}^{t} \frac{\lambda e^{-\lambda|\sigma - s|}}{|x(\sigma) - x(s)|}\, d\sigma\, ds. \end{aligned} \tag{3.8}$$

Since $1/|x|$ is positive definite, there exists a measure μ on R^3 such that $1/|x| = \int e^{i(\xi, x)} \mu(d\xi)$, and so

$$\int_0^{t-1} \int_{t-1}^{t} \frac{\lambda e^{-\lambda|\sigma - s|}}{|x(\sigma) - x(s)|}\, d\sigma\, ds$$

$$= \int_0^{t-1} \lambda\, d\sigma \int_{t-1}^{t} \frac{e^{-\lambda(s-\sigma)}}{|x(\sigma) - x(s)|}\, ds$$

$$= \int \mu(d\xi) \int_0^{t-1} \lambda e^{\lambda\sigma}\, d\sigma \int_{t-1}^{t} e^{-\lambda s} e^{i(\xi, x(\sigma) - x(s))}\, ds$$

$$
\begin{aligned}
(3.9)\qquad &= \int \mu(d\xi) \int_0^{t-1} \lambda\, e^{\lambda(\sigma-t+1)} e^{i\langle \xi, x(\sigma)\rangle}\, d\sigma \int_{t-1}^{t} e^{\lambda(t-1-s)} e^{-i\langle \xi, x(s)\rangle}\, ds \\
&\leqq \left[\int \mu(d\xi) \left| \int_0^{t-1} \lambda\, e^{\lambda(\sigma-t+1)} e^{i\langle \xi, x(\sigma)\rangle}\, d\sigma \right|^2\right]^{1/2} \\
&\quad \times \left[\int \mu(d\xi) \left| \int_{t-1}^{t} e^{\lambda(t-1-s)} e^{-i\langle \xi, x(s)\rangle}\, ds \right|^2\right]^{1/2} \\
&= \left[\int \mu(d\xi) \int_0^{t-1}\int_0^{t-1} \lambda^2 e^{\lambda[\sigma+\tau-2(t-1)]} e^{i\langle \xi, x(\sigma)-x(\tau)\rangle}\, d\sigma\, d\tau\right]^{1/2} \\
&\quad \times \left[\int \mu(d\xi) \int_{t-1}^{t}\int_{t-1}^{t} e^{\lambda[2(t-1)-s-u]} e^{-i\langle \xi, x(s)-x(u)\rangle}\, ds\, du\right]^{1/2} \\
&= \left[\int_0^{t-1}\int_0^{t-1} \frac{\lambda\, e^{[\sigma+\tau-2(t-1)]}}{|x(\sigma)-x(\tau)|}\, d\sigma\, d\tau\right]^{1/2} \\
&\quad \times \left[\int_{t-1}^{t}\int_{t-1}^{t} \frac{\lambda\, e^{\lambda[2(t-1)-s-u]}}{|x(s)-x(u)|}\, ds\, du\right]^{1/2}.
\end{aligned}
$$

Now, $\sigma \leqq t-1$ and $\tau \leqq t-1$ imply $\sigma+\tau-2(t-1) \leqq -|\sigma-\tau|$. Similarly, $t-1 \leqq s$ and $t-1 \leqq u$ imply $2(t-1)-s-u \leqq -|s-u|$. Thus, from (3.9) we conclude that

$$
\int_0^{t-1}\int_{t-1}^{t} \frac{\lambda\, e^{-\lambda|\sigma-s|}}{|x(\sigma)-x(s)|}\, d\sigma\, ds \leqq \left[\int_0^{t-1}\int_0^{t-1} \frac{\lambda\, e^{-\lambda|\sigma-\tau|}}{|x(\sigma)-x(\tau)|}\, d\sigma\, d\tau\right]^{1/2} \times \left[\int_{t-1}^{t}\int_{t-1}^{t} \frac{\lambda\, e^{-\lambda|s-u|}}{|x(s)-x(u)|}\, ds\, du\right]^{1/2}.
$$

Using this in (3.8), we have

$$
\begin{aligned}
z(\alpha, \lambda, 0, t) &\leqq z(\alpha, \lambda, 0, t-1) + z(\lambda, \lambda, t-1, t) \\
&\quad + 2[z(\alpha, \lambda, 0, t-1) z(\alpha, \lambda, t-1, t)]^{1/2},
\end{aligned}
$$

and since $2\sqrt{xy} = 2\sqrt{\theta x \cdot y/\theta} \leqq \theta x + (1/\theta) y$ for any $\theta > 0$, we get

$$
(3.10)\qquad z(\alpha, \lambda, 0, t) \leqq (1+\theta) z(\alpha, \lambda, 0, t-1) + (1+1/\theta) z(\alpha, \lambda, t-1, t),
$$

for any $\theta > 0$.

Using (3.10) and the Hölder inequality, we get for $p>1$, $q>1$, such that $1/p+1/q=1$,

$$
\begin{aligned}
&E^{P_1}\left\{\exp\left\{\alpha \int_0^t\int_0^t \frac{\lambda\, e^{-\lambda|\sigma-s|}}{|x(\sigma)-x(s)|}\, d\sigma\, ds\right\}\right\} \\
&\quad \leqq E^{P_1}\left\{\exp\left\{\alpha(1+\theta) \int_0^{t-1}\int_0^{t-1} \frac{\lambda\, e^{-\lambda|\sigma-\tau|}}{|x(\sigma)-x(\tau)|}\, d\sigma\, d\tau\right\}\right.
\end{aligned}
$$

$$\times \exp\left\{\alpha\left(1+\frac{1}{\theta}\right)\int_{t-1}^{t}\int_{t-1}^{t}\frac{\lambda\, e^{-\lambda|s-u|}}{|x(s)-x(u)|}\,ds\,du\right\}\right\}$$

$$\leqq\left[E^{P_t}\left\{\exp\left\{p\alpha(1+\theta)\int_{0}^{t-1}\int_{0}^{t-1}\frac{\lambda\, e^{-\lambda|\sigma-\tau|}}{|x(\sigma)-x(\tau)|}\,d\sigma\,d\tau\right\}\right\}\right]^{1/p}$$

$$\times\left[E^{P_t}\left\{\exp\left\{q\alpha\left(1+\frac{1}{\theta}\right)\int_{t-1}^{t}\int_{t-1}^{t}\frac{\lambda\, e^{-\lambda|s-u|}}{|x(s)-x(u)|}\,ds\,du\right\}\right\}\right]^{1/q}.$$

For tied-down Brownian motion, the measure P_t is reversible in time, i.e., the probability distribution of

$$\int_{t-1}^{t}\int_{t-1}^{t}\frac{\lambda\, e^{-\lambda|s-u|}}{|x(s)-x(u)|}\,ds\,du$$

is the same as the probability distribution of

$$\int_{0}^{1}\int_{0}^{1}\frac{\lambda\, e^{-\lambda|s-u|}}{|x(s)-x(u)|}\,ds\,du$$

so that the last inequality can be rewritten as

$$E^{P_t}\left\{\exp\left\{\alpha\int_{0}^{t}\int_{0}^{t}\frac{\lambda\, e^{-\lambda|\sigma-s|}}{|x(\sigma)-x(s)|}\,d\sigma\,ds\right\}\right\}$$

$$\leqq\left[E^{P_t}\left\{\exp\left\{p\alpha(1+\theta)\int_{0}^{t-1}\int_{0}^{t-1}\frac{\lambda\, e^{-\lambda|\sigma-\tau|}}{|x(\sigma)-x(\tau)|}\,d\sigma\,d\tau\right\}\right\}\right]^{1/p}$$
$$\times\left[E^{P_t}\left\{\exp\left\{q\alpha\left(1+\frac{1}{\theta}\right)\int_{0}^{1}\int_{0}^{1}\frac{\lambda\, e^{-\lambda|s-u|}}{|x(s)-x(u)|}\,ds\,du\right\}\right\}\right]^{1/q}. \tag{3.11}$$

For any $t'<t$, the measure P_t is absolutely continuous with respect to the measure P on the σ-field up to time t' which we shall denote by $\mathcal{F}_{t'}$. Moreover,

$$\left.\frac{dP_t}{dP}\right|_{\mathcal{F}_{t'}}=\left(\frac{t}{t-t'}\right)^{3/2}\exp\left\{-\frac{\|x(t')\|^2}{2(t-t')}\right\}\leqq\left(\frac{t}{t-t'}\right)^{3/2}.$$

Thus, for any functional G measurable with respect to $\mathcal{F}_{t'}$, $E^{P_t}\{G[x]\}\leqq(t/(t-t'))^{3/2}E^{P}\{G[x]\}$. Using this in (3.11), we obtain

$$E^{P_t}\left\{\exp\left\{\alpha\int_{0}^{t}\int_{0}^{t}\frac{\lambda\, e^{-\lambda|\sigma-s|}}{|x(\sigma)-x(s)|}\,d\sigma\,ds\right\}\right\}$$

$$\leqq t^{3/2p}\left[E^{P}\left\{\exp\left\{p\alpha(1+\theta)\int_{0}^{t-1}\int_{0}^{t-1}\frac{\lambda\, e^{-\lambda|\sigma-\tau|}}{|x(\sigma)-x(\tau)|}\,d\sigma\,d\tau\right\}\right\}\right]^{1/p}$$
$$\times\left(\frac{t}{t-1}\right)^{3/2q}\left[E^{P}\left\{\exp\left\{q\alpha\left(1+\frac{1}{\theta}\right)\int_{0}^{1}\int_{0}^{1}\frac{\lambda\, e^{-\lambda|s-u|}}{|x(s)-x(u)|}\,ds\,du\right\}\right\}\right]^{1/q}. \tag{3.12}$$

Hence, from (3.3) and (3.12) we get, for each $\theta > 0$ and $p > 1$,

$$\text{(3.13)}\quad \overline{\lim_{t\to\infty}} \frac{1}{t} \log E^{P_t}\left\{\exp\left\{\alpha \int_0^t \int_0^t \frac{\lambda e^{-\lambda|\sigma - s|}}{|x(\sigma) - x(s)|} d\sigma\, ds\right\}\right\} \leqq \frac{1}{p} \Phi(p\alpha(1+\theta), \lambda).$$

Now, Φ is a finite convex function of α and therefore continuous, which means that (3.13) implies (3.7) by letting $p \to 1$ and $\theta \to 0$. This completes the proof of Theorem 3.3.

Before proving Theorem 3.2 we need some lemmas.

LEMMA 3.7. *For any $\delta > 0$, let*

$$V_\delta(x) = \frac{1}{|x|} - \frac{1}{(\delta^2 + |x|^2)^{1/2}}.$$

Let P_x denote Brownian motion measure in R^3 for paths starting at x. Then,

$$\text{(3.14)}\qquad \lim_{\delta\to 0} \sup_x E^{P_x}\left\{\int_0^1 V_\delta(x(\sigma))\, d\sigma\right\} = 0.$$

Proof: Since

$$V_\delta(x) \leqq \frac{1}{|x|} - \frac{1}{\delta + |x|} = \frac{\delta}{|x|(\delta + |x|)},$$

it suffices to show that

$$\lim_{\delta\to 0} \sup_x E^{P_x}\left\{\int_0^1 \frac{\delta}{|x(\sigma)|(\delta + |x(\sigma)|)} d\sigma\right\} = 0.$$

Using Brownian scaling, we have, for any $0 < \varepsilon < 1$,

$$\begin{aligned}\int_0^1 E^{P_x}\left\{\frac{\delta}{|x(\sigma)|(\delta + |x(\sigma)|)}\right\} d\sigma &= \int_0^1 \frac{d\sigma}{\sqrt{\sigma}} E^{P_{x/\sqrt{\sigma}}}\left\{\frac{\delta}{|x(1)|(\delta + \sqrt{\sigma}|x(1)|)}\right\} \\ &\leqq \left(\int_0^\varepsilon \frac{d\sigma}{\sqrt{\sigma}}\right) \sup_x E^{P_x}\left\{\frac{1}{|x(1)|}\right\} \\ &\quad + \left(\int_\varepsilon^1 \frac{d\sigma}{\sqrt{\sigma}}\right) \sup_x E^{P_x}\left\{\frac{\delta}{\sqrt{\varepsilon}|x(1)|^2}\right\} \\ &= I_1 + I_2.\end{aligned}$$

But

$$\begin{aligned}E^{P_x}\left\{\frac{1}{|x(1)|}\right\} &= \frac{1}{(2\pi)^{3/2}} \int \frac{\exp\{-\frac{1}{2}|y-x|^2\}}{|y|} dy \\ &\leqq c_1\left(\int_{|y|\leqq 1} \frac{dy}{|y|} + \int_{|y|\geqq 1} \frac{\exp\{-\frac{1}{2}|y-x|^2\}}{2} dy\right) \\ &\leqq c_1\left(\int_{|y|\leqq 1} \frac{dy}{|y|} + \int \exp\{-\tfrac{1}{2}|y|^2\}/dy\right).\end{aligned}$$

Thus, $I_1 \leqq c_2\sqrt{\varepsilon}$. Similarly $I_2 \leqq c_3\delta/\sqrt{\varepsilon}$, and hence $\overline{\lim}_{\delta\to 0}(I_1+I_2) \leqq c_2\sqrt{\varepsilon}$ which implies the lemma.

LEMMA 3.8. *With $V_\delta(x)$ as defined in the preceding lemma, for any $\alpha > 0$ and $\lambda > 0$,*

$$\lim_{\delta\to 0}\overline{\lim_{t\to\infty}}\frac{1}{t}\log E^P\left\{\exp\left\{\alpha\int_0^t\int_0^t V_\delta(x(\sigma)-x(s))\,e^{-\lambda|\sigma-s|}\lambda\,d\sigma\,ds\right\}\right\}=0. \tag{3.15}$$

Proof: For simplicity of exposition we set $\alpha = \frac{1}{2}$ and $\lambda = 1$. Let $G(\omega) = \int_0^\infty e^{-u}V_\delta(x(u)-x(0))\,du$, and let

$$F(\omega)=\tfrac{1}{2}\int_0^t\int_0^t V_\delta(x(\sigma)-x(s))\,e^{-|\sigma-s|}\,d\sigma\,ds$$
$$=\int_0^t ds\int_s^t e^{-(\sigma-s)}V_\delta(x(\sigma)-x(s))\,d\sigma.$$

Using the notation introduced in Section 2,

$$G(\theta_s\omega)=\int_0^\infty e^{-u}V_\delta(x(u+s)-x(s))\,du=\int_s^\infty e^{-(\sigma-s)}V_\delta(x(\sigma)-x(s))\,d\sigma,$$

and so $F(\omega) \leqq \int_0^t G(\theta_s\omega)\,ds$. Let $G_n(\omega) = \int_n^{n+1} V_\delta(x(u)-x(0))\,du$ and note that $G(\omega) \leqq \sum_{n=0}^\infty e^{-n}G_n(\omega)$. Using these inequalities and Hölder's inequality, we get

$$E^P\{\exp\{F(\omega)\}\}\leqq E^P\left\{\exp\left\{\sum_{n=0}^\infty e^{-n}\int_0^t G_n(\theta_s\omega)\,ds\right\}\right\}$$
$$\leqq\prod_{n=0}^\infty\left(E^P\left\{\exp\left\{2^{n+1}e^{-n}\int_0^t G_n(\theta_s\omega)\,ds\right\}\right\}\right)^{1/2^{n+1}}.$$

Thus, letting $\beta = \sup_n (n+1)2^{n+1}e^{-n} < \infty$, we have

$$\log E^P\{\exp\{F(\omega)\}\}\leqq\sum_{n=0}^\infty\frac{1}{2^{n+1}}\log E^P\left\{\exp\left\{2^{n+1}e^{-n}\int_0^t G_n(\theta_s\omega)\,ds\right\}\right\} \tag{3.16}$$
$$\leqq\sup_n\log E^P\left\{\exp\left\{\frac{\beta}{n+1}\int_0^t G_n(\theta_s\omega)\,ds\right\}\right\}.$$

It follows from Lemma 4.1 of [4] that

$$\log E^P\left\{\exp\left\{\frac{\beta}{n+1}\int_0^t G_n(\theta_s\omega)\,ds\right\}\right\}\leqq t\log\sup_x E^{P_x}\{\exp\{\beta G_n(\omega)\}\}, \tag{3.17}$$

where P_x denotes ordinary Brownian motion in R^3 starting from x. By Lemma 3.7, we can take $\delta > 0$ so small that $\beta\sup_x E^{P_x}\{\int_0^1 V_\delta(x(\sigma))\,d\sigma\} < 1$. For such

$\delta > 0$, we have

$$\sup_x E^{P_x}\{\exp\{\beta G_n(\omega)\}\} = \sup_x E^{P_x}\left\{\exp\left\{\beta \int_n^{n+1} V_\delta(x(\sigma)-x(0))\,d\sigma\right\}\right\}$$

$$\text{(3.18)} \qquad \leqq \sup_x E^{P_x}\left\{\exp\left\{\beta \int_0^1 V_\delta(x(\sigma))\,d\sigma\right\}\right\}$$

$$\leqq \frac{1}{1-\beta\sup_x E^{P_x}\{\int_0^1 V_\delta(x(\sigma))\,d\sigma\}}.$$

From (3.16), (3.17), and (3.18) we obtain, for sufficiently small $\delta > 0$,

$$\varlimsup_{t\to\infty} \frac{1}{t}\log E^P\{\exp\{F(\omega)\}\} \leqq \log\left(\frac{1}{1-\beta\sup_x E^{P_x}\{\int_0^1 V_\delta(x(\sigma))\,d\sigma\}}\right).$$

Now, (3.15) follows by letting $\delta \to 0$ and applying Lemma 3.7.

LEMMA 3.9. *Let R_1 and R_2 be two Gaussian processes with mean 0 and covariance functions ρ_1 and ρ_2, respectively. If $\mathcal{F}_T^0$ denotes the σ-field up to time T and if $dR_2/dR_1|_{\mathcal{F}_T^0}$ is bounded for each T, then $\rho_2 \leqq \rho_1$ in the sense of quadratic forms.*

Proof: For any function $f(s)$ we have

$$E^{R_2}\left\{\exp\left\{\int_0^T f(s)x(s)\,ds\right\}\right\} = E^{R_1}\left\{\exp\left\{\int_0^T f(s)x(s)\,ds\right\}\frac{dR_2}{dR_1}\bigg|_{\mathcal{F}_T^0}\right\}$$

$$\leqq c_T E^{R_1}\left\{\exp\left\{\int_0^T f(s)x(s)\,ds\right\}\right\}$$

for some constant c_T. This implies that

$$\exp\{\tfrac{1}{2}\langle\rho_2 f, f\rangle_T\} \leqq c_T \exp\{\tfrac{1}{2}\langle\rho_1 f, f\rangle_T\}.$$

Since f is arbitrary, for any a,

$$\exp\{\tfrac{1}{2}a^2\langle\rho_2 f, f\rangle_T\} \leqq c_T \exp\{\tfrac{1}{2}a^2\langle\rho_1 f, f\rangle_T\},$$

and hence

$$\exp\{\tfrac{1}{2}\langle\rho_2 f, f\rangle_T\} \leqq c_T^{1/a^2} \exp\{\tfrac{1}{2}\langle\rho_1 f, f\rangle_T\}.$$

Letting $a \to \infty$, we obtain $\langle\rho_2 f, f\rangle_T \leqq \langle\rho_1 f, f\rangle_T$ for any T.

LEMMA 3.10. *Let P^ε denote the probability measure corresponding to the Ornstein–Uhlenbeck* (O-U) *process in R^3 starting from the origin with generator $\frac{1}{2}\Delta - \varepsilon x \cdot \nabla$. Then,*

$$\text{(3.19)} \qquad E^P\{\exp\{z_\delta(\alpha, \lambda, 0, t)\}\} \leqq E^{P^\varepsilon}\{\exp\{z_\delta(\alpha, \lambda, 0, t)\}\},$$

where P, as before, is ordinary Brownian motion in R^3 *starting from the origin and* $z_\delta(\alpha, \lambda, 0, t)$ *is defined in Corollary* 3.5.

Proof: By the Cameron–Martin–Girsanov formula, we have, for any $T<\infty$,

$$\begin{aligned}\frac{dP^\varepsilon}{dP}\bigg|_{\mathcal{F}_T^0} &= \exp\left\{-\varepsilon\int_0^T x(s)\,dx(s)-\tfrac{1}{2}\varepsilon^2\int_0^T |x(s)|^2\,ds\right\}\\ &= \exp\left\{-\tfrac{1}{2}\varepsilon|x(T)|^2+\tfrac{3}{2}\varepsilon T-\tfrac{1}{2}\varepsilon^2\int_0^T |x(s)|^2\,ds\right\}\\ &\leqq \exp\{\tfrac{3}{2}\varepsilon T\}.\end{aligned}$$

From Lemma 3.9 we conclude that $\rho_{P_\varepsilon}\leqq\rho_P$ in the sense of quadratic forms and (3.19) then follows from Corollary 3.5.

With entropy as defined in Section 2 we now state a lemma the proof of which is given in [4] (Theorem 7.1 of that paper).

LEMMA 3.11. *Let* $H_\varepsilon(Q)$ *be the entropy of the stationary process* Q *(in* R^3*) relative to the Ornstein–Uhlenbeck process* P^ε*, and let* $H(Q)$ *be the entropy of* Q *relative to the Brownian motion P. Then,*

$$H_\varepsilon(Q)=H(Q)+\tfrac{1}{2}\varepsilon^2\int|\omega(0)|^2Q(d\omega)-\tfrac{3}{2}\varepsilon. \tag{3.20}$$

LEMMA 3.12. *For each* $\varepsilon>0$,

$$\begin{aligned}&\overline{\lim_{t\to\infty}}\frac{1}{t}\log E^P\{\exp\{z_\delta(\alpha,\lambda,0,t)\}\}\\ &\leqq\sup_Q\left[E^Q\left\{2\alpha\int_0^\infty\frac{\lambda e^{-\lambda\sigma}}{(\delta^2+|x(\sigma)-x(0)|^2)^{1/2}}\,d\sigma\right\}-H_\varepsilon(Q)\right],\end{aligned}\tag{3.21}$$

where the supremum is taken over all stationary processes Q in R^3 *and where* $H_\varepsilon(Q)$ *is the entropy of Q relative to the* O–U *process* P^ε *defined in Lemma* 3.10. *Moreover, for each* $\alpha>0$, $\lambda>0$, *and* $\delta>0$,

$$\overline{\lim_{t\to\infty}}\frac{1}{t}\log E^P\{\exp\{z_\delta(\alpha,\lambda,0,t)\}\}\leqq\Phi(\alpha,\lambda), \tag{3.22}$$

where $\Phi(\alpha,\lambda)$ *is given by* (3.2).

Proof: In order to show (3.21), it suffices, in view of Lemma 3.10, to show that

$$\begin{aligned}&\overline{\lim_{t\to\infty}}\frac{1}{t}\log E^{P^\varepsilon}\{\exp\{z_\delta(\alpha,\lambda,0,t)\}\}\\ &\leqq\sup_Q\left[E^Q\left\{2\alpha\int_0^\infty\frac{\lambda e^{-\lambda\sigma}}{(\delta^2+|x(\sigma)-x(0)|^2)^{1/2}}\,d\sigma\right\}-H_\varepsilon(Q)\right].\end{aligned}\tag{3.23}$$

Let

$$F_\delta(\omega)=2\alpha\int_0^\infty \frac{\lambda\, e^{-\lambda\sigma}}{(\delta^2+|\omega(\sigma)-\omega(0)|^2)^{1/2}}\,d\sigma.$$

Since stationary processes have no fixed discontinuities, $\int F_\delta(\omega)Q(d\omega)$ is a bounded continuous function from $\mathcal{M}_S(\Omega)$ to R^1. Using the notation of [4], we observe that

$$\begin{aligned}\int F_\delta(\omega')R_{t,\omega}(d\omega') &= \frac{1}{t}\int_0^t F_\delta(\theta_s\omega_t)\,ds\\ &= \frac{2\alpha}{t}\int_0^t ds\int_0^\infty \frac{\lambda e^{-\lambda\sigma}}{(\delta^2+|\omega_t(\sigma+s)-\omega_t(s)|^2)^{1/2}}\,d\sigma\\ &= \frac{2\alpha}{t}\int_0^t ds\left[\int_0^{t-s} \frac{\lambda\, e^{-\lambda\sigma}}{(\delta^2+|x(\sigma+s)-x(s)|^2)^{1/2}}\,d\sigma\right.\\ &\qquad \left.+\int_{t-s}^\infty \frac{\lambda\, e^{-\lambda\sigma}}{(\delta^2+|\omega_t(\sigma+s)-\omega_t(s)|^2)^{1/2}}\,d\sigma\right].\end{aligned}$$

Therefore,

$$\left|t\int F_\delta(\omega')R_{t,\omega}(d\omega')-z_\delta(\alpha,\lambda,0,t)\right|\leqq \frac{2\alpha}{\delta}\int_0^t ds\int_{t-s}^\infty \lambda\, e^{-\lambda\sigma}\,d\sigma \leqq \frac{2\alpha}{\delta\lambda}, \tag{3.24}$$

and hence,

$$E^{P^\varepsilon}\{\exp\{z_\delta(\alpha,\lambda,0,t)\}\}\leqq E^{P^\varepsilon}\left\{\exp\left\{t\int F_\delta(\omega')R_{t,\omega}(d\omega')\right\}\right\}e^{2\alpha/\delta\lambda}. \tag{3.25}$$

The Ornstein–Uhlenbeck process satisfies the hypotheses (i)–(v) for Theorem 4.6 of [4] (see e.g. Section 7 of [4]) and thus it follows from (3.25) that

$$\varlimsup_{t\to\infty}\frac{1}{t}\log E^{P^\varepsilon}\{\exp\{z_\delta(\alpha,\lambda,0,t)\}\}\leqq \sup_Q\,[E^Q\{F_\delta(\omega)\}-H_\varepsilon(Q)],$$

which is (3.23). Hence, we have (3.21) for every $\varepsilon>0$ and so, using Lemma 3.11, we get

$$\begin{aligned}&\varlimsup_{t\to\infty}\frac{1}{t}\log E^{P}\{\exp\{z_\delta(\alpha,\lambda,0,t)\}\}\\ &\leqq \inf_{\varepsilon>0}\sup_Q\left[E^Q\left\{2\alpha\int_0^\infty \frac{\lambda\, e^{-\lambda\sigma}}{|x(\sigma)-x(0)|}\,d\sigma\right\}-H(Q)+\tfrac{3}{2}\varepsilon\right].\\ &-\Phi(\alpha,\lambda).\end{aligned}$$

We are now ready to prove Theorem 3.2.

Proof of Theorem 3.2: First of all, let $U_\delta=1/(\delta^2+|x|^2)^{1/2}$ and, as before,

$V_\delta = 1/|x| - 1/(\delta^2+|x|^2)^{1/2}$, so that $1/|x| = U_\delta + V_\delta$. Using Hölder's inequality and noting that

$$z_\delta(\alpha, \lambda, 0, t) = \alpha \int_0^t \int_0^t U_\delta(x(\sigma)-x(s))\lambda\, e^{-\lambda|\sigma-s|}\, d\sigma\, ds,$$

we have

$$\begin{aligned} &E^P\{\exp\{z(\alpha,\lambda,0,t)\}\} \\ &\leqq \left(E^P\left\{\exp\left\{p\alpha \int_0^t\int_0^t U_\delta(x(\sigma)-x(s))\, e^{-\lambda|\sigma-s|}\lambda\, d\sigma\, ds\right\}\right\}\right)^{1/p} \\ &\quad \times \left(E^P\left\{\exp\left\{q\alpha \int_0^t\int_0^t V_\delta(x(\sigma)-x(s))\, e^{-\lambda|\sigma-s|}\lambda\, d\sigma\, ds\right\}\right\}\right)^{1/q}, \end{aligned}$$

and

$$\begin{aligned} &\overline{\lim_{t\to\infty}} \frac{1}{t} \log E^P\{\exp\{z(\alpha.\lambda,0,t)\}\} \\ (3.26)\qquad &\leqq \frac{1}{p} \overline{\lim_{t\to\infty}} \frac{1}{t} \log E^P\{\exp\{z_\delta(p\alpha,\lambda,0,t)\}\} \\ &\quad + \frac{1}{q} \overline{\lim_{t\to\infty}} \frac{1}{t} \log E^P\left\{\exp\left\{q\alpha\int_0^t\int_0^t V_\delta(x(\sigma)-x(s))\, e^{-\lambda|\sigma-s|}\lambda\, d\sigma\, ds\right\}\right\}. \end{aligned}$$

From (3.26) and (3.22) we then get

$$\begin{aligned} \overline{\lim_{t\to\infty}} \frac{1}{t} \log E^P\{\exp\{z(\alpha,\lambda,0,t)\}\} &\leqq \frac{1}{p}\,\Phi(p\alpha,\lambda) \\ &\quad + \frac{1}{q} \overline{\lim_{t\to\infty}} \frac{1}{t} \log E^P\left\{\exp\left\{q\alpha\int_0^t\int_0^t V_\delta(x(\sigma)-x(s))\, e^{-\lambda|\sigma-s|}\lambda\, d\sigma\, ds\right\}\right\}, \end{aligned}$$

which, recalling Lemma 3.8 and letting $\delta\to 0$, yields

$$(3.27)\qquad \overline{\lim_{t\to\infty}} \frac{1}{t} \log E^P\{\exp\{z(\alpha,\lambda,0,t)\}\} \leqq \frac{1}{p}\,\Phi(p\alpha,\lambda).$$

As noted earlier, $\Phi(\alpha,\lambda)$ is continuous in α and therefore if we let $p\to 1$ we obtain from (3.27) the desired upper estimate.

For the corresponding lower estimate we use again (3.24) to get

$$\begin{aligned} &E^P\{\exp\{z(\alpha,\lambda,0,t)\}\} \geqq E^P\{\exp\{z_\delta(\alpha,\lambda,0,t)\}\} \\ (3.28)\qquad &\geqq E^P\left\{\exp\left\{t\int F_\delta(\omega')R_{t,\omega}(d\omega')\right\}\right\} e^{-2\alpha/\delta\lambda}. \end{aligned}$$

Now Brownian motion satisfies hypotheses I–II of Theorem 5.5 in [4] and thus we obtain from (3.28)

$$\varliminf_{t\to\infty}\frac{1}{t}\log E^{P}\{\exp\{z(\alpha,\lambda,0,t)\}\}\geqq\varliminf_{t\to\infty}\frac{1}{t}\log E^{P}\Big\{\exp\Big\{t\int F_{\delta}(\omega')R_{t,\omega}(d\omega')\Big\}\Big\}$$
$$\geqq\sup_{Q}[E^{Q}\{F_{\delta}(\omega)\}-H(Q)]. \tag{3.29}$$

Letting $\delta\to 0$ in (3.29) we get the desired lower estimate and hence complete the proof of Theorem 3.2.

4. Proof of Pekar's Conjecture

In Section 3 we proved that if

$$G(\alpha,t)=E^{P_t}\Big\{\exp\Big\{\alpha\int_0^t\int_0^t\frac{e^{-|\sigma-s|}}{|x(\sigma)-x(s)|}\,d\sigma\,ds\Big\}\Big\},$$

then

$$\lim_{t\to\infty}\frac{1}{t}\log G(\alpha,t)=g(\alpha)\quad\text{exists,}$$

and in fact,

$$g(\alpha)=\Phi(\alpha,1)=\sup_{Q}\Big[E^{Q}\Big\{2\alpha\int_0^{\infty}\frac{e^{-\sigma}}{|x(\sigma)-x(0)|}\,d\sigma\Big\}-H(Q)\Big].$$

In this section we want to prove Pekar's conjecture that

$$\lim_{\alpha\to\infty}\frac{g(\alpha)}{\alpha^2}=g_0\quad\text{exists}$$

and

$$g_0=\sup_{\substack{\phi\in L_2(R^3)\\ \|\phi\|=1}}\Big[2\iint\frac{\phi^2(x)\phi^2(y)}{|x-y|}\,dx\,dy-\tfrac{1}{2}\int|\nabla\phi|^2\Big].$$

First of all, if we let

$$K(\alpha,\lambda,t)=E^{P}\{\exp\{z(\alpha,\lambda,0,t\}\},$$

then using Brownian scaling we observe that, for any $\tau>0$, $K(\alpha,\lambda,t)=K(\alpha\sqrt{\tau},\lambda\tau,t/\tau)$. Thus, from Theorem 3.2 we conclude that, for any $\tau>0$, $\Phi(\alpha,\lambda)=(1/\tau)\Phi(\alpha\sqrt{\tau},\lambda\tau)$. In particular, this means $\Phi(\alpha,\lambda)=\alpha^2\Phi(1,\lambda/\alpha^2)$, and so $g(\alpha)/\alpha^2=1/\alpha^2\Phi(\alpha,1)=\Phi(1,1/\alpha^2)$. Hence, to prove Pekar's conjecture, it suffices to show

$$\lim_{\lambda\to 0}\Phi(1,\lambda)=g_0=\sup_{\substack{\phi\in L_2(R^3)\\ \|\phi\|=1}}\Big[2\iint\frac{\phi^2(x)\phi^2(y)}{|x-y|}\,dx\,dy-\tfrac{1}{2}\int|\nabla\phi|^2\Big], \tag{4.1}$$

where

$$\Phi(1,\lambda)=\sup_Q\left[E^Q\left\{2\int_0^\infty \frac{\lambda\, e^{-\lambda\sigma}}{|x(\sigma)-x(0)|}\,d\sigma\right\}-H(Q)\right].$$

We first prove a theorem (Theorem 4.1 below) in an abstract setting which we shall use in the proof of (4.1).

Let X be a Polish space and let T_t, $t \geqq 0$, be a one-parameter family of homeomorphisms of X. Let $\mathcal{M}$ be the space of probability measures on X and $\mathcal{M}_I \subset \mathcal{M}$ be the measures on X invariant under T_t.

For each $\lambda > 0$, define a map $\alpha_{\lambda,x}: X \to \mathcal{M}$ by

$$\alpha_{\lambda,x}(A)=\lambda\int_0^\infty e^{-\lambda t}\chi_A(T_t x)\,dt,$$

where $A \subset X$. For $P \in \mathcal{M}_I$, define $\tilde{P}_\lambda = P\alpha_{\lambda,x}^{-1}$ so that $\tilde{P}_\lambda$ is a probability measure on $\mathcal{M}$, i.e., if $G \subset \mathcal{M}$, then

$$\tilde{P}_\lambda(G)=P\{x\in X: \alpha_{\lambda,x}\in G\}.$$

We will have several occasions to use an elementary identity implied by these definitions. Let $A \subset X$, $P \in \mathcal{M}_I$, and $\lambda > 0$, then

$$\tag{4.2} \begin{aligned} E^P\{\alpha_{\lambda,x}(A)\} &= \lambda\int_0^\infty e^{-\lambda t}E^P\{\chi_A(T_t x)\}\,dt \\ &= \lambda\int_0^\infty e^{-\lambda t}E^P\{\chi_A(x)\}\,dt = \lambda\int_0^\infty e^{-\lambda t}P(A)\,dt = P(A). \end{aligned}$$

In terms of the $\tilde{P}_\lambda$ measure on $\mathcal{M}$ we can express (4.2) as

$$\tag{4.3} E^{\tilde{P}_\lambda}\{Q(A)\}=P(A).$$

In what follows let $\phi(x, y) \in C(X\times X)$ and assume further that $\phi(x, y)$ is symmetric.

Theorem 4.1. *Let $P_n \in M_I$ for $n = 1, 2, \cdots$ and assume $P_n \Rightarrow P$. Let $\lambda_n \to 0$ as $n \to \infty$ and define $\tilde{P}_{n,\lambda_n} = P_n\alpha_{\lambda_n,x}^{-1}$,* i.e., *if $G \subset \mathcal{M}$, $\tilde{P}_{n,\lambda_n}(G) = P_n\{x \in X: \alpha_{\lambda_n,x} \in G\}$. Then:*

(i) *$\{\tilde{P}_{n,\lambda_n}\}$ is a tight family of measures on $\mathcal{M}$.*

(ii) *If we choose a subsequence n_j, $j = 1, 2, \cdots$, such that $\tilde{P}_{n_j,\lambda_{n_j}} \Rightarrow R$, then*

$$\text{(a)}\quad R(\mathcal{M}_I)=1,$$

and, moreover,

$$\text{(b)}\quad \int_{\mathcal{M}_I} Q R(dQ)=P.$$

(iii) $\lim_{j\to\infty} E^{P_{n_j}}\{\lambda_{n_j}\int_0^\infty \phi(x, T_t x)\exp\{-\lambda_{n_j}t\}\,dt\}$
$= \int_{\mathcal{M}_I}\iint \phi(x,y)Q(dx)Q(dy)R(dQ).$

Proof: since $\{P_n\}$ is a tight family of measures on X, there exists for every $l>0$ a compact set $K_l \subset X$ such that $P_n(K_l) \geqq 1-1/2^l$ for all $n=1,2,\cdots$. Thus, using (4.3), we see that, for every $l>0$,

$$E^{\tilde{P}_{n,\lambda_n}}\{Q(K_l)\} \geqq 1-\frac{1}{2^l}, \qquad n=1,2,3,\cdots, \tag{4.4}$$

and from this we obtain in the usual way

$$\tilde{P}_{n,\lambda_n}\{Q: Q(K_l) \geqq 1-1/2^{l/2}\} \geqq 1-\frac{1}{2^{l/2}}, \tag{4.5}$$

for $n=1,2,\cdots$. From (4.5) we have, for any $l_0>0$,

$$\tilde{P}_{n,\lambda_n}\{Q: Q(K_l) \geqq 1-1/2^{l/2} \text{ for all } l \geqq l_0\} \geqq 1-\sum_{l \geqq l_0} \frac{1}{2^{l/2}}. \tag{4.6}$$

But, for any $l_0>0$, the set of Q such that $Q(K_l) \geqq 1-1/2^{l/2}$ for all $l \geqq l_0$ is a compact subset of $\mathcal{M}$. We have therefore shown that $\{\tilde{P}_{n,\lambda_n}\}$ is a tight family of measures on $\mathcal{M}$.

For $s>0$ and $f \in c(X)$ we note that

$$\begin{aligned}
&\int_X f(y)\alpha_{\lambda,x}(dy) - \int_X f(T_s y)\alpha_{\lambda,x}(dy) \\
&\quad = \lambda \int_0^\infty e^{-\lambda t} f(T_t x)\,dt - \lambda \int_0^\infty e^{-\lambda t} f(T_{t+s} x)\,dt \\
&\quad = \lambda \int_0^\infty e^{-\lambda t} f(T_t x)\,dt - \lambda \int_s^\infty e^{-\lambda(t-s)} f(T_t x)\,dt \\
&\quad = \lambda \int_0^s e^{-\lambda t} f(T_t x)\,dt - \lambda \int_s^\infty f(T_t x)[e^{-\lambda(t-s)} - e^{-\lambda t}]\,dt,
\end{aligned}$$

and therefore,

$$\left| \int_X f(y)\alpha_{\lambda,x}(dy) - \int_X f(T_s y)\alpha_{\lambda,x}(dy) \right| \leqq \lambda s \|f\| + \|f\|(1-e^{-\lambda s}).$$

Thus, for $s>0$, $f \in c(X)$, and $n=1,2,\cdots$;

$$\int_X \left| \int_X f(y)\alpha_{\lambda_n,x}(dy) - \int_X f(T_s y)\alpha_{\lambda_n,x}(dy) \right| P_n(dx) \leqq \lambda_n s \|f\| + \|f\|(1-\exp\{-\lambda_n s\}),$$

or equivalently,

$$\begin{aligned}
&\int_{\mathcal{M}} \left| \int_X f(x)Q(dx) - \int_X f(T_s x)Q(dx) \right| \tilde{P}_{n,\lambda_n}(dQ) \\
&\qquad \leqq \lambda_n s \|f\| + \|f\|(1-\exp\{-\lambda_n s\}).
\end{aligned} \tag{4.7}$$

Now we choose a subsequence n_j, $j=1, 2, \cdots$, such that $\tilde{P}_{n_j,\lambda_{n_j}} \Rightarrow R$ as $j\to\infty$. Since both $\int_X f(x)Q(dx)$ and $\int_X f(T_s x)Q(dx)$ are bounded continuous functions of Q on $\mathcal{M}$ and since $\lambda_n \to 0$, we have from (4.7), replacing n by n_j and letting $j\to\infty$,

$$\int_{\mathcal{M}} \left|\int_X f(x)Q(dx) - \int_X f(T_s x)Q(dx)\right| R(dQ) = 0. \tag{4.8}$$

Thus, for each $s>0$ and each $f\in c(X)$, (4.8) implies

$$R\left\{Q\colon \int_X f(x)Q(dx) = \int_X f(T_s x)Q(dx)\right\} = 1.$$

By the usual measure-theoretic arguments we can then conclude that $R(\mathcal{M}_I)=1$.

From (4.3) we see that

$$\int_{\mathcal{M}} Q\tilde{P}_{n_j,\lambda_{n_j}}(dQ) = P_{n_j}$$

and thus if we let $j\to\infty$ we get $\int_{\mathcal{M}} QR(dQ) = P$. We have just proved that $R(\mathcal{M}_I)=1$ and hence we conclude that in fact $\int_{\mathcal{M}_I} QR(dQ)=P$.

Finally, to prove statement (iii) of the theorem, let $P\in\mathcal{M}_I$ and using the symmetry of ϕ observe that

$$\begin{aligned} E^P\left\{\lambda\int_0^\infty \phi(x, T_t x)\, e^{-\lambda t}\, dt\right\} &= E^P\left\{\lambda\int_0^\infty \phi(T_t x, x)\, e^{-\lambda t}\, dt\right\} \\ &= E^P\left\{\lambda\int_0^\infty \phi(x, T_{-t} x)\, e^{-\lambda t}\, dt\right\} \\ &= E^P\left\{\lambda\int_{-\infty}^0 \phi(x, T_t x)\, e^{\lambda t}\, dt\right\}, \end{aligned}$$

i.e.,

$$E^P\left\{\lambda\int_0^\infty \phi(x, T_t x) e^{-\lambda t}\, dt\right\} = E^P\left\{\tfrac{1}{2}\lambda\int_{-\infty}^\infty \phi(x, T_t x)\, e^{-\lambda|t|}\, dt\right\}. \tag{4.9}$$

Let ξ_1 and ξ_2 be independent random variables each having probability density $\lambda e^{-\lambda t}$ for $t\geqq 0$. It is then trivial to prove that $\xi_1-\xi_2$ has probability density $\frac{1}{2}\lambda e^{-\lambda|t|}$, $-\infty<t<\infty$. Using this remark and the fact that $P\in\mathcal{M}_I$, we get

$$\begin{aligned} E^P\left\{\tfrac{1}{2}\lambda\int_{-\infty}^\infty \phi(x, T_t x)\, e^{-\lambda|t|}\, dt\right\} &= E^P\left\{\lambda^2\int_0^\infty\int_0^\infty \phi(x, T_{\sigma-t} x)\, e^{-\lambda\sigma-\lambda t}\, d\sigma\, dt\right\} \\ &= E^P\left\{\lambda^2\int_0^\infty\int_0^\infty \phi(T_t x, T_\sigma x)\, e^{-\lambda\sigma-\lambda t}\, d\sigma\, dt\right\}. \end{aligned} \tag{4.10}$$

From (4.9), (4.10), and the assumption that n_j is a sequence such that $\tilde{P}_{n_j,\lambda_{n_j}} \Rightarrow R$ as $j \to \infty$ we obtain

$$\lim_{j\to\infty} E^{P_{n_j}}\left\{\lambda_{n_j}\int_0^\infty \phi(x, T_t x)\exp\{-\lambda_{n_j}t\}\,dt\right\}$$

$$= \lim_{j\to\infty} E^{P_{n_j}}\left\{\lambda_{n_j}^2\int_0^\infty\int_0^\infty \phi(T_t x, T_\sigma x)\exp\{-\lambda_{n_j}\sigma - \lambda_{n_j}t\}\,d\sigma\,dt\right\}$$

$$= \lim_{j\to\infty} E^{P_{n_j}}\left\{\int_X\int_X \phi(y, z)\alpha_{\lambda_{n_j},x}(dy)\alpha_{\lambda_{n_j},x}(dz)\right\}$$

$$= \lim_{j\to\infty} E^{\tilde{P}_{n_j,\lambda_{n_j}}}\left\{\int_X\int_X \phi(x, y)Q(dx)Q(dy)\right\}$$

$$= E^R\left\{\int_X\int_X \phi(x, y)Q(dx)Q(dy)\right\}$$

$$= \int_{\mathcal{M}_I}\int_X\int_X \phi(x, y)Q(dx)Q(dy)R(dQ).$$

This completes the proof of Theorem 4.1.

LEMMA 4.2. *Let $U_\delta(x) = 1/(\delta^2 + |x|^2)^{1/2}$ and let $H_\varepsilon(Q)$ be the entropy of the stationary process Q with respect to the* O–U *process (defined in Lemma* 3.10). *Then, for each $\varepsilon > 0$,*

$$\overline{\lim_{\lambda\to 0}}\sup_Q\left[2E^Q\left\{\lambda\int_0^\infty e^{-\lambda t}U_\delta(x(t)-x(0))\,dt\right\} - H_\varepsilon(Q)\right]$$

(4.11)

$$\leqq \sup_{\substack{\phi\in L_2(R^3)\\ \|\phi\|=1}}\left[2\iint \phi^2(x)\phi^2(y)U_\delta(x-y)\,dx\,dy - \tfrac{1}{2}\int|\nabla\phi|^2\,dx + \varepsilon\right].$$

Proof: Let

$$N(\lambda) = \sup_Q\left[2E^Q\left\{\lambda\int_0^\infty e^{-\lambda t}U_\delta(x(t)-x(0))\,dt\right\} - H_\varepsilon(Q)\right],$$

and, for each $\lambda > 0$, choose Q_λ such that

$$(4.12)\qquad 2E^{Q_\lambda}\left\{\lambda\int_0^\infty e^{-\lambda t}U_\delta(x(t)-x(0))\,dt\right\} - H_\varepsilon(Q_\lambda) \geqq N(\lambda) - \lambda.$$

From (4.12) we see that $\sup_{\lambda\leq 1} H_\varepsilon(Q_\lambda) < \infty$. Now the O–U process satisfies hypotheses (i)–(v) (see Section 2 of this paper); moreover, in [4] (Theorem 6.4) we showed that a family of stationary measures having uniformly bounded entropies with respect to a Markov process satisfying (i)–(v) is a tight family of measures. Thus, $\{Q_\lambda\}_{\lambda\leq 1}$ is tight. Let $\overline{\lim}_{\lambda\to 0} N(\lambda)=N$. Choosing subsequences, using the tightness just established, and denoting Q_{λ_n} by Q_n, we conclude that, for some sequence $\lambda_n \to 0$,

$$\lim_{n\to\infty}\left[2E^{Q_n}\left\{\lambda_n\int_0^\infty \exp\{-\lambda_n t\}U_\delta(x(t)-x(0))\,dt\right\}-H_\varepsilon(Q_n)\right]=N \tag{4.13}$$

and $Q_n \Rightarrow Q'$.

Now we want to use Theorem 4.1. To do so, let the X of that theorem be the space of trajectories $x(\cdot)$ and let $\phi(x(\cdot), y(\cdot)) = U_\delta(x(0)-y(0))$ which is then a symmetric, bounded, and continuous function on $X\times X$. Let $\tilde{P}_{n,\lambda_n}$ be defined as in Theorem 4.1 (with Q_n playing the role of P_n); we then conclude that $\{\tilde{P}_{n,\lambda_n}\}$ is a tight family of measures on $\mathcal{M}$ so that, choosing another subsequence if necessary,

$$\tilde{P}_{n,\lambda_n} \Rightarrow R, \tag{4.14}$$

where R is some probability measure on $\mathcal{M}$. From (iii) of Theorem 4.1 and (4.13) we then conclude that

$$N=2\int_{\mathcal{M}_I}\iint U_\delta(x(0)-y(0))Q(dx(\cdot))Q(dy(\cdot))R(dQ)-\lim_{n\to\infty}H_\varepsilon(Q_n). \tag{4.15}$$

But $H_\varepsilon(Q)$ is lower semicontinuous (see [4], Theorem 3.3) and $Q_n \Rightarrow Q'$ so that from (4.15) we get

$$N\leqq 2\int_{\mathcal{M}_I}\iint U_\delta(x(0)-y(0))Q(dx(\cdot))Q(dy(\cdot))R(dQ)-H_\varepsilon(Q'). \tag{4.16}$$

From (iib) of Theorem 4.1 we know that $Q'=\int_{\mathcal{M}_I} QR(dQ)$ and therefore we get

$$\begin{aligned} N\leqq 2\int_{\mathcal{M}_I}\iint & U_\delta(x(0)-y(0))Q(dx(\cdot))Q(dy(\cdot))R(dQ)\\ &-H_\varepsilon\left(\int_{\mathcal{M}_I} QR(dQ)\right). \end{aligned} \tag{4.17}$$

In [4] (Theorem 3.5) we proved that $H(Q)$ is linear in Q and therefore, from (4.17),

$$N\leqq\int_{\mathcal{M}_I}\left[2\iint U_\delta(x(0)-y(0))Q(dx(\cdot))Q(dy(\cdot))-H_\varepsilon(Q)\right]R(dQ). \tag{4.18}$$

By part (iia) of Theorem 4.1 we know that $R(\mathcal{M}_I)=1$ and so (4.18) implies

$$N\leqq\sup_Q\left[2\iint U_\delta(x(0)-y(0))Q(dx(\cdot))Q(dy(\cdot))-H_\varepsilon(Q)\right]. \tag{4.19}$$

Let μ be the marginal distribution of the stationary measure Q. In [4] (Section 6) we established the contraction principle which states

$$\inf_{Q:q(Q)=\mu} H(Q)=I(\mu),$$

where $q(Q)$ denotes the marginal distribution of Q and $I(\mu)$ is the I-function introduced by the authors in [3]. Thus, (4.19) implies

$$N \leqq \sup_{\mu}\left[2\iint U_\delta(x-y)\mu(dx)\mu(dy)-I_\varepsilon(\mu)\right], \tag{4.20}$$

where $I_\varepsilon(\mu)$ is the I-function corresponding to the O–U process with parameter ε. If we let $I(\mu)$ be the I-function corresponding to standard Brownian motion, then $I_\varepsilon(\mu)\geqq I(\mu)-\varepsilon$. Moreover, in our earlier papers, we showed that $I(\mu)$ is finite if and only if μ has a density ϕ^2 in which case $I(\mu)=I(\phi)=\frac{1}{2}\int|\nabla\phi|^2\,dx$. Thus, from (4.20), for all $\varepsilon>0$ and $\delta>0$,

$$N \leqq \sup_{\substack{\phi\in L_2(R^3)\\ \|\phi\|=1}}\left[2\iint\frac{\phi^2(x)\,\phi^2(y)}{(|x-y|^2+\delta^2)^{1/2}}\,dx\,dy-\frac{1}{2}\int|\nabla\phi|^2\,dx+\varepsilon\right], \tag{4.21}$$

which is (4.11).

LEMMA 4.3. *We have*

$$\overline{\lim_{\lambda\to 0}}\,\Phi(1,\lambda)\leqq \sup_{\substack{\phi\in L_2(R^3)\\ \|\phi\|=1}}\left[2\iint\frac{\phi^2(x)\phi^2(y)}{|x-y|}\,dx\,dy-\frac{1}{2}\int|\nabla\phi|^2\,dx\right]. \tag{4.22}$$

Proof: By Theorem 3.2,

$$\Phi(1,\lambda)=\lim_{t\to\infty}\frac{1}{t}\log E^P\{\exp\{z(1,\lambda,0,t)\}\}. \tag{4.23}$$

But from inequality (3.26) we have, for any $0<p<1$ and q such that $1/p+1/q=1$, and any $\delta>0$,

$$\overline{\lim_{t\to\infty}}\,\frac{1}{t}\log E^P\{\exp\{z(1,\lambda,0,t)\}\}$$

$$\leqq\frac{1}{p}\,\overline{\lim_{t\to\infty}}\,\frac{1}{t}\log E^P\{\exp\{z_\delta(p,\lambda,0,t)\}\} \tag{4.24}$$

$$+\frac{1}{q}\,\overline{\lim_{t\to\infty}}\,\frac{1}{t}\log E^P\left\{\exp\left\{q\int_0^t\int_0^t V_\delta(x(\sigma)-x(s))\lambda\, e^{-\lambda|\sigma-s|}\,d\sigma\,ds\right\}\right\}.$$

From (3.21) we have, for each $\varepsilon > 0$, and each $\delta > 0$,

$$\begin{aligned}(4.25)\qquad &\overline{\lim_{t\to\infty}} \frac{1}{t} \log E^P\{\exp\{z_\delta(p,\lambda,0,t)\}\}\\ &\leqq \sup_Q \left[E^Q\left\{2p\int_0^\infty \frac{\lambda e^{-\lambda\sigma}}{(\delta^2+|x(\sigma)-x(0)|^2)^{1/2}}\,d\sigma\right\}-H_\varepsilon(Q)\right].\end{aligned}$$

Thus, letting $\delta\to 0$ and $p\to 1$, we see from (4.23), (4.24), (4.25), and Lemma 3.8 that

$$(4.26)\qquad \Phi(1,\lambda)\leqq \sup_Q\left[E^Q\left\{2\int_0^\infty \frac{\lambda e^{-\lambda\sigma}}{|x(\sigma)-x(0)|}\,d\sigma\right\}-H_\varepsilon(Q)\right].$$

If we apply Lemma 4.2 (with $\delta = 0$) to (4.26) and then let $\varepsilon\to 0$ we obtain (4.22).

LEMMA 4.4.

$$(4.27)\qquad \underline{\lim_{\lambda\to 0}}\,\Phi(1,\lambda)\geqq \sup_{\substack{\phi\in L_2(R^3)\\ \|\phi\|=1}}\left\{2\iint \frac{\phi^2(x)\phi^2(y)}{|x-y|}\,dx\,dy-\tfrac{1}{2}\int|\nabla\phi|^2\,dx\right\}.$$

Proof: Let $\phi\in L_2(R^3)$ with $\|\phi\|=1$ and consider the stationary Markov process Q_ϕ whose generator is $\frac{1}{2}\Delta+(\nabla\phi/\phi)\cdot\nabla$. The invariant measure for this process is ϕ^2 and the process is ergodic (in fact, mixing). Therefore,

$$\begin{aligned}(4.28)\qquad \Phi(1,\lambda)&=\sup_Q\left[E^Q\left\{2\int_0^\infty \frac{\lambda e^{-\lambda\sigma}}{|x(\sigma)-x(0)|}\,d\sigma\right\}-H(Q)\right]\\ &\geqq E^{Q_\phi}\left\{2\int_0^\infty \frac{\lambda e^{-\lambda\sigma}}{|x(\sigma)-x(0)|}\,d\sigma\right\}-H(Q_\phi).\end{aligned}$$

The ergodic theorem yields

$$\begin{aligned}(4.29)\qquad \lim_{\lambda\to 0} E^{Q_\phi}\left\{2\int_0^\infty \frac{\lambda e^{-\lambda\sigma}}{|x(\sigma)-x(p)|}\,d\sigma\right\}&=\lim_{\lambda\to 0} W^{Q_\phi}\left\{2\int_0^\infty \frac{e^{-u}}{|x(u/\lambda)-x(0)|}\,du\right\}.\\ &=2\iint\frac{\phi^2(x)\phi^2(y)}{|x-y|}\,dx\,dy.\end{aligned}$$

Moreover, for the process Q_ϕ, $H(Q_\phi)=\frac{1}{2}\int|\nabla\phi|^2\,dx$ (see [4], Section 6). Hence,

$$(4.30)\qquad \underline{\lim_{\lambda\to 0}}\,\Phi(1,\lambda)\geqq 2\iint\frac{\phi^2(x)\phi^2(y)}{|x-y|}\,dx\,dy-\tfrac{1}{2}\int|\nabla\phi|^2\,dx,$$

and the lemma follows since (4.30) holds for any $\phi\in L^2(R^3)$ with $\|\phi\|=1$.

Combining (4.22) and (4.27) we have proved (4.1) which as pointed out at the beginning of this section implies Pekar's conjecture.

Acknowledgment The work in this paper was supported in part by the National Science Foundation under Grant No. NSF MCS-80-02568.

Bibliography

[1] Feynman, R. P., *Statistical Mechanics*, W. A. Benjamin, Reading, MA, 1972.
[2] Pekar, S. I., *Theory of Polarons*, Zh. Eksperim. i Teor. Fiz. 19, 1949.
[3] Donsker, M. D., and Varadhan, S. R. S., *Asymptotic evaluation of certain Markov process expectations for large time*, I, II, III, Comm. Pure Appl. Math. 28, 1975, pp. 1–47; 28, 1975, pp. 279–301; 29, 1976, pp. 389–461.
[4] Donsker, M. D., and Varadhan, S. R. S., *Asymptotic evaluation of certain Markov process expectations for large time*, IV, Comm. Pure Appl. Math. 36, 1983, pp. 183–212.
[5] Adamowski, J., Gerlach, B., and Leschke, H., *Strong coupling limit of polaron energy, revisited*, Physical Letter, Sept. 1980, pp. 249–251.
[6] Lieb, E. H. *Existence and uniqueness of the minimizing solution of Choquard's nonlinear equation*, Studies in Applied Mathematics 57, 1977, pp. 93–105.

Received November, 1982.

Commun. Math. Phys. 97, 187–210 (1985)

Communications in Mathematical Physics

Large Deviations for Stationary Gaussian Processes*

M. D. Donsker and S. R. S. Varadhan

Courant Institute of Mathematical Sciences, New York University, New York, NY 10012, USA

Abstract. In their previous work on large deviations the authors always assumed the base process to be Markovian whereas here they consider the base process to be stationary Gaussian. Similar large deviation results are obtained under natural hypotheses on the spectral density function of the base process. A rather explicit formula for the entropy involved is also obtained.

1. Introduction

Let $\{X_k\}$, $-\infty<k<\infty$, be a stationary Gaussian process with $E\{X_k\}=0$ and $E\{X_0X_j\}=\varrho_j=\frac{1}{2\pi}\int\limits_0^{2\pi} e^{ij\theta}f(\theta)d\theta$. We assume that the spectral density function $f(\theta)$ is continuous on $[0,2\pi]$, $f(0)=f(2\pi)$, and

$$\int\limits_0^{2\pi}\log f(\theta)d\theta>-\infty\,. \tag{1.1}$$

Let $\Omega=\prod\limits_{j=-\infty}^{\infty}\mathbb{R}_j$ where, for each j, $\mathbb{R}_j$ is the real line, i.e., Ω is the space of doubly infinite sequences of real numbers. We specify a point $\omega\in\Omega$ by $\omega=\{x_k\}$, $-\infty<k<\infty$, and let $\omega(j)=x_j$ for $-\infty<j<\infty$. The process $\{X_k\}$ induces a probability measure P on Ω. We will denote integration over Ω with respect to P measure by $E^P\{\}$.

For each positive integer n and each $\omega\in\Omega$, let $\omega^{(n)}$ be the point in Ω obtained by the periodic extension in both directions of the elements $x_1,x_2,\ldots,x_n$ of ω, i.e., if $\omega=\{x_k\}$, $-\infty<k<\infty$, then $\omega^{(n)}$ is the point

$$\ldots,x_1,\ldots,x_{n-1},x_n,x_1,x_2,\ldots,x_n,x_1,x_2,\ldots,x_n,\ldots\,.$$

* The research in this paper was supported by the National Science Foundation, Grant No. MCS-80-02568

Let T be the shift operator acting on the points of Ω. For each $\omega \in \Omega$ and each positive integer n, we define a probability measure on Ω, call it $\pi_n(\omega)$, as follows:

$$\pi_n(\omega) = \frac{1}{n}(\delta_{\omega^{(n)}} + \delta_{T\omega^{(n)}} + \ldots + \delta_{T^{n-1}\omega^{(n)}}). \tag{1.2}$$

In other words, given $\omega \in \Omega$ and $n \geqq 1$, $\pi_n(\omega)$ is the probability measure on Ω which assigns mass $1/n$ to $\omega^{(n)}$ and to each of its $(n-1)$ translates. We note that $\pi_n(\omega)$ is a translation invariant measure on Ω, i.e., a stationary measure on Ω. Let $\underline{M}_S$ be the space of all stationary measures on Ω and impose on $\underline{M}_S$ the topology of weak convergence. We can use the mapping $\pi_n : \Omega \to \underline{M}_S$ and the measure P on Ω to construct a probability measure on $\underline{M}_S$ by defining, for each n, $Q_n = \pi_n P^{-1}$, i.e., if A is a set of stationary measures in $\underline{M}_S$, then

$$Q_n(A) = P\{\omega \in \Omega : \pi_n(\omega) \in A\}. \tag{1.3}$$

The assumptions we made on the Gaussian process $\{X_k\}$ imply that it is an ergodic process, and it then follows from the ergodic theorem that, for almost all ω (P-measure), the measure $\pi_n(\omega)$ converges weakly to the measure P as $n \to \infty$, i.e.,

$$P\{\omega \in \Omega : \pi_n(\omega) \Rightarrow P\} = 1. \tag{1.4}$$

Thus, with reference to the Q_n measure on $\underline{M}_S$ given by (1.3), we expect that if the set A contains the measure P, then $Q_n(A) \to 1$ as $n \to \infty$, whereas if P is not in the closure of A, we expect $Q_n(A) \to 0$ as $n \to \infty$. In this paper we show that in this latter case $Q_n(A)$ approaches zero exponentially fast as $n \to \infty$ and we determine the constant in the exponential rate. To be more specific, let R denote a general element of $\underline{M}_S$, i.e., a stationary measure on Ω, and recall that f is the spectral density of our basic stationary Gaussian process $\{X_k\}$. We define a functional $H_f(R)$ which is, in fact, the entropy of the stationary process R with respect to the stationary Gaussian process $\{X_k\}$, and $H_f(R)$: $\underline{M}_S \to [0, \infty]$ is such that:

If $C \subset M_S$ is closed,

$$\overline{\lim_{n \to \infty}} \frac{1}{n} \log Q_n(C) \leqq - \inf_{R \in C} H_f(R), \tag{1.5}$$

and if $G \subset \underline{M}_S$ is open,

$$\underline{\lim_{n \to \infty}} \frac{1}{n} \log Q_n(G) \geqq - \inf_{R \in G} H_f(R). \tag{1.6}$$

Consistent with remarks made above it will indeed be true that $H_f(P) = 0$. We give an explicit formula for $H_f(R)$ in (1.9) below.

Let $\Phi : \underline{M}_S \to \mathbb{R}$ be bounded and continuous, and let $E^{Q_n}\{\}$ denote integration over $\underline{M}_S$ with respect to Q_n measure. From (1.5) and (1.6) it follows easily that

$$\lim_{n \to \infty} \frac{1}{n} \log E^{Q_n}\{e^{n\Phi(R)}\} = \sup_{R \in \underline{M}_S} [\Phi(R) - H_f(R)]. \tag{1.7}$$

Since $Q_n = \pi_n P^{-1}$, equivalent to (1.7) is

$$\lim_{n \to \infty} \frac{1}{n} \log E^P\{e^{n\Phi(\pi_n(\omega))}\} = \sup_{R \in M_S} [\Phi(R) - H_f(R)]. \tag{1.8}$$

The main results of this paper then are (1.5), (1.6), their implication (1.8), and the formula (1.9) for $H_f(R)$ which we discuss now.

Let $R \in \underline{M}_S$ and let, for $A \subset \mathbb{R}$, $R(A|\omega) = R(X_0 \in A | X_{-1}, X_{-2}, \ldots)$ be the regular conditional probability distribution of X_0 given the entire past. Denote by $r(y|\omega)$ the corresponding density. If $G(\theta)$ is the spectral measure of the stationary process R, then the formula for $H_f(R)$ is

$$H_f(R) = E^R\left\{\int_{-\infty}^{\infty} r(y|\omega)\log r(y|\omega)dy\right\} + \frac{1}{2}\log 2\pi + \frac{1}{4\pi}\int_0^{2\pi}\frac{dG(\theta)}{f(\theta)} + \frac{1}{4\pi}\int_0^{2\pi}\log f(\theta)d\theta, \tag{1.9}$$

where it is understood that $H_f(R)$ is defined to be $+\infty$ if for any reason we cannot define any of the ingredients in (1.9).

The authors have developed a theory of large deviations for Markov processes (see [1, 2], for theoretical results and e.g. [3–5] for some applications thereof). In theory, our methods should apply to rather general stationary processes, but it is difficult to see in any great generality what the natural hypotheses to impose on a stationary process are in order to obtain the analogues of (1.5), (1.6), and (1.8), much less to obtain the analogue of (1.9) in anything like explicit form. For recent work in this direction see Orey [6]. In this paper we consider the special case of a Gaussian stationary process because as seen above the hypotheses needed are natural and we obtain an explicit formula for the corresponding entropy.

In Sect. 2 of this paper we prove (1.5) and (1.6) but in doing so we must use properties of $H_f(R)$ and other entropies which occur in the proofs. So as not to interrupt the argument we prove these properties in Sect. 3.

2. Proofs of the Main Theorems

As in the introduction, let $\{X_k\}$ be a stationary Gaussian process and $E\{X_0X_j\} = \varrho_j = \frac{1}{2\pi}\int_0^{2\pi} e^{ij\theta}f(\theta)d\theta$, where we assume that the spectral density function $f(\theta)$ is continuous on $[0, 2\pi]$, $f(0) = f(2\pi)$, and $\int_0^{2\pi}\log f(\theta)d\theta > -\infty$.

Let $\{\xi_k\}$, $-\infty < k < \infty$, be a sequence of independent Gaussian random variables each with mean 0 and variance 1. If we let $\{a_n\}$ be the Fourier coefficients of $\sqrt{f(\theta)}$, i.e., $\sqrt{f(\theta)} = \sum_{n=-\infty}^{\infty} a_n e^{in\theta}$, then we can write

$$X_k = \sum_{n=-\infty}^{\infty} a_{n-k}\xi_k = \sum_{n=-\infty}^{\infty} a_n\xi_{n+k}. \tag{2.1}$$

Let N be a positive integer and define

$$h_N(\theta) = \sum_{|j|<N} a_j\left(1 - \frac{|j|}{N}\right)e^{ij\theta}, \tag{2.2}$$

where the a_j's are defined in the preceding paragraph. Since $f(\theta)$ is continuous and $f(0)=f(2\pi)$, we have by Fejer's theorem,

$$\lim_{N\to\infty}\sup_{0\leq\theta\leq 2\pi}|h_N(\theta)-\sqrt{f(\theta)}|=0. \tag{2.3}$$

For each N, define the process

$$X_k^N=\sum_{|j|<N} a_j\left(1-\frac{|j|}{N}\right)\xi_{j+k}. \tag{2.4}$$

If we let $f_N(\theta)=h_N^2(\theta)$, then $f_N(\theta)$ is the spectral density of the process $\{X_k^N\}$.

In the introduction we noted that the process $\{X_k\}$ induces a probability measure P on Ω. Similarly the process $\{X_k^N\}$ just defined induces a probability measure P^N on Ω. Indeed, the family $\{\xi_k\}$ of independent, Gaussian, mean 0, variance 1, random variables itself induces a probability measure on Ω which we will denote by μ. Each of these three stationary measures on Ω together with the mapping $\pi_n:\Omega\to \underline{M}_S$ of the introduction generates a probability measure on $\underline{M}_S$. We have already introduced $Q_n=P\pi_n^{-1}$. Analogously, we define $\nu_n=\mu\pi_n^{-1}$ and for each N, $Q_n^N=P^N\Pi_n^{-1}$.

For convenience let $b_k=a_k\left(1-\frac{|k|}{N}\right)$ and define the mapping $F_N:\Omega\to\Omega$ by

$$(F_N(\omega))(j)=\sum_{|k|<N} b_k x_{j+k}, \tag{2.5}$$

so that F_N is a moving average map of Ω into itself. The mapping $F_N:\Omega\to\Omega$ induces a corresponding map $\tilde{F}_N:\underline{M}_S\to\underline{M}_S$.

We also have the mapping $F:\Omega\to\Omega$ given by

$$(F(\omega))(j)=\sum_{k=-\infty}^{\infty} a_k x_{j+k}, \tag{2.6}$$

which exists for almost all ω (μ measure on Ω), since $\sum_{n=-\infty}^{\infty} a_n^2<\infty$. We note that for sets $A\subset\underline{M}_S$:

$$\begin{aligned} Q_n(A)&=P\{\omega\in\Omega:\pi_n(\omega)\in A\}=\mu\{\omega\in\Omega:\pi_n\cdot F(\omega)\in A\},\\ Q_n^N(A)&=P^N\{\omega\in\Omega:\pi_n(\omega)\in A\}=\mu\{\omega\in\Omega:\pi_n\cdot F_N(\omega)\in A\},\\ \nu_n(A)&=\mu\{\omega\in\Omega:\pi_n(\omega)\in A\}. \end{aligned} \tag{2.7}$$

It will be important to introduce also the probability measure $\tilde{Q}_n^N$ on $\underline{M}_S$ given by $\tilde{Q}_n^N=\nu_n\tilde{F}_n^{-1}$ so that

$$\tilde{Q}_n^N(A)-\nu_n\{R\in\underline{M}_S:\tilde{F}_N(R)\in A\}-\mu\{\omega\in\Omega:\tilde{F}_N\cdot\pi_n(\omega)\in A\}. \tag{2.8}$$

To deduce the asymptotic behavior of Q_n as $n\to\infty$, i.e., to prove (1.5) and (1.6), we first determine the asymptotic behavior of Q_n^N as $n\to\infty$ and then let $N\to\infty$. To obtain the asymptotics of Q_n^N we compare it with $\tilde{Q}_n^N$ the asymptotic behavior of which as $n\to\infty$ follows from our earlier work and which we state below as Theorem 2.2. Looking at (2.8) and (2.7) we see that to compare $\tilde{Q}_n^N$ and Q_n^N means comparing $\pi_n \cdot F_N(\omega)$ with $\tilde{F}_N \cdot \pi_n(\omega)$. Now,

$$\tilde{F}_N \cdot \pi_n(\omega) = \frac{1}{n}[\delta_{F_N(\omega^{(n)})} + \delta_{F_N(T\omega^{(n)})} + \ldots + \delta_{F_N(T^{n-1}\omega^{(n)})}], \tag{2.9}$$

and

$$\pi_n \cdot F_N(\omega) = \frac{1}{n}[\delta_{(F_N(\omega))^{(n)}} + \delta_{T(F_N(\omega))^{(n)}} + \ldots + \delta_{T^{n-1}(F_N(\omega))^{(n)}}]. \tag{2.10}$$

Thus, to compare the two measures $\tilde{F}_N \cdot \pi_n$ and $\pi_n \cdot F_N$, we need to determine when $T^{j-1}(F_N(\omega))^{(n)} = F_N(T^{j-1}\omega^{(n)})$ with $1 \leqq j \leqq n$. It is clear that the difference of the two measures $\pi_n \cdot F_N$ and $\tilde{F}_N \cdot \pi_n$ can be estimated in the variation norm over the σ-field generated by the first L components. We obtain

$$\|\pi_n \cdot F_N - \tilde{F}_N \cdot \pi_n\|_{F'_L} \leqq \frac{2N+L}{n}. \tag{2.11}$$

In the space $\underline{M}_S$ we have the topology of weak convergence. This topology can be induced by the following metric. Let R_1 and R_2 be in $\underline{M}_S$ and let L be a positive integer. Denote by R_1^L and R_2^L the respective L-dimensional measures one obtains by considering the joint distribution of L successive random variables in the processes R_1 and R_2. We then define the distance between R_1 and R_2 in $\underline{M}_S$ by

$$d(R_1, R_2) = \sum_{L=1}^{\infty} \frac{1}{2^L} d_L(R_1^L, R_2^L), \tag{2.12}$$

where d_L is the Prohorov distance for L-dimensional measures. This metric in $\underline{M}_S$ yields the topology of weak convergence in $\underline{M}_S$. We now can prove the following lemma.

Lemma 2.1. *For any $\delta > 0$ and large enough n,*

$$\mu\{\omega \in \Omega : d(\pi_n \cdot F_N, \tilde{F}_N \cdot \pi_n) \geqq \varepsilon\} = 0. \tag{2.13}$$

Proof. From (2.12), (2.11), and the fact that the variation norm dominates the Prohorov norm in L-dimensions, we have for $n > N$,

$$d(\pi_n \cdot F_N, \tilde{F}_n \cdot \pi_n) \leqq \sum_{L=1}^{\infty} \frac{1}{2^L} \|\pi_n \cdot F_N - \tilde{F}_N \cdot \pi_n\|_{\underline{F}'_L} \leqq \frac{1}{n} \sum_{L=1}^{\infty} \frac{2N+L}{2^L}.$$

Since the infinite series on the right converges, we can make the distance on the left as small as we please by taking n large enough which implies (2.13).

As noted in (2.8),

$$\tilde{Q}_n^N(A) = \nu_n\{R \in \underline{M}_S : \tilde{F}_N(R) \in A\} = \mu\{\omega \in \Omega : \tilde{F}_N \cdot \pi_n(\omega) \in A\}.$$

Since μ measure on Ω comes from the sequence $\{\xi_k\}$ of independent Gaussian variables with mean 0 and variance 1, the results of [2-IV] give us the asymptotics of ν_n measure on $\underline{M}_S$ and hence of $\tilde{Q}_n^N$ measure on $\underline{M}_S$.

In the introduction we defined for $R \in \underline{M}_S$ the density $r(y|\omega)$ of the regular condition probability distribution of X_0 given the entire past. For $R \in \underline{M}_S$ define

$$I(R) = E^R \left\{ \int_{-\infty}^{\infty} r(y|\omega) \log \left(\frac{r(y|\omega)}{\frac{1}{\sqrt{2\pi}} e^{-y^2/2}} \right) dy \right\}, \tag{2.14}$$

and define for $R \in \underline{M}_S$

$$H_{f_N}(R) = \inf_{R' \in \underline{M}_S : R' \tilde{F}_N^{-1} = R} I(R'). \tag{2.15}$$

Then in Sect. 3, formula (3.21), we show

$$H_{f_N}(R) = E^R \left\{ \int_{-\infty}^{\infty} r(y|\omega) \log r(y|\omega) dy \right\} + \frac{1}{2} \log 2\pi + \frac{1}{4\pi} \int_0^{2\pi} \frac{dG(\theta)}{f_N(\theta)} + \frac{1}{4\pi} \int_0^{2\pi} \log f_N(\theta) d\theta,$$

where $G(\theta)$ is the spectral measure of R and $f_N(\theta)$ (see just after (2.4)) is the spectral density of the $\{X_R^N\}$ process.

Thus, from [2-IV] and the contraction principle discussed in that paper, we have

Theorem 2.2. *If $C \subset \underline{M}_S$ is closed,*

$$\overline{\lim_{n\to\infty}} \frac{1}{n} \log \tilde{Q}_n^N(C) \leqq - \inf_{R \in C} H_{f_N}(R), \tag{2.16}$$

and if $G \subset \underline{M}_S$ is open,

$$\underline{\lim_{n\to\infty}} \frac{1}{n} \log \tilde{Q}_n^N(G) \geqq - \inf_{R \in G} H_{f_N}(R). \tag{2.17}$$

Using this and Lemma 2.1 we can prove the following theorem.

Theorem 2.3. *If $C \subset \underline{M}_S$ is closed,*

$$\overline{\lim_{n\to\infty}} \frac{1}{n} \log Q_n^N(C) \leqq - \inf_{R \in C} H_{f_N}(R), \tag{2.18}$$

and if $G \subset \underline{M}_S$ is open,

$$\underline{\lim_{n\to\infty}} \frac{1}{n} \log Q_n^N(G) \geqq - \inf_{R \in G} H_{f_N}(R). \tag{2.19}$$

Proof. Let $\delta > 0$ and $C^\delta = \bigcup_{R \in C} S(R, \delta)$ where $S(R, \delta)$ is the ball in $\underline{M}_S$ of radius δ centered at R. From (2.7) and (2.8) we see that

$$\begin{aligned} Q_n^N(C) &= \mu\{\omega \in \Omega : \pi_n \cdot F_N(\omega) \in C\} \\ &\leqq \mu\{\omega \in \Omega : \tilde{F}_N \cdot \pi_n(\omega) \in C^\delta\} + \mu\{\omega \in \Omega : d(\pi_n \cdot F_N, \tilde{F}_n \cdot \pi_n) \geqq \delta\} \\ &= \tilde{Q}_n^N(C^\delta) + \mu\{\omega \in \Omega : d(\pi_n \cdot F_N, \tilde{F}_N \cdot \pi_n) \geqq \delta\}. \end{aligned}$$

Let $\bar{C}^\delta$ be the closure of C^δ. Then, from Lemma 2.1 and Theorem 2.2,

$$\overline{\lim_{n\to\infty}} \frac{1}{n} \log Q_n^N(C) \leqq \overline{\lim_{n\to\infty}} \frac{1}{n} \log \tilde{Q}_n^N(\bar{C}^\delta) \leqq -\inf_{R\in\bar{C}^\delta} H_{f_N}(R). \tag{2.20}$$

Since C is closed, $H_{f_N}(R)$ is lower semicontinuous, and $\{R: H_{f_N}(R) \leqq K\}$ is tight, (cf. Lemma 3.10) we obtain (2.18) from (2.20) on letting $\delta\to 0$.

For the lower bound, let G be open in $\underline{M}_S$ and let $R\in G$. Pick $\delta>0$ and an open set V such that $R\in V\subset V^\delta\subset G$. Then, again noting (2.7) and (2.8),

$$\begin{aligned} Q_n^N(G) &\geqq Q_n^N(V^\delta) = \mu\{\omega\in\Omega: \pi_n\cdot F_N(\omega)\in V^\delta\} \\ &\geqq \mu\{\omega\in\Omega: \tilde{F}_N\cdot\pi_n(\omega)\in V\} - \mu\{\omega\in\Omega: d(\pi_n\cdot F_N, \tilde{F}_N\cdot\pi_n)\geqq\delta\} \\ &= \tilde{Q}_n^N(V) - \mu\{\omega\in\Omega: d(\pi_n\cdot F_N, \tilde{F}_N\cdot\pi_n)\geqq\delta\}. \end{aligned}$$

Using again Lemma 2.1 and Theorem 2.2,

$$\underline{\lim_{n\to\infty}} \frac{1}{n}\log Q_n^N(G) \geqq \underline{\lim_{n\to\infty}} \frac{1}{n}\log \tilde{Q}_n^N(V) \geqq -\inf_{R'\in V} H_{f_N}(R') \geqq -H_{f_N}(R). \tag{2.21}$$

Since inequality (2.21) holds for any $R\in G$ we obtain (2.19).

Before proving the next theorem we need some preliminary lemmas.

Lemma 2.4. *Let $\{X_k\}$ and $\{X_k^N\}$ be defined respectively by* (2.1) *and* (2.4). *There exists a function $\varepsilon(N)=o(1)$ as $N\to\infty$ such that if $0<\lambda<\frac{1}{2\varepsilon(N)}$, then for all n,*

$$\frac{1}{n}\log E^\mu\left\{\exp\left\{\lambda\sum_{k=1}^n (X_k - X_k^N)^2\right\}\right\} \leqq -\frac{1}{2}\log(1-2\lambda\varepsilon(N)). \tag{2.22}$$

Proof. First of all, let $\{Y_k\}$ be any stationary Gaussian process with mean 0, covariance $\gamma_j = \frac{1}{2\pi}\int_0^{2\pi} e^{ij\theta} g(\theta)d\theta$, and strictly positive definite covariance matrix

$$A_n = \begin{pmatrix} \gamma_0 & \gamma_1 & \cdots & \gamma_{n-1} \\ \gamma_1 & \gamma_0 & \cdots & \gamma_{n-2} \\ \cdots & \cdots & \cdots & \\ \gamma_{n-1} & \cdots & & \gamma_0 \end{pmatrix}.$$

Let $c = \sup_{0\leqq\theta\leqq 2\pi} |g(\theta)|$. We show now that if $\lambda<\frac{1}{2c}$, then

$$E\left\{\exp\left\{\lambda\sum_{k=1}^n Y_k^2\right\}\right\} \leqq (1-2\lambda c)^{-n/2}. \tag{2.23}$$

To see this, first observe that if c_n is the maximum eigenvalue of A_n,

$$\begin{aligned} E\left\{\exp\left\{\lambda\sum_{k=1}^n Y_k^2\right\}\right\} &= \frac{1}{\sqrt{|A_n|}}\frac{1}{(2\pi)^{n/2}}\int_{-\infty}^{\infty}\!\!\cdots\!\int \exp\left\{\lambda\sum_{k=1}^n y_k^2 - \frac{1}{2}Y'A_n^{-1}Y\right\}dy_1\ldots dy_n \\ &= \frac{1}{\sqrt{|A_n|}}\frac{1}{\sqrt{|A_n^{-1}-2\lambda I|}} = \frac{1}{\sqrt{|I-2\lambda A_n|}} \leqq (1-2\lambda c_n)^{-n/2}, \end{aligned}$$

providing $\lambda < 1/2c_n$. To estimate c_n, consider

$$\sum_{j,k=1}^{n} \gamma_{k-j}\eta_k\bar{\eta}_j = \frac{1}{2\pi}\int_0^{2\pi}\left|\sum_{k=1}^{n}\eta_k e^{ik\theta}\right|^2 g(\theta)d\theta \leqq \sup_{0\leqq\theta\leqq 2\pi}|g(\theta)|\sum_{j=1}^{n}|\eta_j|^2,$$

which implies $c_n \leqq \sup_{0\leqq\theta\leqq 2\pi}|g(\theta)| = c$. Thus, $(1-2\lambda c_n)^{-n/2} \leqq (1-2\lambda c)^{-n/2}$ and $\lambda < 1/2c$ implies $\lambda < 1/2c_n$, so we have (2.23). Now, in particular let

$$y_k = X_k^N - X_k = \sum_{|j|<N} a_j\left(1-\frac{|j|}{N}\right)\xi_{j+k} - \sum_{j=-\infty}^{\infty} a_j\xi_{j+k}.$$

The spectral density $g(\theta)$ for the process $\{Y_k\}$ is now given by

$$g(\theta) = \left|\sum_{j=-\infty}^{\infty} a_j e^{ij\theta} - \sum_{|j|\leqq N} a_j\left(1-\frac{|j|}{N}\right)e^{ij\theta}\right|^2 = |\sqrt{f(\theta)} - h_N(\theta)|^2.$$

Let $\varepsilon(N) = \sup_{0\leqq\theta\leqq 2\pi}|g(\theta)| = \sup_{0\leqq\theta\leqq 2\pi}|\sqrt{f(\theta)} - h_N(\theta)|^2$. From (2.3) we see $\varepsilon(N) = o(1)$ as $N\to\infty$, and from (2.23) if $0<\lambda<1/2\varepsilon(N)$, then we have for all n,

$$\frac{1}{n}\log E\left\{\exp\left\{\lambda\sum_{k=1}^{n}(x_k - x_k^N)^2\right\}\right\} \leqq -\frac{1}{2}\log(1-2\lambda\varepsilon(N)),$$

which is the lemma.

Lemma 2.5. *Let $\omega\in\Omega$ be given by $\omega=\{x_k\}$, $-\infty<k<\infty$ and let $\omega'\in\Omega$ be given by $\omega'=\{x_k+\varepsilon_k\}$, $-\infty<k<\infty$. Then, there is a constant c such that*

$$d(\pi_n(\omega), \pi_n(\omega')) \leqq c\sqrt{\frac{\varepsilon_1^2+\varepsilon_2^2+\ldots+\varepsilon_n^2}{n}}. \tag{2.24}$$

Proof. By definition [cf. (2.12)],

$$d(\pi_n(\omega), \pi_n(\omega')) = \sum_{L=1}^{\infty}\frac{1}{2^L} d_L([\pi_n(\omega)]_L^1, [\pi_n(\omega')]_L^1). \tag{2.25}$$

Let α and β be two L-dimensional prob distributions and $\hat{\alpha}$ and $\hat{\beta}$ their characteristic functions. Since uniform convergence of characteristic functions on compact sets in L-dimensions is equivalent to weak convergence in L-dimensions which is equivalent to convergence in the Prohorov metric d_L in L-dimensions, we can write

$$d_L(\alpha,\beta) = \sum_{k=1}^{\infty}\frac{1}{2^k}\sup_{|\eta|\leqq k}|\hat{\alpha}(\eta) - \hat{\beta}(\eta)|. \tag{2.26}$$

In particular, if $\alpha = \delta_{(x_1,x_2,\ldots,x_L)}$ and $\beta = \delta_{(x_1+\varepsilon_1,\ldots,x_L+\varepsilon_L)}$, we have from (2.26),

$$\begin{aligned} d_L(\alpha,\beta) &= \sum_{k=1}^{\infty}\frac{1}{2^k}\sup_{|\eta|\leqq k}|e^{i\langle x,\eta\rangle} - e^{i\langle x+\varepsilon,\eta\rangle}| \\ &= \sum_{k=1}^{\infty}\frac{1}{2^k}\sup_{|\eta|\leqq k}|e^{i\langle\varepsilon,\eta\rangle} - 1| \\ &\leqq \|\varepsilon\|\sum_{k=1}^{\infty}\frac{k}{2^k} = 2\|\varepsilon\|. \end{aligned}$$

Thus, from the definition of π_n,

$$d_L([\pi_n(\omega)]_L^1, [\pi_n(\omega')]_L^1)$$

$$\leqq \frac{2}{n}(\sqrt{\varepsilon_1^2+\varepsilon_2^2+\ldots+\varepsilon_L^2}+\sqrt{\varepsilon_2^2+\ldots+\varepsilon_{1+L}^2}+\ldots+\sqrt{\varepsilon_n^2+\ldots+\varepsilon_{n+L-1}^2}),$$

where $\varepsilon_{n+j}=\varepsilon_j$. Using Minkowski's inequality on this last, we get

$$d_L([\pi_n(\omega)]_L^1, [\pi_n(\omega')]_L^1)$$

$$\leqq 2\sqrt{\frac{1}{n}[(\varepsilon_1^2+\ldots+\varepsilon_L^2)+(\varepsilon_2^2+\ldots+\varepsilon_{2+L}^2)+\ldots+(\varepsilon_n^2+\ldots+\varepsilon_{n+L}^2)]}$$

$$=2\sqrt{\frac{L}{n}(\varepsilon_1^2+\varepsilon_2^2+\ldots+\varepsilon_n^2)}\,.$$

Finally, from (2.25) and this last inequality,

$$d(\pi_n(\omega), \pi_n(\omega'))\leqq \sum_{L=1}^{\infty}\frac{1}{2^L}\cdot 2\sqrt{\frac{L(\varepsilon_1^2+\ldots+\varepsilon_n^2)}{n}}=c\sqrt{\frac{\varepsilon_1^2+\ldots+\varepsilon_n^2}{n}}\,.$$

Lemma 2.6. *Let F_N and F be defined by* (2.5) *and* (2.6) *respectively. For any $\delta>0$,*

$$\mu\{\omega\in\Omega: d(\pi_n\cdot F_N, \pi_n\cdot F)\geqq\delta\}\leqq\mu\left\{\omega\in\Omega:\sum_{k=1}^{n}\frac{(X_k^N-X_k)^2}{n}\geqq\frac{\delta^2}{c^2}\right\}, \quad (2.27)$$

where c is the constant in Lemma 2.5.

Proof. From (2.1) and (2.4), $X_k=\sum_{n=-\infty}^{\infty}a_n\xi_{n+k}$ and $x_k^N=\sum_{|j|<N}a_j\left(1-\frac{|j|}{N}\right)\xi_{j+k}$. If we let $\varepsilon_k=X_k-X_k^N$, then $X_k=X_k^N+\varepsilon_k$. Thus, from Lemma 2.5,

$$d(\pi_n\cdot F_N, \pi_n\cdot F)\leqq c\sqrt{\frac{1}{n}\sum_{k=1}^{n}(X_k-x_k^N)^2}\,,$$

and (2.27) follows.

Lemma 2.7. *For every $\delta>0$,*

$$\lim_{N\to\infty}\overline{\lim_{n\to\infty}}\frac{1}{n}\log\mu\{\omega\varepsilon\,\Omega: d(\pi_n\cdot F_N, \pi_n\cdot F)\geqq\delta\}=-\infty\,. \quad (2.28)$$

Proof. For fixed N, and $\lambda<1/2\varepsilon(N)$, where $\varepsilon(N)$ is defined above in Lemma 2.4, we have from (2.27) and (2.24) for any $\delta>0$,

$$\mu\{\omega\in\Omega: d(\pi_n\cdot F_N, \pi_n\cdot F)\geqq\delta\}\leqq\mu\left\{\omega\in\Omega:\frac{1}{n}\sum_{k=1}^{n}(X_k^N-X_k)^2\geqq\frac{\delta^2}{c^2}\right\}$$

$$\leqq E^\mu\left\{\exp n\lambda\left(\frac{1}{n}\sum_{k=1}^{n}(X_k^N-X_k)^2\right)\right\}e^{-n\delta^2\lambda/c^2}\leqq e^{-n\delta^2\lambda/c^2}e^{-\frac{n}{2}\log(1-2\lambda\varepsilon(N))}\,.$$

Therefore, for any $\delta>0$ and $\lambda<1/2\varepsilon(N)$ with N fixed,

$$\overline{\lim_{n\to\infty}}\frac{1}{n}\log\mu\{\omega\in\Omega: d(\pi_n\cdot F_N, \pi_n\cdot F)\geqq\delta\}\leqq-\frac{\delta^2\lambda}{c^2}-\frac{1}{2}\log(1-2\lambda\varepsilon(N)). \quad (2.29)$$

Since $\varepsilon(N)=o(1)$ as $N\to\infty$, we get from (2.29) that for all $\delta>0$ and all $\lambda>0$,

$$\overline{\lim_{N\to\infty}}\ \overline{\lim_{n\to\infty}}\ \frac{1}{n}\log\mu\{\omega\in\Omega: d(\pi_n\cdot F_N,\pi_n\cdot F)\geqq\delta\}\leqq-\frac{\delta^2\lambda}{c^2}.$$

In this last if we let $\lambda\to\infty$ we obtain (2.28).

We are now ready to prove (1.5) and (1.6) of the introduction.

Theorem 2.8. *If $C\subset\underline{M}_S$ is closed, then*

$$\overline{\lim_{n\to\infty}}\ \frac{1}{n}\log Q_n(C)\leqq-\inf_{R\in C}H_f(R), \tag{2.30}$$

and if $G\subset\underline{M}_S$ is open, then

$$\underline{\lim}_{n\to\infty}\ \frac{1}{n}\log Q_n(G)\geqq-\inf_{R\in G}H_f(R), \tag{2.31}$$

where $H_f(R)$ is defined by (1.9).

Proof. Let C be closed in $\underline{M}_S$. Then,

$$Q_n(C)=P\{\omega\in\Omega:\pi_n(\omega)\in C\}=\mu\{\omega\in\Omega:\pi_n\cdot F(\omega)\in C\}.$$

For any $\delta>0$, we have

$$\begin{aligned}Q_n(C)&\leqq\mu\{\omega\in\Omega:\pi_n\cdot F_n\in C^\delta\}+\mu\{\omega\in\Omega:d(\pi_n\cdot F_N,\pi_n\cdot F)\geqq\delta\}\\&=P^N\{\omega\in\Omega:\pi_n(\omega)\in C^\delta\}+\mu\{\omega\in\Omega:d(\pi_n\cdot F_N,\pi_n\cdot F)\geqq\delta\}\\&=Q_n^N(C^\delta)+\mu\{\omega\in\Omega:d(\pi_n\cdot F_N,\pi_n\cdot F)\geqq\delta\}\\&\leqq Q_n^N(\bar{C}^\delta)+\mu\{\omega\in\Omega:d(\pi_n\cdot F_N,\pi_n\cdot F)\geqq\delta\}.\end{aligned}$$

From Theorem 2.3, this last inequality implies

$$\overline{\lim_{n\to\infty}}\ \frac{1}{n}\log Q_n(C)\leqq\max\Big(-\inf_{R\in\bar{C}^\delta}H_{f_N}(R),A_N(\delta)\Big),$$

where $A_N(\delta)=\overline{\lim_{n\to\infty}}\ \frac{1}{n}\log\mu\{\omega\in\Omega:d(\pi_n\cdot F_N,\pi_n\cdot F)\geqq\delta\}$. From Lemma 2.7 we then obtain,

$$\overline{\lim_{n\to\infty}}\ \frac{1}{n}\log Q_n(C)\leqq-\underline{\lim}_{N\to\infty}\ \inf_{R\in\bar{C}^\delta}H_{f_N}(R). \tag{2.32}$$

From Lemma 3.10 below it follows that

$$\underline{\lim}_{N\to\infty}\ \inf_{R\in\bar{C}^\delta}H_{f_N}(R)=\inf_{R\in\bar{C}^\delta}H_f(R), \tag{2.33}$$

where $H_f(R)$ is given by (1.9). Using (2.33) in (2.32) we get

$$\overline{\lim_{n\to\infty}}\ \frac{1}{n}\log Q_n(C)\leqq-\inf_{R\in\bar{C}^\delta}H_f(R). \tag{2.34}$$

Since C is closed and $H_f(R)$ is lower semicontinuous (2.34) implies (2.30) on letting $\delta\to 0$.

For the lower bound, let G be open in $\underline{M}_S$ and let R be any element of G. Pick $\delta>0$ and an open set V such that $R\in V\subset V^\delta\subset G$. Then

$$\begin{aligned} Q_n(G) &\geqq Q_n(V^\delta)=\mu\{\omega\in\Omega:\pi_n\cdot F(\omega)\in V^\delta\}\\ &\geqq \mu\{\omega\in\Omega:\pi_n\cdot F_N\in V\}-\mu\{\omega\in\Omega:d(\pi_n\cdot F_N,\pi_n\cdot F)\geqq\delta\}\\ &=Q_n^N(V)-\mu\{\omega\in\Omega:d(\pi_n\cdot F_N,\pi_n\cdot F)\geqq\delta\}\,. \end{aligned} \tag{2.35}$$

Without loss of generality we can assume that $H_f(R)<\infty$ for the $R\in G$ we picked.

As before, let

$$A_N(\delta)=\overline{\lim_{n\to\infty}}\frac{1}{n}\log\mu\{\omega\in\Omega:d(\pi_n\cdot F_N,\pi_n\cdot F)\geqq\delta\}\,.$$

From Lemma 2.7 we can choose N so large that $A_N(\delta)\leqq-(H_f(R)+3)$, i.e., we can choose N so large that

$$\mu\{\omega\in\Omega:d(\pi_n\cdot F_N,\pi_n\cdot F)\geqq\delta\}\leqq e^{-n(H_f(R)+3)+o(n)}\,. \tag{2.36}$$

Also, since V is open we obtain from Theorem 2.3,

$$\underline{\lim_{n\to\infty}}\frac{1}{n}\log Q_n^N(V)\geqq-\inf_{R'\in V}H_{f_N}(R')\,. \tag{2.37}$$

In Lemma 3.9 below we prove that for an $R\in\underline{M}_S$ such that $H_f(R)<\infty$, there exists a sequence $\{R_N\}$ in $\underline{M}_S$ which converges weakly to R and for which $\lim_{N\to\infty}H_{f_N}(R_N)=H_f(R)$. If we let $0<\eta<1$ be given, then from (2.37) and Lemma 3, since V is a neighborhood of R, we can take N so large that

$$\underline{\lim_{n\to\infty}}\frac{1}{n}\log Q_n^N(V)\geqq-\inf_{R'\in V}H_{f_N}(R')\geqq-H_{f_N}(R_N)\geqq-H_f(R)-\eta\,.$$

In other words, for N large enough,

$$Q_n^N(V)\geqq e^{-n(H_f(R)+\eta)+o(n)}\,. \tag{2.38}$$

Using (2.38) and (2.36) in (2.35) we see that for N large enough

$$Q_n(G)\geqq e^{-n(H_f(R)+\eta)+o(n)}-e^{-n(H_f(R)+3)+o(n)}\,.$$

Since $\eta<1$ the first expression on the right is the controlling asymptotic term and we get,

$$\underline{\lim_{n\to\infty}}\frac{1}{n}\log Q_n(G)\geqq-H_f(R)-\eta\,. \tag{2.39}$$

Since η is arbitrary and R is an arbitrary element of G, we conclude finally that

$$\underline{\lim_{n\to\infty}}\frac{1}{n}\log Q_n(G)\geqq-\inf_{R\in G}H_f(R)\,,$$

which completes the proof of the theorem.

3. Properties of Entropy

In this section we derive the formula for $H_{f_N}(R)$ occurring in Theorem 2.2. We also prove those properties of $H_{f_N}(R)$ and $H_f(R)$ which we called upon in obtaining the upper and lower bounds in Theorem 2.8.

We will be somewhat more general at first and start with $I(R)$ as defined by (2.16). Note, that for $R \in \underline{M}_S$, and with $\phi(y) = \frac{1}{\sqrt{2\pi}} e^{-y^2/2}$,

$$\begin{aligned} I(R) &= E^R\left\{\int_{-\infty}^{\infty} r(y|\omega)\log\left(\frac{r(y|\omega)}{\phi(y)}\right)dy\right\} \\ &= E^R\left\{\int_{-\infty}^{\infty} r(y|\omega)\log r(y|\omega)dy\right\} + \frac{1}{2}\log 2\pi + \frac{1}{2}E^R\{\omega^2(0)\}. \end{aligned} \tag{3.1}$$

Let $\{a_k\}$, $-\infty < k < \infty$ be any sequence of reals such that $\sum_{j=1}^{\infty} |a_j| < \infty$ and $|h(\theta)| = \left|\sum_{j=-\infty}^{\infty} a_j e^{ij\theta}\right| \neq 0$ for all $\theta \in [0, 2\pi]$. Consider the mapping $\tau : \Omega \to \Omega$ based on such a sequence $\{a_k\}$ given by $(\tau\omega)(k) = \sum_{j=-\infty}^{\infty} a_j \omega(j+k)$. We will call such mappings τ "admissible."

For τ such an admissible mapping, define for $R \in \underline{M}_S$

$$H^\tau(R) = \inf_{R' : R'\tau^{-1} = R} I(R'). \tag{3.2}$$

We prove below that if R has spectral measure $G(\theta)$, then

$$H^\tau(R) = I(R) + \frac{1}{4\pi}\int_0^{2\pi} \frac{dG(\theta)}{|h(\theta)|^2} - \frac{1}{4\pi}\int_0^{2\pi} dG(\theta) + \frac{1}{2\pi}\int_0^{2\pi} \log|h(\theta)d\theta. \tag{3.3}$$

Before proving (3.3) we need to examine how $I(R\tau^{-1})$ is related to $I(R)$ for an admissible map τ. We start with some preliminary lemmas. The first is easy and we omit the proof. The second is a classic result found in standard references on matrices.

Lemma 3.1. *Let $R \in \underline{M}_S$ and let $R_n(y_1, y_2, \ldots, y_n)$ be the n-dimensional joint density for the R process, i.e., $R_n(y_1, y_2, \ldots, y_n)dy_1 \ldots dy_n = R\{X_1 \in dy_1, \ldots, X_n \in dy_n\}$. Define, with $\phi(y) = \frac{1}{\sqrt{2\pi}} e^{-y^2/2}$,*

$$I_n(R) = \int_{-\infty}^{\infty}\cdots\int R_n(y_1, y_2, \ldots, y_n) \log\left(\frac{R_n(y_1, y_2, \ldots, y_n)}{\prod_{j=1}^{n} \phi(y_j)}\right) dy_1 \ldots dy_n .$$

Then, $I_n(R)-I_{n-1}(R)$ *is monotone increasing in n,* $\lim_{n\to\infty}(I_n(R)-I_{n-1}(R))=I(R)$ *so that also*

$$\lim_{n\to\infty}\frac{I_n(R)}{n}=I(R).$$

Lemma 3.2. *Let D be the determinant of a circulant matrix of order n, i.e.,*

$$D=\begin{vmatrix} d_1 & d_2 & \dots & d_n & \\ d_n & d_1 & d_2 \dots & d_{n-1} & \\ \vdots & & & & \\ d_2 & d_3 & \dots & d_n & d_1 \end{vmatrix}.$$

Then

$$D=\prod_{\xi\in Z_n}(d_1+d_2\xi+\ldots+d_n\xi^{n-1}), \tag{3.4}$$

where Z_n *is the set of* n^{th} *roots of unity.*

Lemma 3.3. *Let* $f(x)$ *be a probability density in d-dimensions and let* $g(x)$ *be the d-dimensional Gaussian density,* $g(x)=\frac{1}{(2\pi)^{d/2}}e^{-(1/2)\|x\|^2}$. *Let A be a nonsingular matrix. Then,*

$$\int_{R^d}\left(\frac{f(A^{-1}x)}{|A|}\right)\log\left(\frac{f(A^{-1}x)/|A|}{g(x)}\right)dx=\int_{R^d}f(y)\log\frac{f(y)}{g(y)}dy$$
$$+\frac{1}{2}\int_{R^d}(\|Ay\|^2-\|y\|^2)f(y)dy-\log|A|. \tag{3.5}$$

Proof. We observe that if $\alpha(dx)=f(x)dx$, then $(\alpha A^{-1})(dx)=\frac{f(A^{-1}x)}{|A|}dx$. Now, letting $x=Ay$,

$$\int_{R^d}\left(\frac{f(A^{-1}x)}{|A|}\right)\log\left(\frac{f(A^{-1}x)/|A|}{g(x)}\right)dx=\int_{R^d}f(y)\log\left(\frac{f(y)}{g(Ay)}\right)dx-\log|A|$$
$$=\int_{R^d}f(y)\log\left(\frac{f(y)}{g(y)}\right)dy+\int_{R^d}f(y)\log\left(\frac{g(y)}{g(Ay)}\right)dy-\log|A|$$
$$=\int_{R^d}f(y)\log\frac{f(y)}{g(y)}dy+\frac{1}{2}\int_{R^d}(\|Ay\|^2-\|y\|^2)f(y)dy-\log|A|.$$

As a preparatory lemma we consider admissible maps of a special kind – where only a finite number of the elements in $\{a_k\}$ are different from 0.

Lemma 3.4. *For any* $R\in \underline{M}_S$, *let* $I(R)$ *be defined by* (3.1) *and let* $\tau:\Omega\to\Omega$ *be defined by* $(\tau\omega)(j)=\sum_{k=-\infty}^{\infty}a_k\omega(k+j)$, *where, for some K,* $a_k=0$ *if* $|k|>K$ *and* $|h(\theta)|=\left\|\sum_{k=-\infty}^{\infty}a_ke^{ik\theta}\right\|\neq 0$ *for all* $\theta\in[0,2\pi]$. *Then,*

$$I(R\tau^{-1})\leq I(R)+\frac{1}{2}[E^{R\tau^{-1}}\{\omega^2(0)\}-E^R\{\omega^2(0)\}-\frac{1}{2\pi}\int_0^{2\pi}\log|h(\theta)|d\theta. \tag{3.6}$$

Proof. Let $n > K$ where K is such that $a_k = 0$ if $|k| > K$. Consider the $(2n+1) \times (2n+1)$ circulant matrix

$$A_{n+1} = \begin{vmatrix} a_0 & a_1 \dots a_n & a_{-n} \dots a_{-1} \\ a_{-1} & a_0 \dots a_{n-1} & \dots a_{-2} \\ a_1 & a_2 \quad a_{-n} & a_0 \end{vmatrix}.$$

Let E_{2n+1} be $(2n+1)$ dimensional space and define a mapping $E_{2n+1} \to E_{2n+1}$ by $x \to A_{2n+1}x$. If $\omega \in \Omega$, let $\Gamma_{2n+1}\omega = (\omega(j))_{|j| \leq n}$, i.e., if $\omega = \{x_k\}$, $-\infty < k < \infty$, then Γ_{2n+1} is the section:

$$x_{-n}, x_{-n+1}, \dots, x_{-1}, x_0, x_1, \dots, x_{n-1}, x_n.$$

It is easy to see that if $|i| \leqq n - K$, then

$$(\tau\omega)(i) = (A_{2n+1}\Gamma_{2n+1}\omega)(i). \tag{3.7}$$

Define the density $R'_{2n+1}(y_{-n}, \dots, y_0, \dots, y_n)$ by

$$R'_{2n+1}(y_{-n}, \dots, y_0, \dots, y_n) dy_{-n} \dots dy_0 \dots dy_n = R\tau^{-1}\Gamma^{-1}_{2n+1},$$

and similarly define the density $\hat{R}'_{2n+1}$ by

$$\hat{R}'_{2n+1}(y_{-n}, \dots, y_0, \dots, y_n) dy_{-n}, \dots, dy_0 \dots dy_n = R\Gamma^{-1}_{2n+1}A^{-1}_{2n+1}.$$

From the definition of $I_n(R)$ in Lemma 3.1,

$$I_{2k+1}(R\tau^{-1}) = \int_{-\infty}^{\infty}\cdots\int R'_{2k+1}(y_{-k}, \dots, y_0, \dots, y_k) \cdot \log\left(\frac{R'_{2k+1}(y_{-k}, \dots, y_0, \dots, y_k)}{\left(\frac{1}{\sqrt{2\pi}}\right)^{2k+1} \exp\left\{-1\frac{1}{2}\sum_{j=-k}^{k} y_j^2\right\}}\right) dy_{-k} dy_0 \dots dy_k. \tag{3.8}$$

From (3.7) we see that $R\tau^{-1}\Gamma^{-1}_{2n+1} = R\Gamma^{-1}_{2l+1}A^{-1}_{2l+1}\Gamma^{-1}_{2n+1}$ provided $l \geqq n + K$. In particular, if we choose $n = k$, $l = k + K$, we have

$$R\tau^{-1}\Gamma^{-1}_{2k+1} = R\Gamma^{-1}_{2(k+K)+1}A^{-1}_{2(k+K)+1}\Gamma^{-1}_{2k+1}.$$

Therefore, from (3.8) we obtain

$$I_{2k+1}(R\tau^{-1}) \leqq \int_{-\infty}^{\infty}\cdots\int \hat{R}'_{2(k+K)+1}(y_{-k-k}, \dots, y_0, \dots, y_{k+K}) \cdot \log \frac{\hat{R}'_{2(k+K)+1}(y_{-K+k}, \dots, y_0, \dots, y_{k+K})}{\left(\frac{1}{\sqrt{2\pi}}\right)^{2(k+K)+1} \exp\left\{-\frac{1}{2}\sum_{j=-k-K}^{k+K} y_j^2\right\}} dy_{-k-K} \dots dy_{k+K}. \tag{3.9}$$

From Lemma 3.3 and (3.9) we get the estimate

$$I_{2k+1}(R\tau^{-1}) \leqq I_{2(k+K)+1}(R) - \log|A_{2(k+K)+1}| + \frac{1}{2}E^R\{\|A_{2(k+K)+1}\Gamma_{2(k+K)+1}\omega\|^2 - \|\Gamma_{2(k+K)+1}\omega\|^2\}. \tag{3.10}$$

Since R is stationary and $a_k=0$ if $|k|>K$,

$$\lim_{n\to\infty}\frac{1}{2n+1}E^R\{\|A_{2n+1}\Gamma_{2n+1}\omega\|^2\}=E^R\left\{\left|\sum_{j=-\infty}^{\infty}a_j\omega(j)\right|^2\right\}=E^{R\tau^{-1}}\{\omega^2(0)\},$$

and
$$\lim_{n\to\infty}\frac{1}{2n+1}E^R\{\|\Gamma_{2n+1}\omega\|^2\}=E^R\{\omega^2(0)\}.$$

Let $c=\lim_{n\to\infty}\frac{1}{2n+1}\log|A_{2n+1}|$, divide both sides of (3.10) by $2k+1$, let $k\to\infty$, and use Lemma 3.1 to obtain

$$I(R\tau^{-1})\leqq I(R)-c+\frac{1}{2}[E^{R\tau^{-1}}\{\omega^2(0)\}-E^R\{\omega^2(0)\}]. \tag{3.11}$$

Thus, it remains to show that

$$c=\frac{1}{2\pi}\int_0^{2\pi}\log|h(\theta)|d\theta,$$

where $h(\theta)=\sum_{k=-\infty}^{\infty}a_k e^{ik\theta}$. Since A_{2n+1} is a circulant matrix, Lemma 3.2 applies. We note that if $\xi\in Z_{2n+1}$ then $\xi^{-j}=\xi^{2n+1-j}$. Since $a_k=0$ if $|k|>K$, we get

$$\begin{aligned}\lim_{n\to\infty}\frac{1}{2n+1}\log|A_{2n+1}|&=\lim_{n\to\infty}\frac{1}{2n+1}\log\prod_{\xi\in Z_{2n+1}}\left(\sum_{|j|\leqq n}a_j\xi^j\right)\\&=\lim_{n\to\infty}\frac{1}{2n+1}\sum_{\xi\in Z_{2n+1}}\log\sum_{j=-\infty}^{\infty}a_j\xi^j=\frac{1}{2\pi}\int_0^{2\pi}\log\left|\sum_{j=-\infty}^{\infty}a_je^{ij\theta}\right|d\theta,\end{aligned}$$

where, of course, in the last equality we used the properties of roots of unity and that the last expression is finite since $|h(\theta)|\neq 0$ for all $\theta\in[0,2\pi]$ by hypothesis.

In the preceding lemma we showed inequality (3.6) under the hypothesis that the admissible mapping τ was in fact based on coefficients $\{a_k\}$ only a finite number of which were different from zero. Using that result we can now show that for any admissible mapping equality actually holds in (3.6).

Lemma 3.5. *Let τ be any admissible mapping. Then, for any $R\in\underline{M}_S$,*

$$I(R\tau^{-1})=I(R)+\frac{1}{2}[E^{R\tau^{-1}}\{\omega^2(0)\}-E^R\{\omega^2(0)\}]-\frac{1}{2\pi}\int_0^{2\pi}\log\left|\sum_{k=-\infty}^{\infty}a_ke^{ik\theta}\right|d\theta. \tag{3.12}$$

Proof. First define the map τ_n by $(\tau_n\omega)(j)=\sum_{k=-n}^{n}a_k\omega(k+j)$ with the same a_k's as in the statement of the lemma. We apply Lemma 3.4 to τ_n so that for any $R\in\underline{M}_S$,

$$\begin{aligned}I(R\tau_n^{-1})\leqq I(R)&+\frac{1}{2}[E^{R\tau_n^{-1}}\{\omega^2(0)\}-E^R\{\omega^2(0)\}]\\&-\frac{1}{2\pi}\int_0^{2\pi}\log\left|\sum_{k=-n}^{n}a_ke^{ik\theta}\right|d\theta.\end{aligned} \tag{3.13}$$

Since $I(R)$ is lower semicontinuous on $\underline{M}_S$, $I(R\tau^{-1}) \leqq \varliminf_{n\to\infty} I(R\tau_n^{-1})$. Moreover, $\lim_{n\to\infty} E^{R\tau_n^{-1}}\{\omega^2(0)\} = E^{R\tau^{-1}}\{\omega^2(0)\}$, and

$$\lim_{n\to\infty} \frac{1}{2\pi}\int_0^{2\pi} \log\left|\sum_{k=-n}^{n} a_k e^{ik\theta}\right| d\theta = \frac{1}{2\pi}\int_0^{2\pi} \log\left|\sum_{k=-\infty}^{\infty} a_k e^{ik\theta}\right| d\theta,$$

since $\left|\sum_{k=-\infty}^{\infty} a_k e^{ik\theta}\right| \neq 0$ for all $\theta \in [0, 2\pi]$. Thus, from (3.13) we conclude, for $R \in \underline{M}_S$,

$$I(R\tau^{-1}) \leqq I(R) + \frac{1}{2}[E^{R\tau^{-1}}\{\omega^2(0)\} - E^R\{\omega^2(0)\}] - \frac{1}{2\pi}\int_0^{2\pi} \log\left|\sum_{k=-\infty}^{\infty} a_k e^{ik\theta}\right| d\theta \tag{3.14}$$

for any admissible mapping τ. Let $g(\theta) = \sum_{j=-\infty}^{\infty} a_j e^{ij\theta}$. By our hypothesis this is an absolutely convergent Fourier series and $g(\theta) \neq 0$ for any $\theta \in [0, 2\pi]$. Wiener's theorem says that the L^2-function $1/g(\theta)$ has a Fourier series $\frac{1}{g(\theta)} = \sum_{j=-\infty}^{\infty} b_j e^{ij\theta}$ and $\sum_{j=-\infty}^{\infty} |b_j| < \infty$. Of course, $\left|\sum_{j=-\infty}^{\infty} b_j e^{ij\theta}\right| \neq 0$ for all $\theta \in [0, 2\pi]$.

Consider then the admissible map σ given by $(\sigma\omega)(j) = \sum_{k=-\infty}^{\infty} b_k \omega(k+j)$ and note that $\sigma = \tau^{-1}$. Apply (3.14) to σ and we get

$$I(R\sigma^{-1}) \leqq I(R) + \frac{1}{2}[E^{R\sigma^{-1}}\{\omega^2(0)\} - E^R\{\omega^2(0)\}] + \frac{1}{2\pi}\int_0^{2\pi} \log\left|\sum_{k=-\infty}^{\infty} a_k e^{ik\theta}\right| d\theta, \tag{3.15}$$

where the sign in the last term on the right is now positive since $\left|\sum_{k=-\infty}^{\infty} a_k e^{ik\theta}\right| = \frac{1}{\left|\sum_{k=-\infty}^{\infty} b_k e^{ik\theta}\right|}$. But (3.15) is true for all $R \in \underline{M}_S$ and, therefore, true in particular if we replace R in (3.15) by $R\tau^{-1}$ yielding, since $\sigma = \tau^{-1}$,

$$I(R\tau^{-1}\sigma^{-1}) = I(R) \leqq I(R\tau^{-1}) + \frac{1}{2}[E^R\{\omega^2(0)\} - E^{R\tau^{-1}}\{\omega^2(0)\}] + \frac{1}{2\pi}\int_0^{2\pi} \log\left|\sum_{k=-\infty}^{\infty} a_k e^{ik\theta}\right| d\theta,$$

or

$$I(R\tau^{-1}) \geqq I(R) + \frac{1}{2}[E^{R\tau^{-1}}\{\omega^2(0)\} - E^R\{\omega^2(0)\}] - \frac{1}{2\pi}\int_0^{2\pi} \log\left|\sum_{k=-\infty}^{\infty} a_k e^{ik\theta}\right| d\theta. \tag{3.16}$$

Inequalities (3.14) and (3.16) give us (3.12) for any admissible map τ which is the lemma.

From the preceding lemma we see that for any admissible mapping τ, $I(R)<\infty$ if and only if $I(R\tau^{-1})<\infty$. Since an admissible mapping τ is $1-1$ we observe that for any τ in this class, and with $\sigma=\tau^{-1}$,

$$H^{\tau}(R)=\inf_{R':R'\tau^{-1}=R} I(R')=I(R\sigma^{-1})$$

$$=I(R)+\frac{1}{2}[E^{R\sigma^{-1}}\{\omega^2(0)\}-E^{R}\{\omega^2(0)\}]+\frac{1}{2\pi}\int_0^{2\pi}\log|h(\theta)|d\theta\,, \quad (3.17)$$

where $h(\theta)=\sum_{j=-\infty}^{\infty} a_j e^{ij\theta}$, and τ is given by $(\tau\omega)(j)=\sum_{k=-\infty}^{\infty} a_k\omega(k+j)$. Note, that in (3.17) the sign before the last term on the right is + [cf. (3.12)] because $\sigma=\tau^{-1}$ and the sequence $\{a_k\}$ belongs to τ.

Moreover if $R\in \underline{M}_S$ has spectral measure $G(\theta)$, then $\frac{1}{2\pi}\int_0^{2\pi} dG(\theta)=E^R\{\omega^2(0)\}$ and also since $\sigma=\tau^{-1}$ we have $E^{R\sigma^{-1}}\{\omega^2(0)\}=\frac{1}{2\pi}\int_0^{2\pi}\frac{dG(\theta)}{|h(\theta)|^2}$. Hence, we can rewrite (3.17) in the form

$$H^{\tau}(R)=I(R)+\frac{1}{4\pi}\int_0^{2\pi}\frac{dG(\theta)}{|h(\theta)|^2}$$

$$-\frac{1}{4\pi}\int_0^{2\pi} dG(\theta)+\frac{1}{2\pi}\int_0^{2\pi}\log|h(\theta)|d\theta\,, \quad (3.18)$$

which is formula (3.3) at the beginning of this section.

In formula (3.18) which holds for any admissible τ, replace $I(R)$ by the last line in (3.1), using again that $E^R\{w^2(0)\}=\frac{1}{2\pi}\int_0^{2\pi} dG(\theta)$, so

$$H^{\tau}(R)=E^R\left\{\int_{-\infty}^{\infty} r(y|\omega)\log r(y|\omega)dy\right\}$$

$$+\frac{1}{2}\log 2\pi+\frac{1}{4\pi}\int_0^{2\pi}\frac{dG(\theta)}{|h(\theta)|^2}+\frac{1}{2\pi}\int_0^{2\pi}\log|h(\theta)|d\theta\,. \quad (3.19)$$

Now, in particular, consider the mapping $F_N:\Omega\to\Omega$ given by (2.5) or more precisely, its induced map $\tilde{F}_N$ in $\underline{M}_S$. This is an admissible map since

$$(F_N\omega)(j)=\sum_{|k|<N} a_k\left(1-\frac{|k|}{N}\right)\omega(k+j)\,,$$

and therefore at most a finite number of the coefficients defining this map are different from 0. Moreover, $h_N(\theta)=\sum_{|k|<N} a_k\left(1-\frac{|k|}{N}\right)e^{ik\theta}$, defined in (2.2), never vanishes since by Fejer's theorem,

$$h_N(\theta)=\int_0^{2\pi}\sqrt{f(\phi)}\,\psi_N(\theta-\phi)d\phi\,,$$

where

$$\psi_N(\theta)=\sum_{|k|<N}\left(1-\frac{|k|}{N}\right)e^{ik\theta},$$

is strictly positive. If as before we let $f_N(\theta)=|h_N(\theta)|^2$, then $f_N(\theta)$ is the spectral density of the process $\{X_k^N\}$, and so with $\tau=F_N$ and $H_{f_N}(R)=H^\tau(R)$, we get from (3.19)

$$H_{f_N}(R)=E^R\left\{\int_{-\infty}^{\infty} r(y|\omega)\log r(y|\omega)dy\right\}+\frac{1}{2}\log 2\pi$$
$$+\frac{1}{4\pi}\int_0^{2\pi}\frac{dG(\theta)}{f_N(\theta)}+\frac{1}{4\pi}\int_0^{2\pi}\log f_N(\theta)d\theta. \tag{3.20}$$

Formula (3.20) is the same as in Sect. 2 just after (2.17).

Finally, we must see in what sense $H_{f_N}(R)$ approximates $H_f(R)$ as defined by formula (1.9). We begin with the following well known lemma.

Lemma 3.6. *Let $\phi(x)$ be a probability density on $\mathbb{R}$ with two moments and let $\sigma^2(\phi)$ be the variance of ϕ. Then, for any $\alpha>0$,*

$$\int_{-\infty}^{\infty}\phi(y)\log\phi(y)dy\geqq-\frac{1}{2}\log 2\pi\alpha-\frac{1}{2\alpha}\sigma^2(\phi). \tag{3.21}$$

Proof. Let $\phi_{\sigma,\mu}=\frac{1}{\sigma\sqrt{2\pi}}\exp\left\{-\frac{(y-\mu)^2}{2\sigma^2}\right\}$. We note

$$\int_{-\infty}^{\infty}\phi(y)\log\phi(y)dy=\int_{-\infty}^{\infty}\phi(y)\log\left(\frac{\phi(y)}{\phi_{\sigma,\mu}(y)}\right)dy+\int_{-\infty}^{\infty}\phi(y)\log\phi_{\sigma,\mu}(y)dy.$$

The first term on the right is nonnegative being the entropy of one probability density with respect to another. Now, $\log\phi_{\sigma,\mu}(y)=-\log(\sigma\sqrt{2\pi})-\frac{1}{2\sigma^2}(y-\mu)^2$, and therefore

$$\int_{-\infty}^{\infty}\phi(y)\log\phi(y)dy\geqq-\frac{1}{2}\log(2\pi\sigma^2)-\frac{1}{2\sigma^2}\int_{-\infty}^{\infty}(y-\mu)^2\phi(y)dy.$$

If, in particular, we choose $\mu=\int_{-\infty}^{\infty}y\phi(y)dy$ and $\sigma^2=\alpha$ we obtain (3.21).

Lemma 3.7. *Let τ be an admissible map and for $R\in \underline{M}_S$ let*

$$e(R)=\int_{-\infty}^{\infty}r(y|\omega)\log r(y|\omega)dy.$$

If $R'=R\tau^{-1}$, then

$$e(R')=e(R)-\frac{1}{2\pi}\int_0^{2\pi}\log|h(\theta)|d\theta, \tag{3.22}$$

where $h(\theta)=\sum_{j=-\infty}^{\infty} a_j e^{ij\theta}$ *and* $\{a_j\}$, $-\infty<j<\infty$ *are the coefficients defining the map* τ.

Proof. From (3.12) and (3.1)

$$I(R')=I(R)+\frac{1}{2}[E^{R'}\{\omega^2(0)\}-E^R\{\omega^2(0)\}]-\frac{1}{2\pi}\int_0^{2\pi}\log|h(\theta)|d\theta$$
$$=e(R)+\frac{1}{2}\log 2\pi+\frac{1}{2}E^{R'}\{\omega^2(0)\}-\frac{1}{2\pi}\int_0^{2\pi}\log|h(\theta)|d\theta.$$

Using (3.1) on the left of this last equation we see that

$$e(R')+\frac{1}{2}\log 2\pi+\frac{1}{2}E^{R'}\{\omega^2(0)\}$$
$$=e(R)+\frac{1}{2}\log 2\pi+\frac{1}{2}E^{R'}\{\omega^2(0)\}-\frac{1}{2\pi}\int_0^{2\pi}\log|h(\theta)|d\theta,$$

from which (3.22) follows.

Lemma 3.8. *With* $f_N(\theta)$ *and* $f(\theta)$ *as in Sect. 2,*

$$\lim_{N\to\infty}\int_0^{2\pi}\log f_N(\theta)d\theta=\int_0^{2\pi}\log f(\theta)d\theta. \tag{3.23}$$

Proof. It suffices to show, since $f_N(\theta)=h_N^2(\theta)$, that

$$\lim_{N\to\infty}\int_0^{2\pi}\log h_N(\theta)d\theta=\int_0^{2\pi}\log\sqrt{f(\theta)}d\theta.$$

Now, $h_N(\theta)=\sqrt{f}*\psi_N$, where ψ_N is Fejer's kernel. Therefore, by Jensen's inequality,

$$\log h_N(\theta)=\log(\sqrt{f}*\psi_N)\geqq\log\sqrt{f}*\psi_N,$$

so that

$$\int_0^{2\pi}\log h_N(\theta)d\theta\geqq\int_0^{2\pi}(\log\sqrt{f}*\psi_N)d\theta=\int_0^{2\pi}\log\sqrt{f(\theta)}d\theta,$$

and hence

$$\lim_{N\to\infty}\int_0^{2\pi}\log h_N(\theta)d\theta\geqq\int_0^{2\pi}\log\sqrt{f(\theta)}d\theta.$$

On the other hand, from (2.3) we have that $\{h_N(\theta)\}$ are uniformly bounded and $\log h_N(\theta)\to\log\sqrt{f}$, so that from Fatou's lemma

$$\overline{\lim_{N\to\infty}}\int_0^{2\pi}\log h_N(\theta)d\theta\leqq\int_0^{2\pi}\log\sqrt{f(\theta)}d\theta.$$

Lemma 3.9. *Let $R \in \underline{M}_S$ such that $H_f(R) < \infty$; then there exists a sequence $\{R_N\} \in \underline{M}_S$ such that $\{R_N\}$ converges weakly to R and*

$$\lim_{N\to\infty} H_{f_N}(R_N) = H_f(R). \tag{3.24}$$

Proof. From its definition (1.9),

$$H_f(R) = e(R) + \frac{1}{2}\log 2\pi + \frac{1}{4\pi}\int_0^{2\pi}\frac{dG(\theta)}{f(\theta)} + \frac{1}{4\pi}\int_0^{2\pi}\log f(\theta)d\theta, \tag{3.25}$$

where $G(\theta)$ is the spectral measure of R and $e(R)$ is as defined in Lemma 3.7. From (3.20), for any $R' \in \underline{M}_S$,

$$H_{f_N}(R') = e(R') + \frac{1}{2}\log 2\pi + \frac{1}{4\pi}\int_0^{2\pi}\frac{dG'(\theta)}{f_N(\theta)} + \frac{1}{4\pi}\int_0^{2\pi}\log f_N(\theta)d\theta, \tag{3.26}$$

where $G'(\theta)$ is the spectral measure of R'. The way we construct the desired sequence $\{R_N\}$ is to define $R_N = R\tau_N^{-1}$, where $\{\tau_N\}$ is a suitably chosen sequence of admissible mappings which first of all will approach the identity map as $N\to\infty$, so that R_N will converge weaky to R. Secondly, $\{\tau_N\}$ will have to be selected so that (3.24) holds. Choosing τ_N means choosing a suitable sequence of coefficients. Let $g_N(\theta) = \sum_{j=-\infty}^{\infty} c_k^{(N)} e^{ik\theta}$, where $\{c_k^{(N)}\}$ are the defining coefficients of the admissible map τ_N. Taking into account Lemma 3.7 and 3.8, we see that for τ_N to satisfy the necessary requirements, $g_N(\theta)$ must satisfy:

(1) $$\lim_{N\to\infty}\int_0^{2\pi}\log|g_N(\theta)|d\theta = 0,$$

(2) $$\lim_{N\to\infty}\int_0^{2\pi}\frac{|g_N(\theta)|^2}{f_N(\theta)}dG(\theta) = \int_0^{2\pi}\frac{dG(\theta)}{f(\theta)},$$

(3) $$\lim_{N\to\infty}\int_0^{2\pi}|g_N(\theta)-1|^2 dG(\theta) = 0.$$

Condition (1), because of (3.22), guarantees $e(R_N)\to e(R)$. Since the spectral measure of R_N, call it $dG_N(\theta) = |g_N(\theta)|^2 dG(\theta)$, we see that (2), (1) and Lemma 3.8 insure that $H_{f_N}(R_N)\to H_f(R)$. Condition (3) insures that τ_N approaches the identity because it is the same as requiring $E^R\{|\omega(0) - (\tau_N\omega)(0)|^2\}\to 0$ as $N\to\infty$.

Let $g_{N,\varepsilon}(\theta) = \dfrac{\sqrt{f_N(\theta)}}{\sqrt{f_N(\theta)+\varepsilon}}$. We will show

(1)′ $$\lim_{\varepsilon\to 0}\lim_{N\to\infty}\left|\int_0^{2\pi}\log|g_{N,\varepsilon}(\theta)|d\theta\right| = 0,$$

(2)′ $$\lim_{\varepsilon\to 0}\lim_{N\to\infty}\left|\int_0^{2\pi}\frac{|g_{N,\varepsilon}(\theta)|^2 dG(\theta)}{f_N(\theta)} - \int_0^{2\pi}\frac{dG(\theta)}{f(\theta)}\right| = 0,$$

and

(3)′ $$\lim_{\varepsilon\to 0}\lim_{N\to\infty}\int_0^{2\pi}|g_{N,\varepsilon}(\theta)-1|^2 dG(\theta) = 0.$$

This will mean there exists a sequence $\varepsilon_N \to 0$ as $N \to \infty$ such that if we define $g_N(\theta) = g_{N,\varepsilon_N}(\theta)$, then (1)′, (2)′, and (3)′ imply (1), (2), (3) and hence the lemma. Because of (2.3), $g_{N,\varepsilon}(\theta) \to \sqrt{f(\theta)}/(\sqrt{f(\theta)}+\varepsilon)$ uniformly for $\theta \in [0,2\pi]$. Also $\frac{|g_{N,\varepsilon}(\theta)|^2}{f_N(\theta)} \to \frac{1}{(\sqrt{f(\theta)}+\varepsilon)^2}$ uniformly for $\theta \in [0,2\pi]$. Thus,

$$\lim_{N\to\infty} \int_0^{2\pi} \log|g_{N,\varepsilon}(\theta) d\theta = \int_0^{2\pi} \log\left(\frac{\sqrt{f(\theta)}}{\sqrt{f(\theta)}+\varepsilon}\right) d\theta , \tag{3.27}$$

$$\lim_{N\to\infty} \int_0^{2\pi} \frac{|g_{N,\varepsilon}(\theta)|^2 dG(\theta)}{f_N(\theta)} = \int_0^{2\pi} \frac{dG(\theta)}{(\sqrt{f(\theta)}+\varepsilon)^2}, \tag{3.28}$$

and

$$\lim_{N\to\infty} \int_0^{2\pi} |g_{N,\varepsilon}(\theta) - 1|^2 dG(\theta) = \int_0^{2\pi} \left|\frac{\sqrt{f(\theta)}}{\sqrt{f(\theta)}+\varepsilon} - 1\right|^2 dG(\theta). \tag{3.29}$$

In (3.27), since by hypothesis $\int_0^{2\pi} \log f(\theta) d\theta > -\infty$, we see $f(\theta) > 0$ almost everywhere on $[0,2\pi]$. Thus

$$\lim_{\varepsilon\to 0} \int_0^{2\pi} \log\left(\frac{\sqrt{f(\theta)}}{(\sqrt{f(\theta)})+\varepsilon}\right) d\theta = 0 ,$$

which shows (1)′. By the monotone convergence theorem

$$\lim_{\varepsilon\to 0} \int_0^{2\pi} \frac{dG(\theta)}{((\sqrt{f(\theta)})+\varepsilon)^2} = \int_0^{2\pi} \frac{dG(\theta)}{f(\theta)} ,$$

which shows (2)′. Finally by the bounded convergence theorem

$$\lim_{\varepsilon\to 0} \int_0^{2\pi} \left|\frac{\sqrt{f(\theta)}}{\sqrt{f(\theta)}+\varepsilon} - 1\right|^2 dG(\theta) = 0 ,$$

which shows (3)′ and completes the proof.

The preceding lemma was needed to complete the proof of the lower bound in Theorem 2.8. We now prove a lemma providing what was needed for the upper bound.

Lemma 3.10. *Let $H_{f_N}(R)$ be given by (3.20) and $H_f(R)$ be given by (1.9). Then for any $l > 0$,*

$$\bigcup_{N=1}^{\infty} \{R \in \underline{M}_S : H_{f_N}(R) \leq l\} \tag{3.30}$$

is tight in $\underline{M}_S$.

If $\{R_N\}$ is any sequence in $\underline{M}_S$ converging weakly to $R \in \underline{M}_S$, then

$$\varliminf_{N\to\infty} H_{f_N}(R_N) \geq H_f(R) . \tag{3.31}$$

Proof. From (2.3) we have that $f_N(\theta)\to f(\theta)$ uniformly in $[0,2\pi]$. From Lemma 3.8 we have that $\lim_{N\to\infty}\int_0^{2\pi}\log f_N(\theta)d\theta=\int_0^{2\pi}\log f(\theta)d\theta$. Since

$$H_{f_N}(R)=E^R\left\{\int_{-\infty}^{\infty}r(y|\omega)\log r(y|\omega)dy\right\}+\frac{1}{2}\log 2\pi+\frac{1}{4\pi}\int_0^{2\pi}\frac{dG(\theta)}{f_N(\theta)}+\frac{1}{4\pi}\int_0^{2\pi}\log f_N(\theta)d\theta,$$

and

$$H_f(R)=E^R\left\{\int_{-\infty}^{\infty}r(y|\omega)\log r(y|\omega)dy\right\}+\frac{1}{2}\log 2\pi+\frac{1}{4\pi}\int_0^{2\pi}\frac{dG(\theta)}{f(\theta)}+\frac{1}{4\pi}\int_0^{2\pi}\log f(\theta)d\theta,$$

it suffices, in order to show (3.30) and (3.31), to show

$$\bigcup_{N=1}^{\infty}\{R\in \underline{M}_S:\Phi_N(R)\leqq l\}\tag{3.32}$$

is tight in $\underline{M}_S$, and, for any sequence $\{R_N\}$ converging to R,

$$\varliminf_{N\to\infty}\Phi_N(R_N)\geqq\Phi(R),\tag{3.33}$$

where

$$\Phi_N(R)=E^R\left\{\int_{-\infty}^{\infty}r(y|\omega)\log r(y|\omega)dy\right\}+\frac{1}{4\pi}\int_0^{2\pi}\frac{dG(\theta)}{f_N(\theta)},$$

and

$$\Phi(R)=E^R\left\{\int_{-\infty}^{\infty}r(y|\omega)\log r(y|\omega)dy\right\}+\frac{1}{4\pi}\int_0^{2\pi}\frac{dG(\theta)}{f(\theta)}.$$

We start with Lemma 3.6 where the ϕ of that lemma is $r(y|\omega)$, giving us for any $\alpha>0$,

$$E^R\left\{\int_{-\infty}^{\infty}r(y|\omega)\log r(y|\omega)dy\right\}\geqq-\frac{1}{2}\log 2\pi\alpha-\frac{1}{2\alpha}\cdot\frac{1}{2\pi}\int_0^{2\pi}dG(\theta).\tag{3.34}$$

Since $f_N(\theta)\to f(\theta)$, there exists a c such that $f_N(\theta)\leqq c$ on $[0,2\pi]$. Thus, from (3.34)

$$\begin{aligned}E^R\left\{\int_{-\infty}^{\infty}r(y|\omega)\log r(y|\omega)dy\right\}&\geqq-\frac{1}{2}\log 2\pi\alpha-\frac{1}{2\alpha}\cdot\frac{1}{2\pi}\int_0^{2\pi}\frac{f_N(\theta)dG(\theta)}{f_N(\theta)}\\&\geqq-\frac{1}{2}\log 2\pi\alpha-\frac{1}{2\alpha}\frac{c}{2\pi}\int_0^{2\pi}\frac{dG(0)}{f_N(\theta)}.\end{aligned}$$

In this last if we let $\alpha=c/\varepsilon$ for $0<\varepsilon<1$, we get

$$E^R\left\{\int_{-\infty}^{\infty}r(y|\omega)\log r(y|\omega)dy\right\}\geqq-\frac{1}{2}\log\frac{2\pi c}{\varepsilon}-\frac{\varepsilon}{4\pi}\int_0^{2\pi}\frac{dG(\theta)}{f_N(\theta)},$$

so that if $\Phi_N(R) \leqq l$, then

$$\frac{1}{4\pi}\int_0^{2\pi}\frac{dG(\theta)}{f_N(\theta)} \leqq \frac{l+\frac{1}{2}\log\frac{2\pi c}{\varepsilon}}{1-\varepsilon}.$$

Since $f_N(\theta)$ is bounded by c on $[0,2\pi]$, this last implies

$$\frac{1}{2\pi}\int_0^{2\pi} dG(\theta) \leqq c\left(\frac{l+\frac{1}{2}\log\frac{2\pi c}{\varepsilon}}{1-\varepsilon}\right).$$

This means

$$E^R\{\omega^2(0)\} \leqq c\left(\frac{l+\frac{1}{2}\log\frac{2\pi c}{\varepsilon}}{1-\varepsilon}\right),$$

and since the constant on the right depends only on l and not on N or R, we see that $\bigcup_{N=1}^{\infty}\{R\in \underline{M}_S : \Phi_N(R)\leqq l\}$ is tight, which is (3.22).

To prove (3.23) we want to show that if $\{R_N\}$ converges weakly to R and $\Phi_N(R_N) \leqq M$, then $\Phi(R) \leqq M$.

We just proved that if $\Phi_N(R)\leqq l$ then $E^R\{\omega^2(0)\}$ is bounded by a constant independent of N. Thus if $\Phi_N(R_N)\leqq M$, we conclude that the second moments of $\{R_N\}$ are uniformly bounded which in turn implies the total mass of the spectral measure of R_N is uniformly bounded, i.e., $\int_0^{2\pi} dG_N(\theta)$ is uniformly bounded. Hence, by Helly's theorem (choosing a subsequence if necessary) $\{G_N\}$ converges weakly to $G' = G + G_1$, where G is the spectral measure of R and G_1 is nonnegative.

By hypothesis

$$\Phi_N(R_N) = E^{R_N}\left\{\int_{-\infty}^{\infty} r_N(y|\omega)\log r_N(y|\omega)dy\right\} + \frac{1}{4\pi}\int_0^{2\pi}\frac{dG_N(\theta)}{f_N(\theta)} \leqq M. \tag{3.35}$$

In (3.1) if, instead of entropy of R with respect to the stationary process generated by independent Gaussian, mean 0 variance 1 variables, we let $I_\lambda(R)$ be entropy of R with respect to the same stationary Gaussian process but with variance λ for $\lambda > 0$, then, of course,

$$I_\lambda(R) = E^R\left\{\int_{-\infty}^{\infty} r(y|\omega)\log r(y|\omega)dy\right\} + \frac{1}{2}\log 2\pi\lambda + \frac{1}{2\lambda}E^R\{\omega^2(0)\}.$$

From (3.35), we see that for any $\lambda > 0$,

$$I_\lambda(R_N) + \frac{1}{4\pi}\int_0^{2\pi}\frac{dG_N(\theta)}{f_N(\theta)} \leqq M + \frac{1}{2}\log 2\pi\lambda + \frac{1}{2\lambda}E^{R_N}\{\omega^2(0)\}. \tag{3.36}$$

Now, $I_\lambda(R)$ is lower semicontinuous on $\underline{M}_S$, and so since $R_N \Rightarrow R$, $f_N(\theta)\to f(\theta)$, $G_N \Rightarrow G' = G + G_1$, we conclude that

$$I_\lambda(R) + \frac{1}{4\pi}\int_0^{2\pi}\frac{dG'(\theta)}{f(\theta)} \leqq M + \frac{1}{2}\log 2\pi\lambda + \frac{1}{4\pi\lambda}\int_0^{2\pi} dG'(\theta). \tag{3.37}$$

Therefore, using the expression for $I_\lambda(R)$,

$$E^R\left\{\int_{-\infty}^{\infty} r(y|\omega)\log r(y|\omega)\right\}$$

$$+\frac{1}{2}\log 2\pi\lambda+\frac{1}{4\pi\lambda}\int_0^{2\pi} dG(\theta)+\frac{1}{4\pi}\int_0^{2\pi}\frac{dG'(\theta)}{f(\theta)}$$

$$\leqq M+\frac{1}{2}\log 2\pi\lambda+\frac{1}{4\pi\lambda}\int_0^{2\pi} dG'(\theta).$$

Since this last is true for any $\lambda>0$, we let $\lambda\to\infty$, after cancelling $\frac{1}{2}\log 2\pi\lambda$ from both sides, to obtain

$$E^R\left\{\int_{-\infty}^{\infty} r(y|\omega)\log r(y|\omega)dy\right\}+\frac{1}{4\pi}\int_0^{2\pi}\frac{dG'(\theta)}{f(\theta)}\leqq M.$$

Since $G'(\theta)=G(\theta)+G_1(\theta)$ and $G_1(\theta)$ is nonnegative, we conclude that in fact

$$\Phi(R)=E^R\left\{\int_{-\infty}^{\infty} r(y|\omega)\log r(y|\omega)dy\right\}+\frac{1}{4\pi}\int_0^{2\pi}\frac{dG(\theta)}{f(\theta)}\leqq M,$$

which completes the proof.

References

1. Donsker, M.D., Varadhan, S.R.S.: Asymptotic evaluation of certain Wiener integrals for large time. Functional Integration and its Applications, Proceedings of the International Conference held at the Cumberland Lodge, Windsor Great Park, London, in April 1974, A. M. Arthurs, (ed.). Oxford: Clarendon 1975
2. Donsker, M.D., Varadhan, S.R.S.: Asymptotic evaluation of certain Markov process expectations for large time, I. Commun. Pure Appl. Math. **28**, 1–47 (1976); II. Commun. Pure Appl. Math. **28**, 279–301 (1975); III. Commun. Pure Appl. Math. **29**, 389–461 (1976); IV. Commun. Pure Appl. Math. **36**, 183–212 (1983)
3. Donsker, M.D., Varadhan, S.R.S.: Asymptotics for the Wiener Sausage. Commun. Pure Appl. Math. **28**, 525–565 (1975)
4. Donsker, M.D., Varadhan, S.R.S.: On laws of the iterated logarithm for local times. Commun. Pure Appl. Math. **30**, 707–753 (1977)
5. Donsker, M.D., Varadhan, S.R.S.: Asymptotics for the Polaron. Commun. Pure Appl. Math. **36**, 505–528 (1983)
6. Orey, S.: Large deviations and Shanon-McMillan theorems (preprint)

Communicated by A. Jaffe

Received July 6, 1984

Expositiones Mathematicae

Expo. Math. 3 (1985), 251–272

Large deviations and applications

S.R.S. Varadhan

1. Introduction

Let $V(x)$ be a periodic continuous function of period 2π on the line. We consider the solution $u(t, x)$ of the initial value problem

$$\frac{\partial u}{\partial t}=\frac{1}{2}\frac{\partial^2 u}{\partial x^2}+V(x)\,u \tag{1.1}$$
$$u(0, x)=1$$

The Feynman-Kac formula expresses the solution of (1.1) as a function space integral

$$u(t, x)=E_x\left\{\exp\left[\int_0^t V(x(s))\,ds\right]\right\} \tag{1.2}$$

where E_x refers to the expectation with respect to the standard Brownian motion starting from the point $x \in R$ at time $t=0$. Since the function $V(x)$ as well as the solution $u(t, x)$ are both periodic in x of period 2π the problem could be considered on the circle Z rather than the real line R.

$$\frac{\partial u}{\partial t}=\frac{1}{2}\frac{\partial^2 u}{\partial \theta^2}+V(\theta)\,u \qquad \text{for } t>0,\ \theta \in Z \tag{1.3}$$
$$u(0, \theta)=1$$

If we denote by $\theta(t)$ the Brownian motion on the circle then the solution $u(t, \theta)$ has again a representation by Feynman-Kac's formula has

$$u(t, \theta)=E_\theta\left[\exp\left\{\int_0^t V(\theta(s))\,ds\right\}\right] \tag{1.4}$$

where E_θ is the expectation with respect to Brownian motion on Z starting from $\theta \in Z$ at time $t=0$.

The operator $\frac{1}{2}\frac{\partial}{\partial \theta^2}+V(\theta)$ is an operator on $L_2(Z)$ with a compact resolvent. Moreover the operator is self adjoint and bounded above. It therefore has a sequence $\{\lambda_j\}$ of eigenvalues tending to $-\infty$ and a sequence $\phi_j(\cdot)$ of eigenfunctions. The largest eigenvalue λ_1 is of multiplicity one and the corresponding eigenfunction is strictly positive. Denoting by $\langle\ ,\ \rangle$ the inner product in $L_2(Z)$ we can obtain an expansion for the solution $u(t,\theta)$ in terms of the eigenvalues and eigenfunctions.

$$u(t,\theta)=\sum_{j=1}^{\infty} e^{\lambda_j t}\,\phi_j(\theta)\,\langle\phi_j,1\rangle \tag{1.5}$$

From the formula it is not difficult to see that

$$\lim_{t\to\infty}\frac{1}{t}\log u(t,\theta)=\lambda_1 \quad \text{for all } \theta\in Z. \tag{1.6}$$

For the largest eigenvalue λ_1 we also have the classical variational formula

$$\lambda_1=\sup_{\substack{\phi\in L_2(Z)\\ \|\phi\|_2=1}}\{\textstyle\int V(\theta)\,\phi^2(\theta)\,d\theta-\frac{1}{2}\int[\phi'(\theta)]^2\,d\theta\} \tag{1.7}$$

since the supremum is attained at a positive ϕ we can take $\phi(\theta)=\sqrt{f(\theta)}$ and vary $f(\cdot)$ over all L_1 functions, that are nonnegative and have total integral 1. The formula then becomes

$$\lambda_1=\sup_{\substack{f\in L_1(Z)\\ f\geq 0,\ \int f\,d\theta=1}}\left[\int V(\theta)\,f(\theta)\,d\theta-\frac{1}{8}\int\frac{[f'(\theta)]^2}{f(\theta)}\,d\theta\right] \tag{1.8}$$

We would like to derive (1.8) directly from (1.4) without passing through the differential equation (1.3) or the interpretation of the limit in (1.4) as an eigenvalue. The reasons for doing this are many. We would like to replace $\exp\left[\int_0^t V(\theta(s))\,ds\right]$ by a more general functional $F_t(\omega)$ depending on the trajectory $\theta(\cdot)$. We aim to calculate limits of the form

$$\lim_{t\to\infty}\frac{1}{t}\log E^P\{\exp[F_t(\omega)]\} \tag{1.9}$$

for suitable functionals $F_t(\omega)$. We would like to generalize formula (1.4) by taking expectations with respect to processes other than the Brownian motion. If one replaces the Brownian motion by a more general reversible Markov process we still have a connection with a self adjoint operator for the special functional $\exp\left[\int_0^t V(\theta(s))\,ds\right]$. If the process is Markov but not reversible, we lose the self adjointness of the operator and hence the analogue of the variational formula does not seem to be valid in any obvious sense. If the process P is a stationary process without any Markov property there is no differential operator in the background at all. A direct derivation of formula (1.8) from (1.4) would therefore give us a clue as to how to generalize the formula in situations where there are no differential equations floating around.

Let us fix the starting point $\theta=0$ for simplicity and denote by P the corresponding Brownian motion measure. Let us define

$$\lambda(t,A)=P\left\{\theta(\cdot): \frac{1}{t}\int_0^t V(\theta(s))\,ds\in A\right\}$$

Then

$$u(t,0)=\int_{-\infty}^{\infty} \exp[t\,y]\,\lambda(t,dy) \tag{1.10}$$

If we denote by

$$m=\frac{1}{2\pi}\int_0^{2\pi} V(y)\,dy$$

then by the ergodic theorem for the Brownian motion on the circle, which has normalized Lebesgue measure as the invariant measure

$$\lambda(t,A)\to 0 \quad \text{as } t\to\infty.$$

for any closed set A, not containing the number m.

Suppose there exists a function $h(y)$ such that decay rate of the probability $\lambda(t,dy)$ locally near the point y is given by $\exp[-t\,h(y)]$ for every $y\in R$. We will make this precise later. If we think of the integral (1.10) as a sum

$$u(t,0)=\sum_y \exp[t\,y]\,\lambda(t,dy)$$

then approximately

$$u(t,0)\sim\sum_y \exp[t\,y]\exp[-t\,h(y)] \tag{1.11}$$

It is now obvious that

$$\lim_{t\to\infty}\frac{1}{t}\log u(t,0)=\sup_y\,[y-h(y)]. \tag{1.12}$$

although we have demonstrated the possibility of a variational formula (1.12) for the rate of growth of $u(t,0)$ we have lost the formula (1.8) in the process. The potential V is hidden in the function $h(y)$.

To remedy this situation we go up one level. We introduce the local time or the occupation time

$$L_t(\omega,A)=\frac{1}{t}\int_0^t \chi_A(\theta(s))\,ds \tag{1.13}$$

$L_t(\omega, A)$ is the proportion of time the Brownian trajectory $\theta(\cdot)$ spends in the set A during the time period $[0, t]$. $L_t(\omega, \cdot)$ is a random probability measure on Z and since we have a measure P on the space of trajectories, for each $t>0$ we get a probability measure Q_t on the space M of all probability measures on Z. Q_t is nothing but the distribution of $L_t(\omega, \cdot)$ where ω is distributed according to P. We can now rewrite $u(t, 0)$ in the form

$$u(t,0) = E^{Q_t}\{\exp[t \int V(\theta)\, \mu(d\theta)]\} \tag{1.14}$$

Here μ is the variable of integration for Q_t. By the ergodic theorem the Q_t measure approaches the δ-measure at the point $\mu_0 \in M$ where μ_0 is the invariant measure i.e. the normalized Lebesgue measure. If we have a rate of decay $\exp[-t\, I(\mu)]$ for the probability $Q_t(A)$, locally around $\mu \in M$ then an argument similar to (1.11) leads to

$$\lim_{t\to\infty} \frac{1}{t} \log u(t,0) = \sup_{\mu\in M} [\int V(\theta)\, \mu(d\theta) - I(\mu)] \tag{1.15}$$

Suppose it were also true that

$$I(\mu) = \frac{1}{8} \int \frac{[f'(\theta)]^2}{f(\theta)}\, d\theta$$

if μ has a nice density $f(\cdot)$ and

$$I(\mu) = \infty \quad \text{otherwise.}$$

Then (1.5) reduces to (1.8). The same sort of reasoning will also lead to a formula for integrals $\alpha(t)$ of the form

$$\alpha(t) = E^P\{\exp[t\, F(L(t, \cdot))]\} = E^{Q_t}\{\exp[t F(\mu)]\}$$

which will look like

$$\lim_{t\to\infty} \frac{1}{t} \log \alpha(t) = \sup_{\mu\in M} [F(\mu) - I(\mu)]. \tag{1.16}$$

Clearly if we take $F(\mu) = \int V(\theta)\, \mu(d\theta)$, a linear functional, we are back to (1.15) or (1.8). But we could take more general F and we would get formulae that cannot be obtained by differential equations.

The variational formula (1.16) is very similar to the energy-entropy formulae in statistical mechanics. See for instance [13].

Let us consider an example that is slightly different. Let

$$a(t) = E^P\left[\exp\left[\int_0^t V(\theta(s), \theta(s+1))\, ds\right]\right]$$

where $V(\theta_1, \theta_2)$ is a function on $Z \times Z$. The function $\int_0^t V(\theta(s), \theta(s+1))\, ds$ cannot be expressed in terms of the local time $L_t(\omega, \cdot)$ alone. We need to build a theory that is at one level higher than the one that we described just now, in order to study the growth rate of $a(t)$.

Given a segment of a trajectory $\theta(s)$ for $0 \leq s \leq t$, we extend it periodically to obtain a trajectory $\theta_t(s)$ for $-\infty < s < \infty$. We look at the periodic trajectory $\theta_t(\cdot)$ as well as its translates $\theta_t(\cdot + s)$ for $0 \leq s \leq t$. We define a measure R_t^ω on the space of all trajectories by

$$R_t^\omega = \frac{1}{t} \int_0^t \delta_{\theta_t(\cdot + s)}\, ds$$

In other words R_t^ω is the orbital measure on the random periodic orbit of $\theta_t(\cdot)$ obtained by time translations. R_t^ω is a random invariant measure (with respect to time translations) on the space of trajectories. Moreover

$$\int V(\theta(0), \theta(1))\, dR_t^\omega = \frac{1}{t} \int_0^t V(\theta(s), \theta(s+1))\, ds$$

provided the addition $\theta + 1$ on the right hand side is interpreted modulo t. The integration on the left is over the space of trajectories $\theta(\cdot)$ continuous except for jumps introduced by the periodization of $\theta(\cdot)$. Since the effect of periodization is only felt near the end the effect is small as t becomes large. The growth rate of $a(t)$ is therefore the same as that of

$$\hat{a}(t) = E^P\{\exp[t \int V(\theta(0), \theta(1))\, dR_t^\omega]\}$$

If we denote by π_t the distribution of R_t^ω then

$$\hat{a}(t) = E^{\pi_t}\{\exp[t \int V(\theta(0), \theta(1))\, dR]\}$$

Here R is an arbitrary stationary process i.e. a translation invariant measure on the space of trajectories on $(-\infty, \infty)$ with values in Z which are continuous except for jumps and R is the variable of integration with respect to π_t. In other words π_t is a mesare on the space of stationary stochastic processes. If we denote by R_0 the Brownian motion viewed as a stationary stochastic process on the time interval $(-\infty, \infty)$ then the ergodic theorem asserts that π_t is close to the δ-measure at R_0 for large t. If we now postulate once again a local decay rate $H(R)$ for π_t probabilities away from R_0, we clearly could predict a variational formula

$$\lim_{t \to \infty} \frac{1}{t} \log a(t) = \sup_R [E^R\{V(\theta(0), \theta(1))\} - H(R)] \tag{1.17}$$

We have three functions $H(R)$, $I(\mu)$ and $h(y)$ giving local decay rates. There are clearly relations between them. Since the sets

$$\left\{\omega : \frac{1}{t} \int_0^t V(\theta(s))\, ds = y\right\}$$

$$\{\omega: \int V(\theta)\, L_t(\omega, d\theta) = y\}$$
$$\{\omega: \int V(\theta(0))\, dR_t^{\omega} = y\}$$

are all the same we can expect

$$I(\mu) = \inf_{R:\, q(R) = \mu} H(R) \tag{1.18}$$

when $q(R)$ is the common one dimensional marginal of the stationary process R.

$$h(y) = \inf_{\mu:\, \int V(\theta)\, \mu(d\theta) = y} I(\mu) \tag{1.19}$$

Formulae like (1.18) and (1.19) are referred to as the contraction principle.

To avoid technical difficulties we will build the theory first for discrete time stochastic processes and then for continuous time processes. We will deal mostly with Markov processes except for the special case of stationary Gaussian processes in Section 10. In Sections 12–14 we will consider some applications.

2. Mathematical Formulation

We consider a sequence P_n of probability measures on a Polish space X which converges weakly to the degenerate distribution or point mass at some point $x_0 \in X$. For closed sets A disjoint from x_0, $P_n(A) \to 0$ as $n \to \infty$ and we are interested in the rate at which $P_n(A) \to 0$ as $n \to \infty$. The examples we consider will be those for which this rate is exponential in n and our interest will be in the exponential constant which of course will depend on the set A.

Formally if we assume that $P_n(A)$ behaves like $\exp[-n\rho(A)]$ for large n, then from the additivity properties of $P_n(A)$ one can conclude that

$$\rho(A \cup B) = \operatorname{Min}[\rho(A), \rho(B)].$$

A natural guess for the set function $\rho(A)$ is then a function of the form

$$\rho(A) = \inf_{x \in A} I(x)$$

where $I(x)$ is a point function i.e. a map from $X \to [0, \infty]$. The function $I(x)$ controls the local rate of decay of the probability measures P_n at the point x.

For technical reasons it is convenient to formulate the connection between the sequence $\{P_n\}$ and the function $I(x)$ in the following terms.

Definition: We say that the sequence $\{P_n\}$ satisfies the large deviation principle with a rate function $I(x)$ if

$$\limsup_{n \to \infty} \frac{1}{n} \log P_n(A) \le -\inf_{x \in A} I(x) \quad \text{for } A \text{ closed;}$$

$$\liminf_{n \to \infty} \frac{1}{n} \log P_n(G) \ge -\inf_{x \in G} I(x) \quad \text{for } G \text{ open.}$$

The rate function $I(x)$ is supposed to be a function on X with the following properties

(i) $0 \leq I(x) \leq \infty$ for all $x \in X$
(ii) $I(x)$ is lower-semi-continuous.
(iii) $\{x\colon I(x) \leq l\}$ is totally bounded in X for each $l < \infty$.

We will look at several examples of this type of a situation in the following sections.

3. Some Elementary Facts

Suppose $\{P_n\}$ is a sequence of probability measures satisfying a large deviation principle with a rate function $I(x)$. Then it is straightforward to conclude the following implications:

(a) Suppose $A \subset X$ is a set such that

$$\inf_{x \in A^0} I(x) = \inf_{x \in A} I(x) = \inf_{x \in \bar{A}} I(x)$$

where A^0, $\bar{A}$ are the interior and the closure of A. Then

$$\lim_{n \to \infty} \frac{1}{n} \log P_n(A) = -\inf_{x \in A} I(x). \tag{3.1}$$

(b) Suppose $F(x)$ is a bounded continuous function on X, then

$$\lim_{n \to \infty} \frac{1}{n} \log \int \exp[n F(x)]\, dP_n = \sup_{x \in X} [F(x) - I(x)]. \tag{3.2}$$

(c) More generally suppose $F_n(x)$ is a sequence of uniformly bounded continuous functions coverging uniformly on compact sets to a limiting function $F(x)$, then

$$\lim_{n \to \infty} \frac{1}{n} \log \int \exp[n F_n(x)]\, dP_n = \sup_{x \in X} [F(x) - I(x)]. \tag{3.3}$$

(d) Suppose Φ is a continuous map from $X \to Y$ where Y is another Polish space then the family $Q_n = P_n \Phi^{-1}$ satisfies a large deviation principle on Y with a rate function $\hat{I}(y)$ given by

$$\hat{I}(y) = \inf_{x\colon \Phi(x) = y} I(x) \tag{3.4}$$

More generally we can have Q_n given by $Q_n = P_n \Phi_n^{-1}$ where Φ_n tends to Φ as $n \to \infty$ uniformly on compact subsets of X. It is understood here that if $[x\colon \Phi(x) = y]$ is empty then $\hat{I}(y)$ is taken to be $+\infty$. We will refer to this as the Contraction Principle.

These implications (a) through (d) are direct consequences of the definition. They are proved in [15] or [16].

4. Cramér's Theorem

The simplest example of large deviation is to consider the sequence P_n of measures on R defined by

$$P_n(A) = \text{Prob}\left\{\frac{X_1 + \dots X_n}{n} \in A\right\}$$

where $X_1, \dots, X_n \dots$ are an independent identically distributed sequence of random variables with a common distribution α. Cramér in [2] showed that the large deviation principle holds for P_n with a rate function

$$I(x) = \sup_\theta [\theta x - \log M(\theta)] \tag{4.1}$$

where

$$M(\theta) = \int_R e^{\theta x}\, \alpha(dx) \tag{4.2}$$

If we take α to be the standard Gaussian distribution then $M(\theta) = \exp\left[\frac{\theta^2}{2}\right]$ and $I(x) = x^2/2$. P_n in this case is easily seen to be given by

$$dP_n = \sqrt{\frac{n}{2\pi}} \exp\left[-\frac{n x^2}{2}\right] dx$$

from which one can easily deduce the large deviation principle. If we take $F(x) = \theta x$ and try to evaluate

$$\begin{aligned}
\lim_{n\to\infty} \frac{1}{n} \log \int e^{n\theta x}\, dP_n &= \lim_{n\to\infty} \frac{1}{n} \log E \exp[\theta(X_1 + \dots X_n)] \\
&= \lim_{n\to\infty} \frac{1}{n} \log(E \exp[\theta x_1])^n \\
&= \log E \exp[\theta X_1] \\
&= \log M(\theta)
\end{aligned}$$

by formula (3.2) we obtain

$$\log M(\theta) = \sup_\theta [\theta x - I(x)] \tag{4.3}$$

Formulae (4.1) and (4.3) express the duality between the pair of conjugate convex functions $I(x)$ and $\log M(\theta)$.

The extensions of Cramér's theorem to the case of vector valued random variables have been considered. See [8] for the case of a Banach space or [1] for the case of topological vector spaces.

5. Empirical Distribution Functions

If $X_1, X_2, \dots X_n, \dots$ is a sequence of independent identically distributed random variables with a common distribution α and we look at the empirical distribution of the first n observation we obtain the random measure $\frac{1}{n}(\delta_{X_1}+\dots\delta_{X_n})$ on R. If we dente by $\boldsymbol{M}$ the space of probability measures on R, then $\frac{1}{n}(\delta_{X_1}+\dots\delta_{X_n})$ is a random element of $\boldsymbol{M}$ and we can denote by Q_n the probability distribution on $\boldsymbol{M}$ which is the distribution of the empirical distribution. As $n\to\infty$, the Glivenko-Cantelli theorem asserts that $Q_n\Rightarrow\delta_\alpha$ as $n\to\infty$. The principle of large deviation is valid in this situation with a rate function $I_\alpha(\mu)$ defined for $\mu\in\boldsymbol{M}$ as follows:

If $\mu\ll\alpha$ with $f(x)=d\mu/d\alpha$ satisfying $\int|f(x)\log f(x)|\,\alpha(dx)<\infty$ then

$$I_\alpha(\mu)=\int f(x)\log f(x)\,\alpha(dx) \tag{5.1}$$

otherwise $I_\alpha(\mu)=\infty$.

The formula (5.1) is clearly the same as relative entropy or the Kullback-Liebler information number. If we look at the map from $\boldsymbol{M}\to R$ given by $\mu\to\int x\,d\mu$, then the contraction principle of Section 3, gives us that

$$\inf_{\mu:\,\int x\,d\mu=a} I_\alpha(\mu)=\sup_\theta\left[\theta a-\log\int e^{\theta x}\,\alpha(dx)\right]. \tag{5.2}$$

To see this, we note that the mean of the empirical distribution $\frac{1}{n}(\delta_{X_1}+\dots+\delta_{X_n})$ is nothing but $\frac{1}{n}(X_1+\dots+X_n)$. The left side of (5.2) is computed by the contraction principle and the right side by Cramér's theorem.

In some sense the large deviation principle for the empirical distribution is a special case of Cramér's theorem when the random variable takes values in the topological vector space of all signed measures on R. θ then belongs to the dual space of bounded continuous functions

$$\begin{aligned} M(\theta)&=E\exp\langle\theta,\delta_X\rangle\\ &=E\exp[\theta(X)]\\ &=\int\exp[\theta(x)]\,\alpha(dx)\\ I(\mu)&=\sup\left[\int\theta(x)\,\mu(dx)-\log M(\theta)\right]. \end{aligned} \tag{5.2}$$

It is not very hard to compute (5.2) and verify that it yields again (5.1). For some of the properties of $I_\alpha(\mu)$ and its relation to Cramér's theorem see [4] and [8].

An interesting special case is when α is a probability distribution giving mass $\alpha_1,\dots,\alpha_k$ to k points $x_1,\dots,x_k$, then the empirical distribution is the normalized frequency count. What we have then is essentially the large deviation principle for the multinomial probability distribution with cell probabilities $\alpha_1,\dots\alpha_k$. If we denote by $\mu_1,\dots,\mu_k$ the probabilities of another distribution on these same points then

$$\text{Prob}\left[\frac{1}{n}(f_1, \ldots, f_k) \text{ is nearly equal to } (\mu_1, \ldots, \mu_k)\right]$$

can be computed by the multinomial distribution and using Stirling's formula one can see that the log of the above probability is nearly

$$-n \sum_{i=1}^{k} \mu_i \log \frac{\mu_i}{\alpha_i}.$$

6. Empirical Processes

The empirical distribution of the first n members of the sequence $X_1, X_2, \ldots$ does not convey any information about the order in which $X_1, \ldots, X_n$ occur. In order to take care of this we define the empirical process as follows:

Consider for any n-tuple $X_1, \ldots, X_n$ the two sided periodic extension of $X_1, \ldots, X_n$ given by

$$\xi_n = (\ldots, X_1, \ldots, X_n, X_1, \ldots, X_n, \ldots).$$

Although ξ_n is ambiguous because of the location of the starting point X_1, the periodic orbit $\xi_n, T\xi_n, \ldots, T^{n-1}\xi_n$ defined by ξ_n is well defined. Here T denotes the shift on the space of sequences.

Given the orbit one can of course define a measure $\frac{1}{n}(\delta_{\xi_n} + \ldots + \delta_{T^{n-1}\xi_n})$ on the orbit. Since the orbit is periodic we now have a stationary measure or a stationary stochastic process on the space of sequences. The stationary process is random because the n-tuple $X_1 \ldots X_n$ is random. We have now a random stationary process which we will call the empirical process. If R_n is the empirical process then the one dimensional marginal of R is $\frac{1}{n}(\delta_{X_n} + \ldots + \delta_{X_n})$ or the empirical distribution. The two dimensional marginal is easily seen to be $\frac{1}{n}(\delta_{X_1, X_2} + \ldots + \delta_{X_{n-1}, X_n} + \delta_{X_n, X_1})$. Except for the last term it is the distribution of consecutive pairs and for large n the effect of the last term can be ignored. We now have a probability distribution π_n on the space of all stationary processes Ω_S and as $n \to \infty$, $\pi_n \Rightarrow \delta_{P_0}$ where P_0 is the infinite product measure with factor α. Clearly $P_0 \in \Omega_S$ and the ergodic theorems assure us that $\pi_n \Rightarrow \delta_{P_0}$ as $n \to \infty$. Note that weak convergence in Ω_S is essentially convergence of finite dimensional distributions.

One can naturally ask if there is a principle of large deviations for π_n on Ω_S. The answer is yes and the rate function is given by

$$H_{P_0}(Q) = E^Q\{I_\alpha(Q_\omega)\}. \tag{6.1}$$

Here Q_ω is the regular conditional probability of X_0 given the past $X_{-1}, X_{-2}, \ldots$ under Q and is a distribution on R for each ω representing the past information.

For each ω, $I_\alpha(Q_\omega)$ is calculated by formula (5.1). This is a nonnegative function of ω and $H_{P_0}(Q)$ is the integral with respect to Q.

When we replace R by a finite set with k elements and take α to be the uniform distribution on these k elements then $\log k - H_{P_0}(Q)$ is nothing but Shannon's entropy of the stationary process Q and the principle of large deviation is essentially the Shannon-Breimann-McMillan theorem of information theory.

We have a map q from Ω_S to $\boldsymbol{M}$ which assigns to any stationary process Q its common one dimensional marginal distribution μ. The map q applied to the empirical process gives the empiricial distribution. The contraction principle should yield

$$I_\alpha(\mu) = \inf_{Q:\, q(Q)=\mu} H_{P_0}(Q). \tag{6.2}$$

One can verify this formula by noting that infimum on the right is attained when Q is the product measure with factor μ. In such a case $Q_\omega = \mu$ for all ω and one can compute from (6.1) that for this special Q, $H_{P_0}(Q) = I_\alpha(\mu)$.

As an application of these ideas let us try to evaluate

$$\lim_{n\to\infty} \frac{1}{n} \log E \exp\{F(X_1, \ldots, X_k) + \ldots + F(X_n, \ldots, X_{n+k-1})\}$$

where $X_1, \ldots, X_n, \ldots$ is a sequence of independent identically distributed random variables with a common distribution α. We assume that F is a bounded continuous function of $X_1, \ldots, X_k$. If we define a functional $\Phi(Q)$ on Ω_S by

$$\Phi(Q) = E^Q F(X_1, \ldots, X_k)$$

then $\log E\{\exp\{F(X_1, \ldots, X_k) + \ldots + F(X_n, \ldots, X_{n+k-1})\}$ differs from $\log \int \exp[n\Phi(Q)]\, \pi_n(dQ)$ by only the effect of periodization and this is a bounded effect. When we divide by n, this disappears as $n \to \infty$. If we apply formula (3.2) we can conclude that

$$\lim_{n\to\infty} \frac{1}{n} \log E\{\exp[F(X_1, \ldots, X_k) + \ldots + F(X_n, \ldots, X_{n+k-1})]\}$$
$$= \sup_Q [E^Q\{F(X_1, \ldots, X_k)\} - H_{P_0}(Q)].$$

7. The Dependent Case

It is not unreasonable to want to replace the underlying stochastic process of independent random variables $X_1, X_2, \ldots$ with common distribution α, with an ergodic stationary process. In other words we change the product measure P_0 to a stationary ergodic measure P. It also seems reasonable at first glance that since the ergodic theorem assures us that $\pi_n \Rightarrow \delta_P$ there might be a principle of large deviation lurking around. The rate function should be given by

$$H_P(Q)=E^Q[I_{P_\omega}(Q_\omega)] \tag{7.1}$$

where Q_ω, P_ω are respectively the regular conditional probability distributions of X_0 given the past $X_{-1}, \ldots, X_{-k}, \ldots$ under Q and P respectively. One can convince oneself that (7.1) is the logical extension of (6.1) when P is not a product measure.

However there is a proverbial fly in the ointment. Q_ω as a conditional probability is defined only almost everywhere with respect to Q just as P_ω is only defined almost everywhere with respect to P. Since Q and P, being stationary processes want to be orthogonal most of the time formula (7.1) does not make sense. Since the expectation is with repect to Q the problem is to define P_ω almost everywhere with respect to Q.

Clearly one cannot do this under great generality but we have to make special assumptions on the nature of P so that P_ω may be defined almost everywhere with respect to Q for any $Q\in\Omega_S$.

If P is a stationary Markov process with transition probability $\pi(x, dy)$ then

$$P_\omega(A)=\pi(X_{-1}, A)$$

and this is clearly well defined everywhere. It seems therefore likely that if we restrict P to Markov processes we may get somewhere. Another possibility is to take P to be a Gaussian process in which case P_ω is a Gaussian measure with mean

$$m(\omega)=\sum_{j=1}^{\infty} a_{-j}X_{-j}$$

and variance θ_0 equal to the prediction error. One can then see for which processes Q the expression $m(\omega)$ makes reasonable sense. These ideas will be explored in the next two sections.

8. The Markov Case

Let $\pi(x, dy)$ be the transition probabilities of a Markov chain on X. For any initial point $X_0=x$, we can construct the probability measure P_x on the space of sequences $X_0=X_1, X_2, \ldots$.

Irrespective of whether there exists a stationary Markov process with transition probability $\pi(x, dy)$, we can construct the measure π_n^x on Ω_S as in Section 7. The starting point x enters because the probability measure on n-tuples is induced by P_x. If there is no finite invariant measure, we expect π_n^x to dissipate as $n\to\infty$.

If the Markov chain is positive recurrent and ergodic then $\pi_n\Rightarrow\delta_P$ where P is the unique stationary Markov process with $\pi(x, dy)$ as transition probability.

We make the following assumptions on $\pi(x, dy)$.

(i) For each bounded continuous function $f(y)$ on X, $(\pi f)(x)=\int f(y)\,\pi(x,dy)$ is a bounded continuous function of x.

(ii) There exists a function $u_0(x)$ on X such that $u_0(x)\geq 1$, $(\pi u_0)(x)=\int u(y)\,\pi(x,dy)$ is bounded on compact subsets of X and for each $l<\infty$

$$\left\{x: \frac{u_0(x)}{(\pi u_0)(x)}\leq l\right\}$$

is a totally bounded subset of X.

(iii) There exists a reference measure α on X such that $\pi(x,dy)$ has a density $\pi(x,y)$ with respect to α. $\pi(x,y)$ is positive almost everywhere with respect to α and the mapping $x\to\pi(x,y)$ is continuous as a map from X into $L_1(\alpha)$.

Assumption (ii) guarantees the existence of a finite invariant measure and assumption (iii) guarantees the uniqueness of the invariant measure and the ergodicity of the stationary process. Assumption (iii) can be weakened somewhat.

Under these conditions one can prove the principle of large deviations with a rate function given by (7.1). The details for the discrete Markov chain case have not appeared in print. But the details for the continuous time analogue have appeared in [5] and the discrete time situation, if anything, is easier.

9. Continuous Time Markov Processes

We start with a continuous time Markov process on a Polish state space X. The process is specified by transition probabilities $p(t,x,dy)$. These transition probabilities generate a semigroup T_t on the space of bounded measurable functions on X. We will assume that the process is Feller, i.e. that the semigroup maps the space $C(X)$ of bounded continuous functions into itself. We will also assume that the process can be realized in the Skorohod space $D_X[0,\infty)$ of right continuous functions with left limits on $[0,\infty)$ with values in X. For each starting point $x\in X$ we have the measure P_X on $D_X[0,\infty)$ which is again a Polish space and we assume that the map $x\to P_x$ is weakly continuous. Given a realization of the process $x(\cdot)$ in $[0,T]$ i.e. a function $x(s)$ for $0\leq s\leq T$ we can extend the function periodically both sides and construct a periodic family $x(s+t)$, $0\leq t\leq T$ of functions on $(-\infty,\infty)$. Denoting this function by $x_t(s)$ we can construct on the space $D_X(-\infty,\infty)$ of trajectories a random measure $\frac{1}{T}\int_0^T \delta_{x_t}\,dt$. This is a translation invariant random measure and therefore induces a measure π_T^x on the space of all stationary stochastic processes. Here x is the starting point and comes in because the measure on the set of trajectories $D_X[0,T]$ is P_x which is constructed from the transition probabilities and the starting point x.

One can ask if there is a large deviation principle here governed by a rate function similar to (7.1). If Q is any stationary process and Q_ω is the regular conditional probability of Q given the past i.e. given $x(s)$ for $-\infty < s \leq 0$ then we denote by Q_ω^T the restriction of Q_ω to the σ-field generated by $x(s)$ for $0 \leq s \leq T$. We also have the measure $P_{x(0,\omega)}^T$ which is the Markov measure corresponding to our transition probabilities starting from $x = x(0, \omega)$ restricted to the same σ-field: If we define

$$H_P^T(Q) = E^Q[I_{P_{x(0,\omega)}^T}(Q_\omega^T)] \tag{9.1}$$

then it turns out that

$$H_P^T(Q) = T H_P(Q) \quad \text{for} \quad T > 0 \tag{9.2}$$

for some $H_P(Q)$ which may be infinite. The large deviation principle holds for π_T^x with this rate function under suitable conditions: the conditions are

(i) The map $x \to P_x$ is weakly continuous.
(ii) There exists a sequence of functions $u_n(x)$ in the domain of the infinitesimal generator L of the semigroup T_t of the Markov process such that

(a) $u_n(x) \geq 1$

(b) For every compact set $W \subset X$

$$\sup_{x \in W} \sup_n u_n(x) < \infty.$$

(c) For each $x \in X$

$$\lim_{n \to \infty} \left(\frac{L u_n}{u_n}\right)(x) = V(x) \qquad \text{exists.}$$

(d) For some $A < \infty$

$$\sup_n \sup_x \left(\frac{L u_n}{u_n}\right)(x) \leq A.$$

(e) $\{x : V(x) \geq l\}$ is a compact set in X for each $l > -\infty$.

(iii) The transition probability $p(1, x, dy)$ is equivalent to a fixed reference measure α with density $p(1, x, y)$ and the map $x \to p(1, x, y)$ is a continuous map of X into $L_1(\alpha)$.

These are analogues of conditions (i) to (iii) in the Markov chain case introduced in Section 8. The proofs and details appear in [5]. Condition (ii) should be interpreted as a coerciveness condition. It requires the generator L to be such that $(L - V)u = 0$ has a solution $u \geq 1$ for some V satisfying (e).

10. The Gaussian Case

Let $\{X_n\}$ be a Gaussian process with mean zero and covariance $E X_i X_j = \rho_{j-i}$. We assume that the covariance arises from a spectral density

$$\rho_n = \frac{1}{2\pi} \int_0^{2\pi} e^{in\theta} f(\theta)\, d\theta.$$

We assume that the process is purely nondeterministic i.e. $\int_0^{2\pi} \log f(\theta)\, d\theta > -\infty$. We further make the regularity assumption that the spectral density is a continuous periodic function of θ of period 2π. Under these conditions we can show that a principle of large deviation holds and formula (7.1) can be simplified to yield the following explicit expression for the rate functional:

$$H_P(Q) = E^Q \int_{-\infty}^{\infty} q(x,\omega) \log q(x,\omega)\, dx + \frac{1}{2}\log 2\pi + \frac{1}{4\pi} \int_0^{2\pi} \log f(\theta)\, d\theta + \frac{1}{4\pi} \int \frac{dG(\theta)}{f(\theta)}.$$

Here $q(x,\omega)$ is the conditional probability density of X_0, under Q given the past history $X_{-1}, X_{-2}, \ldots$. We take $H_P(Q) = \infty$, unless the stationary process Q has two moments and if Q has two moments then $E^Q \int_{-\infty}^{\infty} q(x,\omega) \log q(x,\omega)\, dx$ is well defined as a real number which could be $+\infty$ but never $-\infty$. Here dG is the spectral measure of the process Q i.e.

$$E^Q X_0 X_n = \frac{1}{2\pi} \int e^{in\theta}\, dG(\theta).$$

The details of this situation will appear in [10].

11. The Case of Brownian Motion on R^d

If we take for our underlying Markov process the standard Brownian motion on R^d it satisfies the conditions (i) and (iii) of Section 9 but not condition (ii). In fact it does not have a finite invariant measure. In terms of the rate functions $I(\mu)$ and $H(P)$ this means that $I(\mu)$ and $H(P)$ are strictly positive and never zero. There is always local decay of probability. The upper estimate for closed sets require putting together local estimates for a global estimate and this cannot be done. However the global lower estimate is really a local estimate and can be derived without assumption (ii). The upper estimate can still be derived for compact sets and this can sometimes be used in special problems. In later sections we will finesse the issue by modifying the Brownian motion process either by considering it on a large torus or by introducing a small restoring drift. Condition (ii) will then be satisfied and we have to remove the effect of the modification by a special argument.

12. Application: Laws of the Iterated Logarithm

If $x(t)$, $0 \leq t < \infty$ is the standard one dimensional Brownian motion it is well known that it has a local time as a continuous function. In other words the random measure

$$L_t(\omega, A) = \frac{1}{t} \int_0^t \chi_A(x(s))\, ds$$

has for almost all ω a continuous density $l_t(\omega, y)$ with respect to the Lebesgue measure.

$$l_t(\omega, y) = \frac{1}{t} \int_0^t \delta(x(s) - y)\, ds$$

and

$$L_t(\omega, A) = \int_A l_t(\omega, y)\, dy.$$

Brownian scaling property implies that the distribution of $\sqrt{t}\, l_t(\omega, \sqrt{t}\, y)$ is independent of t and is the same as that of $l_1(\omega, y)$. However for $t \to \infty$, $l_t(\omega, y)$ tends to be more and more dispersed. It is likely to have a zero limit in the topology of uniform convergence on compact sets but not in the topology of $L_1(R)$.

Let us define

$$\hat{l}_t(\omega, y) = \sqrt{\frac{t}{\log\log t}}\, l_t\left(\omega, \sqrt{\frac{t}{\log\log t}}\, y\right).$$

Statistically this has the same distribution as $l_{\log\log t}(\omega, y)$ and therefore will dissipate slowly. By our local decay rates

$$\text{Prob}\{l_{\log\log t}(\omega, \cdot) \sim f\} \simeq \exp[-\log\log t\, I(f)].$$

If we sample at times $t_n = \rho^n$ for some $\rho > 1$

$$\text{Prob}\{l_{\log\log t_n}(\omega, \cdot) \sim f\} \simeq \exp[-[\log n + \log\log \rho]\, I(f)] \sim \frac{1}{n^{I(f)}}.$$

Then by an application of the Borel Cantelli lemma and by standard arguments used in proving laws of the iterated logarithm one can conclude that the family $\{\hat{l}_t(\omega, \cdot)\}$ is conditionally compact as a family in the space of continuous functions with the topology of uniform convergence on compact sets and its limits points are precisely the set C of subprobability densities $f(x)$ with $\frac{1}{8} \int \frac{(f')^2}{f}\, dx \leq 1$.

In particular if Φ is a reasonable function on the space of nonnegative functions with integral at most one then

$$\limsup_{t \to \infty} \Phi(\hat{l}_t(\omega, \cdot)) = \sup_{f \in C} \phi(f) \quad \text{a.e.}$$

we can try various functionals for $\phi(f)$. For example if we take $\Phi(f)=f(0)$ we get

$$\limsup_{t\to\infty} \frac{1}{\sqrt{t\log\log t}} \int_0^t \delta(x(s))\,ds = \sup_C f(0) = \sqrt{2} \quad \text{a.e.}$$

As another example if we take

$$\phi(f)=\inf\left\{a: \int_{-a}^{a} f(x)\,dx=1\right\}$$

by an additional argument we can show that

$$\liminf_{t\to\infty} \left(\sup_{0\le s\le t} |x(s)|\right)\left(\frac{\sqrt{\log\log t}}{t}\right) = \inf_{f\in C} \phi(f) = \frac{\pi}{2\sqrt{2}} \quad \text{a.e.}$$

for these laws of the iterated logarithms and related references see [7]. For further generalizations regarding laws of iterated logarithm see [12] and [14].

13. Application: Wiener Sausage

The following problem comes up in the study of the behavior of the density of states for the Laplace operator in R^d with a random distribution of holes. The operator has Dirichlet boundary conditions at these holes. The centers of these holes have a Poisson point process for their distribution with some uniform intensity and the holes have some fixed size $\varepsilon>0$.

Let $x(s)$, $0\le s\le t$ be the d-dimensional Brownian motion conditioned to start at time 0 from the origin and end at time t again at the origin. Let C_t be the random set which is the range of this random Brownian trajectory. Let C_t^ε be the sausage around C_t of radius ε. That is to say

$$C_t^\varepsilon=\{y: |y-x|<\varepsilon \text{ for some } x\in C_t\}.$$

The set C_t^ε has some d-dimensional volume which we denote by $|C_t^\varepsilon|$. We let

$$a(t)=E_t\{\exp[-\nu|C_t^\varepsilon|]\} \tag{13.1}$$

where E_t refers to expectation with respect to the conditional Brownian motion in the interval $[0,t]$. The problem is to show that

$$\lim_{t\to\infty} \frac{1}{t^{\frac{d}{d+2}}} \log a(t) = -k_d(\nu) \tag{13.2}$$

exists in each dimension and is finite and nonzero.

As a first step it is straightforward to prove that the conditional Brownian motion can be replaced by the free Brownian motion starting at the origin for large times with negligible error so that it is sufficient to prove that

$$\lim_{t\to\infty} \frac{1}{t^{\frac{d}{d+2}}} \log b(t) = -k_d(v) \tag{13.3}$$

where

$$b(t) = E\{\exp[-v|C_t^\varepsilon|]\} \tag{13.4}$$

where E is expectation with respect to the ordinary Brownian motion starting from the origin.

By the Brownian scaling property one can show that

$$b(t) = E\{\exp[-v t^{d/(d+2)} |C_{t^{d/(d+2)}}^{\varepsilon t^{-1/(d+2)}}|]\}. \tag{13.5}$$

It is therefore sufficient to show that

$$\lim_{t\to\infty} \frac{1}{t} \log c(t) = -k_d(v) \tag{13.6}$$

where

$$c(t) = E\{\exp[-vt|C_t^{\varepsilon t^{-1/d}}|]\}. \tag{13.7}$$

We have just replaced $t^{d/(d+2)}$ by the new parameter t.

Formally we have a family of functionals on probability measures μ on R^d defined by

$$F_t(\mu) = |C_\mu^{\varepsilon t^{-1/\alpha}}| \tag{13.8}$$

where C_μ is the support of the measure μ. The range of a continuous trajectory is the support of its occupation distribution.

$$\lim_{t\to\infty} F_t(\mu) = F(\mu) = |C_\mu|. \tag{13.9}$$

The expression $C(t)$ is given by

$$c(t) = E\{\exp[-vt F_t(L_t(\omega, \cdot))]\}.$$

It is not unreasonable to expect that

$$\lim_{t\to\infty} \frac{1}{t} \log c(t) = \sup_f [-vF(f) - I(f)] = -\inf_f [F(f) + I(f)].$$

Since $\inf_{\text{supp} f \subset A} I(f) = \lambda(A)$ where $\lambda(A)$ is the lowest eigenvalue of $-\frac{1}{2}\Delta$ on A with Dirichlet boundary conditions, we can reduce the variational expression is (13.10) to

$$k_d(v) = \inf_A \{v|A| + \lambda(A)\}. \tag{13.11}$$

The infimum is seen to be attained over spheres by means of an isoperimetric inequality and this reduces to just a variation over the radius of the sphere.

$$k_d(v) = \inf_{v>0} \left\{ v r^d v_1 + \frac{\lambda_1}{r^2} \right\} \tag{13.12}$$

where v_1 is he volume of the unit sphere in d-dimensions and λ_1 is the lowest eigenvalue of $-\frac{1}{2}\Delta$ in the unit sphere with zero boundary conditions.

The analysis however is filled with difficulties. First of all the lower bound is fairly easy. If S_r is the sphere of radius r, then by insisting that the Brownian motion not leave the sphere for time t we can make $|C_t^{\varepsilon t^{-1/\alpha}}|$ be at most $v_1(r+\varepsilon t^{-1/d})^d$. The probability of this is approximately $\exp\left[-t\frac{\lambda_1}{r^2}\right]$ for large t. This gives the lower bound since r can be chosen arbitrarily.

The upper bound is technically hard because although the contribution when the range is fixed to be a given set has the right behavior the totality of such sets is huge.

It is here that we need to use our theory or more precisely some refinements of it. The Brownian motion on infinite space lacks the coerciveness condition imposed for upper bounds. We look instead at the Brownian motion on the d-dimensional torus of size l. Fortunately the volume of any set projected to the torus is at most the volume of the set so that any upper bound for any torus is automatically an upper bound for $c(t)$. We denote by

$$c_l(t) = E_l\{\exp[-vt|C_t^{\varepsilon t^{-1/d}}|]\} \tag{13.12}$$

where E_l is Brownian motion on the torus of size l in d dimensions. If we show

$$\lim_{t\to\infty} \frac{1}{t} \log c_l(t) = -\inf_A [v|A| + \lambda_l(A)] = k_{d,l}(v)$$

then it is not hard to show that

$$\lim_{l\to\infty} k_{d,l}(v) = k_d(v)$$

by standard methods.

Although our results apply to Brownian motion on the d torus, the basic topology we use is weak convergence of probability measures. The functional $F(\mu) = |\operatorname{supp}\mu|$ is very unstable in this topology. It is however stable, at least lower semi continuous in the L_1 topology on probability densities. The sausage of radius $\varepsilon t^{-1/\alpha}$ offers a certain mollification. Consider a mollifier $\phi_t(x) = t\phi(t^{1/d}x)$ on the torus, where ϕ is supported in a sphere of radius ε around the origin. Then

$$|C_\mu^{\varepsilon t^{-1/\alpha}}| = |C_{\mu * \phi_t}|.$$

We now need to show that if we define a random density f_t by $f_t = L_t * \phi_t$ then the large deviation principle holds for the distributions of f_t in the L_1-topology. L_t here stands for the occupation time of a Brownian trajectory for duration t. This has been done in [3]. Once we have the basic large deviation principle in the L_1 topology the lower semi continuity of $F(f)$ as a functional of f in the L_1 topology is enough to give the upper bound. The details appear in [6].

14. Application: The Polaron Problem

In a statistical mechanical problem, see [11] for details, the following question comes up. Let $x(s)$, $0 \le s \le t$ be the 3-dimensional Brownian motion conditioned to start at time 0 and end at time t at the origin. We want to look at the integral

$$a(t)=E_t\left[\exp\left[\alpha\int_0^t\int_0^t \frac{e^{-(\sigma-s)}}{|x(\sigma)-x(s)|}\,d\sigma\,ds\right].\right. \tag{14.1}$$

The problem is to show that

$$\lim_{t\to\infty}\frac{1}{t}\log a(t)=g(\alpha) \tag{4.2}$$

exists and that

$$\lim_{\alpha\to\infty}\frac{g(\alpha)}{\alpha^2}=g_0 \tag{14.3}$$

exists and is given by

$$g_0=\sup_{\substack{\phi\in L_2(R^2)\\ \|\phi\|_2=1}}\left[2\iint\frac{\phi^2(x)\,\phi^2(y)}{|x-y|}\,dx\,dy-\tfrac{1}{2}\int|\nabla\phi|^2\,dx\right]. \tag{14.4}$$

First we replace the conditioned Brownian motion by free Brownian motion.

$$b(t)=E\left\{\exp\left[\alpha\int_0^t\int_0^t\frac{e^{-|\sigma-s|}}{|x(\sigma)-x(s)|}\,d\sigma\,ds\right]\right\} \tag{4.5}$$

where E is the expectation with respect to free Brownian motion. The difference is shown to be negligible for large t.

$$\int_0^t\int_0^t\frac{e^{-(\sigma-s)}}{|x(\sigma)-x(s)|}\,d\sigma\,ds=2\int_0^t ds\int_0^t\frac{e^{-(\sigma-s)}}{|x(\sigma)-x(s)|}\,d\sigma\simeq 2\int_0^t F(\theta_s\sigma)\,ds \tag{14.6}$$

where

$$F(\omega)=\int_0^\infty\frac{e^{-\sigma}}{|x(\sigma)-x(0)|}\,d\sigma$$

and θ_s is the translation of the ω trajectory by s in time.

In view of the exponential factor the second integral can be taken up to ∞ for large t. We therefore look at

$$c(t)=E\left\{\exp\left[2\alpha\int_0^t F(\theta_s\omega)\,ds\right]\right\}. \tag{14.7}$$

Our general theory of Section 9 will tell us that

$$\lim_{t\to\infty}\frac{1}{t}\log c(t)=g(\alpha)=\sup_R\left\{E^R\left[2\alpha\int_0^\infty\frac{e^{-\sigma}}{|x(\sigma)-x(0)|}\right]-H(R)\right\} \tag{14.8}$$

where the supremum is over all stationary processes. By using Brownian scaling

$$\frac{g(\alpha)}{\alpha^2}=\sup_R\left[2E^R\left\{\int_0^\infty\frac{e^{-\sigma}\,d\sigma}{|x(\alpha^2\sigma)-x(0)|}\right\}-H(R)\right]. \tag{14.9}$$

By the ergodic theorem, if R is ergodic

$$\lim_{\alpha\to\infty}E^R\left\{\int_0^\infty\frac{e^{-\sigma}}{|x(\alpha^2\sigma)-x(0)|}\,d\sigma\right\}=\iint\frac{r(dx)\,r(dy)}{|x-y|}$$

where $r(dx)$ is the one dimensional marginal of R. Essentially $x(\alpha^2\sigma)$ and $x(\sigma)$ are nearly independent for large α. It is quite reasonable therefore to expect that

$$\lim_{\alpha\to\infty}\frac{g(\alpha)}{\alpha^2}=g_0=\sup_R\left[2\iint\frac{r(dx)\,r(dy)}{|x-y|}-H(R)\right].$$

By the contraction principle if we fix $r(dx)$ at a density $f(x)\,dx$ then

$$\inf_{R:\,r(dx)=f(x)\,dx}H(R)=I(f)=\frac{1}{8}\int\frac{|\nabla f|^2}{f}\,dx.$$

Therefore taking $f(x)=\phi^2(x)$ for some L_2 function $\phi(x)$ we conclude that

$$g_0=\sup_{\substack{\phi\in L_2(R^3)\\ \|\phi\|_2=1}}\left[2\iint\frac{\phi^2(x)\phi^2(y)}{|x-y|}\,dx\,dy-\frac{1}{2}\int\frac{|\nabla\phi|^2}{\phi}\,dx\right].$$

The derivation of (14.8) still has some difficulties. As usual the infinite volume creates problems with the upper bound. It turns out that replacing Brownian motion by the Ornstein-Uhlenbeck process with generator $\frac{1}{2}\Delta-\varepsilon\langle x\cdot\nabla\rangle$ one can finesse the problem. The details can be found in [9].

References

1. Bahadur, R.R. and Zabell, S., Large Deviation of the Sample Mean in General Vector Spaces, Annals of Prob. 7 (1979), 587–621.

2. Cramér, H., On a new limit theorem in the theory of probability: Colloquium on the Theory of Probability, Hermann, Paris (1937).
3. Donsker, M.D. and Varadhan, S.R.S., Asymptotic evaluation of certain Markov process expectations for large time, II, Comm. Pure Appl. Math. 28 (1975) 279-301.
4. Donsker, M.D., and Varadhan, S.R.S., Asymptotic evaluation of certain Markov process expectations for large time, III, Comm. Pure Appl. Math. 29 (1977), 389-461.
5. Donsker, M.D. and Varadhan, S.R.S., Asymptotic Evaluation of Certain Markov process expectations for large time, IV, Comm. Pure Appl. Math. 36 (1983), 183-212.
6. Donsker, M.D. and Varadhan, S.R.S., Asymptotics for the Wiener sausage, Comm. Pure Appl. Math. 28 (1975), 525-565.
7. Donsker, M.D. and Varadhan, S.R.S., On the laws of the iterated logarithms for local times, Comm. Pure Appl. Math. 30 (1977), 707-753.
8. Donsker, M.D. and Varadhan, S.R.S., some Problems of Large Deviations, Symposia Mathematica, Vol. XXI (1977) 313-318.
9. Donsker, M.D. and Varadhan, S.R.S., Asymptotics for the Polaron, Comm. Pure Appl. Math. 36 (1983), 505-528.
10. Donsker, M.D. and Varadhan, S.R.S., Large Deviations for Gaussian Processes (preprint).
11. Feynman, R.P., Statistical Mechanics, W.A. Benjamin, Reading, MA (1972)
12. Jain, N.C., A. Donsker-Varadhan type of invariance principle, Z. Wahrsch. Verw. Gebiete 59 (1982), 117-138.
13. Lanford, Oscar, Entropy and equilibrium states in classical statistical mechanics, Statistical Mechanics and Mathematical Problems, Lecture notes in Physics, Vol. 20, Springer-Verlag, Berlin and New York (1971), 1-113.
14. Mueller, Carl, Strassen's Law for Local Time, to appear.
15. Varadhan, S.R.S., Asymptotic probabilities and differential equations, Comm. Pure Appl. Math. 19 (1966), 261-286.
16. Varadhan, S.R.S., Large Deviations and Applications, CBMS-NSF Regional Conference Series, Number 46, SIAM (1984).

Received 29.5.84

Courant Institute
of Mathematical Sciences
New York University
251 Mercer St.
N.Y. 10012

Journal of Statistical Physics, Vol. 46, Nos. 5/6, 1987

Large Deviations for Noninteracting Infinite-Particle Systems

M. D. Donsker[1] **and S. R. S. Varadhan**[1]

Received November 25, 1986

A large deviation property is established for noninteracting infinite particle systems. Previous large deviation results obtained by the authors involved a single I-function because the cases treated always involved a unique invariant measure for the process. In the context of this paper there is an infinite family of invariant measures and a corresponding infinite family of I-functions governing the large deviations.

KEY WORDS: Large deviations; infinite particle systems; invariant measures; asymptotics for expectations.

1. INTRODUCTION

Let X be a countable set and let $X_0, X_1, X_2,...$ be a Markov chain with state space X and transition probabilities $\{\pi_{xy}\}$. Probabilities and expectations for this chain starting from a point $x \in X$ will be denoted by $P_x^\pi\{\ \}$ and $E_x^\pi\{\ \}$, respectively. We make five assumptions about this chain: it is irreducible, transient, the matrix $\{\pi_{xy}\}$ is doubly stochastic, and if for any finite set $F \subset X$ we define $\tau_F = \inf_{j \geqslant 1}\{X_j \in F\}$, then we assume $\lim_{x \to \infty} P_x^\pi\{\tau_F < \infty\} = 0$, i.e., for any $\varepsilon > 0$ there exists another finite set F_1 such that $F \subset F_1$ and $x \notin F_1$ implies $P_x^\pi\{\tau_F < \infty\} < \varepsilon$. Finally, we assume that the Green's function $G(x, y)$ for the chain has the property $\sup_{x \in X} G(x, x) < \infty$.

For each $x \in X$, let $n_0(x)$ be a nonnegative integer giving the number of particles at x initially, so that $\{n_0(x), x \in X\}$ is the initial configuration of particles. At time 1 all of these particles move independently according to the transition probabilities $\{\pi_{xy}\}$, giving us a new configuration

[1] Courant Institute of Mathematical Sciences, New York University, New York, New York.

$\{n_1(x), x \in X\}$. There is no interaction here; all the $n_0(x)$ particles at x leave independently of one another and for any other point $y \in X$, the $n_0(y)$ particles at y leave independently not only of each other, but of all the particles at x. At time 2 the process is repeated, giving us a new configuration $\{n_2(x), x \in X\}$ and so on. If we let $N = \{0, 1, 2,...\}$, then $Z = N^X$ is the space of all configurations, and if at time j we have a particular configuration $n(\cdot) \in Z$, we can ask for the transition probability $\hat{\pi}(n(\cdot), n'(\cdot))$ that at time $j+1$ we have configuration $n'(\cdot) \in Z$.

From our assumptions, the transition probability $\hat{\pi}(n(\cdot), n'(\cdot))$ is given through its moment generating function by

$$\int_Z \left\{\exp\left[-\sum_{x \in X} \lambda(x)\, n'(x)\right]\right\} \hat{\pi}(n(\cdot), dn'(\cdot))$$
$$= \prod_{x \in X} \left(\sum_{y \in X} \{\exp[-\lambda(y)]\}\, \pi_{xy}\right)^{n(x)} \tag{1.1}$$

where $\lambda: X \to \mathbb{R}$ vanishes outside some finite subset of X.

Important for us will be the family $\underline{A}$ of σ-finite invariant measures $\alpha(\cdot)$ for π_{xy}, i.e., for all $x \in X$, $0 \leqslant \alpha(x) < \infty$, $\sum_{x \in X} \alpha(x)\, \pi_{xy} = \alpha(y)$, and we assume $\alpha(\cdot) \not\equiv 0$. Since the chain is irreducible, if $\alpha(\cdot)$ vanishes at any point, it vanishes everywhere, so actually $\alpha(x) > 0$ for all $x \in X$.

For each $\alpha \in \underline{A}$ we define a measure P_α on Z as follows: for each $x \in X$ the number of particles at x, i.e., $n(x)$, is Poisson-distributed with mean $\alpha(x)$ and $\{n(x), x \in X\}$ are mutually independent.

It is well known and, using (1.1), easy to show that for each $\alpha \in \underline{A}$, P_α is an invariant measure for $\hat{\pi}(\cdot, \cdot)$. Also, using the Kolmogorov zero-one law, we can show that for each $\alpha \in \underline{A}$, P_α is an ergodic measure on Z.

For each $x \in X$ and any positive integer N, let

$$L_N(x) = \frac{n_0(x) + n_1(x) + \cdots + n_{N-1}(x)}{N}$$

i.e., the average number of particles at site x during the first N steps. Let Σ be the space of sequences of particle configurations on X, i.e., $\{n_j(\cdot), j = 0, 1, 2,...\}$. Any initial configuration $n(\cdot)$ of particles and $\hat{\pi}(\cdot, \cdot)$ generate a probability measure on Σ which we will denote by $\underline{P}_n$ [we will often in the sequel use $n(\cdot)$ instead of $n_0(\cdot)$ as a generic notation for the initial configuration]. Let $\underline{M}$ be the space of σ-finite measures σ on X. Then, for each N, L_N maps Σ into $\underline{M}$, and we use this mapping to define a probability measure $Q_{n,N}$ on $\underline{M}$ by $Q_{n,N} = \underline{P}_n \cdot L_N^{-1}$, i.e., if $A \subset \underline{M}$, then $Q_{n,N}(A) = \underline{P}_n\{L_N \in A\}$.

Since, for each $\alpha \in \underline{A}$, P_α is an invariant measure for $\hat{\pi}(\cdot, \cdot)$ and each P_α

is ergodic, it follows from the ergodic theorem that as $N \to \infty$, $Q_{n,N} \Rightarrow \delta_\alpha$ where, of course, δ_α is the Dirac measure on $\underline{M}$ concentrated at $\alpha(\cdot)$. What is essential is that the statement $Q_{n,N} \Rightarrow \delta_\alpha$ holds *for almost all* $n(\cdot) \in Z$ *with respect to* P_α*-measure.* This makes sense since if $\alpha_1 \not\equiv \alpha_2$, then P_{α_1} and P_{α_2} are mutually singular measures on Z.

In this paper we prove that a large-deviation property holds for the $Q_{n,N}$ measure, i.e., we find a functional $I_\alpha(\sigma)$ that for each $\alpha \in \underline{A}$ maps $\underline{M}$ into $[0, \infty]$ such that for each closed set $C \subset \underline{M}$,

$$\overline{\lim_{N \to \infty}} \frac{1}{N} \log Q_{n,N}(C) \leqslant - \inf_{\sigma \in C} I_\alpha(\sigma) \tag{1.2}$$

and for each open set $G \subset \underline{M}$,

$$\underline{\lim_{N \to \infty}} \frac{1}{N} \log Q_{n,N}(G) \geqslant - \inf_{\sigma \in G} I_\alpha(\sigma) \tag{1.3}$$

where both of the statements (1.2) and (1.3) hold for almost all $n(\cdot)$ with respect to P_α-measure on Z.

This last "almost all" statement emphasizes what is new in this paper relative to our previous work (see Refs. 1–3 for the theory, and Refs. 4–6 for some applications), namely that we have here more than one invariant measure. In our earlier papers we always imposed sufficient hypotheses on the underlying process so that there was a unique invariant measure and hence a single I-function governing large deviations. Here the underlying process has a whole family of invariant measures $\{P_\alpha, \alpha \in \underline{A}\}$, and we have an I-function therefore corresponding to each $\alpha \in \underline{A}$.

We obtain an explicit expression for $I_\alpha(\sigma)$, which we discuss now. Let $\underline{V}$ be the space of functions $V: X \to \mathbb{R}$ that vanish outside some finite subset of X and have the property that for some $x \in X$,

$$u_V(x) = u(x) = E_x^\pi \left\{ \exp\left[\sum_{j=0}^{\infty} V(X_j) \right] \right\} < \infty \tag{1.4}$$

It is an elementary consequence of the irreducibility hypothesis (see first few lines of Lemma 2.1 below) that $u(x) < \infty$ for all $x \in X$ if it is finite for any x.

Now, for $\alpha \in \underline{A}$ and $\sigma \in \underline{M}$, define

$$I_\alpha(\sigma) = \sup_{V \in \underline{V}} \left\{ \sum_{x \in X} \sigma(x)\, V(x) - \sum_{x \in X} \alpha(x)(1 - e^{-V(x)})\, u(x) \right\} \tag{1.5}$$

and this is the I-function that enters into (1.2) and (1.3). For a given $\alpha \in \underline{A}$, $I_\alpha(\sigma)$ turns out to be finite if and only if σ behaves "asymptotically" like α. This is made explicit in Theorem 3.3 below. If $\alpha \in \underline{A}$ and $\beta \in \underline{A}$ and $\sigma \in \underline{M}$ exists such that both $I_\alpha(\sigma) < \infty$ and $I_\beta(\sigma) < \infty$, then $\alpha \equiv \beta$ (Lemma 3.4). In

Section 3 other properties of this I-function are established. In Section 2 we prove the upper bound (1.2) and in Section 4 we prove the lower bound (1.3).

Once such a large-deviation property for $Q_{n,N}$ is established, i.e., once one has an $I_\alpha(\sigma)$ for which (1.2) and (1.3) hold, then one easily obtains[7] an asymptotic result for certain expectations with respect to the $\underline{P}_n$ measure. To make this explicit, let Φ: $\underline{M} \to \mathbb{R}$ be bounded and continuous; then (1.2) and (1.3) imply

$$\lim_{N \to \infty} \frac{1}{N} \log E^{\underline{P}_n}\{e^{N\Phi(L_N(\cdot))}\} = \sup_{\sigma \in \underline{M}} [\Phi(\sigma) - I_\alpha(\sigma)] \tag{1.6}$$

for almost all $n(\cdot) \in Z$ (P_α-measure).

To obtain large-deviation properties where more than one invariant measure is involved, the present context of infinite-particle systems is natural. Here, in the simplest situation of noninteracting particles, we obtain an explicit I-function (1.5) in terms of which (1.2) and (1.3) are proved. In the case of infinite-particle systems with interaction the problem is to find the appropriate I-function in terms of which the analogues of (1.2) and (1.3) can be established.

Some interesting large-deviation results for infinite-particle systems with interaction have been found by Cox and Griffeath.[9] Lee[8] obtained a large-deviation property with explicit I-function for an infinite-particle system with no interaction, but where the particles move according to independent Brownian motions with constant drift.

2. THE UPPER BOUND

Using the definitions and notation of the introduction, let $V \in \underline{V}$ and C be a closed subset of $\underline{M}$. Then,

$$\begin{aligned} E^{\underline{P}_n}&\left\{\exp\left[\sum_{j=0}^{N-1} \sum_{x \in X} n_j(x)\, V(x)\right]\right\} \\ &= E^{\underline{P}_n}\left\{\exp\left[N \sum_{x \in X} L_N(x)\, V(x)\right]\right\} \\ &= E^{Q_{n,N}}\left\{\exp\left[N \sum_{x \in X} \sigma(x)\, V(x)\right]\right\} \\ &\geqslant E^{Q_{n,N}}\left\{\exp\left[N \sum_{x \in X} \sigma(x)\, V(x)\right]; \sigma \in C\right\} \\ &\geqslant \left\{\exp\left[N \inf_{\sigma \in C} \sum_{x \in X} \sigma(x)\, V(x)\right]\right\} Q_{n,N}(C) \end{aligned} \tag{2.1}$$

In Theorem 2.10 below we show

$$\lim_{N\to\infty}\frac{1}{N}\log E^{P_n}\left\{\exp\left[\sum_{j=0}^{N-1}\sum_{x\in X} n_j(x)\,V(x)\right]\right\}$$

$$=\sum_{x\in X}\alpha(x)(1-e^{-V(x)})\,u(x) \tag{2.2}$$

for almost all $n(\cdot)\in Z$ (P_α-measure). We recall from the introduction that

$$u(x)=E_x^\pi\left\{\exp\left[\sum_{j=0}^{\infty}V(X_j)\right]\right\}$$

and that if $\alpha_1\not\equiv\alpha_2$, then P_{α_1} and P_{α_2} are mutually singular measures on Z.

Thus, from (2.1) and (2.2) we conclude

$$\overline{\lim_{N\to\infty}}\frac{1}{N}\log Q_{n,N}(C)$$

$$\leqslant-\inf_{\sigma\in C}\sum_{x\in X}\sigma(x)\,V(x)+\sum_{x\in X}\alpha(x)(1-e^{-V(x)})\,u(x)$$

$$=-\inf_{\sigma\in C}\left[\sum_{x\in X}\sigma(x)\,V(x)-\sum_{x\in X}\alpha(x)(1-e^{-V(x)})\,u(x)\right] \tag{2.3}$$

for almost all $n(\cdot)\in Z$ (P_α-measure). Since (2.3) holds for any $V\in\underline{V}$, we obtain

$$\overline{\lim_{N\to\infty}}\frac{1}{N}\log Q_{n,N}(C)$$

$$\leqslant-\sup_{V\in\underline{V}}\inf_{\sigma\in C}\left[\sum_{x\in X}\sigma(x)\,V(x)-\sum_{x\in X}\alpha(x)(1-e^{-V(x)})\,u(x)\right] \tag{2.4}$$

for almost all $n(\cdot)\in Z$ (P_α-measure). In Lemma 2.11 below we show that, because C is closed, (2.4) implies

$$\overline{\lim_{N\to\infty}}\frac{1}{N}\log Q_{n,N}(C)$$

$$\leqslant-\inf_{\sigma\in C}\sup_{V\in\underline{V}}\left[\sum_{x\in X}\sigma(x)\,V(x)-\sum_{x\in X}\alpha(x)(1-e^{-V(x)})\,u(x)\right]$$

$$=-\inf_{\sigma\in C}I_\alpha(\sigma) \tag{2.5}$$

for almost all $n(\cdot)\in Z$ (P_α-measure).

Now (2.5) is just the upper bound (1.2) of the introduction, and hence what needs to be proved in this section is (2.2) and Lemma 2.11. We prove (2.2) as Theorem 2.10 after a succession of preliminary lemmas.

Let $\underline{V}_1$ be the space of functions $V: X \to \mathbb{R}$ such that V vanishes outside some finite subset of X and

$$u(x) = E_x^\pi \left\{ \exp\left[\sum_{j=0}^{\infty} V(X_j) \right] \right\} < \infty$$

for *all* $x \in X$.

For F a finite subset of X, let $\tau = \tau_F = \inf_{j \geqslant 1} \{X_j \in F\}$, i.e., the first entry time of the Markov chain into F, and let $\pi_{xy}^F = P_x^\pi\{X_\tau = y, \tau < \infty\}$. Let M_F be the matrix $\{e^{V(x)} \pi_{xy}^F, x, y \in F\}$, where V vanishes outside F, and let $\rho(M_F)$ be the spectral radius of M_F. Define $\underline{V}_2$ to be the space of functions $V: X \to \mathbb{R}$ such that V vanishes outside some finite set $F = F_V$ and $\rho(M_F) < 1$.

Lemma 2.1. $\underline{V} = \underline{V}_1 = \underline{V}_2$.

Proof. Let $V \in \underline{V}$ and $x \in X$ such that

$$u(x) = E_x^\pi \left\{ \exp\left[\sum_{j=0}^{\infty} V(X_j) \right] \right\} < \infty$$

Since the chain is irreducible, for any $y \in X$ there is a positive integer k such that $\pi_{xy}^{(k)} > 0$. If we let

$$\|V\| = \sup_{z \in F = \operatorname{supp} V} |V(z)|$$

then

$$\begin{aligned}
\infty &> E_x^\pi \left\{ \exp\left[\sum_{j=0}^{\infty} V(X_j) \right] \right\} \\
&\geqslant [\exp(-k\,\|V\|)]\, E_x^\pi \left\{ \exp\left[\sum_{j=k}^{\infty} V(X_j) \right] \right\} \\
&\geqslant [\exp(-k\,\|V\|)]\, E_x^\pi \left\{ \exp\left[\sum_{j=k}^{\infty} V(X_j) \right]; X_k = y \right\} \\
&= [\exp(-k\,\|V\|)]\, \pi_{xy}^{(k)} E_y^\pi \left\{ \exp\left[\sum_{j=0}^{\infty} V(X_j) \right] \right\}
\end{aligned}$$

so that

$$E_y^\pi \left\{ \exp\left[\sum_{j=0}^{\infty} V(X_j) \right] \right\} < \infty$$

which means $V \in \underline{V}_1$ and hence $\underline{V} = \underline{V}_1$.

To show $\underline{V}_1 \subset \underline{V}_2$, let $V \in \underline{V}_1$, so that

$$u(x) = E_x^\pi \left\{ \exp \left[\sum_{j=0}^{\infty} V(X_j) \right] \right\} < \infty$$

for all $x \in X$. Let $F \subset X$ be the finite set outside of which V vanishes, and with τ and π_{xy}^F as defined just before this lemma, let

$$\eta_x = P_x^\pi\{\tau = \infty\} = 1 - \sum_{y \in F} \pi_{xy}^F$$

We note that, since the chain is transient, $\eta_x > 0$ for at least one $x \in F$.

Now, for any $x \in F$, using the strong Markov property of the chain with respect to τ,

$$\begin{aligned}
u(x) &= E_x^\pi \left\{ \exp \left[\sum_{j=0}^{\infty} V(X_j) \right] \right\} \\
&= [\exp V(x)]\, E_x^\pi \left\{ \exp \left[\sum_{j=1}^{\infty} V(X_j) \right]; \tau < \infty \right\} \\
&\quad + [\exp V(x)]\, E_x^\pi \left\{ \exp \left[\sum_{j=1}^{\infty} V(X_j) \right]; \tau = \infty \right\} \\
&= [\exp V(x)]\, E_x^\pi \left\{ \exp \left[\sum_{j=1}^{\infty} V(X_j) \right]; \tau < \infty \right\} \\
&\quad + [\exp V(x)]\, P_x^\pi\{\tau = \infty\} \\
&= [\exp V(x)] \sum_{y \in F} \pi_{xy}^F E_y^\pi \left\{ \exp \left[\sum_{j=0}^{\infty} V(X_j) \right] \right\} + [\exp V(x)]\, \eta_x \\
&= [\exp V(x)] \sum_{y \in F} \pi_{xy}^F u(y) + [\exp V(x)]\, \eta_x
\end{aligned} \tag{2.6}$$

Let u be the column vector with entries $\{u(x),\ x \in F\}$, $e^V \eta$ be the column vector with entries $\{e^{V(x)}\eta_x,\ x \in F\}$; then, with M_F the matrix defined above, we can rewrite (2.6) as

$$(I - M_F)\, u = e^V \eta \tag{2.7}$$

Since (2.7) has the strictly positive solution

$$u(x) = E_x^\pi \left\{ \exp \left[\sum_{j=0}^{\infty} V(X_j) \right] \right\}, \qquad x \in F$$

we conclude that $\rho(M_F) < 1$, i.e., $V \in \underline{V}_2$.

Finally, to show $\underline{V}_2 \subset \underline{V}_1$, let $V \in \underline{V}_2$ with finite support F. Since we are now assuming $\rho(M_F) < 1$ and, as noted earlier, at least one entry in $e^V \eta$ is strictly positive, the matrix equation $(I - M_F)\, u = e^V \eta$ for u has a solution $\bar{u}(x)$, $x \in F$. Indeed,

$$\bar{u} = (I - M_F)^{-1}(e^V \eta) \tag{2.8}$$

and since the chain is irreducible, $\bar{u} = \{\bar{u}(x), x \in F\}$ has strictly positive entries. We extend the column vector $\bar{u}$ by defining for $y \notin F$,

$$\begin{aligned} \bar{u}(y) &= \sum_{z \in F} \pi^F_{yz} \bar{u}(z) + P^\pi_y\{\tau = \infty\} \\ &= \sum_{z \in F} \pi^F_{yz} \bar{u}(z) + 1 - \sum_{z \in F} \pi^F_{yz} \end{aligned} \tag{2.9}$$

so that now $\bar{u}$ is a column vector having entries for all the elements of X. Thus, for any $x \in X$,

$$\begin{aligned} \sum_{y \in X} \pi_{xy} \bar{u}(y) &= \sum_{y \in F} \pi_{xy} \bar{u}(y) + \sum_{y \notin F} \pi_{xy} \bar{u}(y) \\ &= \sum_{y \in F} \pi_{xy} \bar{u}(y) + \sum_{y \notin F} \pi_{xy} \sum_{z \in F} \pi^F_{yz} \bar{u}(z) + \sum_{y \notin F} \pi_{xy} P^\pi_y\{\tau = \infty\} \end{aligned} \tag{2.10}$$

But,

$$\begin{aligned} \sum_{z \in F} \pi^F_{xz} \bar{u}(z) &= E^\pi_x\{\bar{u}(X_\tau); \tau < \infty\} \\ &= E^\pi_x\{\bar{u}(X_\tau); \tau = 1\} + E^\pi_x\{\bar{u}(X_\tau); 2 \leqslant \tau < \infty\} \\ &= \sum_{y \in F} \pi_{xy} \bar{u}(y) + \sum_{y \notin F} \pi_{xy} \sum_{z \in F} \pi^F_{yz} \bar{u}(z) \end{aligned}$$

and using this in (2.10), we have, for any $x \in X$,

$$\begin{aligned} \sum_{y \in X} \pi_{xy} \bar{u}(y) &= \sum_{z \in F} \pi^F_{xz} \bar{u}(z) + \sum_{y \notin F} \pi_{xy} P^\pi_y\{\tau = \infty\} \\ &= \sum_{z \in F} \pi^F_{xz} \bar{u}(z) + P^\pi_x\{\tau = \infty\} \end{aligned} \tag{2.11}$$

First consider (2.11) for $x \notin F$. Using (2.9), we obtain

$$\sum_{y \in X} \pi_{xy} \bar{u}(y) = \bar{u}(x)$$

On the other hand, if $x \in F$, then (2.11) and (2.8) state that for $V \in \underline{V}_2$,

$$\begin{aligned} e^{V(x)} \sum_{y \in X} \pi_{xy} \bar{u}(y) &= e^{V(x)} \sum_{z \in F} \pi^F_{xy} \bar{u}(z) + e^{V(x)} P^\pi_x\{\tau = \infty\} \\ &= e^{V(x)} \sum_{z \in F} \pi^F_{xz} \bar{u}(z) + e^{V(x)} \eta_x \\ &= (M_F \bar{u})_x + (e^V \eta)_x = \bar{u}(x) \end{aligned}$$

Since V vanishes outside of F, we conclude that for all $x \in X$,

$$e^{V(x)} \sum_{y \in X} \pi_{xy} \bar{u}(y) = \bar{u}(x)$$

or, with the abbreviated notation $(\pi \bar{u})(x) = \sum_{y \in X} \pi_{xy} \bar{u}(y)$, we have for all $x \in X$

$$e^{V(x)} (\pi \bar{u})(x) = \bar{u}(x) \tag{2.12}$$

Identity (2.12) holds for a $V \in \underline{V}_2$ and $\bar{u}$ defined by (2.8) and (2.9). In this connection we should note that for $V \in \underline{V}$ the same identity holds for

$$\begin{aligned} u(x) &= E^\pi_x \left\{ \exp\left[\sum_{j=0}^{\infty} V(X_j) \right] \right\} \\ &= [\exp V(x)]\, E^\pi_x \left\{ \exp\left[\sum_{j=1}^{\infty} V(X_j) \right] \right\} \\ &= [\exp V(x)] \sum_{y \in X} \pi_{xy} E^\pi_y \left\{ \exp\left[\sum_{j=0}^{\infty} V(X_j) \right] \right\} \\ &= [\exp V(x)] (\pi u)(x) \end{aligned} \tag{2.13}$$

Since $\bar{u}(x)$ is strictly positive for $x \in F$, we have

$$0 < \inf_{x \in F} \bar{u}(x) \leqslant \sup_{x \in F} \bar{u}(x) < \infty$$

From this and (2.9) we see that for all $y \in X$

$$0 < \min[1, \inf_{z \in F} \bar{u}(z)] \leqslant \bar{u}(y) \leqslant \max[\sup_{z \in F} \bar{u}(z), 1] < \infty \tag{2.14}$$

Now, (2.12) implies that the sequence of random variables

$$Z_n = \bar{u}(X_n) \exp\left[\sum_{j=0}^{n-1} V(X_j) \right]$$

is a martingale because if $\underline{F}_{n-1}$ is the σ-field up to time $n-1$,

$$E_x^\pi\{Z_n|\underline{F}_{n-1}\}=\left\{\exp\left[\sum_{j=0}^{n-1}V(X_j)\right]\right\}(\pi\bar{u})(X_{n-1})$$

$$=\left\{\exp\left[\sum_{j=0}^{n-2}V(X_j)\right]\right\}\bar{u}(X_{n-1})=Z_{n-1}$$

Thus, for all $x\in X$,

$$E_x^\pi\left\{\left[\exp\sum_{j=0}^{n-1}V(X_j)\right]\bar{u}(X_n)\right\}=\bar{u}(x) \tag{2.15}$$

and from this and (2.14) we conclude, for all $x\in X$,

$$E_x^\pi\left\{\exp\left[\sum_{j=0}^{n-1}V(X_j)\right]\right\}\leqslant\frac{\bar{u}(x)}{\inf_{y\in X}\bar{u}(y)}<\infty$$

Finally, from Fatou's lemma, this last inequality implies

$$u(x)=E_x^\pi\left\{\exp\left[\sum_{j=0}^{\infty}V(X_j)\right]\right\}<\infty$$

for all $x\in X$, i.e., $V\in\underline{V}_1$, and the lemma is proved.

Lemma 2.2. Let F be a finite subset of X and let $\phi(x)=P_x^\pi\{\inf_{j\geqslant 1}(X_j\in F)<\infty\}$. Then for all $x\in X$, $\lim_{n\to\infty}E_x^\pi\{\phi(X_n)\}=0$.

Proof. This is obvious, since

$$E_x^\pi\{\phi(X_n)\}=P_x^\pi\{X_j\in F, j>n\}$$

$$=P_x^\pi\{\text{last visit of chain to } F \text{ occurs after time } n\}$$

and the last probability must go to 0 as $n\to\infty$ because the chain is transient.

Corollary. With $\bar{u}(x)$ defined for $x\in F$ by (2.8) and for $x\notin F$ by (2.9), we have for all $x\in X$,

$$\lim_{n\to\infty}E_x^\pi\{|\bar{u}(x_n)-1|\}=0$$

Proof. With $\phi(x)$ as defined in Lemma 2.2, we have for $x\notin F$,

$$\bar{u}(x)=\sum_{y\in F}\pi_{xy}^F\bar{u}(y)+\left(1-\sum_{y\in F}\pi_{xy}^F\right)$$

$$=\sum_{y\in F}\pi_{xy}^F\bar{u}(y)+[1-\phi(x)]$$

Thus,

$$E_x^\pi\{|\bar{u}(X_n)-1|\} \leqslant \sup_{y\in F} \bar{u}(y)\, E_x^\pi\{\phi(X_n)\} + E_x^\pi\{\phi(X_n)\}$$

and from the lemma, for all $x \in X$,

$$\lim_{n\to\infty} E_x^\pi\{|\bar{u}(X_n)-1|\} = 0$$

In this proof there was no loss in generality by taking $\bar{u}(x)$ as defined for $x \notin F$, since F is finite and the chain is transient.

Lemma 2.3. With $\bar{u}(x)$ defined by (2.8) for $x \in F$ and by (2.9) for $x \notin F$, we have $\bar{u}(x) = u(x)$ for all $x \in X$.

Proof. We showed in (2.15) that for *all* $x \in X$,

$$\bar{u}(x) = E_x^\pi\left\{\left[\exp \sum_{j=0}^{N-1} V(X_j)\right] \bar{u}(X_N)\right\}$$

and, because of the corollary to Lemma 2.2, the present lemma will follow if we can show that

$$\left\{\left[\exp \sum_{j=0}^{N-1} V(X_j)\right] \bar{u}(X_N)\right\}$$

are uniformly integrable. Now, by (2.14),

$$0 < \inf_{y\in X} \bar{u}(y) \leqslant \sup_{y\in X} \bar{u}(y) < \infty$$

so it suffices to show that $\{\exp \sum_{j=0}^{N-1} V(X_j)\}$ are uniformly integrable. This follows if, for some $\lambda > 1$,

$$\sup_N E_x^\pi\left\{\exp\left[\lambda \sum_{j=0}^{N-1} V(X_j)\right]\right\} \leqslant \text{const}$$

But in Lemma 2.1 we showed that if $V \in \underline{V}$, then it is in $\underline{V}_2$, i.e., $\rho(M^F) < 1$, where M^F is the matrix $\{e^{V(x)}\pi_{xy}^F,\ x, y \in F\}$. Let $M_\lambda^F = \{e^{\lambda V(x)}\pi_{xy}^F,\ x, y \in F\}$. Since the matrix is finite, its spectral radius $\rho(M_\lambda^F)$ is a continuous function of λ. Since the spectral radius is strictly less than 1 when $\lambda = 1$, there then exists $\lambda > 1$ such that $\rho(M_\lambda^F) < 1$ also. But, again by Lemma 2.1, this means for that λ, $\lambda V \in \underline{V}$, which implies

$$\sup_N E_x^\pi\left\{\exp\left[\lambda \sum_{j=0}^{N-1} V(X_j)\right]\right\} \leqslant \text{const}$$

For any $\alpha \in \underline{A}$ we introduce the Markov chain time-reversed with respect to α, i.e., for $x, y \in X$, let $\pi_{xy}(\alpha) = \pi_{yx}\alpha(y)/\alpha(x)$. The $\pi_{xy}(\alpha)$ are transition probabilities, since

$$\sum_{y \in X} \pi_{xy}(\alpha) = \frac{1}{\alpha(x)} \sum_{y \in X} \pi_{yx}\alpha(y) = \frac{\alpha(x)}{\alpha(x)} = 1$$

Since $\pi_{xy}^{(k)}(\alpha) = \pi_{yx}^{(k)}\alpha(y)/\alpha(x)$, the time-reversed chain is also irreducible. Since the original chain is transient, it has a finite Green's function, i.e., for all $x, y \in X$, $G(x, y) = \sum_{k=0}^{\infty} \pi_{xy}^{(k)} < \infty$. But the Green's function for the reversed chain $G^{\alpha}(x, y)$ is given by

$$G^{\alpha}(x, y) = \sum_{k=0}^{\infty} \pi_{xy}^{(k)}(\alpha) = \sum_{k=0}^{\infty} \pi_{yx}^{(k)} \frac{\alpha(y)}{\alpha(x)} = \frac{\alpha(y)}{\alpha(x)} G(y, x) < \infty$$

which means the reversed chain is also transient. We denote probabilities and expectations with respect to the reversed chain by $P_x^{\pi(\alpha)}\{\ \}$ and $E_x^{\pi(\alpha)}\{\ \}$, respectively.

Let $\underline{V}^{\alpha}$ be the space of functions $V: X \to \mathbb{R}$ that vanish outside some finite set and such that for some $x \in X$,

$$E_x^{\pi(\alpha)}\left\{\exp\left[\sum_{j=0}^{\infty} V(X_j)\right]\right\} < \infty$$

Now, in the proof of Lemma 2.1, we used only the hypotheses that the original chain was irreducible and transient. Hence, Lemma 2.1 applies also to the time-reversed chain and thus $\underline{V}^{\alpha} = \underline{V}_1^{\alpha} = \underline{V}_2^{\alpha}$.

Lemma 2.4. For any $\alpha \in \underline{A}$,

$$\underline{V} = \underline{V}^{\alpha}$$

Proof. Let $\alpha \in \underline{A}$, F be a finite subset of X, and $x, y \in F$. Then,

$$\pi_{xy}^{F}(\alpha) = \pi_{xy}(\alpha) + \sum_{z_1 \in F^c} \pi_{xz_1}(\alpha)\, \pi_{z_1 y}(\alpha) + \sum_{z_1, z_2 \in F^c} \pi_{xz_1}(\alpha)\, \pi_{z_1 z_2}(\alpha)\, \pi_{z_2 y}(\alpha) + \cdots$$

But

$$\pi_{z_1 z_2}(\alpha) = \pi_{z_2 z_1}\alpha(z_2)/\alpha(z_1)$$

so that

$$\pi^F_{xy}(\alpha) = \pi_{yx}\frac{\alpha(y)}{\alpha(x)} + \sum_{z_1 \in F^c} \pi_{yz_1}\frac{\alpha(y)}{\alpha(z_1)}\pi_{z_1x}\frac{\alpha(z_1)}{\alpha(x)} + \cdots$$

$$= \frac{\alpha(y)}{\alpha(x)}\pi^F_{yx}$$

Hence, for a V vanishing outside F, the matrix

$$M_F(\alpha) = \{e^{V(x)}\pi^F_{xy}(\alpha),\ x,\ y \in F\} = \left\{e^{V(x)}\pi^F_{yx}\frac{\alpha(y)}{\alpha(x)},\ x,\ y \in F\right\}$$

Since $M_F = \{e^{V(x)}\pi^F_{xy},\ x,\ y \in F\}$, we see that if we let D be the diagonal matrix

$$D = (e^{-V(y)}\alpha(y)\,\delta_{xy},\ x,\ y \in F)$$

then

$$M_F(\alpha) = D^{-1}M_F^T D$$

which implies $\rho(M_F(\alpha)) = \rho(M_F)$. From Lemma 2.1 and the remarks just preceding this lemma we conclude that $\underline{V} = \underline{V}^\alpha$.

Lemma 2.5. Let $V \in \underline{V}$ and define

$$u_N(x) = E^\pi_x\left\{\exp\left[\sum_{j=0}^{N-1} V(X_j)\right]\right\}$$

For $\alpha \in \underline{A}$,

$$\sup_N \sum_{x \in X} \alpha(x)\,|u_N(x) - u_{N-1}(x)| < \infty$$

Proof. It suffices to show

$$\sup_N \sum_{x \in X} \alpha(x)\,E^\pi_x\left\{\exp\left[\sum_{j=0}^{N-2} V(X_j)\right]|\exp[V(X_{N-1})] - 1|\right\} < \infty$$

and this will follow, since V has finite support, if we show that for every $y \in X$

$$\sup_N \sum_{x \in X} \alpha(x)\,E^\pi_x\left\{\left[\exp\sum_{j=0}^{N-2} V(X_j)\right]\delta_y(X_{N-1})\right\} < \infty \tag{2.16}$$

From Lemma 2.4, $\underline{V} = \underline{V}^\alpha$, so we know

$$E_y^{\pi(\alpha)}\left\{\exp\left[\sum_{j=0}^{\infty} V(X_j)\right]\right\} < \infty$$

for every $y \in X$. Hence, for every $y \in X$,

$$\alpha(y) \sup_N E_y^{\pi(\alpha)}\left\{\exp\left[\sum_{j=1}^{N-1} V(X_j)\right]\right\}$$

$$= \alpha(y) \sup_N \sum_{x_1, x_2, \ldots, x_{N-1}} \left\{\exp\left[\sum_{j=1}^{N-1} V(x_j)\right]\right\}$$

$$\times \pi_{yx_1}(\alpha)\, \pi_{x_1 x_2}(\alpha) \cdots \pi_{x_{n-2}x_{n-1}}(\alpha) < \infty$$

Since $\pi_{xy}(\alpha) = \pi_{yx}\alpha(y)/\alpha(x)$, this last becomes

$$\sup_N \sum_{x_1, x_2, \ldots, x_{N-1}} \alpha(x_{N-1}) \left\{\exp\left[\sum_{j=1}^{N-1} V(x_j)\right]\right\}$$

$$\times \pi_{x_1 y} \pi_{x_2 x_1} \cdots \pi_{x_{N-1} x_{N-2}} < \infty \tag{2.17}$$

Now, if we let $x_{N-1} = x$ and $x_{N-j} = x_{j-1}$, $j = 2, 3, \ldots, N-1$, (2.17) states that for $y \in X$,

$$\sup_N \sum_{x, x_1, x_2, \ldots, x_{N-2}} \alpha(x)[\exp V(x)] \left\{\exp\left[\sum_{j=1}^{N-2} V(x_j)\right]\right\}$$

$$\times \pi_{xx_1} \pi_{x_1 x_2} \cdots \pi_{x_{N-2} y} < \infty$$

i.e., for all $y \in X$,

$$\sup_N \sum_{x \in X} \alpha(x)\, E_x^\pi\left\{\left[\exp \sum_{j=0}^{N-2} V(X_j)\right] \delta_y(X_{N-1})\right\} < \infty$$

which is (2.16).

Lemma 2.6. $u_N(x) \to 1$ as $x \to \infty$ uniformly in N, i.e., for any $\delta > 0$ there exists a finite set F such that $x \notin F$ implies $|1 - u_N(x)| < \delta$ for all N

Proof. For any $V \in \underline{V}$,

$$E_x^\pi\left\{\left[\exp \sum_{j=0}^{N-1} V(X_j)\right] u(X_N)\right\} = u(x)$$

From Lemma 2.3 and (2.14) we then obtain

$$u_N(x) = E_x^\pi \left\{ \exp \left[\sum_{j=0}^{N-1} V(X_j) \right] \right\}$$
$$\leqslant \frac{\sup_{x \in X} u(x)}{\inf_{x \in X} u(x)} \leqslant \text{const} \quad \text{for all } N \text{ and all } x$$

Since, as noted in the proof of Lemma 2.3, $V \in \underline{V}$ implies $\lambda V \in \underline{V}$ for some $\lambda > 1$, we see that

$$\sup_{\substack{x \in X \\ N}} E_x^\pi \left\{ \exp \left[\lambda \sum_{j=0}^{N-1} V(X_j) \right] \right\} \leqslant \text{const}$$

for some $\lambda > 1$. This shows the uniform integrability of $\{\exp \sum_{j=0}^{N-1} V(X_j)\}$.

Now, one of our assumptions on the underlying chain is that $\lim_{x \to \infty} P_x^\pi\{\tau_F < \infty\} = 0$ for any finite subset $F \in X$. In particular, $\lim_{x \to \infty} P_x^\pi\{\tau_{\{y\}} < \infty\} = 0$, but if $y \neq x$, $P_x^\pi\{\tau_{\{y\}} < \infty\} = G(x, y)/G(y, y)$. Thus, we have $G(x, y) \to 0$ as $x \to \infty$ for every $y \in X$.

Moreover,

$$E_x^\pi \left\{ \left| \sum_{j=0}^{N-1} V(X_j) \right| \right\} \leqslant E_x^\pi \left\{ \sum_{j=0}^{\infty} |V(X_j)| \right\} = \sum_{y \in X} G(x, y) \, |V(y)|$$

and, since V has finite support, we see that

$$\sup_N E_x^\pi \left\{ \left| \sum_{j=0}^{N-1} V(X_j) \right| \right\} \to 0 \qquad \text{as} \quad x \to \infty$$

This, together with the uniform integrability shown above, implies the lemma.

Lemma 2.7. Let $V \in \underline{V}$ and $\alpha \in \underline{A}$. Then,

$$\lim_{N \to \infty} \left| \sum_{x \in X} \alpha(x)[\log u_N(x) - \log u_{N-1}(x)] - \sum_{x \in X} \alpha(x)[u_N(x) - u_{N-1}(x)] \right| = 0 \tag{2.18a}$$

and, for any real λ,

$$\lim_{N \to \infty} \left| \sum_{x \in X} \alpha(x)\{[u_N(x)]^\lambda - [u_{N-1}(x)]^\lambda\} - \lambda \sum_{x \in X} \alpha(x)[u_N(x) - u_{N-1}(x)] \right| = 0 \tag{2.18b}$$

Proof. Let $\varepsilon > 0$ be given. There exists a $\delta > 0$ such that $|1-x| < \delta$ and $|1-y| < \delta$ imply

$$|\log x - \log y - (x-y)| < \varepsilon\, |x-y|$$

For this δ, let F be the finite subset of X guaranteed by Lemma 2.6, so that $|1-u_N(x)| < \delta$ for all N provided $x \notin F$. To show the first part of (2.18), it suffices to show

$$\begin{aligned}
&\overline{\lim_{N\to\infty}} \left| \sum_{x\in F} \alpha(x)[\log u_N(x) - \log u_{N-1}(x)] \right.\\
&\qquad \left. - \sum_{x\in F} \alpha(x)[u_N(x) - u_{N-1}(x)] \right|\\
&\qquad + \overline{\lim_{N\to\infty}} \left| \sum_{x\in F^c} \alpha(x)[\log u_N(x) - \log u_{N-1}(x)] \right.\\
&\qquad \left. - \sum_{x\in F^c} \alpha(x)[u_N(x) - u_{N-1}(x)] \right| = 0
\end{aligned} \tag{2.19}$$

Since $\lim_{N\to\infty} u_N(x) = u(x) > 0$ for all $x \in X$ and since summation over $x \in F$ is just a finite sum, the first term in (2.19) is zero. For the second term in (2.19) we get, from our choice of δ and F made above,

$$\begin{aligned}
&\overline{\lim_{N\to\infty}} \left| \sum_{x\in F^c} \alpha(x)[\log u_N(x) - \log u_{N-1}(x)] \right.\\
&\qquad \left. - \sum_{x\in F^c} \alpha(x)[u_N(x) - u_{N-1}(x)] \right|\\
&\qquad \leqslant \overline{\lim_{N\to\infty}}\ \varepsilon \sum_{x\in F^c} \alpha(x)\, |u_N(x) - u_{N-1}(x)|\\
&\qquad \leqslant \varepsilon \sup_N \sum_{x\in X} \alpha(x)\, |u_N(x) - u_{N-1}(x)|
\end{aligned}$$

In Lemma 2.5 we showed that the multiple of ε in this last inequality is finite for any $V \in \underline{V}$ and any $\alpha \in \underline{A}$, so we have (2.19). The proof of the second statement in (2.18) is similar.

Lemma 2.8. For $V \in \underline{V}$, $\alpha \in \underline{A}$, and all $N = 1, 2, \ldots$,

$$\sum_{x\in X} \alpha(x)[(\pi u_N)(x) - u_N(x)] = 0 \tag{2.20}$$

Proof. If we can justify changing the order of summation, (2.20) follows immediately from the fact that α is an invariant measure, since

$$\sum_{x \in X} \alpha(x)[(\pi u_N)(x) - 1] = \sum_{x \in X} \alpha(x)\left[\sum_{y \in X} \pi_{xy} u_N(y) - 1\right]$$
$$= \sum_{x \in X} \alpha(x)[u_N(x) - 1]$$

To justify the interchange, we show

$$\sum_{x \in X} \alpha(x)\, |(\pi u_N)(x) - 1| < \infty$$

which is implied by

$$\sum_{x \in X} \alpha(x)\, |u_N(x) - 1| < \infty \tag{2.21}$$

To show (2.21), we note that for any $x \in X$ and with $F = \operatorname{supp} V$,

$$|u_N(x) - 1| = \left| E_x^\pi \left\{ \exp\left[\sum_{j=0}^{N-1} V(X_j)\right]\right\} - 1 \right|$$
$$\leqslant [\exp(N \,\|V\|)]\; P_x^\pi\{\text{at least on } X_j \in F,\ j = 0, 1,\dots, N-1\}$$
$$= [\exp(N \,\|V\|)][\delta_x(F) + \pi^{(1)}(x, F)$$
$$+ \pi^{(2)}(x, F) + \cdots + \pi^{(N-1)}(x, F)]$$

where

$$\pi^{(k)}(x, F) = \sum_{y \in F} \pi_{xy}^{(k)}, \qquad k = 1, 2,\dots, N-1$$

Thus,

$$\sum_{x \in X} \alpha(x)\, |u_N(x) - 1|$$
$$\leqslant e^{N\|V\|} \sum_{x \in X} \alpha(x)[\delta_x(F) + \pi^{(1)}(x, F) + \cdots + \pi^{(N-1)}(x, F)]$$

But $\sum_{x \in X} \alpha(x)\, \delta_x(F) = \alpha(F)$ and, for each $k = 1, 2,\dots, N-1$,

$$\sum_{x \in X} \alpha(x)\, \pi^{(k)}(x, F) = \sum_{x \in X} \alpha(x) \sum_{y \in F} \pi_{xy}^{(k)} = \sum_{y \in F} \alpha(y) = \alpha(F)$$

Since F is finite and α is a σ-finite measure,

$$\sum_{x \in X} \alpha(x)\, |u_N(x) - 1| \leqslant e^{N\|V\|} N\alpha(F) < \infty$$

Although we have just shown that for each N, $\sum_{x \in X} \alpha(x)\, |u_N(x) - 1| < \infty$, in fact $\sum_{x \in X} \alpha(x)\, |u(x) - 1|$ may be infinite.

Lemma 2.9. Let $V \in \underline{V}$ and $\alpha \in \underline{A}$. Then,

$$\lim_{N \to \infty} \sum_{x \in X} \alpha(x)[u_N(x) - u_{N-1}(x)] = \sum_{x \in X} \alpha(x)(1 - e^{-V(x)})\, u(x)$$

Proof. First of all,

$$\begin{aligned}
u_N(x) &= E_x^\pi \left\{ \exp\left[\sum_{j=0}^{N-1} V(X_j) \right] \right\} \\
&= [\exp V(x)]\, E_x^\pi \left\{ \exp\left[\sum_{j=1}^{N-1} V(X_j) \right] \right\} \\
&= [\exp V(x)] \sum_{y \in X} \pi_{xy} E_y^\pi \left\{ \exp\left[\sum_{j=0}^{N-2} V(X_j) \right] \right\} \\
&= [\exp V(x)] \sum_{y \in X} \pi_{xy} u_{N-1}(y) \\
&= [\exp V(x)](\pi u_{N-1})(x)
\end{aligned}$$

and therefore

$$\begin{aligned}
&\lim_{N \to \infty} \sum_{x \in X} \alpha(x)[u_N(x) - u_{N-1}(x)] \\
&\quad = \lim_{N \to \infty} \sum_{x \in X} \alpha(x)[e^{V(x)}(\pi u_{N-1})(x) - u_{N-1}(x)] \\
&\quad = \lim_{N \to \infty} \left\{ \sum_{x \in X} \alpha(x)(e^{V(x)} - 1)(\pi u_{N-1})(x) \right. \\
&\qquad \left. + \sum_{x \in X} \alpha(x)[(\pi u_{N-1})(x) - u_{N-1}(x)] \right\} \\
&\quad = \lim_{N \to \infty} \left\{ \sum_{x \in X} \alpha(x)(1 - e^{-V(x)})\, u_N(x) \right. \\
&\qquad \left. + \sum_{x \in X} \alpha(x)[(\pi u_{N-1})(x) - u_{N-1}(x)] \right\}
\end{aligned}$$

The first term in this last expression is only a finite sum, since V vanishes outside a finite set, and so in this first term we can bring the limit inside the summation. Moreover, the second term goes to 0 as $N \to \infty$, by Lemma 2.8. Hence, since $V \in \underline{V}$,

$$\lim_{N \to \infty} \sum_{x \in X} \alpha(x)[u_N(x) - u_{N-1}(x)] = \sum_{x \in X} \alpha(x)(1 - e^{-V(x)})\, u(x)$$

Theorem 2.10. For $V \in \underline{V}$ and $\alpha \in \underline{A}$,

$$\lim_{N \to \infty} \frac{1}{N} \log E^{\underline{P}_n} \left\{ \exp \left[\sum_{j=0}^{N-1} \sum_{x \in X} n_j(x)\, V(x) \right] \right\}$$
$$= \sum_{x \in X} \alpha(x) \{ 1 - \exp[-V(x)] \}\, u(x) \tag{2.22}$$

for almost all $n(\cdot) \in Z$ (P_α-measure).

Remark. Although Theorem 2.10 is stated and proved in a form in which the null set depends on $V \in \underline{V}$, by an appropriate choice of a countable subset of $\underline{V}$ and by using standard arguments one can obtain the existence of a single null set that works simultaneously for all $V \in \underline{V}$.

Proof. From (1.1),

$$E^{\underline{P}_n} \left\{ \exp \left[\sum_{j=0}^{N-1} \sum_{x \in X} n_j(x)\, V(x) \right] \right\}$$
$$= \prod_{x \in X} \left(E_x^{\pi} \left\{ \exp \left[\sum_{j=0}^{N-1} V(X_j) \right] \right\} \right)^{n(x)}$$

and therefore

$$\frac{1}{N} \log E^{\underline{P}_n} \left\{ \exp \left[\sum_{j=0}^{N-1} \sum_{x \in X} n_j(x)\, V(x) \right] \right\}$$
$$= \frac{1}{N} \sum_{x \in X} n(x) \log E_x^{\pi} \left\{ \exp \left[\sum_{j=0}^{N-1} V(X_j) \right] \right\}$$
$$= \frac{1}{N} \sum_{x \in X} n(x) \log u_N(x)$$

To show that (2.22) holds for almost all $n(\cdot) \in Z$ (P_α-measure) means then to show

$$\lim_{N \to \infty} \frac{1}{N} \sum_{x \in X} n(x) \log u_N(x) = \sum_{x \in X} \alpha(x)(1 - e^{-V(x)})\, u(x)$$

for almost all $n(\cdot)\in Z$ (P_α-measure). For this it suffices to show that for all real λ

$$\lim_{N\to\infty}\frac{1}{N}\log E^{P_\alpha}\left\{\exp\left[\lambda\sum_{x\in X} n(x)\log u_N(x)\right]\right\}$$
$$=\lambda\sum_{x\in X}\alpha(x)\{1-\exp[-V(x)]\}\,u(x)$$

But

$$\frac{1}{N}\log E^{P_\alpha}\left\{\exp\left[\lambda\sum_{x\in X} n(x)\log u_N(x)\right]\right\}$$
$$=\frac{1}{N}\log\prod_{x\in X}\exp(\alpha(x)\{\exp[\lambda\log u_N(x)]-1\})$$
$$=\frac{1}{N}\sum_{x\in X}\alpha(x)\{[u_N(x)]^\lambda-1\}$$

and so we must show

$$\lim_{N\to\infty}\frac{1}{N}\sum_{x\in X}\alpha(x)\{[u_N(x)]^\lambda-1\}$$
$$=\lambda\sum_{x\in X}\alpha(x)(1-e^{-V(x)})\,u(x) \tag{2.23}$$

From the usual Caesaro argument, (2.23) is implied by

$$\lim_{N\to\infty}\sum_{x\in X}\alpha(x)\{[u_N(x)]^\lambda-[u_{N-1}(x)]^\lambda\}$$
$$=\lambda\sum_{x\in X}\alpha(x)(1-e^{-V(x)})\,u(x) \tag{2.24}$$

which follows from Lemmas 2.7 and 2.9.

Lemma 2.11. If $C\subset \underline{M}$ is closed and

$$\overline{\lim_{N\to\infty}}\frac{1}{N}\log Q_{n,N}(C)$$
$$\leqslant -\sup_{V\in\underline{V}}\inf_{\sigma\in C}\left[\sum_{x\in X}\sigma(x)\,V(x)-\sum_{x\in X}\alpha(x)(1-e^{-V(x)})\,u(x)\right]$$

then

$$\overline{\lim_{N\to\infty}} \frac{1}{N} \log Q_{n,N}(C)$$

$$\leqslant - \inf_{\sigma\in C} \sup_{V\in \underline{V}} \left[\sum_{x\in X} \sigma(x)\, V(x) - \sum_{x\in X} \alpha(x)(1-e^{-V(x)})\, u(x) \right]$$

$$= - \inf_{\sigma\in C} I_\alpha(\sigma)$$

Proof. First of all, one can obtain the desired result if C is compact, using methods developed in our earlier papers (cf. Ref. 1, for example).

We now describe how one can go from the result for compact sets to closed sets.

We begin by picking a $V>0$ such that

$$\sup_x \sum_y G(x, y)\, V(y) = \theta < 1$$

and

$$\sum_{x\in X} \alpha(x)\, V(x) \leqslant \text{const} < \infty$$

With such a V we then have by Portenko's lemma (see Lemma 3.1 below) that $u(x) = u_V(x) \leqslant 1/(1-\theta)$. This implies

$$\lim_{N\to\infty} \frac{1}{N} \log E^{P_\alpha} E^{\underline{P}_n} \left\{ \exp\left[\sum_{j=0}^{N-1} \sum_{x\in X} V(x)\, n_j(x) \right] \right\}$$

$$= \lim_{N\to\infty} \sum_{x\in X} \alpha(x)[u_N(x) - u_{N-1}(x)]$$

$$= \sum_{x\in X} \alpha(x)(1-e^{-V(x)})\, u(x)$$

$$\leqslant \frac{1}{1-\theta} \cdot \text{const} = K < \infty \tag{2.25}$$

Using a Chebycheff argument, we obtain from (2.25)

$$P_\alpha \left\{ n(\cdot)\colon Q_{n,N} \left\{ \frac{1}{N} \sum_{j=0}^{N-1} \sum_{x\in X} V(x)\, n_j(x) \geqslant L \right\} \geqslant e^{-\beta N} \right\} \leqslant e^{KN+\beta N - LN} \tag{2.26}$$

Now, (2.26) implies

$$\overline{\lim_{L\to\infty}}\ \overline{\lim_{N\to\infty}} \frac{1}{N} \log Q_{n,N} \left\{ \frac{1}{N} \sum_{j=0}^{N-1} \sum_{x\in X} V(x)\, n_j(x) \geqslant L \right\} = -\infty \tag{2.27}$$

and (2.27), together with the fact that $\{\sigma\colon \sum_{x\in X} V(x)\,\sigma(x) \leqslant L\}$ is a compact set in the vague topology, implies the lemma.

3. PROPERTIES OF THE I-FUNCTION

In this section we prove various properties of $I_\alpha(\sigma)$ given by (1.5). Some of these results are necessary for establishing the lower bound (1.3) and will be used in the next section, where the lower bound is proved. Other properties of $I_\alpha(\sigma)$ proved here are of separate interest.

As noted earlier, the Green's function $G(x, y) < \infty$ for all $x, y \in X$, since the $\{\pi_{xy}\}$ chain is transient. If $y \neq x$, $P_x^\pi\{\tau_{\{y\}} < \infty\} = G(x, y)/G(y, y)$, and therefore our hypothesis on $P_x^\pi\{\tau_F < \infty\}$ implies

$$\begin{aligned} G(x, y) &\to 0 \quad \text{as} \quad y \to \infty \quad \text{for any } x \in X \\ G(x, y) &\to 0 \quad \text{as} \quad x \to \infty \quad \text{for any } y \in X \end{aligned} \tag{3.1}$$

We also recall that we are assuming

$$\sup_{x \in X} G(x, x) < \infty \tag{3.2}$$

Both (3.1) and (3.2) will be used in this section.

Lemma 3.1. Let $V: X \to R$ vanish outside some finite subset of X and be such that

$$\sup_{x \in X} \sum_{y \in X} G(x, y)\, |V(y)| \leqslant \theta < 1 \tag{3.3}$$

Then, $V \in \underline{V}$ and $|1 - u_V(x)| \leqslant \theta/(1-\theta)$ for each $x \in X$.

Proof. Let $r \geqslant 1$ and consider

$$\begin{aligned} &E_x^\pi\{[V(X_0) + V(X_1) + \cdots]^r\} \\ &\quad \leqslant E_x^\pi\{[|V(X_0)| + |V(X_1)| + \cdots]^r\} \\ &\quad = E_x^\pi\left\{\sum_{i_1, i_2, \ldots, i_r} |V(X_{i_1})|\, |V(X_{i_2})| \cdots |V(X_{i_r})|\right\} \\ &\quad \leqslant r!\, E_x^\pi\left\{\sum_{i_1 \leqslant i_2 \leqslant \cdots \leqslant i_r} |V(X_{i_1})|\, |V(X_{i_2})| \cdots |V(X_{i_r})|\right\} \\ &\quad \leqslant r! \sum_{i_1 \leqslant i_2 \leqslant \cdots \leqslant i_{r-1}} E_x^\pi\left\{|V(X_{i_1})|\, |V(X_{i_2})|\right. \\ &\qquad \left. \times \cdots |V(X_{i_{r-1}})|\, E_{X_{i_{r-1}}}^\pi\left(\sum_{j=0}^{\infty} |V(X_j)|\right)\right\} \\ &\quad \leqslant \theta r! \sum_{i_1 \leqslant i_2 \leqslant \cdots \leqslant i_{r-1}} E_x^\pi\{|V(X_{i_1})|\, |V(X_{i_2})| \cdots |V(X_{i_{r-1}})| \end{aligned}$$

the last inequality following from (3.3). If we keep repeating this process, we end up with

$$E_x^\pi\{[V(X_0)+V(X_1)+\cdots]^r\}\leqslant\theta^r r!$$

and thus

$$|1-u_V(x)|=\left|1-E_x^\pi\left\{\exp\left[\sum_{j=0}^{\infty}V(X_j)\right]\right\}\right|\leqslant\sum_{r=1}^{\infty}\frac{\theta^r r!}{r!}=\frac{\theta}{1-\theta}$$

which implies $u_V(x)<\infty$, so that $V\in\underline{V}$.

Lemma 3.2. Let $x_1, x_2,\ldots, x_N$ be a finite subset of X for which $G(x_i, x_i)\leqslant C$, $i=1, 2,\ldots, N$, $G(x_i, x_j)\leqslant\varepsilon$ for all $i\neq j$, $i, j=1, 2,\ldots, N$. Let $\gamma>0$ be so small that $\gamma[C+(N-1)\varepsilon]=\theta<1$. Let $\alpha\in\underline{A}$. For any $\sigma\in\underline{M}$ such that $I_\alpha(\sigma)<l$, we have

$$\begin{aligned}\frac{e^\gamma-1}{\gamma}&\left[\sum_{i=1}^{N}\alpha(x_i)\right]\left(1-\frac{\theta}{1-\theta}\right)-\frac{l}{\gamma}\\&\leqslant\sum_{i=1}^{N}\sigma(x_i)\\&\leqslant\frac{1-e^{-\gamma}}{\gamma}\left[\sum_{i=1}^{N}\alpha(x_i)\right]\left(1+\frac{\theta}{1-\theta}\right)+\frac{l}{\gamma}\end{aligned}\tag{3.4}$$

Proof. Let

$$V(x)=\begin{cases}\gamma & \text{if } x=x_i,\quad i=1, 2,\ldots, N\\ 0 & \text{otherwise}\end{cases}$$

Then, using the maximum principle,

$$\begin{aligned}\sup_{x\in X}&\sum_{y\in X}G(x, y)\,V(y)\\&=\sup_{1\leqslant i\leqslant N}\sum_{y\in X}G(x_i, y)\,V(y)\\&=\gamma\sup_{1\leqslant i\leqslant N}\sum_{j=1}^{N}G(x_i, x_j)\\&\leqslant\gamma[C+(N-1)\,\varepsilon]=\theta<1\end{aligned}$$

Hence, from Lemma 3.1, for the V just selected, $u_V(x)\leqslant 1+\theta/(1-\theta)$, and $V\in\underline{V}$. Since $\sigma\in\underline{M}$ is such that $I_\alpha(\sigma)\leqslant l$, we then have for this V,

$$\sum_{x\in X}\sigma(x)\,V(x)-\sum_{x\in X}\alpha(x)(1-e^{-V(x)})\,u_V(x)\leqslant l$$

i.e.,

$$\gamma \sum_{i=1}^{N} \sigma(x_i) - (1-e^{-\gamma})\left[\sum_{i=1}^{N} \alpha(x_i)\right]\left(1+\frac{\theta}{1-\theta}\right) \leqslant l$$

or

$$\sum_{i=1}^{N} \sigma(x_i) \leqslant \frac{1-e^{-\gamma}}{\gamma}\left[\sum_{i=1}^{N} \alpha(x_i)\right]\left(1+\frac{\theta}{1-\theta}\right) + \frac{l}{\gamma}$$

If we choose instead

$$V(x) = \begin{cases} -\gamma, & x = x_i, \quad i = 1, 2,..., N \\ 0, & \text{otherwise} \end{cases}$$

then again Lemma 3.1 implies this $V \in \underline{V}$ also and $u_V(x) \geqslant 1 - \theta/(1-\theta)$, which, by the same argument as above, leads to the inequality on the left of (3.4).

Theorem 3.3. Let $\alpha \in \underline{A}$ and $\sigma \in \underline{M}$ such that $I_\alpha(\sigma) = l < \infty$. Then:

1. If, for some sequence $\{y_n\} \to \infty$, $\alpha(y_n) \to 0$, then $\sigma(y_n) \to 0$.
2. If, for some sequence $\{y_n\} \to \infty$, $\alpha(y_n) \to L > 0$, then $\sigma(y_n) \to L$ also.
3. If, for some sequence $\{y_n\} \to \infty$, $\alpha(y_n) \to \infty$, then $\alpha(y_n)/\sigma(y_n) \to 1$.

These three statements combine into: If $\{y_n\} \to \infty$, then $[1+\alpha(y_n)]/[1+\sigma(y_n)] \to 1$.

Proof. We prove each of the three statements. First assume for some sequence $\{y_n\} \to \infty$, $\alpha(y_n) \to 0$. To prove statement 1, we assume the contrary, i.e., there exists $\delta > 0$ such that $\sigma(y_n) \geqslant \delta$ for all n. From (3.2) let $\sup_{x \in X} G(x, x) < C$ and let $\varepsilon > 0$ be given. Let γ_0 be so small that $\gamma < \gamma_0$ implies

$$\frac{1-e^{-\gamma}}{\gamma}\left(1+\frac{\gamma(C+1)}{1-\gamma(C+1)}\right) < 2$$

Choose $\gamma > 0$, but $\gamma < \min(1/2(C+1), \gamma_0)$. Choose N so large that $l/N\gamma < \varepsilon/2$. Because we are assuming here that $\alpha(y_n) \to 0$ and because of (3.1) we can choose N elements $x_1, x_2,..., x_N$ in X such that $G(x_i, x_j) \leqslant 1/N$ for $i \neq j$, $i, j = 1, 2,..., N$, and such that $\alpha(x_j) \leqslant \varepsilon/4$, $j = 1, 2,..., N$.

Now,

$$\gamma[C + (N-1)(1/N)] \leqslant \gamma[C+1] = \theta < 1/2$$

and hence from Lemma 3.2,

$$N\delta \leqslant \sum_{i=1}^{N} \sigma(x_i) \leqslant \frac{1-e^{-\gamma}}{\gamma}\left(1+\frac{\gamma(C+1)}{1-\gamma(C+1)}\right) N\frac{\varepsilon}{4}+\frac{l}{\gamma} \leqslant \frac{N\varepsilon}{2}+\frac{l}{\gamma}$$

i.e.,

$$\delta \leqslant \frac{\varepsilon}{2}+\frac{l}{N\gamma} \leqslant \varepsilon$$

Since $\varepsilon > 0$ is arbitrary, we have a contradiction and statement 1 is proved.

To prove statement 2, suppose $\{y_n\} \to \infty$ such that $\alpha(y_n) \to L > 0$. Again we assume the contrary, i.e., there is a subsequence of $\{y_n\}$, which we will again label $\{y_n\}$, such that either $\sigma(y_n) \geqslant L' > L$ for all $n = 1, 2,...$, or $\sigma(y_n) \leqslant L' < L$ for all $n = 1, 2,...$. For the first case, let $\varepsilon > 0$ be given and choose $\gamma > 0$ so small that $\gamma[C+1] = \theta < 1/2$ and

$$\frac{1-e^{-\gamma}}{\gamma}\left[1+\frac{\gamma(C+1)}{1-\gamma(C+1)}\right] < 1+\varepsilon$$

Choose N so large that $l/N\gamma < \varepsilon$. For that N we can find points $x_1, x_2,..., x_N$ such that $G(x_i, x_j) \leqslant 1/N$ for all $i \neq j$, $i, j = 1, 2,..., N$, and such that $\alpha(x_j) \leqslant L+\varepsilon$. Thus, from Lemma 3.2,

$$NL' \leqslant \sum_{i=1}^{N} \sigma(x_i) \leqslant N(L+\varepsilon)(1+\varepsilon)+l/\gamma$$

or

$$L' \leqslant (L+\varepsilon)(1+\varepsilon)+\varepsilon$$

Since $\varepsilon > 0$ is arbitrary, this contradicts $L' > L$. A similar argument using the lower estimate in Lemma 3.2 takes care of the second case, $L' < L$, and statement 2 is proved.

Finally, to show statement 3, let $\{y_n\} \to \infty$ be a sequence such that $\alpha(y_n) \to \infty$. Again assuming the contrary, suppose there is a subsequence $\{y_n\}$ such that $\sigma(y_n) \geqslant \lambda\alpha(y_n)$ for some $\lambda > 1$, $n = 1, 2,...$, or $\sigma(y_n) \leqslant \lambda\alpha(y_n)$ for some $\lambda < 1$, $n = 1, 2,...$. Take the first case, and use the upper estimate of Lemma 3.2 with $N = 1$. Choose $\gamma > 0$ so small that $\gamma C = \theta < 1$ and

$$\frac{1-e^{-\gamma}}{\gamma}\left(1+\frac{\gamma C}{1-\gamma C}\right) < 1+\frac{\varepsilon}{2}$$

where $\varepsilon > 0$ is given. Choose $M > 0$ so large that $l/M\gamma < \varepsilon/2$. Now, there is an $x \in X$ for which $\alpha(x) > M$. Hence from Lemma 3.2 for this particular x,

$$\frac{\sigma(x)}{\alpha(x)} \leqslant \frac{1-e^{-\gamma}}{\gamma}\left(1+\frac{\gamma C}{1-\gamma C}\right)+\frac{l}{\alpha(x)\,\gamma}$$
$$\leqslant \left(1+\frac{\varepsilon}{2}\right)+\frac{l}{M\gamma} \leqslant 1+\varepsilon$$

and since $\varepsilon > 0$ is arbitrary, this contradicts $\sigma(x)/\alpha(x) \geqslant \lambda > 1$. Similarly, of course, for the other alternative, and the proof is complete.

Lemma 3.4. Let $\alpha \in \underline{A}$ and $\beta \in \underline{A}$. Assume for some $\sigma \in \underline{M}$ that both $I_\alpha(\sigma) < \infty$ and $I_\beta(\sigma) < \infty$. Then $\alpha \equiv \beta$.

Proof. First we want to prove the following: if $\alpha \in \underline{A}$ and $\beta \in \underline{A}$ and if $h(x) = \beta(x)/\alpha(x) \to 1$ as $x \to \infty$, then $\alpha \equiv \beta$.

As in Section 2, let $\pi_{xy}(\alpha) = \pi_{yx}\alpha(y)/\alpha(x)$ and recall that $\{\pi_{xy}(\alpha)\}$ are transition probabilities for an irreducible chain. Now,

$$\sum_{y \in X} \pi_{xy}(\alpha)\, h(y) = \sum_{y \in X} \pi_{yx} \frac{\alpha(y)\,\beta(y)}{\alpha(x)\,\alpha(y)}$$
$$= \frac{1}{\alpha(x)} \sum_{y \in X} \pi_{yx}\beta(y) = \frac{\beta(x)}{\alpha(x)} = h(x)$$

so that h is a harmonic function of the chain time-reversed with respect to α. By hypothesis, $h(x) \to 1$ at ∞, but, for an irreducible Markov chain, if a bounded harmonic function approaches 1 at infinity, it is then identically 1.

Now, to prove this lemma, we have from Theorem 3.3 that $\{y_n\} \to \infty$ implies $[1+\sigma(y_n)]/[1+\alpha(y_n)] \to 1$ and also $[1+\sigma(y_n)]/[1+\beta(y_n)] \to 1$, i.e., $[1+\alpha(y_n)]/[1+\beta(y_n)] \to 1$. Since, by hypothesis, $\{\pi_{xy}\}$ is doubly stochastic, we see that 1 is an invariant measure for the π-chain, i.e., $1 \in \underline{A}$. Since the sum of two invariant measures is also an invariant measure, the statement at the beginning of this proof implies $1+\alpha(\cdot) \equiv 1+\beta(\cdot)$, i.e., $\alpha(\cdot) \equiv \beta(\cdot)$.

Lemma 3.5. Let $\alpha \in \underline{A}$ and $\sigma \in \underline{M}$ such that $I_\alpha(\sigma) < \infty$. Then, for all $y \in X$,

$$\lim_{k \to \infty} \sum_{x \in X} \sigma(x)\, \pi_{xy}^{(k)} = \alpha(y) \tag{3.5}$$

Proof. As noted in the preceding lemma, since $\{\pi_{xy}\}$ is doubly stochastic, 1 is an invariant measure, and therefore to show (3.5), it suffices to show

$$\lim_{k \to \infty} \sum_{x \in X} [\sigma(x)+1]\, \pi_{xy}^{(k)} = \alpha(y)+1 \tag{3.6}$$

for all $y \in X$. Let

$$h(x) = \frac{\sigma(x)+1}{\alpha(x)+1}, \qquad \tilde{\pi}_{xy} = \pi_{yx} \frac{\alpha(y)+1}{\alpha(x)+1}$$

Now

$$\tilde{\pi}^{(k)}_{xy} = \pi^{(k)}_{yx} \frac{\alpha(y)+1}{\alpha(x)+1}$$

and

$$\begin{aligned}\sum_{x \in X} [\sigma(x)+1]\, \pi^{(k)}_{xy} &= \sum_{x \in X} \frac{\alpha(y)+1}{\alpha(x)+1} \tilde{\pi}^{(k)}_{yx} [\sigma(x)+1] \\ &= [\alpha(y)+1] \sum_{x \in X} h(x)\, \tilde{\pi}^{(k)}_{yx} \\ &= [\alpha(y)+1]\, E^{\tilde{\pi}}_{y}\{h(X_k)\}\end{aligned}$$

As we have argued several times already, the $\tilde{\pi}$-chain is also transient, which implies, for every $y \in X$, $X_k \to \infty$ for almost all paths starting from y ($\tilde{\pi}$-measure). Moreover, from Theorem 3.3, $h(x_k) \to 1$ if $x_k \to \infty$, so by bounded convergence $\lim_{k \to \infty} E^{\tilde{\pi}}_{y}\{h(X_k)\} = 1$ and we have (3.6) for all $y \in X$ and the lemma.

We now prove a succession of lemmas leading up to Theorem 3.11, which will be used in Section 4 in order to prove the lower bound (1.3).

In these lemmas as well as in Theorem 3.11, F will denote a fixed, finite subset of X having elements $x_1, x_2, \ldots, x_k$. Let $\boldsymbol{\theta} = \{\theta_1, \theta_2, \ldots, \theta_k\}$ be a point in R^k and let $C = \{V: X \to \mathbb{R}\}$. Let $T: R^k \to C$ be the mapping given by

$$\begin{aligned} V(x_i) &= \theta_i, \qquad i = 1, 2, \ldots, k \\ V(x) &= 0, \qquad x \neq x_i \end{aligned}$$

Let $\underline{Q} \in R^k$ consist of those $\boldsymbol{\theta} \in R^k$ such that $T\boldsymbol{\theta} \in \underline{V}$. For $\alpha \in \underline{A}$ and $\boldsymbol{\theta} \in \underline{Q}$, define

$$\Phi(\boldsymbol{\theta}) = \Phi_F(\boldsymbol{\theta}) = \sum_{i=1}^{k} \alpha(x_i)(1 - e^{-\theta_i})\, u(\boldsymbol{\theta}; x_i) - \sum_{i=1}^{k} \alpha(x_i)\, \theta_i$$

where, of course,

$$u(\boldsymbol{\theta}; x_i) = u_V(x_i) = E^{\pi}_{x_i}\left\{\exp\left[\sum_{j=0}^{\infty} V(X_j)\right]\right\}$$

with $V = T\boldsymbol{\theta}$.

In the first of these lemmas we use the notation of Section 2. Recall that $\pi_{xy}^F = P_x^\pi\{X_\tau = y; \tau < \infty\}$, where $\tau = \inf_{j \geqslant 1}\{X_j \in F\}$, and

$$\eta_x = 1 - \sum_{y \in F} \pi_{xy}^F = P_x^\pi\{X_j \notin F, j \geqslant 1\}$$

For the time-reversed chain with respect to an $\alpha \in \underline{A}$, i.e., the chain with transition probabilities $\pi_{xy}(\alpha) = \pi_{yx}\alpha(y)/\alpha(x)$, we use the notations

$$\pi_{xy}^F(\alpha) = P_x^{\pi(\alpha)}\{X_\tau = y; \tau < \infty\}$$

$$\eta_x(\alpha) = 1 - \sum_{y \in F} \pi_{xy}^F(\alpha) = P_x^{\pi(\alpha)}\{X_j \notin F, j \geqslant 1\}$$

as before.

Lemma 3.6. For $\alpha \in \underline{A}$ and $\boldsymbol{\theta} \in \underline{O}$,

$$\Phi_F(\boldsymbol{\theta}) = \sum_{i=1}^k \alpha(x_i)\, \eta_{x_i}(\alpha)\, u(\boldsymbol{\theta}; x_i) - \sum_{i=1}^k \alpha(x_i)\, \eta_{x_i} - \sum_{i=1}^k \alpha(x_i)\, \theta_i \tag{3.7}$$

Proof. Since $\boldsymbol{\theta} \in \underline{O}$, $V \in \underline{V}$ and therefore from (2.6) we have for any $x \in X$,

$$u_V(x) = e^{V(x)}\left[\sum_{y \in F} \pi_{xy}^F u_V(y) + \eta_x\right]$$

Thus,

$$\begin{aligned}
&\sum_{x \in X} \alpha(x)(1 - e^{-V(x)})\, u_V(x) \\
&\quad = \sum_{x \in F} \alpha(x)(1 - e^{-V(x)})\, u_V(x) \\
&\quad = \sum_{x \in F} \alpha(x)\, u_V(x) - \sum_{x \in F} \alpha(x)\left[\sum_{y \in F} \pi_{xy}^F u_V(y) + \eta_x\right] \\
&\quad = \sum_{y \in F}\left[\alpha(y) - \sum_{x \in F} \alpha(x)\, \pi_{xy}^F\right] u_V(y) - \sum_{x \in F} \alpha(x)\, \eta_x
\end{aligned} \tag{3.8}$$

For $x, y \in F$,

$$\begin{aligned}
\pi_{xy}^F &= \pi_{xy} + \sum_{z_1 \notin F} \pi_{xz_1}\pi_{z_1y} + \sum_{\substack{z_1 \notin F \\ z_2 \notin F}} \pi_{xz_1}\pi_{z_1z_2}\pi_{z_2y} + \cdots \\
&= \pi_{xy}^{(1)} + \pi_{xy}^{(2)} + \cdots + \pi_{xy}^{(n)} + \cdots
\end{aligned}$$

and therefore

$$\begin{aligned}
\alpha(y) - \sum_{x\in F} \alpha(x)\,\pi^F_{xy}
&= \sum_{x\in X} \alpha(x)\,\pi_{xy} - \sum_{x\in F} \alpha(x)\,\pi^F_{xy} \\
&= \sum_{x\notin F} \alpha(x)\,\pi_{xy} - \sum_{x\in F} \alpha(x)[\pi^F_{xy} - \pi_{xy}] \\
&= \sum_{x\notin F} \alpha(x)\,\pi_{xy} - \sum_{x\in F} \alpha(x) \sum_{n=2}^{\infty} \pi^{(n)}_{xy} \\
&= \sum_{x\notin F} \sum_{z_1\in X} \alpha(z_1)\,\pi_{z_1x}\pi_{xy} - \sum_{x\in F} \alpha(x) \sum_{n=2}^{\infty} \pi^{(n)}_{xy} \\
&= \sum_{x\notin F} \sum_{z_1\notin F} \alpha(z_1)\,\pi_{z_1x}\pi_{xy} \\
&\quad - \sum_{x\in F} \alpha(x)\left[\sum_{n=2}^{\infty} \pi^{(n)}_{xy} - \sum_{z_1\notin F} \pi_{xz_1}\pi_{z_1y}\right] \\
&= \sum_{x\notin F} \sum_{z_1\notin F} \alpha(z_1)\,\pi_{z_1x}\pi_{xy} - \sum_{x\in F} \alpha(x) \sum_{n=3}^{\infty} \pi^{(n)}_{xy}
\end{aligned}$$

If we repeat this process we arrive at the following expression, which holds for any $n = 1, 2,\ldots$:

$$\begin{aligned}
\alpha(y) - \sum_{x\in F} \alpha(x)\,\pi^F_{xy}
&= \sum_{z_1, z_2,\ldots, z_n\notin F} \alpha(z_1)\,\pi_{z_1z_2}\pi_{z_2z_3}\cdots\pi_{z_ny} \\
&\quad - \sum_{x\in F} \alpha(x) \sum_{k=n+1}^{\infty} \pi^{(k)}_{xy}
\end{aligned} \tag{3.9}$$

Letting $n\to\infty$ in (3.9), we see that (cf. proof of Lemma 2.3)

$$\begin{aligned}
\alpha(y) - \sum_{x\in F} \alpha(x)\,\pi^F_{xy}
&= \lim_{n\to\infty} \sum_{z_1, z_2,\ldots, z_n\notin F} \alpha(z_1)\,\pi_{z_1z_2}\pi_{z_2z_3}\cdots\pi_{z_ny} \\
&= \lim_{n\to\infty} \sum_{z_1, z_2, z_3,\ldots, z_n\notin F} \alpha(y)\,\pi_{yz_1}(\alpha)\,\pi_{z_1z_2}(\alpha)\cdots\pi_{z_{n-1}z_n}(\alpha) \\
&= \alpha(y)\,\eta_y(\alpha)
\end{aligned} \tag{3.10}$$

Using (3.10) in (3.8), we get

$$\sum_{x \in X} \alpha(x)(1 - e^{-V(x)})\, u_V(x)$$

$$= \sum_{y \in F} \alpha(y)\, \eta_y(\alpha)\, u_V(y) - \sum_{x \in F} \alpha(x)\, \eta_x$$

Therefore,

$$\Phi_F(\boldsymbol{\theta}) = \sum_{i=1}^{k} \alpha(x_i)\, \eta_{x_i}(\alpha)\, u(\boldsymbol{\theta}; x_i) - \sum_{i=1}^{k} \alpha(x_i)\, \eta_{x_i} - \sum_{i=1}^{k} \alpha(x_i)\, \theta_i$$

which is the lemma.

Lemma 3.7. $\underline{O}$ is an open, convex set in R^k, $\Phi(\boldsymbol{\theta}) \geqslant 0$ for $\boldsymbol{\theta} \in \underline{O}$, and if $\tilde{\boldsymbol{\theta}} \in R^k$ such that $\tilde{\boldsymbol{\theta}} \in \partial \underline{O}$ and $\boldsymbol{\theta}^{(n)} \to \tilde{\boldsymbol{\theta}}$ as $n \to \infty$, then $\Phi(\boldsymbol{\theta}^{(n)}) \to \infty$.

Proof. In Section 2 we noted that if $V \in \underline{V}$, there exists $\lambda > 1$ such that $\lambda V \in \underline{V}$, which implies $\underline{O}$ is open. From Lemma 2.1, $\underline{V} = \underline{V}_2$, so $V \in \underline{V}$ implies $\rho(M_F) < 1$, where M_F is the matrix $\{e^{V(x)} \pi^F_{xy},\ x,\ y \in F\}$. Since here F is fixed, $\rho(M_F) = \rho(\boldsymbol{\theta})$ and $\boldsymbol{\theta} \in \underline{O}$ implies $\rho(\boldsymbol{\theta}) < 1$, i.e., the convex function $\log \rho(\boldsymbol{\theta}) < 0$ for $\boldsymbol{\theta} \in \underline{O}$, which implies $\underline{O}$ is convex. That $\Phi(\boldsymbol{\theta}) \geqslant 0$ for $\boldsymbol{\theta} \in \underline{O}$ follows by applying Jensen's inequality to the definition of $\Phi(\boldsymbol{\theta})$.

Although $\underline{O}$ is open and convex, it is not bounded, and indeed we later must examine the behavior of $\Phi(\boldsymbol{\theta})$ as $\boldsymbol{\theta}$ stays in $\underline{O}$ but approaches ∞. Here the sequence $\{\boldsymbol{\theta}^{(n)}\} \in \underline{O}$ converges to the finite point $\tilde{\boldsymbol{\theta}} \in \partial \underline{O}$. In the definition of $\Phi(\boldsymbol{\theta})$, the term

$$\sum_{i=1}^{k} \alpha(x_i)(1 - e^{-\theta_i})\, u(\boldsymbol{\theta}; x_i)$$

is troublesome because, depending on the signs of the θ_i, some terms are positive and some negative, presenting possible cancellations. This problem is obviated by the decomposition in Lemma 3.6. Thus,

$$\Phi(\boldsymbol{\theta}^{(n)}) = \sum_{i=1}^{k} \alpha(x_i)\, \eta_{x_i}(\alpha)\, u(\boldsymbol{\theta}^{(n)}; x_i) - \sum_{i=1}^{k} \alpha(x_i)\, \eta_{x_i} - \sum_{i=1}^{k} \alpha(x_i)\, \theta_i^{(n)} \tag{3.11}$$

In (3.11) not all of the factors $\eta_{x_i}(\alpha)$, $i = 1, 2, \ldots, k$, can be zero, because the chain time-reversed with respect to α is transient. Since $\underline{O}$ is open, $\tilde{\boldsymbol{\theta}} \notin \underline{O}$, and hence $u(\tilde{\boldsymbol{\theta}}; x_i) = \infty$, $i = 1, 2, \ldots, k$. From Fatou's lemma, we see then that

$u(\boldsymbol{\theta}^{(n)}; x_i) \to \infty$ as $\boldsymbol{\theta}^{(n)} \to \tilde{\boldsymbol{\theta}}$ for $i = 1, 2, \ldots, k$. Thus, the first term in (3.11) becomes infinite as $\boldsymbol{\theta}^{(n)} \to \tilde{\boldsymbol{\theta}}$ and so $\Phi(\boldsymbol{\theta}^{(n)}) \to \infty$ as $\boldsymbol{\theta}^{(n)} \to \tilde{\boldsymbol{\theta}}$, completing the lemma.

Now we must examine the behavior of $\Phi(\boldsymbol{\theta})$ as $\boldsymbol{\theta} \to \infty$. There exist rays along which $\Phi(\boldsymbol{\theta})$ remains bounded and we have to take these into account in what follows. To this end, we define a point $\mathbf{a} \in R^k$ to be *special* if $\sup_{\lambda > 0} \Phi(\lambda \mathbf{a}) \leqslant \text{const}$. Let $S = \{\mathbf{a} \in R^k, \mathbf{a} \text{ special}\}$. Clearly, $S \subset \underline{O}$.

Lemma 3.8. If $\mathbf{a} \in S$, then $-\mathbf{a} \in S$ and $\Phi(\mathbf{a}) = 0$. S is a linear subspace of R^k and if $\mathbf{b} \in \underline{O}$ and $\mathbf{a} \in S$, then $\mathbf{b} + \mathbf{a} \in \underline{O}$ and $\Phi(\mathbf{b} + \mathbf{a}) = \Phi(\mathbf{b})$.

Proof. From the definition of $\Phi(\boldsymbol{\theta})$ we see that it is convex and $\Phi(\mathbf{0}) = 0$. Thus, if $\Phi(\lambda \mathbf{a}) \leqslant c$ for all $\lambda > 0$, then $\Phi(\lambda \mathbf{a}) \equiv 0$ for all $\lambda > 0$. Since Φ is analytic in $\boldsymbol{\theta}$, we have $\Phi(\lambda \mathbf{a}) \equiv 0$ for all real λ. To show that S is a linear subspace, it suffices to show that $\mathbf{a}_1 \in S$ and $\mathbf{a}_2 \in S$ imply $\frac{1}{2}(\mathbf{a}_1 + \mathbf{a}_2) \in S$. But

$$0 \leqslant \Phi\left(\frac{\mathbf{a}_1 + \mathbf{a}_2}{2}\right) \leqslant \frac{1}{2}\left[\Phi(\mathbf{a}_1) + \Phi(\mathbf{a}_2)\right] = 0$$

so not only is S a linear subspace, but $\Phi(\boldsymbol{\theta}) \equiv 0$ for $\boldsymbol{\theta} \in S$.

Let $\mathbf{b} \in \underline{O}$ and $\mathbf{a} \in S$. Since $\underline{O}$ is open, there exists $\varepsilon > 0$ small enough so that $[1/(1-\varepsilon)]\, \mathbf{b} \in \underline{O}$. By definition, $(1/\varepsilon)\, \mathbf{a} \in S \subset \underline{O}$ and since $\underline{O}$ is a convex set,

$$\mathbf{b} + \mathbf{a} = (1 - \varepsilon) \frac{1}{1 - \varepsilon} \mathbf{b} + \varepsilon \left(\frac{1}{\varepsilon}\right) \mathbf{a} \in \underline{O}$$

By the convexity of Φ and since $\Phi(\mathbf{a}) = 0$, we have

$$\Phi(\mathbf{a} + \mathbf{b}) \leqslant (1 - \varepsilon)\, \Phi\left(\frac{\mathbf{b}}{1 - \varepsilon}\right)$$

Convex functions being continuous, we obtain from this that

$$\Phi(\mathbf{a} + \mathbf{b}) \leqslant \Phi(\mathbf{b}) \qquad \text{for all} \quad \mathbf{a} \in S \text{ and } \mathbf{b} \in \underline{O} \tag{3.12}$$

In (3.12) we can replace $\mathbf{a}$ by $-\mathbf{a}$, giving $\Phi(\mathbf{b} - \mathbf{a}) \leqslant \Phi(\mathbf{b})$ for all $\mathbf{a} \in S$ and $\mathbf{b} \in \underline{O}$. But we just showed $\mathbf{a} + \mathbf{b} \in \underline{O}$ and therefore this last inequality gives $\Phi(\mathbf{b}) \leqslant \Phi(\mathbf{a} + \mathbf{b})$, which, together with (3.12), finishes the lemma.

We have noted that $S \subset \underline{O}$. Let $\underline{O}_0 = \{\boldsymbol{\theta}' \in \underline{O}, \boldsymbol{\theta}' \perp S\}$, so $\underline{O}_0$ is a closed subset of $\underline{O}$, and any $\boldsymbol{\theta} \in \underline{O}$ can be written $\boldsymbol{\theta} = \boldsymbol{\theta}' + \mathbf{a}$, where $\boldsymbol{\theta}' \in \underline{O}_0$ and $\mathbf{a} \in S$.

Lemma 3.9. Let $\{\boldsymbol{\theta}^{(n)}\} \in \underline{O}_0$ such that $\|\boldsymbol{\theta}^{(n)}\| \to \infty$. Then, $\Phi(\boldsymbol{\theta}^{(n)}) \to \infty$.

Proof. Assume the contrary; then, there exists a constant C such that $\Phi(\boldsymbol{\theta}^{(n)}) \leqslant C$ for all n. For $\lambda > 0$,

$$\Phi\left(\frac{\lambda\boldsymbol{\theta}^{(n)}}{\|\boldsymbol{\theta}^{(n)}\|}\right) = \Phi\left(\frac{\lambda}{\|\boldsymbol{\theta}^{(n)}\|}\boldsymbol{\theta}^{(n)} + \left(1 - \frac{\lambda}{\|\boldsymbol{\theta}^{(n)}\|}\right)\cdot\mathbf{0}\right)$$
$$\leqslant \frac{\lambda}{\|\boldsymbol{\theta}^{(n)}\|}\Phi(\boldsymbol{\theta}^{(n)})$$
$$\leqslant \frac{C\lambda}{\|\boldsymbol{\theta}^{(n)}\|}$$

which implies $\Phi(\lambda\boldsymbol{\theta}^{(n)}/\|\boldsymbol{\lambda}^{(n)}\|) \to 0$ as $n \to \infty$. Since $\underline{Q}_0$ is closed and, for each n, $\|\boldsymbol{\theta}^{(n)}/\|\boldsymbol{\theta}^{(n)}\|\,\| = 1$, there exists a subsequence, which we will label $\boldsymbol{\theta}^{(n)}/\|\boldsymbol{\theta}^{(n)}\|$, converging to some element $\mathbf{a} \in \underline{Q}_0$ such that $\|\mathbf{a}\| = 1$. Since Φ is continuous, we have for all $\lambda > 0$, $\Phi(\lambda\mathbf{a}) = 0$, which means $\mathbf{a}$ is special. But $\mathbf{a} \in \underline{Q}_0$ means $\mathbf{a} \perp \mathbf{a}$, which is impossible, since $\mathbf{a}$ is a unit vector.

Lemma 3.10. Let $\alpha \in \underline{A}$ and $\sigma \in \underline{M}$ such that $I_\alpha(\sigma) < \infty$. If $\mathbf{a} \in S$, then

$$\sum_{i=1}^{k} \sigma(x_i)\, a_i = \sum_{i=1}^{k} \alpha(x_i)\, a_i$$

Proof.

$$I_\alpha(\sigma) = \sup_{V \in \underline{V}}\left[\sum_{x \in X} \sigma(x)\, V(x) - \sum_{x \in X} \alpha(x)(1 - e^{-V(x)})\, u_V(x)\right]$$
$$\geqslant \sup_{\boldsymbol{\theta} \in \underline{Q}}\left[\sum_{i=1}^{k} \sigma(x_i)\,\theta_i - \sum_{i=1}^{k} \alpha(x_i)(1 - e^{-\theta_i})\, u(\boldsymbol{\theta}; x_i)\right]$$

since the first supremum is over V's vanishing outside any finite subset of X, whereas the second supremum is just over V's vanishing outside our given finite subset F.

Thus, using the fact that $\Phi(\mathbf{a}) = 0$ for $\mathbf{a} \in S$,

$$I_\alpha(\sigma) \geqslant \sup_{\boldsymbol{\theta} \in \underline{Q}}\left[\sum_{i=1}^{k} \sigma(x_i)\,\theta_i - \sum_{i=1}^{k} \alpha(x_i)\,\theta_i - \Phi_F(\boldsymbol{\theta})\right]$$
$$\geqslant \sup_{\mathbf{a} \in S}\left[\sum_{i=1}^{k} \sigma(x_i)\, a_i - \sum_{i=1}^{k} \alpha(x_i)\, a_i - \Phi_F(\mathbf{a})\right]$$
$$= \sup_{\mathbf{a} \in S}\left[\sum_{i=1}^{k} \sigma(x_i)\, a_i - \sum_{i=1}^{k} \alpha(x_i)\, a_i\right]$$

Since S is a linear subspace, this last supremum is either ∞ or 0. The former is ruled out by our hypothesis $I_\alpha(\sigma) < \infty$, and hence we have this lemma, because

$$\sup_{\mathbf{a} \in S} \sum_{i=1}^{k} [\sigma(x_i) - \alpha(x_i)]\, a_i = 0$$

implies

$$\sum_{i=1}^{k} [\sigma(x_i) - \alpha(x_i)]\, a_i = 0 \qquad \text{identically in } S$$

Theorem 3.11. Let $\alpha \in \underline{A}$ and $\sigma \in \underline{M}$ such that $I_\alpha(\sigma) < \infty$. Let $\sigma' = (1-\varepsilon)\,\sigma + \varepsilon\alpha$ for $\varepsilon > 0$ fixed, and let F be a finite subset of X. In the variational problem

$$\sup_{\boldsymbol{\theta} \in \underline{Q}} \left[\sum_{i=1}^{k} \sigma'(x_i)\,\theta_i - \sum_{i=1}^{k} \alpha(x_i)(1 - e^{-\theta_i})\, u(\boldsymbol{\theta}; x_i) \right] \tag{3.13}$$

the supremum is attained at a point $\boldsymbol{\theta}^* \in \underline{Q}$.

Proof. Since $\alpha \in \underline{A}$, $I_\alpha(\alpha) = 0$, and the convexity of the I-function implies

$$I_\alpha(\sigma') \geqslant \sup_{\boldsymbol{\theta} \in \underline{Q}} \left[\sum_{i=1}^{k} \sigma'(x_i)\,\theta_i - \sum_{i=1}^{k} \alpha(x_i)(1 - e^{-\theta_i})\, u(\boldsymbol{\theta}; x_i) \right]$$

we see that the supremum in (3.13) is finite. Also, since $\mathbf{0} \in \underline{Q}$, we see that the supremum in (3.13) is nonnegative.

With $\Phi_F(\boldsymbol{\theta})$ as defined earlier, we can write (3.13) as

$$\sup_{\boldsymbol{\theta} \in \underline{Q}} \left[\sum_{i=1}^{k} \sigma'(x_i)\,\theta_i - \sum_{i=1}^{k} \alpha(x_i)\,\theta_i - \Phi_F(\boldsymbol{\theta}) \right] \tag{3.14}$$

As noted before, for any $\boldsymbol{\theta} \in \underline{Q}$ we can write $\boldsymbol{\theta} = \mathbf{a} + \mathbf{b}$, where $\mathbf{a} \in S$ and $\mathbf{b} \in \underline{Q}_0 = \{\boldsymbol{\theta}' \in \underline{Q}\colon\ \boldsymbol{\theta}' \perp S\}$. From Lemma 3.8, $\Phi(\mathbf{a} + \mathbf{b}) = \Phi(\mathbf{b})$. Since $I_\alpha(\sigma') < \infty$, for this $\mathbf{a} \in S$ we have from Lemma 3.10 that $\sum_{i=1}^{k} \sigma'(x_i)\,a_i = \sum_{i=1}^{k} \alpha(x_i)\,a_i$. Thus, (3.14) becomes

$$\begin{aligned}
&\sup_{\mathbf{b} \in \underline{Q}_0} \left[\sum_{i=1}^{k} \sigma'(x_i)\,b_i - \sum_{i=1}^{k} \alpha(x_i)\,b_i - \Phi_F(\mathbf{b}) \right] \\
&\quad = \sup_{\mathbf{b} \in \underline{Q}_0} \left[(1-\varepsilon) \sum_{i=1}^{k} \sigma(x_i)\,b_i \right. \\
&\qquad \left. - (1-\varepsilon) \sum_{i=1}^{k} \alpha(x_i)\,b_i - \sum_{i=1}^{k} \alpha(x_i)(1 - e^{-b_i})\,u(\mathbf{b}; x_i) + \sum_{i=1}^{k} \alpha(x_i)\,b_i \right] \\
&\quad = \sup_{\mathbf{b} \in \underline{Q}_0} \left\{ (1-\varepsilon) \left[\sum_{i=1}^{k} \sigma(x_i)\,b_i - \sum_{i=1}^{k} \alpha(x_i)(1 - e^{-b_i})\,u(\mathbf{b}; x_i) \right] - \varepsilon\Phi_F(\mathbf{b}) \right\}
\end{aligned} \tag{3.15}$$

Let $\mathbf{b}^{(n)}$ be a maximizing sequence in $\underline{Q}_0$. There are three possibilities: (1) $\|\mathbf{b}^{(n)}\| \to \infty$, (2) $\mathbf{b}^{(n)} \to \mathbf{b} \in \partial \underline{Q}_0$, or (3) $\mathbf{b}^{(n)} \to \mathbf{0}^*$, a point in $\underline{Q}_0 \subset \underline{Q}$.

Assume possibility 1 first. By hypothesis,

$$\infty > I_\alpha(\sigma) \geqslant \sup_{\mathbf{b} \in \underline{Q}_0} \left[\sum_{i=1}^{k} \sigma(x_i)\, b_i - \sum_{i=1}^{k} \alpha(x_i)(1 - e^{-b_i})\, u(\mathbf{b}; x_i) \right]$$

so that in the last line of (3.15) the quantity in square brackets is bounded, whereas Lemma 3.9 gives us $\Phi_F(\mathbf{b}^{(n)}) \to \infty$. Hence, in the last line of (3.15) the quantity in curly braces goes to $-\infty$ as $\|\mathbf{b}^{(n)}\| \to \infty$, which contradicts the supremum in (3.13) being nonnegative.

Next, assume possibility 2 holds. From Lemma 3.7, $\Phi_F(\mathbf{b}^{(n)}) \to \infty$ if $\mathbf{b}^{(n)} \to \mathbf{b} \in \partial \underline{Q}_0 \subset \partial \underline{Q}$, and, just as above, the quantity in square brackets in the last line of (3.15) remains bounded. Thus, the quantity in curly braces in the last line of (3.15) again goes to $-\infty$ as $\mathbf{b}^{(n)} \to \mathbf{b} \in \partial \underline{Q}_0$, which gives the same contradiction, and the proof is complete.

4. THE LOWER BOUND

Let $\alpha \in \underline{A}$. To prove the lower bound (1.3), it suffices, since G is open, to show that for any $\sigma \in \underline{M}$ such that $I_\alpha(\sigma) < \infty$,

$$\varliminf_{N \to \infty} \frac{1}{N} \log Q_{n,N}(N_\sigma) \geqslant -I_\alpha(\sigma) \tag{4.1}$$

for almost all $n(\cdot) \in Z$ (P_α-measure), where N_σ is any $\underline{M}$-neighborhood of σ.

We can choose $\varepsilon > 0$ so small that $\sigma' = (1-\varepsilon)\,\sigma + \varepsilon\alpha$ is an element of N_σ and then we can choose an $\underline{M}$-neighborhood of σ', call it $N_{\sigma'}$, such that $N_{\sigma'} \subset N_\sigma$. Moreover, since the I-function is convex and $I_\alpha(\alpha) = 0$, we see that

$$I_\alpha(\sigma') \leqslant (1-\varepsilon)\, I_\alpha(\sigma) + \varepsilon I_\alpha(\alpha) \leqslant I_\alpha(\sigma)$$

Thus, to show (4.1), it suffices to show for σ' of the form above and any $\underline{M}$-neighborhood $N_{\sigma'}$ of σ' that

$$\varliminf_{N \to \infty} \frac{1}{N} \log Q_{n,N}(N_{\sigma'}) \geqslant -I_\alpha(\sigma') \tag{4.2}$$

for almost all $n(\cdot) \in Z$ (P_α-measure).

Sets in M of the form

$$B_{\sigma'} = \{\lambda(\cdot) \in \underline{M} : |\lambda(x) - \sigma'(x)| < \delta \text{ for all } x \in F, \text{ a finite subset of } X)$$

form a basis for the weak topology in $\underline{M}$, so that, to show (4.2), it suffices to show, for any set $B_{\sigma'}$ in the basis and $\sigma' = (1-\varepsilon)\,\sigma + \varepsilon\alpha$, that

$$\varliminf_{N\to\infty} \frac{1}{N} \log Q_{n,N}(B_{\sigma'}) \geqslant -I_\alpha(\sigma') \tag{4.3}$$

for almost all $n(\cdot) \in Z$ (P_α-measure).

With this in mind, we let F be a finite subset of X having elements $x_1, x_2,..., x_k$. Let $\underline{O} \subset R^k$ be the set of $\boldsymbol{\theta} \in R^k$ having the property that $V: X \to \mathbb{R}$ defined by

$$V(x) = \begin{cases} \theta_i & \text{if} \quad x = x_i, \quad i = 1, 2,..., k \\ 0 & \text{otherwise} \end{cases}$$

is in $\underline{V}$. As noted in Section 3, $\underline{O}$ is an open set in R^k. For each $N = 1, 2,...,$ define a probability measure μ_N on R^k by: if $A \subset R^k$,

$$\mu_N(A) = \underline{P}_n\{(L_N(x_1), L_N(x_2),..., L_N(x_k)) \in A\}$$

Let

$$M_N(\boldsymbol{\theta}) = \int \exp(N\langle\boldsymbol{\theta}, \mathbf{y}\rangle)\, d\mu_N(\mathbf{y}) = E^{\underline{P}_n}\left\{\exp\left[N \sum_{i=1}^{k} \theta_i L_N(x_i)\right]\right\}$$

which exists if $\boldsymbol{\theta} \in \underline{O}$. What is important is that, if $\alpha \in \underline{A}$ and $\boldsymbol{\theta} \in \underline{O}$, we have from Lemma 2.10 that

$$\begin{aligned}
&\lim_{N\to\infty} \frac{1}{N} \log M_N(\boldsymbol{\theta}) \\
&\quad = \lim_{N\to\infty} \frac{1}{N} \log E^{P_n}\left\{\exp\left[N \sum_{i=1}^{k} \theta_i L_N(x_i)\right]\right\} \\
&\quad = \lim_{N\to\infty} \frac{1}{N} \log E^{P_n}\left\{\exp\left[\sum_{j=0}^{N-1} \sum_{i=1}^{k} \theta_i n_j(x_i)\right]\right\} \\
&\quad = \lim_{N\to\infty} \frac{1}{N} \log E^{P_n}\left\{\exp\left[\sum_{j=0}^{N-1} \sum_{x\in X} n_j(x)\, V(x)\right]\right\} \\
&\quad = \sum_{x\in X} \alpha(x)\{1 - \exp[-V(x)\}\, u_V(x) \\
&\quad = \sum_{i=1}^{k} \alpha(x_i)[1 - \exp(-\theta_i)]\, u(\boldsymbol{\theta}; x_i) = \psi(\boldsymbol{\theta})
\end{aligned}$$

for almost all $n(\cdot) \in Z$ (P_α-measure).

We note that $\psi(\boldsymbol{\theta})$ is continuously differentiable on $\underline{O}$. For $\alpha \in \underline{A}$, $\sigma \in \underline{M}$ such that $I_\alpha(\sigma) < \infty$, and with $\sigma' = (1-\varepsilon)\,\sigma + \varepsilon\alpha$, we proved (Theorem 3.11) that in the variational problem,

$$\sup_{\boldsymbol{\theta} \in \underline{O}} \left[\sum_{i=1}^{k} \sigma'(x_i)\,\theta_i - \psi(\boldsymbol{\theta}) \right]$$

the supremum is actually attained at a point $\boldsymbol{\theta}^* \in \underline{O}$. This, of course, implies $(\nabla\psi)(\boldsymbol{\theta}^*) = \boldsymbol{\sigma}'$, where $\boldsymbol{\sigma}' = (\sigma'(x_i),\ i = 1, 2,..., k)$.

All of these observations become important because of the following lemma, which is a standard, general result, the proof of which can be found, for example, in Ref. 10.

Lemma 4.1. Let $\{\mu_N\}$ be a family of probability measures on R^k and assume

$$M_N(\boldsymbol{\theta}) = \int \exp(N\langle \boldsymbol{\theta}, \mathbf{y} \rangle)\, d\mu_N(\mathbf{y})$$

exists for $\boldsymbol{\theta}$ belonging to an open set $O \subset R^k$. Assume also that

$$\lim_{N \to \infty} \frac{1}{N} \log M_N(\boldsymbol{\theta} = \psi(\boldsymbol{\theta})$$

exists for $\boldsymbol{\theta} \in O$ and that $\psi(\boldsymbol{\theta})$ is continuously differentiable in O. Finally, assume $\boldsymbol{\sigma}' \in R^k$ is such that $(\nabla\psi)(\boldsymbol{\theta}^*) = \boldsymbol{\sigma}'$ for some $\boldsymbol{\theta}^* \in O$. Then,

$$\varliminf_{N \to \infty} \frac{1}{N} \log \mu_N(N_{\boldsymbol{\sigma}'}) \geqslant -[\langle \boldsymbol{\sigma}', \boldsymbol{\theta}^* \rangle - \psi(\boldsymbol{\theta}^*)] \tag{4.4}$$

where $N_{\boldsymbol{\sigma}'}$ is an R^k neighborhood of $\boldsymbol{\sigma}'$.

Since, as we just noted, all the hypotheses of Lemma 4.1 are satisfied, we conclude that

$$\begin{aligned}
&\varliminf_{N \to \infty} \frac{1}{N} \log \mu_N(N_{\boldsymbol{\sigma}'}) \\
&\quad = \varliminf_{N \to \infty} \frac{1}{N} \log P_n\{(L_N(x_1), L_N(x_2),..., L_N(x_k)) \in N_{\boldsymbol{\sigma}'}) \\
&\quad = \varliminf_{N \to \infty} \frac{1}{N} \log Q_{n,N}(N_{\boldsymbol{\sigma}'}) \\
&\quad \geqslant -[\langle \boldsymbol{\sigma}', \boldsymbol{\theta}^* \rangle - \psi(\boldsymbol{\theta}^*)] \\
&\quad = -\sup_{\boldsymbol{\theta} \in \underline{O}} \left[\sum_{i=1}^{k} \sigma'(x_i)\,\theta_i - \psi(\boldsymbol{\theta}) \right]
\end{aligned}$$

for almost all $n(\cdot) \in Z$ (P_α-measure). But

$$I_\alpha(\sigma') = \sup_{V \in \underline{V}} \left[\sum_{x \in X} \sigma'(x)\, V(x) - \sum_{x \in X} \alpha(x)(1 - e^{-V(x)})\, u_V(x) \right]$$

$$= \sup_{\theta \in \underline{Q}} \left[\sum_{i=1}^{k} \sigma'(x_i)\, \theta_i - \psi(\boldsymbol{\theta}) \right] \tag{4.5}$$

because the first supremum is over V's vanishing outside any finite subset of X, whereas the second supremum is just over V's vanishing outside a fixed, finite set F. From (4.4) and (4.5) we have

$$\varliminf_{N \to \infty} \frac{1}{N} \log Q_{n,N}(N_{\sigma'}) \geqslant -I_\alpha(\sigma') \tag{4.6}$$

for almost all $n(\cdot) \in Z$ (P_α-measure).

Now, since F was any finite subset of X, and since $N_{\sigma'}$ is any R^k neighborhood of σ', we see that (4.6) implies (4.3) and hence the lower bound (1.3).

ACKNOWLEDGMENT

This work was supported by NSF grants NSF-MCS-83-01364 and NSF-DMS-86-00233.

REFERENCES

1. M. D. Donsker and S. R. S. Varadhan, Asymptotic evaluation of certain Wiener integrals for large time, in *Functional Integration and Its Applications* (Proceedings of the International Conference, London) (Clarendon Press, Oxford, 1975), pp. 15–33.
2. M. D. Donsker and S. R. S. Varadhan, Asymptotic evaluation of certain Wiener integrals for large time, I, *Commun. Pure Appl. Math.* **28**:1–47 (1975); II, *Commun. Pure Appl. Math.* **28**:279–301 (1975); III, *Commun. Pure Appl. Math.* **29**:389–461 (1976); IV, *Commun. Pure Appl. Math.* **36**:183–212 (1983).
3. M. D. Donsker and S. R. S. Varadhan, Large deviations for stationary Gaussian processes, *Commun. Math. Phys.* **97**:187–210 (1985).
4. M. D. Donsker and S. R. S. Varadhan, Asymptotics for the Wiener sausage, *Commun. Pure Appl. Math.* **28**:525–565 (1975).
5. M. D. Donsker and S. R. S. Varadhan, Asymptotics for the polaron, *Commun. Pure Appl. Math.* **36**:505–528 (1983).
6. M. D. Donsker and S. R. S. Varadhan, On laws of the iterated logarithm for local times, *Commun. Pure Appl. Math.* **30**:707–753 (1977).

7. S. R. S. Varadhan, Asymptotic probabilities and differential equations, *Commun. Pure Appl. Math.* **19**:261–286 (1966).
8. T. Y. Lee, Large deviation theory for empirical density of the noninteracting infinite particle systems, Thesis, Courant Institute, New York University (June 1986).
9. J. T. Cox and D. Griffeath, Large deviations for some infinite particle system occupation times, *Contemp. Math.* **41**:43–53 (1985).
10. R. S. Ellis, *Entropy, Large Deviations, and Statistical Mechanics* (Springer-Verlag, New York, 1985).

Comm. Pure Appl. Math. 44 (1991) 911-923

Some Familiar Examples for Which the Large Deviation Principle Does Not Hold

J. R. BAXTER AND N. C. JAIN
University of Minnesota

AND

S. R. S. VARADHAN
Courant Institute

This paper is dedicated to Natascha Brunswick.

Abstract

For a class of Markov processes (in continuous or discrete time) we show that if the full large deviation holds for normalized occupation time measures $L_t(\omega, \cdot)$ with some rate function J, then the lower semicontinuous regularization of J must agree with the rate function I introduced by M. D. Donsker and S. R. S. Varadhan. As a consequence we show that for processes such as Brownian motion the full large deviation principle for $L_t(\omega, \cdot)$ cannot hold with *any* rate function.

1. Introduction

Let S be a complete separable metric space. $\mathcal{M}(S)$ will denote the set of probability measures on the Borel subsets of S. The topology of weak convergence of probability measures in $\mathcal{M}(S)$ is metrizable by the Prokhorov metric, and with this metric $\mathcal{M}(S)$ is itself a complete separable metric space. By a ball in $\mathcal{M}(S)$ we will always mean a ball with respect to this metric. Let $\{\mu_t, t \in T\} \subset \mathcal{M}(S)$, where $T = [o, \infty)$ or $\{1, 2, \cdots\}$. This family is said to satisfy the *full large deviation principle* with rate function $J: S \to [o, \infty]$, if (1.1) and (1.2) hold:

(1.1) For any closed set $C \subset S$,

$$\limsup_{t\to\infty} \frac{1}{t} \log \mu_t(C) \leqq -\inf_{x\in C} J(x);$$

(1.2) for any open $G \subset S$,

$$\limsup_{t\to\infty} \frac{1}{t} \log \mu_t(G) \geqq -\inf_{x\in G} J(x).$$

We will say that $\{\mu_t, t \in T\}$ satisfies the *restricted large deviation principle* with rate function J if (1.1) is assumed to hold only for compact sets C. In what follows, X_t, $t \in T$, will denote a Markov process with state space E, a complete separable metric space, and with transition probability function $\pi(t, x, dy)$. If $T = [o, \infty)$, the normalized occupation time measure will be denoted by ($1_A(x) = 1$ if $x \in A$, o otherwise)

$$L_t(\omega, A) = \frac{1}{t}\int_o^t 1_A(X_s(\omega))\, ds\,, \qquad t > o, A \in \mathscr{E}\,, \tag{1.3}$$

where $\mathscr{E}$ denotes the σ-algebra of Borel subsets of E. The process will be assumed to be Feller, i.e., $\int f(y)\pi(t, x, dy)$ as a function of x belongs to $C_b(E)$, the space of bounded continuous functions on E, whenever $f \in C_b(E)$. Let L denote the infinitesimal generator (on the Banach space $C_b(E)$ with sup norm) of the process, and let $D(L)$ denote its domain. If $\mu \in \mathscr{M}(E)$, $I(\mu)$ will denote the functional introduced in [2]

$$I(\mu) = -\inf_u \int \left(\frac{Lu}{u}\right)(x)\, d\mu(x)\,, \tag{1.4}$$

where the inf is taken over $u \in D(L)$ such that $\inf u > o$.

If $T = \{1, 2, \cdots\}$, then the corresponding definitions are given by

$$L_n(\omega, A) = \frac{1}{n}\sum_{j=o}^{n-1} 1_A(X_j(\omega))\,, \qquad n \geqq 1, A \in \mathscr{E} \tag{1.5}$$

and

$$I(\mu) = -\inf_u \int \log\left(\frac{\pi u}{u}\right)(x)\, d\mu(x)\,, \tag{1.6}$$

where the inf is taken over all $u \in C_b(E)$ such that $\inf u > o$.

For $x \in E$, as usual $P^x(\Lambda)$ will denote the probability of the event Λ when the process starts at x. For a given $x \in E$, A a Borel subset of $\mathscr{M}(E)$, let

$$\mu_t(A) = P^x[L_t(\omega, \cdot) \in A]\,, \qquad t \in T\,. \tag{1.7}$$

In many familiar situations such as when X_t, $t \geqq o$, is a symmetric stable process with $E = \mathbb{R}^d$, or X_n, $n = 1, 2, \cdots$ is a simple random walk, with $E = \mathbb{Z}^d$, the results of Donsker and Varadhan (see [3]) show that for every $x \in E$, $\{\mu_t, t \in T\}$ defined by (1.7) satisfies the *restricted* large deviation principle with rate function I (given by (1.4) or (1.6), as the case may be). Since π is Feller in these cases, I is always lower semicontinuous (l.s.c.) but does not have compact level sets. A natural question arises whether a rate function $J : \mathscr{M}(E) \to [o, \infty]$ can be found such that $\{\mu_t, t \in T\}$ satisfies the full large deviation principle ((1.1) and (1.2)) with rate function J. In this note we show that the answer to this question is negative. To accomplish this, we first prove a general uniqueness theorem (Theorem 2.1) for the rate function in the next section. This theorem says that if the process is Feller, J is a lower semicontinuous rate function such that, for some x, μ_t defined by (1.7) satisfies the large deviation principle, and if (1.2) also holds with I, then $J = I$. If we now want to show that for a certain Feller process the full large deviation principle cannot hold for any rate function, we first show that for some closed set $C \subset \mathcal{M}(E)$ the upper bound (1.1) is not satisfied with rate function I. By the uniqueness theorem this implies that the full large deviation principle cannot hold for $\{\mu_t\}$ with any lower semicontinuous rate function J. Remark 2.4 is then used to show that the full large deviation principle cannot hold with any rate function.

The examples mentioned above are either transient or null recurrent Markov processes. We also give an example of a positive recurrent process (in fact, a Doeblin process—see [4], p. 192, for the definition of Hypothesis (D)) for which the full large deviation principle cannot hold with any rate function. All these applications of the uniqueness theorem are given in Section 3.

2. Uniqueness of the Rate Function

The main results of the section are Theorems 2.1 and 2.4. Theorem 2.4 is the discrete time version of Theorem 2.1. Since the proofs are essentially the same in the two cases, we shall prove only Theorem 2.1. The theorems imply that if a full large deviation principle is to hold with a lower semicontinuous rate function J, and the lower bound holds with rate function I defined by (1.4) or (1.6), then $J = I$.

THEOREM 2.1. *Suppose $\pi(t, x, dy)$ is a Feller transition probability function. Suppose $J : \mathcal{M}(E) \to [o, \infty]$ is lower semicontinuous and I is given by* (1.4). *Assume that for some $x \in E$,*

for every closed ball $F \subset \mathcal{M}(E)$

$$\limsup_{t\to\infty} \frac{1}{t} \log P^x\{L_t(\omega, \cdot) \in F\} \leqq -\inf_{\mu\in F} J(\mu), \tag{2.1}$$

and

for every open ball $G \subset \mathcal{M}(E)$

$$\liminf_{t\to\infty} \frac{1}{t} \log P^x\{L_t(\omega, \cdot) \in G\} \geqq -\inf_{\mu\in G} (J \wedge I)(\mu), \tag{2.2}$$

then $J = I$.

Remark 2.2. In the above theorem we do not assume J to have compact level sets.

Proof: We know from [2], page 40, that for any measurable $C \subset \mathcal{M}(E)$, any $x \in E$

$$\limsup_{t\to\infty} \frac{1}{t} \log P^x\{L_t(\omega, \cdot) \in C\} \leqq \inf_u \sup_{\mu \in C} \int \frac{Lu}{u}\, d\mu\,, \tag{2.3}$$

where the inf is over $u \in D(L)$ such that $\inf u > 0$. Fix $\mu_o \in \mathcal{M}(E)$ and let $a < I(\mu_o)$. Then for some $u_o \in D(L)$ with $\inf u_o > o$, we have $\int (Lu_o/u_o)\, d\mu_o < -a$. Since Lu/u is a bounded continuous function for such u, there exists a neighborhood N of μ_o in $\mathcal{M}(E)$ such that

$$\sup_{\mu \in N} \int (Lu_o/u_o)\, d\mu \leqq -a\,,$$

which implies

$$\inf_u \sup_{\mu \in N} \int (Lu/u)\, d\mu \leqq -a\,.$$

If $C = N$ in (2.3), we conclude the following: given $\mu \in \mathcal{M}(E)$, given $a < I(\mu)$, there exists a neighborhood N of μ such that

$$\limsup_{t\to\infty} \frac{1}{t} \log P^x\{L_t(w, \cdot) \in N\} \leqq -a\,. \tag{2.4}$$

Let $\mu \in \mathcal{M}(E)$ and let $a < I(\mu)$. Then (2.4) and (2.2) immediately imply that $J(\mu) \geqq a$, showing that $I \leqq J$. Since J is assumed to be lower semicontinuous, (2.1) implies that given $\mu \in \mathcal{M}(E)$ and $a < J(\mu)$, there exists a neighborhood N of μ for which (2.4) holds. Since (2.2) is symmetric in I and J, the above argument also shows that $J \leqq I$. This proves the theorem.

The discrete time analogue of Theorem 2.1 is the following theorem.

THEOREM 2.3. *Suppose $\pi(1, x, dy)$ is a Feller transition probability function. Suppose $J : \mathcal{M}(E) \to [o, \infty]$ is lower semicontinuous and I is given by (1.6). Assume that for some $x \in E$,*

for every closed ball $F \subset \mathcal{M}(E)$

$$\limsup_{n\to\infty} \frac{1}{n} \log P^x\{L_n(\omega, \cdot) \in F\} \leqq -\inf_{\mu \in F} J(\mu)\,, \tag{2.5}$$

and

(2.6)
$$\text{for every open ball} \quad G \subset \mathcal{M}(E)$$
$$\liminf_{n\to\infty} \frac{1}{n} \log P^x\{L_n(\omega, \cdot) \in G\} \geqq -\inf_{\mu\in G}(J \wedge I)(\mu),$$

then $J = I$; *here* $a \wedge b = \min(a, b)$.

Remark 2.4. In Theorems 2.1 and 2.4 if J is not assumed to be lower semicontinuous, then we can define $J_1(\mu) = \lim_{\varepsilon\to 0}\inf_{\nu\in B_\varepsilon} J(\nu)$, where B_ε is a ball of radius ε with center μ. Then J_1 is lower semicontinuous, $J_1 \leqq J$ and satisfies the conditions of the theorems. Thus we then have $J_1 = I$.

3. Applications

We shall show that empiricals of symmetric stable processes cannot satisfy the full large deviation principle with any rate function. In discrete time we shall show the same to hold for simple random walk. When the space dimension $d = 1$, these are examples of "null recurrent" processes. We shall also give an example of a positive recurrent Markov chain for which the same conclusion holds. The basic strategy is to show that for μ_t, defined by (1.7), the upper bound in (1.1) does not hold for all closed sets with rate function I given by (1.4) (or (1.6)). Theorem 2.1 (or Theorem 2.3) will then show that the full large deviation principle cannot hold with any rate function.

We shall need the following technical result.

PROPOSITION 3.1. *Suppose* $\{\mu_n\}$ *is a sequence of probability measures on a complete separable metric space* S *such that for some* $\delta > o$, *all compact* $K \subset S$, $\limsup_n \mu_n(K) \leqq 1 - 3\delta$. *If* C *is a closed set such that* $\limsup_n \mu_n(C) \leqq \delta$, *then there exists a closed set* $C_1 \subset C^c$ *such that* $\limsup_n \mu_n(C_1) \geqq \delta$.

Before proving the proposition we observe that S is homeomorphic to a G_δ subset of the Hilbert cube. Therefore without any loss of generality S may be given an equivalent metric which makes it totally bounded, and if S^* = completion of S under this metric then S is a Borel subset of S^*. If ν is a measure on S and A is a Borel subset of S^*, then $\nu(A) := \nu(A \cap S)$ extends ν to S^*. Since S^* is compact, every subsequence of $\{\mu_n\}$ on S^* has a weakly convergent subsequence. We now prove a lemma.

LEMMA 3.2. *Let* S, S^*, *and* $\{\mu_n\}$ *be as above. If* $\nu_k = \mu_{n_k} \to \mu$ *weakly on* S^*, *then* $\mu(S) \leqq 1 - 3\delta$.

Proof: Let $K \subset S$ be a compact set. It is enough to show that $\mu(K) \leqq 1 - 3\delta$. K is also a compact subset of S^*. There exist open $U_j \subset S^*$, $j \geqq 1$, such that $U_j \searrow K$. Since $\nu_k \to \mu$ weakly, we have

$$\mu(K) \leqq \mu(U_j) \leqq \liminf_k \nu_k(U_j), \qquad j \geqq 1. \tag{3.1}$$

Therefore, given $\varepsilon > o$ there exist $k(j) \nearrow \infty$ and compact $K_j \subset U_j \cap S$ such that

$$\nu_{k(j)}(K_j) \geqq \mu(K) - \varepsilon, \qquad j \geqq 1. \tag{3.2}$$

Let $B = \bigcup_{j=1}^{\infty} K_j \cup K$. If U is an open subset of S^* which contains the compact set K, then $U \supset K_j$ for all j sufficiently large. Therefore B is a compact subset of S and by (3.2) we then have $\mu(K) \leqq \limsup_n \mu_n(B) + \varepsilon \leqq 1 - 3\delta + \varepsilon$ for all $\varepsilon > o$. Therefore $\mu(K) \leqq 1 - 3\delta$, and the lemma is proved.

Proof of Proposition 3.1: Let $\nu_k \to \mu$ weakly in S^* as above. By Lemma 3.2 we have

$$\mu(S) \leqq 1 - 3\delta. \tag{3.3}$$

Therefore there exists a compact $H \subset S^* - S$ such that $\mu(H) > 2\delta$. Let

$$h(x) = \frac{1}{\rho(x, H)}, \tag{3.4}$$

where ρ denotes the equivalent metric chosen above and $\rho(x, H) = \inf\{\rho(x, y) : y \in H\}$. For any $\alpha > o$, let $V_\alpha = \{x : h(x) < \alpha\} \cap S$. These are open subsets of S and $V_\alpha \nearrow S$ as $\alpha \nearrow \infty$. By weak convergence we have

$$\begin{aligned}\limsup_k \nu_k(V_\alpha) &\leqq \limsup_k \nu_k\{x \in S^* : h(x) \leqq \alpha\} \\ &\leqq \mu\{x \in S^* : h(x) \leqq \alpha\} \leqq \mu(H^c) < 1 - 2\delta.\end{aligned}$$

There exists $k(1)$ such that $\nu_{k(1)}(C \cup V_1) < 1 - \delta$, so we can find compact $K_1 \subset S - (C \cup V_1)$ such that $\nu_{k(1)}(K_1) > \delta$. There exists $m(1)$ such that $K_1 \subset V_{m(1)}$. Then we can find $k(2)$ such that $\nu_{k(2)}(C \cup V_{m(1)+1}) < 1 - 2\delta$, and then a compact $K_2 \subset S - (C \cup V_{m(1)+1})$, $K_2 \subset V_{m(2)}$, such that $\nu_{k(2)}(K_2) > \delta$. We proceed in this manner, and find $m(o) = 1$, $m(i+1) > m(i) + 1$, $k(i+1) > k(i)$, and compacts $K_{i+1} \subset S - (C \cup V_{m(i)+1})$, $K_i \subset V_{m(i)}$, and $\nu_{k(i)}(K_i) > \delta$ for $i \geqq 1$. Let $C_1 = \bigcup_{i=1}^{\infty} K_i$. Then $C_1 \subset S - C$, $\nu_{k(i)}(C_1) \geqq \delta$ for all i. It remains to check that C_1 is a closed subset of S. Let $\{x_n\} \subset C_1$, $x_n \to x \in S$ as $n \to \infty$. If infinitely many x_n's belong to some K_i, then $x \in K_i \subset C_1$, so we might as well assume that $x_n \in K_n$, $n \geqq 1$. Since h is a real-valued continuous function on S, we have $h(x_n) \to h(x)$.

On the other hand, $h(x_n) \in K_n \subset V_{m(n)} - V_{m(n-1)+1}$, i.e., $|h(x_n) - h(x_{n+1})| \geqq 1$, which is a contradiction and the assumption cannot be valid. This proves the proposition.

We now come to the question of a symmetric stable process. A process X_t, $t \geqq o$, with stationary, independent increments, taking values in $\mathbb{R}^d$, is symmetric stable if

$$E^o\{e^{i\langle X_t, u\rangle}\} = e^{-t|u|^\alpha}, \qquad u \in \mathbb{R}^d,$$

where $\langle \cdot, \cdot \rangle$ denotes the Euclidean inner product and $|\cdot|$ the corresponding norm in $\mathbb{R}^d$ and $o < \alpha \leqq 2$. The case $\alpha = 2$ corresponds to Brownian motion with variance parameter equal to 2.

THEOREM 3.3. *Let X_t, $t \geqq o$, be a symmetric stable process. Then there is no rate function $J : \mathcal{M}(\mathbb{R}^d) \to [o, \infty]$ such that the full large deviation principle holds for μ_t, defined by (1.7), with rate function J.*

Proof: Suppose the full large deviation principle holds with rate function J. If J_1 is defined in terms of J as in Remark 2.5, then J_1 is lower semicontinuous and satisfies the full large deviation principle. By the results in [3], the lower bound result holds for a symmetric stable process with rate function I. Since the transition function of the process satisfies the Feller property, it follows from Theorem 2.1 that $I = J_1$, i.e., I must satisfy the large deviation principle. The theorem will therefore be proved if we show that there exists a closed set $C \subset \mathcal{M}(\mathbb{R}^d)$ for which the upper bound cannot hold with rate function I. Suppose we can show that for any x

$$P^x\{\omega : I(L_1(\omega, \cdot)) = \infty\} = 1, \tag{3.5}$$

then by spatial homogeneity and the scaling property of the symmetric stable process we have $I(L_t(\omega, \cdot)) = \infty$, P^x-a.s. The set $C = \{\mu : I(\mu) \leqq 1\}$ is closed since I is lower semicontinuous, and we have $P^x\{L_t(\omega, \cdot) \in C\} = o$ for all $t > o$. We shall show that for any compact $K \subset \mathcal{M}(\mathbb{R}^d)$, any x

$$\lim_{t \to \infty} P^x\{L_t(\omega, \cdot) \in K\} = o. \tag{3.6}$$

Let $t_n \nearrow \infty$, and let $\mu_n(A) = P^x\{L_{t_n}(\omega, \cdot) \in A\}$, for $A \subset \mathcal{M}(\mathbb{R}^d)$. With $S = \mathcal{M}(\mathbb{R}^d)$, the sequence $\{\mu_n\}$ satisfies the conditions of Proposition 3.1 with $\delta = \frac{1}{3}$ and we conclude that there exists a closed set $C_1 \subset C^c$ such that $\limsup_n \mu_n(C_1) \geqq \frac{1}{3}$. It follows that

$$\limsup_{t \to \infty} \frac{1}{t} \log P^x\{L_t(\omega, \cdot) \in C_1\} \geqq o. \tag{3.7}$$

On the other hand, since $C_1 \subset C^c = \{\mu : I(\mu) > 1\}$, if the full large deviation principle were true with rate function I, we would have the lim sup in (3.7) less than or equal to -1, which is a contradiction. Therefore the full large deviation principle cannot hold with rate function I and the theorem is proved if (3.5) and (3.6) can be established.

The next lemma shows that (3.6) holds. The proof of (3.5) will be accomplished through Propositions 3.5, 3.7, and Theorem 3.8.

LEMMA 3.4. *Let $K \subset \mathcal{M}(\mathbb{R}^d)$ be compact. Then*

$$\lim_{t\to\infty} P^x\{L_t(\omega, \cdot) \in K\} = o, \qquad x \in \mathbb{R}^d.$$

Proof: Note that $L_t(\omega, F) \to o$ as $t \to \infty$, P^x-a.s., for any compact $F \subset \mathbb{R}^d$ and for any $x \in \mathbb{R}^d$ since the process is either transient or null-recurrent. Assume that there exist x, $K \subset \mathcal{M}(\mathbb{R}^d)$ compact, $\eta > o$, and $t_n \nearrow \infty$ such that

$$P^x\{L_{t_n}(\omega, \cdot) \in K\} \geqq \eta, \qquad n \geqq 1.$$

Then $P^x(\Lambda) \geqq \eta$, where $\Lambda = \{L_{t_n}(\omega, \cdot) \in K \text{ i.o.}\}$. Therefore, if $\omega \in \Lambda$, there exists a subsequence $\{t'_n\}$ such that $L_{t'_n}(\omega, \cdot) \in K$, i.e., $\{L_{t'_n}(\omega, \cdot), n \geqq 1\}$ is a tight family, which is a contradiction, so the lemma is proved.

We now proceed to prove (3.5). We first observe that if $\alpha \leqq 1$ or if $d \geqq 2$, then P^x-a.s., $L_t(\omega, \cdot)$ is singular with respect to Lebesgue measure in $\mathbb{R}^d$. Since the Lebesgue measure plays the role of a reference measure in [3], $I(L_t(\omega, \cdot)) < \infty \Rightarrow L_t(\omega, \cdot) \ll m$, where m denotes Lebesgue measure. Therefore without any loss of generality we may assume $d = 1$ and $1 < \alpha \leqq 2$. In this situation it is known (see [1]) that $(dL_t(\omega, \cdot)/dm)(x) = l_t(\omega, x)$ exists and has a jointly continuous version in (t, x).

The next proposition is essentially Theorem 1.3 in [5].

PROPOSITION 3.5. *Let Y_t, $t \geqq o$, be a Markov process, reversible with respect to invariant probability measure π. Assume that P_π is ergodic. Let $g \in L^2(\pi)$ be such that $\int g\, d\pi = o$. Assume that there exists C such that*

$$\left|\int fg\, d\pi\right| \leqq C\left(-\int fLf\, d\pi\right)^{1/2}, \qquad f \in D(L),$$

where $D(L)$ denotes the domain of the generator of the process. Then $t^{-1/2} \int_o^t g(Y_s)\, ds$ converges weakly to $N(o, \sigma^2)$, for almost every x with respect to π, where $\sigma^2 = 2(-L^{-1}g, g)$. In particular, if $Lg = -\lambda g$ for some $\lambda > o$, then $\sigma^2 = (2/\lambda)\|g\|^2$. Here $(\cdot, \cdot)$ and $\|\cdot\|$ denote the inner product and the norm, respectively, in $L^2(\pi)$.

Remark 3.6. In the context of Proposition 3.5, $Lf = o \Rightarrow f =$ constant, and since $(1, g) = o$, the expression for σ^2 is well defined.

We now return to the symmetric stable process X_t. Let n be a positive integer, and let $\theta_1, \theta_2, \cdots, \theta_n$ be real numbers linearly independent over the rationals. We define the process

$$Z_n(t) = (e^{i\theta_1 X_t}, \cdots, e^{i\theta_n X_t}) \tag{3.8}$$

which is a process with stationary independent increments with state space T^n, the n-dimensional torus. The normalized Haar measure, denoted by μ, is clearly the invariant measure for its transition probability function. If $k_1, \cdots, k_n$ are integers, let $\Psi_{k_1,\cdots,k_n} : T^n \to T^1$ denote the group character

$$\Psi_{k_1,\cdots,k_n}(e^{ix_1}, \cdots, e^{ix_n}) = e^{i(k_1x_1 + \cdots + k_nx_n)} . \tag{3.9}$$

If $L_{\theta_1,\cdots,\theta_n}$ denotes the infinitesimal generator of the Markov process $Z_n(t)$, $t \geqq o$, then

$$L_{\theta_1,\cdots,\theta_n}\Psi_{k_1,\cdots,k_n} = -|\theta_1 k_1 + \cdots + \theta_n k_n|^{\alpha}\Psi_{k_1,\cdots,k_n} . \tag{3.10}$$

Therefore $\Psi_{k_1,\cdots,k_n}$ are eigenfunctions for $L_{\theta_1,\cdots,\theta_n}$; $\{\Psi_{k_1,\cdots,k_n} : k_1, \cdots k_n \text{ integers}\}$ are a complete orthonormal system in $L^2(\mu)$. It follows from (3.10) that if $f(e^{ix_1}, \cdots, e^{ix_n}) = \cos x_j$, then

$$(L_{\theta_1,\cdots,\theta_n}f)(e^{ix_1}, \cdots, e^{ix_n}) = -|\theta_j|^{\alpha}\cos x_j , \qquad j = 1, \cdots, n . \tag{3.11}$$

Since $\theta_1, \cdots, \theta_n$ are chosen to be linearly independent over the rationals, (3.10) immediately implies that if $L_{\theta_1,\cdots,\theta_n}f = o$ then $f =$ constant. This implies that $Z_n(t)$ with initial distribution μ is a stationary ergodic process. Let $c_1, c_2, \cdots, c_n$ be real, and define

$$Y_n(t) = c_1\cos(\theta_1 X_t) + \cdots + c_n\cos(\theta_n X_t) . \tag{3.12}$$

We now prove

PROPOSITION 3.7. *Let $Y_n(t)$, $t \geqq o$, be defined by (3.12). Then $t^{-1/2} \times \int_o^t Y_n(s)\, ds$ converges weakly to $N(o, \sigma^2)$, as $t \to \infty$, for almost every $x(\mu)$, where $\sigma^2 = c_1^2|\theta_1|^{-\alpha} + \cdots + c_n^2|\theta_n|^{-\alpha}$. In particular, $t^{-1/2}\int_o^t (\cos(\theta_1 X_s), \cdots, \cos(\theta_n X_s))\, ds$ converges weakly to $N(o, \Sigma)$, as $t \to \infty$, where Σ is a diagonal matrix with the i-th diagonal element equal to $|\theta_i|^{-\alpha}$.*

Proof: We simply apply Proposition 3.5 to the process $Z_n(t)$ with $g(e^{ix_1}, \cdots, e^{ix_n}) = c_1\cos x_1 + \cdots + c_n\cos x_n$, and $\pi = \mu$. Then $g \in \text{Range}((-L)^{1/2})$, so the conditions of the proposition are satisfied. Since $2(-L^{-1}g, g) = c_1^2|\theta_1|^{-\alpha} + \cdots + c_n^2|\theta_n|^{-\alpha}$, the proposition is proved.

THEOREM 3.8. *For a symmetric stable process X_t, $t \geqq o$, we have $I(L_1(\omega, \cdot)) = \infty$, P^x-a.s., for all x.*

Proof: As we observed earlier, it suffices to consider $1 < \alpha \leqq 2$ and $d = 1$. Also, if we know the result for some x, P^x-a.s., then, I being invariant under translations of measures in the present context, we have the result for all x. Let

$$u(x) = \exp\left\{\frac{1}{\lambda^{\alpha/2}} \sum_{j=1}^{n} u_j \cos(\lambda\theta_j x)\right\}, \qquad x \in \mathbb{R}, \tag{3.13}$$

where $\lambda > o$, $u_1, u_2, \cdots, u_n$ are rational. If L denotes the generator of the stable process, then $u \in D(L)$. Let $v(x) = u(x/\lambda)$, then by the scaling property $(Lv)(\lambda x) = \lambda^{-\alpha}(Lu)(x)$. This together with the scaling property gives

$$\int_o^1 \frac{Lu}{u}(X_s)\,ds \stackrel{d}{=} \int_o^{\lambda^\alpha} \frac{Lv}{v}(X_s)\,ds, \tag{3.14}$$

where $Z_1 \stackrel{d}{=} Z_2$ means the random variables Z_1 and Z_2 have the same distribution. Let $h(x) = \sum_{j=1}^n u_j\cos(\theta_j x)$. If A is an upper bound for $|h|$, $|h'|$, and $|h''|$, then it is easy to verify that there exists $C > o$ such that $\|Lh^k\|_\infty \leqq CA^k k(k-1)$ for all positive integers k. Therefore

$$Lv = \lambda^{-\alpha/2}Lh + \tfrac{1}{2}\lambda^{-\alpha}Lh^2 + o(\lambda^{-\alpha})$$

where o does not depend on x. Therefore

$$\frac{Lv}{v} = \lambda^{-\alpha/2}Lh + \frac{1}{2}\lambda^{-\alpha}Lh^2 - \lambda^{-\alpha}hLh + o(\lambda^{-\alpha}), \tag{3.15}$$

and we conclude that

$$\lim_{\lambda\to\infty} \int_o^1 \frac{Lu}{u}(X_s)\,ds \stackrel{d}{=} \lim_{\lambda\to\infty} \int_o^{\lambda^\alpha} \left\{\lambda^{-\alpha/2}Lh + \frac{1}{2}\lambda^{-\alpha}Lh^2 - \lambda^{-\alpha}hLh\right\}(X_s)\,ds, \tag{3.16}$$

where these limits are to be understood in the sense of weak convergence. We will now identify the limit on the right side in (3.16). Since $(L\cos(\theta\cdot))(y) = -|\theta|^\alpha\cos(\theta y)$, by Proposition 3.7 we get

$$\frac{1}{\lambda^{\alpha/2}} \int_o^{\lambda^\alpha} Lh(X_s)\, ds \to N(o, \sigma^2) \tag{3.17}$$

weakly as $\lambda \to \infty$, where $\sigma^2 = \sum_{j=1}^n u_j^2 |\theta_j|^\alpha$. Also

$$h^2 = \frac{1}{2} \sum_{j,k} u_j u_k [\cos(\theta_j + \theta_k)x + \cos(\theta_j - \theta_k)x]$$

and

$$hLh = -\frac{1}{2} \sum_{j,k} u_j u_k |\theta_k|^\alpha [\cos(\theta_j + \theta_k)x + \cos(\theta_j - \theta_k)x] ,$$

hence by (3.17) only the constants in the diagonal terms of hLh contribute to the limit in (3.16) and we get

$$\lim_{\lambda\to\infty} \int_o^1 \frac{-Lu}{u}(X_s)\, ds \stackrel{d}{=} Z - \frac{1}{2} \sum_{j=1}^n u_j^2 |\theta_j|^\alpha , \tag{3.18}$$

where Z is a $N(o, \sum_{j=1}^n u_j^2 |\theta_j|^\alpha)$ random variable.

To finish the argument, let $\theta = \{\theta_j\}$ be a sequence of real numbers such that for any n, $\theta_1, \theta_2, \cdots, \theta_n$ are linearly independent over rationals, and let $\{\xi_j\}$ be a sequence of independent $N(o, |\theta_j|^{-\alpha})$ random variables defined on some probability space. Consider the collection of random variables

$$\zeta_{u_1,u_2,\cdots,u_n;\theta_1,\theta_2,\cdots,\theta_n} := \sum_{j=1}^n u_j |\theta_j|^\alpha \xi_j - \frac{1}{2} \sum_{j=1}^n u_j^2 |\theta_j|^\alpha ,$$

where $u_1, u_2, \cdots, u_n$, $n \geqq 1$, are rational. Let

$$\eta = \sup_{\substack{\lambda > o\\ u_1,u_2,\cdots,u_n}} \int_o^1 \frac{-Lu}{u}(X_s)\, ds ,$$

where u is given by (3.13) and the sup is taken over rational $u_1, \cdots, u_n$, and rational λ, then for any $a > o$, $n \geqq 1$,

$$P^o[\eta \leqq a] \leqq P^o[\sup_{\substack{u_1,u_2,\cdots,u_n\\ \text{rational}}} \zeta_{u_1,u_2,\cdots,u_n;\theta_1,\theta_2,\cdots,\theta_n} \leqq a] .$$

Since $\eta(\omega) \leqq I(L_1(\omega, \cdot))$, and $\sup \zeta = \frac{1}{2} \sum_{j=1}^n |\theta_j|^\alpha \xi_j^2$, we have

$$P^o[I(L_1(\omega, \cdot)) \leqq a] \leqq P^o\left[\frac{1}{2} \sum_{j=1}^n |\theta_j|^\alpha \xi_j^2 \leqq a\right],$$

the latter probability clearly tends to zero as $n \to \infty$. Therefore $P^o[I(L_1(\omega, \cdot)) \leqq a] = o$ for every $a > o$, and the theorem is proved.

We now consider the simple random walk example. Let $X_1, X_2, \cdots$ be a simple random walk on the integer lattice $\mathbb{Z}$, i.e., it is a Markov chain for which X_n is the n-th partial sum of a sequence of independent, identically distributed random variables $\eta_1, \eta_2, \cdots$, and $P[\eta_1 = 1] = P[\eta_1 = -1] = \frac{1}{2}$. Let $\pi(x, dy)$ denote the one-step transition probability function of the chain. If $\mu \in \mathcal{M}(\mathbb{Z})$, then according to (1.6) we have

$$I(\mu) = \sup_{u>o} \sum_{j=-\infty}^{\infty} \left(\log \frac{2u(j)}{u(j-1) + u(j+1)} \right) \mu(\{j\}) .$$

Let

$$C = \left\{ \mu : \mu(\{j\}) = \frac{1}{2m+1}, \quad j = o, 1, \cdots, 2m, \quad \text{for some } m \geqq 1 \right\} .$$

C is a closed subset of $\mathcal{M}(\mathbb{Z})$. If $\mu \in C$, and we take $u(2j+1) = 1$, $u(2j) = k \geqq 2$, for all j, then an elementary computation shows that $I(\mu) \geqq (2m+1)^{-1} \log k$, where μ is distributed over $o, 1, \cdots, 2m$. Therefore $I(\mu) = \infty$ whenever $\mu \in C$. For $j \geqq 1$, $n = 2r + 1$, $r \geqq 1$, we have

$$P^j\{L_n(\omega, \cdot) \in C\} \geqq P^j\{X_o = j, X_1 = j+1, \cdots, X_{n-1} = j + n - 1\} = 2^{-n} .$$

Therefore

$$-\log 2 \leqq \limsup_{n \to \infty} \frac{1}{n} \log P^j\{L_n(\omega, \cdot) \in C\} .$$

Since $\inf\{I(\mu) : \mu \in C\} = \infty$, the upper bound in the large deviation principle cannot hold for C with rate function I. By Theorem 2.4 and Remark 2.5 we conclude as in the proof of Theorem 3.3 that the full large deviation principle for the simple random walk cannot hold with any rate function.

The next example is one of a positive recurrent chain (Doeblin chain). Let the state space $E = \{o, 1, 2, \cdots\}$, and with $o < \rho < 1$ let the transition probability function be given by $\pi(x, x+1) = \rho$, $\pi(x, o) = 1 - \rho$ for all $x \in E$. Then for $\mu \in \mathcal{M}(E)$,

$$I(\mu) = \sup_{u>o} \sum_{j} \left(\log \frac{u(j)}{(1-\rho)u(o) + \rho u(j+1)} \right) \mu(\{j\}) .$$

Let

$$C = \left\{\mu : \mu(\{j\}) = \frac{1}{m}, j = l+1, \cdots, l+m, \quad \text{some } m \geqq 1, \quad \text{and } l \geqq 1\right\}.$$

If we pick u such that $u(j) = 1$ for $j \neq l+1$ and $u(l+1) = k$, then $I(\mu) \geqq m^{-1}\log k$, where $\mu(\{j\}) = m^{-1}$ for $j = l+1, \cdots, l+m$; so $I(\mu) = \infty$ whenever $\mu \in C$. For $j \geqq 1$,

$$P^j\{L_n(\omega, \cdot) \in C\} \geqq P^j\{X_0 = j, X_1 = j+1, \cdots, X_{n-1} = j+n-1\}$$
$$\geqq \rho^{n-1}.$$

We then conclude, as in the simple random walk example, that the full large deviation principle for this chain cannot hold with any rate function.

Acknowledgment. This work was partially supported by National Science Foundation grants.

Bibliography

[1] Boylan, E. S., *Local times for a class of Markov processes*, Ill. Jour. Math. 8, 1969, pp. 19–39.
[2] Donsker, M. D., and Varadhan, S. R. S., *Asymptotic evaluation of certain Markov process expectations for large time I*, Comm. Pure Appl. Math. 28, 1975, pp. 1–47.
[3] Donsker, M. D., and Varadhan, S. R. S., *Asymptotic evaluation of certain Markov process expectations for large time III*, Comm. Pure Appl. Math. 29, 1976, pp. 389–461.
[4] Doob, J. L., *Stochastic Processes*, John Wiley & Sons, New York, 1953.
[5] Kipnis, C., and Varadhan, S. R. S., *Central limit theorem for additive functionals of reversible Markov processes and applications to simple exclusions*, Comm. Math. Phy. 104, 1986, pp. 1–19.

Received March 1990.

European Journal of Combinatorics 32 (2011) 1000–1017

Contents lists available at ScienceDirect

European Journal of Combinatorics

journal homepage: www.elsevier.com/locate/ejc

The large deviation principle for the Erdős-Rényi random graph

Sourav Chatterjee, S.R.S. Varadhan

Courant Institute of Mathematical Sciences, New York University, 251 Mercer Street, New York, NY 10012, United States

ARTICLE INFO

Article history:
Available online 6 May 2011

ABSTRACT

What does an Erdős-Rényi graph look like when a rare event happens? This paper answers this question when p is fixed and n tends to infinity by establishing a large deviation principle under an appropriate topology. The formulation and proof of the main result uses the recent development of the theory of graph limits by Lovász and coauthors and Szemerédi's regularity lemma from graph theory. As a basic application of the general principle, we work out large deviations for the number of triangles in $G(n, p)$. Surprisingly, even this simple example yields an interesting double phase transition.

1. Introduction

1.1. The Erdős–Rényi graph

Let $G(n, p)$ be the random graph on n vertices where each edge is added independently with probability p. This model has been the subject of extensive investigations since the pioneering work of [14], yielding a large body of literature (see [2,18] for partial surveys).

This paper studies the following basic aspect of Erdős–Rényi graphs: What does the graph look like if one knows that some rare event has happened? One way to comprehensively answer this question is to formulate a large deviation principle for the Erdős–Rényi graph, in the same way as Sanov's theorem [27] gives a large deviation principle for an i.i.d. sample.

The setting of Sanov's theorem conforms naturally to the abstract theory of large deviations (see Chapter 6 in [11]) because i.i.d. samples can be thought of as random probability measures, allowing them to be viewed as random elements of a single topological space irrespective of the sample size. The first hurdle in formulating such a program for random graphs is in constructing a single abstract space

E-mail address: sourav@cims.nyu.edu (S. Chatterjee).

in which all graphs can be embedded. Fortunately, this issue has been settled recently. In a sequence of papers [5–7,15,19–25] Laszlo Lovász with coauthors (listed here in order of frequency) Sós, Szegedy, Borgs, Chayes, Vesztergombi, Schrijver and Freedman have developed a beautiful, unifying limit theory (see also the related work of [12] which traces this back to work of [1,17].) This sheds light on topics such as graph homomorphisms, Szemerédi's regularity lemma, quasi-random graphs, graph testing and extremal graph theory, and has even found applications in statistics and related areas (see e.g. [10]). Their theory has been developed for dense graphs (number of edges comparable with the square of number of vertices) but parallel theories for sparse graphs are beginning to emerge [3].

1.2. Graph limits and graphons

The limit of a sequence of dense graphs can be defined as follows. We quote the definition verbatim from [22] (see also [6,7,12]). Let G_n be a sequence of simple graphs whose number of nodes tends to infinity. For every fixed simple graph H, let $|\hom(H,G)|$ denote the number of homomorphisms of H into G (i.e. edge-preserving maps $V(H) \to V(G)$, where $V(H)$ and $V(G)$ are the vertex sets). This number is normalized to get the homomorphism density

$$t(H,G) := \frac{|\hom(H,G)|}{|V(G)|^{|V(H)|}}. \tag{1}$$

This gives the probability that a random mapping $V(H) \to V(G)$ is a homomorphism.

Suppose that the graphs G_n become more and more similar in the sense that $t(H,G_n)$ tends to a limit $t(H)$ for every H. One way to define a limit of the sequence $\{G_n\}$ is to define an appropriate limit object from which the values $t(H)$ can be read off.

The main result of [22] (following the earlier equivalent work of [1,17]) is that indeed there is a natural "limit object" in the form of a function $f \in \mathcal{W}$, where $\mathcal{W}$ is the space of all measurable functions from $[0,1]^2$ into $[0,1]$ that satisfy $f(x,y) = f(y,x)$ for all x, y.

Conversely, every such function arises as the limit of an appropriate graph sequence. This limit object determines all the limits of subgraph densities: if H is a simple graph with $V(H) = [k] = \{1,\ldots,k\}$, then

$$t(H,f) = \int_{[0,1]^k} \prod_{(i,j)\in E(H)} f(x_i,x_j)\,\mathrm{d}x_1\cdots \mathrm{d}x_k. \tag{2}$$

Here $E(H)$ denotes the edge set of H. A sequence of graphs $\{G_n\}_{n\geq 1}$ is said to converge to f if for every finite simple graph H,

$$\lim_{n\to\infty} t(H,G_n) = t(H,f). \tag{3}$$

Intuitively, the interval $[0,1]$ represents a 'continuum' of vertices, and $f(x,y)$ denotes the probability of putting an edge between x and y. For example, for the Erdős–Rényi graph $G(n,p)$, if p is fixed and $n\to\infty$, then the limit graph is represented by the function that is identically equal to p on $[0,1]^2$.

These limit objects, i.e. elements of $\mathcal{W}$, are called 'graphons' in [22,6,7]. A finite simple graph G on $\{1,\ldots,n\}$ can also be represented as a graphon f^G is a natural way, by defining

$$f^G(x,y) = \begin{cases} 1 & \text{if}(\lceil nx\rceil, \lceil ny\rceil)\text{is an edge in } G,\\ 0 & \text{otherwise.}\end{cases} \tag{4}$$

Note that this allows *all* simple graphs, irrespective of the number of vertices, to be represented as elements of a single abstract space, namely $\mathcal{W}$.

1.3. The cut metric

With the above representation, it turns out that the notion of convergence in terms of subgraph densities outlined above can be captured by an explicit metric on $\mathcal{W}$, the so-called 'cut distance' (originally defined for finite graphs by Frieze and Kannan [16]). We start with the space $\mathcal{W}$ of measurable functions $f(x, y)$ on $[0, 1]^2$ that satisfy $0 \le f(x, y) \le 1$ and $f(x, y) = f(y, x)$. We define the cut distance

$$d_\square(f, g) := \sup_{S,T \subseteq [0,1]} \left| \int_{S\times T} [f(x, y) - g(x, y)]\mathrm{d}x\mathrm{d}y \right|. \tag{5}$$

We introduce in $\mathcal{W}$ an equivalence relation. Let Σ be the space of measure preserving bijections $\sigma : [0, 1] \to [0, 1]$. Say that $f(x, y) \sim g(x, y)$ if $f(x, y) = g_\sigma(x, y) := g(\sigma x, \sigma y)$ for some $\sigma \in \Sigma$. Denote by $\widetilde{g}$ the closure in $(\mathcal{W}, d_\square)$ of the orbit $\{g_\sigma\}$. The quotient space is denoted by $\widetilde{\mathcal{W}}$ and τ denotes the natural map $g \to \widetilde{g}$. Since $d_\square$ is invariant under σ one can define on $\widetilde{\mathcal{W}}$, the natural distance $\delta_\square$ by

$$\delta_\square(\widetilde{f}, \widetilde{g}) := \inf_\sigma d_\square(f, g_\sigma) = \inf_\sigma d_\square(f_\sigma, g) = \inf_{\sigma_1,\sigma_2} d_\square(f_{\sigma_1}, g_{\sigma_2})$$

making $(\widetilde{\mathcal{W}}, \delta_\square)$ into a metric space. To any finite graph G, we associate f^G as in (4) and its orbit $\widetilde{G} = \tau f^G = \widetilde{f^G} \in \widetilde{\mathcal{W}}$. One of the key results of [6] is the following:

Theorem 1.1 (*Theorem 3.8 in [6]*). *A sequence of graphs $\{G_n\}_{n\ge1}$ converges to a limit $f \in \mathcal{W}$ in the sense defined in (3) if and only if $\delta_\square(\widetilde{G}_n, \widetilde{f}) \to 0$ as $n \to \infty$.*

Szemerédi's regularity lemma and the related deep results of Lovász and Szegedy will play a crucial role in this paper:

1.4. Szemerédi's lemma

Let $G = (V, E)$ be a simple graph, and let X, Y be subsets of V. Then we denote by $e_G(X, Y)$ the number of X-Y edges of G (edges whose endpoints belong to $X \cap Y$ are counted twice), and call

$$\rho_G(X, Y) := \frac{e_G(X, Y)}{|X||Y|}$$

the *density* of the pair (X, Y). Given some $\epsilon > 0$, we call a pair (A, B) of disjoint sets $A, B \subseteq V$ ϵ-regular if all $X \subseteq A$ and $Y \subseteq B$ with $|X| \ge \epsilon|A|$ and $|Y| \ge \epsilon|B|$ satisfy

$$|\rho_G(X, Y) - \rho_G(A, B)| \le \epsilon.$$

A partition $\{V_0, \ldots, V_K\}$ of V is called an ϵ-regular partition of G if it satisfies the following two conditions:

(i) $|V_0| \le \epsilon n$;
(ii) $|V_1| = |V_2| = \cdots = |V_K|$;
(iii) all but at most ϵK^2 of the pairs (V_i, V_j) with $1 \le i < j \le K$ are ϵ-regular.

Szemerédi's regularity lemma goes as follows.

Theorem 1.2 (*Szemerédi's lemma [28]*). *Given $\epsilon > 0$ and an integer $m \ge 1$ there exists an integer $M = M(\epsilon, m)$ such that every graph of order at least M admits an ϵ-regular partition $\{V_0, \ldots, V_K\}$ for some K in the range $m \le K \le M$.*

This result was proved by Szemerédi [28] in 1976 and has since found numerous applications in combinatorics, number theory and many other areas of discrete mathematics. The version presented above is from [13], Section 7.2.. Lovász and Szegedy proved the following related result.

Theorem 1.3 (*Theorem 5.1 in [23]*). *The metric space $(\widetilde{\mathcal{W}}, \delta_\square)$ is compact.*

2. The main result

2.1. The rate function

The main goal of this paper is to prove a large deviation principle for $G(n, p)$ when p is fixed and $n \to \infty$. The discussion in Section 1 gives a topological space (namely, $\widetilde{\mathcal{W}}$) that is suitable for this purpose. The next step is to define a rate function on this space. Let $I_p : [0, 1] \to \mathbb{R}$ be the function

$$I_p(u) := \frac{1}{2} u \log \frac{u}{p} + \frac{1}{2}(1-u) \log \frac{1-u}{1-p} = \frac{1}{2} \sup_{a,b\in\mathbb{R}} [au + b(1-u) - \log(pe^a + (1-p)e^b)] = \frac{1}{2} \sup_{a\in\mathbb{R}} [au - \log(pe^a + (1-p))]. \tag{6}$$

The domain of the function I_p can be extended to $\mathcal{W}$ as

$$I_p(h) := \int_0^1 \int_0^1 I_p(h(x,y))\, dx\, dy = \frac{1}{2} \int_0^1 \int_0^1 \left[h(x,y) \log \frac{h(x,y)}{p} + (1-h(x,y)) \log \frac{1-h(x,y)}{1-p} \right] dxdy \tag{7}$$

$$= \frac{1}{2} \sup_{a(\cdot,\cdot)} \left[\int\!\!\int a(x,y)h(x,y)\, dx\, dy - \int \log(pe^{a(x,y)} + (1-p))\, dx\, dy \right]. \tag{8}$$

The following property of I_p is crucial.

Lemma 2.1. *The function I_p is well defined on $\widetilde{\mathcal{W}}$ and is lower semicontinuous under the cut metric $\delta_\square$ on $\widetilde{\mathcal{W}}$.*

Proof. The supremum in (8) can be taken over all bounded measurable functions a on $[0, 1]^2$. As the supremum of a family of affine linear functionals continuous in the weak topology, $I_p(h)$ is lower semi-continuous in the weak topology and therefore also in the topology of the metric $d_\square$. If $\sigma : [0, 1] \to [0, 1]$ is a measure preserving bijection then $I_p(h_\sigma) = I(h)$. By lower semi-continuity of $g \in \widetilde{h}$, $I_p(g) \le I_p(h)$. But $g \in \widetilde{h}$ implies $h \in \widetilde{g}$ so that $I_p(h) \le I_p(g)$. Hence, $I_p(g) = I_p(h)$ and $I_p(\cdot)$ is well defined and lower semi-continuous on $\widetilde{\mathcal{W}}$. □

2.2. The large deviation principle

The random graph $G(n, p)$ induces probability distributions $\mathbb{P}_{n,p}$ on the space $\mathcal{W}$ through the map $G \to f^G$ and $\widetilde{\mathbb{P}}_{n,p}$ on $\widetilde{\mathcal{W}}$ through the map $G \to f^G \to \widetilde{f}^G = \tau f^G$. The space $\mathcal{W}$ is compact in the weak topology and a large deviation principle for $\mathbb{P}_{n,p}$ on $\mathcal{W}$ in the weak topology with the lower-semicontinuous rate function $I_p(h)$ given by (7) is elementary but is not of much use since quantities like 'triangle counts' are not stable in the weak topology. We will state it for the record and find a use for it later.

Theorem 2.2. *The sequence $\mathbb{P}_{n,p}$ on $\mathcal{W}$ satisfies a large deviation principle in the weak topology. That is, for every weakly closed set $F \subset \mathcal{W}$*

$$\limsup_{n\to\infty} \frac{1}{n^2} \log \mathbb{P}_{n,p}(F) \le -\inf_{f\in F} I_p(f)$$

and for any open set U (again in the weak topology) in $\mathcal{W}$

$$\liminf_{n\to\infty} \frac{1}{n^2} \log \mathbb{P}_{n,p}(U) \ge -\inf_{f\in U} I_p(f).$$

Proof. The weak topology is defined through an arbitrary but finite number of linear functionals. Therefore the large deviation principle can be reduced to the large deviation behavior of a finite set of linear functionals $\{Z_\phi(f)\}$ given by

$$Z_\phi(f) = \langle \phi, f \rangle := \iint f(x,y)\phi(x,y)\mathrm{d}x\mathrm{d}y$$

under the measure $\mathbb{P}^{n,p}$. The limit

$$\lim_{n\to\infty} \frac{1}{n^2} \log \mathbb{E}^{\mathbb{P}_{n,p}}\left[\exp\left(n^2 \iint f(x,y)\phi(x,y)\mathrm{d}x\mathrm{d}y\right)\right]$$

is easily calculated to yield

$$\frac{1}{2}\iint \log(pe^{2\phi(x,y)} + (1-p))\mathrm{d}x\mathrm{d}y.$$

(Note that this is true only if ϕ is symmetric. However, since f is symmetric, it suffices to restrict attention to symmetric ϕ.) Therefore, an abstract Gärtner–Ellis Theorem (see e.g. Theorem 4.5.3 in [11]) gives the upper bound with rate function

$$I_p(f) = \sup_\phi \left[\langle \phi, f\rangle - \frac{1}{2}\iint \log(pe^{2\phi(x,y)} + (1-p))\mathrm{d}x\mathrm{d}y\right].$$

Note that this is the rate function I_p defined in (7) and (8). The supremum is attained at the function

$$\phi_f(x,y) := \frac{1}{2}\log\frac{f(x,y)}{p} - \frac{1}{2}\log\frac{1-f(x,y)}{1-p}.$$

Note that for any $g \neq f$,

$$\begin{aligned}&(\langle \phi_f, f\rangle - I_p(f)) - (\langle \phi_f, g\rangle - I_p(g))\\ &= \frac{1}{2}\iint \left(g(x,y)\log\frac{g(x,y)}{f(x,y)} + (1-g(x,y))\log\frac{1-g(x,y)}{1-f(x,y)}\right)\mathrm{d}x\,\mathrm{d}y > 0.\end{aligned}$$

This shows that every f is an exposed point of the lower semicontinuous rate function I_p, in the parlance of convex analysis. Therefore by the Gärtner–Ellis Theorem (see e.g. Theorem 4.5.20 in [11]) and the compactness of the weak topology, we get the lower bound. □

The large deviation principle for $\widetilde{\mathbb{P}}_{n,p}$ on $(\widetilde{\mathcal{W}}, \delta_\square)$ is much more useful and is the main result of this article.

Theorem 2.3. *For each fixed $p \in (0,1)$, the sequence $\widetilde{\mathbb{P}}_{n,p}$ obeys a large deviation principle in the space $\widetilde{\mathcal{W}}$ (equipped with the cut metric) with rate function I_p defined by (7). Explicitly, this means that for any closed set $\widetilde{F} \subseteq \widetilde{\mathcal{W}}$,*

$$\limsup_{n\to\infty} \frac{1}{n^2}\log \widetilde{\mathbb{P}}_{n,p}(\widetilde{F}) \le -\inf_{\widetilde{h}\in\widetilde{F}} I_p(\widetilde{h}). \tag{9}$$

and for any open set $\widetilde{U} \subseteq \widetilde{\mathcal{W}}$,

$$\liminf_{n\to\infty} \frac{1}{n^2}\log \widetilde{\mathbb{P}}_{n,p}(\widetilde{U}) \ge -\inf_{\widetilde{h}\in\widetilde{U}} I_p(\widetilde{h}). \tag{10}$$

For the upper bound, because $(\widetilde{\mathcal{W}}, \delta_\square)$ is compact, it is sufficient to prove that for any $\widetilde{h} \in \widetilde{\mathcal{W}}$,

$$\lim_{\eta\to 0}\limsup_{n\to\infty} \frac{1}{n^2}\log \widetilde{\mathbb{P}}_{n,p}(S_\square(\widetilde{h},\eta)) \le -I_p(\widetilde{h}).$$

For the lower bound we need to prove that if $\widetilde{h} \in \widetilde{\mathcal{W}}$ and $\eta > 0$ is arbitrary

$$\liminf_{n\to\infty} \frac{1}{n^2}\log \widetilde{\mathbb{P}}_{n,p}(S_\square(\widetilde{h},\eta)) \ge -I_p(\widetilde{h}),$$

where $S_\square(\widetilde{h},\eta) = \{\widetilde{g} : \delta_\square(\widetilde{g},\widetilde{h}) \le \eta\}$.

2.3. Proof of the upper bound in Theorem 2.3

Let $B(\widetilde{h}, \eta) \subset \mathcal{W}$ be defined as

$$B(\widetilde{h}, \eta) = \tau^{-1} S_{\square}(\widetilde{h}, \eta) \subset \mathcal{W}$$

i.e. the union of all the orbits from $S_{\square}(\widetilde{h}, \eta) \subset \widetilde{\mathcal{W}}$. We need to show that

$$\lim_{\eta \to 0} \limsup_{n \to \infty} \frac{1}{n^2} \log \mathbb{P}_{n,p}[B(\widetilde{h}, \eta)] \leq -I_p(\widetilde{h}). \tag{11}$$

Let the set of n vertices be partitioned into K subsets of size a with a remainder of size b, so that $n = Ka + b$. We assume that $b \leq \epsilon n$. We order the vertices so that $V_0 = \{1, 2, \ldots b\}$ and $V_i = \{b + (i-1)a + 1, \ldots, b + ia\}$ for $i = 1, 2, \ldots, K$. We map the vertices into subintervals of the unit interval, with the vertex r represented by the interval $[\frac{r-1}{n}, \frac{r}{n}]$. The sets V_i of vertices will then correspond to the intervals $E_0 = [0, \frac{b}{n}]$ for $i = 0$ and $E_i = [\frac{b+(i-1)a}{n}, \frac{b+ia}{n}]$ for $1 \leq i \leq K$. Let us denote by $\mathcal{V}_K$ the subset of $\mathcal{W}$ consisting of $g(x, y)$ that are of the form

$$g(x, y) = \sum_{i,j=1}^{K} p_{i,j} \mathbf{1}_{E_i}(x) \mathbf{1}_{E_j}(y)$$

where $\{p_{i,j}\}$, $1 \leq i, j \leq K$ is symmetric and satisfies $0 \leq p_{i,j} \leq 1$. For any pair m, M with $m < M$, we define $\mathcal{V}_{m,M} = \cup_{m \leq K \leq M} \mathcal{V}_K$. The following is a restatement of the Szemerédi regularity lemma.

Lemma 2.4. *Given any $\epsilon > 0$ and $m \geq 1$ such that $2m^{-1} < \epsilon$, there is $M = M(\epsilon, m)$ such that for any graph G, there exists a permutation π, i.e. a relabeling of the vertices of the graph, such that*

$$\inf_{g \in \mathcal{V}_{m,M}} d_{\square}(f^{\pi G}, g) \leq \epsilon.$$

Proof. Let ϵ' and m be given. According to Szemerédi's lemma, there is $M(\epsilon', m)$ such that, for some K in the range $m \leq K \leq M$, we can find a partition $V_0, \ldots V_K$ which is ϵ'-regular. After a permutation we can assume that the ordering of the vertices coincides with the ordering of the partitions. We define $p_{i,i} = p_{0,i} = p_{i,0} = 0$ and for $1 \leq i \neq j$, $p_{i,j} = \rho_G(V_i, V_j)$. This leads to

$$g(x, y) = \sum_{\substack{i,j=1 \\ i \neq j}}^{K} \rho_G(V_i, V_j) \mathbf{1}_{E_i}(x) \mathbf{1}_{E_j}(y)$$

to be compared with f^G when $V_0, \ldots, V_K$ is an ϵ' regular partition of G. Recall that

$$d_{\square}(f^G, g) = \sup_{S, T \subset [0,1]} \left| \iint_{S \times T} [f^G(x, y) - g(x, y)] \mathrm{d}x\, \mathrm{d}y \right|.$$

Since both f^G and g are constant on sets of the form $[\frac{i}{n}, \frac{i+1}{n}] \times [\frac{j}{n}, \frac{j+1}{n}]$ it is easy to see that S and T can be restricted to unions of intervals of the form $[\frac{i}{n}, \frac{i+1}{n}]$ i.e. subsets of $[0, 1]$ that represent subsets of vertices. These subsets will also be denoted by S and T. Now, given two such subsets S and T,

$$\iint_{S \times T} (f^G(x, y) - g(x, y)) \mathrm{d}x\, \mathrm{d}y = \sum_{0 \leq i,j \leq K} \iint_{(S \cap E_i) \times (T \cap E_j)} (f^G(x, y) - g(x, y)) \mathrm{d}x\, \mathrm{d}y.$$

Let A_{ij} denote the (i, j)th term in the above sum. Since f^G and g both take values in $[0, 1]$, therefore for each i and each j,

$$|A_{i0}| \leq |E_i||E_0|, \qquad |A_{0j}| \leq |E_0||E_j|.$$

For the same reason, if $|S \cap E_i| < \epsilon'|E_i|$ or $|T \cap E_j| < \epsilon'|E_j|$, then

$$|A_{ij}| \leq \epsilon'|E_i||E_j|.$$

If $1 \le i \ne j$, $|S \cap E_i| \ge \epsilon'|E_i|$, $|T \cap E_j| \ge \epsilon'|E_j|$ and the pair (V_i, V_j) is ϵ'-regular, then

$$\begin{aligned}|A_{ij}| &= \left|\frac{1}{n^2} e_G(S \cap V_i, T \cap V_j) - \rho_G(V_i, V_j)|S \cap E_i||T \cap E_j|\right| \\ &\le |\rho_G(S \cap V_i, S \cap V_j) - \rho_G(V_i, V_j)||S \cap E_i||T \cap E_j| \\ &\le \epsilon'|E_i||E_j|.\end{aligned}$$

Finally, if either $1 \le i = j$ or (V_i, V_j) is not ϵ'-regular (of which there are at most $K + 2\epsilon' K^2$ cases), then we have the trivial bound $|A_{ij}| \le \frac{a^2}{n^2} \le \frac{1}{K^2}$. A combination of the above estimates gives

$$\begin{aligned}\sum_{0 \le i,j \le K} |A_{ij}| &\le 2|E_0| + \epsilon' + (K + 2\epsilon' K^2)\frac{1}{K^2} \\ &\le (2\epsilon' + 2\epsilon' + \epsilon' + K^{-1}).\end{aligned}$$

Thus,

$$d_\square(f^G, g) \le 5\epsilon' + K^{-1} \le 5\epsilon' + m^{-1}$$

Since $m^{-1} < \epsilon/2$, we can choose ϵ' so that $5\epsilon' + m^{-1} < \epsilon$. □

Lemma 2.5. *Let ϵ, m and M be as in Lemma 2.4.*

$$\mathbb{P}_{n,p}(B(\widetilde{h}, \eta)) \le n!\, \mathbb{P}_{n,p}(B(\widetilde{h}, \eta) \cap B(\mathcal{V}_{m,M}, \epsilon))$$

where $B(\mathcal{V}_{m,M}, \epsilon) = \{g : \inf_{f \in \mathcal{V}_{m,M}} d_\square(g, f) \le \epsilon\}$.

Proof. The orbit under the permutation group has at most $n!$ elements and they all have the same probability under $\mathbb{P}_{n,p}$. Moreover by the Lemma 2.4 every orbit meets $B(\mathcal{V}_{m,M}, \epsilon)$, and $B(\widetilde{h}, \eta)$ is invariant under $\sigma \in \Sigma$ and therefore under π. Consequently

$$B(\widetilde{h}, \eta) \subset \bigcup_\pi \pi^{-1}(B(\widetilde{h}, \eta) \cap B(\mathcal{V}_{m,M}, \epsilon))$$

and the lemma follows.

Lemma 2.6. *There exists a function $\delta(\widetilde{h}, \epsilon)$, depending only on $\widetilde{h}$ and ϵ, with $\delta(\widetilde{h}, \epsilon) \to 0$ as $\epsilon \to 0$, such that for each arbitrary but fixed ϵ, m, M satisfying Lemma 2.4,*

$$\lim_{\eta \to 0} \limsup_{n \to \infty} \frac{1}{n^2} \log \mathbb{P}_{n,p}(B(\widetilde{h}, \eta) \cap B(\mathcal{V}_{m,M}, \epsilon)) \le -I_p(\widetilde{h}) + \delta(\widetilde{h}, \epsilon)$$

Proof. Since $\mathcal{V}_{m,M}$ is a finite union $\bigcup_{m \le K \le M} \mathcal{V}_K$ it is sufficient to prove that for each K

$$\lim_{\eta \to 0} \limsup_{n \to \infty} \frac{1}{n^2} \log \mathbb{P}_{n,p}(B(\widetilde{h}, \eta) \cap B(\mathcal{V}_K, \epsilon)) \le -I_p(\widetilde{h}) + \delta(\widetilde{h}, \epsilon)$$

and $\delta(\widetilde{h}, \epsilon) \to 0$ as $\epsilon \to 0$. For fixed K, $\mathcal{V}_K$ consists of a compact set of functions in $L_1([0, 1]^2)$ and can be covered by a finite number of spheres of radius ϵ in L_1 and therefore in $\mathcal{W}$. It is therefore sufficient to show that for fixed K and $g \in \mathcal{V}_K$

$$\lim_{\eta \to 0} \limsup_{n \to \infty} \frac{1}{n^2} \log \mathbb{P}_{n,p}(B(\widetilde{h}, \eta) \cap B(g, 2\epsilon)) \le -I_p(\widetilde{h}) + \delta(\widetilde{h}, \epsilon)$$

We can assume that $B(\widetilde{h}, \eta) \cap B(g, 2\epsilon) \ne \emptyset$. Therefore $g \in B(\widetilde{h}, \eta + 2\epsilon)$. Since $\eta \to 0$ we can assume $\eta < \epsilon$ so that $g \in B(\widetilde{h}, 3\epsilon)$. By lower semi-continuity of $I_p(\cdot)$, $I_p(f) \ge I_p(\widetilde{h}) - \delta(\widetilde{h}, \epsilon)$ on

$B(g, 2\epsilon) \subset B(\widetilde{h}, 5\epsilon)$ and $\delta(\widetilde{h}, \epsilon) \to 0$ as $\epsilon \to 0$. We note that $B(g, 2\epsilon) \subset \mathcal{W}$ is weakly closed and therefore by the upper bound in Theorem 2.2,

$$\lim_{\eta\to 0}\limsup_{n\to\infty}\frac{1}{n^2}\log\mathbb{P}_{n,p}(B(\widetilde{h},\eta)\cap B(g,2\epsilon)) \le \limsup_{n\to\infty}\frac{1}{n^2}\log\mathbb{P}_{n,p}(B(g,2\epsilon))$$
$$\le -\inf_{f\in B(g,2\epsilon)} I_p(f) \le -\inf_{f\in B(\widetilde{h},5\epsilon)} I_p(f) \le -I_p(\widetilde{h}) + \delta(\widetilde{h},\epsilon)$$

where $\delta(\widetilde{h}, \epsilon) \to 0$ as $\epsilon \to 0$. □

Lemmas 2.5 and 2.6 Eq. (11), which proves the upper bound in Theorem 2.3.

2.4. Proof of the lower bound in Theorem 2.3

Let $h(x, y) \in \mathcal{W}$ be given. We define

$$p_{i,j}^{(n)} = n^2 \int\int_{[\frac{i-1}{n},\frac{i}{n}]\times[\frac{j-1}{n},\frac{j}{n}]} h(x,y)\mathrm{d}x\mathrm{d}y$$

and the corresponding function $h_n(x, y) \in \mathcal{W}$ by

$$h_n(x,y) = \sum_{i,j} p_{i,j}^{(n)} \mathbf{1}_{[\frac{i-1}{n},\frac{i}{n}]}(x)\mathbf{1}_{[\frac{j-1}{n},\frac{j}{n}]}(y).$$

Since $\|h_n - h\|_{L_1([0,1]^2)} \to 0$, it follows that $d_\square(h_n, h) \to 0$. It is therefore sufficient to prove that for any $\epsilon > 0$

$$\liminf_{n\to\infty}\frac{1}{n^2}\log\mathbb{P}(d_\square(f^{G(n,p)}, h_n) \le \epsilon) \ge -I_p(h).$$

We define an inhomogeneous random graph where the edge connecting the vertices i and j is present with probability $p_{i,j}^{(n)}$. Different edges are independent. If $\xi_{i,j} = 1$ when the edge connecting i, j is present and 0 otherwise then $\xi_{i,j}$ are independent Binomial random variables with $\mathbb{P}(\xi_{i,j} = 1) = p_{i,j}^{(n)}$. We denote by $\mathbb{P}_{n,h}$ the measure on $\mathcal{W}$ induced by

$$f_n(x,y) := \sum_{\substack{i,j\\ i\ne j}} \xi_{i,j}\mathbf{1}_{[\frac{i-1}{n},\frac{i}{n}]}(x)\mathbf{1}_{[\frac{j-1}{n},\frac{j}{n}]}(y)$$

If A and B are subsets of $\{1, \dots, n\}$, it is straightforward to calculate

$$\psi_n(\lambda) := \frac{1}{n^2}\log\mathbb{E}^{\mathbb{P}_{n,h}}\left[\exp\left(\lambda\sum_{\substack{i\in A, j\in B\\ i\ne j}}(\xi_{i,j} - p_{i,j}^{(n)})\right)\right]$$
$$= \frac{1}{n^2}\left[\sum_{\substack{i,j\in A\cap B\\ i>j}}\log\mathbb{E}^{\mathbb{P}_{n,h}}[\exp(2\lambda(\xi_{i,j} - p_{i,j}^{(n)}))]\right.$$
$$\left. + \sum_{\substack{i\in A\cap B, j\in B\cap A^c\\ \text{or } i\in A\cap B^c, j\in A\cap B\\ \text{or } i\in A\cap B^c, j\in A^c\cap B}}\log\mathbb{E}^{\mathbb{P}_{n,h}}[\exp(\lambda(\xi_{i,j} - p_{i,j}^{(n)}))]\right].$$

Each term in the sum is easily estimated by $\frac{\lambda^2}{2}$, providing an estimate of the form

$$\mathbb{P}_{n,h}\left(\left|\int\int_{\widetilde{A}\times\widetilde{B}}(f_n - h_n)\mathrm{d}x\mathrm{d}y\right| \ge \epsilon\right) \le \mathrm{e}^{-\frac{n^2\epsilon^2}{2}},$$

where $\widetilde{A} = \cup_{i\in A}[\frac{i-1}{n}, \frac{i}{n}]$ and $\widetilde{B}$ is defined similarly. Since the number of sets like $\widetilde{A} \times \widetilde{B}$ is only 2^{2n} it follows that

$$\mathbb{P}_{n,h}(d_\square(f_n, h_n) \geq \epsilon) \to 0$$

as $n \to \infty$. Now the lower bound is easily established by a simple tilting argument. Denoting by $B_{\epsilon,n}$ the set $\{f : d_\square(f, h_n) \leq \epsilon\}$

$$\begin{aligned}\mathbb{P}_{n,p}(B_{\epsilon,n}) &= \int_{B_{\epsilon,n}} \mathrm{d}\mathbb{P}_{n,p} = \int_{B_{\epsilon,n}} \mathrm{e}^{-\log \frac{\mathrm{d}\mathbb{P}_{n,h}}{\mathrm{d}\mathbb{P}_{n,p}}} \mathrm{d}\mathbb{P}_{n,h} \\ &= \mathbb{P}_{n,h}(B_{\epsilon,n}) \frac{1}{\mathbb{P}_{n,h}(B_{\epsilon,n})} \int_{B_{\epsilon,n}} \mathrm{e}^{-\log \frac{\mathrm{d}\mathbb{P}_{n,h}}{\mathrm{d}\mathbb{P}_{n,p}}} \mathrm{d}\mathbb{P}_{n,h}.\end{aligned}$$

By Jensen's inequality

$$\log \mathbb{P}_{n,p}(B_{\epsilon,n}) \geq \log \mathbb{P}_{n,h}(B_{\epsilon,n}) - \frac{1}{\mathbb{P}_{n,h}(B_{\epsilon,n})} \int_{B_{\epsilon,n}} \log \frac{\mathrm{d}\mathbb{P}_{n,h}}{\mathrm{d}\mathbb{P}_{n,p}} \mathrm{d}\mathbb{P}_{n,h}.$$

Since $\mathbb{P}_{n,h}(B_{\epsilon,n}) \to 1$, it is easy to see that

$$\liminf_{n\to\infty} \frac{1}{n^2} \log \mathbb{P}_{n,p}(B_{\epsilon,n}) \geq - \lim_{n\to\infty} \frac{1}{n^2} \int \log \frac{\mathrm{d}\mathbb{P}_{n,h}}{\mathrm{d}\mathbb{P}_{n,p}} \mathrm{d}\mathbb{P}_{n,h}$$

The entropy cost of tilting (i.e. the integral in the preceding display) is

$$\frac{1}{n^2} \sum_{i>j} \left(p_{i,j}^{(n)} \log \frac{p_{i,j}^{(n)}}{p} + (1 - p_{i,j}^{(n)}) \log \frac{1 - p_{i,j}^{(n)}}{1 - p} \right)$$

which converges to $I_p(h)$ as $n \to \infty$. This proves the lower bound.

3. Conditional distributions

Theorem 2.3 gives estimates of the probabilities of rare events for $G(n, p)$. However, it does not quite answer the following question: given that some particular rare event has occurred, what does the graph look like? Naturally, one might expect that if $G(n, p) \in \widetilde{F}$ for some closed set $\widetilde{F} \subseteq \widetilde{\mathcal{W}}$ satisfying

$$\inf_{\widetilde{h}\in\widetilde{F}^o} I_p(\widetilde{h}) = \inf_{\widetilde{h}\in\widetilde{F}} I_p(\widetilde{h}) > 0, \tag{12}$$

then $G(n, p)$ should resemble one of the minimizers of I_p in $\widetilde{F}$. (Here $\widetilde{F}^o$ denotes the interior of $\widetilde{F}$, as usual.) In other words, given that $G(n, p) \in \widetilde{F}$, one might expect that $\delta_\square(G(n, p), \widetilde{F}^*) \approx 0$, where $\widetilde{F}^*$ is the set of minimizers of I_p in $\widetilde{F}$ and

$$\delta_\square(G(n, p), \widetilde{F}^*) := \inf_{\widetilde{h}\in\widetilde{F}^*} \delta_\square(G(n, p), \widetilde{h}).$$

However, it is not obvious that a minimizer must exist in $\widetilde{F}$. Here is where the compactness of $\widetilde{\mathcal{W}}$ comes to the rescue yet one more time: since the function I_p is lower semicontinuous on $\widetilde{F}$ and $\widetilde{F}$ is closed, therefore a minimizer must necessarily exist. The following theorem formalizes this argument.

Theorem 3.1. *Take any $p \in (0, 1)$. Let $\widetilde{F}$ be a closed subset of $\widetilde{\mathcal{W}}$ satisfying (12). Let $\widetilde{F}^*$ be the subset of $\widetilde{F}$ where I_p is minimized. Then $\widetilde{F}^*$ is non-empty and compact, and for each n, and each $\epsilon > 0$,*

$$\mathbb{P}(\delta_\square(G(n, p), \widetilde{F}^*) \geq \epsilon \mid G(n, p) \in \widetilde{F}) \leq \mathrm{e}^{-C(\epsilon,\widetilde{F})n^2}$$

where $C(\epsilon, \widetilde{F})$ is a positive constant depending only on ϵ and $\widetilde{F}$. In particular, if $\widetilde{F}^*$ contains only one element $\widetilde{h}^*$, then the conditional distribution of $G(n, p)$ given $G(n, p) \in \widetilde{F}$ converges to the point mass at $\widetilde{h}^*$ as $n \to \infty$.

Proof. Since $\widetilde{\mathcal{W}}$ is compact and $\widetilde{F}$ is a closed subset, therefore $\widetilde{F}$ is also compact. Since I_p is a lower semicontinuous function on $\widetilde{F}$ (Lemma 2.1) and $\widetilde{F}$ is compact, it must attain its minimum on $\widetilde{F}$. Thus, $\widetilde{F}^*$ is non-empty. By the lower semicontinuity of I_p, $\widetilde{F}^*$ is closed (and hence compact). Fix $\epsilon > 0$ and let

$$\widetilde{F}_\epsilon := \{\widetilde{h} \in \widetilde{F} : \delta_\square(\widetilde{h}, \widetilde{F}^*) \geq \epsilon\}.$$

Then $\widetilde{F}_\epsilon$ is again a closed subset. Observe that

$$\mathbb{P}(\delta_\square(G(n,p), \widetilde{F}^*) \geq \epsilon \mid G(n,p) \in \widetilde{F}) = \frac{\mathbb{P}(G(n,p) \in \widetilde{F}_\epsilon)}{\mathbb{P}(G(n,p) \in \widetilde{F})}.$$

Thus, with

$$I_1 := \inf_{\widetilde{h}\in\widetilde{F}} I_p(\widetilde{h}), \qquad I_2 := \inf_{\widetilde{h}\in\widetilde{F}_\epsilon} I_p(\widetilde{h}),$$

Theorem 2.3 and condition (12) give

$$\limsup_{n\to\infty} \frac{1}{n^2} \log \mathbb{P}(\delta_\square(G(n,p), \widetilde{F}^*) \geq \epsilon \mid G(n,p) \in \widetilde{F}) \leq I_1 - I_2.$$

The proof will be complete if it is shown that $I_1 < I_2$.

Now clearly, $I_1 \leq I_2$. If $I_1 = I_2$, the compactness of $\widetilde{F}_\epsilon$ implies that there exists $\widetilde{h} \in \widetilde{F}_\epsilon$ satisfying $I_p(\widetilde{h}) = I_2$. However, this means that $\widetilde{h} \in \widetilde{F}^*$ and hence $\widetilde{F}_\epsilon \cap \widetilde{F}^* \neq \emptyset$, which is impossible. □

4. Application to triangle counts

4.1. Brief history of the problem

Let $T_{n,p}$ be the number of triangles in $G(n,p)$. The primary objective of this section is to compute the large deviation rate function for the upper tail of $T_{n,p}$ when p remains fixed and $n \to \infty$. In other words, given $p \in [0,1]$ and $\epsilon > 0$, we wish to evaluate the limit

$$\lim_{n\to\infty} \frac{1}{n^2} \log \mathbb{P}(T_{n,p} \geq (1+\epsilon)\mathbb{E}(T_{n,p})) \tag{13}$$

as a function of p and ϵ.

The problem of estimating tail probabilities like $\mathbb{P}(T_{n,p} \geq (1+\epsilon)\mathbb{E}(T_{n,p}))$ has been studied extensively in the random graphs literature, particularly in the case when p is allowed to tend to zero as $n \to \infty$. Computing upper and lower bounds on such tail probabilities that are sharp up to constants in the exponent was a prominent open problem in this area until it was solved recently in [8]. Let us refer to the paper [8] for a survey of the aforementioned literature.

When p is fixed, computing sharp upper and lower bounds is relatively easy. The difficult problem in this case is the exact evaluation of the limit (13). The first progress in this direction was made in [9] where it was shown that, given $p \in (0,1)$, there exist $p^3/6 < t' \leq t'' < 1/6$ such that for all $t \in (p^3/6, t') \cup (t'', 1/6)$,

$$\lim_{n\to\infty} \frac{1}{n^2} \log \mathbb{P}(T_{n,p} \geq tn^3) = -I_p((6t)^{1/3}), \tag{14}$$

when I_p is the entropy function defined in (6). Explicit formulas for p' and p'' are also given in [9]. Unfortunately, the result does not cover all values of (p,t); and neither is the above formula true for all (p,t), as we shall see below.

There is a related unpublished manuscript by Bolthausen, Comets and Dembo [4] on large deviations for subgraph counts. As of now, to the best of our knowledge, the authors of [4] have only looked at subgraphs that do not complete loops, like 2-stars. Besides [9] and [4], we know of no other papers that attack the exact evaluation of (13) (or equivalently, (14)).

4.2. Exact large deviations for the upper tail

In this subsection, the limit (14) is evaluated for all values of p and t. It comes as the solution of the following variational problem. Let $\mathcal{W}$, $\widetilde{\mathcal{W}}$ and $\delta_\square$ be defined as in Section 1. For each $f \in \mathcal{W}$, let

$$T(f) := \frac{1}{6}\int_0^1\int_0^1\int_0^1 f(x,y)f(y,z)f(z,x)\mathrm{d}x\mathrm{d}y\mathrm{d}z$$

and let $I_p(f)$ be defined as in (7). Note that T can be defined on $\widetilde{\mathcal{W}}$ simply by letting $T(\widetilde{f}) := T(f)$, because T is a continuous map on $\mathcal{W}$ under the $d_\square$ pseudometric (Theorem 3.7 in [6]).

For each $p \in [0, 1]$ and $t \in [0, 1/6)$, let

$$\phi(p,t) := \inf\{I_p(f) : f \in \mathcal{W},\ T(f) \geq t\}. \tag{15}$$

For $t \geq 1/6$, let $\phi(p, t) = \infty$. The following result gives the large deviation rate function for the upper tail of $T_{n,p}$. (Note that this is just an illustrative example. Theorem 2.3 can be used to derive large deviations for any subgraph count, or even joint large deviations for the counts of more than one subgraph.)

Theorem 4.1. *Let $G(n, p)$ be the Erdős–Rényi random graph on n vertices with edge probability p. Let $T_{n,p}$ denote the number of triangles in $G(n, p)$. Let ϕ be defined as above. Then for each $p \in (0, 1)$ and each $t \geq 0$,*

$$\lim_{n\to\infty}\frac{1}{n^2}\log\mathbb{P}(T_{n,p} \geq tn^3) = -\phi(p,t).$$

Next, take any $p \in (0, 1)$ and $t \in (p^3/6, 1/6)$. Let $F^*_{p,t}$ be the set of minimizers for the variational problem (15) and $\widetilde{F}^*_{p,t}$ be its image in $\widetilde{\mathcal{W}}$. Then $\widetilde{F}^*_{p,t}$ is a non-empty compact subset of $\widetilde{\mathcal{W}}$. Moreover, for each $\epsilon > 0$ there exists a positive constant $C(\epsilon, p, t)$ depending only on ϵ, p and t such that for any n,

$$\mathbb{P}(\delta_\square(G(n,p), \widetilde{F}^*_{p,t}) \geq \epsilon \mid T_{n,p} \geq tn^3) \leq \mathrm{e}^{-C(\epsilon,p,t)n^2}.$$

Proof. Let $F := \{f \in \mathcal{W} : T(f) \geq t\}$. By Theorem 3.7 in [6], F is a closed subset of $\mathcal{W}$. Therefore by Theorem 2.3,

$$\begin{aligned}\limsup_{n\to\infty}\frac{1}{n^2}\log\mathbb{P}(T_{n,p} \geq tn^3) &= \limsup_{n\to\infty}\frac{1}{n^2}\log\mathbb{P}(G(n,p) \in F)\\ &\leq -\inf_{h\in F} I_p(h) = -\phi(p,t).\end{aligned}$$

Next, let $U := \{f \in \mathcal{W} : T(f) > t\}$. Again by Theorem 3.7 of [6], U is an open set. Therefore by Theorem 2.3, for each $\epsilon > 0$,

$$\begin{aligned}\liminf_{n\to\infty}\frac{1}{n^2}\log\mathbb{P}(T_{n,p} \geq tn^3) &\geq \liminf_{n\to\infty}\frac{1}{n^2}\log\mathbb{P}(G(n,p) \in U)\\ &\geq -\inf_{h\in U} I_p(h) \geq -\phi(p,t+\epsilon).\end{aligned}$$

In Proposition 4.2 below, it is proved that ϕ is a continuous function of t for every fixed p. This completes the proof of the first assertion of the theorem. The second assertion is merely a corollary of Theorem 3.1. The condition (12) required for Theorem 3.1 can be easily shown to follow from the continuity of ϕ in t, because any $\widetilde{f}$ with $T(\widetilde{f}) > t$ lies in the interior of the set $\{\widetilde{h} : T(\widetilde{h}) \geq t\}$. □

4.3. Properties of the rate function

Given Theorem 2.3, there is a natural desire to understand the rate function ϕ. The following proposition summarizes some basic properties of ϕ. The first property is required in the proof of Theorem 4.1 above.

Proposition 4.2. *For each fixed $p \in (0, 1)$, the following hold:*

(i) *The function $\phi(p, t)$ is continuous in t in the interval $[0, 1/6)$.*

(ii) *As a function of* t, $\phi(p,t) = 0$ *in the interval* $[0, p^{3/6}]$ *and strictly increasing in* $(p^3/6, 1/6)$. *Moreover, for* $p^3/6 < t < s < 1/6$,

$$\phi(p,t) < (t/s)^{1/3}\phi(p,s).$$

(iii) *For* $t \in (p^3/6, 1/6)$, $\phi(p,t)$ *can be alternately represented as*

$$\phi(p,t) := \inf\{I_p(f) : f \in \mathcal{W},\ T(f) = t\}.$$

Moreover, if $\{f_n\}_{n\geq 1}$ *is a sequence in* $\mathcal{W}$ *such that* $T(f_n) \geq t$ *for all* n *and* $I_p(f_n) \to \phi(p,t)$, *then* $T(f_n) \to t$. *In particular, the elements of* $\widetilde{F}^*_{p,t}$ *all satisfy* $T(f) = t$.

Proof. Let us start by proving that ϕ is continuous in t. For each $f \in \mathcal{W}$ and $\delta \in [0, 1]$, let

$$f^\delta := f + \delta(1-f).$$

By the inequality

$$\begin{aligned}
&(a+\delta(1-a))(b+\delta(1-b))(c+\delta(1-c)) \\
&\quad = abc + \delta((1-a)bc + (1-b)ac + (1-c)ab) \\
&\qquad + \delta^2((1-a)(1-b)c + (1-a)(1-c)b + (1-b)(1-c)a) \\
&\qquad + \delta^3(1-a)(1-b)(1-c) \\
&\quad \geq abc + \delta^3((1-a)bc + (1-b)ac + (1-c)ab) \\
&\qquad + \delta^3((1-a)(1-b)c + (1-a)(1-c)b + (1-b)(1-c)a) \\
&\qquad + \delta^3(1-a)(1-b)(1-c) \\
&\quad = abc + \delta^3(1-abc),
\end{aligned}$$

we see that

$$T(f^\delta) \geq T(f)(1-\delta^3) + \frac{\delta^3}{6}. \tag{16}$$

Take any $t \in [0, 1/6)$ and any f such that $T(f) \geq t$. Suppose $t_n \downarrow t$. Let δ_n be the smallest number in $[0, 1]$ such that $T(f^{\delta_n}) \geq t_n$. By (16) it follows that δ_n exists and $\lim_{n\to\infty} \delta_n = 0$. Therefore by the dominated convergence theorem, $\lim_{n\to\infty} I_p(f^{\delta_n}) = I_p(f)$. Thus,

$$\lim_{n\to\infty} \phi(p, t_n) \leq \lim_{n\to\infty} I_p(f^{\delta_n}) = I_p(f).$$

Since this is true for every f such that $T(f) \geq t$ and ϕ is a non-decreasing function in t, this proves the right continuity of ϕ.

Next, take a sequence $t_n \uparrow t$. Let f_n be a sequence of functions such that $T(f_n) \geq t_n$ and

$$\lim_{n\to\infty} I_p(f_n) = \lim_{n\to\infty} \phi(p, t_n).$$

For each n, let δ_n be the smallest number in $(0, 1)$ such that $T(f_n^{\delta_n}) \geq t$. By (16), δ_n exists and $\lim_{n\to\infty} \delta_n = 0$. Now, the function I_p on $[0, 1]$ (defined in (6)) is uniformly continuous on $[0, 1]$. As a consequence,

$$\lim_{\delta\to 0} \sup_{f\in\mathcal{W}} |I_p(f^\delta) - I_p(f)| = 0.$$

In particular,

$$\lim_{n\to\infty} I_p(f_n^{\delta_n}) = \lim_{n\to\infty} \phi(p, t_n).$$

But $\phi(p,t) \leq I_p(f_n^{\delta_n})$ for each n. By the monotonicity of ϕ, this proves left continuity.

Next, note that since $\mathbb{P}(T(G_{n,p}) \geq tn^3) \to 1$ for any $t < p^3/6$, Theorem 4.1 and the continuity of ϕ imply that $\phi(p,t) = 0$ for $t \leq p^3/6$. Let us now show that $\phi(p,t)$ is strictly increasing in t when $t \in (p^3/6, 1/6)$.

Fix $p^3/6 \le t < s < 1/6$. Fix $\epsilon > 0$. Take any $f \in \mathcal{W}$ such that $T(f) \ge s$. For each $\delta \in (0, 1)$, let

$$f_\delta(x, y) := (1-\delta)f(x, y) + \delta p. \tag{17}$$

By the inequality

$$((1-\delta)a + \delta p)((1-\delta)b + \delta p)((1-\delta)c + \delta p) \ge (1-\delta)^3 abc$$

we have

$$T(f_\delta) \ge (1-\delta)^3 T(f) \ge (1-\delta)^3 s.$$

Thus, if we take δ such that $(1-\delta)^3 = t/s$, then $T(f_\delta) \ge t$. By the convexity of the function I_p on $[0, 1]$ defined in (6), we see that for any $x \ge p$,

$$I_p((1-\delta)x + \delta p) \le (1-\delta)I_p(x) + \delta I_p(p) = (1-\delta)I_p(x),$$

and therefore,

$$I_p(f_\delta) \le (1-\delta)I_p(f). \tag{18}$$

Thus, $\phi(p, t) \le I_p(f_\delta) \le (t/s)^{1/3} I_p(f)$. Since this holds for any f with $T(f) \ge s$, it shows that

$$\phi(p, t) \le (t/s)^{1/3}\phi(p, s). \tag{19}$$

To show ϕ is a strictly increasing function of t in the interval $(p^3/6, 1/6)$, it therefore suffices to prove that $\phi(p, t) > 0$ for t in this interval. This follows easily, since the strict convexity of I_p on $[0, 1]$ and Eq. (18) show that equality in (19) can hold only if $f \equiv p$ almost everywhere for some f such that $T(f) \ge s$, which is impossible since $s > p^3/6$.

Next, fix $t \in (p^3/6, 1/6)$ and take any sequence $\{f^{(n)}\}$ in $\mathcal{W}$ such that $T(f^{(n)}) \ge t$ for all n and $I_p(f^{(n)}) \to \phi(p, t)$. Recall the subscript notation introduced in Eq. (17) above. By the continuous mapping theorem, there exist $\delta_n \in [0, 1]$ such that for each n,

$$T(f^{(n)}_{\delta_n}) = t.$$

Therefore by (18),

$$\phi(p, t) \le I_p(f^{(n)}_{\delta_n}) \le (1-\delta_n)I_p(f^{(n)}),$$

which proves that $\delta_n \to 0$. This proves that

$$\phi(p, t) = \inf\{I_p(f) : f \in \mathcal{W},\ T(f) = t\}.$$

Since $\delta_n \to 0$, this also proves that $T(f^{(n)}) \to t$. This completes the proof of Proposition 4.2. □

4.4. The 'Replica Symmetric' phase

Note that there are two "extreme" functions that satisfy $T(f) = t$. First, there is the constant function

$$c_t(x, y) \equiv (6t)^{1/3}. \tag{20}$$

On the opposite extreme, there is the function χ_t, defined as

$$\chi_t(x, y) := \begin{cases} 1 & \text{if } \max\{x, y\} \le (6t)^{1/3}, \\ 0 & \text{otherwise.} \end{cases} \tag{21}$$

In a limiting sense, c_t represents an Erdős–Rényi random graph with edge probability $(6t)^{1/3}$, while χ_t represents the union of a clique of size $n(6t)^{1/3}$ and a set of isolated vertices of size $n(1-(6t)^{1/3})$.

It is simple to see that (14) holds if and only if the infimum in (15) is attained at the constant function c_t. The following theorem gives a sufficient condition for this to happen. This extends the main result of [9]. The methods of [9] are closely related to methods from statistical physics; drawing inspiration from this connection, one may call the region where c_t solves the variational problem (15) the 'replica symmetric phase' of the problem.

Theorem 4.3. *For each $0 < p < 1$ and $t \in (p^3/6, 1/6)$, let $h_p(t) := I_p((6t)^{1/3})$. For $t \in [0, p^3/6]$, let $h_p(t) = 0$. Let $\hat{h}_p$ be the convex minorant of h_p (i.e. the maximum convex function lying below h_p). If t is a point in $(p^3/6, 1/6)$ where $h_p(t) = \hat{h}_p(t)$, then the variational problem (15) for this pair (p, t) is uniquely solved by the constant function $c_t \equiv (6t)^{1/3}$. Consequently, $\phi(p, t) = h_p(t)$. Moreover, for such (p, t), for each $\epsilon > 0$*

$$\lim_{n\to\infty} \mathbb{P}(\delta_\square(G(n, p), c_t) \geq \epsilon \mid T_{n,p} \geq tn^3) = 0.$$

Since c_t is the limit of $G(n, (6t)^{1/3})$, this means that for such (p, t) the conditional distribution of $G(n, p)$ given $T_{n,p} \geq tn^3$ is indistinguishable from the law of $G(n, (6t)^{1/3})$ in the large n limit.

Proof. Since h_p is an increasing function, it is easy to see that $h_p(t) = \hat{h}_p(t)$ for $t \leq p^3/6$ (incidentally, this also shows that $\hat{h}_p$ is an increasing function in $[0, 1/6)$, and strictly increasing in $(p^3/6, 1/6)$).

Suppose t is a point in $(p^3/6, 1/6)$ such that $h_p(t) = \hat{h}_p(t)$. We claim that there exists $\beta > 0$ such that

$$t = \arg \max_{x\in[0,1/6]} (\beta x - h_p(x)). \tag{22}$$

To see this, observe that since $\hat{h}_p$ is convex and strictly increasing in the interval $(p^3/6, 1/6)$, there exists $\beta > 0$ and $c \in \mathbb{R}$ such that the line $y = \beta x + c$ lies below the curve $y = \hat{h}_p(x)$ and touches it at $x = t$. But we also know that h_p lies above $\hat{h}_p$ and the two curves touch at t. Thus,

$$t = \arg \max_x (\beta x + c - h_p(x)) = \arg \max_x (\beta x - h_p(x)).$$

This proves the claim.

Now take any $f \in \mathcal{W}$ such that $T(f) \geq t$. Let c_t be the function that is identically equal to $(6t)^{1/3}$. Then $T(c_t) = t$, and by Hölder's inequality and (22),

$$\begin{aligned}
\beta t - I_p(c_t) &= \iint \left(\frac{\beta c_t(x, y)^3}{6} - I_p(c_t(x, y))\right) dx\, dy \\
&\geq \iint \left(\frac{\beta f(x, y)^3}{6} - I_p(f(x, y))\right) dxdy \\
&\geq \frac{\beta}{6} \iiint f(x, y)f(y, z)f(z, x)dxdydz - \iint I_p(f(x, y))dxdy \\
&\geq \beta t - I_p(f).
\end{aligned}$$

Thus, c_t minimizes $I_p(f)$ among all f such that $T(f) \geq t$. This shows that

$$\phi(p, t) = I_p(c_t) = h_p(t).$$

The uniqueness of the optimizer follows from the Hölder step in the above deduction. Finally, the claim about the conditional distribution follows from Theorem 3.1 and the uniqueness of the minimizer. □

It is easy to show that for any $p > 0$, $h_p(t) = \hat{h}_p(t)$ for all $t \in (p^3/6, t') \cup (t'', 1/6)$ where t' and t'' depend on p. Similarly, given any $t \in (0, 1/6)$, there exists $p' < (6t)^{1/3}$ depending on t such that for all $p \in (p', (6t)^{1/3})$, $h_p(t) = \hat{h}_p(t)$. Thus, there is a nontrivial set of (p, t) where c_t solves the variational problem and consequently $\phi(p, t) = h_p(t)$. As mentioned before, this recovers the main result of [9]. The conclusion about the conditional distribution is a new result.

4.5. Replica symmetry breaking

Given Theorem 4.3, it is quite interesting to note that the variational problem (15) is not solved by constant functions everywhere. From the physical point of view espoused in [9], however, this is not surprising; it is simply the effect of replica symmetry breaking down in the 'low temperature regime'.

The phase transition is very easy to establish using Theorem 4.1, by comparing the performances of c_t and χ_t defined in (20) and (21). A simple computation shows that for any $t \in (0, 1/6)$,

$$\lim_{p\to 0} \frac{I_p(c_t)}{\log(1/p)} = \frac{(6t)^{1/3}}{2} > \frac{(6t)^{2/3}}{2} = \lim_{p\to 0} \frac{I_p(\chi_t)}{\log(1/p)}. \tag{23}$$

Combining the above observation with Theorem 3.1, it follows easily that there are values of (p, t) such that given $T_{n,p} \geq tn^3$, the graph $G(n, p)$ must look different than an Erdős–Rényi graph. Again, compactness is crucial. This is formalized by the following theorem.

Theorem 4.4. *Let $\widetilde{C}$ denote the set of constant functions in $\widetilde{\mathcal{W}}$. For each $t \in (0, 1/6)$, there exists $p' > 0$ such that for all $p < p'$, the variational problem (15) is not solved by the constant function $c_t \equiv (6t)^{1/3}$. Moreover, for such (p, t), $\delta_\square(\widetilde{F}^*_{p,t}, \widetilde{C}) > 0$, where $\widetilde{F}^*_{p,t}$ is the set of minimizers defined in Theorem 4.1. Consequently, there exists $\epsilon > 0$ such that*

$$\lim_{n\to\infty} \mathbb{P}(\delta_\square(G(n, p), \widetilde{C}) > \epsilon \mid T_{n,p} \geq tn^3) = 1.$$

Proof. By (23) we see that for each t, there exists $p' > 0$ such that c_t is not a minimizer for the problem (15) if $p < p'$. Take any such (p, t). By Theorem 4.1, $\widetilde{F}^*_{p,t}$ is non-empty and compact. By part (iii) of Proposition 4.2 and the non-optimality of c_t, it follows that $\widetilde{C}$ and $\widetilde{F}^*_{p,t}$ must be disjoint. But $\widetilde{C}$ and $\widetilde{F}^*_{p,t}$ are both compact subsets of $\widetilde{\mathcal{W}}$. Therefore, $\delta_\square(\widetilde{C}, \widetilde{F}^*_{p,t}) > 0$. The last claim follows by Theorem 4.1. □

4.6. The double phase transition

A combination of Theorems 4.3 and 4.4 shows that for all small enough p, the variational problem (15) has a 'double phase transition'. (Actually, there may be more than two phase transitions, but we show that there is at least two.)

Indeed, for all small enough p, Theorem 4.3 (or the results of [9]) show that there exists $p^3/6 < t' \leq t'' < 1/6$ such that $\phi(p, t) = I_p((6t)^{1/3})$ for all $t \in (p^3/6, t') \cup (t'', 1/6)$. On the other hand by Theorem 4.4, it follows that for all small enough p, the variational problem (15) is not solved by a constant function at the point $(p, 1/2)$. Combining these two observations gives the following theorem.

Theorem 4.5. *There exists $p_0 > 0$ such that if $p \leq p_0$, then there exists $p^3/6 < t' < t'' < 1/6$ such that the variational problem (15) is solved by the constant function $c_t \equiv (6t)^{1/3}$ when $t \in (p^3/6, t') \cup (t'', 1/6)$, but there is a non-empty subset of (t', t'') where all optimizers are non-constant.*

Of course, as shown by Theorems 4.3 and 4.4, the significance of optimizers being constant or non-constant is in whether the conditional behavior of $G(n, p)$ given $T_{n,p} \geq tn^3$ is close to that of an Erdős–Rényi graph or not.

4.7. The small p limit

The last theorem of this paper describes the nature of $\phi(p, t)$ and $\widetilde{F}^*_{p,t}$ when t is fixed and p is very small, tending to zero. The essence of the result, perhaps not surprisingly, is that when t is fixed and $p \to 0$, then conditionally on the event $\{T_{n,p} \geq tn^3\}$ the graph $G(n, p)$ must look like a clique.

Theorem 4.6. *For each $t \in (0, 1/6)$,*

$$\lim_{p\to 0} \frac{\phi(p, t)}{\log(1/p)} = \frac{(6t)^{2/3}}{2}.$$

Moreover, if χ_t is the function defined in (21) and $\widetilde{F}^*_{p,t}$ is defined as in Theorem 4.1, then the set $\widetilde{F}^*_{p,t}$ converges to the point $\widetilde{\chi}_t$ as $p \to 0$, in the sense that

$$\lim_{p\to 0} \sup_{\widetilde{f}\in \widetilde{F}^*_{p,t}} \delta_\square(\widetilde{f}, \widetilde{\chi}_t) = 0.$$

Consequently, for each $\epsilon > 0$,

$$\lim_{p\to 0}\lim_{n\to\infty} \mathbb{P}(\delta_\square(G(n,p), \widetilde{\chi}_t) \geq \epsilon \mid T_{n,p} \geq tn^3) = 0.$$

Proof. In this proof, C will denote any constant that does not depend on anything else. The value of C may change from line to line. All integrals will be over the interval $[0, 1]$.

Fix $t \in (0, 1/6)$. For each $p < (6t)^{1/3}$, choose a function $f_p \in F^*_{p,t}$. From the definition of I_p, observe that if $p \leq 1/2$,

$$\left| I_p(f_p) - \frac{1}{2}\iint f_p(x,y) \log\frac{1}{p}\, dx\, dy \right| \leq C. \tag{24}$$

On the other hand, by the definition of $\widetilde{F}^*_{p,t}$,

$$I_p(f_p) \leq I_p(\chi_t) \leq \frac{(6t)^{2/3}}{2}\log\frac{1}{p} + C.$$

Combining the last two inequalities gives

$$\iint f_p(x,y) dx dy \leq (6t)^{2/3} + \frac{C}{\log(1/p)}. \tag{25}$$

Next, let

$$h_p(x,y) := \int f_p(x,z) f_p(z,y)\, dz.$$

Then by two applications of the Cauchy–Schwarz inequality, the inequality (25), and the fact that $f_p(x, y) \in [0, 1]$, we get the following important sequence of inequalities.

$$\begin{cases} (6t)^2 = (6T(f_p))^2 = \left(\iint h_p(x,y) f_p(x,y)\, dx\, dy\right)^2 \\ \leq \iint h_p^2(x,y)\, dx\, dy \iint f_p^2(x,y)\, dx\, dy \\ \leq \int\left(\int f_p^2(x,z)\, dz \int f_p^2(y,z)\, dz\right) dx\, dy \iint f_p^2(x,y)\, dx\, dy \\ = \left(\iint f_p^2(x,y)\, dx\, dy\right)^3 \\ \leq \left(\iint f_p(x,y)\, dx\, dy\right)^3 \leq \left((6t)^{2/3} + \frac{C}{\log(1/p)}\right)^3. \end{cases} \tag{26}$$

A direct consequence of (26), combined with (24), is that

$$\lim_{p\to 0}\frac{\phi(p,t)}{\log(1/p)} = \lim_{p\to 0}\frac{I_p(f_p)}{\log(1/p)} = \lim_{p\to 0}\frac{1}{2}\iint f_p(x,y)\, dx\, dy = \frac{(6t)^{2/3}}{2},$$

which proves the first assertion of the theorem. A second important consequence of (26), to be useful later, is that

$$\iint f_p(x,y)(1 - f_p(x,y))\, dx\, dy \leq \frac{C}{\log(1/p)}. \tag{27}$$

Next, note that for any x, y,

$$\frac{1}{2}\iint (f_p(x,z)f_p(z',y) - f_p(x,z')f_p(z,y))^2\,dz\,dz' = \iint f_p^2(x,z)f_p^2(z',y)\,dz\,dz' - h_p^2(x,y)$$
$$= \int f_p^2(x,z)dz \int f_p^2(y,z)dz - h_p^2(x,y).$$

It follows from this and (26) that

$$\iiiint (f_p(x,z)f_p(z',y) - f_p(x,z')f_p(z,y))^2\,dz\,dz'\,dx\,dy \iint f_p^2(x,y)\,dx\,dy \le \frac{C}{\log(1/p)}.$$

The above inequality and the lower bound on $\iint f_p^2(x,y)dxdy$ from (26) give

$$\iiiint (f_p(x,z)f_p(z',y) - f_p(x,z')f_p(z,y))^2\,dz\,dz'\,dx\,dy \le \frac{C}{t^{2/3}\log(1/p)}. \tag{28}$$

Let $M_p := \iint f_p(x,y)dxdy$. For each x, let $m_p(x) := M_p^{-1/2}\int f_p(x,y)dy$. An application of Jensen's inequality to (28) gives

$$\iint (f_p(x,z)M_p - M_p m_p(x)m_p(z))^2\,dz\,dx \le \frac{C}{t^{2/3}\log(1/p)}.$$

By (26), $M_p \ge (6t)^{2/3}$. Thus, m_p is bounded by $(6t)^{-1/3}$, and

$$\iint (f_p(x,z) - m_p(x)m_p(z))^2\,dz\,dx \le \frac{C}{t^2\log(1/p)}. \tag{29}$$

For each p, let $n_p : [0,1] \to [0,(6t)^{-1/3}]$ be a step function (i.e. a function that is constant on intervals) such that

$$\int (m_p(x) - n_p(x))^2 dx \le p. \tag{30}$$

(Such functions exist because we can approximate m_p by a continuous function to any degree of accuracy by Lusin's Theorem [26], and then approximate the continuous function by a step function.)

Let σ_p be a measure preserving bijection of $[0,1]$ such that $n_p(\sigma_p x)$ is a non-increasing function. Such a bijection is easy to construct because n_p is a step function. Let $\ell_p(x) := n_p(\sigma_p x)$ and $g_p(x,y) := f_p(\sigma_p x, \sigma_p y)$. By the monotonicity and uniform boundedness of ℓ_p there exists a sequence $\{p_i\}_{i\ge1}$ decreasing to zero such that ℓ_{p_i} converges in L^2 to a limit function ℓ. Therefore by (29) and (30), $g_{p_i} \to g$ in L^2, where

$$g(x,y) := \ell(x)\ell(y).$$

By this and (27), g is a $\{0,1\}$-valued function. It is not difficult to see from this and the non-negativity of ℓ that ℓ must also be $\{0,1\}$-valued. (If $\ell(x) \notin \{0,1\}$ on some set of positive measure, then there may be a set A of positive measure where $\ell(x) \in (0,1)$, or there may be a set A of positive measure where $\ell(x) \in (1,\infty)$. In either case, $g(x,y) \notin \{0,1\}$ on $A \times A$.) Since ℓ is monotone decreasing, it follows that ℓ must be the indicator of an interval of the form $[0,b]$ for some $b \in [0,1]$. Lastly, (26) implies that $\iint g(x,y)dxdy = (6t)^{2/3}$, and therefore $b = (6t)^{1/3}$. Consequently, $g = \chi_t$.

The above argument establishes that for any collection $\{f_p\}_{p>0}$ such that $f_p \in F^*_{p,t}$ for each p, there is a sequence $\{p_i\}_{i\ge1}$ decreasing to zero such that $f_{p_i} \to \chi_t$ in the cut metric. The same argument can be extended to show that for any sequence $\{f_{p_i}\}$ such that $p_i \to 0$ and $f_{p_i} \in F^*_{p_i,t}$ for each i, there is a subsequence converging to χ_t in the cut metric. This proves the second assertion of the theorem. The last claim of the theorem follows from this and Theorem 4.1. □

4.8. Open questions

There are many questions that remain unresolved, even in the simple example of upper tails for triangle counts that has been analyzed in this section. For instance, what is the set of optimal solutions of the variational problem (15) in the broken replica symmetry phase (i.e. where the optimizer is not a constant)? Is the solution unique in the quotient space $\widetilde{\mathcal{W}}$, or can there exist multiple solutions? Is it possible to explicitly compute a nontrivial solution of (15) for at least some value of (p, t)? Is it possible to even numerically evaluate or approximate a solution using a computer? Does Theorem 4.3 characterize the full replica symmetric phase? If not, what is the exact phase transition boundary? What happens in the sparse case where p and t are both allowed to tend to zero? At the time of writing this paper, we do not know how to answer any of these questions.

Acknowledgements

The authors thank Amir Dembo for his crucial role in motivating this research and suggesting the formula for the rate function in the case of subgraph counts, Joel Spencer for hinting that Szemerédi's lemma may be useful, and Persi Diaconis for pointing us to the work of Lovász and coauthors (which helped in cutting down the size of the paper by a half). The authors also thank the referee for a very careful report.

Sourav Chatterjee's research was partially supported by NSF grants DMS-0707054 and DMS-1005312, and a Sloan Research Fellowship

S. R. S. Varadhan's research was partially supported by NSF grants DMS-0904701 and OISE-0730136.

References

[1] D. Aldous, Representations for partially exchangeable arrays of random variables, J. Multivariate Anal. 11 (4) (1981) 581–598.
[2] B. Bollobás, Random Graphs, second edition, Cambridge University Press, Cambridge, 2001.
[3] B. Bollobás, O. Riordan, Metrics for sparse graphs, in: Surveys in Combinatorics, in: LMS Lecture Notes Series, vol. 365, CUP, 2009, pp. 211–287.
[4] E. Bolthausen, F. Comets, A. Dembo, Large deviations for random matrices and random graphs, Private communication (2003).
[5] C. Borgs, J. Chayes, L. Lovász, V.T. Sós, K. Vesztergombi, Counting graph homomorphisms, in: Topics in Discrete Mathematics, in: Algorithms Combin., vol. 26, Springer, Berlin, 2006, pp. 315–371.
[6] C. Borgs, J. Chayes, L. Lovász, V.T. Sós, K. Vesztergombi, Convergent sequences of dense graphs I. subgraph frequencies, metric properties and testing, Adv. Math. 219 (6) (2008) 1801–1851.
[7] C. Borgs, J. Chayes, L. Lovász, V.T. Sós, K. Vesztergombi, Convergent sequences of dense graphs II. Multiway cuts and statistical physics. Preprint. 2007, Available at http://research.microsoft.com/en-us/um/people/borgs/papers/conright.pdf.
[8] S. Chatterjee, The missing log in large deviations for subgraph counts. Preprint. 2010, Available at http://arxiv.org/abs/1003.3498.
[9] S. Chatterjee, P.S. Dey, Applications of Stein's method for concentration inequalities. in Ann. Probab. 2009, Available at http://arxiv.org/abs/0906.1034 (in press).
[10] S. Chatterjee, P. Diaconis, A. Sly, Random graphs with a given degree sequence. Preprint. 2010, Available at.
[11] A. Dembo, O. Zeitouni, Large Deviations Techniques and Applications, second edition, Springer, 1998.
[12] P. Diaconis, S. Janson, Graph limits and exchangeable random graphs, Rend. Mat. Appl. 7 (28) (2008) 33–61. 1.
[13] R. Diestel, Graph Theory, second edition, Springer-Verlag, New York, 2000.
[14] P. Erdős, A. Rényi, On the evolution of random graphs, Publ. Math. Inst. Hungar. Acad. Sci. 5 (1960) 17–61.
[15] M. Freedman, L. Lovász, A. Schrijver, Reflection positivity, rank connectivity, and homomorphism of graphs, J. Amer. Math. Soc. 20 (2007) 37–51.
[16] A. Frieze, R. Kannan, Quick approximation to matrices and applications, Combinatorica 19 (1999) 175–220.
[17] D.N. Hoover, Row-column exchangeability and a generalized model for probability, in: Exchangeability in probability and statistics, North-Holland, Amsterdam-New York, Rome,1981, pp. 281–291.
[18] S. Janson, T. Łuczak, A. Ruciński, Random Graphs, Wiley, New York, 2000.
[19] L. Lovász, The rank of connection matrices and the dimension of graph algebras, Eur. J. Comb. 27 (2006) 962–970.
[20] L. Lovász, Connection matrices, in: G. Grimmett, C. McDiarmid (Eds.), Combinatorics, Complexity, and Chance: A Tribute to Dominic Welsh, Oxford University Press, Oxford, 2007, pp. 179–190.
[21] L. Lovász, V.T. Sós, Generalized quasirandom graphs, J. Comb. Theory B. 98 (1) (2008) 146–163.
[22] L. Lovász, B. Szegedy, Limits of dense graph sequences, J. Combin. Theory Ser. B 96 (6) (2006) 933–957.
[23] L. Lovász, B. Szegedy, Szemerédi's lemma for the analyst, Geom. Funct. Anal. 17 (1) (2007) 252–270.
[24] L. Lovász, B. Szegedy, Testing properties of graphs and functions. in Israel J. Math., 2007 (in press).
[25] L. Lovász, B. Szegedy, Contractors and connectors in graph algebras, J. Graph Theory 60 (1) (2009) 11–30.
[26] W. Rudin, Real and Complex Analysis, third edition, McGraw-Hill Book Co., New York, 1987.
[27] I.N. Sanov, On the probability of large deviations of random variables, Selected Translations in Mathematical Statistics and Probability 1 (1957) 213–224. Russian original: Mat. Sb..
[28] E. Szemerédi, Regular partitions of graphs. Problèmes combinatoires et théorie des graphes, Colloq. Internat. CNRS, Univ. Orsay, Orsay, 1976, Colloq. Internat. CNRS, 260, CNRS, Paris, 1978, pp. 399–401.

Communications on Stochastic Analysis
Vol. 6, No. 1 (2012) 1-13

Serials Publications
www.serialspublications.com

LARGE DEVIATIONS FOR RANDOM MATRICES

SOURAV CHATTERJEE* AND S. R. S. VARADHAN†

Dedicated to Professor K. R. Parthasarathy on the occasion of his 75th birthday

ABSTRACT. We prove a large deviation result for a random symmetric $n \times n$ matrix with independent identically distributed entries to have a few eigenvalues of size n. If the spectrum $\mathcal{S}$ survives when the matrix is rescaled by a factor of n, it can only be the eigenvalues of a Hilbert-Schmidt kernel $k(x, y)$ on $[0, 1] \times [0, 1]$. The rate function for k is $I(k) = \frac{1}{2} \int h(k(x, y)dxdy$ where h is the Cramér rate function for the common distribution of the entries that is assumed to have a tail decaying faster than any Gaussian. The large deviation for $\mathcal{S}$ is then obtained by contraction.

1. Introduction

We are interested in the large deviation behavior of random $n \times n$ symmetric matrices $X(\omega) = \{x_{i,j}(\omega)\}$ where the entries for $j \geq i$ are independent identically distributed real random variables having a common distribution μ. Let $\Lambda_n(\omega) = \{\lambda_j^n(\omega); 1 \leq j \leq n\}$ be the set of n real eigenvalues of this random matrix. If we assume that the mean is 0 and the variance is equal to σ^2, according to a theorem of Wigner [6], when divided by $\sqrt{n}$, the empirical distribution of these eigenvalues converges in probability to the semi-circle law, i.e. for any bounded continuous function $f : \mathbb{R} \to \mathbb{R}$

$$\lim_{n \to \infty} \frac{1}{n} \sum_{j=1}^{n} f\left(\frac{\lambda_j(\omega)}{\sqrt{n}}\right) = \int f(y)\phi_\sigma(y)dy$$

in probability, where

$$\phi_\sigma(y) = \begin{cases} 0 & \text{if } |x| \geq 2\sigma, \\ \frac{1}{2\pi\sigma^2}\sqrt{4\sigma^2 - x^2} & \text{if } |x| \leq 2\sigma. \end{cases}$$

In fact it is known [1] that under suitable additional assumptions, for any $\epsilon > 0$,

$$\mathbb{P}\Big[\max_{1 \leq j \leq n} |\lambda_j^n| \geq (2\sigma + \epsilon)\sqrt{n}\Big] \to 0.$$

Received 2011-6-23; Communicated by K. B. Sinha.
2000 *Mathematics Subject Classification.* 60B20; 15B52.
Key words and phrases. Large deviations, eigenvalues, random matrices.
* Sourav Chatterjee's research was partially supported by NSF grants DMS-0707054 and DMS-1005312, and a Sloan Research Fellowship.
† S. R. S. Varadhan's research was partially supported by NSF grants DMS-0904701 and OISE-0730136.

Large deviation results for the distribution of eigenvalues to be different from the Wigner distribution has been considered by Ben Arous and Guionnet [2]. If we decide to divide by n rather than $\sqrt{n}$, and denote the resulting spectrum by $S_n(\omega) = \frac{1}{n}\Lambda_n(\omega)$, then

$$\mathbb{P}[\sup_{\lambda \in S_n} |\lambda| \geq \epsilon] \to 0.$$

If we drop the assumption that the mean is 0, then there will be one large eigenvalue λ_{max} of size n in Λ_n, with $\frac{\lambda_{max}}{n} \simeq \int x d\mu$ and the remaining eigenvalues will follow the semi-circle law as before when divided by $\sqrt{n}$. It is also possible to consider scaling by n^α for $\frac{1}{2} < \alpha < 1$, but they would likely behave like moderate deviations and exhibit Gaussian behavior with rate $n^{2\alpha}$.

Let $\mathcal{S}$ denote the set of all finite or countable collections $S = \{\lambda_j\}$ of positive as well as negative real numbers, with repetitions allowed, that have the property

$$\sum_{\lambda \in S} |\lambda|^2 = \sum_j |\lambda_j|^2 < \infty.$$

They are all possible spectra of self adjoint Hilbert-Schmidt operators. The topology on $\mathcal{S}$ will be ordinary convergence as real numbers of the corresponding eigenvalues outside any arbitrarily small interval around 0. Equivalently, it is the minimal topology such that for any bounded continuous function f that is 0 in some interval around 0, the sum $\sum_{\lambda \in S} f(\lambda)$ is continuous as a map of $\mathcal{S} \to \mathbb{R}$. One can easily construct a metric for this topology. We enumerate separately the positive and negative values in S in decreasing order of their absolute values as $\{u_j^+\}, \{u_j^-\}$. If one or both of them is only a finite set or empty we augment it by adding 0's. If two points S, S' in $\mathcal{S}$ are enumerated as $\{u_j^+\}, \{u_j^-\}$ and $\{v_j^+\}, \{v_j^-\}$, we define the distance

$$d(S, S') := \sum_{j=1}^{\infty} \frac{|u_j^+ - v_j^+|}{2^{j+1}} + \sum_{j=1}^{\infty} \frac{|u_j^- - v_j^-|}{2^{j+1}}.$$

The random matrix yields a random spectrum and after normalization by n, we obtain a random element of $\mathcal{S}$. Its distribution yields a sequence $\{P_n\}$ of probability measures on $\mathcal{S}$. We will prove a large deviation result for P_n with rate n^2. We assume that for all $\theta > 0$,

$$\int \exp[\theta\, x^2] \mu(dx) < \infty. \tag{1.1}$$

In particular this implies that the moment generating function $M(\theta) = \int e^{\theta\, x} \mu(dx)$ is finite for all θ and satisfies

$$\limsup_{|\theta| \to \infty} \frac{1}{\theta^2} \log M(\theta) = 0$$

and the conjugate rate function of Cramér

$$h(x) = \sup_\theta [\theta\, x - \log M(\theta)] \tag{1.2}$$

satisfies

$$\liminf_{|x| \to \infty} \frac{h(x)}{x^2} = +\infty.$$

The condition (1.1) is important, because an eigenvalue of size n can be produced by a single entry of size n in the random matrix and we would like this to have probability that is super-exponentially small in the scale n^2.

The random symmetric $n \times n$ matrix $X(\omega) = \{x_{i,j}(\omega)\}$ is first mapped into a symmetric kernel $k(x, y, \omega)$

$$k(x, y, \omega) = \sum_{i,j=1}^{n} x_{i,j}(\omega)\mathbf{1}_{J_i^n}(x)\mathbf{1}_{J_j^n}(y) \tag{1.3}$$

where J_i^n is the interval $[\frac{i-1}{n}, \frac{i}{n}]$. This induces a family of probability measures Q_n on the space of symmetric kernels $k(x, y)$ on $D = [0, 1] \times [0, 1]$. We will restrict ourselves to $\mathcal{K} = \{k : \int\int_D |k(x, y)|^2\, dxdy < \infty\}$. Then $k(x, y)$ defines on $L^2[0, 1]$ a Hilbert-Schmidt operator which has a countable spectrum with 0 as the only limit point. Actually for each fixed n the range of the map $X \to k$ is a finite dimensional subspace of simple functions $\mathcal{K}_n \subset \mathcal{K}$. The nonzero spectrum of $k(x, y, \omega)$ is the same as that of $\{\frac{x_{i,j}}{n}\}$ and we can obtain P_n from Q_n through the natural map $\mathcal{K} \to \mathcal{S}$ that takes any k to its set of eigenvalues. We define on $\mathcal{K}$ the following rate function

$$I(k(\cdot, \cdot)) = \frac{1}{2}\int_0^1 \int_0^1 h(k(x, y))dxdy \tag{1.4}$$

with h given by (1.2).

Any permutation $\sigma \in \Pi(n)$ of the rows and columns of X, mapping $\{x_{i,j}\} \to \{x_{\sigma(i),\sigma(j)}\}$ leaves the set of eigen-values of X invariant and the group G of measure preserving transformations σ of $[0, 1]$ onto itself lifts to an action on $\mathcal{K}$ mapping $k \to \sigma k$ where $\sigma k(x, y) := k(\sigma x, \sigma y)$. The map $k \to \sigma k$ leaves $I(k(\cdot, \cdot))$ as well as the spectrum $S(k)$ of k invariant. For establishing the large deviations of P_n on $\mathcal{S}$ it is therefore enough to prove a large deviation principle for the images $\tilde{Q}_n$ of Q_n on $\tilde{\mathcal{K}} = \mathcal{K}/G$.

We will be working with the space $\mathcal{K}$ of symmetric kernels $k(x, y)$ on D. If $I(k) < \infty$, then the operator defined by $k(x, y)$ on $L^2[0, 1]$ is Hilbert-Schmidt and has a countable spectrum with 0 as the only limit point. We need to show some sort of continuity of the map $\tilde{\mathcal{K}} \to \mathcal{S}$ mapping the G-orbit $\tilde{k}$ of k to its spectrum $S(k)$, in order to transfer the large deviation result from $\tilde{\mathcal{K}}$ to $\mathcal{S}$. This requires a topology on $\tilde{\mathcal{K}}$ that will be inherited from $\mathcal{K}$. The weak topology on $\mathcal{K}$ turns out to be too weak and the strong or L^1 topology too strong. What works is the topology induced by the cut metric

$$d_{\square}(k_1, k_2) = \sup_{\substack{|\phi|\le 1\\ |\psi|\le 1}} \left| \int (k_1(x, y) - k_2(x, y))\phi(x)\psi(y)dxdy \right|, \tag{1.5}$$

ϕ, ψ being Borel measurable functions on $[0, 1]$. Equivalently,

$$d_{\square}(k_1, k_2) = \sup_{A\times B} \left| \int_{A\times B} (k_1(x, y) - k_2(x, y))dxdy \right|, \tag{1.6}$$

where the supremum is taken over all Borel subsets A, B of $[0,1]$. The induced metric on $\tilde{\mathcal{K}}$ is

$$d_\Box(k_1, k_2) = \inf_\sigma d_\Box(\sigma k_1, k_2) = \inf_\sigma d_\Box(k_1, \sigma k_2) = \inf_{\sigma_1, \sigma_2} d_\Box(\sigma_1 k_1, \sigma_2 k_2)$$

where $\sigma k(x,y) = k(\sigma x, \sigma y)$. We can define on $\mathcal{S}$ the rate function

$$J(S) = \inf_{k : S(k) = S} I(k).$$

Our main result is the following.

Theorem 1.1. *Under assumption (1.1) the sequence of measures P_n on $\mathcal{S}$ satisfies a large deviation property with rate function $J(S)$, i.e for closed $C \subset \mathcal{S}$*

$$\limsup_{n\to\infty} \frac{1}{n^2} \log P_n(C) \le - \inf_{S \in C} J(S)$$

and for $U \subset \mathcal{S}$ that are open

$$\liminf_{n\to\infty} \frac{1}{n^2} \log P_n(U) \ge - \inf_{S \in U} J(S).$$

This is based on a large deviation principle for $\tilde{Q}_n$ on $\tilde{\mathcal{K}}$ in the cut topology with rate function $I(k)$.

Theorem 1.2. *Under assumption (1.1) the sequence of measures $\tilde{Q}_n$ on $\tilde{\mathcal{K}}$ satisfies a large deviation property with rate function $I(\tilde{k})$, i.e for closed $C \subset \tilde{\mathcal{K}}$*

$$\limsup_{n\to\infty} \frac{1}{n^2} \log \tilde{Q}_n(C) \le - \inf_{\tilde{k} \in C} I(\tilde{k})$$

and for $U \subset \tilde{\mathcal{K}}$ that are open

$$\liminf_{n\to\infty} \frac{1}{n^2} \log \tilde{Q}_n(U) \ge - \inf_{\tilde{k} \in U} I(\tilde{k}).$$

To make the connection we need the map $k \to S(k)$ to be continuous in the cut topology. It is valid, provided we restrict it to sets $\mathcal{K}^\ell$ of the form $\mathcal{K}^\ell = \{k : |k(x,y)| \le \ell\}$ for some $\ell < \infty$. This requires truncation at level ℓ and then removing the cut-off. Condition (1.1) provides super-exponential bounds in the Hilbert-Schmidt norm for the error due to cutoff and that is used to complete the proof.

Similar methods were used in [3] to study the large deviation behavior of the number of triangles or other finite subgraphs in random graphs as the number of vertices goes to ∞ but the probability of an edge being connected remains fixed at some $p > 0$. This will correspond to each $x_{i,j}$ taking the values 0 or 1.

2. Some Useful Lemmas.

We will be working with partitions of the unit interval $[0,1]$ into a finite disjoint union of subintervals. We will not worry about the end points. We can adopt any convention that makes it a true partition. For each partition $\mathcal{P}$ of the unit interval into m subintervals $\{J_i\}$ there is a corresponding partition of D into m^2 sub-squares $\{J_i \times J_j\}$.

For each integer m we have the special partition $\mathcal{P}_m$ of the unit interval into m equal subintervals and they will be denoted by $J_i^m = [\frac{i-1}{m}, \frac{i}{m}]$. We denote by $\mathcal{K}_m \subset \mathcal{K}$ the space of symmetric kernels of the form

$$f(x,y) = \sum_{i,j} f_{i,j} \mathbf{1}_{J_i^m}(x) \mathbf{1}_{J_j^m}(y).$$

This provides a faithful representation of symmetric matrices of size $m \times m$ as elements of $\mathcal{K}_m$. If $|f_{i,j}| \leq \ell$, then the corresponding $f \in \mathcal{K}_m^\ell = \mathcal{K}^\ell \cap \mathcal{K}_m$. The following is a simple, but useful lemma.

Lemma 2.1. *If $f(x,y) = \sum f_{i,j} \mathbf{1}_{J_i}(x) \mathbf{1}_{J_j}(y)$ and $g(x,y) = \sum g_{i,j} \mathbf{1}_{J_i}(x) \mathbf{1}_{J_j}(y)$ are both simple functions with respect to the same partition $J_1, \ldots, J_m$ of $[0,1]$, then in calculating the distance*

$$d_\square(f,g) = \sup_{A \times B} \left| \int_{A \times B} (f(x,y) - g(x,y)) dx dy \right|$$

the sets A, B can be restricted to sets of the form $\bigcup_{i \in \mathcal{N}} J_i$ where $\mathcal{N} \subset \{1, 2, \ldots, m\}$.

Proof. First note that

$$\begin{aligned}
\int_{A \times B} (f(x,y) - g(x,y)) dx dy &= \sum_{i,j} \int_{A \cap J_i \times B \cap J_j} (f(x,y) - g(x,y)) dx dy \\
&= \sum_{i,j} |A \cap J_i| |B \cap J_j| (f_{i,j} - g_{i,j}) \\
&= \sum_{i,j} a_i b_j |J_i| |J_j| (f_{i,j} - g_{i,j}),
\end{aligned}$$

where $a_i = \frac{|A \cap J_i|}{|J_i|}$ and $b_i = \frac{|B \cap J_i|}{|J_i|}$. It is now clear that the supremum of the absolute value is achieved when each a_i and b_j is either 0 or 1. □

Remark 2.2. If f and g are defined in terms of two different partitions of $[0,1]$, they can both be viewed as defined with respect to the finite partition which is their common refinement.

Since our large deviation result on $\mathcal{S}$ is deduced from a large deviation result on $\mathcal{K}$, we need some continuity property of the map $k \to S(k)$ of $\mathcal{K} \to \mathcal{S}$.

Lemma 2.3. *For any $\ell < \infty$, the map $k \to S(k)$ from $\mathcal{K} \to \mathcal{S}$ is continuous in the cut topology when restricted to $\mathcal{K}^\ell = \{k : |k(x,y)| \leq \ell\}$.*

Proof. If $f(\lambda) = 0$ near 0 then $\lambda^{-3} f(\lambda)$ can be approximated uniformly on $[-\ell, \ell]$ by a polynomial in λ. Therefore $f(\lambda)$ is approximated by a polynomial involving only powers λ^m for $m \geq 3$. Since we have a bound on $\sum_{\lambda \in S} |\lambda|^2$ it is enough to show that

$$k \to \sum_{\lambda \subset S(k)} \lambda^m$$

are continuous maps of $\mathcal{K}^\ell \to \mathcal{S}$ for each $m \geq 3$. It is elementary to check that since $|k(x,y)| \leq \ell$, the maps

$$k \to \int_{[0,1]^m} k(x_1, x_2) \cdots k(x_{m-1}, x_m) k(x_m, x_1) dx_1 \ldots dx_m = \sum_{\lambda \in S(k)} \lambda^m$$

are continuous maps in the cut topology from $\mathcal{K}^\ell \to \mathcal{S}$ provided $m \geq 3$. We start with

$$\int_{[0,1]^m} k_n(x_1, x_2) \cdots k_n(x_{m-1}, x_m) k_n(x_m, x_1) dx_1 \dots dx_m$$

which can be written as

$$\int_{[0,1]^{m-2}} \int dx_3 \dots dx_m \int_{[0,1]^2} k_n(x_1, x_2)\phi_n(x_1, x_3, \dots, x_n)\psi_n(x_2, x_3, \dots, x_n) dx_1 dx_2$$

where ϕ_n and ψ_n are uniformly bounded. If $d_\square(k_n, k) \to 0$, we can then replace $k_n(x_1, x_2)$ by $k(x_1, x_2)$. This is repeated for each factor. □

We will also need the following lemmas: a multicolor version of Szemerédi's regularity lemma for graphs that can be found in [4] and its consequence.

Lemma 2.4. *Given any $\epsilon > 0$ and integers r and m, there exists M and n_0 such that if the edges of a graph $\mathcal{G}_n$ of size $n \geq n_0$ are colored with any one of r colors, then the vertex set can be partitioned into sets $V_0, \dots, V_p$ for some p in the range $m \leq p \leq M$ so that $|V_0| \leq \epsilon n$, $|V_1| = |V_2| = \cdots = |V_p| = K$ and all but at most ϵp^2 pairs (V_i, V_j) satisfy the following regularity condition. For any $X \subset V_i, Y \subset V_j$ with $i, j \geq 1$ and $|X|, |Y| \geq \epsilon K$ we have*

$$|d_\nu(X, Y) - d_\nu(V_i, V_j)| < \epsilon,$$

where $d_\nu(X, Y)$ is the proportion of edges between X and Y that are colored with color ν.

Lemma 2.5. *For any $\ell < \infty$ and $\epsilon > 0$, there is a compact set of simple functions $W_{\ell,\epsilon} \subset \mathcal{K}$ and $n_0(\epsilon, \ell)$ such that for $n \geq n_0$ and any $k \in \mathcal{K}_n^\ell$, there exists $f \in W_{\ell,\epsilon}$ and $\sigma \in \Pi(n)$ such that*

$$d_\square(\sigma k, f) < \epsilon.$$

Proof. Let $\epsilon > 0$ be given. Let the integer r be chosen such that $\frac{\ell}{r} < \frac{\epsilon}{2}$. Let $\epsilon' = \frac{\epsilon}{12r\ell}$ and $m = \frac{1}{\epsilon'}$. Let $x_{i,j}$ be the value of k in the rectangle $J_i^n \times J_j^n$. The interval $[-\ell, \ell]$ is divided in to r disjoint equal intervals of length $\frac{2\ell}{r}$ and the edge (i, j) is colored according to the interval into which $x_{i,j}$ falls. The 'color' is defined as the value of the mid point of the interval. The color of the edge (i, j) is then a real number $y_{i,j}$ which can equal any one from the finite set $z_1, z_2, \dots, z_r$ and $|y_{i,j} - x_{i,j}| \leq \frac{\epsilon}{2}$. We apply the multicolor version of Szemerédi's regularity theorem (Lemma 2.4) with parameters (ϵ', m, r) to obtain a partition $V_0, V_1, \dots V_p$ of $\{1, 2, \dots, n\}$ with the following properties. $|V_0| = K' \leq \epsilon' n$. $|V_1| = |V_2| = \cdots = |V_p| = K$. If $a_{s,i,j}$ be the proportion of edges between V_i and V_j that have color z_s, then for all but $\epsilon' p^2$ pairs (V_i, V_j), $i, j \geq 1$ $i \neq j$, for any two subsets $X \subset V_i$, $Y \subset V_j$ with $|X| \geq \epsilon' K, |Y| \geq \epsilon' K$ the proportion $b_{s,i,j}$ of edges of color z_s, between X and Y satisfies

$$\sup_s |b_{s,i,j} - a_{s,i,j}| \leq \epsilon'.$$

Let us divide the unit interval into subintervals $J_0, \ldots, J_p$ where $J_0 = [0, \frac{K'}{n}]$ and for $i \geq 1$, $J_i = [\frac{K'+(i-1)K}{n}, \frac{K'+iK}{n}]$. We construct a function $f \in \mathcal{K}^\ell$ as

$$f(x,y) = \sum_{s=1}^{r} z_s f(s,x,y),$$

where

$$f(s,x,y) = \sum_{i,j=0}^{p} a_{s,i,j} \mathbf{1}_{J_i}(x) \mathbf{1}_{J_j}(y).$$

For fixed ℓ and ϵ as long as p remains bounded such functions vary over a compact subset of $L^1(D)$.

For the permutation $\sigma \in \Pi(n)$ of the vertices, we rearrange the order of the vertices, so that those in V_0 corresponds to the first K' indices and for $1 \leq i \leq p$, those in V_i correspond respectively to indices in the range $(K'+(i-1)K+1, K'+iK)$. We will denote by k' the image of $\{x_{\sigma(i),\sigma(j)}\}$ in $\mathcal{K}_n^\ell$. We define $c(s,i,j) = 1$ if $x_{\sigma(i),\sigma(j)}$ belongs to the interval with mid point z_s. Otherwise it is 0. With

$$k'(s,x,y) = \sum_{i,j=1}^{n} c(s,i,j) J_i^n(x) J_j^n(y)$$

and

$$k'(x,y) = \sum_{i,j=1}^{n} x_{\sigma(i),\sigma(j)} J_i^n(x) J_j^n(y)$$

we have that for each x, y,

$$\left| k'(x,y) - \sum_s z_s k'(s,x,y) \right| \leq \frac{\epsilon}{2}.$$

We need to estimate

$$\left| \int_{A\times B} [k'(x,y) - f(x,y)] dx dy \right| \leq \frac{\epsilon}{2} + r\ell \sup_{1\leq s\leq r} \left| \int_{A\times B} [k'(s,x,y) - f(s,x,y)] dx dy \right|.$$

For each value of s, we will estimate

$$\int_{A\times B} [k'(s,x,y) - f(s,x,y)] dx dy = \sum_{i,j=0}^{p} \int_{A\cap J_i \times B\cap J_j} [k'(s,x,y) - f(s,x,y)] dx dy.$$

The summation over (i,j) will be split into several groups. $F_0 = \{(i,i) : i \geq 1\}$. $F_1 = \{(i,j) : i = 0\} \cup \{(i,j) : j = 0\}$. $F_2 = \{(i,j)\}$ is the collection of at most $\epsilon' p^2$ exceptional pairs from Lemma 2.4. $F_3 = \{(i,j)\}$ for which either $|A \cap J_i| \leq \epsilon' |J_i|$ or $|B \cap J_j| \leq \epsilon' |J_j|$. F_4 will be the rest. Then:

$$\left| \sum_{(i,j)\in F_0} \int_{A\cap J_i \times B\cap J_j} [k'(s,x,y) - f(s,x,y)] dx dy \right| \leq \frac{pK^2}{n^2} \leq \frac{1}{p} \leq \epsilon',$$

$$\left| \sum_{(i,j)\in F_1} \int_{A\cap J_i \times B\cap J_j} [k'(s,x,y) - f(s,x,y)] dx dy \right| \leq 2\epsilon',$$

$$\left|\sum_{(i,j)\in F_2}\int_{A\cap J_i\times B\cap J_j}[k'(s,x,y)-f(s,x,y)]dxdy\right| \le \epsilon' p^2|J_i||J_j| \le \epsilon' p^2\frac{1}{p^2}=\epsilon',$$

$$\left|\sum_{(i,j)\in F_3}\int_{A\cap J_i\times B\cap J_j}[k'(s,x,y)-f(s,x,y)]dxdy\right| \le \epsilon'\sum_{i,j}|J_i||J_j| \le \epsilon'.$$

Finally in the remaining set F_4, since $|A\cap J_i| \ge \epsilon'|J_i|$ and $|B\cap J_j| \ge \epsilon'|J_j|$, it follows that

$$\left|\int_{A\cap J_i\times B\cap J_j}[k'(s,x,y)-f(s,x,y)]dxdy\right| \le \epsilon'|J_i||J_j|$$

and hence

$$\left|\sum_{(i,j)\in F_4}\int_{A\cap J_i\times B\cap J_j}[k'(s,x,y)-f(s,x,y)]dxdy\right| \le \epsilon'\sum_{i,j}|J_i||J_j| \le \epsilon'.$$

Adding them up gives

$$\left|\int_{A\times B}[k'(x,y)-f(x,y)]dxdy\right| \le \frac{\epsilon}{2}+6r\ell\epsilon' \le \epsilon.$$

□

3. Lower Bound

The lower bound for $\tilde{Q}_n$ on $\tilde{\mathcal{K}}$ can be proved by proving a lower bound for Q_n on $\mathcal{K}$ and it can be done without truncation and under the (weaker) assumption that $\int e^{\theta\, x^2}d\mu(x)<\infty$ for some θ. This implies a lower bound $h(x)\ge c|x|^2$ when $|x|$ is large.

Theorem 3.1. *Under the weaker assumption that for some $\theta>0$,*

$$\int e^{\theta\, x^2}d\mu(x)<\infty,$$

it follows that for any $f\in\mathcal{K}$ such that

$$I(f)=\frac{1}{2}\int h(f(x,y))dxdy<\infty$$

and for any $\delta>0$,

$$\liminf_{n\to\infty}\frac{1}{n^2}\log Q_n[d_\Box(k,f)<\delta]\ge -I(f).$$

Proof. Since h is a convex function of its argument, for any integer q we can replace f by a simple function g of its averages over $J_r^q\times J_s^q$ so that

$$g(x,y)=\sum_{r,s=1}^{q} g_{r,s}\mathbf{1}_{J_r^q}(x)\mathbf{1}_{J_s^q}(y),$$

where

$$g_{r,s}=q^2\int_{J_r^q\times J_s^q}f(x,y)dxdy.$$

For any q, $I(g) \le I(f)$ and for large q, $d_\square(f,g) < \frac{\delta}{2}$. It suffices to show that

$$\liminf_{n\to\infty} \frac{1}{n^2} \log Q_n[d_\square(g,k) \le \frac{\delta}{2}] \ge -I(g).$$

We have

$$k(x,y) = \sum_{i,j=1}^{n} x_{i,j} \mathbf{1}_{J_i^n}(x) \mathbf{1}_{J_j^n}(y)$$

and we can view both k and g as members of $\mathcal{K}_{nq}$. The distance $d_\square(g,k)$ can be computed as

$$d_\square(g,k) = \sup_{A,B} \left| \int_{A\times B} (g(x,y) - k(x,y)) dxdy \right|,$$

where A and B are taken to be unions of sub-collections of intervals of the form $\{[\frac{i-1}{nq}, \frac{i}{nq}]\}$. There are exactly $2^{nq} \times 2^{nq}$ such pairs A, B. We now tilt the measure so that $\{x_{i,j}\}$ remains symmetric and $\{x_{i,j}\}$ for $j \ge i$ are still independent but the distribution of $x_{i,j}$ is tilted from μ to $\mu_{i,j}$ given by

$$\mu_{i,j}(dx) = \frac{1}{M(\theta_{r,s})} \exp[\theta_{r,s}\, x] \mu(dx)$$

for $(r-1)n < iq \le rn$ and $(s-1)n < jq \le sn$ where $\theta_{r,s} = h'(g_{r,s})$, or equivalently $g_{r,s} = \frac{M'(\theta_{r,s})}{M(\theta_{r,s})}$ for $1 \le r, s \le q$. Let Q_n^g be the law of the new k. The law of large numbers applies to each such pair A, B with uniform (in A and B) exponential error bounds of $e^{-cn^2 + o(n^2)}$ for some $c > 0$. Therefore,

$$Q_n^g\left[\left|\int_{A\times B} (g(x,y) - k(x,y)) dxdy\right| \ge \frac{\delta}{2}\right] \le e^{-c(\delta)n^2 + o(n^2)}.$$

Since $2^{2nq} \ll e^{cn^2}$, it follows that

$$Q_n^g[d_\square(g,k) \ge \frac{\delta}{2}] \to 0.$$

The relative entropy of Q_n^g with respect to Q_n is easily computed to be $n^2 I(g)$.

The following entropy lower bound using Jensen's inequality, establishes the large deviation lower bound. Suppose α, β are two probability measures and $\beta \ll \alpha$ with $H = \int \phi \log \phi \, d\alpha < \infty$ where $\phi = \frac{d\beta}{d\alpha}$. Since $y \log y \ge -e^{-1}$ for all $y \ge 0$, $\int |\phi \log \phi| d\alpha \le H + 2e^{-1}$. Therefore

$$\begin{aligned}
\alpha(A) &\ge \int_A \phi^{-1} d\beta = \int_A \exp[-\log \phi] d\beta \\
&= \beta(A) \frac{1}{\beta(A)} \int_A \exp[-\log \phi] d\beta \\
&\ge \beta(A) \exp\left[-\frac{1}{\beta(A)} \int_A \phi \log \phi d\alpha\right] \\
&\ge \beta(A) \exp\left[-\frac{1}{\beta(A)} \int |\phi \log \phi| d\alpha\right] \\
&\ge \beta(A) \exp\left[-\frac{1}{\beta(A)} [H + 2e^{-1}]\right].
\end{aligned}$$

Taking $\alpha = Q_n$ and $\beta = Q_n^g$ and $A = \{k : d_\Box(k,g) \le \frac{\delta}{2}\}$ we have $H = n^2 I(g)$ and $\beta(A) \simeq 1$. □

4. Upper Bound

We assume that $|x_{i,j}| \le \ell$. According to a result of Lovász and Szegedy [5], for any finite ℓ, the set $\tilde{\mathcal{K}}^\ell = \{\tilde f : \sup_{x,y} |f(x,y)| \le \ell\}$ is compact in $\tilde K$. Therefore in order to prove the large deviation upper bound for $\tilde Q_n$ with rate function $I(f)$ it is sufficient to prove the local version of the upper bound.

Theorem 4.1. *Let $\tilde f \in \tilde{\mathcal{K}}$. Then*

$$\limsup_{\epsilon\to 0} \limsup_{n\to\infty} \frac{1}{n^2} \log \tilde Q_n[\tilde k : d_\Box(\tilde k, \tilde f) < \epsilon] \le -I(\tilde f).$$

Proof. The theorem is proved in several steps. Closed balls $B(f,\epsilon)$ in $\mathcal{K}$ of the form $\{k : d_\Box(k,f) \le \epsilon\}$ are weakly closed. It is not hard to prove that

$$\limsup_{\epsilon\to 0} \limsup_{n\to\infty} \frac{1}{n^2} \log Q_n[k : d_\Box(k, f) \le \epsilon] \le -I(f). \tag{4.1}$$

The argument goes as follows. Assume $I(f) < \infty$. Given $\delta > 0$ pick a nice g such that

$$\frac{1}{2}\left[\int f(x,y)g(x,y)dxdy - \int \log\left[\int e^{z\,g(x,y)}\mu(dz)\right]dxdy\right] \ge I(f) - \delta$$

and apply Cramér type estimate using the moment generating function for the half space

$$H_{f,g,\epsilon} = \left\{k : \langle k, g\rangle \ge \inf_{k'\in B(f,\epsilon)} \langle k', g\rangle\right\}$$

which contains $B(f,\epsilon)$. This gives

$$Q_n[B(f,\epsilon)] \le Q_n[H_{f,g,\epsilon}] \le \exp\left[-\frac{n^2}{2} \inf_{k\in B(f,\epsilon)} \langle k, g\rangle\right] \mathbb{E}^{Q_n} \exp\left[\frac{n^2}{2}\langle k, g\rangle\right].$$

It is easy to see (because $x_{i,j} = x_{j,i}$) that

$$\limsup_{n\to\infty} \frac{1}{n^2} \log \mathbb{E}^{Q_n} \exp\left[\frac{n^2}{2}\langle k, g\rangle\right] = \frac{1}{2}\int_D \log\left[\int \exp[zg(x,y)]\mu(dz)\right]dxdy$$

and

$$\lim_{\epsilon\to 0} \inf_{k\in B(f,\epsilon)} \langle k, g\rangle = \langle f, g\rangle.$$

We can let $\delta \to 0$ at the end. If $I(f) = \infty$, pick g such that

$$\frac{1}{2}\left[\int f(x,y)g(x,y)dxdy - \int \log\left[\int e^{z\,g(x,y)}\mu(dz)\right]dxdy\right] \ge L$$

and let $L \to \infty$ in the end.

This proves (4.1). But to prove it for balls in $\tilde{\mathcal{K}}$ we need to estimate the probability of the pre-image in $\mathcal{K}$ of a ball in $\tilde{\mathcal{K}}$. This amounts to showing that

$$\limsup_{\epsilon\to 0} \limsup_{n\to\infty} \frac{1}{n^2} \log Q_n[k : \inf_{\sigma\in G} d_\Box(k, \sigma f) < \epsilon] \le -I(\tilde f).$$

We start with the set

$$U(\tilde{f},\epsilon) = \cup_{\sigma\in G} B(\sigma f,\epsilon).$$

We saw that for each $f \in \mathcal{K}$, (4.1) holds. We need to replace the union over $\sigma \in G$ by a union over a finite collection. According to Lemma 2.5 there is a compact set $W_{\ell,\epsilon} \subset \mathcal{K}$ such that its orbit by $\Pi(n)$ nearly covers $\mathcal{K}_n^\ell$ for all sufficiently large n. More precisely for $n \geq n_0(\epsilon,\ell)$,

$$\sup_{k\in\mathcal{K}_n^\ell} \inf_{\sigma\in\Pi(n)} \inf_{g\in W_{\ell,\epsilon}} d_\square(\sigma g,k) < \epsilon.$$

$W_{\ell,\epsilon}$ is compact in the cut topology (and even in L^1) and can be covered by the union of balls $B(g,\epsilon)$ of radius ϵ centered around g from a finite collection $F_{\ell,\epsilon}$. Therefore for sufficiently large n

$$\cup_{\sigma\in\Pi(n)} \cup_{g\in F_{\ell,\epsilon}} B(\sigma g, 2\epsilon) \supset \mathcal{K}_n^\ell.$$

We need to estimate the probability under Q_n of

$$\cup_{\sigma'\in\Pi(n)} \cup_{g\in F_{\ell,\epsilon}} \cup_{\sigma\in G}[B(\sigma' g, 2\epsilon) \cap B(\sigma f,\epsilon)]$$

which is at most $n!$ times

$$\sup_{\sigma'\in\Pi(n)} Q_n[\cup_{g\in F_{\ell,\epsilon}} \cup_{\sigma\in G} [B(\sigma' g, 2\epsilon) \cap B(\sigma f,\epsilon)]].$$

Since $F_{\ell,\epsilon}$ is a finite set independent of n and Q_n is invariant under $\Pi(n) \subset G$, it is enough to show that for each $g \in F_{\ell,\epsilon}$

$$\limsup_{\epsilon\to 0} \limsup_{n\to\infty} \frac{1}{n^2} \log Q_n[B(g,2\epsilon) \cap \cup_{\sigma\in G} B(\sigma f,\epsilon)] \leq -I(f).$$

If the intersection is nonempty, then there is a σ' such that $B(g,2\epsilon) \subset B(\sigma' f, 5\epsilon)$. Thus, from (4.1), the invariance of $I(\sigma f)$ under $\sigma \in G$ and the lower semicontinuity of $I(\cdot)$ on $\tilde{\mathcal{K}}$ we see that there is a $\sigma' \in G$ such that

$$\begin{aligned}
&\limsup_{\epsilon\to 0} \limsup_{n\to\infty} \frac{1}{n^2} \log Q_n[B(g,2\epsilon) \cap \cup_{\sigma\in G} B(\sigma f,\epsilon)] \\
&\qquad \leq \limsup_{\epsilon\to 0} \limsup_{n\to\infty} \frac{1}{n^2} \log Q_n[B(\sigma' f, 5\epsilon)] \\
&\qquad = -I(\sigma' f) = -I(f).
\end{aligned}$$

□

5. Truncation

Given $X = \{x_{i,j}\}$ we truncate it at level ℓ. Let $x = f_\ell(x) + g_\ell(x)$ where

$$f_\ell(x) = \begin{cases} x & \text{if } |x| \leq \ell \\ \ell & \text{if } x \geq \ell \\ -\ell & \text{if } x \leq -\ell \end{cases}$$

and $g_\ell(x) = x - f_\ell(x)$. There is a corresponding decomposition of $k = f_\ell(k) + k - f_\ell(k)$ for $k \in \mathcal{K}$. We have the estimate

$$\limsup_{\ell\to\infty} \limsup_{n\to\infty} \frac{1}{n^2} \log \mathbb{E} \exp\Big[\theta \sum_{i,j} [x_{i,j} - f_\ell(x_{i,j})]^2\Big] = 0$$

for any $\theta > 0$. If $\{f_\ell(x_{i,j})\}$ and $\{x_{i,j}\}$ are mapped respectively into k_ℓ and k, we have super exponential estimates on $\Delta(\ell) = \int_D |f_\ell(k(x,y)) - k(x,y)|^2 dxdy$:

$$Q_n[\Delta(\ell) \geq \epsilon] \leq e^{-n^2\theta\epsilon^2} \mathbb{E} \exp\Big[\theta \sum_{i,j} [x_{i,j} - f_\ell(x_{i,j})]^2\Big].$$

Therefore

$$\limsup_{\ell\to\infty} \limsup_{n\to\infty} \frac{1}{n^2} \log Q_n[\Delta(\ell) \geq \epsilon] \leq -\theta\epsilon^2.$$

Since $\theta > 0$ is arbitrary, for any $\epsilon > 0$,

$$\limsup_{\ell\to\infty} \limsup_{n\to\infty} \frac{1}{n^2} \log Q_n[\Delta(\ell) \geq \epsilon] = -\infty.$$

It is easy to complete the proof of Theorem 1.2 using the above identity and the lower-semicontinuity of I.

The difference between the eigenvalues of two kernels k_1 and k_2 can be easily controlled by the Hilbert-Schmidt norm of the difference $k_1 - k_2$. In fact using the variational formula for successive eigenvalues of a compact self adjoint operator,

$$\lambda_{j+1}(A) = \inf_{\substack{\mathcal{M}: \\ \mathrm{codim}(\mathcal{M})=j}} \sup_{\substack{\|z\|=1 \\ z\in\mathcal{M}}} \langle Az, z\rangle$$

it is easily seen that if $\lambda_j(A)$ and $\lambda_j(B)$ are the jth largest positive eigenvalues of two compact self adjoint operators A and B

$$|\lambda_j(A) - \lambda_j(B)| \leq \|A - B\| \leq \|A - B\|_{HS}$$

and a similar formula holds for negative eigenvalues as well. Clearly if $k_1, k_2 \in \mathcal{K}$, then $d(S(k_1), S(k_2)) \leq \|k_1 - k_2\| \leq \|k_1 - k_2\|_{HS}$.

If $C \in \mathcal{S}$ is a closed set, and we truncate X at level ℓ and denote by k_ℓ and k the two images in $\mathcal{K}$, then

$$P_n(C) = Q_n[S(k) \in C] \leq Q_n[S(k_\ell) \in C^\epsilon] + Q_n[d(S(k_\ell), S(k)) \geq \epsilon].$$

Applying the upper bound for the truncated version,

$$\limsup_{n\to\infty} \frac{1}{n^2} \log P_n[C] \leq \max\{-\inf_{S\in\overline{C^\epsilon}} J_\ell(S), -c(\ell,\epsilon)\},$$

where

$$c(\ell,\epsilon) = -\limsup_{n\to\infty} \frac{1}{n^2} \log Q_n[d(S(k_\ell), S(k)) \geq \epsilon]$$

and for any $\epsilon > 0$,

$$\lim_{\ell\to\infty} c(\ell,\epsilon) = -\infty.$$

Moreover

$$J_\ell(S) = \inf_{k:S(k)=S} I_\ell(k); \quad I_\ell(k) = \int_D h_\ell(k(x,y))dxdy$$

and

$$h_\ell(z) = \sup[z\theta - \log \int \exp[\theta f_\ell(x)]\mu(dx)].$$

The upper bound is now easily established by letting $\ell \to \infty$ and then letting $\epsilon \to 0$. We observe that the superexponential estimate implies $c(\ell,\epsilon) \to \infty$ as $\ell \to \infty$ for any $\epsilon > 0$. This completes the proof of Theorem 1.1.

References

1. Bai, Z. D.: Methodologies in the spectral analysis of large-dimensional random matrices, a review. *Statist. Sinica* **9** (1999), no. 3, 611–677.
2. Ben Arous, G. and Guionnet, A.: Large deviations for Wigners law and Voiculescu's non-commutative entropy, *Prob. Th. Rel. Fields* **108** (1997) 517–542.
3. Chatterjee, S. and Varadhan, S. R. S.: The large deviation principle for the Erdős-Rényi random graph. *European J. Combin.* **32** (2011), no. 7, 1000–1017
4. Komlós, J. and Simonovits, M.: Szemerédi's regularity lemma and its applications in graph theory. *Bolyai Soc. Math. Stud.,* **2** (1996) 295–352.
5. Lovász, L. and Szegedy, B.: Szemerédi's lemma for the analyst. *Geom. Funct. Anal.* **17** (2007), no. 1, 252–270.
6. Wigner, E. P.: On the distribution of the roots of certain symmetric matrices. *Ann. of Math. (2)* **67** (1958) 325–327.

Sourav Chatterjee: Courant Institute, New York University, New York, NY, 10012, USA
E-mail address: `sourav@cims.nyu.edu`
URL: `http://www.cims.nyu.edu/~sourav/`

S. R. S. Varadhan: Courant Institute, New York University, New York, NY, 10012, USA
E-mail address: `varadhan@cims.nyu.edu`
URL: `http://www.cims.nyu.edu/~varadhan/`